SAE Handbook 1981

Published by:

Society of Automotive Engineers, Inc.

400 Commonwealth Drive, Warrendale, Pa. 15096

(412) 776-4841

The Standards, Recommended Practices, and Information Reports in the SAE Handbook apply primarily to surface vehicles and machinery. SAE also publishes Aerospace Standards, (AS), Aerospace Recommended Practices (ARP), Aerospace Information Reports (AIR) and Aerospace Material Specifications (AMS). Indexes listing SAE aerospace publications are available. For further information contact:

Technical Division
SAE
400 Commonwealth Dr.
Warrendale, PA 15096
U.S.A.
(412)776-4841

ISBN 0-89883-002-8
ISSN 0362-8205
Library of Congress Catalog Card Number: 25-16527
Copyright © Society of Automotive Engineers, Inc. 1981

All technical reports, including standards approved and practices recommended, are advisory only. Their use by anyone engaged in industry or trade or their use by governmental agencies is entirely voluntary. There is no agreement to adhere to any SAE Standard or Recommended Practice, and no commitment to conform to or be guided by any technical report. In formulating and approving technical reports, the Technical Board, its councils, and committees will not investigate or consider patents which may apply to the subject matter. Prospective users of the report are responsible for protecting themselves against liability for infringement of patents, trademarks, and copyrights.

—SAE Technical Board Rules and Regulations

PRINTED IN U.S.A.

1981 SAE HANDBOOK CONTENTS
PART 1

NUMERICAL AND SUBJECT INDICES FOR SAE STANDARDS, RECOMMENDED PRACTICES, AND INFORMATION REPORTS appear at the back of this volume.

Each surface vehicle Standard, Recommended Practice, or Information Report has a designation consisting of the letter "J" combined with a number. The letter "J" is combined with a nonsignificant number to eliminate any possible confusion between the report number and the SAE numbers within the reports.

Effective with the 1981 SAE Handbook, subsequent revisions of reports will be indicated by the month and year of revision, for example, J1159 AUG79. This new system will be phased in over a five year period. A lower case "a", "b", etc., appended to the report number indicates successive revisions of older reports.

Because of the publication of the SAE Handbook on an annual basis, at times certain reports and standards in this book may not be the latest issue of the document. Therefore, users are cautioned that if it is important that the latest issue of any given Standard or Recommended Practice be used, contact should be made with the SAE Headquarters to determine the status of specific documents. It is also possible to subscribe to the SAE Standards Subscription Service which provides copies of all individual documents as they are published. Details can be found at the end of this volume.

All new and revised reports contain SI (metric) equivalents of all dimensions. The SAE Metric Advisory Committee will appreciate receiving any comment and/or suggestions regarding the conversions.

The ϕ symbol is for the convenience of the user in locating the areas where technical revisions have been made to the previous issue of the report. If the symbol is next to the report title, it indicates a complete revision of the report. The notation "ed." is used to indicate editorial changes.

THE SOCIETY

SAE Technical Reports Referenced in Government Regulations	viii
SAE Technical Board	x
SAE Standing Administrative Committee	x
SAE Technical Board Councils	xi
SAE Technical Committee Personnel	xv
SAE Representation in Other Technical Organizations	xxxvi
SAE Technical Board Rules and Regulations	xxxvii

1 Rules, Nomenclature and Guideposts

Technical Committee Guideposts—SAE J1271 AUG79	1.01
Preparation of SAE Technical Reports—Surface Vehicles and Machines: Standards, Recommended Practices, Information Reports—SAE J1159 AUG79	1.06
†Rules for SAE Use of SI (Metric) Units—SAE J916 JUN80	1.09
Guidelines for Developing and Revising SAE Nomenclature and Definitions—SAE J1115	1.21
*Universal Symbols for Operator Controls—SAE J1500 JUN80	1.22

FERROUS METALS

2 Numbering System, Chemical Compositions

Numbering Metals and Alloys—SAE J1086	2.01
Selection and Use of Steels—SAE J401	2.05
SAE Numbering System for Wrought or Rolled Steel—SAE J402b	2.06
Chemical Compositions of SAE Carbon Steels—SAE J403h	2.06
Chemical Compositions of SAE Alloy Steels—SAE J404j	2.09
Chemical Compositions of SAE Wrought Stainless Steels—SAE J405d	2.11
§Chemical Compositions of SAE Experimental Steels—SAE J1081 MAY80	2.12
Formerly Standard SAE Alloy Steels—SAE J778a	2.13
Formerly Standard SAE Carbon Steels—SAE J118a	2.14
Methods of Determining Hardenability of Steels—SAE J406c	2.15
*Hardenability Bands for Carbon and Alloy H Steels—SAE J1268 JUN80	2.24
Methods of Sampling Steel for Chemical Analysis—SAE J408c	2.69
Permissible Variations from Specified Ladle Chemical Ranges and Limits for Steels—SAE J409c	2.70
High Strength, Low Alloy Steel—SAE J410c	2.72
High Strength, Quenched, and Tempered Structural Steels—SAE J368a	2.74
Engine Poppet Valve Materials—SAE J775a	2.75

3 General Data on Steels

Carbon and Alloy Steels—SAE J411d	3.01
High Strength Carbon and Alloy Die Drawn Steels—SAE J935	3.03
General Characteristics and Heat Treatments of Steels—SAE J412h	3.05
Mechanical Properties of Heat Treated Wrought Steels—SAE J413b	3.13
Estimated Mechanical Properties and Machinability of Hot Rolled and Cold Drawn Carbon Steel Bars—SAE J414a	3.14
Glossary of Carbon Steel Sheet and Strip Terms—SAE J940	3.18
Aging of Carbon Steel Sheet and Strip—SAE J763	3.20
Definitions of Heat Treating Terms—SAE J415j	3.20
Alloy Steel Machinability Ratings—SAE J770c	3.24

4 Methods of Testing Steels

Tensile Test Specimens—SAE J416b	4.01
Surface Hardness Testing with Files—SAE J864	4.02
Hardness Tests and Hardness Number Conversions—SAE J417b	4.03
Grain Size Determination of Steels—SAE J418a	4.07
Methods of Measuring Decarburization—SAE J419	4.10
Methods of Determining Plastic Deformation in Sheet Metal Stampings—SAE J863c	4.14
Sheet Steel Thickness and Profile—SAE J1058	4.16
Properties of Low Carbon Steel Sheets and Strip and Their Relationship to Formability—SAE J877	4.17
Selecting and Specifying Hot and Cold Rolled Steel Sheet and Strip—SAE J126a	4.19

*New
†Technical revision
§ Editorial change

† Classification of Common Imperfections in Sheet Steel—SAE J810 JUN80	4.24
*Undervehicle Corrosion Test—SAE J1293 JAN80	4.34
Cleanliness Rating of Steels by the Magnetic Particle Method—SAE J421b	4.37
Microscopic Determination of Inclusions in Steels—SAE J422a	4.39
† Detection of Surface Imperfections in Ferrous Rods, Bars, Tubes, and Wires—SAE J349 JUN80	4.42
Methods of Measuring Case Depth—SAE J423a	4.43
Method for Determining Breakage Allowances for Steel Sheets—SAE J424	4.45
§ Nondestructive Tests—SAE J358 JAN80	4.45
Infrared Testing—SAE J359a	4.47
Magnetic Particle Inspection—SAE J420b	4.47
Eddy Current Testing by Electromagnetic Methods—SAE J425b	4.48
Leakage Testing—SAE J1267	4.50
Liquid Penetrant Test Methods—SAE J426c	4.52
Penetrating Radiation Inspection—SAE J427b	4.52
Ultrasonic Inspection—SAE J428b	4.55
Acoustic Emission Test Methods—SAE J1242	4.56
Definition for Particle Size—SAE J391	4.56
Definitions for Macrostrain and Microstrain—SAE J932	4.57
Technical Report on Fatigue Properties—SAE J1099	4.57

5 Steel Fasteners

Mechanical and Material Requirements for Externally Threaded Fasteners—SAE J429 JAN80	5.01
Mechanical and Material Requirements for Metric Externally Threaded Steel Fasteners—SAE J1199	5.06
Test Methods for Metric Threaded Fasteners—SAE J1216	5.11
§ Mechanical and Quality Requirements for Machine Screws—SAE J82 JUN79	5.14
§ Mechanical and Material Requirements for Steel Nuts—SAE J995 JUN79	5.15
Mechanical and Material Requirements for Wheel Bolts—SAE J1102	5.18
Mechanical and Chemical Requirements for Nonthreaded Fasteners—SAE J430	5.18
Mechanical and Quality Requirements for Tapping Screws—SAE J933 JUN79	5.19
Steel Self-Drilling Tapping Screws—SAE J78 JUN79	5.21
§ Thread Rolling Screws—SAE J81 JUN79	5.26
Metric Thread Rolling Screws—SAE J1237	5.29
Surface Discontinuities on General Application Bolts, Screws, and Studs—SAE J1061a	5.33
Surface Discontinuities on Bolts, Screws, and Studs—SAE J123c	5.37
Surface Discontinuities on Nuts—SAE J122a	5.40
Decarburization in Hardened and Tempered Threaded Fasteners—SAE J121a	5.44

6 Spring Wire and Springs

Oil Tempered Carbon Steel Spring Wire and Springs—SAE J316	6.01
Oil Tempered Carbon Steel Valve Spring Quality Wire and Springs—SAE J351	6.02
Oil Tempered Chromium-Vanadium Valve Spring Quality Wire and Springs—SAE J132	6.03
Hard Drawn Mechanical Spring Wire and Springs—SAE J113	6.04
Special Quality High Tensile, Hard Drawn Mechanical Spring Wire and Springs—SAE J271	6.04
Oil Tempered Chromium-Silicon Alloy Steel Wire and Springs—SAE J157	6.05
Hard Drawn Carbon Steel Valve Spring Quality Wire and Springs—SAE J172	6.06
Stainless Steel 17-7 PH Spring Wire and Springs—SAE J217	6.07
Stainless Steel, SAE 30302, Spring Wire and Springs—SAE J230	6.08
Music Steel Spring Wire and Springs—SAE J178	6.09

7 Ferrous Castings

Numbering System for Designating Grades of Cast Ferrous Materials—SAE J859	7.01
† Automotive Gray Iron Castings—SAE J431 AUG79	7.01
Automotive Malleable Iron Castings—SAE J158a	7.04
Automotive Ductile (Nodular) Iron Castings—SAE J434c	7.06
Elevated Temperature Properties of Cast Irons—SAE J125	7.08
Automotive Steel Castings—SAE J435c	7.12

8 Tool and Die Steels

Tool and Die Steels—SAE J438b	8.01
Selection and Heat Treatment of Tool and Die Steels—SAE J437a	8.02
Sintered Carbide Tools—SAE J439a	8.07
Sintered Tool Materials—SAE J1072	8.12

9 Ferrous Materials

Sintered Powder Metal Parts: Ferrous—SAE J471d	9.01
Cut Steel Wire Shot—SAE J441	9.05
§ Test Strip, Holder and Gage for Shot Peening—SAE J442 AUG79	9.05
Procedures for Using Standard Shot Peening Test Strip—SAE J443	9.07
Cast Steel Shot—SAE J827	9.08
Cast Shot and Grit Size Specifications for Peening and Cleaning—SAE J444a	9.09
Metallic Shot and Grit Mechanical Testing—SAE J445a	9.10
Size Classification and Characteristics of Glass Beads for Peening—SAE J1173	9.11

10 Ferrous and Nonferrous—General

Use of the Terms Yield Point and Yield Strength—SAE J450	10.01
Surface Texture Measurement of Cold Rolled Sheet Steel—SAE J911	10.01
Surface Texture—SAE J448a	10.02
Surface Texture Control—SAE J449a	10.04
Preferred Thicknesses for Uncoated Flat Metals (Thru 12 mm)—SAE J446a	10.06
Welding, Brazing, and Soldering—Materials and Practices—SAE J1147	10.06

NONFERROUS METALS

11 Nonferrous Metals

Aluminum Alloys—Fundamentals—SAE J451b	11.01
Alloy and Temper Designation Systems for Aluminum—SAE J993b	11.02
General Information on SAE Aluminum Casting Alloys—SAE J452c	11.05
Chemical Compositions and Mechanical and Physical Properties of SAE Aluminum Casting Alloys—SAE J453c	11.08
General Data on Wrought Aluminum Alloys—SAE J454k	11.11
Chemical Compositions, Mechanical Property Limits, and Dimensional Tolerances of SAE Wrought Aluminum Alloys—SAE J457j	11.27
Anodized Aluminum Automotive Parts—SAE J399a	11.28
Bearing and Bushing Alloys (Standard)—SAE J460e	11.29
Bearing and Bushing Alloys (Information Report)—SAE J459c	11.29
Wrought and Cast Copper Alloys—SAE J461d	11.31
Cast Copper Alloys—SAE J462b	11.60
Wrought Copper and Copper Alloys—SAE J463d	11.63
Magnesium Alloys—SAE J464c	11.102
Magnesium Casting Alloys—SAE J465c	11.102

*New † Technical revision § Editorial change

Magnesium Wrought Alloys—SAE J466c 11.107
Special Purpose Alloys ("Superalloys")—SAE J467b 11.111
Zinc Alloy Ingot and Die Casting Compositions—SAE
  J468b 11.122
Zinc Die Casting Alloys—SAE J469a 11.123
Wrought Nickel and Nickel Related Alloys—SAE J470c . 11.123
Electroplating of Nickel and Chromium on Metal Parts—
  Automotive Ornamentation and Hardware—SAE J207 .. 11.133
Solders—SAE J473a 11.138
Electroplating and Related Finishes—SAE J474b 11.139

NONMETALLIC MATERIALS

12 Nonmetallic Materials

† Classification System for Rubber Materials for Automotive
  Applications—SAE J200 APR80 .... 12.01
Recommended Guidelines for Fatigue Testing of Elastomeric
  Materials and Components—SAE J1183 .... 12.29
Test for Dynamic Properties of Elastomeric Isolators—SAE
  J1085a 12.34
§ Latex Foam Rubbers—SAE J17 OCT79 .... 12.37
† Sponge- and Expanded Cellular-Rubber Products—SAE
  J18 DEC79 .... 12.40
Latex Dipped Goods and Coatings for Automotive Applications—SAE J19 .... 12.45
Synthetic Resin Plastic Sealers, Nondrying Type—SAE
  J250 .... 12.45
Methods of Tests for Automotive-Type Sealers, Adhesives,
  and Deadeners—SAE J243 .... 12.46
Sound Deadeners and Underbody Coatings—SAE J671 .. 12.56
Test Method for Measuring Weight of Organic Trim Materials—SAE J860 .... 12.58
Test Method for Determining Dimensional Stability of Automotive Textile Materials—SAE J883 .... 12.58
Method of Testing Resistance to Crocking of Organic Trim
  Materials—SAE J861 .... 12.58
Test Method for Measuring Thickness of Automotive Textiles and Plastics—SAE J882 .... 12.59
Test Method for Determining Cold Cracking of Flexible Plastic Materials—SAE J323 .... 12.59
Nonmetallic Trim Materials—Test Method for Determining
  the Staining Resistance to Hydrogen Sulfide Gas—SAE
  J322 .... 12.60
Test Method for Determining Window Fogging Resistance
  of Interior Trim Materials—SAE J275 .... 12.61
† Test for Chip Resistance of Surface Coatings—SAE J400
  JUN80 .... 12.63
Florida Exposure of Automotive Finishes—SAE J951
  FEB80 .... 12.66
Glossary of Fiberboard Terminology—SAE J947b 12.67
Fiberboard Test Procedure—SAE J315b 12.68
Test Method for Wicking of Automotive Fabrics and Fibrous
  Materials—SAE J913a 12.69
Method of Testing Resistance to Scuffing of Trim Materials—SAE J365a 12.69
Test Method for Determining Blocking Resistance and Associated Characteristics of Automotive Trim Materials—SAE
  J912a 12.71
Fiberboard Crease Bending Test—SAE J119a 12.72
Definitions of Acoustical Terms—SAE J1184 .... 12.72
Flammability of Automotive Interior Materials—Horizontal
  Test Method—SAE J369a 12.73
Test Method for Determining Stiffness (Modulus of Bending)
  of Fiberboards—SAE J949a 12.74
Test Method for Determining Resistance to Snagging and
  Abrasion of Automotive Bodycloth—SAE J948a 12.75
Test Method for Determining Visual Color Match to Master
  Specimen for Fabrics, Vinyls, Coated Fiberboards, and
  Other Automotive Trim Materials—SAE J361a 12.75
Test Method of Stretch and Set of Textiles and Plastics—
  SAE J855a 12.76

Felts—Wool and Part Wool—SAE J314b 12.78
Nonmetallic Automotive Gasket Materials—SAE J90b .. 12.80
Rubber Rings for Automotive Applications—SAE J120a . 12.87
Engine Coolants—SAE J814c 12.93
Engine Coolant Concentrate—Ethylene-Glycol Type—SAE
  J1034 .... 12.97
Coolant System Hoses—SAE J20e 12.98
Fuel and Oil Hoses—SAE J30d 12.102
Power Steering Pressure Hose—High Volumetric Expansion
  Type—SAE J188 .... 12.107
Power Steering Return Hose—Low Pressure—SAE J189 . 12.108
Power Steering Pressure Hose—Low Volumetric Expansion
  Type—SAE J191 .... 12.110
Power Steering Pressure Hose—Wire Braid—SAE J190 .. 12.111
Windshield Wiper Hose—SAE J50a 12.112
Automotive Air Conditioning Hose—SAE J51b 12.112
Emission Control Hose—SAE J1010 .... 12.114
Windshield Washer Tubing—SAE J1037 .... 12.115

13 Fuels and Lubricants

Effective Dates of Revisions—SAE J301 .... 13.01
† Engine Oil Performance and Engine Service Classification—SAE J183 FEB80 .... 13.01
Engine Oil Viscosity Classification—SAE J300d 13.05
Engine Oil Tests—SAE J304c 13.06
Physical and Chemical Properties of Engine Oils—SAE
  J357a 13.08
Axle and Manual Transmission Lubricants—SAE J308c .. 13.12
§ Axle and Manual Transmission Lubricant Viscosity Classification—SAE J306 OCT79 .... 13.13
§ Automotive Lubricating Greases—SAE J310 MAR80 .... 13.14
Fluid for Passenger Car Type Automatic Transmissions—
  SAE J311b 13.17
*Powershift Transmission Fluid Classification—SAE J1285
  FEB80 .... 13.19
Central Fluid Systems—SAE J71a 13.20
† Automotive Gasolines—SAE J312 JUN80 .... 13.22
† Diesel Fuels—SAE J313 JUN80 .... 13.28
*Alternative Automotive Fuels—SAE J1297 APR80 ..... 13.32
The Engine Oil Performance and Engine Service Classification Maintenance Procedure—SAE J1146 .... 13.34

THREADS, FASTENERS, AND COMMON PARTS

14 Threads

Screw Threads—SAE J475a 14.01
Dryseal Pipe Threads—SAE J476a 14.01

15 Fasteners

Hex Bolts—SAE J105 .... 15.01
Hi-Head Finished Hex Bolts—SAE J871a 15.04
Round Head Bolts—SAE J481 .... 15.04
Slotted and Recessed Head Screws—SAE J478a 15.07
Slotted Headless Setscrews—SAE J479a 15.40
Square Head Setscrews—SAE J102 .... 15.41
Flanged 12-Point Screws—SAE J58 .... 15.42
Lag Screws—SAE J103 .... 15.46
Square and Hex Nuts—SAE J104 .... 15.47
Hexagon High Nuts—SAE J482a 15.53
Alignment of Nut Slots—SAE J484 .... 15.54
Crown (Blind, Acorn) Nuts—SAE J483a 15.54
Spring Nuts—SAE J891a 15.55
Push-On Spring Nuts—SAE J892a 15.63
Steel Stamped Nuts of One Pitch Thread Design—SAE
  J1053 .... 15.64
Plain Washers—SAE J488 .... 15.68
Lock Washers—SAE J489b 15.68
Conical Spring Washers—SAE J773b 15.73
Nut and Conical Spring Washer Assemblies—SAE J238 .. 15.74
Torque-Tension Test Procedure for Steel Threaded Fasteners—SAE J174 .... 15.76

Cotter Pins—SAE J487a 15.77
Holes in Bolt and Capscrew Shanks and Slots in Nuts for
  Cotter Pins—SAE J485 .... 15.79
Straight Pins (Solid)—SAE J495 .... 15.79
Grooved Straight Pins—SAE J494 .... 15.80
Spring Type Straight Pins—SAE J496 .... 15.82
Unhardened Ground Dowel Pins—SAE J497 .... 15.84
Rod Ends and Clevis Pins—SAE J493 .... 15.84
Rivets and Riveting—SAE J492 .... 15.85
Blind Rivets—Break Mandrel Type—SAE J1200 .... 15.95
Bolt and Capscrew Sizes for Use in Construction and Industrial Machinery—SAE J370 .... 15.101
Hose Clamps—SAE J536b 15.101

16 Ball Studs and Joints
Ball Studs and Ball Stud Socket Assemblies—SAE J491b 16.01
Ball Stud and Socket Assembly Test Procedure—SAE
  J193 SEP79 .... 16.06
Ball Joints—SAE J490c 16.08
† Spherical Rod Ends—SAE J1120 SEP79 .... 16.12
* Metric Spherical Rod Ends—SAE J1259 APR80 .... 16.15

17 Splines
Involute Splines and Inspections—SAE J498c 17.01
Parallel Side Splines for Soft Broached Holes in Fittings—
  SAE J499a 17.01
Serrated Shaft Ends—SAE J500 .... 17.03
Shaft Ends—SAE J501 .... 17.04
Woodruff Keys—SAE J502 .... 17.05
Woodruff Key Slots and Keyways—SAE J503 .... 17.06

18 V-Belts
V-Belts and Pulleys—SAE J636c 18.01
Automotive V-Belt Drives—SAE J637b 18.04

19 Springs
Leaf Springs for Motor Vehicle Suspension—Metric Bar
  Sizes—SAE J1123a 19.01
Leaf Springs for Motor Vehicle Suspension—SAE J510c 19.01
Pneumatic Spring Terminology—SAE J511a 19.08
Helical Compression and Extension Spring Terminology—
  SAE J1121 .... 19.09
Helical Springs: Specification Check Lists—SAE J1122 . 19.14
Rated Spring Capacity—SAE J274 .... 19.15

20 Speedometers
Factors Affecting Automotive Odometer-Speedometer Accuracy—SAE J862b 20.01
Speedometers and Tachometers—Automotive—SAE J678e 20.03
Speedometer Test Procedure—SAE J1059 .... 20.07
† Electric Tachometer Specification—On Road—SAE J196
  JUN80 .... 20.08
Electric Tachometer Specification—Off Road—SAE J197 20.10
Automatic Vehicle Speed Control—Motor Vehicles—SAE
  J195 .... 20.12

21 Tubing and Fittings
§ Coding System for Identification of Tube, Pipe, and Hose
  Fittings—SAE J846 MAY80 .... 21.01
† Automotive Tube Fittings—SAE J512 NOV79 .... 21.04
§ Spherical and Flanged Sleeve (Compression) Tube Fittings—SAE J246 JUN80 .... 21.18
Refrigeration Tube Fittings—SAE J513f 21.26
§ Hydraulic Tube Fittings—SAE J514 APR 80 .... 21.46
§ Hydraulic Hose Fittings—SAE J516 SEP79 .... 21.72
Hose Push-On Fittings—SAE J1231 .... 21.89
† Hydraulic Hose—SAE J517 JUN80 .... 21.92

*Selection, Installation, and Maintenance of Hose and Hose
  Assemblies—SAE J1273 SEP79 .... 21.104
§ Tests and Procedures for SAE 100R Series Hydraulic Hose
  and Hose Assemblies—SAE J343 JUN80 .... 21.105
Flex-Impulse Test Procedure for Hydraulic Hose Assemblies—SAE J1405 .... 21.106
Tests and Procedures for High-Temperature Transmission
  Oil Hose, Lubricating Oil Hose and Hose Assemblies—
  SAE J1019 .... 21.107
Hydraulic Flanged Tube, Pipe, and Hose Connections,
  4-Bolt Split Flange Type—SAE J518c 21.107
Hydraulic "O" Ring—SAE J515a 21.112
Formed Tube Ends for Hose Connections—SAE J962b .. 21.112
Flares for Tubing—SAE J533b 21.113
† Seamless Low Carbon Steel Tubing Annealed for Bending
  and Flaring—SAE J524 JAN80 .... 21.115
† Welded and Cold Drawn Low Carbon Steel Tubing Annealed
  for Bending and Flaring—SAE J525 JAN80 .... 21.115
† Welded Low Carbon Steel Tubing—SAE J526 JAN80 .. 21.116
† Brazed Double Wall Low Carbon Steel Tubing—SAE J527
  JAN80 .... 21.117
† Welded Flash Controlled Low Carbon Steel Tubing Normalized for Bending, Double Flaring, and Beading—SAE
  J356 JAN80 .... 21.118
Pressure Ratings for Hydraulic Tubing and Fittings—SAE
  J1065 .... 21.119
† Seamless Copper Tube—SAE J528 JAN80 .... 21.121
Metallic Air Brake System Tubing and Pipe—SAE J1149 21.122
Nonmetallic Air Brake System Tubing—SAE J844d 21.124
Performance Requirements for SAE J844d Nonmetallic Tubing and Fitting Assemblies Used in Automotive Air Brake
  Systems—SAE J1131 .... 21.126
Fuel Injection Tubing—SAE J529b 21.127
Automotive Pipe Fittings—SAE J530a 21.128
Automotive Pipe, Filler, and Drain Plugs—SAE J531a .. 21.134
Automotive Straight Thread Filler and Drain Plugs—SAE
  J532a 21.138
Lubrication Fittings—SAE J534c 21.143

ELECTRICAL EQUIPMENT AND LIGHTING

22 Equipment
Nomenclature—Automotive Electrical Systems—SAE J831 22.01
Ignition System Nomenclature and Terminology—SAE
  J139 .... 22.01
Ignition System Measurements Procedure—SAE J973a . 22.02
Storage Batteries—SAE J537j 22.04
Life Test for Automotive Storage Batteries—SAE J240a . 22.12
Grounding of Storage Batteries—SAE J538a 22.13
Voltage Drop for Starting Motor Circuits—SAE J541a .. 22.13
Voltages for Diesel Electrical Systems—SAE J539a 22.13
Starting Motor Mountings—SAE J542c 22.14
Starting Motor Pinions and Ring Gears—SAE J543c 22.15
Generator Mountings—SAE J545b 22.15
Starting Motor and Generator Curves—SAE J544b 22.17
Low Temperature Cranking Load Requirements of an Engine—SAE J1253 .... 22.18
Electrical Generating System (Alternator Type) Performance
  Curve and Test Procedure—SAE J56 .... 22.19
Magneto Mountings—SAE J546 .... 22.20
Spark Plugs—SAE J548d 22.21
Preignition Rating of Spark Plugs—SAE J549a 22.27
† Limits and Methods of Measurement of Radio Interference
  Characteristics of Vehicles and Devices (20–
  1000 MHz)—SAE J551 JUN79 .... 22.27
External Electromagnetic Radiation Suppressors—SAE
  J552 .... 22.34
Electric Fuses (Cartridge Type)—SAE J554b 22.35
*Blade Type Electric Fuses—SAE J1284 DEC79 .... 22.37
Circuit Breakers—SAE J553c 22.40

*New
†Technical revision   § Editorial change

Circuit Breaker—Internal Mounted—Automatic Reset—SAE
J258 .... 22.42
Ignition Switch—SAE J259 .... 22.42
*Automobile, Truck, Truck-Tractor, Trailer, and Motor Coach
Wiring—SAE J1292 JUN80 .... 22.43
Automotive Printed Circuits—SAE J771c 22.45
High Tension Ignition Cable—SAE J557 .... 22.49
Low Tension Wiring and Cable Terminals and Splice
Clips—SAE J163 .... 22.51
Fusible Links—SAE J156a 22.51
Five Conductor Electrical Connectors for Automotive Type
Trailers—SAE J895 .... 22.52
Four- and Eight-Conductor Rectangular Electrical Connectors
for Automotive Type Trailers—SAE J1239 .... 22.54
Seven-Conductor Electrical Connector for Truck-Trailer
Jumper Cable—SAE J560b 22.57
Seven Conductor Jacketed Cable for Truck-Trailer Connections—SAE J1067 .... 22.58
Electrical Terminals—Blade Type—SAE J858a 22.59
† Electrical Terminals—Pin and Receptacle Type—SAE J928
JUN80 .... 22.60
† Electrical Terminals—Eyelet and Spade Type—SAE J561
JUN80 .... 22.61
Nonmetallic Loom—SAE J562a 22.65
Electric Windshield Wiper Switch—SAE J112a 22.66
Electric Windshield Washer Switch—SAE J234 ...... 22.66
Electric Blower Motor Switch—SAE J235 .... 22.67
6- and 12-Volt Cigar Lighter Receptacles—SAE J563b ... 22.67
Electromagnetic Susceptibility Procedures for Vehicle Components (Except Aircraft)—SAE J1113a 22.68
† Battery Cable—SAE J1127 JUN80 .... 22.79
Low Tension Primary Cable—SAE J1128 .... 22.81
Recommended Environmental Practices for Electronic Equipment Design—SAE J1211 .... 22.86
Glossary of Automotive Electronic Terms—SAE J1213 ... 22.102
*Performance Requirements for the Automotive Audio Cassette—SAE J1274 JUN80 .... 22.107
*Testing Methods for Audio Cassettes—SAE J1275 JUN80 22.108

23 Lighting

Terminology—Motor Vehicle Lighting—SAE J387 ..... 23.01
Lighting Identification Code—SAE J759c 23.01
Headlamp Beam Switching—SAE J564c 23.02
Semiautomatic Headlamp Beam Switching Devices—SAE
J565c 23.03
Lamp Bulb Retention System—SAE J567c 23.04
Connectors and Plugs—SAE J856 .... 23.07
Dimensional Specifications for Sealed Beam Headlamp
Units—SAE J571d 23.09
142 mm x 200 mm Sealed Beam Headlamp Unit—SAE
J1132 .... 23.16
Dimensional Specifications for General Service Sealed Lighting Units—SAE J760a 23.18
Lamp Bulbs and Sealed Beam Headlamp Units—SAE
J573g 23.20
§ Service Performance Requirements and Test Procedures for
Motor Vehicle Lamp Bulbs—SAE J1049 AUG79 ..... 23.23
Tests for Motor Vehicle Lighting Devices and Components—SAE J575 JUN80 .... 23.24
Service Performance Requirements for Motor Vehicle Lighting Devices and Components—SAE J256b 23.26
Plastic Materials for Use in Optical Parts Such as Lenses
and Reflectors of Motor Vehicle Lighting Devices—SAE
J576d 23.28
Plastic Material for Use in Housings of Motor Vehicle Lighting Devices—SAE J29 .... 23.28
Color Specification for Electric Signal Lighting Devices—
SAE J578d 23.29
Sealed Beam Headlamp Units for Motor Vehicles—SAE
J579c 23.31
Service Performance Requirements for Sealed Beam Headlamp Units for Motor Vehicles—SAE J32a 23.32
§ Sealed Beam Headlamp Assembly—SAE J580 AUG 79 .. 23.33
Auxiliary Driving Lamps—SAE J581a 23.35
Fog Lamps—SAE J583d 23.36
Auxiliary Low Beam Lamp—SAE J582a 23.36
Motorcycle Headlamps—SAE J584b 23.37
*Motorcycle Auxiliary Front Lamps—SAE J1306 JUN80 .. 23.39
Motorcycle and Motor Driven Cycle Electrical System (Maintenance of Design Voltage)—SAE J392 .... 23.39
Motorcycle Turn Signal Lamps—SAE J131a 23.40
Tail Lamps (Rear Position Lamps)—SAE J585e 23.40
Stop Lamps—SAE J586d 23.41
Mechanical Stop Lamp Switch—SAE J249 .... 23.42
License Plate Lamps—SAE J587f 23.43
Turn Signal Lamps—SAE J588f 23.44
Headlamp-Turn Signal Spacing—SAE J1221 .... 23.45
Supplemental High Mounted Stop and Rear Turn Signal
Lamps—SAE J186a 23.49
Turn Signal Switch—SAE J589b 23.50
Side Turn Signal Lamps—SAE J914b 23.50
Cornering Lamps—SAE J852b 23.51
Turn Signal Flashers—SAE J590e 23.52
Service Performance Requirements for Turn Signal Flashers—SAE J1055 .... 23.52
Vehicular Hazard Warning Signal Flasher—SAE J945b ... 23.54
Service Performance Requirements for Vehicular Hazard
Warning Flashers—SAE J1056 .... 23.55
Warning Lamp Alternating Flashers—SAE J1054 ...... 23.56
Service Performance Requirements for Warning Lamp Alternating Flashers—SAE J1104 .... 23.56
Spot Lamps—SAE J591b 23.58
Parking Lamps (Front Position Lamps)—SAE J222a 23.59
Clearance, Side Marker, and Identification Lamps—SAE
J592f 23.59
Backup Lamps—SAE J593e 23.60
Backup Lamp Switches—SAE J1076 .... 23.60
Headlamp Switch—SAE J253 .... 23.61
Reflex Reflectors—SAE J594f 23.61
Emergency Warning Device—SAE J774c 23.62
Flashing Warning Lamps for Authorized Emergency, Maintenance and Service Vehicles—SAE J595b 23.63
Hazard Warning Signal Switch—SAE J910b 23.64
360 Deg Emergency Warning Lamp—SAE J845 ...... 23.64
School Bus Red Signal Lamps—SAE J887a 23.65
School Bus Stop Arm—SAE J1133a 23.66
Lighting Inspection Code—SAE J599d 23.66
Headlamp Testing Machines—SAE J600a 23.68
Headlamp Aiming Device for Mechanically Aimable Sealed
Beam Headlamp Units—SAE J602c 23.69
Flasher Test Equipment—SAE J823c 23.72

SAE TECHNICAL REPORTS REFERENCED IN GOVERNMENT REGULATIONS

It is known that the documents listed below are referenced or used in the Government regulatory system. Some of these documents have been revised and reissued since the original Government reference usage. However, the specific issue of the document referenced by regulation is still available through the SAE Publications Division, Society of Automotive Engineers, 400 Commonwealth Drive, Warrendale, PA 15096.

CALIFORNIA AIR RESOURCE BOARD (CARB)

Specifications for Fuel Pipes and Openings of Motor Vehicle Fuel Tanks

CARB Fuel Tank Filler Cap and Cap Retainer . . . . SAE J829b
CARB Fuel Tank Filler Cap and Cap Retainer—Threaded Pressure—Vacuum Type . . . . SAE J1114
CARB Engine Rating Code-Spark Ignition . . . . SAE J245

ENVIRONMENTAL PROTECTION AGENCY

Title 40 CFR (40 CFR 85) Published Part J. Gaseous Emissions, Diesel Heavy Duty Engines— Paragraph 85.974–13, (a), (b), (c).

Measurement of CO_2, CO, NO_x, and Diesel Exhaust . . . . . SAE J177
Continuous HC Analysis of Diesel Exhaust . . . . SAE J215
Measurement of In-Take of Exhaust Flow of Diesel Engines SAE J244

NATIONAL HIGHWAY TRAFFIC SAFETY ADMINISTRATION

Federal Motor Vehicle Safety Standards

FMVSS-103 Passenger Car Windshield Defrosting Systems—August 1964 . . . . SAE J902
FMVSS-103 Passenger Car Windshield Defrosting Systems—March 1967 . . . . SAE J902a
FMVSS-104 Passenger Car Driver's Eye Range— November 1965 . . . . SAE J941
FMVSS-104 Passenger Car Windshield Washer Systems—November 1965 . . . . SAE J942
FMVSS-104 Passenger Car Windshield Wiper Systems— May 1966 . . . . SAE J903a
FMVSS-105-75 Moving Barrier Collision Tests—November 1966 . . . . SAE J972
FMVSS-107 Passenger Car Driver's Eye Range— November 1965 . . . . SAE J941
FMVSS-108 Automotive Turn Signal Flashers—October 1965 . . . . SAE J590b
FMVSS-108 Back Up Lamps—February 1968 . . . . SAE J593c
FMVSS-108 Clearance, Side Marker, Identification, and Parking Lamps—July 1972 . . . . SAE J592e
FMVSS-108 Color Specifications for Electric Signal Lighting Devices—February 1977 . . . . . SAE J578c
FMVSS-108 Dimensional Specifications for Sealed Beam Headlamp Units—April 1965 . . . . SAE J571b
FMVSS-108 Headlamp Aiming Device for Mechanically Aimable Sealed Beam Units—August 1963 SAE J602
FMVSS-108 Headlamp Beam Switching—April 1964 . . . SAE J564a
FMVSS-108 Headlamp Mountings—January 1960; Deleted 1972 . . . . SAE J566
FMVSS-108 Lamp Bulbs and Sealed Beam Headlamp Units . . . . SAE J573d
FMVSS-108 License Plate Lamps—March 1969 . . . . . . SAE J587d
FMVSS-108 Lighting Inspection Code—March 1973 . . . . SAE J599e
FMVSS-108 Motorcycle and Motor Driven Cycle Headlamps—January 1949 . . . . SAE J584
FMVSS-108 Parking Lamps (Position Lamps)—December 1970 . . . . SAE J222
FMVSS-108 Plastic Materials for Use in Optical Parts, Such as Lenses and Reflectors, of Motor Vehicle Lighting Devices—August 1966 . . SAE J576b
FMVSS-108 Plastic Materials for Use in Optical Parts, Such as Lenses and Reflectors, of Motor Vehicle Lighting Devices—May 1970 . . . . SAE J576c
FMVSS-108 Reflex Reflectors—March 1970 . . . . SAE J594e
FMVSS-108 School Bus Red Signal Lamps—July 1964 . . SAE J887
FMVSS-108 Sealed Beam Headlamp—June 1966 . . . . . SAE J580a
FMVSS-108 Sealed Beam Headlamp Units for Motor Vehicles—August 1965 . . . . SAE J579a
FMVSS-108 Sealed Beam Headlamp Units for Motor Vehicles—December 1974 . . . . SAE J579c
FMVSS-108 Semi-Automatic Headlamp Beam Switching Devices—February 1969 . . . . SAE J565b
FMVSS-108 Stop Lamps—August 1970 . . . . SAE J586c
FMVSS-108 Tail Lamps—August 1970 . . . . SAE J585d
FMVSS-108 Test for Motor Vehicle Lighting Devices and Components—August 1967 . . . . SAE J575d
FMVSS-108 Test for Motor Vehicle Lighting Devices and Components—August 1970 . . . . SAE J575e
FMVSS-108 Test for Motor Vehicle Lighting Devices and Components—April 1975 . . . . SAE J575f
FMVSS-108 Turn Signal Lamps—June 1966 . . . . SAE J588d
FMVSS-108 Turn Signal Lamps—August 1970; Edit. Change—September 1970 . . . . SAE J588e
FMVSS-108 Turn Signal Operating Units—April 1964 . . . SAE J589
FMVSS-108 Vehicular Hazard Signal Operating Unit— January 1966 . . . . SAE J910
FMVSS-108 Vehicular Hazard Warning Signal Flasher— February 1966 . . . . SAE J945
FMVSS-108 142 mm x 200 mm Sealed Beam Headlamp Unit—January 1949 . . . . SAE J1132
FMVSS-111 Test Procedure for Determining Reflectivity of Rearview Mirrors—August 1974 . . . . . . SAE J964a
FMVSS-116 Brazed Double Wall Low Carbon Steel Tubing—January 1955 . . . . SAE J527
FMVSS-116 Motor Vehicle Brake Fluid—April 1968 . . . . SAE J1703b
FMVSS-201 Instrument Panel Laboratory Impact Test Procedure—June 1965 . . . . SAE J921
FMVSS-201 Instrumentation for Laboratory Impact Tests—November 1966 . . . . SAE J977
FMVSS-201 Passenger Car Side Door Latch Systems— May 1965; Edit. Change—January 1972 . SAE J839b
FMVSS-202 Manikins for Use in Defining Vehicle Seating Accommodation—November 1962 . . . . . SAE J826
FMVSS-203 Steering Wheel Assembly Laboratory Test Procedure—December 1965 . . . . SAE J944
FMVSS-204 Barrier Collision Tests—February 1963 . . . . SAE J850
FMVSS-205 Automotive Glazing—August 1967 . . . . . . SAE J673a
FMVSS-206 Passenger Car Side Door Latch Systems— May 1965; Edit. Change—January 1972 . SAE J839b
FMVSS-206 Vehicular Passenger Door Hinge System— July 1965   SAE J934
FMVSS-208 Anthropomorphic Test Device for Dynamic Testing—June 1968 . . . . SAE J963
FMVSS-208 Human Tolerance to Impact Conditions as Related to Motor Vehicle Design—October 1966 . . . . SAE J885a
FMVSS-208 Instrumentation for Barrier Collision Tests— October 1970 . . . . SAE J211
FMVSS-209 Air Cleaner Test Code—June 1962 . . . . . SAE J726a
FMVSS-209 Motor Vehicle Seat Belt Assembly Installations—September 1965 . . . . SAE J800b

FMVSS-210 Manikins for Use in Defining Vehicle Seating Accommodation—November 1962 ..... SAE J826
FMVSS-222 Instrumentation for Barrier Collision Tests—December 1971 .... SAE J211a
FMVSS-PART 572 Instrumentation for Barrier Collision Tests—December 1971 .... SAE J211a
FMVSS-PART 581 Headlamp Aiming Device for Mechanically Aimable Sealed Beam Units—August 1963 .... SAE J602a
FMVSS-PART 581 Lighting Inspection Code—July 1970 . SAE J599b
FMVSS-PART 581 Surface Texture Control—June 1963 .. SAE J449a
FMVSS-302 Flammability of Automotive Interior Materials—Horizontal Test Method ..... SAE J369a

OCCUPATIONAL SAFETY AND HEALTH ADMINISTRATION

OSHA No. 29 CFR 1910 Slow Moving Vehicle Emblem .. SAE J943a
OSHA No. 29 CFR 1926.1001 Minimum Performance Criteria for Roll-Over Protective Structures (ROPS) for Designated Scrapers, Loaders, Dozers, Graders and Crawler Tractors ..... SAE J1040b
OSHA No. 29 CFR 1928.51 Agricultural Tractor ROPS .. SAE J1194
OSHA No. 29 CFR 1928.57 Safety of Agricultural Equipment .... SAE J208b

SAE OFFICERS—1981

Phillip J. Mazziotti, President

H. L. Brock, Treasurer
Rodger F. Ringham, Assistant Treasurer

Harold C. MacDonald 1980 President
Lewis E. Fleuelling, 1979 Past President

Joseph Gilbert, Secretary and General Manager

SAE TECHNICAL BOARD FOR 1980 AND 1981

L. H. Hodges, 1980 Chairman
E. A. Green, 1981 Chairman

COMMITTEE AND TERM EXPIRATION DATE (END OF ADMINISTRATIVE YEAR)

1980	1981	1982	1983
R. G. Flagan	D. R. Blundell	B. Ancker-Johnson	A. W. Carey
H. P. Freers	J. B. Codlin	N. B. Chew	W. D. Compton
L. H. Hodges	J. Costantino	J. B. Colletti	H. N. Cotter
D. J. Lloyd-Jones	D. W. Hadden	R. F. German	V. P. Hendrickson
W. F. Paul	J. W. Mohr	E. A. Green	R. W. Hildebrandt
B. H. Pauly	G. C. Nield	R. C. Lunn	W. H. Weltyck
P. G. Scully	I. H. R. Rosen	J. L. Palmer	
T. S. Webb		N. H. Pulling	

(M. L. Stoner, Secretary)

SAE TECHNICAL DIVISION STAFF

M. L. Stoner, Manager

D. R. Bentley   A. G. Salem   R. C. Uhl   L. P. Ziegler, Jr.

SAE Headquarters, 400 Commonwealth Drive, Warrendale, PA 15096

P. Couhig   R. T. Northrup

SAE Detroit Branch Office, 2100 West Big Beaver, Troy, MI 48084

CERTIFICATES OF APPRECIATION COMMITTEE

J. B. CODLIN (Chairman)
N. B. Chew
R. G. Flagan

J. F. Hutchinson
W. F. Paul

(SAE Staff, M. L. Stoner)

TECHNICAL COMMITTEE MILITARY LIAISON-AT-LARGE

H. Handler   A. Nehman

METRIC ADVISORY COMMITTEE

L. C. KISER (Chairman)
J. N. Bagnall
L. E. Barbrow
J. T. Benedict
H. E. Guetzlaff
N. B. Johnston

J. T. Keeley
S. E. Mallen
G. Pilkington
H. R. Steding
L. R. Strang
R. P. Trowbridge

(SAE Staff, P. Couhig)

PUBLICATIONS ADVISORY COMMITTEE

R. P. TROWBRIDGE (Chairman)
J. T. Benedict
J. W. Kourik
D. L. Nordeen
J. V. Prange

J. V. Sprong
S. G. Tilden, Jr.
J. R. Tishkowski
F. C. Walters

(SAE Staff, A. G. Salem)

UNIFIED NUMBERING SYSTEM ADVISORY BOARD

A. G. COOK (Chairman)
L. H. Bennett
H. M. Cobb
A. Cohen
J. A. Conway
C. J. Cooley
L. Falcone

H. B. Fernald, Jr.
J. Gadbut
E. J. Kubel, Jr.
E. F. Parker
B. N. Peak, Jr.
R. D. Thomas, Jr.

(SAE Staff, P. Couhig)

AEROSPACE COUNCIL

R. C. COLLINS (Chairman)
B. R. Aubin
S. S. Baits
M. C. Beard
K. F. Becker
M. W. Eastburn
C. W. Eyres
G. Gilder
W. M. Goodwin
E. A. Green
D. Hanink
W. P. Hannan
D. P. Hettermann
F. Hom
B. L. Koff
F. T. Main

G. R. Makepeace
J. L. Mason
N. R. Parmet
D. P. Passeri
W. F. Paul
B. E. Peterman
M. L. Ramey
R. Schmidt
R. P. Skully
H. H. Slaughter
L. H. Smith
J. W. Steffen
G. P. Townsend
T. S. Webb
W. D. Wise

(SAE Staff, A. G. Salem)

CONSTRUCTION, AGRICULTURAL, AND OFF-ROAD MACHINERY COUNCIL

R. F. HUGHES (Chairman)
F. C. WALTERS (Chairman)
L. D. Bergsten
J. B. Codlin
W. D. Drummond
G. W. Eger
R. W. Fabere

D. W. Hadden
L. H. Hodges
C. L. Kepner
G. L. Klose
P. E. Lockie
G. H. Millar
M. R. North

F. W. Ritchey
D. R. Wielage

F. H. Winters

(SAE Staff, R. C. Uhl)

CONSTRUCTION, AGRICULTURAL, AND OFF-ROAD MACHINERY COUNCIL COMMITTEES

Off-Road Machinery Technical Committee
Agricultural Tractor Technical Committee

Construction, Agricultural, and Off-Road Machinery Sound Level Technical Committee

GENERAL MATERIALS COUNCIL

M. P. SEMENEK (Chairman)
F. H. LIEB (Vice-Chairman)
H. D. Berns
G. F. Bush
J. Codlin
A. O. DeHart
S. Dinda
R. G. Flagan
G. H. Gilmore
G. Greene
W. Hart
A. F. Hegerich
B. Hertel

H. R. Jaeckel
A. S. Kasper
L. J. Lamberg
J. A. Larson
J. L. Palmer
L. G. Pless
I. H. R. Rosen
N. A. Schilke
D. A. Scoble
G. E. Stein
G. W. Tuffnell
J. Valdez
R. A. Wilde

(SAE Staff, R. T. Northrup)

GENERAL MATERIALS COUNCIL COMMITTEES

Fatigue Design and Evaluation Committee
Fuels and Lubricants Technical Committee
Iron and Steel Technical Committee

Nonferrous Metals Committee
Nonmetallic Materials Committee

GENERAL STANDARDS COUNCIL

R. S. PIOTROWSKI (Chairman)
K. NAYLOR (Vice-Chairman)
V. T. Czebatol
D. W. Hadden
R. E. Holmgren

R. G. Michel
W. M. Roll
I. H. R. Rosen
D. W. Vial
J. V. Woolley

(SAE Staff, P. Couhig)

GENERAL STANDARDS COUNCIL COMMITTEES

Ball Joint and Spherical Rod End Committee
Ball Stud and Tie Rod Socket Committee
Fasteners Committee

Speedometer and Tachometer Committee
Fluid Conductors and Connectors Technical Committee
V-Belt Committee

MOTOR VEHICLE COUNCIL

J. VERSACE (Chairman)
B. H. PAULY (Vice-Chairman)
W. G. Agnew
B. Ancker-Johnson
W. T. Birge
B. H. Bouwkamp
H. P. Freers
R. T. Gaskill
R. W. Hildebrandt

J. W. Kourik
D. E. Martin
G. C. Nield
D. L. Nordeen
W. J. Oakley
H. R. Pickford
N. H. Pulling
J. Seidl
W. B. Smyth

T. J. Walsh M. R. Young

(SAE Staff, L. P. Ziegler, Jr.)

MOTOR VEHICLE COUNCIL COMMITTEES

Automotive Emissions Committee
Body Engineering Committee
Brake Committee
Electric Vehicle Technical Committee
Electrical Equipment Committee
Electronic Systems Committee
Engine Committee
Fuel Supply Systems Committee
Highway Tire Committee
Human Factors Engineering Committee
Hydraulic Brake Systems Actuating Committee
Lighting Committee
Motor Vehicle Safety Systems Testing Committee
Nomenclature Advisory Committee
Passenger Car and Light Truck Fuel Economy Measurement Committee
Seal Committee
Technical Advisory Group for ISO/TC 22—Road Vehicles
Transmission and Drivetrain Committee
Vehicle Dynamics Committee
Vehicle Identification Numbers Committee
Vehicle Security Committee
Vehicle Sound Level Committee
Wheel Committee

SPECIALIZED VEHICLE AND EQUIPMENT COUNCIL

D. I. REED (Chairman)
J. Abromavage
M. A. Berk
C. A. Bible
R. Cordill
G. W. Eger
W. S. Fox
G. C. Hardwick
D. R. Hartdegen
L. C. Lake
R. H. Lincoln
T. H. Lohr
R. H. Madison
J. W. Mohr
J. A. Orvis
A. J. Troyan
B. R. Weber
N. Woelffer

(SAE Staff, L. P. Ziegler, Jr.)

SPECIALIZED VEHICLE AND EQUIPMENT COUNCIL COMMITTEES

Lawn, Garden, and Small Engine Powered Equipment Committee
Marine Technical Committee
Motorcycle Committee
On-Highway Recreational Vehicle Committee
Snowmobile Committee
Specialized Vehicle and Equipment Sound Level Committee
Trailer Hitch Committee

TRUCK AND BUS COUNCIL

J. E. ALLEN (Chairman)
N. B. Chew
W. H. Close
J. A. Dick
A. M. Fischer
D. Forester
G. A. Frederiksen
W. L. Giles
R. H. Hinchcliff
R. W. Hildebrandt
A. F. Hulverson
J. MacDougall
J. Malus
P. J. Mazziotti
J. F. Mueller
B. H. Pauly
J. M. Prange
I. Rosen
R. W. Sackett
D. Stanley
W. B. Smyth
J. G. Stieber
D. L. Stephens
L. W. Strawhorn
J. D. Winsor

(SAE Staff, A. G. Salem)

TRUCK AND BUS COUNCIL COMMITTEES

Truck and Bus Cab and Occupant Environment Committee
Truck and Bus Chassis Committee
Truck and Bus Electrical Committee
Truck and Bus Fuel Economy Committee
Truck and Bus Operations and Maintenance Committee
Truck and Bus Powertrain Committee

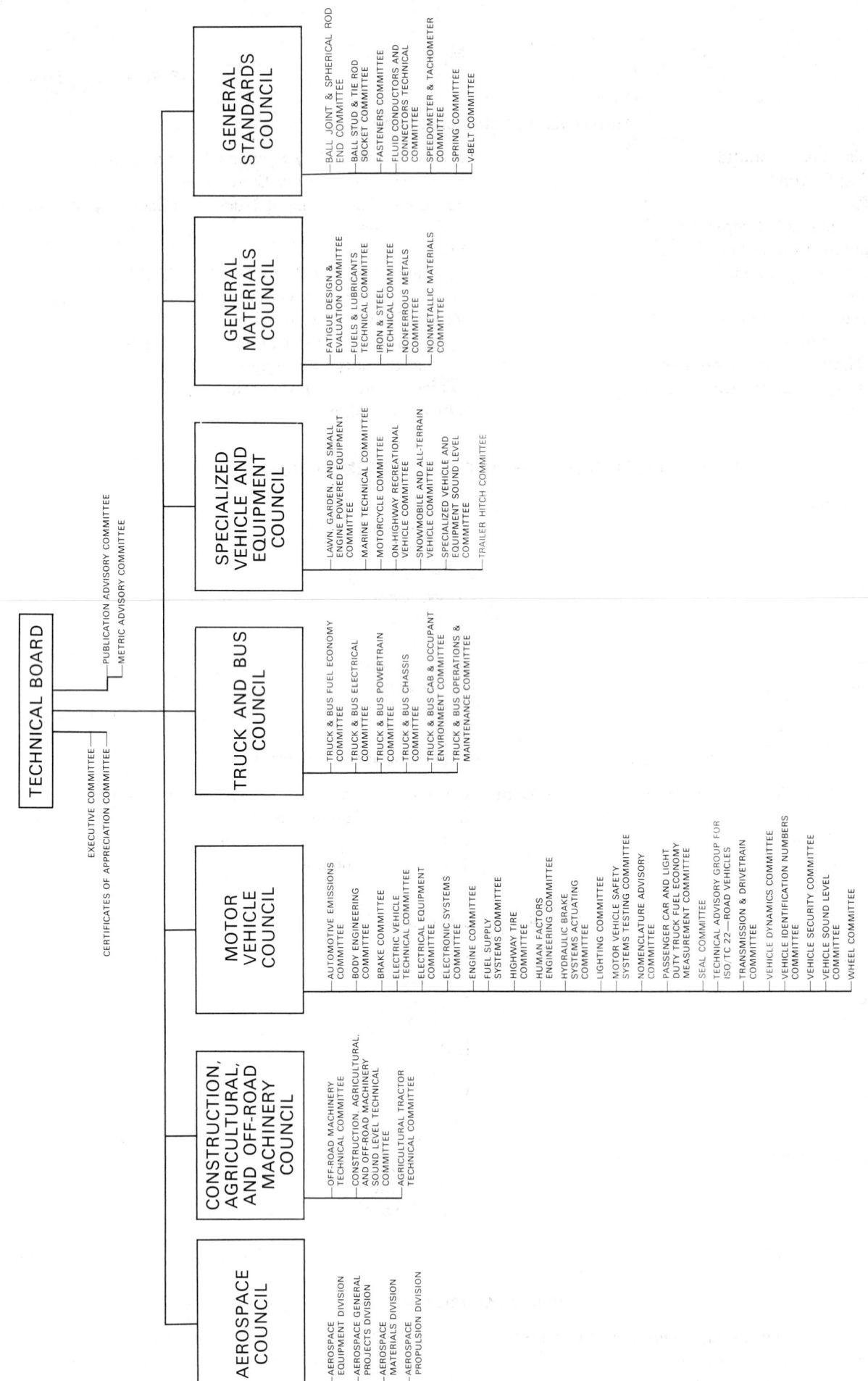

SAE SURFACE VEHICLE TECHNICAL COMMITTEE PERSONNEL

AGRICULTURAL TRACTOR TECHNICAL COMMITTEE

P. E. LOCKIE (Chairman)
R. D. REED (Vice-Chairman)
P. A. Asseff
D. G. Bamford
R. T. Bennett
K. Cheatham

J. B. Liljedahl
C. E. McKeon
M. R. North
M. G. Pribyl
W. E. Splinter
L. F. Stikeleather

J. H. Taylor
M. A. Walck
F. C. Walters
E. J. Zeglen
L. I. Leviticus (Liaison)

(SAE Staff, R. C. Uhl)

AGRICULTURAL TRACTOR TEST CODE SUBCOMMITTEE—R. D. Reed (Chairman), R. H. Becker, W. W. Brixius, E. J. Constien, J. E. Langdon, L. I. Leviticus, B. Nelson, W. E. Splinter, R. E. Turney, P. K. Turnquist, F. C. Walters, D. Withrow, E. J. Zeglen. TRACTOR TIRE SUBCOMMITTEE—F. C. Walters (Chairman), R. W. Ellis (Vice-Chairman), D. G. Bamford, J. C. Boone, K. Cheatham, J. Climstein, T. Crusinberry, M. Denham, J. Evans, W. O. Gipp, C. Hausz, T. D. Kenady, H. Kreb, L. I. Leviticus, D. L. Liedtke, M. North, W. E. Scholich, H. A. Steinkraus, J. Taylor, D. Weber, C. F. Zuege. JOINT SEATING SUBCOMMITTEE OF THE OFF-ROAD MACHINERY TECHNICAL COMMITTEE AND THE AGRICULTURAL TRACTOR TECHNICAL COMMITTEE—J. W. Carter (Chairman), F. L. Wildey (Vice-Chairman), J. C. Barton, J. Evans, G. O. Hall, H. K. Kienzle, B. J. Lindgren, K. Moehle, E. K. Pederson, T. J. Perlinger, D. A. Raab, L. F. Stikeleather, R. F. Swenson, R. C. Wagner, R. Wahls, B. D. Zehr.

AUTOMOTIVE EMISSIONS COMMITTEE

W. G. AGNEW (Sponsor)
K. D. MILLS (Chairman)
W. E. Adams
R. A. Altenkirch
R. C. Bascom
J. L. Bascunana
F. W. Bowditch
W. M. Brehob
J. M. DiBella
W. S. Fagley

C. Ferguson
T. M. Fisher
G. C. Hass
C. Heinen
M. W. Jackson
J. E. A. John
R. Matthews
P. V. Mohan
R. G. Murray
A. H. Pahnke

R. H. Perry
L. Raymond
R. L. Reichlen
K. J. Springer
J. P. Stieger
D. Stock
C. B. Tracy
R. E. Wallace
H. J. Wimette

(SAE Staff, R. T. Northrup)

CARBON MONOXIDE MEASUREMENT SUBCOMMITTEE—M. W. Jackson (Chairman), D. E. Brisbin, T. Dreaver, D. R. Hahnke, D. Hitchcock, S. J. Nowak, Jr., E. Tosolt, V. L. Vickland. DIESEL EMISSIONS SUBCOMMITTEE—C. R. Hudson (Chairman), L. C. Broering, A. J. Burgess, A. W. Cary, D. C. Dowdall, F. J. Hills, J. H. Johnson, M. W. Korth, R. D. McDowell, K. E. Murphy, J. Pearce, J. M. Perez, R. J. Peyla, R. K. Scott, S. M. Shahed, J. P. Steiger, B. Vedak. DIESEL EMISSIONS MEASUREMENT TASK FORCE—K. E. Murphy (Chairman), L. C. Broering, M. S. Englund, J. Hamilton, C. R. Hudson, R. A. Miller. DIESEL PARTICULATE MEASUREMENT TASK FORCE—S. M. Shahed (Chairman), E. E. Gough, F. J. Hills, C. R. Hudson, J. H. Johnson, D. Lenane, K. E. Murphy, K. J. Springer, J. P. Steiger, C. T. Vuk, D. J. Waldham, R. E. Winsor. J177 TASK FORCE—J. M. Perez (Chairman), L. C. Broering, W. Clemens, D. S. Gray, R. LaFrentz, D. Neluis, P. Polhonky, E. J. Sienicki. J255 TASK FORCE—C. C. Everett (Chairman), G. L. Green, J. W. Huber, R. D. McDowell, E. S. Vedak. EMISSION NOMENCLATURE SUBCOMMITTEE—W. E. Adams (Chairman), E. W. Beckman, M. Freeman, G. D. Kittredge, W. A. Kostin, C. D. Shepherd, P. H. Toney. EXHAUST EMISSIONS MEASUREMENT SUBCOMMITTEE—P. V. Mohan (Chairman), N. Baerman, W. J. Brown, J. B. Clark, A. Huff, M. W. Jackson, H. Schlumbohm, J. P. Steiger, A. T. Weibel, J. S. Welstand. PARTICULATES SUBCOMMITTEE—E. F. Fort (Chairman), R. Bauer, M. Beltzer, K. Habibi, R. J. Hames, M. G. Jacko, D. Lenane, D. A. Meyer, D. Pellerito, W. Stobaugh, D. A. Trayser, E. E. Weaver. TAIL PIPE WORKING GROUP—E. F. Fort (Chairman), R. Bauer, M. Beltzer, K. Habibi, R. J. Hames, D. Lenane, D. A. Trayser, E. E. Weaver.

BALL JOINT AND SPHERICAL ROD END COMMITTEE

J. V. WOOLLEY (Sponsor)
J. G. LANGENSTEIN (Chairman)
P. A. D'Anza
S. E. Drenth

A. Henn
D. M. Johnson
W. D. Ross
C. F. Schaening

W. C. Stephens
H. Sterkel
J. Weston

(SAE Staff, P. Couhig)

SUBCOMMITTEE 1—A. Henn (Chairman), M. Gibson, J. Keller, J. Langenstein.
SUBCOMMITTEE 2—J. Keller (Chairman), P. A. D'Anza, A. Henn, N. Smith.

BALL STUD AND TIE ROD SOCKET COMMITTEE

V. T. CZEBATOL (Sponsor)
D. D. PATTON (Chairman)
R. C. KOWALSKE (Vice-Chairman)
W. L. Adkins

R. W. Alexander
J. G. Banjo
J. P. Graham
M. Hassan

R. E. Jayroe
R. D. Perkins
M. A. Schultz

(SAE Staff, P. Couhig)

PERFORMANCE REQUIREMENTS SUBCOMMITTEE—R. W. Alexander (Chairman), J. G. Banjo, M. Hassan, D. D. Patton, C. Regent, M. A. Schultz, R. D. Wisley. J491a SUBCOMMITTEE—M. A. Schultz (Chairman), W. L. Adkins, V. J. Ryszewski, P. E. Yakimowich.

BODY ENGINEERING COMMITTEE

B. H. BOUWKAMP (Sponsor)
D. K. PETTRY (Chairman)
E. ZIPP (Vice-Chairman)
P. Arbogast
W. Atkinson
D. C. Brown

J. J. Cantalupo
J. Colenso
E. Crilley
J. J. Griffin
C. Kaehn
D. K. Haenchen (Corresponding Member)

H. Maruyama
W. M. Miller
R. Morrison
C. Silvester (Corresponding Member)

(SAE Staff, P. Couhig)

DEFROSTING AND INTERIOR CLIMATE CONTROL SUBCOMMITTEE—W. Atkinson (Chairman), L. E. Bischoff, G. T. Boyd, W. J. Christensen, K. E. Cavey, C. A. Copp, D. Haenchen (Corresponding Member), R. H. Jacobsen, Mr. Kaltner (Corresponding Member), R. W. Kelley, J. King, H. W. Kogan, E. Kruse, R. B. Kurre, R. T. Marshall, C. J. McLachlan, L. T. Merritt, R. Michaels, J. Mieras, B. Muchnij, R. J. Olsen, R. M. Premo, R. Proctor, J. D. Puckett, N. A. Rini, C. Scherer, C. Silvester (Corresponding Member), C. D. Sweet, D. L. Tothero, H. Weaver, W. Wiitala, L. E. Windsor (Corresponding Member), D. Wood, R. H. Youmans, K. H. Ziwica (Corresponding Member). HEATING TASK FORCE—G. Boyd (Chairman), L. Bischoff, R. W. Kelley, H. W. Kogan, J. E. Long, B. Muchnij, D. Pritchard, C. Scherer, C. D. Sweet, D. Wood. AIR CONDITIONING TASK FORCE—R. W. Kelley (Chairman), L. E. Bischoff, E. Borowiec, C. A. Copp, K. W. Finch, G. Hoch, Mr. Kaltner (Corresponding Member), H. W. Kogan, E. Kruse, J. E. Long, C. J. McLachlan, L. T. Merritt, J. Mieras, B. Muchnij, R. M. Premo, R. Proctor, J. D. Puckett, C. Scherer, C. D. Sweet. TRUCK HEATING AND DEFROSTING TASK FORCE—R. Michaels (Chairman), R. Ali, W. Atkinson, L. Bischoff, R. Brooks, C. H. Corning, J. F. Dermody, R. B. Dickey, L. Egolf, L. Gillespie, K. Hillers, R. H. Jacobsen, T. Kelly, J. King, J. Kozub, G. Kreaden, M. W. Lankford, C. McLachlan, B. Muchnij, G. R. Palzkill, J. Paul, N. Rini, M. Steiger, W. Strassel, C. D. Sweet, E. Traub, W. S. Trindal. FRONT WINDOW DEFROSTING TASK FORCE—R. T. Marshall (Chairman), L. Bischoff, G. T. Boyd, W. J. Christensen, T. L. Cox, W. M. Elliott, H. Kogan (Corresponding Member), J. E. Long, F. Melin (Corresponding Member), R. Michaels, C. D. Sweet (Corresponding Member), K. H. Ziwica (Corresponding Member). REFRIGERANT 12 STUDY TASK FORCE—W. D. Cooper (Chairman), C. A. Copp, J. E. Holtslag, J. Kellerhouse, H. Kogan, F. Melin, L. T. Merritt, R. Michaels, R. J. Olson, R. Proctor, N. Rini, C. D. Sweet, W. Tisinger, E. J. Winkley. GLAZING MATERIALS SUBCOMMITTEE—R. Morrison (Chairman), E. Edge (Vice-Chairman), H. M. Alexander, A. G. Attryde (Corresponding Member), J. P. Banks, A. J. Bartosic, J. G. Beck, J. G. J. Beke (Corresponding Member), J. W. Blumer, G. Bowen, C. Cameron, J. W. Freiburger, L. R. Gee, L. Haluska, G. M. Hespeler, D. J. Houston, O. Jandeleit (Corresponding Member), P. Knorr, H. Kunert (Corresponding Member), E. Lacey, R. LeBrecque, F. D. Lovett, P. T. Mattimoe, R. Molari, S. Mueller (Corresponding Member), S. Nelson, N. Nitschke, L. W. Parr, L. M. Patrick, P. J. Pennells (Corresponding Member), B. W. Preston, J. D. Ryan, D. J. Schrum, R. Shepard, R. B. Siegler, C. Silvester (Corresponding Member), Z. S. Subotich, F. Takenaka, T. Taormina, W. vonReis, R. Williams. LATCH AND HINGE SUBCOMMITTEE—H. Maruyama (Chairman), C. H. Lee (Vice-Chairman), D. C. Brown, K. W. Finch, A. J. Hammond, J. Kemp, J. W. Kennebeck (Corresponding Member), M. Monk, H. Reizes, D. J. Schrum. REAR VISION SUBCOMMITTEE—D. W. Barkley, D. Baumgardner, G. M. Craig, M. DeVries, R. E. DeWald, R. Donohue, R. Ealba, W. M. Elliott, W. J. Gibson, R. M. Hadley, D. K. Haenchen (Corresponding Member), R. L. Henderson, A. Irving (Corresponding Member), J. Knister, E. Krantz, W. Lim, V. L. Lindberg, F. D. Lovett, R. R. Mourant, C. Neuman, P. Olson, R. Perison, B. Preston, A. W. J. Prince (Corresponding Member), S. L. Reed, C. Silvester (Corresponding Member), K. H. Ziwica (Corresponding Member). DISTORTION TASK FORCE—D. Helder (Chairman), G. Craig, M. DeVries, M. Herman, R. E. LaBrecque, V. Lindberg, R. R. Mourant, S. Reed, R. Snider. TRUCK TASK FORCE—H. Boyce, N. Darmstadter, W. Gillies, J. E. Hadley, M. McHenry, C. Neuman, L. Seiler, P. L. Sherrick (Corresponding Member). RESTRAINT SYSTEMS SUBCOMMITTEE—J. Cantalupo (Chairman), J. F. Meldrum (Vice-Chairman), L. O. Baker, F. C. Booth, D. Braun (Corresponding Member), R. Brocklehurst, R. F. Chandler, F. M. Coffey, P. Cotter, C. G. Dewey, R. H. Fredericks, D. K. Haenchen, H. Hanson, W. E. Hering, G. M. Hespeler, T. Hrynik, H. G. Johannessen, W. A. Kohn, E. R. McKenna, R. J. Neff, V. T. Newton, L. M. Patrick, F. Pepe, F. Peters, G. H. Pope, C. H. Pulley, D. H. Robbins, R. A. Rogers, H. G. Roulea, T. K. Sandberg, D. J. Schrum, M. Schulman, R. J. Settimi, D. M. Severy, W. C. Shelton, R. E. Simpson, N. E. Shoemaker, C. Silvester (Corresponding Member), G. Smith, W. Smith, L. W. Schneider, R. G. Snyder, M. Stacks, J. P. Stapp, G. W. Tannahill, J. Versace. CHILDREN'S RESTRAINT SYSTEMS TASK FORCE—H. Willson (Chairman), E. Berndt, J. Brock, S. Davis, R. Fredericks, A. Frew, E. P. Grenier, V. G. Radovich, J. Smreker, E. Soderwall, M. Stacks, H. Wheeler, G. Wize. COMFORT, FIT, AND CONVENIENCE TASK FORCE—E. R. Berndt, F. Booth, L. Carlson, P. J. Cotter, B. J. Howard, P. Hundemer, H. G. Johannessen, W. A. Kohn, J. F. Meldrum, R. D. Osmond, J. L. Pearson, D. Perry, F. Peters, R. Pilarski, J. Podorsek, L. Schneider, W. Smith, J. Smreker, O. Soderwall. HARDWARE AND ASSEMBLY TASK FORCE—T. K. Sanderg (Chairman), R. J. Amorosi, E. L. Barcus, R. Black, F. Booth, C. G. Dewey, D. M. Dymond, R. P. Fisher, R. H. Fredericks, J. W. Freiburger, D. A. Garstecki, J. E. Glauser, T. Hrynik, G. D. Hunter, A. M. Irving, R. Jasinksi, W. A. Kohn, E. R. McKenna, F. Pepe, F. Razzovoli, L. Romanzi, R. Settimi, R. E. Simpson, R. Stewart, T. H. Vos. INFORMATION REPORT WORKING GROUP—G. J. Howard (Chairman), D. C. Hammond, J. F. Meldrum, J. Podorsek, R. Pilarski, J. Smreker. WEBBING TASK FORCE—R. J. Neff (Chairman), F. Booth, J. G. Brock, J. A. DeAngelis, G. M. Degiorgio, R. H. Dobson, R. M. Kemmerer, J. O. Martino, K. Motte, L. Romanzi, W. C. Shelton, R. Thurber, J. Turnbull, J. T. Wallag. SEATING SUBCOMMITTEE—E. Zipp (Chairman), L. O. Baker, W. Butterfield, D. C. Cazabon, J. Cherry, C. Derrickson, N. D'Souza, W. Eagleson, A. Freehan, D. R. Fruchte, R. T. Hampson, L. Heck, A. Heffelfinger, S. Hendricks, R. Karbowski, W. B. Kelley, H. Klueger, C. Maykuth, R. Murphy, N. Pocock, R. W. Russell, R. P. Sharpe, D. J. Van Kirk, P. Vogt, K. H. Ziwica. SEATING MANUAL TASK FORCE—A. Freehan (Chairman), K. A. Bhatt, W. Butterfield, L. Heck, N. Pocock, C. Reed, R. Schueler, P. Vogt. WINDSHIELD WIPING AND WASHING SUBCOMMITTEE—J. J. Griffin (Chairman), Z. K. Baranowski, L. E. Bischoff, F. Bohannon, K. E. Cavey, K. W. Finch, C. F. Finn, C. E. Fisher, A. L. Fuller, R. F. Jaske, T. C. Jensen, A. J. Kovaleski, R. B. Kurre, J. J. Lafter, C. E. Larson, P. J. Passon, J. Perkins, R. M. Premo, P. J. Roath, C. Silvester (Corresponding Member), C. Strobel, P. Waterman.

BRAKE COMMITTEE

R. W. HILDEBRANDT (Sponsor)
S. G. TILDEN, JR. (Chairman)
R. O. TUEGEL (Vice-Chairman)
E. T. Andrews
K. Aoki
A. G. Beier
T. A. Brandon
F. Brooks, Sr.
R. G. Brown
C. M. Brunhofer
G. L. Buyck
R. L. Doering
J. Fent

A. M. Fischer
R. A. Gallant
P. J. Garthe
J. Getz
F. H. Highley
L. Kiselis
B. W. Klein
H. C. Klein (Corresponding Member)
R. H. Madison
G. P. Mathews
D. W. Morrison
P. A. Myers
R. E. Nelson

W. J. Oakley
P. Oppenheimer (Corresponding Member)
F. W. Petring
P. Raves
J. M. Rowell
B. D. Sibley
G. W. Smith
J. Sorsche (Corresponding Member)
R. F. Stelzer
R. K. Super
R. B. Temple
K. R. Thibo
W. J. Zechel

(SAE Staff, L. P. Ziegler, Jr.)

BRAKE SUBCOMMITTEE 1—IN-SERVICE PERFORMANCE—G. L. Buyck (Chairman), J. Carlisle, W. Harris, R. Horvath, C. O. Jones, G. Jacoby, R. Lampsa, D. H. Mahannah, J. W. Miller, P. A. Myers, F. W. Petring, J. M. Sypniewski, C. R. Walker. PASSENGER CAR TASK FORCE—F. W. Petring (Chairman), B. B. Brombaugh, G. L. Buyck, R. Lampsa. TRUCK TASK FORCE—W. Harris (Chairman), C. Walker (Vice-Chairman), P. W. Baker, J. Carlisle, L. E. Carroll, W. E. French, L. Kelly, J. Miller, P. A. Myers, J. M. Sypniewski, K. R. Thibo, D. B. Wilson. BRAKE SUBCOMMITTEE 2—BRAKE LININGS—P. Raves (Chairman), G. B. Jacoby (Vice-Chairman), A. E. Anderson, K. F. Barnhardt, E. Bayus, T. J. Billings, B. B. Brombaugh, A. K. Carpenter, P. H. Dougherty, E. W. Drislane, A. Dunmore, R. W. Gibson, C. W. Greening, W. Harris, F. H. Highley, M. G. Jacko, E. Kakaley, R. W. Kickel, J. M. Knoll, R. Link, D. H. Mahannah, M. E.

Markert, W. W. Matthes, G. K. McCann, G. E. McMullen, J. Mikaila, R. E. Nelson, J. C. Nordman, D. W. Pitts, C. Rodziewicz, A. A. Stalkis, D. E. Steis, J. Stukenborg, S. G. Tilden, Jr., A. R. Wadum. BRAKE SUBCOMMITTEE 3—DYNAMOMETER TEST CODE—P. A. Myers (Chairman), W. C. Harris (Vice-Chairman), R. D. Bailey, R. H. Bauer, T. C. Bush, R. W. Clawson, O. B. Cruse, L. DelVecchio, R. H. Doan, R. E. Grambo, C. W. Greening, R. W. Griffin, R. E. Hannon, E. Kakaley, R. W. Kickel, J. M. Knoll, D. R. Krebs, R. P. Lord, W. W. Matthes, G. K. McCann, T. A. Michelhaugh, J. Mikaila, J. Moules, C. Newell, P. Raves, C. Sillin, D. E. Steis, E. E. Tuttle, W. J. Williams. BRAKE SUBCOMMITTEE 5—NOMENCLATURE AND METRICATION CONTROL—F. Brooks, Sr. (Chairman), R. E. Dix (Vice-Chairman), T. A. Brandon, J. L. Chalk, R. L. Doering, E. W. Drislane, L. S. Feldheim, C. T. Hoffman, Jr., R. H. Madison, M. Pastore, E. Rysso, R. K. Super, R. T. Smart. BRAKE SUBCOMMITTEE 6—PARKING BRAKES—B. D. Sibley (Chairman), B. G. Mazurek (Vice-Chairman), E. F. Beatty, B. J. Besso, R. Boughton, G. Choinski, J. B. Dale, M. E. Gatt, B. L. Hirsch, D. T. Jendrusch, L. H. Klauer, M. Kluger, J. P. Koenig, E. C. Lipshield, M. J. Locke, J. MacDougall, J. L. Pappin, G. C. Ploeg, R. Pfluger, A. H. Przepiora, J. L. Rust, J. J. Strbik, R. Swanson, J. R. Risley, T. F. Troester. BRAKE SUBCOMMITTEE 7—ROAD TEST PROCEDURES—P. J. Garthe (Chairman), T. F. Troester (Vice-Chairman), P. G. Marting (Secretary), C. Afanador, A. G. Beier, R. Bennett, B. B. Brombaugh, D. D. Brown, J. E. Byers, R. A. Chirakos, D. T. Courtright, O. B. Cruse, P. H. Dougherty, L. B. Duff, J. G. Einhorn, A. D. Foster, R. A. Gallant, D. Gross, W. C. Harris, M. C. Hayward, R. J. Hoffman, A. V. Hogan, J. W. Hulverson, D. W. Hunter, P. Jank, R. A. Joslin, R. E. Kennel, B. Latvala, M. J. Locke, D. H. Mahannah, E. N. Mattausch, W. W. Matthes, J. McKinley, R. E. Nelson, B. C. Ocker, M. A. Pastore, B. R. Prokop, R. W. Radlinski, D. I. Reed, D. Renner, P. M. Rock, R. Schoen, P. J. Soltis, P. B. Smith, T. I. Sheikh, W. L. Sherry, W. Simond, F. C. Skelton, D. E. Steis, R. B. Swanberg, D. L. Swanson, R. B. Temple, P. A. Thesier, J. L. Turak, J. A. Urban, E. L. Volker, C. R. Walker, A. D. Warner, D. B. Weinfurther, D. L. Williams, G. A. Whitcomb, C. R. Wuellner, W. J. Zechel. MOTOR VEHICLES 10 000 LB GVWR AND LIGHTER TASK FORCE 7a—F. Skelton (Chairman), J. L. Turak (Vice-Chairman), A. V. Hogan (Secretary), P. G Anthony, R. W. Bennett, B. B. Brombaugh, P. H. Dougherty, R. A. Gallant, P. J. Garthe, R. J. Hoffmann, J. W. Hulverson, D. W. Hunter, R. A. Joslin, M. J. Locke, M. A. Postore, R. W. Radlinski, D. Renner, P. M. Rock, W. Simon, P. B. Smith D. E. Steis, R. R. Swanberg, R. P. Temple, P. A. Thesier, T. F. Troester, E. L. Volker, D. L. Williams, C. R. Wuellner, W. J. Zechel. BRAKE WEAR TEST TASK GROUP—R. P. Temple (Chairman), P. H. Dougherty, W. H. Gillespie, R. E. Nelson, M. A. Pastore, P. B. Smith, R. E. Swanberg, J. L. Turak. MOTOR VEHICLES OVER 10 000 LB GVWR TASK FORCE 7b—P. A. Thesier (Chairman), E. L. Volker (Vice-Chairman), W. C. Harris (Secretary), D. T. Courtright, O. B. Cruse, L. B. Duff, J. G. Einhorn, D. W. Hunter, P. Jank, R. A. Joslin, R. E. Kennel, W. W. Matthes, J. McKinley, R. W. Radlinski, D. Renner, T. I. Sheikh, W. Simon, D. E. Steis, T. F. Troester, C. R. Walker, G. A. Whitcomb. TRAILER WITH TOW VEHICLE 10 000 LB GVWR AND LIGHTER TASK FORCE 7c—D. Reed (Chairman), W. L. Sherry (Vice-Chairman), J. A. Abromavage, A. G. Beier, S. Borden, D. D. Brown, R. A. Chirakos, E. P. Conradi, A. D. Foster, D. Gross, E. N. Mattausch, B. R. Prokop, D. Renner, R. B. Swanberg, D. L. Swanson, R. G. Tschabrun, A. D. Warner, D. B. Weinfurther, C. R. Wuellner. LOW DUTY TRAFFIC PROCEDURE WORKING GROUP—P. Thesier (Chairman), D. Haugen, B. E. Kirkham, M. Surbrook. BRAKE SUBCOMMITTEE 10—WHEEL SLIP BRAKE CONTROL SYSTEMS—J. Getz (Chairman), R. C. Bueller, R. L. Davis, J. W. Douglas, D. R. Elliott, J. J. Goestenkors, R. A. Grimm, D. Gross, J. B. Hageman, R. Hodges, J. M. Holden, M. Hutchins, H. L. McCord, G. Meggison, D. C. Marriman, W. H. Morse, J. C. Neisch, J. M. Rowell, R. W. Rowell, D. B. Sibley, R. A. Wagner.

COMMITTEE FOR AUTOMOTIVE RUBBER SPECIFICATIONS (CARS)

R. J. SLINGERLEND (Chairman)
R. Alig
R. J. Arabia
F. Barlow
L. Bradford
J. H. Bramley
R. C. Bremer, Jr.
A. B. Calhoun
D. Coz
R. A. Davis
F. N. DeMott
J. R. Dillhoeffer
C. R. Dingess
C. Doney
J. R. Dunn
R. C. Edwards
S. Flanders
E. P. Francis

E. H. Gibbs, Jr.
W. B. Green
C. E. Heckert
P. D. Hinckley
R. F. Hinderer
G. Ilkka
E. N. Ipiotis
D. F. Kruse
J. Laslo
W. G. Levans
J. J. Leyden
F. Lohr, Jr.
M. Lowman
L. D. Malone
L. Marlin
L. R. Mayo
W. J. McCortney
J. T. McIntyre

D. G. McLeod
A. A. McNeish
G. C. Norman
W. W. Paris
L. S. Porter
J. R. Ruff
S. Schafer
E. L. Scheinbart
D. Scheufler
R. P. Schmuckal
D. A. Seil
N. F. Silver
W. J. Snodden
J. Stringfellow
R. Strong
E. W. Thomas
H. F. Trommer
C. D. Wemmers

(SAE Staff, R. T. Northrup)

CONSTRUCTION, AGRICULTURAL, AND OFF-ROAD MACHINERY SOUND LEVEL TECHNICAL COMMITTEE

L. D. BERGSTEN (Chairman)
C. C. GODSEY (Vice-Chairman)
R. Bartholomae
C. W. Farmer
J. D. Harris

R. H. Hoerr
J. Hoeschen
R. F. Holton
F. M. Kessler
B. Lindgren

A. D. Loken
R. I. Myers
P. D. Schomer
F. Walters

(SAE Staff, R. T. Northrup)

AGRICULTURAL MACHINERY SOUND LEVEL SUBCOMMITTEE—J. D. Harris (Chairman), F. Walters (Vice-Chairman), W. R. Bryant, R. N. Coleman (Liasion), M. H. Ihnen, R. Job, G. J. Leech, L. I. Leviticus, R. L. Mann, L. W. Randt, D. Walker. CONSTRUCTION SITE SOUND LEVEL SUBCOMMITTEE—F. Kessler (Chairman), P. D. Schomer (Vice-Chairman), L. D. Bergsten, J. B. Codlin, J. D. Harris, R. F. Holton, G. Kamperman, H. T. Larmore, C. Sanders, S. Stempler, C. E. Warren, G. Wildey, W. A. Wood. EARTHMOVING MACHINERY SOUND LEVEL SUBCOMMITTEE—R. I. Myers (Chairman), A. D. Loken (Vice-Chairman), R. F. Holton (Secretary), L. D. Bergsten, R. M. Blythe (Observer), W. J. Cowan (Observer), J. Gentille (Observer), C. C. Godsey, J. D. Harris, M. Hawke (Observer), J. Hoeschen, J. Imes (Observer), J. Inhofer, G. W. Kamperman, G. J. Leech, E. O'Keefe, N. T. Putman, W. Smith, D. L. Spreitzer (Observer), G. A. Stangl, D. Tesar (Observer), D. Yokich. ISO SOUND LEVEL DOCUMENTS SUBCOMMITTEE—R. F. Holton (Chairman), R. J. Nelissen (Vice-Chairman), L. D. Bergsten, A. D. Loken, G. J. Leech, R. I. Myers, D. E. Wolbers. SOUND LEVEL MEASUREMENT VARIABILITY SUBCOMMITTEE—B. Lindgren (Chairman), M. Alexander, M. Cocking, C. Everett, W. Flint, D. O'Brien. WARNING HORNS SOUND LEVEL SUBCOMMITTEE—R. H. Hoerr (Chairman), G. R. Gray (Vice-Chairman), C. C. Everett, B. Geidraitis, R. I. Myers, J. A. Neese, M. L. Ruther, M. Wieland.

ELECTRIC VEHICLE TECHNICAL COMMITTEE

W. G. AGNEW (Sponsor)
C. C. CHRISTIANSON (Chairman)
F. T. DeWolf
A. F. Dicker

W. H. Koch
J. Lunan
N. A. Richardson
J. T. Salihi

J. Werth
M. C. Yew
F. M. Petler (Liaison)

(SAE Staff, L. P. Ziegler, Jr.)

ELECTRICAL EQUIPMENT COMMITTEE

B. H. BOUWKAMP (Sponsor)
F. BAUER (Chairman)
R. E. KIRK (Vice-Chairman)
C. E. Bates
J. W. Brodhacker
A. J. Burgess
G. L. Cameron
R. J. Craver
D. Droste

W. M. Elliott
J. D. Fobian
J. T. Hardin
N. S. Hatch
R. E. Heller
R. W. MacKay
J. D. Marks
E. W. Meyer
J. W. Mueller

J. W. Quinlan
T. Shewchuck
J. Simeroth
E. J. Szabo
O. Taraborrelli
N. J. Van Halteren
T. Varga

(SAE Staff, P. Couhig)

CIRCUIT PROTECTION AND SWITCHING DEVICES SUBCOMMITTEE—T. Shewchuck (Chairman), R. A. Carlson (Vice-Chairman), J. B. Andrews (Alternate), K. R. Audino, W. D. Bragdon, P. M. Byam (Alternate), J. R. Canaday (Alternate), W. H. Curtis, J. R. Dawson, H. J. Ewald, J. Gulau (Alternate), N. S. Hatch, A. J. Kochanski, M. Koler, J. A. Krehbiel (Alternate), E. A. Lau, R. N. Lentes, W. M. Lorenz, W. A. Mathews (Alternate), J. B. Meister, W. E. Melville (Alternate), O. R. Nichols, L. C. Nyman (Alternate), R. Pribish, R. G. Rhoads, J. Salliotte, G. H. Simpson, (Alternate), G. Timoff (Alternate), M. Uecker, R. Van Kampen (Alternate), R. Young, W. L. Zurbriggen, G. Zulauf. FUSE TASK FORCE—W. H. Curtis (Chairman), H. Anderson (Alternate), K. R. Audino, M. Bezusko, R. W. Crum, N. S. Hatch, W. A. Mathews (Alternate), L. Misewicz, J. W. Mueller (Alternate), D. R. Patchett, J. L. Read (Alternate), R. Walz (Alternate). CRANKING MOTOR SUBCOMMITTEE—G. L. Cameron (Chairman), J. Augustine, A. J. Barber, D. G. Baugher, R. P. Bowcott (Corresponding Member), J. Cerano, K. Grovos, C. M. Hayes, J. L. Kruse, H. R. Mortensen. ELECTRICAL DISTRIBUTION SYSTEMS SUBCOMMITTEE—N. S. Hatch (Chairman), J. W. Mueller (Vice-Chairman), R. Aikins (Alternate), R. L. Blumer, M. B. Brandt, E. Bullis, A. J. Burgess, E. E. Gough (Alternate), J. V. Hemp, L. T. Horvath, Z. Hotra, R. C. Joshi, V. Karabcievschy, F. C. Keller (Alternate), J. M. Keller (Alternate), S. E. Koehl (Alternate), T. F. Langston, W. G. Lippert, K. McBroom (Alternate), C. W. McClain, D. R. Patchett (Alternate), L. R. Phillippi (Alternate), R. H. Rose (Alternate), J. W. Simeroth, L. D. Stone (Alternate), L. Susalla (Alternate), R. Szudarek, F. J. Tavarozzi, T. B. Thomas (Alternate), J. M. Vitale, B. R. Weber, B. C. Williams (Alternate), W. B. Williams, P. Witwer (Alternate), H. L. Woolf, E. L. Yochum, R. B. Young. BATTERY BOOSTER CABLE TASK GROUP—J. M. Vitale (Chairman), R. L. Blumer, R. Casey, J. D. Fobian, N. S. Hatch, J. V. Hemp, R. C. Joshi, J. W. Mueller, R. H. Rose, H. L. Woolf, E. L. Yochum. CONNECTORS TASK FORCE—C. Hurton (Vice-Chairman), W. F. Abbott, T. Cairns, R. Kennedy, M. S. Mamrick, D. O'Neal, D. R. Patchett, G. T. Ritter, L. Wall, R. G. Yorks. LOW TENSION TERMINALS TASK FORCE—F. J. Tavarozzi (Chairman), J. R. Dawson, Z. Hotra, V. Karabcievschy, R. D. Kearns, D. A. King, T. F. Langston, J. W. Mueller, E. L. Yochum. ELECTRICAL GENERATING SYSTEMS SUBCOMMITTEE—T. J. Varga (Chairman), A. Alexander (Alternate), R. Arora, F. J. Berger, J. Cerano, T. D. Drennon, R. J. Entwisle, J. G. Giacobino, R. Hojna (Alternate), C. J. Juhnke, L. Leaf, D. Santis. EMISSIONS RELATED ELECTRICAL DEVICES SUBCOMMITTEE—J. D. Marks (Chairman), J. E. Acker, D. A. Brockman, L. P. Carrion, R. J. Craver, T. E. Evernham, E. E. Gough, W. J. Hallauer, B. Harvey, T. L. Heffelfinger, P. Hubbard, E. R. Larges, G. E. Rigsby. IGNITION SUBCOMMITTEE—J. T. Hardin (Chairman), R. J. Craver, D. O. Crim, K. Denneler, P. J. DePaulis, R. E. Elrod, G. G. Gast, D. G. Guetersloh, W. J. Hallauer, R. Handy, N. S. Hatch, G. B. Herzog, K. D. Kobelentz (Alternate), E. R. Larges, E. Late, J. McKian, R. H. Rose (Alternate), G. E. Roy, L. D. Stone, E. L. Yochum. SPARK PLUG RATING ENGINE STANDARDIZATION PANEL—V. A. Nawrocki (Chairman), L. Campbell, R. Elrod, J. Leptich, R. Taylor, F. Westenkirchner. IGNITION MEASUREMENTS WORKING GROUP—D. O. Crim (Chairman), O. J. Coletti (Alternate), P. J. DePaulis, K. D. Kobelentz, E. R. Larges, E. E. Late, J. R. Lennis, R. Sieja. RADIO FREQUENCY INTERFERENCE SUBCOMMITTEE—F. Bauer (Chairman), P. H. Andersen (Vice-Chairman), P. Tharman (Secretary), J. Adams, C. Bandera, R. E. Bell, R. W. Bensman, N. H. Berry, Jr., N. H. Berry, Sr., J. V. Bienkowski, D. J. Brown, W. J. Clauss, J. Deitz, E. Dydo, N. Fabian, J. M. Gloudemans, G. R. Hofmann, W. A. Kesselman, W. A. King, J. Klouda, G. A. Koch, A. R. Lavis, L. Lodge, E. McBrien, D. McGrew, F. Quail, B. Radin, D. J. Rutkowski, T. Rybak, N. H. Shepherd, J. Smreker, R. L. Sprague, R. R. Steiner, R. Szudarek, H. E. Taggart, D. L. Tribble, J. L. Vajgart, E. L. Williams, D. H. Wood, G. H. Woodward, T. C. Young. STORAGE BATTERY SUBCOMMITTEE—C. E. Bates (Chairman), H. F. Andrews, R. Bennett, O. C. Birt, J. W. Brodhacker, J. R. Duwve, R. E. Eddy, G. G. Gast, H. Harback, C. W. Hill, L. Holden, H. Hutz, F. C. Keller, C. H. Leaf, F. Lee, F. J. Lysaght, R. W. MacKay, J. H. Miller, J. R. Pierson, J. Reinman, C. J. Van Halteren. J240 TASK FORCE—J. R. Duwve (Chairman), P. E. Baird, C. E. Bates, O. C. Birt, H. R. Ellis, G. G. Gast, R. E. Hoffman, A. Howard, H. Hutz, F. Lee, R. W. MacKay, R. MacKenzie, U. W. Peters, E. M. Valeriote, C. J. Van Halteren, W. F. Werzner. TAPE CARTRIDGE SUBCOMMITTEE—O. Taraborrelli (Chairman), A. Magic (Vice-Chairman), A. Bank, B. L. Bezenek, R. Garretson, R. H. Hamilton, E. Hanson, G. Hull (Alternate), J. E. Jackson, S. Kessler, J. W. Weber.

ELECTRONIC SYSTEMS COMMITTEE

D. S. POTTER (Sponsor)
G. B. ANDREWS (Chairman)
J. F. ZIOMEK (Vice-Chairman)
F. Bauer
J. F. Cassidy
M. L. Crawford
W. M. Elliott
R. G. Fenske
M. E. Hartz
G. B. Herzog
T. H. Horrell

J. B. King
M. Kutzin
R. H. Lewis
M. S. Mamrick
O. T. McCarter
E. W. Meyer
L. L. Nagy
K. Niemi
F. F. Oettinger
R. H. Parker
C. F. Raiss

J. G. Rivard
D. E. Robb
T. L. Schaller
D. R. Sendelbach
K. S. Shanmugam
D. Taylor
W. C. Troll
W. G. Wolber
F. Zeisler

(SAE Staff, P. Couhig)

DESIGN INFORMATION GUIDES SUBCOMMITTEE—F. Zeisler (Chairman), D. A. Brockman, R. Cohen, W. F. Datwyler, A. G. Gillund, M. E. Hartz, R. L. Kerrigan, J. B. King, B. Kirk, M. Kutzin, L. A. Lopez, J. W. McNamee, R. L. McReynolds, P. Recupito, H. Roth, J. B. Russell, R. Silco. DIAGNOSTICS SUBCOMMITTEE—R. H. Lewis (Chairman), R. Baudendistel, J. Beck, D. A. Brockman, R. G. Bunnell (Alternate), A. K. Chang, R. E. Citerone, W. L. Doelp, S. Groves, M. E. Hartz, T. F. Hornung, A. Joniec, J. B. King, E. D. Lowell, J. A. Marino, F. Newton, M. F. Quinn, J. L. Sachs, A. Santaruga (Alternate), H. Schmitt, M. J. Sinko, J. G. Sutton (Alternate), R. Volk (Alternate). ELECTRONIC DISPLAYS SUBCOMMITTEE—D. R. Sendelbach (Chairman), E. M. Arakelian, L. Baber, D. I. R. Bitetti (Corresponding Member), K. H. Brunkmann (Corresponding Member), W. H. Daniels, R. DuBois, D. L. Evans, M. Guimond, C. Leady, L. A. Lopez, J. Siegel, C. Slupek, E. Strandt, C. E. Thompson, B. H. Van Vlack, J. Wenzel, R. West, M. H. Westbrook (Corresponding Member). EMI STANDARDS AND TEST METHODS SUBCOMMITTEE—M. L. Crawford (Chairman), C. Lowman (Vice-Chairman), F. Miesterfeld (Secretary), G. B. Andrews, J. A. Bijon, N. R. Brainard, E. L. Bronaugh, J. Cacciatore, P. Fanson, W. R. Faris, P. Franklin (Alternate), V. P. Freay, J. G. Harvey, H. P. Hsu, J. G. Olin, F. Pagano, E. R. Pezon, L. Rockwell, J. Severinsen (Alternate), J. Shragal (Alternate), W. Sperber, G. Vrooman, T. J. Waraksa, S. P. Wetter, ENVIRONMENTAL STANDARDS AND TEST METHODS SUBCOMMITTEE—D. E. Robb (Chairman), P. H. Andersen, J. P. Butler, H. J. Ciolli, K. W. Doversberger, W. C. Fahling, G. G. Gast, C. E. Harrison, R. B. Hood, A. Humble, F. M. Jenkins, L. L. Kent, W. G. Pracejus. MICROPROCESSORS SUBCOMMITTEE—M. Kutzin (Chairman), R. LaJeunesse (Vice-Chairman), G. Boone, D. Carley, W. L. Densham, T. J. Flis, S. Groves, M. E. Hartz, C. Howanski, M. Kitchen, J. McCormich, F. Meisterfeld, J. Olmstead, R. Sabourin, W. Seitz, R. Smith, S. Stonestreet, B. T. Sullivan, F. Thorley, R. G. Yorks. MICROWAVE SYSTEMS SUBCOMMITTEE—L. L. Nagy (Chairman), E. Belohoubek (Vice-Chairman), J. H. Bryant, R. J. Ellis, J. Wewett, L. W. Roberts, R. Spence, F. Sterzer, R. M. Storwick, W. Troll, Y. Wu. RELIABILITY SUBCOMMITTEE—J. B. King (Chairman), C. Barrett, W. Binroth, C. C. Chandler, D. Denton, K. W. Doversberger, W. Fitch, S. Flint, W. Frick, W. Full, G. G. Gast, H. Haerer, J. Howell, J. Jennings, T. Kelly, L. L. Kent, D. S. McAlexander, M. Murray, G. Oltmann, A. Seitz, J. Stanley,

J. Thielmann, E. Thomas, J. E. Thompson, T. Van Presson, D. Van Sicklen, K. Weissert. TRANSDUCERS SUBCOMMITTEE—W. G. Wolber (Chairman), D. McNamara (Secretary), L. Carrion, J. W. Crow (Alternate), W. F. Datwyler, B. H. Drill, R. W. Eshelman, A. C. Fielek, R. K. Frank, R. B. Hood, J. Lappington, R. L. Louisignau (Alternate), J. T. Maupin, D. Peters, J. N. Reddy, M. J. Sinko, W. Sousa, H. Wertheimer.

ENGINE COMMITTEE

C. MARKS (Sponsor)
R. J. KLEIN (Chairman)
C. A. FORDHAM (Vice-Chairman)
N. Alvis
T. W. Asmus
A. Carey
D. Cather
W. B. Clemmens
D. H. Connor
N. Cross
J. Decker
J. Ferguson
R. L. Good
R. J. Griffiths
S. J. Hoot
E. G. Jacobsen
L. Matson
G. Millard
F. Piech
W. E. Schwieder
H. J. Steinhagen
C. D. Wink
R. H. Whiteside
K. Yamaoka
C. S. L. Yee

(SAE Staff, N. T. Northrup)

AIR CLEANER TEST CODE SUBCOMMITTEE—R. Florine (Chairman), D. Belden (Vice-Chairman), B. Reber (Secretary), J. H. Allen, A. H. Anderson, A. Caronia, R. Dobransky, J. Dreznes, F. Egberts, R. L. Foster, R. Hall, R. Hoagland, U. E. Holzhausen, W. Holtzman, L. M. Hough, J. W. Johnson, P. Labadie, G. Lawson, R. Mayer, R. McGillick, J. Mohling, D. Moser, G. Olson, L. G. Ponce, G. Rauscher, B. Reber, C. Reinhardt, W. Risse, A. Ringel, R. E. Sendelback, G. Slater, D. Tortorici, K. Williamson, J. Woidke. AUTOMOTIVE AIR CLEANER TASK FORCE—D. Belden (Automotive Coordinator), J. H. Allen, A. H. Anderson, M. Cadigan, A. Caronia, H. Dahlstrom, R. Florine, R. L. Foster, R. Hall, W. Holtzman, L. M. Hough, D. Moser, W. L. O'Neill, G. Paquette, B. Reber, R. E. Sendelbach, F. Weglarz. INDUSTRIAL AIR CLEANER TASK FORCE—G. Slater (Industrial Coordinator), J. H. Allen, D. W. Beach, C. Broad, J. Dreznes, F. Egberts, W. Holtzman, U. E. Holzhausen, J. W. Johnson, G. Jones, P. Labadie, G. Lawson, P. Lutz, R. Mayer, B. McKernan, J. Mohling, G. Paquette, J. B. Peterson, W. H. Risse, G. Slater, D. Tortorici, J. Tsengouras, J. Workman. BEARINGS SUBCOMMITTEE—A. O. DeHart (Chairman), A. A. Aalto, A. Body, T. E. Cavender, P. Das, R. W. Ewing, M. J. Haddad, D. J. Grasso, J. D. Jones, S. Mori, J. Mydlarz, A. Rike, R. W. Sutliff, E. J. Vargo. CLUTCH, FLYWHEEL, AND HOUSING SUBCOMMITTEE—J. G. Steinhagen (Chairman), J. L. Bair, P. H. Brandes, L. D. Carufel, W. Cross, J. M. Landis, D. F. Linn, D. E. McIntyre, J. T. Mehne, C. Noble, D. Piper, J. Rathburn, L. Traxson, O. B. Turkyilmaz, E. G. Vogl, E. P. Welcher, D. L. Williams. COOLING SYSTEMS SUBCOMMITTEE—G. Millard (Chairman), J. H. Steller (Vice-Chairman), B. Carkin, S. Cavanaugh, M. G. Clingerman, W. Crute, W. P. Hollenbeck, W. L. Howard, R. Jacobsen, W. V. Johansen, T. H. Liedel, D. J. Lineberry, J. H. Miller, J. D. Morse, J. R. Pharis, E. J. Rambie, F. Rising, A. Schaefer, E. J. Schlaffer, R. L. Schreiner, R. E. Seekins, K. K. Spence, D. Thekkanath, W. L. Was, I. D. Woodhull, B. Wong, C. S. L. Yee. COOLING SYSTEM TASK FORCE—E. J. Schlaffer (Chairman), M. G. Clingerman, J. R. Pharis, F. Rising, R. L. Schreiner, J. H. Steller, C. S. L. Yee. DIESEL FUEL INJECTION EQUIPMENT SUBCOMMITTEE—S. L. Gaal (Chairman), V. Lang (Secretary), L. Bailey, E. Bluhm, R. L. Bodnar, C. V. Brack, R. J. Brager, A. J. Burgess, D. Butterfield, R. A. DeKeyser, F. Deluca, A. S. Lamb, S. L. Langlie, D. Niemeyer, W. Parrish, C. Rebhahn, M. Schlichtmann, A. Smith, K. Suzuki, B. Thorpe, G. D. Wolff. DIESEL FUEL INJECTION EQUIPMENT SUBCOMMITTEE WORKING GROUP 1—R. J. Brager (Chairman), L. A. Bailey, C. V. Brack, A. J. Burgess, D. Butterfield, S. L. Gaal, V. Lang, D. Niemeyer, C. Rabhahn, M. Schlichtmann, A. Smith, B. Thorpe. DIESEL FUEL INJECTION EQUIPMENT SUBCOMMITTEE WORKING GROUP 2—D. Niemeyer (Chairman), E. Bluhm, D. Butterfield, R. A. DeKeyser, S. L. Gaal, W. Parrish. DIESEL FUEL INJECTION EQUIPMENT SUBCOMMITTEE WORKING GROUP 3—R. A. DeKeyser (Chairman), L. A. Bailey, A. J. Burgess, D. Butterfield, S. L. Gaal, S. L. Langlie, C. W. Novak, M. Schlichtmann, A. Smith, G. D. Wolff. DIESEL FUEL INJECTION EQUIPMENT SUBCOMMITTEE WORKING GROUP 4—G. D. Wolff (Chairman), S. L. Gaal, A. S. Lamb, S. L. Langlie, A. Smith, B. Thorpe. ELECTRONIC GOVERNING STUDY GROUP—M. Trenne (Chairman), N. Alvis, S. Carter, R. J. Griffiths, R. Tadlock, L. Tomczak, T. Wolanzyk. ENGINE COOLING FAN STRUCTURAL ANALYSIS SUBCOMMITTEE—C. S. L. Yee (Chairman), L. Adams, M. Bennet, R. Jacobsen, F. Krey, B. McKinney, T. Osborn, G. Peterson, R. Rising, N. Robb, J. Smith, L. Thompson, M. Van Galder, R. Willingham. EXHAUST BACKPRESSURE MEASUREMENT TEST PROCEDURE STUDY GROUP—D. K. Smith (Chairman), C. A. Anderson, R. Arbizzani, D. A. Blaser, W. B. Clemmens, G. Danielson, B. Fallon, A. J. Gebeau, J. T. Guillemette, R. K. Hillquist, B. A. Hootman, J. P. Kennedy, D. S. Klonowski, H. E. Koontz, W. D. McBain, R. C. Northrup, C. Shepherd, T. A. Warlick, E. Wyckoff. EXHAUST SYSTEMS SUBCOMMITTEE—C. D. Shepherd (Chairman), R. Arbizzani, D. Bentley, I. (Van) Cudini, F. Hubbell, G. House, C. J. Miller, R. C. Pollack, B. A. Rosa, J. J. Shaughnessy. FILTER TEST METHODS SUBCOMMITTEE—M. Cadigan (Chairman), A. Anderson, A. Caronia, A. Crumrine, W. Dallas, R. L. Foster, J. Gehrking, R. L. Hall, R. M. Hardt, D. Hodgkins, U. E. Holzhausen, L. M. Hough, G. C. Lawson, R. McGillick, D. Moser, Jr. L. F. Niebergall, W. G. Nostrand, L. G. Ponce, G. Puia, G. Rauscher, D. D. Roberson, L. E. Rupert, A. F. Vijlee, F. Weglarz. LOCOMOTIVE SPARK ARRESTERS SUBCOMMITTEE—L. U. DeBernardo (Chairman), R. S. Farr, D. J. Goding, F. N. Harris, C. Jursch, P. Labadie, C. W. Novak. PISTON AND PISTON RING SUBCOMMITTEE—K. W. Thurston (Chairman), R. D. Anderson, H. Becker, R. Burgette, G. Glover, W. Groves, W. P. Holcombe, T. R. Hoopes, B. Martin, D. L. Rex, L. H. Saylor, E. Schwartz, S. Wood. POWER TEST CODE SUBCOMMITTEE—A. Carey (Chairman), L. J. Billington, P. Carlson, A. Hogan, C. J. Hutson, H. MacInnes, C. E. McInerney, O. M. Smith, M. H. Suter. SMALL ENGINE SUBCOMMITTEE—G. C. Hardwick (Chairman), M. G. Adams, I. Barker, L. J. Bernauer, J. A. Donohue, J. A. Gresch, J. Mieritz, D. G. Owens, W. D. Schultz, S. Seidlitz, D. Williamson, N. Woelffer.

FASTENERS COMMITTEE

W. M. ROLL (Sponsor)
H. W. ELLISON (Chairman)
A. G. Baustert
R. B. Belford
A. A. Bien
A. R. Breed
R. M. Byrne
D. B. Carroll
G. L. Cowing
W. J. Derner
W. W. Dodson
T. S. Doppke
A. W. Gair
D. A. Garrison
D. H. Gill
G. A. Gobb
F. E. Graves
R. W. Hayden
T. P. Hurst
J. Karagozian
R. S. Knecht
F. A. Kocian
J. F. Koenigshof
L. M. Lalik
C. S. Larson
S. E. Mallen
H. G. Muenchinger
A. G. Nauta
R. S. Piotrowski
C. F. Schaening
C. J. Schim
G. Schremmer
H. B. Schweppe
J. Passon
L. R. Strang
E. J. Streichert
T. E. Urich
D. W. Vial
R. A. Walker
C. W. Wilson

(SAE Staff, P. Couhig)

BOLT AND NUT SUBCOMMITTEE—R. B. Belford (Chairman), G. L. Cowing, W. J. Derner, G. R. Dorset, A. J. Nauta, R. S. Piotrowski, L. R. Strang. CLEVIS AND CLEVIS PIN SUBCOMMITTEE—C. J. Schim (Chairman), W. Bacon, W. Bodily, R. Brunner, P. Chiesa, C. Donahue, H. G. Fiedler, D. Gill, A. G. Nauta, P. Pelli, C. F. Schaening. HOSE CLAMP SUBCOMMITTEE—V. A. Rusnack (Chairman), C. P. Coldren, C. W. Collins, J. V. Elias, J. Freedman, L. H. Forman, J. F. Kelman, P. Kern, H. P. McKowan, T. H. Miller, C. F. Schaening, R. Thurston. SLOTTED AND RECESSED HEAD SCREW SUBCOMMITTEE—C. F. Schaening (Chairman), E. R. Baugh, R. M. Byrne, D. W. Vial. SPRING NUT SUBCOMMITTEE—D. H. Gill (Chairman), A. A. Bien, G. Clay, R. Derby,

W. Duffy, J. Freedman, G. A. Gobb, G. Lucas, R. A. Munse, W. L. Seitz. STAMPED NUT SUBCOMMITTEE—A. A. Bien (Chairman), W. B. Duffy, D. H. Gill, G. Gobb, R. Schmidt. TECHNICAL PROGRAM SUBCOMMITTEE—C. W. Wilson (Chairman), R. B. Belford, D. R. Bowen, R. M. Byrne. TORQUE-TENSION RELATIONSHIP SUBCOMMITTEE—G. A. Gobb (Chairman), A. G. Baustert, R. B. Belford, A. A. Bien, W. R. Chandler, H. W. Ellison, R. J. Erisman, A. W. Gair, T. P. Hurst, F. R. Kull, D. McCrindle, R. S. Piotrowski, K. G. Roth. BLIND RIVET SUBCOMMITTEE—D. H. Gill (Chairman), Q. C. Achuff, R. B. Belford, A. A. Bien, R. R. Deans, R. G. Hawes, T. Rowley, C. J. Schim, W. D. Wedgewood. WASHER SUBCOMMITTEE—J. Freedman (Chairman), C. Coldren, P. A. D'Anza, E. R. Freisth, W. A. Rossiter, R. Scherler, G. Schremmer, G. Strumbos, D. W. Vial, D. Wagner.

FATIGUE DESIGN AND EVALUATION STEERING COMMITTEE

H. R. JAECKEL (Sponsor)
M. P. SEMENEK (Chairman)
T. Bernard
H. D. Berns
D. Dabell
N. E. Dowling
D. R. Galliart

W. F. Hofmeister
R. W. Landgraf
J. A. Larson
H. Mindlin
J. Morrow
M. D. Morton
D. V. Nelson

T. E. Parker
R. I. Stephens
L. E. Tucker
R. E. Wahlstrom
R. D. Zipp

(SAE Staff, R. T. Northrup)

DESIGN HANDBOOK DIVISION—H. D. Berns (Chairman), C. F. Barrett, B. E. Boardman, W. R. Brose, R. K. Brown, S. L. Bussa, D. A. Campbell, P. E. Cary, C. Confer, A. Conle, B. E. Coursin, B. Dabell, H. Demink, T. Dilger, K. Donaldson, N. E. Dowling, W. P. Evans, H. O. Fuchs, D. R. Galliart, J. A. Graham, W. Hayes, W. F. Hofmeister, R. D. Isaacson, H. R. Jaeckel, K. L. Jerina, H. C. Johnson, K. V. Johnson, T. M. Johnson, R. E. Jones, G. Keller, R. W. Landgraf, J. A. Larson, R. Leever, M. D. Mathers, J. McKittrick, G. A. Miller, H. Mindlin, M. R. Mitchell, J. Morrow, M. D. Morton, M. Naujoks, D. V. Nelson, C. C. Osgood, T. E. Parker, F. D. Richards, R. E. Ricklefs, D. J. Romero, M. P. Semenek, M. R. Shah, S. W. Shin, N. Slack, D. Socie, R. I. Stephens, D. R. Storjohann, R. A. Testin, R. B. Thakkar, A. J. Titus, D. Vukovich, R. E. Wahlstrom, E. D. Walker, G. H. Walter, E. M. Wene, J. Wert, R. A. Wilde, R. B. Wilson, R. D. Zipp. FATIGUE LIFE PREDICTION AND VERIFICATION DIVISION—D. R. Galliart (Chairman), L. E. Alban, L. H. Ames, P. G. Anderson, P. Artwohl, S. A. Avramidis, F. K. Baldauf, T. V. Baughn, R. T. Bennett, H. D. Berns, B. Boardman, C. W. Booth, R. F. Brodrick, W. R. Brose, R. Bucci, R. W. Bunneke, O. Buxbaum, M. H. Camp, H. L. Chesney, T. M. Clarke, C. H. Confer, A. Conle, L. F. Counter, B. Coursin, B. Dabell, E. P. Dahlberg, R. DeWit, T. J. Dolan, K. Donaldson, N. E. Dowling, R. G. Dubensky, H. L. Dunegan, C. Feltner, F. E. Fox, H. O. Fuchs, J. P. Gallagher, J. George, J. A. Graham, J. W. Grant, D. Hall, D. R. Hartdegan, W. C. Herbein, H. Bilberry, W. F. Hofmeister, F. R. Holliday, J. M. Holt, J. A. Horwath, R. Hunter, G. Jacoby, H. Jaeckel, K. L. Jerina, T. M. Johnson, R. E. Jones, G. W. Keller, K. J. Kennely, L. Koch, R. Krzywanos, P. Kurath, H. S. Lamba, R. W. Landgraf, R. Langham, J. Larson, J. A. Larson, A. F. Lawrence, R. C. Leever, J. Lukasik, C. H. Luther, R. A. Mantz, J. Martin, E. Matheny, G. Mauritzson, D. A. Miller, G. A. Miller, H. Mindlin, M. R. Mitchell, J. Morrow, M. D. Morton, D. Nelson, D. V. Nelson, C. Osgood, T. E. Parker, N. C. Pavlinac, J. Potter, H. S. Reemsnyder, R. C. Rice, D. J. Romero, W. L. Ropp, T. Russell, R. H. Sailors, G. Sandoz, M. A. Schober, M. P. Semenek, M. R. Shah, E. C. Sheets, P. N. Sheth, S. Shin, G. Singh, N. L. Slack, G. L. Smith, D. F. Socie, R. I. Stephens, D. Storjohann, R. A. Testin, R. B. Thakkar, W. A. Thomas, A. J. Titus, T. H. Topper, L. E. Tucker, D. Vukovich, G. H. Walter, E. M. Wene, T. Wermimont, G. A. Wilkinson, R. B. Wilson, S. W. Wohlfort, G. Yeager, T. C. Zebehazy, R. D. Zipp. MATERIAL PROPERTIES AND PROCESSING ANALYSIS DIVISION—H. Mindlin (Chairman), D. Aichbhaumik, L. E. Alban, R. Arnett, J. Au, S. A. Avramidis, F. K. Baldauf, C. Barrett, G. Bartlett, D. L. Baughman, R. R. Biederman, D. I. Biehler, B. E. Boardman, E. A. Borch, R. F. Brodrick, R. Bunneke, H. D. Berns, Y. F. Bilimoria, H. Burrier, J. T. Cammett, L. Campana, D. A. Campbell, P. E. Cary, R. A. Cellitti, J. B. Cohan, D. L. Coovert, B. Coursin, M. DiBernardo, N. E. Dowling, R. G. Dubensky, J. R. Eagan, T. M. Elsesser, T. Ericksson, W. P. Evans, P. A. Finn, F. C. Flowers, F. E. Fox, H. O. Fuchs, L. N. Gilbertson, J. W. Grant, R. F. Griswold, D. Hale, E. Hawkinson, R. A. Hesse, W. F. Hofmeister, L. Hrusovsky, C. Jatczak, K. L. Jerina, O. Johari, T. M. Johnson, C. J. Kelly, H. S. Lamba, C. J. Lambright, R. W. Landgraf, J. Larson, G. Lee, R. C. Leever, J. W. Linhart, R. T. Lloyd, C. H. Luther, E. Macherauch, J. L. McCall, G. Miller, M. R. Mitchell, L. Mordfin, J. D. Morrow, D. V. Nelson, J. M. Nielson, T. Parker, K. Pearson, Jr., W. Platko, J. Price, R. C. Rice, F. D. Richards, R. Ricklefs, G. T. Robertson, M. J. Rowney, R. Sailors, M. Schober, M. P. Semenek, S. W. Shin, N. Slack, K. W. Smith, P. Spiegelberg, K. Spray, C. Stanton, R. I. Stephens, P. A. Stoll, D. Straits, S. Taira, R. A. Testin, W. A. Thomas, J. Tripp, W. C. Truckenmiller, L. E. Tucker, D. T. Vukovich, R. E. Wahlstrom, E. D. Walker, N. M. Walter, J. J. Wert, R. M. Westrich, T. Whittow, R. A. Wilde, G. A. Wilkinson, R. B. Wilson, P. M. Winter, W. M. Wood, J. E. Woodilla, T. Zajac, T. C. Zebehazy, R. E. Zinkham, R. D. Zipp. MECHANICAL PRE-STRESSING TASK FORCE—R. E. Wahlstrom (Chairman), S. A. Avramidis, F. K. Baldauf, C. Barrett, G. Bartlett, D. L. Baughman, E. A. Borch, R. F. Brodrick, L. Campana, P. E. Cary, R. A. Cellitti, M. DiBernardo, J. R. Eagan, F. E. Fox, H. O. Fuchs, R. F. Griswold, D. Hale, E. Hawkinson, W. Platko, M. Schober, R. I. Stephens, D. Straits, W. A. Thomas, N. M. Walter, W. M. Wood, T. Zajec. SERVICE DATA ACQUISITION AND ANALYSIS DIVISION—M. D. Morton (Chairman), L. H. Ames, D. H. Anderson, S. A. Avramidis, R. T. Bennett, H. D. Berns, R. F. Brodrick, S. L. Bussa, D. A. Campbell, C. H. Confer, L. F. Counter, B. Coursin, R. W. Diesner, T. J. Dolan, T. M. Elsesser, F. C. Flowers, F. E. Fox, W. C. Frank, H. O. Fuchs, D. R. Galliart, C. B. Goldman, J. A. Graham, W. F. Hofmeister, F. R. Holliday, J. M. Holt, R. Isaacson, H. R. Jaeckel, K. Johnson, T. M. Johnson, R. W. Landgraf, A. F. Lawrence, R. C. Leever, H. Mindlin, J. Morrow, D. Nelson, D. R. Olberts, M. A. Schober, R. N. Spiess, G. Singh, C. Stanton, M. K. Stark, R. I. Stephens, D. Storjohann, C. Struckemeyer, S. R. Swanson, R. B. Thakkar, A. J. Titus, T. H. Topper, R. Trempe, L. E. Tucker, E. M. Wene, R. M. Wetzel, R. A. Wilde, F. Zapp. STRUCTURAL ANALYSIS DIVISION—T. Bernard (Chairman), H. D. Berns, T. Dilger, R. G. Dubensky, G. Leese, R. C. Leever, M. P. Semenek, N. L. Slack.

FLUID CONDUCTORS AND CONNECTORS TECHNICAL COMMITTEE

R G. MICHEL (Sponsor)
W. A. HERTEL (Chairman)
H. D. Berns
H. R. Burns
B. Christensen
W. Coyne
W. E. Currie
V. T. Czebatol
H. Dorman
C. M. Foote
R. C. Gibson
R. C. Harrison

J. S. Hinske
R. E. Holmgren
L. Horvath
J. Inch
N. B. Johnston
J. A. Jones
D. E. Kimmet
R. B. Koch
O. R. Linger
R. J. Lobmeyer
A. J. Maguire
K. Naylor

H. B. Newman
C. F. Schaening
C. J. Shim
A. M. Schmidt
R. F. Sievert
E. J. Streichert
J. Wegmann
R. Wiley
J. C. White
J. Zimmel

(SAE Staff, P. Couhig)

METRIC TUBING TASK FORCE—E. J. Streichert (Chairman), V. T. Czebatol, W. A. Hertel, C. F. Schaening, C. J. Schim, A. M. Schmidt. AIR BRAKE SYSTEMS TUBING NONMETALLIC SUBCOMMITTEE—R. Mayne (Chairman), L. R. Phillippi (Vice-Chairman), B. Arterburn, D. Burge (Alternate), W. S. Busdiecker, G. Hopf, M. Cooke, G. E. Crewson, W. E. Currie, R. Kane, R. B. Koch, A. M. Schmidt, R. Silver, D. E. Washkewicz. AUTOMOTIVE AND HYDRAULIC TUBING SUBCOMMITTEE—V. T. Czebatol (Chairman), J. Hill, J. L. Lytell, R. F. Sievert, E. J. Streichert, A. V. Walsen, L. Witt, J. Zimmel. AUTOMOTIVE PIPE PLUG AND FITTINGS SUBCOMMITTEE—D. W. Vial (Chairman), D. N. Badgley, D. Burge, R. Gurnick, W. A. Hertel, M. Kotsakis, C. F. Schaening, A. M. Schmidt, C. Weicht. AUTOMOTIVE TUBE FITTINGS SUBCOMMITTEE—A. M. Schmidt (Chairman), K. Naylor (Vice-Chairman), D. Burge, V. T. Czebatol,

R. Dickson, R. C. Gibson, W. A. Hertel, A. J. Maguire, M. J. Pizzurro, R. J. Reitz, C. F. Schaening, C. J. Schim, R. Schmidt, R. R. Sjoberg, J. Zimmel. METRIC FITTINGS TASK FORCE—A. M. Schmidt (Chairman), D Burge, R. C. Gibson, W. A. Hertel, R. Holmes, K. Naylor, C. F. Schaening, C. J. Schim. CODING SYSTEMS FOR IDENTIFICATION OF FLUID CONDUCTORS AND CONNECTORS SUBCOMMITTEE—A. M. Schmidt (Chairman), C. M. Foote, R. C. Gibson, J. S. Hinske, H. Smiley, C. F. Schaening. HYDRAULIC HOSE AND HOSE FITTINGS SUBCOMMITTEE—O. R. Linger (Chairman), D. Batzer, H. D. Berns, H. R. Burns, G. C. Burrington, M. Chermak, B. Christensen, H. M. Cooke, W. Coyne, W. E. Currie, J. W. Espy, R. Fischer, W. A. Hertel, R. E. Holmgren, F. Johnson, J. A. Jones, E. M. Kavick, C. R. Leslie, J. T. Lewis, R. Lobmeyer, J. Loder, G. L. Lukavich, H. Lusher, R. Luther, R. J. May, R. McKenna, L. Moreiras, L. P. Muller, W. J. Stafford, H. V. Patel, V. Paul, D. Piccoli, E. Selvaggi, R. R. Sjoberg, J. M. Stomper, J. E. Willis, R. Wooten. J343 TASK FORCE—H. M. Cooke (Chairman), W. E. Currie, E. L. Falendysz, D. Kimmet, O. R. Linger, R. Lobmeyer, J. Loder, E. Selvaggi. HYDRAULIC TUBE FITTINGS SUBCOMMITTEE—J. S. Hinske (Chairman), H. D. Berns, H. M. Cooke, B. Coyne, V. T. Czebatol, B. Coyne, V. T. Czebatol, P. A. D'Anza, H. Dorman, B. Feifarek, R. C. Gibson, W. A. Hermann, W. A. Hertel, A. Huie, R. Jaquette, J. A. Jones, G. T. Lyon, H. Newman, L. B. O'Sickey, A. M. Schmidt, F. J. Schwarzenbart, R. F. Seivert. LUBRICATION FITTINGS SUBCOMMITTEE—J. Wegmann (Chairman), J. Rashid, E. N. Robinson, C. F. Schaening, C. J. Schim, J. M. Stomper. REFRIGERATION FITTINGS SUBCOMMITTEE—R. C. Gibson (Chairman), D. G. Burge, W. A. Hertel, J. Inch, C. F. Schaening, E. Schlotterbeck, A. M. Schmidt, W. C. Yocum.

FUEL SUPPLY SYSTEMS COMMITTEE

D. E. MARTIN (Sponsor)
L. E. ARNETT (Chairman)
G. E. MOORE (Vice-Chairman)
A. M. Bower
T. Carr
T. Detwiler
W. J. Dibsky
J. J. Gares

S. Hasko
J. C. Hoelle
R. W. Jack
D. K. Lawrence
T. O. Mitchell
J. V. Milo
R. L. Murray
M. M. Noble

R. G. Pochert
G. H. Pope
G. D. Robinson, Jr.
B. J. Seger
R. Southers
G. G. Sutcliffe
J. F. Valenzona
P. H. Waddle

(SAE Staff, R. T. Northrup)

FUELS AND LUBRICANTS EXECUTIVE COMMITTEE

J. PALMER (Sponsor)
N. T. BARTHOLOMAEI (Chairman)
A. M. Bierylo
C. C. Colyer
H. E. Deen
J. H. Freeman, Jr.

N. E. Gallopoulos
D. Garland
N. A. Hunstad
R. H. Kabel
J. W. Lane
W. E. MacDonald

R. Mackinven
L. W. Okon
C. H. Ruof
R. B. Sneed
S. R. Sprague
J. B. Stucker

(SAE Staff, R. C. Uhl)

FUELS AND LUBRICANTS TECHNICAL COMMITTEE

N. T. BARTHOLOMAEI (Chairman)
J. A. McLAIN (Vice-Chairman)
B. Adinoff
W. R. Alexander
J. L. Bascunana
P. A. Bennett
W. E. Bettoney
F. J. Blatz
E. H. Brumbaugh
K. Cashmore
A. V. Churchill
C. C. Colyer
D. J. Cornelius
B. E. Council
G. J. Decker
D. W. Dinsmore
D. S. Durham
R. E. Elrod
R. W. Geiser
A. W. Gilbert
J. M. Gottlieb
W. Hart
D. Haseltine
R. A. Hayes
S. H. Hill

P. D. Hobson
W. C. Hollibaugh
G. E. Holman
H. H. Hopkins
P. S. Hossack
N. A. Hunstad
D. R. Jones
R. E. Kay
W. H. Keeber
J. W. Lane
M. E. LePera
R. Q. Little
C. F. Long
W. E. MacDonald
T. F. McDonnell
A. C. McDonough
N. V. Messina
H. H. Mullinger
P. S. Myers
J. L. Newcombe
L. W. Okon
J. L. Palmer
D. P. Pellerito
R. H. Perry
L. G. Pless

R. I. Potter
C. H. Ruof
A. B. Sarkis
C. F. Schwarz
E. H. Scott
R. K. Smith
R. B. Sneed
G. A. Miller
R. E. Streets
B. C. Streigler
J. B. Stucker
R. Teasley, Jr.
J. G. Valdez
W. E. Waddey
J. F. Wagner
D. Wagster, Jr.
A. B. Weinberg
W. Winer
E. Yatsko, Jr.
A. H. Zeitz, Jr.
H. E. Deen (Consultant)
H. H. Donaldson (Consultant)
R. H. Kabel (Consultant)
T. P. Sands (Consultant)
S. R. Sprague (Consultant)

(SAE Staff, R. C. Uhl)

SUBCOMMITTEE 1—FUELS—C. H. Ruof (Chairman), W. E. Bettoney, C. C. Colyer, H. E. Davis, E. H. DeLong, F. B. Fitch, N. E. Gallopoulos, L. M. Gibbs, J. M. Gottlieb, R. Y. Heisler, P. D. Hobson, G. E. Holman, B. J. Krause, M. E. LePera, W. E. MacDonald, H. H. Mullinger, L. T. Murphy, D. P. Pellerito, L. G. Pless, C. F. Schwarz, R. B. Sneed, J. H. Steury, H. A. Toulmin, R. Tupa, W. T. Wotring, E. Yatsko, Jr., A. H. Zeitz, Jr. SUBCOMMITTEE 2—ENGINE OILS—R. H. Kabel (Chairman), J. L. Newcombe (Secretary), N. T. Bartholomaei, A. H. Birke, J. G. Brandes, K. Cashmore, H. E. Davis, E. H. DeLong, F. E. Didot, J. DeJovine, W. Hart, R. C. Hercules, P. D. Hobson, G. E. Holman, G. W. Holmes, R. E. Kay, J. W. Lane, M. E. LePera, W. C. Long, T. F. Lonstrup, H. V. Lowther, R. Mackinven, R. J. McConnell, T. F. McDonnell, N. V. Messina, E. L. Miller, H. H. Mullinger, L. T. Murphy, L. W. Okon, D. P. Pellerito, L. G. Pless, J. D. Plowright, R. J. Pohl, R. I. Potter, R. L. Riedel, L. A. Schaap, J. W. Schulte, C. Schwarz, T. W. Selby, R. L. Stambaugh, R. M. Stewart, J. B. Stucker, S. E. Swedberg, R. Teasley, Jr., W. E. Waddey, A. B. Weinberg. SUBCOMMITTEE 3—GEAR LUBRICANTS—J. B. Stucker (Chairman), B. Adinoff, T. C. Bowen, Jr., S. R. Calish, Jr., K. Cashmore, H. Chambers, C. C. Colyer, E. H. DeLong, J. R. Dickey, D. W. Dinsmore, J. M. Gottlieb, W. Hart, C. F. Long, H. V. Lowther, J. A. McLain, H. H. Mullinger, L. W. Okon, W. F. Olszewski, R. E. Osborne, R. I. Potter, D. L. Powell, H. G. Russell, L. F. Schiemann, L. A. Schaap. SUBCOMMITTEE 4—LUBRICATING GREASES—J. W. Lane (Chairman), B. Adinoff, K. Cashmore, E. H. DeLong, M. Movsesian, R. J. Muller, H. H. Mullinger, A. G. Papay, L. W. Okon, D. P. Pellerito, F. S. Sayles, T. M. Verdura, N. D. Hudecki. SUBCOMMITTEE 5—FLUIDS—H. E. Deen (Chairman), N. T. Bartholomaei, K. Cashmore, E. DeLong, D. Dexter, S. A. Fasone, J. M. Gottlieb, M. L. Haviland, R. A. Hayes, D. G. Jones, R. Q. Little, C. H. McKenney, J. A. McLain, R. S.

McMullen, M. I. Michael, G. Miller, H. Mullinger, E. F. Outten, J. Pandosh, B. A. Pearson, D. P. Pellerito, R. I. Potter, D. H. Rodgers, W. D. Ross, H. G. Russell, D. Wagster, P. Willermet. SUBCOMMITTEE 6—SMALL ENGINE LUBRICANTS—D. H. Garland (Chairman), K. Cashmore, E. H. Delong, B. F. Dorsey, L. Hall, P. D. Hobson, R. M. Kellam, L. Lake, G. G. Mazon, C. R. McCaffree, J. S. McKay II, M. E. McHenry, K. Pike, H. M. Pollari, D. Schultz, A. Weinberg, D. Weinberg.

HIGHWAY TIRE COMMITTEE

R. T. GASKILL (Sponsor)
R. H. SNYDER (Chairman)
R. H. Attenhofer
T. P. Baker
R. L. Brown
K. L. Campbell
A. Casanova
L. Chapin

J. Davis
A. H. Easton
J. B. Gale
W. Heath
R. E. Johnson
L. E. Kiselis
J. M. Lawther
J. L. Lipps

L. Marick
R. L. Marlowe
J. Musteric
A. Neill
R. A. Pepoy
K. G. Peterson
J. N. Schutz
W. J. Woehrle

(SAE Staff, P. Couhig)

TIRE TEST SUBCOMMITTEE—R. L. Marlowe (Chairman), R. H. Attenhofer, D. Anderson, K. D. Bird, S. K. Clark, W. B. Crum, J. W. Davis, R. G. Dunlop, J. B. Gale, H. Hodges, T. Jerney, R. E. Johnson, H. M. Keen, J. E. Ripley, R. H. Snyder, W. J. Woehrle. TIRE PENETRATION RESISTANCE TASK FORCE—J. B. Gale (Chairman), R. H. Attenhofer, W. R. Bartilson, G. Beckwith, J. W. Davis, J. Mishory, H. D. Somerville, N. E. Williams. TIRE TRACTION DEVICE TASK FORCE—L. Chapin (Chairman), M. H. Allmacher, J. D. Andrus, W. L. Cline, T. Herlick, H. J. Karp, M. Klein, S. H. Lieser, H. E. McCarthy, W. McClain, L. C. Miller, E. J. Pawlowski, R. F. Spies, R. M. Vondrasek. TIRE TRACTION TASK FORCE—K. L. Campbell (Chairman), C. V. Allen, R. H. Attenhofer, C. Beauregard, K. D. Bird, A. Y. Casanova, W. B. Crum, J. W. Davis, J. B. Gale, T. King, T. M. Knowles, W. J. LeBlanc, F. E. Matyja, W. S. McDowell, J. E. Ripley, S. R. Sacia, W. B. Straub, E. A. Whitehurst, C. T. Wright. TIRE VALVE SUBCOMMITTEE—J. L. Lipps (Chairman), J. L. Wiseberger (Co-Chairman), C. H. Denker, G. C. Fagert, H. K. Hochschwender, F. Koch, D. M. Magoulick, D. A. Makled, R. Nece, R. Paul, J. W. Shipp, C. Stasiunas, W. Szymaskiewiecz. TRUCK TIRE SUBCOMMITTEE—R. A. Pepoy (Chairman), G. A. Bindon, F. Chikos, L. Cooper, J. B. Gale, R. E. Johnson, L. Marick, W. S. McDowell, J. W. McGrath, T. W. Morsheimer, J. Rodgers, F. D. Smithson, A. J. Tribuzi.

HUMAN FACTORS ENGINEERING COMMITTEE

J. VERSACE (Sponsor)
J. STARKEY (Chairman)
L. L. Baker
J. C. Barton
W. A. Devlin
R. E. DeWald
R. J. Donohue
L. M. Forbes
T. Gage

D. E. Gobuty
J. L. Hovey
C. O. Jones
C. Kaehn
L. T. Katchmar
P. R. Knaff
T. J. Kuechenmeister
J. F. McCabe
A. Mital

L. P. Olson
G. S. Popa
N. H. Pulling
A. Raouf
R. W. Roe
R. G. Snyder
J. F. Stofflet
W. E. Woodson

(SAE Staff, P. Couhig)

DESIGN DEVICES SUBCOMMITTEE—J. L. Hovey (Chairman), J. Banashieski, R. Hapiak, R. P. Vuylsteke. DRIVER VISIBILITY SUBCOMMITTEE—W. A. Devlin (Chairman), D. Cavey, F. Decker, R. Ealba, W. M. Elliott, C. F. Finn, D. C. Hammond, R. L. Henderson, C. O. Jones, W. McAuley, M. J. McKale, J. F. Meldrum, S. L. Reed, T. Taormina, K. H. Ziwica (Consultant). CONTROLS AND DISPLAYS SUBCOMMITTEE—T. J. Kuechenmeister (Chairman), C. E. Austin, F. R. Blair, J. W. Carson, G. A. Cleek, D. D. Colovas, W. H. Daniels, R. DeWald, C. F. Finn, V. J. Geraci, W. Gillies, P. Green, A. Hallen, D. C. Hammond, G. M. Hespeler, J. M. Howard, C. O. Jones, C. McGrew, D. Mitchell, R. M. Nicholson, M. Perel, G. Rabideau, H. B. Rath, J. Sauter, E. Farber (Knowledgeable Contributor), W. A. Green (Knowledgeable Contributor), S. Konz (Knowledgeable Contributor), K. H. Kroemer (Knowledgeable Contributor), J. Starkey (Knowledgeable Contributor), H. W. Stoudt (Knowledgeable Contributor), D. Waterman (Knowledgeable Contributor). CONTROLS OPERABILITY TASK FORCE—C. Austin, W. H. Daniels, D. D. Jack, C. O. Jones, T. J. Kuechenmeister, H. E. Mellinger, M. Perel, G. Urdahl, D. Waterman (Knowledgeable Contributor). CONTROL EFFORTS TASK FORCE—D. C. Hammond (Co-Chairman), S. Z. Konz (Co-Chairman), M. Alexander, H. W. Baxley, G. Rabideau, L. A. Schaefer, K. Aoki (Knowledgeable Contributor), R. Chernikoff (Knowledgeable Contributor), K. H. Kroemer (Knowledgeable Contributor), J. L. Purswell (Knowledgeable Contributor), S. L. Reed (Knowledgeable Contributor), J. D. Weisz (Knowledgeable Contributor), D. J. Wing (Knowledgeable Contributor). HUMAN ACCOMODATIONS PRACTICES SUBCOMMITTEE—W. A. Devlin (Chairman), A. DeSantis, J. Dobek, C. F. Finn, W. Gillies, D. C. Hammond, H. Kazanjian, C. Klann, R. E. Knipper, G. S. Popa, S. Reed, J. N. Rickabaugh, R. W. Roe, T. Taormina, R. P. Vuylsteke. INSTRUMENTS AND DISPLAYS TASK FORCE—J. W. Carson, W. H. Daniels, C. F. Finn, W. Gillies, G. Hespeler, C. O. Jones, T. J. Kuechenmeister, J. M. Miller, R. M. Nicholson, H. Nissley, S. L. Reed, E. Farber (Knowledgeable Contributor), H. E. Mellinger (Knowledgeable Contributor), J. Starkey (Knowledgeable Contributor). SYMBOLS AND IDENTIFICATION TASK FORCE—D. Mitchell (Chairman), F. R. Blair, R. Boike, R. DeWald, S. B. Fawcett, C. F. Finn, V. J. Geraci, P. A. Green, E. Gregory, A. Hallen, D. D. Jack, C. O. Jones, T. J. Kuechenmeister, R. R. Luycky, J. Pappas, R. S. Piotrowski, H. B. Rath, J. Sauter, E. Tomczak, W. A. Green (Knowledgeable Contributor), H. E. Mellinger (Knowledgeable Contributor).

HYDRAULIC BRAKE SYSTEMS ACTUATING COMMITTEE

R. W. HILDEBRANDT (Sponsor)
R. G. BROWN (Chairman)
D. WASMER (Vice-Chairman)
R. L. Coffman
J. Conley
R. L. Doering
E. P. Francis
S. E. Hagerty

C. Harrington
J. L. Harvey
F. Hussey
H. C. Klein
J. MacDougall
A. L. Marshall
V. Meyer
W. J. Oakley

P. Oppenheimer
J. M. Rowell (Alternate)
R. W. Shiffler
M. Tamaki
W. Weisbrod
A. J. Wilson
P. T. Wood
K. E. Yost

(SAE Staff, L. P. Ziegler, Jr.)

AUTOMOTIVE BRAKE AND STEERING HOSE SUBCOMMITTEE—K. E. Yost (Chairman), G. J. Resnik (Vice-Chairman), F. Bumann, M. Carter, L. J. Dreyer, J. H. Elgin, E. P. Francis, J. B. Johnson, R. L. Jones, H. H. Klever, T. A. Knurek, C. G. Mitton, J. Mutzner, M. Newberry, E. P. Pickles, D. A. Pittenger, F. F. Sadr, G. Santoro, A. M. Spisak, D. L. Veit, J. R. Whiston. VACUUM BRAKE HOSE TASK FORCE—K. E. Yost (Chairman), T. A. Knurek (Vice-Chairman), F. Bumann, M. Carter, L. J. Dreyer, E. P. Francis, J. B. Johnson, J. T. McElhose, M. Newberry, D. A. Pittinger, G. J. Resnik, F. F. Sadr, G. Santoro, A. M. Spisak, D. L. Veit, J. R. Whiston, A. E. Williams, A. B. Wood. POWER STEERING HOSE TASK FORCE—K. E. Yost (Chairman), D. L. Veit (Vice-Chairman), W. Behnke, C. R. Brandich, F. Bumann, M. Carter, D. H. Dengler, R. Jenkins, F. J. Lasker, J. Loder, J. T. McElhose, W. Mitzenius, P. Muller, J. Mutzner, P. Piccoli, J. Shea, R. Wiley. HYDRAULIC BRAKE HOSE TASK FORCE—J. R. Whiston (Chairman). HANDLING OF BRAKE FLUIDS SUBCOMMITTEE—W. Weisbrod (Chairman), F. Hussey (Vice-Chairman), J. Bowers, R. B. Broadwater, R. G. Brown, R. J. Donaldson, E. P. Francis, J. L. Harvey, J. MacDougall, A. L. Marshall, R. F. Smith. HYDRAULIC BRAKE COMPONENTS SUBCOMMITTEE—S. E. Hagerty (Chairman), H. W. Baxley, J. Bowers, T. A. Brandon, J. Cavicchia, V. T. Czebatol, S. DeRosa, H. Dubke, C. A. Feeser, A. Folgart, S. Fox, J. W. Getchei, P. S. Gritt, T. Hamilton, C. A. Hanning, C. G. Hockett, S. Hori, K. Jensen, W. R. Kienle, R. P. Kolena, H. Lang, F. Limberg, M. J. Locke, C. Marciniak, J. Martinic, D. McIntyre, W. Melinat, R. A. Merrifield, K. Messersmith, R. W. Miaso, J. Miller, P. J. Miller, B. Newell, W. J. Oakley, E. S. Orzel, R. Poplawski, G. Preston, G. Ploeg, J. Prokopek, P. M. Rock, R. H. Rosback, L. Rossigno, R. Hudy, N. Runkle, P. Ruseckus, C. Schaening, J. Schwarz, P. M. Schwarz, J. Sellinger, J. Smiley, D. Snider, E. Streichert, P. B. Smith, K. Swanson, R. Temple, R. O. Tuegel, J. Vallo, G. Waterman, F. Watkins, R. Winter, R. Horvath (Alternate), J. Kosinski (Alternate), R. Newsock (Alternate), A. R. Wadum (Alternate). HYDRAULIC BRAKE SYSTEMS TUBING, TAPPED HOLES, AND MALE FITTINGS RECOMMENDED PRACTICE TASK GROUP—W. Melinat (Chairman), R. Babich, V. T. Czebatol, H. Dubke, T. Hamilton, M. J. Locke, R. A. Merrifield, K. Messersmith, R. W. Miaso, C. Schaening, A. Schmidt, E. Streichert, J. Vallo. MASTER CYLINDER TASK GROUP—J. C. Cavicchia, S. DeRosa, S. Fowler, J. W. Getchei, P. Gritt, S. Hagerty, C. Hanning, C. G. Hockett, R. Kolena, H. Lang, M. J. Locke, R. Poplawski, P. B. Smith, D. Snider, J. Schwartz, J. Williams, R. Winter. HYDRAULIC BRAKE ACTUATING ELASTOMERIC SUBCOMMITTEE—V. W. Meyer (Chairman), H. H. Anderson, R. G. Brown, R. J. Clifford, L. L. Collins, J. H. Conley, V. T. Cortesi, E. P. Francis, J. L. Harvey, E. L. Hoffmann, D. B. Judsen, R. P. Kolena, A. L. Marshall, A. R. Mellyn, C. G. Mitton, R. C. Pocock, J. Rentschler, R. M. Shane. MOTOR VEHICLE BRAKE FLUIDS SUBCOMMITTEE—A. L. Marshall (Chairman), R. B. Broadwater (Vice-Chairman), E. P. Bobby, J. Bowers, R. G. Brown, I. Burgess (Corresponding Member), J. Burgess (Corresponding Member), L. Cabelkova-Taguchi, D. R. Chapman, R. L. Coffman, J. Conley, E. Crivelli, R. J. Donaldson, E. P. Francis, R. E. Hannon, C. Harrington, J. L. Harvey, F. Haynes (Corresponding Member), G. W. Holbrook, F. Hussey, I. Keddie, I. Liebman, L. Loy, L. Mackowiak, C. J. Marciniak, D. Marshall, R. Roenigk, R. Salvador, R. W. Shiffler, R. F. Smith, M. Tamaki (Corresponding Member), A. Torkelson, J. Van Sloun, W. Weisbrod. SIMULATED SERVICE TASK GROUP—E. L. Hoffmann (Chairman), J. Bowers, J. Conley, E. P. Francis, W. Holbrook, A. L. Marshall. SPECIFICATION DEVELOPMENT TASK GROUP—R. B. Broadwater (Chairman), E. P. Bobby, J. Bowers, R. G. Brown, I. Burgess (Corresponding Member), J. Burgess (Corresponding Member), L. Cabelkova-Taguchi, D. R. Chapman, R. L. Coffman, J. H. Conley, E. Crivelli, R. J. Donaldson, E. P. Francis, C. Harrington, J. L. Harvey, F. Haynes (Corresponding Member), F. Hussey, I. Keddie, A. L. Marshall, R. J. L. Parker (Corresponding Member), R. Roenigk, R. Salvador (Corresponding Member), R. W. Shiffler, M. Tamaki (Corresponding Member), W. Weisbrod. CORROSION TEST DEVELOPMENT WORKING GROUP—F. Hussey (Chairman), E. P. Bobby, R. B. Broadwater, J. H. Conley, E. Crivelli, E. P. Francis, J. L. Harvey, A. L. Marshall, W. Weisbrod. VAPOR LOCK TASK GROUP—R. W. Shiffler (Chairman), E. P. Bobby, R. G. Brown, G. R. Browning, J. H. Conley, C. Harrington, J. L. Harvey, H. F. Hussey, I. Liebman, A. L. Marshall, D. B. Pellerito, M. Tamaki, W. Weisbrod. NON-CONVENTIONAL BRAKE FLUIDS TASK GROUP—J. L. Harvey (Chairman), B. Belier (Observer), V. Bloom, E. P. Bobby, J. Bowers, R. B. Broadwater, R. G. Brown, L. Cabelkova-Taguchi (Corresponding Member), D. R. Chapman, R. L. Coffman, J. H. Conley, E. Crivelli, B. Driscoll, E. P. Francis, R. E. Hannon, C. J. Harrington, G. W. Holbrook, F. Hussey, I. Liebman, A. L. Marshall, R. Salvador (Corresponding Member), R. W. Shiffler, M. Tamaki (Corresponding Member), J. Van Sloun, W. Weisbrod. REFEREE MATERIALS SUBCOMMITTEE—J. L. Harvey (Chairman), W. Weisbrod (Vice-Chairman), W. Amber, E. P. Bobby, J. Bowers, R. G. Brown, L. Cabelkova-Taguchi (Corresponding Member), R. L. Coffman, J. H. Conley, E. P. Francis, C. W. Greening, R. E. Hannon, C. Harrington (Corresponding Member), F. Haynes (Corresponding Member), E. L. Hoffmann, F. Hussey, I. Liebman, L. Mackowiak, A. L. Marshall, V. W. Meyer, R. Salvador (Corresponding Member), R. M. Shane, G. R. Sims (Corresponding Member), M. Tamaki (Corresponding Member), A. Torkelson, D. A. Wasmer, J. Van Sloun.

IRON AND STEEL EXECUTIVE COMMITTEE

D. RUNDLE (Chairman)	W. J. Cormack	C. J. Rhodes
B. WRIGHT (Vice-Chairman)	D. D. Dodge	E. F. Ryntz
F. P. Bens	M. L. Frey	B. A. Smith
R. M. Buck	N. O. Kates	G. W. Tuffnell
G. F. Bush	C. Krubsack	G. Witt

(SAE Staff, P. Couhig)

IRON AND STEEL TECHNICAL COMMITTEE

D. F. RUNDLE (Chairman)	H. A. Doyle	C. J. Rhodes
P. A. DeGIOIA (Vice-Chairman)	G. N. Douglas	G. H. Robinson
F. J. Arabia	J. R. Easterday	E. F. Ryntz
R. J. Belz	A. G. Forrest	S. R. Scales
R. D. Bennett	M. L. Frey	M. P. Semenek
F. P. Bens	R. W. Goetz	D. L. Shangle
E. T. Bittner	W. T. Groves	H. T. Sheppard
H. N. Bogart	J. S. Hanson	B. A. Smith
F. Borik	D. J. Hayes	Y. A. Smith
R. M. Buck	R. H. Hays	C. H. Sperry
G. F. Bush	L. A. Huebner	J. V. Sprong
D. P. Buswell	K. W. Jerwann	R. F. Steigerwald
S. R. Callaway	N. O. Kates	E. J. Streichert
H. R. Chappell	J. H. King	M. Suess
E. F. Chojnowski	C. Krubsack	E. R. Suveges
J. F. Clark	C. D. Loyd	J. E. Tripp
A. G. Cook	A. S. MacDonald	G. W. Tuffnell
W. J. Cormack	D. E. McVicker	P. M. Unterweiser
H. W. Dailey	R. S. Moriarty	E. T. Vitcha
A. J. Dearing	J. H. Nelson	R. F. Webster
M. V. DiBernardo	W. H. B. Newell	T. R. Weins
D. V. Doane	P. K. Patnaik	B. E. Wright
J. M. Dobos	W. J. Ptashnik	
D. D. Dodge	J. E. Rainey	

(SAE Staff, P. Couhig)

PANEL D—TRACTOR AND EARTHMOVING—C. A. Krubsack (Chairman), C. J. Meyer (Vice-Chairman), R. M. Baker, R. D. Bennett, E. T. Bittner, S. R. Callaway, R. Cummins, P. A. DeGioia, H. C. Dill, H. Fischer, A. G. Flores, A. G. Forrest, R. H. Hays, J. A. Hildebrandt, C. McKenney, E. J. Nugent, D. F. Rundle, D. Shangle, C. H. Sperry. DIVISION 1—CARBON AND ALLOY STEELS—D. P. Buswell (Chairman), J. W. Abar, P. L. Arnold, G. A. Beaudoin, C. Beck, R. M.

Buck, J. F. Clark, W. J. Cormack, R. Cummins, M. V. DiBernardo, A. Dunhoff, R. P. Edwards, H. B. Fernald, M. L. Frey, B. M. Glasgal, R. L. Green, R. G. Groen, W. T. Groves, D. J. Hayes, J. A. Hildebrandt, J. C. Holzwarth, R. H. Hudson, J. G. Hutchinson, W. C. Leslie, R. T. Morelli, T. E. Moss, J. Nelson, W. H. B. Newell, J. M. Quigley, R. Richmond, J. R. Ruff, D. L. Shangle, R. H. Smith, P. A. Speer, G. W. Tuffnell, L. J. Vande Walle, T. R. Weins, H. W. White, L. S. Willis, D. J. Wulpi. DIVISION 3—TEST PROCEDURES—C. A. Martini (Chairman), H. B. Aaron, P. Bailey, P. A. DeGioia, D. Dieberg, G. W. Henger, J. Sweet, P. Vernia. DIVISION 8—CARBON AND ALLOY STEEL HARDENABILITY—D. V. Doane (Chairman), J. W. Abar, R. M. Baker, D. P. Buswell, A. Dunhoff, B. M. Glasgal, R. G. Groen, D. J. Hayes, R. H. Hays, G. W. Henger, J. G. Hutchinson, K. W. Jerwann, C. A. Krubsack, M. C. Long, L. Mair, P. L. Mangonon, R. T. Morelli, G. H. Robinson, R. H. Smith, G. H. Walter, L. S. Willis. DIVISION 9—AUTOMOTIVE IRON CASTINGS—E. F. Ryntz (Chairman), D. D. Day (Vice-Chairman), H. N. Bogart, R. T. Brower, E. F. Chojnowski, P. M. Gribbell, W. W. Holden, O. K. Hunsaker, L. R. Jenkins, G. M. Lahr, N. Lillybeck, D. Matter, D. McCarthy, P. J. Mikelonis, J. H. Panasiewicz, P. K. Patnaik, F. Preston, A. H. Rauch, R. Reesman, T. J. Reilly, D. A. Rhoda, D. F. Rundle, C. F. Walton, T. R. Weins, W. E. Wistehuff. DIVISION 10—AUTOMOTIVE STEEL CASTINGS—H. T. Sheppard (Chairman), R. M. Baker, W. J. Cormack, P. J. Neff, J. R. Ruff, W. G. Sholz, C. H. Sperry. DIVISION 12—JOINT COMMITTEE ON DEFINITIONS OF TERMS RELATING TO HEAT TREATMENT OF METALS—M. L. Frey (Chairman), F. J. Arabia, G. F. Bush, W. J. Cormack, D. D. Dodge, R. W. Foreman, D. J. Funk, J. E. Jacoby, N. O. Kates, R. J. Light, J. A. Lincoln, H. D. Monsch, I. A. Rohrig, P. Vernia, C. F. Walton. DIVISION 13—HIGH STRENGTH, LOW ALLOY STEELS—G. J. Hansen (Chairman), G. A. Beaudoin, J. J. Cahill, M. E. Dilley, J. M. Dobos, J. A. Hildebrandt, T. Moss, R. B. Olsen, C. J. Rhodes, P. Speer, T. H. Spencer, C. H. Sperry, J. V. Sprong, R. E. Stitt, J. R. Zanetti. DIVISION 15—CORROSION RESISTANT STEELS—R. F. Steigerwald (Chairman), A. G. Cook, J. P. Gallivan, J. R. Huth, R. T. Morelli, J. G. Tack, G. W. Tuffnell, S. E. Tyson, E. T. Vitcha, W. I. Weed, F. E. Yanko. DIVISION 18—ABRASIVE WEAR OF FERROUS MATERIALS—J. L. Lytell (Chairman), H. S. Avery, K. Budinski, T. J. Bulat, J. F. Dehnke, J. Dodd, W. H. Feilbach, W. E. Fuller, H. W. Lloyd, H. O. McIntire, A. E. Miller, D. Rosenblatt, W. Scholz. DIVISION 25—NON-DESTRUCTIVE TEST METHODS—D. D. Dodge (Chairman), G. N. Douglas, J. R. Frederick, D. J. Hagemaier, R. G. Hentschel, W. C. Hitt, T. W. Judd, P. C. McEleney, L. Mordfin, J. K. Schmitt, K. Skeie, R. W. Thams. DIVISION 27—PREVENTION OF CORROSION OF IRON AND STEEL—G. F. Bush (Chairman), H. D. Baker, G. W. Bush, R. L. Chance, C. O. Durbin, G. H. Ekstrand, J. A. Hamman, G. J. Hansen, E. W. Johnson, A. W. Kennedy, P. H. Lake, K. Lehman, J. L. Livermore, F. M. Loop, E. E. Madion, R. J. Neville, A. D. Squitero. DIVISION 29—MECHANICAL AND CHEMICAL REQUIREMENTS FOR THREADED FASTENERS—D. A. Garrison (Chairman), R. Belford (Secretary), F. J. Arabia, A. R. Breed, R. M. Byrne, R. A. Cellitti, H. R. Chappell, P. J. Chiesa, G. W. DeLoof, T. S. Doppke, R. P. Edwards, R. L. Faunce, F. E. Graves, J. S. Hanson, R. W. Hayden, D. J. Hayes, K. E. Kuschell, C. S. Larson, S. E. Mallen, K. E. McCullough, J. Michael, R. Mitchell, B. E. Olson, J. S. Orlando, L. Polek, W. J. Ptashnik, D. Radmore, C. Reiter, A. E. Schneider, L. J. Skowron, B. A. Smith, W. Suzuki, T. E. Urich, L. Vande Walle, D. Wagner. DIVISION 32—SHEET AND STRIP STEEL—S. Dinda (Chairman), G. Belyea, W. C. Berg, J. J. Cahill, J. Dobos, G. R. Fenrich, R. Banules, G. M. Goodwin, G. J. Hansen, S. Hoskins, G. B. Jacobs, S. A. Kane, R. G. Lang, J. A. Lumm, W. H. McFarland, W. E. McNabb, P. Michel, M. Monaco, J. E. Morgan, R. Moriarty, R. Neville, A. Ruitta, E. Schuerer, C. Scoble, J. Sprong, J. Streichert, R. Twisdom. DIVISION 32a—UNDERVEHICLE CORROSION OF COATED STEELS TASK FORCE—R. J. Neville (Chairman), L. Allegra, G. T. Gira, A. M. Kalson, S. R. Koprich, F. M. Kuhanek, H. H. Lawson, G. G. Levy, E. J. Oles, E. A. Praschan, C. R. Rarey, E. C. Rhea, W. C. Sievert, M. S. Walker. DIVISION 33—GEAR METALLURGY—R. H. Hays (Chairman), I. Binder, D. L. Borden, R. Brown, R. W. Buenneke, H. E. Collins, P. A. DeGioia, D. E. Diesburg, D. M. Donegan, R. C. Ferbitz, D. Fergle, A. G. Forrest, R. L. Green, W. T. Groves, E. E. Hawkinson, G. Hruby, R. L. Hughes, D. Hughson, C. A. Krubsack, D. E. McVicker, W. J. Ptashnik, V. P. Rao, J. B. Saxton, D. L. Shangle, M. M. Shea, J. E. Tripp, G. Tuffnell, D. T. Vukovich, G. H. Walter, R. A. Wilde, R. Zimmerman. DIVISION 34—TOOL MATERIALS—B. E. Wright (Chairman), W. E. Burd, R. A. Johnson, J. H. King, H. T. Sheppard. DIVISION 35—ELEVATED TEMPERATURE PROPERTIES OF FERROUS MATERIALS—C. Clark (Chairman), R. Amadee, J. R. Drake, E. Gratzer, G. M. Lahr, J. Larson, T. E. Lewis, D. I. Morton, B. Royer, H. Tanczyn, G. W. Tuffnell, C. Vigor, E. Vitcha. DIVISION 36—FERROUS POWDERED METALS—D. Glover (Chairman), D. Anderson, G. Gaft, J. R. Gleixner, D. Gustafson, L. A. Heubner, C. T. Kunze, R. A. Powell, A. Reid, W. Roth, H. Sheppard.

LAWN, GARDEN, AND SMALL ENGINE POWERED EQUIPMENT COMMITTEE

GEORGE C. HARDWICK (Chairman)
S. SEIDLITZ (Vice-Chairman)
G. M. Adams
I. Barker
L. J. Bernauer
R. Chhabra

B. Christy
J. A. Donohue
J. A. Gresch
E. Keszthelyi
J. Mieritz
D. G. Owens

G. Ruehlow
W. D. Schultz
D. Williamson
N. Woelffer

(SAE Staff, R. T. Northrup)

LIGHTING COMMITTEE

D. L. NORDEEN (Sponsor)
P. G. SCULLY (Chairman)
R. W. OYLER (Vice-Chairman)
R. Austin
J. E. Bair
A. F. Bleiweiss
A. J. Burgess
A. Cardarelli
R. P. Donely
W. M. Elliott
D. M. Finch
C. E. Granfors
W. M. Heath
E. Hitzemeyer

G. A. Knapp
P. Lawrenz
A. T. Lewry
R. J. Love
P. W. Maurer
W. A. McKinney
G. E. Meese
H. Miyazawa
S. F. Kimball
R. A. Nixon
P. Olson
K. C. Ploeger
J. Simeroth
C. A. Slater

P. Smester
Z. S. Subotich
D. S. Sumple
C. E. Terrill
R. M. Terry
K. Uding
R. L. Vile
P. E. Westlake
R. A. Witt
G. Wright
H. J. T. Young
T. C. Tjader

(SAE Staff, R. C. Uhl)

AUXILIARY DEVICES SUBCOMMITTEE—A. T. Lewry (Chairman), D. L. DeLano, W. Heath, E. Wojcik, M. E. Uecker. FLASHER TASK FORCE—E. Wojcik (Chairman), J. E. Bair, A. F. Bleiweiss, G. Columbo, L. Cyr, D. M. Finch, A. J. Hollis, A. T. Lewry, A. J. Little, W. G. Pracejus, T. Shewchuck. SOCKET TASK FORCE—D. L. Delano (Chairman), R. C. Brownlee, J. H. Dewar, E. M. Kooker, A. T. Lewry, J. M. Packman, R. G. Plyler, P. E. Westlake. SWITCH TASK FORCE—H. J. Ewald, K. Audino, W. D. Bragdon, O. R. Nichols, Jr., T. Drennon, M. Slavin, M. Koler, E. Lau, R. Lentes, R. G. Rhoads, J. Salliotte, W. L. Zurbriggen, T. Loughran, D. Vescio, G. Wright, A. T. Lewry, R. A. Leiper. MATERIALS SUBCOMMITTEE—P. Lawrenz, J. V. Bailey, J. E. Bennett, A. F. Bleiweiss, A. J. Burgess, E. P. DePalma, C. J. Devonshire, W. M. Heath, R. LaPado, W. McMahan, M. Miller, J. F. Mizia, R. A. Nixon, L. C. Owen, C. A. Slater, P. Smester, T. C. Tjader, P. Westlake, G. M. Ziver. REFLECTIVE DEVICES SUBCOMMITTEE—K. D. Uding (Chairman), J. E. Bennett (Vice-Chairman), R. L. Austin, C. J. Devonshire, D. M. Finch, W. M. Heath, S. A. Heenan, D. W. Moore, K. C. Ploeger, P. G. Scully, P. R. Smester, A. F. Bleiweiss. ROAD ILLUMINATION DEVICES SUBCOMMITTEE—J. E. Bair (Chairman), P. E. Westlake (Vice-Chairman), W. M. Elliott, D. M. Finch, S. Kimball, P. Lawrenz, G. E. Meese, R. A. Nixon, R. W. Oyler, E. Pitkjaan, H. L. Kesler, T. G. Tallon, E. L. Hopkins, T. C. Tjader. LIGHTING INSPECTION CODE TASK FORCE—J. M. DePaepe (Chairman), J. E. Bair, A. Cardarelli, E. L. Hopkins, D. King, T. N. Salow, P. Smester, R. M. Terry, R. L. Vile, P. Westlake, R. Woschitz, M. Sugiyama. SIGNALLING AND MARKING DEVICES SUBCOMMITTEE—R. J. Love (Chairman), G. P. Wright (Vice-Chairman), L. Bachynsky, J. E. Bair, V. D. Bhise, A. J. Birkhoff, A. F. Bleiweiss, T. Dawson, C. J. Devonshire, W. M. Elliott, W. M. Heath, C. O. Jones, G. R. Mengel, B. C. Muccioli, C. Newman, R. W. Oyler, L. Poirier, P. G. Scully. GASEOUS DISCHARGE LAMP T. F.—L. Bachynsky (Chairman), T. Dawson, L. W. Fabry, C. Newman. FLASHING AND 360 DEG WARNING LAMPS—

C. Newman (Chairman), T. Dawson, L. Bachynsky, L. W. Fabry. BULB TASK FORCE—R. L. Vile (Chairman), R. J. Love, J. E. Bair, P. S. Rust. CHROMATICITY TASK FORCE—G. Mengel (Chairman). TURN SIGNAL TASK FORCE—R. J. Love, A. F. Bleiwiess, G. P. Wright, P. Lawrenz. LAMP MOUNTING HEIGHT—W. M. Heath (Chairman), C. Finn, E. Hitzemeyer, E. Laginess, C. Terrill. BACK UP LAMP—L. Poirier (Chairman), P. Lawrenz, B. C. Muccioli. REAR FOG LAMP—C. O. Jones (Chairman). LICENSE PLATE—C. Finn (Chairman). TEST METHODS AND EQUIPMENT SUBCOMMITTEE—W. A. McKinney (Chairman), L. Bachynsky, J. Fitzgerald, D. F. King, W. E. L. Pears, F. Skrbina, P. Smester, Z. S. Subotich, G. E. Swierb, A. E. Ackerly. LIGHTING IDENTIFICATION CODE TASK FORCE—C. Slater (Chairman), A. Cardarelli, A. M. Dekovich, R. M. Terry, M. Sugiyama, C. E. Terrill. INTERNATIONAL STANDARDS SUBCOMMITTEE—P. W. Maurer (Chairman), A. F. Bleiwiess, R. A. Ehrhardt, C. Finn, E. Hitzemeyer, M. Elliott, E. M. Kooker, E. J. Laginess, P. Lawrenz, G. E. Meese, B. C. Muccioli, H. J. T. Young, Robert C. Oliver.

LUBRICANTS REVIEW INSTITUTE BOARD

C. C. COLYER (Chairman)
J. B. Bidwell
P. Carl

G. J. Decker
S. Gratch
J. L. Palmer

R. A. Stapf
W. A. Wallace

(SAE Staff, A. G. Salem)

MARINE TECHNICAL COMMITTEE

A. J. TROYAN (Chairman)
R. F. KRESS (Vice-Chairman)
D. A. Armstrong
B. Arnold
D. D. Beach
L. C. Bibow
W. M. Crook
C. Game

L. B. Gray
J. Hodge
E. C. Kiekhaefer
J. A. Langley
J. T. Laskey
N. Leeper
G. J. Lippmann
J. M. McClellan

F. Miszczak
D. I. Reed
W. M. Rosenfeld
G. W. Schulz
W. C. Shanks
R. Sluka
S. Q. Wales
D. H. Wood

(SAE Staff, L. P. Ziegler, Jr.)

ELECTRICAL SYSTEMS SUBCOMMITTEE—D. H. Wood (Chairman), E. Yochum (Vice-Chairman), J. R. Brown, J. R. Draxler, E. C. Game, N. S. Hatch, J. P. Hutcheson, C. J. Juhnke, K. D. Kobelentz, J. M. McClellan, G. Pace, D. I. Reed, J. L. Schaefer. ENGINE FUEL SYSTEMS SUBCOMMITTEE—L. C. Bibow (Chairman), G. Baltz, L. J. Bernauer, R. Erickson, L. Granholm, C. Lysaught, P. McAlindon, D. I. Reed, W. T. Sawyer, R. L. Scott, J. Worley. FUEL FILTRATION TASK FORCE—J. R. Kaiser (Chairman), C. S. Berry, J. J. Fontana, R. W. Heintz, J. W. Hurst, J. Langley, E. H. Lotter, C. Noble, P. Quick. ENGINE POWER RATING SUBCOMMITTEE—J. A. Langley (Chairman), D. A. Armstrong, D. D. Beach, L. C. Bibow, J. R. Branstrator, W. M. Crook, J. Hodge, A. E. Liffengran, D. I. Reed, E. J. Reinelt, W. C. Shanks, R. Sluka, D. Wilcox. POWER UNIT INSTALLATION SUBCOMMITTEE—D. A. Armstrong, D. D. Beach, J. R. Branstrator, A. D. Dudleston, J. A. Langley, D. R. Osborn, R. H. Roy, A. J. Troyan, T. O. Elsner (Liaison). PROPELLER-SHAFT ENDS AND HUBS SUBCOMMITTEE—B. Arnold (Chairman), D. D. Beach, C. Benson, J. E. Brandt, F. Carr, R. Daniels, W. C. Edmundson, R. F. Kress, W. M. Rosenfeld, G. Sullivan, W. Thompson, G. Torgersen, A. J. Troyan, G. Walter, L. E. Yerkovich.

MOTORCYCLE COMMITTEE

C. A. BIBLE (Sponsor)
L. C. LAKE (Chairman)
R. HAGIE (Vice-Chairman)
J. Carson
T. Carter
C. L. Hale
R. Henderson
R. E. Harris
R. Harrison
C. Hartman
R. Huebner

K. Jelinek
P. D. Keller
R. A. Little
S. Munro
H. Neas
R. P. Parker
F. M. Petler
K. Reimers
G. A. Robinson
R. D. Ross
W. J. Ross

B. Shepard
M. R. Stahl
G. Stubbs
I. J. Wagar
J. Walsh
G. Winn
H. Worthington
M. Yoakum
H. Young

(SAE Staff, R. C. Uhl)

ELECTRICAL SYSTEMS SUBCOMMITTEE—F. M. Petler (Chairman), C. L. Hale, B. Henderson, R. E. Harris, M. Iwase, K. D. Kobelentz, L. C. Lake, D. McCormack, H. Miyazawa, Y. Nakajima, R. W. Oyler, D. W. Pedley, G. A. Robison, A. E. Seckinger, B. Shepard, M. Elliott, H. Worthington, G. Winn, H. Young, W. Ross. HEADLIGHT TASK FORCE—P. E. Westlake (Chairman), D. E. David, R. Hagie, C. L. Hale, W. Heath, B. Henderson, M. Iwase, C. H. Kaehn, K. D. Kobelentz, L. C. Lake, Y. Nakajima, R. W. Oyler, F. M. Petler, K. Watanabe, H. Worthington, G. Winn, H. Young, T. W. Hudson, Sr. MOTORCYCLE SOUND LEVEL SUBCOMMITTEE—J. B. Walsh (Chairman), R. Hagie, C. Hale, R. Harrison, R. Hillquist, L. C. Lake, R. Little, R. Parker, B. Shepard.

MOTOR VEHICLE SAFETY SYSTEMS TESTING COMMITTEE

J. VERSACE (Sponsor)
H. G. BRILMYER (Chairman)
R. A. WILSON (Vice-Chairman)
L. E. Baltz
J. Brinn

R. P. Daniel
J. E. Hofferberth
W. Koebnick
D. J. McDowell
W. D. Nelson

G. W. Nyquist
J. Scowcroft
C. H. Lee (Liaison)

(SAE Staff, L. P. Ziegler, Jr.)

ACCIDENT INVESTIGATION PRACTICES SUBCOMMITTEE—W. D. Nelson (Chairman), J. Abromavage, S. P. Baker, R. Clark, P. Cooley, R. Cromack, E. F. Grush, R. Hamilton, D. L. Hendricks, J. Kistle, S. Lee, D. Lischynski, G. M. Mackay, J. Marsh, A. J. McLean, K. L. Morgan, M. Oversby, A. Rininger,

V. Roberts, D. N. Schmidt, H. W. Sherman, A. Siegel, J. D. States, B. Sweigard, A. L. Thompson, C. Whitney, R. L. Wilson. HUMAN BIOMECHANICS AND SIMULATION SUBCOMMITTEE—G. W. Nyquist (Chairman), S. W. Alderson, S. H. Backaitis, J. Brinn, R. P. Daniel, R. Eppinger, V. R. Hodgson, R. P. Hubbard, A. I. King, R. E. Knight, J. O. Kortge, J. B. Lenox, J. Melvin, H. J. Mertz, R. F. Neathery, P. Presad, H. M. Reynolds, D. H. Robbins, J. Smrcka, D. J. Thomas, R. J. Vargovick, R. W. Roe (Liaison). MECHANICAL HUMAN SIMULATION TASK FORCE—R. P. Daniel (Chairman), S. W. Alderson, S. H. Backaitis, T. Colvin, R. E. Knight, J. O. Kortge, R. F. Neathery, J. Smrcka, R. J. Vargovick. DUMMY TESTING EQUIPMENT WORKING GROUP—M. Koga (Chairman), D. Alianello, J. Blaker, T. Colvin, R. P. Daniel, K. Edinger, S. Goldner, T. Hayak, D. Lombardi, G. W. Nyquist, F. M. Peters, M. Walsh, M. Wolanin. ANTHROPOMETRICS TASK FORCE—R. P. Hubbard (Chairman), S. H. Backaitis, E. Churchill, C. E. Clauser, K. W. Kennedy, J. T. McConville, H. M. Reynolds, J. Young, R. W. Roe (Liaison). HUMAN MECHANICAL RESPONSE TASK FORCE—A. I. King (Chairman), R. Eppinger, R. P. Hubbard, J. W. Melvin, R. F. Neathery, D. J. Thomas. HUMAN INJURY CRITERIA TASK FORCE—J. Brinn (Chairman), R. Eppinger, V. R. Hodgson, J. W. Melvin, H. J. Mertz, R. L. Stalnaker. ANALYTICAL HUMAN SIMULATION TASK FORCE—D. H. Robbins (Chairman), B. M. Dowman, H. S. Chan, C. Chou, E. Knight, J. F. Marquardt, P. Prasad, R. J. Vargovick. IMPACT AND ROLLOVER TEST PROCEDURES SUBCOMMITTEE—W. Koebnick (Chairman), S. H. Backaitis, J. J. Cantalupo, D. H. Edwards, R. H. Fredericks, J. W. Freiburger, D. Haenchen, N. J. Nagrant, F. M. Peters, J. A. Searle, C. Teal. PASSENGER COMPARTMENT ENERGY ABSORPTION SUBCOMMITTEE—R. P. Daniel (Chairman), C. Culver (Vice-Chairman), B. Albright, R. Behar, G. E. Dumas, T. P. Eviston, L. G. Heck, J. W. Kennebeck, J. C. King, A. R. Lenkner, F. M. Peters, J. Ruby, J. Searle (Corresponding Member), C. Silvester (Corresponding Member), P. F. Woley. PERFORMANCE CRITERIA SUBCOMMITTEE—J. Brinn (Chairman), S. H. Backaitis, W E. Hering, A. Hirsch, M. S. Koga, W. B. McCormick, J. McElhaney, R. McHenry, L. P. Patrick, F. M. Peters, C. A. Preuss, R. Vargovick. SAFETY TEST INSTRUMENTATION SUBCOMMITTEE—L. E. Baltz (Chairman), R. L. Anderson, S. H. Backaitis, D. Blaisdell, J. W. Kennebeck, M. J. Locke, J. R. Nothstine, E. R. Shull, D. T. Siems, J. Vandenburg. TOWABILITY SUBCOMMITTEE—D. J. McDowell (Chairman), H. Carr, B. Fielek, J. Fobian, G. Holmes, J. F. Jensen, R. Kalbacher, D. Morrow, D. Murray, J. Schultz, K. H. Ziwica.

NOMENCLATURE ADVISORY COMMITTEE

M. J. TAUSCHEK (Sponsor)
D. A. FAY (Chairman)
K. WEBER (Vice-Chairwoman)
L. O. Baker

T. A. Brandon
L. C. DeWeese
J. W. Justusson
M. Petler

C. F. Thelin
S. G. Tilden, Jr.
V. L. Vickland

(SAE Staff, L. P. Ziegler, Jr.)

OWNER'S MANUAL CONSUMER INFORMATION SUBCOMMITTEE—C. F. Thelin (Chairman), M. Allmacher, L. Baker, D. Black, R. A. Darienzo, J. D. Fobian, P. G. Hundt, M. Johnston, F. Louis, J. E. Martens, R. W. Mattutat, G. C. Nield, M. Poirer, E. Robibero, A. G. Spring, Jr., J. E. Steger, G. W. Warren, R. G. Yancheson.

NONFERROUS METALS COMMITTEE

G. HARLAN GILMORE (Sponsor)
ART KASPER (Chairman)
R. S. Busk
C. S. Barker
R. Batra
S. M. Brandt
A. R. Canady
A. Cohen
R. J. Cox
A. O. DeHart

R. E. Doane
W. M. Harris
C. N. Isackson
P. J. Kranz
Z. J. Lansky
D. E. McVicker
N. P. Milano
C. J. Rhodes
H. L. Schmedt
R. Seese

C. D. Skrzypek
R. B. Smith
W. E. Smith
A. A. Staklis
C. R. Straesser
G. W. Tuffnell
A. Wilcox
W. B. Young
R. P. Zerwekh

BEARINGS AND BUSHINGS SUBCOMMITTEE—A. O. DeHart (Chairman), S. M. Brandt, G. R. Conrad, G. R. Kingsbury, G. J. LeBrasse, C. G. Massieon, H. R. Payne, C. C. Pratt, A. W. Rike, C. R. Straesser. CAST ALUMINUM AND ZINC ALLOYS SUBCOMMITTEE—S. M. Brandt, R. P. Dunn, C. Gibbons, J. M. Greer, T. R. Hitchcock, A. Kearney, P. T. Madziar, D. E. McVicker, C. R. Mielke, H. D. Monsch, J. N. Schweitzer, C. D. Skrzypek, R. B. Smith, A. A. Staklis, D. Shah, R. P. Walson, G. W. Wlodgya, W. B. Young, F. J. Zeglen. CAST AND WROUGHT MAGNESIUM SUBCOMMITTEE—R. S. Busk (Chairman), A. H. Braun, S. Erickson, T. R. Hitchcock, J. G. Mezoff, H. J. Proffitt. COPPER AND COPPER BASE ALLOYS SUBCOMMITTEE—P. J. Kranz (Chairman), A. Cohen, R. J. Cox, R. H. Hammer, A. H. Hesse, W. N. Ling, D. C. Marsden, D. Osborne, R. E. Ricksecker, T. E. Rishavy, J. Sawicki, M. D. Scott. ELECTROPLATING AND ANODIZING SUBCOMMITTEE—C. N. Isackson (Chairman), J. W. Backus, F. W. Baker, E. F. Barkman, V. J. Cassidy, R. J. Clauss, J. R. Crain, O. Klingenmaier, T. Leggin, R. McLafferty, D. C. Marsden, D. Montgomery, L. M. Morse, F. E. Odell, P. Smith, A. D. Squitero, G. W. Tuffnell. NONFERROUS POWDER METALS SUBCOMMITTEE—W. E. Smith (Chairman), R. C. Botterman, R. E. Doane, M. F. Hostetler, S. Lawrence, D. Nims, L. F. Pease, III, W. H. Roth, R. R. VanValkenburg, R. P. Walson. SPECIAL PURPOSE ALLOYS SUBCOMMITTEE—A. R. Canady, R. J. Kekervis, A. Roy, E. T. Vitcha. WROUGHT ALUMINUM SUBCOMMITTEE—S. M. Brandt, J. M. Dobos, W. M. Harris, D. L. Graham, A. Kasper, J. H. McCullogh, D. E. McVicker, L. W. Moriarty, Calvin J. Rhodes, D. Shah, L. B. Shives, G. E. Stein, W. E. Swenson, Jr., R. P. Walson, H. W. Zoeller.

NONMETALLIC MATERIALS BALLOTING COMMITTEE

L. J. LAMBERG (Sponsor)
T. J. WALLAG (Chairman)
J. J. MESTDAGH (Vice-Chairman)
R. F. Anderson
H. Baker
K. Balliett
R. C. Bremer, Jr.
A. B. Calhoun
M. Carter, Jr.
F. N. DeMott
C. O. Durbin
M. Gale
A. F. Hegerich

J. E. Hinsch
G. A. Ilkka
R. J. Kane
Z. J. Lansky
F. H. Lieb
L. Malone
R. D. Mayne
J. B. McCallum
W. J. McCortney
D. G. McLeod
E. Melgun
M. Newberry
G. C. Norman

S. Patel
E. Presser
N. R. Roobol
L. Roslinski
J. Sawicki
R. D. Silver
R. W. Siorek
R. J. Slingerlend
A. E. Williams
W. J. Wilson
G. J. Wolf
F. Woolary
J. S. Ziomek

(SAE Staff, R. T. Northrup)

NONMETALLIC MATERIALS EXECUTIVE COMMITTEE

L. J. LAMBERG (Sponsor)
T. J. WALLAG (Chairman)
J. J. MESTDAGH (Vice-Chairman)
H. Baker
K. Balliett
R. C. Bremer, Jr.
M. Carter, Jr.
C. O. Durbin
M. Gale
A. F. Hegerich

J. E. Hinsch
G. A. Ilkka
R. J. Kane
Z. J. Lansky
F. H. Lieb
L. Malone
J. B. McCallum
W. J. McCortney
E. Melgun
M. Newberry

E. Presser
L. M. Roslinski
J. Sawicki
R. D. Silver
R. W. Siorek
R. J. Slingerlend
A. E. Williams
W. J. Wilson
G. M. Wolf
J. S. Ziomek

(SAE Staff, R. T. Northrup)

ADHESIVES AND BODY SEALER SUBCOMMITTEE—H. Baker (Chairman), J. G. Claussen, H. Koski, J. L. Martin, G. Tiedeck, E. K. Williams, J. W. Zimmerman. AUTOMOTIVE FOAM AND SPONGE SUBCOMMITTEE—K. Balliet (Chairman), R. Martindale, R. L. Pleiness, D. A. Seil. AUTOMOTIVE GASKETS SUBCOMMITTEE—J. Sawicki (Chairman), D. Andress, A. Aramian, D. S. Brown, A. W. Gilbert, A. L. Gordon, A. H. Johnson, R. D. Mayne, M. T. Passarella, E. Presser, D. O. Stovall, E. D. Taylor, A. Warniak, D. W. Watson. COOLANT HOSE SUBCOMMITTEE—M. Carter, Jr. (Chairman), F. Arabia, D. Barber, M. A. Boshmer, H. Cox (Liaison), H. G. Deline (Liaison), L. J. Dreyer, J. R. Dunn, R. Endress, J. W. Espy, E. Francis, C. Glomski, J. Golding, J. Hamann, R. H. Hannum, M. B. Johnson, P. J. Jones, J. E. Juntenen, R. Kaizer, R. J. Kane, D. Kimmet, T. Knurek, F. J. Lasker, J. Madden, J. T. McElhose, S. Monthey, M. Movsesian, P. W. Muller, M. Newberry, A. Olcott, A. Patterson, L. R. Phillipi, D. A. Pittenger, E. Presser, H. E. Sears, R. Semin, A. Shah, C. T. Simmons, L. J. Stork, G. Vyse, E. Wilks, A. E. Williams, R. L. Wooten, K. E. Yost, J. J. Young. ENGINE COOLANTS SUBCOMMITTEE—C. O. Durbin (Chairman), R. Christ, J. W. Compton, J. Conley, D. A. Cramer, F. R. Duffey, R. H. Fay, J. Gould, V. R. Graytok, H. J. Hannigan, R. A. Hayes, D. Hudgens, C. J. Korpics, D. Pellerito, E. J. Rambie, L. C. Rowe, C. F. Rueping, T. Yates. FIBERBOARD MATERIALS SUBCOMMITTEE—J. B. McCallum (Chairman), B. Bennett, R. Billingshurst, D. Brown, R. Daugherty, H. Ely, G. C. Fewson, D. Flotow, N. Fringer, R. E. Gostioux, W. H. Hooth, B. Krishnan, R. Lamb, G. R. Lewis, B. Radke, A. Roman, D. Stovall, G. Strass, S. Wallag, W. Walter. TASK FORCE FOR FIBERBOARD CLASSIFICATION SYSTEM—J. B. McCallum (Chairman), B. Brezina, D. Flotow, R. E. Gostioux, B. Hooth, R. Lamb, A. Roman. FUEL, OIL, AND EMISSION HOSE SUBCOMMITTE—M. Gale (Chairman), F. Arabia, R. Briggs, R. A. Brullo, M. Carter, Jr., W. E. Currie, T. Dendinger, J. R. Dunn, G. Evanoff, E. A. Green, S. Gupta, T. L. Jackson, H. H. Klever, O. Linger, J. T. McElhose, J. J. Millard, M. Movsesian, P. W. Muller, M. Newberry, J. T. Pecora, J. E. Putacik, P. Robertson, J. Sawicki, S. L. Schafer, H. E. Sears, R. M. Smith, L. E. Sollberger, J. Spoulding, H. Visser, D. W. Watson, J. Wellbaum, C. D. Wemmers, R. Wiley, A. E. Williams. NON-LEAD FLEET COLORS SUBCOMMITTEE—J. E. Hinsch (Chairman), R. F. Brady, F. Brotherson, R. Brown, P. G. Campbell, N. D. Cassel, J. B. Coxon, P. R. Desai, L. J. Dick, D. Dissonette, E. J. Flewelling, R. Gerear, J. E. Grady, R. H. Groesbeck, A. H. Groth, D. G. Hamaker, D. Hudson, R. J. Kane, K. L. Kelly, F. G. Kostrub, M. Kuo, P. F. Kusy, F. L. Lafferman, C. Livingston, J. Madden, E. J. Masar, E. O. McLaughlin, E. Melgun, G. Muir, H. J. O'Connor, S. Panush, J. O. Peterson, D. Pettigrew, M. Plaza, E. A. Praschan, N. Price, B. Reynolds, J. V. Roddewig, N. R. Roobol, W. A. Russo, R. Sackett, R. J. Sadowski, W. St. John, R. E. Schuster, D. H. Soule, B. Sweet, E. Tysl, A. W. D. Vos, T. J. Wallag, C. Weiman, T. J. Young. PLASTICS SUBCOMMITTEE—J. J. Mestdagh (Chairman), R. Kwapisz (Vice-Chairman), A. B. Calhoun, G. G. Eaton, W. H. Englehart, D. Fedor, J. E. Greenup, S. Hoffman, C. Konya, P. F. Kusy, J. Madden, E. L. Melgun, A. A. Staklis, F. O. Swanson, R. C. Turner, D. W. Watson. SPECIAL TASK FORCE OF THE SAE PLASTICS SUBCOMMITTEE—J. J. Mestdagh (Chairman), C. Konya, P. F. Kusy, R. Kwapisz, F. O. Swanson. SOUND AND HEAT INSULATION MATERIALS SUBCOMMITTEE—F. H. Lieb, (Chairman), T. Kozyra (Vice-Chairman), J. Allen, P. Andries, R. R. Audette, K. S. Bagga, W. J. Banacki, M. J. Banks, B. Bernard, F. Botz, J. Carlson, R. Canfield, V. Cardela, L. Caudill, J. Collins, J. A. D'Amico, C. W. Dietrich, E. F. Franz, C. K. Furton, F. Gordon, H. W. Gorenflo, J. C. Haines, H. Hartley, D. Henrickson, G. Huber, W. Katzer, J. F. Kerscher, G. Kleeman, B. Krishnan, K. Kubofcik, T. L. Lauvray, G. P. Lindaberry, R. W. Livingston, C. K. Mallon, J. Masiak, R. McAdams, A. P. Mueller, T. Nitsch, V. G. Parrill, R. T. Reidenbach, H. Reiher, N. Riley, D. J. Schmitz, A. L. Schichner, D. F. Schumacher, P. H. Scott, W. Smith, Jr., D. Stewart, J. C. White, Jr., E. K. Williams. STRUCTURAL COMPOSITE MATERIALS SUBCOMMITTEE—J. S. Ziomek (Chairman), W. Steuf (Vice-Chairman), T. Barkimer, D. Cedar, J. G. Claussen, A. A. G. Cooper, J. Epel, C. T. Hawk, III, R. R. Hicks, J. C. Hsu, P. Ilitch, K. Kinsman, P. F. Kusy, C. H. Leaf, M. K. McDougall, J. S. McKinley, J. J. Mestdagh, W. C. Phillips, B. Putnam, A. D. Rosenstein, B. A. Sandors, R. E. Vollbach, A. S. Williamson, D. Yates. VIBRATION CONTROL SUBCOMMITTEE—W. J. Wilson (Chairman), D. Allgood, J. H. Bramley, R. C. Bremer, Jr., F. Brouwer, P. Browne, G. Buchaman, T. Bush, P. F. Cain, A. B. Calhoun, W. T. Cummins, F. N. DeMott, T. Dendinger, J. T. Drompp, J. Gira, P. V. Heck, A. F. Hegerich, D. F. Kruse, A. F. Lawrence, M. Lowman, R. D. Mayne, R. D. McFeeters, H. L. Oh, L. R. Oliver, A. F. Pacis, J. Pizarck, G. E. Rudd, F. Sanders, J. Stringfellow, T. Susko, A. Utter, W. Woolar, F. C. Yakes. ROUND ROBIN TEST TASK FORCE ON SAE J1085—W. J. Wilson (Chairman), T. Bush, J. Gerhardt, D. F. Kruse, L. LaRowe, M. Lowman, G. E. Rudd, W. Woolar.

OFF-ROAD MACHINERY TECHNICAL COMMITTEE

G. L. KLOSE (Chairman)
R. G. RUMPF (Vice-Chairman)
D. C. Ager
R. I. Besser
W. L. Black
H. D. Bordeaux
R. R. Brubaker
D. R. Buerschinger
J. E. Carr
D. E. Cheklich
J. B. Codlin
D. W. Driscol
W. T. Estabrook
A. L. Garman

R. F. Griffin
D. Guy
N. F. Hanson
J. H. Hyler
C. L. Kepner
S. T. Kieller
P. Larsen
L. E. Miller
R. Oliver
F. W. Ritchey
A. H. Saele
J. R. Salonimer
R. E. Schwary
P. P. Seabase

D. B. Shore
J. E. Staab
D. J. Thompson
A. J. Tolbert
L. A. Venere
J. P. Walsh
L. L. Williams
J. L. Woodward
D. W. Hadden (Liaison)
P. D. Hopler (Liaison)
A. H. Huebner (Liaison)
M. R. North (Liaison)
L. Watts (Liaison)

(SAE Staff, R. C. Uhl)

SUBCOMMITTEE 1—TRACTORS AND MOUNTED ATTACHMENTS—A. H. Saele (Chairman), S. A. Hyland (Vice-Chairman), J. M. Baylor, H. E. Beams, R. J. Fanslow, F. J. Halterman, E. E. Harvey, R. E. Hauff, R. C. Henneman, J. W. Hoeschen, D. J. Mickus, D. Ribbing, M. G. Watt. SUBCOMMITTEE 2—TIRE AND RIM—J. R. Salonimer (Chairman), R. E. Dagnall (Vice-Chairman), R. F. Boersma, F. M. Burns, W. F. Busbey, H. R. Cozad, D. A. Cucchi, W. T. Estabrook, R. H. Finefrock, J. J. France, F. C. Hausz, J. Jewett, L. Johnson, W. J. Johnson, D. Kuhl, H. D. Larson, R. F. McCarthy, D. B. Meisner, R. L. Ochsenhirt, W. T. Smith, R. E. Snodgrass, W. S. Trindal, C. R. Ward, W. G. Weber, R. N. Wenda, K. Wolfgram, G. Zambelas. SUBCOMMITTEE 4—HYDRAULIC FLUID POWER SYSTEMS AND COMPONENTS—D. B. Shore (Chairman), J. T. Parrett (Vice-Chairman), H. P. Barthe, R. M. Barton, T. Berg, D. F. Carl, W. L. Chichester, L. Claar, J. M. Dillon, J. P. Fairbairn, E. L. Falendysz, A. A. Farris, W. D. Fast, E. C. Fitch, M. P. Gehring, W. F. Heilman, R. Inoue, R. B. Janvrin, W. F. Marshall, D. R. Mattson, R. Messerschmidt, H. O. Mielke, J. V. Miller, E. J. Ratkay, H. B. Reese Jr., S. D. Royle, R. K. Schantz, J. L. Schmitt, H. F. Schultz, R. E. Schwary, W. L. Snyder, D. D. Straight, R. W. White, W. F. Zoller, A. Alameddin (Consultant), R. Deitrich (Consultant), C. W. Henry (Consultant), J. M. Karhnak, Jr. (Consultant), D. MacIntosh (Consultant), P. O'Donnell (Consultant), G. D. Olson (Consultant), H. H. Schmiel (Consultant), J. W. Wilcox (Consultant), J. Feuerstein

(Liaison), W. A. Hertel (Liaison), G. R. Flanagan (Liaison), J. C. White (Liaison). SUBCOMMITTEE 5—ELECTRICAL EQUIPMENT—R. R. Brubaker (Chairman), G. E. Redzinski (Vice-Chairman), R. C. Aigner, R. H. Bartel, P. Bibbee, R. L. Branstetter, A. K. Chaudhuri, R. N. Hagen, J. M. Haggard, K. A. Julian, G. A. Koch, D. W. Mattick, H. O. Mosquera, R. W. Oyler, L. J. Raver, K. L. Schar, J. Scheberle, W. L. Scully, B. E. Shoulders, J. E. Szudy, W. S. Trindal, J. P. Welsh, J. C. Giacobino (Alternator Consultant), J. C. Hardy (Alternator Consultant), C. J. Juhnke (Alternator Consultant), E. Szabo (Alternator Consultant), D. Wickland (Alternator Consultant), N. R. Eisenhut (Battery Consultant), N. Sangmavi (Battery Consultant), A. Fox (Battery Consultant), R. C. Slautterback (Battery Consultant), J. E. Bair (Lighting Consultant), R. Layko (Lighting Consultant), K. A. Degnan (Wiring Consultant), J. Haney (Wiring Consultant), G. M. Giannini (Switch Consultant), D. G. Grier (Gauge Consultant), K. Kehm (Gauge Consultant), E. Lindberg (Gauge Consultant), C. Price (Gauge Consultant), B. K. Reed (Gauge Consultant). SUBCOMMITTEE 6—ENGINES AND POWER UNITS—D. W. Driscol (Chairman), P. H. Brandes, L. D. Carufel, W. W. Cross, T. Dickson, C. R. Freeberg, J. A. Gresch, A. G. Johnson, R. P. Kesl, D. F. Linn, D. McLean, J. T. Mehne, T. I. Schoepel, H. Shipe, H. Steinhagen, D. J. Thompson, O. B. Turkilmaz. SAFETY ITEM SUBGROUP OF SUBCOMMITTEE 6—J. A. Gresch (Chairman), L. D. Carufel, R. Day, D. W. Driscol, C. R. Freeberg, W. Kenyon, D. Linn, J. Mehne, T. I. Schoepel, H. Steinhagen, D. J. Thompson, O. B. Turkyilmaz, A. A. Zagotta. SUBCOMMITTEE 8—GROUND ENGAGING TOOLS—A. Garman (Chairman), R. Montgomery (Vice-Chairman), H. Auter, G. R. Bailey, C. Bole, V. Bowen, M. Carlson, F. Christensen, J. R. Cryder, R. E. Dannan, L. E. Eaton, R. W. Foley, M. R. Goar, R. Goetzke, W. Jones, S. Kabay, A. W. McGraw, Jr., L. R. McLean, J. T. Passarelli, C. L. Reid, Jr., T. B. Russell, M. Shaughnessy, C. Stephens, J. Thompson, P. Tyler, R. E. Wahlstrom, W. Wilkinson, R. D. Lank (Consultant), J. M. Poker (Consultant), W. Webster (Consultant), E. M. Wilson (Consultant). SUBCOMMITTEE 9—LOADERS AND MOUNTED ATTACHMENTS—J. P. Walsh (Chairman), C. G. Termont (Vice-Chairman), M. L. Abbott, A. W. Acker, R. H. Anderson, C. A. Ardelt, O. T. Bentrup, D. G. Castine, J. J. Dunning, C. Frisbee, R. Hayes, E. R. Hodgman, D. A. Holtkamp, F. J. Hoppe, R. M. Jesswein, P. L. Kelsey, L. Lenart, J. M. Mather, L. A. Molby, J. M. Poker, G. B. Western, P. M. Wilke. SUBCOMMITTEE 10—BRAKING—L. A. Venere (Chairman), D. R. Thomas (Vice-Chairman), B. L. Amos, R. I. Besser, W. L. Daniels, L. W. Davis, W. C. Davidson, J. F. Dernovshek, G. J. Dvorznak, L. R. Elliott, R. T. Evans, T. Fouser, R. Fowler, R. L. Grennan, D. W. Hadden, M. J. Hapeman, D. J. Hogberg, O. L. Holcomb, P. L. Kelsey, R. C. Lehman, B. L. McGrew, R. E. Miller, G. Moore, P. A. Myers, R. F. Plantan, P. Rosenberger, R. F. Rought, F. R. Schubert, G. W. Stearns, R. R. Svenson, T. Whitlock, H. W. Whyy, D. Oldenberg (Liaison), F. Staunton, (Liaison). RETARDATION OF OFF-HIGHWAY MACHINES AD HOC COMMITTEE—S. Kershaw, G. M. Bloom, J. L. Eldridge, C. C. Gray, R. J. McClone, T. Milner, E. Thomas, G. Trowbridge, J. R. Wiley. SUBCOMMITTEE 12—MACHINE TEST PROCEDURES—D. C. Ager (Chairman), D. A. Lockie (Vice-Chairman), A. F. Boyer, C. F. Crandall, J. L. Dahle, H. C. Davis, G. Douglas, R. C. Gessel, W. O. Gipp, D. W. Hadden, R. J. Hanna, J. H. Hyler, W. C. Jackson, G. Keller, S. T. Kieller, D. J. Nelson, E. Nyborg, G. L. Parr, J. R. Prosek, G. C. Randall, J. C. Skeel, L. E. Spencer, G. A. Stangl, K. C. Strebig, S. Swan, J. Szudy, J. Thauberger, J. Wagner, J. L. Woodward. AD HOC COMMITTEE ON ROPS FOR ROLLERS AND COMPACTORS—F. Bean, G. Klose, E. Melhuish, B. Ringwelski, J. Williams. ROPS FOR DUMPERS WITH FULL MOUNTED BODIES AD HOC COMMITTEE—G. Douglas, D. Evans, D. Guhl, D. Hadden, F. Hoppe, R. Kress, J. Siewert, V. Suopys, S. Swan, J. Szudy. AD HOC COMMITTEE ON FATIGUE AND CORROSION—S. Adams, H. C. Davis, D. Galliart, G. Gavan, G. Keller, D. Lockie, S. Swann. PLANNING AD HOC COMMITTEE—S. B. Admas, A. F. Boyer, D. J. Hanna, D. A. Lockie, J. C. Thauberger, D. C. Ager, W. O. Gipp, G. L. Klose, G. L. Parr, J. L. Woodward. SUBCOMMITTEE 13—TERMINOLOGY AND DEFINITIONS—C. L. Kepner (Chairman), B. J. Slinger (Vice-Chairman), G. L. Ball, W. Collins, J. Condra, L. L. Lemke, P. J. Sperry, J. E. Staab, H. T. Larmore (Consultant), W. E. Miller (Consultant). SUBCOMMITTEE 15—MACHINE CLIMATIZATION—N. F. Hanson (Chairman), D. G. Batchelor (Vice-Chairman), W. G. Bugelski, G. P. Burton, N. B. Esabrook, A. B. Follansbee, J. G. Gbur, H. R. Grivna, E. G. Hanus, R. Hushower, C. R. Jefferson, L. Kloepfer, A. D. Lewis, M. Margeson, V. J. Matles, J. Rymes, R. Schaefer, R. Shaw, F. Venhoff, L. E. Windsor, D. A. Winer, G. Wolf, R. Ali (Consultant), B. J. Homkes (Consultant), J. Krueger (Consultant). SUBCOMMITTEE 16—SERVICEABILITY—R. J. Oliver (Chairman), P. D. Redenbarger (Vice-Chairman), D. E. Chiklich, C. E. Fenson, S. C. Hammons, J. A. Higginbotham, R. F. Hughes, A. R. Link, S. R. Mayer, J. Miller, R. Miller, D. W. Patton, D. Phillips, D. R. Potter, T. R. Turnbull, C. W. Volland. AD HOC COMMITTEE FOR FAN GUARDS—T. Bachon, C. E. Fenson, R. F. Hughes, D. Kella, D. Miller, A. Sheth. SUBCOMMITTEE 17—CRANES AND EXCAVATORS—R. F. Griffin (Chairman), P. A. Chalupsky (Vice-Chairman), R. L. Bauer, H. Babcock, T. O. Davidson, M. Hamilton, A. B. Hill, E. G. Johnson, K. V. Johnson, D. C. Juergens, J. H. Keller, R. M. Kohner, H. E. Luyties, J. Maytum, T. S. McKosky, J. G. Morrow, R. Paciorek, D. B. Rees, H. I. Shapiro, F. J. Tucek, E. J. Vroonland, R. Wheeler, J. A. Gilligan (Consultant), D. W. Hadden (Consultant). LATTICE BOOM COMPETENCY AD HOC COMMITTEE—E. J. Vroonland (Chairman), M. Anderson, R. Berg, E. G. Johnson, K. V. Johnson, D. Marcotte, G. Rennich, B. Snelling, R. E. Welch, C. Wick. WORKING GROUP A—HYDRAULIC CRANES—K. V. Johnson (Chairman), R. A. Fritsch (Vice-Chairman), J. D. Anderson, R. D. Becker, G. D. Dolezal, T. Holmes, J. Keller, H. Miller, M. Pasierbowicz. WORKING GROUP B—LATTICE CRANES—R. Kohner (Chairman), M. Anderson, J. Ashton, M. Brunet, H. Durscher, T. Holly, R. W. Job, K. V. Johnson, C. A. Larson, M. Pasierbowicz, D. B. Rees, J. P. Rennich, R. Sittner, W. J. Snelling, E. J. Vroonland. WORKING GROUP C—EXCAVATORS—J. Maytum (Chairman), D. Marek (Vice-Chairman), C. R. Clauson, C. Crawshaw, M. Donahue, R. Klitz, P. Moyer, R. A. Plemitscher. WORKING GROUP D—NOMENCLATURE—R. L. Bauer (Chairman), A. Freedy, R. W. Gustafson, J. G. Morrow, P. Cunningham, T. Stoychoff. WORKING GROUP E—INDICATING DEVICES—R. M. Paciorek (Chairman), T. E. McHugh (Vice-Chairman), R. A. Begun, W. C. Bennett, A. Bonnell, R. F. Griffin, E. G. Johnson, H. E. Luyties, D. S. Schaner, R. Wheeler, R. P. Wiener, J. J. Elengo (Consultant), J. W. Six (Consultant). WORKING GROUP G—TOWER CRANES—E. Anderson, M. Anderson, J. Boring, L. Knutson, M. Kohler, T. Olson, C. Rosenberg, E. Vroonland (Alternate). WORKING GROUP H—DYNAMIC DERATING—K. V. Johnson (Chairman), M. C. Anderson, R. F. Bailey, D. A. Davis, E. J. Holler, C. W. Ireland, F. R. Johnson, Jr., T. Greak, R. Kasprzak, R. B. Madden, P. Malone, W. R. McCreight, D. E. McCuaig, F. M. Monteiro, J. G. Morrow, D. B. Rees, C. C. Stevens, R. E. Welch, C. Wick. SUBCOMMITTEE 18—HUMAN FACTORS ENGINEERING—L. L. Williams (Chairman), E. C. Williams (Vice-Chairman), T. Antenucci, J. E. Carr, J. L. Clingerman, D. C. Fuller, R. L. Gerard, G. O. Hall, A. D. Loken, V. M. Pandav, N. H. Pulling, F. Riddle, D. G. Roley, R. L. Sawall, R. L. Sealine, D. L. Steele, R. F. Swenson, S. A. Tennyson, A. O. Radke (Consultant), R. B. Sleight (Consultant), R. J. McCracken (Liaison), G. R. Gaven (Liaison), T. J. Smith (Liaison), K. Conway (Liaison), J. H. Crowley (Liaison). VISIBILITY AD HOC COMMITTEE—R. Gerard, R. Johanningmeier, H. Kerner, D. Roley, N. Ross, S. Tennyson. CONTROLS AD HOC COMMITTEE—J. E. Carr (Chairman), D. C. Fuller, R. L. Gerard, A. D. Loken, V. M. Pandav, S. A. Tennyson, E. C. Williams, L. L. Williams. SUBCOMMITTEE 19—STEERING AND CONTROLS—P. P. Seabase (Chairman), G. Fisher (Vice-Chairman), J. A. Abramczyk, L. Becker, J. Feuerstein, T. O. Goodney, O. W. Johnson, T. Johnson, M. Naft, R. J. Pelouch, G. Redzinski, G. R. Rose, R. G. Fumpf, J. E. Scheberle, D. Servais, J. Siburt, C. A. Treadwell, R. A. Yoder, L. A. Cochran (Consultant), E. L. Collins (Consultant), H. T. Larmore (Consultant), J. Lauck (Consultant). STEERING AD HOC COMMITTEE—R. G. Rumf (Chairman), G. E. Steward (Secretary), T. O. Goodney, M. Naft, R. J. Pelouch, G. R. Rose, P. O. Seabase, C. A. Treadwell, E. J. Ohms, R. N. Wenzel, R. L. Zillman. SUBCOMMITTEE 20—SOIL EFFECTS—L. E. Eaton, G. Hand, J. E. Langdon, K. J. M. Melzer, C. A. Reaves, C. E. Sanders, E. T. Selig, E. V. Semonin, J. Sikkut, R. J. Sullivan, A. J. Tolbert, R. D. Wismer. SUBCOMMITTEE 25—GENERAL PURPOSE TRACTORS (INDUSTRIAL)—P. W. Larsen (Chairman), J. J. Dunning (Vice-Chairman), A. Brudnak, J. G. Gbur, R. A. Gillette, R. D. Moore, J. C. Skroski, D. K. Spring. SUBCOMMITTEE 28—FORESTRY AND LOGGING MACHINERY—D. Guy (Chairman), T. J. Smith (Vice-Chairman), H. Bradshaw, J. F. Briggs, Jr., J. Dikken, H. Hammond, R. C. Henneman, L. Kramer, I. McKenzie, L. E. Miller, D. Oldenburg, D. Paisley, D. Sirois, M. Smyth, T. Stoychoff, J. G. Morrow (Liaison). SUBCOMMITTEE 29—SPECIALIZED MINING MACHINERY—J. L. Woodward (Chairman), B. E. Morgan (Vice-Chairman), C. D. Albright, G. V. Auchard, J. Ault, R. Banks, S. Barker, E. Barrett, G. Bayles, K. Bickel, S. Black, G. R. Bockosh, T. Bodimer, E. Bollinger, D. Boness, P. M. Budzak, D. E. Burkhart, Jr., L. Chabot, A. Christopher, W. R. Cobb, D. Strassel, K. Conswa, M. Cosgrove, L. Crow, J. C. Cumming, C. Cypher, J. L. Dahle, C. Dieter, F. DuBreuil, G. Dvorznak, B. F. Eads, W. P. Edwards, B. Evilsizer, R. Faltsman, J. Fletcher, P. Forsyth, G. R. Frey, D. Freed, W. G. Doepken, C. W. Gardner, C. Garrett, G. R. Gavan, S. G. Gaydos, R. L. Gerard, M. Golben, J. M. Goris, D. J. Gorman, J. Gowins, R. Griffin, D. W. Hadden, W. E. Halopoff, L. Hansson, C. R. Harrison, H. Hartley, D. A Holtkamp, F. J. Hoppe, T. J. Huecker, R. F. Hughes, J. Janes, W. V. Jessee, D. A. Johnson, G. A. Johnson, J. Johnson, J. F. Judeikis, F. Jurczak, J. M. Karhnak, Jr., K. Keeney, V. Kevorkian, J. Klein, R. A. Knopp, E. H. Kurt, M. K. Lebegue, J. P. Mann, C. Marshall, G. C. Marshall, B. Mathies, R. Mazzoni, T. McCormick, R. J. McCracken, M. D. McGuire, R. Medina, R. C. Miles, W. E. Miller, B. Minor, E. Mullen, R. O. Mullin, B. Nelson, B. F. Noble, G. E. O'Keefe, M. Olson, J. Omer, N. Petelski, H. G. Pyles, H. B. Reese, J. Reeves, F. W. Ritchey, K. Rohlfs, R. D. Saltsman, A. Sampson, L. M. Schmelzer, B. Self, D. Shannon, J. L. Sheppard, D. Shutty, B. J. Smith, T. W. Snyder, F. Staunton, J. W. Steelman, R. B. Stevens, W. O. Stewart, K. Strebig, R. F. Sundstrom, C. Switalski, R. Thiele, J. M. Trapp, B. J. Turley, J. K. Van Burnt, J. Van Sloun, F. Van Vleet, G. G. Wattley, B. Waytulonis, D. J. Wehner, W. J. Wiehagen, G. B. Wiley, J. L. Woodward, G. M. Dufresne, A. Bacho. WORKING GROUP 3 OF SUBCOMMITTEE 29—HUMAN FACTORS—W. E. Halopoff (Chairman), C. D. Albright, B. Ashworth, J. Ault, J. Bogus, E. R. Bollinger, E. J. Conway, J. F. Colinet, J. L. Dahle, C. Duncan, G. J. Dvorznak, K. Ewald, C. L. Foster, B. Frey, L. Harding, D. Hardman, M. G. Helander, E. D. Hennen, V. Jessee, G. Marshall, L. B. McDonald, R. Mazzoni, R. C. Miles, C. R. Piro, G. Radomsky, W. D. Roper, W. R. Self, J. L. Sheppard, J. J. Teti, G. G. Wattley, K. Whitehead, W. J. Wiehagen, G. Bockosh (Alternate), W. J. Debevec (Alternate), S. D. Huntsman (Alternate), P. D. Lindahl (Alternate). WORKING GROUP 8 OF SUBCOMMITTEE 29—BRAKES—J. L. Dahel (Chairman), J. Bakos, G. Bockosh (Alternate), J. Bogus, J. Cumming, P. Forsyth, G. R. Frey, J. F. Gowins, T. J. Huecker (Alternate), V. Jessee, J. Judeikis, R. A. Knopp, T. McCormick, R. Medina (Alternate), G. O'Keefe, H. Pyles (Alternate), K. Rohlfs, T. Snyder, F. Staunton, R. Svenson, J. Trapp, J. Van Sloun, D. J. Wehner. SUBCOMMITTEE 30—ELECTRIC PROPULSION—OFF-ROAD MACHINES—D. Buerschinger (Chairman), F. A. Green (Vice-Chairman), F. Bartley, C. Bennett, L. Bowers, R. Brunetti, R. Cheers, T. Eva, S. Faulkner, R. C. Gutberlet, L. Ishler, R. McIntosh, T. Milner, R. Sinor, V. Thompson.

xxix

ON-HIGHWAY RECREATIONAL VEHICLE COMMITTEE

C. A. BIBLE (Sponsor)
R. H. MADISON (Chairman)
J. C. Abromavage
D. Arnold
D. Bartz
L. M. Baulis
A. Chamberlain
R. Cordill
J. Crane

R. Curtis
P. L. Graney
R. Herzler
W. L. Jacobson
E. Kent
G. Lucas
L. W. Moore
M. Nerem
C. J. Owen

D. I. Reed
D. Reeker
W. L. Sherry
B. R. Weber
P. White
C. A. Wilhelm
R. E. Wilkinson
A. E. Zollinger

(SAE Staff, L. P. Ziegler, Jr.)

MOTOR HOME SUBCOMMITTEE—R. Cordill (Chairman), J. Allen, D. Barz, W. Baumgartner, D. Betts, E. Blazek, W. Boehnlein, R. Campbell, R. Curtis, J. Dalton, N. F. Erickson, R. C. Frank, R. Franke, T. Gloor, P. L. Graney, R. J. Haller, R. Herzler, G. B. Hickner, J. C. Holtsberry, W. L. Jacobson, W. Korn, T. Krausch, G. Lucas, S. Lyons, L. W. Moore, D. A. Nelson, H. D. Reeker, R. Reschly, G. Schultz, R. Snowberger, H. L. Steinfeld, G. Stevens, K. Stransky, I. W. Strayer, R. Wilkinson, C. A. Wilhelm, A. E. Zollinger. MOTOR HOME VIN TASK FORCE—M. Nerem (Chairman), D. Armstrong, D. Betts, R. Burns, R. Campbell, M. W. Dixon, N. F. Erickson, R. J. Haller, P. Leverenz, R. H. Madison, W. Stark, G. Stevens, R. Zeller, A. E. Zollinger. CENTER OF GRAVITY AND LOAD DISTRIBUTION TASK FORCE—W. L. Jacobson (Chairman), W. Baumgartner, P. L. Graney, G. Hickner, D. A. Nelson, K. Stransky, R. Zeller. TIEDOWN LOCATION AND METHODS TASK FORCE—G. Lucas (Chairman), S. Lyons. MOUNTING OF AUXILIARY POWER UNITS TASK FORCE—R. Burdick (Co-Chairman), E. Keszthelyi (Co-Chairman). TRAILER SUBCOMMITTEE—B. R. Weber (Chairman), W. L. Sherry (Vice-Chairman), J. C. Abromavage, W. T. Birge, R. Fertitta, C. N. French, M. H. Hines, F. J. Huston, C. E. Klatt, R. H. Madison, W. P. Marshall, M. Nerem, E. V. Olgren, D. I. Reed.

PASSENGER CAR AND LIGHT TRUCK FUEL ECONOMY MEASUREMENT COMMITTEE

H. FREERS (Sponsor)
W. S. FREAS (Acting Chairman)
J. B. Baker
J. L. Bascunana
J. A. Bert
C. J. Daye
R. J. Dennison

R. G. Finnerty
E. C. Klaubert
L. J. LaDouceur
J. Mac Dougall
R. D. Matthews
W. G. Mears
H. McGee

W. L. McNulty
W. J. Most
J. D. Murrell
W. E. Schwieder
J. P. Steiger
G. Thompson

(SAE Staff, R. T. Northrup)

DYNAMOMETER VARIABILITY TASK FORCE—B. Mears (Chairman), P. Bandoian, W. S. Fagley, F. E. Johnson, E. C. Klaubert, C. W. LaPointe, W. J. Most, D. Paulsell, G. F. Pierce-Ruhland, G. Piodkowski, R. E. Rice, F. Sam, D. F. Sougstad, M. L. Straub, H. B. Weaver. FRONTAL AREA MEASUREMENT TASK FORCE—W. Freas (Chairman), D. Bonawitz, C. Chapin, W. Fogerty, H. Kazanjian, T. Takahashi. ROAD LOAD MEASUREMENT TASK FORCE—W. S. Freas (Chairman), D. Bonawitz, C. Chapin, E. C. Klaubert, C. LaPointe, W. L. McNulty, D. Sutherland, T. Takahashi, G. Thompson.

SEAL COMMITTEE

E. D. TAYLOR (Chairman)
D. L. OTTO (Vice-Chairman)
G. L. GREMAUX (Secretary)
J. W. Abar
R. L. Augustine
D. Bainard
A. S. Berens
E. S. Bower
R. V. Brink
B. Butler
R. J. Boyle
O. H. Cannedy
J. J. Carr
D. A. Cather, Jr.
J. Chandler
F. J. Charhut
D. J. Cornelious
R. L. Dega

W. G. Fuetterer
T. J. Gair
G. L. Germaux
L. T. Gilbert
A. Ginn
J. M. Gottlieb
F. Hatch
T. S. Hemenway
K. L. Hoer
L. Horve
E. T. Jagger
C. Kramer
B. F. Kupfert
R. K. MacLaren
W. Maltman
L. R. Marcy
J. Marinik
L. G. Marlin

K. Martek
A. Matsushima
D. L. Otto
J. O. Payne
B. A. Pearson
D. E. Perry
R. S. Rainey
V. D. Raut
T. V. Saether
L. H. Saylor
J. Schaus
J. A. Serio
L. L. Smith
S. N. Smith
J. D. Symons
D. A. Vandeven
M. R. Young

(SAE Staff, R. T. Northrup)

J1245, APP. ENGINE COOLANT PUMP SEALS SECTION—A. Matsushima (Chairman), J. W. Abar, B. A. Butler, J. D. Symons. J946 SEALS, APPLICATION GUIDE TO RADIAL LIP SECTION—D. R. Bainard (Chairman), J. M. Gottlieb, R. K. MacLaren, D. L. Otto. J281, CAST IRON SEALING SECTION—D. C. Hill (Chairman), K. Martek, L. H. Saylor. EDITORIAL SUBCOMMITTEE—J. D. Symons (Chairman), D. R. Bainard, G. L. Gremaux, K. L. Hoer, L. R. Marcy, J. O. Payne, S. N. Smith. J780, ENGINE COOLANT PUMP SEALS SECTION—A. Matsushima (Chairman), J. W. Abar, B. A. Butler, J. D. Symons. J1002, SEALS, EVALUATION OF ELASTOHYDRODYNAMIC SECTION—L. R. Marcy (Chairman). FACE SEAL SUBCOMMITTEE—S. N. Smith (Chairman), J. W. Abar, B. A. Butler, F. H. Charhut, A. Ginn, A. Matsushima, J. D. Symons. GASKETS SUBCOMMITTEE—E. D. Taylor (Chairman), D. E. Czernik, J. M. Gottlieb, D. J. McDowell, J. A. Serio, C. Stout. RADIAL LIP SEAL SUBCOMMITTEE—D. L. Otto (Chairman), R. L. Augustine, D. R. Bainard, A. S. Berens, R. V. Brink, B. Butler, J. Carr, D. A. Cather, Jr., J. Chandler, F. J. Charhut, D. J. Cornelious, E. S. Czekansky, R. L. Dega, L. T. Gilbert, A. Ginn, J. M. Gottlieb, G. L. Gremaux, F. Hatch, K. L. Hoer, L. Horve, T. S. Hemenway, D. J. Jutzi, B. F. Kupfert, R. K. MacLaren, W. Maltman, L. R. Marcy, L. G. Marlin, K. Martek, J. O. Payne, D. Perry, J. E. Schaus, J. A. Serio, L. L. Smith, S. N. Smith, C. Stout, J. D. Symons, E. D. Taylor, D. A. VanDeven. J120, ''O'' RINGS SECTION—E. S. Bower, J. M. Gottlieb, C. R. Kramer, J. A. Serio. SEAL AND FLUID COMPATIBILITY—D. J. Jutzi (Chairman), J. J. Carr, D. J. Cornelious, E. S. Czekansky, L. T. Gilbert, J. M. Gottlieb, L. G. Marlin, D. E. Perry. SEALANTS SUBCOMMITTEE—J. J. Carr (Chairman), D. J. Cornelious, D. E. Czernik, D. J. McDowell, E. D. Taylor. SEAL MANUAL—RADIAL LIP SECTION—R. V. Brink (Chairman), D. R. Bainard, R. L. Dega, K. L. Hoer, L. Horve, D. J. Jutzi, A. Matsushima, D. L. Otto,

S. N. Smith, J. D. Symons. SPECIAL SEALS SUBCOMMITTEE—F. Hatch (Chairman), E. S. Bower, J. J. Carr, D. A. Cather, Jr., J. M. Gottlieb, D. C. Hill, H. J. Knott, C. R. Kramer, L. H. Saylor, J. E. Schaus, L. L. Smith, S. N. Smith, C. Stout, D. A. VanDeven. J654, STATIC AND RECIPROCATING ELASTOMER TRANSMISSION SEALS SECTION—H. J. Knott (Chairman), E. S. Bower, R. V. Brink, A. Ginn, C. R. Kramer, K. Martek, L. L. Smith, TECHNICAL SESSIONS SUBCOMMITTEE—J. A. Serio (Chairman), L. Horve, K. Martek, A. Matsushima.

J111, SEALS, TERMINOLOGY OF RADIAL SECTION—D. A. VanDeven (Chairman), D. A. Cather, Jr., A. Matsushima, S. N. Smith, J. D. Symons. J110, SEALS, TESTING OF RADIAL LIP SECTION—J. Chandler (Chairman), R. L. Augustine, E. S. Czekansky, J. M. Gottlieb, B. F. Kupfert, R. L. MacLaren, A. Matsushima, J. E. Schaus. TFE SEALS SECTION—D. A. VanDeven (Chairman), D. R. Bainard, B. A. Butler, D. A. Cather, Jr., J. M. Gottlieb, F. Hatch, K. L. Hoer, D. E. Perry, J. E. Schaus.

SEAL EXECUTIVE COMMITTEE

M. R. YOUNG (Sponsor)
D. L. OTTO (Chairman)
E. D. TAYLOR (Vice-Chairman)
S. N. SMITH (Secretary)

J. Carr
L. T. Gilbert
F. Hatch
K. Martek

A. Matsushima
J. A. Serio
J. D. Symons
D. A. VanDeven

(SAE Staff, R. T. Northrup)

SNOWMOBILE COMMITTEE

T. H. LOHR (Chairman)
N. Berg
M. A. Berk
R. R. Cote
W. Fox
L. C. Lake

G. Marier
R. W. Muth
D. Nelson
K. K. Prasad
S. Quick
C. Walsh

A. R. Erickson (Honorary Member)
J. Fandy (Liaison)
W. L. Severson (Liaison)
R. Harrison (Consultant)
W. L. Konickson (Consultant)

(SAE Staff, R. C. Uhl)

SNOWMOBILE PERSONAL SAFETY SUBCOMMITTEE—K. K. Prasad (Chairman), R. Lietzow (Vice-Chairman), M. A. Berk, R. R. Cote, W. Fox, L. E. Haas, L. C. Lake, T. H. Lohr, T. Lykken, D. Nelson, C. Picard, S. Quick, C. Walsh, J. Fandy (Liaison), W. L. Konickson (Consultant). SNOWMOBILE ELECTRICAL, LIGHTING AND SIGNALLING SUBCOMMITTEE—R. R. Cote (Chairman), D. Nelson (Vice-Chairman), F. Flathman, L. C. Lake, G. Marier, B. Muise, J. A. Orvis, W. L. Scully, A. Streicher, L. Swanson, M. Van Overbeke, C. Walsh, K. Denneler (Consultant), B. G. House (Consultant), V. Lunde (Consultant), K. F. Marsh (Consultant), D. Pedley (Consultant), E. Yochum (Consultant), J. Fandy (Liaison). TERMINOLOGY AND PERFORMANCE SUBCOMMITTEE—R. Niemchick (Chairman), N. Berg, S. Braun, G. Marier, S. Quick, C. Walsh.

SPECIALIZED VEHICLE AND EQUIPMENT SOUND LEVEL COMMITTEE

D. R. HARTDEGEN (Chairman)
G. M. Adams
L. J. Eriksson
R. D. Hellweg, Jr.

R. A. Lanpheer
R. H. Lincoln
J. W. Moore
W. E. Roper

W. Snyder
J. Walsh
J. Wootten

(SAE Staff, R. T. Northrup)

MARINE SOUND LEVEL SUBCOMMITTEE—R. A. Lanpheer (Chairman), P. C. Ball, R. B. Passmore, D. Piper, J. J. Pok, D. I. Reed, W. M. Sanford, D. Thornburg, J. R. Winberg. MOTORIZED SNOW VEHICLE SUBCOMMITTEE—J. W. Moore (Chairman), N. Berg, M. Basham, C. Barrows, J. Blass, T. Brammer, L. E. Haas, R. Harrison, M. Khosropour, R. Kostecki, T. Kozyra, L. C. Lake, R. W. Muth, J. Nicolas, K. F. Nowak, I. N. Swanson, J. Thomas, R. Trapp, G. Tremblay. SMALL ENGINE POWERED EQUIPMENT SUBCOMMITTEE—G. M. Adams (Chairman), M. Curtis, R. DeRuyter, W. Durant, R. Elmy, R. Faenger, L. Fahning, D. Gordon, L. Hall, I. Kamlukin, M. M. Khosropour, J. T. Nadolny, R. Peterson, L. Presnall, M. Prevost, R. Siegrist, H. Trefz, J. R. Winberg, N. Woelffer.

SPEEDOMETER AND TACHOMETER COMMITTEE

K. F. NAYLOR (Sponsor)
J. G. ZELENKA (Chairman)
R. Abel
D. Barberi
W. W. Bischoff (Corresponding Member)

R. J. Broadman
W. C. Ellis
J. S. Harley
S. Hasko
S. R. Holland

J. Marlotte
H. B. Rath
R. D. Straszheim
W. C. Subluskey

(SAE Staff, P. Couhig)

ELECTRIC SPEEDOMETER AND TACHOMETER SUBCOMMITTEE—K. F. Naylor (Sponsor), R. D. Straszheim (Chairman), C. T. Aquilino, E. Baez, R. J. Broadman, G. J. Crowdes, R. M. Currie, T. DeCook, T. G. Faria, R. Jenks, C. N. Lansdown, W. M. Maki, D. Neill, R. Pitchford, H. B. Rath, G. T. Rini, C. Sebold, J. R. Shaffer, H. J. Simms.

SPRING COMMITTEE

H. M. REIGNER (Sponsor)
K. CAMPBELL (Chairman)
T. A. Bank

J. J. Bozyk
J. Brannan
R. N. Darash

G. W. Folland
J. Gibbs
L. A. Habrle

R. E. Hanslip
W. J. Jarae
E. H. Judd
J. F. Kelly
W. Mayers
M. W. Mericle

G. W. Myrick
E. C. Oldfield
W. Platko
F. T. Rowland
H. L. Schmedt
G. Schremmer

K. E. Siler
J. E. Silvis
B. Sterne
E. J. Streichert
F. Waksmundzki

(SAE Staff, P. Couhig)

BELLEVILLE WASHER SUBCOMMITTEE—W. M. Wood (Chairman), H. H. Hall, S. Q. Jafri, E. H. Judd, W. Mayers, G. Schremmer. COIL SPRING SUBCOMMITTEE—W. Mayers (Chairman), A. M. Peach (Vice-Chairman), M. L. Banta, W. J. Behnke, G. C. Cuff, H. O. Fuchs, D. E. Gartner, L. A. Habrle, E. H. Judd, F. J. Korpics, D. LaDuke, L. Lombardo, J. L. Luttinen, C. Meyer, I. Neimanis, T. L. Satchell, K. E. Siler, B. Sterne, E. J. Streichert, M. C. Turkish, R. Van Eerden, W. M. Wood. LEAF SPRING SUBCOMMITTEE—F. T. Rowland (Chairman), G. W. Follant (Vice-Chairman), K. Campbell, D. Curtin, R. E. Hanslip, J. F. Kelly, E. C. Oldfield, K. E. Siler, B. Sterne, E. J. Streichert, F. Waksmundzki. PNEUMATIC SPRING SUBCOMMITTEE—T. A. Bank (Chairman), T. Burkley, E. L. Harrod, R. E. Houser, J. R. Hughlett, W. S. Locke, J. M. Mann, W. C. Pierce, D. Strader, C. Wreford, M. C. Yew, W. J. Young. TORSION BAR SPRING SUBCOMMITTEE—K. Campbell (Chairman), R. Siorek (Vice-Chairman), W. Allison, J. J. Bozyk, G. Dentel, R. E. Hanslip, J. Marsland, W. Platko, D. W. Schumann, A. F. Skover, B. Sterne, D. Tuttle, R. Walstrom, W. J. Young.

SYMBOLS COMMITTEE

D. J. SCHWARTZ (Chairman)
D. D. Beach
M. Belzer
M. A. Berk
H. Bream

T. F. Crusinberry
R. Hagie
D. Mitchell
V. K. Rajpaul
S. Tennyson

J. R. Tishkowski
C. W. Volland
F. C. Walters
B. R. Weber
B. Wood

(SAE Staff, L. P. Ziegler, Jr.)

TECHNICAL ADVISORY GROUP FOR ISO/TC 22—ROAD VEHICLES

M. J. TAUSCHEK (Sponsor)
R. J. SZYDLOWSKI (Chairman)
J. L. MAPLEBACK (Vice-Chairman)
R. L. Atkin
A. W. Carey, Jr.
T. J. Carr

N. Chew
J. P. DeKany
J. R. Farron
E. D. Heins
J. W. Kourik
F. Krall

A. C. Malliaris
J. E. Mitchell
E. Newman
F. M. Petler
J. Starkey
R. Steen

(SAE Staff, L. P. Ziegler, Jr.)

TRAILER HITCH COMMITTEE

J. C. ABROMAVAGE (Chairman)
S. J. Borden
G. Conradi
E. DeAngelis
W. E. Dotterweich
R. Fertitta
C. N. French

E. M. Hermansen
F. J. Houston
R. Klein
M. C. Maryonovich
M. Nerem
D. Pearch
D. I. Reed

A. Roberts
W. L. Sherry
J. Smith
D. Swanson
B. R. Weber
K. J. Williams
D. A. Young

(SAE Staff, L. P. Ziegler, Jr.)

TRANSMISSION AND DRIVETRAIN STEERING COMMITTEE

E. UPTON (Chairman)
C. E. COONEY (Chairman)
B. W. Cartwright
E. L. Clary
M. G. Gabriel
C. W. Greening

D. E. Hobson
J. W. Holdeman
G. E. Huffaker
V. J. Jandasek
D. L. Otto
C. E. Shellman

G. R. Smith
E. D. Taylor
R. L. Thomas
F. H. Walker

(SAE Staff, R. T. Northrup)

TRANSMISSION AND DRIVETRAIN TECHNICAL COMMITTEE

M. R. YOUNG (Sponsor)
E. W. UPTON (Chairman)
C. E. COONEY (Vice-Chairman)
F. Abar
A. Adams
R. E. Annis
R. A. Armstrong
P. Ashburn
W. A. Bartkowiak
J. L. Bascunana
A. P. Blomquist
H. H. Bloom

E. E. Bower
C. E. Brady
D. K. Cameron
B. W. Cartwright
E. L. Clary
R. Cleveland
R. Colello
R. R. Cornish
R. W. Craig
W. E. Daggett
H. E. Deen
D. W. Dundore

E. L. Egbert
R. Emmadi
P. E. Fadow
M. G. Gabriel
R. B. Gibson
L. E. Green
C. W. Greening
K. B. Harmon
G. B. Haueter
S. J. Haydock
R. N. Hazzard
J. A. Heck

E. T. Hendzel
D. C. Hill
D. E. Hobson
J. W. Holdeman
G. E. Huffaker
A. C. Huevel
J. G. Jelinek
E. L. Jones
R. D. Jones
C. E. Juntunen
K. R. Kaza
R. A. Kobe
J. Koinis
K. E. Kritsch
A. LaCroix
T. R. Leonard
F. A. Lloyd
W. E. Loney
J. E. Mahoney

L. R. Marcy
K. Martek
A. Matsushima
W. McCall
E. C. Maki
R. A. Mercure
H. Miner
R. E. Mitchell
F. C. Nagele
D. L. Otto
B. A. Pearson
D. P. Pellerito
C. R. Potter
S. V. Puidokas
L. H. Saylor
M. M. Schall
H. L. Sharp
C. E. Shellman
P. L. Silbert

E. Y. Sing
D. M. Slaubaugh
G. R. Smith
J. B. Snoy
R. J. Socin
R. W. Stapleton
L. G. Steinl
E. D. Taylor
R. L. Thomas
J. J. Thornton
G. E. Tozer
D. A. VanDeven
R. J. S. Vodicka
E. R. Wagner
F. H. Walker
T. M. Wang
R. W. Wayman
F. H. Whitmyer

(SAE Staff, R. T. Northrup)

AXLE SUBCOMMITTEE—E. E. Hobson (Chairman), K. E. Kritsch, E. C. Maki, J. Mueller, R. J. Niemiec, S. V. Puidokas, R. A. Shriver. BEARINGS, SHAFT, AND ONE-WAY SUBCOMMITTEE—H. Fisher, C. E. Juntunen, D. Slaubaugh, R. W. Stapleton. DRIVESHAFT SUBCOMMITTEE—C. E. Cooney (Chairman), E. R. Wagner (Vice-Chairman), R. J. Brown, R. T. Evans, R. N. Hazzard, D. W. Holzinger, J. Koinis, E. E. Sowers, T. Trojanowski, C. M. Ziegler. EDITORIAL AND PUBLICATIONS SUBCOMMITTEE—E. L. Clary (Chairman), P. D. Cruise, D. J. Drayton, M. G. Gabriel, N. Hull, E. L. Jones, J. E. Mahoney, O. E. Philips, G. R. Smith, E. R. Wagner, K. L. Westercamp. FLUIDS SUBCOMMITTEE—B. A. Pearson (Chairman), H. Deen, M. Haviland, W. J. Lendener, H. L. Miner, R. E. Osborne, D. Rogers, M. M. Schall. FRICTION SUBCOMMITTEE—C. W. Greening (Chairman), B. A. Clay, H. E. Deen, L. DuHaime, R. Fish, J. E. Gray, R. P. Katz, W. E. Loney, B. Martin, J. A. McLain, D. C. Mitchell, C. Potter, D. H. Rodgers, W. D. Ross. GEARS AND SPLINE SUBCOMMITTEE—D. K. Cameron (Chairman), R. E. Cleveland, R. E. Mitchell, J. J. Thornton, E. W. Upton, H. Van Lent. HYDRAULIC CONTROLS SUBCOMMITTEE—C. E. Shellman (Chairman), R. W. Craig, W. E. Daggett, R. J. Socin, E. W. Upton. POWER TAKE-OFF SUBCOMMITTEE—R. L. Thomas (Chairman), H. M. Ball, C. H. Clauson, G. Huffaker, A. Huevel, R. Kennard, D. Toner, P. Weis, E. Zahn. TERMINOLOGY SUBCOMMITTEE—M. G. Gabriel (Chairman), E. F. Bowie, M. W. Dundore, E. L. Egbert, K. R. Kaza, W. L. Schulz, L. G. Steinl, D. H. Wade, D. R. Weilant. TRANSMISSION TEST AND PERFORMANCE SUBCOMMITTEE—F. H. Walker (Chairman), P. Ashburn, S. Luchter, D. Maddock, W. L. McNulty, R. A. Mercure, R. D. Moan, P. J. Ross, F. Slocum, O. K. Thiel, M. J. Waclawek. TRUCK, BUS, AND INDUSTRIAL DRIVETRAIN SUBCOMMITTEE—G. E. Huffaker (Chairman), D. K. Bell, C. E. Cooney, D. L. Cutts, R. Drewes, E. T. Hendzel, A. C. Huevel, R. G. Joyner, D. F. Linn, E. E. Londt, J. Owens, J. Rathburn, R. Schimmel, O. Webser, R. W. Wolfe, E. W. Zahn.

TRUCK AND BUS CAB AND OCCUPANT ENVIRONMENT COMMITTEE

L. W. STRAWHORN (Chairman)
R. J. Brooks
N. A. Bundra
F. M. Callahan
J. H. Culbertson
R. M. Clarke
G. C. Crane
R. E. Didion
T. E. Dobbs

A. M. Fischer
T. D. Gillespie
R. E. Heglund
B. Klingenberg
B. Koepke
F. Krall
S. Kraus
S. Larsson
J. W. Lawrence

C. M. Overbey
V. E. Pound
W. J. Rheaume
G. Rossow
R. H. Singer
H. Sullivan
W. E. Whitmer
J. Winslow
W. J. Young

(SAE Staff, A. G. Salem)

OCCUPANT DIMENSIONAL PARAMETERS SUBCOMMITTEE—N. A. Bundra, (Chairperson), C. D. Bishop, F. M. Callahan, T. Carr, W. H. Daniels, L. J. Dupuis, Jr., S. B. Fawcett, T. Gage, D. C. Hammond, R. E. Heglund, C. G. Holstein, C. Jones, S. Kraus, S. Larsson, C. Overbey, W. Rheaume, N. A. Rini, E. F. Saxman, R. G. Snyder, H. Sullivan, N. L. Thomas, K. Thompson. TRUCK RIDE SUBCOMMITTEE—G. C. Crane (Chairman), N. C. Mehta (Vice-Chairman), M. L. Baum, H. Cook, J. Elhbeck, T. Gillespie, N. Goodwin, R. R. Hegmon, R. Jable, T. Larson, P. Levering, J. Lommel, E. C. Maki, L. Novoa, C. Overbey, P. R. Pierce, W. Presley, B. S. Repa, G. M. Reynolds, J. Rodgers, D. Roper, G. Rosso, R. W. Sobson, J. Stadler, L. W. Strawhorn, R. VanSteenkiste, F. Waksmundski, R. M. Clarke.

TRUCK AND BUS CHASSIS COMMITTEE

D. L. STEPHENS (Chairman)
R. W. HILDEBRANDT (Vice-Chairman)

P. C. Bertelson
R. S. Graham

R. W. Hildebrandt
W. L. Reiersgaard

(SAE Staff, A. G. Salem)

BRAKE SUBCOMMITTEE—R. W. Hildebrandt (Chairman). AIR BRAKE SYSTEM COMPONENTS—C. F. Smith (Chairman), J. Fent (Vice-Chairman), W. E. Anthes, E. F. Beatty, A. F. Beier, H. R. Burns, M. L. Carton, G. Choinski, R. Clinefelter, D. Cooper, T. E. Donley, R. Evans, A. M. Fischer, C. R. Herring, J. M. Holden, G. Hopkins, R. J. Hornacek, P. G. Hykes, G. W. Kurasz, H. Linkner, G. P. Mathews, M. Newberry, R. Plantan, A. H. Przepiora, B. D. Sibley, R. F. Stelzer, H. L. Stringer, W. N. Tazelaar, C. Thornton, D. Wasmer. AIR BRAKE HOSE SPECIFICATION TASK FORCE—H. R. Burns (Chairman), A. G. Beier, J. M. Holden, L. C. Huneke, W. T. Johnson, D. E. Piccoli, R. E. Ritchie, R. Ruda, E. F. Stanton, R. F. Stelzer, W. N. Tazelaar. AIR BRAKE RESERVOIR TEST CODE TASK FORCE—A. G. Beier (Chairman), E. Franke, J. M. Holden, G. Pfeifer, V. L. Schumacher, G. J. Snyder. AIR BRAKE SYSTEM VALVE PERFORMANCE TASK FORCE—J. Fent (Chairman), A. G. Beier, M. L. Carton, G. Crewson, C. R. Herring, G. Hopkins, R. J. Hornacek, W. N. Tanzelaar, D. Wasmer. TASK FORCE TO REVISE J982—J. Fent, D. Merriman. AIR BRAKE COMPONENTS TIME LAG MEASUREMENT TASK FORCE—D. A. Hoffman (Chairman), D. Brooks, J. Fent, R. J. Hornacek, D. Merriman. AIR BRAKE ACTUATOR TEST PROCEDURE TASK FORCE—M. L. Carton (Chairman), E. Beatty, J. Fent, D. Hoffman, G. W. Kurasz, G. P. Mathews, B. D. Sibley. SLACK ADJUSTER TEST PROCEDURE TASK FORCE—B. D. Sibley (Chairman), J. Fent, A. Holmes, P. Johnston, G. P. Mathews, J. A. Urban. CHASSIS SUBCOMMITTEE—W. L. Reiersgaard (Chairman), R. Morrison. STEERING SUBCOMMITTEE—D. C. Shropshire, S. E.

Dukes, R. E. Knight, R. S. Button, L. A. Meacock, R. J. Parker, J. P. Smith, R. B. Schwartz, J. M. Madonio, S. S. Mazur. SUSPENSION SUBCOMMITTEE—R. S. Graham, E. Volker, J. D. Brannan, H. C. Owen, C. Small, J. F. Kelly, W. C. Pierce, R. Pierman, R. Hickson, A. Riggs, G. Montgomery. WHEEL SUBCOMMITTEE—P. C. Bertelson.

TRUCK AND BUS ELECTRICAL COMMITTEE

R. W. SACKETT (Chairman)
E. SZABO (Vice-Chairman)
J. E. Allen
H. R. Brink, Jr.
A. Burgess
R. Dickey

C. Granfors
T. Irick
A. Lesesky
V. Miles
C. Owen
L. R. Phillippi

D. O. Rupert
G. H. Schlensker
P. J. Speece
R. Speth
J. Sypniewski
E. Tolnar

(SAE Staff, A. G. Salem)

TRUCK AND BUS OPERATIONS AND MAINTENANCE COMMITTEE

J. MacDOUGALL (Chairman)
J. A. DICK (Vice-Chairman)

R. J. Deierlein
F. Eaton

G. E. Trotte
P. Hinds

(SAE Staff, A. G. Salem)

TRUCK AND BUS POWERTRAIN COMMITTEE

P. J. MAZZIOTTI (Chairman)
G. FREDERIKSEN (Vice-Chairman)

R. V. Gorman

A. Demien

(SAE Staff, A. G. Salem)

V-BELT ADVISORY COMMITTEE

L. W. VIRTUE (Chairman)
A. E. Beaty

E. Bonow
R. D. Hoback

F. A. Shedd
B. L. Speer

(SAE Staff, P. Couhig)

LAB TESTING FOR V-BELTS SUBCOMMITTEE—J. Shepherd (Chairman), F. Beyer, C. Eaglin, G. Kirshberg, R. D. Mueller, W. H. Reichardt, P. F. Stoeck, W. Waanders. SYNCHRONOUS BELT SUBCOMMITTEE—J. J. Zaiss (Chairman), E. C. Adams, P. Burke, J. J. Doyle, W. Erickson, W. G. Fuetterer, R. Goodwin, M. Gravel, R. Hoback, T. Kowalski, G. Krolls (Corresponding Member), F. LaManna, J. Madden, J. Matthews, C. A. Mudge, R. Nelson, J. Rowe, J. Rushmore, G. Swenson, L. Wedding, R. Wetzel. V-BELT TEST SUBCOMMITTEE—J. Shepherd (Chairman), S. Bemis, W. G. Fuetterer, I. A. Groothuis, R. Hoback, W. L. Johnson, L. Mehelich, R. J. Nelson, L. Nye, H. J. Schlamadinger, B. Speer, R. F. Wallace, R. Wetzel, J. J. Zaiss. V-RIBBED BELT SUBCOMMITTEE—I. A. Groothuis (Chairman), E. Bonow, S. K. Fan, J. Kent, L. Mehelich, D. Pittman, R. Putnam, F. Shedd, J. Shepherd, B. L. Speer, L. Virtue.

V-BELT COMMITTEE

R. S. PIOTROWSKI
(Sponsor)
L. VIRTUE (Chairman)
J. T. Alden
A. Beaty
E. Bonow
P. D. Boos
D. J. Clifford
W. P. Coyne
P. Davidson (Alternate)
W. D. Erickson (Alternate)
S. K. Fan
D. E. Gardner (Alternate)
M. D. Gayer (Alternate)

I. Groothuis
D. D. Hall (Alternate)
G. Hitchcock (Alternate)
R. D. Hoback (Alternate)
J. E. Kent
L. J. Leaf
L. Mehelich
E. J. McCarthy
C. Mudge
S. Naida
R. J. Nelson (Alternate)
D. Pittman (Alternate)
R. Putnam
W. H. Reichardt (Alternate)

J. R. Ruff
G. T. Russell
S. Sabharwal
J. S. Sears
M. Shadday (Alternate)
F. A. Shedd (Alternate)
J. Shepherd
B. Speer
T. Sulkowski (Alternate)
R. L. Turner
R. Wallace
C. M. Williams
L. Williams
J. J. Zaiss

(SAE Staff, P. Couhig)

VEHICLE DYNAMICS COMMITTEE

R. T. GASKILL (Sponsor)
W. F. MILLIKEN (Chairman)
R. T. BUNDORF (Vice-Chairman)
W. Bergman
W. B. Carlson
D. Cox
D. A. Glemming
J. T. Hamilton
L. Howell
C. O. Jones

P. G. Keith
J. W. Kent, Jr.
R. H. Klein
W. F. LeFevre
S. A. Lippmann
T. Maeda
J. M. Mann
D. T. McRuer
S. D. Ogden
D. A. Perrin

R. E. Rasmussen
B. Repa
R. Rice
H. K. Sachs
F. K. Schenkel
L. Segel
S. Shadle
L. Sweet
C. F. Thelin
R. W. Topping

D. H. Weir
R. A. Rider

F. Winsor

P. Riede, Jr.

(SAE Staff, R. C. Uhl)

DIRECTIONAL CONTROL SUBCOMMITTEE—R. S. Rice (Chairman), C. Beauregard, T. D. Gillespie, J. T. Hamilton, R. H. Klein, E. Kramer, R. L. Leffert, L. C. Miller. MOTORCYCLE DYNAMICS SUBCOMMITTEE—D. H. Weir (Chairman), H. H. Hurt, J. S. McKibben, R. J. Miennert, R. S. Rice, Y. Watanabe, R. Maiman. VEHICLE AERODYNAMICS SUBCOMMITTEE—F. K. Schenkel (Chairman), J. B. Barlow, W. H. Bettes, R. Buchheim, G. W. Carr, A. Costelli, A. Cogotti, H. J. Emmelmann, M. Gleason, R. A. Hunter, K. B. Kelly, L. U. Nilsson, Y. Nishimura, L. Janssen, G. F. Rombert, G. Sovran, C. V. Williams, J. E. Williams. VEHICLE AND SUSPENSION PARAMETER MEASUREMENT SUBCOMMITTEE—R. W. Topping (Chairman), R. L. Anderson, R. Mosrie, C. Thatcher, M. Bethell, R. L. Henry, E. G. Klosterhaus, B. Los, K. J. O'Neil, R. D. Queary, J. Sahakian, C. B. Winkler, R. A. Cripe. U. S. A. COMMITTEE FOR ISO/TC22/SC9—VEHICLE DYNAMICS AND ROAD HOLDING ABILITY—R. T. Bundorf (Chairman), W. Bergman, L. M. Forbes, W. F. Milliken, R. E. Rasmussen, L. Segel, D. H. Weir, W. B. Carlson, D. A. Perrin, R. Rice. TECHNICAL ADVISORY GROUP—WORKING GROUP 3—B. M. Gallaway, R. R. Hegmon, A. G. Veith, F. D. Smithson, F. C. Brenner, E. A. Whitehurst. ERDA AERODYNAMICS RESEARCH ADVISORY COMMITTEE—W. F. Milliken (Chairman), F. H. Abernathy, T. Barrows, J. L. Bascunana, R. T. Bundorf, R. T. Gaskill, W. H. Hucho, R. Hunter, K. B. Kelly, C. LaPointe, P. Lissaman, J. Marti, N. P. Moore, A. Roshko, G. F. Romberg, G. Sovran, C. Van Schayk, M. Zlotnick. WIND TUNNEL BOUNDARY CORRECTION WORKING PANEL—J. E. Williams (Chairman), P. Beebe, G. Sovran, J. G. LaFond, D. Wilsden, L. Butz, K. R. Cooper, A. Farahan.

VEHICLE IDENTIFICATION NUMBERS COMMITTEE

C. MARKS (Sponsor)
M. W. DIXON (Chairman)
R. H. Brushwood
E. Cerrelli
H. G. Cook
D. Costa
J. R. Doto
R. Durbin
N. F. Erickson
C. Finn
J. E. Forss

F. C. Funk
G. Gardner
P. Gilliland
R. J. Haller
R. F. Ingegneri
J. H. Leverenz
R. H. Madison
T. S. Owens
P. Perry
B. J. Riley
B. Schiff

D. M. Schwentker
R. Sjoberg
R. Sostkowski
G. O. Stevens
J. R. Tishkowski
S. M. Weglian
G. R. Williams
T. F. Williams
R. Wilson
D. R. Wolfslayer

(SAE Staff, L. P. Ziegler, Jr.)

VEHICLE COMPONENTS PARTS IDENTIFICATION SUBCOMMITTEE—P. Gilliland (Chairman), C. O. Brickey, H. G. Cook, D. Costa, M. W. Dixon, J. R. Doto, R. Durbin, N. F. Erickson, R. L. Haller, R. W. Harris, G. H. Hespeler, T. J. Horrigan, R. F. Ingegneri, J. H. Leverenz, G. Milliron, T. S. Owens, P. Perry, F. M. Petler, D. M. Schwentker, R. Sostkowski, J. E. Timmerman, M. F. Von Leer, S. M. Weglian, D. R. Wolfslayer. VEHICLE INFORMATION RECORD SUBCOMMITTEE—B. Schiff (Chairman), D. Costa, M. W. Dixon, J. R. Doto, N. F. Erickson, D. A. Frisco, G. Gardner, R. J. Haller, F. M. Petler, D. M. Schwentker, G. O. Stevens, J. R. Tishkowski, M. F. Von Leer, S. M. Weglian, T. F. Williams.

VEHICLE SECURITY COMMITTEE

D. E. MARTIN (Sponsor)
D. R. WOLFSLAYER (Chairman)
H. G. Cook
D. Costa
J. R. Doto
N. F. Erickson
C. F. Finn

G. Gardner
P. W. Gilliland
C. W. Oliver, Jr.
R. Planton
B. J. Riley
W. Riley
D. M. Schwentker

R. H. Sostkowski
J. R. Tishkowski
M. F. Von Leer
S. M. Weglian
G. R. Williams
J. Wilson
R. L. Wilson

(SAE Staff, L. P. Ziegler, Jr.)

RELATED ACTIVITIES SUBCOMMITTEE—J. R. Doto (Chairman). RELATED LAWS SUBCOMMITTEE—G. R. Williams (Chairman).

VEHICLE SOUND LEVEL COMMITTEE

C. MARKS (Sponsor)
R. K. HILLQUIST (Chairman)
L. J. ERIKSSON (Vice-Chairman)
R. D. Bruce
E. P. Davies
T. M. Howell
F. R. Kishline

R. M. Law
S. A. Lippmann
J. T. Lisbon
C. Marks
N. Mehta
N. A. Miller
C. A. Preuss

J. F. Shaffer
D. G. Thomas
J. A. Thomas
J. H. Venema
E. R. Welbourne
E. Willette
G. P. Wilson

(SAE Staff, R. T. Northrup)

COACH SOUND LEVEL SUBCOMMITTEE—J. F. Shaffer (Chairman), P. Jettinghoff, K. M. Jordan (Alternate), L. K. Mikalonis (Alternate), R. N. Pawlysz, J. J. Pisarski, M. E. Rumbaugh, Jr., G. E. Swetnam, D. Whitney. DATA HANDLING PRACTICES SUBCOMMITTEE—N. C. Mehta (Chairman), J. Hendal, C. A. Preuss. ENGINE SOUND LEVEL MEASUREMENT SUBCOMMITTEE—R. M. Law (Chairman), R. E. Canfield, J. R. Crowe, D. Gorgon, J. W. Johnson, K. K. Kuehner, J. McCormack, J. T. Nadolny, T. J. Pearsall, D. K. Stephenson, D. D. Tiede, G. A. Weinert, T. Wu. EXHAUST AND INTAKE SILENCER SUBCOMMITTEE—L. Eriksson (Chairman), J. Cahill (Secretary), P. Cheng, J. Dreznes, F. Egberts, K. Ligot, K. F. Nowak, W. L. O'Neill, R. G. Palmer, D. G. Thomas,

D. W. Rowley, D. VanWyck. HEAVY TRUCK SUBCOMMITTEE—N. A. Miller (Chairman), E. A. Bacsanyi, D. Gerhardt, D. Gray, R. Jenkins, R. L. Kesler, F. W. Krey, R. Kunicki, R. M. Law, P. J. Miller, C. L. Moon, D. Olree, D. Verhoff, D. R. Whitney, R. K. Witwer. INSTRUMENTATION SUBCOMMITTEE—J. T. Lisbon (Chairman), R. Schumacher (Co-Chairman), J. Bair, W. H. Flint, T. J. Henry, J. S. Mutziger. LIGHT VEHICLE EXTERIOR SOUND LEVEL TEST METHODS SUBCOMMITTEE—T. M. Howell (Chairman), D. Bass, P. R. Donavan, R. Hanson, R. D. Hellweg, R. K. Hillquist, D. Klonowski, R. H. Paddy, C. A. Preuss, J. Schreiber, W. A. Smith, J. Swing, J. A. Thomas, R. A. Vorpagel, D. R. Whitney. LIGHT VEHICLE INTERIOR SOUND LEVEL TEST METHODS SUBCOMMITTEE—F. R. Kishline (Chairman), K. F. Bagga, R. E. Canfield, J. D. Haines, R. K. Hillquist, T. W. Kozyra, J. J. Pisarski, J. R. Rutledge, D. Schmitz, J. H. Venema. PRACTICES AND PROTOCOL SUBCOMMITTEE—D. G. Thomas (Chairman). RAIL RAPID TRANSIT SUBCOMMITTEE—G. P. Wilson (Chairman), C. S. Caccavari, L. Kurzweil, R. J. Murray, A. Paolillo. J986b REVIEW TASK FORCE—R. K. Hillquist (Convener), T. M. Howell, F. R. Kishline, R. F. Schumacher, R. A. Vorpagel. SUBCOMMITTEE TO REVISE SAE J377—J. H. Venema (Chairman), R. A. Little, J. A. Neese, G. L. Papow, C. E. Sloan.

WHEEL COMMITTEE

H. PICKFORD (Sponsor)
J. R. AURELIA (Chairman)
M. D. WEBER (Vice-Chairman)
R. N. Archer
P. C. Bertelson

R. A. DeRegnaucourt
J. Guzek
E. J. Hayes
H. S. Karzun
L. E. Kiselis

F. C. Koch
D. D. MacIntyre
K. B. O'Neil
J. N. Schweitzer
N. Zorka

(SAE Staff, P. Couhig)

PASSENGER CAR WHEEL SUBCOMMITTEE—D. D. MacIntyre (Chairman), J. R. Aurelia, M. Bailey, R. G. Falzon, J. P. Gerstner, J. Guzek, F. C. Koch, A. L. Lapsys, L. Malwitz, D. J. Orban, N. Zorka. TRUCK WHEEL SUBCOMMITTEE—R. A. DeRegnaucourt (Chairman), J. R. Aurelia, R. J. Baroni, P. C. Bertelson, S. W. Blate, P. E. DeVergilio, J. J. Duggan, P. A. Fensel, G. L. Goldberg, D. R. Grejda, D. L. Kerr, J. R. Kinstler, P. Levering, W. C. Long, D. D. MacIntyre, L. Palmer, R. G. Schaefer, M. D. Weber, P. G. Willer, R. K. Williamson, R. A. Winter. BOLT PATTERN TASK FORCE—J. R. Kinstler (Chairman), J. R. Aurelia, S. W. Blate, R. A. DeRegnaucourt, J. J. Duggan, A. Erickson, L. Fox, D. Grejda, J. E. Haug, D. Jendrusch, D. L. Kerr, P. Levering, W. C. Long, D. D. MacIntyre, G. J. Montgomery, R. A. Pete, R. W. Sackett, R. G. Schaefer, P. W. Tomlinson, M. D. Weber, P. G. Willer. HUB TASK FORCE—J. J. Duggan (Chairman), R. Baroni, A. Erickson, F. A. Fensel, L. Fox, R. Hogan, R. E. Kennel, D. L. Kerr, J. T. Kosinski, P. Levering, W. C. Long, R. W. Murphy, R. A. Pete, J. Petras, D. Sandberg, R. G. Schaefer, D. L. Stephens.

SAE ADVISORY COMMITTEE FOR THE SAE/DOT TRUCK AND BUS FUEL ECONOMY MEASUREMENT STUDY

W. B. SMYTH (Chairman)
R. S. JOHNSON (Vice-Chairman)
J. E. Allen
T. P. Baker
M. Balban
R. Belcer
W. H. Bettes
R. G. Cadwell
W. H. Close
K. R. Cooper
J. H. Culbertson
W. Dixon
R. Ehrlich
C. H. Ek

D. D. Forester
C. J. France
T. Franquist
T. Gelinas
B. Gibson
G. P. Hanley
J. E. Haworth
R. E. Hoffmeister
R. A. Hunter
D. Jesmantas
J. D. Lineberry
B. Mason
W. R. Minning
R. T. Northrup

J. M. Prange
W. R. Rodger
G. W. Rossow
F. K. Schenkel
J. B. Schnell
H. E. Seiff
D. Stattenfield
J. G. Stieber
L. Strawhorn
C. J. Travis, Jr.
E. Upton
J. C. Walter
T. Young

(SAE Staff, R. T. Northrup)

DATA ANALYSIS TASK FORCE—G. W. Rossow (Chairman), J. E. Allen, M. Balban, P. Bracht, C. H. Ek, W. J. K. Gibson, R. A. Hunter, C. Lafferty, J. Lange, C. Lunsford, R. H. Richardson, H. E. Seiff, N. J. Sheth, W. B. Smyth, C. J. Travis. PROGRAM INFORMATION TASK FORCE—T. Gelinas (Chairman), J. Allen, W. H. Close, E. Coch, E. J. Niederbuehl, G. Rossow. AERODYNAMICS SUBCOMMITTEE—W. H. Bettes (Chairman), B. Bowen, W. Bowman, F. T. Buckley, K. R. Cooper, M. Gleason, R. A. Hunter, R. S. Johnson, T. Kangas, J. Kettinger, G. C. Madzsar, B. Mason, A. T. McDonald, R. W. Murphy, G. M. Palmer, E. Peterson. AERODYNAMICS TASK FORCE—W. H. Bettes (Chairman), K. R. Cooper, R. A. Hunter, B. Mason, J. B. Barlow. BASIC ENGINE MODIFICATION SUBCOMMITTEE—R. Cadwell (Chairman), G. D. Aravosis, J. G. Barker, D. E. Cole, C. W. Coon, Jr., D. A. Fay, R. J. Kakoczki, D. W. Knopf, R. A. Mercure, D. F. Merrion, P. E. Phelan, C. C. Young. BUS ADVISORY GROUP (BAG)—G. P. Hanley (Chairman), J. E. Baker, M. S. Balban, G. Davis, J. G. DeRoos, C. H. Ek, D. D. Forester, C. J. France, E. R. Gerlach, D. Guthrie, B. Harrison, E. T. Hendzel, A. C. Huevel, G. Hunt, R. A. Hunter, R. Ingle, E. Kravitz, B. S. Mallhi, H. G. Miller, E. J. Niederbuehl, A. L. Neumann, M. Prange, J. R. Prior, A. E. Prunka, W. Raithel, B. Reynolds, G. Rossow, G. H. Roziewski, J. B. Schnell, J. G. Stieber, W. J. Sulak, W. P. Tell, G. Thur, H. B. Tyson, J. Wetzel. SCHOOL BUS TASK FORCE—B. Harrison (Chairman), G. Davis, D. Guthrie, E. T. Hendzel, A. C. Huevel, E. J. Niederbuehl. TRANSIT ENERGY ACTIVITY TASK FORCE—W. J. Sulak (Chairman), J. E. Allen, J. E. Baker, E. R. Gerlach, D. H. Guthrie, G. P. Hanley, G. K. Hussong, A. L. Neumann, E. J. Niederbuehl, R. T. Northrup, J. R. Prior, G. Prytula, J. B. Schnell, H. E. Seiff, J. F. Shaffer, B. W. Skinner, J. Zakotnik. DRIVELINE COMPONENTS AND MODIFICATIONS SUBCOMMITTEE—E. W. Upton (Chairman), E. C. Maki (Vice-Chairman), J. Arzoian, P. Ashburn, W. F. Baugh, C. E. Cooney, Jr., R. L. Earl, M. G. Gabriel, K. W. Gordon, C. Greening, S. J. Haydock, D. E. Hobson, J. W. Holdeman, G. E. Huffaker, C. R. Jones, D. F. Linn, C. F. Lundbom, J. F. Mueller, R. K. Nelson, D. L. Otto, B. A. Pearson, G. R. Smith, J. Vandervort, F. H. Whitmyer, R. W. Wolf. DRIVING CYCLE SUBCOMMITTEE—G. W. Rossow (Chairman), J. B. Feiten, C. J. France, R. Gallivan, R. S. Johnson, C. Laferty, E. F. Saxman, H. E. Seiff. FANS AND ACCESSORIES SUBCOMMITTEE—J. D. Lineberry (Chairman), E. G. Blair, B. M. Bonifant, K. A. Boyd, E. D. Davis, W. J. Fuetterer, Jr., R. L. Goff, D. Gross, J. G. Hammond, G. P. Hanley, D. W. Lawson, R. J. Lipski, J. Mall, J. M. Olson, D. Parsons, F. G. Rising, M. E. Rumbaugh, Jr., C. W. Selby, R. B. Spokas, G. E. Ternent, G. W. Trabbic. IN-SERVICE TEST PROCEDURES SUBCOMMITTEE—J. Allen (Chairman), B. Bolstad, W. H. Close, F. Eaton, R. A. Hunter, R. Huter, J. Lumb, T. Mannix, W. A. Mareneck, R. D. Marksland, V. Miller, A. R. Pagnotto, J. C. Paterson, J. M. Salvaggio, D. Strout, L. Swenson, C. J. Travis, Jr., H. Watts (Alternate), M. White, C. Wiley. SAE/TMC TASK FORCE FOR IN-SERVICE TEST PROCEDURES—J. E. Allen (Co-Chairman), C. J. Travis, Jr. (Co-Chairman), M. M. Alderfer, J. Bartkiewicz, A. Carlson, W. C. Chisholm, K. Frassa, W. J. K. Gibson, T. Hollingsworth, R. A. Hunter, L. Kibbee, D. Kitchell, M. Langford, J. Lumb, W. A. Mareneck, V. Miller, J. Moriarity, J. C. Patterson, K. Penaluna, R. S. Rosenthal, K. VanLiew, B. Wessels, M. White, C. Wiley. OEM VEHICLE TEST PROCEDURES SUBCOMMITTEE—D. Jesmantas (Chairman), R. Beckmann, C. H. Ek, C. J. France, W. S. Freas, R. Gallivan, L. L. Hieber, A. L. Huevel, R. S. Johnson, M. S. Lantz, N. A. Miller, L. Orr, J. A. Rockwell, R. S. Rosenthal, D. R. Schomaker, W. B. Smyth. ROLLING RESISTANCE SUBCOMMITTEE—T. P. Baker (Chairman), R. H. Attenhofer, F. C. Brenner, C. A. Brunot, S. K. Clark, L. W. DeRaad, L. T. Dorsch, C. H. Ek, B. Gapco, I. Gusakov, R. E. Knight, W. A. Leasure, Jr., J. R. Luchini, K. Moehring, R. C. Moore, J. Moules, A. S. Myint, G. E. Pollard, G. R. Potts, L. Ragsdale, M. S. Schramm, D. J. Schuring, R. N. Simmons, A. Stiebel, G. D. Thompson, T. C. Warholic. VEHICLE CATEGORIZATION SUBCOMMITTEE—W. G. Dixon (Chairman). VEHICLE CORRELATION AND SIMULATION SUBCOMMITTEE—C. H. Ek (Chairman), R. Buck, G. P. Hanley, N. A. Miller, W. Moody, W. E. Schilke, D. Stattenfield.

The following Subcommittees are responsible to the Advisory Committee and Standing Technical Committees. Detailed membership is shown under Technical Committee Listings:

Aerodynamics—Subcommittee of the Truck and Bus Committee
Basic Engine Modification—Subcommittee of the Engine Committee
Driveline Components and Modifications—Subcommittee of the Transmission and Drivetrain Technical Committee
Driving Cycle—Subcommittee of the Transportation and Maintenance and Truck and Bus Committees
Fans and Accessories—Subcommittee of the Engine Committee
In-Service Test Procedures—Subcommittee of the Transportation and Maintenance Committee
OEM Vehicle Test Procedure—Subcommittee of the Truck and Bus Committee
Rolling Resistance—Subcommittee of the Highway Tire Committee
Vehicle Categorization—Subcommittee of the Truck and Bus Committee
Vehicle Correlation and Simulation—Subcommittee of the Truck and Bus Committee

SAE REPRESENTATION IN OTHER TECHNICAL ORGANIZATIONS

AMERICAN NATIONAL STANDARDS INSTITUTE

ORGANIZATIONAL MEMBER COUNCIL
J. Gilbert (M. L. Stoner, Alt.)

ACOUSTICAL STANDARDS MANAGEMENT BOARD
R. K. Hillquist (R. T. Northrup, Alt.)

AMERICAN NATIONAL METRIC COUNCIL
ELECTRONIC EQUIPMENT AND COMPONENTS SECTOR
R. G. Fenske

EXECUTIVE STANDARDS COUNCIL
M. L. Stoner

GRAPHIC TECHNICAL ADVISORY BOARD
A. G. Salem

MECHANICAL STANDARDS MANAGEMENT BOARD
M. L. Stoner (R. C. Uhl, Alt.)

SAFETY AND HEALTH STANDARDS MANAGEMENT BOARD
Leo P. Ziegler, Jr. (P. Couhig, Alt.)

U. S. NATIONAL COMMITTEE FOR AMERICAN-BRITISH-CANADIAN CONFERENCE ON THE UNIFICATION OF ENGINEERING STANDARDS
R. P. Trowbridge

VISUAL ALERTING SYSTEMS COMMITTEE
VASCOM INTERNATIONAL TECHNOLOGY SUBCOMMITTEE
Leo P. Ziegler, Jr.

STANDARDS FOR SAFETY IN THE CONSTRUCTION INDUSTRY (A10)
R. C. Uhl

STANDARDIZATION AND UNIFICATION OF SCREW THREADS (B1)
H. W. Ellison, George Garcina, L. R. Strang

PIPE AND HOSE COUPLING THREADS
J. Hinske

ALLOWANCES AND TOLERANCES FOR CYLINDRICAL PARTS AND LIMIT GAGES (B4)
K. O. Kverneland

MACHINE TOOLS, COMPONENTS, ELEMENTS, PERFORMANCE AND EQUIPMENT (B5)
Robert W. Kynast

STANDARDIZATION OF BOLTS, NUTS, RIVETS, SCREWS, AND SIMILAR FASTENERS (B18)
S. E. Mallen, H. W. Ellison, R. S. Piotrowski, Ralph R. Sjoberg, David W. Vial

STANDARDIZATION OF WASHERS AND MACHINE RINGS (B27)

TRANSMISSION CHAINS AND SPROCKET TEETH (B29)
M. Love

SAFETY STANDARD FOR CRANES, DERRICKS AND HOISTS, JACKS AND SLINGS (B30)
F. J. Strnad (H. R. Cozad and Paul A. Chalupsky, Alts.)

WIRE DIAMETERS AND METAL THICKNESSES (B32)
E. J. Streichert

CLASSIFICATION AND DESIGNATION OF SURFACE QUALITIES (B46)
P. E. McKim

SAFETY CODE FOR POWERED INDUSTRIAL TRUCKS (B56)
Arthur H. Pickford, J. E. Staab, R. C. Uhl

SAFETY STANDARDS FOR LAWN MOWERS, SNOW THROWERS, POWER EDGERS AND TRIMMERS, GARDEN TRACTORS AND RELATED EQUIPMENT AND ATTACHMENTS (B71)
I. Kamlukin (G. E. Buske, Alt)

SPLINES AND SPLINED SHAFTS (B92)
C. L. Carlisle

FLUID POWER SYSTEMS AND COMPONENTS (B93)
E. L. Falendysz, R. T. Northrup

AGRICULTURAL TRACTOR AND AGRICULTURAL MACHINERY (B114)
C. F. Walters, R. C. Uhl

NATIONAL ELECTRIC CODE (C1) CODE MAKING PANEL NO. 19
C. J. Owen

RADIO ELECTRICAL COORDINATION (C63)
F. Bauer, H. K. Mertel

COMPONENTS FOR ELECTRONIC EQUIPMENT (C83)
E. U. Thomas

RADIO FREQUENCY RADIATION HAZARDS (C95)
D. A. Weber

TEMPERATURE MEASUREMENT THERMOCOUPLES (C96)
R. B. Clark

INSPECTION REQUIREMENTS FOR MOTOR VEHICLES (D7)
Harold G. Brillmyer

METHOD OF RECORDING AND MEASURING MOTOR VEHICLE FLEET ACCIDENT EXPERIENCE (D15)
W. D. Nelson

STANDARDIZATION OF FREIGHT CONTAINERS (MH5)
(M. L. Stoner, Alt.)

ACOUSTICS (S1)
R. K. Hillquist (R. T. Northrup, Alt.)

MECHANICAL SHOCK AND VIBRATION (S2)
S. Rubin (M. L. Stoner, Alt.)

BIOACOUSTICS (S3)
R. K. Hillquist (R. T. Northrup. Alt.)

ABBREVIATIONS (Y1)
H. E. Guetzlaff

STANDARDS FOR DRAWINGS AND DRAFTING PRACTICE (EXCLUSIVE OF ARCHITECTURAL DRAWINGS (Y14)
H. Guetzlaff, G. Garcina, E. Kardos (L. Porter, Alt.)

SPECIFICATIONS AND METHODS OF TEST FOR SAFETY GLAZING MATERIAL (Z26)
G. Beck (D. W. Palmeter, P. Couhig, Alts.)

SUBCOMMITTEE ON SYMBOLS (Z535.3), James E. Carr

GLOSSARY OF ENVIRONMENTAL TERMINOLOGY (Z84)

AMERICAN SOCIETY FOR TESTING AND MATERIALS

COMMITTEE A1 ON STEEL
A. G. Cook

ADVISORY COMMITTEE
A. G. Cook

COMMITTEE B2 ON NONFERROUS METALS AND ALLOYS
R. J. Cox

SUBCOMMITTEE I ON REFINED COPPER, R. J. Cox
SUBCOMMITTEE VII ON REFINED NICKEL AND COBALT AND CONTAINING NICKEL AND/OR COBALT ALLOYS, CAST & WROUGHT, R. J. Cox

COMMITTEE B6 ON DIE-CAST METALS AND ALLOYS
R. B. Smith

SUBCOMMITTEE I ON ALUMINUM BASE DIE-CASTING ALLOYS, R. B. Smith
SUBCOMMITTEE II ON ZINC BASE DIE-CASTING ALLOYS, R. B. Smith

COMMITTEE B7 ON LIGHT METALS AND ALLOYS CAST & WROUGHT
R. T. Northrup, S. M. Brandt

ADVISORY SUBCOMMITTEE
SUBCOMMITTEE I ON ALUMINUM AND ALUMINUM ALLOY INGOTS, S. M. Brandt
SUBCOMMITTEE III ON WROUGHT ALUMINUM & WROUGHT ALUMINUM ALLOYS, R. B. Smith

COMMITTEE B8 ON ELECTRODEPOSITED METALLIC COATING AND RELATED FINISHES
L. M. Morse

COMMITTEE D2 ON PETROLEUM PRODUCTS AND LUBRICANTS
C. C. Colyer

COMMITTEE D11 ON RUBBER PRODUCTS
G. C. Norman

SUBCOMMITTEE XXX U.S.A. NATIONAL COMMITTEE FOR ISO/TC45, S. K. Flanders

COMMITTEE E20 ON TEMPERATURE MEASUREMENT
R. B. Clark

COMMITTEE F3 ON GASKETS
E. P. Francis

COMMITTEE G1 ON CORROSION OF METALS
G. F. Bush

AMERICAN WELDING SOCIETY

AUTOMOTIVE WELDING COMMITTEE
T. M. Hudson

COORDINATING RESEARCH COUNCIL, INC.

SAE and the American Petroleum Institute jointly organized the Coordinating Research Council, Inc. in 1942. Object of the CRC is to encourage and promote the arts and sciences by directing to scientific cooperative research in developing the best combinations of fuels, lubricants, and equipment powered by internal-combustion engines; and to afford means of cooperation with the government on matters of national interest within this field.

In the course of research sponsored by the three-technical committees of the Council (the Air Pollution Research Advisory Committee, Coordinating Fuel and Equipment Research Committee, and Coordinating Lubricants and Equipment Research Committee) a number of research techniques have been developed. These considered suitable for standardization have been referred to the American Society for Testing and Materials.

Information on the work of the technical committees, may be secured from the Coordinating Research Council, Inc., 219 Perimeter Center Parkway, Atlanta, GA 30346.

JOINT ASM, ASTM, AFA AND SAE COMMITTEE ON DEFINITION OF HEAT-TREATING TERMS

SAE Reps: M. L. Frey (Chairman), R. D. Chapman, H. D. Monsch, E. S. Rowland

NATIONAL FIRE PROTECTION ASSOCIATION

COMMITTEE ON AVIATION AND AIRPORT FIRE PROTECTION
A. G. Salem

COMMITTEE ON MOTOR CRAFT AND MARINAS
R. H. Roy

TRANSPORTATION RESEARCH COUNCIL

R. S. Shackson (W. M. Spreitzer, M. L. Stoner, Alts.)

U. S. NATIONAL COMMITTEE OF THE INTERNATIONAL COMMISSION ON ILLUMINATION

D. M. Finch, G. E. Meese, P. W. Maurer

U. S. TECHNICAL COMMITTEE TO INTERNATIONAL (CIE) COMMITTEE 4.8, AIRBORNE LIGHTING

(SAE COMMITTEE A20, AIRCRAFT LIGHTING)
Paul H. Greenlee

COMMITTEE ON COLORS OF SIGNAL LIGHTS
(D. M. Finch, Alt.)

SAE TECHNICAL BOARD RULES AND REGULATIONS

1. SAE TECHNICAL BOARD: The Technical Board (hereinafter called Board) is the agent of the SAE Board of Directors (hereinafter called Directors) with authority to direct and supervise all SAE cooperative engineering programs, including standardization and research, subject only to the right of appeal to the Directors by anyone in disagreement with action of the Board.

The objective of the Board is to make the technical knowledge, experience and skill of engineers effectively useful to the public, industry, government, and educational institutions, through cooperative engineering action; and to enhance the value of Society membership through technical committee activities.

The Board will determine engineering relationships with government, industry, other technical societies, educational institutions, or civic organizations. It will authorize all programs to be undertaken for, or in cooperation with, other organizations, as well as SAE participation in their technical committees.

2. EXECUTIVE AND ADMINISTRATIVE COMMITTEES OF THE TECHNICAL BOARD:

2.1 At the Board's first meeting in each administrative year, the Chairman, with the approval of the Board shall appoint an Executive Committee from the membership of the Board to serve for one year. The Chairman of the Board shall be the Chairman of the Executive Committee.

2.2 The Executive Committee shall be the executive and administrative agent of the Board, and, on matters requiring prompt disposition which arise between meetings of the Board, shall exercise all powers of the Board, except the approval of standards, recommended practices and information reports. The Executive Committee shall notify the Board of all actions taken. All actions of the Executive Committee shall be subject to review and confirmation of the Board.

2.3 Committees may make no commitments involving expenditures of Society funds without the approval of the Executive Committee.

2.4 The Executive Committee may approve the payment of necessary and reasonable travel expenses to meetings of the Board or of its committees, of an officer or regular employee of the government, or an instructor of a recognized educational institution, either as a member of such groups or as an invited consultant.

2.5 In intervals between meetings of the Board, the Executive Committee shall place before the Board any question on which a Board member desires the opinion of or action by the entire Board.

2.6 The Board may appoint such administrative committees as are necessary to carry on its work.

3. COUNCILS:

3.1 The Board may organize Councils and delegate to them authority to provide for, promote, direct and supervise the development of SAE standards, recommended practices, information reports and conduct research within defined areas of the Society's interests. The Board retains the authority for final review, approval, rejection or referral of actions taken when dissenting views (of members or nonmembers who were consulted in committee work or the preparation of documents) cannot be substantially reconciled.

Each Council shall organize committees to carry on the various phases of its assignment. Such committees may organize subgroups as they find necessary. These subgroups may be designated as divisions, subcommittees, task forces, or panels (hereinafter these groups are referred to as committees).

3.2 Each Council will determine that a consensus exists of those substantially concerned with the provisions of a proposed standard or recommended practice. The appropriate Council shall verify that the following requirements have been met with respect to each proposed standard or recommended practice:

3.2.1 All substantially concerned parties who are technically competent shall have had an opportunity to participate and their views shall have been given due consideration.

3.2.2 There shall be evidence of use or of potential use of a proposed standard or recommended practice.

3.2.3 Due consideration shall have been given to the existence of other comparable standards having national acceptance in the given field.

3.2.4 There shall be no unfair discrimination inherent in the proposed standard or recommended practice.

3.2.5 There shall be assurance by those concerned of a satisfactory level of technical quality of the proposed standard or recommended practice.

3.3 The Councils may establish operational procedures within their scope and in compliance with these rules and regulations, subject to approval by the Board.

3.4 The Board will appoint the initial membership and chairmen of its Councils. Thereafter, each year each Council shall nominate a chairman for Board approval. The Chairmen of Councils shall be Board members unless otherwise authorized by the Executive Committee. Each new Chairman of the Technical Board will request Board member(s) to serve as sponsoring or liaison member(s) with the various Councils of the Board as may be needed to carry on communications between the Board and its Councils. After the initial membership of each Council has been approved by the Board, the chairman of a Council is authorized to effect changes in non-Board member personnel of the Council.

4. COMMITTEES:

4.1 Selection of participants of SAE technical committees shall be based upon the following:

4.1.1 That a substantial balance of voting members with respect to technical background and experience shall exist within the committee organization. On technical committees dealing with the standardization of parts, products or materials, the organization and operating procedures of the committee, which includes its subcommittees and panels, shall provide an opportunity for qualified individuals from the substantially interested producer, consumer and general interest groups to participate and vote.

4.1.2 That a committee so chosen is competent and authoritative in its field.

4.1.3 Subordinate Committee Structure: Working groups such as task force ad hoc committees, panels or other groups may be organized under operating committees with membership having specific specialized expertise for the purpose of drafting or writing proposed documents. Such groups may or may not be balanced in accordance with the requirements specified in paragraph 4.1.1. Such groups may be specifically organized to develop recommendations for standards for military use or for use by other governmental agencies. The work of all such subordinate committee groups shall be subject to review and approval by parent committee(s) or council(s) wherein balanced structure as noted in paragraph 4.1.1 above is required.

4.2 Committee participants shall be designated as member, liaison member or consultant member. Liaison and consultant members are not eligible to vote on committee actions.

All participants are appointed by the chairmen on the basis of need for their particular services. Liaison members provide coordination with paralleling activities of other committees and organizations. Consultant members supply advice on the specific program for which they have been appointed. Governmental or other agency employees may be appointed as members, liaison members or consultant members of the committee with aforementioned responsibilities and privileges.

4.3 Council chairmen, with the advice of Council members, shall designate chairmen of newly formed committees reporting directly to their Council. Existing committees, reporting directly to the Council, shall nominate a chairman annually for Council approval. Rotation of the chairmen is encouraged where practical. Renominations of chairmen who have served five or more consecutive years shall be subject to review and approval by the Council. Committee chairmen at all levels may appoint chairmen of their subordinate committees.

4.4 A committee may initiate a project within its scope. In cases where such projects overlap areas of activity of another Council's committee, the Board shall be asked for prior review and approval.

4.5 Except as provided in paragraph 7.1, action by committees shall be by vote of the majority of their members.

5. RESPONSIBILITIES AND QUALIFICATIONS OF MEMBERS OF THE BOARD, COUNCILS AND COMMITTEES: In discharging their responsibilities, members of the Board, Councils, and Committees function as individuals and not as agents or representatives of any organization with which they may be associated. Members are appointed to SAE technical committees on the basis of their individual qualifications which enable them to contribute to the work of these committees. Members of the Board shall be members of the Society. SAE membership is not a prerequisite for membership on Councils or technical committees established by the Board or its Councils.

6. STANDARDS, RECOMMENDED PRACTICES, AND INFORMATION REPORTS: Standards, recommended practices and information reports are known collectively as SAE technical reports.

6.1 Reports of committees recommending the approval of a standard or recommended practice shall represent a consensus of the committee. In standardization practice a consensus is achieved when substantial agreement is reached by concerned interests according to the judgment of the Technical Board and/or one of its Councils.

6.1.1 Consensus implies that all dissenting viewpoints have been considered, and that an objective effort has been made toward their resolution.

6.1.2 Substantial agreement means much more than simple majority, but not necessarily unanimity.

6.2 Standards are documentations of sound, established, broadly accepted engineering practices.

6.3 Recommended Practices are documentations of data that are intended as guides toward standard engineering practice. Their content may be of a more general nature or they may propound data that have not yet gained broad acceptance.

6.3.1 Committees may, at their discretion, add an introductory note to any recommended practice stating that "This SAE Recommended Practice is intended as a guide toward standard practice, but may be subject to frequent change to keep pace with experience and technical advances. Hence, its use where flexibility of revision is impractical, is not recommended."

6.4 Information Reports are compilations of engineering reference data or educational material which are useful to the technical community. Cooperative committee development action is a significant technical feature of such reports.

6.5 SAE committees are encouraged to develop test procedures and may develop and define performance levels[1] where they are appropriate and according to the rules as described in Paragraphs 6.5.1 through 6.5.4.

6.5.1 Performance levels should be separated from test procedures either in separate technical reports or in different sections of the same report. Where they are separated, the two reports shall be cross-referenced. The scopes of reports containing performance levels shall clearly state the applicability and limitations of such levels.

6.5.1.1 Where physical, dimensional or performance characteristics are necessary to identify items and assure compatibility or interchangeability of replaceable components (e.g., material specifications, screw threads, oil grates, hoses and fittings, and flange dimensions), the defining characteristics, dimensions, identifying markings, properties, or performance characteristics may be included in SAE technical reports.

6.5.2 The Councils of the Technical Board are responsible for developing specific policies concerning the advisability of their committees developing performance levels for specific procedures or subject areas. Such policies will direct development of performance values for documents involving dimensional requirements or grade, class or type identification and differentiation. When SAE technical reports concern major systems or complete machines or vehicles, the necessity and desirability of developing performance levels are to be directed to and accepted by the appropriate Council before development work is commenced by a Technical Committee.

6.5.3 Where documents prepared to standardize test procedures and equipment are for the purpose of assuring consistent results between laboratories in the measurement of significant characteristics and properties, such documents should not contain performance levels. In such cases where the performance level is of interest, an information report outlining repeatability or accuracy of methods or data obtained from typical range of tests to show state-of-the-art should be prepared. An alternative is to present such information in a technical paper published through an engineering activities program.

6.5.4 If a Technical Committee concludes that performance levels are appropriate, the rationale used in selecting their applicability and limits must be clearly defined for consideration by and submitted for approval by its Council. The Technical Committee's rationale shall include information giving reason for identified differences between the submitted document and other existing standards or regulations. Values that exceed the state-of-the-art are to be avoided.

6.6 For specific details on preparation of documents, one should refer to guideline preparation documents; i.e., J1159–SAE Recommended Practice for Preparation of SAE Technical Reports—Surface Vehicles: Standards, Recommended Practices, Information Reports; Technical Committee Guideposts—An SAE Technical Committee Guide Recommended by the SAE Technical Board; and the Aerospace Organization and Operating Guide and appendices thereto.

7. PROCEDURE FOR APPROVAL OF STANDARDS, RECOMMENDED PRACTICES AND INFORMATION REPORTS[2]

7.1 Documents submitted to a Council for approval, in general, should

---

[1] Performance levels are the specified minimum, maximum, mean or average performance requirements.

[2] Hereinafter, Standards, Recommended Practices and Information Reports are referred to as documents.

have the unanimous approval of the committee making such a submittal. Where unanimous approval cannot be achieved, documents shall have the approval of at least three-quarters of the responding committee members who have not waived their vote. Unresolved dissenting views, including those of liaison and consultant members, as well as non-members, shall accompany the document when submitted to the Council.

7.1.1 Any person (member or non-member of a committee group) submitting an unresolved dissenting view on a document shall have the right to appeal to the Board the approval of the document by a Council. To take his appeal, such person must file with the Council a notice in writing within twenty (20) days after receipt of notification that the document has been approved by the Council, and the notice shall state the person's desire to appeal, shall set forth the reasons for his appeal, identifying all portions of the document to which exception is taken, and shall include supporting information, data and material in support of his position.

7.2 Committee documents shall normally require confirmation by letter ballot except when they are submitted for final voice vote approval. In such instances the documents shall be distributed to the members of the voting group at least two weeks prior to the meeting.

7.3 Every approved document shall carry the following statements: "All technical reports, including standards approved and practices recommended, are advisory only. Their use by anyone engaged in industry or trade or their use by governmental agencies is entirely voluntary. There is no agreement to adhere to any SAE standard or recommended practice, and no commitment to conform to or be guided by any technical report. In formulating and approving technical reports, The Technical Board, its Councils and committees will not investigate or consider patents which may apply to the subject matter. Prospective users of the report are responsible for protecting themselves against liability for infringement of patents, trademarks and copyright."

7.4 Council members will review documents for technical content, for policy implications and for impact of the documents on users and the public in terms of broad technological considerations. Approval by Council will also be based on the committee record of voting and the consensus attained from all participants.

7.5 Councils will strive for unanimous approval and in no case will they approve a document which has not been found acceptable by three-quarters of its responding members. The Chairman of the Board is required to approve all documents. He may declare a document submitted to letter ballot by a Council to have been approved after three weeks from date of circulation of the document, provided ballots have been returned by at least three-quarters of the members, and provided there are no negative ballots. A written negative ballot within three weeks after the circulation of a document will require the proposal to be reconsidered by Council before final action. A Council will approve the document, reject the document, or refer the document with dissenting views to the Board for action. Council will refer to the Board any approved document where a notice of appeal has been filed as provided in paragraph 7.1.1.

7.6 Neither the Council nor the Board will alter the technical content of a document without first referring it back to the responsible Council or committee. All approved documents referred to the Board by a Council where a notice of appeal has been filed shall be handled by either

7.6.1 referring the document to the responsible committee with instructions to attempt to resolve the matters appealed and to provide the person appealing an opportunity to present to the committee such evidence as is set forth in his notice of appeal, or

7.6.2 setting a time at any Board meeting scheduled after the filing of the notice of appeal where the person appealing shall have an opportunity to present to the Board such evidence as is set forth in his notice of appeal. Upon hearing the evidence, the Board shall determine the merits of the appeal and shall approve, reject or refer back to the committee (as in subparagraph (a) above) the document. The Board's decision shall be transmitted to the person appealing within ten (10) days after the decision and it shall briefly set forth the reasons therefor.

Failure of any person appealing to appear at a scheduled Board hearing shall be deemed a waiver or a withdrawal of his appeal.

7.7 The effective date of a document shall be the date it is approved by the Council or by the Board, unless otherwise stipulated by the Council, the Board or the Board's Chairman.

8. TECHNICAL BOARD MEETINGS:

8.1 Only members of the Board have the right to attend and vote at Board meetings, and to cast letter ballots on matters under consideration by the Board. Others may attend at the discretion of the Board Chairman.

8.2 The Chairman of the Board shall preside at meetings of the Board and, if he is not present, a member of the Executive Committee designated by the Chairman or by that Committee shall preside.

8.3 One-half of the members of the Board shall constitute a quorum.

8.4 Action by the Board shall be by majority vote of those present; provided that any member may call for a letter ballot on action taken, and when a letter ballot is taken, action by the Board shall be by three-quarters vote of the entire Board; documents referred to the Board by the Councils for approval shall require approval by three-quarters of the entire membership of the Board.

8.5 The first order of business shall be an executive session.

8.6 Questions of parliamentqry procedure shall be determined by Roberts Rules of Order.

8.7 The last order of business shall be the determination of the time and place of the next meeting of the Board.

9. ELECTION OF ONE DELEGATE AND TWO ALTERNATES TO THE ANNUAL NOMINATING COMMITTEE:
At the Technical Board's first meeting of the administrative year, the Executive Committee will submit for the election of the Board, recommendations for one delegate and two alternates to serve on the Society's Annual Nominating Committee. All delegates and alternates shall be voting members of the Society. They need not be members of the Technical Board. These names must be submitted to the Secretary of the Society not later than September 30. (It is the duty of the Annual Nominating Committee, each year, to select nominees for President, Treasurer and four Directors for inclusion in the ballot for election of Officers to be submitted to the voting members.)

10. ELECTION OF TECHNICAL BOARD NOMINATING COMMITTEE:
At the Technical Board's first meeting of the administrative year, the Board shall elect, from its membership, seven members to serve as a Nominating Committee to nominate a voting member of the Society to serve as a Director on the Society's Board of Directors. The first member elected to the Nominating Committee shall serve as Temporary Chairman and shall assume the responsibility of calling the first meeting of the Committee. A quorum shall consist of five members. The nomination made by the Committee shall be approved by at least four of these members. It shall be the duty of this Nominating Committee to submit to the Society's Secretary, not later than June 15, each year, the name of a consenting nominee to serve as a Director for a term of three years. This nominee shall be listed on the ballot for election of Officers submitted to the voting members.

11. SECRETARY:

11.1 The Secretary of the Board and his staff assistants shall be designated by the General Manager of the Society.

11.2 The Secretary shall be responsible for the performance of such staff functions as the Board or its Executive Committee may direct, and shall provide for the recording and distribution of minutes of meetings of the Board, Councils and committees, subject to the financial limitations imposed by the Directors for the operations of the Board.

11.3 The Secretary shall have authority to release information and publicity with respect to the work of the Board and its committees, unless the Board specifically directs otherwise.

12. RECORDS: The records of the Board and its committees, shall be maintained for a reasonable time in the offices of the Society where they will be available for inspection by members of the Society except as the Board or the staff security officer (in the case of classified material) directs otherwise.

13. AMENDMENTS: Amendments to these rules and regulations shall be approved by not less than three-quarters of the members of the Board, subject to final approval by the Directors.

Adopted January 10, 1946
Revised January 14, 1972
Revised October 20, 1976
Revised June 8, 1977
Revised May 25, 1978

1 Rules, Nomenclature and Guideposts

**TECHNICAL COMMITTEE GUIDEPOSTS—
SAE J1271 AUG79**

SAE Recommended Practice

Report of Publications Advisory Committee approved August 1979.

1. Introduction

1.1 These Guideposts were written to provide information needed by SAE technical committee members. Subject matter covers relations of technical committees to the SAE organization and, in broad terms, committee operating procedures.

1.2 The *Guideposts* are the outgrowth of the principles and policies of the Society, and they reflect the philosophies, traditions, and methodology that have emerged from years of successful operations of SAE technical committees.

1.3 These Guideposts are necessarily brief and presented in an outline form. For additional information refer to the latest issue of the SAE Technical Board Rules and Regulations and appropriate Council Operating Practices.

2. SAE Objective

2.1 The objective of the Society is to promote the Arts, Sciences, Standards, and Engineering Practices connected with the design, construction, and utilization of self-propelled mechanisms, prime movers, components thereof, and related equipment. SAE serves its members and the General Public through meetings and programs developed by its various Engineering Activities and Sections, through its Placement Committee, and through its publications; *it serves industry, government, and the public through the development of technical reports*[1] *including engineering standards and recommended practices, and distributing these documents.*

3. SAE Technical Board

3.1 **Organization**—The Technical Board is the agent of the SAE Board of Directors with authority to direct and supervise all SAE Cooperative Engineering Programs, including standardization, research, and the participation in technical committee activity of other organizations. Fig. 1 shows its position in the Society's structure, the councils, and a few examples of technical committees it administers.

3.2 **Philosophies**

3.2.1 The Technical Board will consider those projects for which industry, government, the public, or other responsible agencies have expressed a need and which lend themselves to cooperative solution. Within their own operations, technical committees frequently generate projects meeting the above-noted criteria.

3.2.2 The Technical Board expects technical committees to set up their own organizations, procedures, and programs within their scopes and the limits of the Technical Board's *Rules and Regulations* and individual council guidelines.

3.3 **Recognition of Achievements**—Annually the Board awards a maximum of 30 *Certificates of Appreciation* to technical committee members and to individuals representing SAE in other organizations. Nominations for awards are submitted through the councils to the Certificates of Appreciation Committee by the various technical committees. Supporting data outlining the basis for nomination is required.

4. Councils of the SAE Technical Board

4.1 The Technical Board delegates to its councils the authority to direct and approve (see paragraph 6.2) SAE Standards, Recommended Practices, and Information Reports (subject to the right of anyone to appeal a decision to the Board). The councils are authorized to establish committees that may be needed to accomplish this assignment.

4.2 **Committee Sponsor**—It is intended that the chairman of each council appoint annually council members to act as sponsors for committees functioning directly under the council. The council chairman may appoint the council technical committee chairmen as members of the council and as sponsors for their committees. The committee sponsor shall represent the committee to the council, and serve as liaison between the committee and the council. During periods when a committee is without a sponsor, the council chairman will perform such functions.

4.2.1 Committee sponsor's responsibilities shall include:

(a) Providing regular communication of significant committee activity to the council and council actions to the technical committees. To facilitate communication, a sponsor report should be included as an agenda item at each technical committee meeting. Each technical committee chairman should prepare an annual status report for submission to the council through his sponsor. (The date for this report to be selected to be convenient to council meetings and national meetings.) Time should be provided on the council agenda for discussion of these reports with the sponsors.

(b) Reviewing and counseling with committee chairman on committee programs and membership and providing council policy guidance.

(c) Providing coordination among sponsor's assigned committees and with other sponsors.

(d) Reviewing technical committee organization and proposing appropriate changes to the council.

5. SAE Technical Committees

5.1 **Objectives**—The objectives of a technical committee are to coordinate and utilize the knowledge, experience, and skill of engineers and other qualified individuals on technical problems within the scope of its activities to:

(a) Conduct necessary investigations and develop technical reports.

(b) Review technical reports periodically, revise as necessary, and maintain content abreast of latest technology.

(c) Advise, consult, and cooperate with industry, government, educational institutions, the public, other standardizing bodies, and other SAE committees and members.

(d) Assist committees of the SAE Engineering Activity Board (see Fig. 1) in the preparation and presentation of papers at national meetings and specialty conferences.

5.2 **Principles**—The end products of the committee's work are offered as the best judgment of a group technically competent to deal with the problems covered and do *not* represent an industry-trade position. Employers of committee members are not committed by an action of an SAE committee. Over many years, the extensive use of SAE technical reports clearly indicates that committee members, working as individuals, do produce results that are practical and useful to industry, government, and the public.

5.3 **Scope**—A technical committee shall be responsible for a field of endeavor, as defined by its scope. In cases where projects overlap areas of interest of another council's committee, the originating committee shall submit the project(s) for review and approval to the other concerned committee and its council prior to issuance. A committee is established when a new major project area is to be undertaken and no existing commitee is available. A committee is discharged when the assigned work is completed and there is no further need for its services. The councils retain responsibility for periodic review of technical reports developed by their disbanded committees.

5.4 **Membership**

5.4.1 QUALIFICATIONS—All participants are appointed to SAE technical

---

[1] The term "technical reports" as used in these Guideposts, stands for the end product of a committee's efforts and may consist of an SAE Standard, Recommended Practice, Information Report, or Aerospace Material Specification.

1.02

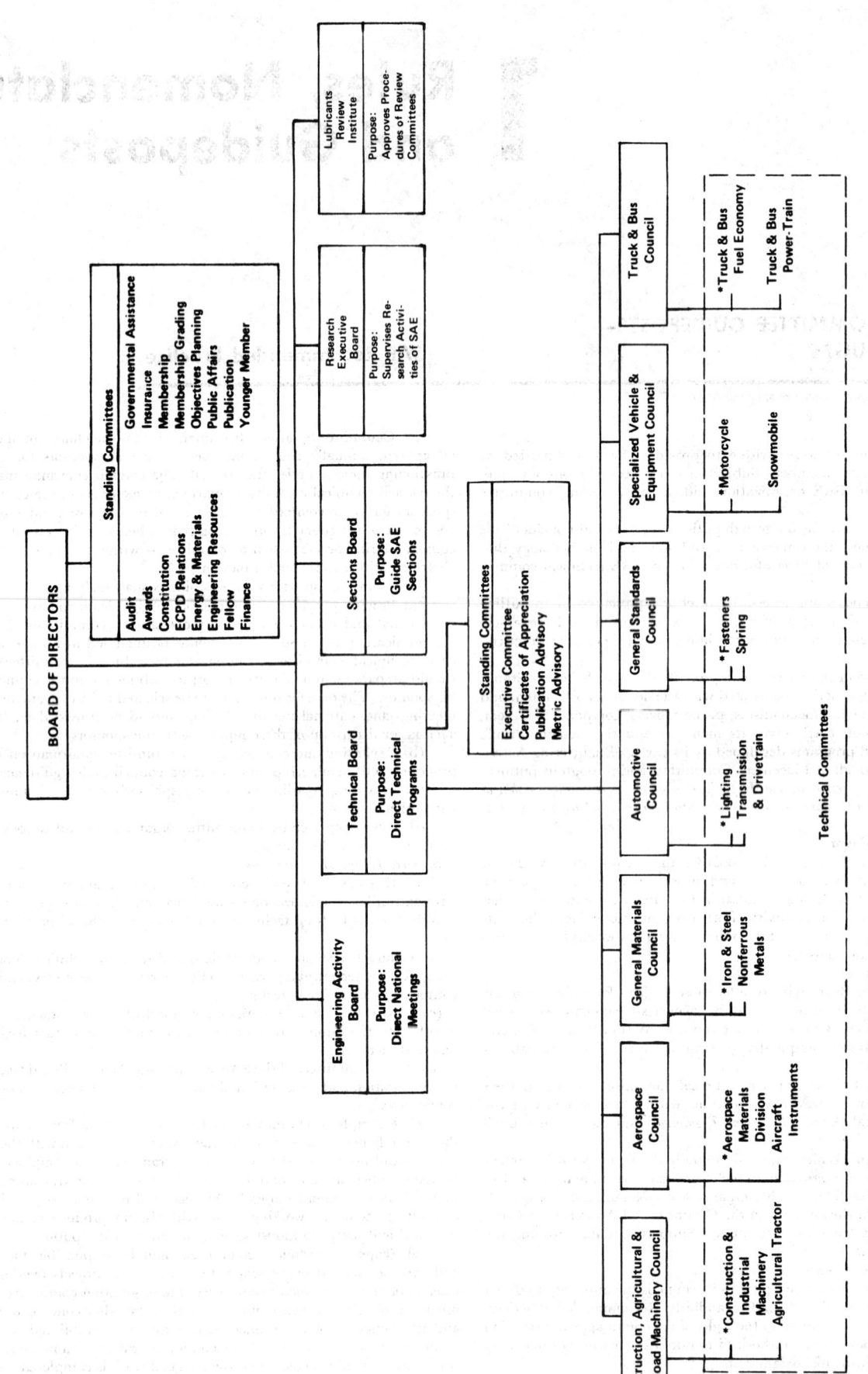

FIG. 1—SAE ORGANIZATION

*Only typical examples of committees reporting to the Councils are shown.

committees by the committee chairman on the basis of their individual qualifications which enable them to contribute to the work of these committees. Overall, committee membership should attain an equitable balance of representation by knowledgeable *parties at interest*. All relevant points of view should be invited to participate. (Ref. paragraph 4.1.1 Technical Board Rules and Regulations.) SAE membership is not a prerequisite for committee membership.

5.4.2 The policy of the Society is that SAE technical committee members act as *individuals* and not as agents or representatives of their employers. Their actions are accepted as personal actions and do not necessarily represent their employers' attitudes or views.

5.4.3 GRADES—In addition to committee officers (see paragraph 5.5.1), committee participants shall be classified as *member, liaison member,* and *consultant member*.

5.4.4 *Liaison and consultant members* are appointed by the chairman on the basis of need for their particular services. *Liaison members* relay information to and from paralleling activities of other committees and organizations. *Consultant members* supply advice on the specific program for which they have been appointed. *Liaison* and *consultant members* are not eligible to vote on committee actions except at the request of the committee chairman.

5.4.5 Governmental agency employees may be appointed as *members, liaison members,* or *consultant members* of the committee with aforementioned responsibilities and privileges.

5.5 Organization—The number of members on a technical committee may vary depending on the specific needs. Typical organization patterns are shown in Figs. 2 and 3.

5.5.1 OFFICERS—The committee shall have a chairman and may have a vice-chairman and/or a secretary.

5.5.1.1 The chairman and vice-chairman of a newly formed committee shall be appointed by the council chairman with advice of council members. Existing committees shall nominate a chairman and vice-chairman annually for council approval. (See paragraph 4.3 of SAE Technical Board Rules and Regulations). The chairman may become a member of a council by appointment by the council chairman.

5.5.1.2 The secretary shall be appointed annually by the committee chairman.

5.5.1.3 Reasonably frequent rotation of committee chairmen, where practical and desirable, is encouraged. However, renominations of chairmen who have served five or more consecutive years shall be reviewed and approved by the council.

5.5.1.4 It is the duty of the chairman to:
(a) Plan and conduct committee meetings.
(b) Establish subcommittees, appoint their chairmen, and supervise their operation.
(c) Establish working panels, including appointment of their chairmen and/or individual member work assignments, and supervise their operation.
(d) Assign projects so as to balance and expedite the committee's work.
(e) Act for the committee between meetings, subject to confirmation at the next meeting.
(f) Supervise and report voting on all committee reports. (See paragraph 5.5.2.1 for responsibilities of SAE staff representative.)
(g) Review the membership annually to maintain an active and balanced committee.
(h) Recommend revisions of committee procedures as needed.
(i) Arrange for the nomination of candidates for the Technical Board Certificate of Appreciation Award. Candidates must be nominated by April 1 (see paragraph 3.3).
(j) Serve as chairman of the steering committee (or executive committee), if applicable.
(k) For additional duties of the chairman, refer to SAE J1159, Preparation of SAE Technical Reports—Surface Vehicles and Machines: Standards, Recommended Practices, Information Reports.

5.5.2 SAE STAFF REPRESENTATIVE—An SAE staff representative will advise the committee officers on procedures and assist the committee in its organization and operation, attend meetings, and assure that meeting minutes are prepared and properly distributed.[2]

5.5.2.1 SAE staff representative performs tally of committee voting and disseminates results of balloting.

5.5.3 EXECUTIVE COMMITTEES (Fig. 2)—When a technical committee has numerous subcommittees, projects, or is so large as to make meetings of the entire group impractical, an executive committee may be established to organize and manage the affairs of the committee. The executive committee shall include all committee officers, and may include all subcommittee chairmen and such additional members as may be desirable to form an efficient working group. The technical committee officers shall be the officers of the executive committee.

5.5.4 SUBCOMMITTEES—Subcommittees are organized to carry out specific continuing technical segments of the committee's scope. The original chairman shall be appointed by the parent technical committee chairman; thereafter, the chairman may be nominated by the subcommittee, subject to review and approval by the parent technical committee chairman. It is desirable that the subcommittee chairman be a member of the parent committee.

5.5.4.1 The original membership will be appointed by either the parent committee chairman or by the subcommittee chairman. Thereafter, membership matters are handled by the subcommittee chairman.

5.5.4.2 Duties of subcommittee officers in relation to their subcommittees are the same as those of the technical committee officers in relation to the technical committee, except for paragraph 5.5.1.4, items b, i, and j.

5.5.5 WORKING TECHNICAL PANELS (Fig. 3)—When a committee or subcommittee wishes to have several of its members work together in preparation of a draft technical report, a temporary working panel may be formed. After completion of its task, responsibility for review and maintenance of resulting technical reports reverts to the committee or subcommittee, and the panel is discharged.

5.6 Committee Relationships with SAE and Non-SAE Groups

5.6.1 INTRA-SAE RELATIONSHIPS—As a primary principle, each SAE technical report should be reasonably self-contained or cross-reference other SAE documents. Development of a draft technical report will often require use of data which falls within the scope of another SAE committee. In these instances, liaison should be established by formation of a joint sub-group, by membership on or from that committee, or through the SAE staff. In any event, comments and/or approval by the consultant committee should be solicited by the committee preparing the draft technical report. Adherence to this principle will avoid duplication of effort and will insure against conflicts and ambiguities. Because of his intimate knowledge of SAE activities, the SAE staff representative can help the technical committee in its relationships with other Society groups. See paragraph 6.6 in SAE J1159.

5.6.2 LIAISON WITH OTHER ORGANIZATIONS—Technical committees should coordinate their efforts with parallel activities in other organizations such as the American National Standards Institute, American Society for Testing and Materials, American Iron and Steel Institute, American Petroleum Institute, and the Aerospace Industries Association. To maintain this liaison, the Technical Board appoints individuals to represent the Society. It is the duty of these representatives to report developments to the appropriate SAE committee, and to present SAE views which are the consensus of the concerned SAE committee(s) to these organizations.

5.6.2.1 Representatives of SAE on American National Standards committees or standards committees of other standards writing organizations shall be appointed by the Chairman of the SAE Technical Board.

5.6.2.2 Representatives of SAE on non-SAE standards committees shall report to a technical committee or, if none exists, to the appropriate council.

5.6.2.3 Representatives of SAE shall, where feasible, develop SAE positions regarding draft technical reports developed by such non-SAE committees through consultation with the appropriate SAE technical committee or council.

5.6.2.4 The representative of SAE shall report activities of the non SAE committee annually and shall advise the technical committee or council of the SAE vote on approval or disapproval of standards or other substantive matters coming before the non-SAE committee. The SAE representative may seek additional SAE support in backing up the SAE position through the appropriate SAE committees and councils.

5.6.2.5 Approval of American National Standards by American National Standards committees for which SAE is sponsor, co-sponsor, or secretariat shall be by the appropriate SAE council on either recommendation of the appropriate SAE committee or the SAE representatives on the American National Standards comitee.

5.6.2.6 Upon approval by SAE of an American National Standard for which SAE is *sponsor, co-sponsor,* or *secretariat,* the subject standard shall be considered an approved SAE Standard. In the event that SAE is not the publisher of the subject standard as an American National Standard, but does provide coverage of the subject matter contained in the standard in the SAE handbook or other SAE publications, such SAE publications will be revised as soon as possible after SAE approval of the American National Standard but not later than one year after such approval. When SAE coverage of the subject matter is approved in advance of a revision to the American National Standard, the Society will immediately initiate a revision of the American National Standard.

5.6.3 JOINT SPONSORSHIP WITH OTHER ORGANIZATIONS—SAE joint sponsorship of committees with outside organizations is discouraged, unless such joint sponsorship is of direct benefit to SAE committees. Where possible, SAE

---

[2] When the chairman or his appointed secretary records the minutes, a copy (or copies, as appropriate) should be forwarded to the SAE office where they will be reviewed and distributed and maintained for at least five years and made available for inspection and for distribution as appropriate.

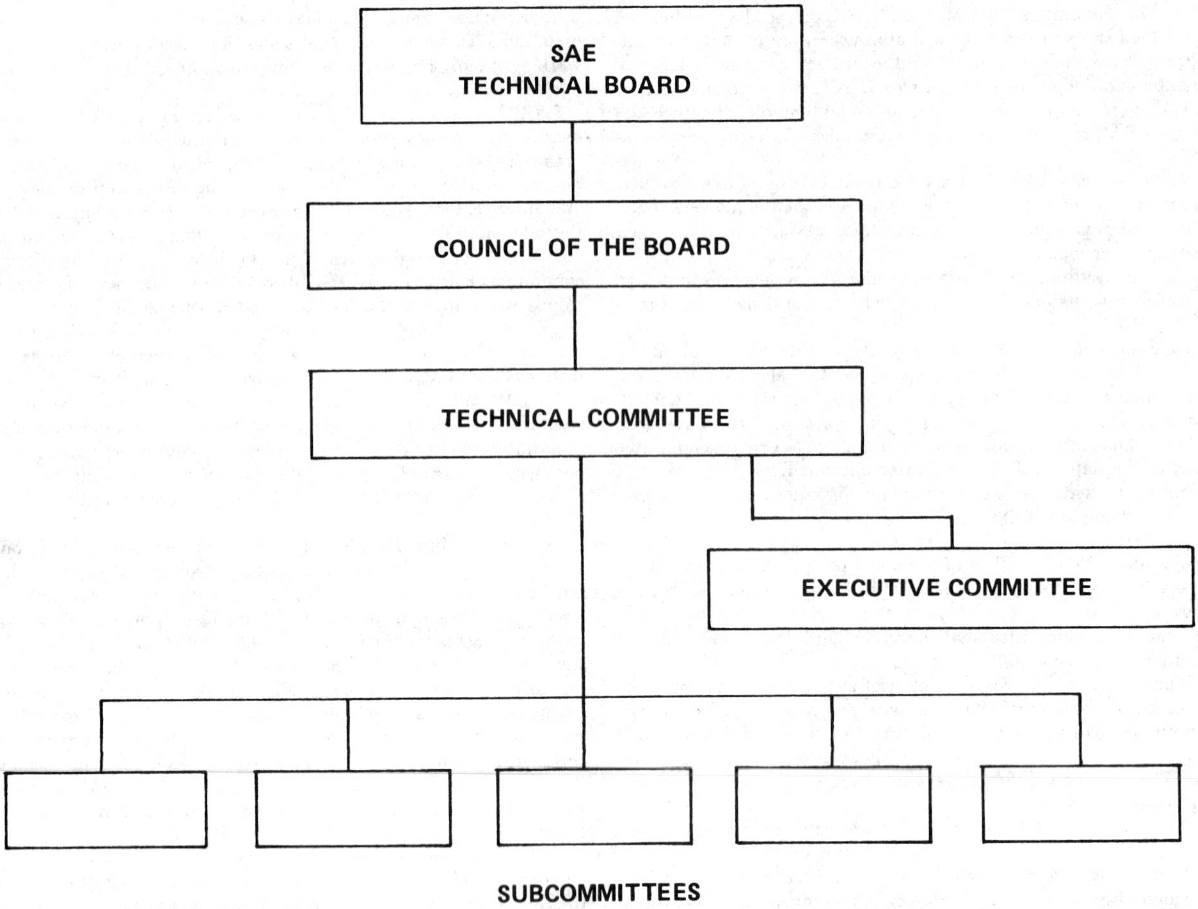

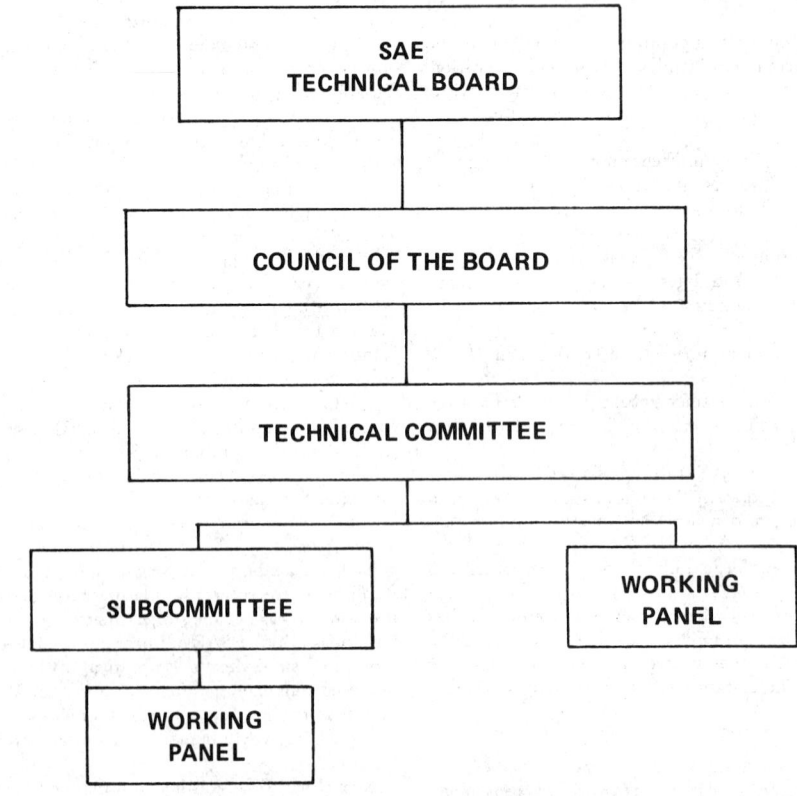

FIG. 2—EXAMPLE OF EXECUTIVE COMMITTEE OF LARGE TECHNICAL COMMITTEE

FIG. 3—TECHNICAL COMMITTEE

should perform its standardization and cooperative engineering functions without the establishment of jointly sponsored groups.

5.6.3.1 In cases where SAE technical committee work or technical projects are of major interest outside of SAE (for example, splines and screw threads), cooperation is encouraged in established SAE sponsored organizations such as the American National Standards Institute (ANSI) and the Coordinating Research Council (CRC).

5.6.3.2 Where joint sponsorship of a project with an outside organization is proposed, specific Technical Board approval is required.

5.6.3.3 Any technical reports resulting from such cooperative activity will be subject to normal SAE review and approval procedures. In such technical reports recognition of the participation of outside groups is appropriate.

5.6.4 COOPERATION WITH GOVERNMENT AGENCIES—Technical committee cooperation with government agencies in developing technical reports of mutual interest is encouraged. Such technical reports will be identified by normal SAE numbering systems when published by the SAE, and appropriately cross-referenced when issued in some other manner.

5.6.4.1 Where there is a divergence of technical opinion on a proposed technical report between a committee and interested governmental agency, the committee may offer the government its technical opinion in the form of comments upon a proposed government prepared specification or standard.

5.6.4.2 If the committee so chooses, and with review and approval by the appropriate council, the committee's technical opinion can be published in the form of an SAE technical report.

5.6.5. SAE PARTICIPATION IN INTERNATIONAL STANDARDS

5.6.5.1 SAE may serve as the U. S. technical secretariat for ANSI in International Organization for Standards (ISO) and International Electro-Technical Commission (IEC) technical committees, subcommittees, and working groups only when approved by the Technical Board.

5.6.5.2 In administering such ISO and IEC technical secretariats, SAE shall form a U. S. Technical Advisory Group (TAG) consistent with procedures of the American National Standards Institute. Membership on the TAG shall follow the rules governing membership on SAE technical committees. Appointments to the TAG shall be subject to confirmation by the appropriate SAE council. The TAG will call upon cognizant SAE technical committees and outside activities, if appropriate, to assist in developing U. S. positions on proposed ISO or IEC standards and in the development of draft standards.

5.6.5.3 SAE shall follow the provisions of the ANSI Guidelines for ISO Standards Activities.

6. Technical Committee Technical Reports—The major effort of technical committee activity is the development of technical reports for publication by the SAE.

6.1 Development—The initial work on a draft technical report is usually handled by a task force which presents its work to a parent group, preferably well in advance of a meeting date. Corrections to the proposal are officially recognized at the meeting of the parent group and documented in the minutes. Depending on procedures established for each group, mailing of draft technical reports may be handled directly by the chairman, the SAE staff representative, or by a delegated member.

6.1.1 GUIDES FOR PREPARATION OF TECHNICAL REPORTS—Annually, a large number of documents are developed by the technical committee for publication. It is not practical for the SAE staff to restyle them. Technical committees will use the following SAE publications.

(a) *Preparation of SAE Technical Reports*—Surface Vehicles and Machines: Standards, Recommended Practices, Information Reports—SAE J1159.

(b) *Rules for SAE Use of SI (Metric) Units*—SAE J916—establishes the rules for the use of Système International (SI) units in SAE documents including specifications and standards.

(c) *AMS Editorial Procedure and Form* for the preparation of Aerospace Material Specifications (AMS) and other Aerospace Material Documents.

(d) *Aerospace Council's Organization and Operating Guide for Aerospace Cooperative Engineering Program* for the preparation of Aerospace Standards, Military Standards, Aerospace Recommended Practices, and Aerospace Information Reports.

6.1.2 COMMITTEE CORRESPONDENCE—It is required that all correspondence within and between committees be classified by subject so that it may be readily identified. Copies of committee correspondence should be sent to the chairman and SAE staff representative. Committees shall use technical committee correspondence forms which are available, upon request, from the SAE staff representative. Committees shall not use stationery with a company or business letterhead.

6.2 Approval—Draft technical reports submitted to a council for approval, in general, should have the unanimous approval of the committee making such a submittal. Where unanimous approval cannot be achieved, draft technical reports shall have the approval of at least three-quarters of the responding committee members who have not waived their vote. Dissenting views, including those of liaison and consulting members shall accompany draft technical reports when they are circulated to the council for final review and approval prior to publication. The committee's reasons for not accepting the dissenting views should be included.

6.2.1 Committee draft technical reports shall normally require confirmation by letter ballot, except when they are submitted for final voice vote approval. In such instances the draft technical reports shall be distributed to the members of the voting group at least two weeks prior to the meeting. Where a single draft technical report is a joint project of two committees reporting to separate councils, it shall be submitted to both councils for review and approval.

6.2.2 The Technical Board retains the authority for final review and approval when dissenting views cannot be reconciled.

6.3 Publication and Timing—After approval by the council, the technical report will be published at the earliest opportunity.

6.3.1 The preparation of technical reports intended for publication in the SAE Handbook should be scheduled so that council approval can be obtained prior to the closing date set by the Publications Advisory Committee. At least three weeks should be allowed for circulating drafts to the councils. Timing on technical reports which are to be released in loose-leaf or pamphlet form is not as critical with respect to publication date. The SAE staff representative should be consulted as required, to determine target dates.

6.4 Distribution and Use—A basic tenet of SAE technical committee operating policy is that technical reports produced by technical committees are advisory in nature. The use of such technical reports by industry, government, or other responsible agencies is entirely voluntary.

6.4.1 Early recognition by the SAE membership of new or revised technical reports is highly desirable. This provides better service for members, government, and the public, and may result in beneficial comment leading to further improvement. For these reasons, information should be submitted by the technical committee chairman to SAE *AUTOMOTIVE ENGINEERING* as a news item, or, if the technical report has wide appeal, it may be given more extensive treatment. With a view to providing increased service, notice of all new and revised technical reports will appear as soon as possible after council approval in *AUTOMOTIVE ENGINEERING* mazazine.

6.5 Review—Every technical report shall be reviewed at least every five years. The staff advisor shall initiate such reviews. At such reviews, the technical report may be reaffirmed, revised, or canceled. If reaffirmed, no formal ballot of the affected council is needed, but the council should be informed of the action. Regular balloting of the council is required for a revision or cancellation.

7. Some General Considerations for Technical Reports—SAE technical reports are to be limited to technical and engineering considerations. They are not to include provisions that are of a commercial nature such as prices, warranties, allocation of risk of loss or conditions of acceptance or rejection, nor are such considerations to be a basis for SAE documents.

7.1 Minimum Requirements—SAE technical reports should be written in terms of performance rather than design so as not to exclude any technically adequate equipment, product, design, material, or process. Where technical requirements are established to achieve a stated purpose, such requirements should be the minimum required to achieve such purpose. In terms of standardization or interchangeability of products, only that portion of the product necessary to accomplish such standard or interchangeability should be specified in the document. When a specific product, design, material, or the like is known not to conform to the requirements or conditions of an SAE technical report applicable to the same class of products, designs, materials, or the like, the reasons (in terms of performance characteristics) for such failure are to be set forth in the minutes or files of the appropriate SAE committee together with all data supporting the conclusions of the committee.

7.2 Two Users—An SAE technical report for a particular product, design, material, or process should only be undertaken when there are two or more interested users, unless the only user in the government and the activity is undertaken at the request of an appropriate agency of the government.

7.3 Source of Supply—It is desirable that technical reports not contain a reference to sources of supply of parts or products, or the identity of manufacturers. Where a committee finds it necessary to specify a particular brand of product, such specification is to be accompanied by the statement *or equivalent.*

7.4 Other Society or Association Product Listings—An SAE technical report may reference a list of products that has been developed by other recognized organizations; however, in such case the source of the list is to be clearly identified. A statement is to be made that the listing is included only for the convenience of the user and does not indicate either approval by SAE or the technical committee or its fitness for purposes specified.

7.4.1 An example of a completely referenced technical report is the SAE Information Report, Special Purpose Alloys ("Superalloys")—SAE J467.

7.4.2 An example of a partially referenced technical report is the SAE Standard, Road Vehicle—Hydraulic Brake Hose Assemblies for Use with Non-Petroleum Base Hydraulic Fluids—SAE J1401.

7.5 Test Materials—A particular product or material may be identified by name when it is essential to uniformity in testing. In such cases, an *or* *equivalent* statement should be added to the company product or material referenced.

7.6 Patents and Copyrights—The committees in developing a technical report are not to consider whether the subject matter set forth is patented or copyrighted. However, if the committee is aware of any copyrights applicable to published material then such material shall not be used in the technical report. In the event it is known by the committee that following the teachings of a technical report will probably result in the infringement of a patent, the committee is to set forth criteria which will permit the user to conform to the technical report without infringing such patent.

7.6.1 NOTICE ON ALL TECHNICAL REPORTS—Every approved technical report shall carry the following statements: *All technical reports, including standards approved and practices recommended, are advisory only. Their use by anyone engaged in industry or trade or their use by governmental agencies is entirely voluntary. There is no agreement to adhere to any SAE Standard or Recommended Practice, and no commitment to conform to or be guided by any technical report. In formulating and approving technical reports, the Technical Board, its councils, and committees will not investigate or consider patents which may apply to the subject matter. Prospective users of the report are responsible for protecting themselves against liability for infringement of patents, trademarks, and copyrights.*

PREPARATION OF SAE TECHNICAL REPORTS—SURFACE VEHICLES AND MACHINES: STANDARDS, RECOMMENDED PRACTICES, INFORMATION REPORTS—SAE J1159 AUG79

SAE Recommended Practice

Report of Publications Advisory Committee approved July 1976 and completely revised August 1979.

1. Foreword—This SAE Recommended Practice has been developed by the Publications Advisory Committee of the SAE Technical Board. The Publications Advisory Committee was formed as the Publications Policy Committee of the Technical Board in 1956. The objectives of the committee are (1) to guide and promote efficient dissemination of material produced under the Technical Board and (2) to recommend immediate and long-range policies to the Technical Board to assure that Technical Board information will be available to those who need it in a form suitable for their work.

2. Purpose—The purpose of this recommended practice is to establish a uniform practice for technical committees for the preparation of technical reports.

3. Scope—It applies to reports of Surface Vehicles and Machinery Technical Committees only. Aerospace technical reports are covered by editorial practices of the Aerospace Council. Close adherence to this recommended practice by technical committees of SAE will help to assure uniform technical reports. Should questions on format, style, or other matters pertaining to the organization and editorial practices of technical reports be raised within technical committees of the Technical Board, they should be referred to the Chairman of the Publications Advisory Committee for interpretation or for discussion by the full Publications Advisory Committee.

4. Technical Committee Chairman Responsibilities

4.1 A technical committee chairman is responsible for seeing that his committee and all subcommittees or working group members understand their responsibilities relative to publications, particularly:

(a) Accuracy of technical content and references,
(b) Conformance to policies and guidelines outlined in this report, and
(c) Clearance through technical committee chairman of all subcommittee or committee work, whether major or minor (including editorial changes and corrections). It is the chairman's responsibility to see that the following actions are taken relative to technical reports.

4.1.1 CLASSIFICATION—This shall be recommended by the time technical agreement has been reached on the content of the report. See Section 5 of this report.

4.1.2 LEGAL ASPECTS—The report shall be checked against rules prepared by the Society's legal counsel. See Technical Committee Guideposts, SAE J1271, Section 7.

4.1.3 PUBLICATION METHOD—This shall be recommended based on the following methods.

4.1.3.1 *SAE Handbook*—The Handbook is used for technical reports (Standards, Recommended Practices, and Information Reports) of value to a substantial number of SAE members.

4.1.3.2 *Handbook Supplements*—Supplements are used for descriptive or educational material of broad interest to SAE members and other engineers in the automotive and allied industries. They are also used to provide groups of related technical reports from the SAE Handbook.

4.1.3.3 *Separate Reports*—A properly approved technical report may be issued separately in addition to publishing in the Handbook, either because the subject is timely and must not wait for the next Handbook issue, or because of major interest.

4.1.4 ORGANIZATION AND FORMAT—The draft of the report shall be prepared using details given in Section 6 of this report.

4.1.5 METRICATION—The report shall be checked to make sure that metric units have been included or used as the basis for the report. Details of use shall be checked for conformance with Rules for SAE Use of SI (Metric) Units, SAE J916.

4.1.6 KEY INDEX WORDS—Along with the report, the committee shall submit a list of key words to provide a basis for index preparation by staff. A list of 3–10 words is suggested, depending on the nature of the report. Instructions are given in Section 7 of this report.

4.1.7 EXPLANATION OF PROPOSED REPORT—When the report is submitted to a council of the Technical Board, a statement shall be included outlining:

(a) Significance of report,
(b) Background information (rationale),
(c) Reason for choice of classification, and
(d) Recommendation for method of distribution.

The statement will be retained in the committee file by the SAE Technical Division.

4.1.8 CUT-OFF DATE—When beginning the final approval process, the approved committee draft report should be submitted to SAE staff in time to meet the deadline for publication of the Handbook. Time must be allowed for approval by a cognizant council of the Technical Board and submittal of final report draft and transmittal to the Publications Division by June 30 of each year. Staff should be consulted on adequate lead time.

5. Classification and Numbering of Technical Reports[1]

5.1 Classification—Technical reports are approved for publication by the Technical Board, and must be based on sound technology and cooperative engineering work. Before publication, a report must be classified into one of the following three classifications: (See paragraph 4.1.1 of this report.)

5.1.1 SAE STANDARDS—These reports are a documentation of broadly accepted engineering practices or requirements for a material, product, process, procedure, or test method.

5.1.1.1 A product standard may be primarily a descriptive standard covering dimensions, composition, and other details or it may be a functional or performance standard, or both.

5.1.1.2 Performance standards involve requirements or levels against which the functions can be evaluated. This frequently involves the need to define test methods by which these requirements are measured. Preferably, performance standards and test procedure standards should be in separate reports. If this is not practical, they should be in separate sections of the same report. Where performance standards are given, it is desirable to publish the rationale simultaneously as an SAE Information Report in order to provide all users with the basis for selection of performance levels.

5.1.2 SAE RECOMMENDED PRACTICES—These reports are documentation of data that are intended as guides to standard engineering practice. Their content may be of a more general nature, or they may propound data that have not yet gained broad acceptance.

5.1.2.1 A technical committee preparing such a report may add an introductory note stating, "This SAE Recommended Practice is intended as a guide toward standard practice but may be subject to frequent change to keep

---

[1] The Society also approves and issues reports for the aerospace industry. These are called Aerospace Standards (AS), Aerospace Recommended Practices (ARP), Aerospace Information Reports (AIR), and Aerospace Material Specifications (AMS). Their definitions are similar to the above.

pace with experience and technical advances, and this should be kept in mind when considering its use."

5.1.3 SAE INFORMATION REPORTS—These reports are compilations of engineering reference data or educational material useful to the technical community.

5.1.4 EXAMPLES

5.1.4.1 *Standard*—Automotive Carburetor Flanges, SAE J623. A product standard based on dimensions.

5.1.4.2 *Recommended Practice*—Surface Texture Control, SAE J449. Description of the techniques of control, including preparation of standards, etc.

5.1.4.3 *Information Report*—Mechanical Properties of Heat Treated Steels, SAE J413. General information giving guidance on the relationship of various properties.

5.2 Numbering—Prior to submission to the appropriate council of the Technical Board, SAE staff will assign a "J" number to all reports.

5.2.1 SUFFIXES—When a technical report is approved by a council or the Technical Board, the report "J" number will be supplemented by a date suffix (e.g., SAE J1159 AUG79).[2] This date suffix should be shown in all indexes and should be used as appropriate in references. (See paragraph 6.6 of this report.) The date suffix will be advanced for each technical or editorial revision.

5.2.2 INTEGRITY OF SAE "J" NUMBERS—Changes to an SAE Standard or Recommended Practice which alter it sufficiently to affect its interchangeability or interchangeable application shall require a new "J" number identity. The superseded "J" number shall continue to exist unless retired. When an SAE technical report is retired, its "J" number shall continue in the index with its date suffix, classification, a cross reference to any superseding "J" number, and an indication of its last date and method of publication.

6. Guide for Preparation of Technical Reports

6.1 Organization of Report—The following recommendations for items to be included in SAE technical reports are all-inclusive. Accordingly, all items may not be required for every type of report. In addition, the sequence in which they are presented herein may be changed, for good reason, by the responsible technical committee.

6.1.1 TITLE—Each report shall have a title which does not duplicate an existing title. It should be as short, concise, and descriptive as possible.

6.1.2 REPORT CLASSIFICATION—See Section 5 of this report.

6.1.3 APPROVAL NOTE—Includes credit to originating committee, original approval date, and date of last revision or reaffirmation. Latest editorial revision, if any, is included. For example: Report of the Iron and Steel Technical Committee, approved April 1963, last revised June 1975, editorial change March 1979.

6.1.4 INTRODUCTION (as applicable)—The introduction shall provide the basis for data or information in the report, background or general description of the report, and a brief explanation of changes from previous version of the report.

6.1.5 PURPOSE AND/OR SCOPE (as applicable)—If both are used, the purpose shall precede the scope. The purpose will explain the objectives or end to be obtained by use of the report. The scope will briefly give the extent of treatment and applicability of the report.

6.1.6 DEFINITIONS, GLOSSARY OF TERMS, TERMINOLOGY, AND DESIGNATIONS—All definitions, glossary, terminology, and designations should be reviewed for consistency with SAE reports; SAE Motor Vehicle and Machinery, Safety and Environmental Terminology, HS215, and Guidelines for Developing and Revising SAE Nomenclature and Definitions, SAE J1115.

6.1.7 Test procedures including:

6.1.7.1 Description of facilities and environment.

6.1.7.2 Listing of instrumentation and equipment. (Where instruments must be identified by brand name, the phrase *or equivalent* must be used.)

6.1.7.3 Details for installation of instrumentation and equipment and preparation required for vehicles or machinery, test samples or components.

6.1.7.4 Forms for data sheets, graphs, and reports.

6.1.8 Dimensional data, including tables and charts.

6.1.9 General specifications which augment dimensional data.

6.1.10 Illustrations, photographs.

6.1.11 Performance requirements (if not included in a separate document)—See paragraph 6.5 of SAE Technical Board Rules and Regulations for reservations on performance requirements.

6.1.12 Component materials and mechanical and physical properties.

6.1.13 Inspection requirements.

6.1.14 APPENDICES—As applicable, to add supplementary engineering reference data or educational material not an integral part of the basic technical report. They may also be issued as a separate SAE Information Report.

6.1.15 RATIONALE—The rationale which accompanies a technical report when being sent to upper committees for approval may be included as an Appendix or may be the subject of a separate SAE Information Report. The rationale will provide an expanded explanation of the purpose and scope of the technical report; an explanation of the reasons for decisions, conclusions, and recommendations; and a report on actual tests made which support conclusions and recommendations or which provide the basis for performance criteria.

6.1.16 REFERENCES—List all SAE technical reports and other standards referenced in the report.

6.1.17 BIBLIOGRAPHY (as required)—A list of publications from which authoritative information was gathered for inclusion in the report.

6.2 Preparation of Technical Reports

6.2.1 PARAGRAPH NUMBERING—A decimal numbering system should be used where practical to aid organization in long or complicated reports. Use decimal point to indicate successive subheadings (Example: 1., 1.1, 1.1.1).

6.2.2 NEW REPORTS—Double space drafts of new reports. Mark and date each successive draft legibly.

6.2.3 REVISIONS TO EXISTING REPORTS[3]—Revisions may be handled in several ways. The best method depends upon the extent of the revisions. A major revision, for example, would probably best be handled by complete retyping as in paragraph 6.2.2. Other approaches in handling are:

(a) Cut and paste or mark copies of printed report, indicating deletions and insertions clearly.

(b) Double space new copy, describing clearly where it is to be inserted in existing report.

6.3 Artwork—These instructions are based on a 50% reduction for final size. A one-column figure has a final maximum width of 92 mm (3.62 in). Largest maximum final width is 178 mm (7.00 in).

(a) Typewriter lettering and pencil drawings are unacceptable as both fade and drop out in reproduction process. Please use ink.

(b) Clear, sharp glossy prints of specified original artwork are acceptable if originals must be kept in committee's files.

6.3.1 LINE DRAWINGS (excluding graphs)

Main lines—Equiv. O Leroy pen (0.3 mm width)
Inside lines
Dimension line leaders } Equiv. 00 Leroy pen (0.2 mm width)
Phantom line, etc.

6.3.2 GRAPHS—The heaviest line weight used on graphs should be the curves.

Curves—Equiv. 1 Leroy pen (0.4 mm width)
Ordinate and Abscissa—Equiv. 0 Leroy pen (0.3 mm width)
Grid lines }
Tic marks } Equiv. 00 Leroy pen (0.2 mm width)

6.3.3 LETTERING (excluding section and reference letters)—All lettering to be capitals unless lower case letters are necessary for a specific term. Lettering shall be vertical except for quantity symbols which shall be in italics. Use only Roman alphabet, except where letters are recognized standard symbols.

6.3.3.1 All lettering should be placed outside visible outline of part. Label, with line and arrowhead to area being identified, should be kept reasonably close to figure.

6.3.3.2 Lettering, including Greek, numbers, fractions: 120 Leroy—Equiv. 00 pen (0.120 in [3.05 mm] letter height).

6.3.4 SECTION AND REFERENCE LETTERS—All section and reference letters: 140 Leroy—Equiv. 0 pen (0.140 in [3.56 mm] letter height).

6.3.5 NUMBERS—Align column of numbers on decimal point.

6.3.6 ABBREVIATIONS—Do NOT use ", ', or ° (angular). Use in, ft, deg.

6.4 Tables—Tables shall be numbered consecutively throughout the report, and referred to in the text. Each shall be titled.

(a) Concise descriptions, measurement units, and letter symbols shall be included in column headings.

(b) Be concise in numerical ranges. Do not overlap ranges or leave gaps in ranges. An example of good practice is:

0.75 thru 1.25 mm
Over 1.25 thru 2.00 mm
Over 2.00 thru 3.25 mm

6.5 Decimal Dimensioning—The dimensions in all new and revised SAE technical reports shall be expressed in decimal units. Nominal sizes shall be expressed as decimals or fractions, as determined by their design basis or historic use. Where these considerations are not decisive, decimal nominal sizes will be used.

6.5.1 The number of significant digits used in a dimension should relate to the precision of the quantity stated. This is particularly important in decimalizing dimensions previously expressed as fractions. A dimension of $1\tfrac{3}{16}$ with an intended precision of about ±0.01 shall be decimalized as 1.19, not 1.1875.

---

[2] The practice of using date suffixes for SAE technical reports is initiated with the issue of J1159 AUG79. The previous practice of using suffix letters will be phased out as SAE technical reports are revised. (NOTE: Date suffixes will be phased in in the 1981 SAE Handbook.)

[3] See paragraph 6.9.

A discussion of the precision of a value, and the number of decimals proper to retain, is given in Rules for SAE Use of SI (Metric) Units, SAE J916.

6.5.2 ROUNDING OFF—When it is necessary to reduce the number of decimals by rounding off, the method shown in SAE J916 shall be used.

6.5.3 ZEROS—Where decimal values less than 1 appear, a zero shall be placed to the left of the decimal point.

6.6 Cross-Referencing—As necessary, other SAE reports or reports of other organizations may be referenced.

6.6.1 References shall be made to other standards or technical reports by name of the standards writing organization, number of the standard, and optimally the date of issue (e.g., SAE J804 FEB79; ASTM D1405 JUN78). If date is included, it is assumed that a specific report is being referenced even though it may be obsolete.

6.6.2 Where references are made, the referencing committee shall notify the committee responsible for maintaining the referenced technical report and request that the referencing committee be furnished with drafts of any proposed changes to the referenced report. Such drafts shall be sent to the chairman and SAE staff advisor of the referencing committee.

6.6.3 If the proposed changes are acceptable to the referencing committee, the chairman of the referencing committee should notify the committee responsible for the referenced technical report. The referencing technical report should then be revised to reflect the latest date of the referenced technical report.

6.6.4 If the changes are unacceptable to the referencing committee, the chairman of the referencing committee should notify the committee responsible for the referenced technical report that the proposed changes have affected applicability of the referenced report and a determination should be made between the two committees whether or not a new document number is required.

6.6.5 As an aid to committees and users of SAE Technical Reports, SAE staff shall maintain a listing of cross-referenced technical reports. The listing should give access to both the referenced and referencing technical reports.

6.6.6 If the SAE technical report corresponds to but is not identical to a technical report of another organization, it should be stated in the approval note. Example: This report conforms essentially to American National Standard B18.2.

6.6.7 Joint development of a technical report should be indicated in the approval note. Example: This is a joint report of SAE and ASTM . . .

6.7 Use of Basic Terms

6.7.1 SURFACE VEHICLE OR MACHINE—The term *surface vehicle* or *machine* is preferred to *automotive* for use in identifying technical reports which do not apply to the aerospace industry.

6.7.1.1 *Vehicle*—The term *vehicle* pertains to self-propelled devices for carrying passengers, goods, or equipment . . . a car, bus, truck, or boat.

6.7.1.2 *Machine*—The term *machine* pertains to self-propelled or mobile devices designed to alter or transmit energy and force for the performance of useful work . . . tractor, loader, grader, ditcher, combine.

6.7.2 MECHANICAL PROPERTIES—Mechanical properties are those properties of a material that pertain to its elastic and plastic behavior when force is applied: For example, yield strength, ultimate strength, elongation, hardness, etc.

6.7.3 PHYSICAL PROPERTIES—Physical properties are those properties other than mechanical properties that pertain to the physics of a material: For example, density, electrical conductivity, thermal expansion, etc., often improperly used to express mechanical properties.

6.7.4 USE OF *Shall* OR *Should*—The use of *should* or *shall* has no bearing on the voluntary nature of SAE technical reports. Inclusion of an SAE technical report in a document, standard, or contract by a company or agency is a voluntary act. When a technical report is so cited, the report becomes a requirement within the limitations set forth by the document, standard, or contract. The following shall apply to use of these words:

Shall—Shall is to be used wherever the criterion for conformance with the specific recommendation requires that there be no deviation. Its use shall not be avoided on the grounds that compliance with the report is considered voluntary.

Should—Should is to be used wherever noncompliance with the specific recommendation is permissible. *Should* shall not be substituted for *shall* on the grounds that compliance with the report is considered voluntary.

6.7.5 USE OF *Safe* AND *Safety*—The words *safe* and *safety* shall be used in SAE technical reports only when they are in whole or in part commonly used engineering terms, such as: fail-safe, factor of safety, safety glass. To preclude any misinterpretation of the words *safe* and *safety*, more definite descriptive words shall be used, such as:

"lock wiring" rather than "safety wiring" . . .

"lock nut" rather than "safety nut" . . .

"relief valve" rather than "safety valve" . . .

"the integrity of the painted surface" rather than "the safety of the painted surface" . . .

"to provide protection against shock" rather than "to provide safety against shock" . . .

"will provide for reliable transportation of the component" rather than "will provide for safe transportation of the component."

If circumstances arise which strongly indicate a need for the use of the words *safe* or *safety*, the Publications Advisory Committee shall be consulted.

6.8 Measurement Unit Symbols and Abbreviations—These may be found in Rules for SAE Use of SI (Metric) Units, SAE J916.

6.9 Indicating Revisions to SAE Standards, Recommended Practices, and Information Reports

6.9.1 INDICATIONS OF TECHNICAL REVISIONS—In drafts of revisions to existing "J" reports, the symbol ϕ shall be used to indicate technical changes.[4] A technical change requires a change in the date suffix, that is, JXXX JUN75 to JXXX JUN79. This indication of change shall be put on the draft revision at the earliest circulation.

6.9.1.1 The ϕ symbol is always placed in the left margin for single column copy and in the left- or right-hand margin, respectively, for double column copy. It is used to indicate change in text, tables, or figures. In the case of text, a separate symbol is used for each paragraph. In the case of tables or figures, the symbol is used once for each table or figure. In no case should it be located where it might be confused with the international symbol for *diameter*, which has a similar appearance.

6.9.1.2 Should the revision be so extensive that most of the report is changed, the symbol is put next to the title of the report, and the history note under the title shall read "completely revised (date)".

6.9.1.3 The ϕ symbol will be carried on the published document. If the document is again revised, the old symbols are deleted and new ones appropriate to the new technical revision are added.

6.9.1.4 The SAE Handbook and any separately published "J" reports will carry the following explanatory note: "The ϕ symbol is for the convenience of the user in locating the areas where technical revisions have been made to the previous issue of the report. If the symbol is next to the report title, it indicates a complete revision of the report."

6.9.2 EDITORIAL REVISIONS—Editorial revisions shall be indicated in both draft and published technical reports by the symbol *ed*. The history note under the report title will also call out editorial revisions by the following: "Editorial change (date)". The ϕ symbol indicating last technical revisions shall be retained.

7. Indexing Information—Following are instructions to assist in providing key words for use by SAE staff in preparing SAE Handbook index.

7.1 Selection—Terms chosen should be taken directly from or derived from the material being indexed on the basis of their relative significance and effectiveness in later retrieval of needed information. Terms should be used consistently. Terms should be as specific as the nature of the material and user's requirements allow.

7.2 Clarity—Subject terms should be nouns. Avoid terms that will not be recognized, coined terms, jargon, and slang. Generally the plural form should be used. Terms should be clarified if they have more than one accepted meaning, or must show distinction from other subject terms. This may be done by compound terms such as metal tubing, plastic tubing, or a parenthetical qualifying expression appended to distinguish the homographs as Tolerances (Mechanics) and Tolerances (Physiology).

7.3 Word Order—Terms consisting of two or more words should be listed in their natural word order, that is: Lighting equipment not Equipment, Lighting. When two or more candidate terms are true synonyms, one should be selected as the preferred index term, the other(s) entered as a see reference, that is, air cleaners use air filters, cloth use fabrics.

7.4 Punctuation—Punctuation marks should be used in index terms only when essential.

7.5 Abbreviation—Abbreviated word forms, including acronyms, should be used only when meanings are well established or when significant convenience results.

7.6 Inversions—For committee consideration, inverted entries, that is Engines, Aircraft, or Engines, Passenger Car, etc., make it possible to find most members of a related class together for ease of retrieval. The terms to be inverted must be carefully selected and care taken that another useful grouping is not scattered by inversion.

8. References—Available from the Society of Automotive Engineers, Inc., 400 Commonwealth Drive, Warrendale, Pennsylvania 15096:

Technical Committee Guideposts, SAE J1271 AUG79.

SAE Handbook.

Rules for SAE Use of SI (Metric) Units, SAE J916.

Standard—Automotive Carburetor Flanges, SAE J623.

Recommended Practice—Surface Texture Control, SAE J449.

Information Report—Mechanical Properties of Heat Treated Steels, SAE J413.

---

[4] Symbol made on a typewriter by a combination of a zero and a slash.

SAE Motor Vehicle, Safety and Environmental Terminology, HS 215.
Guidelines for Developing and Revising SAE Nomenclature and Definitions, SAE J1115.
SAE Technical Board Rules and Regulations.
SAE J804.
Available from the American Society for Testing and Materials, 1916 Race Street, Philadelphia, Pennsylvania 19103:
ASTM D 1405—Estimation of Net Heat of Combustion of Aviation Fuels.
Available from the American National Standards Institute, 1430 Broadway, New York, New York 10018:
American National Standard B18.2.

RULES FOR SAE USE OF SI (METRIC) UNITS—
SAE J916 JUN80

SAE Recommended Practice

Report of the Publication Policy Committee, approved June 1965, last revised by Metric Advisory Committee June 1980.

1. Introduction—In the spring of 1969 the SAE Board of Directors issued a statement that "SAE will include SI[1] units in SAE Standards and other technical reports." Much investigation has attended the determination of units of measure for use, since measurement practice all over the world is to some degree in a state of transition. Engineering use of measurement units in nearly every metric country of the world, and in all of those nations adopting metric units, is confronted with the struggle between the noncoherent technical metric units, such as kilogram-force and calorie, and the SI units, such as newton and joule.

This document establishes the rules for the use of SI units in SAE reports, including specifications and standards. It must be remembered that a technical committee may produce its reports in any units it feels are proper for the users—U.S. inch-pound, SI, or other metric. However, if the units used do not conform to the Units Approved for SAE Use (see paragraph 2), they must be followed by SI units in parentheses.

Throughout this document, SI is intended to include recognized SI units as established by the international General Conference on Weights and Measures,[2] (CGPM) and a limited number of other units that are not formal SI units.

These other units are all included in the American National Standard Z210.1, "Standard for Metric Practice" in "The Metric System of Measurement" issued by the Secretary of Commerce in the 10-26-77 Federal Register, and in ISO 1000, the worldwide document for use by all ISO[3] committees.

By careful contact with other countries, the General Conference, and ISO, this document will be updated as often as necessary to keep the use of SI units in SAE reports as nearly as possible in harmony with the units that will be adopted for United States and world use.

2. Units Approved for SAE Use—All SAE documents produced under the Board of Directors' directive to "include SI units" must utilize as applicable:

2.1 Base Units of SI

Quantity	Unit (symbol)
length	—meter[4] (m)
mass	—kilogram (kg)
time	—second (s)
electric current	—ampere (A)
thermodynamic temperature	—kelvin (K)
amount of substance	—mole (mol)
luminous intensity	—candela (cd)

2.2 Supplementary Units of SI

Quantity	Unit (symbol)
plane angle	—radian (rad)
spherical angle	—steradian (sr)

2.3 Recognized Derived Units of SI with Special Names

Quantity	Unit (symbol)	Formula
absorbed dose	—gray (Gy)	J/kg
activity (of a radionuclide)	—becquerel (Bq)	$1/s$, s^{-1}
Celsius temperature	—degree Celsius (°C)	[5]
dose equivalent	—sievert[6] (Sv)	J/kg
electric capacitance	—farad (F)	C/V
electric conductance	—siemens (S)	A/V
electric inductance	—henry (H)	Wb/A
electric potential diff.	—volt (V)	W/A
electric resistance	—ohm (Ω)	V/A
energy, work	—joule (J)	N·m
force	—newton (N)	kg·m/s²
frequency	—hertz (Hz)	$1/s$, s^{-1}
illuminance	—lux (lx)	lm/m²
luminous flux	—lumen (lm)	cd·sr
magnetic flux	—weber (Wb)	V·s
magnetic flux density	—tesla (T)	Wb/m²
power	—watt (W)	J/s
pressure or stress	—pascal (Pa)	N/m²
quantity of electricity	—coulomb (C)	A·s

See Z210.1 paragraph 2, for more complete description.

2.4 Other Units that May be Used with SI

Quantity	Unit (symbol)
plane angle	—degree (°) (decimal divisions preferred)
time	—minute (min), hour (h), day (d), week, and year
mass	—metric ton (t)
area	—hectare (ha)
sound pressure level	—decibel (dB)
volume	—liter[4] (L)[7]
navigation velocity	—knot (kn)[8]
distance	—nautical mile (nmi)[8]

When these units are used, they need not be followed by SI units unless it suits the purpose of the document.

The liter which the General Conference established as a special name for the cubic decimeter, is approved for SAE use, normally for fluid measurement only, and the only prefixed use allowed is mL.

In the case of time, committees are urged to use the second and its multiples, but the units given above are permitted.

The unit metric ton (exactly 1 Mg) is in wide use but should be limited to commercial description of vehicle mass, or freight mass, and no prefix is permitted.

---

[1] SI—The International System of Units (Système International) officially abbreviated "SI" in all languages—the modern metric system.
[2] CGPM Resolutions and Recommendations are published in NBS Special Publication 330, "The International System of Units (SI)."
[3] The International Organization for Standardization.
[4] "re" spelling is also used.
[5] In 1976 the CIPM decided that the degree Celsius is a special name for the kelvin, to be used to express Celsius temperature. For formula see paragraph 8.
[6] Approved by CGPM in 1979.
[7] In 1979 the CGPM approved the symbol "L" for liter and it is recommended for North American use. The alternative symbol "l" will also be used during a transition period.
[8] Abbreviation, not a symbol.

The unit hectare (exactly 1 hm²) is restricted to land and water area measurement.

2.5 Other derived units that are formed from those units and derived units indicated above are also acceptable. For example, the SI unit designation for electric field strength is V/m; however, it is also expressed in terms of base units as $kg \cdot m/(s^3 \cdot A)$ or $kg \cdot m \cdot s^{-3} \cdot A^{-1}$. Likewise, torque and bending moment $(N \cdot m)$ may also be expressed as $kg \cdot m^2/s^2$ or $kg \cdot m^2 \cdot s^{-2}$.

3. Units Not Approved for Use as SI—Gravimetric force units, such as kilogram-force, or kilogram-force per square millimeter, which have been common in some countries, must not be used in SAE reports. Similarly, calorie, bar, angstrom, and dyne are not SI units and are not to be used. However, as stated in Section 1, this restriction does not preclude use of these units where a committee considers them to be the proper units for the users of the report, and provided they are followed with approved SI units in parentheses.

4. Multiplying Prefixes—Table 1 lists the prefixes to be used with SI units, observing the rules given in Section 5.

TABLE 1—SI UNIT PREFIXES

Multiples and Submultiples	Prefixes	Symbols	Pronunciations
10^{18}	exa	E	ex'a
10^{15}	peta	P	pet'a
10^{12}	tera	T	ter'a
10^{9}	giga	G	ji'ga
10^{6}	mega	M	meg'a
10^{3}	kilo	k	kil'o
10^{2}	hecto	h	hek'to
10	deka	da	dek'a
10^{-1}	deci	d	des'i
10^{-2}	centi	c	sen'ti
10^{-3}	milli	m	mil'i
10^{-6}	micro	μ	mi'kra
10^{-9}	nano	n	nan'o
10^{-12}	pico	p	pe'co
10^{-15}	femto	f	fem'to
10^{-18}	atto	a	at'to

5. Rules for Use of Units

5.1 Requirements of this document establish the use of SI units in one of the following manners:

5.1.1 As regular units followed by other units in parentheses.

5.1.2 In parentheses following other units.

5.1.3 As regular units where presently usable by the user, in which case no units need be added in parentheses.

5.1.4 Under special circumstances it is permissible to deviate from these rules. See Appendix B.

5.2 SI units must be those shown in Appendix A or their decimal multiples, except as covered in paragraph 6.2. In case of need for other units the Metric Advisory Committee of the SAE Technical Board should be consulted. If units for quantities not included in Appendix A and not clearly covered by paragraph 6.2 are required, the above committee should be contacted for guidance.

An apparent anomaly exists in the use of the joule for work $(J = N \cdot m)$ and the use of $N \cdot m$ for torque or bending moment. These are, however, entirely different units. In the former, the unit of work results from unit force moving through unit distance. In the latter, there is no implication of movement, and unit force acts at right angles to the lever arm of unit length. This would be readily seen if vectors were incorporated in the unit symbols. For these reasons, it is important to express work or energy in joules and moment of force or torque in newton meters, not joules.

5.3 Use of Prefixes

5.3.1 Use of prefixes representing 10 raised to a power which is a multiple of 3 is recommended. In the case of prefixed units which carry exponents, such as units of area and volume, this may not be practical, however, and any listed prefix may be used.

5.3.2 Compound prefixes, such as milli-micro, are never used.

5.3.3 In general, prefixes in the denominator of a compound unit should be avoided except for established usage. (Since the kilogram is a base unit of SI, use of kg in the denominator is not contrary to this guidance.)

5.3.4 When expressing a quantity by a numerical value and a unit, prefixes should preferably be chosen so that the numerical value lies between 0.1 and 1000. This is, of course, not true where certain multiples and units have been agreed to for particular use, such as kPa for pressure, or where tabular use requires the same unit in a series, even though this means exceeding the preferred range of 0.1–1000.

5.3.5 The prefix becomes a part of the symbol or name with no separation (meganewton, MN).

5.3.6 Errors in *calculations* can be minimized if all quantities are expressed in SI units, and prefixes are replaced by powers of 10.

5.3.7 With SI units of higher order, such as m^2 or m^3, the prefix is also raised to the same order; for example, $1\ mm^3$ is $(10^{-3}\ m)^3$ or $10^{-9}\ m^3$.

5.4 Symbols and Abbreviations

5.4.1 DISTINCTION—The distinction between unit symbols and unit abbreviations is not always recognized, particularly with certain U. S. inch-pound units of measurement. The symbols for some U. S. units are also abbreviations (ft, in, yd). In many cases the unit symbol and the abbreviation are not the same (such as unit symbol ft^3/min and abbreviation cfm; unit symbol A and abbreviation amp; unit symbol in^3 and abbreviation cu in). A positive distinction can be made between unit symbols and unit abbreviations. The SI unit symbol designation is the same in all languages. Abbreviations are conventional representations of words or names in a particular language; they may be different in different languages.

5.4.2 UNIT SYMBOL COMPOSITION—Unit symbols are letters or groups of letters predominantly from the English alphabet representing the units in which physical quantities are measured (m for meter, $W \cdot h$ for watthour). Non-English alphabet unit symbols are (Ω) for ohm, (°) for the plane angle degree or used with the Celsius (°C) temperature scale, and (μ) for the prefix micro. All unit symbols are printed in Roman (upright) type.

5.4.3 UNIT SYMBOL STYLE[9]—Unit symbols are, in general, shown as lower case letters. If, however, the symbol is derived from a proper name, it or the

[9] Handling of Unit Names—Names of units are never capitalized except at the beginning of sentences or in titles. (Modifiers used in unit names are capitalized if proper names; for example, degree Fahrenheit.) Compound unit names are formed with a space for product and the word "per" for quotient. Prefixes become part of the word: ampere (A), milliampere (mA), ampere second $(A \cdot s)$, meter per second (m/s).

TABLE 2—ABBREVIATIONS AND SYMBOLS FOR UNITS OTHER THAN SI

Unit Name	Symbol	Abbreviation	Unit Name	Symbol	Abbreviation
brake horsepower		bhp	inch pound—force	$in \cdot lbf$	
Brinell hardness number		Bhn	kilocycle	kc	
British thermal unit	Btu		kilogram—force	kgf	
calorie	cal		mile	mi	
candlepower		cp	mile per hour	mi/h	mph
cubic foot per minute	ft^3/min	cfm	minute (angle)	'	min
cubic foot per second	ft^3/s	cfs	ounce	oz	
cycle per minute	c/min	cpm	ounce—force	ozf	
cycle per second	c/s	cps	part per gallon		ppg
cycle	c		pint	pt	
degree Fahrenheit	°F		pound	lb	
degree Rankine	°R		poundal	pdl	
dram	dr		pound—force	lbf	
foot	ft		pound—force per		
footcandle	fc		square inch	lbf/in^2	psi
foot per minute	ft/min		pound—force per		
foot per second	ft/s		square inch absolute		psia
foot pound—force	$ft \cdot lbf$		pound—force per		
friction horsepower		fhp	square inch gage		psig
gallon	gal		quart	qt	
gallon per minute	gal/min	gpm	revolution per minute	r/min	rpm
gallon per second	gal/s	gps	revolution per second	r/s	rps
horsepower	hp		Saybolt universal second		SUS
inch	in				
inch of mercury	in Hg		second (angle)	"	sec
inch of water	in H_2O		yard	yd	

first letter (where more than one) is an upper case letter (Hz, Wb, Pa). An exception to the above permits the upper case (L) to represent the unit liter because of the confusion that can occur between the lower case unit symbol (l) and the number one (1).

The letter style must be followed for SI unit symbols and prefixes even in applications where all other lettering is upper case (such as technical drawings). The only exception allowed is for computer and machine displays with limited character sets. For symbols for use in systems with limited character sets, refer to ANSI X3.50 or ISO 2955. The symbols for limited character sets must never be used when the available character set permits the use of the proper symbols as given herein.

ϕ 5.4.4 QUANTITY SYMBOLS—Unit symbols must not be confused with quantity symbols. Quantity symbols are single letters representing the magnitude of physical quantities (I for electric current, e for charge of an electron) and are established in upper or lower case that must always be maintained (f—frequency, F—force, m—mass, M—moment of force).

Quantity symbols are single letters of the English or Greek alphabet, and are printed in italic (slanting) type.

ϕ 5.4.5 ABBREVIATIONS—Abbreviations are shortened forms of words or phrases formed in various ways that have been accepted and established (ANSI Y1.1). They are generally letters from the word being abbreviated, except where the abbreviation is taken from another language (no for number, lb for pound). Abbreviations are never to be used when a mathematical operation sign is involved, unless the abbreviation is also the symbol.

ϕ 5.4.6 SYMBOLIZED COMPOUND (DERIVED) UNITS[9]—Compound (derived) units constitute a mathematical expression. Where compound units include the solidus (/), it must not be repeated in the same expression. In complicated cases, negative powers or parentheses should be used. For example, write: m/s^2 or m·s^{-2} but not m/s/s; or write kg·m/(s^3·A) or kg·m·s^{-3}·A^{-1} but not kg·m/s^3/A.

5.4.7 PLURAL—The form of symbols and abbreviations is the same for singular or plural (1 in, 10 in, 1 s, 27 s).

5.4.8 Periods are not used after symbols or abbreviations. The same abbreviation is used for related noun, verb, adverb, etc. (inclusion, include, inclusive are all abbreviated incl). When these rules would cause confusion, spell out the word. Words of four letters or less are not abbreviated.

5.4.9 When writing a quantity, a space should be left between the numerical value and a unit symbol—for example, write 35 mm, not 35mm. An exception occurs when the symbols for degree of plane angle or degree Celsius are used, in which case the space is omitted (25°C). *ed.*

5.5 Miscellaneous

5.5.1 With nominal sizes that are not measurements but are names for items, no conversion should be made: for example, 1/4-20 UNC thread, 1 in pipe, 2 x 4 lumber.

ϕ 5.5.2 The decimal marker used by SAE is the dot on the line (.) for quantities in either U. S. customary or SI units.

To facilitate the reading of numbers having five or more digits, the digits should be placed in groups of three separated by a space instead of a comma, counting both to the left and to the right of the decimal point. In the case of four digits, the spacing is optional. This style also avoids confusion caused by the use elsewhere of the comma to express the decimal marker.

For example, use:

1 532 or 1532 instead of 1,532
132 541 816 instead of 132,541,816
983 769.788 16 instead of 983,769.78816

5.5.3 Surface roughness expressed in microinches should be converted to micrometers (μm).

5.5.4 Linear dimensions on engineering drawings will customarily be given in millimeters regardless of length.

6. General

6.1 The principal departure of SI from the gravimetric form of metric engineering units is the separate and distinct units for mass and force. The kilogram is restricted to the unit of mass. The newton is the unit of force and should be used in place of the kilogram-force. The newton instead of the kilogram-force should be used in combination units which include force; for example, pressure or stress (N/m^2 = Pa), energy (N·m = J), and power (N·m/s = W).

Considerable confusion exists in the use of the term weight to mean either force or mass.

In scientific use, the term *weight* of a body usually means a force related to gravity,[10] which varies in time and space. Weight, if used to mean force also varies. Observed values differ by over 0.5% at various points on the earth's surface. *ed.*

In commercial and everyday use, the term weight is nearly always synonymous with mass. Thus, in speaking of a person's weight, the quantity referred to is mass.

Because of this dual use, it is wise to avoid the term weight, except under circumstances in which its meaning is completely clear. When the term is used, it is important to know whether mass or force is intended, and to use SI units properly as clarified in the first paragraph of this section, using kilograms for mass and newtons for force.

6.2 Many units for rates are not shown in Appendix A, but should be derived from approved units. For example, the proper unit for mass per unit time is kg/s.

6.3 Expressions that can be stated as a ratio of the same unit, such as 0.006 inch per inch, should be changed to a designation of a ratio such as 0.006:1. Where an expression might be shown in two different units one of which is a multiple of the other, reduce the expression to a common unit and show it as a ratio. Example: 1.50 in per ft = 0.125 ft per ft. Express as a ratio 0.125:1.

6.4 It has been internationally recommended that pressure units themselves should not be modified to indicate whether the pressure is *absolute* (that is, above zero) or *gage* (that is, above atmospheric pressure). If, therefore, the context leaves any doubt as to which is meant, the word *pressure* must be qualified appropriately.

For example:

". . . at a gage pressure of 200 kPa" or *ed.*
". . . at an absolute pressure of 95 kPa" or
". . . reached an absolute pressure of 95 kPa",
etc.

7. Conversion Techniques

—Conversion of quantities between systems of units involves careful determination of the number of significant digits to be retained. To convert "1 quart of oil" to "0.9463529 liter of oil" is, of course, nonsense because the intended accuracy of the value does not warrant expressing the conversion in this fashion.

This section provides information to be used as a guide in the conversion of quantities specified in SAE Standards. In certain circumstances, reasons may exist for using other guidance. For example, in the case of interchangeable dimensions on engineering drawings, a more specific approach is outlined in SAE J390, Dual Dimensioning, although the methods given here will usually produce the same results.

All conversions, to be logically established, *must* depend upon an intended precision of the original quantity—either implied by a specific tolerance, or by the nature of the quantity. The first step in conversion is to establish this precision.

7.1 Precision of a Value—It is absolutely necessary to determine the intended precision of a value before converting.

The intended precision of a value *should* relate to the number of significant digits shown. The implied precision is plus or minus one-half unit of the last significant digit in which the value is stated. This is true because it may be assumed to have been rounded from a greater number of digits, and one-half unit of the last significant digit retained is the limit of error resulting from rounding. For example, the number 2.14 may have been rounded from any number between 2.135 and 2.145. Whether rounded or not, a quantity should always be expressed with this implication of precision in mind. For instance, 2.14 in implies a precision of ±0.005 in, since the last significant digit is in units of 0.01 in.

Two problems interfere with this, however:

(a) Quantities *may* be expressed in digits which are not intended to be significant. The dimension 1.1875 in may be a very precise one in which the digit in the fourth place is significant, or it may in some cases be an exact decimalization of a rough dimension 1 3/16 in, in which case the dimension is given with too many decimal places relative to its intended precision.

(b) Quantities may be expressed omitting significant zeros. The dimension 2 in may mean "about 2 in", or it may, in fact, mean a very precise expression which should be written 2.0000 in. In the latter case, while the added zeros are not significant in establishing the value, they are very significant in expressing the proper intended precision.

Therefore, it is necessary to determine an approximate implied precision before converting. This can usually be done by using knowledge of the circumstances or information on the accuracy of measuring equipment.

If accuracy of measurement is known, this will provide a convenient lower limit to the precision of the dimension, and in some cases may be the only basis for establishing it. The implied precision should never be smaller than the accuracy of measurement.

A tolerance on a dimension will give a good indication of the intended precision, although the precision will, of course, be much smaller than the tolerance. A dimension of 1.635 ± 0.003 in obviously is intended to be quite precise, and the precision implied by the number of significant digits is correct (±0.0005 in, total 0.001 in). A dimension of 4.625 ± 0.125 in is obviously a different matter. The use of thousandths of an inch to express a tolerance of

---

[10] The force which if applied to the body would give it acceleration equal to the local acceleration of free fall.

0.25 in is probably the result of decimalization of fractions, and the expression is probably better written 4.62 ± 0.12, with an implied precision of ±0.005 (total implied precision 0.01 in). The circumstances, however, should be examined and judgment applied.

A rule of thumb often helpful for determining implied precision of a toleranced value is to assume it is one-tenth of the tolerance. Since the implied precision of the converted value should be no greater than that of the original, the total tolerance should be divided by 10, converted, and the proper significant digits retained in both the converted value and converted tolerance such that total implied precision is not reduced—that is, such that the last significant digit retained is in units no larger than one-tenth the converted total tolerance.

EXAMPLE: 200 ± 15 psi. Tolerance is 30 psi, divided by 10 is 3 psi, converted is about 20.7 kPa. The value (200 psi) converted is 1 378.9514 ± 103.421 355 kPa which should be rounded to units of 10 kPa, since 10 kPa is the largest unit smaller than one-tenth the converted tolerance. The conversion should be 1380 ± 100 kPa.

EXAMPLE: 25 ± 0.1 oz of alcohol. Tolerance is 0.2 oz, one-tenth of tolerance is 0.02 oz, converted is about 0.6 cm^3. The converted value (739.34 ± 2.957 cm^3) should be rounded to units of 0.1 cm^3 and becomes 739.3 ± 3.0 cm^3.

7.2 Conversion Procedure—In the sections that follow, the "total implied precision" discussed in paragraph 7.1 is referred to as "TIP".

7.2.1 First determine TIP.

7.2.2 Convert the dimension, TIP, and the tolerance if any, by the accurate conversion factor given in this document or ANSI Z210.1.

7.2.3 Choose the smallest number of decimals to retain, such that the last digit retained is in units equal to or smaller than the converted TIP.

7.2.4 Round off to this number of decimals by the following rules:

7.2.4.1 Where the digit next beyond the last digit to be retained is less than 5, the last digit retained should not be changed. Example: 4.46325 if rounded to three places would be 4.463.

7.2.4.2 Where the digits beyond the last digit to be retained amount to more than 5 followed by zeros, the last digit retained should be increased by one. Example: 8.37652 if rounded to three places would be 8.377.

7.2.4.3 Where the digit next beyond the last digit to be retained is exactly 5, the last digit retained, if even, is unchanged; but if odd, the last digit is increased by one. Example: 4.36500 becomes 4.36 when rounded to two places. 4.35500 also becomes 4.36 when rounded.

7.2.5 EXAMPLES

7.2.5.1 Test pressure 200 ± 15 psi
TIP not evident in this case
Total tolerance 30 psi, divided by 10 is 3 psi converted equals 20.68 kPa, for TIP use 10 kPa
Units to use, 10 kPa
200 ± 15 psi equals 1 378.9514 ± 103.421 355 kPa, round to 1380 ± 100 kPa

7.2.5.2 A stirring rod 6 in long
Estimate of TIP. Assume intended precision ±1/16 in, TIP = 1/8 in
Converted TIP 1/8 × 25.4 = 3.17 mm
Units to use, 1 mm
6 in equals 152.4 mm, round to 152 mm

7.2.5.3 50 000 psi tensile strength
Estimate of TIP 400 psi from nature of use and precision of measuring equipment
Converted TIP 2.8 MPa
Units to use, 1 MPa
50 000 psi equals 344.737 85 MPa, round to 345 MPa

7.2.5.4 5.163 in length
Estimate of TIP 0.001 in (significant digits judged correct)
Converted TIP 0.0254 mm
Units to use, 0.01 mm
5.163 in equals 131.1402 mm, round to 131.14 mm

7.2.5.5 12.125 in length
Estimate of TIP 0.06 in from nature of use
Converted TIP 1.524 mm
Units to use 1 mm
12.125 in equals 307.975 mm, round to 308 mm

7.2.6 In dealing with toleranced quantities or quantities that establish limits, the rounding may be required in one direction only. When *maximum* or *minimum* are specified and judgment shows that these terms are mandatory, a maximum quantity must be rounded downward and a minimum rounded upward. The following illustrations show rounding of a dimension to two decimal places under different circumstances.

Dimension converted to 131.7625 mm
Round to two decimal places

(a) Normal dimension, untoleranced
Round to 131.76 mm (closest to original)

(b) Dimension stated as *minimum*
Round to 131.77 mm (rounded *up*)

(c) Dimension stated as *maximum*
Round to 131.76 mm (rounded *down*)

Similarly, a toleranced quantity may be rounded as in item (a). However, if critical it may be first converted to limits and each limit rounded in the appropriate fashion depending on the nature of the individual limit. For absolute maintenance of the original limits, the upper limit should be rounded down and the lower limit rounded up. (This is method B described in ANSI Z210.1.)

8. Temperature Conversion—The SI unit for temperature is the kelvin. SAE will use kelvins principally for thermodynamics, but the Celsius[11] temperature scale will also be commonly used.

The Celsius scale is related to the kelvin scale as follows:
One degree Celsius equals one kelvin exactly. Celsius temperature ($t_{°C}$) is related to kelvin temperature (T_K) as follows:

$$T_K = 273.15 + t_{°C}$$

The Celsius scale is related to the Fahrenheit scale as follows:
One degree Celsius equals 9/5 of a degree Fahrenheit, exactly. Celsius temperature ($t_{°C}$) is related to Fahrenheit temperature ($t_{°F}$) as follows:

$$t_{°C} = 5/9(t_{°F} - 32)$$

General guidance for converting tolerances from degrees Fahrenheit to kelvins or degrees Celsius is given below:

Conversion of Temperature Tolerance Requirements	
Tolerance, °F	Tolerance, K or °C
±1	±0.5
±2	±1
±5	±3
±10	±5.5
±15	±8.5
±20	±11
±25	±14

Normally, temperatures expressed in a whole number of degrees Fahrenheit should be converted to the nearest 0.5 kelvin (or degree Celsius). As with other quantities, the number of significant digits to retain will depend upon implied accuracy of the original dimension, for example:

100 ± 5°F: implied accuracy estimated to be 2°F.
37.7777 ± 2.7777°C rounds to 38 ± 3°C.
1000 ± 50°F: implied accuracy estimated to be 20°F.
537.7777 ± 27.7777°C rounds to 540 ± 30°C.

9. Bibliography

American Society for Testing and Materials, 1916 Race St., Philadelphia, PA 19103.

American National Standards Institute (ANSI), 1430 Broadway, New York, NY 10018.

Standard for Metric Practice (ANSI Z210.1 and ASTM E380). American National Standards Institute.

ANSI X3.50, Representations for U. S. Customary, SI, and Other Units to be Used in Systems with Limited Character Sets.

International Organization for Standardization,[12] Geneva, Switzerland:

ISO 1000, SI Units and Recommendations for the Use of their Multiples and of Certain Other Units.

ISO 2955, Information Processing—Representations of SI and Other Units for Use in Systems with Limited Character Sets.

Superintendent of Documents, U. S. Government Printing Office, Washington, DC 20402.

National Bureau of Standards, NBS Special Publication 330, The International System of Units (SI).

National Bureau of Standards, Washington, DC 20234.

Federal Register Notice of 10-26-77, NBS Letter Circular LC1078 The Metric System of Measurement as issued by the Secretary of Commerce.

---

[11] The term "Celsius" officially replaced "Centigrade" to eliminate confusion with French metric decimalized angular measurement (a "grad" or "grade" is 1% of a right angle, and a "centigrad" or "centigrade" is 1% of a "grad").

[12] Available in U. S. from American National Standards Institute.

APPENDIX A
Application of SI Units
(including conversion factors)

The following table illustrates recommended SI use for applications in the industries served by SAE. The particular recommendations are not mandatory, but should be followed in all SAE effort unless other use conforming to SAE J916 is strongly preferred.

1. Arrangement—The unit applications are arranged in alphabetical order of quantities, by principal nouns. Thus to find SI use for Surface Tension look under Tension, Surface, and for Specific Energy look under Energy, Specific.

2. Rates and Other Derived Quantities—It is of course not practical to list all possible applications but others such as rates can be readily derived. For example, if guidance is desired for Heat Energy per Unit Volume, looking up Energy and Volume will show the recommendation kJ/m^3 (or other prefix, depending on factors discussed in paragraph 5.3).

3. Conversion Factors—Conversion factors are shown from Old Units to Metric Units to seven significant digits, unless the precision with which the factor is known does not warrant seven digits.

Exact conversion factors are indicated by *.

For conversion from Metric Units to Old Units, divide rather than multiply ϕ by the factor. For example, to convert 16.3 lb/yd^3 to kg/m^3 multiply by 0.593 276 3. The answer is 9.670 403 6 kg/m^3 which should be rounded properly according to the precision of the 16.3 lb/yd^3, probably to 9.7 kg/m^3. To convert 9.7 kg/m^3 to lb/yd^3 divide by 0.593 276 3. The answer is 16.349 886 lb/yd^3 which also should be rounded, probably to 16.3 lb/yd^3.

φ TABLE A.1

Quantity	Typical Application	From Old Units	To Metric Units	Multiply by
Acceleration, angular	General	rad/s²	rad/s²	1*
Acceleration, linear	Vehicle General (includes acceleration of gravity)†	(mile/h)/s ft/s²	(km/h)/s m/s²	1.609 344* 0.304 8*
Angle, plane	Rotational calculations	r (revolution) rad	r (revolution) rad	1* 1*
	Geometric and general	° (deg) ′ (min) ″ (sec)	° ° (decimalized) ° (decimalized)	1* 1/60* 1/3600*
Angle, solid	Illumination calculations	sr	sr	1*
Area	Cargo platforms, frontal areas, fabrics, roof and floor areas, general	in² ft²	m² m²	0.000 645 16* 0.092 903 04*
	Pipe, conduit	in² ft²	cm² m²	6.451 6* 0.092 903 04*
	Small areas, orifices	in²	mm²	645.16*
	Brake & clutch contact area, glass, radiators, agricultural	in²	cm²	6.451 6*
	Land and water areas (Small)	ft²	m²	0.092 903 04*
	(Large)	acre	ha	0.404 687 3⁽ᶜ⁾
	(Very Large)	mile²	km²	2.589 998⁽ᶜ⁾
Area per time	Field operations (agricultural)	acre/h	ha/h	0.404 687 3⁽ᶜ⁾
	Auger sweeps, silo unloader	ft²/s	m²/s	0.092 903 04*
Bending moment	(See Moment of force)			
Capacitance, electric	Capacitors	μF	μF	1*
Capacity, electric	Battery rating	A·h	A·h	1*
Capacity, heat	General	Btu/°F⁽ᵃ⁾	kJ/K⁽ᵇ⁾	1.899 101
Capacity, heat, specific	General	Btu/(lb·°F)⁽ᵃ⁾	kJ/(kg·K)⁽ᵇ⁾	4.186 8*
Capacity, volume	(See volume)			
Coefficient of heat transfer	General	Btu/(h·ft²·°F)⁽ᵃ⁾	W/(m²·K)⁽ᵇ⁾	5.678 263
Coefficient of linear expansion	Shrink fit, general	°F⁻¹, (1/°F)	K⁻¹, (1/K)⁽ᵇ⁾	1.8*
Conductance, electric	General	mho	S	1*
Conductance, thermal	(See Coefficient of heat transfer)			
Conductivity, electric	Material property	mho/ft	S/m	3.280 840
Conductivity, thermal	General	Btu·ft/(h·ft²·°F)⁽ᵃ⁾	W/(m·K)⁽ᵇ⁾	1.730 735
Consumption, fuel	(See Efficiency, fuel)			
Consumption, oil	Vehicle performance testing	qt/1000 miles	L/1000 km	0.588 036 4
Consumption, specific, fuel	(See Efficiency, fuel)			
Consumption, specific, oil	Engine testing	lb/(hp·h) lb/(hp·h)	g/(kW·h) g/MJ	608.277 4 168.965 9
Current, electric	General	A	A	1*
Damping coefficient		lbf·s/ft	N·s/m	14.593 90
Density, current	General	A/in² A/ft²	kA/m² A/m²	1.550 003 10.763 91

For footnotes see end of Table.
† Standard acceleration of gravity 9.806 650 m/s² exactly.

◊ TABLE A.1 (continued)

Quantity	Typical Application	From Old Units	To Metric Units	Multiply by
Density, magnetic flux	General	kilogauss	T	0.1*
Density, (mass)	Solid	lb/yd^3 lb/in^3 lb/ft^3 ton (short)/yd^3 ton (long)/yd^3	kg/m^3 kg/m^3 kg/m^3 kg/m^3 kg/m^3	0.593 276 3 27 679.90 16.018 46 1 186.553 1 328.939
	Liquid	lb/gal	kg/L	0.119 826 4
	Gas	lb/ft^3	kg/m^3	16.018 46
Density of heat flow rate	Irradiance, general	Btu/(h·ft^2)[a]	W/m^2	3.154 591
Diffusivity, thermal	Heat transfer	ft^2/h	m^2/h	0.092 903 04*
Drag	(See Force)			
Economy, fuel	(See Efficiency, fuel)			

Efficiency, fuel	Highway vehicles
	  economy
	  consumption
	  specific fuel consumption
		mile/gal
		—
		lb/(hp·h)
			km/L
			L/100 km
			g/MJ
				0.425 143 7
				**
				168.965 9

	Off-highway equipment
	  economy
	  consumption
	  specific fuel consumption
	  specific fuel consumption
		hp·h/gal
		gal/h
		lb/(hp·h)
		lb/(hp·h)
			kW·h/L
			L/h
			g/(kW·h)
			g/MJ
				0.196 993 1
				3.785 412
				608.277 4
				168.965 9

	Aircraft gas turbine engines
	  Thrust specific fuel consumption (Turbo-jet)
	  Shaft specific fuel consumption (Turbo-shaft)
		lb/(lbf·h)
		lb/(hp·h)
			kg/(kN·h)
			kg/(kW·h)
				101.971 6
				0.608 277 4

Energy, work, enthalpy, quantity of heat	Impact strength	ft·lbf	J	1.355 818
	Heat[a]	Btu kcal	kJ kJ	1.055 056 4.186 8*
	Electrical	kW·h kW·h	kW·h MJ	1* 3.6*
	Mechanical, hydraulic, general	ft·lbf ft·pdl hp·h	J J MJ	1.355 818 0.042 140 11 2.684 520
Energy per area	Solar radiation	Btu/ft$^{2[a]}$	MJ/m^2	0.011 356 53
Energy, specific	General[a]	cal/g[d] Btu/lb	J/g kJ/kg	4.186 8* 2.326*
Enthalpy	(See Energy)			
Entropy	(See Capacity, heat)			
Entropy, specific	(See Capacity, heat specific)			
Floor loading	(See Mass per area)			
Flow, heat, rate	(See Power)			
Flow, mass, rate	General	lb/min lb/s	kg/min kg/s	0.453 592 4 0.453 592 4
	Dust flow	g/min	g/min	1*
Flow, volume	Air, gas, general	ft^3/s ft^3/s	m^3/s m^3/min	0.028 316 85 1.699 011
	Liquid flow, pump capacity	gal/s gal/s gal/min	L/s m^3/s L/min	3.785 412 0.003 785 412 3.785 412
	Seal and packing leakage, sprayer flow	oz/s oz/min	mL/s mL/min	29.573 53 29.573 53

For footnotes see end of Table.
**Convenient conversion: 235.215 ÷ (mile/gal) = L/100 km

φ TABLE A.1 (continued)

Quantity	Typical Application	From Old Units	To Metric Units	Multiply by
Flux, luminous	Light bulbs	lm	lm	1*
Flux, magnetic	Coil rating	maxwell	Wb	0.000 000 01*
Force, thrust, drag	Pedal, spring, belt, hand lever, general	lbf ozf	N N	4.448 222 0.278 013 9
	Drawbar, breakout, rim pull, winch line pull, general[e]	lbf lbf	N kN	4.448 222 0.004 448 222
	General	pdl kgf dyne	N N N	0.138 255 0 9.806 650 0.000 01*
Force per length	Beam loading (See also Spring rate)	lbf/ft	N/m	14.593 90
Frequency	System, sound and electrical	Mc/s kc/s Hz, c/s	MHz kHz Hz	1* 1* 1*
	Mechanical events, rotational (See Velocity, rotational)			
Hardness	Conventional hardness numbers, BHN, R, etc., not affected by change to SI			
Heat	(See Energy)			
Heat capacity	(See Capacity, heat)			
Heat capacity, specific	(See Capacity, heat specific)			
Heat flow rate	(See Power)			
Heat flow-density of	(See Density of heat flow)			
Heat, specific	General[a]	cal/g[d] Btu/lb	kJ/kg kJ/kg	4.186 8* 2.326
Heat transfer coefficient	(See Coefficient of heat transfer)			
Illuminance, illumination	General	fc	lx	10.763 91
Impact strength	(See Strength, impact)			
Impedance, mechanical	(See Damping coefficient)			
Inductance, electric	Filters and chokes, permeance	H	H	1*
Intensity, luminous	Light bulbs	candlepower	cd	1*
Intensity, radiant	General	W/sr	W/sr	1*
Leakage	(See Flow, volume)			
Length	Land distances, maps, odometers	mile	km	1.609 344*[c]
	Field size, turning circle, braking distance, cargo platforms, water depth, land levelling (cut and fill)	rod yd ft	m m m	5.029 210[c] 0.914 4* 0.304 8*
	Engineering drawings, engineering part specifications, motor vehicle dimensions, general	in	mm	25.4*
	Field drainage (runoff), evaporation, irrigation depth, rain and snowfall	in	cm	2.54*
	Coating thickness, filter rating	mil μin micron	μm μm μm	25.4* 0.025 4* 1*

	Surface texture
	  Roughness, average
	  Roughness sampling length, waviness height and spacing
		μin
		in
			μm
			mm
				0.025 4*
				25.4*

	Radiation wavelengths, optical measurements, (interference)	μin	nm	25.4*

For footnotes see end of Table.

φ TABLE A.1 (continued)

Quantity	Typical Application	From Old Units	To Metric Units	Multiply by
Load	(See Mass)			
Luminance	Brightness	footlambert	cd/m^2	3.426 259
Magnetization	Coil field strength	A/in	A/m	30.370 08
Mass	Vehicle mass (weight), axle rating, rated load, tire load, lifting capacity, tipping load, load, general	ton (long)	Mg, t	1.016 047
		ton (short)	Mg, t	0.907 184 7
		lb	kg	0.453 592 4
		slug	kg	14.593 90
	Small mass	oz	g	28.349 52
Mass per area	Fabric, surface coatings	oz/yd^2	g/m^2	33.905 75
		lb/ft^2	kg/m^2	4.882 428
		oz/ft^2	g/m^2	305.151 7
	Floor loading	lb/ft^2	kg/m^2	4.882 428
	Application rate, fertilizer, pesticide	lb/acre	kg/ha	1.120 851$^{(c)}$
	Crop yield, soil erosion	ton (short)/acre	t/ha	2.241 702$^{(c)}$
Mass per length	General	lb/ft	kg/m	1.488 164
		lb/yd	kg/m	0.496 054 7
Mass per time	Machine work capacity, harvesting, materials handling	ton (short)/h	t/h, Mg/h	0.907 184 7
Modulus, bulk	(See Pressure)			
Modulus of elasticity	General	lbf/in^2	MPa	0.006 894 757
Modulus of rigidity	(See Modulus of elasticity)			
Modulus, section	General	in^3	mm^3	16 387.06
		in^3	cm^3	16.387 06
Moment, bending	(See Moment of force)			
Moment of area, second	General	in^4	mm^4	416 231.4
		in^4	cm^4	41.623 14
Moment of force, torque, bending moment	General, engine torque, fasteners	lbf·in	N·m	0.112 984 8
		lbf·ft	N·m	1.355 818
		kgf·cm	N·m	0.098 066 5*
	Locks, light torque	ozf·in	mN·m	7.061 552
Moment of inertia	Flywheel, general	lb·ft^2	kg·m^2	0.042 140 11
Moment of mass	Unbalance	oz·in	kg·mm	0.720 077 8
Moment of momentum	(See Momentum, angular)			
Moment of section	(See Moment of area, second)			
Momentum	General	lb·ft/s	kg·m/s	0.138 255 0
Momentum, angular	Torsional vibration	lb·ft^2/s	kg·m^2/s	0.042 140 11
Permeability	Magnetic core properties	H/ft	H/m	3.280 840
Permeance	(See Inductance)			
Potential, electric	General	V	V	1*
Power	General, light bulbs	W	W	1*
	Air conditioning, heating	Btu/min$^{(a)}$	W	17.584 27
		Btu/h$^{(a)}$	W	0.293 071 1
	Engine, alternator, drawbar, power take-off, general	hp (550 ft·lbf/s)	kW	0.745 699 9
Power per area	Solar radiation	Btu/(ft^2·h)$^{(a)}$	W/m^2	3.154 591

For footnotes see end of Table.

φ TABLE A.1 (continued)

Quantity	Typical Application	From Old Units	To Metric Units	Multiply by
Pressure	All pressure except very small, bulk modulus	lbf/in² lbf/ft² in Hg (60°F) in H₂O (60°F) mm Hg (0°C) kgf/cm² bar atm (standard = 760 torr)	kPa kPa kPa kPa kPa kPa kPa kPa	6.894 757 0.047 880 26 3.376 85 0.248 84 0.133 322 98.066 5* 100* 101.325*
	Very small pressures (high vacuum)	lbf/in²	Pa	6 894.757
Pressure, sound, level	Acoustical measurements[f]	dB	dB	1*
Quantity of electricity	General	C	C	1*
Radiant intensity	(See Intensity, radiant)			
Resistance, electric	General	Ω	Ω	1*
Resistivity, electric	General	Ω·ft Ω·ft	Ω·m Ω·cm	0.304 8* 30.48*
Sound pressure level	(See Pressure, sound level)			
Speed	(See Velocity)			
Spring rate, linear	General spring properties	lbf/in	N/mm	0.175 126 8
Spring rate, torsional	General	lbf·ft/deg	N·m/deg	1.355 818
Strength, field, electric	General	V/ft	V/m	3.280 840
Strength, field, magnetic	General	oersted	A/m	79.577 47
Strength, impact	Materials testing	ft·lbf	J	1.355 818
Stress	General	lbf/in²	MPa	0.006 894 757
Surface tension	(See Tension, surface)			
Temperature	General use	°F	°C	$t_{°C} = (t_{°F} - 32)/1.8$*
	Absolute temperature, thermodynamics, gas cycles	°R	K	$T_K = T_{°R}/1.8$*
Temperature interval	General use	°F	K[b]	1 K = 1°C = 1.8°F*
Tension, surface	General	lbf/in dyne/cm	mN/m mN/m	175 126.8 1*
Thrust	(See Force)			
Time	General	s h min	s h min	1* 1* 1*
Torque	(See Moment of Force)			
Toughness, fracture	Metal properties	ksi √in	MPa·m^{1/2}	1.098 843
Vacuum	(See Pressure)			
Velocity, angular	(See Velocity, rotational)			
Velocity, linear	Vehicle	mile/h knot (international)	km/h km/h	1.609 344* 1.851 999 8
	General	ft/s ft/min in/s	m/s m/min mm/s	0.304 8* 0.304 8* 25.4*
Velocity, rotational	Mechanical events (rotational) and general	rad/s r/s r/min	rad/s r/s, r·s⁻¹ r/min, r·min⁻¹	1* 1* 1*
Viscosity, dynamic	General liquids	centipoise	mPa·s	1*

For footnotes see end of Table.

φ TABLE A.1 (continued)

Quantity	Typical Application	From Old Units	To Metric Units	Multiply by
Viscosity, kinematic	General liquids	centistokes	mm^2/s	1*
Volume	Truck body, shipping or freight, bucket capacity, grain bins and tanks, general	yd^3	m^3	0.764 554 9
		ft^3	m^3	0.028 316 85
		bushel	m^3	0.035 239 07
	Automobile luggage capacity	ft^3	L	28.316 85
	Gas pump displacement, air compressor, small gaseous, air reservoir	in^3	cm^3	16.387 06
	Engine displacement			
	large engines	in^3	L	0.016 387 06
	small engines	in^3	cm^3	16.387 06
	Liquid—fuel, lubricant, etc.	gal	L	3.785 412
		qt	L	0.946 352 9
		pt	L	0.473 176 5
	Small liquid	oz	mL	29.573 53
	Irrigation, reservoir	acre·ft	m^3	1 233.489$^{(c)}$
			dam^3	1.233 489$^{(c)}$
Volume per area	Application rate, pesticide	gal/acre	L/ha	9.353 958$^{(c)}$
Weight	May mean either mass or force—see paragraph 6.1			
Work	(See Energy)			
Young's modulus	(See Modulus of elasticity)			

φ ^{a}Conversions of Btu and calorie are based on the International Table Btu and calorie.
φ ^{b}In these expressions K indicates temperature interval. Therefore K may be replaced with °C if desired without changing the value or affecting the conversion factor, for example: kJ/(kg·K) = kJ/(kg·°C).
φ ^{c}Official use in surveys and cartography involves the U.S. statute mile based on the U.S. survey foot, which is longer than the international foot by two parts per million. The factors used in this standard for acre, acre foot, U.S. statute mile, and rod are based on the U.S. survey foot. Factors for all other old length related units are based on the international foot. For detail, see ASTM E 380.

φ ^{d}Not to be confused with kcal/g. The kcal is often called "calorie" in the nutritional field.
φ ^{e}Lift capacity ratings for cranes, hoists, and related components such as rope, cables, chains, etc. should be rated in mass units. Those items such as winches, which can be used for pulling as well as lifting, shall be rated in both force and mass units.
φ ^{f}When weighting is specified, show weighting level in parentheses following the symbol, for example: dB (A).

APPENDIX B

As covered in paragraph 5.1, SI units are required in SAE reports. To assist committees in carrying out this requirement in special circumstances, some qualifying rules are covered here.

B1. In standards that have alternative or optional procedures based on apparatus calibrated in either U. S. inch-pound or SI units, converted values need not be included. If optional procedures or dimensions produce equally acceptable results, the options may be shown by using the word *or* rather than parentheses: for example, in a 2-in gage length metal tension test specimen, the gage length may be shown as "2 in or 50 mm".

B2. A specific equivalent, for example 1.00 in (25.4 mm), need be inserted only the first time it occurs in each paragraph.

B3. Special instructions cover the use of tabular material.

Case 1. Limited Tabular Material—Provide SI equivalents in tables in parentheses or in separate columns.

STRAIGHT WHEEL GRINDERS

H	L		R
	in	(mm)	
3/8-24 UNF-2A	1-1/8	(28.58)	
1/2-13 UNC-2A	1-3/4	(44.45)	
5/8-11 UNC-2A	2-1/8	(53.98)	Governed by thickness
5/8-11 UNC-2A	3-1/8	(79.38)	of wheel used
3/4-10 UNC-2A	3-1/4	(82.55)	

Case 2. One or Two Large Tables—When the size of a table and limitations of space (on the printed page) make it impractical to expand the table to include SI equivalents, the table should be duplicated in U. S. inch-pound units and in SI units.

φ DIMENSIONS IN U.S. INCH-POUND UNITS

Chain No.	H60	H74	H75	H78	H82	H124
P (in)	2.308	2.609	2.609	2.609	3.075	4.000
A (in)	0.312	0.375	0.312	0.500	0.562	0.750
F (in)	0.73	1.00	0.75	1.12	1.25	1.56
H (in)	0.75	0.88	0.72	0.88	1.19	1.44
Proof test load (lb)						
Class M	2.800	4.000	2.800	6.400	8.000	12.000
Class P	3.500	5.000	3.500	8.000	10.000	15.000
No. of pitches per nominal 120 in strand	52	46	46	46	39	30
Theoretical length of nominal 120 in strand	120.02	120.01	120.01	120.01	119.92	120.00
Measuring load (lb)	190	270	190	130	510	810

DIMENSIONS IN SI UNITS

Chain No.	H60	H74	H75	H78	H82	H124
P (mm)	58.62	66.27	66.27	66.27	78.10	101.60
A (mm)	7.92	9.52	7.92	12.70	11.27	19.05
F (mm)	18.5	25.4	19.0	28.4	31.75	39.62
H (mm)	18.5	22.3	18.3	22.3	30.2	36.6
Proof test load (kN)						
Class M	12.50	17.80	12.50	28.50	35.60	53.40
Class P	15.60	22.20	15.60	35.60	44.50	66.80
No. of pitches per nominal 3048 mm strand	52	46	46	46	39	30
Theoretical length of nominal 3048 mm strand	3048.5	3048.2	3048.2	3048.2	3046.0	3048.0
Measuring load (N)	850	1200	850	1010	2400	3600

If the above Cases 1 and 2 would still result in major increase in the size of the standard, consideration must be given to other methods. SAE staff should first be consulted on techniques of arranging column spacing, etc., to accomplish addition of SI as shown in Cases 1 and 2.

Cases 3 and 4 are two approaches to reduce the number of pages involved in adding SI to reports with extensive tabular data. They should be used only in extreme cases since they do not accomplish the intent of SAE policy. Also, these approaches should not be considered when the users of the report are judged to need SI units for its use.

Case 3. Extensive Tabular Material—When the tabulated data is extensive and the above procedures would require an impractical addition to the standard, a summary appendix may be prepared listing all of the values appearing in the tables, along with the conversion of each, as follows:

TABLE B.1—SI EQUIVALENTS

φ Inches to Millimeters

in	mm	in	mm	in	mm
0.015	0.38	0.350	8.89	0.987	25.07
0.020	0.51	0.375	9.52	1.000	25.40
0.028	0.71	0.383	9.73	1.128	28.65
0.038	0.97	0.431	10.95	1.178	29.92
0.044	1.12	0.437	11.10	1.270	32.26
0.050	1.27	0.487	12.37	1.410	35.81
0.056	1.42	0.500	12.70	1.571	39.90
0.064	1.63	0.540	13.72	1.963	49.86
0.071	1.80	0.612	15.55	2.356	59.84
0.143	3.63	0.625	15.88	2.749	69.82
0.191	4.85	0.700	17.78	3.142	79.81
0.239	6.07	0.750	19.05	3.544	90.02
0.262	6.65	0.790	20.07	3.990	101.35
0.286	7.26	0.875	22.22	4.430	112.52
0.334	8.48	0.889	22.58		

φ Square Inches to Square Centimeters

in²	cm²	in²	cm²	in²	cm²
0.11	0.71	0.44	2.84	1.00	6.45
0.20	1.29	0.60	3.87	1.27	8.19
0.31	2.00	0.79	5.10	1.56	10.06

φ Pounds per Foot to Kilograms per Meter

lb/ft	kg/m	lb/ft	kg/m	lb/ft	kg/m
0.376	0.560	1.502	2.235	3.33	4.96
0.668	0.994	2.044	3.042	4.303	6.403
1.043	1.552	2.670	3.973	5.313	7.906

Pounds per Square Inch to Megapascals

psi	MPa	psi	MPa
50 000	345	80 000	550
60 000	415	90 000	620

Case 4.—In extreme cases when all the above approaches do not apply because of the size and number of tables, conversion factors may be placed in a footnote under each table, as in the following example.

TABLE B.2

Nominal Size, in	Outside Diameter, in[a]	Wall Thickness, in[a]	Nominal Mass per ft, Plain End, lb/ft[b]	Weight Class	Schedule No.	Test Pressure,[c] psi Butt-Welded	Test Pressure,[c] psi Grade A	Test Pressure,[c] psi Grade B
20	20.000	0.250	52.73	—	10	—	450	500
		0.281	59.18	—	—	—	500	600
		0.312	65.60	—	—	—	550	650
		0.344	72.21	—	—	—	600	700
		0.375	78.60	STD	20	—	700	800
		0.406	84.96	—	—	—	750	850
		0.438	91.51	—	—	—	800	900
		0.469	97.83	—	—	—	850	950
		0.500	104.13	XS	30	—	900	1000
		0.594	123.11	—	40	—	1100	1200
		0.812	166.40	—	60	—	1500	1700
		1.031	208.87	—	80	—	1900	2200
		1.281	256.10	—	100	—	2300	2700
		1.500	296.37	—	120	—	2700	2800
		1.750	341.10	—	140	—	2800	2800
		1.969	379.17	—	160	—	2800	2800
24	24.000	0.250	63.41	—	10	—	400	450
		0.281	71.18	—	—	—	400	500
		0.312	78.93	—	—	—	450	550
		0.344	86.91	—	—	—	500	600
		0.375	94.62	STD	20	—	550	650
		0.406	102.31	—	—	—	600	700
		0.438	110.22	—	—	—	650	750
		0.469	117.86	—	—	—	700	825
		0.500	125.49	XS	—	—	750	900
		0.562	140.68	—	30	—	850	1000
		0.688	171.29	—	40	—	1000	1200
		0.938	231.03	—	—	—	1400	1600
		0.969	238.85	—	60	—	1500	1700
		1.219	296.58	—	90	—	1800	2100
		1.531	367.39	—	100	—	2300	2700
		1.812	429.39	—	120	—	2700	2800
		2.062	483.12	—	140	—	2800	2800
		2.344	542.14	—	160	—	2800	2800
26	26.000	0.250	68.75	—	—	—	350	400
		0.281	77.18	—	—	—	390	450
		0.312	85.60	—	10	—	430	500
		0.344	94.26	—	—	—	480	560
		0.375	102.63	STD	—	—	520	610
		0.406	110.98	—	—	—	560	660
		0.438	119.57	—	—	—	610	710
		0.469	127.88	—	—	—	650	760
		0.500	136.17	XS	20	—	690	810
		0.562	152.68	—	—	—	780	910

[a] 1 in = 25.4 mm
[b] 1 lb/ft = 1.49 kg/m
[c] 1 psi = 6.9 kPa

B4. Graphs and charts may be handled in several ways depending on the circumstances. In adding SI units to a graphic presentation of data, the practice of specific addition of metric conversions to existing ordinate or abscissa values should be avoided.

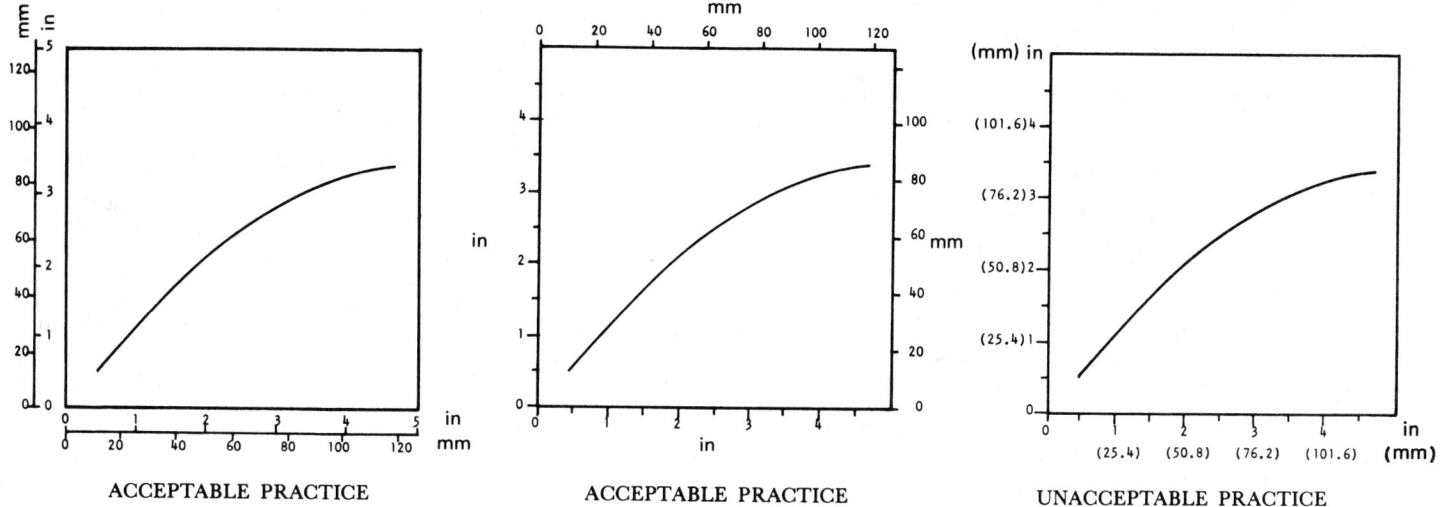

ACCEPTABLE PRACTICE   ACCEPTABLE PRACTICE   UNACCEPTABLE PRACTICE

GUIDELINES FOR DEVELOPING AND REVISING SAE NOMENCLATURE AND DEFINITIONS—SAE J1115

SAE Information Report

Report of Nomenclature Advisory Committee of SAE Automotive Council approved June 1975. Editorial change February 1976.

Introduction—Historically SAE has been concerned with nomenclature as an integral part of the standards development process. Guidelines for automotive nomenclature were written in 1916, were last revised in 1941, and were included in the SAE Handbook until 1962. The present diversity of groups working on nomenclature in the various ground vehicle committees led to the organization of the Nomenclature Advisory Committee under SAE Automotive Council.

Objective—The objective of the Committee is to promote understandable and precise communication relating to the engineering aspects of on-highway vehicles, their components, their design, and their evaluation. In order to reach this objective, the Committee is primarily concerned with the definition or redefinition of needed terms considering (a) current usage, (b) changing needs, and (c) the interactive use of a particular term by various SAE committees, government agencies, and other national and international organizations. In order to facilitate and encourage the use of generally accepted terminology by SAE committees and other organizations, the Nomenclature Advisory Committee plans to prepare and maintain a glossary of terms appearing in SAE technical reports.

Guidelines—Since the basic approach of the Committee is one of advice and coordination, the following guidelines for developing and revising SAE nomenclature and definitions are recommended:

1. Before developing and revising nomenclature, check for similar terms already defined in existing SAE Standards, Recommended Practices, and Information Reports and in Federal motor vehicle standards, in order to minimize duplication, to avoid conflict, and to achieve uniformity of format. In addition to the SAE Handbook, consult the following SAE and DOT reports:

   Vehicle Dynamics Terminology, SAE J670d
   Seating Manual, SAE J782a
   Motor Vehicle Dimensions, SAE J1100a
   Recommendations for Writing SAE Technical Reports
   49 CFR 571.3 and appropriate FMVSS

2. If dictionary definitions of common generic terms can be used, they need not be included in nomenclature listings; for example, "Acceleration."

3. Develop general definitions for general terms; for example, define

"Fully Latched" generically and not specifically as applied to doors, hoods or trunk lids. General definitions must be valid for all possible situations or contingencies, not just for the situation under immediate consideration.

4. Specific concepts or components should be identified by correspondingly specific terms when defined in a document, so that these terms can stand alone when extracted from that document and integrated with other terms in a glossary. For example, use "Tire Valve Core" rather than "Core" alone, so that the term will not be confused with another type of core, such as "Radiator Core."

5. The abbreviation for a defined term, when included, should follow it and be placed in parentheses; for example, "Decibel (dB)."

6. Terminology should follow normal word order; for example, use "Lighting Device," not "Device, Lighting," and "Brake Cylinder," not "Cylinder (Brake)." The glossary will index the defined term under each significant word in that term. For example, "Accelerator Heel Point" will appear in the index in the following permuted forms:

Accelerator Heel Point
Heel Point, Accelerator
Point, Accelerator Heel

7. Term definitions should be directed at concise statements of the items being defined, rather than at specifications, performance requirements, or test procedures. As an example, the following description includes both a definition and a test procedure:

Windshield Slope Angle—the angle between the vertical reference line and a chord of the windshield arc running from the lower DLO to the upper DLO at the car centerline, when such chord is no longer than 18.0 in. If the windshield is longer than 18.0 in, the angle to be measured will be formed by a chord 18.0 in long, drawn from the lower DLO to the intersecting point on the windshield.

8. The opposite of a term already defined within a document need not also be defined; for example, the definition of "Asymmetrical Beam" is implied by the definition of "Symmetrical Beam."

9. Nomenclature which refers to a diagram should have a sufficient written definition to make the term understandable without the diagram.

10. Explanatory or historical notes should be stated separately from and should follow the base definition and be so identified. The following example illustrates this usage:

Static Loaded Tire Radius—the loaded radius of a stationary tire inflated to normal recommended pressure.

Note. In general, the static loaded radius is different from the radius of a slowly rolling tire; and the static radius of a tire rolled into position may be different from that of a tire loaded without being rolled.

11. When two or more terms have the same definition and are used interchangeably, a preferred term should be chosen and defined. Synonymous terms may also be listed but with only a reference to the preferred item term. Example:

Barrel Gasket—the cylindrical sleeve of rubber-like material, etc.
Barrel Seal—use Barrel Gasket.

12. Definitions should be clear and useful to all who use the SAE Handbook as an engineering or technical reference.

UNIVERSAL SYMBOLS FOR OPERATOR CONTROLS—SAE J1500 JUN80

SAE Standard

Report of the Symbols Committee of the SAE Technical Board, approved June 1980.

1. Introduction—There is an increasing use of graphic symbols throughout the world in different technical fields to communicate information and identify operation or function.

Vehicles as compared to equipment which have like functions should have like graphic symbol identification.

2. Purpose—The compilation of symbols in this document are for reference use by all current and future committees responsible for the development of symbols.

3. Scope—This standard is to delineate the symbols used to identify controls, indicators, and tell-tales for automotive vehicles, trucks, off-the-road vehicles, construction equipment, industrial and recreational transportation and is for reference purposes only. The symbol application is to be found within the appropriate standards listed.

	UPPER BEAM	LOWER BEAM	HEADLAMP CLEANER	HEADLAMP LEVELING CONTROL	FRONT FOG LIGHT	REAR FOG LIGHT	PARKING LIGHTS
ISO 2575/3 REFERENCE	⊟D	≡D	⊟D̥	⇕D	⧧D	D⧧	P⩽
FMVSS 101/80	⊟D						
CMVSS 101	⊟D	≡D					
EEC 78/316	⊟D	≡D	⊟D̥		⧧D	D⧧	P⩽
SAE J1048 AUTOMOTIVE	⊟D	≡D	⊟D̥		⧧D	D⧧	P⩽
SAE TRUCKS							
SAE J389b AGRICULTURAL	⊟D	≡D					
SAE J298 INDUSTRIAL	⊟D	≡D					≡P
SAE J107 MOTORCYCLES	⊟D	≡D					
SAE MARINE							
SAE SNOWMOBILES							

		MASTER LIGHTING SWITCH	TURN SIGNALS	HAZARD WARNING	SIDE LAMPS	CLEARANCE LAMPS	WINDSHIELD WIPER	WINDSHIELD WASHER
	ISO 2575/3 REFERENCE	☼	⇔ ⇨	△			⌐	⌐
REGULATORY	FMVSS 101/80	≡D	⇔ ⇨	△		≡D D≡	⌐	⌐
REGULATORY	CMVSS 101	≡D ☼	⇔ ⇨	△		≡D D≡	⌐	⌐
REGULATORY	EEC 78/316	☼	⇔ ⇨	△	≡D D≡		⌐	⌐
STANDARDS	SAE J1048 AUTOMOTIVE	☼	⇔ ⇨	△			⌐	⌐
STANDARDS	SAE TRUCKS							
STANDARDS	SAE J389b AGRICULTURAL		⇔ ⇨	⚠			⌐	⌐
STANDARDS	SAE J298 INDUSTRIAL		↰ ↱	⚠			⌐	
STANDARDS	SAE J107 MOTORCYCLES		⇔ ⇨	△				
STANDARDS	SAE MARINE							
STANDARDS	SAE SNOWMOBILES							

	WINDSHIELD WASHER & WIPER	WINDSCREEN DEMISTER DEFOGGER	REAR WINDOW WIPER	REAR WINDOW WASHER	REAR WINDOW WIPER & WASHER	REAR WINDOW DEMISTER DEFOGGER	VENTILATING FAN
ISO 2575/3 REFERENCE	⛆	⛆	▭	▭	▭	▭	✿
FMVSS 101/80	⛆	⛆				▭	✿
CMVSS 101	⛆						✿
EEC 78/316	⛆	⛆				▭	✿
SAE J1048 AUTOMOTIVE	⛆	⛆				▭	✿
SAE TRUCKS							
SAE J389b AGRICULTURAL	⛆	⛆					✿
SAE J298 INDUSTRIAL		⛆					
SAE J107 MOTORCYCLES							
SAE MARINE							
SAE SNOWMOBILES							

		BATTERY CHARGING	ENGINE COOLANT	ENGINE OIL	FUEL	UNLEADED FUEL	CHOKE	HORN
	ISO 2575/3 REFERENCE	🔋	🌡️	🛢️	⛽	⛽	choke	📯
REGULATORY	FMVSS 101/80	🔋	🌡️	🛢️	⛽			
REGULATORY	CMVSS 101	🔋		🛢️	⛽			
REGULATORY	EEC 78/316	🔋	🌡️	🛢️	⛽		choke	📯
STANDARDS	SAE J1048 AUTOMOTIVE	🔋	🌡️	🛢️	⛽	⛽	choke	📯
STANDARDS	SAE TRUCKS							
STANDARDS	SAE J389b AGRICULTURAL	🔋	🌡️	🛢️	⛽		choke	📯
STANDARDS	SAE J298 INDUSTRIAL	⚡	🌡️	🛢️	⛽		choke	📯
STANDARDS	SAE J107 MOTORCYCLES	🔋	🌡️	🛢️	⛽		choke	📯
STANDARDS	SAE MARINE							
STANDARDS	SAE SNOWMOBILES							

	LIGHTER	SEAT BELT	PARKING BRAKE	BRAKE FAILURE	FRONT HOOD (BONNET)	REAR HOOD (BOOT)	TRANSMISSION
ISO 2575/3 REFERENCE	🚬	🧍‍♂️	(P)	(!)	🚗	🚗	
FMVSS 101/80		🧍‍♂️					
CMVSS 101		🧍‍♂️		(!)			
EEC 78/316		🧍‍♂️	(P)	(!)	🚗	🚗	
SAE J1048 AUTOMOTIVE	🚬	🧍‍♂️			🚗	🚗	
SAE TRUCKS							
SAE J389b AGRICULTURAL	🚬	👐	P				⚙
SAE J298 INDUSTRIAL	☀		[P]				⚙
SAE J107 MOTORCYCLES							
SAE MARINE							
SAE SNOWMOBILES							

		TRANSMISSION OIL PRESSURE	TRANSMISSION OIL TEMPERATURE	TRANSMISSION OIL FILTER	AIR FILTER	GREASE LUBRICANT	OIL LUBRICANT	OIL LEVEL
	ISO 2575/3 REFERENCE							
REGULATORY	**FMVSS 101/80**							
	CMVSS 101							
	EEC 78/316							
STANDARDS	**SAE J1048** AUTOMOTIVE							
	SAE TRUCKS							
	SAE J389b AGRICULTURAL	⊙	🌡	⊙	▽	10	20/50	—
	SAE J298 INDUSTRIAL	⊙	🌡	⊙	▽	10	20/50	—
	SAE J107 MOTORCYCLES							
	SAE MARINE							
	SAE SNOWMOBILES							

	DIPSTICK	FIRING ORDER	ENGINE R.P.M.	FUEL SHUT-OFF	HAND BRAKE	FORWARD	REVERSE
ISO 2575/3 REFERENCE							
FMVSS 101/80							
CMVSS 101							
EEC 78/316							
SAE J1048 AUTOMOTIVE							
SAE TRUCKS							
SAE J389b AGRICULTURAL	🖉	1·3·2·4	⛽		┣○ ○┫	↑	▯ R ↓
SAE J298 INDUSTRIAL	🖉	1·3·2·4	⛽	⊘	┣○ ○┫ Engaged Disengaged	↑ ↑	↕ ↕
SAE J107 MOTORCYCLES							
SAE MARINE							
SAE SNOWMOBILES							

		NEUTRAL	UP DOWN	ADD INCREASE	DECREASE	CONTROL LEVER	VOLUME LEVEL	TEMPERATURE
REGULATORY	ISO 2575/3 REFERENCE							
REGULATORY	FMVSS 101/80							
REGULATORY	CMVSS 101							
REGULATORY	EEC 78/316							
STANDARDS	SAE J1048 AUTOMOTIVE							
STANDARDS	SAE J680b TRUCKS							
STANDARDS	SAE J389b AGRICULTURAL	N	↓/↑	+	−	↕	●◐○	🌡🌡
STANDARDS	SAE J298 INDUSTRIAL	(N)	⬆ Up ⬇ Down	+		↕	● Full ◐ 1/2 ○ Empty	🌡🌡
STANDARDS	SAE J107 MOTORCYCLES							
STANDARDS	SAE MARINE							
STANDARDS	SAE SNOWMOBILES							

	TEMPERATURE CONTROL	DIFFERENTIAL LOCK	ROCK SHAFT	CONNECTION	AXLE DISCONNECT	TOW	POWER TAKE-OFF
ISO 2575/3 REFERENCE							
FMVSS 101/80							
CMVSS 101							
EEC 78/316							
SAE J1048 AUTOMOTIVE							
SAE TRUCKS							
SAE J389b AGRICULTURAL	↑↓	⊕	↗ ↘	⇇ ⇉	🚗 🚗	⚙	⚙
SAE J298 INDUSTRIAL	🌡🌡	⊕	Raised Lowered	Engage Disengage	Connect Disconnect	⚙	On Off
SAE J107 MOTORCYCLES							
SAE MARINE							
SAE SNOWMOBILES							

		REMOTE CYLINDER	HEAVY LIGHT	BACKHOE DIPPERSTICK CONTROL	BACKHOE BOOM CONTROL	BACKHOE SWING CONTROL	BACKHOE BUCKET CONTROL	LOADER LIFT ARM CONTROL
REGULATORY	ISO 2575/3 REFERENCE							
REGULATORY	FMVSS 101/80							
REGULATORY	CMVSS 101							
REGULATORY	EEC 78/316							
STANDARDS	SAE J1048 AUTOMOTIVE							
STANDARDS	SAE TRUCKS							
STANDARDS	SAE J389b AGRICULTURAL	■	■	■	■	■	■	■
STANDARDS	SAE J298 INDUSTRIAL	■	■	■	■	■	■	■
STANDARDS	SAE J107 MOTORCYCLES							
STANDARDS	SAE MARINE							
STANDARDS	SAE SNOWMOBILES							

	LOADER BOOM CONTROL	LOADER BUCKET CONTROL	STABILIZER CONTROL	UNLOADING AUGER	HEADER	REEL SPEED	REEL HEIGHT
ISO 2575/3 REFERENCE							
FMVSS 101/80							
CMVSS 101							
EEC 78/316							
SAE J1048 AUTOMOTIVE							
SAE TRUCKS							
SAE J389b AGRICULTURAL	◨	◨	◨	◨	◨	◨	◨
SAE J298 INDUSTRIAL	Float / Level Raise	◨	◨				
SAE J107 MOTORCYCLES							
SAE MARINE							
SAE SNOWMOBILES							

		PLATFORM HEIGHT	CYLINDER SPEED (COMBINE)	CONCAVE CLEARANCE (COMBINE)	GROUND SPEED	SPEED RANGE	ALL MECHANISMS	PRESSURIZE OPEN SLOWLY
	ISO 2575/3 REFERENCE							
REGULATORY	FMVSS 101/80							
REGULATORY	CMVSS 101							
REGULATORY	EEC 78/316							
STANDARDS	SAE J1048 AUTOMOTIVE							
STANDARDS	SAE TRUCKS							
STANDARDS	SAE J389b AGRICULTURAL	✦	⟳	✦⊙	🚜	🐇 CONTINUOUSLY VARIABLE 🐢	🚜	⚠
STANDARDS	SAE J298 INDUSTRIAL					🐇 Fast 🐢 Slow		⚠
STANDARDS	SAE J107 MOTORCYCLES							
STANDARDS	SAE MARINE							
STANDARDS	SAE SNOWMOBILES							

	BASKET LIFT	HOURS	READ OPERATOR'S MANUAL	REVERSING LAMP	HAND ACCELERATOR THROTTLE	DIESEL ENGINE CUT-OFF	
REGULATORY ISO 2575/3 REFERENCE							
FMVSS 101/80							
CMVSS 101							
EEC 78/316				®⇐	)•(	⊗	
STANDARDS SAE J1048 AUTOMOTIVE							
SAE TRUCKS							
SAE J389b AGRICULTURAL	⬒	⧖	📖				
SAE J298 INDUSTRIAL		⧖					
SAE J107 MOTORCYCLES							
SAE MARINE							
SAE SNOWMOBILES							

FERROUS METALS

FERROUS METALS

2 Numbering System, Chemical Compositions

Numbering Metals and Alloys—SAE J1086	2.01
Selection and Use of Steels—SAE J401	2.05
SAE Numbering System for Wrought or Rolled Steel—SAE J402b	2.06
Chemical Compositions of SAE Carbon Steels—SAE J403h	2.06
Chemical Compositions of SAE Alloy Steels—SAE J404j	2.09
Chemical Compositions of SAE Wrought Stainless Steels—SAE J405d	2.11
§ Chemical Compositions of SAE Experimental Steels—SAE J1081 MAY80	2.12
Formerly Standard SAE Alloy Steels—SAE J778a	2.13
Formerly Standard SAE Carbon Steels—SAE J118a	2.14
Methods of Determining Hardenability of Steels—SAE J406c	2.15
* Hardenability Bands for Carbon and Alloy H Steels—SAE J1268 JUN80	2.24
Methods of Sampling Steel for Chemical Analysis—SAE J408c	2.69
Permissible Variations from Specified Ladle Chemical Ranges and Limits for Steels—SAE J409c	2.70
High Strength, Low Alloy Steel—SAE J410c	2.72
High Strength, Quenched, and Tempered Structural Steels—SAE J368a	2.74
Engine Poppet Valve Materials—SAE J775a	2.75

3 General Data on Steels

Carbon and Alloy Steels—SAE J411d	3.01
High Strength Carbon and Alloy Die Drawn Steels—SAE J935	3.03
General Characteristics and Heat Treatments of Steels—SAE J412h	3.05
Mechanical Properties of Heat Treated Wrought Steels—SAE J413b	3.13
Estimated Mechanical Properties and Machinability of Hot Rolled and Cold Drawn Carbon Steel Bars—SAE J414a	3.14
Glossary of Carbon Steel Sheet and Strip Terms—SAE J940	3.18
Aging of Carbon Steel Sheet and Strip—SAE J763	3.20
Definitions of Heat Treating Terms—SAE J415j	3.20
Alloy Steel Machinability Ratings—SAE J770c	3.24

4 Methods of Testing Steels

Tensile Test Specimens—SAE J416b	4.01
Surface Hardness Testing with Files—SAE J864	4.02
Hardness Tests and Hardness Number Conversions—SAE J417b	4.03
Grain Size Determination of Steels—SAE J418a	4.07
Methods of Measuring Decarburization—SAE J419	4.10
Methods of Determining Plastic Deformation in Sheet Metal Stampings—SAE J863c	4.14
Sheet Steel Thickness and Profile—SAE J1058	4.16
Properties of Low Carbon Steel Sheets and Strip and Their Relationship to Formability—SAE J877	4.17
Selecting and Specifying Hot and Cold Rolled Steel Sheet and Strip—SAE J126a	4.19
† Classification of Common Imperfections in Sheet Steel—SAE J810 JUN80	4.24
* Undervehicle Corrosion Test—SAE J1293 JAN80	4.34
Cleanliness Rating of Steels by the Magnetic Particle Method—SAE J421b	4.37
Microscopic Determination of Inclusions in Steels—SAE J422a	4.39

* New
† Technical revision
§ Editorial change

† Detection of Surface Imperfections in Ferrous Rods, Bars, Tubes, and Wires—SAE J349
  JUN80.. 4.42
Methods of Measuring Case Depth—SAE J423a 4.43
Method for Determining Breakage Allowances for Steel Sheets—SAE J424 .... 4.45
§ Nondestructive Tests—SAE J358 JAN80 .... 4.45
Infrared Testing—SAE J359a 4.47
Magnetic Particle Inspection—SAE J420b 4.47
Eddy Current Testing by Electromagnetic Methods—SAE J425b 4.48
Leakage Testing—SAE J1267 .... 4.50
Liquid Penetrant Test Methods—SAE J426c 4.52
Penetrating Radiation Inspection—SAE J427b 4.52
Ultrasonic Inspection—SAE J428b 4.55
Acoustic Emission Test Methods—SAE J1242 .... 4.56
Definition for Particle Size—SAE J391 .... 4.56
Definitions for Macrostrain and Microstrain—SAE J932 .... 4.57
Technical Report on Fatigue Properties—SAE J1099 .... 4.57

5 Steel Fasteners

Mechanical and Material Requirements for Externally Threaded Fasteners—SAE J429
  JAN80.. 5.01
Mechanical and Material Requirements for Metric Externally Threaded Steel
  Fasteners—SAE J1199... 5.06
Test Methods for Metric Threaded Fasteners—SAE J1216 .... 5.11
§ Mechanical and Quality Requirements for Machine Screws—SAE J82 JUN79 .. 5.14
§ Mechanical and Material Requirements for Steel Nuts—SAE J995 JUN79 .... 5.15
Mechanical and Material Requirements for Wheel Bolts—SAE J1102 .... 5.18
Mechanical and Chemical Requirements for Nonthreaded Fasteners—SAE J430.. 5.18
Mechanical and Quality Requirements for Tapping Screws—SAE J933 JUN79 .. 5.19
Steel Self-Drilling Tapping Screws—SAE J78 JUN79 .... 5.21
§ Thread Rolling Screws—SAE J81 JUN79 .... 5.26
Metric Thread Rolling Screws—SAE J1237 .... 5.29
Surface Discontinuities on General Application Bolts, Screws, and Studs—SAE J1061a 5.33
Surface Discontinuities on Bolts, Screws, and Studs—SAE J123c 5.37
Surface Discontinuities on Nuts—SAE J122a 5.40
Decarburization in Hardened and Tempered Threaded Fasteners—SAE J121a .. 5.44

6 Spring Wire and Springs

Oil Tempered Carbon Steel Spring Wire and Springs—SAE J316 .... 6.01
Oil Tempered Carbon Steel Valve Spring Quality Wire and Springs—SAE J351 .. 6.02
Oil Tempered Chromium-Vanadium Valve Spring Quality Wire and Springs—SAE J132 6.03
Hard Drawn Mechanical Spring Wire and Springs—SAE J113 .... 6.04
Special Quality High Tensile, Hard Drawn Mechanical Spring Wire and Springs—SAE
  J271... 6.04
Oil Tempered Chromium-Silicon Alloy Steel Wire and Springs—SAE J157 ... 6.05
Hard Drawn Carbon Steel Valve Spring Quality Wire and Springs—SAE J172 . 6.06
Stainless Steel 17-7 PH Spring Wire and Springs—SAE J217 .... 6.07
Stainless Steel, SAE 30302, Spring Wire and Springs—SAE J230 .... 6.08
Music Steel Spring Wire and Springs—SAE J178 .... 6.09

7 Ferrous Castings

Numbering System for Designating Grades of Cast Ferrous Materials—SAE J859 . 7.01
† Automotive Gray Iron Castings—SAE J431 AUG79 .... 7.01
Automotive Malleable Iron Castings—SAE J158a 7.04
Automotive Ductile (Nodular) Iron Castings—SAE J434c 7.06
Elevated Temperature Properties of Cast Irons—SAE J125 .... 7.08
Automotive Steel Castings—SAE J435c 7.12

8 Tool and Die Steels

Tool and Die Steels—SAE J438b 8.01
Selection and Heat Treatment of Tool and Die Steels—SAE J437a 8.02
Sintered Carbide Tools—SAE J439a 8.07
Sintered Tool Materials—SAE J1072 .... 8.12

9 Ferrous Materials

Sintered Powder Metal Parts: Ferrous—SAE J471d 9.01
Cut Steel Wire Shot—SAE J441 .... 9.05
§ Test Strip, Holder and Gage for Shot Peening—SAE J442 AUG79 .... 9.05
Procedures for Using Standard Shot Peening Test Strip—SAE J443 .... 9.07
Cast Steel Shot—SAE J827 .... 9.08
Cast Shot and Grit Size Specifications for Peening and Cleaning—SAE J444a .. 9.09

Metallic Shot and Grit Mechanical Testing—SAE J445a 9.10
Size Classification and Characteristics of Glass Beads for Peening—SAE J1173 .... 9.11

10 Ferrous and Nonferrous—General
Use of the Terms Yield Point and Yield Strength—SAE J450 .... 10.01
Surface Texture Measurement of Cold Rolled Sheet Steel—SAE J911 .... 10.01
Surface Texture—SAE J448a 10.02
Surface Texture Control—SAE J449a 10.04
Preferred Thicknesses for Uncoated Flat Metals (Thru 12 mm)—SAE J446a 10.06
Welding, Brazing, and Soldering—Materials and Practices—SAE J1147 .... 10.06

2 Numbering System, Chemical Compositions

NUMBERING METALS AND ALLOYS—
SAE J1086 AND ASTM E527

SAE and ASTM Recommended Practice

Report of Unified Numbering System Advisory Board approved August 1974. Editorial change June 1977.

UNS designations shall not be used for metals and alloys which are not registered under the system described herein, or for any metal or alloy whose composition differs from those registered.

1. Scope

1.1 This recommended practice describes a unified numbering system (UNS) for metals and alloys which have a "commercial standing" (see Note 1), and covers the procedure by which such numbers are assigned.

Section 2 describes the system of alphanumeric numbers (or codes) established for each family of metals and alloys.

Section 3 outlines the organization established for administering the system.

Section 4 describes the procedure for requesting number assignment to metals and alloys for which UNS numbers have not previously been assigned.

1.2 The UNS provides a means of correlating many nationally used numbering systems currently administered by societies, trade associations, and individual users and producers of metals and alloys, thereby avoiding confusion caused by use of more than one identification number for the same material; and by the opposite situation of having the same number assigned to two or more entirely different materials. It provides, also, the uniformity necessary for efficient indexing, record keeping, data storage and retrieval, and cross referencing.

1.3 *A UNS number is not in itself a specification,* since it establishes no requirements for form, condition, quality, etc. It is a unified identification of metals and alloys for which controlling limits have been established in specifications published elsewhere. (See Note 2.)

2. Description of Numbers (or Codes) Established for Metals and Alloys

2.1 The unified numbering system (UNS) establishes 15 series of numbers for metals and alloys, as shown in Table 1. Each UNS number consists of a single letter-prefix followed by five digits. In most cases the letter is suggestive of the family of metals identified, for example, A = aluminum, P = precious metals, H = H-steels. Table 2 shows the secondary division of some primary series of numbers.

2.2 Whereas some of the digits in certain of the UNS number groups have special assigned meaning, each series is independent of the others in such significance; this practice permits greater flexibility and avoids complicated and lengthy UNS numbers. (See Note 3.)

2.3 Wherever feasible, identification "numbers" from existing systems are incorporated into the UNS numbers. For example: The carbon steel which is presently identified by "AISI 1020" (American Iron & Steel Institute), is covered by "UNS G10200" and the free cutting brass, which is presently identified by "CDA 360" (Copper Development Association), is covered by "UNS C36000."

(Readers are cautioned *not* to make their own "assignments" of numbers from such listings, as this can result in unintended and unexpected duplication and conflict.)

2.4 The ASTM and the SAE periodically publish up-to-date listings of all UNS numbers assigned to specific metals and alloys, with appropriate reference information on each. (See Note 6.) Many trade associations also publish similar listings related to materials of primary interest to their organizations.

3. Organization for Administering Unified Numbering System for Metals and Alloys

3.1 The organization for administering the UNS consists of: (1) an advisory board, (2) several number-assigning offices, (3) a corps of volunteer consultants, and (4) staffs at ASTM and SAE. In addition, SAE and ASTM committees dealing with various groups of materials may be consulted.

3.1.1 The Advisory Board has approximately 20 volunteer members who are affiliated with major producing and using industries, trade associations, government agencies, and standards societies, and who have extensive experience with identification, classification, and specification of materials. The Board is the administrative arm of SAE and ASTM on all matters pertaining to the UNS. It coordinates thinking on the format of each series of numbers and the administration of each by selected experts. It sets up ground rules for determining eligibility of any material for a UNS number, for requesting such numbers, and for appealing unfavorable rulings. It is the final referee on matters of disagreement between requesters and assigners.

3.1.2 UNS number assigners for certain materials are set up at trade associations which have successfully administered their own numbering systems; for other materials, assigners are located at the offices of SAE and ASTM. Each of these assigners has the responsibility for administering a

ϕTABLE 1—PRIMARY SERIES OF NUMBERS

UNS Series	Metal
Nonferrous metals and alloys	
A00001–A99999	Aluminum and aluminum alloys
C00001–C99999	Copper and copper alloys
E00001–E99999	Rare earth and rare earth-like metals and alloys (18 items, see Table 2)
L00001–L99999	Low melting metals and alloys (14 Items, see Table 2)
M00001–M99999	Miscellaneous nonferrous metals and alloys (12 Items, see Table 2)
N00001–N99999	Nickel and nickel alloys
P00001–P99999	Precious metals and alloys (8 Items, see Table 2)
R00001–R99999	Reactive and refractory metals and alloys (14 Items, see Table 2)
Z00001–Z99999	Zinc and zinc alloys
Ferrous metals and alloys	
D00001–D99999	Specified mechanical properties steels
F00001–F99999	Cast irons
G00001–G99999	AISI and SAE carbon and alloy steels (except tool steels)
H00001–H99999	AISI H-steels
J00001–J99999	Cast steels (except tool steels)
K00001–K99999	Miscellaneous steels and ferrous alloys
S00001–S99999	Heat and corrosion resistant (stainless) steels
T00001–T99999	Tool steels

specific series of numbers, as shown in Table 3. Each considers requests for assignment of new UNS numbers, and informs applicants of the action taken. Trade association UNS number assigners also report immediately to both SAE and ASTM details of each number assignment. ASTM and SAE assigners collaborate with designated consultants when considering requests for assignment of new numbers.

3.1.3 Consultants are selected by the Advisory Board to provide expert knowledge of a specific field of materials. Since they are utilized primarily by the Board and the SAE and ASTM number assigners, they are not listed in this recommended practice. At the request of the ASTM or SAE number assigner, a consultant considers a request for a new number in the light of the ground rules established for the material involved, decides whether a new number is justified, and informs the ASTM or SAE number assigner accordingly.

This utilization of experts (consultants and number assigners) is intended to insure prompt and fair consideration of all requests. It permits each decision to be based on current knowledge of the needs of a specific industry of producers and users.

3.1.4 Staff members at SAE and ASTM maintain duplicate master listings of all UNS numbers assigned.

3.1.5 Established SAE and ASTM committees which normally deal with standards and specifications for the materials covered by the UNS, and other knowledgeable persons, are called upon by the Advisory Board for advice when considering appeals from unfavorable rulings in the matter of UNS number assignments.

4. Procedure for Requesting Number Assignment to Metals and Alloys Not Already Covered by UNS Numbers (or Codes)

4.1 UNS numbers are assigned only to metals and alloys which have a commercial standing (as defined in Note 1).

4.2 The need for a new number should always be verified by determining from the latest complete listing of already assigned UNS numbers that a usable number is or is not available. (See Note 4.)

4.3 For a new UNS number to be assigned, the composition (or other properties, as applicable) must be significantly different from those of any metal or alloy which has already been assigned a UNS number.

4.3.1 In the case of metals or alloys that are normally identified or specified by chemical composition, the chemical composition limits must be reported.

4.3.2 In the case of metals or alloys which are normally identified or specified by mechanical (or other) properties, such properties and limits thereof must be reported. Only those chemical elements and limits, if any, which are significant in defining such materials need be reported.

4.4 Requests for new numbers shall be submitted on "Application for UNS Number Assignment" forms (Fig. 1). Copies of these are available from any UNS number assigning office (Table 3) or facsimiles may be made of the one herein.

4.5 All instructions on the printed application form should be read carefully and all information provided as indicated. (See Note 5.)

4.6 To further assist in assigning UNS numbers, the requester is encouraged to suggest a possible UNS number in each request, giving appropriate consideration to any existing number presently used by a trade association, standards society, producer, or user.

4.7 Each completed application form shall be sent to the UNS number assigning office having responsibility for the series of numbers which appears to most closely relate to the material described on the form (Table 3).

Note 1. The terms "commercial standing," "production usage," and others, are intended to portray a material in active industrial use, although the actual amount of such use will depend, among other things, upon the type of materials. (Obviously gold will not be used in the same "tonnages" as hot rolled steel.)

Different standardizing groups use different criteria to define the status that a material has to attain before a standard number will be assigned to it. For instance, the American Iron and Steel Institute requires for stainless steels "two or more producers with combined production of 200 tons per year for at

φ **TABLE 2—SECONDARY DIVISION OF SOME SERIES OF NUMBERS**

UNS Series	Metal	UNS Series	Metal
E00001–E99999 Rare earth and rare earth-like metals and alloys		M00001–M99999 Miscellaneous nonferrous metals and alloys	
E00000–E00999	Actinium	M00001–M00999	Antimony
E01000–E20999	Cerium	M01001–M01999	Arsenic
E21000–E45999	Mixed rare earths[a]	M02001–M02999	Barium
E46000–E47999	Dysprosium	M03001–M03999	Calcium
E48000–E49999	Erbium	M04001–M04999	Germanium
E50000–E51999	Europium	M05001–M05999	Plutonium
E52000–E55999	Gadolinium	M06001–M06999	Strontium
E56000–E57999	Holmium	M07001–M07999	Tellurium
E58000–E67999	Lanthanum	M08001–M08999	Uranium
E68000–E68999	Lutetium	M10001–M19999	Magnesium
E69000–E73999	Neodymium	M20001–M29999	Manganese
E74000–E77999	Praseodymium	M30001–M39999	Silicon
E78000–E78999	Promethium		
E79000–E82999	Samarium	P00001–P99999 Precious metals and alloys	
E83000–E84999	Scandium	P00001–P00999	Gold
E85000–E86999	Terbium	P01001–P01999	Iridium
E87000–E87999	Thulium	P02001–P02999	Osmium
E88000–E89999	Ytterbium	P03001–P03999	Palladium
E90000–E99999	Yttrium	P04001–P04999	Platinum
		P05001–P05999	Rhodium
F00001–F99999 Cast irons	Gray, malleable, pearlitic malleable, and ductile (nodular) cast irons	P06001–P06999	Ruthenium
		P07001–P07999	Silver
K00001–K99999 Miscellaneous steels and ferrous alloys		R00001–R99999 Reactive and refractory metals and alloys	
		R01001–R01999	Boron
L00001–L99999 Low-melting metals and alloys		R02001–R02999	Hafnium
L00001–L00999	Bismuth	R03001–R03999	Molybdenum
L01001–L01999	Cadmium	R04001–R04999	Rhenium
L02001–L02999	Cesium	R05001–R05999	Tantalum
L03001–L03999	Gallium	R06001–R06999	Thorium
L04001–L04999	Indium	R07001–R07999	Tungsten
L05001–L05999	Lead	R08001–R08999	Vanadium
L06001–L06999	Lithium	R10001–R19999	Beryllium
L07001–L07999	Mercury	R20001–R29999	Chromium
L08001–L08999	Potassium	R30001–R39999	Cobalt
L09001–L09999	Rubidium	R40001–R49999	Columbium
L10001–L10999	Selenium	R50001–R59999	Titanium
L11001–L11999	Sodium	R60001–R69999	Zirconium
L12001–L12999	Thallium		
L13001–L13999	Tin	Z00001–Z99999 Zinc and zinc alloys	Zinc

[a] Alloys in which the rare earths are used in the ratio of their natural occurrence (that is, unseparated rare earths). In this mixture, cerium is the most abundant of the rare earth elements.

FIG. 1—APPLICATION FORM FOR UNS NUMBER ASSIGNMENT (FRONT)

APPLICATION FOR UNS NUMBER ASSIGNMENT
and
**Data Input Sheet for Entering a Specific Material in the
SAE-ASTM Unified Numbering System for Metals and Alloys**
(See Reverse Side for Instructions for Completing This Form)

Material Description _____

_____

_____

_____ Suggested UNS No. _____

*UNS Assigned Description _____

_____

_____ *UNS Assigned No. _____

*Chemical Composition   (percent by wt.)

Aluminum	Al	_____	Indium	In	_____	Selenium	Se	_____
Antimony	Sb	_____	Iridium	Ir	_____	Silicon	Si	_____
Arsenic	As	_____	Iron	Fe	_____	Silver	Ag	_____
Beryllium	Be	_____	Lead	Pb	_____	Sulfur	S	_____
Bismuth	Bi	_____	Lithium	Li	_____	Tantalum	Ta	_____
Boron	B	_____	Magnesium	Mg	_____	Tellurium	Te	_____
Cadmium	Cd	_____	Manganese	Mn	_____	Thorium	Th	_____
Carbon	C	_____	Mercury	Hg	_____	Tin	Sn	_____
Chromium	Cr	_____	Molybdenum	Mo	_____	Titanium	Ti	_____
Cobalt	Co	_____	Nickel	Ni	_____	Tungsten	W	_____
Columbium	Cb	_____	Nitrogen	N	_____	Uranium	U	_____
Copper	Cu	_____	Oxygen	O	_____	Vanadium	V	_____
Germanium	Ge	_____	Phosphorus	P	_____	Zinc	Zn	_____
Gold	Au	_____	Platinum	Pt	_____	Zirconium	Zr	_____
Hafnium	Hf	_____	Rhenium	Re	_____			_____
Hydrogen	H	_____	Rhodium	Rh	_____			_____

Other_____

_____

_____

*Cross References

AA _____

ACI _____

AISI _____

ANSI _____

AMS _____

ASME _____

ASTM _____

AWS _____

CDA _____

FED _____

MIL SPEC _____

SAE _____

OTHER _____

Requesting Person and Organization (full address) _____

_____ Date of Request _____

*Assigning Org. _____ *Date of UNS Assignment _____

*Assigner's Name and Office _____

Applicant: DO NOT write in shaded areas.   * These items for Computer Operator

FIG. 1—APPLICATION FORM FOR UNS NUMBER ASSIGNMENT (BACK)

GENERAL

Before attempting to complete this form, the applicant should be thoroughly familiar with the objectives of the UNS and the "ground rules" for assigning numbers, as stated in Section 4 of SAE J1086 and ASTM E 527.

MATERIAL DESCRIPTION

Identify the base element; the single alloying element that constitutes 50% or more of the total alloy content; other distinguishing predominant characteristics (such as "casting"); and common or generic names if any (such as "ounce metal" or "Waspalloy"). When no single element makes up 50% or more of the total alloy content, list in decreasing order of abundance, the two alloying elements which together constitute the largest portion of the total alloy contents; except that if no two elements make up at least 50% of the total alloy content, list the three most abundant, and so on. Instead of "iron", use "steel" to identify the base element of those iron-low-carbon alloys commonly known as steels.

When mechanical properties or physical characteristics are the primary defining criteria and chemical composition is secondary or nonsignificant, enter such properties and characteristics with the appropriate values or limits for each.

SUGGESTED UNS NO.

While applicant's suggestion may or may not be the one finally assigned, it will assist proper identification of the material by the UNS Number Assigner.

CHEMICAL COMPOSITION

φ Enter limits such as 0.13–0.18 (not .13–.18 or 0.13 to 0.18), 1.5 max, 0.040 min, and balance. In space designated "other," enter information such as "Each 0.05 max, Total 0.15 max" and "Sn plus Pb 2.0 min". Additional specific elements not included in the list on this form may be entered in the spaces provided at the end of the list.

CROSS REFERENCES

Letter symbols listed indicate widely known trade associations and standards issuing organizations. Enter after appropriate symbols any known specification numbers or identification numbers issued by such groups to cover material equivalent to, similar to, or closely resembling the subject material.

Examples ; SAE J404 (50B44), AISI 415, ASTM A638 (660)

In space designated "other" enter any pertinent numbers issued by groups not listed above. In these instances, the full name and address of the issuing group shall be included.

SUBMIT COMPLETED FORM TO APPROPRIATE UNS NUMBER ASSIGNER, AS LISTED IN SAE J1086 AND ASTM E527

ϕ TABLE 3—NUMBER ASSIGNERS AND AREAS OF RESPONSIBILITY

1. The Aluminum Association 750 Third Avenue New York, New York 10017 Attention: Office for Unified Numbering System for Metals Telephone: (212) 972-1800, ext. 32	Aluminum and aluminum alloys UNS Number Series: A00001–A99999 Carbon and alloy steels UNS Number Series: G00001–G99999
2. American Iron and Steel Institute 1000 16th Street N.W. Washington, D.C. 20036 Attention: Office for Unified Numbering System for Metals Telephone: (202) 223-9040	H-steels UNS Number Series: H00001–H99999 Heat and corrosion resistant (stainless) steels UNS Number Series: S00001–S99999 Tool steels UNS Number Series: T00001–T99999 Miscellaneous steels and ferrous alloys UNS Number Series: K00001–K99999 Copper and copper alloys UNS Number Series: C00001–C99999
3. Copper Development Association 405 Lexington Avenue New York, New York 10017 Attention: Office for Unified Numbering System for Metals Telephone: (212) 953-7321	Rare earth and rare earth-like metals and alloys UNS Number Series: E00001–E99999 Low melting metals and alloys UNS Number Series: L00001–L99999 Miscellaneous nonferrous metals and alloys UNS Number Series: M00001–M99999
4. American Society for Testing and Materials 1916 Race Street Philadelphia, Pa. 19103 Attention: Office for Unified Numbering System for Metals Telephone: (215) 299-5521	Cast steels UNS Number Series: J00001–J99999 Zinc and zinc alloys UNS Number Series: Z00001–Z99999 Precious metals and alloys UNS Number Series: P00001–P99999 Cast irons and cast steels UNS Number Series: F00001–F99999
5. Society of Automotive Engineers 400 Commonwealth Drive Warrendale, Pa. 15096 Attention: Office for Unified Numbering System for Metals Telephone: (412) 776-4841	Nickel and nickel alloys UNS Number Series: N00001–N99999 Steels specified by mechanical properties UNS Number Series: D00001–D99999 Reactive and refractory metals and alloys UNS Number Series: R00001–R99999

least two years;" the Copper Development Association requires that the material be "in commercial use (without tonnage limits);" the Aluminum Association requires that the alloy must be "offered for sale (not necessarily in commercial use);" the SAE Aerospace Materials Division calls for "repetitive procurement by at least two users."

While it is apparent that no hard and fast usage definition can be set up for an all-encompassing system, the UNS numbers are intended to identify metals and alloys that are in more or less regular production and use.

A UNS number will not ordinarily be issued for a material which has just been conceived or which is still in only experimental trial.

Note 2. Organizations that issue specifications should report to appropriate UNS number assigning offices (paragraph 3.1.2), any specification changes which affect descriptions shown in published UNS listings.

Note 3. This arrangement of alpha-numeric six character numbers is a compromise of the thinking that identification numbers should indicate many characteristics of the material, and the thinking that numbers should be short and uncomplicated to be widely accepted and used.

Note 4. In assigning UNS numbers, and consequently in searching complete listings of numbers, the predominant element of the metal or alloy usually determines the prefix letter of the series to which it is assigned. In certain instances where no one element predominates, arbitrary decisions are made as to what prefix letter to use, depending upon the producing industry and other factors.

Note 5. The application form is designed to serve also as a data input sheet to facilitate processing each request through to final printout of the data on electronic data processing equipment and to minimize transcription errors at number-assigning offices and data processing centers.

Note 6. One such listing is ASTM Publication No. DS-56 and SAE Handbook Supplement HS 1086 (a joint ASTM-SAE publication).

SELECTION AND USE OF STEELS—SAE J401   SAE Information Report

Report of Iron and Steel Division approved June 1911 and last revised by Iron and Steel Technical Committee November 1959. Reaffirmed without change January 1966. Editorial change November 1977.

[SAE Specifications for automotive steels are developed cooperatively by steel producers and users.]

Introduction—Steel compositions included in SAE Handbook are considered adequate for practically all parts made of wrought ferrous materials that are necessary for the production of automotive apparatus, and include grades that have been found commercially available and technically adequate for the service required of such parts. (SAE steel compositions should not be confused with specifications, of which composition is only a part.)

Only general applications of the SAE steels are indicated; definite applications are not shown, because—as it will be readily appreciated—the selection of a steel for a given part must be based upon an intimate knowledge of a number of important factors such as the severity of the service to be imposed, the detailed design of the part, the method of fabrication, its machinability, and the cost and availability of the material. It is only after a careful consideration of these factors that steels for the great variety of automotive and other mechanical parts can be selected properly.

Heat Treating—It is important to bear in mind that the information pertaining to typical heat treatments is based on the tests and experience of steel manufacturers and users and is intended only as a guide in selecting steels and their heat treatment. Care has been taken to provide reliable data, but steel users should supplement these data with suitable inspection tests and practices in order to be assured that each lot of material purchased will respond in a normal manner to the heat treatment generally applied in their particular practices.

Variations in the effect of the usual forms of heat treatment may be due to personnel, variations in the steel composition or manufacture, and to changes in local conditions such as control and precision of heat treating equipment. In order to minimize the effects of such variations, steel users should keep the product of each mill heat of steel separate in the stock room and during processing so that necessary adjustments of treatment can be made.

Frequently the necessity for drastic quenching can be avoided by the proper selection of a steel from a hardenability, or depth of hardening, standpoint. Plain carbon steels, unless their section size is very small and uniform, normally will not develop their maximum hardness throughout the section even with the most drastic quench. If depth hardness is an important factor, the user should consider an alloy steel of the proper carbon and alloy level which will meet his hardenability requirement with a less drastic quench and less chance of distortion and/or cracking. When selecting a steel, the user should bear in mind that the surface hardness obtainable by quenching is mainly a function of the carbon content and that depth hardness, or hardenability, is dependent upon the presence of alloying elements in addition to the carbon content of the steel.

Filleting—Another detail of greatest importance to the designer and the producer of heat treated steel parts is correct filleting. Sharp corners and inadequate fillets are the primary cause of most fatigue failures. In parts subjected to dynamic stresses, sharp corners and inadequate fillets produce serious stress concentration, causing the actual service stresses to build up to a point where they may amount to several times the normal working stress calculated by the designer. The use of generous fillets is especially desirable with all high strength alloy steels.

While adequate fillets may reduce the weakening effect of abrupt changes in section, they will not appreciably reduce the serious distortion which inevitably develops when parts of widely varying section are quenched during heat treatment. Whether oil or water quenching is required, heavy sections cool less rapidly than light sections, and, therefore, variations in section size tend to produce distortions. Consequently it is desirable to minimize variations in section as much as practicable.

Elastic Deflection—The steel user should also remember that the elastic deflection under load of a given part is a function of the section of the part rather than of the composition, heat treatment, or hardness of the particular steel that may be used. In other words, the modulus of elasticity of all the commercial steels, both plain carbon and alloy types, is the same so far as practical designing is concerned. Consequently if a part deflects excessively within the elastic range, the remedy lies in the field of design, not in the field of metallurgy. Either a heavier section must be used, the points of support must be increased, or some similar change made, since under the same conditions of loading all steels deflect the same amount within the elastic limit.

High Tensile Low Alloy Steels—In addition to the standard list of SAE steels, there is a specific class of High Strength Low Alloy Steels. In this category, materials are normally furnished to mechanical property requirements rather than chemical ranges, and various steel makers produce these grades under specific trade names. These steels have high mechanical properties, higher abrasion resistance and, in some cases, better resistance to atmospheric corrosion than plain structural or copper bearing steels. These steels are not intended for quenching and tempering. They have better welding properties than quenched and tempered structurals of similar strength. Applications are usually for untreated high strength parts made from hot or cold rolled structural shapes, plates, bars, sheets, and strip.

SAE NUMBERING SYSTEM FOR WROUGHT OR ROLLED STEEL—SAE J402b

SAE Standard

Report of Iron and Steel Division approved January 1912 and last revised by Iron and Steel Technical Committee May 1969. Editorial change November 1977.

This SAE Standard is intended to supply a uniform means of designating wrought ferrous materials reported in SAE Standards and Recommended Practices.

[History of standardized automotive steels goes back to Bulletin No. 9 of the Mechanical Branch of the Association of Licensed Automobile Manufacturers, dated Dec. 23, 1905, which have ALAM Specifications Nos. 1 through 6 for steels. ALAM Bulletin No. 13 soon followed with specifications for steel and gray iron castings, notes on malleable iron castings, and a standard or arbitration bars and test specimens for the iron castings.

In 1910 SAE took over ALAM's standardization work, setting up the SAE Standards Committee and, under it, the SAE Iron and Steel Division (now the SAE Iron and Steel Technical Committee). SAE Standards and SAE Recommended Practices developed by this group have been of such engineering and commercial value that for many years they have been used in nearly all fields of mechanical manufacture in the United States and many other countries.]

Only compositions which conform to the SAE compositions given in the current SAE Handbook should bear the prefix "SAE."

A numeral index system is used to identify the compositions of the SAE steels, which system makes possible use of numerals on shop drawings and blueprints to describe partially the composition of the material.

The first digit indicates the type to which the steel belongs, that is, "1" indicates a carbon steel; "2" a nickel steel; and "3" a nickel-chromium steel. In the case of the simple alloy steels, the second digit generally indicates an alloy or alloy combination, and sometimes the approximate percentage of the predominant alloying element. Usually the last two or three digits indicate the approximate carbon content in "points" or hundredths of one percent. Thus, "SAE 5135" indicates a chromium steel of approximately 1% chromium (0.80 to 1.05%) and 0.35% carbon (0.33 to 0.38%).

In some instances, in order to avoid confusion, it has been found necessary to depart from this system of identifying the approximate alloy composition of a steel by varying the second and third digits of the number. Instances of such departure are the steel numbers selected for several of the corrosion and heat resisting alloys and the triple alloy steels.

The basic numerals of the various types of SAE steel are given in the table.

BASIC NUMBERING SYSTEM FOR SAE STEELS

Numerals and Digits	Type of Steel and Average Chemical Contents, %	Numerals and Digits	Type of Steel and Average Chemical Contents, %	Numerals and Digits	Type of Steel and Average Chemical Contents, %
	CARBON STEELS	43BVXX	Ni 1.82; Cr 0.50; Mo 0.12 and 0.25; V 0.03 minimum	72XX	W 1.75; Cr 0.75
10XX	Plain Carbon (Mn 1.00% max)	47XX	Ni 1.05; Cr 0.45; Mo 0.20 and 0.35		SILICON MANGANESE STEELS
11XX	Resulphurized	81XX	Ni 0.30; Cr 0.40; Mo 0.12	92XX	Si 1.40 and 2.00; Mn 0.65, 0.82 and 0.85 Cr 0.00 and 0.65
12XX	Resulphurized and Rephosphorized	86XX	Ni 0.55; Cr 0.50; Mo 0.20		
15XX	Plain Carbon (max Mn range—over 1.00–1.65%)	87XX	Ni 0.55; Cr 0.50; Mo 0.25		LOW ALLOY HIGH TENSILE STEELS
		88XX	Ni 0.55; Cr 0.50; Mo 0.35	9XX	Various
	MANGANESE STEELS	93XX	Ni 3.25; Cr 1.20; Mo 0.12		
13XX	Mn 1.75	94XX	Ni 0.45; Cr 0.40; Mo 0.12		STAINLESS STEELS
		97XX	Ni 0.55; Cr 0.20; Mo 0.20		(Chromium-Manganese-Nickel)
	NICKEL STEELS	98XX	Ni 1.00; Cr 0.80; Mo 0.25	302XX	Cr 17.00 and 18.00; Mn 6.50 and 8.75; Ni 4.50 and 5.00
23XX	Ni 3.50		NICKEL-MOLYBDENUM STEELS		(Chromium-Nickel)
25XX	Ni 5.00	46XX	Ni 0.85 and 1.82; Mo 0.20 and 0.25	303XX	Cr 8.50, 15.50, 17.00, 18.00, 19.00, 20.00, 20.50, 23.00, 25.00
	NICKEL-CHROMIUM STEELS	48XX	Ni 3.50; Mo 0.25		Ni 7.00, 9.00, 10.00, 10.50, 11.00, 11.50, 12.00, 13.00, 13.50, 20.50, 21.00, 35.00
31XX	Ni 1.25; Cr 0.65 and 0.80		CHROMIUM STEELS		(Chromium)
32XX	Ni 1.75; Cr 1.07	50XX	Cr 0.27, 0.40, 0.50 and 0.65		
33XX	Ni 3.50; Cr 1.50 and 1.57	51XX	Cr 0.80, 0.87, 0.92, 0.95, 1.00 and 1.05	514XX	Cr 11.12, 12.25, 12.50, 13.00, 16.00, 17.00, 20.50 and 25.00
34XX	Ni 3.00; Cr 0.77	501XX	Cr 0.50		
	MOLYBDENUM STEELS	511XX	Cr 1.02	515XX	Cr 5.00
40XX	Mo 0.20 and 0.25	521XX	Cr 1.45		
44XX	Mo 0.40 and 0.52		CHROMIUM VANADIUM STEELS		BORON INTENSIFIED STEELS
	CHROMIUM-MOLYBDENUM STEELS	61XX	Cr 0.60, 0.80 and 0.95; V 0.10 and 0.15 minimum	XXBXX	B denotes Boron Steel
41XX	Cr 0.50, 0.80 and 0.95; Mo 0.12, 0.20, 0.25 and 0.30		TUNGSTEN CHROMIUM STEELS		LEADED STEELS
	NICKEL-CHROMIUM-MOLYBDENUM STEELS	71XXX	W 13.50 and 16.50; Cr 3.50	XXLXX	L denotes Leaded Steel
43XX	Ni 1.82; Cr 0.50 and 0.80; Mo 0.25				

CHEMICAL COMPOSITIONS OF SAE CARBON STEELS—SAE J403h

SAE Standard

Report of Iron and Steel Division approved June 1911 and last revised by Iron and Steel Technical Committee November 1977.

(In 1941, the SAE Iron and Steel Division in collaboration with the American Iron and Steel Institute made a major change in the method of expressing composition ranges for the SAE steels. The plan, as now applied, is based in general on narrower ladle analysis ranges plus certain check analysis allowances on individual samples, in place of the fixed ranges and limits without tolerances formerly provided for carbon and other elements in SAE steels. To avoid the possibility of confusion and conflict between SAE and AISI steel designations, all proposed changes in compositions or additions or deletions of numbers will be coordinated between the two organizations.)

For years the variety of chemical compositions of steel has been a matter of concern in the steel industry. It was recognized that production of fewer grades of steel could result in improved deliveries and provide a better opportunity to achieve advances in technology, manufacturing practices and quality; and thus develop more fully the possibilities of application inherent in those grades.

Comprehensive and impartial studies were directed toward determining which of the many grades being specified were the ones in most common demand, and the feasibility of combining compositions having like requirements. From these studies, the most common grades of steel have been selected and kept in current revision. The ladle chemical composition limits or ranges of these grades are given in Tables 1A, 1B, 2, and 3. These ladle limits or ranges are subject to standard variations for check analysis as given in SAE J409.

It is recognized that chemical compositions other than those listed in Tables 1A, 1B, 2, and 3 will at times be needed for specialized applications or processing. When such a steel is required, the elements comprising the desired chemical composition are specified in one of three ways: (1) by a minimum limit; (2) by a maximum limit; or (3) by minimum and maximum limits, termed a range.

Standard ladle analysis limits and ranges for the various elements of carbon steels are given in Tables 4 and 5. In these tables, "range" is the arithmetical difference between the minimum and maximum limits (that is, 0.19–0.25 is a 0.06 range). These ladle limits and ranges are also subject to standard variations for check analysis as given in SAE J409.

It will be noted that certain grade numbers in Table 1B have slightly wider ranges for carbon and manganese than the same grade number in Table 1A.

These wider ranges are necessary when producing steel for structural shapes, plates, strip, sheets and welded tubing because of a combination of factors involving complex problems arising from the use of larger mill equipment due generally to the larger sizes of these products. Chemical composition demands for these products are also of a different pattern and offer less flexibility of steel application. These differences are also reflected in Tables 4 and 5.

TABLE 1A—CARBON STEEL COMPOSITIONS APPLICABLE ONLY TO SEMIFINISHED PRODUCTS FOR FORGING, TO HOT ROLLED AND COLD FINISHED BARS, TO WIRE RODS, AND TO SEAMLESS TUBING

SAE No.	Ladle Chemical Composition[a,c] Limits, %				Corresponding AISI No.	SAE No.	Ladle Chemical Composition[a,c] Limits, %				Corresponding AISI No.
	C	Mn	P, max	S, max			C	Mn	P, max	S, max	
1005	0.06 max	0.35 max	0.040	0.050	—	1042	0.40–0.47	0.60–0.90	0.040	0.050	1042
1006	0.08 max	0.25–0.40	0.040	0.050	1006	1043	0.40–0.47	0.70–1.00	0.040	0.050	1043
1008	0.10 max	0.30–0.50	0.040	0.050	1008	1044	0.43–0.50	0.30–0.60	0.040	0.050	—
1010	0.08–0.13	0.30–0.60	0.040	0.050	1010	1045	0.43–0.50	0.60–0.90	0.040	0.050	1045
1012	0.10–0.15	0.30–0.60	0.040	0.050	1012	1046	0.43–0.50	0.70–1.00	0.040	0.050	1046
1013	0.11–0.16	0.50–0.80	0.040	0.050	—	1047[b]	0.43–0.51	1.35–1.65	0.040	0.050	—
1015	0.13–0.18	0.30–0.60	0.040	0.050	1015	1048[b]	0.44–0.52	1.10–1.40	0.040	0.050	1048
1016	0.13–0.18	0.60–0.90	0.040	0.050	1016	1049	0.46–0.53	0.60–0.90	0.040	0.050	1049
1017	0.15–0.20	0.30–0.60	0.040	0.050	1017	1050	0.48–0.55	0.60–0.90	0.040	0.050	1050
1018	0.15–0.20	0.60–0.90	0.040	0.050	1018	1051[b]	0.45–0.56	0.85–1.15	0.040	0.050	—
1019	0.15–0.20	0.70–1.00	0.040	0.050	1019	1052[b]	0.47–0.55	1.20–1.50	0.040	0.050	1052
1020	0.18–0.23	0.30–0.60	0.040	0.050	1020	1053	0.48–0.55	0.70–1.00	0.040	0.050	—
1021	0.18–0.23	0.60–0.90	0.040	0.050	1021	1055	0.50–0.60	0.60–0.90	0.040	0.050	1055
1022	0.18–0.23	0.70–1.00	0.040	0.050	1022	1060	0.55–0.65	0.60–0.90	0.040	0.050	1060
1023	0.20–0.25	0.30–0.60	0.040	0.050	1023	1061[b]	0.55–0.65	0.77–1.05	0.040	0.050	—
1024[b]	0.19–0.25	1.35–1.65	0.040	0.050	1024	1064	0.60–0.70	0.50–0.80	0.040	0.050	1064
1025	0.22–0.28	0.30–0.60	0.040	0.050	1025	1065	0.60–0.70	0.60–0.90	0.040	0.050	1065
1026	0.22–0.28	0.60–0.90	0.040	0.050	1026	1066[b]	0.60–0.71	0.85–1.15	0.040	0.050	—
1027[b]	0.22–0.29	1.20–1.50	0.040	0.050	1027	1069	0.65–0.75	0.40–0.70	0.040	0.050	—
1029	0.25–0.31	0.60–0.90	0.040	0.050	—	1070	0.65–0.75	0.60–0.90	0.040	0.050	1070
1030	0.28–0.34	0.60–0.90	0.040	0.050	1030	1072[b]	0.65–0.76	1.00–1.30	0.040	0.050	—
1035	0.32–0.38	0.60–0.90	0.040	0.050	1035	1074	0.70–0.80	0.50–0.80	0.040	0.050	1074
1036[b]	0.30–0.37	1.20–1.50	0.040	0.050	1036	1075	0.70–0.80	0.40–0.70	0.040	0.050	—
1037	0.32–0.38	0.70–1.00	0.040	0.050	1037	1078	0.72–0.85	0.30–0.60	0.040	0.050	1078
1038	0.35–0.42	0.60–0.90	0.040	0.050	1038	1080	0.75–0.88	0.60–0.90	0.040	0.050	1080
1039	0.37–0.44	0.70–1.00	0.040	0.050	1039	1084	0.80–0.93	0.60–0.90	0.040	0.050	1084
1040	0.37–0.44	0.60–0.90	0.040	0.050	1040	1085	0.80–0.93	0.70–1.00	0.040	0.050	—
1041[b]	0.36–0.44	1.35–1.65	0.040	0.050	1041	1086	0.80–0.93	0.30–0.50	0.040	0.050	1086
						1090	0.85–0.98	0.60–0.90	0.040	0.050	1090
						1095	0.90–1.03	0.30–0.50	0.040	0.050	1095

Note: LEAD: When lead is required as an added element to a standard steel, a range of 0.15 to 0.35%, inclusive, is generally used. Such a steel is identified by inserting the letter "L" between the second and third numerals of the grade number, that is, 10L45.

[a] These steels may be produced by the basic open hearth, the basic oxygen, or the basic electric steelmaking process (see SAE J411). Where silicon is required, the following limits and ranges are commonly used: for steel designations up to but excluding SAE 1015, 0.10% max; for SAE 1015 to SAE 1025, 0.10% max or ranges of 0.10–0.20%, 0.15–0.30%, 0.20–0.40%, or 0.30–0.60%; for over SAE 1025, ranges of 0.10–0.20%, 0.15–0.30%, 0.20–0.40%, or 0.30–0.60%.

[b] These grades, with max Mn in excess of 1%, have been renumbered 1500 series. See Table 3. Present 10XX standard designation will be discontinued after 1970.

[c] Certain qualities and commodities are customarily produced to lower limits of phosphorus and sulfur. (See SAE J411, Table 1 for listing.)

TABLE 1B—CARBON STEEL COMPOSITIONS APPLICABLE ONLY TO STRUCTURAL SHAPES, PLATES, STRIP, SHEETS[a] AND WELDED TUBING

SAE No.	Ladle Chemical Composition[b,c,d] Limits, %				SAE No.	Ladle Chemical Composition[b,c,d] Limits, %			
	C	Mn	P, max	S, max		C	Mn	P, max	S, max
1006	0.08 max	0.25–0.45	0.040	0.050	1040	0.36–0.44	0.60–0.90	0.040	0.050
1008	0.10 max	0.25–0.50	0.040	0.050	1041	0.36–0.45	1.30–1.65	0.040	0.050
1009	0.15 max	0.60 max	0.040	0.050	1042	0.39–0.47	0.60–0.90	0.040	0.050
1010	0.08–0.13	0.30–0.60	0.040	0.050	1043	0.39–0.47	0.70–1.00	0.040	0.050
1012	0.10–0.15	0.30–0.60	0.040	0.050	—	—	—	—	—
1015	0.12–0.18	0.30–0.60	0.040	0.050	1045	0.42–0.50	0.60–0.90	0.040	0.050
1016	0.12–0.18	0.60–0.90	0.040	0.050	1046	0.42–0.50	0.70–1.00	0.040	0.050
1017	0.14–0.20	0.30–0.60	0.040	0.050	—	—	—	—	—
1018	0.14–0.20	0.60–0.90	0.040	0.050	1048	0.43–0.52	1.05–1.40	0.040	0.050
1019	0.14–0.20	0.70–1.00	0.040	0.050	1049	0.45–0.53	0.60–0.90	0.040	0.050
1020	0.17–0.23	0.30–0.60	0.040	0.050	1050	0.47–0.55	0.60–0.90	0.040	0.050
1021	0.17–0.23	0.60–0.90	0.040	0.050	1052	0.46–0.55	1.20–1.55	0.040	0.050
1022	0.17–0.23	0.70–1.00	0.040	0.050	1055	0.52–0.60	0.60–0.90	0.040	0.050
1023	0.19–0.25	0.30–0.60	0.040	0.050	1060	0.55–0.66	0.60–0.90	0.040	0.050
1024	0.18–0.25	1.30–1.65	0.040	0.050	1064	0.59–0.70	0.50–0.80	0.040	0.050
1025	0.22–0.28	0.30–0.60	0.040	0.050	1065	0.59–0.70	0.60–0.90	0.040	0.050
1026	0.22–0.28	0.60–0.90	0.040	0.050	1070	0.65–0.76	0.60–0.90	0.040	0.050
1027	0.22–0.29	1.20–1.55	0.040	0.050	1074	0.69–0.80	0.50–0.80	0.040	0.050
1030	0.27–0.34	0.60–0.90	0.040	0.050	1078	0.72–0.86	0.30–0.60	0.040	0.050
1033	0.29–0.36	0.70–1.00	0.040	0.050	1080	0.74–0.88	0.60–0.90	0.040	0.050
1035	0.31–0.38	0.60–0.90	0.040	0.050	1084	0.80–0.94	0.60–0.90	0.040	0.050
1036	0.30–0.38	1.20–1.55	0.040	0.050	1085	0.80–0.94	0.70–1.00	0.040	0.050
1037	0.31–0.38	0.70–1.00	0.040	0.050	1090	0.84–0.98	0.60–0.90	0.040	0.050
1038	0.34–0.42	0.60–0.90	0.040	0.050	1095	0.90–1.04	0.30–0.50	0.040	0.050
1039	0.36–0.44	0.70–1.00	0.040	0.050					

[a] Sheet steel surface requirements may be broadly identified as to the end use by the suffix letter "E" for exposed parts requiring a good painted surface and suffix "U" for unexposed parts for which surface finish is unimportant.

[b] These steels may be produced by the basic open hearth, the basic oxygen, or the basic electric steelmaking process (see SAE J411, Carbon and Alloy Steels). Where silicon is required, the following limits and ranges are commonly used: for steel designations up to but excluding SAE 1015, 0.10% max; for SAE 1015 to SAE 1025, 0.10% max or ranges of 0.10–0.20% or 0.15–0.30%; for over SAE 1025, ranges of 0.10–0.20% or 0.15–0.30%.

[c] Certain qualities and commodities are customarily produced to lower limits of phosphorus and sulfur. (See SAE J411, Table 1 for listing.)

[d] For sheet and strip products in grades 1006 and 1008, only the maximum of the manganese range applies.

TABLE 2—FREE CUTTING CARBON STEEL COMPOSITIONS APPLICABLE ONLY TO SEMIFINISHED PRODUCTS FOR FORGING, HOT ROLLED AND COLD FINISHED BARS, WIRE RODS AND SEAMLESS TUBING

SAE No.	Ladle Chemical Composition[a] Limits, %				Corresponding AISI No.	SAE No.	Ladle Chemical Composition[a] Limits, %				Corresponding AISI No.
	C	Mn	P	S			C	Mn	P	S	
1110	0.08–0.13	0.30–0.60	0.04 max	0.08–0.13	1110				Max		
1211	0.13 max	0.60–0.90	0.07–0.12	0.10–0.15	1211	1108	0.08–0.13	0.50–0.80	0.040	0.08–0.13	1108
1212	0.13 max	0.70–1.00	0.07–0.12	0.16–0.23	1212	1117	0.14–0.20	1.00–1.30	0.040	0.08–0.13	1117
						1118	0.14–0.20	1.30–1.60	0.040	0.08–0.13	1118
1213	0.13 max	0.70–1.00	0.07–0.12	0.24–0.33	1213	1137	0.32–0.39	1.35–1.65	0.040	0.08–0.13	1137
						1139	0.35–0.43	1.35–1.65	0.040	0.13–0.20	1139
1215	0.09 max	0.75–1.05	0.04–0.09	0.26–0.35	1215	1140	0.37–0.44	0.70–1.00	0.040	0.08–0.13	1140
						1141	0.37–0.45	1.35–1.65	0.040	0.08–0.13	1141
12L13[b]	0.13 max	0.70–1.00	0.07–0.12	0.24–0.33		1144	0.40–0.48	1.35–1.65	0.040	0.24–0.33	1144
12L14[b]	0.15 max	0.85–1.15	0.04–0.09	0.26–0.35	12L14	1146	0.42–0.49	0.70–1.00	0.040	0.08–0.13	1146
						1151	0.48–0.55	0.70–1.00	0.040	0.08–0.13	1151

Note: SAE Grades 1110, 1211, 1212, 1213, 12L14, 1108, and 1109 are customarily furnished without specified silicon content.
When silicon is required, the following limits and ranges are commonly used: for steel designations 1108, 1109, and 1110, 0.10% max; for 1116 and over, 0.10% max, or ranges 0.10–0.20%, 0.15–0.30%, 0.20–0.40%, or 0.30–0.60%.

[a] These steels may be produced by the basic open hearth, the basic oxygen, or the basic electric steelmaking process (see SAE J411).
[b] When lead is required as an added element to other grades of steel, a range of 0.15–0.35% inclusive, is generally used. Such a steel is identified by inserting the letter "L" between the second and third numerals of the grade number, that is, 11L17.

TABLE 3—HIGH MANGANESE CARBON STEEL COMPOSITIONS APPLICABLE ONLY TO SEMIFINISHED PRODUCTS FOR FORGING, TO HOT ROLLED AND COLD FINISHED BARS, TO WIRE RODS, AND TO SEAMLESS TUBING

SAE No.	Former SAE No.	Ladle Chemical Composition[a] Limits, %			
		C	Mn	P, Max	S, Max
1513	—	0.10–0.16	1.10–1.40	0.040	0.050
1522	—	0.18–0.24	1.10–1.40	0.040	0.050
1524	1024	0.19–0.25	1.35–1.65	0.040	0.050
1526	—	0.22–0.29	1.10–1.40	0.040	0.050
1527	1027	0.22–0.29	1.20–1.50	0.040	0.050
1536	1036	0.30–0.37	1.20–1.50	0.040	0.050
1541	1041	0.36–0.44	1.35–1.65	0.040	0.050
1548	1048	0.44–0.52	1.10–1.40	0.040	0.050
1551	1051	0.45–0.56	0.85–1.15	0.040	0.050
1552	1052	0.47–0.55	1.20–1.50	0.040	0.050
1561	1061	0.55–0.65	0.75–1.05	0.040	0.050
1566	1066	0.60–0.71	0.85–1.15	0.040	0.050

NOTE: When lead is required as an added element to a standard steel, a range of 0.15–0.35%, inclusive, is generally used. Such a steel is identified by inserting the letter "L" between the second and third numerals of the grade number, that is, 15L41.

[a] Where silicon is required, the following limits and ranges are commonly used: for steel designations up to and including SAE 1524, 0.10% max or ranges of 0.10–0.20%, 0.15–0.30%, 0.20–0.40%, or 0.30–0.60%; for SAE 1525 and over, ranges of 0.10–0.20%, 0.15–0.30%, 0.20–0.40%, or 0.30–0.60%.

TABLE 4—CARBON STEEL LADLE CHEMICAL LIMITS AND RANGES APPLICABLE ONLY TO SEMIFINISHED PRODUCTS FOR FORGING, HOT ROLLED AND COLD FINISHED BARS, WIRE RODS, AND SEAMLESS TUBING

Element	Standard Chemical Ranges and Limits, %		
	Limit or Max of Specified Range %	Range	Lowest Max
Carbon[a]			0.06
	To 0.12 incl	—	
	Over 0.12 to 0.25 incl	0.05	
	Over 0.25 to 0.40 incl	0.06	
	Over 0.40 to 0.55 incl	0.07	
	Over 0.55 to 0.80 incl	0.10	
	Over 0.80	0.13	
Manganese			0.35
	To 0.40 incl	0.15	
	Over 0.40 to 0.50 incl	0.20	
	Over 0.50 to 1.65 incl	0.30	
Phosphorus	Basic Steels:		0.040
	To 0.040 incl		
	Over 0.040 to 0.08 incl	0.03	
	Over 0.08 to 0.13 incl	0.05	
	Acid Bessemer Steel:		0.12
	To 0.12 incl		
	Over 0.08 to 0.13	0.05	
Sulfur	Basic Steels:		0.05
	To 0.050 incl		
	Acid Bessemer Steel:		0.060
	To 0.060 incl		
	Basic Steels and Acid Bessemer Steel		
	Over 0.050 to 0.09 incl	0.03	
	Over 0.09 to 0.15 incl	0.05	
	Over 0.15 to 0.23 incl	0.07	
	Over 0.23 to 0.35 incl	0.09	
Silicon[b]			0.10
	To 0.10 incl	—	
	Over 0.10 to 0.15 incl	0.08	
	Over 0.15 to 0.20 incl	0.10	
	Over 0.20 to 0.30 incl	0.15	
	Over 0.30 to 0.60 incl	0.20	
Copper	When copper is required, 0.20 minimum is generally used.		
Lead[c]	When lead is required, a range of 0.15 to 0.35 is generally used.		

[a] Carbon: The carbon ranges shown in the column headed range apply when the specified maximum limit for manganese does not exceed 1.10%. When the maximum manganese limit exceeds 1.10%, add 0.01 to the carbon ranges shown in the table.
[b] Silicon: Because of the technological nature of the process, acid bessemer steels are not produced with specified silicon content. It is not common practice to produce a rephosphorized and resulfurized carbon steel to specified limits for silicon because of its adverse effect on machinability.
[c] Lead: A ladle analysis for lead is not determinable, since lead is added to the ladle stream while each ingot is poured. Check analysis tolerances for lead are shown in SAE J409.

TABLE 5—CARBON STEEL LADLE CHEMICAL LIMITS AND RANGES APPLICABLE ONLY TO STRUCTURAL SHAPES, PLATES, STRIP, SHEETS AND WELDED TUBING

Element	Standard Chemical Ranges and Limit, %		
	Limit or Max of Specified Range	Range	Lowest Max
Carbon[a]			0.08[b]
	To 0.15 incl	0.05	
	Over 0.15 to 0.30 incl	0.06	
	Over 0.30 to 0.40 incl	0.07	
	Over 0.40 to 0.60 incl	0.08	
	Over 0.60 to 0.80 incl	0.11	
	Over 0.80 to 1.35 incl	0.14	
Manganese			0.40
	To 0.50 incl	0.20	
	Over 0.50 to 1.15 incl	0.30	
	Over 1.15 to 1.65 incl	0.35	
Phosphorus			0.04
	To 0.08 incl	0.03	
	Over 0.08 to 0.15 incl	0.05	
Sulfur			0.05
	To 0.08 incl	0.03	
	Over 0.08 to 0.15 incl	0.05	
	Over 0.15 to 0.23 incl	0.07	
	Over 0.23 to 0.33 incl	0.10	
Silicon			0.10
	To 0.15 incl	0.08	
	Over 0.15 to 0.30 incl	0.15	
	Over 0.30 to 0.60 incl	0.30	
Copper	When copper is required 0.20 minimum is commonly specified.		

[a] The carbon ranges shown in the column headed Range apply when the specified maximum limit for manganese does not exceed 1.00%. When the maximum manganese limit exceeds 1.00%, add 0.01 to the carbon ranges shown in the table.
[b] 0.12 carbon maximum for structural shapes and plates.

CHEMICAL COMPOSITIONS OF SAE ALLOY STEELS—SAE J404j

SAE Standard

Report of Iron and Steel Division approved June 1911 and last revised by Iron and Steel Technical Committee November 1977.

In 1941, the SAE Iron and Steel Division in collaboration with the American Iron and Steel Institute made a major change in the method of expressing composition ranges for the SAE steels. The plan, as now applied, is based in general on narrower ladle analysis ranges plus certain product (check) analysis allowances on individual samples, in place of the fixed ranges and limits without tolerances formerly provided for carbon and other elements in SAE steels (Ref. J408). To avoid the possibility of confusion and conflict between SAE and AISI steel designations, all proposed changes in compositions or additions or deletions of numbers will be coordinated between the two organizations.

The compositions in this SAE Standard may apply to open hearth and basic oxygen, or electric furnace steels. Where they apply to electric furnace steels, the maximum phosphorus and maximum sulfur shall each be 0.025%. The nominal chemical limits or ranges in the compositions given in Table 1A are subject to standard variations in check analysis given in SAE J409.

Table 1A is applicable to billets, blooms, slabs, and hot rolled and cold finished bars.

Cross Index to Equivalent Grades and Government Specifications—Attention is called to the SAE Aeronautical Material Specifications (AMS) Index which is published twice a year. This index gives a cross reference to AMS grades, SAE grades, AISI grades, and Government Specifications (MIL, QQS, and so on) for metals, alloys, and nonmetallic materials.

TABLE 1A—ALLOY STEEL COMPOSITIONS[a]

UNS No.	SAE No.	Ladle Chemical Composition Limits, %									Corresponding AISI No.
		C	Mn	P	S	Si	Ni	Cr	Mo	V	
G13300	1330	0.28–0.33	1.60–1.90	0.035	0.040	0.15–0.30	—	—	—	—	1330
G13350	1335	0.33–0.38	1.60–1.90	0.035	0.040	0.15–0.30	—	—	—	—	1335
G13400	1340	0.38–0.43	1.60–1.90	0.035	0.040	0.15–0.30	—	—	—	—	1340
G13450	1345	0.43–0.48	1.60–1.90	0.035	0.040	0.15–0.30	—	—	—	—	1345
G40230	4023	0.20–0.25	0.70–0.90	0.035	0.040	0.15–0.30	—	—	0.20–0.30	—	4023
G40240	4024	0.20–0.25	0.70–0.90	0.035	0.035–0.050	0.15–0.30	—	—	0.20–0.30	—	4024
G40270	4027	0.25–0.30	0.70–0.90	0.035	0.040	0.15–0.30	—	—	0.20–0.30	—	4027
G40280	4028	0.25–0.30	0.70–0.90	0.035	0.035–0.050	0.15–0.30	—	—	0.20–0.30	—	4028
G40320	4032	0.30–0.35	0.70–0.90	0.035	0.040	0.15–0.30	—	—	0.20–0.30	—	—
G40370	4037	0.35–0.40	0.70–0.90	0.035	0.040	0.15–0.30	—	—	0.20–0.30	—	4037
G40420	4042	0.40–0.45	0.70–0.90	0.035	0.040	0.15–0.30	—	—	0.20–0.30	—	—
G40470	4047	0.45–0.50	0.70–0.90	0.035	0.040	0.15–0.30	—	—	0.20–0.30	—	4047
G41180	4118	0.18–0.23	0.70–0.90	0.035	0.040	0.15–0.30	—	0.40–0.60	0.08–0.15	—	4118
G41300	4130	0.28–0.33	0.40–0.60	0.035	0.040	0.15–0.30	—	0.80–1.10	0.15–0.25	—	4130
G41350	4135	0.33–0.38	0.70–0.90	0.035	0.040	0.15–0.30	—	0.80–1.10	0.15–0.25	—	—
G41370	4137	0.35–0.40	0.70–0.90	0.035	0.040	0.15–0.30	—	0.80–1.10	0.15–0.25	—	4137
G41400	4140	0.38–0.43	0.75–1.00	0.035	0.040	0.15–0.30	—	0.80–1.10	0.15–0.25	—	4140
G41420	4142	0.40–0.45	0.75–1.00	0.035	0.040	0.15–0.30	—	0.80–1.10	0.15–0.25	—	4142
G41450	4145	0.43–0.48	0.75–1.00	0.035	0.040	0.15–0.30	—	0.80–1.10	0.15–0.25	—	4145
G41470	4147	0.45–0.50	0.75–1.00	0.035	0.040	0.15–0.30	—	0.80–1.10	0.15–0.25	—	4147
G41500	4150	0.48–0.53	0.75–1.00	0.035	0.040	0.15–0.30	—	0.80–1.10	0.15–0.25	—	4150
G41610	4161	0.56–0.64	0.75–1.00	0.035	0.040	0.15–0.30	—	0.70–0.90	0.25–0.35	—	4161
G43200	4320	0.17–0.22	0.45–0.65	0.035	0.040	0.15–0.30	1.65–2.00	0.40–0.60	0.20–0.30	—	4320
G43400	4340	0.38–0.43	0.60–0.80	0.035	0.040	0.15–0.30	1.65–2.00	0.70–0.90	0.20–0.30	—	4340
G43406	E4340[b]	0.38–0.43	0.65–0.85	0.025	0.025	0.15–0.30	1.65–2.00	0.70–0.90	0.20–0.30	—	E4340
G44220	4422	0.20–0.25	0.70–0.90	0.035	0.040	0.15–0.30	—	—	0.35–0.45	—	—
G44270	4427	0.24–0.29	0.70–0.90	0.035	0.040	0.15–0.30	—	—	0.35–0.45	—	—
G46150	4615	0.13–0.18	0.45–0.65	0.035	0.040	0.15–0.30	1.65–2.00	—	0.20–0.30	—	4615
G46170	4617	0.15–0.20	0.45–0.65	0.035	0.040	0.15–0.30	1.65–2.00	—	0.20–0.30	—	—
G46200	4620	0.17–0.22	0.45–0.65	0.035	0.040	0.15–0.30	1.65–2.00	—	0.20–0.30	—	4620
G46260	4626	0.24–0.29	0.45–0.65	0.035	0.04 max	0.15–0.30	0.70–1.00	—	0.15–0.25	—	4626
G47180	4718	0.16–0.21	0.70–0.90	—	—	—	0.90–1.20	0.35–0.55	0.30–0.40	—	4718
G47200	4720	0.17–0.22	0.50–0.70	0.035	0.040	0.15–0.30	0.90–1.20	0.35–0.55	0.15–0.25	—	4720
G48150	4815	0.13–0.18	0.40–0.60	0.035	0.040	0.15–0.30	3.25–3.75	—	0.20–0.30	—	4815
G48170	4817	0.15–0.20	0.40–0.60	0.035	0.040	0.15–0.30	3.25–3.75	—	0.20–0.30	—	4817
G48200	4820	0.18–0.23	0.50–0.70	0.035	0.040	0.15–0.30	3.25–3.75	—	0.20–0.30	—	4820
G50401	50B40	0.38–0.43	0.75–1.00	0.035	0.040	0.15–0.30	—	0.40–0.60	—	—	—
G50441	50B44	0.43–0.48	0.75–1.00	0.035	0.040	0.15–0.30	—	0.40–0.60	—	—	50B44
G50460	5046	0.43–0.48	0.75–1.00	0.035	0.040	0.15–0.30	—	0.20–0.35	—	—	—
G50461	50B46[c]	0.44–0.49	0.75–1.00	0.035	0.040	0.15–0.30	—	0.20–0.35	—	—	50B46
G50501	50B50[c]	0.48–0.53	0.75–1.00	0.035	0.040	0.15–0.30	—	0.40–0.60	—	—	50B50
G50600	5060	0.56–0.64	0.75–1.00	0.035	0.040	0.15–0.30	—	0.40–0.60	—	—	—
G50601	50B60[c]	0.56–0.64	0.75–1.00	0.035	0.040	0.15–0.30	—	0.40–0.60	—	—	50B60
G51150	5115	0.13–0.18	0.70–0.90	0.035	0.040	0.15–0.30	—	0.70–0.90	—	—	—
G51170	5117	0.15–0.20	0.70–0.90	0.035	0.040	0.20–0.35	—	0.70–0.90	—	—	5117
G51200	5120	0.17–0.22	0.70–0.90	0.035	0.040	0.15–0.30	—	0.70–0.90	—	—	5120
G51300	5130	0.28–0.33	0.70–0.90	0.035	0.040	0.15–0.30	—	0.80–1.10	—	—	5130
G51320	5132	0.30–0.35	0.60–0.80	0.035	0.040	0.15–0.30	—	0.75–1.00	—	—	5132
G51350	5135	0.33–0.38	0.60–0.80	0.035	0.040	0.15–0.30	—	0.80–1.05	—	—	5135
G51400	5140	0.38–0.43	0.70–0.90	0.035	0.040	0.15–0.30	—	0.70–0.90	—	—	5140
G51470	5147	0.46–0.51	0.70–0.95	0.035	0.040	0.15–0.30	—	0.85–1.15	—	—	5147
G51500	5150	0.48–0.53	0.70–0.90	0.035	0.040	0.15–0.30	—	0.70–0.90	—	—	5150
G51550	5155	0.51–0.59	0.70–0.90	0.035	0.040	0.15–0.30	—	0.70–0.90	—	—	5155
G51600	5160	0.56–0.64	0.75–1.00	0.035	0.040	0.15–0.30	—	0.70–0.90	—	—	5160
G51601	51B60[c]	0.56–0.64	0.75–1.00	0.035	0.040	0.15–0.30	—	0.70–0.90	—	—	51B60
G50986	50100[b]	0.98–1.10	0.25–0.45	0.025	0.025	0.15–0.30	—	0.40–0.60	—	—	—
G51986	51100[b]	0.98–1.10	0.25–0.45	0.025	0.025	0.15–0.30	—	0.90–1.15	—	—	E51100
G52986	52100[b]	0.98–1.10	0.25–0.45	0.025	0.025	0.15–0.30	—	1.30–1.60	—	—	E52100
G61180	6118	0.16–0.21	0.50–0.70	0.035	0.040	0.15–0.30	—	0.50–0.70	—	0.10–0.15	6118
G61500	6150	0.48–0.53	0.70–0.90	0.035	0.040	0.15–0.30	—	0.80–1.10	—	0.15 min	6150
G81150	8115	0.13–0.18	0.70–0.90	0.035	0.040	0.15–0.30	0.20–0.40	0.30–0.50	0.08–0.15	—	8115
G81451	81B45[c]	0.43–0.48	0.75–1.00	0.035	0.040	0.15–0.30	0.20–0.40	0.35–0.55	0.08–0.15	—	81B45

(Table continued on next page)

TABLE 1A—ALLOY STEEL COMPOSITIONS[a] (continued)

UNS No.	SAE No.	C	Mn	P	S	Si	Ni	Cr	Mo	V	Corresponding AISI No.
G86150	8615	0.13–0.18	0.70–0.90	0.035	0.040	0.15–0.30	0.40–0.70	0.40–0.60	0.15–0.25	—	8615
G86170	8617	0.15–0.20	0.70–0.90	0.035	0.040	0.15–0.30	0.40–0.70	0.40–0.60	0.15–0.25	—	8617
G86200	8620	0.18–0.23	0.70–0.90	0.035	0.040	0.15–0.30	0.40–0.70	0.40–0.60	0.15–0.25	—	8620
G86220	8622	0.20–0.25	0.70–0.90	0.035	0.040	0.15–0.30	0.40–0.70	0.40–0.60	0.15–0.25	—	8622
G86250	8625	0.23–0.28	0.70–0.90	0.035	0.040	0.15–0.30	0.40–0.70	0.40–0.60	0.15–0.25	—	8625
G86270	8627	0.25–0.30	0.70–0.90	0.035	0.040	0.15–0.30	0.40–0.70	0.40–0.60	0.15–0.25	—	8627
G86300	8630	0.28–0.33	0.70–0.90	0.035	0.040	0.15–0.30	0.40–0.70	0.40–0.60	0.15–0.25	—	8630
G86370	8637	0.35–0.40	0.75–1.00	0.035	0.040	0.15–0.30	0.40–0.70	0.40–0.60	0.15–0.25	—	8637
G86400	8640	0.38–0.43	0.75–1.00	0.035	0.040	0.15–0.30	0.40–0.70	0.40–0.60	0.15–0.25	—	8640
G86420	8642	0.40–0.45	0.75–1.00	0.035	0.040	0.15–0.30	0.40–0.70	0.40–0.60	0.15–0.25	—	8642
G86450	8645	0.43–0.48	0.75–1.00	0.035	0.040	0.15–0.30	0.40–0.70	0.40–0.60	0.15–0.25	—	8645
G86451	86B45[e]	0.43–0.48	0.75–1.00	0.035	0.040	0.15–0.30	0.40–0.70	0.40–0.60	0.15–0.25	—	—
G86500	8650	0.48–0.53	0.75–1.00	0.035	0.040	0.15–0.30	0.40–0.70	0.40–0.60	0.15–0.25	—	—
G86550	8655	0.51–0.59	0.75–1.00	0.035	0.040	0.15–0.30	0.40–0.70	0.40–0.60	0.15–0.25	—	8655
G86600	8660	0.56–0.64	0.75–1.00	0.035	0.040	0.15–0.30	0.40–0.70	0.40–0.60	0.15–0.25	—	—
G87200	8720	0.18–0.23	0.70–0.90	0.035	0.040	0.15–0.30	0.40–0.70	0.40–0.60	0.20–0.30	—	8720
G87400	8740	0.38–0.43	0.75–1.00	0.035	0.040	0.15–0.30	0.40–0.70	0.40–0.60	0.20–0.30	—	8740
G88220	8822	0.20–0.25	0.75–1.00	0.035	0.040	0.15–0.30	0.40–0.70	0.40–0.60	0.30–0.40	—	8822
G92540	9254	0.51–0.59	0.60–0.80	0.035	0.040	1.20–1.60	—	0.60–0.80	—	—	—
G92600	9260	0.56–0.64	0.75–1.00	0.035	0.040	1.80–2.20	—	—	—	—	9260
G93106	9310[b]	0.08–0.13	0.45–0.65	0.025	0.025	0.15–0.30	3.00–3.50	1.00–1.40	0.08–0.15	—	—
G94151	94B15[c]	0.13–0.18	0.75–1.00	0.035	0.040	0.15–0.30	0.30–0.60	0.30–0.50	0.08–0.15	—	—
G94171	94B17[c]	0.15–0.20	0.75–1.00	0.035	0.040	0.15–0.30	0.30–0.60	0.30–0.50	0.08–0.15	—	94B17
G94301	94B30	0.28–0.33	0.75–1.00	0.035	0.040	0.15–0.30	0.30–0.60	0.30–0.50	0.08–0.15	—	94B30

[a] For standard variations in composition limits, see Table 4 of SAE J409. Small quantities of certain elements which are not specified or required may be found in alloy steels. These elements are to be considered as incidental and are acceptable to the following maximum amount: copper to 0.35%, nickel to 0.25%, chromium to 0.20%, and molybdenum to 0.06%.
[b] Electric furnace steel.
[c] Boron content is 0.0005% min.

TABLE 1B—ALLOY STEEL PLATE COMPOSITIONS[a,b,c]
(OPEN HEARTH AND BASIC OXYGEN)

UNS No.	SAE No.	C	Mn	P max	S max	Si[d]	Ni	Cr	Mo	V
G13300	1330	0.27–0.34	1.50–1.90	0.035	0.040	0.15–0.30	—	—	—	—
G13350	1335	0.32–0.39	1.50–1.90	0.035	0.040	0.15–0.30	—	—	—	—
G13400	1340	0.36–0.44	1.50–1.90	0.035	0.040	0.15–0.30	—	—	—	—
G13450	1345	0.41–0.49	1.50–1.90	0.035	0.040	0.15–0.30	—	—	—	—
G41180	4118	0.17–0.23	0.60–0.90	0.035	0.040	0.15–0.30	—	0.40–0.65	0.08–0.15	—
G41300	4130	0.27–0.34	0.35–0.60	0.035	0.040	0.15–0.30	—	0.80–1.15	0.15–0.25	—
G41350	4135	0.32–0.39	0.65–0.95	0.035	0.040	0.15–0.30	—	0.80–1.15	0.15–0.25	—
G41370	4137	0.33–0.40	0.65–0.95	0.035	0.040	0.15–0.30	—	0.80–1.15	0.15–0.25	—
G41400	4140	0.36–0.44	0.70–1.00	0.035	0.040	0.15–0.30	—	0.80–1.15	0.15–0.25	—
G41420	4142	0.38–0.46	0.70–1.00	0.035	0.040	0.15–0.30	—	0.80–1.15	0.15–0.25	—
G41450	4145	0.41–0.49	0.70–1.00	0.035	0.040	0.15–0.30	—	0.80–1.15	0.15–0.25	—
G43400	4340	0.36–0.44	0.55–0.80	0.035	0.040	0.15–0.30	1.65–2.00	0.60–0.90	0.20–0.30	—
G43406	E4340	0.37–0.44	0.60–0.85	0.025	0.025	0.15–0.30	1.65–2.00	0.65–0.90	0.20–0.30	—
G46150	4615	0.12–0.18	0.40–0.65	0.035	0.040	0.15–0.30	1.65–2.00	—	0.20–0.30	—
G46170	4617	0.15–0.21	0.40–0.65	0.035	0.040	0.15–0.30	1.65–2.00	—	0.20–0.30	—
G46200	4620	0.16–0.22	0.40–0.65	0.035	0.040	0.15–0.30	1.65–2.00	—	0.20–0.30	—
G51600	5160	0.54–0.65	0.70–1.00	0.035	0.040	0.15–0.30	—	0.60–0.90	—	—
G61500	6150	0.46–0.54	0.60–0.90	0.035	0.040	0.15–0.30	—	0.80–1.15	—	0.15 min
G86150	8615	0.12–0.18	0.60–0.90	0.035	0.040	0.15–0.30	0.40–0.70	0.35–0.60	0.15–0.25	—
G86170	8617	0.15–0.21	0.60–0.90	0.035	0.040	0.15–0.30	0.40–0.70	0.35–0.60	0.15–0.25	—
G86200	8620	0.17–0.23	0.60–0.90	0.035	0.040	0.15–0.30	0.40–0.70	0.35–0.60	0.15–0.25	—
G86220	8622	0.19–0.25	0.60–0.90	0.035	0.040	0.15–0.30	0.40–0.70	0.35–0.60	0.15–0.25	—
G86250	8625	0.22–0.29	0.60–0.90	0.035	0.040	0.15–0.30	0.40–0.70	0.35–0.60	0.15–0.25	—
G86270	8627	0.24–0.31	0.60–0.90	0.035	0.040	0.15–0.30	0.40–0.70	0.35–0.60	0.15–0.25	—
G86300	8630	0.27–0.34	0.60–0.90	0.035	0.040	0.15–0.30	0.40–0.70	0.35–0.60	0.15–0.25	—
G86370	8637	0.33–0.40	0.70–1.00	0.035	0.040	0.15–0.30	0.40–0.70	0.35–0.60	0.15–0.25	—
G86400	8640	0.36–0.44	0.70–1.00	0.035	0.040	0.15–0.30	0.40–0.70	0.35–0.60	0.15–0.25	—
G86550	8655	0.49–0.60	0.70–1.00	0.035	0.040	0.15–0.30	0.40–0.70	0.35–0.60	0.15–0.25	—
G87420	8742	0.38–0.46	0.70–1.00	0.035	0.040	0.15–0.30	0.40–0.70	0.35–0.60	0.20–0.30	—

[a] Small quantities of certain elements not required may be found. These elements are to be considered as incidental and are acceptable to the following maximum amounts: copper to 0.35%, nickel to 0.25%, chromium to 0.20%, and molybdenum to 0.06%.
[b] When electric furnace steel is ordered, the carbon range is restricted 0.01%, manganese 0.05%, chromium 0.05% up to 1.25% incl. and 0.10% over 1.25%. The maximum phosphorus and sulfur is 0.025% each.
[c] Boron or lead may be added to these compositions.
[d] Silicon available in ranges of 0.10–0.20%, 0.20–0.30%, and 0.35% maximum (when carbon deoxidized) when so specified by the purchaser.

CHEMICAL COMPOSITIONS OF SAE WROUGHT STAINLESS STEELS—SAE J405d

SAE Standard

Report of Iron and Steel Division approved June 1911 and last revised by Iron and Steel Technical Committee August 1970. Editorial change May 1971.

The chemical composition of standard types of wrought stainless steels are listed in Tables 1, 2, and 3. The 302XX series designates nickel-chromium manganese, corrosion resistant types, nonhardenable by thermal treatment (Table 1). The 303XX series are nickel-chromium, corrosion resistant steels, nonhardenable by thermal treatment (Table 1). The 514XX series, however, include both a hardenable, martensitic chromium steel (Table 2), and non-hardenable, ferritic, chromium steel (Table 3). Table 4 lists proprietary and modifications of standard types to which elements have been added to provide special machinability characteristics. In order to avoid confusion with the AISI designations, the use of a suffix to denote a free machining steel or a variation in the carbon or silicon range has been retained. Therefore, the footnotes should be carefully followed. Reference to SAE J412 is suggested for general information and usage of these types of materials.

TABLE 1—WROUGHT CHROMIUM-NICKEL AUSTENITIC STEELS (NOT HARDENABLE BY THERMAL TREATMENT)

SAE No.[a]	C Max	Mn Max	Si Max	P Max	S Max	Cr Range	Ni Range	Other Elements	AISI Type
30201	0.15	5.5 – 7.5	1.00	0.060	0.030	16.00–18.00	3.50– 5.50	N, 0.25 max	201
30202	0.15	7.5 –10.0	1.00	0.060	0.030	17.00–19.00	4.00– 6.00	N, 0.25 max	202
30301	0.15	2.00	1.00	0.045	0.030	16.00–18.00	6.00– 8.00	—	301
30302	0.15	2.00	1.00	0.045	0.030	17.00–19.00	8.00–10.00	—	302
30302B	0.15	2.00	2.00–3.00	0.045	0.030	17.00–19.00	8.00–10.00	—	302B
303C3	0.15	2.00	1.00	0.20	0.15 min	17.00–19.00	8.00–10.00	Zr or Mo, 0.60 max[c]	303
30303 Se	0.15	2.00	1.00	0.20	0.06	17.00–19.00	8.00–10.00	Se, 0.15 min	303 Se
30304	0.08	2.00	1.00	0.045	0.030	18.00–20.00	8.00–10.50	—	304
30304L	0.03	2.00	1.00	0.045	0.030	18.00–20.00	8.00–12.00	—	304L
30305	0.12	2.00	1.00	0.045	0.030	17.00–19.00	10.50–13.00	—	305
30308	0.08	2.00	1.00	0.045	0.030	19.00–21.00	10.00 12.00	—	308
30309	0.20	2.00	1.00	0.045	0.030	22.00–24.00	12.00–15.00	—	309
30309S	0.08	2.00	1.00	0.045	0.030	22.00–24.00	12.00–15.00	—	309S
30310	0.25	2.00	1.50	0.045	0.030	24.00–26.00	19.00–22.00	—	310
30310S	0.08	2.00	1.50	0.045	0.030	24.00–26.00	19.00–22.00	—	310S
30314	0.25	2.00	1.50–3.00	0.045	0.030	23.00–26.00	19.00–22.00	—	314
30316	0.08	2.00	1.00	0.045	0.030	16.00–18.00	10.00–14.00	Mo, 2.00–3.00	316
30316L[d]	0.03	2.00	1.00	0.045	0.030	16.00–18.00	10.00–14.00	Mo, 2.00–3.00	316L
30317	0.08	2.00	1.00	0.045	0.030	18.00–20.00	11.00–15.00	Mo, 3.00–4.00	317
30321[e]	0.08	2.00	1.00	0.045	0.030	17.00–19.00	9.00–12.00	Ti, 5 × C min	321
30330	0.15	2.00	1.50[b]	0.045	0.04	14.00–17.00	33.00–37.00	—	—
30347	0.08	2.00	1.00	0.045	0.030	17.00–19.00	9.00–13.00	Cb-Ta, 10 × C min	347
30348	0.08	2.00	1.00	0.045	0.030	17.00–19.00	9.00–13.00	Cb-Ta, 10 × C min; Ta, 0.10 max	348
30384	0.08	2.00	1.00	0.045	0.030	15.00–17.00	17.00–19.00	—	384
30385	0.08	2.00	1.00	0.045	0.030	11.50–13.50	14.00–16.00	—	385

[a] The suffixes with grade numbers denote: B—2.00-3.00 silicon range; Se—a free machining steel with selenium addition; L—extra low carbon grade; S—lower carbon grade.
[b] To minimize carbon or nitrogen pick-up 0.75-1.50 Si is recommended for high temperature application involving carbon or nitrogen atmosphere.
[c] At producer's option; reported only when intentionally added.
[d] 10.0-15.0 Ni permitted for tubular products.
[e] 9.0-13.0 Ni permitted for tubular products.

TABLE 2—WROUGHT STAINLESS MARTENSITIC CHROMIUM STEELS (HARDENABLE BY THERMAL TREATMENT)

SAE No.[a]	C Max	Mn Max	Si Max	P Max	S Max	Cr Range	Ni Range	Other Elements	AISI Type[a]
51403	0.15	1.00	0.50	0.040	0.030	11.50–13.00	—	—	403
51410	0.15	1.00	1.00	0.040	0.030	11.50–13.50	—	—	410
51414	0.15	1.00	1.00	0.040	0.030	11.50–13.50	1.25–2.50	—	414
51416	0.15	1.25	1.00	0.06	0.15 min	12.00–14.00	—	Zr or Mo, 0.60 max[b]	416
51416 Se	0.15	1.25	1.00	0.06	0.06	12.00–14.00	—	Se, 0.15 min	416 Se
51420	Over 0.15	1.00	1.00	0.040	0.030	12.00–14.00	—	—	420
51420F	0.15 min	1.25	1.00	0.06	0.15 min	12.00–14.00	—	Mo, 0.60 max[b]	420F
51420F Se	0.30–0.40	1.25	1.00	0.06	0.06	12.00–14.00	—	Se, 0.15 min	—
51431	0.20	1.00	1.00	0.040	0.030	15.00–17.00	1.25–2.50	—	431
51440A	0.60–0.75	1.00	1.00	0.040	0.030	16.00–18.00	—	Mo, 0.75 max	440A
51440B	0.75–0.95	1.00	1.00	0.040	0.030	16.00–18.00	—	Mo, 0.75 max	440B
51440C	0.95–1.20	1.00	1.00	0.040	0.030	16.00–18.00	—	Mo, 0.75 max	440C
51440F	0.95–1.20	1.25	1.00	0.06	0.15 min	16.00–18.00	—	Zr or Mo, 0.75 max[b]	—
51440F Se	0.95–1.20	1.25	1.00	0.06	0.06	16.00–18.00	—	Se, 0.15 min	—
51501	Over 0.10	1.00	1.00	0.040	0.030	4.00–6.00	—	Mo, 0.40–0.65	501
51502	0.10	1.00	1.00	0.04	0.030	4.00–6.00	—	Mo, 0.40–0.65	502

[a] Suffixes A, B, and C denote differing carbon ranges for the same grade; F—a free machining steel; Se—a free machining steel with selenium addition.
[b] At producer's option; reported only when intentionally added.

2.11

TABLE 3—WROUGHT STAINLESS FERRITIC CHROMIUM STEELS (NOT HARDENABLE BY THERMAL TREATMENT)

SAE No.[a]	Chemical Composition Limits, %								AISI Type[a]
	C Max	Mn Max	Si Max	P Max	S Max	Cr Range	Ni Range	Other Elements	
51405[b]	0.08	1.00	1.00	0.040	0.030	11.50–14.50	—	Al, 0.10–0.30	405
51409	0.08	1.00	1.00	0.045	0.045	10.50–11.75	0.50 max	Ti, 6 × C or max of 0.75, Fe, rem	
51429	0.12	1.00	1.00	0.040	0.030	14.00–16.00	—	—	429
51430	0.12	1.00	1.00	0.040	0.030	16.00–18.00	—	—	430
51430F	0.12	1.25	1.00	0.060	0.15 min	16.00–18.00	—	Mo, 0.60 max[c]	430F
51430F Se	0.12	1.25	1.00	0.060	0.060	16.00–18.00	—	Se, 0.15 min	430F Se
51434	0.12	1.00	1.00	0.040	0.030	16.00–18.00	—	Mo, 0.75–1.25	434
51436	0.12	1.00	1.00	0.040	0.030	16.00–18.00	—	Mo, 0.75–1.25; Cb + Ta, 5 × C − 0.70	436
51442	0.20	1.00	1.00	0.04	0.035	18.00–23.00	—	—	—
51446	0.20	1.50	1.00	0.04	0.030	23.00–27.00	—	N, 0.25 max	446

[a] Suffix F—denotes a free machining steel; Se—denotes a free machining steel with selenium addition.
[b] Essentially nonhardenable by heat treatment.
[c] At producer's option; reported only when intentionally added.

TABLE 4—WROUGHT STAINLESS STEELS FOR SPECIAL MACHINABILITY

Proprietary Designation	Chemical Composition Limits, %							
	C Max	Mn Max	Si Max	P Max	S Max	Cr Range	Ni Range	Other Elements
203 EZ[a]	0.08	5.00–6.50	1.00	0.04	0.18–0.35	16.00–18.00	5.00–6.50	Mo, 0.50 max; Cu, 1.75/2.25
303 MA[a]	0.15	2.00	1.00	0.05	0.11–0.16	17.00–19.00	8.00–10.00	Mo, 0.40–0.60; Al, 0.60–1.00
303 Pb[a]	0.15	2.00	1.00	0.04	0.12–0.25	17.00–19.00	8.00–10.00	Mo, 0.60 max; Pb, 0.12–0.30
303 Cu[a]	0.15	2.00	1.00	0.15	0.10 min	17.00–19.00	6.00–10.00	Se, 0.10 max; Cu, 2.50–4.00
303 Plus X[a]	0.15	2.50–4.50	1.00	0.20	0.15 min	17.00–19.00	7.00–10.00	Mo, 0.60 max
416 Plus X[b]	0.15	1.50–2.50	1.00	0.06	0.15 min	12.00–14.00	—	Mo, 0.60 max

[a] Not hardenable by thermal treatment.
[b] Hardenable by thermal treatment.

CHEMICAL COMPOSITIONS OF SAE EXPERIMENTAL STEELS—SAE J1081 MAY80

SAE Information Report

Report of the Iron and Steel Technical Committee, approved April 1974, last revised February 1977, editorial change May 1980.

This SAE Information Report provides a uniform means of designating wrought steels during a period of usage prior to the time they meet the requirements for SAE standard steel designation. The numbers consist of the prefix *EX* followed by a sequential number starting with 1. A number once assigned is never assigned to any other composition.

An EX number may be obtained for an experimental steel composition by submitting a written request to SAE Staff, indicating the chemical composition and other pertinent characteristics of the material. If the request is approved according to established procedures, SAE Staff will assign an EX number to the grade. This number will remain in effect until the experimental grade meets the requirements for an SAE standard steel or the grade is discontinued according to established procedures.

Table 1 is a listing of the chemical composition limits of experimental steels which were considered active on the date of the last survey prior to the date of this report. These ladle limits are subject to standard variations for check analysis as given in SAE J409.

TABLE 1—SAE EXPERIMENTAL STEEL COMPOSITIONS

EX No.	Ladle Chemical Composition Limits, % by weight								
	C	Mn	P, max	S, max	Si	Ni	Cr	Mo	B
EX 10	0.19–0.24	0.95–1.25	0.035	0.040	0.20–0.35	0.20–0.40	0.25–0.40	0.05–0.10	—
EX 15	0.18–0.23	0.90–1.20	0.035	0.040	0.20–0.35	—	0.40–0.60	0.13–0.20	—
EX 16	0.20–0.25	0.90–1.20	0.035	0.040	0.20–0.35	—	0.40–0.60	0.13–0.20	—
EX 17	0.23–0.28	0.90–1.20	0.035	0.040	0.20–0.35	—	0.40–0.60	0.13–0.20	—
EX 18	0.25–0.30	0.90–1.20	0.035	0.030	0.20–0.35	—	0.40–0.60	0.13–0.20	—
EX 19	0.18–0.23	0.90–1.20	0.035	0.040	0.20–0.35	—	0.40–0.60	0.08–0.15	0.0005 min
EX 20	0.13–0.18	0.90–1.20	0.035	0.040	0.20–0.35	—	0.40–0.60	0.13–0.20	—
EX 21	0.15–0.20	0.90–1.20	0.035	0.040	0.20–0.35	—	0.40–0.60	0.13–0.20	—
EX 24	0.18–0.23	0.75–1.00	0.035	0.040	0.20–0.35	—	0.45–0.65	0.20–0.30	—
EX 30	0.13–0.18	0.70–0.90	0.035	0.040	0.20–0.35	0.70–1.00	0.45–0.65	0.45–0.60	—
EX 31	0.15–0.20	0.70–0.90	0.035	0.040	0.20–0.35	0.70–1.00	0.45–0.65	0.45–0.60	—
EX 32	0.18–0.23	0.70–0.90	0.035	0.040	0.20–0.35	0.70–1.00	0.45–0.65	0.45–0.60	—
EX 33	0.17–0.24	0.85–1.25	0.035	0.040	0.20–0.35	0.20 min	0.20 min	0.05 min	—
EX 34	0.28–0.33	0.90–1.20	0.035	0.040	0.20–0.35	—	0.40–0.60	0.13–0.20	—
EX 36	0.38–0.43	0.90–1.20	0.035	0.040	0.20–0.35	—	0.45–0.65	0.13–0.20	—
EX 38	0.43–0.48	0.90–1.20	0.035	0.040	0.20–0.35	—	0.45–0.65	0.13–0.20	—
EX 39	0.48–0.53	0.90–1.20	0.035	0.040	0.20–0.35	—	0.45–0.65	0.13–0.20	—
EX 40	0.51–0.59	0.90–1.20	0.035	0.040	0.20–0.35	—	0.45–0.65	0.13–0.20	—
EX 54	0.19–0.25	0.70–1.05	0.035	0.040	0.35 max	—	0.40–0.70	0.05 min	—
EX 55	0.15–0.20	0.70–1.00	0.035	0.040	0.20–0.35	1.65–2.00	0.45–0.65	0.65–0.80	—
EX 56	0.080–0.13	0.70–1.00	0.035	0.040	0.20–0.35	1.65–2.00	0.45–0.65	0.65–0.80	—
EX 57	0.08 max	1.25 max	0.040	0.15–0.35	1.00 max	—	17.00–19.00	1.75–2.25	—
EX 58	0.16–0.21	1.00–1.30	0.035	0.040	0.15–0.30	—	0.45–0.65	—	—
EX 59	0.18–0.23	1.00–1.30	0.035	0.040	0.15–0.30	—	0.70–0.90	—	—
EX 60	0.20–0.25	1.00–1.30	0.035	0.040	0.15–0.30	—	0.70–0.90	—	—
EX 61	0.23–0.28	1.00–1.30	0.035	0.040	0.15–0.30	—	0.70–0.90	—	—
EX 62	0.25–0.30	1.00–1.30	0.035	0.040	0.15–0.30	—	0.70–0.90	—	—
EX 63	0.31–0.38	0.75–1.10	0.035	0.040	0.15–0.35	—	0.45–0.65	—	0.0005–0.003
EX 64	0.16–0.21	1.00–1.30	0.035	0.040	0.15–0.35	—	0.70–0.90	—	—
EX 65	0.21–0.26	1.00–1.30	0.035	0.040	0.15–0.35	—	0.70–0.90	—	—

APPENDIX A
DESIGNATION AND NUMBERING OF EXPERIMENTAL COMPOSITIONS

A. Scope—This Appendix establishes a system for designating and numbering new or experimental compositions for steel products during a period of limited use in which technical and commercial desirability is evaluated.

A.1 Designation and Numbering—Such materials shall be designated by the prefix EX and numbered by assigning a sequential, non-significant number beginning with 1 in the order of approval for listing.

A.1.1 Application for listing may be made by any person, acting in his own behalf or for his company or association by written application to the SAE technical staff listing the range of chemical composition and other pertinent characteristics of the proposed material. This request shall be forwarded to the Chairman of the appropriate Division of ISTC for action.

A.1.2 ASSIGNMENT OF NUMBER—If the Division Chairman approves, he shall inform the appropriate member of the SAE Technical Staff who will assign the number, add the material to the list, and inform the applicant of the number assigned.

A.2 Publication—The current numbers, their range of chemical composition, and other pertinent characteristics, shall be published yearly in the SAE Handbook in SAE J1081. Publication of interim bulletins on newly listed numbers may be authorized by the ISTC Executive Committee.

A.3 Discontinuance—A material may be removed from the list by any of the following:

A.3.1 Adoption of the material as an SAE standard steel and assignment of a permanent number.

A.3.2 Formal action to delist by the cognizant Division.

A.3.2.1 Every two years each Division shall determine the status of the steels within its jurisdiction and delete those which an ISTC ballot shows to be of insufficient interest to warrant further consideration.

A.3.3 A request to delist by the applicant who originally requested listing, followed by action prescribed in A.3.2.

A.4 Reassignment of a number once assigned may be done only if the material to which it was originally assigned is relisted.

FORMERLY STANDARD SAE ALLOY STEELS—SAE J778a

SAE Information Report

Report of Iron and Steel Technical Committee approved June 1961 and last revised November 1977.

This SAE Information Report provides a list of those SAE Alloy Steels which, because of decreased usage, have been deleted from the SAE Handbook since 1936. Information on SAE steels prior to 1936 may be obtained from SAE offices on request.

While the steels in the list below are no longer considered as standard steels, they may be purchased as nonstandard steels which call for broader chemical ranges for most of their elements. The producer, in such cases, should be consulted for chemical limits.

In Table 1, the last column lists the date a steel was last listed in the SAE Handbook. When applicable, the corresponding AISI standard steel appears in the next to the last column.

TABLE 1—FORMERLY STANDARD SAE ALLOY STEELS

SAE No.	C	Mn	P, Max	S, Max	Si	Cr	Ni	Mo	V, Min	AISI No.	DATE
1320	0.18–0.23	1.60–1.90	0.040	0.040	0.20–0.35	—	—	—	—	A1320	1956
2317	0.15–0.20	0.40–0.60	0.040	0.040	0.20–0.35	—	3.25–3.75	—	—	A2317	1956
2330	0.28–0.33	0.60–0.80	0.040	0.040	0.20–0.35	—	3.25–3.75	—	—	A2330	1953
2340	0.38–0.43	0.70–0.90	0.040	0.040	0.20–0.35	—	3.25–3.75	—	—	A2340	1953
2345	0.43–0.48	0.70–0.90	0.040	0.040	0.20–0.35	—	3.25–3.75	—	—	A2345	1952
2512	0.09–0.14	0.45–0.60	0.025	0.025	0.20–0.35	—	4.75–5.25	—	—	E2512	1953
2515	0.12–0.17	0.40–0.60	0.040	0.040	0.20–0.35	—	4.75–5.25	—	—	A2515	1956
2517	0.15–0.20	0.45–0.60	0.025	0.025	0.20–0.35	—	4.75–5.25	—	—	E2517	1959
3115	0.13–0.18	0.40–0.60	0.040	0.040	0.20–0.35	0.55–0.75	1.10–1.40	—	—	A3115	1953
3120	0.17–0.22	0.60–0.80	0.040	0.040	0.20–0.35	0.55–0.75	1.10–1.40	—	—	A3120	1956
3130	0.28–0.33	0.60–0.80	0.040	0.040	0.20–0.35	0.55–0.75	1.10–1.40	—	—	A3130	1956
3135	0.33–0.38	0.60–0.80	0.040	0.040	0.20–0.35	0.55–0.75	1.10–1.40	—	—	3135	1960
X3140	0.38–0.43	0.70–0.90	0.040	0.040	0.20–0.35	0.70–0.90	1.10–1.40	—	—	A3141	1947
3140	0.38–0.43	0.70–0.90	0.040	0.040	0.20–0.35	0.55–0.75	1.10–1.40	—	—	3140	1964
3145	0.43–0.48	0.70–0.90	0.040	0.040	0.20–0.35	0.70–0.90	1.10–1.40	—	—	A3145	1952
3150	0.48–0.53	0.70–0.90	0.040	0.040	0.20–0.35	0.70–0.90	1.10–1.40	—	—	A3150	1952
3215	0.10–0.20	0.30–0.60	0.040	0.050	0.15–0.30	0.90–1.25	1.50–2.00	—	—	—	1941
3220	0.15–0.25	0.30–0.60	0.040	0.050	0.15–0.30	0.90–1.25	1.50–2.00	—	—	—	1941
3230	0.25–0.35	0.30–0.60	0.040	0.050	0.15–0.30	0.90–1.25	1.50–2.00	—	—	—	1941
3240	0.35–0.45	0.30–0.60	0.040	0.040	0.15–0.30	0.90–1.25	1.50–2.00	—	—	A3240	1941
3245	0.40–0.50	0.30–0.60	0.040	0.040	0.15–0.30	0.90–1.25	1.50–2.00	—	—	—	1941
3250	0.45–0.55	0.30–0.60	0.040	0.040	0.15–0.30	0.90–1.25	1.50–2.00	—	—	—	1941
3310	0.08–0.13	0.45–0.60	0.025	0.025	0.20–0.35	1.40–1.75	3.25–3.75	—	—	E3310	1964
3312	0.08–0.13	0.45–0.60	0.025	0.025	0.20–0.35	1.40–1.75	3.25–3.75	—	—	—	1948
3316	0.14–0.19	0.45–0.60	0.025	0.025	0.20–0.35	1.40–1.75	3.25–3.75	—	—	E3316	1956
3325	20–30	0.30–0.60	0.040	0.050	0.15–0.30	1.25–1.75	3.25–3.75	—	—	—	1936
3335	30–40	0.30–0.60	0.040	0.050	0.15–0.30	1.25–1.75	3.25–3.75	—	—	—	1936
3340	35–45	0.30–0.60	0.040	0.050	0.15–0.30	1.25–1.75	3.25–3.75	—	—	—	1936
3415	0.10–0.20	0.30–0.60	0.040	0.050	0.15–0.30	0.60–0.95	2.75–3.25	—	—	—	1941
3435	0.30–0.40	0.30–0.60	0.040	0.050	0.15–0.30	0.60–0.95	2.75–3.25	—	—	—	1936
3450	0.45–0.55	0.30–0.60	0.040	0.050	0.15–0.30	0.60–0.95	2.75–3.25	—	—	—	1936
4012	0.09–0.14	0.75–1.00	0.035	0.040	0.15–0.30	—	—	0.15–0.25	—	4012	1977
4053	0.50–0.56	0.75–1.00	0.040	0.040	0.20–0.35	—	—	0.20–0.30	—	4053	1956
4063	0.60–0.67	0.75–1.00	0.040	0.040	0.20–0.35	—	—	0.20–0.30	—	4063	1964
4068	0.63–0.70	0.75–1.00	0.040	0.040	0.20–0.35	—	—	0.20–0.30	—	A4068	1957
4119	0.17–0.22	0.70–0.90	0.040	0.040	0.20–0.35	0.40–0.60	—	0.20–0.30	—	A4119	1956
4125	0.23–0.28	0.70–0.90	0.040	0.040	0.20–0.35	0.40–0.60	—	0.20–0.30	—	A4125	1950
4317	0.15–0.20	0.45–0.65	0.040	0.040	0.20–0.35	0.40–0.60	1.65–2.00	0.20–0.30	—	4317	1953
4337	0.35–0.40	0.60–0.80	0.040	0.040	0.20–0.35	0.70–0.90	1.65–2.00	0.20–0.30	—	4337	1964
4419	0.18–0.23	0.45–0.65	0.035	0.040	0.15–0.30	—	—	0.45–0.60	—	4520	1977
4419H	0.17–0.23	0.35–0.75	0.035	0.040	0.15–0.30	—	—	0.45–0.60	—	4419H	1977
4608	0.06–0.11	0.25–0.45	0.040	0.040	0.025 Max	—	1.40–1.75	0.15–0.25	—	4608	1956
46B12	0.10–0.15	0.45–0.65	0.040	0.040	0.20–0.35	—	1.65–2.00	0.20–0.30	—	46B12[a]	1957
X4620	0.18–0.23	0.50–0.70	0.040	0.040	0.20–0.35	—	1.65–2.00	0.20–0.30	—	X4620	1956
4621	0.18–0.23	0.70–0.90	0.035	0.040	0.15–0.30	—	1.65–2.00	0.20–0.30	—	4621	1977
4621H	0.17–0.23	0.60–1.00	0.035	0.040	0.15–0.30	—	1.55–2.00	0.20–0.30	—	4621H	1977
4640	0.38–0.43	0.60–0.80	0.040	0.040	0.20–0.35	—	1.65–2.00	0.20–0.30	—	A4640	1952
4812	0.10–0.15	0.40–0.60	0.040	0.040	0.20–0.35	—	3.25–3.75	0.20–0.30	—	4817	1956

(Table continued on next page)

TABLE 1—FORMERLY STANDARD SAE ALLOY STEELS (continued)

SAE No.	C	Mn	P, Max	S, Max	Si	Cr	Ni	Mo	V, Min	AISI No.	DATE
5015	0.12–0.17	0.30–0.50	0.035	0.040	0.15–0.30	0.30–0.50	—	—	—	5015	1977
5045	0.43–0.48	0.70–0.90	0.040	0.040	0.20–0.35	0.55–0.75	—	—	—	—	1953
5145	0.43–0.48	0.70–0.90	0.035	0.040	0.15–0.30	0.70–0.90	—	—	—	5145	1977
5145H	0.42–0.49	0.60–1.00	0.035	0.040	0.15–0.30	0.60–1.00	—	—	—	5145H	1977
5152	0.48–0.55	0.70–0.90	0.040	0.040	0.20–0.35	0.90–1.20	—	—	—	5152	1956
6115	0.10–0.20	0.30–0.60	0.040	0.050	0.15–0.30	0.80–1.10	—	—	0.15	—	1936
6117	0.15–0.20	0.70–0.90	0.040	0.040	0.20–0.35	0.70–0.90	—	—	0.10	6117	1956
6120	0.17–0.22	0.70–0.90	0.040	0.040	0.20–0.35	0.70–0.90	—	—	0.10	6120	1961
6125	0.20–0.30	0.60–0.90	0.040	0.050	0.15–0.30	0.80–1.10	—	—	0.15	—	1936
6130	0.25–0.35	0.60–0.90	0.040	0.050	0.15–0.30	0.80–1.10	—	—	0.15	—	1936
6135	0.30–0.40	0.60–0.90	0.040	0.050	0.15–0.30	0.80–1.10	—	—	0.15	—	1941
6140	0.35–0.45	0.60–0.90	0.040	0.050	0.15–0.30	0.80–1.10	—	—	0.15	—	1936
6145	0.43–0.48	0.70–0.90	0.040	0.050	0.20–0.35	0.80–1.10	—	—	0.15	6145	1956
6195	0.90–1.05	0.20–0.45	0.030	0.035	0.15–0.30	0.80–1.10	—	—	0.15	—	1936
71360	0.50–0.70	0.30 Max	0.035	0.040	0.15–0.30	3.00–4.00	12.00–15.00 W	—	—	—	1936
71660	0.50–0.70	0.30 Max	0.035	0.040	0.15–0.30	3.00–4.00	15.00–18.00 W	—	—	—	1936
7260	0.50–0.70	0.30 Max	0.035	0.040	0.15–0.30	0.50–1.00	1.50–2.00 W	—	—	—	1936
8632	0.30–0.35	0.70–0.90	0.040	0.040	0.20–0.35	0.40–0.60	0.40–0.70	0.15–0.25	—	8632	1951
8635	0.33–0.38	0.75–1.00	0.040	0.040	0.20–0.35	0.40–0.60	0.40–0.70	0.15–0.25	—	8635	1956
8641	0.38–0.43	0.75–1.00	0.040	0.040–0.060	0.20–0.35	0.40–0.60	0.40–0.70	0.15–0.25	—	8641	1956
8653	0.50–0.56	0.75–1.00	0.040	0.040	0.20–0.35	0.50–0.80	0.40–0.70	0.15–0.25	—	8653	1956
8647	0.45–0.50	0.75–1.00	0.040	0.040	0.20–0.35	0.40–0.60	0.40–0.70	0.15–0.25	—	8647	1948
8715	0.13–0.18	0.70–0.90	0.040	0.040	0.20–0.35	0.40–0.60	0.40–0.70	0.20–0.30	—	8715	1956
8717	0.15–0.20	0.70–0.90	0.040	0.040	0.20–0.35	0.40–0.60	0.40–0.70	0.20–0.30	—	8717	1956
8719	0.18–0.23	0.60–0.80	0.040	0.040	0.20–0.35	0.40–0.60	0.40–0.70	0.20–0.30	—	8719	1952
8735	0.33–0.38	0.75–1.00	0.040	0.040	0.20–0.35	0.40–0.60	0.40–0.70	0.20–0.30	—	8735	1952
8742	0.40–0.45	0.75–1.00	0.040	0.040	0.20–0.35	0.40–0.60	0.40–0.70	0.20–0.30	—	8742	1964
8745	0.43–0.48	0.75–1.00	0.040	0.040	0.20–0.35	0.40–0.60	0.40–0.70	0.20–0.30	—	8745	1953
8750	0.48–0.53	0.75–1.00	0.040	0.040	0.20–0.35	0.40–0.60	0.40–0.70	0.20–0.30	—	8750	1956
9250	0.45–0.55	0.60–0.90	0.040	0.040	1.80–2.20	—	—	—	—	9250	1941
9255	0.51–0.59	0.70–0.95	0.035	0.040	1.80–2.20	—	—	—	—	9255	1977
9261	0.55–0.65	0.75–1.00	0.040	0.040	1.80–2.20	0.10–0.25	—	—	—	9261	1956
9262	0.55–0.65	0.75–1.00	0.040	0.040	1.80–2.20	0.25–0.40	—	—	—	9262	1961
9315	0.13–0.18	0.45–0.65	0.025	0.025	0.20–0.35	1.00–1.40	3.00–3.50	0.08–0.15	—	E9315	1959
9317	0.15–0.20	0.45–0.65	0.025	0.025	0.20–0.35	1.00–1.40	3.00–3.50	0.08–0.15	—	E9317	1959
9437	0.35–0.40	0.90–1.20	0.040	0.040	0.20–0.35	0.30–0.50	0.30–0.60	0.08–0.15	—	9437	1950
9440	0.38–0.43	0.90–1.20	0.040	0.040	0.20–0.35	0.30–0.50	0.30–0.60	0.08–0.15	—	9440	1950
94B40	0.38–0.43	0.75–1.00	0.040	0.040	0.20–0.35	0.30–0.50	0.30–0.60	0.08–0.15	—	94B40	1964
9442	0.40–0.45	0.90–1.20	0.040	0.040	0.20–0.35	0.30–0.50	0.30–0.60	0.08–0.15	—	9442	1950
9445	0.43–0.48	0.90–1.20	0.040	0.040	0.20–0.35	0.30–0.50	0.30–0.60	0.08–0.15	—	9445	1950
9447	0.45–0.50	0.90–1.20	0.040	0.040	0.20–0.35	0.30–0.50	0.30–0.60	0.08–0.15	—	9447	1950
9747	0.45–0.50	0.50–0.80	0.040	0.040	0.20–0.35	0.10–0.25	0.40–0.70	0.15–0.25	—	9747	1950
9763	0.60–0.67	0.50–0.80	0.040	0.040	0.20–0.35	0.10–0.25	0.40–0.70	0.15–0.25	—	9763	1950
9840	0.38–0.43	0.70–0.90	0.040	0.040	0.20–0.35	0.70–0.90	0.85–1.15	0.20–0.30	—	9840	1964
9845	0.43–0.48	0.70–0.90	0.040	0.040	0.20–0.35	0.70–0.90	0.85–1.15	0.20–0.30	—	9845	1950
9850	0.48–0.53	0.70–0.90	0.040	0.040	0.20–0.35	0.70–0.90	0.85–1.15	0.20–0.30	—	9850	1961
43BV12	0.08–0.13	0.75–1.00	—	—	0.20–0.35	0.40–0.60	1.65–2.00	0.20–0.30	0.03	—	—
43BV14	0.10–0.15	0.45–0.65	—	—	0.20–0.35	0.40–0.60	1.65–2.00	0.08–0.15	0.03	—	—

[a] Boron content 0.0005 min.

FORMERLY STANDARD SAE CARBON STEELS—SAE J118a

SAE Information Report

Report of Iron and Steel Technical Committee approved August 1969 and last revised November 1977.

This SAE Information Report provides a list of those SAE Carbon Steels which, because of decreased usage, have been deleted from the SAE Handbook since 1952. Information on SAE steels prior to 1952 may be obtained from SAE offices on request.

While the steels listed in Table 1 are no longer considered standard steels, they may be purchased as nonstandard steels which call for similar chemical ranges. The producer, in such cases, should be consulted for chemical limits.

In Table 1, the last column lists the date a steel was last listed in the SAE Handbook. When applicable, the corresponding AISI standard steel appears in the next to the last column.

TABLE 1—FORMERLY STANDARD SAE CARBON STEELS
(Apply to only semi-finished products for forgings, bars, wire rods, and seamless tubing)

SAE No.	C	Mn	P	S	Corresponding AISI No.	Date Last Listed in SAE Handbook
1009	0.15 max	0.60 max	0.040 max	0.050 max	1009	1965
1011	0.08-0.13	0.60-0.90	0.040 max	0.050 max	—	1977
1033	0.30-0.36	0.70-1.00	0.040 max	0.050 max	1033	1965
1034	0.32-0.38	0.50-0.80	0.040 max	0.050 max	C1034	1968
1059	0.55-0.65	0.50-0.80	0.040 max	0.050 max	—	1968
1062	0.54-0.65	0.85-1.15	0.040 max	0.050 max	C1062	1953
1086	0.80-0.94	0.30-0.50	0.040 max	0.050 max	—	1977
1109	0.08-0.13	0.60-0.90	0.040 max	0.08-0.13	1109	1977
1111	0.13 max	0.60-0.90	0.07-0.12	0.10-0.15	B1111	1969
1112	0.13 max	0.70-1.00	0.07-0.12	0.16-0.23	B1112	1969
1113	0.13 max	0.70-1.00	0.07-0.12	0.24-0.33	B1113	1969
1114	0.10-0.16	1.00-1.30	0.040 max	0.08-0.13	C1114	1952
1115	0.13-0.18	0.60-0.90	0.040 max	0.08-0.13	1115	1965
1116	0.14-0.20	1.10-1.40	0.040 max	0.16-0.23	C1116	1952
1119	0.14-0.20	1.00-1.30	0.040 max	0.24-0.33	1119	1977
1120	0.18-0.23	0.70-1.00	0.040 max	0.08-0.13	1120	1965
1126	0.23-0.29	0.70-1.00	0.040 max	0.08-0.13	1126	1965
1132	0.27-0.34	1.35-1.65	0.040 max	0.08-0.13	1132	1977
1138	0.34-0.40	0.70-1.00	0.040 max	0.08-0.13	1138	1965
1145	0.42-0.49	0.70-1.00	0.040 max	0.04-0.07	1145	1977
1518	0.15-0.21	1.10-1.40	0.040 max	0.050 max	—	1977
1525	0.23-0.29	0.80-1.10	0.040 max	0.050 max	—	1977
1547	0.43-0.51	1.35-1.65	0.040 max	0.050 max	—	1977
1572	0.65-0.76	1.00-1.30	0.040 max	0.050 max	—	1977

When silicon is required, the following limits and ranges were commonly used: for SAE 1009 and steels designations up to but excluding SAE 1114, 0.10% max; for SAE 1114 and over, 0.10% max or the ranges of 0.10-0.20% or 0.15-0.30%; for over SAE 1033 ranges of 0.10-0.20% or 0.15-0.30%. Grades 1111, 1112, and 1113 were customarily furnished without specified silicon content.

These steels were produced by the basic open hearth, the basic oxygen, or the basic electric steel making process; except for SAE 1111, 1112, and 1113 which were processed by the acid bessemer process.

METHODS OF DETERMINING HARDENABILITY OF STEELS—SAE J406c

SAE Standard

Report of Iron and Steel Division approved January 1942 and completely revised by Iron and Steel Technical Committee December 1977.

Scope—This SAE Standard prescribes the procedure for making hardenability tests and recording results on shallow and medium hardening steels, but not deep hardening steels that will normally air harden.

Included are procedures using the 1 in (25 mm) standard hardenability end quench specimen for both medium and shallow hardening steels, Surface-Area-Center (SAC) method for shallow hardening steels, subsize method for bars less than 1¼ in (32 mm) in diameter, and methods for determining carburized hardenability.

Any hardenability tests made under other conditions than those given in this SAE Standard will not be deemed standard and will be subject to agreement between supplier and user. Whenever check tests are made, all laboratories concerned must arrange to use the same alternate procedure with reference to test specimen and method of grinding for hardness testing.

For routine testing of the hardenability of successive heats of steel required to have hardenability within certain limits, it is sufficient to designate hardenability simply in terms of distance from the quenched end to the point at which a certain hardness is obtained. This designation is also adequate for comparing steels of different compositions to see whether they have similar hardenability.

Hardenability limits for specifying steel in this manner are obtained by measuring the hardenability of a steel which has proved satisfactory for the use intended. The hardenability test may be used in this way as an empirical test.

Hardenability data may be used to estimate hardnesses obtainable with any steel in new machine parts not yet in production and not similar to any parts on which production experience is available. Various hardenability application methods are described in the selected references in the Bibliography. It appears none of these methods are precise, but these are often useful for estimation purposes. Final correlation on actual parts is necessary.

HARDENABILITY TEST FOR MEDIUM HARDENING STEELS

Introduction—This method covers the procedure for determining the hardenability of steel by the end quench test for both the 1 in (25 mm) standard specimen and the sub-size test specimen. Also included are charts for plotting hardenability test results and for predicting hardness U curves in various sizes of rounds.

Test Specimen—The test specimen is a 1 in (25 mm) diameter cylinder 4 in (102 mm) long with means for hanging it in a vertical position for end quenching. Fig. 1 shows a test specimen in the fixture ready for quenching illustrating the preferred form of specimen. Fig. 2 gives the details of the preferred test specimen. Fig. 3 is an example of an optional specimen which provides the same diameter and approximately the same length and will provide satisfactory heat transfer characteristics.

The bar from which the specimen is machined shall be a forged or rolled 1¼ in (32 mm) round representing the full cross-section of the product. A cast specimen may be used in lieu of a rolled or forged specimen, except in the case of boron-treated steels: experience has shown that cast specimens of boron-treated steels give erratic results. The option of using as-cast specimens for non-boron steels, deletion of normalizing prior to heating for end-quenching or modification of other testing details shall be negotiated between supplier and user. It is of primary importance that the specimen represent the full cross section of the ingot since test specimens from a portion of the bloom, billet, or bar may introduce factors tending to affect the reproducibility of results. The condition of this hot formed bar shall be such that there is no decarburization on the 1 in (25 mm) specimen machined from it. If any test specimen shows obvious defects or flaws, the specimen should be discarded and a new specimen obtained.

Optional Specimen Preparation—The following method is satisfactory for most purposes, but for check testing against specifications, the method in the preceding paragraph is mandatory.

The test specimen shall be machined from the center of the bar in the case of sections from 1¼ to 2 in (32-51 mm) round or square. In sections over 2 in (51 mm), the test specimen shall be machined from one-half of the section with the axis of the specimen located at a point halfway between the center and surface of the bar and marked to identify the position of the test bar with reference to the original bar. The hardness readings shall be made on the two sides of the test specimen corresponding to a position in the bar approximately halfway between the center and the surface.

Normalizing Prior to Heating for End Quenching—The forged or rolled round shall be normalized prior to machining the test specimen. This is of importance since the structure of material before the final austenitizing treatment may materially affect the hardening characteristics. In order that variations in prior structure may be controlled as much as possible, the normalizing temperature listed in Table 1 should be used. The steel shall be held at such

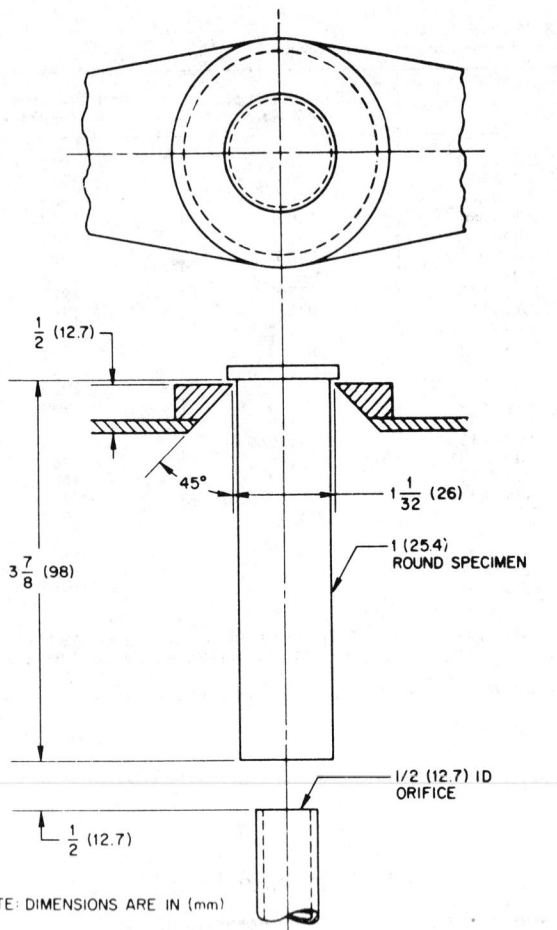

FIG. 1—HARDENABILITY TEST SPECIMENS IN FIXTURE FOR WATER QUENCHING

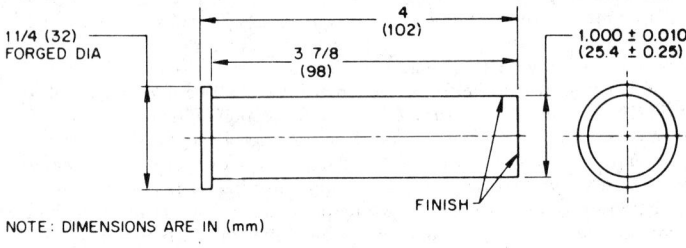

FIG. 2—PREFERRED TEST SPECIMEN

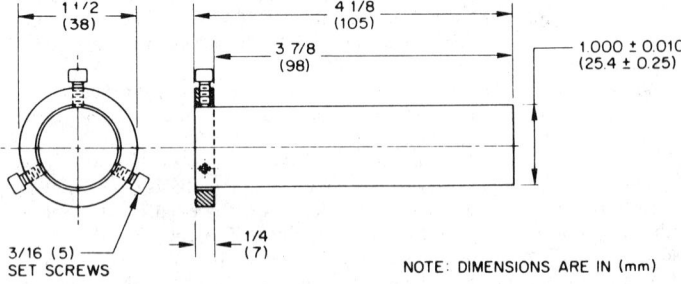

FIG. 3—OPTIONAL TEST SPECIMEN

TABLE 1—NORMALIZING AND QUENCHING TEMPERATURES[a,b] APPLICABLE TO STEEL ORDERED TO END-QUENCH HARDENABILITY REQUIREMENTS

Max Ordered Carbon Content, %	Normalizing Temperature		Austenitizing Temperature	
	°F	°C	°F	°C
Steel Series 1000, 1300, 1500, 4000, 4100, 4300, 4600, 5000, 5100, 6100, 8600, 8700, 9400				
Up to 0.25 incl	1700	925	1700	925
0.26 to 0.36 incl	1650	900	1600	870
0.37 and over[c]	1600	870	1550	845
Steel Series 4800, 9200				
Up to 0.25 incl	1700	925	1550	845
0.26 to 0.36 incl	1650	900	1500	815
0.37 and over	1600	870	1475	800
0.50 and over (9200)	1650	900	1600	870

[a] A variation of ±10°F (±5°C) from the above temperature is permissible.
[b] When testing H steels, the normalizing and austenitizing temperatures should be the same as for the equivalent standard steels. EXAMPLES: For 8622 H, the normalizing and austenitizing temperature should be the same as for SAE 8622; for 4032 H (carbon 0.30/0.37), the temperature should be the same as for SAE 4032 (carbon 0.30/0.35).
[c] Normalizing and austenitizing temperatures shall be 50°F (30°C) higher for the 6100 series.

tizing temperature shown in Table 1. The specimen shall be placed in a furnace which is at the specified temperature and shall be held at this temperature for 30 min.[1] It is necessary to determine by means of a thermocouple the time required for a test specimen to come to temperature to be sure that the above heating time and temperature requirements are met.

It is important that while heating the test specimen care be taken that its environment is such that practically no scaling or decarburization takes place on the end to be quenched. An adequately protective atmosphere in the furnace is suitable for meeting the above requirements. In the absence of such atmospheres, the specimen shall be inserted in a suitable container which maintains a nonoxidizing atmosphere. Placing fine graphite powder or cast iron chips in the base of the container are two methods of preventing oxidation of the quenched end.

Fig. 4 illustrates a type of container which has been used with success. However, any similar type will be satisfactory.

Quenching—The test piece shall be placed on a fixture so that a column of water at a temperature of 40–85°F (5–30°C) may be directed against the bottom face of the piece. The column of water passing through an opening ½ in (13 mm) in diameter shall rise to a free height of 2½ in (63 mm) above the opening. The fixture shall be dry at the beginning of each test.

In performing the test, the water supply shall be shut off with a quick-opening valve and the hot specimen placed over the water pipe so that the bottom of the specimen is ½ in (13 mm) from the opening of the water pipe and the water shall then be turned on. The time between removal of the specimen from the furnace and the beginning of the quench shall be not more than 5 s. The sample shall remain on the fixture for at leat 10 min. A condition of still air shall be maintained around the piece during cooling.

[1] In production testing, slightly longer times up to 35 min may be used without appreciably affecting results.

temperature for 1 h and cooled in still air. If the normalized specimen is too hard, it may be given a short time temper at about 100°F (55°C) below the Ac_1 to improve machinability. Cast specimens usually are not normalized before quenching. The record of hardenability test results must always state the prior thermal history of the specimen tested.

Heating for End Quenching—The specimen shall be heated to the austeni-

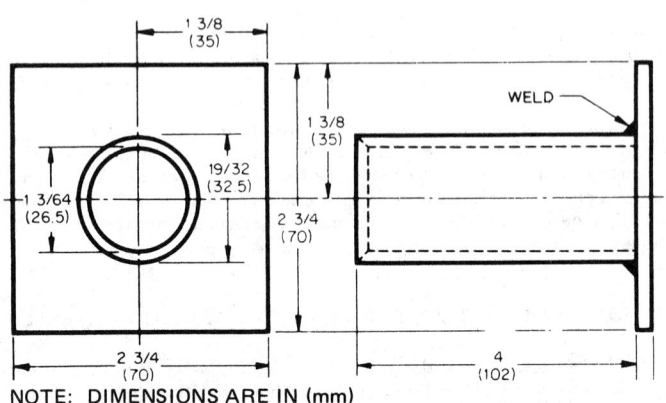

FIG. 4—HARDENING BAR PROTECTING FIXTURE TO BE CONSTRUCTED OF HEAT RESISTING ALLOY

FIG. 5—COMMERCIALLY AVAILABLE FIXTURE FOR POSITIONING SPECIMEN FOR HARDNESS INDENTATIONS

Hardness Measurement—Two flats 180 deg apart shall be ground to a minimum depth of 0.015 in (0.38 mm) along the entire length of the bar and Rockwell C hardness measurements made along the length of the bar. Deviation from the standard depth can affect reproducibility of results, and correlation with cooling rates in quenched bars.

The preparation of the two flats must be carried out with considerable care. They should be mutually parallel and the grinding done in such a manner that no change of the quenched structure takes place. Very light cuts (less than 0.0005 in (0.013 mm) with water cooling and a coarse, soft grinding wheel are recommended to avoid heating the specimen. In order to detect tempering due to grinding, the flat shall be etched as follows.

Two etchant solutions are used:
No. 1—5% nitric acid (concentrated) and 95% water by volume.
No. 2—50% hydrochloric acid (concentrated) and 50% water by volume.

Wash the sample in hot water. Etch in solution No. 1 until black. Wash in hot water. Immerse in solution No. 2 for 3 s and wash in hot water. Dry in air blast.

The presence of lighter or darker areas indicates that hardness and structure have been altered in grinding. All structural changes caused by grinding shall be removed before hardness tests are made. This may be accomplished by resurfacing and again etching, or new flats may be prepared.

When hardness readings are made, the test specimen rests on one of its flats on an anvil firmly attached to the hardness machine. It is important that no vertical movement be allowed when the major load is applied. The anvil must be constructed to move the test specimen past the penetrator in accurate steps of $\frac{1}{16}$ in (1.6 mm). Fig. 5 is an example of a fixture commercially available which provides for the controlled movement of the specimens.

The Rockwell tester should be checked against standard test blocks before testing the hardenability bar. It is recommended that the test block be interposed between the specimen and the indenter to check the seating of the indenter and the specimen simultaneously.

Care must be exercised in registering the point of the indenter with the hardened end of the specimen, as well as providing for accurate spacing between indentations. A low power measuring microscope is suitable for use in determining the distance from the quenched end to the center of the first impression and in checking the distance from center to center of the succeeding impressions. It has been found that with reasonable operating care and a well-built fixture, it is practical to locate the center of the first impression 0.0625 ± 0.003 in (1.6 ± 0.075 mm) from the quenched end. The variations between spacings should be even smaller. Obviously, it is more important to position the indenter accurately when testing low hardenability steels than when testing high hardenability steels. The positioning of the indenter should be checked with sufficient frequency to provide assurance that accuracy requirements are being met. In cases of lack of reproducibility or of differences between laboratories, indenter spacing should be measured immediately.

Readings shall be taken in steps of $\frac{1}{16}$ in (1.6 mm) for the first 1 in (25 mm). Distance between readings for the last 2 in (51 mm) may be at the discretion of the tester. When a flat on which readings have been made is used as a base, the burrs around the indentation shall be removed by grinding unless a fixture is used which has been relieved to accommodate the irregularities due to the indentations.

Hardness readings should be made on one flat, or preferably, two flats 180 deg apart. Testing on two flats will assist in the detection of errors in specimen preparation and hardness measurement. If the two probes on opposite sides differ by more than 4 HRC points at any one position, the test should be repeated on new flats, 90 deg from first two flats. If the retest also has greater than 4 HRC points spread, a new specimen should be tested.

For reporting purposes, hardness readings should be recorded to the nearest integer, with 0.5 HRC values rounded to the next higher integer.

Plotting of Tests—Tests should be plotted on a standard chart prepared for this purpose (Fig. 6) in which the ordinates represent hardness and the

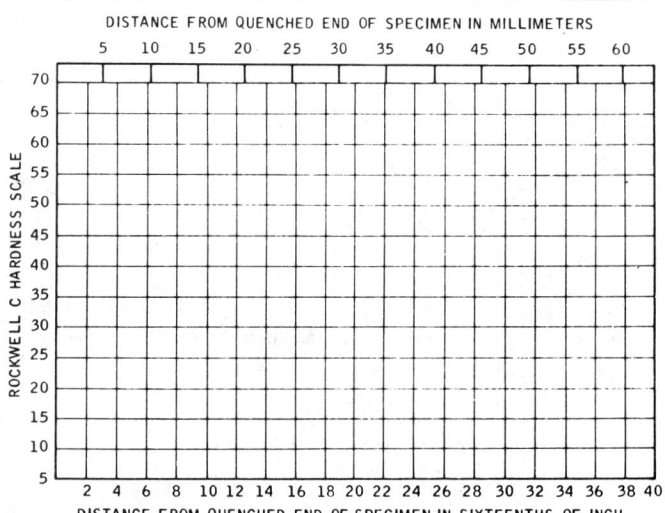

FIG. 6—STANDARD FORM FOR PLOTTING HARDENABILITY CURVES

abscissas represent distance from the quenched end. Readings at identical distances should be averaged and the resultant values used for plotting.

Fig. 6 shows the Standard Form for Plotting Hardenability Curves.

Construction of Hardness U Curves—A chart is also provided for using the hardenability curve to predict hardness U curves in various sized rounds when oil or water quenched. Fig. 7 shows this chart. The curves show the locations in various sizes of rounds where the cooling rates are the same as at various positions along the end-quenched hardenability test bar. It should be noted that these curves assume good heat treatment practice: separation of pieces in the quench and good control of temperature and cleanliness of the quench medium. The ranges given reflect variations found under experimental conditions; under production conditions even wider variations may be found.

Index of Hardenability—The hardenability of steel is usually reported as a series of values of hardness versus distance from the quenched end of the test bar, either in graphical or tabular form. Hardenability may also be designated by the use of either one or the other of the two following codes, the first of which is preferred.

(a) That code which designates a minimum hardness value at a distance specified as J 36 min = $^{8}/_{16}$ in (13 mm) or in case of both minimum and maximum limitations may also be specified in terms of Rockwell C hardness at the required distance from the quenched end.

As an example of this method, a hardenability requirement could be specified as J 36 min = $^{8}/_{16}$ in (13 mm) or in case of both minimum and maximum restrictions, could be specified as J 36-50 = $^{8}/_{16}$ in (13 mm). This means that at the specified distance of $^{8}/_{16}$ in (13 mm) from the quenched end the Rockwell C hardness should be a minimum of 36 and a maximum of 50.

(b) The alternate method would be a code which indicates the distance from the quenched end where the following hardness reference numbers occur. The requirement may be specified as a minimum distance only or as a minimum and a maximum distance at the hardness reference number which applies. Table 2 indicates the hardness reference numbers in terms of Rockwell C values for various carbon contents.

As an example, an alloy steel of 0.34 mean carbon content could be specified to have a hardenability index of J 45 = $^{4}/_{16}$ in (6.3 mm) min which means

TABLE 2—HARDNESS REFERENCE NUMBERS FOR STEELS OF VARIOUS CARBON CONTENTS

Mean of Ordered Carbon Range	Hardness Reference No., HRC	
	Alloy Steels	Carbon Steels
0.08-0.17	25	—
0.18-0.22	30	25
0.23-0.27	35	30
0.28-0.32	40	35
0.33-0.42	45	40
0.43-0.52	50	45
0.53-0.62	55	50

that the minimum requirements of this steel would be 45 HRC at a distance of $^{4}/_{16}$ in (6.3 mm) from the quenched end.

If this steel were one having both minimum and maximum curves or limits, the index of hardenability might be specified as J 45 = $^{4}/_{16}$–$^{11}/_{16}$ in (6.3–17 mm).

In addition to the specification in accordance with either of the above two codes, the minimum and maximum hardness at the $^{1}/_{16}$ in (1.6 mm) position may be specified. These hardness values should be in agreement with the maximum and minimum carbon content specified.

Subsize Test Specimen—For determining hardenability of steel received in bars less than $1^{1}/_{4}$ in (32 mm) in diameter, the test bar may be made $^{3}/_{4}$, $^{1}/_{2}$, or $^{1}/_{4}$ in (19, 13, or 6.3 mm) diameter, as desired, and end quenched as prescribed for the 1 in (25 mm) round. Modifications in the water orifice are required for quenching cylinders of less than 1 in (25 mm) diameter. The details of orifices for quenching specimens less than 1 in (25 mm) diameter are given in Table 3.

Because of the greater aircooling effect on test bars less than 1 in (25 mm) diameter and especially in bars smaller than $^{3}/_{4}$ in (19 mm) diameter, the cooling rates at various distances from the quenched end will not be the same as in the standard test bar.

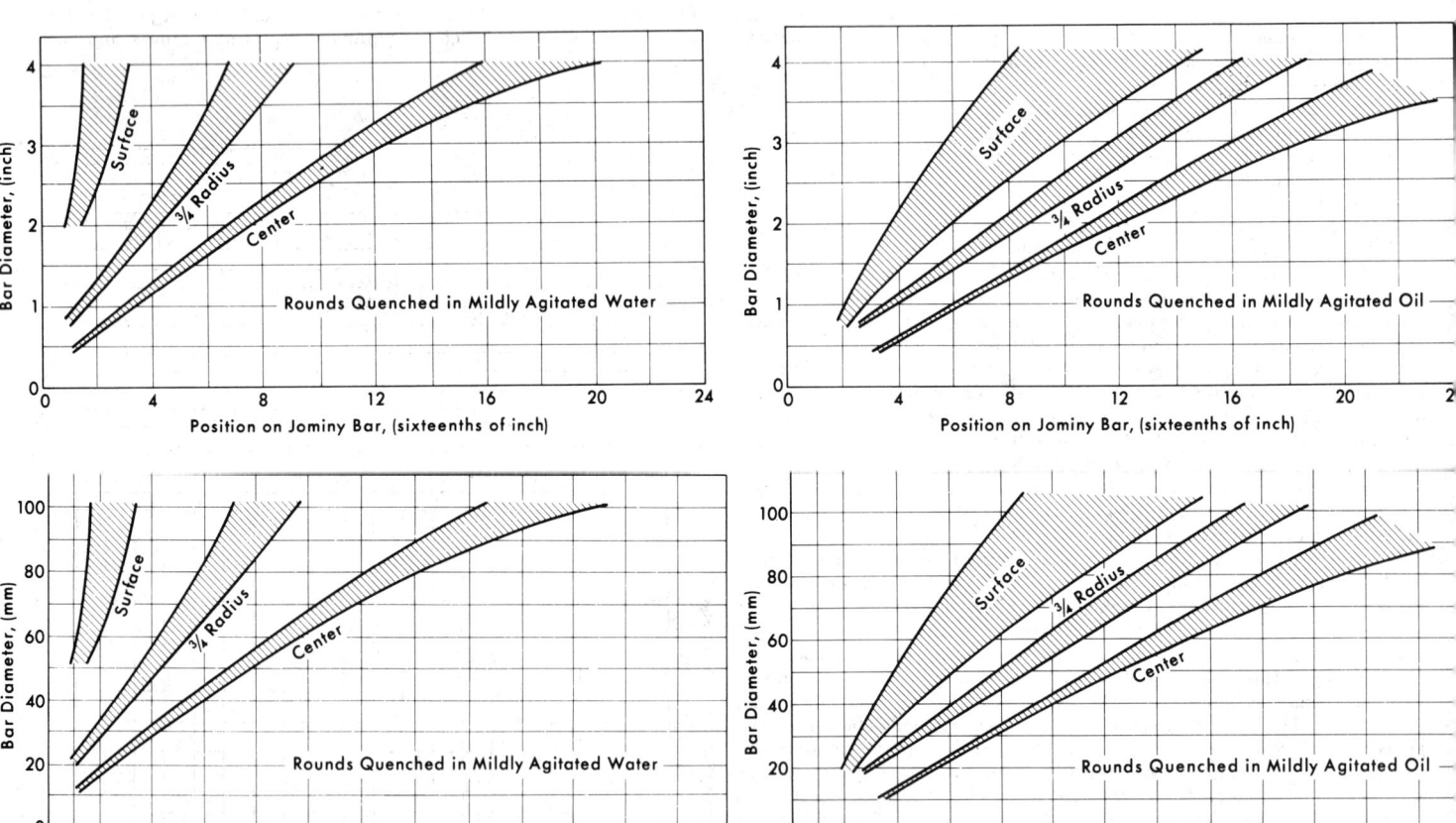

FIG. 7—CORRELATION OF COOLING RATES IN JOMINY BAR AND QUENCHED ROUND BARS

These charts have been revised by a Division 8 task force, to reflect probable variations in actual cooling rates of round bars.

TABLE 3—ORIFICES FOR QUENCHING SMALL SPECIMENS

Test Bar Dia		Orifice Size		Distance from Orifice to Quenched End of Specimen		Free Height of Water Column	
in	mm	in	mm	in	mm	in	mm
3/4	19	1/2	13	1/2	13	2-1/2	63
1/2	13	1/4	6.3	3/8	9.5	4	102
1/4	6.3	1/8	3.2	1/4	6.3	8	203

Hardness curves from these smaller bars are not comparable with curves from the 1 in (25 mm) diameter specimen. If the standard hardness curve from subsize specimens is needed, it becomes necessary to make actual cooling rate determination on the subsize specimen in question.

HARDENABILITY TESTS FOR SHALLOW HARDENING STEELS

Introduction—This method covers two different tests: the 1 in (25 mm) standard hardenability specimen method, and the SAC hardenability test method. These methods are applicable to carbon and low alloy steels, other than carbon tool steels, and are suitable only for shallow hardening steels which will not harden completely through in 1 in (25 mm) and larger diameters with a water quench.

The 1 in (25 mm) Standard Hardenability Specimen Method

Procedure—The 1 in (25 mm) standard hardenability specimen may be used to determine the hardenability of shallow hardening steels other than the carbon tool steels by a modification in the hardness survey. The procedure in preparing the specimen prior to hardness measurements is that given in the section on Test Procedure for 1 in (25 mm) Standard Hardenability Specimen. An anvil providing a means of very accurately measuring the distance from the quenched end is essential.

Hardness values are obtained from $1/16$–$1/2$ in (1.6–13 mm) from the quenched end in intervals of $1/32$ in (0.8 mm). Beyond $1/2$ in (13 mm) from the quenched end, the intervals are the same as for the 1 in (25 mm) standard hardenability specimen. For readings to $1/2$ in (13 mm) from the quenched end, two hardness traverses are made, both with readings $1/16$ in (1.6 mm) apart; one starting at $1/16$ in (1.6 mm) and being completed at $1/2$ in (13 mm) from the quenched end, and the other starting at $3/32$ in (2.4 mm) and being completed at $15/32$ in (12 mm) from the quenched end.

Only two flats 180 deg apart need be ground if the mechanical fixture has a grooved bed which will accommodate the indentations on the flat surveyed first. The second hardness traverse is made after turning the bar over. If the fixture does not have such a grooved bed, two pairs of flats should be ground, the flats of each pair being 180 deg apart. The two hardness surveys are made on adjacent flats.

For plotting test results, the Standard Form for Plotting Hardenability Curves (Fig. 6) should be used. Distances for the odd number $1/32$ in (0.8 mm) should be estimated with care.

SAC Hardenability Test Method

Introduction—This method is referred to as the SAC test. The designation SAC means Surface-Area-Center, the area being determined and calculated from a cross section hardness survey of a suitably quenched 1 in (25 mm) diameter cylindrical test specimen.

Sampling—The bar from which the specimen is machined shall be a forged or rolled $1\frac{1}{4}$ in (32 mm) diameter round representing the full cross section of the product. Specimen locations representing other than the full cross section may introduce factors tending to offset reproducibility of results and shall be subject to agreement between steel producer and user.

In all cases, specimens shall be machined from bars of sufficient diameter to ensure freedom from decarburization.

Normalizing Prior to Heating for Quench—The material from which the test specimen is to be machined shall be normalized, holding for 1 h at the appropriate temperature designated in Table 1 followed by cooling in air. The record of hardenability test results must always state the prior thermal history of the specimen tested.

Test Specimens—The test specimen is a 1 in (25 mm) diameter cylinder 4 in (102 mm) long, as shown in Fig. 8. Specimen A is for use with quenching fixture A, Fig. 9; and specimen B is for use with quenching fixture B, Fig. 9.

Heating for Quenching—The specimen shall be heated to the austenitizing temperature shown in Table 1. The specimen shall be held at the specified temperature for 30 min. The furnace chamber shall be muffle type, and the amount of scaling controlled only by standardizing the time of exposure.

Quenching—Frequently SAC test specimens are hand quenched by laboratories that do not have the special quenching fixture shown in Fig. 9. In most

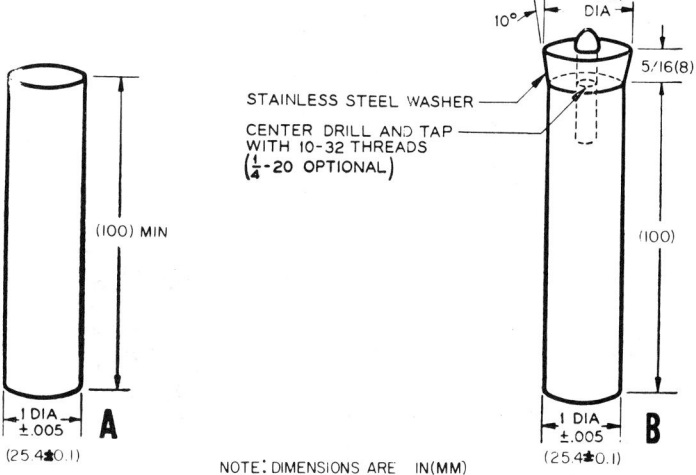

FIG. 8—SAC TEST SPECIMENS

cases, hand quenching yields satisfactory results provided the specimen is well agitated during quenching. However, it is recommended that one of the quenching fixtures shown in Fig. 9 be used. The apparatus shall be adjusted so that the water flows upward through the top opening to a free height of $2\frac{1}{2}$ in (63 mm) without specimen, guide tube, or basket in place. The water temperature shall be 65–85°F (18–30°C).

The specimen shall be removed from the furnace by gripping near one end with tongs, and it shall then be dropped into the basket or guide tube. Time to remove the specimen from the furnace and drop it into the quenching fixture shall be 5 s max. The specimen shall remain in the quench for 60 s and shall then be removed by lifting out the basket or guide tube and inverting.

Preparation for Hardness Testing—The quenched specimen shall be cut transversely so that the top portion is 2 in (50 mm) long as illustrated in Fig. 10. The test slug can be $1/2$–1 in (13–25 mm) long. A suitable soft cutoff wheel shall be used with precaution against heating of the specimen.

The faces of the test slug shall be ground flat and parallel and finished with a grit sufficiently fine to permit satisfactory hardness determination. Four flats shall be ground 0.005–0.008 in (0.12–0.20 mm) deep at 90 deg spacing on the cylindrical surface of the test slug.

The test surfaces shall be checked for freedom from tempering by etching as recommended in the section on Hardness Measurement on medium hardening steels. The presence of darkened areas indicates that tempering has taken place and grinding shall be continued until tempering effects are removed.

Hardness Testing—Surface hardness shall be taken on the surfaces of the four ground flats. The average of these readings shall represent the surface hardness.

Face readings shall be obtained by testing at $1/16$ in (1.6 mm) increments across two diameters 90 deg removed. A fixture suitable for indexing the specimen is illustrated in Fig. 11. Readings for each comparable distance from the surface shall be averaged. Since only one center reading can be taken, the average center hardness shall be the average of five readings; the center reading and the four readings taken at $1/16$ in (1.6 mm) from the center.

If an indexing fixture is not available, the surface to be tested may be lightly scribed with two diameters 90 deg removed, and with seven circles $1/16$ in (1.6 mm) removed. Test locations shall be on each circle at the intersection of each of the four radii.

Estimation of Area—Area shall be computed by Rockwell Inch (Rockwell millimetre) units, as shown in Fig. 12. Ordinates shall be Rockwell C units starting with Rockwell C of zero. Abscissas shall constitute 1 in (25 mm) max. Areas shall be Rockwell Inches (Rockwell millimetres) under the curve S-s', computed by the total of the trapezoids.

Index of Hardenability—The hardenability (and hardness) of a steel shall be designated by a code known as the SAC number. The code shall consist of a set of three numbers, including first, the surface hardness; second, the Rockwell Inch Rockwell millimetre) area; and last, the center hardness, each of which shall be determined as herein described. Values shall be rounded to the nearest whole number.

Example: SAC Number 63-52-42 indicates surface hardness of 63 HRC, Rockwell Inch area of 52, and center hardness of 42 HRC. The equivalent example using Rockwell millimetre units for area would be 63-1320-42.

Record of Test—Test results shall be recorded on a suitable form as illustrated in Fig. 13.

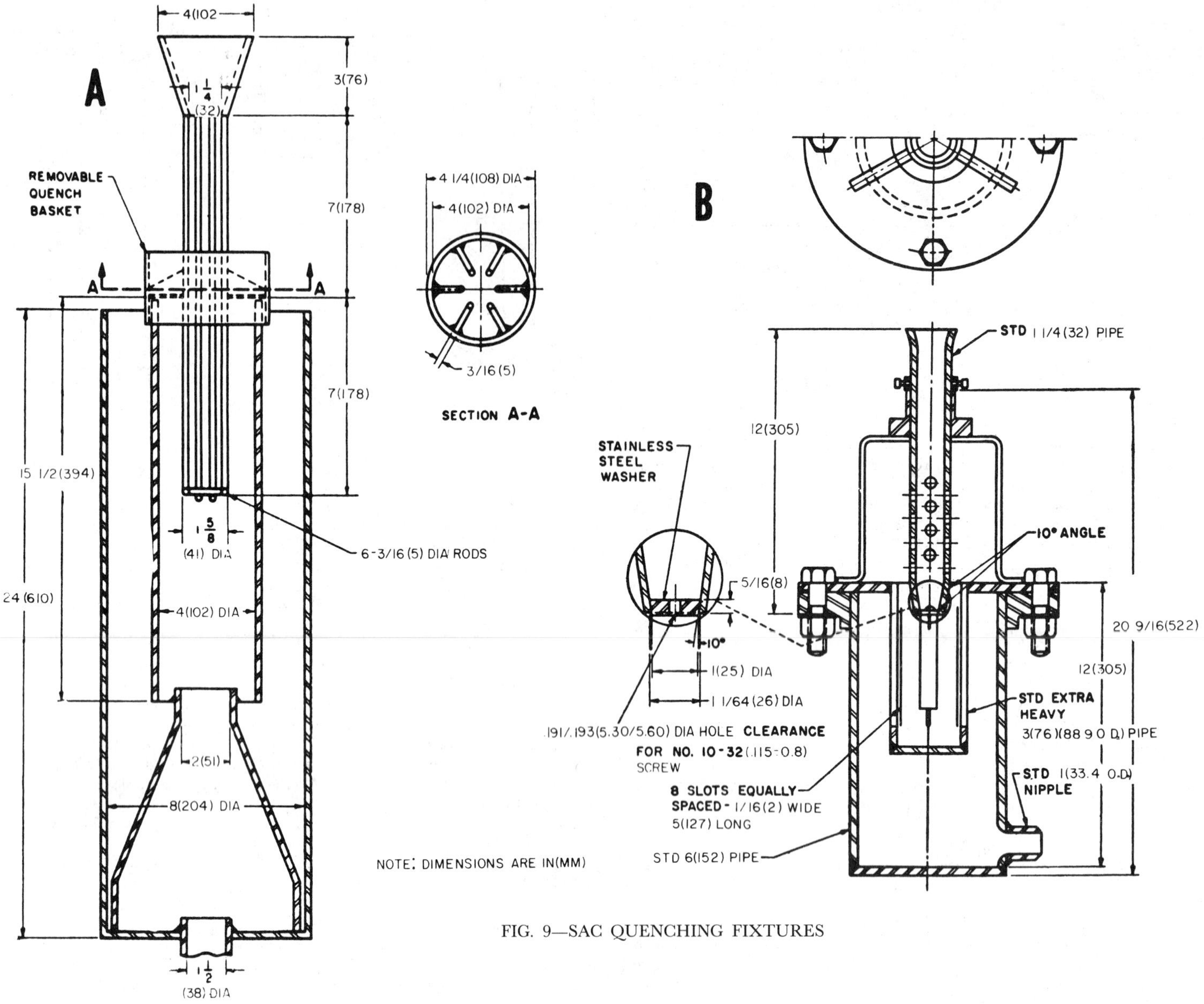

FIG. 9—SAC QUENCHING FIXTURES

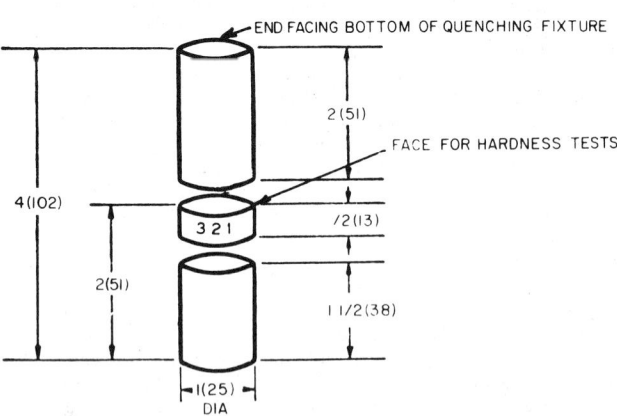

FIG. 10—LOCATION OF TEST SLUG AND OF STAMPED IDENTIFICATION

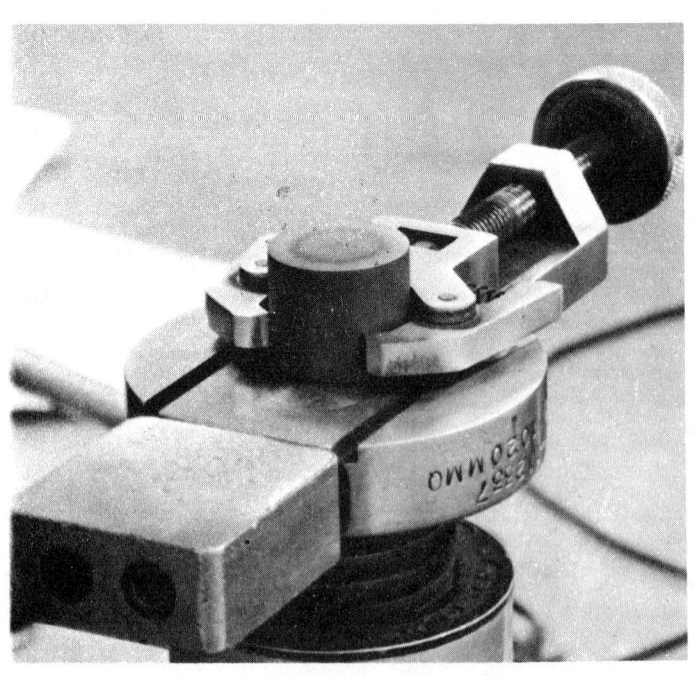

FIG. 11—INDEX FIXTURE

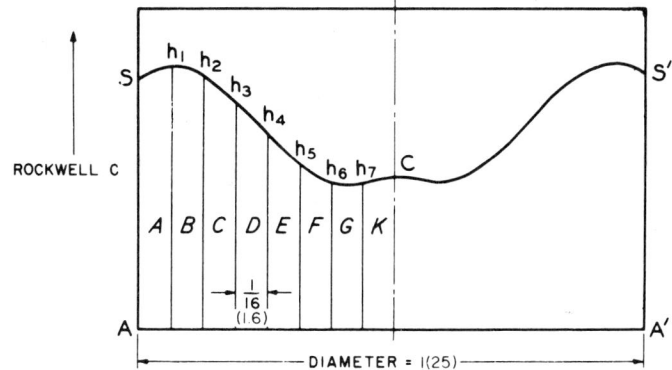

Let S = average surface hardness
h_1, h_2, h_3, and so forth = average hardness at depths indicated
C = average center hardness

Then: area of $A = \dfrac{S + h_1}{2} \times 1/16\ (1.6)$ area of $B = \dfrac{h_1 + h_2}{2} \times 1/16\ (1.6)$

total area $= 2\,(A + B + C + D + E + F + G + K)$

$= 1/8\,(3.2)\left(\dfrac{S}{2} + h_1 + h_2 + h_3 + h_4 + h_5 + h_6 + h_7 + \dfrac{C}{2}\right)$

NOTE: DIMENSIONS ARE IN (mm)

FIG. 12—SAC ESTIMATION OF AREA

APPENDIX
METHODS FOR DETERMINING CARBURIZED HARDENABILITY

Scope—This method prescribes procedures for determining the hardenability of steels after carburizing and for subsequently recording the results. It is of interest to note that such a procedure was used by Walter Jominy when he first introduced the end-quench test. Information on carburized hardenability is important in controlling carburizing and quenching practice, and in determining the ability of a specific steel to meet the hardness-case depth requirements for a specific part.

Carburized parts are used chiefly where high surface stresses are imposed. Failures generally originate in the surface layers where the service stresses are most severe. Generally, therefore, high case strength and high endurance limit are critical factors. It has been proved that compressive stresses in the case improve the fatigue durability and that high case hardness is associated with high durability. These compressive stresses are created by developing a high-hardness high-carbon case on a low-hardness low-carbon core. The core must have adequate hardness and strength to support the hard case, but increasing core hardness over the necessary minimum reduces compressive stresses at the surface and therefore decreases the fatigue resistance.

Originally, it was considered that core hardenability alone was needed for steel selection and heat treatment of carburized parts, as adequate core hardenability. Equal additions of carbon, however, do not have the same effect on the hardenability of all base compositions; conclusions reached on the basis of core hardenability may not be correct for the case. It has, therefore, become clear that adequate hardenability of both case and core is essential for proper selection of the optimum grade and control of its processing to a specific part.

Test Procedure—The end-quench specimens and quenching and testing procedures, described in detail in previous sections of this standard, are used. When evaluating the carburized hardenability characteristics of a steel, high-carbon-potential pack carburizing procedures are employed as described in the next paragraph. Results using this practice have been reported previously.[2,3]

Direct Quench—In the determination of case hardenability, a standard end-quench hardenability specimen and a carbon-gradient specimen, 1 in diameter by 6 in (25 x 152 mm) long, prepared from the same bar, are simultaneously carburized in a covered alloy steel box for 9 h at 1700°F (925°C). The composition of the carburizing medium is to be: charcoal 50%, coke 30%, barium carbonate 12%, sodium carbonate 3%, calcium carbonate 3%, molasses binder 2%.

All new carburizer is used for each batch to provide uniform carburizing

[2] J. A. Halgren and E. A. Solecki, "Case Hardenability of SAE 4028, 8620, 4620, and 4815 Steels." SAE Transactions, Vol. 69 (1961), p. 662.
[3] Atlas "Hardenability of Carburized Steels." New York: Climax Molybdenum Co., 1960.

SAC TEST REPORT

DATE_____   TEST NO._____
LABORATORY_____   SOURCE_____

GRADE	HEAT NO.	GRAIN	ANALYSIS							
			C	Mn	S	P	Si	Ni	Cr	Mo

NORMALIZING TEMP (F)_____   QUENCHING TEMP (F)_____
REMARKS_____
_____
_____

	INDIVIDUAL READINGS	AVERAGE ORDINATES		INDIVIDUAL READINGS	AVERAGE ORDINATES
S		S/2	S		S/2
1		1	1		1
2		2	2		2
3		3	3		3
4		4	4		4
5		5	5		5
6		6	6		6
7		7	7		7
C		C/2	C		C/2
TOTAL ORDINATES			TOTAL ORDINATES		
AREA = TOTAL ORDINATES DIVIDED BY 8			AREA = TOTAL ORDINATES DIVIDED BY 8		
HARDENABILITY RATING	S | A | C		HARDENABILITY RATING	S | A | C	

FIG. 13—SAC TEST REPORT

conditions and to overcarburize so the highest carbon level to be investigated (1.10%) will be sufficiently subsurface to permit accurate location.

The hardenability specimen is end quenched, and the carbon-gradient bar is either cooled in loose hydrated lime or immersion quenched in oil. If oil quenched, the carbon-gradient bar is tempered at 1200°F (650°C) for 10 min in lead or salt to soften it for machining. Samples for carbon analysis are lathe turning machined in radial increments 0.005 in (0.13 mm) deep. The carbon-gradient curve is obtained by plotting the carbon content for each radial increment against the average depth of the increment below the surface.

On the assumption that the distribution of carbon in the end-quench specimen is the same as in the carbon-gradient bar, parallel flats are ground on the end-quench specimen to levels corresponding to carbon concentrations of 1.10, 1.00, 0.90% and, in some cases, lower carbon levels.

To minimize the effect of softer underlying layers, Rockwell A hardness values are determined with impressions along the centerline of each flat. The A values are converted to C values using conversion tables given in SAE J417. The hardness value at the $1/16$ in (1.6 mm) position is affected by carburizing the end of the bar, therefore this reading is discarded. If hardness values at the $1/16$ in (1.6 mm) position are desired, the quenched end can be copper-plated to prevent carburizing.

A pictorial representation of the procedure, giving an example of a carbon-gradient curve, and the sequence of operations is shown in Fig. A.1. Grinding the end-quenched hardenability bar is critical. Extreme care should be exercised to avoid tempering. See the section of this standard entitled Hardness Measurement.

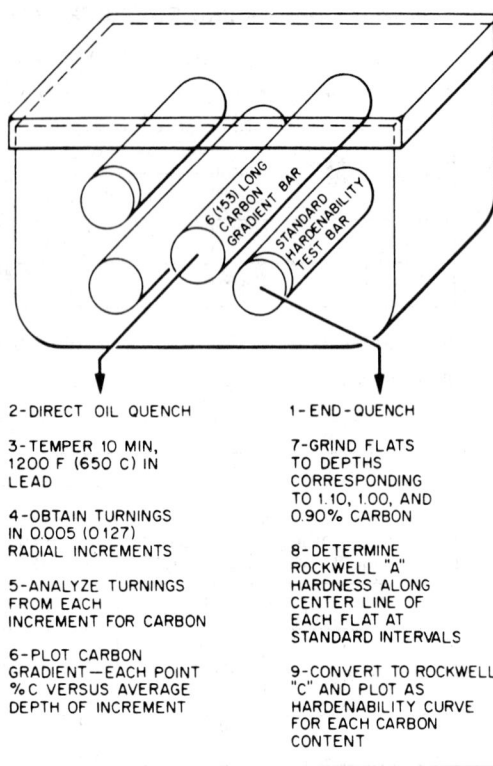
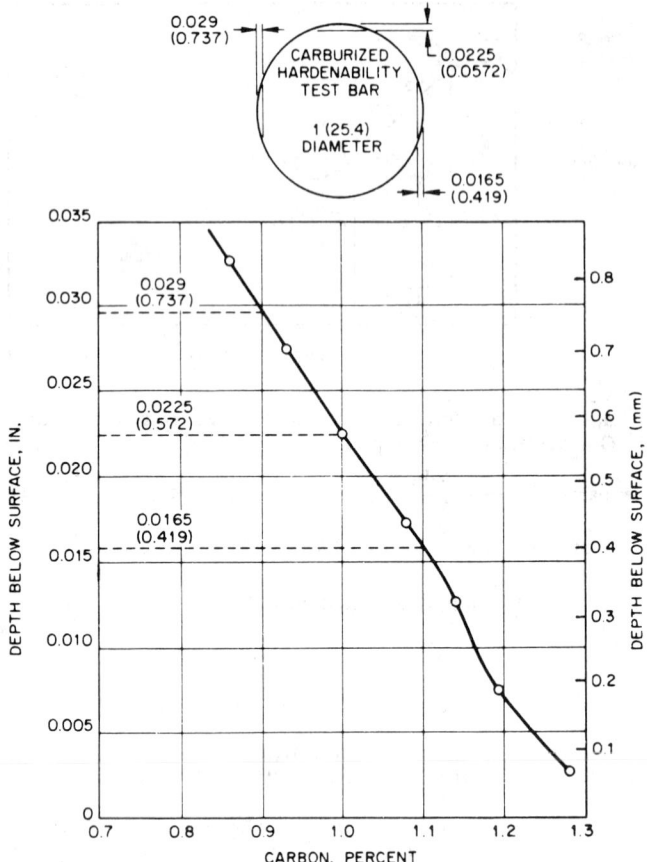

FIG. A-1—STANDARD CARBURIZING PROCEDURE, PACK CARBURIZE 9 h AT 1700°F (925°C).

Reheat and Quench—Steels may be tested under reheat and quench conditions by modifying the practice slightly. The end-quench specimens and the corresponding carbon-gradient bars are pack-carburized as described above, then the bars are removed from the box and either cooled in still air or oil-quenched, depending upon the proposed plant practice. The specimens are then reheated in a controlled atmosphere furnace held at 1550°F (845°C), for a total time of 55 min in the furnace to approximate the specified 30 min at furnace temperature. The hardenability specimen is then end quenched and the carbon-gradient bar is either cooled in lime or oil quenched and tempered as described above.

Alternate Procedures—It is apparent that the test can be tailored to suit individual plant practice, but the procedures described above should be used when comparing results with other laboratories.

An example of an alternate procedure would be to grind flats on the end-quench specimen before carburizing, then carburize the specimen in the same carburizing furnace with parts, end quenching the specimen after carburizing. Surface hardenability may be determined on the preground flats, and hardenability as a function of case depth can be determined by grinding flats to specified depths.

Reproducibility—The method described for direct quenching provides good reproducibility, as indicated by two tests. In the first test, four carbon-gradient bars and four end-quench bars machined from the same normalized bar stock were simultaneously carburized and sequentially quenched.

In the second test, the case hardenability of one heat was determined on three separate occasions with carburizing temperatures between 1700 and 1750°F (925 and 955°C).

The results of the reproducibility tests are given in Fig. A.2.

BIBLIOGRAPHY

Selected References on Hardenability Testing

1. W. E. Jominy and A. L. Boegehold, "A Hardenability Test for Carburizing Steel." ASM Transactions, Vol. 26 (1938), No. 2, pp. 574–599.
2. J. L. Burns, T. L. Moore, and R. S. Archer, "Quantitative Hardenability." ASM Transactions, Vol. 26 (1938), No. 1, pp. 1–33.
3. W. E. Jominy, "A Hardenability Test for Shallow Hardening Steels." ASM Transactions, Vol. 27 (1939), pp. 1072–1085.
4. "Symposium on Hardenability of Alloy Steels." ASM 1939.
5. M. Asimow and M. A. Grossmann, "Hardening Characteristics of Various Shapes." ASM Transactions, Vol. 28 (1940), pp. 949–977.
6. "Standardization Sought in Determining the Hardenability of Steels" (a symposium). SAE Journal, Vol. 49, No. 1 (July 1941), pp. 266–293.
7. A. E. Focke, "Hardenability of Steel." Iron Age, Aug. 20, 1942, pp. 37–40; Aug. 27, 1942, pp. 43–51; Sept. 3, 1942, pp. 56–59.
8. Morse Hill, "The End-Quench Test: Reproducibility." ASM Transactions, Vol. 31 (1943), p. 923 ff.
9. "Symposium on the Hardenability of Steel." Special Report No. 36, British Iron and Steel Institute, 1946.
10. G. K. Manning, "End Quench Hardenability Versus Hardness of Quenched Rounds." Metal Progress, Vol. 50, No. 4 (October 1946), pp. 647–650.
11. E. W. Wienman, R. F. Thomson, and A. L. Boegehold, "Correlation of End Quenched Test Bars and Rounds in Terms of Hardness and Cooling Characteristics." ASM Transactions, Vol. 44 (1952), pp. 802–834.
12. G. K. Manning, "Comparison of Tests for Hardenability of Shallow Hardening Steels." SAE Journal, Vol. 61, July 1953, pp. 30–36.
13. D. J. Carney, "Another Look at Quenchants, Cooling Rates and Hardenability." ASM Transactions, Vol. 46 (1954), pp. 882–925.
14. John Birtalan, R. G. Henley, Jr., and A. L. Christenson, "Thermal Reproducibility of the End-Quench Test." ASM Transactions, Vol. 46 (1954), p. 928 ff.

Selected References on Calculating Hardenability

1. M. A. Grossmann and R. L. Stephenson, "The Effect of Grain Size on Hardenability." ASM Transactions, Vol. 29 (1941), pp. 1–19.
2. M. A. Grossmann, "Hardenability Calculated from Chemical Compositions." AIME Transactions, Vol. 150 (1942), pp. 227–259.
3. I. R. Kramer, S. Siegel, and J. Brooks, "Factors for the Calculation of Hardenability." AIME Transactions, Vol. 163 (1946), p. 670 ff.
4. C. F. Jatczak and D. J. Girardi, "Multiplying Factors for the Calculation of Hardenability of Hypereutectoid Steels Hardened from 1700 F." ASM Transactions, Vol. 51 (1960), p. 335 ff.
5. A. F. deRetana and D. V. Doane, "Predicting the Hardenability of Carburizing Steels." Metal Progress, September 1971, p. 65.
6. C. F. Jatczak, "Determining Hardenability from Composition." Metal Progress, Vol. 100, No. 3 (September 1971), p. 60.
7. D. H. Breen, G. H. Walter, C. J. Keith, and J. T. Sponzilli, "Computer-Based System Selects Optimum Cost Steels." Metal Progress, I: Dec. 1972, p. 42; II: Feb. 1973, p. 76; III: April 1973, p. 105; IV: June 1973, p. 83; V: Nov. 1973, p. 43.

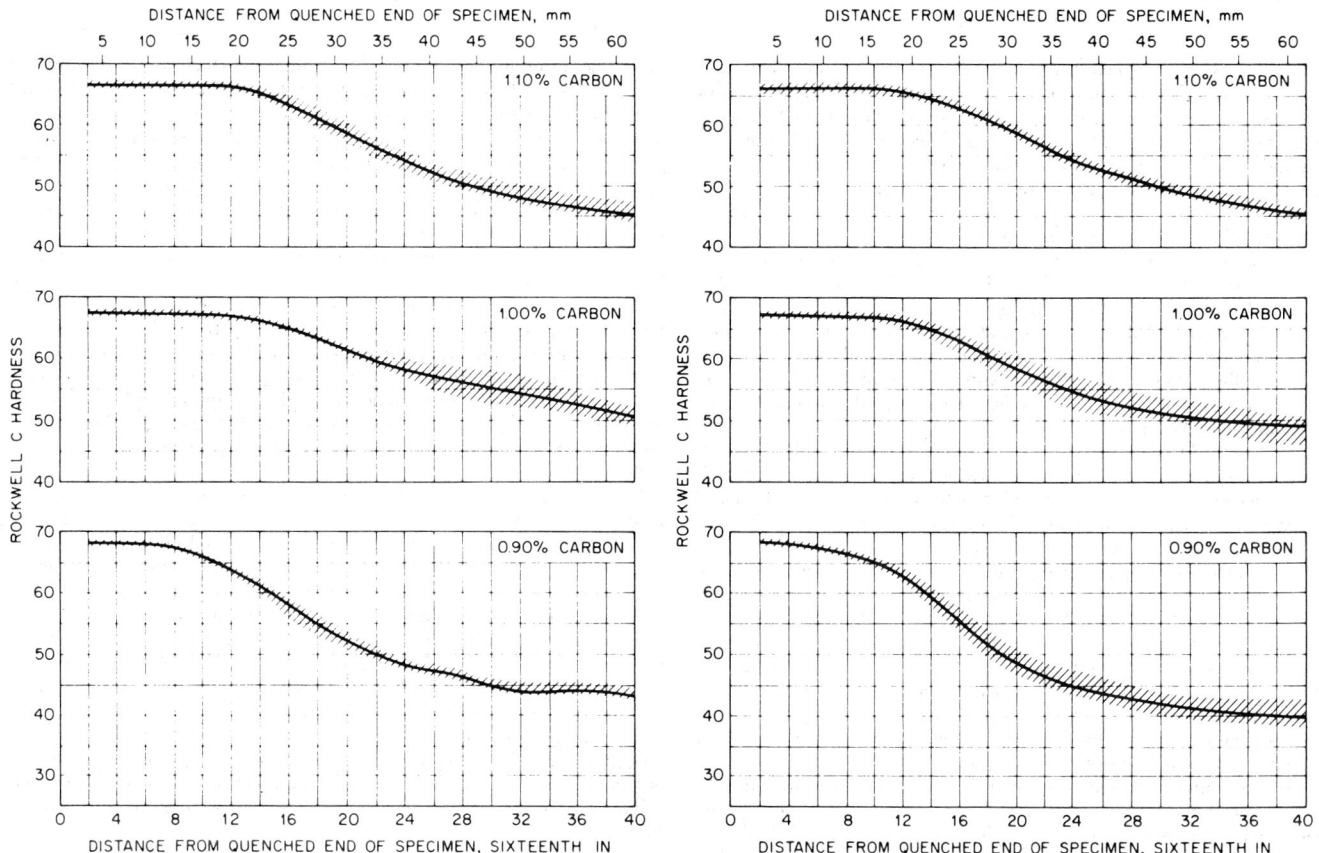

FIG. A-2—REPRODUCIBILITY OF CASE HARDENABILITY

HARDENABILITY BANDS FOR CARBON AND ALLOY H STEELS—SAE J1268 JUN80

SAE Standard

Report of the Iron and Steel Technical Committee, approved June 1980.

[The SAE Iron and Steel Technical Committee—Division 14 in cooperation with the Technical Committee on Alloy Steel Bars of the American Iron and Steel Institute devised hardenability bands for the constructional alloy steels. The initial SAE Standard for H Steels was published in 1948 and numerous revisions and additions have been issued in the intervening years. The SAE Iron and Steel Technical Committee established Division 8 in 1960 to devise hardenability bands for carbon steels. As a result of their efforts, the SAE Recommended Practice J776 came into being.

Since that time, Division 14 has been dissolved and Division 1 came into being and is responsible for Carbon and Alloy Steel Bars, and Division 8 is now responsible for Hardenability for Carbon and Alloy Steels. This trend has also been followed by the American Iron and Steel Institute which now has a technical committee for bars which covers both carbon and alloy steels.

Thus, with this revision, the SAE Recommended Practice J776 and the SAE Standard J407 will be retired and the carbon and alloy H-Band Steels will be combined in this new standard.]

Grades of Steel—H steels and their corresponding minimum and maximum hardenability limits are shown for all of the carbon and alloy steels for which there are sufficient hardenability data and for which grades the standard end quench test can be used. As information is accumulated on other grades, this standard will be revised to include such grades.

Chemical Composition Limits—To permit steel producers the necessary latitude to meet a common standard of hardenability limits, the chemical composition limits of these steels have been modified somewhat from those limits applicable to the same grades when specified by chemical composition only. These modifications permit adjustments in manufacturing ranges of chemistry to correct for individual plant melting characteristics which might otherwise influence the levels and widths of the bands. The modifications have not been great enough to influence the general characteristics of the original compositions of the series under consideration.

The chemical composition limits for electric furnace, open hearth, and BOF steels are outlined in Tables 1 and 2 and are subject to the permissible variations for product analysis outlined in Tables 1 and 3 of SAE J409.

Identification—As a means of identifying steels specified to hardenability band limits, the suffix letter *H* has been added to the conventional series number. In the Unified Numbering System, the H appears as a prefix. It is important that steel consumers use this letter in specification requirements, as there is no other means of determining when hardenability band limits apply. When the letter is used, all conditions pertaining to chemical composition limits, restrictions, testing technique, and so forth, as outlined herein apply.

Grain Size—The limits set forth for bands are intended to apply to steels exhibiting austenitic grain size 5 or finer (see SAE J418). In cases where coarse grain steel is desired, the hardenability limits shall be a matter of agreement between producer and consumer.

Use of Hardenability Limits—Band limits are shown graphically and are so depicted for convenience in estimating the hardness value obtainable at various locations on the end quench test bar and for quick comparisons of the various H grades. The values of *Diameter of Rounds with Same As-Quenched Hardness* shown above each H-band, are approximate and were selected from the ranges appearing in Fig. 7 of SAE J406. It should be noted that starting with this issue of J1268 that Standard H Steels are presented graphically in both U. S. customary units and also in metric (SI) units. The metric hardenability bands were prepared by careful conversion from existing bands in U. S. customary units.

In either case, for specification purposes, the tabulated values of Rockwell hardness (HRC) are used. Values below 20 HRC are not specified because such values are invalid.

Two points from the tabulated values are commonly designated according to one of methods A, B, C, D, or E, which are defined in the following paragraphs. Those various methods are illustrated graphically in Figs. 1 and 2.

A. The minimum and maximum hardness values at any desired distance. This method is illustrated in Figs. 1 and 2 as points A–A.

B. The minimum and maximum distance at which any desired hardness value occurs. This method is illustrated in Figs. 1 and 2 as points B–B. If the desired hardness does not fall on an exact sixteenth (or 1.5 mm) position, the minimum distance selected should be the nearest sixteenth (or 1.5 mm) position toward the quenched end and the maximum should be the nearest sixteenth (or 1.5 mm) position away from the quenched end.

C. Two maximum hardness values at two desired distances, illustrated in Figs. 1 and 2 as points C–C.

D. Two minimum hardness values at two desired distances, illustrated in Figs. 1 and 2 as points D–D.

E. Any minimum hardness plus any maximum hardness, illustrated in Figs. 1 and 2 as points E–E.

When hardenability is specified according to one of the aforementioned methods A to E, the balance of the hardenability band is not applicable.

In cases when it is considered desirable, the maximum and minimum limits at a distance of $\frac{1}{16}$ in (or 1.5 mm) from the quenched end can be specified in addition to the other two points as previously described in paragraphs A to E, inclusive.

When the full H-band is specified the hardenability can be reported by listing hardness values from the quenched end to the test specimen for each $\frac{1}{16}-\frac{16}{16}$ in and $\frac{1}{4}$ in increments from there to $\frac{32}{16}$ in. In the case of the metric end quench test, hardness values would be reported starting at the quenched end, for each 1.5 mm out to 24 mm and in 6 mm increments from there to 51 mm. Except at the $\frac{1}{16}$ in or 1.5 mm position, it is customary to accept a tolerance of two points HRC for a $\frac{3}{16}$ in or 5 mm portion of the curve. This tolerance is necessary because curves of individual heats vary somewhat in shape from the standard band limits and thus deviate slightly at one or more positions in the full length of the curves.

For shallow hardening carbon H steels, distances from the quenched end may be reported by listing hardness values for each $\frac{1}{16}$ in or 1.5 mm or $\frac{1}{32}$ in or 0.75 mm (rather than full sixteenths or 1.5 mm only as with alloy steels).

The technique of testing for acceptance should be in accordance with SAE J406.

General—The hardenability limits set forth have been formulated to conform with steel production practices which represent, at this time, the narrowest limits that can be produced with regular quality steel. Those limits may be used in conjunction with higher quality classifications when desired. In either case, all conditions and quality features outlined in AISI Steel Products Manual Section—Alloy, Carbon and High Strength Low Alloy Steels: Semifinished for Forging, Hot Rolled Bars, Cold Finished Bars 1977 apply except for the conditions concerning chemical compositions, size, and shape of products which are modified herein.

TABLE 1—CARBON AND CARBON BORON H STEELS COMPOSITION

UNS No.	SAE or AISI Steel No.	Chemical Composition, %				
		C	Mn	Si	P, max[b]	S, max[b]
H10380	1038H	0.34/0.43	0.50/1.00	0.15/0.35	0.040	0.050
H10450	1045H	0.42/0.51	0.50/1.00	0.15/0.35	0.040	0.050
H15220	1522H	0.17/0.25	1.00/1.50	0.15/0.35	0.040	0.050
H15240	1524H	0.18/0.26	1.25/1.75	0.15/0.35	0.040	0.050
H15260	1526H	0.21/0.30	1.00/1.50	0.15/0.35	0.040	0.050
H15410	1541H	0.35/0.45	1.25/1.75	0.15/0.35	0.040	0.050
H15211	15B21H[a]	0.17/0.24	0.70/1.20	0.15/0.35	0.040	0.050
H15351	15B35H[a]	0.31/0.39	0.70/1.20	0.15/0.35	0.040	0.050
H15371	15B37H[a]	0.30/0.39	1.00/1.50	0.15/0.35	0.040	0.050
H15411	15B41H[a]	0.35/0.45	1.25/1.75	0.15/0.35	0.040	0.050
H15481	15B48H[a]	0.43/0.53	1.00/1.50	0.15/0.35	0.040	0.050
H15621	15B62H[a]	0.54/0.67	1.00/1.50	0.40/0.60	0.040	0.050

[a] These steels contain 0.0005–0.003% boron.
[b] If electric furnace practice is specified or required, the limits for phosphorus and sulfur are 0.025%, respectively and the prefix E is added to the SAE or AISI number.

TABLE 2[a]—STANDARD ALLOY H STEEL COMPOSITIONS

UNS No.	SAE or AISI Steel No.	Chemical Composition, %[b,c]						
		C	Mn	Si	Ni	Cr	Mo	V
H13300	1330H	0.27/0.33	1.45/2.05	0.15/0.35	—	—	—	—
H13350	1335H	0.32/0.38	1.45/2.05	0.15/0.35	—	—	—	—
H13400	1340H	0.37/0.44	1.45/2.05	0.15/0.35	—	—	—	—
H13450	1345H	0.42/0.49	1.45/2.05	0.15/0.35	—	—	—	—
H40270	4027H	0.24/0.30	0.60/1.00	0.15/0.35	—	—	0.20/0.30	—
H40280[d]	4028H[d]	0.24/0.30	0.60/1.00	0.15/0.35	—	—	0.20/0.30	—
H40320	4032H	0.29/0.35	0.60/1.00	0.15/0.35	—	—	0.20/0.30	—
H40370	4037H	0.34/0.41	0.60/1.00	0.15/0.35	—	—	0.20/0.30	—
H40420	4042H	0.39/0.46	0.60/1.00	0.15/0.35	—	—	0.20/0.30	—
H40470	4047H	0.44/0.51	0.60/1.00	0.15/0.35	—	—	0.20/0.30	—
H41180	4118H	0.17/0.23	0.60/1.00	0.15/0.35	—	0.30/0.07	0.08/0.15	—
H41300	4130H	0.27/0.33	0.30/0.70	0.15/0.35	—	0.75/1.20	0.15/0.25	—
H41350	4135H	0.32/0.38	0.60/1.00	0.15/0.35	—	0.75/1.20	0.15/0.25	—
H41370	4137H	0.34/0.41	0.60/1.00	0.15/0.35	—	0.75/1.20	0.15/0.25	—
H41400	4140H	0.37/0.44	0.65/1.10	0.15/0.35	—	0.75/1.20	0.15/0.25	—
H41420	4142H	0.39/0.46	0.65/1.10	0.15/0.35	—	0.75/1.20	0.15/0.25	—
H41450	4145H	0.42/0.49	0.65/1.10	0.15/0.35	—	0.75/1.20	0.15/0.25	—
H41470	4147H	0.44/0.51	0.65/1.10	0.15/0.35	—	0.75/1.20	0.15/0.25	—
H41500	4150H	0.47/0.54	0.65/1.10	0.15/0.35	—	0.75/1.20	0.15/0.25	—
H41610	4161H	0.55/0.65	0.65/1.10	0.15/0.35	—	0.65/0.95	0.25/0.35	—
H43200	4320H	0.17/0.23	0.40/0.70	0.15/0.35	1.55/2.00	0.35/0.65	0.20/0.30	—
H43400	4340H	0.37/0.44	0.55/0.44	0.15/0.35	1.55/2.00	0.65/0.95	0.20/0.30	—
H43406[f]	E4340H[f]	0.37/0.44	0.60/0.95	0.15/0.35	1.55/2.00	0.65/0.95	0.20/0.30	—
H46200	4620H	0.17/0.23	0.35/0.75	0.15/0.35	1.55/2.00	—	0.20/0.30	—
H47180	4718H	0.15/0.21	0.60/0.95	0.15/0.35	0.85/1.25	0.30/0.60	0.30/0.40	—
H47200	4720H	0.17/0.23	0.45/0.75	0.15/0.35	0.85/1.25	0.30/0.60	0.15/0.25	—
H48150	4815H	0.12/0.18	0.30/0.70	0.15/0.35	3.20/3.80	—	0.20/0.30	—
H48170	4817H	0.14/0.20	0.30/0.70	0.15/0.35	3.20/3.80	—	0.20/0.30	—
H48200	4820H	0.17/0.23	0.40/0.80	0.15/0.35	3.20/3.80	—	0.20/0.30	—
H50401[e]	50B40H[e]	0.37/0.44	0.65/1.10	0.15/0.35	—	0.30/0.70	—	—
H50441[e]	50B44H[e]	0.42/0.49	0.65/1.10	0.15/0.35	—	0.30/0.70	—	—
H50460	5046H	0.43/0.50	0.65/1.10	0.15/0.35	—	0.13/0.43	—	—
H50461[e]	50B46H[e]	0.43/0.50	0.65/1.10	0.15/0.35	—	0.13/0.43	—	—
H50501[e]	50B50H[e]	0.47/0.54	0.65/1.10	0.15/0.35	—	0.30/0.70	—	—
H50601[e]	50B60H[e]	0.55/0.65	0.65/1.10	0.15/0.35	—	0.30/0.70	—	—
H51200	5120H	0.17/0.23	0.60/1.00	0.15/0.35	—	0.60/1.00	—	—
H51300	5130H	0.27/0.33	0.60/1.10	0.15/0.35	—	0.75/1.20	—	—
H51320	5132H	0.29/0.35	0.50/0.90	0.15/0.35	—	0.65/1.10	—	—
H51350	5135H	0.32/0.38	0.50/0.90	0.15/0.35	—	0.70/1.15	—	—
H51400	5140H	0.37/0.44	0.60/1.00	0.15/0.35	—	0.60/1.00	—	—
H51470	5147H	0.45/0.52	0.60/1.05	0.15/0.35	—	0.80/1.25	—	—
H51500	5150H	0.47/0.54	0.60/1.00	0.15/0.35	—	0.60/1.00	—	—
H51550	5155H	0.50/0.60	0.60/1.00	0.15/0.35	—	0.60/1.00	—	—
H51600	5160H	0.55/0.65	0.65/1.10	0.15/0.35	—	0.60/1.00	—	—
H51601[e]	51B60H[e]	0.55/0.65	0.65/1.10	0.15/0.35	—	0.60/1.00	—	—
H61180	6118H	0.15/0.21	0.40/0.80	0.15/0.35	—	0.40/0.80	—	0.10/0.15
H61500	6150H	0.47/0.54	0.60/1.00	0.15/0.35	—	—	—	0.15
H81451[e]	81B45H[e]	0.42/0.49	0.70/1.05	0.15/0.35	0.15/0.45	0.30/0.60	0.08/0.15	—
H86170	8617H	0.14/0.20	0.60/0.95	0.15/0.35	0.35/0.75	0.35/0.65	0.15/0.25	—
H86200	8620H	0.17/0.23	0.60/0.95	0.15/0.35	0.35/0.75	0.35/0.65	0.15/0.25	—
H86220	8622H	0.19/0.25	0.60/0.95	0.15/0.35	0.35/0.75	0.35/0.65	0.15/0.25	—
H86250	8625H	0.22/0.28	0.60/0.95	0.15/0.35	0.35/0.75	0.35/0.65	0.15/0.25	—
H86270	8627H	0.24/0.30	0.60/0.95	0.15/0.35	0.35/0.75	0.35/0.65	0.15/0.25	—
H86300	8630H	0.27/0.33	0.60/0.95	0.15/0.35	0.35/0.75	0.35/0.65	0.15/0.25	—
H86301[e]	86B30H[e]	0.27/0.33	0.60/0.95	0.15/0.35	0.35/0.75	0.35/0.65	0.15/0.25	—
H86370	8637H	0.34/0.41	0.70/1.05	0.15/0.35	0.35/0.75	0.35/0.65	0.15/0.25	—
H86400	8640H	0.37/0.44	0.70/1.05	0.15/0.35	0.35/0.75	0.35/0.65	0.15/0.25	—
H86420	8642H	0.39/0.46	0.70/1.05	0.15/0.35	0.35/0.75	0.35/0.65	0.15/0.25	—
H86450	8645H	0.42/0.49	0.70/1.05	0.15/0.35	0.35/0.75	0.35/0.65	0.15/0.25	—
H86451[e]	86B45H[e]	0.42/0.49	0.70/1.05	0.15/0.35	0.35/0.75	0.35/0.65	0.15/0.25	—
H86500	8650H	0.47/0.54	0.70/1.05	0.15/0.35	0.35/0.70	0.35/0.65	0.15/0.25	—
H86550	8655H	0.50/0.60	0.70/1.05	0.15/0.35	0.35/0.75	0.35/0.65	0.15/0.25	—
H86600	8660H	0.55/0.65	0.70/1.05	0.15/0.35	0.35/0.75	0.35/0.65	0.15/0.25	—
H87200	8720H	0.17/0.23	0.60/0.95	0.15/0.35	0.35/0.75	0.35/0.65	0.20/0.30	—
H87400	8740H	0.37/0.44	0.70/1.05	0.15/0.35	0.35/0.75	0.35/0.65	0.20/0.30	—
H88220	8822H	0.19/0.25	0.70/1.05	0.15/0.35	0.35/0.75	0.35/0.65	0.30/0.40	—
H92600	9260H	0.55/0.65	0.65/1.10	1.70/2.20	—	—	—	—
H93100[f]	9310H[f]	0.07/0.13	0.40/0.70	0.15/0.35	2.95/3.55	1.00/1.45	0.08/0.15	—
H94151[e]	94B15H[e]	0.12/0.18	0.70/1.05	0.15/0.35	0.25/0.65	0.25/0.55	0.08/0.15	—
H94171[e]	94B17H[e]	0.14/0.20	0.70/1.05	0.15/0.35	0.25/0.65	0.25/0.55	0.08/0.15	—
H94301[e]	94B30H[e]	0.27/0.33	0.70/1.05	0.15/0.35	0.25/0.65	0.25/0.55	0.08/0.15	—

[a] The ranges and limits on this table apply only to material not exceeding 200 in² (0.13 m²) in cross sectional area, 18 in (460 mm) in width, or 10 000 lb (4.5 tonne) per piece in weight. Ranges and limits are subject to the permissible variations for product analysis shown in Table 4 of SAE J409.

[b] Small quantities of certain elements may be found in alloy steel which are not specified or required. These elements are to be considered as incidental and acceptable to the following maximum amounts: copper to 0.35%, nickel to 0.25%, chromium to 0.20%, and molybdenum to 0.06%.

[c] For open hearth and basic oxygen steels maximum sulfur content is to be 0.040% and maximum phosphorus content is to be 0.035%. Maximum phosphorus and sulfur in basic electric furnace steels are to be 0.025% each.

[d] Sulfur content range is 0.035/0.050%.

[e] These steels contain 0.0005–0.003% boron.

[f] Electric furnace steel.

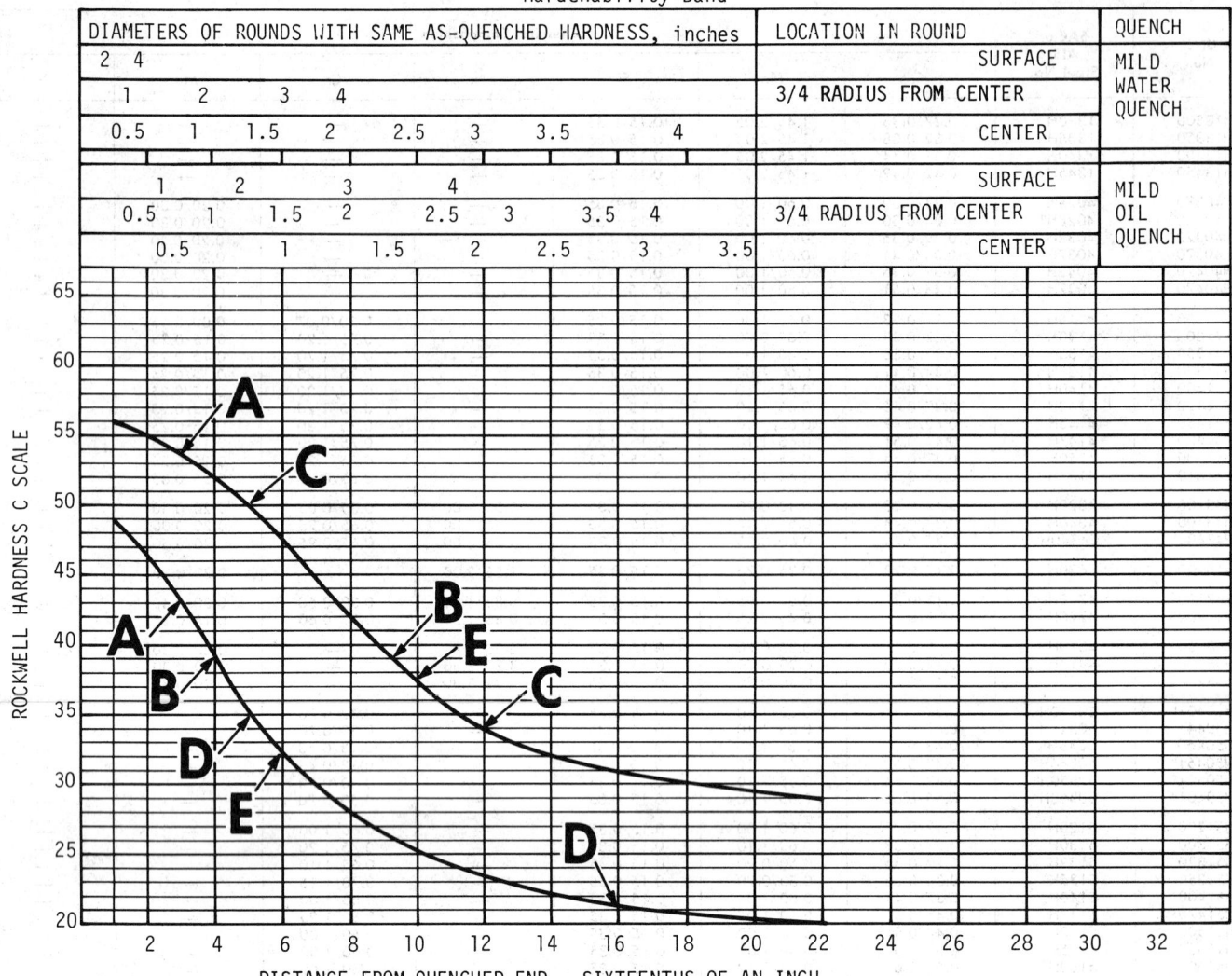

FIG. 1—EXAMPLES ILLUSTRATING ALTERNATE METHODS OF SPECIFYING HARDENABILITY REQUIREMENTS IN U.S. CUSTOMARY UNITS

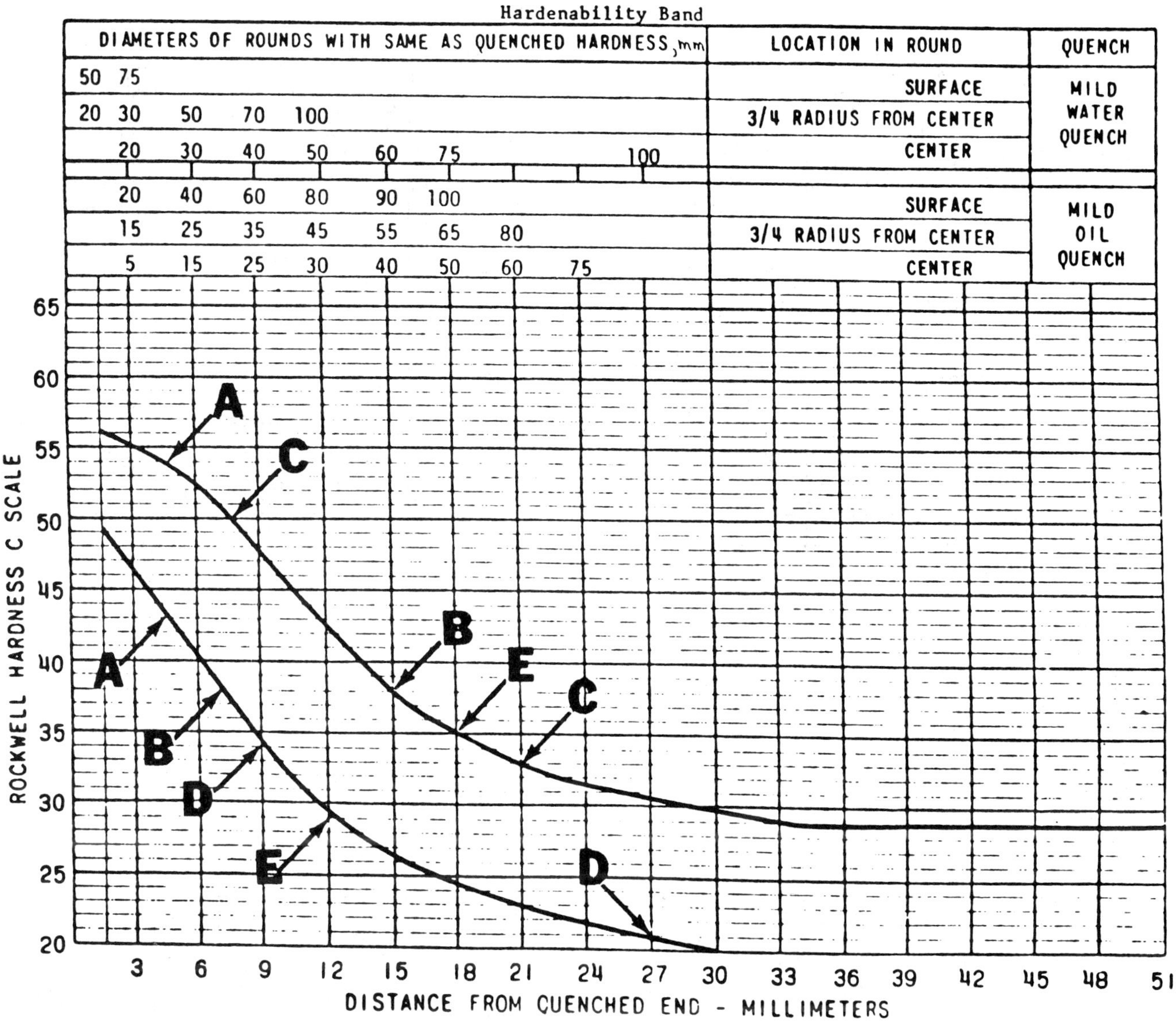

Hardenability Band
DIAMETERS OF ROUNDS WITH SAME AS QUENCHED HARDNESS, mm
LOCATION IN ROUND
QUENCH
50 75
SURFACE
20 30 50 70 100
3/4 RADIUS FROM CENTER
20 30 40 50 60 75 100
CENTER
MILD WATER QUENCH
20 40 60 80 90 100
SURFACE
15 25 35 45 55 65 80
3/4 RADIUS FROM CENTER
5 15 25 30 40 50 60 75
CENTER
MILD OIL QUENCH
ROCKWELL HARDNESS C SCALE
65 60 55 50 45 40 35 30 25 20
3 6 9 12 15 18 21 24 27 30 33 36 39 42 45 48 51
DISTANCE FROM QUENCHED END - MILLIMETERS
A
C
B
E
C
A
B
D
E
D

Method	Example
A Minimum and maximum hardness values at a designated distance	A–A 43 to 54 HRC at J 4.5 mm
B A hardness value at minimum and maximum distances	B–B 38 HRC at J 7.5 mm min 38 HRC at J 15 mm max
C Two maximum hardness values at two designated distances	C–C 50 HRC at J 7.5 mm max 33 HRC at J 21 mm max
D Two minimum hardness values at two distances	D–D 34 HRC at J 9 mm min 21 HRC at J 27 mm min
E Any minimum hardness plus any maximum hardness	E–E 29 HRC at J 12 mm min 35 HRC at J 18 mm max

FIG. 2—EXAMPLES ILLUSTRATING ALTERNATE METHODS OF SPECIFYING HARDENABILITY REQUIREMENTS IN METRIC (SI) UNITS

SERIES A
H-Bands in U. S. Customary Units

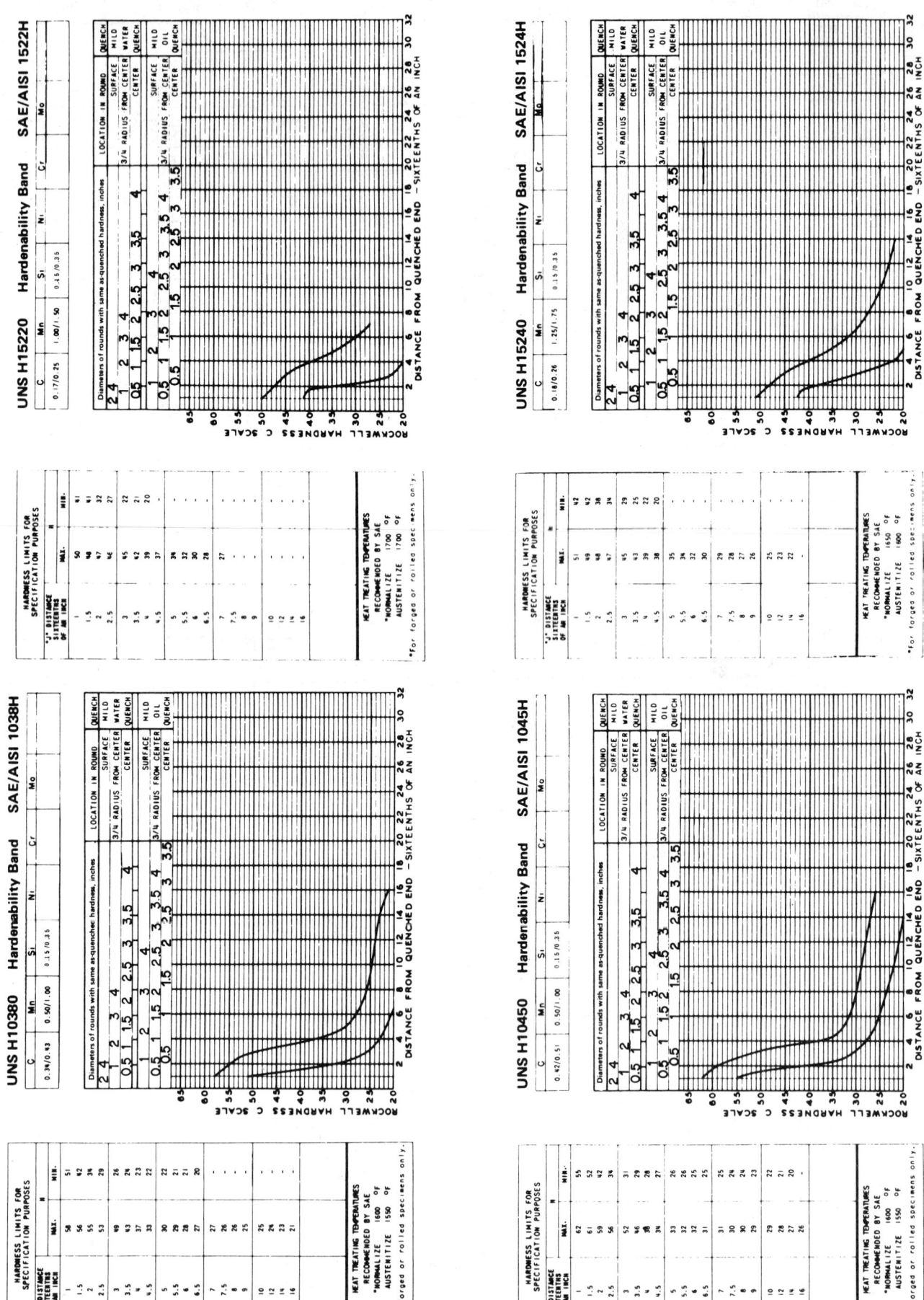

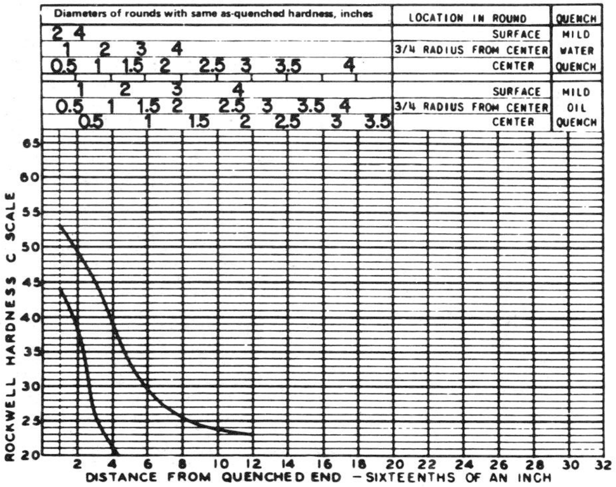

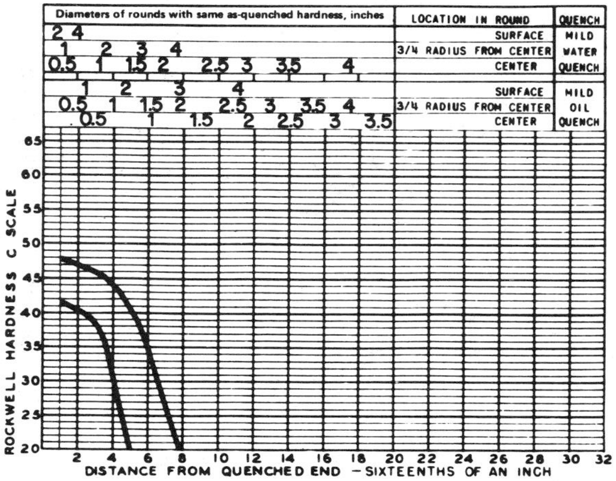

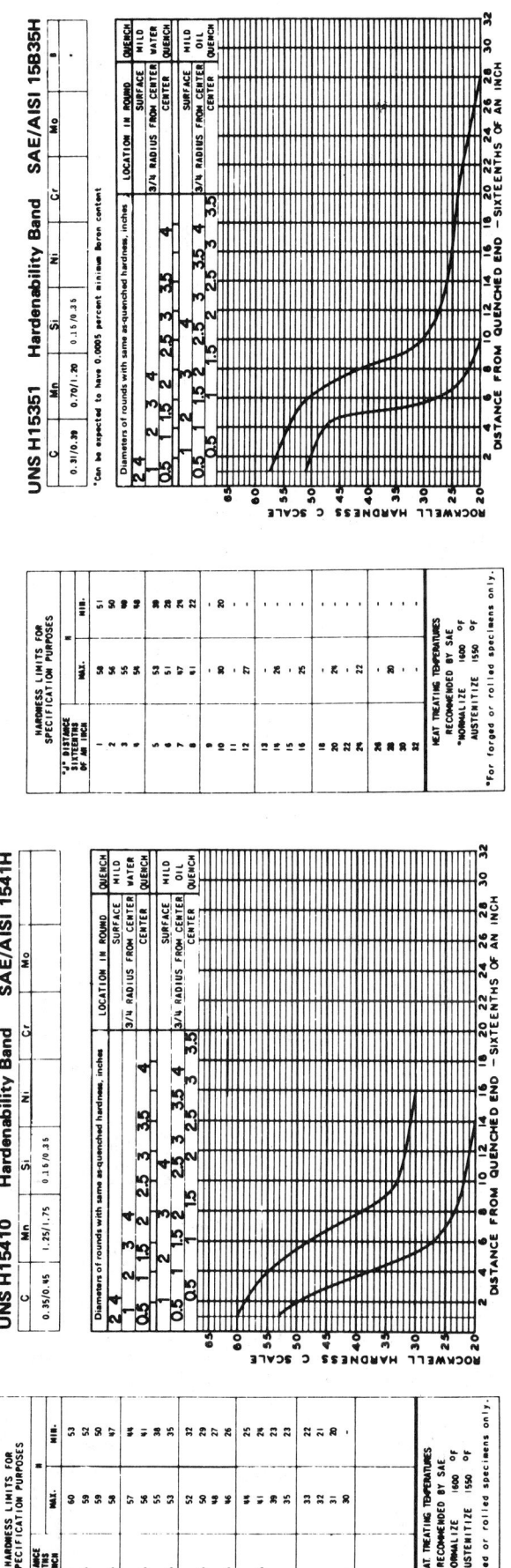

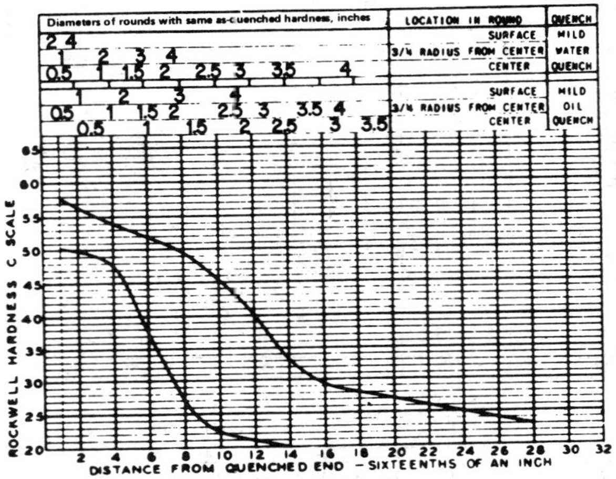

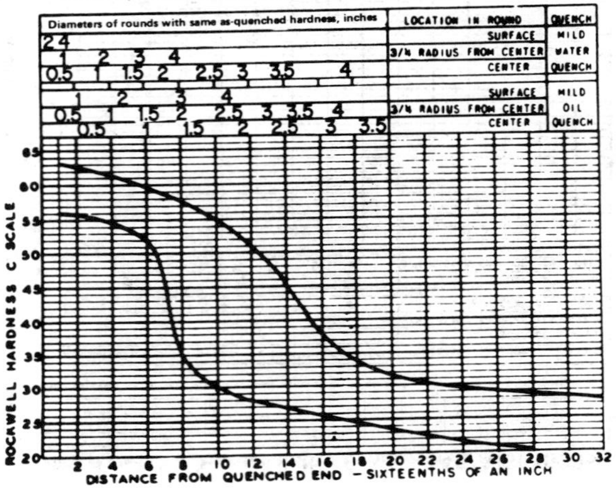

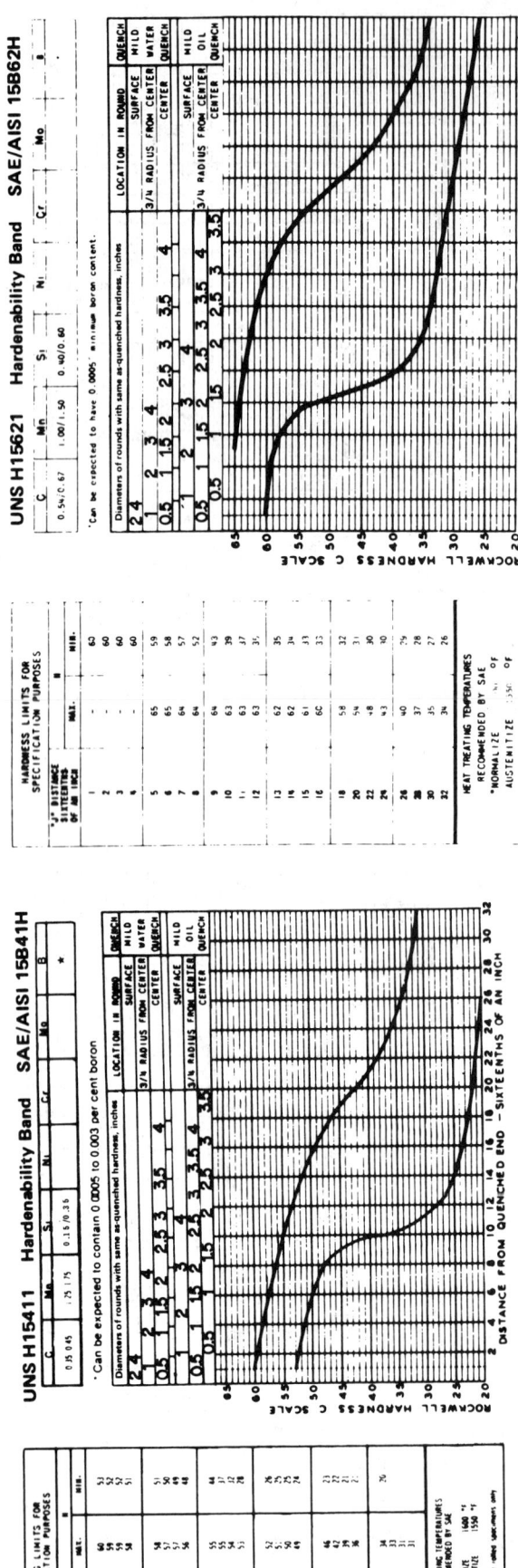

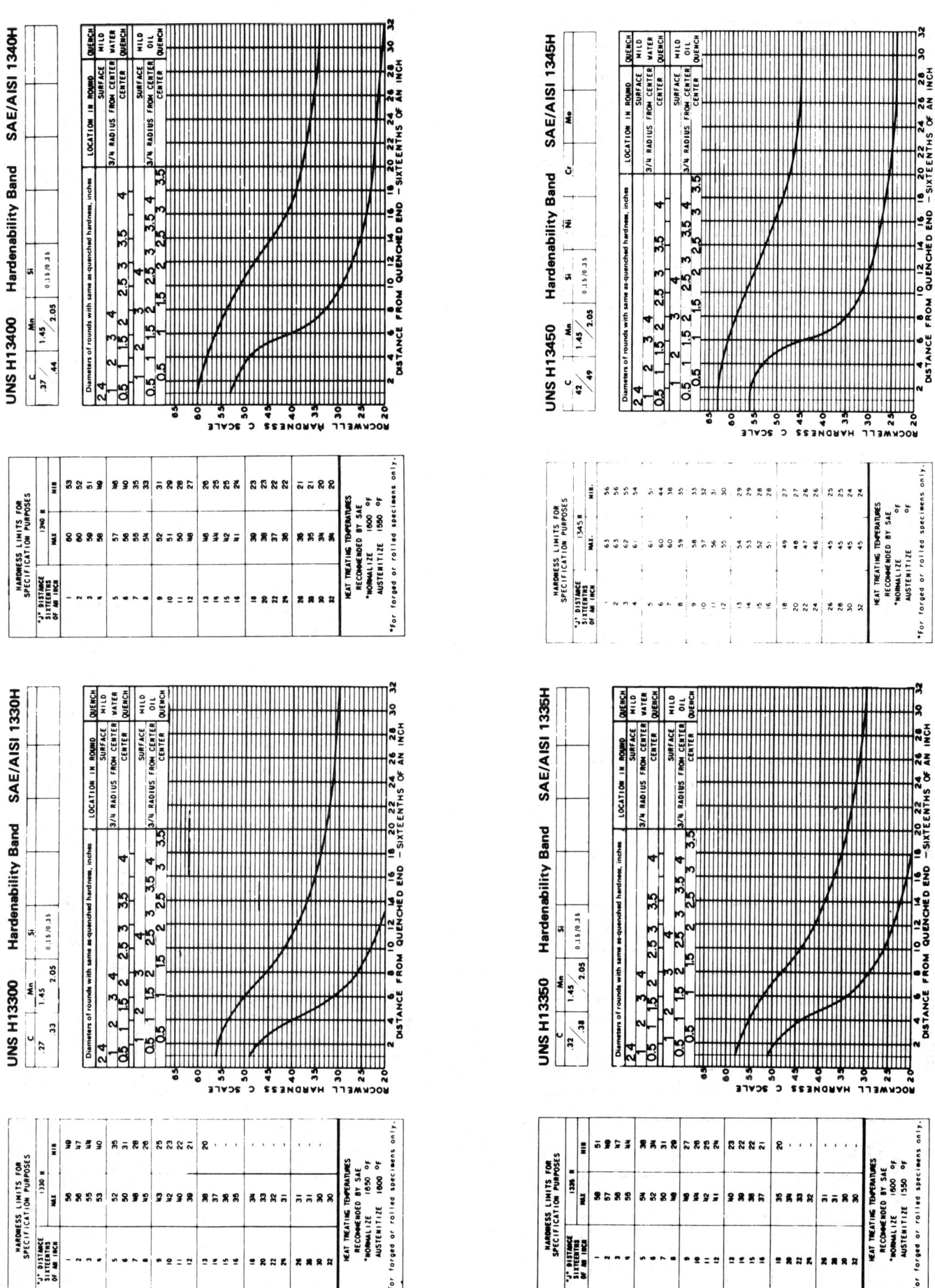

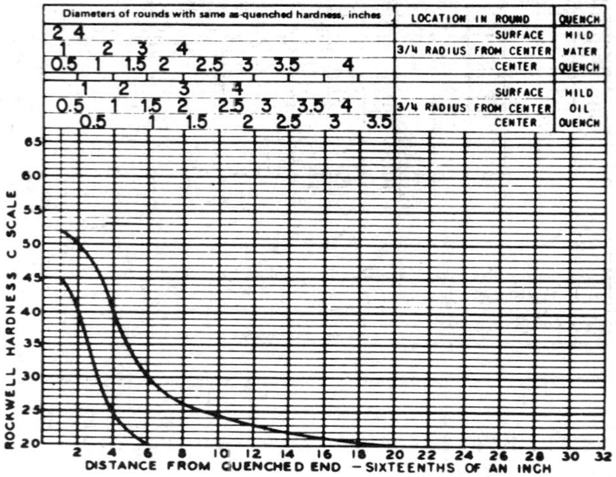

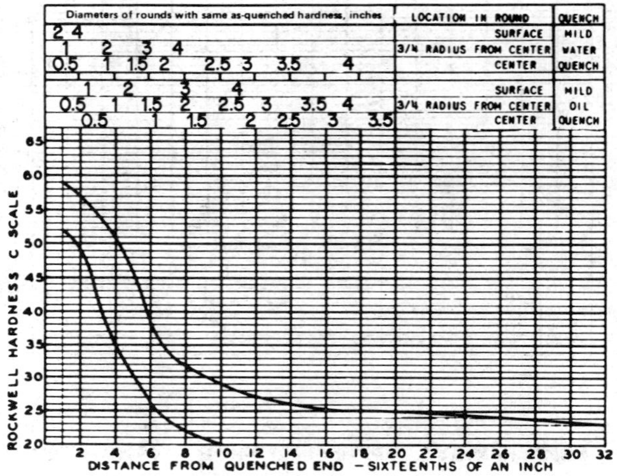

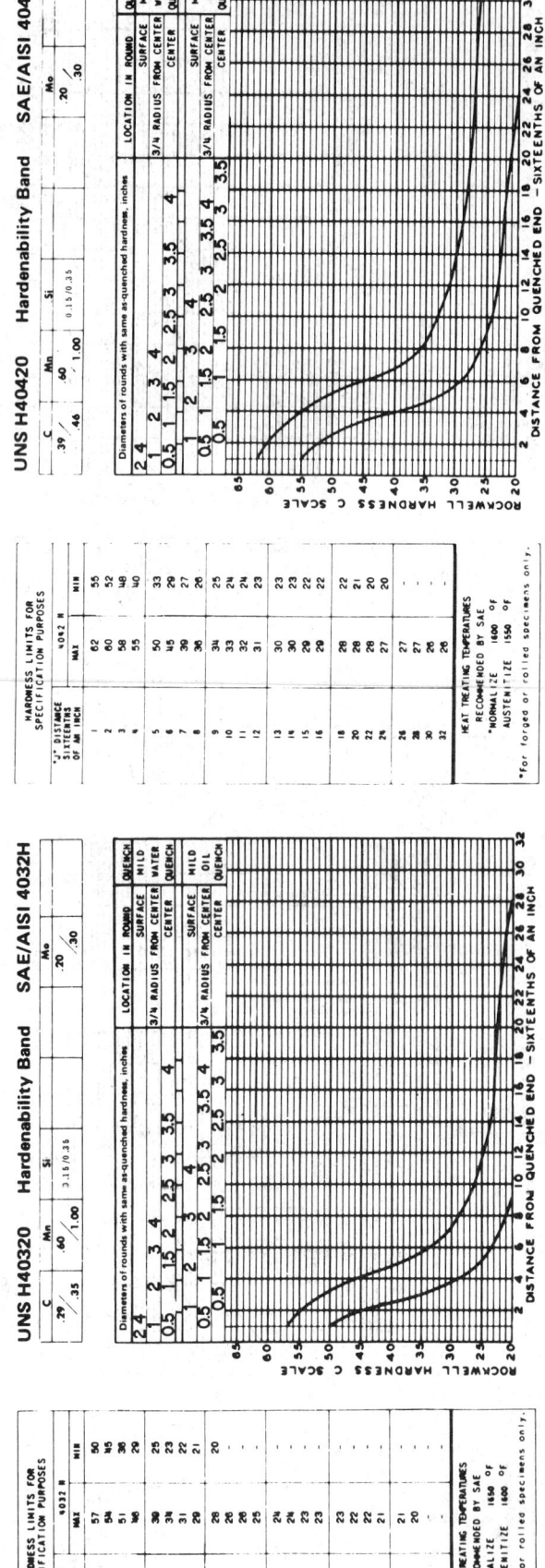

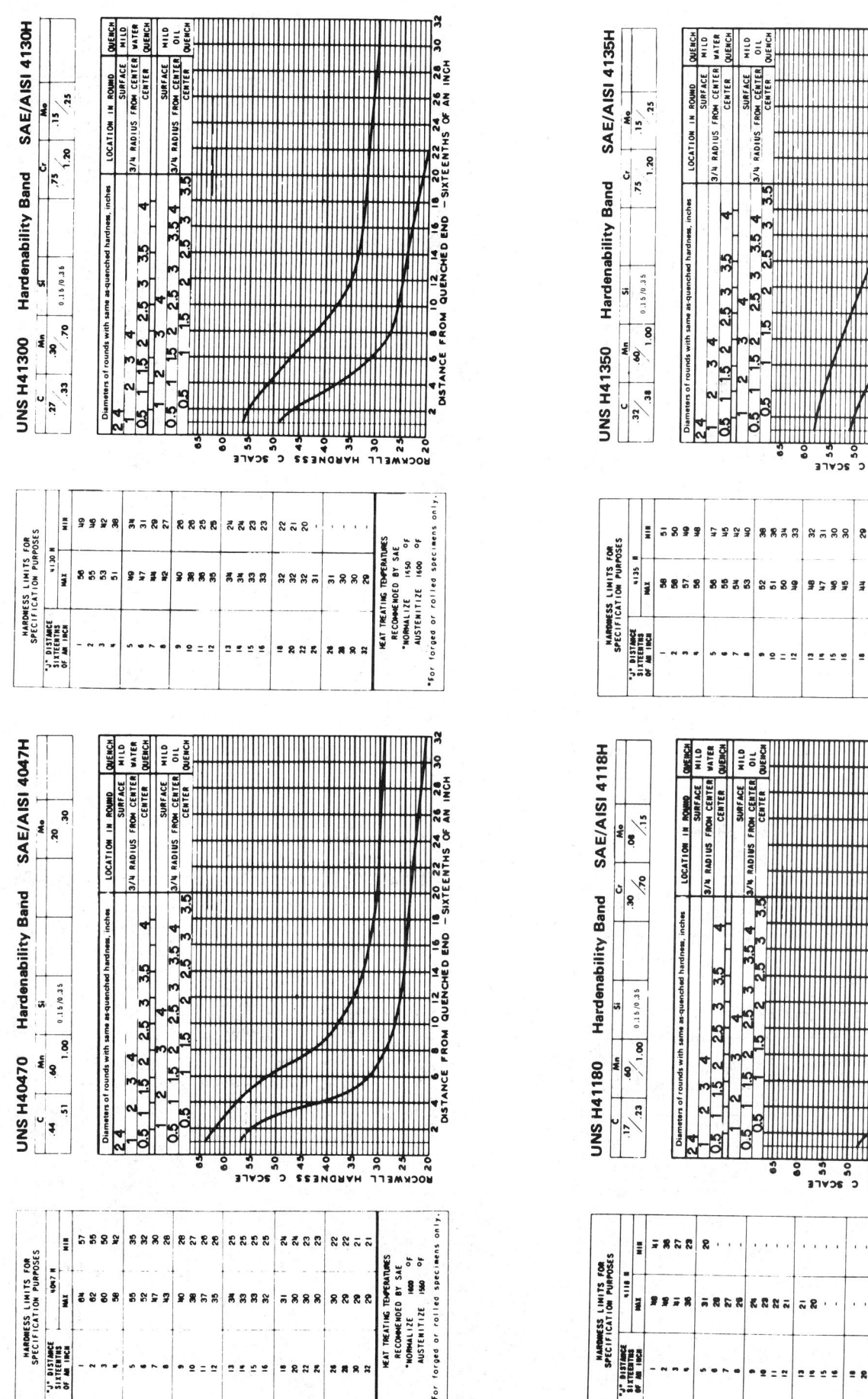

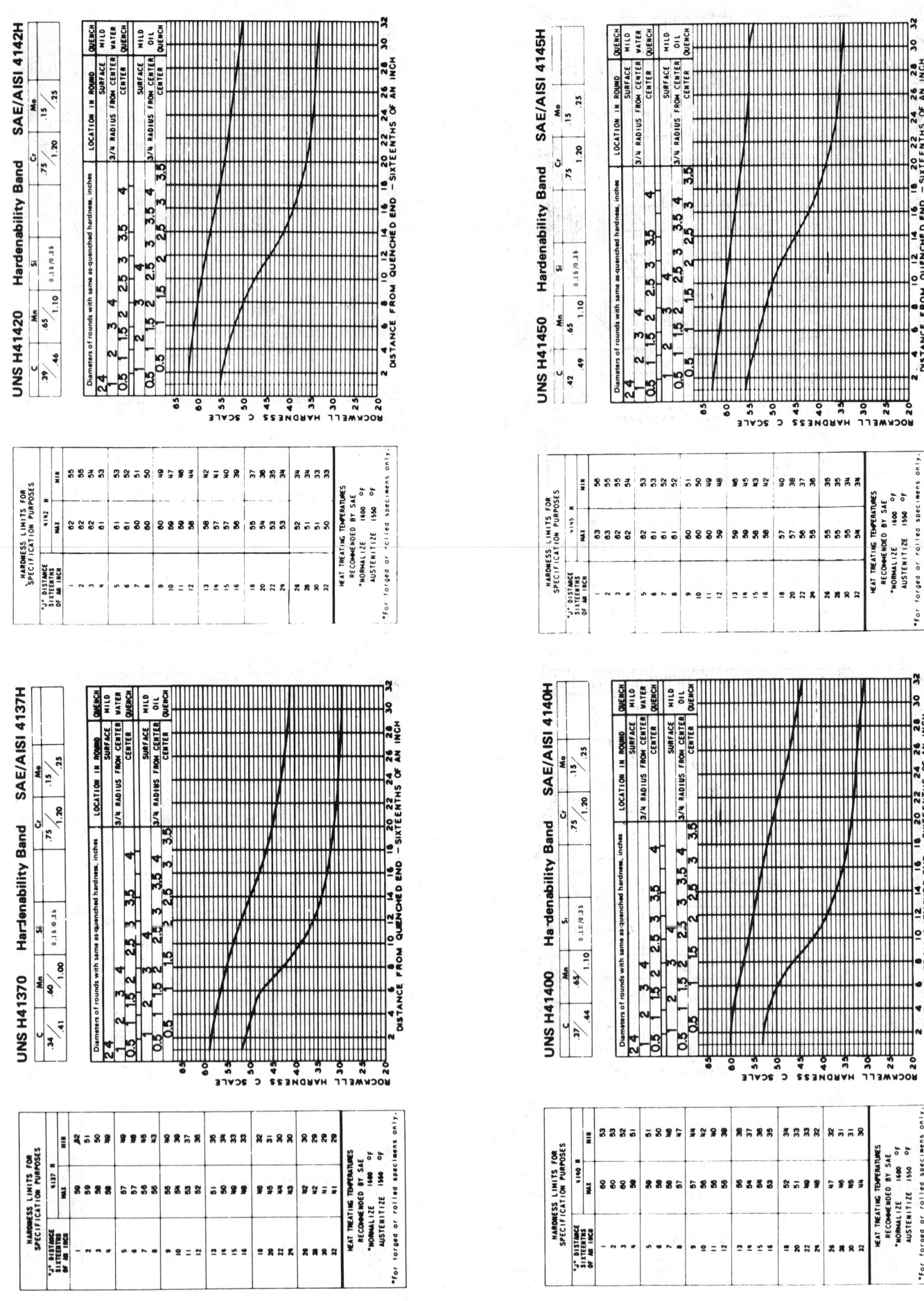

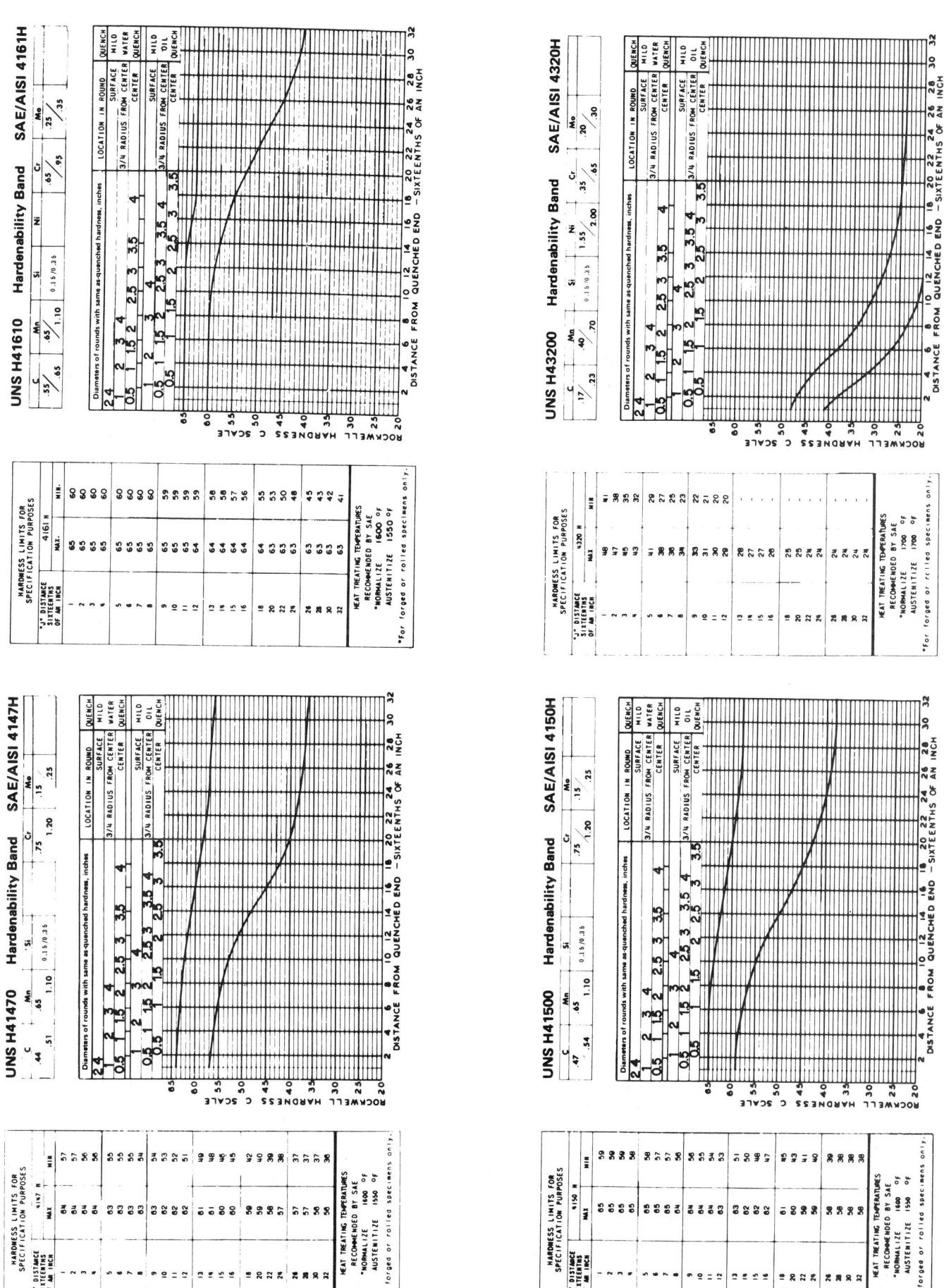

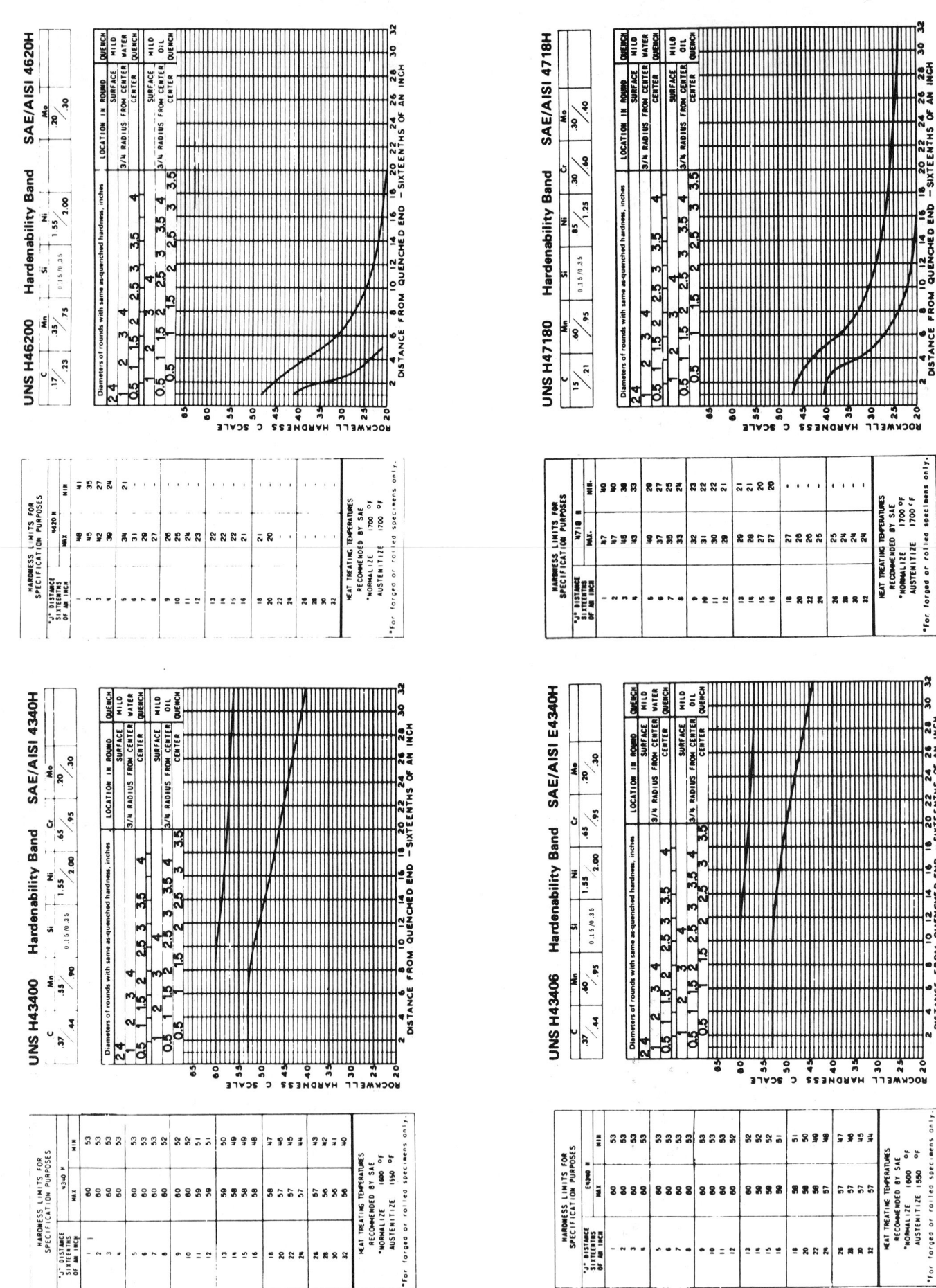

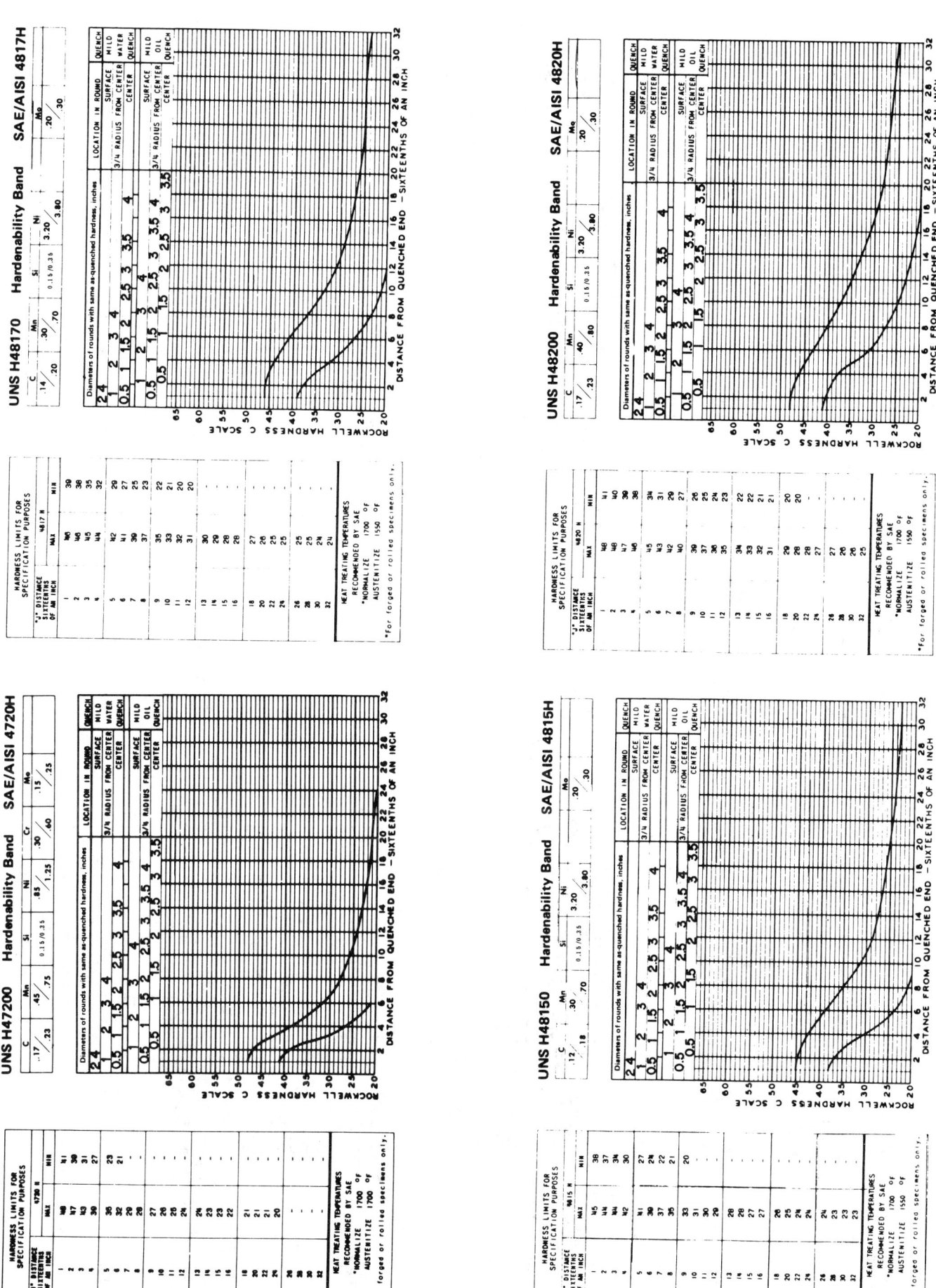

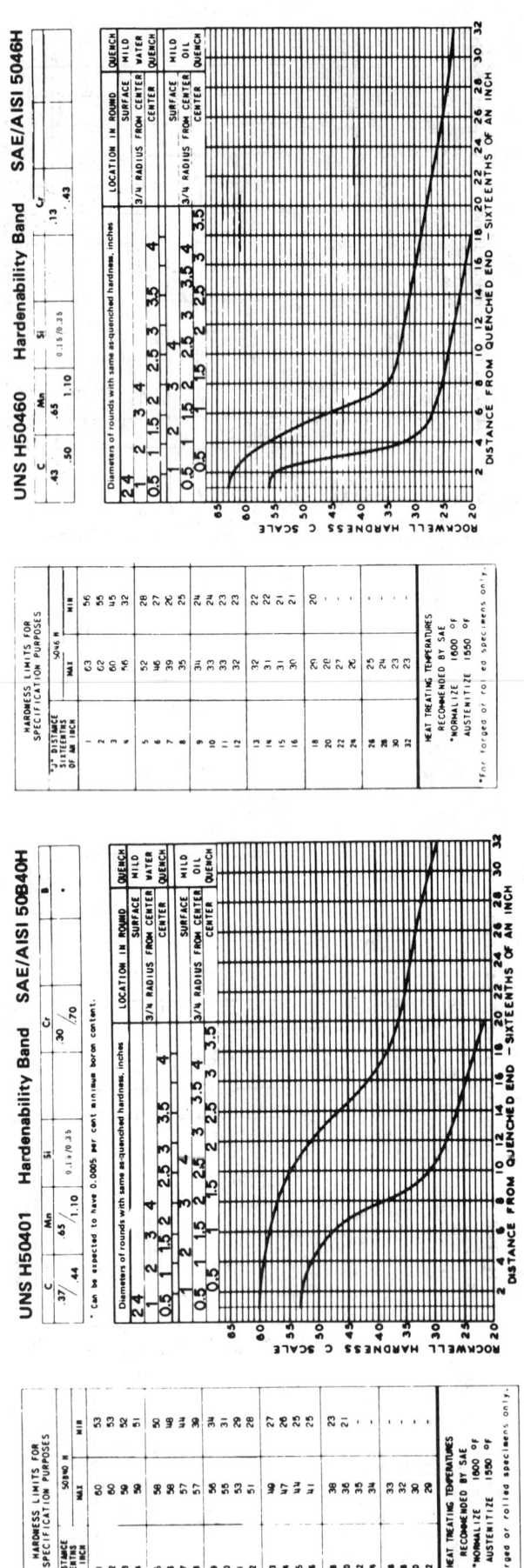

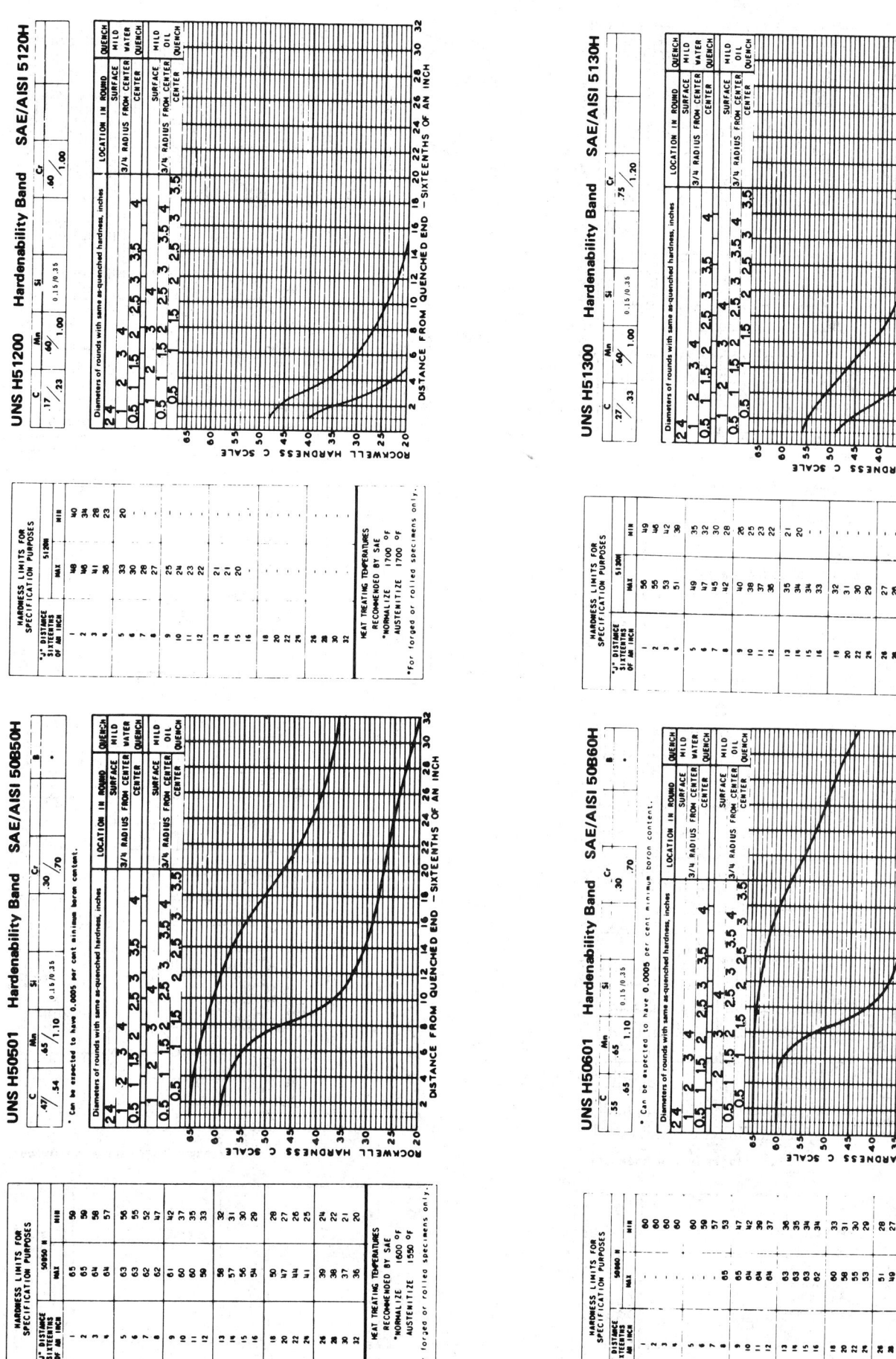

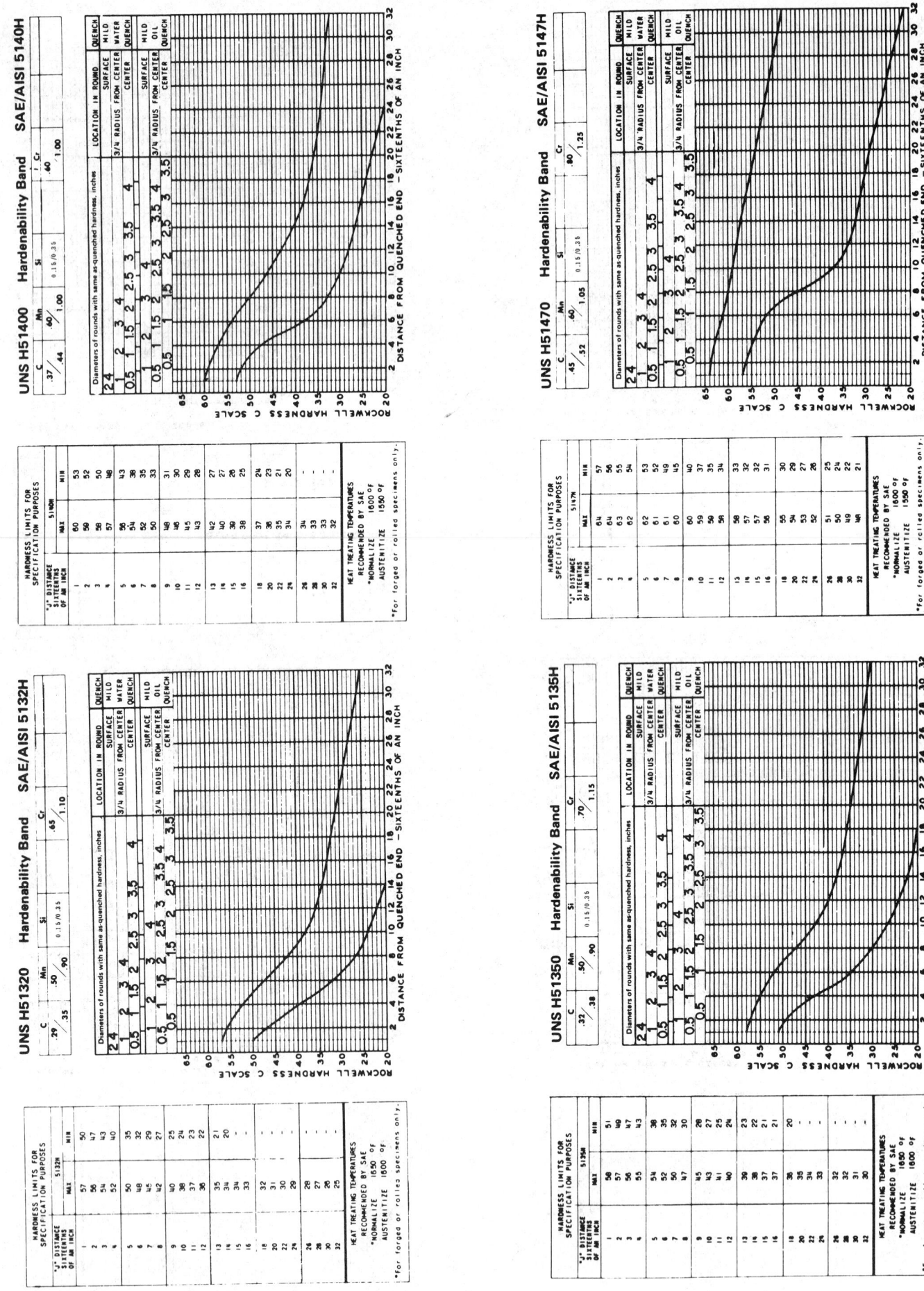

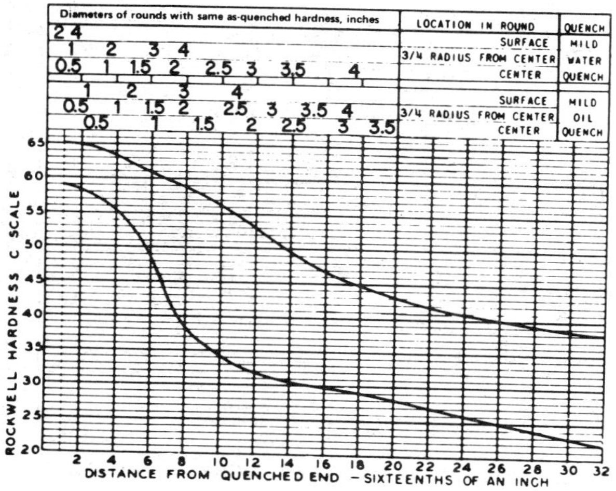

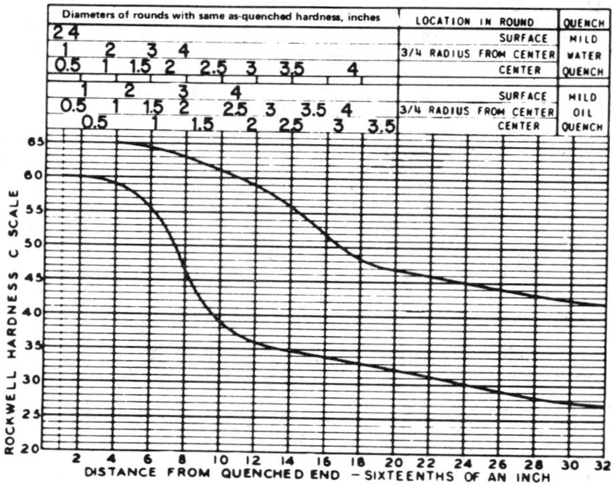

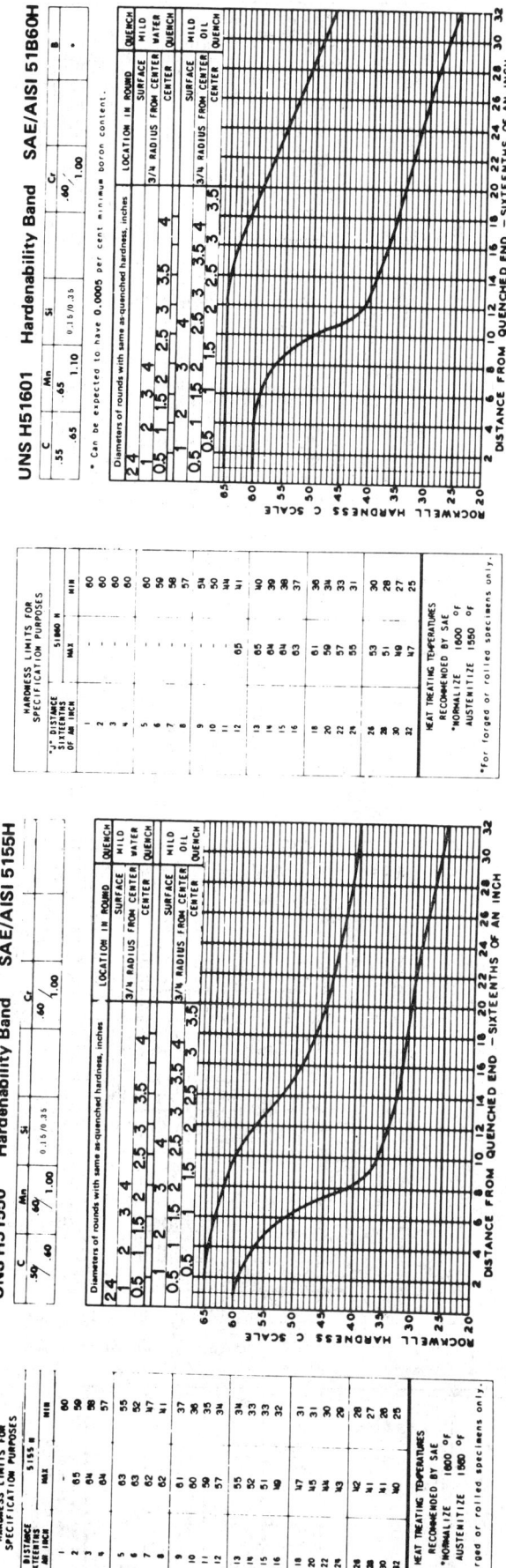

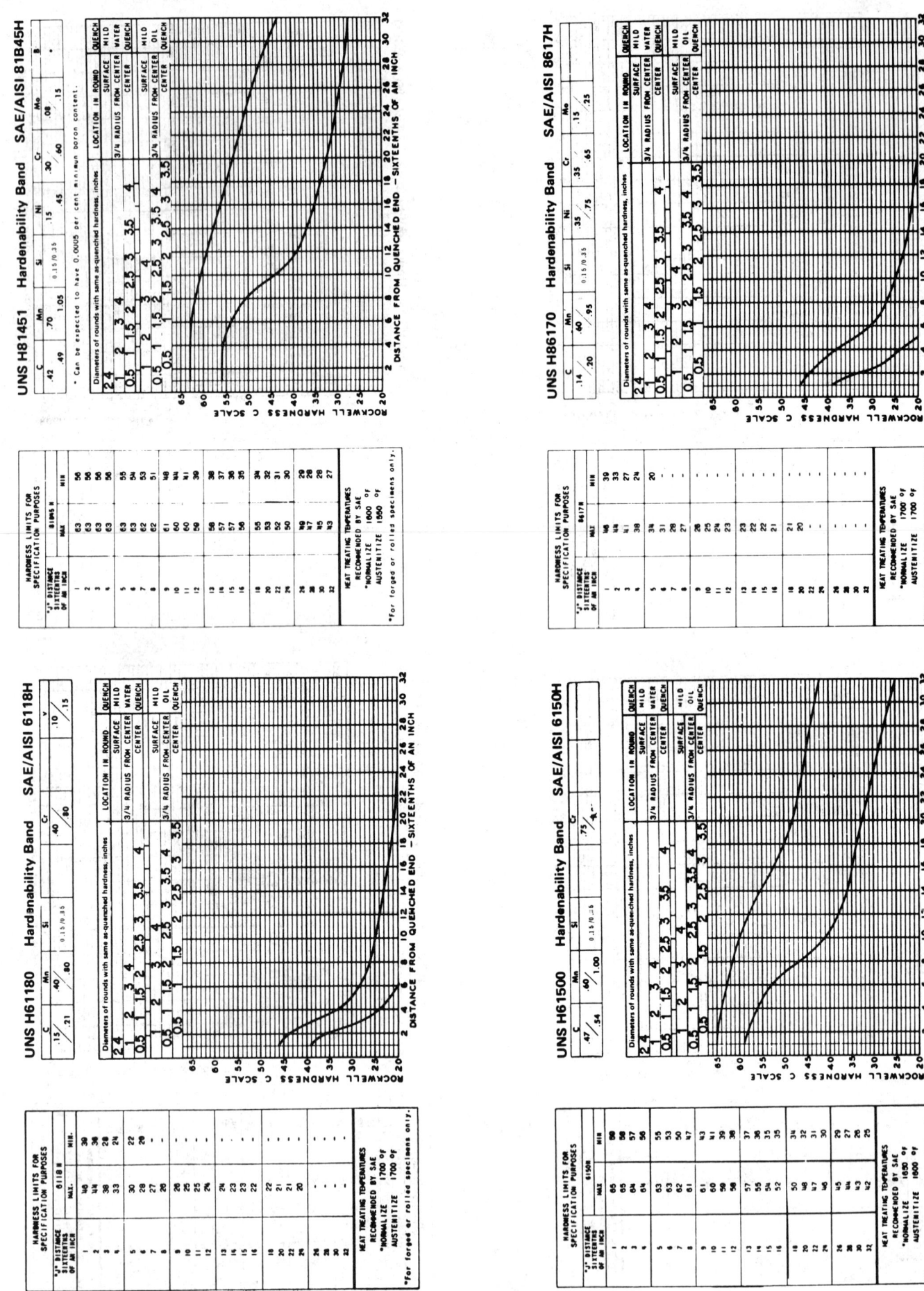

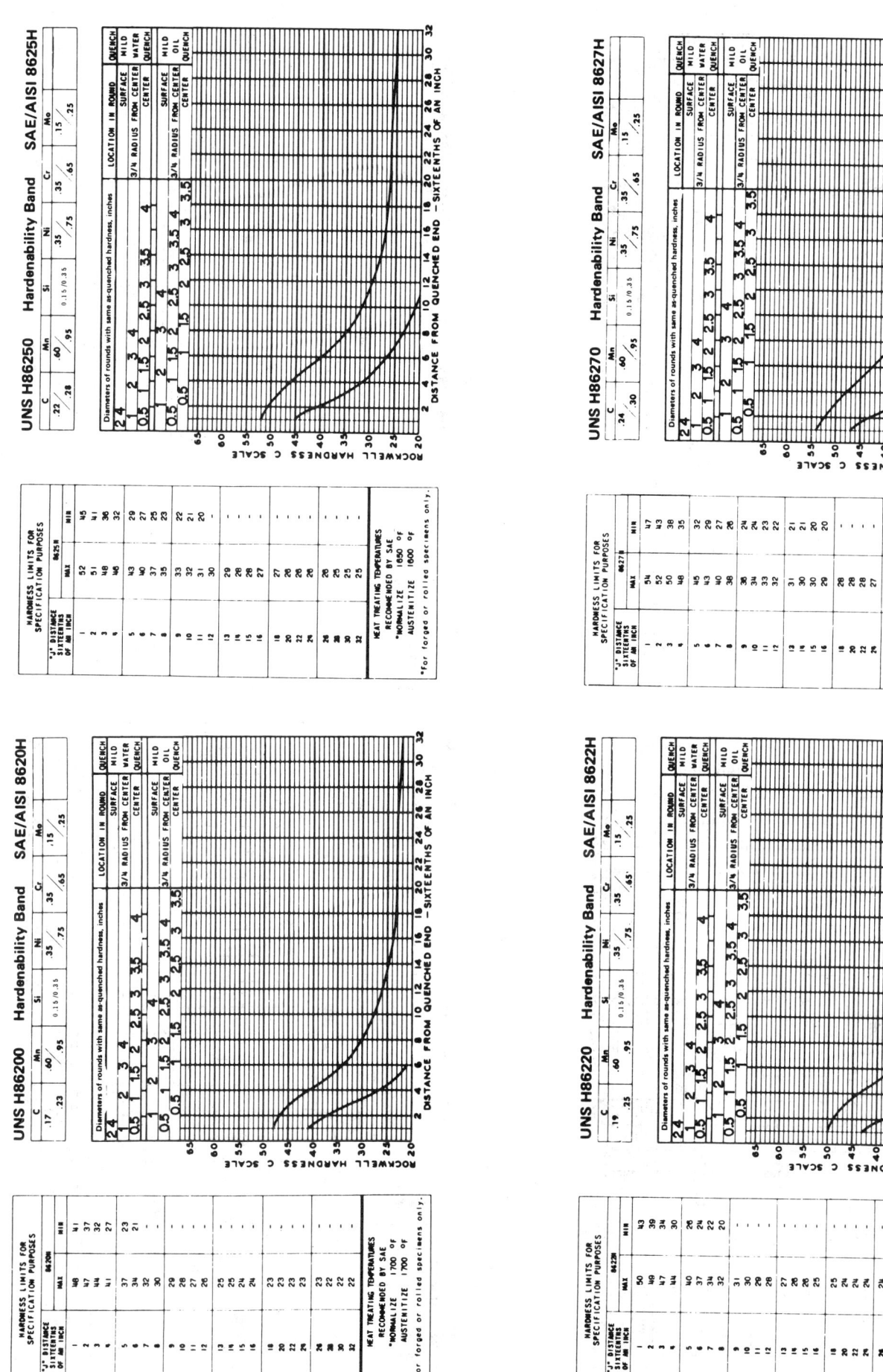

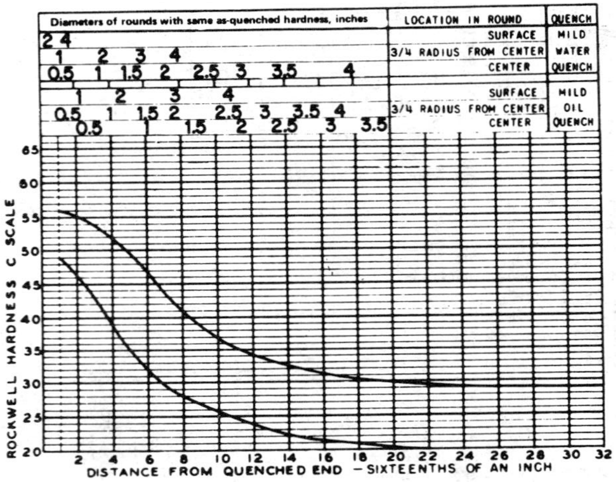

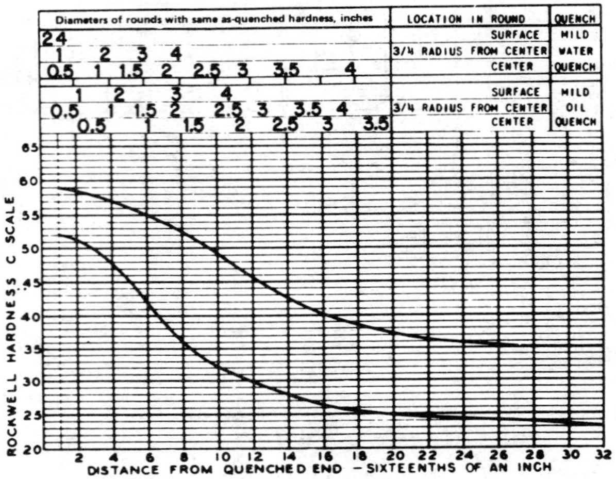

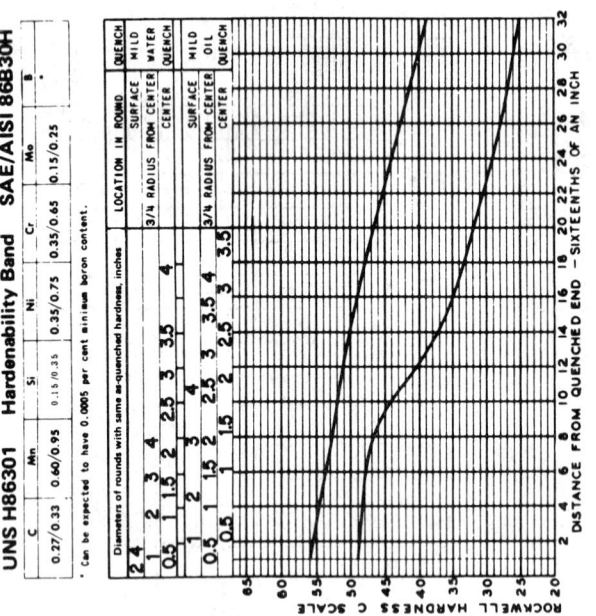

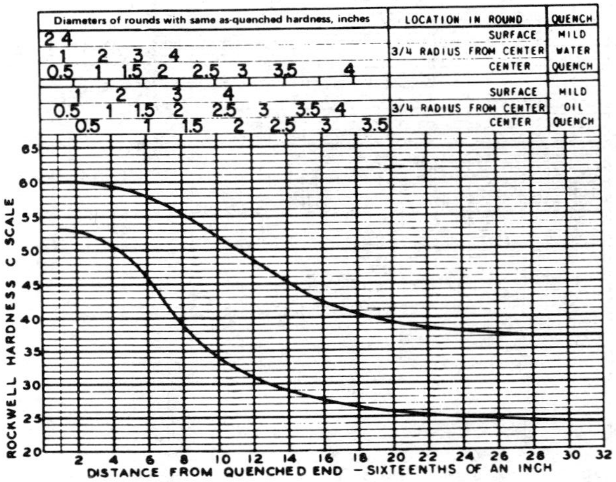

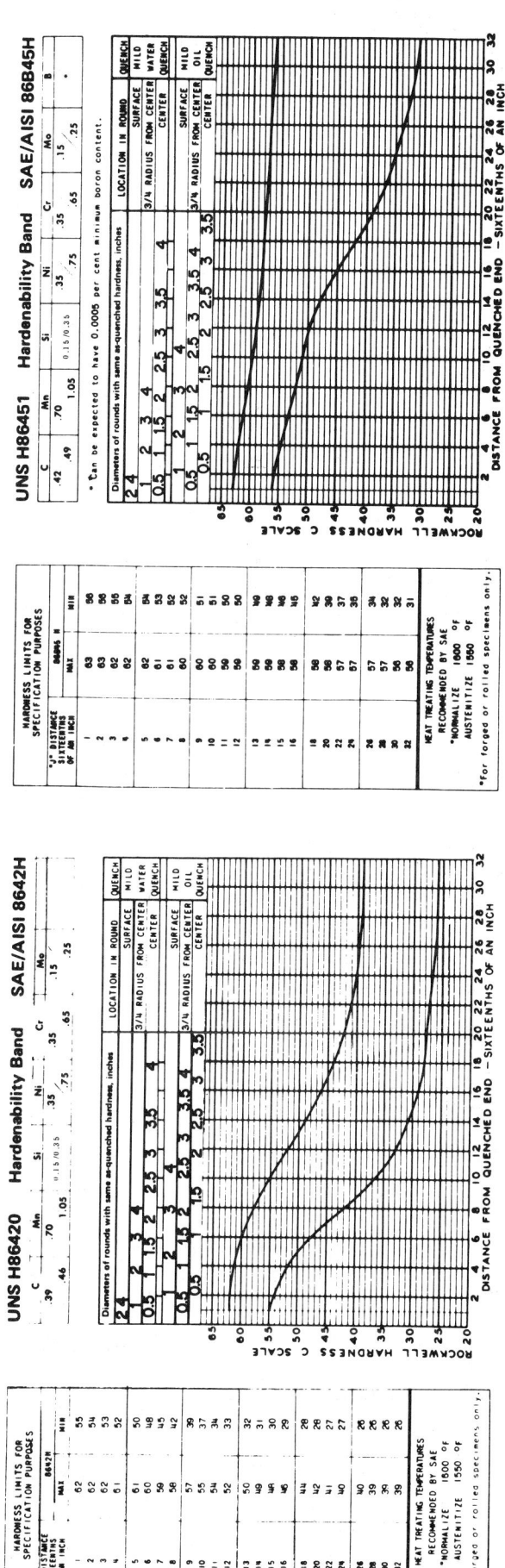
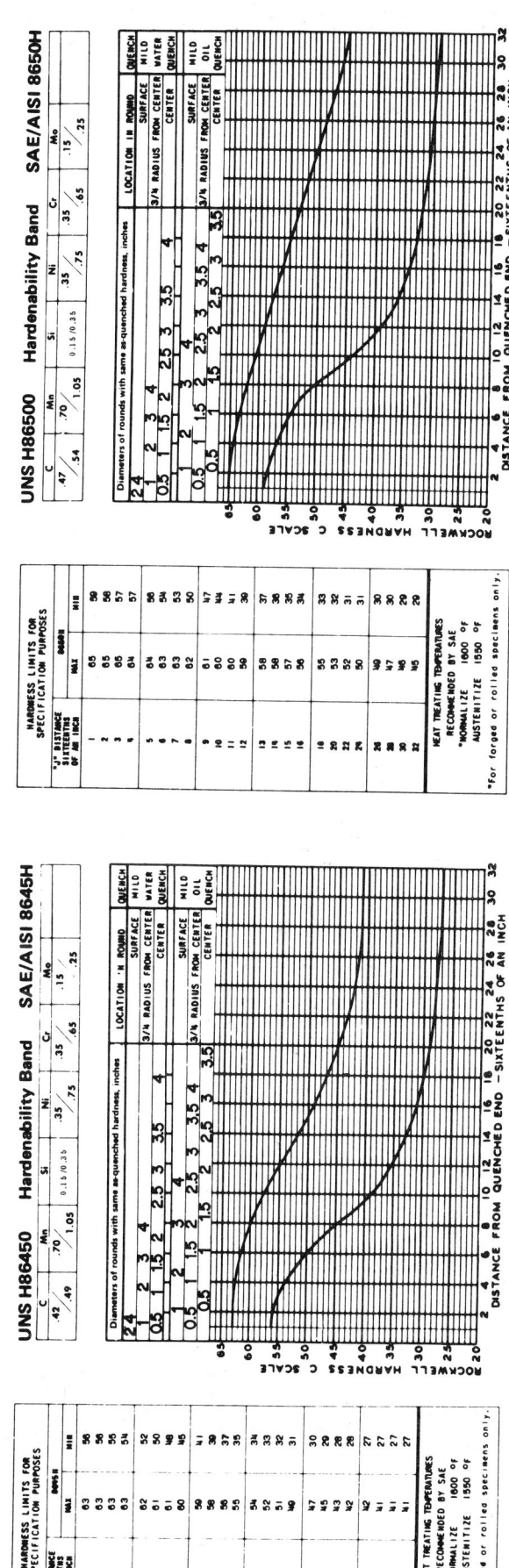

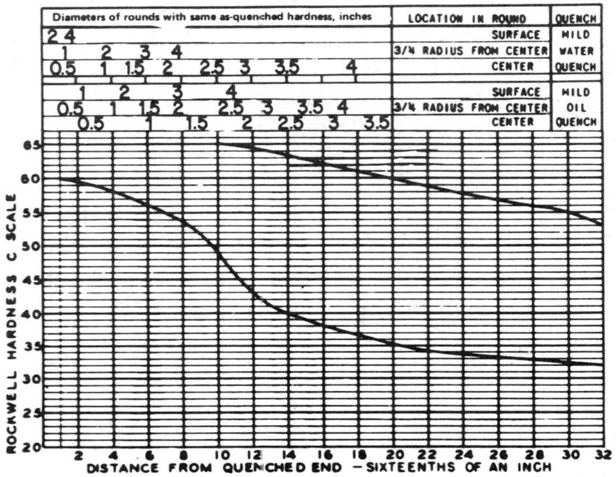

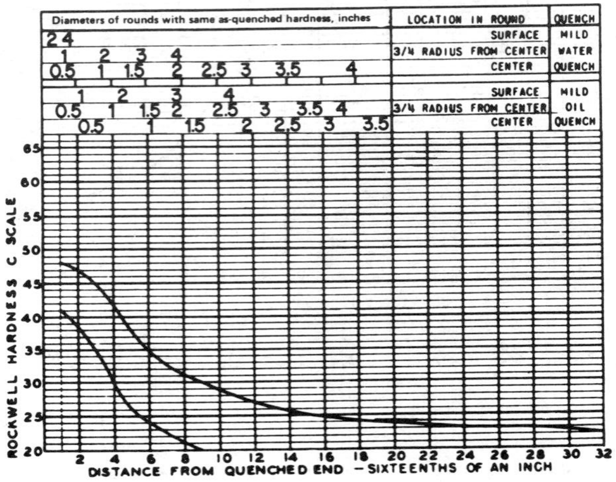

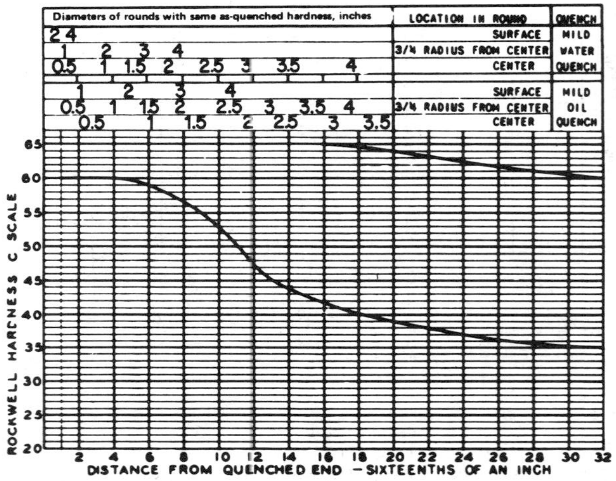

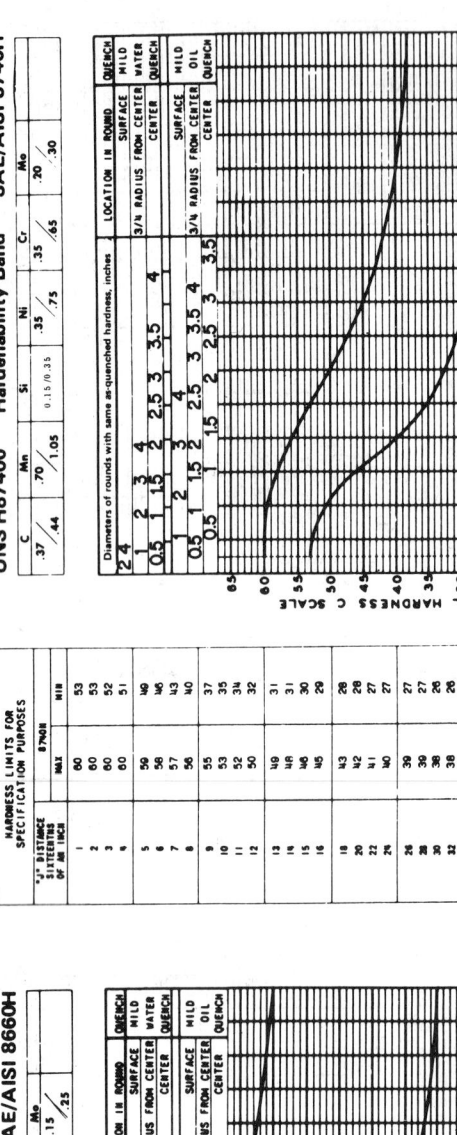

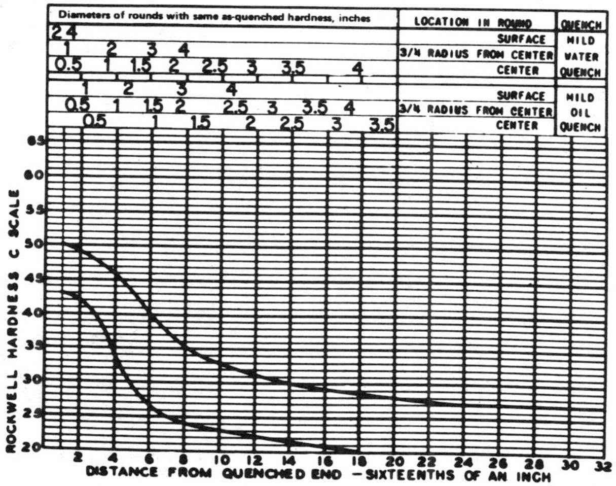

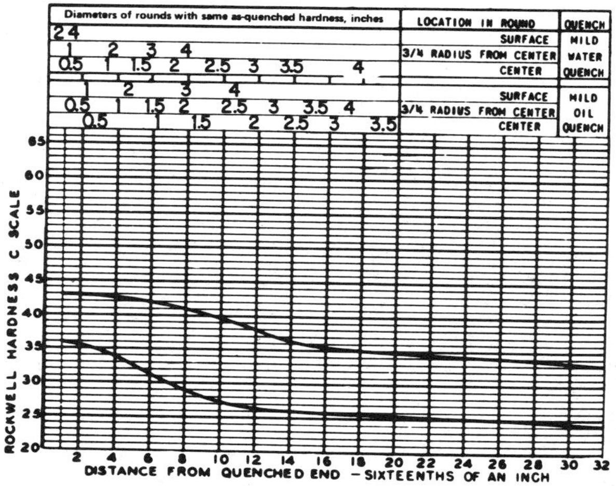

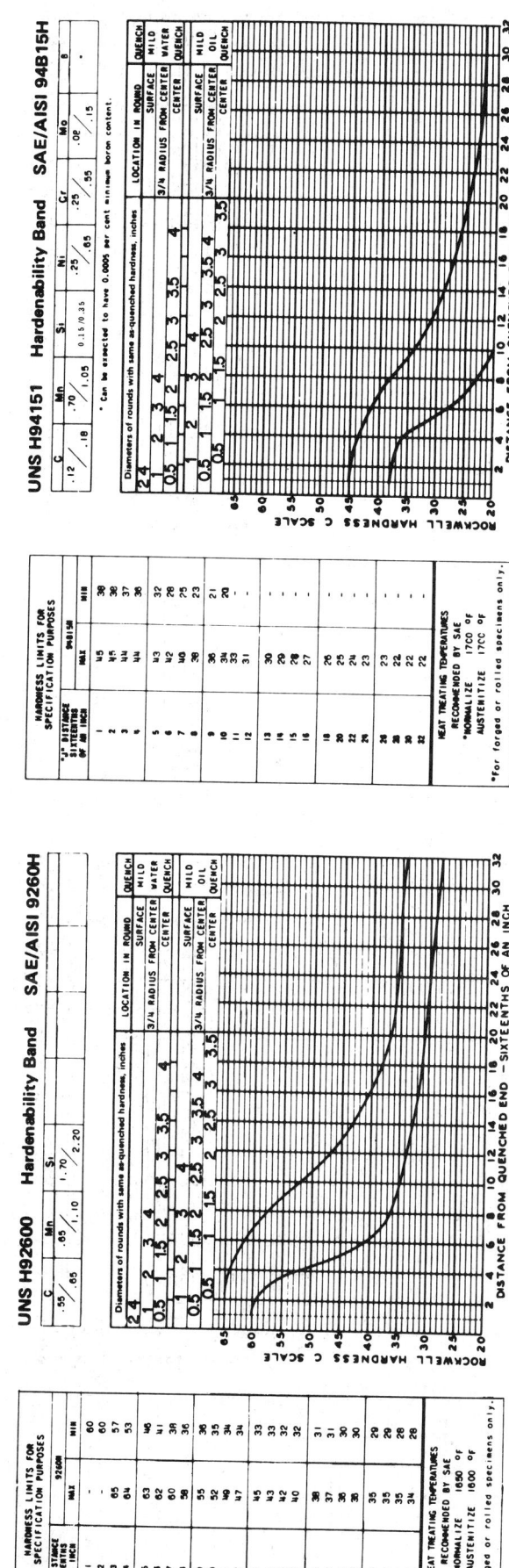

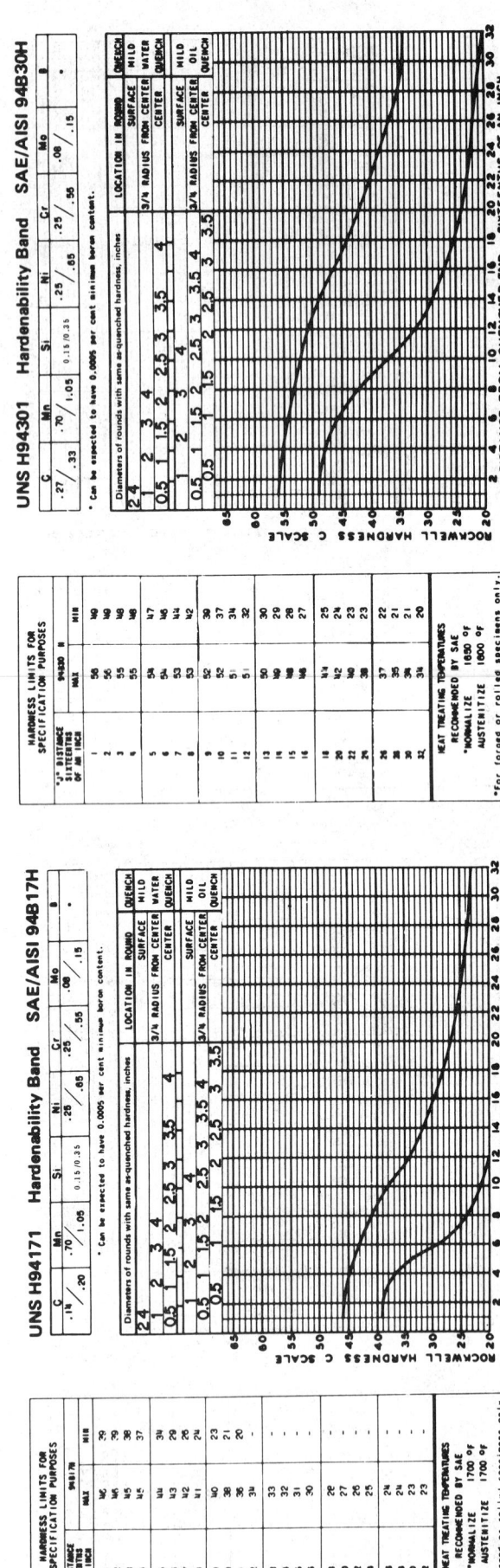

SERIES B
H-Bands in Metric (SI) Units

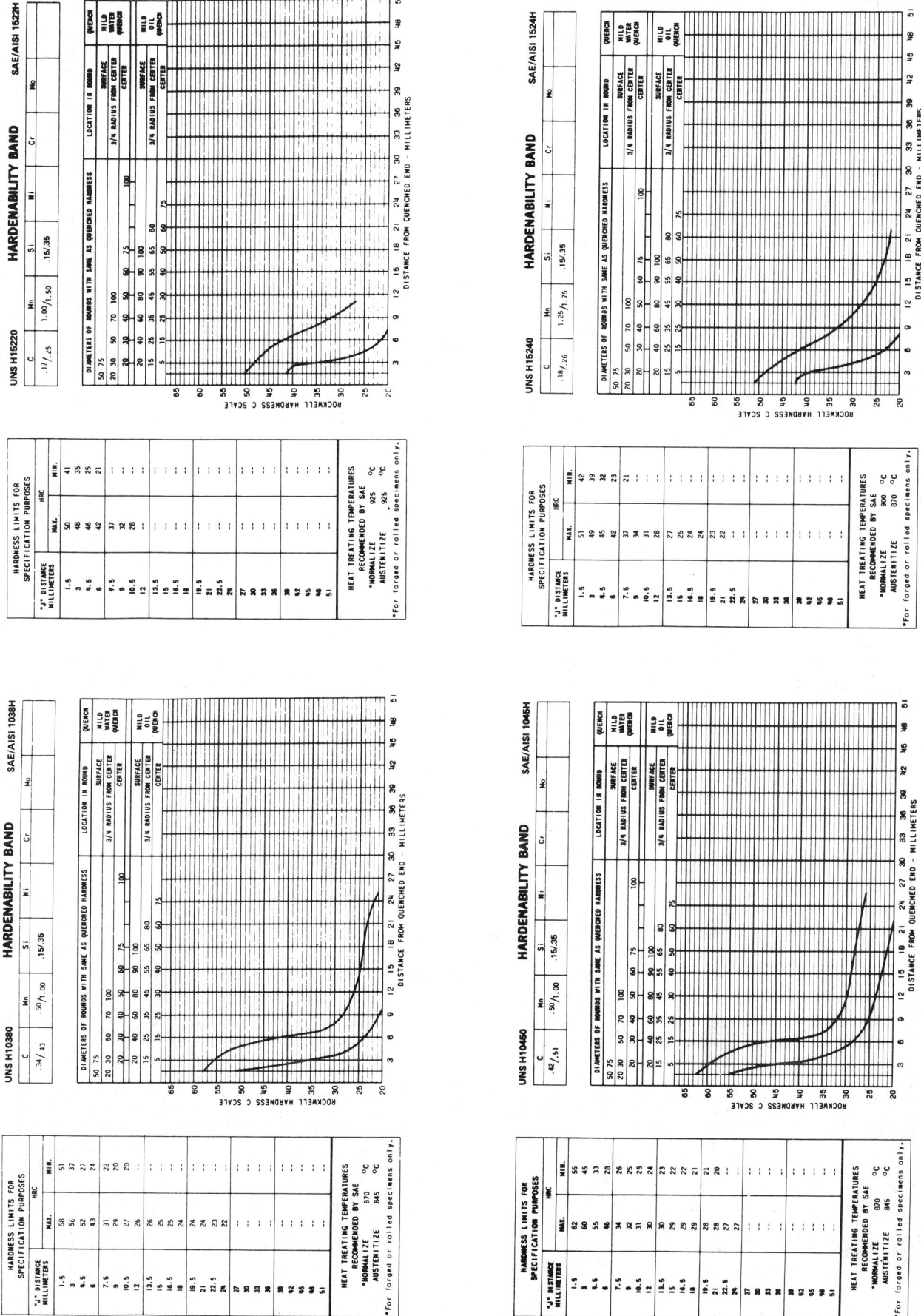

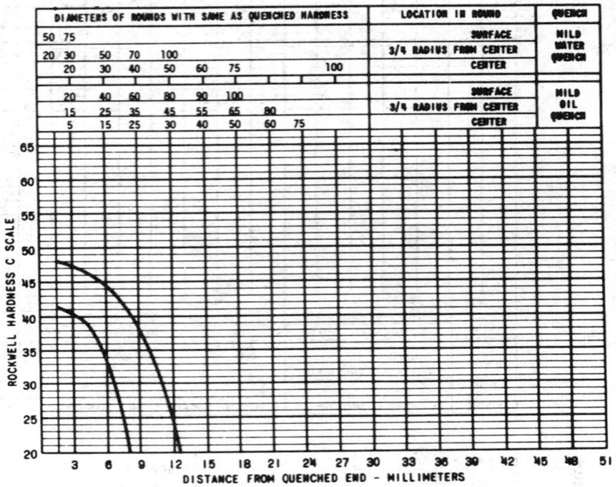

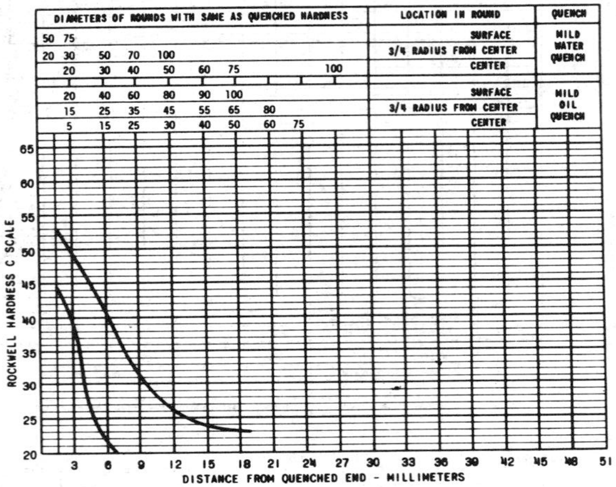

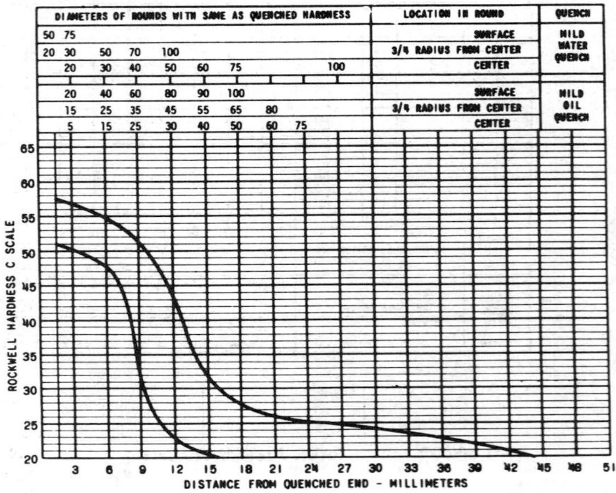

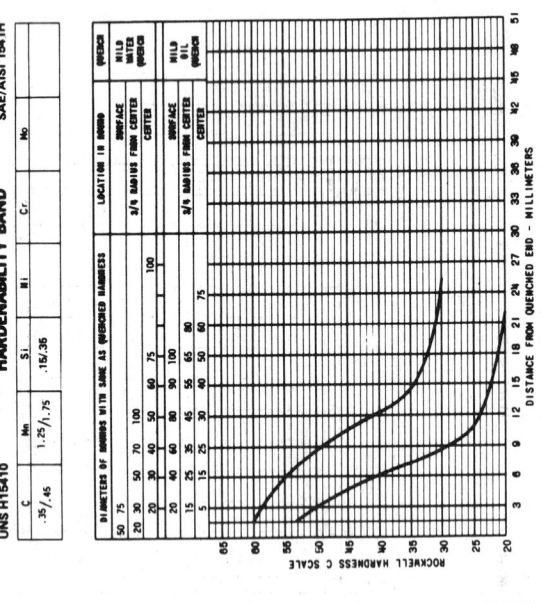

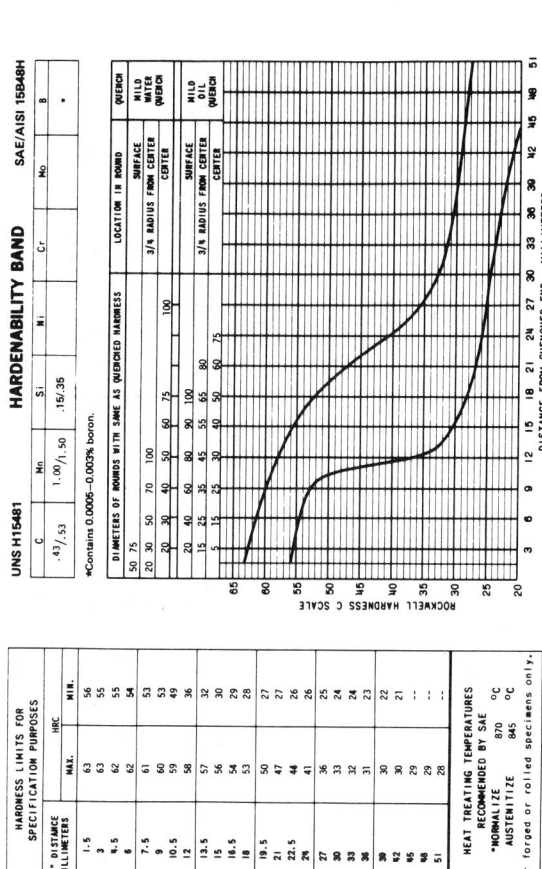

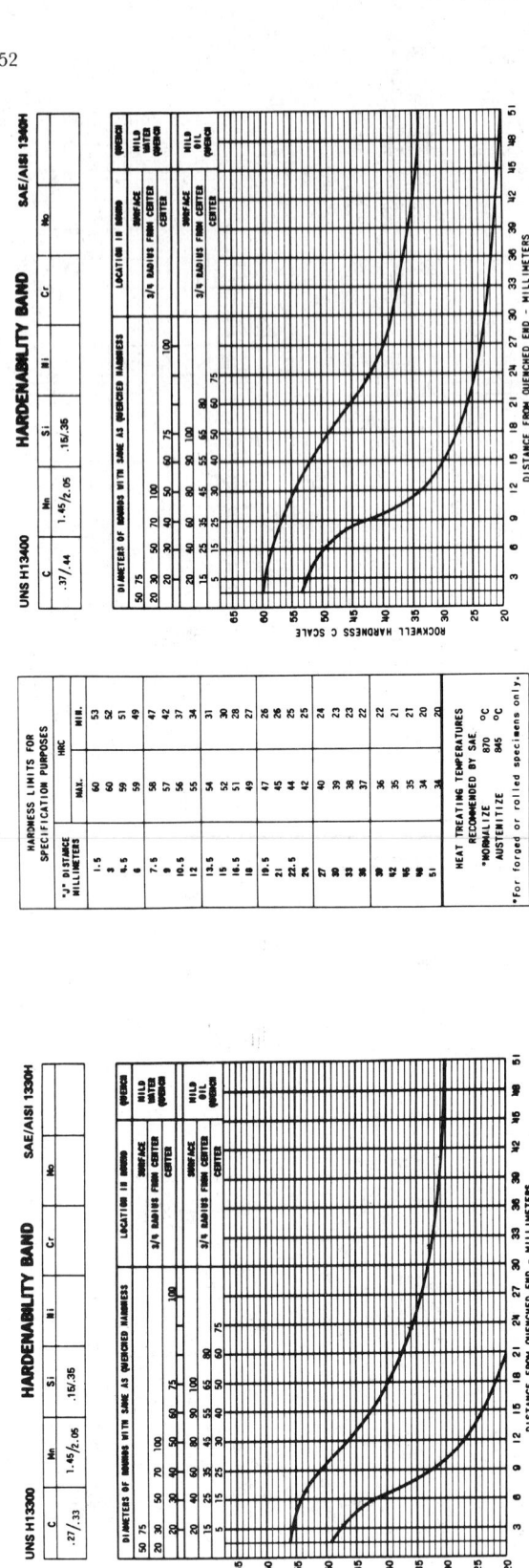

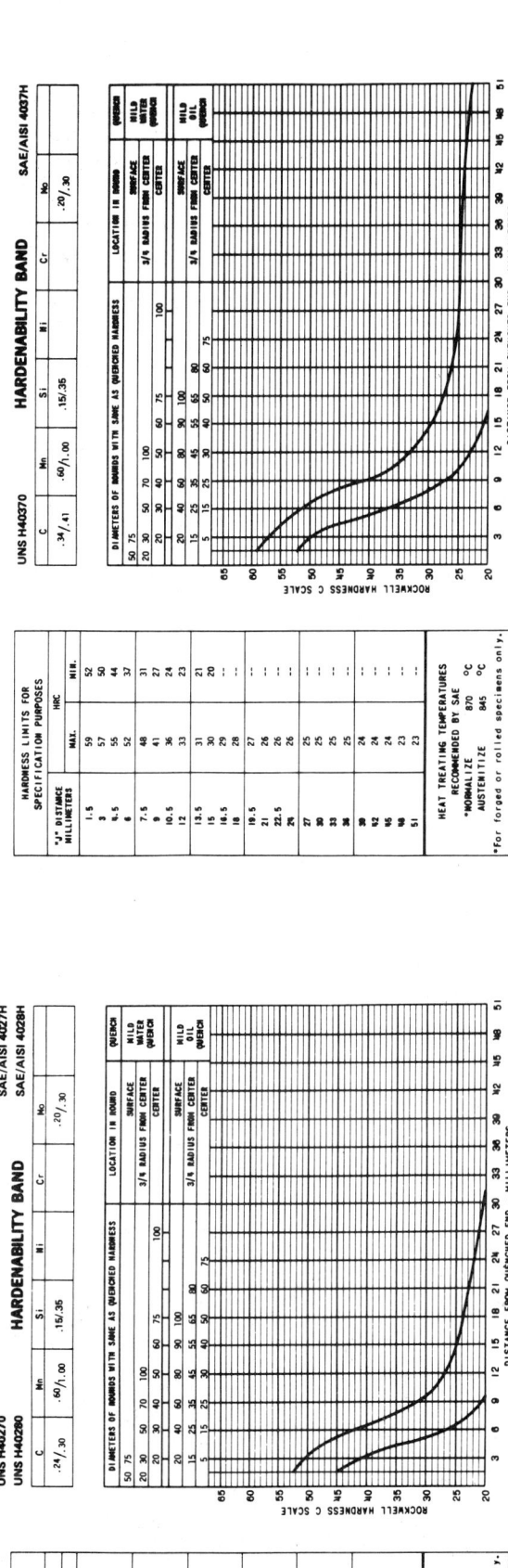

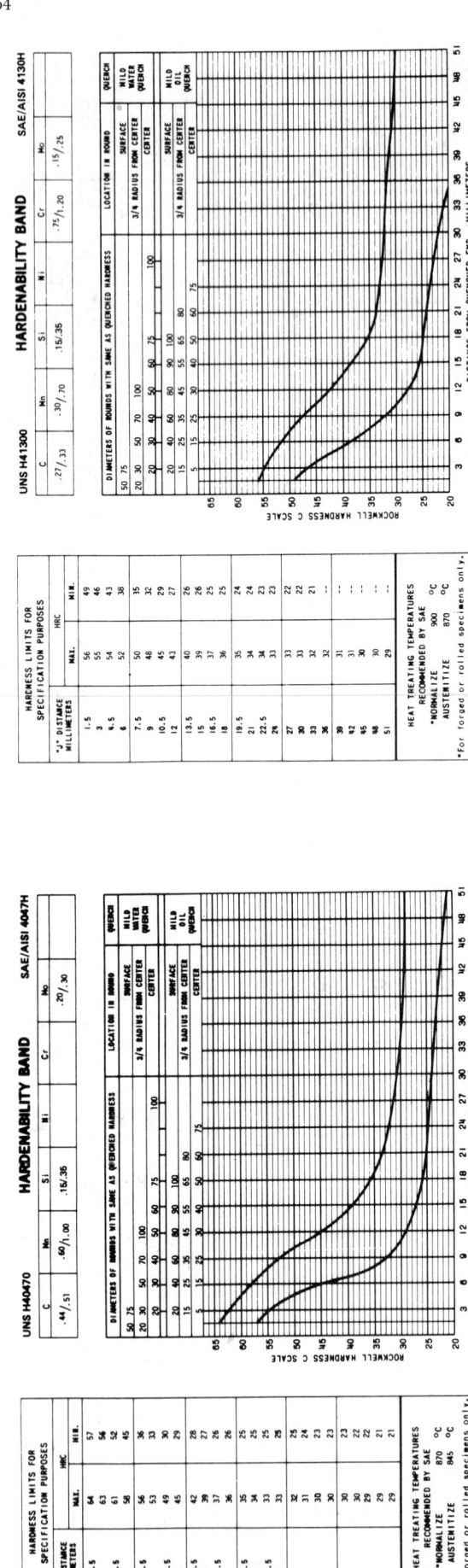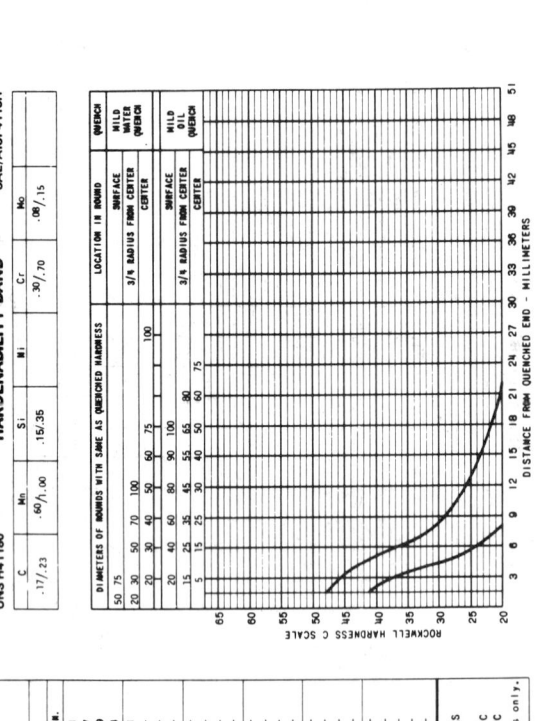

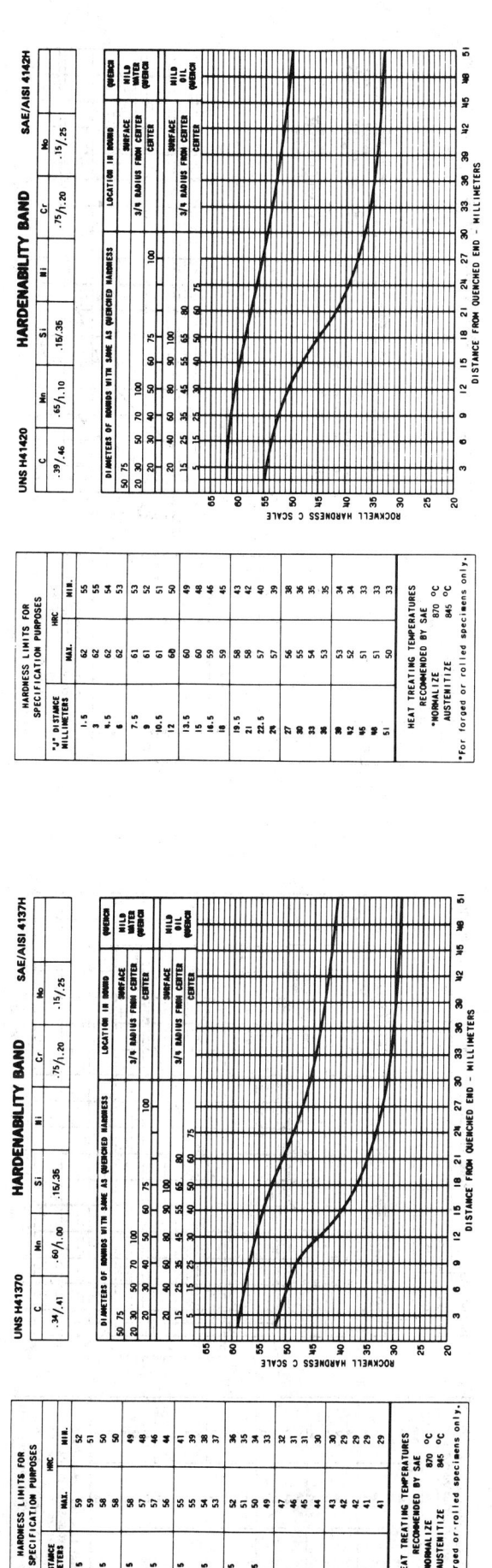
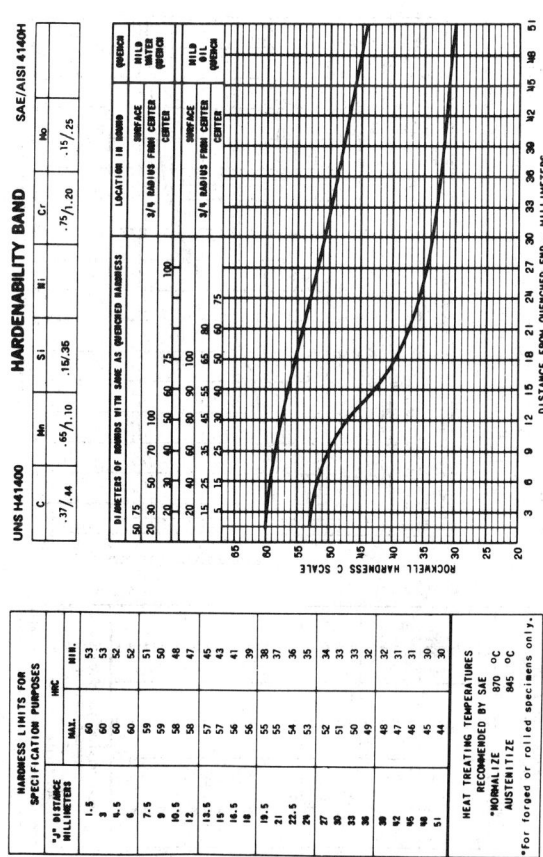

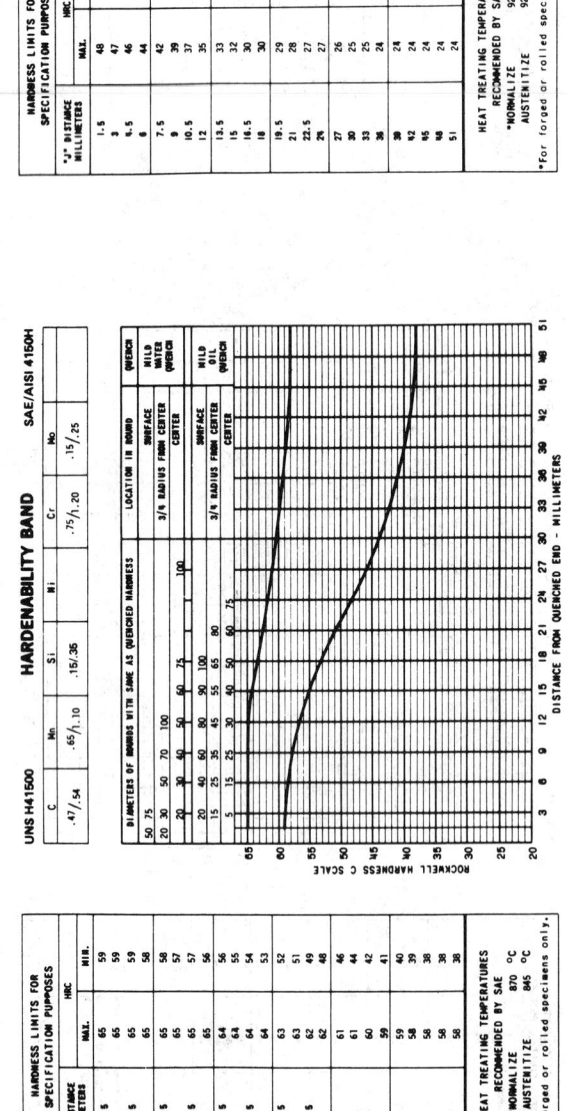

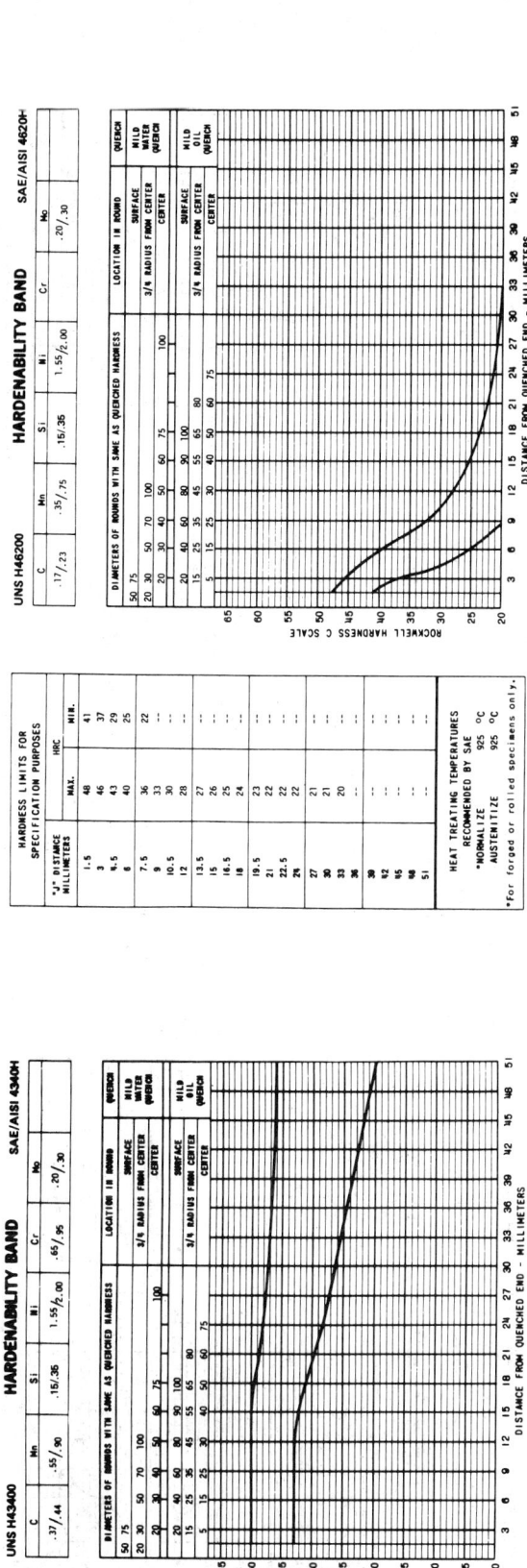

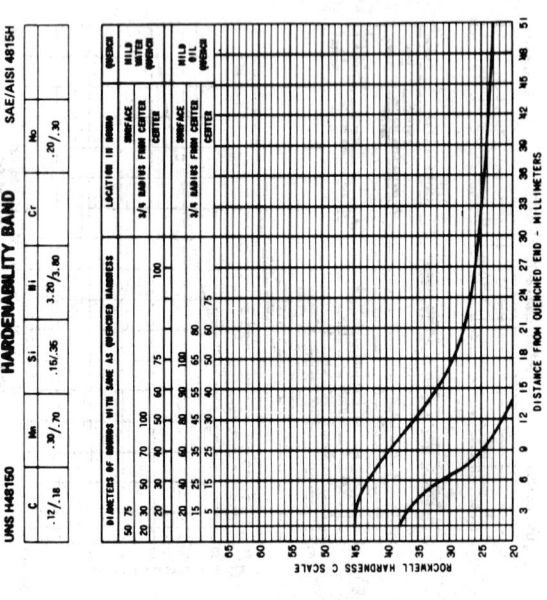

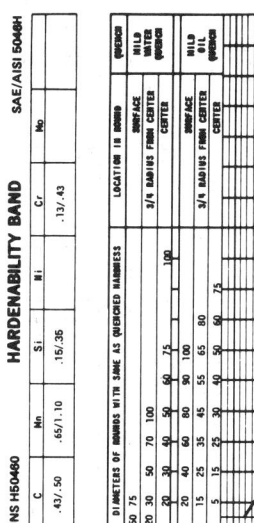
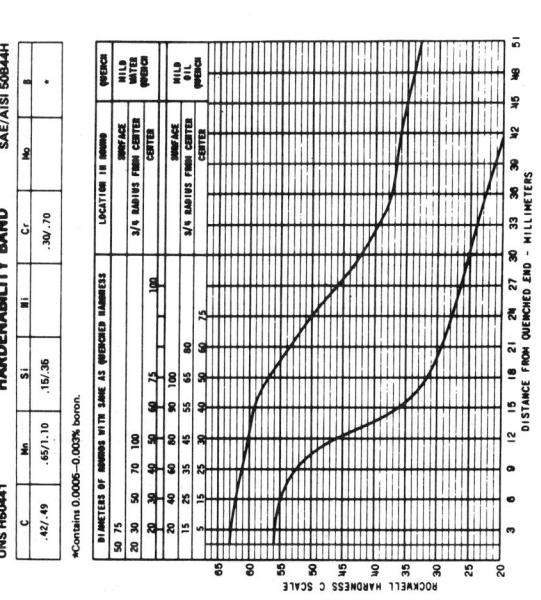

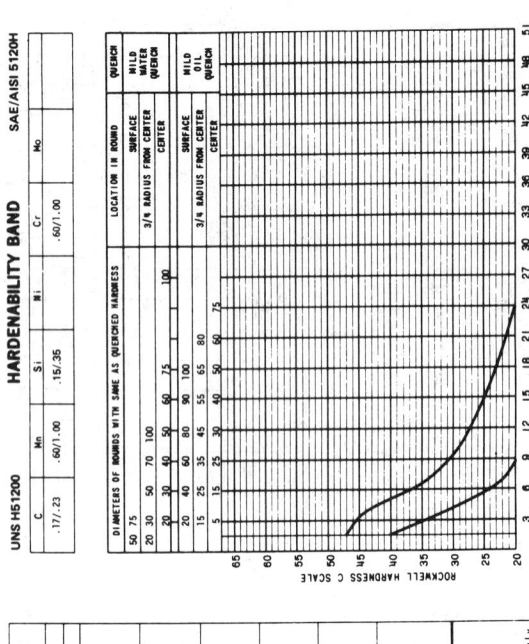
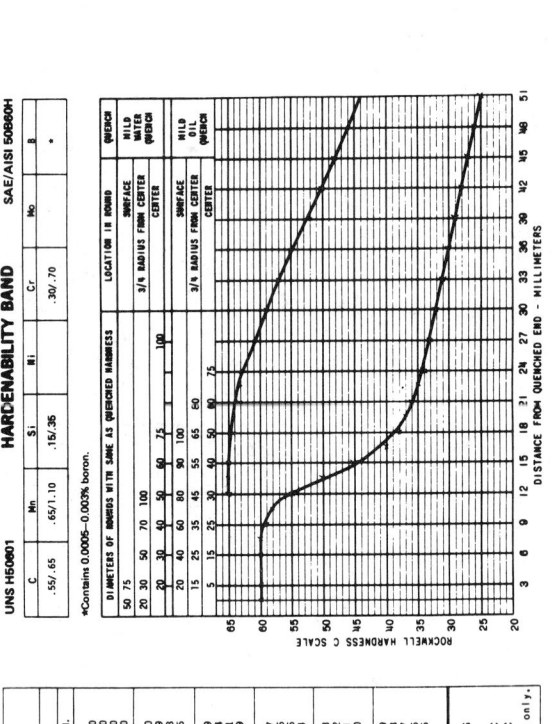

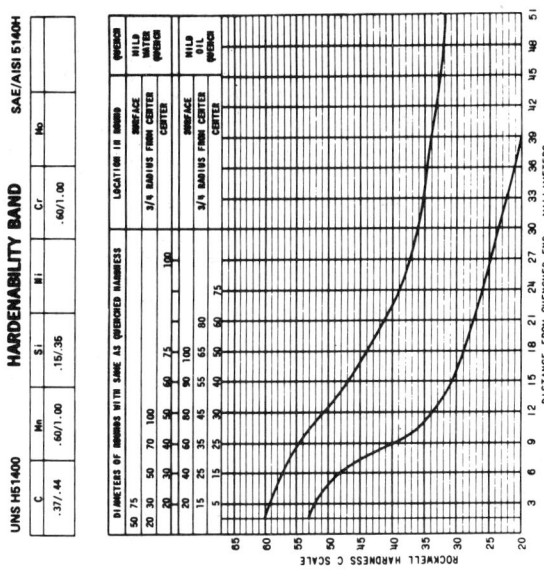

2.62

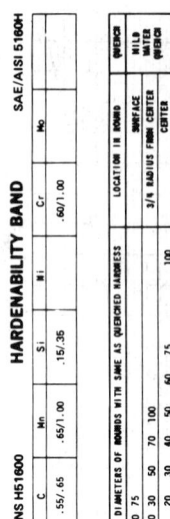
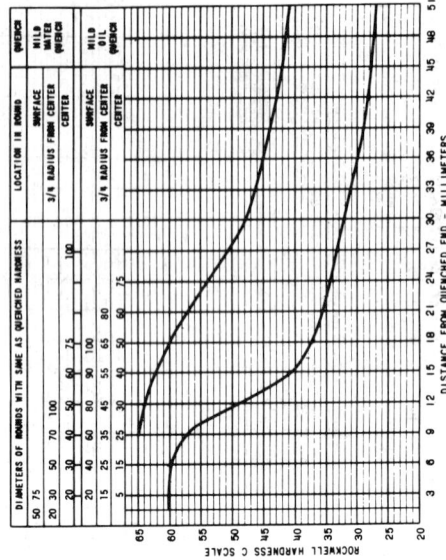
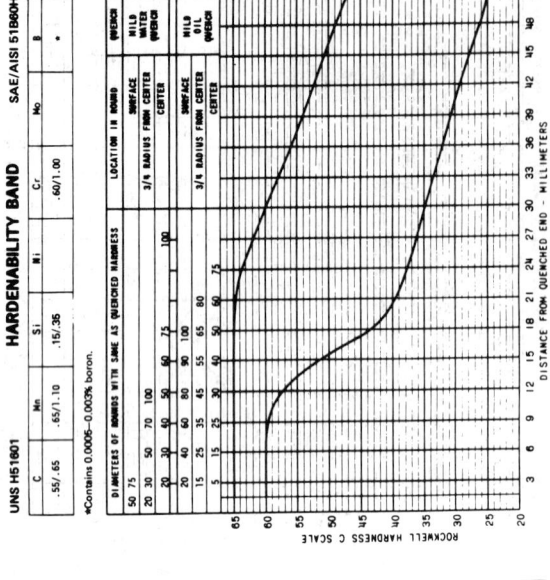
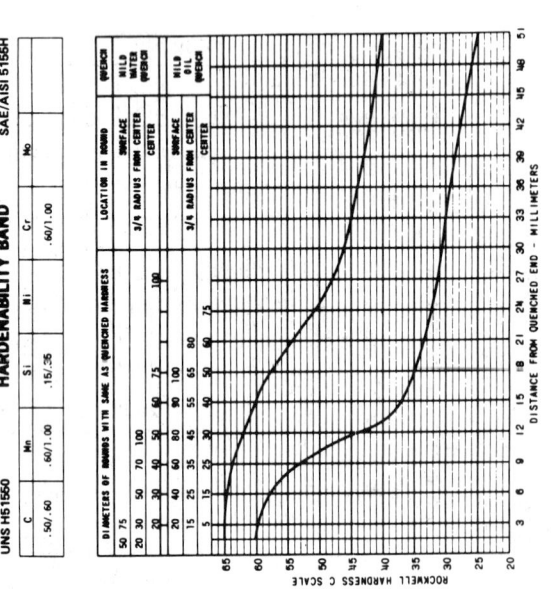

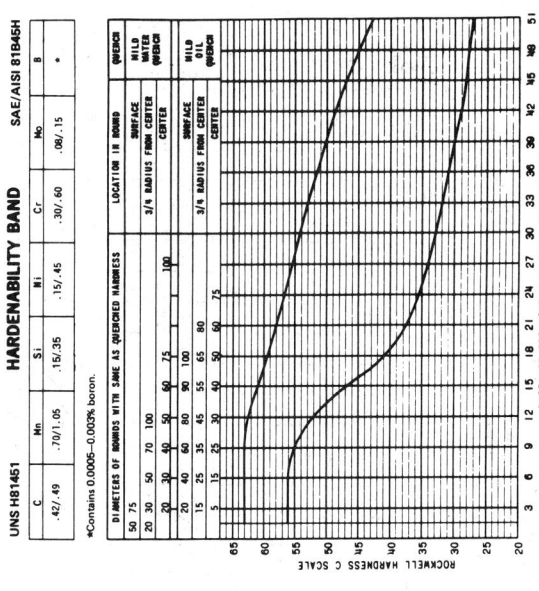

SAE/AISI 8620H — UNS H86200

HARDENABILITY BAND

C	Mn	Si	Ni	Cr	Mo
.17/.23	.60/.95	.15/.35	.35/.75	.35/.65	.15/.25

"J" DISTANCE MILLIMETERS	HARDNESS LIMITS FOR SPECIFICATION PURPOSES HRC MAX.	MIN.
1.5	48	41
3	47	37
4.5	45	33
6	42	28
7.5	38	24
9	35	22
10.5	33	20
12	31	—
13.5	30	—
15	29	—
16.5	28	—
18	27	—
19.5	26	—
21	25	—
22.5	25	—
24	24	—
27	24	—
30	23	—
33	23	—
36	23	—
39	23	—
42	23	—
45	22	—
48	22	—
51	22	—

HEAT TREATING TEMPERATURES RECOMMENDED BY SAE
*NORMALIZE 930 °C
AUSTENITIZE 930 °C
*For forged or rolled specimens only.

SAE/AISI 8622H — UNS H86220

HARDENABILITY BAND

C	Mn	Si	Ni	Cr	Mo
.19/.25	.60/.95	.15/.35	.35/.75	.35/.65	.15/.25

"J" DISTANCE MILLIMETERS	HRC MAX.	MIN.
1.5	50	43
3	50	39
4.5	48	35
6	45	31
7.5	42	26
9	39	25
10.5	35	23
12	33	21
13.5	31	20
15	31	—
16.5	30	—
18	29	—
19.5	28	—
21	27	—
22.5	25	—
24	26	—
27	25	—
30	24	—
33	24	—
36	24	—
39	24	—
42	24	—
45	24	—
48	24	—
51	24	—

HEAT TREATING TEMPERATURES RECOMMENDED BY SAE
*NORMALIZE 930 °C
AUSTENITIZE 930 °C
*For forged or rolled specimens only.

SAE/AISI 8625H — UNS H86250

HARDENABILITY BAND

C	Mn	Si	Ni	Cr	Mo
.22/.28	.60/.95	.15/.35	.35/.75	.35/.65	.15/.25

"J" DISTANCE MILLIMETERS	HRC MAX.	MIN.
1.5	52	45
3	51	41
4.5	49	37
6	47	33
7.5	44	30
9	41	28
10.5	39	26
12	36	24
13.5	34	22
15	33	21
16.5	31	20
18	30	—
19.5	29	—
21	29	—
22.5	28	—
24	28	—
27	27	—
30	27	—
33	26	—
36	26	—
39	26	—
42	26	—
45	25	—
48	25	—
51	25	—

HEAT TREATING TEMPERATURES RECOMMENDED BY SAE
*NORMALIZE 900 °C
AUSTENITIZE 870 °C
*For forged or rolled specimens only.

SAE/AISI 8627H — UNS H86270

HARDENABILITY BAND

C	Mn	Si	Ni	Cr	Mo
.24/.30	.60/.95	.15/.35	.35/.75	.35/.65	.15/.25

"J" DISTANCE MILLIMETERS	HRC MAX.	MIN.
1.5	54	47
3	53	43
4.5	51	39
6	49	36
7.5	46	33
9	44	31
10.5	42	28
12	39	26
13.5	37	25
15	35	24
16.5	34	23
18	33	22
19.5	32	21
21	31	20
22.5	30	20
24	30	—
27	29	—
30	28	—
33	28	—
36	27	—
39	27	—
42	27	—
45	27	—
48	27	—
51	27	—

HEAT TREATING TEMPERATURES RECOMMENDED BY SAE
*NORMALIZE 900 °C
AUSTENITIZE 870 °C
*For forged or rolled specimens only.

2.65

UNS H86370 — HARDENABILITY BAND — SAE/AISI 8637H

C	Mn	Si	Ni	Cr	Mo
.34/.41	.70/1.05	.15/.35	.35/.75	.35/.65	.15/.25

Hardness Limits for Specification Purposes

"J" DISTANCE MILLIMETERS	HRC MAX.	HRC MIN.
1.5	59	52
3	59	51
4.5	58	50
6	57	48
7.5	56	46
9	55	43
10.5	54	40
12	53	37
13.5	52	35
15	50	31
16.5	49	31
18	47	30
19.5	46	29
21	44	28
22.5	43	27
24	41	27
27	39	26
30	38	25
33	37	24
36	36	24
39	36	24
42	35	23
45	35	23
48	35	23
51	35	23

HEAT TREATING TEMPERATURES RECOMMENDED BY SAE
*NORMALIZE 870 °C
AUSTENITIZE 845 °C
*For forged or rolled specimens only.

UNS H86400 — HARDENABILITY BAND — SAE/AISI 8640H

C	Mn	Si	Ni	Cr	Mo
.37/.44	.70/1.05	.15/.35	.35/.75	.35/.65	.15/.25

Hardness Limits for Specification Purposes

"J" DISTANCE MILLIMETERS	HRC MAX.	HRC MIN.
1.5	60	53
3	60	53
4.5	60	52
6	60	51
7.5	59	49
9	58	47
10.5	57	43
12	56	40
13.5	54	37
15	53	35
16.5	51	31
18	50	31
19.5	48	31
21	47	30
22.5	45	29
24	44	28
27	42	26
30	40	26
33	39	25
36	38	25
39	38	24
42	37	24
45	37	24
48	37	24
51	37	24

HEAT TREATING TEMPERATURES RECOMMENDED BY SAE
*NORMALIZE 870 °C
AUSTENITIZE 845 °C
*For forged or rolled specimens only.

UNS H86300 — HARDENABILITY BAND — SAE/AISI 8630H

C	Mn	Si	Ni	Cr	Mo
.27/.33	.60/.95	.15/.35	.35/.75	.35/.65	.15/.25

Hardness Limits for Specification Purposes

"J" DISTANCE MILLIMETERS	HRC MAX.	HRC MIN.
1.5	56	49
3	55	46
4.5	55	43
6	53	40
7.5	50	37
9	48	33
10.5	45	30
12	43	28
13.5	40	27
15	38	26
16.5	37	25
18	35	24
19.5	34	23
21	33	22
22.5	32	22
24	32	21
27	31	21
30	30	20
33	29	—
36	29	—
39	29	—
42	29	—
45	29	—
48	29	—
51	29	—

HEAT TREATING TEMPERATURES RECOMMENDED BY SAE
*NORMALIZE 900 °C
AUSTENITIZE 870 °C
*For forged or rolled specimens only.

UNS H86301 — HARDENABILITY BAND — SAE/AISI 86B30H

C	Mn	Si	Ni	Cr	Mo	B
.27/.33	.60/.95	.15/.35	.35/.75	.35/.65	.15/.25	*

*Contains 0.0005–0.003% boron.

Hardness Limits for Specification Purposes

"J" DISTANCE MILLIMETERS	HRC MAX.	HRC MIN.
1.5	56	49
3	56	48
4.5	56	48
6	55	48
7.5	55	48
9	54	48
10.5	54	48
12	53	47
13.5	53	46
15	52	44
16.5	52	43
18	52	41
19.5	51	40
21	51	38
22.5	50	37
24	50	36
27	49	34
30	48	32
33	46	31
36	45	30
39	44	29
42	42	28
45	41	27
48	40	26
51	39	25

HEAT TREATING TEMPERATURES RECOMMENDED BY SAE
*NORMALIZE 900 °C
AUSTENITIZE 870 °C
*For forged or rolled specimens only.

SAE/AISI 8720H — UNS H87200

HARDENABILITY BAND

C	Mn	Si	Ni	Cr	Mo
.17/.23	.60/.95	.15/.35	.35/.75	.35/.65	.20/.30

Hardness Limits for Specification Purposes

"J" Distance Millimeters	HRC Max.	HRC Min.
1.5	48	41
3	47	39
4.5	45	37
6	43	32
7.5	40	28
9	37	25
10.5	34	23
12	32	21
13.5	30	20
16.5	29	—
18	28	—
19.5	27	—
21	26	—
22.5	26	—
24	25	—
27	25	—
30	24	—
33	24	—
36	23	—
39	23	—
42	23	—
45	22	—
48	22	—
51	22	—

Heat Treating Temperatures Recommended by SAE
*Normalize 925 °C
Austenitize 925 °C
*For forged or rolled specimens only.

SAE/AISI 8740H — UNS H87400

HARDENABILITY BAND

C	Mn	Si	Ni	Cr	Mo
.37/.44	.70/1.05	.15/.35	.35/.75	.35/.65	.20/.30

Hardness Limits for Specification Purposes

"J" Distance Millimeters	HRC Max.	HRC Min.
1.5	60	53
3	60	53
4.5	60	53
6	60	51
7.5	60	49
9	59	46
10.5	58	44
12	57	41
13.5	56	39
16.5	54	34
18	52	33
19.5	50	32
21	49	31
22.5	47	30
24	46	29
27	44	28
30	43	27
33	42	27
36	41	27
39	40	27
42	39	26
45	39	26
48	38	26
51	38	26

Heat Treating Temperatures Recommended by SAE
*Normalize 870 °C
Austenitize 845 °C
*For forged or rolled specimens only.

SAE/AISI 8655H — UNS H86550

HARDENABILITY BAND

C	Mn	Si	Ni	Cr	Mo
.50/.60	.70/1.05	.15/.35	.35/.75	.35/.65	.15/.25

Hardness Limits for Specification Purposes

"J" Distance Millimeters	HRC Max.	HRC Min.
1.5	—	60
3	—	60
4.5	—	60
6	—	58
7.5	65	57
9	65	56
10.5	65	55
12	65	54
13.5	65	53
16.5	65	51
18	65	48
—	—	45
19.5	64	43
21	64	41
22.5	63	40
24	63	39
27	62	37
30	61	36
33	60	35
36	59	34
39	58	34
42	57	33
45	56	33
48	55	32
51	53	32

Heat Treating Temperatures Recommended by SAE
*Normalize 870 °C
Austenitize 845 °C
*For forged or rolled specimens only.

SAE/AISI 8660H — UNS H86600

HARDENABILITY BAND

C	Mn	Si	Ni	Cr	Mo
.55/.65	.70/1.05	.15/.35	.35/.75	.35/.65	.15/.25

Hardness Limits for Specification Purposes

"J" Distance Millimeters	HRC Max.	HRC Min.
1.5	—	60
3	—	60
4.5	—	59
6	—	58
7.5	—	57
9	—	55
10.5	—	53
12	—	51
13.5	—	49
16.5	—	47
18	—	45
19.5	—	44
21	—	43
22.5	—	43
24	—	41
27	65	39
30	64	38
33	64	37
36	63	37
39	62	37
42	61	36
45	60	36
48	60	35
51	60	35

Heat Treating Temperatures Recommended by SAE
*Normalize 870 °C
Austenitize 845 °C
*For forged or rolled specimens only.

2.68

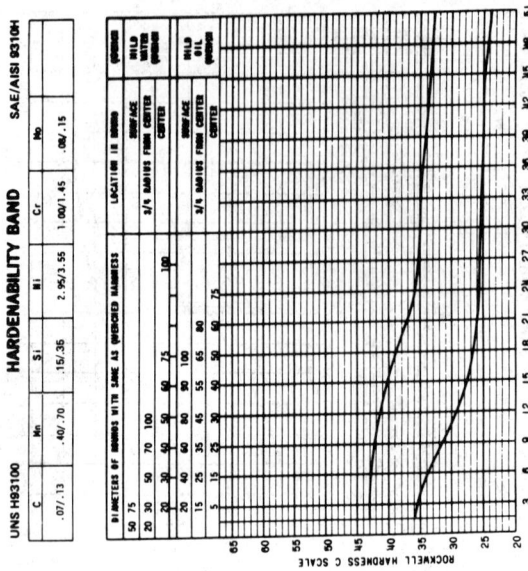
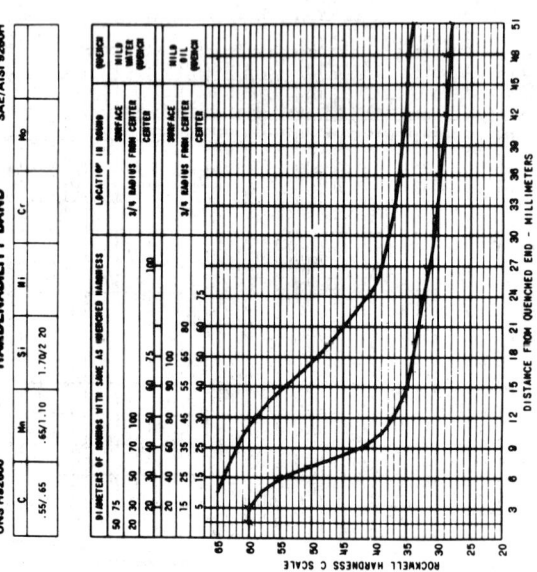

UNS H94171 — HARDENABILITY BAND — SAE/AISI 94B17H

C	Mn	Si	Ni	Cr	Mo	B
.14/.20	.70/1.05	.15/.35	.25/.65	.25/.55	.08/.15	*

*Contains 0.0005–0.003% boron.

HARDNESS LIMITS FOR SPECIFICATION PURPOSES

"J" DISTANCE MILLIMETERS	HRC MAX.	HRC MIN.
1.5	46	39
3	46	39
4.5	46	38
6	45	37
7.5	45	35
9	44	31
10.5	43	27
12	42	25
13.5	41	23
15	39	22
16.5	37	21
18	35	20
19.5	34	--
21	33	--
22.5	32	--
24	31	--
27	29	--
30	28	--
33	27	--
36	26	--
39	25	--
42	24	--
45	24	--
48	23	--
51	23	--

HEAT TREATING TEMPERATURES RECOMMENDED BY SAE
- *NORMALIZE 925 °C
- AUSTENITIZE 925 °C

*For forged or rolled specimens only.

UNS H94301 — HARDENABILITY BAND — SAE/AISI 94B30H

C	Mn	Si	Ni	Cr	Mo	B
.27/.33	.70/1.05	.15/.35	.25/.65	.25/.55	.08/.15	*

*Contains 0.0005–0.003% boron.

HARDNESS LIMITS FOR SPECIFICATION PURPOSES

"J" DISTANCE MILLIMETERS	HRC MAX.	HRC MIN.
1.5	56	49
3	56	49
4.5	56	48
6	55	48
7.5	55	47
9	54	46
10.5	54	45
12	53	43
13.5	53	41
15	52	38
16.5	52	36
18	51	34
19.5	51	32
21	50	30
22.5	49	28
24	48	27
27	45	26
30	43	24
33	41	23
36	39	23
39	38	22
42	36	21
45	35	21
48	34	20
51	34	20

HEAT TREATING TEMPERATURES RECOMMENDED BY SAE
- *NORMALIZE 900 °C
- AUSTENITIZE 870 °C

*For forged or rolled specimens only.

METHODS OF SAMPLING STEEL FOR CHEMICAL ANALYSIS—SAE J408c

SAE Recommended Practice

Report of Iron and Steel Division approved January 1912 and last revised by Iron and Steel Technical Committee January 1971.

Definitions

Ladle Analysis (heat analysis) represents the chemical analysis of a heat of steel as reported to the purchaser. It is the analysis for all the specified elements and is determined by analyzing one or more test samples obtained during the pouring of the steel. When such samples are unobtainable or if it is evident that the samples do not represent the analysis of the melt, additional samples are taken from the solid steel. An analysis based on these representative samples may be used.

In **Ladle Sampling** it is common practice in most melting operations to obtain more than one test sample; often three or more are taken representing the first, middle, and last portions of the heat. These samples are used to survey the uniformity of the heat and for control purposes.

Check Analysis, as used in the steel industry, means an analysis of the metal after it has been rolled or forged into semifinished or finished forms, and is either for the purpose of verifying the average composition of a heat or lot as represented by the ladle analysis, or to determine variations in the composition of a heat or lot. It is not used, as the term might imply, for a duplicate determination made to confirm a previous result. The results of analyses representing different locations in the same piece or taken from different pieces of a lot may differ from each other and from the ladle analysis due to segregation.

Segregation is the result of that natural phenomenon in the solidification of a steel ingot in which various components of the steel having the lowest freezing points are concentrated in parts of the ingot last to solidify. This concentration at different locations results in such a distribution of elements in the ingot that certain areas contain more, while others contain less, of a given element than the average composition of the ingot as a whole.

Segregation in varying degrees is found in all types of steel ingots. The principal factors affecting the amount of segregation are the type and composition of steel, the casting temperature, the ingot shape and size, and the inherent segregating characteristics of the elements being considered.

Rimmed or capped steels are characterized by a lack of uniformity in their chemical composition, especially for the elements carbon, phosphorus, and sulfur, and for this reason check analysis is not considered appropriate for these elements unless misapplication is clearly indicated.

Certain qualities of some commodities are not subject to check analysis requirements except when misapplication is clearly indicated. Examples of these qualities are merchant quality in carbon steel bars and regular quality in carbon steel plates.

In all types of steel, because of the degree to which phosphorus and sulfur tend to segregate, check analysis for these elements is equally inappropriate for rephosphorized or resulfurized steels unless indications of misapplication clearly exist.

The effect of segregation makes the ladle analysis more representative of the average composition of a melt of steel than the analysis of a single sample from the finished material. However check analysis of properly located samples will afford a reasonable comparison with the ladle analysis if taken from a sufficient number of pieces of the finished material to constitute a fair average.

Methods of Sampling for Check Analysis—Each heat of steel in a lot or shipment is considered separately. To indicate adequately the representative composition of a heat or lot, samples selected to represent the melt as fairly as possible are taken from a minimum number of pieces as follows:

> Four pieces for lots up to 15 tons, inclusive
> Six pieces for lots over 15 tons

If the number of pieces in a melt is less than the number of samples specified above, one sample shall be taken from each piece.

Preparation of Samples—Steel subjected to check analysis should be in the condition as received from the steel producer because if it has been subjected to subsequent heating operations it may not yield analytical results which properly represent its original composition.

Drillings or chips are taken without the application of water, oil, or other lubricant, and must be free from scale, grease, dirt, or other foreign substances. Provisions should be made to discard the surface metal before obtaining samples for check analysis to assure freedom from scale, decarburization, and so on, and thus obtain representative results. They should not be overheated during cutting to the extent of causing decarburization. Chips must be well mixed, and those too coarse to pass a No. 10 (2000 μm) sieve or too fine to remain on a No. 30 (590 μm) sieve are not suitable for proper analysis. Sieve size numbers are ASTM designations.

When chips are taken by drilling, for pieces having a cross sectional area up to 16 in.^{2} inclusive, the diameter of the drill is approximately $\frac{1}{2}$ in.; for steel over 16 in.^{2} cross sectional area, the diameter of the drill is approximately 1 in. Each sample commonly consists of not less than 2 oz of drillings.

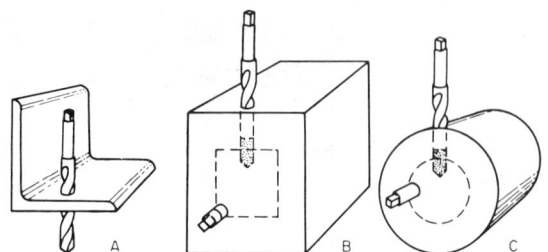

FIG. 1—LOCATION OF SAMPLES FOR CHECK ANALYSIS

In referring samples to other analysts for check analyses, pieces of the original full size section, when possible, should be submitted rather than cuttings unless the latter are specifically requested.

Sampling Steel Products, Rolled and Forged

Large Sections—For large sections, including blooms, billets, rounds, squares, shapes, and so forth, samples shall be taken at any point midway between the outside and the center of the piece by drilling parallel to the axis. In cases where this is not practicable, the piece shall be drilled on the side (see Figs. 1B and 1C), but the drillings shall not be collected until they represent the portion midway between the outside and the center. The tensile test specimen may be used for sampling if it conforms to the above conditions.

Small Sections—For material of small cross section, if the method described in the section for large sections is not applicable, the sample shall be taken by machining off the entire cross section, or, if this is not possible, by drilling entirely through the material at a point midway between the outside and the center (Fig. 1A).

Bored Forgings—Forged, Turned, and Bored Pipe—For bored forgings, samples shall be taken midway between the inner and outer surface of the wall. For forged, turned, and bored pipe, samples shall be taken by drilling through the pipe wall, or cuttings shall be taken across the end of the tube, or millings shall be taken from a broken tensile test specimen cut from the wall of the tube.

Plates—For plates, samples shall be taken at a point midway between the center and edge of the plate. For thicknesses 2 in. and less, samples are customarily taken by drilling through the thickness of the plate. For thicknesses over 2 in., samples are taken by drilling the edge of the plate at a point midway between the rolled surface and the midthickness.

Sheets Rolled Longitudinally—For sheets rolled from slabs or bars longitudinally, the specimen for sampling shall be cut 2 in. in width and across the full width of the sheet as rolled. The specimen shall be cleaned by pickling or grinding and then folded once or more by bringing the ends together and closing the bend. The sample for analysis shall be taken in the middle of this length by milling the inside sheared edges or drilling entirely through from the flat surface. Sampling by milling is preferable. For sheets of a light gage, more than one specimen may be taken and stacked together before folding.

Sheets Rolled Transversely—For sheets rolled from slabs or bars transversely, the specimen shall be cut from the side of the sheet, halfway between the middle and end as rolled 2 in. in width and 18 in. in length. If the sheet is 0.037 in. in thickness or lighter, the specimen shall be cut from the full length of the sheet as rolled. The specimen selected shall be cleaned by pickling or grinding and then folded once or more by bringing the ends together and closing the bend. The sample for analysis shall be taken in the middle of this length by milling the inside sheared edges or drilling entirely through from the flat surface. Sampling by milling is preferable.

Sheets Not of the Full Size Rolled—Sheets cut from larger sheets and not of the full size rolled shall be sampled by milling or drilling the sheet in a sufficient number of places so that the sample is representative of the entire sheet. The sampling may be facilitated by folding the sheet both ways.

PERMISSIBLE VARIATIONS FROM SPECIFIED LADLE CHEMICAL RANGES AND LIMITS FOR STEELS—SAE J409c — SAE Standard

Report of Iron and Steel Division approved January 1942 and last revised by Iron and Steel Technical Committee March 1972.

Supplementary to the ladle chemical analysis, a check analysis may be made on steel in the semifinished or finished form. For definitions and methods of sampling steel for chemical check analysis, refer to SAE J408.

Check analysis is applicable where a high degree of uniformity in composition is essential—for example, material that is to be heat treated. Individual determinations may vary from the specified ladle (heat or cast) analysis ranges or limits to the extent shown in Tables 1–4, but the several determinations of any element in a heat may not vary both above and below the specified range, except for lead, which may vary both above and below the specified range.

In the carbon steel series, rimming steel or capped steel is sometimes specified or required because of characteristics inherent in these types of steel. In these applications, check analysis results are not essential as a measure of the desired properties. Rimming steel and capped steel are characterized by a lack of homogeneity in their chemical composition, especially for the elements

TABLE 1—PERMISSIBLE VARIATIONS FROM SPECIFIED CHEMICAL RANGES AND LIMITS FOR CARBON STEEL IN HOT ROLLED AND COLD FINISHED BARS AND SEMIFINISHED FOR FORGING, WIRE ROD, AND SEAMLESS TUBING

Element	Limit or Max of Specified Range, %	Variation, %, Over Max Limit or Under Min Limit			
		Bars, Wire Rod, Seamless Tubing and Semifinished for Forging to 100 in² (0.065 m²) incl	Semifinished Products for Forging		
			Over 100 to 200 in² (0.065 to 0.129 m²) incl	Over 200 to 400 in² (0.129 to 0.258 m²) incl	Over 400 to 800 in² (0.258 to 0.516 m²) incl
Carbon	To 0.25 incl	0.02	0.03	0.04	0.05
	Over 0.25 to 0.55 incl	0.03	0.04	0.05	0.06
	Over 0.55	0.04	0.05	0.06	0.07
Manganese	To 0.90 incl	0.03	0.04	0.06	0.07
	Over 0.90 to 1.65 incl	0.06	0.06	0.07	0.08
Phosphorus	Over max only to 0.040 incl	0.008	0.008	0.010	0.015
Sulfur	Over max only to 0.050 incl	0.008	0.010	0.010	0.015
Silicon	To 0.35 incl	0.02	0.02	0.03	0.04
	Over 0.35 to 0.60	0.05	—	—	—
Copper	Under min only for copper bearing steels	0.02	0.03	—	—
Lead[a]	0.15 to 0.35 incl	0.03	0.03	—	—

[a]Check analysis tolerance for lead applies, both over and under, to a range of 0.15-0.35% lead.

TABLE 2—PERMISSIBLE VARIATIONS FROM SPECIFIED CHEMICAL RANGES AND LIMITS FOR CARBON STEEL STRUCTURAL SHAPES AND PLATES, SHEETS, STRIP, AND WELDED TUBING

Element	Limit or Max of Specified Range, %	Variation, %, over Max Limit or under Min Limit	
		Under Min Limit	Over Max Limit
Carbon	To 0.15 incl	0.02	0.03
	Over 0.15 to 0.40 incl	0.03	0.04
	Over 0.40 to 0.80 incl	0.03	0.05
	Over 0.80	0.03	0.06
Manganese	To 0.60 incl	0.03	0.03
	Over 0.60 to 1.15 incl	0.04	0.04
	Over 1.15 to 1.65 incl	0.05	0.05
Phosphorus	—	—	0.01
Sulfur	—	—	0.01
Silicon	To 0.30 incl	0.02	0.03
	Over 0.30 to 0.60	0.05	0.05
Copper	Under min only for copper bearing steels	0.02	—
Lead[a]	0.15 to 0.35 incl	0.03	0.03

[a]Check analysis tolerance for lead applies both over and under a range of 0.15-0.35% lead.

TABLE 3—PERMISSIBLE VARIATIONS FROM SPECIFIED CHEMICAL RANGES AND LIMITS FOR ALLOY STEELS

Element	Limit or Max of Specified Range, %	Variation, %, Over Max Limit or Under Min Limit				
		Bars, Sheets, Strip and Semifinished Products to 100 in² (0.065 m²) incl	Semifinished Products			Plates
			Over 100 to 200 in² (0.065 to 0.129 m²) incl	Over 200 to 400 in² (0.129 to 0.258 m²) incl	Over 400 to 800 in² (0.258 to 0.516 m²) incl	
Carbon	To 0.30 incl	0.01	0.02	0.03	0.04	0.02
	Over 0.30 to 0.75 incl	0.02	0.03	0.04	0.05	0.03
	Over 0.75	0.03	0.04	0.05	0.06	0.04
Manganese	To 0.90 incl	0.03	0.04	0.05	0.06	0.04
	Over 0.90 to 2.10 incl	0.04	0.05	0.06	0.07	0.05
Phosphorus	Over max only	0.005	0.010	0.010	0.010	0.010
Sulfur	To 0.060 incl[a]	0.005	0.010	0.010	0.010	0.010
Silicon	To 0.40 incl	0.02	0.02	0.03	0.04	0.02
	Over 0.40 to 2.20 incl	0.05	0.06	0.06	0.07	0.06
Nickel	To 1.00 incl	0.03	0.03	0.03	0.03	0.03
	Over 1.00 to 2.00 incl	0.05	0.05	0.05	0.05	0.05
	Over 2.00 to 5.30 incl	0.07	0.07	0.07	0.07	0.07
	Over 5.30 to 10.00 incl	0.10	0.10	0.10	0.10	0.10
Chromium	To 0.90 incl	0.03	0.04	0.04	0.05	0.04
	Over 0.90 to 2.10 incl	0.05	0.06	0.06	0.07	0.06
	Over 2.10 to 3.99	0.10	0.10	0.12	0.14	0.10
Molybdenum	To 0.20 incl	0.01	0.01	0.02	0.03	0.01
	Over 0.20 to 0.40 incl	0.02	0.03	0.03	0.04	0.03
	Over 0.40 to 1.15 incl	0.03	0.04	0.05	0.06	0.04
Tungsten	To 1.00 incl	0.04	0.05	0.05	0.06	0.05
	Over 1.00 to 4.00 incl	0.08	0.09	0.10	0.12	0.09
Vanadium	To 0.10 incl	0.01	0.01	0.01	0.01	0.01
	Over 0.10 to 0.25 incl	0.02	0.02	0.02	0.02	0.02
	Over 0.25 to 0.50 incl	0.03	0.03	0.03	0.03	0.03
	Min value specified check under min limit	0.01	0.01	0.01	0.01	0.01
Aluminum[b]	Up to 0.10 incl	0.03	—	—	—	0.03
	Over 0.10 to 0.20 incl	0.04	—	—	—	0.04
	Over 0.20 to 0.30 incl	0.05	—	—	—	0.05
	Over 0.30 to 0.80 incl	0.07	—	—	—	0.07
	Over 0.80 to 1.80 incl	0.10	—	—	—	0.10
Lead[b]	0.15 to 0.35 incl	0.03[c]	—	—	—	0.03[c]
Copper[b]	To 1.00 incl	0.03	—	—	—	0.03
	Over 1.00 to 2.00 incl	0.05	—	—	—	0.05

[a]Sulfur over 0.060% is not subject to check analysis.
[b]Tolerances shown apply only to 100 in² or less and to all plates.
[c]Tolerance is over and under.

TABLE 4—PERMISSIBLE VARIATIONS FROM SPECIFIED CHEMICAL RANGES AND LIMITS FOR STAINLESS STEELS

Element	Limit or Max of Specified Range, %	Variation, %, over Max Limit or Under Min Limit
Carbon	To 0.030 incl Over 0.030 to 0.20 incl Over 0.20 to 0.60 incl Over 0.60 to 1.20 incl	0.005 0.01 0.02 0.03
Manganese	To 1.00 incl Over 1.00 to 3.00 incl Over 3.00 to 6.00 incl Over 6.00 to 10.00 incl	0.03 0.04 0.05 0.06
Phosphorus	To 0.040 incl Over 0.040 to 0.20 incl	0.005 0.010
Sulfur	To 0.040 incl Over 0.040 to 0.20 incl Over 0.20 to 0.50 incl	0.005 0.010 0.020
Silicon	To 1.00 incl Over 1.00 to 3.00 incl	0.05 0.10
Chromium	Over 4.00 to 10.00 incl Over 10.00 to 15.00 incl Over 15.00 to 20.00 incl Over 20.00 to 27.00 incl	0.10 0.15 0.20 0.25
Nickel	To 1.00 incl Over 1.00 to 5.00 incl Over 5.00 to 10.00 incl Over 10.00 to 20.00 incl Over 20.00 to 22.00 incl	0.03 0.07 0.10 0.15 0.20
Molybdenum	Over 0.20 to 0.60 incl Over 0.60 to 1.75 incl Over 1.75 to 4.00 incl	0.03 0.05 0.10
Titanium	All ranges	0.05
Columbium-Tantalum	All ranges	0.05
Tantalum	To 0.10 incl	0.02
Cobalt	0.05 to 0.20 incl	0.01
Aluminum	0.10 to 0.30 incl	0.05
Selenium	All ranges	0.03
Nitrogen	To 0.19 incl Over 0.19 to 0.25 incl	0.01 0.02

carbon, phosphorus, and sulfur. Therefore, check analysis is not applicable for elements other than manganese and copper.

Rephosphorized steel is not subject to check analysis for phosphorus, and resulfurized steel is not subject to check analysis for sulfur unless misapplication is clearly indicated.

Boron is not subject to check analysis tolerances.

HIGH STRENGTH, LOW ALLOY STEEL—SAE J410c

SAE Recommended Practice

Report of Iron and Steel Technical Committee approved January 1947 and last revised October 1969.

Introduction—High strength, low alloy steel represents a specific group of steels in which enhanced mechanical properties and, in some cases, resistance to atmospheric corrosion are obtained by the addition of moderate amounts of one or more alloying elements other than carbon. Different types are available, some of which are carbon-manganese steels and others which contain further alloy additions, governed by special requirements for weldability, formability, toughness, strength and economics. Descriptions of the different grades are given in the paragraph entitled General Information. These steels may be obtained in the form of sheet, strip, plates, structural shapes, bars, and bar size sections.

These steels are especially characterized by their mechanical properties, obtained in the as-rolled condition. They are not intended for quenching and tempering, and the purchaser should not subject them to such treatment without assuming responsibility for the ensuing mechanical properties. For certain applications these steels are sometimes annealed, normalized, or stress relieved with some influence on mechanical properties.

Where these steels are used for fabrication by welding, care must be exercised in selection of grade and in the details of the welding process. Certain grades may be welded without preheat or postheat.

Application—These steels, because of their high strength-to-weight ratio, abrasion resistance and, in the case of certain compositions, improved atmospheric corrosion resistance, are adapted particularly for use in mobile equipment and other structures where substantial weight savings are generally desirable. Typical applications are truck bodies, frames, structural members, scrapers, truck wheels, cranes, shovels, booms, chutes, and conveyors.

Mechanical Properties—The mechanical properties of these steels are shown in Table 1. For larger sizes than shown in the table, consultation with suppliers on availability and characteristics is suggested.

NOTE: PERMISSIBLE USE OF YIELD POINT—Present steel industry practice is to express the yield of these materials as yield point rather than yield strength, and determination is by drop of beam, dividers, or other method, covered in ASTM A 370, Methods and Definitions for Mechanical Testing of Steel Products. Unless otherwise specified, this recommended practice is acceptable for material supplied in this report, and use of an extensometer is not required. The yield point shall meet the requirements shown for yield strength in Table 1A.

High strength, low alloy steel may be specified annealed or normalized, or otherwise specially prepared for special forming properties. In those cases mechanical properties should be agreed upon between supplier and purchaser.

Bend Test—Longitudinal bend test specimens shall stand being bent at room temperature through 180 deg without cracking on the outside of the bent portion. The inside diameter of the bend shall have a relationship to the thickness of the specimen as prescribed in Table 1B.

The bend test requirements in Table 1B are for mill acceptance purposes and are made on prepared test specimens. They have no relationship to shop bending practices (see General Information) and are not to be used as a basis for fabricating procedures.

Dimensional Tolerances—Standard manufacturing tolerances for dimensions as given in the AISI Steel Products Manual for High Strength-Low Alloy Steel shall apply.

TABLE 1A—MINIMUM MECHANICAL PROPERTIES[a]

Grade	Form	Yield Strength,[b] psi min	Tensile Strength, psi min	Elongation, % min 2 in.	Elongation, % min 8 in.
942X	Plates, shapes, bars to 4 in. incl	42,000	60,000	24	20
945A, C	Sheet and strip	45,000	60,000	22	—
	Plates, shapes, bars				
	To 1/2 in. incl	45,000	65,000	22	18
	1/2–1-1/2 in. incl	42,000	62,000	24	19
	1-1/2–3 in. incl	40,000	62,000	24	19
945X	Sheet and strip	45,000	60,000	25	—
	Plates, shapes, bars				
	To 1-1/2 in. incl	45,000	60,000	22	19
950A, B, C, D	Sheet and strip	50,000	70,000	22	—
	Plates, shapes, bars				
	To 1/2 in. incl	50,000	70,000	22	18
	1/2–1-1/2 in. incl	45,000	67,000	24	19
	1-1/2–3 in. incl	42,000	63,000	24	19
950X	Sheet and strip	50,000	65,000	22	—
	Plates, shapes, bars				
	To 1-1/2 in. incl	50,000	65,000	—	18
955X	Sheet and strip	55,000	70,000	20	—
	Plates, shapes, bars				
	To 1-1/2 in. incl	55,000	70,000	—	17
960X	Sheet and strip	60,000	75,000	18	—
	Plates, shapes, bars				
	To 1-1/2 in. incl	60,000	75,000	—	16
965X	Sheet and strip	65,000	80,000	16	—
	Plates, shapes, bars				
	To 3/4 in. incl	65,000	80,000	—	15
970X	Sheet and strip	70,000	85,000	14	—
	Plates, shapes, bars				
	To 3/4 in. incl	70,000	85,000	—	14
980X	Sheet and strip	80,000	95,000	12	—
	Plates to 3/8 in. incl	80,000	95,000	—	10

[a] Mechanical properties to be determined in accordance with ASTM A 370.
[b] Yield strength to be measured at 0.2% offset.

TABLE 2—CHEMICAL COMPOSITION [a,b] LADLE ANALYSIS, MAX, %

Grade	C	Mn	P
942X	0.21	1.35	0.04
945A	0.15	1.00	0.04
945C	0.23	1.40	0.04
945X	0.22	1.35	0.04
950A	0.15	1.30	0.04
950B	0.22	1.30	0.04
950C	0.25	1.60	0.04
950D	0.15	1.00	0.15
950X	0.23	1.35	0.04
955X	0.25	1.35	0.04
960X	0.26	1.45	0.04
965X	0.26	1.45	0.04
970X	0.26	1.65	0.04
980X	0.26	1.65	0.04

[a] Sulfur, 0.05% max; silicon, 0.90% max.
[b] It is generally believed that atmospheric corrosion resistance can be enhanced by the addition of moderate amounts of one or more alloying elements, other than carbon. If this characteristic is a requirement, the addition of such elements within the limits of this specification shall be negotiated.

TABLE 3—SUGGESTED MINIMUM INSIDE RADII FOR COLD BENDING—RATIO OF BEND RADIUS TO THICKNESS

Grade	Thickness of Material, in.		
	To 0.180	0.180–0.250	0.250–0.500
942X	—	1	2
945A, C	1	2	2-1/2
945X	1	1	2
950A, B, C, D	1	2	3
950X	1-1/2	2-1/2	2-1/2
955X	2	3	3
960X	2-1/2	3-1/2	3-1/2
965X	3	4	4
970X	3-1/2	4-1/2	4-1/2
980X	3-1/2	4-1/2	4-1/2[a]

[a] Available to 0.375 in. inclusive.

Hot forming is recommended for bending thicknesses over 1/2 in. While the recommendations in Table 3 for inside radii for cold bending are minimum, it may be found necessary to increase the radii, depending on the strength level of the material and the application.

Description of Grades

Grade 942X—This is a columbium or vanadium treated carbon manganese high strength steel similar to 945X and 945C except for somewhat improved welding and forming properties.

Chemical Composition—Chemical composition of steel furnished to this specification shall conform to Table 2.

NOTE: Fully killed steel made to fine-grain practice may be specified by the use of suffix "K". EXAMPLE: 945AK. Each of these grades made to "K" practice may not be available from some suppliers, and this practice should only be specified when improved low temperature notch toughness is important.

The steels designated by the suffix "X" contain strengthening elements such as columbium, vanadium, or nitrogen, added singly or in combination. These steels are usually made semikilled; however, if killed steel is desired, it may be specified by use of the "K" suffix, as SAE 950XK.

Choice of, and limits for, additional elements, other than those specified in Table 2 necessary to attain required or desirable properties, shall be made known to the purchaser and not changed without his full knowledge and consent.

General Information—**Recommended Bending Practice** The recommended fabricating practice for cold forming these steels is to avoid bends in any direction that are more severe than shown in Table 3.

TABLE 1B—180 DEG BEND TEST REQUIREMENTS, RATIO OF BEND DIAMETER TO THICKNESS

Grade	Sheet and Strip	Plates, Shapes, Bars—Thickness of Material, in.						
		To 3/8 incl	3/8–1/2 incl	1/2–3/4 incl	3/4–1 incl	1–1-1/2 incl	1-1/2–2 incl	2–4 incl
942X	—	1	1	1	1-1/2	2	2-1/2	3
945A	1	1	1	2	2	2	3	3
945C	1	1	1	2	2	2	3	3
945X	1	1	1	1	1-1/2	2	2-1/2	—
950A	1	1	1	2	2	2	3	3[a]
950B	1	1	1	2	2	2	3	3[a]
950C	1	1	1	2	2	2	3	3[a]
950D	1	1	1	2	2	2	3	3[a]
950X	1	1	1	1	1-1/2	2-1/2	3	—
955X	1	1	1-1/2	1-1/2	2	3	3-1/2[b]	—
960X	1-1/2	1-1/2	2	2	2-1/2	3	—	—
965X	2	2	2-1/2	3	—	—	—	—
970X	3	3	—	—	—	—	—	—
980X	3	3	—	—	—	—	—	—

[a] Thickness 2–3 in. inclusive.
[b] Applicable to webs of structural shapes.

Grade 945A—This is a high strength, low alloy steel with excellent welding characteristics, both arc and resistance, and the best formability, weldability, and low temperature notch toughness of the high strength steels. It is generally used in sheet, strip, and light plate thickness.

Grade 945C—This is a carbon-manganese high strength steel with satisfactory arc welding properties, if adequate precautions are observed. It is similar to grade 950C, except that lower carbon and manganese improve arc welding characteristics, formability, and low temperature notch toughness at some sacrifice in strength.

Grade 945X—This is a columbium or vanadium treated carbon-manganese high strength steel similar to 945C, except for somewhat improved welding and forming properties.

Grade 950A—This is a high strength, low alloy steel with good weldability, both arc and resistance, with good low temperature notch toughness, and good formability. It is generally used in sheet, strip, and light plate thickness.

Grade 950B—This is a high strength, low alloy steel with satisfactory arc welding properties and fairly good low temperature notch toughness and formability.

Grade 950C—This is a carbon-manganese high strength steel which can be arc welded with special precautions, unsuitable for resistance welding. The formability and toughness are fair.

Grade 950D—This is a high strength, low alloy steel with good weldability, both arc and resistance, and fairly good formability. Where low temperature properties are important, the effect of phosphorus in conjunction with other elements present should be considered.

Grade 950X—This is a columbium or vanadium treated carbon-manganese high strength steel similar to 950C, except for somewhat improved welding and forming properties.

Grade 955X, 960X, 965X, 970X, 980X—These are steels similar to 945X and 950X with higher strength obtained by increased amounts of strengthening elements, such as carbon or manganese or by the addition of nitrogen up to about 0.015%. This increased strength involves reduced formability and usually decreased weldability. Toughness will vary considerably with composition and mill practice.

TABLE 4—GRADES IN APPROXIMATE ORDER OF INCREASING EXCELLENCE

Weldability	Formability	Toughness
980X	980X	980X
970X	970X	970X
965X	965X	965X
960X	960X	960X
955X, 950C, 942X	955X	955X
945C	950C	945C, 950C, 942X
950B, 950X	950D	945X, 950X
945X	950B, 950X, 942X	950D
950D	945C, 945X	950B
950A	950A	950A
945A	945A	945A

HIGH STRENGTH, QUENCHED, AND TEMPERED STRUCTURAL STEELS—SAE J368a

SAE Recommended Practice

Report of Iron and Steel Technical Committee approved November 1968 and last revised August 1974.

Introduction—The steels covered by this SAE Recommended Practice have enhanced mechanical properties obtained by quench and temper treatment. Grade Q980 is a carbon-manganese steel, while grades Q980B, Q990B, and Q9100B are carbon-manganese boron steels. Other grades (designated by suffix A) represent steels containing one or more additional alloying elements as required to achieve higher strengths and to accommodate greater thicknesses. These steels are produced fully deoxidized and to fine grain practice.

Since these steels are characterized by their mechanical properties, care must be exercised in the selection of grade, especially where fabrication by welding or forming is required. Special procedures may be applicable to varying compositions and section sizes, as produced by a given supplier; therefore, the purchaser should consult with the producer in order to be aware of these variables.

Application—These steels, because of their high strength-to-weight ratio, abrasion resistance and, in the case of certain compositions, improved atmospheric corrosion resistance, are adapted particularly for use in mobile equipment and other structures where substantial weight savings are desirable. Typical applications are truck bodies, frames, structural members, scrapers, truck wheels, cranes, shovels, booms, chutes, and conveyors.

General Requirements—Material furnished under this report shall conform to the applicable requirements of the current edition of ASTM A 6, Specification for General Requirements for Delivery of Rolled Steel Plates, Shapes, Sheet Piling and Bars for Structural Use.

Mechanical Properties—The mechanical properties of these steels shall be as shown in Table 1. Producers should be consulted concerning requirements for plate thicknesses greater than those shown in the table and for all sheets and

TABLE 1—MECHANICAL PROPERTIES[a]

Grade and Thickness	Strength				Elongation in 2 in or 50 mm[c], min, %	Bend Test[d]	Typical Bhn Range
	Yield,[b] min		Tensile				
	10³ psi	MPa	10³ psi	MPa			
Q980							
To 3/4 in (19.05 mm) incl	80	550	95–115	655–795	18	2T	200–235
Q980B							
To 1-1/4 in (31.75 mm) incl	80	550	95–115	655–795	18	2T	200–235
Q980A							
To 1-1/2 in (38.1 mm) incl	80	550	95–115	655–795	18	2T	200–235
Q990B							
To 1-1/2 in (38.1 mm) incl	90	620	100–130	690–895	18	3T	217–285
Q990A							
To 1 in (25.4 mm) incl	90	620	100–130	690–895	18	2T	217–285
Over 1 to 2 (25.4 to 50.8 mm) incl	90	620	100–130	690–895	18	3T	217–285
Over 2 to 4 in (50.8 to 101.6 mm) incl	90	620	100–130	690–895	18	4T	217–285
Q9100B							
To 1 in (25.4 mm) incl	100	690	110 min	760 min	18	2T	223 min
Over 1 to 1-1/4 in (25.4 to 31.75 mm) incl	100	690	110 min	760 min	18	3T	223 min
Q9100A							
To 1 in (25.4 mm) incl	100	690	110–130	760–895	18	2T	223–285
1 to 2-1/2 in (25.4 to 63.5 mm) incl	100	690	110–130	760–895	18	2T	223–285
Q9110A							
To 1 in (25.4 mm) incl	110	760	115–135	795–930	18	2T	241–285
1 to 2-1/2 in (25.4 to 63.5 mm) incl	110	760	115–135	790–930	18	2T	241–285

[a] Properties such as fatigue resistance and fracture mechanics should be considered when making a material selection.
[b] Yield strength to be measured at 0.2% offset or 0.5% extension under load.
[c] Minimum percent elongation for all material under 1/4 in (6.35 mm) thick shall be 15%.
[d] Ratios of bend diameter to thickness of specimens. T = Thickness of specimen. Longitudinal bend specimens at room temperature shall bend through 180 deg without cracking on the outside of the bent portion. The inside diameter of the bend shall have a relationship to the thickness of the specimen as described in Table 1.

rolled shapes. Consultation with suppliers on availability and characteristics is also suggested.

Chemical Compositions—Chemical composition of steels furnished to this specification shall conform to Table 2.

General Information

Formability—The high strength level of these steels reduces the formability; however, moderate forming may be performed by using proper radii, preferably by forming transverse to the final rolling direction. Recommended *minimum* inside radii for cold bending (in any direction) are:

Thickness		Inside Radii, min
in	mm	
To 1 incl	To 25.4 incl	2T
1 thru 2	25.4 thru 50.8	3T
Over 2	Over 50.8	Cold bending not recommended

For some grades of material and for some applications, it may be necessary to increase the minimum radii shown.

Bend Test—The bend test is performed by the producer and is intended for product qualification purposes. The bend test has no relationship to shop

TABLE 2—CHEMICAL COMPOSITION (HEAT ANALYSIS), MAX, % BY WT[a]

Grade	C	Mn	P	S	B
Q980	0.20	1.35	0.04	0.05	—
Q980B	0.20	1.50	0.04	0.05	0.0005 min
Q980A	0.20	1.10	0.04	0.05	—
Q990B	0.20	1.50	0.04	0.05	0.0005 min
Q990A	0.20	1.50	0.04	0.05	—
Q9100B	0.20	1.50	0.04	0.05	0.0005 min
Q9100A	0.21	1.50	0.04	0.05	—
Q9110A	0.21	1.50	0.04	0.05	—

[a]Where thickness, fabricating, and service requirements dictate the need of additional chemical elements or of a change in the limits shown to produce the mechanical properties shown in Table 1, the supplier shall obtain approval of the purchaser.

bending practices (see above paragraph) and its requirements are not to be used as a basis for fabricating procedures.

Recommended Welding Practices—These high strength steels are weldable. It is common practice to use low hydrogen metal arc processes for welding these quenched and tempered steels. (Proper consideration must be given to any structure and/or property changes which might occur in the base metal as a result of the welding process.) The producer should be consulted for those special procedures applicable to the material which will be supplied against any given requirement.

ENGINE POPPET VALVE MATERIALS—SAE J775a

SAE Information Report

Report of Iron and Steel Technical Committee approved June 1961 and last revised July 1969.

Definition—Poppet valve materials are defined as those metallic alloy compositions which have been developed to meet special requirements for durability and function in the presence of gases inducted to and rejected from the combustion chamber of internal combustion engines.

This definition may include iron, nickel, or cobalt base alloys which are suitable for end use other than engine valve applications. It will also include materials which are suited to engine valve applications exclusively, and which have not been described previously in the SAE Handbook.

Purpose and Scope—This SAE report on valve materials is intended to provide engineers, designers, and metallurgists with the following information:

1. A rough classification of valve materials according to type, severity, and durability requirements of end use.
2. A general description of valve design and construction and its relation to valve materials.
3. General descriptive information on the type of analyses of valve materials.
4. General information on valve material physical and mechanical properties.

This report is intended to summarize pertinent information on engine valve materials which is now scattered among many publications.

Classification of Valve Materials—Except for low alloy intake valve steels, valve materials are similar in many respects to other corrosion resistant high-hot-strength metallic materials. Because of their unique end use, certain properties (such as resistance to corrosive attack by products of combustion) are emphasized, while others (such as ductility) are de-emphasized. Valve materials are chosen for good high temperature corrosion resistance and for good high temperature fatigue and creep strength. Other properties of importance in a valve material are hot hardness, wear resistance at ordinary temperatures, and, of course, the requisite properties that permit their economic manufacture.

Valve materials are classified either according to end use or according to type. The type classifications follow (see also Table 1):

1. Martensitic Steels—These are steels with carbon contents up to 1.0% and chromium contents up to 22.0%. Other elements may be varied to influence fabrication or heat treat characteristics or to control oxidation resistance. These steels are hardenable by quenching and develop desirable wear characteristics. The martensitic materials are almost always used as inlet valves and are used as exhaust valves where the maximum valve operating temperature is limited to approximately 1200 F.

2. Austenitic Steels—Steels of this group are substantially fully austenitic at ordinary temperatures and the temperatures of end use. The austenitizing elements commonly used are nickel, manganese, nitrogen and carbon. Some of the austenitics, particularly those high in nitrogen and carbon, can be hardened by solution treating plus aging to precipitate carbides and nitrides.

These types are useful in one-piece valve construction. Others cannot be hardened significantly, and find their field of usage only in composite valve structures.

Alloys of the austenitic steel group are used as exhaust valves in medium to heavy duty service. These steels are useful up to a maximum operating temperature of approximately 1600 F. Austenitic steels, by virtue of their physical and mechanical properties, are frequently chosen as the base material for faced valves where extreme durability is required.

3. Superalloys—This classification covers nickel or cobalt base alloys, and some highly alloyed iron base materials. Many of these materials were originally developed for the extreme requirements of gas turbine bucket applications. Because of the similarity of application, these alloys, or modifications of them, may be useful for exhaust valves also. These alloys are only used where the ultimate of dependability or durability is required.

4. Facing Alloys—This classification describes high alloy grades mainly used as overlays at critical points of wear and corrosion, and sometimes made into valve castings.

Designation of SAE Valve Steels—The majority of valve material applications can be divided into functional groups which make possible an orderly system of identification. The SAE numbers used for valve materials throughout this report have a somewhat obvious relation to the types of valves in which the materials are commonly used. Numbers have been assigned on the following bases:

1. Letter Prefix—Intake valves may be made of constructional low alloy steel or heat and corrosion resistant high alloy steels. The prefix letters NV designate low alloy constructional steel. HNV designates high alloy intake valve material. [In this report, the HNV designation has been retained for two instances (HNV3 and HNV6) where these materials also are used for exhaust valve applications.]

Exhaust valves may be made of hardenable martensitic steels, austenitic steels, or superalloys. They may be iron, nickel, or cobalt base alloys. The prefix letters EV are used to designate austenitic exhaust valve steels, and the letters HEV for the extremely high strength alloy used in very severe diesel and gasoline engine service.

High alloy, welded overlays used at critical points of wear or corrosion have been designated in the material tables with the prefix VF.

2. Number Suffix—A number is arbitrarily assigned and based on the order in which the material appears in the various categories of Table 1. Future grades will be assigned numbers in the order in which they are added to a category.

3. SAE Number—Where an SAE alloy or stainless steel number presently exists for a valve steel, the SAE number will appear in parentheses following the valve steel number.

Valve Design and Application as Related to Materials—The design of a valve

TABLE 1—NOMINAL ANALYSES OF TYPICAL VALVE STEELS, %[a]

SAE No.	C	Mn	Si	Cr	Ni	Co	W	Mo	Fe, Max	Other	Similar Commercial Designations
Active Intake Valves											
Constructional											
NV 1 (SAE 1541)	0.41	1.50	0.25	—	—	—	—	—	—	—	—
NV 2 (SAE 1547)	0.47	1.50	0.25	—	—	—	—	—	—	—	—
NV 3	0.50	0.80	0.30	0.40	0.30	—	—	0.15	—	—	NE 8150
NV 4 (SAE 3140)	0.40	0.80	0.30	0.65	1.25	—	—	—	—	—	—
NV 5 (SAE 8645)	0.45	0.90	0.30	0.50	0.55	—	—	0.20	—	—	—
NV 6 (SAE 5150)	0.50	0.80	0.30	0.80	—	—	—	—	—	—	—
High Alloy											
HNV 1	0.55	0.40	1.50	8.0	—	—	—	0.75	—	—	Sil 2
HNV 2	0.40	0.30	3.9	2.2	—	—	—	—	—	—	Sil F
HNV 3	0.45	0.40	3.3	8.5	—	—	—	—	—	—	Sil 1
HNV 4	0.45	0.40	3.3	7.0	1.0	—	—	—	—	—	731
HNV 5	0.35	0.40	2.5	13.0	8.0	—	—	0.50	—	—	CNS
HNV 6	0.80	0.40	2.3	20.0	1.3	—	—	—	—	—	XB
HNV 7 (SAE 71360)	0.55	0.20	0.20	3.5	—	—	14.00	—	—	—	—
Active Exhaust Valves											
Martensitic											
HNV 3	0.45	0.40	3.3	8.5	—	—	—	—	—	—	Sil 1
HNV 6	0.80	0.40	2.3	20.0	1.3	—	—	—	—	—	XB
Austenitic											
EV 3	0.20	1.3	1.0	21.0	11.5	—	—	—	—	—	21-12
EV 4	0.20	1.3	1.0	21.0	11.5	—	—	—	—	N-0.15-0.25	21-12N
EV 5	0.38	1.0	3.0	19.0	8.0	—	—	—	—	—	Sil 10
EV 6	0.38	1.0	3.0	19.00	8.0	—	—	—	—	N-0.15-0.25	Sil 10N
EV 7	0.20	5.00	0.50	21.00	4.50	—	—	—	—	N-0.30	2155N
EV 8	0.53	9.00	0.15	21.00	3.75	—	—	—	—	N-0.42, S-0.07	21-4N
EV 9	0.45	0.50	0.60	14.00	14.00	—	2.40	0.35	—	—	TPA
EV 10	1.00	0.8	3.0	14.5	14.5	—	—	—	—	—	CAST 14-14
EV 11	0.70	6.3	0.55	21.0	1.9max	—	—	—	—	N-0.23	Sil 746
EV 12	0.55	8.3	0.15	21.0	2.2	—	—	—	—	N-0.30, S-0.07	21-2N
EV 13	0.53	11.5	2.6	21.0	—	—	—	—	—	N-0.40 min	Gamon H
Superalloys											
HEV 1	0.10	1.50	0.50	21.30	20.00	20.00	2.5	3.00	—	N-0.15 Cb and Ta-1.00	N-155
HEV 2	0.05	2.30	0.05	16.00	Base	0.50	—	—	—	Al-0.05, Ti-3.10	TPM
HEV 3	0.05	0.60	0.30	15.00	Base	0.70	—	—	—	Al-0.70, Ti-2.50 Cb and Ta-1.00	Inconel X-750
HEV 5	0.05	1.0	0.6	20.00	Base	2.0max	—	—	5.0	Al-1.2, Ti-2.5	Nimonic 80A
HEV 6	0.05	1.0	1.5 max	20.00	Base	18.00	—	—	—	Al-1.2, Ti-2.5	Nimonic 90
Active Facings											
Alloy Overlays											
VF 1	0.20	0.80	0.20	20.00	Base	—	—	—	1.0	—	80-20 Ni-Cr
VF 2	1.20	0.5	1.20	28.00	3.00	Base	4.50	0.50	3.0	—	Stellite 6
VF 3	2.40	—	0.70	29.00	39.00	10.00	15.00	—	6.5	—	Eatonite
VF 4	2.00	0.30	0.30	26.00	Base	0.50	8.75	—	4.0	—	X-782
VF 5	1.75	0.30	1.00	25.00	22.00	Base	12.00	—	2.0	—	Stellite F
VF 6	2.5	0.5	1.3	30.0	1.5	Base	13.0	0.5	3.0 min	—	Stellite 1
VF 7	1.4	2.5	0.7	30.0	1.5	Base	8.3	—	3.0 min	—	Stellite 12
Noncurrent Valves											
EV 1	0.45	0.50	0.50	23.5	4.8	—	—	2.8	—	—	XCR
EV 2	0.40	4.3	0.80	24.0	3.8	—	—	1.4	—	—	TXCR
Experimental Valves											
						V		Cb	N		
XEV-A	0.5	11.5	0.25	20.5	—	0.5	2.2	1.0	0.5	—	DV2A
XEV-B	0.5	11.5	1.00	20.5	—	0.5	2.2	1.0	0.5	—	DV2B
XEV-C	0.2	6.5	0.4	21.0	5.5	—	—	—	0.2	—	Rep. 2157

[a]Table 1 is intended to supply information on typical valves and is not a complete list of all materials used for valve steels.

and its application to the engine are as important as, and frequently much more important than, the selection of the valve material. Valve durability is limited by the operating temperature of the valve and by the stresses imposed upon it. It is sometimes, but not always, possible to select a stronger or a more temperature resistant material to overcome defects in valve design and application.

Good cooling is paramount to satisfactory valve durability and this implies proper seating of the valve in the engine and proper cooling of the immediate area around the valve seat. There is no known material which will deliver satisfactory durability under conditions of blowby, excessive valve seat distortion and/or inadequate seat cooling. As a rough rule of thumb, a reduction of temperature of 25 F will about double the burning durability of any given valve material. Again, valve stresses arise generally from the dynamics of the valve train and from the manner in which the valve closes against its seat, thus implying the necessity for careful analysis of the valve train kinematics.

In certain instances, where valve materials of requisite properties to obtain the desired durability are not available or cannot be used economically, various means of valve fortification are employed. The most important of these are as follows:

1. **Face Coatings**—These are welded overlays applied to valve faces and intended to develop optimum corrosion and wear resistance at the valve seating surface. Cobalt and nickel base alloys are usually chosen for this purpose.

2. **Head Coatings**—These coatings are applied to the tops of the heads of exhaust valves to inhibit corrosion. Because hot hardness is not required, nickel-chromium alloys are most frequently chosen for this purpose.

3. **Aluminizing**—This is a special case of a protective coating. It comprises a thin layer of aluminum applied to the valve face and sometimes to the top of the valve head. Aluminum, when diffused, alloys with the base material and provides a thin hard corrosion resistant coating.

4. **Sodium Cooling**—This is obtained by the use of hollow valves partially filled with metallic sodium which transfers heat by convection from the hot head end of the valve to the stem. Sodium filling reduces maximum valve temperatures from 150 to 350 F.

5. **Stem and Tip Welding**—Occasionally, the valve tip, and more often the entire stem, is made of a martensitic steel. The stem material is welded to the head section of the valve and heat treated to provide more desirable wear properties.

6. **Valve Rotators**—These are mechanical devices which cause the valve to turn in service, thus causing deposits to be wiped off the valve face and stem. Rotation tends to ensure good heat transfer from the valve face to the seat. Several types are in use. Valve rotators have been responsible for durability improvements of from two to ten times that of the same material without rotation.

7. **Automatic Tappets**—These devices function by maintaining a small predetermined clearance in the valve train when the valve is closed, thus ensuring positive seating. Automatic tappets compensate for wear and for thermal changes in the valve train. Their use tends to avoid excessive stresses resulting from too large clearance, as well as excessive valve temperatures resulting from too small or negative clearance.

Mechanical and Physical Properties—The selection of a valve material on the basis of physical properties requires careful consideration. The properties which are of most importance in the valve application (corrosion resistance, hot fatigue strength, and so forth) are, unfortunately, the most difficult properties to determine in the laboratory, and consequently complete information on these desired properties is not available. Therefore, other data (for example, alloy content, stretch or creep data, hot tensile strength, and so forth) may be used as a guide, but final confirmation should always come from engine testing done in a manner that will control as many variables as possible.

In attempting to improve valve durability, it is self-evident that the property being explored should have some relation to the failures manifested in the engine. For example, it would be inappropriate to try to improve valve burning resistance by the selection of a material with a higher creep strength, unless the burning could be directly related to LACK of creep strength in the valve material. In the selection of the new valve material, the first step should be a thorough analysis of valve performance in the engine in question, including operating temperatures and in particular, a thorough analysis of any failures which may be recurrent. With this background, valve material physical and mechanical data can be used as a guide to an intelligent choice of materials for further engine testing. (See Table 2.)

Corrosion Resistance—A most fundamental property of a valve material in controlling valve durability is its corrosion or burning resistance. Many attempts have been made to develop a satisfactory method for measuring the corrosion resistance of valve materials, but unfortunately none of these tests have been completely successful. Final recourse is usually made to engine testing.

The corrosion of valves is accelerated by virtue of both their temperature and environment. Corrosion occurs through three mechanisms:

1. Oxidation.
2. Attack by various metal oxides.
3. Attack by fuel and lubricant additives or contaminants.

Valve alloys are high in chromium and sometimes contain silicon as well to give them inherent oxidation stability. Silicon is sometimes intentionally removed from exhaust valve alloys to prevent certain surface reactions with metal oxides. Diesel engine valve alloys are frequently made with high chromium contents which afford protection from sulfur attack even in alloys containing substantial nickel contents.

TABLE 2—MECHANICAL AND PHYSICAL PROPERTIES OF TYPICAL VALVE STEELS

	RUPTURE AND CREEP							TENSILE PROPERTIES						PHYSICAL CONSTANTS			
SAE No	Stress for 1% Stretch at 1350 F in 100 Hr, psi		High Temperature Stress for Rupture at 1350 F in 100 Hr, psi		Brinell Hot Hardness			SAE No.	Temperature F	Tensile Strength, psi	Yield Strength, psi	Elongation, %	Reduction of Area, %	SAE No.	Specific Gravity gm/cc	Density lb/cu in.	Coefficient of Thermal Exp., 70-1400 F, in/in F × 10⁶
					Mutual Indentation		Cold Ball										
	Forge and Age	Solution Treat and Age Harden	Forge and Age	Solution Treat and Age Harden	at 1200 F	at 1400 F	at 1400 F										
Martensitic								HNV 3	Room	133,000	100,000	22	50	HNV 3	7.75	0.280	7.2
HNV 6	3,000		5,000		85	40	70		1400	20,000	—	stretches					
								HNV 6	Room	156,000	111,000	15.5	25	HNV 6	7.80	0.284	7.4
									1400	13,500	11,500	72	85				
Austenitic								EV 3 and EV 4	Room	119,000	65,000	26.2	19.7	EV 3 and EV 4	7.90	0.288	10.3
EV 3	5-7,000	Not Used	9-12,000	Not Used	—	70	—		1400	42,800	31,300	13.3	18.0				
EV 4	6-8,000	15-17,000	9-12,000	16-19,000	—	80	115	EV 5 and EV 6	Room	156,000	74,650	34.2	25.5	EV 5 and EV 6	7.70	0.278	9.9
EV 5	6-8,000	14-16,000	9-12,000	16-18,000	145	80	120		1400	41,200	28,400	18	30.5				
EV 6	6-9,000	14-17,000	9-12,000	16-19,000	—	90	162	EV 7	Room	125,000	70,000	50	35	EV 7	7.85	0.284	10.4
EV 7	8-10,000	18-20,000	—	—	—	100	121		1400	44,000	31,000	27	23				
EV 8	6-8,000	18-20,000	—	26-30,000	185	100	193	EV 8	Room	163,000	104,000	9	9	EV 8	7.73	0.279	10.25
EV 9	6-8,000	Not Used	8-12,000	Not Used	—	70	108		1400	62,000	37,000	18	25				
EV 10	As Cast	14,000			—	100	—	EV 11	Room	150,000	75,000	20	30	EV 11	7.75	0.280	10.3
EV 11	6-8,000	19-21,000	20,000	26-30,000	165	100	—		1400	50,000	30,000	8	10	HEV 1	8.20	0.296	9.7
EV 12	6-8,000	18-20,000	—	26-30,000	185	100	193	HEV 1	Room	108,000	53,100	63	70.7	HEV 2 and HEV 3	8.30	0.300	8.3
EV 13	—	19-20,000	—	25,000	220	130	—		1400	65,000	30,000	33	33				
XEV-A	7-9,000	21-22,000	—	—	200	120	—	HEV 2	Room	149,500	96,000	33.5	36.5	HEV 4	9.13	0.330	8.6
XEV-B	7-9,000	21-22,000	—	—	200	120	—		1400	78,500	68,500	14.0	14.0	HEV 5	8.20	0.296	8.3
XEV-C	8-10,000	18-20,000	—	—	180	100	—	HEV 3	Room	160,000	90,000	25	30	HEV 6	8.27	0.299	8.3
									1400	80,000	65,000	10	25				

SAE No.	Stress for 1% Stretch of 1350 F in 100 Hr, psi	Stress for Rupture at 1350 F in 100 Hr, psi	Stress for Rupture at 1350 F in 1000 Hr, psi	Brinell Hot Hardness Mutual Indentation		SAE No.	Temperature F	Tensile Strength, psi	Yield Strength, psi	Elongation, %	Reduction of Area, %
				at 1200 F	at 1350 F						
Superalloys						HEV 4	Room	150,000	70,000	65	35
HEV 1	24,000	28,000	20,000	120	105		1400	66,000	38,000	12	26
HEV 2	29,000	32,000	—	—	210	HEV 5	Room	155,000	90,000	39	38
HEV 3	48,000	50,000	38,000	225	195		1400	80,000	68,000	20	20
HEV 4	—	36,000	30,250	—	125	HEV 6	Room	166,000	100,000	40	20
HEV 5	38,000	41,000	26,500	235	225		1400	100,000	70,000	15	5
HEV 6	42,000	45,500	35,500	260	245						

3 General Data on Steels

CARBON AND ALLOY STEELS—SAE J411d

SAE Information Report

Report of the Iron and Steel Technical Committee approved February 1948 and last revised August 1973.

1. Steel—Steel is a malleable alloy of iron and carbon which has been made molten in the process of manufacture and which contains approximately 0.05–2.0% carbon, as well as some manganese and sometimes other alloying elements.

1.1 Carbon Steel—Steel is considered to be carbon steel when no minimum content is specified or required for aluminum, chromium, cobalt, columbium, molybdenum, nickel, titanium, tungsten, vanadium, or zirconium, or any other element added to obtain a desired alloying effect; when the specified minimum for copper does not exceed 0.40%; or when the maximum content specified for any of the following elements does not exceed the following percentage: manganese, 1.65%; silicon, 0.60%; copper, 0.60%. Boron may be added to killed fine grain carbon steel to improve hardenability.

In all carbon steels, small quantities of certain residual elements, such as copper, nickel, molybdenum, chromium, etc., are unavoidably retained from raw materials. Those elements are considered incidental. However, if any of these elements are considered detrimental for special applications, the maximum acceptable content of these incidental elements should be specified by the purchaser.

1.2 Alloy Steel—Steel is considered to be alloy steel when the maximum of the range given for the content of alloying elements exceeds one or more of the following limits: manganese, 1.65%; silicon, 0.60%; copper, 0.60%; or in which a definite range or definite minimum quantity for any of the following elements is specified or required within the limits of the recognized field of constructional alloy steels: aluminum and chromium up to 3.99%; cobalt, columbium, molybdenum, nickel, titanium, tungsten, vanadium, zirconium, or any other alloying element added to obtain a desired alloying effect.

2. Steelmaking Processes—These fall into two general groups, acid or basic, according to the character of the furnace lining. Thus, open hearth or electric processes may be either acid or basic. Basic oxygen is a relatively recent addition to the list of steelmaking processes and, as the name implies, is exclusively basic. The choice of an acid or basic furnace is usually determined mainly by the phosphorus in the available raw materials and the content of phosphorus permissible in the finished steel.

Phosphorus is an acid-forming element and, in its oxide form, will react with any suitable base to form a slag in the steelmaking furnace. In basic processes, the metallurgist and steelmaker take advantage of this chemical behavior by oxidizing the phosphorus with iron oxide, which yields up its oxygen to the phosphorus. This permits the iron to remain as part of the steelmaking bath, while the acid phosphoric oxide is separated by floating up into the molten basic lime slag. In acid processes, furnaces are generally lined with silica, which is acid in nature and will not tolerate the use of basic materials for fluxes. Since an acid slag has no affinity for impurities such as phosphorus, the steel cannot be dephosphorized by fluxing and the content of this element remains at the level contained in the raw material, or may be concentrated somewhat in the finished steel due to loss of other materials from the original metallic charge.

Most iron ores in the United States are of a phosphorus content suitable only for basic steelmaking processes; hence, all of the nation's wrought steel is so made. A small proportion is low enough in phosphorus that steel could be made from it by acid processes, although such steels invariably would have higher residual phosphorus levels than steels made in basic lined furnaces.

The following are the principal steelmaking processes used in the United States with approximate percentage of total raw steel production in 1972.

Basic oxygen: 56.0%
Basic open hearth: 26.2%
Basic electric: 17.8%

2.1 Basic Open Hearth—Because of its flexibility to accommodate iron produced from available ores and basic fluxes for removal of objectionable elements such as phosphorus and sulfur, the basic open hearth, until recently, has been the most widely used steelmaking process. It is used to produce all SAE carbon and alloy steels, except compositions designated as electric furnace grades.

2.2 Basic Electric—The principal advantage of this process is optional control in the furnace permitting steel to be treated under oxidizing, reducing, or neutral slags, and pouring off and replacement of slags during the process. In this manner, and depending upon specified requirements, objectionable elements may be substantially reduced and a high degree of refinement obtained in the steel bath. Practically all grades of steel can be made by the basic electric furnace; and the process is used exclusively for producing SAE wrought stainless steels.

2.3 Basic Oxygen—Steelmaking capacity by this process was first introduced in the United States and Canada in 1954 and has been growing at a substantial rate. It is now the most popular method of steel production. The prime advantage of this process is the rate at which steel can be produced. The nature of the process is such that large quantities of molten iron must be readily available, since refining is accomplished by the exothermic reactions of high-purity oxygen with the various elements contained in the molten iron. In respect to its essential chemical composition and metallurgical characteristics, steel of a specified grade which has been made by the basic oxygen process is similar to basic open hearth steel of the same grade. The differences between these two processes are chiefly in the design of the furnace employed and in the relative extent to which high-purity oxygen is used as a refining agent.

2.4 Vacuum Degassing Process—The use of vacuum degassing in the United States has grown rapidly since the first production installation. It can be employed with any of the principal steelmaking processes and is adaptable to all grades of carbon and alloy steel. The advantages of this process lie in its ability to reduce substantially the steel's soluble gas content and minimize the formation of oxide-type inclusions.

One of the basic mechanisms involved in vacuum degassing is carbon deoxidation at pressures lower than atmospheric (generally under 10 mm Hg). This process removes a substantial portion of the oxygen, nearly all of the hydrogen, and some nitrogen. The reduction in oxygen content generally results in fewer nonmetallic inclusions with corresponding improvement in mechanical properties. The reduced hydrogen content provides steel with improved internal soundness and resistance to internal rupturing or "flaking."

This process is particularly adaptable to steels which are used in critical stress applications, such as large generator rotors, bearing and aircraft components, etc.

2.5 Strand Casting—This process involves the direct casting of steel from the ladle into slabs, blooms, or billets. In strand casting, a heat of steel is tapped into a ladle in the conventional manner. The liquid steel is then teemed into a tundish which acts as a reservoir to provide for a controlled casting rate. The steel flows from the tundish into the casting machine and rapid solidification begins in the open-ended molds. The partially solidified slab, bloom, or billet is continuously extracted from the mold. Solidification is completed by cooling the moving steel surface. Several strands may be cast simultaneously, depending upon the heat size and section size. A reduction in size may be carried out by hot working the product prior to cutting the strand into lengths. Chemical segregation is minimized, due to the rapid solidification rate of the strand cast product.

When two or more heats are cast without interruption, the process is called continuous strand casting.

3. Quality Classifications—Technically, quality, as the term relates to steel products, may be indicative of many conditions such as the degree of internal soundness, relative uniformity of composition, relative freedom from injurious surface imperfections, and finish. Steel quality also relates to general suitability for particular applications. Sheet steel surface requirements may be broadly identified as to the end use by the suffix E for exposed parts requiring a good painted surface, and suffix U for unexposed parts for which surface finish is unimportant.

Carbon steel may be obtained in a number of fundamental qualities which reflect various degrees of the quality conditions mentioned above. The quality designations for various products are listed in Table 1. Some of those qualities may be modified by such requirements as limited austenitic grain size, special discard, macroetch test, special hardenability, maximum incidental alloy elements, restricted chemical composition, and nonmetallic inclusions. In addition, several of the products have special qualities which are intended for specific end uses or fabricating practices, that is, scrapless nut quality, axle shaft quality, gun barrel, shell quality.

Alloy steels also may be obtained in special qualities, some of which are listed in Table 1. Superimposed upon some of these qualities may be such requirements as extensometer test, fracture test, impact test, macroetch test, nonmetallic inclusion tests, special hardenability test, and grain size test.

For complete descriptions of the qualities and supplementary requirements for carbon and alloy steels, reference should be made to the latest applicable AISI Steel Products Manual Section. Manual titles are listed at the end of this SAE Information Report.

4. Types of Steel—In most steelmaking processes, the primary reaction is the combination of carbon and oxygen to form a gas. If the oxygen available for this reaction is not removed prior to or during casting, the gaseous products continue to evolve during solidification. Proper control of the evolution of gas determines the type of steel.

4.1 Killed steel is a type of steel from which there may be only a slight evolution of gases during solidification of the metal. Killed steels have more uniform chemical composition and properties than the other types. However, there may be variations in composition, depending upon the steelmaking practices used. Alloy steels are of the killed type, while carbon steels may be killed or may be of the following types.

4.2 Rimmed steels have marked differences in chemical composition across the section. The typical structure of rimmed steel results from a marked gas evolution during solidification of the outer rim, caused by a reaction between the carbon in the solidifying metal and dissolved oxygen. The outer rim is lower in carbon, phosphorus, and sulfur than the average composition, whereas the inner portion, or core, is higher than the average in those elements. The technology of manufacturing rimmed steels limits the maximum contents of carbon and manganese and those maximum contents vary among producers. Rimmed steels do not retain any significant percentages of highly oxidizable elements, such as aluminum, silicon, or titanium.

Rimmed steel products, because of their chemical composition and their surface and other characteristics, may be used advantageously for the manufacture of finished articles involving cold bending, cold forming, deep drawing, and, in some cases, cold heading applications.

4.3 Semikilled steels have characteristics intermediate between those of killed and rimmed steels. During the solidification of semikilled steel, some gas is evolved and entrapped within the body of the ingot. This tends to compensate for the shrinkage which accompanies solidification.

4.4 Capped steels have characteristics which combine some features of rimmed and semikilled steels. After pouring, the rimming action is stopped after a brief interval by means of mechanical or chemical capping. The thin lower carbon rim has surface and forming properties comparable to those of rimmed steel, whereas the uniformity of composition and properties more nearly approaches that of semikilled steels. Capped steel products, because of their chemical composition, surface, and other characteristics, may be used to

TABLE 1—FUNDAMENTAL QUALITY DESCRIPTION[a] OF CARBON AND ALLOY STEELS[b]

Carbon Steels				Alloy Steels

SEMIFINISHED FOR FORGING
  Forging Quality
    Special Hardenability
    Special Internal Soundness
    Nonmetallic Inclusion Requirement
    Special Surface
 	
CARBON STEEL STRUCTURAL SECTIONS
  Structural Quality
 	
CARBON STEEL PLATES
  Regular Quality
  Structural Quality
  Cold Drawing Quality
  Cold Pressing Quality
  Cold Flanging Quality
  Forging Quality
  Pressure Vessel Quality
  Marine Quality
 	
HOT ROLLED CARBON STEEL BARS
  Merchant Quality
  Special Quality
    Special Hardenability
    Special Internal Soundness
    Nonmetallic Inclusion Requirement
    Special Surface
  Scrapless Nut Quality
  Axle Shaft Quality
  Cold Extrusion Quality
  Cold Heading and Cold Forging Quality
 	
COLD FINISHED CARBON STEEL BARS
  Standard Quality
    Special Hardenability
    Special Internal Soundness
    Nonmetallic Inclusion Requirement
    Special Surface
  Cold Heading and Cold Forging Quality
  Cold Extrusion Quality
	HOT ROLLED SHEETS
	  Commercial Quality
	  Drawing Quality
	  Drawing Quality Special Killed
	  Physical Quality
	
	COLD ROLLED SHEETS
	  Commercial Quality
	  Drawing Quality
	  Drawing Quality Special Killed
	  Physical Quality
	
	PORCELAIN ENAMELING SHEETS
	  Commercial Quality
	  Drawing Quality
	
	LONG TERNE SHEETS
	  Commercial Quality
	  Drawing Quality
	  Drawing Quality Special Killed
	  Physical Quality
	
	GALVANIZED SHEETS
	  Commercial Quality
	  Drawing Quality
	  Drawing Quality Special Killed
	  Physical Quality
	  Lock Forming Quality
	
	ELECTROLYTIC ZINC COATED SHEETS
	  Commercial Quality
	  Drawing Quality
	  Drawing Quality Special Killed
	  Physical Quality
	
	HOT ROLLED STRIP
	  Commercial Quality
	  Drawing Quality
	  Drawing Quality Special Killed
	  Physical Quality
	
	COLD ROLLED STRIP
	  Specific quality descriptions are not provided in cold rolled strip, since this product is largely produced for specific end use.
		TIN MILL PRODUCTS
		  Specific quality descriptions are not applicable to tin mill products.
		
		CARBON STEEL WIRE
		  Industrial Quality Wire
		  Cold Extrusion Wires
		  Heading, Forging and Roll Threading Wires
		  Mechanical Spring Wires
		  Upholstery Spring Construction Wires
		  Welding Wire
		
		CARBON STEEL FLAT WIRE
		  Stitching Wire
		  Stapling Wire
		
		CARBON STEEL PIPE
		
		STRUCTURAL TUBING
		
		LINE PIPE
		
		OIL COUNTRY TUBULAR GOODS
		
		STEEL SPECIALTY TUBULAR PRODUCTS
		  Pressure Tubing
		  Mechanical Tubing
		  Aircraft Tubing
		
		HOT ROLLED CARBON STEEL WIRE RODS
		  Industrial Quality
		  Rods for Manufacture of Wire Intended for Electric Welded Chain
		  Rods for Heading, Forging, and Roll Threading Wire
		  Rods for Lock Washer Wire
		  Rods for Scrapless Nut Wire
		  Rods for Upholstery Spring Wire
		  Rods for Welding Wire
				ALLOY STEEL PLATES
				  Regular Quality or Structural Quality
				  Drawing Quality
				  Pressure Vessel Quality
				  Structural Quality
				  Aircraft Quality
				  Aircraft Physical Quality
				
				HOT ROLLED ALLOY STEEL BARS
				  Regular Quality
				  Aircraft Quality or Steel Subject to Magnetic Particle Inspection
				  Axle Shaft Quality
				  Bearing Quality
				  Cold Heading Quality
				  Special Cold Heading Quality
				  Rifle Barrel Quality, Gun Quality, Shell or A.P. Shot Quality
				
				ALLOY STEEL WIRE
				  Aircraft Quality
				  Bearing Quality
				  Special Surface Quality
				
				COLD FINISHED ALLOY STEEL BARS
				  Regular Quality
				  Aircraft Quality or Steel Subject to Magnetic Particle Inspection
				  Axle Shaft Quality
				  Bearing Shaft Quality
				  Cold Heading Quality
				  Special Cold Heading Quality
				  Rifle Barrel Quality, Gun Quality, Shell or A.P. Shot Quality
				
				LINE PIPE
				
				OIL COUNTRY TUBULAR GOODS
				
				STEEL SPECIALTY TUBULAR GOODS
				  Pressure Tubing
				  Mechanical Tubing
				  Stainless and Heat Resisting Pipe, Pressure Tubing, and Mechanical Tubing
				  Aircraft Tubing
				  Pipe

[a] In the case of certain qualities, phosphorus and sulfur are ordinarily furnished to lower limits than the specified maximum. For details, refer to the appropriate AISI Manual.

[b] Detailed description of many of the categories listed in this table appear in an appropriate section of the AISI manual.

advantage when the material is to withstand cold bending, cold forming, or cold heading.

5. Commonly Specified Elements—The effect of a single commonly specified element on steelmaking practice and carbon and alloy steel properties is dependent upon the effect of other elements. These interrelations, frequently of a complex nature, should be considered when evaluating a change in specified composition. It should be noted also that as the number of elements specified increases, and as restrictive requirements increase, availability generally decreases, to the ultimate end that special heats may become a necessity and material must be ordered in heat lots.

5.1 **Carbon**—The amount of carbon required in the finished steel limits the type of steel that can be made. As the carbon content of rimmed steels increases, surface quality becomes impaired. Killed steels, by comparison, have poorer surfaces in the lower carbon grades. Carbon has a moderate tendency to segregate, and because of its major effect on properties, carbon segregation is frequently of more significant importance than the segregation of other elements. It is the principal hardening element in all steel. Tensile strength in the as-rolled condition increases as the carbon increases up to about 0.85% carbon. Ductility and weldability decrease with increasing carbon.

5.2 **Manganese** has a lesser tendency to segregate than any of the common elements. Steels above 0.60% manganese cannot be readily rimmed.

Manganese is beneficial to surface quality in all carbon ranges (with the exception of extremely low carbon rimmed steels) and is particularly beneficial in high sulfur steels. It contributes to strength and hardness, but to a lesser degree than does carbon, the amount of increase being dependent upon the carbon content. Increasing the manganese content decreases ductility and weldability, but to a lesser extent than does carbon. Manganese has a moderate effect on increasing the hardenability of a steel.

5.3 **Phosphorus** segregates, but to a lesser degree than carbon and sulfur. Increasing phosphorus increases strength and hardness and decreases ductility and notch-impact toughness in the as-rolled condition. The latter adverse effects are greater in quenched and tempered higher carbon steels. Higher phosphorus is often specified in low carbon free-machining steels to improve machinability.

5.4 **Sulfur**—Increased sulfur content lowers transverse ductility and notched impact toughness, but has only a slight effect on longitudinal mechanical properties. Weldability decreases with increasing sulfur content. This element is very detrimental to surface quality, particularly in the lower carbon and lower manganese steels. For these reasons, only a maximum limit is specified for most steels. The only exception is the group of free-machining steels, where sulfur is added to improve machinability, in which case a range is specified.

Sulfur has a greater segregation tendency than any of the other common elements.

Sulfur occurs in steel, principally in the form of sulfide inclusions. Obviously, greater frequency of such inclusions is to be expected in the resulfurized grades.

5.5 **Silicon** is one of the principal deoxidizers used in steelmaking, and therefore the amount of silicon present is related to the type of steel. Rimmed and capped steels contain no significant amounts of silicon. Semikilled steels may contain moderate amounts of silicon, although there is a definite maximum amount that can be tolerated in such steels. Killed carbon steels may contain any amount of silicon up to 0.60% maximum.

Silicon is somewhat less effective than manganese in increasing as-rolled strength and hardness. Silicon has only a slight tendency to segregate. In low carbon steels, silicon is usually detrimental to surface quality, and this condition is more pronounced in low carbon resulfurized grades.

5.6 **Copper** has a moderate tendency to segregate. Copper, in appreciable amounts, is detrimental to hot working operations. Copper adversely affects forge welding, but it does not seriously affect arc or acetylene welding. Copper is detrimental to surface quality and exaggerates the surface defects inherent in resulfurized steels. Copper is, however, beneficial to atmospheric corrosion resistance when present in amounts exceeding 0.20%.

5.7 **Lead** is an element sometimes added to carbon and alloy steels through mechanical dispersion during teeming, for the purpose of improving the machining characteristics of such steels. When so added, the range is generally 0.15–0.35%.

5.8 **Boron** is added to fully killed steel to improve hardenability. Boron-treated steels will usually have a boron content in the range of 0.0005–0.003%. Whenever boron is substituted in part for other alloys, it should be done so only with hardenability in mind, because the lowered alloy content may be harmful on some applications. Boron is most effective in lower carbon steels.

5.9 **Chromium** is generally added to steel to increase resistance to corrosion and oxidation, increase hardenability, improve high-temperature strength, or improve abrasion resistance in high carbon compositions. Chromium is a strong carbide former. Complex chromium-iron carbides go into solution in austenite slowly; therefore, a sufficient heating time before quenching is necessary. If the chromium is obtained. Chromium is essentially a hardening element, and is frequently used with a toughening element such as nickel to produce superior mechanical properties. At higher temperatures, chromium contributes increased strength, but is ordinarily used for applications of this nature in conjunction with molybdenum.

5.10 **Nickel,** when used as an alloying element in constructional steels, is a ferrite strengthener. Since nickel does not form any carbide compounds in steel, it remains in solution in the ferrite, thus strengthening and toughening the ferrite phase. Nickel steels are easily heat treated because nickel lowers the critical cooling rate. In combination with chromium, nickel produces alloy steels with greater hardenability, higher impact strength, and greater fatigue resistance than are possible with carbon steels.

5.11 **Molybdenum** is added to constructional steels in the normal amounts of 0.10–0.60%. When molybdenum is in solid solution in austenite prior to quenching, the reaction rates for transformation become considerably slower as compared with carbon steel. Molybdenum steels in the quenched condition require higher tempering temperatures to obtain the same degree of softness as comparable carbon and alloy steels. Alloy steels which contain 0.15–0.30% molybdenum show a minimized susceptibility to temper embrittlement.

6. AISI Steel Products Manuals—The American Iron and Steel Institute's Technical Committee cooperates with the SAE Iron and Steel Technical Committee on standardization of compositions and related data on steels used in the automotive industries. AISI publishes Steel Products Manual Sections as listed below. Copies are available to steel users from the American Iron and Steel Institute, 1000 16th St., NW, Washington, DC 20036. (The latest revision dates are available from AISI.)

Carbon Steel: Semifinished for Forging; Hot Rolled and Cold Finished Bars; Hot Rolled Deformed Concrete Reinforcing Bars	Carbon Steel Flat Wire
	Steel Tubular Products
	Railway Track Materials
Carbon Steel: Plates, Structural Sections; Rolled Floor Plates; Steel Sheet Piling	Wrought Steel Wheels
	Forged Railway Axles
Alloy Steel Plates	Alloy Steel Sheets and Strip
Alloy Steel: Semifinished; Hot Rolled and Cold Finished Bars	Stainless and Heat Resisting Steels
	Tool Steels
Carbon Steel Sheets	Flat Rolled Electrical Steel
Carbon Steel Strip	Alloy Steel Wire
Tin Mill Products	High Strength Low Alloy Steel (out of print)
Wire and Rods, Carbon Steel	Carbon Steel Pipe, Structural Tubing, Line Pipe, Oil Country Tubular Goods

HIGH STRENGTH CARBON AND ALLOY DIE DRAWN STEELS—SAE J935

SAE Recommended Practice

Report of Iron and Steel Technical Committee approved September 1965. Editorial change September 1971.

Scope—This SAE Recommended Practice is intended to provide basic information on properties and characteristics of high strength carbon and alloy steels which have been subjected to special die drawing. This includes both cold drawing with heavier-than-normal drafts and die drawing at elevated temperatures.

Introduction—Die drawing of hot rolled bars increases the strength and hardness. At the same time, the ratio of yield strength to tensile strength is increased and the notched bar impact values are reduced. Various factors control the degree of change in the mechanical properties. The final properties are dependent upon chemical composition, hot rolled microstructure (except in the case of alloy steel where a normalize treatment is used prior to drawing), size, shape and the amount of reduction in cross-sectional area, die geometry, straightening procedures, and manner or temperature level of the stress relieving operation.

As noted in Table 1, plain carbon and alloy steels of medium carbon content respond readily to this special processing. Compositional additives may be employed to improve machinability.

In the production of these products, drafts of approximately 10–35% reduction in cross-sectional area are employed at either room or elevated temperatures depending on the practices and facilities of the individual producer. Stress relieving temperatures vary over a similarly wide range, depending on producer facilities and the end product requirements.

Die drawn and stress relieved bars are employed instead of quenched and tempered bars because of their unique combinations of properties. The more

TABLE 1—MINIMUM MECHANICAL PROPERTIES

Tensile Strength, psi	Yield Strength, psi	Elongation in 2 in., %[a]	Reduction in Area, %[a]	Brinell Hardness	Grades	Size Range, in.	Tolerances
Carbon Steels							
120,000	100,000	10.0	25.0	241/321[b]	1541 1045	Up to 3 (round)	See Table 2.
					1052 1141	1/4 thru 3-1/2 (round)	
				248/321[b]	1144 1151	1/4 thru 4-1/2 (round) and 1/4 thru 2 (hexagon)	See Table 2.
140,000	125,000	5.0	15.0	280	1144 1151	1/4 thru 2-1/2 (round) and 1/4 thru 1-1/2 (hexagon)	See Table 3.
Alloy Steels							
125,000	105,000	14.0	45.0	269	41XX 51XX[c]		See Table 4.
150,000[d]	130,000	10.0	35.0	302	41XX[c] 51XX[c]	7/16 thru 3-1/2 (round)	
180,000	165,000	5.0	20.0	363	41XX[c]		

[a] Typical minimum.
[b] Typical hardness ranges, subject to negotiation. Hardness to be taken on a flat below decarb or on the mid-radius. In case of disagreement between hardness and tensile or yield strength, the latter properties govern.
[c] May contain Pb or Te or other additives for improved machinability.
[d] See SAE J429.

TABLE 2 — SIZE TOLERANCES FOR CARBON STEELS

	Size Range, in.	Tolerance, in.[e]
Rounds	1/4 to 1-1/2 incl	0.004
	Over 1-1/2 to 2-1/2 incl	0.005
	Over 2-1/2 to 4 incl	0.006
	Over 4 to 4-1/2 incl	0.007
Hexagons	1/4 to 3/4 incl	0.004
	Over 3/4 to 1-1/2 incl	0.005
	Over 1-1/2 to 2 incl	0.006

[e] All tolerances are minus.

TABLE 3 — SIZE TOLERANCES FOR CARBON STEELS

	Size Range, in.	Tolerance, in.[e]
Rounds	5/16 to less than 7/16	0.004
	7/16 to 1-1/2 incl	0.005
	Over 1-1/2 to 2-1/2 incl	0.006
Hexagons	1/4 to less than 3/8	0.004
	3/8 to less than 7/16	0.005
	7/16 to 1-1/2 incl	0.006

[e] All tolerances are minus.

TABLE 4 — SIZE TOLERANCES FOR ALLOY STEELS (ROUNDS)

Size Range, in.	Tolerance, in.[e]
7/16 to 1-1/2 incl	0.005
Over 1-1/2 to 2-1/2 incl	0.006
Over 2-1/2 to 3-1/2 incl	0.007

[e] All tolerances are minus.

important of these are a pearlitic structure which machines more readily and a greater uniformity of hardness throughout the cross section, except when high hardenability grades are involved. When dimensional stability is critical during or after machining, or after cold forming operations, the individual producer should be consulted for special processing to meet such conditions.

The torsional strength and endurance limit are similar to those of quenched and tempered grades at the same strength level. The wear resistance of these special processed steels is approximately equal to that of quenched and tempered bars of the same surface hardness.

Hardness—The hardness values for all grades are shown in Table 1. The typical hardness ranges indicated for the 120,000 psi tensile strength steels are subject to negotiation between producer and consumer. Hardness determinations are commonly made on a flat ground on the outside diameter or on a cross section from the mid-radius to within $\frac{1}{4}$ in. of the surface. If, when testing the finished product, there is disagreement between the typical hardness and tensile or yield strength values, the latter properties shall govern.

Impact Characteristics—The impact test values of special die drawn bars, as measured by the Izod or Charpy notched bar test, are lower than those of quenched and tempered carbon bars and they are significantly lower than those of quenched and tempered alloy steels. Failures of machine components usually result from fatigue, corrosion, wear or shock loading. With the possible exception of the latter, there is no known correlation between the cases of failure and the notched bar impact test. In the case of shock loading, whatever relation exists must be derived empirically, that is, from experience. When low temperatures or high pressures are involved and where doubt exists as to the suitability of these steels, the design of the part should be reviewed.

Surface Finish—A number of surface finishes are available depending on producers' facilities and end use requirements. Bars can be supplied in the die drawn condition turned and polished, or ground and polished from die drawn or turned bars. The bars frequently have a dark appearance when the last operation is stress relieving. Surface finishes are subject to negotiation with each producer. The following ranges of Arithmetical Average (AA) values are considered normal for each condition.

Cold Drawn	50/125 AA
Turned and Polished	15/40 AA
Cold Drawn-Ground and Polished	8/20 AA
Turned-Ground and Polished	8/20 AA

Machinability—Machinability values for any given grade or condition will vary considerably from shop to shop as a function of equipment, tooling grade and design, set up conditions, lubrication, and personnel. The following ratings which are considered typical and which are offered only for purposes of comparison are based on a value of 100% for SAE 1112.

Typical Machinability Ratings

SAE Grade	Heavy Drafted, Stress Relieved, %
1045	56
1050	54
1141	67
1144	85
High Tensile 1144	80
41XX 150,000 psi, TS	75 } with free
51XX 180,000 psi, TS	60 } machining additives

GENERAL CHARACTERISTICS AND HEAT TREATMENTS OF STEELS—SAE J412h

SAE Information Report

Report of Iron and Steel Division approved January 1912, editorial change April 1970 and last revised by Iron and Steel Technical Committee July 1976.

Introduction—The information and data contained in this report are intended as a guide in the selection of steel grades for various purposes. SAE steels are generally purchased on the basis of chemical composition requirements (SAE J403, J404, and J405). This information report can be used as a reference for determining the general characteristics of commonly used SAE steels. The use of the typical heat treatments listed in Tables 1 and 2 is recommended.

Normalizing—The normalizing heat treatment consists of:

1. Uniformly heating steel to a temperature high enough to obtain complete transformation to austenite.
2. Holding at the austenitizing temperature until the mass is of equal temperature throughout.
3. Air cooling, allowing free air circulation to give uniform cooling. Normalizing temperatures are dependent upon the steel grade, while holding time at temperature will vary with the mass being heat treated. Hence, normalizing cycles and subsequent steel properties may vary considerably with steel grade, part size, individual furnace conditions and cooling facilities.

Normalizing is generally performed to obtain desired mechanical properties but is also used for the following functions:

1. Modify and refine coarse as-rolled or forged structures.
2. Improve hardening characteristics by refining grain size and homogenizing microstructure.
3. Improve machining characteristics. This treatment is especially beneficial for 0.15–0.40% carbon steels.

Annealing—When the term "annealing" is applied without qualification, the term implies full annealing. Full annealing consists of austenitizing [the process of forming austenite by heating a ferrous alloy into the transformation range (partial austenitizing) or above the transformation range (complete austenitizing)] and then cooling uniformly and slowly, through the transformation range. In isothermal annealing, the heating is the same as used for a full anneal, but the steel is held for a given time at a constant temperature in the transformation range before proceeding to a constant uniform cooling. These practices produce a coarse pearlitic structure which greatly improves machinability of medium carbon steels. Spheroidizing is an annealing process which under suitable conditions of temperature and time produces a spheroidal or globular form of carbide in steel and is recommended prior to machining steels higher than 0.60% carbon.

In addition to producing desired mechanical properties, improving machinability, and obtaining the desired microstructure, the various forms of annealing are also used to improve the cold forming properties of steels.

Time-temperature cycle used and microstructure obtained for the annealing process have a broad meaning to metallurgists. For example, the following terms are all means of identifying a type of annealing: bright anneal, black anneal, box anneal, flame anneal, isothermal anneal, process anneal, recrystallization anneal, spheroidizing, and full anneal. Definitions for the various annealing processes are given in SAE J415.

Carbon Restoration—Carbon restoration or carbon correction is a special thermal treatment for restoring carbon to the decarburized skin normally found on all grades of hot rolled, cold drawn, or cold drawn and annealed steel products. The process was originally applied to medium carbon steels where substantial differences in carbon content can occur between the base metal and the decarburized zone. The intent was to adjust the carbon potential of the furnace atmosphere to the carbon content of the steel being treated and, thus, the term "carbon restoration" came into use. More recent usage has found the process applied to low and high carbon steels as well as the medium carbon grades. Current practice also shows some application for products with an enrichment of the surface carbon content slightly above the base analysis. The principal application of the carbon restored product is in those cases in which the full hardness is desired on the as-rolled or drawn surface after a heat treating operation.

This special treatment consists of a gas carburizing process usually carried out at the steel mill or cold finishing plant in which time, temperature, and carbon potential of the atmosphere are so controlled that carbon content at the surface is at least as high as the minimum product analysis limit for the grade of steel being treated. Carbon restoration is normally limited to a depth of approximately 0.015 in (0.38 mm). Requirements and limits are generally agreed upon between producer and user and are usually determined by chemical analysis, microscopic examination or hardness tests. Dependent upon requirements and method of production, a shallow outer layer of essentially carbon free material may be present up to approximately 0.001 in (0.025 mm) deep. Because of its shallowness, this condition is normally not considered to be detrimental. Subsequent heat treating by austenitizing and quenching will usually obliterate this condition. The producer should be consulted if this condition is not acceptable.

The process is carried out in either a batch or continuous furnace, suitably instrumented to provide the necessary control. As in any process of this type, some leeway or tolerance is necessary to make the operation commercial. Therefore, the carbon content at the surface of the treated stock is often higher than that of the base stock. One of the important controls in this process is to avoid excessive carbon content in the restored layer, as it is a most important consideration in the application of the product. The higher the carbon in the restored layer, the greater the sensitivity of the product to cooling rate after restoration. Excessive surface hardness may result unless suitable precautions are taken. If the product is to be machined, this higher carbon content may also result in an excess of carbides in the restored layer when the material is controlled cooled for best machinability of the unrestored

TABLE 1—TYPICAL TREATMENTS FOR CASE HARDENING GRADES OF CARBON STEELS

SAE Steels[a]	Carbon Temperature, F	Cooling Method	Reheat Temperature, F	Cooling Medium	Carbonitriding, Temperature, F[b]	Cooling Method	Temper, F[c]
1010	—	—	—	—	1450-1650	Oil	250-400
1015	—	—	—	—	1450-1650	Oil	250-400
1016	1650-1700	Water or Caustic	—	—	1450-1650	Oil	250-400
1018	1650-1700	Water or Caustic	1450	Water or Caustic[d]	1450-1650	Oil	250-400
1019	1650-1700	Water or Caustic	1450	Water or Caustic[d]	1450-1650	Oil	250-400
1020	1650-1700	Water or Caustic	1450	Water or Caustic[d]	1450-1650	Oil	250-400
1022	1650-1700	Water or Caustic	1450	Water or Caustic[d]	1450-1650	Oil	250-400
1026	1650-1700	Water or Caustic	1450	Water or Caustic[d]	1450-1650	Oil	250-400
1030	1650-1700	Water or Caustic	1450	Water or Caustic[d]	1450-1650	Oil	250-400
1109	1650-1700	Water or Oil	1400-1450	Water or Caustic[d]	—	—	250-400
1117	1650-1700	Water or Oil	1450-1600	Water or Caustic[d]	1450-1650	Oil	250-400
1118	1650-1700	Oil	1450-1600	Oil	—	—	250-400
1513	1650-1700	Oil	1450	Oil	—	—	250-400
1518	1650-1700	Oil	1450	Oil	—	—	250-400
1522	1650-1700	Oil	1450	Oil	—	—	250-400
1524 (1024)	1650-1700	Oil	1450	Oil	—	—	250-400
1525	1650-1700	Oil	1450	Oil	—	—	250-400
1526	1650-1700	Oil	1450	Oil	—	—	250-400
1527 (1027)	1650-1700	Oil	1450	Oil	—	—	250-400

[a] Generally, it is not necessary to normalize the carbon grades for fulfilling either dimensional or machinability requirements of parts made from the steel grades listed in the table, although where dimension is of vital importance normalizing temperatures of at least 50 F above the carburizing temperatures are sometimes required.

[b] The higher manganese steels such as 1118 and the 1500 series are not usually carbonitrided. If carbonitriding is performed, care must be taken to limit the nitrogen content because high nitrogen will increase their tendency to retain austenite.

[c] Even where recommended draw temperatures are shown, the draw is not mandatory on many applications. Tempering is generally employed for a partial stress relief and improves resistance to cracking from grinding operations. Higher temperatures than those shown may be employed where the hardness specification on the finished parts permits.

[d] 3% sodium hydroxide.

TABLE 2—TYPICAL TREATMENTS FOR HEAT TREATING GRADES OF CARBON STEELS

SAE Steels	Normalizing Temperature, F	Annealing Temperature, F	Hardening Temperature, F	Quenching Medium	Temper[a]
1030	—	—	1575–1600	Water or Caustic	To desired hardness
1035	—	—	1550–1600	Water or Caustic	To desired hardness
1037	—	—	1525–1575	Water or Caustic	To desired hardness
1038[b]	—	—	1525–1575	Water or Caustic	To desired hardness
1039[b]	—	—	1525–1575	Water or Caustic	To desired hardness
1040[b]	—	—	1525–1575	Water or Caustic	To desired hardness
1042	—	—	1500–1550	Water or Caustic	To desired hardness
1043[b]	—	—	1500–1550	Water or Caustic	To desired hardness
1045[b]	—	—	1500–1550	Water or Caustic	To desired hardness
1046[b]	—	—	1500–1550	Water or Caustic	To desired hardness
1050[b]	1600–1700	—	1500–1550	Water or Caustic	To desired hardness
1053	1600–1700	—	1500–1550	Water or Caustic	To desired hardness
1060	1600–1700	1400–1500	1575–1625	Oil	To desired hardness
1074	1550–1650	1400–1500	1575–1625	Oil	To desired hardness
1080	1550–1650	1400–1500[c]	1575–1625	Oil[d]	To desired hardness
1084	1550–1650	1400–1500[c]	1575–1625	Oil[d]	To desired hardness
1085	1550–1650	1400–1500[c]	1575–1625	Oil[d]	To desired hardness
1090	1550–1650	1400–1500[c]	1575–1625	Oil[d]	To desired hardness
1095	1550–1650	1400–1500[c]	1575–1625	Water and Oil	To desired hardness
1137	—	—	1550–1600	Oil	To desired hardness
1141	—	1400–1500	1500–1550	Oil	To desired hardness
1144	1600–1700	1400–1500	1500–1550	Oil	To desired hardness
1145	—	—	1475–1500	Water or Oil	To desired hardness
1146	—	—	1475–1500	Water or Oil	To desired hardness
1151	1600–1700	—	1475–1500	Water or Oil	To desired hardness
1536	1600–1700	—	1500–1550	Water or Oil	To desired hardness
1541 (1041)	1600–1700	1400–1500	1500–1550	Water or Oil	To desired hardness
1548 (1048)	1600–1700	—	1500–1550	Oil	To desired hardness
1552 (1052)	1600–1700	—	1500–1550	Oil	To desired hardness
1566 (1066)	1600–1700	—	1575–1625	Oil	To desired hardness

[a] Even where recommended draw temperatures are shown, the draw is not mandatory on many applications. Tempering is generally employed for a partial stress relief and improves resistance to cracking from grinding operations. Higher temperatures than those shown may be employed where the hardness specification on the finished parts permits.

[b] Commonly used on parts where induction hardening is employed. However, all steels from SAE 1030 up may have induction hardening applications.

[c] Spheroidal structures are often required for machining purposes and should be cooled very slowly or be isothermally transformed to produce the desired structure.

[d] May be water or brine quenched by special techniques such as partial immersion or time quenched; otherwise they are subject to quench cracking.

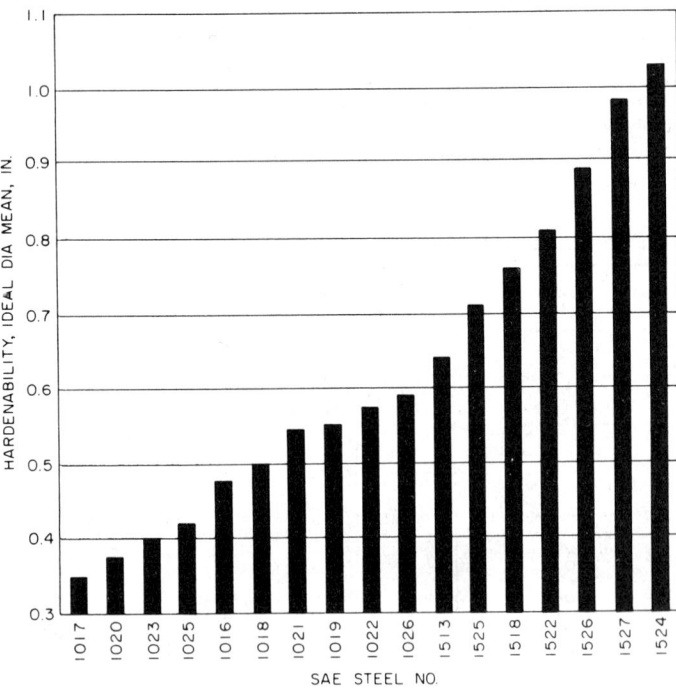

BASIS OF CALCULATION:
NO. 7 GRAIN SIZE
MEAN CARBON OF GRADE
MEAN MANGANESE OF GRADE

FIG. 1—SELECTION OF CARBURIZING GRADES OF CARBON STEEL ON RELATIVE HARDENABILITY BASIS

portion of the product. It is sometimes necessary, therefore, to perform a second thermal treatment to generate the optimum structure for subsequent operations; machining, cold extrusion, etc.

Although carbon restoration can be applied to any product that the available equipment can accommodate, its usual application is to bars and rods in either coils or cut lengths. The carbon restored product will approximate the mechanical properties of annealed bars or rods of the base carbon level. If they are cold drawn after carbon restoration, the product will have the appearance, characteristics, and mechanical properties of cold drawn material as given in SAE J414, except that it is essentially free of decarburization. In any application, it is well to keep in mind that the surface condition of the carbon restored product with respect to seams is the same as that of the hot rolled or cold drawn stock with which the process started.

Case Hardening—The process as considered in this report refers to the various dry or pack carburizing methods as well as to the processes utilizing gases or molten baths including nitriding and carbonitriding.

There are two methods of proceeding after carburizing: (1) quenching directly and (2) cooling slowly or box cooling.

The first method involves removal of the work from the furnace or from the carburizing box and quenching the parts while they are at or slightly below the carburizing temperature or by direct quenching from gas carburizing furnaces.

The second method allows the work to cool slowly without any quenching, in the box or container or in a cooling chamber provided in the furnace. The work is subsequently austenitized and quenched in either water or oil.

The relative value of these two methods is dependent upon the type of steel treated, the method of carburizing, the kind of furnace installation, and the mechanical properties desired.

Tempering of parts after carburizing, cyaniding, and activated bath treatments is sometimes omitted in commercial practice, but is included in the accompanying recommendations as being in accord with good heat treating practice.

Parts carburized in activated baths should be treated similarly to other carburized work and may be given any of the hardening treatments shown under the specific steels.

CHARACTERISTICS OF PLAIN CARBON STEELS

SAE 1005, 1006, 1008, 1010, 1011, 1012, 1013, 1015 *Group I*—These steels are the lowest carbon steels of the plain carbon type and are selected when cold formability is the primary requisite. They are produced as rimmed, semikilled or capped, and killed steels. Rimmed steel is used for sheet and strip where excellent surface finish or good drawing qualities are required. Rimmed steel is also used for cold headed wire products. Semikilled and capped steels represent intermediate deoxidizing practices. Capped steels are used for sheet, strip and wire products, and semikilled steels are used for structural steel applications. Aluminum killed steel is used for some difficult stampings or where nonaging properties are needed. Killed steels should be used for hot forging or heat treating applications.

These steels have relatively low tensile values. Within the carbon range of the group, strength and hardness will increase with increase in carbon and with cold work. Such increases in strength are at the sacrifice of ductility or the ability to withstand cold deformation.

When under 0.15% carbon, the steels are susceptible to grain growth, causing brittleness and surface roughness in subsequent forming which may occur as the result of cold work followed by heating to elevated temperatures. Consequently, if cold worked parts formed from these steels are to be later heated to temperatures in excess of 1100 F, the user should exercise care to avoid brittleness. If the coarse grain condition develops, it can be overcome by heating the parts to a temperature well in excess of the upper critical point, approximately 1750 F minimum to 1850 F maximum.

The machinability of bar, rod, and wire products in this group is improved by cold drawing. In general, these steels are considered suitable for welding or brazing.

SAE 1016, 1017, 1018, 1019, 1020, 1021, 1022, 1023, 1025, 1026, 1029, 1030 *Group II*—Steels in this group have increased strength and hardness and reduced cold formability compared to the lowest carbon group. For heat treating purposes, they are commonly known as carburizing or case hardening grades. Killed steel is preferred for forgings and when uniform response to heat treatment is required. Semikilled or rimmed steel may be indicated for other uses depending on the combination of properties desired. Rimmed steels can ordinarily be supplied up to 0.25% carbon.

Selection of one of these steels for carburizing applications depends on the nature of the part, the properties desired, and the processing practices preferred. Increase in carbon content of the base steel results in greater core hardness with a given

1513	quench. Increase in manganese improves the hardenability of both the core and the case; in carbon steels, this is the only change in composition that will increase case hardenability. The information in Fig. 1 indicates the calculated hardenability of this group and can be used as a guide for the selection of a steel grade of appropriate hardenability.
1518	
1522	
1524(1024)	
1525	
1526	
1527(1027)	

In this group, the intermediate manganese grades (0.60 to 1.00) machine better than the lower manganese grades. For carburizing applications, SAE 1016, 1018, and 1019 are widely used for water quenched parts. SAE 1022 and the 1500 series in this group, are used for heavier sections or with thin sections where oil quenching is desired. SAE 1527 (1027) is used for parts given a light case to obtain satisfactory core properties, without drastic quenching. SAE 1025, 1029, and 1030, while not usually regarded as carburizing types, are sometimes used in this manner for larger sections or where greater core hardness is needed.

In cold formed or headed parts, the lowest manganese grades offer the best formability at their carbon level. The next higher manganese types (SAE 1018, 1021, and 1026) provide increased strength.

In general, these steels may be considered as suitable for welding or brazing. These steels are used for numerous forged parts. Forgings usually machine better in the as-forged or normalized condition compared with an annealed condition.

SAE	
1030	*Group III*—Steels of the medium carbon type are selected for uses where their higher mechanical properties are needed, and are frequently further hardened and strengthened by heat treatment or by cold work. These grades are produced as semi-killed or capped, and killed steels. Steels in this group are suitable for a wide variety of automotive applications. Selection of the particular carbon and manganese level is governed by a number of factors. Increase in mechanical properties required, section thickness, or depth of hardening, ordinarily necessitates either higher carbon or a higher manganese, or both. The heat treating practice used, especially the quenching medium, also has a great effect on the steels selected. In general, any of the grades over 0.30% carbon may be induction or flame hardened.
1034	
1035	
1037	
1038	
1039	
1040	
1042	
1043	
1044	
1045	
1046	
1049	The lower carbon and manganese steels in this group find wide usage for certain types of cold formed parts. In practically all cases, the parts cold formed from these steels are generally annealed or normalized prior to use. Stampings are usually limited to flat parts or simple bends. The higher carbon grades are frequently cold drawn to specified mechanical properties for use without heat treatment for some applications.
1050	
1053	
1536(1036)	
1541(1041)	
1547(1047)	
1548(1048)	
1551(1051)	
1552(1052)	

All of these steels can be used for forgings, the selection being governed by the section size and the mechanical properties desired after heat treatment. Thus, SAE 1030 and 1035 are used for many small forgings where moderate properties are desired. SAE 1536 (1036) is used for more critical parts where a higher strength level and better uniformity is essential. The SAE 1038, 1052, 1053, and the 1500 groups are used for larger forgings. They are also used for small forgings where high hardness after oil quenching is desired. Suitable heat treatment is necessary on forgings from this group to provide machinability.

These steels are also widely used for parts machined from bar stock. They are used both with and without heat treatment, depending upon the application and the level of properties needed. As a class, they are considered good for normal machining operations. It is also possible to weld these steels by most commercial methods, but precautions should be taken to avoid cracking from rapid heating or cooling.

SAE	
1055	*Group IV*—Steels in this group are of the high carbon type which are used for applications where the higher carbon is needed to improve wear characteristics and where strength levels required are higher than those attainable with the lower carbon groups.
1059	
1060	
1064	
1065	
1069	In general, cold forming methods are not practical with this group of steels, being limited to flat stampings and springs coiled from small diameter wire. Practically all parts from these steels are heat treated before use. Variations in heat treating methods are required to obtain optimum properties for particular composition and application.
1070	
1074	
1075	
1078	
1080	Typical uses in the spring industry include SAE 1065 for pretempered wire, SAE 1064 for small washers and thin stamped parts, SAE 1074 for light, flat springs formed from annealed stock, and SAE 1080 and 1085 for thicker flat springs. SAE 1085 is also used for heavier coiled springs.
1084	
1085	
1086	
1090	

1095	Because of good wearing properties when properly heat treated, the high carbon steels find wide usage in the farm implement industry. Typical applications are plow beams, plow shares, scraper blades, discs, mower knives, and harrow teeth.
1561(1061)	
1566(1066)	
1572(1072)	

CHARACTERISTICS OF FREE CUTTING CARBON STEELS

This class of steels is intended for those uses where improved machinability is desired as compared with carbon steels of similar carbon and manganese content. Machinability refers to the effects of hardness, strength, ductility, grain size, microstructure, and chemical composition upon the cutting tool wear, chip formation, ease of metal removal and surface finish quality of the steel being cut. Sulfur, sulfur and phosphorus combination, and lead additions to carbon steels are made for the sole purpose of decreasing machining costs. Lower costs are achieved either by increased production through greater machining speeds and improved tool life, or by eliminating secondary operations through an improvement in finish. Sulfur and phosphorus additions result in some sacrifice of cold forming properties, weldability, and forging characteristics, whereas lead additions have very little effect on these characteristics. The addition of bismuth, selenium, or tellurium are known to further enhance the machinability of free-cutting steels, but SAE standard grade designations have not been assigned to these steels.

SAE	
1110	These steels are nonkilled free-machining screw stock grades. They have excellent machining characteristics and are used for a wide variety of machined parts. While of excellent strength in the cold drawn condition, they have an unfavorable property of cold shortness. They are also subject to nonuniformity, particularly when made as Bessemer steels. These steels may be cyanided or carburized, but when best response to heat treating is necessary, killed open hearth or killed electric furnace steels are recommended, although killed steels are not recommended for best machinability. Machinability improves within the 1100 series groups as sulfur goes up. Sulfur combines mostly with the manganese in steel and precipitates as sulfide inclusions. These inclusions favor machining by causing the formation of a broken chip and by providing a built-in lubricant that prevents the chips from sticking to the tool and undermining the cutting edge. By minimizing this adherence, less power is required, finish is improved, and the speed of machining may often be doubled as compared with a similar nonresulfurized grade. The 1200 series steels are both rephosphorized and resulfurized. Phosphorus is soluble in iron and promotes chip breakage in cutting operations through increased hardness and brittleness. Like carbon, an excessive amount of phosphorus could raise strength and hardness levels so high as to impair machinability. Hence, the 1200 series phosphorus content is limited to either 0.04–0.09% or 0.07–0.12% range and carbon is limited to 0.13% max for the same reason. Lead is insoluble in steel, dispersed microscopically in the rolled product. These lead particles act as a lubricant helping to prevent tool buildup during machining and to serve as chip breakers in a manner similar to sulfide inclusions. The lead addition in 12L13 and 12L14 augments the effect of sulfur, permitting a further increase of machining speed and better finish. Economic reasons usually limit leaded resulfurized steels to high speed screw machining products wherein the superior machinability of the steel can be used to the fullest extent.
1111	
1112	
1113	
12L13	
12L14	
1215	

SAE	
1108	Steels in this group are used where a combination of good machinability and response to heat treatment is needed. The lower carbon varieties are used for small parts which are to be cyanided or carbonitrided. SAE 1117, 1118, and 1119 carry more manganese for better hardenability, permitting oil quenching after case hardening heat treatments in many instances. These steels are available with lead added to further enhance machinability.
1109	
1116	
1117	
1118	
1119	

SAE	This group of steels has characteristics comparable to carbon steels of the same carbon level, except for changes as noted previously.
1132	They are widely used for parts where a large amount of machining is necessary, or where threads, splines, or other operations offer special tooling problems. SAE 1137, for example, is widely used for nuts, bolts, and studs with machined threads. The higher manganese SAE 1132, 1137, 1141, and 1144 offer greater hardenability, the higher carbon types being suitable for oil quenching for many parts. All these steels may be selectively hardened by induction or flame heating if desired. These steels are available with lead added to further enhance the machinability.
1137	
1140	
1141	
1144	
1145	
1146	
1151	

COMMON CONSTRUCTIONAL ALLOY STEELS (OTHER THAN STAINLESS AND AUSTENITIC)

A steel is classified as an alloy steel when the maximum of the range given for the content of alloying elements exceeds one or more of the following limits: manganese 1.65%, silicon 0.60%, copper 0.60%; or, when there is specified a definite range or a minimum quantity of chromium (up to 3.99%), nickel, molybdenum, aluminum, cobalt, columbium, titanium, tungsten, vanadium, or zirconium.

The principal uses for alloying elements in the common construction steels are:

1. To develop maximum mechanical properties with minimum distortion and cracking.
2. To develop to a lesser degree special qualities such as resistance to tempering, increased toughness, low notch sensitivity, better machinability in the hardened and tempered condition than possessed by a carbon steel of equivalent carbon content similarly heat treated.

Alloy steels are generally not specified for use without appropriate heat treatment. Some special properties which certain alloying elements influence are: retardation of transformation, lowering of transformation temperature, resistance to creep at elevated temperatures, retention of toughness at subzero temperature, resistance to wear, and effect on hardness and machinability. Other special properties may be imparted by the proper choice of alloying elements.

The HARDENABILITY or response to heat treatment is probably the most important single criterion for the selection of steel. Hardenability is that property of steels which determines the depth and distribution of hardness induced by quenching from above the transformation range. For alloy steels, hardenability is usually measured by the end quench test described in SAE J406. This test provides a comparison of the hardenability of various steel compositions and permits the determination of variations from one heat to another of the same composition or of the variations in different parts of the same heat. In other words, the end quench hardenability is a convenient summation of all of the characteristics which determine the depth and distribution of hardness in a particular material. By tempering the end-quench test bar, the final hardness of the product produced by quenching and tempering may be predicted.

Hardenability should not be confused with hardness per se or with maximum hardness. The maximum hardness obtainable with any steel quenched at the necessary minimum cooling rate depends only on the carbon content. That is to say, the maximum surface hardness obtainable in hardened steels is governed entirely by the carbon content at the surface. It has been established that, under the conditions of scale free heating, complete solution of excess constituents, either ferrite or carbides, and achievement of critical cooling rate, maximum surface hardness is attained at about 0.60% carbon. Further increase of carbon, speed of quenching, or alloy content does not increase the hardness.

Decarburized, scaled, or overheated material or surfaces that have been quenched at less than the critical cooling rate may not be expected to attain full hardness. Mechanical properties such as yield strength, fatigue strength, and toughness are related to the thoroughness with which the quenching is done. It must be kept in mind that for lower carbon material, or for sections above a certain limiting size (see SAE J406, Fig. 10) the minimum cooling rate necessary for production of maximum surface hardness may be greater than that obtainable by oil quenching and a more drastic quench must be used to obtain expected results.

The term hardening implies that the hardness of the material is increased by suitable treatment, usually involving heating to a suitable austenitizing temperature followed by cooling at a certain minimum rate which depends upon the alloy content. If quenching is complete, the resulting structure is martensite. As indicated previously, its hardness depends upon carbon content of the steel. If the quenching conditions are such that a minimum of 90% of martensite is produced, followed by proper tempering, it may reasonably be expected that the surface hardness and cross-sectional hardness are commercially satisfactory measures of the mechanical properties. If the quenching

TABLE 3—TYPICAL HEAT TREATMENTS FOR CARBURIZING GRADES OF ALLOY STEELS

SAE Steels[a]	Pretreatments			Carburizing Temperature, °F	Cooling Method	Reheat Temperature, F	Quenching Medium	Tempering[f] Temperature, F
	Normalize[b]	Normalize and Temper[c]	Cycle Anneal[d]					
4012, 4023, 4024, 4027, 4028, 4032	Yes	—	—	1650–1700	Quench in oil[g]	—	—	250–350
4118	Yes	—	—	1650–1700	Quench in oil[g]	—	—	250–350
4320	Yes	—	Yes	1650–1700 / 1650–1700	Quench in oil[g] / Cool slowly	— / 1525–1550[i]	— / Oil	250–350 / 250–350
4419, 4422, 4427	Yes	—	Yes	1650–1700	Quench in oil[g]	—	—	250–350
4615, 4617, 4620, 4621, 4626, 4718	Yes	—	Yes	1650–1700 / 1650–1700 / 1650–1700	Quench in oil[g] / Cool slowly / Quench in oil	— / 1500–1550[i] / 1500–1550[h]	— / Oil / Oil	250–350 / 250–350 / 250–350
4720	Yes	—	Yes	1650–1700	Quench in oil	1500–1550[h]	Oil	250–350
4815, 4817, 4820	—	Yes	Yes	1650–1700 / 1650–1700 / 1650–1700	Quench in oil[g] / Cool slowly / Quench in oil	— / 1475–1525[i] / 1475–1525[h]	— / Oil / Oil	250–325 / 250–325 / 250–325
5015, 5115, 5120	Yes	—	—	1650–1700	Quench in oil[g]	—	—	250–350
6118	Yes	—	—	1650	Quench in oil[g]	—	—	325
8115, 8615, 8617, 8620, 8622, 8625, 8627, 8720, 8822	Yes / Yes	—	— / Yes	1650–1700 / 1650–1700 / 1650–1700	Quench in oil[g] / Cool slowly / Quench in oil	— / 1550–1600[i] / 1550–1600[h]	— / Oil / Oil	250–350 / 250–350 / 250–350
9310	—	Yes	—	1600–1700 / 1600–1700	Quench in oil / Cool slowly	1450–1525[h] / 1450–1525[i]	Oil / Oil	250–325 / 250–325
94B15, 94B17	Yes	—	—	1650–1700	Quench in oil[g]	—	—	250–350

[a] These steels are fine grain. Heat treatments are not necessarily correct for coarse grain.
[b] Normalizing temperature should be at least as high as the carburizing temperature followed by air cooling.
[c] After normalizing, reheat to temperature of 1100–1200 F and hold at temperature approximately 1 hr per in. of maximum section or 4 hr minimum time.
[d] Where cycle annealing is desired, heat to at least as high as the carburizing temperature, hold for uniformity, cool rapidly to 1000–1250 F, hold 1 to 3 hr, then air cool or furnace cool to obtain a structure suitable for machining and finish.
[e] It is general practice to reduce carburizing temperatures to approximately 1550 F before quenching to minimize distortion and retained austenite. For 4800 series steels, the carburizing temperature is reduced to approximately 1500 F before quenching.
[f] Tempering treatment is optional. Tempering is generally employed for partial stress relief and improved resistance to cracking from grinding operations. Temperatures higher than those shown are used in some instances where application requires.
[g] This treatment is most commonly used and generally produces a minimum of distortion.
[h] This treatment is used where the maximum grain refinement is required and/or where parts are subsequently ground on critical dimensions. A combination of good case and core properties is secured with somewhat greater distortion than is obtained by a single quench from the carburizing treatment.
[i] In this treatment the parts are slowly cooled, preferably under a protective atmosphere. They are then reheated and oil quenched. A tempering operation follows as required. This treatment is used when machining must be done between carburizing and hardening or when facilities for quenching from the carburizing cycle are not available. Distortion is at least equal to that obtained by a single quench from the carburizing cycle, as described in note e.

rate is such that a smaller percentage of martensite is produced (and a correspondingly larger percentage of other transformation products) a corresponding reduction in mechanical properties may be expected. This may be summarized by saying that steels that have the same hardenability may, after quenching and tempering, be expected to have very nearly the same mechanical properties and that, as a result, many different compositions may be used interchangeably to obtain a similar range of mechanical properties in the finished part.

Properties such as case of annealing, degree of hardening from cold working, ease of machining, and some other characteristics may be peculiar to either composition or heat treatment or both. The choice of steel for a given application is more often dictated by the overall economic consideration than by any other factor.

The commonly used alloy steels may be divided into two grades:
1. The low carbon carburizing grade.
2. The higher carbon directly hardenable grade.

Carburizing Grades of Alloy Steels

Properties of the Case—The properties of carburized and hardened cases depend upon the carbon and alloy content, the depth of case, the structure of the case, and the degree and distribution of residual stresses. The carbon content of the case depends upon the details of the carburizing process and the response of iron and the alloying elements present to carburization. The original carbon content of the steel has little or no effect upon the carbon content produced in the case. The hardenability in the case, therefore, depends upon the alloy content of the steel and the FINAL carbon content produced by carburizing, but not upon the INITIAL carbon content of the steel.

When heating for hardening results in complete carbide solution in the case, the effect of alloying elements upon the hardenability of the case will in general be the same as the effect of these elements upon the hardenability of the core. An exception is that boron increases significantly the hardenability of the low carbon core, but has little effect upon the hardenability of the higher carbon case. It is also true that some elements which raise the hardenability of the core may tend to produce more retained austenite and consequently somewhat lower indentation hardness in the case.

Alloy steels are frequently used for case hardening because the required surface hardness can be obtained by moderate rates of cooling such as may be secured with an oil quench. This may mean less distortion than would be encountered with water quenching. It is usually desirable to select a steel which will attain a minimum surface hardness of 58 Rockwell C (RC) after carburizing and oil quenching. Where section sizes are large, a high hardenability alloy steel may be necessary, while for medium and light sections, a low hardenability steel will suffice.

In general, the case hardening alloy steels may be divided into three classes so far as the hardenability of the case is concerned. The three classes are: low hardenability such as the 4000, 5000, 5100, 6100, and 8100 series: intermediate hardenability such as the 4300, 4400, 4500, 4600, 4700, 8600, 8800,[1] and 94B00 series; high hardenability such as 4800 and 9300 series. Since the carbon content or hardenability of the case, there is no appreciable difference in the case hardenability of 4815 steel compared with 4820.

The steels having high case hardenability generally have reasonably high core hardenabilities; although, the core hardenability is dependent upon the carbon content of the basic steel as well as the alloy content. These steels are used particularly for carburized parts having thick sections, such as heavy duty truck bevel drive pinions and gears and large roller bearings. Good case properties can be obtained by oil quenching. These steels are likely to have substantial amounts of retained austenite in the case after carburizing and quenching. The amount of retained austenite may be held to reasonable limits by controlling the carbon content of the case to produce near eutectoid case, by refrigerating the parts, or by reheating and quenching after carburizing. Lower case hardenability steels are used in smaller parts which are less heavily loaded. Steels with intermediate case hardenability are used for tractor and automotive gears, piston pins, ball studs, universal crosses, and roller bearings. Satisfactory case hardness should be produced in most cases by oil quenching.

Core Properties—The core properties of the case hardened steels depend upon the carbon and alloy content of the original steel and the severity of the quench. Many of the generally used types of alloy case hardening steels are made with two or more carbon contents so as to provide a choice of core hardness. The most desirable hardness for the core depends upon the design and function of the individual part. In general, where high compressive loads are encountered, relatively high core hardness is beneficial in supporting the case. Low core hardnesses may be desirable where more than normal toughness is essential.

The case hardening steels may be divided into three general classes with respect to hardenability of the core. (For hardenability of individual steels, see SAE J407.) Due to the fact that H bands have not been established for all steels, it is impossible to give an accurate comparative rating of hardenability of all of the steels in any one group. Low hardenability core steels include SAE 4012, 4023, 4024, 4027[2], 4028[2], 4118[2], 4419[2], 4422[2], 4615, 4617, 4626[2], 5015, 5115, 5120[2], 6118[2], and 8615.

Medium hardenability core steels include SAE 4032, 4427, 4620, 4621, 4720, 4815[2], 8617, 8620, 8622 and 8720.

High hardenability core steels include SAE 4320, 4718, 4817, 4820, 8625, 8627, 8822, 9310, 94B15 and 94B17. 94B15 and 94B17 have been classed as high hardenability steels in the core due to the marked effect of boron upon the hardenability of low carbon steels.

Heat Treatment—With few exceptions the alloy carburizing steels are made of fine grain, and most are therefore suitable for direct quenching from the carburizing temperature (1700 F) or from a reduced temperature of 1500–1600 F. If the carburizing is to be done at temperatures above 1700 F and the parts are direct quenched, careful studies should be made of the suitability of the products so treated. Several other types of heat treatment involving single and double quenching are also used for some of these steels. See Table 3.

Directly Hardenable Grades of Alloy Steel—These steels may be considered in five groups on the basis of approximate mean carbon content of the SAE specification. In general, the last two figures of the specification agree with the mean carbon content. Consequently, the heading "0.30–0.37 Mean Carbon Content of SAE Specification" includes steel such as SAE 1330 and 4137.

Mean Carbon Content of SAE Specification, %	Common Applications (See also more detailed discussion below)
0.30–0.37	Heat treated parts requiring moderate strength and great toughness.
0.40–0.42	Heat treated parts requiring higher strength and good toughness.
0.45–0.50	Heat treated parts requiring fairly high hardness and strength with moderate toughness.
0.50–0.60	Springs and hand tools.
1.02	Ball and roller bearings.

It is necessary to deviate from the preceeding plan in the classification of the carbon-molybdenum steels. When carbon-molybdenum steels are used, it is customary to specify higher carbon content for any given application than would be specified for other alloy steels, due to the low alloy content of these steels. For example, SAE 4063 is used for the same applications as SAE 4140, 4145, and 5150. Consequently, in the following tables and discussion, the carbon-molybdenum steels have been shown in the groups where they belong on the basis of applications rather than carbon content.

For the present discussion, steels of each carbon content are divided into two or three groups on the basis of hardenability. Transformation ranges and consequently heat treating practices vary somewhat with different alloying elements even though the hardenability is not changed.

0.30–0.37 Mean Carbon Content—These steels are frequently used for water quenched parts of moderate section size and for oil quenched parts of small section size. Typical applications of these steels are connecting rods, steering arms and steering knuckles, axle shafts, bolts, studs, screws, and other parts requiring strength and toughness where section size is small enough to permit obtaining the desired mechanical properties with the customary heat treatment. Steels falling in this classification may be subdivided into two groups on the basis of the hardenability:

Low hardenability steels in the 0.30–0.37 mean carbon content classification include SAE 1330, 1335, 4037, 4130, 5130, 5132, 5135, and 8630. Medium hardenability steels in the 0.30–0.37 mean carbon content classification include SAE 4135, 4137, 8637, and 94B30.

0.40–0.42 Mean Carbon Content—In general, these steels are used for medium and large size parts requiring a high degree of strength and toughness. The choice of the proper steel depends upon the section size and the mechanical properties which must be produced. The low and medium hardenability steels are used for average size automotive parts such as steering knuckles, axle shafts, and propeller shafts. The high hardenability steels are used particularly for large axles and shafts and for large aircraft parts. These steels are usually considered as oil quenching, although some large parts made of the low and medium hardenability classifications may be quenched in water under properly controlled conditions. These steels may be roughly divided into three groups on the basis of hardenability:

1. Low hardenability steels in the 0.40–0.42 mean carbon content classification include SAE 1340, 4047, and 5140.
2. Medium hardenability steels in the 0.40–0.42 mean carbon content classification include 4140, 4142, 50B40, 8640, 8642, and 8740.
3. High hardenability steels in the 0.40–0.42 mean carbon content classification include SAE 4340.

0.45–0.50 Mean Carbon Content—These steels are used primarily for gears and other parts requiring fairly high hardness as well as strength and

---

[1] These are borderline cases which might be considered in the group with high case hardenability.

[2] These are borderline cases which might be considered in the next higher hardenability group.

TABLE 4—TYPICAL HEAT TREATMENTS FOR DIRECTLY HARDENABLE GRADES OF ALLOY STEELS

SAE Steels[a]	Normalizing Temperature, F	Annealing[d] Temperature, F	Hardening[e] Temperature, F	Quenching Medium	Temper
1330	1600–1700[b]	1550–1650	1525–1575	Water or oil	To desired hardness
1335, 1340, 1345	1600–1700[b]	1550–1650	1500–1550	Oil	To desired hardness
4037, 4042	—	1500–1575	1525–1575	Oil	To desired hardness
4047	—	1450–1550	1500–1575	Oil	To desired hardness
4130	1600–1700[b]	1450–1550	1500–1600	Water or oil	To desired hardness
4135, 4137, 4140, 4142	—	1450–1550	1550–1600	Oil	To desired hardness
4145, 4147, 4150	—	1450–1550	1500–1550	Oil	To desired hardness
4161	—	1450–1550	1500–1550	Oil	To desired hardness, 700 F, min
4340	1600–1700[b,c]	1450–1550	1500–1550	Oil	To desired hardness
50B40, 50B44, 5046, 50B46	1600–1700[b]	1500–1600	1500–1550	Oil	To desired hardness
50B50, 5060, 50B60	1600–1700[b]	1500–1600	1475–1550	Oil	To desired hardness
5130, 5132	1600–1700[b]	1450–1550	1525–1575	Water, caustic solution, or oil	To desired hardness
5135, 5140, 5145	1600–1700[b]	1500–1600	1500–1550	Oil	To desired hardness
5147, 5150, 5155, 5160, 51B60	1600–1700[b]	1500–1600	1475–1550	Oil	To desired hardness
50100, 51100, 52100	—	1350–1450	1425–1475 / 1500–1600	Water / Oil	To desired hardness
6150	—	1550–1650	1550–1625	Oil	To desired hardness
81B45	1600–1700[b]	1550–1650	1500–1575	Oil	To desired hardness
8630	1600–1700[b]	1450–1550	1525–1600	Water or oil	To desired hardness
8637, 8640	—	1500–1600	1525–1575	Oil	To desired hardness
8642, 8645, 86B45, 8650	—	1500–1600	1500–1575	Oil	To desired hardness
8655, 8660	—	1500–1600	1475–1550	Oil	To desired hardness
8740	—	1500–1600	1525–1575	Oil	To desired hardness
9254, 9255, 9260	—	—	1500–1650	Oil	To desired hardness
94B30	1600–1700[b]	1450–1550	1550–1625	Oil	To desired hardness

[a] These steels are fine grain unless otherwise specified.
[b] These steels should be either normalized or annealed for optimum machinability.
[c] Temper at 1100–1225.
[d] The specific annealing cycle is dependent upon the alloy content of the steel, the type of subsequent machining operations and desired surface finish.
[e] Frequently, these steels, with the exception of 4340, 50100, 51100, and 52100, are hardened and tempered to a final machinable hardness without preliminary heat treatment.

toughness. Such parts are usually oil quenched, and a minimum of 90% martensite in the as-quenched condition is desirable.

1. Low hardenability steels in the 0.45–0.50 mean carbon content classification include SAE 5046, 50B44, 50B46, and 5145.
2. Medium hardenability steels in the 0.45–0.50 mean carbon content classification include SAE 4145, 5147, 5150, 81B45, 8645 and 8650.
3. High hardenability steels in the 0.45–0.50 mean carbon content classification include SAE 4150 and 86B45.

0.50–0.60 Mean Carbon Content—These steels are used primarily for springs and hand tools. The hardenability necessary depends upon the thickness of the material and the quenching practice.

1. Medium hardenability steels in the 0.50–0.60 mean carbon content classification include SAE 50B50, 5060, 50B60, 5150, 5155, 51B60, 6150, 8650, 9254, 9255, and 9260.
2. High hardenability steels in the 0.50–0.60 mean carbon content classification include SAE 4161, 8655, and 8660.

1.02 Mean Carbon Content—These are straight chromium electric furnace steels used primarily for the races and balls or rollers of antifriction bearings. They are also used for other parts requiring high hardness and wear resistance. The compositions of the three steels are identical except for a variation in chromium with a corresponding variation in hardenability.

1. The low hardenability steel in the 1.02 mean carbon content classification is SAE 50100.
2. The medium hardenability steels in the 1.02 mean carbon content classification are SAE 51100 and 52100.

Heat Treatments—Typical treatments are given in Table 4.

Resulfurized Steel—Some of the alloy steels (SAE 4024 and 4028) are resulfurized to give better machinability at a relatively high hardness.

CHARACTERISTICS OF WROUGHT STAINLESS STEELS

The composition and corresponding physical characteristics of these steels divides them into three broad groups or types:

1. Stainless Chromium-Nickel Austenitic Steels (*Not Hardenable*)—The first group are high nickel-chromium alloys. They are austenitic at room temperature and higher and cannot be hardened by thermal treatment.

Table 5 gives typical heat treatments for these steels.

TABLE 5—TYPICAL HEAT TREATMENTS FOR GRADES OF CHROMIUM-NICKEL AUSTENITIC STEELS NOT HARDENABLE BY THERMAL TREATMENT

SAE Steels	AISI No.	Treatment No.	Normalizing Temperature, F	Annealing[a] Temperature, F	Hardening Temperature, F	Quenching Medium	Temper
30201	201	1	—	1850-2050	—	Water or air	—
30202	202	1	—	1850-2050	—	Water or air	—
30301	301	1	—	1800-2100	—	Water or air	—
30302	302	1	—	1800-2100	—	Water or air	—
30303F[b]	303	1	—	1800-2100	—	Water or air	—
30304	304	1	—	1800-2100	—	Water or air	—
30305	305	1	—	1800-2100	—	Water or air	—
30309	309	1	—	1800-2100	—	Water or air	—
30310	310	1	—	1800-2100	—	Water or air	—
30316	316	1	—	1800-2100	—	Water or air	—
30317	317	1	—	1800-2100	—	Water or air	—
30321	321	1	—	1800-2100	—	Water or air	—
30325	325	1	—	1800-2100	—	Water or air	—
30330	—	1	—	2050-2250	—	Air	—
30347	347	1	—	1800-2100	—	Water or air	—

[a] Quench to produce full austenitic structure using water or air in accordance with thickness of section. Annealing temperatures given cover process and full annealing as already established and used by industry, the lower end of the range being used for process annealing.
[b] Suffix A denotes steel differing only in carbon content. Suffix F denotes a free machining steel.

SAE 30201 is an austenitic chromium-nickel-manganese stainless steel usually required in flat products. In the annealed condition it exhibits higher strength values than the corresponding chromium-nickel stainless steel (SAE 30301). It is nonmagnetic in the annealed condition but may be magnetic when cold worked. SAE 30201 is used to obtain high strength by work hardening and is well suited for corrosion resistant structural members requiring high strength with low weight. It has excellent resistance to a wide variety of corrosive media, showing behavior comparable to stainless grade SAE 30301. It has high ductility and excellent forming properties. Due to this steel's work hardening rate and yield strength, tools for forming must be designed to allow for a higher springback or recovery rate. It is used for automotive trim, automotive wheel covers, railroad passenger car bodies and structural members, truck trailer bodies.

SAE 30202, like its corresponding chromium-nickel stainless steel SAE 30302, is a general purpose stainless steel. It has excellent corrosion resistance and deep drawing qualities. It is nonhardenable by thermal treatments but may be cold worked to high tensile strengths. In the annealed condition it is nonmagnetic but slightly magnetic when cold worked. Applications for this stainless steel are hub cap, railcar and truck trailer bodies, and spring wire.

SAE 30301 is capable of developing high tensile strength, while retaining high ductility, by moderate to severe cold working. It is used largely in the cold rolled or cold drawn condition in the form of sheet, strip, and wire. It is nonmagnetic when annealed but is magnetic when cold worked. Its corrosion resistance is not quite equal to SAE 30302. This steel is used for applications requiring a combination of high strength and excellent forming properties such as in structural members, automotive trim, wheel discs, and rings. It is used for flat and wire springs, windshield wiper arms, grills, steering wheel spokes, and similar applications. It is also used for cream separators and milking machine parts and other such products where a combination of formability and resistance to corrosion by food products is needed.

SAE 30302 is the general purpose stainless steel of this type. Its corrosion resistance is superior to that of SAE 30301, and it is the most widely used of all the chromium-nickel stainless and heat resisting steels. It is used for deep drawing largely in the softer tempers. It can be worked to high tensile strength but with lower ductility than SAE 30301. It is nonmagnetic when annealed but is magnetic when cold worked. This steel is used on automotive parts where excellent corrosion resistance and high physical properties or good forming and drawing properties are required. It is used for hub caps, radiator grills, windshield wiper parts such as tension bars and binder strips, hose clamps, antennas, control cables, fender guards, fire walls, and hydraulic tubing. It is used for other similar parts which have severe forming requirements combined with a need for resistance to rusting or tarnishing.

SAE 30303F has elements added to improve its machining and non-seizing characteristics. This steel, the free machining modification of SAE 30302, is recommended for the manufacture of parts produced on automatic machines. It can be forged but requires much more care than is necessary with SAE 30302. Its corrosion resistance is slightly inferior to that of SAE 30302. It is nonmagnetic when annealed but is slightly magnetic when cold worked. It is used for screws, nuts, carburetor parts, aircraft fittings, water pump shafts, and other machined parts requiring some corrosion resistance. It is not recommended for applications involving severe cold working, cold upsetting, or welding.

SAE 30304 is a low carbon steel similar to SAE 30302 but somewhat superior in corrosion resistance and having superior welding properties for certain types of equipment. It is nonmagnetic when annealed but is slightly magnetic when cold worked. It is used for diesel injection pump valve springs, roller chains, parachute hardware, and welded parts that can be heat treated after welding or that are not liable to damage by intergranular corrosion if heat treating after welding is not performed. This steel is also available with 0.03/0.05% carbon for certain applications.

SAE 30305 is similar to SAE 30302 and 30304 but it does not harden as rapidly with cold working as do either of those grades. It also has much less change in magnetic permeability when cold worked. Because of its lower work hardening tendency, it is better suited for spun parts, multiple drawing operations, severe cold heading, and parts requiring large amounts of cold deformation.

SAE 30309 has higher corrosion and oxidation resistance than SAE 30304. It is resistant to oxidation at temperatures up to about 2000 F. It is nonmagnetic when annealed but may be very slightly magnetic when cold worked. It is used primarily in high temperature applications such as thermocouple wells, heat exchangers, glass lehr belts, and aircraft cabin heaters.

SAE 30310 has very high corrosion and heat resisting properties. Like SAE 30309, it resists oxidation at temperatures up to about 2000 F. It is more stable and somewhat stronger at high temperature and is more safely hot worked than SAE 30309. It is nonmagnetic when annealed or cold worked. It is used in such applications as diesel injector cup wipers, jet engine burner liners, and nozzle vanes.

SAE 30316 is similar to SAE 30304 in fabricating qualities and general corrosion resistance. However, it has superior corrosion resistance to other chromium-nickel steels when exposed to sea water and many types of chemical corrodents, especially those of a reducing nature. It also has superior strength at elevated temperatures. It is nonmagnetic when annealed but is slightly magnetic when cold worked. It is used in applications such as wire screens, dye making and chemical processing equipment, and in elevated temperature service, especially where strength is important, up to about 1500 F. This steel is also available with 0.03/0.05% carbon for certain applications.

SAE 30317 is similar to SAE 30316 but with greater corrosion resistance in many environments and with somewhat greater high temperature strength. It is primarily used for paper making equipment.

SAE 30321 has a specified titanium content. Its properties are similar to those of SAE 30304 except that it can be recommended for use in the manufacture of welded parts requiring immunity to intergranular attack and where heat treating after welding is not feasible. It may also be used where temperatures in the range of about 800 to 1650 F are encountered in fabrication or service and where the possibility of intergranular corrosion exists. It is nonmagnetic when annealed but is slightly magnetic when cold worked. It is used for exhaust manifolds, manifold flanges, and high temperature bolts and locknuts.

SAE 30325 is primarily used in hot caustic solutions. It has been used for valve trim, and pump shafting.

SAE 30330 alloy, in the wrought and cast form is used for high temperature oxidation resistance essentially over 1650 F and utilized in the construction of heat treating baskets, similar items, and heat treating furnace parts.

SAE 30347 is similar to SAE 30321 except it contains columbium instead of titanium. This columbian bearing alloy is used in the same applications as SAE 30321 except that it does not cold form as satisfactorily.

2. Stainless Martensitic Chromium Steels (Hardenable)—The second group are chromium alloys that may contain small amounts (up to about 3%) of nickel. They are ferritic at room temperature but become austenitic at elevated temperature and can be rapidly cooled to produce a hard, martensitic structure in the same manner as other hardenable steels are heat treated. As they can be heat treated to produce martensite, they are commonly known as martensitic stainless steels.

SAE 51409 is an 11% chromium alloy developed especially for automotive mufflers and tailpipes. Resistance to corrosion and oxidation is very similar to SAE 51410. It is nonhardenable and has good forming and welding characteristics. This alloy is recommended for mildly corrosive applications where surface appearance is not critical.

SAE 51410 is the general purpose steel of this type. It can be hardened by heat treating to develop a wide range of mechanical properties. 39/41 RC is about the maximum useful hardness obtainable. It has fair machining properties and corrosion resistance, although in this respect, it is inferior to SAE 51430. Best corrosion resistance is obtained in the heat treated condition. It is magnetic in all conditions. It is used in applications requiring high strength combined with moderate resistance to corrosion. Possessing fair strength and good oxidation resistance to about 1200 F, it is used in manifold stud bolts, heat control shafts, steam valves, bourdon tubes, and gun mounts.

SAE 51414 has somewhat better corrosion resistance than SAE 51410. It will attain slightly higher mechanical properties when heat treated than SAE 51410 and will develop a maximum useful hardness of about 41/43 RC. It is magnetic in all conditions. It is used in the form of tempered strip and in bars and forgings for heat treated parts. It is also used for valve trim and stems.

SAE 51416F is similar to SAE 51410. It can be heat treated to a maxi-

TABLE 6—TYPICAL HEAT TREATMENTS FOR STAINLESS CHROMIUM STEELS

SAE Steels	AISI No.	Treatment No.	Normalizing Temperature, F	Subcritical Annealing Temperature, F	Full Annealing[a] Temperature, F	Hardening Temperature, F	Quenching Medium	Temper
51409	—	I	—	—	1550-1650	—	air	—
51410	410	II / I	—	1300-1350[b]	1550-1650	1750-1850	Oil or air	To desired hardness
51414	414	II / I	—	1200-1250[b]	—	1750-1850	Oil or air	To desired hardness
51416	416	II	—	1300-1350[b]	1550-1650	1750-1850	Oil or air	To desired hardness
51420	420	II	—	1350-1450[b]	1550-1650	1800-1850	Oil or air	To desired hardness
51420F[c]	—	II	—	1350-1450[b]	1550-1650	1800-1850	Oil or air	To desired hardness
51430	430	I	—	1400-1500[d]	—	—	—	—
51430F[c]	—	I	—	1250-1500[d]	—	—	—	—
51431	431	I	—	1150-1225[b]	—	1800-1900	Oil or air	To desired hardness
51434		I	—	1400-1500[d]	—	—	—	—
51436								
51440A[c]	440A[c]							
51440B[c]	440B[c]		—	1350-1440[b]	1550-1650	1850-1950	Oil or air	To desired hardness
51440C[c]	440C[c]							
51440F[c]			—	—	—	—	—	—
51442	442	I	—	1400-1500[d]	—	—	—	—
51446	446	I	—	1500-1650[b]	—	—	—	—
51501	501		—	1325-1375[d]	1525-1600	1600-1700	Oil or air	To desired hardness

[a] Cool slowly in furnace.
[b] Usually air-cooled but may be furnace cooled.
[c] Suffixes A, B, and C denote three types of steel differing only in carbon content. Suffix F denotes a free machining steel.
[d] Cool rapidly in air.

mum of about 39/41 RC. Elements have been added to improve its machining and nonseizing characteristics at some sacrifice in corrosion resistance and weldability. It is the most readily machinable of all the stainless steels and is suited for use on automatic screw machines. It is magnetic in all conditions. Available in bars, wire, and forgings, it is used for water pump shafts, carburetor needle valves, heat control valve shafts, manifold parts, and other parts requiring a hardenable, free machining, corrosion, and heat resisting steel. Free machining types are not recommended for welding.

SAE 51420 is capable of being hardened to a wide range of mechanical properties depending upon the actual carbon and chromium contents. With low side of carbon range and chromium, the grade behaves about like SAE 51410. With about 12.5% or higher chromium and 0.30% carbon, the maximum useful hardness expected is about 49/53 RC. It offers its maximum corrosion resistant properties only in fully hardened condition. It is magnetic in all conditions. It is used in cutlery, hardened pump shafts, water pump parts, glass and plastic molds, bomb shackle parts, and drive screws.

SAE 51420F is similar to SAE 51420 except that elements have been added to improve its machinability.

SAE 51431 is a nickel bearing chromium steel capable of being heat treated to a maximum useful hardness of about 42/44 RC. Its corrosion resistance is superior to that of the other hardenable grades SAE 51410, 51420, 51430, and 51440. It is magnetic in all conditions. It is used for aircraft bolting, cable terminals, bomb shackle parts, and other parts requiring a hardenable steel with high mechanical properties and superior corrosion resistance.

SAE 51434 and 51436 are special purpose nonhardenable ferritic stainless steels. These two alloys are modifications of the basic SAE 51430 alloy and were designed especially for use as materials for automotive trim. Their superior resistance to atmospheric corrosion in the presence of salt spray and road dirt (due to winter road conditioning and dust laying chemicals) is their main virtue.

These alloys may be fabricated in the same manner as SAE 51430.

SAE 51440A is hardenable to a greater quenched hardness than SAE 51420 and with greater toughness than SAE 51440B or 51440C. It can be hardened to a maximum of about 51/56 RC. Maximum corrosion resistance is obtained only from a polished surface on fully hardened material. It is magnetic in all conditions. It is used for paint spray nozzles and some types of bearings.

SAE 51440B is hardenable to a greater quenched hardness than SAE 51440A and greater toughness than SAE 51440C. Maximum corrosion resistance is obtained only from a polished surface on fully hardened material. Depending upon carbon content, it can be hardened from 53/58 RC. It is magnetic in all conditions. It is used for balls and races.

SAE 51440C acquires upon heat treatment the highest quenched hardness and greatest wear resistance of any corrosion or heat resistant steel. It can be hardened to 55/60 RC and is corrosion resisting only in the fully hardened and polished condition. It is magnetic in all conditions. It is used for diesel engine pump parts, instrument parts, crankshaft counterweight pins, valve trim, ball bearings, races, and other parts requiring a hard wear and corrosion resisting surface.

SAE 51440F is similar to SAE 51440C except that elements have been added to improve its machinability and nonseizing characteristics. It is used for carburetor parts.

SAE 51501 is used for its heat and corrosion resistance and good mechanical properties at elevated temperatures. It can be produced with about 0.5% molybdenum to improve its toughness. It can be heat treated to various hardnesses depending upon its carbon content. It is magnetic in all conditions. It is used in service up to about 1200 F. A 0.15% maximum carbon type is used for tubing in oil stills and heat exchangers. Higher carbon types are used as valve stems, valves, and hot or cold work dies and mandrels.

3. Stainless Ferritic Chromium Steels (Not Hardenable)—The third group contains more chromium and less carbon than the second group. Nickel, if present, is incidental. By virtue of this high chromium and low carbon content, these steels are ferritic at room and elevated temperatures. As they do not transform to austenitic, they cannot be hardened by heat treatment and are known as ferritic stainless steels.

Table 6 gives typical heat treatments for these steels.

SAE 51430 has superior corrosion and heat resistance as compared with SAE 51410. It is not normally considered hardenable by heat treatment. Its ductility is only fair, but it is resistant to destructive oxidation up to about 1500 F. It is magnetic in all conditions. It is applied in parts requiring only a moderate draw, such as moldings, windshield wiper yokes, heat control valves, shafts and bushings, magneto drive covers, fasteners of all types, wire strainer screens, and fender guards.

SAE 51430F is similar to SAE 51430 except that elements have been added to improve its machinability and nonseizing characteristics. It can be machined at speeds approximating 85% of those of Bessemer screw stock. It is less amenable to both cold and hot work than SAE 51430. It is used for oil burner nozzles and other machined parts requiring good corrosion of heat resistance. This type is not recommended for welding.

SAE 51442 is somewhat superior in corrosion and heat resistance to SAE 51430. It has good resistance to oxidation up to 1600 F. It is magnetic in all conditions.

SAE 51446 has the maximum amount of chromium consistent with commercial malleability. It is used principally for the manufacture of parts which must resist high temperatures in service without scaling but which are not highly stressed. It resists destructive oxidation to a temperature of about 2000 F. It is used for glass seals, salt bath electrodes, thermocouple wells, combustion chambers and soot blowers, and other parts where resistance to oxidation is important but where the need to carry a load is negligible.

4. Stainless Steels Possessing Special Machinability Features—This group represents proprietary modifications of standard SAE steels to provide steels of special machining characteristics in comparison to the standard free machining counterparts. These steels are particularly suited for parts made in automatic screw machines. Table 7 gives typical heat treatments for these steels.

Type 203-EZ is a chromium-nickel-manganese free machining stainless steel. It is austenitic and does not respond to thermal treatments. This steel is nonmagnetic when annealed, but is slightly magnetic when cold worked.

Types 303 Ma, 303 Pb, 303 Cu, and 303 Plus X are modifications of SAE 30303. They are austenitic and do not respond to thermal treatments. These steels are nonmagnetic when annealed, but are slightly magnetic when cold worked.

Type 416 Plus X is a modification of SAE 51416. This steel is martensitic and hardenable by thermal treatments to RC 40 minimum. It is magnetic in all conditions.

TABLE 7—TYPICAL HEAT TREATMENTS FOR WROUGHT STAINLESS STEELS OF SPECIAL MACHINABILITY

Proprietary Designation	Treatment Number	Subcritical Annealing Temperature, F	Full Annealing Temperature, F	Hardening Temperature, F	Quenching Medium	Temper
203-EZ	1	—	1850–2050a	—	Water or air	—
303 Ma	1	—	1850–2050a	—	Water or air	—
303 Pb	1	—	1850–2050a	—	Water or air	—
303 Cu	1	—	1850–2050a	—	Water or air	—
303 Plus X	1	—	1850–2050a	—	Water or air	—
416 Plus X	1	1300–1350b	1550–1650c	—	—	—
	11	—	—	1750–1850	Oil or air	To desired hardness

aQuench to produce full austenitic structure using water or air in accordance with thickness of section. Annealing temperatures given cover process and full annealing as already established and used by industry, the lower end of the range being used for process annealing.
bUsually air-cooled but may be furnace-cooled.
cCool slowly in the furnace.

MECHANICAL PROPERTIES OF HEAT TREATED WROUGHT STEELS—SAE J413b

SAE Information Report

Report of Iron and Steel Division approved January 1932 and completely revised by Iron and Steel Technical Committee November 1978.

The charts in this report illustrate the principle that, regardless of composition, steels of the same cross sectional hardness produced by tempering after through hardening, will have approximately the same longitudinal[1] tensile strength at room temperature.

Chart 1 shows the relation between hardness and longitudinal tensile strength of 0.30–0.50% carbon steels in the fully hardened and tempered, as rolled, normalized, and annealed conditions. Chart 2 showing the relation between longitudinal tensile strength and yield strength, and Chart 3 illustrating longitudinal tensile strength versus reduction of area, are typical of steels in the quenched and tempered condition. Chart 3 shows the direct relationship between ductility and hardness and illustrates the fact that the reduction of area decreases as hardness increases, and that, for a given hardness, the reduction of area is generally higher for alloy steels than for plain carbon steels.

It is evident from these curves that steels of the same cross sectional hardness have about the same strength characteristics, so that any one of several different compositions would yield the same results. For some specific application then, the first thing to be determined is what composition is required to obtain proper hardening in the size section involved. This information is not contained in mechanical property charts, but can be determined from published data or by means of a hardenability test. Methods of making this hardenability test and interpretation of the test results are provided in SAE J406b.

Having selected a steel that will through harden in the size section under consideration, the engineer must decide from the service stresses imposed on the finished part what tensile properties are required in the part. These tensile properties may then be converted to hardness values from the charts given here; and from Chart 4 showing the effect of tempering temperature on hardness, the appropriate tempering temperature to obtain this hardness can be selected. In Chart 4 the curves are approximate values to be used as a guide. Carbon steels and lean alloy steels, when fully hardened, will fall

[1] Longitudinal means parallel to rolling direction.

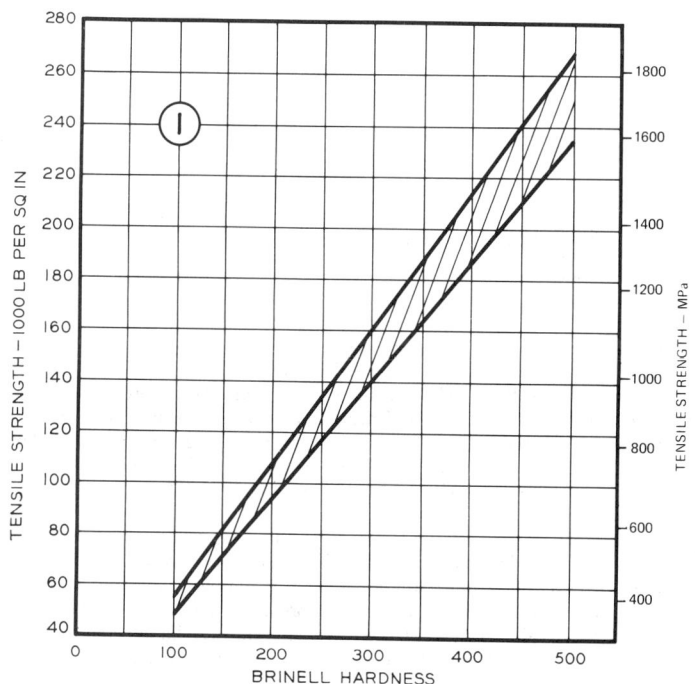

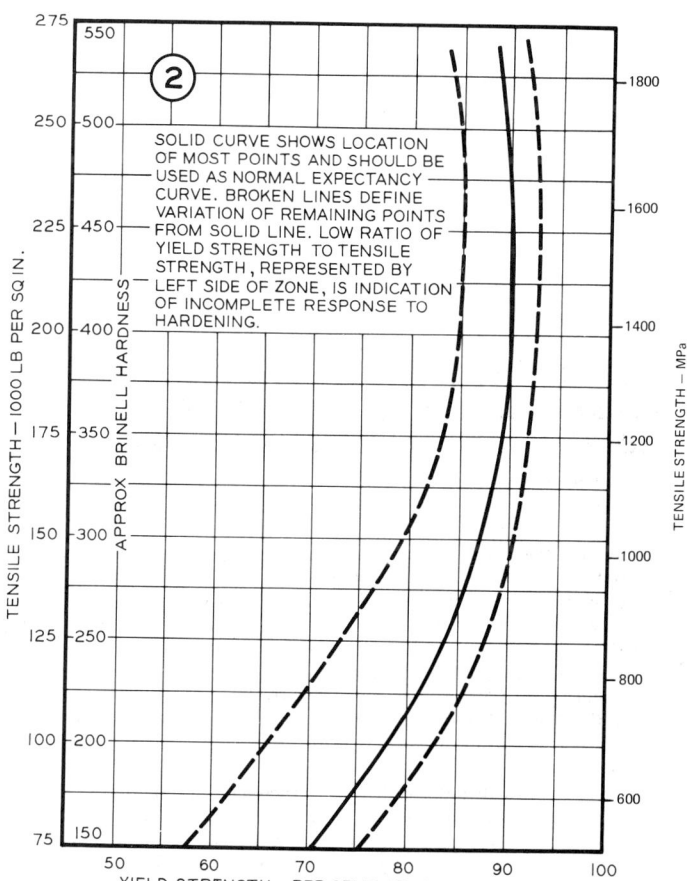

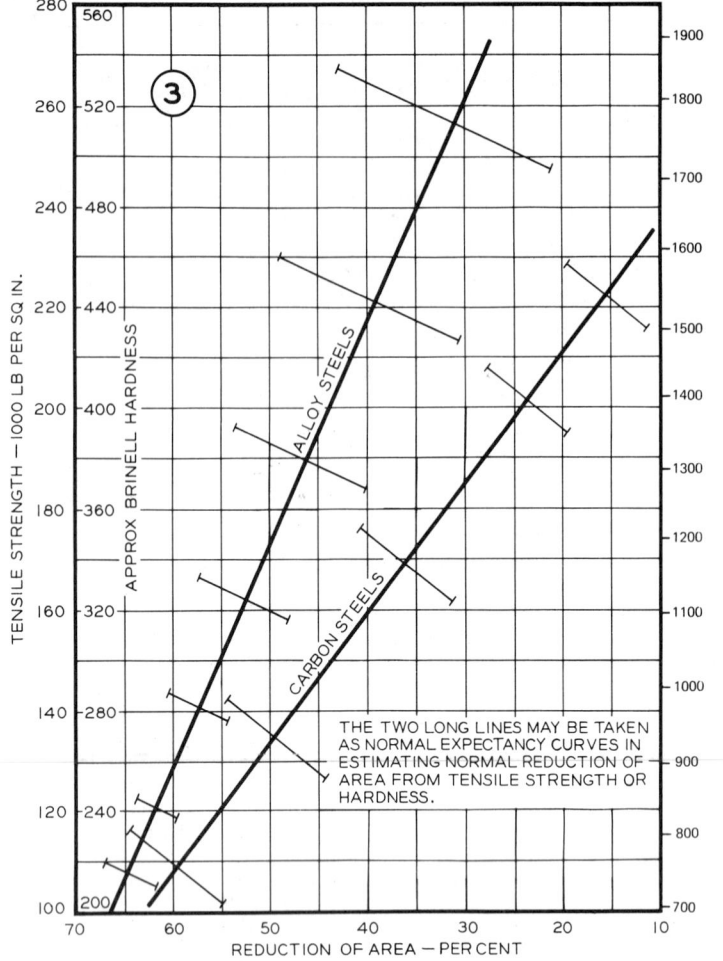

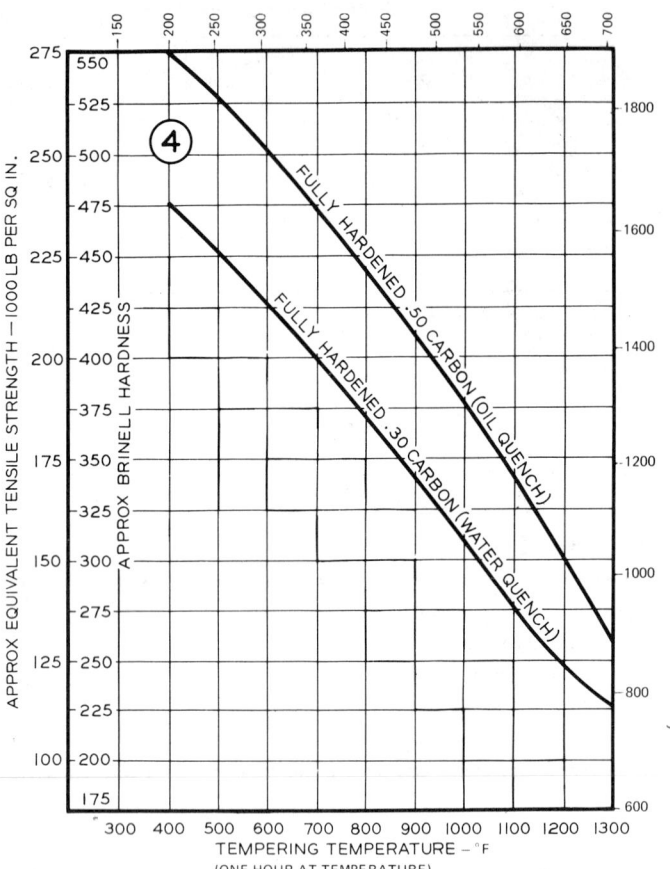

slightly below the curves and strongly alloyed steels will fall slightly above the curves.

Chart 4 showing the effect of tempering temperature on hardness is a summary of information contained in a large number of mechanical property charts published by steel companies, alloy suppliers and users, and represents, as do the charts on tensile and yield strengths, and reduction of area, data on all SAE alloy and carbon steels with carbon contents of 0.30–0.50%.

Mechanical property values obtained from these few summary charts will be as accurate as the information formerly available in a large number of charts, each representing an individual type of steel. For more exact information it would be necessary to make tests on samples from individual heats of steel.

NOTE: Mechanical properties in this report are monotonic and do not represent cyclic test loading conditions. Cyclic loading and cyclic material properties are described in SAE J1099, Technical Report on Fatigue Properties.

ESTIMATED MECHANICAL PROPERTIES AND MACHINABILITY OF HOT ROLLED AND COLD DRAWN CARBON STEEL BARS—SAE J414a

SAE Information Report

Report of Iron and Steel Technical Committee approved January 1950 and completely revised August 1977.

Mechanical Properties—The mechanical properties listed in Tables 1, 2, and 3 are given as a matter of general information. They do not form a part or requirement of any specification unless each instance is approved by the source of supply. The properties in Tables 1, 2, and 3 can generally be expected from bars in sizes ranging from 3/4–1 1/4 in (20–30 mm) based on the standard round tensile test specimen with 2-in (50-mm) gage length.

Sizes under 3/4 in (20 mm) will show slightly higher strength than those shown in the Tables. The mass effect of larger sections has a direct influence on mechanical properties and results in slightly lower values as the section increases.

Properties of turned and polished or turned and ground types of cold finished material will correspond to the hot rolled values.

The cold drawn properties are based on conventional production from hot rolled bars. When required, these properties may be varied by modified cold drawing practices or a combination of cold drawing practice plus thermal treatment for grades SAE 1050 and lower in carbon. Grades higher in carbon than SAE 1050 are commonly annealed before cold drawing.

Machinability Ratings—The machinability ratings listed are based on a value of 100% for SAE 1212 cold drawn. This value involves turning at a cutting speed of 180 surface feet (55 m) per min for feeds up to 0.007 in (0.18 mm) per revolution and depths of cut up to 0.250 in (6.4 mm) using appropriate cutting fluids with high speed steel tools, SAE Grade T-1 (18-4-1) hardened to 63/65 RC.

Relative machinability data shown in the Tables represent results obtained from various experimental data and actual shop production information obtained from results of machining cold drawn bars on single and multiple spindle automatic machines. Various factors influence machinability and, therefore, results shown in Tables are average and may be affected to some degree by amount of cold reduction, mechanical properties, grain size, and microstructure.

TABLE 1—ESTIMATED MECHANICAL PROPERTIES AND MACHINABILITY RATINGS OF NONRESULFURIZED CARBON STEEL BARS, MANGANESE 1.00% MAXIMUM

UNS No.	SAE and/or AISI No.	Type of Processing	Tensile Strength, psi	MPa	Estimated Minimum Values Yield Strength, psi	MPa	Elongation in 2 in, %	Reduction in Area, %	Brinell Hardness	Average Machinability Rating (Cold Drawn 1212 = 100%)
G10060	1006	Hot Rolled	43 000	300	24 000	170	30	55	86	
		Cold Drawn	48 000	330	41 000	280	20	45	95	50
G10080	1008	Hot Rolled	44 000	303	24 500	170	30	55	86	
		Cold Drawn	49 000	340	41 500	290	20	45	95	55
G10100	1010	Hot Rolled	47 000	320	26 000	180	28	50	95	
		Cold Drawn	53 000	370	44 000	300	20	40	105	55
G10120	1012	Hot Rolled	48 000	330	26 500	180	28	50	95	
		Cold Drawn	54 000	370	45 000	310	19	40	105	55
G10150	1015	Hot Rolled	50 000	340	27 500	190	28	50	101	
		Cold Drawn	56 000	390	47 000	320	18	40	111	60
G10160	1016	Hot Rolled	55 000	380	30 000	210	25	50	111	
		Cold Drawn	61 000	420	51 000	350	18	40	121	70
G10170	1017	Hot Rolled	53 000	370	29 000	200	26	50	105	
		Cold Drawn	59 000	410	49 000	340	18	40	116	65
G10180	1018	Hot Rolled	58 000	400	32 000	220	25	50	116	
		Cold Drawn	64 000	440	54 000	370	15	40	126	70
G10190	1019	Hot Rolled	59 000	410	32 500	220	25	50	116	
		Cold Drawn	66 000	460	55 000	380	15	40	131	70
G10200	1020	Hot Rolled	55 000	380	30 000	210	25	50	111	
		Cold Drawn	61 000	420	51 000	350	15	40	121	65
G10210	1021	Hot Rolled	61 000	420	33 000	230	24	48	116	
		Cold Drawn	68 000	470	57 000	390	15	40	131	70
G10220	1022	Hot Rolled	62 000	430	34 000	230	23	47	151	
		Cold Drawn	69 000	480	58 000	400	15	40	137	70
G10230	1023	Hot Rolled	56 000	370	31 000	210	25	50	111	
		Cold Drawn	62 000	430	52 500	360	15	40	121	65
G10250	1025	Hot Rolled	58 000	400	32 000	220	25	50	116	
		Cold Drawn	64 000	440	54 000	370	15	40	126	65
G10260	1026	Hot Rolled	64 000	440	35 000	240	24	49	126	
		Cold Drawn	71 000	490	60 000	410	15	40	143	75
G10300	1030	Hot Rolled	68 000	470	37 500	260	20	42	137	
		Cold Drawn	76 000	520	64 000	440	12	35	149	70
G10350	1035	Hot Rolled	72 000	500	39 500	270	18	40	143	
		Cold Drawn	80 000	550	67 000	460	12	35	163	65
G10370	1037	Hot Rolled	74 000	510	40 500	280	18	40	143	
		Cold Drawn	82 000	570	69 000	480	12	35	167	65
G10380	1038	Hot Rolled	75 000	520	41 000	280	18	40	149	
		Cold Drawn	83 000	570	70 000	480	12	35	163	65
G10390	1039	Hot Rolled	79 000	540	43 500	300	16	40	156	
		Cold Drawn	88 000	610	74 000	510	12	35	179	60
G10400	1040	Hot Rolled	76 000	520	42 000	290	18	40	149	
		Cold Drawn	85 000	590	71 000	490	12	35	170	60
G10420	1042	Hot Rolled	80 000	550	44 000	300	16	40	163	
		Cold Drawn	89 000	610	75 000	520	12	35	179	60
		NCD[b]	85 000	590	73 000	500	12	45	179	70
G10430	1043	Hot Rolled	82 000	570	45 000	310	16	40	163	
		Cold Drawn	91 000	630	77 000	530	12	35	179	60
		NCD[b]	87 000	600	75 000	520	12	45	179	70
G10440	1044	Hot Rolled	80 000	550	44 000	300	16	40	163	
G10450	1045	Hot Rolled	82 000	570	45 000	310	16	40	163	
		Cold Drawn	91 000	630	77 000	530	12	35	179	55
		ACD[a]	85 000	590	73 000	500	12	45	170	65
G10460	1046	Hot Rolled	85 000	590	47 000	320	15	40	170	
		Cold Drawn	94 000	650	79 000	540	12	35	187	55
		ACD[a]	90 000	620	75 000	520	12	45	179	65
G10490	1049	Hot Rolled	87 000	600	48 000	330	15	35	179	
		Cold Drawn	97 000	670	81 500	560	10	30	197	45
		ACD[a]	92 000	630	77 000	530	10	40	187	55
G10500	1050	Hot Rolled	90 000	620	49 500	340	15	35	179	
		Cold Drawn	100 000	690	84 000	580	10	30	197	45
		ACD[a]	95 000	660	80 000	550	10	40	189	55
G10550	1055	Hot Rolled	94 000	650	51 500	360	12	30	192	
		ACD[a]	96 000	660	81 000	560	10	40	197	55

[a] ACD represents annealed cold drawn.
[b] NCD represents normalized cold drawn.
[c] SACD represents spheroidized annealed cold drawn.

TABLE 1—ESTIMATED MECHANICAL PROPERTIES AND MACHINABILITY RATINGS OF NONRESULFURIZED CARBON STEEL BARS, MANGANESE 1.00% MAXIMUM (continued)

UNS No.	SAE and/or AISI No.	Type of Processing	Tensile Strength, psi	Tensile Strength, MPa	Estimated Minimum Values Yield Strength, psi	Yield Strength, MPa	Elongation in 2 in, %	Reduction in Area, %	Brinell Hardness	Average Machinability Rating (Cold Drawn 1212 = 100%)
G10600	1060	Hot Rolled	98 000	680	54 000	370	12	30	201	
		SACD[c]	90 000	620	70 000	480	10	45	183	60
G10640	1064	Hot Rolled	97 000	670	53 500	370	12	30	201	
		SACD[c]	89 000	610	69 000	480	10	45	183	60
G10650	1065	Hot Rolled	100 000	690	55 000	380	12	30	207	
		SACD[c]	92 000	630	71 000	490	10	45	187	60
G10700	1070	Hot Rolled	102 000	700	56 000	390	12	30	212	
		SACD[c]	93 000	640	72 000	500	10	45	192	55
G10740	1074	Hot Rolled	105 000	720	58 000	400	12	30	217	
		SACD[c]	94 500	650	73 000	500	10	40	192	55
G10780	1078	Hot Rolled	100 000	690	55 000	380	12	30	207	
		SACD[c]	94 000	650	72 500	500	10	40	192	55
G10800	1080	Hot Rolled	112 000	770	61 500	420	10	25	229	
		SACD[c]	98 000	680	75 000	520	10	40	192	45
G10840	1084	Hot Rolled	119 000	820	65 500	450	10	25	241	
		SACD[c]	100 000	690	77 000	530	10	40	192	45
G10850	1085	Hot Rolled	121 000	830	66 500	460	10	25	248	
		SACD[c]	100 500	690	78 000	540	10	40	192	45
G10860	1086	Hot Rolled	112 000	770	61 500	420	10	25	229	
		SACD[c]	97 000	670	74 000	510	10	40	192	45
G10900	1090	Hot Rolled	122 000	840	67 000	460	10	25	248	
		SACD[c]	101 000	700	78 000	540	10	40	197	45
G10950	1095	Hot Rolled	120 000	830	66 000	460	10	25	248	
		SACD[c]	99 000	680	76 000	520	10	40	197	45

[a] ACD represents annealed cold drawn.
[b] NCD represents normalized cold drawn.
[c] SACD represents spheroidized annealed cold drawn.

TABLE 2—ESTIMATED MECHANICAL PROPERTIES AND MACHINABILITY RATINGS OF NONRESULFURIZED CARBON STEEL BARS, MANGANESE MAXIMUM OVER 1.00%

UNS No.	SAE and/or AISI No.	Type of Processing	Tensile Strength, psi	Tensile Strength, MPa	Estimated Minimum Values Yield Strength, psi	Yield Strength, MPA	Elongation in 2 in, %	Reduction in Area, %	Brinell Hardness	Average Machinability Rating (Cold Drawn 1212 = 100%)
G15240	1524	Hot Rolled	74 000	510	41 000	280	20	42	149	
		Cold Drawn	82 000	570	69 000	480	12	35	163	60
G15270	1527	Hot Rolled	75 000	520	41 000	280	18	40	149	
		Cold Drawn	83 000	570	70 000	480	12	35	163	65
G15360	1536	Hot Rolled	83 000	570	45 500	310	16	40	163	
		Cold Drawn	92 000	630	77 500	530	12	35	187	55
G15410	1541	Hot Rolled	92 000	630	51 000	350	15	40	187	
		Cold Drawn	102 500	710	87 000	600	10	30	207	45
		ACD[a]	94 000	650	80 000	550	10	45	184	60
G15470	1547	Hot Rolled	94 000	650	52 000	360	15	30	192	
		Cold Drawn	103 000	710	88 000	610	10	28	207	40
		ACD[a]	95 000	660	85 000	590	10	35	187	45
G15520	1552	Hot Rolled	108 000	740	59 500	410	12	30	217	
		ACD[a]	98 000	680	83 000	570	10	40	193	50
G15480	1548	Hot Rolled	96 000	660	53 000	370	14	33	197	
		Cold Drawn	106 500	730	89 500	620	10	28	217	45
		ACD[a]	93 500	640	78 500	540	10	35	192	50

[a] ACD represents annealed cold drawn.

TABLE 3—ESTIMATED MECHANICAL PROPERTIES AND MACHINABILITY RATINGS OF RESULFURIZED CARBON STEEL BARS[a]

UNS No.	SAE and/or AISI No.	Type of Processing	Tensile Strength, psi	Tensile Strength, MPa	Yield Strength, psi	Yield Strength, MPa	Estimated Minimum Values Elongation in 2 in, %	Reduction in Area, %	Brinell Hardness	Average Machinability Rating (Cold Drawn 1212 = 100%)
G11080	08	Hot Rolled	50 000	340	27 500	190	30	50	101	
		Cold Drawn	56 000	390	47 000	320	20	40	121	80
G11090	9	Hot Rolled	50 000	340	27 500	190	30	50	101	
		Cold Drawn	56 000	390	47 000	320	20	40	121	80
G11170		Hot Rolled	62 000	430	34 000	230	23	47	121	
		Cold Drawn	69 000	480	58 000	400	15	40	137	90
G11180		Hot Rolled	65 000	450	36 000	250	23	47	131	
		Cold Drawn	72 000	500	61 000	420	15	40	143	85
G11190	9	Hot Rolled	62 000	430	34 000	230	23	47	121	
		Cold Drawn	69 000	480	58 000	400	15	40	137	100
G11320	11	Hot Rolled	83 000	570	45 500	310	16	40	167	
		Cold Drawn	92 000	630	77 000	530	12	35	183	75
G11370	1137	Hot Rolled	88 000	610	48 000	330	15	35	179	
		Cold Drawn	98 000	680	82 000	570	10	30	197	70
G11400	1140	Hot Rolled	79 000	540	43 500	300	16	40	156	
		Cold Drawn	88 000	610	74 000	510	12	35	170	70
G11410	1141	Hot Rolled	94 000	650	51 500	360	15	35	187	
		Cold Drawn	105 100	720	88 000	610	10	30	212	70
G11440	1144	Hot Rolled	97 000	670	53 000	370	15	35	197	
		Cold Drawn	108 000	740	90 000	620	10	30	217	80
G11450	1145	Hot Rolled	85 000	590	47 000	320	15	40	170	
		Cold Drawn	94 000	650	80 000	550	12	35	187	65
G11460	1146	Hot Rolled	85 000	590	47 000	320	15	40	170	
		Cold Drawn	94 000	650	80 000	550	12	35	187	70
G11510	1151	Hot Rolled	92 000	630	50 500	340	15	35	187	
		Cold Drawn	102 000	700	86 000	590	10	30	207	65
G12110	1211	Hot Rolled	55 000	380	33 000	230	25	45	121	
		Cold Drawn	75 000	520	58 000	400	10	35	163	95
G12120	1212	Hot Rolled	56 000	390	33 500	230	25	45	121	
		Cold Drawn	78 000	540	60 000	410	10	35	167	100
G12130	1213	Hot Rolled	56 000	390	33 500	230	25	45	121	
		Cold Drawn	78 000	540	60 000	410	10	35	167	135
G12144	12L14	Hot Rolled	57 000	390	34 000	230	22	45	121	
		Cold Drawn	78 000	540	60 000	410	10	35	163	160

[a] All 1100 and 1200 series steels are rated on basis of 0.10% max. silicon or coarse grain melting practice.

GLOSSARY OF CARBON STEEL SHEET AND STRIP TERMS—SAE J940

SAE Information Report

Report of Iron and Steel Technical Committee approved November 1965. Reaffirmed without change March 1973.

Scope—This glossary is intended to provide engineers, metallurgists, and production personnel with uniform definitions of commonly used carbon sheet and strip terms. The glossary serves to supplement information and photographs reported in SAE J810, J763, J877, J863, and J403.

Many of the terms listed apply only to hot-dipped zinc-coated products or to uncoated products. The letter C following the term identifies a term applying to coated materials, while the letters NC identify a term applying to uncoated materials. Where no identification is provided, the term is common to both.

Glossary—

Aging—A term applied to changes in physical and mechanical properties of low carbon steel that occur with the passage of time. See SAE J763.

Annealing Border (NC)—See Oxidized Surface. See also SAE J810.

Annealing Stain (NC)—A discoloration on annealed material which may occur anywhere on the sheet (usually lighter than annealing border). It results from residue, or oxidation, during annealing.

Arc (C)—A narrow curved pencil-like line in the coating running transversely approximately 2 in. from each edge of the strip.

Band Mark—An indentation caused by the packaging band resulting from external pressure on cut lengths or coils and may occur in handling, transit, and storing.

Beads (C)—Small lumps in the coating surface. Particles of dross picked up in the coating or iron oxide particles imbedded in the strip surface from the furnace hearth rolls.

Black Spots (NC)—Carbonaceous deposits caused by tightly wound areas in a coil not being exposed to the circulating gases during open coil type annealing. This condition is aggravated by poor strip shape.

Blank—A flat piece of sheet steel produced in blanking dies or by shearing for an identified part. The blank is usually subjected to further press operations.

Blanking—Cutting desired shapes out of flat sheet.

Blister—A small raised area on the surface resulting from the expansion of gas concentrated at a subsurface inclusion. May occur as isolated spots, but often found in longitudinal streaks.

Box Annealing—The term given to the process of softening steel by heating it in a closed container through which, in most cases, a controlled atmosphere is circulated to prevent oxidation during the heating and cooling cycle.

Breaks—Creases or ridges usually in "untempered" or in aged material where the yield point has been exceeded. Depending on the origin of the break, it might be termed a crossbreak, a coil break, an edge break, a sticker break, etc. See SAE J810.

Bright Annealing (NC)—Annealing in a protective atmosphere to prevent discoloration of the surface.

Bright Finish (NC)—A high quality finish produced on ground and polished rolls. Suitable for electroplating.

Buckles—A series of waves in sheets which are ordinarily transverse to the direction of rolling; see SAE J810. In formed panels, "excess metal" in the form of wrinkles, kinks, or folds.

Buildup (C)—Localized lineal areas showing a difference in cross-sectional contour during coiling. Usually occurs on the edges of the strip.

Camber—Deviation from a straight edge, usually referring to the greatest deviation of the concave side edge of a sheet or strip from a straight line.

Capped Steel—This is a type of steel with characteristics similar to those of rimmed steels, but to a degree intermediate between those of rimmed and semikilled steels. It can be either mechanically capped or chemically capped when the ingot is cast, but in either case the full rimming action is stopped, resulting in a more uniform composition than rimmed steel.

Carbon Edge (NC)—Carbonaceous deposits in a wavy pattern along the edges of the sheet or coil. (See also Snaky Edges).

Chatter—A series of lines uniformly spaced appearing transverse to the rolling direction usually resulting from material being rolled on units having loose bearings. Results in a slight thickness variation where lines appear.

Checked Edges—Sawtooth edges seen after hot rolling and/or cold rolling. See also Ragged Edge in SAE J810 and Sawtooth Edge in this report.

Coil Breaks—Creases or ridges which appear as parallel lines across the direction of rolling and which generally extend across the width of the sheet. See SAE J810.

Coil Weld—A joint between two lengths of metal within a coil. See J810.

Cold Rolled Sheets (NC)—A product produced from a hot rolled pickled coil which has been given substantial cold reduction at room temperature. The resulting product usually requires further processing to make it suitable for most common applications. The usual end product is characterized by improved surface, greater uniformity in thickness and improved mechanical properties compared to hot rolled sheet.

Corrugations—Transverse ripples caused by a variation in strip shape during hot or cold reduction.

Cross Breaks—See Coil Breaks.

Curtains (C)—An uneven pattern of the coating resulting from runback of the applied material.

Cut Edge—A mechanically sheared edge obtained by slitting, shearing, or blanking.

Deep Drawing—An extreme condition of drawing. The term "deep drawing" is commonly used to describe metal stamping operations which are a combination of drawing and severe stretching.

Developed Blank—See Finished Blank.

Differential Coating (C)—A coated product having a specified coating on one surface and a significantly lighter coating on the other surface, usually hot-dipped galvanized.

Dimple—A defect resulting from foreign matter being mechanically impressed into the sheet surface.

Dings—Accidental impact damage, similar in appearance to dimples.

Drawability—The ability of the sheet steel to be formed or drawn into the intended end product without fracturing.

Drawing—The shaping of a flat blank into the desired contour by causing the metal to flow over a draw ring and round a punch. The flow of metal is restrained by sufficient blank holder pressure to prevent buckling.

Drawing Compound / Drawing Lubricant—A substance applied to minimize metal to metal contact between the sheet metal and the die. Proper application of the proper lubricant can improve flow characteristics of the metal and prevent scoring, galling, and pickup on the dies or part.

Ductility—The ability of a metal to be deformed plastically without fracturing. See Formability.

"E" Finish—This designation indicates that the material is to be used for an exposed part requiring a good painted surface. See SAE J403.

Earing (Scalloping)—The formation of scallops (ears or marked unevenness) around the top edge of a drawn cup caused by differences in the directional properties of the sheet metal used.

Edge Break (Side Strain)—See Edge Strain. See also SAE J810.

Edge Strain—Transverse strain lines of Lüder's Lines ranging from 1 to 12 in. in from the edges of the sheet. See SAE J810.

Elastic Ratio—The yield point divided by the tensile strength.

Electrogalvanizing (C)—The electroplating of zinc upon iron or steel.

End Mark—A roll mark caused by the end of a sheet marking the roll during hot or cold rolling.

Entry Mark (Exit Mark)—A slight corrugation caused by the entry or exit rolls of a roller leveling unit.

Finished Blank (Developed Blank)—A blank that requires little or no trimming after being formed.

Flex Roll—The movable jump roll designed to push up against the sheets as it passes through the roller leveler. The roll can be adjusted to produce varying amounts of deflection of the sheet up to the diameter of the roll.

Flex Rolling—Passing sheets through a flex roll unit to minimize yield point elongation so as to reduce the tendency for stretcher strains to appear in forming.

Floppers—Lines or ridges which are transverse to the direction of rolling and generally confined to the section midway between the edges of the coil as rolled. They are somewhat irregular and tend toward a flat arc shape. See SAE J810.

Fluting—A series of sharp parallel kinks or creases occurring in the arc when sheet steel is rolled formed into a cylindrical shape. See SAE J810.

Formability—The degree to which a metal can be shaped through plastic deformation. See Ductility.

Forming—The shaping of sheet metal by bending, uniaxial stretching, biaxial stretching, compression, or by a combination thereof.

Friction Gouges or Scratches—A series of relatively short scratches variable in form or severity. See SAE J810.

Full Center—A "fullness" in the center portion of the sheet or strip.

Galling—Scratches caused by localized cold welding of particles to the tool during the forming operation.

Galvannealed Coating (C)—Galvannealed sheets are hot dipped zinc-coated sheets which have been processed to produce a zinc-iron alloy coating. This product does not have a spangle and is suitable for painting after cleaning. The alloy produced lacks ductility and powdering of the coating can occur during forming.

Ghost Lines (NC) (Ghost Welt Lines)—Lines running parallel to the rolling direction that appear in a panel when it is stretched. These lines may not be evident unless panel has been sanded or painted. (Not to be confused with leveler lines.)

Handling Breaks—Irregular breaks caused by improper handling of sheets during processing. These breaks result from the bending or sagging of the sheets while being handled.

Healed Over Scratch (NC)—A scratch that occurred in an earlier mill operation and was partially masked in subsequent rolling. It may open up during forming. See SAE J810.

Hot Rolled—Hot rolled sheets are those that are reduced to required thickness at temperatures at which scale is formed and, therefore, carry hot mill oxide.

Hot Rolled, Pickled—The hot rolled product which has been pickled to remove the hot mill oxide.

Impact or Recoil Line—The line on a drawn panel where a change in thickness occurs. Caused by: (1) the transfer of the panel from the die to the punch; (2) the reaction from the blank being pulled sharply through the draw ring (recoil); (3) the impact of the punch contacting the blank.

Inclusions—Nonmetallic materials in a solid metallic matrix.

Ironing—Thinning the walls of deep drawn articles by reducing the clearance between punch and die.

Killed Steel—Steel deoxidized with certain deoxidizing elements, such as aluminum, silicon, etc. The term "killed" is used because such additions cause the steel to lie quietly in the molds during solidification. See SAE J877.

Laminations—Defects aligned parallel to the worked surface of the sheet resulting from the presence of inclusions. See SAE J810.

Loose Metal—Refers to an area in a formed panel that is not stiff enough to hold its shape, may be confused with Oil Canning.

Lüder's Lines—See Stretcher Strains.

Luster Finish (NC)—A finish produced on ground rolls suitable for decorative painting and plating with additional surface preparation after forming.

Matte Finish—The texture produced on sheets by rolls which have been blasted to various degrees of roughness depending upon the end use.

Mechanical Properties—The properties of a material that reveal its elastic and plastic behavior when force is applied, for example, yield strength, ultimate strength, elongation, hardness, etc.

Mill Edge—The normal edge produced in hot rolling. This edge is customarily removed when hot rolled sheets are further processed into cold rolled sheets.

Minimized Spangle (C)—Minimized spangle galvanized sheet has very small spangles which are obtained by treating the galvanized sheet during the solidification of the zinc to restrict the normal zinc spangle formation.

Necking—Reducing the thickness of a sheet in a localized area by stretching.

Normalizing—Heating steel to a suitable temperature above the transformation range and then cooling in air to a temperature substantially below the transformation range. See SAE J415.

Offal—The material trimmed from blanks or formed panels.

Oil Can (Oil Canning)—Refers to an area in a formed panel that when depressed slightly will recover its original contour after the depressing force is removed.

Orange Peel—A course textured or pebbly surface condition which becomes evident during forming. See SAE J810.

Oxide Border (NC)—See Oxidized Surface.

Oxidized Surface (NC)—Surface having a thin, tightly adhering, (discolored from straw to blue) oxidized skin extending in from the edge of the coil or sheet. Sometimes called "Annealing Border." See SAE J810.

Physical Properties—The properties other than mechanical properties, that pertain to the physics of a material; for example, density, electrical conductivity, thermal expansion, etc. Often improperly used to express mechanical properties.

Pickle Patch (NC)—A tightly adhering oxide or scale not removed during the pickling process. See SAE J810.

Pickle Stain (NC)—Discoloration present after pickling.

Pickling—The removal of surface oxides from sheets by chemical or electrochemical reaction.

Pickup—Metal particles adhering to a work roll or tool which cause a series of dents, scratches, or pits on a sheet or part.

Pinchers—Fernlike ripples or creases usually diagonal to the rolling direction. See SAE J810.

Pipe—In sheets it appears as a separation midway between the surfaces containing oxide inclusions. See SAE J810.

Pits—Small cavities in the surface of the sheet. See SAE J810.

Reel Breaks (Reel Kinks)—Transverse breaks or ridges on successive inner laps of a coil which are the result of crimping the lead end of the coil into a gripping segmented mandrel.

Ridge—A longitudinal line where the thickness of the metal is slightly greater than the thickness adjacent.

Rimmed Steel—A type of steel characterized by a gaseous effervescence when cooling in the mold. This results in a relatively pure iron outer rim. Rimmed steel is subject to aging. See SAE J877.

Roller Leveler—A series of small diameter staggered rolls used primarily to improve flatness and/or to remove yield point elongation.

Roller Leveler Breaks—Obvious transverse breaks usually $\frac{1}{8}$ to $\frac{1}{4}$ in. apart caused by the sheet fluting during roller leveling. These will not be removed by stretching.

Roller Leveler Lines—Lines running transverse to the direction of leveling. These may be seen upon stoning or light sanding after leveling and before drawing. Moderate stretching will usually remove them.

Rosebuds (C)—Noted only on minimized spangle. Concentric rings of distorted coating, giving the effect of an opened rosebud.

Rough Developed Blank—A blank that will require trimming after being formed.

Saw Tooth Edge—See Checked Edges.

Scabs—Elongated patches of loosened metal which have been rolled into the surface of the sheet or strip.

Scale—Oxides of iron which form on the surface of hot steel.

Scoring—Marring or scratching of a formed part by metal pickup on the punch or die. Also see Galling.

Scratches—Lines caused by the abrasion of one surface against another during rolling, processing, or shipping.

Scribed Square Test—A method to determine the percent increase in unit area of selected regions of a formed panel. See SAE J863.

Seam Lines (C)—A continuous line of small beads.

Seams—Open, broken surface running in straight longitudinal lines caused by the presence of oxides near the surface of the sheet.

Segregation—The variation in chemical composition resulting from natural phenomena in the solidification of a steel ingot. The various elements of the steel having lowest freezing points are concentrated in parts of the ingot last to solidify.

Semikilled Steel—Steel that is partially deoxidized so that there is greater degree of gas evolution than in killed steel, but less than in capped or rimmed steel. The uniformity in composition lies between that of killed steel and rimmed steel.

Side Strain—See Edge Strain.

Skin Lamination—Subsurface separation which usually results in surface rupture. See SAE J810.

Skin Pass—See Temper Rolling.

Slivers—Surface ruptures somewhat similar in appearance to skin laminations, but usually more prominent. See SAE J810.

Smudge (NC)—A dark residue on the surface of sheet steel. See Smut.

Smut (NC)—A reaction product sometimes left on the surface of the sheet after pickling or annealing. See Smudge.

Snaky Edges (NC)—Carbonaceous deposits in a wavy pattern along the edges of the annealed strip. See Carbon Edge.

Spangle (C)—The characteristic crystalline form in which the hot dipped zinc coating solidifies on steel strip.

Spinning—The shaping of flat cicular blanks by forcing the blank against a chuck or form block while it is rotating.

Springback—The tendency of metal to partially return to its original shape after cold forming.

Sticker Breaks (NC)—Arc-shaped breaks usually located near the middle of the sheet. See SAE J810.

Stiffness—The ability of a metal or shape to resist deflection.

Strain Hardening—An increase in hardness and strength caused by plastic deformation at temperatures lower than the recrystallization range. See SAE J877.

Strain Hardening Exponent—A measure of the rate of strain hardening. The constant 'n' in the expression.

$$\delta = \delta_o \, \delta^n$$

where:
δ = True stress
δ_o = True stress at unit strain
δ^n = True strain

See SAE J877.

Strain Ratio—This is expressed as 'r' value. It is the ratio of width to thickness strain determined in uniform elongation portion of a tension test. It is a good measure of the crystallographic directionality of the material. See SAE J877.

Stretchability—The ability of a metal to be stretched over a punch without splitting.

Stretch Forming—Shaping of a sheet or part, usually of uniform cross section by applying suitable tension or stretch and forming it around or over a die of the desired shape.

Stretching—The operation where the blank is stretched around the punch with no metal flow over the draw ring. The metal thickness is reduced.

Stretcher Leveling—Leveling where a piece of metal is gripped at each end and subjected to a stress higher than its yield strength to obtain a high degree of flatness. Sometimes called patent leveling.

Stretcher Strain (Lüder's Lines)—Irregular surface patterns of ridges and valleys which develop during forming of annealed last or temper rolled, aged steel. See SAE J810.

Surface Texture—The finish of the surface of sheet steel presently described by the roughness (peak) height in micro inches and the peaks per inch. (See Matte Finish and SAE J448).

Synthetic Cold Rolled (NC)—A hot rolled pickled sheet given a sufficient final temper pass to impart a surface approximating that of cold rolled steel.

Temper Rolling—Light cold rolling of sheet steel. This operation is performed to improve flatness, minimize the tendency to stretcher strain and flute, obtain the desired texture and mechanical properties.

Tensile Strength—The unit stress at the highest load reached during the tension test. See SAE J877.

Tiger Stripes (C)—Continuous bright lines in the rolling direction.

Total Elongation—Percent elongation measured after fracture in a tension test. See SAE J877.

Traverse Lines—Lines closely spaced across the full width of the sheet and running in the direction of rolling.

"U" Finish—This designation indicates that the material is to be used for an unexposed part for which surface finish is unimportant. See SAE J403.

Uniform Elongation (Uniform Strain E_u)—The percent elongation at the onset of necking, usually taken as the strain to maximum load in the tension test.

Wiped Coat (C)—A hot dipped galvanized coating where virtually all the free zinc is removed by wiping prior to solidification leaving only a thin zinc iron alloy layer.

Work Hardening—Same as Strain Hardening.

Wrinkling—Small buckles which occur in drawing sheet metal as it passes over the drawing ring radius.

Yield Point—The stress beyond which the metal is permanently deformed. See SAE J877 and J450.

Yield Point Elongation—Percent elongation at the end of nonhomogeneous yielding in a tension test.

AGING OF CARBON STEEL SHEET AND STRIP—SAE J763

SAE Information Report

Report of Iron and Steel Technical Committee approved May 1961. Reaffirmed without change March 1973.

Scope and Purpose—This report briefly covers the aging of hot rolled, cold rolled, and coated carbon steel sheet and strip. Its purpose is to provide general information concerning the phenomenon of aging so that associated problems may be recognized.

Terminology

Aging—A term applied to changes in physical and mechanical properties of low carbon steel that occur with the passage of time. These changes adversely affect formability. Aging accelerates as the temperature is raised.

Strain Aging—A term applied to changes in properties which occur after cold working. Rimmed steel is particularly susceptible to strain aging.

Quench Aging—A term applied to those changes in properties which occur after rapid cooling of the product.

NOTE: Rapid cooling as a process is normally performed by the producer and is not a factor for consideration by the fabricator; therefore, this explanation is for reference purposes only and quench aging is not further considered in this report.

Temper Rolling or Skin Passing denotes cold rolling of sheets to improve flatness, impart a desired surface finish or reduce the tendency to stretcher strain and flute. Temper rolling should not be confused with the term "cold reduction", which involves a substantial reduction in thickness.

Effects—Among the easily recognizable effects resulting from strain aging of temper rolled carbon steel sheet and strip are the following:

1. **Surface Irregularities**—Strain aging is manifested during forming by visible surface irregularities commonly known as stretcher strains (Lüder's Lines) or fluting. These may render the material unsuitable for use on exposed or external parts.
2. **Ductility Reduction**
3. **Hardness Increase**
4. **Tensile Test Properties Changes:**
   (a) Increase in yield point
   (b) Increase in yield point elongation
   (c) Decrease in total elongation

Guides—The adverse effects of strain aging may be controlled by:
1. The use of special killed or stabilized steels.
2. The use of material which has not been temper passed (for unexposed parts only).
3. Effective roller leveling of temper passed steel immediately prior to fabrication. This will minimize the tendency of the sheets to stretcher strain and to flute but such leveling will not restore softness and ductility. In fact roller leveling further work hardens the sheet and hence further reduces ductility.
4. Rotation of stock by application in order of receipt (inventory control by "first-in, first-out").

BIBLIOGRAPHY

It is recognized that this report cannot cover all of the aspects of aging of carbon steel sheet and strip. For those desiring additional information, the following basic bibliography is provided:

1. Kenyon, R. L. and Burns, R. S., "Aging in Iron and Steel," ASM PREPRINT 40, 1939 Meeting Iron and Steel, 13, April 1940, p. 227; May 1940, p. 260.
2. Epstein, A., "Aging of Iron and Steel," METALS HANDBOOK, 1948, p. 438.
3. Shoenberger, L. R. and Paliwoda, E. J., "Accelerated Strain Aging of Commercial Sheet Steels," TRANSACTIONS ASM, 45, 1953, p. 344.
4. Leslie, W. C. and Rickett, R. L., "Influence of Aluminum and Silicon Deoxidation on the Strain Aging of Low-Carbon Steels," TRANSACTIONS AIME, 197, 1953, p. 1021.
5. Morgan, E. R. and Shyne, J. C., "The Strain Aging of Boron-Treated Low-Carbon Steels," JOURNAL IRON AND STEEL INSTITUTE, 185, 1957, p. 156.
6. Morgan, Eric R., "Can an Improved Non-Aging Steel Be Produced Commercially," METAL PROGRESS, 73, June 1958, p. 88.
7. "Strain Aging," METALS HANDBOOK, 1961, pp. 81, 327.
8. "Stretcher Strains," METALS HANDBOOK, 1961, p. 326.

DEFINITIONS OF HEAT TREATING TERMS—SAE J415j

SAE Information Report

Report of Iron and Steel Division approved June 1911 and last revised by Iron and Steel Technical Committee Division 12 March 1977.

(These definitions were prepared by the Joint Committee on Definitions of Terms Relating to Heat Treatment appointed by the American Society for Testing and Materials, The American Society for Metals, the American Foundrymen's Association, and the SAE.)

This glossary is not intended to be a specification, and it should not be interpreted as such. Since this is intended to be strictly a set of definitions, temperatures have been omitted purposely.

Ac_{cm}, Ac_1, Ac_3, Ac_4—Defined under Transformation Temperature.
Ae_{cm}, Ae_1, Ae_3, Ae_4—Defined under Transformation Temperature.

Age Hardening—Hardening by aging, usually after rapid cooling or cold working. See Aging.

Age Softening—Aluminum Alloys—Spontaneous decrease of strength and hardness which takes place at room temperature in certain strain hardened alloys.

Aging—A change in the properties of certain metals and alloys that occurs at ambient or moderately elevated temperatures after hot working or a heat treatment (quench aging in ferrous alloys, natural or artificial aging in ferrous and nonferrous alloys) or after a cold working operation (strain aging). The change in properties is often, but not always, due to a phase change (precipitation), but never involves a change in chemical composition of the metal or alloys. See also Age Hardening, Artificial Aging, Natural Aging, Overaging, Precipitation Hardening, Precipitation Heat Treatment, Progressive Aging, Quench Aging, Strain Aging.

Anneal to Temper—Copper and Copper Alloys—A final anneal used to produce specified mechanical properties in a material.

Annealing—A generic term denoting a treatment, consisting of heating to and holding at a suitable temperature followed by cooling at a suitable rate, used primarily to soften metallic materials, but also to simultaneously produce desired changes in other properties or in microstructure. The purpose of such changes may be, but is not confined to, one or more of: Improvement of machinability; facilitation of cold work; improvement of mechanical or electrical properties or increase in stability of dimensions.

Annealing—Aluminum and Aluminum Alloys—Annealing cycles are designed to: (1) remove part or all of the effects of cold working. (Recrystallization may or may not be involved); (2) cause substantially complete coalescence of precipitates from solid solution in relatively coarse form; or (3) both, depending on the composition and condition of the material. When the term is used without qualification, full annealing is implied. Specific process names in commercial use are: Final Annealing, Full Annealing, Intermediate Annealing, Partial Annealing, Recrystallization Annealing, Stress Relief Annealing.

Annealing—Ferrous—The time-temperature cycles used vary widely in both maximum temperature attained and in cooling rate employed, depending on the composition of the material, its condition and the results desired. When applicable, the following more specific commercial process names should be used: Black Annealing, Blue Annealing, Box Annealing, Bright Annealing, Cycle Annealing, Flame Annealing, Full Annealing, Graphitizing, In-Process Annealing, Isothermal Annealing, Malleablizing, Orientation Annealing, Process Annealing, Quench Annealing, Spheroidizing. When the term is used without qualification full annealing is implied. When applied only for the relief of stress, the process is properly called stress relieving.

Artificial Aging—Aging above room temperature. See Aging and Precipitation Heat Treatment. Compare with Natural Aging.

Austempering—Quenching a ferrous alloy from a temperature above the transformation range, in a medium having a rate of heat abstraction high enough to prevent the formation of high temperature transformation products, and then holding the alloy, until transformation is complete, at a temperature below that of pearlite formation and above that of martensite formation.

Austenitizing—Forming austenite by heating a ferrous alloy into the transformation range (partial austenitizing) or above the transformation range (complete austenitizing). When used without qualification, the term implies complete austenitizing.

Baking—Heating to a low temperature in order to remove gases.

Betatizing—Forming beta constituent by heating a non-ferrous alloy into the temperature region in which the constituent forms.

Black Annealing—Box annealing or pot annealing ferrous alloy sheet, strip or wire. See Box Annealing.

Blank Carburizing—Simulating the carburizing operation without introducing carbon. This is usually accomplished by using an inert material in place of the carburizing agent, or by applying a suitable protective coating to the ferrous alloy.

Blank Nitriding—Simulating the nitriding operation without introducing nitrogen. This is usually accomplished by using an inert material in place of the nitriding agent, or by applying a suitable protective coating to the ferrous alloy.

Blue Annealing—Heating hot rolled ferrous sheet in an open furnace to a temperature within the transformation range and then cooling in air, in order to soften the metal. The formation of a bluish oxide on the surface is incidental.

Bluing—Subjecting the scale free surface of a ferrous alloy to the action of air, steam, or other agents at a suitable temperature, thus forming a thin blue film of oxide and improving the appearance and resistance to corrosion.

NOTE: This term is ordinarily applied to sheet, strip, or finished parts. It is used also to denote the heating of springs after fabrication, in order to improve their properties.

Box Annealing—Annealing a metal or alloy in a sealed container under conditions that minimize oxidation. In box annealing a ferrous alloy, the charge is usually heated slowly to a temperature below the transformation range, but sometimes above or within it, and is then cooled slowly; this process is also called "close annealing" or "pot annealing." See Black Annealing.

Bright Annealing—Annealing in a protective medium to prevent discoloration of the bright surface.

Burning—Permanently damaging a metal or alloy by heating to cause either incipient melting or intergranular oxidation. See Overheating.

Carbon Potential—A measure of the ability of an environment containing active carbon to alter or maintain, under prescribed conditions, the carbon content of the steel exposed to it.

NOTE: In any particular environment, the carbon level attained will depend on such factors as temperature, time and steel composition.

Carbon Restoration—Replacing the carbon lost in the surface layer from previous processing by carburizing this layer to substantially the original carbon level.

Carbonitriding—A case hardening process in which a suitable ferrous material is heated above the lower transformation temperature in a gaseous atmosphere of such composition as to cause simultaneous absorption of carbon and nitrogen by the surface and, by diffusion, create a concentration gradient. The process is completed by cooling at a rate which produces the desired properties in the workpiece.

Carburizing—A process in which an austenitized ferrous material is brought into contact with a carbonaceous atmosphere of sufficient carbon potential to cause absorption of carbon at the surface and, by diffusion, create a concentration gradient.

Case—In a ferrous alloy, the outer portion that has been made harder than the inner portion or Core by Case Hardening.

Case Hardening—A generic term covering several processes applicable to steel that change the chemical composition of the surface layer by absorption of carbon, nitrogen, or a mixture of the two and, by diffusion, create a concentration gradient. The processes commonly used are: carburizing and quench hardening, cyaniding, nitriding, carbonitriding. The use of the applicable specific process name is preferred.

Cementation—The introduction of one or more elements into the outer portion of a metal object by means of diffusion at high temperature.

Close Annealing—See Box Annealing.

Cold Treatment—Exposing to suitable sub-zero temperatures for the purpose of obtaining desired conditions or properties, such as dimensional or microstructural stability. When the treatment involves the transformation of retained austenite, it is usually followed by a tempering treatment.

Conditioning Heat Treatment—A preliminary heat treatment used to prepare a material for a desired reaction to a subsequent heat treatment. For the term to be meaningful, the treatment used must be specified.

Controlled Cooling—Cooling from an elevated temperature in a predetermined manner, to avoid hardening, cracking, or internal damage, or to produce a desired microstructure or mechanical properties. The term applies to cooling following hot working.

Core—In a case hardened or surface hardened ferrous alloy, the inner portion that is softer than the outer portion or Case.

Critical Cooling Rate—The minimum rate of continuous cooling to prevent undesirable transformations. For steel it is the minimum rate at which austenite must be continuously cooled to suppress transformations above the M_s temperature.

Critical Temperature Range—Synonymous with Transformation range, which is preferred.

Cyaniding—A case hardening process in which a ferrous material is heated above the lower transformation range in a molten salt containing cyanide to cause simultaneous absorption of carbon and nitrogen at the surface and, by diffusion, create a concentration gradient. Quench hardening completes the process.

Cycle Annealing—An annealing process employing a predetermined and closely controlled time-temperature cycle to produce specific properties or microstructure.

Decarburization—The loss of carbon from the surface of a ferrous alloy as a result of heating in a medium that reacts with the carbon.

Differential Heating—Heating that intentionally produces a temperature gradient within an object such that, after cooling, a desired stress distribution or variation in properties is present within the object.

Diffusion Coating—Any process whereby a basis metal or alloy is either: (1) coated with another metal or alloy and heated to a sufficient temperature in a suitable environment or (2) exposed to a gaseous or liquid medium containing the other metal or alloy, thus causing diffusion of the coating or of the other metal or alloy into the basis metal with resultant change in the composition and properties of its surface.

Direct Quenching—Quenching carburized parts directly from the carburizing operation.

Double Aging—Employment of two different aging treatments to control the type of precipitate formed from a super-saturated alloy matrix in order to obtain the desired properties. The first aging treatment, sometimes referred to as intermediate or stabilizing, is usually carried out at a higher temperature than the second.

Double Tempering—A treatment in which quench hardened steel is given two complete tempering cycles at substantially the same temperature for the purpose of assuring completion of the tempering reaction and promoting stability of the resulting microstructure.

Drawing—A misnomer for Tempering.

ɸ ***Ductile Nitriding***—See nitriding.

Ferritizing Anneal—A treatment given as-cast gray or ductile (nodular) iron to produce an essentially ferritic matrix. For the term to be meaningful, the final microstructure desired or the time-temperature cycle used must be specified.

Final Annealing—Nonferrous—An imprecise term used to denote the anneal used to prepare a material for shipment to the user.

Flame Annealing—Annealing in which the heat is applied directly by a flame.

Flame Hardening—A surface hardening process in which only the surface layer of a suitable workpiece is heated by a suitably intense flame to above the upper transformation temperature and immediately quenched.

Fog Quenching—Quenching in a mist.

Full Annealing—Aluminum and Aluminum Alloys—An imprecise term used to denote the annealing cycle required to produce minimum strength. For the term to be meaningful, the composition and condition of the material and the time-temperature cycle used must be stated.

Full Annealing—Ferrous—Austenitizing and then cooling at such a rate that the hardness of the product approaches a minimum.

Gas Cyaniding—A misnomer for Carbonitriding.

Grain Growth—An increase in the average size of the grains (see Notes 1 and 2) in polycrystalline metal, usually as a result of heating at elevated temperature.

NOTES: (1) A grain is an individual crystal in a polycrystalline metal and includes twined regions and subgrains when present.

(2) Grain size is a measure of the mean diameter, area, or volume of all individual grains observed in a polycrystalline metal. In metals containing two or more phases, the grain size refers to that of the matrix unless otherwise specified. For further information on grain size and its measurement, see ASTM E 112, Methods for Estimating the Average Grain Size of Metals.

Graphitizing—Annealing a ferrous alloy in such a way that some or all of the carbon is precipitated as graphite.

Hardenability—In a ferrous alloy, the property that determines the depth and distribution of hardness induced by quenching.

Hardening—Increasing the hardness by suitable treatment, usually involving heating and cooling. When applicable, the following more specific terms should be used: Age Hardening, Case Hardening, Precipitation Hardening, Quench Hardening, Surface Hardening.

Heat Treatment—Heating and cooling a solid metal or alloy in such a way as to produce desired conditions or properties. Heating for the sole purpose of hot working is excluded from the meaning of this definition.

Homogeneous Carburizing—A process that converts a low carbon ferrous alloy to one of substantially uniform and higher carbon content throughout the section, so that a specific response to hardening may be obtained.

Homogenizing—Holding at high temperature to reduce or eliminate chemical segregation by diffusion.

Hot-Cold Working—Mechanical deformation of austenitic and precipitation hardening alloys at a temperature just below the recrystallization range to increase the yield strength and hardness by either plastic deformation or precipitation hardening effects induced by plastic deformation or both.

Hot Quenching—An imprecise term used to cover a variety of quenching procedures in which a quenching medium is maintained at a prescribed temperature above 160°F (71°C).

Induction Hardening—A surface hardening process in which only the surface layer of a suitable ferrous workpiece is heated by electrical induction to above the upper transformation temperature and immediately quenched.

Induction Heating—Heating by electrical induction.

Intermediate Annealing—Annealing wrought metals at one or more stages during manufacture and before final thermal treatment.

Intermediate Annealing—Aluminum and Aluminum Alloys—An imprecise term used to denote annealing of wrought products at one or more stages during processing but before final heat treatment. For the term to be meaningful, the type and condition of the material and the time-temperature cycle used must be stated.

Interrupted Aging—Aging at two or more temperatures, by steps, and cooling to room temperature after each step. See Aging and compare with Progressive Aging.

Interrupted Quenching—A quenching procedure in which the work piece is removed from the first quench at a temperature substantially higher than that of the quenchant and is then subjected to a second quenching system having a different cooling rate than the first.

Isothermal Annealing—Austenitizing a ferrous alloy and then cooling to and holding at a temperature at which austenite transforms to a relatively soft ferrite-carbide aggregate.

Isothermal Transformation—A change in phase at constant temperature.

Malleablizing—A process in which the as-cast malleable-type (white) iron is thermally treated for the purpose of converting most of all of the carbon in Fe_3C to graphite (temper carbon) to produce a family of products with improved ductility.

Maraging—A precipitation hardening treatment applied to a special group of iron base alloys to precipitate one or more intermetallic compounds in matrix of essentially carbon-free martensite.

NOTE: The first developed series of maraging steels contained, in addition to iron, more than 10% nickel and one or more supplemental hardening elements. In this series, the aging is done at about 900°F.

Marquenching—See martempering (2).

Martempering—(1) A hardening procedure in which an austenitized ferrous workpiece is quenched into an appropriate medium whose temperature is maintained substantially at the M_s of the workpiece, held in the medium until its temperature is uniform throughout but not long enough to permit bainite to form and then cooled in air. The treatment is frequently followed by tempering. (2) When the process is applied to carburized material, the controlling M_s temperature is that of the case. This variation of the process is frequently called marquenching.

Martensite Range—The temperature interval between M_s and M_f.

M_f—Defined under Transformation Temperature.

M_s—Defined under Transformation Temperature.

Natural Aging—Spontaneous aging of a supersaturated solid solution at room temperature. See Aging and compare with Artificial Aging.

ɸ ***Nitriding***—A case hardening process in which a ferrous-base material is heated to approximately the iron-nitrogen eutectoid temperature in either a gaseous or a liquid medium containing active nitrogen, thus causing absorption of nitrogen at the surface and, by diffusion, creating a concentration gradient. Within the capabilities of the particular material, slow cooling produces full hardness of the case.

In *conventional nitriding* a hardened and tempered alloy steel or tool steel is treated for sufficient time to produce highly saturated nitrides in the case.

In an important variation of the process, sometimes called *ductile nitriding*, applied to any ferrous-base material, the amount of active nitrogen and the time of exposure are so controlled as to produce a case of lower nitrogen content which, within the capabilities of the material, is fully hard on a micro scale but lower in hardness on a macro scale and relatively ductile.

Normalizing—Heating a ferrous alloy to a suitable temperature above the transformation range and then cooling in air to a temperature substantially below the transformation range.

Overaging—Aging under conditions of time and temperature greater than those required to obtain maximum change in a certain property, so that the property is altered in the direction of the initial value. See Aging.

Overheating—Heating a metal or alloy to such a high temperature that its properties are impaired. When the original properties cannot be restored by further heat treating, by mechanical working, or by a combination of working and heat treating, the overheating is known as Burning.

Partial Annealing—Aluminum and Aluminum Alloys—An imprecise term used to denote a treatment given cold worked material to reduce the strength to a controlled level or to effect stress relief. To be meaningful, the type of material, the degree of cold work it had undergone and the time-temperature cycle used must be stated.

Patenting—In wire making, a heat treatment applied to medium carbon or high carbon steel before the drawing of wire or between drafts. This process consists of heating to a temperature above the transformation range and then cooling to a temperature below Ae_1 in air or in a bath of molten lead or salt.

Postheating—Heating weldments immediately after welding, for tempering, for stress relieving, or for providing a controlled rate of cooling to prevent formation of a hard or brittle structure.

Pot Annealing—See Box Annealing.

Precipitation Hardening—Hardening caused by the precipitation of a constituent from a supersaturated solid solution. See also Age Hardening and Aging.

Precipitation Heat Treatment—Artificial aging in which a constituent precipitates from a supersaturated solid solution. See Artificial Aging, Interrupted Aging, and Progressive Aging.

Preheating—Heating before some further thermal or mechanical treatment. For tool steel, heating to an intermediate temperature immediately before final austenitizing. For some nonferrous alloys, heating to a high temperature for a long time in order to homogenize the structure before working.

Process Annealing—Ferrous and Copper and Copper Alloys—An imprecise term used to denote various treatments used to improve workability. For the term to be meaningful, the condition of the material and the time-temperature cycle used must be stated.

Progressive Aging—Aging by increasing the temperature in steps or continuously during the aging cycle. See Aging and compare with Interrupted Aging and Step Aging.

Pseudocarburizing—See Blank Carburizing.

Pseudonitriding—See Blank Nitriding.

Quench Aging—Aging induced by rapid cooling after Solution Heat Treatment.

Quench Annealing—Annealing an austenitic ferrous alloy by Solution Heat Treatment.

Quench Hardening—Copper Alloys—Hardening suitable alloys by betatizing and quenching to develop a martensite-like structure.

Quench Hardening—Ferrous—Hardening a suitable ferrous alloy by austenitizing and then cooling at a rate such that a substantial amount of austenite transforms to martensite.

Quenching—Rapid cooling. When applicable, the following more specific terms should be used: Direct Quenching, Fog Quenching, Hot Quenching, Interrupted Quenching, Selective Quenching, Spray Quenching, and Time Quenching.

Recrystallization—(1) The change from one crystal structure to another, as occurs on heating or cooling through a transformation temperature. (2) The formation of a new, strain-free grain structure from that existing in cold worked metal, usually accomplished by heating.

Recrystallization Annealing—Annealing cold worked metal to produce a new grain structure without phase change.

Recrystallization Temperature—The approximate minimum temperature at which complete recrystallization of a cold worked metal occurs within a specified time.

Secondary Hardening—The hardening phenomenon that occurs during high temperature tempering of certain steels containing one or more carbide forming alloying elements. Up to an optimum combination of tempering time and temperature, the reaction results either in the retention of hardness or an actual increase in hardness.

Selective Carburizing—Carburizing only selected surfaces of a workpiece by preventing absorption of carbon by all other surfaces.

Selective Case Hardening—Case hardening only selected surfaces of a workpiece.

Selective Heating—Intentional heating of only certain portions of a workpiece.

Selective Quenching—Quenching only certain portions of a workpiece.

Shell Hardening—A surface hardening process in which a suitable steel workpiece, when heated through and quench hardened, develops a martensitic layer or shell that closely follows the contour of the piece and surrounds a core of essentially pearlitic transformation product. This result is accomplished by a proper balance between section size, steel hardenability, and severity of quench.

Slack Quenching—The incomplete hardening of steel due to quenching from the austenitizing temperature at a rate slower than the critical cooling rate for the particular steel, resulting in the formation of one or more transformation products in addition to martensite.

Snap Temper—A precautionary interim stress-relieving treatment applied to high hardenability steels immediately after quenching to prevent cracking because of delay in tempering them at the prescribed higher temperature.

Soaking—Prolonged holding at a selected temperature.

Solution Heat Treatment—Heating an alloy to a suitable temperature, holding at that temperature long enough to cause one or more constituents to enter into solid solution, and then cooling rapidly enough to hold these constituents in solution.

ɸ *Soft Nitriding*—A misnomer for ductile nitriding.

ɸ *Solutionizing*—Another name for solution heat treatment, used principally in copper-beryllium technology.

Spheroidizing—Heating and cooling to produce a spheroidal or globular form of carbide in steel. Spheroidizing methods frequently used are:
1. Prolong holding at a temperature just below Ae_1.
2. Heating and cooling alternately between temperatures that are just above and just below Ae_1.
3. Heating to a temperature above Ae_1 or Ae_3 and then cooling very slowly in the furnace or holding at a temperature just below Ae_1.
4. Cooling at a suitable rate from the minimum temperature at which all carbide is dissolved, to prevent the re-formation of a carbide network and then reheating in accordance with Method 1 or 2 above. (Applicable to hypereutectoid steel containing a carbide network.)

Spray Quenching—Quenching in a spray of liquid.

Stabilizing Treatment—Ferrous—A treatment applied for the purpose of stabilizing the dimensions of a workpiece or the structure of a material such as (1) before finishing to final dimensions, heating a workpiece to or somewhat beyond its operating temperature and then cooling to room temperature a sufficient number of times to insure stability of dimensions in service (2) transforming retained austenite in those materials which retain substantial amounts when quench hardened (see cold treatment). (3) heating a solution treated austenitic stainless steel that contains controlled amounts of titanium or columbium plus tantalum to a temperature below the solution heat treating temperature to cause precipitation of finely divided, uniformly distributed carbides of those elements, thereby substantially reducing the amount of carbon available for the formation of chromium carbides in the grain boundaries upon subsequent exposure to temperatures in the sensitizing range.

Step Aging—Aluminum Alloys—Employment of two different aging treatments to control the type of precipitate formed from a super-saturated alloy matrix in order to obtain the desired properties. The first aging treatment, sometimes referred to as intermediate or stabilizing, is usually carried out at a higher temperature than the second.

Strain Aging—Aging induced by cold working. See Aging.

Stress Relieving—Heating to a suitable temperature, holding long enough to reduce residual stresses and then cooling slowly enough to minimize the development of new residual stresses.

Subcritical Annealing—Ferrous—A process anneal performed at a temperature below Ac_1.

Surface Hardening—A generic term covering several processes applicable to a suitable ferrous alloy that produces by quench hardening only, a surface layer that is harder or more wear resistant than the core. There is no significant alteration of the chemical composition of the surface layer. The processes commonly used are induction hardening, flame hardening, and shell hardening. Use of the applicable specific process name is preferred.

Temper Brittleness—Brittleness that results when certain steels are held within, or are cooled slowly through, a certain range of temperature below the transformation range. The brittleness is revealed by notched bar impact tests at or below room temperature.

Tempering—(1) Reheating a quench hardened or normalized ferrous alloy to a temperature below the transformation range (Ac_1) and then cooling at any desired rate. (2) A term used in conjunction with a qualifying adjective to designate the relative properties of a particular metal or alloy induced by cold work or heat treatment, or both.

Tempering—Copper Alloys—Heating quench hardened material to a temperature below the solution treatment temperature to produce desired changes in properties.

Tempering—Ferrous—Heating a quench hardened or normalized ferrous alloy to a temperature below the transformation range to produce desired changes in properties.

Time Quenching—Interrupted quenching in which the duration of holding in the quenching medium is controlled.

Transformation Ranges or Transformation Temperature Ranges—Those ranges of temperature within which austenite forms during heating and transforms during cooling. The two ranges are distinct, sometimes overlapping but never coinciding. The limiting temperatures of the ranges depend on the composition of the alloy and on the rate of change of temperature, particularly during cooling. See Transformation Temperature.

Transformation Temperature—The temperature at which a change in phase occurs. The term is sometimes used to denote the limiting temperature of a transformation range. The following symbols are used for iron and steels:

Ac_{cm}—In hypereutectoid steel, the temperature at which the solution of cementite in austenite is completed during heating.

Ac_1—The temperature at which austenite begins to form during heating.

Ac_3—The temperature at which transformation of ferrite to austenite is completed during heating.

Ac_4—The temperature at which austenite transforms to delta ferrite during heating.

Ae_1, Ae_3, Ae_{cm}, Ae_4—The temperatures of phase changes at equilibrium.

Ar_{cm}—In hypereutectoid steel, the temperature at which precipitation of cementite starts during cooling.

Ar_1—The temperature at which transformation of austenite to ferrite or to ferrite plus cementite is completed during cooling.

Ar_3—The temperature at which austenite begins to transform to ferrite during cooling.

Ar_4—The temperature at which delta ferrite transforms to austenite during cooling.

M_s—The temperature at which transformation of austenite to martensite starts during cooling.

M_f—The temperature, during cooling, at which transformation of austenite to martensite is substantially completed.

NOTE: All these changes except the formation of martensite occur at lower temperatures during cooling than during heating, and depend on the rate of change of temperature.

ALLOY STEEL
MACHINABILITY RATINGS—SAE J770c

SAE Information Report

Report of Iron and Steel Technical Committee approved June 1961 and last revised April 1966. Editorial change March 1971.

Introduction—This SAE Information Report is intended to provide a guide to the machinability characteristics of SAE alloy steel grades. The ratings and properties shown in Table 1 are provided as general information and not as requirements for specifications unless each instance is approved by the source of supply.

Machinability Ratings—While it is recognized that there is a considerable difference of views regarding alloy steel machinability, it is believed that the ratings contained in this report reflect current industry experience. The data on which they were based was obtained by a detailed survey of both producers and users and summarizes the combined experience of both groups. Various factors influence machinability and, therefore, results shown in Table 1 are average and may be affected to some degree by amount of cold reduction, mechanical properties, grain size, and microstructure. The machinability ratings listed are based on a value of 100% for SAE 1212 cold drawn. This value involves turning at a cutting speed of 180 surface feet per minute for feeds up to 0.007 in. per revolution and depths of cut up to 0.250 in., using appropriate cutting fluids with high speed steel tools, SAE Grade T1 (18-4-1) hardened to 63-65 RC.

Most low carbon alloy steels are machined in the as-rolled or as-rolled and cold drawn or cold finished condition. Higher carbon alloy steels and high hardenability low carbon steels, such as SAE 9310, may be conditioned for machining by subcritical annealing, annealing for softening to no specified structure, annealing to a specified structure such as lamellar pearlite or a percentage of lamellar pearlite and spheroidization, or to a fully spheroidized condition.

The structures imparted to the bars are evaluated in the machining operation by the tooling setup and type of tool used. It is possible to use widely diverging hardnesses and structures with different tooling setups and obtain satisfactory results both as to finish and parts per hr.

The above indicates the necessity for a systematic tabulation of expected machinability ratings by type, condition, and microstructure.

Microstructure—The microstructures shown in Table 1 are identified as follows:

Type A—Predominantly lamellar pearlite and ferrite.

Type B—Predominantly spheroidized.

Type C—This is a hot rolled structure which depends upon grade, size, and rolling conditions of the producing mill. The structure may be coarse or fine pearlite or bainite. The pearlite at low magnification may be blocky or acicular. For descriptive information, see U. S. Air Force Machinability Report, Volume 2, 1951, published by Curtiss-Wright Corporation.

Type D—This is a structure resulting from a subcritical anneal or temper anneal. It is usually a granular or spheroidized carbide condition confined to the hot rolled grain pattern, which may be blocky or acicular.

TABLE 1—MACHINABILITY RATINGS FOR ALLOYS STEEL

SAE No.	Machinability Rating	Condition	Range of Typical Hardnesses (Brinell)	Micro-Structure Type No.	SAE No.	Machinability Rating	Condition	Range of Typical Hardnesses (Brinell)	Micro-Structure Type No.
1330	55	Annealed and Cold Drawn	179-235	A	5115	65	Hot Rolled and Cold Drawn	163-201	C
1335	55	Annealed and Cold Drawn	179-235	A	5120	70	Hot Rolled and Cold Drawn	163-201	C
1340	50	Annealed and Cold Drawn	183-241	A	5130	70	Annealed and Cold Drawn	174-212	A
1345	45	Annealed and Cold Drawn	183-241	A	5132	70	Annealed and Cold Drawn	174-212	A
					5135	70		179-217	A
4012	70	Hot Rolled and Cold Drawn	149-196	C	5140	65	Annealed and Cold Drawn	179-217	A
4023	70	Hot Rolled and Cold Drawn	156-207	C	5145	65	Annealed and Cold Drawn	179-229	A
4024	75	Hot Rolled and Cold Drawn	156-207	C	5147	65	Annealed and Cold Drawn	179-229	A
4027	70	Annealed and Cold Drawn	167-212	A	5150	60	Annealed and Cold Drawn	183-235	A, B
4028	75	Annealed and Cold Drawn	167-212	A	5155	55	Annealed and Cold Drawn	183-235	A, B
4032	70	Annealed and Cold Drawn	174-217	A	5160	55	Spheroidized Annealed and Cold Drawn	179-217	B
4037	70	Annealed and Cold Drawn	174-217	A	51B60	55	Spheroidized Annealed and Cold Drawn	179-217	B
4042	65	Annealed and Cold Drawn	179-229	A					
4047	65	Annealed and Cold Drawn	179-229	A	50100	40	Spheroidized Annealed and Cold Drawn	183-241	B
					51100	40	Spheroidized Annealed and Cold Drawn	183-241	B
4118	60	Hot Rolled and Cold Drawn	170-207	C	52100	40	Spheroidized Annealed and Cold Drawn	183-241	B
4130	70	Annealed and Cold Drawn	187-229	A					
4135	70	Annealed and Cold Drawn	187-229	A	6118	60	Hot Rolled and Cold Drawn	179-217	C
4137	70	Annealed and Cold Drawn	187-229	A	6150	55	Annealed and Cold Drawn	183-241	B, A
4140	65	Annealed and Cold Drawn	187-229	A					
4142	65	Annealed and Cold Drawn	187-229	A	8115	65	Hot Rolled and Cold Drawn	163-202	C
4145	60	Annealed and Cold Drawn	187-229	A	81B45	65	Annealed and Cold Drawn	179-223	A
4147	60	Annealed and Cold Drawn	187-235	A					
4150	55	Annealed and Cold Drawn	187-241	A, B	8615	70	Hot Rolled and Cold Drawn	179-235	C
4161	50	Spheroidized and Cold Drawn	187-241	B-A	8617	70	Hot Rolled and Cold Drawn	179-235	C
					8620	65	Hot Rolled and Cold Drawn	179-235	C
4320	60	Annealed and Cold Drawn	187-229	D, B, A	8622	65	Hot Rolled and Cold Drawn	179-235	C
4340	50	Annealed and Cold Drawn	187-241	B, A	8625	60	Annealed and Cold Drawn	179-223	A
E4340	50	Annealed and Cold Drawn	187-241	B, A	8627	60	Annealed and Cold Drawn	179-223	A
					8630	70	Annealed and Cold Drawn	179-229	A
4419	65	Annealed and Cold Drawn	170-212	A	8637	65	Annealed and Cold Drawn	179-229	A
4422	65	Hot Rolled and Cold Drawn	170-212	C	8640	65	Annealed and Cold Drawn	184-229	A
4427	65	Annealed and Cold Drawn	170-212	A					
					8642	65	Annealed and Cold Drawn	184-229	A
4615	65	Hot Rolled and Cold Drawn	174-223	C	8645	65	Annealed and Cold Drawn	184-235	A
4617	65	Hot Rolled and Cold Drawn	174-223	C	86B45	65	Annealed and Cold Drawn	184-235	A
4620	65	Hot Rolled and Cold Drawn	183-229	C	8650	60	Annealed and Cold Drawn	187-248	A, B
4621	60	Hot Rolled and Cold Drawn	183-229	C	8655	55	Annealed and Cold Drawn	187-248	A, B
4626	70	Hot Rolled and Cold Drawn	170-212	C	8660	55	Spheroidized Annealed and Cold Drawn	179-217	B
4718	60	Hot Rolled and Cold Drawn	187-229	C	8720	65	Hot Rolled and Cold Drawn	179-235	C
4720	65	Hot Rolled and Cold Drawn	187-229	C	8740	65	Annealed and Cold Drawn	184-235	A
4815	50	Annealed and Cold Drawn	187-229	D, B					
4817	50	Annealed and Cold Drawn	187-229	D, B	8822	55	Hot Rolled and Cold Drawn	179-223	B
4820	50	Annealed and Cold Drawn	187-229	D, B					
					9254	45	Spheroidized Annealed and Cold Drawn	187-241	B
5015	65	Hot Rolled and Cold Drawn	156-196	C	9255	40	Spheroidized Annealed and Cold Drawn	179-229	B
50B40	65	Annealed and Cold Drawn	174-223	A	9260	40	Spheroidized Annealed and Cold Drawn	184-235	B
50B44	65	Annealed and Cold Drawn	174-223	A					
5046	60	Annealed and Cold Drawn	174-223	A	9310	50	Annealed and Cold Drawn	184-229	D
50B46	60	Annealed and Cold Drawn	174-223	A					
50B50	55	Annealed and Cold Drawn	183-235	A	94B15	70	Hot Rolled and Cold Drawn	163-202	C
5060	55	Spheroidized Annealed and Cold Drawn	170-212	B	94B17	70	Hot Rolled and Cold Drawn	163-202	C
50B60	55	Spheroidized Annealed and Cold Drawn	170-212	B	94B30	70	Annealed and Cold Drawn	170-223	A

4 Methods of Testing Steels

TENSILE TEST SPECIMENS—SAE J416b
SAE Recommended Practice

Report of Iron and Steel Division approved June 1911 and last revised by Iron and Steel Technical Committee June 1963. Reaffirmed without change June 1979. Information in this SAE Recommended Practice conforms to ASTM A 370 and ASTM E 8, except that the 0.160-in and 0.113-in diameter specimens have been added. Information on these conforms to Federal Test Method Standard No. 151, Method 211.

When required, unless otherwise specified in the SAE Standards or Recommended Practices, tensile test specimens for metals shall be selected and prepared in accordance with this report. ASTM E 8, Methods of Tension Testing of Metallic Materials, gives more detailed information on tensile testing procedure, and ASTM E 4, Methods of Verification of Testing Machines, provides information on testing equipment calibration.

In recommending these specimens for use in tensile tests it is not intended to exclude entirely the use of other test specimens for special materials or for special forms of material. It is, however, recommended that these specimens be used wherever it is feasible.

Machining of specimens shall be done in such a manner as to avoid leaving severe machining strains in the material. Specimens shall be finished so that the surfaces are smooth and free from nicks and tool marks. All ragged edges shall be smoothed.

Full Section Test Specimens—For wire, rod, and bars less than 3/4-in. diameter or distance between flats, specimens having the full cross section of the material are recommended. It is permissible to reduce the cross section slightly by grinding or machining throughout the test section to insure fracture between gage marks. This may be done either without changing shape of cross section or, on squares, hexagons, or octagons by turning to a round. The following limits apply to such reduction:

1. If the same cross sectional shape is retained the final area must be not less than 90% of the original area and the diameter or distance between flats must not be reduced more than 0.01 in.
2. If the rod is turned to a round, the final area must be not less than 90% of the area of the maximum inscribed circle and the final diameter must not be less than the original distance between flats —0.01 in.

Fillets must be used at the ends of the reduced section. The fillet radius should be not less than the section diameter.

Flat Test Specimens—The tensile test specimen shown in Fig. 1 (8 in. gage length) is recommended for plates, shapes, and flat material having a thickness 3/16 in. or over.

The tensile test specimen shown in Fig. 1 (2 in. gage length) is recommended for sheet, plate, flat wire, strip, band, and hoop ranging in thickness from 0.005 to 5/8 in. in either case. Where size of material permits, one of the specimens shown in Fig. 2 may also be used.

Round Test Specimens—The 1/2-in. diameter round test specimen shown in Fig. 2 is considered standard and is recommended for general testing of metals. Small size specimens proportional to the standard specimen may be used when it is necessary to test material from which the standard specimen or the specimens shown in Fig. 1 cannot be prepared.

Smaller miniature specimens for so-called microtensile tests may be used on agreement between supplier and user.

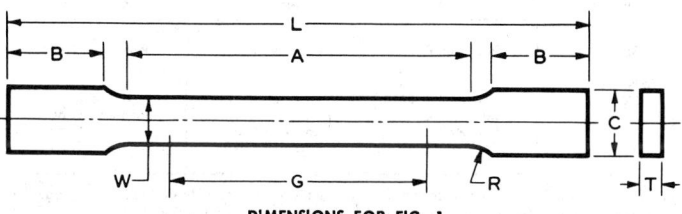

DIMENSIONS FOR FIG. 1

Dimensions	2 in. Gage Length	8 in. Gage Length
G, Gage length	2.000 ±0.005 in.	8.00 ±0.01 in.[a]
W, Width[b]	0.500 ±0.010 in.[c]	1-1/2 +1/8 in.[d] −1/4 in.
T, Thickness	Thickness of material	Thickness of material
R, Radius of fillet	1/2 in. min	1 in. min[e]
L, Overall length	8 in. min	18 in. min
A, Length of reduced section	2-1/4 in. min	9 in. min
B, Length of grip section[f]	2 in. min[g]	3 in. min
C, Width of grip section[b]	3/4 in. approx[h]	2 in. approx[i]

[a] Punch mark for measuring elongation after fracture shall be made on the flat or on the edge of the specimen and within the reduced section. Either a set of nine or more punch marks 1-in. apart, or one or more pairs of punch marks 8-in. apart may be used.
[b] When necessary, a narrower specimen may be used. In such case the width should be as great as the width of the material being tested permits. In such cases, the sides may be parallel throughout the length of the specimen.
[c] The ends of the reduced section shall not differ in width by more than 0.002 in. There may be a gradual taper in width from the ends to the center, but the width at either end shall not be more than 0.005 in. greater than the width at the center.
[d] The ends of the reduced section shall not differ in width by more than 0.004 in. There may be a gradual taper in width from the ends to the center, but the width at either end shall not be more than 0.015 in. greater than the width at the center.
[e] A 1/2-in. min radius at the ends of the reduced section is permitted for steel specimens under 100,000 psi tensile strength when a profile cutter is used to machine the reduced section.
[f] It is desirable, if possible, to make the length of the grip section great enough to allow the specimen to extend into the grips a distance equal to two-thirds or more of the length of the grips.
[g] If the thickness of the specimen is over 3/8 in. longer grips and correspondingly longer grip sections of the specimen may be necessary to prevent failure in the grip section.
[h] The ends of the specimen shall be symmetrical with the centerline of the reduced section within 0.01 in. However, for steel if the ends are symmetrical within 0.05 in., a specimen may be considered satisfactory for all but referee testing.
[i] The ends of the specimen shall be symmetrical with the centerline of the reduced section within 0.10 in.

FIG. 1—STANDARD RECTANGULAR TENSILE TEST SPECIMEN

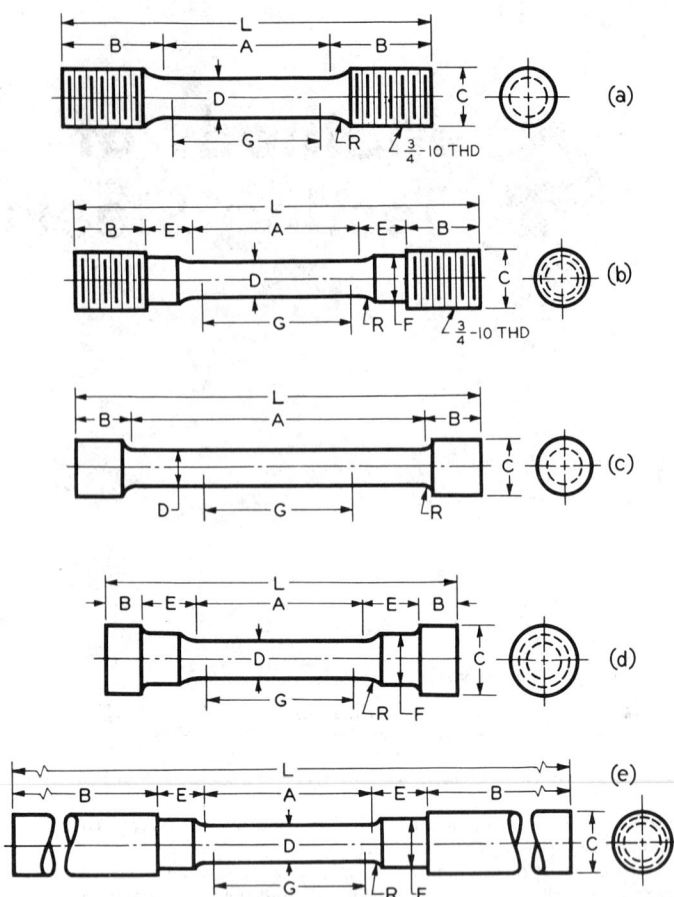

DIMENSIONS FOR FIG. 2

Dimensions, in.	Standard 1/2-in. Round Tensile Test Specimen				
	(a)	(b)	(c)	(d)	(e)
G, Gage length[a]	2.000 ±0.005	2.000 ±0.005	2.000 ±0.005	2.000 ±0.005	2.000 ±0.005
D, Dia[b]	0.500 ±0.010	0.500 ±0.010	0.500 ±0.010	0.500 ±0.010	0.500 ±0.010
R, Radius of fillet[a]	3/8 min	3/8 min	1/16 min	3/8 min	3/8 min
A, Length of reduced section	2-1/4 min	2-1/4 min	4 approx	2-1/4 min	2-1/4 min
L, Overall length	5 approx	5-1/2 approx	5-1/2 approx	4-3/4 approx	9-1/2 approx
B, Length of end section[c]	1-3/8 approx	1 approx	3/4 approx	1/2 approx	3 min
C, Dia of end section	3/4	3/4	23/32	7/8	3/4
E, Length of shoulder and fillet section	—	5/8 approx	—	3/4 approx	5/8 approx
F, Dia of shoulder		5/8		5/8	19/32

Dimensions, in.	Examples of Small Size Specimens Proportional to Standard			
	0.350-in. Round	0.250-in. Round	0.160-in. Round	0.113-in. Round
G, Gage length	1.400 ±0.005	1.000 ±0.005	0.64 ±0.005	0.45 ±0.005
D, Dia[d]	0.350 ±0.007	0.250 ±0.005	0.160 ±0.003	0.113 ±0.002
R, Radius of fillet	1/4 min	3/16 min	0.15 min	0.09 min
A, Length of reduced section[e]	1-3/4 min	1-1/4 min	3/4 min	5/8 min

[a] The gage length and fillets shall be as shown, but the ends may be of any form to fit the holders of the testing machine in such a way that the load shall be axial. If the ends are to be held in grips it is desirable, if possible, to make the length of the grip section great enough to allow the specimen to extend into the grips a distance equal to two-thirds or more of the length of the grips.

[b] The reduced section may have a gradual taper from the ends toward the center with the ends not more than 0.005 in. larger in diameter than the center.

[c] On specimen (e) it is desirable, if possible, to make the length of the grip section great enough to allow the specimen to extend into the grips a distance equal to two-thirds or more of the length of the grips.

[d] The reduced section may have a gradual taper from the ends toward the center, with the ends not more than 0.003 in. larger in diameter than the center.

[e] If desired, on the small size specimens the length of the reduced section may be increased to accommodate an extensometer. However, reference marks for the measurement of elongation should nevertheless be spaced at the indicated gage length.

FIG. 2—ROUND TENSILE TEST SPECIMENS WITH VARIOUS TYPES OF ENDS

SURFACE HARDNESS TESTING WITH FILES—SAE J864

SAE Recommended Practice

Report of Iron and Steel Technical Committee approved June 1963. Reaffirmed without change January 1968.

Application—This procedure describes the technique of using a file for testing the surface hardness of miscellaneous iron and steel parts as designated by engineering specifications. In presenting this procedure, it is recognized that it is subjective and that it must be used with considerable judgment on the part of the operator. File hardness tests should be used only when specified on the engineering drawing and not in lieu of conventional hardness methods.

Scope—Hardness testing with files consists essentially of cutting or abrading the surface of metal parts, and approximating the hardness by the feel, or extent to which, the file bites into the surface. The term file hard means that the surface hardness of the parts tested is such that new hardness files will not cut the surface of the material being tested.

Apparatus Required

Standard Files—Hand files meeting the following requirements:
- 6 or 8 in. pillar
- No. 1 Swiss double cut
- 66 cuts per in.
- Hardness of File:
  - Rockwell C 65–68 designated No. 65
  - Rockwell C 61–63 designated No. 62
  - Rockwell C 57–59 designated No. 58
  - Rockwell C 54–56 designated No. 55
  - Rockwell C 49–51 designated No. 50
- Chemical Composition:
  - Carbon 1.20–1.40%
  - Manganese 0.20–0.40%
  - Phosphorus 0.04 max
  - Sulfur 0.05 max
  - Silicon 0.10–0.20%

Standard Prover—Standard steel or iron provers, 2 in. dia and approximately $\frac{1}{4}$ in. thick, hardened and tempered to the Rockwell of the lower limit of each standard file range, are required for testing the standard files.

The prover shall be filed with a discarded test file to remove any hard or soft skin that will interfere with the accuracy of the test. Similar provers hardened to ranges below and above the medians can be used to prove the file will cut below and not above the designated file hardness.

CHART OF FILES AND PROVERS

File No.	Standard Prover Hardness	Prover Cutting Hardness	Prover Noncutting Hardness
Untempered 65	Rockwell C 65	Rockwell C 63	Rockwell C 67
Tempered[a]			
62	Rockwell C 61	Rockwell C 60	Rockwell C 64
58	Rockwell C 57	Rockwell C 56	Rockwell C 60
55	Rockwell C 54	Rockwell C 53	Rockwell C 57
50	Rockwell C 49	Rockwell C 48	Rockwell C 52

[a] Using tempered files below Rockwell C 65 is less accurate and, therefore, more judgment on the part of the operator must be exercised as the hardness of the file decreases.

Testing of files is performed by passing the test file across the $\frac{1}{4}$ in. thick face of the prover.

A $\frac{1}{4}$ in. prover is specified because at the higher hardnesses, the surface area contacted affects the cutting area considerably. Narrow areas can be cut more readily than wide areas due to the concentration of pressure that may be obtained.

Standard Test Pieces—Standard test pieces (of the same contour, steel or iron composition and heat treatment) varying in hardness by small increments, with which the parts being inspected can be compared, are recommended. In the case of steels, each family of SAE steels, namely, 10XX,

20XX, 30XX, and the like should be used. They can be heat treated to compare with the operation, such as carburized, quench and temper or carbonitrided, quenched and temper.

These pieces will enable the operator to learn, with fair accuracy, the feel of the file as it cuts, or does not cut, in relation to the Rockwell hardness.

Surface Condition—In testing high surface hardnesses with a file, experience has shown that surface condition is important, very smoothly ground surfaces cannot be touched with a file as readily as surfaces that have been filed. Sometimes a testing file will cut a prover of a certain steel at, for example, C 64, yet when the file is applied to a smoothly ground part made of the same steel and at the same hardness, the part feels harder. It is important when comparisons are made that the surface smoothness of the parts being tested be the same as the standard test piece. The standard test pieces can be made in a series of microfinish of 20, 60, 125, and 200.

On surfaces with a microfinish, the direction of the test filing across or parallel with the finish direction affects results. When filing, the direction of the file in relation to the microfinish should always be the same as the standard test piece.

Testing Hardness
1. Check file against standard prover.
2. Apply the file to the surface of the part being tested at such an angle that only a few teeth will engage the surface at once. Use slow, firm strokes in an effort to feel the manner in which the file cuts or does not cut. To prolong the life of the file, use as short a stroke as possible.

NOTE: To standardize pressure, attach specimen to a balance scale platform and measure the file effort against specimen in pounds. Application should be between 10 and 12 lb.

3. Compare the parts with a standard test piece with a known hardness range. This will assist in determining whether the part falls within the range specified by the engineering specifications. File hardness should not replace conventional methods where penetrators will not break through surface hardened areas.

NOTE: During the testing of a batch of parts, the files must not be allowed to become so dull as to cause difficulty in discriminating between parts within specification and those below specification. This can be prevented by frequent checking against the standard part or prover.

4. The hardness should be specified according to the Rockwell reading of the file as to the surface requirements of an iron or steel part. The designation should be "file hard—65" for Rockwell C 65 surface; "file hard—62" for Rockwell C 62, and so on.

HARDNESS TESTS AND HARDNESS NUMBER CONVERSIONS—SAE J417b

SAE Information Report

Report of Iron and Steel Technical Committee approved January 1946 and last revised May 1970. Reaffirmed without change June 1979.

Scope—This report lists approximate hardness conversion values; test methods for Diamond Pyramid Hardness (Vickers DPH), Brinell Hardness, Rockwell Hardness, Rockwell Superficial Hardness, Shore Hardness; and information regarding surface preparation, specimen thickness, effect of curved surfaces, and recommendations for Rockwell surface hardness testing for case hardened parts.

The tables in this report give the approximate relationship of DPH (Vickers), Brinell, Rockwell, and Scleroscope hardness values and corresponding approximate tensile strengths of steels. It is impossible to give exact relationships because of the inevitable influence of size, mass, composition, and method of heat treatment. Where more precise conversions are required, they should be developed specially for each steel composition, heat treatment, and part.

The accompanying conversion tables for steel hardness numbers are based on extensive tests on carbon and alloy steels, mostly in the heat treated condition, but have been found to be reliable on practically all constructional alloy steels and tool steels in the as-forged, annealed, normalized, and quenched and tempered conditions, provided they are homogeneous. Such special cases as high manganese steel, 18% chromium—8% nickel steel and other austenitic steels, and nickel base alloys, as well as constructional alloy steels and tool steels in the cold worked condition, may not conform to the relationships given with the same degree of accuracy as the steels for which the tables are intended.

All numbers in these tables given in bold face type were prepared jointly by the American Society for Testing and Materials, the American Society for Metals, and SAE from carefully checked data. The values given regular face type were taken from the Army-Navy Approximate Hardness Tensile Strength Relationship of Carbon and Low Alloy Steels (AN-QQ-H-201) published in the 1943 SAE Handbook, with only minor adjustments.

Use of Conversion Tables—The conversions given in the accompanying Tables 1, 2, and 3 are recommended for use in converting the results of one form of hardness test to another only on flat surfaces and only when the specific test procedures and precautions outlined in the several hardness test methods are followed. Attention is called to the limitation in ASTM E 10 on the use of the standard steel ball to hardnesses under 500 Brinell. The Rockwell Superficial and Diamond Pyramid Hardness tests require especially smooth surfaces for accurate results. In all tests, a specimen should be of sufficient thickness to avoid anvil effect—which thickness is roughly 10 times the depth of the indentation. It is important that conversions from Brinell Hardness to shallow impression type tests, such as Rockwell Superficial and Diamond Pyramid Hardness tests, be made only on materials that are of uniform hardness to a depth at least 10 times that of the indentation. Such hardness conversions should not be made on surface hardened, coated, or decarburized surfaces. Although the Rockwell Hardness and the Rockwell Superficial Hardness values in the tables are given to tenths of a point in order to maintain exact relationships between the various scales, it is customary to report these values to the nearest point. Experience has shown that even under carefully controlled conditions, some deviations from the conversion relationships will occur.

The numbers given in parentheses in the tables are values beyond the practical range of usefulness of the type of test under which they appear and have no strict application. They are included in the tables as a matter of information only.

Diamond Pyramid Hardness (Vickers, DPH), Table 1—Diamond Pyramid Hardness is determined by forcing a square base diamond pyramid having an apex angle of 136 deg into the test specimen under loads usually of 5 to 50 kg and measuring the diagonals of the recovered indentations. The Diamond Pyramid Hardness is defined as the load per unit area of surface contact in kilograms per square millimeter as calculated from the average diagonal as follows:

$$DPH = \frac{2L \sin \frac{a}{2}}{d^2}$$

where DPH = diamond pyramid hardness
  d = length of average diagonal in millimeters
  a = apex angle = 136 deg
  L = load in kilograms

For further information on standard methods of Diamond Pyramid Hardness Testing, refer to ASTM E 92-57.

Brinell Hardness—Tables 2 and 3

Test Ball—The diameter of the ball shall be 10.00 ± 0.01 mm (0.3937 ± 0.0004 in.). The load applied shall be 3000 kg (6614 lb) for at least 15 sec on iron and steel.

Test Impression—The average diameter of the impression shall be obtained from two measurements at right angles to each other, made with an instrument having a reading error not over 0.01 mm (0.0004 in.).

Test Specimen—The surface of the specimen should be flat and reasonably free from scratches. The specimen shall be taken deep enough to represent the true composition of the material to be tested, and shall be maintained in a plane normal to the direction of the testing load.

Exceptions—This test should not be used on soft steels less than $\frac{3}{8}$ in. thick or on areas small enough to permit deflection of the edges of the specimen owing to the flow from the ball depression.

For further information on standard methods of Brinell Hardness testing, refer to ASTM E 10. For Brinell Hardness Numbers for Various Loads, see Table 3.

Rockwell Hardness—Table 4

Principle of Test—The Rockwell Hardness tester is essentially a machine that measures hardness by determining the depth of penetration of a penetrator into the specimen under certain arbitrarily fixed conditions of test. The penetrator may be either a steel ball or a diamond sphero-conical penetrator. The hardness value as read from the dial is an arbitrary number which is related to the depth of indentation, and since the scales are reversed, the

TABLE 1—APPROXIMATE EQUIVALENT HARDNESS NUMBERS[a] FOR DIAMOND PYRAMID HARDNESS NUMBERS (VICKERS, DPH), FOR STEEL

Diamond Pyramid Hardness No.	Brinell Hardness No. 10-mm Ball, 3000-kg Load			Rockwell Hardness No.[b]				Rockwell Superficial Hardness No., Superficial Brale Penetrator			Shore Scleroscope Hardness No.	Tensile Strength (Approximate) in 1000 psi	Diamond Pyramid Hardness No.
	Standard Ball	Hultgren Ball	Tungsten-Carbide Ball	A-Scale, 60-kg Load, Brale Penetrator	B-Scale, 100-kg Load, 1/16-in. Dia Ball	C-Scale, 150-kg Load, Brale Penetrator	D-Scale, 100-kg Load, Brale Penetrator	15-N Scale, 15-kg Load	30-N Scale, 30-kg Load	45-N Scale, 45-kg Load			
Col. 1	Col. 2	Col. 3	Col. 4	Col. 5	Col. 6	Col. 7	Col. 8	Col. 9	Col. 10	Col. 11	Col. 12	Col. 13	Col. 14
940	—	—	—	85.6	—	68.0	76.9	93.2	84.4	75.4	97	—	940
920	—	—	—	85.3	—	67.5	76.5	93.0	84.0	74.8	96	—	920
900	—	—	—	85.0	—	67.0	76.1	92.9	83.6	74.2	95	—	900
880	—	—	767	84.7	—	66.4	75.7	92.7	83.1	73.6	93	—	880
860	—	—	757	84.4	—	65.9	75.3	92.5	82.7	73.1	92	—	860
840	—	—	745	84.1	—	65.3	74.8	92.3	82.2	72.2	91	—	840
820	—	—	733	83.8	—	64.7	74.3	92.1	81.7	71.8	90	—	820
800	—	—	722	83.4	—	64.0	73.8	91.8	81.1	71.0	88	—	800
780	—	—	710	83.0	—	63.3	73.3	91.5	80.4	70.2	87	—	780
760	—	—	698	82.6	—	62.5	72.6	91.2	79.7	69.4	86	—	760
740	—	—	684	82.2	—	61.8	72.1	91.0	79.1	68.6	84	—	740
720	—	—	670	81.8	—	61.0	71.5	90.7	78.4	67.7	83	—	720
700	—	615	656	81.3	—	60.1	70.8	90.3	77.6	66.7	81	—	700
690	—	610	647	81.1	—	59.7	70.5	90.1	77.2	66.2	—	—	690
680	—	603	638	80.8	—	59.2	70.1	89.8	76.8	65.7	80	—	680
670	—	597	630	80.6	—	58.8	69.8	89.7	76.4	65.3	—	—	670
660	—	590	620	80.3	—	58.3	69.4	89.5	75.9	64.7	79	—	660
650	—	585	611	80.0	—	57.8	69.0	89.2	75.5	64.1	—	—	650
640	—	578	601	79.8	—	57.3	68.7	89.0	75.1	63.5	77	—	640
630	—	571	591	79.5	—	56.8	68.3	88.8	74.6	63.0	—	—	630
620	—	564	582	79.2	—	56.3	67.9	88.5	74.2	62.4	75	—	620
610	—	557	573	78.9	—	55.7	67.5	88.2	73.6	61.7	—	—	610
600	—	550	564	78.6	—	55.2	67.0	88.0	73.2	61.2	74	—	600
590	—	542	554	78.4	—	54.7	66.7	87.8	72.7	60.5	—	298	590
580	—	535	545	78.0	—	54.1	66.2	87.5	72.1	59.9	72	293	580
570	—	527	535	77.8	—	53.6	65.8	87.2	71.7	59.3	—	288	570
560	—	519	525	77.4	—	53.0	65.4	86.9	71.2	58.6	71	283	560
550	505	512	517	77.0	—	52.3	64.8	86.6	70.5	57.8	—	276	550
540	496	503	507	76.7	—	51.7	64.4	86.3	70.0	57.0	69	270	540
530	488	495	497	76.4	—	51.1	63.9	86.0	69.5	56.2	—	265	530
520	480	487	488	76.1	—	50.5	63.5	85.7	69.0	55.6	67	260	520
510	473	479	479	75.7	—	49.8	62.9	85.4	68.3	54.7	—	254	510
500	465	471	471	75.3	—	49.1	62.2	85.0	67.7	53.9	66	247	500
490	456	460	460	74.9	—	48.4	61.6	84.7	67.1	53.1	—	241	490
480	448	452	452	74.5	—	47.7	61.3	84.3	66.4	52.2	64	235	480
470	441	442	442	74.1	—	46.9	60.7	83.9	65.7	51.3	—	228	470
460	433	433	433	73.6	—	46.1	60.1	83.6	64.9	50.4	62	222	460
450	425	425	425	73.3	—	45.3	59.4	83.2	64.3	49.4	—	217	450
440	415	415	415	72.8	—	44.5	58.8	82.8	63.5	48.4	59	212	440
430	405	405	405	72.3	—	43.6	58.2	82.3	62.7	47.4	—	205	430
420	397	397	397	71.8	—	42.7	57.5	81.8	61.9	46.4	57	199	420
410	388	388	388	71.4	—	41.8	56.8	81.4	61.1	45.3	—	193	410
400	379	379	379	70.8	—	40.8	56.0	81.0	60.2	44.1	55	187	400
390	369	369	369	70.3	—	39.8	55.2	80.3	59.3	42.9	—	180	390
380	360	360	360	69.8	(110.0)	38.8	54.4	79.8	58.4	41.7	52	175	380
370	350	350	350	69.2	—	37.7	53.6	79.2	57.4	40.4	—	170	370
360	341	341	341	68.7	(109.0)	36.6	52.8	78.6	56.4	39.1	50	164	360
350	331	331	331	68.1	—	35.5	51.9	78.0	55.4	37.8	—	159	350
340	322	322	322	67.6	(108.0)	34.4	51.1	77.4	54.4	36.5	47	155	340
330	313	313	313	67.0	—	33.3	50.2	76.8	53.6	35.2	—	150	330
320	303	303	303	66.4	(107.0)	32.2	49.4	76.2	52.3	33.9	45	146	320
310	294	294	294	65.8	—	31.0	48.4	75.6	51.3	32.5	—	142	310
300	284	284	284	65.2	(105.5)	29.8	47.5	74.9	50.2	31.1	42	138	300
295	280	280	280	64.8	—	29.2	47.1	74.6	49.7	30.4	—	136	295
290	275	275	275	64.5	(104.5)	28.5	46.5	74.2	49.0	29.5	41	133	290
285	270	270	270	64.2	—	27.8	46.0	73.8	48.4	28.7	—	131	285
280	265	265	265	63.8	(103.5)	27.1	45.3	73.4	47.8	27.9	40	129	280
275	261	261	261	63.5	—	26.4	44.9	73.0	47.2	27.1	—	127	275
270	256	256	256	63.1	(102.0)	25.6	44.3	72.6	46.4	26.2	38	124	270
265	252	252	252	62.7	—	24.8	43.7	72.1	45.7	25.2	—	122	265
260	247	247	247	62.4	(101.0)	24.0	43.1	71.6	45.0	24.3	37	120	260
255	243	243	243	62.0	—	23.1	42.2	71.1	44.2	23.2	—	117	255
250	238	238	238	61.6	99.5	22.2	41.7	70.6	43.4	22.2	36	115	250
245	233	233	233	61.2	—	21.3	41.1	70.1	42.5	21.1	—	113	245
240	228	228	228	60.7	98.1	20.3	40.3	69.6	41.7	19.9	34	111	240
230	219	219	219	—	96.7	(18.0)	—	—	—	—	33	106	230
220	209	209	209	—	95.0	(15.7)	—	—	—	—	32	101	220
210	200	200	200	—	93.4	(13.4)	—	—	—	—	30	97	210
200	190	190	190	—	91.5	(11.0)	—	—	—	—	29	92	200
190	181	181	181	—	89.5	(8.5)	—	—	—	—	28	88	190
180	171	171	171	—	87.1	(6.0)	—	—	—	—	26	84	180
170	162	162	162	—	85.0	(3.0)	—	—	—	—	25	79	170
160	152	152	152	—	81.7	(0.0)	—	—	—	—	24	75	160
150	143	143	143	—	78.7	—	—	—	—	—	22	71	150
140	133	133	133	—	75.0	—	—	—	—	—	21	66	140
130	124	124	124	—	71.2	—	—	—	—	—	20	62	130
120	114	114	114	—	66.7	—	—	—	—	—	—	57	120
110	105	105	105	—	62.3	—	—	—	—	—	—	—	110
100	95	95	95	—	56.2	—	—	—	—	—	—	—	100
95	90	90	90	—	52.0	—	—	—	—	—	—	—	95
90	86	86	86	—	48.0	—	—	—	—	—	—	—	90
85	81	81	81	—	41.0	—	—	—	—	—	—	—	85

[a] The values in this table shown in **bold face type** correspond to the values shown in the corresponding joint SAE-ASM-ASTM Committee on Hardness Conversions as printed in ASTM E 140, Table 1.

[b] Values in ( ) are beyond normal range and are given for information only.

TABLE 2—APPROXIMATE EQUIVALENT HARDNESS NUMBERS[a] FOR BRINELL HARDNESS NUMBERS[b], FOR STEEL

Brinell Indentation Dia, mm	Brinell Hardness No.[b] 10-mm Ball, 3000-kg Load			Diamond Pyramid Hardness No.	Rockwell Hardness No.[c]				Rockwell Superficial Hardness No. Superficial Brale Penetrator			Shore Sclero-scope Hardness No.	Tensile Strength (Approxi-mate) in 1000 psi	Brinell Indentation Dia, mm
	Standard Ball	Hultgren Ball	Tungsten-Carbide Ball		A-Scale, 60-kg Load, Brale Penetrator	B-Scale, 100-kg Load 1/16-in. Dia Ball	C-Scale, 150-kg Load Brale Penetrator	D-Scale, 100-kg Load Brale Penetrator	15-N Scale, 15-kg Load	30-N Scale, 30-kg Load	45-N Scale, 45-kg Load			
Col. 1	Col. 2	Col. 3	Col. 4	Col. 5	Col. 6	Col. 7	Col. 8	Col. 9	Col. 10	Col. 11	Col. 12	Col. 13	Col. 14	Col. 15
—	—	—	—	940	85.6	—	68.0	76.9	93.2	84.4	75.4	97	—	—
—	—	—	—	920	85.3	—	67.5	76.5	93.0	84.0	74.8	96	—	—
—	—	—	—	900	85.0	—	67.0	76.1	92.9	83.6	74.2	95	—	—
—	—	—	767	880	84.7	—	66.4	75.7	92.7	83.1	73.6	93	—	—
—	—	—	757	860	84.4	—	65.9	75.3	92.5	82.7	73.1	92	—	—
2.25	—	—	745	840	84.1	—	65.3	74.8	92.3	82.2	72.2	91	—	2.25
—	—	—	733	820	83.8	—	64.7	74.3	92.1	81.7	71.8	90	—	—
—	—	—	722	800	83.4	—	64.0	73.8	91.8	81.1	71.0	88	—	—
2.30	—	—	712	—	—	—	—	—	—	—	—	—	—	2.30
—	—	—	710	780	83.0	—	63.3	73.3	91.5	80.4	70.2	87	—	—
—	—	—	698	760	82.6	—	62.5	72.6	91.2	79.7	69.4	86	—	—
—	—	—	684	740	82.2	—	61.8	72.1	91.0	79.1	68.6	—	—	—
2.35	—	—	682	737	82.2	—	61.7	72.0	91.0	79.0	68.5	84	—	2.35
—	—	—	670	720	81.8	—	61.0	71.5	90.7	78.4	67.7	83	—	—
—	—	—	656	700	81.3	—	60.1	70.8	90.3	77.6	66.7	—	—	—
2.40	—	—	653	697	81.2	—	60.0	70.7	90.2	77.5	66.5	81	—	2.40
—	—	—	647	690	81.1	—	59.7	70.5	90.1	77.2	66.2	—	—	—
—	—	—	638	680	80.8	—	59.2	70.1	89.8	76.8	65.7	80	—	—
—	—	—	630	670	80.6	—	58.8	69.8	89.7	76.4	65.3	—	—	—
2.45	—	—	627	667	80.5	—	58.7	69.7	89.6	76.3	65.1	79	—	2.45
2.50	—	601	—	677	80.7	—	59.1	70.0	89.8	76.8	65.7	—	—	2.50
	—	—	601	640	79.8	—	57.3	68.7	89.0	75.1	63.5	77	—	
2.55	—	578	—	640	79.8	—	57.3	68.7	89.0	75.1	63.5	—	—	2.55
	—	—	578	615	79.1	—	56.0	67.7	88.4	73.9	62.1	75	—	
2.60	—	555	—	607	78.8	—	55.6	67.4	88.1	73.5	61.6	—	—	2.60
	—	—	555	591	78.4	—	54.7	66.7	87.8	72.7	60.6	73	298	
2.65	—	534	—	579	78.0	—	54.0	66.1	87.5	72.0	59.8	—	292	2.65
	—	—	534	569	77.8	—	53.5	65.8	87.2	71.6	59.2	71	288	
2.70	—	514	—	553	77.1	—	52.5	65.0	86.9	70.7	58.0	—	278	2.70
	—	—	514	547	76.9	—	52.1	64.7	86.5	70.3	57.6	70	274	
2.75	495	—	—	539	76.7	—	51.6	64.3	86.3	69.9	56.9	—	269	2.75
	—	495	—	530	76.4	—	51.1	63.9	86.0	69.5	56.2	—	265	
	—	—	495	528	76.3	—	51.0	63.8	85.9	69.4	56.1	68	264	
2.80	477	—	—	516	75.9	—	50.3	63.2	85.6	68.7	55.2	—	258	2.80
	—	477	—	508	75.6	—	49.6	62.7	85.3	68.2	54.5	—	252	
	—	—	477	508	75.6	—	49.6	62.7	85.3	68.2	54.5	66	252	
2.85	461	—	—	495	75.1	—	48.8	61.9	84.9	67.4	53.5	—	244	2.85
	—	461	—	491	74.9	—	48.5	61.7	84.7	67.2	53.2	—	242	
	—	—	461	491	74.9	—	48.5	61.7	84.7	67.2	53.2	65	242	
2.90	444	—	—	474	74.3	—	47.2	61.0	84.1	66.0	51.7	—	231	2.90
	—	444	—	472	74.2	—	47.1	60.8	84.0	65.8	51.5	—	230	
	—	—	444	472	74.2	—	47.1	60.8	84.0	65.8	51.5	63	230	
2.95	429	429	429	455	73.4	—	45.7	59.7	83.4	64.6	49.9	61	219	2.95
3.00	415	415	415	440	72.8	—	44.5	58.8	82.8	63.5	48.4	59	212	3.00
3.05	401	401	401	425	72.0	—	43.1	57.8	82.0	62.3	46.9	58	202	3.05
3.10	388	388	388	410	71.4	—	41.8	56.8	81.4	61.1	45.3	56	193	3.10
3.15	375	375	375	396	70.6	—	40.4	55.7	80.6	59.9	43.6	54	184	3.15
3.20	363	363	363	383	70.0	—	39.1	54.6	80.0	58.7	42.0	52	177	3.20
3.25	352	352	352	372	69.3	(110.0)	37.9	53.8	79.3	57.6	40.5	51	171	3.25
3.30	341	341	341	360	68.7	(109.0)	36.6	52.8	78.6	56.4	39.1	50	164	3.30
3.35	331	331	331	350	68.1	(108.5)	35.5	51.9	78.0	55.4	37.8	48	159	3.35
3.40	321	321	321	339	67.5	(108)	34.3	51.0	77.3	54.3	36.4	47	154	3.40
3.45	311	311	311	328	66.9	(107.5)	33.1	50.0	76.7	53.3	34.4	46	149	3.45
3.50	302	302	302	319	66.3	(107)	32.1	49.3	76.1	52.2	33.8	45	146	3.50
3.55	293	293	293	309	65.7	(106.0)	30.9	48.3	75.5	51.2	32.4	43	141	3.55
3.60	285	285	285	301	65.3	(105.5)	29.9	47.6	75.0	50.3	31.2	—	138	3.60
3.65	277	277	277	292	64.6	(104.5)	28.8	46.7	74.4	49.3	29.9	41	134	3.65
3.70	269	269	269	284	64.1	(104.0)	27.6	45.9	73.7	48.3	28.5	40	130	3.70
3.75	262	262	262	276	63.6	(103.0)	26.6	45.0	73.1	47.3	27.3	39	127	3.75
3.80	255	255	255	269	63.0	(102.0)	25.4	44.2	72.5	46.2	26.0	38	123	3.80
3.85	248	248	248	261	62.5	(101.0)	24.2	43.2	71.7	45.1	24.5	37	120	3.85
3.90	241	241	241	253	61.8	100.0	22.8	42.0	70.9	43.9	22.8	36	116	3.90
3.95	235	235	235	247	61.4	99.0	21.7	41.4	70.3	42.9	21.5	35	114	3.95
4.00	229	229	229	241	60.8	98.2	20.5	40.5	69.7	41.9	20.1	34	111	4.00
4.05	223	223	223	234	—	97.3	(18.8)	—	—	—	—	—	—	4.05
4.10	217	217	217	228	—	96.4	(17.5)	—	—	—	—	33	105	4.10
4.15	212	212	212	222	—	95.5	(16.0)	—	—	—	—	—	102	4.15
4.20	207	207	207	218	—	94.6	(15.2)	—	—	—	—	32	100	4.20
4.25	201	201	201	212	—	93.8	(13.8)	—	—	—	—	31	98	4.25
4.30	197	197	197	207	—	92.8	(12.7)	—	—	—	—	30	95	4.30
4.35	192	192	192	202	—	91.9	(11.5)	—	—	—	—	29	93	4.35
4.40	187	187	187	196	—	90.7	(10.0)	—	—	—	—	—	90	4.40
4.45	183	183	183	192	—	90.0	(9.0)	—	—	—	—	28	89	4.45
4.50	179	179	179	188	—	89.0	(8.0)	—	—	—	—	27	87	4.50
4.55	174	174	174	182	—	87.8	(6.4)	—	—	—	—	—	85	4.55
4.60	170	170	170	178	—	86.8	(5.4)	—	—	—	—	26	83	4.60
4.65	167	167	167	175	—	86.0	(4.4)	—	—	—	—	—	81	4.65
4.70	163	163	163	171	—	85.0	(3.3)	—	—	—	—	25	79	4.70
4.80	156	156	156	163	—	82.9	(0.9)	—	—	—	—	—	76	4.80
4.90	149	149	149	156	—	80.8	—	—	—	—	—	23	73	4.90
5.00	143	143	143	150	—	78.7	—	—	—	—	—	22	71	5.00
5.10	137	137	137	143	—	76.4	—	—	—	—	—	21	67	5.10
5.20	131	131	131	137	—	74.0	—	—	—	—	—	—	65	5.20
5.30	126	126	126	132	—	72.0	—	—	—	—	—	20	63	5.30
5.40	121	121	121	127	—	69.8	—	—	—	—	—	19	60	5.40
5.50	116	116	116	122	—	67.6	—	—	—	—	—	18	58	5.50
5.60	111	111	111	117	—	65.7	—	—	—	—	—	15	56	5.60

[a] The values in this table shown in **bold face type** correspond to the values shown in the corresponding joint SAE-ASM-ASTM Committee on Hardness Conversions as printed in ASTM E 140, Table 3.

[b] Brinell numbers are based on the diameter of impressed indentation. If the ball distorts (flattens) during test, Brinell numbers will vary in accordance with the degree of such distortion when related to hardnesses determined with a Vickers Diamond Pyramid, Rockwell Brale, or other penetrator which does not sensibly distort. At high hardnesses, therefore, the relationship between Brinell and Vickers or Rockwell scales is affected by the type of ball used. Steel balls (Standard or Hultgren) tend to flatten slightly more than carbide balls, resulting in larger indentation and lower Brinell number than shown by a carbide ball. Thus, on a specimen of 640 Vickers, a Hultgren ball will leave a 2.55 mm impression (578 Bhn), and the carbide ball a 2.50 mm impression (601 Bhn). Conversely, identical impression diameters for both types of ball will correspond to different Vickers or Rockwell values. Thus, if both impressions are 2.55 mm (578 Bhn), material tested with a Hultgren ball has a Vickers Hardness 640, while material tested with a carbide ball has a Vickers Hardness 615.

[c] Values in ( ) are beyond normal range and are given for information only.

TABLE 3—BRINELL HARDNESS NUMBERS (10-MM BALL DIAMETER)

Dia of Indentation, mm	Loads, kg						Dia of Indentation, mm	Loads, kg					
	500	1000	1500	2000	2500	3000		500	1000	1500	2000	2500	3000
2.00	158	316	473	632	788	945	4.25	33.6	67.2	101	134	167	201
2.05	150	300	450	600	750	899	4.30	32.8	65.6	98.5	131	164	197
2.10	143	286	428	572	714	856	4.35	32.0	64.0	96.0	128	160	192
2.15	136	272	409	544	681	817	4.40	31.2	62.4	93.5	125	156	187
2.20	130	260	390	520	650	780	4.45	30.5	61.0	91.5	122	153	183
2.25	124	248	373	496	621	745	4.50	29.8	59.6	89.5	119	149	179
2.30	119	238	356	476	593	712	4.55	29.1	58.2	87.0	116	145	174
2.35	114	228	341	456	568	682	4.60	28.4	56.8	85.0	114	142	170
2.40	109	218	327	436	545	653	4.65	27.8	55.6	83.5	111	139	167
2.45	104	208	314	416	522	627	4.70	27.1	54.2	81.5	108	136	163
2.50	100	200	301	400	500	601	4.75	26.5	53.0	79.5	106	133	159
2.55	96.3	193	289	385	482	578	4.80	25.9	51.8	78.0	104	130	156
2.60	92.6	185	278	370	462	555	4.85	25.4	50.8	76.0	102	127	152
2.65	89.0	178	267	356	445	534	4.90	24.8	49.6	74.5	99.2	124	149
2.70	85.7	171	257	343	429	514	4.95	24.3	48.6	73.0	97.2	122	146
2.75	82.6	165	248	330	413	495	5.00	23.8	47.6	71.5	95.2	119	143
2.80	79.6	159	239	318	398	477	5.05	23.3	46.6	70.0	93.2	117	140
2.85	76.8	154	231	307	384	461	5.10	22.8	45.6	68.5	91.2	114	137
2.90	74.1	148	222	296	371	444	5.15	22.3	44.6	67.0	89.2	112	134
2.95	71.5	143	215	286	358	429	5.20	21.8	43.6	65.5	87.2	109	131
3.00	69.1	138	208	276	346	415	5.25	21.4	42.8	64.0	85.6	107	128
3.05	66.8	134	201	267	334	401	5.30	20.9	41.8	63.0	83.6	105	126
3.10	64.6	129	194	258	324	388	5.35	20.5	41.0	61.5	82.0	103	123
3.15	62.5	125	188	250	313	375	5.40	20.1	40.2	60.5	80.4	101	121
3.20	60.5	121	182	242	303	363	5.45	19.7	39.4	59.0	78.8	98.5	118
3.25	58.6	117	176	234	293	352	5.50	19.3	38.6	58.0	77.2	96.5	116
3.30	56.8	114	171	227	284	341	5.55	18.9	37.8	57.0	75.6	95.0	114
3.35	55.1	110	166	220	276	331	5.60	18.6	37.2	55.5	74.4	92.5	111
3.40	53.4	107	161	214	267	321	5.65	18.2	36.4	54.5	72.8	90.8	109
3.45	51.8	104	156	207	259	311	5.70	17.8	35.6	53.5	71.2	89.2	107
3.50	50.3	101	151	201	252	302	5.75	17.5	35.0	52.5	70.0	87.5	105
3.55	48.9	97.8	147	196	244	293	5.80	17.2	34.4	51.5	68.8	85.8	103
3.60	47.5	95.0	143	190	238	285	5.85	16.8	33.6	50.5	67.2	84.2	101
3.65	46.1	92.2	139	184	231	277	5.90	16.5	33.0	49.6	66.0	82.5	99.2
3.70	44.9	89.8	135	180	225	269	5.95	16.2	32.4	48.7	64.8	81.2	97.3
3.75	43.6	87.2	131	174	218	262	6.00	15.9	31.8	47.8	63.6	79.5	95.5
3.80	42.4	84.8	128	170	212	255	6.05	15.6	31.2	46.9	62.4	78.0	93.7
3.85	41.3	82.6	124	165	207	248	6.10	15.3	30.6	46.0	61.2	76.7	92.0
3.90	40.2	80.4	121	161	201	241	6.15	15.0	30.0	45.2	60.4	75.3	90.3
3.95	39.1	78.2	118	156	196	235	6.20	14.8	29.6	44.4	59.2	73.8	88.7
4.00	38.1	76.2	115	152	191	229	6.25	14.5	29.0	43.6	58.0	72.6	87.1
4.05	37.1	74.2	112	148	186	223	6.30	14.2	28.4	42.8	56.8	71.3	85.5
4.10	36.2	72.4	109	145	181	217	6.35	14.0	28.0	42.0	56.0	70.0	84.0
4.15	35.3	70.6	106	141	177	212	6.40	13.7	27.4	41.3	54.8	68.8	82.5
4.20	34.4	68.8	104	138	172	207	6.45	13.5	27.0	40.5	54.0	67.5	81.0

Thickness of Specimens—The minimum allowable thickness of any specimen varies according to the hardness, the load applied, and the kind of test point or penetrator used. See Tables 2 and 3 of ASTM E 18 for selection of Rockwell scales for a given hardness and thickness of specimen.

Curved Surfaces—Data for hardness tests on a highly curved surface should be accompanied by a statement of the radius of curvature. In testing small rounds, the effect of curvature can be eliminated by making a small flat spot on the specimen. See Tables 5 and 6 of ASTM E 18 for corrections for tests on cylindrical specimens.

Case Hardened Parts—The following information defines the minimum effective case depths which will allow the accurate determination of indentation surface hardness measurements for standard and superficial hardness tests. These practices are for fully hardened cases either as quenched or with low (approximately 350 F) temperature temper. Tempering to lower hardness levels may require less indention load than described.

Effective case is defined as the depth to RC 50 or its equivalent (see SAE J423). These practices will not avoid errors caused by surface metal of reduced hardness resulting from decarburization, retained austenite, grinding damage, etc. These recommendations may be used for all levels of core hardness.

It is recommended that surface hardness be *specified and measured* with a scale which has indentation loads no greater than the following:

Minimum Effective Case Depth on Parts	Scale
0.007	H R15N
0.010	H R30N
0.012	H R45N
0.015	H RA
0.018	H RD
0.021	H RC

Rockwell Scales—The black figures are used only for the diamond brale penetrator with various loads. Scale A applies when the major load is 60 kg, scale D when it is 100 kg, and scale C when the load is 150 kg. The red figures are used for readings obtained with ball penetrators regardless of size or magnitude of major load; scale B applies when the major load of 100 kg is applied to the 1/16-in. steel ball penetrator. All data should be prefixed by a letter showing whether the values are on the A, B, C, or D scale.

Testing Cast Iron—Materials such as cast iron with graphite particles and some nonferrous materials whose crystalline aggregates are comparatively large must be tested with a penetrator of sufficient size to overcome local or grain hardness in order to secure mass hardness.

Superficial Hardness Tester—The Rockwell Superficial Hardness tester utilizes the same principle as the regular Rockwell tester, but employs a light minor load of 3 kg and a light major load of 15, 30, or 45 kg in conjunction with a more sensitive depth measuring system. It is recommended for use on thin strip or sheet material, nitrided or lightly carburized pieces, finished pieces on which large test marks would be undesirable, areas near edges, extremely small parts or sections, and shapes that would collapse under the comparatively heavy test loads of the regular Rockwell tester. When the 120-deg diamond cone penetrator is used, readings are designated by the letter N prefixed by the major load (that is, 15-N, 30-N, or 45-N). Similarly, the letter T prefixed by the major load is applied to readings taken with the 1/16-in. steel ball. Special penetrators for very soft metals or nonmetallic materials include 1/8-in., 1/4-in. and 1/2-in. steel balls, designated by the letters W, X, and Y, respectively. In using the Rockwell Superficial Hardness tester, the general methods prescribed for the regular Rockwell tester should be observed.

number is higher the harder the material. A minor load of 10 kg is first applied which causes an initial penetration which sets the penetrator on the material and holds it in position. The dial is set at zero on the black figure scale and the major load is applied. After the major load is applied and removed, according to standard procedure, the reading is taken while the minor load is still in position.

Preparation of Surfaces—Concordant results are dependent on surface roughness being much less than the size of the impression. Surfaces that are ridged perceptibly to the eye by rough grinding or machining offer unequal support to the penetrator. The degree of surface preparation then depends to some extent on the requirements of testing, whether they be production or research.

TABLE 4—APPROXIMATE EQUIVALENT HARDNESS NUMBERS[a] FOR ROCKWELL C HARDNESS NUMBERS, FOR STEEL

Rockwell C-Scale Hardness No.[b]	Diamond Pyramid Hardness No.	Brinell Hardness No. 10-mm Ball, 3000-kg Load			Rockwell Hardness No.[b]			Rockwell, Superficial Hardness No., Superficial Brale Penetrator			Shore Scleroscope Hardness No.	Tensile Strength (Approximate) in 1000 psi	Rockwell C-Scale Hardness No.[b]
		Standard Ball	Hultgren Ball	Tungsten-Carbide Ball	A-Scale, 60-kg Load, Brale Penetrator	B-Scale, 100-kg Load, 1/16-in. Dia Ball	D-Scale, 100-kg Load, Brale Penetrator	15-N Scale, 15-kg Load	30-N Scale, 30-kg Load	45-N Scale, 45-kg Load			
Col. 1	Col. 2	Col. 3	Col. 4	Col. 5	Col. 6	Col. 7	Col. 8	Col. 9	Col. 10	Col. 11	Col. 12	Col. 13	Col. 14
68	940	—	—	—	85.6	—	76.9	93.2	84.4	75.4	97	—	68
67	900	—	—	—	85.0	—	76.1	92.9	83.6	74.2	95	—	67
66	865	—	—	—	84.5	—	75.4	92.5	82.8	73.3	92	—	66
65	832	—	—	739	83.9	—	74.5	92.2	81.9	72.0	91	—	65
64	800	—	—	722	83.4	—	73.8	91.8	81.1	71.0	88	—	64
63	772	—	—	705	82.8	—	73.0	91.4	80.1	69.9	87	—	63
62	746	—	—	688	82.3	—	72.2	91.1	79.3	68.8	85	—	62
61	720	—	—	670	81.8	—	71.5	90.7	78.4	67.7	83	—	61

TABLE 4—APPROXIMATE EQUIVALENT HARDNESS NUMBERS[a] FOR ROCKWELL C HARDNESS NUMBERS, FOR STEEL (continued)

Rockwell C-Scale Hardness No.[b]	Diamond Pyramid Hardness No.	Brinell Hardness No. 10-mm Ball, 3000-kg Load			Rockwell Hardness No.[b]			Rockwell, Superficial Hardness No., Superficial Brale Penetrator			Shore Scleroscope Hardness No.	Tensile Strength (Approximate) in 1000 psi	Rockwell C-Scale Hardness No.[b]
		Standard Ball	Hultgren Ball	Tungsten-Carbide Ball	A-Scale, 60-kg Load, Brale Penetrator	B-Scale, 100-kg Load, 1/16-in. Dia Ball	D-Scale, 100-kg Load, Brale Penetrator	15-N Scale, 15-kg Load	30-N Scale, 30-kg Load	45-N Scale, 45-kg Load			
Col. 1	Col. 2	Col. 3	Col. 4	Col. 5	Col. 6	Col. 7	Col. 8	Col. 9	Col. 10	Col. 11	Col. 12	Col. 13	Col. 14
60	697	—	613	654	81.2	—	70.7	90.2	77.5	66.6	81	—	60
59	674	—	599	634	80.7	—	69.9	89.8	76.6	65.5	80	—	59
58	653	—	587	615	80.1	—	69.2	89.3	75.7	64.3	78	—	58
57	633	—	575	595	79.6	—	68.5	88.9	74.8	63.2	76	—	57
56	613	—	561	577	79.0	—	67.7	88.3	73.9	62.0	75	—	56
55	595	—	546	560	78.5	—	66.9	87.9	73.0	60.9	74	301	55
54	577	—	534	543	78.0	—	66.1	87.4	72.0	59.8	72	292	54
53	560	—	519	525	77.4	—	65.4	86.9	71.2	58.6	71	283	53
52	544	500	508	512	76.8	—	64.6	86.4	70.2	57.4	69	273	52
51	528	487	494	496	76.3	—	63.8	85.9	69.4	56.1	68	264	51
50	513	475	481	481	75.9	—	63.1	85.5	68.5	55.0	67	255	50
49	498	464	469	469	75.2	—	62.1	85.0	67.6	53.8	66	246	49
48	484	451	455	455	74.7	—	61.4	84.5	66.7	52.5	64	237	48
47	471	442	443	443	74.1	—	60.8	83.9	65.8	51.4	63	229	47
46	458	432	432	432	73.6	—	60.0	83.5	64.8	50.3	62	222	46
45	446	421	421	421	73.1	—	59.2	83.0	64.0	49.0	60	215	45
44	434	409	409	409	72.5	—	58.5	82.5	63.1	47.8	58	208	44
43	423	400	400	400	72.0	—	57.7	82.0	62.2	46.7	57	201	43
42	412	390	390	390	71.5	—	56.9	81.5	61.3	45.5	56	194	42
41	402	381	381	381	70.9	—	56.2	80.9	60.4	44.3	55	188	41
40	392	371	371	371	70.4	—	55.4	80.4	59.5	43.1	54	181	40
39	382	362	362	362	69.9	—	54.6	79.9	58.6	41.9	52	176	39
38	372	353	353	353	69.4	—	53.8	79.4	57.7	40.8	51	171	38
37	363	344	344	344	68.9	—	53.1	78.8	56.8	39.6	50	168	37
36	354	336	336	336	68.4	(109.0)	52.3	78.3	55.9	38.4	49	162	36
35	345	327	327	327	67.9	(108.5)	51.5	77.7	55.0	37.2	48	157	35
34	336	319	319	319	67.4	(108.0)	50.8	77.2	54.2	36.1	47	153	34
33	327	311	311	311	66.8	(107.5)	50.0	76.6	53.3	34.9	46	149	33
32	318	301	301	301	66.3	(107.0)	49.2	76.1	52.1	33.7	44	145	32
31	310	294	294	294	65.8	(106.0)	48.4	75.6	51.3	32.5	43	142	31
30	302	286	286	286	65.3	(105.5)	47.7	75.0	50.4	31.3	42	138	30
29	294	279	279	279	64.7	(104.5)	47.0	74.5	49.5	30.1	41	135	29
28	286	271	271	271	64.3	(104.0)	46.1	73.9	48.6	28.9	41	132	28
27	279	264	264	264	63.8	(103.0)	45.2	73.3	47.7	27.8	40	128	27
26	272	258	258	258	63.3	(102.5)	44.6	72.8	46.8	26.7	38	125	26
25	266	253	253	253	62.8	(101.5)	43.8	72.2	45.9	25.5	38	122	25
24	260	247	247	247	62.4	(101.0)	43.1	71.6	45.0	24.3	37	120	24
23	254	243	243	243	62.0	100.0	42.1	71.0	44.0	23.1	36	117	23
22	248	237	237	237	61.5	99.0	41.6	70.5	43.2	22.0	35	114	22
21	243	231	231	231	61.0	98.5	40.9	69.9	42.3	20.7	35	112	21
20	238	226	226	226	60.5	97.8	40.1	69.4	41.5	19.6	34	110	20
(18)	230	219	219	219	—	96.7	—	—	—	—	33	106	(18)
(16)	222	212	212	212	—	95.5	—	—	—	—	32	102	(16)
(14)	213	203	203	203	—	93.9	—	—	—	—	31	98	(14)
(12)	204	194	194	194	—	92.3	—	—	—	—	29	94	(12)
(10)	196	187	187	187	—	90.7	—	—	—	—	28	90	(10)
(8)	188	179	179	179	—	89.5	—	—	—	—	27	87	(8)
(6)	180	171	171	171	—	87.1	—	—	—	—	26	84	(6)
(4)	173	165	165	165	—	85.5	—	—	—	—	25	80	(4)
(2)	166	158	158	158	—	83.5	—	—	—	—	24	77	(2)
(0)	160	152	152	152	—	81.7	—	—	—	—	24	75	(0)

[a] The values in this table shown in **bold face type** correspond to the values shown in the corresponding joint SAE-ASM-ASTM Committee on Hardness Conversions as printed in ASTM E 140, Table 2.
[b] Values in ( ) are beyond normal range and are given for information only.

For further information on standard methods of Rockwell hardness testing of metallic materials, refer to ASTM E 18.

Shore Hardness—The Shore hardness number is the reading obtained on an arbitrary scale ranging from 0 to 120 by the rebound of a small diamond pointed hammer dropped from a fixed height. Two types of instrument are in common use, one in which the rebound is read directly on a vertical scale and the other on which the reading is registered by the instrument on a recording dial.

CAUTION: Shore readings are affected by variations in mass, form, surface, composition, and physical condition of different specimens being tested.

GRAIN SIZE DETERMINATION OF STEELS—SAE J418a

SAE Recommended Practice

Report of Iron and Steel Division approved June 1934 and last revised by Iron and Steel Technical Committee June 1964. Editorial change April 1968. Reaffirmed without change June 1979.

[This SAE Recommended Practice is based on ASTM E 112, Standard Methods for Estimating the Average Grain Size in Metals. It is published here with permission of the ASTM for convenience of SAE Handbook users.]

Scope—This classification for grain size comprises three sets of comparison charts to be used for determining grain size. These charts are presented in three categories as follows:[1]

Plate I—Untwinned grains (flat etch)

Plate II—Twinned grains (flat etch)

Plate IV—Austenite grains in steel (McQuaid-Ehn test or other test)

[1] Plates I, II, and IV may be obtained from ASTM Headquarters, 1916 Race Street, Philadelphia, Pennsylvania at a nominal cost. (For ordering purposes, Plates I, II, and IV of ASTM E 112 should be requested.) Examples of grain size standards from Plates I, II and IV are shown in Figs. 1, 2, and 3.

TABLE 1—SUGGESTED COMPARISON CHARTS

Material	Plate No.	Basic Magnification
Austenitic	II or IV	100x
Ferritic	I	100x
Carburized	IV	100x
Stainless	II	100x
Super-Strength Alloys	I or II	100x

TABLE 2—MICRO-GRAIN SIZE RELATIONSHIPS

ASTM Micro-Grain Size No.	Calculated Dia of Average Grain		Calculated Area of Average Grain Section		Nominal Grains per sq in. at 100 ×
	mm	in., × 10^{-3}	sq mm, × 10^{-3}	sq in., × 10^{-6}	
00[a]	0.508	20.0	258	400	0.250
0	0.359	14.1	129	200	0.50
0.5	0.302	11.9	91.2	141	0.707
1.0	0.254	10.0	64.5	100	1.0
—	0.250	9.84	62.5	96.9	1.03
1.5	0.214	8.41	45.6	70.7	1.41
—	0.200	7.87	40.0	62.0	1.61
—	0.180	7.09	32.4	50.2	1.99
2.0	0.180	7.07	32.3	50.0	2.0
2.5	0.151	5.95	22.8	35.4	2.83
—	0.150	5.91	22.5	34.9	2.87
3.0	0.127	5.00	16.1	25.0	4.0
—	0.120	4.72	14.4	22.3	4.48
3.5	0.107	4.20	11.4	17.7	5.66
—	0.900	3.54	8.10	12.6	7.97
4.0	0.0898	3.54	8.06	12.5	8.0
4.5	0.076	2.97	5.70	8.84	11.3
—	0.070	2.76	4.90	7.59	13.2
5.0	0.064	2.50	4.03	6.25	16.0
—	0.060	2.36	3.60	5.58	17.9
5.5	0.0534	2.10	2.85	4.42	22.6
—	0.050	1.97	2.50	3.88	25.8
6.0	0.045	1.77	2.02	3.13	32.0
—	0.040	1.58	1.60	2.48	40.3
6.5	0.038	1.49	1.43	2.21	45.3
—	0.035	1.38	1.23	1.90	52.7
7.0	0.032	1.25	1.01	1.56	64.0
—	0.030	1.18	0.90	1.40	71.7
7.5	0.027	1.05	0.713	1.10	90.5
—	0.025	0.984	0.625	0.969	103
8.0	0.0224	0.884	0.504	0.781	128
—	0.0200	0.787	0.40	0.620	161

[a] The use of 00 is recommended instead of "−1" or "minus 1" to avoid confusion.

Table 1 lists a number of materials and the comparison charts that are suggested for use in estimating their grain size by the comparison method.

NOTE: The suggestions in Table 1 are based upon the customary practices in industry. For specimens prepared according to special techniques, the appropriate comparison chart should be selected on a structural appearance basis as described in the Scope.

Establishing Ferrite Grain Size—Ferrite grain size is already established in the sample to be examined by prior processing.

In hot worked material, a specimen representing a plane transverse to the direction of working is generally suitable. However, on flat rolled material, or any other in which elongation is likely to be encountered, both a transverse plane and one parallel to the direction of working should be examined.

Revealing the Ferrite Grain Size—The specimen may be prepared by appropriate polishing and etching with a 5% nital solution for 10 sec which generally produces good grain boundary delineation.

Establishing Austenite Grain Size—Numerous methods are in use for establishing austenite grain size, and a knowledge of grain coarsening behavior is helpful in deciding which method to use. The size of austenite grains, in any particular steel, depends primarily on the temperature to which that steel is heated and the time it is held at the temperature. It should be remembered that the atmosphere in heating may affect the grain growth behavior at the outside of the piece.

Austenite grain size is influenced by most previous treatments to which steel may have been subjected, as, for example, quenching, normalizing, hot working, and cold working. It is therefore advisable, when testing for austenite grain size, to consider the effects of prior or subsequent treatments on the precise piece (or typical piece) which is under consideration.

As may be agreed upon between the manufacturer and the purchaser, austenite grain size shall in this classification be established by either of the following:

1. Carburizing at 1700 F (925 C) for 8 hr (the McQuaid-Ehn test), which is recommended for carburizing grades, and often employed for other grades as well.

2. Heating at a temperature not over 50 F (28C) above the normal heat treating temperature and for not over 50% more than the normal heat treating time, the normal values being those mutually agreed upon.

The rate of cooling depends on the method of treatment, as will be evident from the methods described in the following paragraphs.

Revealing the Austenite Grain Size—For revealing austenite grain size the following methods are generally used:

1. Outlining the grains with cementite, as in carburizing (McQuaid-Ehn test) or as in high carbon steels. In the hypereutectoid zone of a McQuaid-Ehn test, or in hypereutectoid steels cooled from the austenitic condition, the austenite grain size is outlined by the cementite which precipitated in the grain boundaries. It is therefore possible to read the grain size by etching the micrographic specimen with a suitable etchant, such as nital or picral, or alkaline sodium picrate.

2. Outlining the grains with ferrite, as in the hypoeutectoid zone in carburizing, or in medium carbon steels generally, or by an interrupted cooling or gradient quench on low carbon steels. Ferrite precipitates in the austenite grain boundaries, thus indicating the austenite grain size in the hypoeutectoid zone in a McQuaid-Ehn test (see Plate Series II). Ferrite similarly outlines the former austenite grains in a medium carbon steel (say 0.50% carbon) when it has been cooled slowly from the austenite range. In low carbon steels (say 0.20% carbon) cooled slowly from the austenite range to room temperature, the amount of ferrite is so large that the former austenite grain size is masked. In this case, the steel may be cooled slowly to an intermediate temperature, to allow only a small amount of ferrite to precipitate, followed by quenching in water; an example would be a piece previously heated to 1675 F, transferred to a furnace at perhaps 1350–1450 F, held at this temperature for perhaps 3–5 minutes, and then quenched in water; the austenite grain size would be revealed by small ferrite grains outlining low carbon martensite grains.

3. Fine pearlite outlining of martensite grains, as in eutectoid steels at a not quite fully hardened zone. A method applicable particularly to eutectoid steels, which cannot be judged so readily by some other methods, is either (a) to harden a bar of such size that it is fully hardened at the outside, but not quite fully hardened in the interior, or (b) to employ a "gradient quench" in which the heated piece is for a portion of its length immersed in water and therefore fully hardened, the remainder of the piece projecting above the quenching bath, being therefore not hardened. With either method, there will be a small zone which is almost but not quite fully hardened. In this zone the former austenite grains will consist of martensite grains surrounded by small amounts of fine pearlitic ("nodular troostite"), thus revealing the grain size. These methods are also applicable to steels somewhat higher and lower than the eutectoid composition.

4. Appropriate etching of fully hardened martensite. The former austenite grain size may be revealed in steels fully hardened to martensite by using an etching reagent which develops contrast between the martensite grains. A reagent which has been recommended is 1 gr of picric acid, 5 ml of HCl

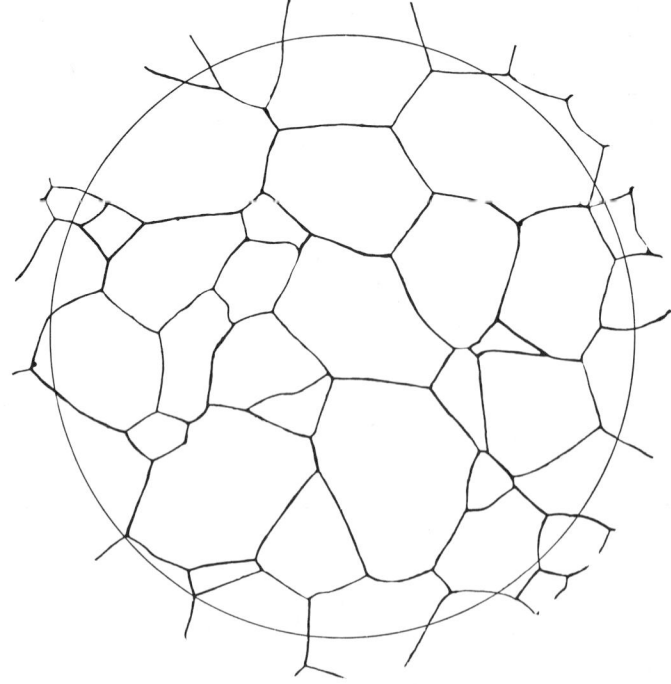

FIG. 1—EXAMPLE OF UNTWINNED GRAINS (FLAT ETCH) FROM PLATE I. GRAIN SIZE NO. 3 AT 100X

(specific gravity 1.19), and 95 ml of ethyl alcohol. Tempering for 15 minutes at 450 F prior to etching distinctly improves the contrast.

5. *Oxidation method*—The oxidation method depends on the fact that when steels are heated in an oxidizing atmosphere, oxidation takes place in part preferentially along the grain boundaries. A common procedure, therefore, is to polish the test specimen to a metallographic polish, heat it in air at the desired temperature for the desired length of time, and then repolish the specimen lightly so as merely to remove scale; whereupon the austenite grain boundaries are visible as outlined by oxide.

Estimating the Grain Size—The estimation of micro-grain size should be made by direct comparison at 100 diameters with the appropriate chart and selecting the standard which most nearly matches the image of the test specimen or interpolating between two standards. This estimated grain size should be reported to the nearest appropriate unit listed in Table 2. When the grains are of a size outside the range covered by the comparison charts or when a magnification of 100 is not satisfactory, reference should be made to ASTM E 112.

Where greater accuracy than that obtainable by the comparison method is required, a quantitative grain count may be made either by the Jeffries' planimetric or Heyn's intercept method. Both methods are more accurate for a given microscopic field but are more laborious, particularly where a number of fields must be viewed because of variations in grain size within the specimen. Heyn's intercept method is particularly suitable where the grains are not equiaxed. (See ASTM E 112.)

Report—In reporting grain size, the test conditions should be stated, including the temperature and time used in establishing the austenite grain size, and the method of revealing the grain size.

Fracture Method—There are sets of fracture standards in which the grain size is judged from the appearance of the fracture. It has been found that the arbitrarily numbered fracture grain sizes agree very well with the arbitrarily numbered grain sizes presented in Plate Series I. This coincidence makes the fracture grain sizes interchangeable with the austenite grain sizes determined microscopically (except that "duplexing" or mixed grain size is not readily discernible in fractures). The sizes observed microscopically shall be considered the primary standard, since they can be determined with measuring instruments.

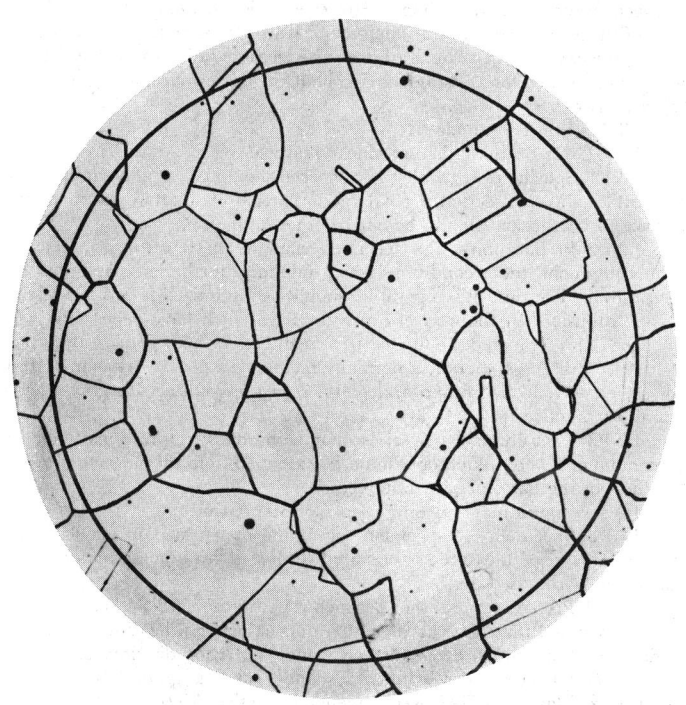

FIG. 2—EXAMPLE OF TWIN GRAINS (FLAT ETCH) FROM PLATE II. GRAIN SIZE NO. 3 AT 100X

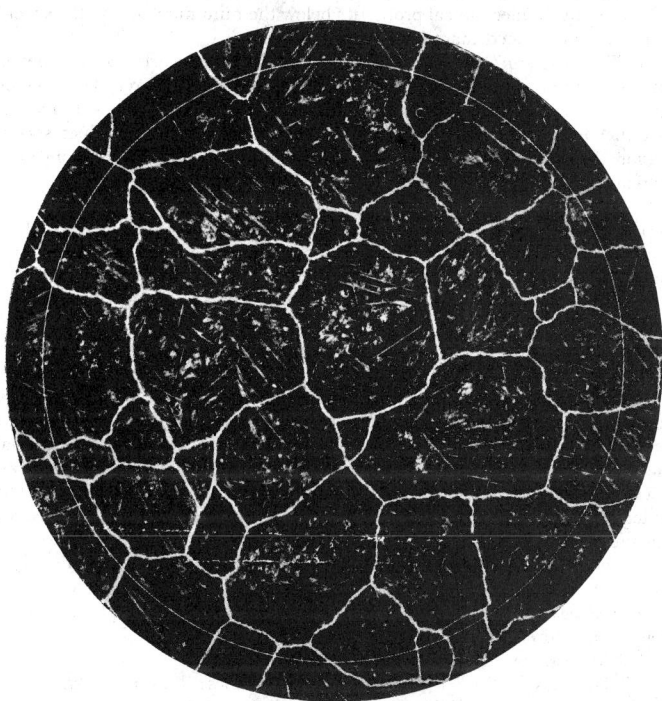

FIG. 3—EXAMPLE OF AUSTENITE GRAINS IN STEEL FROM PLATE IV. GRAIN SIZE NO. 3 AT 100X

METHODS OF MEASURING DECARBURIZATION—SAE J419

SAE Recommended Practice

Report of Iron and Steel Technical Committee approved May 1959. Reaffirmed without change June 1979.

1. Scope—This report covers the recommended practice for the evaluation and measurement of decarburization in ferrous material. Included are definitions of types with charts and micrographs and methods most commonly used for the measurement of decarburization.

2. Definition—Decarburization is the loss of carbon at the surface of commercial ferrous materials which have been heated for fabrication or when heated to modify mechanical properties.

2.1 Complete Decarburization—Complete loss of carbon as determined by examination.

2.2 Partial Decarburization—Any measurable loss of carbon content, less than complete, with respect to carbon level of base material.

2.3 Effective Decarburization—Any measurable loss of carbon content which results in mechanical properties below the minimum acceptable specifications for hardened material.

3. Types of Decarburization—Three general types of decarburization may be prevalent in ferrous materials dependent on manner and degree of carbon loss from the material. Classifying decarburization into 3 types may aid in selecting the process necessary to utilize the material to meet a product specification. Accompanying photomicrographs are illustrations of typical conditions which may be encountered.

3.1 Type 1 Decarburization—indicated by the curve and photomicrographs in Fig. 1, covers that condition in which carbon free ferrite exists for a measurable distance below the surface. Underneath the ferrite will exist varying degrees of partial decarburization.

3.2 Type 2 Decarburization—indicated by the curve and photomicrographs in Fig. 2, covers that condition in which there is a loss of more than 50% of the base carbon at the surface but where no measurable depth of complete decarburization is evident.

3.3 Type 3 Decarburization—indicated by the curve and photomicrographs in Fig. 3, covers that condition where some loss of carbon at the surface is evident but to a degree less than 50% of the base carbon of the material.

3.3.1 Further subdividing of Type 3 Decarburization may be necessary for highly stressed members such as spring or high strength materials. In this category the effective decarburization may be determined by microhardness testing for materials lower than 0.6% base carbon.

Chemical analysis procedures may be required when examining high carbon materials.

4. Methods of Measuring Decarburization—The methods used to any extent for the measurement of decarburization are:

4.1 Microscopic
4.2 Hardness: 4.2.1 Cross Section Microhardness Traverse; 4.2.2 Longitudinal Traverse; 4.2.3 File
4.3 Chemical Analysis

The accuracy of the method to be used is dependent on the degree of decarburization, microstructure, and base carbon content of the steel. The metallographic method is sufficiently accurate for most annealed and hot rolled materials but inaccurate for small amounts of decarburization in high carbon (above 0.60%), high hardness steels. The hardness method is also insensitive in this latter case and recourse must be taken to chemical analysis.

File method is often suitable for detecting decarburization of hardened materials during shop processing but not for accurate measurement.

It is fundamental that true measure of decarburization lies in chemical analysis for carbon content. This method is normally used only in research investigations or to check accuracy of other methods. Analysis is difficult and slow in application because of limitations of size and section of material. Method of procuring sample itself depends upon shape and hardness of test piece. Parts and/or test specimens too hard to machine should be tempered at 1100 to 1200 F to permit machining of surface layers into chips for subsequent carbon analysis. Obviously a sample which is annealed to permit milling of chips may be modified in its condition of decarburization. Standard methods for carbon determination are described in textbooks of analytical chemistry.

4.1 Microscopic Method

4.1.1 *Specimen*—The area to be examined should be cut at right angles to the surface. Samples are preferably taken when the material is in full annealed or in hot rolled condition. Other conditions, such as spheroidized annealed, hardened, or cold worked material, may be examined but care must be used in interpretation. For sections up to $\frac{1}{2}$ in., the entire cross section is normally mounted for examination. For larger sections, a specimen should be cut to include about $\frac{3}{4}$ in. of the surface to be examined. Corners of straight sided sections should not be included, since they are not considered representative.

4.1.2 *Preparation*—In mounting the specimen for grinding and polishing, protection from rounding the surface to be examined is essential. The specimen should be mounted in a clamp or in a plastic mount, the latter being the preferred method. An additional method of protection is to electroplate a metallic coating of 0.001–0.003 in. on the specimen before mounting.

After mounting the surface should be ground and polished in accordance with good metallographic practice.

Etching in a 3% nital (concentrated nitric acid in alcohol) is usually suitable for showing changes in microstructure caused by decarburization.

4.1.3 *Measurement*—Magnification for examination can be agreed on between purchaser and producer. However, it is recommended that 100× magnification be used. If the microscope is of a type with a ground glass screen, the extent of decarburization can be measured directly with a scale. If an eyepiece is used for measurement, it should be an appropriate type containing a cross hair or a scale.

4.2 Hardness Methods

4.2.1 Cross Section Microhardness Traverse

4.2.1.1 *Specimen*—Sample to be checked should be cut at right angles to the surface. If cross section is too large, a portion of suitable size including surface to be checked should be cut before examination.

4.2.1.2 *Preparation*—The specimen shall be hardened by quenching from equipment under conditions which minimize further change in carbon distribution. The time at temperature should be minimized to avoid excessive carbon diffusion. In the case of finished parts, which have been previously quenched and tempered, no further treatment is necessary. For sections up to $\frac{1}{2}$ in. the entire cross section is normally mounted in plastic. After mounting the surface should be ground and polished in accordance with good metallographic practice.

4.2.1.3 *Measurement*—A series of microhardness impressions made by pyramidal or Knoop indentors should be extended from the surface until the hardness of the base metal is obtained.

4.2.2 Longitudinal Traverse (Taper or Step Grind)

4.2.2.1 *Specimen*—A specimen containing the surface on which decarburization is to be measured is prepared so that it can be manipulated on a superficial hardness tester.

4.2.2.2 *Preparation*—If the specimen is not in the hardened condition, it is recommended that it be hardened by quenching from heating equipment under conditions which avoid further change in carbon distribution.

For the taper grind specimen, a shallow taper is ground through the decarburized layer, see SAE Recommended Practice, Methods of Measuring Case Depth—SAE J423. The angle is chosen so that hardness readings spaced equal distances apart will represent the hardness at the desired increments below the surface. Unless special anvils are used on the hardness tester, a parallel section should be prepared so that indentations will be at right angles to the tapered surface.

For the step grind procedure, flats are ground at predetermined intervals below the original surface. These flats should have sufficient area to allow several hardness readings to be taken on each flat.

4.2.2.3 *Measurement*—A superficial hardness tester such as a Rockwell Superficial or Vickers Tester using a light load should be employed in making the hardness measurements. The depth of decarburization is defined as the distance measured from the nearest original surface to the point at which no increase in hardness is found.

4.2.3 File Method

4.2.3.1 *Specimen*—A specimen of suitable size is obtained from the desired location.

4.2.3.2 *Preparation*—The specimen shall be hardened by quenching from heating equipment under conditions which avoid further decarburization.

4.2.3.3 *Measurement*—After hardening, the sample is filed. Base metals expected to harden to above RC 60 and found to be file soft are probably decarburized. Decarburization of base metals that will not harden to RC 60 cannot be detected by this method unless specially prepared files are used. The extent and severity of any decarburization detected by this method should be verified by either of the other two methods.

4.3 Chemical Analysis—Procedure is the same as SAE J423.

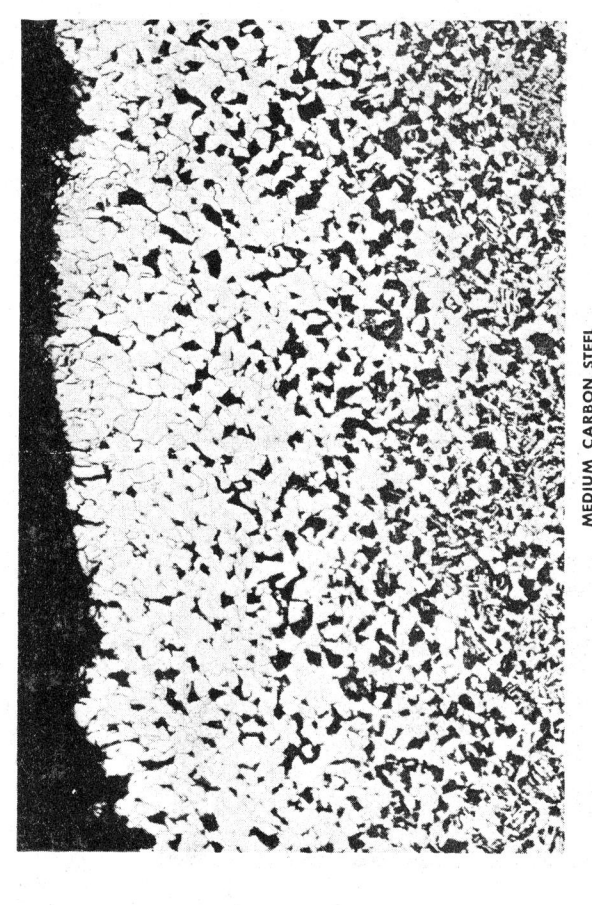

MEDIUM CARBON STEEL

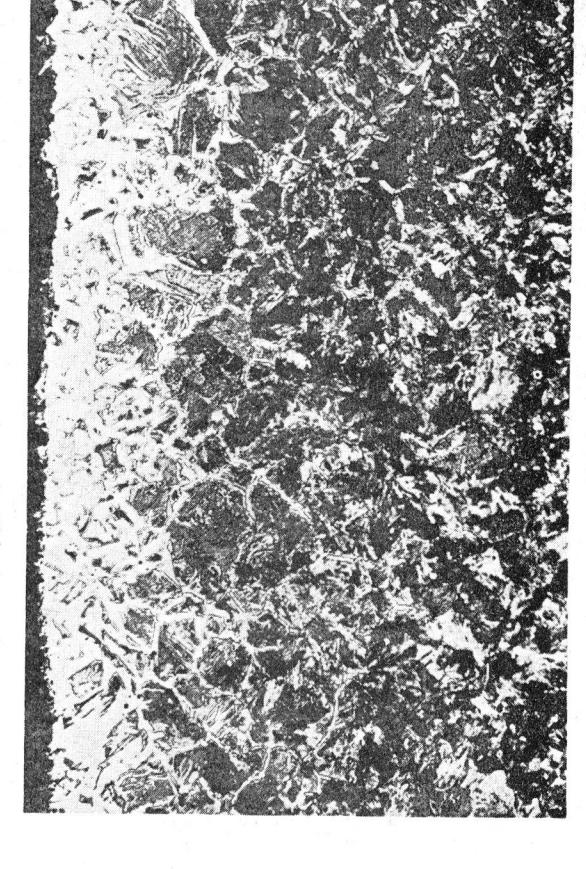

HIGH CARBON STEEL

Magnification 100 X—Nital Etch

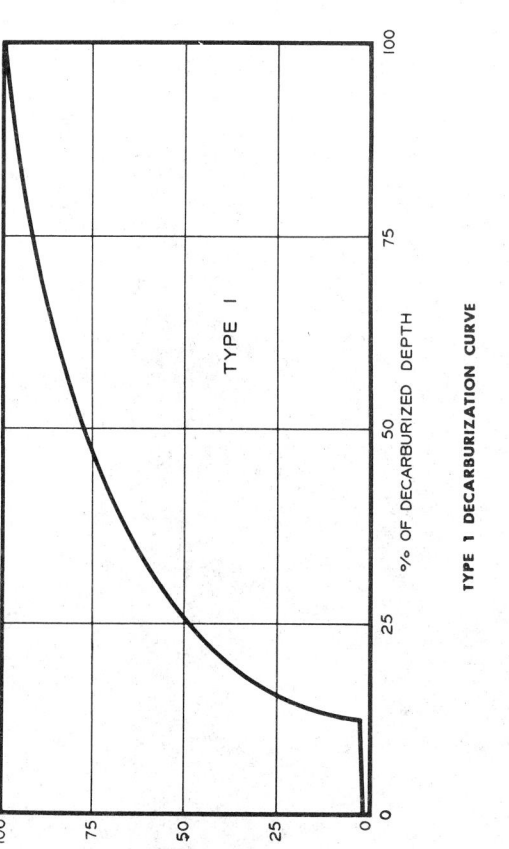

LOW CARBON STEEL

TYPE 1 DECARBURIZATION CURVE

FIG. 1—TYPE 1 DECARBURIZATION

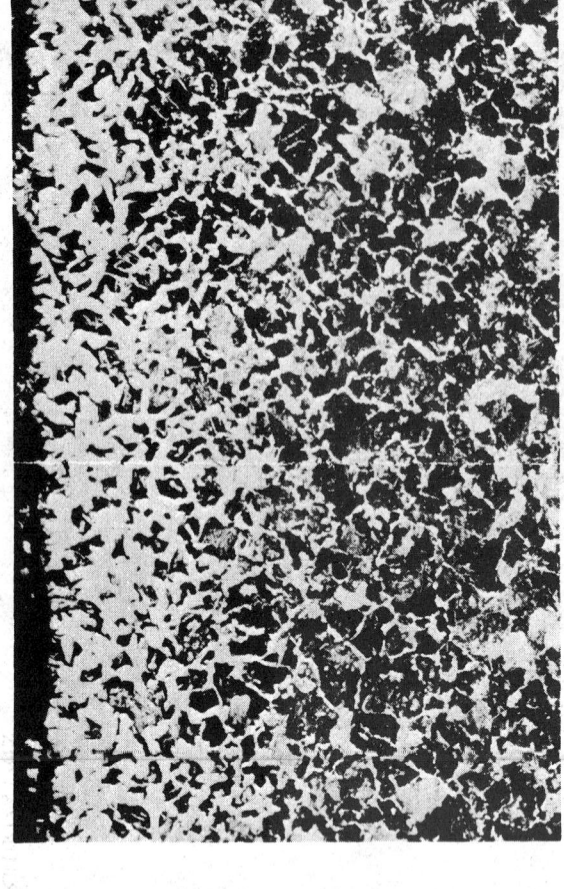
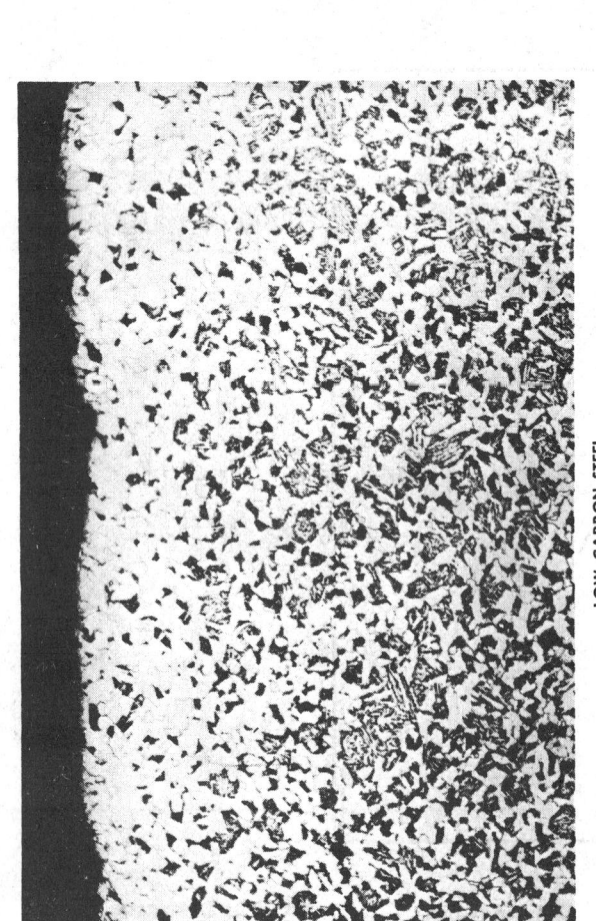

FIG. 2—TYPE 2 DECARBURIZATION

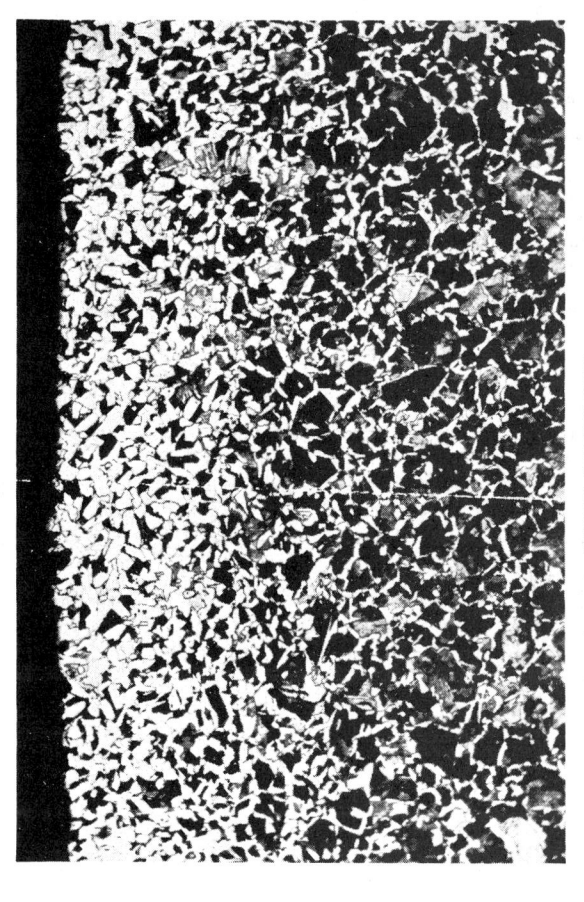

TYPE 3 DECARBURIZATION CURVE

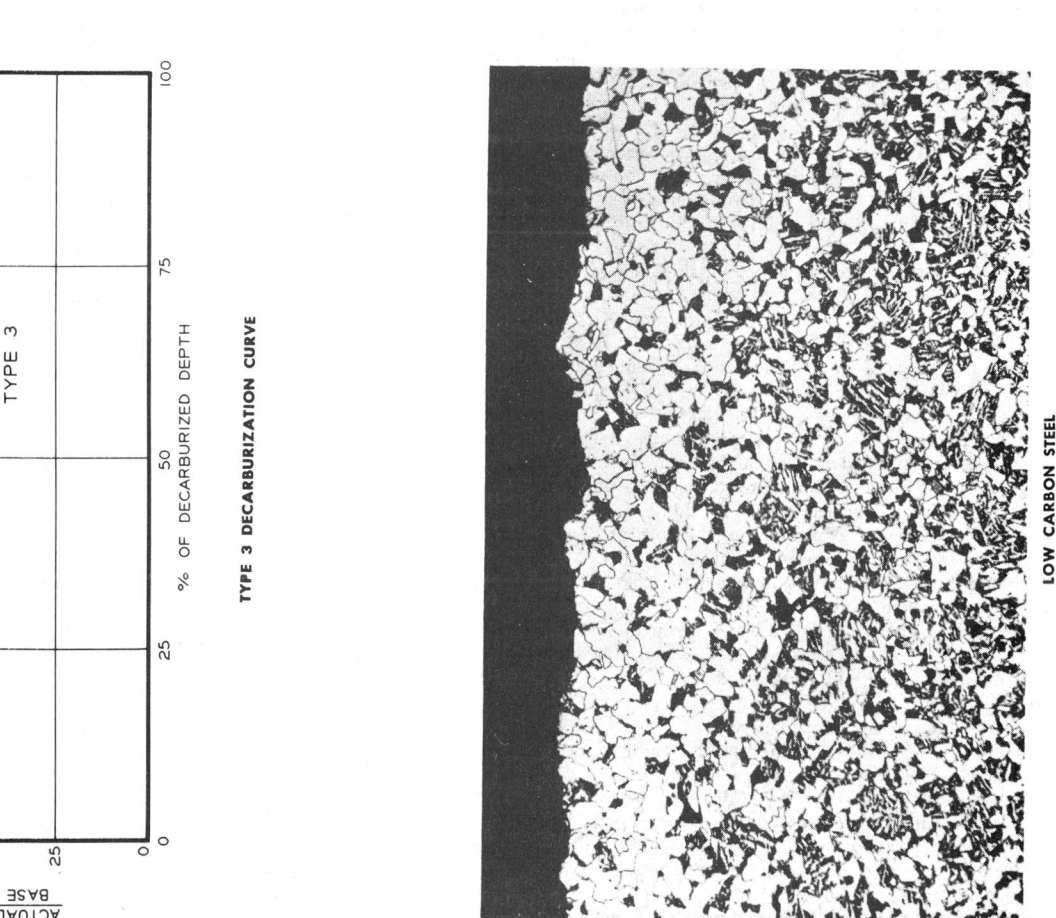

MEDIUM CARBON STEEL

HIGH CARBON STEEL

LOW CARBON STEEL

Magnification 100 X—Nital Etch

FIG 3—TYPE 3 DECARBURIZATION

METHODS OF DETERMINING PLASTIC DEFORMATION IN SHEET METAL STAMPINGS—SAE J863c

SAE Recommended Practice

Report of Iron and Steel Technical Committee approved June 1963 and last revised July 1972.

Scope—This SAE Recommended Practice describes methods for determining plastic deformation encountered in the forming or drawing of sheet steel.

Introduction—The preferred method for determining plastic strain is the circle grid and the severity curve. The scribed square and change in thickness methods may also be used to evaluate deformation during the forming of a flat sheet into the desired shape.

Methods

Circle Grid Method—The test system employs electrochemically etched circle patterns on the surface of a sheet metal blank and a severity curve for the evaluation of strains developed by forming in press operations. It is useful in the laboratory and in the press room. Selection from the various steels which are commercially available can be done effectively by employing this technique. In addition, corrective action in die or part design to improve performance is often indicated.

The severity curve in Fig. 1 has been developed from actual measurements of the major (e_1) and associated minor (e_2) strains found in critical areas of production type stampings. Strain combinations which locate below this curve are safe, while those which fall above the curve are critical. The left of zero portion of the curve (tension-compression) represents 25% change in unit area. The right side (tension-tension) defines a severity limit since no constant percent change in area will be found to be critical.

PROCEDURE

1. Obtain or prepare a stencil with selected circles in a uniform pattern. The circles may be 0.10–1.0 in. (2.5–25.4 mm) in diameter; the most convenient diameter is 0.20 in. (5.1 mm) because it is easy to read and the gage spacing is short enough to show the maximum strain in a specific location on the part.

2. The sheet metal blanks should be cleaned to remove excess oil and dirt; however, some precoated sheets can be etched without removing the coating. The area(s) to be etched should be determined from observation of panels previously formed; generally, the area which has a split problem is selected for etching. Normally, the convex side of the radius is gridded. If sufficient time is available, the entire blank may be etched, since valuable information can be obtained about the movement of metal in stamping a part when strains can be evaluated in what may appear to be noncritical areas. Additionally, for complex shapes it may be desirable to etch both surfaces of blanks so that the strains which occur in reverse draws can be determined.

3. The etch pad is saturated with an appropriate electrolyte. Various electrolytes are available from suppliers of the etching equipment. Some electrolytes are more effective than others for etching certain surfaces, such as terne plate and other metallic coated steels. A rust inhibiting solution is preferred for steel sheets.

4. A ground clamp from the transformer of suitable amperage (10–50 A is usually used) is fastened to the blank and the second lead is attached to the etch pad. Although the current may be turned on at this time, caution should be taken not to lay the pad on the etch blank as it will arc. It is advisable to refrain from touching the metal of the etch pad and the grounded sheet blank.

5. The stencil is placed with the plastic coating against the sheet surface in the area to be etched. Wetting the stencil with a minimum amount of electrolyte will assist in smoothing out the wrinkles and gives a more uniform etch. The etch pad is now positioned on the stencil and the current turned on, if it is not already on. Apply suitable pressure to the pad. Only the minimum time necessary to produce a clear etched pattern should be used. The etching time will vary with the amperage available from the power source and the stencil area, as well as the pad area in contact with the stencil. Rocker type etch pads give good prints and require less amperage than flat surfaced pads. Excessive current causes stencil damage.

6. The etching solution activates the surface of the metal and may cause rusting unless it is inhibited. After the desired area has been etched, the blank should be wiped or rinsed, dried, and neutralized.

7. The etched blank is now ready for forming. The lubricants and press conditions should simulate production situations.

8. If a sequence of operations is used in forming a part, it is desirable to etch sufficient blanks so that each operation can be studied.

MEASUREMENT OF STRAIN AFTER FORMING—After forming, the circles are generally distorted into elliptical shapes (Fig. 2). These ellipses have major and minor strain axes. The major strain (e_1) is always defined to be the direction in which the greatest positive strain has occurred without regard to original blank edges or the sheet rolling direction.

The minor strain (e_2) is defined to be 90 deg to the major strain direction.

There are several methods for determining the major and minor strains of the formed panel. Typical tools are a pair of dividers and a scale ruled in 50ths of an inch (0.5 mm). For sharp radii, a thin plastic scale which can follow the contour of the stamping can be used to determine the dimensions of the ellipses. (Scales are available to read the percent strain directly.)

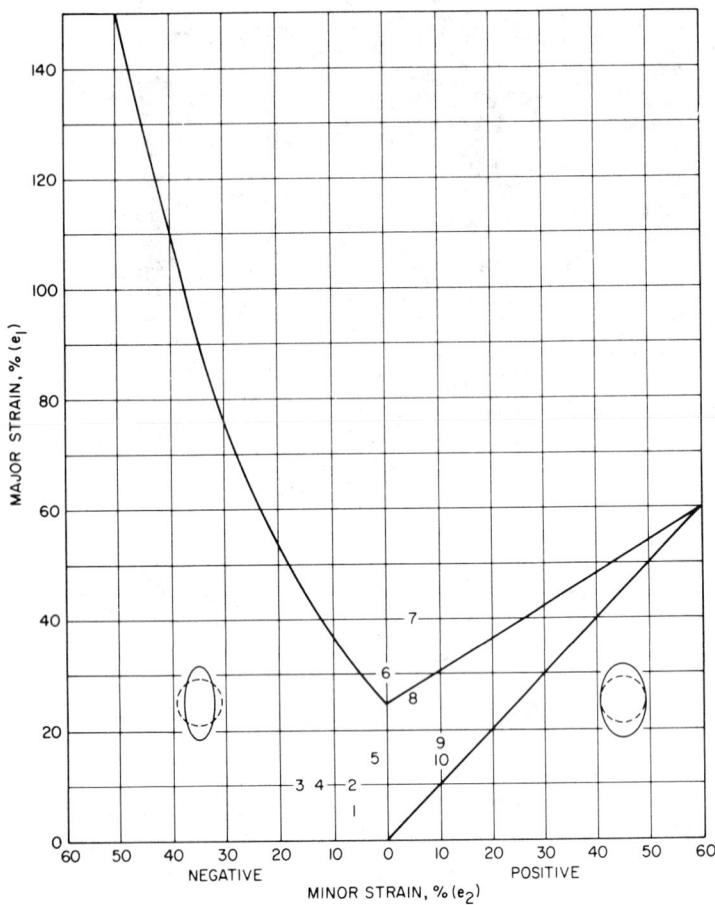

FIG. 1—SEVERITY CURVE

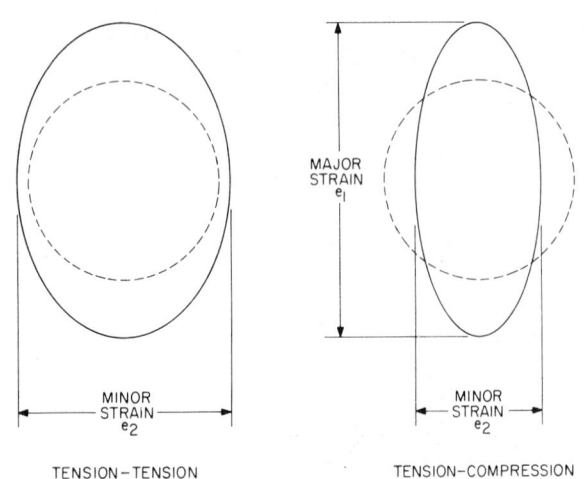

FIG. 2

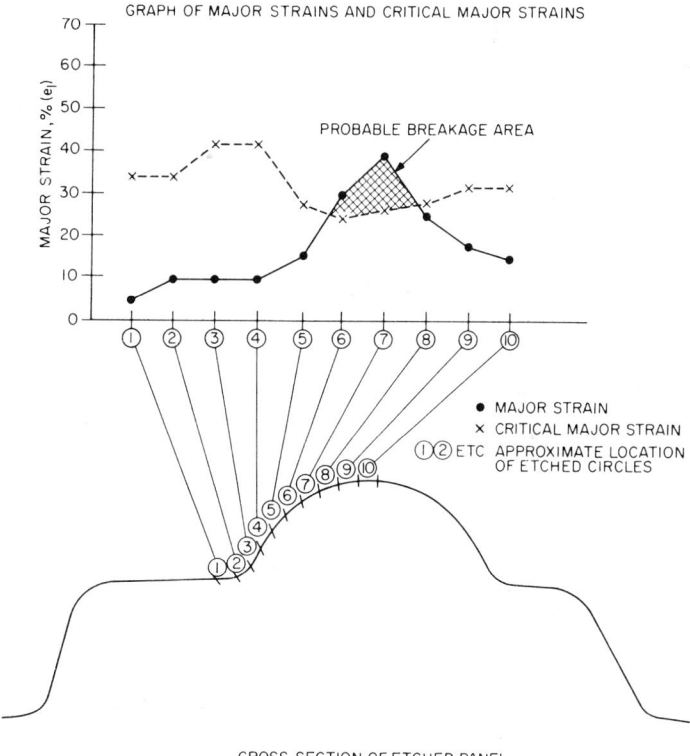

FIG. 3

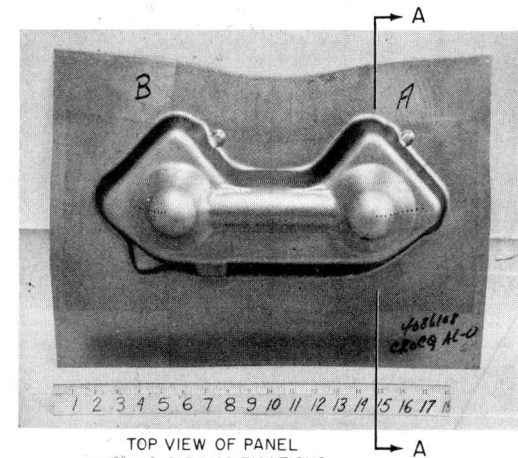

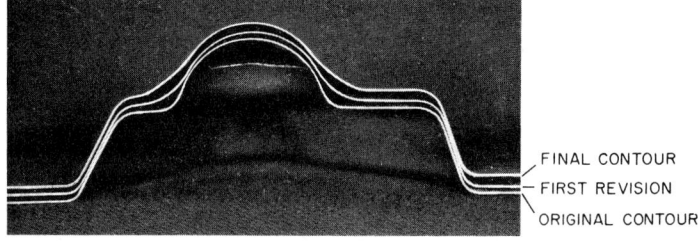

FIG. 4

EVALUATION OF STRAIN MEASUREMENTS—The e_1 strain is always positive while the e_2 strain may be zero, positive, or negative, as indicated on the severity curve chart (Fig. 1). The maximum e_1 and associated e_2 values measured in critical areas on the formed part are plotted on the graph paper containing the severity curve by locating the point of intersection of the e_1, e_2 strains.

If this point is on or below the severity curve, the strain should not cause breakage. Points further below the curve indicate that a less ductile material of a lower grade may be applied. Points above the severity curve show that fabrication has induced strains which could result in breakage. Therefore, in evaluations on stampings exhibiting these strains, efforts should be made to provide and e_1, e_2 strain combination which would lie on or below the severity curve. A different e_1, e_2 strain combination can be obtained through changes of one or more of the forming variables such as die conditions, lubricants, blank size, thickness, or material grade.

When attempting to change the relationship of e_1 and e_2 strains, it should be noted that on the severity curve the most severe condition for a given e_1 strain is at 0% e_2 strain. This means the metal works best when it is allowed to deform in two dimensions, e_1 and e_2, rather than being restricted in one dimension. A change in e_2 to decrease the severity can be made by changing one of the previously mentioned forming variables or the die design, for example, improving lubrication on the tension-tension side will increase e_2 and decrease the severity.

In addition to the severity curve, the e_1, e_2 strain measurements may be used to evaluate the material requirements on the basis of strain gradients, as illustrated in Fig. 3, or by plotting contours of equivalent strain levels on the surface of the formed part. Even when the level of strain is relatively low, parts in which the e_1 strain is changing rapidly either in magnitude or direction over a short span on the surface may require more ductile grades of sheet metal, change in lubrication, or change in part design.

EXAMPLE OF MAJOR AND MINOR STRAIN DISTRIBUTION—A formed panel (Fig. 4) with a cross section as shown in Fig. 3 is used to illustrate major and minor strain combinations. A plot of the major strain distribution should be made by finding the ellipse with the largest major strain (circle 7) and measuring both the major and minor strains in the row of ellipses running in the direction of the major strain. The solid dots (Fig. 3) are the measured major strains for each ellipse. The Xs are the critical major strains as determined from the severity curve at the corresponding minor strain (intersection of the measured major strain and the severity curve).

Usually a single row of ellipses will suffice to determine the most severe strain distribution. The resulting strain distribution plot (Fig. 4) illustrates both severity of the strain compared to the critical strain limits and the concentration of strain in the stamping. Steep strain gradients should be avoided because they are inherent fracture sites.

EXAMPLE FOR REDUCING SPLITTING TENDENCY—In an area such as represented in Fig. 3, the splitting tendency can be reduced as follows:

If the radius of the part in the region of circle 1 is increased, some strain can be induced to take place in this area which will allow the major strain in circle 7 to be reduced sufficiently to bring the strain combination below the critical limit. This course of action requires no building nor reshaping of the punch, only grinding in the radius.

The total average major strain required to make this formation is only 17.5%; yet in a 0.2 in. (5.1 mm) circle the strain is as high as 40%. The strain distribution curve puts forth graphically the need to distribute the strain over the length of the line by some means as described above.

Change in lubrication can also improve the strain distribution of a stamping. If the strain over the punch is critical, the amount of stretch (strain) required to make the shape can be reduced by allowing metal to flow in over the punch by decreasing the friction through the use of a more effective lubricant in the hold-down area.

If the part is critical, a change in material may help. That is, a material having a better uniform elongation will distribute the strain more uniformly or a material having a higher "r" value will make it possible to "draw" in more metal from the hold-down area so that less stretch is necessary to form the part.

Scribed Square Method—The basic technique is to draw a panel from a blank which has been scribed both longitudinally and transversely with a series of parallel lines spaced at 1 in. (25.4 mm) intervals. The lines on the panel are measured after drawing and the stretch or draw calculated as the percent increase in area of a 1 in. (25.4 mm) square. This is a fairly simple procedure for panels having generous radii and fairly even stretch or draw. A great many major panels fall in this category and in these instances it is quite easy to pick out the square area exhibiting the greatest increase.

If the square or line to be measured is no longer a flat surface, place a narrow strip of masking (or other suitable tape) on the formed surface and mark the points which are to be measured. Remove the tape, place on a plane surface, and determine the distance between the points with a steel scale.

There will be cases of minor increase in area with major elongation in the one direction. In these instances, the percent elongation should be recorded.

Thickness Method—There are instances when the maximum stretch is confined to an area smaller than 1 in.^{2} (25.4 mm^2) or the shape of the square has been distorted irregularly, making measurement difficult and calculation inaccurate. When either of these conditions exists, an electronic thickness gage may be used at the area in question or this area may be sectioned and the decrease in metal thickness measured with a ball point micrometer. The increase in unit area can be calculated by dividing the original thickness by the final thickness.

EXAMPLE—Assuming the blank thickness to be 0.035 in. (0.889 mm) and the final thickness to be 0.028 in. (0.7112 mm), the increase in unit area would be 0.035/0.028 = 1.25 (0.7112/0.889 = 1.25) or 25% increase in unit area.

SHEET STEEL THICKNESS AND PROFILE—SAE J1058

SAE Information Report

Report of Iron and Steel Technical Committee approved February 1976.

1. Scope—This report provides information regarding methods for specifying thickness of sheet steel and how thickness tolerances apply. This report also explains the profile of as-rolled sheet steel.

2. Specifying Thickness

2.1 Method I—Ordered to Minimum Thickness—This is the most common method used today because it provides the design engineer with a specification system where he can select a minimum thickness without regard to tolerance due to ordered width. When material is ordered to minimum thickness, the tolerance is all plus and can be determined by doubling the tabular values shown in the ASTM A568—Carbon Sheet Steel thickness tolerances tables. An example of this method is as follows: 0.031 in (7.9 mm) x 36.5 in (930 mm) coil. Using Table 23 (Cold Rolled Sheet), attached, tolerance would be plus 0.006 in (0.15 mm).

2.2 Method II—Ordered to Nominal Thickness—Sheet steel can be ordered to a specified nominal thickness, with the tolerances being plus and minus as shown in the ASTM A568, Table 23 (Cold Rolled Sheet—thickness tolerances). An example of this method is shown below.

0.034 in (8.6 mm) x 36.5 in (930 mm)

Tolerance would be plus and minus 0.003 in (0.08 mm).

3. Defining Profile

3.1 Feather Edge[1]—Feather edge is generally understood to be the thickness deviation between a location 3/8 in (9.5 mm) from the mill trimmed edge of sheet steel and a position 1 in (25 mm) to 2 in (50 mm) in from the mill trimmed edge.

3.2 Crown[1]—Crown is generally understood to be the difference in thickness between a point 3/8 in (9.5 mm) in from the mill trimmed edge and the center area of the sheet across the width as rolled. A more correct interpretation of crown would be the difference in thickness 1 in (25 mm) to 2 in (50 mm) in from the mill trimmed edge and the thickness at the center of the sheet width as rolled.

[1] To illustrate the phenomenon of feather edge and crown, profiles of typical hot rolled and cold rolled sheet are illustrated below. Actually, however, no such "classic profile" exists.

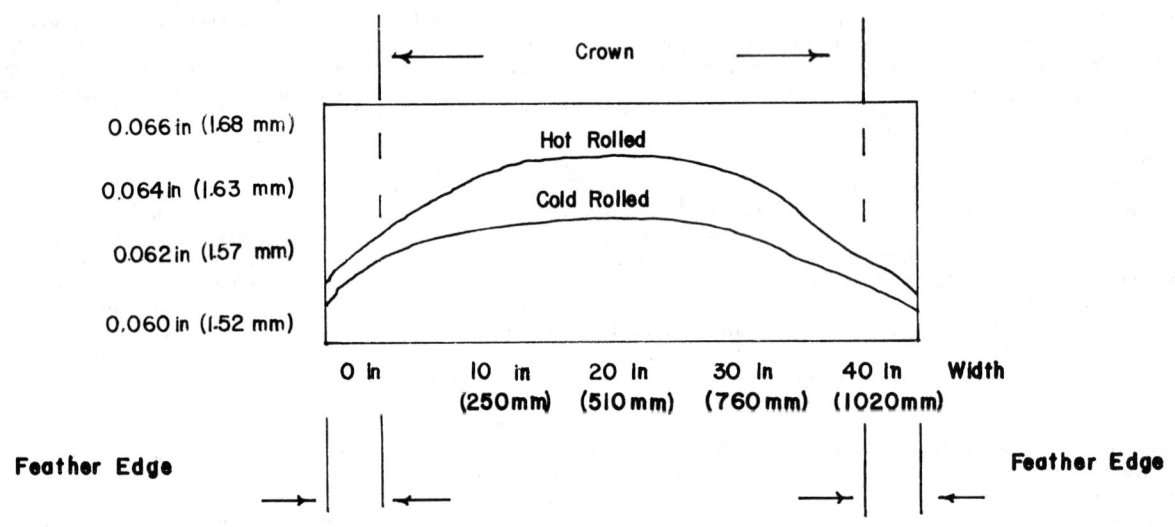

THICKNESS TOLERANCES FOR WIDTH AND THICKNESSES OVER AND UNDER, in (mm)

Specified Width, in (mm)	Specified Thickness, in (mm)					
	0.1419 (3.60 mm) 0.0972 (2.47 mm)	0.0971 (2.47 mm) 0.0710 (1.80 mm)	0.0709 (1.80 mm) 0.0568 (1.44 mm)	0.0567 (1.44 mm) 0.0389 (0.99 mm)	0.0388 (0.98 mm) 0.0195 (0.50 mm)	0.0194 (0.49 mm) 0.0142 (0.36 mm)
Over 12 (300 mm) to 15 (380 mm) incl.	0.005 (0.13 mm)	0.005 (0.13 mm)	0.005 (0.13 mm)	0.004 (0.10 mm)	0.003 (0.08 mm)	0.002 (0.05 mm)
Over 15 (380 mm) to 72 (1830 mm) incl.	0.006 (0.15 mm)	0.005 (0.13 mm)	0.005 (0.13 mm)	0.004 (0.10 mm)	0.003 (0.08 mm)	0.002 (0.05 mm)
Over 72 (1830 mm)	0.007 (0.18 mm)	0.006 (0.15 mm)	0.005 (0.13 mm)	0.004 (0.10 mm)	0.003 (0.08 mm)	NA

NOTE 1: Thickness is measured at any point across the width not less than 3/8 in (9.5 mm) from a side edge.

NOTE 2: Regardless of whether thickness tolerance is specified equally or unequally, over and under, the total tolerance should be equal to twice the tabular tolerances.

PROPERTIES OF LOW CARBON STEEL SHEETS AND STRIP AND THEIR RELATIONSHIP TO FORMABILITY—SAE J877

SAE Information Report

Report of Iron and Steel Technical Committee approved October 1963. Editorial change June 1967. Reaffirmed without change November 1968.

Problems associated with evaluation of formability or drawability of sheet steel have, for many years, received a great deal of attention and creative thought on the part of scientific metallurgists, sheet steel producers, and fabricators. By nature, these problems are complex and difficult to solve because of the number of involved variables.

In the mid-1930's, when drawability started to become an important factor in the performance of sheet steel, the tests available were Rockwell hardness, Olsen or Erichsen ductility, and standard tension tests. Investigators accumulated masses of data based on the available tests and had varying degrees of success in attempting to correlate them with the fabricating performance. Sheet processing at this time was not as sophisticated as we think it is today. Sheets were either hot rolled; normalized and second annealed; or pickled, cold reduced, and box annealed. The metallurgists were further burdened with calibration of testing equipment and reproducibility of results on rimmed steel, a material of known nonuniformity.

As long ago as 1940 the AISI Technical Committee on Sheet Steel reviewed this problem extensively. The ASM Committee on Formability of Steel published the results of an excellent severity classification in *Metal Progress*, August 1955, and has since produced sections in the 1961 edition of Metals Handbook on "The Selection of Low Carbon Steel Sheets for Deep Drawing," "The Selection of Low Carbon Steel Sheets for Formability," and "Low Carbon Steel Sheet." The last is replete with mechanical properties and their expected ranges for the various types and qualities of sheet steel.

Traditional Mechanical Properties—The mechanical properties of low carbon steel sheets are not accurately related to their performance in fabrication, and are not ordinarily used in specifications unless special strength properties are required in the fabricated product. As a matter of general interest, Fig. 1 (which is abstracted from "Low Carbon Steel Sheet") gives typical ranges of mechanical properties of sheets manufactured by three representative mills. With this amount of overlap in the bar charts, it is obviously difficult to separate even killed steel from rimmed steel on the basis of mechanical property data from a limited testing program. It will be noted that the ranges are broader for hot rolled sheets than for the more tailored cold rolled sheets and that cold rolled, special killed has the narrowest range. It is true also that the range covered by each bar in Fig. 1 is considerably restricted by eliminating the effects of gage or thickness and segregation. In hot rolled sheets the heavier gages, that is 10 and 12 gage, will be considerably softer and more ductile than 16 gage. This is because the method of manufacture necessitates finishing and coiling at higher temperatures; hence, a tendency towards more complete annealing after coiling or piling. Likewise, cold rolled sheets in the heavier gages are softer than lighter gage sheets because, in general, they are cold reduced less before annealing.

In rimmed steel, to a large extent, and in killed steel, to a somewhat smaller degree, segregation plays a large part in the distribution curves or charts of mechanical test data when sampling is done at random. The metalloids—carbon, sulfur, and phosphorus—are the elements which are most prone to segregate in a rimmed steel ingot. These elements are from two to five times greater in the top center (core) of the ingot than they are in the skin material which solidifies first. The segregation of these elements causes increased hardness and strength and decreased ductility. For instance, in hot rolled sheets there may be as much as five, or even seven, points variation in Rockwell B from edge to center of a sheet taken from the top portion of the ingot and the same difference between center readings on sheets taken from the top and bottom of the same ingot. In cold rolled sheets this effect of segregation may amount to as much as 12 points Rockwell B. If we superimpose on these the effects of variation from heat to heat through chemistry and variations in finishing and annealing temperatures, it becomes obvious that the selection of the test specimen becomes very important.

During the production of both hot rolled and cold rolled sheets, sampling procedures include prescribed test locations. This enables the steel producer to prejudge material with a greater degree of accuracy than can possibly be done with random sampling.

Test Specimen—The selection of the test specimen is as important as the test itself. The specimen should be as representative of the sheet or lift as possible. While it is not possible to establish precise sampling procedures because each situation is somewhat different from any other, some guides may be suggested:

1. Because the severe stretch or draw in a stamping usually takes place well in from the edge of the sheet, test specimens should be taken from the central three-fourths of the width.

2. Because segregation affects the properties in the center of the sheet and has no effect on the edges, any attempt to determine the range or distribution of properties throughout a lift or a coil should include specimens from the center of the sheet.

3. Because the ends of coils are usually not representative of the body, specimens should be taken only after uncoiling 50–100 ft. Coil ends in the hot rolled sheets cool faster after coiling than the center. In cold rolled sheets the outside laps during annealing may be as much as 100 F hotter than the balance of the coil. These conditions produce variations in addition to segregation.

Standard Tests of Formability—The Rockwell hardness and the Olsen cup tests are usually the first, and frequently the only, mechanical tests made to indicate formability. These tests are not an exact measure of formability but they may serve as a guide and are useful for broad applications. In respect to these test values, 0.400 Olsen may be related to about 35–40 Rockwell B for a 20 gage sheet and these are indicative of good drawing quality. Likewise, 0.375 Olsen may be related to about 45–50 Rockwell B and average drawing quality, while 0.360 Olsen and 55–60 Rockwell B indicate below average drawing quality or commercial quality level. These tests are easily and quickly made and require a minimum quantity of steel.

Olsen Cup Test—The Olsen cup test value recorded is the height of the cup in thousandths of an inch at the instant the punch load starts to drop. Load values are not otherwise significant. The higher cup draws usually indicate better ductility.

Some laboratories read the cup height at visible fracture which results in recorded values higher by 0.010–0.020 in. Those experienced in using this test can predict the approximate coarsening behavior of the steel after forming by examination of the sides of the cup. With ASTM grain sizes 8 and 9, the sides and top of the cup are smooth; with grain size 7 and larger, there is moderate

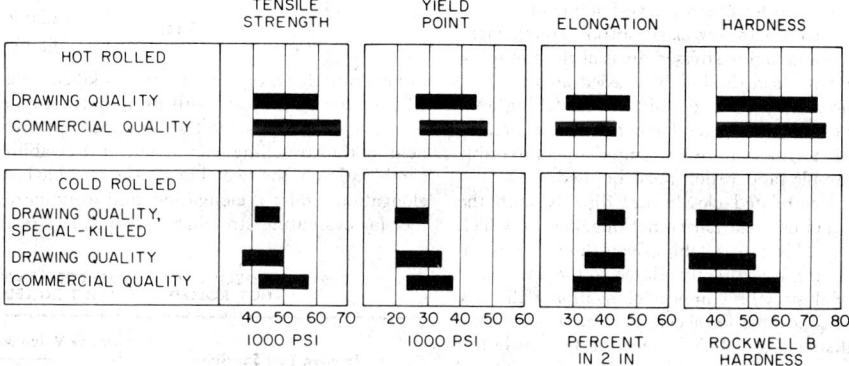

FIG. 1—TYPICAL RANGE OF MECHANICAL PROPERTIES OF 1008 STEEL FURNISHED BY THREE MILLS. HOT ROLLED SHEET THICKNESSES FROM 0.0598 TO 0.1345 IN. (16 TO 10 GAGE INCLUSIVE); COLD ROLLED FROM 0.0299 TO 0.0598 IN. (22 TO 16 GAGE INCLUSIVE). ALL COLD ROLLED GRADES INCLUDE A TEMPER ROLLING OR SKIN PASSING. ALL GRADES ROLLED FROM RIMMED STEEL EXCEPT THE ONE LABELED SPECIAL KILLED.

to heavy roughening which becomes progressively coarser as the grain size increases. Ridges across the top of an Olsen cup will indicate the presence of roller leveler or coil breaks previous to testing.

Directionality and brittle fracture may also be detected by examination of the cup fracture.

Rockwell Hardness Test—This test is used as one of the controls in sheet steel production and, as indicated previously, if testing position within the ingot is known, Rockwell hardness is a fairly good indicator of formability. This does not mean that rigid limits of inspection and rejection can be set up on the basis of Rockwell testing even with controlled testing positions, but it does provide the sheet producer with broad limits by which he may prejudge applicability for severe forming as compared to relatively simple forming applications.

There are some restrictions which must be adhered to if hardness testing is to be performed properly:

1. There are minimum thicknesses for various hardness levels which can be tested on the Rockwell B scale. Twenty gage (0.036 in.) can be tested on the B scale if it is no softer than B 40. Thinner and softer sheets must be tested on the F scale or even the proper T scale. These limits are well delineated in ASTM E 18, Methods of Test for Rockwell Hardness and Rockwell Superficial Hardness of Metallic Materials.

2. To secure the most accurate data for control or investigational work the surfaces of the sheet to be tested must be prepared by touching them to a 180 grit belt sander. Heavy or coarse surface texture (40 μ in. or rougher) may produce hardness readings as much as 5 Rockwell B points lower than the true hardness of the sheet.

Tensile Tests—When properly interpreted, the full data obtained from properly prepared and performed tensile tests give a somewhat more accurate measure of formability, but even they must be considered inadequate when samples are taken without regard to the nonuniformity in properties across the width of a sheet or length of a coil.

Tension tests should be made on a standard $\frac{1}{2}$ in. wide, 2 in. gage length ($2\frac{1}{4}$ in. parallel section) specimen if elongation comparisons are to be made. The edges of the gage length should be carefully milled. Blanked or sheared specimens without proper preparation by machining give inaccurate results, particularly elongation and tensile strength. Both parallel milled and center taper milled specimens are used. The center tapered specimen insures center breaks, but at a slight sacrifice in elongation compared to a parallel milled specimen. See ASTM E 8, Methods of Tension Testing of Metallic Materials, Sections 5(b), 7, and 9(a).

Yield Point—The determination of the yield point is complicated by the variety of load extension contours, which depend upon the condition of the steel and the conditions under which it is loaded. If the yield point is sharp and shows yield point elongation, the value recorded when testing cold rolled sheets is known as the lower yield point, and is the lowest stress reached during yield point extension. When tests are made against minimum specification values, the upper yield point is recorded. If the yield point is indefinite, as indicated by a smooth stress strain diagram, the yield point is recorded as the stress at 0.5% extension under load. The yield point is very significant. A low yield point is best if other properties are favorable and if it is not caused by abnormal grain growth. Steels with low yield point deform easily in compression; hence, there is less tendency for buckles to form in tension-compression types of shell or cup drawing.

Tensile Strength—Tensile strength is recorded as the stress at the highest load reached during the tensile test.

Elastic Ratio—This is the yield point divided by the tensile strength and is an important consideration in severe stretching. The lower this value, the greater the spread between the yield point and the tensile strength. With low elastic ratios, the sheet starts to deform at lower stress levels and continues to deform over a wider stress range. Steels with low elastic ratios stretch more evenly between moderately stressed and highly stressed areas of the stamping rather than by concentrating the stretch in the highly stressed areas.

Total Elongation—This is direct measure of ductility in stretching and, therefore, the steel with high elongation will stretch farther before failure. As pointed out previously, if elongation values are to be compared, comparable specimen sizes and methods of sample preparation must be used.

Uniform Elongation—This is correlated closely and directly with the amount of deformation that can occur before localized necking, beyond which point forming operations are not readily controllable. Even though uniform elongation is a unidirectional property, it is directly related to the amount of stretch possible in multidirectional stretch. Uniform elongation will vary between 20 and 28%; however, it is possible for sheets to stretch more than 40% in one direction in a biaxial stress field. This is because the tensile test specimen is not restrained and, therefore, contracts laterally.

Yield Point Elongation—This is direct measure of the depth or intensity of stretcher strains that will develop in lightly formed areas. Even a small amount of yield point elongation is undesirable in exposed parts and must be minimized by properly roller leveling immediately before drawing.

New Testing Methods—Recently, scientific experimenters have made significant progress in understanding the behavior of sheet metal in a drawing operation. They have even devised some new tests whose applications may reveal some of the specifics we need to know about formability and drawability.

Plastic strain ratio, work hardening exponent, anisotropy, stretch forming, and scribed square (see SAE J863) are already becoming widely accepted terms describing some areas of drawability determination.

Plastic Strain Ratio, r—When a tensile test specimen from a sheet of ductile metal having isotropic mechanical properties is stretched 20%, the width will contract 10% and the thickness will also contract 10%. This is essentially true for a hot rolled or a normalized low carbon steel sheet. If the sheet has been cold reduced and annealed subcritically by conventional methods, it will have a certain degree of preferred crystallographic orientation, resulting in anisotropic mechanical properties. In this case the width contraction of the tension test specimen may be 12% and the thickness contraction 8%. Since the plastic strain ratio, r, is defined as the width strain divided by the thickness strain, $r = 1.5$ for cold reduced, annealed specimen and $r = 1.0$ for the normalized specimen.

In the case of an anisotropic sheet the direction of sampling for the tensile test specimen also influences the plastic strain ratio. The results for rimmed and aluminum killed steel sheets, shown in Table 1, may be considered typical.

Aluminum killed steel sheets processed for flattened or "pancake" grain characteristically have higher r values in all test directions than rimmed steels. For both grades the maximum r normally occurs in the transverse direction and the minimum in the diagonal direction. As a class, the aluminum killed steels processed to obtain high average r values are capable of producing the most difficult stampings.

Work Hardening Exponent—One of the properties of ductile metals is to increase in hardness and strength as they are plastically deformed. This is demonstrated in a tension test in which the first sections of the test specimen are strengthened sufficiently to stretch plastically by work hardening to prevent further extension until all other sections are equally extended. Local necking begins when the work hardening rate is not great enough to compensate for the increased stress associated with the reduction in area of the cross section. A high work hardening rate is conducive to a high uniform elongation, and usually to a high total elongation. This characteristic is important in drawn parts in which the metal undergoes uniaxial or biaxial stretching.

Cup Drawing Tests—Olsen and Erichsen cup drawing or ductility tests take their place along with hardness tests among the "old-line" methods of evaluating drawing sheets.

A few years ago more sophisticated cup drawing tools and procedures appeared in Europe and Japan, and the development of these tests continues. Foremost among these is the Swift cup drawing test. The punch shape may be varied from essentially flat to hemispherical or elliptical. Although less widely used, the Fukui conical cup drawing test has also been carefully developed. The punch head used in this test may be essentially flat or hemispherical, as in the Swift test. The 30 deg conical entry section of the Fukui die controls wrinkling of the blank, so that no hold-down is required, provided the proper relationship is maintained between sheet thickness and blank diameter. In both the Fukui and Swift tests, the dies are modified for testing different sheet thicknesses.

These cup drawing tests differ in principle from Erichsen and Olsen ductility tests in that the hold-down and lubrication are designed to permit passage of the blank, without wrinkling, over the die radius and into the drawn cup. The limiting size of blank which can be drawn without breakage occurring in the cylinder wall is expressed as the limiting drawing ratio:

$$LDR = \frac{\text{Blank diameter}}{\text{Cup diameter}}$$

This ratio for a typical aluminum killed steel is 2.22 for the Swift test and 2.75 for the Fukui test, both using flat bottomed punches.

Stretchability—Cup tests can be adapted to measure the drawing component or the stretching component of drawability, and some types measure a combination of the two. The tensile test, having capability of measuring total elongation, uniform elongation, and work hardening exponent, is a powerful tool for evaluating stretchability. Biaxial stretchability can be measured by

TABLE 1—MECHANICAL PROPERTIES AND CHARACTERISTICS OF HOT ROLLED AND COLD ROLLED SHEET STEEL

Tension Test Specimen	r Value Angle with Direction of Rolling		
	0 deg	45 deg	90 deg
Rimmed Steel, Annealed	1.22	0.95	1.55
Aluminum Killed, Annealed	1.61	1.21	1.88
Rimmed or Aluminum Killed, Normalized	0.92	0.95	0.98

the hydraulic bulge test, but at considerable cost in time, effort, and sample material. The bulge test, like the tensile test, stretches the sheet without introducing surface frictional forces. This is not realistic. The results of uniaxial stretch tests with surface contact between the punch and the strip (in other words, stretch forming test) seem to be more meaningful. The advantages of this test are the relatively large area of contact between punch and strip and the opportunity it affords to test the directionality of the sheet material. Lubrication and deformation speed effects can also be studied. This test effectively measures the effects of temper rolling and of aging.

Sheet Steel Quality and Expected Properties—Table 2 indicates mechanical properties and characteristics of some of the types of hot rolled and cold rolled sheet steel. The exactness of this table is indicated by showing a range for Rockwell B and typical properties for yield point, tensile strength, and elongation. Such ranges and values are not recommended for specification use. This table is simply intended to indicate what is available with respect to thickness, chemistry, and mechanical properties.

TABLE 2—MECHANICAL PROPERTIES OF HOT ROLLED AND COLD ROLLED SHEET STEEL

Sheet Quality	Normally Available Thickness Range, in.	Standard Chemistry (Ladle)	For Exposed or Unexposed Parts	Usual Rockwell B Range	Strain Aging Propensity	Strain or Flute	Typical Properties[a]		
							Yield Point, psi	Tensile Strength, psi	Elongation, % (1/2 × 2 in.)
Hot Rolled and Hot Rolled Pickled, Commercial Quality	0.059–0.229	1008–1010 1012	U	50–75	Yes	Yes	38,000	52,000	30
Drawing Quality	0.075–0.187	1006–1008	U	45–65	Yes	Yes	35,000	50,000	36
Drawing Quality, Special Killed	0.075–0.187	1006	U	48–68	Yes	Yes	35,000	50,000	40
Cold Rolled									
Commercial Quality, Annealed—Last	0.025–0.110	1008–1010	U	35–60	No	Yes	34,000	46,000	37
Commercial Quality, Annealed—Temper Rolled	0.025–0.110	1008–1010	E	38–60	Yes	No	34,000	46,000	35
Drawing Quality, Annealed—Last	0.025–0.110	1006	U	35–50	No	Yes	32,000	45,000	41
Drawing Quality, Annealed—Temper Rolled	0.025–0.110	1006	E	35–55	Yes	No	30,000	45,000	39
Drawing Quality, Special Killed, Annealed—Last	0.025–0.110	1006	U	35–48	No	Yes	30,000	43,000	42
Drawing Quality, Special Killed, Annealed—Temper Rolled	0.025–0.110	1006	E	35–48	No	No	25,000	43,000	41

NOTE: Cold rolled sheets are usually ordered temper rolled for various strain and fluting hazards, but for unexposed parts they may be ordered annealed last. If the consumer needs to avoid strain and fluting and does not have effective roller leveling equipment, he should order special killed steel temper rolled.

[a] These values are given as information only and are not intended as criteria for acceptance or rejection.

ϕ SELECTING AND SPECIFYING HOT AND COLD ROLLED STEEL SHEET AND STRIP—SAE J126a

SAE Recommended Practice

Report of Iron and Steel Technical Committee approved September 1969 and completely revised February 1979.

1. Scope—This SAE Recommended Practice outlines a procedure for selecting the proper specification for carbon steel sheet and strip which are purchased *to make an identified part*. Specifications considered are:

ASTM A 109, Steel, Carbon, Cold Rolled Strip.

ASTM A 569, Steel, Carbon (0.15 maximum percent), Hot Rolled Sheet, Commercial Quality (HRCQ).

ASTM A 621, Steel, Sheet, Carbon, Hot Rolled, Drawing Quality (HRDQ).

ASTM A 622, Steel, Sheet, Carbon, Hot Rolled, Drawing Quality, Special Killed (HRDQSK).

ASTM A 568, Steel, Carbon and High-Strength Low-Alloy Hot Rolled Sheet, and Cold Rolled Sheet, General Requirements.

ASTM A 366, Steel, Carbon, Cold Rolled Sheet, Commercial Quality (CRCQ).

ASTM A 619, Steel, Sheet, Carbon, Cold Rolled, Drawing Quality, (CRDQ).

ASTM A 620, Steel, Sheet, Carbon, Cold Rolled, Drawing Quality, Special Killed (CRDQSK).

ASTM A 749M, Steel, Carbon and High-Strength Low-Alloy, Hot Rolled Strip, General Requirements.

ASTM A 635, Steel Sheet and Strip, Carbon, Hot Rolled Commercial Quality, Heavy Thickness Coils.

(Metric ASTM documents are designated by suffix *M*)

It also describes how codes or symbols for specifying certain characteristics may be used in electronic data processing systems. Characteristics covered are:

(A) Hot or cold rolled.
(B) Sheet or strip.
(C) Severity of draw (*quality* of steel).
(D) Surface condition (finish, etc.).
(E) Edge condition.
(F) Dimensions.

It is intended that other characteristics and part identification be covered by a supplement to the specification, as necessary.

2. Procedure—Evaluate the part to determine the requirements for characteristics A–F, as follows:

A—Hot or Cold Rolled Product—Normally the finish or thickness of the metal required for a part will determine whether *hot-rolled* or *cold-rolled* product should be specified. (See Table 1 and Table 4.)

B—Sheet or Strip—Principal factors to consider in determining whether sheet or strip should be specified are:

TABLE 1A—STEEL SHEET OR STRIP PRODUCT CLASSIFICATION BY SIZE, INCH-POUND UNITS

Product	Thickness, in	Width, in	Other Limitations	Specification Symbol (ASTM No.)
Hot Rolled Sheet	0.0449 thru 0.1799	Over 12.00	a	A 569, A 619, or A 622
	0.1800 thru 0.5000	Over 12.00 thru 48.00	a	
		Over 48.00	b	A 635
	0.2300 thru 0.5000	Over 12.00 thru 48.00	b	
Hot Rolled Strip	0.0255 thru 0.2030	0.50 thru 3.50	a	A 569, A 621, or A 622
	0.0344 thru 0.2020	3.50 thru 6.00		
	0.0449 thru 0.2299	3.50 thru 12.00		
	0.2300 thru 0.5000	Over 8.00 thru 12.00	b	
	0.2300 thru 0.5000	Over 6.00 thru 12.00	b	A 635
Cold Rolled Sheet	0.0142 thru 0.0821	2.00 thru 12.00	c	A 366, A 621, or A 620
		Over 12.00 thru 23.94	d	
		Over 23.94		
	Over 0.0821	Over 12.00 thru 23.94	d	
		Over 23.94		
Cold Rolled Strip	Thru 0.2499	Over 0.50 thru 23.94	e	A 109
		Over 12.00 thru 23.94	f	

[a] Cut lengths may not be available in all combinations of width and thickness.
[b] Coils only.
[c] Cold rolled sheet, coils, and cut lengths, slit from wider coils with cut edge (only) and in thicknesses 0.0142 thru 0.0821, carbon 0.25% maximum by cast or heat analysis.
[d] When no special edge or finish (other than matte or luster) is required and/or single strand rolling of widths under 24 in is not required.
[e] Widths 2.00 thru 12.00 in, in thicknesses, 0.0142 thru 0.0821 are classified as "sheet" when slit from wider coils having a cut edge only and carbon of 0.25% maximum.
[f] Over 12.00 thru 23.94 in is classified as strip when temper class or special properties are specified.

TABLE 1B—STEEL SHEET OR STRIP PRODUCT CLASSIFICATION BY SIZE, METRIC UNITS

Product	Thickness, mm	Width, mm	Other Limitations	Specification Symbol (ASTM No.)
Hot Rolled Sheet	1.2 thru 6.00 1.2 thru 4.5	Over 300 thru 1200 Over 1200		A 569M, A 621M, or A 622M
	6.00 thru 12.5	Over 300 thru 1200		
	Over 4.5 thru 12.5	Over 1200	b	A 635M
Hot Rolled Strip	Over 0.65 thru 5 Over 0.9 thru 5 Over 1.2 thru 6	Thru 100 Over 100 thru 200 Over 200 thru 300	a	A 569M, A 621M, or A 622M
	Over 6.0 thru 12.5	Over 200 thru 300	b	A 635M
Cold Rolled Sheet	0.35 thru 2.00	50 thru 300	c	A 366M, A 619M, or A 620M
	0.35 and Over	Over 300	d	
Cold Rolled Strip	Thru 6.00	Thru 600	e	A 109M

[a] Cut lengths and coils, (cut lengths may not be available in all combinations of width and thicknesses).
[b] Coils only.
[c] Cold rolled sheet, coils, and cut lengths, slit from wider coils with cut edge (only) and in thicknesses 0.35 thru 2.00 mm, carbon 0.25% maximum by cast or heat analysis.
[d] When no special edge or finish (other than matte or luster) is required and/or single strand rolling of widths through 600 mm is not required.
[e] Widths thru 300 mm with thicknesses from 0.35 thru 2.00 mm are classified as "sheet" when slit from wider coils having a cut edge only and carbon of 0.25% maximum.

Size of part, or more specifically, size of flat steel required to develop part.
Thickness of metal required for the part.
Hot or cold rolled steel.
Selection and specification of temper for cold rolled strip. (See Table 3.)
Equipment on which the metal will be handled and fabricated.
Steel industry product classification by size. (See Table 1.)

C—Selection of Sheet Steel for Formability—For Cold Rolled Carbon Sheet and Hot Rolled Carbon Sheet and Strip, three levels of formability or drawability (called *quality* in the steel industry) are available as indicated in the Scope. They are commonly referred to as CQ, DQ, and DQSK.

The following procedure is based on a Forming Severity Index (FSI) which has been developed through experience in production forming of sheet metal stampings. The procedure is recommended for determining the quality needed for a specific part and fabrication operation.

1. Form a sample or prototype part from a specimen of known quality of steel using the gridding procedure outlined in SAE J863. (For the most accurate description of the quality of steel used for the gridded blank, the mechanical properties of a sample taken from material adjacent to the blank used for the gridded part should be known. This can be from the same blank, or the sheet preceding or following the sheet to be gridded.) If the specimen fractures, form another sample from the next better quality of steel.

2. Measure the e_1 and e_2 strains as described in SAE J863. (This should be done on a sample part which has not fractured during forming.)

3. Calculate the Forming Severity Index (FSI), for the critical area or areas of the sample part, using the following formulae:

NOTE: The e_2 strain is the associated strain (minor) measured perpendicular to the major strain, e_1.

When the e_2 strain as measured on the grid in the maximum stretch area is 0 to +30%.

$$FSI = (0.6e_2 + 15 + 350t) - e_1$$

When the e_2 strain is 0 to -30% (biaxial tension compression forming).

$$FSI = (1.5e_2 + 15 + 350t) - e_1$$

where: t = thickness of gridded panel in inches.
e_1 = major strain expressed as a percent (not a decimal value).
e_2 = minor strain expressed as a percent (not a decimal value).

The sign of e_2 is disregarded (an e_2 of -30 is expressed as 30 in the formula).

NOTE:
(a) The thickness t in the above formulae is only a correction factor for calculation of the FSI. Material formability may not be dependent on the sheet metal thickness.
(b) If t is expressed in millimeters, the multiplier will be 13.8 instead of 350.

(c) The reliability of the equations for stock over 0.125 in or 3 mm has not been established.

(d) The constant 15 in the above formulae can be modified by mutual consent of the supplier and user to provide the desired degree of risk of breakage. A constant of 15 approximates a safety factor of ten percentage of strain points for die, lubricant, press, and material variance of e_1 strain (major strain). A constant of 20 gives a safety factor of 5, and a constant of 10 gives a safety factor of 15.

4. Select the quality of steel sheet needed for the part from Table 2. When a change in material is indicated by Table 2, the selection should be verified using the new quality of sheet metal to form another sample part.

5. After a sufficient production history has been obtained through the use of the SAE Recommended Practice J424, Determination of Breakage Allowance for Steel Sheets, the quality of steel being used should be reviewed. For a given part an unusually high scrap rate indicates either a tool, lubricant, or material quality selection problem. If material quality is found to be the problem, upgrading should be considered. Conversely, an unusually low scrap rate indicates a less expensive quality should be considered. SAE Recommended Practice J863 should be followed to determine the most beneficial material change.

Examples: In the case of 0.080 in or 2.00 mm thick sheet steel with e_1 strain of 33% and an e_2 strain of +10%, the Forming Severity Index (FSI) would be:

(in)
$$FSI = [0.6(10) + 15 + 350(0.080)] - 33$$
$$FSI = 16$$

(mm)
$$FSI = [0.6(10) + 15 + 13.8(2.0)] - 33$$
$$FSI = 16$$

If the gridded panel was CRCQ steel, the FSI indicates the selection of a lower quality steel such as HRDQ (A621). Note that in cases such as this where the FSI is greater than +6 there are other factors than material which should be investigated to obtain the most economical production.

If the gridded panel was HRDQ, the indicated selection would be HRCQ (A569).

In the case of 0.036 in or 0.9 mm thick stock with an e_1 strain of 55% and an e_2 strain of -15%, the FSI would be:

(in)
$$FSI = [1.5(15) + 15 + 350(0.036)] - 55$$
$$FSI = -5$$

(mm)
$$FSI = [1.5(15) + 15 + 13.8(0.09)] - 55$$
$$FSI = -5$$

If the gridded panel was CRCQ and trouble was being encountered in forming production parts, the indicated selection would be to upgrade to CRDQ (A619). If the gridded panel was CRDQ, the indicated selection would be to upgrade to CRDQSK (A620) and if the panel was CRDQSK some other changes may be required in the forming operations or in the design of the part (see comments in Table 2).

D—Surface Condition (Finish, etc.)—Consider any surface conditions required for the part. Consult Table 4 for a description of surfaces applicable to the product selected in A, B, and C above. Designate surface, finish, coating, etc. by symbol shown.

E—Edge Condition—Consider any necesary edge conditions required for the part. Consult Table 5 for a description of edges applicable to the product

TABLE 2—SELECTION OF QUALITY OF STEEL SHEET AND STRIP BASED ON THE FORMING SEVERITY INDEX (FSI)

Quality of Steel Used for the Gridded Sample Panel	Quality Suggested When the Forming Severity Index is Within the Range Given Below			
	−20 to −11	−10 to −2	0 to +5	+6 and Greater
DQSK	DQSK[a]	DQSK[b]	DQSK	DQ
DQ	DQSK[b]	DQSK	DQ	CQ
CQ	DQSK	DQ	CQ	CQ[c]

[a] This indicates too much is being expected of the sheet metal, a part redesign or breakdown into separate components may be necessary in addition to the factors below ([b]).
[b] This represents a part which is difficult to form, other factors, such as tools, design, drawing compound application, blank development, etc., should be considered to provide an adequate Forming Severity Index because further upgrading is not possible. (Premium quality grades of steel with a high plastic strain ratio (r_m) values may be necessary if the above factors cannot be modified to produce a more favorable Forming Severity Index).
[c] The part should have no forming problems and if further economics are desired, factors other than material quality should be investigated, such as, thickness reduction, drawing compound change, or simplified tooling. In the appropriate thickness range HR should be considered in place of CR.

selected in A, B, and C above. Designate the required edge condition by the symbol shown.

F—Dimensions—Consider all dimensions required for the part. Refer to ASTM A 568 for thickness tolerance tables.

Designate dimensions in the following order: thickness, width, and length. (Note: Use the symbol C for length of material purchased in coil form.) When the measuring unit is inches, all fractions thereof shall be expressed as decimals and not as common fractions (for example, 1.25 in, not $1\frac{1}{4}$ in). State the thickness to three decimal places and width and length to two decimal places.

When metric units are used for dimensions, the thickness, width, and length should be expressed in millimeters. State thickness to one decimal place and width and length in whole numbers.

Example:

	(in)	(mm)
For cut length:	0.035 x 36.25 x 84.75	0.9 x 900 x 2153
For coil stock:	0.047 x 47.37 x C	1.2 x 1200 x C

(These units are not intended to indicate conversion of inch to metric units.)

G—For a more complete explanation of industry nomenclature for Cold Rolled or Hot Rolled Steel Sheet, the AISI Steel Customer Communication Handbooks are recommended.

TABLE 3A—(INCH-POUND UNITS)—SELECTION AND SPECIFICATION OF THE AMOUNT OF TEMPER (COLD WORK)—(APPLICABLE TO COLD ROLLED STEEL STRIP ONLY)

Requirement of Part (Relative to Hardness and Maximum Severity of Bend Involved In Forming the Part)

Stock Thickness, in		Rockwell Hardness		Bend Test Requirement	Temper of Strip Normally Required
Over	Through	Minimum	Maximum (Approx.)		
0.070	—	B84	—	No bending in any direction.	No. 1 (Hard)
0.040	0.070	B90	—		
0.025	0.040	30T76	—		
—	0.025	15T89	—		
0.040	—	B70	B85[b]	Bend 90 deg across rolling direction around a radius equal to that of the metal thickness.[a]	No. 2 (Half Hard)
0.025	0.040	30T64	30T74		
—	0.025	15T81	15T86		
0.040	—	B60	B75[b]	Bend 180 deg across rolling direction over one thickness of the strip and 90 deg in the direction of rolling around a radius equal to the thickness.	No. 3 (Quarter Hard)[a]
0.025	0.040	30T58	30T68		
—	0.025	15T83	15T88		
0.040	—	—	B65[b]	Bend flat upon itself in any direction.	No. 4 (Skin Rolled)[c]
0.025	0.040	—	30T55		
—	0.025	—	15T82.5		
0.040	—	—	B55[b]	Bend flat upon itself in any direction.	No. 5 (Dead Soft)[c]
0.025	0.040	—	30T55		
—	0.025	—	15T79.5		

[a] To bend across the rolling direction means that the crease formed by the bend shall be at right angles to the length of the strip. To bend along the rolling direction means that the crease formed by the bend shall be parallel with the length of the strip.
[b] Rockwell hardness values apply to special killed steels and also rimmed or semi-killed steels at time of shipment only. Aging of these latter steels may result in slightly higher values when tested at a later date.
[c] Number 4 and 5 tempers may sometimes be ordered with a carbon range of 0.15–0.25%. In each instance the maximum hardness requirement is established by agreement.

TABLE 3B—(METRIC UNITS)—SELECTION AND SPECIFICATION OF THE AMOUNT OF TEMPER (COLD WORK)—(APPLICABLE TO COLD ROLLED STEEL STRIP ONLY)

Requirement of Part (Relative to Hardness and Maximum Severity of Bend Involved In Forming the Part)

Stock Thickness, mm		Rockwell Hardness		Bend Test Requirement	Temper of Strip Normally Required
Over	Under	Minimum	Maximum (Approx.)		
1.8	—	B84	—	No bending in any direction.	No. 1 (Hard)
1.0	1.8	B90	—		
0.6	1.0	30T76.0	—		
—	0.6	15T90.0	—		
1.0	—	B70	B85[b]	Bend 90 deg across rolling direction, around a radius equal to that of the metal thickness.[a]	No. 2 (Half Hard)
0.6	1.0	30T63.5	30T73.5		
—	0.6	15T83.5	15T88.5		
1.0	—	B60	B75[b]	Bend 180 deg across rolling direction over one thickness of the strip and 90 deg in the direction of rolling around a radius equal to the thickness.[a]	No. 3 (Quarter Hard)
0.6	1.0	30T56.5	30T67.0		
—	0.6	15T80.0	15T85.0		
1.0	—	—	B65[b]	Bend flat upon itself in any direction.	No. 4 (Skin Rolled)[c]
0.6	1.0	—	30T60.0		
—	0.6	—	15T82.0		
1.0	—	—	B55[b]	Bend flat upon itself in any direction.	No. 5 (Dead Soft)[c]
0.6	1.0	—	30T53.0		
—	0.6	—	15T78.5		

[a] To bend across the rolling direction means that the crease formed by the bend shall be at right angles to the length of the strip. To bend along the rolling direction means that the crease formed by the bend shall be parallel with the length of the strip.
[b] Rockwell hardness values apply to special killed steels and also rimmed or semi-killed steels at time of shipment only. Aging of these latter steels may result in slightly higher values when tested at a later date.
[c] Number 4 and 5 tempers may sometimes be ordered with a carbon range of 0.15–0.25%. In each instance the maximum hardness requirement is established by agreement.

TABLE 4—SELECTION AND SPECIFICATION OF SURFACE CONDITION OF STEEL SHEET AND STRIP

Description of Surface	Surface Described Applicable To	Specification Symbol
Surface finish as normally used for unexposed automotive parts. Matte (dull) appearance. Normally annealed last.	Cold rolled sheet	U[a]
Surface finish as normally used for exposed automotive parts which require a good painted surface. Free from strain markings and fluting. Matte (dull) appearance. Temper rolled.	Cold rolled sheet	E[b]
Same as above, except commercial bright appearance.	Cold rolled sheet	B
Same as above, except luster appearance.	Cold rolled sheet	L
No. 1 or dull finish (no luster). Especially suitable for lacquer or paint adhesion. Facilitates drawing by reducing the contact friction between the die and the metal.	Cold rolled strip	1
No. 2 or regular bright finish (moderately smooth). Suitable for many applications, but not generally applicable for parts to be plated, unless polished and buffed.	Cold rolled strip	2
No. 3 or best bright finish (relatively high luster). Particularly suitable for parts to be plated.	Cold rolled strip	3
As rolled or black (oxide or scale not removed).	Hot rolled sheet and strip	A
Pickled (scale removed), not oiled.	Hot rolled sheet and strip	P
Same as above, except oiled.	Hot rolled sheet and strip	O

[a] U—unexposed is presently also designated as Class 2, Cold Rolled Sheet.
[b] E—exposed is presently also designated as Class 1, Cold Rolled Sheet.

TABLE 5—SELECTION AND SPECIFICATION OF EDGE CONDITION OF STEEL SHEET AND STRIP

Description of Edge	Edge Described Applicable To	Specification Symbol
Cut Edge.	Cold rolled sheet	None required
No. 1 Edge is a prepared edge of a specified contour (round, square, or beveled) supplied when a very accurate width is required, or where the finish of the edge is required to be suitable for electroplating, or both.	Cold rolled strip	1
No. 2 Edge is a natural mill edge carried through the cold rolling from the hot rolled strip without additional processing of the edge.	Cold rolled strip	2
No. 3 Edge is an approximately square edge produced by slitting, on which the burn is not eliminated.	Cold rolled strip	3
No. 4 Edge is a rounded edge produced by edge rolling the natural edge of hot rolled strip or slit-edge strip. This edge is produced when the width tolerance and edge condition are not as exacting as for No. 1 Edge.	Cold rolled strip	4
No. 5 Edge is an approximately square edge produced by rolling or filing of a slit-edge to remove burr only.	Cold rolled strip	5
No. 6 Edge is a square edge produced by edge rolling the natural edge of hot rolled strip or slit-edge strip, where the width tolerance and finish required are not as exacting as for No. 1 Edge.	Cold rolled strip	6
Mill Edge.	Hot rolled sheet and strip	M
Cut Edge.	Hot rolled sheet and strip	C
Square Edge (square and smooth, corners slightly rounded). Produced by rolling through vertical edging rolls during the hot rolling operation.	Hot rolled strip	S

TABLE 6—SUMMARY OF AVAILABLE TYPES OF LOW CARBON STEEL SHEET AND STRIP

Product Name	"Quality" or Temper[a]	Applicable Basic Specification Number	Surface Finish, etc.		Edge	
			Description	Symbol	Description	Symbol
Cold Rolled Carbon Steel Sheet	Commercial quality	A 366	For exposed parts: Temper rolled For unexposed parts: Annealed last		Cut	See Note[b]
	Drawing quality	A 619	Matte (dull) — — Matte (dull) Commercial bright — Luster —	E U B L		
	Drawing quality, special killed	A 620				
Hot Rolled Carbon Steel Sheet	Commercial quality	A 569 A 635			Mill	M
	Drawing quality	A 621	As rolled (black) Pickled—dry Pickled and oiled	A P O	Cut	C
	Drawing quality, special killed	A 622				
Hot Rolled Carbon Steel Strip	Commercial quality	A 569			Mill	M
	Drawing quality	A 621	As rolled (black) Pickled—dry Pickled and oiled	A P O	Square	S
	Drawing quality, special killed	A 622			Cut	C
Cold Rolled Carbon Steel Strip	Temper[a] Description No. 1 No. 2 No. 3 No. 4 No. 5	A 109	Matte (dull) Regular bright Best bright	1 2 3	See Table 4	1 2 3 4 5 6

[a] Temper designation applicable to cold rolled strip only, "quality" not applicable. (Metric ASTM documents are designated by suffix "M".)
[b] No symbol necessary; cut edge is "standard".

CLASSIFICATION OF COMMON IMPERFECTIONS IN SHEET STEEL—SAE J810 JUN80

SAE Information Report

Report of the Iron and Steel Technical Committee, approved April 1956, completely revised June 1980.

Common or obvious surface imperfections, which sometimes occur in sheet steel, are normally visible to the naked eye before or after fabrication. Illustrations and definitions of these imperfections are contained in this SAE Information Report. The identifying names are those commonly used throughout the steel industry. These imperfections are variable in appearance and severity. Extreme conditions have been selected in some instances in order to obtain suitable photographs.

Photographs are courtesy of the American Iron and Steel Institute, Steel Products Manual covering Sheet Steel; Carbon, High Strength Low Alloy, and Alloy; Copyright 1979.

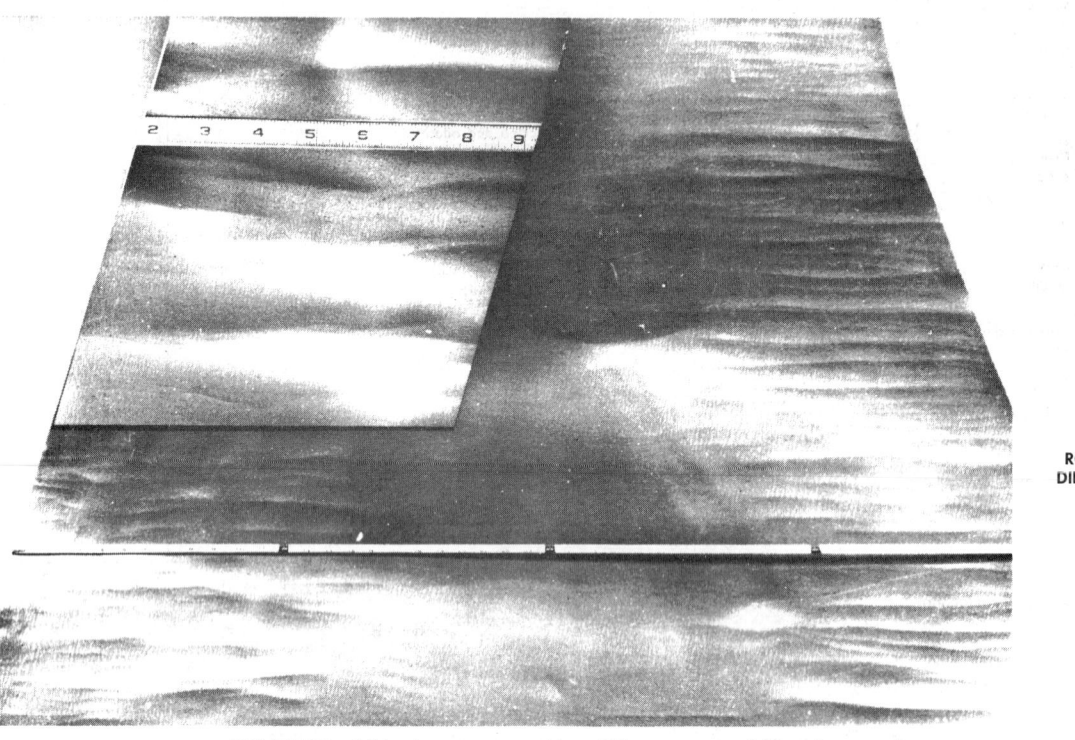

ROLLING DIRECTION

COIL BREAKS—Coil breaks are creases or ridges which appear as parallel lines, transverse to the direction of rolling, and which generally extend across the width of the sheet.

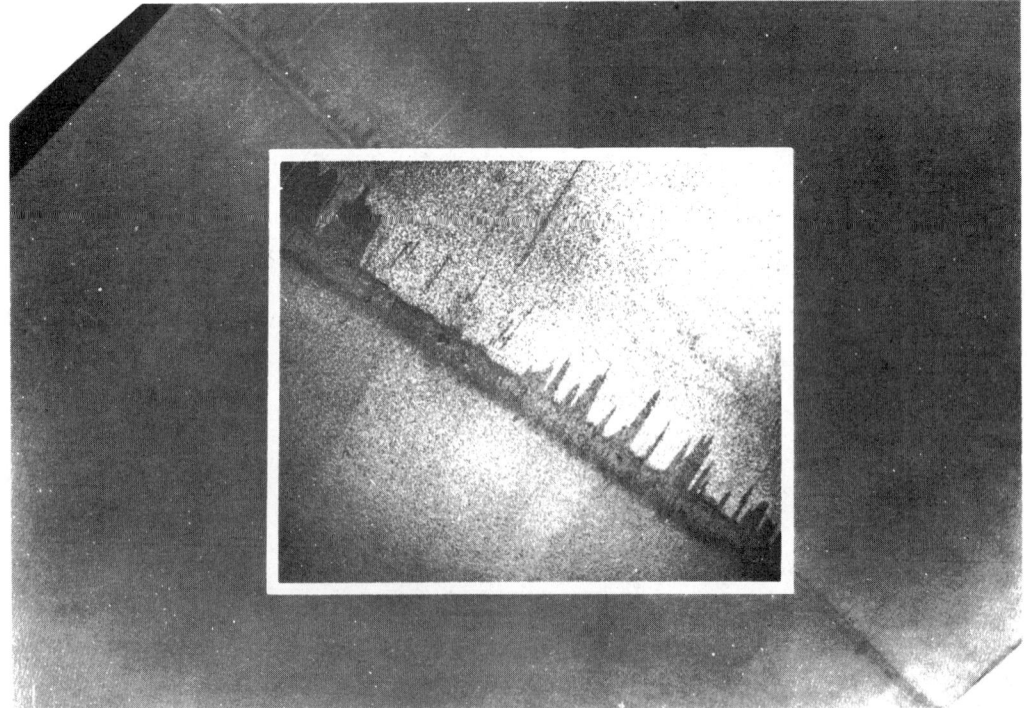

ROLLING DIRECTION

COIL WELD—A coil weld is a joint between two lengths of metal within a coil. Coil welds are not always visible in the cold reduced product.

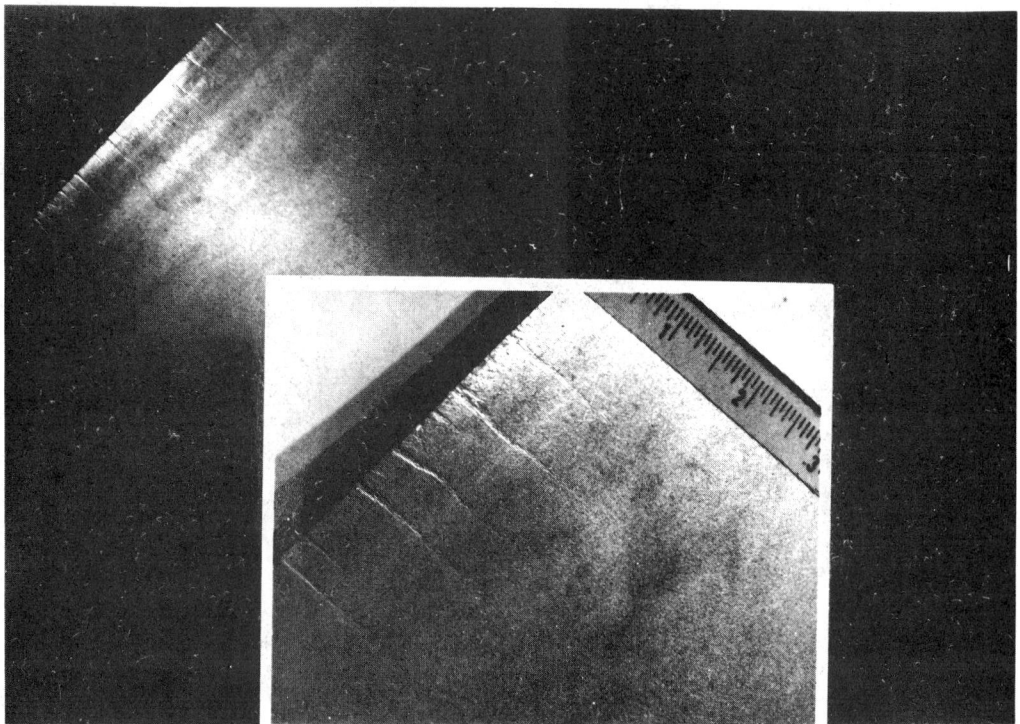

ROLLING DIRECTION

EDGE BREAKS—Edge breaks are short creases which extend in varying distances from the side edge of the temper rolled sheet.

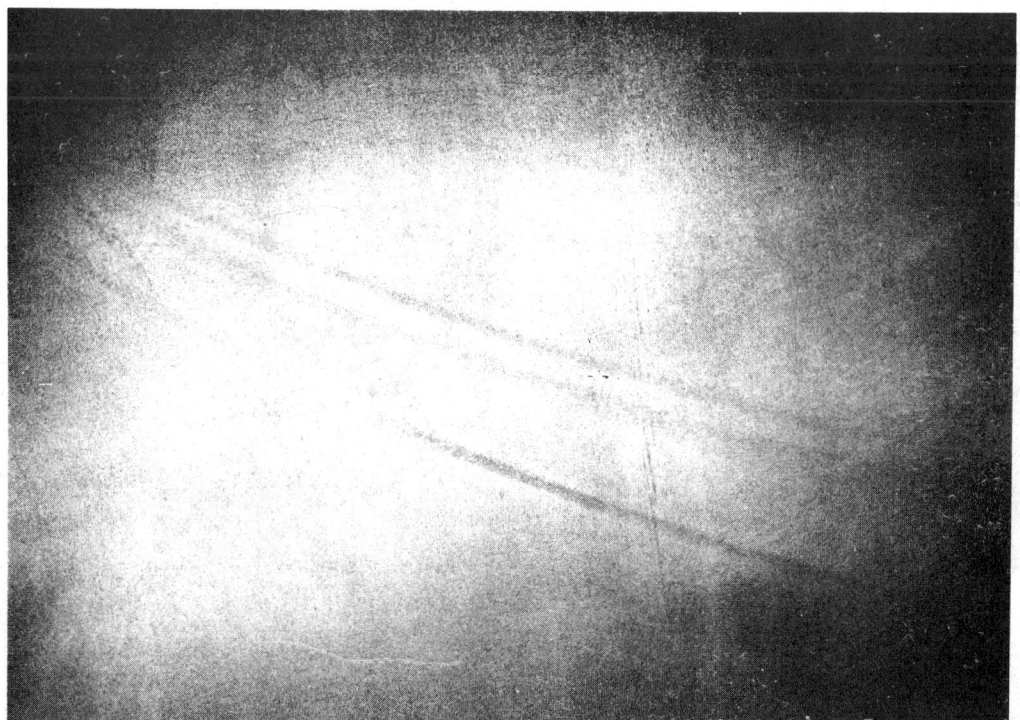

ROLLING DIRECTION

FLOPPERS—Floppers are lines or ridges which are diagonally transverse to the direction of rolling and generally confined to the section midway between the edges of a coil as rolled. They are somewhat irregular and tend toward a flat arc shape.

FLUTING—Fluting is a series of sharp parallel kinks or creases occurring in the arc when sheet steel is formed cylindrically. Photograph shown above is from a test specimen.

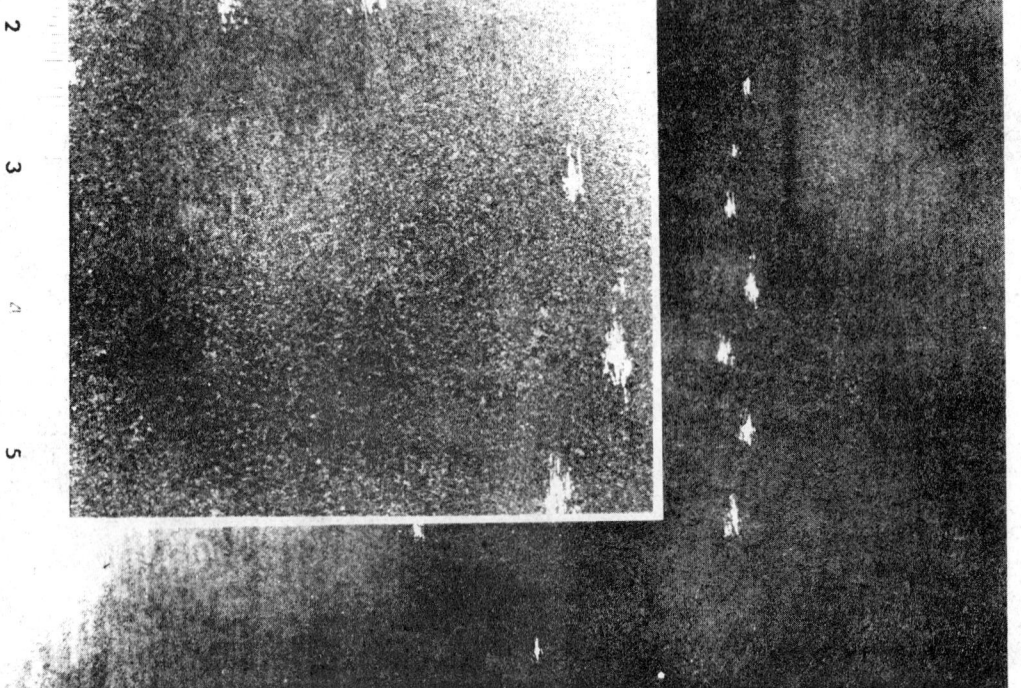

ROLLING DIRECTION

FRICTION DIGS—Friction digs are a series of relatively short scratches variable in form and severity.

HEALED-OVER SCRATCH—Healed-over scratch is a scratch that occurred in an earlier mill operation and was partially masked in subsequent rolling. It might open up during forming.

ROLLING DIRECTION ⟷

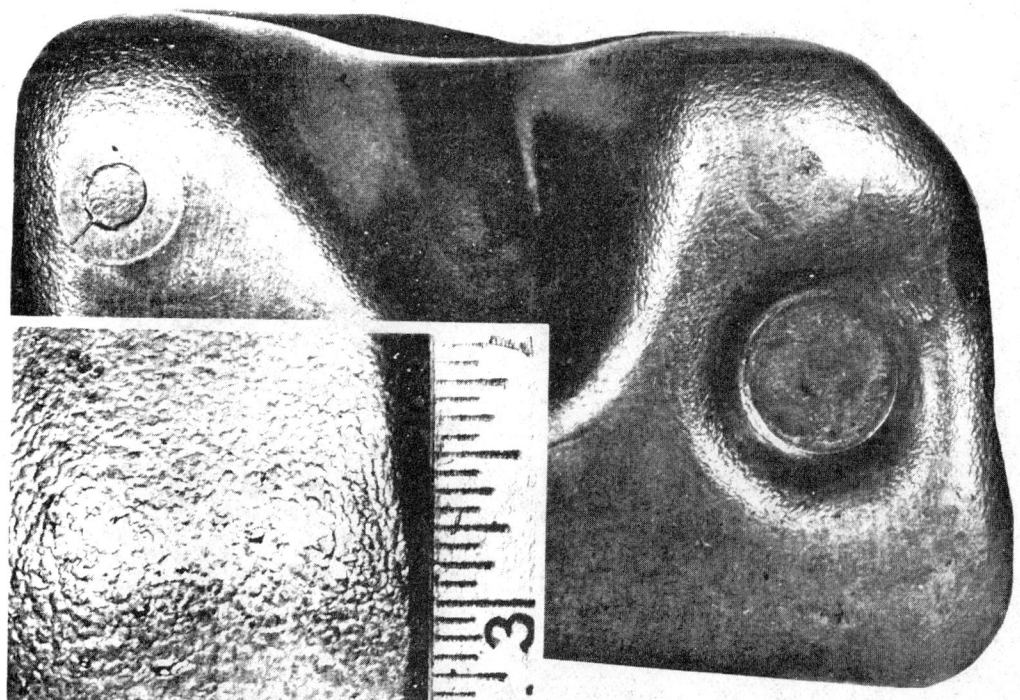

ORANGE PEEL—Orange peel is a coarse grain condition which becomes evident during drawing.

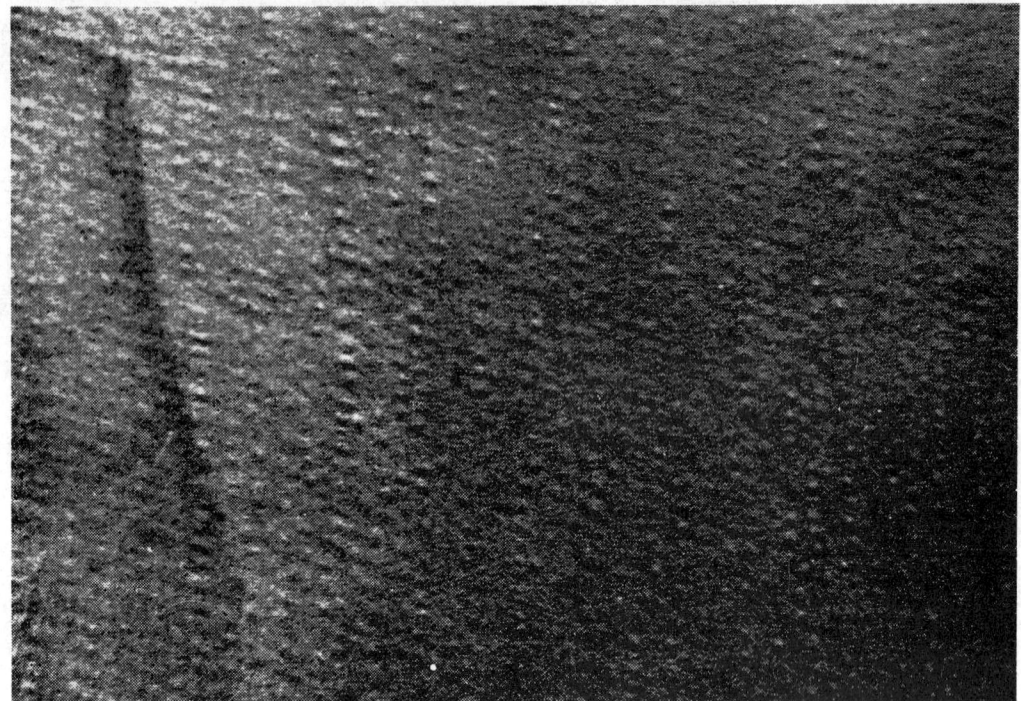

ORANGE PEEL STRAIN—Orange peel strain is a pebbly surface condition which develops during drawing.

ROLLING DIRECTION

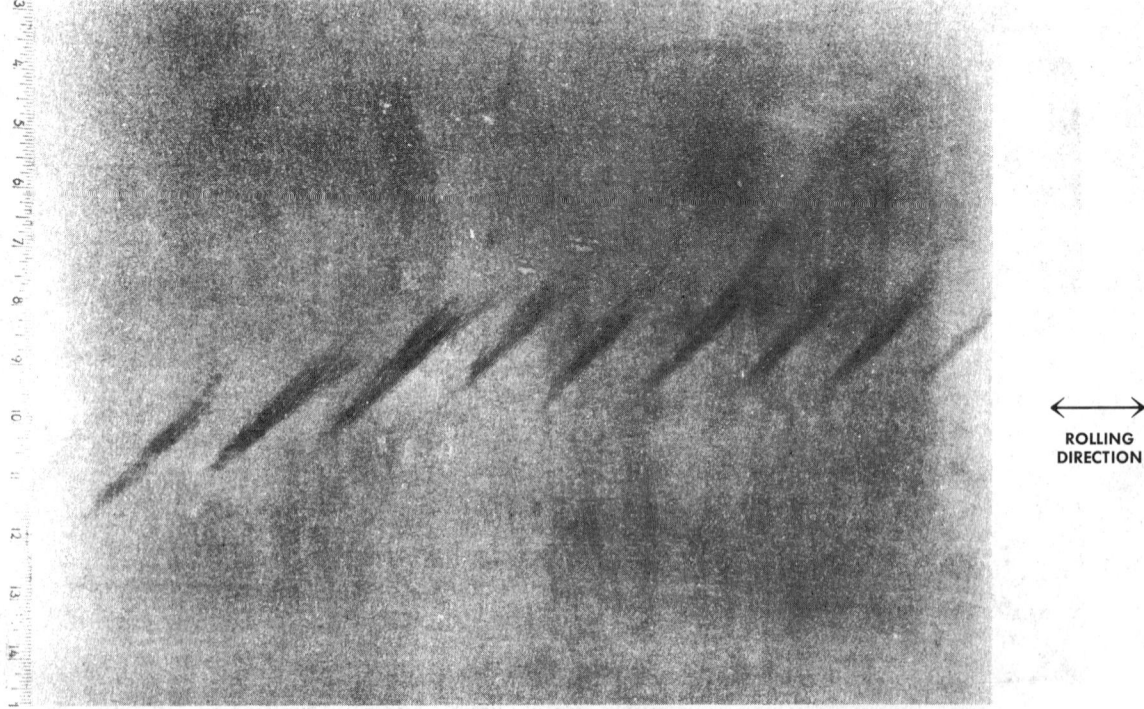

PINCHERS—Pinchers are fern-like ripples or creases usually diagonal to the rolling direction.

ROLLING DIRECTION

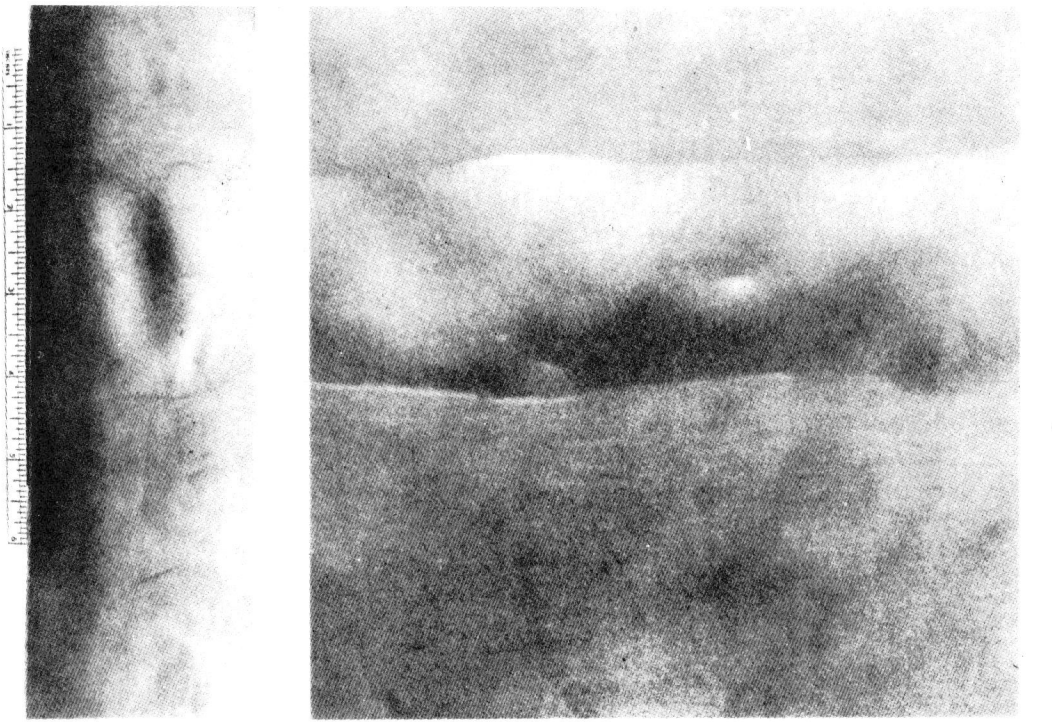

PIPE LAMINATION—Pipe lamination is a separation midway between the surfaces containing oxide inclusions.

ROLLED-IN DIRT—Rolled-in dirt is extraneous matter rolled into the surface of the sheet.

ROLLED-IN SCALE—Rolled-in scale consists of scale partially rolled into the surface of the sheet.

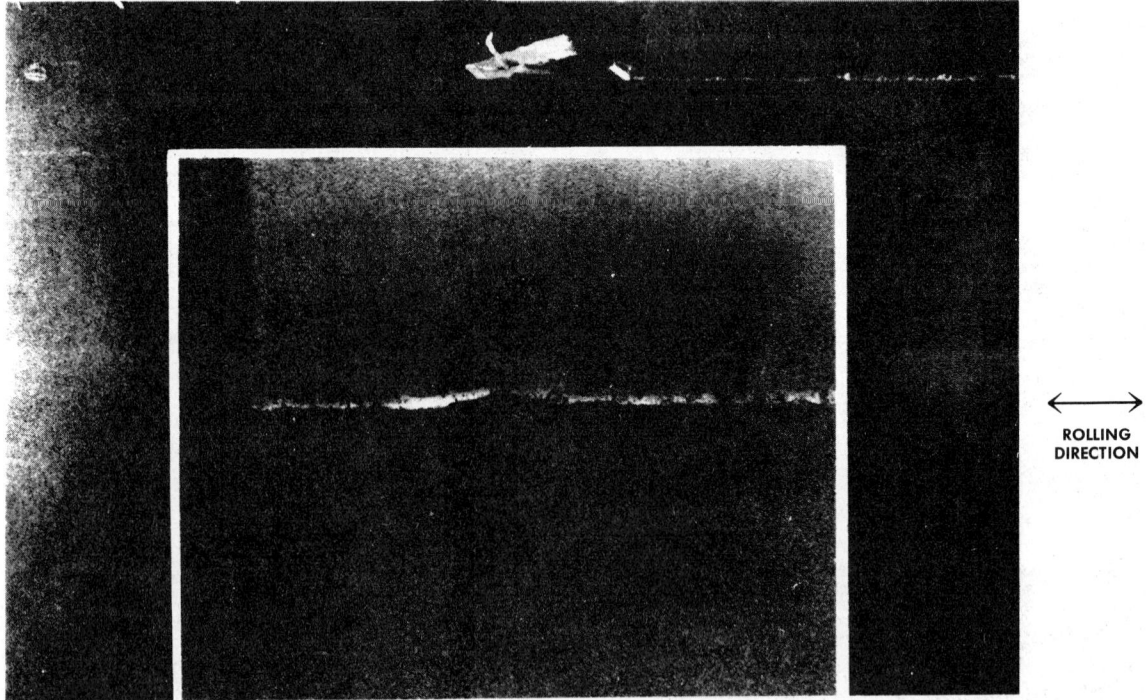

SKIN LAMINATION—Skin lamination is a sub-surface separation which usually results in a surface rupture.

SLIVERS—Slivers are surface ruptures somewhat similar in appearance to skin laminations but usually more prominent.

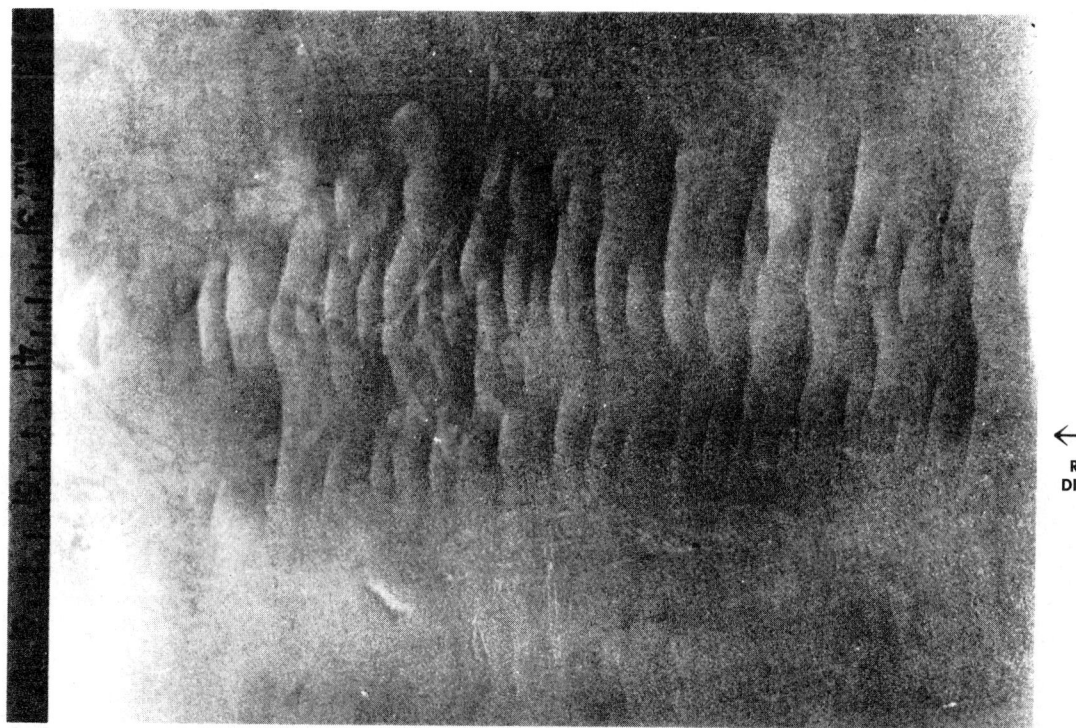

STICKER BREAKS—Sticker breaks are arc-shaped types of coil breaks usually located near the middle of the sheet.

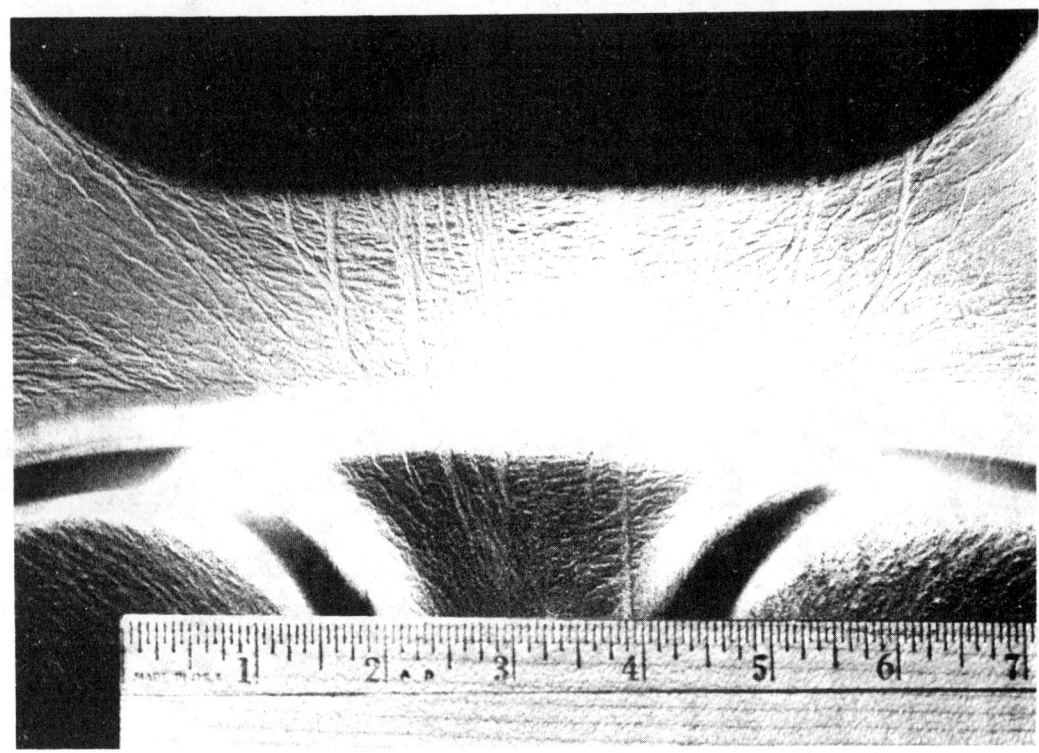

STRETCHER STRAINS (LUDER'S LINES)—Stretcher strains are irregular surface patterns of ridges and valleys which develop during drawing.

ROLLING DIRECTION

GHOST LINES—Ghost lines are lineal irregularities in the surface which develop in drawing. They are parallel to the direction of rolling.

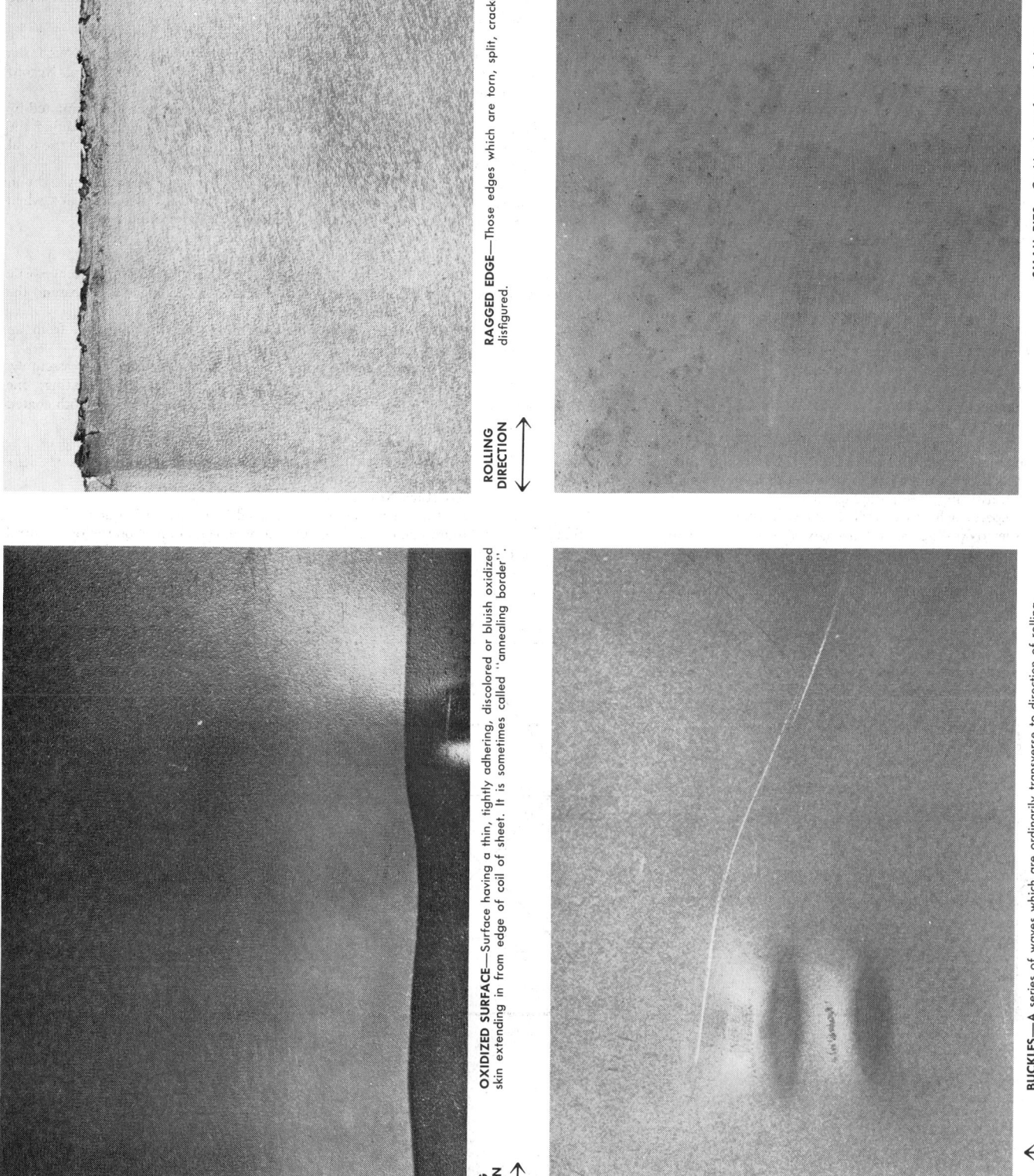

RAGGED EDGE—Those edges which are torn, split, cracked, ragged, burred, or otherwise disfigured.

↕ ROLLING DIRECTION

SMALL PITS—Cavities in surface of sheet.

OXIDIZED SURFACE—Surface having a thin, tightly adhering, discolored or bluish oxidized skin extending in from edge of coil of sheet. It is sometimes called "annealing border".

↕ ROLLING DIRECTION

BUCKLES—A series of waves which are ordinarily transverse to direction of rolling.

↔ ROLLING DIRECTION

UNDERVEHICLE CORROSION TEST—SAE J1293 JAN80

SAE Recommended Practice

Report of the Iron and Steel Technical Committee, approved January 1980.

1. Purpose—This recommended practice is a road test procedure for comparing the corrosion resistance of coated steels in an undervehicle deicing salt environment. This test can also be used for uncoated steels.

2. Background—A single result from one test material exposed on a given rack should not be compared to a single result from a test material exposed on another rack. A suggestion for an evaluation would be to use a minimum of two vehicles with random placement of the panels on each test rack and then averaging the results. However, the ranking of test materials from rack to rack should be similar.

The key features of this test are:

(a) Measurement of steel substrate corrosion rather than coating breakdown.

(b) Both exposed and crevice test conditions.

(c) Measurement of both the area and depth of attack.

(d) Realistic comparison of precoated and post-applied coated materials.

The reliability and reproducibility of the test are demonstrated in SAE Technical Paper 800144.

3. Method—The test method described herein should be followed carefully and any deviations reported, as they may influence the results.

3.1 Selection of Test Materials

3.1.1 Future coatings could be evaluated against one of the following coatings used in the development of this test, namely: hot dip galvanized, zinc-rich primer, anodic electro-deposited primer, and metallo-organic petroleum-based (M.O.P.B.) coating. Carbon steel is suggested as a standard to indicate the severity of the environment.

3.1.2 The number of coatings is limited by the size of the test rack. Six is a typical number.

3.2 Test Coupon Preparation

3.2.1 Gage is optional but should be similar in a given test where galvanic edge protection is a consideration. 50 x 125 mm (2 x 5 in) steel coupons can be sheared directly from prefinished sheet. Post-applied coatings are applied after test coupon assembly (see paragraph 3.3).

3.2.2 Prepare each test coupon as shown in Fig. 1.

3.2.3 (OPTIONAL) If gage and mechanical properties are appropriate, a Ball Punch Deformation Test (ASTM E 643-78) may be drawn so that the coupons nest properly, as follows: ball diameter, 22.2 mm ($7/8$ in); upper die diameter, 25.4 mm (1.0 in); and depth, 6.35 mm (0.250 in). The cup provides an indication of leading surface abrasion and formability.

3.2.4 (OPTIONAL) An X may be scribed between the cup and mounting holes on exposed surfaces only and should penetrate to the substrate metal (ASTM D 1654-74). The X-scribe can provide a measure of the undercutting resistance of the coatings after test exposure.

3.2.5 Locate coupon identification as shown in Fig. 1 by XX.

3.2.6 Before assembly of coupons for post-applied coatings, weigh the coupons to allow for corrosion weight loss determination after testing. Record in milligrams.

3.2.7 Bare steel coupon weights of precoated materials can be estimated by using a nondestructive coating thickness measurement and subtracting the weight calculated from this thickness (knowing coating density) from the total coupon weight. Record in milligrams.

3.2.8 Weigh metallic coated coupons and record total coupon weight in milligrams. This data is required for coating weight loss determination when the steel substrate is not attacked during the exposure interval.

3.3 Test Coupon Assembly

3.3.1 Assemble the test coupons as shown in Fig. 2.

3.3.2 Cut 10 mm I.D. x 15 mm O.D. x 0.25 mm thick shims from plastic sheet. Select a temperature resistant material, such as mica, to withstand the high temperature bakes for post-applied coatings.

3.3.3 6.35 mm ($1/4$ in) I.D. rubber grommets can be obtained from an electrical supply house.

3.3.4 Apply post-applied coatings to the coupon assembly after assembly.

3.3.5 Assemble two coupon assemblies for each one-sided coating; one assembly, with both coated surfaces exposed, and the other, with both coated surfaces in the crevice (see Evaluation, paragraph 4.1.5).

3.4 Test Rack Assembly

3.4.1 Assemble the test rack as shown in Figs. 3A and 3B.

3.4.2 Locate the test coupon assemblies in random order.

3.5 Test Rack Mounting

3.5.1 Select a test vehicle with expected high driving frequency.

3.5.2 Mount one test rack per vehicle as shown in Fig. 4 on the rear control arm on the side opposite from the exhaust.

Minimize U-bolt protrusions. If the test vehicle does not have a rear control

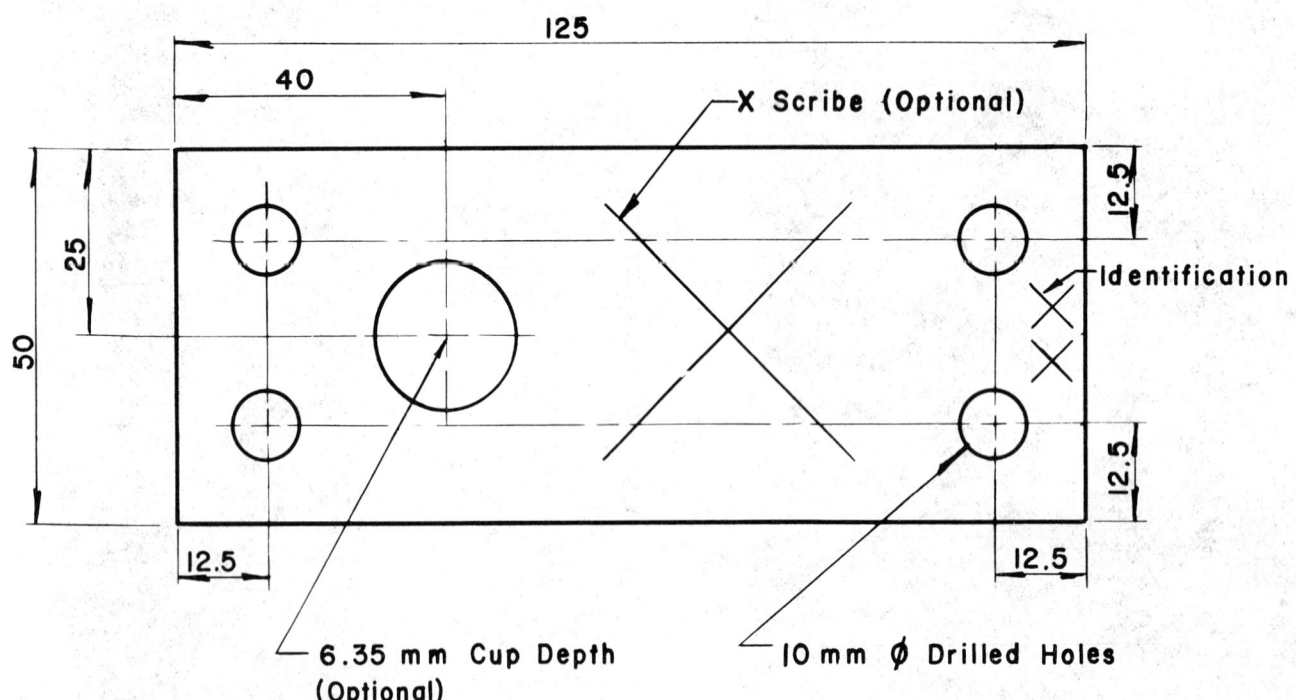

DIMENSIONS ARE IN MILLIMETERS

FIG. 1—TEST COUPON DESIGN

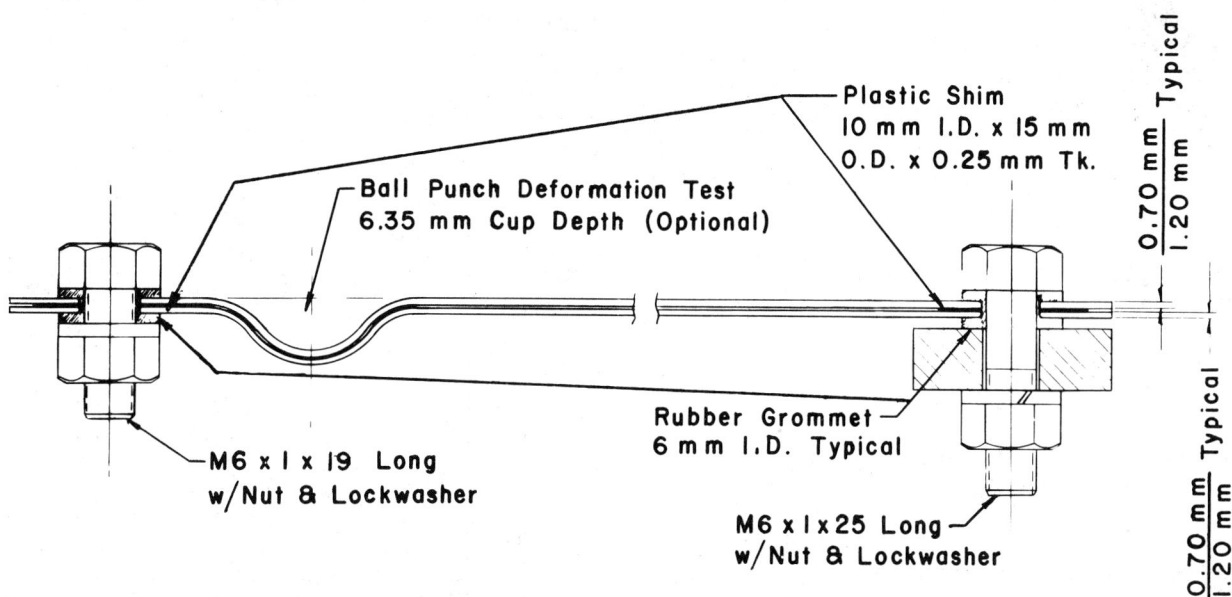

FIG. 2—TEST COUPON ASSEMBLY

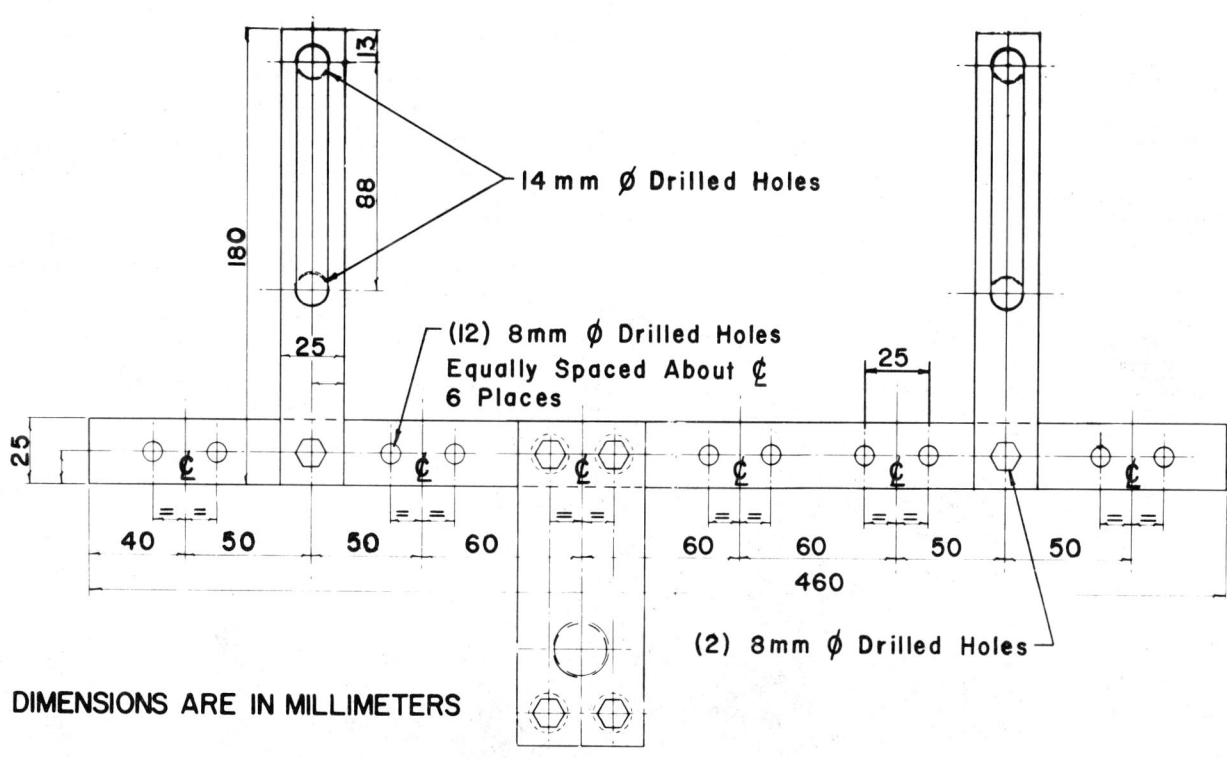

DIMENSIONS ARE IN MILLIMETERS

FIG. 3A—TEST RACK ASSEMBLY—TOP VIEW

arm, select a mounting location in the general vicinity such that the test coupons are horizontal, away from the exhaust, and not likely to receive mechanical damage.

3.5.3 Record mounting date and odometer reading.

3.6 Exposure

3.6.1 Try to begin the test before a winter season.

3.6.2 The test duration is optional but should be sufficient for ranking of materials. In the development of the test, one and two year exposures were evaluated.

3.6.3 Test results should include a notation of the geographical area where the mileage was accumulated, e.g., city and state or province.

4. Evaluation

4.1 General Inspection Before Disassembly

4.1.1 Record removal date, calculate total exposure time on vehicle in days and total distance travelled.

4.1.2 Measure gap and record in millimeters. If variable, record minimum and maximum in millimeters.

4.1.3 Inspect for mechanical damage or other unusual occurrences, such as foreign matter, and report.

4.1.4 Report rack and sample position relative to the geometry of the car.

4.1.5 The evaluations that follow apply to each of the four surfaces of each coupon assembly for two-sided coatings (see Fig. 5).

For one-sided coatings, two coupon assemblies are required for complete evaluation (see Fig. 6). The evaluations apply to only the two-coated surfaces of each coupon assembly.

4.2 Initial Cleaning—Wash each panel with warm water using a nonmetallic brush, sponge, or cheesecloth, and wipe or blow dry.

4.3 Carbon Steel Evaluation

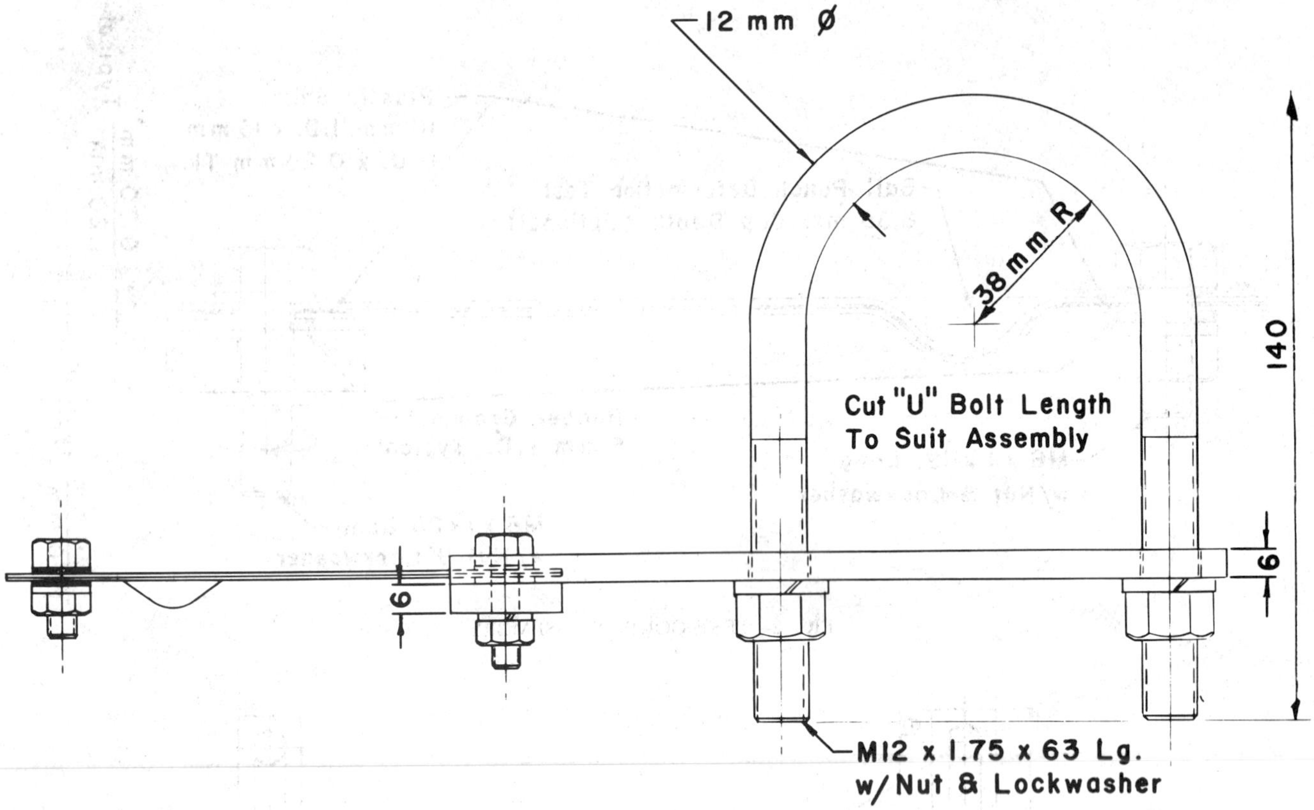

FIG. 3B—TEST RACK ASSEMBLY—END VIEW

4.3.1 Examine for edge abrasion and note if present. Estimate percent red rust, ignoring the bolted area, and describe tightness and color of scale.

4.3.2 Immerse into Clarke's solution (ASTM G1-72) until clean. Wipe or blow dry. Record time required for use in paragraph 4.3.6.

4.3.3 Reweigh and record weight loss in milligrams and calculate thickness loss in micrometers using a nominal area of 50 x 125 mm (ASTM G1-72).

4.3.4 Estimate visually or by the use of a low power microscope the percent area of base metal attack as indicated by surface roughening or pitting. Treat the cup the same as the flat surface for a single evaluation on all samples.

4.3.5 Note the occurrence of perforation. If perforated, no further measurements are required. If not perforated, examine for pitting using a microscope at 10X and measure the 10 deepest pits at 200X by focusing at the top and bottom of the pits. Ignore the cup. Report the average depth and range of the 10 deepest pits in micrometers.

If insufficient number of pits are present (i.e., less than 10), report the average depth and range of the pits present and note with an asterisk the number of pits involved.

4.3.6 Reclean sample in Clarke's solution for same length of time as recorded in paragraph 4.3.2 for blank weight loss correction.

4.4 Metallic Coatings Evaluation

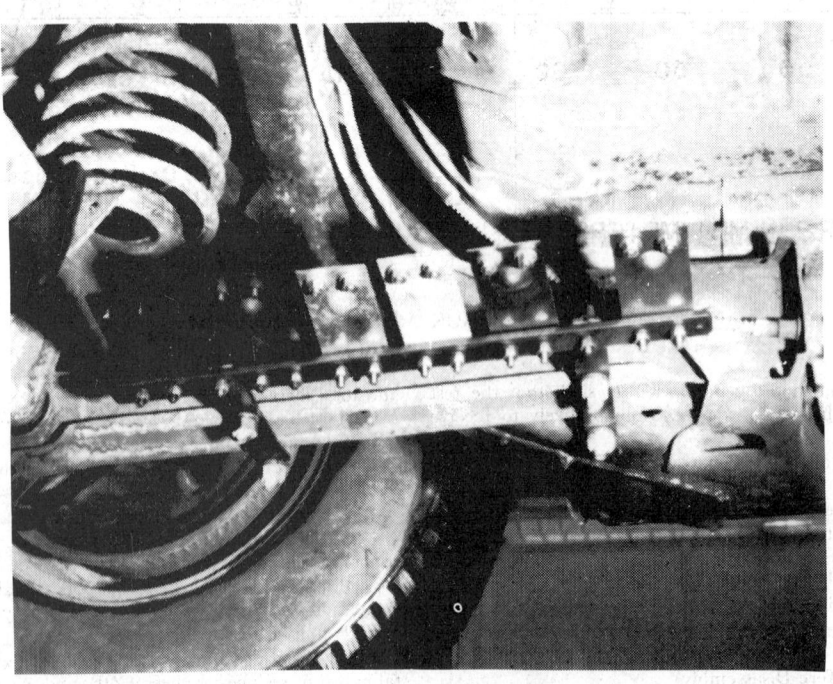

FIG. 4—TEST RACK MOUNTING

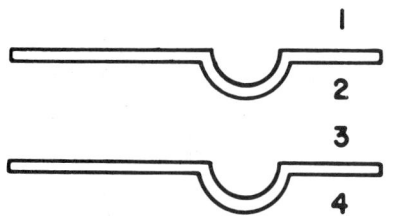

FIG. 5—SURFACE IDENTIFICATION FOR TWO-SIDED COATING

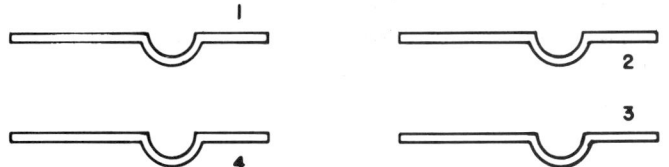

FIG. 6—SURFACE IDENTIFICATION FOR ONE-SIDED COATING

4.4.1 Evaluate qualitatively as for carbon steel in paragraph 4.3.1.

4.4.2 Immerse into appropriate cleaning solution (ASTM G1-72) until clean, then wipe or blow dry. For example, clean galvanized coupons in 15% ammonium hydroxide at room temperature followed by a dip in boiling 5% chromic acid containing silver nitrate until clean.

4.4.3 Weigh and report metallic coating weight loss in milligrams and calculate thickness loss in micrometers using a nominal area 50 x 125 mm (ASTM G1-72). Omit this step if a significant amount of red rust is present.

4.4.4 Estimate visually or by the use of a low power microscope the red rust, surface roughening, or pitting.

4.4.5 Strip for coating weight determination. Galvanized coatings can be stripped according to ASTM A90-69. Omit this step if a significant amount of red rust is present.

4.4.6 Reconfirm base metal attack and examine for pitting as outlined in paragraph 4.3.5.

4.5 Organic Coatings Evaluation

4.5.1 Evaluate qualitatively as for carbon steel in paragraph 4.3.1.

4.5.2 If the coating was scribed, measure total width of creepback from the scribe and record half of the maximum width in millimeters. Note the presence of filiform corrosion.

4.5.3 Remove organic coating using an appropriate solvent or stripping solution. For example, zinc-rich primers and metallo-organic petroleum-based coatings can be removed with MEK, and electro-deposited primer, with hot 10% sodium hydroxide. All stripping is enhanced by using a soft bristle brush.

4.5.4 Chemically clean as for carbon steel in paragraph 4.3.2.

4.5.5 Calculate steel coupon thickness loss as in paragraph 4.3.3.

4.5.6 Estimate visually or by the use of a low power microscope the percent base metal attack as indicated by roughening or pitting.

4.5.7 Examine for pitting as in paragraph 4.3.5.

4.6 Ranking

4.6.1 Rank each surface of the test materials separately according to percent base metal attack. This ranking is based on measurements obtained from paragraphs 4.3.4, 4.4.4, and 4.5.6 and include base metal attack resulting from creepback.

4.6.2 Rank each surface of the test materials separately according to average pit depth and show range of pit depths.

4.6.3 Rank each material based on thickness loss as calculated from weight loss if available.

BIBLIOGRAPHY

1. Robert J. Neville, "Task Force on Undervehicle Corrosion of Coated Steels (SAE Iron and Steel Technical Committee, Division 32)." Paper 770364 presented at SAE Congress and Exposition, Detroit, February 1977.

2. ANSI/ASTM E643-78, "Method of Conducting a Ball Punch Deformation Test for Metallic Sheet Material."

3. ASTM D1654-74, "Evaluation of Painted or Coated Specimens Subjected to Corrosive Environments."

4. ANSI/ASTM G1-72, "Recommended Practice for Preparing, Cleaning, and Evaluating Corrosion Test Specimens."

5. ANSI/ASTM A90-69, "Tests for Weight of Coating on Zinc-Coated (Galvanized) Iron or Steel Articles."

6. Robert J. Neville, "Results of a Test for Undervehicle Corrosion Resistance (SAE Task Force, Iron and Steel Technical Committee, Division 32)." Paper 800144 presented at SAE Congress and Exposition, Detroit, February 1980.

CLEANLINESS RATING OF STEELS BY THE MAGNETIC PARTICLE METHOD—SAE J421b

SAE Recommended Practice

Report of Iron and Steel Division approved January 1941 and last revised by Iron and Steel Technical Committee November 1977.

1. Scope—This SAE Recommended Practice provides a rating procedure for the cleanliness rating of steels by the magnetic particle method. The procedure is based on counting the number of indications (frequency) and employs a weighted value to obtain a severity factor. The method outlined is similar to that described in SAE Aerospace Material Specification AMS 2301.

2. Sampling—Experience has shown that sampling of steel is primarily a matter of agreement between producer and consumer. For heat qualification, unless otherwise agreed, a sample shall be taken from product representing the top and bottom of the first ingot and last usable ingot from heats having not over 10 ingots or not over 30 tons (2700 kg) or from portions of heats within these limits; and from the top and bottom of the first ingot, middle ingot, and last usable ingot of heats having more than 10 ingots or over 30 tons (2700 kg). Where ingot identity has not been maintained, specimens are taken from not less than 10% of the steel to be evaluated.

3. Test Specimens

3.1 The specimens shall be prepared in accordance with the details given in paragraph 3.2. The recommended procedure for developing the specimens from blooms, billets, and bars, in round or square sections is as follows:

3.1.1 CROSS SECTION OVER 36 IN^2 (230 CM^2)—Cut a quarter section as shown in Figs. 1 or 2 and develop the specimen by machining, or forging and machining, to a straight cylinder of a diameter between $2\frac{1}{2}$ and 6 in. An alternative method is to forge or roll the full section to 6 in square or round and machine the quarter section in accordance with Item 3.1.2.

3.1.2 CROSS SECTION 16–36 IN^2 (100–230 CM^2) INCLUSIVE—Cut a quarter section as shown in Figs. 1 or 2 and develop the specimen by machining, or forging and machining, to a straight cylinder of the largest possible diameter.

3.1.3 CROSS SECTION LESS THAN 16 IN^2 (100 CM^2)—Machine the specimen to a straight cylinder. An alternative method is to use a three-step step-down specimen, each step having 3 inches in length. The diameter, D, of the first step is the stock size less minimum stock removal shown in paragraph 3.2.2; the diameter of the second step is $\frac{3}{4}$ D; and the diameter of the third step is $\frac{1}{2}$ D.

3.2 The specimens shall conform to the following requirements unless specified otherwise in paragraph 3.1.

3.2.1 The length of the rated surface is nominally 5 in (13 cm). A 1 in extension for holding is usually employed.

3.2.2 The minimum amount of stock removed from the surface shall be as follows:

Nominal Size of Specimen, Round or Square		Minimum Stock Removal from Surface	
in	mm	in	mm
To 1/2	13	0.030	0.76
Over 1/2 to 3/4	13–19	0.045	1.1
Over 3/4 to 1	19–25	0.060	1.5
Over 1 to 1-1/2	25–38	0.075	1.9
Over 1-1/2 to 2	38–51	0.090	2.3
Over 2 to 2-1/2	51–64	0.125	3.2
Over 2-1/2 to 3-1/2	64–89	0.156	4.0
Over 3-1/2 to 4-1/2	89–114	0.187	4.7
Over 4-1/2 to 6	114–152	0.250	6.4

3.2.3 All quarter sections shall be cut oversize as shown in Figs. 1 and 2 so that the center of the original stock will be approximately on the surface of the test specimen. The location of the center of the original stock shall be identified on the test specimen.

4. Preparation of Specimen

4.1 After the specimen is rough turned, it shall be heat treated to a hardness of 250 Bhn minimum for carbon 0.25% or over, 200 Bhn minimum

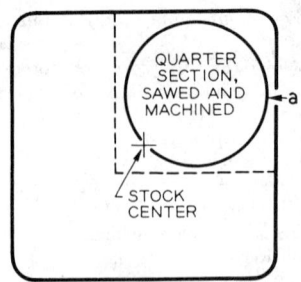

a, DENOTES SURFACE REMOVAL
NOTE: THIS METHOD IS ALSO APPLICABLE TO ROUND SECTIONS

FIG. 1—QUARTER SECTION SPECIMEN FROM SQUARE SECTION FOR MAGNETIC PARTICLE TEST, MACHINE ONLY

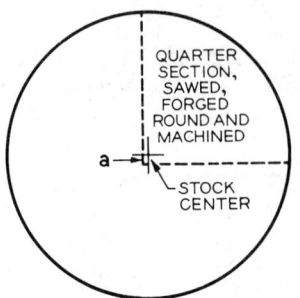

a, DENOTES DISTANCE EQUAL TO SURFACE REMOVAL
NOTE: METHOD ALSO APPLICABLE TO SQUARE SECTIONS

FIG. 2—QUARTER SECTION SPECIMEN FROM ROUND SECTION FOR MAGNETIC PARTICLE TEST, FORGING AND MACHINING

for carbons less than 0.25% by oil or water quenching from well above the critical temperature and tempering within the range of 400–1200 F (200–650 C), depending upon the composition of the steel. Care shall be taken to avoid quenching cracks. The heat treatment tends to develop a more uniform structure hard enough to retain some residual magnetism, thus helping to hold the magnetic powder in place after test.

4.2 After heat treatment, the specimen shall be ground, and the ends of the specimen shall also be ground or otherwise cleaned to insure good contact for magnetizing. Care shall be taken to avoid grinding checks during the grinding. The grinding shall be transverse to the length of the specimen. Longitudinal grinding scratches may be deep enough to retain the magnetic powder and confuse the inclusion determination.

4.3 Before magnetizing, the specimen shall be thoroughly washed with some quick-drying solvent in order to remove all grease and finger marks.

5. Procedure

5.1 Magnetization shall be accomplished by passing direct current through the specimen in the longitudinal direction for $\frac{1}{5}$–$\frac{1}{2}$ s. The magnitude of the current shall be 800–1200 amp per in (300–500 amp per cm) of diameter of specimen.

5.2 In general the wet continuous method that has the specimen covered with magnetic particle suspension during magnetization shall be used. Hardened steel specimens (Rockwell C 50 or higher) may be tested using the wet residual method, applying the suspension after magnetization. Care shall be taken not to disturb indications before inspection is completed.

5.3 The finely divided magnetic particles are usually suspended in kerosene or other light oil of about 40 s Saybolt Universal viscosity. About 1 oz (30 mL) of nonfluorescent magnetic particles to 1 gal (3.8 L) of oil should be used. The suspension concentration of nonfluorescent particles shall be 1.0–2.0% by volume when tested by demagnetizing and allowing to settle 30–45 min in an ASTM 100-mL pear shaped or cone shaped graduated centrifuge tube.

6. Examination of Specimen

6.1 The specimen must be examined under a well diffused light. The standard white fluorescent light is satisfactory. In order to obtain the best dispersion, the longitudinal axis of the light should be at right angles to longitudinal axis of the specimen. The larger inclusions will be plainly visible and the relatively small inclusions may also be detected. If inclusions of $\frac{1}{32}$ in (0.8 mm) or smaller are of interest, it will be helpful to use a low-power hand magnifying glass. The magnetic powder indications produced by inclusions can be distinguished by a practiced operator from indications due to other causes, such as cracks, flow lines, and other related indications. The size of each inclusion appearing on the surface of the specimen shall be recorded.

6.2 The indications representing inclusions may be recorded by photographing, contact printing, or by drawing of a diagram. Transfers of the results of the magnetic particle test may be made by pressing transparent cellulose tape over the areas containing indications and mounting these transfers on heavy paper or cardboard. The transfers are sufficiently accurate to be examined under the binocular or low-power magnifying glass. The transfer may be more rapidly made and is more accurate on curved surfaces than photography and is particularly useful in showing the different distribution of indications near the surface and outside the billet. Specially prepared absorbent paper, known as imbibition paper, may also be used for retaining a permanent record of the magnetic particle inspection.

7. Expression of Results

7.1 Magnetic particle test results are normally expressed in terms of frequency and severity.

7.1.1 Frequency is the total number of indications in a given area. A common area has been 40 in^2 (260 cm^2). Frequency may also be expressed in terms of number of indications per square inch of surface examined.[1]

7.1.2 Severity is the weighted value of the magnetic particle indications according to the following table:

Length of Inclusion		Weight Factor
in	mm	
Over 1/16 to 1/8	2–3	0.5
Over 1/8 to 1/4	3–6	1
Over 1/4 to 1/2	6–13	2
Over 1/2 to 3/4	13–19	4
Over 3/4 to 1	19–25	8
Over 1	25	16

The severity value is obtained by multiplying the number of indications of a given length by the weight factor and adding these results. Severity should be expressed as the weighted value for a given area. A common area has been 40 in^2 (260 cm^2). Severity may also be expressed as the weighted value per square inch of surface examined.[1]

7.2 The average of the frequency and severity values for all the specimens in a heat may be used to express the magnetic particle results for the heat.

7.3 The frequency and severity values for one heat may be readily compared with the values of another heat. In making such comparisons between heats, however, care should be taken to compare only results obtained on billets or bars of approximately the same size.

7.4 If a step-down test is used, results should be related to the individual diameters.

7.5 Magnetic particle results may also be expressed as the total length of indications for a stated area or per unit area.

7.6 When aircraft quality steel is specified, the following maximum average frequency and severity rating per unit area can be expected:

Test Bar	Carbon Content, %	Frequency		Severity	
		psi	pscm	psi	pscm
Individual Test Bar	Under 0.25	0.75	4.8	0.75	4.8
	0.25 and over	0.67	4.3	0.55	3.5
Average of All Test Bars from a Heat	Under 0.25	0.37	2.4	0.28	1.8
	0.25 and over	0.34	2.2	0.25	1.6

7.7 For applications other than aircraft quality steel, the frequency and severity rating should be an arrangement between the individual producer and consumer and related to the particular end product quality involved.

[1] The method of evaluating inclusions in terms of per square inch for frequency and severity has been adopted by the Society of Automotive Engineers, Aeronautical Materials Specification AMS 2301.

MICROSCOPIC DETERMINATION OF INCLUSIONS IN STEELS—SAE J422a

SAE Recommended Practice

Report of Iron and Steel Division approved January 1941 and last revised by Iron and Steel Technical Committee May 1958. Editorial change January 1967. Reaffirmed without change January 1970.

This recommended microscopic practice for evaluating the inclusion content in steel has been developed as a practical method of quantitatively determining the degree of cleanliness of steel. This method has been established as a reasonable control for steel mill operations and acceptance for production manufacturing. It has been widely accepted for carbon and alloy steel bars, billets, and slabs. Exceptions are resulfurized grades which are outside the limits of these photomicrographs and the high carbon bearing quality steels which are generally classified using ASTM E 45-60T, Method A, Jernkontoret Charts.

Preparation of Samples—This microscopic method is based on examination of specimens approximately $\frac{1}{4}$ in.^{2} in area ($\frac{3}{8}$ x $\frac{3}{4}$ in.). The exact dimensions are not of prime importance since the area examined represents an extremely small part of the bar, billet, or heat being evaluated. For bars $1\frac{1}{2}$ in. and smaller, the face obtained by cutting from surface to center with the short dimension parallel to the rolling direction is polished and examined. If one-half the diameter is more than 1 in., the specimen shall be taken midway between the outside and center. The manner of cutting a specimen from a $1\frac{1}{2}$ in. round bar is shown in Fig. 1. A disk, $\frac{3}{8}$ in. in thickness should be sliced from the bar, the section indicated in Fig. 1 cut out of the disk and the shaded area polished parallel to the direction of rolling.

Bars and billets over 4 to 6 in. are normally forged to 4 in. sq before specimens are obtained from a midway position as described above for bars over 2 in. This is illustrated in Fig. 2. The area that shall be polished is shown shaded and extends $\frac{3}{8}$ in. parallel to the length of the bar or billet and $\frac{3}{4}$ in. in the longitudinal center plane normal to the longitudinal axis, so that the polished face is midway between the outside and center of the bar or billet.

It is generally desirable to facilitate polishing by hardening the specimen. Polishing may be done by any desired technique. One generally followed is:

Step 1—Grind.

Step 2—Rough polish, going successively from Nos. 240, 320, 400 grits and Nos. 0, 00, and 000 emery papers.

Step 3—Fine polish, employing some medium such as alumina or other powders having a uniform particle size of 0.3 μm to less than 0.1 μm.

Step 4—Wash in hot water and follow by rinsing in alcohol.

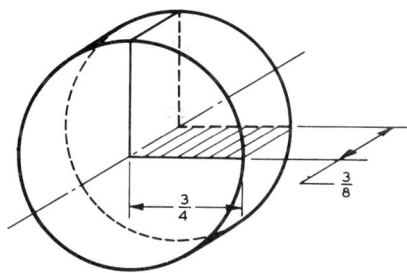

FIG. 1—SPECIMEN FROM 1½ IN. ROUND SECTION FOR MICROSCOPIC TEST

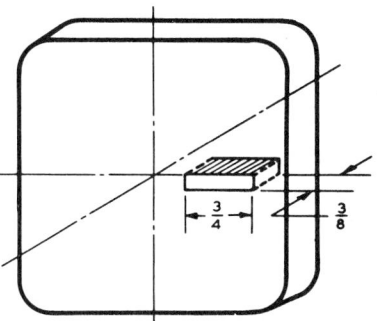

FIG. 2—SPECIMEN FROM LARGE BAR OR BILLET FOR MICROSCOPIC TEST

Polishing scratches in the direction of rolling tends to confuse the appearance of the specimen. It is of utmost importance that the polished surface not be pitted or the inclusions distorted.

The entire polished surface of the prepared specimen is examined at 100 diameters. The examination may be made using the eyepiece or by projecting the field on a ground glass screen. In practice, visual observation of the prepared sample is often used to locate critical areas for microscopic examination.

Classification—The inclusions observed are compared with the accompanying series of photomicrographs of oxides and silicates classified from 1 to 8 inclusive. The length of the field shown is represented as 0.045 in., and the classification is based on length with consideration given to width in the photomicrographs over class 6. The maximum length of each type of inclusion oxide or silicate, is generally used to evaluate a specimen. The silicate photomicrographs are used for all slag or fluid type inclusions and the oxide photomicrographs for all oxide or hard type inclusions. For example, a specimen may be classified 5-O (oxide) 4-S (silicate) to indicate that the longest oxide inclusion noted was comparable to photomicrograph 5 and the longest silicate inclusion noted was comparable to photomicrograph 4.

Modifications may be used such as suffix numerals to indicate the number of long inclusions noted or the exact length of a particular inclusion in thousandths of an inch when over the maximum length indicated by the photomicrographs.

In evaluating steel cleanliness it is important to recognize that the value obtained applies directly to that area being examined. For proper inclusion determination, adequate sampling is of prime importance. Inclusions vary from heat to heat, ingot to ingot, and in different portions of the same ingot product. The accompanying standard series of photomicrographs is designed for use in evaluating the severity of the most common types of inclusions and it should be recognized that they do not represent a complete metallographic study of steel cleanliness.

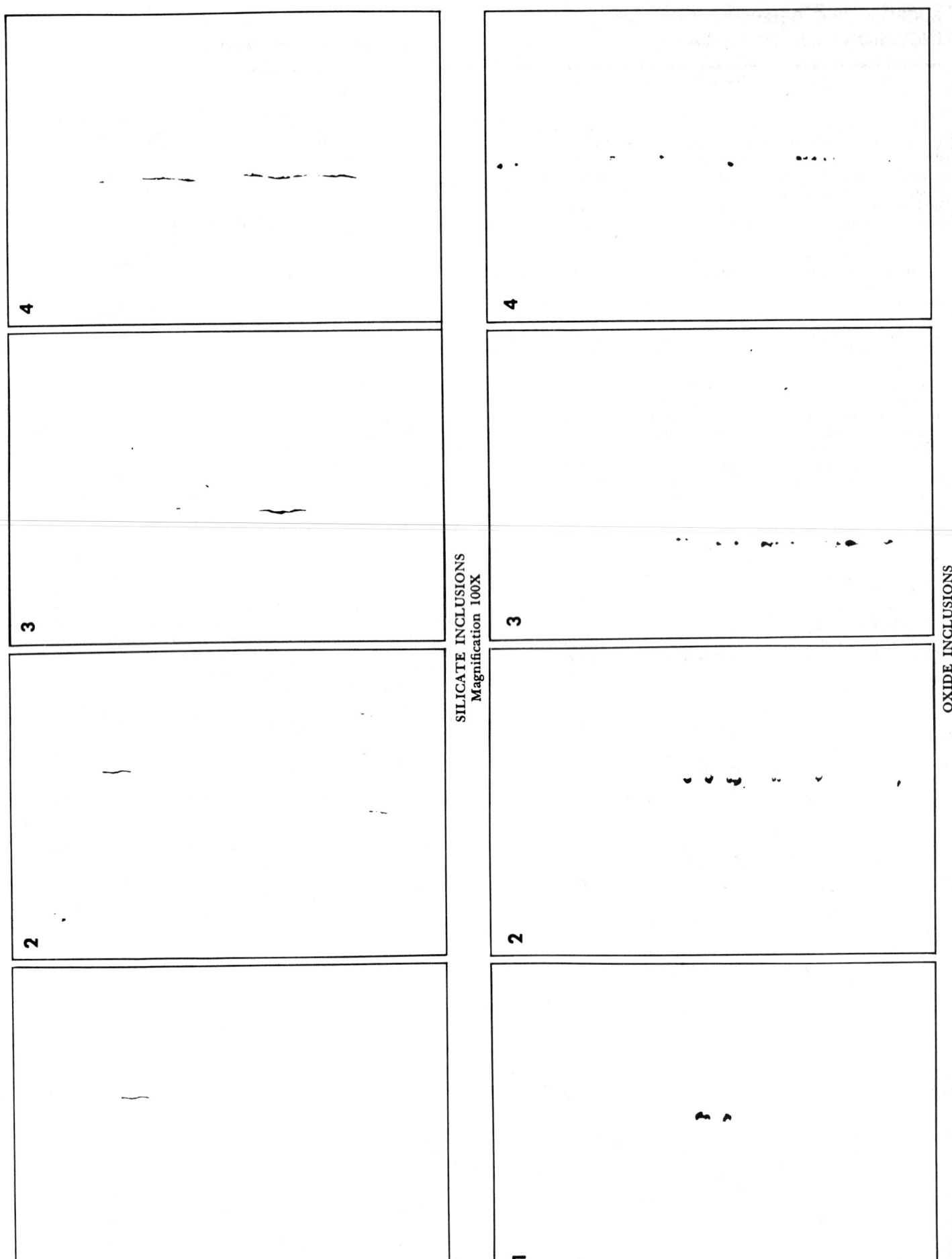

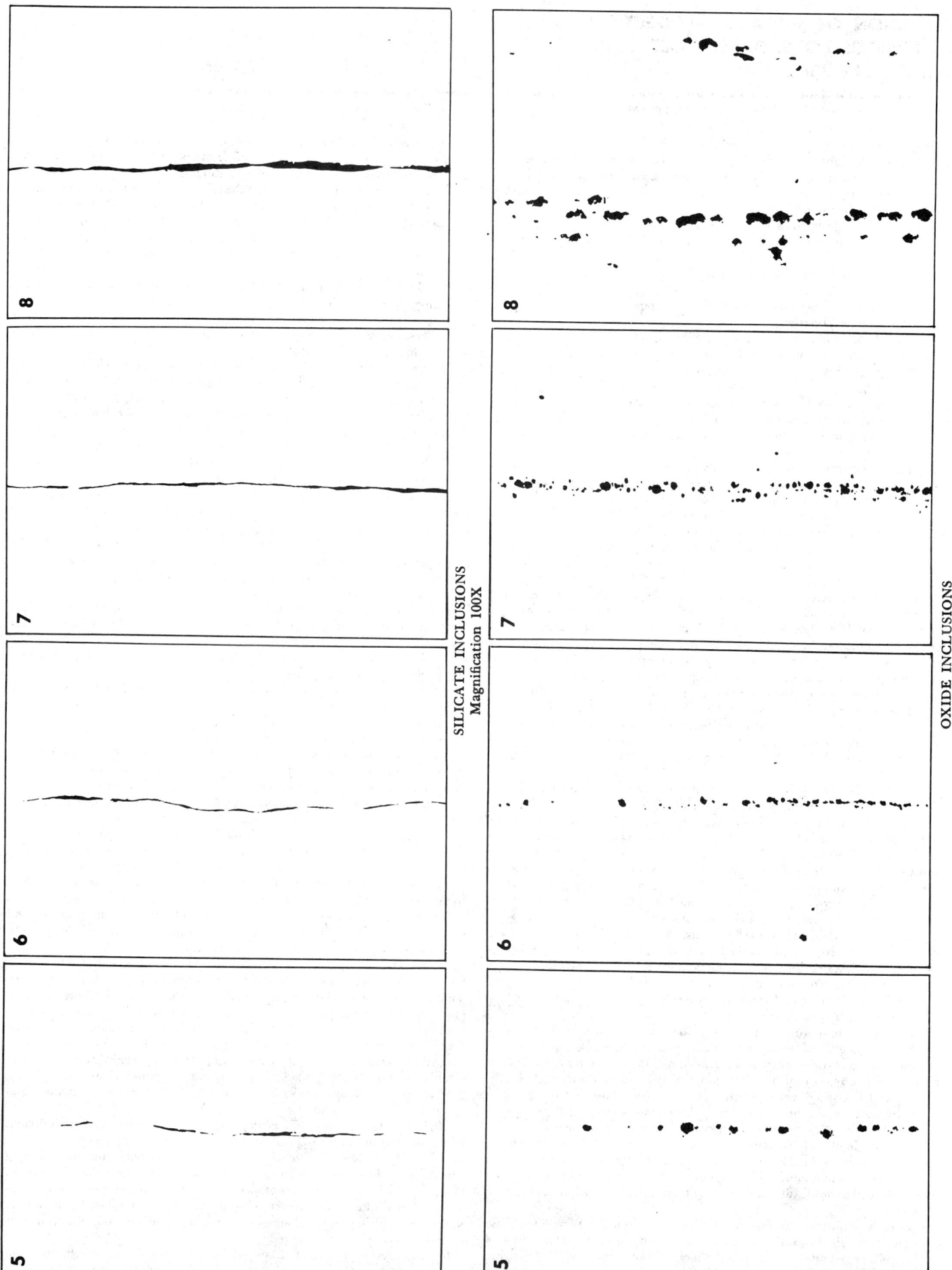

DETECTION OF SURFACE IMPERFECTIONS IN FERROUS RODS, BARS, TUBES, AND WIRES—SAE J349 JUN80

SAE Information Report

Report of the Iron and Steel Technical Committee, approved July 1968, completely revised June 1980.

Scope—This SAE Information Report provides a summary of several methods that are available for detecting, and in some instances detecting and measuring, surface imperfections in ferrous rods, bars, tubes, and wires. References relating to detailed technical information and to specific applications are enumerated in the bibliography.

Limitations—Imperfections which are open to the surface of ferrous rods, bars, tubes, or wires are the only types considered. Such imperfections include:

1. Longitudinal types (seams and laps)
2. Point types (pits)
3. Mechanical types (scratches, nicks, and gouges)

Test Methods for Detection

Visual Examination—Depending upon end product requirements, visual inspection, with or without the aid of magnification, is in some cases adequate to detect the surface imperfections under consideration. Conditions which limit the size and nature of the imperfections as well as surface qualities are factors to be considered in its application. Visual examination may be aided by using surface preparations such as buffing, light grinding, pickling, or blast cleaning using a small particle grit or sand. Sometimes pickling is used in conjunction with blast cleaning.

Liquid Penetrant—SAE J426 briefly describes the equipment and techniques for detecting surface imperfections using a liquid penetrant. The method is generally used for inspecting nonmagnetic materials, but it can also be used on ferrous materials. Being a test involving a penetrating liquid, it is insensitive to the directional aspects of surface flaws; consequently, it will detect laps, seams, cracks, and similar surface imperfections without regard to their orientation. Good indications, however, are dependent upon surface cleanliness.

Magnetic Particle Inspection—The methods available, recommended usage for types of surface discontinuities, and inspection techniques are described in SAE J420. The sensitivity level to be achieved is dependent upon the system employed. Magnetic particle inspection is especially useful to find laps, seams, cracks, inclusions, and some mechanical flaws in ferromagnetic materials, but it has limited value when inspecting for gouges and pits that are circular in nature or too broad to induce a magnetic leakage field. Although a clean surface is important for satisfactory indications, the shape of the cross section and straightness of the test specimens are inconsequential in obtaining satisfactory results. However, if magnetic particle inspection is to be applied to coiled materials, it must be performed on representative samples cut from the coils.

Electromagnetic—Electromagnetic methods are used for the detection of surface imperfections. Eddy current and fringe flux techniques are discussed here. SAE J425 gives general information relative to the nature and use of eddy currents in the broad field of nondestructive testing. This discussion gives additional information on electromagnetic methods as they apply to the detection of surface imperfections in ferrous bars, tubes, rods, and wires.

A distinct advantage of electromagnetic testing is that it can give information as to the severity of surface imperfections. This makes it possible to establish a quality level of the material being tested by accepting that with imperfections not detrimental to the end use of the material and by rejecting that with more severe imperfections. The minimum surface imperfection which can be detected by electromagnetic methods is determined by:

1. The surface condition of the material
2. The type and size of the test coil used
3. The test frequency used
4. The discriminating capabilities of the test instrumentation
5. The smoothness of operation of the material handling equipment

EDDY CURRENT—Eddy current testing is a method of electromagnetic testing in which eddy current flow is induced in the material under test by an exciting coil energized with an alternating current. Changes in the flow caused by variations in the material are reflected into a sensing coil or coils for subsequent analysis by suitable instrumentation and techniques.

The principle of eddy current testing, simply stated, is mutual induction. Mutual induction is the development of an induced emf in one circuit by the change of current in another. Thus, if a piece of metal is placed in the field of an exciting coil carrying alternating current, eddy currents will be induced in the metal.

Testing is performed by passing the bar, rod, tube, or wire lengthwise through or near the inspection coil, which may contain separate exciting and sensing coils or a single coil that may be used for both purposes. The exciting coil is energized with alternating current of one or more frequencies. The electrical impedance of the sensing coil is modified by the proximity of the material under test. The extent of this modification is determined by the distance between the coil and the material, and the electrical conductivity and magnetic permeability of the material. The presence of metallurgical or mechanical discontinuities on the surface of the material will alter the apparent electrical impedance of the coil. During passage of the material being tested, the test coil induces eddy currents in the material and senses changes in amplitude and/or phase of these eddy currents. These changes produce electrical signals which are amplified and modified so as to actuate a suitable signalling device. If variations in magnetic permeability exist in the test material, they may cause spurious signals with some types of eddy current tests. These signals are generally eliminated by saturating the test piece with a uniform magnetic field at the test coil.

Two general coil types will be discussed here. One coil type is the encircling or feed-through type where the coil or coils are stationary while the material is fed through by means of a suitable transport mechanism. Either absolute or differential coil arrangements can be used. The differential coil arrangement is particularly sensitive to short imperfections such as pits, slivers, or nicks. Longitudinal imperfections, such as cracks or seams, may be indicated if they are variable. The absolute arrangement is sensitive to variables such as material properties, size, shape, and imperfections.

The other type is the probe coil. This type can be made to rotate around the material, or the coil can be held stationary while the material is rotated and traversed longitudinally in close proximity to the coil. The probe coil type is reliable and lends itself to mechanization of round product testing. The advantages are that no material saturation is necessary; that it is sensitive to continuous, uniform, longitudinal type imperfections; and that very shallow surface imperfections can be detected.

FRINGE FLUX—Fringe (or leakage) flux testing is a nondestructive method for detecting cracks and other discontinuities at or near the surface in ferromagnetic materials. The method consists of the following steps:

1. The part is magnetized immediately prior to or during the test to a proper level approaching saturation.
2. A flux sensor containing magnetic transducers is placed on the surface in the magnetized area.
3. The part or the magnetic flux sensor is moved progressively at a constant speed so the entire surface is scanned by the sensor.
4. Each magnetic transducer in the flux sensor is connected to an electronic console which amplifies, filters, and electronically processes the signals such that significant discontinuities are indicated (visually and audibly), then marked with paint or automatically removed from the production line, or both.

The fringe flux test is somewhat similar to a magnetic particle test with the flux sensor replacing the magnetic particles. It is somewhat similar to eddy current testing in the scanning and capability. The severity of the discontinuity can be estimated and a rejection level set with respect to the magnitude of the electromagnetic indication produced by the discontinuity.

If properly applied, this method is capable of detecting the presence and location of significant discontinuities such as pits, scabs, slivers, gouges, roll-ins, laps, seams, cracks, holes, and imperfections in welds.

Ultrasonic—Ultrasonic test methods, as described in SAE J428b, can be used for the detection of surface discontinuities in bar and tube products. Various adaptations of the basic method are employed. The choice is influenced by factors such as cross-sectional area of the bar and the size and nature of the imperfections sought. Under proper conditions, ultrasonic waves can be propagated on and just below the surface of the bar. This test mode is particularly well suited to surface inspection. However, surface roughness and cleanliness must be controlled to prevent false determinations.

In general, ultrasonic inspection is limited to bars and wires greater than 2.5 mm (0.1 in) in diameter.

Methods of Measurement—Electromagnetic (eddy current or fringe flux) and ultrasonic testing may be considered quantitative in that acceptance standards can be established, and the equipment set to reject materials having surface imperfections exceeding the predetermined acceptable conditions. Actual deviations from an acceptance standard can be interpreted quantitatively, after acquiring experience with the material being tested and gaining familiarity with the signal changes resulting from the type of imperfection or imperfections being investigated.

Only the surface length of an imperfection can usually be determined from liquid penetrant testing, magnetic particle testing, and visual examination and are usually interpreted qualitatively. However, some indication of the depth of surface discontinuities is sometimes possible on hot rolled products if a fluorescent powder is used when inspecting by the magnetic particle method. (Ref. 9, p. 354.)

Methods of actual depth measurement that are easy to use will either destroy the initial evidence or are a destructive test method. Those commonly used are:

File or Grind and Inspect—Depth is often determined by merely filing or grinding the imperfection until it disappears visually. When using magnetic particle inspection, the material is ground or filed and then magnetic particle inspected again to determine if the imperfection is completely removed. The depth of the resulting groove, to the point of complete removal of the discontinuity, can then be measured.

Macroexamination—Depth can be determined by cutting and grinding a section perpendicular to the direction of the imperfection and macroetching the sample. The depth is measured by a suitable means which could be a scale, Brinell glass, or low power microscope.

Microexamination—A very accurate method of measuring depth is by cutting and metallographically polishing a section perpendicular to the direction of the imperfection and measuring the depth microscopically.

Macroetching—If all conditions, including acid concentration, temperature, and time are controlled, a surface discontinuity of a section can be exaggerated by macroetching. The depth of the etched imperfection can be estimated and a rough approximation made of the original imperfection depth since the amount of material removed can be determined by measuring the cross section before and after etching. Subtracting this difference from the estimated depth of the etched surface imperfection gives the rough estimate.

Bibliography

1. J. M. Mandula and E. S. Monk, "NDT Systems for Steel Billets, Bars, and Tubes." Materials Evaluation, Vol. 34 (10), October 1976, pp. 230–236.
2. W. A. Black, "Evaluation of Surface Defects by Nondestructive Testing—A Progress Report." Journal of Metals, October 1965, pp. 1136–1140.
3. J. M. Mandula, "Billetscan—A New Eddy Current Device for Total Surface Inspection of Square Billets." Materials Evaluation, Vol. 30 (3), March 1972, pp. 49–54.
4. T. W. Judd, "Orbitest for Round Tubes." Materials Evaluation, Vol. 28 (1), Jan. 1970, pp. 8–12.
5. C. E. Betz, "Principles of Penetrants." Second Edition, Magnaflux Corporation, Chicago, IL, 1969.
6. ASM Handbook, Vol. 11, "Nondestructive Inspection and Quality Control." American Society for Metals, Metals Park, OH 44073, 1976.
7. W. J. McGonnagle, "Nondestructive Testing." Second Edition, Gordon and Breach, Science Publishers, Inc., New York, 1969.
8. R. C. McMaster, "Nondestructive Testing Handbook." Vol. II, The Ronald Press Co., New York, 1959.
9. C. E. Betz, "Principles of Magnetic Particle Testing." Magnaflux Corporation, Chicago, IL, 1966.
10. H. L. Libby, "Introduction to Electromagnetic Nondestructive Test Methods." John Wiley & Sons, New York, 1971.
11. "Nondestructive Testing." (NASA SP-5113), National Aeronautics and Space Administration, Washington, DC, 1973.

METHODS OF MEASURING CASE DEPTH—SAE J423a

SAE Recommended Practice

Report of Iron and Steel Technical Committee approved January 1950 and last revised June 1963. Editorial change January 1968. Reaffirmed without change June 1979.

Scope—Case hardening may be defined as a process for hardening a ferrous material in such a manner that the surface layer, known as the case, is substantially harder than the remaining material, known as the core. The process embraces carburizing, nitriding, carbonitriding, cyaniding, induction and flame hardening. In every instance, chemical composition, mechanical properties, or both are affected by such practice.

This testing procedure describes various methods for measuring the depth to which change has been made in either chemical composition or mechanical properties. Each procedure has its own area of application established through proved practice, and no single method is advocated for all purposes.

Methods employed for determining the depth of case are either Chemical, Mechanical, or Visual, and the specimens or parts may be subjected to the described test either in the soft or hardened condition. The measured case depth may then be reported as either effective or total case depth on hardened specimens, and as total case depth on unhardened specimens.

It should be recognized that the relationship between case depths as determined by the different methods can vary extensively. Factors affecting this relationship include case characteristics, parent steel composition, quenching conditions, and others. It is not possible to predict, in some instances for example, effective case depth by chemical or visual means. It is important, therefore, that the method of case depth determination be carefully selected on the basis of specific requirements, consistent with economy.

Definitions for Case Depth

Effective Case Depth is the perpendicular distance from the surface of a hardened case to the furthest point where a specified level of hardness is maintained. The hardness criterion is 50 RC, except where otherwise specified. Effective case depth should always be determined on the part itself, or on samples or specimens having a heat treated condition representative of the part under consideration.

Total Case Depth is the distance (measured perpendicularly) from the surface of the hardened or unhardened case to a point where differences in chemical or physical properties of the case and core no longer can be distinguished.

Chemical Methods

General—This method is generally applicable only to carburized cases, but may be used for cyanided or carbonitrided cases. The procedure consists in determining the carbon content (and nitrogen when applicable) at various depths below the surface of a test specimen. This method is considered the most accurate for measuring total case depth on carburized cases.

Procedure for Carburized Cases—Test specimens shall normally be of the same grade of steel as parts being carburized. Test specimens may be actual parts, rings, or bars and should be straight or otherwise suitable for accurate machining of surface layers into chips for subsequent carbon analysis.

Test specimens shall be carburized with parts or in a manner representative of the procedure to be used for parts in question. Care should be exercised to avoid distortion and decarburization in cooling test specimens after carburizing. In cases where parts and test specimens are quenched after carburizing, such specimens should be tempered at approximately 1100–1200 F and straightened to 0.0015 max total indicator reading (TIR) before machining is attempted. The time at temperature should be minimized to avoid excessive carbon diffusion.

Test specimens must have clean surfaces and shall be machined dry in increments of predetermined depth. The analysis of machined chips will then accurately reveal the depth of carbon penetration. Chosen increments usually vary between 0.002 and 0.010 in. depending upon the accuracy desired and expected depth of case.

Chips from each increment shall be kept separate and analyzed individually for carbon content by an accepted method. Total case depth is considered to be the distance from the surface equivalent to the depth of the last increment of machining whose chips analyze to a carbon content 0.04% higher than that of the established carbon content of the core.

Mechanical Methods

General—This method is considered to be one of the most useful and accurate of the case depth measuring methods. It can be effectively used on all types of hardened cases, and is the preferred method for determination of effective case depth. The use of this method requires the obtaining and recording of hardness values at known intervals through the case. For determination of effective case depth, the 50 RC criterion is generally used. The sample or part is considered to be through hardened when the hardness level does not drop below the effective case depth hardness value. In some instances involving flame and induction hardened cases, it is desirable to use a lower hardness criterion. Suggested hardness levels are tabulated below for various nominal carbon levels.

Carbon Content	Effective Case Depth Hardness
0.28-0.32% C	35 RC
0.33-0.42% C	40 RC
0.43-0.52% C	45 RC
0.53% and over	50 RC

A plot of hardness versus depth from the surface will facilitate this reading. Figs. 1, 2, 3, and 4 illustrate the recommended procedures.

Hardness testers which produce small shallow impressions should be used for all of the following procedures, in order that the hardness values obtained will be representative of the surface or area being tested. Those testers which are used to produce Diamond Pyramid or Knoop Hardness Numbers are recommended, although testers using heavier loads, such as the Rockwell superficial, A or C scales, can be used in some instances on flame and induction hardened cases.

Considerable care should be exercised during preparation of samples for case depth determination by any of the mechanical methods, to insure against grinding or cutting burn. The use of an etchant for burn detection is recommended as a general precaution, because of the serious error which can be introduced by its presence.

Hardness Traverse Procedure—Cut specimens perpendicular to hardened surface at critical location being careful to avoid any cutting or grinding practice which would affect the original hardness.

Grind and polish specimen. Surface finish of the area to be traversed shall be polished finely enough so the hardness impressions are unaffected—that is, the lighter the indentor load, the finer the polish necessary.

The procedure illustrated by Fig. 2 is recommended for the measurement of light and medium cases. The alternate procedure illustrated in Fig. 3 is recommended for medium and heavier cases.

The hardness traverse should be started far enough below the surface to ensure proper support from the metal between the center of the impression and the surface. Subsequent impressions are spaced far enough apart so as not to distort hardness values. The distance from the surface of the case to the center of the impression is measured on a calibrated optical instrument, micrometer stage, or other suitable means.

Taper Grind Procedure—This procedure, illustrated by Fig. 1, is recommended for measurement of light and medium cases.

A shallow taper is ground through the case, and hardness measurements are made along the surface thus prepared. The angle is chosen so that readings spaced equal distances apart will represent the hardness at the desired increments below the surface of the case.

Unless special anvils are used, a parallel section should be prepared so that readings are taken at right angles to the surface. Care should be exercised in grinding to prevent tempering or rehardening.

Step Grind Procedure—This procedure illustrated by Fig. 4 is recommended for measurement of medium and heavy cases.

It is essentially the same as the taper grind section method with the exception that hardness readings are made on steps which are known distances below the surface.

A variation in this procedure is the step grind method where two predetermined depths are ground to insure that the effective case depth is within specified limits.

Visual Methods

General—This method employs any visual procedure with or without the aid of magnification for reading the depth of case produced by any of the various processes. Samples may be prepared by combinations of fracturing, cutting, grinding, and polishing methods. Etching with a suitable reagent is normally required to produce a contrast between the case and core. Nital (concentrated nitric acid in alcohol) of various strengths is frequently used for this purpose.

Macroscopic—Magnification methods for determination of case depth measurement are recommended for routine process control, primarily because of the short time required for determinations, and the minimum of specialized equipment and trained personnel needed. They have the added advantage of being applicable to the measurement of all types of cases. However, the accuracy can be improved by correlation with other methods more in keeping with engineering specifications for the parts being processed. These methods are applied normally to hardened specimens, and while a variety of etchants may be employed with equal success, the following procedures are typical and widely used.

1. *Fracture*—Prepare product or sample by fracturing. Examine at a magnification not to exceed 20 diameters with no further preparation.

2. *Fracture and Etch*—Water quench product or samples directly from the carburizing temperature. Fracture and etch in 20% nitric acid in water for a time established to develop maximum contrast. Rinse in water and read while wet.

3. *Fracture or Cut, and Rough Grind*—Prepare specimen by either fracturing, or cutting and rough grinding. Etch in 10% nital for a period of time established to provide a sharp line of demarcation between case and core. Examine at magnification not to exceed 20 dia (Brinell glass) and read all the darkened area for approximate total case depth.

4. *Fracture or Cut, and Polish or Grind*—Prepare specimen by fracturing or cutting. Polish or grind through No. 000 or finer metallographic emery paper or both. Etch in 5% nital for approximately 1 min. Rinse in two clean alcohol or water rinses. Examine at magnification not to exceed 20 dia (Brinell glass) and read all of the darkened zone. After correlation effective case depth can be determined by reading from external surface of specimen to a selected line of the darkened zone.

Microscopic—Microscopic methods are generally for laboratory determination and require a complete metallographic polish and an etch suitable for the material and the process. The examination is made most commonly at 100 diameters.

1. *Carburized Cases*—The microscopic method may be used for laboratory determinations of total case and effective case depths in the hardened condition. When the specimen is annealed properly the total case depth and the depth of the various zones—hypereutectoid, eutectoid and hypoeutectoid—also can be determined quite precisely.[1]

(a) Hardened Condition

Fracture or cut specimen at right angles to the surface.

Prepare specimen for microscope and etch in 2 to 5% nital (concentrated nitric acid in alcohol).

For effective case depth, read from surface to metallographic structures which have been shown to be equivalent to 50 RC.

For total case depth, read to the line of demarcation between the case and core. In alloy steels quenched from a high temperature the line of demarcation is not sharp. Read all the darkened zone that indicates a difference in carbon from the uniform core structure.

(b) Annealed Condition

For specimens previously hardened or not cooled under controlled conditions.

The specimen to be annealed may be protected by copper plate or any suitable means for preventing loss of carbon.

Pack in a small thin wall container with a suitable material such as charcoal.

Place container in furnace at 75-150 F above the upper critical temperature (AC_3) for the core. (Generally an annealing temperature of 1600-1700 F is satisfactory.)

Leave in furnace long enough for specimen to reach furnace temperature, but not for an excessive time at temperature as carbon diffusion will increase total case depth.

(c) Cooling Rates

Carbon Steels—A satisfactory cooling rate is obtained by cooling the con-

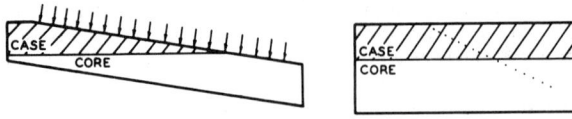

FIG. 1—SPECIMEN FOR TAPER GRIND PROCEDURE

FIG. 2—SPECIMEN FOR CROSS SECTION PROCEDURE

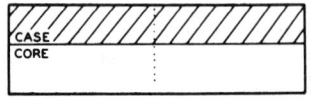

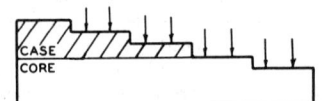

FIG. 3—SPECIMEN FOR ALTERNATE CROSS SECTION PROCEDURE

FIG. 4—SPECIMEN FOR STEP GRIND PROCEDURE

---

[1] For certain applications involving moderate to high hardenability alloy steels in the 0.4 to 0.8% carbon range, the M_s method of case depth determination to specific carbon level has been found to be effective. In this method, the specimen is austenitized at the time and temperature sufficient to more than take into solution the alloy and carbon at the desired level of measurement. It is then quenched into salt at the M_s temperature of the carbon level desired, held just long enough to temper the martensite at all lower carbon levels and water quenched. Subsequent polishing and etching disclose a sharp line of demarcation between tempered and untempered martensite, which can be read with a Brinell glass to a precision of ±0.002. Additional information on this technique can be obtained by reference to "The Application of M_s Points to Case Depth Measurement," by E. S. Rowland and S. R. Lyle, ASM Transactions, Vol. 37 (1946) pp. 26-47.

tainer in mica, lime or other insulating material at a rate which will reduce the temperature to 800 F in 2½ to 3 hr. Cool as desired below 800 F.

Alloy Steels—Slower cooling rates or isothermal transformations are required. If martensite is retained in the structure, better contrast after etching may be obtained by tempering the specimens at 1000–1100 F. Cool as desired after tempering.

Section, prepare and etch specimen as desired under (a) Hardened Condition. Etching time is usually longer.

For total case depth measurement, read the depth of carbon enrichment.

For specimens cooled slowly after carburizing. If the production carburizing cycle provides the proper cooling rate or the cooling rate is otherwise controlled as described for the annealed condition, specimens may be prepared and examined without reheating after carburizing. This is often possible when the parts are cooled in solid compound when the boxes are not too small.

2. Carbonitrided Cases—Carbonitrided cases are measured for total case depth in the hardened condition. High quenching temperatures, high alloy content of the steel and high carbon content of the cone decrease the accuracy of readings obtained by this method.

Section, prepare, etch and read as described in Carburized Cases, (a) Hardened Condition.

3. Cyanided Cases—Cyanided cases are thin, and only the microscopic method is recommended for accurate case depth measurements. The usual cyanide case contains a light etching layer followed by a totally martensitic constituent, which in turn is followed by martensite with increasing networks of other constituents, depending on the type of steel which has been cyanided. Cyanided cases are read in the hardened condition only and results reported as total case depth.

Section, prepare and etch specimen as described in Carburized Cases, (a) Hardened Condition.

Read to the line of demarcation between the case and core.

(When a sharp line of demarcation does not exist, the use of a hardness test such as described under Mechanical Methods is recommended.)

4. Nitrided Cases—The microscopic method is used chiefly in those situations where the available sample cannot readily be prepared for the more desirable hardness traverse method. It may be difficult to read the case depth because the nitride network gradually diminishes.

Section and prepare the specimen as described in Carburized Cases, (a) Hardened Condition.

Etch in 10% nital.

Read all darkened zone for total case depth.

5. Flame or Induction Hardened Cases—Since no chemical change occurs in flame or induction hardening, readings must be made in the hardened or hardened and tempered condition only. A procedure for reading effective case depth may be established by correlating microstructures with a hardness traverse method. A minimum hardness of 50 RC is used commonly but some other point may be selected or required, for example, in lower carbon steels that do not reach 50 RC when fully hardened. See Carbon Content table, under Mechanical Methods, General. The microstructure at the selected location will differ depending on steel composition, prior treatment (annealed, heat treated, or other treatments) and on the hardness level chosen.

Section, prepare and etch specimen as described in Carburized Cases, (a) Hardened Condition.

For total case depth, read the entire zone containing structures hardened by the process.

For effective case depth, read to selected microstructure correlated with specified hardness.

METHOD FOR DETERMINING BREAKAGE ALLOWANCES FOR STEEL SHEETS—SAE J424

SAE Recommended Practice

Report of Iron and Steel Technical Committee approved October 1959. Editorial change December 1967. Reaffirmed without change March 1973.

Scope—This method for establishing breakage allowances applies to parts fabricated from sheets and is recommended as equitable to both the sheet producer and the fabricator. This recommended practice is not intended to replace laboratory and control procedures.

Definition—Breakage for the purpose of this proposal is defined as unrepairable parts, broken during drawing and classed as scrap. Parts showing laminations, resulting from pipe, may be excluded provided they are separately identified. Broken parts which can be salvaged are not covered in this proposed method.

Method—This procedure is intended to establish a breakage allowance without the need for re-inspection of each broken stamping. It will apply to overall breakage on a given part (as calculated by the method outlined below) in excess of 1% up to and including 8%. Inherent variations in steel sheets and normal variables in the stamping operation preclude 100% satisfactory performance. Therefore, it is accepted that practical perfection is attained when 99% of the stampings are produced without breakage. When the overall breakage is in excess of 8%, it is considered to be the result of abnormal stamping conditions, and this method does not apply.

Two or More Suppliers—When there are two or more suppliers, the recommended procedure for determining a breakage allowance for an identified part is based on the average percentage of breakage of AT LEAST 75% OF THE BLANKS RUN on that part, on one set of dies, during at least one month (3000 piece minimum). The total production of all suppliers used to obtain this 75% minimum is to be included in the calculation starting with the best performance. The average breakage thus determined shall be considered the allowance for the part.

Example:

Vendor	Parts Produced	Parts Scrap	% Scrap
A	32,466	630	1.94
B	27,856	579	2.08
C	67,120	1477	2.20
D	56,200	1349	2.40
E	40,900	1125	2.75
F	850	60	7.05
	Total 225,392	Total 5220	Average 2.32

75% of 225,392 equals 169,044; therefore it is necessary to include the total production of vendors A, B, C and D, since the total A, B and C is only 127,442, which is less than 75% of the total. Total production of 183,642 parts (A + B + C + D) with 4035 parts being rejected, results in a percentage allowance of 2.20%. On this basis, vendors D, E and F exceed the allowance.

One Supplier—When there is only one supplier in any one month, the recommended procedure for determining a breakage allowance for an identified part is based on the average percentage of breakage on that part, on one set of dies, during at least two consecutive months (5000 piece minimum). This applies whether the supplier in the two consecutive months is the same or a different one. The average breakage thus determined shall be considered the allowance for the part.

Exceptions—Individual lifts or coils which exhibit unreasonably high or unusually variable breakage will not be considered in determining the allowance. Such material should be set aside and the supplier notified.

NONDESTRUCTIVE TESTS—SAE J358 JAN80

SAE Information Report

Report of the Iron and Steel Technical Committee, approved September 1968, last revised June 1978, editorial change January 1980.

Nondestructive tests are those tests which detect factors related to the serviceability or quality of a part or material without limiting its usefulness. Material defects such as surface cracks, laps, pits, internal inclusions, bursts, shrink, seam, hot tears, and composition analysis can be detected. Sometimes their dimensions and exact location can be determined. Such tests can usually be made rapidly. Processing results such as hardness, case depth, wall thickness, ductility, decarburization, cracks, apparent tensile strength, grain size, and lack of weld penetration or fusion may be detectable and measurable. Service results such as corrosion and fatigue cracking may be detected and measured by nondestructive test methods. In many cases, imperfections can be automatically detected so that parts or materials can be classified.

The SAE Handbook describes the following nondestructive test methods:

TABLE 1—FEATURES OF NONDESTRUCTIVE TESTS

Method	Principle	Material	Applications	Advantages	Limitations
Magnetic particle (SAE J420)	Magnetic particles attracted by leakage flux at surface flaws of magnetic object aid visual inspection.	Magnetic materials	Surface flaws such as cracks, laps, and seams. Some subsurface flaws.	Easy to interpret, fast, simple to perform	Parts must be relatively clean. Usually requires high current source. Parts sometimes must be demagnetized. Standards difficult to establish.
Electromagnetic (eddy current) (SAE J425)	Alternating current coil induces eddy currents in test object. Flaws and material properties affect flow of currents. Information derived from meter or cathode ray tube indications.	Metals	Material composition, structure, hardness changes, cracks, case depth, voids, large inclusions, tubing weld defects, laminations, coating thickness, porosity.	Intimate contact between coil and material not required. Versatile. Special coils easily made. Fast operation can be automated. Electric circuit design variations permit selective sensitivity and function. Sensitive to surface and near surface inhomogeneities.	Sensitive to many variables. Sensitivity varies with depth. Reference standards needed. Response often comparative.
Liquid penetrant (SAE J426)	Liquid penetrant is drawn into surface flaws by capillary action, then revealed by developer material to aid in visual inspection.	Nonporous material, metals, plastics, glazed ceramics	Surface flaws such as cracks, porosity, pits, seams, and laps.	Simple to perform, applicable to complex shapes, on site inspection.	Only surface flaws detected. Surfaces must be clean. Penetrant washes out of wide defects. Standards difficult to establish.
Penetrating Radiation (SAE J427)	General—Penetrating radiation is differentially absorbed by materials, depending upon thickness and type of material.	Most materials	Internal defects such as inclusions, porosity, shrink, hot tears, cracks, cold shuts, and coarse structure in cast metals; lack of fusion and penetration in welds. Thickness measurement. Detection of missing internal parts in an assembly.	More standards established than for other methods. Internal defects detected. Permanent film record. Automatic thickness gaging.	Health precautions necessary. Defect must be at least 2% of total section thickness. Film processing requires time, facilities, and care. Difficulty with complex shapes. Most costly nondestructive test method.
	X-ray source produces radiation electrically, by deceleration of electrons.			Versatile—energy adjustable. Fluoroscopy available. Image intensification available. Thickness up to 24 in (600 mm) of steel.	Electric power and water required. Equipment heavy and costly. Shielded area usually required.
	Gamma source produces radiation in decay of radioactive material.			More portable than x-ray. Lower cost than x-ray. Thickness up to 10 in (250 mm) steel can be tested.	Government license required. Energy cannot be adjusted or turned off. Source must be replaced. Orientation affects the test.
	Neutron source produces radiation by nuclear reactors, accelerators, or decay of radioactive material.			Penetrates dense metals but is attenuated by light elements such as in water, plastics, and oil. Usable on radioactive objects.	Government license required. Less portable and more expensive than x-ray.
Ultrasonic (SAE J428)	Mechanical vibrational waves (frequency range 0.1–25 MHz) are introduced into a test object. This energy is reflected and scattered by inhomogeneities or becomes resonant. Information is interpreted from cathode ray tube or read from meter.	Metals, plastics, ceramics, glass, rubber, graphite, concrete	Inclusions, cracks, porosity, bursts, laminations, structure, lack of bond, thickness measurement, weld defects.	Variety of inspection elements and circuitry permits selective test. High speed test. Can be automated and recorded. Penetrates up to 60 ft (18 m) steel. Indicates flaw location. Access to only one surface usually needed.	Difficulty with complex shapes. Surface roughness may affect test. Defect orientation affects test. Comparative standards only. Requires couplant.
Infrared (SAE J359)	Electromagnetic radiation from test objects above a temperature of absolute zero is detected and correlated to quality. Information is displayed by meter, recorder, photograph, or CRT.	Most materials	Discontinuities that interrupt heat flow: flaws, voids, inclusions, lack of bond. Higher or lower than normal resistances in circuitry.	High sensitivity. One-sided inspection possible. Applicable to complex shapes and assemblies of dissimilar components. Active or passive specimens.	Emissivity variations in materials, coatings and colors must be considered. In multilayer assemblies, hot spots can be hidden behind cool surface component. Relatively slow.
Acoustic Emission (SAE J1242)	Acoustic emission is a transient elastic wave generated by rapid release of energy from a localized source within a solid material. Rate and amplitude of high frequency (0.1–1 MHz) acoustic emissions are noted and correlated to structure or object characteristics.	Most solid materials	Determine or monitor integrity of structures such as weldments or castings.	Remote and continuous real time surveillance of structures is possible. Inaccessible flaws can be detected. Location of flaws can be determined. Permanent record can be made.	Part must be stressed. Non-propagating flaws cannot be detected. Non-relevant noise must be filtered out. Transducers must be placed upon the object.
ed. Leakage Testing (SAE J1267)	Material flows across an interface at a leak site. Rate of flow is pressure, time, and leak size dependent. Detection of the trans interface migration is done in one of eight or more ways.	Totally independent of materials.	Any vessel containing a product at a pressure different from ambient or a vessel in which a pressure different from ambient can be created for evaluation.	Provides assurance that the vessel will retain contents as designed. Advantages vary for the individual methods.	Vary from method to method.

SAE J359   Infrared
SAE J420   Magnetic Particle
SAE J425   Eddy Current
SAE J426   Liquid Penetrant
SAE J427   Penetrating Radiation
SAE J428   Ultrasonic
SAE J1267   Leakage Testing

Table 1 summarizes the principal features of most of these tests. In addition to the tests described, other nondestructive tests exist which are less well established, but whose use is expanding. Among these are microwave tests, holography, and sonic tests. Microwaves are used to locate defects in nonmetallic substances and to determine some physical characteristics of those materials. Optical holography uses coherent light from a laser beam to detect strains and defects in materials by means of three-dimensional imaging and interferometry techniques. Acoustical holography uses ultrasonic waves to image discontinuities in the interior of solids. Recent refinements in sonic testing permit more objective determination of the physical properties of cast iron. Complete information concerning each nondestructive test can be obtained from books listed in the bibliographies of the aforementioned reports.

Increasing consumer demand for product quality at reasonable cost has resulted in development of nondestructive tests which can be applied to materials and manufactured parts. Although a variety of complementary nondestructive methods is available, development time is generally required for application to specific materials or products. The effect of part contour, surface condition, heat treatment, composition variation, and other variables may limit the ability of certain tests to detect imperfections with desired accuracy.

Nondestructive tests properly applied to basic material can add greater assurance of performance to design strengths, thereby affecting material and manufacturing economy. In addition, parts can be tested after each basic operation which is critical to service performance of the finished part. In-process nondestructive tests can also serve as basic components of feedback process control systems since all tests are based upon measurements which do not damage the material or part being inspected.

INFRARED TESTING—SAE J359a

SAE Information Report

Report of Iron and Steel Technical Committee approved July 1973. Reaffirmed without change October 1976.

Purpose—The purpose of this report is to provide general information relative to the nature and use of infrared techniques for nondestructive testing. The report is not intended to provide detailed technical information, but will serve as an introduction to the theory and capabilities of infrared testing and as a guide to more extensive references.

General—Infrared (IR) nondestructive testing can be defined as a method by which objects (raw materials, in-process, or finished items) can be evaluated by detecting reading out, and interpreting the infrared emissions which are a function of the physical, electrical, mechanical, and thermal properties that may generate a temperature differential or influence heat transfer. As an independent system, this relatively new method of testing bridges existing gaps in nondestructive testing technology and is useful in supplementing or verifying other methods. It is being successfully applied as a process control technique where it can monitor extremely high temperatures in minute areas, using either focused optics or optical fibers, and it can be coupled by feedback to automated systems. Its precise measurement lends itself to smelting of high purity materials. IR testing equipment now has a sufficiently broad production base to assure the availability of options to suit a particular problem or for versatile operations. IR testing has been used successfully in detecting delaminations in solid propellant missile motors, ply separations in automobile tires, and effectiveness of vacuum, batt, or foamed-in-place insulation; and it can detect flaws, voids, and lack of bond in welds and solder joints, castings, etc. It has found high acceptance in the form of various infrared "cameras" producing in real-time thermal images of items ranging in size from very large missiles to electronic microcircuits and displaying actual and potential defects.

Principle—Infrared testing is concerned with the electromagnetic radiations in that part of the spectrum between 0.7 and 100 μm. All objects at a temperature greater than absolute zero radiate electromagnetic radiation and, as a result of increased molecular motion with temperature rise, the intensity and frequency increase. Thus, temperature can be measured by measuring the intensity of radiation. When the temperature of an object is high enough to radiate in the visible wavelengths of about 0.4–0.7 μm, ordinary photography will record shadings corresponding to any localized heat changes. As the radiation changes upon cooling to the longer red wavelengths where visible detection of the light begins to fail, some photographic red films can still provide a record. Beyond this narrow threshold, detection of radiation is made possible by the use of infrared detectors and systems.

Infrared detectors fall into two general types:

1. Photodetectors which produce a signal from a semiconductor, the signal being proportional to the impinging radiation. These detectors include photoelectromagnetic, photovoltaic, and photoconductive types.

2. Thermal detectors, which undergo a physical change in response to thermal change. These detectors include thermistors, thermocouples, bolometers, oil film evaporation and radiometer types.

An optical system of one type or another is usually introduced for magnification and/or focusing, a controlled black body for reference in alternation with the emission from the object, and the appropriate electronic circuitry and display.

Systems can be selected for measuring infrared emissions as low as $-150\,°$C to as high as desired, for detecting gradients as small as $0.05\,°$C. for focusing from $\frac{1}{2}$ in. (12.7 mm) to infinity, and with resolution as small as 0.00015 in. (0.0038 mm).

Procedure—Most of the systems are portable and have negligible but external power requirements. Readout is rapid and the system, being remote from the specimens, is inherently nondestructive. However, a specimen could be energized or heated to destruction if so desired, while the system recorded points or elements of failure. Operation of most systems requires minimum training. Experience required for evaluation of the readout is totally dependent on the type information desired; that is, common sense would allow the interpretation of the "Polaroid" display of a heat leak on a foundry furnace, while evaluation of the same display of relative temperatures of an electronic microcircuit taken by an infrared microscope would require a knowledge of the operating thermal characteristics of the components of the specimen.

In a more complex single specimen, such as a printed circuit board, a good knowledge is needed of the theory of heat transfer as well as the construction of the board. Lateral transfer of heat to another component functioning as a sink could mask a defect which would otherwise be indicated by an abnormal temperature.

Because infrared radiates the same as visible light and because the detector measures the surface radiation, a defect in a multilayer specimen could be concealed behind a surface component at normal operating temperature.

In evaluation of thermographs or thermoplots of complex specimens, consideration must be given to the wide variations in emmissivity resulting from dissimilar materials, coatings, etc.

None of the foregoing present insurmountable problems. It would be expected that in production testing a good standard sample would be made available, then one of many techniques such as "flicker" comparison of thermographs, overlays, etc., would quickly distinguish between go and no-go.

BIBLIOGRAPHY

1. Kruse, McGlaucklin, McQuistan, ELEMENTS OF INFRARED TECHNOLOGY. New York: John Wiley & Sons, 1963.
2. Jamieson, McFee, Plass, Grube, Richards, INFRARED PHYSICS AND ENGINEERING. New York: McGraw-Hill Co., 1963.
3. Hackforth, INFRARED RADIATION. New York: McGraw-Hill Book Co., 1960.
4. "Infrared." EDN, Vol. 10, No. 5, May 1965.
5. Riccardo Vanzetti, "Infrared Radiation: A New Dimension for Production Reliability and Maintainability." Raytheon Co., Wayland, Mass.
6. Glenn W. Carter, "A State-of-the-Art Evaluation of Infrared in Heat Transfer Engineering." IBM, Owego, N. Y., IBM No. 65-825-1434.
7. TRANSACTIONS OF THE INFRARED SESSIONS. SNT Convention, February 1965. Society for Nondestructive Testing, 914 Chicago Ave., Evanston, Ill.
8. Riccardo Vanzetti, PRACTICAL APPLICATIONS OF INFRARED TECHNOLOGY. New York: John Wiley & Sons, 1972.
9. *Applied Optics,* Vol. 7, No. 9 (September 1968). (Special edition on infrared containing 23 papers.)
10. P. Vogel, "Thermal Fingerprint." *Army Research and Development News Magazine,* May–June 1972.

MAGNETIC PARTICLE INSPECTION—SAE J420b

SAE Information Report

Report of Iron and Steel Technical Committee approved June 1952 and last revised August 1973. Reaffirmed without change October 1976.

Purpose—The purpose of this report is to provide general information relative to the nature and use of magnetic particles for nondestructive testing. The report is not intended to provide detailed technical information, but will serve as an introduction to the theory and capabilities of magnetic particle testing, and as a guide to more extensive references.

General—Magnetic particle inspection is a nondestructive means of inspecting ferromagnetic materials such as iron and steel for discontinuities (cracks, seams, near surface inclusions) by the detection of leakage fields through the use of magnetic particles.

Magnetic particle inspection is an aid to visual inspection of objects. Surface discontinuities that might not be seen with the aid of optical magnification are often dependably detected in manufacturing operations of maintenance. It is not applicable to nonmagnetic materials. The usual basic steps in magnetic particle inspection of an object are: clean, magnetize, apply magnetic particles, inspect, and demagnetize. Magnetic particle inspection is a relatively simple procedure. It is most effective when the various factors, such as types of magnetization, current, particles, equipment, and method, are properly selected for the application.

Principle—The principle of magnetic particle inspection, simply stated, is bipolar magnetism. The material subjected to the inspection is magnetized in a fashion which will produce north and south poles on opposite edges of a discontinuity. Finely divided magnetic particles are introduced into the leakage field between the poles, and are held there by the magnetic flux. The visible accumulation of these particles indicates the discontinuity.

Procedure—A magnetic field is induced in the part to be tested by the application of an electric current through the part, or through a central conductor inserted through a hole in the part, or by means of a yoke, prods, or coil. The type of magnetization selected is determined primarily by the need to establish magnetic flux lines perpendicular to the direction of anticipated surface imperfections. Any discontinuity at or near the surface of the part will interrupt the magnetic flux induced in the part and a leakage field will be fomed at the surface of the part. Magnetic particles in the vicinity of this leakage field will be attracted to it, forming a visible indication which, to experienced interpreters, expresses the characteristics of the discontinuity. Following the creation of the indication, the interpretation of the indication, and the evaluation of the discontinuity, the part is suitably demagnetized.

Adequate light must be provided for the quick and sure detection of the indications of dicontinuities. Lights should be adjusted to give broad highlights on finish machined parts. If fluorescent lighting is used, the tubes should be located transverse to the long axis of the parts being inspected. A nominal illumination level of 100 ft-c (9.3 lx) of white light should be present on the part surface in the case of nonfluorescent inspection. Personnel should have eyesight, corrected or uncorrected, capable of distant vision of 20/30 in at least one eye and should be able to read Jaeger Type No. 2 with both eyes at 12 in. (305 mm).

An adequate source of ultraviolet black light must be provided for inspection when using the fluorescent magnetic particle inspection method. A filtered high-pressure mercury vapor source is generally recommended. The emitted light should have an intensity of 90 ft-c (8.36 lx) at a 15 in. (380 mm) distance from the source, or no less than 900 mW/in.^{2} (5806 mW/cm^2) on the part surface. For detection of certain fine indications, illumination at the part surface may need to be as high as 250 ft-c (23.2 lx). Personnel vision requirements are the same as for nonfluorescent inspection, but in addition, visual acuity in the green-yellow spectrum must be satisfactory.

Demagnetization consists of removing objectional residual magnetic fields from parts which have been subjected to magnetic particle inspection. This must be done to prevent the deflection of adjacent sensitive instruments and to prevent the attraction of small magnetic chips, or the like, which could cause damage to contacting surfaces. The most common type of demagnetization consists of drawing the magnetized part through a high intensity alternating current solenoid. Another type, often used on heavier parts, consists of passing an alternating current or reversing direct current through the part or through a surrounding solenoid, and then gradually reducing the current value to zero.

Irons and steels exhibit various magnetic characteristics with variations in hardness and composition. Continuous magnetization during particle application is used on relatively soft steels since they usually do not retain sufficient magnetism to allow the use of the residual method. These steels are processed for inspection by introducing the magnetic particles into the leakage fields created at the discontinuities while the magnetizing force is in operation. Parts processed through the use of the continuous magnetizing force are said to be processed by the "continuous" method. Use of the continuous method makes possible the successful inspection of irons and steels which do not retain sufficient magnetism for processing by the residual method. In addition to this, certain subsurface discontinuities are easily detected in both hardened and unhardened parts by this method when direct current magnetization is employed.

The residual magnetization test method may be applied to hardened steels, since they will retain magnetism after the force has been removed. These remaining magnetic fields will produce leakage fields adjacent to discontinuities strong enough to hold magnetic particles and produce indications. Parts processed through the use of these retained fields are said to be processed by the "residual" method. Use of the residual method often eliminates nonrelevant indications. It is especially useful for the detection of surface discontinuities in suitably hardened parts. An adequate level of magnetization should be assured by a suitable method.

Wet particles used in suspension liquid consist of finely ground magnetic oxide of iron. These particles are coated so they can be easily dispersed in a liquid vehicle. They are generally available in a concentrated paste in red or black nonfluorescent colors. They are also available coated with a material which fluoresces under near ultraviolet (black) light. Wet particles are commonly used in maintenance, process, and finish inspection of machine and engine parts. The wet process offers the advantage of ease of application of the particles, sensitivity in locating the finest discontinuities, and, especially with the fluorescent particle, rapid inspection rates.

Dry particles consist of finely divided magnetic material in powder form. These particles are coated so as to be easily conveyed by air to the part being inspected. They are generally available in red, black, and gray colors for maximum contrast with the test object. Dry particles are commonly used for the maintenance, process, and finish inspection of heavy weldments, heavy castings, and heavy forgings. Dry particles are superior for the inspection of very rough surfaces and for the location of subsurface discontinuities in rough castings, forgings, and weldments.

Circular magnetization consists of inducing a circular magnetic field in a part so that the magnetic lines of force take the form of concentric rings about the axis of the current. This is accomplished by passing the current directly through the part, or by passing the current though a conductor which passes through a hole in the part, sometimes by use of prods. The circular method is used chiefly to indicate discontinuities radiating from and parallel to the axis of the current flow.

Longitudinal magnetization consists of inducing a longitudinal magnetic field in a part by making it the core of a solenoid, such as placing it in a coil or by making it a link in a magnetic circuit through use of a yoke. In a part so magnetized, the lines of force will be parallel to the axis of the solenoid, and the part will exhibit the properties of a bar magnet. The longitudinal method is used to indicate discontinuities transverse or circumferential to the long axis of a part.

Moving field magnetization consists of inducing fields in a part in more than one direction almost simultaneously. The fields induced may be a combination of circular and longitudinal or may be a combination of either type. The moving field method may be used on many parts ordinarily requiring two or more distinct magnetization and inspection operations. The moving field method, because of the rapidly changing field directions, makes possible the location of all detectable discontinuities after only one processing. This may, in some cases, eliminate a great percentage of the time required for the inspection if the parts were processed by more conventional methods.

Alternating current magnetization is commonly used on moderately stressed parts in production and for the detection of fatigue discontinuities in maintenance. Alternating current magnetization is always equal to, and often superior to, direct current magnetization for the detection of surface discontinuities. Subsurface discontinuities are not revealed when alternating current is used. In moderately stressed parts, this greatly simplifies inspection.

Direct current magnetization is commonly used on highly stressed parts. It is also able to disclose certain subsurface discontinuities in addition to surface discontinuities.

Half-wave direct current is commonly used in the inspection of heavy weldments, heavy castings, and heavy forgings, in conjunction with dry magnetic particles. Half-wave direct current is essentially a pulsating direct current. The pulsations impart mobility to the magnetic particles, thereby assisting in aligning them in the weaker leakage fields produced by subsurface discontinuities. Subsurface discontinuities are best revealed by the use of this type current.

BIBLIOGRAPHY

1. Nondestructive Testing Handbook, R. C. McMaster, ed. Evanston, Ill.: Society for Nondestructive Testing, 1963, Vol. 2, Sections 30–34.
2. C. E. Betz, "Principles of Magnaflux." Magnaflux Corp., Chicago, Ill., 1966.
3. ASTM E 269, Standard Definitions of Terms Relating to Magnetic Particle Inspection. American Society for Testing and Materials, Philadelphia, Pennsylvania 19103.
4. MIL-M-6867, Magnetic Inspection Units. Department of Defense.
5. MIL-I-6868, Inspection Process, Magnetic Particle. Department of Defense.
6. MIL STD-410, Qualification of Inspection Personnel. Department of Defense.
7. AMS-2640, Magnetic Particle Inspection, SAE, New York, 1969.
8. Programmed Instruction Handbook PI-4-3, Magnetic Particle Testing. Convair Div., General Dynamics Corp., 1967.

EDDY CURRENT TESTING BY ELECTROMAGNETIC METHODS—SAE J425b

SAE Information Report

Report of Iron and Steel Technical Committee approved June 1960 and last revised April 1976.

1. Purpose—The purpose of this report is to provide general information relative to the nature and use of eddy current techniques for nondestructive testing. The report is not intended to provide detailed technical information but to serve as an introduction to the principles and capabilities of eddy current testing, and as a guide to more extensive references listed in the Bibliography.

2. General—Eddy current testing is a method of electromagnetic testing which uses induced electrical currents to indicate or measure certain characteristics of electrically conducting bodies (ferrous and nonferrous). Applications are in one of three general categories: metal sorting, surface discontinuity detection, or thickness measurement. Under appropriate conditions and with proper instrumentation, eddy current testing has been used to:

(a) Detect discontinuities such as, but not limited to, seams, laps, slivers, scabs, pits, cracks, voids, inclusions, and cold shuts.

(b) Sort for chemical composition on a qualitative basis.

(c) Sort for physical properties such as hardness, case depth, and heat damage.

(d) Measure conductivity and related properties.

(e) Measure dimensions such as the thickness of metallic coatings, plating, cladding, wall thickness or outside diameter of tubing, corrosion depth, and wear.

(f) Measure the thickness of nonmetals, when a metallic backing sheet can be employed.

3. *Principle* —Eddy currents are induced in a test piece by a time varying magnetic field generated by an alternating current flowing in a coil. The coil configuration may assume a wide variety of shapes, sizes, and arrangements. The coil may surround the test piece or may be placed on or near the surface.

Eddy currents are influenced by many characteristics of the metal: conductivity, magnetic permeability, geometry, and homogeneity. This fact makes it possible to evaluate many different characteristics of the test piece with appropriate test procedures.

In electromagnetic testing, energy is dissipated in the test piece by two separate processes: (a) magnetic hysteresis and (b) eddy current flow. In magnetic materials, both effects are present. In non-magnetic and magnetically saturated materials, the hysteresis effect is absent or suppressed; and the prevalent losses are due to eddy currents.

Saturation is a term used generally to describe the condition of a ferromagnetic material at its maximum values of magnetization. To provide saturation, a direct current magnetic field or a permanent magnet of sufficient magnitude is applied to bring the material to a point where the ratio force approaches unity. In this condition, the material behaves as if it were non-magnetic. Theoretically, magnetic saturation should not be necessary for non-ferromagnetic material, but some non-magnetic materials contain small amounts of ferromagnetic material which can generate electrical noise during testing. This noise can usually be eliminated by the use of a saturating field.

4. *Procedure* —The effect of the characteristics of the test piece on the eddy currents may be studied in a number of different ways. A characteristic to be studied is related to a change in the amplitude, distribution, or phase of the eddy currents, or some combination of these three. These changes are reflected as changes in the exciting coil or in auxiliary coils located to be sensitive to the eddy currents. These changes may be measured as voltage differences, current differences, phase differences, or changes in the impedance of the coil or coils.

The coils and the instrumentation can be arranged to measure a given characteristic directly, or they may be used as a comparator. In the latter case, the measurement is the difference between the characteristics of the test piece and a similar piece of known or acceptable characteristics. Such measurements can also be made to determine differences between various segments of the same test piece.

With the best instrumentation, it is sometimes difficult to separate effects of the characteristics to be measured from effects of other characteristics. The success of an eddy current test depends on:

(a) proper coil design and arrangement,
(b) selection of the proper test frequency,
(c) selection of the proper analysis circuit,
(d) use of proper magnetic field strength,
(e) optimization and maintenance of electromagnetic coupling between the coil and test piece, and
(f) selection of the most suitable stage in the manufacturing process for the inspection procedure.

Eddy current effects are most pronounced near the surface, with sensitivity for detecting irregularities of composition or structure falling off as depth below the surface increases. Depth of eddy current penetration of an object decreases as test frequency increases. Ferromagnetic metals, such as steel, are generally tested with low frequencies in the range of 1 Hz to 10,000 Hz (10 kHz). Non-magnetic metals with higher conductivity, such as aluminum, are generally tested with frequencies around 100 kHz, while those with lower conductivity, such as titanium, are generally tested with frequencies in the range of 1 MHz to 10 MHz. There are numerous exceptions to these generalities.

5. *Test Coil Methods*

5.1 Single Coil—In this method, a single coil is used. It may have one, two, or three windings for excitation and detection. A winding is excited from an alternating current source within the test instrument. The amplitude and phase of the voltage across a winding is a function of the effect of the test piece on the coil.

5.2 Differential Coil—An arrangement where two separate detector coils are used to compare two different test pieces, or two different portions of the same test piece. A voltage appears at the output terminal of the coils when the effective permeability, conductivity, mass, geometry, or homogeneity of the metal in the two coils differ.

6. *Method of Analysis*

6.1 Lumped Impedance—In the lumped impedance analysis, a single coil is employed. A characteristic of the test piece is correlated to the amplitude and phase coil voltage.

6.2 Impedance Plane Analysis

6.2.1 MAGNETIC PARAMETER AMPLITUDE—The single coil, the two-coil, or the differential coil method may be employed in this analysis. The variation in amplitude and phase of the detector coil voltage is measured and plotted in an impedance plane. The coil parameters are correlated to a test piece characteristic. Some variation in chemistry and size can be tolerated in this system providing the proper test frequency is employed.

6.2.2 PHASE ANGLE ANALYSIS—A two-coil method is more suited to this type of analysis. The phase angle between the voltage at the driving coil and that at the detector coil is measured and related to a test piece characteristic.

7. *Equipment*—One of the advantages of electromagnetic equipment is that it lends itself to automatic operation for regularly shaped parts. Electromagnetic equipment can be large, elaborate, and expensive when multiple stations and materials handling sections are included, such as used on sheets and plates. Manual systems which are small, simple, and inexpensive are also common in other instances such as used with large or irregularly shaped objects.

The electronic apparatus is capable of energizing an encircling coil or probe with alternating currents of suitable frequency and amplitude and is capable of sensing the electromagnetic response of the sensors. Equipment may include a detector phase discriminator, filter circuits, modulation circuits, magnetic saturation devices, recorders, and signaling devices as required by the application.

The encircling or probe coil assembly is capable of inducing current in the part and sensing changes in the electric and magnetic characteristics of the part.

A mechanical device capable of passing a part (such as a tube) through the encircling coil or past the probe may be used. It generally operates at uniform speed with minimum vibration of the coil, probe or part, and maintains the article to be inspected in proper register or concentricity with the probe or encircling coil. A mechanism capable of uniformly rotating or moving the part or the probe may be required.

An end effect suppression device, a means capable of suppressing the signals produced at the ends of tubes or bars, may be used.

Reference standards are generally required to adjust the sensitivity of the electronic apparatus.

7.1 Typical Examples of Equipment Variations for Different Applications:

7.1.1 Equipment using impedance plane analysis and operable over a range of test frequencies from 1 Hz to 10 kHz has been used to sort carbon steel mixtures involving different compositions and/or different heat treat conditions. A unique advantage of this instrument is that it is possible to quickly determine the optimum frequency for performing a given test. Similar equipment has been calibrated to indicate conductivity, hardness, case depth, and dimensions.

7.1.2 Equipment using a single coil to scan the surface has been used to detect and indicate the depth of seams, cracks, laps, slivers, and similar surface imperfections in bars, rounds, billets and tubular products. The sensitivity of this equipment depends on the surface condition of the product under test. On a hot-rolled surface with thin, tightly adherent scale, seams as shallow as 0.010 in (0.25 mm) are reliably evaluated. Product with heavy or broken scale should be cleaned by grit blasting prior to testing. Under more favorable (smoother, less scale) surface conditions seams as shallow as 0.005 in (0.13 mm) have been evaluated. On polished (ground) surfaces, seams and cracks as shallow as 0.001 in (0.025 mm) have been detected.

7.1.3 Equipment using differential test coils has been used to detect imperfections in carbon steel tubular and bar products. Testing frequencies ranging from 400 Hz to over 20 kHz have been used. At the lowest testing frequencies, and with the use of magnetic saturation, defects have been reliably detected (OD, ID, or subsurface) in the wall of tubular products with wall thicknesses as great as 0.62 in (15.9 mm). When testing at frequencies as low as 400 Hz, the testing speed is limited to about 100 ft/min (30.5 m/min). When higher testing frequencies are used, the testing speed can be correspondingly increased. Higher testing frequencies can be used for testing product with thinner walls and higher resistivity.

7.1.4 Vector sensitive instruments operate on the impedance plane principle. The frequency range of these instruments is from 100 Hz to 6 MHz. This type of operation considers both the amplitude and phase of the eddy currents. This allows one to optimize the instrument response for a selected material variable, while minimizing response to another variable, such as probe spacing.

BIBLIOGRAPHY

1. Robert C. McMaster, ed., NONDESTRUCTIVE TESTING HANDBOOK, New York: The Ronald Press Co., 1959, Vol. 2, Chapters 36–42.
2. PROGRAMMED INSTRUCTION HANDBOOKS, PI-4-5, Eddy Current Testing, 1971. CLASSROOM TRAINING HANDBOOK, CT-6-5, Eddy Current Testing, 1971. The above prepared by General Dynamics and available from American Society for Nondestructive Testing.
3. Hugo L. Libby, INTRODUCTION TO ELECTROMAGNETIC NONDESTRUCTIVE TEST METHODS, New York: John Wiley and Sons, Inc., 1971.
4. ASTM ANNUAL STANDARDS, Part 11, Articles E215, E243, E268, E309, E376, E426.

LEAKAGE TESTING—SAE J1267

SAE Information Report

Report of Iron and Steel Technical Committee approved May 1979.

Purpose—This information report provides basic information on leakage testing, as applied to nondestructive testing, and affords the user sufficient information so that he may decide whether leakage testing methods apply to his particular need. Detailed references are listed in the Bibliography.

General—Leakage testing is a form of nondestructive testing capable of determining the existence of leak sites and, under proper conditions, measuring the quantity of material passing through these sites. The word *leak* means the hole through which fluid (liquid or gas) passes in either a pressurized or evacuated system, while *leakage* is the term used to connote the mass flow of fluid regardless of the size of the leak. Leakage rate is the quantity of fluid per unit time that flows through the leak at a given temperature as a result of a specified pressure difference across the leak. The ASTM accepted unit of leakage rate is standard cubic centimeters per second (std. cm^3/s), frequently referred to as atmosphere cubic centimeters per second (atm. cm^3/s). The SI terminology is Pascal cubic meters per second (Pa m^3/s). (1 Pa m^3/s = 10 atm. cm^3/s, approximately.)

There is no container in which a differential pressure exists (either pressurized or vacuum) that does not leak to some extent. Absolute leak tightness is an absolute impossibility. Any container must, therefore, have a maximum leakage rate specified. In considering the leakage rate that can be tolerated, one must decide whether the rate represents the total leakage from the system or the maximum leakage from a single leak. Additional factors to be considered include shelf life, product contained, toxicity, legal requirements, consequences of excessive leakage, cost of product, cost of testing, and customer requirements. Once a leakage rate has been specified then a leak test procedure describing the operating and test conditions needs to be detailed. Since leakage rate relates pressure, volume, and time, more than one set of procedural values can yield the same leakage rate. In general, the pressure used should reflect pressures that the item would see in service, however, this is not a requirement. In some isolated cases, using markedly different pressures can cause leaks to pass grossly different rates of fluid due to elastic deformation of the item being tested. Regardless of the type of leakage testing being done, safety considerations for the personnel performing these tests must be a paramount consideration.

Principles—There are eight or more primary methods which may be employed to detect, locate, and/or measure leakage. The following paragraphs identify these methods and describe their principles, as well as their capabilities and limitations.

Mass Loss and Pressure Change—These are two related methods. Traditionally, these are used for sizable leakage rates, and provide no information as to the leak site. Mass loss is calculated on the basis of change in mass at two times; accordingly, extremely accurate weighing is a requirement of this method. Pressure change methods operate in a similar fashion, except that a change in pressure is the signaling mechanism. Pressure change systems usually measure change of the gaseous systems. Since pressure is temperature dependent, the temperature of the system must either remain constant or be compensated for by use of ideal gas laws. Mass loss and pressure change methods, using most techniques, are time consuming and thus are limited in leakage testing applications.

Theoretically, these methods are very accurate if one has sufficient time to conduct the test.

Ultrasonic Leak Testing—This is a method valuable for detecting leakage great enough to produce turbulent flow. Turbulent flow in a gas occurs when the velocity approaches the speed of sound in that gas; this is approximately 10^{-1} to 10^{-2} std. cm^3/s. This method takes advantage of the fact that turbulent flow generates sound frequencies from audible upward to 60 kHz. In using only the ultrasonic component generated, fewer false signals are detected because there are fewer other sources of ambient ultrasound. Because of the highly directional nature of ultrasound, the leak can usually be accurately located. Output of ultrasonic leak detectors is an audible signal or a meter deflection, the strength of which is a function of the leakage rate. Advantages of the ultrasonic method are that the equipment is simple to operate, it can be done with the probe removed from the leak, and it does not require any material which could clog a leak and require cleaning. Its primary disadvantage lies in its lack of sensitivity to small leakage rates (less than 10^{-2} std. cm^3/s).

Chemical Penetrant Leak Tests—These are incapable of providing leakage rate information, but do clearly point out sites for repair. Sensitivities are generally conceded to be in the range of 10^{-3} std. cm^3/s, although greater sensitivities have been achieved. Two basic forms of chemical penetrants are available, liquid tracers (quite similar to liquid penetrants, see SAE Information Report J426) and gaseous tracers. Hydrostatic testing with water alone is not a substitute for leakage testing.

Liquid Tracers—Liquid tracers are usually a solution of a tracer dye and a liquid in which it is soluble. It is essential to determine the coloring power of the tracer solution in the concentration being used as this relates to the sensitivity, as does the wettability of the tracer solution. As a general rule in white light systems, basic dyes work best in a water solution and solvent dyes are better suited to an oil based system. White light liquid tracer systems are generally inferior to fluorescent liquid tracer systems. This sensitivity inferiority is due, in part, to the increased visibility of fluorescent dyes, and the inherent contrast of the dyes to the near black background used in testing. By dissolving a fluorescent tracer in a volatile liquid, very small leaks can be found, because the liquid which evaporates leaves behind a concentrated dye which is more visible. Advantages of liquid tracers lie in their cost, sensitivity, and ease of use. Foremost among their disadvantages is that they use material which could temporarily clog a leak. Also, liquid tracers require cleaning of the parts after use and care in their application so as not to create false signals. In addition, one may experience occasional difficulty in tracing large leakages to their source due to liquid spread.

Gaseous Tracers—Gaseous tracers are gases which color indicating media, thereby denoting the location of a leak. The most widely used gas for this application is ammonia. Indicating media for ammonia gas are:
1. Phenolphthalein which turns from clear to pink.
2. Bromocresol purple which turns from a yellow-green to purple.
3. Bromothymol blue, which turns from yellow to blue.

Carbon dioxide gas can be used for leak testing with an indicating medium of sodium carbonate and phenolphthalein in an agar-agar solution. This bright red indicator will turn white at a leak site.

There is another medium, which is much less widely used, due to the inherent danger of its chemicals. Pressurizing a component with ammonia and then allowing hydrogen chloride to be brought near, will produce a white cloud of ammonium chloride vapor which is clearly visible. These gases are highly corrosive and dangerous to human tissue. Extreme care and a high level of ventilation are needed, as well as consideration for the safety of the personnel performing the test.

There is little difference in the level of sensitivity for gaseous tracers when compared to liquid tracers; both are typically at 10^{-3} std. cm^3/s. Rates as low as 10^{-6} std. cm^3/s have been reported for gaseous tracers. Primary among their advantages is their low cost of operation since no instrumentation is needed. Disadvantages are that some gases could corrode the test object, be hazardous to personnel, require cleanup, and clog leaks.

Bubble Leak Test—These methods are widely used. They possess sensitivities to a commercial range as small as 10^{-4} std. cm^3/s (10^{-2} is a practical value for an unskilled operator). In the laboratory, under ideal conditions with special combinations of liquid and gas, rates as low as 10^{-7} std. cm^3/s have been detected. The method operates on the basis of a differential pressure at a leak creating a flow of gas. This gas, upon escaping, will produce one or more bubbles in the test liquid. These bubbles mark the location of the leak and the frequency and size can be used to estimate the leakage rate.

Procedurally, the test object is fixtured and pressurized, and then the indicating liquid (not a soap or detergent solution) is brought into contact with the component. This precludes the liquid from temporarily blocking a small leak which could cause the acceptance of a leaking component. Precleaning of the test object is necessary because surface contaminants also may cause a temporary blockage of a leak. From a practical standpoint, any gas may be used to pressurize the object. Should air be used, it must be very clean, again to preclude temporary blockage of a leak. Shop air is generally too dirty, wet, and oily to use for leak testing.

Ample illumination must be provided to permit the inspector to be able to see a stream of bubbles; 1000 lm/m^2 (100 fc) is recommended as a minimum level.

Indication of leakage may be accomplished by the use of:
1. A liquid in which the test object is immersed.
2. A liquid film which produces bubbles when a leakage passes. (A vacuum box which surrounds the test area may be used to create the pressure differential.)

Liquids used in bubble testing must not corrode the object being tested. Frequently, it is desired to enhance the sensitivity of a bubble test. Enhancement can be done by increasing the time for testing or increasing the pressure. In some instances neither of these approaches is practical. Changing the gas to one of a lower molecular weight and/or lowering the surface tension of the liquid will also enhance the sensitivity. Visual inspection should be conducted at distances less than 0.6 m (2 ft) for best results.

A vacuum box places an area to be tested under a sub-atmospheric pressure. A clear window through which observations are made and a liquid in which leakage appears are necessary for the vacuum box technique. When used, (usually for welds in large vessels) adjacent testing locations must be overlapped to assure full coverage.

The advantages of bubble testing lie in its simplicity, cost, and relative sensitivity. Disadvantages include the need for cleanup, the fact that fine leaks may not be detected due to a lack of time, the possibility of clogging, and finally that bubble testing is a visual inspection, and as such, bubble testing is limited by the performance of an operator.

Thermal Conductivity Leak Testing—These methods have a minimum leakage rate detectability of 10^{-5} std. cm^3/s. They are based on the principle that certain gases have a markedly different thermal conductivity when compared to air. Equipment for this method consists of two heated filaments in a bridge circuit. One filament is cooled by air and the other by the test gas. Any differences unbalance the bridge and can be related to leakage. The two gases with the greatest difference in thermal conductivity are hydrogen and helium. Most thermal conductivity leak testing is done with argon, CO_2, neon, or R-12 (freon). Advantages include cost of equipment, reduced sensitivity to contaminants in the ambient atmosphere than other instrumented leak detectors, and simplicity of operation. Disadvantages include the limited gases which can be used.

Halogen-Based Leak Detectors—These use a halogen gas as the pressurizing medium and may take several forms, including the halide torch, the heated anode detector, and the electron capture detector. The upper limit of sensitivity is 10^{-9} std. cm^3/s. Halogen leak detector tests are normally not conducted using elemental halogens as a detector gas. Halogen leak detector tests are conducted using a chlorinated, fluorinated, or chlori-fluorinated hydrocarbon as the tracer gas.

Simplest and least expensive in the halogen family of leak detectors is the *halide torch*. It consists of a halide free source of gas, frequently acetylene, and a search tube to look for leaks, both of which feed a burner with a copper plate. In operation, the flame of the torch is blue when no halogens are present. The flame turns green when small leaks are detected, and turns violet when exposed to larger leaks. Search rates are approximately 6 mm ($\frac{1}{4}$ in)/s. Halide torches have a leakage detectability of 10^{-4} std. cm^3/s. Since torches generate toxic vapors they must be used only in areas with adequate ventilation and cannot be used in flammable environments.

Due to the widespread use of the *heated anode halogen detector* in the refrigeration industry, this instrumentation is the most widely used of the halogen leak detectors. Operationally, ions are emitted from a hot plate to a collector. These positive ions increase in proportion to the amount of halogen present. Sensitivities of 10^{-9} std. cm^3/s are obtainable. This detector has the advantages of high sensitivity, and the ability to operate in air. Its disadvantages include responding to halogen containing suspended particles from sources like cigarette smoke and chlorinated hydrocarbons used in cleaning compounds, and that the decomposed products are toxic and corrosive. Further, the anode operates at 900°C (1650°F) which makes it unusable in a flammable environment, and there is a need to recalibrate the unit regularly as the calibration changes with use.

The *electron capture leak test* method uses a non-electron capturing gas (argon or nitrogen) as a background gas. The electron capture test gas is ionized producing tritium. In operation, the halogens drawn through the sensor reduce the ion content which produces a current. This current is proportional to the amount of halogen. Electron capture is frequently used with sulphur hexafluoride as a tracer. Sensitivities of 10^{-10} std. cm^3/s or better have been achieved. Advantages include very good calibration sensitivity, the absence of a heated element, and non-production of toxic or corrosive gases. Cost is the primary disadvantage of this system.

One of the most sensitive types of leakage testing equipment is the *mass spectrometer*. Leakage rates of 10^{-11} std. cm^3/s are achievable under ideal conditions. This method is the most accurate form for vacuum testing. A mass spectrometer operates on the principle of sorting gaseous ions with respect to their molecular weight. In a helium mass spectrometer, baffles with slits allow He^+ ions to pass through to the detector while all other ions are blocked. The number of He^+ ions which arrive at the collector per unit time is a measure of the leakage rate. Rates are typically displayed on a calibrated meter. As in any tracer gas system, care should be exercised to keep false signals from being sensed and displayed as leakage. Grease, oil, rubber, and other materials can act as storage reservoirs for helium.

Sensitivity is usually considered to be the greatest advantage of the mass spectrometer, also the fact that it is not affected by background contamination (other than He) is a great asset. Using helium provides inherent safety when compared with other gases which are toxic.

Cost is the greatest disadvantage of the mass spectrometer; however, many thousands are currently in use.

Applications—Any attempt to list the more common products evaluated by these test methods would be cumbersome and fail to serve the user. Briefly, any product containing a pressure different from atmospheric can be leak tested. The decision to leak test or not to leak test should be based on economic considerations and applicable legal requirements.

With the capability to sense leakage rates to 10^{-11} std. cm^3/s, there is no reasonable leakage rate that cannot be detected using available leakage testing technology.

Table 1 is presented to give the reader a better understanding of leakage rates.

TABLE 1—COMPARISON OF LEAKAGE RATES

Leakage Rate (Std. cm^3/s)	Approximate Time to Fill	
	$1 cm^3$	$1 in^3$
10^{-1}	10 s	3 min
10^{-2}	2 min	27 min
10^{-3}	17 min	4 h
10^{-4}	3 h	2 days
10^{-5}	28 h	19 days
10^{-6}	12 days	6 months
10^{-7}	4 months	5 years
10^{-8}	3 years	52 years
10^{-9}	32 years	520 years
10^{-10}	320 years	5200 years
10^{-11}	3200 years	52 000 years

Bibliography

Nondestructive Testing Handbook, 2nd Edition, Volume 1, 1979, American Society for Nondestructive Testing, Columbus, OH 43221.

J. W. Marr, Leakage Testing Handbook, 1968, NASA CR-952, Jet Propulsion Laboratory, National Aeronautics and Space Administration, Washington, DC.

N. I. Sax, Dangerous Properties of Industrial Materials, 2nd Edition, 1963, Reinhold Publishing Corp., New York, NY.

J. W. Perry, Chemical Engineering Handbook, 3rd Edition, 1950, McGraw-Hill Book Co., New York, NY.

Metals Handbook, 8th Edition, Volume 11, 1976, pp. 260–270, American Society for Metals, Metals Park, OH 44073.

ASME Boiler and Pressure Vessel Code, latest addenda, American Society for Mechanical Engineers, New York, NY.

Qualification and Certification of Personnel, Recommended Practice SNT-TC-1A, Supplement G (Leak Testing), American Society for Nondestructive Testing, Columbus, OH 43221.

R. J. Roehrs, Leak Testing of Welded Vessels, 2nd Annual Symposium on Nondestructive Testing of Welds, sponsored by IITRI, Chicago, IL, 1967.

ASTM E 425, "Standard Definitions of Terms Relating to Leak Testing." ASTM Annual Standards, Part 11, American Society for Testing and Materials, Philadelphia, PA 19103.

ASTM E 427, "Recommended Practice for Testing for Leaks Using the Halogen Leak Detector (Alkali-Ion Diode)." ASTM Annual Standards, Part 11, American Society for Testing and Materials, Philadelphia, PA 19103.

ASTM E 432, "Standard Recommended Guide for the Selection of a Leak Testing Method." ASTM Annual Standards, Part 11, American Society for Testing and Materials, Philadelphia, PA 19103.

ASTM E 479, "Standard Recommended Guide for Preparation of a Leak Test Specification." ASTM Annual Standards, Part 11, American Society for Testing and Materials, Philadelphia, PA 19103.

ASTM E 493, "Standard Test Methods for Leaks Using the Mass Spectrometer Leak Detector in the Inside-Out Testing Mode." ASTM Annual Standards, Part 11, American Society for Testing and Materials, Philadelphia, PA 19103.

ASTM E 498, "Standard Methods of Testing for Leaks Using the Mass Spectrometer Leak Detector or Residual Gas Analyzer in Tracer Probe Mode." ASTM Annual Standards, Part 11, American Society for Testing and Materials, Philadelphia, PA 19103.

ASTM E 499, "Standard Methods of Testing for Leaks Using the Mass Spectrometer Leak Detector in the Detector Probe Mode." ASTM Annual Standards, Part 11, American Society for Testing and Materials, Philadelphia, PA 19103.

ASTM E 515, "Standard Methods of Testing for Leaks Using Bubble Emission Techniques." ASTM Annual Standards, Part 11, American Society for Testing and Materials, Philadelphia, PA 19103.

LIQUID PENETRANT TEST METHODS—SAE J426c

SAE Information Report

Report of Iron and Steel Technical Committee approved June 1960 and last revised June 1978.

Purpose—The purpose of this report is to supply the user with sufficient information so that he may decide whether liquid penetrant test methods apply to his particular inspection problem. Detailed technical information can be obtained by referring to the Bibliography at the end of this report.

General—Liquid penetrant testing is a sensitive, nondestructive inspection method suitable for the detection of very small discontinuities that are open to the surface. It is generally used on materials such as metals, plastics, and ceramics. However, the magnetic particle method is generally preferred for ferromagnetic materials. Specific applications include detection of cold shuts, seams, shrinkage, porosity, cracks, and other imperfections which are open to the surface of nonporous objects.

Principle—The liquid penetrant test method is based upon capillary action, using low surface tension liquids. The liquid penetrant is applied to the surface to be inspected by dipping, spraying, or brushing. The excess penetrant is removed and a developer is applied. The bleeding out of penetrant from the discontinuity into the developer yields indications which can be observed and evaluated. This is done under ultraviolet or white light, depending upon the type of liquid penetrant used—fluorescent or nonfluorescent (visible).

Procedure
1. Clean parts by washing, degreasing, or etching.
2. Apply penetrant to the surface to be inspected.
3. Allow adequate time for penetration.
4. Remove excess penetrant from the surface.
5. Dry the surface to be inspected. (Perform after the next step if a wet developer is used.)
6. Apply a developer when applicable. The developer is a material which acts like a blotter and draws penetrant from the defect. Dry or wet (aqueous or nonaqueous) developers are used.
7. Locate imperfections by observing penetrant bleed-out from the discontinuity.
8. Post clean parts. Remove residual penetrant and developer.

NOTE: Excessive part temperatures can degrade penetrants.

Characteristics—There are a number of different classifications of penetrant test materials. These are discussed in the following paragraphs.

1. Penetrants are classified into two types. One type of penetrant employs fluorescent dyes to make surface imperfections visible under ultraviolet light. The other type of penetrant employs red nonfluorescent dyes which are visible under white light.

2. The penetrants are further classified according to the method of excess penetrant removal:

Method A—The penetrants contain an emulsifier which makes them water washable.

Method B—The penetrants require that an emulsifier be applied over the penetrants to make them water washable. Hence, they are called post-emulsifiable penetrants.

Method C—The penetrants are solvent removable.

Generally, there are several recognized sensitivity levels of penetrant performance relating primarily to the width of discontinuity that must be detected. Selection of the method (that is, A, B, or C) and sensitivity levels will be made based upon the following factors: surface roughness, surface treatment, size of discontinuity to be disclosed, environment, production required, equipment available, type of material to be inspected, subsequent use of the part, disposal restrictions, cost, and others. Where minimal sensitivity with respect to size of discontinuities is needed, the color contrast or visible dye penetrants are usually employed. Either of the three methods above can be used.

Fluorescent penetrants make discontinuities more discernable. Fluorescent penetrants may be selected for use with method A, B, or C. Two principal factors affect performance: the amount of fluorescent dye that is dissolved, and the ability of the penetrant to be retained in the surface apertures after surface excess is removed.

Generally, the least effective method is method C, solvent removable: Because the process is difficult to control accurately, results may vary widely. Either the self-emulsified, water-washable, or the post-emulsified method will yield equivalent results if proper selection of materials is made. For ultra-high performance, a very low activity emulsifier mixture with water is used to clean off the surface excess to a measured depth.

Developers—One of three types of developers is used to draw the penetrant from the discontinuities. The first, called dry developer, consists of a dry, light-colored, powdery material. Dry developer is applied to the surface of the parts after removal of the excess penetrant and drying of the part. Dry developer is applied by immersing the parts in a tank containing powder, by brushing it on with a paintbrush (usually not a desirable technique), or by blowing the powder onto the surface of the part.

The second type is aqueous wet developer and consists of powdered material suspended in water. The use of the wet developers permits rapid coverage of a large number of parts or of parts that have complicated shapes. After application of the wet developer, the part is dried and then inspected for penetrant indications.

The third developer is called nonaqueous wet developer. The powdered developer is suspended in a suitable solvent and sprayed onto the surface of the dry part. The solvent evaporates quickly, leaving a fine coating of developer on the surface of the part. The nonaqueous wet developer produces the greatest sensitivity when inspecting parts with small, tight defects.

Selection of the type of development method or material used for an application is important to the achievement of reliable inspections. The three types of developers vary widely in the degree of enhancement of indications. Nonaqueous wet developer, aqueous wet developer, or dry powder may be preferred depending upon the application. Surface finish may affect the degree of enhancement of developers. An aqueous developer, either soluble or particulate, should not generally be used with water-washable penetrants.

BIBLIOGRAPHY

1. C. E. Betz, "Principle of Penetrants." Magnaflux Corp., 1963.
2. Nondestructive Testing Handbook, R. C. McMaster, ed., Columbus, OH 43221. American Society for Nondestructive Testing, 1959, Vol. 1, Sections 6–8.
3. AMS 2645G, Fluorescent Penetrant Inspection. May 1, 1969.
4. AMS 2646B, Contrast Dye Penetrant Inspection. May 1, 1969.
5. AMS 3155A, Oil, Fluorescent Penetrant Solvent Soluble. June 30, 1964.
6. AMS 3156A, Oil, Fluorescent Penetrant Water Soluble. June 30, 1964.
7. AMS 3158, Solution, Fluorescent Penetrant Water Base. June 30, 1964.
8. ASTM E 165, Methods for Penetrant Inspection.
9. ASTM E 270, Definitions of Terms Relating to Liquid Penetrant Inspection.
10. MIL-I-6866B (ASG), Inspection, Penetrant Method of.
11. MIL-I-25135C (ASG), Inspection Materials, Penetrant.
12. Programmed Instruction Handbook PI-4-2. Convair Div., General Dynamics Corp., 1967.
13. Recommended Practice SNT-TC-1A, Supplement D-Liquid Penetrant, Qualification and Certification of Personnel. American Society for Nondestructive Testing, Columbus, OH 43221.
14. Metals Handbook, Volume 11, 1976, pp. 20–44. American Society for Metals, Metals Park, OH 44073.

PENETRATING RADIATION INSPECTION—SAE J427b

SAE Information Report

Report of Iron and Steel Technical Committee approved June 1960 and last revised June 1978.

Purpose—The purpose of this report is to provide basic information on penetrating radiation, as applied in the field of nondestructive testing, and to supply the user with sufficient information so that he may decide whether penetrating radiation methods apply to his particular inspection need. Detailed information references are listed in the Bibliography.

General—Penetrating radiation is a versatile nondestructive test method used in modern industry. The use of penetrating x-rays, gamma rays, thermal neutrons, and other forms of radiation which do not affect the material being inspected, provide the basic information by which soundness can be determined. Radiography provides a permanent record on film of internal conditions. Fluoroscopy differs from radiography in that the radiation image is projected on a fluorescent screen or other readout monitor and observed visually rather than recorded on a film. Penetrating radiation enables industry to monitor a variety of products for a number of types of imperfections. Objects inspected range in size from microminiature electronic parts to very large components in a wide range of manufactured forms (for example; castings, weldments, assemblies).

Principles—X-rays, gamma rays, and neutrons possess the capability of penetrating materials, even those that are opaque to light. In passing through matter, some of these rays are absorbed or scattered. Materials absorb x-rays and gamma rays in proportion to their mass. Neutron absorption, on the other hand, is not related directly to atomic number or mass; neighboring elements can differ in neutron absorption by factors of 100 or more. Differential absorption of the radiant energy passing through the object due to the presence of voids, discontinuities, or density variations caused by inhomogeneity or internal construction is recorded on radiographic film or observed directly by fluoroscopic methods. With acceptable conditions of technology and equipment, it is generally agreed that discontinuities can be detected which present to the axis of radiation a minimum dimension of 1-2% of the thickness of the object undergoing radiographic examination, or 2-6% for fluoroscopic examination. Two-dimensional imperfections, such as cracks and cold shuts, are not detectable unless they present an effective thickness difference of the above magnitude, or greater.

Procedure

1. Radiographic Film Technique—A radiographic film is a photographic record produced by the passage of x-rays, gamma rays, or neutrons through an object onto a film. When film is exposed to a radiation source or light, an invisible change is produced in the film emulsion. The areas so exposed become dark when the film is immersed in a developing solution; the amount of darkening depends upon the degree of exposure. Image formation is usually enhanced through use of thin metal screens in intimate contact with the film. Lead screens are used in x-ray exposures made with energy above 100 kV and in gamma ray exposures. Screens are necessary for film detection of thermal neutrons. Gadolinium metal screens are normally used for this purpose. The developing, fixing, and washing of exposed film may be done either manually or in an automatic film processor. The exposed, processed, and dried radiographic film is examined under transmitted light. Interpretation of the image is performed in accordance with established codes, specifications, or acceptance criteria.

The finished radiograph should be viewed under conditions which provide for the best visualization of detail combined with maximum comfort and minimum fatigue for the observer. A high-intensity illuminator with adjustable intensity is almost a necessity for optimum radiographic observation and interpretation. Penetrameters are used to indicate the image quality which exists in a radiograph. The type generally used in the United States is a small rectangular plate of the same material as the object being x-rayed. It is uniform in thickness (usually 2% of the object thickness) and has holes drilled through it. ASTM specifies hole diameters 1, 2, and 4 times the thickness of the penetrameter. Step, wire, and bead penetrameters are also used. (See ASTM E 94, Recommended Practice for Radiographic Testing.) For neutron radiography, image quality indicators provide a measure of the relative exposure due to gamma rays, higher energy neutrons, and scattered neutrons. Additional image quality indicators are suggested to provide measures of contrast and resolution capability. (See ASTM E 545.)

ADVANTAGES—Film radiography provides a permanent, visible record of the internal condition of the subject. Preservation of films is a common practice in industry.

DISADVANTAGES—High cost is the chief objection to film radiography. One-half of the average inspection cost is the radiographic film cost. X-ray paper products reduce this advantage when maximum performance capability is not required.

Inspection results are not available until radiographic film has been exposed, processed, and interpreted.

2. Fluoroscopic Inspection Technique—Fluoroscopy is the process of examining an object by direct observation of the fluorescence of a screen caused by radiation transmitted through an object. The arrangement of the x-ray source, object, and imaging plane is identical to that used in radiography. The fluorescent screen, image intensifier tube, television camera, and similar electronic imaging devices convert x-rays to visible light for further signal processing and operator interpretation.

ADVANTAGES—Production line inspection systems are available. These can result in low cost per part inspected and can meet the inspection requirements of high-volume production.

DISADVANTAGES—The sensitivity of the fluoroscopic process is not as great as that of radiography, 2-6% being routine. The lack of a permanent record of the examination may be a disadvantage. For systems employing television detection, however, magnetic recording can be used, or photographs may be taken of the television image.

Application—The ability of high energy radiation to penetrate all engineering materials and the differential rates of absorption for different materials is responsible for the extensive use of this nondestructive inspection technique throughout industry. Accordingly, penetrating radiation inspection methods are extensively used for flaw detection in the following areas:

1. Castings—The increasingly wide use that penetrating radiation inspection methods enjoy in the castings field result from the fact that most of the flaws and discontinuities inherent in ferrous and nonferrous castings can be readily detected by this inspection medium. Shrinkage, gas porosity, inclusions, hot tears, cold cracks and shuts, core shifts, and major surface irregularities may be detectable by radiographic or fluoroscopic inspection techniques. In addition, the following discontinuities which are peculiar to light metal (aluminum and magnesium) castings are detectable: gas holes, dross inclusions, segregation, microshrinkage, hydrogen porosity, microporosity, shrinkage, sponge, cold shuts, and other discontinuities common to light metal castings.

2. Weldments—Penetrating radiation inspection of weldments is an accepted procedure for the detection of internal discontinuities. It is used in the establishment of welding procedures to qualify welders and especially to control quality of welded joints in finished products. The following imperfections or discontinuities are detectable by radiography: porosity, cracks, poor penetration and fusion, inclusions, and other discontinuities common in welded joints.

3. Finished Assemblies—Penetrating radiation techniques are applicable to the inspection of fabricated assemblies relative to placement of internal components, such as electronic devices, mufflers, fuel tanks, bonded honeycomb, and tires. Electrical connections as well as the position of bolts and nuts in finished enclosures are frequently checked by radiography. Neutron radiography of assemblies provides a capability to verify proper placement of hydrogen-containing materials in metal assemblies. By this method rubber *O* rings, plastic parts, propellants, fluid levels, and similar materials can be visualized even when these objects are inside metallic containers.

4. Miscellaneous Applications—Less frequent use is made of radiographic techniques in the inspection of forgings, powder metal parts, and of nonmetallic materials such as plastic, rubber, ceramic, and solid propellant. The limited use of this inspection medium for forgings is explained by the fact that forging defects are smaller in size and unsuitably oriented for good detection.

Equipment—There are a number of factors which affect the use of penetrating radiation to varying extents. These factors can be grouped into three general categories: source of radiation, object or material to be examined, and detecting or recording medium.

Sources for neutron radiography include nuclear reactors, accelerators, and radioactive isotopes. These sources can be moved (in a truck, for example) but most neutron radiographic inspection is done by bringing the inspection object to the source. Radiation sources for other types of radiography involve either x-ray generators or one of several radio isotopes. X-rays are produced when high-velocity electrons impinge upon target atoms. The energy of the x-radiation produced is a function of the velocity of the impinged electrons, which in turn is dependent upon the applied anode voltage (kV or MeV). The practical thickness range of steel which can be inspected by x-ray units is proportional to their radiation energy, as shown in Fig. 1. The usefulness of Fig. 1 can be extended to other materials by referring to Table 1, which gives equivalence factors for various other materials as compared to steel.

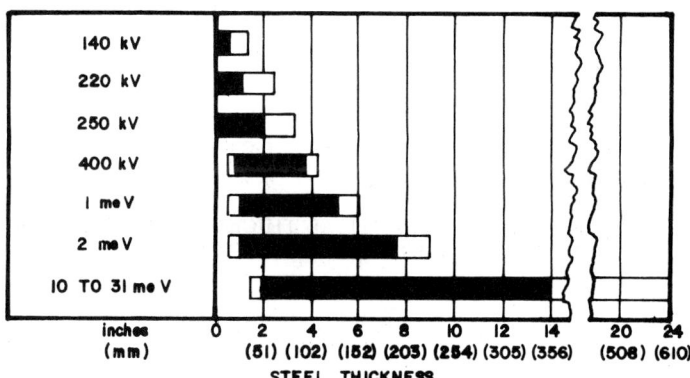

THE SHADED PORTIONS OF THE BAR REPRESENT THICKNESSES WHERE 1% SENSITIVITY IS OBTAINED. THE UNSHADED PORTION AT BAR ENDS REPRESENT 2% SENSITIVITY.

FIG. 1—APPROXIMATE PRACTICAL THICKNESS RANGES OF STEEL FOR VARIOUS X-RAY KILOVOLTAGES

TABLE 1—APPROXIMATE RADIOGRAPHIC EQUIVALENCE FACTORS FOR SEVERAL METALS IN RELATION TO STEEL[a] (ADAPTED FROM ASTM E 94)

Metal	Density	Radiographic Equivalence Factor					
		140 kV	220 kV	250 kV	400 kV	1 meV	2 meV
Aluminum	2.7	0.083	0.24	0.24	—	—	—
Magnesium	1.7	0.05	0.08	0.08	—	—	—
Steel	7.8	1.0	1.0	1.0	1.0	1.0	1.0
Stainless (18.8)	7.9	1.0	1.0	1.0	1.0	1.0	1.0
Copper	8.9	1.8	1.4	1.4	1.4	—	—
Zinc	7.1	—	1.3	1.3	1.3	—	—
Brass	8.4	—	1.3	1.3	1.3	1.2	—
Lead	11.3	—	11.0	—	—	5.0	2.5

[a] Note: To determine upper practical limit for materials listed other than steel, divide the value given for steel by the proper equivalence factor. Table 1 may be extended to apply to radioisotopes by taking the average of the energy values given in Fig. 2, and using the nearest size x-ray unit in the table.

The radiographic isotopes emit gamma radiation of known energy levels. The approximate practical thickness range of the most commonly used radioisotopes for steel is included in Fig. 2. The energy level of the gamma radiation for each isotope determines its equivalence factor for materials other than steel (included in Fig. 2). Table 1 can be utilized to approximate these equivalence factors by averaging the energy values for a given source and using the closest energy level column in the table.

Other factors such as economics, flexibility, sensitivity, maintenance costs, and portability must of necessity be considered when deciding the type of unit to be used.

Generally, x-ray film is used as the detecting medium. Various types of film are commercially available. These differ in speed, grain, and contrast. The selection of a film is interrelated with the type and energy of the radiation, and the material and thickness of the object to be inspected. Factors such as sensitivity required and exposure time are also considerations. Industrial x-ray paper may be used as a detecting medium. This stabilization paper offers several advantages: lower material cost, increased processing speed, darkroom simplicity, and space savings. Consideration should be given to this process if maximum sensitivity and long periods of radiographic print storage are not required. Other detecting media are available, such as Polaroid film, and Xeroradiography.

Fluoroscopic systems are available for instantaneous radiographic inspections. The sensitivity of this type of inspecting medium is somewhat less than the photographic method. This, coupled with the fact that no record is maintained as with film, may limit the usefulness of this inspection technique.

Protection—Personnel protection from all forms of radiation is an essential requirement in the field of penetrating radiation. It is a fact that scattered as well as direct rays have a biological and physical effect on all living matter. It is recommended (and is generally a legal requirement) that all persons operating or working near any source of radiation keep a record of the radiation dosage received weekly and at no time exceed the limits allowed by the Nuclear Regulatory Commission, or licensing state.

BIBLIOGRAPHY

1. "Radiography in Modern Industry." Eastman Kodak Co., Rochester, NY, 1969.
2. John R. Bradford, ed., "Radioisotopes in Industry." 1953.
3. R. C. McMaster, ed., Nondestructive Testing Handbook, Vol. I, Sections 13–27, 1959. American Society for Nondestructive Testing, Columbus, OH 43321.
4. H. Berger, Neutron Radiography. New York: American Elsevier Publishing Co., 1965.
5. W. J. McGonnagle, Nondestructive Testing. New York: McGraw Hill Book Co., 1961.
6. R. Halmshaw, ed., Industrial Radiology Techniques. New York: American Elsevier Publishing Co., 1971.
7. E. T. Clarke, "Gamma Radiography of Light Metals." Nondestructive Testing, Vol. 16, May–June 1958, p. 265.
8. "Qualification and Certification of Personnel." Recommended Practice No. SNT-TC-1A, Supplement A (Radiography), American Society for Nondestructive Testing, Columbus, OH 43321.
9. "ASME Boiler and Pressure Vessel Code." American Society of Mechanical Engineers, New York.
10. Justin G. Schneeman, Industrial X-Ray Interpretation. Evanston, IL: Intex Publishing Co., 1968.
11. "Radiographic Testing." Programmed Instruction Handbook PI-4-6, Convair Div., General Dynamics Corp., 1967.
12. "Radiographic Inspection." AMS 2635B, SAE, 1967.
13. "Military Standard Inspection—Radiographic." MIL-STD-453, U. S. Department of Defense.
14. Metals Handbook, Vol. 11, 1976, pp. 105–161. American Society for Metals, Metals Park, OH 44073.
15. ASTM E 94, "Recommended Practice for Radiographic Testing," American Society for Testing and Materials, Philadelphia, PA 19103.
16. ASTM Annual Standards, Part II, Article E 545, "Determining Image Quality in Thermal Neutron Radiographic Testing," American Society for Testing and Materials, Philadelphia, PA 19103.
17. H. Berger, ed., "Practical Applications of Neutron Radiography and Gaging," ASTM STP 586, American Society for Testing and Materials, Philadelphia, PA 19103.
18. M. R. Hawkesworth, ed., "Radiography with Neutrons," British Nuclear Energy Society, London, 1975.
19. Neutron Radiography Issue, Atomic Energy Review, Vol. 15, No. 2, International Atomic Energy Agency, Vienna, 1977.

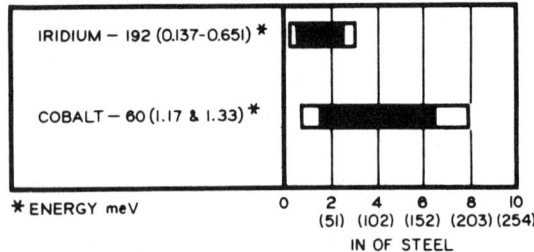

NOTE: DIMENSIONS ARE IN (mm)

THE SHADED PORTIONS OF THE BAR REPRESENT THICKNESSES WHERE 2% SENSITIVITY IS OBTAINED. THE UNSHADED PORTION AT BAR ENDS REPRESENT POORER SENSITIVITY IN THE PRACTICAL RANGE.

FIG. 2—APPROXIMATE PRACTICAL THICKNESS OF STEEL FOR VARIOUS RADIOACTIVE ISOTOPE SOURCES

ULTRASONIC INSPECTION—SAE J428b

SAE Information Report

Report of Iron and Steel Technical Committee approved June 1960 and last revised May 1978.

Purpose—The purpose of this report is to provide basic information on ultrasonics, as applied in the field of nondestructive inspection. References to detailed information are listed in the Bibliography.

General—Ultrasonic testing is a versatile nondestructive inspection method which is applicable to most solid materials, metallic or nonmetallic. Materials inspected include steel, aluminum, cast iron, concrete, rubber, glass, and plastics. Through these tests, surface and internal discontinuities such as laps, seams, voids, cracks, blow holes, inclusions, lack of bond, and porosity can be detected. Material thickness can be accurately measured from one side. Under certain conditions, materials at elevated temperatures can be inspected.

Totally automatic systems are in use in heavy industry. Location of defects can be marked on parts, or graphic recordings made of parts. Parts can be automatically removed from a processing line when defect severity exceeds established limits.

In many cases the extent of discontinuities can be determined. The minimum size discontinuity which can be located by ultrasonics in a given application is determined by:

1. The sensitivity of the test equipment.
   (a) Physical characteristics of the transducer.
   (b) Gain/band width characteristics of the instrument.
2. The material inspected.
   (a) Physical properties (modulus, grain size).
   (b) Surface condition (rough, smooth, wavy, and so forth).
3. The test frequency used. In general, higher test frequencies permit detection of smaller discontinuities. Lower frequencies permit penetration of greater thickness of material, or of coarse grained material that cannot be inspected with the higher frequencies.
4. Orientation of discontinuity and its distance from the ultrasound entrant surface.
5. Type of defect and acoustic impedance mismatch.

In addition to discontinuity detection, ultrasonic energy is also used to gage the thickness of materials from one side. Yield or tensile strength of nodular cast irons can be estimated through its relationship to the velocity of ultrasonic energy in the metal.

Principles—Ultrasonic inspection is made possible by the ability of most solid materials to support the transmission of high frequency sound waves. This ability to support these mechanical vibrations varies in extent for different materials, and depends upon certain physical properties of each material, such as density, modulus, grain structure, etc.

All ultrasonic tests involve introducing controlled ultrasonic energy into the object under test, and observing how the passage of sound is affected in transit. Any discontinuity in the material can reflect, disperse, or attenuate the energy. The ultrasonic energy used for testing is usually generated in short bursts or pulses by piezoelectric transducers driven by appropriate electronic circuitry. Test frequencies used are usually between 1–25 MHz, and the pulse repetition rates from a few Hz to thousands of Hz. Since air will not support these ultrasonic signals, a liquid such as water or oil is used to couple the energy from the transducer into the material under test.

Types of Tests—There are three basic types of ultrasonic test techniques. The Pulse Echo technique is by far the most commonly used.

1. Pulse Echo—A pulse of ultrasonic energy is transmitted into the part. The time required for the reflected energy to return to the transducer is observed. A discontinuity is usually indicated by:

(a) Reflections received from locations where no physical discontinuities (such as end faces, grooves, or holes) are known to exist.
(b) Loss of the reflection from the known physical discontinuity.

Advantages:
(a) Single transducer operation permits inspection with access to only one side of the material.
(b) The resolution and sensitivity of this method, in most applications, is superior to other ultrasonic methods.

Disadvantages: The minimum thickness of material which can be inspected is about 0.01 in (0.254 mm) with present-day equipment.

2. Through Testing—Either a pulsed or continuous beam of energy is coupled into the material from one transducer. A second transducer, placed in a position to receive the transmitted energy, receives the energy leaving the material. Changes in the amplitude of the received energy indicate discontinuities in the part.

Advantages:
(a) The energy passes through the part only one time, permitting this test to be used on materials difficult to penetrate.
(b) Very thin materials can be tested.

Disadvantages:
(a) Precision fixturing for two transducers and preparation of two test surfaces are required.
(b) The accuracy is inherently less than in Pulse Echo testing.
(c) Depth of discontinuity cannot be determined.

3. Resonance Testing—A swept frequency beam of ultrasonic energy is coupled into the part from the transducer. When the resonant frequency of the material under test is reached, a thickness indication is obtained.

Advantages: Material down to a few thousandths of an inch can be measured or inspected from one side.

Disadvantages:
(a) Interpretation and calibration of instrument can be troublesome.
(b) For flaw detection, the sensitivity is generally less than in Pulse Echo testing.

Procedure—Two techniques are used in the three types of ultrasonic inspection, contact testing and immersed testing. In any application the material under test should be cleaned to remove any loose particles or scale prior to inspection.

1. Contact Testing—The transducer is placed directly against the material under test, with a film of liquid couplant (for example, water, oil, glycerine) between.

Advantages:
(a) Precision positioning apparatus requirements are reduced.
(b) Surface defects can be detected by transmitting a surface wave along the outside contour of most parts. Results are a direct function of surface smoothness, improving with better surface.
(c) Good sound penetration.
(d) Portable battery operated equipment available.
(e) Relatively low cost equipment.

Disadvantages:
(a) Good surface is required.
(b) The energy cannot be readily focused to obtain increased resolution and sensitivity in a given area.
(c) Difficult to control shape and direction of beam.
(d) The transducer is subject to wear thus requiring replacement or wear/shoes in some applications.

2. Immersion Testing—The material to be inspected is placed in a reservoir of couplant liquid. The transducer is immersed in the reservoir and accurately positioned relative to the material under test. Water columns may also be used where immersion is undesirable.

Advantages:
(a) The energy can be focused or shaped for the part, permitting increased resolution and sensitivity.
(b) Immersion coupling facilitates the inspection of nonuniformly contoured parts.
(c) Better close-to-surface resolution than other ultrasonic techniques.
(d) Automatic inspection—uniform couplant.
(e) Transducer wear is minimized.

Disadvantages:
(a) The requirement of immersing the sample.
(b) The necessity of accurate positioning of the material and transducer(s).
(c) The sample (or object) size is limited by the size of the immersion reservoir.

BIBLIOGRAPHY

1. J. & H. Krautkramer, "Ultrasonic Testing of Materials." New York: Springer-Verlag, 1969.
2. T. F. Hueter and R. H. Bolt, "Sonics" (Fifth Edition). New York: John Wiley & Sons, Inc., 1966.
3. R. Goldman, "Ultrasonic Technology." New York: Reinhold Publishing Corp., 1962.
4. J. Frederick, "Ultrasonic Engineering." New York: John Wiley & Sons, Inc., 1965.
5. R. C. McMaster, ed., "Nondestructive Testing Handbook", Vol. II, Section 43-50, 1959. American Society for Nondestructive Testing, Columbus, OH 43321.
6. A. L. Phillips, ed., "Welding Handbook" (Sixth Edition), pp. 6.54-60. New York: American Welding Society, 1968.
7. Metals Handbook, Vol. 11, 1976, pp. 161-198. American Society for Metals, Metals Park, OH 44073.

ACOUSTIC EMISSION TEST METHODS—SAE J1242

SAE Information Report

Report of Iron and Steel Technical Committee approved June 1978.

Purpose—The purpose of this report is to supply the user with sufficient information so that he may decide whether acoustic emission test methods apply to his particular inspection problem. Detailed technical information can be obtained by referring to the Bibliography at the end of this report.

General—Acoustic emission is defined as *a transient elastic wave generated by the rapid release of energy from a localized source or sources within a material.* The emission may be the result of any of several changes taking place in the material. A crack may be growing, the material may be undergoing permanent deformation, the internal structure may be changing due to heat treatment, or, in the case of composite materials, the fibers that strengthen the material may be breaking.

Some metals produce audible acoustic emission when they are bent. This is due to a deformation process called twinning. Tin, magnesium, and zinc show this effect.

Acoustic emission technology is applicable to many nondestructive inspection problems. These include detection and growth monitoring of fatigue cracks and stress-corrosion cracking, in-process determination of weld quality, measurement of adhesive bond strength, and in certain cases, the detection of loose parts in assembled components. Acoustic emission is particularly useful for monitoring the growth of a crack in order to give warning of impending failure, and to detect subsurface deformation.

There are several advantages of acoustic emission as a nondestructive test method when compared with more common methods such as radiography, ultrasonics, or magnetic particle techniques. Some of these are as follows:
 1. It is capable of monitoring a complete structure in real time.
 2. It is very sensitive to the presence of active flaws when compared to other nondestructive test methods, but usually requires these other methods to characterize the flaws.
 3. It can detect discontinuities that may be inaccessible to other nondestructive test methods.
 4. It is suitable for use during proof testing in those structures that will be stressed sufficiently to produce local plastic deformation during the test.

Limitations of acoustic emission testing include:
 1. Inactive non-propagating flaws cannot be detected.
 2. The significance of a detected source of emission cannot be assessed unambiguously.
 3. As with many other nondestructive tests, acoustic emission tests are best used in conjunction with other nondestructive methods, such as ultrasonics and radiography.
 4. The part or system under test must be stressed by an external force.

Principle—There are two types of acoustic emission: *burst* and *continuous*. A single burst of emission lasts from a few microseconds to several milliseconds. Continuous emission consists of a series of closely spaced noise peaks of random amplitude that occur without interruption. Burst emissions usually have a larger amplitude than continuous emissions.

A force must be applied to a specimen to cause it to give off acoustic emission. Emission usually occurs only while the stress is increasing. When the stress stops increasing the emission stops. When the force is reapplied it must exceed the previous stress level before the specimen will emit again.

Most of the acoustic emission signals that are useful in nondestructive testing are usually of low amplitude and have frequencies that are above the audible range. Ordinarily they are between 100 kHz and 1 MHz, depending upon the application. Low frequencies are filtered out in order to avoid interference from unwanted sources of noise such as machines or electrical equipment. The maximum distance that the signals will travel in a structure and still be detectable depends on the type of material and on the range of frequencies in the signal. In steel pressure vessels the acoustic emission caused by crack growth in welds can travel 10 m or more from the source of the emission to the transducer that is detecting it.

The location of a source of the emission is determined by triangulation methods. These are based on the differences in the times required for the signals to reach the various elements in an array of transducers.

Procedure—Specially designed transducers are used for detecting the acoustic emission in a test specimen or structure. These must be coupled to the test specimen with a suitable liquid or grease, or by means of an epoxy cement or other adhesive. The output of the transducer is amplified and the low frequencies filtered out. Processing of the signal is usually very desirable. The simplest method for monitoring an acoustic emission test is to electronically convert the high frequency acoustic emission signals to lower frequencies that can be heard with the human ear. The most common methods, however, use chart recorders or cathode ray oscilloscopes to display the test results. Magnetic tape is used for storing larger amounts of data for later processing or display. Specialized equipment for the detection and processing of acoustic emission signals is available from several manufacturers. Data processing as applied to acoustic emission tests is limited only by the creativity and sophistication of the user and the data processing facility.

BIBLIOGRAPHY

1. Metals Handbook, Eighth Edition, American Society for Metals, 1976, Vol. 11, "Nondestructive Inspection and Quality Control", pp. 234–243.
2. "Acoustic Emission", A Symposium presented December 7–8, 1971, at Bal Harbor, Florida. American Society for Testing and Materials, STP 505, (1972).
3. Monitoring Structural Integrity by Acoustic Emission, A Symposium presented at Fort Lauderdale, Florida, January 17–18, 1974. American Society for Testing and Materials, STP 571, (1975).
4. R. G. Liptai and D. O. Harris, "Acoustic Emission—An Introductory Review." Materials Research and Standards, Vol. 11, No. 3, March 1971, pp. 8–10.
5. C. R. Horak and A. F. Weyhreter, "Acoustic Emission System for Monitoring Components and Structures in a Severe Fatigue Noise Environment", Materials Evaluation, Vol. 35, No. 5, May 1977, pp. 59–63.
6. J. C. Spanner, "Acoustic Emission Techniques and Applications", Intex Publishing Co., 1974.
7. ASTM E 569, "Standard Recommended Practice for Acoustic Emission Monitoring of Structures During Controlled Stimulation", ASTM Annual Standards, Part 11, American Society for Testing and Materials, Philadelphia, PA 19103.
8. ASTM E 610, "Standard Definitions of Terms Relating to Acoustic Emission", ASTM Annual Standards, Part 11, American Society for Testing and Materials, Philadelphia, PA 19103.

DEFINITION FOR PARTICLE SIZE—SAE J391

SAE Recommended Practice

Report of Fatigue Design and Evaluation Committee approved May 1969.

Scope—"Effective particle or domain size" is a phrase used in X-ray diffraction literature to describe the size of the coherent regions within a material which are diffracting. Coherency in this sense means diffracting as a unit. Small particle size causes X-ray line broadening and as such can be measured. It has been shown related to substructure as observed in transmission electron microscopy. Particle size is affected by hardening, cold working, and fatigue; conversely, there is increasing evidence that particle size, per se, affects both static and dynamic strength.

Definition

Effective Particle or Domain Size, as determined by diffraction, is a one-dimensional measure of the average size of essentially perfect regions within a material.

NOTE: Such regions are related to the substructure seen in transmission electron microscopy. Small particle size contributes to diffraction line broadening.

DEFINITIONS FOR MACROSTRAIN AND MICROSTRAIN—SAE J932

SAE Recommended Practice

Report of Iron and Steel Technical Committee approved July 1965.

Scope—In work on analysis and measurement of residual stresses of materials, it has been noted that there are frequently differences in interpretation of the terms "macrostress" and "microstrain." To assist communications among research personnel in this area, basic definitions for these two terms have been developed by Division 4, Residual Stresses and Fatigue, of the SAE Iron and Steel Technical Committee. Since "macrostress" is nearly always computed from "macrostrain" in residual stress analysis, to be consistent, the definitions given are for "macrostrain" and "microstrain."

Definitions

Macrostrain is the mean strain over any finite gage length of measurement large in comparison with interatomic distances.

NOTE: Macrostrain can be measured by several methods, including electrical resistance strain gages and mechanical or optical extensometers. Elastic macrostrain can be measured by X-ray diffraction.

Microstrain is the strain over a gage length comparable to interatomic distances.

NOTE: These are the strains being averaged by the macrostrain measurement. Microstrain is not measurable with existing techniques. Variance of the microstrain distribution can, however, be measured by X-ray diffraction.

TECHNICAL REPORT ON FATIGUE PROPERTIES—SAE J1099

SAE Information Report

Report of the Fatigue Design and Evaluation Steering Committee approved February 1975.

1. Scope

1.1 Pertinent information to provide design guidance in avoiding fatigue failures is outlined in this SAE Information Report. Of necessity, it is brief, but it does provide a basis for approaching complex fatigue problems. The information provided here can be used in preliminary design estimates of fatigue life, the selection of materials and the analysis of service load and/or strain data.

2. Introduction

2.1 Designing to avoid fatigue failures of a component of a vehicle (or machine or structure) is one of the more difficult tasks an engineer faces. Many factors are involved with the relationships between the factors only partially established and largely empirical. Fatigue failure is caused by repeated loading with the number of cycles of loading to failure varying with the load range. "Damage" accrues gradually over the life span of cyclical loading, with small changes taking place during the total life. Crack initiation occurs at varying percentages of the final life, and crack propagation continues until final fracture takes place.

2.2 Designing to avoid fatigue failures requires knowledge of the following:

2.2.1 The expected load-time history (or better, the local stress-time and strain-time history at the most critical locations).

2.2.2 The nature of the environment in which the component is operated (wet, dry, corrosive, temperature, etc.)

2.2.3 The properties of the material as it exists in the finished component at the most critically stressed locations ("inherent" fatigue properties, residual stress effects, surface effects, sensitivity to corrosion, "cleanliness," variability, etc.).

2.2.4 The geometry of the component and its notches (stress concentrations, surface finish, manufacturing variability, etc.).

2.3 Scatter in fatigue life is another aspect of fatigue life evaluation and prediction which must be considered. This often calls for statistically based analyses. Circumstances dictate the degree of sophistication required in all aspects of an evaluation or prediction.

2.4 In the decades that have followed the first recorded observation of the fatigue phenomenon, well over 100 years ago, much has been done to enable progressively better fatigue life predictions. The concept of the fatigue limit—the stress level below which fatigue failure is highly improbable—has been used extensively, probably because it is simple to apply and because it has an empirical relationship to the ultimate tensile strength.

2.5 Early in automotive history, the "proving ground approach" to evaluating performance of "systems" came into being. Weak components failed and had to be strengthened; it was that simple. System tests of this kind still play a predominant role in the industry and are used together with laboratory component and system testing today. A vast amount of specific experience and data have been acquired in this way, permitting comparative testing, in which a new component or system is compared with those made previously. This provides another valuable approach to the problem of measuring fatigue performance and some guidance in design. More general approaches are necessary, however, for many applications.

2.6 In the past 15–20 years, substantial advances have been made in the understanding of fatigue phenomenon, design engineering concepts and testing methods and equipment. Considerable attention has been devoted to the subject of fracture mechanics. In many design situations it will be necessary to consider the mechanism of crack propagation and fracture as well as the initiation of the crack. Fracture mechanics may be suited to the task of predicting the life of a cracked component. The yield strength of a material determines the minimum size crack to which fracture mechanics can be applied. With the proper information, the rate at which a given crack will grow slowly to a specific size as well as the size at which it will propagate catastrophically can be estimated. Also, instances when a given crack will not propagate can be estimated. At the present time, only a small quantity of data on fracture mechanics properties of steels used by the ground vehicle industry have been published. As larger amounts of data become available, definitions and tables will be added to this report. More information on this subject is given in Refs. 5–11.

3. Material Property Tables

3.1 Table 1 is a listing of monotonic stress-strain properties for selected metals. Table 2 is a listing of cyclic and fatigue properties. Steels are listed first, followed by the aluminums. Within these broad classifications, the metals are listed in order of SAE specification then increasing true fracture strength. Table 3 lists the non-standard abbreviations necessitated by computer printed tables in this report. A brief introduction, definitions and discussion follow the tables. An example application is given at the end of the report.

3.2 Material properties given in the current tables represent only the beginning for this report. In order that the quantity and quality of this data may grow, it is necessary that additional information be contributed to the existing data bank. As data become available, this report will be updated. Persons wishing to contribute may contact the Fatigue Design & Evaluation Committee through the SAE Detroit Branch Office, 18121 East Eight Mile Road, East Detroit, Michigan 48021.

3.3 The majority of the properties listed in the tables have been contributed by members of the SAE Fatigue Design & Evaluation Committee. Values listed in the tables were obtained by testing a sample of the metal specified. The size of the sample varied from a very few test bars cut from a single sample of the metal to numerous bars from more than one source. A data reference number is given to identify the original source of the properties. As defined, these properties are measured on smooth polished specimens and therefore do not include such influences as environmental effects (wet or corrosive conditions, elevated temperature, etc.), surface roughness effects, mean or residual stress effects, notch effects, etc.

3.4 There are many procedures for using this information for the above-mentioned purposes. They are too lengthy to be included in this report; however, there are a number of publications which discuss these procedures. One of them is the SAE Fatigue Design Handbook (1)[1] which discusses fatigue properties, methods of determining fatigue properties, and illustrates the use of this data for making design decisions. Other useful references are listed at the end of this report.

---

[1] Numbers in parentheses refer to References listed at the end of the report.

TABLE 1—MONOTONIC STRESS-STRAIN PROPERTIES OF SELECTED METALS (sort: steel, A1, SAE spec., increasing true fracture strength)

SAE Spec	BHn	Data Ref†	Grain Dir	Process Description	Ult Str ksi (MPa)	Yield Str ksi (MPa)	True Fract Str ksi (MPa)	%RA	True Fract Ductility	Strain Hard'G Exponent	Str Cof ksi (MPa)
A-538-A[2]	405	6	L	Sol Tr & Aged	220 (1517)	215 (1482)	275 (1896)	67	1.10	0.030	
A-538-B[2]	460	6	L	Sol Tr & Aged	270 (1862)	260 (1793)	310 (2137)	56	0.82	0.020	
A-538-C[2]	480	6	L	Sol Tr & Aged	290 (1999)	280 (1931)	325 (2241)	55	0.81	0.015	
AM-350[5]		1	L	HR & Annealed	191 (1317)	64 (441)	298 (2055)	52	0.74		
AM-350[5]	496	1	L	CD	276 (1903)	270 (1862)	316 (2179)	20	0.23		
Gainex[3]		7	LT	HR Sheet	77 (531)	58 (400)	117 (807)	58	0.86	0.20	
Gainex[3]		7	L	HR Sheet	74 (510)	57 (393)	118 (814)	64	1.02	0.20	
H-11	660	6	L	Ausformed	375 (2586)	295 (2034)	460 (3172)	33	0.40	0.120	
R-100[4]	236	11	LT	As Rec Plate	177 (1220)	117 (807)	214 (1475)				
R-100[4]	236	11	L	As Rec Plate	169 (1165)	112 (772)	236 (1627)				
RQC-100[1]	290	10	LT	HR Plate	136 (938)	130 (896)	155 (1069)	43	0.56	0.06	170 (1172)
RQC-100[1]	290	10	L	HR Plate	135 (931)	128 (883)	193 (1331)	67	1.02	0.06	170 (1172)
10B62	430	7	L	Q&T	238 (1641)	219 (1510)	258 (1779)	38	0.89	0.042	260 (1793)
1005-1009	90	7	LT	HR Sheet	52 (359)	39 (269)	104 (717)	73	1.3	0.12	73 (503)
1005-1009	125	7	LT	CD Sheet	68 (469)	65 (448)	108 (745)	66	1.09	0.029	78 (538)
1005-1009	125	7	L	CD Sheet	60 (414)	58 (400)	122 (841)	64	1.02	0.049	76 (524)
1005-1009	90	7	L	HR Sheet	50 (345)	38 (262)	123 (848)	80	1.6	0.16	77 (531)
1015	80	4	L	Normalized	60 (414)	33 (228)	105 (724)	68	1.14	0.26	
1020	108	12	L	HR Plate	64 (441)	38 (262)	103 (710)	62	0.96	0.19	107 (738)
1040	225	13	L	As Forged	90 (621)	50 (345)	152 (1048)	60	0.93	0.22	
1045	225	7	L	Q&T	105 (724)	92 (634)	178 (1227)	65	1.04	0.13	166 (1145)
1045	410	7	L	Q&T	210 (1448)	198 (1365)	270 (1862)	51	0.72	0.076	302 (2082)
1045	390	7	L	Q&T	195 (1344)	185 (1276)	270 (1862)	59	0.89	0.044	
1045	450	7	L	Q&T	230 (1586)	220 (1517)	305 (2103)	55	0.81	0.041	
1045	500	7	L	Q&T	265 (1827)	245 (1689)	330 (2275)	51	0.71	0.047	
1045	595	7	L	Q&T	325 (2241)	270 (1862)	395 (2723)	41	0.52	0.071	
1080 +Mn	326	14	L	HR Plate	162 (1117)	92 (634)	181 (1248)	17	0.17		
1080 +Mn	375	14	L	Q&T	189 (1303)	166 (1145)	235 (1620)	31	0.37		
1080 +Mn	415	14	L	Q&T	206 (1420)	180 (1241)	243 (1675)	31	0.36		
1080 +Mn	505	14	L	Q&T	265 (1827)	235 (1620)	295 (2034)	30	0.36		
1080 +Mn	555	14	L	Q&T	309 (2130)	273 (1882)	339 (2337)	17	0.18		
1144	265	16	L	CD Strain Rel	135 (931)	104 (717)	168 (1158)	33	0.51		
1144	305	16	L	Drawn at Temp	150 (1034)	148 (1020)	220 (1517)	25	0.29		
1541F	290	15	L	Q&T Forging	138 (951)	129 (889)	185 (1276)	49	0.68	0.12	
1541F	260	15	L	Q&T Forging	129 (889)	114 (786)	185 (1276)	60	0.93	0.13	
30304	160	1	L	HR & Annealed	108 (745)	37 (255)	228 (1572)	74	1.37		
30304	327	1	L	CD	138 (951)	108 (745)	246 (1696)	69	1.16		
30310	145	1	L	HR & Annealed	93 (641)	32 (221)	168 (1158)	64	1.01		
4130	258	1	L	Q&T	130 (896)	113 (779)	206 (1420)	67	1.12		
4130	365	1	L	Q&T	207 (1427)	197 (1358)	264 (1820)	55	0.79		
4140	310	16	L	Q&T Drawn at Temp	156 (1076)	140 (965)	221 (1524)	60	0.69		
4142	310	16	L	Drawn at Temp	154 (1062)	152 (1048)	162 (1117)	29	0.35		
4142	335	16	L	Drawn at Temp	181 (1248)	179 (1234)	246 (1696)	28	0.34		
4142	380	6	L	Q&T	205 (1413)	200 (1379)	265 (1827)	48	0.66	0.051	
4142	400	6	L	Q&T and Deformed	225 (1551)	210 (1448)	275 (1896)	47	0.63	0.032	
4142	450	6	L	Q&T	255 (1758)	230 (1586)	290 (1999)	42	0.54	0.043	
4142	475	6	L	Q&T and Deformed	295 (2034)	275 (1896)	300 (2068)	20	0.22	0.01	
4142	450	6	L	Q&T and Deformed	280 (1931)	270 (1862)	305 (2103)	37	0.46	0.016	
4142	475	6	L	Q&T	280 (1931)	250 (1724)	315 (2172)	35	0.43	0.048	
4142	670	6	L	As Quenched	355 (2448)	235 (1620)	375 (2586)	6	0.06	0.136	
4142	560	6	L	Q&T	325 (2241)	245 (1689)	385 (2654)	27	0.31	0.091	
4340	243	1	L	HR & Annealed	120 (827)	92 (634)	158 (1089)	43	0.57		
4340	409	1	L	Q&T	213 (1469)	199 (1372)	226 (1558)	38	0.48		
4340	350	2	L	Q&T	180 (1241)	170 (1172)	240 (1655)	57	0.84	0.066	229 (1579)
5160	430	7	L	Q&T	242 (1669)	222 (1531)	280 (1931)	42	0.87	0.055	308 (2124)
52100	518	1	L	Sol Heat Q&T	292 (2013)	279 (1924)	318 (2193)	11	0.12		
9262	260	7	L	Annealed	134 (924)	66 (455)	151 (1041)	14	0.16	0.22	253 (1744)
9262	280	7	L	Q&T	145 (1000)	114 (786)	177 (1220)	33	0.41	0.14	
9262	410	7	L	Q&T	227 (1565)	200 (1379)	269 (1855)	32	0.38	0.06	283 (1951)
950C	159	8	LT	HR Plate	82 (565)	46 (317)	135 (931)	64	1.03	0.19	134 (924)
950C	150	3	L	HR Bar	82 (565)	47 (324)	145 (1000)	69	1.19	0.21	
950X	150	5	L	Plate Channel	64 (441)	50 (345)	109 (752)	65	1.06	0.16	98 (676)
950X	156	9	L	HR Plate	77 (531)	48 (331)	145 (1000)	72	1.24	0.19	131 (903)
980X	225	5	L	Plate Channel	101 (696)	82 (565)	177 (1220)	68	1.15	0.13	181 (1248)
1100 Al	26	1	L	As Received	16 (110)	14 (97)		88	2.09		
2014-T6	155	1	L	Sol Tr & Artif Age	74 (510)	67 (462)	87 (600)	25	0.29		
2024-T351		7	L	Sol Tr Strn Harden	68 (469)	55 (379)	81 (558)	25	0.28	0.032	66 (455)
2024-T4		2	L	Sol Tr & RT Age	69 (476)	44 (303)	92 (634)	35	0.43	0.20	117 (807)
5456-H311	95	1	L	Strain Hardened	58 (400)	34 (234)	76 (524)	35	0.42		
7075-T6		2	L	Sol Tr & Artif Age	84 (579)	68 (469)	108 (745)	33	0.41	0.113	120 (827)

† See list of Data References
[1] Tradename—Bethlehem Steel Corp.
[2] ASTM Specification
[3] Tradename—Armco Steel Corp.
[4] Tradename—Republic Steel Corp.
[5] Grade Number—Allegheny Ludlum Steel Corp.

DATA REFERENCES

1. R. W. Smith, M. H. Hirschberg, and S. S. Manson, "Fatigue Behavior of Materials in Low and Intermediate Life Range." NASA Tech. Note D-1574, April 1963.
2. T. Endo and JoDean Morrow, "Cyclic Stress-Strain and Fatigue Behavior of Representative Aircraft Metals." Journal of Materials, Vol. 4 (1969), p. 159.
3. P. C. Rosenberger, "Fatigue Behavior of Smooth and Notched Specimens of Man-Ten Steel." M.S. Thesis, Dept. of Theoretical and Applied Mechanics, University of Illinois, 1968.
4. A. Keshavan, "Some Studies on the Deformation and Fracture of Normalized Steel under Cyclic Conditions." Ph.D. Thesis, University of Waterloo, 1967.
5. P. Watson and T. H. Topper, "An Evaluation of the Fatigue Performance of Automotive Steels." Paper 710597 presented at Mid-Year Meeting, Chicago, May 1971.
6. R. W. Landgraf, "Cyclic Deformation and Fatigue of Hardened Steels." T & AM Report 320, Dept. of Theoretical and Applied Mechanics, University of Illinois, Urbana, 1968.
7. R. W. Landgraf, M. R. Mitchell, and N. R. LaPointe, "Monotonic & Cyclic Properties of Engineering Materials." Metallurgical Dept., Scientific Research Staff, Ford Motor Co., Dearborn, Michigan, June 1972.

TABLE 2—CYCLIC STRESS-STRAIN AND FATIGUE PROPERTIES OF SELECTED METALS (sort: steel, Al, SAE spec., increasing true fracture strength)

SAE Spec	BHn	Data Ref†	Grain Dir	Process Description	Mod of Elas ksi (GPa)	Cyc Yld ksi (MPa)	Cyc Strain Hard'g Exp	Cyc Str Cof ksi (MPa)	Fat Str Cof ksi (MPa)	Fat Str Exp	Fat Duc Cof	Fat Duc Exp
A-538-A[2]	405	6	L	Sol Tr & Aged	27000 (186)	150 (1034)	0.09		240 (1655)	−0.065	0.30	−0.62
A-538-B[2]	460	6	L	Sol Tr & Aged	27000 (186)	195 (1344)	0.075		310 (2137)	−0.071	0.80	−0.71
A-538-C[2]	480	6	L	Sol Tr & Aged	26000 (179)	215 (1482)	0.08		325 (2241)	−0.07	0.60	−0.75
AM-350[5]		1	L	HR & Annealed	28000 (193)	196 (1351)	0.13		406 (2799)	−0.14	0.33	−0.84
AM-350[5]	496	1	L	CD	26000 (179)	235 (1620)	0.21		390 (2689)	−0.102	0.10	−0.42
Gainex[3]		7	LT	HR Sheet	29200 (201)	58 (400)	0.11	114 (786)	117 (807)	−0.07	0.86	−0.65
Gainex[3]		7	L	HR Sheet	29200 (201)	54 (372)	0.11	114 (786)	117 (807)	−0.071	0.86	−0.65
H-11	660	6	L	Ausformed	30000 (207)	340 (2344)	0.07		460 (3172)	−0.077	0.08	−0.74
R-100[4]	236	11	LT	As Rec Plate	28500 (197)							
R-100[4]	236	11	L	As Rec Plate	28000 (193)							
RQC-100[1]	290	10	LT	HR Plate	30000 (207)	87 (600)	0.14	208 (1434)	180 (1241)	−0.07	0.66	−0.69
RQC-100[1]	290	10	L	HR Plate	30000 (207)	87 (600)	0.14	208 (1434)	180 (1241)	−0.07	0.66	−0.69
10B62	430	7	L	Q&T	28000 (193)	140 (965)	0.16	309 (2130)	258 (1779)	−0.067	0.32	−0.56
1005-1009	90	7	LT	HR Sheet	30000 (207)	35 (241)	0.12	71 (490)	84 (579)	−0.09	0.15	−0.43
1005-1009	125	7	LT	CD Sheet	30000 (207)	41 (283)	0.11	83 (572)	75 (517)	−0.059	0.30	−0.51
1005-1009	125	7	L	CD Sheet	29000 (200)	36 (248)	0.11	71 (490)	78 (538)	−0.073	0.11	−0.41
1005-1009	90	7	L	HR Sheet	29000 (200)	33 (228)	0.12	67 (462)	93 (641)	−0.109	0.10	−0.39
1015	80	4	L	Normalized	30000 (207)	35 (241)	0.22	137 (945)	120 (827)	−0.11	0.95	−0.64
1020	108	12	L	HR Plate Plate	29500 (203)	35 (241)	0.18	112 (772)	130 (896)	−0.12	0.41	−0.51
1040	225	13	L	As Forged	29000 (200)	56 (386)	0.18		223 (1538)	−0.14	0.61	−0.57
1045	225	7	L	Q&T	29000 (200)	60 (414)	0.18	195 (1344)	178 (1227)	−0.095	1.00	−0.66
1045	410	7	L	Q&T	29000 (200)	120 (827)	0.146	335 (2310)	270 (1862)	−0.073	0.60	−0.70
1045	390	7	L	Q&T	30000 (207)	110 (758)	0.17		230 (1586)	−0.074	0.45	−0.68
1045	450	7	L	Q&T	30000 (207)	140 (965)	0.15		260 (1793)	−0.07	0.35	−0.69
1045	500	7	L	Q&T	30000 (207)	185 (1276)	0.12		330 (2275)	−0.08	0.25	−0.68
1045	595	7	L	Q&T	30000 (207)	250 (1724)	0.13		395 (2723)	−0.081	0.07	−0.60
1080 +Mn	326	14	L	HR Plate	30000 (207)							
1080 +Mn	375	14	L	Q&T	30400 (210)							
1080 +Mn	415	14	L	Q&T	29300 (202)							
1080 +Mn	505	14	L	Q&T	29700 (205)							
1080 +Mn	555	14	L	Q&T	30400 (210)							
1144	265	16	L	CD Strain Rel	28500 (197)	80 (552)	0.15		145 (1000)	−0.08	0.32	−0.58
1144	305	16	L	Drawn at Temp	28800 (199)	82 (565)	0.18		230 (1586)	−0.09	0.27	−0.53
1541F	290	15	L	Q&T Forging	29900 (206)	95 (655)	0.17	255 (1758)	185 (1276)	−0.076	0.68	−0.65
1541F	260	15	L	Q&T Forging	29900 (206)	85 (586)	0.16	235 (1620)	185 (1276)	−0.071	0.93	−0.65
30304	160	1	L	HR & Annealed	27000 (186)	104 (717)	0.36		350 (2413)	−0.15	1.02	−0.69
30304	327	1	L	CD	25000 (172)	127 (876)	0.17		330 (2275)	−0.12	0.89	−0.77
30310	145	1	L	HR & Annealed	28000 (193)	50 (345)	0.26		240 (1655)	−0.15	0.60	−0.57
4130	258	1	L	Q&T	32000 (221)	82 (565)	0.13		185 (1276)	−0.083	0.92	−0.63
4130	365	1	L	Q&T	29000 (200)	120 (827)	0.12		246 (1696)	−0.081	0.89	−0.69
4140	310	16	L	Q&T Drawn at Temp	29200 (201)	90 (621)	0.14		265 (1827)	−0.08	1.2	−0.59
4142	310	16	L	Drawn at Temp	29000 (200)	108 (745)	0.18		210 (1448)	−0.10	0.22	−0.51
4142	335	16	L	Drawn at Temp	28900 (199)	181 (1248)	0.14		181 (1248)	−0.08	0.06	−0.62
4142	380	6	L	Q&T	30000 (207)	120 (827)	0.17		265 (1827)	−0.08	0.45	−0.75
4142	400	6	L	Q&T and Deformed	29000 (200)	130 (896)	0.16		275 (1896)	−0.09	0.50	−0.75
4142	450	6	L	Q&T	30000 (207)	155 (1069)	0.15		290 (1999)	−0.08	0.40	−0.73
4142	475	6	L	Q&T and Deformed	29000 (200)	160 (1103)	0.15		300 (2068)	−0.082	0.20	−0.77
4142	450	6	L	Q&T and Deformed	29000 (200)	155 (1069)	0.16		305 (2103)	−0.09	0.60	−0.76
4142	475	6	L	Q&T	30000 (207)	195 (1344)	0.13		315 (2172)	−0.081	0.09	−0.61
4142	670	6	L	As Quenched	29000 (200)	320 (2206)	0.05		375 (2586)	−0.075		
4142	560	6	L	Q&T	30000 (207)	250 (1724)	0.12		385 (2654)	−0.089	0.07	−0.76
4340	243	1	L	HR & Annealed	28000 (193)	66 (455)	0.18		174 (1200)	−0.095	0.45	−0.54
4340	409	1	L	Q&T	29000 (200)	120 (827)	0.15		290 (1999)	−0.091	0.48	−0.60
4340	350	2	L	Q&T	28000 (193)	110 (758)	0.14		240 (1655)	−0.076	0.73	−0.62
5160	430	7	L	Q&T	28000 (193)	145 (1000)	0.15	335 (2310)	280 (1931)	−0.071	0.40	−0.57
52100	518	1	L	Sol Heat Q&T	30000 (207)	192 (1324)	0.16		375 (2586)	−0.09	0.18	−0.56
9262	260	7	L	Annealed	30000 (207)	76 (524)	0.15	200 (1379)	151 (1041)	−0.071	0.16	−0.47
9262	280	7	L	Q&T	28000 (193)	94 (648)	0.12	197 (1358)	177 (1220)	−0.073	0.41	−0.60
9262	410	7	L	Q&T	29000 (200)	152 (1048)	0.089	292 (2013)	269 (1855)	−0.057	0.38	−0.65
950C	159	8	LT	HR Plate	29600 (204)	50 (345)	0.15		170 (1172)	−0.12	0.95	−0.61
950C	150	3	L	HR Bar	30000 (207)	45 (310)	0.185		141 (972)	−0.11	0.85	−0.59
950X	150	5	L	Plate Channel	30000 (207)	49 (338)	0.134	115 (793)	91 (627)	−0.075	0.35	−0.54
950X	156	9	L	HR Plate	29500 (203)	56 (386)	0.114	134 (924)	146 (1007)	−0.10	0.85	−0.61
980X	225	5	L	Plate Channel	28200 (194)	81 (558)	0.134	181 (1248)	153 (1055)	−0.08	0.21	−0.53
1100 AL	26	1	L	As Received	10000 (69)	9 (62)	0.15		28 (193)	−0.106	1.8	−0.69
2014-T6	155	1	L	Sol Tr & Artif Age	10000 (69)	60 (414)	0.16		123 (848)	−0.106	0.42	−0.65
2024-T351		7	L	Sol Tr Strn Harden	10600 (73)	62 (427)	0.065	95 (655)	160 (1103)	−0.124	0.22	−0.59
2024-T4		2	L	Sol Tr & RT Age	10200 (70)	64 (441)	0.08		147 (1014)	−0.11	0.21	−0.52
5456-H311	95	1	L	Strain Hardened	10000 (69)	52 (359)	0.16		105 (724)	−0.11	0.46	−0.67
7075-T6		2	L	Sol Tr & Artif Age	10300 (71)	76 (524)	0.146		191 (1317)	−0.52	0.19	−0.52

† See Data References, Table 1
[1] Tradename—Bethlehem Steel Corp.
[2] ASTM Specification
[3] Tradename—Armco Steel Corp.
[4] Tradename—Republic Steel Corp.
[5] Grade Number—Allegheny Ludlum Steel Corp.

8. L. E. Tucker, "Monotonic, Cyclic & Fatigue Properties of A13C Steel." Materials Engineering Dept., Report No. 0 1NOV70 DCM01001, Deere & Co., Moline, Illinois, September 1971.

9. L. E. Tucker, Materials Engineering Dept. Report Nos. 22SEP70 TA51 006, 007, and 008 Deere & Co., Moline, Illinois, 1970–1973.

10. G. Ihm, "Monotonic, Cyclic and Fatigue Properties of RQC-100 Steel." Materials Engineering Dept. Report No. 01NOV70 DCM01010, Deere & Co., Moline, Illinois, February 1972.

11. L. E. Tucker, "Monotonic Tension Properties of Republic R-100 Steel." Materials Engineering Dept. Report No. 15JUN71 AM51001, Deere & Co., Moline, Illinois, June 1971.

12. L. E. Tucker, Materials Engineering Dept. Report Nos. 22SEP70 TA51 003, 004, and 005, Deere & Co., Moline, Illinois, 1970–1973.

13. T. Zebehazy, "Material Properties." Chevrolet Engineering Lab Report TWO 839-4, Chevrolet Engineering Lab., Warren, Michigan, 1973.

14. L. E. Tucker, "Monotonic Stress-Strain Properties of 1080 Mod. (Mn 0.8–1.1)." Materials Engineering Dept. Report No. 26APR72 AA51001, Deere & Co., Moline, Illinois, June 1972.

15. L. E. Tucker, "Monotonic and Cyclic Properties of 1541F Normalized Quenched and Tempered to Rc 25–31." Materials Engineering Dept. Report No. 09AUG71 RA51003, Deere & Co., Moline Illinois, November 1971.

TABLE 3—LIST OF NON-STANDARD ABBREVIATIONS OF MATERIAL PROPERTY NAMES

Abbreviation	Name
CYC STRAIN HARD'G EXP, n'	Cyclic Strain Hardening Exponent
CYC STR COF, K'	Cyclic Strength Coefficient
CYC YLD, S_{ys}, σ'_{ys}	Cyclic Yield Strength
DIR	Direction
FAT DUC COF, ϵ'_f	Fatigue Ductility Coefficient
FAT DUC EXP, c	Fatigue Ductility Exponent
FAT STR COF, σ'_f	Fatigue Strength Coefficient
FAT STR EXP, b	Fatigue Strength Exponent
L	Longitudinal Grain Direction
LT	Long Transverse Grain Direction
MOD OF ELAS, E	Modulus of Elasticity
STRAIN HARD'G EXP, n	Strain Hardening Exponent
STR COF, K	Strength Coefficient
TRUE FRACT DUCT, ϵ_f	True Fracture Ductility
TRUE FRACT STR, σ_f	True Fracture Strength
ULT STR, S_u	Ultimate Strength
YIELD STR, S_{ys}, σ_{ys}	Yield Strength
% RA	Percent Reduction of Area

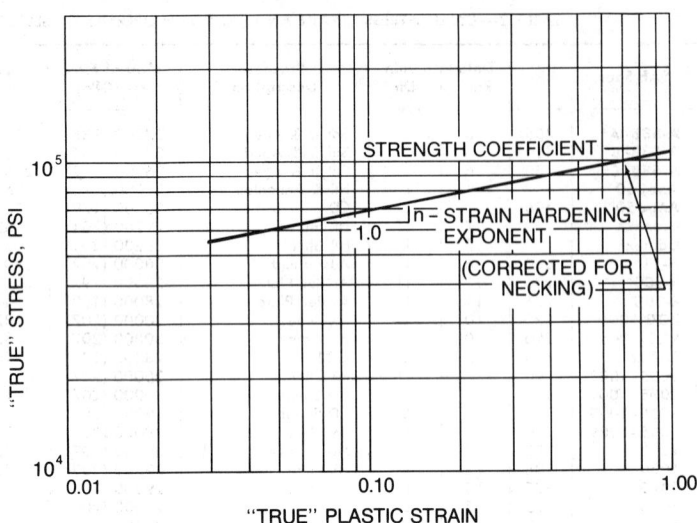

FIG. 2—"TRUE" STRESS-PLASTIC STRAIN PLOT, 1020 H.R. STEEL

16. Y. F. Bilimoria, personal communication with L. E. Tucker, Deere & Co., Materials Engineering Dept. File No. 09MAR AFA51009, Moline, Illinois, February 1973.

4. Monotonic Stress-Strain Properties

4.1 Monotonic[2] stress-strain properties are generally determined by testing a smooth polished specimen under axial loading.

4.2 The load, diameter and/or strain on the uniform test section is measured during the test in order to determine the materials stress-strain response as illustrated in Figs. 1 and 2. Properties, most of which are discrete points on the stress-strain curve, can be defined to describe the behavior of a material.

4.3 Definitions:

4.3.1 ULTIMATE TENSILE STRENGTH (S_u) is the engineering stress at maximum load. In a ductile material it is governed by necking of the specimen.

$$S_u = P_{max}/A_o \qquad (1)$$

where

P_{max} = maximum load
A_o = original cross sectional area

[2] The term monotonic is used in preference to static since the test is usually conducted by continuously increasing with time the distance between the cross heads of the test machine (or better still, the strain on the specimen) until fracture occurs.

4.3.2 TRUE FRACTURE STRENGTH (σ_f) is the "true" tensile stress required to cause fracture.

$$\sigma_f = P_f/A_f \qquad (2)$$

where

P_f = load at failure
A_f = minimum cross sectional area after failure

The value must be corrected for the effect of triaxial stress present due to necking. One such correction suggested by Bridgman (2) is illustrated in Fig. 3. In this figure, the ratio of the corrected value to the uncorrected value ($\sigma_f/(P_f/A_f)$) is plotted against true tensile strain.

4.3.3 TENSILE YIELD STRENGTH (S_{ys}, σ_{ys}) is the stress to cause a specified amount of inelastic strain, usually 0.2%. It is usually determined by constructing a line of slope E through 0.2% strain and zero stress. The stress where the constructed line intercepts the stress-strain curve is taken as the yield strength (E = modulus of elasticity, see Fig. 1).

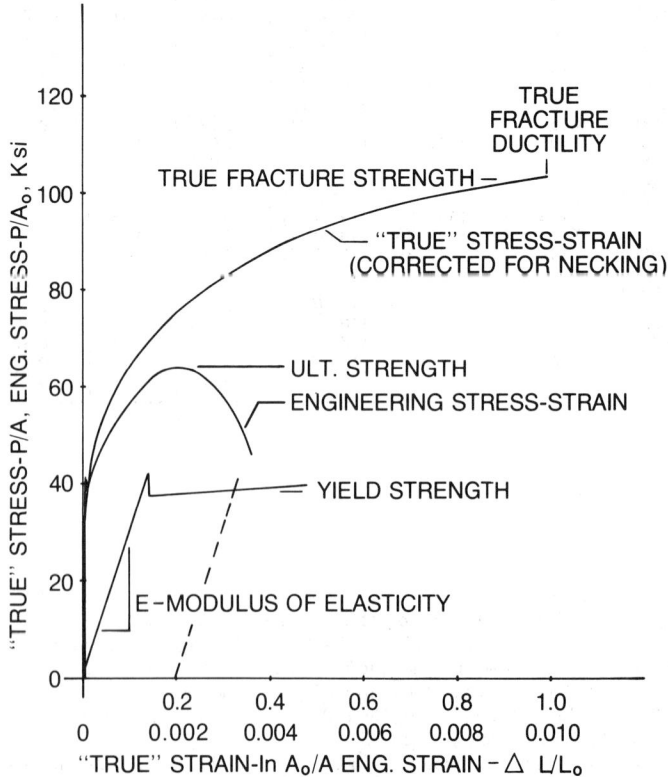

FIG. 1—ENG. AND "TRUE" STRESS-STRAIN PLOT, 1020 H.R. STEEL

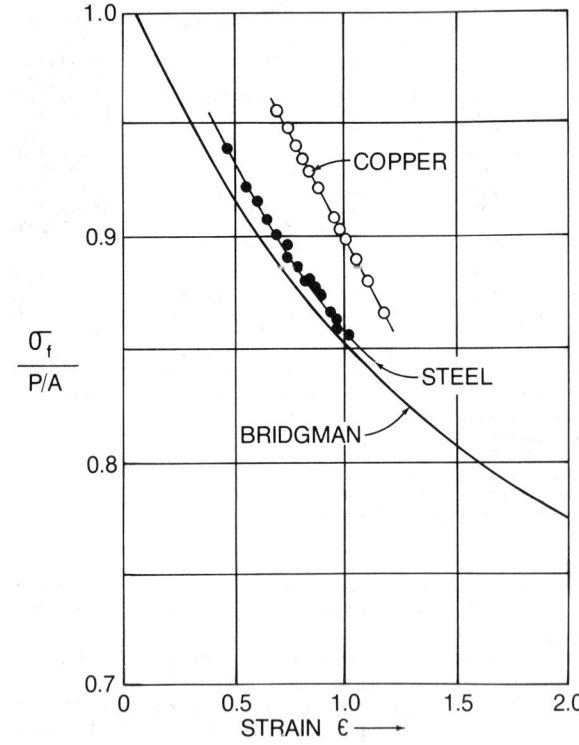

FIG. 3—RELATIONSHIP BETWEEN BRIDGMAN CORRECTION FACTOR $\sigma_f/(P/A)$ AND TRUE TENSILE STRAIN

4.3.4 Percent Reduction of Area (% RA) is the percentage of reduction in cross sectional area after fracture.

$$\% \text{RA} = 100 \left(\frac{A_o - A_f}{A_o} \right) \quad (3)$$

4.3.5 True Fracture Ductility (ε_f) is the "true" plastic strain after fracture.

$$\varepsilon_f = \ln(A_o/A_f) = \ln(100/(100-\%\text{RA})) \quad (4)$$

4.3.6 Monotonic Strain Hardening Exponent (n) is the power to which the "true" plastic strain must be raised to be proportional to "true" stress. It is generally taken as the slope of log σ_p versus log ε_p plot as shown in Fig. 2.

$$\sigma = K\varepsilon_p^n \quad (5)$$

4.3.7 Monotonic Strength Coefficient (K) is the "true" stress at a "true" plastic strain of unity as shown in Fig. 2. If fracture ductiliity is less than 1.0, it is necessary to extrapolate. (see Eq. 5)

4.4 Discussion

4.4.1 Monotonic tension properties of a material can be classed into two groups, engineering stress-strain properties and "true" stress-strain properties. Engineering properties are associated with the original cross sectional area of the test specimen and "true" values relate to actual area while the specimen is under load. The difference between "true" and engineering values is insignificant in the low strain region, up to 1–2% strain.

4.4.2 Until the Test Bar Begins to Locally Neck, some simple relationships exist between engineering and "true" stress-strain values. Eq. 6 gives the relationship between engineering and true strain.

$$\varepsilon = \ln(1 + e) \quad (6)$$

where

ε = "true" strain
e = engineering strain

Similarly, Eq. 7 relates true stress to engineering stress.

$$\sigma = S(1 + e) \quad (7)$$

where

σ = "true" stress
S = engineering stress

4.4.3 A more detailed discussion and derivation of monotonic stress-strain properties can be found in ASTM STP 465 (3). Figs. 1 and 2 graphically illustrate a majority of these properties.

5. Cyclic Stress-Strain Properties

5.1 Cyclic stress-strain properties are determined by testing smooth polished specimens under axial cyclic strain control. The cyclic stress-strain curve is defined as the locus of tips of stable "true" stress-strain hysteresis loops obtained from companion test specimens. A typical stable hysteresis loop is illustrated in Fig. 4 and a set of stable loops with a cyclic stress-strain curve down through the loop tips is illustrated in Fig. 5. As illustrated, the height of the loop from tip-to-tip is defined as the stress range ($\Delta\sigma$). For completely reversed testing one-half of the stress range is generally equal to

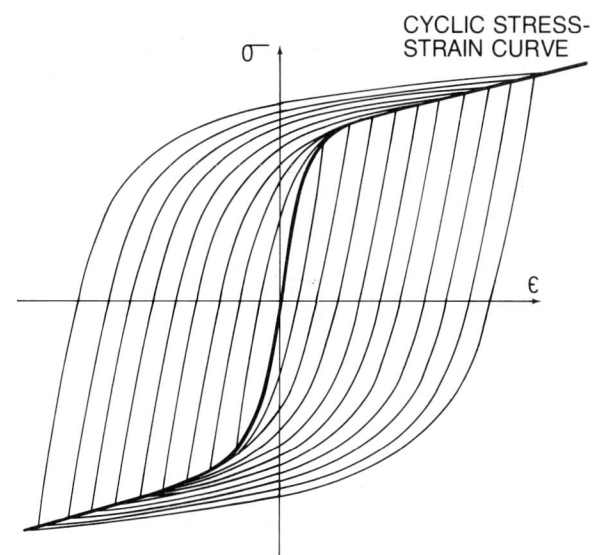

FIG. 5—CYCLIC STRESS-STRAIN CURVE DRAWN THROUGH STABLE LOOP TIPS

the stress amplitude while one-half of the width from tip-to-tip is defined as the strain amplitude ($\Delta\varepsilon/2$). Plastic strain amplitude is found by subtracting the elastic strain amplitude ($\Delta\varepsilon_e/2$) from the strain amplitude as indicated in Eqs. 8–10.

$$\Delta\varepsilon_p/2 = \Delta\varepsilon/2 - \Delta\varepsilon_e/2 \quad (8)$$

According to Hooke's law,

$$\Delta\varepsilon_e/2 = \Delta\sigma/2E \quad (9)$$

where E = modulus of elasticity,

$$\Delta\varepsilon_p/2 = \Delta\varepsilon/2 - \Delta\sigma/2E \quad (10)$$

5.2 A more complete discussion of the cyclic stress-strain curve and other methods of obtaining the curve are given in STP 465 (3) and in Ref. 4

5.3 Definitions

5.3.1 Cyclic Yield Strength (0.2% σ_{ys}) is the stress to cause 0.2% inelastic strain as measured on a cyclic stress-strain curve. It is usually determined by constructing a line parallel to the slope of the cyclic stress-strain curve at zero stress through 0.2% strain and zero stress. The stress where the constructed line intercepts the cyclic stress-strain curve is taken as the 0.2% cyclic yield strength.

5.3.2 Cyclic Strain Hardening Exponent (n') is the power to which "true" plastic strain amplitude must be raised to be proportional to "true" stress amplitude. It is taken as the slope of the log $\Delta\varepsilon_p/2$ and $\Delta\sigma/2$ versus log $\Delta\sigma/2$ plot, where $\Delta\varepsilon_p/2$ and $\Delta\sigma/2$ are measured from cyclically stable hysteresis loops.

$$\Delta\sigma/2 = K'(\Delta\varepsilon_p/2)^{n'} \quad (11)$$

where $\Delta\varepsilon_p/2$ = "true" plastic strain amplitude
The line defined by this equation is illustrated in Fig. 6.

5.3.3 Cyclic Strength Coefficient (K') is the "true" stress at a "true" plastic strain of unity in Eq. 11. It may be necessary to extrapolate as indicated in Fig. 6.

5.4 Discussion

5.4.1 Stress-strain response of some steels can change significantly when subjected to inelastic strains such as can occur at notch roots due to cyclic loading. When fatigue failure occurs, particularly low cycle fatigue, such inelastic straining is generally present. Hence, the cyclic stress-strain curve may better represent the steel's stress-strain response than the monotonic stress-strain curve.

5.4.2 In many field test situations it may be desirable to convert measured strains to stress in order to estimate fatigue life. The cyclic stress-strain curve can be described with an equation using the cyclic properties. Eq. 10 can be rewritten rearranging the terms as shown in Eq. 12.

$$\Delta\varepsilon/2 = \Delta\sigma/2E + \Delta\varepsilon_p/2 \quad (12)$$

Rearranging the terms in Eq. 11 indicates the relationship between plastic strain amplitude and stress amplitude.

$$\Delta\varepsilon_p/2 = (\Delta\sigma/2K')^{1/n'} \quad (13)$$

Substituting Eq. 13 into Eq. 12 yields an equation relating cyclic strain

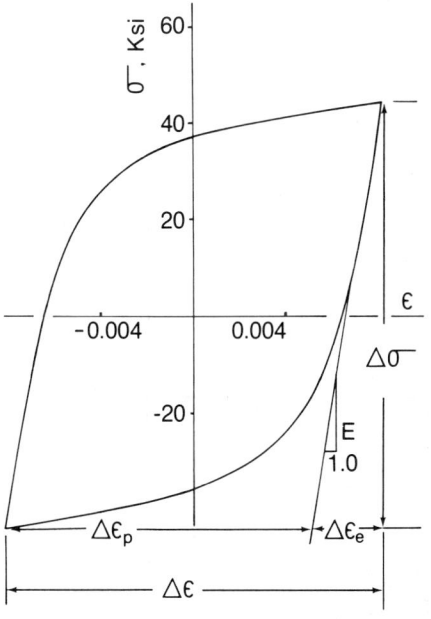

FIG. 4—STABLE STRESS-STRAIN HYSTERESIS LOOP

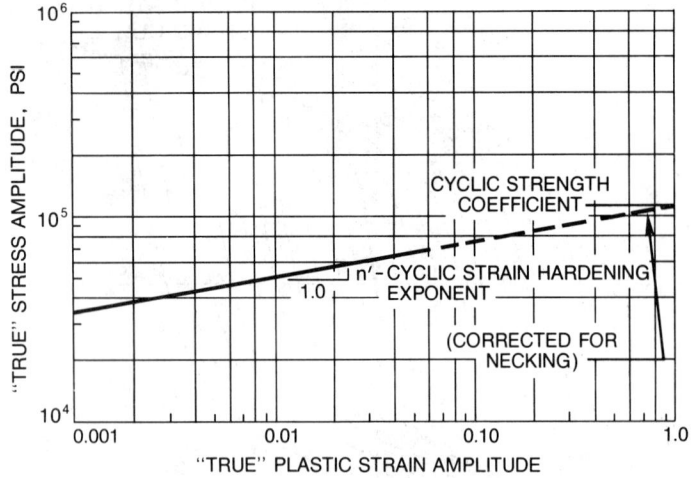

FIG. 6—CYCLIC STRESS-PLASTIC STRAIN PLOT, 1020 H.R. STEEL

amplitude to cyclic stress amplitude in terms of the previously defined properties and the modulus of elasticity.

$$\Delta\varepsilon/2 = \Delta\sigma/2E + (\Delta\sigma/2K')^{1/n'} \qquad (14)$$

5.4.3 For a more detailed discussion see STP 465 (3).

6. Fatigue Properties

6.1 Fatigue resistance of metals is generally described in terms of the number of constant amplitude stress or strain reversals[3] required to cause failure. The properties defined in this section are determined on smooth polished axial specimens tested under strain control. Stress amplitude, strain amplitude and plastic strain amplitude can each be plotted against reversals to failure. The plot of log "true" plastic strain amplitude versus log reversals to failure are typically straight lines as illustrated in Figs. 7 and 8. The intercept at one reversal and the slope of these straight lines can be described as fatigue properties.

6.2 Definitions

6.2.1 FATIGUE DUCTILITY EXPONENT (c) is the power to which the life in reversals must be raised to be proportional to the "true" strain amplitude. It is taken as the slope of the log ($\Delta\varepsilon_p/2$) versus log ($2N_f$) plot.

---

[3] A reversal is counted each time the stress or strain-time signal changes direction. In constant amplitude testing, one cycle is equal to two reversals.

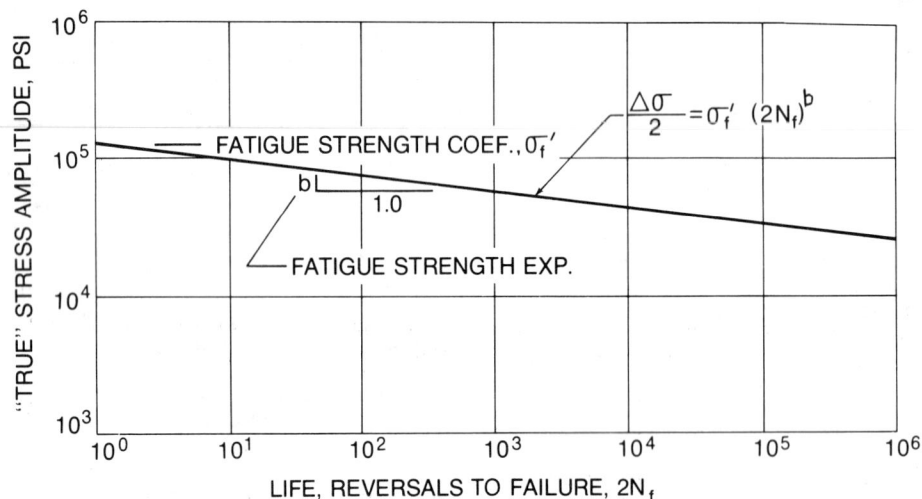

FIG. 7—STRESS AMPLITUDE VERSUS REVERSALS TO FAILURE, 1020 H.R. STEEL

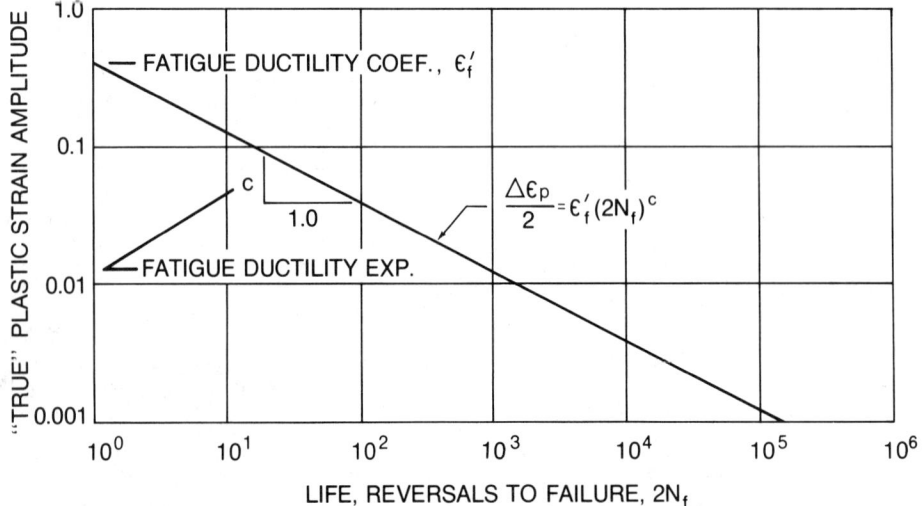

FIG. 8—PLASTIC STRAIN AMPLITUDE VERSUS REVERSALS TO FAILURE, 1020 H.R. STEEL

6.2.2 FATIGUE DUCTILITY COEFFICIENT (ϵ'_f) is the "true" strain required to cause failure in one reversal. It is taken as the intercept of the log ($\Delta\epsilon_p/2$) versus log ($2N_f$) plot at $2N_f = 1$.

6.2.3 FATIGUE STRENGTH EXPONENT (b) is the power to which life in reversals must be raised to be proportional to "true" stress amplitude. It is taken as the slope of the log $\Delta\sigma/2$ versus log ($2N_f$) plot.

6.2.4 FATIGUE STRENGTH COEFFICIENT (σ'_f) is the "true" stress required to cause failure in one reversal. It is taken as the intercept of the log $\Delta\sigma/2$ versus log ($2N_f$) plot at $2N_f = 1$.

6.2.5 TRANSITION FATIGUE LIFE ($2N_t$) is the life where elastic and plastic components of the total strain are equal. It is the life at which the plastic and elastic strain-life lines cross.

6.3 Discussion

6.3.1 A metals resistance to strain cycling can be considered as the summation of the elastic and plastic resistance as indicated by Eq. 15.

$$\Delta\epsilon/2 = (\Delta\epsilon_e/2) + (\Delta\epsilon_p/2) \qquad (15)$$

An equation of the "true" plastic strain-life relationship can be written in terms of the above fatigue properties (Fig. 8).

$$\Delta\epsilon_p/2 = \epsilon'_f (2N_f)^c \qquad (16)$$

where $2N_f$ is reversal to failure. The "true" elastic strain-life relationship is simply the stress-life relationship divided by the modulus of elasticity (Fig. 7).

$$\Delta\epsilon_e/2 = (\sigma'_f/E)(2N_f)^b \qquad (17)$$

Substituting Eq. 16 and Eq. 17 into Eq. 15 gives an equation between "true" strain amplitude and reversals to failure in terms of the fatigue properties.

$$\Delta\epsilon/2 = (\sigma'_f/E)(2N_f)^b + \epsilon'_f(2N_f)^c \qquad (18)$$

The above equation is illustrated in Fig. 9.

7. Illustrative Example

7.1 The purpose of this section is to illustrate some of the possible applications of cyclic stress-strain and fatigue properties. One example is presented illustrating how the data in this report might be utilized. It by no means covers the many possible ways in which this data can or will be used in design considerations. Other writings such as Sandor's recent book (17) and other references cover this subject in much greater detail.

7.2 Consider the problem of estimating fatigue life for a given strain history. Three different metals will be considered: 1020 hot rolled (HR) steel, 1045 quenched and tempered (Q&T) steel at 390 BHN, and 2024-T4 aluminum. A strain-time history at the point of possible failure is given in the following sequence of peaks and valleys: $e_0 = 0.0, e_1 = 0.0045, e_2 = -0.002, e_3 = 0.004, e_4 = -0.0045, e_5 = 0.003, e_6 = -0.0045$, and $e_7 = 0$. (see Fig. 10.) To simplify the example, it is assumed that the strains given include the influence of any stress concentration that may be present. For example, they might represent the strains at a notch root. Further details concerning notch effects and a method for converting applied loads or strains measured near a notch to notch root strains can be found in Refs. 12–14.

7.3 First, a check should be made to determine if yielding takes place during the first block of straining. This is done by comparing the monotonic yield strength with the largest strain multiplied by the modulus of elasticity. Since yielding may occur in compression as well as tension, the peak or valley with the largest absolute value should be used in the comparison. (see Table 4.) Clearly, the 1020 yields while the 1045 remains elastic at least for the first block of load. The 2024 yields a small amount.

7.4 The next question to ask is whether the chosen metals will cylically harden or soften. That is, will the metals stress-strain response change during cycling such that yielding may occur later or possibly stop? The first indication can be obtained by comparing the monotonic and cyclic strain hardening exponents of the metals. If the cyclic is less than the monotonic strain hardening exponent, the metal will generally cyclically harden. If it is greater it will usually cyclically soften. (see Table 5.)

7.5 Little change should be expected for the 1020 HR, considerable cyclic softening for the 1045 Q&T, and considerable cyclic hardening for the 2024. A check can be made to see if yielding occurs after the metal has been cycled for some time by comparing twice the cyclic yield strength to the largest strain range in the history multiplied by the modulus of elasticity. For this comparison, the largest strain range is defined as the maximum strain minus the minimum strain, regardless of whether they occur adjacent in the load sequence or not (Table 6).

7.6 The 1020 HR is still in a yield condition; 1045 Q&T is now yielding, whereas it did not yield in the first reversal; and the 2024 T4 is no longer yielding. Note, that if the question as to whether yield would occur had been based upon the monotonic value, an incorrect answer would have been arrived at in two out of the three choices.

7.7 Most important is the consideration of fatigue life of the component. Life can be estimated from the strain data alone. For simplicity, the effect of mean stress will be ignored here, although it may be an influential and even the deciding factor in a more detailed analysis. The first step is to separate the strain signal into individual cycles via a cycle counting routine. Numerous methods for counting have been suggested. However, the most promising is the rainflow method (15). This method can be shown to separate the history into stress-strain hysteresis loops which are comparable to those found in the constant amplitude material properties testing. As illustrated in Fig. 10, rainflow counting breaks this service history into three closed loops: 0.0045 to -0.0045, 0.004 to -0.002, and -0.0045 to 0.0025 strain.

7.8 Fatigue life will be estimated using a linear damage rule:

$$\text{Fatigue life in blocks to failure} = \frac{1}{\sum \frac{n}{N_f}} \qquad (19)$$

where:

Block = one time through strain sequence
n = cycles counted in sequence at given amplitude
N_f = cycles to crack initiation

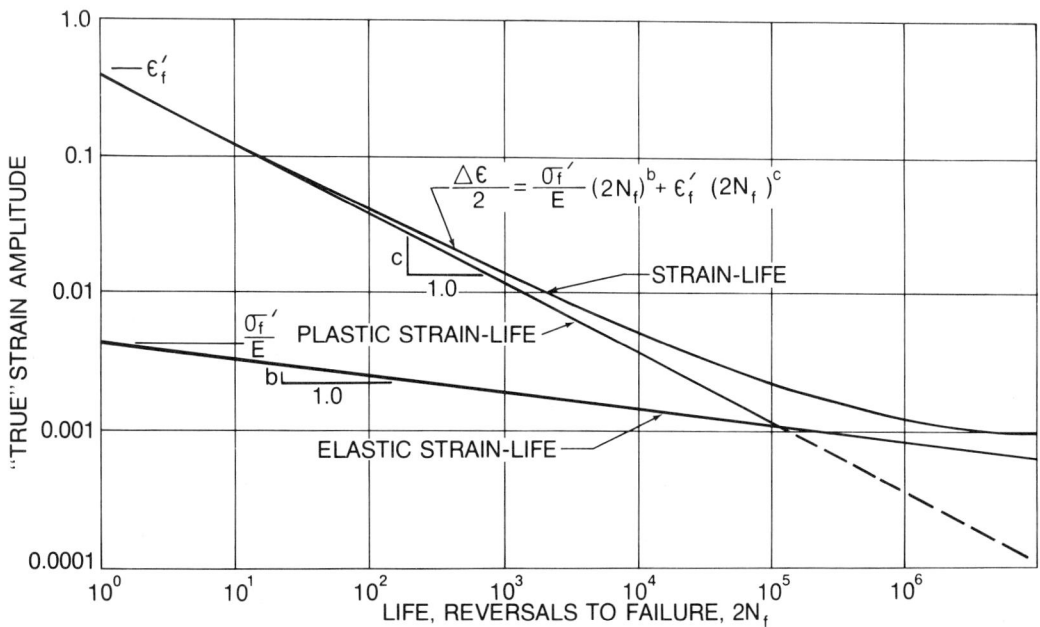

FIG. 9—STRAIN AMPLITUDE VERSUS REVERSALS TO FAILURE, 1020 H.R. STEEL

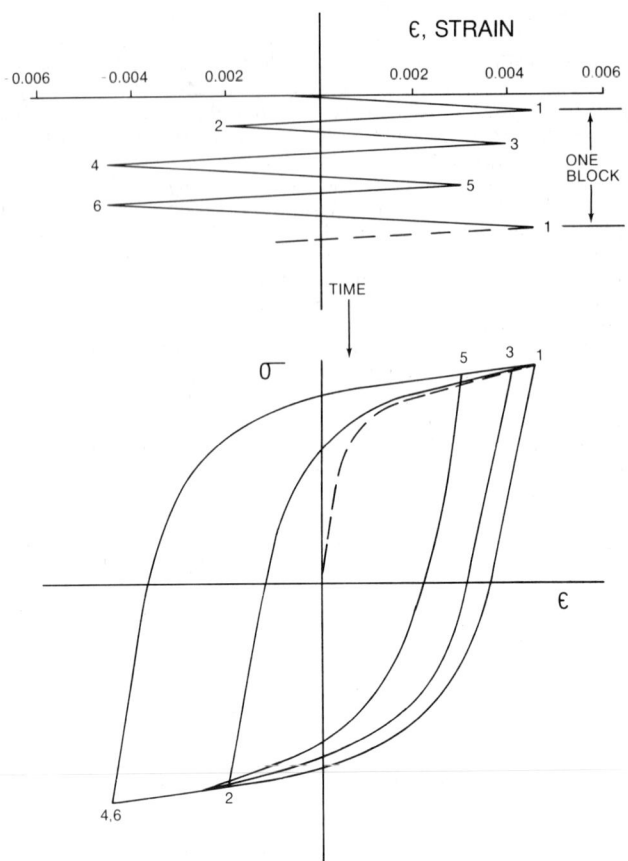

FIG. 10—STRAIN-TIME PLOT AND SCHEMATIC OF RESULTING STRESS-STRAIN RESPONSE

7.9 The life, N_f, is found by substituting the four fatigue properties and the modulus of elasticity as given in Table 2 into Eq. 18 (Table 7):

$$\text{for 1020 HR } \frac{\Delta\varepsilon}{2} = \frac{130}{29,500}(2N_f)^{-0.12} + 0.41\,(2N_f)^{-0.51}$$

$$\text{for 1045 Q\&T } \frac{\Delta\varepsilon}{2} = \frac{230}{30,000}(2N_f)^{-0.074} + 0.45\,(2N_f)^{-0.68}$$

$$\text{for 2024 T4 } \frac{\Delta\varepsilon}{2} = \frac{147}{10,200}(2N_f)^{-0.11} + 0.21\,(2N_f)^{-0.52}$$

7.10 Cycles-to-failure, N_f, for the strain ranges counted, found by solving the above equations, is given in Table 8.

Fatigue life for each metal is estimated below:

$$\text{Life 1020 HR steel} = \frac{1}{1/7,100 + 1/14,000 + 1/21,500} = 3864 \text{ blocks}$$

$$\text{Life 1045 Q\&T steel} = \frac{1}{1/6,700 + 1/43,500 + 1/2,150,000} = 5650 \text{ blocks}$$

$$\text{Life 2024-T4} = \frac{1}{1/57,000 + 1/330,000 + 1/1,090,000} = 46,500 \text{ blocks}$$

7.11 For the given condition of equal strain histories, it is clear that the 2024-T4 would have the longest fatigue life. The above prediction is strain based and does not account for any mean stress effects. Also, the load-carrying capacity of an aluminum member would be smaller than for the steel members because of the differences in elastic modulus.

7.12 In order to include mean stress, the stresses which correspond to the given strains must be determined. Since inelastic material behavior is occurring, it will be necessary to follow cyclic stress-strain response instead of simply multiplying strain times modulus of elasticity to obtain stress.

7.13 To simplify the analysis, since hardening or softening generally takes place in the early portion of the life, the stress-strain response will be assumed to be equal to the stable cyclic condition. Stable cyclic response is described by Eq. 14 involving the modulus of elasticity, cyclic strain hardening exponent, and cyclic strength coefficient. These properties are listed in Table 2 except for the cyclic strength coefficient of 1045 Q&T and 2024-T4. In the case where the cyclic strength coefficient is not reported, it can be estimated knowing the fatigue strength and ductility coefficient and the cyclic strain hardening exponent. The cyclic strength and ductility coefficient represent a point on the plastic strain amplitude versus stress amplitude plot and the cyclic strain hardening exponent is the slope of the line. The cyclic strength coefficient can be calculated from Eq. 11 as follows:

$$\Delta\sigma/2 = K'(\Delta\varepsilon_p/2)^{n'}$$
$$K' = (\Delta\sigma/2)/(\Delta\varepsilon_p/2)^{n'}$$
$$K' = \sigma'_f/(\varepsilon'_f)^{n'}$$

See Table 9.

7.14 for a repeated sequence, the first half cycle is unique starting at zero stress and zero strain; for this example, it is assumed to follow the stable cyclic stress-strain response as indicated in Eq. 20.

$$\varepsilon = \frac{\sigma}{E} + \left(\frac{\sigma}{K'}\right)^{1/n'} \qquad (20)$$

TABLE 4

SAE Spec	Modulus of Elasticity, E, ksi	$e_{max} \cdot E$, ksi	Monotonic Yield Strength, ksi
1020 HR	29,500	133	38
1045 Q&T	30,000	135	185
2024-T4	10,200	46	44

TABLE 5

SAE Spec	Monotonic Strain Hardening Exponent, n	Cyclic Strain Hardening Exponent, n'
1020 HR	0.19	0.18
1045 Q&T	0.044	0.17
2024-T4	0.20	0.08

TABLE 6

SAE Spec	Modulus of Elasticity, E, ksi	$\Delta e_{max} \cdot E$, ksi	2·Cyclic Yield Strength, ksi
1020 HR	29,500	266	70
1045 Q&T	30,000	270	220
2024-T4	10,200	92	128

TABLE 7

SAE Spec	E, Modulus of Elasticity, ksi	σ'_f Fatigue Strength Coefficient, ksi	b, Fatigue Strength Exponent	ε'_f, Fatigue Ductility Coefficient	c, Fatigue Ductility Exponent
1020 HR	29,500	130	−0.12	0.41	−0.51
1045 Q&T	30,000	230	−0.074	0.45	−0.68
2024-T4	10,200	147	−0.11	0.21	−0.52

TABLE 8

Hysteresis Loop	Strain Range	Fatigue Life, N_f		
		1020 HR	1045 Q&T	2024-T4
1–6	0.0090	7,100	6,700	57,000
4–5	0.0070	14,000	43,500	330,000
2–3	0.0060	21,500	215,000	1,090,000

7.15 The next reversal follows the stress-strain outer loop curve. The outer loop shape has been found to be different from the cyclic curve by approximately a factor of two (16). Hence, the outer loop equation is as follows:

$$\frac{\Delta \varepsilon}{2} = \frac{\Delta \sigma}{2E} + \left(\frac{\Delta \sigma}{2K'}\right)^{1/n'} \qquad (21)$$

7.16 Substituting the properties from Table 9 into Eq. 21 the stress range for each strain range may be calculated, as shown in Table 10.

7.17 Starting with the stress determined by following the cyclic stress-strain curve to the first point in the history, the above ranges can be algebraically added to each succeeding point to determine the sequential stress peaks. With stable cyclic response, each block of the strain sequence will result in a repeating stress sequence. The repeating sequence for the three materials is indicated in Table 11.

7.18 Rainflow counting identifies the cycles as three closed loops with tips at points 1 and 6, 2 and 3, and 4 and 5. The strain range and mean stress of each cycle is tabulated in Table 12.

7.19 Numerous methods have been suggested to account for mean stress effect. For this example, a method suggested by Morrow (1) will be used. This results in Eq. 18 being modified as shown below:

$$\frac{\Delta \varepsilon}{2} = \frac{\sigma'_f - \sigma_o}{E}(2N_f)^b + \varepsilon'_f(2N_f)^c \qquad (22)$$

7.20 Cycles-to-failure is found by solving Eq. 22 for the combinations of mean stress and strain ranges from Table 12. This is shown in Table 13. Fatigue life for each metal is estimated below:

$$\text{Life 1020 HR} = \frac{1}{1/7,100 + 1/21,000 + 1/14,200} = 3862 \text{ blocks}$$

$$\text{Life 1045 Q\&T} = \frac{1}{1/6,700 + 1/145,000 + 1/60,500} = 5790 \text{ blocks}$$

$$\text{Life 2024-T4 aluminum} = \frac{1}{1/57,000 + 1/635,000 + 1/535,000}$$
$$= 47,600 \text{ blocks}$$

7.21 The addition of a mean stress correction in this example had little influence on the resulting estimates of life. This is generally true when the damage is principally due to a few large cycles-low cycle fatigue. In cases where failure is due to a larger number of small cycles, mean stress may have a much greater influence.

TABLE 9

SAE Spec	E, Modulus of Elasticity ksi	n', Cyclic Strain Hardening Exponent	K', Cyclic Strength Coefficient, ksi
1020 HR	29,500	0.18	112
1045 Q&T	30,000	0.17	263
2024-T4	10,200	0.08	166

TABLE 10

Hysteresis Loop Tips*	Strain Range	Stress Range, ksi		
		2024-T4	1045 Q&T	1020 HR
1-2	0.0065	66.3	153.0	73.2
2-3	0.006	61.2	147.0	71.7
1-4	0.009	91.8	175.6	79.4
4-5	0.007	71.4	158.5	74.7

*Associated hysteresis loop tips determined by rainflow counting.

TABLE 11

Sequence Point	Strain Sequence	Stable Cyclic Stress Sequence, ksi		
		2024-T4	1045 Q&T	1020 HR
1	0.0045	45.9	87.8	39.7
2	−0.002	−20.4	−65.2	−33.5
3	0.004	40.8	81.8	38.2
4	−0.0045	−45.9	−87.8	−39.7
5	0.0025	25.5	70.7	35.0
6	−0.0045	−45.9	−87.8	−39.7

TABLE 12

Hysteresis Loop	Strain Range	Mean Stress, ksi		
		2024-T4	1045 Q&T	1020 HR
1-6	0.009	0.01	0.0	0.0
2-3	0.006	10.2	8.3	2.35
4-5	0.007	−10.2	−8.55	−2.35

TABLE 13

Hysteresis Loop	Fatigue Life, N_f		
	2024-T4	1045 Q&T	1020 HR
1-6	57,000	6,700	7,100
2-3	635,000	145,000	21,000
4-5	535,000	60,500	14,200

8. References

8.1 Graham, J. A., Editor, *Fatigue Design Handbook*, Society of Automotive Engineers, Inc., 400 Commonwealth Drive, Warrendale, Pa. 15096.

8.2 Bridgeman, P. W., Transactions of ASM, American Society for Metals, Vol. 32, p. 553, 1944; (also Dieter, G. E. *Mechanical Metallurgy*, McGraw-Hill Book Co., Inc., 1961, New York, New York, pp. 250-254.

8.3 Raske, D. T. and Morrow, JoDean, Mechanics of Materials in Low Cycle Fatigue Testing, *Manual on Low Cycle Fatigue Testing*, ASTM STP 465, American Society for Testing and Materials, 1969, pp. 1-25.

8.4 Landgraf, R. W., Morrow, JoDean, and Endo, T., Determination of the Cyclic Stress-Strain Curve. Journal of Materials, ASTM, Vol. 4 No. 1, March 1969, pp. 176-188.

8.5 Gallagher, J. P., What the Designer Should Know About Fracture Mechanics Fundamentals. Paper 710151 presented at SAE Automotive Engineering Congress, Detroit, January, 1971.

8.6 Sinclair, G. M., What the Designer Should Know About Fracture Mechanics Testing. Paper 710152 presented at SAE Automotive Engineering Congress, Detroit, January, 1971.

8.7 Ripling, E. J., How Fracture Mechanics Can Help the Designer. Paper 710153 presented at SAE Automotive Engineering Congress, Detroit, January 1971.

8.8 Campbell, J. E., Berry, W. E., and Fedderson, C. E., *Damage Tolerant Design Handbook*, MCIC HB-01, Metal and Ceramics Information Center, Battalle Columbus Laboratories, Columbus, Ohio.

8.9 Jaske, C. E., Fedderson, C. E., Davies, K. B., Rice, R. C., *Analysis of Fatigue, Fatigue Crack Propagation and Fracture Data*, NASACR 132332, Battelle Columbus Laboratories, Columbus, Ohio, November 1973.

8.10 Moore, T. D., *Structural Alloys Handbook*, Mechanical Properties Data Center, BelFour Stulen, Inc., Traverse City, Michigan.

8.11 Wolf, J., Brown, W. F., Jr., *Aerospace Structural Metals Handbook*, Vol. 1-4, Mechanical Properties Data Center, BelFour Stulen, Inc., Traverse City, Michigan.

8.12 Raske, D. T., Review of Methods for Relating the Fatigue Notch Factor to the Theoretical Stress Concentration Factor, Simulation of the Fatigue Behavior of the Notch Root in Spectrum Loaded Notched Members, Chapter II, TAM Report No. 333—Department of Theoretical and Applied Mechanics, University of Illinois, Urbana, January 1970.

8.13 Topper, T. H., Wetzel, R. M. and Morrow, JoDean, Neuber's Rule Applied to Fatigue of Notched Specimens, Journal of Materials, ASTM, Vol. 4, No. 1, March 1969, pp. 200-209.

8.14 Tucker, Lee E., A Procedure for Designing Against Fatigue Failure of Notched Parts, SAE Paper No. 720265, Society of Automotive Engineers, New York, New York 10001.

8.15 Dowling, N. E., Fatigue Failure Predictions for Complicated Stress-Strain Histories. J. Materials, ASTM, March 1972; (see also: Fatigue Failure Predictions for Complicated Stress-Strain Histories. TAM Report No. 337, Theoretical and Applied Mechanics Dept., University of Illinois, Urbana 1970).

8.16 Morrow, JoDean, Cyclic Plastic Strain Energy and Fatigue of Metals, *Internal Friction, Damping, and Cyclic Plasticity*, ASTM STP 378, American Society for Testing and Materials, 1965, pp. 45-87.

8.17 Sandor, B. I., *Fundamentals of Cyclic Stress and Strain*, University of Wisconsin Press, Madison, Wisconsin.

5 Steel Fasteners

MECHANICAL AND MATERIAL REQUIREMENTS FOR EXTERNALLY THREADED FASTENERS—SAE J429 JAN80

SAE Standard

Report of the Iron and Steel Technical Committee, approved January 1949, last revised May 1979.

1. Scope—This SAE Standard covers the mechanical and material requirements for steel bolts, screws, studs, sems,[1] and U-bolts[2] used in automotive and related industries in sizes to 1½ in, inclusive.

[1] Sems: Screw and washer assemblies.
[2] U-bolts covered by this SAE Standard are those used primarily in the suspension and related areas of vehicles. For specification purposes, this standard treats U-bolts as studs. Thus, wherever the word "studs" appears, "U-bolts" is also implied. (Designers should recognize that the "U" configuration may not sustain a load equivalent to two bolts or studs of the same size and grade; thus, actual load carrying capacity of U-bolts should be determined by saddle load tests.)

NOTE: Previous issues of this standard also covered nuts, now covered separately in SAE J995 (August, 1967).

2. Designations

2.1 Designation System—Grades are designated by numbers where increasing numbers represent increasing tensile strength, and by decimals of whole numbers where decimals represent variations at the same strength level. The grade designations are given in Table 1.

2.2 Grades—Bolts and screws are normally available only in Grades 1, 2, 5, 5.2, 7, 8, and 8.2. Studs are normally available only in Grades 1, 2, 4, 5, 8, and 8.1. Grade 5.1 is applicable to sems which are heat treated following

TABLE 1—MECHANICAL REQUIREMENTS AND IDENTIFICATION MARKING FOR BOLTS, SCREWS, STUDS, SEMS, AND U-BOLTS[j]

Grade Designation	Products	Nominal Size Dia, in	Full Size Bolts, Screws, Studs, Sems		Machine Test Specimens of Bolts, Screws, and Studs				Surface Hardness	Core Hardness		Grade Identification Marking[i]
			Proof Load (Stress), psi	Tensile Strength (Stress) Min, psi	Yield[a] Strength (Stress) Min, psi	Tensile Strength (Stress) Min, psi	Elongation[f] Min, %	Reduction of Area Min, %	Rockwell 30N Max	Rockwell Min	Rockwell Max	
1	Bolts, Screws, Studs	1/4 thru 1-1/2	33,000[k]	60,000	36,000[b]	60,000	18	35	—	B70	B100	None
2	Bolts, Screws, Studs	1/4 thru 3/4[c]	55,000[k]	74,000	57,000	74,000	18	35	—	B80	B100	
		Over 3/4 to 1-1/2	33,000	60,000	36,000[b]	60,000	18	35	—	B70	B100	None
4	Studs	1/4 thru 1-1/2	65,000	115,000	100,000	115,000	10	35	—	C22	C32	None
5	Bolts, Screws, Studs	1/4 thru 1	85,000	120,000	92,000	120,000	14	35	54	C25	C34	
		Over 1 to 1-1/2	74,000	105,000	81,000	105,000	14	35	50	C19	C30	
5.1[d]	Sems[h] Bolts Screws	No. 6 thru 5/8 No. 6 thru 1/2	85,000	120,000	—	—	—	—	59.5[g]	C25	C40[g]	
5.2	Bolts Screws	1/4 thru 1	85,000	120,000	92,000	120,000	14	35	56	C26	C36	
7[e]	Bolts Screws	1/4 thru 1-1/2	105,000	133,000	115,000	133,000	12	35	54	C28	C34	
8	Bolts, Screws, Studs	1/4 thru 1-1/2	120,000	150,000	130,000	150,000	12	35	58.6	C33	C39	
8.1	Studs	1/4 thru 1-1/2	120,000	150,000	130,000	150,000	10	35	—	C32	C38	None
8.2	Bolts Screws	1/4 thru 1	120,000	150,000	130,000	150,000	10	35	61	C35	C42	

[a] Yield strength is stress at which a permanent set of 0.2% of gage length occurs.
[b] Yield point shall apply instead of yield strength at 0.2% offset.
[c] Grade 2 requirements for sizes 1/4 through 3/4 apply only to bolts and screws 6 in and shorter in length, and to studs of all lengths. For bolts and screws longer than 6 in, Grade 1 requirements shall apply.
[d] Grade 5 material heat treated before assembly with a hardened washer is an acceptable substitute.
[e] Grade 7 bolts and screws are roll threaded after heat treatment.
[f] See Table 6 for gage length.
[g] Hex washer head and hex flange products without assembled washers shall have a core hardness not exceeding Rockwell C38 and a surface hardness not exceeding Rockwell 30N 57.5.
[h] Sems and similar products without washers.
[i] See footnote 2 of text.
[j] Not applicable to studs or slotted and cross recess head products.
[k] Proof load test: Requirements in these grades only apply to stress relieved products.

assembly of the washer on the screw, and to products without assembled washer.

3. Materials and Processes

3.1 Steel Characteristics—Bolts, screws, studs, and sems shall be made of steel conforming to the description and chemical composition requirements specified in Table 2 for the applicable grade.

3.2 Heading Practice—Methods other than upsetting and/or extrusion are permitted only by special agreement between purchaser and supplier.

Grade 1 bolts and screws may be hot or cold headed, at option of the manufacturer.

Grades 2, 5, 5.2, 7, 8, and 8.2 bolts and screws in sizes up to 3/4 in, inclusive, and in lengths up to 6 in, inclusive, shall be cold headed, except that by special agreement they may be hot headed. Larger sizes and longer lengths may be hot or cold headed, at option of the manufacturer.

Grade 5.1 bolts, screws, and sems shall be cold headed.

3.3 Threading Practice—Grades 2, 5, 5.2, 8, and 8.2 bolts and screws in sizes up to 3/4 in, inclusive, and lengths up to 6 in, inclusive, shall be roll threaded, except by special agreement. Grade 7 bolts and screws shall be roll threaded after heat treatment. Grade 5.1 bolts, screws, and sems shall be roll threaded. Threads of all sizes of Grade 1 bolts and screws, and Grades 2, 5, 5.2, 8, and 8.2 bolts and screws in sizes over 3/4 in and/or lengths longer than 6 in, may be rolled, cut, or ground, at option of the manufacturer. Threads of all grades and sizes of studs may be rolled, cut, or ground, at option of the manufacturer.

3.4 Heat Treatment Practice—Grade 1 bolts and screws and Grades 1 and 2 studs need not be heat treated. When specified by purchaser, Grade 2 cold headed bolts and screws shall be stress relieved at a minimum stress relief temperature of 875°F (468°C). Grades 4 and 8.1 studs are manufactured from pretreated material and the studs, as manufactured, need no further heat treatment. Grades 5 and 5.2 bolts, screws, and studs shall be heat treated, oil or water quenched, at option of manufacturer, and tempered at a minimum tempering temperature of 800°F (427°C). Grade 5.1 bolts, screws, and sems shall be heat treated, quenched, and tempered at a minimum tempering temperature of 650°F (343°C). For Grade 5.1 sems, quenchants whose principal constituent is water shall not be used, unless specifically approved by the user. Grades 7 and 8 bolts and screws and Grade 8 studs shall be heat treated, oil quenched, and tempered at a minimum tempering temperature of 800°F (427°C). Grade 8.2 bolts and screws shall be fully austenitized, quenched in oil or water, and tempered at a minimum temperature of 650°F (340°C).

3.5 Decarburization—Unless otherwise specified, Grades 5, 5.1, 5.2, 7, 8, and 8.2 bolts, screws, and studs shall conform to Class C unless Class B is specified as described in SAE J121 (September, 1969).

3.6 Surface Discontinuities—Grades 5, 5.1, 5.2, 7, 8, 8.1, and 8.2 bolts, screws, and studs in sizes up to 1 in inclusive, and lengths up to 6 in inclusive shall be in conformity with the requirements of SAE J1061 (September, 1973).

When the engineering requirements of the application necessitate that surface discontinuities of bolts, screws, and studs should be more closely controlled, the purchaser shall specify the applicable limits in the original inquiry and purchase order. For certain fasteners, this may be done by reference to SAE J123 (September, 1973).

4. Mechanical Requirements

Bolts, screws, studs, and sems shall be tested in accordance with the mechanical testing requirements for the applicable type, grade, size, and length of product as specified in Table 3, and shall meet the mechanical requirements specified for that product in Table 1.

In the case of U-bolts having thread length equal to 3D or longer, cut stud-like specimens from either leg of the "U" (utilizing the maximum available thread length) and test as shown for studs. Where thread length is less than 3D, test for hardness only as shown for "short studs." (Applicable mechanical tests are shown in Table 3 and requirements in Table 1.)

5. Methods of Test

5.1 Hardness—The hardness of bolts, screws, studs, and sems shall be determined at mid-radius of a transverse section through the threaded portion of the product taken at a distance of one diameter from the end of the product. The reported hardness shall be the average of four hardness readings located at 90 deg to one another. The preparation of test specimens and the performance of hardness tests shall be in conformity with the requirements of SAE J417 (January, 1946).

To meet the requirements of paragraph 4, the hardness shall not exceed the maximum hardness specified in Table 1 for the applicable grade. In addition, as required in paragraph 4 and Table 3, the hardness shall be not less than the minimum hardness specified in Table 1 for the applicable grade.

5.2 Surface Hardness—Tests to determine surface hardness conditions shall be conducted on the ends, hexagon flats, or unthreaded shanks which have been prepared by lightly grinding or polishing to insure accurate reproducible readings in accordance with SAE J417 (January, 1946). Proper correction factors shall be used when hardness tests are made on curved surfaces, per ASTM E 18.

Depending on the location and individual surface upon which the test is conducted, some increase in hardness above that specified in Table 1, when measured on the Rockwell 30N scale, may occur for reasons other than carburization. To insure that lots of products not considered acceptable for this cause are in fact carburized, the metallographic and hardness checking technique described in SAE J121 (September, 1969) shall be used.

In applying the J121 (September, 1969) procedure, a difference between Knoop and Rockwell 30N readings by conversion may occur. This difference is disregarded since the primary purpose of the Knoop traverse in J121 (September, 1969) is to establish the existence of carburization.

5.3 Proof Load—The proof load test consists of stressing the bolt, screw, stud, or sem with a specified load which the product must withstand without permanent set.

The overall length of the specimen shall be measured between conical or ball centers on the centerline of the specimen, using mating centers on the measuring anvils. The specimen shall be marked so that it can be placed in the measuring fixture in the same position for all measurements. The measurement instrument shall be capable of measurement to 0.0001 in. In the case of sems, the washer may be removed from the screw prior to assembly in the testing machine; however, for referee testing, the washer shall be removed. For bolts, screws, and sems, 3D or longer, the specimen shall be assembled in the fixture of the tensile machine so that six complete threads are exposed between the grips. This is obtained by freely running the nut or fixture to the thread runout of the specimen and then unscrewing the specimen six full turns. Short bolts, 2 1/4–3D in length, threaded to within 2 1/2 pitches of the bearing surface shall be assembled finger tight in the fixture and unscrewed two full turns. When proof load testing studs, one end of the stud shall be assembled in a threaded fixture to the thread runout. For studs having unlike threads, this shall be the end with the finer pitch thread. The other end of the stud shall likewise be assembled in a threaded fixture, as above for bolts. The bolt, screw, stud, or sem shall then be axially loaded to the proof load specified for the applicable size, thread series, and grade in Table 5, the load retained for a period of 10 s, the load removed, and the overall length again measured.

TABLE 2—CHEMICAL COMPOSITION REQUIREMENTS[a]

Grade	Material and Treatment	C Min	C Max	Mn Min	P Max	S Max	B Min
1	Low or medium carbon steel	—	0.55	—	0.048	0.058	—
2	Low or medium carbon steel	—	0.55	—	0.048	0.058[b]	—
4	Medium carbon cold drawn steel	—	0.55	—	0.048	0.058	—
5	Medium carbon steel, quenched and tempered	0.28	0.55	—	0.048	0.058[c]	—
5.1	Low or medium carbon steel, quenched and tempered[e]	0.15	0.30	—	0.048	0.058	—
5.2	Low carbon martensite steel, fully killed, fine grain, quenched and tempered	0.15	0.25	0.74	0.048	0.058	0.0005
7	Medium carbon alloy steel, quenched and tempered[d]	0.28	0.55	—	0.040	0.045	—
8	Medium carbon alloy steel, quenched and tempered[d]	0.28	0.55	—	0.040	0.045	—
8.1	Elevated temperature drawn steel—medium carbon alloy or SAE 1541 (or 1541H steel)	0.28	0.55	—	0.048	0.058	—
8.2	Low carbon martensite steel, fully killed, fine grain, quenched and tempered[f]	0.15	0.25	0.74	0.048	0.058	0.0005

[a] All values are for product analysis (percent by weight). For cast or heat analysis, use standard permissible variations as shown in SAE J409 (January, 1942).

[b] For studs only, sulfur content may be 0.33% max.

[c] For studs only, sulfur content may be 0.13% max.

[d] Steel shall be fine grain, with hardenability that will produce a minimum hardness of Rockwell C47 at the center of a transverse section one diameter from the threaded end of the bolt, screw, or stud after oil quenching (see SAE J407 (August, 1947)). Carbon steel may be used by agreement between producer and consumer, for sizes 1/4–3/4 in diameter products. SAE 1541 (or 1541H) steel, oil quenched and tempered, may be used at the option of the producer for products 7/16 in nominal diameter and smaller.

[e] For sems only, sizes 7/16–5/8 in diameter, low carbon martensite steel (as specified for Grade 5.2) may be used.

[f] Steel with hardenability that will produce a minimum hardness of Rockwell C38 at the center of a transverse section one diameter from the threaded end of the bolt or screw after quenching.

TABLE 3—MECHANICAL TESTING REQUIREMENTS FOR BOLTS, SCREWS, STUDS, AND SEMS

Product	Grade	Specified Min Tensile Strength of Product, lb	Length of Product	Hardness[a]		Tests Conducted Using Full Size Products[a]			Tests Conducted Using Machine Test Specimens[a]			
				Max	Min	Proof Load	Wedge Tensile Strength	Axial Tensile Strength	Yield Strength	Axial Tensile Strength	Elongation	Reduction of Area
Short Bolts and Screws	1, 2, 5, 5.2, 7, 8, 8.2	All	Less than 2-1/4D[b]	*	*	—	—	—	—	—	—	—
Special Head Bolts and Screws	1, 2, 5, 5.2, 7, 8, 8.2	All	All	*	*	—	—	—	—	—	—	—
Square and Hex Bolts and Screws	1, 2, 5, 5.2, 7, 8, 8.2	100,000 and less	2-1/4D to 8D or 8 in, whichever is greater	*	—	*	*	—	—	—	—	—
			Over 8D or 8 in, whichever is greater, thru and including 12 in	*	—	Option C	*	—	Option B	Option B	Option B	Option B
			Over 12 in	*	—	Option C	Option A	—	Option B	Option B	Option B	Option B
		Over 100,000	2-1/4D and longer	*	—	Option C	Option A	—	Option B	Option B	Option B	Option B
All Other Bolts and Screws	1, 2, 5, 5.2, 7, 8, 8.2	100,000 and less	2-1/4 to 8D or 8 in, whichever is greater	*	—	*	—	—	*	—	—	—
			Over 8D or 8 in, whichever is greater	*	—	Option C	—	Option A	Option B	Option B	Option B	Option B
		Over 100,000	2-1/4D and longer	*	—	Option C	—	Option A	Option B	Option B	Option B	Option B
Short Studs	1, 2, 4, 5, 8, 8.1	All	Less than 3D	*	*	—	—	—	—	—	—	—
All Other Studs	1, 2, 4, 5, 8, 8.1	100,000 and less	3D to 8D or 8 in, whichever is greater	*	—	*	*	—	—	—	—	—
			Over 8D or 8 in, whichever is greater	*	—	Option C	Option A	—	Option B	Option B	Option B	Option B
		Over 100,000	3D and longer	*	—	Option C	Option A	—	Option B	Option B	Option B	Option B
Short Bolts, Screws, and Sems	5.1	All	Less than 2-1/4D	*	*	—	—	—	—	—	—	—
Hex Head Bolts, Screws, and Sems	5.1	All	2-1/4D and longer	*	—	*	*	—	—	—	—	—
Other Bolts, Screws, and Sems	5.1	All	2-1/4D and longer	*	—	*	—	*	—	—	—	—
Tests to be performed in accordance with paragraph				5.1		5.3	5.5	5.4	5.6			

[a] Asterisks (*) denote mandatory tests. Where options are indicated, all Option A tests (which apply to full size products) or all Option B tests (which apply to machined specimens) shall be performed. Option C tests (which apply to full size products) are not mandatory unless specified in the original inquiry and purchase order. Option A and Option C tests shall be performed in case arbitration is necessary. Dashes (—) denote tests which are not required.
[b] D equals nominal diameter of the product.
[c] Special head bolts and screws are those with special configurations or with drilled heads which are weaker than the threaded section.
[d] For purposes of Table 3 requirements, "length of product" is the nominal length including point chamfer as defined in SAE J105 (June, 1911), and all special point products shall be measured from the bearing surface to the crest of the last complete thread form.

The speed of testing, as determined with a free running cross head, shall not exceed 0.12 in/min.

To meet the requirements of paragraph 4, the length of the bolt, screw, stud, or sem after loading shall be the same as before loading within a tolerance of ±0.0005 in allowed for measurement error.

Variables, such as straightness and thread alignment (plus measurement error), may result in apparent elongation of the fasteners when the proof load is initially applied. In such cases, the fastener may be retested using a 3% greater load, and may be considered satisfactory if the length after this loading is the same as before this loading (within the 0.0005 in tolerance for measurement error).

5.4 Axial Tensile Strength—Following proof load testing, the same bolt, screw, stud, or sem shall be reassembled in the testing machine per paragraph 5.3 and axial loading applied until failure. Typical fixturing is illustrated in Fig. 1. The speed of testing, as determined with a free running cross head, shall not exceed 1 in/min.

To meet the requirements of paragraph 4, the bolt, screw, stud, or sems shall not fracture before having withstood the minimum tensile load specified for the applicable size, thread series, and grade in Table 5. In addition, for bolts, screws, and sems with regular style heads, the ultimate failure location shall occur in the body or threaded section and not at the junction of the head and shank. (See footnote c under Table 3.)

5.5 Wedge Tensile Strength

5.5.1 BOLTS AND SCREWS—Following proof load testing, the same bolt or screw shall be assembled with a wedge inserted under the head, as illustrated in Fig. 2, installed in the testing machine and tensile tested to failure, as described in paragraph 5.3. The angle of the wedge for the bolt or screw size and grade is specified in Table 4. The wedge shall be so placed that no corner of the square or hexagon bolt or screw head takes the bearing load; that is, a flat of the head shall be aligned with the direction of uniform thickness of the wedge. The wedge shall have a thickness of one-half the bolt or screw diameter measured at the thin side of the hole. The hole in the wedge shall have the following clearance over the nominal size of the bolt or screw, and its top

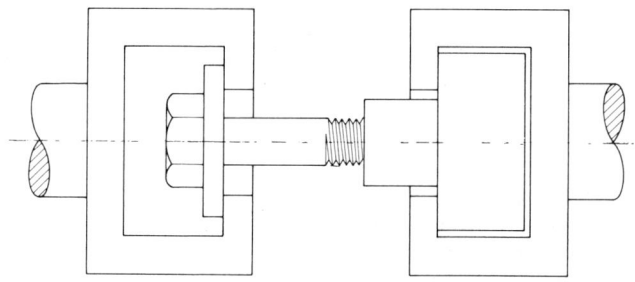

FIG. 1—TENSILE TESTING OF FULL SIZE BOLT OR SCREW

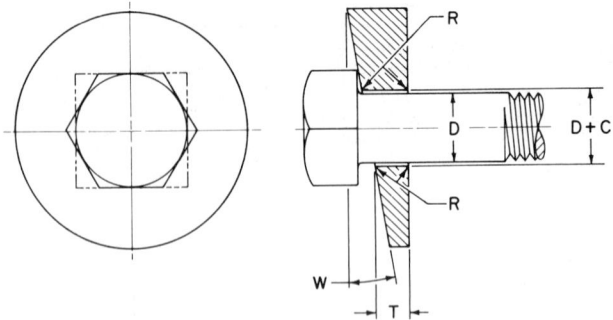

C = CLEARANCE OF HOLE (SEE PARA. 5.4.1.)
D = DIAMETER OF BOLT OR SCREW
R = RADIUS OR CHAMFER (SEE PARA. 5.4.1.)
T = THICKNESS OF WEDGE AT THIN SIDE OF HOLE EQUALS ONE HALF DIAMETER OF BOLT OR SCREW
W = WEDGE ANGLE (SEE TABLE 4)

FIG. 2—WEDGE TEST DETAILS—BOLTS AND SCREWS

φ TABLE 4—TENSILE TEST WEDGE ANGLES

Product	Grade	Nominal Size of Product, in	Wedge Angle, deg
Bolts and Screws[b]	1, 2	1/4 thru 1	10
		Over 1 to 1-1/2	6
	5, 5.2, 7, 8, 8.2[a]	1/4 thru 1	10
	5, 7, 8[a]	Over 1 to 1-1/2	6
Hex Head Bolts[b] Screws and Sems	5.1	No. 6 thru 5/8	6
Studs	1, 2, 5, 8, 8.1	1/4 thru 3/4	6
		Over 3/4 to 1-1/2	4

[a] For Grades 5, 5.2, 7, 8, and 8.2 bolts and screws which are threaded 1 dia and closer to the underside of head, wedge angle shall be 6 deg for sizes 1/4 through 3/4 in, and 4 deg for sizes over 3/4 in.
[b] For hex flange and hex washer head product, the wedge angle shall be 6 deg.

and bottom edges shall be rounded or chamfered 45 deg to the following dimensions:

Nominal Bolt or Screw Size, in	Clearance in Hole, in	Radius or Depth of Chamfer, in
No. 6 thru 12	0.020	0.020
1/4 thru 1/2	0.030	0.030
9/16 thru 3/4	0.050	0.060
7/8 and 1	0.060	0.060
1-1/8 and 1-1/4	0.060	0.125
1-3/8 and 1-1/2	0.094	0.125

To meet the requirements of paragraph 4, the bolt, screw, stud, or sems shall not fracture before having withstood the minimum tensile load specified for the applicable size, thread series, and grade in Table 5. In addition, the ultimate failure location shall occur in the body or threaded section and not at the junction of the head and shank. (See footnote c under Table 3.)

5.5.2 Studs—Following proof load testing, the stud shall be assembled per paragraph 5.3 except with a threaded wedge, as illustrated in Fig. 3. The angle of the wedge for the stud size and grade shall be as specified in Table 4. The stud shall be assembled in the testing machine and tensile tested to failure, as described in paragraph 5.3.

The length of the threaded section of the wedge shall be equal to the diameter of the stud. To facilitate removal of the broken stud, the wedge shall be counterbored. The thickness of the wedge at the thin side of the hole shall equal the diameter of the stud plus the depth of counterbore. The supporting fixture, as shown in Fig. 3, shall have a hole clearance over the nominal size of the stud, and shall have its top and bottom edges rounded or chamfered to the same limits specified for the hardened wedge in paragraph 5.5.1.

To meet the requirements of paragraph 4, the stud shall not fracture before having withstood the minimum tensile load specified for the applicable size, thread series, and grade in Table 5.

5.6 Testing of Machined Test Specimens—Where bolts, screws, and studs cannot be tested in full size for proof load and tensile strength requirements, tests shall be conducted using test specimens machined from the bolt, screw, or stud.

For $1\frac{1}{2}$ in diameter bolts, screws, and studs, a standard 0.500 in round 2 in gage length test specimen shall be turned from the bolt, screw, or stud with the axis of the specimen located midway between the center and outside surface of the bolt, screw, or stud shank, as shown in Fig. 4. Bolts, screws, and studs $\frac{3}{4}$ through $1\frac{3}{8}$ in diameter shall have their shanks machined to the dimensions of a standard 0.500 in round 2 in gage length test specimen concentric with the axis of the bolt, screw, or stud, leaving the bolt or screw head and threaded sections intact, as shown in Fig. 5. Bolts, screws, and studs $\frac{1}{4}$ through $\frac{5}{8}$ in diameter shall have their shanks machined to subsize specimens having dimensions shown in Fig. 5 and Table 6.

The test specimen shall be tensile tested as described in paragraph 5.3; and the yield strength, tensile strength, elongation, and reduction of area determined.

To meet the requirements of paragraph 4, the test specimen must have a yield strength, tensile strength, elongation, and reduction of area equal to or greater than the values for these properties specified for the applicable product size and grade in Table 1.

5.7 Common Test Fixture Details—The grips of the tensile testing machine shall be self-aligning to avoid side thrust on the specimen.

The wedge shall have a minimum hardness of Rockwell C45.

The hole in the fixture or washer used under the head of bolts and screws during proof load and tensile testing shall have the same clearance as that specified for wedges (paragraph 5.5.1).

Wedges, nuts, and fixtures into which bolts, screws, and studs are threaded for proof load, tensile strength, and wedge tensile testing shall have threads which are of the same size, pitch, and tolerance class as the product being tested. (For *standard products*, Class 3B tolerances are normally applicable.) For studs having interference fit threads, wedges shall be threaded to provide a *finger-free fit*.

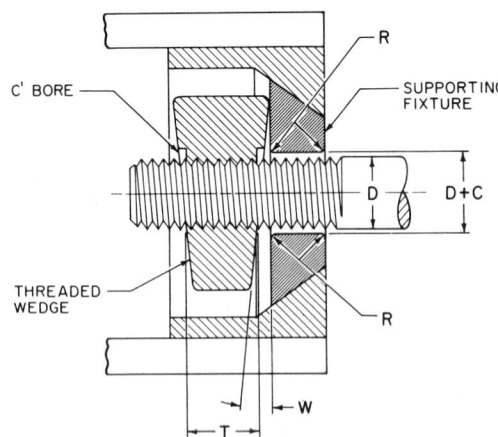

C = CLEARANCE OF HOLE (SEE PARA. 5.4.1.)
D = DIAMETER OF STUD
R = RADIUS OR CHAMFER (SEE PARA. 5.4.1.)
T = D PLUS DEPTH OF COUNTERBORE
W = WEDGE ANGLE (SEE TABLE 4)

FIG. 3—WEDGE TEST DETAILS—STUDS

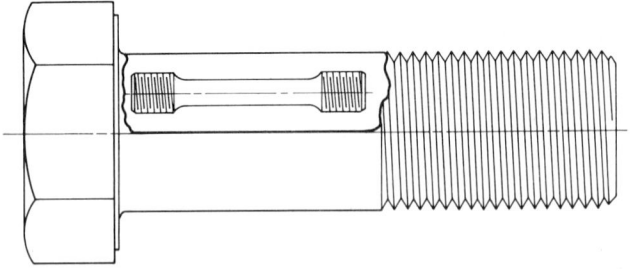

FIG. 4—LOCATION OF STANDARD ROUND 2 IN GAGE LENGTH TENSILE TEST SPECIMEN WHEN TURNED FROM LARGE SIZE BOLTS OR SCREWS

TABLE 5—PROOF LOAD AND TENSILE STRENGTH REQUIREMENTS[a]

Nominal Dia of Product and Threads per in	Stress Area, in²	Grade 1		Grade 2		Grade 4		Grades 5 and 5.2[b]		Grade 5.1		Grade 7		Grades 8, 8.1, 8.2[b]	
		Proof Load, lb	Tensile Strength Min, lb	Proof Load, lb	Tensile Strength Min, lb	Proof Load, lb	Tensile Strength Min, lb	Proof Load, lb	Tensile Strength Min, lb	Proof Load, lb	Tensile Strength Min, lb	Proof Load, lb	Tensile Strength Min, lb	Proof Load, lb	Tensile Strength Min, lb
Coarse Thread Series—UNC															
No. 6–32	0.00909	—	—	—	—	—	—	—	—	750	1,100	—	—	—	—
8–32	0.0140	—	—	—	—	—	—	—	—	1,200	1,700	—	—	—	—
10–24	0.0175	—	—	—	—	—	—	—	—	1,500	2,100	—	—	—	—
12–24	0.0242	—	—	—	—	—	—	—	—	2,050	2,900	—	—	—	—
1/4 –20	0.0318	1,050	1,900	1,750	2,350	2,050	3,650	2,700	3,800	2,700	3,800	3,350	4,250	3,800	4,750
5/16–18	0.0524	1,750	3,150	2,900	3,900	3,400	6,000	4,450	6,300	4,450	6,300	5,500	6,950	6,300	7,850
3/8 –16	0.0775	2,550	4,650	4,250	5,750	5,050	8,400	6,600	9,300	6,600	9,300	8,150	10,300	9,300	11,600
7/16–14	0.1063	3,500	6,400	5,850	7,850	6,900	12,200	9,050	12,800	9,050	12,800	11,200	14,100	12,800	15,900
1/2 –13	0.1419	4,700	8,500	7,800	10,500	9,200	16,300	12,100	17,000	12,100	17,000	14,900	18,900	17,000	21,300
9/16–12	0.182	6,000	10,900	10,000	13,500	11,800	20,900	15,500	21,800	15,500	21,800	19,100	24,200	21,800	27,300
5/8 –11	0.226	7,450	13,600	12,400	16,700	14,700	25,400	19,200	27,100	19,200	27,100	23,700	30,100	27,100	33,900
3/4 –10	0.334	11,000	20,000	18,400	24,400	21,700	38,400	28,400	40,100	—	—	35,100	44,400	40,100	50,100
7/8 – 9	0.462	15,200	27,700	15,200	27,700	30,000	53,100	39,300	55,400	—	—	48,500	61,400	55,400	69,300
1 – 8	0.606	20,000	36,400	20,000	36,400	39,400	69,700	51,500	72,700	—	—	63,600	80,600	72,700	90,900
1-1/8 – 7	0.763	25,200	45,800	25,200	45,800	49,600	87,700	56,500	80,100	—	—	80,100	101,500	91,600	114,400
1-1/4 – 7	0.969	32,000	58,100	32,000	58,100	63,000	111,400	71,700	101,700	—	—	101,700	127,700	116,300	145,400
1-3/8 – 6	1.155	38,100	69,300	38,100	69,300	75,100	132,800	85,500	121,300	—	—	121,300	153,600	138,600	173,200
1-1/2 – 6	1.405	46,400	84,300	46,400	84,300	91,300	161,600	104,000	147,500	—	—	147,500	186,900	168,600	210,800
Fine Thread Series—UNF															
No. 6–40	0.01015	—	—	—	—	—	—	—	—	850	1,200	—	—	—	—
8–36	0.01474	—	—	—	—	—	—	—	—	1,250	1,750	—	—	—	—
10–32	0.0200	—	—	—	—	—	—	—	—	1,700	2,400	—	—	—	—
12–28	0.0258	—	—	—	—	—	—	—	—	2,200	3,100	—	—	—	—
1/4 –28	0.0364	1,200	2,200	2,000	2,700	2,350	4,200	3,100	4,350	3,100	4,350	3,800	4,850	4,350	5,450
5/16–24	0.0580	1,900	3,500	3,200	4,300	3,750	6,700	4,900	6,950	4,900	6,950	6,100	7,700	6,950	8,700
3/8 –24	0.0878	2,900	5,250	4,800	6,500	5,700	10,100	7,450	10,500	7,450	10,500	9,200	11,700	10,500	13,200
7/16–20	0.1187	3,900	7,100	6,550	8,800	7,700	13,650	10,100	14,200	10,100	14,200	12,500	15,800	14,200	17,800
1/2 –20	0.1599	5,300	9,600	8,800	11,800	10,400	18,400	13,600	19,200	13,600	19,200	16,800	21,300	19,200	24,000
9/16–18	0.203	6,700	12,200	11,200	15,000	13,200	23,300	17,300	24,400	17,300	24,400	21,300	27,000	24,400	30,400
5/8 –18	0.256	8,450	15,400	14,100	18,900	16,600	29,400	21,800	30,700	21,800	30,700	26,900	34,000	30,700	38,400
3/4 –16	0.373	12,300	22,400	20,500	27,600	24,200	42,900	31,700	44,800	—	—	39,200	49,600	44,800	56,000
7/8 –14	0.509	16,800	30,500	16,800	30,500	33,100	58,500	43,300	61,100	—	—	53,400	67,700	61,100	76,400
1 –12	0.663	21,900	39,800	21,900	39,800	43,100	76,200	56,400	79,600	—	—	69,600	88,200	79,600	99,400
1 –14 uns	0.679	22,400	40,700	22,400	40,700	44,100	78,100	57,700	81,500	—	—	71,300	90,300	81,500	101,900
1-1/8 –12	0.856	28,200	51,400	28,200	51,400	55,600	98,400	63,300	89,900	—	—	89,900	113,800	102,700	128,400
1-1/4 –12	1.073	35,400	64,400	35,400	64,400	69,700	123,400	79,400	112,700	—	—	112,700	142,700	128,800	161,000
1-3/8 –12	1.315	43,400	78,900	43,400	78,900	85,500	151,200	97,300	138,100	—	—	138,100	174,900	157,800	197,200
1-1/2 –12	1.581	52,200	94,900	52,200	94,900	102,800	181,800	17,000	166,000	—	—	166,000	210,300	189,700	237,200

[a] Proof loads and tensile strengths are computed by multiplying the proof load stresses and tensile strength stresses given in Table 1 by the stress area of the thread. The stress area of sizes and thread series not included in Table 5 may be computed from the formula: $A_s = 0.7854 \left[D - \dfrac{0.9743}{n} \right]^2$ where D equals nominal diameter in inch, and n equals threads per inch.

[b] Grades 5.2 and 8.2 applicable to sizes 1/4 through 1 in.

6. Marking—Unslotted bolts, screws, and hex head sems shall be marked with the grade identification symbol shown in Table 1. In addition, bolts and screws shall be marked with the manufacturer's identification symbol. Markings shall be located on the top of the head, and may be either raised or depressed, at option of the manufacturer.

Studs need not be marked.

7. Testing Requirements

7.1 Manufacturer's Responsibility—During the manufacture of products to the requirements of this specification, the manufacturer shall make periodic tests to ensure that the properties of the product are being maintained within specified limits. Such tests shall be conducted in accordance with a sampling plan, preferably the sampling plan given in paragraph 7.3, and the test results shall be recorded in a test report. When requested in writing by the purchaser, the manufacturer shall furnish a copy of the test report certified to be a report of the results of the last completed set of tests for the specific type, size, length, and grade of product.

Additional tests of products in individual shipments are not normally contemplated. Unless otherwise agreed at time of original inquiry and purchase order, individual heats of steel need not be identified in the finished product.

7.2 Purchaser's Options—If the purchaser requires that additional tests be performed by the manufacturer to determine that the properties of prod-

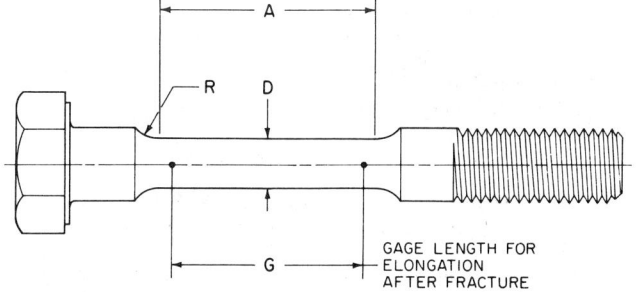

FIG. 5—TENSILE TEST SPECIMEN FOR BOLTS OR SCREWS WITH TURNED DOWN SHANK

TABLE 6—DIMENSIONS OF MACHINED TEST SPECIMENS (SEE FIG. 5 AND PARAGRAPH 5.5)

Nominal Dia of Product	Gage Length G	Dia Parallel Section, D	Length Parallel Section, Min, A	Fillet Radius, Min, R
3/4 thru 1-1/2	2.000 ± 0.005	0.500 ± 0.010	2.25	0.38[a]
1/4 thru 5/8	1.400 ± 0.005	0.350 ± 0.007	1.75	0.25
	1.000 ± 0.005	0.250 ± 0.005	1.25	0.19

[a] Minimum radius recommended 0.38 in; 0.12 minimum permitted.

ucts in an individual shipment are within specified limits, or if the purchaser requires that a sampling plan different from that given in paragraph 7.3 shall be used when determining the acceptability of a lot, or shipment, of products, the purchaser shall specify the complete testing requirements, including sampling plan and basis of acceptance, in the original inquiry and purchase order.

7.3 General—An acceptable sampling plan is outlined below:

Number of Pieces in Lot	Minimum Number of Specimens To Be Tested
50 and under	2
51 to 500	3
501 to 35,000	5
35,001 and over	8

A lot, for purposes of selecting test specimens, shall consist of all products offered for inspection and testing at one time that are of the same type, grade, size, length, and thread series and are manufactured essentially at one time and under the same process conditions.

The same test specimen may be used for different tests wherever practical.

When tested in accordance with this sampling plan, a lot shall be subject to rejection if any of the test specimens fail to meet the applicable test requirements. If the failure of a test specimen is due to improper preparation of the specimen or to incorrect testing technique, the specimen shall be discarded and another specimen substituted.

APPENDIX

(Relative to 150,000 psi tensile bolts and screws produced from low carbon martensite steels and designated as Grade 8.2)

Coverage for 150,000 psi tensile bolts and screws produced from low carbon martensite steels is included in SAE J429 (January, 1949) because several large steel and bolt producers and users have reported highly favorable results with such products over a period of more than three years. This coverage is designated by a separate grade number (Grade 8.2) to distinguish such fasteners from Grade 8 made of medium carbon and medium carbon alloy steels.

Limited data available concerning room temperature ductility and low temperature impact characteristics indicate that fasteners made to Grade 8.2 requirements may have advantages compared to alloy steels historically used for Grade 8 fasteners.

Heat treatment control for elements such as decarburization or carburization and quench medium heat transfer are more critical for Grade 8.2 than for Grade 8 steels. Thus, more attention should be given to verification of the use of proved practices. (It is suggested that users initially require details of heat treatment practices from the fastener producer until a broad spectrum of suppliers are familiar with the closer controls necessary.)

Users should recognize the difference in stress relaxation characteristics of various steels between the temperature range of 650°F, minimum, specified for Grade 8.2 and 800°F, minimum, specified for Grade 8, when considering bolts and screws that may be exposed to such temperature range. The data available on elevated temperature properties of Grade 8.2 indicates that performance testing is desirable in applications where the operating temperature exceeds 500°F (as may also be the case with Grade 8 fasteners).

The requirements stated, herein, limit the use of steels to those which have been used on a production basis with highly favorable results. There is much evidence that other steels are satisfactory also, but these are excluded from the standard until more widespread experience is had with them.

MECHANICAL AND MATERIAL REQUIREMENTS FOR METRIC EXTERNALLY THREADED STEEL FASTENERS—SAE J1199

SAE Standard

Report of Iron and Steel Technical Committee approved February 1978.

1. Scope

1.1 This standard covers the mechanical and material requirements for eight property classes of steel, externally threaded metric fasteners in sizes M1.6 thru M36, inclusive, and suitable for use in automotive and related applications.

1.2 Products included are bolts, screws, studs, U-bolts, pre-assembled screw and washer assemblies (sems), and products manufactured the same as sems except without washer.

1.3 Products not covered are tapping screws, thread rolling screws, and self-drilling screws. Mechanical and material requirements for these products are covered in other SAE documents.

1.4 The term *stud* as referred to herein applies to a cylindrical rod of moderate length, threaded on either one or both ends or throughout its entire length. It does not apply to headed, collared, or similar products which are more closely characterized by requirements shown herein for bolts.

1.5 For specification purposes, this standard treats U-bolts as studs. Thus, wherever the word *studs* appears, *U-bolts* is also implied. U-bolts covered by this standard are those used primarily in the suspension and related areas of vehicles. (Designers should recognize that the U configuration may not sustain a load equivalent to two bolts or studs of the same size and grade; thus actual load carrying capacity of U-bolts should be determined by saddle load tests.)

2. Designations

2.1 Property classes are designated by numbers where increasing numbers generally represent increasing tensile strengths. The designation symbol consists of two parts:

(a) The first numeral of a two-digit symbol or the first two numerals of a three-digit symbol approximates $1/100$ of the minimum tensile strength in MPa.

(b) The last numeral approximates $1/10$ of the ratio expressed as a percentage between minimum yield stress and minimum tensile stress.

2.2 For specification purposes (on engineering drawings, purchase orders, etc.) all property class designations are used in combination with a single basic specification number as follows:

SAE J1199 (4.6) SAE J1199 (9.8)
SAE J1199 (4.8) SAE J1199 (10.9)
SAE J1199 (5.8) SAE J1199 (12.8)
SAE J1199 (8.8) SAE J1199 (12.9)

2.3 Property Classes

2.3.1 Machine screws are normally available only in classes 4.8 and 9.8; other bolts, screws, and studs are available in all classes within the specified product size limitations given in Table 1.

2.3.2 Screw and washer assemblies (sems) are covered by classes 4.8 and 9.8, and allowable deviations from normal 9.8 requirements are stated in footnotes throughout the standard.

2.3.3 At the option of the manufacturer, class 5.8 may be supplied when either class 4.6 or 4.8 is ordered, and class 4.8 may be supplied when class 4.6 is ordered.

2.4 Conversion Guidance

2.4.1 For guidance purposes only, to assist designers in selecting a property class:

(a) Class 4.6 is approximately equivalent to SAE J429, Grade 1 and ASTM A307, Grade A.

(b) Class 5.8 is approximately equivalent to SAE J429, Grade 2.

(c) Class 8.8 is approximately equivalent to SAE J429, Grade 5, and ASTM A449.

(d) Class 9.8 has properties approximately 9% stronger than SAE J429, Grade 5, and ASTM A449.

(e) Class 10.9 is approximately equivalent to SAE J429, Grade 8, and ASTM A354, Grade BD.

TABLE 1—MECHANICAL REQUIREMENTS FOR BOLTS, SCREWS, AND STUDS

Property Class	Nominal Dia	Full Size Bolts, Screws, and Studs		Machined Test Specimens of Bolts, Screws, and Studs (Sizes larger than M24)				Surface Hardness	Product Hardness			
									Rockwell		Vickers	
		Proof Load (Stress) MPa (4)	Tensile Strength (Stress) Min MPa (4)	Yield Strength (Stress) Min(1) MPa	Tensile Strength (Stress) Min MPa	Elongation Min %	Reduction of Area Min %	Rockwell 30N Max	Min (8)	Max	Min (8)	Max
4.6	M5 thru M36	225	400	240(2)	400	22	35	—	B67	B87	120	180
4.8	M1.6 thru M16	310	420	—	—	—	—	—	B71	B87	130	180
5.8	M5 thru M24(3)	380	520	—	—	—	—	—	B82	B95	160	220
8.8	M17 thru M36	600	830	660	830	12	35	(6)	C23	C34	254	336
9.8	M1.6 thru M16(5)	650	900	—	—	—	—	(6)	C27	C36	279	354
10.9	M6 thru M36	830	1040	940	1040	9	35	(6)	C33	C39	327	382
12.8	See Appendix											
12.9(7)	M1.6 thru M36	970	1220	1100	1220	8	35	(6)	C38	C44	372	434

NOTES:
1. Yield strength is stress at which a permanent set of 0.2% of gage length occurs.
2. Yield point shall apply instead of yield strength at 0.2% offset for class 4.6 products.
3. Class 5.8 requirements apply to bolts and screws with lengths 150 mm and shorter, and to studs of all lengths.
4. Proof load and tensile strength values for full size products of each property class are given in Table 5.
5. For class 9.8 screw and washer assemblies (sems), base metal hardness may be 25-40 HRC (270-390 HV) and surface hardness shall not exceed 60 HR30N. This requirement applicable also to products manufactured same as sems except without washer, sizes M1.6 thru M12.
6. Surface hardness shall not exceed base metal hardness by more than 2 points (Rockwell C equivalent), and in the case of classes 10.9 and 12.9 shall also not exceed 59 HR30N and 63 HR30N, respectively.
7. Caution is advised when considering the use of class 12.9 bolts and screws. Capability of the bolt manufacturer, as well as the anticipated in-use environment, should be considered. High strength products such as class 12.9, require rigid control of heat treating operations and careful monitoring of as-quenched hardness, surface discontinuities, depth of partial decarburization, and freedom from carburization. Some environments may cause stress corrosion cracking of non-plated as well as electroplated products.
8. Minimum hardness requirement is waived if minimum tensile strength is met.

(f) Class 12.9 has properties approximately equivalent to ASTM A574.

2.4.2 Note that class 8.8 is applicable to sizes above 16 mm, and class 9.8 is applicable to sizes 16 mm and smaller.

3. Materials and Processes

3.1 Steel Characteristics—Bolts, screws, and studs shall be made of steel conforming to the description and chemical composition requirements specified in Table 2 for the applicable property class.

3.2 Heading Practice

3.2.1 Methods other than upsetting and/or extrusion are permitted only by special agreement between purchaser and manufacturer.

3.2.2 Class 4.6 may be hot or cold headed at option of the manufacturer.

3.2.3 Class 4.8, 5.8, 8.8, 9.8, 10.9, and 12.9 bolts and screws in sizes up to M20 inclusive, and lengths up to 10 times the nominal product size or 150 mm, whichever is shorter, shall be cold headed, except that they may be hot headed by special agreement of the purchaser. Larger sizes and longer lengths may be cold or hot headed at option of the manufacturer.

3.3 Threading Practice—Class 4.8, 5.8, 8.8, 9.8, 10.9, and 12.9 bolts and screws in sizes up to M20 inclusive, and lengths up to 150 mm inclusive, shall be roll threaded, except by special agreement. Threads of all sizes of class 4.6 bolts and screws and class 4.8, 5.8, 8.8, 9.8, 10.9, and 12.9 bolts and screws in sizes over M20 and/or lengths longer than 150 mm may be rolled, cut or ground, at option of the manufacturer. Threads of all classes and sizes of studs may be rolled, cut, or ground at option of the manufacturer.

3.4 Heat Treatment Practice

3.4.1 Class 4.6 bolts and screws and class 4.6, 4.8, and 5.8 studs need not be heat treated. Class 4.8 and 5.8 bolts and screws shall be stress relieved if necessary to assure the soundness of the head to shank junction. When specified by the purchaser, class 5.8 bolts and screws shall be stress relieved at a minimum stress relief temperature of 470°C. Where higher temperatures are necessary to relieve stresses in severely upset heads, mechanical requirements shall be agreed upon by manufacturer and purchaser.

3.4.2 Class 8.8 and 9.8 bolts, screws, and studs shall be heat treated, quenched in oil or water-base quenchant at the option of the manufacturer, and tempered at a minimum tempering temperature of 425°C for class 8.8 and 410°C for class 9.8. For class 9.8 screw and washer assemblies (sems), quenchants whose principal constituent is water shall NOT be used, and tempering temperature shall be no less than 340°C. See also Section 3.4.6.

3.4.3 Medium carbon alloy steel class 10.9 bolts, screws, and studs shall be heat treated, oil quenched, and tempered at a minimum tempering temperature of 425°C. Low carbon martensite steel class 10.9 bolts, screws, and studs shall be heat treated, quenched in oil or water-base quenchant at the option of the manufacturer, and tempered at a minimum tempering temperature of 340°C. See also Section 3.4.6.

3.4.4 Class 12.9 bolts, screws, and studs shall be heat treated, oil quenched, and tempered at a minimum tempering temperature of 380°C. See also Section 3.4.6.

3.4.5 Under no circumstances should heat treatment or carbon restoration be accomplished in the presence of nitrogen compounds, such as carbonitriding or cyaniding.

3.4.6 Tempering Temperature Audit Test (for checking whether products have been tempered at specified temperature). Conduct hardness test (SAE J1216, Section 3.1) on one or more bolts, screws, or studs; retemper the product(s) at a temperature 10°C less than the specified minimum tempering temperature for 30 min; repeat product hardness test. The difference between the mean hardnesses (before and after retempering) shall be no greater than 2 points Rockwell C (approximately 20 Vicker points).

4. Mechanical and Physical Properties

4.1 Mechanical—Bolts, screws, and studs shall be tested in accordance with the mechanical testing requirements for the applicable type, property class, size, and length of product as specified in Table 3, and shall meet the mechanical requirements specified for that product in Table 1.

4.2 Decarburization—Unless otherwise specified, class 8.8, 9.8, and 10.9 products shall conform to decarburization class 1/2H, and class 12.9 products shall conform to decarburization class 3/4H as specified in SAE J121. In addition for class 12.9, no gross decarburization is permitted on product threads.

4.3 Surface Discontinuities

4.3.1 Bolts, screws, and studs of classes 8.8, 9.8, and 10.9 in sizes up to M24 inclusive, and lengths up to 150 mm inclusive, shall not have surface discontinuities exceeding the limits specified in SAE J1061.

Surface discontinuities for sizes and lengths of products not covered in the scope of SAE J1061 shall be within limits specified by purchaser.

4.3.2 When the engineering requirements of the application necessitate that surface discontinuities must be more closely controlled, such as for class 12.9 products, the purchaser shall specify the applicable limits in the original inquiry and purchase order. For certain fasteners, this may be done by reference to SAE J123.

5. Methods of Test

5.1 General—Procedures for conducting the tests to determine the mechanical properties as specified in Table 3 for the applicable product, property class, size, and length are given in SAE J1216. Table 3 specifies the applicable test method to be followed when determining each mechanical property.

6. Marking

6.1 Bolts and Screws—Slotted and cross-recessed screws of all sizes and other screws and bolts of sizes smaller than M5 need not be marked. All other bolts and screws of sizes M5 and larger shall be marked permanently and

TABLE 2—CHEMICAL COMPOSITION REQUIREMENTS
Product Analysis (% by mass)

Property Classes 4.6 and 4.8
    Manufacturer's option—
        Low or medium carbon steels (for all sizes), within following limits:
            C 0.55 max, P 0.048 max, S 0.058 max

Property Class 5.8
    Manufacturer's option—
        Low or medium carbon steels (for all sizes), within following limits:
            C 0.13—0.55, P 0.048 max, S 0.058 max
            For studs only, sulfur content may be 0.33 max

Property Class 8.8
    Manufacturer's option—
        Medium carbon steels (for all sizes), within following limits:
            C 0.28—0.55, P 0.048 max, S 0.058 max
            For studs only, sulfur content may be 0.13 max
        Medium carbon alloy steels (for sizes <u>over</u> M24), within following limits:
            C 0.28—0.55, P 0.040 max, S 0.045 max

    When authorized by purchaser—
        Low carbon martensite steels (for sizes thru M20), within following limits:
            C 0.15—0.27, Mn 0.74—1.46, P 0.038 max, S 0.048 max, B 0.0005—0.003 (See Note 1 below)
        Medium carbon boron steels (for sizes thru M24), within following limits:
            C 0.25—0.40, Mn 0.74 min, P 0.048 max, S 0.058 max, B 0.0005—0.003 (See Note 1 below)

Property Class 9.8
    Medium carbon steels (for all sizes), within following limits:
        C 0.28—0.55, P 0.048 max, S 0.058 max
            For studs only, sulfur content may be 0.13 max
            For screw and washer assemblies (sems) and for products manufactured
            same as sems except without washer, sizes thru M12 only, carbon content
            may be 0.15—0.40 (See Note 2 Table 6)

    When authorized by purchaser—
        Low carbon martensite steels (for sizes thru M20), within following limits:
            C 0.15—0.27, Mn 0.74—1.46, P 0.038 max, S 0.048 max, B 0.0005—0.003 (See Note 1 below)
        Medium carbon boron steels (for sizes thru M24), within following limits:
            C 0.25—0.40, Mn 0.74 min, P 0.048 max, S 0.058 max, B 0.0005—0.003 (See Note 1 below)

Property Class 10.9
    Manufacturer's option—
        Medium carbon alloy steels (for all sizes), within following limits:
            C 0.28—0.55, P 0.040 max, S 0.045 max
            Fine grain
            Hardenability—47 min HRC (See Note 2 below)
        SAE 1541 or SAE 1541H (for sizes thru M12)
            Fine grain
    When authorized by purchaser—
        Carbon steels (for sizes thru M20); Fine grain
        Low carbon martensite steels (for sizes thru M20), within following limits:
            C 0.15—0.27, Mn 0.74—1.46, P 0.038 max, S 0.048 max, B 0.0005—0.003 (See Note 1 below)
            Hardenability—40 min HRC (See Note 2 below)
        Medium carbon boron steels (for sizes thru M24), within following limits:
            C 0.25—0.40, Mn 0.74 min, P 0.048 max, S 0.058 max, B 0.0005—0.003 (See Note 1 below)
            Hardenability—47 min HRC (See Note 2 below)

Property Class 12.9
    Alloy steels (for all sizes), within following limits:
        C 0.31—0.65, P 0.045 max, S 0.045 max
            One or more of the alloying elements chromium,
            nickel, molybdenum, or vandium shall be present in
            sufficient quantity to insure that the specified
            strength properties are met after quenching and
            tempering.
        Fine grain
        Hardenability—47 min HRC (See Note 2 below)

NOTES: 1. Products made from low carbon martensite steels and medium carbon boron steels shall be identified as specified in Table 6, Note (1).
    2. Steels shall have hardenability that is capable of producing the minimum hardness (Rockwell C) shown at the center of a transverse section one nominal diameter from the threaded end of bolt, screw, or stud (after quenching).

TABLE 3—MECHANICAL TESTING REQUIREMENTS FOR BOLTS, SCREWS, AND STUDS

Product	Property Class	Specified Min Tensile Strength of Product (See Table 5) kN	Length of Product (2)	Product Hardness Max	Product Hardness Min	Surface Hardness (4) Max	Tests Conducted Using Full Size Products — Proof Load	Tests Conducted Using Full Size Products — Wedge Tensile Strength (5)	Tests Conducted Using Full Size Products — Axial Tensile Strength	Tests Conducted Using Machined Test Specimens — Yield Strength	Tests Conducted Using Machined Test Specimens — Tensile Strength	Tests Conducted Using Machined Test Specimens — Elongation	Tests Conducted Using Machined Test Specimens — Reduction of Area	Decarburization in Threaded Section (4)
Short bolts and screws	all	all	less than 2 1/4D	•	•	•	—	—	—	—	—	—	—	○
Special head bolts and screws (3)	all	all	all	•	•	•	—	—	—	—	—	—	—	○
Hex bolts and screws (6) (7)	all	450 and less	2 1/4D to 8D or 200 mm, whichever is greater	•	—	•	○	•	—	—	—	—	—	○
Hex bolts and screws (6) (7)	all	450 and less	over 8D or 200 mm, whichever is greater thru and incl 300 mm	•	—	•	○	•	—	—	—	—	—	○
Hex bolts and screws (6) (7)	all	450 and less	over 300 mm	•	—	•	○	A	—	B	B	B	B	○
Hex bolts and screws (6) (7)	all	over 450	2 1/4D and longer	•	—	•	○	A	—	B	B	B	B	○
All other bolts and screws	all	450 and less	2 1/4D to 8D or 200 mm, whichever is greater	•	—	•	○	—	•	—	—	—	—	○
All other bolts and screws	all	450 and less	over 8D or 200 mm, whichever is greater	•	—	•	○	—	A	B	B	B	B	○
All other bolts and screws	all	over 450	2 1/4D and longer	•	—	•	○	—	A	B	B	B	B	○
Short studs	all	all	less than 2 1/4D	•	•	•	—	—	—	—	—	—	—	○
All other studs	all	450 and less	2 1/4D to 8D or 200 mm, whichever is greater	•	—	•	○	—	•	—	—	—	—	○
All other studs	all	450 and less	over 8D or 200 mm, whichever is greater	•	—	•	○	—	A	B	B	B	B	○
All other studs	all	over 450	2 1/4D and longer	•	—	•	○	—	A	B	B	B	B	○
Tests to be conducted in accordance with paragraph				3.1	3.2	3.3	3.6	3.5	See SAE J1216	3.7				See SAE J121

NOTES:
1. • denotes a mandatory test. For each product all mandatory tests (•) shall be performed. In addition, either all tests denoted A (which apply to full size products) or all tests denoted B (which apply to machined test specimens) shall be performed; except optional B tests are not applicable to products M24 and smaller. ○ denotes tests to be performed when specifically required in the original inquiry and purchase order. In case arbitration is necessary, A tests shall be performed. Dashes (—) indicate tests which are not required.
2. D equals nominal diameter of product. For purposes of Table 3 requirements, ''length of product'' is the nominal length including point chamfer as defined in SAE J105, and all special point products shall be measured from the bearing surface to the crest of the last complete thread form.
3. Special head bolts and screws are those with special configurations or with drilled heads which are weaker than the threaded section.
4. Surface hardness and decarburization requirements apply only to property classes 8.8, 9.8, 10.9, and 12.9.
5. Tensile test wedge angles are specified in Table 4.
6. Includes flange, washer, and other hex head configurations which are not weaker than the threaded section.
7. Includes class 9.8 sems and 9.8 products manufactured same as sems except without washer (sizes M1.6 thru M12). For purposes of determining applicability of tensile testing, length of sems is the distance measured from the underside of bearing plane of the unflattened washer to the last full thread of the screw.

TABLE 4—TENSILE TEST WEDGE ANGLES

Product	Property Class	Nominal Dia	Wedge Angle Deg
Hex and hex washer head machine screws	4.8, 9.8	thru M10	6
Hex bolts and screws threaded 2D and closer to underside of head	12.9	thru M20	4
Hex bolts and screws threaded 2D and closer to underside of head	12.9	over M20 to M36	4
All other hex bolts and screws	12.9	thru M20	6
All other hex bolts and screws	12.9	over M20 to M36	4
Hex bolts and screws threaded 1D and closer to underside of head	8.8, 9.8, 10.9	thru M20	6
Hex bolts and screws threaded 1D and closer to underside of head	8.8, 9.8, 10.9	over M20 to M36	4
Hex flange and hex washer head bolts and screws	4.6, 4.8, 5.8, 8.8, 9.8, 10.9	thru M36	6
All other hex bolts and screws	4.6, 4.8, 5.8, 8.8, 9.8, 10.9	thru M24	10
All other hex bolts and screws	4.6, 4.8, 5.8, 8.8, 9.8, 10.9	over M24 to M36	6
Studs	All	thru M20	6
Studs	All	Over M20 to M36	4

TABLE 5—PROOF LOAD AND TENSILE STRENGTH VALUES (1)

Nominal Thread Dia and Thread Pitch	Stress Area (2) mm²	Class 4.6		Class 4.8		Class 5.8		Class 8.8		Class 9.8		Class 10.9		Class 12.9	
		Proof Load kN	Tensile Strength Min kN	Proof Load kN	Tensile Strength Min kN	Proof Load kN	Tensile Strength Min kN	Proof Load kN	Tensile Strength Min kN	Proof Load kN	Tensile Strength Min kN	Proof Load kN	Tensile Strength Min kN	Proof Load kN	Tensile Strength Min kN
M1.6 x 0.35	1.27			0.39	0.53					0.83	1.14			1.23	1.55
M2 x 0.4	2.07			0.64	0.87					1.35	1.86			2.01	2.53
M2.5 x 0.45	3.39			1.05	1.42					2.20	3.05			3.29	4.14
M3 x 0.5	5.03			1.56	2.11					3.27	4.53			4.88	6.14
M3.5 x 0.6	6.78			2.10	2.85					4.41	6.10			6.58	8.27
M4 x 0.7	8.78			2.72	3.69					5.71	7.90			8.52	10.7
M5 x 0.8	14.2	3.20	5.68	4.40	5.96	5.40	7.38			9.23	12.8	11.8	14.8	13.8	17.3
M6 x 1	20.1	4.52	8.04	6.23	8.44	7.64	10.4			13.1	18.1	16.7	20.9	19.5	24.5
M8 x 1.25	36.6	8.24	14.6	11.3	15.4	13.9	19.0			23.8	32.9	30.4	38.1	35.5	44.6
M10 x 1.5	58.0	13.1	23.2	18.0	24.4	22.0	30.2			37.7	52.2	48.1	60.3	56.3	70.8
M12 x 1.75	84.3	19.0	33.7	26.1	35.4	32.0	43.8			54.8	75.9	70.0	87.7	81.8	103
M14 x 2	115	25.9	46.0	35.7	48.3	43.7	59.8			74.8	104	95.4	120	112	140
M16 x 2	157	35.3	62.8	48.7	65.9	59.7	81.6			102	141	130	163	152	192
M20 x 2.5	245	55.1	98.0			93.1	127	147	203			203	255	238	299
M24 x 3	353	79.4	141			134	184	212	293			293	367	342	431
M30 x 3.5	561	126	224					337	466			466	583	544	684
M36 x 4	817	184	327					490	678			678	850	792	997

NOTES: 1. Proof loads and tensile strengths are computed by multiplying the stresses given in Table 1 by the stress area of the thread.
2. Stress area = $0.7854(D-0.9382P)^2$ where D is nominal thread diameter in mm and P is thread pitch in mm.

clearly to identify the property class and the manufacturer. The property class symbols shall be as given in Table 6; the symbol for the manufacturer's identification shall be at his option. Markings shall be located on the top of the head of bolts and screws, and may be either raised or depressed at option of the manufacturer. Alternatively, for hex head products, the markings may be indented on the side of the head.

Property class marking shall conform to the following:

Bolt or Screw Size mm	Height of Symbol mm
5 thru 6	1.5 min
8 thru 10	2.3 min
12 and 14	3.2 min
16 and larger	4.0 min

Metric bolts and screws shall not be marked with radial line symbols.

6.2 Studs—All studs of sizes M5 and larger shall be marked permanently and clearly to identify the property class. The symbols used shall be as given in Table 6. Markings shall be located on the extreme end of the stud, and may be raised or depressed at the option of the manufacturer. For studs with an interference fit thread, the markings shall be located at the nut end. Studs of sizes smaller than M12 may be marked using the property class symbols given in Table 6.

TABLE 6—PROPERTY CLASS IDENTIFICATION SYMBOLS

Property Class	Identification Symbol	
	Bolts, Screws, and Studs Sizes M5 and Larger	Optional for Studs Sizes M5 thru M11
4.6	4.6	
4.8	4.8	
5.8	5.8	
8.8 (1)	8.8	○
9.8 (1) (2)	9.8	+
10.9 (1)	10.9	□
12.9	12.9	△

NOTES: 1. Products made of low carbon martensite steel shall be additionally identified by underlining the numerals.
2. Products manufactured same as sems except without washer, from steel having optional carbon content permitted in Table 2, shall be additionally identified with an inverted T located between the numerals 9.8 as follows: 9|8.

APPENDIX

A1. Preface—This Appendix covers the mechanical and material requirements for high strength, high ductility, heat treated boron steel bolts, screws, and studs. The range of steels permitted is limited to those which have been used on a production basis with highly successful results. There is much evidence that other steels are satisfactory also, but these are excluded until more widespread experience is had with them. The data available on elevated temperature properties of low carbon martensite steels (and of many other steels) indicates that performance testing is desirable in applications where the operating temperature exceeds 260°C.

A2. Requirements

Property Class Designation —————— 12.8

Applicable to ————— Bolts, screws, and studs; sizes thru 20 mm

Mechanical and Physical Requirements (See Section 4.1 and below)
Proof Load ————— 940 MPa
Tensile Strength (Stress), Min ————— 1220 MPa
Product Hardness, Rockwell ————— 38-43 HRC
  Vickers ————— 372-423 HV
Surface Hardness ————— See Note
Decarburization (See Section 4.2) ————— Same as Class 10.9
Surface Discontinuities (See Section 4.3) ————— Same as Class 10.9
Tensile Strength, kN (See Table 5) ————— Same as Class 12.9
Proof Load Strength (See Below)

Nom Thd Dia and Pitch	kN	Nom Thd Dia and Pitch	kN
M1.6 x 0.35	1.19	M6 x 1	18.9
M2 x 0.4	1.95	M8 x 1.25	34.4
M2.5 x 0.45	3.19	M10 x 1.5	54.5
M3 x 0.5	4.73	M12 x 1.75	79
M3.5 x 0.6	6.37	M14 x 2	108
M4 x 0.7	8.25	M16 x 2	148
M5 x 0.8	13.4	M20 x 2.5	230

Materials and Processes
  Heading Practice (See Sections 3.2 and 3.2.3) — Same as Class 10.9
  Threading Practice (See Section 3.3) ————— Same as Class 10.9
  Heat Treatment Practice
    Class 12.8 products shall be heat treated in a *continuous type* furnace having a protective atmosphere, quenched in a suitable medium, and tempered at a minimum temperature of 200°C. Under no circumstances should heat treatment or carbon restoration be accomplished in the presence of nitrogen compounds, such as carbonitriding or cyaniding.
Steel Characteristics
    Class 12.8 products shall be made from low carbon martensite (boron)

steel conforming to the following description and chemical composition (product analysis, per cent mass):

Carbon	0.16–0.27
Manganese	0.74–1.46
Silicon	0.32 max
Phosphorus	0.038 max
Sulfur	0.048 max
Boron	0.0005–0.003

Steel shall be fine grain, fully killed, with hardenability that is capable of producing a low carbon martensite with minimum hardness of 40 HRC at the center of a transverse section one nominal diameter from the threaded end of bolt, screw, or stud (after quenching).

Marking (See Sections 6.1 and 6.2)

Identification symbols for class 12.8 shall be **12.8**.

NOTE: Surface hardness shall not exceed base metal hardness by more than 2 points (HRC equivalent) and shall not exceed 82 HR15N (43 HRC equivalent).

TEST METHODS FOR METRIC THREADED FASTENERS—SAE J1216

SAE Standard

Report of Iron and Steel Technical Committee approved March 1978.

1. Scope

1.1 This standard establishes procedures for conducting tests to determine the mechanical properties of externally and internally threaded fasteners.

1.2 Property requirements and the applicable tests for their determination are specified in individual product standards. In those instances where the testing requirements are unique or at variance with these standard procedures, the product standard shall specify the controlling testing requirements.

1.3 This standard describes mechanical tests for determining the following properties:

For externally threaded fasteners:
Product hardness (Section 3.1)
Surface hardness (Section 3.2)
Proof load (Section 3.3)
Yield strength (Section 3.4)
Axial tensile strength (Section 3.5)
Wedge tensile strength (Section 3.6)
Elongation (Section 3.7)
Reduction of area (Section 3.7)

For internally threaded fasteners:
Product hardness (Section 4.1)
Proof load (Section 4.2)

2. General Precautions

2.1 Improper machining or preparation of test specimens may give erroneous results. Care should be exercised to assure good workmanship in machining. Improperly machined specimens should be discarded and other specimens substituted.

2.2 If any test specimen fails because of mechanical reasons such as failure of testing equipment or improper specimen preparation, it shall be discarded and another test specimen substituted.

3. Test Methods for Externally Threaded Fasteners

3.1 Product Hardness—For routine inspection, hardness of bolts, screws, and studs may be determined on head, end, or shank after removal of any plating or other coating and after suitable preparation of the specimen. For referee purposes, the hardness shall be determined at mid-radius of a transverse section through the threaded portion of the product taken at a distance of one diameter from the end of the product. For sizes smaller than M6 the reported hardness shall be the average of two hardness readings. For sizes M6 and larger, the reported hardness shall be the average of four hardness readings located at 90 deg to one another. The preparation of test specimens and the performance of hardness tests shall be in conformity with the requirements of SAE J417.

3.2 Surface Hardness—Tests to determine surface hardness conditions shall be conducted on the ends, hexagon flats, or unthreaded shanks which have been prepared by lightly grinding or polishing to ensure accurate reproducible readings in accordance with SAE J417. Proper correction factors shall be used when hardness tests are made on curved surfaces, per ASTM E18.

3.3 Proof Load

3.3.1 The proof load of a bolt, screw, or stud is the axially applied load which the product must withstand without permanent set.

3.3.2 The overall length of the specimen shall be measured between conical or ball centers on the centerline of the specimen, using mating centers on the measuring anvils. The specimen shall be marked so that it can be placed in the measuring fixture in the same position for all measurements. The measurement instrument shall be capable of measurement to 2.5 μm.

3.3.3 The grips of the testing machine shall be self-aligning to avoid side thrust on the specimen. For bolts and screws the specimen shall be assembled in the fixture of the tensile machine so that six complete threads are exposed between the grips. This is obtained by freely running the nut or fixture to the thread runout of the specimen and then unscrewing the specimen six full turns (two turns for products less than 3 D in length and threaded to within 2.5 thread pitches of the bearing surface). Typical fixturing is illustrated in Fig. 1. When proof load testing sems, the washer may be removed prior to testing; however, for referee testing the washer shall be removed. When proof load testing studs one end of the specimen shall be assembled in a threaded fixture to the thread runout. For studs having unlike threads, this shall be the end with the finer pitch thread. For studs having unlike diameters (step studs), this shall be the end with the larger minor diameter. The other end of the stud shall likewise be assembled in a threaded fixture, except it shall be unscrewed six full turns from the thread runout, thus leaving six complete threads exposed between the grips (two turns for products less than 3 D in length and threaded to within 2.5 thread pitches of the bearing surface).

3.3.4 The bolt, screw, or stud shall then be axially loaded to the proof load specified for the applicable size and class in the product standard, the load retained for a period of 15 s, the load removed, and the overall length again measured. The speed of testing, as determined with a free running cross head, shall not exceed 3 mm/min.

3.3.5 To meet requirements, the length of the specimen after loading shall be the same as before loading within a tolerance of 12.5 μm allowed for measurement error.

3.3.6 Variables, such as straightness and thread alignment (plus measurement error), may result in apparent elongation of the specimen when the proof load is initially applied. In such cases, the specimen may be retested using a 3% greater load, and may be considered satisfactory if the length after this loading is the same as before this loading within the 12.5 μm tolerance for measurement error.

3.4 Yield Strength

3.4.1 The yield strength of a bolt, screw, or stud is the stress at which the fastener exhibits a specified limiting deviation from the proportionality of stress to strain. The deviation is expressed in terms of strain.

3.4.2 The yield strength test shall be set up as outlined in Section 3.3.3 using, however a machined product as shown in Fig. 4. The speed of testing (determined with a free running cross head), shall not exceed 3 mm/min. Load shall be applied axially and as the load is applied the total elongation of the specimen or any part of the specimen which includes the six exposed threads shall be measured. Yield strength shall be determined by the offset method.

3.4.3 It is necessary to obtain data (autographic or numerical) from which a stress-strain diagram may be drawn. On the stress-strain diagram an offset line is then drawn parallel to the modulus line at a distance equal to the specified offset from the modulus line (for the specified offset refer to the product

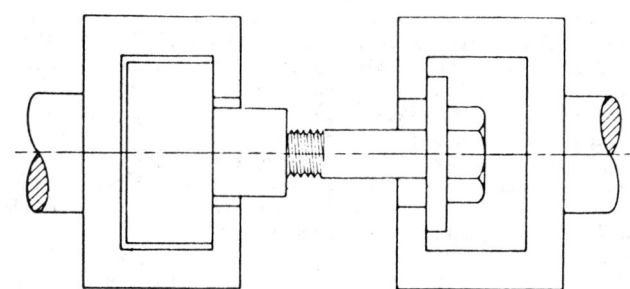

FIG. 1—TENSILE TESTING OF FULL SIZE BOLT OR SCREW

standard). The yield strength load is the load at which the offset line intersects the stress-strain diagram. Yield strength is then obtained by dividing the yield strength load by the original nominal tensile stress area of the exposed threaded section.

3.4.4 To meet requirements, the yield strength shall not be less than the minimum yield strength specified in the product standard.

3.5 Axial Tensile Strength

3.5.1 The axial tensile strength of a bolt, screw, or stud is the maximum tensile stress which the fastener is capable of sustaining when axially loaded. Tensile strength is calculated from the original nominal tensile stress area of the threaded section and the maximum load occurring during a tensile test.

3.5.2 When bolt, screw, or stud is to be proof load or yield strength as well as axial tensile tested, both tests may be conducted on the same specimen.

3.5.3 The axial tensile test shall be set-up as outlined in Section 3.3.3. Loading shall be continued until fracture occurs. The speed of testing, as determined with a free running cross head, shall not exceed 25 mm/min.

3.5.4 When tensile testing sems, the washer may be removed prior to testing; however, for referee testing the washer shall be removed.

3.5.5 To meet requirements, the specimen shall support a load, prior to fracture, not less than the minimum tensile strength specified for the applicable size and class in the product standard. In addition, for bolts and screws, (except flat and oval head machine screws) the fracture shall occur in the body or threaded section with no failure at the junction of the head and shank.

3.6 Wedge Tensile Strength—The wedge tensile strength of a bolt, screw, or stud is the maximum tensile stress which the fastener is capable of sustaining when loaded eccentrically. Wedge tensile strength is calculated as specified in Section 3.5.1.

3.6.1 BOLTS AND SCREWS

3.6.1.1 When bolt or screw is to be proof load tested as well as wedge tensile tested the same specimen may be used for both tests. This shall be assembled with a wedge inserted under the head, as illustrated in Fig. 2, installed in the testing machine and tensile tested to failure, as described in Section 3.3.3. The angle of the wedge for the bolt or screw type, size and class is specified in the product standard. The wedge shall be so placed that no corner of a hexagon head takes the bearing load; that is, a flat of the head shall be aligned with the direction of uniform thickness of the wedge. When wedge tensile testing sems, the washer may be removed prior to testing; however, for referee testing the washer shall be removed. The wedge shall have a minimum hardness of Rockwell C48. The wedge shall have a thickness of one-half the bolt or screw diameter measured at the thin side of the hole. The hole in the wedge shall have the following clearance over the nominal size of the bolt or screw, and its top and bottom edges shall be rounded or chamfered 45 deg to the following dimensions:

Nominal Product Size	Clearance in Hole approx mm	Radius or Depth of Chamfer approx mm
thru M6	0.5	0.7
Over M6 thru M12	0.8	0.8
Over M12 thru M20	1.6	1.3
Over M20 thru M39	3.2	1.6

3.6.1.2 To meet requirements, the specimen shall support a load, prior to fracture, not less than the minimum tensile strength specified for the applicable size and class in the product standard. In addition, the fracture shall occur in the body or threaded section with no failure at the junction of head and shank.

3.6.2 STUDS

3.6.2.1 When stud is to be proof load tested as well as wedge tensile tested the same specimen may be used for both tests. One end of stud shall be assembled in a threaded fixture to the thread runout. For studs having unlike threads, this shall be the end with the finer pitch thread. For studs having unlike diameters (step studs), this shall be the end with the larger minor diameter. The other end of the specimen shall be assembled in a threaded wedge to the runout and then unscrewed six full turns (two turns for products less than 3 D in length and threaded to within 2.5 thread pitches of the bearing surface), thus leaving six (or two) complete threads exposed between the grips, as illustrated in Fig. 3. The angle of the wedge for the stud size and class shall be as specified in the product standard. The stud shall be assembled in the testing machine and tensile tested to failure, as described in Section 3.3.3.

3.6.2.2 The minimum hardness of the threaded wedge shall be 48 HRC. The length of the threaded section of the wedge shall be equal to the diameter of the stud. To facilitate removal of the broken stud, the wedge shall be

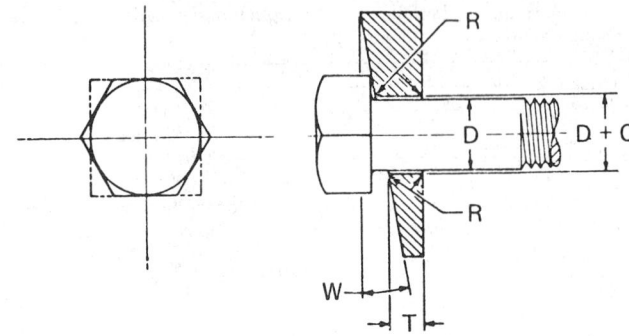

C = Clearance of hole (see para. 3.6.1.1)
D = Diameter of bolt or screw
R = Radius or chamfer (see para. 3.6.1.1)
T = Thickness of wedge at thin side of hole equals one half diameter of bolt or screw
W = Wedge angle (given in product standard)

FIG. 2—WEDGE TEST DETAILS—BOLTS AND SCREWS

counterbored. The thickness of the wedge at the thin side of the hole shall equal the diameter of the stud plus the depth of counterbore. The thread in the wedge shall have Grade 4H tolerances in accordance with ISO 965/1, except when testing studs having an interference fit thread, in which case the wedge shall be threaded to provide a finger free fit. The supporting fixture, as shown in Fig. 3, shall have a hole clearance over the nominal size of the stud, and shall have its top and bottom edges rounded or chamfered to the same limits specified for the hardened wedge in Section 3.6.1.1.

3.6.2.3 To meet requirements, the specimen shall support a load, prior to fracture, not less than the minimum tensile strength specified for the applicable size and class in the product standard.

3.7 Testing of Machined Test Specimens

3.7.1 Where product specifications specify or authorize the testing of specimens machined from bolts, screws, or studs (normally nominal diameters over M24), standard specimens shall be turned from the product, as shown in Fig. 4.

3.7.2 The test specimen shall be axially tensile tested and the yield strength, tensile strength, elongation, and reduction of area shall be determined.

3.7.3 To meet requirements, the test specimen must have a yield strength, tensile strength, elongation, and reduction of area equal to or greater than the values for these properties specified for the applicable size and class in the product standard.

4. Test Methods for Internally Threaded Fasteners

4.1 Product Hardness

4.1.1 Hardness shall be determined on a polished surface located on the nut face halfway between the major diameter of the thread and a corner, or, when

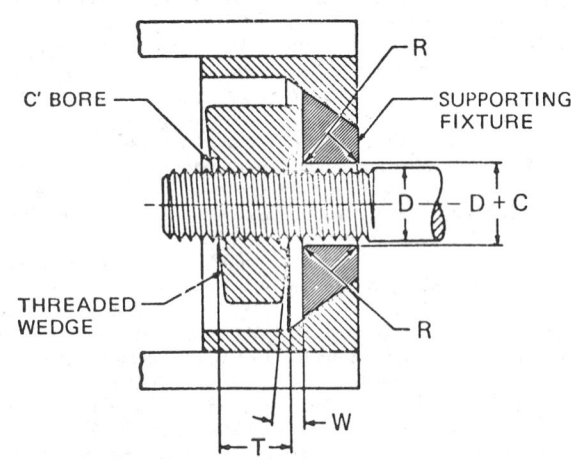

C = Clearance of hole (see para. 3.6.1.1)
D = Diameter of stud
R = Radius or chamfer (see para. 3.6.1.1)
T = D plus depth of counterbore
W = Wedge angle (given in product standard)

FIG. 3—WEDGE TEST DETAILS—STUDS

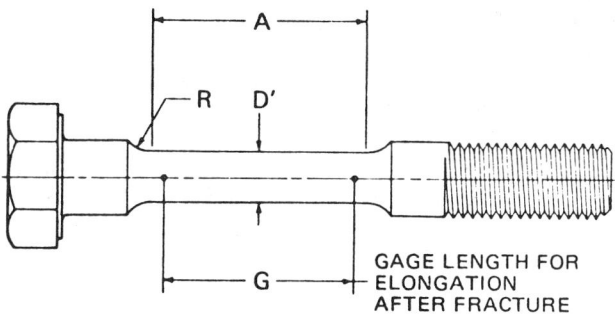

- D' = Diameter of turned down shank (D' < minor dia of product thread).
- G = Gage length = 5D'
- A = Length of turned down shank = G + D'
- R = Fillet radius (R ≥ 4 mm)

The reduction of the shank diameter of bolts, screws or studs should not exceed 25 percent of the original nominal diameter of the product. This results in a cross sectional area of the turned down section of approximately 56 percent of the original nominal diameter cross sectional area.

FIG. 4—TENSILE TEST SPECIMEN FOR BOLTS OR SCREWS WITH TURNED DOWN SHANK

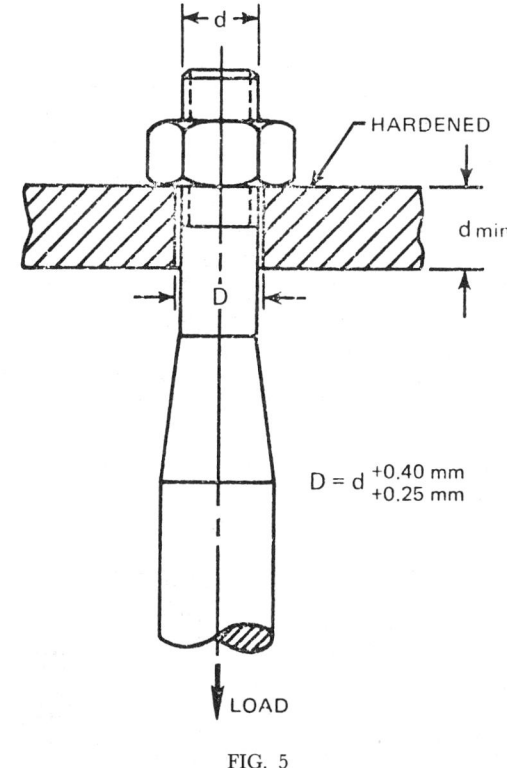

FIG. 5

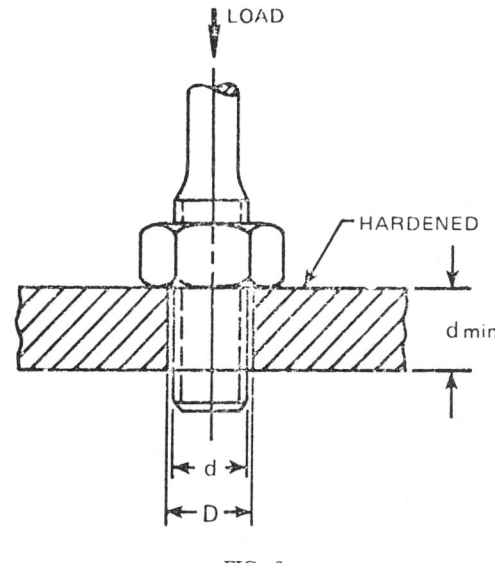

FIG. 6

practicable, on a wrench face one-third of the distance from a corner to the center of the wrench face. In preparing the surface, sufficient material shall be removed to assure elimination of any decarburization or other surface irregularities. Rockwell hardness testing shall be performed in accordance with the provisions of ISO R80, and Vickers hardness in accordance with the provisions of ISO R81.

4.2 Proof Load

4.2.1 The proof load of a nut is the axially applied load the nut must withstand without thread stripping or rupture.

4.2.2 The nut shall be assembled on a test bolt or on a hardened and threaded mandrel, as specified below and as illustrated in Fig. 5 or Fig. 6. The proof load as specified for the nut in the product standard shall be applied against the nut in an axial direction. For referee testing, the mandrel shall be used with tension loading. If the threads of the test bolt or mandrel are damaged during the test, the test shall be discarded.

4.2.3 To meet requirements, the nuts shall resist this load without failure by stripping or rupture, and shall be removable from the test bolt or mandrel by the fingers after the load is released. (NOTE: Occasionally it may be necessary to use a manual wrench or other means to start the nut in motion. Use of such means is permissible, providing the nut is removable by the fingers following an initial loosening of not more than one-half turn of the nut.)

4.2.4 Test bolts shall have 6 g threads, as specified in ISO 965, Part I, and shall have a yield strength in excess of the specified proof load of the nut being tested.

4.2.5 Mandrels shall have a hardness of Rockwell C45 minimum; and shall have 4 g 6 g threads, as specified in ISO 965, Part I, except the minimum major diameter shall be the specified minimum major diameter, and the maximum major diameter, shall be minimum major diameter plus 0.25 times the major diameter tolerance.

MECHANICAL AND QUALITY REQUIREMENTS FOR MACHINE SCREWS—SAE J82 JUN79

SAE Standard

Report of the Iron and Steel Technical Committee and Fasteners Committee, approved July 1972, editorial change June 1979.

1. **Scope**—This SAE Standard covers the mechanical and quality requirements for two grades of carbon steel, slotted and recessed, 82 deg flat countersunk, 82 deg oval countersunk, pan, fillister, hex, and hex washer head machine screws in sizes No. 4 through 3/4 in. for use in automotive and related industries. The dimensions of these screws are covered in SAE J478.

2. **Designations**—The two grades of machine screws are designated Grade 60M and Grade 120M, indicating 60,000 and 120,000 psi minimum tensile strength, respectively.

3. **Materials and Processes**

    3.1 Steel Characteristics—Machine screws shall be made of steel conforming to the description and chemical composition requirements specified in Table 1 for the applicable grade.

    3.2 Heading Practice—Machine screws shall be cold headed and/or extruded, unless other methods are permitted by special agreement of the purchaser.

    3.3 Threading Practice—Machine screws shall be roll threaded, except by special agreement of purchaser.

    3.4 Heat Treatment Practice—Grade 60M machine screws need not be heat treated. When specified by purchaser, Grade 60M screws shall be stress relieved. Grade 120M machine screws shall be heat treated, oil or water quenched and tempered at a minimum tempering temperature of 650 F (343 C).

    3.5 Finish—Unless otherwise specified, machine screws shall be supplied with a natural (as processed) finish unplated or uncoated. Plated and coated finishes shall be supplied in accordance with requirements of the purchaser.

    NOTE: Class 2A allowance in sizes No. 8 and smaller may not accommodate a commercial thickness of 0.00015 in minimum. To accommodate this commercial thickness on these smaller size screws, the before-plating size may have to be reduced. Any such reduction will affect strength properties. When necessary to maintain Class 2A limits after plating of any size screw, Class 2AG shall be specified.

4. **Mechanical Requirements**

    4.1 Hardness—Machine screws shall have a hardness not in excess of the maximum specified in Table 2. Screws which are excepted from tensile testing in accordance with paragraphs 4.2.1 and 4.2.2 shall have a hardness not less than the minimum and not more than the maximum specified in Table 2.

    4.2 Tensile Strength

    4.2.1 HEX AND HEX WASHER HEAD MACHINE SCREWS—No. 4 and No. 5 hex and hex washer head machine screws which are shorter than 0.50 in. are not subject to tensile testing. No. 4 and No. 5 hex and hex washer head machine screws 0.50 in. and longer shall meet the tensile load requirements specified in Table 3 when axially tensile tested in accordance with paragraph 5.2.

    Hex and hex washer head machine screws in sizes No. 6 to 3/4 in. inclusive, which are shorter than either 0.50 in. or 3D (where D is nominal screw size in inches) are not subject to tensile testing. Hex and hex washer head machine screws in these sizes and with a length that is both equal to or longer than 0.50 in., and also at least 3D, shall meet the tensile load requirements specified in Table 3 when wedge tensile tested in accordance with paragraph 5.3.

    4.2.2 OTHER MACHINE SCREWS—Machine screws with head styles other than hex or hex washer head which are shorter than 0.50 in., are not subject to tensile testing. Such machine screws 0.50 in. and longer shall meet the tensile load requirements specified in Table 3 when axially tensile tested in accordance with paragraph 5.2.

5. **Methods of Test**

    5.1 Hardness—The hardness shall be determined at mid-radius of a transverse section through the screw taken at a distance of one diameter from the end of the screw. For screws smaller than No. 10 size, a referee test may be made at mid-radius using microhardness measurement techniques.

    5.2 Axial Tensile Strength—Screws shall be assembled in a tensile testing machine with a minimum of six threads exposed, and an axial load applied against the bearing surface until failure occurs. The speed of testing as determined with a free-running cross head shall not exceed 1 in./min. The grips of the testing machine shall be self-aligning to avoid side thrust on the specimen.

    To meet the requirements of paragraphs 4.2.1 and 4.2.2, the load at failure shall not be less than the tensile load given in Table 3 for the applicable size and grade.

    5.3 Wedge Tensile Strength—Screws shall be installed in a tensile testing machine with a 6 deg wedge inserted under the head, as illustrated in Fig. 1 and tensile tested to failure as described in paragraph 5.2. The wedge shall be so placed that no corner of the hexagon screw head takes the bearing load; that is, a flat of the head shall be aligned with the direction of uniform thickness of the wedge.

    To meet the requirements of paragraph 4.2.1, the load at failure should not be less than the tensile load given in Table 3 for the applicable size and grade. In addition, failure shall occur in the body or threads with no fracture at the junction of the head and shank or failure due to any portion of the shank being pulled out of the head.

    5.3.1 Wedge may be either circular or square. See Fig. 1. Recommended outside dimension is 1.25 in. for screw sizes No. 4 through No. 12 and 1.75 in. for larger sizes. Thickness of wedge at thin size of hole shall be equivalent to one-half the nominal diameter of the screw but not less than 0.12 in. Hole shall be 0.020 in. over the nominal diameter of the screw for sizes No. 4 through No. 12, 0.030 in. over for sizes 1/4 through 1/2 in., and 0.050 in. over for sizes 9/16 through 3/4 in. Top and bottom edges of the hole shall be rounded or chamfered, with radius and depth of chamfer as follows: 0.020 in. for sizes No. 4 through No. 12, 0.030 in. for sizes 1/4 through 1/2 in., and 0.060 in. for sizes 9/16 through 3/4 in. The wedge shall have a minimum hardness of Rockwell C45.

6. **Marking**—Machine screws need not be marked to identify grade or manufacturer.

7. **Testing Requirements**

    7.1 Manufacturer's Responsibility—During the manufacture of machine

TABLE 2—MECHANICAL REQUIREMENTS

Grade	Tensile Strength, min, psi	Hardness Rockwell	
		min	max
60M	60,000	B70	B100
120M	120,000	C25	C38

TABLE 1—CHEMICAL COMPOSITION REQUIREMENTS[a]

Grade	Material and Treatment	Element, %			
		C min	C max	P, max	S, max
60M	Carbon steel	—	0.30	0.048	0.058
120M	Carbon steel, quenched and tempered	0.15	0.55	0.048	0.058

[a] All values are for check analysis (percent by weight). For ladle analysis, use standard permissible variations as shown in SAE J409.

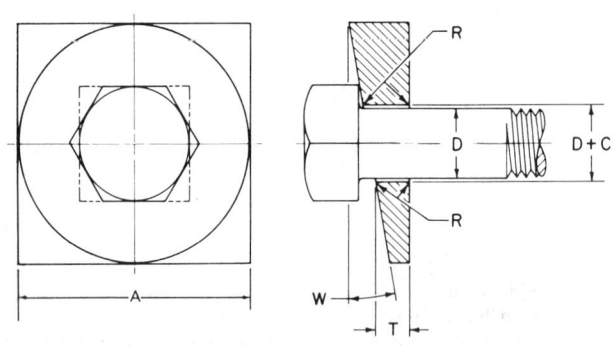

A = WEDGE OUTSIDE DIMENSIONS (SEE PARAGRAPH 5.3.1)
C = CLEARANCE HOLE (SEE PARAGRAPH 5.3.1)
D = DIAMETER OF SCREW
R = RADIUS OR CHAMFER (SEE PARAGRAPH 5.3.1)
T = THICKNESS AT THIN SIDE OF HOLE (SEE PARAGRAPH 5.3.1)
W = WEDGE ANGLE (6 DEG)

FIG. 1—WEDGE TEST DETAILS

TABLE 3—TENSILE LOAD REQUIREMENTS FOR MACHINE SCREWS

Nominal Size or Basic Major Dia of Thread and Threads per in		Stress Area, in²	Tensile Strength,[a] lb, min		Nominal Size or Basic Major Dia of Thread and Threads per in		Stress Area, in²	Tensile Strength,[a] lb, min	
			Grade 60M	Grade 120M				Grade 60M	Grade 120M
No.	4-40 0.112	0.00604	360	720	No.	5/16-18 0.312	0.0524	3,150	6,300
	4-48 0.112	0.00661	390	780		5/16-24 0.312	0.0580	3,500	6,950
	5-40 0.125	0.00796	470	940		3/8-16 0.375	0.0775	4,650	9,300
	5-44 0.125	0.00830	490	980		3/8-24 0.375	0.0878	5,250	10,500
	6-32 0.138	0.00909	550	1100		7/16-14 0.438	0.1063	6,400	12,800
	6-40 0.138	0.01015	600	1200		7/16-20 0.438	0.1187	7,100	14,200
	8-32 0.164	0.0140	850	1700		1/2-13 0.500	0.1419	8,500	17,000
	8-36 0.164	0.01474	880	1750		1/2-20 0.500	0.1599	9,600	19,200
	10-24 0.190	0.0175	1050	2100		9/16-12 0.562	0.182	10,900	21,800
	10-32 0.190	0.0200	1200	2400		9/16-18 0.562	0.203	12,200	24,400
	12-24 0.216	0.0242	1450	2900		5/8-11 0.625	0.226	13,600	27,100
	12-28 0.216	0.0258	1550	3100		5/8-18 0.625	0.256	15,400	30,700
	1/4-20 0.250	0.0318	1900	3800		3/4-10 0.750	0.334	20,100	40,100
	1/4-28 0.250	0.0364	2200	4350		3/4-16 0.750	0.373	22,400	44,800

[a]Tensile strength values for Grade 60M and Grade 120M are based on 60,000 and 120,000 psi, respectively.

screws to the requirements of this standard, the manufacturer shall make periodic tests to ensure that the properties of the screw are being maintained within the specified limits. Such tests shall be conducted in accordance with a sampling plan, preferably the sampling plan given in Table 4.

Additional tests of screws in individual shipments are not normally contemplated. Unless otherwise agreed at time of original inquiry and purchase order, individual heats of steel need not be identified.

7.2 Purchaser's Options—If the purchaser requires that additional tests be performed by the manufacturer to determine that the properties of screws in an individual shipment are within specified limits, or if the purchaser requires that a specific (that shown in Table 4 or other) sampling plan shall be used when determining the acceptability of a lot or shipment of screws, or if the purchaser requires test data to be recorded and furnished, the purchaser shall specify the complete testing requirements, including sampling plan and basis of acceptance, in the original inquiry and purchase order.

7.3 General—An acceptable sampling plan is outlined in Table 4.

A lot, for purposes of selecting test specimens, shall consist of all screws offered for inspection and testing at one time that are of the same head style, grade, size, length, and thread series and are manufactured essentially at one time and under the same process conditions.

TABLE 4

Number of Pieces In Lot	Minimum Number of Specimens To be Tested
50 and under	2
51 to 500	3
501 to 35,000	5
35,001 and over	8

The same test specimen may be used for different tests wherever practical.

When tested in accordance with this sampling plan, a lot shall be subject to rejection if any of the test specimens fail to meet the applicable test requirements. If the failure of a test specimen is due to improper preparation of the specimen or to incorrect testing technique, the specimen shall be discarded and another specimen substituted.

MECHANICAL AND MATERIAL REQUIREMENTS FOR STEEL NUTS—SAE J995 JUN79

SAE Standard

Report of the Iron and Steel Technical Committee, approved August 1967, last revised September 1974, editorial change June 1979.

1. Scope—This SAE Standard covers the mechanical and material requirements for three grades of steel nuts suitable for use in automotive and related engineering applications, in sizes 1/4 to 1 1/2 in., inclusive, and with dimensions conforming with the requirements of the latest issue of SAE J104.

1.1 This standard does not include limits for surface discontinuities. Where usage requires such control, limits may be specified separately. For sizes 1/4 through 1 in., this may be done by the statement: "Surface discontinuities shall not exceed the limits specified in SAE J122."

2. Designation—The three grades of nuts are designated Grades 2, 5, and 8.

3. Material—Nuts shall be made of steel conforming to the chemical composition limits specified in Table 1.

4. Mechanical Requirements

4.1 Proof Load—Nuts described in this standard shall withstand the proof load stress specified for the applicable nut grade, size, and thread series shown in Table 2.

4.2 Hardness—Nuts shall have a hardness not in excess of that specified in Table 2.

5. Test Methods

5.1 Proof Load Test—The nut shall be assembled on a test bolt or on a hardened and threaded mandrel, as described below and illustrated in Fig. 1. The specified proof load for the nut shall be applied against the nut in an axial direction. (See footnote b of Table 2 for method for computing the proof

TABLE 1—CHEMICAL COMPOSITION REQUIREMENTS

Nut Grade No.	C Max	Mn Min	P Max	S Max
2	0.47	—	0.12[a]	0.15[b]
5	0.55	0.30	0.05[c,d]	0.15[b,d]
8	0.55	0.30	0.04	0.05[e]

NOTE: All values are for ladle analysis (per cent by weight) and are subject to standard variations for check analysis as given in SAE J409.

[a] Resulfurized and rephosphorized material is not subject to rejection based on check analysis for sulfur.
[b] If agreed between purchaser and producer, sulfur content may be 0.23 max.
[c] Phosphorus content may be 0.13 max for acid bessemer steel only.
[d] If agreed between purchaser and producer, sulfur content may be 0.35 max and phosphorus content may be 0.12 max provided that manganese content is 0.70 min.
[e] If agreed between purchaser and producer, sulfur content may be 0.33 max provided that manganese content is 1.35 min.

load in pounds for a nut.) The nut shall resist this load without failure by stripping or rupture, and shall be removable from the test bolt or mandrel by the fingers after the load is released. (NOTE: Occasionally it may be necessary to use a manual wrench or other means to start the nut in motion. Use of such means is permissible, providing the nut is removable by the fingers following an initial loosening of not more than one-half turn of the nut.) If the threads of the test bolt or mandrel are damaged during the test, the test shall be discarded. (See Fig. 1.)

Test bolts shall have threads conforming to Class 2A tolerances and shall have a yield strength in excess of the specified proof load of the nut being tested.

Mandrels shall have a hardness of Rockwell C45 minimum; and shall be threaded to Class 3A tolerance, except that the major diameter shall be the minimum major diameter with a tolerance of +0.002 in.

For referee purposes, the proof load test shall be conducted using a hardened mandrel.

5.2 Hardness Test—Rockwell hardness shall be determined on the top or bottom of the nut.

Hardness determinations shall be made on a polished surface located on the nut face halfway between the major diameter of the thread and the one corner, or, if applicable, on a wrench face one-third of the distance from a corner to the center of the wrench face. In preparing the surface, sufficient material shall be removed to assure elimination of any decarburization or other surface irregularities.

Hardness tests shall be conducted in accordance with SAE J417.

6. Marking

6.1 Grade 2 nuts are not required to be marked for grade identification. All grades of hex jam, heavy hex jam, hex slotted, heavy hex slotted, hex thick and heavy hex nuts are not required to be marked for grade identification, unless specified by the purchaser.

6.2 Grade 5 and Grade 8 hex nuts, sizes $\frac{1}{4}$ thru $1\frac{1}{2}$, shall be marked for grade identification. Three "styles" of grade marking are acceptable: Style A is applicable to all sizes of nuts. Style B may be used at the supplier's option for sizes $\frac{5}{8}$ and larger; and may be used for smaller sizes only when authorized by the purchaser. Style C is applicable to nuts which are fabricated by cutting from hex bar.

TABLE 2—PROOF LOAD AND HARDNESS REQUIREMENTS FOR NUTS[a]

Nut Grade	Nut Size Dia, in.	Proof Load Stress, psi[b]		Rockwell Hardness
		Thread Series		
		UNC 8 UN	UNF, 12 UN and Finer	
2[c]	1/4 thru 1-1/2	90,000	90,000	C32 max
5	1/4 thru 1	120,000	109,000	C32 max
	Over 1 thru 1-1/2	105,000	94,000	C32 max
8	1/4 thru 5/8	150,000	150,000	C24-C32
	Over 5/8 thru 1			C26-C34
	Over 1 thru 1-1/2			C26-C36

[a] Values listed are not normally applicable to jam, slotted, castle, heavy, or thick nuts (see Appendix).
[b] The proof load in pounds for a nut is computed by multiplying the proof load stress for the nut grade, size, and thread series, as shown in Table 2, and the stress area for the applicable size and thread series shown in Table 3. (See Appendix, Table 6, for computed values for some products.)
[c] Normally applicable to square nuts only. Also, square nuts normally available in Grade 2 only.

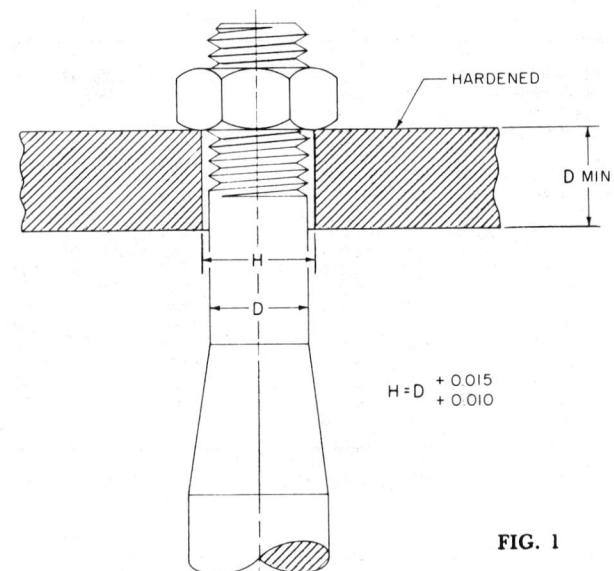

FIG. 1

$$H = D \begin{array}{c} +0.015 \\ +0.010 \end{array}$$

Unless otherwise specified by the purchaser, marking for manufacturer identification shall be at supplier's option. No more than 10% of the nut top surface area may be used for grade and manufacturer markings. In the case of double chamfer nuts, one face only is considered a top surface.

Style A markings shall consist of a dot and a radial or circumferential line at 120 deg counterclockwise from the dot for Grade 5 nuts, and 60 deg counterclockwise from the dot for Grade 8 nuts. These marks shall be located on the top surface of the nut on a circular line or path approximately midway between hole diameter and hex flat diameter, and in a manner which precludes any metal projecting above the top surface. Radial and circumferential lines shall conform to the following dimensions (inch, nominal):

Nut Size	Width (W)	Length (L)	Deep (D)
1/4 and 5/16	0.015	0.05	0.010
3/8 thru 9/16	0.020	0.06	0.010
5/8 thru 7/8	0.030	0.08	0.010
1 and larger	0.030	0.12	0.010

Dot may be round (diameter = W) or rectangular (maximum length = L/2).

Style B markings shall consist of a dot at one corner of the nut and a radial line at the corner 120 deg counterclockwise from the dot for Grade 5 nuts, and 60 deg counterclockwise from the dot for Grade 8 nuts. These marks shall be located on the chamfer surface of the top of the nut, and may be raised or depressed at the option of the supplier. Raised marks shall not project beyond the height or width of the nut.

Style C markings shall consist of notches at the hexagon corners, one notch at each corner for Grade 5, and two notches at each corner for Grade 8. Note: While SAE J995 and SAE J104 do not cover flange nuts, it is appropriate to recommend that Grade 5 and Grade 8 products be marked as described above for hex nuts, with the following additional supplier options: (1) raised or depressed Style A markings on top of flange or on top surface of nut, (2) lines up to 2 times the dimensions specified above.

7. Testing Requirements

7.1 Manufacturer's Responsibility—During the manufacture of products to the requirements of this specification, the manufacturer shall make periodic tests to ensure that the properties of the product are being maintained within specified limits. Such tests shall be conducted in accordance with a sampling plan, preferably the sampling plan given in paragraph 7.3, and the test results shall be recorded in a test report. When requested in writing by the purchaser, the manufacturer shall furnish a copy of the test report certified to be a report of the results of the last completed set of tests for the specific type, size, length, and grade of product.

Additional tests of products in individual shipments are not normally contemplated. Unless otherwise agreed at time of original inquiry and purchase order, individual heats of steel need not be identified in the finished product.

7.2 Purchaser's Options—If the purchaser requires that additional tests be performed by the manufacturer to determine that the properties of products in an individual shipment are within specified limits, or if the purchaser requires that a sampling plan different from that given in paragraph 7.3 shall be used when determining the acceptability of a lot, or shipment, of products, the purchaser shall specify the complete testing requirements, including sampling plan and basis of acceptance, in the original inquiry and purchase order.

7.3 General—An acceptable sampling plan is outlined below:

Number of Pieces in Lot	Minimum Number of Specimens To Be Tested
50 and under	2
51 to 500	3
501 to 35,000	5
35,001 and over	8

A lot, for purposes of selecting test specimens, shall consist of all products offered for inspection and testing at one one time that are of the same type, grade, size, length, and thread series and are manufactured essentially at one time and under the same process conditions.

The same test specimens may be used for different tests wherever practical.

If the failure of a test specimen is due to improper preparation of the specimen or to incorrect testing technique, the specimen shall be discarded and another specimen substituted.

TABLE 3 — TENSILE STRESS AREAS (TEST BOLT OR MANDREL)

Coarse Thread Series UNC		Fine Thread Series UNF		8-Thread Series 8 UN	
Nominal Size and Threads Per Inch	Tensile Stress Area, sq in.	Nominal Size and Threads Per Inch	Tensile Stress Area, sq in.	Nominal Size and Threads Per Inch	Tensile Stress Area, sq in.
1/4 —20	0.0318	1/4 —28	0.0364	—	—
5/16—18	0.0524	5/16—24	0.0580	—	—
3/8 —16	0.0775	3/8 —24	0.0878	—	—
7/16—14	0.1063	7/16—20	0.1187	—	—
1/2 —13	0.1419	1/2 —20	0.1599	—	—
9/16—12	0.182	9/16—18	0.203	—	—
5/8 —11	0.226	5/8 —18	0.256	—	—
3/4 —10	0.334	3/4 —16	0.373	—	—
7/8 — 9	0.462	7/8 —14	0.509	—	—
1 — 8	0.606	1 —12	0.663	1 —8	0.606
1-1/8 — 7	0.763	1-1/8 —12	0.856	1-1/8—8	0.790
1-1/4 — 7	0.969	1-1/4 —12	1.073	1-1/4—8	1.000
1-3/8 — 6	1.155	1-3/8 —12	1.315	1-3/8—8	1.233
1-1/2 — 6	1.405	1-1/2 —12	1.581	1-1/2—8	1.492

APPENDIX

ed. Grade designations in this standard apply only to hexagon nuts and square nuts as indicated in Table 2. Tables 4 and 5 give proof load stress values for these types of nuts, but grade designations have not been established for these. Until these are established, products may be described by notes, such as the following: "SAE J995; Proof Load Stress: 96,000 psi; Hardness: Rockwell C32 max."

ed. TABLE 4 — PROOF LOAD AND HARDNESS REQUIREMENTS FOR MISCELLANEOUS NUTS 120,000 BASIC PROOF LOAD STRESS

Nut Size & Thread Series	Proof Load Stress, psi				Rockwell Hardness, Max
	1/4 thru 1 in. Dia		Over 1 thru 1-1/2 in. Dia		
Type of Nut	UNC and 8 UN	UNF and 12 UN and Finer	UNC and 8 UN	UNF and 12 UN and Finer	
Hex Jam	72,000	65,000	63,000	57,000	C32
Hex Slotted	96,000	87,000	84,000	75,000	C32
Heavy Hex	133,000	120,000	116,000	105,000	C32
Heavy Hex Jam	72,000	65,000	63,000	57,000	C32
Heavy Hex Slotted	105,000	96,000	92,000	84,000	C32
Hex Thick	133,000	120,000	116,000	105,000	C32
Hex Thick Slotted	105,000	96,000	92,000	84,000	C32

ed. TABLE 5 — PROOF LOAD AND HARDNESS REQUIREMENTS FOR MISCELLANEOUS NUTS 150,000 BASIC PROOF LOAD STRESS

Nut Size & Thread Series	Proof Load Stress, psi		Rockwell Hardness, Max
	1/4 to 1-1/2 in. Dia		
Type of Nut	UNC 8 UN	UNF, 12 UN and Finer Pitch Series	
Hex Jam	90,000	90,000	
Hex Slotted	120,000	120,000	
Heavy Hex	165,000	150,000	See Table 2
Heavy Hex Jam	90,000	90,000	
Heavy Hex Slotted	132,000	120,000	
Hex Thick	165,000	150,000	
Hex Thick Slotted	132,000	120,000	

ed. TABLE 6 — PROOF LOAD FOR MISCELLANEOUS NUTS,[a,b] LB (UNC THREADS ONLY)

Nut Size and Threads Per Inch	Square Nuts Grade 2	Hex Nuts Grade 5	Hex Nuts Grade 8	Hex Jam and Heavy Hex Jam Nuts X	Hex Jam and Heavy Hex Jam Nuts Y	Hex Slotted Nuts X	Hex Slotted Nuts Y	Heavy Hex and Hex Thick Nuts X	Heavy Hex and Hex Thick Nuts Y	Heavy Hex Slotted and Hex Thick Slotted X	Heavy Hex Slotted and Hex Thick Slotted Y
1/4—20	2,850	3,800	4,750	2,300	2,850	3,050	3,800	4,250	5,250	3,350	4,200
5/16—18	4,700	6,300	7,850	3,750	4,700	5,050	6,300	6,950	8,650	5,500	6,900
3/8—16	7,000	9,300	11,600	5,600	7,000	7,450	9,300	10,300	12,800	8,150	10,200
7/16—14	9,550	12,800	15,900	7,650	9,550	10,200	12,800	14,100	17,500	11,200	14,000
1/2—13	12,800	17,000	21,300	10,200	12,800	13,600	17,000	18,900	23,400	14,900	18,700
9/16—12	16,400	21,800	27,300	13,100	16,400	17,500	21,800	24,200	30,000	19,100	24,000
5/8—11	20,300	27,100	33,900	16,300	20,300	21,700	27,100	30,100	37,300	23,700	29,800
3/4—10	30,100	40,100	50,100	24,000	30,100	32,100	40,100	44,400	55,100	35,100	44,100
7/8—9	41,600	55,400	69,300	33,300	41,600	44,400	55,400	61,400	76,200	48,500	61,000
1—8	54,500	72,700	90,900	43,600	54,500	58,200	72,700	80,600	100,000	63,600	80,000
1-1/8—7	68,700	80,100	114,000	48,100	68,700	64,100	91,600	88,500	126,000	70,200	101,000
1-1/4—7	87,200	102,000	145,000	61,000	87,200	81,400	116,000	112,000	160,000	89,100	128,000
1-3/8—6	104,000	121,000	173,000	72,800	104,000	97,000	139,000	134,000	191,000	106,000	152,000
1-1/2—6	126,000	148,000	211,000	88,500	126,000	118,000	169,000	163,000	232,000	129,000	185,000

[a] Proof load stress values in Table 4 and 5 are based on requirements shown in Table 2 for Grades 5 and 8 nuts. Primarily, each value is derived from the ratio of the minimum thickness of the product involved to the minimum thickness of square machine screw nuts and hex nuts (see SAE J107) of the same size—and "correction factors" added or subtracted to compensate for differences in width across flats, width and depth of slots, and depth of countersink.

[b] Computed according to Table 2, Footnote b, using psi values shown in Tables 2, 4, and 5. Values in "X" columns are related to Table 4; those in "Y" columns to Table 5.

MECHANICAL AND MATERIAL REQUIREMENTS FOR WHEEL BOLTS—SAE J1102

SAE Standard

Report of the Iron and Steel Technical Committee approved October 1974.

1. Scope—This SAE Standard covers the chemical, metallurgical and mechanical requirements for two types of passenger car and truck wheel bolts, as follows:
(a) Nonserrated shank bolts which are heat treated
(b) Serrated shank bolts which are case hardened

2. Materials and Processes

2.1 Steel Characteristics

2.1.1 Bolts in sizes through 14 mm or 9/16 inch diameter shall be made of killed steel with carbon content of 0.28 to 0.47, sulfur 0.058 max., and phosphorus 0.048 max.

2.1.2 Bolts in sizes over 14 mm or 9/16 inch diameter shall be made of a medium carbon alloy steel (carbon content 0.28 to 0.55), fine grain with hardenability that will produce a minimum hardness of Rockwell C47 at the center of a transverse section one diameter from the threaded end of the bolt after oil quenching.

2.1.3 The preceding analyses are product analyses (percent by wt) and refer to individual determinations on uncarburized or core portion of bolts.

2.2 Heading Practice—Bolts in sizes through 20 mm or 3/4 inch shall be cold headed. Larger sizes may be hot or cold headed, at option of the manufacturer.

2.3 Threading Practice—All bolts, regardless of size shall be roll threaded.

2.4 Heat Treatment Practice

2.4.1 Nonserrated shank bolts shall be heat treated, quenched in a liquid medium and tempered at a minimum tempering temperature of 425°C (800°F).

2.4.2 Nonserrated shank bolts shall conform to Class C Decarburization (as described in SAE J121), unless otherwise specified.

2.4.3 Serrated shank bolts shall be carburized in a non-nitriding atmosphere to a total case depth of 0.10–0.30 mm (0.004–0.012 in.), oil quenched and tempered at a minimum tempering temperature of 450°C (850°F). Case depth shall be measured on the body or head of the bolt.

2.5 Surface Discontinuities—Bolts in sizes up to 24 mm (1 in.) inclusive, and lengths up to 150 mm (6 in.) inclusive shall not have surface discontinuities exceeding the limits specified in SAE J1061. Surface discontinuities for other sizes and lengths shall be within limits specified by purchaser.

3. Mechanical and Performance Requirements

3.1 Proof Load—600 MPa for metric bolts, 85 ksi for inch bolts

3.2 Axial Tensile Strength—830 MPa for metric bolts, 120 ksi minimum for inch bolts

3.3 Core Hardness—Rockwell C25-34.

3.4 Surface Hardness (Serrated Shank Bolts Only)—Rockwell 15N77 minimum.

3.5 Bend Test—Bolts shall withstand a 10 deg bend without breaking into separate parts.

3.6 Serration Test (Serrated Shank Bolts Only)—Bolts shall assemble in a test plate without visual evidence of surface stripping of the serrations.

4. Test Methods

4.1 Proof Load—Same as defined in SAE J429.

4.2 Axial Tensile Strength—Same as defined in SAE J429.

4.3 Core Hardness—Same as defined in SAE J429.

4.4 Surface Hardness—Tests to determine surface hardness shall be conducted on the ends, hexagon flats or unthreaded and nonserrated portion of shanks, in accordance with SAE J417. Proper correction factors shall be used when hardness tests are made on curved surfaces, per ASTM E18.

4.5 Bend Test—The test bolt shall be threaded or clamped into a hardened block or other suitable device with three threads exposed. A force perpendicular to the centerline of the bolt shall be applied against the bolt head and continued until the plane of the bearing surface under the bolt head is permanently bend through 10 deg.

4.6 Serration Test (Serrated Shank Bolts Only)—The test bolt shall be pressed into a hole in a steel plate or appropriate wheel hub or axle flange until the head is seated by applying an axial compression load to the head of the bolt. The bolt shall then be removed and visually examined for evidence of serration stripping. If a wheel hub or axle flange is used it shall be of the same material and hardness required for the part into which the bolt is normally assembled in production. If plate is used, it shall be 12.7 mm (0.5 in.) thick with hardness of Brinell 269–285. The hole size shall be as specified by the purchaser; however, if not specified the diameter shall be the average of the mean major and mean minor serration diameters.

5. Marking

5.1 Nonserrated shank bolts shall be marked with three radial lines 120 deg apart.

5.2 Serrated shank bolts shall be marked with three radial lines spaced 165-30-165 deg apart.

5.3 Bolts with metric series threads shall be marked also with the letters "METRIC".

5.4 Markings shall be located on the top of the head and may be either raised or depressed at the option of the manufacturers.

MECHANICAL AND CHEMICAL REQUIREMENTS FOR NONTHREADED FASTENERS—SAE J430

SAE Recommended Practice

Report of Iron and Steel Technical Committee approved March 1960. Reaffirmed without change October 1968.

CARBON STEEL SOLID RIVETS

Scope—These specifications cover the mechanical and chemical requirements for carbon steel solid rivets used in automotive and other related industries.

Grouping—Rivets in Grades 0 and 1 fall in two groups; namely, those of small diameter—7/16 in. and less, usually driven cold, and over 7/16 in. usually driven hot. It is recommended that the rivets for cold driving in sizes over 7/16 in. be ordered annealed.

General Data—Rivets for cold driving are specified so as to provide the necessary ductility for the application.

The properties of rivets intended for hot driving are not necessarily those found in the driven rivet. Therefore, the specifications for hot driven product are designed to furnish satisfactory properties after cooling from the driving heat.

Steel Process—Steel shall be produced by any suitable process to conform to the chemistry specified.

Quality—Surface shall be commercially free of injurious cracks and seams, and steel shall be free of injurious pipe and excessive segregation.

Annealed rivets for cold driving should be free of loose scale.

Number of Tests on Annealed Rivets

A. Samples shall be selected for hardness test from each lot of rivets as follows:

Items in Lot	Number of Samples
To 800	3
801—8000	6
8001—22000	9
over 22000	15

B. All samples must meet the hardness requirements of the specification for acceptance, but retests are permitted as stated in paragraph C. Retests.

C. Retests—If any sample from the same lot fails to meet the specified requirement, double the number of samples shall be tested, in which case for acceptance all of the additional samples shall meet the specification.

TABLE 1—GRADES AND PROPERTIES

Grade	Tensile Properties of Hot Rolled Rod or Bar from which Rivets are Produced		Chemical Composition Ladle Analysis, % Max	Heat Treatment of Rivets
0	Tensile strength, psi Yield point, min, psi Elongation in 8 in., Min, %	40,000—55,000 23,000 27	P —0.040 S —0.050	7/16 in. dia and under are furnished annealed to Rockwell B 65 max[a]
1	Tensile strength, psi Yield point, Min, psi Elongation in 8 in., Min, %	52,000—62,000 28,000 24	P —0.040 S —0.050	7/16 in. dia and under are furnished annealed to Rockwell B 85 max[a]
2[b]	Tensile strength, psi Yield point, Min, psi Elongation in 8 in., Min, %	55,000—70,000 29,000 22	C —0.28 Mn —0.30—0.90 P —0.040 S —0.050 Si —0.25	Not specified
3[b]	Tensile strength, psi Yield point, Min, psi Elongation in 8 in., Min, %	68,000—82,000 38,000 20	C —0.30 Mn —1.65 P —0.040 S —0.050	Not specified

[a]Hardness to be taken in accordance with SAE Information Report, Hardness Tests and Hardness Number Conversions—SAE J417.

[b]Grades 2 and 3 intended for hot driving only. The tensile requirements of grade 3 are met by heating to 1450 F, holding at this temperature for not less than 30 min and cooling slowly in the furnace.

MECHANICAL AND QUALITY REQUIREMENTS FOR TAPPING SCREWS—SAE J933 JUN79

SAE Recommended Practice

Report of the Iron and Steel Technical Committee, approved July 1965, last revised September 1969, reaffirmed with editorial change June 1979.

1. Scope

1.1 This SAE Recommended Practice covers the mechanical and quality requirements for steel tapping screws used in automotive and related industries. It does not apply to corrosion resistant (stainless) steel screws. (Dimensional requirements for most types of screws mentioned herein are covered in SAE J478.)

1.2 The primary objective of the specification is to insure that screws form or cut mating threads in materials of construction into which they are normally driven, without deforming their own thread and without breaking during assembly or service.

NOTE: Certain limitations on basic material and manufacturing processes have been incorporated because the size and configuration of the parts under consideration make them vulnerable to relatively small variations in chemistry, heat treatment, etc., and because experience has shown that in processing it is difficult to keep these variables consistently "in balance." Until improved performance tests are developed, these limitations will supplement the "performance" features of the specification.

2. Performance Requirements

2.1 **General**—In cases where screws are plated subsequent to delivery to the purchaser (or where plating of screws is otherwise under the control of the purchaser), the screw producer is not responsible for failures due to plating. In such cases, additional screws from the same lot shall be stripped of plating, baked, lubricated with machine oil, and retested in the plain finish condition.

2.2 **Drive Test for Types A, B, C, D, F, G, T, AB, and BP**[1]—Sample screws (coated or uncoated, as received) shall, without deforming their own thread, form a mating thread in test plate described below until a thread of full diameter is completely through the test plate.

The test plate shall be made of low carbon cold rolled steel, having hardness of Rockwell B70–85 or equivalent, and thickness as specified in Table 1. Test holes shall be drilled or punched and redrilled, or reamed, to ±0.001 in. of nominal diameter specified in Table 1 for type and size screw being tested.

2.3 **Torsional Strength Test**—Shank of sample screw (coated or uncoated, as received) shall be securely clamped in a mating, split, blind-hole die (Fig. 1) or other means, such that the clamped portion of the threads is not damaged and at least two full threads project above the clamping device and at least two full form threads exclusive of point, flute (s), or end slot are held within the clamping device. (A blind hole may be used in place of the clamping device, providing the hole depth is such as to insure that breakage will occur beyond the point, or the full length of the flute (s) or end slot.) By means of a suitably calibrated torque measuring device, apply torque to the screw until failure occurs. The torque required to cause failure shall equal or exceed the minimum value shown in Table 2 for the type and size of screw being tested.

2.4 **Ductility Test**—Not required at this time; under development.

3. Material and Processing Requirements

3.1 **Material**—Screws shall be made from cold heading quality, killed steel wire, conforming to the composition limits shown below:

Tapping Screw Size (dia)	Analysis[a]	Chemical Composition, % by wgt	
		Carbon	Manganese
No. 4 and smaller	Ladle Check	0.13–0.25 0.11–0.27	0.60–1.65 0.57–1.71
No. 5 thru 1/2 in.	Ladle Check	0.15–0.25 0.13–0.27	0.70–1.65 0.64–1.71

[a]Ladle analyses are shown for informational purposes. Check analyses are mandatory and refer to individual determinations on uncarburized or core portion of screws.

3.2 **Heat Treatment**—Shall be in carbonitriding or gas carburizing system. Screws shall be quenched in a liquid medium and then tempered by reheating to 650 F min.

Cyaniding systems may be approved by a purchaser when the producer shows that a continuous flow (no batch) quenching process is employed which consistently produces uniform case and core.

3.3 **Total Case Depth**—Shall conform to the following, as measured at thread flank midpoint between crest and root:

Size	Thickness, in.
No. 4 thru 6	0.002–0.007
No. 8 thru 12	0.004–0.009
1/4 and larger	0.005–0.011

3.4 **Surface Hardness After Tempering**—Shall be equivalent to Rockwell C45 minimum. For routine quality control purposes (where case depth and geometry of screws permit), measurements may be made on end, shank,

---

[1]This test does not apply to Types BF, BG, and BT screws.

TABLE 1—STANDARD TEST PLATE THICKNESSES AND HOLE SIZES FOR DRIVE TEST INSPECTION OF TAPPING SCREWS[a]

Nominal Screw Size	Thickness					Hole Size									
	Types AB, A, B, BP and C			Types D, F, G and T		Type A		Types AB, B and BP		Type C				Types D, F, G, and T			
										Coarse Thread		Fine Thread		Coarse Thread		Fine Thread	
	Gage	Max	Min	Max	Min	Drill Size	Hole Dia	Drill Size	Hole Dia	Drill Size	Hole Dia	Drill Size	Hole Dia	Drill Size	Hole Dia	Drill Size	Hole Dia
2	18	0.0500	0.0460	0.0800	0.0760	No. 48	0.0760	No. 48	0.0760	No. 48	0.0760	No. 48	0.0760	No. 49	0.0730	—	—
3	18	0.0500	0.0460	0.0960	0.0920	No. 46	0.0810	No. 46	0.0810	No. 44	0.0860	No. 43	0.0890	No. 46	0.0810	—	—
4	18	0.0500	0.0460	0.1110	0.1070	No. 44	0.0860	No. 44	0.0860	No. 41	0.0960	No. 40	0.0980	No. 41	0.0960	—	—
5	18	0.0500	0.0460	0.1110	0.1070	No. 36	0.1065	No. 36	0.1065	No. 35	0.1100	No. 35	0.1100	No. 37	0.1040	—	—
6	14	0.0770	0.0730	0.1425	0.1385	No. 32	0.1160	No. 32	0.1160	No. 31	0.1200	1/8	0.1250	No. 31	0.1200	—	—
7	14	0.0770	0.0730	—	—	No. 30	0.1285	No. 30	0.1285	—	—	—	—	—	—	—	—
8	14	0.0770	0.0730	0.1420	0.1380	No. 29	0.1360	No. 29	0.1360	No. 27	0.1440	No. 26	0.1470	No. 26	0.1470	—	—
10	1/8	0.1270	0.1230	0.1905	0.1845	No. 21	0.1590	No. 21	0.1590	No. 19	0.1660	11/64	0.1719	No. 17	0.1730	No. 16	0.1770
12	1/8	0.1270	0.1230	0.1905	0.1845	3/16	0.1875	3/16	0.1875	No. 11	0.1910	No. 10	0.1935	No. 8	0.1990	—	—
14	1/8	0.1270	0.1230	—	—	5.5mm	0.2165	—	—	—	—	—	—	—	—	—	—
1/4	3/16	0.1905	0.1845	0.2530	0.2470	—	—	5.5mm	0.2165	7/32	0.2188	1	0.2280	1	0.2280	A	0.2340
16	3/16	0.1905	0.1845	—	—	B	0.2380	—	—	—	—	—	—	—	—	—	—
18	3/16	0.1905	0.1845	—	—	G	0.2610	—	—	—	—	—	—	—	—	—	—
5/16	3/16	0.1905	0.1845	0.3155	0.3095	I	0.2720	—	—	J	0.2770	L	0.2900	L	0.2900	M	0.2950
20	3/16	0.1905	0.1845	—	—	L	0.2900	—	—	—	—	—	—	—	—	—	—
24	3/16	0.1905	0.1845	—	—	11/32	0.3438	—	—	—	—	—	—	—	—	—	—
3/8	3/16	0.1905	0.1845	0.3780	0.3720	—	—	21/64	0.3281	R	0.3390	11/32	0.3438	T	0.3580	T	0.3580
7/16	3/16	0.1905	0.1845	—	—	—	—	13/32	0.4062	10mm	0.3937	—	—	—	—	—	—
1/2	3/16	0.1905	0.1845	—	—	—	—	15/32	0.4688	29/64	0.4531	—	—	—	—	—	—

[a] Requirements shown in each column of Tables 1 and 2 are applicable also to screws which have thread forming characteristics similar to the type(s) designated in the column heading.

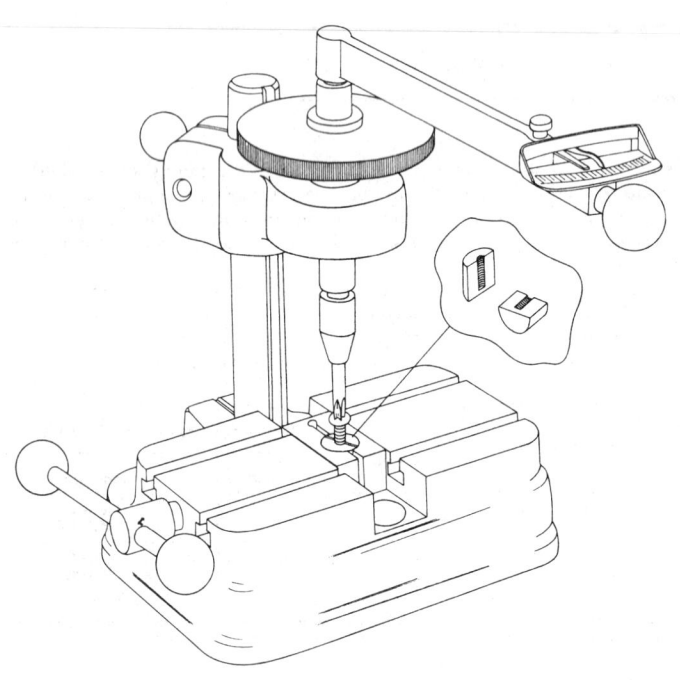

FIG. 1

TABLE 2—TORSIONAL STRENGTH REQUIREMENTS FOR TAPPING SCREWS[a]

Nominal Screw Size	Minimum Torsional Strength, lb-in.					
	Type A	Types AB, B, BF, BG, BP and BT	Type C		Types D, F, G and T	
			Coarse Thread	Fine Thread	Coarse Thread	Fine Thread
2	4	4	5	6	5	6
3	9	9	9	10	9	10
4	12	13	13	15	13	15
5	18	18	18	20	18	20
6	24	24	23	27	23	27
7	30	30	—	—	—	—
8	39	39	42	47	42	47
10	48	56	56	74	56	74
12	83	88	93	108	93	108
14	125	—	—	—	—	—
1/4	—	142	140	179	140	179
16	152	—	—	—	—	—
18	196	—	—	—	—	—
5/16	—	290	306	370	306	370
20	250	—	—	—	—	—
24	492	—	—	—	—	—
3/8	—	590	560	710	560	710
7/16	—	—	—	—	—	—
1/2	—	—	—	—	—	—

[a] Requirements shown in each column of Tables 1 and 2 are applicable also to screws which have thread forming characteristics similar to the type(s) designated in the column heading.

or head using Rockwell 15 N. As an alternate, or where this method is not applicable, a microhardness instrument with a Knoop or diamond pyramid indenter and a 500 gram load may be used. In such cases, measurements shall be made on the thread profile of a properly prepared longitudinal metallographic specimen.

3.5 Core Hardness After Tempering—Shall be Rockwell C28–38[2], as determined at mid-radius of a transverse section through the screw taken at a distance sufficiently behind the point of the screw to be through the full minor diameter.

3.6 Microstructure—Shall show no band of free ferrite between case and core, as determined by metallographic examination.

[2] Hardness shall not exceed maximum shown and preferably should be no higher than Rockwell C36 to insure against failure in assembly and service.

4. Testing Requirements

4.1 The requirements of this specification shall be met in continuous mass production for stock, and the producer shall make sample inspections to insure that the product is controlled within the specified limits. Additional tests of individual shipments are not ordinarily contemplated. Individual heats of steel are not identified in the finished product, and testing on a heat basis is not feasible.

When specified on purchase order or engineering drawing, the manufacturer shall furnish a report certified to be the latest set of test results for the following tests for each stock size in each shipment: Core Hardness, Torsional Strength Test, and Drive Test (in case of self-drilling screws, both Drill and Drive Tests).

Strength Test, and Drive Test.

Chemical and metallurgical tests (material analysis, surface hardness, case depth, metallographic examination of cross section) are not inspected frequently during normal processing, and consequently, such testing shall be considered as special and shall only be done by the manufacturer when specifically required in the original inquiry and purchase order.

4.2 When testing on a lot basis is specified, a lot shall consist of all screws of a single type, size and length, manufactured under essentially the same conditions and submitted for inspection at one time. Unless otherwise agreed between purchaser and supplier the following number of tests shall be conducted to determine acceptance of each of the mechanical requirements (core hardness, torsional strength, and driving characteristics):

Lot Size (pieces)	No. of Tests	Lot Size (pieces)	No. of Tests
50 and less	2	501 to 35,000	5
51 to 500	3	Over 35,000	8

4.3 If failure of a test specimen is due to improper preparation of the specimen or to incorrect testing technique, the specimen shall be discarded and another specimen substituted.

STEEL SELF-DRILLING TAPPING SCREWS—SAE J78 JUN79 — SAE Standard

Report of the Iron and Steel Technical Committee and Fasteners Committee, approved July 1972, last revised October 1973, reaffirmed without change June 1979.

1. Scope

1.1 General—This standard covers the dimensional and general specifications, including performance requirements, for carbon steel self-drilling tapping screws suitable for use in general applications.

It is the objective of this standard to insure that carbon steel self-drilling tapping screws, by meeting the mechanical and performance requirements specified, shall drill a hole and form or cut mating threads in materials into which they are driven without deforming their own thread and without breaking during assembly.

An Appendix is included to provide a recommended technique for measuring the case depth on the screws.

1.2 Screw Types and Application—The two types of self-drilling tapping screws covered by this standard are designated and described as follows:

1.2.1 TYPE BSD—Type BSD screws shall have spaced threads with drill points of varying configuration, designated Style 2 and Style 3, designed to accommodate different panel thickness conditions as delineated in Table 5.

1.2.2 TYPE CSD—Type CSD screws shall have threads of machine screw diameter-pitch combinations approximating Unified Form with drill points of varying configuration, designated Style 2 and Style 3, designed to accommodate different panel thickness conditions as delineated in Table 5. Type CSD screws are not subject to thread gaging but shall meet dimensions specified in this standard. They are intended for application where the use of a machine screw pitch thread is preferred over the spaced thread.

1.3 Head Types—The head types applicable to self-drilling tapping screws covered by this standard shall include those specified in SAE J478 and ANSI B18.6.4, Slotted and Recessed Head Tapping Screws and Metallic Drive Screws, except for slotted head and hex (nonwasher) head designs which are not recommended for self-drilling screws.

2. Dimensional Requirements

2.1 General Dimensions—Dimensions and general specifications applicable to heads, body, and screw length for Type BSD and Type CSD screws shall conform to those specified for Type B and Type C tapping screws, respectively, as specified in SAE J478 or ANSI B18.6.4, except as specified in paragraphs 2.2–2.4.

2.2 Heads—The underside on all noncountersunk styles of heads on milled point self-drilling screws may be chamfered at the periphery of head in accordance with the dimensions specified in Fig. 1 and Table 1.

2.3 Eccentricity—Eccentricity is defined as one-half of the full or total indicator reading.

2.3.1 ECCENTRICITY OF HEX AND HEX WASHER HEADS—Hex and hex washer heads shall not be eccentric with the axis of screw by an amount equal to more than 4% of the basic screw diameter.

2.3.2 ECCENTRICITY OF RECESS—The recess in recessed head screws shall not be eccentric with the axis of screw by an amount equal to more than 4% of the basic screw diameter.

2.4 Length of Thread

2.4.1 TYPE BSD SCREWS—For screws of nominal lengths equal to or shorter than 1.50 in., the full form threads shall extend close to the head such that the specified minor diameter limits are maintained to within one pitch (thread), or closer if practicable, of the underside of the head. See Fig. 2. For screws of nominal lengths longer than 1.50 in., the length of full form thread shall be as specified by the purchaser.

2.4.2 TYPE CSD SCREWS—For screws of nominal lengths equal to or shorter than 1.50 in., the full form threads shall extend close to the head such that the specified major diameter limits are maintained to within two pitches (threads), or closer if practicable, of the underside of the head. See Fig. 3. For screws of nominal lengths longer than 1.50 in., the length of full form thread shall be as specified by the purchaser.

2.5 Threads and Points—The threads and points applicable to screws covered by this standard are generally described under paragraph 1.2. They shall conform to the dimensions specified in Table 2.

3. Material and Process Requirements

3.1 Material and Chemistry—Screws shall be made from cold heading quality, killed steel wire conforming to the following chemical composition:

TABLE 1—HEAD CHAMFER DIMENSIONS FOR MILLED POINT SELF-DRILLING TAPPING SCREWS, IN (FIG. 1)

Nominal Screw Size	U Washer Thickness		R_1 Chamfer Height Hex Washer Heads	R_2 Chamfer Height Recessed Heads
	Max	Min	Ref	Ref
4	0.030	0.020	0.015	0.015
6	0.040	0.025	0.015	0.015
8	0.050	0.035	0.020	0.015
10	0.050	0.035	0.020	0.020
12	0.050	0.035	0.020	0.020
1/4	0.060	0.040	0.025	0.020

Analysis[a]	Composition Limits, % by weight			
	Carbon		Manganese	
	Max	Min	Max	Min
Ladle	0.25	0.15	1.65	0.70
Check	0.27	0.13	1.71	0.64

[a]Ladle analyses are shown for informational purposes. Check analyses are mandatory and refer to individual determinations on uncarburized or core portions of screws.

3.2 Heat Treatment—Screws shall be heat treated in a carbonitriding or gas carburizing system. Cyaniding systems may be approved by the purchaser when it is shown that a continuous flow (no batch) quenching process which consistently produces uniform case and core hardnesses is employed.

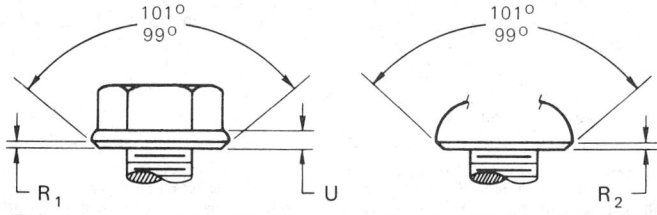

FIG. 1A—HEX WASHER HEAD FIG. 1B—RECESSED HEADS

FIG. 1—HEAD CHAMFERS ON MILLED POINT SCREWS

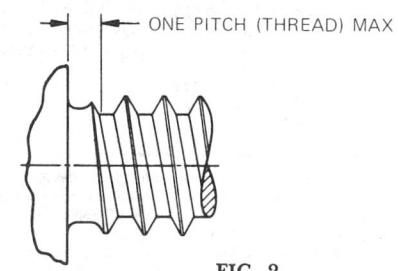

FIG. 2

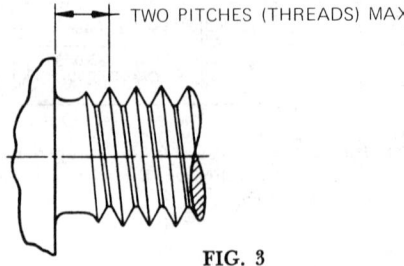

FIG. 3

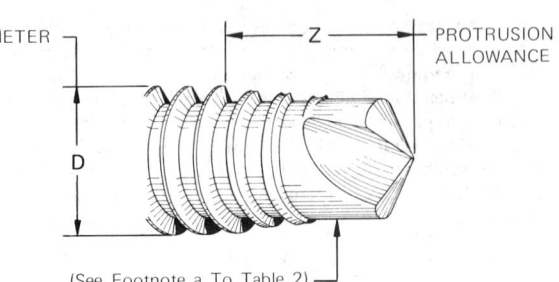

(See Footnote a To Table 2)

FIG. 4—TYPICAL SELF-DRILLING TAPPING SCREW POINT

3.2.1 Tempering Temperature—Minimum tempering temperature shall be 625°F.

When cyaniding systems are approved, the minimum tempering temperature shall be 450°F.

3.2.2 Case Depth—Screws shall have a case depth conforming to the tabulation below:

Nominal Screw Size	Case Depth, in	
	Max	Min
4 and 6	0.007	0.002
8 thru 12	0.009	0.004
1/4	0.011	0.005

Case depth shall be measured at a midpoint between crest and root on the thread flank. A recommended technique for measuring case depth is given in the Appendix.

3.2.3 Case Hardness—Screws shall have a case hardness equivalent to Rockwell C 52–58. For routine quality control purposes (where case depth and geometry of screw permit), case hardness may be measured on end, shank, or head using Rockwell 15 N. As an alternate, or where this method is not applicable, a microhardness instrument with a Knoop or diamond pyramid indenter and a 500 g load may be used. In such cases, measurements shall be made on the thread profile of a properly prepared longitudinal metallographic specimen.

3.2.4 Core Hardness—Screws shall have a core hardness equivalent to Rockwell C 32–40, when measured at mid-radius of a transverse section through the screw taken at a distance sufficiently behind the point of the screw to be through the full minor diameter.

3.3 Ductility—Heads of screws shall not separate completely from the shank when a permanent deformation of 5 deg is induced between the plane of the under head bearing surface and a plane normal to the axis of the screw, when tested in accordance with paragraph 3.3.1.

3.3.1 Ductility Test—The sample screw shall be inserted into a drilled hole in a hardened wedge block, or other suitable device, and an axial compressive (or impact) load applied against the top of the screw head. Loading shall be continued until the plane of the under head bearing surface is bent permanently through 5 deg with respect to a plane normal to the axis of the screw.

3.4 Finish—Unless otherwise specified, screws shall be supplied with a natural (as processed) finish, unplated or uncoated. Where corrosion preventative or decorative finishes are required, screws shall be plated or coated as specified by the user. However, where steel screws are plated or coated and subject to hydrogen embrittlement, they shall be suitably treated subsequent to the plating or coating operation to obviate such embrittlement. Cadmium or zinc electroplated screws shall be subjected to the hydrogen embrittlement test in paragraph 3.4.1.

3.4.1 Hydrogen Embrittlement Test—Cadmium and zinc electroplated screws shall drill their own hole and form a thread in a steel test plate with a thickness equal to the maximum specified for the applicable screw type and size in Table 5. The head of the screw shall be seated against one or more ANSI B27.2 Standard Type B Plain Washers, Narrow Series (size corresponding to screw size and minimum stack thickness corresponding to maximum unthreaded length under the head), or an equivalent spacer, and tightened with a torque equal to the hydrogen embrittlement torque specified in Table 3. The assembly shall remain in this tightened state for 24 h. The original hydrogen embrittlement torque shall then be reapplied, following which the screw shall be removed by the application of removal torque. There shall be no evidence of failure of the screws.

3.4.2 In cases where screws are plated or coated following delivery to the purchaser (or where plating or coating of screws is otherwise under the control of the purchaser), the screw producer shall not be responsible for failures of screws to meet mechanical or performance requirements due to plating or coating. In such cases, additional screws from the same lot shall be stripped of plating or coating, baked, lubricated with machine oil, and retested in the natural finish.

4. Performance Requirements and Tests

4.1 Torsional Strength—Screws shall not fail with the application of a torque less than the torsional strength torque specified in Table 3, when tested in accordance with paragraph 4.1.1.

4.1.1 Torsional Strength Test—The sample screw shall be securely clamped by suitable means (Fig. 5) such that the threads in the clamped length are not damaged, and that at least two full threads project above the clamping device, and that at least two full threads exclusive of point, flutes, or thread cutting slot, are held within the clamping device. By means of a suitably calibrated torque measuring device, torque shall be applied to the

TABLE 2—DIMENSIONS OF THREADS AND POINTS FOR TYPES BSD AND CSD SELF-DRILLING TAPPING SCREWS (FIG. 4)

Nominal Size[b] or Basic Screw Diameter		Type BSD							Type CSD					Types BSD and CSD							
		Threads Per Inch	D Major Diameter		d Minor Diameter		Z[c] Protrusion Allowance (Ref)		Threads Per Inch	D Major Diameter		Z[c] Protrusion Allowance (Ref)		L							
														Minimum Practical Nominal Screw Lengths (Ref)							
														Style 2 Points				Style 3 Points			
														Formed		Milled		Formed		Milled	
			Max	Min	Max	Min	Style 2 Point	Style 3 Point		Max	Min	Style 2 Point	Style 3 Point	90 deg Heads	Csk Heads	90 deg Heads	Csk Heads	90 deg Heads	Csk Heads	90 deg Heads	Csk Heads
4	0.1120	24	0.114	0.110	0.086	0.082	0.163	—	40	0.1120	0.1072	0.130	—	5/16	3/8	3/8	7/16	—	3/8	7/16	—
6	0.1380	20	0.139	0.135	0.104	0.099	0.190	0.220	32	0.1380	0.1326	0.152	0.172	5/16	3/8	7/16	7/16	3/8	7/16	7/16	1/2
8	0.1640	18	0.166	0.161	0.122	0.116	0.211	0.251	32	0.1640	0.1586	0.162	0.202	3/8	7/16	7/16	1/2	7/16	1/2	1/2	9/16
10	0.1900	16	0.189	0.183	0.141	0.135	0.235	0.300	24	0.1900	0.1834	0.193	0.258	7/16	1/2	15/32	19/32	1/2	9/16	9/16	21/32
12	0.2160	14	0.215	0.209	0.164	0.157	0.283	0.353	24	0.2160	0.2094	0.223	0.293	1/2	5/8	17/32	21/32	1/2	5/8	21/32	25/32
1/4	0.2500	14	0.246	0.240	0.192	0.185	0.318	0.393	20	0.2500	0.2428	0.275	0.350	1/2	5/8	17/32	11/16	1/2	5/8	11/16	27/32

[a] Drill portion of points may be milled and/or cold formed and details of point taper and flute design shall be optional with the manufacturer, provided the screws meet the performance requirements specified in this standard and are capable of drilling the maximum panel thicknesses shown in Table 5 prior to thread pickup.

[b] Where specifying nominal size in decimals, zeros preceding decimal and in fourth decimal place shall be omitted.

[c] Protrusion allowance Z is the distance, measured parallel to the axis of screw, from the extreme end of the point to the first full form thread beyond the point and encompasses the length of drill point and the tapered incomplete threads. It is intended for use in calculating the maximum effective design grip length Y on the screw in accordance with the following:

$$Y = L_{min} - Z$$

TABLE 3—MECHANICAL AND PERFORMANCE REQUIREMENTS FOR
TYPES BSD AND CSD SELF-DRILLING TAPPING SCREWS

Nominal Screw Size	Minimum Torsional Strength, lb-in		Hydrogen Embrittlement Test Torque, lb-in	
	Type BSD	Type CSD	Cadmium Plated Screws Types BSD and CSD	Zinc Plated Screws Types BSD and CSD
4	14	14	10.5	12
6	24	24	18	20
8	42	48	36	41
10	61	65	49	55
12	92	100	72	85
1/4	150	156	114	132

TABLE 4—DRILL-DRIVE TEST CONDITIONS AND REQUIREMENTS FOR
TYPES BSD AND CSD SELF-DRILLING TAPPING SCREWS

Nominal Screw Size	Test Plate Thickness,[a] in		Axial Loading,[b] lb			Time to Drill and Form Thread,[c] s
	Max	Min	A Max	B Max	C Max	Max
4	0.068	0.062	25	30	40	2.0
6	0.068	0.062	30	35	45	2.5
8	0.068	0.062	30	35	45	3.0
10	0.068	0.062	35	40	50	3.5
12	0.068	0.062	45	50	60	4.0
1/4	0.068	0.062	45	50	60	5.0

[a] Test plates shall be low carbon, cold rolled steel having a hardness of Rockwell B60-85.
[b] Axial loads are varied to offset the detrimental effects on drilling capability created by finishes applied to screws in accordance with the following:
   Column A — Axial loads tabulated shall apply to plain, oiled, and commercial phosphate coating and cadmium and zinc platings up to 0.0003 in thickness.
   Column B — Axial loads tabulated shall apply special electroplated finishes exceeding 0.0003 in thickness and to special coatings, such as thread sealing hot melts, etc.
   Column C — Axial loads tabulated shall apply to chromium finish.
[c] Tool speed shall be 2500 rpm for screw sizes No. 4 through No. 10. Tool speed of 1800 rpm is recommended for screw sizes No. 12 and 1/4; however, 2500 rpm may be used provided care is exercised to minimize influence of high heat buildup due to surface speed.

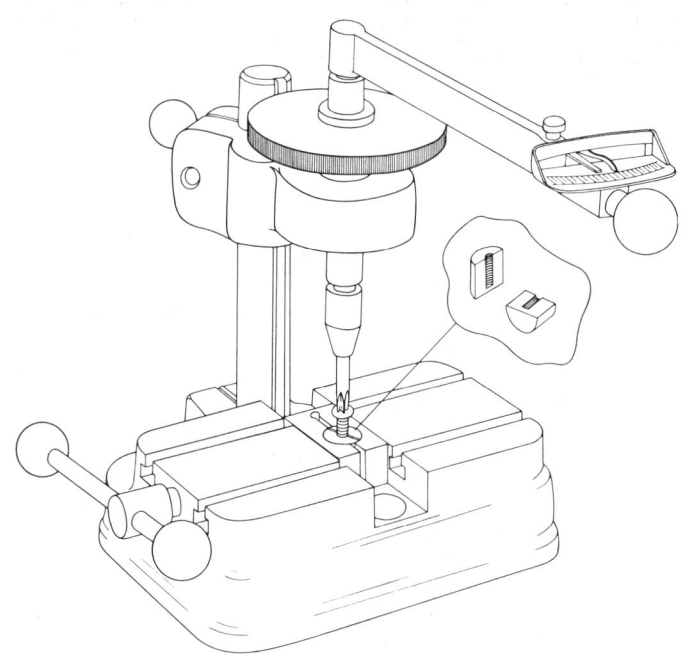

FIG. 5—TYPICAL TORSIONAL STRENGTH TEST FIXTURE

screw until failure of the screw occurs. The torque required to cause failure shall be recorded as the torsional strength torque.

4.2 Drill-Drive Test—Sample screws shall be selected at random from the lot and shall be used to drill holes and form or cut mating threads in a test plate. The time in seconds for the screw to drill and thread a hole completely through the test plate shall be recorded. The test plate material and thickness, and load applied against the screw during drilling and threading, and the other test conditions are specified in Table 4. Each screw shall be used to drill and thread only one hole. A typical drill drive test fixture is depicted in Fig. 6.

The drill-drive test shall be conducted in accordance with the following sampling plan:

Lot Size[a]	Sample Size
Up to 5,000	6
5,001 to 15,000	12
15,001 to 50,000	18
50,001 and over	25

[a] Lot size is defined as a quantity submitted for inspection.

If the actual time for each of the sample screws to drill and thread a hole does not exceed the maximum time specified in Table 4, the lot shall be acceptable. If one or more of the test times exceed the maximum specified in Table 4, a retest shall be made using twice the original sample size. The lot shall then be acceptable in accordance with the following:

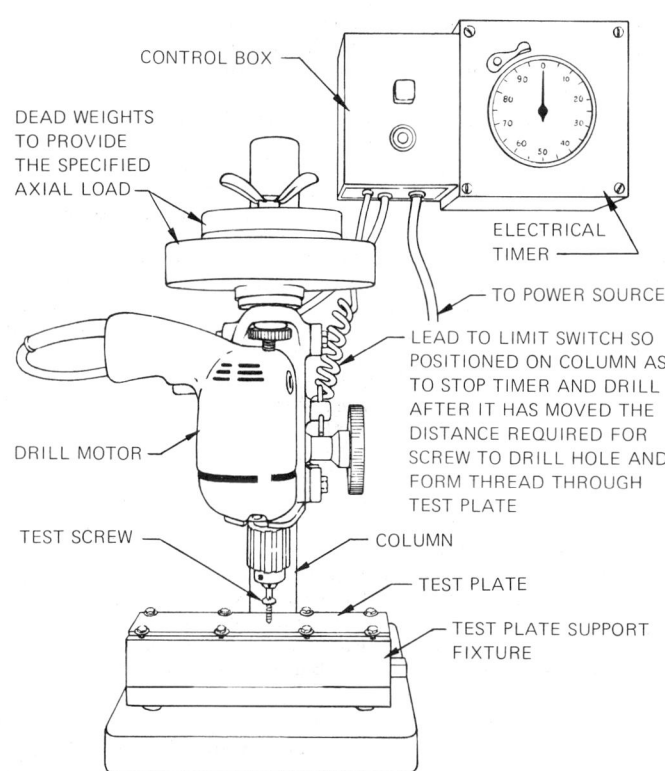

FIG. 6—TYPICAL DRILL-DRIVE TEST FIXTURE

Sample Size	Slow Drive[a]	Excessive Drive[b]
12	1	0
24	1	0
36	2	1
50	3	1

[a] A "slow drive" is defined as a screw having a drilling and threading time in excess of, but less than, twice the specified maximum.
[b] An "excessive drive" is defined as a screw having a drilling and threading time twice the specified maximum or greater.

4.3 Drive to Failure Test—There shall be a satisfactory difference between starting torque and failure torque. The difference may be expressed as a ratio or range of torques.

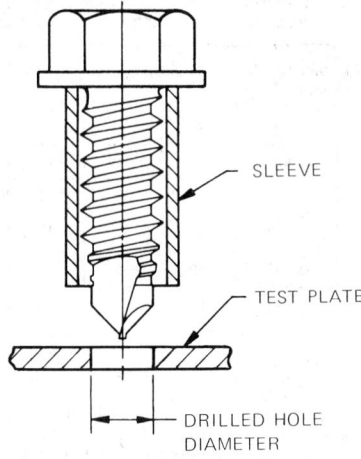

FIG. 7—DRILL HOLE SIZE TEST

TABLE 5—SELF-DRILLING TAPPING SCREW SELECTION CHART

Screw Type	Point Style	Nominal Screw Size	Pa Recommended Panel Thickness, in
BSD and CSD	2	4	0.080 Max
		6	0.090 Max
		8	0.100 Max
		10	0.110 Max
		12	0.140 Max
		1/4	0.175 Max
	3	6	0.090-0.110
		8	0.100-0.140
		10	0.110-0.175
		12	0.110-0.210
		1/4	0.110-0.210

ᵃIf the panel to be drilled is comprised of two or more layers (see Figs. 8B and 8C), the gap between the layers (which might consist of a sealing strip, airspace caused by warpage, etc., or just the separation caused by the pressure exerted by the driver) must be considered in determining the point style for the particular fastener. Using a self-drilling tapping screw as covered in this standard in a multilayer application with an excessive gap could result in point breakage since the tapping in one layer begins before completion of the drilling of the other layers and since the advancement of the screw in the tapping operation is much faster than in the drilling operation.

(Test conditions and performance ratios or torque ranges are to be developed.)

4.4 Drill Hole Size—When desired to determine that the drill point does not drill an oversize hole that would cause a loss of thread engagement and result in premature stripping of the mating thread, a drill hole size test may be conducted in accordance with paragraph 4.4.1. The diameter of the hole drilled by the screw shall not exceed the point diameter of the test screw by more than 0.005 in.

4.4.1 DRILL HOLE SIZE TEST—The sample screw shall be inserted through a sleeve or collar (Fig. 7) having an inside diameter of approximately 0.010 in. greater than the major diameter of the screw. The length of sleeve or collar should be such that sufficient unthreaded point length extends through the sleeve or collar to drill a hole through the minimum thickness material specified in Table 4 without thread pickup. After the hole is drilled in the test plate the screw shall be removed and the diameter of the drilled hole gaged.

5. Screw Selection and Installation Considerations—Screw point style selection should be made on the basis of the recommended panel thicknesses specified in Table 5. For multipanel applications which exceed the thickness tabulated, clearance holes should be provided in the uppermost panel or panels to reduce the thickness to be drilled by the screw.

Driving tools which operate between 1800 and 3000 rpm are commonly used for self-drilling tapping screw applications.

Fig. A-1 illustrates comparisons between the structure of case and core produced by the method recommended herein and a regular quenched and tempered structure. Case depths were measured on each of three screws after carbonitriding and microhardness traverses were run. The same parts were then water-quenched from 1430 F and case depths were again measured. Results of each method appear under the photographs.

5.1 Optional Head Marking—For the purpose of identifying self-drilling tapping screws in assembled components, the consumer, at his option, may specify identifying head markings. Heads of self-drilling tapping screws, when specified by the consumer, shall be marked as shown in Fig. 9.

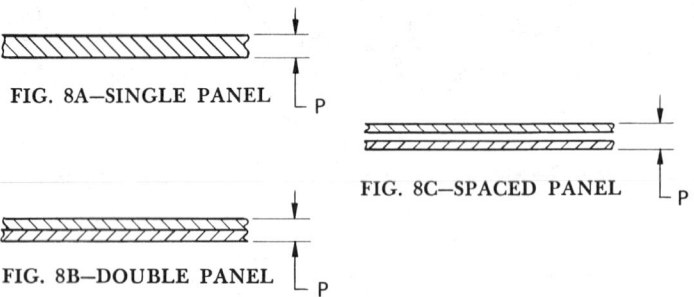

FIG. 8A—SINGLE PANEL

FIG. 8C—SPACED PANEL

FIG. 8B—DOUBLE PANEL

FIG. 8—TYPICAL PANEL CONFIGURATIONS

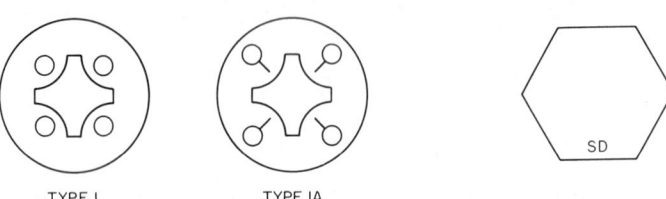

TYPE I TYPE IA
ROUND DEPRESSIONS TO BE LOCATED AS SHOWN
RECESS TYPES

HEX AND HEX WASHER TYPES

MARKS MAY BE RAISED OR DEPRESSED AT MANUFACTURERS OPTION

FIG. 9

APPENDIX
RECOMMENDED TECHNIQUE FOR MEASURING CASE DEPTH

The accurate measurement of case depth of fasteners which have been carburized or carbonitrided is often affected by conflicting results obtained by different laboratories. The conventional microscopic method relies on the ability of the technician to distinguish the line of demarcation between the case and the core; and with annealed or tempered structures, this line is often not very sharply defined. Consequently, it is common for different people to come up with varying results on the same sample. The following recommended technique, however, greatly reduces the element of visual interpretation inherent in microscopic examination.

Samples are prepared by grinding to approximately one-half diameter on a longitudinal plane. They are heated for 7 min. at 1430°F and water quenched. Further grinding on 240 and 600 grit papers, followed by a 30 s etch in $2\frac{1}{2}$ nital, rinse in methanol, and dry in an air blast, results in a structure as seen in Fig. A-1. The austenizing temperature of 1430°F is sufficient to completely transform and harden the hypereutectoid case, whereas the core will not harden. The polishing, while not up to full metallographic standards, is sufficient to reveal the structure clearly when etched. It is not necessary to mount the specimens in plastic, which is a time-saving factor.

The following is a guide for case depth measurement:

A. Standard Method

1. Prepare and quench samples as described in the preceding paragraph.
2. Measure total case at thread flank, midpoint between crest and root. The depth of the case is taken as the line of demarcation between the hardened hypereutectoid zone and the unhardened core. Recommend magnification 100X.
3. Take readings on four separate threads and average the results to obtain the case depth of the screw. (The average figure rather than the minimum depth found at any one point should be used as the criterion for case depth, as a single low reading represents only a very small proportion of the total area involved.)

B. Referee Method

1. Take a sample representing the conventional heat treatment of the part and mount and polish for microscopic examination.
2. Measure case with Tukon or equivalent microhardness instrument. Start at surface at midpoint of screw flank between root and crest, measure hardness at 0.001 in increments. Total case depth will be perpendicular distance from surface to a point of Rc 45 minimum.

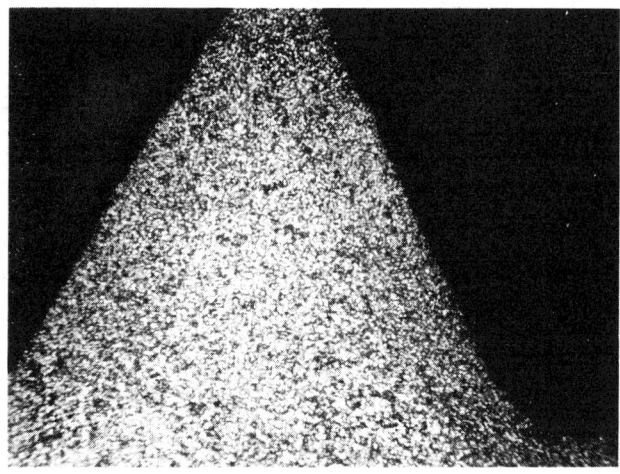

Structure of case and core after anneal at 1600°F for 7 min, nital etch (100X)

0.002 in case depth, water quench from 1430°F, nital etch (100X)

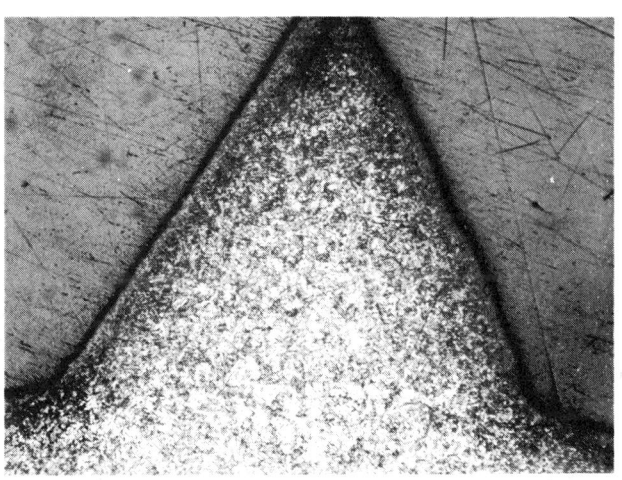

0.0015/0.002 in case depth, 0.002 in to Rc 45, oil quenched and tempered, nital etch (100X)

0.003 in case depth, water quench from 1430°F, nital etch (100X)

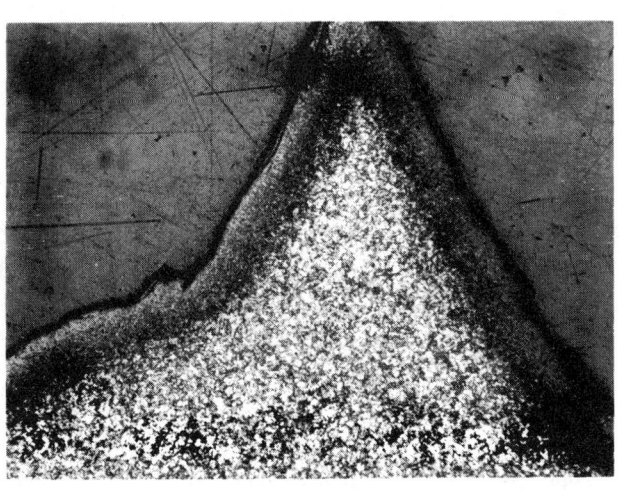

0.0025/0.003 in case depth, 0.0025 in to Rc 45, oil quenched and tempered, nital etch (100X)

0.005/0.0055 in case depth, water quench from 1430°F, nital etch (100X)

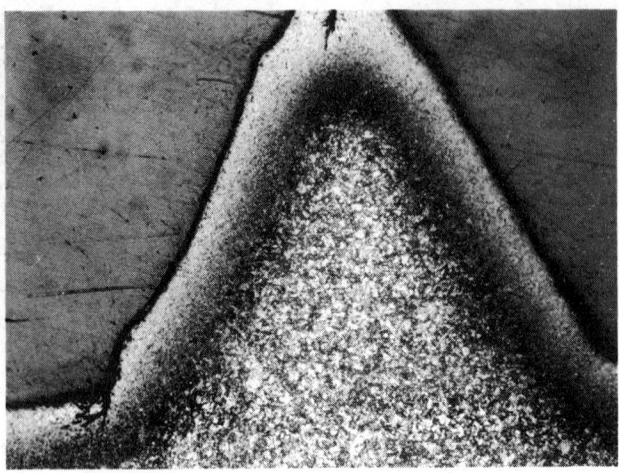

0.004/0.005 in case depth, 0.0045 in to Rc 45, oil quenched and tempered, nital etch (100X)

FIG. A-1—COMPARISONS OF STRUCTURE

THREAD ROLLING SCREWS—SAE J81 JUN79   SAE Standard

Report of the Iron and Steel Technical Committee and Fasteners Committee, approved July 1972, editorial change June 1979. Note: SAE Standard J1237 is available for Metric Thread Rolling Screws.

1. Scope—This standard covers requirements for thread rolling screws suitable for use in general engineering applications. (It is intended that "thread rolling" screws have performance capabilities beyond those normally expected of other standard types of tapping screws.)

NOTE: The performance requirements covered in this standard apply only to the combination of laboratory conditions described in the testing procedures. If other conditions are encountered in an actual service application (such as different materials, thicknesses, hole sizes, etc.), values shown herein for drive torque, torque-to-clamp load, and proof torque may require adjustment.

2. Requirements

2.1 Material and Process Requirements

2.1.1 MATERIAL AND CHEMISTRY—Screws shall be made from cold heading quality, killed steel wire conforming to the following chemical composition requirements:

Analysis	Composition Limits, % by weight			
	Carbon		Manganese	
	Min	Max	Min	Max
Ladle	0.15	0.25	0.70	1.65
Check	0.13	0.27	0.64	1.71

2.1.2 HEAT TREATMENT—Screws shall be heat treated in a gas carburizing system. Cyaniding systems may be approved by the purchaser when it is shown that a continuous flow (no batch) quenching process which consistently produces uniform case and core hardnesses is employed. Carbonitriding systems may also be used when approved by the purchaser.

2.1.4 TEMPERING TEMPERATURE—Minimum tempering temperature shall be 650 F.

2.1.5 FINISH—Screws shall be cadmium or zinc plated with a coating thickness of 0.0002–0.0004 in., or have a phosphate and oil coating, as specified by the purchaser. At the option of the manufacturer, screws may be provided with an additional supplementary lubricant as necessary to meet the performance requirements.

Electroplated screws shall be baked for a minimum of 1 h within the temperature range 375–450 F as soon as practicable after plating to avoid hydrogen embrittlement. (In batch type processing, a minimum of 4 h is normally required to insure that all parts in the batch receive this treatment.)

In cases where screws are plated or coated following delivery to the purchaser (or where plating or coating of screws is otherwise under the control of the purchaser), the screw producer shall not be responsible for failures of the screw to meet mechanical or performance requirements due to plating or coating.

2.2 Dimensional Requirements

2.2.1 HEAD DIMENSIONS—Head dimensions shall conform to those specified in SAE J478 and ANSI B18.6.4, Slotted and Recessed Head Tapping Screws and Metallic Drive Screws.

2.2.2 THREAD AND POINT DIMENSIONS—Thread and point dimensions shall conform to the values shown in Table 1. Threads shall conform to a 60 deg basic thread form, but are not subject to thread gaging. Details of point configurations shall be optional with the manufacturer, provided all dimensions specified are maintained and the screws meet the performance requirements set forth in this standard.

2.2.3 THREAD LENGTH—For screws of nominal lengths equal to or shorter than the nominal lengths tabulated below, the full form threads shall extend close to the head such that the specified thread major diameter limits are maintained to within two pitches (threads), or closer if practicable, from the underside of the head. See Fig. 1. Screws of nominal lengths longer than those tabulated shall, unless otherwise specified, have a minimum length of full form thread equivalent to six times the basic screw diameter of 1.50 in., whichever is shorter.

Nominal Screw Size	Nominal Screw Length	Full Form Thread Length[a] Min	Two Pitches Length[b] Coarse Thread
2	5/8	0.52	0.036
3	5/8	0.59	0.042
4	3/4	0.67	0.050
5	7/8	0.75	0.050
6	7/8	0.83	0.062
8	1	0.98	0.062
10	1-1/4	1.14	0.083
1/4	1-1/2	1.50	0.100
5/16	1-1/2	1.50	0.111
3/8	1-1/2	1.50	0.125
7/16	1-1/2	1.50	0.143
1/2	1-1/2	1.50	0.154

[a]Tabulated values through No. 10 size are 6 times the basic screw diameter, rounded off to two decimal places.
[b]Values are tabulated for convenient reference.

2.3 Mechanical and Performance Requirements

2.3.1 HARDNESS

2.3.1.1 *Core Hardness*—Screws shall have a core hardness of Rockwell C28-38, when tested as specified in paragraph 3.1.

2.3.1.2 *Case Hardness*—Screws shall have a case hardness equivalent to Rockwell C45 minimum, when tested as specified in paragraph 3.2.

2.3.1.3 *Case Depth*—Screws shall have a case depth conforming to the following, when tested as specified in paragraph 3.3.

Nominal Size	Case Depth, in	
	Min	Max
No. 2 thru 6	0.002	0.007
No. 8 and 10	0.004	0.009
1/4 and larger	0.005	0.011

2.3.2 TENSILE STRENGTH—Hex and hex washer head screws which have nominal lengths equal to or longer than 1/2 in. or three times the nominal screw diameter, whichever is longer, shall have tensile strengths not less than those specified in Table 2, when tested in accordance with paragraph 3.4. Screws with shorter lengths or screws with other head styles are not subject to tensile testing.

2.3.3 TORSIONAL STRENGTH—Screws shall not fail with the application of a torque less than the torsional strength torque specified in Table 2, when tested in accordance with paragraph 3.5.

2.3.4 DRIVE TORQUE—Screws shall, without deforming their own thread, form a mating internal thread in a test plate with the application of a torque not exceeding the drive torque specified in Table 2 for the applicable screw size and finish, when tested in accordance with paragraph 3.6.

2.3.5 TORQUE-TO-CLAMP LOAD—Hex and hex washer head screws, in sizes No. 6 and larger, shall develop the clamp load specified in Table 2 with the application of a torque not exceeding the clamp load torque specified in Table 2 for the applicable screw size and finish, when tested in accordance with paragraph 3.7. Smaller sizes of screws and screws with other head styles are not subject to torque-to-clamp load requirements.

2.3.6 PROOF TORQUE—Hex and hex washer head screws shall withstand without failure the proof torque and shall be capable of being removed from the test plate following application of the proof torque specified in Table 2 for the applicable screw size and finish, when tested in accordance with paragraph 3.7. Screws with other head styles are not subject to proof torque requirements.

2.3.7 DUCTILITY—Heads of screws shall not separate from the shank when a permanent deformation of 7 deg is induced between the plane of the under head bearing surface and a plane normal to the axis of the screw, when tested in accordance with paragraph 3.8.

2.3.8 HYDROGEN EMBRITTLEMENT—Cadmium and zinc electroplated screws shall withstand without failure the hydrogen embrittlement torque specified in Table 2 for the applicable screw size and finish, when tested in accordance with paragraph 3.9.

3. Test Methods

3.1 Core Hardness—Core hardness shall be determined at mid-radius of a transverse section through the screw taken at a distance sufficiently behind the point of the screw to be through the full minor diameter.

3.2 Case Hardness—For routine quality control purposes (where case depth and geometry of screws permit), case hardness may be measured on end, shank, or head using Rockwell 15 N. Hardness tests shall be made on plain finish or plated screws after removal of finish. As an alternate, or where this method is not applicable, a microhardness instrument with a Knoop indenter and a 500 g load may be used. In such cases, measurements shall be made on the thread profile of a properly prepared longitudinal metallographic specimen.

3.3 Case Depth—Case depth shall be measured at the midpoint between crest and root on the thread flank. A recommended technique for measuring case depth is given in the Appendix of SAE J78.

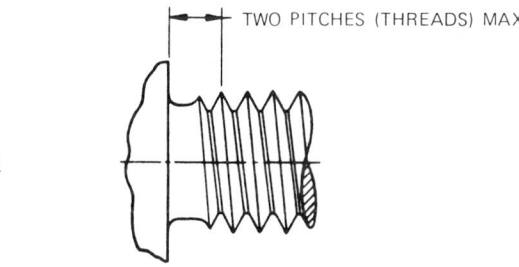

FIG. 1

TABLE 1—THREAD AND POINT DIMENSIONS OF THREAD ROLLING SCREWS

Nominal Screw Size and Threads per Inch[a]	Major Diameter[b] Max	Point Diameter[b] Max	Dia of Circumscribing Circle[c] Max	Circumscribing Circle (Point)[c] Max	Point Length Max[d]	Point Length Min[e]	Min Practical Nominal Screw Lengths 90 deg Heads	Min Practical Nominal Screw Lengths Csk Heads
No. 2-56	0.086	—	0.088	0.070	0.062	0.036	5/32	3/16
3-48	0.099	—	0.101	0.081	0.073	0.042	3/16	7/32
4-40	0.112	0.086	0.115	0.090	0.088	0.050	3/16	1/4
5-40	0.125	0.099	0.128	0.103	0.088	0.050	7/32	1/4
6-32	0.138	0.106	0.141	0.111	0.109	0.062	1/4	5/16
8-32	0.164	0.132	0.167	0.137	0.109	0.062	1/4	11/32
10-24	0.190	0.147	0.194	0.153	0.146	0.083	5/16	13/32
1/4-20	0.250	0.198	0.255	0.206	0.175	0.100	13/32	1/2
5/16-18	0.313	0.255	0.318	0.264	0.194	0.111	15/32	5/8
3/8-16	0.375	0.310	0.281	0.320	0.219	0.125	9/16	23/32
7/16-14	0.438	0.361	0.445	0.375	0.250	0.143	21/32	13/16
1/2-13	0.500	0.416	0.508	0.433	0.269	0.154	23/32	7/9

[a]Fine thread series screws are also available.
[b]These dimensions are applicable to screw blanks prior to thread rolling and to types of screws where the periphery of the thread approximates a circle.
[c]These dimensions are applicable to types of screws where some portions of the periphery of the thread are further from the screw axis than others (lobular, elliptical, tri-roundular, etc).
[d]These values are equal to 3.5 times the pitch distance rounded off to three decimal places.
[e]These values are equal to 2 times the pitch distance rounded off to three decimal places.

TABLE 2—MECHANICAL AND PERFORMANCE REQUIREMENTS FOR THREAD ROLLING SCREWS

Nominal Screw Size and Threads per Inch	Tensile Strength, Min lb	Torsional Strength		Drive Torque				Clamp Load, lb	Clamp Load Torque				Proof Torque				Hydrogen Embrittlement Torque			
				For PC and CP Screws		For ZP Screws			For PC and CP Screws		For ZP Screws		For PC and CP Screws		For ZP Screws		For CP Screws		For ZP Screws	
		Min in-lb	Min ft-lb	Max in-lb	Max ft-lb	Max in-lb	Max ft-lb		Max in-lb	Max ft-lb	Max in-lb	Max ft-lb	in-lb	ft-lb	in-lb	ft-lb	in-lb	ft-lb	in-lb	ft-lb
No. 2-56	500	6	—	4.5	—	6	—	—	—	—	—	—	7	—	8	—	4.5	—	5	—
3-48	660	10	—	7.5	—	9.5	—	—	—	—	—	—	12	—	13.5	—	7.5	—	8.5	—
4-40	810	14	—	9	—	13	—	—	—	—	—	—	17	—	19	—	10.5	—	12	—
5-40	1,100	22	—	12	—	16	—	—	—	—	—	—	25	—	28	—	17	—	19	—
6-32	1,250	24	—	14	—	20	—	460	19	—	25	—	28	—	33	—	18	—	20	—
8-32	1,900	48	—	25	—	32	—	700	37	—	48	—	50	—	57	—	36	—	41	—
10-24	2,350	65	—	35	—	52	—	900	55	—	68	—	68	—	77	—	49	—	55	—
1/4-20	4,300	156	13	90	7.5	120	10	1600	120	10	144	12	162	13.5	186	15.5	114	9.5	132	11
5/16-18	7,100	330	27.5	180	15	240	20	2600	252	21	312	26	342	28.5	372	31	252	21	276	23
3/8-16	10,500	600	50	240	20	300	25	4000	480	40	612	51	636	53	690	57.5	456	38	510	42.5
7/16-14	14,400	840	70	360	30	480	40	5400	744	62	900	75	888	74	960	80	630	52.5	720	60
1/2-13	19,100	1080	90	540	45	660	55	7200	996	83	1140	95	1170	97.5	1260	105	816	68	930	77.5

Legend: CP — cadmium plated
ZP — zinc plated
PC — phosphate coated

NOTE: Values shown in Table 2 are intended for specification purposes and for acceptability of screws to the requirements of the specification. These values are not valid for use in design or assembly unless all conditions of the application are identical with those specified for the inspection tests.

3.4 Tensile Strength Test—Screws shall be assembled in a tensile testing machine with a minimum of six threads exposed, and an axial load applied against the under head bearing surface until screw failure occurs. The speed of testing as determined with a free-running cross head shall not exceed 1 in./min. The grips of the testing machine shall be self-aligning to avoid side thrust on the specimen. The tensile strength of the screw shall be the maximum load in pounds occurring coincident with or prior to screw fracture (that is, screw breakage into two or more parts).

3.5 Torsional Strength Test—The sample screw shall be securely clamped by suitable means (Fig. 2) such that the threads in the clamped length are not damaged, and that at least two full threads project above the clamping device, and that at least two full threads exclusive of point (2 to 3½ thread pitches) are held within the clamping device. A blind hole may be used in place of a threaded clamping device, provided the hole depths is such as to insure that breakage will occur beyond the point (2 to 3½ thread pitches). By means of a suitably calibrated torque measuring device, torque shall be applied to the screw until failure of the screw occurs. The torque required to cause failure shall be recorded as the torsional strength torque.

3.6 Drive Test—The sample screw shall be driven into the hole in a test plate (paragraph 3.11) until an internal thread of full major diameter is formed completely through the full thickness of the plate or until the screw head comes into contact with the plate, whichever occurs first. Speed of driving shall not exceed 500 rpm. For referee purposes, speed of driving shall not exceed 30 rpm. The maximum torque occurring during the test shall be recorded as the drive torque.

3.7 Clamp Load and Proof Torque Test—The test shall be conducted using a load indicating type washer, or other load measuring device, capable of measuring the actual tension induced in the screw as the screw is tightened. The device shall be accurate within ±5% of the test clamp load to be induced in the screw.

Place plain washer, or equivalent punched or drilled steel strip (paragraph 3.12), and then load-indicating-type washer (paragraph 3.7) on sample screw and position this assembly for driving into prescribed hole in test plate (paragraph 3.11). Drive the screw into the test plate until the screw is seated and continue tightening until a tensile load equal to the clamp load as specified in Table 2 is developed. Restrain plain washer from turning to prevent damage to load-indicating-type washer. The torque necessary to develop the clamp load shall be recorded as the clamp load torque.

Tightening shall be continued until a torque equal to the proof torque as specified in Table 2 has been applied to the screw. The assembly shall remain in this tightened state for 10 s, following which the screw shall be removed from the test plate by the application of removal torque.

If convenient, the clamp load and proof torque test may be conducted in conjunction with the drive test.

3.8 Ductility Test—The sample screw shall be inserted into a hole in a hardened 7 deg wedge block, or other suitable device, and an axial compressive load applied against the top of the screw head. The hole shall be 0.020–0.040 in. larger than the nominal screw diameter. Loading shall be continued until the plane of the under head bearing surface of protruding-type heads or the plane through the largest diameter of the head of countersunk-type heads is bent permanently through 7 deg with respect to a plane normal to the axis of the screw.

3.9 Hydrogen Embrittlement Test—Screws shall be threaded into a tapped hole or free running nut (paragraph 3.12) having thickness of at least 1.5 times the nominal screw size and tightened with a torque equal to the hydrogen embrittlement torque specified in Table 2 for the applicable screw size and finish. Spacers should be used for screws with unthreaded shanks and may be used with other lengths providing full thread engagement is maintained within the test nut or tapped hole. The assembly shall remain in this tightened state for 24 h. The original hydrogen embrittlement torque shall then be reapplied, following which the screw shall be removed by the application of removal torque. Nuts may be hardened to permit reusability.

3.10 Torque Wrenches—Torque wrenches used in all tests shall be accurate within ±2% of the maximum of the specified torque range of the wrench.

Alternatively, a torque sensing power device of equivalent accuracy may be used.

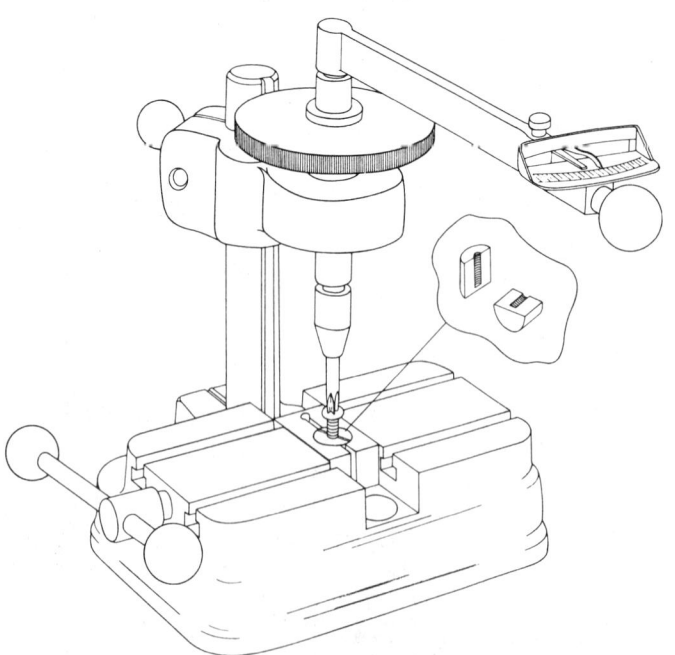

FIG. 2

3.11 Test Plate—Test plates shall be low carbon cold rolled steel having a hardness of Rockwell B 70–85. Test plate thicknesses and hole sizes are given in Table 3. Test holes shall be drilled or punched and redrilled, or reamed, to ±0.001 in. of the hole sizes specified in Table 3.

3.12 Under Head Bearing Test Surface—The surface condition of plain commercially available flat washers, free running nuts, and cold rolled steel is normally suitable for tests specified in paragraphs 3.6, 3.7, and 3.9. For referee purposes, however, the surface shall conform to 20–30 μin. (AA roughness range).

4. Inspection

4.1 Inspection Procedure—Screws shall be inspected to determine conformance with the requirements of this standard. Inspection shall be performed in accordance with sampling plans given in MIL-STD-105D. Alternate inspection procedures may be specified by the purchaser on the purchase order or engineering drawing.

TABLE 3—TEST PLATE THICKNESSES AND HOLE SIZES FOR DRIVE CLAMP LOAD AND PROOF TORQUE TESTS

Nominal Screw Size and Threads per Inch	Thickness, in		Hole Dia, in
	Max	Min	
No. 2-56	0.1270	0.1230	0.075
3-48	0.1270	0.1230	0.087
4-40	0.1270	0.1230	0.098
5-40	0.1270	0.1230	0.110
6-32	0.1270	0.1230	0.120
8-32	0.1905	0.1845	0.147
10-24	0.1905	0.1845	0.166
1/4-20	0.2540	0.2460	0.219
5/16-18	0.3175	0.3075	0.277
3/8-16	0.3800	0.3700	0.339
7/16-14	0.4425	0.4325	0.394
1/2-13	0.5050	0.4950	0.456

NOTE: Values shown in Table 3 are intended for specification purposes and for acceptability of screws to the requirements of the specification. These values are not valid for use in design or assembly unless all conditions of the application are identical with those specified for the inspection tests.

METRIC THREAD ROLLING SCREWS—SAE J1237 — SAE Standard

Report of Iron and Steel Technical Committee approved June 1978. Editorial change May 1979. NOTE: SAE Standard J81 is available for Thread Rolling Screws dimensioned in the inch system.

1. Scope—This standard covers requirements for metric thread rolling screws suitable for use in general engineering applications. (It is intended that *thread rolling* screws have performance capabilities beyond those normally expected of other standard types of tapping screws.)

1.1 Requirements for three material-process options are stated:

(a) Screws (in sizes M2 x 0.4 thru M12 x 1.75) manufactured from low carbon steel, carburized, and tempered. These screws are designated SAE J1237 Type 2 ●.

(b) Screws (in sizes M2 x 0.4 thru M12 x 1.75) manufactured from medium carbon alloy steel, heat treated to achieve properties comparable to SAE J1199 class 9.8 screws, and additionally, with the point selectively hardened. These screws are designated SAE J1237 Type 9 ●.

(c) Screws (in sizes M5 x 0.8 thru M12 x 1.75) manufactured from medium carbon alloy steel, heat treated to achieve properties comparable to SAE J1199 class 10.9 screws, and additionally, with the point selectively hardened. These screws are designated SAE J1237 Type 10 ●.

1.2 When SAE J1237 is specified without Type designation, either Type 2 ● or Type 9 ● may be supplied.

NOTE: The performance requirements covered in this standard apply only to the combination of laboratory conditions decribed in the testing procedures. If other conditions are encountered in an actual service application (such as different materials, thicknesses, hole sizes, etc.), values shown herein for drive torque, torque-to-clamp load, and proof torque may require adjustment.

2. Requirements

2.1 Material and Process Requirements—Type 2 ●

2.1.1 MATERIAL AND CHEMISTRY—Type 2 ● screws shall be made from cold heading quality, killed steel wire conforming to the following chemical composition requirements:

Analysis	Composition Limits[a], % by Mass			
	Carbon		Manganese	
	Min	Max	Min	Max
Cast or Heat	0.15	0.25	0.70	1.65
Product	0.13	0.27	0.64	1.71

[a] Boron permitted in the range of 0.0005–0.003

2.1.2 HEAT TREATMENT—Type 2 ● screws shall be heat treated in a continuous carbonitriding or gas carburizing system. Cyaniding systems may be approved by the purchaser when it is shown that a continuous flow (no batch) quenching process which consistently produces uniform case and core hardnesses is employed.

2.1.3 TEMPERING TEMPERATURES—Minimum tempering temperatures shall be 340°C.

2.2 Material and Process Requirements—Types 9 ● and 10 ●

2.2.1 MATERIAL AND CHEMISTRY—Unless otherwise specified by purchaser, Type 9 ● and 10 ● screws shall be made from cold heading quality, killed alloy steel wire conforming to the following chemical composition requirements (SAE 4037):

	Cast or Heat Analysis	Product Analysis
	% by Mass	
Carbon	0.35–0.40	0.33–0.42
Manganese	0.70–0.90	0.67–0.93
Phosphorus	0.035 max	0.040 max
Sulfur	0.040 max	0.045 max
Silicon	0.15–0.30	0.13–0.32
Molybdenum	0.20–0.30	0.18–0.32

2.2.2 HEAT TREATMENT—Type 9 ● and 10 ● screws shall be heat treated in a continuous non-carburizing system operated under fine grain practice, oil quenched, and tempered at a minimum tempering temperature of 460°C for Type 9 ● and 425°C for Type 10 ●.

2.2.3 Lead threads on Types 9 ● and 10 ● shall be induction hardened to achieve a minimum hardness equivalent to 45 RC (Rockwell C45) on 1 to 3 full threads and one or more lead threads, as shown in Fig. 1. (File test according to SAE J864.)

2.3 Finish—Screws shall be cadmium or zinc electroplated with a coating thickness of 5–10 μm, or have a zinc phosphate and oil coating as specified by the purchaser. Unless otherwise specified, screws may be provided with an additional supplementary lubricant as necessary to meet the performance requirements. Other finishes are available, however, it is the intent of this standard that the mechanical and performance requirements shall apply only to those screws having one of the three finishes specified above. When other finishes are required, the purchaser and manufacturer may agree on performance values other than those of Section 2.7.

Electroplated screws shall be baked within the temperature range 190–230°C as soon as practicable after plating to avoid hydrogen embrittlement. In continuous type processing, a minimum of 1 h is required. In batch type processing, a minimum of 4 h is normally required to insure that all parts in the batch receive this treatment.

In cases where screws are plated or coated following delivery to the purchaser (or where plating or coating of screws is otherwise under the control of the purchaser) the screw producer shall not be responsible for failures of the screw to meet mechanical or performance requirements due to plating or coating.

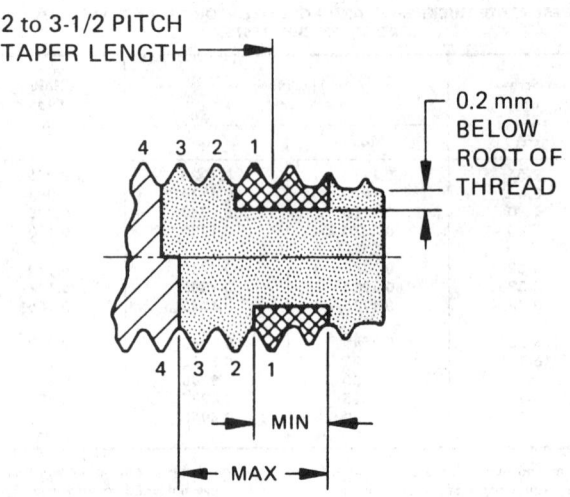

2 to 3-1/2 PITCH TAPER LENGTH

0.2 mm BELOW ROOT OF THREAD

▨ = REQUIRED HARDENED AREA SHALL ENCOMPASS FIRST FULL THREAD

▦ = PERMISSIBLE HARDENED AND TRANSITION ZONE MAY EXTEND TO ENCOMPASS SECOND AND THIRD FULL FORM THREAD

NOTE TO USER—When selecting length of Type 9 • and 10 • screws for any application, one objective should be location of the induction hardened zone beyond the nut anchorage, or a minimum of six full form threads in the threaded hole.

FIG. 1

2.4 Dimensional Requirements

2.4.1 HEAD DIMENSIONS—Standard head styles for thread rolling screws are flat countersunk, oval countersunk, pan, hex, hex washer, and hex flange head. Flat countersunk, oval countersunk, and pan head screws are available slotted or with Type 1 or Type 1A cross recess drives. Head, slot, and cross recess dimensions shall be specified by purchaser.

2.4.2 THREAD AND POINT DIMENSIONS—Thread and point dimensions shall conform to those given in Table 2. Threads shall conform to a 60 deg basic thread form. Threads are not subject to thread gaging. Details of point configuration shall be optional with the manufacturer providing all specified dimensions are maintained and screws meet the performance requirements of this standard.

2.4.3 THREAD LENGTH—For screws of nominal lengths within the ranges listed under column Y of Table 1, the full form threads shall extend close to the head such that the specified thread major diameter limits are maintained to within the respective Y distance from the underside of the head, or closer if practicable. See Fig. 2. Screws of longer nominal lengths, unless otherwise specified by the purchaser, shall have a minimum length of full form thread as specified in column L_T.

2.5 Mechanical Requirements—Type 2 • Screws

2.5.1 HARDNESS

2.5.1.1 *Core Hardness*—Type 2 • screws shall have a core hardness of 28–38 RC (Rockwell C) when tested as specified in Section 3.1. Core hardness shall not exceed maximum shown and preferably should be no higher than 36 RC (Rockwell C) on electroplated parts.

2.5.1.2 *Case Hardness*—Type 2 • screws shall have a case hardness of 45 RC (Rockwell C), minimum, when tested as specified in Section 3.2.

2.5.1.3 *Case Depth*—Type 2 • screws shall have a total case depth conforming to the following, when tested as specified in Section 3.3.

Nominal Size	Case Depth, mm	
	Min	Max
2 thru 3.5	0.05	0.18
4 and 5	0.10	0.23
6 thru 12	0.13	0.28

2.5.2 TENSILE STRENGTH—Type 2 • screws with hex head, hex washer head, and hex flange head, which have lengths equal to or longer than 12 mm or 3 times the nominal screw diameter, whichever is longer, shall have tensile strengths not less than those specified in Table 3, when tested in accordance with Section 3.4. Screws with shorter lengths or screws with other head styles which are weaker than the threaded section are not subject to tensile testing.

2.5.3 TORSIONAL STRENGTH—Type 2 • screws shall not fail with the application of a torque less than the torsional strength torque specified in Table 3 when tested in accordance with Section 3.5.

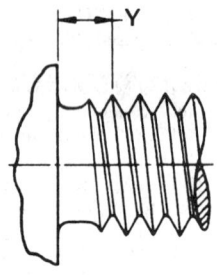

FIG. 2

TABLE 1

Nominal Screw Size and Thread Pitch	L_T Full Form Thread Length			Y Unthreaded Length Under Head			
	For Nominal Screw Lengths >Than	Min (1)	For Nominal Screw Lengths ≤Than	Max (2)	For Nominal Screw Lengths >Than	≤Than	Max (3)
M2 × 0.4	16	12.0	6	0.40	6	16	0.8
M2.5 × 0.45	20	15.0	8	0.45	8	20	0.9
M3 × 0.5	25	18.0	9	0.50	9	25	1.0
M3.5 × 0.6	30	21.0	10	0.60	10	30	1.2
M4 × 0.7	35	24.0	12	0.70	12	35	1.4
M5 × 0.8	40	30.0	15	0.80	15	40	1.6
M6 × 1	45	38.0	18	1.00	18	45	2.0
M8 × 1.25	45	38.0	24	1.25	24	45	2.5
M10 × 1.5	45	38.0	30	1.50	30	45	3.0
M12 × 1.75	50	38.0	36	1.75	36	50	3.5

NOTES: 1. Tabulated values thru 5 mm size are equal to 6 times basic screw diameter rounded to nearest millimetre.
2. Tabulated values are equal to 1 times thread pitch.
3. Tabulated values are equal to 2 times thread pitch.
4. All dimensions are millimetres.

TABLE 2—THREAD AND POINT DIMENSIONS OF THREAD ROLLING SCREWS

Nominal Screw and Thread Pitch	P Major Dia (1) Max	Point Dia (1) Max	C Dia of Circumscribing Circle (2) Max	Cp Circumscribing Circle (Point) (2) Max	Point Length Max (3)	Point Length Min (4)	L Min Practical Nominal Screw Length Pan, Hex, Hex Washer Heads	L Min Practical Nominal Screw Length Flat and Oval Ctsk Heads
M2 × 0.4	2.00	1.6	—	—	1.4	0.8	4	5
M2.5 × 0.45	2.50	2.1	2.57	2.13	1.6	0.9	4	6
M3 × 0.5	3.00	2.5	3.07	2.58	1.8	1.0	5	8
M3.5 × 0.6	3.50	2.9	3.58	2.99	2.1	1.2	6	8
M4 × 0.7	4.00	3.4	4.08	3.40	2.4	1.4	8	10
M5 × 0.8	5.00	4.4	5.09	4.31	2.8	1.6	8	10
M6 × 1	6.00	5.3	6.10	5.12	3.5	2.0	10	12
M8 × 1.25	8.00	7.1	8.13	6.92	4.4	2.5	10	16
M10 × 1.5	10.00	9.0	10.15	8.69	5.2	3.0	13	16
M12 × 1.75	12.00	10.5	12.18	10.48	6.1	3.5	16	20

NOTES: 1. These dimensions are applicable to types of screws where periphery of the thread approximates a circle.
2. These dimensions are applicable to types of screws where some portions of the periphery of the thread are farther from the screw axis than others (lobular, triroundular, etc.).
3. These values are equal to 3.5 times the pitch distance rounded off to 1 decimal place.
4. These values are equal to 2 times the pitch distance rounded off to 1 decimal place.
5. All dimensions are millimetres.

2.5.4 DUCTILITY—Heads of screws shall not separate from the shank when a permanent deformation of 7 deg is induced between the plane of the under head bearing surface and a plane normal to the axis of the screw, when tested in accordance with Section 3.8.

2.6 Mechanical Requirements—Type 9 • and 10 • Screws

2.6.1 Type 9 • and 10 • screws shall conform to the mechanical requirements specified in Table 4 when tested according to wedge tensile and hardness procedures published in SAE J1216, Test Methods for Metric Threaded Fasteners.

2.6.2 Type 9 • and 10 • screws shall conform to Class 3/4H decarburization limits as described in SAE J121.

2.6.3 DUCTILITY—Heads of screws shall not separate from the shank when a permanent deformation of 10 deg is induced between the plane of the under head bearing surface and a plane normal to the axis of the screw, when tested in accordance with Section 3.8.

2.7 Performance Requirements—Types 2 •, 9 •, and 10 •

2.7.1 DRIVE TORQUE—Screws shall, without deforming their own thread, form a mating internal thread in a test plate with the application of a torque not exceeding the drive torque specified in Table 6 for the applicable screw size and finish, when tested in accordance with Section 3.6.

2.7.2 TORQUE-TO-CLAMP LOAD—Screws subject to tensile test, M4 × 0.7 size and larger, shall develop the clamp load specified in Table 6 with the application of a torque not exceeding the clamp load torque specified in Table 6 for the applicable screw size and finish, when tested in accordance with Section

TABLE 3—MECHANICAL REQUIREMENTS FOR TYPE 2 • THREAD ROLLING SCREWS

Basic Dia and Thread Pitch (millimetres)	Min Tensile Strength kN	Min Torsional Strength N·m
M2 × 0.4	1.9	0.7
M2.5 × 0.45	3.15	1.2
M3 × 0.5	4.68	2.2
M3.5 × 0.6	6.3	3.5
M4 × 0.7	8.17	5.2
M5 × 0.8	13.2	10.5
M6 × 1	18.7	17.7
M8 × 1.25	34.0	43.0
M10 × 1.5	53.9	87.0
M12 × 1.75	78.4	152.0

TABLE 5—TENSILE STRENGTH VALUES—TYPE 9 • AND 10 • SCREWS

Nominal Thread Dia and Thread Pitch	Type 9 • Min Tensile Strength kN	Type 10 • Min Tensile Strength kN
M2 × 0.4	1.86	
M2.5 × 0.45	3.05	
M3 × 0.5	4.53	
M3.5 × 0.6	6.10	
M4 × 0.7	7.90	
M5 × 0.8	12.8	14.8
M6 × 1	18.1	20.9
M8 × 1.25	32.9	38.1
M10 × 1.5	52.2	60.3
M12 × 1.75	75.9	87.7

TABLE 4—MECHANICAL REQUIREMENTS TYPE 9 • AND 10 • SCREWS

Type	Wedge Tensile Strength (Stress) MPa (1) (2)	Surface Hardness Rockwell 15N, max	Product Hardness Rockwell (3)
9 •	900 min	(4)	C27–36
10 •	1040 min	(4)	C33–39

NOTES: 1. Wedge tensile strength values for full size products are specified in Table 5. Wedge tensile strengths are applicable only to screws which have lengths equal to or longer than 12 mm or 3 times the nominal diameter, whichever is longer. Screws with shorter lengths or screws with head styles which are weaker than the threaded section are not subject to wedge tensile testing.
2. Tensile wedge angles: 6 deg when screws are threaded one diameter or closer to the head; 10 deg on all others.
3. Minimum product hardness values applicable to screws not subject to tensile tests, and these hardness requirements exclude induction hardened zone Section 2.2.3 (Fig. 1).
4. Surface hardness shall not exceed product hardness by more than 3 points Rockwell C equivalent, and in the case of Type 10 • shall also not exceed Rockwell 15N 80, except as noted in Section 2.2.3 (Fig. 1).

TABLE 6—PERFORMANCE REQUIREMENTS—TYPES 2 •, 9 •, AND 10 •

Basic Dia And Thread Pitch (millimetres)	Test Plate		Drive Torque		Clamp Load	Clamp Load Torque		Proof Torque		Hydrogen Embrittlement Torque	
	Thickness	Pilot Hole Dia	For ZPC and CP Screws	For ZP Screws		For ZPC and CP Screws	For ZP Screws	For ZPC and CP Screws	For ZP Screws	For CP Screws	For ZP Screws
	mm	mm	max N·m	max N·m	kN	max N·m	max N·m	N·m	N·m	N·m	N·m
M2 x 0.4	3	1.77	0.4	0.6				0.6	0.7	0.4	0.5
M2.5 x 0.45	3	2.25	0.8	1.0				1.3	1.4	0.9	1.0
M3 x 0.5	3	2.7	1.3	1.7				2.4	2.5	1.7	1.9
M3.5 x 0.6	3	3.15	1.9	2.4				3.7	4.0	2.6	3.0
M4 x 0.7	5	3.6	2.6	3.4	3.1	4.2	4.8	5.4	5.8	3.8	4.4
M5 x 0.8	5	4.55	4.8	6.0	5.0	8.0	10.0	11.0	12.0	7.8	9.0
M6 x 1	6	5.4	7.5	9.2	6.9	15.0	16.0	19.0	20.0	13.0	15.0
M8 x 1.25	8	7.3	16.0	20.0	12.6	34.0	40.0	46.0	48.0	32.0	36.0
M10 x 1.5	10	9.2	28.0	35.0	20.0	68.0	81.0	92.0	96.0	65.0	74.0
M12 x 1.75	12	11.0	46.0	55.0	29.5	110.0	130.0	160.0	170.0	110.0	130.0

Legend: CP—Cadmium Electroplated
ZP—Zinc Electroplated
ZPC—Zinc Phosphate Coated—(Commonly known as Phosphate and oil)

NOTE: Values shown in Table 6 are intended for specification purposes and for determination of acceptability of screws to the requirements of this standard. These values are not valid for use in design or assembly unless all conditions of the application are identical with those specified for the inspection tests.

3.7. Smaller sizes of screws and screws not subject to tensile test are not subject to torque-to-clamp load requirements.

2.7.3 PROOF TORQUE—Screws with head styles subject to tensile test shall withstand without failure the proof torque and shall be capable of being removed from the test plate following application of the proof torque specified in Table 6 for the applicable screw size and finish, when tested in accordance with Section 3.7. Screws not subject to tensile test are not subject to proof torque requirements.

2.7.4 HYDROGEN EMBRITTLEMENT—Cadmium and zinc electroplated screws shall withstand without failure the hydrogen embrittlement torque specified in Table 6 for the applicable screw size and finish, when tested in accordance with Section 3.9.

3. Test Methods

3.1 Core Hardness—Core hardness shall be determined at mid-radius of a transverse section through the screw taken at a distance sufficiently behind the point of the screw to be through the full minor diameter.

3.2 Case Hardness—For routine quality control purposes (where case depth and geometry of screws permit), case hardness may be measured on end, shank, or head using Rockwell 15 N. Hardness tests shall be made on plain finish or plated screws after removal of finish. As an alternate, or where this method is not applicable, a microhardness instrument with a Knoop or DPH indenter and a 500 g load may be used. In such cases, measurements shall be made on the thread profile of a properly prepared longitudinal metallographic specimen. Due to normal hardness gradients in a case structure, microhardness values shall be taken at the center of the total case.

3.3 Case Depth—Total case depth shall be measured at the midpoint between crest and root on the thread flank. A recommended technique for measuring total case by microscopic methods is given in SAE J423.

NOTE: An effective case depth of Rockwell C45 equivalent as measured by microhardness methods is normally less than the total case depth measured by microscopic methods. The purchaser and supplier may agree on effective case depth values.

3.4 Tensile Strength Test—Screws shall be assembled in a tensile testing machine with a minimum of six threads exposed, and an axial load applied against the under head bearing surface until screw failure occurs. The speed of testing as determined with a free running cross head shall not exceed 25 mm/min. The grips of the testing machine shall be self-aligning to avoid side thrust on the specimen. The tensile strength of the screw shall be the maximum load in Newtons occurring coincident with or prior to screw fracture (such as, screw breakage into two or more parts).

3.5 Torsional Strength Test—The sample screw shall be securely clamped by suitable means (Fig. 3) such that the threads in the clamped length are not damaged, and that at least two full threads project above the clamping device, and that at least two full threads exclusive of point (2-3.5 thread pitches), are held within the clamping device. A blind hole may be used in place of a threaded clamping device, provided the hole depth is such as to insure that breakage will occur beyond the point (2-3.5 thread pitches). By means of a suitably calibrated torque measuring device, torque shall be applied to the screw until failure of the screw occurs. The torque required to cause failure shall be recorded as the torsional strength torque.

3.6 Drive Test—The sample screw shall be driven into the hole in a test plate (Section 3.6.1) until an internal thread of full major diameter is formed completely through the full thickness of the plate or until the screw head comes into contact with the plate, whichever occurs first. Speed of driving shall not exceed 500 rpm. For referee purposes, speed of driving shall not exceed 30 rpm. The maximum torque occurring during the test shall be recorded as the drive torque.

3.6.1 TEST PLATE—Test plates shall be low carbon cold rolled steel having a hardness of Rockwell B75-90. Test plate thicknesses and hole sizes are given in Table 6. Test holes shall be drilled or punched and redrilled, or reamed, to ±0.025 mm of the hole sizes specified in Table 6.

3.7 Clamp Load and Proof Torque Test—The test shall be conducted using a load indicating type washer, or other load measuring device, capable of measuring the actual tension induced in the screw as the screw is tightened. The device shall be accurate within ±5% of the test clamp load to be induced in the screw.

Assemble a plain washer, or equivalent punched or drilled steel strip as specified in Section 3.7.1, and then the load-indicating type washer on the sample screw and position this assembly for driving into the test plate (Section 3.6.1). The screw shall be driven into the test plate until the screw is seated. Tightening shall be continued until a tensile load equal to the clamp load as

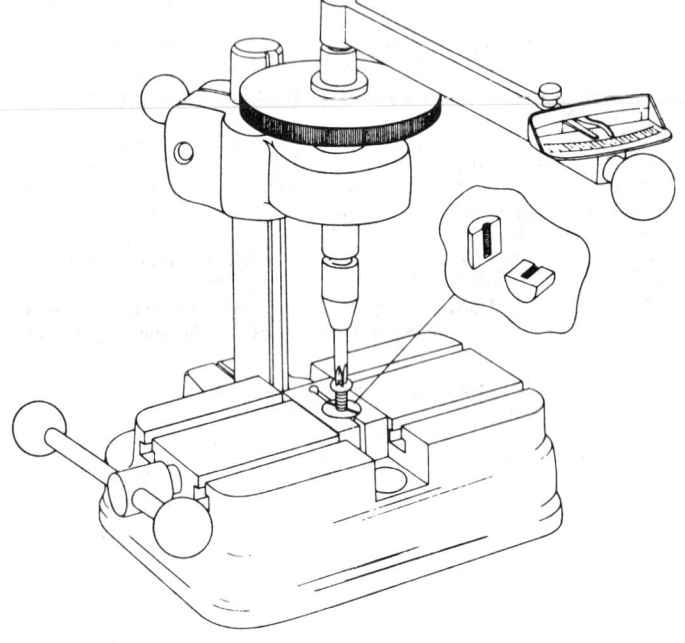

FIG. 3

specified in Table 6 is developed. Care shall be taken to prevent the under head bearing surface from turning during tightening. The torque necessary to develop the clamp load shall be recorded as the clamp load torque.

Tightening shall be continued until a torque equal to the proof torque as specified in Table 6 has been applied to the screw. The assembly shall remain in this tightened state for 10 s following which the screw shall be removed from the test plate by the application of removal torque.

If convenient, the clamp load and proof torque test may be conducted in conjunction with the drive test.

3.7.1 UNDER HEAD BEARING TEST SURFACE—The surface condition of plain commercially available flat washers, free running nuts, and cold rolled steel is normally suitable for tests specified in Sections 3.6, 3.7, and 3.9. For referee purposes, the surface shall conform to 0.50–0.75 μm (AA roughness range).

3.8 Ductility Test—The sample screw shall be inserted into a drilled hole in a hardened wedge block, or other suitable device, and an axial compressive load applied against the top of the screw head. The hole shall be 0.50–1.00 mm larger than the nominal screw diameter. Loading shall be continued until the plane of the under head bearing surface is bent permanently through the angle specified for the screw type with respect to a plane normal to the axis of the screw.

3.9 Hydrogen Embrittlement Test—Screws shall be threaded into a tapped hole or free running nut having thicknesses of at least 1.5 times the nominal screw size and tightened with a torque equal to the hydrogen embrittlement torque specified in Table 6 for the applicable screw size and finish. Spacers should be used for screws with unthreaded shanks and may be used with other lengths providing full thread engagement is maintained within the test nut or tapped hole. The assembly shall remain in this tightened state for 24 h. The original hydrogen embrittlement torque shall then be reapplied following which the screw shall be removed by the application of removal torque. Nuts may be hardened to permit reusability.

3.10 Torque Wrenches—Torque wrenches used in all tests shall be accurate within ±2% of the maximum of the specified torque range of the wrench.

Alternatively, a torque sensing power device of equivalent accuracy may be used.

4. Inspection

4.1 Inspection Procedure—Screws shall be inspected to determine conformance with the requirements of this standard. Inspection shall be performed in accordance with sampling plans given in MIL-STD-105D. Alternate inspection procedures may be specified by the purchaser on the purchase order or engineering drawing.

5. Identification—The following head identification markings may be specified as an option by the purchaser on hex or similar non-recessed or non-slotted head styles:

Type	Identification Mark
2 •	2 •
9 •	9 •
10 •	10 •

SURFACE DISCONTINUITIES ON GENERAL APPLICATION BOLTS, SCREWS, AND STUDS—SAE J1061a

SAE Recommended Practice

Report of Iron and Steel Technical Committee approved September 1973. Editorial change; metric products added July 1975.

1. Scope—This recommended practice defines, illustrates, and specifies allowable limits for various types of surface discontinuities that may occur during the manufacture and processing of bolts, screws, and studs in sizes through 24 mm or 1 in. diameter inclusive with lengths to 150 mm or 6 in. inclusive, having specified minimum tensile strengths of 900 MPa or 120,000 psi and greater, which are primarily intended for use in automotive assemblies.

1.1 The basic recommended practice does not include inspection sampling requirements. It is intended that the purchaser shall specify, in the original inquiry and purchase order, the inspection sampling requirements which the producer must satisfy to demonstrate the acceptability of bolts and screws with respect to surface discontinuities.

2. Types of Surface Discontinuities—For the purpose of this recommended practice, surface discontinuities on bolts, screws, and studs are divided into nine "types", defined as follows:

2.1 Crack—A crack is a clean (crystalline) fracture passing through or across the grain boundaries without inclusion of foreign elements. Cracks are normally caused by overstressing the metal during forging or other forming operation, or during heat treatment. Where parts are subjected to significant reheating, cracks usually are discolored by scale.

2.1.1 QUENCH CRACKS—Quench cracks may occur during heat treatment due to excessively high thermal and transformation stresses. They usually traverse an irregular and erratic course on the surface of the fastener. Typical quench cracks are shown in Fig. 1.

2.1.2 FORGING CRACKS—Forging cracks may occur during the cutoff or forging operations and are located on the top of the heads of screws and bolts. Typical forging cracks are shown in Fig. 2.

2.2 Seam—Seams are generally inherent in the raw material from which fasteners are manufactured. They are narrow, generally straight or smooth-curved line discontinuities, running longitudinally, on the shank and/or

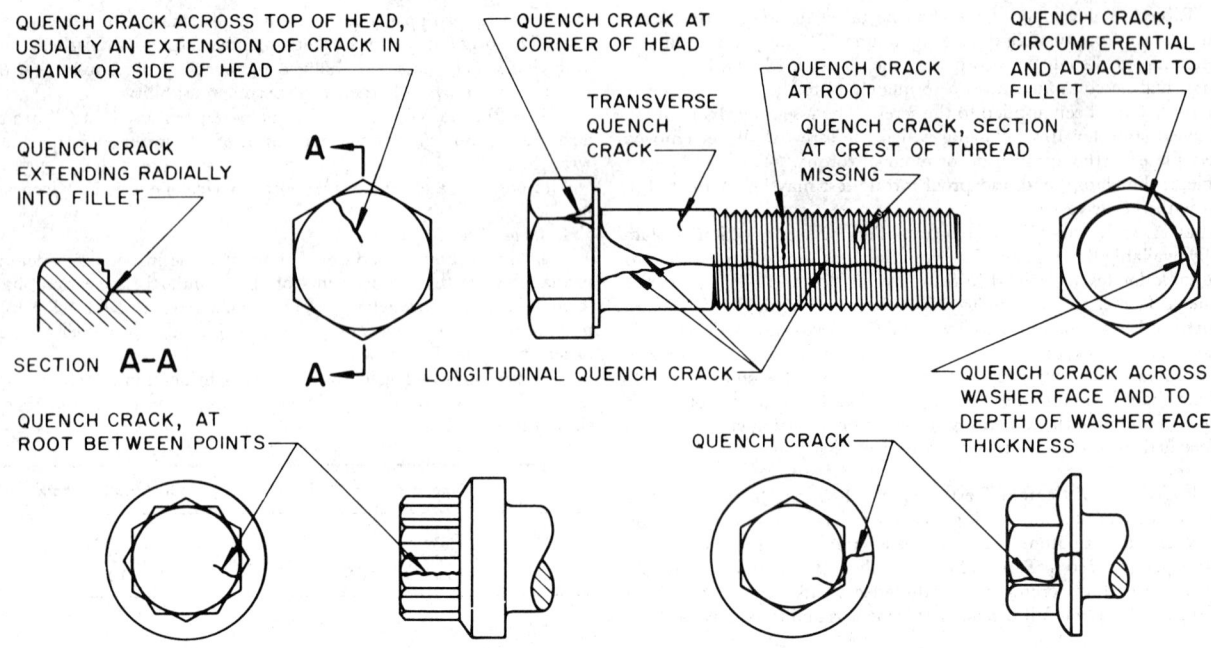

NOTE: Quench cracks of any depth, any length, or in any location are not permissible discontinuities.

FIG. 1—TYPICAL QUENCH CRACKS

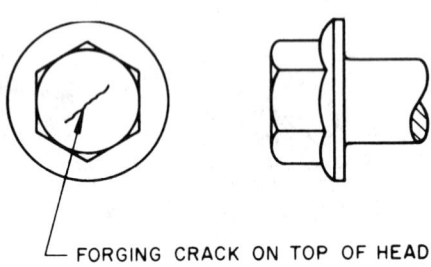

NOTE: Forging cracks are permissible discontinuities if with the limits specified in paragraph 3.4.

FIG. 2—TYPICAL FORGING CRACKS

thread. Seams may extend onto the tops of the heads of circular head products as well as being present at the periphery of the head. Seams may also extend into the chamfer circle, washer face, and wrenching flats of hex head products. Typical seams are shown in Fig. 3.

2.3 Burst—A burst is an open break in the metal (material). Bursts may occur on the flats or corners of the heads of bolts and screws, at the periphery of flanged or circular head products, or on the raised periphery of indented head bolts and screws. Typical bursts are shown in Fig. 4.

2.4 Shear Burst—A shear burst is an open break in the metal, occurring most frequently at the periphery of products having circular or flanged heads and are generally located at approximately 45 deg to the product axis. Shear bursts may also occur on the sides of hex head products. Typical discontinuities of this type are shown in Fig. 4.

2.5 Void—A void is a shallow pocket or hollow on the surface of the bolt or screw due to nonfilling of metal during forging or upsetting. Typical voids are shown in Fig. 5.

2.6 Fold—A fold is a doubling over of metal which may occur during the forging operation. Folds may occur at or near the intersection of diameter

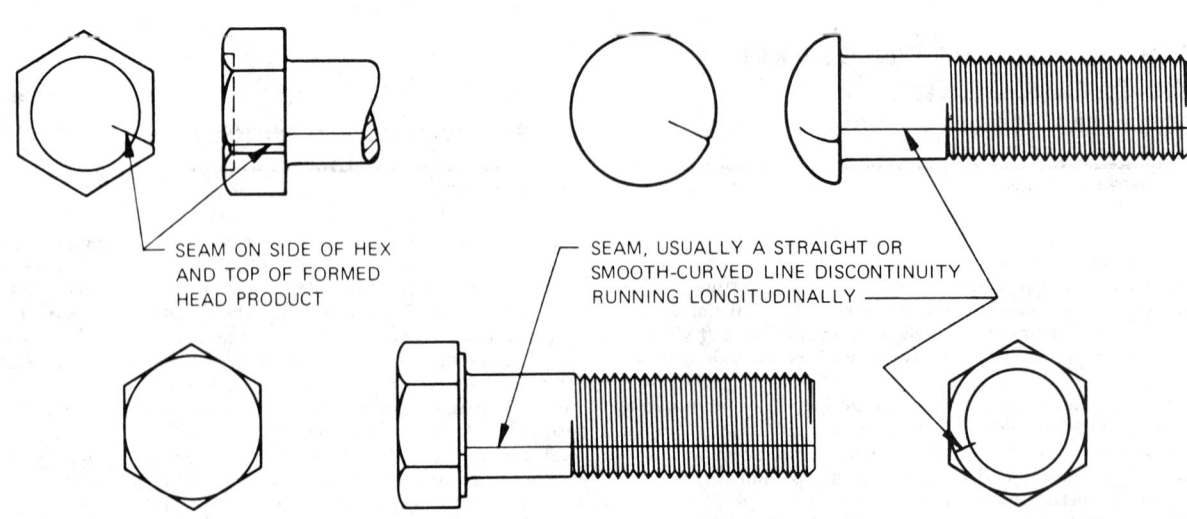

NOTE: Seams are permissible discontinuities if within the limits specified in paragraph 3.5.

FIG. 3—TYPICAL SEAMS

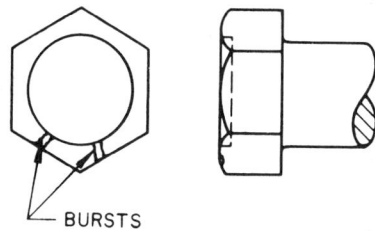

NOTE: Bursts in raised periphery of indented head bolts and screws are permissible discontinuities if within the limits specified in paragraph 3.6.3.

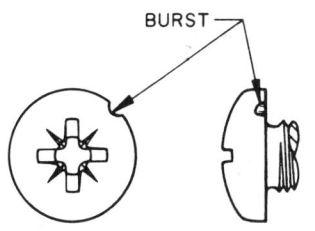

NOTE: Bursts in circular head products, with or without recess, are permissible discontinuities if within the limits specified in paragraph 3.6.2.

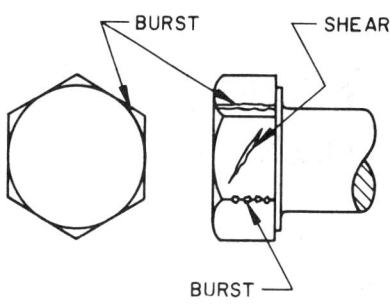
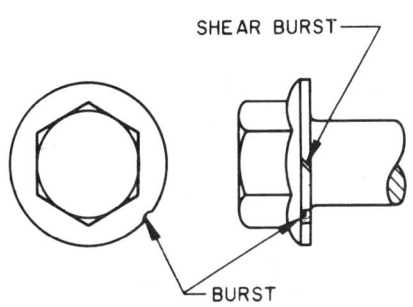

NOTE: Bursts and shear bursts are permissible discontinuities if within the limits specified in paragraph 3.6.

FIG. 4—TYPICAL BURSTS AND SHEAR BURSTS

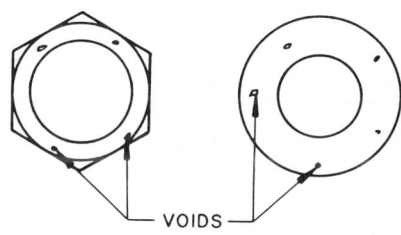

NOTE: Voids are permissible discontinuities if within the limits specified in paragraph 3.7.

FIG. 5—TYPICAL VOIDS ON BEARING SURFACE

changes, and are especially prevalent with noncircular necks, shoulders, and heads. Typical folds are shown in Fig. 6.

2.7 Tool Marks—Tool marks are longitudinal or circumferential grooves of shallow depth produced by the movement of manufacturing tools over the surface of the fastener. Typical tool marks are shown in Fig. 7.

2.8 Nick or Gouge—A nick or gouge is an indentation on the surface of the fastener, produced by forceful abrasion or the impact of product coming into contact with other product or manufacturing equipment during manufacture.

3. Limits for Surface Discontinuities

3.1 Letter Definitions—Throughout the following requirements D designates the nominal size (basic major diameter of thread) of bolts, screws, and studs, except for bolts and screws with shoulders, in which case D designates the largest shoulder diameter. F designates the nominal flange diameter or head diameter of products having circular heads. For metric-series products, use millimeter; for inch-series products use inch.

3.2 Quench Cracks—Quench cracks of any depth, any length, or in any location, are not permitted. (See paragraph 2.1.1 and Fig. 1.)

3.3 Folds—Folds located in internal corners at or below the bearing surface, for example, in the fillet at the junction of head and shank, are not permitted. (See paragraph 2.6 and Fig. 6.)

3.4 Forging Cracks—Forging cracks on the top of heads of bolts and screws shall not exceed a length of 1 D or a width or depth of 0.040 D or 0.25 mm (0.010 in.), whichever is greater.

3.5 Seams—Seams in the shanks of bolts, screws, and studs shall not exceed a depth of 0.030 D or 0.20 mm (0.008 in.), whichever is greater. Seams extending into the heads and flanges of fasteners which do not open beyond the limits specified for bursts are acceptable. (See paragraph 2.2 and Fig. 3.)

3.6 Bursts and Shear Bursts

3.6.1 No burst in the flats of hex bolts and screws shall extend into the top crown of head surface (chamfer circle) or the under head bearing surface. In addition, bursts occurring at the intersection of two wrenching flats shall not reduce the width across corners below the specified minimum.

3.6.2 Flanges of bolts and screws and peripheries of circular head products may have two or more bursts or shear bursts, providing that only one has a width greater than 0.040 F; in addition, this one burst shall not have a width exceeding 0.080 F.

3.6.3 Bursts in the raised periphery of indented head bolts and screws shall not exceed a width of 0.060 D or 0.40 mm (0.015 in.), whichever is greater, or have a depth extending into the indented portion. (See paragraph 2.3 and Fig. 4.)

3.7 Voids on Bearing Surface—Voids on the bearing surface of bolts and screws shall not exceed a depth of 0.25 mm (0.010 in.), and the combined area of all voids shall not exceed 10% of the specified minimum area of the bearing surface. The method for determining area of voids shall be as agreed upon by purchaser and producer. (See paragraph 2.5 and Fig. 5.)

3.8 Tool Marks—Tool marks on the bearing surface shall not exceed surface roughness measurement of 2.8 μm (110 μ in.) determined as the arithmetic average deviation from the mean surface. (See paragraph 2.7 and Fig. 7.)

3.9 Nicks and Gouges—Nicks and gouges located in the threaded length shall not be of size and number which will interfere with assembly of the proper GO thread gage on the thread with the application of not more than 0.05 times DN·m (12 times D in.-lb) of torque, where D is the nominal bolt, screw, or stud size in inches. The manufacturer shall exercise due care during the manufacture and handling of parts to minimize the number and magnitude of nicks and gouges.

4. Inspection Procedure—Bolts, screws, and studs shall be inspected. For referee purposes, unless otherwise specified by purchaser, inspection shall be in accordance with the procedures outlined in paragraphs 4.1 and 4.2.

4.1 Visual Inspection—A representative sample with a size as given in Table 1 shall be picked at random from the lot. The sample shall be examined visually for quench cracks, bursts, shear bursts, forging cracks, folds, tool marks, seams, voids on the bearing surface, and nicks and gouges.

4.1.1 If any part is found with quench cracks or with folds at internal corners at or below the bearing surface, the lot shall be subject to rejection.

4.1.2 If any part is found with seams, bursts, shear bursts, forging cracks, tool marks, voids, or nicks and gouges which exceed the allowable limits as specified for the applicable type of discontinuity under paragraph 3, the lot shall be subject to rejection.

4.2 Seam Inspection—A representative sample with a size as given in Table 1 (if acceptable by visual inspection) shall then be examined for seams by deep (surface) acid etch or magnetic inspection techniques (Magna-glo, Magna-flux, eddy current, etc.). (NOTE: Other examining procedures may be used providing they have an ability to detect discontinuities of the size specified in paragraph 3.5.)

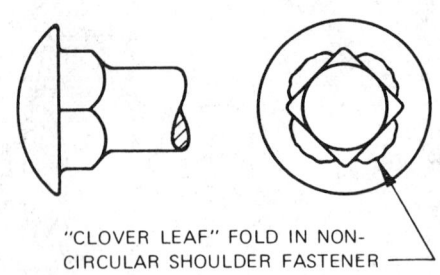

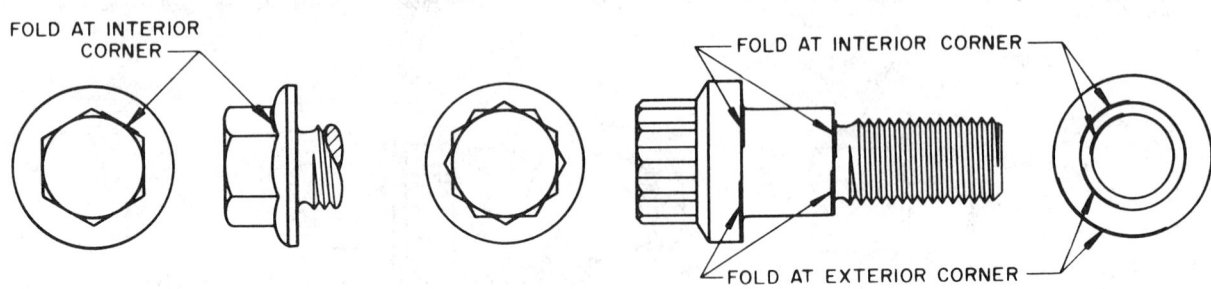

NOTE: Folds in interior corners at or below the bearing surface are not permissible discontinuities. Folds at exterior corners are permissible discontinuities if within the limits specified in paragraph 3.3

FIG. 6—TYPICAL FOLDS

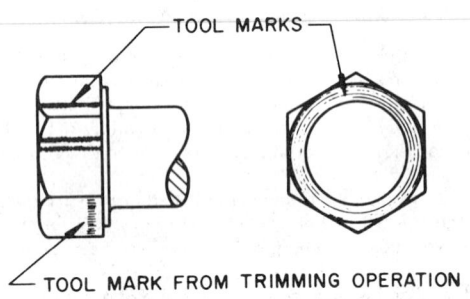

NOTE: Tool marks are permissible discontinuities if within the limits specified in paragraph 3.8.

FIG. 7—TYPICAL TOOL MARKS

TABLE 1—SAMPLE SIZE FOR VISUAL AND SEAM INSPECTION

Lot Size	Sample Size Visual and Nondestructive Techniques	Sample Size Destructive Techniques
Up to 1,500, incl	15	2
1,501 to 5,000	25	3
5,001 to 15,000	50	5
15,001 to 50,000	75	8
50,001 and over	100	10

(NOTE: During this inspection, attention should also be given to examining the product for indications showing transverse quench cracks or folds in internal corners as these discontinuities are sometimes not readily seen in visual examination.)

4.2.1 All parts showing indications which could be interpreted as seams shall be set aside. From this group, a secondary sample with a size as given in Table 2 shall be picked at random and each part in this sample sectioned through the shank perpendicular to the axis and etched for microscopic examination. The section should be through the unthreaded body adjacent to the thread runout. For bolts and screws which are threaded to the head, the section should be taken where the seam indication intersects the root of the thread at a distance of approximately 1 D from the underside of the head, where D is nominal size of bolt or screw. (A recommended procedure is outlined in ASTM E 3, Methods of Preparing Metallographic Specimens.) If during the microscopic examination any part is found having a seam with a depth in excess of the limit specified in paragraph 3.5, the lot shall be subject to rejection.

TABLE 2—SAMPLE SIZE FOR MICROSCOPIC EXAMINATION OF PRODUCTS WITH SEAM INDICATIONS

Number of Products Showing Seam Indications	Sample Size
1	1
2 to 8	2
9 to 15	3
16 to 25	5
26 to 50	8
51 to 90	13
91 to 100	20

SURFACE DISCONTINUITIES ON BOLTS, SCREWS, AND STUDS—SAE J123c

SAE Recommended Practice

Report of Iron and Steel Technical Committee approved September 1973 and last revised June 1976.

1. Scope—This recommended practice defines, illustrates, and specifies allowable limits for various types of surface discontinuities that may occur or become apparent during the manufacture and processing of bolts, screws, and studs which are primarily intended for use in automotive assemblies subjected to severe dynamic stresses and necessitating use of high strength fasteners having appropriate fatigue resistant properties.

1.1 The basic recommended practice does not include inspection sampling requirements. It is intended that the purchaser shall specify, in the original inquiry and purchase order, the inspection sampling requirements which the producer must satisfy to demonstrate the acceptability of bolts and screws with respect to surface discontinuities. Appendix outlines inspection sampling plans applicable when such requirements are not specified by the purchaser in the original inquiry, purchase order, or in related specifications.

2. Types of Surface Discontinuities—For the purpose of this recommended practice, surface discontinuities on bolts, screws, and studs are divided into 10 "types," defined as follows:

2.1 Crack—A crack is a clean (crystalline) fracture passing through or across the grain boundaries without inclusion of foreign elements. Cracks are normally caused by overstressing the metal during forging or other forming operation, or during heat treatment. Where parts are subjected to significant reheating, cracks usually are discolored by scale.

2.1.1 QUENCH CRACKS—Quench cracks may occur during heat treatment due to excessively high thermal and transformation stresses. They usually traverse an irregular and erratic course on the surface of the fastener. Typical quench cracks are shown in Fig. 1.

2.1.2 FORGING CRACKS—Forging cracks may occur during the cutoff or forging operations and are located on the top of the heads of screws and bolts. Typical forging cracks are shown in Fig. 2.

2.2 Seam—Seams are generally inherent in the raw material from which fasteners are manufactured. They are narrow, generally straight or smooth-curved line discontinuities, running longitudinally on the shank and/or thread. Seams may extend onto the tops of the heads of circular head products as well as being present at the periphery of the head. Seams may also extend into the chamfer circle, washer face, and wrenching flats of hex head products. Typical seams are shown in Fig. 3.

2.3 Burst—A burst is an open break in the metal (material). Bursts may occur on the flats or corners of the heads of bolts and screws, at the periphery of flanged or circular head products, or on the raised periphery of indented head bolts and screws. Typical bursts are shown in Fig. 4.

2.4 Shear Burst—A shear burst is an open break in the metal, occurring most frequently at the periphery of products having circular or flanged heads and are generally located at approximately 45 deg to the product axis. Shear bursts may also occur on the sides of hex head products. Typical discontinuities of this type are shown in Fig. 4.

2.5 Void—A void is a shallow pocket or hollow on the surface of the bolt or screw due to nonfilling of metal during forging or upsetting. Typical voids are shown in Fig. 5.

2.6 Lap—A lap is a fold-over of metal in the threads of screws, bolts, and studs. If laps occur, they generally show a pattern of consistency between the product, that is, laps will be identically located and with the same direction of traverse between all product. Typical laps in external threads are shown in Fig. 6A.

2.7 Fold—A fold is a doubling over of metal which may occur during the forging operation. Folds may occur at or near the intersection of diameter changes and are especially prevalent with noncircular necks, shoulders, and heads. Typical folds are shown in Fig. 7.

2.8 Tool Marks—Tool marks are longitudinal or circumferential grooves of shallow depth produced by the movement of manufacturing tools over the surface of the fastener. Typical tool marks are shown in Fig. 8.

2.9 Nick or Gouge—A nick or gouge is an indentation on the surface of the fastener, produced by forceful abrasion or the impact of product coming into contact with other product or manufacturing equipment during manufacture.

3. Limits for Surface Discontinuities

3.1 Letter Definitions—Throughout the following requirements, D designates the nominal size (basic major diameter of thread) of bolts, screws, and studs, except for bolts and screws with shoulders, in which case D designates

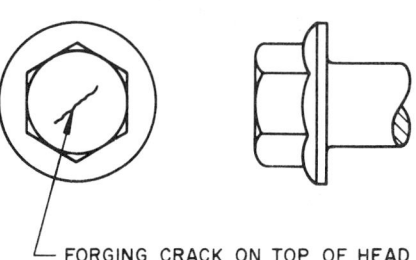

FORGING CRACK ON TOP OF HEAD

NOTE: Forging cracks are permissible discontinuities if within the limits specified in paragraph 3.5.

FIG. 2—TYPICAL FORGING CRACKS

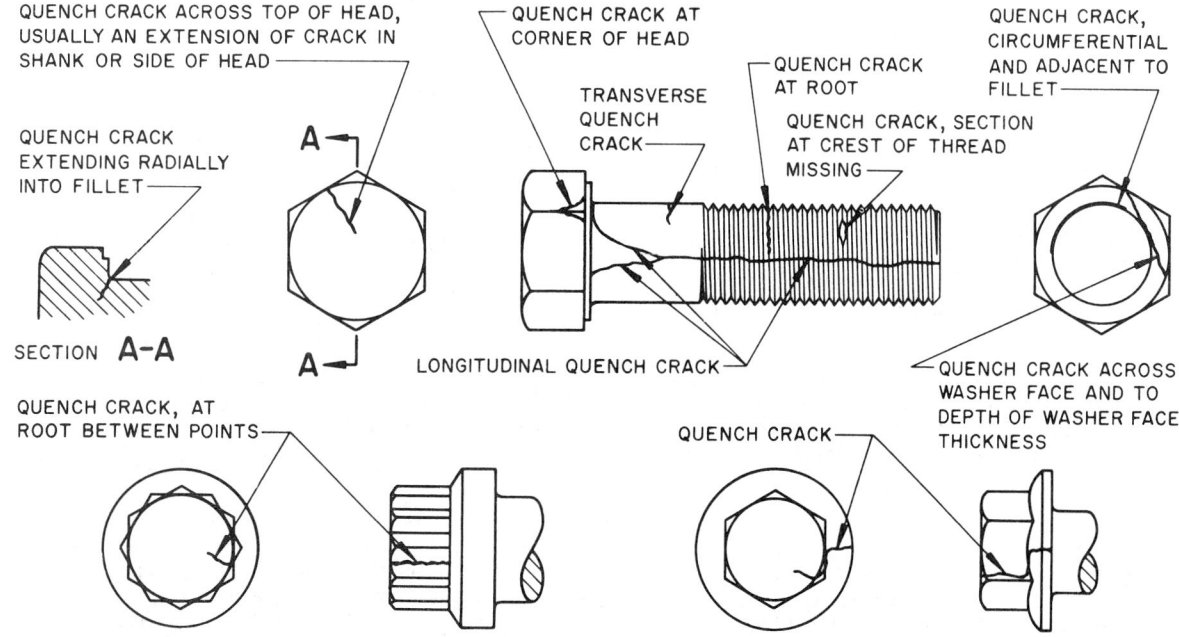

NOTE: Quench cracks of any depth, any length, or in any location are not permissible discontinuities.

FIG. 1—TYPICAL QUENCH CRACKS

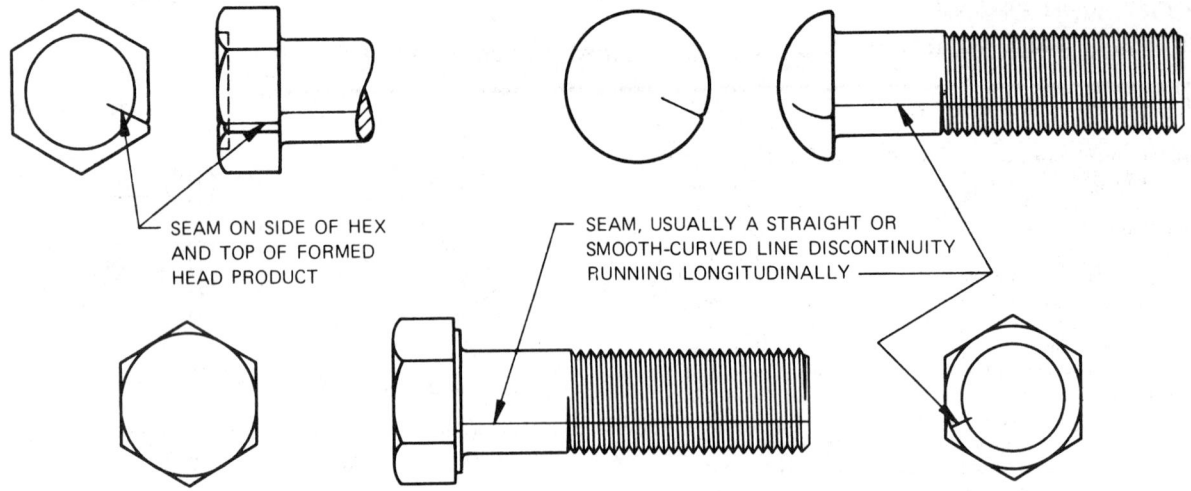

FIG. 3—TYPICAL SEAMS

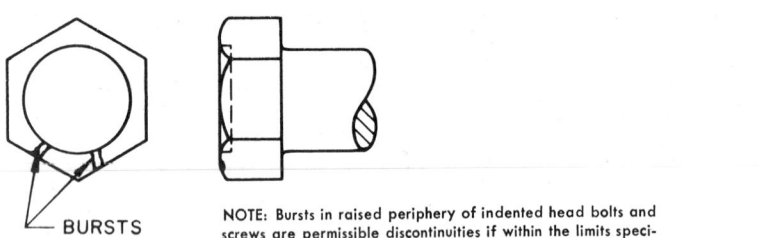

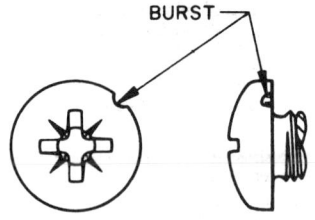

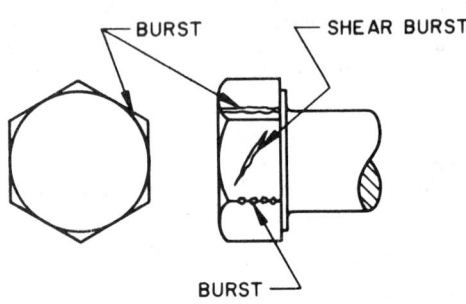

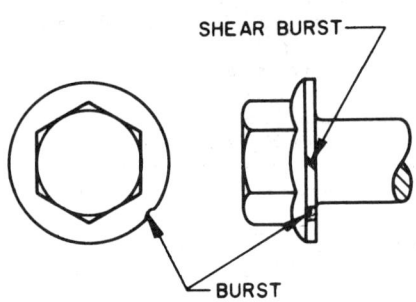

NOTE: Bursts and shear bursts are permissible discontinuities if within the limits specified in paragraph 3.7.

FIG. 4—TYPICAL BURSTS AND SHEAR BURSTS

the largest shoulder diameter. F designates the nominal flange diameter or head diameter of products having circular heads. For metric-series products, use millimeter; for inch-series products, use inch.

3.2 Quench Cracks—Quench cracks of any depth, any length, or in any location, are not permitted. (See paragraph 2.1.1 and Fig. 1).

3.3 Laps in Screw Threads—Laps of any depth and any length which (a) are present in the root of the screw thread, or (b) originate on the flank, traverse toward the interior, and extend in depth below the pitch line of the bolt, screw, or stud, or (c) originate below the pitch line on the pressure flank and traverse toward the major diameter, are not permitted. (This requirement is not applicable to tapping screws having spaced threads.) (See paragraph 2.6 and Fig. 6A.) When approved by the purchaser, marks on threads caused by the serrations on thread rolling dies (see Fig. 6B) shall be excluded from these requirements.

3.4 Folds

3.4.1 Folds located in internal corners at or below the bearing surface, for example, in the fillet at the junction of head and shank, are not permitted.

3.4.2 Folds located at the intersection of the flange periphery and bearing surface, are not permitted. (See paragraph 2.7 and Fig. 7.)

3.5 Forging Cracks—Forging cracks on the top of head bolts and screws shall not exceed a length of 1 D or a width or depth of 0.20 mm (0.008 in) +0.010 D.

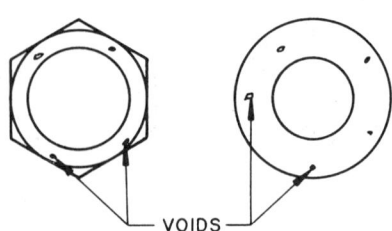

NOTE: Voids are permissible discontinuities if within the limits specified in paragraph 3.8.

FIG. 5—TYPICAL VOIDS ON BEARING SURFACE

3.6 Seams—For metric-series bolts, screws and studs, seams in the shanks shall not exceed (a) an open width at the surface of 0.13 mm for sizes 6.3 to 12 mm, inclusive, and 0.25 mm for sizes 14 mm and larger, and (b) a depth of 0.015 D +0.10 mm for sizes 6.2 to 16 mm, inclusive, and 0.030 D for sizes over 16 mm. (See paragraph 2.2 and Fig. 3.)

For inch-series bolts, screws and studs, seams in the shanks shall not exceed (a) an open width at the surface of 0.005 in for sizes $\frac{1}{4}$ to $\frac{7}{16}$ in, inclusive, and 0.010 in for sizes $\frac{1}{2}$ in and larger, and (b) a depth of 0.015 D +0.004 in for

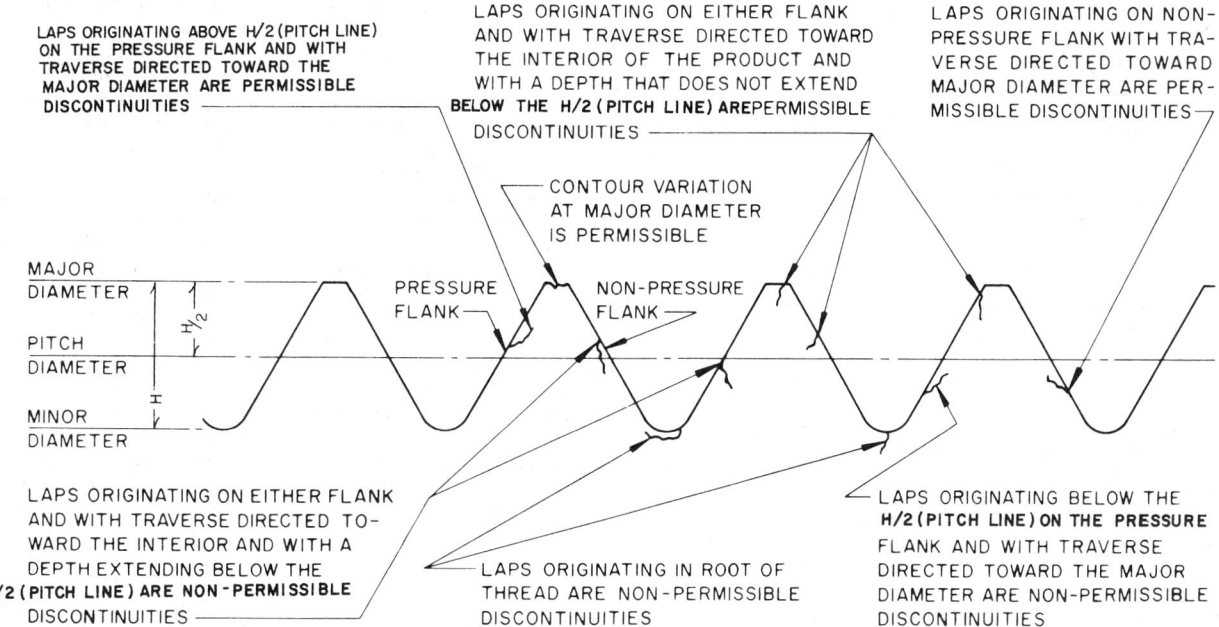

FIG. 6A—SURFACE DISCONTINUITIES IN EXTERNAL SCREW THREADS

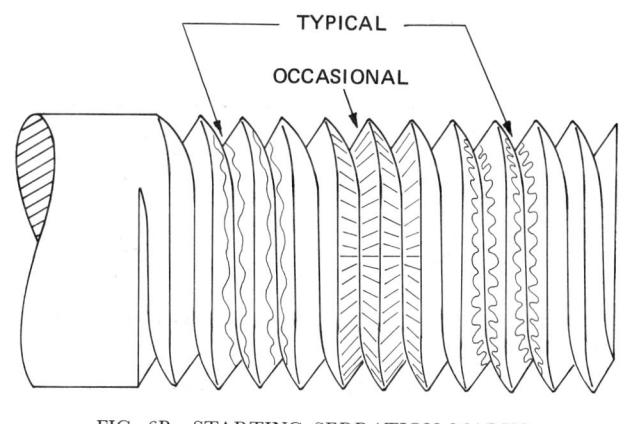

FIG. 6B—STARTING SERRATION MARKS

sizes 1/4 to 5/8 in, inclusive, and 0.030 D for sizes over 5/8 in. (See paragraph 2.2 and Fig. 3.)

Seams extending into the heads and flanges of fasteners which do not open beyond the limits specified for bursts are acceptable.

3.7 Bursts and Shear Bursts

3.7.1 Bursts in the flats of hex bolts and screws shall not exceed a width or an open depth of 0.25 mm (0.010 in) + 0.025 D. In addition, no burst shall extend into the bearing surface, nor shall any burst occurring at the intersection of two wrenching flats reduce the width across corners below the specified minimum. (See paragraphs 2.3 and 2.4 and Fig. 4.)

3.7.2 Flanges of bolts and screws and peripheries of circular head products may have two or more bursts or shear bursts providing that only one has a width greater than 0.13 mm (0.005 in) + 0.020 F or an open depth greater than 0.08 mm (0.003 in) + 0.012 F; in addition, this one burst shall not have a width exceeding 0.25 mm (0.010 in) + 0.040 F or an open depth of 0.15 mm (0.006 in) + 0.024 F.

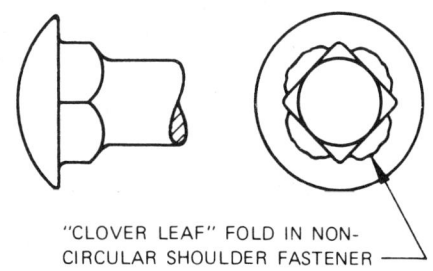

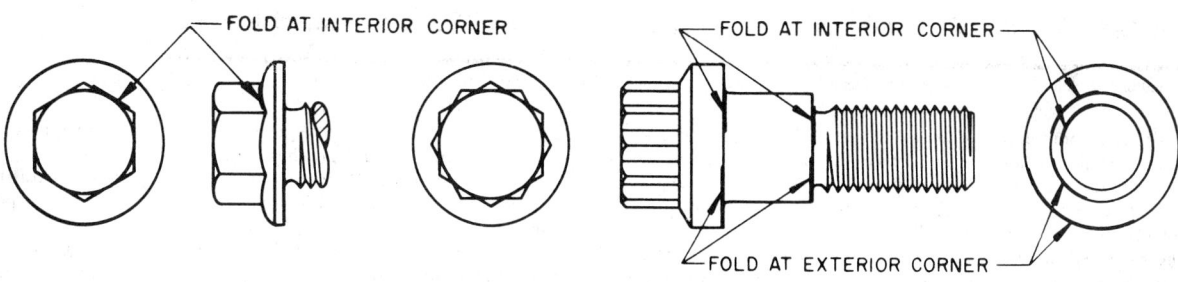

FIG. 7—TYPICAL FOLDS

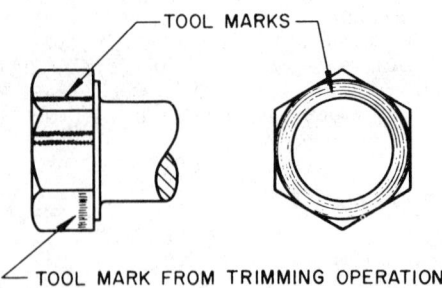

NOTE: Tool marks are permissible discontinuities if within the limits specified in paragraph 3.9.

FIG. 8—TYPICAL TOOL MARKS

3.7.3 Bursts in the raised periphery of indented head bolts and screws shall not exceed a width of 0.25 mm (0.010 in) + 0.020 D, or have a depth greater than the height of the raised periphery. (See paragraph 2.3 and Fig. 4.)

3.8 Voids on Bearing Surface—Voids on the bearing surface of bolts and screws shall not exceed a depth of 0.25 mm (0.010 in) and the combined area of all voids shall not exceed 5% of the specified minimum area of the bearing surface. The method for determining area of voids shall be as agreed upon by purchaser and producer.

3.9 Tool Marks—Tool marks on the bearing surface shall not exceed surface roughness measurement of 2.8 μm (110 μin) determined as the arithmetic average deviation from the mean surface. (See paragraph 2.8 and Fig. 8.)

3.10 Nicks and Gouges—Nicks and gouges located in the threaded length shall not be of size and number which will interfere with assembly of the proper GO thread gage on the thread with the application of not more than 0.06 times DN · m (12 times D in-lb) of torque, where D is the nominal bolt, screw, or stud size in inches. The manufacturer shall exercise due care during the manufacture and handling of parts to minimize the number and magnitude of nicks and gouges.

4. Inspection Procedure—Bolts, screws and studs shall be inspected in accordance with the procedures outlined in paragraphs 4.1 and 4.2, unless otherwise specified by purchaser.

4.1 Visual Inspection—A representative sample[1] shall be picked at random from the lot. The sample shall be examined visually for quench cracks, bursts, shear bursts, forging cracks, folds, tool marks, seams, voids on the bearing surface, and nicks and gouges.

4.1.1 If any part is found with quench cracks or with folds at internal corners at or below the bearing surface, the lot shall be subject to rejection.

4.1.2 If any part is found with seams, bursts, shear bursts, forging cracks, tool marks, voids, or nicks and gouges which exceed the allowable limits as specified for the applicable type of discontinuity under paragraph 3, the lot shall be subject to rejection.

4.2 Seam and Lap Inspection—The same sample (if acceptable by visual inspection) shall then be further examined for laps in threads and seams by deep (surface) acid etch or magnetic inspection techniques (Magna-glo, Magna-flux, eddy current, etc.). (NOTE: Other examining procedures may be used providing they have an equivalent ability to detect discontinuities of the size specified in paragraphs 3.3 and 3.6.)

4.2.1 All parts showing indications which could be interpreted as seams shall be set aside. From this group, a secondary sample[1] shall be picked at random and each part in this sample sectioned through the shank perpendicular to the axis and etched for microscopic examination. The section should be through the unthreaded body adjacent to the thread runout. For bolts and screws which are threaded to the head, the section should be taken where the seam indication intersects the root of the thread at a distance of approximately one D from the underside of the head, where D is nominal size of bolt or screw. (A recommended procedure is outlined in ASTM E 3, Methods of Preparing Metallographic Specimens.) If during the microscopic examination any part is found having a seam with a depth in excess of the limit specified in paragraph 3.6, the lot shall be subject to rejection.

4.2.2 Laps in Screw Threads—All products showing indications of laps in the threads shall be set aside. If the original sample examined in paragraph 4.2 was inspected using magnetic techniques, this same sample (if acceptable by inspection in paragraph 4.1) shall again be examined by magnetic inspection techniques but with the direction of flow of magnetization current changed so as to detect discontinuities located transverse to the bolt or screw axis. (NOTE: During this inspection, attention should also be given to examining the product for indications showing transverse quench cracks or folds in internal corners as these discontinuities are sometimes not readily seen in visual examination.) From the products set aside, five specimens shall be selected at random, surface etched, and examined visually to determine that the pattern of laps is reasonably consistent between products. One of the five specimens shall be sectioned longitudinally on the center line of the part and on a plane passing through the point at which any lap extends closest to the minor diameter of the thread and the section etched for microscopic examination. If during the microscopic examination any lap is found with a location and direction of traverse which classifies the lap as nonpermissible, the lot shall be subject to rejection.

APPENDIX—SAMPLING PLAN

A1. Scope—This Appendix outlines inspection sampling plans applicable when the purchaser has not specified a plan or plans in the original inquiry or purchase order, or in related specifications.

A2. Sample size for visual and magnetic analysis inspection of surface discontinuities is:

Lot Size	Sample Size
Up to 1,500 incl	15
1,501 to 5,000	25
5,001 to 15,000	50
15,001 to 50,000	75
50,001 and over	100

A3. Sample size for microscopic examination of products with seam indications is:

Number of Products Showing Seam Indications	Sample Size
1	1
2 to 8	2
9 to 15	3
16 to 25	5
26 to 50	8
51 to 90	13
91 to 100	20

[1] See Appendix.

SURFACE DISCONTINUITIES ON NUTS—J122a   SAE Recommended Practice

Report of Iron and Steel Technical Committee approved September 1969. Editorial change; metric products added July 1975.

1. Scope—This recommended practice defines, illustrates, and specifies allowable limits for the various types of surface discontinuities that may occur during the manufacture and processing of metric-series nuts, in sizes 6.3 to 25 mm and inch-series nuts in sizes $\frac{1}{4}$ to 1 in., inclusive, which are primarily intended for use in automotive assemblies.

1.1 The basic recommended practice does not include inspection sampling requirements. It is intended that the purchaser shall specify in the original inquiry and purchase order the inspection sampling requirements which the producer must satisfy to demonstrate the acceptability of nuts with respect to surface discontinuities. Appendix outlines inspection sampling plans applicable when such requirements are not specified by the purchaser in the original inquiry and purchase order.

2. Types of Surface Discontinuities—For the purpose of this recommended practice, surface discontinuities on nuts are divided into 11 "types," defined as follows:

2.1 Cracks—A crack is a clean (crystalline) fracture passing through or across the grain boundaries without inclusion of foreign elements. Cracks are normally caused by overstressing the metal during forging or other forming operation, or during heat treatment. Where parts are subjected to significant reheating, cracks usually are discolored by scale.

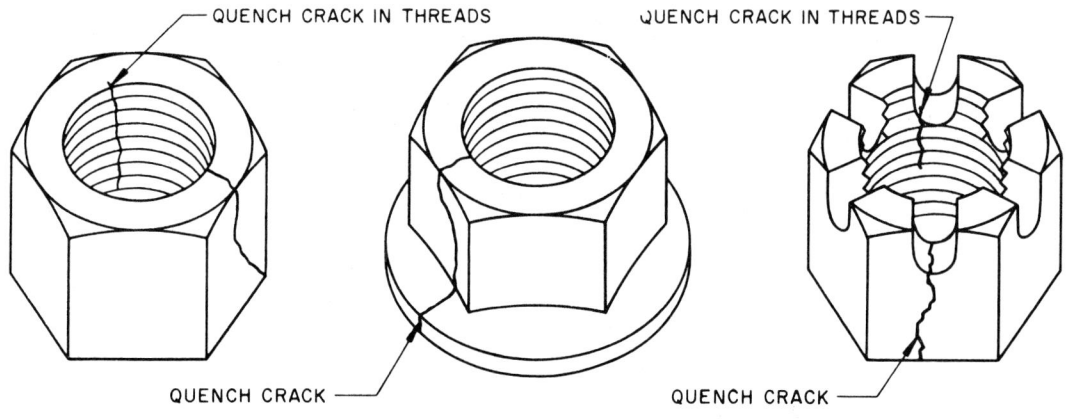

NOTE: Quench cracks of any depth, any length, or in any location on a nut are not permissible discontinuities.

FIG. 1—TYPICAL QUENCH CRACKS

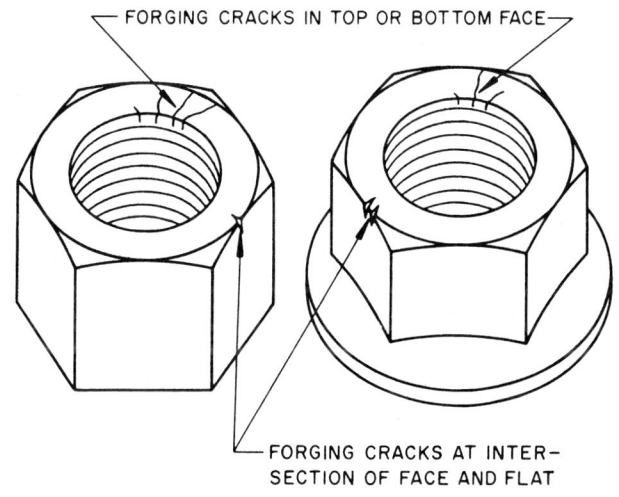

NOTE: Forging cracks are permissible discontinuities if within the limits specified in paragraph 3.2.2.2.

FIG. 2—TYPICAL FORGING CRACKS

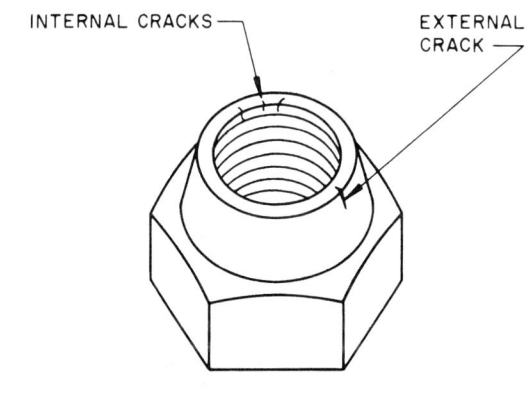

NOTE: Locking element formation cracks located on the external surface of the nut are not permissible discontinuities.
    Locking element formation cracks located on the internal surface of the nut are permissible discontinuities if within the limits specified in paragraph 3.2.3.1.

FIG. 3—TYPICAL LOCKING ELEMENT FORMATION CRACKS ON PREVAILING-TORQUE NUTS

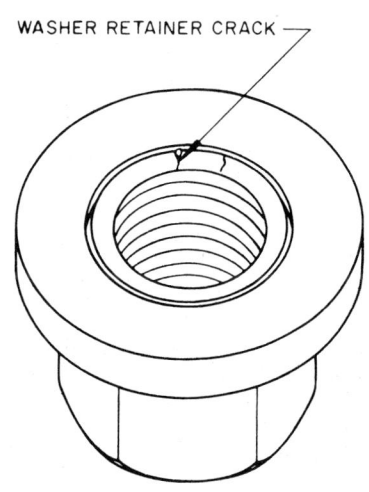

NOTE: Washer retainer cracks are permissible discontinuities.

FIG. 4—TYPICAL WASHER RETAINER CRACKS

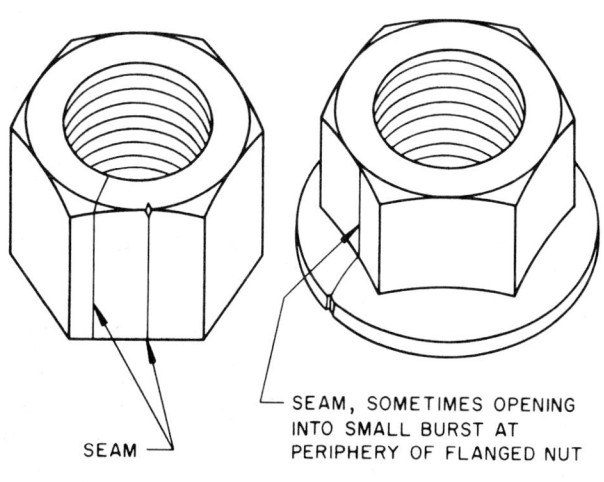

NOTE: Seams are permissible discontinuities **only** if within the limits specified in paragraph 3.3.1.

FIG. 5—TYPICAL SEAMS

2.1.1 QUENCH CRACKS—Quench cracks may occur during heat treatment due to excessively high thermal and transformation stresses. They usually traverse an irregular and erratic course on the surface of the nut. Typical quench cracks are shown in Fig. 1.

2.1.2 FORGING CRACKS—Forging cracks may occur during the cutoff or forging operations and are located on the top and bottom face of the nut, and at the intersection of the face and flat. Typical forging cracks are shown in Fig. 2.

2.1.3 LOCKING ELEMENT FORMATION CRACKS—These cracks may occur due to the application of pressure against the nut when introducing the locking element into prevailing torque nuts. Cracks of this type are usually located in the vicinity of the locking element and may either be on the internal or external surface. Typical locking element formation cracks are shown in Fig. 3.

2.1.4 WASHER RETAINER CRACK—A washer retainer crack is an opening in a lip or hub of metal used for captivating a washer on a nut. Washer retainer cracks may occur when pressure is applied to the lip or hub during assembly of the washer. Such cracks are limited to the contour of the hub or lip used for retaining purposes. Typical washer retainer cracks are shown in Fig. 4.

2.2 Seams—A seam is a narrow continuous discontinuity in the metal. Seams are generally inherent in the raw material from which the nut is made. Seams in nuts are usually straight or smooth curved line discontinuities running generally parallel to the nut axis. Typical seams are shown in Fig. 5.

2.3 Bursts—A burst is an open break in the metal. Bursts may occur at the periphery of flanged nuts. A typical burst is shown in Fig. 6.

2.4 Shear Failures—A shear failure is an open break in the metal, located at the periphery of flanged nuts and approximately at a 45 deg angle to the nut axis. Shear failures occur most frequently with flanged nuts and are due to overstressing the metal during forging. A typical shear failure is shown in Fig. 6.

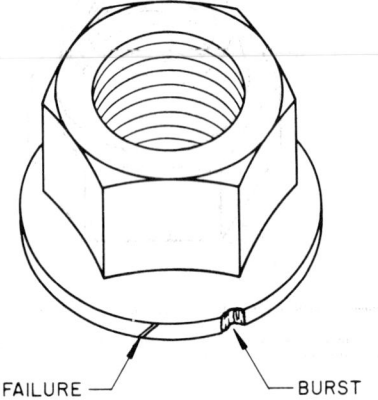

NOTE: Bursts are permissible discontinuities if within the limits specified in paragraph 3.4. Shear failures are permissible discontinuities if within the limits specified in paragraph 3.5.

FIG. 6—TYPICAL BURST AND SHEAR FAILURES ON FLANGED NUTS

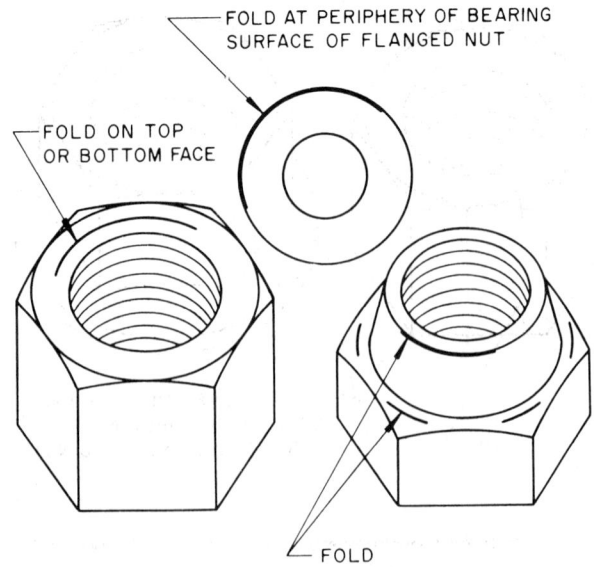

NOTE: Folds are permissible discontinuities if within the limits specified in paragraph 3.6.

FIG. 7—TYPICAL FOLDS

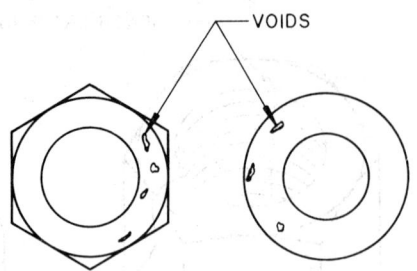

NOTE: Voids on the bearing surface of nuts are permissible if within the limits specified in paragraph 3.7.

FIG. 8—TYPICAL VOIDS ON BEARING SURFACES

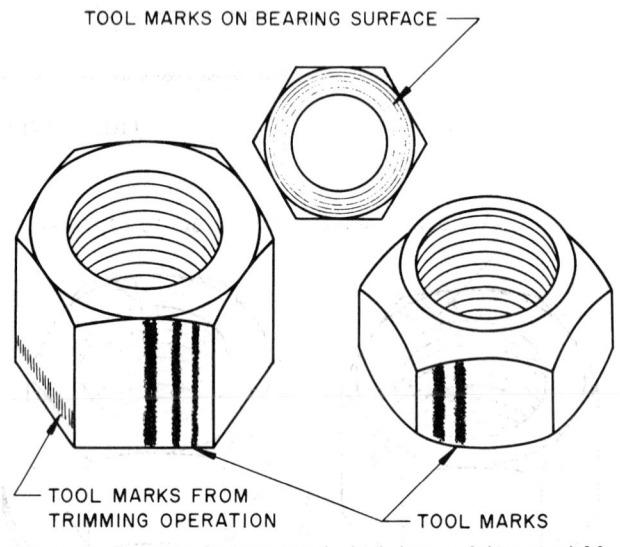

NOTE: Tool marks are permissible discontinuities if within the limits specified in paragraph 3.8.

FIG. 9—TYPICAL TOOL MARKS

2.5 Folds—A fold is a doubling over of metal which may occur during the forging operation. Folds in nuts may occur at or near the intersection of diameter changes or on the top or bottom face of the nut. Typical folds are shown in Fig. 7.

2.6 Voids—A void is a shallow pocket or hollow on the surface of the nut due to nonfilling of metal during forging. Typical voids are shown in Fig. 8.

2.7 Tool Marks—Tool marks are longitudinal or circumferential grooves of shallow depth, produced by the movement of manufacturing tools over the surface of the nut. Typical tool marks are shown in Fig. 9.

2.8 Nicks or Gouges—A nick or gouge is an indentation on the surface of a nut, produced by forceful abrasion or the impact of product coming into contact with other product or manufacturing equipment during manufacture.

3. Limits for Surface Discontinuities

3.1 Letter Definitions—Throughout the following requirements, D designates the nominal size (basic major diameter of thread of nuts). F designates the nominal flange diameter. For metric-series nuts use millimeter; for inch-series nuts use inch.

3.2 Cracks

3.2.1 QUENCH CRACKS of any depth, any length, or in any location, are not permitted. (See paragraph 2.1.1 and Fig. 1.)

3.2.2 FORGING CRACKS are permissible discontinuities providing that nuts with forging cracks, in addition to meeting the dimensional requirements detailed below, can meet the requirements of the cone proof load test described in paragraph 4.3.

3.2.2.1 Forging cracks located in the top and bottom face of nuts are permissible discontinuities providing that (a) there are no more than two forging cracks which extend from the tapped hole across the full width of the faces; (b) no forging crack extends into the tapped hole beyond the first full thread; and (c) the width of any forging crack does not exceed 0.20 mm (0.008 in.) +0.010 D.

3.2.2.2 Forging cracks located at the intersection of top or bottom face with the flat (these discontinuities are sometimes interpreted as bursts) shall not exceed 0.25 mm (0.010 in.) +0.020 D. (See paragraph 2.1.2 and Fig. 2.)

3.2.3 LOCKING ELEMENT FORMATION CRACKS of any length or any depth, located on the external surface of prevailing torque nuts, are not permitted.

3.2.3.1 Locking element formation cracks located on the internal surface of prevailing torque nuts shall not exceed two thread pitches in length; shall not extend into the thread root; and the width of any crack shall not exceed the following: For metric-series nuts, 0.18 mm for sizes 6.3 to 11 mm, inclusive, and 0.25 mm for sizes 12 to 25 mm, inclusive. For inch-series nuts, 0.007 in. for sizes ¼ to ⁷⁄₁₆ in., inclusive, and 0.010 in. for sizes ½ to 1 in., inclusive (See paragraph 2.1.3 and Fig. 3.)

3.2.4 WASHER RETAINER CRACKS are permissible discontinuities. (See paragraph 2.1.4 and Fig. 4.)

3.3 **Seams**—Seams are permissible discontinuities providing that nuts with seams, in addition to meeting the dimensional requirements detailed below, can meet the requirements of the cone proof load test described in paragraph 4.3.

3.3.1 Seams shall not exceed the following open width at the surface; for metric-series nuts, 0.13 mm for sizes 6.3 to 11 mm, inclusive, and 0.25 mm for sizes 12 to 25 mm, inclusive. For inch-series nuts, 0.005 in. for nut sizes ¼ to ⁷⁄₁₆ in., inclusive, and 0.010 in. for nut sizes ½ to 1 in., inclusive. (See paragraph 2.2 and Fig. 5.)

3.4 **Bursts**—Bursts in flanged nuts shall not exceed a width of 0.13 mm (0.005 in.) +0.020 F or an open depth of 0.08 mm (0.003 in.) +0.012 F, except that one burst may have a width no greater than 0.25 mm (0.010 in.) +0.040 F, or an open depth no greater than 0.15 mm (0.006 in.) +0.024 F. (See paragraph 2.3 and Fig. 6.)

3.5 **Shear Failures**—Shear failures on flanged nuts shall not exceed a width of 0.020 F or a depth of 0.030 F. (See paragraph 2.4 and Fig. 6.)

3.6 **Folds**—Folds located at the intersection of the flange periphery and bearing surface of flanged nuts shall not project below the bearing surface. (See paragraph 2.5 and Fig. 7.)

3.7 **Voids**—Voids on the bearing surface of nuts shall not exceed a depth of (0.25 mm (0.010 in.), and the combined area of all voids on the bearing surface shall not exceed 5% of the specified minimum area of the bearing surface. (See paragraph 2.6 and Fig. 8.)

3.8 **Tool Marks**—Tool marks on the bearing surface shall not exceed a surface roughness measurement of 2.8 μm (110 μ in.) determined as the arithmetic average deviation from the mean surface. Tool marks on other surfaces are permissible discontinuities. (See paragraph 2.7 and Fig. 9.)

3.9 **Nicks and Gouges**—Nicks and gouges are permissible discontinuities; however, the manufacturer shall exercise due care during the manufacture and handling of nuts to minimize the number and magnitude of nicks and gouges. (See paragraph 2.8.)

4. *Inspection Procedure*—Nuts shall be inspected in accordance with the procedures outlined in paragraphs 4.1, 4.2, and 4.3, unless otherwise specified by purchaser.

4.1 **Visual Inspection**—A representative sample[1] shall be picked at random from the lot. The sample shall be examined visually for quench cracks, locking element formation cracks, width of seams, bursts, shear failures, forging cracks, folds, voids on the bearing surface, and tool marks on the bearing surface.

4.1.1 If any nuts are found with quench cracks, or if any prevailing torque nuts are found with locking element formation cracks located on the external surface, the lot shall be subject to rejection.

4.1.2 If any nuts are found with locking element formation cracks located on the internal surface, seams, bursts, shear failures, forging cracks, folds, voids, or tool marks which exceed the allowable limits as specified for the applicable type of discontinuity under paragraph 3, the lot shall be subject to rejection.

4.2 **Magnetic Analysis Inspection**—The same sample (if acceptable by visual inspection) shall then be further examined by magnetic inspection techniques (Magna-glo, Magna-flux, eddy current, etc.). (NOTE: Other examining procedures may be used providing they have an equivalent ability to detect discontinuities of the size specified under paragraph 3.) All nuts showing indications that could be interpreted as seams, and all nuts showing strong indications of forging cracks with potentially significant depth, shall be set aside. From this group a secondary sample[1] shall be picked at random. Each nut in this sample shall be cone proof load tested as outlined in paragraph 4.3. If any nut fails to meet the requirements of this test, the lot shall be subject to rejection.

4.3 **Cone Proof Load Test**—The purpose of the test[2] is to detect the presence of detrimental seams or forging cracks. The use of a conical washer and threaded mandrel, as illustrated in Fig. 10, exaggerates the influence of such discontinuities on the load carrying ability of the nut by introducing a simultaneous dilation and stripping action on the nut.

4.3.1 The mandrel shall have a hardness of Rockwell C45 minimum. For metric-series nuts, the mandrel shall be threaded to class 5 g 6 g tolerances, except the minimum major diameter shall be the specified minimum major

[1] See Appendix.
[2] Same as test specified in ASTM A 194, Steel Nuts for High Pressure and High Temperature Service.

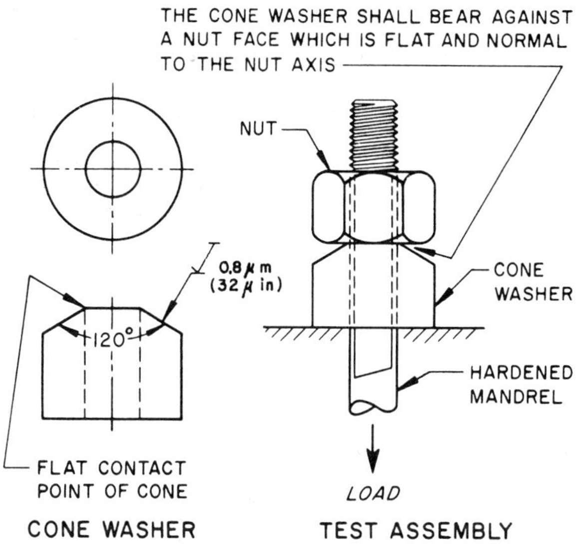

FIG. 10—USE OF CONICAL WASHER AND THREADED MANDREL

diameter for class 6 g, and the maximum major diameter shall be minimum major diameter plus 0.25 times the class 6 g major diameter tolerance; for inch-series nuts, the mandrel shall be threaded to Class 3A tolerance except that the major diameter shall be minimum major diameter with a tolerance of +0.002 in.

4.3.2 The cone washer shall have a hardness of Rockwell C57 minimum; a hole diameter equivalent to the nominal diameter of the mandrel, +0.05, −0 mm (+0.002, −0 in.); and a flat contact point as follows: For metric-series nuts, 0.10–0.15 mm for sizes 6.3 through 12 mm, and 0.35–0.41 mm for sizes 14 through 25 mm diameter; for inch-series nuts, 0.004–0.006 for sizes ¼ through ½ in. diameter, and 0.014–0.016 for sizes ½ through 1 in. diameter.

4.3.3 The nut and the cone washer shall be assembled on mandrel, and the cone proof load for the nut shall be applied against the nut through the cone washer. (The speed of testing, as determined with free running cross head, shall not exceed 3 mm/minute (0.125 in./minute). Loading shall be applied for 10 sec.) The cone proof load of a nut shall be computed using the following formula:

$$PL_c = PL_a (1 - 0.30D)$$

where:
- PL_c = cone proof load, N(lb)
- PL_a = specified axial proof load, N(lb)
- D = nominal size of nut, mm(in.)

4.3.4 To meet the requirements of the cone proof load test, the nut shall support its specified cone proof load.

APPENDIX—SAMPLING PLAN

A1. Scope—This appendix outlines inspection sampling plans applicable when the purchaser has not specified a plan or plans in the original inquiry or purchase order, or in related specifications.

A2. Sample size for visual and magnetic analysis inspection of surface discontinuities is:

Lot Size	Sample Size
Up to 1,500 incl	15
1,501 to 5,000	25
5,001 to 15,000	50
15,001 to 50,000	75
50,001 and over	100

A3. Sample size for cone proof load testing of products with seam and/or forging crack indications is:

Number of Products Showing Seam Indications	Sample Size
1	1
2 to 8	2
9 to 15	3
16 to 25	5
26 to 50	8
51 to 90	13
91 to 100	20

DECARBURIZATION IN HARDENED AND TEMPERED THREADED FASTENERS—SAE J121a

SAE Recommended Practice

Report of Iron and Steel Technical Committee approved September 1969 and last revised July 1976.

1. **Scope**—This recommended practice covers methods for measuring, classifying, and specifying decarburization in the threaded section of hardened and tempered steel bolts, screws, studs, and similar parts. It is not intended to cover products which are specifically carburized to achieve special properties.

2. **Definitions**—According to SAE J419, "decarburization" is the loss of carbon at the surface of commercial ferrous materials which have been heated to facilitate fabrication or heated to modify mechanical properties. SAE J419 defines also "complete decarburization," "partial decarburization," and "effective decarburization," as related to unhardened steels. This standard extends these definitions, as follows, to cover more specifically hardened and tempered steel bolts, screws, studs, and similar products.

   2.1 **Partial Decarburization**—Decarburization with loss of carbon sufficient to cause a lighter shade of tempered martensite than that of the immediately adjacent base metal, when examined metallographically by the method outlined in paragraph 4.1, but insufficient carbon loss to show clearly defined ferrite grains. (The hardness traverse method, outlined in paragraph 4.2, is the referee method for determining that partial decarburization is not present at a point below that shown in Fig. 1 for each classification.)

   2.2 **Gross Decarburization**—Decarburization with sufficient carbon loss to show only clearly defined ferrite grains under metallographic examination by the method outlined in paragraph 4.1. This is sometimes called "Complete Decarburization."

   2.3 **Carbon Restoration**—A process of restoring surface carbon loss by heat treating in an atmosphere furnace of properly controlled carbon potential. (This process is generally required to produce products having decarburization as defined herein.)

   2.4 **Carburization**—A darker shade of tempered martensite than that of the immediately adjacent base metal, when examined metallographically by the method outlined in paragraph 4.1, and harder by at least 30 points (Knoop or Vickers DPH) than the hardness at root diameter when checked by the method outlined in paragraph 4.2. (The limits established by this standard exclude this condition.)

   2.5 **Base Metal Hardness**—For purposes of this report, hardness at root diameter on a line bisecting the included angle of the thread (Position 1 in Fig. 2) is considered "base metal hardness."

3. **Classes of Decarburization**—This report establishes two classes of decarburization for inch series threaded products and three classes for SI (metric) series threaded products. Each class is characterized by dimensional limits for decarburized zone, gross decarburized zone, and/or base carbon zone (as applied to longitudinal sections through the thread axis). Decarburization limits applicable to the more commonly used ISO Modified Threads (mm) and Unified Threads (in) are shown in Fig. 1. Limits applicable to other threaded products are as follows:

   For Class 1/2H ——— N = 0.50H; G = 0.015 mm (0.0006 in) max
   For Class 2/3H ——— N = 0.67H; G = 0.015 mm (0.0006 in) max
   For Class 3/4H ——— N = 0.75H; G = 0.015 mm (0.0006 in) max

NOTE: This recommended practice recognizes that the surface may vary in carbon content from the base metal carbon content, and stipulates that this variation shall be either "partial decarburization" or "gross decarburization" or a "carbon restored surface" to the extent allowed in Fig. 1 for the different classes. "Carburization" is not permitted in the surface zone.

4. **Methods for Measuring Decarburization**—Two methods for measuring decarburization are provided. the microscopic method is intended primarily for routine inspection purposes. The hardness method is intended primarily for referee purposes. In the case of gross decarburization, however, only the microscopic method is applicable.

   4.1 **Microscopic Method**[1]:

   4.1.1 SPECIMENS—Use longitudinal sections taken through the thread axis of the bolt, screw, or stud, after all heat treating operations have been performed on the product.

   4.1.2 PREPARATION

   (a) Mount specimen for grinding and polishing. Protection from rounding the surface to be examined is essential. The specimen should be mounted in a clamp or in a plastic mount, the latter being the preferred method.

   (b) After mounting, grind and polish the surface in accordance with good metallographic practice.

   (c) Etching in a 3% nital (concentrated nitric acid) or picral (saturated picric acid) is usually suitable for showing changes in microstructure caused by decarburization.

   4.1.3 MEASUREMENT—Unless otherwise agreed on between purchaser and producer, examine at 100X magnification. Compare with Fig. 1 and definitions in paragraphs 2.1, 2.2, and 2.3.

   If the microscope is of a type with a ground glass screen, the extent of decarburization can be measured directly with a scale. If an eye-piece is used for measurement, it should be an appropriate type containing a cross hair or a scale.

   4.2 **Hardness Method**[2]

   4.2.1 Prepare specimens as outlined in paragraphs 4.1.1 and 4.1.2.

   4.2.2 Determine hardness at three positions, as shown in Fig. 2, using a Knoop indenter with a 500 g load or Vickers DPH with 300 g load.

   4.2.3 Interpret hardness readings as follows:

   (a) A decrease of more than 30 hardness points from Position 1 to Position 2 indicates that the part does not conform to the classification specified.

   (b) An increase of more than 30 points between Position 1 and Position 3 is regarded as "carburization," and indicates that part does not conform to the classification specified.

   NOTE: Careful differentiation must be made between an increase in hardness caused by carburization or by cold working the surface (such as from thread rolling).

[1] Same as outlined in SAE J419, except "Specimen."
[2] Hardness Method applicable only for threads with pitches 1.25 mm and larger, and for all sizes of threaded products having base metal hardness Rockwell C40 and higher and surface hardness Rockwell 30N 60 and higher.

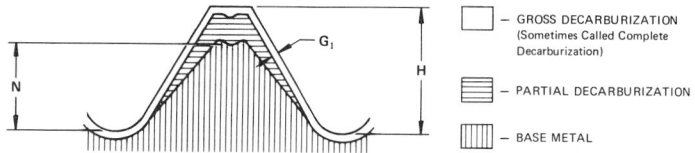

	Limits for ISO Modified (OMFS) Metric Threads[e]—mm						Limits for Unified Threads[e]—in			
Thread Pitch P	Thread Height H[a]	Decarburization Class 1/2 H[b]	Decarburization Class 2/3 H[b]	Decarburization Class 3/4 H[b]	G[d]	Threads Per in	Thread Height H[a]	Decarburization Class C (1/2 H)[c]	Decarburization Class B (2/3 H)[c]	G[d]
		N	N	N				N	N	
		min	min	min	max			min	min	max
0.35	0.202	0.10	0.13	0.15	↑	28	0.02191	0.011	0.015	↑
0.40	0.232	0.12	0.15	0.17		24	0.02556	0.013	0.017	
0.45	0.262	0.13	0.17	0.20		20	0.03067	0.015	0.020	
0.50	0.292	0.15	0.19	0.22						
						18	0.03408	0.017	0.023	
0.60	0.352	0.18	0.23	0.26		16	0.03834	0.019	0.025	
0.70	0.411	0.21	0.27	0.31		14	0.04382	0.022	0.029	
0.80	0.471	0.24	0.31	0.35						
1.00	0.591	0.30	0.39	0.44		13	0.04719	0.024	0.032	
					0.015	12	0.05112	0.026	0.035	0.0006
1.25	0.740	0.37	0.49	0.56		11	0.05577	0.028	0.037	
1.50	0.890	0.44	0.59	0.67						
1.75	1.040	0.52	0.69	0.78		10	0.06134	0.031	0.041	
2.00	1.189	0.59	0.79	0.89		9	0.06816	0.034	0.045	
						8	0.07668	0.038	0.051	
2.50	1.488	0.74	0.99	1.12						
3.00	1.787	0.89	1.19	1.34		7	0.08763	0.044	0.059	
3.50	2.086	1.04	1.39	1.56		6	0.10224	0.051	0.068	
4.00	2.386	1.19	1.59	1.79	↓					↓

All dimensions above are millimetres

All dimensions above are inch.

[a] H is height of external thread at its maximum boundary non-plated condition.
[b] Decarburization Class 1/2 H is commonly specified for Class 8.8 and Class 9.8 threaded fasteners; Class 2/3 H for Class 10.9 fasteners; Class 3/4 H for Class 12.9 fasteners.
[c] For unified inch threads, 1/2 H and 2/3 H were formerly referred to as X and 4/3 X, respectively (where X equals one-half thread height).
[d] G shall be measured perpendicular to the flank of the thread midway between crest and root. (The additional depth of gross decarburization shown at thread crest is due to "thread enfoliation" caused by thread rolling.)
[e] See paragraph 3 for limits applicable to other threaded products.

FIG. 1

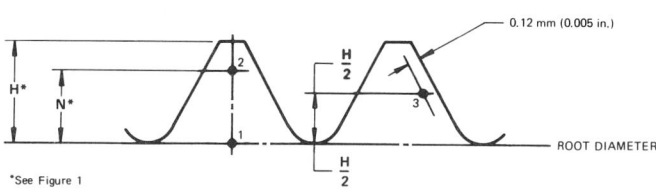

FIG. 2

6 Spring Wire and Springs

OIL TEMPERED CARBON STEEL SPRING WIRE AND SPRINGS—SAE J316

SAE Recommended Practice

Report of Iron and Steel Technical Committee approved November 1967. Editorial change July 1970.

Scope—This specification covers the mechanical, chemical, and dimensional requirements of oil tempered carbon steel spring wire used in the automotive and related industries. It is especially intended for the manufacture of mechanical springs and wire forms which are not subjected to a large number of high stress cycles. Class I wire is intended for moderate stress and Class II for higher stress level applications. This specification also covers the basic material and heat treat requirements for springs fabricated from this wire.

Manufacture—The steel shall be made by the electric furnace, open hearth, or basic oxygen process, and shall be free from pipe and undue segregation. The wire shall be properly drawn, properly austenitized, oil quenched, and tempered to produce the specified mechanical properties.

Chemical Composition—The steel shall conform to the chemical composition[1] as follows:

	Grade A	Grade B
Carbon	0.60–0.85%	0.60–0.75%
Manganese	0.90–1.20%	0.60–0.90%
Phosphorus	0.040% max	0.040% max
Sulfur	0.050% max	0.050% max
Silicon	0.15–0.30%	0.15–0.30%

Unless otherwise specified by the purchaser, Grade A shall be supplied for all wire 0.192 in. diameter and larger, and Grade B shall be supplied for all wire less than 0.192 in. diameter.

[1] Subject to limits shown for check analysis in SAE J409.

Mechanical Properties—The tensile strength of the wire shall conform to the requirements in Table 1 for the various sizes. Hardness ranges indicated in Table 1 apply to finished springs and are subject to normal variations found in standard hardness testing procedures.

Wrap Test—Wire 0.162 in. and smaller in diameter shall wind on itself as an arbor without breaking or checking of the surface. Larger diameter wire, up to and including 0.312 in., shall wind without breaking or checking the surface on a mandrel twice the wire diameter. Wrap test on wires over 0.312 in. diameter is not applicable.

Microstructure—A longitudinal section shall show a fine homogeneous tempered martensitic structure. Decarburization shall be determined by etching a polished transverse section of the wire in nital and examining the entire periphery at 100X magnification, measuring the worst area present, but not measuring decarburization which is directly associated with a seam or other surface defect. Carbon free depth of decarburization shall not exceed 0.5% of the wire diameter. Combined depth of carbon free and partial decarburization shall not exceed 2% of the wire diameter on sizes less than 0.250 in. or 0.005 in. on sizes 0.250 in. or larger.

Surface Condition—The surface of the wire shall be smooth and free from rust, scale, die marks, deep scratches, slivers, seams, laps, pits, or cracks which would affect the fabrication of the finished parts or their serviceability. Unless otherwise specified by the purchaser, seams shall not exceed 3.5% of the wire diameter or 0.010 in., whichever is the smaller as measured on a transverse section.

Workmanship—The wire shall be uniform in quality and in temper and shall not be wavy or crooked. It shall be homogeneous and free from injurious imperfections caused in its manufacture, whether such imperfections are

TABLE 1

Decimal Size, in.	Class I Wire				Class II Wire				Decimal Size, in.	Class I Wire				Class II Wire			
	Tensile Strength,[a] 10³ psi		Hardness		Tensile Strength,[a] 10³ psi		Hardness			Tensile Strength,[a] 10³ psi		Hardness		Tensile Strength,[a] 10³ psi		Hardness	
	Min	Max	Min	Max	Min	Max	Min	Max		Min	Max	Min	Max	Min	Max	Min	Max
			R15N				R15N					RC				RC	
0.020	293	323	88.0	90.0	324	354	90.0	92.0	0.135	215	240	45	50	241	266	50	55
0.023	289	319	88.0	90.0	320	350	90.0	92.0	0.148	210	235	44	49	236	261	49	54
0.026	286	316	88.0	90.0	317	347	89.0	91.0	0.162	205	230	43	48	231	256	48	53
0.029	283	313	87.5	89.5	314	344	89.0	91.0	0.177	200	225	42	48	226	251	48	53
0.032	280	310	87.5	89.5	311	341	88.5	90.5	0.192	195	220	41	47	221	246	47	52
0.035	274	304	87.0	89.0	305	335	88.0	90.0	0.207	190	215	40	46	216	241	46	51
0.041	266	296	86.5	88.5	297	327	87.5	89.5	0.225	188	213	40	45	214	239	45	50
0.047	259	289	86.0	88.0	290	320	87.0	89.0	0.244	187	212	40	45	213	238	45	50
0.054	253	283	86.0	88.0	284	314	87.0	89.0	0.250	185	210	40	45	211	236	45	50
									0.312	183	208	39	44	209	234	44	49
			R45N				R45N		0.375	180	205	39	44	206	231	44	49
0.062	247	277	55.0	60.0	278	308	58.0	63.0	0.437	175	200	37	43	201	226	43	48
0.072	241	271	54.0	59.0	272	302	57.0	62.0	0.500	170	195	36	42	196	221	42	47
0.080	235	265	53.0	58.0	266	296	56.0	61.0	0.562	165	190	35	41	191	216	41	46
0.091	230	260	52.0	57.0	261	291	55.5	60.5	0.625	165	190	35	41	191	216	41	46
0.105	225	255	51.5	56.5	256	286	55.0	60.0									
0.120	220	250	50.5	55.5	251	281	54.5	59.5									

[a] Examination of the tensile fracture shall not show a coarse or cuppy condition.

NOTE: Tensile strength values for intermediate diameters may be interpolated.

apparent at the time of receiving inspection or while the wire is processed by the manufacturer. Each coil shall be one continuous length of wire, properly coiled and firmly tied.

Permissible Variations in Dimensions—The diameter of the wire shall not vary from the specified by more than the following:

Diameter, in.	Permissible Variations, ± in.	Permissible Out-Of-Round, in.
0.020 to 0.027	0.0008	0.0008
Over 0.027 to 0.072	0.0010	0.0010
Over 0.072 to 0.375	0.0020	0.0020
Over 0.375 to 0.625	0.0030	0.0030

Finished Parts

Heat Treatment—Springs coiled from this wire shall be stress relieved for a minimum of 30 minutes at temperature. Normally, the temperature used will be the maximum which will leave the original hardness of the wire essentially unchanged. Typical temperatures are:

450 F—small diameter (0.020 to 0.054 in.)
550 F—medium diameter (over 0.054 to 0.120 in.)
650 F—large diameter (over 0.120 in.)

It should be recognized that in certain applications, such as extension springs with initial tension requirements, lower than the typical stress relieving temperature may be used. This is also true for thin flexible spring designs to minimize distortion. Springs requiring maximum resistance to relaxation at moderately elevated temperatures may be stress relieved at higher than the typical temperatures with some loss of hardness.

Hardness—Hardness of springs shall be measured on suitably ground flats on wire sizes of 0.062 in. and larger, or on ground mounted sections for wire sizes of less than 0.062 in.

Surface Condition—The surface condition of the finished parts shall be as described for the wire except in certain instances where shot peening might be used. In addition, there shall be no excessive coiling marks, nicks, or gouges which would impair the serviceability of the parts.

OIL TEMPERED CARBON STEEL VALVE SPRING QUALITY WIRE AND SPRINGS—SAE J351

SAE Recommended Practice

Report of Iron and Steel Technical Committee approved July 1968.

Scope—This specification covers the physical and chemical requirements of oil tempered carbon steel valve spring quality wire used for the manufacture of engine valve springs and other springs requiring high-fatigue properties. This specification also covers the basic material and processing requirements of springs fabricated from this wire.

Manufacture and Workmanship—The steel shall be made by the electric furnace, open hearth, or basic oxygen process. Sufficient discard must be made to insure freedom from all pipe and undue segregation. The wire shall be properly drawn, austenitized, oil quenched, and tempered to produce the specified mechanical properties.

The wire shall be uniform in quality and in temper and shall not be wavy or crooked. It shall be homogeneous and free from injurious imperfections caused in its manufacture, whether such imperfections are apparent at the time of receiving inspection or while the wire is being processed by the user. Each coil shall be one continuous length of wire properly coiled and firmly tied. Welds in coils are not permitted.

Chemical Composition—The steel shall conform to the chemical composition[1] as follows:

Carbon	0.60-0.75%[2]
Manganese	0.60-0.90%[2]
Phosphorus	0.025% max
Sulphur	0.030% max
Silicon	0.15-0.30%

Mechanical Properties—The tensile properties of the wire shall conform to the requirements in Table 1 for the various sizes. Hardness ranges indicated in Table 1 apply to finished springs and are subject to normal variations found in standard hardness testing procedure.

Permissible Variations in Dimensions—The diameter of the wire shall not vary from that specified by more than that shown in Table 2.

Wrap Test—Wire 0.162 in. and smaller in diameter shall wind on itself without breaking or checking the surface. Larger diameter wire up to and including 0.250 in. shall wind without breaking or checking on a mandrel twice the diameter of the wire.

Microstructure—A longitudinal section shall show a fine homogeneous tempered martensitic structure. Decarburization shall be determined by etching a polished *transverse* section of the tempered wire in nital and examining the entire periphery at 100X magnification, measuring the worst area present.

Inspection at 100X magnification shall show no completely decarburized (carbon free) areas and partial decarburization shall not exceed a depth of 0.001 in. on wire 0.192 in. and smaller or 0.0015 in. on wire larger than 0.192 in.

Surface Conditions—The surface of the wire specimens shall be examined after etching in a solution of equal parts of hydrochloric acid and water at a temperature of 165-175 F for a sufficient time to remove approximately 1% of the wire diameter. This examination shall be made using a binocular microscope at a magnification not to exceed 10X. The surface of the wire shall be free from imperfections such as seams, pits, die marks, scratches, and other defects tending to impair the fatigue value of the springs.

Twist Test—A 10 in. specimen of wire slowly twisted four revolutions in one direction and then twisted in the opposite direction until failure shall show a square break normal to the axis of the wire without splits or cracks.

Electromagnetic Inspection—When specified, the entire length of every coil used by engine valve spring manufacturers must be inspected for surface imperfections with a magnetic and/or eddy current defect analyzer. Springs made from wire containing surface defects 0.002 in. deep and greater must be rejected.

Finished Parts

1. There shall be no excessive coiling marks, nicks, or gouges which would affect the specified wire diameter by more than 1% or would impair the serviceability of the parts.

TABLE 1

Wire Diameter, in.	Tensile Strength, 10³ psi		Hardness		Reduction of Area,[a] min, %
	Min	Max	Min	Max	
			R45N		
0.062 to 0.092 incl	240	260	52	57	N/A[b]
Over 0.092 to 0.128 incl	235	255	51	56	45
			RC		
Over 0.128 to 0.162 incl	230	250	46	51	45
Over 0.162 to 0.192 incl	225	245	45	50	45
Over 0.192 to 0.225 incl	220	240	44	49	45
Over 0.225 to 0.250 incl	215	235	44	49	45

Note: Examination of the tensile fracture shall not show a coarse or cuppy condition.
[a] The 45% minimum value is for as received wire. A 40% minimum value is acceptable for wire produced at the mill when tested immediately after tempering.
[b] N/A—Reduction of area does not apply to 0.092 and below.

[1] For permissible variations from specified chemical ranges and limits for steel, refer to SAE J409.
[2] Carbon and manganese may be varied from the specified ranges by agreement between the manufacturer and the purchaser, provided the mechanical properties specified are maintained.

2. Springs coiled from this wire shall be stress relieved for a minimum of 30 minutes at heat. Normally the temperature used will be the maximum which will leave the original hardness of the wire essentially unchanged. Typical temperatures are 650–750 F.

3. Hardness of springs shall be measured on suitably ground flats. Hardness values shall conform to Table 1.

4. Engine valve springs shall be shot peened to a minimum of 90% coverage on the ID of the spring (reference: SAE Manual on Shot Peening—SAE J808, HS-84). After shot peening the springs shall be stress relieved at 400–450 F.

TABLE 2

Dia, in.	Permissible Variations, ± in.	Permissible Out of Round, in.
0.062 to 0.092 incl	0.0008	0.0008
Over 0.092 to 0.148 incl	0.0010	0.0010
Over 0.148 to 0.177 incl	0.0015	0.0015
Over 0.177 to 0.250 incl	0.0020	0.0020

OIL TEMPERED CHROMIUM-VANADIUM VALVE SPRING QUALITY WIRE AND SPRINGS—SAE J132

SAE Recommended Practice

Report of Iron and Steel Technical Committee approved October 1969. Editorial change April 1974.

Scope—This SAE Recommended Practice covers the mechanical and chemical requirements of oil tempered chromium-vanadium valve spring quality wire used for the manufacture of engine valve springs and other springs used at moderately elevated temperatures and requiring high fatigue properties. It also covers the basic material and processing requirements of spring fabricated from this wire.

Manufacture and Workmanship—The steel shall be made by the electric furnace, open hearth, or basic oxygen process. Sufficient discard shall be made to insure freedom from all pipe and undue segregation. The wire shall be properly drawn, austenitized, oil quenched, and tempered to produce the specified mechanical properties.

The wire shall be uniform in quality and in temper and shall not be wavy or crooked. It shall be homogeneous and free from injurious imperfections caused in its manufacture, whether such imperfections are apparent at the time of receiving inspection or while the wire is being processed by the user. Each coil shall be one continuous length of wire properly coiled and firmly tied. Welds in coils are not permitted.

Chemical Composition—The steel shall conform to the following chemical composition[1] (percentage by weight):

Carbon	0.48–0.53
Manganese	0.70–0.90
Phosphorus	0.020 max
Sulfur	0.035 max
Silicon	0.20–0.35
Chromium	0.80–1.10
Vanadium	0.15 min

Mechanical Properties—The tensile properties of the wire shall conform to the requirements in Table 1 for the various sizes.

Permissible Variations in Dimensions—The diameter of the wire, prior to forming springs, shall not vary from that specified by more than the values in Table 2.

Wrap Test—Wire 0.162 in. and smaller in diameter shall wind on itself as a mandrel without breaking or cracking the surface. Larger diameter wire up to and including 0.312 in. shall wind without breaking or cracking of the surface when wound on a mandrel twice the diameter of the wire. The wrap test is not applicable to wire over 0.312 in. in diameter.

Microstructure—A longitudinal section shall show a fine homogeneous tempered martensitic structure. Decarburization shall be determined by etching a polished *transverse* section of the tempered wire in nital and by examining the entire periphery at 100X magnification, measuring the worst area present. Inspection at 100X magnification shall show no completely decarburized (carbon free) areas and partial decarburization shall not exceed a depth of 0.001 in. on wire 0.192 in. and smaller or 0.0015 in. on wire larger than 0.192 in. See SAE J419.

Surface Conditions—The surface of the wire specimens shall be examined after etching in a solution of equal parts of hydrochloric acid and water at a temperature of 165-175°F for a sufficient time to remove approximately 1% of the wire diameter. This examination shall be made using a binocular microscope at a magnification not to exceed 10X. The surface of the wire shall be free from imperfections such as seams, pits, die marks, scratches, and other defects tending to impair the fatigue value of the springs.

Special Surface Inspection—When specified, the entire length of every coil used by engine valve spring manufacturers shall be inspected for surface imperfections with a magnetic and/or eddy current defect analyzer or equivalent. Springs made from wire containing surface defects 0.002 in. deep and greater must be rejected.

Finished Parts

1. Coiling marks or nicks which flatten but do not gouge the specified wire diameter by more than 1% or that would not impair the serviceability of the parts are permitted.

2. Springs coiled from this wire shall be stress relieved for a minimum of 30 minutes at heat unless otherwise agreed upon by purchaser and supplier. Normally the temperature used will be the maximum which will leave the original hardness of the wire essentially unchanged. Typical temperatures are 700–800°F.

3. Hardness of springs shall be measured on suitable ground flats for wire sizes 0.062 in. and larger, or on ground mounted sections for wire sizes of less than 0.062 in. Hardness values shall conform to Table 1.

4. When specified, valve and other high fatigue requirement springs shall be shot peened to a minimum of 90% coverage on the inside diameter of the springs (reference SAE J808). After shot peening the springs shall be stress relieved at a temperature of 400–475°F.

TABLE 1—TENSILE PROPERTIES

Wire Diameter, In	Tensile Strength, 10^3 psi		Hardness[a]		Reduction of Area[b] Min %
	Min	Max	Min	Max	
			Rockwell 15N		
0.020	300	325	88.5	89.7	N/A[c]
0.031	290	315	88.0	89.3	N/A[c]
0.041	280	305	87.5	88.8	N/A[c]
0.054	270	295	86.9	88.3	N/A[c]
			Rockwell 45N		
0.062	265	290	57.9	61.4	N/A[c]
0.080	255	275	56.4	59.4	N/A[c]
0.106	245	265	55.0	57.9	45
			Rockwell C		
0.135	235	255	48	51	45
0.162	225	245	47	50	40
0.192	220	240	46	49	40
0.244	210	230	45	48	40
0.283	205	225	44	47	40
0.312	203	223	43	47	40
0.375	200	220	43	46	40
0.438	195	215	42	45	40
0.500	190	210	41	45	40

Values for intermediate sizes may be interpolated.
[a]Hardness ranges indicated apply to finished springs and are subject to normal variations found in standard hardness testing procedures.
[b]The 45% and 40% minimum values are for as received wire. These values may be decreased by five points when tested immediately after tempering.
[c]N/A—Reduction of area does not apply to wire under 0.106 in in diameter.

TABLE 2—PERMISSIBLE VARIATIONS

Diameter, in	Permissible Plus Variations in	Permissible Out of Round, ±in
0.020 to 0.075 incl	0.0008	0.0008
Over 0.075 to 0.148 incl	0.0010	0.0010
Over 0.148 to 0.375 incl	0.0015	0.0015
Over 0.375 to 0.500 incl	0.0020	0.0020

[1]For permissible variations from specified chemical ranges and limits for steel, refer to SAE J409.

HARD DRAWN MECHANICAL SPRING WIRE AND SPRINGS—SAE J113

SAE Recommended Practice

Report of Iron and Steel Technical Committee approved August 1969.

Scope—This specification covers the mechanical and chemical requirements of hard drawn carbon steel spring wire in two classes used for the manufacture of mechanical springs and wire forms generally employed for applications subject to static loads or infrequent stress repetitions. This specification also covers basic material and processing requirements of the springs and forms fabricated therefrom. Class 2 is a higher tensile strength product and is furnished only when specified.

Manufacture and Workmanship—The steel shall be made by the electric furnace, open hearth, or basic oxygen process. Sufficient discard must be made to insure freedom from all pipe and undue segregation. Rolling practice shall be controlled to insure that the finished wire shall have no seams greater than 3.5% of the wire diameter or 0.010 in. whichever is the smaller, as measured on a transverse section. The material shall be properly patented or controlled cooled and cold drawn to produce the specified mechanical properties and shall be uniform in quality, not kinked or improperly cast. Each coil shall be one continuous length of wire properly coiled and firmly tied. Welds prior to cold drawing are permitted.

Chemical Composition—The steel shall conform to the chemical composition as follows:

Element	Class 1	Class 2
Carbon, %	0.45–0.75[a]	0.50–0.85[a]
Manganese, %	0.60–1.30[b]	0.60–1.30[b]
Phosphorous, % (max)	0.040	0.040
Sulfur, % (max)	0.050	0.050
Silicon, %	0.10–0.30	0.10–0.30

[a] Carbon in any one lot should not vary more than 0.20%.
[b] Manganese in any one lot should not vary more than 0.30%.

Mechanical Properties—The tensile properties of the wire shall conform to the requirements of Table 1 for the various sizes.

Wrap Test—Class 1 wire up to 0.162 in. inclusive shall wind on itself as an arbor, and over 0.162 in. to 0.312 in. inclusive shall wind on a mandrel twice the wire diameter without the surface breaking or checking. Class 2 wire up to 0.162 in. inclusive shall wind on an arbor twice the diameter, and over 0.162 in. to 0.312 in. dia. inclusive shall wind on a mandrel four times the wire diameter without the surface breaking or checking. Wrap test is not applicable to wire over 0.312 in.

Surface Condition—The surface of the wire shall be smooth and free from rust, scale, die marks, deep scratches, slivers, seams, laps, pits, or cracks which would affect the fabrication of the finished parts or their serviceability.

Permissible Variations in Dimensions—The diameter of the wire shall not vary from that specified by more than shown in Table 2.

Finished Parts—Tension and compression springs coiled from this wire shall be stress relieved for a minimum of 30 minutes at a temperature of 450–550 F unless otherwise agreed upon between purchaser and supplier. The surface condition of the finished parts shall be as described for the wire except in certain instances where shot peening might be used. In addition, there shall be no excessive coiling marks, nicks, or gouges which would affect the serviceability of the part.

To relieve hydrogen embrittlement, electroplated parts may be heated at 350–375 F for a minimum of 2 hr immediately after plating.

TABLE 1—MECHANICAL PROPERTIES

Wire Dia, in.	Class 1 Tensile Strength,[a] 10^3 psi		Class 2 Tensile Strength,[a] 10^3 psi	
	Minimum	Maximum	Minimum	Maximum
0.020	283	323	324	364
0.023	279	319	320	360
0.026	275	315	316	356
0.029	271	311	312	352
0.032	266	306	307	347
0.035	261	301	302	342
0.041	255	293	294	332
0.048	248	286	287	325
0.054	243	279	280	316
0.062	237	272	273	308
0.072	232	266	267	301
0.080	227	261	262	296
0.092	220	253	254	287
0.106	216	248	249	281
0.120	210	241	242	273
0.135	206	237	238	269
0.148	203	234	235	266
0.162	200	230	231	261
0.177	195	225	226	256
0.192	192	221	222	251
0.207	190	218	219	247
0.225	186	214	215	243
0.250	182	210	211	239
0.312	174	200	201	227
0.375	167	193	194	220
0.438	165	190	191	216
0.500	156	180	181	205
0.531	154	178	179	203

[a] The tensile strength for intermediate sizes can be interpolated. The tensile fracture shall not show a coarse or cuppy condition.

TABLE 2—PERMISSIBLE VARIATIONS

Diameters, in.	Permissible Variations, ± in.	Permissible Out of Round, in.
0.020 to 0.027 incl	0.0008	0.0008
Over 0.027 to 0.072 incl	0.0010	0.0010
Over 0.072 to 0.375 incl	0.0020	0.0020
Over 0.375	0.0030	0.0030

SPECIAL QUALITY HIGH TENSILE, HARD DRAWN MECHANICAL SPRING WIRE AND SPRINGS—SAE J271

SAE Recommended Practice

Report of Iron and Steel Technical Committee approved June 1972.

Scope—This recommended practice covers the mechanical and chemical requirements of special quality high tensile, hard drawn carbon steel spring wire with restricted size tolerances. This material is used where such restricted dimensional requirements are necessary for the manufacture of highly stressed mechanical springs and wire forms. It is generally employed for applications subject to static loads or infrequent stress repetitions. This recommended practice also covers basic materials and processing requirements of springs and forms fabricated therefrom.

Manufacture and Workmanship—The steel shall be made by electric furnace, open hearth, or basic oxygen process. Sufficient discard must be made to insure freedom from pipe and undue segregation. Rolling practice shall be controlled to insure that the finished wire shall have no seams greater than 3.5% of the wire diameter or 0.010 in. (0.25 mm), whichever is the smaller, as measured on a transverse section.

Using either properly patented or uniformly control cooled rods, the wire shall be cold drawn to produce the specified mechanical properties. The wire shall be uniform in quality, not kinked or improperly cast. To test for cast, one or more convolutions of wire shall be cut from the coil and placed on a flat surface. The wire shall lay substantially flat on itself and not show a wavy condition. Each finished coil shall be one continuous length of wire, properly coiled and firmly tied. Welds prior to cold drawing are permitted.

Chemical Composition—The steel shall conform to the following chemical composition (percent by weight):

Carbon[a]	0.70–1.00
Manganese[a]	0.20–1.30
Phosphorus	0.040 max
Sulfur	0.040 max
Silicon	0.10–0.30

NOTE: For permissible variations from specified chemical ranges and limits for steel, refer to SAE J409.

[a] Carbon in any one lot shall not vary more than 0.20% and Manganese not more than 0.30%. Normally, lower Mn is used for the smaller size wire, and higher Mn is used for the larger sizes.

Mechanical Properties—The tensile properties of the wire shall conform to the requirements shown in Table 1 by sizes.

Permissible Variations in Dimensions—The diameter of the wire shall not vary from that specified by more than shown in Table 2.

Wrap Test Requirements—Wire shall withstand winding (at least five turns closely wound) without fracture over an arbor of diameter as indicated below. The test shall be conducted on wire prior to stress relieving.

Wire Dia, in (mm)	Mandrel Size
0.020 to 0.162 (0.51 to 4.11) incl	2X dia
Over 0.162 to 0.312 (4.11 to 7.92) incl	4X dia
Over 0.312 (7.92)	Wrap test not applicable

Surface Condition—The surface of the wire shall be smooth and free from rust, scale, die marks, deep scratches, slivers, seams, laps, pits, or cracks which would affect the fabrication of the finished parts of their serviceability.

Finished Parts

1. The surface of the finished springs shall be as described for the wire. In addition, there shall be no excessive coiling marks, nicks, or gouges which would impair the serviceability of the parts.

2. Unless otherwise agreed upon by purchaser and supplier, tension and compression springs coiled from this wire shall be stress relieved for a minimum of 30 min at heat. The normal temperature range is 450–550 F (232–288 C).

3. Electroplating of parts made from special quality high strength, hard drawn wire is not recommended because of susceptibility to hydrogen embrittlement. If plating is necessary, parts should be heated at a temperature of not less than 350 F (177 C) for a minimum of 2 h immediately after plating. Other temperatures and times may be necessary.

TABLE 1—TENSILE PROPERTIES

Wire Dia		Tensile Strength[a]			
		Min		Max	
in	mm	ksi	MPa	ksi	MPa
0.020	0.508	350	2415	387	2670
0.023	0.584	346	2385	381	2630
0.026	0.660	337	2325	373	2570
0.029	0.737	332	2290	367	2530
0.032	0.813	327	2255	361	2490
0.035	0.889	325	2245	359	2480
0.041	1.041	314	2165	348	2400
0.048	1.219	306	2110	339	2335
0.054	1.377	301	2075	332	2290
0.062	1.575	295	2035	325	2245
0.072	1.829	287	1980	317	2185
0.080	2.032	282	1945	312	2150
0.092	2.337	276	1905	305	2100
0.106	2.692	269	1855	297	2050
0.120	3.048	263	1815	291	2005
0.135	3.429	258	1780	285	1965
0.148	3.760	253	1745	280	1930
0.162	4.115	249	1715	275	1895
0.177	4.496	245	1690	270	1860
0.192	4.877	241	1660	267	1840
0.207	5.258	238	1640	264	1820
0.225	5.715	235	1620	260	1795
0.250	6.350	230	1585	255	1760
0.312	7.925	217	1495	242	1670
0.375	9.525	210	1450	235	1620
0.438	11.125	206	1420	231	1590
0.500	12.700	195	1345	220	1515
0.531	13.487	190	1310	210	1450

[a] Tensile strength for intermediate sizes shall be interpolated.

TABLE 2—PERMISSIBLE VARIATIONS

Diameters		Permissible Variations		Permissible Out-of-Round	
in	mm	±in	±mm	in	mm
0.020 to 0.026 incl	0.51 to 0.66 incl	0.0003	0.008	0.0003	0.008
Over 0.026 to 0.063 incl	Over 0.66 to 1.60 incl	0.0005	0.013	0.0005	0.013
Over 0.063 to 0.207 incl	Over 1.60 to 5.26 incl	0.0010	0.025	0.0010	0.025
Over 0.207 to 0.375 incl	Over 5.26 to 9.52 incl	0.0020	0.05	0.0020	0.05
Over 0.375	Over 9.52	0.0030	0.08	.00030	0.08

OIL TEMPERED CHROMIUM—SILICON ALLOY STEEL WIRE AND SPRINGS—SAE J157

SAE Recommended Practice

Report of Iron and Steel Technical Committee approved March 1970.

Scope—This SAE Recommended Practice covers the mechanical and chemical requirements of oil tempered chromium silicon alloy steel wire used for the manufacture of springs requiring resistance to set when used at moderately elevated temperatures. It also covers the basic material and processing requirements of springs fabricated from this wire.

Manufacture and Workmanship—The steel shall be made by the electric furnace, open hearth, or basic oxygen process. Sufficient discard shall be made to insure freedom from all pipe and undue segregation. The wire shall be properly drawn, austenitized, oil quenched, and tempered to produce the specified mechanical properties.

The wire shall be uniform in quality and in temper and shall not be wavy or crooked. It shall be homogeneous and free from injurious imperfections caused in its manufacture, whether such imperfections are apparent at the time of receiving inspection or while the wire is processed by the spring manufacturer. Each coil shall be one continuous length of wire, properly coiled and firmly tied.

Chemical Composition—The steel shall conform to the chemical composition[1] as follows:

Carbon	0.51–0.59%
Manganese	0.60–0.80%
Phosphorus	0.035% max
Sulfur	0.040% max
Silicon	1.20–1.60%
Chromium	0.60–0.80%

Mechanical Properties—The tensile strength of the wire shall conform to the requirements in Table 1 for the various sizes. Hardness ranges indicated in Table 1 apply to finished springs and are subject to normal variations found in standard hardness testing procedures.

[1] For permissible variations from specified ranges and limits for steel, refer to SAE J409.

On Rockwell C, normal variations are considered ±1 point RC or equivalent by conversions.

Permissible Variations in Dimensions—The diameter of the wire shall not vary from that specified by more than the following:

Diameter, in.	Permissible Variations, ±in.	Permissible Out of Round, in.
0.032–0.072 incl	0.0010	0.0010
Over 0.072–0.438 incl	0.0020	0.0020

Wrap Test—Wire 0.162 in. and smaller in diameter shall wind on itself without breaking or cracking the surface. Larger diameter wire, up to and including 0.312 in., shall wind without breaking or cracking the surface on a mandrel twice the wire diameter. The wrap test on wires over 0.312 in. diameter is not applicable.

Microstructure—A longitudinal section shall show a fine homogeneous tempered martensitic structure. Decarburization shall be determined by etching a polished transverse section of wire in nital and examining the entire periphery at 100X magnification, measuring the worst area present, but not measuring decarburization which is directly associated with a seam or other surface defect. Depth of carbon free decarburization shall not exceed 0.5% of the wire diameter.

Combined depth of carbon free and partial decarburization shall not exceed 2% of the wire diameter on all sizes of wire.

Surface Condition—The surface of the wire shall be smooth and free from rust, scale, die marks, deep scratches, slivers, seams, laps, pits, or cracks which would affect the fabrication of the finished parts. Unless otherwise specified by the purchaser, seams shall not exceed 3.5% of the wire diameter or 0.010 in., whichever is the smaller as measured on a transverse section.

Finished Parts—Springs coiled from this wire shall be stress relieved immediately after coiling for a minimum of 30 min at heat unless otherwise agreed upon between purchaser and supplier. The temperature used will be the maximum which will leave the original hardness of the wire essentially unchanged. Typical temperatures are 650–800 F.

It should be recognized that in certain applications, such as some torsional or extension springs, lower than typical stress relieving temperatures may be used. Hardness of springs shall be measured on a suitable ground flat on wire sizes of 0.062 in. and larger or on ground mounted sections for wire sizes less than 0.062 in. The hardness scale and values shall conform to the requirements of Table 1 for the respective wire diameters.

The surface conditions on the finished parts shall be as described for the wire, except certain instances where shot peening might be used. In addition, there shall be no excessive coiling marks, nicks, or gouges which would impair the serviceability of the parts.

When springs are shot peened, the surface appearance will be altered. Because of a resulting decrease in the spring resistance to relaxation, shot peening is permitted only when agreed upon by the purchaser. After shot peening, the springs shall be stress relieved at 400–475 F for a minimum of 30 min at heat.

TABLE 1

Wire Diameter, in.	Tensile Strength, 10^3 psi		Hardness		Reduction of Area[a] min %
	Min	Max	Min	Max	
			R 15N		
0.032	300	325	88.5	90.0	b
0.041	298	323	88.5	90.0	b
0.054	292	317	88.0	89.5	b
			R 45N		
0.062	290	315	59.5	63.0	b
0.080	285	310	59.0	62.0	b
0.092	280	305	58.5	61.5	45
0.120	275	300	57.5	61.0	45
			RC		
0.135	270	295	51.5	54.0	45
0.162	265	290	51.0	53.5	40
0.177	260	285	50.5	53.0	40
0.192	260	283	50.5	53.0	40
0.218	255	278	50.0	52.5	40
0.250	250	275	50.0	52.5	40
0.312	245	270	49.0	52.0	40
0.375	240	265	48.5	51.5	40
0.438	235	260	48.0	51.0	40

Values for intermediate sizes may be interpolated.

[a] The reduction of area values are for as-received wire. These values may be decreased by 5 points when tested immediately after tempering.

[b] Reduction of area does not apply to wire under 0.092 in. in diameter.

HARD DRAWN CARBON STEEL VALVE SPRING QUALITY WIRE AND SPRINGS—SAE J172

SAE Recommended Practice

Report of Iron and Steel Technical Committee approved April 1970.

Scope—This SAE Recommended Practice covers the mechanical and chemical requirements of the best quality hard drawn carbon steel spring wire used for the manufacture of engine valve springs and other springs requiring high fatigue properties. It also covers the basic material and processing requirements of springs fabricated from this wire.

Manufacture and Workmanship—The steel shall be made by the electric furnace, open hearth, or basic oxygen process. Sufficient discard must be made to insure freedom from all pipe and undue segregation. The material shall be properly patented or control cooled and cold drawn to produce the specified mechanical properties. The wire shall be uniform in quality and shall not be kinked, wavy, or crooked. It shall be free from injurious imperfections caused in its manufacture, whether such imperfections are apparent at the time of receiving inspection or while the wire is being processed by the user. Each coil shall be one continuous length of wire, properly coiled and firmly tied. Welds in coils are not permitted.

Chemical Composition—The steel shall conform to the chemical composition[1] as follows:

[1] For permissible variations from specified chemical ranges and limits for steel, refer to SAE J409.

Carbon	0.60–0.75%[a]
Manganese	0.60–0.90%[a]
Phosphorus	0.025% max
Sulfur	0.030% max
Silicon	0.15–0.30%

[a] Carbon and manganese may be varied from the above specified ranges, by the manufacturer, by agreement with the purchaser, provided the mechanical properties specified are maintained.

Mechanical Properties—The tensile properties of the wire shall conform to the requirements in Table 1 for the various sizes.

Permissible Variations in Dimensions—The diameter of the wire shall not vary from that specified by more than the dimensions specified in Table 2.

Microstructure—Decarburization shall be determined by etching a polished transverse section of the wire in nital and examining the entire periphery at 100X magnification, measuring the worst area present. Inspection shall show no completely decarburized (carbon free) areas. Partial decarburization shall not exceed a depth of 0.001 in. on wire 0.192 in. and smaller or 0.0015 in. on wire larger than 0.192 in.

Surface Conditions—The surface of the as-received wire specimens, which have been stress relieved at approximately 800 F for $\frac{1}{2}$ hr, shall be examined after etching in a solution of equal parts of hydrochloric acid and water at a temperature of 165–175 F for a sufficient time to remove approximately 1% of the wire diameter. This examination shall be made using a binocular microscope at a magnification not to exceed 10X. The surface of the wire shall be free from imperfections such as seams, pits, die marks, scratches, and other defects tending to impair the fatigue value of the springs.

Special Surface Inspection—When specified, the entire length of every coil used by engine valve spring manufacturers must be inspected for surface imperfections with a magnetic and/or eddy current defect analyzer. Springs made from this wire containing surface defects 0.002 in. deep and greater must be rejected.

Finished Parts

1. There shall be no excessive coiling marks, nicks, or gouges which would affect the specified wire diameter by more than 1% or would impair the serviceability of the parts.

2. Springs coiled from this wire shall be stress relieved for a minimum of 30 min at heat unless otherwise agreed upon between purchaser and supplier. Typical temperatures are 650–750 F.

3. Engine valve springs shall be shot peened to a minimum of 90% coverage on the ID of the spring (reference SAE J808). After shot peening, the springs shall be stress relieved at 400–450 F.

TABLE 1

Wire Diameter, in.	Tensile Strength, 10^3 psi		Reduction of Area, Min, %
	Min	Max	
Over 0.092 to 0.128 incl	235	255	40
Over 0.128 to 0.162 incl	230	250	40
Over 0.162 to 0.192 incl	225	245	40
Over 0.192 to 0.225 incl	220	240	40
Over 0.225 to 0.250 incl	215	235	40

TABLE 2

Diameter, in.	Permissible Variations, ±in.	Permissible Out of Round, in.
Over 0.092 to 0.148 incl	0.0010	0.0010
Over 0.148 to 0.177 incl	0.0015	0.0015
Over 0.177 to 0.250 incl	0.0020	0.0020

STAINLESS STEEL 17-7 PH SPRING WIRE AND SPRINGS—SAE J217

SAE Recommended Practice

Report of Iron and Steel Technical Committee approved November 1970.

1. Scope—This recommended practice covers a high quality corrosion resisting steel wire, cold drawn, formed, and heat treated to produce uniform mechanical properties. It is magnetic in all conditions. It is intended for the manufacture of springs and wire forms that are to be heat treated after forming to enhance the spring properties. This recommended practice covers basic materials and processing requirements of springs and forms fabricated therefrom.

2. Manufacturing and Workmanship—The steel shall be made by the electric arc, electric induction, or other suitable commercial processes, using proper controls to prevent injurious segregation or inclusions. The wire shall be properly annealed and cold drawn to produce the specified mechanical properties after heat treatment. The wire shall be uniform in quality and shall not be kinked or improperly cast. To test for cast, a few convolutions of wire shall be cut and placed on a flat surface. The wire shall lie substantially flat on itself and not show a wavy condition. Each bundle shall be one continuous length of wire properly coiled and firmly tied. Welds are permitted prior to the final drawing operations. Care should be exercised in spring coiling and forming to avoid injurious tool and die marks and small radii in coiling and bending.

3. Finish—This wire is usually supplied with a thin surface film to assist in preventing galling or seizure of the wire by the coiling or forming tools. This wire is available with the following finishes: lead, oxide, copper, and others.

4. Chemical Composition—This wire shall conform to the chemical composition (percent by weight) as follows:

Carbon	0.090 max
Manganese	1.00 max
Silicon	1.00 max
Phosphorous	0.040 max
Sulfur	0.030 max
Chromium	16.00–18.00
Nickel	6.50–7.75
Aluminum	0.75–1.50

5. Mechanical Properties—The tensile strength of the wire shall conform to the requirements in Table 1 for the various sizes after specimens have been heat treated at 900 ±10 F for 1 hr and air cooled. Prior to heat treating, the samples will be cleaned to remove all drawing lubricants and metallic or nonmetallic coating by immersing the wire sample in 15–25% nitric acid at room temperature for 5 min followed by a thorough water wash.

6. Permissible Variation in Diameter—Permissible variations in the diameter of the wire shall be as specified in Table 2.

7. Wrap Test—The cold drawn wire shall withstand, without cracking, wrapping at room temperature, around an arbor equal to the nominal diameter of the wire.

8. Coiling Test—The wire as cold drawn shall show a uniform pitch with no splits, cracks, or fractures when wound in a tightly closed coil on an arbor of the size shown in Table 3, and the resultant coil stretched to a permanent set of four times its as-wound length. This requirement shall apply only to wire having a nominal diameter of 0.016–0.125 in.

TABLE 1—HEAT TREATED WIRE TENSILE STRENGTHS

Wire Diameter, in	Tensile Strength after 1 h at 900 ± 10 F (air cooled), ksi	
	Min	Max
0.016 to 0.020 incl	335	365
Over 0.020 to 0.025 incl	330	360
Over 0.025 to 0.030 incl	325	355
Over 0.030 to 0.041 incl	320	350
Over 0.041 to 0.051 incl	310	340
Over 0.051 to 0.061 incl	305	335
Over 0.061 to 0.071 incl	297	327
Over 0.071 to 0.086 incl	292	322
Over 0.086 to 0.090 incl	282	312
Over 0.090 to 0.100 incl	279	309
Over 0.100 to 0.106 incl	274	304
Over 0.106 to 0.130 incl	272	302
Over 0.130 to 0.138 incl	260	290
Over 0.138 to 0.146 incl	258	288
Over 0.146 to 0.162 incl	256	286
Over 0.162 to 0.180 incl	254	284
Over 0.180 to 0.207 incl	252	282
Over 0.207 to 0.225 incl	248	278
Over 0.225 to 0.306 incl	242	272
Over 0.306 to 0.440 incl	235	265

TABLE 2

Nominal Diameter, in	Variation, ± in	Out-of Round, in
0.016 to under 0.024	0.0004	0.0004
0.024 to under 0.033	0.0005	0.0005
0.033 to under 0.044	0.00075	0.00075
0.044 to under 0.313	0.0010	0.0010
0.313–0.0440	0.0015	0.0015

TABLE 3

Nominal Wire Diameter, in	Arbor Diameter, in
0.016 to 0.024 incl	0.067
Over 0.024 to 0.034 incl	0.102
Over 0.034 to 0.045 incl	0.145
Over 0.045 to 0.055 incl	0.212
Over 0.055 to 0.078 incl	0.250
Over 0.078 to 0.125 incl	0.328

9. Surface Condition—The surface of wire specimens shall be prepared in accordance with paragraph 10.3. The prepared specimens shall have a surface free from injurious imperfections, such as seams, pits, die scratches, and other defects which will impair the serviceability of the part.

10. Finished Parts

10.1 The surface conditions on the finished parts shall be as described for the wire, except certain instances where shot peening might be used. In addition, there shall be no excessive coiling marks, nicks, or gouges which would impair the serviceability of the part. When springs are shot peened, the surface appearance will be altered. Because of a resulting decrease in the spring resistance to relaxation, shot peening is permitted only when agreed upon by the purchaser. After shot peening, the springs shall be stress relieved at 450–500 F for a minimum of 30 min at heat.

10.2 All forming shall be done on the wire in the as-drawn condition.

10.3 Springs made from this wire must be cleaned and passivated after coiling to insure maximum corrosion resistance of the stainless steel. All metallic coatings, must be removed prior to heat treatment. One procedure is as follows:

(a) Remove drawing compounds from the wire surface by a 5 min dip in alkaline cleaner at approximately 190 F, followed by a water rinse.

(b) Remove metallic and most nonmetallic coatings from the wire surface and passivate the surface by immersing parts in a nitric acid solution of 15–25% at 140–160 F for 5 min or until clean. Follow with a water rinse.

10.4 After passivating, springs and forms made from this wire must be heated at 900 ± 10 F for 1 hr and air cooled. No forming should be done to the wire or parts after heat treating.

STAINLESS STEEL, SAE 30302, SPRING WIRE AND SPRINGS—SAE J230

SAE Recommended Practice

Report of Iron and Steel Technical Committee approved February 1971.

1. Scope—This recommended practice covers a high-strength corrosion-resisting steel wire, uniform in mechanical properties, intended for the manufacture of springs and wire forms. It covers basic materials and processing requirements of springs and forms fabricated therefrom.

2. Manufacture and Workmanship—The steel shall be made by the electric arc, electric induction, or other suitable commercial processes, using proper controls to prevent injurious segregation or inclusions. The wire shall be properly annealed and cold drawn to produce the specified mechanical properties. The wire shall be uniform in quality and shall not be kinked or improperly cast. To test for cast, a few convolutions of wire shall be cut from the bundle and placed on a flat surface. The wire shall lie substantially flat on itself and not show a wavy condition. Each unit shall be a continuous length with welds being permitted before final drawing. Welds are not permitted at finished size except by negotiation between manufacturer and user.

3. Finish—This wire is usually supplied with a thin surface film that will prevent galling or seizure of the wire by the coiling or forming tools. This wire is available with the following finishes: lead, oxide, bright, copper, and other.

4. Chemical Composition—This wire shall conform to the chemical composition (percent by weight) as follows:

Carbon	0.15 max
Manganese	2.00 max
Silicon	1.00 max
Phosphorus	0.045 max
Sulfur	0.030 max
Chromium	17.00–19.00
Nickel	8.00–10.00

5. Mechanical Properties—The tensile properties of the wire shall conform to the requirements in Table 1 for the various sizes. Both lower and higher tensile strengths are available upon mutual agreement between supplier and purchaser.

6. Permissible Variations in Diameter—Permissible variations in the wire diameter shall be as specified in Table 2.

7. Wrap Test Requirement—For sizes 0.162 in. diameter and smaller, the wire shall wind on itself as an arbor without fracture. For sizes over 0.162 in., the wire shall wind without cracking on an arbor equal to twice its diameter. This test shall be conducted prior to any stress relieving or passivation.

8. Coiling Test—The as cold drawn wire shall show a uniform pitch with no splits, cracks, or fractures when wound in a tightly closed coil on an arbor three to four times the wire diameter and the resultant coil stretched to a permanent set of four times its as-wound length. This requirement shall apply only to wire sizes 0.105 in. diameter and smaller.

9. Surface Condition—Surface of the wire shall be free from injurious imperfections, such as seams, pits, die scratches, and other defects which impair the serviceability of the part. (Visually examine at 10X magnification.)

10. Finished Parts

10.1 The surface condition of the finished parts shall be as described for the wire, except in certain instances where shot peening might be used. In addition, there shall be no excessive coiling marks, nicks, or gouges which would impair the serviceability of the part. When the springs are shot peened, the surface appearance will be altered. Because of a resulting decrease in the spring resistance to relaxation, shot peening is permitted only when agreed upon by the purchaser. After shot peening, the springs shall be stress relieved at 450–500 F for a minimum of 30 min at heat.

TABLE 1

Wire Diameter, in	Tensile Strength, min, ksi	Tensile Strength, max, ksi
0.009 and smaller	325	355
0.010	320	350
0.011	318	348
0.012	316	346
0.013	314	344
0.014	312	342
0.015	310	340
0.016	308	338
0.017	306	336
0.018	304	334
0.019 to 0.020 incl	300	330
Over 0.020 to 0.022 incl	296	326
Over 0.022 to 0.024 incl	292	322
Over 0.024 to 0.026 incl	289	319
Over 0.026 to 0.028 incl	286	316
Over 0.028 to 0.032 incl	277	307
Over 0.032 to 0.036 incl	273	303
Over 0.036 to 0.041 incl	269	299
Over 0.041 to 0.047 incl	262	292
Over 0.047 to 0.054 incl	260	290
Over 0.054 to 0.062 incl	255	285
Over 0.062 to 0.072 incl	250	280
Over 0.072 to 0.080 incl	245	275
Over 0.080 to 0.092 incl	240	270
Over 0.092 to 0.105 incl	232	262
Over 0.105 to 0.120 incl	225	255
Over 0.120 to 0.148 incl	210	240
Over 0.148 to 0.162 incl	205	235
Over 0.162 to 0.177 incl	195	225
Over 0.177 to 0.207 incl	185	215
Over 0.207 to 0.225 incl	180	210
Over 0.225 to 0.250 incl	175	205
Over 0.250 to 0.312 incl	160	190
Over 0.312 to 0.375 incl	140	170

10.2 Lead coatings shall be removed from springs prior to stress relieving when a temperature of 550 F or above is required.

10.3 Springs made from this wire are normally stress relieved for a minimum of 30 min. Typical temperatures are 550–600 F. It should be recognized that other than typical stress relieving temperatures may be used or omitted completely, depending upon the spring design and application.

10.4 Springs made from this wire must be cleaned and passivated after coiling to insure maximum corrosion resistance of the stainless steel. All metallic coatings must be removed prior to heat treatment. One procedure is as follows:

(a) Remove drawing compounds from the wire surface by a 5 min dip in alkaline cleaner at approximately 190 F followed by a water rinse.

(b) Remove metallic and most nonmetallic coatings from the wire surface and passivate the surface by immersing parts in a nitric acid solution of 15–25% at 140–160 F for 5 min or until clean. Follow with a water rinse.

TABLE 2

Diameter, in	Permissible Variations, ± in	Permissible Out of Round, in
0.003 to 0.0046 incl	0.0001	0.0001
Over 0.0046 to 0.007 incl	0.00015	0.00015
Over 0.007 to 0.008 incl	0.0002	0.0002
Over 0.008 to 0.012 incl	0.00025	0.00025
Over 0.012 to 0.024 incl	0.0004	0.0004
Over 0.024 to 0.033 incl	0.0005	0.0005
Over 0.033 to 0.044 incl	0.00075	0.00075
Over 0.044 to 0.312 incl	0.0010	0.0010
Over 0.312 to 0.375 incl	0.0015	0.0015

MUSIC STEEL SPRING WIRE AND SPRINGS—SAE J178

SAE Recommended Practice

Report of Iron and Steel Technical Committee approved June 1970.

Scope—This SAE Recommended Practice covers a high quality, hard drawn, steel spring wire, uniform in mechanical properties, intended for the manufacturer of spring and wire forms subjected to high stresses or requiring good fatigue properties. It covers basic materials and processing requirements of springs and forms fabricated therefrom.

Manufacture and Workmanship—The steel shall be made by the electric furnace, open hearth, or basic oxygen process. Sufficient discard must be made to insure freedom from all pipe and undue segregation. The wire shall be properly patented and cold drawn to produce the specified mechanical properties. The wire shall be uniform in quality and shall not be kinked or improperly cast. To test for cast, a few convolutions of wire shall be cut from the bundle and placed on a flat surface. The wire shall lie substantially flat on itself and not show a wavy condition. Each unit shall be one continuous length of wire, properly coiled and firmly tied. Welds are not permitted at finished size except by negotiation between manufacturer and user.

Finish—Music steel spring wire is available in the following finishes:
1. Bright (white liquor).
2. Phosphate.
3. Tin.
4. Cadmium alloy.
5. Other finishes.

Chemical Composition—The steel shall conform to the following chemical composition[1]:

Element	%, by wgt.
Carbon	0.80–1.00[a]
Manganese	0.20–0.60[a]
Phosphorus	0.025 max
Sulfur	0.030 max
Silicon	0.10–0.30

[a] Carbon and manganese may be varied from the above specified ranges by the manufacturer, by agreement with the purchaser, provided the mechanical properties specified are maintained.

Mechanical Properties—The tensile strength of the wire shall conform to the requirements in Table 1.

Permissible Variations in Dimensions—Permissible variations in the wire diameter shall be as specified in Table 2.

Wrap Test Requirement—The wire shall wind on itself as an arbor without fracture. This test shall be conducted on wire prior to any stress relieving.

Decarburization—Transverse sections of the wire properly mounted, polished, and etched shall show a maximum affected depth of partial decarburization to be no more than 1% of the wire diameter, but not to exceed 0.0015 in. when examined at a magnification of 100X.

Coiling Test—The coiling test shall be applied only to sizes 0.105 in. and smaller in diameter. The wire shall be of good uniform mechanical properties. A length of tightly closed coil shall be wound on an arbor 3–3½ times the diameter of the wire, and a 5 in. length of coil shall be stretched so that it sets to approximately three times its original length. The wire so tested shall show a uniform pitch with no splits or fracture. This test should be conducted on wire prior to any stress relieving.

Surface Condition—The surface of the wire specimens, which has been stress relieved at approximately 800 F for ½ hr, shall be examined after etching in a solution of equal parts of commercial hydrochloric acid and water at a temperature of 165–175 F, for a sufficient time to remove approximately 1% of the wire diameter. This examination shall be made using a binocular microscope at a magnification not to exceed 10X for sizes 0.020 in. and larger. For sizes under 0.020 in., the magnification used is subject to negotiation.

The surface of the wire shall be free from imperfections such as seams, pits, die marks, scratches, and other defects tending to impair the fatigue properties of the springs.

TABLE 1—TENSILE STRENGTH OF WIRE

Wire Diameter, in.	Tensile Strength,[a] 10^3 psi		Wire Diameter, in.	Tensile Strength,[a] 10^3 psi	
	Min	Max		Min	Max
0.004	439	485	0.055	300	331
0.005	426	471	0.059	296	327
0.006	415	459	0.063	293	324
0.007	407	449	0.067	290	321
0.008	399	441	0.072	287	317
0.009	393	434	0.076	284	314
0.010	387	428	0.080	282	312
0.011	382	422	0.085	279	308
0.012	377	417	0.090	276	305
0.013	373	412	0.095	274	303
0.014	369	408	0.100	271	300
0.015	365	404	0.102	270	299
0.016	362	400	0.107	268	296
0.018	356	393	0.110	267	295
0.020	350	387	0.112	266	294
0.022	345	382	0.121	263	290
0.024	341	377	0.125	261	288
0.026	337	373	0.130	259	286
0.028	333	368	0.135	258	285
0.030	330	365	0.140	256	283
0.032	327	361	0.145	254	281
0.034	324	358	0.150	253	279
0.036	321	355	0.156	251	277
0.038	318	352	0.162	249	275
0.040	315	349	0.177	245	270
0.042	313	346	0.192	241	267
0.045	309	342	0.207	238	264
0.048	306	339	0.225	235	260
0.051	303	335	0.250	230	255

[a] The tensile strength may be interpolated for intermediate sizes. Higher tensile strength music spring wire is available.

[1] For permissible variations from specified chemical ranges and limits for steel, refer to SAE J409.

TABLE 2

Diameter, in.	Permissible Variations, ±in.	Permissible Out of Round, in.
0.004 to 0.010 incl	0.0002	0.0002
Over 0.010 to 0.028 incl	0.0003	0.0003
Over 0.028 to 0.063 incl	0.0004	0.0004
Over 0.063 to 0.080 incl	0.0005	0.0005
Over 0.080 to 0.250 incl	0.0010	0.0010

Finished Parts

1. The springs shall be free from rust and there shall be no marks, nicks, cracks, or gouges which will impair the serviceability of the part.

2. Springs made from this wire shall normally be stress relieved for a minimum of 30 min at heat. Typical temperatures are 450–500 F. It should be recognized that in certain applications lower than typical stress relieving temperatures may be used or omitted completely. This is also true for thin flexible spring designs to minimize distortion.

3. To relieve hydrogen embrittlement, parts which are electroplated after coiling shall be heated at 350 F minimum for a minimum of 2 hr immediately after plating. Higher minimum temperatures and time may be necessary.

7 Ferrous Castings

NUMBERING SYSTEM FOR DESIGNATING GRADES OF CAST FERROUS MATERIALS—SAE J859

SAE Recommended Practice

Report of Iron and Steel Technical Committee approved June 1963. Editorial change October 1964.

Scope—This SAE Recommended Practice is intended to supply a uniform means of designating cast ferrous materials reported in SAE Standards and Recommended Practices. The system outlined in this report is intended to facilitate the addition of new, widely used casting materials and also the deletion of now obsolete grades.

Designation System

Basic Number—The method of identifying ferrous cast materials will be based on a designation consisting of a letter prefix followed by an integral four digit number arranged in the following manner:

1. **Prefix Letter**—G —Gray Iron
   M—Malleable Iron
   D —Nodular Iron

2. **First Two Digits**—For G —Tensile Strength
   For M—Yield Strength
   For D —Yield Strength (10 will be used for 100,000 psi)

3. **Second Two Digits**—For G use 00
   For M—Elongation
   For D —Elongation

Suffix Letters—Suffix letters may be used to indicate special conditions within each category of materials. Suffix letters shall be assigned in the sequence in which the conditions or requirements are adopted for a particular category of material.

EXAMPLE: SAE D5506 is a ductile iron having a minimum yield strength of 55,000 psi and a minimum elongation of 6%.

ɸ AUTOMOTIVE GRAY IRON CASTINGS—SAE J431 AUG79

SAE Standard

Report of the Iron and Steel Division, approved January 1935, completely revised by the Iron and Steel Technical Committee August 1979.

1. Scope—This standard covers the hardness and microstructural requirements for gray iron sand mold castings used in automotive and allied industries. The chemical requirements for alloy gray iron automotive camshafts are included in the standard under *casting for special application with controlled composition and microstructure*. The Appendix provides general information on application of gray iron in automotive castings and chemical composition to meet hardness, microstructural, and other properties needed for particular service conditions. The mechanical properties in the Appendix are provided for design purposes.

NOTE: This document was rewritten in June 1970. The materials described by the grade numbers are different than in the former writing.

2. Grades—The specified grades, hardness ranges, and metallurgical description are shown in Tables 1 and 2 and in Sections 5 and 6.

3. Hardness

3.1 The area or areas on the casting where hardness is to be checked shall be established by agreement between supplier and purchaser.

3.2 The foundry shall exercise the necessary controls and inspection techniques to insure compliance with the specified hardness range. Brinell hardness shall be determined according to ASTM E 10, Test for Brinell Hardness of Metallic Materials, after sufficient material has been removed from the casting surface to insure representative hardness readings. The 10 mm ball and 3000 kg load shall be used unless otherwise agreed upon.

4. Heat Treatment

4.1 Unless otherwise specified, castings of grades G1800 and G2500 may be annealed in order to meet the desired hardness range.

4.2 Appropriate heat treatment for removal of residual stresses, or to

TABLE 1—GRADES OF GRAY IRON

Grade	Casting Hardness Range at Specified Locations	Description
G1800	187 HB max or as agreed (4.4 BID max)	Ferritic-pearlitic
G2500	170–229 HB or as agreed (4.6–4.0 BID)	Pearlitic-ferritic
G3000	187–241 HB or as agreed (4.4–3.9 BID)	Pearlitic
G3500	207–255 HB or as agreed (4.2–3.8 BID)	Pearlitic
G4000	217–269 HB or as agreed (4.1–3.7 BID)	Pearlitic

NOTE: Brinell impression diameter (BID) is the diameter in millimeters (mm) of the impression of a 10 mm ball at 3000 kg load.

TABLE 2—BRAKE DRUMS AND CLUTCH PLATES FOR SPECIAL SERVICE

Grade	C %, min	Casting Hardness at Specified Locations	Microstructure	
			Graphite[b]	Matrix[c]
G2500a	3.40	170–229 HB or as agreed (4.6–4.0 BID)	Type VII A, distribution size 2–4	Lamellar pearlite. Ferrite if present not to exceed 15%
G3500b	3.40[a]	207–255 HB or as agreed (4.2–3.8 BID)	Type VII A, distribution size 3–5	Lamellar pearlite. Ferrite or carbide if present not to exceed 5%
G3500c	3.50[a]	207–255 HB or as agreed (4.2–3.8 BID)	Type VII A, distribution size 3–5	Lamellar pearlite. Ferrite or carbide if present not to exceed 5%

[a] Grades G3500b and G3500c normally require alloying to obtain the specified hardness at the high carbon levels specified.
[b] See ASTM A 247, Recommended Practice for Evaluating the Microstructure of Graphite in Iron Castings.
[c] See ASTM E 562, Determining Volume Fraction by Systematic Manual Point Count.

NOTE: Brinell impression diameter (BID) is the diameter in millimeters (mm) of the impression of a 10 mm ball at 3000 kg load.

improve machinability or wear resistance, may be specified by agreement between supplier and purchaser.

5. Microstructure

5.1 The microstructure of gray iron shall consist of flake graphite in a matrix of ferrite or pearlite or mixtures thereof.

5.2 As graphite size and shape somewhat affect hardness-strength ratio and other properties, the type size and distribution of the graphite flakes at a designated location on the casting may be specified by agreement between supplier and purchaser in accordance with ASTM A 247, Recommended Practice for Evaluating the Microstructure of Graphite in Iron Castings.

5.3 Unless otherwise specified, the matrix microstructure of castings covered by this standard shall be substantially free of primary cementite and/or massive steadite. Castings in grades G1800 and G2500 may have a matrix of ferrite and/or pearlite. Grades G3000, G3500, and G4000 shall be substantially pearlitic in matrix structure.

6. Casting for Special Application with Controlled Composition and Microstructure

6.1 Heavy Duty Brake Drums and Clutch Plates

6.1.1 These castings are considered as special cases and are covered in Table 2.

6.2 Alloy Gray Iron Automotive Camshafts

6.2.1 These castings are considered as special cases.

6.2.2 GRADE DESIGNATION—G4000d.

6.2.3 CHEMICAL COMPOSITION—Alloy gray iron camshafts shall contain alloys within the following range or as agreed upon by supplier and purchaser:

Chromium	0.85–1.50%
Molybdenum	0.40–0.60%
Nickel	As agreed

6.2.4 CASTING HARDNESS—241–321 HB determined on a bearing surface or as agreed by supplier and purchaser.

6.2.5 MICROSTRUCTURE—Extending 45 deg on both sides of the centerline of the cam nose and to a minimum depth of $\frac{1}{8}$ in (3.2 mm) the surface shall consist of primary carbides (of acicular or cellular form or a mixture thereof) and graphite in a fine pearlitic matrix. The graphite shall be type VII, A and E distribution, 4–7 flake size according to ASTM A 247, Recommended Practice for Evaluating the Microstructure of Graphite in Gray Iron Castings. The amount of primary carbides and location at which the structure is checked shall be a matter of agreement between the supplier and the purchaser.

6.2.6 SELECTIVE HARDENING—The cam areas of camshaft castings are usually selectively hardened by flame or induction hardening by the supplier. The depth and surface hardness of the hardened case shall be as agreed by supplier and purchaser.

7. Quality Assurance
Sampling plans are a matter of agreement between supplier and purchaser. The supplier shall employ adequate equipment and controls to insure that parts conform to agreed upon requirements.

8. General

8.1 Castings furnished to this standard shall be representative of good foundry practice and shall conform to dimensions and tolerances specified on the casting drawing.

8.2 Minor imperfections usually not associated with the structural functioning may occur in castings. These imperfections often are repairable; however, repairs should be made only in areas and by methods approved by the purchaser.

8.3 Additional casting requirements such as vendor identification, other casting information, and special testing, may be agreed upon by purchaser and supplier. These should appear as product specifications on the casting or part drawing.

APPENDIX—GRAY IRON
(A material description and not a part of the standard)

A1. Definition—Gray iron is a cast iron in which the graphite is present as flakes instead of temper carbon nodules as in malleable iron or small spherulites as in ductile iron.

A2. Chemical Composition—The ranges in composition generally employed in producing the various grades of most automotive gray iron castings are shown in Table A-1. The composition ranges for such special applications as heavy duty brake drums and clutch plates and camshafts are shown in Table A-3 and Table A-5, respectively. The contents of certain elements for these applications are critical in terms of service requirements and the ranges are specified in the standard.

A2.1 The specific composition range for a given grade may vary according to the prevailing or governing section size of the castings being produced.

A2.2 Alloying elements such as chromium, copper, nickel, tin, molybdenum, or other elements may be employed to meet the specified hardness or microstructural requirements or to provide the properties needed for particular service conditions.

A3. Microstructure

A3.1 The microstructure of the various grades of gray iron are generally a mixture of flake graphite in a matrix of ferrite, pearlite, or tempered pearlite. The relative amounts of each of these constituents depends on the analysis of the iron, casting design, and foundry techniques as they affect solidification and subsequent cooling rate and heat treatments if any.

A3.2 The distribution and size of graphite flakes, like the matrix structure of gray iron, depends greatly on the solidification rate and cooling rate of the casting. If a section solidifies very rapidly, an appreciable amount of carbide causing a mottled fracture or chilled corners can be present. If a section cools slowly, as in a massive heavy section casting, an appreciable amount of ferrite may be present. In like manner, light sections will contain small graphite flakes while graphite will form in much larger flakes if the same iron is poured into a heavy casting.

A3.3 For these reasons the strength and hardness of gray iron are greatly influenced by the rate of cooling during and after solidification, the design and nature of the mold and the casting, and by other factors such as inoculation practice, in addition to the composition of the iron.

A3.4 Alloying with nickel, chromium, molybdenum, tin, copper, or other alloys usually promotes a more stable pearlitic structure and is often done to obtain increased hardness, strength, and wear resistance, especially in heavy sections subjected to severe service.

A3.5 Alloying is sometimes used to obtain structures containing a controlled percentage of carbides, as in camshaft or valve lifter castings.

A3.6 Primary carbides and/or pearlite can be decomposed by appropriate heat treatment. Gray irons of suitable composition and structure can be hardened by liquid quenching, or by flame or induction selective hardening.

A4. Mechanical Properties—The mechanical properties listed in Table A-2 can be used for design purposes. However, the stability of a particular grade for an intended application is best determined by laboratory or service tests. Typical mechanical properties for such specialized applications as heavy duty brake drums and clutch plates are shown in Table A-4.

A5. Application of Gray Iron in Automotive Castings

A5.1 The graphite flakes in gray iron give this metal many desirable

TABLE A-1—TYPICAL BASE COMPOSITIONS, %

Grade	Carbon	Silicon	Manganese	Sulfur, max[a]	Phosphorus, max	Approximate Carbon Equiv.[b]
G1800	3.40 3.70	2.80 2.30	0.50–0.80	0.15	0.25	4.25–4.5
G2500	3.20 3.50	2.40 2.00	0.60–0.90	0.15	0.20	4.0–4.25
G3000	3.10 3.40	2.30 1.90	0.60–0.90	0.15	0.15	3.9–4.15
G3500	3.00 3.30	2.20 1.80	0.60–0.90	0.15	0.12	3.7–3.9
G4000	3.00 3.30	2.10 1.80	0.70–1.00	0.15	0.10	3.7–3.9 (usually alloyed)

[a] Typical value.
[b] $CE = \% C + 1/3\% Si$.

TABLE A-2—TYPICAL MECHANICAL PROPERTIES FOR DESIGN PURPOSES[a]

Grade	Hardness Range	Tensile Strength, min, psi (MPa)	Transverse Strength,[b] min, lb (kN)	Deflection,[b] min, in (mm)
G1800	187 HB max (4.4 BID, min)	18 000 (124)	1720 (7.65)	0.14 (3.6)
G2500	170–229 HB (4.6–4.0 BID)	25 000 (173)	2000 (8.90)	0.17 (4.3)
G3000	187–241 HB (4.4–3.9 BID)	30 000 (207)	2200 (9.79)	0.20 (5.1)
G3500	207–255 HB (4.2–3.8 BID)	35 000 (241)	2450 (10.90)	0.24 (6.1)
G4000	217–269 HB (4.1–3.7 BID)	40 000 (276)	2600 (11.56)	0.27 (6.9)

[a] The reported mechanical properties were determined on separately cast 1.2 in D (30 mm D) test bars and may vary throughout the casting depending on composition and cooling rate. Heavy wall (slowly cooled) castings may exhibit lower properties.
[b] See ASTM A 438, Transverse Testing of Gray Cast Iron, for information concerning the B transverse test bar and the transverse test.

NOTE: Brinell impression diameter (BID) is the diameter in millimeters (mm) of the impression of a 10 mm ball at 3000 kg load.

TABLE A-3—TYPICAL BASE COMPOSITION OF BRAKE DRUMS AND CLUTCH PLATES FOR SPECIAL SERVICE, %

Grade	Carbon[a]	Silicon[b]	Manganese[b]	Sulfur[c]	Phosphorus, max
G2500a	3.40 min	1.60–2.10	0.60–0.90	0.12	0.15
G3500b	3.40 min	1.30–1.80	0.60–0.90	0.12	0.15
G3500c	3.50 min	1.30–1.80	0.60–0.90	0.12	0.15

Alloy may be added as required.
[a] Mandatory.
[b] As required.
[c] Typical value.

TABLE A-4—MECHANICAL PROPERTIES FOR DESIGN PURPOSES FOR BRAKE DRUMS, CLUTCH PLATES, AND SPECIAL SERVICE[a]

Grade	Hardness Range	Tensile Strength, min, psi (MPa)	Transverse Strength,[b] min, lb (kN)	Deflection,[b] min, in (mm)
G2500a	170–229 HB (4.6–4.0 BID)	25 000 (173)	2000 (8.90)	0.17 (4.3)
G3500b	207–255 HB (4.2–3.8 BID)	35 000 (241)	2400 (10.68)	0.24 (6.1)
G3500c	207–255 HB (4.2–3.8 BID)	35 000 (241)	2400 (10.68)	0.24 (6.1)

[a] The reported mechanical properties were determined on separately cast 1.2 in D (30 mm D) test bars and may vary throughout the casting depending on composition and cooling rate. Heavy wall (slowly cooled) castings may exhibit lower properties.
[b] See ASTM A 438, Transverse Testing of Gray Cast Iron, for information concerning the B transverse test bar and the transverse test.

NOTE: Brinell impression diameter (BID) is the diameter in millimeters (mm) of the impression of a 10 mm ball at 3000 kg load.

TABLE A-5—TYPICAL CHEMICAL COMPOSITION OF ALLOY GRAY IRON AUTOMOTIVE CAMSHAFTS, GRADE G4000d, %

Total Carbon	3.10–3.60
Silicon	1.95–2.40
Manganese	0.60–0.90
Phosphorus	0.10 max
Sulfur	0.15 max
Chromium	0.85–1.50
Molybdenum	0.40–0.60
Nickel	0.20–0.45 optional
Copper	Residual

properties. These include excellent machinability, high thermal conductivity, vibration dampening properties, and resistance to wear or scuffing. Due to its low freezing temperature for a ferrous alloy, high fluidity, and low shrinkage properties, it is more readily cast in complex shapes than other ferrous metals.

A5.2 Gray iron castings of the lower strength grades G1800 and G2500 are characterized by excellent machinability, high damping capacity, low modulus of elasticity, and comparative ease of manufacture. When higher strength is obtained by a reduction in the carbon or carbon equivalent, castings are more difficult to machine, have lower damping capacity and higher modulus of elasticity, and may be more difficult to manufacture.

TYPICAL APPLICATIONS OF GRAY IRON FOR AUTOMOTIVE CASTINGS

Grade	General Data
G1800	Miscellaneous soft iron castings (as cast or annealed) in which strength is not of primary consideration. Exhaust manifolds may be made of this grade of iron alloyed or unalloyed. These may be annealed casting for exhaust manifolds in order to avoid growth cracking due to heat.
G2500	Small cylinder blocks, cylinder heads, aircooled cylinders, pistons, clutch plates, oil pump bodies, transmission cases, gear boxes, clutch housings, and light duty brake drums.
G3000	Automobile and diesel cylinder blocks, cylinder heads, flywheels, differential carrier castings, pistons, medium duty brake drums, and clutch plates.
G3500	Diesel engine blocks, truck and tractor cylinder blocks and heads, heavy flywheels, tractor transmission cases, and heavy gear boxes.
G4000	Diesel engine castings, liners, cylinders, and pistons.

A6. Special Application of Gray Iron

A6.1 Heavy Duty Brake Drums and Clutch Plates—Automotive brake drums and clutch plates for heavy duty service are considered as special cases. Typical chemical analyses and mechanical properties are listed in Tables A-3 and A-4. Heavy duty irons for such service require high carbon contents for resistance to thermal shock and to minimize heat checking. To maintain strength levels specified for grades G3500b and G3500c normally requires alloying due to their high carbon contents.

A6.1.1 MICROSTRUCTURE—See Table 2 for microstructure requirements.

A6.1.2 SUGGESTED USAGE—Following are suggested grades for brake drums and clutch plates according to types of service.

Grade	Suggested Usage
G2500a	Brake drums and clutch plates for moderate service requirements, where high carbon iron is desired to minimize heat checking. (See paragraph 6.)
G3500b	Brake drums and clutch plates for heavy duty service where both resistance to heat checking and higher strength are definite requirements. (See paragraph 6.)
G3500c	Extra heavy duty service brake drums. (See paragraph 6.)

A6.2 Automotive Camshafts—Alloy gray iron automotive camshafts are also considered as special cases. The chemical composition of such castings is usually within the range given in Table A-5 but may be modified by mutual agreement.

A6.2.1 In casting hardenable iron for camshafts, the aim is to obtain a suitable microstructure in critical locations of the casting and balance the composition to obtain response to induction or flame hardening treatment. These depend not only on the chemistry of the iron but even more on the cross section of the casting and details of melting practice. In making a given casting, it is recognized that the foundry will find it necessary to adjust the chemistry to narrow limits within the range of analysis in Table A-5.

A6.2.2 As the performance of an automotive camshaft is determined by the microstructure and hardness, producers do not normally use tensile or transverse tests for quality control purposes. Camshaft iron with chemistry as given in Table A-5 would be expected to have the following minimum mechanical properties shown in Table A-6:

TABLE A-6—MECHANICAL PROPERTIES FOR AUTOMOTIVE GRAY IRON CAMSHAFTS, GRADE G4000d

Tensile Strength, min psi (MPa)	40 000 (276)
Transverse Strength,[a] min lb (kN)	2600 (11.56)
Deflection,[a] min in (mm)	0.27 (6.9)
Hardness, HB (BID)	241–321 (3.9–3.4)

[a] See ASTM A 438, Transverse Testing of Gray Cast Iron, for information concerning the B transverse test bar and the transverse test.

NOTE: Brinell impression diameter (BID) is the diameter in millimeters (mm) of the impression of a 10 mm ball at 3000 kg load.

A6.2.3 MICROSTRUCTURE—See SAE J431 for microstructure requirements for grade G4000d alloy cast iron camshafts.

A7. Additional Information—Additional information concerning gray iron castings, their properties and uses can be obtained from:

1. *Metals Handbook,* Vol. 1, 8th Edition. American Society for Metals, Metals Park, OH.
2. *Cast Metals Handbook.* American Foundrymen's Society, Des Plaines, IL.
3. 1971 *Gray Iron Castings Handbook.* Gray and Ductile Iron Founders Society, Cleveland, OH.
4. H. D. Angus, *Physical and Engineering Properties of Cast Iron.* British Cast Iron Research Association, Birmingham, England, 1960.
5. *Gray, Ductile, and Malleable Iron Castings Current Capabilities.* STP-455, American Society for Testing and Materials, 1916 Race Street, Philadelphia, PA 19103.
6. G.N.I. Gilbert, *Engineering Data on Cast Iron.* British Cast Iron Association (1968), Alvechurch, Birmingham, England.

AUTOMOTIVE MALLEABLE IRON CASTINGS—SAE J158a — SAE Standard

Report of Iron and Steel Technical Committee approved June 1970 and last revised January 1975. SAE J432 and J433 have been discontinued and replaced by this report.

1. Scope—This SAE Standard covers castings of ferritic, pearlitic, tempered pearlitic, and tempered martensitic grades of malleable iron used in the products of the automotive and allied industries. Castings shall be heat treated to meet this SAE Standard.

2. Grades—The specified grades, hardness range and final heat treatment are shown in Table 1.

TABLE 1—GRADES OF MALLEABLE IRON

Grade	Casting Hardness Range[a]	Heat Treatment
M3210	156 Bhn, max 4.8 BID, min or as agreed	Annealed
M4504	163–217 Bhn 4.7–4.1 BID or as agreed	Air quenched and tempered
M5003	187–241 Bhn 4.4–3.9 BID or as agreed	Air quenched and tempered
M5503	187–241 Bhn 4.4–3.9 BID or as agreed	Liquid quenched and tempered
M7002	229–269 Bhn 4.0–3.7 BID or as agreed	Liquid quenched and tempered
M8501	269–302 Bhn 3.7–3.5 BID or as agreed	Liquid quenched and tempered

[a] Brinell impression diameter (BID) is the diameter in millimeters (mm) of the impression of a 10 mm ball at 3000 kg load.

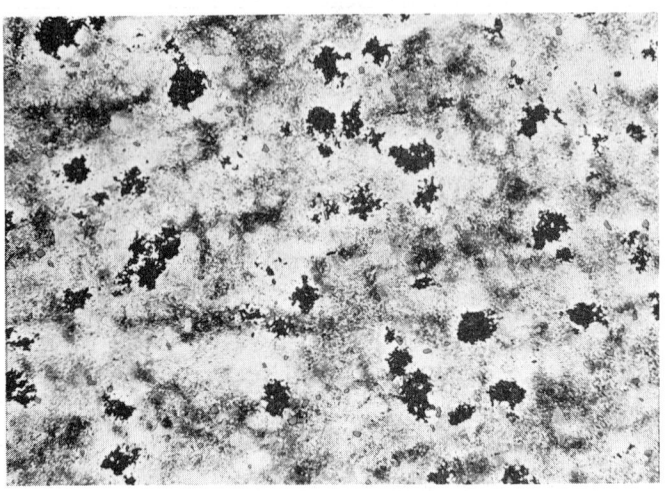

FIG. 3—M4504, APPROXIMATE 207 BHN (100X), TYPICAL MICROSTRUCTURES

FIG. 1—REFERENCE PHOTOMICROGRAPH OF ALLOWABLE MAXIMUM PEARLITE IN GRADE M3210 IRON (100X, 2% NITAL ETCH), TYPICAL MICROSTRUCTURES

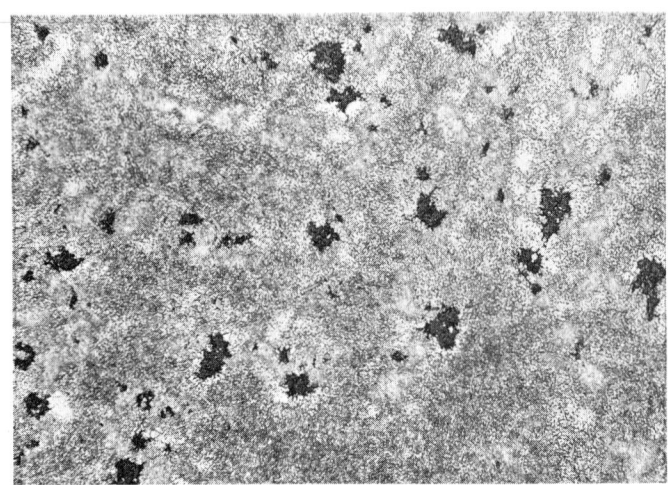

FIG. 4—M5003, APPROXIMATE 229 BHN (100X), TYPICAL MICROSTRUCTURES

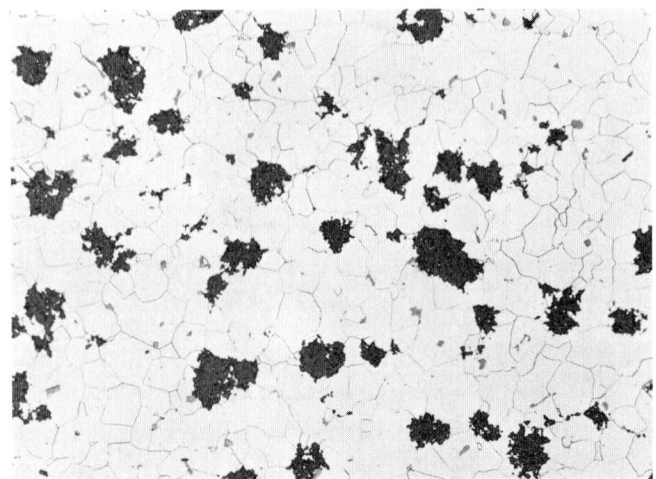

FIG. 2—M3210, APPROXIMATE 143 BHN (100X), TYPICAL MICROSTRUCTURES

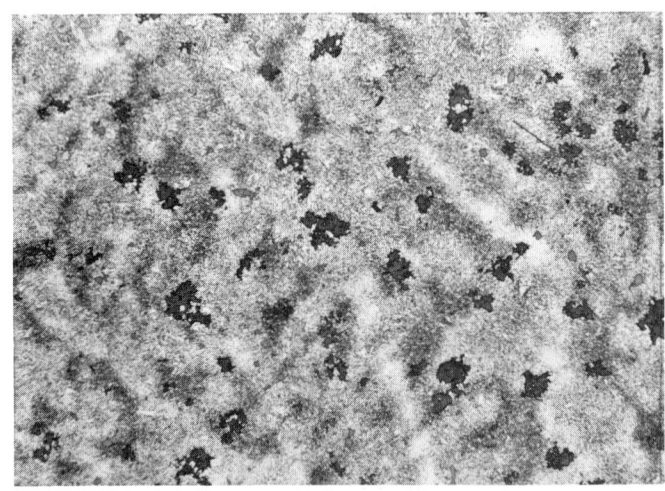

FIG. 5—M5503, APPROXIMATE 229 BHN (100X), TYPICAL MICROSTRUCTURES

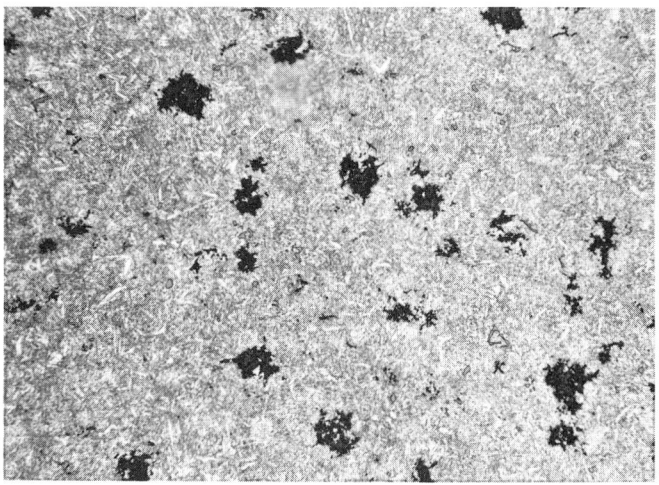

FIG. 6—M7002, APPROXIMATE 262 BHN (100X), TYPICAL MICROSTRUCTURES

FIG. 7—M8501, APPROXIMATE 285 BHN (100X), TYPICAL MICROSTRUCTURES

The foundry may also produce grades M4504 and M5003 by liquid quenching and tempering and/or alloying.

3. Hardness—The foundry shall exercise the necessary controls and inspection procedures to insure compliance with the specified hardness range. Hardness readings shall be taken according to ASTM E 10, Test for Brinell Hardness of Metallic Materials, after sufficient material has been removed from the casting surface to insure representative hardness readings. The area or areas on the casting where hardness is to be checked shall be established by agreement between supplier and purchaser and shown on the drawing.

4. Microstructure

4.1 The microstructure of Grade M3210 malleable iron shall consist of temper carbon nodules distributed in a matrix of ferrite. The rim or surface layer (see paragraph A.3.3) shall not exceed 0.050 in. (1.27 mm).

Unless otherwise specified, the material below the rim can contain some pearlite; however, it shall not exceed the amount shown in Fig. 1.

4.2 The microstructure of the other grades shall consist of temper carbon nodules distributed in a matrix of ferrite and tempered pearlite in air quenched castings and in a matrix of tempered martensite in liquid quenched castings. All grades shall be substantially free of primary graphite or primary cementite.

5. Quality Assurance—Sampling plans are a matter of agreement between supplier and purchaser. The supplier shall employ adequate equipment and controls to insure that parts conform to the agreed upon requirements.

6. General

6.1 Castings furnished to this standard shall be representative of good foundry practice and shall conform to dimensions and tolerances specified on the casting drawing.

6.2 Minor imperfections usually not associated with the structural function may occur in castings. These imperfections often are repairable; however, repairs shall be made only in areas allowed by the purchaser and only by approved methods.

6.3 Additional casting requirements may be agreed upon by the purchaser and supplier. These should appear as additional product requirements on the casting drawing.

APPENDIX—MALLEABLE IRON
(A material description not a part of the standard)

7. Definition and Classification—Malleable iron is a cast iron in which the graphite is present as temper carbon nodules, instead of flakes as in gray iron or small spherulites as in ductile iron.

The term malleable iron includes all grades of malleable iron, those with a ferritic, pearlitic, tempered pearlite, or tempered martensite matrix.

8. Chemical Composition—The chemical composition range of malleable iron generally conforms to the following range:

Total carbon	2.20–2.90%
Silicon	0.90–1.90%
Manganese	0.15–1.25%
Sulfur	0.02–0.20%
Phosphorus	0.02–0.15%

Individual foundries will produce to narrower ranges than those shown above. The composition is controlled such that the molten iron solidifies with all the carbon in the combined form, producing a "white iron" structure which is heat treated to specifications.

9. Microstructure

9.1 The microstructure of malleable irons covered in this standard consist of temper carbon nodules in a matrix of ferrite, pearlite, tempered pearlite, or tempered martensite or combinations of these. The structure of the matrix is controlled by heat treatment and/or composition.

9.2 The matrix of the M3210 grade of malleable iron is essentially ferritic but a small amount of pearlite is permitted. The matrices of the other grades of malleable iron contain combined carbon as pearlite, tempered pearlite, or tempered martensite.

9.3 Because of reaction with the annealing furnace atmosphere, some depletion of carbon and silicon occurs at the surface of the castings. This usually results in a rim which, if excessive, can result in poor machinability. The rim on M3210 malleable iron can consist of coarse pearlite underlying a graphite-free layer sometimes containing more or less combined carbon than the underlying material.

9.4 Typical microstructure of the various grades of malleable iron are shown in Figs. 2–7.

10. Mechanical Properties—The mechanical properties listed in Table A-1 can be used for design purposes, but the suitability of a particular metal for an intended use is best determined by laboratory or service tests.

The mechanical properties vary with microstructure and hardness. For optimum mechanical properties, especially in the liquid quenched and tempered grade, section size should be limited to $3/4$ in. (19.05 mm) to insure a uniform structure.

11. Typical Applications

11.1 M3210 is used where good machinability is important, such as steering gear housings, carriers, and mounting brackets.

TABLE A-1—TYPICAL MECHANICAL PROPERTIES FOR DESIGN PURPOSES

Grade	Hardness Range[a]	Heat Treatment	Tensile Strength, psi (MPa)	Yield Strength, 0.2% Off-set, psi (MPa)	Elongation, % in 2 in.	Modulus of Elasticity, 10^6 psi (GPa)
M3210	156 Bhn max 4.8 BID min	Annealed	50,000 (345)	32,000 (224)	10	25 (172)
M4504	163–217 Bhn 4.7–4.1 BID	Air quenched and tempered	65,000 (448)	45,000 (310)	4	26 (179)
M5003	187–241 Bhn 4.4–3.9 BID	Air quenched and tempered	75,000 (517)	50,000 (345)	3	26 (179)
M5503	187–241 Bhn 4.4–3.9 BID	Liquid quenched and tempered	75,000 (517)	55,000 (379)	3	26 (179)
M7002	229–269 Bhn 4.0–3.7 BID	Liquid quenched and tempered	90,000 (621)	70,000 (483)	2	26 (179)
M8501	269–302 Bhn 3.7–3.5 BID	Liquid quenched and tempered	105,000 (724)	85,000 (586)	1	26 (179)

[a] Brinell impression diameter (BID) is the diameter in millimeters (mm) of the impression of a 10 mm ball at 3000 kg load.

11.2 M4504 is used where slightly higher strength and hardness than M3210 is required, such as certain compressor crankshafts and hubs.

11.3 M5003 is used where moderate strength and/or selective hardening are required for parts such as planet carriers, certain transmission gears, and differential cases.

11.4 M5503 is used where better machinability than M5003 and/or improved response to induction hardening is necessary for parts requiring moderate strength.

11.5 M7002 is used for parts where high strength is required such as connecting rods and universal joint yokes.

11.6 M8501 is used where high strength and wear resistance is required, such as certain gears.

12. Additional Information
 1. *Cast Metals Handbook.* American Foundrymen's Society, Des Plaines, Ill.
 2. *Malleable Iron Castings.* Malleable Founders Society, Cleveland, Ohio.
 3. *Metals Handbook.* Vol. 1, 8th Edition. American Society for Metals, Metals Park, Ohio.
 4. *Modern Pearlitic Malleable Castings Handbook.* Malleable Research and Development Foundation, Dayton, Ohio.
 5. H. D. Angus, *Physical and Engineering Properties of Cast Iron.* British Cast Iron Research Association, Birmingham, England, 1960.
 6. G. N. J. Gilbert, *Engineering Data on Malleable Cast Iron.* British Cast Iron Research Association, Birmingham, England, 1968.
 7. *Gray, Ductile, and Malleable Iron Castings Current Capabilities.* STP-455, American Society for Testing and Materials, 1916 Race Street, Philadelphia, Pennsylvania 19103.

AUTOMOTIVE DUCTILE (NODULAR) IRON CASTINGS—SAE J434c   SAE Standard

Report of Iron and Steel Technical Committee approved September 1956 and last revised January 1975.

1. Scope—This SAE Standard applies to ductile iron castings products such as those used in the automotive and allied industries. Castings may be specified in the as-cast or heat treated condition. NOTE: This document was rewritten in June 1970. The materials described by the grade numbers are different than in the former writing.

2. Grades—The specified grades, hardness range and metallurgical description are shown in Table 1.

3. Hardness

3.1 The foundry shall exercise the necessary controls and inspection procedures to insure compliance with the specified hardness range. Brinell hardness readings shall be determined according to ASTM method E 10 after sufficient material has been removed from the casting surface to insure representative hardness readings. The area or areas on the casting where hardness is to be checked shall be established by agreement between supplier and purchaser and shown on the drawing.

3.2 Unless otherwise specified, castings may be heat treated to the appropriate hardness range.

4. Microstructure—The graphite component of the microstructure shall consist of at least 80% spheroidal graphite conforming to Types I and II in Fig. 1. The matrix microstructure shall consist of either ferrite, ferrite and pearlite, pearlite, tempered pearlite, or tempered martensite or a combination of these. The microstructure shall be substantially free of primary cementite.

5. Quality Assurance—Sampling plans are a matter of agreement between supplier and purchaser. The supplier shall employ adequate equipment and controls to insure that parts conform to the agreed upon requirements.

6. General

6.1 Castings furnished to this standard shall be representative of good

TABLE 1—GRADES OF DUCTILE IRON

Grade	Casting Hardness Range[a]	Description
D4018	170 Bhn, max 4.6 BID, min (or as agreed)	Ferritic
D4512	156–217 Bhn 4.80–4.10 BID (or as agreed)	Ferritic-pearlitic
D5506	187–255 Bhn 4.4–3.8 BID (or as agreed)	Ferritic-pearlitic
D7003	241–302 Bhn 3.90–3.50 BID (or as agreed)	Pearlitic
DQ&T	Range specific	Martensitic

[a] Brinell impression diameter (BID) is the diameter in millimeters (mm) of the impression of a 10 mm ball at 3000 kg load.

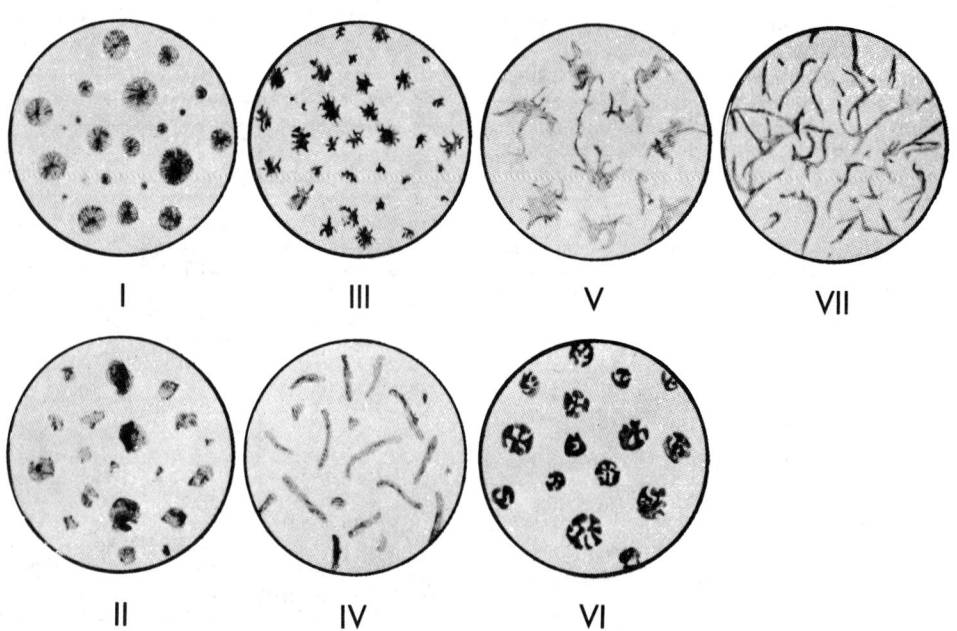

FIG. 1—CLASSIFICATION OF GRAPHITE SHAPE IN CAST IRONS (FROM ASTM A 247)

foundry practice and shall conform to dimensions and tolerances specified on the casting drawing.

6.2 Minor imperfections usually not associated with the structural functioning may occur in castings. These imperfections are often repairable; however, repairs should be made only in areas allowed by the purchaser and only by approved methods.

APPENDIX—DUCTILE (NODULAR) IRON
(A material description not a part of the standard)

7. Definition and Classification—Ductile (nodular) iron, also known as spheroidal graphite iron, is cast iron in which the graphite is present as spheroids, instead of flakes as in gray iron or temper carbon nodules as in malleable iron.

Ductile iron castings may be used in the as-cast condition, or may be heat treated.

8. Chemical Composition—The typical chemical composition of unalloyed iron generally conforms to the following ranges:

Total carbon	3.20–4.10%
Silicon	1.80–3.00%
Manganese	0.10–1.00%
Phosphorus	0.015–0.10%
Sulfur	0.005–0.035%

Individual foundries will produce to narrower ranges than those shown. The spheroidal graphite structure is produced by alloying the molten iron with small amounts of one or more elements such as magnesium or cerium.

9. Microstructure

9.1 The microstructure of the various grades of ductile iron consists of spheroidal graphite in a matrix of either ferrite, pearlite, tempered pearlite, tempered martensite, or a combination of these. The relative amounts of each of these constituents is dependent upon the grade of material specified, casting design as it affects cooling rate, and heat treatments, if any.

9.2 The matrix microstructure of as-cast ductile iron depends to a great extent on the solidification rate and cooling rate of the casting, as shown in Fig. 2. If a section solidifies rapidly, especially sections of 0.25 in. or less, an appreciable amount of carbide may be present in the casting. If a section cools slowly, as in a massive, heavy casting, a largely ferrite matrix may result.

9.3 Alloying elements can also alter the microstructure usually resulting in increased amounts of pearlite. Large variations in structure can be eliminated or minimized by modifying the casting design or the runner system or both, or by controlled cooling, or any combination of these. Primary carbides, and/or pearlite can be decomposed by appropriate heat treatments.

9.4 A rim may occur on heat treated castings consisting of a graphite-free layer sometimes containing more or less combined carbon than the underlying material.

9.5 Typical microstructure of the grades of ductile iron are as follows:

D4018 is annealed ferritic ductile iron. The annealing time and temperature cycle is such that primary carbides, if present in the as-cast structure, are decomposed, and the resulting matrix is ferritic as shown in Fig. 3.

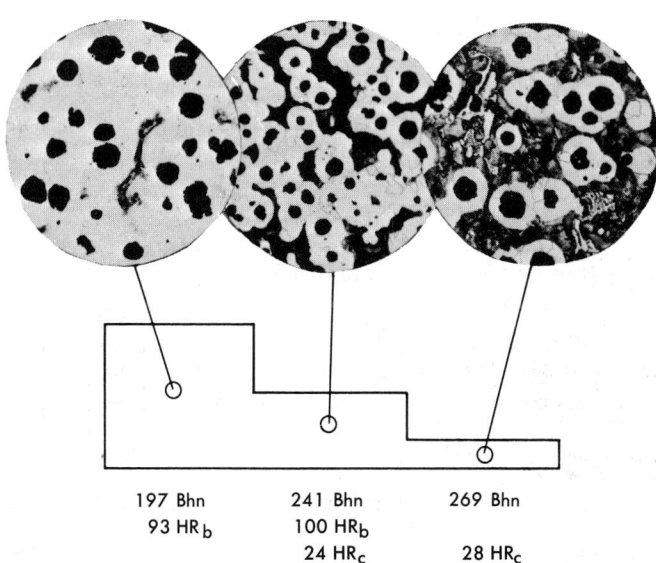

FIG. 2—EXAMPLE OF MICROSTRUCTURAL VARIATION WHICH MAY OCCUR IN AS-CAST CONDITION AS FUNCTION OF METAL THICKNESS (THAT IS, SOLIDIFICATION RATE)

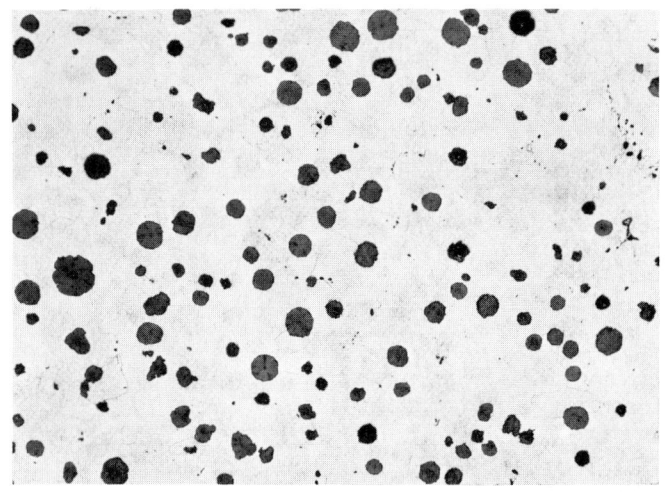

FIG. 3—D4018, APPROXIMATE 156 BHN (100X) TYPICAL MICROSTRUCTURES

D4512 is ferritic ductile iron supplied either as cast or heat treated. The matrix, shown in Fig. 4, is essentially ferrite but this grade can contain pearlite, depending on the section size.

D5506 is ferritic-pearlitic ductile iron supplied either as-cast or heat treated. The matrix, shown in Fig. 5, is essentially pearlite. This grade may contain substantially more ferrite.

D7003 (not shown) is generally air or liquid quenched and tempered to a specified hardness range. The resulting matrix is tempered pearlite or tempered martensite. Time and temperature before hardening can be such that primary carbides are decomposed.

DQ&T is a liquid quenched and tempered grade. The resulting matrix is tempered martensite. The Brinell hardness range is a matter of agreement between supplier and purchaser.

10. Mechanical Properties

10.1 The mechanical properties shown in Table 2 can be used for design purposes but the suitability of a particular grade for an intended use is best determined by laboratory or service tests.

10.2 The mechanical properties will vary with the microstructure which, especially in the as-cast condition, is dependent on section size as well as chemical composition and some foundry processes.

10.3 For optimum mechanical properties in the quenched and tempered grade, section size for unalloyed iron should be limited to approximately $3/4$ in. to insure a uniform, through hardened structure.

11. Typical Applications

11.1 D4018 is used in moderately stressed parts requiring high ductility and good machinability, such as automotive suspension parts.

11.2 D4512 is used for moderately stressed parts where machinability is less important, such as differential cases and carriers.

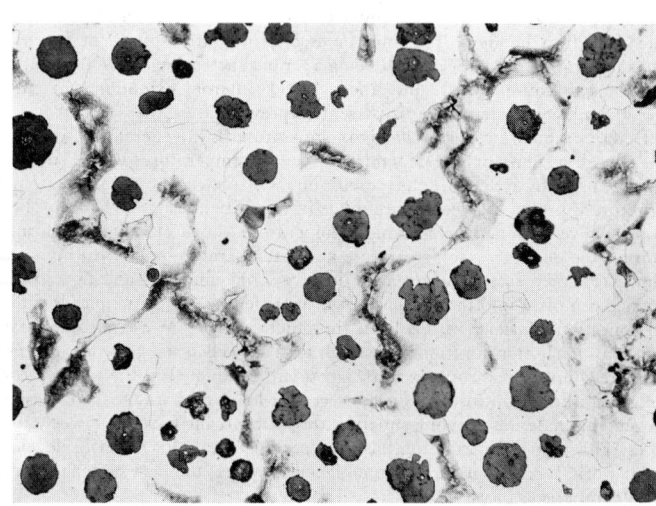

FIG. 4—D4512, APPROXIMATE 179 BHN (100X) TYPICAL MICROSTRUCTURES

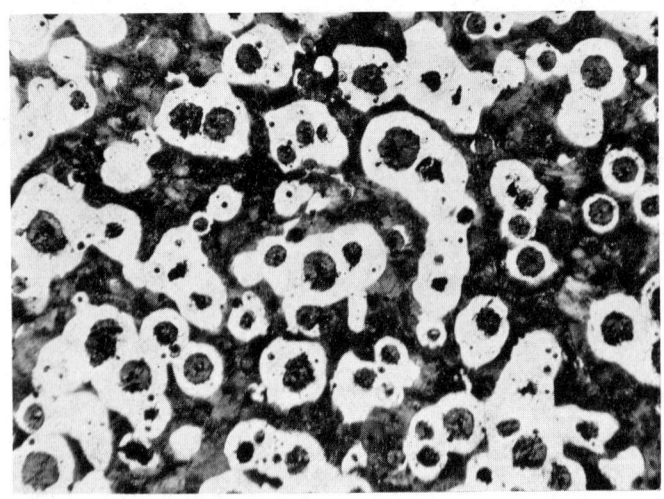

FIG. 5—D5506, APPROXIMATE 235 BHN (100X) TYPICAL MICROSTRUCTURES

11.3 D5506 is used for more highly stressed parts, such as automotive crankshafts.

11.4 D7003 is used where high strength or improved wear resistance is required and where selective hardening is to be employed.

11.5 DQ&T is used where the uniformity of a heat treated material is required to control the range of mechanical properties or machinability.

TABLE 2—TYPICAL MECHANICAL PROPERTIES FOR DESIGN PURPOSES

Grade	Hardness Range[a]	Description	Tensile Strength, psi (MPa)	Yield Strength, 0.2% Offset, psi (MPa)	Elongation, % in 2 in	Modulus of Elasticity, 10⁶ psi (GPa)
D4018	170 Bhn max 4.6 BID min	Ferritic	60,000 (414)	40,000 (276)	18	22 (152)
D4512	156–217 Bhn 4.80–4.10 BID	Ferritic-pearlitic	65,000 (448)	45,000 (310)	12	22 (152)
D5506	187–255 Bhn 4.4–3.8 BID	Ferritic-pearlitic	80,000 (552)	55,000 (379)	6	22 (152)
D7003	241–302 Bhn 3.9–3.5 BID	Pearlitic	100,000 (689)	70,000 (483)	3	22 (152)
DQ&T	Range specified by agreement	Martensitic	A wide variety of desirable properties will result from liquid quenching and tempering			22 (152)

[a] Brinell impression diameter (BID) is the diameter in millimeters (mm) of the impression of a 10 mm ball at 3000 kg load.

12. Additional Information

1. *Metals Handbook,* Vol. 1, 2, and 5, 8th Edition. American Society for Metals, Metals Park, Ohio.
2. *Gray and Ductile Iron Castings Handbook.* Gray and Ductile Iron Founders Society, Cleveland, Ohio.
3. H. D. Angus, *Physical Engineering Properties of Cast Iron.* British Cast Iron Research Association, Birmingham, England.
4. *Gray, Ductile, and Malleable Iron Castings Current Capabilities,* STP-455, American Society for Testing and Materials, 1916 Race Street, Philadelphia, Pennsylvania 19103.

ELEVATED TEMPERATURE PROPERTIES OF CAST IRONS—SAE J125

SAE Information Report

Report of Iron and Steel Technical Committee approved September 1969.

Scope—The purpose of this SAE Information Report is to provide automotive engineers and designers with a concise statement of the basic characteristics of cast iron under elevated temperature conditions. As such, the report concentrates on general statements regarding these properties with limited illustrative data, anticipating that those who may be interested in more detail will want to use the bibliography provided at the conclusion of the report.

Introduction—Cast irons, like steels and other metals, lose strength as operating temperatures increase. Composition is of importance not only because of its effect on the basic properties of materials at elevated temperatures, but also because in cast irons it influences growth resulting from oxidation and microstructural changes. Irons may be used in most atmospheres at temperatures up to 750 F without growth being a serious factor. Beyond 900 F graphitization can cause growth and above 1200 F internal oxidation can cause growth unless sufficient alloy is present to prevent it.

Deterioration of properties at high temperatures is in general time-dependent as well as temperature-dependent. Even at temperatures where strength has been greatly reduced, many useful hours of life can be obtained from a structure if proper allowances are made in the initial design. Where applications involve sustained stress at high temperature, the most valuable information for the designer is the creep rate at the temperature and stress involved. However, creep rate data generally involve long time tests and as a consequence complete information has not been generated for all materials under all conditions. Instead, it has been the practice for many years to compare materials in shorter duration tests. Such tests are called stress-to-rupture tests or more simply stress-rupture tests. These are conducted in the temperature ranges of interest but usually at much greater loads than any realistic design. In general, materials showing superior stress-rupture life have the lowest creep rates. This type of information has been used primarily for material development work. It can be used by the designer, however, to select better material on a comparison basis.

Several types of iron are included in this section to show trends; they are representative of broad classes of irons used commercially, and for which thermal data are available in the literature.

TABLE 1—CHEMICAL COMPOSITION FOR CAST IRONS SHOWN IN FIGS. 1–7

Material	Alloying, %								
	T.C.	C	Si	Mn	P	S	Cr	Ni	Mo	Mg
Alloy gray cast iron (ASTM A 48, No. 60) (2)[a]	3.06	—	1.79	0.70	0.04	0.09	0.61	0.04	0.84	—
Alloy gray cast iron (SAE G4500) (2)	3.31	—	1.56	0.68	0.19	0.114	0.08	0.08	0.73	—
Gray cast iron SAE G4000 (2)	3.27	—	1.74	0.72	0.26	0.156	0.08	0.15	0.07	—
Ferritic malleable (3) (Creep and stress rupture data)	2.16	—	1.01	0.29	0.11	0.074	0.017	—	—	—
	2.29		1.17	0.38	0.148	0.095	0.000			
Pearlitic malleable (3)	2.27	—	1.01	0.89	0.135	0.098	0.019	—	—	—
	2.29		1.15	0.75	0.110	0.086	0.000			
Ferritic malleable (3) (Elevated temperature tensile strength data)	2.30	—	0.98	0.30	0.162	0.078	—	—	—	—
	2.33		1.05	0.34	0.168	0.084				
Gray cast iron SAE G4500 (4)	2.84	—	1.52	1.05	0.07	0.124	0.31	—	0.20	—
Ferritic ductile iron (4)	3.7	—	2.6	0.40	—	—	—	1.0	—	—
Low carbon steel (7)	—	0.08	0.25	0.30	0.045	0.060	—	—	—	—
		0.20		0.80						
Austenitic ductile + Cr and Mo (9)	2.38	—	1.99	0.62	—	—	3.05	30.18	0.95	0.13
Austenitic ductile + Cr (9)	2.98	—	2.20	1.15	—	—	2.36	20.48	—	0.085
Ferritic ductile iron (14)	3.64	—	2.66	0.46	0.032	0.014	—	0.66	—	0.076
Austenitic cast iron (17)	2.63	—	2.14	1.23	0.16	0.059	2.09	14.9	—	—

[a] Parenthetical numbers indicate source of data in bibliography.

Effect of Elevated Temperature on Mechanical Properties

Tensile Strength—The tensile strength of ferrous materials generally shows small changes from room temperature up to 600–800 F, at higher temperature the strengths usually fall rather rapidly. The presence of alloying elements which affect the stability of the higher strength microstructures tends to delay this effect or raise the temperature at which rapid loss of strength occurs. In some ferrous alloys, changes in microstructure occur at temperatures between room temperature and 800 F which may cause small changes in strength and, in fact, may cause reversals in the strength versus temperature curve. In Fig. 1, examples of tensile strength versus temperature for some typical cast irons are illustrated in comparison with the behavior of low carbon steel. Generally, the changes in structure which occur over this temperature range are associated with tempering after hardening. These changes are irreversible.

Stress Rupture Properties—Where metals are required to sustain loads over long periods of time at elevated temperatures, the stress-rupture test is used as an indication of the relative load-carrying ability at the test temperature.

The material is stressed in tension under a constant load at a constant temperature and the time that it takes the sample to rupture under these conditions is recorded. Separate samples of the material are stressed under a number of different loads at the same temperature and the rupture times are plotted against load to give a stress-rupture curve for the material. Typical stress-rupture curves for a number of SAE cast irons and alloyed irons at 800 F are plotted in Fig. 2. Stress-rupture curves for these and other cast irons at 1000 F are plotted in Fig. 3, which includes a stress-rupture curve for low carbon wrought steel for comparison.

Creep Properties of Irons—Another important temperature effect on cast irons is the effect of creep or elongation per hour at a given stress and temperature. These tests are difficult to run on cast irons because of the growth and oxidation phenomena. Fig. 4 shows typical creep curves for some SAE and alloyed cast irons at 800 and 1000 F. A creep curve for a plain carbon steel at 1000 F is included for comparison. The values shown on the creep rate curves in Fig. 4 include any growth that occurred during the tests. This allows these figures to be used as design parameters.

Growth of Cast Iron—Growth in irons is generally defined as the permanent increase in volume which occurs after prolonged exposure to constant elevated temperatures or after repeated heating and cooling.

The mechanism of growth is rather complex in irons owing to the fact that it results from several different and independent phenomena; which phenomena occur in any given case is dependent upon the temperature involved, the environment, the chemical composition and structure of the iron, and the severity of cycling, if any. The main phenomena which cause growth in irons at temperature follow:

1. Oxidation—This may progress into the body of an iron along graphite flakes or cracks, resulting in a greater dimensional change than in materials where oxidation is limited to the surfaces.

2. Graphitization—If the iron has carbides in its structure and is heated to

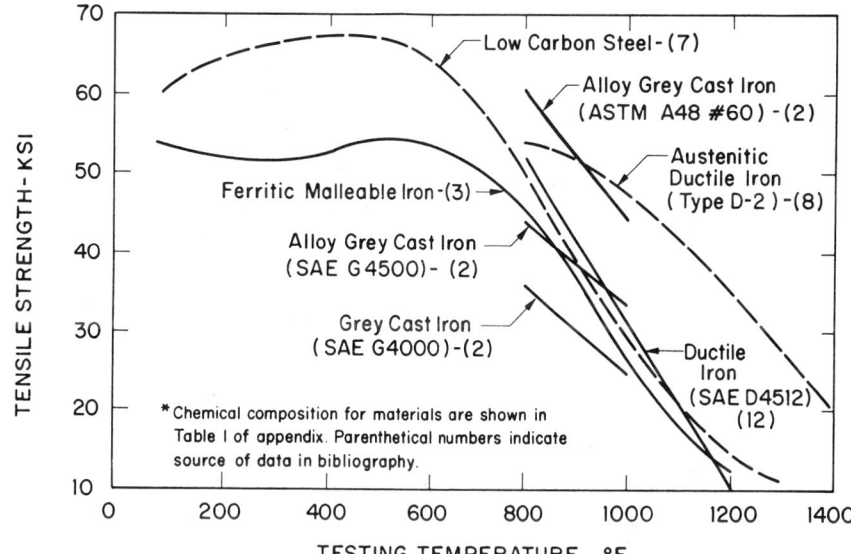

FIG. 1—ELEVATED TEMPERATURE TENSILE STRENGTH

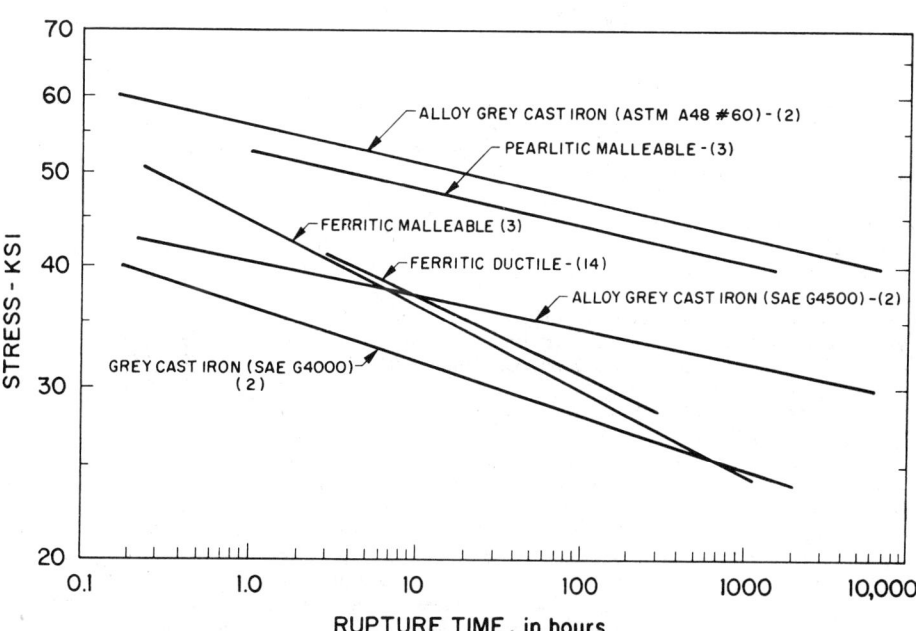

FIG. 2—STRESS RUPTURE PROPERTIES OF CAST IRONS AT 800 F

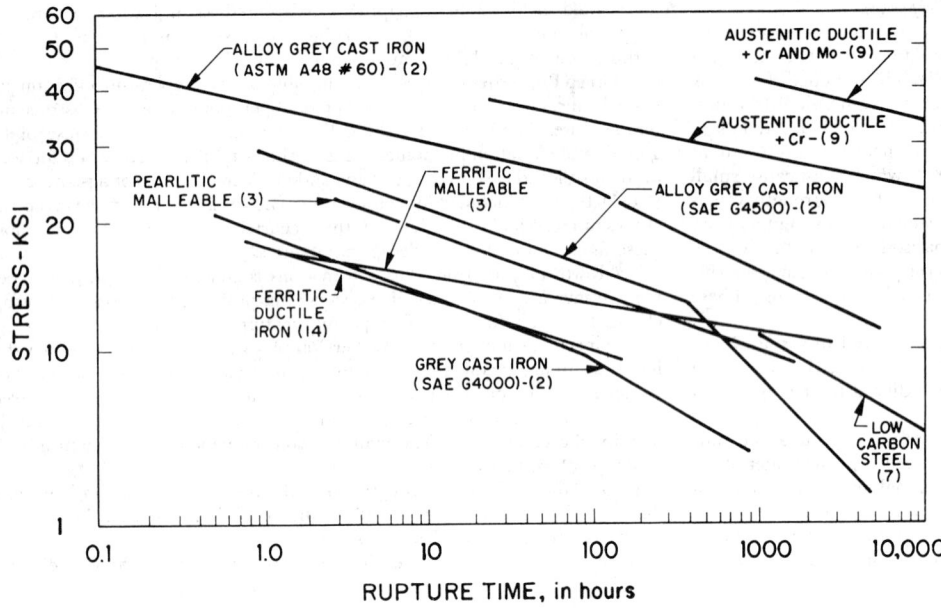

FIG. 3—STRESS RUPTURE PROPERTIES OF CAST IRONS AT 1000 F

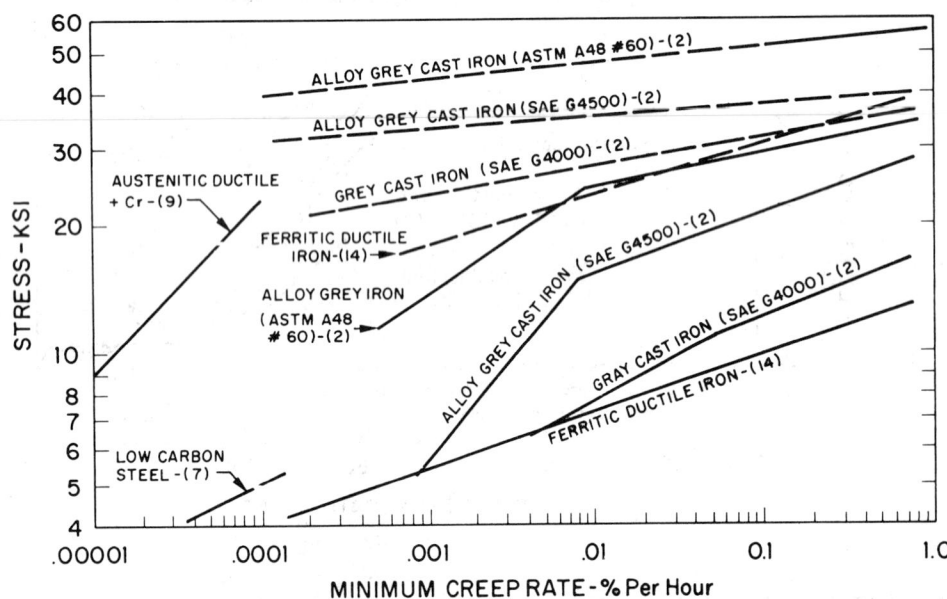

FIG. 4—CREEP PROPERTIES OF CAST IRONS

a temperature which will decompose the carbide structures (this temperature will vary considerably with chemistry), the resulting ferrite and graphite will occupy considerably greater volume than the original carbide.

3. Crazing or Thermal Cracking—When an iron is repeatedly heated and cooled through a transformation range, the stresses, imposed by the expansion and contraction resulting from the transformation, will cause crazing.

Ductile and malleable irons are less affected by oxidation than gray iron. Some investigators have indicated that this difference is a result of graphite carbon shapes. Alloyed irons are usually less susceptible to growth, and this can be either due to an increased stability of the structure at the temperature involved, improved oxidation resistance, or a change of the critical temperature so the part does not experience a transformation in the temperature range. Examples of the growth experienced by some common SAE irons are shown in Fig. 5.

Endurance Limit—Relatively little data are available on the endurance limit of cast irons at elevated temperatures. The curves in Fig. 6 show the relationship between the tensile strength and the endurance limit of a low carbon equivalent low alloy gray cast iron up to a temperature of 1100 F. It can be seen that the endurance ratio is nearly constant from room temperature to 1100 F for this iron. It is probable that for most gray cast irons heated in air, the endurance ratio would remain nearly constant up to a temperature at which changes occur in the structure or severe oxidation takes place. A second set of curves for an austenitic gray iron are included in Fig. 6. This shows how the endurance limit is affected by structure.

Effect of Temperature on Physical Properties

Modulus of Elasticity—The modulus of elasticity at room temperature varies considerably for the various types of iron. It is difficult to give representative figures; but, in general, it can be said that the cast irons do not show a marked drop in modulus up to 800 F.

Specific Heat—of ferritic malleable iron:

Temperature Range		Mean Specific Heat	
F	C	Btu/lb-F	Cal/g-C
70- 210	20-100	0.122	0.122
70- 570	20-300	0.128	0.128
70- 750	20-400	0.139	0.139
70-1300	20-700	0.159	0.159

The specific heat of pearlitic malleable iron is substantially the same as that of ferritic malleable.

In gray cast iron the specific heat varies with the temperature and the structure; therefore, a temperature range which causes a change in microstructure may result in a reversal in the specific heat versus temperature curve which will result in a curve such as Fig. 7. A commonly used figure for average cast irons is 0.13 cal/g-C.

Thermal Conductivity—The thermal conductivity of gray cast iron varies considerably with both temperature and composition. Most plain and

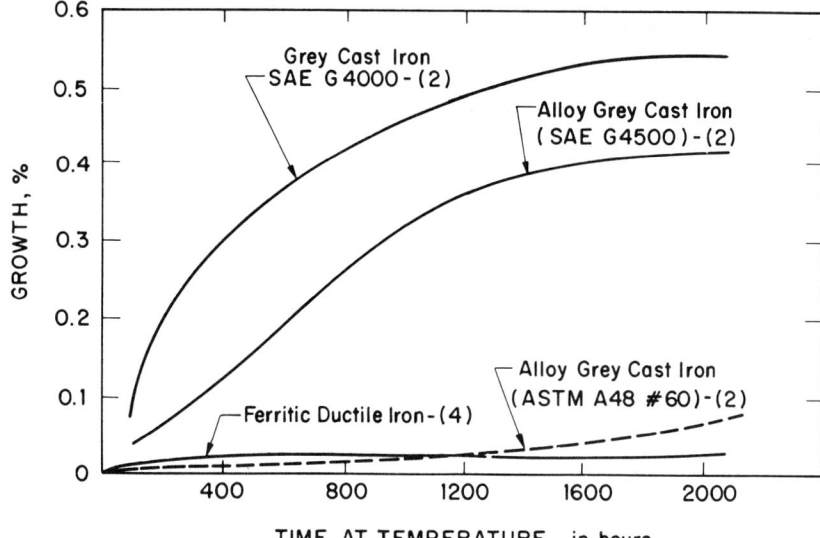

FIG. 5—CAST IRON GROWTH AT 1000 F

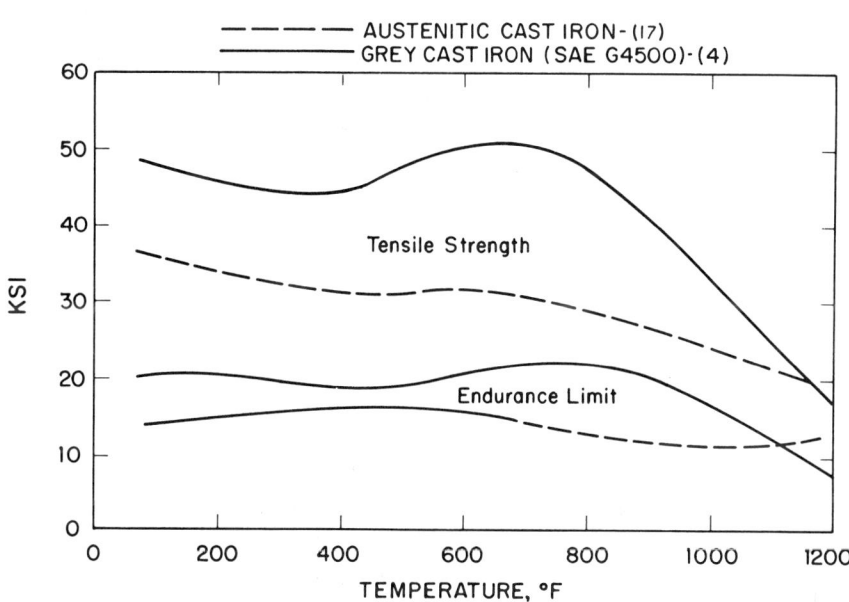

FIG. 6—EFFECT OF TEMPERATURE ON STRENGTH AND ENDURANCE LIMIT

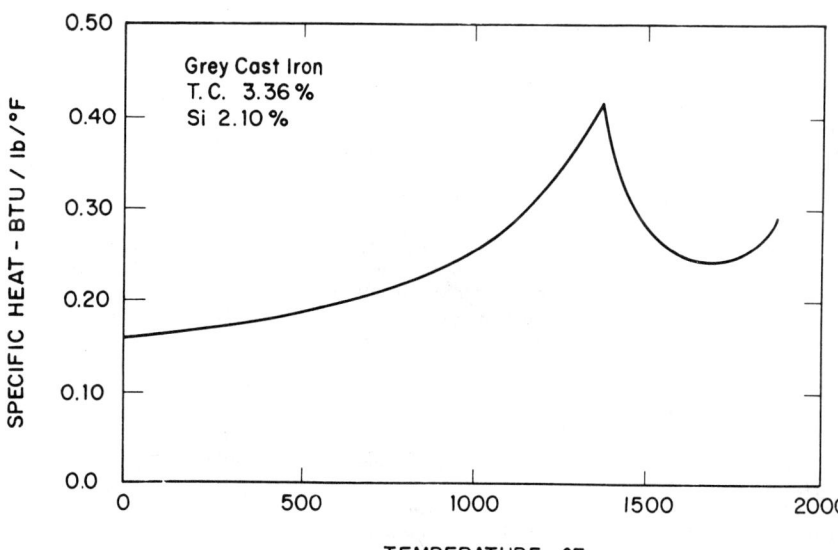

FIG. 7—SPECIFIC HEAT VERSUS TEMPERATURE FOR A GRAY CAST IRON (4)

alloy gray irons will have a thermal conductivity of 0.10 to 0.135 cal-cm/sec-cm²-C at 100 C and this will drop to about 0.09-0.115 cal-cm/sec-cm²-C at 400 C. Some specific cases are given in the following table:

Grade or Type	Cal-cm/sec-cm²-C		Btu-in./hr-ft²-F	
	At 100C	At 400C	At 212F	At 750F
SAE G4000 + Low S	0.135	0.114	391.5	330.6
SAE G5000	0.110	0.100	319	290
Mo alloyed (ASTM A 48, No. 55)	0.118	0.108	342.2	313.2
Mo + Ni alloyed (ASTM A 48, No. 55)	0.106	0.097	307.4	281.3

Malleable iron has a thermal conductivity of about 0.151 cal-cm/sec-cm²-C at 100 C (437.9 Btu-in./hr-ft²-F at 212 F) and 0.139 cal-cm/sec-cm²-C at 400 C (403.1 Btu-in./hr-ft²-F at 750 F).

Ductile cast iron has the following conductivity:

Type	Cal-cm/sec-cm²-C		Btu-in./hr-ft²-F	
	At 100C	At 400C	At 212F	At 750F
Ferritic	0.0845	0.0821	239.3	238.1
Pearlitic	0.075	0.0761	217.5	220.7
Austenitic	At Room Temperature		At Room Temperature	
	0.032		92.8	

Average Coefficient of Expansion—The coefficient of expansion of cast irons varies with temperature, and to a lesser degree with alloy content and structure or heat treatment. The following table gives values commonly used:

Type of Iron	Mean Coefficient of Linear Expansion	
	cm/cm-C 0–500C	in./in.-F 32–932F
Gray iron	12.96×10^{-6}	7.2×10^{-6}
Ferritic malleable	11.88×10^{-6}	6.6×10^{-6}
Pearlitic malleable	13.5×10^{-6}	7.5×10^{-6}
	20–200C	68–392F
Ferritic ductile	$11.85\text{–}12.65 \times 10^{-6}$	$6.6\text{–}7.0 \times 10^{-6}$
Pearlitic ductile	$11.7\text{–}11.85 \times 10^{-6}$	$6.5\text{–}6.6 \times 10^{-6}$
Austenitic ductile	$3.98\text{–}18.9 \times 10^{-6}$	$2.2\text{–}10.5 \times 10^{-6}$

BIBLIOGRAPHY

1. "Mechanical Properties of Metals and Alloys." U. S. Dept. of Commerce Circular C-447, National Bureau of Standards, 1943.
2. Kattus and McPherson, "Properties of Cast Iron at Elevated Temperatures." ASTM Special Technical Publication No. 248.
3. Malleable Iron Casting Handbook. Malleable Founders Society, 1960.
4. Gray Iron Castings Handbook. Gray Iron Founders' Society Inc., 1958.
5. Cast Metals Handbook. American Foundrymens Society, 1957.
6. Colin J. Smithell, "Metals Reference Book." Washington Butterworths, 1962.
7. Metals Handbook, 8th Edition. American Society for Metals, 1961.
8. "Engineering Properties of Ductile Ni-Resist Austenitic Irons." International Nickel Co., 1955.
9. Schelleng and Eash, "Effect of Composition on the Elevated-Temperature Properties of Ductile Iron." Proceedings of ASTM, Vol. 57, 1957.
10. Greene and Sefing, "Cast Irons in High Temperature Service." *Corrosion*, Vol. 11, No. 7, July 1955.
11. Turnbull and Wallace, "Molybdenum Effect on Gray Iron Elevated Temperature Properties." Transactions AFS, Vol. 67, 1959.
12. F. B. Foley, "Mechanical Properties at Temperature of Ductile Cast Iron." Preprint No. 55-A-204, ASME, 1955.
13. "Engineering Properties of Ni-Resist Ductile Irons." International Nickel Co., 1958.
14. "Elevated Temperature Properties of Ductile Cast Irons." ASM Transactions, Vol. 47, 1955.
15. Scholz, Doane, and Timmons, "Effects of Molybdenum on Stability and High Temperature Properties of Pearlitic Malleable Iron." AFS Transactions, Vol. 63, 1955.
16. D. A. Pearson, "Stress-Rupture and Elongation of Malleable Iron at Elevated Temperatures." AFS Transactions, Vol. 74, 1966.
17. W. L. Collins, "Fatigue and Static Load Tests of an Austenitic Cast Iron at Elevated Temperatures." ASTM Proceeding, Vol. 48, 1948.

AUTOMOTIVE STEEL CASTINGS—SAE J435c

SAE Standard

Report of Iron and Steel Technical Committee approved October 1946 and last revised July 1974.

1. Scope—These specifications cover steel castings used in the automotive and allied industries.

2. General Data—In order to cover adequately the varying requirements encountered, the cast steels are classified into the following groups:

(1) GRADE 00XX STEEL CASTINGS—These are specified by chemical composition and minimum mechanical properties (except the carburizing grade, which is based on chemical composition only).

(2) GRADE 0XX STEEL CASTINGS—These are specified by minimum mechanical properties, for miscellaneous uses where the requirements do not justify hardenability control.

(3) GRADE HX STEEL CASTINGS—These are specified by hardenability requirements.

2.1 All castings shall be furnished in the heat treated condition by one of the following heat treatments unless otherwise specified:

(1) Annealed (3) Normalized and tempered
(2) Normalized (4) Quenched and tempered

2.2 The choice of alloying elements used in the manufacture of the 0XX grades of steel castings shall be at the option of the foundry. Certain specific cases may arise where chemistry is a controlling factor in obtaining desired properties, such as wear resistance. Under these circumstances, chemistry may form part of this specification by specific agreement between the producer and purchaser, in inquiry and purchase order.

3. Grades of Steel Castings—Grades and requirements of automotive cast steels are shown in Table 1. A typical material specification is: Steel casting, SAE J435 (0120).

4. Mechanical Properties—When steel castings are specified to minimum mechanical properties, the values shown for the appropriate grade in Table 1 shall apply to the test bars. When test samples are to be taken from the castings, their location and the required mechanical properties shall be agreed upon in writing by the producer and user at the time of bidding.

4.1 The mechanical property requirements pertain to the properties of standard test bars that have been poured in accordance with ASTM A 370, Methods and Definitions for Mechanical Testing of Steel Products. They shall be poured with metal which has received deoxidation treatment identical to that metal from which the castings are poured. All test coupons from which specimens are to be prepared shall be given the same heat treatment as the castings, except that the tempering temperature for the castings are to be adjusted for equivalent Brinell hardness.

5. Tensile Test—Tensile test specimens shall be machined to the dimensions shown in SAE J416 or in ASTM E 8, Methods of Tension Testing of Metallic Materials. One tensile test shall be made from each heat or lot[1]. If any specimen shows defective machining or develops flaws, it may be discarded and another specimen substituted for the same lot.

6. Quality Control—After meeting acceptance tests for ten consecutive heats used for making castings of any one grade, the manufacturer may assemble the castings from succeeding lots in groups of five heats each. The castings in each such group shall be accepted on the basis of one test specimen taken at random from one of the five heats provided that the chemical analysis of all heats in the group falls within the range established by the first ten consecutive acceptable heats as determined by both chemistry and mechanical testing.

---

[1] The term "lot" shall be considered as all castings in a heat subjected to the same heat treating procedure, for example, a heat of castings could be divided into several lots by virtue of different heat treatments, but all of the same chemistry.

TABLE 1—GRADES OF CAST STEEL AND REQUIREMENTS

SAE Grade	Approx Equivalent		Description	Chemical Composition, % by wt[a]					Test Bar Minimum Mechanical Properties						
	ASTM	QQ-S-681d		C	Mn	Si	P, Max	S, Max	Tensile Strength		Yield Strength		Elongation in 2 in, %	Reduction in Area, %	Brinell Hardness[c]
									psi	MPa	psi	MPa			
0022	A 27	—	Low carbon, suitable for carburizing	0.12–0.22	0.50–0.90	0.60 max	0.040	0.045	—	—	—	—	—	—	187 max
0025	A 27	—	Carbon steel welding grades	0.25 max	0.75 max[b]	0.80 max	0.040	0.045	60 000	413.7	30 000	206.8	22	30	187 max
0030	A 27	65-35		0.30 max	0.70 max[b]	0.80 max	0.040	0.045	65 000	448.2	35 000	241.3	24	35	131–187
0050A	—	0050A	Carbon steel medium strength grades	0.40–0.50	0.50–0.90	0.80 max	0.040	0.045	85 000	586.0	45 000	310.3	16	24	170–229
0050B[d]	—	0050B		0.40–0.50	0.50–0.90	0.80 max	0.040	0.045	100 000	689.5	70 000	482.6	10	15	207–255
080	A 148	80-50	Medium strength low alloy grades	Optional	Optional	Optional	0.040	0.045	80 000	551.6	50 000	344.7	22	35	163–207
090	A 148	90-60		Optional	Optional	Optional	0.040	0.045	90 000	620.5	60 000	413.7	20	40	187–241
0105	A 148	105-85	High strength alloy grades	Optional	Optional	Optional	0.040	0.045	105 000	723.9	85 000	586.0	17	35	217–248
0120	A 148	120-95		Optional	Optional	Optional	0.040	0.045	120 000	827.4	95 000	655.0	14	30	248–311
0150	A 148	150-125		Optional	Optional	Optional	0.040	0.045	150 000	1034.2	125 000	861.8	9	22	311–363
0175	—	175-145		Optional	Optional	Optional	0.040	0.045	175 000	1206.6	145 000	999.7	6	21	363–415
HA, HB, HC	Hardenability Grades. See Figs. 1, 2, and 3.														

[a] These analyses are mandatory, unless otherwise specified by the purchaser. The optional and alloying elements may be used at the producer's discretion to obtain the mechanical properties or hardenability specified, unless otherwise specified. Boron in the range of 0.003–0.007% with 1.35% manganese maximum may be used as an optional alloying element by agreement between producer and purchaser.

[b] For each reduction of 0.01% carbon below the maximum specified, an increase of 0.04% manganese above the maximum specified will be permitted to a maximum of 1% manganese.
[c] Obtain from parts, not test bars, in location not over 3 in. in thickness.
[d] Properties require a liquid quench and temper. Casting section should be 1 in (25.4 mm) or less.

6.1 After meeting acceptance tests for ten consecutive heats, by mutual agreement between the producer and purchaser, subsequent castings may be accepted by hardness checking of the castings in lieu of mechanical properties of test bars, to the hardness ranges shown in Table 1, provided that the chemistry falls within the range of the ten accepted heats.

6.2 *Retests*—If retest is required, the heat may be requalified by using another specimen, and the four other heats in the group shall be tested individually. The same heat treating procedure used for the first ten consecutive heats shall be used for all subsequent heats. This procedure shall be established for each grade separately.

7. **Hardenability**—Whenever hardenability is specified on alloy steel castings that are to be heat treated by the purchaser or producer, the limits shown for grades HA, HB or HC (Figs. 1, 2, and 3) whichever is specified, shall apply. Alloy steels may thus be specified on drawings to hardenability in the following way: Steel Casting SAE HB.

7.1 Castings which are shipped to the buyer under hardenability requirements, and on which the buyer will do the heat treating, should not have a hardness greater than 248 Bhn for Grades HA and HB, and 269 Bhn for Grade HC when they leave the foundry.

7.2 The hardenability of the steel used shall be determined in accordance with SAE J406, or ASTM A 255, Method of End-Quench Test for Hardenability of Steel. Calculated hardenability may be used in lieu of actual test by agreement between producer and user. The specimens may be cast to size in molds or prepared from the standard test bar coupon. Unless otherwise agreed to between the producer and purchaser, the producer shall keep a record of the hardenability data representing all heats in question for a period of three months.

7.3 The hardenability of cast steels is similar to that of wrought steels of the same chemical analysis. Details as discussed for wrought steels in SAE J407, apply equally to cast steels. The bands for cast steels given in Figs. 1, 2, and 3 are based on a nominal carbon content of 0.30%. Bands for steel of other carbon contents shall be established by mutual consent of the purchaser and producer. It is suggested that H bands for wrought steels as shown in SAE J407 be used as guides.

8. **Chemical Analysis**—All castings based on chemical analysis shall conform to the ranges given for each grade in Table 1. The only chemical requirements for the balance of the castings in these specifications will be maximum phosphorus and sulfur content as indicated. Chemical analysis of each melt of steel shall be made from a test specimen obtained by the producer during the pouring of the heat. Chemical analysis samples shall be taken from metal not less than $\frac{1}{4}$ in. beneath the surface. The chemical composition thus determined shall conform to the requirements prescribed in Table 1.

8.1 When high strength cast steels are to be used for fabrication by fusion welding, it is recommended that the carbon content be limited to 0.35% carbon, max.

8.2 *Product Analysis*—When so desired, an analysis may be made by the purchaser from a test specimen or casting representing each melt. The procedure previously outlined shall be followed, and the chemical composition as determined shall conform with the requirements specified for the particular grade of steel in question.

8.3 *Residual Alloy Limitations*—When required, residual alloy limitations are subject to the specific agreement between producer and purchaser in inquiry and order. The following additional special conditions are noted:

Grade 0022—Maximum hardness in Table 1 is before carburizing and provides a limit to residual alloys.

Grade 0025 and 0030—Residual alloys shall be limited by the ability of these grades to be readily welded.

8.4 If close control of residual alloys is required, the maximum DI (ideal diameter) to control hardenability should be negotiated between purchaser and producer.

9. **Finish Requirement**—When so specified in the inquiry and purchase order and mutually agreed upon by the producer and purchaser, castings made to any SAE grade listed in Table 1 may be subject to surface finish requirements. Such requirements shall be based upon comparison with the ACI Surface Indicator Scale (available from Steel Founders' Society of America, Cast Metals Federation Building, 20611 Center Ridge Road, Rocky River, Ohio 44116. See Fig. 4.

9.1 The SIS Scale is a comparator intended for visual inspection. Optical magnifiers and measuring instruments are not to be used, but it is permissible to judge similar surfaces by touch, that is, rubbing a finger across the casting and the scale, in addition to visual observation. Specifying one of the four SIS numbers is sufficient for purposes of designating general surface smoothness in a given area of a casting or over its entire surface. If control of localized surface irregularities is desired, it should be mutually agreed between producer and purchaser.

9.2 The surface finishes represented by the SIS numbers are typical of the major processes employed in the production of steel castings. The molding material and molding technique together are the principal factors in determining the as-cast smoothness. Surface as smooth as or smoother than SIS-1 can be attained by ceramic molding or, in some instances, by shell molding. Usually, shell molding surfaces will fall between SIS-1 and SIS-2. Fine dry sand molding can produce surfaces equal to SIS-2. Green sand molding will generally produce casting surfaces equivalent to SIS-3. Castings of large size are usually made in molds using coarse sand and may be as rough as SIS-4.

9.3 Different parts of the same casting may have different surface smoothness. Variations can result from casting design, position in the mold, and other factors. Therefore, critical surfaces, where smoothness must be equal to or better than the designated SIS number should be noted on the drawing. A rougher surface may be acceptable in noncritical areas of the same casting and the appropriate SIS number should be so noted on the drawing. Engineers should recognize that smoother surface finishes are obtained, generally, by going to a more costly molding process and that the practice of specifying finishes in excess of those required for the application should be avoided.

9.4 This standard does not include the use of the raised bars (on the ACI

HARDENABILITY BAND — GRADE HA

C	Mn	Si	Ni	Cr	Mo
.25 / .34		ANALYSIS TO BE SELECTED BY FOUNDRY			

DIAMETERS OF ROUNDS WITH SAME AS QUENCHED HARDNESS						LOCATION IN ROUND	QUENCH
3.8						SURFACE	MILD WATER QUENCH
1.1	2.0	2.9	3.8	4.8	5.8 6.7	3/4 RADIUS FROM CENTER	
0.7	1.2	1.6	2.0	2.4	2.8 3.2 3.6 3.9	CENTER	
					3.8	SURFACE	MILD OIL QUENCH
0.8	1.8	2.5	3.0	3.4	3.8	3/4 RADIUS FROM CENTER	
0.2	0.6	1.0	1.4	1.7	2.0 2.4 2.8 3.1	CENTER	

FIG. 1

Hardness Limits for Specification Purposes — Grade HA

"J" DISTANCE SIXTEENTHS OF AN INCH	MAX	MIN
1	54	46
2	53	45
3	51	40
4	48	35
5	44	30
6	41	27
7	39	25
8	37	22
9	34	21
10	32	20
11	31	
12	29	
13	28	
14	27	
15	26	
16	25	
18	24	
20	23	
22	21	
24	20	
26		
28		
30		
32		

HEAT TREATING TEMPERATURES RECOMMENDED BY SAE
NORMALIZE 1650 °F
AUSTENITIZE 1600 °F

HARDENABILITY BAND — GRADE HB

C	Mn	Si	Ni	Cr	Mo
.25 / .34		ANALYSIS TO BE SELECTED BY FOUNDRY			

FIG. 2

Hardness Limits for Specification Purposes — Grade HB

"J" DISTANCE SIXTEENTHS OF AN INCH	MAX	MIN
1	56	46
2	55	46
3	55	44
4	54	42
5	53	40
6	52	38
7	51	36
8	50	35
9	48	33
10	46	32
11	44	31
12	42	30
13	41	29
14	40	28
15	39	27
16	38	26
18	37	25
20	35	24
22	34	21
24	34	20
26	33	
28	33	
30	32	
32	32	

HEAT TREATING TEMPERATURES RECOMMENDED BY SAE
NORMALIZE 1650 °F
AUSTENITIZE 1600 °F

HARDENABILITY BAND — GRADE HC

C	Mn	Si	Ni	Cr	Mo
.25 / .34		ANALYSIS TO BE SELECTED BY FOUNDRY			

FIG. 3

Hardness Limits for Specification Purposes — Grade HC

"J" DISTANCE SIXTEENTHS OF AN INCH	MAX	MIN
1	57	49
2	56	48
3	55	48
4	55	47
5	55	46
6	55	46
7	55	45
8	55	44
9	55	43
10	55	42
11	55	41
12	55	42
13	55	42
14	54	41
15	54	41
16	54	40
18	54	40
20	54	39
22	54	39
24	53	38
26	53	38
28	53	38
30	52	37
32	51	37

HEAT TREATING TEMPERATURES RECOMMENDED BY SAE
NORMALIZE 1600 °F
AUSTENITIZE 1550 °F

FIG. 4

ACI SURFACE INDICATOR SCALE — SIS-1, SIS-2, SIS-3, SIS-4

NOTES FOR FIGS. 1, 2, AND 3: THE EXPERIENCE OF USERS OF STEEL CASTINGS OF THESE GENERAL TYPES INDICATES THAT GRADE HA CAN BE USED FOR STEEL CASTINGS WITH CRITICAL SECTIONS UP TO 2½ IN (63.5 MM), GRADE HB FOR CASTINGS WITH CRITICAL SECTIONS FROM 2½ TO 4 IN (63.5-102 MM), AND GRADE HC FOR CASTINGS WITH CRITICAL SECTIONS FROM 4 TO 7 IN (102-178 MM). IT SHOULD BE POINTED OUT THAT THE ABOVE RECOMMENDATIONS WILL NOT RESULT IN THOROUGH HARDENING ON THE QUENCH. THEY ARE BASED ON A COMPROMISE BETWEEN IDEAL MICROSTRUCTURE AND PRODUCTION DIFFICULTIES, PARTICULARLY QUENCH CRACKING. THE PARTICULAR APPLICATION REQUIRES THAT THE CASTING SO TREATED WILL NOT SHOW UNIFORM PROPERTIES THROUGHOUT THE SECTION.

IF THE PARTICULAR APPLICATION REQUIRES THAT THE CASTING POSSESS THE ULTIMATE IN MECHANICAL PROPERTIES, PARTICULARLY NOTCHED IMPACT STRENGTH AT THE CENTER OF THE SECTION, THE FOLLOWING LIMITATIONS SHOULD BE OBSERVED IN SELECTING THE HARDENABILITY BANDS. GRADE HA, LOWER BAND FOR ¾ IN (19 MM) MAXIMUM SECTION, UPPER BAND FOR 1¼ IN (32 MM) MAXIMUM SECTION; GRADE HB, LOWER BAND 1¼ IN (32 MM) MAXIMUM SECTION, UPPER BAND 1¾ IN (44 MM) MAXIMUM SECTION; GRADE HC, LOWER BAND 2¾ IN (70 MM) MAXIMUM SECTION, UPPER BAND 4 IN (102 MM) MAXIMUM SECTION. THESE VALUES FOR SECTION THICKNESSES ARE APPLICABLE ONLY IF WATER QUENCHING IS USED. IF OIL QUENCHING IS USED REFERENCE SHOULD BE MADE TO A TEST ON HARDENABILITY. SUCH A TEXT SHOULD ALSO BE CONSULTED FOR DETAILS AS TO THE APPLICATION OF HARDENABILITY DATA.

SIS replica) which indicate the height and depth of localized irregularities extending beyond the range of general variations.

10. Inspection of Castings

10.1 General—This section covers methods for qualification and routine control of characteristics and properties indigenous to the dimensions, shape, and molding practice of the casting, as opposed to methods used to qualify and control the mechanical properties of the steel used to produce the casting. Such methods are primarily nondestructive in nature, but simulated service loading and destructive tests on the casting may also be part of initial qualification and routine quality control procedures. All testing procedures and tooling which are to be used should be decided upon at the time an inquiry is submitted and must be on the order and agreed upon by both parties prior to the start of bidding on any item.

10.2 Test Procedures—Areas to be inspected, minimum number of pieces from each lot which shall be inspected, and standards of acceptability must be agreed upon by producer and purchaser. Proper application and interpretation of all test procedures must be understood by all involved in the inspection.

10.2.1 CASTING STRENGTH

10.2.1.1 *Proof Testing*—Simulated service loading may constitute one basis for acceptance of individual castings.

10.2.1.2 *Destructive Tests*—Loading to destruction may be used and may be based on loads, fracture mode, soundness, and so on.

10.2.2 INTERNAL QUALITY—Specifications for internal quality, when agreed upon by purchaser and producer, require the use of radiography, ultrasonics, or destructive testing methods.

10.2.2.1 *Destructive Testing*—Usually involves cross-sectioning or machining the casting at critical areas with visual inspection for defects.

10.2.2.2 *Radiographic Tests*—May be made with either x-ray or gamma ray sources, and specifications should follow standards such as those listed:

ASTM E 52, Industrial Radiographic Terminology for Use in Radiographic Inspection of Castings and Weldments

ASTM E 94, Recommended Practice for Radiographic Testing

ASTM E 142, Method for Controlling Quality of Radiographic Testing

ASTM E 186, Industrial Radiographic Reference for Steel Castings

ASTM E 71, Industrial Radiographic Standards for Steel Castings

ASTM E 99, Reference Radiographs for Steel Welds

10.2.2.3 *Ultrasonic Testing*—May be used for checking dimensional variations in areas inaccessible to conventional tools, as well as testing for internal soundness. This method is usually specified only for critically stressed castings and requires equipment and techniques of a specialized nature. Therefore, there are not, as yet, many standardized specifications to be used as acceptance or rejection references. Some of those available are:

ASTM E 113, Recommended Practice for Ultrasonic Testing by the Resonance Method

ASTM E 114, Recommended Practice for Ultrasonic Testing by the Reflection Method, Using Pulsed Longitudinal Waves Induced by Direct Contact

ASTM A 609, Specification for Longitudinal Beam Ultrasonic Inspection

10.2.3 SURFACE QUALITY—Specifications may require visual inspection, acid etching, magnetic particle inspection, or liquid penetrant inspection. Whenever possible, specifications should be based on standards such as are listed.

SAE J420, Magnetic Particle Inspection

ASTM E 109, Method for Dry Powder Magnetic Particle Inspection

ASTM E 138, Method for Wet Magnetic Particle Inspection

ASTM E 165, Methods for Liquid Penetrant Inspection

ASTM E 125, Methods for Reference Photographs for Magnetic Particle Indications on Ferrous Castings

10.3 Workmanship—All castings shall be made in a workmanlike manner and shall conform substantially to the dimensions on drawings furnished by the purchaser before manufacture is started, or to the dimensions predicated by the pattern supplied by the purchaser, if no drawings are provided. In the absence of adequate limiting dimensions, the weight of each casting shall be within the following percentages of the established weight, which is the average weight of ten acceptable castings. The castings shall be free from injurious defects and shall be satisfactorily cleaned for the intended use when offered for inspection.

Established Weight		% Tolerance
lb	kg	
Up to 100	45	+5
Over 100 to 500	45–225	+4, −3
Over 500 to 10,000	225–4500	+3, −2-1/2
Over 10,000	4500	+2-1/2, −2

10.4 Marking—The manufacturer's identification mark and pattern number shall appear on all castings except those of such small size as to make such marking impractical. An alloy castings melt number may be required at option of purchaser, and if agreed upon by the producer.

10.5 Inspection by Purchaser—The manufacturer shall afford the purchaser's inspector all reasonable facilities necessary to satisfy him that the material is being produced and furnished in accordance with this specification. Foundry inspection by the purchaser shall not interfere unnecessarily with the manufacturer's operations. All tests and inspections (except product analysis) shall be made at the place of manufacture unless otherwise agreed to.

10.6 Rejection—Unless otherwise specified, a rejection for any reason shall be reported to the producer within six months from the receipt of castings by the purchaser.

10.7 Rehearing—Tested samples representing rejected material shall be held for two weeks from the date of the test report. In case of dissatisfaction with the results of the tests, the producer may make claim for a rehearing within that time.

11. Repair of Defective Steel Castings—Repair of defects in steel castings may be performed in: (1) the foundry or (2) the machine shop. Welding practice shall conform to recommended practices for welding as published by the American Welding Society; "Recommended Practice for Repair Welding and Fabrication Welding of Steel Castings," published by Steel Founders Society of America; or SAE Handbook Supplement 22, Repair of Ferrous Castings—SAE J804.

Repair procedures other than the welding of castings, specified as follows, may not be used by the foundry without prior knowledge and consent of the purchaser.

11.1 Minor Defects—Minor defects in nonmachined areas which are removed at a depth not exceeding 10% of the wall thickness, but in no case greater than 1/8 in. (3.2 mm), may be blended smooth by grinding for good appearance, by specific agreement between supplier and purchaser.

11.2 Peening, Plugging, or Impregnating—The repair of defects in castings shall not be accomplished by peening, plugging, or impregnating, unless agreed to by purchaser.

11.3 Repair Welding—Unless otherwise specified by the procuring agency, defects in castings may be removed and repaired by welding at the discretion of the producer provided the weld repair area has a degree of hardness and other mechanical properties comparable to that of the parent metal. Repair welds shall be subjected to the same inspection standards required of the casting and the area shall be suitably marked to facilitate inspection.

8 Tool and Die Steels

TOOL AND DIE STEELS—SAE J438b

SAE Standard

Report of Iron and Steel Technical Committee approved January 1949 and last revised May 1970.

Purpose and Scope—This standard covers the identification, classification, and chemical composition of tool and die steels for use by engineers, metallurgists, tool designers, tool room supervisors, heat treaters, and tool makers.

Definition—Tool and die steels are defined as certain carbon or alloy steels, capable of being hardened and tempered. They are usually melted in electric furnaces and produced to meet special requirements. They may be used in

TABLE 1—CHEMICAL COMPOSITIONS OF TOOL AND DIE STEELS[a]

SAE Steel Designation	C	Mn	Si	Cr	V	W	Mo	Co
Water Hardening Tool Steels								
W108[b]	0.70–0.85	—[b]	—[b]	—[b]	—	—	—	—
W109[b]	0.85–0.95	—[b]	—[b]	—[b]	—	—	—	—
W110[b]	0.95–1.10	—[b]	—[b]	—[b]	—	—	—	—
W112[b]	1.10–1.30	—[b]	—[b]	—[b]	—	—	—	—
W209	0.85–0.95	—[b]	—[b]	—[b]	0.15–0.35	—	—	—
W210	0.95–1.10	—[b]	—[b]	—[b]	0.15–0.35	—	—	—
W310	0.95–1.10	—[b]	—[b]	—[b]	0.35–0.50	—	—	—
Shock Resisting Tool Steels								
S1—Chromium-Tungsten	0.45–0.55	0.20–0.40	0.25–0.45[c]	1.25–1.75	0.15–0.30	1.00–3.00	0.40[d]	—
S2—Silicon-Molybdenum	0.45–0.55	0.30–0.50	0.80–1.20	—	0.25[d]	—	0.40–0.60	—
S5—Silicon-Manganese	0.50–0.60	0.60–0.90	1.80–2.20	0.30[d]	0.25[d]	—	0.30–0.50	—
Cold Work Tool Steels								
Oil Hardening Types								
O1—Low Manganese	0.85–0.95	1.00–1.30	0.20–0.40	0.40–0.60	0.20[d]	0.40–0.60	—	—
O2—High Manganese	0.85–0.95	1.40–1.80	0.20–0.40	0.35[d]	0.20[d]	—	0.30[d]	—
O6—Molybdenum Graphitic	1.35–1.55	0.30–1.00	0.80–1.20	—	—	—	0.20–0.30	—
Medium Alloy Air Hardening Types								
A2—5% Chromium Air Hard	0.95–1.05	0.45–0.75	0.20–0.40	4.75–5.50	0.40[d]	—	0.90–1.40	—
High Carbon-High Chromium Types								
D2—High Carbon-High Chromium (Air)	1.40–1.60	0.30–0.50	0.30–0.50	11.00–13.00	0.80[d]	—	0.70–1.20	0.60[d]
D3—High Carbon-High Chromium (Oil)	2.00–2.35	0.24–0.45[c]	0.25–0.45	11.00–13.00	0.80[d]	0.75[d]	0.80[d]	—
D5—High Carbon-High Chromium (Cobalt)	1.40–1.60	0.30–0.50	0.30–0.50	11.00–13.00	0.80[d]	—	0.70–1.20	2.50–3.50
D7—High Carbon-High Chromium-High Vanadium	2.15–2.50	0.30–0.50	0.30–0.50	11.50–13.50	3.80–4.40	—	0.70–1.20	—
Hot Work Tool Steels								
Chromium Base Types								
H11—Chromium-Molybdenum-V	0.30–0.40	0.20–0.40	0.80–1.20	4.75–5.50	0.30–0.50	—	1.25–1.75	—
H12—Chromium-Molybdenum-Tungsten	0.30–0.40	0.20–0.40	0.80–1.20	4.75–5.50	0.10–0.50	1.00–1.70	1.25–1.75	—
H13—Chromium-Molybdenum-VV	0.30–0.40	0.20–0.40	0.80–1.20	4.75–5.50	0.80–1.20	—	1.25–1.75	—
Tungsten Base Types								
H21—Tungsten	0.30–0.40	0.20–0.40	0.15–0.30	3.00–3.75	0.30–0.50	8.75–10.00	—	—
High Speed Tool Steels								
Tungsten Base Types								
T1—Tungsten 18-4-1	0.65–0.75	0.20–0.40	0.20–0.40	3.75–4.50	0.90–1.30	17.25–18.75	—	—
T2—Tungsten 18-4-2	0.75–0.85	0.20–0.40	0.20–0.40	3.75–4.50	1.80–2.40	17.50–19.00	—	—
T4—Cobalt-Tungsten 18-4-1-5	0.70–0.80	0.20–0.40	0.20–0.40	3.75–4.50	0.80–1.20	17.25–18.75	0.70–1.00	4.25–5.75
T5—Cobalt-Tungsten 18-4-2-8	0.75–0.85	0.20–0.40	0.20–0.40	3.75–4.50	1.80–2.40	17.50–19.00	0.70–1.00	7.00–9.00
T8—Cobalt-Tungsten 14-4-2-5	0.75–0.85	0.20–0.40	0.20–0.40	3.75–4.50	1.80–2.40	13.25–14.75	0.70–1.00	4.25–5.75
Molybdenum Base Types								
M1—Molybdenum 8-2-1	0.75–0.85	0.20–0.40	0.20–0.40	3.75–4.50	0.90–1.30	1.15–1.85	7.75–9.25	—
M2—Molybdenum-Tungsten 6-6-2	0.78–0.88	0.20–0.40	0.20–0.40	3.75–4.50	1.60–2.20	5.50–6.75	4.50–5.50	—
M3—Molybdenum-Tungsten 6-6-3	1.00–1.25	0.20–0.40	0.20–0.40	3.75–4.50	2.25–3.25	5.50–6.75	4.75–6.25	—
M4—Molybdenum-Tungsten 6-6-4	1.25–1.40	0.20–0.40	0.20–0.40	4.00–4.75	3.90–4.50	5.25–6.50	4.50–5.50	—
Special Purpose Tool Steels								
Low Alloy Types								
L6—Nickel-Chromium[e]	0.65–0.75	0.55–0.85[c]	0.20–0.40	0.65–0.85	0.25[d]	—	0.25[d]	—
L7—Chromium	0.95–1.05	0.25–0.45	0.20–0.40	1.25–1.75	—	—	0.30–0.50	—

[a] These compositions are not intended for forging die steels.
[b] Water hardening steels listed herein are usually available in four grades or qualities as follows:

Special (Grade 1)—The highest quality water hardening carbon tool steel, controlled for hardenability, chemistry held to closest limits, and subject to most rigid tests to insure maximum uniformity in performance.
Extra (Grade 2)—A high quality water hardening carbon tool steel, controlled for hardenability, subject to tests to insure good service for general application.
Standard (Grade 3)—A good quality water hardening carbon tool steel, not controlled for hardenability, recommended for application where some latitude with respect to uniformity is permissible.
Commercial (Grade 4)—A commercial quality water hardening carbon tool steel, not controlled for hardenability, not subject to special tests.

On Special and Extra Grades, limits on manganese, silicon, and chromium are not generally required in lieu of the following Shepherd hardenability limits:

0.70–0.85 C and 0.85–0.95 C

	Hardenability, 64ths In. Penetration	Fracture Grain Size, min
Shallow	10 max	8
Regular	9 to 13	8
Deep	12 min	8

0.95–1.10 C and 1.10–1.30 C

	Hardenability, 64ths In. Penetration	Fracture Grain Size, min
Shallow	8 max	9
Regular	7 to 11	9
Deep	10 to 16	8

On Standard and Commercial Grades, the following limits on composition are generally required:

	Mn	Si	Cr
Standard, max	0.35	0.35	0.15
Commercial, max	0.35	0.35	0.20

Total of manganese, silicon, and chromium not to exceed 0.75%.

[c] May be present in percentages other than shown.
[d] Optional element. Steels have found satisfactory application either with or without the element present.
[e] Nickel content 1.25–1.75.

certain hand tools, precision gages, or in mechanical fixtures for cutting, shaping, forming, and blanking of materials at either cold or elevated temperatures.

This definition is not intended to include that type of tonnage production open hearth steel used in the manufacture of ordinary mechanics' hand tools, nor steel used in the manufacture of such products as hammers, picks, files, hollow drill steel, mining bits and cutters, large rolling mill rolls, and low alloy medium carbon forging die blocks. These exceptions are stated as a matter of guidance only and are not inclusive.

Identification and Classification of Tool Steels—This method of identification and classification of tools steels was designed to follow the most commonly used and generally accepted terminology of tool steel types of classes. It includes such basic principles as method of quenching, applications, special characteristic, and steels for particular industries. The method is believed to be as simplified as possible and aims to avoid complications in details composition or metallurgical specifications. The method provides appropriate symbols for generally accepted types of tool steel. It also provides for the addition of new products as they may be developed. See Table 1.

The present commonly used tool steels have been grouped into 6 major headings and each commonly accepted group of tool steels under these headings has been assigned an alphabetical letter symbol. Each major group identified by a letter symbol may contain a number of individual types of tool steels. These types are identified by a suffix number which follows the letter symbol. For water hardening tool steels this number suffix consists of three digits, the last two digits representing the approximate mean of the carbon content in tenths of one per cent. To the above may be added after a dash (-) a suffix to further designate the grade and hardenability of W1 steels. (Examples: W110-2R would indicate a Grade 2 with regular hardenability. W110-3 would indicate a Grade 3 not controlled for hardenability.)

Water Hardening Tool Steels: W
Shock Resisting Tool Steels: S
Cold Work Tool Steels: O—Oil Hardening Types
                           A—Medium Alloy Air Hardening Types
                           D—High Carbon High Chromium Types
Hot Work Tool Steels: H—H1-H19 incl—Chromium Base Types
                     H20-H39 incl—Tungsten Base Types
                     H40-H59 incl—Molybdenum Base Types
High Speed Tool Steels: T—Tungsten Base Types
                     M—Molybdenum Base Types
Special Purpose Tools Steels: L—Low Alloy Type

The chemical composition[1] of each type is given only as a representative type analysis. The carbon content is shown only in those cases where it is considered an identifying element to the steel.

Standards for Austenitic Grain Size of Tool and Die Steels—It is recommended that the following method be used for the determination of austenitic grain size of hardened tool steels:

The Shepherd Penetration Fracture Test—This test is used to determine hardenability and fracture grain size on tool and die steels. It is generally applied to carbon tool steels. To perform the test, a sample is machined to ¾ in. diameter x 3 in. long and pretreated by holding at 1600 F for 30 min and quenching in oil, followed by a retreatment by holding for 30 min at 1450 F, then quenching in brine. The specimen is notched midway and fractured by impact.

The penetration of hardening is measured in 64ths of an inch on half of the fractured sample after grinding and etching lightly in hot 50% hydrochloric acid solution.

The grain size is judged by comparing the surface of the fracture of the hardened case with Shepherd fracture grain size standards.

[1] In cooperation with the American Iron and Steel Institute.

SELECTION AND HEAT TREATMENT OF TOOL AND DIE STEELS—SAE J437a

SAE Information Report

Report of Iron and Steel Technical Committee approved January 1949 and last revised May 1969. Editorial change April 1970.

Purpose and Scope—The information in this report covers data relating to SAE J438, Tool and Die Steels, and is intended as a guide to the selection of the steel best suited for the intended purpose and to provide recommended heat treatments and other data pertinent to their use.

Specific requirements as to physical properties are not included because the majority of tool and die steels are either worked or given special heat treatments by the purchaser. The purchaser may or may not elect to use the accompanying data for specification purposes.

The Selection of Tool and Die Steels[1] Simplification of the problems connected with the selection of tool steels has long been an aim of both producers and consumers. This article is restricted to a discussion of the general principles involved in selection and will include a tabulation of the metallurgical characteristics of the principal tool steel types as an aid in selection. A correlation of these metallurgical characteristics with the requirements of the tool in operation should form the basis of a sound approach to the selection of a steel for any application. See Table 1.

Practical experience indicates that in the majority of instances the choice is not limited to a single type of tool steel or even to a particular family of tool steels for a workable solution to an individual tooling problem. Because it is desirable to select the steel that will give the most economical overall performance, the tool life obtained with each steel under consideration should be judged by weighing such factors as expected productivity, ease of fabrication, and cost.

The majority of tool steel applications can be divided into a small number of groups or types of operations: cutting, shearing, forming, drawing, extrusion, rolling, and battering. Cutting tools include drills, taps, broaches, hobs, lathe tools, and the like. Shearing tools include shears, blanking and trimming dies, punches, and such. Forming tools include draw, forging, cold heading, and die casting dies. Battering tools include chisels and all forms of tools involving heavy shock. Many of these classifications can be further divided into cold and hot working tools.

For each of these groups, certain metallurgical characteristics are of utmost importance. Most cutting tools require high hardness, high resistance to the softening effect of heat, and high wear resistance. Shearing tools require high wear resistance combined with fair toughness, and these characteristics must be properly balanced depending on the tool design, thickness of stock being sheared, and temperature of the shearing operation. Forming tools must possess high wear resistance or high toughness and high strength, and many require maximum resistance to heat softening. In battering tools high toughness is most important.

Hardness, strength, toughness, wear resistance, and resistance to heat softening are, therefore, prime selective factors for tool steel applications. Many other properties must be seriously considered in individual applications; these include permissible distortion in hardening, permissible surface decarburization, hardenability or depth of hardness desired, resistance to heat checking, machinability and grindability, as well as heat treating requirements, including temperatures, atmospheres, and equipment.

Table 1 lists those properties which merit special consideration when selecting steels for any application, from the list shown. For compositions of these steels, see Table 1 of SAE J438.

Table 2 is presented as an aid in the relative evaluation of those properties which must be considered for the proper heat treatment of the steels.

Relation of Design to Heat Treatment—The design bears, in many ways, upon the serviceability of the tool or machine part, and unsatisfactory performance may frequently be traced directly to faulty design. This discussion is concerned only with design as it affects the heat treating operation and, through the heat treatment, the serviceability of the finished part. It is the purpose of this discussion to bring about a better mutual understanding

[1] Condensed from the ASM Handbook, 1948 edition, pp. 658–659, with the permission of the American Society for Metals.

TABLE 1—COMPARISON OF TOOL STEELS ON BASIS OF PROPERTIES AFFECTING SELECTION

SAE Steel Designation	Nondeforming Properties	Safety in Hardening	Depth of Hardening[a]	Toughness	Resistance to Softening Effect of Heat	Wear Resistance	Machinability
Water Hardening Tool Steels							
W108	Poor	Fair	Shallow	Good[b]	Poor	Fair	Best
W109	Poor	Fair	Shallow	Good[b]	Poor	Fair	Best
W110	Poor	Fair	Shallow	Good[b]	Poor	Good	Best
W112	Poor	Fair	Shallow	Good[b]	Poor	Good	Best
W209	Poor	Fair	Shallow	Good	Poor	Fair	Best
W210	Poor	Fair	Shallow	Good	Poor	Good	Best
W310	Poor	Fair	Shallow	Good	Poor	Good	Best
Shock Resisting Tool Steels							
S1—Chromium-Tungsten	Fair	Good	Medium	Good	Fair	Fair	Fair
S2—Silicon-Molybdenum	W Poor[c] O Fair[c]	W Poor[c] O Good[c]	Medium	Best	Fair	Fair	Good
S5—Silicon-Manganese	W Poor[c] O Fair[c]	W Poor[c] O Good[c]	Medium	Best	Fair	Fair	Fair
Cold Work Tool Steels							
Oil Hardening Types							
O1—Low Manganese	Good	Good	Medium	Fair	Poor	Good	Good
O2—High Manganese	Good	Good	Medium	Fair	Poor	Good	Good
O6—Molybdenum Graphitic	Fair	Good	Medium	Fair	Poor	Good	Best
Medium Alloy Air Hardening Types							
A2—5% Chromium Air Hard	Best	Best	Deep	Fair	Fair	Good	Fair
High Carbon-High Chromium Types							
D2—High Carbon-High Chromium (Air)	Best	Best	Deep	Fair	Fair	Best	Poor
D3—High Carbon-High Chromium (Oil)	Good	Good	Deep	Poor	Fair	Best	Poor
D5—High Carbon-High Chromium-Cobalt	Best	Best	Deep	Fair	Fair	Best	Poor
D7—High Carbon-High Chromium-High Vanadium	Best	Best	Deep	Poor	Fair	Best	Poor
Hot Work Tool Steels							
Chromium Base Types							
H11—Chromium-Molybdenum-V	Good	Good	Deep	Good	Good	Fair	Fair
H12—Chromium-Molybdenum-Tungsten	Good	Good	Deep	Good	Good	Fair	Fair
H13—Chromium-Molybdenum-VV	Good	Good	Deep	Good	Good	Fair	Fair
Tungsten Base Types							
H21—Tungsten	Good	Good	Deep	Good	Good	Fair	Fair
High Speed Tool Steels							
Tungsten Base Types							
T1—Tungsten 18-4-1	Good	Good	Deep	Poor	Good	Good	Fair
T2—Tungsten 18-4-2	Good	Good	Deep	Poor	Good	Good	Fair
T4—Cobalt-Tungsten 18-4-1-5	Good	Fair	Deep	Poor	Best	Good	Fair
T5—Cobalt-Tungsten 18-4-2-8	Good	Fair	Deep	Poor	Best	Good	Fair
T8—Cobalt-Tungsten 14-4-2-5	Good	Fair	Deep	Poor	Best	Good	Fair
Molybdenum Base Types							
M1—Molybdenum 8-2-1	Good	Fair	Deep	Poor	Good	Good	Fair
M2—Molybdenum-Tungsten 6-6-2	Good	Fair	Deep	Poor	Good	Good	Fair
M3—Molybdenum-Tungsten 6-6-3	Good	Fair	Deep	Poor	Good	Best	Fair
M4—Molybdenum-Tungsten 6-6-4	Good	Fair	Deep	Poor	Good	Best	Fair
Special Purpose Tool Steels							
Low Alloy Types							
L6—Nickel-Chromium	Fair	Good	Medium	Fair	Poor	Fair	Fair
L7—Chromium	Fair	Good	Medium	Fair	Poor	Good	Fair

[a]These are intended to emphasize major differences between the groups of steels and do not account for the minor differences in depths of hardening that exist between steels of the same group. This is particularly true of the Water Hardening W Steels which are frequently furnished with varying degrees of hardenability as listed in Table 1.
[b]Toughness decreases somewhat with increasing depth of hardening.
[c]W as shown here indicates water quench. O as shown here indicates oil quench.

between the designer and the steel treater so that faulty design which may cause cracking or distorting during heat treating can be avoided.

The fundamental principles of good design from a heat treatment standpoint are quite simple. Heat treated steel has a certain strength depending upon the analysis of the steel, the quality of the metal, and the heat treatment which it has received. When subjected to a combination of forces its ultimate strength, the steel cracks or fails. There are 2 types of force combining to break steel, which are:

1. The internal stress set up during fabrication and heat treatment of the tool.
2. The external force of service.

Sometimes the internal stresses alone exceed the strength of the metal, and the parts crack in hardening. Again, the internal stresses may equal 90% or more of the total strength, in which case failure will develop in service under relatively light loads. It therefore appears that the useful strength of a part decreases in proportion as the internal stresses increase.

Internal stresses arise from many causes, but the most serious by far are those developed by differential cooling resulting from quenching. This differential cooling is largely a function of the size and shape of the piece being quenched; in other words, the design. Here, then, is the relation of design to heat treatment, and the basic principle of successful design is to plan shapes which will allow the piece to cool as uniformly as possible during quenching.

Some shapes are almost impossible to harden because of the abruptness in the change of sections, but a certain latitude in design is recognized when using an oil hardening or air hardening steel.

Errors in design reach further than merely affecting the internal stress of hardening. A sharp angle serves to concentrate greatly the stresses of service. The design of the part may be entirely responsible for concentrating the service stresses at a point already weakened by internal stresses produced during hardening.

Reducing all the foregoing to a single statement, a part is properly designed from the standpoint of heat treatment when the entire piece may be heated and cooled at approximately the same rate during the heat treating operation. Perfection in this regard is unattainable because, even in a sphere, the surface cools more rapidly than the interior. The designer should, however, attempt to so shape his parts that they will heat and cool as uniformly as possible. The greater the temperature difference between any two points on a given part during quenching and the closer these two points are together, the greater will be the internal stress and, therefore, the poorer the design.

The principles described in this article are illustrated in Fig. 1.

Heat Treat Data—The thermal treatments listed in Table 3 cover the generally used treatments for the forging, normalizing, and annealing of tool and die steels.

The thermal treatments listed in Table 2, under selection, cover the usual ranges of temperature for hardening and tempering tool and die steels.

The information listed in Tables 2 and 3 is not intended for specification because of the need for altering treatments for specific applications.

Allowance for Machining of Tool Steel Bars[2]—Tool and die steels should be ordered oversize with sufficient material to be removed from all surfaces by machining or grinding to allow for:

(a) Surface decarburization.
(b) Surface defects such as slivers, seams, laps, scale marks, and the like.
(c) Undersize tolerance as given in Tables 6, 7, 8, and 9.

Table 4 lists the minimum allowance per side over finish size for machining or grinding rounds, squares, hexagons, and octagons.

Polished or ground tool steel quality round drill rod is free from decarburization or any surface defects requiring surface removal.

[2]In cooperation with the American Iron and Steel Institute.

TABLE 2—APPROXIMATE COMPARISON OF TOOL AND DIE STEELS ON BASIS OF SOME HEAT TREATING CHARACTERISTICS

SAE Steel Designation	Quench Medium	Preheat Temperature, F	Hardening Temperature Range,[a] F	Hardness after Quenching, Rockwell C	Tempering Temperature Range,[a] F	Hardness after Tempering, Rockwell C	Decarburization (Prevention of During Heat Treatment)
Water Hardening Tool Steel							
W108	Water	—[b]	1420-1450	65-67	350-525	65-56	—[c]
W109	Water	—[b]	1420-1450	65-67	350-525	65-56	—[c]
W110	Water	—[b]	1420-1450	65-67	350-525	65-56	—[c]
W112	Water	—[b]	1420-1500	65-67	350-525	65-56	—[c]
W209	Water	—[b]	1420-1500	65-67	350-525	65-56	—[c]
W210	Water	—[b]	1420-1500	65-67	350-525	65-56	—[c]
W310	Water	—[b]	1420-1500	65-67	350-525	65-56	—[c]
Shock Resisting Tool Steels							
S1—Chromium-Tungsten	Oil	1200-1300	1650-1800	57-59	300-1000	57-45	—[d]
S2—Silicon-Molybdenum	Water / Oil	—[b]	1550-1575 / 1600-1625	60-62 / 58-60	300-500 / 300-500	60-54 / 58-54	—[c] / —[c]
S5—Silicon-Manganese	Water / Oil	—[b]	1550-1600 / 1600-1675	60-62 / 58-60	300-650 / 300-650	60-54 / 58-54	—[c] / —[c]
Cold Work Tool Steels							
Oil Hardening Types							
O1—Low Manganese	Oil	—[b]	1450-1500	63-65	300-800	62-50	—[c]
O2—High Manganese	Oil	—[b]	1420-1450	63-65	375-500	62-57	—[c]
O6—Molybdenum Graphitic	Oil	—[b]	1450-1500	63-65	300-800	63-50	—[c]
Medium Alloy Air Hardening Types							
A2—5% Chromium Air Hard	Air	1200-1300	1725-1775	61-63	400-700	60-57	—[d]
High Carbon-High Chromium Types							
D2—High Carbon-High Chromium	Air	1200-1300	1800-1875	61-63	400-700	60-58	—[d]
D3—High Carbon-High Chromium	Oil	1200-1300	1750-1800	62-64	400-700	62-58	—[d]
D5—High Carbon-High Chromium-Cobalt	Air	1200-1300	1800-1875	60-62	400-700	59-57	—[d]
D7—High Carbon-High Chromium-High Vanadium	Air	1200-1300	1850-1950	63-65	300-500 / 850-1000	65-63 / 62-58	—[d]
Hot Work Tool Steels							
Chromium Base Types							
H11—Chromium-Molybdenum-V	Air	1450-1500	1825-1875	53-55	1000-1100	51-43	—[d]
H12—Chromium-Molybdenum-Tungsten	Oil, Air	1450-1500	1800-1900	53-55	1000-1100	51-43	—[d]
H13—Chromium-Molybdenum-VV	Air	1400-1450	1825-1875	53-55	1000-1100	51-43	—[d]
Tungsten Base Types							
H21—Tungsten	Oil, Air	1500-1550	2100-2150	50-52	950-1150	50-47	—[d]
High Speed Tool Steels							
Tungsten Base Types							
T1—Tungsten 18-4-1	Oil, Air, Salt	1500-1550	2300-2375	63-65	1025-1100	65-63	—[d]
T2—Tungsten 18-4-2	Oil, Air, Salt	1500-1550	2300-2375	63-65	1025-1100	65-63	—[d]
T4—Cobalt-Tungsten 18-4-1-5	Oil, Air, Salt	1500-1550	2300-2375	63-65	1025-1100	65-63	—[d]
T5—Cobalt-Tungsten 18-4-2-8	Oil, Air, Salt	1500-1550	2300-2400	63-65	1050-1100	65-63	—[d]
T8—Cobalt-Tungsten 14-4-2-5	Oil, Air, Salt	1500-1550	2300-2375	63-65	1025-1100	65-63	—[d]
Molybdenum Base Types							
M1—Molybdenum 8-2-1	Oil, Air, Salt	1400-1500	2150-2250	63-65	1025-1050	65-63	—[d]
M2—Molybdenum-Tungsten 6-6-2	Oil, Air, Salt	1450-1500	2175-2250	63-65	1025-1075	65-63	—[d]
M3—Molybdenum-Tungsten 6-6-3	Oil, Air, Salt	1450-1500	2150-2225	63-65	1025-1075	65-63	—[d]
M4—Molybdenum-Tungsten 6-6-4	Oil, Air, Salt	1450-1500	2150-2225	63-65	1025-1075	65-63	—[d]
Special Purpose Tool Steels							
Low Alloy Types							
L6—Nickel-Chromium	Oil	—[b]	1500-1600	62-64	400-800	62-48	—[c]
L7—Chromium	Oil	—[b]	1525-1550	63-65	350-500	62-60	—[c]

[a] The purpose of these columns is to show the usual ranges of temperature employed in hardening and tempering and is not to be used as a specification.
[b] For large tools and tools having intricate sections, preheating at 1050 to 1200 F is recommended.
[c] Use moderately oxidizing atmosphere in furnace or a suitable neutral salt bath.
[d] Use protective pack from which volatile matter has been removed, carefully balanced neutral salt bath, or atmosphere controlled furnaces. In the latter case, the furnace atmosphere should be in equilibrium with the carbon content of the steel being treated. Furnace atmosphere dew point is considered a reliable method for measuring and controlling this equilibrium.

TABLE 3—FORGING, NORMALIZING, AND ANNEALING TREATMENTS OF TOOL AND DIE STEELS

SAE Steel Designation[a]	Forging[b]			Normalizing[c]		Annealing[d]			
	Heat Slowly to	Start Forging at	Do Not Forge below	Heat Slowly to	Hold at	Temperature	Maximum Rate of Cooling, F/hr	Approximate Brinell Hardness	Approximate Rockwell B
Water Hardening Tool Steels									
W108	1450	1800-1950	1500	1450	1500	1400-1450	75	159-202	84-94
W109	1450	1800-1950	1500	1450	1500	1375-1425	75	159-202	84-94
W110	1450	1800-1900	1500	1450	1550	1400-1450	75	159-202	84-94
W112	1450	1800-1900	1500	1450	1625	1400-1450	75	159-202	84-94
W209	1450	1800-1900	1500	1450	1500	1375-1425	75	159-202	84-94
W210	1450	1800-1900	1500	1450	1550	1400-1450	75	159-202	84-94
W310	1450	1800-1900	1500	1450	1550	1400-1450	75	159-202	84-94
Shock Resisting Tool Steels									
S1—Chromium-Tungsten	1500	1800-2000	1600	Do not normalize		1450-1500	50	192-235	92-99
S2—Silicon-Molybdenum	1500	1900-2100	1600	1500	1650	1400-1450	50	192-229	92-98
S5—Silicon-Manganese	1500	1900-2050	1600	1500	1600	1400-1450	50	192-229	92-98
Cold Work Tool Steels									
Oil Hardening Types									
O1—Low Manganese	1500	1750-1900	1550	1500	1600	1425-1475	50	183-212	90-96
O2—High Manganese	1500	1750-1900	1550	1500	1500	1375-1425	50	183-212	90-96
O6—Molybdenum Graphitic	1500	1750-1900	1500	1500	1625	1425-1475	20	183-217	90-96
Medium Alloy Air Hardening Types									
A2—5% Chromium Air Hard	1600	1850-2000	1650	Do not normalize		1550-1600	40	202-229	94-98
High Carbon-High Chromium Types									
D2—High Carbon-High Chromium (Air)	1650	1850-2000	1650	Do not normalize		1600-1650	40	207-255	95-102
D3—High Carbon-High Chromium (Oil)	1650	1850-2000	1650	Do not normalize		1600-1650	50	212-255	96-102
D5—High Carbon-High Chromium-Cobalt	1600	1850-2000	1650	Do not normalize		1600-1650	40	207-255	95-102
D7—High Carbon-High Chromium-High Vanadium	1650	2050-2125	1800	Do not normalize		1600-1650	50	235-262	99-103

(Table continued on next page)

8.05

TABLE 3—FORGING, NORMALIZING, AND ANNEALING TREATMENTS OF TOOL AND DIE STEELS (continued)

SAE Steel Designation [a]	Forging [b]			Normalizing [c]		Annealing [d]			
	Heat Slowly to	Start Forging at	Do Not Forge below	Heat Slowly to	Hold at	Temperature	Maximum Rate of Cooling, F/hr	Approximate Brinell Hardness	Approximate Rockwell B
Hot Work Tool Steels									
Chromium Base Types									
H11—Chromium-Molybdenum-V	1650	1950–2100	1650	Do not normalize		1550–1600	50	192–229	92–98
H12—Chromium-Molybdenum-Tungsten	1650	1950–2100	1650	Do not normalize		1600–1650	50	192–229	92–98
H13—Chromium-Molybdenum-VV	1650	1950–2100	1650	Do not normalize		1550–1600	50	192–229	92–98
Tungsten Base Types									
H21—Tungsten	1600	2000–2150	1650	Do not normalize		1600–1650	50	202–235	94–99
High Speed Tool Steels									
Tungsten Base Types									
T1—Tungsten 18-4-1	1600	1950–2100	1750	Do not normalize		1600–1650	50	217–255	96–102
T2—Tungsten 18-4-2	1600	2000–2150	1750	Do not normalize		1600–1650	50	223–255	97–102
T4—Cobalt-Tungsten 18-4-1-5	1600	2000–2150	1750	Do not normalize		1600–1650	50	229–255	98–102
T5—Cobalt-Tungsten 18-4-2-8	1600	2000–2150	1800	Do not normalize		1600–1650	50	248–293	102–106
T8—Cobalt-Tungsten 14-4-2-5	1600	2000–2150	1750	Do not normalize		1600–1650	50	229–255	98–102
Molybdenum Base Types									
M1—Molybdenum 8-2-1	1500	1900–2050	1700	Do not normalize		1525–1600	50	207–248	95–102
M2—Moybdenum-Tungsten 6-6-2	1500	1950–2100	1700	Do not normalize		1550–1625	50	217–248	96–102
M3—Molybdenum-Tungsten 6-6-3	1500	2000–2150	1700	Do not normalize		1550–1625	50	223–255	97–102
M4—Molybdenum-Tungsten 6-6-4	1500	2000–2150	1700	Do not normalize		1550–1625	50	229–255	98–102
Special Purpose Tool Steels									
Low Alloy Types									
L6—Nickel-Chromium	1500	1800–2000	1600	1550	1650	1400–1450	50	183–212	90–96
L7—Chromium	1500	1800–2000	1550	1550	1650	1450–1500	50	174–212	88–96

[a] These tool and die steels are the same as those listed in Table 1 of this report.
[b] The temperature at which to start forging is given as a range, the higher side of which should be used for large sections and heavy or rapid reductions and the lower side for smaller sections and lighter reductions. As the alloy content of the steel increases, the time of soaking at forging temperature increases proportionately. Likewise, as the alloy content increases, it becomes more necessary to cool slowly from the forging temperature. With very high alloy steels, such as high speed or air hardening steels, this slow cooling is imperative in order to prevent cracking and to leave the steel in a semisoft condition. Either furnace cooling or burying in an insulating medium, such as lime, mica, or silocel, is satisfactory.
[c] The length of time the steel is held after being uniformly heated through at the normalizing temperature, varies from about 15 min for a small section to about 1 hr for large sizes. Cooling from the normalizing temperature is done in still air. The purpose of normalizing after forging is to refine the grain structure and to produce a uniform structure throughout the forging. Normalizing should not be confused with low temperature (about 1200 F) annealing used for the relief of residual stresses resulting from heavy machining, bending, and forming.
[d] The annealing temperature is given as a range, the upper limit of which should be used for large sections and the lower limit for smaller sections. The length of time the steel is held after being uniformly heated through at the annealing temperature varies from about 1 hr for light sections and small furnace charges of carbon or low alloy steel to about 4 hr for heavy sections and large furnace charges of high alloy steel.

For information on the forging and heat treating of tool steels, see ASM Handbook, 1948 edition, pp. 653–655.

TABLE 4—MINIMUM ALLOWANCES FOR MACHINING AND MAXIMUM DECARBURIZATION LIMITS (ROUNDS, HEXAGONS, AND OCTAGONS)

Ordered Size, in.	Minimum Allowance Per Side for Machining Prior to Heat Treatment, in.			
	Hot Rolled	Forged	Rounds Rough Turned	Cold Drawn
Up to 0.5, incl	0.016	—	—	0.016
Over 0.5 to 1, incl	0.031	—	—	0.031
Over 1 to 2, incl	0.048	0.072	—	0.048
Over 2 to 3, incl	0.063	0.094	0.020	0.063
Over 3 to 4, incl	0.088	0.120	0.024	0.088
Over 4 to 5, incl	0.112	0.145	0.032	—
Over 5 to 6, incl	0.150	0.170	0.040	—
Over 6 to 8, incl	0.200	0.200	0.048	—
Over 8	—	0.200	0.072	—

Maximum Decarburization Limits
80% of above allowances per side

NOTE: Rounds 1/4 in. and over of high speed steel are normally furnished free of scale and decarburization.

TABLE 5A—SIZE TOLERANCES FOR HOT ROLLED BARS (ROUNDS,* SQUARES, OCTAGONS, QUARTER OCTAGONS, HEXAGONS)

Specified Sizes, in.	Size Tolerances, in.	
	Under	Over
To 0.5, incl	0.005	0.012
Over 0.5 to 1, incl	0.005	0.016
Over 1 to 1.5, incl	0.006	0.020
Over 1.5 to 2, incl	0.008	0.025
Over 2 to 2.5, incl	0.010	0.030
Over 2.5 to 3, incl	0.010	0.040
Over 3 to 4, incl	0.012	0.050
Over 4 to 5.5, incl	0.015	0.060
Over 5.5 to 6.5, incl	0.018	0.100
Over 6.5 to 8, incl	0.020	0.150

*For high speed steel rounds free of scale and decarburization, see Table 4.

TABLE 5B—WIDTH AND THICKNESS TOLERANCES FOR HOT ROLLED FLATS

Specified Widths, in.	Width Tolerances, in.	
	Under	Over
To 1, incl	0.016	0.031
Over 1 to 3, incl	0.031	0.047
Over 3 to 5, incl	0.047	0.063
Over 5	0.063	0.094

Specified Widths, in.	Thickness Tolerances for Thicknesses Given, in.							
	To 0.25 Incl		Over 0.25 to 0.5, Incl		Over 0.5 to 1, Incl		Over 1 to 2, Incl	
	Under	Over	Under	Over	Under	Over	Under	Over
To 1, incl	0.006	0.010	0.008	0.012	0.010	0.016	—	—
Over 1 to 2, incl	0.006	0.014	0.008	0.016	0.010	0.020	0.020	0.024
Over 2 to 3, incl	0.006	0.018	0.008	0.020	0.010	0.024	0.020	0.027
Over 3 to 4, incl	0.008	0.020	0.010	0.022	0.013	0.024	0.024	0.030
Over 4 to 5, incl	0.010	0.020	0.012	0.024	0.015	0.030	0.027	0.035
Over 5 to 6, incl	0.012	0.020	0.014	0.030	0.018	0.030	0.030	0.035

TABLE 6A—WIDTH AND TOLERANCES FOR FORGED FLATS

Specified Widths, in.	Width Tolerances, in.	
	Under	Over
Over 1 to 3, incl	0.031	0.078
Over 3 to 5, incl	0.062	0.125
Over 5 to 7, incl	0.125	0.187
Over 7	0.187	0.312

Specified Widths, in.	Thickness Tolerances for Thicknesses Given, in.									
	To 1, Incl		Over 1 to 3, Incl		Over 3 to 5, Incl		Over 5 to 7, Incl		Over 7, Incl	
	Under	Over	Under	Over	Under	Over	Under	Over	Under	Over
Over 1 to 3, incl	0.016	0.031	0.031	0.078	—	—	—	—	—	—
Over 3 to 5, incl	0.031	0.062	0.047	0.094	0.062	0.125	—	—	—	—
Over 5 to 7, incl	0.047	0.094	0.062	0.125	0.078	0.156	0.125	0.187	—	—
Over 7	0.062	0.125	0.078	0.156	0.094	0.187	0.156	0.219	0.187	0.312

TABLE 6B—SIZE TOLERANCES FOR FORGED BARS (ROUNDS, SQUARES, OCTAGONS, HEXAGONS)

Specified Sizes, in.	Size Tolerances, in.	
	Under	Over
Over 1 to 2 incl	0.030	0.060
Over 2 to 3, incl	0.030	0.080
Over 3 to 5, incl	0.060	0.125
Over 5 to 7, incl	0.125	0.187
Over 7	0.187	0.312

NOTE: Refer to Table 4 for diameter tolerances on rounds of high speed steels free of scale and decarburization.

TABLE 7—SIZE TOLERANCES FOR COLD DRAWN BARS

Rounds, Octagons, Quarter Octagons, and Hexagons		Squares and Flats	
Size Range, in.	Tolerance, ± in.	Size Range, in.	Tolerance, ± in.
0.25 to 0.50, excl	0.002	0.25 to 0.75, incl	0.002
0.50 to 1, excl	0.0025	Over 0.75 to 1.50, incl	0.003
1 to 2.75, incl	0.003	Over 1.50	0.004

TABLE 9A—SIZE TOLERANCE FOR DRILL ROD ROUNDS (POLISHED OR GROUND)

Size Range, in.	Standard Manufacturing Tolerance, ± in.	Closer Tolerance, ± in.
Up to 0.124, incl	0.0003	0.0002
0.125 to 0.499, incl	0.0005	0.00025
0.500 to 1.500, incl	0.001	0.0005

TABLE 8—DIAMETER TOLERANCES FOR CENTERLESS GROUND BARS (ROUNDS)

Diameter Range, in.	Tolerance, in.	
	Under	Over
0.25 to 0.50 excl	0.0015	0.0015
0.50 to 3.0625 excl	0.002	0.002

TABLE 9B—SIZE TOLERANCES FOR DRILL ROD SHAPES OTHER THAN ROUNDS (COLD DRAWN)

Size Range, in.	Tolerance, ± in.
Up to 0.25, excl	0.0005
0.25 to 0.75 excl	0.001
0.75 to 1, incl	0.0015

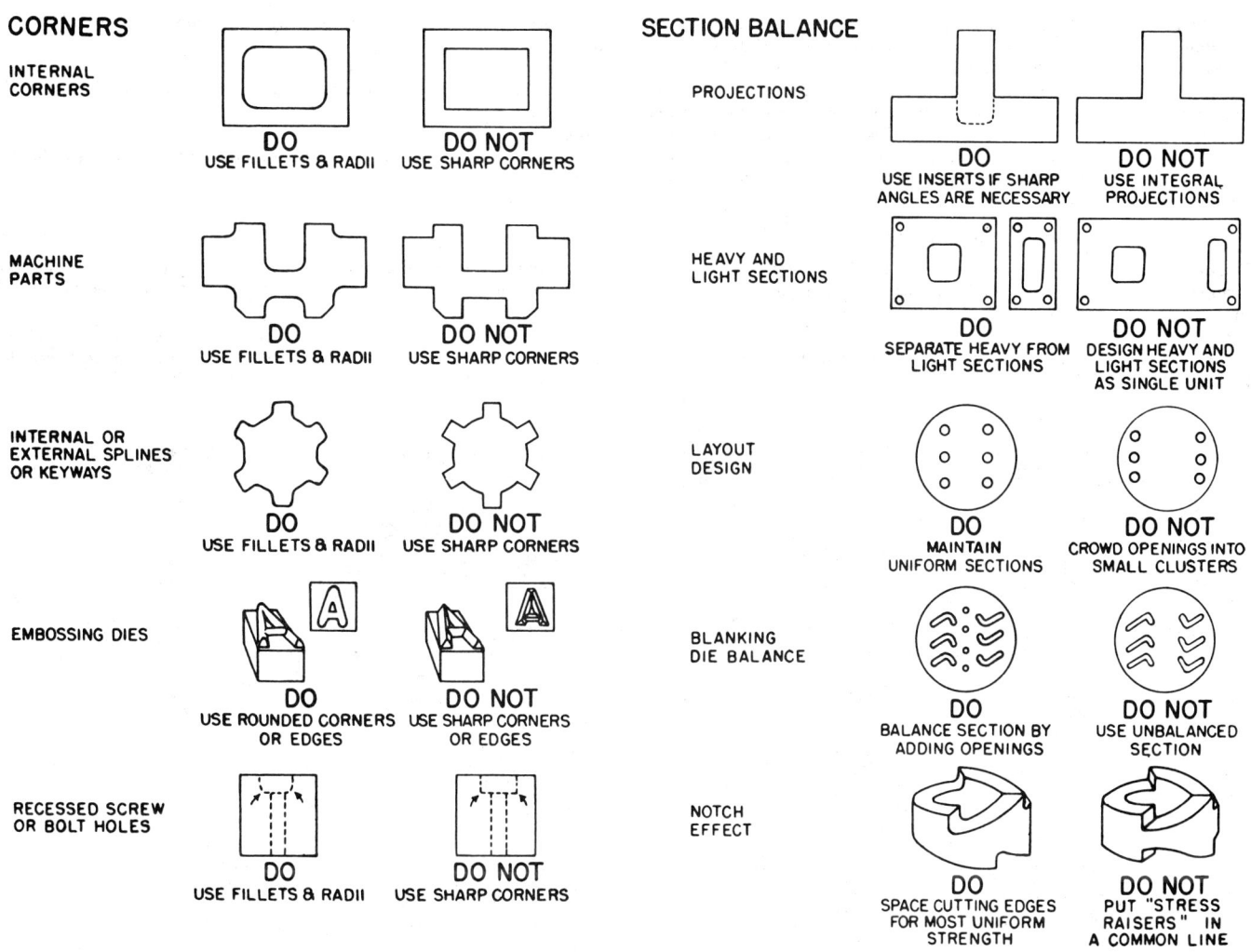

FIG. 1—TOOL AND DIE DESIGN TIPS (TO REDUCE BREAKAGE IN HEAT TREATING)

SINTERED CARBIDE TOOLS—SAE J439a

SAE Recommended Practice

Report of Iron and Steel Technical Committee approved April 1956 and last revised April 1969. Editorial change February 1977.

Scope—This recommended practice covers methods for measuring or evaluating five properties or characteristics of sintered carbide which contribute significantly to the performance of sintered carbide tools. These properties are: hardness, specific gravity, apparent porosity, structure, and grain size. They are covered under separate headings below.

HARDNESS

General—The Rockwell hardness tester provides a simple, rapid, and reliable means of measuring the hardness of sintered carbide tools. A hardness value is easily obtained, but is subject to error if precautionary measures are not taken in making this test. Hardness determinations, therefore, shall be made according to the requirements outlined below and in ASTM E 18, Methods of Test for Rockwell Hardness and Rockwell Superficial Hardness of Metallic Materials.

Apparatus

1. Rockwell hardness testing machine with 60 kg load and diamond brale penetrator for use with the A scale.[1]

[1] It is recommended that a diamond brale especially selected for use with the Rockwell "A" scale be used for this type of testing. This type of penetrator is of higher quality, free from chips and other imperfections, and should be specified for Rockwell "A" scale use. Slowing down of the rate of speed at which the major load is applied during testing will aid in increasing the life of the diamond brale. This change in load application does not affect the accuracy of the hardness reading. The use of the superficial scale is not recommended for hardness testing of sintered carbides unless extreme care is exercised with regard to parallelism and smoothness during surface preparation of the specimen.

2. Two tungsten carbide test blocks with a hardness of 90.0 and 92.0 Rockwell A (RA) respectively.

Material—Sample—Preparation of the surface of the specimen prior to making the hardness test is of major importance. It is recommended that a finish equivalent to that produced with a 220 grit diamond grinding wheel be obtained on the surface which is to be checked for hardness. Because of the shallow penetration of the diamond penetrator used in making this test, the surface being tested for hardness must be parallel to the surface opposite of that being tested. Both surfaces must be smooth and devoid of any bulge or other irregularity affecting parallelism. If these two surfaces are only slightly out of parallel, an error will be obtained in the hardness reading.

It is important that the Rockwell hardness testing machine is located in such a manner and area that it is free from vibration while hardness tests are being performed. Vibration is detected by the bouncing effect transmitted through the needle of the indicator after the major load has been applied.

Procedure—The hardness test shall be made using the RA scale. This reading is obtained by observing the deflection of the needle pointer on the black scale with a 60 kg load and the diamond brale penetrator.[1]

Before making the hardness test on the carbide material, the Rockwell testing machine shall be checked for accuracy, using a tungsten carbide test block of known hardness. Two check blocks of different hardness values are recommended to assure accurate hardness readings over the general range of hardness of the common grades of sintered carbides. The check blocks should have a hardness of 90.0 and 92.0 RA respectively. The check block having the hardness closest to the expected hardness of the carbide material to be checked shall be selected for calibrating the Rockwell tester. The average of five readings should check within ±0.2 of a hardness number. If the Rockwell tester varies appreciably from the hardness number of the test block, the dial of the machine must be adjusted so that the correct reading is obtained. The amount of variation is noted and this correction plus or minus is applied when taking the hardness reading on the specimens of sintered carbide being tested. This dial adjustment will be made just before the major load is applied. With careful manipulations, hardness readings can be accurately duplicated when the hardness tester is calibrated in this manner.

SPECIFIC GRAVITY

General—The specific gravity of sintered carbide tool materials shall be determined by the immersion method, using as a basis the difference in weight of the carbide in air and in water.

Apparatus

1. A standard analytical balance of 200 g capacity and 0.1 mg sensitivity at full load.

2. A 150- or 250-ml beaker, depending upon the size of the carbide specimen.

3. Small diameter nonferrous wire.

4. Thermometer 0–100 C for room temperatures capable of being read to nearest 0.5 C.

Materials

1. The specimen shall be surface ground all over with a 100 grit diamond wheel before testing.

2. Distilled water.

Procedure

1. Weigh the specimen, to the nearest 0.5 mg.

2. Support a beaker of distilled water[2] over the pan of the balance by a suitable bridge. Water level should be high enough to cover the specimen by at least $\frac{1}{4}$ in.

[2] Care should be used to see that no air bubbles are present on the sample after immersion, and that the wire twist on the sample is completely submerged. Several drops of a suitable wetting agent will aid in eliminating air bubbles.

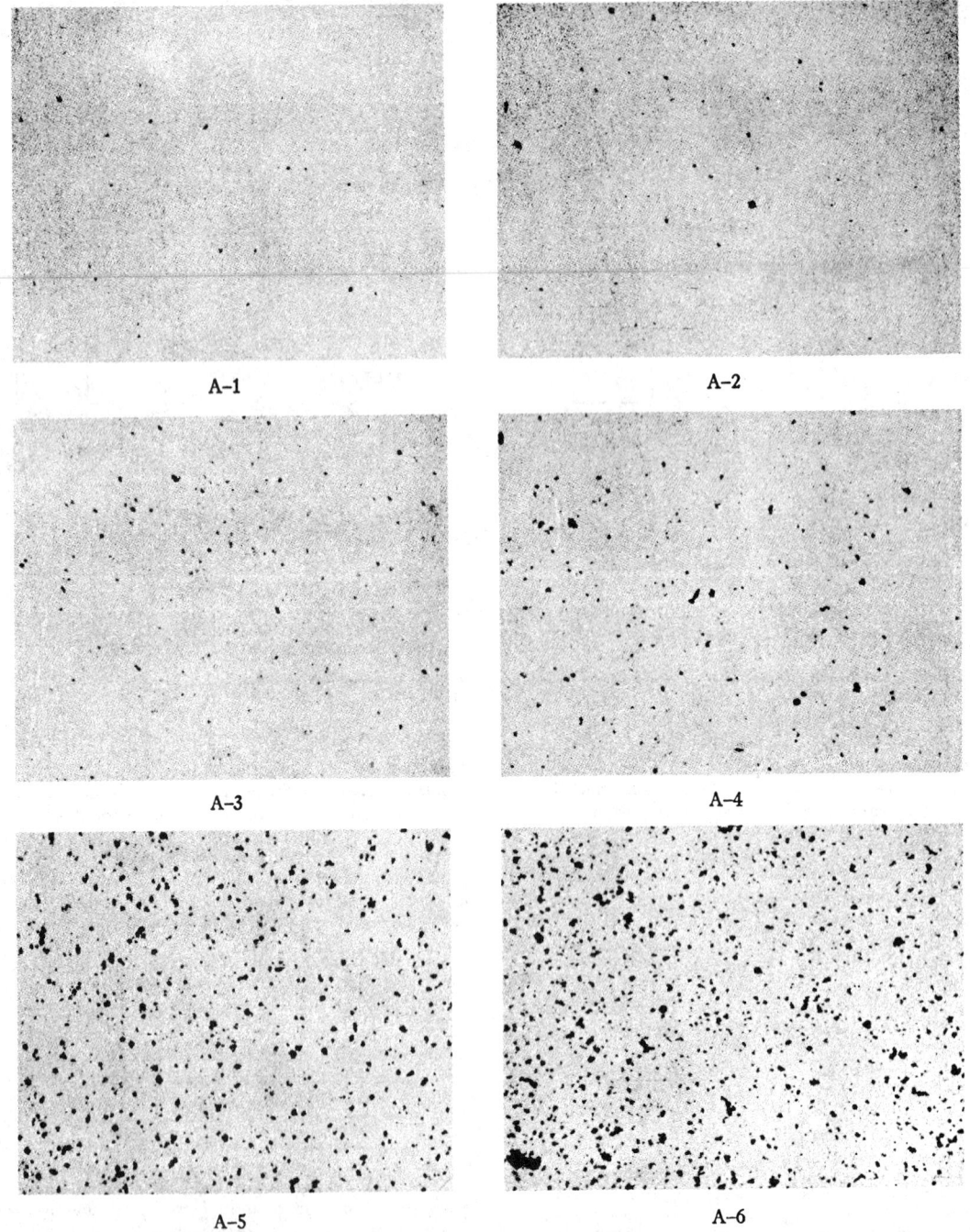

FIG. 1—TYPE A—APPARENT POROSITY MICROSTRUCTURE OF CEMENTED CARBIDES (×200) (B 276)

3. Suspend the specimen and the wire from the beam hook, placing the specimen in the water, and weigh to the nearest 0.5 mg.

4. Remove the specimen from the wire and weigh the wire alone in water. Subtract this weight from the total weight found in step 3.

5. Observe the temperature of the water to the nearest 1.0 C.

Calculations

W_a = Weight of Specimen in Air
W_w = Weight of Specimen in Water
D = Relative Density of Water at Test Temperature
(Density relative to that of water at 4 C)

$$\text{Specific Gravity} = \frac{W_a \times D}{W_a - W_w}$$

APPARENT POROSITY, STRUCTURE, AND GRAIN SIZE

General—Apparent porosity, structure, and grain size shall be evaluated by metallographic examination, as outlined below.

Apparent porosity is the term applied to the inherent porosity, nonmetallic inclusions, and uncombined carbon as observed in the microstructure of the properly prepared surface of sintered carbides.

Structure refers to the type and distribution of the metal carbides and binder material observed in the microstructure of the properly prepared surface of sintered carbides.

Grain size is the term applied to the predominating particle sizes, in microns, of the metal carbides observed in the microstructure.

Sample Preparation—Select a specimen approximately 1/2 in. square from the area of particular interest of the sample to be tested. Sectioning should be done with a diamond cutoff wheel. Mount unwieldy specimens in hard bakelite or its equivalent, then grind as follows:

1. Rough grind using a green silicon carbide wheel.
2. Fine grind using a 320 grit diamond wheel running at a speed of approximately 5500 surface fpm.

Samples should be polished using the ordinary metallographic polishing equipment. Impregnate a paper polishing disc, properly attached to the bronze disc of the polishing lap, with a light (SAE 10) oil. Apply diamond paste to the oiled paper and work it in with the fingertip. At least two

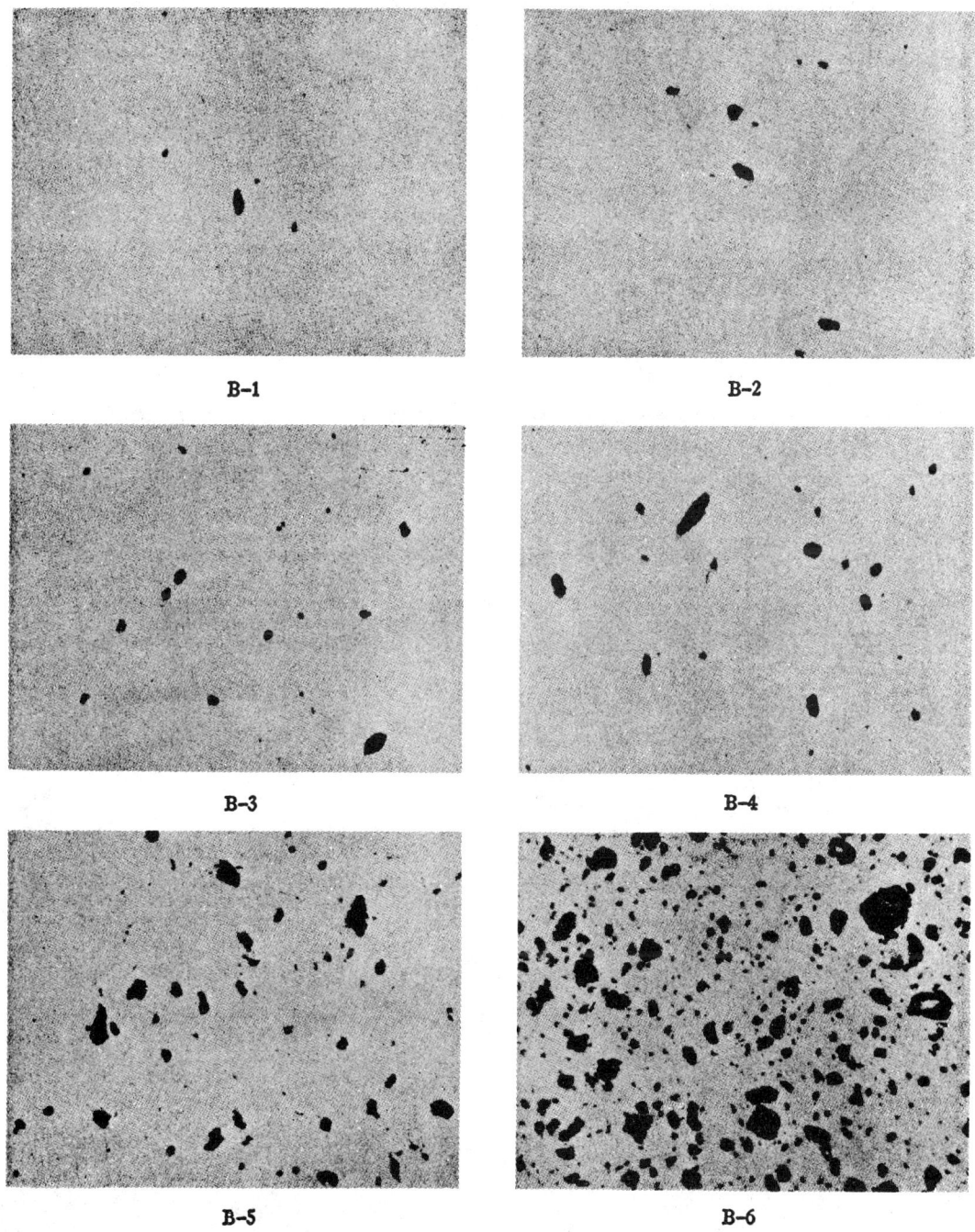

FIG. 2—TYPE B—APPARENT POROSITY MICROSTRUCTURE OF CEMENTED CARBIDES (×200) (B 276)

polishing laps should be used in the following order:
1. A diamond lap using a 10 μm maximum diamond powder.
2. A diamond lap using a 1 μm maximum diamond powder.

Hold the specimen 1–3 in. from the center of the lap running at approximately 1150 rpm. Considerable pressure should be exerted on the specimen while polishing in intervals of approximately 10 sec. Rotating the specimen 90 deg between each interval is recommended. (CAUTION: Light pressure and too much polishing may cause pitting of the specimen.)

Polishing is ineffective when the specimen is above approximately 150 F; therefore, polishing time should be carefully watched to keep the temperature of the specimen below this point. Extreme cleanliness is necessary to prevent contamination of the diamond laps. The specimen should be washed thoroughly with a suitable solvent after each polishing operation.

Apparent Porosity Evaluation—After the prescribed sample preparation, the sample shall be examined in the unetched condition at a magnification of 200X.

A porosity rating shall be made by comparing the observed field with the porosity charts of Figs. 1–3.

The rating charts depict both the type of porosity, designated alphabetically, and the quantity of porosity, designated numerically. Type A classifies porosity sizes under 10 μm in diameter; Type B classifies porosity sizes between 10 and 40 μm in diameter; Type C classifies cluster porosity or that developed by the presence of uncombined carbon, and is considered the type most detrimental to tool performance.

Structure Evaluation—After the prescribed sample preparation the sample shall be etched and examined at a magnification of 1500X. The etchant shall consist of a fresh solution having equal parts of 10% potassium hydroxide and 10% potassium ferricyanide. The sample shall be immersed in the etchant for 2 minutes, then rinsed with water and the polished surface swabbed with wet cotton. A second immersion for approximately another 2 minutes shall be made to delineate the structure. The sample shall then be washed with water and dried with alcohol and air to prevent staining. Examination of the prepared surface shall be made with a metallographic microscope utilizing an oil immersion objective. Figs. 4A and 4B are typical photomicrographs of tungsten carbide (WC) with 6% cobalt and 13% cobalt, respectively. Figs. 4C and 4D are typical photomicrographs of tungsten carbide (WC) plus solid

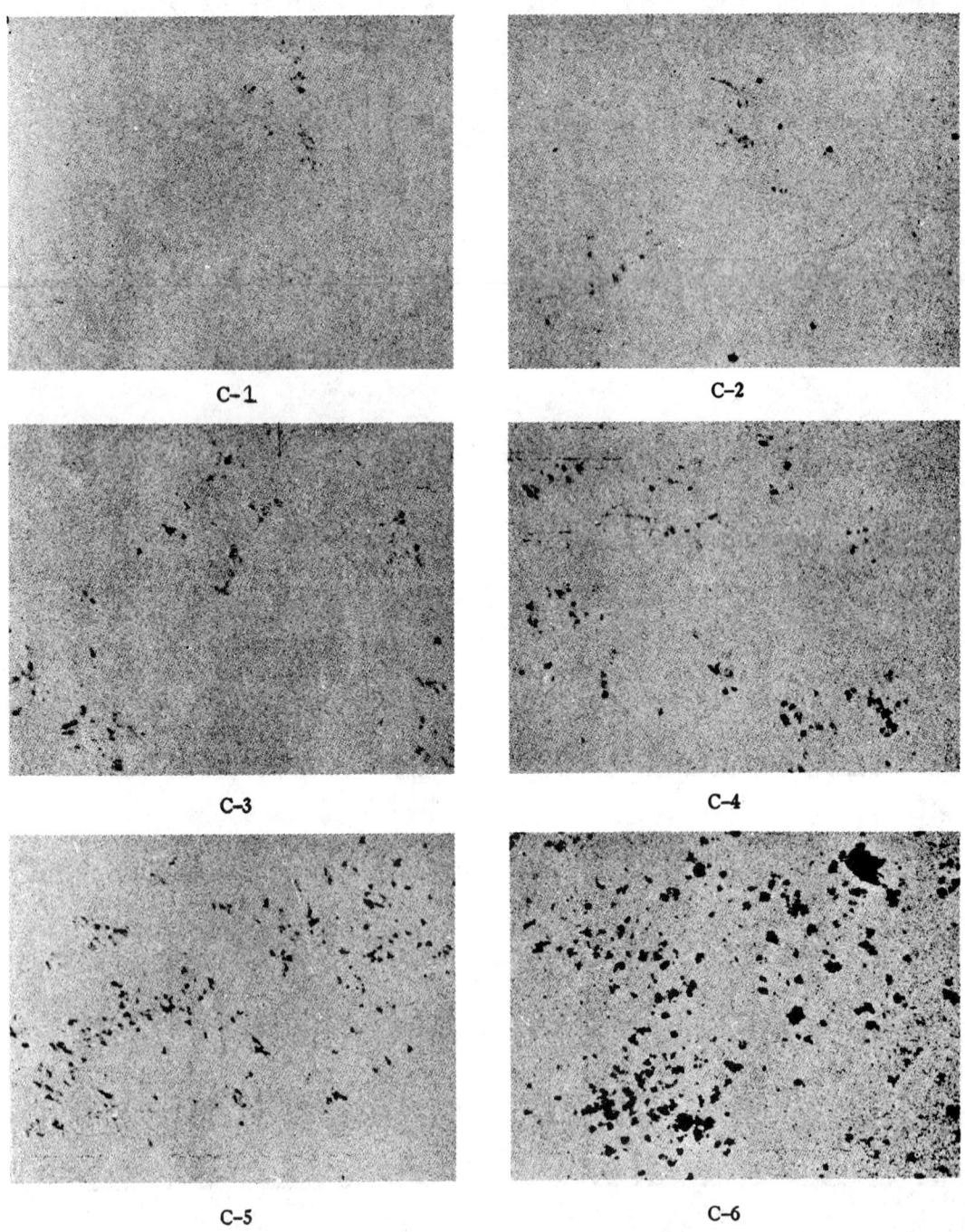

FIG. 3—TYPE C—APPARENT POROSITY MICROSTRUCTURE OF CEMENTED CARBIDES (×200) (B 276)

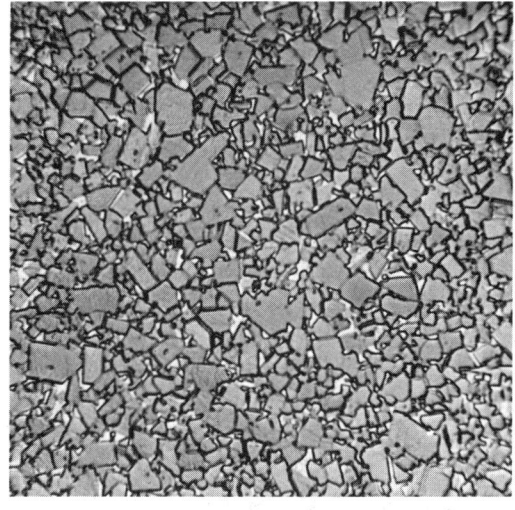

A - TUNGSTEN CARBIDE WITH 6% COBALT

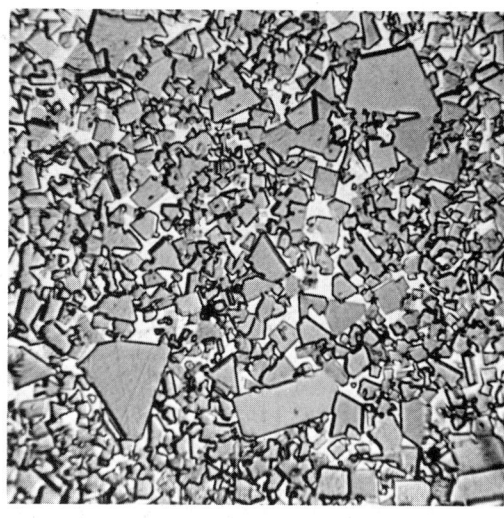

B - TUNGSTEN CARBIDE WITH 13% COBALT

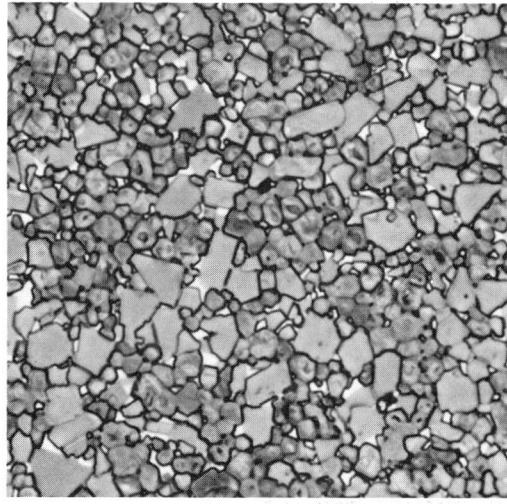

C - TUNGSTEN CARBIDE + SOLID SOLUTION CARBIDE WITH 4.5% COBALT

D - TUNGSTEN CARBIDE + SOLID SOLUTION CARBIDE WITH 11% COBALT

FIG. 4—TYPICAL MICROSTRUCTURES OF SINTERED CARBIDES (1500×, Murakami's reagent)

solution carbide (WC-TiC-TaC) with 4.5% cobalt and 11% cobalt, respectively. The tungsten carbide particles are angular and gray in appearance, the solid solution particles, where present, are rounded and usually darker gray, and the cobalt binder appears white. The abnormal "eta phase" carbide is not depicted by the photomicrographs. It is a brittle, carbon deficient carbide detrimental to tool performance but is readily detected as a very rapid etching, black constituent.

The data provided by this test are an excellent indicator for identifying a particular producer's product and its uniformity.

Grain Size Evaluation—Sample preparation, etching technique, and equipment shall be the same as described for structure evaluation.

The grain size shall be determined by comparing representative areas of the observed sample field with the carbide grain size chart (Fig. 5). This chart illustrates the relationship of particle sizes from 1 to 10 μm as observed at a magnification of 1500X.

The grain size rating shall consist of a sequence of numbers such as 231. Each number refers to a carbide particle size range; that is, a "1" includes all particles which are 1 μm or finer, a "2" includes all particles over 1 through 2 μm, a "3" includes all particles over 2 through 3 μm, etc., as illustrated by the carbide grain size chart. The sequence of the numbers shall be in the order of the sample area they represent, with the first number representing the greatest area. A minimum of 80% of the representative sample area shall be included in the rating.

Grain size and distribution has considerable influence on the mechanical properties of sintered carbide. Thus, materials having similar composition but different grain size and distribution may have very different performance characteristics.

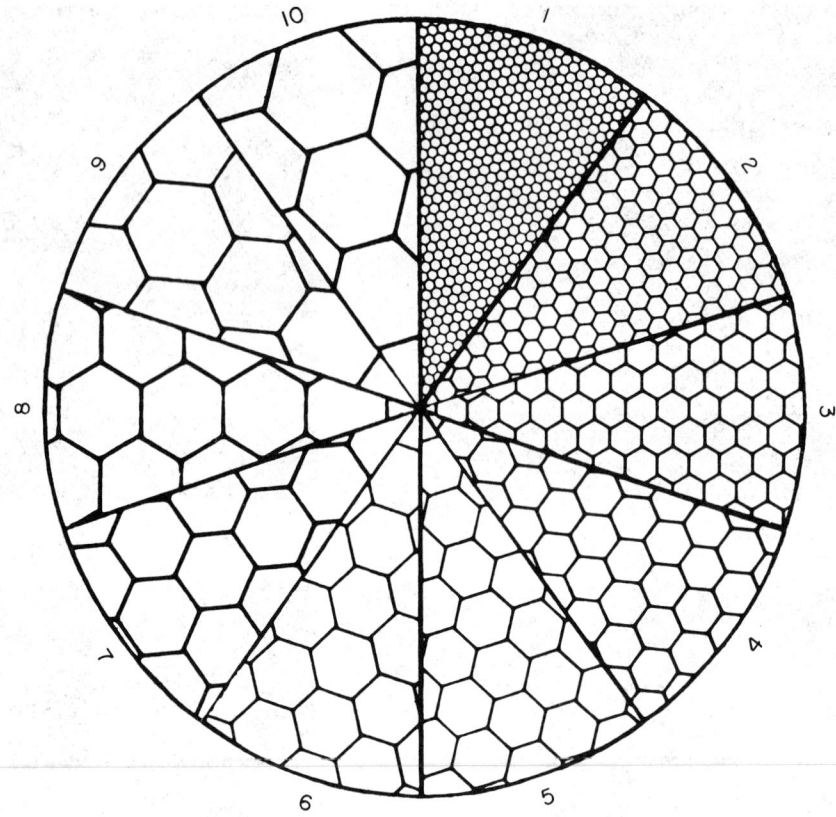

FIG. 5—CARBIDE GRAIN SIZE CHART, 1–10 μm AT 1500×
(1 μm = 0.00003937 in)

SINTERED TOOL MATERIALS—SAE J1072

SAE Recommended Practice

Report of Iron and Steel Technical Committee approved January 1974. Editorial change February 1977.

Scope—This SAE Recommended Practice covers the identification and classification of ceramic, sintered carbide, and other cermet tool products.

Its purpose is to provide a standard method for designating the characteristics and properties of sintered tool materials.

Description—The method is a typical letter-number "line call out" system which consists of two parts:

1. Basic Classification—which identifies the type of sintered tool product.
2. Suffix Requirements—which provide quantitative values for the properties or characteristics of a specific material.

(Note: Only pertinent properties or characteristics should be included. Alpha-numeric combinations which describe commercially unavailable products must be avoided.)

Basic Classification—A sequence of five digits is used to type each sintered tool product. The first digit indicates the type of compound. The second digit indicates the binder metal. The third digit indicates the *predominating* base metal, while the fourth and fifth digits indicate *other* base metals. See Table 1.

Example: SAE J1072 23200—tungsten carbide with a cobalt binder
SAE J1072 30600—aluminum oxide ceramic
SAE J1072 23234—tungsten, titanium, tantalum carbide with a cobalt binder

Suffix Requirements—A letter designating the property or characteristic to be specified is followed by numbers which designate the level of the property or characteristic. (Note: Suffix Z shall be used to describe any property or characteristic for which an appropriate suffix letter has not been assigned.)

Suffix A—Binder Metal Quantity—A three-digit number following the letter "A" designates the specified percentage by weight of binder metal to the nearest 0.1%. The product range shall be ±5% of the specified quantity.

Example: A045 specifies 4.5% of binder metal described by the basic classification or a product range of 4.3–4.7%.

Suffix B—Base Metal Quantity—A three-digit number following the letter "B" designates the specified percentage by weight of base metal to the nearest 0.1%. The product range shall be ±5% of the specified quantity. Each base metal described by the fourth and fifth digits of the basic classification shall be included.

Example: B096B047 specifies 9.6% of base metal described by the fourth digit of the basic classification and 4.7% of base metal described by the fifth digit of the basic classification. The product ranges would be 9.1–10.1 and 4.5–4.9%, respectively.

(Note: The Basic Classification provides for two base metals other than the predominating base metal. Any additional base metal and its quantity must be described by suffix Z.)

Suffix C—Hardness per SAE J439—A three-digit number following the letter "C" designates the specified Rockwell A hardness to the nearest 0.1 Rockwell A. The hardness range of the product shall be ±0.3 Rockwell A for products specified 89.9 Rockwell A or softer and ±0.2 Rockwell A for products specified 90.0 Rockwell A or harder.

Example: C922 specifies 92.2 Rockwell A or a product hardness range of 92.0–92.4 Rockwell A.

Suffix D—Specific Gravity per SAE J439—A three-digit number following the letter "D" designates the specified specific gravity or density in g/cm^3 to the nearest 0.1 g/cm^3. The density range of the product shall be ±0.1 g/cm^3 from that specified.

Example: D126 specifies a density of 12.6 g/cm^3 and a product density range of 12.5–12.7 g/cm^3.

Suffix E—Grain Size per SAE J439—One to three digits following the letter "E" designate the specified grain size.

Example: A grain size rating of 231 per SAE J439 shall be specified E231.

Suffix F—Apparent Porosity per SAE J439—Three digits following the letter "F" specify the maximum amount of each type porosity allowable, as depicted by the photographs of SAE J439. The first digit indicates the amount of type A porosity, the second digit indicates the amount of type B porosity, and the third digit the amount of type C porosity.

Example: F421 specifies a maximum apparent porosity rating of A-4, B-2, and C-1.

Suffix G—Transverse Rupture Strength per ASTM B 406—A three-digit number following the letter "G" designates the specified minimum transverse rupture strength in psi x 1000 (MPa x 6.9).

Example: G095 specifies a minimum transverse rupture strength of 95,000 psi (656 MPa).

Suffix Z—Other Properties—Properties not described by listed suffix designations and those which require more explicit description shall be written out.

Examples: Z1—thermal expansion from room temperature to 1500°F (816°C) shall be 3.46 x 10^{-6}/°F (6.23 x 10^{-6}/°C).

Z2—titanium carbide coated.

TABLE 1—BASIC CLASSIFICATION

Material Compound	Binder Metal	Base Metal
1–Nitride	0–None	0–None
2–Carbide	1–Nickel	1–Columbium
3–Oxide	2–Iron	2–Tungsten
9–Other[a]	3–Cobalt	3–Titanium
	9–Other[a]	4–Tantalum
		5–Chromium
		6–Aluminum
		9–Other[a]

[a] Shall be described by suffix Z.

9 Ferrous Materials

SINTERED POWDER METAL PARTS: FERROUS—SAE J471d — SAE Standard

Report of Nonferrous Metals Division approved January 1939, revised by Nonferrous Metals Committee June 1966, and last revised by Iron and Steel Technical Committee August 1973.

Powder metal (P/M) parts are manufactured by pressing metal powders to the required shape in a precision die and sintering to produce metallurgical bonds between the particles, thus generating the appropriate mechanical properties. The shape and mechanical properties of the part may be subsequently modified by repressing or by conventional methods such as machining and/or heat treating.

While powder metallurgy embraces a number of fields wherein metal powders may be used as raw materials, this standard is concerned primarily with information relating to mechanical components and bearings produced from iron-base materials.

1. Bearings—Powder metal bearings are classified broadly in two groups: ferrous and nonferrous. While much of the basic information is common to both types, this standard is concerned only with the former. Information relating to copper- and aluminum-base materials is under development.

1.1 Chemical Composition—The chemical composition shall be determined on an oil-free basis and shall conform to the limits set out in Table 1. The analysis shall be performed in accordance with ASTM procedure, or any other approved method agreed upon by the manufacturer and the purchaser.

Subject to agreement between purchaser and manufacturer metallographic estimates of combined carbon values may be used.

In cases of disagreement in respect of composition, samples shall be submitted to independent umpire analysis.

1.2 Physical and Mechanical Properties—A most important characteristic of oil impregnated sintered bearings is their property of self-lubrication resulting from the internal oil reservoir created by the interconnected pore structure. The quantity of oil available is thus directly proportional to the pore volume of the bearing. The mechanical strength of bearings of the same composition produced under similar manufacturing conditions is inversely proportional to the pore volume. Although a tensile bar pressed and sintered under the same conditions as the bearing is sometimes used to evaluate materials, the generally accepted test is a radial crush test in which the load required to break the bearing is related to its physical dimensions via a constant, K, specified for each material.

1.2.1 DENSITY—The density of the bearing, fully impregnated with lubricant (see Appendix B), shall conform to the limits set out in Table 1. If in one bearing the variation of density from any one section to any other is less than 0.3 g/cm³, the density of the bearing as a whole shall fall within the limits prescribed in Table 1. If this point-to-point variation exceeds 0.3 g/cm³, the manufacturer and purchaser shall agree upon a critical section of the part in which the density requirements of the specification must be fulfilled.

1.2.2 OIL CONTENT—The oil content of the bearing shall not be less than that specified in Table 1. (See Appendix B.)

1.2.3 RADIAL CRUSHING STRENGTH—Radial crushing strength (see Appendix A) shall not be less than the value calculated as follows:

$$P = \frac{KLT^2}{D - T}$$

where:
- P = radial crushing load, lb (N)
- D = outside diameter of bearing, in. (mm)
- T = wall thickness of bearing, in. (mm)
- L = length of bearing, in. (mm)
- K = strength constant shown in Table 1

1.2.4 PERMISSIBLE LOADS—In calculating permissible loads, the operating conditions, housing conditions, and construction should be considered. Permissible bearing loads for various operating conditions are shown in Table 2. These are intended only as a general guide.

Certain conditions will increase the permissible loads, such as additional lubrication, pressure lubrication, hardening of the shaft, loads of short duration.

Certain conditions will tend to reduce the load-carrying capacity of bearings regardless of type or make: continued start-stop operation, oscillating and reciprocating motion, extremely high or low temperatures; excessively close or loose bearing clearances; deflection or misalignment of shaft; dust, grit, corrosive fumes, or poor shaft finish.

1.3 Dimensional Characteristics

1.3.1 TOLERANCES—Dimensional tolerances allowed shall conform to the limits prescribed in Tables 3 and 4, unless otherwise agreed between supplier and purchaser.

TABLE 1—PROPERTIES OF FERROUS P/M BEARINGS

SAE No.	Density, g/cm³	Chemical Composition, %				Minimum Oil Content by Volume, %	Strength Constant	
		Cu	C	Others	Fe		psi	MPa
850	5.7–6.1	—	0.25 max	2.0 max	Bal	18	25,000	172
851	5.7–6.1	—	0.25–0.60	2.0 max	Bal	18	30,000	207
862	5.8–6.2	7–11	0.30 max	2.0 max	Bal	18	40,000	276
863	5.8–6.2	18–22	0.30 max	2.0 max	Bal	18	40,000	276

TABLE 2—PERMISSIBLE BEARING LOADS

Shaft Velocity		Permissible Loads			
		SAE 850/851		SAE 862/863	
ft/min	m/min	psi	MPa	psi	MPa
Static (0)	0	7500	52	15,000	103
Slow and intermittent (25)	7.6	3600	25	8,000	55
50–100	15.2–30.4	1800	12	3,000	21
100–150	30.4–45.7	450	3.1	700	4.8
150–200	45.7–61	300	2.1	400	2.8
Over 200	61	225	1.6	300	2.1

For shaft velocities in excess of 200 ft/min (61 m/min), the permissible load may be calculated as follows:

$$P = 50,000/V$$

where:
- P = safe load per square inch of projected area, psi
- V = shaft velocity, ft/min

or:

$$P = 105/V$$

- P = safe load per square metre of projected area, MPa
- V = shaft velocity, m/min

TABLE 3—COMMERCIAL DIMENSIONAL TOLERANCES

Note: This table is intended for bearings with a 3:1 maximum length to inside diameter ratio and a 20:1 maximum length to wall thickness ratio. Bearings having greater ratios than these are not covered by the table.

Inside Diameter and Outside Diameter		Total Diameter Tolerance[a]			
		Inside Diameter		Outside Diameter	
in	mm	in	mm	in	mm
Up to 0.760	Up to 19.31	0.001	0.025	0.001	0.025
0.761 to 1.510	19.32 to 38.36	0.0015	0.025	0.0015	0.04
1.511 to 2.510	38.37 to 63.76	0.002	0.05	0.002	0.05
2.511 to 3.010	63.77 to 76.46	0.003	0.08	0.002	0.05
3.011 to 4.010	76.47 to 101.86	0.004	0.10	0.004	0.10
4.011 to 5.010	101.87 to 127.26	0.005	0.13	0.005	0.13
5.011 to 6.010	127.27 to 152.65	0.006	0.15	0.006	0.15

Length		Total Length Tolerance[b]	
in	mm	in	mm
Up to 1.495	Up to 37.97	0.010	0.25
1.496 to 1.990	37.98 to 50.54	0.015	0.38
1.991 to 2.990	50.55 to 75.96	0.020	0.51
2.991 to 4.985	75.97 to 126.61	0.030	0.76

Outside Diameter		Wall Thickness, max		Concentricity Tolerance[c]	
in	mm	in	mm	in	mm
Up to 1.510	Up to 38.36	Up to 0.355	9.02	0.003	0.08
1.511 to 2.010	38.37 to 51.06	Up to 0.505	12.83	0.004	0.10
2.011 to 4.010	51.07 to 101.86	Up to 1.010	25.65	0.005	0.13
4.011 to 5.010	101.87 to 127.26	Up to 1.510	38.35	0.006	0.15
5.011 to 6.010	127.27 to 152.65	Up to 2.010	51.05	0.007	0.18

[a]Total tolerance on the inside diameter and outside diameter is a minus tolerance only.
[b]Total tolerance is split into plus and minus.
[c]Total indicator reading.

1.3.2 RECOMMENDED PRESS FITS—Plain cylindrical journal bearings are commonly installed by press fitting the bearing into a housing using an insertion arbor. For housings rigid enough to withstand the press fit without appreciable distortion and for bearings with wall thickness approximately one-eighth of the bearing outside diameter, the press fits shown in Table 5 are recommended.

1.3.3 RUNNING CLEARANCES—Proper running clearances for sintered bearings depend to a great extent upon the particular application. Therefore, only minimum recommended clearances are listed in Table 6. It is assumed that ground steel shafting will be used and that all bearings will be oil impregnated.

2. Mechanical Components

2.1 General Information—This section of the standard relates to mechanical or structural components such as cams, gears, levers, shock absorber parts, transmission parts, etc., which are produced by powder metallurgy methods. Many of these parts are used in the "as-sintered" or "as-sized" condition; however, in a large number of applications, additional processing of the parts is required. Additional processes include machining, heat treatment, sealing, or surface treatments. These notes are intended to provide a general guide on the application and use of some of these processes.

2.1.1 HEAT TREATMENT—P/M parts are porous and thus provide more surface area in any metal/gas reactions proceeding during heat treatment. In any given set of heat treating circumstances, the depth of carburization or decarburization will increase with decreasing density. Provided that the proper care is taken to maintain the appropriate carbon potential, carbon-bearing iron-base P/M parts can be heat treated by conventional quench-hardening methods. It should be noted that the porous material will, on cooling, absorb some of the quench medium, perhaps resulting in some minor problems during tempering or further treatment.

The absorption of fluids by the porous materials usually precludes the use of liquid salt bath treatments.

2.1.2 STEAM TREATMENT—This process consists of heating ferrous parts to 1000–1100°F (540–600°C) and subjecting them to superheated steam under pressure. A layer of black iron oxide is formed on all external and internal (interconnected porosity) surfaces. This oxide layer improves wear resistance,

TABLE 4—FLANGE AND THRUST BEARINGS DIAMETER AND THICKNESS TOLERANCES[a]

Flange Bearings, Flange Diameter Tolerances

Diameter Range		Standard		Special	
in	mm	in	mm	in	mm
0 to 1-1/2	0 to 38	±0.005	±0.13	±0.0025	±0.06
Over 1-1/2 to 3	39 to 76	±0.010	±0.25	±0.005	±0.13
Over 3 to 6	77 to 152	±0.025	±0.64	±0.010	±0.25

Flange Bearings, Flange Thickness Tolerances

Diameter Range		Standard		Special	
in	mm	in	mm	in	mm
0 to 1-1/2	0 to 38	±0.005	±0.13	±0.0025	±0.06
Over 1-1/2 to 3	39 to 76	±0.010	±0.25	±0.007	±0.18
Over 3 to 6	77 to 152	±0.015	±0.38	±0.010	±0.25

Thrust Bearings (1/4 in (6.35 mm) Thickness, max), Thickness Tolerances, All Diameters[b]

Standard		Special	
in	mm	in	mm
±0.005	±0.13	±0.0025	±0.06

Parallelism on Faces, max

Diameter Range		Standard		Special	
in	mm	in	mm	in	mm
0 to 1-1/2	0 to 38	0.005	0.13	0.003	0.03
Over 1-1/2 to 3	39 to 76	0.007	0.18	0.005	0.13
Over 3 to 6	77 to 152	0.010	0.25	0.007	0.18

[a]Standard and special tolerances are specified for diameters, thickness, and parallelism. Special tolerances should not be specified unless required since they require additional or secondary operations and, therefore, are costlier.
[b]Outside diameter tolerances same as for flange bearings.

TABLE 5—RECOMMENDED PRESS FITS

Outside Diameter of Bearing		Press Fit			
		Min		Max	
in	mm	in	mm	in	mm
Up to 0.760	Up to 19.31	0.001	0.025	0.003	0.03
0.761 to 1.510	19.32 to 38.36	0.0015	0.04	0.004	0.10
1.511 to 2.510	38.37 to 63.76	0.002	0.05	0.005	0.13
2.511 to 3.010	63.77 to 76.45	0.002	0.05	0.006	0.15
Over 3.010	Over 76.45	0.002	0.05	0.007	0.18

TABLE 6—RUNNING CLEARANCES

Shaft Size		Total Clearance, min	
in	mm	in	mm
Up to 0.760	Up to 19.31	0.0005	0.01
0.761 to 1.510	19.32 to 38.36	0.001	0.025
1.511 to 2.510	38.37 to 63.76	0.0015	0.04
Over 2.510	Over 63.76	0.002	0.05

surface hardness, compressive strength and, under some conditions, corrosion resistance. The presence of oxide within the pores tends to close these channels, reducing the volume of interconnected porosity and providing a measure of pressure tightness. Steam treatment usually results in a decrease in impact resistance. It should also be noted that oxidation can lead to the generation of internal stresses with a general degradation of mechanical properties.

2.1.3 PLATING—P/M parts can be electroplated by conventional techniques providing certain precautions are taken to prevent the absorption of the plating solution into the porous body. Trapped electrolyte will eventually exude, causing corrosion and flaking of the plate. The degree of surface preparation required is governed by the part density. Infiltrated parts and parts with a density in excess of 7.0 g/cm^3 can be plated by procedures normally employed for wrought materials. At lower densities the parts must be sealed by resin impregnation if the plating is to be deposited from a liquid electrolyte. Certain types of mechanical plating can be applied to porous materials without difficulty.

2.1.4 INFILTRATION—Infiltration is a process in which the residual interconnected porosity in an iron-base P/M part is filled with a metal of lower melting point. The infiltrant, normally copper or a copper-base alloy, is placed in contact with the part and the two are heated above the melting point of the infiltrant. In the liquid state the infiltrant is drawn into the interconnected porosity of the part by capillary action. The major disadvantage of the process is that it may result in some loss of dimensional accuracy.

The process has the following advantages:

(a) Improved mechanical properties. Higher tensile strength and hardness values, together with improved impact and fatigue resistance, are obtained as a result of infiltrating the part.

(b) Elimination of porosity. The sealing effect resulting from the filling of interconnected porosity eliminates problems associated with electrolyte entrapment in plating or gas permeation in heat treatment. Infiltrated parts can usually be used in most applications requiring pressure tightness.

2.1.5 IMPREGNATION—Impregnation is the process of filling the pores of a part with oil or a plastic resin. Oil is used primarily for self-lubricating parts or bearings; plastic resins may be used

(a) To effect pressure tightness.

(b) To seal porosity as a pretreatment prior to plating.

(c) To provide an uninterrupted surface for machining. Impregnation improves tool life and surface finish.

Two basic techniques are in use for oil impregnation:

(a) The parts are immersed in hot oil for a period varying between 30 min and several hours depending upon the size, shape, and type of part.

(b) The parts are immersed in oil under vacuum in some suitable vessel.

The latter method ensures the removal of air pockets from within the component.

In the case of plastic impregnation, only the vacuum technique is employed.

2.1.6 MACHINING—It is not possible to give many useful rules or principles for the machining of P/M materials because of the diversity of materials, machining techniques, and objectives. In general, the machining characteristics of P/M materials are different from those of wrought materials of similar hardness or composition. It is obvious that a machining operation may close surface porosity and hence interfere with the intended function of a bearing surface. If possible, machining operations should be carried out dry since coolants may be retained in the pores subsequently leading to corrosion or act as an adulterant to the impregnating lubricant. Wet machining can be used without difficulty on infiltrated or impregnated parts. It is necessary to examine each individual application in detail to devise the optimum method and conditions for machining.

2.2 Properties

2.2.1 CHEMICAL COMPOSITION—The chemical composition shall be determined on an oil-free basis and shall conform to the limits prescribed in Table 7. The analysis shall be carried out in accordance with ASTM procedure or by any approved method agreed upon by the manufacturer and purchaser.

Subject to agreement between purchaser and manufacturer, metallographic estimates of combined carbon values may be used.

In cases of disagreement in respect of composition, samples must be submitted to independent umpire analysis.

2.2.2 DENSITY—In structural parts of complex shape, there may be variation in density from one section of the part to another. If this variation is less than 0.3 g/cm^3, the overall density of the part as a whole shall fall within the limits prescribed in Table 7. If the variation exceeds 0.3 g/cm^3, the manufacturer and purchaser shall agree upon a critical section of the part in which the density requirements of the specification must be fulfilled. This critical section would ordinarily be that at which the stresses are highest.

Density shall be determined on a dry basis, that is, on the unimpregnated part. (See Appendix B.)

2.2.3 OTHER MECHANICAL PROPERTIES—The properties given in Table 7 are typical of materials within the specified density ranges and properly sintered.

Most P/M parts are too small to allow tensile bars to be cut from the actual component; thus, it is a common practice for the manufacturer and purchaser to agree upon an empirical acceptance test based upon the conditions of service of the part. This test may be an axial or radial crushing test, an impact test in which a weight is allowed to fall a specific distance onto a specified area of the part, a bending test, etc. The method of carrying out this test must be agreed upon, including the method of holding the specimen, the rate of application of the load, etc. Hardness tests are often used in conjunction with tests of this type. The actual metal hardness may be obscured by the collapse of the pore structure under the localized load of such a test. Consequently, it is usually impossible to correlate hardness measurements carried out by different methods as is often possible in wrought materials. In general, a particular hardness specification for a part should be developed by agreement of the supplier and customer. The familiar Rockwell scales such as B, C, and 15T are used to advantage.

The hardness values obtained for porous sintered materials on any scale should not be compared with the hardness readings yielded by wrought metals of similar composition because of the "pore effect" on hardness readings obtained on sintered materials and discussed earlier in this passage.

APPENDIX A
RADIAL CRUSHING STRENGTH

Radial crushing strength shall be determined by compressing the test specimen between two flat surfaces at a no load speed of 0.1 in./min (2.54 mm/min), the direction of the load being normal to the longitudinal axis of the specimen. The point at which the load drops due to the first crack shall be considered the crushing strength.

In the case of flanged bearings, the flange shall be cut off and the two parts tested separately. Each section shall meet the minimum requirement as calculated by the formula given in paragraph 1.2.3.

APPENDIX B
DENSITY AND OIL CONTENT OF SINTERED STRUCTURAL PARTS AND OIL-IMPREGNATED BEARINGS

B1. Scope—This appendix covers test procedures for determining the density and oil content of sintered structural parts and bearings.

B2. Preparation of Test Specimens

B2.1 Weight—The specimen weight must be a minimum of 2.0 g. Several specimens can be used to reach the minimum weight.

B2.2 Impregnation—Either of the following two methods may be used to impregnate the test specimen for the purpose of determining weights of oil-impregnated specimens in air, B, or in water, C; however, the vacuum method is preferred.

B2.2.1 Reduce the pressure over the specimen immersed in oil held at room temperature to not more than 2 in. (50.8 mm) of mercury pressure for 30 min by a suitable evacuating method, after which permit the pressure to increase to atmospheric pressure and the specimen to remain immersed in oil at room temperature and atmospheric pressure for 10 min.

B2.2.2 Immerse the specimen for at least 4 h in oil (viscosity of approximately 200 sus at 100°F (46 μm^2/s at 38°C)), held at a temperature of 180 ±10°F (82 ±5°C), and then cool to room temperature by immersion in oil at room temperature.

B2.3 Oil Removal—Samples which are delivered to the purchaser with oil shall be freed from lubricant for determining weight A by extracting the lubricant in Soxhlet apparatus of suitable size using toluol or petroleum ether as a solvent. After extraction, the residual solvent shall be removed by heating samples for 1 h at 250°F (120°C). Alternate extraction and drying shall be continued until the dry weight in air is constant to 0.1%.

NOTE: A practical and fast method of oil removal is to heat the specimen in a reducing atmosphere in the temperature range of 1400–1600°F (760–871°C). This method, which is in close agreement with the Soxhlet apparatus, can be used if agreed upon by both parties.

B3. Procedure

B3.1 Using an analytical balance, obtain the dry and impregnated weights of the test specimen in air. These are weights A and B, respectively. These, and all subsequent weighings, should be to the nearest 0.001 g.

B3.2 Select a fine wire, less than 0.010 in. (2.54 mm) in diameter, for supporting the specimen in a beaker of distilled water when suspended from the beam hook of the balance. A wetting agent (in the amount of 0.1–0.2% by weight) is to be used to reduce the effects of surface tension.

B3.3 Support the beaker of water over the pan of the balance, using a suitable bridge.

B3.4 Twist the wire around the specimen and suspend it from the beam hook so that the specimen is completely immersed in the water. The water should cover the specimen by at least $\frac{1}{4}$ in. (6 mm), and the wire twist should be completely submerged. Care must be taken to ensure that no air bubbles adhere to the specimen or to the wire.

B3.5 Weigh the specimen and wire in water. This is weight C.

TABLE 7—PROPERTIES OF STRUCTURAL COMPONENTS

SAE No.	Grade	Class	Type	Chemical Composition, %					Density g/cm³	Ultimate Yield Strength (typical)		Yield Strength in Compression (typical)		Ultimate Yield Strength After Heat Treatment (typical)	
				C	Cu	Ni	Others	Fe		psi	MPa	psi	MPa	psi	MPa
853	1	1	1	0.25 max	—	—	2.0 max	Bal	5.6–6.0	16,000	110	12,000	83	—	—
853	1	1	2						6.0–6.4	20,000	138	17,000	117	—	—
853	1	1	3						6.4–6.8	26,000	179	20,000	138	—	—
853	1	1	4						6.8–7.2	30,000	207	25,000	172	—	—
853	1	1	5						7.2 min	40,000	276	30,000	207	—	—
853	2	1	1	0.25–0.60	—	—	2.0 max	Bal	5.6–6.0	20,000	138	18,000	124	30,000	207
853	2	1	2						6.0–6.4	26,000	179	22,000	152	40,000	276
853	2	1	3						6.4–6.8	34,000	234	24,000	166	50,000	345
853	2	1	4						6.8–7.2	50,000	345	35,000	241	70,000	483
853	2	1	5						7.2 min	60,000	414	42,000	290	80,000	552
853	3	1	1	0.60–0.90	—	—	2.0 max	Bal	5.6–6.0	26,000	179	22,000	152	40,000	276
853	3	1	2						6.0–6.4	34,000	234	26,000	179	50,000	345
853	3	1	3						6.4–6.8	40,000	276	28,000	193	64,000	441
853	3	1	4						6.8–7.2	60,000	414	42,000	290	80,000	552
853	3	1	5						7.2 min						
864	1	3	1	0.60–0.90	1–3	—	2.0 max	Bal	5.6–6.0	30,000	207	—	—	40,000	276
864	1	3	2						6.0–6.4	41,000	283	—	—	49,000	338
864	1	3	3						6.4–6.8	58,000	400	—	—	89,000	614
864	1	3	4						6.8–7.2	76,000	524	—	—	124,000	855
864	1	3	5						7.2 min	105,000	724	—	—	154,000	1062
864	2	3	1	0.60–0.90	3–6	—	2.0 max	Bal	5.6–6.0	34,000	234	—	—	45,000	310
864	2	3	2						6.0–6.4	48,000	331	—	—	54,000	372
864	2	3	3						6.4–6.8	68,000	469	—	—	92,000	634
864	3	3	1	0.60–0.90	6–11	—	2.0 max	Bal	5.6–6.0	36,000	248	—	—	—	—
864	3	3	2						6.0–6.4	51,000	352	—	—	—	—
864	4	3	1	0.60–0.90	18–22	—	2.0 max	Bal	5.6–6.0	33,000	228	—	—	—	—
864	4	3	2						6.0–6.4	47,000	324	—	—	—	—
?	1	1	3	0.3 max	2.5 max	1–3	2.0 max	Bal	6.4–6.8	28,000	193	—	—	—	—
	1	1	4						6.8–7.2	38,000	262	—	—	—	—
	1	1	5						7.2 min	45,000	310	—	—	—	—
?	1	2	3	0.3–0.6	2.5 max	1–3	2.0 max	Bal	6.4–6.8	37,000	255	—	—	82,000	565
	1	2	4						6.8–7.2	50,000	345	—	—	110,000	689
	1	2	5						7.2 min	61,000	421	—	—	134,000	924
?	1	3	3	0.6–0.9	2.5 max	1–3	2.0 max	Bal	6.4–6.8	48,000	331	—	—	100,000	689
	1	3	4						6.8–7.2	65,000	448	—	—	135,000	931
	1	3	5						7.2 min	79,000	545	—	—	160,000	1103
?	2	1	3	0.3 max	2.0 max	3–5.5	2.0 max	Bal	6.4–6.8	36,000	248	—	—	—	—
	2	1	4						6.8–7.2	49,000	338	—	—	—	—
	2	1	5						7.2 min	58,000	400	—	—	—	—
?	2	2	3	0.3–0.6	2.0 max	3–5.5	2.0 max	Bal	6.4–6.8	45,000	310	—	—	112,000	772
	2	2	4						6.8–7.2	62,000	428	—	—	154,000	1062
	2	2	5						7.2 min	74,000	510	—	—	180,000	1241
?	2	3	3	0.6–0.9	2.0 max	3–5.5	2.0 max	Bal	6.4–6.8	57,000	393	—	—	—	—
	2	3	4						6.8–7.2	77,000	531	—	—	—	—
	2	3	5						7.2 min	93,000	641	—	—	—	—
Infiltrated Materials															
870				0.25 max	15–25	—	4.5	Bal	7.1 min	65,000	448	70,000	483	—	—
872				0.6–0.9	15–25	—	4.5	Bal	7.1 min	85,000	586	90,000	621	120,000	827

NOTE: All properties given above are typical of materials produced from elemental powder mixes as distinct from prealloyed powders.

B3.6 Remove the specimen and reweigh the wire in water, immersed to the same point as before. This is weight E.

B3.7 For interconnected porosity determinations, measure the temperature during the test to the nearest whole degree and determine the specific gravity of the impregnant, which is S.

B4. Calculation

B4.1 The density of structural parts shall be calculated as follows:

$$D = \frac{A}{B - C + E} = \frac{A}{B - (C - E)}$$

where: D = density, g/cm³
A = weight in air of the oil-free specimen, g
B = weight in air of the oil-impregnated specimen, g
C = weight of the oil-impregnated specimen and wire in water, g
E = weight of wire, g

B4.2 The wet density of bearings supplied fully impregnated with lubricant shall be calculated as follows:

$$D = \frac{B}{B - C + E}$$

where: D = density, g/cm³
B = weight of oil-impregnated specimen, g
C = weight of oil-impregnated specimen and wire in water, g
E = weight of wire in water, g

B4.3 The interconnected porosity or oil content by volume shall be calculated as follows:

$$P = \frac{B - A}{(B - C + E) \times S} \times 100 = \frac{B - A}{(B - (C - E)) \times S} \times 100$$

where: P = oil content by volume, %
A = weight in air of oil-free specimen, g
B = weight in air of oil-impregnated specimen, g
C = weight of oil-impregnated sample immersed in water, g
E = weight of wire in water, g
S = specific gravity of impregnant at the temperature of test

For faster results, the alternate procedure given below gives a close approximation of density. In cases of dispute, however, the foregoing procedure should be used.

ALTERNATE PROCEDURE—Weigh the part in air, then coat the entire part with an air-drying transparent acrylic lacquer. The part is subsequently weighed again in air, then in water.

Density is calculated as follows:

$$D = \frac{A}{B - C}$$

where: A = weight of the original part in air, g
B = weight of the part in air after coating with lacquer, g
C = weight of the part immersed in water after coating with lacquer, g
D = density, g/cm^3

NOTE: The foregoing methods give the density of the part in relation to the density of water at the testing temperature, that is, specific gravity. Although it is common practice to assume density and specific gravity to be equal, this is in fact not true since the maximum density of pure water is 0.999972 g/cm^3 at 39.16°F (3.98°C) and decreases with increasing temperature.

The resulting error increases to 0.5% above 90°F (32°C) and to 2.5% at 210°F (99°C). It is therefore suggested that the test temperature be held below 80°F (26°C) in order to minimize the error.

CUT STEEL WIRE SHOT—SAE J441 — SAE Recommended Practice

Report of Iron and Steel Technical Committee approved January 1952. Reaffirmed with editorial change January 1969.

[This SAE Recommended Practice is considered to be tentative and is subject to modification to meet new developments or requirements. It is offered as a guide in the selection and use of cut steel wire shot.]

Description—Cut steel wire shot shall be the product of carbon steel wire cut into the form of cylinders with lengths approximately equal to the wire diameter. Conditioned cut steel wire shot with cut edges prerounded may be specified when required for special applications.

Identification—All cut steel wire shot shall be classified according to the wire size from which it is obtained. It shall be identified by the prefix letters CW meaning cut wires. This designation shall be followed by a suffix number equivalent to the mean diameter of the wire from which the shot is produced.

Chemical Composition—The chemical composition shall conform in general to the following specification:

Carbon, %	0.45–0.75
Manganese, %	0.60–1.20
Phosphorus, %	0.045 max
Sulfur, %	0.050 max
Silicon, %	0.10–0.30

Tensile Properties—Shot shall be made from wire conforming to the tensile strengths shown in Table 1.

Hardness—The hardness of the shot particles (as cut) shall have the minimum values, given in Table 2, as determined by any of the various methods applicable to small sections, such as a Tukon tester with Vickers indenter, at loads determined to provide a reliable conversion to Rockwell.

Size Classification—Cut steel wire shot shall be made from wire of the diameters shown in Table 3. Shot sizes varying from those shown are available and may be obtained by arrangement between the shot manufacturer and user.

Inspection Procedure—Shot particles to be checked for length and hardness are to be mounted and ground and polished to the centerline of the cylinder longitudinal cross section. The combined length of 10 random particles shall be within the tolerances of Table 4. As an alternate method, the total weight of 50 random particles shall be within the limits of Table 4.

Soundness—Shot particles shall be free of shear cracks and laps and shall not contain excessive seams or burrs.

Packaging—This material shall be packaged to prevent loss.

TABLE 1—TENSILE PROPERTIES OF CUT STEEL WIRE SHOT

Shot Size	Mean Wire Dia, in.	Tensile Strength, psi
CW-62	0.0625	237,000–272,000
CW-54	0.054	243,000–279,000
CW-47	0.047	248,000–286,000
CW-41	0.041	255,000–293,000
CW-35	0.035	261,000–301,000
CW-32	0.032	265,000–305,000
CW-28	0.028	271,000–311,000
CW-23	0.023	275,000–314,000
CW-20	0.020	283,000–320,000

TABLE 2—HARDNESS OF CUT STEEL WIRE SHOT

Shot Size	Min Hardness, RC
CW-62	36
CW-54	39
CW-47	41
CW-41	42
CW-35	44
CW-32	45
CW-28	46
CW-23 and finer	48

TABLE 3—SIZE CLASSIFICATION FOR CUT STEEL WIRE SHOT

Shot Size	Wire Dia, in.
CW-62	0.0625 ± 0.002
CW-54	0.054 ± 0.002
CW-47	0.047 ± 0.002
CW-41	0.041 ± 0.002
CW-35	0.035 ± 0.001
CW-32	0.032 ± 0.001
CW-28	0.028 ± 0.001
CW-23	0.023 ± 0.001
CW-20	0.020 ± 0.001

TABLE 4—LENGTH TOLERANCES AND WEIGHT LIMITS FOR CUT STEEL WIRE SHOT

Shot Size	Length of 10 Pieces, in.	Weight of 50 Pieces, G
CW-62	0.620 ± 0.040	1.09–1.33
CW-54	0.540 ± 0.040	0.72–0.88
CW-47	0.470 ± 0.040	0.48–0.58
CW-41	0.410 ± 0.040	0.31–0.39
CW-35	0.350 ± 0.030	0.20–0.24
CW-32	0.320 ± 0.030	0.14–0.18
CW-28	0.280 ± 0.030	0.10–0.12
CW-23	0.230 ± 0.020	0.05–0.07
CW-20	0.200 ± 0.020	0.04–0.05

TEST STRIP, HOLDER AND GAGE FOR SHOT PEENING—SAE J442 AUG79 — SAE Standard

Report of Iron and Steel Technical Committee, approved January 1952, last revised by Fatigue Design and Evaluation Steering Committee November 1977, editorial change August 1979.

This SAE Standard is supplemented by an SAE Recommended Practice, Procedures for Using Standard Shot Peening Test Strip, SAE J443.

Outline of Method of Control—The control of a peening machine operation is primarily a matter of the control of the properties of a blast of shot in its relation to the work being peened. The basis of measurement of these properties is as follows: If a flat piece of steel is clamped to a solid block and exposed to a blast of shot, it will be curved upon removal from the block. The curvature will be convex on the peened side. The extent of this curvature on a standard sample serves as a means of measurement of the blast. The degree of curvature depends upon the properties of the blast, the properties of the test strip, and the nature of exposure to the blast, as described below.

Properties of the blast are the velocity, size, shape, density, kind of material, and hardness of the shot.

The properties of exposure to the blast are the length of time, angle of impact, and shot flow rate.

The properties of the test strip depend upon the physical dimensions and mechanical properties of the strip.

Based on these principles, the SAE has adopted the following standards: test strips, holding block, and gage. Specifications of these parts, the method of use, and a standard designation are presented herein.

Specifications of Intensity Measuring Equipment

Test Strips and Holding Fixtures—Standard test strips, N, A, and C are shown in Fig. 1, and test strip holder is shown in Fig. 2. The relationship between test strips N, A, and C are shown in Fig. 3. This curve shows N, A, and C strip readings for conditions of identical blast and exposure.

Gage—The gage for determining the curvature of the test strip is shown in Fig. 4. The curvature of the strip is determined by a measurement of the height of the combined longitudinal and transverse arcs across standard chords. This arc height is obtained by measuring the displacement of a central point on the nonpeened surface from the plane of four balls forming the corners of a particular rectangle. To use this gage, the test strip is located so that the indicator stem bears against the NONPEENED surface.

Designation Standard of Intensity Measurement—The standard designation of intensity measurement includes the gage reading and the test strip used. It may be explained by the following example:

```
     13A
      ↑↑
      │└─ Test Strip
      └── Gage Reading
```

This example signifies that the gage reading of the peened test strip as measured on the gage is 13. This can be considered a dimensionless number relating to the number of graduations read on the dial indicator of the Almen gage.

Another example is:

6–8C

This signifies gage readings on the C size test strip measured with the same gage. This example is typical of the method used for specifying a gage reading tolerance for an application.

As shown in both of the examples, the gage reading is given first and is followed by the test strip designation.

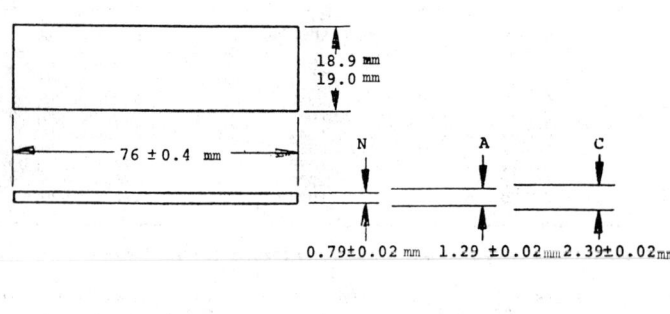

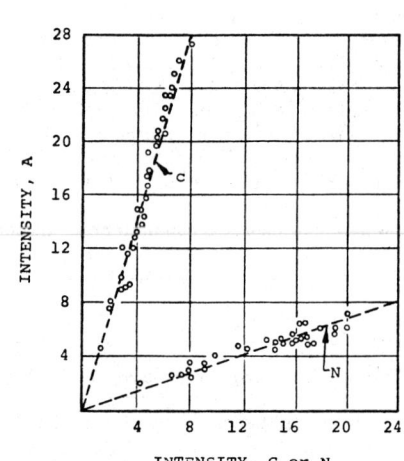

ed. FIG. 3—CORRELATION OF A, N, AND C STRIPS AS CHECKED ON AN ALMEN GAGE

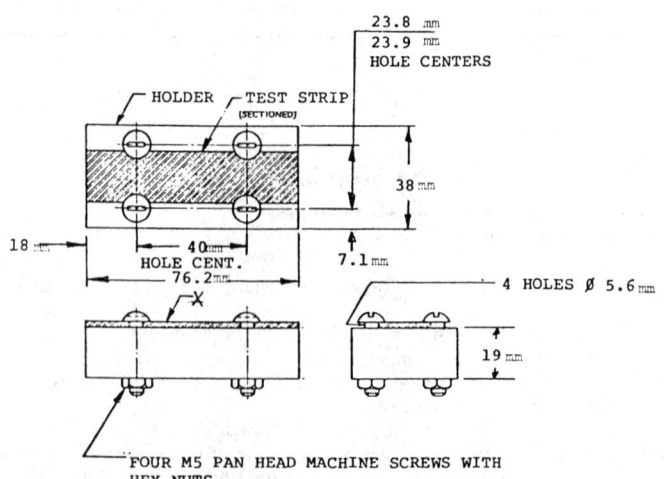

FIG. 1—TEST STRIP SPECIFICATIONS

FIG. 2—ASSEMBLED TEST STRIP AND HOLDER

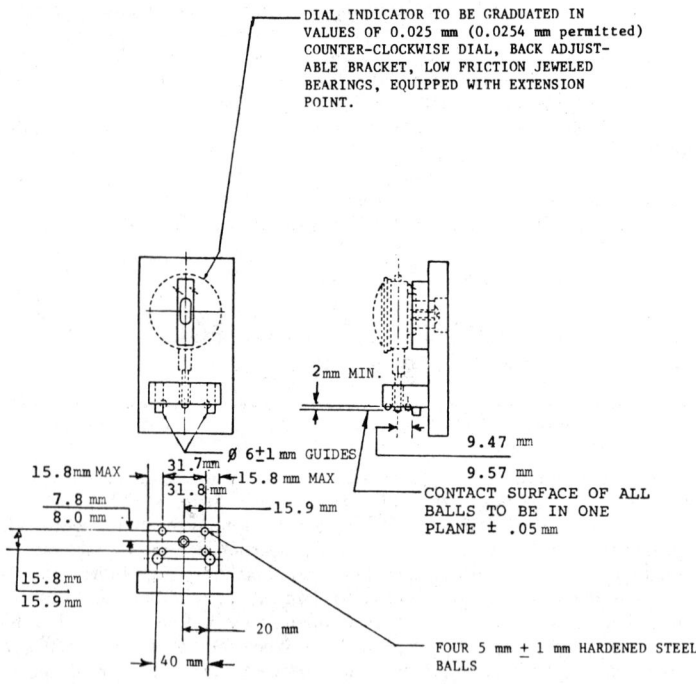

ed. FIG. 4—ALMEN GAGE

PROCEDURES FOR USING STANDARD SHOT PEENING TEST STRIP—SAE J443

SAE Recommended Practice

Report of Iron and Steel Technical Committee approved January 1952 and last revised June 1961. Reaffirmed without change May 1968.

This SAE Recommended Practice is intended to provide uniform procedures for using the standard shot peening test strips reported in the SAE Standard, Test Strip, Holder and Gage for Shot Peening—SAE J442.

In the procedures reported in this recommended practice, it is recommended that test strip A be used for intensities that produce arc heights of 0.006A to 0.024A. For lesser intensities of peening, the N strip is recommended, and for greater intensities of peening, the C strip is recommended.

Procedure Based on Arc Height Exposure Time Relationship

General—1. Fasten the test strip tightly and centrally to the test strip holder.

2. Expose the surface X (Fig. 2 of SAE Standard, Test Strip, Holder, and Gage for Shot Peening—SAE J442) of the strip to the blast to be measured. Record the time of exposure or its equivalent.

3. Remove the strip from the holder and measure the arc height on the gage. The zero position of the gage must be frequently checked and, if necessary, adjusted.

4. Using different exposure times, repeat Steps 1, 2, and 3 sufficiently to determine a curve similar to Fig. 1.

5. The gage reading corresponding with the point A where the curve flattens out is generally taken as the measurement of the intensity of that particular peening. In some cases, this point is difficult to pick out and requires some judgment.

Production Setup Procedure—Blast Measurement—The procedure to be used in making a production setup in which a setting of the machine is to be determined for a desired arc height and shot size may be described as follows:

1. Provide a fixture to support the test strip in a manner to simulate the most critical surface of the part to be peened. In cases where more than one critical surface is to be peened, the fixture should provide for the mounting of the required additional test strips.

2. With an estimated setting of the machine (shot flow rate, shot velocity, and type of shot), a series of test strips should be exposed to the blast of shot, each for a different exposure time so that a curve such as shown by Fig. 1 may be established.

3. If the intensity measurement obtained from the curve does not fall within the desired limit, machine settings must be changed. If a higher arc height is desired, either higher shot velocity or larger shot is necessary, assuming a given type of shot. If lower arc height is desired, a lower shot velocity or smaller shot is needed. These velocity changes may be made by changing wheel speed or air pressure. In certain cases, an adjustment may be made in the direction of the shot stream, but the most efficient peening is obtained with the direction of the main part of the blast stream normal to the critical section of the part being peened.

4. After new settings are made, arc heights are again determined as prescribed in Step 2.

5. Suppose with the first trial, the curve B of Fig. 2 was obtained, and the desired arc height is as indicated by the horizontal broken line. The shot velocity or shot size is accordingly too great and one or both must be reduced. Suppose the second trial resulted in the curve C. Here the shot velocity or shot size is too small. Perhaps the third trial would result in curve D, which is the correct one for the required arc height.

6. When the machine settings are found that yield the desired arc height, the time of exposure of the part is also indicated. For example, from curve D in Fig. 2, the time of exposure T, corresponding with point Q on the curve, is that which would ordinarily be used.

Alternate Procedure Based on Coverage Measurement

General—1. Fasten the test strip tightly and centrally to the test strip holder.

2. Expose the surface X (Fig. 2 of SAE Standard, Test Strip, Holder, and Gage for Shot Peening) of the test strip to the blast to be measured. Record time of exposure or its equivalent.

3. Remove the strip from the holder and measure the arc height on the gage. The zero position of the gage must be frequently checked and, if necessary, adjusted.

Coverage—The degree of coverage can be determined as follows:

1. Polish each strip to obtain a reflecting surface by means of metallurgical polishing cloths or their equivalent.

2. Fasten to the test strip holder.

3. Expose the polished surface to the blast under conditions identical to that used in determining the arc height or Almen gage reading.

4. Remove the strip from the holder and place it in the field of a metallurgical camera.

5. Using a piece of transparent paper as a ground glass, and with a magnification of approximately 50 diameters, trace the indented areas with a sharp pencil. The indented areas can be identified by the contrast of the polished surface and the inclined surfaces of the indentations.

6. Measure with a planimeter the area of all of the indentations enclosed by a circle of known diameter. The ratio of the indented areas to the total area is the percentage coverage.

Relationship of Coverage to Exposure Time—There is a definite and quantitative relationship between coverage and exposure time. This relationship may be expressed as follows:

$$C_n = 1 - (1 - C_1)^n$$

where C_1 = % coverage (decimal) after 1 cycle
C_n = % coverage (decimal) after n cycles
n = number of cycles

As this expression indicates, coverage approaches 100% as a limit. It is difficult to obtain accurate measurements of coverage above 98%, but a measurement at a lower degree of coverage will serve as a means of determining the exposure time or equivalent required to obtain any desired coverage. Since coverage approaches 100% as a limit, and since actual measurement can be made up to and including 98%, 98% is arbitrarily chosen to represent full coverage. Beyond this value, the coverage is expressed as a multiple of the exposure time required to produce 98%. For example, 1.5 coverage represents

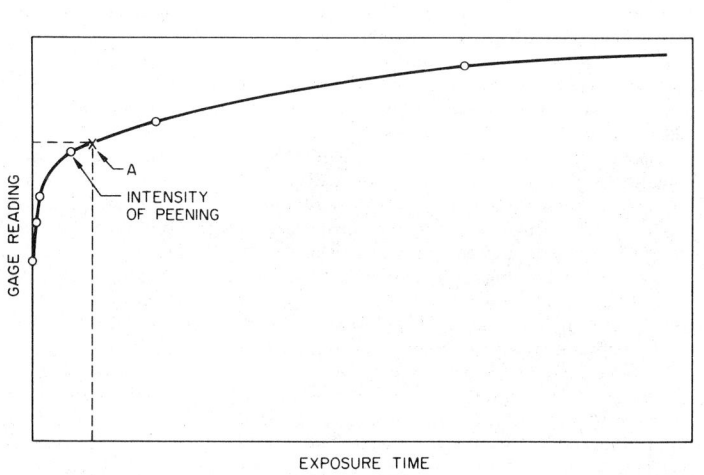

FIG. 1—INTENSITY DETERMINATION CURVE

FIG. 2—INTENSITY DETERMINATION CURVES B, C, AND D

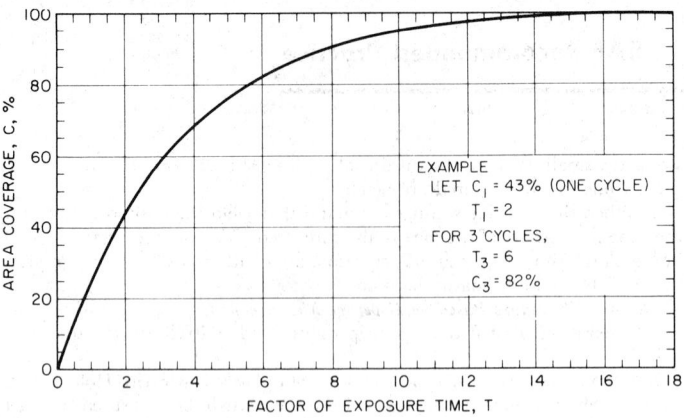

FIG. 3—RELATIONSHIP OF COVERAGE TO EXPOSURE TIME

a condition in which the specimen has been exposed to the blast 1.5 times the exposure required to obtain 98% coverage. A chart plotted to a convenient exposure time scale is shown in Fig. 3.

Production Setup Procedure—Blast Measurement—The procedure to be used in making a production setup in which a setting of the machine is to be determined for a desired arc height and shot size may be described as follows:

1. Provide a fixture to support the test strip in a manner to simulate the most critical surface of the part to be peened. In cases where more than one critical surface is to be peened, the fixture should provide for the mounting of the required additional test strips.

2. With an estimated setting of the machine (shot flow rate, shot velocity, and type of shot), under which a low degree of coverage is expected, a polished strip should be exposed to the blast of shot for a definite exposure time or its equivalent.

3. The strip is then removed from the block and the coverage measured.

4. From this measurement of coverage, the required exposure time is calculated to produce the desired coverage.

5. A regular test strip (not polished) is then exposed to the blast for a time indicated by the coverage calculation.

6. If, under these conditions, the blast measurement thus obtained does not fall within the desired limits, machine settings must be changed. If a higher arc height is desired, either higher shot velocity or larger shot is necessary, assuming a given type of shot. If a lower arc height is desired, a lower shot velocity or smaller shot is needed. These velocity changes may be made by changing wheel speed or air pressure. In certain cases, an adjustment may be made in the direction of the shot stream but the most efficient peening is obtained with the direction of the main part of the blast stream normal to critical section of part being peened.

7. Suppose for example, the desired conditions are 0.010 A and 98% coverage. Suppose further that the coverage as measured in the first trial was 76%. Referring to the chart of Fig. 3, the exposure time used in this test is equivalent to five units. At fourteen units, 98% would be obtained. Therefore, the exposure time must be increased in the ratio of fourteen to five, or 2.8 times the exposure used in the first trial. This is the exposure time to be used in determining the arc height.

8. If the arc height does not fall within the desired limits, the above process is repeated with blast conditions changed as described in Step 6.

CAST STEEL SHOT—SAE J827

SAE Recommended Practice

Report of Iron and Steel Technical Committee approved June 1962. Reaffirmed with editorial change January 1969.

Description and Scope—Cast steel shot is the product obtained by atomizing molten steel into random sizes and quenching, with subsequent screening and heat treating, to the hardness desired. It includes all shot sizes from No. 70 to No. 1320 as described in SAE J444.

Classification—Cast steel shot shall be identified by prefix letters CS followed by the appropriate shot number in accordance with SAE J444.

EXAMPLE: CS 330 indicates a cast steel shot identified by a nominal screen aperture of 0.0331 in.

Chemical Composition—The chemical composition shall conform in general to the following:

Carbon	0.85-1.20
Manganese	0.60-1.20
Phosphorus, max	0.050
Sulfur, max	0.050
Silicon, min	0.40

Hardness—Ninety per cent of the hardness checks performed on a representative sample shall fall within the range of Rockwell C 40 to 50. The hardness may be determined by any of the various methods applicable to small sections, such as a Tukon tester with Vickers indenter, at loads determined to provide a reliable conversion to Rockwell C.

Microstructure—The microstructure of cast steel shot shall be uniformly tempered martensite with fine, well distributed carbides, if any. Carbide networks, transformation products, decarburized surfaces, inclusions, porosity, and quench cracks are undesirable.

Density—Cast steel shot shall not weigh less than 7 gm per cc nor contain more than 10% hollows. The method of determining density may be a displacement method or an actual count of the hollows in a mounted, polished specimen.

General Appearance—The cast steel shot shall be as nearly spherical as is commercially possible with a minimum of elongated or compound particles, tails, hollows, broken pieces, slag, and dirt.

Mechanical Tests—Several designs of shot testing machines are available commercially for application to routine acceptance procedures. See SAE J445, for methods of checking uniformity of shipments of shot or to determine the relative fatigue life of different types of shot.

Inspection Procedure—Samples for chemical analysis, hardness and microstructure, and density—if checked by a displacement method—shall be at least 100 gm each as obtained from a representative sample of each shipment.

Shot particles to be checked for hardness and microstructure are to be mounted in plastic and ground and polished to the centerline with care to prevent work hardening of the polished surface. At least 10 hardness readings are to be taken at random.

A count of hollows may be taken from a representative portion or all of a polished specimen prior to etching for examining microstructure.

A simple method for determining density by displacement is as follows:

Using a 100 ml graduate, fill graduate with water to 50 ml mark. Add 100 gm of shot. Note rise in water level. Divide volume of shot as shown by water displaced into the weight of shot. This will give the apparent density of the shot.

For critical density measurements, a pycnometer method for determining true density is recommended.

CAST SHOT AND GRIT SIZE SPECIFICATIONS FOR PEENING AND CLEANING—SAE J444a

SAE Recommended Practice

Report of Production Division approved January 1946 and last revised by Mechanical Prestressing of Metals Division November 1976.

[This SAE Recommended Practice pertains to blast cleaning and shot peening and provides for standard cast shot and grit size numbers. It supersedes the previous SAE Recommended Practice, Shot for Peening. For shot, this number corresponds with the aperture size of the nominal screen. For grit, this number corresponds with the number of the nominal screen with the prefix G added. These screens are in accordance with the National Bureau of Standards series as given in ASTM E 11, Specification for Sieves for Testing Purposes.

The accompanying shot and grit classifications and size designations were formulated by representatives of shot and grit suppliers, equipment manufacturers, and automotive users who constituted the Shot Peening Division of the Iron and Steel Technical Committee.]

Shot should be round and solid. When used for peening or cleaning round, solid shot withstands breakage better than irregularly shaped or porous particles.

As used in the actual peening process, it is highly desirable that the shot be reasonably spherical and reasonably solid. It must be within the desired size limits.

Since it may be more economical for the user to mechanically *season* the shot, the degree of control of shape and quality of shot may be established by agreement between user and shot supplier.

Testing Procedure

1. (a) A rotating and tapping type of testing machine shall be used.
   (b) The shaking speed shall be 275–295 rpm.
   (c) The taps per minute shall be 145–160, when tapping machines are used.
2. The size of the sample shot shall be 100 g, to be obtained from a representative quantity.
3. Screening sieves shall be in accordance with the National Bureau of Standards series as given in ASTM E 11. They shall be of the 8 in (203 mm) diameter series, of either standard 1 in (25 mm) or 2 in (51 mm) height.
4. The time of test shall be 5 min ± 5 s for shot size up to and including U.S. Standard Screen size 35; and 10 min ± 5 s for finer screen sizes.
5. Any alternate method agreed upon between the supplier and user which gives equivalent results will be acceptable.

Grit for Cleaning—See Table 2.

Cross References:
SAE J827—Cast Steel Shot: for information on Composition and Shapes.
SAE J445—Metallic Shot and Grit Mechanical Testing: for information on Shot Quality Determination.

TABLE 1—CAST SHOT SPECIFICATIONS FOR SHOT PEENING OR BLAST CLEANING

NBS Screen No.	Standard[a] mm	Screen Size (in)	S1320	S1110	S930	S780	S660	S550	S460	S390	S330	S280	S230	S170	S110	S70
4	4.75	(0.187)	All Pass	—	—	—	—	—	—	—	—	—	—	—	—	—
5	4.00	(0.157)	—	All Pass	—	—	—	—	—	—	—	—	—	—	—	—
6	3.35	(0.132)	90% min	—	All Pass	—	—	—	—	—	—	—	—	—	—	—
7	2.80	(0.111)	97% min	90% min	—	All Pass	—	—	—	—	—	—	—	—	—	—
8	2.36	(0.0937)	—	97% min	90% min	—	All Pass	—	—	—	—	—	—	—	—	—
10	2.00	(0.0787)	—	—	97% min	85% min	—	All Pass	—	—	—	—	—	—	—	—
12	1.70	(0.0661)	—	—	—	97% min	85% min	—	All Pass	5% max	All Pass	—	—	—	—	—
14	1.40	(0.0555)	—	—	—	—	97% min	85% min	—	All Pass	5% max	All Pass	—	—	—	—
16	1.18	(0.0469)	—	—	—	—	—	97% min	85% min	—	96% min	85% min	5% max	All Pass	—	—
18	1.00	(0.0394)	—	—	—	—	—	—	96% min	85% min	—	5% max	All Pass	—	—	—
20	0.850	(0.0331)	—	—	—	—	—	—	—	96% min	85% min	—	10% max	All Pass	—	—
25	0.710	(0.0278)	—	—	—	—	—	—	—	—	96% min	85% min	—	10% max	All Pass	—
30	0.600	(0.0234)	—	—	—	—	—	—	—	—	—	96% min	85% min	—	10% max	All Pass
35	0.500	(0.0197)	—	—	—	—	—	—	—	—	—	—	97% min	—	10% max	—
40	0.425	(0.0165)	—	—	—	—	—	—	—	—	—	—	—	85% min	—	10% max
45	0.355	(0.0139)	—	—	—	—	—	—	—	—	—	—	—	97% min	—	—
50	0.300	(0.0117)	—	—	—	—	—	—	—	—	—	—	—	—	80% min	—
80	0.180	(0.007)	—	—	—	—	—	—	—	—	—	—	—	—	90% min	80% min
120	0.125	(0.0049)	—	—	—	—	—	—	—	—	—	—	—	—	—	90% min
200	0.075	(0.0029)	—	—	—	—	—	—	—	—	—	—	—	—	—	—

[a] Corresponds to ISO Recommendations.

TABLE 2—CAST GRIT SPECIFICATIONS FOR BLAST CLEANING

NBS Screen No.	Standard[a] mm	Screen Size (in)	G10	G12	G14	G16	G18	G25	G40	G50	G80	G120	G200	G325
4	4.75	(0.187)	—	—	—	—	—	—	—	—	—	—	—	—
5	4.00	(0.157)	—	—	—	—	—	—	—	—	—	—	—	—
6	3.35	(0.132)	—	—	—	—	—	—	—	—	—	—	—	—
7	2.80	(0.111)	All Pass	—	—	—	—	—	—	—	—	—	—	—
8	2.36	(0.0937)	—	All Pass	—	—	—	—	—	—	—	—	—	—
10	2.00	(0.0787)	80%	—	All Pass	—	—	—	—	—	—	—	—	—
12	1.70	(0.0661)	90%	80%	—	All Pass	—	—	—	—	—	—	—	—
14	1.40	(0.0555)	—	90%	80%	—	All Pass	—	—	—	—	—	—	—
16	1.18	(0.0469)	—	—	90%	75%	75%	All Pass	—	—	—	—	—	—
18	1.00	(0.0394)	—	—	—	85%	75%	—	All Pass	—	—	—	—	—
20	0.850	(0.0331)	—	—	—	—	—	—	—	—	—	—	—	—
25	0.710	(0.0278)	—	—	—	—	85%	70%	—	All Pass	—	—	—	—
30	0.600	(0.0234)	—	—	—	—	—	—	—	—	—	—	—	—
35	0.500	(0.0197)	—	—	—	—	—	—	—	—	—	—	—	—
40	0.425	(0.0165)	—	—	—	—	—	80%	70%	—	All Pass	—	—	—
45	0.355	(0.0139)	—	—	—	—	—	—	—	—	—	—	—	—
50	0.300	(0.0117)	—	—	—	—	—	—	80%	65%	—	All Pass	—	—
80	0.180	(0.007)	—	—	—	—	—	—	—	75%	65%	—	All Pass	—
120	0.125	(0.0049)	—	—	—	—	—	—	—	—	75%	60%	—	All Pass
200	0.075	(0.0029)	—	—	—	—	—	—	—	—	—	70%	55%	—
325	0.045	(0.0017)	—	—	—	—	—	—	—	—	—	—	65%	20%

[a] Corresponds to ISO Recommendation.

METALLIC SHOT AND GRIT MECHANICAL TESTING—SAE J445a

SAE Information Report

Report of Iron and Steel Technical Committee approved January 1957 and last revised June 1962. Reaffirmed without change January 1969.

Scope—This report is intended to provide users and producers of metallic shot and grit[1] with general information on methods of mechanically testing metal abrasives in the laboratory.

Introduction—The introduction of metallic shot for cleaning and peening operations created a demand for a laboratory device to measure the quality of incoming material. Ideally, such a testing machine would not only compare various shipments of the same type of shot, but could also be used to compare the various types of shot.

The results obtained from testing machines generally are not used to establish shot consumption or cost in production machines, because of other considerations not duplicated in the laboratory installations. However, the test machines can be used to check incoming shot for conformity with previous shipments of the same type of shot from the same manufacturer and to obtain data on comparative life of various types of shot under laboratory conditions.

Various shot testing machines have been developed and are available commercially. As might be expected, the machines differ in detail, but are alike in the fundamental principle that a sample of shot is subjected to repeated impacts on a hardened target. The percentage of breakdown is readily determined by means of a screen analysis. These data can be used to check the uniformity of shipments or to determine the relative fatigue life.

Sampling—In any such test the results obtained are no better than the sample tested. For this reason, it is extremely important that a representative sample be selected. It is recommended that about ½ lb of shot be taken from each of several bags in the shipment to be tested. These samples should be combined and quartered to select the approximate amount desired for screening in accordance with the SAE Recommended Practice, Cast Shot and Grit for Peening and Cleaning—SAE J444.

Calibration—Because results can be influenced by the condition of a test machine, it is recommended that the machine be recalibrated periodically. This may be accomplished by reserving an adequate amount of shot of known life characteristics and comparing the results obtained on occasional tests of the "standard" shot. Through the subsequent evaluation of the condition of the machine, it can be adjusted as necessary when off-standard conditions are observed.

Examples of Test Procedures

1. Average Life by Measurement of the Area Under the Breakdown Curve—If a representative sample of shot is observed as it is broken down in a testing machine, and the percent of the sample retained on a control screen is plotted on rectangular coordinate paper against the number of cycles, a breakdown curve typical of the abrasive is obtained. The area under this curve is a measure of the average number of cycles required to reduce the size of the abrasive pellets to fragments which pass through the control screen. This average number of cycles, commonly referred to as the average life of the abrasive, is a complete evaluation of the life of the abrasive under the conditions of the test.

Average life data may be used for control checking and for comparing abrasives of different types and from different sources if the control screen aperture is approximately equal to the removal size in the blast operation, and if the test machine is ADJUSTED to permit correlation with the blast equipment. The reader is cautioned against inferring that the peripheral speed of the abrasive impeller is an indication of correlation.

Procedure

(a) Place 50 to 100 gr of the shot to be tested into the testing machine.
(b) Run until about 20% passes through the control screen.
(c) Screen, and plot the percent retained on the control screen against the number of cycles, using rectangular coordinate paper.
(d) Return the sample to the machine and continue running.
(e) Repeat steps (c) and (d) at intervals dictated by the rapidity of breakdown of the sample, until the breakdown curve can be drawn.
(f) Measure the area under the curve and record the value as average life, in cycles.

EXAMPLE: A sample of S550 steel shot tested in accordance with the described procedure yielded the breakdown curve shown on the graph, Fig. 1. The area under the curve measured 2025 cycles—the average life of the sample under the conditions of the test.

2. Complete Breakdown to Stabilized Loss Method—A constant loss rate for a given shot material can be determined from breakdown results of a series of repetitive tests. This requires that the tests be repeated on a constant time or cycle interval, with the nonusable, pulverized shot removed periodically,

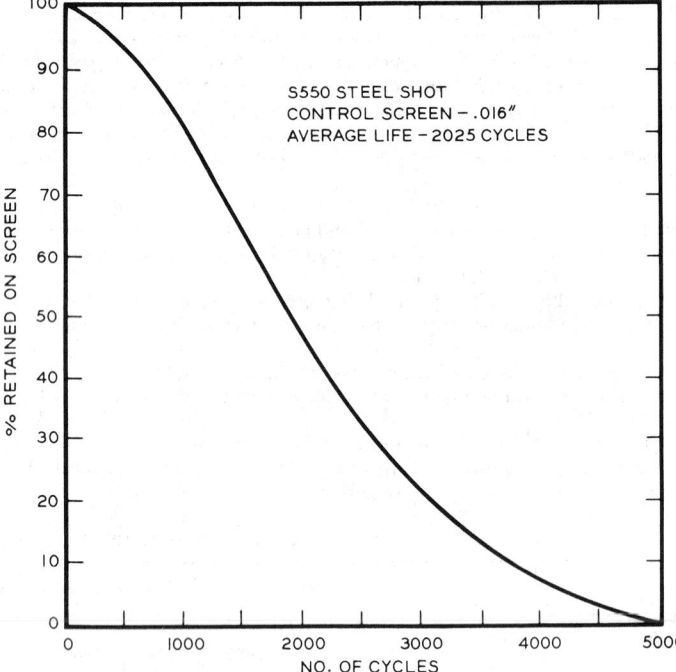

FIG. 1—SHOT LIFE TEST

weighed, and replaced with new shot. The loss rate should approach a constant value, as a given weight for a given time interval or number of cycles.

Procedure

(a) Weigh out the required sample of material.
(b) Run in test machine for a given number of cycles or given time interval.
(c) Screen and weigh the shot from the test machine, discarding the portion which passes through a specific throwout screen.
(d) Weigh an equivalent amount of new shot to restore sample to original weight.
(e) Repeat procedure until the average loss approaches a constant value.

EXAMPLE:

Charge in machine . . . . .100 gr
Number of passes 100
Loss of material below a specific screen loss per 100 passes . . . . 40 gr
Average amount in machine (based on ½ the weight loss for the last cycle period) . . . . 80 gr
% loss per 100 passes $\frac{40}{80} \times 100$ . . . . =50%
= 1/25 per pass

3. Short Method of Comparing Shipment Samples—When a pattern of results has been compiled for a particular shot material by using either of the foregoing procedures, tests of incoming shipments may be made more quickly by eliminating much of the work described. A sample may be tested for a definite time or a definite number of cycles, and screened to make a direct numerical comparison of shipment sample breakdown characteristics.

EXAMPLE: It is desired to check an SAE No. 390 shot.

(a) From a representative sample, weigh out 50 gr.
(b) Place shot in machine and run exactly 15 minutes.
(c) Remove shot from machine and run screen analysis on screen Nos. 18 (0.0394), 20 (0.0331), 25 (0.028), 30 (0.0232), and 35 (0.0197). (Analysis may be run using only one of the finer screens representative of the finest usable size for the production application of the shot—if prior experience with the material warrants this simplified procedure.)
(d) Compare results with results obtained on previous shipments of the same material. If results show the breakdown to be equal to, or less than, the established reference material, the shot is acceptable, while if it shows greater breakdown than normal, it is subject to further testing, and possible rejection.

---

[1] Shot and/or grit will hereafter be referred to as Shot.

SIZE CLASSIFICATION AND CHARACTERISTICS OF GLASS BEADS FOR PEENING—SAE J1173

SAE Recommended Practice

Report of Mechanical Prestressing Division approved January 1977.

Scope—This SAE Recommended Practice pertains to peening with glass beads, and provides for standard glass bead size numbers.[1]

Introduction—The glass bead classification number is the approximate nominal diameter of the glass spheres in that classification, in hundredths of a millimetre, with the prefix *GB* added.

Glass beads used for peening shall be made from high quality glass of the *soda-lime* type. They should be as resistant as possible to breakage from shock-impact, or by abrasion during shipment and handling. The particles should be substantially round, and, free from inherent chemical impurities or contaminants that might be detrimental to the workpiece.

Selection of Sample—A representative sample of the shipment shall be selected for evaluation. This can be accomplished by:

(a) Splitting the entire large quantity by repeated passes through the sample reducer (riffle or splitter) as described in ASTM Method 271 or

(b) Randomly selecting the cube root of the number of containers which can then be reduced in order to obtain a representative sample. Other sampling techniques may be used if agreed upon between the supplier and vendor. Representative samples of the whole should result in 50 g test quantities which may be sealed in properly labeled containers for the required tests.

Sieve Analysis for Size Classification

1. This test shall be performed on a 50 g representative sample *prior* to the performance of roundness or other tests on that sample.
2. The sieve analysis shall be performed in accordance with ASTM Method D-1241, "Sieve Analysis of Glass Spheres."
3. The screens are in accordance with the U.S. Standard Series sieves described in ASTM Specification E-11, "Wire Cloth Sieves for Testing Purposes."
4. Classification limits shall be as shown in Table 1.

Roundness Test

1. The roundness test shall be performed in accordance with ASTM Method D-1115, "Roundness of Glass Spheres."
2. Acceptable limits of roundness for peening purposes shall be as shown in Table 2.

[1] The accompanying Glass Bead Size Classification was formulated by representatives of glass bead suppliers, equipment manufacturers, and automotive users who constituted the Mechanical Prestressing Division of the SAE Fatigue Design and Evaluation Committee.

TABLE 1—GLASS BEAD STANDARD SIZE NUMBERS—FOR PEENING

U.S. Standard Screen No.	Nominal Sieve Aperature size mm	in	GB 100	GB 70	GB 50	GB 35	GB 25	GB 20	GB 18	GB 15	GB 12	GB 10	GB 8	GB 7	GB 6
a14	1.40	0.0555	all pass												
16	1.18	0.0469	5% max												
a18	1.00	0.0394		all pass											
20	0.850	0.0031	90% min	5% max											
a25	0.710	0.0278			all pass										
30	0.600	0.0234	5% max	90% min	5% max										
a35	0.500	0.0197				all pass									
40	0.425	0.0165			90% min	5% max									
a45	0.355	0.0139		5% max			all pass								
50	0.300	0.0117				90% min	5% max	all pass							
a60	0.250	0.0098				5% max		5% max	all pass						
70	0.212	0.0083					5% max	90% min	5% max	all pass					
a80	0.180	0.0070							90% min	5% max	all pass				
100	0.150	0.0059						5% max		90% min	5% max	all pass			
a120	0.125	0.0049							5% max		90% min	5% max	all pass		
140	0.106	0.0041							5% max			90% min	5% max	all pass	
a170	0.090	0.0035								5% max			90% min	5% max	all pass
200	0.075	0.0029												90% min	5% max
a230	0.063	0.0025									5% max				90% min
270	0.053	0.0021										5% max			90% min
a325	0.045	0.0017										5% max			
400	0.038	0.0015											5% max		
a−400															15% max

a Corresponds to ISO Recommendation R 565.

TABLE 2—ROUNDNESS LIMITS FOR GLASS BEADS FOR PEENING

SAE Standard Bead Size No.	Percent Round	SAE Standard Bead Size No.	Percent Round
GB 330	65% min	GB 50	80% min
GB 230	65% min	GB 40	85% min
GB 170	70% min	GB 35	90% min
GB 120	70% min	GB 30	90% min
GB 80	80% min	GB 25	90% min
GB 70	80% min	GB 20	90% min
GB 60	80% min		

Coatings

1. The beads shall not be coated with silicone or any other coating.
2. Method of testing for silicone coating shall be as follows:

Slowly pour 50 g of the sample beads into a 250 mL beaker containing 200 mL of distilled water. A small number of beads floating on the water is acceptable, but no coagulation (which is an indication of silicone coating) is permitted.

Chemical Composition

1. The method of analysis for silica shall be in accordance with ASTM Method C-169, "Chemical Analysis of Soda-Lime Glass, (for silicon dioxide)."
2. Silica content shall not be less than 67% in order to provide the highest chemical stability.

Specific Gravity

1. The density of the glass particles is determined by a specific gravity measurement as follows:

   (a) Dry a quantity of the beads by placing them in an open dish in an oven at 105–110°C until a constant weight is achieved.

   (b) Place a 60 g sample of the beads in a 100 mL graduated cylinder containing 50 mL of distilled water.

   (c) The total volume -50 represents the volume of the glass particles.

   (d) Specific gravity is computed as follows:

$$\text{Sp Gr} = \frac{\text{Weight of original sample of dried beads (g)}}{\text{Final total volume (mL)} - \text{Original volume of water (mL)}}$$

2. Specific gravity shall be not less than 2.3.

Hardness

1. Hardness shall be taken:

   (a) By Moh's Scratch Hardness or

   (b) By Knoop penetrator using 100 g load or

   (c) By diamond pyramid penetrator using 50 g load.

2. Hardness shall be as follows:

   #5 to #6 Moh's Scale
   525 to 575 KHN
   500 to 550 KPH

Free Iron Content

1. Magnetic particles shall not exceed 0.1% of the original sample, by weight.
2. Iron particle content is determined by slowly sprinkling 500 g of the sample bead material on an inclined tray made at 0.62 in (1.6 mm) aluminum 6 in (152 mm) wide by 12 in (305 mm) long. The tray is supported by a non-magnetic frame so that it is inclined with a 9 in (229 mm) rise in 12 in (305 mm) horizontal distance. Four 1 in x 1 in x 6 in (25 x 25 x 152 mm) bar magnets are positioned against the under surface and crosswise of the inclined tray about the middle of its length.
3. The magnetic particles (Iron) that accumulate on the tray as the beads roll down are carefully brushed into a tared dish. The procedure is repeated until all magnetic particles are collected.
4. The tared dish is then reweighed and the magnetic content is calculated as percent of the total original sample.

Air Inclusions—Not more than 10% of the beads shall show air inclusions of more than 25% of their area. Test microscopically while immersed in 1.5 Refractive Index Fluid.

10 Ferrous and Nonferrous—General

USE OF THE TERMS YIELD POINT AND YIELD STRENGTH—SAE J450

SAE Recommended Practice

Report developed by Publication Policy Committee approved by Nonferrous Metals Committee and Iron and Steel Technical Committee June 1960. Reaffirmed without change by Nonferrous Metals Committee March 1970.

DEFINITIONS AND APPLICATION

Yield Strength is defined as the stress at which a material exhibits a specified limiting deviation from the proportionality of stress to strain. The deviation is expressed in terms of strain. The term is applicable to materials whose stress-strain diagram in the region of yield is a smooth curve as well as to those whose diagram at a similar location is sharp kneed; that is, it is applicable to the tensile testing of all metallic materials.

Yield Point is defined as the first stress in a material, less than the maximum attainable stress, at which an increase in strain occurs without an increase in stress. This behavior is known as yielding. The stress-strain diagram is characterized by a sharp knee which determines the yield point. Strictly construed, the term is applicable only to materials which exhibit these characteristics.

Since in their commercial form only ferrous metals exhibit the phenomenon of yielding and then only under some circumstances, it follows that the term yield point has only limited application to the results of tensile testing of ferrous materials and is not applicable to the testing of nonferrous metals.

RECOMMENDED USAGE

Nonferrous Metals—Only the term yield strength is applicable. Specifications and test reports must always state the method of test, that is offset or extension under load and the percent offset or the total strain under load, whichever is applicable.

Ferrous Metals—Yield strength is the general term. It is applicable to stress-strain curves of both the smooth, rounded type, and the sharp kneed type. Specifications and test reports must state the method of test and the limiting values of strain, the same as for nonferrous metals.

Strictly interpreted, the term "yield point" is intended for application only in those cases in which the material exhibits the unique characteristic of yielding as defined previously under yield point. However, an important exception exists in the case of some materials with smooth stress-strain curves. When the specification of such a material prescribes a yield point, a value equivalent to the yield point in practical significance may be determined by the use of dividers or by the extension under load method. When a test value so determined is 80,000 psi or less, it may be recorded as yield point but when the test value is more than 80,000 psi, use of the term yield strength is mandatory.

For a more detailed discussion of the terms involved and a description of the applicable methods of test, refer to the following:

ASTM E 6 —Definitions Relating to Methods of Testing
ASTM E 8 —Tension Testing of Metallic Materials
ASTM A 370—Mechanical Testing of Steel Products

SURFACE TEXTURE MEASUREMENT OF COLD ROLLED SHEET STEEL—SAE J911

SAE Recommended Practice

Report of Iron and Steel Technical Committee approved January 1965. Reaffirmed without change October 1969.

1. Scope

1.1 This SAE Recommended Practice describes a method for measuring the surface texture of cold rolled, matte finish, sheet steel product with a roughness height rating of 20–80 μ in.

2. Samples

2.1 Samples shall be selected to be representative of the surface of the material as produced. Surfaces near the rolled edges should be avoided. Samples for sheet examination shall be 6 in. long by 4 in. wide, with the longer dimension parallel to the direction of rolling.

3. Tracing Procedure

3.1 Measurements shall be made with the pickup traverse parallel to the direction of rolling (longitudinal) and at 90 deg to the direction of rolling (transverse), and shall be identified accordingly.

3.2 The numerical rating for each direction of traverse shall be the average of the readings from three separate parallel traces spaced not less than 1 in. apart.

4. Surface Roughness Measurement—Microinch (Vertical Component)

4.1 Instrumentation shall comply with specifications for stylus type instruments as detailed in Section 4 of American Standard-Surface Texture-ASA B46.1-1962 and SAE J449.

4.2 The instrumentation shall be prepared for use and connected in accordance with manufacturer's instructions.

4.3 The location of the instrumentation shall be quiet and vibration free. Suitability of location can be checked by observing the meter of the instrument with the pickup stationary, but in contact with the test sample. A meter reading of 2 (or less) μ in. indicates a suitable location.

4.4 The instrument shall be adjusted to have a roughness width cutoff of 0.030 in.

4.5 The instrument shall be calibrated against the 125 μ in. precision reference standard prior to each set of measurements. Calibration shall be carried out in accordance with manufacturer's instructions. Calibration shall be in terms of an arithmetic average.

4.6 The motor drive shall be adjusted so as to position the one-half-mil-radius diamond stylus perpendicular to the surface of the test sample.

4.7 The motor drive shall be adjusted in accordance with manufacturer's instructions to provide a sampling length of 1 in. It is important that the 1 in. length be held as nearly exact as possible.

4.8 All surface roughness measurements shall be expressed in arithmetic average microinch values as observed on the meter of the indicating instrument. So that uniform interpretation may be made of the readings, it should be understood that the reading which is considered significant is the mean reading around which the needle tends to dwell or fluctuate under small amplitude. This reading is to be determined after consideration of the entire 1 in. trace.

5. Automatic Peaks Per Inch Count

5.1 The procedure for automatic peaks per inch counting shall be as stated in Sections 1, 2, 3, and 4, except as amended below.

5.2 For purposes of this section the term "peak" shall be defined as a vertical irregularity having a peak to valley magnitude of 50.0 μ in. or greater.

5.3 For purposes of this procedure, the term "peaks per inch" shall be defined as the total number of peaks occurring over 1 in. of sampling length.

5.4 Instrumentation shall be as defined in paragraph 4.1 and capable of counting the peaks defined in paragraph 5.2.

5.5 Instrumentation shall be prepared for use connected in accordance with manufacturer's instructions.

5.6 The motor drive shall be adjusted in accordance with manufacturer's instructions to provide a peak count sampling length of 1 in. It is important that the 1 in. length be held as nearly exact as possible.

5.7 The instrumentation shall be calibrated in accordance with manufacturer's instructions prior to each set of measurements.

SURFACE TEXTURE—SAE J448a   SAE Standard

Report of Surface Finish Committee approved March 1949 and last revised June 1963. Conforms in general with American Standard for Surface Texture, ASA B46. 1-1962.

1. General Data—This SAE Standard is concerned with the geometrical irregularities of surfaces of solid materials. It establishes definite classifications for various degrees of roughness and waviness and for several varieties of lay. It also provides a set of symbols for use on drawings, and in specifications, reports, and the like. The ranges for roughness and waviness are divided into a number of steps, and the general types of lay are established by type characteristics.

This standard does not define what degrees of surface roughness and waviness or what type of lay are suitable for any specific purpose. It does not specify the means by which any degree of such irregularities may be obtained or produced. Neither is it concerned with the other surface qualities such as luster, appearance, color, corrosion resistance, wear resistance, hardness, microstructure, and absorption characteristics any of which may be governing considerations in specific applications.

Surfaces, in general, are very complex in character. Although the height, width, length, shape, and direction of surface irregularities may all be of practical importance in specific applications, this standard deals only with their height, width, and direction.

2. Precision Reference Specimens—Surface roughness designation by this standard is based on instrument readings of surfaces to be rated in comparison with those of precision reference specimens having known roughness values and having a wide distribution of replicas. Surfaces described in the specifications for these specimens are designed primarily to serve for calibration of instruments used for measuring surface roughness height. They are not intended to have the appearance or characteristics of commonly produced surfaces, nor are they intended for use in visual or tactual comparisons.

Specifications are given for surface contour, material, accuracy, uniformity, and rating that will be satisfactory for the purpose.

2.1 **Surface Contour**—The normal surface profile of precision reference specimens of roughness height shall consist of a series of isosceles triangles having included angles of 150 deg. Such a profile is shown in Fig. 1.

A departure from this triangular profile is permitted at the bottom of the grooves, provided that the deviated portion does not exceed 0.000130 in. in width and that there shall be no solid material at any point beyond a line corresponding to a flat of this width. This departure shall not affect the portion above this flat, which portion shall meet the allowed tolerance for accuracy.

2.2 **Material**—The material from which precision reference specimens are made shall be such that repeated measurements on these specimens can be made without significant loss of accuracy.

2.3 **Accuracy**—Average roughness values of precision reference specimens shall not vary from the designated value by more than ± 1 Mu in. or $\pm 3\%$, whichever is the larger. The average spacing of the grooves of precision reference specimens shall be within 2% or 20 Mu in. (whichever is the smaller) of the theoretical spacing corresponding to the nominal roughness height.

2.4 **Uniformity**—The average deviation of roughness height of individual grooves of any precision reference specimen shall not exceed 4% of the total roughness height. The average deviation of the groove spacings on a given precision reference specimen shall not exceed 3% of the average spacing.

2.5 **Rating**—Precision reference specimens shall be rated for roughness height and roughness width as provided in the section on Recommended Values of Roughness and Waviness of this Standard. With tracer type instruments having a finite tracer tip radius, it is impossible to bottom the ideally sharp grooves as described for the ideal triangular profile. Accordingly, the proper reading of a tracer type instrument on the precision reference specimens will depend on the tracer tip radius. Ratings of the specimens for checking the calibration of such instruments shall be supplied with the specimens.[1]

3. Definitions—(See Fig. 3.)

3.1 **Surface Texture**—Repetitive or random deviations from the nominal surface which form the pattern of the surface. Surface texture includes roughness, waviness, lay, and flaws.

3.2 **Surface**—The surface of an object is the boundary which separates that object from another object, substance, or space. Surfaces with which this standard is concerned shall be those requiring control of roughness or other surface characteristics.

3.2.1 NOMINAL SURFACE—Nominal surface is the intended surface contour, the shape and extent of which is usually shown and dimensioned on a drawing or descriptive specification.

3.2.2 MEASURED SURFACE—The measured surface is a representation of the surface obtained by instrumentation or other means.

3.3 **Profile**—The profile is the contour of a surface in a plane perpendicular to the surface, unless some other angle is specified.

3.3.1 NOMINAL PROFILE—The nominal profile is the profile disregarding surface texture.

3.3.2 MEASURED PROFILE—The measured profile is a representation of the profile obtained by instrumental or other means. (See Fig. 2.)

3.4 **Centerline (Roughness)**—The centerline is the line about which roughness is measured and is a line parallel to the general direction of the profile within the limits of the roughness—width cutoff, such that the sums of the areas contained between it and those parts of the profile which lie on either side of it are equal.[2]

3.5 **Microinch**—One millionth of a linear inch (0.000001 in.). This is the unit of height for roughness. Microinches may be abbreviated as Mu in.

3.6 **Roughness**—Roughness consists of the finer irregularities in the surface texture usually including those irregularities which result from the inherent action of the production process. These are considered to include traverse feed marks and other irregularities within the limits of the roughness—width cutoff. (See Fig. 3.)

3.7 **Waviness**—Waviness is the usually widely spaced component of surface texture and is generally of wider spacing than the roughness—width cutoff. Waviness may result from such factors as machine or work reflections, vibration, chatter, heat treatment, or warping strains. Roughness may be considered as superposed on a wavy surface. Their directions are not necessarily related.

3.8 **Lay**—The direction of the predominant surface pattern, ordinarily determined by the production method used.

[1] See also Appendix C in ASA B46.1-1962.
[2] Centerline, as defined above, is also known mathematically as the median line.

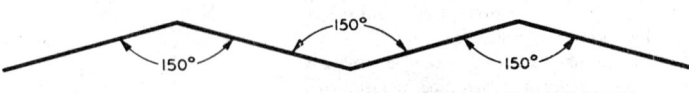

FIG. 1—SURFACE PROFILE OF PRECISION REFERENCE SPECIMEN

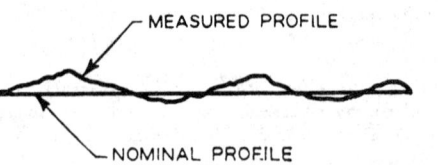

FIG. 2—MEASURED PROFILE

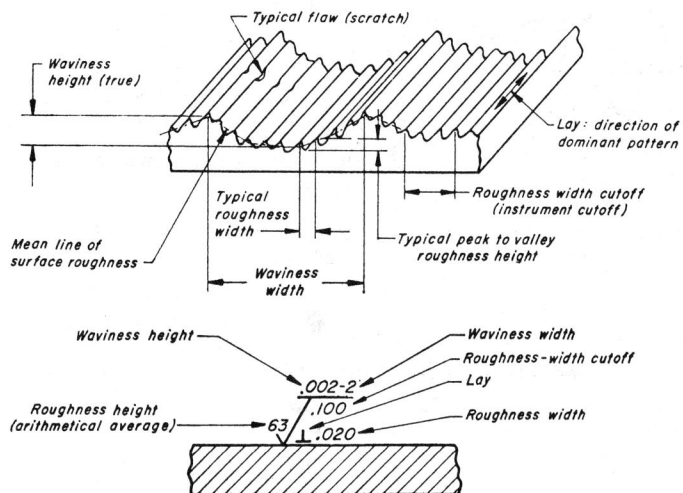

FIG. 3—MEANING OF EACH PART OF SYMBOL DEFINED

3.9 Flaws—Flaws are irregularities which occur at one place or at relatively infrequent or widely varying intervals in a surface. Flaws include such defects as cracks, blow holes, checks, ridges, and scratches. Unless otherwise specified, the effect of flaws shall not be included in the roughness height measurements.

4. Specification and Rating

4.1 Roughness Height Rating—The height of the roughness shall be specified in microinches as the arithmetical average of the absolute deviations from the mean surface. This value will be identified as a roughness number; for example, 16 means that the surface has an arithmetical average absolute deviation from the mean surface of 16 Mu in.[3]

4.2 Roughness Width Rating—The maximum permissible spacing of repetitive units of the dominant surface pattern. It may be specified in inches adjacent to the lay symbol. Irregularities having spacings up to and including the maximum specified are rated as roughness width and are to be included in the measurement of roughness height. When no maximum dimension is specified, spacings up to and including the width of the irregularities due to machine feed are rated as roughness width and are to be included in the measurement of roughness height.

4.3 Roughness—Width Cutoff—The greatest spacing of repetitive surface irregularities to be included in the measurement of average roundness height. Roughness—width cutoff is rated in inches. Roughness—width cutoff must always be greater than the roughness width in order to obtain the total roughness height rating.

Standard roughness—width cutoff values (inches) are:

0.003   0.010   0.030   0.100   0.300   1.000

When no value is specified, the value 0.030 is assumed. Refer to SAE J449, Surface Texture Control.

4.4 Waviness Height Rating—Waviness heights may be specified directly in inches as the vertical distance from peaks to valleys of waves.

4.5 Waviness Width Rating—Waviness widths may be specified directly in inches as the distance from peak to peak of the waves.

4.6 Lay Specifications—The lay of a surface shall be specified by the lay symbol indicating direction of dominant visible surface marks.

5. Measurement or Evaluation

For compliance with specified ratings, surfaces are to be evaluated by comparison with specified reference standards or by direct instrument measurements as described below.

5.1 Roughness—Roughness height values may be measured by any acceptable method, for instance, sight, feel, or instrument. For routine measurements, comparison may be made with a master surface that satisfactorily meets the requirements of the surface being measured. In making comparisons care should be exercised to avoid errors due to differences in material, contour, and type of operation represented by the reference surface and the work.

In using instruments for comparison or for direct measurement, care should be exercised to insure that the specified quality or characteristic of the surface is measured.[4]

Roughness measurements, unless otherwise specified, are taken in the direction which gives the maximum value of the reading normally across the lay.

5.2 Waviness—Waviness values for height and width may be measured by any suitable device for linear measurement.

[3] Instruments calibrated in rms (root mean square) average will read approximately 11% higher on a given surace than those calibrated for arithmetic average (aa).
[4] See ASA B46. 1-1962 for instrument specifications.

ROUGHNESS HEIGHT VALUE IS PLACED ADJACENT TO AND ON THE INSIDE OF THE LONG LEG.

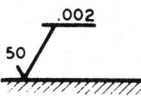
WAVINESS HEIGHT VALUE, WHEN REQUIRED, IS PLACED ABOVE THE EXTENSION LINE.

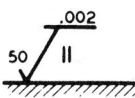
LAY DESIGNATION, WHEN REQUIRED, IS INDICATED BY THE LAY SYMBOL, PLACED UNDER THE EXTENSION TO THE RIGHT OF THE LONG LEG LINE.

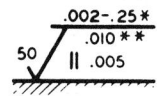
ROUGHNESS WIDTH VALUE, WHEN REQUIRED, IS PLACED TO THE RIGHT OF THE LAY SYMBOL.

* WHEN WAVINESS WIDTH VALUE IS REQUIRED, THE VALUE MAY BE PLACED TO THE RIGHT OF THE WAVINESS HEIGHT VALUE.
** ROUGHNESS WIDTH CUTOFF VALUE, WHEN REQUIRED, IS PLACED IMMEDIATELY BELOW THE RIGHT-HAND EXTENSION.

FIG. 4—SURFACE SYMBOL

6. Recommended Values of Roughness and Waviness—The use of only one number shall indicate the maximum value of either the height or the width of irregularities. Any less degree shall be satisfactory. When two numbers are used, they shall specify the maximum and minimum permissible values.

	SAE Roughness Height Values, Mu in.					
	3	8	20	50	125	320	800
1	4	10	25	63	160	400	1000
	5	13	32	80	200	500	
2	6	16	40	100	250	600	

SAE Waviness Height Values, in.					
0.00002	0.00008	0.0003	0.001	0.005	0.015
0.00003	0.0001	0.0005	0.002	0.008	0.020
0.00005	0.0002	0.0008	0.003	0.010	0.030

7. Surface Symbol—The symbol used to designate surface irregularities is the check mark and extension as shown in Fig. 4.

The point of the symbol may be on the line indicating the surface, on a witness line, or on a leader pointing to the surface. The long leg and extension shall preferably be to the right and erect, as the drawing is read. For preferred proportions see SAE Aerospace-Automotive Drawing Standard on Surface Texture—Roughness, Waviness, and Lay.

8. Symbols Indicating Direction of Lay—A lay symbol used with a surface symbol shall specify the direction of the visible pattern of the marks on the surface. (See Fig. 5.)

Typical examples would be the use of the symbols, as in Figs. 6 and 7, to express the given specifications.[5]

[5] For more complete discussion of application, see SAE Aerospace-Automotive Drawing Standard on Surface Texture—Roughness, Waviness, and Lay.

∥   PARALLEL TO THE SURFACE BOUNDARY LINE INDICATED BY THE SYMBOL.
EXAMPLE: PARALLEL SHAPING, END VIEW OF TURN AND OD GRIND.

⊥   PERPENDICULAR TO THE SURFACE BOUNDARY LINE INDICATED BY THE SYMBOL.
EXAMPLE: END VIEW OF SHAPING, LONGITUDINAL VIEW OF TURN AND OD GRIND.

X   ANGULAR IN BOTH DIRECTIONS TO THE SURFACE BOUNDARY LINE INDICATED BY THE SYMBOL.
EXAMPLE: SIDE WHEEL GRIND, HONE, AND TRAVERSED END MILL.

M   MULTIDIRECTIONAL.
EXAMPLE: LAP, SUPERFINISH.

C   APPROXIMATELY CIRCULAR RELATIVE TO THE CENTER OF THE SURFACE INDICATED BY THE SYMBOL.
EXAMPLE: FACING.

R   APPROXIMATELY RADIAL RELATIVE TO THE CENTER OF THE SURFACE INDICATED BY THE SYMBOL.

FIG. 5—SYMBOLS INDICATING DIRECTION OF LAY

10.04

FIG. 6—SPECIMEN APPLICATION

ROUGHNESS HEIGHT = 16 MU IN.

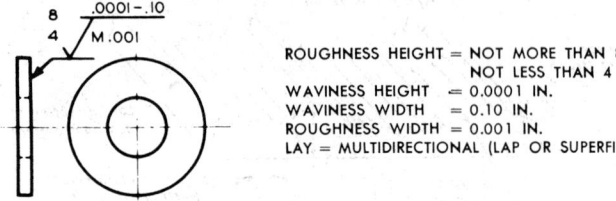

ROUGHNESS HEIGHT = NOT MORE THAN 8 MU IN.
                   NOT LESS THAN 4 MU IN.
WAVINESS HEIGHT = 0.0001 IN.
WAVINESS WIDTH   = 0.10 IN.
ROUGHNESS WIDTH = 0.001 IN.
LAY = MULTIDIRECTIONAL (LAP OR SUPERFINISH)

FIG. 7—SPECIMEN APPLICATION

SURFACE TEXTURE CONTROL—SAE J449a — SAE Recommended Practice

Report of Surface Finish Committee approved January 1953 and last revised June 1963. Conforms to Precision Reference Specimens of American Standard for Surface Texture, ASA B46. 1-1962.

1. Purpose—SAE J448, Surface Texture, has been set up for precision reference specimens using a controlled surface profile to obtain reproducible roughness values. These specimens are for instrument calibration. Appropriate symbols for roughness, waviness, and lay have also been standardized (ASA B46.1-1962 and SAE J448).

For production control, especially from one geographical location to another, means are required to facilitate the inspection of surface characteristics called for by specifications which include not only roughness but profile waviness and lay. In order to integrate the requirements of the designer with the actual production of surfaces, a second grade of control standards must be adopted which will be functional in nature for the specific product being manufactured. These control standards may be Calibrated Pilot Specimens (actual parts with satisfactory texture) or Roughness Comparison Specimens (ASA B46.1-1962). This SAE Recommended Practice describes the usage of these control standards.

2. Roughness Comparison Specimens—In order to comply with a specific type of lay with required roughness and waviness values, a number of roughness machined comparison specimens are available commercially. These specimens are intended to have the appearance and feel of typical machine surfaces and are made to cover both a range of roughness and a variety of methods of surface preparation.

Roughness comparison specimens are well adapted for use by designers and draftsmen to relate numerical specifications of surface roughness and lay to general experience of appearance and texture of machined surfaces. They may also be used for visual and tactual comparison with production surfaces. Care should be taken when comparing specimens that the effect of shape, curvature, material, lay, and spectral characteristics do not produce misleading results. Samples of specific parts are usually the best control specimens.

3. Designation of Surface Texture—Surface texture should be specified for production parts only on those surfaces which must be under functional control. For all other surfaces the finish resulting from the machining method required to obtain dimensional accuracy is generally satisfactory, unless the appearance of the surface is of prime concern.

The recommended degree of functional roughness, direction of lay, and waviness for any specific surface cannot be accurately foretold because of many factors influencing optimum performance in any one application. The choice of surface characteristics will be determined by such factors as loading, speed and direction of movement, physical characteristics of both materials in contact, type and amount of lubrication, contaminants, and temperature.

The primary reasons for designation of surface finish control are to improve performance, increase service life, or reduce cost. The required data for this control comes from past experience of field service or experimental results. All significant variables should be considered when establishing test methods and analyzing results.

Under conditions of complete lubrication, it would appear axiomatic that the finer the surface roughness with complete lubrication, the more efficient will be the performance. Most new moving parts do not attain a condition of complete lubrication due to imperfect geometry and running clearances, and they must therefore wear-in by actual removal of metal. In certain instances, experience may indicate that a specific degree of roughness or a specific degree of lay is necessary to accommodate this wear-in process which lessens the chance of galling, seizure, or excessive wear.

The surface chosen for a specific application will be determined by its required function and a compromise made between sufficient roughness to allow proper wearing-in and resulting smoothness for expected service life. In certain cases, a roughness number in itself may not adequately define the character of the surface found by experience to give the best results. Special reference samples may be made to give manufacturing, inspection, and engineering samples for comparison with the manufactured parts. In general, with lower dimensional tolerances and better manufacturing practices, with adequate lubrication and compatible surfaces, finer finishes would be expected to give optimum results. Frequently, cases where surfaces are not compatible, such as hardened parts running together, a certain degree of roughness or character of surface may assist lubrication in obtaining satisfactory wear-in. However, where hardened parts run against soft materials, the hardened parts must have a fine finish to avoid distress on the soft parts.

Typical normal ranges of surface roughness applications on functional parts are shown in Fig. 1. Specific applications may require finer or coarser roughness values than those indicated, especially for gears and bearings.

The designation of surface texture requirements on drawings should conform with SAE J448, and with the Surface Texture section of the SAE Aerospace-Automotive Drawing Standards. The designer should be sure that those surfaces calling for control are of sufficient importance to warrant the expenditure of time and money necessary for this control. Profuse and loose usage of controlled finishes, where not essential, detracts from the emphasis that should be given to important surfaces. Where properly used, designation and control of a surface in accordance with SAE J448, can eliminate much confusion and many rejects.

4. Production of Required Surface Texture—Unless service or experimental results have indicated that only one process method will give completely satisfactory performance, the method of machining to obtain the desired finish should be left to the discretion of the processing shop supervisory staff. They will have more intimate knowledge of the desirable machines to produce economical parts under required schedules. It is important, therefore, that production engineers and master mechanics become thoroughly familiar with surface texture as defined and rated by SAE J448.

Fig. 2 is a reproduction from the section on Surface Texture in the SAE Aerospace-Automotive Drawing Standards, and it shows typical surface roughness values obtained by various production methods. This chart indicates surface roughness values up to 2000 Mu in., although in automotive practice controlled surfaces rarely exceed 100 Mu in. roughness. Fine surface finishes may require more operations and greater care in production; but if quantities

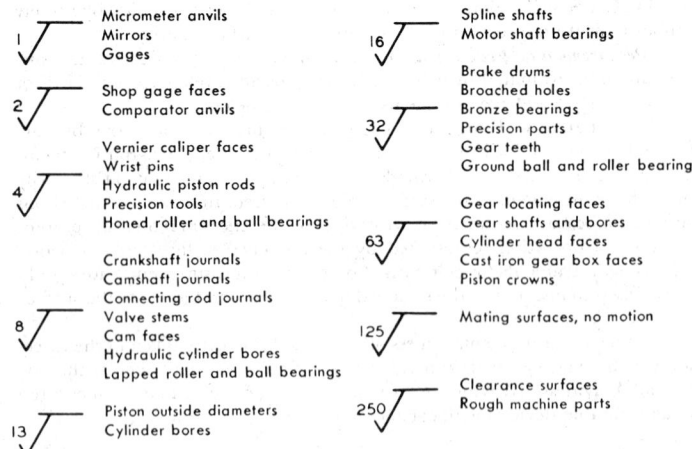

FIG. 1—TYPICAL VALUES OF SURFACE ROUGHNESS FINISHES

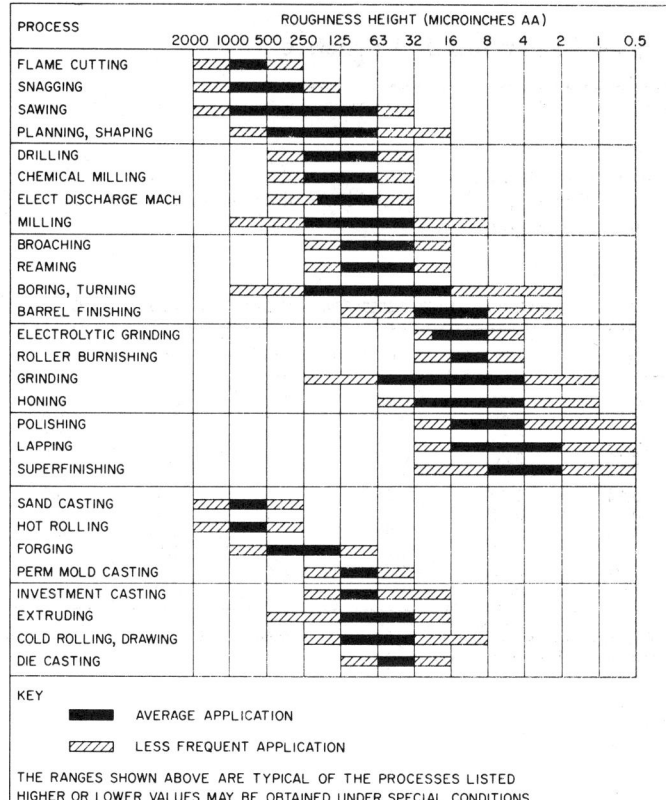

FIG. 2—SURFACE ROUGHNESS AVAILABLE BY COMMON PRODUCTION METHODS

are large, special tooling with honing, lapping, or high speed grinding can produce better finishes more economically than with older methods of production.

5. Inspection and Control—A specific surface roughness depends on reproducible production techniques. The surface in question may be inspected by use of instruments or by visual or tactual comparison. Instruments should be calibrated by use of precision reference specimens. Approved pilot specimens or replicas may be used for comparison with the surface in question. Instruments may be used as a final check of the pilot specimen with the production machined surface. Roughness comparison specimens may be used for visual or tactual control but are not recommended for instrument comparison with manufactured part.

Production machined surfaces are composed of irregular peaks and valleys having a variety of grooves, angles, and variable roughness widths. The precision reference specimens have uniform roughness heights, groove angles, widths. Instrument readings will, therefore, reflect the effect of the irregular production surface character so that the readings will be relative rather than absolute.

The geometry of the precision reference specimens being uniform, instruments may be checked for worn or chipped stylus points by a comparison of corrected readings for two widely separated reference surfaces. The correction factors and methods of checking instrument accuracy are available in the literature accompanying the reference specimens. These factors are required because of the specific stylus tip radius used on the instrument. It is impossible to contact the bottom of the grooves, so this factor is comparatively small for coarse surfaces but large for fine ones.

Readings of stylus tip instruments are also affected by the roughness width because of low and high frequency response limits beyond which the instrument will not give reliable readings. In some instances the roughness width or wave length will be specified on the drawing, therefore requiring that the instrument have a definite width cutoff value. If no width is specified, it is necessary that the frequency response of the instrument does not limit its sensitivity to any significant roughness of the surface to be inspected; that is, that the significant roughness width is not greater or smaller than the instrument is capable of measuring and that the frequency response is correct for the roughness width range being measured. Technical data on roughness width cutoff of instruments is available from the various instrument manufacturers.

When continuously averaging stylus type instruments are used, the length of trace (sampling length) used should be not less than 20 times the roughness width cutoff value. For instruments having meters which indicate integrated roughness over a fixed length of trace, the sampling length shall preferably be at least five times the roughness width cutoff value.

Where the continuously-averaging type instrument is used, it is not necessary for the traversing length to be traversed continuously in one direction provided that the time required to reverse the direction of trace is short compared to the time the tracer is in motion. For this type of operation, the minimum length of travel shall be not less than five times the roughness-width cutoff. Where surfaces are not large enough to permit the recommended minimum traversing length, the readings may not be the actual roughness of a surface but may be useful for comparative purposes.

For proper surface control of production parts, the process should be completely specified and should include depth of cut, cutting speed, feed, grit size, lubricant, and so forth. Selection of process methods should be based on surface inspection of production specimens. Production may be controlled at the machine by visual, tactual, or instrument comparison of production pieces with sample specimens. If control is required at more than one station, sample specimens may be cut into the required number of pieces; or if a large quantity is required, electroformed or plastic reference specimens may be satisfactorily employed in many instances.

Final inspection of the production pieces may be by visual, tactual, or instrument comparison with the sample specimen or by instrument comparison with the precision reference specimen depending on the roughness variation allowed and on past experience with surfaces for a similar function. For disputed surfaces, instruments calibrated with the precision reference specimens should be used. A 100% inspection for all parts is necessary only for highly critical surfaces where failure to meet the surface requirements might result in costly delays. Normal sampling inspection should prove adequate for most production parts.

The thoroughness of surface roughness inspection should depend on the judgment of the inspector. He should take into account the roughness value tolerances allowed and the physical proportions of the surface.

Instrument readings are subject to the skill of interpretation of the inspector. Readings of stylus tip instruments fluctuate because of the roughness irregularities of a machined surface. All meters are damped to minimize acute fluctuations; nevertheless, extremely high and low momentary readings do occur. The reading which should be recorded as representing the roughness value of the surface should be a mean reading around which the needle tends to dwell or fluctuate under a small amplitude. Occasional extreme fluctuations represent flaws or defects rather than average surface conditions and should not be used in determining average roughness. If in the opinion of the inspector, the extreme fluctuations are too frequent, indicating excessive lack of uniformity in the surface, the manufacturing cause, such as loading of cutting edges, overheating, too rapid feeds, should be investigated.

6. Steps for Control of Surface Texture—The following outline summarizes briefly the steps necessary for proper control of surface texture. Some of these steps may be eliminated as experience is gained in the requirements for specific surface applications.

6.1 Establish need for control by:
    (a) Field experience.
    (b) Improvement of design.
    (c) Past experience with similar designs.
6.2 Experimental steps:
    (a) Designate tentative degrees of texture.
    (b) Trial manufacture.
    (c) Performance tests.
    (d) Choose most desirable texture.
6.3 Designate texture on drawing.
6.4 Processing steps:
    (a) Specify processing methods.
    (b) Prepare pilot specimen.
    (c) Obtain engineering approval.
    (d) Produce in quantity.
6.5 Inspection procedures at machines or final inspection, or both (Usually only one procedure is followed for a given surface):
    (a) Instrument calibration and comparison with:
        1. Precision reference specimens.
    (b) Instrument comparison with:
        1. Calibrated pilot specimens.
    (c) Visual and tactual comparison with one of the following:
        1. Calibrated pilot specimens.
        2. Replicas of calibrated pilot specimens.
        3. Roughness comparison specimens.
6.6 Correct processing when irregular texture is detected.
6.7 Revise texture requirements if field service records indicate need.

PREFERRED THICKNESSES FOR UNCOATED FLAT METALS (THRU 12 MM)—SAE J446a

SAE Recommended Practice

Report of Iron and Steel Technical Committee approved January 1954 and completely revised May 1978. Conforms to ANSI B32.3-1977, American National Standard Preferred Metric Sizes for Flat Metal Products.

1. Scope—The preferred thicknesses in Table 1 provide an orderly metric (SI) series for designating the thickness of uncoated flat metal products of rectangular cross section. The thicknesses shown are also applicable to base metals which may be coated in later operations.

2. General—Requirements of industry permit leeway in the choice of thickness in some instances, but it is recognized that for many applications, particularly the tonnage requirements of the mass production industries, thicknesses are normally determined by critical engineering design or manufacturing considerations.

2.1 However, for general applications or where requirements permit some latitude in the selection of thickness, the preferred thicknesses given in Table 1 will facilitate interchangeability of different metals in design, reduce inventory, and increase the availability in warehouse stocks of thicknesses commonly required for general applications.

2.2 All of the sizes listed are not necessarily produced in all metals and grades. Producers or distributors must be consulted to determine availability of a particular thickness for a given metal product.

3. Basis of Tables—Most of the thicknesses in Table 1 were derived from preferred numbers listed in ANSI Z17.1. Some deviations from this listing occur as the result of rounding and as the result of demonstrated or anticipated need for sizes other than those which occur in the above document.

NOTE: This recommended practice is under the jurisdiction of SAE Iron and Steel Technical Committee, and is the direct responsibility of Division 32 on Sheet and Strip Steel.

TABLE 1—PREFERRED THICKNESSES FOR UNCOATED METAL AND ALLOYS[a]

mm	mm	mm	mm	mm	mm
0.05	0.28	0.80	1.6	3.0	5.5
0.06	0.30	0.85	1.7	3.2	6.0
0.08	0.35	0.90	1.8	3.4	6.5
0.10	0.40	0.95	1.9	3.5	7.0
0.12	0.45	1.00	2.0	3.6	7.5
0.14	0.50	1.05	2.1	3.8	8.0
0.16	0.55	1.1	2.2	4.0	9.0
0.18	0.60	1.2	2.4	4.2	10
0.20	0.65	1.3	2.5	4.5	11
0.22	0.70	1.4	2.6	4.8	12
0.25	0.75	1.5	2.8	5.0	

[a] Bold type sizes shown are first preference sizes from ANSI B32.3-1977. All others are second and third preference sizes.

WELDING, BRAZING, AND SOLDERING—MATERIALS AND PRACTICES—SAE J1147

SAE Information Report

Report of AWS/SAE Joint Committee on Automotive Welding approved July 1976. Editorial change April 1978.

1. The Joint AWS/SAE Committee on Automotive Welding was organized on January 16, 1974, for the primary purpose of facilitating the development and publication of various documents related to the selection, specification, testing, and use of welding materials and practices, particularly for the automotive and related industries. A secondary purpose is the dissemination of technical information.

1.1 SAE participation in this activity is intended to insure that the needs and thinking of the industries mentioned are adequately considered prior to publication of selected Standards, Recommended Practices, and Information Reports. To this end, such documents are subject to the approval of the SAE General Materials Council, as well as the Joint AWS/SAE Committee and the AWS Board of Directors.

1.2 Documents which are approved by the Council are printed as separate sheets or booklets, with both AWS D number and SAE HS J number identifications. They are not published in the SAE Handbook. SAE numbered items approved to date of this Information Report are:

(A) SAE HS J1156 (AWS D8.6): Automotive Resistance Spot Welding Electrodes. This document recommends a "standard" for spot welding electrodes suitable for use by smaller suppliers to the automotive industry. Typical electrode cap "standards" of the major automobile manufacturers are included for information purposes.

(B) SAE HS J1188 (AWS D8.7—78): Specifications for Automotive Weld Quality—Resistance Spot Welding.

(C) SAE HS J1196 (AWS D8.8—78): Specifications for Automotive Frame Weld Quality—Arc Welding.

2. The SAE Metal Joining Subcommittee (disbanded prior to organization of the Joint AWS/SAE Committee on Automotive Welding) developed the following document. This, also, is published only as a handbook supplement.

(A) SAE HS J836: Automotive Metallurgical Joining. This document is an abbreviated summary of metallurgical joining, brazing, and soldering, intended to reflect usage in the automotive industry.

3. Copies of SAE and AWS documents are available from the following addresses:

Society of Automotive Engineers, Inc.
Dept 325
400 Commonwealth Drive
Warrendale, PA 15096

American Welding Society
2501 NW 7th St.
Miami, FL 33125

NONFERROUS METALS

NONFERROUS METALS

11 Nonferrous Metals

Aluminum Alloys—Fundamentals—SAE J451b	11.01
Alloy and Temper Designation Systems for Aluminum—SAE J993b	11.02
General Information on SAE Aluminum Casting Alloys—SAE J452c	11.05
Chemical Compositions and Mechanical and Physical Properties of SAE Aluminum Casting Alloys—SAE J453c	11.08
General Data on Wrought Aluminum Alloys—SAE J454k	11.11
Chemical Compositions, Mechanical Property Limits, and Dimensional Tolerances of SAE Wrought Aluminum Alloys—SAE J457j	11.27
Anodized Aluminum Automotive Parts—SAE J399a	11.28
Bearing and Bushing Alloys (Standard)—SAE J460e	11.29
Bearing and Bushing Alloys (Information Report)—SAE J459c	11.29
Wrought and Cast Copper Alloys—SAE J461d	11.31
Cast Copper Alloys—SAE J462b	11.60
Wrought Copper and Copper Alloys—SAE J463d	11.63
Magnesium Alloys—SAE J464c	11.102
Magnesium Casting Alloys—SAE J465c	11.102
Magnesium Wrought Alloys—SAE J466c	11.107
Special Purpose Alloys ("Superalloys")—SAE J467b	11.111
Zinc Alloy Ingot and Die Casting Compositions—SAE J468b	11.122
Zinc Die Casting Alloys—SAE J469a	11.123
Wrought Nickel and Nickel Related Alloys—SAE J470c	11.123
Electroplating of Nickel and Chromium on Metal Parts—Automotive Ornamentation and Hardware—SAE J207	11.133
Solders—SAE J473a	11.138
Electroplating and Related Finishes—SAE J474b	11.139

* New
† Technical revision
§ Editorial change

11 Nonferrous Metals

ALUMINUM ALLOYS—FUNDAMENTALS—SAE J451b SAE Information Report

Report of Nonferrous Metals Division approved January 1934 and last revised by Nonferrous Metals Committee June 1971. Editorial change March 1976.

Purpose—This information report is intended to give general data on the properties of aluminum and information on working, joining, forming, machining, finishing, and heat treating of aluminum.

Properties—Commercially pure aluminum is a face-centered cubic metal with a specific gravity of about 2.71 (0.098 lb/in.^{3}), a thermal conductivity of about 0.52 cgs units (at 25 C), and a melting point of approximately 1215 F. Its coefficient of thermal expansion (approximately 0.000013 per degree F) is about twice that of steel or cast iron and about one-third greater than that of copper or brass. The electrical conductivity of pure aluminum is about 62% of the International Annealed Copper Standard. In the form of cast test bars, the commercially pure metal has a typical tensile strength of 14,000 psi and a typical elongation of 30% in 2 in., while sheet in the annealed temper has a typical tensile strength of about 13,000 psi and a typical elongation of about 45% in 2 in. The modulus of elasticity, for all practical purposes, is 10,000,000 psi. The commercially pure metal and many of its alloys are highly resistant to atmospheric corrosion and to attack by many chemicals, with the notable exception of strong alkalis. Because it is so high in the electrochemical series, however, it is subject to galvanic attack if coupled with metals such as the copper alloys in the presence of an electrolyte.

Alloying Elements—Additions of alloying elements usually increase the specific gravity (silicon and magnesium lower it), decrease the electrical and thermal conductivity and the melting point, increase the strength, and have a rather slight effect on the coefficient of thermal expansion and the modulus of elasticity. Some alloying elements, alone or in combination, produce alloys that respond to heat treatment. The addition of alloying elements can increase or decrease corrosion resistance, depending on the alloying element, heat treatment, and service environment. Aluminum alloys which are adversely affected by such additions are often protected by metallurgical cladding with a sacrificial alloy. The alloying elements commonly used in this country are copper, silicon, magnesium, manganese, and zinc.

Working and Heat Treating—Aluminum and its commercial alloys, being rather ductile materials, can be hot or cold worked into most of the common manufactured forms. The commercially pure metal and some of the alloys are not heat treatable compositions, and attain their strengths either by virtue of the alloy content or because of strain hardening resulting from cold work. The strength of many of the alloys, however, can be further increased by suitable heat treatments.

The response of an aluminum alloy to heat treatment depends on the presence of one or more alloying elements substantially more soluble in aluminum at temperatures of about 900–1000 F than at room temperature. By heating the material for a sufficient time at the proper solution treating temperature, the alloying elements are substantially dissolved by the aluminum; and by quenching rapidly from the solution treating temperature, the elements are retained in solid solution. Longer heating times are required for castings than for wrought products, and for heavy as compared to light sections. Alloys which are susceptible to intergranular corrosion should be quickly quenched after solution heat treatment to prevent reprecipitation along grain boundaries.

Certain of the heat treatable alloys, notably the so-called duralumin (Cu, Mg, Si) type alloys, age harden considerably at room temperature within a few days after quenching; the others, although they harden slowly at room temperature, must be heated to about 300 F for a few hours to attain their maximum strengths. With a few exceptions, most alloys which age harden substantially at room temperature can be made to develop even greater strength by a precipitation treatment at 300–500 F. It is generally agreed that precipitation treatments or age hardening result from lattice strains and the precipitation of alloying elements or compounds from the supersaturated solid solution in the form of minute particles. Recent studies indicate that the strengthening of heat treatable aluminum alloy by aging is due to both the uniform dispersion of a finely dispensed submicroscopic precipitate and the distortion of the lattice structure by these particles before they reach a visible size. It is believed that these particles, because of their critical size and location in the crystal structure, impede or prevent slip and thus increase the strength of the metal. Because of this phenomenon, these aging treatments are normally referred to as precipitation treatments. Room temperature aging, on the other hand, is believed to be the result of zone hardening. In this connection, it is interesting to note that the better workability of the as-quenched material can be retained in those alloys which age at room temperature by the simple expedient of storing the quenched material at about 0 F.

The effects of either cold work or heat treatment on the strength and workability of the materials can be removed by annealing at temperatures of about 600–800 F, depending on the alloy and temper. It must be remembered, however, that the strength of a non-heat treatable alloy can be regained, after annealing, only by the introduction of additional cold work.

Joining—Aluminum and its alloys can be joined by fusion welding, resistance welding, soldering, brazing, and adhesive bonding. The choice of process is dependent on alloy composition, material thickness, joint configuration, and expected service environment. The inert gas shielded metal arc process (MIG) and inert gas shielded tungsten arc process (TIG) are the most widely used fusion welding processes. Oxygas and coated electrode welding techniques are sometimes used, but the fluxes required with these processes, if not completely removed after welding, can promote corrosion. Brazing techniques now in common use include torch, dip, and furnace brazing.

All aluminum alloys can be joined by one or more of the available processes. Heat treated aluminum alloys (like the ferrous base alloys) are subject to reductions in strength after welding. Heat treating after welding will restore most of the prewelded mechanical properties. Work hardened aluminum alloys provide good as-welded mechanical properties and are used for applications such as storage tanks, boats, ships, and railroad cars.

Forming—Aluminum and its alloys can be formed hot or cold with considerable ease, although the bend radii for cold forming and the allowance for spring-back must be increased as the strength of the material increases. For severe forming, very deep drawing, or spinning, the annealed (0) temper usually is employed; while for the less drastic operations, the intermediate, cold-worked temper (H12, H22, or H32; or H14, H24, or H34), or the T3 or T4 type temper immediately after quenching usually is selected. The full hard (H18, H28, or H38) or the heat treated and aged (T6) tempers are not usually used where more than slight forming is required. Heat treatable alloys, however, can often be formed in the annealed or the as-quenched tempers and subsequently heat treated to the desired temper.

Machining—The aluminum alloys can generally be machined easily, if suitable practices and proper tools are used. Substantial tonnages of aluminum alloy rods and bars are regularly used for making screw-machine products.

Finishing and Coating—The aluminum alloys can be given a wide variety of mechanical, chemical, electrochemical, or paint finishes. The more common mechanical finishes include sand or grit blasting, scratch brushing, and buffing, while the chemical finishes may be a simple dip coating or an etching treatment. The possibility of generating an explosive mixture of finely powdered metal and air should be borne in mind in connection with mechanical finishing operations. Paint coatings may be either a clear lacquer or a pigmented coating and may be applied to secure either decoration or protection, or both. Paint adhesion is generally enhanced by the application of chemical conversion coatings prior to painting. Electroplating, although not extensively practiced in the past, is now gaining increased commercial use.

Anodic coatings can be produced to provide good protection against corrosion and are also good bases for subsequent paint coatings. These coatings can be dyed, and they make possible a variety of colored surfaces suitable for many decorative applications. Their hard, wear resistant surfaces are made use of in many applications.

The appearance of automotive bright anodized trim parts produced from 5252 or the 5X57 type sheet or a 6463 extrusion is dependent upon the alloy, the temper, the finishing procedure, the aluminum producers' controls of their fabrication procedures, and the metal handling and forming techniques used. Strength requirements and formability considerations generally dictate alloy selections. Variations of temper within the bright sheet trim alloys offer further opportunity to adjust mechanical and formability properties. However, the relationship between alloy, temper, and appearance must be given careful consideration. Alloy 5457-0, widely used, has excellent formability associated with the annealed temper. It offers a good and acceptable finish for many decorative trim parts, but lacks the image clarity or brightness of the less workable strain hardened tempers, such as the H25 and H28 tempers of all 5X57 type automotive trim alloys. Alloy 5657, when supplied in a modified strain hardened temper to achieve a higher minimum elongation, may have formability and finishing capabilities intermediate between the annealed and H25 tempers. Partially recrystallized structures, which may be experienced when material is produced to significantly higher minimum elongation requirements than those specified herein for the H25 temper, may give, under some conditions, an undesirable appearance after forming and finishing. Adequate control of finishing procedures is required to provide the highly lustrous and good image clarity possible using the 5X57 type decorative aluminum trim alloys. "Out-of-control" finishing procedures used after forming can produce trim parts having an unfavorable appearance or corrosion resistance. Improper handling and forming techniques can also contribute to an undesirable appearance (scratches, gouges, strains, etc.) of the final automotive trim part.

To simplify presentation of information about the aluminum alloys, the materials have been grouped under the general headings of casting alloys and wrought alloys. Generally speaking, a given composition is not used commercially for both wrought and cast products, and the casting alloys usually contain a somewhat greater total alloy content than the wrought alloys. When yield strength is specified, it is that stress at which the stress-strain curve deviates 0.2% from the modulus line (normally referred to as 0.2% offset).

Additional information on aluminum alloys and commercially available forms can be found in SAE J454.

ALLOY AND TEMPER DESIGNATION SYSTEMS FOR ALUMINUM—SAE J993b SAE Standard

Report of Nonferrous Metals Committee approved July 1967 and last revised September 1973. Conforms to American National Standard H35.1-1972.

1. Scope—This standard provides systems for designating wrought aluminum and wrought aluminum alloys, aluminum and aluminum alloys in the form of castings and foundry ingot, and the tempers in which aluminum and aluminum alloy wrought products and aluminum alloy castings are produced.

2. Wrought Aluminum and Aluminum Alloy Designation System (see Note 5.1)—A system of four-digit numerical designations is used to identify wrought aluminum and wrought aluminum alloys. The first digit indicates the alloy group as shown in Table 1. The last two digits identify the aluminum alloy or indicate the aluminum purity. The second digit indicates modifications of the original alloy or impurity limits.

2.1 Aluminum—In the 1xxx group for minimum aluminum purities of 99.00% and greater, the last two of the four digits in the designation indicate the minimum aluminum percentage (Note 5.2). These digits are the same as the two digits to the right of the decimal point in the minimum aluminum percentage when it is expressed to the nearest 0.01%. The second digit in the designation indicates modifications in impurity limits. If the second digit in the designation is zero, it indicates that there is no special control on individual impurities; integers 1 through 9, which are assigned consecutively as needed, indicate special control of one or more individual impurities or alloying elements.

2.2 Aluminum Alloys—In the 2xxx through 8xxx alloy groups, the last two of the four digits in the designation have no special significance but serve only to identify the different aluminum alloys in the group. The second digit in the alloy designation indicates alloy modifications (Note 5.3). If the second digit in the designation is zero, it indicates the original alloy; integers 1 through 9, which are assigned consecutively, indicate alloy modifications.

2.3 Experimental Alloys—Experimental alloys are also designated in accordance with this system, but they are indicated by the prefix X. The prefix is dropped when the alloy is no longer experimental. During development and before they are designated as experimental, new alloys are identified by serial numbers assigned by their originators. Use of the serial number is discontinued when the X number is assigned.

2.4 National Variations—National variations (Note 5.4) of wrought aluminum and wrought aluminum alloys registered by another country in accordance with this system are identified by a serial letter (Note 5) before the numerical designation.

3. Cast Aluminum and Aluminum Alloy Designation System[1] (see Note 5.1)—A system of four-digit numerical designations is used to identify aluminum and aluminum alloys in the form of castings and foundry ingot. The first digit indicates the alloy group, as shown in Table 2. The second two digits identify the aluminum alloy or indicate the aluminum purity. The last digit, which is separated from the others by a decimal point, indicates the product form, that is, castings or ingot. A modification of the original alloy or impurity limits is indicated by a serial letter (Note 5.6) before the numerical designation.

3.1 Aluminum Castings and Ingot—In the 1xx.x group for minimum aluminum purities of 99.00% and greater, the second two of the four digits in the designation indicate the minimum aluminum percentage (Note 5.2).

TABLE 1—DESIGNATION SYSTEM FOR WROUGHT ALUMINUM AND ALUMINUM ALLOY

Composition	Alloy No.
Aluminum, 99.0% min and greater	1xxx
Aluminum alloys grouped by major alloying element[a,b,c]	
Copper	2xxx
Manganese	3xxx
Silicon	4xxx
Magnesium	5xxx
Magnesium and silicon	6xxx
Zinc	7xxx
Other element	8xxx
Unused series	9xxx

[a] For codification purposes, an alloying element is any element which is intentionally added for any purpose other than grain refinement and for which minimum and maximum limits are specified.

[b] Standard limits for alloying elements and impurities are expressed to the following places:

Less than 1/1000%	0.000X
1/1000 up to 1/100%	0.00X
1/100 up to 1/10%	
Unalloyed aluminum made by a refining process	0.0XX
Alloys and unalloyed aluminum not made by a refining process	0.0X
1/10 through 1/2%	0.XX
Over 1/2%	0.X, X.X, etc.

[c] Standard limits for alloying elements and impurities are expressed in the following sequence: silicon; iron; copper; manganese; magnesium; chromium; nickel; zinc (Note 1); titanium; other elements (each); other elements (Total); aluminum (Note 2).

Note 1—Additional specified elements having limits are inserted in alphabetical order of their chemical symbols between zinc and titanium, or are specified in footnotes.

Note 2—Aluminum is specified as minimum for unalloyed aluminum, and as a remainder for aluminum alloys.

[1] The castings and ingot alloy designation system described herein is not currently in use for some SAE cast aluminum alloys. It is applicable to Aluminum Association (AA) and American National Standard Institute (ANSI), and other, specification systems. Although the chemical composition limits shown in most SAE reports conform to the limits shown for comparable castings and ingots covered in AA and ANSI publications, the designation system described herein is not currently used in SAE Standards and Information Reports.

TABLE 2—DESIGNATION SYSTEM FOR CAST ALUMINUM AND ALUMINUM ALLOY

Composition	Alloy No.
Aluminum, 99.00% min and greater	1xx.x
Aluminum alloy group by major alloying element[a, b, c]	
Copper	2xx.x
Silicon, with added copper and/or magnesium	3xx.x
Silicon	4xx.x
Magnesium	5xx.x
Zinc	7xx.x
Tin	8xx.x
Other element	9xx.x
Unused series	6xx.x

[a]For codification purposes, an alloying element is any element which is intentionally added for any purpose other than grain refinement and for which minimum and maximum limits are specified.

[b]Standard limits for alloying elements and impurities are expressed to the following places:

Less than 1/1000%	0.000X
1/1000 up to 1/100%	0.00X
1/100 up to 1/10%	
Unalloyed aluminum made by a refining process	0.0XX
Alloys and unalloyed aluminum not made by a refining process	0.0X
1/10 through 1/2%	0.XX
Over 1/2%	0.X, X.X, etc.

[c]Standard limits for alloying elements and impurities are expressed in the following sequence: silicon; iron; copper; manganese; magnesium; chromium; nickel; zinc (Note 1); titanium; other elements (each); other elements (Total); aluminum (Note 2).

Note 1—Additional specified elements having limits are inserted in alphabetical order of their chemical symbols between zinc and titanium, or are specified in footnotes.

Note 2—Aluminum is specified as minimum for unalloyed aluminum, and as a remainder for aluminum alloys.

These digits are the same as the two digits to the right of the decimal point in the minimum aluminum percentage when it is expressed to the nearest 0.01%. The last digit, which is to the right of the decimal point, indicates the product form: 1xx.0 indicates castings, and 1xx.1 indicates ingot. Special control of one or more individual elements other than aluminum is indicated by a serial letter (Note 5.6) before the numerical designation.

3.2 Aluminum Alloy Castings and Ingot—In the 2xx.x through 9xx.x alloy groups, the second two of the four digits in the designation have no special significance but serve only to identify the different aluminum alloys in the group. The last digit, which is to the right of the decimal point, indicates the product form: xxx.0 indicates castings, xxx.1 indicates ingot which has chemical composition limits conforming to paragraph 3.2.1, and xxx.2 indicates ingot which has chemical composition limits that differ but fall within the limits for xxx.1 ingot. Alloy modifications (Note 5.3) are indicated by a serial letter (Note 5.9) before the numerical designation.

3.2.1 Limits for alloying elements and impurities for xxx.1 ingot are the same as for the alloy in the form of castings, except for the limits noted in Table 3.

3.3 Experimental Alloys—Experimental alloys are also designated in accordance with this system, but they are indicated by the prefix X. The prefix is dropped when the alloy is no longer experimental. During development and before they are designated as experimental, new alloys are identified by serial numbers assigned by their originators. Use of the serial number is discontinued when the X number is assigned.

4. *Temper Designation System*—The temper designation system is used for all forms of wrought and cast aluminum and aluminum alloys except ingot. It is based on the sequences of basic treatments used to produce the various tempers. The temper designation follows the alloy designation, the two being separated by a hyphen. Basic temper designations consist of letters. Subdivisions of the basic tempers, where required, are indicated by one or more digits following the letter. These designate specific sequences of basic treatments, but only operations recognized as significantly influencing the characteristics of the product are indicated. Should some other variation of the same sequence of basic operations be applied to the same alloy, resulting in different characteristics, then additional digits are added to the designation.

4.1 Basic Temper Designations

F As Fabricated—Applies to the products of shaping processes in which no special control over thermal conditions or strain-hardening is employed. For wrought products, there are no mechanical property limits.

O Annealed (Wrought Products Only)—Applies to wrought products which are fully annealed to obtain the lowest strength condition.

H Strain Hardened (Wrought Products Only)—Applies to products which have their strength increased by strain-hardening, with or without supplementary thermal treatments to produce some reduction in strength. The H is always followed by two or more digits.

W Solution Heat-Treated—An unstable temper applicable only to alloys which spontaneously age at room temperature after solution heat-treatment. This designation is specific only when the period of natural aging is indicated; for example, W $\frac{1}{2}$ h.

T Thermally Treated to Produce Stable Tempers Other Than F, O, or H—Applies to products which are thermally treated, with or without supplementary strain-hardening, to produce stable tempers. The T is always followed by one or more digits.

4.2 Subdivisions of Basic Tempers

4.2.1 Subdivisions of H Temper: Strain Hardened

4.2.1.1 The first digit following the H indicates the specific combination of basic operations, as follows:

H1 Strain Hardened Only—Applies to products which are strain hardened to obtain the desired strength without supplementary thermal treatment. The number following this designation indicates the degree of strain hardening.

H2 Strain Hardened and Partially Annealed—Applies to products which are strain hardened more than the desired final amount and then reduced in strength to the desired level by partial annealing. For alloys that age soften at room temperature, the H2 tempers have the same minimum ultimate tensile strength as the corresponding H3 tempers. For other alloys, the H2 tempers have the same minimum ultimate tensile strength as the corresponding H1 tempers and slightly higher elongation. The number following this designation indicates the degree of strain hardening remaining after the product has been partially annealed.

H3 Strain Hardened and Stabilized—Applies to products which are strain hardened and whose mechanical properties are stabilized by a low-temperature thermal treatment which results in slightly lowered tensile strength and improved ductility. This designation is applicable only to those alloys which, unless stabilized, gradually age soften at room temperature. The number following this designation indicates the degree of strain hardening before the stabilization treatment.

4.2.1.2 The digit following the designations H1, H2, and H3 indicates the degree of strain hardening. Numeral 8 has been assigned to indicate tempers having an ultimate tensile strength equivalent to that achieved by a cold reduction (temperature during reduction not to exceed 120°F (49°C)) of approximately 75% following a full anneal. Tempers between 0 (annealed) and 8 are designated by numerals 1 through 7. Material having an ultimate tensile strength about midway between that of the 0 temper and that of the 8 temper is designated by the numeral 4; about midway between the 0 and 4 tempers by the numeral 2; and about midway between the 4 and 8 tempers by the numeral 6. Numeral 9 designates tempers whose minimum ultimate tensile strength exceeds that of the 8 temper by 2.0 ksi (14 MPa) or more. For two-digit H tempers whose second digit is odd, the standard limits for ultimate tensile strength are exactly midway between those of the adjacent two-digit H tempers whose second digits are even.

Note: For alloys which cannot be cold reduced, an amount sufficient to establish an ultimate tensile strength applicable to the 8 temper (75% cold reduction after full anneal), the 6 temper tensile strength may be established by a cold reduction of approximately 55% following a full anneal, or the 4

TABLE 3

Element, %	For Castings	For Ingot
Iron, max	Sand and permanent mold:	
	Up thru 0.15	0.03 less than castings
	Over 0.15 thru 0.25	0.05 less than castings
	Over 0.25 thru 0.6	0.10 less than castings
	Over 0.6 thru 1.0	0.2 less than castings
	Over 1.0	0.3 less than castings
	Die	
	Up thru 1.3	0.3 less than castings
	Over 1.3	1.1 maximum
Magnesium, min	All	
	Less than 0.50	0.05 more than castings[a]
	0.5 and greater	0.1 more than castings[a]
Zinc, max	Die	
	Over 0.25 thru 0.6	0.10 less than castings
	Over 0.6	0.1 less than castings

[a]Applicable only when the specified magnesium range for castings is greater than 0.15%.

temper tensile strength may be established by a cold reduction of approximately 35% after a full anneal.

4.2.1.3 The third digit (Note 10), when used, indicates a variation of a two-digit temper. It is used when the degree of control of temper or the mechanical properties are different from, but close to, those for the two-digit H temper designation to which it is added, or when some other characteristic is significantly affected. (See Appendix for three-digit H tempers.)

NOTE: The minimum ultimate tensile strength of a three-digit H temper is at least as close to that of the corresponding two-digit H temper as it is to the adjacent two-digit H tempers.

4.2.2 SUBDIVISIONS OF T TEMPER: THERMALLY TREATED

4.2.2.1 Numerals 1 through 10 following the T indicate specific sequences of basic treatments, as follows (Note 5.8):

T1 COOLED FROM AN ELEVATED TEMPERATURE SHAPING PROCESS AND NATURALLY AGED TO A SUBSTANTIALLY STABLE CONDITION—Applies to products for which the rate of cooling from an elevated temperature shaping process, such as casting or extrusion, is such that their strength is increased by room temperature aging.

T2 ANNEALED (CAST PRODUCTS ONLY)—Applies to cast products which are annealed to improve ductility and dimensional stability.

T3 SOLUTION HEAT TREATED AND THEN COLD WORKED—Applies to products which are cold worked to improve strength, or in which the effect of cold work in flattening or straightening is recognized in mechanical property limits.

T4 SOLUTION HEAT TREATED AND NATURALLY AGED TO A SUBSTANTIALLY STABLE CONDITION—Applies to products which are not cold worked after solution heat treatment, or in which the effect of cold work in flattening or straightening may not be recognized in mechanical property limits.

T5 COOLED FROM AN ELEVATED TEMPERATURE SHAPING PROCESS AND THEN ARTIFICIALLY AGED—Applies to products which are cooled from an elevated temperature shaping process, such as casting or extrusion, and then artificially aged to improve mechanical properties or dimensional stability or both.

T6 SOLUTION HEAT TREATED AND THEN ARTIFICALLY AGED—Applies to products which are not cold worked after solution heat treatment, or in which the effect of cold work in flattening or straightening may not be recognized in mechanical property limits.

T7 SOLUTION HEAT TREATED AND THEN STABILIZED—Applies to products which are stabilized to carry them beyond the point of maximum strength to provide control of some special characteristics.

T8 SOLUTION HEAT TREATED, COLD WORKED, AND THEN ARTIFICIALLY AGED—Applies to products which are cold worked to improve strength, or in which the effect of cold work in flattening or straightening is recognized in mechanical property limits.

T9 SOLUTION HEAT TREATED, ARTIFICIALLY AGED, AND THEN COLD WORKED—Applies to products which are cold worked to improve strength.

T10 COOLED FROM AN ELEVATED TEMPERATURE SHAPING PROCESS, ARTIFICIALLY AGED, AND THEN COLD WORKED—Applies to products which are artificially aged after cooling from an elevated temperature shaping process, such as casting or extrusion, and then cold worked to improve strength further.

4.2.2.2 Additional digits (Note 5.9), the first of which shall not be zero, may be added to designations T1 through T10 to indicate a variation in treatment which significantly alters the characteristics of the product. (See Appendix for specific additional digits for T tempers.)

5. Notes

5.1 Producers of wrought aluminum and wrought aluminum alloys, and aluminum and aluminum alloy castings and foundry ingot, may register chemical composition limits and designations conforming to this standard with the Aluminum Association (AA) provided the aluminum or aluminum alloy is offered for sale, the complete chemical composition limits are registered, and the composition is significantly different from that of any aluminum or aluminum alloy for which a numerical designation already has been assigned. A numerical designation assigned in conformance with this standard should be used only to indicate an aluminum or aluminum alloy having chemical composition limits identical to those registered with AA for that aluminum or aluminum alloy.

5.2 The aluminum content for unalloyed aluminum made by a refining process is the difference between 100.000% and the sum of all other metallic elements present in amounts of 0.0010% or more each, expressed to the third decimal; for unalloyed aluminum not made by a refining process, it is the difference between 100.00% and the sum of all other metallic elements present in amounts of 0.010% or more each, expressed to the second decimal.

5.3 A modification of the original alloy is limited to any one or a combination of the following:

(a) Change of not more than the following amounts in the arithmetic mean of the limits for an alloying element:

Arithmetic Mean of Limits for Alloying Elements in Original Alloy, %	Maximum Change, %
Up thru 1.0	0.15
Over 1.0 thru 2.0	0.20
Over 2.0 thru 3.0	0.25
Over 3.0 thru 4.0	0.30
Over 4.0 thru 5.0	0.35
Over 5.0 thru 6.0	0.40
Over 6.0	0.50

To determine compliance when limits are specified for a combination of two or more elements in one alloy composition, the mean of such a combination should be compared to the sum of the mean values of the same individual elements, or any combination thereof, in another alloy composition.

(b) Addition or deletion of not more than one alloying element with limits having an arithmetic mean of not more than 0.30%.

(c) Substitution of one alloying element for another element serving the same purpose.

(d) Change in limits for impurities.

(e) Change in limits for grain refining elements.

(f) Distinctive iron or silicon limits, or both, reflecting high purity base metal.

An alloy shall not be registered as a modification if it meets the requirements for a national variation.

5.4 A national variation has composition limits which are similar but not identical to those registered by another country, with differences such as:

(a) Differences in the arithmetic mean of limits for alloying elements not exceeding the following amounts:

Arithmetic Mean of Limits for Alloying Elements in Original Alloy or Modification, %	Maximum Difference, %
Up thru 1.0	0.15
Over 1.0 thru 2.0	0.20
Over 2.0 thru 3.0	0.25
Over 3.0 thru 4.0	0.30
Over 4.0 thru 5.0	0.35
Over 5.0 thru 6.0	0.40
Over 6.0	0.50

To determine compliance when limits are specified for a combination of two or more elements in one alloy composition, the mean of such a combination should be compared to the sum of the mean values of the same individual elements, or any combination thereof, in another alloy composition.

(b) Substitution of one alloying element for another element serving the same purpose.

(c) Different limits on impurities except for low iron. Low iron, reflecting high purity base metal, should be considered an alloy modification. See paragraph 5.3 (f).

(d) Different limits on grain refining elements.

(e) Inclusion of a minimum limit for iron or silicon, or both.

Wrought aluminum and wrought aluminum alloys meeting these requirements shall not be registered as a new alloy or alloy modification.

5.5 The serial letters are assigned internationally in alphabetical sequence with A but omitting I, O, and Q.

5.6 The serial letters are assigned in alphabetical sequence starting with A but omitting I, O, Q, and X, the X being reserved for experimental alloys.

5.7 Numerals 1 through 9 may be arbitrarily assigned as the third digit and registered with AA for an alloy and product to indicate a variation of a two-digit H temper provided the temper is used or is available for use by more than one user, mechanical property limits are registered, the characteristics of the temper are significantly different from those of all other tempers which have the same sequence of basic treatments and for which designations already have been assigned for the same alloy and product, and the following are also registered if characteristics other than mechanical properties are considered significant: (a) test methods and limits for the characteristics, or (b) the specific practices used to produce the temper. Zero has been assigned to indicate variations negotiated between the manufacturer and purchaser which are not used widely enough to justify registration.

5.8 A period of natural aging at room temperature may occur between or after the operations listed for tempers T3 through T10. Control of this period is exercised when it is metallurgically important.

5.9 Additional digits may be arbitrarily assigned and registered with AA for an alloy and product to indicate a variation of tempers T1 through T10 provided the temper is used or is available for use by more than one user, mechanical property limits are registered, the characteristics of the temper are significantly different from those of all other tempers which have the same sequence of basic treatments and for which designations already have been assigned for the same alloy and product, and the following are also registered if characteristics other than mechanical properties are considered significant: (a) test methods and limits for the characteristics, or (b) the specific practices used to provide the temper. Variations in treatment which do not alter the characteristics of the product are considered alternate treatments for which additional digits are not assigned.

APPENDIX

A1. Three-Digit H Tempers

A1.1 The following three-digit H temper designations have been assigned for wrought products in all alloys:

H111 Applies to products which are strain hardened less than the amount required for a controlled H11 temper.

H112 Applies to products which acquire some temper from shaping processes not having special control over the amount of strain hardening or thermal treatment, but for which there are mechanical property limits.

A1.2 The following three-digit H temper designations have been assigned for wrought products in alloys containing over a nominal 4% magnesium.

H311 Applies to products which are strain hardened less than the amount required for a controlled H31 temper.

H321 Applies to products which are strain hardened less than the amount required for a controlled H32 temper.

H323 Applies to products which are specially fabricated to have acceptable resistance to stress corrosion cracking.
H343

A1.3 The following three-digit H temper designations have been assigned for:

Patterned or Embossed Sheet	Fabricated from
H114	O temper
H124, H224, H324	H11, H21, H31 temper, respectively
H134, H234, H334	H12, H22, H32 temper, respectively
H144, H244, H344	H13, H23, H33 temper, respectively
H154, H254, H354	H14, H24, H34 temper, respectively
H164, H264, H364	H15, H25, H35 temper, respectively
H174, H274, H374	H16, H26, H36 temper, respectively
H184, H284, H384	H17, H27, H37 temper, respectively
H194, H294, H394	H18, H28, H38 temper, respectively
H195, H295, H395	H19, H29, H39 temper, respectively

A2. Additional Digits for T Tempers

A2.1 The following specific additional digits have been assigned for stress-relieved tempers of wrought products:

T51 STRESS RELIEVED BY STRETCHING—Applies to the following products when stretched the indicated amounts after solution heat treatment or cooling from an elevated temperature shaping process.

Product	Stretch, Permanent Set, %
Plate	1.5–3
Rod, bar, shapes, extruded tube	1–3
Drawn tube	0.5–3

Applies directly to plate and rolled or cold-finished rod and bar. These products receive no further straightening after stretching.

Applies to extruded rod, bar, shapes, and tube and to drawn tube when designated as follows:

T510 Products that receive no further straightening after stretching.

T511 Products that may receive minor straightening after stretching to comply with standard tolerances.

T52 STRESS RELIEVED BY COMPRESSING—Applies to products which are stress relieved by compressing after solution heat treatment, or cooling from an elevated temperature shaping process to produce a permanent set of 1–5%.

T54 STRESS RELIEVED BY COMBINED STRETCHING AND COMPRESSING—Applies to die forgings which are stress relieved by restriking cold in the finish die.

A2.2 The following temper designations have been assigned for wrought products heat treated from O or F temper to demonstrate response to heat-treatment.

T42 SOLUTION HEAT TREATED FROM THE O OR F TEMPER—To demonstrate response to heat treatment, and naturally aged to a substantially stable condition.

T62 SOLUTION HEAT TREATED FROM THE O OR F TEMPER—To demonstrate response to heat treatment, and artificially aged.

Temper designations T42 and T62 may also be applied to wrought products heat treated from any temper by the user, when such heat treatment results in the mechanical properties are applicable to these tempers.

GENERAL INFORMATION ON SAE ALUMINUM CASTING ALLOYS—SAE J452c

SAE Information Report

Report of Nonferrous Metals Division approved January 1934 and last revised by Nonferrous Metals Committee July 1973.

The SAE Standards for aluminum casting alloys cover a wide range of castings for general and special use, but do not include all the alloys in commercial use. The castings are made principally by sand cast, permanent mold, or die cast methods; however, shell molding, investment casting, plaster cast, and other less common foundry methods may also be used. If the alloys listed do not have the desired characteristics, it is recommended that the manufacturers of aluminum castings be consulted.

There are two general types of cast aluminum alloys: nonheat treatable and heat treatable. The nonheat treatable alloys normally are used in the as-cast condition (F), but may be annealed (T2) to relieve casting stresses or to reduce the possibility of distortion during machining. The heat treatable alloys usually are used in a heat treated condition because of the increased strengths resulting from the heat treatment. These treatments generally consist of a high temperature solution treatment, followed by quenching in water, and a low temperature aging treatment (T6).

By aging the solution treated castings at higher temperature (T7 or T71), a product having more stable properties in service at elevated temperatures and less likely to distort during machining is obtained. Occasionally, the aging treatment is omitted and the castings are used in the quenched condition (T4); at other times (especially in castings to be used at elevated temperatures) the solution treatment is omitted and the castings are merely stabilized or aged (T5). Various combinations of properties can be secured by adjusting the thermal treatments, but only the commonly used conditions form a part of SAE J453.

The choice among the alloys depends upon the service requirements of the casting and, in some cases, upon the ability to produce the desired part. Some of the alloys are adapted for use at elevated temperatures; some are particularly resistant to corrosion under severe conditions of exposure; and some are more easily cast into complicated shapes that require pressure tightness. Others may be chosen primarily because of their lower cost when mechanical property requirements are of secondary importance. Specifications for the individual alloys indicate typical applications which will assist in the choice of alloy for any purpose.

The mechanical properties and dimensional tolerances required, the number of castings to be made, the surface finish desired, and the amount of machining necessary are factors in deciding which to use: sand, permanent mold, or die casting.

Sand castings usually are used when the number required is small, when a smooth surface is not required or can be secured economically by machining, and when close tolerances on as-cast dimensions are not essential.

Permanent mold castings usually have a smoother finish and may thus require less machining, can be cast to closer tolerances, and, for the same chemistry, generally have slightly higher mechanical properties.

Die castings can be held to much closer tolerances than either sand or permanent mold castings. In addition, practically all holes can be included, and the castings have a smooth surface requiring a minimum of preparation for final finishings. The uniformity of the castings permits simpler tooling and lower costs of subsequent machining operations. The saving which results from these lower machining and finishing costs is the determining factor in deciding the volume necessary to justify the cost of the die.

Producers of castings should be consulted to ensure the selection of the most suitable and economical casting method and the alloy best suited for conditions under which the casting will be used.

In the design of patterns for the production of aluminum alloy sand castings, a shrinkage is usually allowed (see Table 2) and may vary slightly depending upon the form and size of the casting. Producers of castings should also be consulted concerning the design of the pattern so that the best results may be obtained with the alloy to be used. The information provided in Table 2 is based on a study made by the American Foundrymen's Society.

TABLE 1—TYPICAL USES OF SAE ALUMINUM CASTING ALLOYS AND SIMILAR SPECIFICATIONS

SAE No.	AA Designation	Former ASTM Designation[a,b]	Former Commercial[a] Designation	Type of Casting[c]	Federal	AMS	Typical Uses and General Data
34	222.0	CG100A	122	S / PM	QQ-A-601 Class 7 / QQ-A-596 Class 2	—	Primarily a piston alloy, but also used for aircooled cylinder heads and valve tappet guides.
35	443.0	S5B	43	S / PM	QQ-A-601 Class 2 / QQ-A-596 Class 2	—	Used for intricate castings having thin sections; good resistance to corrosion; fair strength but good ductility.
38	295.0	C4A	195	S	QQ-A-601 Class 4	4230, 4231	General structural castings requiring high strength and shock resistance.
39	242.0	CN42A	142	S / PM	QQ-A-601 Class 6 / QQ-A-596 Class 3	4222	Used primarily for aircooled cylinder heads, but also used for pistons in high performance gasoline engines.
303	384.0	SC114A	384	D	QQ-A-591 Alloy 384	—	General purpose alloy with high fluidity; used for thin-walled castings or castings with large areas.
304	C443.0	S5C	43	D	QQ-A-591 Alloy 43	—	Good casting characteristics and resistance to corrosion.
305	A413.0	S12A	A13	D	QQ-A-591 Alloy A13	—	Highly resistant to corrosion; excellent casting characteristics; used for complicated castings of thin section.
306	A380.0	SC84A	A380	D	QQ-A-591 Alloy A380	4291 Comp. 2	Good casting characteristics and fair resistance to corrosion; not especially suited for thin sections; limited to cold-chamber machines.
308	380.0	SC84B	380	D	QQ-A-591 Alloy 380	—	Same as SAE 306 but suitable for use in either cold-chamber or gooseneck machines.
309	A360.0	SG100A	A360	D	QQ-A-591 Alloy A360	—	Excellent casting characteristics; suited for use in thin-walled or intricate castings produced in cold-chamber casting machine; corrosion resistance high.
310	D712.0	ZG61A	40E	S	QQ-A-601 Class 17	—	General purpose structural castings developing strengths equivalent to SAE 38 without requiring heat treatment.
311	705.0	ZG32A	Ternalloy 5	S / PM	QQ-A-601 Class 24 / QQ-A-596 Class 13	—	High strength, general purpose alloy; excellent machinability and dimensional stability; very good corrosion resistance; can be anodized.
312	707.0	ZG42A	Ternalloy 7	S / PM	QQ-A-601 Class 25 / QQ-A-596 Class 14	—	High strength, general purpose alloy; excellent machinability and dimensional stability; very good corrosion resistance; can be anodized.
313	A712.0	ZG61B	A612	S	QQ-A-601 Class 23	—	High strength, general purpose alloy; excellent machinability; easily polished; very good corrosion resistance; can be anodized.
314	C712.0	ZC60A	C612	PM	—	—	High strength, general purpose alloy; excellent machinability; easily polished; very good corrosion resistance; can be anodized.
315	713.0	ZC81A / ZC81B	Tenzalloy / Tenzalloy	S / PM	QQ-A-601 Class 22 / QQ-A-596 Class 12	—	High strength, general purpose alloy; excellent machinability; easily polished; very good corrosion resistance; can be anodized.
320	514.0	G4A	214	S	QQ-A-601 Class 5	—	Moderate strength; high resistance to corrosion.
321	A332.0	SN122A	A132	PM	QQ-A-596 Class 9	4210, 4212, 4214	Pistons, low expansion.
322	355.0	SC51A	355	S / PM	QQ-A-601 Class 10 / QQ-A-596 Class 6	4280, 4281	General use where high strength and pressure tightness are required, such as pump bodies and liquid-cooled cylinder heads.
323	356.0	SG70A	356	S / PM	QQ-A-601 Class 3 / QQ-A-596 Class 8	4217 / 4284, 4286	For intricate castings requiring good strength and ductility.
324	520.0	G10A	220	S	QQ-A-601 Class 16	4240	High strength and ductility; structural use; required special foundry practice.
326	319.0	SC64D	Allcast	S / PM	QQ-A-601 Class 18 / QQ-A-596 Class 11	—	General purpose alloy.
327	328.0	SC82A	Red X-8	S	QQ-A-601 Class 20	—	Similar to SAE 322 and 323.
329	—	—	319	S / PM	QQ-A-601 Class 18	—	General purpose, low-cost alloy; good foundry characteristics; good machinability; similar to SAE 326.
331	333.0	SC94A	333	PM	QQ-A-596D Class 17	—	General purpose permanent mold alloy used for engine parts, meter housings, flywheel housings, and regulator parts.
332	F332.0	SC103A	F132	PM	—	—	Primarily intended for automotive use in pistons.
334	Z332.0	—	E132	PM	—	—	Pistons.
335	C355.0	SC51B	C355	PM	—	—	Similar to SAE 322, but has greater strength and ductility.
336	A356.0	SG70B	A356	PM	—	—	Similar to SAE 323, but has greater strength and ductility.
380	B295.0	CS43A	B195	PM	QQ-A-596 Class 4	4282, 4283	Modification of SAE 38 suitable for use in permanent molds.
382	201.0	CQ51A	KO1	S / PM	—	4228 / 4229	Very high strength at room and elevated temperature; good impact strength and ductility; premium cast alloy.
383	383.0	SC102A	—	D	—	—	Similar to SAE 308 with better castability.

[a] Former ASTM and commercial alloy designations listed for information only.
[b] ASTM B 26 covers all sand casting specifications, ASTM B 85 all die casting specifications, and ASTM B 108 all permanent mold casting specifications.
[c] S—sand cast, PM—permanent mold, D—die cast.

TABLE 2—SAE ALUMINUM CASTING ALLOYS CHARACTERISTICS[a]

SAE No.	AA Designation	Type Casting	Pattern Shrinkage[b] Allowance		Resistance to Hot Cracking[c]	Pressure Tightness	Fluidity[d]	Solidification Shrinkage Tendency[e]
			in/ft	%				
34	222.0	S	5/32	1.30	3	3	3	3
		PM	—b	—b	4	4	3	4
35	443.0	S	5/32	1.30	1	1	1	1
		PM	—b	—b	1	1	1	2
38	295.0	S	5/32	1.30	4	4	3	3
39	242.0	S	5/32	1.30	4	3	3	4
		PM	—b	—b	4	4	3	4
303	384.0	D	—b	—b	2	2	1	—
304	C443.0	D	—b	—b	1	3	3	—
305	A413.0	D	—b	—b	1	2	1	—
306	A380.0	D	—b	—b	2	2	2	—
308	380.0	D	—b	—b	2	2	2	—
309	A360.0	D	—b	—b	1	1	1	—
310	D712.0	S	3/16	1.56	5	3	4	4
311	705.0	S	3/16	1.56	5	4	4	4
		PM	—b	—b	5	4	4	5
312	707.0	S	3/16	1.56	5	4	4	4
		PM	—b	—b	5	4	4	5
313	A712.0	S	3/16	1.56	5	3	4	4
314	C712.0	PM	—b	—b	5	4	4	4
315	713.0	S	3/16	1.56	5	3	4	4
		PM	—b	—b	5	4	4	5
320	514.0	S	5/32	1.30	4	5	5	5
321	A312.0	PM	—b	—b	1	2	1	3
322	355.0	S	5/32	1.30	1	1	1	1
		PM	—b	—b	1	1	2	2
323	356.0	S	5/32	1.30	1	1	1	1
		PM	—b	—b	1	1	2	1
324	520.0	S	1/10	0.83	2	5	4	5
326	319.0	S	5/32	1.30	2	2	2	2
		PM	—b	—b	2	2	2	3
327	328.0	S	5/32	1.30	1	2	1	2
329	—	S	5/32	1.30	2	2	2	2
331	333.0	PM	—b	—b	2	2	1	3
		S	5/32	1.30	4	3	3	4
332	F332.0	PM	—b	—b	1	2	1	2
334	Z332.0	PM	—b	—b	1	2	1	2
335	C355.0	PM	—b	—b	1	1	2	2
336	A356.0	PM	—b	—b	1	1	2	1
380	B295.0	PM	—b	—b	4	3	3	3
382	201.0	PM	—b	—b	4	3	3	4
383	383.0	D	—b	—b	1	1	1	—

Other Characteristics

SAE No.	Normally Heat Treated	Resistance to Corrosion[f]	Machining[g]	Polishing[h]	Electro-plating[i]	Anodize Appearance[j]	Chemical Oxide Coating[k] (Protection)	Strength at Elevated Temperature[l]	Suitability for Welding[m]	Suitable for Brazing[n]
34	Yes	4	1	2	1	3	4	1	4	No
	Yes	5	1	2	1	3	4	1	4	No
35	No	3	5	5	2	5	2	4	1	Ltd.
	No	3	5	4	2	4	2	4	1	Ltd.
38	Yes	3	2	2	1	2	3	3	3	No
39	Yes	4	2	2	1	3	4	1	4	No
	Yes	4	2	2	1	2	3	1	4	No
303	No	5	3	3	2	4	4	2	No	No
304	No	2	5	4	2	4	3	5	No	No
305	No	3	4	5	3	5	3	3	No	No
306	No	5	3	3	1	3	5	2	No	No
308	No	5	3	3	1	3	5	2	No	No
309	No	3	3	3	1	3	3	2	No	No
310	Aged Only	2	1	1	2	2	3	5	4	Yes
311	Aged Only	2	1	1	3	2	2	5	4	Yes
	Aged Only	2	1	1	3	1	2	5	4	Yes
312	Yes	2	1	1	3	2	2	5	4	Yes
	Yes	2	1	1	3	1	2	5	4	Yes
313	Aged Only	2	1	1	2	2	3	5	4	Yes
314	Aged Only	2	1	1	2	1	3	5	4	Yes
315	Aged Only	2	1	1	2	2	3	5	4	Yes
	Aged Only	2	1	1	2	2	2	5	4	Yes
320	No	1	1	1	5	1	1	2	5	No
321	Yes	3	4	5	4	5	2	2	2	No
322	Yes	3	3	3	1	4	2	2	2	No
	Yes	3	3	3	2	4	2	2	2	No
323	Yes	2	4	5	2	4	2	3	2	No
	Yes	2	3	3	1	4	2	2	2	No
324	Yes	1	1	1	4	1	1	—	5	No
326	Yes	3	3	4	2	4	3	3	2	No
	Yes	3	3	3	2	4	2	3	2	No
327	Yes	3	3	4	5	2	4	2	2	No
329	Yes	3	3	3	2	4	2	3	2	No
331	Yes	3	3	2	3	4	3	2	3	No
	Yes	4	1	1	1	2	2	1	4	No
332	Aged Only	3	3	4	3	5	3	3	2	No
334	Aged Only	3	3	4	3	5	3	3	2	No
335	Yes	3	3	3	3	4	2	2	2	No
336	Yes	2	3	3	2	4	2	3	2	No
380	Yes	4	3	2	1	3	2	2	4	No
382	Yes	4	1	1	1	2	2	1	4	No
383	No	4	3	3	1	3	4	2	No	No

[a] 1 indicates best of group; 5 indicates poorest of group.
[b] Not applicable to permanent mold and die castings. Allowances are for average sand castings. Shrinkage requirements will vary with intricacy of design and dimensions.
[c] Ability of alloy to withstand contraction stresses while cooling through hot-short or brittle temperature range.
[d] Ability of liquid alloy to flow readily in mold and fill thin sections.
[e] Decrease in volume accompanying freezing of alloy and measure of amount of compensating feed metal required in form of risers.
[f] Based on alloy resistance in 5% salt spray test (ASTM B117).
[g] Composite rating based on ease of cutting, chip characteristics, quality of finishing, and tool life. Ratings, in the case of heat treatable alloys, based on T6 temper. Other tempers, particularly the annealed temper, may have lower rating.
[h] Composite rating based on ease and speed of polishing and quality of finish provided by typical polishing procedure.
[i] Ability of casting to take and hold an electroplate applied by present standard methods.
[j] Rated on lightness of color, brightness, and uniformity of clear anodized coating applied in sulfuric acid electrolyte.
[k] Rated on combined resistance of coating and base alloy to corrosion.
[l] Rating based on tensile and yield strengths of temperature up to 500 °F (260 °C), after prolonged heating at testing temperatures.
[m] Based on ability of material to be fusion welded with filler rod of same alloy.
[n] Refers to suitability of alloy to withstand brazing temperatures without excessive distortion or melting.
[o] Not recommended for service at temperatures exceeding 200 °F (93 °C).

TABLE 3—TYPICAL THERMAL TREATMENTS

SAE No.	Temper	AA Designation	Sand Castings					Permanent Mold Castings				
			Solution Heat Treatment			Precipitation Heat Treatment			Solution Heat Treatment			Precipitation Heat Treatment		
			Temperature		Hours	Temperature		Hours	Temperature		Hours	Temperature		Hours
			±10 °F	±6 °C		±10 °F	±6 °C		±10 °F	±6 °C		±10 °F	±6 °C	
34	T2	222.0	—	—	—	600	316	2–4	—	—	—	—	—	—
	T551		—	—	—	—	—	—	—	—	—	340	171	18–22
	T61		950	510	8–12	310	154	10–12	—	—	—	—	—	—
	T65		—	—	—	—	—	—	950	510	8–12	340	171	7–9
38	T4	295.0	960	516	12	—	—	—	—	—	—	—	—	—
	T6		960	516	12	310	154	3–5	—	—	—	—	—	—
	T62		960	516	12	310	154	12–16	—	—	—	—	—	—
	T7		960	516	12	500	260	4–6	—	—	—	—	—	—
	T21		—	—	—	650	343	2–4	—	—	—	—	—	—
39	T571	242.0	—	—	—	—	—	—	—	—	—	340	171	20–26
	T61		960	516	6a	450	232	1–3	960	516	6	400	204	3–5
310	T5	D712.0	—	—	—	356	180	10	—	—	—	—	—	—
311	T5	705.0	—	—	—	210	99	8	—	—	—	210	99	8
312	T5	707.0	—	—	—	210	99	8	—	—	—	210	99	8
	T7		990	532	8–16	350	177	4–8	990	532	8–16	350	177	4–8
313	T5	A712.0	—	—	—	Room	Room	21 d	—	—	—	—	—	—
314	T5	C712.0	—	—	—	—	—	—	—	—	—	210	99	8
315	T5c	713.0	—	—	—	Room	Room	21 d	—	—	—	—	—	—
	T5c		—	—	—	—	—	—	—	—	—	Room	Room	21 d
321	T551	A332.0	—	—	—	—	—	—	—	—	—	400	204	8–12
	T65		—	—	—	—	—	—	960	516	8	340	171	14–18
322	T51	355.0	—	—	—	440	227	7–9	—	—	—	—	—	—
	T6		980	527	8–12	310	154	3–5	980	527	8–12	310	154	3–5
	T71		980	527	8–12	475	246	4–6	—	—	—	—	—	—
323	T51	356.0	—	—	—	440	227	7–9	—	—	—	—	—	—
	T6		1000	538	8–12	310	154	2–3	1000	538	8–12	310	154	2–3
	T71		1000	538	12	475	246	2–4	1000	538	10	425	217	2–4
324	T4b	520.0	810	432	16	—	—	—	—	—	—	—	—	—
326	T6	319.0	940	504	6–12	315	157	3–5	—	—	—	—	—	—
	T61		—	—	—	—	—	—	940	504	6–12	315	157	8–12
327	T6	328.0	940	504	8–12	310	154	3–5	—	—	—	—	—	—
331	T5	330.0	—	—	—	—	—	—	—	—	—	400	204	7–9
	T6		—	—	—	—	—	—	940	504	8	310	154	2–5
	T7		—	—	—	—	—	—	940	504	8	500	260	4–6
332	T5	F332.0	—	—	—	—	—	—	—	—	—	400	204	8
334	T551	Z332.0	—	—	—	—	—	—	—	—	—	400	204	8–12
335	T61	C355.0	—	—	—	—	—	—	980	527	8–12	310	154	10–12
336	T61	A356.0	—	—	—	—	—	—	1000	538	8–12	310	154	6–10
380	T4	B295.0	—	—	—	—	—	—	950	510	8	—	—	—
	T6		—	—	—	—	—	—	950	510	8	310	154	5–7
	T7		—	—	—	—	—	—	950	510	8	500	260	4–6
382	T6	201.0	985	529	14–20	310	154	20	985	929	14–20	310	154	20
	T7		985	529	14–20	370	188	5	985	929	14–20	370	188	5

aQuench in air blast.   bRequires special delayed quench in boiling water.   cMay be precipitation heat treated, 10–16 h at 250 °F (121 °C).

CHEMICAL COMPOSITIONS AND MECHANICAL AND PHYSICAL PROPERTIES OF SAE ALUMINUM CASTING ALLOYS—SAE J453c

SAE Standard

Report of Nonferrous Metals Division approved January 1939 and last revised by Nonferrous Metals Committee July 1973.

Chemical Compositions—Chemical analysis shall be made in accordance with ASTM E 34, Standard Methods for Chemical Analysis of Aluminum and Aluminum Base Alloys, or any other approved method agreed upon by the manufacturer and the purchaser. The analysis may be made spectrographically, provided that, in case of dispute, the results secured by the ASTM E 34 methods shall be the basis for acceptance.

For purposes of determining conformance to limits indicated in the tables, an observed or a calculated value obtained from analysis is rounded off to the nearest unit in the last righthand place of figures used in expressing the specified limit in accordance with ASTM E 29, Recommended Practices for Designating Significant Places in Specified Limiting Values.

Mechanical and Physical Properties—The specified mechanical properties shown in this SAE Standard are the values that should be obtained from standard test specimens, separately cast under conditions that duplicate, as closely as possible, the conditions of solidification of the casting, and tested without machining, except to adapt the ends to the grips of the testing equipment. The specified properties for sand casting alloys (see Table 5) are for $\frac{1}{2}$ in. diameter standard test bars cast without chills in green sand molds, and the specified properties for the permanent mold alloys (see Table 6) are for $\frac{1}{2}$ in. (12.7 mm) diameter standard test bars cast in a permanent mold.

The typical tensile properties given for die casting alloys (see Table 7) are for $\frac{1}{4}$ in. (6.4 mm) diameter standard die cast test bars as shown in ASTM E 8, Methods of Tension Testing of Metallic Materials.

The properties obtained from test specimens machined from castings will vary, depending upon the location from which the bar is taken. Specimens taken from thin sections may have properties higher than those of separately cast test bars, while specimens taken from heavy sections or from locations near gates or risers may show lower properties. These relations are not peculiar to aluminum alloy castings but are the same in the castings of other metals. In general, when test bars machined from a casting are used as the basis for acceptance or rejection, the mechanical properties of these test bars cut from the castings shall be agreed upon between the purchaser and supplier.

The separately cast test specimen serves as a control of the metal quality, and in the case of heat treated alloys, serves also as a control of the heat treatment process, since such test bars must be heat treated with the castings they represent. Factors of safety used in design cover the variations of commercial castings from the properties specified for the alloy, which are based on tests of separately cast test specimens.

TABLE 1—CHEMICAL COMPOSITIONS OF SAE ALUMINUM SAND CASTING ALLOYS[a]

SAE No.	AA Designation	Si	Fe	Cu	Mn	Mg	Cr	Ni	Zn	Ti	Ag	Other Elements Each	Other Elements Total
34	222.0	2.0	1.5	9.2–10.8	0.50	0.15–0.35	—	0.50	0.8	0.25	—	—	0.35
35	443.0	4.5–6.0	0.8	0.6	0.50	0.05	0.25	—	0.50	0.25	—	—	0.35
38	295.0	1.5	1.0	4.0–5.0	0.35	0.03	—	—	0.35	0.25	—	0.05	0.15
39	242.0	0.7	1.0	3.5–4.5	0.35	1.2–1.8	0.25	1.7–2.3	0.35	0.25	—	0.05	0.15
310	D712.0	0.25	0.50	0.25	0.10	0.50–0.65	0.40–0.6	—	5.0–6.5	0.15–0.25	—	0.05	0.20
311	705.0	0.20	0.8	0.20	0.40–0.6	1.4–1.8	0.20–0.40	—	2.7–3.3	0.25	—	0.05	—
312	707.0	0.20	0.8	0.20	0.40–0.6	1.8–2.4	0.20–0.40	—	4.0–4.5	0.25	—	0.05	—
313	A712.0	0.15	0.50	0.35–0.65	0.05	0.6–0.8	—	—	6.0–7.0	0.25	—	0.05	0.15
315	713.0	0.25	1.0	0.40–1.0	0.6	0.20–0.50	0.35	0.15	7.0–8.0	0.25	—	0.10	0.25
320	514.0	0.35	0.50	0.15	0.35	3.5–4.5	—	—	0.15	0.25	—	0.05	0.15
322	355.0	4.5–5.5	0.8[b]	1.0–1.5	0.50[b]	0.40–0.6	0.25	—	0.35	0.25	—	0.05	0.15
323	356.0	6.5–7.5	0.6	0.25	0.35	0.20–0.40	—	—	0.35	0.25	—	0.05	0.15
324	520.0	0.25	0.30	0.25	0.15	9.5–10.6	—	—	0.15	0.25	—	0.05	0.15
326	319.0	5.5–6.5	1.0	3.0–4.0	0.50	0.10	—	0.35	1.0	0.25	—	—	0.50
327	328.0	7.0–8.6	1.0	1.0–2.0	0.20–0.6	0.20–0.6	0.35	0.25	1.5	0.25	—	—	0.50
329	—	5.5–6.5	1.2	3.0–4.0	0.8	0.10–0.50	—	0.50	1.0	0.25	—	—	0.50
382	201.0	0.10	0.15	4.0–5.2	0.20–0.50	0.15–0.55	—	—	—	0.15–0.35	0.40–1.2	0.05	0.10

[a]Values are maximum except where indicated as a range. Aluminum is remainder.
[b]If Fe exceeds 0.45%, it is desirable to have Mn present in an amount equal to one-half Fe.

TABLE 2—CHEMICAL COMPOSITIONS OF SAE ALUMINUM PERMANENT MOLD CASTING ALLOYS[a]

SAE No.	AA Designation	Si	Fe	Cu	Mn	Mg	Cr	Ni	Zn	Ti	Ag	Other Elements Each	Other Elements Total
34	222.0	2.0	1.5	9.2–10.8	0.50	0.15–0.35	—	0.50	0.8	0.25	—	—	0.35
35	443.0	4.5–6.0	0.8	0.6	0.50	0.05	0.25	—	0.50	0.25	—	—	0.35
39	242.0	0.7	1.0	3.5–4.5	0.35	1.2–1.8	0.25	1.7–2.3	0.35	0.25	—	0.05	0.15
312	707.0	0.20	0.8	0.20	0.40–0.6	1.8–2.4	0.20–0.40	—	4.0–4.5	0.25	—	0.05	—
314	C712.0	0.35	1.4	0.35–0.65	0.05	0.25–0.45	—	—	6.0–7.0	0.25	—	0.05	0.15
315	713.0	0.25	1.3	0.40–1.0	0.6	0.20–0.50	0.35	0.15	7.0–8.0	0.25	—	0.10	0.25
321	A332.0	11.0–12.5	1.3	0.50–1.5	0.35	0.7–1.3	—	2.0–3.0	0.35	0.25	—	0.05	—
322	355.0	4.5–5.5	0.8[b]	1.0–1.5	0.50[b]	0.40–0.6	0.25	—	0.35	0.25	—	0.05	0.15
323	356.0	6.5–7.5	0.6	0.25	0.35	0.20–0.40	—	—	0.35	0.25	—	0.05	0.15
326	319.0	5.5–6.5	1.0	3.0–4.0	0.50	0.10	—	0.35	1.0	0.25	—	—	0.50
329	—	5.5–6.5	1.2	3.0–4.0	0.8	0.10–0.50	—	0.50	1.0	0.25	—	—	0.50
331	333.1	8.0–10.0	1.0	3.0–4.0	0.50	0.05–0.50	—	0.50	1.0	0.25	—	—	0.50
332	F332.0	8.5–10.5	1.2	2.0–4.0	0.50	0.50–1.5	—	0.50	1.0	0.25	—	—	0.50
334	—	11.0–13.0	1.0	1.8–2.8	0.50	0.7–1.3	—	1.0	1.0	0.25	—	—	0.50
335	C355.0	4.5–5.5	0.20	1.0–1.5	0.10	0.40–0.6	—	—	0.10	0.20	—	0.05	0.15
336	A356.0	6.5–7.5	0.20	0.20	0.10	0.20–0.40	—	—	0.10	0.20	—	0.05	0.15
380	B295.0	2.5–3.5	1.2	3.5–4.5	0.50	0.10	—	0.35	1.0	0.25	—	—	0.50
382	201.0	0.10	0.15	4.0–5.2	0.20–0.50	0.15–0.55	—	—	—	0.15–0.35	0.40–1.2	0.05	0.10

[a]Values are maximum except where indicated as a range. Aluminum is remainder.
[b]If Fe exceeds 0.45%, it is desirable to have Mn present in an amount equal to one-half Fe.

TABLE 3—CHEMICAL COMPOSITIONS OF SAE ALUMINUM DIE CASTING ALLOYS[a]

SAE No.	AA Designation	Si	Fe	Cu	Mn	Mg	Cr	Ni	Zn	Ti	Sn	Other Elements Each	Other Elements Total
303	384.0	10.5–12.0	1.3	3.0–4.5	0.50	0.10	—	0.50	3.0[b]	—	0.35	—	0.50
304	C443.0	4.5–6.0	2.0	0.6	0.35	0.10	—	0.50	0.50	—	0.15	—	0.25
305	A413.0	11.0–13.0	1.3	0.6	0.35	0.10	—	0.50	0.50	—	0.15	—	0.25
306	A380.0	7.5–9.5	1.3	3.0–4.0	0.50	0.10	—	0.50	3.0[b]	—	0.35	—	0.50
308	380.0	7.5–9.5	2.0	3.0–4.0	0.50	0.10	—	0.50	3.0[b]	—	0.35	—	0.50
309	A360.0	9.0–10.0	1.3	0.6	0.35	0.40–0.6	—	0.50	0.50	—	0.15	—	0.25
383	383.0	9.5–11.5	1.3	2.0–3.0	0.50	0.10	—	0.30	3.0	—	0.15	—	0.50

[a]Values are maximum except where indicated as a range. Aluminum is remainder.
[b]Lower Zn content may be specified on agreement between purchaser and supplier.

TABLE 4—CHEMICAL COMPOSITION OF SAE ALUMINUM CASTING ALLOYS INGOTS[a]

SAE No.	AA Designation	Si	Fe	Cu	Mn	Mg	Cr	Ni	Zn	Ti	Sn	Other Elements Each	Other Elements Total
34	222.1	2.0	1.2	9.2–10.8	0.50	0.20–0.35	—	0.50	0.8	0.25	—	—	0.35
35	443.1	4.5–6.0	0.6	0.6	0.50	0.05	0.25	—	0.50	0.25	—	—	0.35
38	295.1	0.7–1.2	0.8	4.0–5.0	0.30	0.03	—	—	0.30	0.20	—	0.05	0.15
39	242.1	0.7	0.8	3.5–4.5	0.35	1.3–1.8	0.25	1.7–2.3	0.35	0.25	—	0.05	0.15
303	384.1	10.5–12.0	1.0	3.0–4.5	0.50	0.10	—	0.50	2.9[b]	—	0.20	—	0.50
304	C443.1	4.5–6.0	1.0	0.6	0.35	0.10	—	0.50	0.35	—	0.15	—	0.25
305	A413.1	11.0–13.0	1.0	0.6	0.35	0.10	—	0.50	0.35	—	0.15	—	0.25
306	A380.1	7.5–9.5	1.0	3.0–4.0	0.50	0.10	—	0.50	2.9[b]	—	0.20	—	0.50
308	380.1												
309	A360.1	9.0–10.0	1.0	0.6	0.35	0.45–0.6	—	0.50	0.35	—	0.15	—	0.25
310	D712.1	0.25	0.40	0.25	0.10	0.50–0.65	0.40–0.6	—	5.0–6.5	0.15–0.25	—	0.05	0.20
311	705.1	0.20	0.6	0.20	0.40–0.6	1.5–1.8	0.20–0.40	—	2.7–3.3	0.25	—	0.05	—
312	707.1	0.20	0.6	0.20	0.40–0.6	1.9–2.4	0.20–0.40	—	4.0–4.5	0.25	—	0.05	—
313	A712.1	0.15	0.40	0.35–0.65	0.05	0.65–0.8	—	—	6.0–7.0	0.25	—	0.05	0.15
314	C712.1	0.35	1.2	0.35–0.65	0.05	0.30–0.45	—	—	6.0–7.0	0.25	—	0.05	0.15
315	713.1	0.25	0.8[c]	0.40–1.0	0.6	0.25–0.50	0.35	0.15	7.0–8.0	0.25	—	0.10	0.25
320	514.1	0.35	0.40	0.15	0.35	3.6–4.5	—	—	0.15	0.25	—	0.05	0.15
321	A332.1	11.0–12.5	1.1	0.50–1.5	0.35	0.8–1.3	—	2.0–3.0	0.35	0.25	—	0.05	—
322	355.1	4.5–5.5	0.6	1.0–1.5	0.50[d]	0.45–0.6	0.25	—	0.35	0.25	—	0.05	0.15
323	356.1	6.5–7.5	0.50	0.25	0.35	0.25–0.40	—	—	0.35	0.25	—	0.05	0.15
324	520.1	0.20	0.20	0.20	0.10	9.6–10.6	—	—	0.10	0.20	—	0.05	0.15
326	319.1	5.5–6.5	0.8	3.0–4.0	0.50	0.10	—	0.35	1.0	0.25	—	—	0.50
327	328.1	7.0–8.6	0.8	1.0–2.0	0.20–0.6	0.25–0.6	0.35	0.25	1.5	0.25	—	—	0.50
329	—	5.5–6.5	1.0	3.0–4.0	0.8	0.10–0.50	—	0.50	1.0	0.25	—	—	0.50
331	333.1	8.0–10.0	0.8	3.0–4.0	0.50	0.10–0.50	—	0.50	1.0	0.25	—	—	0.50
332	F332.1	8.5–10.5	1.0	2.0–4.0	0.50	0.6–1.5	—	0.50	1.0	0.25	—	—	0.50
334	Z332.1	11.0–13.0	0.8	1.8–2.8	0.50	0.8–1.3	—	1.0	1.0	0.25	—	—	0.50
335	C355.2	4.5–5.5	0.15	1.0–1.5	0.05	0.45–0.6	—	—	0.05	0.20	—	0.05	0.15
336	A356.2	6.5–7.5	0.12	0.10	0.05	0.25–0.40	—	—	0.05	0.20	—	0.05	0.15
380	B291.1	2.5–3.5	1.0	3.5–4.5	0.50	0.10	—	0.35	1.0	0.25	—	0.05	0.15
382	201.2	0.10	0.10	4.0–5.2	0.20–0.50	0.20–0.55	—	—	—	0.15–0.55	Ag 0.40–1.2	0.05	0.10
383	383.1	9.5–11.5	0.6–1.0	2.0–3.0	0.50	0.10	—	0.30	2.9	—	0.15	—	0.50

[a] Percentages are maximum except where noted as a range. Aluminum is remainder.
[b] Lower Zn content may be specified by agreement between purchaser and supplier.
[c] When used for permanent mold castings (ASTM ZC81B), 1.1% Fe is permissible.
[d] If Fe exceeds 0.45%, it is desirable to have Mn present in an amount equal to one-half Fe.

TABLE 5—MECHANICAL AND PHYSICAL PROPERTIES OF SAE ALUMINUM SAND CASTING ALLOYS

SAE No.	Temper	AA Designation	Tensile Strength, Min (psi)	Tensile Strength, Min (kPa)	Yield Strength (0.2% offset), Min (psi)	Yield Strength (0.2% offset), Min (kPa)	Elongation in 2 in (51 mm) Min, %	Hardness (500 kg) Bhn	Endurance Limit (R. R. Moore Specimen) 5 × 10^8 Cycles (psi)	Endurance Limit (kPa)	Specific Gravity	Density (lb/in³)	Density (kg/m³)	Thermal Conductivity at 25 °C cgs units	Coefficient of Thermal Expansion 68–392 °F (20–200 °C) × 10^−6 per °F	per °C
34	T2	222.0	23 000	158 579	—	—	—[a]	—	—	—	2.95	0.107	2962	—	12.9	23.2
	T61		30 000	206 843	—	—	—[a]	115	8 500	58 605	—	—	—	0.31	—	—
35	F	443.0	17 000	117 210	7 000	48 263	3.0	40	8 000	55 158	2.69	0.097	2685	0.34	12.9	23.2
38	T4	295.0	29 000	199 948	13 000	89 632	6.0	60	7 000	48 263	2.81	0.102	2823	0.33	13.2	23.8
	T6		32 000	220 632	20 000	137 895	3.0	75	7 500	51 711	—	—	—	—	—	—
	T62		36 000	248 211	28 000	193 056	—[a]	90	8 000	55 158	—	—	—	0.33	—	—
	T7		29 000	199 948	16 000	110 316	3.0	—	—	—	—	—	—	—	—	—
39	T21	242.0	23 000	158 579	—	—	—[a]	—	—	—	—	—	—	—	—	—
	T61		32 000	220 632	20 000	137 895	—[a]	—	—	—	2.81	0.102	2823	0.31	13.1	23.6
310	T5[b]	D712.0	34 000	234 421	25 000	172 369	4.0	80	—	—	—	0.102	2823	—	—	—
311	T5[c]	705.0	30 000	206 843	17 000	117 211	5.0	65	—	—	—	0.101	2796	—	—	—
312	T5[c]	707.0	33 000	227 527	22 000	151 685	2.0	75	—	—	—	0.101	2796	—	—	—
	T7		37 000	255 106	30 000	206 843	1.0	80	—	—	—	—	—	—	—	—
313	T5[d]	A712.0	32 000	220 632	20 000	137 895	2.0	75	8 000	55 158	2.81	0.102	2823	0.33	13.8	24.8
315	T5[d]	713.0	30 000	206 843	22 000	151 685	3.0	75	9 000	62 053	2.81	0.102	2768	0.33	13.4	24.2
320	F	514.0	22 000	151 685	9 000	62 053	6.0	50	7 000	48 263	2.65	0.096	2657	0.33	13.9	25.0
322	T51	355.0	25 000	172 369	18 000	124 106	—[a]	65	8 000	55 158	2.71	0.098	2713	0.40	13.0	23.4
	T6		32 000	220 632	20 000	137 895	2.0	80	9 000	62 053	—	—	—	0.34	—	—
	T71		30 000	206 843	22 000	151 685	—[a]	75	10 000	68 948	—	—	—	—	—	—
323	T51	356.0	23 000	158 579	16 000	110 316	—[a]	60	8 000	55 158	2.68	0.097	2685	0.40	—	—
	T6		30 000	206 843	20 000	137 895	3.0	70	8 500	58 605	—	—	—	0.36	12.5	22.5
324	T4	520.0	42 000	289 580	22 000	151 685	12.0	75	8 000	55 158	2.57	0.093	2574	0.21	14.2	25.6
326	F	319.0	23 000	158 579	13 000	89 632	1.5	70	—	—	2.79	0.101	2796	0.26	12.4	22.3
	T6		32 000	220 632	20 000	137 895	2.5	80	—	—	—	—	—	—	—	—
327	F	328.0	25 000	172 369	14 000	96 527	1.0	60	—	—	2.71	0.098	2713	—	—	—
	T6		34 000	234 421	21 000	144 790	1.0	85	—	—	—	—	—	—	—	—
382	T6	201.0	60 000	413 685	50 000	344 738	5.0	130	14 000	96 527	2.80	0.101	2796	0.29	12.6	22.7
	T7		60 000	413 685	50 000	344 738	3.0	130	14 000	96 527	2.80	0.101	2796	0.29	12.6	22.7

[a] Not required. The error in determining elongation of 1% or less is comparable with the value being measured.
[b] Aged 21 d at room temperature or artificially aged at 350 ± 10 °F (177 ± 6 °C) for 10 h.
[c] Aged 21 d at room temperature or artificially aged at 210 ± 10 °F (99 ± 6 °C) for 8 h.
[d] Aged 21 d at room temperature.
[e] The properties shown are obtainable in castings of this alloy and heat treat condition when produced under the quality assurance provisions of MIL-A-21180, Aluminum Alloy Castings, High Strength. Strict adherence to these requirements is mandatory.

TABLE 6—MECHANICAL AND PHYSICAL PROPERTIES OF SAE ALUMINUM PERMANENT MOLD CASTING ALLOYS

SAE No.	Temper	AA Designation	Specified Limits					Typical Properties								
			Tensile Strength, Min		Yield Strength (0.2% offset), Min		Elongation in 2 in (51 mm) Min, %	Hardness (500 kg) Bhn	Endurance Limit (R. R. Moore Specimen) 5 x 10^8 Cycles		Specific Gravity	Density		Thermal Conductivity at 25 °C cgs units	Coefficient of Thermal Expansion 68–392°F (20–200 °C) x 10^{-6}	
			psi	kPa	psi	kPa			psi	kPa		lb/in^3	kg/m^3		per °F	per °C
34	T551	222.0	30 000	206 843	—	—	—[a]	115	8 500	58 605	2.95	1.107	2962	0.31	12.9	23.2
	T65		40 000	275 790	—	—	—[a]	140	9 000	62 053						
35	F	443.0	21 000	144 790	7 000	48 263	2.0	45	8 000	55 158	2.69	0.097	2685	0.34	12.9	23.2
39	T571	242.0	34 000	234 422	—	—	—[a]	105	10 500	72 395	2.81	0.102	2823	0.32	13.1	23.6
	T61		40 000	275 790	—	—	—[a]	110	9 500	65 500				0.31		
311	T5[b]	705.0	37 000	255 106	17 000	117 211	10.0	70	—	—	2.76	0.100	2768	—	—	—
312	T5[b]	707.0	42 000	289 580	25 000	172 369	4.0	95	—	—	2.77	0.100	2768	—	—	—
	T7		45 000	310 264	35 000	241 317	3.0	95	—	—						
314	T5[b]	C712.0	28 000	193 053	18 000	124 106	7.0	70	11 000	75 842	2.84	0.103	2851	0.37	13.6	24.5
315	T5[c]	713.0	32 000	220 632	22 000	151 685	4.0	75	—	—	2.74	0.099	2740	—	—	—
321	T551	A332.0	31 000	213 737	—	—	—[a]	105	13 500	93 079	2.72	0.098	2713	0.28	11.5	20.7
	T65		40 000	275 790	—	—	—[a]	125	—	—						
322	T6	355.0	37 000	255 106	23 000	158 579	1.5	90	10 000	68 948	—	0.098	2713	0.34	13.0	23.4
323	F	356.0	21 000	144 790	—	—	3.0	—	—	—	—	—	—	—	—	—
	T6		33 000	227 527	22 000	151 685	3.0	80	13 000	89 632	2.68	0.097	2685	0.36	12.5	22.5
	T71		25 000	172 369	—	—	3.0	—	—	—						
326	F	319.0	27 000	186 158	14 000	96 526	2.5	85	—	—	2.74	0.100	2768	0.26	12.4	22.3
	T61		40 000	275 790	24 000	165 474	2.0	95	—	—						
331	F	333.0	28 000	193 053	—	—	—[a]	90	14 500	99 974	2.77	0.100	2768	0.25	—	—
	T5		30 000	206 843	—	—	—[a]	100	12 000	82 737	2.77	0.100	2768	0.28	—	—
	T6		35 000	241 317	—	—	—[a]	105	15 000	103 421	2.77	0.100	2768	0.28	—	—
	T7		31 000	213 737	—	—	—[a]	90	12 000	82 737	2.77	0.100	2768	0.33	—	—
332	T5	F332.0	31 000	213 737	—	—	—[a]	105	—	—	2.77	0.100	2768	0.25	12.0	21.6
334	T551	Z332.0	31 000	213 737	—	—	—[a]	—	—	—	2.72	0.098	2713	0.28	11.5	20.7
335	T61	C355.0	40 000	275 790	30 000	206 843	3.0	90	10 000	68 948	2.71	0.098	2713	0.34	13.0	23.4
336	T61	A356.0	38 000	262 001	26 000	179 264	5.0	80	13 000	89 632	2.68	0.097	2685	0.36	12.5	22.5
380	T4	B295.0	33 000	227 527	15 000	103 421	4.5	75	9 500	65 500	2.80	0.101	2796	0.31	13.0	23.4
	T6		35 000	241 317	22 000	151 685	2.0	90	10 000	68 948						
	T7		33 000	227 527	16 000	110 316	3.0	80	9 000	62 053						
382	T6	201.0	60 000[d]	413 685	50 000[d]	344 738	5.0[d]	130	14 000	96 527	2.80	0.101	2768	0.29	12.6	22.7
	T7		60 000[d]	413 685	50 000[d]	344 738	3.0[d]	130	14 000	96 527	2.80	0.101	2768	0.29	12.6	22.7

[a] Not required. The error in determining elongation of 1% or less is comparable with the value being measured.
[b] Aged 21 d at room temperature or artificially aged at 210 ± 10 °F (99 ± 6 °C) for 8 h.
[c] Aged 21 d at room temperature or one day at room temperature plus 250 ± 10 °F (121 ± 6 °C) for 16 h.
[d] The properties shown are obtainable in castings of this alloy and heat treat condition when produced under the quality assurance provisions of MIL-A-21180 Aluminum Alloy Castings, High Strength. Strict adherence to the requirements is mandatory.

TABLE 7—TYPICAL MECHANICAL AND PHYSICAL PROPERTIES OF SAE ALUMINUM DIE CASTING ALLOYS[a]

SAE No.	AA Designation	Tensile Strength		Yield Strength, (0.2% offset)		Elongation in 2 in (51 mm), %	Endurance Limit (R. R. Moore Specimen) 5 x 10^8 Cycles		Shear Strength		Specific Gravity	Density		Thermal Conductivity at 25°C cgs units	Coefficient of Thermal Expansion 68–392 °F (20–200 °C) x 10^{-6}	
		psi	kPa	psi	kPa		psi	kPa	psi	kPa		lb/in^3	kg/m^3		per °F	per °C
303	384.0	48 000	330 948	24 000	165 474	2.5	20 000	137 895	29 000	199 948	2.70	0.098	2740	0.23	11.8	21.2
304	C443.0	33 000	227 527	14 000	96 526	9.0	17 000	117 211	19 000	131 000	2.69	0.097	2685	0.34	12.9	23.2
305	A413.0	42 000	289 580	19 000	131 000	3.5	19 000	131 000	25 000	172 369	2.65	0.096	2657	0.29	12.0	21.6
306	A380.0	47 000	324 054	23 000	158 579	3.5	20 000	137 895	27 000	186 158	2.71	0.098	2740	0.24	12.2	22.0
308	380.0	46 000	317 159	23 000	158 579	2.5	20 000	137 895	28 000	193 053	2.72	0.098	2740	0.23	12.1	21.8
309	A360.0	46 000	317 159	24 000	165 474	3.5	18 000	124 106	26 000	179 264	2.64	0.095	2630	0.27	12.2	22.0
383	383.0	45 000	310 264	22 000	151 685	—	—	—	—	—	—	—	—	—	—	—

[a] The data in Table 3 do not constitute a part of the SAE Standard. Tensile properties are determined from test specimens cast in a die and conforming to chemical composition specified. Test bars machined from castings do not provide a reliable measure of the strength properties of the casting and this method should not be used to determine conformance to data shown in Table 3. Hardness values have not been shown because they are considered too unreliable on die castings.

GENERAL DATA ON WROUGHT ALUMINUM ALLOYS—SAE J454k

SAE Information Report

Report of Nonferrous Metals Division approved June 1911 and last revised by Nonferrous Metals Committee June 1974.

The SAE Standards for wrought aluminum alloys cover materials with a considerable range of properties and other characteristics, but do not include all of the commercially available materials. If none of the materials listed provides the characteristics required by a particular application, users may find it helpful to consult with the suppliers of aluminum alloy products.

Wrought aluminum alloys are generally classified as heat treatable and nonheat treatable. The heat treatable group includes alloys in which higher strengths are produced by heat treatment processes. The nonheat treatable group are alloys in which increased strengths are obtained by strain hardening developed by cold working methods.

Temper designation on the nonheat treatable alloys indicates whether the material has been brought to the final level of strength by cold working only (designation H1X), or by cold working followed by partial annealing (designation H2X), or by cold working followed by stabilizing (designation H3X). Basic temper designation and subdivision applicable to alloys described later in this section are shown in SAE J993. In addition, the following list describes special tempers that appear in the tabular data for the alloys listed:

Alloy 2218-T72—Die Forgings—Solution heat treated and stabilized by precipitation heat treatment.

2618-T61—Die Forgings—Solution heat treated and precipitation heat treated for good elevated temperature strength properties.

2024-T72—Sheet and Plate—Solution heat treated and artificially aged by the user to develop increased resistance to stress-corrosion cracking.

2024-T361—Sheet and Plate—Solution heat treated, cold-worked by reduction of approximately 6%.

2024-T861—Sheet and Plate—Solution heat treated, cold-worked by reduction of approximately 6% and artificially aged.

2219-T31—Sheet and Plate—Solution heat treated, cold-worked by flattening or straightening operations.

2219-T81—Sheet and Plate—T31 Temper material precipitation heat treated.

2219-T37—Sheet and Plate—Solution heat treated and cold-worked by reduction of approximately 8%.

2219-T87—Sheet and Plate—T37 material precipitation heat treated.

6063-T83, T831 and T832 Tempers—Drawn Tube—Solution heat treated, cold-worked with varying reductions and subsequently precipitation heat treated to develop the different levels of specified mechanical properties.

7075-T73, T7351, T73510, T73511 and T7352—Wrought Products—T73 temper products in any wrought form are solution heat treated and stabilized to develop the specified mechanical properties and high resistance to stress-corrosion cracking. The T7351 temper for plate, rolled rod or bar, or cold finished rod and bar and drawn tube is stress-relieved by straightening after solution heat treatment and prior to artificial aging.

Data on typical heat treatments for aluminum alloy mill products have been deleted, and instead the reader is referred to the section on Heat Treatments/Fabrication in the Aluminum Association Manual, Aluminum Standards and Data. Data are given for the recommended time and temperature for satisfactorily performing solution heat treatments, precipitation heat treatments, and annealing of aluminum alloy mill products used in automotive applications.

Data on Similar Specifications of SAE Wrought Aluminum Alloys has been deleted, and instead the reader is referred to the section on Specification Cross Reference/General Information in the Aluminum Association Manual, Aluminum Standards and Data, specifically, the tables Aluminum Mill Product Specifications and Specifications Covering Aluminum Mill Products.

The following data, presented in Tables 1–7, are reprinted from the AA Manual, Aluminum Standards Data, with the permission of the Aluminum Association:

1. Typical Characteristics and Applications of Wrought Aluminum Alloys (Table 1)
2. Wrought Aluminum Alloy Products and Tempers (Table 2)
3. Wrought Aluminum Alloy Specialty Mill Products and Tempers (Table 3)
4. Typical Physical Properties of Wrought Aluminum Alloys (Table 4)
5. Typical Mechanical Properties of Wrought Aluminum Alloys (Table 5)
6. Typical Tensile Properties at Various Temperatures (Table 6)
7. Nominal Chemical Composition—Wrought Alloys (Table 7)

TABLE 1—TYPICAL CHARACTERISTICS AND APPLICATIONS OF WROUGHT ALUMINUM

Alloy and Temper	Resistance to Corrosion		Workability (Cold)[e]	Machineability[e]	Brazeability[f]	Weldability[f]			Some Applications of Alloy
	General[a]	Stress-Corrosion Cracking[b]				Gas	Arc	Resistance Spot and Seam	
EC-0	A	A	A	E	A	A	A	B	Electrical conductors
H12, H111	A	A	A	E	A	A	A	A	
H14, H24	A	A	A	D	A	A	A	A	
H16, H26	A	A	B	D	A	A	A	A	
H18	A	A	B	D	A	A	A	A	
1060-0	A	A	A	E	A	A	A	B	Chemical equipment, railroad tank cars
H12	A	A	A	E	A	A	A	A	
H14	A	A	A	D	A	A	A	A	
H16	A	A	B	D	A	A	A	A	
H18	A	A	B	D	A	A	A	A	
1100-0	A	A	A	E	A	A	A	B	Sheet metal work, spun hollowware, fin stock
H12	A	A	A	E	A	A	A	A	
H14	A	A	A	D	A	A	A	A	
H16	A	A	B	D	A	A	A	A	
H18	A	A	C	D	A	A	A	A	
2011-T3	D[c]	D	C	A	D	D	D	D	Screw machine products
T4, T451	D[c]	D	B	A	D	D	D	D	
T8	D	B	D	A	D	D	D	D	
2014-0	—	—	—	D	D	D	D	B	Truck frames, aircraft structures
T3, T4, T451	D[c]	C	C	B	D	D	B	B	
T6, T651, T6510, T6511	D	C	D	B	D	D	B	B	
2017-T4, T451	D[c]	C	C	B	D	D	B	B	Screw machine products, fittings
2018-T61	—	—	—	B	—	—	—	—	Aircraft engine cylinders, heads and pistons
2024-0	—	—	—	D	D	D	D	D	Truck wheels, screw machine products, aircraft structures
T4, T3, T351, T3510, T3511	D[c]	C	C	B	D	C	B	B	
T361	D[c]	C	D	B	D	D	C	B	
T6	D	B	C	B	D	D	C	B	
T861, T81, T851, T8510, T8511	D	B	D	B	D	D	C	B	
T72	—	—	—	B	—	—	—	—	
2025-T6	D	C	—	B	D	D	B	B	Forgings, aircraft propellers
2117-T4	C	A	B	C	D	D	B	B	
2218-T61	D	C	—	—	—	—	—	C	Jet engine impellers and rings
T72	D	C	—	B	D	D	C	B	

(Table continued on next page)

TABLE 1—TYPICAL CHARACTERISTICS AND APPLICATIONS OF WROUGHT ALUMINUM (CONTINUED)

Alloy and Temper	Resistance to Corrosion		Work-ability (Cold)[e]	Machine-ability[e]	Braze-ability[f]	Weldability			Some Applications of Alloy
	General[a]	Stress-Corrosion Cracking[b]				Gas	Arc	Resistance Spot and Seam	
2219-0	—	—	—	—	D	D	A	B	Structural uses at high temperatures (to 600 °F (316 °C)), high strength weldments
T31, T351, T3510, T3511	D[c]	C	C	B	D	A	A	A	
T37	D[c]	C	D	B	D	A	A	A	
T81, T851, T8510, T8511	D	B	D	B	D	A	A	A	
T87	D	B	D	B	D	A	A	A	
2618-T61	D	C	—	B	D	D	C	B	Aircraft engines
3003-0	A	A	A	E	A	A	A	A	Cooking utensils, chemical equipment, pressure vessels, sheet metal work, builder's hardware, storage tanks
H12	A	A	A	E	A	A	A	A	
H14	A	A	B	D	A	A	A	A	
H16	A	A	C	D	A	A	A	A	
H18	A	A	C	D	A	A	A	A	
H25	A	A	B	D	A	A	A	A	
3004-0	A	A	A	D	B	B	A	B	Sheet metal work, storage tanks
H32	A	A	B	D	B	B	A	A	
H34	A	A	B	C	B	B	A	A	
H36	A	A	C	C	B	B	A	A	
H38	A	A	C	C	B	B	A	A	
3105-0	A	A	A	E	B	B	A	B	Residential siding, mobile homes, rain carrying goods, sheet metal work
H12	A	A	B	E	B	B	A	A	
H14	A	A	B	D	B	B	A	A	
H16	A	A	C	D	B	B	A	A	
H18	A	A	C	D	B	B	A	A	
H25	A	A	B	D	B	B	A	A	
4032-T6	C	B	—	B	D	D	B	C	Pistons
5005-0	A	A	A	E	B	A	A	B	Appliances, utensils, architectural, electrical conductor
H12	A	A	A	E	B	A	A	A	
H14	A	A	B	D	B	A	A	A	
H16	A	A	C	D	B	A	A	A	
H18	A	A	C	D	B	A	A	A	
H32	A	A	A	E	B	A	A	A	
H34	A	A	B	D	B	A	A	A	
H36	A	A	C	D	B	A	A	A	
H38	A	A	C	D	B	A	A	A	
5050-0	A	A	A	E	B	A	A	B	Builder's hardware, refrigerator trim, coiled tubes
H32	A	A	B	D	B	A	A	A	
H34	A	A	B	D	B	A	A	A	
H36	A	A	C	C	B	A	A	A	
H38	A	A	C	C	B	A	A	A	
5052-0	A	A	A	D	C	A	A	B	Sheet metal work, hydraulic tube, appliances
H32	A	A	B	D	C	A	A	A	
H34	A	A	B	C	C	A	A	A	
H36	A	A	C	C	C	A	A	A	
H38	A	A	C	C	C	A	A	A	
5056-0	A[d]	B[d]	A	D	D	C	A	B	Cable sheathing, rivets for magnesium, screen wire, zippers
H111	A[d]	B[d]	A	D	D	C	A	A	
H12, H32	A[d]	B[d]	B	D	D	C	A	A	
H14, H34	A[d]	B[d]	C	D	D	C	A	A	
H18, H38	A[d]	C[d]	C	C	D	C	A	A	
H192	B[d]	D[d]	D	B	D	C	A	A	
H392	B[d]	D[d]	D	B	D	C	A	A	
5083-0	A[d]	B[d]	B	D	D	C	A	B	Unfired, welded pressure vessels, marine, auto aircraft cryogenics, TV towers, drilling rigs, transportation equipment, missile components
H321	A[d]	B[d]	C	D	D	C	A	A	
H323	A[d]	B[d]	C	D	D	C	A	A	
H343	A[d]	B[d]	C	C	D	C	A	A	
H111	A[d]	B[d]	C	D	D	C	A	A	
5086-0	A[d]	A[d]	A	D	D	C	A	B	
H32, H116, H117	A[d]	A[d]	B	D	D	C	A	A	
H34	A[d]	B[d]	B	C	D	C	A	A	
H36	A[d]	B[d]	C	C	D	C	A	A	
H38	A[d]	B[d]	C	C	D	C	A	A	
H111	A[d]	A[d]	B	D	D	C	A	A	
5154-0	A[d]	A[d]	A	D	D	C	A	B	Welded structures, storage tanks, pressure vessels, salt water service
H32	A[d]	A[d]	B	D	D	C	A	A	
H34	A[d]	A[d]	B	C	D	C	A	A	
H36	A[d]	A[d]	C	C	D	C	A	A	
H38	A[d]	A[d]	C	C	D	C	A	A	
5252-H24	A	A	B	D	C	A	A	A	Automotive and appliance trim
H25	A	A	B	C	C	A	A	A	
H28	A	A	C	C	C	A	A	A	
5254-0	A[d]	A[d]	A	D	D	C	A	B	Hydrogen peroxide and chemical storage vessels
H32	A[d]	A[d]	B	D	D	C	A	A	
H34	A[d]	A[d]	B	C	D	C	A	A	
H36	A[d]	A[d]	C	C	D	C	A	A	
H38	A[d]	A[d]	C	C	D	C	A	A	

(Table continued on next page)

TABLE 1—TYPICAL CHARACTERISTICS AND APPLICATIONS OF WROUGHT ALUMINUM (CONTINUED)

Alloy and Temper	Resistance to Corrosion		Work-ability (Cold)[c]	Machine-ability[e]	Braze-ability[f]	Weldability			Some Applications of Alloy
	General[a]	Stress-Corrosion Cracking[b]				Gas	Arc	Resistance Spot and Seam	
5454-0	A	A	A	D	D	C	A	B	Welded structures, pressure vessels, marine service
H32	A	A	B	D	D	C	A	A	
H34	A	A	B	C	D	C	A	A	
H111	A	A	B	D	D	C	A	A	
5456-0	A[d]	B[d]	B	D	D	C	A	B	High strength welded structures, storage tanks, pressure vessels, marine applications
H111	A[d]	B[d]	C	D	D	C	A	A	
H321[g], H116, H117	A[d]	B[d]	C	D	D	C	A	A	
H323	A[d]	B[d]	C	D	D	C	A	A	
H343	A[d]	B[d]	C	C	D	C	A	A	
5457-0	A	A	A	E	B	A	A	B	
5652-0	A	A	A	D	C	A	A	B	Hydrogen peroxide and chemical storage vessels
H32	A	A	B	D	C	A	A	A	
H34	A	A	B	C	C	A	A	A	
H36	A	A	C	C	C	A	A	A	
H38	A	A	C	C	C	A	A	A	
5657-H241	A	A	A	D	B	A	A	A	Anodized automotive and appliance trim
H25	A	A	B	D	B	A	A	A	
H26	A	A	B	D	B	A	A	A	
H28	A	A	C	D	B	A	A	A	
6005-T5	B	A	C	C	A	A	A	A	Heavy-duty structures requiring good corrosion resistance,—truck and marine, railroad cars, furniture, pipelines
6053-0	—	—	—	E	A	A	A	B	Wire and rod for rivets
T6, T61	A	A	—	C	A	A	A	A	
6061-0	B	A	A	D	A	A	A	B	Heavy-duty structures requiring good corrosion resistance, truck and marine, railroad cars, furniture, pipelines
T4, T451, T4510, T4511	B	B	B	C	A	A	A	A	
T6, T651, T652, T6510, T6511	B	A	C	C	A	A	A	A	
6063-T1	A	A	B	D	A	A	A	A	Pipe railing, furniture, architectural extrusions
T4	A	A	B	D	A	A	A	A	
T5, T52	A	A	B	C	A	A	A	A	
T6	A	A	C	C	A	A	A	A	
T83, T831, T832	A	A	C	C	A	A	A	A	
6066-0	C	A	B	D	D	D	B	B	Forgings and extrusions for welded structures
T4, T4510, T4511	C	B	C	C	D	D	B	B	
T6, T6510, T6511	C	B	C	B	D	D	B	B	
6070-T4, T4511	B	B	B	C	B	A	A	A	Heavy-duty welded structures, pipelines
T6	B	B	C	C	B	A	A	A	
6101-T6, T63	A	A	C	C	A	A	A	A	High strength bus conductors
T61, T64	A	A	B	D	A	A	A	A	
6151-T6, T652	—	—	—	—	—	—	—	—	Moderate strength intricate forgings for machine and automotive parts
6201-T81	A	A	—	C	A	A	A	A	High strength electric conductor wire
6262-T6, T651, T6510, T6511	B	A	C	B	A	A	A	A	Screw machine products
T9	B	A	D	B	A	A	A	A	
6463-T1	A	A	B	D	A	A	A	A	Extruded architectural and trim sections
T5	A	A	B	C	A	A	A	A	
T6	A	A	C	C	A	A	A	A	
7001-0	C[c]	C	D	B	D	D	D	B	High strength structures
T6, T6510, T6511	—	—	—	—	—	—	—	—	
7075-0	—	—	—	D	D	D	C	B	Aircraft and other structures
T6, T651, T652, T6510, T6511	C[c]	C	D	B	D	D	C	B	
T73, T7351	C	B	D	B	D	D	C	B	
7079-0	—	—	—	—	D	D	C	B	Structural parts for aircraft
T6, T651, T652, T6510, T6511	C[c]	C	D	B	D	D	C	B	
7178-0	—	—	—	—	D	D	C	B	Aircraft and other structures
T6, T651, T6510, T6511	C[c]	C	D	B	D	D	C	B	

[a]Ratings A through E are relative ratings in decreasing order of merit, based on exposures to sodium chloride solution by intermittent spraying or immersion. Alloys with A and B ratings can be used in industrial and seacoast atmospheres without protection. Alloys with C, D, and E ratings generally should be protected at least on faying surfaces.

[b]Stress-corrosion cracking ratings are based on service experience and on laboratory tests of specimens exposed to the 3.5% sodium chloride alternate immersion test.
  A = No known instance of failure in service or in laboratory tests.
  B = No known instance of failure in service; limited failures in laboratory tests of short transverse specimens.
  C = Service failures with sustained tension stress acting in short transverse direction relative to grain structure; limited failures in laboratory tests of long transverse specimens.
  D = Limited service failures with sustained longitudinal or long transverse stress.

[c]In relatively thick sections the rating would be E.
[d]This rating may be different for material held at elevated temperature for long periods.
[e]Ratings A through D for Workability (cold), and A through E for Machinability, are relative ratings in decreasing order of merit.
[f]Ratings A through D for Weldability and Brazeability are relative ratings defined as follows:
[g]Material in this temper is not recommended for, and should not be used in, applications requiring exposure to sea water.
  A = Generally weldable by all commercial procedures and methods.
  B = Weldable with special techniques or for specific applications which justify preliminary trials or testing to develop welding procedure and weld performance.
  C = Limited weldability because of crack sensitivity or loss in resistance to corrosion and mechanical properties.
  D = No commonly used welding methods have been developed.

TABLE 2—WROUGHT ALUMINUM ALLOY PRODUCTS AND TEMPERS

Alloy	Sheet	Plate	Tube Drawn	Tube Extruded	Pipe	Structural Shapes[a]	Extruded Wire, Rod, Bar and Shapes	Rolled or Cold Finished Rod	Rolled or Cold Finished Bar	Rolled or Cold Finished Wire	Rivets	Forgings and Forging Stock
EC[b]	O H12 H14 H16 H18 H112	O H12 H14 H112		H111	H111	H111	H111	O H12 H14 H16 H22 H24 H26 H111	H12 H111 H112	O H12 H14 H16 H19 H22 H24 H26		
1060	O H12 H14 H16 H18 H112	O H12 H14 H112	O H12 H14 H18 H112	O H112								
1100	O H12 H14 H16 H18 H112	O H12 H14 H112	O H12 H14 H16 H18	O H112			O H112	O H112 F	O H112 F	O H112 H12 H14 H16 H18	O H14	H112
1345								O H12 H14 H16 H18 H19	O H12 H14 H16 H18 H19	O H12 H14 H16 H18 H19		
2011								T3 T4 T451 T8	T3 T4 T451 T8	T3 T4 T451 T8		
2014	O T3 T4 T6	O T451 T651	O T4 T6	O T4 T4510 T4511 T6 T6510 T6511			O T4 T4510 T4511 T6510 T6511	O T4 T451 T6 T651	O T4 T451 T6 T651			T4 T6 T652
Alclad 2014	O T3 T4 T6	O T451 T651										
2017								O T4 T451	O T4 T451	O H13 T4	O H13 T4	
2018												T61
2024	O T3 T361 T4 T81 T861	O T361 T351 T851 T861	O T3	O T4 T3510 T3511 T8510 T8511			O T4 T3510 T3511 T81 T8510 T8511	O H13 T4 T351 T81 T851	O T351 T6 T81 T851	O H13 T36 T4	O H13 T4	
Alclad 2024	O T3 T361 T4 T81 T861	O T361 T351 T851 T861										
2025												T6
2117								O H13 H15 T4		O H13 H15 T4	O H13 H15 T4	
2124		T851										
2218												T61 T72
2219	O T31 T37 T81 T87	O T351 T37 T851 T87		O T31 T3510 T3511 T81 T8510 T8511			O T31 T3510 T3511 T81 T8510 T8511	O T31 T81 T851	O T31 T81 T851			T6 T852
Alclad 2219	O T31 T37 T81 T87	O T351 T37 T851 T87										

(Table continued on next page)

TABLE 2—WROUGHT ALUMINUM ALLOY PRODUCTS AND TEMPERS (CONTINUED)

Alloy	Sheet	Plate	Tube		Pipe	Structural Shapes[a]	Extruded Wire, Rod, Bar and Shapes	Rolled or Cold Finished			Rivets	Forgings and Forging Stock
			Drawn	Extruded				Rod	Bar	Wire		
2618												T61
3003	O H12 H14 H16 H18	O H12 H14 H112	O H12 H14 H16 H18 H25	O H112	H18 H112		O H112	O H112 F	O H112 F	O H112 H12 H14 H16 H18	O H14	H112
Alclad 3003	O H12 H14 H16 H18	O H12 H14 H112	O H14 H18 H25	O H112								
3004	O H32 H34 H36 H38	O H32 H34 H112	O H34 H36 H38	O								
Alclad 3004	O H32 H34 H36 H38	O H32 H34 H112										
3005	O H12 H14 H16 H18 H26 H28											
3105	O H12 H14 H16 H18 H25											
4032												T6
4043										O		
5005	O H12 H14 H16 H18 H32 H34 H36 H38	O H12 H14 H32 H34 H112						O H32		O H12 H14 H16 H19 H22 H24 H26 H32	O H32	
5050	O H32 H34 H36 H38	O H112	O H32 H34 H36 H38					O F	O F	O H32 H34 H36 H38		
5052	O H32 H34 H36 H38	O H32 H34 H112	O H32 H34 H36 H38					O F	O F	O H32 H34 H36 H38	O H32	
5056								O F	O F	O H111 H12 H14 H18 H32 H34 H38 H192 H392	O H32	
Alclad 5056								H393		H192 H392		
5083	O H321 H323 H343	O H112 H321		O H111 H112			O H111 H112	O H111 H112	O H111 H112			H111 H112
5086	O H32 H34 H36 H38 H112	O H32 H34 H112	O H32 H34 H36	O H111 H112			O H111 H112	O H111 H112	O H111 H112			

(Table continued on next page)

TABLE 2—WROUGHT ALUMINUM ALLOY PRODUCTS AND TEMPERS (CONTINUED)

Alloy	Sheet	Plate	Tube Drawn	Tube Extruded	Pipe	Structural Shapes[a]	Extruded Wire, Rod, Bar, and Shapes	Rolled or Cold Finished Rod	Rolled or Cold Finished Bar	Rolled or Cold Finished Wire	Rivets	Forgings and Forging Stock
5154	O H32 H34 H36 H38	O H32 H34 H112	O H34 H38	O H112			O H112	O H112 F	O H112 F	O H112 H32 H34 H36 H38		
5252	H24 H25 H28											
5254	O H32 H34 H36 H38	O H32 H34 H112										
5454	O H32 H34	O H32 H34 H112	H32 H34	O H111 H112			O H111 H112	O H111 H112	O H111 H112			
5456	O H321 H323 H343	O H112 H321		O H111 H112			O H111 H112	O H111 H112	O H111 H112			H112
5457	O											
5652	O H32 H34 H36 H38	O H32 H34 H112										
5657	H241 H25 H26 H28											
6005							T1 T5					
6053								O H13 T61		O H13 T61	O H13 T61	T6
6061	O T4 T6	O T451 T6 T651	O T4 T6	O T4 T4510 T4511 T6 T6510 T6511	T6	T4 T6	O T4 T4510 T4511 T6 T6510 T6511	O T4 T4510 T4511 T6 T651	O T4 T451 T6 T651	O H13 T6 T89 T93 T913 T94	O H13 T6	T6 T652
Alclad 6061	O T4 T6	O T451 T651										
6063			O T4 T6 T83 T831 T832	O T1 T4 T5 T52 T6	T6		O T1 T4 T5 T52 T6					
6066			O T4 T6	O T4 T4510 T4511 T6 T6510 T6511			O T4 T4510 T4511 T6 T6510 T6511	O T4 T6	O T4 T6			T6
6070				T6			T6					
6101[b]				T6 T61 T63 T64 H111	T6 T61 T63 T64 H111	T6 T61 T63 T64 H111	T6 T61 T63 T64 H111					
6151												T6 T652
6201[b]										T81		
6262			T6 T9	T6 T6510 T6511			T6 T6510 T6511	T6 T651 T9	T6 T651 T9	T6 T651 T9		
6463							T1 T5 T6 T62					

(Table continued on next page)

TABLE 2—WROUGHT ALUMINUM ALLOY PRODUCTS AND TEMPERS (CONTINUED)

Alloy	Sheet	Plate	Tube Drawn	Tube Extruded	Pipe	Structural Shapes[a]	Extruded Wire, Rod, Bar, and Shapes	Rolled or Cold Finished Rod	Rolled or Cold Finished Bar	Rolled or Cold Finished Wire	Rivets	Forgings and Forging Stock
7001				O T6 T6510 T6511			O T6 T6510 T6511	O T6	O T6			
7075	O T6 T73	O T651 T73 T7351	O T6 T73	O T6 T6510 T6511 T73			O T6 T6510 T6511 T73	O H13 T6 T651 T73 T7351	O T6 T651 T73 T7351	O H13	O H13 T6	T6 T652 T73
Alclad 7075	O T6	O T651										
Alclad one side 7075	O T6	O T651										
7079	O T6	T651					O T6 T6510 T6511					T6 T652
Alclad 7079	O T6											
Alclad one side 7079	O T6											
7178	O T6 T76	O T651 T7651					O T6 T6510 T6511 T76 T76510 T76511	O H13 T6 T76 T7651	O T6 T76 T7651	O H13	O H13 T6	
Alclad 7178	O T6	O T6 T651										

[a] Rolled or extruded.
[b] Products listed for these alloys are for electric conductors only.

Note: Additional alloys, tempers, and products are obtainable from some suppliers. See SAE J993. Suppliers should be consulted for current availability of alloys, tempers, and products.

TABLE 3—SPECIALTY MILL WROUGHT ALUMINUM PRODUCTS

Specialty Product Designation	Specialty Product Description Alloy	Temper	Form	Specialty Product Designation	Specialty Product Description Alloy	Temper	Form
Brazing sheet				Commercial roofing and siding			
Nos. 11 and 12	3003 clad with 4343 on one side (No. 11) or both (No. 12)	O H12 H14	Sheet	Corrugated roofing and siding V-beam roofing and siding Ribbed roofing Ribbed siding	Alclad 3004 Alclad 3004 Alclad 3004 Alclad 3004		Sheet Sheet Sheet Sheet
Nos. 21 and 22	6951 clad with 4343 on one side (No. 21) or both (No. 22)	O T42 T62	Sheet				
Nos. 23 and 24	6951 clad with 4045 on one side (No. 23) or both (No. 24)	O T42 T62	Sheet	Duct sheet	Alloy and temper with min tensile strength of 16.0 ksi		Coiled or flat sheet
Reflector sheet							
Clad 1100	1100 clad with 1175 on one or both sides		Sheet	Tread plate	6061	O T4 T6	Sheet and plate with raised pattern on one surface
Clad 3003	3003 clad with 1175 on one or both sides		Sheet				
Painted sheet	1100 3003	O H12 H14 H16 H18	Coiled sheet	Heat exchanger tube	1060 3003 Alclad 3003 5052 5454 6061	H14 H14 H25 H14 H25 H32 H34 H32 H34 T4 T6	Tube
	3105	O H12 H14 H16 H18 H25	Coiled sheet				
	5005 5050 5052	O H32 H34 H36 H38	Coiled sheet	Rigid electrical conduit	3003 6063	H12 T1	Tube

NOTE: Other alloys and tempers may be available from individual producers for some of these products.

TABLE 4A—TYPICAL PHYSICAL PROPERTIES OF WROUGHT ALUMINUM

The following typical properties are not guaranteed since in most cases they are averages for various sizes, product forms and methods of manufacture and may not be exactly representative of any particular product or size. These data are intended only as a basis for comparing alloys and tempers and should not be specified as engineering requirements or used for design purposes.

Alloy	Temper	Thermal Conductivity at 25 °C (77 °F)		Electrical Conductivity at 20 °C (68 °F), % IACS		Electrical Resistivity at 20 °C (68 °F)	
		cal/cm/cm²/°C/s	Btu/in/ft²/°F/h	Equal Volume	Equal Weight	$\mu\Omega$-cm	Ω-cir mil, ft
EC[a]	All	0.56	1625	62	204	2.8	17
1060	O	0.56	1625	62	204	2.8	17
	H18	0.55	1600	61	201	2.8	17
1100	O	0.53	1540	59	194	2.9	18
	H18	0.52	1510	57	187	3.0	18
2011	T3	0.36	1050	39	123	4.4	27
	T8	0.41	1190	45	142	3.8	23
2014	O	0.46	1340	50	159	3.4	21
	T4	0.32	930	34	108	5.1	31
	T6	0.37	1070	40	127	4.3	26
2017	O	0.46	1340	50	159	3.4	21
	T4	0.32	930	34	108	5.1	31
2018	T61	0.37	1070	40	127	4.3	26
2024	O	0.46	1340	50	160	3.4	21
	T3, T4, T361[b]	0.29	840	30	96	5.7	35
	T6, T81, T861[b]	0.36	1050	38	122	4.5	27
2025	T6	0.37	1070	40	128	4.3	26
2117	T4	0.37	1070	40	130	4.3	26
2218	T72	0.37	1070	40	126	4.3	26
2219	O	0.41	1190	44	138	3.9	24
	T31, T37	0.27	780	28	88	6.2	37
	T62, T81, T87	0.29	870	30	94	5.7	35
3003	O	0.46	1340	50	163	3.4	21
	H12	0.39	1130	42	137	4.1	25
	H14	0.38	1100	41	134	4.2	25
	H18	0.37	1070	40	130	4.3	26
3004	All	0.39	1130	42	137	4.1	25
3105	All	0.41	1190	45	148	3.8	23
4032	O	0.37	1070	40	132	4.3	26
	T6	0.33	960	35	116	4.9	30
4043	O	0.39	1130	42	140	4.1	25
5005	All	0.48	1390	52	172	3.3	20
5050	All	0.46	1340	50	165	3.4	21
5052	All	0.33	960	35	116	4.9	30
5056	O	0.28	810	29	98	5.9	36
	H38	0.26	750	27	91	6.4	38
5083	O	0.28	810	29	98	5.9	36
5086	All	0.30	870	31	104	5.6	33
5154	All	0.30	870	32	107	5.4	32
5252	All	0.33	960	35	116	4.9	30
5254	All	0.30	870	32	107	5.4	32
5454	O	0.32	930	34	113	5.1	31
	H38	0.32	930	34	113	5.1	31
5456	O	0.28	810	29	98	5.9	36
5652	All	0.33	960	35	116	4.9	30
6005	T5	0.40	1160	43	142	4.0	24
6053	O	0.41	1190	45	148	3.8	23
	T4	0.37	1070	40	132	4.3	26
	T6	0.39	1130	42	139	4.1	25
6061	O	0.43	1250	47	155	3.7	22
	T4	0.37	1070	40	132	4.3	26
	T6	0.40	1160	43	142	4.0	24
6063	O	0.52	1510	58	191	3.0	18
	T1[c]	0.46	1340	50	165	3.4	21
	T5	0.50	1450	55	181	3.1	19
	T6, T83	0.48	1390	53	175	3.3	20
6066	O	0.37	1070	40	132	4.3	26
	T6	0.35	1020	37	122	4.7	28
6070	T6	0.41	1190	44	145	3.9	24
6101	T6	0.52	1510	57	188	3.0	18
	T61	0.53	1540	59	194	2.9	18
	T63[d]	0.52	1510	58	191	3.0	18
	T64	0.54	1570	60	198	2.9	17
	T65	0.52	1510	58	191	3.0	18
6151	O	0.49	1420	54	178	3.2	19
	T4	0.39	1130	42	138	4.1	25
	T6	0.41	1190	45	148	3.8	23
6262	T9	0.41	1190	44	145	3.9	24
6463	T1[c]	0.46	1340	50	165	3.4	21
	T5	0.50	1450	55	181	3.1	19
	T6	0.48	1390	53	175	3.3	20
7001	T6	0.30	870	31	98	5.6	33
7072	O	0.53	1540	59	193	2.9	18
7075	T6	0.31	900	33	105	5.2	31
7079	T6	0.30	870	32	104	5.4	32
7178	T6	0.30	870	31	98	5.6	33

[a] Electric conductor grades, 99.45% min aluminum.
[b] Tempers T361 and T861 were formerly designated T36 and T86, respectively.
[c] Formerly designated T42.
[d] Formerly designated T62.

TABLE 4B—TYPICAL PHYSICAL PROPERTIES OF WROUGHT ALUMINUM

The following typical properties are not guaranteed since in most cases they are averages for various sizes, product forms, and methods of manufacture and may not be exactly representative of any particular product or size. These data are intended only as a basis for comparing alloys and tempers and should not be specified as engineering requirements.

Alloy	Density		Specific Gravity	Average Coefficient[a] of Thermal Expansion		Melting Range[b,c] Approx	
	lb/in^3	lb/ft^3		68-212 °F, °F	20-100 °C, °C	°F	°C
EC[d]	0.098	169	2.70	13.2	23.8	1195-1215	646-657
1060	0.098	169	2.70	13.1	23.6	1195-1215	646-657
1100	0.098	169	2.71	13.1	23.6	1190-1215	643-657
2011	0.102	176	2.82	12.7	22.9	1005-1190	541-643[e]
2014	0.101	175	2.80	12.8	23.0	945-1180	507-638[e]
2017	0.101	174	2.79	13.1	23.6	955-1185	513-641[e]
2018	0.101	175	2.80	12.4	22.3	945-1180	507-638[f]
2024	0.100	173	2.77	12.9	23.2	935-1180	502-638[e]
2025	0.101	175	2.79	12.6	22.7	970-1185	521-641[e]
2117	0.099	171	2.74	13.2	23.8	1030-1200	554-649[f]
2218	0.102	176	2.81	12.4	22.3	940-1175	504-635[f]
2219	0.103	178	2.84	12.4	22.3	1010-1190	543-643[e]
2618	0.100	173	2.76	12.4	22.3	1040-1200	560-649
3003	0.099	170	2.73	12.9	23.2	1190-1210	643-654
3004	0.098	170	2.72	13.3	23.9	1165-1210	629-654
3105	0.098	169	2.71	13.1	23.6	1175-1210	635-654
4032	0.097	167	2.69	10.8	19.4	990-1060	532-571[e]
4043	0.097	167	2.69	—	—	1065-1170	574-632
4045	—	—	—	—	—	1065-1110	574-599
4343	—	—	—	—	—	1065-1135	574-613
5005	0.098	169	2.70	13.2	23.8	1170-1210	632-654
5050	0.097	168	2.69	13.2	23.8	1155-1205	624-652
5052	0.097	167	2.68	13.2	23.8	1125-1200	607-649
5056	0.095	165	2.64	13.4	24.1	1060-1180	571-638
5083	0.096	166	2.66	13.2	23.8	1075-1185	579-641
5086	0.096	166	2.66	13.2	23.8	1085-1185	585-641
5154	0.096	166	2.66	13.3	23.9	1100-1190	593-643
5252	0.097	167	2.68	13.2	23.8	1125-1200	607-649
5254	0.096	166	2.66	13.3	23.9	1100-1190	593-643
5356	0.096	166	2.64	—	—	1065-1180	574-638
5454	0.097	167	2.68	13.1	23.6	1115-1195	602-646
5456	0.096	166	2.65	13.3	23.9	1060-1180	571-638
5457	0.097	167	2.68	13.1	23.6	1165-1210	629-654
5652	0.097	167	2.68	13.2	23.8	1100-1200	593-649
5657	0.098	169	2.70	13.1	23.6	1180-1215	638-657
6005	0.098	169	2.70	13.1	23.6	1180-1205	638-652
6053	0.097	167	2.69	12.8	23.0	1070-1205	577-652[f]
6061	0.098	169	2.70	13.1	23.6	1080-1205	582-652[f]
6063	0.098	169	2.70	13.0	23.4	1140-1210	616-654
6066	0.098	169	2.70	12.9	23.2	1045-1195	563-648[e]
6070	0.098	169	2.71	—	—	1050-1200	566-649[e]
6101	0.098	169	2.70	13.0	23.4	1150-1210	621-654
6151	0.098	169	2.70	12.9	23.2	1090-1200	588-649[f]
6201	0.097	167	2.69	13.0	23.4	1080-1210	582-654
6253	—	—	—	—	—	1075-1205	579-652
6262	0.098	170	2.72	13.0	23.4	1080-1205	582-652[f]
6463	0.098	169	2.70	13.0	23.4	1140-1210	616-654
6951	0.098	169	2.70	13.0	23.4	1140-1210	616-654
7001	0.102	176	2.81	13.0	23.4	890-1160	477-627[e]
7072	0.098	169	2.72	13.1	23.6	1195-1215	646-657
7075	0.101	175	2.80	13.1	23.6	890-1175	477-635[g]
7079	0.099	171	2.74	13.1	23.6	900-1180	482-638[f]
7178	0.102	175	2.81	13.0	23.4	890-1165	477-629[g]

[a]Coefficient to be multiplied by 10^6. Example: 12.2 × 10^6 0.0000122.
[b]Melting ranges shown apply to wrought products of 1/4 in (6.35 mm) thickness or greater.
[c]Based on typical composition of the indicated alloys.
[d]Electric conductor grade: 99.45% minimum aluminum.
[e]Eutectic melting is not eliminated by homogenization.
[f]Eutectic melting can be completely eliminated by homogenization.
[g]Homogenization may raise eutectic melting temperature 10-20 deg, but usually does not eliminate eutectic melting.

TABLE 5—TYPICAL MECHANICAL PROPERTIES OF WROUGHT ALUMINUM (NOT FOR DESIGN PURPOSES)

Note: Typical mechanical properties are averages which take into account the variations introduced by type of wrought product, size, shape, and method of manufacture. Typical mechanical properties for all except 0 temper material are higher than specified minimum properties. For 0 temper products, typical ultimate and yield values are slightly lower than specified (maximum) values. The following typical properties are not guaranteed since in most cases they are averages for various sizes, product forms, and methods of manufacture and may not be exactly representative of any particular product or size. These data are intended only as a basis for comparing alloys and tempers and should not be specified as engineering requirements or used for design purposes.

Alloy and Temper	Tension						Hardness	Shear		Fatigue		Modulus	
	Strength				Elongation, % in 2 in		Brinell No., 500 kg load 10 mm ball	Ultimate Shearing Strength		Endurance[b] Limit		Modulus[c] of Elasticity	
	Ultimate		Yield		1/16 in Thick Specimen	1/2 in Dia Specimen							
	ksi	MPa	ksi	MPa				ksi	MPa	ksi	MPa	ksi x 10³	MPa x 10³
EC-0[d]	12	83	4	28	—	—[e]	—	8	55	—	—	10.0	69
–H12[d]	14	97	12	83	—	—	—	9	62	—	—	10.0	69
–H14[d]	16	110	14	97	—	—	—	10	69	—	—	10.0	69
–H16[d]	18	124	16	110	—	—	—	11	76	—	—	10.0	69
–H19[d]	27	186	24	165	—	—[f]	—	15	103	7	48	10.0	69
1060-0	10	69	4	28	43	—	19	7	48	3	21	10.0	69
–H12	12	83	11	76	16	—	23	8	55	4	28	10.0	69
–H14	14	97	13	90	12	—	26	9	62	5	34	10.0	69
–H16	16	110	15	103	8	—	30	10	69	6.5	45	10.0	69
–H18	19	131	18	124	6	—	35	11	76	6.5	45	10.0	69
1100-0	13	90	5	34	35	45	23	9	62	5	34	10.0	69
–H12	16	110	15	103	12	25	28	10	69	6	41	10.0	69
–H14	18	124	17	117	9	20	32	11	76	7	48	10.0	69
–H16	21	145	20	138	6	17	38	12	83	9	62	10.0	69
–H18	24	165	22	152	5	15	44	13	90	9	62	10.0	69
2011-T3	55	379	43	296	—	15	95	32	221	18	124	10.2	70
–T8	59	407	45	310	—	12	100	35	241	18	124	10.2	70
2014-0	27	186	14	97	—	18	45	18	124	13	90	10.6	73
–T4,–T451	62	427	42	290	—	20	105	38	262	20	138	10.6	73
–T6,–T651	70	483	60	414	—	13	135	42	290	18	124	10.6	73
Alclad 2014-0	25	172	10	69	21	—	—	18	124	—	—	10.5	72
–T3	63	434	40	276	20	—	—	37	255	—	—	10.5	72
–T4,–T451	61	421	37	255	22	—	—	37	255	—	—	10.5	72
–T6,–T651	68	469	60	414	10	—	—	41	283	—	—	10.5	72
2017-0	26	179	10	69	—	22	45	18	124	13	90	10.5	72
–T4,–T451	62	427	40	276	—	22	105	38	262	18	124	10.5	72
2018-T61	61	421	46	317	—	12	120	39	269	17	117	10.8	74
2024-0	27	186	11	76	20	22	47	18	124	13	90	10.6	73
–T3	70	483	50	345	18	—	120	41	283	20	138	10.6	73
–T4,–T351	68	469	47	324	20	19	120	41	283	20	138	10.6	73
–T361[g]	72	496	57	393	13	—	130	42	290	18	124	10.6	73
Alclad 2024-0	26	179	11	76	20	—	—	18	124	—	—	10.6	73
–T3	65	448	45	310	18	—	—	40	276	—	—	10.6	73
–T4,–T351	64	441	42	290	19	—	—	40	276	—	—	10.6	73
–T361[g]	67	462	53	365	11	—	—	41	283	—	—	10.6	73
–T81,–T851	65	448	60	414	6	—	—	40	276	—	—	10.6	73
–T861[g]	70	483	66	455	6	—	—	42	290	—	—	10.6	73
2025-T6	58	400	37	255	—	19	110	35	241	18	124	10.4	72
2117-T4	43	296	24	165	—	27	70	28	193	14	97	10.3	71
2218-T72	48	331	37	255	—	11	95	30	207	—	—	10.8	74
2219-0	25	172	11	76	18	—	—	—	—	—	—	10.6	73
–T42	52	359	27	186	20	—	—	—	—	—	—	10.6	73
–T31,–T351	52	359	36	248	17	—	—	—	—	—	—	10.6	73
–T37	57	393	46	317	11	—	—	—	—	—	—	10.6	73
–T62	60	414	42	290	10	—	—	—	—	15	103	10.6	73
–T81,–T851	66	455	51	352	10	—	—	—	—	15	103	10.6	73
–T87	69	476	57	393	10	—	—	—	—	15	103	10.6	73
3003-0	16	110	6	41	30	40	28	11	76	7	48	10.0	69
–H12	19	131	18	124	10	20	35	12	83	8	55	10.0	69
–H14	22	152	21	145	8	16	40	14	97	9	62	10.0	69
–H16	26	179	25	172	5	14	47	15	103	10	69	10.0	69
–H18	29	200	27	186	4	10	55	16	110	10	69	10.0	69
Alclad 3003-0	16	110	6	41	30	40	—	11	76	—	—	10.0	69
–H12	19	131	18	124	10	20	—	12	83	—	—	10.0	69
–H14	22	152	21	145	8	16	—	14	97	—	—	10.0	69
–H16	26	179	25	172	5	14	—	15	103	—	—	10.0	69
–H18	29	200	27	186	4	10	—	16	110	—	—	10.0	69
3004-0	26	179	10	69	20	25	45	16	110	14	97	10.0	69
–H32	31	214	25	172	10	17	52	17	117	15	103	10.0	69
–H34	35	241	29	200	9	12	63	18	124	15	103	10.0	69
–H36	38	262	33	228	5	9	70	20	138	16	110	10.0	69
–H38	41	283	36	248	5	6	77	21	145	16	110	10.0	69
Alclad 3004-0	26	179	10	69	20	25	—	16	110	—	—	10.0	69
–H32	31	214	25	172	10	17	—	17	117	—	—	10.0	69
–H34	35	241	29	200	9	12	—	18	124	—	—	10.0	69
–H36	38	262	33	228	5	9	—	20	138	—	—	10.0	69
–H38	41	283	36	248	5	6	—	21	145	—	—	10.0	69
3105-0	17	117	8	55	24	—	—	12	83	—	—	10.0	69
–H12	22	152	19	131	7	—	—	14	97	—	—	10.0	69
–H14	25	172	22	152	5	—	—	15	103	—	—	10.0	69
–H16	28	193	25	172	4	—	—	16	110	—	—	10.0	69
–H18	31	214	28	193	3	—	—	17	117	—	—	10.0	69
–H25	26	179	23	159	8	—	—	15	103	—	—	10.0	69

(Table continued on next page)

TABLE 5—TYPICAL MECHANICAL PROPERTIES OF WROUGHT ALUMINUM[a] (NOT FOR DESIGN PURPOSES) (CONTINUED)

Alloy and Temper	Tension				Elongation, % in 2 in		Hardness Brinell No., 500 kg load 10 mm ball	Shear Ultimate Shearing Strength		Fatigue Endurance[b] Limit		Modulus[c] of Elasticity	
	Ultimate		Yield		1/16 in Thick Specimen	1/2 in Dia Specimen							
	ksi	MPa	ksi	MPa				ksi	MPa	ksi	MPa	ksi x 10³	MPa x 10³
4032-T6	55	379	46	317	—	9	120	38	262	16	110	11.4	79
5005-O	18	124	6	41	25	—	28	11	76	—	—	10.0	69
-H12	20	138	19	131	10	—	—	14	97	—	—	10.0	69
-H14	23	159	22	152	6	—	—	14	97	—	—	10.0	69
-H16	26	179	25	172	5	—	—	15	103	—	—	10.0	69
-H18	29	200	28	193	4	—	—	16	110	—	—	10.0	69
-H32	20	138	17	117	11	—	36	14	97	—	—	10.0	69
-H34	23	159	20	138	8	—	41	14	97	—	—	10.0	69
-H36	26	179	24	165	6	—	46	15	103	—	—	10.0	69
-H38	29	200	27	186	5	—	51	16	110	—	—	10.0	69
5050-O	21	145	8	55	24	—	36	15	103	12	83	10.0	69
-H32	25	172	21	145	9	—	46	17	117	13	90	10.0	69
-H34	28	193	24	165	8	—	53	18	124	13	90	10.0	69
-H36	30	207	26	179	7	—	58	19	131	14	97	10.0	69
-H38	32	221	29	200	6	—	63	20	138	14	97	10.0	69
5052-O	28	193	13	90	25	30	47	18	124	16	110	10.2	70
-H32	33	228	28	193	12	18	60	20	138	17	117	10.2	70
-H34	38	262	31	214	10	14	68	21	145	18	124	10.2	70
-H36	40	276	35	241	8	10	73	23	159	19	131	10.2	70
-H38	42	290	37	255	7	8	77	24	165	20	138	10.2	70
5056-O	42	290	22	152	—	35	65	26	179	20	138	10.3	71
-H18	63	434	59	407	—	10	105	34	234	22	152	10.3	71
-H38	60	414	50	345	—	15	100	32	221	22	152	10.3	71
5083-O	42	290	21	145	—	22	—	25	172	—	—	10.3	71
-H321	46	317	33	228	—	16	—	—	—	23	159	10.3	71
5086-O	38	262	17	117	22	—	—	23	159	—	—	10.3	71
-H32, -H116, -H117	42	290	30	207	12	—	—	—	—	—	—	10.3	71
-H34	47	324	37	255	10	—	—	27	186	—	—	10.3	71
-H112	39	269	19	131	14	—	—	—	—	—	—	10.3	71
5154-O	35	241	17	117	27	—	58	22	152	17	117	10.2	70
-H32	39	269	30	207	15	—	67	22	152	18	124	10.2	70
-H34	42	290	33	228	13	—	73	24	165	19	131	10.2	70
-H36	45	310	36	248	12	—	78	26	179	20	138	10.2	70
-H38	48	331	39	269	10	—	80	28	193	21	145	10.2	70
-H112	35	241	17	117	25	—	63	—	—	17	117	10.2	70
5252-H25	34	234	25	172	11	—	68	21	145	—	—	10.0	69
-H38,-H28	41	283	35	241	5	—	75	23	159	—	—	10.0	69
5254-O	35	241	17	117	27	—	58	22	152	17	117	10.2	70
-H32	39	269	30	207	15	—	67	22	152	18	124	10.2	70
-H34	42	290	33	228	13	—	73	24	165	19	131	10.2	70
-H36	45	310	36	248	12	—	78	26	179	20	138	10.2	70
-H38	48	331	39	269	10	—	80	28	193	21	145	10.2	70
-H112	35	241	17	117	25	—	63	—	—	17	117	10.2	70
5454-O	36	248	17	117	22	—	62	23	159	—	—	10.2	70
-H32	40	276	30	207	10	—	73	24	165	—	—	10.2	70
-H34	44	303	35	241	10	—	81	26	179	—	—	10.2	70
-H111	38	262	26	179	14	—	70	23	159	—	—	10.2	70
-H112	36	248	18	124	18	—	62	23	159	—	—	10.2	70
5456-O	45	310	23	159	—	24	—	—	—	—	—	10.3	71
-H111	47	324	33	228	—	18	—	—	—	—	—	10.3	71
-H112	45	310	24	165	—	22	—	—	—	—	—	10.3	71
-H321,[i] -H116, -H117	51	352	37	255	—	16	90	30	207	—	—	10.3	71
5457-O	19	131	7	48	22	—	32	12	83	—	—	10.0	69
-H25	26	179	23	159	12	—	48	16	110	—	—	10.0	69
-H38,-H28	30	207	27	186	6	—	55	18	124	—	—	10.0	69
5652-O	28	193	13	90	25	30	47	18	124	16	110	10.2	70
-H32	33	228	28	193	12	18	60	20	138	17	117	10.2	70
-H34	38	262	31	214	10	14	68	21	145	18	124	10.2	70
-H36	40	276	35	241	8	10	73	23	159	19	131	10.2	70
-H38	42	290	37	255	7	8	77	24	165	20	138	10.2	70
5657-H25	23	159	20	138	12	—	40	14	97	—	—	10.0	69
-H38,-H28	28	193	24	165	7	—	50	15	103	—	—	10.0	69
6061-O	18	124	8	55	25	30	30	12	83	9	62	10.0	69
-T4,-T451	35	241	21	145	22	25	65	24	165	14	97	10.0	69
-T6,-T651	45	310	40	276	12	17	95	30	207	14	97	10.0	69
Alclad 6061-O	17	117	7	48	25	—	—	11	76	—	—	10.0	69
-T4,-T451	33	228	19	131	22	—	—	22	152	—	—	10.0	69
-T6,-T651	42	290	37	255	12	—	—	27	186	—	—	10.0	69
6063-O	13	90	7	48	—	—	25	10	69	8	55	10.0	69
-T1[h]	22	152	13	90	20	—	42	14	97	9	62	10.0	69
-T4	25	172	13	90	22	—	—	—	—	—	—	10.0	69
-T5	27	186	21	145	12	—	60	17	117	10	69	10.0	69
-T6	35	241	31	214	12	—	73	22	152	10	69	10.0	69
-T83	37	255	35	241	9	—	82	22	152	—	—	10.0	69
-T831	30	207	27	186	10	—	70	18	124	—	—	10.0	69
-T832	42	290	39	269	12	—	95	27	186	—	—	10.0	69

(Table continued on next page)

TABLE 5—TYPICAL MECHANICAL PROPERTIES OF WROUGHT ALUMINUM[a] (NOT FOR DESIGN PURPOSES) (CONTINUED)

Alloy and Temper	Tension						Hardness	Shear		Fatigue		Modulus	
	Strength				Elongation, % in 2 in		Brinell No., 500 kg load 10 mm ball	Ultimate Shearing Strength		Endurance[b] Limit		Modulus[c] of Elasticity	
	Ultimate		Yield		1/16 in Thick Specimen	1/2 in Dia Specimen							
	ksi	MPa	ksi	MPa				ksi	MPa	ksi	MPa	ksi ×10³	MPa ×10³
6066-0	22	152	12	83	—	18	43	14	97	—	—	10.0	69
-T4,-T451	52	359	30	207	—	18	90	29	200	—	—	10.0	69
-T6,-T651	57	393	52	359	—	12	120	34	234	16	110	10.0	69
6070-T6	55	379	51	352	10	—	—	34	234	14	97	10.0	69
6101-H111	14	97	11	76	—	—	—	—	—	—	—	10.0	69
-T6	32	221	28	193	15[i]	—	71	20	138	—	—	10.0	69
6262-T9	58	400	55	379	—	10	120	35	241	13	90	10.0	69
6463-T1[h]	22	152	13	90	20	—	42	14	97	10	69	10.0	69
-T5	27	186	21	145	12	—	60	17	117	10	69	10.0	69
-T6	35	241	31	214	12	—	74	22	152	10	69	10.0	69
7001-0	37	255	22	152	—	14	60	—	—	—	—	10.3	71
-T6,-T651	98	676	91	627	—	9	160	—	—	22	152	10.3	71
7075-0	33	228	15	103	17	16	60	22	152	—	—	10.4	72
-T6,-T651	83	572	73	503	11	11	150	48	331	23	159	10.4	72
Alclad 7075-0	32	221	14	97	17	—	—	22	152	—	—	10.4	72
-T6,-T651	76	524	67	462	11	—	—	46	317	—	—	10.4	72
7079-0	33	228	15	103	17	16	—	—	—	—	—	10.4	72
-T6,-T651	78	538	68	469	—	14	145	45	310	23	159	10.4	72
Alclad 7079-0	32	221	14	97	16	—	—	—	—	—	—	10.4	72
-T6	71	490	62	427	10	—	—	42	290	—	—	10.4	72
7178-0	33	228	15	103	15	16	—	—	—	—	—	10.4	72
-T6,-T651	88	607	78	538	10	11	—	—	—	—	—	10.4	72
-T76,-T7651	83	572	73	503	—	11	—	—	—	—	—	10.3	71
Alclad 7178-0	32	221	14	97	16	—	—	—	—	—	—	10.4	72
-T6,-T651	81	558	71	490	10	—	—	—	—	—	—	10.4	72

[b]Based on 500,000,000 cycles of completely reversed stress using the R. R. Moore type of machine and specimen.
[c]Average of tension and compression moduli. Compression modulus is about 2% greater than tension modulus.
[d]Electrical conductor grade, 99.45% min aluminum.
[e]EC-0 wire will have an elongation of approximately 23% in 10 in.
[f]EC-H19 wire will have an elongation of approximately 1-1/2% in 10 in.
[g]Tempers T361 and T861 were formerly designated T36 and T86, respectively.
[h]Formerly designated T42.
[i]Based on 1/4 in thick specimen.
[j]Material in this temper is not recommended for, and should not be used in, applications requiring exposure to sea water.

TABLE 6—TYPICAL TENSILE PROPERTIES AT VARIOUS TEMPERATURES[a] (NOT FOR DESIGN PURPOSES)

The following typical properties are not guaranteed since in most cases they are averages for various sizes, product forms, and methods of manufacture and may not be exactly representative of any particular product or size. These data are intended only as a basis for comparing alloys and tempers and should not be specified as engineering requirements or used for design purposes.

Alloy and Temper	Temperature		Tensile Strength				Elongation in 2 in, %	Alloy and Temper	Temperature		Tensile Strength				Elongation in 2 in, %
	°F	°C	Ultimate		Yield[b]				°F	°C	Ultimate		Yield[b]		
			ksi	MPa	ksi	MPa					ksi	MPa	ksi	MPa	
1100-0	−320	−196	25	172	6	41	50	2024-T6, -T651	−320	−196	84	579	68	469	11
	−112	−80	15	103	5.5	38	43		−112	−80	72	496	59	408	10
	−18	−28	14	97	5	34	40		−18	−28	70	483	58	400	10
	75	24	13	90	5	34	40		75	24	69	476	57	393	10
	212	100	10	69	4.6	32	45		212	100	65	448	54	372	10
	300	149	8	55	4.2	29	55		300	149	45	310	36	248	17
	400	204	6	41	3.5	24	65		400	204	26	179	19	131	27
	500	260	4	28	2.6	18	75		500	260	11	76	9	62	55
	600	316	2.9	20	2	14	80		600	316	7.5	52	6	41	75
	700	371	2.1	14	1.6	11	85		700	371	5	34	4	28	100
1100-H14	−320	−196	30	207	20	138	45	2024-T81, -T851	−320	−196	85	586	78	538	8
	−112	−80	20	138	18	124	24		−112	−80	74	510	69	476	7
	−18	−28	19	131	17	117	20		−18	−28	73	503	68	469	7
	75	24	18	124	17	117	20		75	24	70	482	65	448	7
	212	100	16	110	15	103	20		212	100	66	455	62	427	8
	300	149	14	97	12	83	23		300	149	55	379	49	338	11
	400	204	10	69	7.5	52	26		400	204	27	186	20	138	23
	500	260	4	28	2.6	18	75		500	260	11	76	9	62	55
	600	316	2.9	20	2	14	80		600	316	7.5	52	6	41	75
	700	371	2.1	14	1.6	11	85		700	371	5	34	4	28	100
1100-H18	−320	−196	34	234	26	179	30	2024-T861[c]	−320	−196	92	634	85	586	5
	−112	−80	26	179	23	158	16		−112	−80	81	558	77	531	5
	−18	−28	25	172	23	158	15		−18	−28	78	538	74	510	5
	75	24	24	165	22	152	15		75	24	75	517	71	490	5
	212	100	21	145	19	131	15		212	100	70	483	67	462	6
	300	149	18	124	14	97	20		300	149	54	372	48	331	11
	400	204	6	41	3.5	24	65		400	204	21	145	17	117	28
	500	260	4	28	2.6	18	75		500	260	11	76	9	62	55
	600	316	2.9	20	2	14	80		600	316	7.5	52	6	41	75
	700	371	2.1	14	1.6	11	85		700	371	5	34	4	28	100

(Table continued on next page)

TABLE 6—TYPICAL TENSILE PROPERTIES AT VARIOUS TEMPERATURES (NOT FOR DESIGN PURPOSES) (CONTINUED)

Alloy and Temper	Temperature		Tensile Strength				Elongation in 2 in, %	Alloy and Temper	Temperature		Tensile Strength				Elongation in 2 in, %
	°F	°C	Ultimate		Yield[b]				°F	°C	Ultimate		Yield[b]		
			ksi	MPa	ksi	MPa					ksi	MPa	ksi	MPa	
2011-T3	75	24	55	379	43	296	15	2117-T4	−320	−196	56	386	33	228	30
	212	110	47	324	34	234	16		−112	−80	45	310	25	172	29
	300	149	23	193	19	131	25		−18	−28	44	303	24	165	28
	400	204	16	110	11	76	35		75	24	43	296	24	165	27
	500	260	6.5	45	3.8	26	45		212	100	36	248	21	145	16
	600	316	3.1	21	1.8	12	90		300	149	30	207	17	117	20
	700	371	2.3	16	1.4	10	125		400	204	16	110	12	83	35
2014-T6, -T651	−320	−196	84	579	72	496	14		500	260	7.5	52	5.5	38	55
	−112	−80	74	510	65	448	13		600	316	4.7	32	3.3	23	80
	−18	−28	72	496	62	427	13		700	371	2.9	20	2	14	110
	75	24	70	482	60	414	13	2219-T62	−320	−196	73	503	49	338	16
	212	100	63	434	57	393	15		−112	−80	63	434	44	303	13
	300	149	40	275	35	241	20		−18	−28	60	414	42	290	12
	400	204	16	110	13	90	38		75	24	58	400	40	275	12
	500	260	9.5	66	7.5	52	52		212	100	54	372	37	255	14
	600	316	6.5	45	5	34	65		300	149	45	310	33	228	17
	700	371	4.3	30	3.5	24	72		400	204	34	234	25	172	20
2017-T4, -T451	−320	−196	80	551	53	365	28		500	260	27	186	20	138	21
	−112	−80	65	448	42	290	24		600	316	10	69	8	55	40
	−18	−28	64	441	41	283	23		700	371	4.4	30	3.7	26	75
	75	24	62	427	40	276	22	2219-T81, -T851	−320	−196	83	572	61	420	15
	212	100	57	393	39	269	18		−112	−80	71	490	54	372	13
	300	149	40	276	30	207	15		−18	−28	69	476	52	358	12
	400	204	16	110	13	90	35		75	24	66	455	50	345	12
	500	260	9	62	7.5	52	45		212	100	60	414	47	324	15
	600	316	6	41	5	34	65		300	149	49	338	40	276	17
	700	371	4.3	30	3.5	24	70		400	204	36	248	29	200	20
2024-T3 (sheet)	−320	−196	85	586	62	427	18		500	260	29	200	23	158	21
	−112	−80	73	503	52	358	17		600	316	7	48	6	41	55
	−18	−28	72	496	51	352	17		700	371	4.4	30	3.7	26	75
	75	24	70	482	50	345	17	2618-T61	−320	−196	78	538	61	420	12
	212	100	66	455	48	331	16		−112	−80	67	462	55	379	11
	300	149	55	379	45	310	11		−18	−28	64	441	54	372	10
	400	204	27	186	20	138	23		75	24	64	441	54	372	10
	500	260	11	76	9	62	55		212	100	62	427	54	372	10
	600	316	7.5	52	6	41	75		300	149	50	345	44	303	14
	700	371	5	34	4	28	100		400	204	32	221	26	179	24
2024-T4, -T351 (plate)	−320	−196	84	579	61	420	19		500	260	13	90	9	62	50
	−112	−80	71	490	49	338	19		600	316	7.5	52	4.5	31	80
	−18	−28	69	476	47	324	19		700	371	5	34	3.5	24	120
	75	24	68	469	47	324	19	3003-O	−320	−196	33	227	8.5	59	46
	212	100	63	434	45	310	19		−112	−80	20	138	7	48	42
	300	149	45	310	36	248	17		−18	−28	17	117	6.5	45	41
	400	204	26	179	19	131	27		75	24	16	110	6	41	40
	500	260	11	76	9	62	55		212	100	13	90	5.5	38	43
	600	316	7.5	52	6	41	75		300	149	11	76	5	34	47
	700	371	5	34	4	28	100		400	204	8.5	59	4.3	30	60
									500	260	6	41	3.4	23	65
									600	316	4	28	2.4	16	70
									700	371	2.8	19	1.8	14	70
3003-H14	−320	−196	35	241	25	172	30	5050-H38	−320	−196	46	317	36	248	—
	−112	−80	24	165	22	152	18		−112	−80	34	234	30	207	—
	−18	−28	22	152	21	145	16		−18	−28	32	221	29	200	—
	75	24	22	152	21	145	16		75	24	32	221	29	200	—
	212	100	21	145	19	131	16		212	100	31	214	29	200	—
	300	149	18	124	16	110	16		300	149	27	186	25	172	—
	400	204	14	96	9	62	20		400	204	14	96	7.5	52	—
	500	260	7.5	52	4	28	60		500	260	9	62	6	41	—
	600	316	4	28	2.4	16	70		600	316	6	41	4.2	29	—
	700	371	2.8	19	1.8	12	70		700	371	3.9	27	2.6	18	—
3003-H18	−320	−196	41	283	33	228	23	5052-O	−320	−196	44	303	16	110	46
	−112	−80	32	221	29	200	11		−112	−80	29	200	13	90	35
	−18	−28	30	207	28	193	10		−18	−28	28	193	13	90	32
	75	24	29	200	27	186	10		75	24	28	193	13	90	30
	212	100	26	179	21	145	10		212	100	28	193	13	90	36
	300	149	23	158	16	110	11		300	149	23	158	13	90	50
	400	204	14	96	9	62	18		400	204	17	117	11	76	60
	500	260	7.5	52	4	28	60		500	260	12	83	7.5	52	80
	600	316	4	28	2.4	16	70		600	316	7.5	52	5.5	38	110
	700	371	2.8	19	1.8	12	70		700	371	5	34	3.1	21	130
3004-O	−320	−196	42	290	13	90	38	5052-H34	−320	−196	55	379	36	248	28
	−112	−80	28	193	11	76	30		−112	−80	40	276	32	221	21
	−18	−28	26	179	10	69	26		−18	−28	38	262	31	214	18
	75	24	26	179	10	69	25		75	24	38	262	31	214	16
	212	100	26	179	10	69	25		212	100	38	262	31	214	18
	300	149	22	152	10	69	35		300	149	30	207	27	186	27
	400	204	14	96	9.5	66	55		400	204	24	165	15	103	45
	500	260	10	69	7.5	52	70		500	260	12	83	7.5	52	80
	600	316	7.5	52	5	34	80		600	316	7.5	52	5.5	38	110
	700	371	5	34	3	21	90		700	371	5	34	3.1	21	130

(Table continued on next page)

TABLE 6—TYPICAL TENSILE PROPERTIES AT VARIOUS TEMPERATURES (NOT FOR DESIGN PURPOSES) (CONTINUED)

Alloy and Temper	Temperature		Tensile Strength				Elongation in 2 in, %	Alloy and Temper	Temperature		Tensile Strength				Elongation in 2 in, %
			Ultimate		Yield[b]						Ultimate		Yield[b]		
	°F	°C	ksi	MPa	ksi	MPa			°F	°C	ksi	MPa	ksi	MPa	
3004-H34	−320	−196	52	358	34	234	26	5052-H38	−320	−196	60	414	44	303	25
	−112	−80	38	262	30	207	16		−112	−80	44	303	38	262	18
	−18	−28	36	248	29	200	13		−18	−28	42	290	37	255	15
	75	24	35	241	29	200	12		75	24	42	290	37	255	14
	212	100	34	234	29	200	13		212	100	40	276	36	248	16
	300	149	28	193	25	172	22		300	149	34	234	28	193	24
	400	204	21	145	15	103	35		400	204	25	172	15	103	45
	500	260	14	96	7.5	52	55		500	260	12	83	7.5	52	80
	600	316	7.5	52	5	34	80		600	316	7.5	52	5.5	38	110
	700	371	5	34	3	21	90		700	371	5	34	3.1	21	130
3004-H38	−320	−196	58	400	43	296	20	5083-0	−320	−196	59	407	24	165	36
	−112	−80	44	303	38	262	10		−112	−80	43	296	21	145	30
	−18	−28	42	290	36	248	7		−18	−28	42	290	21	145	27
	75	24	41	283	36	248	6		75	24	42	290	21	145	25
	212	100	40	276	36	248	7		212	100	40	276	21	145	36
	300	149	31	214	27	186	15		300	149	31	214	19	131	50
	400	204	22	152	15	103	30		400	204	22	152	17	117	60
	500	260	12	83	7.5	52	50		500	260	17	117	11	76	80
	600	316	7.5	52	5	34	80		600	316	11	76	7.5	52	110
	700	371	5	34	3	21	90		700	371	6	41	4.2	29	130
4032-T6	−320	−196	66	455	48	331	11	5086-0	−320	−196	55	379	19	131	46
	−112	−80	58	400	46	317	10		−112	−80	39	269	17	117	35
	−18	−28	56	386	46	317	9		−18	−28	38	262	17	117	32
	75	24	55	379	46	317	9		75	24	38	262	17	117	30
	212	100	50	345	44	303	9		212	100	38	262	17	117	36
	300	149	37	255	33	228	9		300	149	29	200	16	110	50
	400	204	13	90	9	62	30		400	204	22	152	15	103	60
	500	260	8	55	5.5	28	50		500	260	17	117	11	76	80
	600	316	5	34	3.2	22	70		600	316	11	76	7.5	52	110
	700	371	3.4	23	2	14	90		700	371	6	41	4.2	29	130
5050-0	−320	−196	37	255	10	69	—	5154-0	−320	−196	52	358	19	131	46
	−112	−80	22	152	8.5	59	—		−112	−80	36	248	17	117	35
	−18	−28	21	145	8	55	—		−18	−28	35	241	17	117	32
	75	24	21	145	8	55	—		75	24	35	241	17	117	30
	212	100	21	145	8	55	—		212	100	35	241	17	117	36
	300	149	19	131	8	55	—		300	149	29	200	16	110	50
	400	204	14	96	7.5	52	—		400	204	22	152	15	103	60
	500	260	9	62	6	41	—		500	260	17	117	11	76	80
	600	316	6	41	4.2	29	—		600	316	11	76	7.5	52	110
	700	371	3.9	27	2.6	18	—		700	371	6	41	4.2	29	130
5050-H34	−320	−196	44	303	30	207	—	5254-0	−320	−196	52	358	19	131	46
	−112	−80	30	207	25	172	—		−112	−80	36	248	17	117	35
	−18	−28	28	193	24	165	—		−18	−28	35	241	17	117	32
	75	24	28	193	24	165	—		75	24	35	241	17	117	30
	212	100	28	193	24	165	—		212	100	35	241	17	117	36
	300	149	25	172	22	152	—		300	149	29	200	16	110	50
	400	204	14	96	7.5	52	—		400	204	22	152	15	103	60
	500	260	9	62	6	41	—		500	260	17	117	11	76	80
	600	316	6	41	4.2	29	—		600	316	11	76	7.5	52	110
	700	371	3.9	27	2.6	18	—		700	371	6	41	4.2	29	130
5454-0	−320	−196	54	372	19	131	39	6061-T6, -T651	−320	−196	60	414	47	324	22
	−112	−80	37	255	17	117	30		−112	−80	49	338	42	290	18
	−18	−28	36	248	17	117	27		−18	−28	47	324	41	283	17
	75	24	36	248	17	117	25		75	24	45	310	40	276	17
	212	100	36	248	17	117	31		212	100	42	290	38	262	18
	300	149	29	200	16	110	50		300	149	34	234	31	214	20
	400	204	22	152	15	103	60		400	204	19	131	15	103	28
	500	260	17	117	11	76	80		500	260	7.5	52	5	34	60
	600	316	11	76	7.5	52	110		600	316	4.6	32	2.7	19	85
	700	371	6	41	4.2	29	130		700	371	3	21	1.8	12	95
5454-H32	−320	−196	59	407	36	248	32	6063-T1	−320	−196	34	234	16	110	44
	−112	−80	42	290	31	214	23		−112	−80	26	179	15	103	36
	−18	−28	41	283	30	207	20		−18	−28	24	165	14	96	34
	75	24	40	276	30	207	18		75	24	22	152	13	90	33
	212	100	39	269	29	200	20		212	100	22	152	14	96	18
	300	149	32	221	26	179	37		300	149	21	145	15	103	20
	400	204	25	172	19	131	45		400	204	9	62	6.5	45	40
	500	260	17	117	11	76	80		500	260	4.5	31	3.5	24	75
	600	316	11	76	7.5	52	110		600	316	3.2	22	2.5	17	80
	700	371	6	41	4.2	29	130		700	371	2.3	16	2	14	105
5454-H34	−320	−196	63	434	41	283	30	6063-T5	−320	−196	37	255	24	165	28
	−112	−80	46	317	36	248	21		−112	−80	29	200	22	152	24
	−18	−28	44	303	35	241	18		−18	−28	28	193	22	152	23
	75	24	44	303	35	241	16		75	24	27	186	21	145	22
	212	100	43	296	34	234	18		212	100	24	165	20	138	18
	300	149	34	234	28	193	32		300	149	20	138	18	124	20
	400	204	26	179	19	131	45		400	204	9	62	6.5	45	40
	500	260	17	117	11	76	80		500	260	4.5	31	3.5	24	75
	600	316	11	76	7.5	52	110		600	316	3.2	22	2.5	17	80
	700	371	6	41	4.2	29	130		700	371	2.3	16	2	14	105

(Table continued on next page)

TABLE 6—TYPICAL TENSILE PROPERTIES AT VARIOUS TEMPERATURES (NOT FOR DESIGN PURPOSES) (CONTINUED)

Alloy and Temper	Temperature		Tensile Strength				Elongation in 2 in, %	Alloy and Temper	Temperature		Tensile Strength				Elongation in 2 in, %
			Ultimate		Yield[b]						Ultimate		Yield[b]		
	°F	°C	ksi	MPa	ksi	MPa			°F	°C	ksi	MPa	ksi	MPa	
5456-0	−320	−196	62	427	26	179	32	6063-T6	−320	−196	47	324	36	248	24
	−112	−80	46	317	23	158	25		−112	−80	38	262	33	228	20
	−18	−28	45	310	23	158	22		−18	−28	36	248	32	221	19
	75	24	45	310	23	158	20		75	24	35	241	31	214	18
	212	100	42	290	22	152	31		212	100	31	214	28	193	15
	300	149	31	214	20	137	50		300	149	21	145	20	138	20
	400	204	22	152	17	117	60		400	204	9	62	6.5	45	40
	500	260	17	117	11	76	80		500	260	4.5	31	3.5	24	75
	600	316	11	76	7.5	52	110		600	316	3.3	23	2.5	17	80
	700	371	6	41	4.2	29	130		700	371	2.3	16	2	14	105
5652-0	−320	−196	44	303	16	110	46	6101-T6	−320	−196	43	296	33	228	24
	−112	−80	29	200	13	90	35		−112	−80	36	248	30	207	20
	−18	−28	28	193	13	90	32		−18	−28	34	234	29	200	19
	75	24	28	193	13	90	30		75	24	32	221	28	193	19
	212	100	28	193	13	90	30		212	100	28	193	25	172	20
	300	149	23	158	13	90	50		300	149	21	145	19	131	20
	400	204	17	117	11	76	60		400	204	10	69	7	48	40
	500	260	12	83	7.5	52	80		500	260	4.8	33	3.3	23	80
	600	316	7.5	52	5.5	38	110		600	316	3	21	2.3	16	100
	700	371	5	34	3.1	21	130		700	371	2.5	17	1.8	12	105
5652-H34	−320	−196	55	379	36	248	28	6151-T6	−320	−196	57	393	50	345	20
	−112	−80	40	276	32	221	21		−112	−80	50	345	46	317	17
	−18	−28	38	262	31	214	18		−18	−28	49	338	45	310	17
	75	24	38	262	31	214	16		75	24	48	331	43	296	17
	212	100	38	262	31	214	18		212	100	43	296	40	276	17
	300	149	30	207	27	186	27		300	149	28	193	27	186	20
	400	204	24	165	15	103	45		400	204	14	96	12	83	30
	500	260	12	83	7.5	52	80		500	260	6.5	45	5	103	50
	600	316	7.5	52	5.5	38	110		600	316	5	34	3.9	27	43
	700	371	5	34	3.1	21	130		700	371	4	28	3.2	22	35
5652-H38	−320	−196	60	414	44	303	25	6262-T651	−320	−196	60	414	47	324	22
	−112	−80	44	303	38	262	18		−112	−80	49	338	42	290	18
	−18	−28	42	290	37	255	15		−18	−28	47	324	41	283	17
	75	24	42	290	37	255	14		75	24	45	310	40	276	17
	212	100	40	276	36	248	16		212	100	42	290	38	262	18
	300	149	34	234	28	193	24		300	149	34	234	31	214	20
	400	204	25	172	15	103	45								
	500	260	12	83	7.5	52	80	6262-T9	−320	−196	74	510	67	462	14
	600	316	7.5	52	5.5	38	110		−112	−80	62	427	58	400	10
	700	371	5	34	3.1	21	130		−18	−28	60	414	56	386	10
6053-T6, -T651	75	24	37	255	32	221	13		75	24	58	400	55	379	10
	212	100	32	221	28	193	13		212	100	53	365	52	358	10
	300	149	25	172	24	165	13		300	149	38	262	37	255	14
	400	204	13	90	12	83	25		400	204	15	103	13	90	34
	500	260	5.5	38	4	28	70		500	260	8.5	59	6	41	48
	600	316	4	28	2.7	19	80		600	316	4.6	32	2.7	19	85
	700	371	2.9	20	2	14	90		700	371	3	21	1.8	12	95
7075-T6, -T651	−320	−196	102	703	92	634	9	7079-T6, -T651 (cont.)	300	149	33	228	28	193	37
	−112	−80	90	620	79	545	11		400	204	16	110	13	90	60
	−18	−28	86	523	75	517	11		500	260	11	76	8.5	59	100
	75	24	83	572	73	503	11		600	316	7.5	52	6	41	175
	212	100	70	483	65	448	14		700	371	5.5	38	4.3	30	175
	300	149	31	214	27	186	30								
	400	204	16	110	13	90	55	7178-T6, -T651	−320	−196	106	731	94	648	5
	500	260	11	76	9	62	65		−112	−80	94	648	84	579	8
	600	316	8	55	6.5	45	70		−18	−28	91	627	81	558	9
	700	371	6	41	4.6	32	70		75	24	88	607	78	538	11
									212	100	73	503	68	469	14
7075-T73, -T7351	−320	−196	92	634	72	496	14		300	149	31	214	27	186	40
	−112	−80	79	545	67	492	14		400	204	15	103	12	83	70
	−18	−28	76	524	65	448	13		500	260	11	76	9	62	76
	75	24	73	503	63	434	13		600	316	8.5	59	7	48	80
	212	100	63	434	58	400	15		700	371	6.5	45	5.5	38	80
	300	149	31	214	27	186	30								
	400	204	16	110	13	90	55	7178-T76, -T7651	−320	−196	106	731	89	614	10
	500	260	11	76	9	62	65		−112	−80	91	627	78	538	10
	600	316	8	55	6.5	45	70		−18	−28	88	607	76	524	10
	700	371	6	41	4.6	32	70		75	24	83	572	73	503	11
									212	100	69	476	64	441	17
7079-T6, -T651	−320	−196	92	634	80	552	12		300	149	31	214	27	186	40
	−112	−80	82	565	70	483	14		400	204	15	103	12	83	70
	−18	−28	79	545	68	469	14		500	260	11	76	9	62	76
	75	24	78	538	68	469	14		600	316	8.5	59	7	48	80
	212	100	67	462	60	414	18		700	371	6.5	45	5.5	38	80

[a]These data are based on a limited amount of testing and represent the lowest strength during 10,000 h of exposure at testing temperature under no load; stress applied at 5000 psi/min (34 MPa) to yield strength and then at strain rate of 0.05 in/in/min (1.3 mm/mm/min) to failure. Under some conditions of temperature and time, the application of heat will adversely affect certain other properties of some alloys.
[b]Offset equals 0.2%.
[c]Temper T861 was formerly designated T86.

TABLE 7—NOMINAL CHEMICAL COMPOSITION—WROUGHT ALLOYS

The following values are shown as a basis for general comparison of alloys and are not guaranteed. The information in this table does not include the Alclad materials.

Percent of Alloying Elements—Aluminum and Normal Impurities Constitute Remainder

Alloy	Si	Cu	Mn	Mg	Cr	Ni	Zn	Pb	Bi
EC	—	—		99.45% min aluminum		—	—	—	—
1050	—	—		99.50% min aluminum		—	—	—	—
1060	—	—		99.60% min aluminum		—	—	—	—
1100	—	0.12		99.00% min aluminum		—	—	—	—
1145	—	—		99.45% min aluminum		—	—	—	—
1175	—	—		99.75% min aluminum		—	—	—	—
1200	—	—		99.00% min aluminum		—	—	—	—
1230	—	—		99.30% min aluminum		—	—	—	—
1235	—	—		99.35% min aluminum		—	—	—	—
1345	—	—		99.45% min aluminum		—	—	—	—
2011	—	5.5	—	—	—	—	—	0.40	0.40
2014	0.8	4.4	0.8	0.50	—	—	—	—	—
2017	—	4.0	0.7	0.60	—	—	—	—	—
2018	—	4.0	—	0.7	—	2.0	—	—	—
2024	—	4.4	0.6	1.5	—	—	—	—	—
2025	0.8	4.5	0.8	—	—	—	—	—	—
2117	—	2.6	—	0.35	—	—	—	—	—
2124	—	4.4	0.6	1.5	—	—	—	—	—
2218	—	4.0	—	1.5	—	2.0	—	—	—
2219[a]	—	6.3	0.30	—	—	—	—	—	—
2618[b]	—	2.3	—	1.6	—	1.0	—	—	—
3003	—	0.12	1.2	—	—	—	—	—	—
3004	—	—	1.2	1.0	—	—	—	—	—
3005	—	—	1.2	0.40	—	—	—	—	—
3105	—	—	0.50	0.50	—	—	—	—	—
4032	12.2	0.9	—	1.1	—	0.9	—	—	—
4043	5.2	—	—	—	—	—	—	—	—
4045	10.0	—	—	—	—	—	—	—	—
4343	7.5	—	—	—	—	—	—	—	—
5005	—	—	—	0.8	—	—	—	—	—
5050	—	—	—	1.4	—	—	—	—	—
5052	—	—	—	2.5	0.25	—	—	—	—
5056	—	—	0.12	5.1	0.12	—	—	—	—
5083	—	—	0.7	4.45	0.15	—	—	—	—
5086	—	—	0.45	4.0	0.15	—	—	—	—
5154	—	—	—	3.5	0.25	—	—	—	—
5252	—	—	—	2.5	—	—	—	—	—
5254	—	—	—	3.5	0.25	—	—	—	—
5356[c]	—	—	0.12	5.0	0.12	—	—	—	—
5454	—	—	0.8	2.7	0.12	—	—	—	—
5456	—	—	0.8	5.1	0.12	—	—	—	—
5457	—	—	0.30	1.0	—	—	—	—	—
5652	—	—	—	2.5	0.25	—	—	—	—
5657	—	—	—	0.8	—	—	—	—	—
6003	0.7	—	—	1.2	—	—	—	—	—
6053	0.7	—	—	1.3	0.25	—	—	—	—
6061	0.6	0.27	—	1.0	0.20	—	—	—	—
6063	0.40	—	—	0.7	—	—	—	—	—
6066	1.3	0.9	0.8	1.1	—	—	—	—	—
6070	1.4	0.3	0.7	0.8	—	—	—	—	—
6101	0.50	—	—	0.6	—	—	—	—	—
6151	0.9	—	—	0.6	0.25	—	—	—	—
6201	0.7	—	—	0.8	—	—	—	—	—
6253	—	—	—	1.2	0.25	—	2.0	—	—
6262	0.6	0.27	—	1.0	0.09	—	—	0.55	0.55
6463	0.40	—	—	0.7	—	—	—	—	—
6951	0.30	0.25	—	0.6	—	—	—	—	—
7001	—	2.1	—	3.0	0.30	—	7.4	—	—
7072	—	—	—	—	—	—	1.0	—	—
7075	—	1.6	—	2.5	0.30	—	5.6	—	—
7079	—	0.6	0.20	3.3	0.20	—	4.3	—	—
7178	—	2.0	—	2.9	0.30	—	6.8	—	—

[a]Titanium, 0.06%; vanadium, 0.10%, zirconium, 0.18%.   [b]Iron, 1.1%; titanium, 0.07%.   [c]Titanium, 0.13%.

CHEMICAL COMPOSITIONS, MECHANICAL PROPERTY LIMITS, AND DIMENSIONAL TOLERANCES OF SAE WROUGHT ALUMINUM ALLOYS—SAE J457j

SAE Standard

Report of Nonferrous Metals Division approved June 1911 and last revised by Nonferrous Metals Committee May 1972.

Specification of Wrought Aluminum Alloys—The Aluminum Association publishes a manual entitled Aluminum Standards and Data which includes the specified limits registered for the chemical composition and mechanical properties of standard aluminum alloys commercially available in the wrought product forms shown in Tables 2 and 3 of SAE J454.

NOTE: The Aluminum Association standards may be obtained at no cost to users of aluminum from The Aluminum Association, 750 Third Avenue, New York, New York 10017.

Chemical Composition Limits—The chemical composition limits for each wrought aluminum alloy are registered in accordance with the designation system presented in ANSI H35.1, Alloy and Temper Designation Systems for Aluminum, and SAE J993. The chemical composition limits are tabulated by alloy and are the same for all wrought product forms in which the alloy is available. In addition, the tabulation is by alloy components in the case of alclad sheet, plate, and tubular products.

Mechanical Property Limits—The specified mechanical property limits are established for each wrought product; that is, sheet, plate, drawn tube, extruded shapes, pipe and extruded tube, rolled shapes, rod, bar, wire, and die forgings.

Dimensional Tolerances—The dimensional tolerances to which aluminum alloy mill products are commercially produced are in accordance with the latest revision of the ANSI H35.2, Dimensional Tolerances for Aluminum Mill Products. For sizes outside commercial limits, the tolerance shall be as agreed upon by the supplier and purchaser.

Standards for Chemical Composition and Mechanical Property Limits—The following American Society for Testing and Materials Standards and related American National Standards contain the specified chemical composition limits and mechanical property limits for each of the wrought products listed:

ASTM	ANSI	Title
B 209	H38.2	Specification for Aluminum-Alloy Sheet and Plate
B 210	H38.3	Specification for Aluminum-Alloy Drawn Seamless Tubes
B 211	H38.4	Specification for Aluminum-Alloy Bars, Rods, and Wire
B 221	H38.5	Specification for Aluminum-Alloy Extruded Bars, Rods, Shapes, and Tubes
B 234	H38.6	Specification for Aluminum-Alloy Drawn Seamless Tubes for Condensers and Heat Exchangers
B 236	C7.27	Specification for Aluminum Bar for Electrical Purposes (Bus Bar)
B 241	H38.7	Specification for Aluminum-Alloy Seamless Pipe and Seamless Extruded Tube
B 247	H38.8	Specification for Aluminum-Alloy Die and Hand Forgings
B 308	H38.10	Specification for Aluminum-Alloy Standard Structural Shapes, Rolled or Extruded

ANODIZED ALUMINUM AUTOMOTIVE PARTS—SAE J399a

SAE Information Report

Report of Nonferrous Metals Committee approved June 1969 and last revised April 1976.

Automotive parts can be fabricated from either coiled sheet, flat sheet or extruded shapes. Alloy selection is governed by finish requirements, forming characteristics, and mechanical properties.

Bright anodizing alloys 5657 and 5252[1] sheet provide a high luster and are preferred for trim which can be formed from an intermediate temper, such as H25. Bright anodizing alloy 5457 is used for parts which require high elongation and a fully annealed ("0") temper. Alloy 6463 is a medium strength bright anodizing extrusion alloy; Alloy X7016 is a high strength bright anodizing extrusion alloy primarily suited for bumper applications.

To satisfy anti-glare requirements for certain trim applications, sheet alloy 5205 and extrusion alloy 6063 are capable of providing the desired low-gloss anodized finish.

Bright anodizing alloys require control of the chemical composition of the alloy to enhance response to chemical brightening and to result in the formation of anodic coatings that are essentially transparent. Additionally, aluminum producers employ fabricating practices to minimize other metallurgical factors that adversely affect response to bright anodizing procedures. For non-heat-treatable alloys, a highly fragmented grain structure is preferred. Fully annealed, recrystallized grain structures are not optimum for bright anodizing. Where high elongations are required with intermediate tempers, fabricating practices are selected to minimize grain recrystallization.

Another factor to be considered for trim application is the type of mill surface finish that is required. When the metal working treatments do not mar the mill produced surface appreciably, the smooth, bright rolled, "automotive trim" surface is desirable since it often eliminates the need for expensive mechanical buffing operations. Where trim fabricating procedures might be expected to damage a bright-rolled surface, duller mill finishes can be used and parts are buffed after forming. Bright rolled mill surfaces occasionally are protected with a removable tape or water soluble film.

Selection of anodic coating required to protect aluminum parts is influenced by the required corrosion performance and appearance characteristics. Generally, anodic coatings 0.0003–0.0005 in (0.0076–0.0127 mm) thick are used for exterior trim application. Thinner anodic coatings 0.0001 to 0.0003 in (0.0025–0.0076 mm) are sufficient for interior trim components. Anodic coatings can be dyed to impart color, painted, or inlaid with vinyl or other plastics for aesthetic and/or functional purposes.

The Aluminum Association's "Designation System for Finishes" is a recommended guide to assist in specifying anodic coatings for automotive trim.

The American Society for Testing and Materials (ASTM) offers several test methods which are commonly used as the basis for many user specifications. These are:

ASTM B110, Dielectric Strength of Anodically Coated Aluminum
ASTM B457, Measuring Impedance of Anodic Coatings of Aluminum
ASTM B244, Measuring Thickness of Anodic Coatings on Aluminum with Eddy Current Instruments
ASTM B136, Resistance of Anodically Coated Aluminum to Staining by Dyes
ASTM B137, Weight of Coating on Anodically Coated Aluminum
ASTM B368, Copper-Accelerated Acetic Acid Salt Spray (Fog) Testing (CASS Test)
ASTM B538, Fact (Ford Anodized Aluminum Corrosion Test) Testing
ASTM B580, Guide to the Specification of Anodic Oxide Coatings on Aluminum
ASTM B429, Measurement and Calculation of Reflecting Characteristics of Metallic Surfaces Using Integrating Sphere Instruments
ASTM E430, Measurement of Gloss of High Gloss Metal Surfaces Using Abridged Goniophotometer or Goniophotometer

[1] Details of alloy numbers shown are published in Aluminum Association Standards.

BEARING AND BUSHING ALLOYS—SAE J460e — SAE Standard

Report of Nonferrous Metals Division approved June 1911. Editorial change July 1974. Last revised by Nonferrous Metals Committee October 1974. Editorial change April 1978.

CHEMICAL COMPOSITION OF SAE BEARING AND BUSHING ALLOYS

	SAE No.	Elements,[a] %								
Tin Base		Sn, min	Sb	Pb	Cu	Fe	As	Bi	Zn	Al	Others, Total
Bearing	11	86.0	6.0–7.5	0.50	5.0–6.5	0.08	0.10	0.08	0.005	0.005	0.20
Bearing	12[b]	88.0	7.0–8.0	0.50	3.0–4.0	0.08	0.10	0.08	0.005	0.005	0.20
Lead Base		Pb	Sn	Sb	Cu	As	Bi	Zn	Al	Cd	Others, Total
Bearing	13	Remainder	5.0–7.0	9.0–11.0	0.50	0.25	0.10	0.005	0.005	0.05	0.20
Bearing	14	Remainder	9.2–10.7	14.0–16.0	0.50	0.6	0.10	0.005	0.005	0.05	0.20
Bearing	15	Remainder	0.9–1.3	14.0–15.5	0.50	0.8–1.2	0.10	0.005	0.005	0.02	0.20
Bearing	16[c]	Remainder	3.5–4.7	3.0–4.0	0.10	0.05	0.10	0.005	0.005	0.005	0.40
Plated Overlay[d]		Pb			Sn			Others, Total[h]			
	19	Remainder			8.0–12.0			3.5			
	190	Remainder			5.0–9.0			3.5			
Copper-Lead[d]		Cu	Pb	Sn	Ag	Zn	P	Fe	Others, Total		
	48	67.0–74.0	25.0–32.0	0.25	1.5	0.10	0.02	0.35	0.15		
	480	60.0–70.0	30.0–40.0	0.50	1.5	—	—	0.35	0.30		
	481	56.5–62.5[e]	37.5–42.5	0.25	5.5[f]	0.10	0.02	0.35	0.30		
	49	73.0–79.0	21.0–27.0	0.50[g]	—	—	—	0.35	0.45		
		Cu	Pb	Sn	Fe	Others, Total		Others, Each			
	482	Remainder	24.0–32.0	4.0–7.0	0.35	0.45		0.15			
	484	Remainder	40.0–44.0	1.0–5.0	0.35	0.45		0.15			
	485	Remainder	44.0–58.0	1.0–5.0	0.35	0.45		0.15			
Aluminum Base		Al	Sn	Cd	Si	Cu	Fe	Mn	Ni	Ti	Others, Total	Others, Each
	770	Remainder	5.5–7.0	—	0.7	0.7–1.3	0.7	0.10	0.7–1.3	0.20	0.30	—
	780	Remainder	5.5–7.0	—	1.0–2.0	0.7–1.3	0.7	0.10	0.20–0.7	0.10	0.15	—
	781[i]	Remainder	—	0.8–1.4	3.5–4.5	0.05–0.15	0.35	0.03	—	0.02	0.25[j]	0.05
	782	Remainder	—	2.7–3.5	0.30	0.7–1.3	0.30	—	0.7–1.3	0.10	0.15	0.05
	783	Remainder	17.5–22.5	—	0.50 max	0.7–1.3	0.50 max	0.10 max	0.10 max	0.10 max	0.15	—
Other Copper Base		Cu	Pb	Sn	Zn	Sb	Ni	Fe	Others, Total		
	791[k]	Remainder	3.5–4.5	3.5–4.5	1.5–4.0	—	—	0.10	0.20		
	792 or 797	77.0 min	9.0–11.0	9.0–11.0	0.7	0.50	0.50	0.35	0.40		
	793 or 798	83.0 min	7.0–9.0	3.5–4.5	4.0 max	0.50	0.50	0.35	0.30		
	794 or 799	68.5–75.5	21.0–25.0	3.0–4.0	3.0 max	0.50	0.50	0.35	0.40		
	795	88.0–92.0	—	0.25–0.7	Remainder	—	—	0.10	0.20		

[a] All values not given as ranges are maximum except as shown otherwise.
[b] Formerly SAE 110.
[c] SAE 16 is cast into and on a porous sintered matrix, usually copper-nickel, bonded to steel. The surface layer is 0.001–0.005 in. in thickness.
[d] A corrosion-resistant overlay such as SAE 19 or 190 may be used with alloy SAE 48, 480, and 481 and is recommended for alloy SAE 49.
[e] Copper plus silver.
[f] This specification, covering cast copper-lead linings on steel backs, is designed to permit adjustment of lead and silver contents according to the hardness of the mating shafts. Exact chemical composition shall be as agreed between purchaser and supplier.
[g] The tin maximum may be raised to 1.25% upon agreement between purchaser and supplier.
[h] A copper content of 1–3% is often specified by agreement between purchaser and supplier.
[i] SAE 781 includes a magnesium content specified at 0.05–0.15%.
[j] A 0.15% max Zn content is permissible within this range.
[k] SAE 791 is similar to SAE CA544, as shown in SAE J463d (June, 1976).

BEARING AND BUSHING ALLOYS—SAE J459c — SAE Information Report

Report of Nonferrous Metals Committee approved February 1947 and last revised August 1966. Editorial change July 1974.

The choice of the material to use for main and connecting rod bearings depends upon a number of factors among which are fatigue resistance, corrosion resistance, conformability, embedability, wear resistance, and abrading tendency.

Relative importance of the above factors will depend on each particular application. The resistance to fatigue of bearing materials depends to a great extent upon the design of the bearing, the strength and rigidity of the supporting structure, the thickness of the backing metal (ordinarily SAE 1010 steel), the thickness of the bearing material, and the physical properties of the bond between the bearing material and the backing. Resistance to corrosion depends upon chemical composition and characteristics of both lubricant and bearing alloy, and upon temperature and other operating conditions.

The classification and uses of the SAE bearing and bushing alloys are given in Table 1.

The tin- and lead-base babbitts exhibit a minimum tendency to abrade or wear the bearing journal and also are resistant to corrosion from organic acid of the type normal to lubricating oil. Bearings of SAE 15 should not be subjected to the action of break-in oils containing oleic or similar acid. Bearings of any of these alloys show the greatest life when used in thicknesses of 0.005 in. or less.

Copper-lead bearings are inferior to tin- and lead-base babbitts in non-abrasive properties, but when properly made they have superior resistance to fatigue. Used as insert bearings, they may be made by casting on a formed steel backing or continuous steel strip, preferably with provision for rapid cooling or chilling from the molten condition, or they may be made by sintering the powdered metals on a steel strip. Alloying of the copper-lead with a small per cent of silver is desirable for some applications. The general performance of copper-lead bearings when used under extreme service conditions may be improved by plating the bearings with a corrosion resistant coating of lead-tin or other suitable alloy up to 0.002 in. thickness.

The recently developed higher-tin copper-leads are generally made by sintering a copper or copper-lead or other copper alloy matrix, which is usually filled with a corrosion resistant lead-tin alloy. This generally eliminates the need for overlay, either for corrosion resistance or for compatibility. The higher-lead compositions in this group will run on either hard or soft shafts. In SAE 482, 484, and 485, the tin content will in general be mainly present in the lead-rich phase to prevent corrosion, but a small amount of the tin may be used in the copper-rich phase to increase its fatigue strength, and a further small amount will normally enter the copper by diffusion.

Composite bearings consisting of thin lead babbitt coatings on a matrix of copper and nickel powders sintered to a steel backing show very desirable characteristics as regards resistance to both fatigue and corrosion.

Steel backed aluminum alloy bearings are available for highly loaded, high speed, high temperature applications. Cast solid aluminum alloy bearings are sometimes used in heavily loaded moderate to high speed applications such as diesel main and connecting rod bearings. As compared with solid aluminum

TABLE 1—CLASSIFICATION AND USES OF BEARING AND BUSHING ALLOYS

SAE No.	Nominal Composition, %						Method of Manufacture	Characteristics	Applications
Tin Base Alloys	Sn	Sb	Cu						
11	87.5	6.75	5.75				Cast on steel, bronze or brass backs, or directly in the bearing housing.	Soft, corrosion resistant, moderately fatigue resistant.	Main and connecting rod bearings; motor bushings. Operates with either hard or soft journal.
12	89	7.5	3.5						
Lead Base Alloys	Pb	Sn	Sb	As					
13	rem	6	10	—			SAE 15 is cast on steel; others are cast on steel, bronze, or brass, or in the bearing housing. SAE 16 is cast into and on a porous sintered matrix, usually copper-nickel, bonded to steel.	Soft, moderately fatigue resistant, corrosion resistant. (See text accompanying this table.)	Main and connecting rod bearings. Operates with hard or soft journal. Requires good journal finish.
14	rem	10	15	—					
15	rem	1	15	1					
16	rem	4.5	3.5	—					
Lead-Tin Overlays	Pb		Sn						
19	90		10				Electrodeposited as a thin layer on copper-lead or silver bearing faces.	Soft, corrosion resistant. Bearings so coated run satisfactorily against soft shafts throughout the life of the coating.	Heavy duty, high speed main and connecting rod bearings.
190	93		7						
Copper-Lead Alloys	Cu		Pb					Moderately hard. Somewhat subject to oil corrosion. Some oils minimize this; protection with overlay may be desirable. Fatigue resistance good to fairly good. Listed in order of decreasing hardness and fatigue resistance.	Main and connecting rod bearings. The higher lead alloys can be used unplated against a soft shaft, although an overlay is helpful. The lower lead alloys may be used against a hard shaft, or with an overlay against a soft one.
49	76		24				Cast or sintered on steel back.		
48	70		30						
480	65		35						
481	60		40				Cast on steel back.		
	Cu	Pb	Sn					Moderately hard. Corrosion resistance improved over copper-leads of equal lead content without tin. Fatigue resistance fairly good. Listed in order of decreasing hardness and fatigue resistance.	Main and connecting rod bearings. Generally used without overlay. SAE 484 and 485 may be used with hard or soft shaft, while a hardened or cast shaft is recommended for SAE 482.
482	67	28	5				Steel back, lined with a structure combining sintered copper alloy matrix with corrosion-resistant lead alloy.		
484	55	42	3						
485	46	51	3						
Aluminum Base Alloys	Sn	Cu	Ni	Al	Si	Cd			
770	6.25	1	1	rem	—	—	Cast in permanent molds; work-hardened to improve physical properties.	Hard. Extremely fatigue resistant. Resistant to oil corrosion.	Main and connecting rod bearings. Generally used with suitable overlay.
780	6	1	0.5	rem	1.5	—	Bonded to steel back.[a]	Same as 770.	Same as 770.
781	—	—	—	rem	4	1	Bonded to steel back.[b]	Same as 770.	Main and connecting rod bearings generally used with overlay. Bushings and thrust bearings with or without overlay.
782	—	1	1	rem	—	3	Bonded to steel back.[a]	Same as 770.	Same as 781.
Other Copper Base Alloys	Cu	Sn	Pb	Zn					
795	90	0.5	—	rem			Wrought solid bronze.	Hard, strong, good fatigue resistance.	Intermediate load oscillating motion such as tie-rods, brake shafts, and so forth.
791	rem	4	4	4			Wrought solid bronze.	General purpose bearing material, good shock and load capacity. High temperature resistant. Hard shaft desirable. Less score resistant than higher lead alloys.	Moderate to high loads. Transmission bushings and thrust washers, piston pin bushings, and so forth.
793	rem	4	8	4 max			Cast on steel back.	General purpose bearing material, good shock and load capacity. High temperature resistant. Hard shaft desirable. Less score resistant than higher lead alloys.	Medium to high loading, typical applications, transmission and chassis bushings, and thrust washers
798							Sintered on steel back.		
792	80	10	10	—			Cast on steel back.	Has maximum shock and load carrying capacity of conventional cast bearing alloys; hard, both fatigue and corrosion resistant. Hard shaft desirable.	Heavy loads with oscillating or rotating motion. Typical applications, piston pin steering knuckle, differential axles, thrust washers, wear plates, and so forth.
797							Sintered on steel back.		
794	73.5	3.5	23	3 max			Cast on steel back.	Higher lead content gives improved surface action for higher speeds but results in somewhat less corrosion resistance.	Intermediate load application for both oscillating and rotating shafts, that is rocker-arm bushings, transmissions, and farm implements.
799							Sintered on steel back.		

[a] This alloy is procurable in strip form without steel backing. [b] This alloy can be produced as castings or wrought strip without steel back.

alloy bearings, steel backed aluminum alloy bearings have the following advantages:

If adequately bonded, they operate satisfactorily under higher unit loading, due to their greater fatigue resistance. Differential expansion is minimized, permitting closer assembly clearances and helping to maintain tight fit of the bearing in the housing. For either type of bearing, shaft hardness is less important when a suitable overlay is used.

Bearings made of pure silver on a steel backing, plated with lead and indium, or lead and tin, have been used for very heavy duty service where the high cost is warranted. For bearings of this type, see AMS 4815.

Rolled split bushings are made from wrought alloys or composite strip having suitable bearing and fabricating properties. Composite strip can be made by casting, sintering, or cladding the lining to a steel back. Solid bushings, without a split, can be made by machining cast or wrought material. Depending on the intended service of bushings, their mechanical properties, particularly hardness, can be varied by annealing or other heat treatment, or in the case of rolled split bushings, by rolling the strip. Measurement of hardness is sometimes used as a rapid method of inspection.

In Rockwell hardness testing of a given bearing material, it is advisable to choose a scale that will usually give a reading of 50 to 80, recognizing that for thin walled bearings or thin linings on harder backs, where there is an anvil effect, it is desirable to lean toward superficial Rockwell scales. For porous or coarsely multiphase materials, balls of comparatively large diameter are preferred.

Chemical composition limits for commonly used bearing and bushing alloys are published in SAE J460. Dimensions, tolerances, and terminology for commonly used sleeve type half bearings are published in SAE J506.

For further information on bearing materials, the reader is referred to the following: ASM Handbook (1948) pp. 745–755 and ASM Handbook, 8th Edition, Vol. 1, pp. 843–863.

WROUGHT AND CAST COPPER ALLOYS—SAE J461d   SAE Information Report

Report of Nonferrous Metals Division approved January 1934 and last revised by Nonferrous Metals Committee March 1976.

General—For convenience, this SAE Information Report is presented in two parts as shown below. To avoid repetition, however, data applicable to both wrought and cast alloys is included only in Part I.

Part I—Wrought Copper and Copper Alloys
Types of Copper (Table 1)
General Characteristics (Table 3)
Electrical Conductivity
Thermal Conductivity
General Mechanical Properties (Table 10)
Yield Strength
Fatigue Strength
Physical Properties (Table 2)
General Fabricating Properties (Table 3)
Formability
Bending
Hot Forming
Machinability
Joining
Surface Finishing
Color
Corrosion Resistance
Effect of Temperature
Typical Uses (Table 3)

Part II—Cast Copper Alloys
Types of Casting Alloys
Effects of Alloy Elements and Impurities
General Characteristics (Table 11)
Physical Properties (Table 12)
Typical Uses (Table 11)

PART I—WROUGHT COPPER AND COPPER ALLOYS

Factors influencing the uses of wrought copper and copper alloys concern electrical conductivity, thermal conductivity, machinability, formability, fatigue characteristics, strength, corrosion resistance, the ease with which alloys can be joined, and the fact that these materials are nonmagnetic. Copper and its alloys also have a wide range of rich, pleasing colors. The only other metal with such distinctive coloring is gold. These materials are all easily finished by buffing, scratch brushing, plating or chemically coloring, or clear protective coating systems.

When it is desired to improve one or more of the important properties of copper, alloying often solves the problem. A wide range of alloys, therefore, has been developed and commercially employed, such as the high copper alloys, brasses, leaded brasses, tin bronzes, heat treatable alloys, copper-nickel alloys, nickel silvers, and special bronzes.

The various types of copper and the principal alloys are listed in Tables 1 and 3, along with information describing composition, fabricating properties, and applications.

1. Types of Wrought Copper—C11000, C11100, C11300, C11400, C11500, and C11600 coppers[1] are either electrolytically or fire-refined, cast in the form of refinery shapes, containing a controlled amount of oxygen for the purpose of obtaining a level set on the top of the casting. It generally contains 0.01–0.04% oxygen, which exists as a copper-cuprous oxide eutectic surrounding the crystals of copper. Within these limits, the oxygen has only a very slight effect on the electrical, mechanical, and physical properties of copper. Because of the oxidizing effect of oxygen on impurities, its presence in copper indicates a reduction or elimination of certain impurities which would otherwise have adverse effects on conductivity.

C10200 copper is electrolytically refined and specially produced to be free from cuprous oxide although it is made without the use of residual metallic or metalloidal deoxidizers. Because of its freedom from residual deoxidizers, it has high electrical conductivity.

C12000 and C12200 copper is cast in the form of refinery shapes, free from cuprous oxide, produced through the use of metallic or metalloidal deoxidizers. Because it is necessary to use some excess of reducing agent, the electrical and thermal conductivity of the copper is lowered, and this fact should be considered when high conductivity is needed.

C10200, C12000 and C12200 coppers possess only slightly different mechanical properties from the C11XXX types. They differ little in respect to tensile strength when cold worked to similar extents, but do have somewhat higher ductility and also are not normally subject to hydrogen embrittlement.

2. Electrical Conductivity—The greatest single area of use for copper itself results from the high electrical conductivity of the metal. The combination of the property of high electrical conductivity with ease of forming and high corrosion resistance makes copper the preferred material for current-carrying members. The conductivity of copper for electrical conductors is 101% IACS (see Table 2) in the annealed or soft condition. The tensile strength of the soft copper, 220 MPa (32 ksi) can be increased to 345/380 MPa (50/55 ksi) by cold rolling, in which condition the electrical conductivity is decreased to about 97%. Heating such copper above 200°C for an extended period of time will soften it to a tensile strength of 205/240 MPa (30/35 ksi).

Silver is added to copper to increase its resistance to softening at elevated temperature without decreasing the electrical conductivity. Cold worked silver-bearing copper (see Table 4) can be heated to about 350°C for short periods of time without appreciable softening, and is less susceptible to creep rupture in highly stressed situations. Rolling mill practice and amount of silver have an effect upon the softening of such materials.

Cadmium added in small amounts (0.10%) to copper results in an alloy having superior resistance to softening at temperatures used in forming automotive radiators. Resistance to softening is retained even after the application of large amounts of cold work. The application of this material permits higher strength solders to be used and allows for the increase of soldering temperature range to a point not feasible with other high conductivity materials. Electrical and thermal conductivities are not appreciably different than for silver bearing copper.

Fig. 1 illustrates the softening characteristics of electrolytic copper and silver bearing and cadmium bearing copper alloys in terms of tensile strength for the times and temperatures indicated.

The 0.85% silver-bearing alloy is the best, of the three commonly available alloys, to resist creep rupture. The silver-bearing coppers find use in radiator construction where the material is subjected to slightly elevated temperature during soldering operations, also for commutators which are baked to set mica between the copper segments. Copper must not be softened by these treatments.

---
[1] Since the nomenclature used in the nonferrous metals trade is not always consistent, copper and copper base alloys are referenced by specification numbers described in SAE J463.

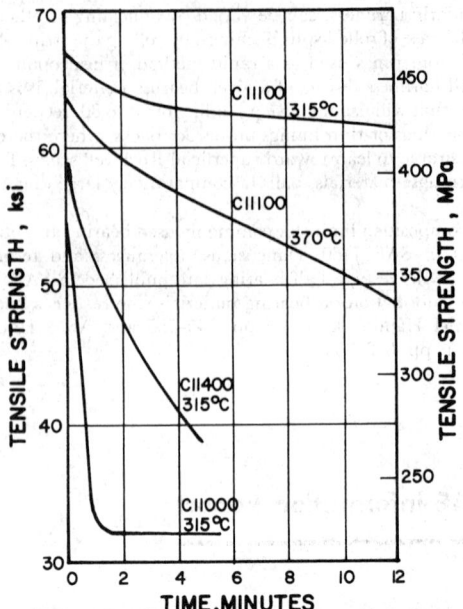

FIG. 1—SOFTENING CHARACTERISTICS OF THREE COPPER ALLOYS

To prevent embrittlement which takes place with copper should be specified if the material is to be heated much above 425°C in an atmosphere of reducing gases such as hydrogen. Embrittlement results from the action of the reducing gases with the copper oxide normally present in all C11XXX types.

The addition of chromium to copper produces an alloy with a combination of high tensile strength (485 MPa [70 ksi]) and electrical conductivity (80% IACS). This alloy C18400 has the ability to retain its mechanical properties and wear resistance to a high degree at elevated temperatures. The copper chromium alloys have found considerable use as fabricated into welding tips and seam welding wheels. Zirconium bearing copper (C15000) is also finding wide use in high temperature-high strength applications.

Heat treated beryllium bearing copper alloys having tensile strengths up to 1345 MPa (195 ksi) and fatigue strengths up to 345 MPa (50 ksi) are available; however, a drop in electrical conductivity to about 50% and high cost must be considered. Where repetitive or cycling operations must be performed, such properties have made application of these alloys economical.

Conducting contacts, springs or other stressed parts that are manufactured by forming may employ chromium or beryllium bearing coppers. The parts are formed by cold working and then strengthened by heat treatment.

The high degree of ductility and toughness of commercially pure copper usually make it unsuited for cutting or machining operations. Coppers with lead, tellurium and sulfur were developed to combine the properties of copper with improved machinability. Parts that must be formed by extensive machining and be highly conductive are made from the free machining coppers. Tellurium copper has a 95% electrical conductivity and a machinability rating of 80-90. Sulfur bearing copper has a 95% IACS electrical conductivity and the same machinability rating, whereas lead copper has an electrical conductivity of 98% IACS and a machinability rating of 80. The machinability rating for copper is 20.

Where higher tensile strength 620 MPa (90 ksi) is required along with good machinability (60) and lower electrical conductivity (10%) can be tolerated, aluminum silicon alloys may be used to advantage.

For applications requiring good fatigue properties, the nickel-silver, phosphor, or beryllium alloys will serve. These alloys, however, have relatively low electrical conductivity ranging from 5 to 50%.

C12000 is also a good choice in the selection of a conductor to be used where creep strength is to be considered, as may be the case when the material is to operate at slightly elevated temperature.

3. Thermal Conductivity—For the alpha solid solutions of copper alloys, at least, the thermal conductivity is a nearly linear function of the electrical conductivity multiplied by the absolute temperature. Good conductors of electricity are also good conductors of heat and poor conductors of electricity are poor conductors of heat.

When high thermal conductivity is of principal importance, the same considerations given electrical conductivity apply.

4. Mechanical Properties—Except for the heat treatable alloys, strength is determined mainly by composition and degree of cold work. Mechanical properties of the most important alloys are to be found in Table 10.

Copper and copper alloys containing aluminum, silicon, tin, iron, and manganese, in various combinations and concentrations, are much stronger by virtue of their chemistry than the other coppers or alloys. For heavy sections or parts requiring high strength, inherently stronger alloys should be specified. For lighter or smaller sections which can be made adequately from stronger tempers, other alloys are successful. For example, the tensile strength of C26000 used in the production of radiator tanks can be increased by adjusting the rolling mill procedure from 310 to 365 MPa (45 to 53 ksi) without a harmful reduction in ductility. Similarly C26000 strip 0.11 mm (.0045 in) intended for fabrication into lockseam tube, used in radiator construction, is available in an annealed temper having a tensile strength of about 440 MPa (64 ksi) and an elongation of 32% in 50 mm (2 in). This represents an 18.5% increase in tensile strength without any sacrifice of ductility, compared to material produced by rolled-to-temper methods.

The tensile strength of the copper-zinc series of alloys, the most widely used group in the industry, increases in general for any specific temper as the copper content decreases. The alloys also are characterized by extremely high ductility, excellent forming characteristics and ease of finishing. The relationship of the increase of properties with zinc content is shown in Fig. 2.

A series of heat treatable alloys are commercially available and have strength as high as 1380 MPa (200 ksi). These alloys are produced with carefully controlled compositions and contain such elements as chromium, beryllium, nickel, phosphorus and silicon. The attractive diversity of properties obtainable in heat treatable copper alloys can be observed in Table 10.

The zirconium alloy copper (C15000) might be included in this group because it does respond to heat treatment; however, its strength is developed primarily through the application of cold working. Heat treatment primarily restores high electrical conductivity and ductility and increases surface hardness. The alloy has found use in the production of welding tips and wheels, stud bases for rectifiers, commutators for motors and electrical switch parts.

One outstanding characteristic of the heat treatable alloys is that they may be formed into articles, such as complex springs, while in the soft or partially work hardened state, and the mechanical properties subsequently improved to their maximum by heat treatment.

5. Yield Strength—Yield strength is the stress at which a material exhibits a specified limiting deviation of strain. Ordinarily the yield strength of copper and copper alloys is taken at 0.5% extension under load (strain) although for some design purposes values taken at 0.1 or 0.2% offset may be used.

Where residual stresses, due to forming, approach or exceed the yield strength, stress corrosion cracking may occur. Also, stresses may reach levels high enough to cause elastic drift in springs. In either case, it may be advisable to apply a low temperature, stress relieving treatment. Suggested temperatures for accomplishing a stress relief are listed in Table 5.

The tendency of a formed part to stress crack can be determined by the application of ASTM B 154, Method of Mercurous Nitrate Test for Copper

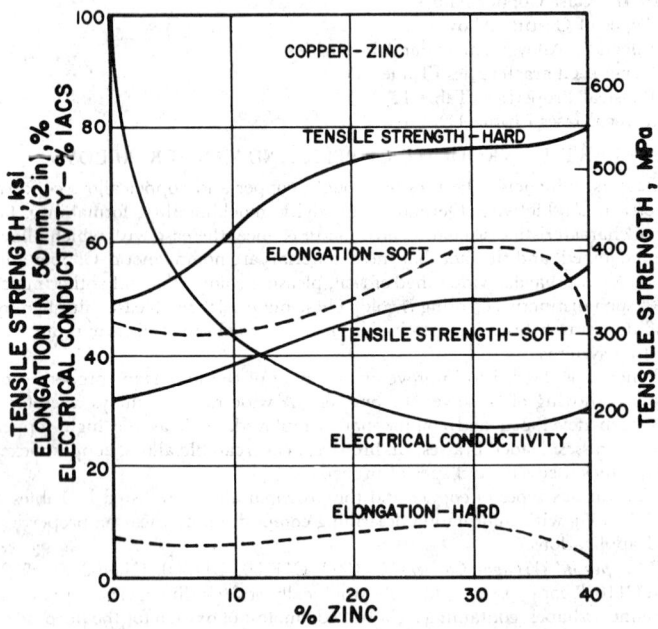

FIG. 2—RELATIONSHIP OF INCREASE OF PROPERTIES WITH ZINC CONTENT

and Copper Alloys. The effectiveness of thermal stress relieving treatment can be judged by the same test.

6. Fatigue Strength—Materials subjected to alternating tensile and compressive stress, or shear stress alternating in direction, will fail by "fatigue" fracture at much lower values for maximum stress than if subjected to steady loads. The same effect applies, but to a lesser degree if the stresses are constant in direction but vary in magnitude. Springs, diaphragms, bellows, flexible hose and similar applications are frequently exposed to such conditions, and when this is the case, the maximum stress used in design calculations must be less than the safe tensile or shear stress applicable when the load is constant.

Generally, the harder the material the higher its fatigue strength, although fatigue strengths vary with surface and temper conditions and corrosion. The heat treatable alloys such as beryllium alloys, copper containing either nickel and silicon or nickel and phosphorus can be hardened by heat treatment and, therefore, optimum spring characteristics can be realized by such treatment after the spring has been formed. Phosphor alloys, nickel-silver alloys and other nonheat treatable alloys must always be used in a condition sufficiently soft to successfully cold form into the desired shape. Since the fatigue strength increases with the hardness of the material, the highest values for the precipitation hardening alloys must be compared with less than the highest values for other materials. Such a comparison indicates a marked superiority of the former alloys.

A comparison of the fatigue strengths of several alloys follows: C17000, 275 MPa (40 ksi); C67400, 215 MPa (31 ksi); C52100, 185 MPa (27 ksi); C65500, 160 MPa (23 ksi); C77000, 155 MPa (22.5 ksi); C51000, 150 MPa (22 ksi); C26800, 140 MPa (20 ksi); C11000, 70 MPa (10 ksi).

7. Formability—All coppers form easily and readily and work harden slowly. Generally, C10200 and C12200 may be best for severe cold forming, although for thin gage material, mill practices may be adjusted to develop extreme ductilities in C11000.

Best results in cold forming operations are obtained through the use of nonleaded alloys, including copper, silver-bearing alloys, and all nonleaded alloys containing more than 63% copper. Included in this list are the phosphor alloys, nickel-silver alloys (65-18) nickel alloys, and the age hardenable alloys in the solution treated condition. Some of the age hardenable alloys have excellent ductility even after fully heat treated and are capable of being cold formed to a high degree.

C26000 is the most widely used for operations such as cold drawing, upsetting, stamping, and bending. C23000, C22000 and C21000 are not as strong as C26000 but harden at a slower rate when cold worked, thereby allowing successive operations without intermediate annealing.

Forming operations such as deep drawing, eyelet work, coining, flanging, spinning, or similar cold working, all require annealed material usually ordered by specifying grain size. Care should be exercised to specify the grain size most suitable for the part to be made. Depending upon the severity of the cold forming operation during the production of a specific part, it is sometimes necessary for the fabricator to perform an anneal or soften the material before further press operations can be successfully applied. The grain size of any of these anneals must be controlled. Grain size that is too small can lead to breakage during forming, whereas excessively large grain sizes may also lead to breakage because of lowering tensile strength of the alloy, or, if the part does not fail, its surface may become excessively rough (a condition known as orange peel) and require costly finishing operations where polishing and plating are required. Table 6 suggests approximate annealing temperature ranges to apply for intermediate annealing. The alloy should be annealed at the lowest temperature experimentally in seeking the proper grain size and smoothest polishing surface. It should be remembered that insufficiently annealed work can always be reannealed at higher temperature. Overannealed damage is beyond salvage.

Currently the trend is toward the use of thinner gages of strip for fabrication purposes. Reduction of grain size to less than 0.010 mm makes available material of sufficient strength to withstand deeper drawing without intermediate annealing processes.

Coining operations demand metal of large grain size for maximum sharpness of impression. Cold upsetting, particularly of screws, rivets and bolts should be performed on metal lightly cold drawn to develop some strength in unsupported sections to resist bending as the parts are being fabricated.

Where machining is an important factor in making the finished part and cold forming is part of the fabricating process, leaded alloys in light drawn tempers are logical choices. The presence of lead results in easier machining. Control of the lead content and temper allows the alloy to be cold worked as, for example, in thread rolling.

Table 3 lists cold workability ratings for various alloys. The ratings are arbitrary for their approximate relative suitability for being cold worked; the ratings being excellent, good, fair, or not recommended. It must be realized that such arbitrary ratings cannot be too precise, due to the multiplicity of cold working operations that must be considered. Operations taken into account in assigning the ratings include drawing, forming, stamping, spinning, bending, and heading. The ratings given take into account not only the relative power required to cold work the alloy, but also the amount of deformation which is possible without fracture.

8. Bending—Bending is often the controlling factor in selection of temper for strip products. For a particular alloy and thickness, the harder the temper the more generous the bending radius must be for successful bending. Bending characteristics of strip are more favorable when the axis of the bend is at a right angle to the rolling direction. Bending problems may be prevented when sharp or difficult bends must be made in more than one direction by designing the bending tools to accommodate blanks cut on some axis other than parallel to the direction of rolling. Table 7 recommends radii for forming 90 deg bends in respect to rolling direction, gage and temper for various alloys.

9. Hot Forming—Copper and a series of copper alloys lend themselves well to production by hot forming, die-pressed forging and extrusion. Where sufficient support is not provided by the tooling during hot working operations, the higher leaded alloys become susceptible to cracking.

Alloys specified for hot forming include many coppers, zinc alloys containing 58–63% copper, tin alloys, aluminum alloys, silicon alloys, and nickel-silver alloys. Table 8 lists the relative forgeability of the various alloys and takes into account such variables as pressure, die wear, and hot plasticity.

10. Machinability—The addition of lead to copper alloys greatly improves their machinability. The greater the amount of lead, the easier the alloy machines or cuts. Lead also improves the blanking quality in strip alloys by reducing their ductility, thereby providing a sharp, clean shear.

Lead does not dissolve in copper or its alloys and is finely dispersed throughout the alloy. During a cutting operation the presence of lead produces short or broken chips which are easily flushed away by lubricants. Excellent finishes can be attained with the use of proper tools and feeds, and machining rates are frequently as high as maximum machine capabilities. Screw machines often utilize speeds as high as 10,000 rpm producing parts from free cutting zinc alloy rod. The Copper Base Alloy Rod Handbook, published by the Copper Development Association (CDA), is recommended for information on tool shapes, feeds, speeds, and so forth.

Ordinarily, half hard C36000 containing about 3.25% lead is preferred for machining. For knurling or thread rolling operations demanding greater ductility, softer or lower leaded alloys should be specified.

Lead, tellurium, or sulfur added to copper combine the properties of pure copper with improved machinability, and all three alloys may be used where the basic properties of copper itself are required. Lead is insoluble in the copper and both tellurium and sulfur form insoluble compounds with copper thereby acting much in the same manner as lead in producing chips during machining operations. Table 3 lists the arbitrary relative machinability rating of many alloys. The numerical rating is a reasonable indication of the amount of power required for any given type and degree of cutting operation, and tool life will be found to vary in proportion to such a rating. The type of chip also plays an important part, for in certain operations almost any type of chip can be tolerated; whereas in others, for example deep drilling, box milling and tapping, long stringy chips may cause scoring of the stack and tool breakage. The table also lists the type of chip expected for each alloy by designating with the letters L, M, or S, indicating that the chips are long, medium, or short.

11. Joining—Copper and most of its alloys are readily joined by soldering and brazing and by most of the commonly used welding processes. Table 3 indicates the approximate relative suitability of alloys for being joined by various processes. The choice of method depends on shape of the work, composition of the metal, and the end use of the product. Thus, where welded joints of maximum strength are required, it is necessary to use C10200, C12000, C12200 or C14500 coppers instead of the C11XXX types. Arc welding of zinc-bearing alloys is hampered by the vaporization of zinc; so, if oxyacetylene welding is not feasible, the parts are designed for brazing or soldering. On aluminum alloys, soldering is impossible, and brazing is difficult because interfering oxides form even under very active fluxes; oxyacetylene welding is also impossible because of interfering oxides; therefore, arc welding processes are the only practicable joining methods. Brazing and soldering are the preferred methods for joining leaded alloys, because welding develops increasing porosity and cracking as lead content increases.

Methods most commonly used are:

11.1 Resistance Welding—Applicable on all nonleaded alloys. Flash butt welding successful on all. Spot and seam welding practicable on those with conductivities below 30%. Copper-silicon and copper-nickel alloys are the most weldable and coppers are the least weldable.

11.2. Gas Shielded Arc Welding—Widely employed on all. Silicon and aluminum alloys are readily welded by these processes. So are the nickel alloys if done with the specially alloyed filler metal developed for these processes.

11.3 Coated Metal Arc Welding—Excellent where good flux-coated electrodes are available, as for the nickel alloys. Coated electrodes are also available for aluminum and phosphor alloys. Process is not suitable for copper because of high heat requirements.

11.4. Carbon Arc Welding—Less costly than gas shielded arc welding and produces good results on silicon alloys. Also useful for welding copper with silicon or phosphor alloy rods.

11.5. Oxyacetylene Welding—Good results on deoxidized copper using specially alloyed welding rods, and on silicon alloys. Excellent for zinc alloys if low fuming rods are used. Not suitable for aluminum alloys.

11.6. Brazing—Generally useful with either silver alloys or phosphorus alloys. Latter are less costly. They are also considered self-fluxing on copper but best results require use of flux. Phosphorus alloys are often used to join the 90-10 nickel alloy but are not recommended for use on alloys with higher nickel contents. Tough pitch coppers are readily "gassed" and embrittled by exposure to hydrogen at high temperatures, therefore they are not suitable for parts to be furnace brazed in hydrogen-bearing atmospheres, and they cannot be safely brazed by flame processes if heating time is prolonged. Special fluxes are available to help in brazing of aluminum alloys.

11.7. Soldering—Readily done on all copper base metals except the aluminum alloys. Suitable fluxes are not available for use on these alloys.

12. Effect of Temperature—Copper and its alloys are not harmed by temperatures as low as $-185°C$, rather, a gain in mechanical properties is noticed with decreasing temperature.

Most copper alloys do not find application above 200°C since, dependent upon the amount of cold work applied during fabrication, most of the alloys soften between 200 and 425°C. Further, oxidation also must be considered above these temperatures.

13. Color—Copper alloys are the only large tonnage metals that have a wide red and yellow color range. Red and pink for the copper rich materials, gold shades for C21000, C22000 and C23000 although sometimes these alloys are also a pleasing red because of the formation of a superficial copper oxide on the surface. The series of alloys becomes more yellow at 80% copper, 20% zinc (C24000) and develop the familiar yellow at the 70-30 (C26000) composition. The color reverses at about 55% copper, 45% zinc.

14. Surface Finishing—A number of types of mechanical finishes and treatments can be applied rather easily to the copper alloys. Among these are deburring, bright rolling, ball burnishing, wheel polishing and buffing, belt polishing, scratch brushing, and sand blasting.

The highest luster that can be produced on copper alloys is by the combination of wheel polishing and buffing. Both manually operated and automatic buffing equipment together with an assortment of polishing and buffing wheels are available to accomplish this job.

C26000 lends itself extremely well to finishing such as just described and because of this and its excellent corrosion resistance, is often used for such items as automobile wheel covers or hub caps.

The ease with which a part can be polished can depend upon surface roughness termed orange peel which can develop on cold drawn parts if the base stock before forming has a grain size over 0.050 mm and the forming operation is severe. The degree of orange peel depends upon grain size and degree of forming. Therefore, when surface finish is important after the forming of a given part, attention must be directed to the grain size of the starting stock or to parts given an intermediate anneal in the fabrication sequence. Table 9 is a guide to specifying annealed tempers for strip in relation to type of operation being performed, gage of the material, and grain size.

Copper alloys in general offer a noncorroding surface for electroplating. A base electroplate of copper is usually not necessary under nickel or chromium used as decorative electroplates. Therefore, copper alloys, in general, allow thinner plates of such metals as tin, nickel, chromium, or silver, than do other metals.

Fused enamels are applied on C21000, C22000 and C23000 with very beautiful effects. Alloys containing much more than 7% zinc should not be used with transparent enamels as cloudiness or color change can result.

Tarnishing or discoloration of copper alloys may be retarded and, in many cases, delayed indefinitely by application of a lacquer selected with consideration to the service environment in which the object is to exist.

There are literally hundreds of transparent coatings which can be applied to copper articles. Perhaps in no other unit process of metal finishing is such a wide variety of materials available for use.

Lacquer or protective coating systems, which have very effectively protected the surface of copper alloys for both interior and exterior exposure for a number of years, have been developed and time tested.

15. Corrosion Resistance—Copper and copper alloys have been extensively and successfully used for many years in a variety of corrosive conditions. Copper is highly resistant to the effects of atmosphere, naturally occurring fresh and salt waters, alkaline solutions (except those containing ammonia) and many organic chemicals. The severity of oxidizing conditions controls its behavior in acidic media. Many salt solutions are successfully handled. Sulfur and its sulfide compounds do combine with copper to produce copper sulfide as a corrosion product. As the zinc content of the copper alloys is increased over 15%, resistance to corrosion from sulfide compounds is markedly increased. This fact is important when radiator materials are selected for possible use on farm equipment that might come into contact with insecticide sprays that contain sulfur.

The commercial copper alloys vary widely in chemical composition; therefore, there is considerable variation in their resistance to corrosion. Many of the alloying elements improve corrosion resistance of the parent metal as well as enhance its mechanical properties.

Extensive use indicates the suitability and, often, superiority of copper and its alloys for many applications, including the following broad classifications: atmospheric exposures, such as hardware, building fronts, automotive radiators, and hub caps; fresh water supply lines, including those buried in soil; sea water applications; heat exchanges; and industrial and chemical plant equipment handling a variety of products.

TABLE 1—GENERAL INFORMATION—NAME, NOMINAL COMPOSITION, AND COMPARABLE STANDARDS OF WROUGHT COPPER ALLOYS

UNS No.[a]	Name[b]	Nominal Composition percent by weight		SAE No.	ASTM No.[c]	Former SAE No.
		Cu	Other			
C10200	Oxygen free copper (OF)	99.9	—	CA102	B75, B152, B280	—
C11000	Electrolytic tough pitch copper (ETP)	99.9	—	CA110	B3, B133, B152, B283	71, 83
C11100	Electrolytic tough pitch, anneal resistant copper	99.9	(Trace elements)	CA111	—	71
C11300	Tough pitch copper with Ag (STP)	99.9	0.03 Ag	CA113	B152	71
C11400	Tough pitch copper with Ag (STP)	99.9	0.04 Ag	CA114	B152	71
φ C11500	Tough pitch copper with Ag (STP)	99.9	0.06 Ag	CA115	B152	—
C11600	Tough pitch copper with Ag (STP)	99.9	0.09 Ag	CA116	B152	71
C12000	Phosphorus deoxidized copper (DLP)	99.9	0.0008 P	CA120	B68, B75, B152, B280	75
C12200	Phosphorus deoxidized copper (DHP)	99.9	0.02 P	CA122	B68, B75 B152, B280	—
C14500	Phosphorus deoxidized tellurium copper (DPTE)	99.5	0.5 Te, 0.008 P	CA145	B283, B301	—
C14700	Sulfur bearing copper	99.7	0.3 S	CA147	B301	—
C15000	Zirconium copper	99.8	0.15 Zn	CA150		—
C16200	Cadmium copper	99.0	1 Cd	CA162		—
C17000	Beryllium copper	98.0	1.7 Be	CA170	B194	—
C17200	Beryllium copper	98.0	1.9 Be	CA172	B194, B196	—
C17500	Beryllium copper	97.0	0.5 Be, 2.5 Co	CA175	B441, B534	—
C17600	Beryllium copper	97.0	0.4 Be, 1.5 Co, 1 Ag	CA176	B441	—
C18400	Chromium copper	99.0	0.8 Cr	CA184		—
C18700	Leaded copper	99.0	1 Pb	CA187	B301	—
φ C19200	High copper alloy	99.0	1 Fe, 0.03 P	CA192	B111	—
C21000	Gilding, 95%	95.0	5 Zn	CA210	B36	—
C22000	Commercial bronze, 90%	90.0	10 Zn	CA220	B36, B135	—
C23000	Red brass, 85%	85.0	15 Zn	CA230	B36, B135	74D, 79A
C24000	Low brass, 80%	80.0	20 Zn	CA240	B36	79B
C26000	Cartridge brass, 70%	70.0	30 Zn	CA260	B36, B134 B135	70A, 74C, 80A
C26800	Yellow brass, 66%	66.0	34 Zn	CA268	B36	70C
C27000	Yellow brass, 65%	65.0	35 Zn	CA270	B134	80B
C33000	Low leaded brass, (tube)	66.0	34 Zn, 0.5 Pb	CA330	B135	74B
C33100	Leaded brass	66.0	33 Zn, 1 Pb	CA331		
C34200	High leaded brass	65.0	33 Zn, 2 Pb	CA342	B121	
C34500	Leaded brass	63.0	35 Zn, 2 Pb	CA345	B453	
C35000	Medium leaded brass, 62%	63.0	36 Zn, 1 Pb	CA350	B121, B453	
C36000	Free cutting brass	62.0	35 Zn, 3 Pb	CA360	B16	72
C37700	Forging brass	60.0	38 Zn, 2 Pb	CA377	B283	88
C46400	Naval brass, unhibited	60.0	39 Zn, 0.8 Sn	CA464	B21, B283	73
φ C46500	Naval brass, arsenical	60.0	40 Zn, 0.5 As	CA465		
φ C46600	Naval brass, antimonial	60.0	40 Zn, 0.5 Sb	CA466		
φ C46700	Naval brass, phosphorized	60.0	40 Zn, 0.5 P	CA467		
C51000	Phosphor bronze, 5% A	95.0	5 Sn, 0.2 P	CA510	B103, B139, B159	77A, 81
φ C51100	Phosphor bronze	96.0	4 Sn, 0.2 P	CA511	B103	
C52100	Phosphor bronze, 8% C	92.0	8 Sn, 0.2 P	CA521	B103	77C
φ C52400	Phosphor bronze, 10% D	90.0	10 Sn, 0.2 P	CA524	B103	
C54400	Phosphor bronze, B-2	88.0	4 Sn, 4 Zn, 4 Pb	CA544	B103, B139	—
C60800	Aluminum bronze	95.0	5 Al	CA608	B111	
φ C61400	Aluminum bronze, D	91.0	7 Al, 2 Fe	CA614	B150, B169	701D
φ C61800	Aluminum bronze	89.0	10 Al, 1 Fe	CA618		
C62300	Aluminum bronze	88.0	9 Al, 3 Fe	CA623	B150, B283	701B
C62400	Aluminum bronze	86.0	11 Al, 3 Fe	CA624		701B
C63000	Aluminum bronze	82.0	10 Al, 3 Fe, 5 Ni	CA630	B150, B283	701C
φ C64200	Aluminum silicon bronze	91.0	7 Al, 2 Si	CA642	B150, B283	
C65500	High silicon bronze, A	97.0	3 Si	CA655	B97, B98, B283	—
C67000	Manganese bronze, B	65.0	24 Zn, 4 Mn, 4 Al, 3 Fe	CA670	B138	—
C67300	Manganese bronze	60.0	34 Zn, 3 Mn, 2 Pb, 1 Si	CA673		—
C67400	Manganese bronze	58.0	37 Zn, 3 Mn, 1 Al, 1 Si	CA674		—
C67500	Manganese bronze, A	58.0	40 Zn, 0.3 Mn, 1 Fe, 1 Sn	CA675	B138	—

TABLE 1—GENERAL INFORMATION—NAME, NOMINAL COMPOSITION, AND COMPARABLE STANDARDS OF WROUGHT COPPER ALLOYS (continued)

UNS No.[a]	Name[b]	Nominal Composition percent by weight		SAE No.	ASTM No.[c]	Former SAE No.
		Cu	Other			
C70600	Copper nickel, 10%	90.0	10 Ni	CA706	B111, B171	—
C71000	Copper nickel, 20%	80.0	20 Ni	CA710	B111, B122	—
C71500	Copper nickel, 30%	70.0	30 Ni	CA715	B111, B122, B171	—
C75200	Nickel silver, 65-18	65.0	18 Ni, 17 Zn	CA752	B122, B151	—
C77000	Nickel silver, 55-18	55.0	18 Ni, 27 Zn	CA770	B122, B151	—

[a] Unified numbering system.
[b] Alloy names are shown for information only, and should not be used. Use the appropriate designation only. (Example: UNS C21000 Copper Alloy).
[c] ASTM numbers listed are only those forms or shapes covered in the specification for wrought copper alloy.

TABLE 2A—TYPICAL PHYSICAL PROPERTIES OF WROUGHT COPPER ALLOYS

Metric (SI) Units

UNS No.	Melting Point °C Liquidus	Melting Point °C Solidus	Density g/cm³ at 20°C	Coefficient of Thermal Expansion/°C × 10⁻⁵ 20–100°C	20–200°C	20–300°C	Thermal Conductivity W/m·K	Electrical Resistivity nΩ·m[a]	Specific Heat J/kg·K	Modulus GPa Elastic	Modulus GPa Rigid
C10200	1083	—	8.94	1.70	1.73	1.77	391	17.1	385	117	44
C11000	1083	1065	8.91	1.70	1.73	1.77	391	17.1	385	117	44
C11100	1083	1065	8.91	1.70	1.73	1.77	388	17.2	385	117	44
C11300	1082	—	8.91	1.70	1.73	1.77	388	17.2	385	117	44
C11400	1082	—	8.91	1.70	1.73	1.77	388	17.2	385	117	44
C11500	1082	—	8.91	1.70	1.73	1.77	388	17.2	385	117	44
C11600	1082	—	8.91	1.70	1.73	1.77	388	17.2	385	117	44
C12000	1083	—	8.94	1.70	1.73	1.77	386	17.6	385	117	44
C12200	1083	—	8.94	1.70	1.73	1.77	339	20.3	385	117	44
C14500	1075	1051[b]	8.94	1.71	1.74	1.78	355	18.6	385	117	44
C14700	1076	1067	8.94	1.70	1.73	1.77	374	18.1	385	117	44
C15000	1080	980	8.89	1.63	1.80	2.01	367[d]	18.6[d]	385	117	44
C16200	1076	1030	8.89	1.70	1.73	1.77	360	19.2	385	117	44
C17000	980	865	8.26	1.67	1.70	1.78	118	76.8	420	131	50
C17200	980	865	8.26	1.67	1.70	1.78	118	76.8	420	131	50
C17500	1075	1070	8.75	—	1.76	—	234	37.9	420	124	47
C17600	1054	1010	8.75	—	—	—	—	31.6	—	124	47
C18400	1075	1070	8.89	1.76	—	—	324[d]	21.6[d]	385	131	50
C18700	1080	953[c]	8.94	—	—	1.76	377	17.9	385	117	44
C19200	1084	—	8.87	1.62	—	—	216	34.5	385	117	44
C21000	1065	1050	8.86	—	—	1.81	234	30.8	377	117	44
C22000	1045	1020	8.80	—	—	1.84	189	39.2	377	117	44
C23000	1025	990	8.75	—	—	1.87	159	46.6	377	117	44
C24000	1000	965	8.67	—	—	1.91	140	53.9	377	110	41
C26000	955	915	8.53	—	—	1.99	121	61.6	377	110	41
C26800	930	905	8.47	—	—	2.03	116	63.9	377	103	39
C27000	930	905	8.47	—	—	2.03	116	63.9	377	103	39
C33000	940	905	8.50	—	—	2.02	116	66.3	377	103	39
C33100	940	905	8.50	—	—	2.02	116	66.3	377	103	39
C34200	910	885	8.47	—	—	2.02	116	66.3	377	103	39
C34500	915	885	8.45	—	—	2.03	119	66.3	—	69	—
C35000	915	895	8.47	—	—	2.03	116	66.3	377	97	37
C36000	900	885	8.50	—	—	2.05	116	66.3	377	97	37
C37700	895	880	8.44	—	—	2.07	119	63.9	377	103	39
C46400	900	885	8.41	—	—	2.12	116	66.3	377	103	39
C46500	900	885	8.41	—	—	2.12	116	66.3	377	103	39
C46600	900	885	8.41	—	—	2.12	116	66.3	377	103	39
C46700	900	885	8.41	—	—	2.12	116	66.3	377	103	39
C51000	1050	950	8.86	—	—	1.78	69	115.0	377	110	41
C51100	1060	975	8.86	—	—	1.78	83	87.0	377	110	41
C52100	1020	880	8.80	—	—	1.82	62	133.0	377	110	41
C52400	1000	845	8.78	—	—	1.84	50	157.0	377	110	41
C54400	1000	930	8.89	—	—	1.73	87	90.7	377	103	39
C60800	1063	1050	8.17	—	—	1.81	80	100.0	377	121	46
C61400	1045	1040	7.89	—	—	1.62	67	123.0	377	117	44
C61800	1045	1040	7.53	—	—	1.62	64	133.0	377	117	44
C62300	1045	1040	7.66	—	—	1.62	61	144.0	377	117	44
C62400	1040	1025	7.45	—	—	1.92	54	144.0	377	117	44
C63000	1054	1035	7.58	—	—	1.62	38	192.0	377	117	44
C64200	1005	985	7.69	—	—	1.81	45	186.0	377	110	41
C65500	1025	970	8.53	—	—	1.80	36	246.0	377	103	39
C67000	930	905	7.81	—	—	2.00	24	75.0	—	103	39
C67300	890	845	8.28	—	—	2.00	—	75.0	—	103	39
C67400	885	865	8.08	—	—	2.00	100	75.0	377	97	37
C67500	890	865	8.36	—	—	2.12	106	71.8	377	103	39
C70600	1150	1100	8.94	—	—	1.71	45	191.0	377	124	47
C71000	1200	1150	8.94	—	—	1.64	36	266.0	377	138	52
C71500	1240	1170	8.94	—	—	1.62	29	375.0	377	152	57
C75200	1110	1070	8.73	—	—	1.62	33	287.0	377	124	47
C77000	1055	—	8.70	—	—	1.67	29	314.0	377	124	47

[a] See Table 2B for percent IACS electrical conductivity.
[b] Small amounts of tellurium-rich constituent remains liquid down to 490°C.
[c] Small amounts of lead-rich constituent remains liquid down to 325°C.
[d] After precipitation-hardening treatment.

TABLE 2B—TYPICAL PHYSICAL PROPERTIES OF WROUGHT COPPER ALLOYS

Customary Units

UNS No.	Melting Point, °F Liquidus	Melting Point, °F Solidus	Density[a]	Coefficient of Thermal Expansion[b] 68–212°F	Coefficient of Thermal Expansion[b] 68–392°F	Coefficient of Thermal Expansion[b] 68–572°F	Thermal Conductivity[c]	Electrical Resistivity[d]	Electrical Conductivity[e]	Thermal Capacity[f]	Modulus Elastic[g]	Modulus Rigid[h]
C10200	1981	—	.323	9.4	9.6	9.8	226	10.3	101	.092	17	6.4
C11000	1981	1949	.322	9.4	9.6	9.8	226	10.3	101	.092	17	6.4
C11100	1981	—	.322	9.4	9.6	9.8	224	10.3	101	.092	17	6.4
C11300	1981	—	.322	9.4	9.6	9.8	224	10.3	100	.092	17	6.4
C11400	1981	—	.322	9.4	9.6	9.8	224	10.4	100	.092	17	6.4
ϕ C11500	1981	—	.322	9.4	9.6	9.8	224	10.4	100	.092	17	6.4
C11600	1981	—	.322	9.4	9.6	9.8	224	10.4	99	.092	17	6.4
C12000	1981	—	.323	9.4	9.6	9.8	223	10.7	97	.092	17	6.4
C12200	1981	—	.323	9.4	9.5	9.8	196	12.2	85	.092	17	6.4
C14500	1960	1931[i]	.323	9.4	9.6	9.8	205	10.9	95	.092	17	6.4
C14700	1970	1953	.323	9.4	9.6	9.8	216	10.9	95	.092	17	6.4
C15000	1979	—	.323	9.4	9.6	9.8	212[k]	11.2[k]	93[k]	.092	17	6.4
C16200	1969	—	.321	9.4	9.6	9.8	208	11.9	87	.092	17	6.4
C17000	1800	1600	.298	9.3	9.4	9.9	—	47.2	22	—	19	7.3
C17200	1800	1600	.298	9.3	9.4	9.9	—	47.2	22	.100	19	7.3
C17500	1955	1885	.316	—	9.8	—	—	23.1	45	—	18	6.8
C17600	1930	1850	.316	—	9.8	—	—	19.0	50	—	18	6.8
C18400	1967	—	.321	9.4	9.6	9.8	187[k]	13.0[k]	80[k]	.092	19	7.2
C18700	1976	1747[j]	.323	9.4	9.6	9.8	218	10.6	98	.092	17	6.4
ϕ C19200	1983	—	.320	9.0	—	—	125	20.8	50	.092	17	6.4
C21000	1950	1920	.320	—	—	10.0	135	18.5	00	.090	17	6.4
C22000	1910	1870	.318	—	—	10.2	109	23.6	44	.090	17	6.4
C23000	1880	1810	.316	—	—	10.4	92	28.0	37	.090	17	6.4
C24000	1830	1770	.313	—	—	10.6	81	32.4	32	.090	16	6.0
C26000	1750	1680	.308	—	—	11.1	70	37.0	28	.090	16	6.0
C26800	1710	1660	.306	—	—	11.3	67	38.4	27	.090	15	5.6
C27000	1710	1660	.306	—	—	11.3	67	38.4	27	.090	15	5.6
C33000	1720	1660	.307	—	—	11.2	67	39.9	26	.090	15	5.6
C33100	1720	1660	.307	—	—	11.2	67	39.9	26	.090	15	5.6
C34200	1670	1630	.306	—	—	11.3	67	39.9	26	.090	15	5.6
C34500	1650	1625	.305	—	—	11.4	69	39.9	26	—	10	—
C35000	1650	1630	.305	—	—	11.4	67	39.9	26	.090	14	5.3
C36000	1650	1630	.307	—	—	11.4	67	39.9	26	.090	14	5.3
C37700	1640	1620	.305	—	—	11.5	69	38.4	27	.090	15	5.6
C46400	1650	1630	.304	—	—	11.8	67	39.9	26	.090	15	5.6
ϕ C46500	1650	1630	.304	—	—	11.8	67	39.9	26	.090	15	5.6
ϕ C46600	1650	1630	.304	—	—	11.8	67	39.9	26	.090	15	5.6
ϕ C46700	1650	1630	.304	—	—	11.8	67	39.9	26	.090	15	5.6
C51000	1920	1750	.320	—	—	9.9	40	69.1	15	.090	16	6.0
ϕ C51100	1945	1785	.320	—	—	9.9	48	52.0	20	.090	16	6.0
C52100	1880	1620	.318	—	—	10.1	36	79.8	13	.090	16	6.0
ϕ C52400	1830	1550	.317	—	—	10.2	29	94.3	11	.090	16	6.0
C54400	1830	1700	.321	—	—	9.6	50	54.6	19	.090	15	5.6
C60800	1945	1920	.295	—	—	10.0	46	60.0	17	.090	17.5	6.6
ϕ C61400	1915	1905	.285	—	—	9.0	39	74.1	14	.090	17	6.4
ϕ C61800	1910	1900	.274	—	—	9.0	37	79.8	13	—	17	—
C62300	1910	1890	.274	—	9.0	9.4	31	79.8	13	—	16	—
C62400	1910	1895	.274	—	9.0	9.2	34	79.8	13	—	16	—
C63000	1930	1890	.274	—	9.0	9.4	22	138.0	8	.090	17	6.4
ϕ C64200	1840	1800	.278	—	—	10.0	26	113.0	8	.090	16	6.0
C65500	1880	1780	.308	—	—	10.0	21	148.0	7	.090	15	5.6
C67000	1710	1665	.282	—	—	11.0	14	86.4	12	—	15	—
C67300	1620	1555	.299	—	—	11.0	—	86.4	12	—	15	—
C67400	1625	1550	.292	—	—	11.0	58	86.4	12	—	14	—
C67500	1630	1590	.302	—	—	11.8	61	43.2	24	.090	15	5.6
C70600	2100	2010	.323	—	—	9.5	26	115.0	9	.090	18	6.8
C71000	2192	2066	.323	—	—	9.1	21	160.0	6	.090	20	7.5
C71500	2260	2140	.323	—	—	9.0	17	225.0	5	.090	22	8.3
C75200	2030	1960	.316	—	—	9.0	19	173.0	6	.090	18	6.8
C77000	1930	—	.314	—	—	9.3	17	189.0	6	.090	18	6.8

[a] lb/in³ at 68°F. See Table 2A for specific gravity (g/cm³ at 20°C).
[b] Per F at temperature range indicated (multiply factor given by 10⁻⁶).
[c] Btu/ft²/ft h F at 68°F.
[d] (Annealed) ohms (circular mil/ft) at 68°F.
[e] (Annealed) percent IACS at 68°F (volume basis).
[f] (Specific heat) Btu/lb/F at 68°F.
[g] (Tension) psi (multiply factor given by 10⁶).
[h] Psi (multiply factor given by 10⁶).
[i] Small amount of tellurium-rich constituent remains liquid down to 1575°F.
[j] Small amount of lead-rich constituent remains liquid down to 619°F.
[k] After precipitation-harding heat treatment.

TABLE 3—FABRICATION PROPERTIES, OTHER CHARACTERISTICS AND TYPICAL USES

UNS No.	Approximate Relative Suitability For Being, Worked[a]		Best Temperature For Hot Working, °C	Approximate Relative Suitability[a] For Being Joined By						Resistance Welding			Machinability[c]	Type of Chip[b]	Typical Uses	Characteristics
	Cold	Hot		Soldering	Brazing	Oxyacetylene Welding	Carbon Arc Welding	Gas Shielded Arc Welding	Coated Metal Arc Welding	Spot	Seam	Butt				
C10200	E	E	760–870	E	E	F	F	G	NR	NR	NR	G	20	L	Thermal and electrical conductors, electronic parts, glass-to-metal seals.	Oxygen-free 100% minimum electrical conductivity, excellent ductility, high purity, no gassing. Not subject to hydrogen embrittlement. Designated for use where processing involves heating in a reducing atmosphere.
C11000	E	E	760–870	E	G	NR	F	F	NR	NR	NR	G	20	L	Electrical wiring and components, radiator fins, gaskets, washers, cold heading wire, water deflectors, heat plugs, clock cases, plating anodes, screen wire.	Minimum electrical conductivity 100%, highest electrical conductivity of any metal except silver, has very high ductility. Will embrittle when heated to redness in a reducing atmosphere.
C11100	E	E	760–870	E	G	NR	F	F	NR	NR	NR	G	20	L	Radiator fins.	Has a softening temperature higher than the silver bearing coppers and electrolytic tough pitch copper.
C11300	E	E	760–780	E	G	NR	F	F	NR	NR	NR	G	20	L	Commutator bars, segments, collector rings and contacts, core and fin stock for radiators.	Minimum electrical conductivity 98%. Resistance to softening increased by presence of silver. Effect increases with increased silver added. Higher silver-content copper used for continued exposure to somewhat higher temperature.
C11400	E	E	760–780	E	G	NR	F	F	NR	NR	NR	G	20	L		
φ C11500	E	E	760–870	E	G	NR	F	F	NR	NR	NR	G	20	L		
C11600	E	E	760–870	E	G	NR	F	F	NR	NR	NR	G	20	L		
C12000	E	E	760–870	E	E	G	G	E	NR	NR	NR	G	20	L	Electrical conductors, applications involving welding or brazing.	Regarded as an alternate to C10200. More resistant to embrittlement than C11000 high electrical conductivity.
C12200	E	E	760–870	E	E	G	G	E	NR	NR	NR	G	20	L	Tube, all types of hydraulic systems, fuel lines, vacuum lines, air conditioning, heat exchangers, anodes, air, gasoline, hydraulic and oil lines, oil coolers, gage lines.	Slightly improved mechanical properties. Electrical conductivity about 85%. Not subject to hydrogen embrittlement.
C14500	E	E	760–845	E	E	F	F	G	NR	NR	NR	G	80	S	Forgings and screw machine parts requiring high electrical and thermal conductivity, furnace brazing. Electrical connectors, motor and switch parts, soldering coppers and welding torch tips.	Free machining copper, combined with high electrical conductivity (90–96%).
C14700	E	E	760–870	E	E	NR	NR	NR	NR	NR	NR	G	80	S	Transformer and circuit-breaker terminals, studs, bolts, nuts and current-carrying parts requiring fine machining.	Free machining copper, combined with high electrical conductivity (90–96%).
C15000	E	E	760–870	E	G	F	F	F	NR	NR	NR	G	20	L	Resistance welding electrodes. Miscellaneous current-carrying components at elevated temperatures.	Precipitation hardened. Combined high strength and conductivity, and resistance to softening at elevated temperatures.
C16200	E	G	760–870	E	E	F	F	G	NR	NR	NR	G	20	L	Electrical contacts and terminals, signal relays. Hard temper used for spring contact in small apparatus and resistance welding electrodes.	Moderate high strength and high electrical conductivity.
C17000	G	G	705–775	G	G	NR	F	G	NR	NR	NR	G	20	L	Leaf springs, electrical contacts, coil springs and bellows requiring severe forming distributor breaker arm, welding tips and welding wheels.	C17000, C17200, C17500, and C17600 can develop the highest mechanical properties by heat treatment. Complete range of properties.
C17200	G	G	705–775	G	G	NR	F	G	NR	NR	NR	G	20	L		
C17500	G	G	760–925	G	G	NR	F	G	NR	NR	NR	G	20	L		
C17600	G	G	760–925	G	G	NR	F	G	NR	NR	NR	G	20	L	Clips, welder tips and wheels.	
C18400	E	E	900–925	G	G	NR	F	G	NR	NR	NR	G	20	L	Spot welding electrodes and wheels, flash welding dies and commutator segments.	Precipitation hardened. Fairly high electrical conductivity. Resistance to softening at elevated temperatures.
C18700	G	NR	—	E	G	NR	NR	NR	NR	NR	NR	F	80	S	Screw machine parts requiring high electrical and thermal conductivity.	Free machining copper, high electrical conductivity. Unsuited for hot working.
φ C19200	E	E	815–950	E	E	G		E	NR	NR	NR	G	20		Flexible hose, electrical terminals, fuse clips, gaskets, air-conditioning and heat exchanger tubing.	Resistance to softening and also stress corrosion.
C21000	E	G	760–870	E	E	G	F	G	NR	NR	NR	G	20	L	Emblems, vitreous enamel base, ornamental trim and jewelry.	C21000, C22000 and C23000 are generally reddish in color, soft and malleable, higher annealing point than copper and slightly stronger and similar in corrosion resistance. Good for drawing and forming. Resistance to dezincification and season cracking is excellent.
C22000	E	G	760–870	E	E	G	F	G	NR	NR	NR	G	20	L	Emblems, vitreous enamel base, ornamental trim, jewelry, expansion plugs, valve parts, escutcheon fasteners and spring clips.	
C23000	E	G	790–900	E	E	G	F	G	NR	F	NR	G	30	L	Radiator ports, heat exchanger tubes, tube bends.	
C24000	E	F	815–900	E	E	G	F	G	NR	F	NR	G	30	L	Bellows and water temperature switch housing, flexible hose, pump lines.	Color is light golden, strength and ductility continue to increase.

TABLE 3—FABRICATION PROPERTIES, OTHER CHARACTERISTICS AND TYPICAL USES (continued)

UNS No.	Approximate Relative Suitability For Being Worked[a]		Best Temperature For Hot Working, °C	Approximate Relative Suitability[a] For Being Joined By							Resistance Welding			Machinability[c]	Type of Chip[b]	Typical Uses	Characteristics
	Cold	Hot		Soldering	Brazing	Oxyacetylene Welding	Carbon Arc Welding	Gas Shielded Arc Welding	Coated Metal Arc Welding		Spot	Seam	Butt				
C26000	E	F	730–845	E	E	G	F	F	NR		G	NR	G	30	L	Radiator tanks and lockseam tubes, header plates, reflectors, lamp bases, terminals, ground straps, baffles, ammeter shells and speedometer counterweights, washers, wheel covers, trim, carburetor parts.	Color is brass yellow. Greatest ductility of the copper-zinc series. Strength is higher than any of the preceding copper-zinc alloys.
C26800	E	NR	—	E	E	G	F	F	NR		G	NR	G	30	L	Radiator cores and tanks, lamp fixtures, socket shells, eyelets, fasteners and grommets, hinges, locks, pins, rivets, screws and springs.	Strength increases and ductility decreases, but is still very good.
C27000	E	NR	—	E	E	G	F	F	NR		G	NR	G	30	L		
C33000	E	NR	—	E	G	F	F	F	NR		F	NR	F	60	M	Tube carburetor parts, oil cooler tube, radiator and ornamental work, pump and power cylinders and liners.	Provides some degree of machinability, together with moderate cold working properties.
C33100	E	NR	—	E	G	NR	NR	NR	NR		NR	NR	F	70	M	Keys.	Intended for blanking, piercing, and machining.
C34200	E	NR	—	E	G	NR	NR	NR	NR		NR	NR	F	90	S	Clock plates and nuts, clock and watch backs, keys, gears and wheels.	Provides increased machinability with moderate cold working properties.
C34500	F	F	705–790	E	G	NR	NR	NR	NR		NR	NR	F	90	S	Screw machine parts requiring roll threads, knurls or staking operations.	Best combination of machinability and cold working properties.
C35000	F	NR	—	E	G	NR	NR	NR	NR		NR	NR	F	70	M	Keys.	Intended for blanking, piercing, and machining.
C36000	NR	F	709–790	E	G	NR	NR	NR	NR		NR	NR	F	100	S	Automatic screw machine parts and carburetor, magneto parts, radiator drums and other fittings, plugs, inserts, gears, pinions, locks.	The standard free cutting brass and its machinability has become the standard by which other alloys are rated.
C37700	NR	E	650–815	E	G	NR	NR	NR	NR		NR	NR	F	80	S	Forgings and pressings of all kinds. Headings, air conditioning tube fittings, convertible top hardware (latches, hinges, etc.) forged valve bodies.	Excellent hot working properties and widely used as forging rod. At ordinary temperatures it is strong, hard and free cutting.
C46400	F	E	650–815	E	E	G	F	F	NR		G	F	G	30	L	Aircraft turnbuckle barrels and balls, cold headed parts, forgings, screw machine parts, marine hardware, condenser plates, welding rod, nozzles and fittings.	Excellent hot and fair cold working properties of somewhat higher strength, good salt water corrosion resistance.
φ C46500	F	E	650–815	E	E	G	F	F	NR		G	F	G	30	L		
φ C46600	F	E	650–815	E	E	G	F	F	NR		G	F	G	30	L		
φ C46700	F	E	650–815	E	E	G	F	F	NR		G	F	G	30	L		
C51000	E	NR	—	E	E	F	G	G	F		G	F	E	20	L	Springs, bearings, clips, contacts, switch parts, diaphragms, welding rod, thermostats, bellows, clutch disks, lock washers, fasteners.	C51000 and C52100 have a remarkable combination of strength, ductility and resilience, and fatigue resistance.
φ C51100	E	NR	—	E	E	F	G	G	F		G	F	E	20	L		
C52100	G	NR	—	E	E	F	G	G	F		G	F	E	20	L	Springs, clips, contacts, terminal wire and bushings, diaphragms and bellows.	
φ C52400	G	NR	—	E	E	F	G	G	F		G	F	E	20	L		
C54400	G	NR	—	E	G	NR	NR	NR	NR		NR	NR	F	80	S	Bearings, bushings, gears, pinions, shafts, thrust washers, valve parts.	Free cutting, good cold working properties also suitable for blanking, forming and bending.
C60800	G	F	790–870	NR	F	NR	—	G	G		G	G	G	20	—	Condenser, evaporator and heat exchanger tubes, ferrules.	
φ C61400	G	G	785–925	NR	F	NR	G	See Note d	G		G	G	G	20	L	Gibs, wear strips, gears, bushings, nuts, bolts and threaded members.	Good cold working properties and corrosion resistance. High strength and ductility.
φ C61800	F	G	760–885	F	G	NR	—	G	G		G	G	G	40	—	Bushings, bearings, corrosion applications, welding rod.	
C62300	NR	G	730–815	NR	F	NR	G	See Note d	G		G	G	G	30	L	Valve guides, spark plug inserts, gears, valve seat inserts, oil plugs and shifter forks.	Good hot working properties; high strength retained well at elevated temperatures; acid and oxidation resistant.
C62400	NR	E	720–775	NR	F	NR	G	See Note d	G		G	G	G	30	L	Valve guides, spark plug inserts, gears, valve seat inserts, oil plugs, shifter forks, wear strips, ball bearings and hydraulic valve components.	Excellent hot working, poor cold working properties; heat treated for high mechanical properties.
C63000	NR	G	705–760	NR	F	NR	G	See Note d	G		G	G	G	20	L	Retractable landing gear, propeller gears, large valve seat inserts, spacer bearings, high pressure pump components.	Very high mechanical properties in the heat treated condition; difficult to cold work; good hot working properties, excellent corrosion resistance.
φ C64200	NR	E	705–760	NR	F	NR	—	F	F		F	F	F	60	—	Valve stems, gears, bolts, nuts, valve bodies and components.	Free machining, high strength, high corrosion resistance.
C65500	E	E	705–760	G	E	G	G	E	E		E	E	E	30	L	Hydraulic pressure lines, bolts, clamps, piston rings, rivets and shafting.	Relatively high strength, marked ductility and capability for being both hot and cold worked and joined by all procedures. Excellent corrosion resistance.

TABLE 3—FABRICATION PROPERTIES, OTHER CHARACTERISTICS AND TYPICAL USES (continued)

UNS No.	Approximate Relative Suitability For Being, Worked[a] Cold	Approximate Relative Suitability For Being, Worked[a] Hot	Best Temperature For Hot Working, °C	Approximate Relative Suitability[a] For Being Joined By Soldering	Brazing	Oxyacetylene Welding	Carbon Arc Welding	Gas Shielded Arc Welding	Coated Metal Arc Welding	Resistance Welding Spot	Resistance Welding Seam	Resistance Welding Butt	Machinability[c]	Type of Chip[b]	Typical Uses	Characteristics
C67000	NR	E	565–745	NR	F	NR	NR	See Note d	G	G	F	G	30	S	Diesel injector nozzles; high pressure hydraulic applications, cams, pistons and other components involving high mechanical loads and sliding contact.	High strength and good wear resistant properties.
C67300	F	E	625–745	NR	G	NR	NR	NR	NR	NR	NR	F	70	S	Forged water pump impellers; gears, axial piston pump components, bushings and bearings.	Hot forgeable free cutting alloy having fairly high strength and good corrosion resistant properties.
C67400	F	E	565–745	NR	F	NR	NR	See Note d	G	G	F	G	30	L	Connecting rods, transmission synchronizing stop ring, door striker plates, shifter shoes, differential idler pins, forged water pump impellers, axial piston pump parts, bushings and bearings.	Hot forgeable; high strength alloy with good wear resistant properties and good corrosion resistance.
C67500	NR	E	625–790	E	E	G	F	F	NR	G	F	G	30	L	Clutch disks, pump rods, shafting, balls, valve stems and bodies.	Strong, rigid and abrasion resistant; adopted to hot forging and pressing, hot-heading and upsetting.
C70600	G	G	760–980	E	E	F	NR	E	G	G	G	E	20	L	Condenser and heat exchanger tubes.	Used where requirements are severe. Strong, tough and very resistant to general corrosion as well as stress corrosion cracking; also serviceable at higher temperatures than copper and brasses. Well suited for condenser and heat exchanger tube.
C71000	G	G	760–980	E	E	G	NR	E	E	E	E	E	20	L	Condenser and heat exchanger tubes, ferrules.	C71000 and C71500 are used where requirements are severe. Strong, tough and very resistant to general corrosion as well as stress corrosion cracking; also; serviceable at higher temperatures than copper and brasses. Well suited for condenser and heat exchanger tube.
C71500	G	G	925–1035	E	E	G	NR	E	E	E	E	E	20	L	Automatic oil coolers, heat exchanger tube.	
C75200	E	NR	—	E	E	G	NR	F	NR	G	F	G	20	L	Rivets, screws, name plates, radio dials, etching stock, trim.	C75200 and C77000 are manufactured in a wide range of nickel contents. Higher the nickel the more silver white the alloy. 65% copper alloys have good cold working properties and are used for cold drawing, spinning, forming and stamping. The lower copper content alloys (55% Cu) are used for spring application.
C77000	G	NR	—	E	E	G	NR	F	NR	G	F	G	30	L	Springs, resistance wire.	

[a] E = Excellent; G = Good; F = Fair; NR = Not Recommended.
[b] S = Short; M = Medium; L = Long.
[c] Approximate relative machinability rating (Free Cutting Brass = 100).
[d] Consumable electrode excellent. Tungsten are good, with AC preferred.

TABLE 4—TYPICAL SOFTENING TEMPERATURE

UNS No.	Temperature °C
C11000	230
C10200	280
C11400	315
C12200	340
C11100	355

TABLE 5—TYPICAL THERMAL STRESS RELIEVING TREATMENTS

SAE No.	Temperature °C	Time, h
C24000	260	1
C26000	260	1
C26800		
C27000	245	1
C51000	190	1
C71500	480	1

TABLE 6—APPROXIMATE ANNEALING TEMPERATURE RANGES FOR INTERMEDIATE ANNEALING OF FABRICATED PARTS

UNS No.	Metal Temperature °C	Average Grain Size, mm
C11000	400–480	0.025
	345–400	0.020
C22000	405–635	0.040
	440–565	0.025
C23000	480–610	0.040
	400–480	0.025
C26000	455–595	0.040
	370–480	0.025
C26800	455–595	0.040
	370–455	0.025
C51000	565–650	0.030
	480–565	0.015
C75200	595–705	0.035
	480–620	0.020

TABLE 7—RECOMMENDED RADII FOR FORMING 90 DEG BENDS

UNS No.	Temper	Thickness mm	Thickness in	Bend Perpendicular To Rolling Direction mm	Bend Perpendicular To Rolling Direction in	Bend At 45 Deg To Rolling Direction mm	Bend At 45 Deg To Rolling Direction in	Bend Parallel To Rolling Direction mm	Bend Parallel To Rolling Direction in
C11000	Half Hard	0.50	.020	Sharp		Sharp		1.6	.06
	Extra Hard			1.2	.05	1.2	.05	1.6	.06
C17200	Annealed	0.10	.004					0.4	.02
	Quarter Hard							0.4	.02
	Half Hard							0.8	.03
	Hard							0.8	.03
	Annealed	0.25	.010					1.2	.05
	Quarter Hard							1.6	.06
	Half Hard							2.4	.09
	Hard							4.0	.16
	Annealed	0.50	.020					2.8	.11
	Quarter Hard							3.2	.12
	Half Hard							4.8	.19
	Hard							8.0	.31
C23000	Soft	0.13/1.60	.005/.063	Sharp		Sharp		Sharp	
	Half Hard	0.50/1.30	.020/.050	Sharp		Sharp		Sharp	
	Hard	0.65	.025	Sharp		0.4	.02	1.2	.05
		1.00	.040	0.4	.02	.12	.05	1.6	.06
		1.60	.063	1.2	.05	2.4	.09	3.2	.12
	Extra Hard	0.30	.012	0.4	.02	1.2	.05	1.6	.06
		0.65	.025	0.8	.03	1.6	.06	3.2	.12
		1.00	.040	1.6	.06	3.2	.12	4.8	.19
	Spring	0.30	.012	1.2	.05	2.4	.09	4.8	.19
		0.65	.025	1.6	.06	4.8	.19	>6.4	>.25
		1.00	.040	2.4	.09	6.4	.25	>6.4	>.25
C24000	Hard	0.50	.020	Sharp		0.8	.03	1.2	.05
	Spring	0.50	.020	1.6	.06	4.8	.19	6.4	.25
C26000	Half Hard	0.13/0.80	.005/.032	Sharp		Sharp		Sharp	
		0.81/1.30	.033/.050	Sharp		Sharp		0.4	.02
	Hard	1.31/0.48	.005/.019	Sharp		0.4	.02	0.8	.03
		0.65	.025	0.4	.02	0.8	.03	1.6	.06
		1.00	.040	0.8	.03	1.6	.06	2.4	.09
		1.60	.063	1.6	.06	3.2	.12	4.8	.19
	Extra Hard	0.40	.106	0.8	.03	1.6	.06	3.2	.12
		0.65	.025	1.6	.06	2.4	.09	4.8	.19
		1.00	.040	2.4	.09	3.2	.12	6.4	.25
	Spring	0.25	.010	1.6	.06	3.2	.12	4.8	.19
		0.65	.025	2.4	.09	4.8	.19	>6.4	>.25
		1.00	.040	3.2	.12	>6.4	>.25	>6.4	>.25
	Extra Spring	0.25	.010	1.6	.06	3.2	.12	6.4	.25
		0.65	.025	2.4	.09	>6.4	>.25	>6.4	>.25
		1.00	.040	3.2	.12	>6.4	>.25	>6.4	>.25
C26800	Half Hard	0.13/0.80	.005/.032	Sharp		Sharp		Sharp	
		0.81/1.30	.033/.050	Sharp		Sharp		0.4	.02
	Hard	0.13/0.30	.005/.012	Sharp		0.4	.02	0.8	.03
		1.00	.040	0.8	.03	1.6	.06	2.4	.09
		1.60	.063	1.6	.06	3.2	.12	4.8	.19
C51000	Half Hard	0.50/0.80	.020/.032	Sharp		Sharp		0.4	.02
	Hard	1.00	.040	1.2	.05	2.4	.09	4.8	.19
	Extra Hard	1.00	.040	1.6	.06	3.2	.12	6.4	.25
	Spring	0.30	.012	0.8	.03	1.6	.06	4.8	.19
		0.65	.025	1.6	.06	3.2	.12	6.4	.25
		1.00	.040	2.4	.09	4.8	.19	>6.4	>.25
	Extra Spring	0.30	.012	1.2	.05	2.4	.09	6.4	.25
		0.50	.020	2.4	.09	4.8	.19	>6.4	>.25
C52100	Half Hard	0.13/0.48	.005/.019	Sharp		Sharp		0.4	.02
	Hard	0.50	.020	0.4	.02	1.2	.05	2.4	.09
		1.00	.040	1.2	.05	2.4	.09	4.8	.19
		1.60	.063	2.4	.09	4.8	.19	>6.4	>.25
	Extra Hard	1.00	.040	2.4	.09	4.8	.19	>6.4	>.25
	Spring	1.00	.040	3.2	.12	6.4	.25	>6.4	>.25
	Extra Spring	1.60	.063	6.4	.25	>6.4	>.25	>6.4	>.25
C75200	Half Hard	1.00	.040			0.4	.02	0.8	.03
	Hard	1.00	.040	1.6	.06	1.6	.06	1.6	.06
	Extra Hard	1.00	.040	3.2	.12	3.2	.12	4.8	.19
	Spring	1.00	.040	4.0	.16	4.8	.19	5.6	.22
	Extra Spring	1.00	.040	4.0	.16	5.6	.22	6.4	.25

TABLE 8—RELATIVE FORGEABILITY RATING OF TYPICAL ALLOYS AS HOT PRESSED

UNS No.	Relative Forgeability Rating
C11000	65
C12200	65
C37700	100
C46400	85
C67500	80
C65500	40
C63700	70
C62400	75
C67400	90

TABLE 9—ANNEALED TEMPERS OF STRIP NOMINAL GRAIN SIZE, mm

UNS No.	Thickness mm Over	Thickness mm Thru	Thickness in Over	Thickness in Thru	Eyelet Type Forming	Deep Drawing, Spinning	Embossing, Severe Forming	Coining, Extremely Severe Drawing
C24000, C26000, and C26800	—	0.50	—	.020	0.015	0.025	0.035	—
	0.50	1.30	.020	.050	0.025	0.035	0.050	0.050
	1.30	2.30	.050	.090	0.035	0.050	0.070	0.070
	2.30	4.60	.090	.180	0.050	0.070	0.070	0.120
	4.60	—	.180	—	0.070	0.120	0.120	0.120
C21000, C22000, and C23000	—	0.50	—	.020	0.015	0.025	0.025	—
	0.50	1.30	.020	.050	0.025	0.035	0.050	0.050
	1.30	4.60	.050	.180	0.035	0.035	0.050	0.050
	4.60	—	.180	—	0.035	0.050	0.050	0.070
C10200, C11000, C11300 thru C12000, and C12200	—	0.50	—	.020	0.015	0.025	0.025	—
	0.50	1.30	.020	.050	0.015	0.025	0.025	0.035
	1.30	4.60	.050	.180	0.015	0.025	0.035	0.035
	4.60	—	.180	—	0.025	0.035	0.035	0.035

TABLE 10A—TYPICAL MECHANICAL PROPERTIES OF WROUGHT COPPER AND COPPER ALLOYS

Metric (SI) Units

UNS No.	Form	Temper	Size Section, mm	Tensile Strength, MPa	Yield Strength, 0.5% Ext Under Load, MPa	Elongation In 50 mm, %	Reduction of Area, %	Hardness RF	Hardness RB	Hardness R30T	Shear Strength, MPa	Fatigue Strength MPa	Fatigue Strength Million Cycles
C10200 C11000 C11100 C11300 C11400 C11500 C11600 C12000 C12200	Plate, Sheet, Strip, and Rolled Bar	Soft Anneal	1.0	220	70	45	—	40	—	—	150	—	—
		Deep-Drawing Anneal	1.0	235	75	45	—	45	—	—	160	75	100
		Light Cold Rolled	1.0	250	195	30	—	60	10	25	170	—	—
		1/2 Hard	1.0	290	250	14	—	84	40	50	180	90	100
		Hard	1.0	345	310	6	—	90	50	57	195	90	100
		Spring	1.0	380	345	4	—	94	60	63	200	90	100
		Extra Spring	1.0	385	365	4	—	95	62	64	200	—	—
		Hot Rolled	1.0	235	70	45	—	45	—	—	160	—	—
		Hot Rolled and Annealed	1.0	230	70	45	—	42	—	—	150	—	—
C10200 C11000 C12000 C12200	Rod, Bar and Shapes	Soft Anneal	25.0	220	70	55	70	40	—	—	150	—	—
		Hard	6.0[g]	380	345	10	—	94	60	—	260	—	—
			25.0[g]	330	305	16	55	87	47	—	185	17	300
			50.0[g]	310	275	20	—	85	45	—	180	—	—
			12.0[h]	330	305	16	—	87	47	—	185	—	—
			12.0[i]	275	220	30	—	—	35	—	180	—	—
C10200 C12000 C12200	Tube	Soft Anneal	25.0 x 1.6	220	70	45	—	40	—	—	150	75[a]	20
		Light Anneal	25.0 x 1.6	235	75	45	—	45	—	—	160	—	—
		Light Drawn	25.0 x 1.6	275	220	25	—	77	35	45	180	95[a]	20
		Drawn	25.0 x 1.6	275	220	25	—	77	35	45	—	—	—
		Hard Drawn	25.0 x 1.6	380	345	8	—	95	60	63	200	130[a]	20
			50.0 x 1.6	380	345	8	—	95	60	63	—	—	—
			100.0 x 1.6	380	345	8	—	95	60	63	—	—	—
C10200 C11000	Wire	Annealed	2.0	240	—	35[f]	—	—	—	—	165	—	—
C14500 and C14700	Rod	1/2 Hard	6.0	295	260	20	—	—	38	—	170	—	—
			25.0	290	250	25	—	—	38	—	165	—	—
			50.0	285	240	25	—	—	35	—	165	—	—
		Hard	6.0	365	330	10	—	—	52	—	200	—	—
			25.0	330	290	15	—	—	50	—	185	—	—
C15000	Round Rod	Drawn and Heat Treated	25.0	415	345	12	—	—	67	—	—	195	100
C16200	Round Rod	Drawn	25.0	450	345	25	—	—	70	—	—	—	—
			50.0	415	240	30	—	—	65	—	—	—	—
			75.0	380	170	35	—	—	60	—	—	—	—
	Square, Rectangular and Hex Rod and Bar	Drawn	25.0	450	345	25	—	—	70	—	—	—	—
			Over 25.0	380	240	30	—	—	65	—	—	—	—
	Forging	As Forged	25.0	450	—	25	—	—	60	—	—	—	—
			50.0	380	—	30	—	—	55	—	—	—	—
			Over 50.0	380	—	30	—	—	55	—	—	—	—
C17000	Strip	A Soft[d]	—	475	220[j]	48	—	—	62	56	—	—	—
		1/4 Hard[d]	—	565	485[j]	22	—	—	79	68	—	—	—
		1/2 Hard[d]	—	635	565[j]	15	—	—	92	76	—	—	—
		Hard[d]	—	760	715[j]	5	—	—	99	81	—	—	—
								RC	R30N				
		AT (3h at 315°C)[k]	—	1140	1000[j]	3	—	36	56	—	—	260	100
		1/4 HT (2h at 315°C)[k]	—	1185	1035[j]	2	—	37	57	—	—	275	100
		1/2 HT (2h at 315°C)[k]	—	1255	1105[j]	1	—	38	58	—	—	290	100
		HT (2h at 315°C)[k]	—	1310	1170[j]	1	—	40	60	—	—	295	100
		AM[l]	—	745	565[j]	20	—	20	40	—	—	—	—
		1/4 HM[l]	—	815	635[j]	17	—	24	44	—	—	—	—
		1/2 HM[l]	—	850	725[j]	14	—	28	48	—	—	—	—
		HM[l]	—	985	840[j]	11	—	32	52	—	—	—	—
		XHM[l]	—	1160	1020[j]	4	—	34	54	—	—	—	—
C17200	Strip	A (Soft)[d]	—	475	220[j]	48	—	—	62	56	—	—	—
		1/4 Hard[d]	—	565	485[j]	22	—	—	79	68	—	—	—
		1/2 Hard[d]	—	635	550[j]	15	—	—	92	76	—	—	—
		Hard[d]	—	760	715[j]	5	—	—	99	81	—	—	—
								RC	R30N				
		AT (3h at 315°C)[k]	—	1230	1070[j]	6	—	38	58	—	—	260	100
		1/4 H (2h at 315°C)[k]	—	1295	1140[j]	4	—	40	60	—	—	275	100
		1/2 H (2h at 315°C)[k]	—	1365	1205[j]	3	—	42	62	—	—	295	100
		HT (2h at 315°C)[k]	—	1390	1240[j]	2	—	43	63	—	—	305	100

TABLE 10A—TYPICAL MECHANICAL PROPERTIES OF WROUGHT COPPER AND COPPER ALLOYS (continued)

Metric (SI) Units

UNS No.	Form	Temper	Size Section, mm	Tensile Strength, MPa	Yield Strength, 0.5% Ext Under Load, MPa	Elongation In 50 mm, %	Reduction of area, %	Hardness RF	Hardness RB	Hardness R30T	Shear Strength, MPa	Fatigue Strength MPa	Fatigue Strength Million Cycles
C17200 continued	Rod and Bar	A (Soft)[d]	—	495	170[j]	48	—	—	RB 62	R30T —	—	—	—
		1/2 Hard[d]	Under 25.0	760	620[j]	15	—	—	96	—	—	—	—
		1/2 Hard[d]	Over 25.0	690	620[j]	15	—	—	96	—	—	—	—
		AT (3h at 315°C)[k]	—	1230	1105[j]	6	—	—	RC 38	—	—	345	100
		1/2 HT (2h at 315°C)[k]	Under 25.0	1380	1255[j]	4	—	—	42	—	—	350	100
		1/2 HT (2h at 315°C)[k]	Over 25.0	1325	1205[j]	4	—	—	42	—	—	350	100
C17500	Strip and Plate	A (Soft)[d]	—	310	170[j]	28	—	—	RB 32	37	—	—	—
		1/2 Hard[d]	—	470	385[j]	8	—	—	70	64	—	—	—
		Hard[d]	—	540	485[j]	5	—	—	83	72	—	—	—
		AT (3h at 480°C)[k]	—	760	620[j]	12	—	—	96	80	—	230	100
		1/2 HT (2h at 480°C)[k]	—	825	745[j]	8	—	—	98	81	—	240	100
		HT (2h at 480°C)[k]	—	825	745[j]	8	—	—	98	81	—	240	100
C17500 C17600	Rod, Bar Shapes and Tubing	A (Soft)[d]	—	310	170[j]	28	—	—	35	—	—	—	—
		1/2 Hard[d]	—	495	450[j]	12	—	—	68	—	—	—	—
		AT (3h at 480°C)[k]	—	760	620[j]	18	—	—	96	—	—	275	100
		1/2 HT (2h at 480°C)[k]	—	825	760[j]	14	—	—	98	—	—	275	100
C18400	Round Rod	Drawn	25.0	485	—	20	—	—	80	—	—	—	—
			50.0	450	—	20	—	—	75	—	—	—	—
			75.0	415	—	20	—	—	70	—	—	—	—
	Hex Rod and Bar	Drawn	25.0	485	—	20	—	—	75	—	—	—	—
			Over 25.0	415	—	20	—	—	70	—	—	—	—
	Forgings	As Forged	25.0	485	—	20	—	—	80	—	—	—	—
			50.0	415	—	20	—	—	75	—	—	—	—
			Over 50.0	415	—	20	—	—	70	—	—	—	—
C18700	Rod	Hard	6.0	415	380	10	—	—	55	—	220	—	—
			12.0	380	345	11	—	—	50	—	205	—	—
			18.0	365	330	12	—	—	50	—	200	—	—
			25.0	350	315	14	—	—	50	—	195	—	—
C19200	Tube	A (Soft)	48.0 x 2.4	255	80	40	—	—	—	—	—	—	—
		Light Drawn	4.8 x 0.8	290	215	3	—	—	—	—	—	—	—
C21000	Plate, Sheet, Strip and Rolled Bar[e]	Annealed 0.050 mm	1.0	235	70	45	—	46	—	—	185	—	—
		0.035 mm	1.0	240	75	45	—	52	—	4	195	—	—
		0.025 mm	1.0	250	85	43	—	56	—	9	200	—	—
		0.015 mm	1.0	260	95	42	—	60	—	15	205	—	—
		1/4 Hard	1.0	290	220	25	—	—	38	44	220	—	—
		1/2 Hard	1.0	330	275	12	—	—	52	54	235	—	—
		3/4 Hard	1.0	350	310	8	—	—	59	58	240	—	—
		Hard	1.0	385	345	5	—	—	64	60	255	—	—
		Extra Hard	1.0	420	380	4	—	—	70	64	270	—	—
		Springs	1.0	440	400	4	—	—	73	66	275	—	—
		Extra Spring	1.0	450	—	—	—	—	74	68	—	—	—
C22000	Plate, Sheet Strip and Rolled Bar[e]	Annealed 0.050 mm	1.0	255	70	45	—	53	—	6	195	—	—
		0.035 mm	1.0	260	85	45	—	57	—	12	205	—	—
		0.025 mm	1.0	270	95	44	—	60	—	16	215	—	—
		0.015 mm	1.0	285	105	42	—	65	—	26	220	—	—
		1/4 Hard	1.0	310	240	25	—	—	42	44	230	—	—
		1/2 Hard	1.0	360	310	11	—	—	58	56	240	—	—
		3/4 Hard	1.0	395	345	8	—	—	67	63	250	—	—
		Hard	1.0	420	370	5	—	—	70	63	260	—	—
		Extra Hard	1.0	460	400	4	—	—	75	67	275	—	—
		Spring	1.0	495	425	3	—	—	78	69	275	145	15
		Extra Spring	1.0	525	—	—	—	—	82	73	290	—	—
	Tube	Soft Anneal	25.0 x 1.6	260	70	52	—	55	—	10	—	—	—
		Light Anneal	25.0 x 1.6	285	105	42	—	65	—	35	—	—	—
		Drawn (General Purpose)	25.0 x 1.6	310	240	25	—	—	50	50	—	—	—
		Hard Drawn	25.0 x 1.6	415	365	6	—	—	69	62	—	—	—

TABLE 10A—TYPICAL MECHANICAL PROPERTIES OF WROUGHT COPPER AND COPPER ALLOYS (continued)

Metric (SI) Units

UNS No.	Form	Temper	Size Section, mm	Tensile Strength, MPa	Yield Strength, 0.5% Ext Under Load, MPa	Elongation In 50 mm, %	Reduction of Area, %	Hardness RF	Hardness RB	Hardness R30T	Shear Strength, MPa	Fatigue Strength MPa	Fatigue Strength Million Cycles
C23000	Sheet and Strip	Annealed 0.070 mm	1.0	270	70	48	—	56	—	10	215	—	—
		0.050 mm	1.0	275	85	47	—	59	—	14	215	—	—
		0.035 mm	1.0	285	95	46	—	63	—	22	215	—	—
		0.025 mm	1.0	295	110	44	—	66	—	28	220	—	—
		0.015 mm	1.0	310	125	42	—	71	—	38	230	—	—
		1/4 Hard	1.0	345	270	25	—	—	55	54	240	—	—
		1/2 Hard	1.0	395	340	12	—	—	65	60	255	—	—
		3/4 Hard	1.0	425	360	8	—	—	73	67	270	—	—
		Hard	1.0	495	395	5	—	—	77	68	290	—	—
		Extra Hard	1.0	540	420	4	—	—	83	72	305	—	—
		Spring	1.0	580	435	3	—	—	86	74	315	—	—
		Extra Spring	1.0	595	—	—	—	—	88	77	—	—	—
	Tube	Soft Anneal	25.0 x 1.6	275	85	55	—	60	—	15	—	—	—
		Light Anneal	25.0 x 1.6	305	125	45	—	71	—	38	—	—	—
		Light Drawn	25.0 x 1.6	345	275	30	—	—	55	54	—	—	—
		Drawn (General Purpose)	25.0 x 1.6	385	315	25	—	—	62	60	—	—	—
		Hard Drawn	25.0 x 1.6	455	400	8	—	—	77	68	—	—	—
C24000	Sheet and Strip	Annealed 0.070 mm	1.0	290	85	52	—	57	—	8	—	—	—
		0.050 mm	1.0	305	95	50	—	61	—	16	220	—	—
		0.035 mm	1.0	315	105	48	—	66	—	28	—	—	—
		0.025 mm	1.0	330	115	47	—	69	—	32	—	—	—
		0.015 mm	1.0	345	140	46	—	75	—	42	230	—	—
		1/4 Hard	1.0	365	275	30	—	—	55	54	250	—	—
		1/2 Hard	1.0	420	345	18	—	—	70	64	270	—	—
		3/4 Hard	1.0	455	370	12	—	—	76	68	285	—	—
		Hard	1.0	510	405	7	—	—	82	71	295	—	—
		Extra Hard	1.0	570	425	5	—	—	87	75	310	—	—
		Spring	1.0	625	450	3	—	—	91	77	330	165	20
		Extra Spring	1.0	640	—	—	—	—	92	78	—	—	—
C26000	Plate, Sheet, Strip, Rolled Bar and Wire[e]	Annealed 0.120 mm	1.0	305	75	66	—	54	—	11	—	90	100
		0.070 mm	1.0	315	95	65	—	58	—	15	220	90	100
		0.050 mm	1.0	325	105	62	—	64	—	26	—	—	—
		0.035 mm	1.0	340	115	57	—	68	—	31	235	95	100
		0.025 mm	1.0	350	130	55	—	72	—	36	—	—	—
		0.015 mm	1.0	365	150	54	—	78	—	43	240	105	100
		1/4 Hard	1.0	370	275	43	—	—	55	54	250	—	—
		1/2 Hard	1.0	425	360	23	—	—	70	65	275	125	100
		3/4 Hard	1.0	475	395	15	—	—	79	70	290	—	—
		Hard	1.0	525	435	8	—	—	82	73	305	145	100
		Extra Hard	1.0	595	450	5	—	—	88	76	315	—	—
		Spring	1.0	650	450	3	—	—	91	77	330	160	100
		Extra Spring	1.0	695	450	3	—	—	93	78	—	—	—
	Tube	Soft Anneal	25.0 x 1.6	325	105	65	—	64	—	26	—	—	—
		Light Anneal	25.0 x 1.6	360	140	55	—	75	—	40	—	—	—
		Hard Drawn	25.0 x 1.6	540	440	8	—	—	82	73	—	—	—
C26800	Plate, Sheet, Strip and Rolled Bar[e]	Annealed 0.120 mm	1.0	—	—	—	—	56	—	5	—	—	—
		0.070 mm	1.0	315	95	65	—	58	—	15	220	80	100
		0.050 mm	1.0	325	105	62	—	64	—	26	230	105	100
		0.035 mm	1.0	340	115	57	—	68	—	31	235	—	—
		0.025 mm	1.0	350	130	55	—	72	—	36	240	—	—
		0.015 mm	1.0	365	150	54	—	78	—	43	240	—	—
		1/4 Hard	1.0	370	275	43	—	—	55	54	250	—	—
		1/2 Hard	1.0	420	345	23	—	—	70	65	275	—	—
		3/4 Hard	1.0	460	380	15	—	—	77	69	285	—	—

11.45

TABLE 10A—TYPICAL MECHANICAL PROPERTIES OF WROUGHT COPPER AND COPPER ALLOYS (continued)

Metric (SI) Units

UNS No.	Form	Temper	Size Section, mm	Tensile Strength, MPa	Yield Strength, 0.5% Ext Under Load, MPa	Elongation In 50 mm, %	Reduction of Area, %	Hardness RF	Hardness RB	Hardness R30T	Shear Strength, MPa	Fatigue Strength MPa	Fatigue Strength Million Cycles
C26800 continued	Plate, Sheet, Strip and Rolled Bar[e]	Hard	1.0	510	415	8	—	—	80	70	295	95	100
		Extra Hard	1.0	585	425	5	—	—	87	74	310	—	—
		Spring	1.0	625	425	3	—	—	90	76	325	140	100
		Extra Spring	1.0	655	—	—	—	—	92	78	—	140	100
C27000	Wire	Annealed 0.035 mm	2.0	345	—	60	—	—	—	—	235	—	—
		1/8 Hard	2.0	400	—	35	—	—	—	—	260	—	—
		1/4 Hard	2.0	485	—	20	—	—	—	—	290	150[a]	300
		1/2 Hard	2.0	605	—	15	—	—	—	—	315	—	—
		3/4 Hard	2.0	690	—	12	—	—	—	—	345	—	—
		Hard	2.0	760	—	8	—	—	—	—	380	—	—
		Extra Hard	2.0	825	—	4	—	—	—	—	400	—	—
		Spring	2.0	835	—	3	—	—	—	—	415	—	—
C33000 and C33100	Tube	Soft Anneal	25.0 x 1.6	325	105	60	—	64	—	26	—	—	—
		Light Anneal	25.0 x 1.6	360	140	50	—	75	—	37	—	—	—
		Hard Drawn	25.0 x 1.6	515	415	7	—	—	80	69	—	—	—
C34200 C35000	Plate, Sheet, Strip and Rolled Bar	Annealed 0.035 mm	1.0	340	115	52	—	68	—	31	235	—	—
		1/4 Hard	1.0	370	275	38	—	—	55	54	250	—	—
		1/2 Hard	1.0	420	345	20	—	—	70	65	275	—	—
		Hard	1.0	510	415	7	—	—	80	69	295	—	—
		Extra Hard	1.0	585	425	5	—	—	87	74	310	—	—
C34500 C35000	Rod	1/2 Hard	12.0	395	205	20	—	—	68	—	250	—	—
			25.0	380	205	25	—	—	68	—	240	—	—
			50.0	345	170	30	—	—	65	—	220	—	—
C36000	Rod	Soft	25.0	340	125	53	58	68	—	—	205	—	—
		1/2 Hard	6.0	470	360	18	48	—	80	—	260	—	—
			25.0	400	310	25	50	—	78	—	235	140[b]	100
			50.0	380	305	32	52	—	75	—	220	95[b]	300
		Hard	3.0	585	345	—	—	—	—	—	—	—	—
			6.0	515	275	5[c]	—	—	—	—	—	—	—
			18.0	485	240	10[c]	—	—	—	—	—	—	—
	Flat Products	Soft	25.0 x 150.0	330	140	25[c]	—	—	—	—	205	—	—
			Over 25.0 x 150.0	310	125	30[c]	—	—	—	—	195	—	—
		1/2 Hard	6.0 x 25.0	385	310	20[c]	—	—	62	—	230	—	—
			12.0 x 150.0	345	205	20[c]	—	—	—	—	205	—	—
			50.0 x 50.0	340	205	25[c]	—	—	—	—	205	—	—
			50.0 x 150.0	310	170	25[c]	—	—	—	—	185	—	—
			Over 50.0 x 100.0	310	170	25[c]	—	—	—	—	185	—	—
C37700	Die Forgings	As Extruded	1.0	360	140	45[c]	65	78	—	—	—	—	—
		As Forged	4 kg	400	160	40[c]	—	—	—	—	—	—	—
C46400 C46500 C46600 C46700	Rod and Bar	Soft	6.0	400	185	45	60	—	56	—	275	—	—
			25.0	395	170	47	60	—	56	—	275	—	—
			50.0	385	170	47	60	—	55	—	275	—	—
			Over 50.0	375	150	—	—	—	—	—	—	—	—
		1/2 Hard or Light Annealed	6.0	435	205	40	55	—	60	—	290	—	—
			25.0	435	205	40	55	—	60	—	290	—	—
			50.0	425	195	43	55	—	60	—	290	—	—
			75.0	395	180	—	—	—	—	—	—	—	—
			Over 75.0	395	165	—	—	—	—	—	—	—	—
		Hard	6.0	550	395	20	45	—	85	—	310	—	—
			25.0	515	365	20	45	—	82	—	305	—	—
			50.0	460	275	35	50	—	75	—	295	—	—
	Shapes	As Extruded	—	400	170	40	—	—	—	—	275	—	—
C51000	Sheet and Strip	Annealed 0.050 mm	1.0	325	130	64	—	73	26	—	250	—	—
		0.035 mm	1.0	340	140	58	—	75	28	—	255	—	—
		0.025 mm	1.0	345	145	52	—	77	30	—	260	—	—
		0.015 mm	1.0	365	150	50	—	79	34	—	275	—	—
		1/2 Hard	1.0	470	380	28	—	—	78	69	340	—	—
		Hard	1.0	560	515	10	—	—	87	75	—	170	100
		Extra Hard	1.0	635	550	6	—	—	93	78	—	—	—
		Spring	1.0	690	550	4	—	—	95	79	—	150	100
		Extra Spring	1.0	740	550	3	—	—	97	80	—	—	—

TABLE 10A—TYPICAL MECHANICAL PROPERTIES OF WROUGHT COPPER AND COPPER ALLOYS (continued)

Metric (SI) Units

UNS No.	Form	Temper	Size Section, mm	Tensile Strength, MPa	Yield Strength, 0.5% Ext Under Load, MPa	Elongation In 50 mm, %	Reduction of Area, %	Hardness RF	Hardness RB	Hardness R30T	Shear Strength, MPa	Fatigue Strength MPa	Fatigue Strength Million Cycles
C51000 continued	Rod	Hard	12.0 25.0	515 485	450 400	25 25	— —	— —	80 78	— —	370 345	— 200	— 300
	Wire	Soft, 0.035 mm	2.0	345	140	58	—	—	38	—	—	—	—
		1/4 Hard	2.0	470	415	24	—	—	49	—	—	—	—
		1/2 Hard	2.0	585	550	8	—	—	—	—	—	—	—
		Hard	2.0	760	—	5	—	—	—	—	—	185	100
		Extra Hard	2.0	895	—	3	—	—	—	—	—	205	100
		Spring	2.0	905	—	2	—	—	—	—	—	—	—
C51100 C54400	Sheet and Strip	Annealed 0.050 mm 0.035 mm 0.025 mm 0.015 mm	1.0 1.0 1.0 1.0	315 330 345 350	— — — —	48 47 46 46	— — — —	70 73 75 76	— — — —	— — — —	— — — —	— — — —	— — — —
		1/2 Hard	1.0	425	370	19	—	—	70	65	—	—	—
		Hard	1.0	550	510	7	—	—	86	74	—	—	—
		Extra Hard	1.0	635	—	4	—	—	91	78	—	—	—
		Spring	1.0	675	550	3	—	—	93	79	—	—	—
		Extra Spring	1.0	710	550	2	—	—	95	80	—	—	—
C52100	Sheet and Strip	Soft	1.0	475	165	63	—	82	50	—	—	—	—
		1/2 Hard	1.0	525	380	32	—	—	84	73	350	—	—
		Hard	1.0	640	495	10	—	—	93	78	—	150	100
		Extra Hard	1.0	730	550	4	—	—	96	80	—	—	—
		Spring	1.0	770	—	3	—	—	98	81	—	—	—
		Extra Spring	1.0	825	—	2	—	—	100	82	—	—	—
C52400	Sheet and Strip	Soft 0.035 mm	1.0	455	195	68	—	—	55	—	—	—	—
		1/2 Hard	1.0	570	—	32	—	—	92	—	—	—	—
		Hard	1.0	690	—	13	—	—	97	—	—	—	—
		Extra Hard	1.0	795	—	7	—	—	100	—	—	—	—
		Spring	1.0	840	—	4	—	—	101	—	—	—	—
		Extra Spring	1.0	845	—	3	—	—	103	—	—	—	—
C54400	Sheet and Strip	See C51100											
	Rod	Hard	12.0 25.0	515 470	435 385	15 20	— —	— —	83 80	— —	— —	— —	— —
	Flat Products	Hard	8.0 18.0	415 380	310 240	20 25	— —	— —	70 —	— —	— —	— —	— —
C60800	Tube	Annealed	25.0 x 1.6	415	185	—	—	77	—	—	—	—	—
C61400	Plate, Sheet, Strip and Rolled Bar	Soft	3.0 8.0 12.0 25.0	560 550 540 525	310 275 240 230	40 40 42 45	35 35 40 40	— — — —	84 83 82 81	— — — —	310 290 275 275	205 195 180 170	100 100 100 100
		Hard	3.0 8.0 12.0 25.0	615 585 550 540	415 400 370 310	32 35 38 40	— — — —	— — — —	87 86 85 84	— — — —	345 330 275 260	— — — —	— — — —
	Rod and Bar	—	12.0 25.0 50.0	585 565 550	310 275 240	35 35 35	55 55 60	— — —	91 90 88	— — —	330 310 275	— — —	— — —
C61800	Rod	1/2 Hard	25.0 50.0 75.0	585 570 550	295 270 270	23[c] 25[c] 28[c]	— — —	— — —	88 88 88	— — —	325 310 295	195 195 180	100 100 100

BHN (1000 kg)

UNS No.	Form	Temper	Size Section, mm	Tensile Strength, MPa	Yield Strength, 0.5% Ext Under Load, MPa	Elongation In 50 mm, %	Reduction of Area, %	BHN			Shear Strength, MPa	Fatigue MPa	Fatigue Million Cycles
C62300	Rod and Bar	Drawn	12.0 25.0 75.0	655 635 620	415 380 310	15[c] 20[c] 25[c]	— — —	180 — —			— — —	— — —	— — —
	Shapes	—	—	585	230	25[c]	—	170			—	—	—
	Forgings	As Forged	Thru 40.0 Over 40.0	585 565	230 215	25[c] 30[c]	— —	140			—	—	—

BHN (3000 kg)

UNS No.	Form	Temper	Size Section, mm	Tensile Strength, MPa	Yield Strength, MPa	Elongation %		BHN					
C62400	Rod	As Extruded	All Sizes	655	345	12[c]	—	200			—	—	—
	Forgings	As Forged	—	655	345	12[c]	—	200			—	—	—

TABLE 10A—TYPICAL MECHANICAL PROPERTIES OF WROUGHT COPPER AND COPPER ALLOYS (continued)

Metric (SI) Units

UNS No.	Form	Temper	Size Section, mm	Tensile Strength, MPa	Yield Strength, 0.5% Ext Under Load, MPa	Elongation In 50 mm, %	Reduction of Area, %	Hardness RF	Hardness RB	Hardness R30T	Shear Strength, MPa	Fatigue Strength MPa	Fatigue Strength Million Cycles
C62400 continued	Forgings	Hardened and H₂O Quenched	—	655	345	—			265		—	—	—
		Hardened, H₂O Quenched and Tempered	—	690	—	12c			210				
C63000	Rod and Bar	Annealed	25.0 50.0 100.0	725 655 620	415 345 310	12c 15c 17c	— — —	— — —	100 — —	— — —	— — —	— — —	— — —
	Shapes	As Extruded	All Sizes	620	310	17c	—	—	—	—	—	—	—
C64200	Rod and Bar	Drawn	18.0 40.0	705 640	470 415	22c 26c	— —	— —	94 90	— —	405 —	345 —	100 —
C65500	Plate, Sheet, Strip and Rolled Bare	Annealed 0.070 mm 0.040 mm	1.0 1.0	385 415	145 170	63 60	— —	76 85	40 62	— —	290 295	— 110	— 100
		1/4 Hard 1/2 Hard Hard Extra Hard	1.0 1.0 1.0 1.0	470 540 650 715	240 310 400 415	30 17 8 6	— — — —	— — — —	75 87 93 96	67 75 78 80	325 345 395 415	— — 160 —	— — 100 —
		Spring	1.0	760	425	4	—	—	97	81	435	—	—
		Hot Rolled	1.0	435	170	—	—	—	—	—	—	—	—
		Hot Rolled and Cold Rolled Finish	1.0	450	205	—	—	—	70	—	—	—	—
	Rod, Bar and Shapes	Soft 0.050 mm	25.0	400	150	60	80	—	60	—	295	130	300
		1/4 Hard	—	415	170	55	—	—	—	—	—	—	—
		1/2 Hard	25.0	540	310	35	65	—	85	—	360	—	—
		Hard	12.0 25.0	670 635	— 380	— 22	— 62	— —	— 90	— —	— 400	230 —	300 —
		Extra Hard	25.0	745	415	13	60	—	95	—	425	—	—
	Rod	Soft	—	400	150	50	—	—	60	—	295	—	—
		Hard	25.0 40.0 75.0	485 450 415	275 240 195	25 30 30	— — —	— — —	— — —	— — —	345 — —	— — —	— — —
C67000	Rod and Bar	Soft	All Sizes	620	345	20	—	—	—	—	—	—	—
		1/2 Hard Hard	All Sizes All Sizes	770 815	460 485	13 12	— —	— —	— —	— —	— —	— —	— —
	Forgings	Soft	—	620	345	20	—	—	—	—	—	—	—
		1/2 Hard Hard	— —	770 815	460 485	13 12	— —	— —	— —	— —	— —	— —	— —
C67300	Rod and Bar	As Extruded	All Sizes	485	275	25	—	—	70	—	—	—	—
		Soft	All Sizes	450	240	25	—	—	60	—	—	—	—
		1/2 Hardg	25.0 75.0 Over 75.0	495 485 485	360 345 310	18 22 25	— — —	— — —	82 80 75	— — —	— — —	215 215 —	100 100 —
		1/2 Hardh	All Sizes	485	310	25	—	—	75	—	—	—	—
		Hardg	25.0 50.0	550 495	415 345	15 19	— —	— —	86 84	— —	— —	— —	— —
	Shapes	As Extruded	All Sizes	485	275	25	—	—	70	—	—	—	—
	Forgings	As Forged	—	485	275	25	—	—	70	—	—	—	—
		Forged and Heat Treated	—	515	310	20	—	—	85	—	—	—	—
C67400	Rod and Bar	As Extruded	All Sizes	505	260	16	—	—	80	—	—	—	—
		Extruded and Drawn	25.0 50.0 75.0	675 635 605	425 385 345	15 18 22	— — —	— — —	90 88 85	— — —	— — —	215 — —	100 — —
	Shapes	As Extruded	All Sizes	505	260	16	—	—	80	—	—	—	—
	Forgings	As Forged	—	470	235	18	—	—	75	—	—	—	—
C67500	Rod and Bar	Soft	25.0	450	205	33	—	—	65	—	290	—	—
		1/2 Hard	25.0 50.0 65.0	530 495 485	310 290 240	23 27 30	— — —	— — —	83 77 —	— — —	325 305 —	— — —	— — —
		Hard	25.0 40.0 65.0	590 550 515	415 380 345	10 12 15	— — —	— — —	90 — —	— — —	330 — —	— — —	— — —
	Shapes	Soft	—	450	205	33	—	—	65	—	290	—	—

TABLE 10A—TYPICAL MECHANICAL PROPERTIES OF WROUGHT COPPER AND COPPER ALLOYS (continued)

Metric (SI) Units

UNS No.	Form	Temper	Size Section, mm	Tensile Strength, MPa	Yield Strength, 0.5% Ext Under Load MPa	Elongation In 50 mm, %	Reduction of Area, %	Hardness			Shear Strength, MPa	Fatigue Strength	
								RF	RB	R30T		MPa	Million Cycles
C70600	Condenser Tube Plate	—	25.0	290	125	35	—	—	15	—	—	—	—
	Tube	Annealed	25.0 x 1.6	305	110	42	—	65	15	26	—	—	—
		Light Drawn	25.0 x 1.6	415	395	10	—	100	72	70	—	—	—
C71000	Tube	Annealed	25.0 x 1.6	345	150	45	—	—	35	—	—	—	—
		Annealed 0.035 mm	1.0	305	95	37	—	72	27	34	—	—	—
		0.015 mm	1.0	325	115	38	—	83	47	48	—	—	—
		1/4 Hard	1.0	380	—	—	—	—	59	56	—	—	—
		1/2 Hard	1.0	435	395	7	—	—	71	64	—	—	—
		Hard	1.0	505	—	—	—	—	80	70	—	—	—
		Extra Hard	1.0	540	—	—	—	—	83	72	—	—	—
		Spring	1.0	565	—	—	—	—	85	73	—	—	—
C71500	Condenser Tube Plate	—	25.0	380	140	45	—	—	35	—	—	—	—
	Tube	Annealed	25.0 x 1.6	415	170	45	—	80	45	—	—	—	—
		Drawn and Stress Relieved	25.0 x 1.6	515	380	20	—	—	—	—	—	—	—
	Plate, Sheet, Strip and Rolled Bar[e]	Annealed 0.035 mm	1.0	310	190	45	—	78	34	39	—	—	—
		0.015 mm	1.0	415	150	40	—	84	50	49	—	—	—
		1/4 Hard	1.0	450	345	20	—	—	74	66	—	—	—
		1/2 Hard	1.0	505	470	12	—	—	81	71	—	—	—
		Hard	1.0	565	510	4	—	—	86	74	—	—	—
		Extra Hard	1.0	595	545	2	—	—	88	75	—	—	—
		Spring	1.0	615	565	2	—	—	89	76	—	—	—
C75200	Plate, Sheet, Strip and Rolled Bar	Annealed 0.035 mm	1.0	400	170	40	—	85	40	47	—	—	—
		0.015 mm	1.0	415	205	32	—	90	55	55	—	—	—
		1/4 Hard	1.0	450	345	20	—	—	73	65	—	—	—
		1/2 Hard	1.0	510	425	8	—	—	83	72	—	—	—
		Hard	1.0	535	510	3	—	—	87	75	—	—	—
		Extra Hard	1.0	635	—	—	—	—	91	77	—	—	—
		Spring	1.0	660	—	—	—	—	93	78	—	—	—
	Rod and Bar	Annealed 0.035 mm	12.0	385	170	42	—	—	—	—	—	—	—
		0.015 mm	12.0	400	180	35	—	—	—	—	—	—	—
		1/4 Hard	12.0	485	415	20	—	—	78	—	—	—	—
		Hard	6.0[g]	620	515	10	—	—	—	—	—	—	—
			12.0[g]	550	470	12	—	—	—	—	—	—	—
			25.0	515	425	15	—	—	—	—	—	160	300
			Over 25.0[g]	485	380	20	—	—	—	—	—	—	—
C77000	Plate, Sheet, Strip and Rolled Bar	Annealed 0.035 mm	1.0	415	185	40	—	90	55	49	—	110	100
		0.015 mm	1.0	—	—	—	—	91	60	56	—	—	—
								RG					
		1/4 Hard	1.0	540	—	—	—	43	79	69	—	—	—
		1/2 Hard	1.0	600	—	—	—	60	87	75	—	—	—
		Hard	1.0	690	585	3	—	72	91	77	—	145	100
		Extra Hard	1.0	745	620	2	—	77	96	80	—	—	—
		Spring	1.0	795	—	2	—	80	99	81	—	—	—
	Rod and Bar	Annealed 0.035 mm	12.0	415	185	45	—	—	—	—	—	—	—
		1/4 Hard	12.0[m]	585	—	—	—	—	—	—	—	—	—
		Hard	6.0[g]	690	—	—	—	—	—	—	—	—	—
			12.0[g]	620	—	—	—	—	—	—	—	—	—
			25.0	585	—	—	—	—	—	—	—	—	—
			Over 25.0[g]	550	—	—	—	—	—	—	—	—	—

[a] Rotating beam tests on rod.
[b] Independent rotating beam tests, diameter of test sections 8.89 mm.
[c] Elongation in 4X diameter or thickness of specimen.
[d] Capable of being hardened by further heat treatment.
[e] Plate generally available in only annealed, 1/4 hard, and 1/2 hard.
[f] Elongation in 250 mm.
[g] Rods only.
[h] Bars only.
[i] Shapes only.
[j] Yield strength measured at 0.2% offset.
[k] After heat treatment.
[l] Mill heat treated.
[m] Rounds only.

TABLE 10B—TYPICAL MECHANICAL PROPERTIES OF WROUGHT COPPER AND COPPER ALLOYS

Customary Units

UNS No.	Form	Temper	Size Section, in	Tensile Strength, ksi	Yield Strength, 0.5% Ext Under Load, ksi	Elongation in 2 in, %	Reduction of Area, %	Hardness RF	Hardness RB	Hardness R30T	Shear Strength, ksi	Fatigue Strength ksi	Fatigue Strength Million Cycles
C10200 C11000 C11100 C11300 C11400 φ C11500 C11600 C12000 C12200	Plate, Sheet, Strip, and Rolled Bar	Soft, Anneal	0.04	32	10	45	—	40	—	—	22	—	—
		Deep Drawing Anneal	0.04	34	11	45	—	45	—	—	23	11	100
		Light Cold Rolled	0.04	36	28	30	—	60	10	25	25	—	—
		1/2 Hard	0.04	42	36	14	—	84	40	50	26	13	100
		Hard	0.04	50	45	6	—	90	50	57	28	13	100
		Spring	0.04	55	50	4	—	94	60	63	29	14	100
		Extra Spring	0.04	57	53	4	—	95	62	64	29	—	—
		Hot Rolled	0.04	34	10	45	—	45	—	—	23	—	—
		Hot Rolled and Annealed	0.04	33	10	45	—	42	—	—	22	—	—
C10200 C11000 C12000 C12200	Rod, Bar and Shapes	Soft Anneal	1.0	32	10	55	70	40	—	—	22	—	—
		Hard	0.25[g]	55	50	10	—	94	60	—	29	—	—
			1.00[g]	48	44	16	55	87	47	—	27	17	300
			2.00[g]	45	40	20	—	85	45	—	26	—	—
			0.50[h]	48	44	16	—	87	47	—	27	—	—
			0.50[i]	40	32	30	—	—	35	—	28	—	—
C10200 C12000 C12200	Tube	Soft Anneal	1 x 0.06	32	10	45	—	40	—	—	22	11[a]	20
		Light Anneal	1 x 0.06	34	11	45	—	45	—	—	23	—	—
		Light Drawn	1 x 0.06	40	32	25	—	77	35	45	26	14[a]	20
		Drawn	1 x 0.06	40	32	25	—	77	35	45	—	—	—
		Hard Drawn	1 x 0.06	55	50	8	—	95	60	63	29	19[a]	20
			1 x 0.06	55	50	8	—	95	60	63	—	—	—
			1 x 0.06	55	50	8	—	95	60	63	—	—	—
C10200 C11000	Wire	Annealed	0.08	35	—	35[f]	—	—	—	—	24	—	—
C14500 and C14700	Rod	1/2 Hard	0.25	43	38	20	—	—	38	—	25	—	—
			1.00	42	36	25	—	—	38	—	24	—	—
			2.00	41	35	25	—	—	35	—	24	—	—
		Hard	0.25	53	48	10	—	—	52	—	29	—	—
			1.00	48	42	15	—	—	50	—	27	—	—
C15000	Round Rod	Drawn and Heat Treated	1.00	60	50	12	—	—	67	—	—	28	100
C16200	Round Rod	Drawn	1.00	65	50	25	—	—	70	—	—	—	—
			2.00	60	35	30	—	—	65	—	—	—	—
			3.00	55	25	35	—	—	60	—	—	—	—
	Square, Rectangular and Hex Rod and Bar	Drawn	1.00	65	50	25	—	—	70	—	—	—	—
			Over 1.00	55	35	30	—	—	65	—	—	—	—
	Forging	As Forged	1.00	65	—	25	—	—	60	—	—	—	—
			2.00	55	—	30	—	—	55	—	—	—	—
			Over 2.00	55	—	30	—	—	55	—	—	—	—
C17000	Strip	A Soft[d]	—	69	32[j]	48	—	—	62	56	—	—	—
		1/4 Hard[d]	—	82	70[j]	22	—	—	79	68	—	—	—
		1/2 Hard[d]	—	92	82[j]	15	—	—	92	76	—	—	—
		Hard[d]	—	110	104[j]	5	—	—	99	81	—	—	—
									RC	R30N			
		AT (3 h at 600°F)[k]	—	165	145[j]	3	—	—	36	56	—	38	100
		1/4 HT (2 h at 600°F)[k]	—	172	150[j]	2	—	—	37	57	—	40	100
		1/2 HT (2 h at 600°F)[k]	—	182	160[j]	1	—	—	38	58	—	42	100
		HT (2 h at 600°F)[k]	—	190	170[j]	1	—	—	40	60	—	43	100
		AM[l]	—	108	82[j]	20	—	—	20	40	—	—	—
		1/4 HM[l]	—	118	92[j]	17	—	—	24	44	—	—	—
		1/2 HM[l]	—	123	105[j]	14	—	—	28	48	—	—	—
		HM[l]	—	143	122[j]	11	—	—	32	52	—	—	—
		XHM[l]	—	168	148[j]	4	—	—	34	54	—	—	—
C17200	Strip	A (Soft)[d]	—	69	32[j]	48	—	—	62	56	—	—	—
		1/4 Hard[d]	—	82	70[j]	22	—	—	79	68	—	—	—
		1/2 Hard[d]	—	92	82[j]	15	—	—	92	76	—	—	—
		Hard[d]	—	110	104[j]	5	—	—	99	81	—	—	—
									RC	R30N			
		AT (3 h at 600°F)[k]	—	178	155[j]	6	—	—	38	58	—	38	100
		1/4 H (2 h at 600°F)[k]	—	188	165[j]	4	—	—	40	60	—	40	100
		1/2 H (2 h at 600°F)[k]	—	198	175[j]	3	—	—	42	62	—	43	100
		HT (2 h at 600°F)[k]	—	202	180[j]	2	—	—	43	63	—	44	100
	Rod and Bar	A (Soft)[d]	—	72	25[j]	48	—	—	RB 62	R30T —	—	—	—
		1/2 Hard[d]	Under 1.00	110	90[j]	15	—	—	96	—	—	—	—
		1/2 Hard[d]	Over 1.00	100	90[j]	15	—	—	96	—	—	—	—

TABLE 10B—TYPICAL MECHANICAL PROPERTIES OF WROUGHT COPPER AND COPPER ALLOYS (continued)

Customary Units

UNS No.	Form	Temper	Size Section, in	Tensile Strength, ksi	Yield Strength, 0.5% Ext Under Load ksi	Elongation in 2 in, %	Reduction of Area, %	Hardness RF	Hardness RB	Hardness R30T	Shear Strength, ksi	Fatigue Strength ksi	Fatigue Strength Million Cycles
C17200 continued	Rod and Bar	AT (3 h at 600°F)[k]	—	178	160[j]	6	—	—	RC 38	—	—	50	100
		1/2 HT (2 h at 600°F)[k]	Under 1.00	200	182[j]	4	—	—	42	—	—	51	100
		1/2 HT (2 h at 600°F)[k]	Over 1.00	192	175[j]	4	—	—	42	—	—	51	100
C17500	Strip and Plate	A (Soft)[d]	—	45	25[j]	28	—	—	RB 32	37	—	—	—
		1/2 Hard[d]	—	68	56[j]	8	—	—	70	64	—	—	—
		Hard[d]	—	78	70[j]	5	—	—	83	72	—	—	—
		AT (3 h at 900°F)[k]	—	110	90[j]	12	—	—	96	80	—	33	100
		1/2 HT (2 h at 900°F)[k]	—	120	108[j]	8	—	—	98	81	—	35	100
		HT (2 h at 900°F)[k]	—	120	108[j]	8	—	—	98	81	—	35	100
C17500 C17600	Rod, Bar Shapes and Tubing	A (Soft)[d]	—	45	25[j]	28	—	—	35	—	—	—	—
		1/2 Hard[d]	—	72	65[j]	12	—	—	68	—	—	—	—
		AT (3 h at 900°F)[k]	—	110	90[j]	18	—	—	96	—	—	40	100
		1/2 HT (2 h at 900°F)[k]	—	120	110[j]	14	—	—	98	—	—	40	100
C18400	Round Rod	Drawn	1.00	70	—	20	—	—	80	—	—	—	—
			2.00	65	—	20	—	—	75	—	—	—	—
			3.00	60	—	20	—	—	70	—	—	—	—
	Hex Rod and Bar	Drawn	1.00	70	—	20	—	—	75	—	—	—	—
			Over 1.00	60	—	20	—	—	70	—	—	—	—
	Forgings	As Forged	1.00	70	—	20	—	—	80	—	—	—	—
			2.00	60	—	20	—	—	75	—	—	—	—
			Over 2.00	60	—	20	—	—	70	—	—	—	—
C18700	Rod	Hard	0.25	60	55	10	—	—	55	—	32	—	—
			0.50	55	50	11	—	—	50	—	30	—	—
			0.75	53	48	12	—	—	50	—	29	—	—
			1.00	51	46	14	—	—	50	—	28	—	—
φ C19200	Tube	A (Soft)	1.88 x 0.09	37	12	40	—	—	—	—	—	—	—
		Light Drawn	1.88 x 0.09	42	31	3	—	—	—	—	—	—	—
C21000	Plate, Sheet, Strip and Rolled Bar[e]	Annealed 0.050 mm	0.04	34	10	45	—	46	—	—	27	—	—
		0.035 mm	0.04	35	11	45	—	52	—	4	28	—	—
		0.025 mm	0.04	36	12	43	—	56	—	9	29	—	—
		0.015 mm	0.04	38	14	42	—	60	—	15	30	—	—
		1/4 Hard	0.04	42	32	25	—	—	38	44	32	—	—
		1/2 Hard	0.04	48	40	12	—	—	52	54	34	—	—
		3/4 Hard	0.04	51	45	8	—	—	59	58	35	—	—
		Hard	0.04	56	50	5	—	—	64	60	37	—	—
		Extra Hard	0.04	61	55	4	—	—	70	64	39	—	—
		Springs	0.04	64	58	4	—	—	73	66	40	—	—
		Extra Spring	0.04	65	—	—	—	—	74	68	—	—	—
C22000	Plate, Sheet Strip and Rolled Bar[e]	Annealed 0.050 mm	0.04	37	10	45	—	53	—	6	28	—	—
		0.035 mm	0.04	38	12	45	—	57	—	12	30	—	—
		0.025 mm	0.04	39	14	44	—	60	—	16	31	—	—
		0.015 mm	0.04	41	15	42	—	65	—	26	32	—	—
		1/4 Hard	0.04	45	35	25	—	—	42	44	33	—	—
		1/2 Hard	0.04	52	45	11	—	—	58	56	35	—	—
		3/4 Hard	0.04	57	50	8	—	—	67	63	36	—	—
		Hard	0.04	61	54	5	—	—	70	63	38	—	—
		Extra Hard	0.04	67	58	4	—	—	75	67	40	—	—
		Spring	0.04	72	62	3	—	—	78	69	42	21	15
		Extra Spring	0.04	76	—	—	—	—	82	73	—	—	—
	Tube	Soft Anneal	1 x 0.06	38	10	52	—	55	—	10	—	—	—
		Light Anneal	1 x 0.06	41	15	42	—	65	—	35	—	—	—
		Drawn (General Purpose)	1 x 0.06	45	35	25	—	—	50	50	—	—	—
		Hard Drawn	1 x 0.06	60	53	6	—	—	69	62	—	—	—
C23000	Sheet and Strip	Annealed 0.070 mm	0.04	39	10	48	—	56	—	10	—	—	—
		0.050 mm	0.04	40	12	47	—	59	—	14	—	—	—
		0.035 mm	0.04	41	14	46	—	63	—	22	—	—	—
		0.025 mm	0.04	43	16	44	—	66	—	28	—	—	—
		0.015 mm	0.04	45	18	42	—	71	—	38	—	—	—
		1/4 Hard	0.04	50	39	25	—	—	55	54	—	—	—

TABLE 10B—TYPICAL MECHANICAL PROPERTIES OF WROUGHT COPPER AND COPPER ALLOYS (continued)

Customary Units

UNS No.	Form	Temper	Size Section, in	Tensile Strength, ksi	Yield Strength, 0.5% Ext Under Load, ksi	Elongation in 2 in, %	Reduction of Area, %	Hardness RF	Hardness RB	Hardness R30T	Shear Strength, ksi	Fatigue Strength ksi	Fatigue Strength Million Cycles
C23000 continued	Sheet and Strip	1/2 Hard	0.04	57	49	12	—	—	65	60	—	—	—
		3/4 Hard	0.04	62	52	8	—	—	73	67	—	—	—
		Hard	0.04	70	57	5	—	—	77	68	—	—	—
		Extra Hard	0.04	78	61	4	—	—	83	72	44	—	—
		Spring	0.04	84	63	3	—	—	86	74	46	—	—
		Extra Spring	0.04	86	—	—	—	—	88	77	—	—	—
	Tube	Soft Anneal	1 x 0.06	40	12	55	—	60	—	15	—	—	—
		Light Anneal	1 x 0.06	44	18	45	—	71	—	38	—	—	—
		Light Drawn	1 x 0.06	50	40	30	—	—	55	54	—	—	—
		Drawn (General Purpose)	1 x 0.06	56	46	25	—	—	62	60	—	—	—
		Hard Drawn	1 x 0.06	70	58	8	—	—	77	68	—	—	—
C24000	Sheet and Strip	Annealed 0.070 mm	0.04	42	12	52	—	57	—	8	—	—	—
		0.050 mm	0.04	44	14	50	—	61	—	16	32	—	—
		0.035 mm	0.04	46	15	48	—	66	—	28	—	—	—
		0.025 mm	0.04	48	17	47	—	69	—	32	—	—	—
		0.015 mm	0.04	50	20	46	—	75	—	42	33	—	—
		1/4 Hard	0.04	53	40	30	—	—	55	54	36	—	—
		1/2 Hard	0.04	61	50	18	—	—	70	64	39	—	—
		3/4 Hard	0.04	66	54	12	—	—	76	68	41	—	—
		Hard	0.04	74	59	7	—	—	82	71	43	—	—
		Extra Hard	0.04	83	62	5	—	—	87	75	45	—	—
		Spring	0.04	91	65	3	—	—	91	77	48	24	20
		Extra Spring	0.04	93	—	—	—	—	92	78	—	—	—
C26000	Plate, Sheet, Strip, Rolled Bar and Wire[e]	Annealed 0.120 mm	0.04	44	11	66	—	54	—	11	—	13	100
		0.070 mm	0.04	46	14	65	—	58	—	15	32	13	100
		0.050 mm	0.04	47	15	62	—	64	—	26	—	—	—
		0.035 mm	0.04	49	17	57	—	68	—	31	34	14	100
		0.025 mm	0.04	51	19	55	—	72	—	36	—	—	—
		0.015 mm	0.04	53	22	54	—	78	—	43	35	15	100
		1/4 Hard	0.04	54	40	43	—	—	55	54	36	—	—
		1/2 Hard	0.04	62	52	23	—	—	70	65	40	18	100
		3/4 Hard	0.04	69	57	15	—	—	79	70	42	—	—
		Hard	0.04	76	63	8	—	—	82	73	44	21	100
		Extra Hard	0.04	86	65	5	—	—	88	76	46	—	—
		Spring	0.04	94	65	3	—	—	91	77	48	23	100
		Extra Spring	0.04	99	65	3	—	—	93	78	—	—	—
	Tube	Soft Anneal	1 x 0.06	42	15	65	—	64	—	26	—	—	—
		Light Anneal	1 x 0.06	57	20	55	—	75	—	40	—	—	—
		Hard Drawn	1 x 0.06	78	64	8	—	—	82	73	—	—	—
C26800	Plate, Sheet, Strip and Rolled Bar[e]	Annealed 0.120 mm	0.04	—	—	—	—	56	—	5	—	—	—
		0.070 mm	0.04	46	14	65	—	58	—	15	32	12	100
		0.050 mm	0.04	47	15	62	—	64	—	26	33	15	100
		0.035 mm	0.04	49	17	57	—	68	—	31	34	—	—
		0.025 mm	0.04	51	19	55	—	72	—	36	35	—	—
		0.015 mm	0.04	53	22	54	—	78	—	43	35	—	—
		1/4 Hard	0.04	54	40	43	—	—	55	54	35	—	—
		1/2 Hard	0.04	61	50	23	—	—	70	65	40	—	—
		3/4 Hard	0.04	67	55	15	—	—	77	69	41	—	—
		Hard	0.04	74	60	8	—	—	80	70	43	14	100
		Extra Hard	0.04	85	62	5	—	—	87	74	45	—	—
		Spring	0.04	91	62	3	—	—	90	76	47	20	100
		Extra Spring	0.04	95	—	—	—	—	92	78	—	20	100
C27000	Wire	Annealed 0.035 mm	0.08	50	—	60	—	—	—	—	34	—	—
		1/8 Hard	0.08	58	—	35	—	—	—	—	38	—	—
		1/4 Hard	0.08	70	—	20	—	—	—	—	42	22[a]	300
		1/2 Hard	0.08	88	—	15	—	—	—	—	46	—	—
		3/4 Hard	0.08	100	—	12	—	—	—	—	50	—	—
		Hard	0.08	110	—	8	—	—	—	—	55	—	—
		Extra Hard	0.08	120	—	4	—	—	—	—	58	—	—
		Spring	0.08	128	—	3	—	—	—	—	60	—	—

TABLE 10B—TYPICAL MECHANICAL PROPERTIES OF WROUGHT COPPER AND COPPER ALLOYS (continued)

Customary Units

UNS No.	Form	Temper	Size Section, in	Tensile Strength, ksi	Yield Strength, 0.5% Ext Under Load, ksi	Elongation in 2 in, %	Reduction of Area, %	Hardness RF	Hardness RB	Hardness R30T	Shear Strength, ksi	Fatigue Strength ksi	Fatigue Strength Million Cycles
C33000 and C33100	Tube	Soft Anneal	1 x 0.06	47	15	60	—	64	—	26	—	—	—
		Light Anneal	1 x 0.06	52	20	50	—	75	—	37	—	—	—
		Hard Drawn	1 x 0.06	75	60	7	—	—	80	69	—	—	—
C34200 C35000	Plate, Sheet, Strip and Rolled Bar	Annealed 0.035 mm	0.04	49	17	52	—	68	—	31	34	—	—
		1/4 Hard	0.04	54	40	38	—	—	55	54	36	—	—
		1/2 Hard	0.04	61	50	20	—	—	70	65	40	—	—
		Hard	0.04	74	60	7	—	—	80	69	43	—	—
		Extra Hard	0.04	85	62	5	—	—	87	74	45	—	—
C34500 C35000	Rod	1/2 Hard	0.50	57	30	20	—	—	68	—	36	—	—
			1.00	55	30	25	—	—	68	—	35	—	—
			2.00	50	25	30	—	—	65	—	32	—	—
C36000	Rod	Soft	1.00	49	18	53	58	68	—	—	30	—	—
		1/2 Hard	0.25	68	52	18	48	—	80	—	38	—	—
			1.00	58	45	25	50	—	78	—	34	—	—
			2.00	55	44	32	52	—	75	—	32	20[b] / 14[b]	100 / 300
		Hard	0.12	85	50	—	—	—	—	—	—	—	—
			0.25	75	40	5[c]	—	—	—	—	—	—	—
			0.75	70	35	10[c]	—	—	—	—	—	—	—
	Flat Products	Soft	1 x 6	48	20	25[c]	—	—	—	—	30	—	—
			Over 1 x 6	45	18	30[c]	—	—	—	—	28	—	—
		1/2 Hard	0.25 x 1	56	45	20[c]	—	—	62	—	33	—	—
			0.50 x 6	50	30	20[c]	—	—	—	—	30	—	—
			2.00 x 2	50	30	25[c]	—	—	—	—	30	—	—
			2.00 x 6	45	25	25[c]	—	—	—	—	27	—	—
			Over 2.00 x 4	45	25	25[c]	—	—	—	—	27	—	—
C37700	Die Forgings	As Extruded	0.04	52	20	45[c]	65	78	—	—	—	—	—
		As Forged	2 lbs	58	23	40[c]	—	—	—	—	—	—	—
C46400 φ C46500 φ C46600 φ C46700	Rod and Bar	Soft	0.25	58	27	45	60	—	56	—	40	—	—
			1.00	57	25	47	60	—	56	—	40	—	—
			2.00	56	25	47	60	—	55	—	40	—	—
			Over 2.00	54	22	—	—	—	—	—	—	—	—
		1/2 Hard or Light Annealed	0.25	63	30	40	55	—	60	—	42	—	—
			1.00	63	30	40	55	—	60	—	42	—	—
			2.00	62	28	43	55	—	60	—	42	—	—
			3.00	57	26	—	—	—	—	—	—	—	—
			Over 3.00	57	24	—	—	—	—	—	—	—	—
		Hard	0.25	80	57	20	45	—	85	—	45	—	—
			1.00	75	53	20	45	—	82	—	44	—	—
			2.00	67	40	35	50	—	75	—	43	—	—
	Shapes	As Extruded	—	58	25	40	—	—	—	—	40	—	—
C51000	Sheet and Strip	Annealed 0.050 mm	0.04	47	19	64	—	73	26	—	36	—	—
		0.035 mm	0.04	49	20	58	—	75	28	—	37	—	—
		0.025 mm	0.04	50	21	52	—	77	30	—	38	—	—
		0.015 mm	0.04	53	22	50	—	79	34	—	40	—	—
		1/2 Hard	0.04	68	55	28	—	—	78	69	49	—	—
		Hard	0.04	81	75	10	—	—	87	75	—	25	100
		Extra Hard	0.04	92	80	6	—	—	93	78	—	—	—
		Spring	0.04	100	80	4	—	—	95	79	—	22	100
		Extra Spring	0.04	107	80	3	—	—	97	80	—	—	—
	Rod	Hard	0.50	75	65	25	—	—	80	—	54	—	—
			1.00	70	58	25	—	—	78	—	50	29	300
	Wire	Soft, 0.035 mm	0.08	50	20	58	—	—	38	—	—	—	—
		1/4 Hard	0.08	68	60	24	—	—	49	—	—	—	—
		1/2 Hard	0.08	85	80	8	—	—	—	—	—	—	—
		Hard	0.08	110	—	5	—	—	—	—	—	27	100
		Extra Hard	0.08	130	—	3	—	—	—	—	—	30	100
		Spring	0.08	140	—	2	—	—	—	—	—	—	—
φ C51100 C54400	Sheet and Strip	Annealed 0.050 mm	0.04	46	—	48	—	70	—	—	—	—	—
		0.035 mm	0.04	48	—	47	—	73	—	—	—	—	—
		0.025 mm	0.04	50	—	46	—	75	—	—	—	—	—
		0.015 mm	0.04	51	—	46	—	76	—	—	—	—	—
		1/2 Hard	0.04	62	54	19	—	—	70	65	—	—	—
		Hard	0.04	80	74	7	—	—	86	74	—	—	—
		Extra Hard	0.04	92	—	4	—	—	91	78	—	—	—
		Spring	0.04	98	80	3	—	—	93	79	—	—	—
		Extra Spring	0.04	103	80	2	—	—	95	80	—	—	—

TABLE 10B—TYPICAL MECHANICAL PROPERTIES OF WROUGHT COPPER AND COPPER ALLOYS (continued)

Customary Units

UNS No.	Form	Temper	Size Section, in	Tensile Strength, ksi	Yield Strength, 0.5% Ext Under Load, ksi	Elongation in 2 in, %	Reduction of Area, %	Hardness RF	Hardness RB	Hardness R30T	Shear Strength, ksi	Fatigue Strength ksi	Fatigue Strength Million Cycles
C52100	Sheet and Strip	Soft	0.04	60	24	63	—	82	50	—	—	—	—
		1/2 Hard	0.04	76	55	32	—	—	84	73	51	—	—
		Hard	0.04	93	72	10	—	—	93	78	—	22	100
		Extra Hard	0.04	106	80	4	—	—	96	80	—	—	—
		Spring	0.04	112	—	3	—	—	98	81	—	—	—
		Extra Spring	0.04	120	—	2	—	—	100	82	—	—	—
φ C52400	Sheet and Strip	Soft, 0.035 mm	0.04	66	28	68	—	—	55	—	—	—	—
		1/2 Hard	0.04	83	—	32	—	—	92	—	—	—	—
		Hard	0.04	100	—	13	—	—	97	—	—	—	—
		Extra Hard	0.04	115	—	7	—	—	100	—	—	—	—
		Spring	0.04	122	—	4	—	—	101	—	—	—	—
		Extra Spring	0.04	128	—	3	—	—	103	—	—	—	—
C54400	Sheet and Strip	See C51100											
	Rod	Hard	0.50	75	63	15	—	—	83	—	—	—	—
			1.00	68	57	20	—	—	80	—	—	—	—
	Flat Products	Hard	0.38	60	45	20	—	—	70	—	—	—	—
			0.75	55	35	25	—	—	—	—	—	—	—
C60800	Tube	Annealed	1 x 0.06	60	27	—	—	77	—	—	—	—	—
φ C61400	Plate, Sheet, Strip and Rolled Bar	Soft	0.12	82	45	40	35	—	84	—	45	30	100
			0.31	80	40	40	35	—	83	—	42	28	100
			0.50	78	35	42	40	—	82	—	40	26	100
			1.00	76	33	45	40	—	81	—	40	25	100
		Hard	0.12	89	60	32	—	—	87	—	50	—	—
			0.31	85	58	35	—	—	86	—	48	—	—
			0.50	80	54	38	—	—	85	—	40	—	—
			1.00	78	45	40	—	—	84	—	38	—	—
	Rod and Bar	—	0.50	85	45	35	55	—	91	—	48	—	—
			1.00	82	40	35	55	—	90	—	45	—	—
			2.00	80	35	35	60	—	88	—	40	—	—
φ C61800	Rod	1/2 Hard	1.00	85	42	23^c	—	—	88	—	47	28	100
			2.00	82	39	25^c	—	—	88	—	45	28	100
			3.00	80	39	28^c	—	—	88	—	43	26	100
C62300	Rod and Bar	Drawn	0.50	95	60	15^c	—	BHN(1000 kg) 180			—	—	—
			1.00	92	55	20^c	—	—			—	—	—
			3.00	90	45	25^c	—	—			—	—	—
	Shapes	—	—	85	33	25^c	—	170				—	—
	Forgings	As Forged	Thru 1.5	85	33	25^c	—	140			—	—	—
			Over 1.5	82	31	30^c	—						
C62400	Rod	As Extruded	All Sizes	95	50	12^c	—	BHN(3000 kg) 200			—	—	—
	Forgings	As Forged	—	95	50	12^c	—	200			—	—	—
		Hardened and H$_2$O Quenched	—	—	—	—	—	265			—	—	—
		Hardened, H$_2$O Quenched and Tempered	—	100	—	12^c	—	210			—	—	—
C63000	Rod and Bar	Annealed	1.00	105	60	12^c	—	—	100	—	—	—	—
			2.00	95	50	15^c	—	—	—	—	—	—	—
			4.00	90	45	17^c	—	—	—	—	—	—	—
	Shapes	As Extruded	All Sizes	90	45	17^c	—	—	—	—	—	—	—
φ C64200	Rod and Bar	Drawn	0.75	102	68	22^c	—	—	94	—	59	50	100
			1.50	93	60	26^c	—	—	90	—	—	—	—
C65500	Plate, Sheet, Strip and Rolled Bar^e	Annealed 0.070 mm	0.04	56	21	63	—	76	40	—	42	—	—
		0.040 mm	0.04	60	25	60	—	85	62	—	43	16	100
		1/4 Hard	0.04	68	35	30	—	—	75	67	47	—	—
		1/2 Hard	0.04	78	45	17	—	—	87	75	50	—	—
		Hard	0.04	94	58	8	—	—	93	78	57	23	100
		Extra Hard	0.04	104	60	6	—	—	96	80	60	—	—
		Spring	0.04	110	62	4	—	—	97	81	63	—	—
		Hot Rolled	0.04	63	25	—	—	—	—	—	—	—	—
		Hot Rolled and Cold Rolled Finish	0.04	65	30	—	—	—	70	—	—	—	—

TABLE 10B—TYPICAL MECHANICAL PROPERTIES OF WROUGHT COPPER AND COPPER ALLOYS (continued)

Customary Units

UNS No.	Form	Temper	Size Section, in	Tensile Strength, ksi	Yield Strength, 0.5% Ext Under Load, ksi	Elongation in 2 in, %	Reduction of Area, %	Hardness RF	Hardness RB	Hardness R30T	Shear Strength, ksi	Fatigue Strength ksi	Fatigue Strength Million Cycles
C65500 continued	Rod, Bar and Shapes	Soft, 0.050 mm	1.00	58	22	60	80	—	60	—	43	19	300
		1/4 Hard	—	60	25	55	—	—	—	—	—	—	—
		1/2 Hard	1.00	78	45	35	65	—	85	—	52	—	—
		Hard	0.50	97	—	—	—	—	—	—	—	34	300
			1.00	92	55	22	62	—	90	—	58	—	—
		Extra Hard	1.00	108	60	13	60	—	95	—	62	—	—
	Rod	Soft	—	58	22	50	—	—	60	—	43	—	—
		Hard	1.00	70	40	25	—	—	—	—	50	—	—
			1.50	65	35	30	—	—	—	—	—	—	—
			3.00	60	28	30	—	—	—	—	—	—	—
C67000	Rod and Bar	Soft	All Sizes	90	50	20	—	—	—	—	—	—	—
		1/2 Hard	All Sizes	112	67	13	—	—	—	—	—	—	—
		Hard	All Sizes	118	70	12	—	—	—	—	—	—	—
	Forgings	Soft	—	90	50	20	—	—	—	—	—	—	—
		1/2 Hard	—	112	67	13	—	—	—	—	—	—	—
		Hard	—	118	70	12	—	—	—	—	—	—	—
C67300	Rod and Bar	As Extruded	All Sizes	70	40	25	—	—	70	—	—	—	—
		Soft	All Sizes	65	35	25	—	—	60	—	—	—	—
		1/2 Hard[g]	1.00	72	52	18	—	—	82	—	—	31	100
			3.00	70	50	22	—	—	80	—	—	31	100
			Over 3.00	70	45	25	—	—	75	—	—	—	—
		1/2 Hard[h]	All Sizes	70	45	25	—	—	75	—	—	—	—
		Hard[g]	1.00	80	60	15	—	—	86	—	—	—	—
			2.00	72	50	19	—	—	84	—	—	—	—
	Shapes	As Extruded	All Sizes	70	40	25	—	—	70	—	—	—	—
	Forgings	As Forged	—	70	40	25	—	—	70	—	—	—	—
		Forged and Heat Treated	—	75	45	20	—	—	85	—	—	—	—
C67400	Rod and Bar	As Extruded	All Sizes	73	38	16	—	—	80	—	—	—	—
		Extruded and Drawn	1.00	98	62	15	—	—	90	—	—	31	100
			2.00	92	56	18	—	—	88	—	—	—	—
			3.00	88	50	22	—	—	85	—	—	—	—
	Shapes	As Extruded	All Sizes	73	38	16	—	—	80	—	—	—	—
	Forgings	As Forged	—	68	34	18	—	—	75	—	—	—	—
C67500	Rod and Bar	Soft	1.00	65	30	33	—	—	65	—	42	—	—
		1/2 Hard	1.00	77	45	23	—	—	83	—	47	—	—
			2.00	72	42	27	—	—	77	—	44	—	—
			2.50	70	35	30	—	—	—	—	—	—	—
		Hard	1.00	84	60	10	—	—	90	—	48	—	—
			1.50	80	55	12	—	—	—	—	—	—	—
			2.50	75	50	15	—	—	—	—	—	—	—
	Shapes	Soft	—	65	30	33	—	—	65	—	42	—	—
C70600	Condenser Tube Plate	—	1.00	42	18	35	—	—	15	—	—	—	—
	Tube	Annealed	1 x 0.06	44	16	42	—	65	15	26	—	—	—
		Light Drawn	1 x 0.06	60	57	10	—	110	72	70	—	—	—
C71000	Tube	Annealed	1 x 0.06	50	22	45	—	—	35	—	—	—	—
	Plate, Sheet, Strip and Rolled Bar[e]	Annealed 0.035 mm	0.04	44	14	37	—	72	27	34	—	—	—
		0.015 mm	0.04	47	17	38	—	83	47	48	—	—	—
		1/4 Hard	0.04	55	—	—	—	—	59	56	—	—	—
		1/2 Hard	0.04	63	57	7	—	—	71	64	—	—	—
		Hard	0.04	73	—	—	—	—	80	70	—	—	—
		Extra Hard	0.04	78	—	—	—	—	83	72	—	—	—
		Spring	0.04	82	—	—	—	—	85	73	—	—	—
C71500	Condenser Tube Plate	—	1.00	55	20	45	—	—	35	—	—	—	—
	Tube	Annealed	1 x 0.06	60	25	45	—	80	45	—	—	—	—
		Drawn and Stress Relieved	1 x 0.06	75	55	20	—	—	—	—	—	—	—
	Plate, Sheet, Strip and Rolled Bar[e]	Annealed 0.035 mm	0.04	54	20	45	—	78	34	39	—	—	—
		0.015 mm	0.04	60	22	40	—	84	50	49	—	—	—
		1/4 Hard	0.04	65	50	20	—	—	74	66	—	—	—
		1/2 Hard	0.04	73	68	12	—	—	81	71	—	—	—
		Hard	0.04	82	74	4	—	—	86	74	—	—	—
		Extra Hard	0.04	86	79	2	—	—	88	75	—	—	—

TABLE 10B—TYPICAL MECHANICAL PROPERTIES OF WROUGHT COPPER AND COPPER ALLOYS (continued)

Customary Units

UNS No.	Form	Temper	Size Section, in	Tensile Strength, ksi	Yield Strength, 0.5% Ext Under Load, ksi	Elongation in 2 in, %	Reduction of Area, %	Hardness RF	Hardness RB	Hardness R30T	Shear Strength, ksi	Fatigue Strength ksi	Fatigue Strength Million Cycles
C71500 continued	Plate, Sheet, Strip and Rolled Bar[e]	Spring	0.04	89	82	2	—	—	89	76	—	—	—
C75200	Plate, Sheet, Strip and Rolled Bar	Annealed 0.035 mm	0.04	58	25	40	—	85	40	47	—	—	—
		0.015 mm	0.04	60	30	32	—	90	55	55	—	—	—
		1/4 Hard	0.04	65	50	20	—	—	73	65	—	—	—
		1/2 Hard	0.04	74	62	8	—	—	83	72	—	—	—
		Hard	0.04	85	74	3	—	—	87	75	—	—	—
		Extra Hard	0.04	92	—	—	—	—	91	77	—	—	—
		Spring	0.04	96	—	—	—	—	93	78	—	—	—
	Rod and Bar	Annealed 0.035 mm	0.50	56	25	42	—	—	—	—	—	—	—
		0.015 mm	0.50	58	26	35	—	—	—	—	—	—	—
		1/4 Hard	0.50	70	60	20	—	—	78	—	—	—	—
		Hard	0.25[g]	90	75	10	—	—	—	—	—	—	—
			0.50[g]	80	68	12	—	—	—	—	—	—	—
			1.00	75	62	15	—	—	—	—	—	23	300
			Over 1.00[g]	70	55	20	—	—	—	—	—	—	—
C77000	Plate, Sheet, Strip and Rolled Bar	Annealed 0.035 mm	0.04	60	27	40	—	90	55	49	—	16	100
		0.015 mm	0.04	—	—	—	—	91	60	56	—	—	—
								RG					
		1/4 Hard	0.04	78	—	—	—	43	79	69	—	—	—
		1/2 Hard	0.04	87	—	—	—	60	87	75	—	—	—
		Hard	0.04	100	85	3	—	72	91	77	—	21	100
		Extra Hard	0.04	108	90	2	—	77	96	80	—	—	—
		Spring	0.04	115	—	2	—	80	99	81	—	—	—
	Rod and Bar	Annealed 0.035 mm	0.50	60	27	45	—	—	—	—	—	—	—
		1/4 Hard	0.50[m]	85	—	—	—	—	—	—	—	—	—
		Hard	0.25[g]	100	—	—	—	—	—	—	—	—	—
			0.50[g]	90	—	—	—	—	—	—	—	—	—
			1.00[g]	85	—	—	—	—	—	—	—	—	—
			Over 1.00[g]	80	—	—	—	—	—	—	—	—	—

[a] Rotating beam tests on rod.
[b] Independent rotating beam tests, diameter of test sections 0.350 in.
[c] Elongation in 4X diameter or thickness of specimen.
[d] Capable of being hardened by further heat treatment.
[e] Plate generally available in only annealed, 1/4 hard, and 1/2 hard.
[f] Elongation in 10 in.
[g] Rods only.
[h] Bars only.
[i] Shapes only.
[j] Yield strength measured at 0.2% offset.
[k] After heat treatment.
[l] Mill heat treated.
[m] Round only.

PART II—CAST COPPER ALLOYS

1. General—The cast copper base alloys consist of a relatively few families or alloy types which have become standard through the years because of their excellent attributes for particular applications. Within each alloy type, many commercial modifications exist. Those most commonly used by the automotive and related industries are shown in Table 11 with the general characteristics and typical uses of each. Table 12 lists the typical physical properties of these same alloys.

2. Types of Cast Alloys

2.1 Tin bronzes are predominantly copper plus tin. Variations containing up to 4% of zinc and 2% lead have been used for pumps handling sea water, some acids, salt solution, and oils. Excellent worm gears are made from tin bronze containing 8% or more tin. Up to 2% nickel may be added for other types of gears. An alloy of 88 Cu, 5 Sn, 5 Ni, 2 Zn is heat treatable to provide higher strength. (Typical properties: 585 MPa [85 ksi] tensile strength, 415 MPa [60 ksi] yield strength, 8% elongation, and Brinell 185.)

2.2 High lead tin bronzes are produced by adding lead to amounts equal to or more than the tin content. These are used for bearing applications where a combination of wear resistance and good anti-friction properties are desired.

2.3 Lead red brasses are alloys of copper, tin, lead, and zinc. They are the most widely used of all cast copper alloys and are satisfactory for a great many applications, including water pumps, small gears, fittings, and valve bodies.

2.4 Aluminum bronzes (alloys of copper, aluminum, iron, and, in some modifications, nickel) are used extensively for structural applications. Their excellent resistance to corrosion leads to their use in sulfide bearing environments and sea water. Some aluminum bronzes may be heat treated to quite high strength. (Typical properties: 725 MPa [105 ksi] tensile strength, 415 MPa [60 ksi] yield strength, 5% elongation, and Brinell 220.)

2.5 Leaded yellow brasses, containing copper, lead, and more than 20% zinc, are inexpensive free machining general purpose alloys.

2.6 High tensile brasses, also known as manganese bronzes, are alloys of copper, aluminum, manganese, iron, and zinc. They are higher in strength than most of the cast copper alloys. In addition, they are readily cast and possess fairly good corrosion resistance.

2.7 Special purpose alloys with exceptional corrosion resistance include silicon bronze, silicon brass and copper-nickel.

3. Effect of Alloying Elements and Impurities

3.1 Zinc—Added to copper as a predominating alloying constituent, in amounts of 5–40%, to form alloys known as brasses. These groups are called leaded red and semired, silicon, yellow, and high strength yellow brasses. Zinc imparts strength. It is completely soluble in copper, forming solid solution except in such cases as in high strength yellow brasses in which a duplex type of structure is obtained. Smaller amounts of zinc up to 5% are used in tin bronzes to tighten up the structure and aid in producing sound castings for pressure work. Zinc is not considered a very detrimental impurity in most alloys. However, it is generally kept to below 5% in bearing bronzes because large amounts would tend to impair bearing qualities.

3.2 Tin—Added to copper in amounts of 5–20% to form a series of alloys known as tin bronzes and leaded tin bronzes. While the copper-tin constitution diagram shows that it is possible to have approximately 16% tin in solid solution at 520°C, the presence of a hard constituent (alpha-delta copper tin eutectoid) develops in the range of 6–8% tin because of deviation from true equilibrium conditions. The tin strengthens and hardens copper, making it tough and resistant to wear and increases its corrosion resistance. Smaller amounts of tin are used in leaded red and semired brasses for increasing the strength of such general utility alloys. Tin is not generally harmful as an impurity except in high tensile manganese bronze, where it is limited to 0.2%. It is generally felt that in this alloy, tin lowers the strength and ductility.

3.3 Lead—Added alone to copper in large amounts of around 35% for automotive bearings and agricultural and aircraft gear pumps. In practically all other cases, it is added to copper base alloys as an additional alloying element. Small amounts of lead up to 1.5% increase machinability without important decreases in strength. Larger amounts of 5–25% increase machinability greatly, and, in tin-containing alloys, increase antifrictional qualities, however, with reduction of strength.

3.4 Aluminum—Added to copper as a predominating alloying constituent to form a series of high strength alloys known as aluminum bronzes. It is soluble in copper to the extent of about 9.5%. It is added to high strength yellow brasses in varying amounts, being a very necessary part of the high tensile alloy. Aluminum, when present as an impurity, has very detrimental effects upon high leaded bronzes, causing lead sweating and unsoundness during solidification. It is also considered detrimental in the nonleaded tin bronzes, causing unsoundness.

3.5 Iron—Added to copper alloys as a strengthening constituent for silicon, aluminum, and manganese bronzes. It combines with aluminum or manganese or both to form hard compounds. These compounds imbed themselves into the matrix to give the alloys wear resistance. Iron, when present as an impurity, is not desirable since it forms hard spots and is detrimental to machining.

3.6 Phosphorus—Added to copper and copper alloys principally as a deoxidizer. It is added to bronzes in greater amounts than necessary for purely deoxidization considerations to improve hardness and wear resistance, particularly in chill mold castings.

3.7 Nickel—Added to bronzes as an alloying constituent for refining the grain and toughening the alloy. It is also used in amounts up to 15% for nickel brasses to displace that amount of zinc. In this alloy, it promotes strength, corrosion resistance, and whiteness. Nickel is added to some of the high tin gear bronzes to provide improved wear characteristics. When present as an impurity, it does not have detrimental effects; and most specifications permit approximately 1%.

3.8 Silicon—Added to copper as an alloying constituent to form copper-silicon alloys. These alloys have high corrosion resistance, high strength, and toughness. Small amounts of silicon are used as deoxidizing elements. When silicon is present as an impurity, it is extremely detrimental in leaded tin bronzes, promoting unsoundness and lead sweating.

3.9 Beryllium—Added to copper together with small amounts of cobalt or nickel as an alloying constituent to form a series of precipitation-hardenable beryllium-copper alloys. When hardened, these are the strongest of the known copper alloys. They are used for plastic molds, resistance welding electrodes, welding gun components and nonsparking tools. Beryllium, though rarely present as an impurity, has the affect of increasing fluidity and decreasing electrical conductivity in most of the copper alloys.

3.10 Manganese—Used primarily as an alloying constituent for high-strength alloy brasses, where it forms compounds with other alloying elements such as iron and aluminum. It is also used, to some extent, for deoxidizing. It is not considered very detrimental as an impurity.

3.11 Chromium—Added to copper as an alloying constituent to produce a precipitation-hardening type alloy, which in the heat treated condition, has mechanical properties far exceeding that of copper at a slight sacrifice of the electrical conductivity. Heat treatment develops a nominal hardness of 120 Brinell and a nominal electrical conductivity of 80% IACS. The alloy in the heat treated condition is useful for resistance welding electrodes where high electrical conductivity coupled with strength and hardness values superior to copper are desired. Chromium is generally not present as an impurity.

3.12 Antimony—Rarely added to copper alloys. When present as an impurity, it is not considered very detrimental in amounts up to 0.5%. If present in greater amounts, it does tend to decrease physical properties.

TABLE 11—CHARACTERISTICS AND USES OF CAST COPPER ALLOYS

UNS No.[a]	SAE No.	Former SAE No.	ASTM No.	Former ASTM No.	Former Name[b]	General Characteristics	Typical Uses
C83600	CA836	40	B271, B505, B584	B145	Leaded red brass	General utility alloys with reasonable corrosion resistance and strength. Good casting and machining properties. Hydrostatic tightness.	Water pump fittings, valve bodies, and general plumbing hardware. C83600 used for bearing backs.
C83800	CA838	—	B271, B505, B584	B145	Leaded red brass		
C85200	CA852	—	B271, B584	B146	Leaded yellow brass	Inexpensive, good machining casting alloy.	Radiator parts, fittings for water-cooling systems, and battery terminals.
C85400	CA854	41	B271, B584	B146	Leaded yellow brass		
C85800	CA858	—	B176	—	Brass die castings	General purpose low-cost yellow brass alloy with good machining and soldering characteristics.	Plumbing hardware, lock mechanisms, window hardware, gear shift forks, bevel gears.
C87800	CA878	—	B176	—	Brass die castings	Higher in mechanical strength, hardness, and wear resistance than C85800 and C87900, but is more difficult to machine. It is used only for highest strength and wear resistance.	
C87900	CA879	—	B176	—	Brass die castings	General purpose alloy with higher strength than C87800. It is somewhat easier to die cast, but slightly more difficult to machine.	
C86200	CA862	430A	B271, B505, B584	B147	High strength yellow brass	Excellent strength, corrosion resistance, and casting properties.	Brackets, shafts, gears, and structural applications.
C86300	CA863	430B	B271, B505, B584	B147	High strength yellow brass		
C86500	CA865	43	B271, B505, B584	B147	High strength yellow brass		
C87200	CA872	—	B271, B584	B198	Silicon bronze	Good strength, toughness, and corrosion resistance. Good casting qualities.	Pump parts, gears, and shafts.
C87400	CA874	—	B271, B585	B198	Silicon bronze		
C87500	CA875	—	B271, B584	B198	Silicon brass		
C90300	CA903	620	B271, B505, B584	B143	Tin bronze	Hard, strong, tough, resistant to wear, fine grained. Good machinability and corrosion resistant to sea water.	Worm wheels, gears, bushings for heavy loads and low speeds, zinc-containing alloys for pressure castings. Good resistance to pounding.
C90500	CA905	62	B271, B505, B584	B143	Tin bronze		
C90700	CA907	65	B505, B584	—	Phosphor bronze		
C92200	CA922	622	B271, B505, B584	B143	Leaded tin bronze		
C92300	CA923	—	B271, B505, B584	B143	Leaded tin bronze		
C92500	CA925	640	B505, B584	—	Nickel phosphor bronze		
C92700	CA927	63	B505, B584	—	Leaded tin bronze		
φ C92900	CA929	—	B427, B505	—	Leaded nickel tin bronze		
C93200	CA932	660	B271, B505, B584	B144	High leaded tin bronze	Excellent antifriction qualities and castings and machining properties. Resistant to wear. Antifriction qualities increase with lead content. Corrosion resistant.	C93200—General bearing and bushing applications. C93500—High speeds and light loads on bearing backs. C93700—High speeds and heavy loads.
C93500	CA935	66	B271, B505, B584	B144	High leaded tin bronze		
C93700	CA937	64	B271, B505, B584	B144	High leaded tin bronze		
C93800	CA938	67	B271, B505, B584	B144	High leaded tin bronze		C93800—General service bearing material for moderate pressures and high speeds. C94300—Light to moderate loads and high speeds.
C94300	CA943	—	B271, B505, B584	B144	High leaded tin bronze		
C94700	CA947	—	B505, B584	B292	Nickel tin bronze	High strength constructural castings. Easy to cast Hydrostatic tightness. Bearing applications. Corrosion and wear resistant. C94700 is heat treatable.	Worm gears, valve stems and nuts, impellers, screw conveyors, roller bearing cages, and railway electrification hardware.
C94800	CA948	—	B505, B584	B292	Leaded nickel tin bronze		
C95200	CA952	68a	B148, B271, B505	—	Aluminum bronze	Good strength. Wear and corrosion resistant. C95300, C95400, C95500 and C95800 are heat treatable.	Gears, worm wheels, valve guides and seats, and structural applications.
C95300	CA953	68b	B148, B271, B505	—	Aluminum bronze		
C95400	CA954	—	B148, B271, B505	—	Aluminum bronze		
C95500	CA955	—	B148, B271, B505	—	Aluminum bronze		
φ C95800	CA958	—	B148, B271, B505	—	Nickel aluminum bronze		
C96200	CA962	—	B369	—	Copper nickel	Good corrosion resistance and toughness. Low-temperature strength.	Pipe couplings, coastal power-plants, air conditioning, and saline water corrosion.

[a] Unified Numbering System
[b] Alloy names are shown for information only, and should not be used. Use appropriate designation only. (Example: C87800 Copper Alloy.)

TABLE 12A—TYPICAL PHYSICAL PROPERTIES OF CAST COPPER ALLOYS (METRIC [SI] UNITS)[a]

UNS No.	Melting Point °C		Density g/cm³ at 20°C	Coefficient of Thermal Expansion per °C × 10⁻⁵ 20–200°C	Thermal Conductivity $\frac{W}{m \cdot K}$	Electrical Resistivity[b] nΩ·m	Modulus of Elasticity GPa
	Liquidus	Solidus					
C83600	1010	855	8.83	1.80	72	115.0	96
C83800	1005	844	8.60	1.80	73	115.0	90
C85200	941	927	8.50	2.07	84	96.2	76
C85400	940	927	8.45	2.02	88	88.5	83
C85800	899	871	8.40	2.16	—	86.2	103
C86200	941	899	7.58	2.16	35	227.3	103
C86300	923	885	7.84	2.16	35	217.4	97
C86500	880	862	8.30	2.03	87	78.1	103
C87200	971	860	8.40	1.66	28	285.7	103
C87400	916	821	8.27	1.96	28	256.4	103
C87500	916	821	8.27	1.96	28	256.4	103
C87800	916	821	8.27	1.96	28	256.4	138
C87900	926	899	8.50	2.16	—	114.9	103
C90300	1000	854	8.70	1.80	75	144.9	103
C90500	999	854	8.72	1.98	75	156.3	103
C90700	1000	832	8.78	1.84	71	178.6	97
C92200	988	825	8.65	1.80	70	120.5	97
C92300	1000	854	8.80	1.80	75	142.9	97
C92500	1000	854	8.85	1.80	—	—	90
C92700	982	848	8.80	1.82	47	156.3	90
C92900	1030	857	8.79	1.71	58	188.7	97
C93200	977	854	8.93	1.80	58	142.9	97
C93500	1000	844	8.87	1.80	71	113.6	100
C93700	928	762	8.95	1.85	47	169.5	76
C93800	943	854	9.25	1.85	52	151.5	69
C94300	954	899	9.29	1.80	63	188.7	76
C94700	1027	904	8.80	1.98	54	142.9	103
C94700 (HT)	1027	904	8.80	1.98	59	113.6	103
C94800	1027	904	8.80	1.98	39	142.9	103
C95200	1045	1042	7.64	1.62	50	156.3	103
C95300	1045	1040	7.53	1.62	63	—	110
C95300 (HT)	1045	1040	7.53	1.62	63	133.3	103
C95400	1037	1027	7.45	1.62	59	—	107
C95400 (HT)	1037	1027	7.45	1.66	59	133.3	110
C95500	1054	1038	7.53	1.62	42	—	110
C95500 (HT)	1054	1038	7.53	1.62	42	204.1	117
C95800	1063	1043	7.64	1.62	36	243.9	114
C96200	1149	1099	8.94	1.71	45	156.3	124

[a] Specific heat, for all alloys listed is 377 J/kg·K except for the following: C95400 and C95500: 420 J/kg·K C95800: 440 J/kg·K.

[b] See Table 12B for % IACS Electrical Conductivity.

TABLE 12B—TYPICAL PHYSICAL PROPERTIES OF CAST COPPER ALLOYS (CUSTOMARY UNITS)

UNS No.	Melting Point °F Liquid	Melting Point °F Solid	Density lb/in^3	Specific Gravity	Coefficient of Thermal Expansion 10^6 in/in/°F (20–400°F)	Thermal Conductivity % of Cu[a]	Electrical Conductivity % IACS[b]	Modulus of Elasticity 10^6 psi
C83600	1840	1570	0.318	8.83	10.0	18	15	14
C83800	1840	1550	0.312	8.60	10.0	18	15	13
C85200	1725	1700	0.307	8.50	11.5	21	18	11
C85400	1725	1700	0.305	8.45	11.2	23	20	12
C85800	1650	1600	0.305	8.40	12.0	—	20	15
C86200	1725	1650	0.288	7.85	12.0	9	8	15
C86300	1690	1625	0.283	7.84	12.0	9	8	14
C86500	1620	1585	0.301	8.30	11.3	22	22	15
C87200	1780	1580	0.302	8.40	9.2	7	6	15
C87400	1680	1510	0.300	8.27	10.9	7	7	15
C87500	1680	1510	0.300	8.27	10.9	7	7	15
C87800	1680	1510	0.300	8.27	10.9	7	7	20
C87900	1700	1650	0.308	8.50	12.0	—	15	15
C90300	1830	1570	0.318	8.70	10.0	19	12	15
C90500	1830	1570	0.315	8.72	11.0	19	11	15
C90700	1830	1528	0.317	8.78	10.2	19	10	14
C92200	1810	1520	0.312	8.65	10.0	18	14	14
C92300	1830	1570	0.317	8.80	10.0	19	12	14
C92500	1830	1570	0.317	8.85	10.0	—	11	13
C92700	1800	1550	0.317	8.80	10.1	12	11	13
C92900	1880	1575	0.320	8.79	9.5	15	9	14
C93200	1800	1570	0.322	8.93	10.0	15	12	14
C93500	1830	1570	0.320	8.87	10.0	18	15	14.5
C93700	1705	1400	0.320	8.95	10.3	12	10	11
C93800	1730	1570	0.334	9.25	10.3	13	12	10
C94300	1750	1650	0.336	9.29	10.0	16	9	11
C94700	1880	1660	0.320	8.80	11.0	14	12	15
C94700 (HT)	1880	1660	0.320	8.80	11.0	15	15	15
C94800	1880	1660	0.320	8.80	11.0	10	12	15
C95200	1913	1907	0.276	7.64	9.0	13	11	15
C95300	1913	1904	0.272	7.53	9.0	16	15	16
C95300 (HT)	1913	1904	0.272	7.53	9.0	16	13	15
C95400	1900	1880	0.269	7.45	9.0	15	13	15.5
C95400 (HT)	1900	1880	0.269	7.45	9.2	15	12	16
C95500	1930	1900	0.272	7.53	9.0	11	8.5	16
C95500 (HT)	1930	1900	0.272	7.53	9.0	11	8	17
C95800	1940	1910	0.276	7.64	9.0	9	7	16.5
C96200	2100	2010	0.323	8.94	9.5	11	11	18

[a] Cu = 226 Btu/ft^2/ft/h/F at 68°F.

[b] International Annealed Copper Standard.

CAST COPPER ALLOYS—SAE J462b

SAE Standard

Report of Nonferrous Metals Division approved June 1911 and last revised by Nonferrous Metals Committee May 1976.

1. *Scope*—This standard prescribes the chemical and mechanical requirements for a wide range of copper base casting alloys used in the automotive industry. It is not intended to cover ingot. (ASTM B30 is suggested for this purpose.)

2. *Chemical and Mechanical Properties*—The chemical composition and mechanical properties of products identified by UNS designations shall conform to the limits shown in Tables 1 and 2. Chemical analyses obtained by use of instruments, such as spectrograph, x-ray, and atomic absorption, the copper (%) may be reported as "calculated by difference." Mechanical property values are applicable to standard specimens (Sand Cast ASTM B208, Centrifugal Cast ASTM B271, Continuous Cast ASTM B505) cast under production conditions used for casting the part(s) identified by the UNS designation. Samples for chemical analysis should be taken from test bars where practical to do so.

3. *Workmanship*—Castings shall be of uniform quality, free from blowholes, porosity, hard spots, shrinkage defects or cracks, or other injurious defects.

TABLE 1—CHEMICAL COMPOSITION OF CAST COPPER ALLOYS[o]

UNS NO.[b]	Cu[c]	Sn	Pb	Zn[c]	Fe	Sb	Ni	Mn	As	S	P	Al	Si
C83600	84.0–86.0[d]	4.0–6.0	4.0–6.0	4.0–6.0	0.30	0.25	1.0[d]	—	—	0.08	0.05[e]	0.005	0.005
C83800	82.0–83.8[d]	3.3–4.2	5.0–7.0	5.0–8.0	0.30	0.25	1.0[d]	—	—	0.08	0.03[e]	0.005	0.005
C85200	70.0–74.0	0.7–2.0	1.5–3.8	20.0–27.0	0.6	0.20	1.0	—	—	0.05	0.02	0.005	0.05
C85400	65.0–70.0	0.50–1.5	1.5–3.8	24.0–32.0	0.7	—	1.0	—	—	—	—	0.35	0.05
C85800	57.0 min[f]	1.5	1.5	31.0–34.0	0.50	0.05	0.50	0.25	0.05	0.05	0.01	0.50	0.25
C86200	60.0–66.0	0.20	0.20	22.0–28.0	2.0–4.0	—	1.0	2.5–5.0	—	—	—	3.0–4.9	—
C86300	60.0–66.0	0.20	0.20	22.0–28.0	2.0–4.0	—	1.0	2.5–5.0	—	—	—	5.0–7.5	—
C86500	55.0–60.0	1.0	0.40	36.0–42.0	0.40–2.0	—	1.0	0.10–1.5	—	—	—	0.50–1.5	—
C87200	89.0 min[f]	1.0	0.50	5.0	2.5	—	—	1.5	—	—	—	1.5	1.0–5.0
C87400	79.0 min[f]	—	1.0	12.0–16.0	—	—	—	—	—	—	—	0.8	2.5–4.0
C87500	79.0 min[f]	—	0.50	12.0–16.0	—	—	—	—	—	—	—	0.5	3.0–5.0
C87800[g]	80.0 min[h]	0.25	0.15	12.0–16.0	0.15	0.05	0.20	0.15	0.05	0.05	0.01	0.15	3.8–4.2
C87900	63.0 min[f]	0.25	0.25	30.0–60.0	0.40	0.05	0.50	0.15	0.05	0.05	0.01	0.15	0.8–1.2
C90300	86.0–89.0	7.5–9.0	0.30	3.0–5.0	0.20	0.20	1.0[d]	—	—	0.05	0.05[e]	0.005	0.005
C90500	86.0–89.0[d]	9.0–11.0	0.30	1.0–3.0	0.20	0.20	1.0[d]	—	—	0.05	0.05[e]	0.005	0.005
C90700	88.0–90.0[i]	10.0–12.0	0.50	—	0.15	0.20	0.50	—	—	0.05	0.30[e]	0.005	0.005
C92200	86.0–90.0	5.5–6.5	1.0–2.0	3.0–5.0	0.25	0.25	1.0[d]	—	—	0.05	0.05[e]	0.005	0.005
C92300	85.0–89.0[d]	7.5–9.0	0.30–1.0	2.5–5.0	0.25	0.25	1.0[d]	—	—	0.05	0.05[e]	0.005	0.005
C92500	85.0–88.0	10.0–12.0	1.0–1.5	0.50	0.30	0.25	0.8–1.5	—	—	0.05	0.30[e]	0.005	0.005
C92700	86.0–89.0	9.0–11.0	1.0–2.5	0.7	0.25	0.25	1.0	—	—	0.05	0.25[e]	0.005	0.005
C92900	82.0–86.0[i]	9.0–11.0	2.0–3.2	0.25	0.20	0.25	2.8–4.0	—	—	0.05	0.50[e]	0.005	0.005
C93200	81.0–85.0[d]	6.3–7.5	6.0–8.0	2.0–4.0	0.20	0.35	1.0[d]	—	—	0.08	0.15[e]	0.005	0.005
C93500	83.0–86.0	4.3–6.0	8.0–11.0	2.0	0.20	0.30	1.0[d]	—	—	0.08	0.05[e]	0.005	0.005
C93700	78.0–82.0[d]	9.0–11.0	8.0–11.0	0.8	0.15[j]	0.55	1.0[d]	—	—	0.08	0.15[e]	0.005	0.005
C93800	75.0–79.0[d]	6.3–7.5	13.0–16.0	0.8	0.15	0.8	1.0[d]	—	—	0.08	0.05[e]	0.005	0.005
C94300	68.5–73.5[d]	4.5–6.0	22.0–25.0	0.8	0.15	0.8	1.0[d]	—	—	0.08	0.05[e]	0.005	0.005
C94700	85.0–90.0	4.5–6.0	0.10[k]	1.0–2.5	0.25	0.15	4.5–6.0	0.20	—	0.05	0.05	0.005	0.005
C94800	84.0–89.0	4.5–6.0	0.30–1.0	1.0–2.5	0.25	0.15	4.5–6.0	0.20	—	0.05	0.05	0.005	0.005
C95200	86.0 min[l]	—	—	—	2.5–4.0	—	—	—	—	—	—	8.5–9.5	—
C95300	86.0 min[l]	—	—	—	0.8–1.5	—	—	—	—	—	—	9.0–11.0	—
C95400	83.0 min[f]	—	—	—	3.0–5.0	—	2.5	0.50	—	—	—	10.0–11.5	—
C95500	78.0 min[f]	—	—	—	3.0–5.0	—	3.0–5.0	3.5	—	—	—	10.0–11.5	—
C95800	79.0 min[f]	—	0.03	—	3.5–4.5[m]	—	4.0–5.0[m]	0.8–1.5	—	—	—	8.5–9.5	0.10

| | | | | | | | | | | Nb | C | | |
| C96200 | 84.5–87.0 | — | 0.03 | — | 1.0–1.8 | — | — | 1.5 | — | 1.0 | 0.15 | 9.0–11.0 | 0.30 |

[a] These specification limits do not preclude the possible presence of other unnamed elements. However, analysis shall regularly be made only for the minor elements listed in the table plus all major elements except one. The major element which is not analyzed shall be determined by difference between the sum of those elements analyzed and 100%. By agreement between producer and consumer, analysis may be required and limits established for elements not specified.

[b] Unified Numbering System. For cross reference to SAE, former SAE, ASTM, former ASTM, and former trade names, see SAE Information Report for Wrought and Cast Copper Alloys, SAE J461.

[c] In reporting chemical analyses by the use of instruments such as spectrograph, X-ray and atomic absorption, copper may be indicated as "remainder." In reporting chemical analyses obtained by wet methods, zinc may be indicated as "remainder" on those alloys with over 2% zinc.

[d] In determining copper minimum, copper may be calculated at Cu + Ni.

[e] For continuous castings, phosphorus shall be 1.5% maximum.

[f] Total named elements shall be 99.5% minimum.

[g] Magnesium requirement is 0.01% maximum.

[h] Total named elements shall be 99.8% minimum.

[i] Cu + Sn + Pb + Ni + P shall be 99.5% minimum.

[j] The iron shall be 0.35% maximum when used for steel backed bearings.

[k] The mechanical properties of C94700 (heat treated) may not be attained if the lead content exceeds 0.01%.

[l] Total named elements shall be 99.0% minimum.

[m] Iron content shall not exceed nickel content.

[n] For welding grades, lead may not exceed 0.01%.

[o] Percent by mass (weight); maximum, unless otherwise shown.

TABLE 2—MECHANICAL PROPERTIES OF CAST COPPER ALLOYS

UNS NO.[a]	SAE Suffix[b,d]	ASTM	Casting Method[c,d] and Condition	Tensile Strength, min MPa	Tensile Strength, min ksi	Yield Strength, min MPa	Yield Strength, min ksi	Elongation, min, % in 50 mm (2 in)
						0.5% Ext. Under Load		
C83600	A	B271, B584	Sand, Centrifugal	205	30	95	14	20
C83600	B	B505	Continuous	250	36	130	19	15
C83600	C	B505	Continuous	345	50	170	15	12
C83800	A	B271, B584	Sand, Centrifugal	205	30	90	13	20
C83800	B	B505	Continuous	205	30	95	15	16
C85200		B271, B584	Sand, Centrifugal	240	35	85	12	25
C85400		B271, B584	Sand, Centrifugal	205	30	75	11	20
						0.2% Offset		
C85800		B176	Die[e]	380	55	205	30	15
C86200		B271, B505, B584	Sand, Centrifugal, Cont.	620	90	310	45	18
C86300	A	B271, B584	Sand, Centrifugal	760	110	415	60	12
C86300	B	B505	Continuous	760	110	425	62	14
C86500	A	B271, B584	Sand, Centrifugal	450	65	170	25	20
C86500	B	B505	Continuous	485	70	170	25	25
						0.5% Ext. Under Load		
C87200		B271, B584	Sand, Centrifugal	310	45	125	18	20
C87400		B271, B584	Sand, Centrifugal	345	50	145	21	18

TABLE 2—MECHANICAL PROPERTIES OF CAST COPPER ALLOYS

UNS NO.[a]	SAE Suffix[b,d]	ASTM	Casting Method[c,d] and Condition	Tensile Strength, min MPa	Tensile Strength, min ksi	Yield Strength, min MPa	Yield Strength, min ksi	Elongation, min, % in 50 mm (2 in)
						0.5% Ext. Under Load		
C87500		B271, B584	Sand, Centrifugal	415	60	165	24	16
						0.2% Offset		
C87800		B176	Die[e]	585	85	345	50	25
C87900		B176	Die[e]	485	70	240	35	25
						0.5% Ext. Under Load		
C90300	A	B271, B584	Sand, Centrifugal	275	40	125	18	20
C90300	B	B505	Continuous	305	44	150	22	18
C90500	A	B271, B584	Sand, Centrifugal	275	40	125	18	20
C90500	B	B505	Continuous	305	44	170	25	10
C90700	A		Sand	240	35	125	18	10
C90700	B	B505	Continuous	275	40	170	25	10
C92200	A	B271, B584	Sand, Centrifugal	235	34	110	16	24
C92200	B	B505	Continuous	260	38	130	19	18
C92300	A	B271, B584	Sand, Centrifugal	250	36	110	16	18
C92300	B	B505	Continuous	275	40	130	19	16
C92500	A		Sand	240	35	125	18	10
C92500	B	B505	Continuous	275	40	165	24	10
C92700	A		Sand	240	35	125	18	10
C92700	B	B505	Continuous	260	38	140	20	8
C92900		B427, B505	Sand, Continuous	310	45	170	25	8
C93200	A	B271, B584	Sand, Centrifugal	205	30	95	14	15
C93200	B	B505	Continuous	240	35	140	20	10
C93500	A	B271, B584	Sand, Centrifugal	195	28	85	12	15
C93500	B	B505	Continuous	205	30	110	16	12
C93700	A	B271, B584	Sand, Centrifugal	205	30	85	12	15
C93700	B	B505	Continuous	240	35	140	20	6
C93700	C		Continuous	275	40	170	25	6
C93800	A	B271, B584	Sand, Centrifugal	180	26	95	14	12
C93800	B	B505	Continuous	170	25	110	16	5
C94300	A	B271, B584	Sand, Centrifugal	145	21	—	—	10
C94300	B	B505	Continuous	145	21	95	15	7
C94700	A	B505, B584	Sand, Continuous	310	45	140	20	25
C94700	B	B505, B584	Sand, Continuous (HT)	515	75	345	50	5
C94800		B505, B584	Sand, Continuous	275	40	140	20	20
C95200	A	B148, B271	Sand, Centrifugal	450	65	170	25	20
C95200	B	B505	Continuous	470	68	180	26	20
C95300	A	B148, B271	Sand, Centrifugal	450	65	170	25	20
C95300	B	B505	Continuous	485	70	180	26	25
C95300	C	B148, B271, B505	Sand, Centrifugal, Cont. (HT)	550	80	275	40	12
C95400	A	B148, B271	Sand, Centrifugal	515	75	205	30	12
C95400	B	B505	Continuous	585	85	220	32	12
C95400	C	B148, B271	Sand, Centrifugal (HT)	620	90	310	45	6
C95400	D	B505	Continuous (HT)	655	95	310	45	10
C95500	A	B148, B271	Sand, Centrifugal	620	90	275	40	6
C95500	B	B505	Continuous	655	95	290	42	10
C95500	C	B148, B271	Sand, Centrifugal (HT)	760	110	415	60	5
C95500	D	B505	Continuous (HT)	760	110	425	62	8
C95800	A	B148, B271	Sand, Centrifugal	585	85	240	35	15
C95800	B	B505	Continuous (3)	620	90	260	38	18
C96200		B369	Sand	310	45	170	25	20

[a] UNIFIED NUMBERING SYSTEM. For cross reference to SAE, former SAE, former ASTM, and former trade names, see SAE Information Report for Wrought and Cast Copper Alloys, SAE J461.

[b] Suffix symbols may be specified to distinguish between two or more sets of mechanical properties, heat treatment, conditions, etc. as applicable.

[c] All alloys listed are in the "as cast" condition except those designated as heat treated (HT) and C95800 which is temper annealed.

[d] Most commonly used method of casting is shown for each alloy. However, unless the purchaser specifies the method of casting or the mechanical properties by supplement to the UNS number, the supplier may use any method which will develop the properties indicated.

[e] Mechanical properties shown for die castings are typical, not minimum.

WROUGHT COPPER AND COPPER ALLOYS—SAE J463d   SAE Standard

Report of Nonferrous Metals Division approved June 1911 and last revised by Nonferrous Metals Committee June 1976.

1. Scope—This standard* describes the chemical, mechanical, and dimensional requirements for a wide range of wrought copper and copper alloys used in the automotive and related industries.

1.1 Wrought forms covered by this standard include sheet, strip, bar, plate, rod, wire, tube, and shapes; however, form required must be specified by purchaser.

2. Chemical and Mechanical Properties—The chemical composition of products identified by the UNS designations shall conform to the limits shown in Table 1. Mechanical properties shall conform to limits shown in Table 2A (metric [SI] units) or 2B (customary units).

2.1 Products shall be of uniform quality and free from defects (such as desegregation, pipes, nonmetallic inclusions, cracks, seams, laps, buckles, and die or roll marks) detrimental to their appearance, fabrication and/or performance in service.

*Note: If none of the alloys listed herein include the characteristics required for a particular application, users are encouraged to consider alloy specifications listed in CDA Publication "Standards Handbook for Copper Alloy Wrought Mill Products," published by the Copper Development Association, 405 Lexington Avenue, New York, New York 10017, before creating specifications of their own.

2.2 Both inside and outside surfaces of tubing shall be clean and smooth.

2.3 Forgings shall not be brazed, soldered, welded, or ground to hide defects or to salvage defective products, unless specifically approved by the purchaser.

2.4 Necessary brazes in soft annealed copper wire shall be in accordance with best commercial practice.

3. Testing—Unless otherwise specified all properties stated herein are based on latest methods of test published in the ASTM Standards.

4. Dimensional Tolerances—Standard forms of products identified by the UNS designations shall conform to the dimensions specified by the purchaser, within the tolerance limits shown in Tables 4–11, the "key" for which is Table 3, "Index to Standard Product Tolerance Tables." Specified dimensions not covered by these tables shall be within the tolerance limits shown in ASTM B248 (plate, sheet, strip, and rolled bar), ASTM B249 (rod, bar, and shapes), ASTM B250 (wire), and ASTM B251 (pipe and tube). (Note: The terms "refractory" and "nonrefractory" used in Table 3 are common in the copper industry, the first applying to alloys which, because of their hardness on abrasiveness, require dimensional tolerances greater than those established for nonrefractory alloys.)

TABLE 1—CHEMICAL COMPOSITIONS OF WROUGHT COPPER ALLOYS[a]

UNS NO.[b]	% by Weight, Maximum (Except where otherwise noted)											
	Cu	Fe	Zn	Pb	Sn	Mn	Ni	Al	Si	P	Be	Other Named Elements
C10200[c]	99.9 min	—	—	—	—	—	—	—	—	—	—	—
C11000[c]	99.9 min	—	—	—	—	—	—	—	—	—	—	—
C11100[c]	99.9 min	—	—	—	—	—	—	—	—	—	—	See Note d
C11300[c,e]	99.9 min[f]	—	—	—	—	—	—	—	—	—	—	Ag, .027 min (8)[g]
C11400[c,e]	99.9 min[f]	—	—	—	—	—	—	—	—	—	—	Ag, .034 min (10)[g]
C11500[c,e]	99.9 min[f]	—	—	—	—	—	—	—	—	—	—	Ag, .054 min (16)[g]
C11600[c,e]	99.9 min[f]	—	—	—	—	—	—	—	—	—	—	Ag, .085 min (25)[g]
C12000	99.9 min	—	—	—	—	—	—	—	—	.004–.012	—	—
C12200[h]	99.9 min	—	—	—	—	—	—	—	—	.015–.040	—	—
C14500[i]	99.9 min[j]	—	—	—	—	—	—	—	—	.004–.012[k]	—	Te, .40–.60
C14700	99.9 min[l]	—	—	—	—	—	—	—	—	—	—	S, .2–.5
C15000	99.8 min	—	—	—	—	—	—	—	—	—	—	Zr, .10–.20
C16200	99.8 min	.02	—	—	—	—	—	—	—	—	—	Cd, .7–1.2
C17000	99.5 min[m]	Note n	—	—	—	—	Note n	—	—	—	1.6–1.8	Co[n]
C17200	99.5 min[m]	Note n	—	—	—	—	Note n	—	—	—	1.8–2.0	Co[n]
C17500	99.5 min[m]	.10	—	—	—	—	—	—	—	—	.40–.70	Co, 2.4–2.7
C17600	99.5 min[m]	—	—	—	—	—	—	—	—	—	.25–.50	Co, 1.4–1.7 Ag, .9–1.1
C18400	99.8 min[o]	.15	.70	—	—	—	—	—	.10	.05	—	As, .005 Cr, .40–1.2 Li, .05 Ca, .005
C18700	99.9 min[o]	—	—	.8–1.5	—	—	—	—	—	—	—	—
C19200	98.7 min	.8–1.2	—	—	—	—	—	—	—	.01–.04	—	—
C21000	94.0–96.0	.05	rem	.05	—	—	—	—	—	—	—	—
C22000	89.0–91.0	.05	rem	.05	—	—	—	—	—	—	—	—
C23000	84.0–86.0[p]	.05	rem	.05[p]	—	—	—	—	—	—	—	—
C24000	78.5–81.5	.05	rem	.05	—	—	—	—	—	—	—	—
C26000	68.5–71.5	.05	rem	.07	—	—	—	—	—	—	—	—
C26800	64.0–68.5	.05	rem	.15	—	—	—	—	—	—	—	—
C27000	63.0–68.5	.07	rem	.10	—	—	—	—	—	—	—	—
C33000	65.0–68.0	.07	rem	.20–.8[q]	—	—	—	—	—	—	—	—
C33100	65.0–68.0	.06	rem	.70–1.2	—	—	—	—	—	—	—	—
C34200	62.5–66.5	.10	rem	1.5–2.5	—	—	—	—	—	—	—	—
C34500	62.0–64.0	.10	rem	1.5–2.8	—	—	—	—	—	—	—	—
C35000	59.0–64.0[r]	.10	rem	.8–1.4	—	—	—	—	—	—	—	—
C36000	60.0–63.0	.35	rem	2.5–3.7	—	—	—	—	—	—	—	—
C37700	58.0–62.0	.30	rem	1.5–2.5	—	—	—	—	—	—	—	—
C46400	59.0–62.0	.10	rem	.20	.50–1.0	—	—	—	—	—	—	—
C46500	59.0–62.0	.10	rem	.20	.50–1.0	—	—	—	—	—	—	As, .02–.10

TABLE 1—CHEMICAL COMPOSITIONS OF WROUGHT COPPER ALLOYS[a] (continued)

UNS NO.[b]	% by Weight, Maximum (Except where otherwise noted)											
	Cu	Fe	Zn	Pb	Sn	Mn	Ni	Al	Si	P	Be	Other Named Elements
C46600	59.0–62.0	.10	rem	.20	.50–1.0	—	—	—	—	—	—	Sb, .02–.10
C46700	59.0–62.0	.10	rem	.20	.50–1.0	—	—	—	.02–.10	—	—	—
C51000	99.5 min[s]	.10	.30	.05	4.2–5.8	—	—	—	—	.03–.35	—	—
C51100	99.5 min[s]	.10	.30	.05	3.5–4.9	—	—	—	—	.03–.35	—	—
C52100	99.5 min[s]	.10	.20	.05	7.0–9.0	—	—	—	—	.03–.35	—	—
C52400	99.5 min[s]	.10	.20	.05	9.0–11.0	—	—	—	—	.03–.35	—	—
C54400	99.5 min[t]	.10	1.5–4.5	3.5–4.5	3.5–4.5	—	—	—	—	.01–.50	—	—
C60800	99.5 min[o]	.10	—	.10	—	—	—	5.0–6.5	—	—	—	As, .2–.35
C61400	99.5 min[o]	1.5–3.5	.20	.10	—	1.0	—	6.0–8.0	—	.015	—	—
C61800	99.5 min[o]	.50–1.5	.02	.02	—	—	—	8.5–11.0	.10	—	—	—
C62300	99.5 min[o]	2.0–4.0	—	—	.60	.50	1.0	8.5–11.0	.25	—	—	—
C62400	99.5 min[o]	2.0–4.5	—	—	.20	.30	—	10.0–11.5	.25	—	—	—
C63000	99.5 min[o]	2.0–4.0	.30	—	.20	1.5	4.0–5.5	9.0–11.0	.25	—	—	—
C64200	99.5 min[o]	.30	.50	.05	.20	.10	.25	6.3–7.6	1.5–2.2	—	—	As, .15
C65500	99.5 min[o]	.8	1.5	.05	—	1.5	.60	—	2.8–3.8	—	—	—
C67000	63.0–68.0	2.0–4.0	rem	.20	.50	2.5–5.0	—	3.0–6.0	—	—	—	—
C67300	58.0–63.0	.50	rem	.4–3.0	.30	2.0–3.5	.25	.25	.50–1.5	—	—	—
C67400	57.0–60.0	.35	rem	.50	.30	2.0–3.5	.25	.50–2.0	.50–1.5	—	—	—
C67500	57.0–60.0	.8–2.0	rem	.20	.50–1.5	.05–.50	—	.25	—	—	—	—
C70600	99.5 min[o]	1.0–1.8	1.0[u]	.05[u]	—	1.0	9.0–11.0	—	—	—	—	See Note u
C71000	99.5 min[o]	1.0	1.0	.05	—	1.0	19.0–23.0	—	—	—	—	—
C71500	99.5 min[o]	.40–.70	1.0[u]	.05[u]	—	1.0	29.0–33.0	—	—	—	—	See Note u
C75200	63.0–68.5	.25	rem	.10	—	.50	16.5–19.5	—	—	—	—	—
C77000	53.5–56.5	.25	rem	.10	—	.50	16.5–19.5	—	—	—	—	—

[a] These specification limits do not preclude the possible presence of other unnamed elements. However, analysis shall regularly be made only for the minor elements listed in the table, plus all major elements except one. The major element which is not analyzed shall be determined by difference between the sum of those elements analyzed and 100%. By agreement between manufacturer and purchaser, analysis may be required and limits established for elements not specified.
[b] Unified Numbering System. For cross reference to SAE, Former SAE, ASTM, and Former Trade Names, see SAE J461.
[c] These are high conductivity coppers which have in the annealed condition a minimum conductivity of 100% IACS.
[d] Small amounts of Cd or other elements may be added by agreement to improve resistance to softening at elevated temperatures.
[e] This includes Low Resistance Lake Copper and Electrolytic Copper.
[f] This includes Cu + Ag.
[g] Figures in parentheses are troy ounces per avoirdupois ton.
[h] This includes Oxygen-Free Copper which contains P in an amount agreed upon.
[i] This includes Oxygen-Free Tellurium Bearing Copper which contains P in an amount agreed upon.
[j] This includes Cu + Ag + Te.
[k] Other deoxidizers may be used as agreed upon, in which case P need not be present.
[l] This includes Cu + Ag + S.
[m] The value of Cu is exclusive of Ag.
[n] Ni + Co, 0.20% min.
Ni + Fe, + Co, 0.6% max.
[o] This includes copper plus elements with specified limits. UNS C70600 (CA706), Cu + Ag, 86.5% min and UNS C71500 (CA715), Cu + Ag, 65% min. Specific limits are defined as any numerical values, whether maximum only, minimum only or ranges.
[p] For pipe and tube, the Cu limit may be 83.0% minimum and the Pb 0.06% max.
[q] For tube over 5 in O.D., the Pb may be less than 0.20%.
[r] Copper 61.0% min for rod.
[s] This includes Cu + Sn + P.
[t] This includes Cu + Sn + P + Pb + Zn.
[u] When the product is for welding applications and so specified by the purchaser, Zn shall be 0.50% max, Pb 0.02% max, P 0.02% max, S 0.02% max, and C 0.05% max.

TABLE 2A—MINIMUM MECHANICAL PROPERTIES OF WROUGHT COPPER ALLOYS

Metric (SI) Units

UNS No.[ee]	Form	Temper	Size Section, mm Over/Thru		Tensile Strength, MPa Min	Tensile Strength, MPa Max	Yield Strength, Min MPa 0.5% Ext Under Load	Elongation, Min %[c] In 4 x Dia or Thickness of Specimen	Hardness RF[b] Min	Hardness RF[b] Max	Hardness R30T[b] Min	Hardness R30T[b] Max	Grain Size, mm Min	Grain Size, mm Max
C10200 C11000 C11100 C11300 C11400 C11500 C11600 C12000 C12200	Plate, Sheet, Strip, and Rolled Bar	Soft Anneal	—		—	—	—	—	—	65	—	—	Note a	—
		Deep-Drawing Anneal	—		—	—	—	—	30	75	—	—	Note a	0.050
		Light Cold Rolled	—		220	275	—	—	40	82	—	49	—	—
		1/2 Hard[r]	—		225	315	—	—	77	80	43	57	—	—
		Hard[r]	—		285	360	—	—	86	93	54	62	—	—
		Spring[r]	—		345	400	—	—	91	97	60	66	—	—
		Extra Spring[r]	—		360	—	—	—	92	—	61	—	—	—
		Hot Rolled	—		205	260	—	—	—	75	—	41	—	—
		Hot Rolled and Annealed	—		205	260	—	—	—	65	—	31	—	—

UNS No.	Form	Temper	Size Section, mm Over/Thru	Tensile Strength, MPa Type B Matl[u] Min	Tensile Strength Max	Yield Strength	Elongation Type B[u] Min %	Hardness Type A[s,u] Min	Hardness Max					
C10200 C11000 C12000 C12200	Rod, Bar and Shapes	Soft Anneal	All Sizes[m]	—	255	—	25	—	65	—	—	—	—	
		Hard	—/6.5[w]	345	—	—	—	68	95	—	—	—	—	
			6.5/9.5[w]	310	—	—	10	68	95	—	—	—	—	
			9.5/25[w]	275	—	—	12	68	95	—	—	—	—	
			25/50[w]	240	—	—	15	68	95	—	—	—	—	
			50/75[w]	230	—	—	15	68	95	—	—	—	—	
			4.8/9.5[x]	290	—	—	12	68	95	—	—	—	—	
			9.5/13[x]	275	—	—	12	68	95	—	—	—	—	
			13/50[x]	230	—	—	15	68	95	—	—	—	—	
			50/100[x]	220	—	—	15	68	95	—	—	—	—	
			All Sizes[y]	220	—	—	15	—	—	—	—	—	—	

UNS No.	Form	Temper	OD	Wall	Tensile Min	Tensile Max	Yield	Elongation In 50 mm			R15T[d] Min	R15T[d] Max	Grain Min	Grain Max
C10200 C12000 C12200	Tube	Soft Anneal	All Sizes	0.4/0.9	—	—	—	40[e]	—	—	—	60	0.040	—
			All Sizes	0.9/—	—	—	—	—	—	50[d]	—	—	0.040	—
		Light Anneal	All Sizes	0.4/0.9	—	—	—	—	—	—	—	65	—	0.040
			All Sizes	0.9/—	—	—	—	—	—	55[d]	—	—	—	0.040
		Light Drawn	All Sizes	All Sizes	250	325	—	—	—	—	30	60	—	—
		Drawn	All Sizes	All Sizes	250	—	—	—	—	—	30	—	—	—
		Hard Drawn	—/25	0.5/3.0	310	—	—	—	—	—	55	—	—	—
			25/50	0.4/4.5	310	—	—	—	—	—	55	—	—	—
			50/100	1.5/6.5	310	—	—	—	—	—	55	—	—	—

UNS No.	Form	Temper	Size Section, mm Over/Thru	Tensile Min	Tensile Max	Yield	Elongation In 250 mm	RF[b] Min	RF[b] Max	R30T[b] Min	R30T[b] Max	Grain Min	Grain Max
C10200 C11000	Wire	Annealed	0.08/0.25	—	—	—	15	—	—	—	—	—	—
			0.25/0.50	—	—	—	20	—	—	—	—	—	—
			0.50/2.5	—	—	—	25	—	—	—	—	—	—
			2.5/7.5	—	—	—	30	—	—	—	—	—	—
			7.5/12	—	—	—	35	—	—	—	—	—	—

UNS No.	Form	Temper	Size Section, mm Over/Thru	Tensile Min	Tensile Max	Yield	Elongation In 4 x Dia						
C14500	Rod	1/2 Hard[f]	1.6/6.5	260	—	205	8	—	—	—	—	—	—
			6.5/65	260	—	205	12	—	—	—	—	—	—
		Hard	1.6/6.5	330	—	275	4	—	—	—	—	—	—
			6.5/30	305	—	260	8	—	—	—	—	—	—
			30/50	275	—	240	8	—	—	—	—	—	—
C14700	Rod	1/2 Hard[f]	1.6/6.5		—	205	8	—	—	—	—	—	—
			6.5/65		—	205	12	—	—	—	—	—	—
		Hard[g]	1.6/6.5		—	275	4	—	—	—	—	—	—
			6.5/30		—	260	8	—	—	—	—	—	—
			30/50		—	240	8	—	—	—	—	—	—
C15000	Round Rod												

TABLE 2A—MINIMUM MECHANICAL PROPERTIES OF WROUGHT COPPER ALLOYS (continued)

Metric (SI) Units

UNS No.[ee]	Form	Temper	Size Section, mm Over/Thru	Tensile Strength, MPa Min	Tensile Strength, MPa Max	Yield Strength, Min MPa 0.5% Ext Under Load	Elongation, Min %[c] In 50 mm	Hardness Min	Hardness Max	Hardness Min	Hardness Max	Grain Size, mm Min	Grain Size, mm Max
				Type B Matl[u]				RB		R30T[b]			
C16200	Round Rod	Drawn	—/25 25/50 50/75	415 380 345	— — —	— — —	20 25 25	65 60 55	— — —	— — —	— — —	— — —	— — —
	Square, Rectangular and Hex Rod and Bar	Drawn	—/25 25/—	415 345	— —	— —	20 20	55 50	— —	— —	— —	— —	— —
	Forging	As Forged	—/25 25/—	415 345	— —	— —	20 25	55 50	— —	— —	— —	— —	— —

UNS No.[ee]	Form	Temper	Size Section, mm Over/Thru	Tensile Strength, MPa Min	Tensile Strength, MPa Max	Yield Strength, Min MPa 0.2% Offset	Elongation, Min %[c] In 50 mm[n]	Hardness Min	Hardness Max	Hardness Min	Hardness Max	Elec Cond % IACS Min	Heat Treat h at 315°C
								RB[o,q]		R30T[p,q]			
C17000	Strip	A Soft[l]	All Sizes	415	540	—	35	45	78	46	67	17	—
		1/4 Hard[l] 1/2 Hard[l] Hard[l]	All Sizes All Sizes All Sizes	515 585 690	605 690 825	— — —	10 5 2	68 88 96	90 96 102	62 74 79	75 79 83	16 15 15	— — —
								RC[o,q]		R30N[p,q]			
		AT[y] 1/4 HT[y] 1/2 HT[y] HT[y]	All Sizes All Sizes All Sizes All Sizes	1035 1100 1170 1240	— — — —	885 930 1000 1070	3 2.5 1 1	33 35 37 39	— — — —	53 55 56 59	— — — —	22 22 22 22	3 2 2 2
		AM[z]	All Sizes	690	795	485 min 655 max	18	18	23	37	44	23	—
		1/4 HM[z]	All Sizes	760	860	550 min 725 max	15	21	26	42	47	23	—
		1/2 HM[z]	All Sizes	830	930	655 min 795 max	12	25	30	46	50	24	—
		HM[z]	All Sizes	930	1030	760 min 930 max	9	30	35	50	55	25	—
		XHM[z]	All Sizes	1100	1200	930 min 1100 max	2	32	36	52	56	24	—
								RB[o,q]		R30T[p,q]			
C17200	Strip	A (Soft)[l]	All Sizes	415	535	—	35	45	78	46	67	17	—
		1/4 Hard[l] 1/2 Hard[l] Hard[l]	All Sizes All Sizes All Sizes	515 585 690	605 690 825	— — —	10 5 2	68 88 96	90 96 102	62 74 79	75 79 83	16 15 15	— — —
								RC[o,q]		R30N[p,q]			
		AT[y]	All Sizes	1140	1350	—	—	36	—	56	—	22	3
		1/4 Hard[y] 1/2 Hard[y] HT[y]	All Sizes All Sizes All Sizes	1200 1770 1310	1410 1480 1520	— — —	— — —	38 39 40	— — —	58 59 60	— — —	22 22 22	2 2 2
	Rod and Bar							RB[o,q]					
		A (Soft)[l]	All Sizes	415	585	—	—	45	85	—	—	17	—
		Hard[l]	—/9.5 9.5/25 25/—	655 620 585	895 825 795	— — —	— — —	92 91 88	103 102 104	— — —	— — —	15 15 15	— — —
								RC[o,q]					
		AT[y]	All Sizes	1140	1310	—	—	36	40	—	—	22	3
		HT[y]	—/9.5 9.5/25 25/—	1280 1240 1200	1480 1450 1410	— — —	— — —	39 38 37	45 44 43	— — —	— — —	22 22 22	3 2 2

TABLE 2A—MINIMUM MECHANICAL PROPERTIES OF WROUGHT COPPER ALLOYS (continued)

Metric (SI) Units

UNS No.[ee]	Form	Temper	Size Section, mm Over/Thru	Tensile Strength, MPa Min	Tensile Strength, MPa Max	Yield Strength, Min MPa 0.2% Offset	Elongation, Min %[c] In 50 mm	Hardness Min RB[o,q]	Hardness Max RB[o,q]	Hardness Min R30T[p,q]	Hardness Max R30T[p,q]	Elec Cond, %IACS Min	Heat Treat h at 315°C
C17500	Strip and Plate	A (Soft)[l]	All Sizes	—	380	140 min 205 max	20	20	45	29	45	20	—
		1/2 Hard[l]	All Sizes	415	515	345 min 485 max	5	65	76	60	67	25	—
		Hard[l]	All Sizes	485	585	415 min 550 max	2	78	88	69	75	25	—
		AT[y]	All Sizes	690	825	550 min 690 max	8	92	100	77	82	45	—
		1/2 HT[y] HT[y]	All Sizes	760	895	655 min 825 max	5	95	102	79	83	48	—
	Hot Worked Sizes, Forgings	A (Soft)[l]	All Sizes	—	380	140 min 205 max	20	20	45	—	—	20	—
		AT[y]	All Sizes	690	825	550 min 690 max	10	92	100	—	—	45	—
C17500 C17600	Rod, Bar Shapes and Tubing	A (Soft)[l]	All Sizes	240	380	140 min 205 max	20	25	45	—	—	20	—
		1/2 Hard	All Sizes	450	585	380 min 515 max	10	60	75	—	—	20	—
		AT[y]	All Sizes	690	825	550 min 690 max	10	92	100	—	—	45	—
		1/2 HT[y]	All Sizes	760	895	690 min 825 max	8	92	102	—	—	48	—

UNS No.[ee]	Form	Temper	Size Section, mm Over/Thru	Tensile Strength, MPa Min	Tensile Strength, MPa Max	Yield Strength, Min MPa 0.5% Ext Under Load	Elongation, Min %[c] In 50 mm	Hardness Min RB	Hardness Max RB	Hardness Min R30T[b]	Hardness Max R30T[b]	Grain Size, mm Min	Grain Size, mm Max
C18400	Round Rod	Drawn	—/25 25/50 50/75	450 415 380	— — —	— — —	15 15 15	75 70 65	— — —	— — —	— — —	— — —	— — —
	Hex Rod and Bar	Drawn	—/25 25/—	450 380	— —	— —	15 15	70 65	— —	— —	— —	— —	— —
	Forgings	As Forged	—/25 25/50 50/—	450 380 380	— — —	— — —	15 15 15	72 70 65	— — —	— — —	— — —	— — —	— — —
C18700	Rod	1/2 Hard[f]	1.6/6.5 6.5/65	260 260	— —	205 205	8 12	— —	— —	— —	— —	— —	— —
		Hard	1.6/6.5 6.5/30 30/50	330 305 275	— — —	275 260 240	4 8 8	— — —	— — —	— — —	— — —	— — —	— — —
C19200	Tube	A (Soft) Light Drawn	All Sizes All Sizes	260 275	— —	80 240	— —	— —	— —	— —	— —	— —	— —
C21000	Plate, Sheet, Strip and Rolled Bar[t]	Annealed 0.050 mm 0.035 mm 0.025 mm 0.015 mm		— — — —	— — — —	— — — —	— — — —	RF[b] 40 47 50 54	52 54 61 65	— — 1 7	4 7 17 23	0.035 0.025 0.015 Note 1	0.090 0.050 0.035 0.025
		1/4 Hard	0.50/0.90 0.90/— 0.30/0.70 0.70/—	255	325	—	—	RB[b] 20 24 — —	48 52 — —	— — 34 37	— — 51 54	— — — —	— — — —

TABLE 2A—MINIMUM MECHANICAL PROPERTIES OF WROUGHT COPPER ALLOYS (continued)

Metric (SI) Units

UNS No.[ee]	Form	Temper	Size Section, mm Over/Thru	Tensile Strength, MPa Min	Tensile Strength, MPa Max	Yield Strength, Min MPa 0.5% Ext Under Load	Elongation, Min %[c] In 50 mm	Hardness Min	Hardness Max	Hardness Min	Hardness Max	Grain Size, mm Min	Grain Size, mm Max
								RB[b]	RB[b]	R30T[b]	R30T[b]		
C21000	Plate, Sheet, Strip and Rolled Bar[t]	1/2 Hard	0.50/0.90 0.90/— 0.30/0.70 0.70/—	290	360	—	—	40 44 — —	56 60 — —	— — 46 48	— — 57 59	— — — —	— — — —
		3/4 Hard	0.50/0.90 0.90/—	315	385	—	—	50 53	61 64	— —	— —	— —	— —
		3/4 Hard	0.30/0.70 0.70/—	315	385	—	In 4 x Dia or Thickness of Specimen —	— —	— —	52 54	60 62	— —	— —
		Hard	0.50/0.90 0.90/— 0.30/0.70 0.70/—	345	405	—	—	57 60 — —	64 67 — —	— — 57 59	— — 62 64	— — — —	— — — —
		Extra Hard	0.50/0.90 0.90/— 0.30/0.70 0.70/—	385	440	—	—	64 66 — —	70 72 — —	— — 62 63	— — 66 67	— — — —	— — — —
		Springs	0.50/0.90 0.90/— 0.30/0.70 0.70/—	415	470	—	—	68 70 — —	73 75 — —	— — 64 65	— — 68 69	— — — —	— — — —
		Extra Spring	0.50/0.90 0.90/— 0.30/0.70 0.70/—	420	475	—	—	69 71 — —	74 76 — —	— — 65 66	— — 69 70	— — — —	— — — —
C22000	Plate, Sheet Strip and Rolled Bar[t]							RF[b]	RF[b]				
		Annealed 0.050 mm 0.035 mm 0.025 mm 0.015 mm		—	—	—	—	50 54 58 62	60 64 70 75	1 7 13 19	16 21 31 39	0.035 0.025 0.015 Note 1	0.090 0.050 0.035 0.025
								RB[b]	RB[b]				
		1/4 Hard	0.50/0.90 0.90/— 0.30/0.70 0.70/—	275	345	—	—	27 31 — —	52 56 — —	— — 38 41	— — 53 56	— — — —	— — — —
		1/2 Hard	0.50/0.90 0.90/— 0.30/0.70 0.70/—	325	395	—	—	50 53 — —	63 66 — —	— — 52 54	— — 61 63	— — — —	— — — —
		3/4 Hard	0.50/0.90 0.90/— 0.30/0.70 0.70/—	360	425	—	—	59 62 — —	68 71 — —	— — 58 60	— — 64 66	— — — —	— — — —
		Hard	0.50/0.90 0.90/— 0.30/0.70 0.70/—	395	455	—	—	65 68 — —	72 75 — —	— — 62 64	— — 66 68	— — — —	— — — —
		Extra Hard	0.50/0.90 0.90/— 0.30/0.70 0.70/—	440	495	—	—	72 74 — —	77 79 — —	— — 67 68	— — 71 72	— — — —	— — — —
		Spring	0.50/0.90 0.90/— 0.30/0.70 0.70/—	475	530	—	—	76 78 — —	79 81 — —	— — 70 71	— — 72 73	— — — —	— — — —
		Extra Spring	0.50/0.90 0.90/— 0.30/0.70 0.70/—	495	550	—	—	78 80 — —	81 83 — —	— — 71 72	— — 73 74	— — — —	— — — —

TABLE 2A—MINIMUM MECHANICAL PROPERTIES OF WROUGHT COPPER ALLOYS (continued)

Metric (SI) Units

UNS No.[ee]	Form	Temper	Size Section, mm Over/Thru		Tensile Strength, MPa Min	Tensile Strength, MPa Max	Yield Strength, Min MPa 0.5% Ext Under Load	Elongation, Min %[c] In 4 x Dia or Thickness of Specimen	Hardness Min	Hardness Max	Hardness Min	Hardness Max	Grain Size, mm Min	Grain Size, mm Max
C22000	Tube		Wall Thickness						RF[b]	RF[b]	R30T[b]	R30T[b]		
		Soft Anneal	—/1.1		—	—	—	—	—	—	—	30	0.025	0.060
			1.1/—		—	—	—	—	—	70	—	—	0.025	0.060
		Light Anneal	—/1.1		—	—	—	—	—	—	—	37	Note a	0.035
			1.1/—		—	—	—	—	—	78	—	—	Note a	0.035
		Drawn (General Purpose)	All Sizes		275	—	—	—	—	—	38	—	—	—
			OD	Wall										
		Hard Drawn	—/100	0.5/6.5	360	—	—	—	—	—	55	—	—	—
C23000	Sheet and Strip		Size Section, mm Over/Thru											
		Annealed												
		0.070 mm	—		—	—	—	—	53	60	6	16	0.050	0.100
		0.050 mm	—		—	—	—	—	56	63	10	20	0.035	0.070
		0.035 mm	—		—	—	—	—	58	66	13	24	0.025	0.050
		0.025 mm	—		—	—	—	—	60	72	16	34	0.015	0.035
		0.015 mm	—		—	—	—	—	62	79	19	48	Note a	0.025
									RB[b]	RB[b]				
		1/4 Hard	0.50/0.90		305	370	—	—	33	58	—	—	—	—
			0.90/—						37	62	—	—	—	—
			0.30/0.70						—	—	42	57	—	—
			0.70/—						—	—	45	60	—	—
		1/2 Hard	0.50/0.90		350	420	—	—	56	68	—	—	—	—
			0.90/—						59	71	—	—	—	—
			0.30/0.70						—	—	56	64	—	—
			0.70/—						—	—	58	66	—	—
		3/4 Hard	0.50/0.90		395	460	—	—	66	73	—	—	—	—
			0.90/—						69	76	—	—	—	—
			0.30/0.70						—	—	63	68	—	—
			0.70/—						—	—	65	70	—	—
		Hard	0.50/0.90		435	495	—	—	72	78	—	—	—	—
			0.90/—						74	80	—	—	—	—
			0.30/0.70						—	—	67	71	—	—
			0.70/—						—	—	68	72	—	—
		Extra Hard	0.50/0.90		495	550	—	—	78	83	—	—	—	—
		Extra Hard	0.90/—		495	550	—	—	80	85	—	—	—	—
			0.30/0.70						—	—	70	74	—	—
			0.70/—						—	—	71	75	—	—
		Spring	0.50/0.90		540	595	—	—	82	85	—	—	—	—
			0.90/—						84	87	—	—	—	—
			0.30/0.70						—	—	74	76	—	—
			0.70/—						—	—	75	77	—	—
		Extra Spring	0.50/0.90		565	620	—	—	84	87	—	—	—	—
			0.90/—						86	89	—	—	—	—
			0.30/0.70						—	—	75	77	—	—
			0.70/—						—	—	76	78	—	—
	Tube		Wall Thickness						RF[d]	RF[d]	R30T[d]	R30T[d]		
		Soft Anneal	—/1.1		—	—	—	—	—	—	—	36	0.025	0.060
			1.1/—		—	—	—	—	—	75	—	—	0.025	0.060
		Light Anneal	—/1.1		—	—	—	—	—	—	—	39	Note a	0.035
			1.1/—		—	—	—	—	—	85	—	—	Note a	0.035
		Light Drawn[g]	All Sizes		305	400	—	—	—	—	43	75	—	—
		Drawn (General Purpose)	All Sizes		305	—	—	—	—	—	43	—	—	—

TABLE 2A—MINIMUM MECHANICAL PROPERTIES OF WROUGHT COPPER ALLOYS (continued)

Metric (SI) Units

UNS No.[ee]	Form	Temper	Size Section, mm Over/Thru		Tensile Strength, MPa Min	Tensile Strength, MPa Max	Yield Strength, Min MPa 0.5% Ext Under Load	Elongation, Min %[c] In 4 × Dia or Thickness of Specimen	Hardness Min	Hardness Max	Hardness Min	Hardness Max	Grain Size, mm Min	Grain Size, mm Max
			OD	Wall					RF[b]		R30T[d]			
C23000	Tube	Hard Drawn	—/100	0.5/6.5	395	—	—	—	—	—	65	—	—	—
			Size Section, mm Over/Thru											
C24000	Sheet and Strip	Annealed 0.070 mm 0.050 mm 0.035 mm 0.025 mm 0.015 mm	— — — — —		— — — — —	— — — — —	— — — — —	— — — — —	53 57 61 63 66	64 67 72 77 83	2 8 16 20 25	21 27 35 42 50	0.050 0.035 0.025 0.015 Note a	0.120 0.070 0.050 0.035 0.025
									RB[b]					
		1/4 Hard	0.50/0.90 0.90/— 0.30/0.70 0.70/—		330	400	—	—	38 42 — —	61 65 — —	— — 42 45	— — 57 60	— — — —	— — — —
		1/2 Hard	0.50/0.90 0.90/— 0.30/0.70 0.70/—		380	450	—	—	59 62 — —	70 73 — —	— — 56 58	— — 64 66	— — — —	— — — —
		3/4 Hard	0.50/0.90 0.90/— 0.30/0.70 0.70/—		420	490	—	—	69 72 — —	76 79 — —	— — 63 65	— — 68 70	— — — —	— — — —
		Hard	0.50/0.90 0.90/— 0.30/0.70 0.70/—		470	530	—	—	76 78 — —	82 84 — —	— — 68 69	— — 72 73	— — — —	— — — —
		Extra Hard	0.50/0.90 0.90/— 0.30/0.70 0.70/—		540	600	—	—	83 85 — —	87 89 — —	— — 72 73	— — 75 76	— — — —	— — — —
		Spring	0.50/0.90 0.90/— 0.30/0.70 0.70/—		565	640	—	—	87 89 — —	90 92 — —	— — 75 76	— — 77 78	— — — —	— — — —
		Extra Spring	0.50/0.90 0.90/— 0.30/0.70 0.70/—		615	670	—	—	88 90 — —	91 93 — —	— — 76 77	— — 78 79	— — — —	— — — —
									RF[d]					
C26000	Plate[t], Sheet, Strip, Rolled Bar and Wire	Annealed 0.120 mm 0.070 mm 0.050 mm 0.035 mm 0.025 mm 0.015 mm	— — — — — —		— — — — — —	— — — — — —	— — — — — —	— — — — — —	50 52 61 65 67 72	62 67 73 76 79 85	— 3 20 25 27 33	21 27 35 38 42 50	0.070 0.050 0.035 0.025 0.015 Note 1	— 0.120 0.070 0.050 0.035 0.025
									RB[b]					
		1/4 Hard	0.50/0.90 0.90/— 0.30/0.70 0.70/—		340	405	—	—	40 44 — —	61 65 — —	— — 43 46	— — 57 60	— — — —	— — — —
		1/2 Hard	0.50/0.90 0.90/— 0.30/0.70 0.70/—		395	460	—	—	60 63 — —	74 77 — —	— — 56 58	— — 66 68	— — — —	— — — —
		3/4 Hard	0.50/0.90 0.90/— 0.30/0.70 0.70/—		440	510	—	—	72 75 — —	70 82 — —	— — 65 67	— — 70 72	— — — —	— — — —

TABLE 2A—MINIMUM MECHANICAL PROPERTIES OF WROUGHT COPPER ALLOYS (continued)

Metric (SI) Units

UNS No.[ee]	Form	Temper	Size Section, mm Over/Thru	Tensile Strength, MPa Min	Tensile Strength, MPa Max	Yield Strength, Min MPa 0.5% Ext Under Load	Elongation, Min %[c] In 4 x Dia or Thickness of Specimen	Hardness Min	Hardness Max	Hardness Min	Hardness Max	Grain Size, mm Min	Grain Size, mm Max
								RB[b]	RB[b]	R30T[d]	R30T[d]		
C26000	Plate[t], Sheet, Strip, Rolled Bar and Wire	Hard	0.50/0.90 0.90/— 0.30/0.70 0.70/—	490	560	—	—	70 81 — —	84 86 — —	— — 70 71	— — 73 74	— — — —	— — — —
		Extra Hard	0.50/0.90 0.90/— 0.30/0.70 0.70/—	570	635	—	—	85 87 — —	89 91 — —	— — 74 75	— — 76 77	— — — —	— — — —
		Spring	0.50/0.90 0.90/— 0.30/0.70 0.70/—	625	690	—	—	89 90 — —	92 93 — —	— — 76 76	— — 78 78	— — — —	— — — —
		Extra Spring	0.50/0.90 0.90/— 0.30/0.70 0.70/—	655	715	—	—	91 92 — —	94 95 — —	— — 77 77	— — 79 79	— — — —	— — — —
	Tube		Wall Thickness					RF[b]	RF[b]				
		Soft Anneal	—/0.75 0.75/—	—	—	—	—	— —	— 80	— —	40 —	0.025 0.025	0.060 0.060
		Light Anneal	—/0.75 0.75/—	—	—	—	—	— —	— 90	— —	60 —	Note a Note a	0.035 0.035
		Drawn (General Purpose)	All Sizes	370	—	—	—	—	—	53	—	—	—
			OD / Wall										
		Hard Drawn	—/100 / 0.5/6.5	455	—	—	—	—	—	70	—	—	—
C26800	Plate[t], Sheet, Strip and Rolled Bar		Size Section, mm Over/Thru					RF[d]	RF[d]	R30T[b]	R30T[b]		
		Annealed 0.120 mm 0.070 mm 0.050 mm 0.035 mm 0.025 mm 0.015 mm	—	—	—	—	—	50 52 61 65 67 72	62 67 73 76 79 85	— 3 20 25 27 33	21 27 35 38 42 50	0.070 0.050 0.035 0.025 0.015 Note a	— 0.120 0.070 0.050 0.035 0.025
								RB[b]	RB[b]				
		1/4 Hard	0.50/0.90 0.90/— 0.30/0.70 0.70/—	340	405	—	—	40 44 — —	61 65 — —	— — 43 46	— — 57 60	— — — —	— — — —
		1/2 Hard	0.50/0.90 0.90/— 0.30/0.70 0.70/—	380	450	—	—	57 60 — —	71 74 — —	— — 54 56	— — 64 66	— — — —	— — — —
		3/4 Hard	0.50/0.90 0.90/— 0.30/0.70 0.70/—	425	495	—	—	70 73 — —	77 80 — —	— — 65 67	— — 69 71	— — — —	— — — —
		Hard	0.50/0.90 0.90/— 0.30/0.70 0.70/—	470	540	—	—	76 78 — —	82 84 — —	— — 68 69	— — 72 73	— — — —	— — — —
								RF[b]	RF[b]				
		Extra Hard	0.50/0.90 0.90/— 0.30/0.70 0.70/—	545	615	—	—	83 85 — —	87 90 — —	— — 73 74	— — 75 76	— — — —	— — — —

TABLE 2A—MINIMUM MECHANICAL PROPERTIES OF WROUGHT COPPER ALLOYS (continued)

Metric (SI) Units

UNS No.[ee]	Form	Temper	Size Section, mm Over/Thru	Tensile Strength, MPa Min	Tensile Strength, MPa Max	Yield Strength, Min MPa 0.5% Ext Under Load	Elongation, Min %[c] In 4 × Dia or Thickness of Specimen	Hardness Min	Hardness Max	Hardness Min	Hardness Max	Grain Size, mm Min	Grain Size, mm Max
								RF[b]	RF[b]	R30T[b]	R30T[b]		
C26800	Plate, Sheet, Strip and Rolled Bar	Spring	0.50/0.90 0.90/— 0.30/0.70 0.70/—	595	655	—	—	87 89 — —	90 92 — —	— — 75 76	— — 77 78	—	—
		Extra Spring	0.50/0.90 0.90/— 0.30/0.70 0.70/—	620	685	—	—	88 90 — —	91 93 — —	— — 76 77	— — 78 79	—	—
C27000	Wire	Annealed 0.100 mm 0.070 mm 0.050 mm 0.035 mm 0.025 mm 0.015 mm		— — — — — —	— — — — — —	— — — — — —	— — — — — —	— — — — — —	— — — — — —	— — — — — —	— — — — — —	0.070 0.050 0.035 0.025 0.015 Note a	— 0.100 0.070 0.050 0.035 0.025
		1/8 Hard 1/4 Hard 1/2 Hard 3/4 Hard Hard[h] Extra Hard[i]	—	345 425 545 635 705 795	450 530 650 740 805 890	—	—	—	—	—	—	—	—
		Spring[j]	—	825	—	—	—	—	—	—	—	—	—
			Wall Thickness							R30T[d]	R30T[d]		
		Soft Anneal	—/0.75 0.75/—	— —	— —	— —	— —	— —	— 80	— —	40 40	0.025 0.025	0.060 0.060
C33000	Tube	Light Anneal	—/0.75 0.70/—	— —	— —	— —	— —	— —	— 90	— —	60 —	Note a Note a	0.035 0.035
		Drawn (General Purpose)	All Sizes	370	—	—	—	—	—	53	—	—	—
			OD	Wall				RB[b]					
		Hard Drawn[g]	—/100	0.50/6.5	455	—	—	—	—	—	70	—	—	—
			Wall Thickness					RF[d]	RF[d]				
		Soft Anneal	—/0.75 0.75/—	— —	— —	— —	— —	— —	— 80	— —	40 40	0.025 0.025	0.060 0.060
C33100	Tube	Light Anneal	—/0.75 0.75/—	— —	— —	— —	— —	— —	— 90	— —	60 —	Note a Note a	0.035 0.035
		Drawn (General Purpose)	All Sizes	370	—	—	—	—	—	53	—	—	—
			OD	Wall									
		Hard Drawn	—/100	0.50/6.5	455	—	—	—	—	—	70	—	—	—
			Over/Thru					RF[b]	RF[b]	R30T[b]	R30T[b]		
C34200 C35000	Plate, Sheet, Strip and Rolled Bar	Annealed 0.070 mm 0.050 mm 0.035 mm 0.025 mm	All Sizes All Sizes All Sizes All Sizes	— — — —	— — — —	— — — —	— — — —	54 61 65 67	67 73 76 79	12 20 25 27	27 35 38 42	0.050 0.035 0.025 0.015	0.100 0.070 0.050 0.035
								RB[b]	RB[b]				
		1/4 Hard 1/2 Hard Hard Extra Hard	All Sizes All Sizes All Sizes All Sizes	340 380 470 545	405 450 540 615	— — — —	— — — —	40 57 76 83	65 74 84 89	43 54 68 73	60 66 73 76	— — — —	— — — —
C34500 C35000	Rod	Soft	—/12.5 12.5/25 25/50	315 305 275	— — —	110 105 105	20 25 30	— — —	— 45 45	— — —	— — —	— — —	— — —

TABLE 2A—MINIMUM MECHANICAL PROPERTIES OF WROUGHT COPPER ALLOYS (continued)

Metric (SI) Units

UNS No.[ee]	Form	Temper	Size Section, mm Over/Thru		Tensile Strength, MPa Min	Tensile Strength, MPa Max	Yield Strength, Min MPa 0.5% Ext Under Load	Elongation, Min %[c] In 4 x Dia or Thickness of Specimen	Hardness RB[b] Min	Hardness RB[b] Max	Hardness R30T[b] Min	Hardness R30T[b] Max	Grain Size, mm Min	Grain Size, mm Max
C34500 C35000	Rod	1/4 Hard	—/12.5 12.5/25 25/50		360 345 290	— — —	170 140 105	10 15 20	50 40 35	75 70 65	— — —	— — —	— — —	— — —
		1/2 Hard	—/12.5 12.5/25 25/50		395 380 345	— — —	170 170 140	7 10 15	60 55 40	80 75 70	— — —	— — —	— — —	— — —
C36000	Rod	Soft	—/25 25/50 50/—		330 305 275	— — —	140 125 105	15 20 25	— — —	— — —	— — —	— — —	— — —	— — —
		1/2 Hard	—/12.5 12.5/25 25/50 50/100 100/—		395 380 345 310 275	— — — — —	170 170 140 105 105	7 10 15 20 20	— — — — —	— — — — —	— — — — —	— — — — —	— — — — —	— — — — —
		Hard	1.5/5.0 5.0/12.5 12.5/20		550 480 450	— — —	310 240 205	— 4 6	— — —	— — —	— — —	— — —	— — —	— — —
	Flat Products		Thickness	Width										
		Soft	—/25 25/—	—/150 —/150	305 275	— —	125 105	20 25	— —	— —	— —	— —	— —	— —
		1/2 Hard	—/12.5 —/12.5 12.5/50 12.5/50 50/—	—/25 25/150 —/50 50/150 50/100	345 310 310 275 275	— — — — —	170 115 115 105 105	10 15 20 20 20	— — — — —	— — — — —	— — — — —	— — — — —	— — — — —	— — — — —
C37700	Die Forgings	As Forged	Over/Thru —/38 38/—		345 315	— —	125 105	25 50	— —	— —	— —	— —	— —	— —
C46400 C46500 C46600 C46700	Rod and Bar	Soft	—/25 25/50 50/—		370 360 345	— — —	140 140 140	30 30 30	— — —	— — —	— — —	— — —	— — —	— — —
		1/2 Hard or Light Annealed	—/12.5 12.5/25 25/50 50/75 75/100 100/—		415 415 400 270 270 270	— — — — — —	185 185 180 170 150 150	22 25 25 25 27 30	— — — — — —	— — — — — —	— — — — — —	— — — — — —	— — — — — —	— — — — — —
		Hard	—/25 25/50		460 425	— —	310 255	13 18	— —	— —	— —	— —	— —	— —
	Shapes	As Extruded	All Sizes		360	—	140	30	—	—	—	—	—	—
C51000	Sheet and Strip	Soft	1.0/— 0.75/— 0.50/1.0 0.25/0.75		295	400	—	—	16 — 12 —	64 — 60 —	32 — — 24	59 — — 53	— — — —	— — — —
		1/2 Hard	1.0/— 0.75/— 0.50/1.0 0.25/0.75		400	505	—	—	64 — 60 —	85 — 82 —	— 59 — 53	— 73 — 69	— — — —	— — — —
		Hard	1.0/— 0.75/— 0.50/1.0 0.25/0.75		525	625	—	—	86 — 84 —	93 — 91 —	— 73 — 71	— 78 — 75	— — — —	— — — —
		Extra Hard	1.0/— 0.75/— 0.50/1.0 0.25/0.75		605	710	—	—	92 — 89 —	96 — 95 —	— 77 — 74	— 81 — 78	— — — —	— — — —
		Spring	1.0/— 0.75/— 0.50/1.0 0.25/0.75		655	760	—	—	94 — 92 —	98 — 97 —	— 79 — 76	— 82 — 80	— — — —	— — — —

TABLE 2A—MINIMUM MECHANICAL PROPERTIES OF WROUGHT COPPER ALLOYS (continued)

Metric (SI) Units

UNS No.[ee]	Form	Temper	Size Section, mm Over/Thru	Tensile Strength, MPa Min	Tensile Strength, MPa Max	Yield Strength, Min MPa 0.5% Ext Under Load	Elongation, Min %[c] In 4 x Dia or Thickness of Specimen	Hardness RB[b] Min	Hardness RB[b] Max	Hardness R30T[b] Min	Hardness R30T[b] Max	Grain Size, mm Min	Grain Size, mm Max
C51000	Sheet and Strip	Extra Spring	1.0/— 0.75/— 0.50/1.0 0.25/0.75	690	785	—	—	95 — 94 —	99 — 98 —	— 80 — 77	— 83 — 81	— — — —	— — — —
	Rod	Soft	—/6.5	275	400	—	—	—	—	—	—	—	—
		Hard	—/6.5 6.5/12.5 12.5/25 25/— 6.5/9.5 9.5/—	550 485 415 380 415 380	885 — — — — —	—	— 13 15 18 10 15	—	—	—	—	—	—
		Spring	—/0.65[aa]	860	—	—	—	—	—	—	—	—	—
		Spring	0.65/1.6[aa] 1.6/3.2[aa] 3.2/6.5[aa] 6.5/9.5[aa] 9.5/12.5[aa]	795 760 725 690 620	— — — — —	—	— — 3.5 5.0 9.0	—	—	—	—	—	—
	Wire for General Purposes	Soft	—	295	400	—	—	—	—	—	—	—	—
		1/4 Hard 1/2 Hard 3/4 Hard Hard	— — — —	415 550 660 745	525 670 795 885	—	—	—	—	—	—	—	—
	Wire for Spring Purposes	—	—/0.65 0.65/1.6 1.6/3.2	1000 930 845	— — —	—	In 50 mm — — —	—	—	—	—	—	—
		—	3.2/6.5 6.5/9.5 9.5/12.5	860 825 720	— — —	—	— 5.0 9.0	—	—	—	—	—	—
C51100 C54400	Sheet and Strip	Soft	1.0/— 0.75/— 0.50/1.0 0.25/0.75	275	380	—	—	7 — 0 —	50 — 45 —	— 24 — 16	— 50 — 46	— — — —	— — — —
		1/2 Hard	1.0/— 0.75/— 0.50/1.0 0.25/0.75	380	485	—	—	60 — 53 —	81 — 78 —	— 57 — 52	— 73 — 71	— — — —	— — — —
		Hard	1.0/— 0.75/— 0.50/1.0 0.25/0.75	495	600	—	—	82 — 80 —	90 — 88 —	— 71 — 69	— 77 — 75	— — — —	— — — —
		Extra Hard	1.0/— 0.75/— 0.50/1.0 0.25/0.75	580	685	—	—	88 — 86 —	94 — 92 —	— 75 — 73	— 80 — 78	— — — —	— — — —
		Spring	1.0/— 0.75/— 0.50/1.0 0.25/0.75	625	725	—	—	90 — 88 —	96 — 94 —	— 77 — 75	— 81 — 79	— — — —	— — — —
		Extra Spring	1.0/— 0.75/— 0.50/1.0 0.25/0.75	660	750	—	—	92 — 89 —	97 — 94 —	— 78 — 76	— 82 — 80	— — — —	— — — —
C52100	Sheet and Strip	Soft	1.0/— 0.75/—	365	460	—	—	29 —	70 —	— 38	— 68	—	—
		Soft	0.50/1.0 0.25/0.75	365	460	—	In 4 x Dia or Thickness of Specimen	20 —	66 —	— 27	— 62	—	—

TABLE 2A—MINIMUM MECHANICAL PROPERTIES OF WROUGHT COPPER ALLOYS (continued)

Metric (SI) Units

UNS No.[ee]	Form	Temper	Size Section, mm Over/Thru	Tensile Strength, MPa Min	Tensile Strength, MPa Max	Yield Strength, Min MPa 0.5% Ext Under Load	Elongation, Min %[c] In 4 × Dia or Thickness of Specimen	Hardness RB[b] Min	Hardness RB[b] Max	Hardness R30T[b] Min	Hardness R30T[b] Max	Grain Size, mm Min	Grain Size, mm Max
C52100	Sheet and Strip	1/2 Hard	1.0/— 0.75/— 0.50/1.0 0.25/0.75	475	580	—	—	76 — 69 —	91 — 88 —	67 — 63	78 — 75	—	—
		Hard	1.0/— 0.75/— 0.50/1.0 0.25/0.75	585	690	—	—	91 — 89 —	97 — 95 —	76 — 73	81 — 80	—	—
		Extra Hard	1.0/— 0.75/— 0.50/1.0 0.25/0.75	670	770	—	—	95 — 93 —	100 — 98 —	78 — 77	83 — 82	—	—
		Spring	1.0/— 0.75/— 0.50/1.0 0.25/0.75	725	820	—	—	97 — 95 —	102 — 100 —	79 — 78	84 — 83	—	—
		Extra Spring	1.0/— 0.75/— 0.50/1.0 0.25/0.75	760	840	—	—	98 — 96 —	103 — 101 —	80 — 79	84 — 83	—	—
C52400	Sheet and Strip	Soft	1.0/— 0.75/— 0.50/1.0 0.25/0.75	400	505	—	—	35 — 25 —	75 — — —	40 71 29	78 — 84	—	—
		1/2 Hard	1.0/— 0.75/— 0.50/1.0 0.25/0.75	525	625	—	—	78 — 74 —	95 — 93 —	67 — 63	80 — 77	—	—
		Hard	1.0/— 0.75/— 0.50/1.0 0.25/0.75	650	750	—	—	94 — 92 —	101 — 100 —	78 — 75	82 — 81	—	—
		Extra Hard	1.0/— 0.75/— 0.50/1.0 0.25/0.75	740	840	—	—	93 — 97 —	103 — 102 —	80 — 79	84 — 83	—	—
		Spring	1.0/— 0.75/— 0.50/1.0 0.25/0.75	795	890	—	—	99 — 98 —	104 — 103 —	81 — 80	85 — 84	—	—
		Extra Spring	1.0/— 0.75/—	825	915	—	—	100 —	105 —	82	86	—	—
		Extra Spring	0.50/1.0 0.25/0.75	825	915	—	—	99 —	104 —	81	85	—	—
C54400	Sheet and Strip	See C51100											
	Rod	Hard	1.5/6.5[bb] 6.5/12.5[bb] 12.5/25[bb] 25/—[bb]	450 415 380 345	— — — —	— — — —	8 10 12 15	—	—	—	—	—	—
	Flat Products	Hard	6.5/9.5[cc] 9.5/—[cc]	380 345	— —	— —	10 15	—	—	—	—	—	—
C60800	Tube	Annealed	All Sizes	345	—	130	—	—	—	—	—	0.010	0.045

UNS No.[ee]	Form	Temper	Thickness	Width	Tensile Strength, MPa Min	Tensile Strength, MPa Max	Yield Strength, Min MPa 0.5% Ext Under Load	Elongation, Min %[c] In 50 mm	BHN 1000 kg Min	BHN 1000 kg Max			Grain Size, mm Min	Grain Size, mm Max
C61400	Plate[f], Sheet, Strip and Rolled Bar	Soft	—/12.5 12.5/50 50/125	All All All	495 485 450	— — —	220 205 195	35 35 35	—	—			—	—
		Hard	—/12.5 12.5/25	All All	550 485	— —	310 275	25 30	—	—			—	—

TABLE 2A—MINIMUM MECHANICAL PROPERTIES OF WROUGHT COPPER ALLOYS (continued)

Metric (SI) Units

UNS No.[ee]	Form	Temper	Size Section, mm Over/Thru	Tensile Strength, MPa Min	Tensile Strength, MPa Max	Yield Strength, Min MPa 0.5% Ext Under Load	Elongation, Min %[c] In 4 x Dia or Thickness of Specimen	Hardness Min	Hardness Max	Hardness Min	Hardness Max	Grain Size, mm Min	Grain Size, mm Max
								RB[b]		BHN 1000 kg			
C61400	Rod and Bar	—	—/12.5 12.5/25 25/50 50/75	550 515 485 485	— — — —	275 240 220 205	30 30 30 30	— — — —	— — — —	— — — —	— — — —	— — — —	— — — —
C61800	Rod	1/2 Hard	—/75	—	—	—	—	—	—	—	—	—	—
C62300	Rod and Bar	Drawn	—/12.5 12.5/25 25/— 50/25 75/—	620 605 580 525 515	— — — — —	310 305 275 255 205	12 15 15 20 20	90 — — — —	— — — — —	155 — — — —	— — — — —	— — — — —	— — — — —
	Shapes	—	All Sizes	—	—	205	20	—	—	—	—	—	—
										3000 kg			
	Forgings	As Forged	—/38 38/—	515 495	— —	205 195	20 25	— —	— —	— —	— —	— —	— —
C62400	Rod	As Extruded	All Sizes	550	—	310	7.0	—	—	175	—	—	—
	Forgings	As Forged	—	550	—	310	7.0	—	—	175	—	—	—
		Hardened and H₂O Quenched	—	—	—	—	—	—	—	240	—	—	—
		Hardened, H₂O Quenched and Tempered	—	600	—	—	8.0	—	—	179	—	—	—
										R30T[b]			
C63000	Rod and Bar	Annealed	12.5/25 25/50 50/100	690 620 585	— — —	345 310 295	5 6 10	— — —	— — —	— — —	— — —	— — —	— — —
	Shapes	As Extruded	All Sizes	585	—	295	10	—	—	—	—	—	—
C64200	Rod and Bar	Drawn	—/12.5 12.5/25 25/50 50/75	620 585 550 515	— — — —	310 310 290 240	9 12 12 15	— — — —	— — — —	— — — —	— — — —	— — — —	— — — —
										RF[b]			
C65500	Plate[t], Sheet, Strip and Rolled Bar	Annealed 0.070 mm 0.040 mm	— —	360 380	400 440	— —	— —	— —	— —	70 76	82 93	0.050 Note 1	0.110 0.055
		1/4 Hard 1/2 Hard[k] Hard[k] Extra Hard[k]	— — — —	425 490 600 685	495 560 670 745	— — — —	— — — —	65 79 88 93	60 91 96 98	— — — —	— — — —	— — — —	— — — —
		Spring[k]	—	725	780	—	—	94	99	—	—	—	—
		Hot Rolled	—	380	495	—	—	—	—	72	—	—	—
		Hot Rolled and Cold Rolled Finish	—	400	495	—	—	60	80	—	—	—	—
										R30T[b]			
	Rod, Bar and Shapes	Soft	All Forms and Sizes	360	—	105	35	—	—	—	—	—	—
		1/4 Hard 1/2 Hard	All Forms and Sizes —/50[dd]	380 485	— —	165 260	25 17	— —	— —	— —	— —	— —	— —
		Hard	—/6.5[dd] 6.5/38[dd]	585 585	— —	345 345	8 13	— —	— —	— —	— —	— —	— —

TABLE 2A—MINIMUM MECHANICAL PROPERTIES OF WROUGHT COPPER ALLOYS (continued)

Metric (SI) Units

UNS No.[ee]	Form	Temper	Size Section, mm Over/Thru	Tensile Strength, MPa Min	Tensile Strength, MPa Max	Yield Strength, Min MPa 0.5% Ext Under Load	Elongation, Min %[c] In 4 × Dia or Thickness of Specimen	Hardness RB[b] Min	Hardness RB[b] Max	Hardness R30T[b] Min	Hardness R30T[b] Max	Grain Size, mm Min	Grain Size, mm Max
C65500	Rod, Bar and Shapes	Extra Hard	—/12.5[w]	690	—	380	7	—	—	—	—	—	—
	Rod	Soft	All Sizes	360	—	105	35	—	—	—	—	—	—
		Hard	—/25	450	—	265	20	—	—	—	—	—	—
			25/38	415	—	205	25	—	—	—	—	—	—
			38/75	380	—	165	27	—	—	—	—	—	—
C67000	Rod and Bar	Soft	All Sizes	585	—	310	10[v]	—	—	—	—	—	—
		1/2 Hard	All Sizes	725	—	415	7[v]	—	—	—	—	—	—
		Hard	All Sizes	795	—	470	5[v]	—	—	—	—	—	—
	Forgings	Soft	—	585	—	310	10	—	—	—	—	—	—
		1/2 Hard	—	675	—	345	7	—	—	—	—	—	—
		Hard	—	690	—	380	5	—	—	—	—	—	—
C67300	Rod and Bar	As Extruded	All Sizes	360	—	170	20	60	—	—	—	—	—
		Soft	All Sizes	360	—	170	20	50	—	—	—	—	—
		1/2 Hard[w]	—/25	450	—	275	12	70	—	—	—	—	—
			25/75	400	—	240	15	70	—	—	—	—	—
			75/—	360	—	205	18	65	—	—	—	—	—
		1/2 Hard[x]	All Sizes	415	—	205	20	70	—	—	—	—	—
		Hard[w]	—/25	485	—	345	10	70	—	—	—	—	—
			25/50	425	—	290	15	70	—	—	—	—	—
	Shapes	As Extruded	All Sizes	360	—	170	20	60	—	—	—	—	—
	Forgings	As Forged	—	360	—	170	20	60	—	—	—	—	—
		Forged and Heat Treated	—	475	—	240	12	70	—	—	—	—	—
C67400	Rod and Bar	As Extruded	All Sizes	485	—	235	12	75	—	—	—	—	—
		Extruded and Drawn	—/25	540	—	275	8	84	—	—	—	—	—
			25/50	515	—	275	10	80	—	—	—	—	—
			50/75	485	—	250	12	78	—	—	—	—	—
	Shapes	As Extruded	All Sizes	485	—	235	12	75	—	—	—	—	—
	Forgings	As Forged	—	450	—	205	15	75	—	—	—	—	—
C67500	Rod	Soft	All Sizes	380	—	150	20	—	—	—	—	—	—
		1/2 Hard	—/25	495	—	250	13	—	—	—	—	—	—
			25/65	485	—	240	15	—	—	—	—	—	—
			65/—	450	—	220	17	—	—	—	—	—	—
		Hard	—/25	550	—	385	8	—	—	—	—	—	—
			25/38	525	—	360	10	—	—	—	—	—	—
			38/65	505	—	330	12	—	—	—	—	—	—
			65/—	470	—	310	16	—	—	—	—	—	—
	Bar	Soft	All Sizes	380	—	150	20	—	—	—	—	—	—
		1/2 Hard	—/25	495	—	250	13	—	—	—	—	—	—
			25/65	485	—	240	15	—	—	—	—	—	—
			65/—	450	—	220	17	—	—	—	—	—	—
		Hard	—/25	550	—	385	8	—	—	—	—	—	—
			25/65	525	—	360	12	—	—	—	—	—	—
			65/—	505	—	330	16	—	—	—	—	—	—
	Shapes	Soft	All Sizes	470	—	310	20	—	—	—	—	—	—

TABLE 2A—MINIMUM MECHANICAL PROPERTIES OF WROUGHT COPPER ALLOYS (continued)

Metric (SI) Units

UNS No.[ee]	Form	Temper	Size Section, mm Over/Thru	Tensile Strength, MPa Min	Tensile Strength, MPa Max	Yield Strength, Min MPa 0.5% Ext Under Load	Elongation, Min %[c] In 50 mm		Hardness				Grain Size, mm	
									RB[b] Min	RB[b] Max	R30T[b] Min	R30T[b] Max	Min	Max
C70600	Condenser Tube Plate	—	—/65	275	—	105	30		—	—	—	—	—	—
	Tube	Annealed	—	275	—	105	—		—	—	—	—	—	—
		Light Drawn	—	310	—	240	—		—	—	—	—	—	—

(Hardness Min/Max columns for C71000 Plate Annealed rows are RF[b])

UNS No.	Form	Temper	Size	Tensile Min	Tensile Max	Yield Min	Hardness RF Min	Hardness RF Max	RB Min	RB Max	R30T Min	R30T Max	Grain Min	Grain Max
C71000	Tube	Annealed	—	310	—	110	—	—	—	—	—	—	—	—
	Plate[t], Sheet, Strip and Rolled Bar	Annealed 0.035 mm	—	—	—	—	67	76	18	35	28	40	0.025	0.050
		0.015 mm	—	—	—	—	76	90	35	58	40	55	Note a	0.020
		1/4 Hard	—	325	435	—	—	—	45	72	46	65	—	—
		1/2 Hard	—	385	485	—	—	—	64	78	59	69	—	—
		Hard	—	460	545	—	—	—	76	84	67	73	—	—
		Extra Hard	—	495	580	—	—	—	79	87	69	75	—	—
		Spring	—	525	600	—	—	—	82	88	71	75	—	—
C71500	Condenser Tube Plate	—	—/65	345	—	140	35		—	—	—	—	—	—
			65/125	310	—	125	35		—	—	—	—	—	—
	Tube	Annealed	—	360	—	125	—		—	—	—	—	—	—
		Drawn and Stress Relieved	—/1.2 Wall	495	—	345	12		—	—	—	—	—	—
			1.2/— Wall	495	—	345	15		—	—	—	—	—	—
	Plate[t], Sheet, Strip and Rolled Bar	Annealed 0.035 mm	—	—	—	—	70	85	23	45	31	46	0.025	0.050
		0.015 mm	—	—	—	—	74	93	37	63	40	58	Note a	0.025
		1/4 Hard	—	400	485	—	—	—	67	81	61	71	—	—
		1/2 Hard	—	455	550	—	—	—	76	85	67	74	—	—
		Hard	—	515	605	—	—	—	83	89	72	76	—	—
		Extra Hard	—	550	635	—	—	—	85	91	73	77	—	—
		Spring	—	580	650	—	—	—	87	91	74	77	—	—
C75200	Plate[t], Sheet, Strip and Rolled Bar	Annealed 0.070 mm	—	—	—	—	70	80	25	40	32	43	0.050	0.100
		0.035 mm	—	—	—	—	75	88	35	55	40	53	0.025	0.050
		0.015 mm	—	—	—	—	83	93	45	70	46	64	Note a	0.025
		1/4 Hard	—	400	495	—	—	—	50	75	49	67	—	—
		1/2 Hard	—	455	550	—	—	—	68	82	62	72	—	—
		Hard	—	540	625	—	—	—	80	90	70	76	—	—
		Extra Hard	—	595	675	—	—	—	87	94	74	79	—	—
		Spring	—	620	695	—	—	—	89	96	75	80	—	—
	Rod and Bar	Annealed 0.070 mm	—	—	—	—	—	—	—	—	—	—	0.050	0.100
		0.035 mm	—	—	—	—	—	—	—	—	—	—	0.025	0.050
		0.015 mm	—	—	—	—	—	—	—	—	—	—		0.030
		1/4 Hard	0.50/12.5[y]	415	550	—	—	—	—	—	—	—	—	—

TABLE 2A—MINIMUM MECHANICAL PROPERTIES OF WROUGHT COPPER ALLOYS (continued)

Metric (SI) Units

UNS No.[ee]	Form	Temper	Size Section, mm Over/Thru	Tensile Strength, MPa Min	Tensile Strength, MPa Max	Yield Strength, Min MPa 0.5% Ext Under Load	Hardness RF[b] Min	Hardness RF[b] Max	Hardness RB[b] Min	Hardness RB[b] Max	Hardness R30T[b] Min	Hardness R30T[b] Max	Grain Size, mm Min	Grain Size, mm Max
C75200	Rod and Bar	Hard	0.5/6.5[w]	550	690	—	—	—	—	—	—	—	—	—
			6.5/12.5[w]	480	620	—	—	—	—	—	—	—	—	—
			12.5/25[w]	450	555	—	—	—	—	—	—	—	—	—
			25/—[w]	415	550	—	—	—	—	—	—	—	—	—
			All Sizes[x]	470	605	—	—	—	—	—	—	—	—	—
C77000	Plate[t], Sheet, Strip and Rolled Bar	Annealed 0.070 mm	—	—	—	—	72	83	29	45	35	46	0.050	0.100
		0.035 mm	—	—	—	—	76	91	37	60	41	57	0.025	0.050
		0.015 mm	—	—	—	—	84	98	47	73	47	65	Note a	0.025
							RG[b]							
		1/4 Hard	—	475	600	—	23	62	70	88	63	75	—	—
		1/2 Hard	—	540	655	—	51	69	81	92	71	78	—	—
		Hard	—	635	740	—	67	76	90	96	76	80	—	—
		Extra Hard	—	705	795	—	73	80	95	99	79	82	—	—
		Spring	—	745	825	—	77	83	97	100	80	—	—	—
	Rod and Bar	Annealed 0.070 mm	—	—	—	—	—	—	—	—	—	—	0.050	0.100
		0.035 mm	—	—	—	—	—	—	—	—	—	—	0.025	0.050
		0.015 mm	—	—	—	—	—	—	—	—	—	—	—	0.030
		1/4 Hard	0.50/12.5[y]	515	655	—	—	—	—	—	—	—	—	—
		Hard	0.50/6.5[w]	620	700	—	—	—	—	—	—	—	—	—
			0.50/6.5[w]	550	690	—	—	—	—	—	—	—	—	—
			12.5/25[w]	515	655	—	—	—	—	—	—	—	—	—
			25/—[w]	485	620	—	—	—	—	—	—	—	—	—
			All Sizes[x]	515	655	—	—	—	—	—	—	—	—	—

[a] Although no minimum grain size is required, this material must be fully recrystallized.
[b] Values are approximate. F and B scales for metal 0.50 mm and over in thickness. 30T scale for metal 0.30 mm and over in thickness. (0.40 mm for annealed material to ASTM B36 and B122).
[c] In any case, a minimum gage length of 25 mm shall be used.
[d] Hardness values shall apply only to tubes having a wall thickness of 0.40 mm or over for annealed temper and 0.50 mm or over for drawn temper (0.30 mm for drawn temper to ASTM B135), to round tubes having an inside diameter of 8.0 mm or over, and to rectangular, including square, tubes having an inside major distance between parallel surfaces of 4.8 mm or over. For all other tubes, no Rockwell values shall apply. Hardness tests shall be made on the inside surface of the tube. When suitable equipment is not available for determining the specified hardness, other Rockwell scales and values may be specified, subject to agreement between purchaser and supplier.
[e] 3.2 to 22.0 mm outside diameter 0.75 to 1.1 mm wall.
[f] Generally available in round, hexagonal, and octagonal.
[g] Normally available in round only.
[h] Not generally available in sizes over 12.5 mm in diameter.
[i] Not generally available in sizes over 9.5 mm in diameter. Square and rectangular wire not generally available.
[j] Not generally available in sizes over 6.5 mm in diameter. Square and rectangular wire not generally available.
[k] Commercially supplied only as strip. Manufacturer should be consulted for sheet or plate.
[l] Capable of being hardened by further heat treatment.
[m] Rods and bars.
[n] Applicable to material 0.10 mm and over.
[o] Applicable to material 0.80 mm and over.
[p] Applicable to material 0.40 mm and over.
[q] When stated on contract or order, tension test shall be waived provided the strip meets the hardness requirement. In case of dispute, tension test shall be the basis for acceptance.
[r] Commonly supplied only as strip.
[s] 6.5 mm and over.
[t] Plate generally available only in annealed, quarter hard, and half hard.
[u] Type A material to listed hardness limits supplied unless otherwise specified.
[v] Cold finished.
[w] Rods only.
[x] Bars only.
[y] After heat treatment.
[z] After mill heat treatment.
[aa] Rounds only.
[bb] Rounds and hexagons.
[cc] Squares and rectangles.
[dd] Rods and square bars.
[ee] Unified Number System. For cross reference to SAE, former SAE, ASTM and former trade names, see SAE J461.

TABLE 2B—MINIMUM MECHANICAL PROPERTIES OF WROUGHT COPPER ALLOYS

Customary Units

UNS No.[ee]	Form	Temper	Size Section, in Over/Thru		Tensile Strength, ksi Min	Tensile Strength, ksi Max	Yield Strength, Min ksi 0.5% Ext Under Load	Elongation, Min %[c] In 4 x Dia or Thickness of Specimen	Hardness Min	Hardness Max	Hardness Min	Hardness Max	Grain Size, mm Min	Grain Size, mm Max
									RF[b]	RF[b]	R30T[b]	R30T[b]		
C10200 C11000 C11100 C11300 C11400 φC11500 C11600 C12000 C12200	Plate, Sheet, Strip and Rolled Bar	Soft Anneal	—		—	—	—	—	—	65	—	—	Note a	—
		Deep-Drawing Anneal	—		—	—	—	—	30	75	—	—	Note a	0.050
		Light Cold Rolled	—		32	40	—	—	40	82	—	49	—	—
		1/2 Hard[r]	—		37	46	—	—	77	89	43	57	—	—
		Hard[r]	—		43	52	—	—	86	93	54	62	—	—
		Spring[r]	—		50	58	—	—	91	97	60	66	—	—
		Extra Spring[r]	—		52	—	—	—	92	—	61	—	—	—
		Hot Rolled	—		30	38	—	—	—	75	—	41	—	—
		Hot Rolled and Annealed	—		30	38	—	—	—	65	—	31	—	—
					Type B Matl[u]		Type B[u]	Type A[s,u]						
C10200 C11000 C12000 C12200	Rod, Bar and Shapes	Soft Anneal	All Sizes[m]		—	37	25	—	65					
		Hard	—/.250[w]		50	—	—	—	68	95	—	—	—	—
			.250/.375[w]		45	—	—	10	68	95	—	—	—	—
			.375/1.0[w]		40	—	—	12	68	95	—	—	—	—
			1.0/2.0[w]		35	—	—	15	68	95	—	—	—	—
			2.0/3.0[w]		33	—	—	15	68	95	—	—	—	—
			.188/.375[x]		42	—	—	12	68	95	—	—	—	—
			.375/.500[x]		40	—	—	12	68	95	—	—	—	—
			.500/2.0[x]		33	—	—	15	68	95	—	—	—	—
			2.0/4.0[x]		32	—	—	15	68	95	—	—	—	—
			All Sizes[y]		32	—	—	15	—	—	—	—	—	—
			OD	Wall				In 2 in			R15T[d]	R15T[d]		
C10200 C12000 C12200	Tube	Soft Anneal	All Sizes	.014/.034	30	—	—	40[e]	—	—	—	60	0.040	—
			All Sizes	.034/—	—	—	—	—	—	50[d]	—	—	0.040	—
		Light Anneal	All Sizes	.014/.034	—	—	—	—	—	—	—	65	—	0.040
			All Sizes	.034/—	—	—	—	—	—	55[d]	—	—	—	0.040
											R30T[d]	R30T[d]		
		Light Drawn	All Sizes	All Sizes	36	47	—	—	—	—	30	60	—	—
		Drawn	All Sizes	All Sizes	36	—	—	—	—	—	30	—	—	—
		Hard Drawn	—/1.0	.019/.120	45	—	—	—	—	—	55	—	—	—
			1.0/2.0	.034/.180	45	—	—	—	—	—	55	—	—	—
			2.0/4.0	.059/.250	45	—	—	—	—	—	55	—	—	—
			Size Section, in					In 10 in	RF[b]		R30T[b]			
C10200 C11000	Wire	Annealed	.0029/.0100 dia		—	—	—	15	—	—	—	—	—	—
			.0100/.0201 dia		—	—	—	20	—	—	—	—	—	—
			.0201/.1019 dia		—	—	—	25	—	—	—	—	—	—
			.1019/.2893 dia		—	—	—	30	—	—	—	—	—	—
			.2893/.4600 dia		—	—	—	35	—	—	—	—	—	—
								In 4 x Dia						
C14500	Rod	1/2 Hard[f]	.062/.250		38	—	30	8	—	—	—	—	—	—
			.250/2.625		38	—	30	12	—	—	—	—	—	—
		Hard	.062/.250		48	—	40	4	—	—	—	—	—	—
			.250/1.250		44	—	38	8	—	—	—	—	—	—
			1.250/2.000		40	—	35	8	—	—	—	—	—	—
C14700	Rod	1/2 Hard[f]	.062/.250		38	—	30	8	—	—	—	—	—	—
			.250/2.625		38	—	30	12	—	—	—	—	—	—
		Hard[g]	.062/.250		48	—	40	4	—	—	—	—	—	—
			.250/1.250		44	—	38	8	—	—	—	—	—	—
			1.250/2.000		40	—	35	8	—	—	—	—	—	—
C15000	Round Rod													

TABLE 2B—MINIMUM MECHANICAL PROPERTIES OF WROUGHT COPPER ALLOYS (continued)

Customary Units

UNS No.[ee]	Form	Temper	Size Section, in Over/Thru	Tensile Strength, ksi Min	Tensile Strength, ksi Max	Yield Strength, Min ksi 0.5% Ext Under Load	Elongation, Min %[c] In 2 in	Hardness Min	Hardness Max	Hardness Min	Hardness Max	Grain Size, mm Min	Grain Size, mm Max
				Type B Matl[u]				RB		R30T[b]			
C16200	Round Rod	Drawn	—/1.0 1.0/2.0 2.0/3.0	60 55 50	— — —	— — —	20 25 25	65 60 55	— — —	— — —	— — —	— — —	— — —
	Square, Rectangular and Hex Rod and Bar	Drawn	—/1.0 Thick 1.0/— Thick	60 50	— —	— —	20 25	55 50	— —	— —	— —	— —	— —
	Forging	As Forged	—/1.0 Thick 1.0/— Thick	60 50	— —	— —	20 25	55 50	— —	— —	— —	— —	— —
						0.2% Offset	In 2 in[n]	RB[o,q]		R30T[p,q]		Elec Cond, % IACS Min	Heat Treat h at 600°F
C17000	Strip	A (Soft)[l]	All Sizes	60	78	—	35	45	78	46	67	17	—
		1/4 Hard[l] 1/2 Hard[l] Hard[l]	All Sizes All Sizes All Sizes	75 85 100	88 100 120	— — —	10 5 2	68 88 96	90 96 102	62 74 79	75 79 83	16 15 15	— — —
								RC[o,q]		R30N[p,q]			
		AT[y]	All Sizes	150	—	130 min	3	33	—	53	—	22	3
		1/4 HT[y] 1/2 HT[y] HT[y]	All Sizes All Sizes All Sizes	160 170 180	— — —	135 min 145 min 155 min	2.5 1 1	35 37 39	— — —	55 56 59	— — —	22 22 22	2 2 2
		AM[z]	All Sizes	100	115	70 min 95 max	18	18	23	37	44	23	—
		1/4 HM[z]	All Sizes	110	125	80 min 105 max	15	21	26	42	47	23	—
		1/2 HM[z]	All Sizes	120	135	95 min 115 max	12	25	30	46	50	24	—
		HM[z]	All Sizes	135	150	110 min 135 max	9	30	35	50	55	25	—
		XHM[z]	All Sizes	160	175	135 min 160 max	2	32	36	52	56	24	—
								RB[o,q]		R30T[p,q]			
	Strip	A (Soft)[l]	All Sizes	60	78	—	35	45	78	46	67	17	—
		1/4 Hard[l] 1/2 Hard[l] Hard[l]	All Sizes All Sizes All Sizes	75 85 100	88 100 120	— — —	10 5 5	68 88 96	90 96 102	62 74 79	75 79 83	16 15 15	— — —
								RC[o,q]		R30N[p,q]			
C17200		AT[y]	All Sizes	165	195	—	—	36	—	56	—	22	3
		1/4 HT[y] 1/2 HT[y] HT[y]	All Sizes All Sizes All Sizes	175 185 195	205 215 220	— — —	— — —	38 39 40	— — —	58 59 60	— — —	22 22 22	2 2 2
								RB[o,q]					
	Rod and Bar	A (Soft)[l]	All Sizes	60	85	—	—	45	85	—	—	17	—
		Hard[l]	—/.375 .375/1.0 1.0/—	95 90 85	130 120 115	— — —	— — —	92 91 88	103 102 104	— — —	— — —	15 15 15	— — —
								RC[o,q]					
		AT[y]	All Sizes	165	190	—	—	36	40	—	—	22	3

TABLE 2B—MINIMUM MECHANICAL PROPERTIES OF WROUGHT COPPER ALLOYS (continued)

Customary Units

UNS No.[ee]	Form	Temper	Size Section, in Over/Thru	Tensile Strength, ksi Min	Tensile Strength, ksi Max	Yield Strength, Min ksi 0.2% Offset	Elongation, Min %[c] In 2 in[n]	Hardness Min	Hardness Max	Hardness Min	Hardness Max	Elec Cond, % IACS Min	Heat Treat h at 600°F
				Type B Matl[u]				RC[o,q]		R30N[p,q]			
C17200	Rod and Bar	HT[y]	—/.375 .375/1.0 1.0/—	185 180 175	215 210 205	— — —	— — —	39 38 37	45 44 43	— — —	— — —	22 22 22	3 2 2
				Min	Max		In 2 in	RB[o,q]		R30T[p,q]			
C17500	Strip and Plate	A (Soft)[l]	All Sizes	—	55	20 min 30 max	20	20	45	29	45	20	—
		1/2 Hard[l]	All Sizes	60	75	50 min 70 max	5	65	76	60	67	25	—
		Hard[l]	All Sizes	70	85	60 min 80 max	2	78	88	69	75	25	—
		AT[y]	All Sizes	100	120	80 min 100 max	8	92	100	77	82	45	—
		1/2 HT[y]	All Sizes	110	130	95 min 120 max	5	95	102	79	83	48	—
		HT[y]	All Sizes	110	130	95 min 120 max	5	95	102	79	83	48	—
	Hot Worked Sizes, Forgings	A (Soft)[l]	All Sizes	—	55	20 min 30 max	20	20	45	—	—	20	—
		AT[y]	All Sizes	100	120	80 min 100 max	10	92	100	—	—	45	—
										R30T[o,q]			
C17500 C17600	Rod, Bar Shapes and Tubing	A (Soft)[l]	All Sizes	35	55	20 min 30 max	20	25	45	—	—	20	—
		1/2 Hard	All Sizes	65	85	55 min 75 max	10	60	75	—	—	20	—
		AT[y]	All Sizes	100	120	80 min 100 max	10	92	100	—	—	45	—
		1/2 HT[y]	All Sizes	110	130	100 min 120 max	8	92	102	—	—	48	—
						0.5% Ext Under Load		RB		R30T[b]		Grain Size, mm Min	Grain Size, mm Max
C18400	Round Rod	Drawn	—/1.0 1.0/2.0 2.0/3.0	65 60 55	— — —	— — —	15 15 15	75 70 65	— — —	— — —	— — —	— — —	— — —
	Hex Rod and Bars	Drawn	—/1.0 Thick 1.0/— Thick	65 55	— —	— —	15 15	70 65	— —	— —	— —	— —	— —
	Forgings	As Forged	—/1.0 1.0/2.0 2.0/—	65 55 55	— — —	— — —	15 15 15	72 70 65	— — —	— — —	— — —	— — —	— — —
C18700	Rod	1/2 Hard[f]	.062/.250 .250/2.625	38 38	— —	30 30	8 12	— —	— —	— —	— —	— —	— —
		Hard	0.62/.250 .250/1.250 1.250/2.000	48 44 40	— — —	40 38 35	4 8 8	— — —	— — —	— — —	— — —	— — —	— — —
ФC19200	Tube	A (Soft) Light Drawn	All Sizes All Sizes	38 40	— —	12 35	— —	— —	— —	— —	— —	— —	— —

TABLE 2B—MINIMUM MECHANICAL PROPERTIES OF WROUGHT COPPER ALLOYS (continued)

Customary Units

UNS No.[ee]	Form	Temper	Size Section, in Over/Thru	Tensile Strength, ksi Min	Tensile Strength, ksi Max	Yield Strength, Min ksi 0.5% Ext Under Load	Elongation, Min %[c] In 2 in	Hardness Min	Hardness Max	Hardness Min	Hardness Max	Grain Size, mm Min	Grain Size, mm Max
								RF[b]	RF[b]	R30T[b]	R30T[b]		
C21000	Plate, Sheet, Strip and Rolled Bar[t]	Annealed 0.050 mm 0.035 mm 0.025 mm 0.015 mm	—	—	—	—	—	40 47 50 54	52 54 61 65	— — 1 7	4 7 17 23	0.035 0.025 0.015 Note a	0.090 0.050 0.035 0.025
								RB[b]	RB[b]				
		1/4 Hard	.019/.036 .036/— .011/.028 .028/—	37	47	—	—	20 24 — —	48 52 — —	— — 34 37	— — 51 54	—	—
		1/2 Hard	.019/.036 .036/— .011/.028 .028/—	42	52	—	—	40 44 — —	56 60 — —	— — 46 48	— — 57 59	—	—
		3/4 Hard	.019/.036 .036/—	46	56	—	—	50 53	61 64	— —	— —	—	—
							In 4 x Dia or Thickness of Specimen						
		3/4 Hard	.011/.028 .028/—	46	56	—	—	— —	— —	52 54	60 62	—	—
		Hard	.019/.036 .036/— .011/.028 .028/—	50	59	—	—	57 60 — —	64 67 — —	— — 57 59	— — 62 64	—	—
		Extra Hard	.019/.036 .036/— .011/.028 .028/—	56	64	—	—	64 66 — —	70 72 — —	— — 62 63	— — 66 67	—	—
		Spring	.019/.036 .036/— .011/.028 .028/—	60	68	—	—	68 70 — —	73 75 — —	— — 64 65	— — 68 69	—	—
		Extra Spring	.019/.036 .036/— .011/.028 .028/—	61	69	—	—	69 71 — —	74 76 — —	— — 65 66	— — 69 70	—	—
C22000	Plate, Sheet, Strip and Rolled Bar[t]	Annealed 0.050 mm 0.035 mm 0.025 mm 0.015 mm	—	—	—	—	—	RF[b] 50 54 58 62	RF[b] 60 64 70 75	1 7 13 19	16 21 31 39	0.035 0.025 0.015 Note a	0.090 0.050 0.035 0.025
								RB[b]	RB[b]				
		1/4 Hard	.019/.036 .036/— .011/.028 .028/—	40	50	—	—	27 31 — —	52 56 — —	— — 38 41	— — 53 56	—	—
		1/2 Hard	.019/.036 .036/— .011/.028 .028/—	47	57	—	—	50 53 — —	63 66 — —	— — 52 54	— — 61 63	—	—
		3/4 Hard	.019/.036 .036/— .011/.028 .028/—	52	62	—	—	59 62 — —	68 71 — —	— — 58 60	— — 64 66	—	—
		Hard	.019/.036 .036/— .011/.028 .028/—	57	66	—	—	65 68 — —	72 75 — —	— — 62 64	— — 66 68	—	—

TABLE 2B—MINIMUM MECHANICAL PROPERTIES OF WROUGHT COPPER ALLOYS (continued)

Customary Units

UNS No.[ee]	Form	Temper	Size Section, in Over/Thru		Tensile Strength, ksi Min	Tensile Strength, ksi Max	Yield Strength, Min ksi 0.5% Ext Under Load	Elongation, Min %[c] In 4 x Dia or Thickness of Specimen	Hardness Min	Hardness Max	Hardness Min	Hardness Max	Grain Size, mm Min	Grain Size, mm Max
									RB[b]	RB[b]	R30T[b]	R30T[b]		
C22000	Plate, Sheet, Strip and Rolled Bars[t]	Extra Hard	.019/.036 .036/— .011/.028 .028/—		64	72	—	—	72 74 — —	77 79 — —	— — 67 68	— — 71 72	— — — —	— — — —
		Spring	.019/.036 .036/— .011/.028 .028/—		69	77	—	—	76 78 — —	79 81 — —	— — 70 71	— — 72 73	— — — —	— — — —
		Extra Spring	.019/.036 .036/— .011/.028 .028/—		72	80	—	—	78 80 — —	81 83 — —	— — 71 72	— — 73 74	— — — —	— — — —
	Tube		Wall Thickness						RF[b]	RF[b]				
		Soft Anneal	—/.045 .045/—		— —	— —	— —	— —	— —	— 70	— —	30 —	0.025 0.025	0.060 0.060
		Light Anneal	—/.045 .045/—		— —	— —	— —	— —	— —	— 78	— —	37 —	Note a Note a	0.035 0.035
		Drawn (General Purpose)	All Sizes		40	—	—	—	—	—	38	—	—	—
			OD	Wall										
		Hard Drawn	—/4.0	.019/.250	52	—	—	—	—	—	55	—	—	—
C23000	Sheet and Strip		Size Section, in											
		Annealed 0.070 mm 0.050 mm 0.035 mm 0.025 mm 0.015 mm	— — — — —		— — — — —	— — — — —	— — — — —	— — — — —	53 56 58 60 62	60 63 66 72 79	6 10 13 16 19	16 20 24 34 48	0.050 0.035 0.025 0.015 Note a	0.100 0.070 0.050 0.035 0.025
									RB[b]	RB[b]				
		1/4 Hard	.019/.036 .036/— .011/.028 .028/—		44	54	—	—	33 37 — —	58 62 — —	— — 42 45	— — 57 60	— — — —	— — — —
		1/2 Hard	.019/.036 .036/— .011/.028 .028/—		51	61	—	—	56 59 — —	68 71 — —	— — 56 58	— — 64 66	— — — —	— — — —
		3/4 Hard	.019/.036 .036/— .011/.028 .028/—		57	67	—	—	66 69 — —	73 76 — —	— — 63 65	— — 68 70	— — — —	— — — —
		Hard	.019/.036 .036/— .011/.028 .028/—		63	72	—	—	72 74 — —	78 80 — —	— — 67 68	— — 71 72	— — — —	— — — —
		Extra Hard	.019/.036 .036/—		72	80	—	—	78 80	83 85	— —	— —	— —	— —
		Extra Hard	.011/.028 .028/—		72	80	—	—	— —	— —	70 71	74 75	— —	— —
		Spring	.019/.036 .036/— .011/.028 .028/—		78	86	—	—	82 84 — —	85 87 — —	— — 74 75	— — 76 77	— — — —	— — — —
		Extra Spring	.019/.036 .036/— .011/.028 .028/—		82	90	—	—	84 86 — —	87 89 — —	— — 75 76	— — 77 78	— — — —	— — — —

TABLE 2B—MINIMUM MECHANICAL PROPERTIES OF WROUGHT COPPER ALLOYS (continued)

Customary Units

UNS No.[ee]	Form	Temper	Size Section, in Over/Thru		Tensile Strength, ksi Min	Tensile Strength, ksi Max	Yield Strength, Min ksi 0.5% Ext Under Load	Elongation, Min %[c] In 4 x Dia or Thickness of Specimen	Hardness Min	Hardness Max	Hardness Min	Hardness Max	Grain Size, mm Min	Grain Size, mm Max
C23000	Tube		Wall Thickness						RF[d]	RF[d]	R30T[d]	R30T[d]		
		Soft Anneal	—/.045		—	—	—	—	—	—	—	36	0.025	0.060
			.045/—		—	—	—	—	—	75	—	—	0.025	0.060
		Light Anneal	—/.045		—	—	—	—	—	—	—	39	Note a	0.035
			.045/—		—	—	—	—	—	85	—	—	Note a	0.035
		Light Drawn[g]	All Sizes		44	58	—	—	—	—	43	75	—	—
		Drawn (General Purpose)	All Sizes		44	—	—	—	—	—	43	—	—	—
			OD	Wall					RF[b]					
		Hard Drawn	—/1.0	.019/.120	57	—	—	—	—	—	65	—	—	—
			1.0/2.0	.034/.180					—	—	65	—	—	—
			2.0/4.0	.059/.250					—	—	65	—	—	—
C24000	Sheet and Strip		Size Section, in											
		Annealed												
		0.070 mm	—		—	—	—	—	53	64	2	21	0.050	0.120
		0.050 mm	—		—	—	—	—	57	67	8	27	0.035	0.070
		0.035 mm	—		—	—	—	—	61	72	16	35	0.025	0.050
		0.025 mm	—		—	—	—	—	63	77	20	42	0.015	0.035
		0.015 mm	—		—	—	—	—	66	83	25	50	Note a	0.025
									RB[b]					
		1/4 Hard	.019/.036		48	58	—	—	38	61	—	—	—	—
			.036/—						42	65	—	—	—	—
			.011/.028						—	—	42	57	—	—
			.028/—						—	—	45	60	—	—
		1/2 Hard	.019/.036		55	65	—	—	59	70	—	—	—	—
			.036/—						62	73	—	—	—	—
			.011/.028						—	—	56	64	—	—
			.028/—						—	—	58	66	—	—
		3/4 Hard	.019/.036		61	71	—	—	69	76	—	—	—	—
			.036/—						72	79	—	—	—	—
			.011/.028						—	—	63	68	—	—
			.028/—						—	—	65	70	—	—
		Hard	.019/.036		68	77	—	—	76	82	—	—	—	—
		Hard	.036/—		68	77	—	—	78	84	—	—	—	—
			.011/.028						—	—	68	72	—	—
			.028/—						—	—	69	73	—	—
		Extra Hard	.019/.036		78	87	—	—	83	87	—	—	—	—
			.036/—						85	89	—	—	—	—
			.011/.028						—	—	72	75	—	—
			.028/—						—	—	73	76	—	—
		Spring	.019/.036		85	93	—	—	87	90	—	—	—	—
			.036/—						89	92	—	—	—	—
			.011/.028						—	—	75	77	—	—
			.028/—						—	—	76	78	—	—
		Extra Spring	.019/.036		89	97	—	—	88	91	—	—	—	—
			.036/—						90	93	—	—	—	—
			.011/.028						—	—	76	78	—	—
			.028/—						—	—	77	79	—	—
C26000	Plate,[t] Sheet, Strip, Rolled Bar and Wire	Annealed							RF[d]					
		0.120 mm	—		—	—	—	—	50	62	—	21	0.070	—
		0.070 mm	—		—	—	—	—	52	67	3	27	0.050	0.120
		0.050 mm	—		—	—	—	—	61	73	20	35	0.035	0.070
		0.035 mm	—		—	—	—	—	65	76	25	38	0.025	0.050
		0.025 mm	—		—	—	—	—	67	79	27	42	0.015	0.035
		0.015 mm	—		—	—	—	—	72	85	33	50	Note a	0.025

TABLE 2B—MINIMUM MECHANICAL PROPERTIES OF WROUGHT COPPER ALLOYS (continued)

Customary Units

UNS No.[ee]	Form	Temper	Size Section, in Over/Thru	Tensile Strength, ksi Min	Tensile Strength, ksi Max	Yield Strength, Min ksi 0.5% Ext Under Load	Elongation, Min %[c] In 4 x Dia or Thickness of Specimen	Hardness Min	Hardness Max	Hardness Min	Hardness Max	Grain Size, mm Min	Grain Size, mm Max
C26000	Plate,[t] Sheet, Strip, Rolled Bar and Wire							RB[b]	RB[b]	R30T[d]	R30T[d]		
		1/4 Hard	.019/.036 .036/— .011/.028 .028/—	49	59	—	—	40 44 — —	61 65 — —	— — 43 46	— — 57 60	— — — —	— — — —
		1/2 Hard	.019/.036 .036/— .011/.028 .028/—	57	67	—	—	60 63 — —	74 77 — —	— — 56 58	— — 66 68	— — — —	— — — —
		3/4 Hard	.019/.036 .036/— .011/.028 .028/—	64	74	—	—	72 75 — —	79 82 — —	— — 65 67	— — 70 72	— — — —	— — — —
		Hard	.019/.036 .036/— .011/.028 .028/—	71	81	—	—	79 81 — —	84 86 — —	— — 70 71	— — 73 74	— — — —	— — — —
		Extra Hard	.019/.036 .036/— .011/.028 .028/—	83	92	—	—	85 87 — —	89 91 — —	— — 74 75	— — 76 77	— — — —	— — — —
		Spring	.019/.036	91	100	—	—	89	92	—	—	—	—
		Spring	.036/— .011/.028 .028/—	91	100	—	—	90 — —	93 — —	— 76 76	— 78 78	— — —	— — —
		Extra Spring	.019/.036 .036/— .011/.028 .028/—	95	104	—	—	91 92 — —	94 95 — —	— — 77 77	— — 79 79	— — — —	— — — —
	Tube		Wall Thickness					RF[b]	RF[b]				
		Soft Anneal	—/.030 .030/—	— —	— —	—	—	— —	— 80	— —	40 —	0.025 0.025	0.060 0.060
		Light Anneal	—/.030 .030/—	— —	— —	—	—	— —	— 90	— —	60 —	Note a Note a	0.035 0.035
		Drawn (General Purpose)	All Sizes	54	—	—	—	—	—	53	—	—	—
			OD / Wall										
		Hard Drawn	—/4.0 / .019/.250	66	—	—	—	—	—	70	—	—	—
C26800	Plate,[t] Sheet, Strip and Rolled Bar		Size Section, in					RF[d]	RF[d]	R30T[b]	R30T[b]		
		Annealed 0.120 mm 0.070 mm 0.050 mm 0.035 mm 0.025 mm 0.015 mm	— — — — — —	— — — — — —	— — — — — —	—	—	50 52 61 65 67 72	62 67 73 76 79 85	— 3 20 25 27 33	21 27 35 38 42 50	0.070 0.050 0.035 0.025 0.015 Note a	— 0.120 0.070 0.050 0.035 0.025
								RB[b]	RB[b]				
		1/4 Hard	.019/.036 .036/— .011/.028 .028/—	49	59	—	—	40 44 — —	61 65 — —	— — 43 46	— — 57 60	— — — —	— — — —
		1/2 Hard	.019/.036 .036/— .011/.028 .028/—	55	65	—	—	57 60 — —	71 74 — —	— — 54 56	— — 64 66	— — — —	— — — —
		3/4 Hard	.019/.036 .036/— .011/.028 .028/—	62	72	—	—	70 73 — —	77 80 — —	— — 65 67	— — 69 71	— — — —	— — — —

TABLE 2B—MINIMUM MECHANICAL PROPERTIES OF WROUGHT COPPER ALLOYS (continued)

Customary Units

UNS No.[ee]	Form	Temper	Size Section, in Over/Thru	Tensile Strength, ksi Min	Tensile Strength, ksi Max	Yield Strength, Min ksi 0.5% Ext Under Load	Elongation, Min %[c] In 4 x Dia or Thickness of Specimen	Hardness Min	Hardness Max	Hardness Min	Hardness Max	Grain Size, mm Min	Grain Size, mm Max
								RB[b]	RB[b]	R30T[b]	R30T[b]		
C26800	Plate,[t] Sheet, Strip and Rolled Bar	Hard	.019/.036 .036/— .011/.028 .028/—	68	78	—	—	76 78 — —	82 84 — —	— — 68 69	— — 72 73	—	—
		Extra Hard	.019/.036 .036/—	79	89	—	—	83 85	87 90	— —	— —	—	—
		Extra Hard	.011/.028 .028/—	79	89	—	—	— —	— —	73 74	75 76	—	—
		Spring	.019/.036 .036/— .011/.028 .028/—	86	95	—	—	87 89 — —	90 92 — —	— — 75 76	— — 77 78	—	—
		Extra Spring	.019/.036 .036/— .011/.028 .028/—	90	99	—	—	88 90 — —	91 93 — —	— — 76 77	— — 78 79	—	—
C27000	Wire	Annealed 0.100 mm 0.070 mm 0.050 mm 0.035 mm 0.025 mm 0.015 mm		— — — — — —	— — — — — —	— — — — — —	— — — — — —	— — — — — —	— — — — — —	— — — — — —	— — — — — —	0.070 0.050 0.035 0.025 0.015 Note a	— 0.100 0.070 0.050 0.035 0.025
		1/8 Hard 1/4 Hard 1/2 Hard 3/4 Hard Hard[h] Extra Hard[i]	—	50 67 79 92 102 115	65 77 94 107 117 129	—	—	—	—	—	—	—	—
		Spring[j]	—	120	—	—	—	—	—	—	—	—	—
			Wall Thickness					RF[d]	RF[d]	R30T[d]	R30T[d]		
C33000	Tube	Soft Anneal	—/.030 .030/—	— —	— —	— —	— —	— —	— 80	— —	40 —	0.025 0.025	0.060 0.060
		Light Anneal	—/.030 .030/—	— —	— —	— —	— —	— —	— 90	— —	60 —	Note a Note a	0.035 0.035
		Drawn (General Purpose)	All Sizes	54	—	—	—	—	—	53	—	—	—
			OD / Wall										
		Hard Drawn[g]	—/4.0 / .019/.250	66	—	—	—	—	—	70	—	—	—
			Wall Thickness										
C33100	Tube	Soft Anneal	—/.030 .030/—	— —	— —	— —	— —	— —	— 80	— —	40 —	0.025 0.025	0.060 0.060
		Light Anneal	—/.030 .030/—	— —	— —	— —	— —	— —	— 90	— —	60 —	Note a Note a	0.035 0.035
		Drawn (General Purpose)	All Sizes	54	—	—	—	—	—	53	—	—	—
			OD / Wall										
		Hard Drawn	—/4.0 / .019/.250	66	—	—	—	—	—	70	—	—	—
								RF[b]	RF[b]	R30T[b]	R30T[b]		
C34200 C35000	Plate, Sheet, Strip and Rolled Bar	Annealed 0.070 mm 0.050 mm 0.035 mm 0.025 mm	All Sizes All Sizes All Sizes All Sizes	—	—	—	—	54 61 65 67	67 73 76 79	12 20 25 27	27 35 38 42	0.050 0.035 0.025 0.015	0.100 0.070 0.050 0.035

TABLE 2B—MINIMUM MECHANICAL PROPERTIES OF WROUGHT COPPER ALLOYS (continued)

Customary Units

UNS No.[ee]	Form	Temper	Size Section, in Over/Thru		Tensile Strength, ksi Min	Tensile Strength, ksi Max	Yield Strength, Min ksi 0.5% Ext Under Load	Elongation, Min %[c] In 4 x Dia or Thickness of Specimen	Hardness RB[b] Min	Hardness RB[b] Max	Hardness R30T[b] Min	Hardness R30T[b] Max	Grain Size, mm Min	Grain Size, mm Max
C34200 C35000	Plate, Sheet, Strip and Rolled Bar	1/4 Hard	All Sizes		49	59	—	—	40	65	43	60	—	—
		1/2 Hard	All Sizes		55	65	—	—	57	74	54	66	—	—
		Hard	All Sizes		68	78	—	—	76	84	68	73	—	—
		Extra Hard	All Sizes		79	89	—	—	83	89	73	76	—	—
C34500 C35000	Rod	Soft	—/.50		46	—	16	20	—	—	—	—	—	—
			.50/1.0		44	—	15	25	—	45	—	—	—	—
			1.0/2.0		40	—	15	30	—	45	—	—	—	—
		1/4 Hard	—/.50		52	—	25	10	50	75	—	—	—	—
			.50/1.0		50	—	20	15	40	70	—	—	—	—
			1.0/2.0		42	—	15	20	35	65	—	—	—	—
		1/2 Hard	—/.50		57	—	25	7	60	80	—	—	—	—
			.50/1.0		55	—	25	10	55	75	—	—	—	—
			1.0/2.0		50	—	20	15	40	70	—	—	—	—
C36000	Rod	Soft	—/1.0		48	—	20	15	—	—	—	—	—	—
			1.0/2.0		44	—	18	20	—	—	—	—	—	—
			2.0/—		40	—	15	25	—	—	—	—	—	—
		1/2 Hard	—/.50		57	—	25	7	—	—	—	—	—	—
			.50/1.0		55	—	25	10	—	—	—	—	—	—
			1.0/2.0		50	—	20	15	—	—	—	—	—	—
			2.0/4.0		45	—	15	20	—	—	—	—	—	—
			4.0/—		42	—	15	20	—	—	—	—	—	—
		Hard	.061/.088		80	—	45	—	—	—	—	—	—	—
			.088/.500		70	—	35	4	—	—	—	—	—	—
			.500/.750		65	—	30	6	—	—	—	—	—	—
	Flat Products		Thickness	Width										
		Soft	—/1.0	—/6.0	44	—	18	20	—	—	—	—	—	—
			1.0/—	6.0/—	40	—	15	25	—	—	—	—	—	—
		1/2 Hard	—/.50	—/1.0	50	—	25	10	—	—	—	—	—	—
			—/.50	1.0/6.0	45	—	17	15	—	—	—	—	—	—
			.50/2.0	—/2.0	45	—	17	20	—	—	—	—	—	—
			.50/2.0	2.0/6.0	40	—	15	20	—	—	—	—	—	—
			2.0/—	2.0/4.0	40	—	15	20	—	—	—	—	—	—
C37700	Die Forgings		Over/Thru											
		As Forged	—/1.50		50	—	18	25	—	—	—	—	—	—
			1.50/—		46	—	15	20	—	—	—	—	—	—
C46400 C46500 C46600 C46700	Rod and Bar	Soft	—/1.0		54	—	20	30	—	—	—	—	—	—
			1.0/2.0		52	—	20	30	—	—	—	—	—	—
			2.0/—		50	—	20	30	—	—	—	—	—	—
		1/2 Hard or Light Annealed	—/.50		60	—	27	22	—	—	—	—	—	—
			.50/1.0		60	—	27	25	—	—	—	—	—	—
			1.0/2.0		58	—	26	25	—	—	—	—	—	—
			2.0/3.0		54	—	25	25	—	—	—	—	—	—
			3.0/4.0		54	—	22	27	—	—	—	—	—	—
			4.0/—		54	—	22	30	—	—	—	—	—	—
		Hard	—/1.0		67	—	45	13	—	—	—	—	—	—
			1.0/2.0		62	—	37	18	—	—	—	—	—	—
	Shapes	As Extruded	All Sizes		52	—	20	30	—	—	—	—	—	—
C51000	Sheet and Strip	Soft	.039/—		43	58	—	—	16	64	—	—	—	—
			.029/—						—	—	32	59	—	—
			.019/.039						12	60	—	—	—	—
			.009/.029						—	—	24	53	—	—
		1/2 Hard	.039/—		58	73	—	—	64	85	—	—	—	—
			.029/—						—	—	59	73	—	—
			.019/.039						60	82	—	—	—	—
			.009/.029						—	—	53	69	—	—
		Hard	.039/—		76	91	—	—	86	93	—	—	—	—
			.029/—						—	—	73	78	—	—
			.019/.039						84	91	—	—	—	—
			.009/.029						—	—	71	75	—	—

TABLE 2B—MINIMUM MECHANICAL PROPERTIES OF WROUGHT COPPER ALLOYS (continued)

Customary Units

UNS No.[ee]	Form	Temper	Size Section, in Over/Thru	Tensile Strength, ksi Min	Tensile Strength, ksi Max	Yield Strength, Min ksi 0.5% Ext Under Load	Elongation, Min %[c] In 4 × Dia or Thickness of Specimen	Hardness RB[b] Min	Hardness RB[b] Max	Hardness R30T[b] Min	Hardness R30T[b] Max	Grain Size, mm Min	Grain Size, mm Max
C51000	Sheet and Strip	Extra Hard	.039/— .029/— .019/.039 .009/.029	88	103	—	—	92 — 89 —	96 — 95 —	— 77 — 74	— 81 — 78	—	—
		Spring	.039/— .029/— .019/.039 .009/.029	95	110	—	—	94 — 92 —	98 — 97 —	— 79 — 76	— 82 — 80	—	—
		Extra Spring	.039/— .029/— .019/.039 .009/.029	100	114	—	—	95 — 94 —	99 — 98 —	— 80 — 77	— 83 — 81	—	—
	Rod	Soft	—/.249[aa]	40	58	—	—	—	—	—	—	—	—
		Hard	—/.249[bb] .249/.500[bb] .500/1.00[bb] 1.00/—[bb] .249/.375[cc] .375/—[cc]	80 70 60 55 60 55	128 — — — — —	—	— 13 15 18 10 15	—	—	—	—	—	—
		Spring	—/.025[s]	125	—	—	—	—	—	—	—	—	—
		Spring	.025/.062[aa] .062/.125[aa] .125/.250[aa] .250/.375[aa] .375/.500[aa]	115 110 105 100 90	— — — — —	—	— — 3.5 5.0 9.0	—	—	—	—	—	—
	Wire for General Purposes	Soft	—	43	58	—	—	—	—	—	—	—	—
		1/4 Hard 1/2 Hard 3/4 Hard Hard	— — — —	60 80 96 108	76 97 115 128	—	—	—	—	—	—	—	—
	Wire for Spring Purposes	—	—/.025 .025/.062 .062/.125	145 135 130	— — —	—	In 2 in — — —	—	—	—	—	—	—
		—	.125/.250 .250/.375 .375/.500	125 120 105	— — —	—	— 5.0 9.0	—	—	—	—	—	—
φ C51100 C54400	Sheet and Strip	Soft	.039/— .029/— .019/.039 .009/.029	40	50	—	—	7 — 0 —	50 — 45 —	— 24 — 16	— 50 — 46	—	—
		1/2 Hard	.039/— .029/— .019/.039 .009/.029	55	70	—	—	60 — 53 —	81 — 78 —	— 57 — 52	— 73 — 71	—	—
		Hard	.039/— .029/— .019/.039 .009/.029	72	87	—	—	82 — 80 —	90 — 88 —	— 71 — 69	— 77 — 75	—	—
		Extra Hard	.039/— .029/— .019/.039 .009/.029	84	99	—	—	88 — 86 —	94 — 92 —	— 75 — 73	— 80 — 78	—	—
		Spring	.039/— .029/— .019/.039 .009/.029	91	105	—	—	90 — 88 —	96 — 94 —	— 77 — 75	— 81 — 79	—	—
		Extra Spring	.039/— .029/— .019/.039 .009/.029	96	109	—	—	92 — 89 —	97 — 94 —	— 78 — 76	— 82 — 80	—	—

TABLE 2B—MINIMUM MECHANICAL PROPERTIES OF WROUGHT COPPER ALLOYS (continued)

Customary Units

UNS No.[ee]	Form	Temper	Size Section, in Over/Thru	Tensile Strength, ksi Min	Tensile Strength, ksi Max	Yield Strength, Min ksi 0.5% Ext Under Load	Elongation, Min %[c] In 2 in	Hardness RB[b] Min	Hardness RB[b] Max	Hardness R30T[b] Min	Hardness R30T[b] Max	Grain Size, mm Min	Grain Size, mm Max
C52100	Sheet and Strip	Soft	.039/— .029/—	53	67	—	—	29 —	70 —	— 38	— 68	—	—
		Soft	.019/.039 .009/.029	53	67	—	—	20 —	66 —	— 27	— 62	—	—
		1/2 Hard	.039/— .029/— .019/.039 .009/.029	69	84	—	—	76 — 69 —	91 — 88 —	— 67 — 63	— 78 — 75	—	—
		Hard	.039/— .029/— .019/.039 .009/.029	85	100	—	—	91 — 89 —	97 — 95 —	— 76 — 73	— 81 — 80	—	—
		Extra Hard	.039/— .029/— .019/.039 .009/.029	97	112	—	—	95 — 93 —	100 — 98 —	— 78 — 77	— 83 — 82	—	—
		Spring	.039/— .029/— .019/.039 .009/.029	105	119	—	—	97 — 95 —	102 — 100 —	— 79 — 78	— 84 — 83	—	—
		Extra Spring	.039/— .029/— .019/.039 .009/.029	110	122	—	—	98 — 96 —	103 — 101 —	— 80 — 79	— 84 — 83	—	—
φC52400	Sheet and Strip	Soft	.039/— .029/— .019/.039 .009/.039	58	73	—	—	35 — 25 —	75 — 71 —	— 40 — 29	— 78 — 84	—	—
		1/2 Hard	.039/— .029/— .019/.039 .009/.029	76	91	—	—	78 — 74 —	95 — 93 —	— 67 — 63	— 80 — 77	—	—
		Hard	.039/— .029/— .019/.039 .009/.029	94	109	—	—	94 — 92 —	101 — 100 —	— 78 — 75	— 82 — 81	—	—
		Extra Hard	.039/— .029/— .019/.039 .009/.029	107	122	—	—	93 — 97 —	103 — 102 —	— 80 — 79	— 84 — 83	—	—
		Spring	.039/— .029/— .019/.039 .009/.029	115	129	—	—	99 — 98 —	104 — 103 —	— 81 — 80	— 85 — 84	—	—
		Extra Spring	.039/— .029/—	120	133	—	—	100 —	105 —	— 82	— 86	—	—
		Extra Spring	.019/.039 .009/.029	120	133	—	—	99 —	104 —	— 81	— 85	—	—
	Sheet and Strip	See C51100											
C54400	Rod	Hard	.061/.250[bb] .249/.500[bb] .500/1.0[bb] 1.0/—[bb]	65 60 55 50	— — — —	— — — —	In 4 x Dia or Thickness of Specimen 8 10 12 15	—	—	—	—	—	—
	Flat Products	Hard	.249/.375[cc] .375/—[cc]	55 50	— —	— —	10 15	—	—	—	—	—	—
C60800	Tube	Annealed	All Sizes	50	—	19	—	—	—	—	—	0.010	0.045

TABLE 2B—MINIMUM MECHANICAL PROPERTIES OF WROUGHT COPPER ALLOYS (continued)

Customary Units

UNS No.[ee]	Form	Temper	Size Section, in Over/Thru		Tensile Strength, ksi Min	Tensile Strength, ksi Max	Yield Strength, Min ksi 0.5% Ext Under Load	Elongation, Min %[c] In 2 in	Hardness RB[b] Min	Hardness RB[b] Max	Hardness BHN 1000 kg Min	Hardness BHN 1000 kg Max	Grain Size, mm Min	Grain Size, mm Max
			Thickness	Width										
φC61400	Plate,[t] Sheet, Strip and Rolled Bar	Soft	—/.500 .500/2.0 2.0/5.0	All Widths All Widths All Widths	72 70 65	— — —	32 30 28	35 35 35	— — —	— — —	— — —	— — —	— — —	— — —
		Hard	—/.500 .500/1.0	All Widths All Widths	80 70	— —	45 40	25 30	— —	— —	— —	— —	— —	— —
			Size Section, in					In 4 x Dia						
	Rod and Bar	—	—/.500 .500/1.0 1.0/2.0 2.0/3.0		80 75 70 70	— — — —	40 35 32 30	30 30 30 30	— — — —	— — — —	— — — —	— — — —	— — — —	— — — —
φC61800	Rod	1/2 Hard	—/3.0		—	—	—	—	—	—	—	—	—	—
C62300	Rod and Bar	Drawn	—/.500 .500/1.0 1.0/2.0 2.0/3.0 3.0/—		90 88 84 76 75	— — — — —	45 44 40 37 30	12 15 15 20 20	90 — — — —	— — — — —	155 — — — —	— — — — —	— — — — —	— — — — —
	Shapes	—	All Sizes		75	—	30	20	—	—	—	—	—	—
	Forgings	As Forged	—/1.5 1.5/—		75 72	— —	30 28	20 25	— —	— —	— —	— —	— —	— —
											3000 kg			
C62400	Rod	As Extruded	All Sizes		80	—	45	7.0	—	—	175	—	—	—
		As Forged	—		80	—	45	7.0	—	—	175	—	—	—
	Forgings	Hardened and H₂O Quenched	—		—	—	—	—	—	—	240	—	—	—
		Hardened, H₂O Quenched and Tempered	—		87	—	—	8.0	—	—	179	—	—	—
									RF[b]					
C63000	Rod and Bar	Annealed	.499/1.0 1.0/2.0 2.0/4.0		100 90 85	— — —	50 45 42.5	5 6 10	— — —	— — —	— — —	— — —	— — —	— — —
	Shapes	As Extruded	All Sizes		85	—	42.5	10	—	—	—	—	—	—
φC64200	Rod and Bar	Drawn	—/.500 .500/1.0 1.0/2.0 2.0/3.0		90 85 80 75	— — — —	45 45 42 35	9 12 12 15	— — — —	— — — —	— — — —	— — — —	— — — —	— — — —
C65500	Plate,[t] Sheet, Strip and Rolled Bar	Annealed 0.070 mm 0.040 mm	—		52 55	58 64	— —	— —	— —	— —	70 76	82 93	0.050 Note a	0.110 0.055
		1/4 Hard 1/2 Hard[k] Hard[k] Extra Hard[k]	—		62 71 87 99	72 81 97 108	— — — —	— — — —	65 79 88 93	80 91 96 98	— — — —	— — — —	— — — —	— — — —
		Spring[k]	—		105	113	—	—	94	99	—	—	—	—
		Hot Rolled	—		55	72	—	—	—	—	72	—	—	—
		Hot Rolled and Cold Rolled Finish	—		58	72	—	—	60	80	—	—	—	—
	Rod, Bar and Shapes	Soft	All Forms and Sizes		52	—	15	35	—	—	—	—	—	—
		1/4 Hard	All Forms and Sizes		55	—	24	25	—	—	—	—	—	—
		1/2 Hard	—/2.0[dd]		70	—	38	17	—	—	—	—	—	—

TABLE 2B—MINIMUM MECHANICAL PROPERTIES OF WROUGHT COPPER ALLOYS (continued)

Customary Units

UNS No.[ee]	Form	Temper	Size Section, in Over/Thru	Tensile Strength, ksi Min	Tensile Strength, ksi Max	Yield Strength, Min ksi 0.5% Ext Under Load	Elongation, Min %[c] In 4 x Dia or Thickness of Specimen	Hardness Min	Hardness Max	Hardness Min	Hardness Max	Grain Size, mm Min	Grain Size, mm Max
								RB[b]		RF[b]			
C65500	Rod, Bar and Shapes	Hard	—/.250[dd] .250/1.500[dd]	85 85	— —	50 50	8 13	— —	— —	— —	— —	— —	— —
		Extra Hard	—/.500[w]	100	—	55	7	—	—	—	—	—	—
	Bar	Soft	All Sizes	52	—	15	35	—	—	—	—	—	—
		Hard	—/1.0 1.0/1.5 1.5/3.0	65 60 55	— — —	38 30 24	20 25 27	— — —	— — —	— — —	— — —	— — —	— — —
C67000	Rod and Bar	Soft	All Sizes	85	—	45	10[v]	—	—	—	—	—	—
		1/2 Hard Hard	All Sizes All Sizes	105 115	— —	60 68	7[v] 5[v]	— —	— —	— —	— —	— —	— —
	Forgings	Soft	—	85	—	45	10	—	—	—	—	—	—
		1/2 Hard Hard	— —	93 100	— —	50 55	7 5	— —	— —	— —	— —	— —	— —
C67300	Rod and Bar	As Extruded	All Sizes	55	—	25	20	60	—	—	—	—	—
		Soft	All Sizes	55	—	25	20	50	—	—	—	—	—
		1/2 Hard[w]	—/1.0	65	—	40	12	70	—	—	—	—	—
								R30T[b]					
		1/2 Hard[w]	1.0/3.0 3.0/—	58 55	— —	35 30	15 18	70 65	— —	— —	— —	— —	— —
		1/2 Hard[x]	All Sizes	60	—	30	20	70	—	—	—	—	—
		Hard[w]	—/1.0 1.0/2.0	70 62	— —	50 42	10 15	70 70	— —	— —	— —	— —	— —
	Shapes	As Extruded	All Sizes	55	—	25	20	60	—	—	—	—	—
	Forgings	As Forged	—	55	—	25	20	60	—	—	—	—	—
		Forged and Heat Treated	—	69	—	35	12	70	—	—	—	—	—
C67400	Rod and Bar	As Extruded	All Sizes	70	—	34	12	75	—	—	—	—	—
		Extruded and Drawn	—/1.0 1.0/2.0 2.0/3.0	78 75 70	— — —	40 40 36	8 10 12	84 80 78	— — —	— — —	— — —	— — —	— — —
	Shapes	As Extruded	All Sizes	70	—	34	12	75	—	—	—	—	—
	Forgings	As Forged	—	65	—	30	15	75	—	—	—	—	—
C67500	Rod	Soft	All Sizes	55	—	22	20	—	—	—	—	—	—
		1/2 Hard	—/1.0 1.0/2.5 2.5/—	72 70 65	— — —	36 35 32	13 15 17	— — —	— — —	— — —	— — —	— — —	— — —
		Hard	—/1.0 1.0/1.5 1.5/2.5 2.5/—	80 76 73 68	— — — —	56 52 48 45	8 10 12 16	— — — —	— — — —	— — — —	— — — —	— — — —	— — — —
	Bar	Soft	All Sizes	55	—	22	20	—	—	—	—	—	—
		1/2 Hard	—/1.0 1.0/2.5 2.5/—	72 70 65	— — —	36 35 32	13 15 17	— — —	— — —	— — —	— — —	— — —	— — —
		Hard	—/1.0 1.0/2.5 2.5/—	76 72 68	— — —	52 47 45	8 12 16	— — —	— — —	— — —	— — —	— — —	— — —

TABLE 2B—MINIMUM MECHANICAL PROPERTIES OF WROUGHT COPPER ALLOYS (continued)

Customary Units

UNS No.[ee]	Form	Temper	Size Section, in Over/Thru	Tensile Strength, ksi Min	Tensile Strength, ksi Max	Yield Strength, Min ksi 0.5% Ext Under Load	Elongation, Min %[c] In 4 x Dia or Thickness of Specimen	Hardness RB[b] Min	Hardness RB[b] Max	Hardness R30T[b] Min	Hardness R30T[b] Max	Grain Size, mm Min	Grain Size, mm Max
C67500	Shapes	Soft	All Sizes	55	—	22	20	—	—	—	—	—	—
C70600	Condenser Tube Plate	—	—/2.5	40	—	15	In 2 in 30	—	—	—	—	—	—
	Tube	Annealed Light Drawn	— —	40 45	— —	15 35	— —	— —	— —	— —	— —	— —	— —

							Hardness Min RF[b]	Hardness Max RF[b]					
C71000	Tube	Annealed	—	45	—	16	—	—	—	—	—	—	—
	Plate,[t] Sheet, Strip and Rolled Bar	Annealed 0.035 mm 0.015 mm	— —	— —	— —	— —	67 76	76 90	18 35	35 58	28 40	40 55	0.025 Note a	0.050 0.020
		1/4 Hard 1/2 Hard Hard Extra Hard	— — — —	47 56 67 72	63 70 79 84	— — — —	— — — —	— — — —	45 64 76 79	72 78 84 87	46 59 67 69	65 69 73 75	— — — —	— — — —
		Spring	—	76	87	—	—	—	82	88	71	75	—	—

							Elongation, Min %[c] In 2 in						
C71500	Condenser Tube Plate	—	—/2.5 2.5/5.0	50 45	— —	20 18	35 35	— —	— —	— —	— —	— —	— —
	Tube	Annealed	—	52	—	18	—	—	—	—	—	—	—
		Drawn and Stress Relieved	—/.048 Wall .048/— Wall	72 72	— —	50 50	12 15	— —	— —	— —	— —	— —	— —

							Hardness Min RF[b]	Hardness Max RF[b]					
	Plate,[t] Sheet, Strip and Rolled Bar	Annealed 0.035 mm 0.015 mm	— —	— —	— —	— —	70 74	85 93	23 37	45 63	31 40	46 58	0.025 Note a	0.050 0.025
		1/4 Hard 1/2 Hard Hard Extra Hard	— — — —	58 66 75 80	72 80 88 92	— — — —	— — — —	— — — —	67 76 83 85	81 85 89 91	61 67 72 73	71 74 76 77	— — — —	— — — —
		Spring	—	84	94	—	—	—	87	91	74	77	—	—
C75200	Plate,[t] Sheet, Strip and Rolled Bar	Annealed 0.070 mm 0.035 mm 0.015 mm	— — —	— — —	— — —	— — —	70 75 83	80 88 93	25 35 45	40 55 70	32 40 46	43 53 64	0.050 0.025 Note a	0.100 0.050 0.025
		1/4 Hard 1/2 Hard Hard Extra Hard	— — — —	58 66 78 86	72 80 91 96	— — — —	— — — —	— — — —	50 68 80 87	75 82 90 94	49 62 70 74	67 72 76 79	— — — —	— — — —
		Spring	—	90	101	—	—	—	89	96	75	80	—	—

TABLE 2B—MINIMUM MECHANICAL PROPERTIES OF WROUGHT COPPER ALLOYS (continued)

Customary Units

UNS No.[ee]	Form	Temper	Size Section, in Over/Thru	Tensile Strength, ksi Min	Tensile Strength, ksi Max	Yield Strength, Min ksi 0.5% Ext Under Load	Hardness Min	Hardness Max	Hardness Min	Hardness Max	Hardness Min	Hardness Max	Grain Size, mm Min	Grain Size, mm Max
							RF[b]		RB[b]		R30T[b]			
C75200	Rod and Bar	Annealed 0.070 mm 0.035 mm 0.015 mm	— — —	— — —	— — —	— — —	— — —	— — —	— — —	— — —	— — —	— — —	0.050 0.025 —	0.100 0.050 0.030
		1/4 Hard	.049/.500[y]	60	80	—	—	—	—	—	—	—	—	—
		Hard	.019/.250[w] .250/.500[w] .500/1.0[w] 1.0/—[w]	80 70 65 60	100 90 85 80	— — — —	— — — —	— — — —	— — — —	— — — —	— — — —	— — — —	— — — —	— — — —
			All Sizes[x]	68	88	—	—	—	—	—	—	—	—	—
C77000	Plate,[t] Sheet, Strip and Rolled Bar	Annealed 0.070 mm 0.035 mm 0.015 mm	— — —	— — —	— — —	— — —	72 76 84	83 91 98	29 37 47	45 60 73	35 41 47	46 57 65	0.050 0.025 Note a	0.100 0.050 0.025
							RG[b]							
		1/4 Hard 1/2 Hard Hard Extra Hard	— — — —	69 78 92 102	87 95 107 115	— — — —	23 51 67 73	62 69 76 80	70 81 90 95	88 92 96 99	63 71 76 79	75 78 80 82	— — — —	— — — —
		Spring	—	108	120	—	77	83	97	100	80	—	—	—
	Rod and Bar	Annealed 0.070 mm 0.035 mm 0.015 mm	— — —	— — —	— — —	— — —	— — —	— — —	— — —	— — —	— — —	— — —	0.050 0.025 —	0.100 0.050 0.030
		1/4 Hard	.019/.500[y]	75	95	—	—	—	—	—	—	—	—	—
		Hard	.019/.050[w] .250/.500[w] .500/1.0[w] 1.0/—[w]	90 80 75 70	110 100 95 90	— — — —	— — — —	— — — —	— — — —	— — — —	— — — —	— — — —	— — — —	— — — —
			All Sizes[x]	75	95	—	—	—	—	—	—	—	—	—

[a] Although no minimum grain size is required, this material must be fully recrystallized.
[b] Values are approximate. F and B scales for metal .020 inch and over in thickness. 30T scale for metal .012 inch and over in thickness. (.015 inch for annealed material to ASTM B36 and B122).
[c] In any case, a minimum gage length of 1 inch shall be used.
[d] Hardness values shall apply only to tubes having a wall thickness of .015 inch or over for annealed temper and .020 inch or over for drawn temper (.012 inch for drawn temper to ASTM B135), to round tubes having an inside diameter of .312 inch or over, and to rectangular, including square, tubes having an inside major distance between parallel surfaces of .188 inch or over. For all other tubes, no Rockwell values shall apply. Hardness tests shall be made on the inside surface of the tube. When suitable equipment is not available for determining the specified hardness, other Rockwell scales and values may be specified, subject to agreement between purchaser and supplier.
[e] .125 to .875 inch outside diameter; .030 to .045 inch wall.
[f] Generally available in round, hexagonal, and octagonal.
[g] Normally available in round only.
[h] Not generally available in sizes over .500 inch in diameter.
[i] Not generally available in sizes over .375 inch in diameter. Square and rectangular wire not generally available.
[j] Not generally available in sizes over .250 inch in diameter. Square and rectangular wire not generally available.
[k] Commercially supplied only as strip. Manufacturer should be consulted for sheet or plate.
[l] Capable of being hardened by further heat treatment.
[m] Rods and bars.
[n] Applicable to material .004 inch and over.
[o] Applicable to material .032 inch and over.
[p] Applicable to material .015 inch and over.
[q] When stated on contract or order, tension test shall be waived provided the strip meets the hardness requirement. In case of dispute, tension test shall be the basis for acceptance.
[r] Commonly supplied only as strip.
[s] .250 inch and over.
[t] Plate generally available only in annealed, quarter hard, and half hard.
[u] Type A material to listed hardness limits supplied unless otherwise specified.
[v] Cold finished.
[w] Rods only.
[x] Bars only.
[y] After heat treatment.
[z] After mill heat treatment.
[aa] Rounds only.
[bb] Rounds and hexagons.
[cc] Squares and rectangles.
[dd] Rods and square bars.
[ee] Unified Number System. For cross reference to SAE, former SAE, ASTM and former trade names, see SAE J461.

TABLE 3—INDEX TO STANDARD PRODUCT TOLERANCE TABLES

UNS No.[a]	Classification[b]	UNS No.[a]	Classification[b]	UNS No.[a]	Classification[b]	UNS No.[a]	Classification[b]
C10200	Nonrefractory	C17500	Refractory	C34500	Nonrefractory	C61800	Refractory
C11000	Nonrefractory	C17600	Refractory	C35000	Nonrefractory	C62300	Refractory
C11100	Nonrefractory	C18400	Refractory[d]	C36000	Nonrefractory	C62400	Refractory
C11300	Nonrefractory	C18700	Nonrefractory	C37700	Refractory[e]	C63000	Refractory
C11400	Nonrefractory	C19200	Nonrefractory	C46400	Refractory[e,f]	C64200	Refractory
C11500	Nonrefractory	C21000	Nonrefractory	C46500	Refractory[e]	C65500	Refractory[e]
C11600	Nonrefractory	C22000	Nonrefractory	C46600	Refractory[e]	C67000	Refractory[e]
C12000	Nonrefractory	C23000	Nonrefractory	C46700	Refractory[e]	C67300	Refractory[e]
C12200	Nonrefractory	C24000	Nonrefractory	C51000	Refractory[e]	C67400	Refractory[e]
C14500	Nonrefractory	C26000	Nonrefractory	C51100	Refractory[e]	C67500	Refractory[e]
C14700	Nonrefractory	C26800	Nonrefractory	C52100	Refractory[e]	C70600	Refractory
C15000	Nonrefractory[c]	C27000	Nonrefractory	C52400	Refractory[e]	C71000	Refractory
C16200	Nonrefractory[c]	C33000	Nonrefractory	C54400	Refractory[e]	C71500	Refractory
C17000	Refractory	C33100	Nonrefractory	C60800	Refractory	C75200	Refractory[e]
C17200	Refractory	C34200	Nonrefractory	C61400	Refractory	C77000	Refractory

[a] Unified Numbering System. For cross reference to SAE, former SAE, ASTM, and former trade names, see SAE J461.
[b] To determine tolerances, use the applicable portion of the standard product tolerance Tables 4–11. This index does not imply that all wrought forms of each of the listed alloys are readily available.
[c] Refractory for hot rolled or as extruded rod and bar.
[d] Nonrefractory for round wire and round tube.
[e] Nonrefractory for hot rolled or as extruded rod and bar.
[f] Nonrefractory for rectangular bar and wire.

TABLE 4A—THICKNESS TOLERANCES—FLAT PRODUCTS
COLD ROLLED WITH SLIT, SLIT AND EDGE ROLLED, SHEARED, SAWED, OR MACHINED EDGES

Metric (SI) Units

Thickness, mm		Thickness Tolerances, Plus and Minus,[a] mm For Specified Width, mm															
		Nonrefractory								Refractory							
Over	Thru	Over–Thru 200	200 300	300 350	350 500	500 700	700 900	900 1200	1200 1500	— 200	200 300	300 350	350 500	500 700	700 900	900 1200	1200 1500
		Strip				Sheet				Strip				Sheet			
—	0.1	0.008	0.015	0.015	—	—	—	—	—	0.010	0.02	0.02	—	—	—	—	—
0.1	0.2	0.015	0.025	0.025	0.040	—	—	—	—	0.020	0.03	0.03	0.05	—	—	—	—
0.2	0.3	0.020	0.030	0.030	0.045	0.06	0.08	0.09	0.10	0.025	0.03	0.03	0.06	—	—	—	—
0.3	0.4	0.025	0.040	0.040	0.050	0.06	0.08	0.09	0.12	0.030	0.05	0.05	0.06	—	—	—	—
0.4	0.5	0.03	0.045	0.045	0.050	0.08	0.09	0.10	0.13	0.035	0.06	0.06	0.08	—	—	—	—
0.5	0.6	0.04	0.05	0.05	0.06	0.08	0.09	0.10	0.13	0.05	0.06	0.06	0.08	0.10	0.13	0.15	0.18
0.6	0.8	0.05	0.05	0.05	0.06	0.09	0.10	0.13	0.15	0.06	0.08	0.08	0.09	0.13	0.15	0.18	0.20
0.8	1.0	0.05	0.06	0.06	0.08	0.10	0.13	0.15	0.18	0.08	0.09	0.09	0.10	0.15	0.18	0.20	0.25
1.0	1.5	0.06	0.08	0.08	0.09	0.13	0.15	0.18	0.20	0.09	0.10	0.10	0.12	0.18	0.20	0.25	0.30
1.5	3.0	0.08	0.09	0.09	0.10	0.15	0.18	0.20	0.25	0.10	0.12	0.12	0.13	0.20	0.25	0.30	0.35
3.0	5.0	0.09	0.10	0.10	0.12	0.18	0.20	0.25	0.30	0.12	0.13	0.13	0.15	0.25	0.30	0.35	0.40
		Bar				Plate				Bar				Plate			
5.0	7.5	0.10	0.12	0.12	0.13	0.23	0.25	0.30	0.35	0.13	0.15	0.15	0.18	0.30	0.35	0.40	0.45
7.5	12.0	0.12	0.13	0.13	0.15	0.30	0.33	0.38	0.45	0.15	0.18	0.18	0.20	0.38	0.43	0.48	0.58
12.0	20.0	0.14	0.18	0.18	0.23	0.38	0.43	0.48	0.58	0.20	0.25	0.25	0.30	0.48	0.53	0.60	0.73
20.0	30.0	0.18	0.23	0.23	0.28	0.45	0.53	0.60	0.73	0.25	0.30	0.30	0.38	0.58	0.65	0.75	0.93
30.0	40.0	0.55	0.55	0.55	0.55	0.63	0.73	0.90	0.70	0.70	0.70	0.70	0.70	0.80	0.93	1.13	0.93
40.0	50.0	0.65	0.65	0.65	0.65	0.65	0.75	0.90	1.10	0.83	0.83	0.83	0.83	0.83	0.95	1.13	1.28

[a] NOTE to users: If tolerances are desired all plus or all minus, it is normal practice to specify double the values given herein.

TABLE 4B—THICKNESS TOLERANCES OF STRIP, SHEET, PLATE, AND ROLLED BAR

Customary Units

Thickness, in		Thickness Tolerances, Plus and Minus,[a] in For Specified Width, in															
		Nonrefractory								Refractory							
Over	Thru	Over–Thru 8.00	8.00 12.00	12.00 14.00	14.00 20.00	20.00 28.00	28.00 36.00	36.00 48.00	48.00 60.00	— 8.00	8.00 12.00	12.00 14.00	14.00 20.00	20.00 28.00	28.00 36.00	36.00 48.00	48.00 60.00
		Strip				Sheet				Strip				Sheet			
—	.004	.0003	.0006	.0006	—	—	—	—	—	.0004	.0008	.0008	—	—	—	—	—
.004	.006	.0004	.0008	.0008	.0013	—	—	—	—	.0006	.0010	.0010	.0015	—	—	—	—
.006	.009	.0006	.0010	.0010	.0015	—	—	—	—	.0008	.0013	.0013	.0020	—	—	—	—
.009	.013	.0008	.0013	.0013	.0018	.0025	.0030	.0035	.0040	.0010	.0015	.0015	.0025	—	—	—	—
.013	.017	.0010	.0015	.0015	.0020	.0025	.0030	.0035	.0045	.0013	.0020	.0020	.0025	—	—	—	—
.017	.021	.0013	.0018	.0018	.0020	.0030	.0035	.0040	.0050	.0015	.0025	.0025	.0030	—	—	—	—
.021	.026	.0015	.0020	.0020	.0025	.0030	.0035	.0040	.0050	.0020	.0025	.0025	.0030	.0040	.0050	.0060	.0070
.026	.037	.0020	.0020	.0020	.0025	.0035	.0040	.0050	.0060	.0025	.0030	.0030	.0035	.0050	.0060	.0070	.0080
.037	.050	.0020	.0025	.0025	.0030	.0040	.0050	.0060	.0070	.0030	.0035	.0035	.0040	.0060	.0070	.0080	.0100
.050	.073	.0025	.0030	.0030	.0035	.0050	.0060	.0070	.0080	.0035	.0040	.0040	.0045	.0070	.0080	.0100	.0120
.073	.130	.0030	.0035	.0035	.0040	.0060	.0070	.0080	.0100	.0040	.0045	.0045	.0050	.0080	.0100	.0120	.0140
.130	.188	.0035	.0040	.0040	.0045	.0070	.0080	.0100	.0120	.0045	.0050	.0050	.0060	.0100	.0120	.0140	.0160
		Rolled bar				Plate				Rolled bar				Plate			
.188	.205	.0035	.0040	.0040	.0045	.0070	.0080	.0100	.0120	.0045	.0050	.0050	.0060	.0100	.0120	.0140	.0160
.205	.300	.0040	.0045	.0045	.0050	.0090	.0100	.0120	.0140	.0050	.0060	.0060	.0070	.0120	.0140	.0160	.0180
.300	.500	.0045	.0050	.0050	.0060	.0120	.0130	.0150	.0180	.0060	.0070	.0070	.0080	.0150	.0170	.0190	.0230
.500	.750	.0055	.0070	.0070	.0090	.0150	.0170	.0190	.0230	.0080	.0100	.0100	.0120	.0190	.0210	.0240	.0290
.750	1.000	.0070	.0090	.0090	.0110	.0180	.0210	.0240	.0290	.0100	.0120	.0120	.0150	.0230	.0260	.0300	.0370
1.000	1.500	.0220	.0220	.0220	.0220	.0220	.0250	.0290	.0360	.0280	.0280	.0280	.0280	.0280	.0320	.0370	.0450
1.500	2.000	.0260	.0260	.0260	.0260	.0260	.0300	.0360	.0440	.0330	.0330	.0330	.0330	.0330	.0380	.0450	.0550

[a] NOTE to user: If tolerances are desired all plus or all minus, it is normal practice to specify double the values given herein.

TABLE 5A—WIDTH TOLERANCES SLIT METAL AND SLIT METAL WITH ROLLED EDGES

Metric (SI) Units

Width, mm		Width Tolerances,[a] Plus and Minus, mm For Specified Thickness, mm			
Over	Thru	Over 0.1 Thru 1.0	1.0 3.0	3.0 5.0	5.0 13.0
—	50	0.13	0.25	0.30	0.38
50	200	0.20	0.33	0.38	0.38
200	600	0.40	0.40	0.40	0.80

[a] NOTE to users: If tolerances are specified as all plus or all minus, double the values given.

TABLE 5B—WIDTH TOLERANCES STRIP METAL WITH SLIT EDGES AND SLIT METAL WITH ROLLED EDGES

Customary Units

Width, in		Width Tolerances,[a] Plus and Minus, in For Specified Thickness, in			
Over	Thru	Over .003 Thru .032	.032 .125	.125 .188	.188 .500
—	2.0	.005	.010	.012	.015
2.0	8.0	.008	.013	.015	.015
8.0	20.0	.016	.016	.016	.031

[a] If tolerances are specified as all plus or all minus, double the values given.

TABLE 6A—THICKNESS TOLERANCES FOR FLAT PRODUCTS—WITH ROLLED OR DRAWN EDGES INCLUDING SQUARES AND BUS BAR STOCK—WITH ANY STANDARD EDGE CONTOUR

Metric (SI) Units

Nonrefractory								Refractory							
Thickness, mm		Thickness Tolerances, Plus and Minus,[a] mm For Specified Width, mm						Thickness, mm		Thickness Tolerances, Plus and Minus,[a] mm For Specified Width, mm					
Over	Thru	Over–Thru 12	12 30	30 50	50 100	100 200	200 300	Over	Thru	Over–Thru 12	12 30	30 50	50 100	100 200	200 300
		Flat Wire		Strip						Flat Wire		Strip			
—	0.5	0.2	0.03	—	—	—	—	—	1.5	0.04	0.05	—	—	—	—
0.5	1.5	0.03	0.04	0.05	—	—	—	1.5	2.5	0.05	0.08	0.10	0.13	—	—
1.5	2.5	0.04	0.05	0.08	0.09	—	—	2.5	3.5	0.08	0.10	0.12	0.15	—	—
2.5	3.5	0.05	0.06	0.09	0.10	—	—	3.5	5.0	0.10	0.12	0.13	0.18	0.23	0.30
3.5	5.0	0.08	0.09	0.10	0.12	0.15	0.20								
		Bar								Bar					
5.0	15.0	0.09	0.10	0.12	0.12	0.15	0.20	5.0	15.0	0.13	0.13	0.15	0.18	0.23	0.30
15.0	25.0	—	0.12	0.13	0.13	0.18	0.23	15.0	25.0	—	0.15	0.18	0.20	0.25	0.33
25.0	50.0	—	0.13	0.13	0.15	0.20	—	25.0	50.0	—	—	0.15	0.18	0.23	0.28
50.0	100.0	—	—	—	0.60%[b]	—	—	50.0	100.0	—	—	—	1.00%[b]	—	—

[a] NOTE to users: If tolerances are desired all plus or all minus, it is normal practice to specify double the values given herein.
[b] Expressed to the nearest 0.03 mm.

TABLE 6B—THICKNESS TOLERANCES FOR RECTANGULAR AND SQUARE BAR AND WIRE

Customary Units

Thickness, in		Thickness Tolerances, Plus and Minus,[a] in For Specified Width, in							
		Bar						Wire	
Over	Thru	Over–Thru .500	.500 1.250	1.250 2.000	2.000 4.000	4.000 8.000	8.000 12.000	— .500	.500 1.250
		Nonrefractory							
—	.013	—	—	—	—	—	—	.0010	.0013
.013	.050	—	—	—	—	—	—	.0013	.0015
.050	.090	—	—	—	—	—	—	.0015	.0020
.090	.130	—	—	—	—	—	—	.0020	.0025
.130	.188	—	—	—	—	—	—	.0030	.0035
.188	.500	.0035	.0040	.0045	.0045	.0060	.0080	—	—
.500	1.000	—	.0045	.0050	.0050	.0070	.0090	—	—
1.000	2.000	—	.0050	.0050	.0060	.0080	—	—	—
2.000	4.000	—	—	—	.30%[b]	—	—	—	—
		Refractory							
—	.050	—	—	—	—	—	—	.0015	.0020
.050	.090	—	—	—	—	—	—	.0020	.0030
.090	.130	—	—	—	—	—	—	.0030	.0040
.130	.188	—	—	—	—	—	—	.0040	.0045
.188	.500	.0050	.0050	.0060	.0070	.0090	.0120	—	—
.500	1.000	—	.0060	.0070	.0080	.0100	.0130	—	—
1.000	2.000	—	.0060	.0070	.0090	.0110	—	—	—
2.000	4.000	—	—	—	.50%[b]	—	—	—	—

[a] NOTE to users: If tolerances are desired all plus or all minus, it is normal practice to specify double the values given herein.
[b] Expressed to the nearest 0.001 in.

TABLE 7—WIDTH TOLERANCES FOR RECTANGLES (NOT INCLUDING SQUARES) FLAT WIRE AND BAR

TABLE 7A—METRIC (SI) UNITS

Width, mm		Width Tolerance Plus and Minus, mm		Width, mm		Width Tolerance Plus and Minus, mm	
Over	Thru	Nonrefractory	Refractory	Over	Thru	Nonrefractory	Refractory
—	1.5	0.03	0.04	15.0	30.0	0.13	0.18
1.5	2.5	0.04	0.05	30.0	50.0	0.20	0.25
2.5	3.5	0.05	0.08	50.0	100.0	0.30	0.38
3.5	5.0	0.08	0.10	100.0	300.0	0.60%[b]	1.00%[b]
5.0	15.0	0.09	0.13				

TABLE 7B—CUSTOMARY UNITS

Width, in		Width Tolerance[a] Plus and Minus, in		Width, in		Width Tolerance Plus and Minus, in	
Over	Thru	Nonrefractory	Refractory	Over	Thru	Nonrefractory	Refractory
—	.050	.0013	.0015	.500	1.250	.0050	.0070
.050	.090	.0015	.0020	1.250	2.000	.0080	.0100
.090	.130	.0020	.0030	2.000	4.000	.0120	.0150
.130	.188	.0030	.0040	4.000	12.00	.30%[c]	.50%[c]
.188	.500	.0035	.0050				

[a] NOTE to users: If tolerances are desired all plus or all minus, it is normal practice to specify double the values given herein.
[b] Expressed to the nearest 0.03 mm.
[c] Expressed to the nearest .001 in.

TABLE 8A—CROSS SECTION TOLERANCES

Metric (SI) Units

TABLE 8A1—WIRE,[a] BARE AND METALLIC COATED, DRAWN TO SIZE

Diameter or Distance Between Parallel Surfaces, mm		Tolerances, Plus and Minus, mm				Diameter or Distance Between Parallel Surfaces, mm		Tolerances, Plus and Minus, mm			
		Nonrefractory		Refractory				Nonrefractory		Refractory	
Over	Thru	Round	Hexagon Octagon	Round	Hexagon Octagon	Over	Thru	Round	Hexagon Octagon	Round	Hexagon Octagon
—	0.25	0.003	—	0.005	—	1.25	1.50	0.015	0.03	0.03	0.08
0.25	0.50	0.005	—	0.008	—	1.50	2.00	0.020	0.04	0.040	0.10
0.50	0.75	0.008	—	0.013	—	2.00	4.00	0.025	0.05	0.05	0.10
0.75	1.00	0.010	0.020	0.018	0.05	4.00	12.00	0.040	0.08	0.05	0.10
1.00	1.25	0.013	0.025	0.020	0.08	12.00	19.00	0.050	0.10	0.08	0.13

TABLE 8A2—ROD[b] AND WIRE

Diameter or Distance Between Parallel Surfaces, mm		Tolerances, Plus and Minus, mm				Diameter or Distance Between Parallel Surfaces, mm		Tolerances, Plus and Minus, mm			
		Nonrefractory		Refractory				Nonrefractory		Refractory	
Over	Thru	Round	Hexagon Octagon	Round	Hexagon Octagon	Over	Thru	Round	Hexagon Octagon	Round	Hexagon Octagon
—	5.0	0.033	0.06	0.05	—	25.0	50.0	0.06	0.13	0.10	0.15
5.0	15.0	0.038	0.08	0.05	0.10	50.0	—	0.30%[c]	0.60%[c]	0.40%[c]	0.80%[c]
15.0	25.0	0.050	0.10	0.08	0.13						

TABLE 8A3—EXTRUDED ROD, BAR AND WIRE—FINISHED AS EXTRUDED

Diameter or Distance Between Parallel Surfaces mm		Tolerances, Plus and Minus, mm	
		Round, Square, Rectangular, Hexagon and Octagon	
Over	Thru	Nonrefractory	Refractory
—	25.0	0.25	0.50
25.00	50.0	0.38	0.75
50.0	75.0	0.63	1.25
75.0	90.0	0.88	1.75
90.0	100.0	1.50	3.00

TABLE 8A4—DIAMETER TOLERANCES HOT ROLLED ROUND ROD AND WIRE

Diameter, mm		Tolerances Plus and Minus, mm
Over	Thru	
6.3	20.0	0.38
20.0	30.0	0.50
30.0	40.0	0.75
40.0	75.0	1.60
75.0	—	3.15

[a] Wire for redrawing or rerolling, double the following tolerances.
[b] Rod-Cold drawn to size—except piston finish rod.
[c] Expressed to the nearest 0.03 mm.

TABLE 8B—DIAMETER (Across Flats) TOLERANCES OF WIRE, ROD AND BAR

Customary Units

Dia or Dist Between Parallel Surfaces, in		Tolerances, Plus and Minus,[a] in										
		Wire				Cold Drawn Rod				As Extruded Rod and Bar		Hot Rolled Round Rod
		Nonrefractory		Refractory		Nonrefractory		Refractory		Nonrefractory	Refractory	Nonrefractory and Refractory
Over	Thru	Round	Hexagon, Octagon	Round	Hexagon, Octagon	Round	Hexagon, Octagon	Round	Hexagon, Octagon	Rounds, Squares, Rectangles, Hexagon, Octagon	Rounds, Squares, Rectangles, Hexagon, Octagon	Rounds
—	.010	.0001	—	.0002	—	—	—	—	—	—	—	—
.010	.020	.0002	—	.0003	—	—	—	—	—	—	—	—
.020	.030	.0003	—	.0005	—	—	—	—	—	—	—	—
.030	.040	.0004	.0008	.0007	.0020	—	—	—	—	—	—	—
.040	.050	.0005	.0010	.0008	.0030	—	—	—	—	—	—	—
.050	.060	.0006	.0012	.0010	.0030	—	—	—	—	—	—	—
.060	.080	.0008	.0016	.0015	.0040	Note c	Note c	Note c	—	—	—	—
.080	.150	.0010	.0020	.0020	.0040	.0013	.0025	.0020	—	—	—	—
.150	.249	.0015	.0030	.0020	.0040	.0015	.0030	.0020	.0040	—	—	—
.250 only		—	—	—	—	—	—	—	—	—	—	+ .020–.010
.250	.500	.0015	.0030	.0020	.0040	.0015	.0030	.0020	.0040	—	—	.015
.500	.750	.0020	.0040	.0030	—	.0020	.0040	.0030	.0050	—	—	.015
.750	1.000	—	—	—	—	.0020	.0040	.0030	.0050	.010[d]	.020[d]	.020
1.000	1.250	—	—	—	—	.0025	.0050	.0040	.0060	.015	.030	.020
1.250	1.500	—	—	—	—	.0025	.0050	.0040	.0060	.015	.030	.030
1.500	2.000	—	—	—	—	.0025	.0050	.0040	.0060	.015	.030	.062
2:000	3.000	—	—	—	—	.15%[b]	.30%[b]	.20%[b]	.40%[b]	.025	.050	.062
3.000	3.500	—	—	—	—	.15%[b]	.30%[b]	.20%[b]	.40%[b]	.035	.070	.125
3.500	4.000	—	—	—	—	.15%[b]	.30%[b]	.20%[b]	.40%[b]	.060	.120	.125
4.000	—	—	—	—	—	.15%[b]	.30%[b]	.20%[b]	.40%[b]	—	—	.125

[a] If tolerances are desired all plus or all minus, double the values given.
[b] Expressed to the nearest .001 in.
[c] Up to and including .150 in.
[d] Up to and including 1.000 in.

TABLE 9A—WALL THICKNESS TOLERANCES[a,b] OF ROUND TUBE

Metric (SI) Units

Wall Thickness mm		Tolerances, Plus and Minus, mm For Specified Outside Diameters,[c] mm													
		Over 0.8 Thru 3.2		3.2 15.9		15.9 25.4		25.4 50.8		50.8 101.6		101.6 177.8		177.8 254.0	
Over	Thru	Nonre-fractory	Refrac-tory	Nonre-fractory	Refrac-tory	Nonre-fractory	Refrac-tory	Nonre-fractory	Refrac-tory	Nonre-fractory	Refrac-tory	Nonre-fractory	Refrac-tory	Nonre-fractory	Refrac-tory
—	0.46	0.05	0.06	0.03	0.04	0.04	0.05	0.05	0.06	—	—	—	—	—	—
0.46	0.64	0.08	0.10	0.05	0.06	0.05	0.06	0.06	0.08	—	—	—	—	—	—
0.64	0.89	0.08	0.10	0.06	0.08	0.06	0.08	0.08	0.10	0.10	0.13	—	—	—	—
0.89	1.47	0.08	0.10	0.08	0.10	0.09	0.11	0.09	0.11	0.13	0.17	0.18	0.23	—	—
1.47	2.11	—	—	0.09	0.11	0.10	0.13	0.10	0.13	0.15	0.19	0.20	0.25	0.25	0.33
2.11	3.05	—	—	0.10	0.13	0.13	0.17	0.13	0.17	0.18	0.23	0.23	0.28	0.28	0.36
3.05	4.19	—	—	0.13	0.18	0.15	0.18	0.15	0.19	0.20	0.25	0.25	0.33	0.30	0.38
4.19	5.59	—	—	0.18	—	0.19	0.23	0.20	0.25	0.25	0.33	0.30	0.38	0.36	0.46
5.59	7.21	—	—	—	—	0.23	0.30	0.25	0.33	0.30	0.38	0.36	0.46	0.41	0.51
7.21	9.65	—	—	—	—	0.28	—	0.30	0.38	0.36	0.46	0.41	0.51	0.46	0.58
9.65	—	—	—	—	—	—	—	5%	6%	5%	6%	6%	8%	6%	8%

[a] Maximum deviation at any point, if tolerances all plus or all minus are desired, double the values given.
[b] Tolerances of following dimensions may be specified for any two, but not all three: outside diameter, inside diameter, wall thickness.
[c] When round tube is ordered by outside and inside diameters, the maximum plus and minus deviation of the wall thickness from the mean at any point shall not exceed the values given in this table by more than 50%.

TABLE 9B—WALL THICKNESS TOLERANCES OF ROUND TUBE

Customary Units

Wall Thickness in		Tolerances, Plus and Minus,[a] in For Specified Outside Diameter,[b] in												
		Over .030 Thru .125		.125 .625		.625 1.00		1.00 2.00		2.00 4.00		4.00 7.00		7.00 10.00	
Over	Thru	Nonrefractory	Refractory	Nonrefractory	Refractory	Nonrefractory	Refractory	Nonrefractory	Refractory	Nonrefractory	Refractory	Nonrefractory	Refractory	Nonrefractory	Refractory
—	.017	.0020	.0025	.0010	.0015	.0015	.0020	.0020	.0025	—	—	—	—	—	—
.017	.024	.0030	.0040	.0020	.0025	.0020	.0025	.0025	.0030	—	—	—	—	—	—
.024	.034	.0030	.0040	.0025	.0030	.0025	.0030	.0030	.0040	.0040	.0050	—	—	—	—
.034	.057	.0030	.0040	.0030	.0040	.0035	.0045	.0035	.0045	.0050	.0065	.0070	.0090	—	—
.057	.082	—	—	.0035	.0045	.0040	.0050	.0040	.0050	.0060	.0075	.0080	.0100	.0100	.0130
.082	.119	—	—	.0040	.0050	.0050	.0065	.0050	.0065	.0070	.0090	.0090	.0110	.0110	.0140
.119	.164	—	—	.0050	.0070	.0060	.0070	.0060	.0075	.0080	.0100	.0100	.0130	.0120	.0150
.164	.219	—	—	.0070	—	.0075	.0090	.0080	.0100	.0100	.0130	.0120	.0150	.0140	.0180
.219	.283	—	—	—	—	.0090	.0120	.0100	.0130	.0120	.0150	.0140	.0180	.0160	.0200
.283	.379	—	—	—	—	.0110	—	.0120	.0150	.0140	.0180	.0160	.0200	.0180	.0230
.379	—	—	—	—	—	—	—	5%	6%	5%	6%	6%	8%	6%	8%

[a] Maximum deviation at any point. Note to Users: If tolerances are desired all plus or all minus, it is normal practice to specify double the values given herein.
[b] When round tube is ordered by outside and inside diameters, the maximum plus and minus deviation of the wall thickness from the nominal at any pont shall not exceed the values given in this table by more than 50 percent.

TABLE 10—AVERAGE DIAMETER TOLERANCES OF ROUND TUBE[a,b]

TABLE 10A—METRIC (SI) UNITS

Diameter, mm		Application of Tolerance	Diameter Tolerances Plus and Minus, mm	
Over	Thru		Nonrefractory	Refractory
—	3.2	ID	0.05	0.08
—	3.2	OD	0.05	0.06
3.2	15.9	ID or OD	0.05	0.06
15.9	25.4	ID or OD	0.06	0.08
25.4	50.8	ID or OD	0.08	0.10
50.8	76.2	ID or OD	0.10	0.13
76.2	101.6	ID or OD	0.13	0.15
101.6	127.0	ID or OD	0.15	0.20
127.0	152.4	ID or OD	0.18	0.23
152.4	203.2	ID or OD	0.20	0.25
203.2	254.0	ID or OD	0.25	0.33

TABLE 10B—CUSTOMARY UNITS

Diameter, in		Application of Tolerance	Diameter Tolerances Plus and Minus, in	
Over	Thru		Nonrefractory	Refractory
—	.125	ID	.0020	.0030
—	.125	OD	.0020	.0025
.125	.625	ID or OD	.0020	.0025
.625	1.00	ID or OD	.0025	.0030
1.00	2.00	ID or OD	.0030	.0040
2.00	3.00	ID or OD	.0040	.0050
3.00	4.00	ID or OD	.0050	.0060
4.00	5.00	ID or OD	.0060	.0080
5.00	6.00	ID or OD	.0070	.0090
6.00	8.00	ID or OD	.0080	.0100
8.00	10.00	ID or OD	.0100	.0130

[a] The average outside or inside diameter of a tube is the average of the maximum and minimum outside diameters, or of the maximum and minimum inside diameters, whichever is applicable, as determined at any cross-section of the tube.
[b] Tube tolerances—tolerances of the following dimensions may be specified for any two, but not all three: outside diameter, inside diameter, wall thickness.

TABLE 11—ROUNDNESS TOLERANCES OF ROUND TUBE

T/D Ratio of Nominal Wall Thickness to Nominal Outside Diameter		Roundness Tolerances,[a] % Nominal Outside Diameter Expressed to Nearest 0.03 mm (.001 in)
Over	Thru	
0.01	0.03	1.5%
0.03	0.05	1.0%
0.05	0.10	0.8% or 0.05 mm (.002 in)[b]
0.10	—	0.7% or 0.05 mm (.002 in)[b]

[a] The deviation from roundness is measured as the difference between major and minor outside diameters, as determined at any one cross-section of the tube.
[b] Whichever value is greater.

NOTE: For tube in any drawn temper in straight lengths. Not applicable to as extruded tube, redraw tube, annealed tube or any tube furnished in coils, or drawn tube whose wall thickness is under 0.40 mm (.016 in). Compliance with the roundness tolerance shall be determined by taking measurements on the outside diameter only, irrespective of the manner in which the tube dimensions are specified; whether outside diameter and wall thickness; outside diameter and inside diameter; or inside diameter and wall thickness.

MAGNESIUM ALLOYS—SAE J464c

SAE Information Report

Report of Nonferrous Metals Division approved January 1940 and last revised by Nonferrous Metals Committee April 1979.

1. Purpose—This report on magnesium alloys covers those alloys which have been more commonly used in the United States for automotive, aircraft, and missile applications. Basic information on nomenclature and temper designation is given. Design data and many characteristics covered by a purchase specification are not included.

2. Sources of Magnesium—Magnesium is the third most abundant structural element in the earth's crust, and considered inexhaustible. Common sources are sea water, natural brines, magnesite, and dolomite. Three methods of extraction are used in the United States. One method involves treating sea water with a source of alkalinity (lime or caustic soda) to precipitate the magnesium as hydroxide, which is then mixed with hydrochloric acid to produce magnesium chloride. The magnesium chloride is reduced electrolytically to produce magnesium metal and a mixture of chlorine and hydrochloric acid. A second method produces coproduct magnesium metal and pure chlorine in the electrolytic cell by the reduction of anhydrous magnesium chloride. The anhydrous cell feed results from the dehydration of natural brines. Another method of extraction which is also used in the United States and in other countries is a thermal reduction method, generally referred to as the ferro-silicon process, employing an alloy of iron and silicon to reduce magnesium oxide. Most of the magnesium ingot sold is of 99.80% purity. Grades of magnesium of 99.90, 99.95, and 99.98% purity are also available. The higher purity grades are used mostly in nuclear applications and for reduction purposes.

ϕ *3. Properties*—Magnesium is extremely light with the common alloys having a specific gravity of about 1.8 compared to 2.7 for aluminum. The heavier structural metals like iron, copper, and zinc are approximately four times as heavy as magnesium. Magnesium melts at 1202°F (650°C). The coefficient of thermal expansion between 68–212°F (20–100°C) is approximately 0.0000145/°F (0.0000261/°C) and is slightly higher than for aluminum, 0.000013/°F (0.000023/°C), and over twice that of steel. The thermal and electrical conductivities of magnesium are relatively high and some alloys approach values comparable to aluminum alloys. The modulus of elasticity is approximately 6 500 000 psi (45 GPa). The pure metal is not used for structural applications. A number of alloys have been developed with good strength-to-weight ratios.

4. Alloying Elements—Common alloying elements used in magnesium alloys are aluminum, manganese, rare earths, silver, thorium, zinc, and zirconium. Alloys are stronger than the pure metal but have lower electrical and thermal conductivites. Certain of the alloys respond to heat treatment with an increase in strength and hardness. Another means used to increase the strength of magnesium is by cold work. Most commercial alloys are stable at room temperature. Certain alloying elements such as the rare earths and thorium give better high temperature strength than can be obtained with the more common alloying elements aluminum and zinc.

5. Alloy Nomenclature—A designation system for magnesium alloys used commercially and described in ASTM B 275, Recommended Practice for Codification of Light Metals and Alloys, Cast and Wrought, was adopted by SAE in 1971. The initial letter(s) represent the major alloying elements(s) with the following numeral(s) representing the nominal percent by weight of each element. The final letter is assigned arbitrarily.

6. Temper Designation—The same temper designation system is used for both aluminum base and magnesium base alloys. It is described in detail under the aluminum alloy section of this book and in ASTM B 296, Recommended Practice for Temper Designation of Magnesium Alloys, Cast and Wrought.

7. Working—Magnesium alloys are available in most commercial forms such as die castings, investment, sand and permanent mold castings, extrusion, forgings, sheet, and plate. It can be formed by drawing, spinning, and pressing. The working is done best at elevated temperatures because of improved workability and freedom from springback. Magnesium can be joined by adhesive bonding, bolting, riveting, and welding. Arc welding, using an inert gas shield, is the most commonly used method of fusion welding. Spot welding is used extensively. Magnesium, in all its forms, can be readily machined with exceptional speed and tool life.

8. Finishing and Coating—Bare magnesium is suitable for many applications. Protective finishes may be required to prevent tarnishing or for protection from corrosion in humid industrial or marine atmospheres. It is subject to galvanic attack when coupled to most other metals, and such connections should be adequately protected if moisture will be present. Magnesium can be finished by plating and painting for either protection or decoration.

9. Testing—Magnesium alloys are tested like other metals using standard ϕ ASTM methods. The tensile and compressive yield strengths are defined as the stress at which the stress-strain curve deviates 0.2% from the initial modulus line.

ϕ MAGNESIUM CASTING ALLOYS—SAE J465c

SAE Standard

Report of Nonferrous Metals Division approved January 1940 and completely revised by Nonferrous Metals Committee April 1979.

1. Purpose—This SAE Standard covers the most commonly used magnesium alloys suitable for casting by the various commercial processes. The chemical composition limits and minimum mechanical properties are shown. Other equally important characteristics such as surface finish, and dimensional tolerances are not covered by this standard.

2. Introduction—Magnesium alloys are cast by all casting methods, the most common being pressure die casting, investment casting, sand casting, and permanent mold casting. Several alloys are available for use as sand, investment, and permanent mold castings to give the desired end use and production characteristics. Many of these are not suitable for use in the pressure die casting process. Most of the alloys used for sand, investment, and permanent mold castings may be heat treated to increase strength or improve stability. Die castings, while in the same composition range as some of the sand castings, are not heat treated because of undesirable effects such as blistering. Magnesium alloy sand, investment, and permanent mold castings are generally sold in the solution heat treated (-T4) condition for best ductility. Artificial aging after solution heat treatment (-T6) increases the yield strength considerably but decreases the ductility. Many times an artificial age (-T5) from the as-cast condition (-F) is sufficient to give the desired strength and stability.

3. Characteristics—The characteristics of the most commonly used alloys for sand casting are compared in Table 2 which was compiled by the American Foundrymen's Society.

Approximately the same ratings shown in Table 2 would apply for the same alloys when used for permanent mold and investment castings. Not all sand casting alloys are suitable for use in permanent molds.

The mechanical properties given in Table 5 are those obtained from separately cast test specimens. These test specimens are cast and heat treated under conditions that duplicate as closely as possible the conditions under which the castings they represent are made. The test bars are not machined except to fit the grips of the testing machine. The properties given for the die casting alloys are from die-cast test bars similar to Fig. 13 of ASTM B 557-74, Methods of Tension Testing Wrought and Cast Aluminum- and Magnesium-Alloy Products.

The mechanical properties of test specimens machined from castings will depend upon the type and size of casting and the location from which the specimen is taken. Specimens from thin sections or heavily chilled sections may have properties comparable to or superior to those from separately cast test specimens. Specimens from sections near gates and risers generally have lower properties. Separately cast test bars serve as a control on the metal quality and the heat treating process, if such are used. Minimum properties of test specimens cut from castings are generally guaranteed on the basis of an average of not less than three specimens from the thickest, the thinnest, and an average cross-section. Minimum mechanical properties for designated areas are sometimes specified.

All magnesium casting alloys are readily machined. Those in the as-cast and aged (-T5) and solution heat treated and aged (-T6) conditions are more stable and give less trouble from warping and growth.

4. Sand Castings—Sand castings are used when a small number of castings are required or the casting is large or complicated. In many cases sand cores are used with permanent mold castings or die castings. Dimensional tolerances, on the whole, are greater for sand castings than for permanent mold castings and the surface is not as smooth.

In the design of patterns, a shrinkage factor of 5/32 in/ft (13 mm/m) is generally used, but this may be reduced to 1/8 in/ft (10 mm/m) or less if free shrinkage is restrained by bosses, internal cores, or gates and risers. Walls as thin as 0.150 in (3.80 mm) can be readily made in large size castings. Thinner walls are possible for smaller areas. For example, a 0.120 in (3.05 mm) thick wall can be cast covering an area of about 1 ft^2 (0.1 m^2).

In order to obtain the best results from castings, the foundry should be

TABLE 1—TYPICAL MECHANICAL PROPERTIES OF SEPARATELY CAST TEST BARS MAGNESIUM DIE CASTING ALLOYS

Alloy Designation			Tensile Strength, psi (MPa)	Yield Strength 0.2% Offset, psi (MPa)	Elongation in 2 in (50.8 mm), %
UNS	ASTM and SAE	Old SAE			
M11910	AZ91A	501	34 000 (234)	23 000 (159)	3
M11912	AZ91B	501A	34 000 (234)	23 000 (159)	3
M10600	AM60A		32 000 (221)	19 000 (131)	8
M10410	AS41A		31 000 (214)	20 000 (138)	6

No minimum properties are required of die castings. Table 3 lists ASTM, AMS, and Federal specifications covering AZ91A and AZ91B.

consulted on the design of the casting, choice of alloy, heat treatment, and properties attainable. The selection of the alloy and heat treatment is governed by the characteristics desired in the casting and the limitations of the casting process.

Considerations of cost and secondary characteristics such as finishing, welding, and pressure tightness may be the deciding factor on which alloy to use. Table 3 lists similar ASTM, AMS, Federal, and Military specifications covering the SAE sand casting alloys given in this standard. The composition of the various alloys is shown in Table 4. Minimum properties of separately cast test bars are given in Table 5.

The concept of *premium quality castings* has been introduced by research workers and the foundry industry. The most important feature of premium quality is higher integrity of the product, and the reliability of properties in designated areas of each and every single casting. Table 6 shows the minimum requirements for mechanical properties in designated and other areas of premium quality castings.

4.1 General Data—Alloys M10100, M11630, M11810, M11914, and M11920 are used for most commercial applications. With the exception of M10100, which is a binary magnesium-aluminum alloy, they contain aluminum and zinc as alloying elements. This alloy family is used where moderately high strength at room temperature is desired. These alloys generally have good castability and are the lowest in cost of the commercial alloys. Individual differences in strength, ductility, and pressure tightness exist in this family of alloys. M11630 has the best toughness but has a tendency to microporosity in complex designs. M11920 has the highest tensile yield strength of the Mg-Al-Zn alloys. It has been used extensively in aircraft engines. M10100 has good castability and pressure tightness. Alloys M11914 and M11810 have better pressure tightness than M11630 and have good weldability. Both M11914 and M11810 have been used extensively in aircraft and racing car wheels. The upper operating limit for the Mg-Al-Zn casting alloys is generally considered to be about 300°F (149°C).

A second series of alloys is based upon the Mg-Zn-Zr alloy system. These alloys are also generally used at service temperatures below 300°F (149°C), although the addition of rare earth metals (alloy M16410) and thorium (alloy M16620) somewhat improves their ability to withstand exposure to more elevated temperatures. Alloys M16410 and M16620 have improved foundry characteristics and weldability over M16510 and M16610. Alloy M16610-T6 has a high strength-to-weight ratio compared to most commercial casting alloys, but shows less favorable foundry characteristics. Alloy M16630-T6 has a high strength-to-weight ratio, is readily castable, and shows little or no tendency to microporosity. It is designed to take advantage of a new principle of heat treatment involving the inward diffusion of hydrogen and formation of hydrides. M13010 is a low-strength casting alloy intended for applications requiring exceptionally good damping characteristics.

A third group of alloys is based on the Mg-Re-Zr system. These alloys are used in applications for operation at temperatures up to 550°F (288°C) where tensile or creep strength is a requirement. Alloy M12390 also is excellent where pressure tightness is a requirement. It rates second to M13010 in damping capacity. Alloy M18220 has the best short-time strength properties, up to 400°F (205°C), of all magnesium alloys.

The fourth group of alloys consists of Mg-Th-Zr alloys with or without zinc, which find applications in parts operating at temperatures up to 650°F (343°C).

5. Permanent Mold Castings—Any of the alloys listed in Table 4 as sand casting alloys can be used for permanent mold castings. Cracking tendencies limit the usefulness of many of the alloys since they cannot be cast in large sizes or complicated shapes. Permanent mold castings are used for economy of production when the number of pieces required justifies the increased mold cost. Permanent mold casting permits the production of more uniform castings, with closer dimensional tolerances and superior surface finish than with sand casting. The minimum wall thickness possible to obtain is somewhat greater on permanent mold castings than for sand castings because of the chilling effect of the mold. However, thicknesses down to 0.150 in (3.80 mm) covering large areas may sometimes be cast. Thinner walls can be cast covering smaller areas. Complex parts which cannot be made entirely as a permanent mold casting can often be produced in semi-permanent molds using sand cores. The characteristics of the various magnesium base alloys are typical of the alloy, whether cast in sand, investment, or permanent molds. Some of the characteristics, such as hot shortness, limit the usefulness of some of the alloys to such an extent that they are seldom used for permanent mold castings. In the Mg-Al-Zn alloy group, M11920, M10100, M11914, and M11810 are most commonly used as permanent mold castings. M12390, M13310, and M18220 alloys can be cast in permanent molds quite readily.

The minimum properties of separately cast test bars and test specimens cut from castings are generally the same for a given alloy, whether cast in sand or permanent molds. The same minimum mechanical properties are used for both sand and permanent mold castings of the same alloy. Hence, those shown in Table 5 for sand castings are used for permanent mold castings. Table 3 lists UNS, ASTM, AMS, Federal, and Military specifications covering the SAE alloys given in this standard. Applications for permanent mold castings are the same as for sand castings. Producibility, cost, surface, and tolerances should be considered in deciding the process to be used.

TABLE 2—PHYSICAL PROPERTIES AND CHARACTERISTICS OF MAGNESIUM SAND-CASTING ALLOYS

Alloy Designation			Approximate Melting Range, °F (°C)			Pattern Shrinkage Allowance, in/ft (mm/m)[b]	Foundry Characteristics[c]				Alloy Designation			Other Characteristics				
UNS	ASTM and SAE	Old SAE	Non-Equilibrium Solidus[a]	Solidus	Liquidus		Pressure Tightness	Fluidity[d]	Microporosity Tendency[e]	Normally Heat Treated	UNS	ASTM and SAE	Old SAE	Machining[f]	Electroplating[g]	Surface Treatment[h]	Suitability to Brazing[i]	Suitability to Welding[j]
M10100[m]	AM100A	502	810 (432)	867 (464)	1100 (593)	5/32 (13.0)	2	1	2	Yes	M10100[m]	AM100A	502	1	2	2	No	1
M11630	AZ63A	50	685 (363)	850 (454)	1130 (610)	5/32 (13.0)	3	1	3	Yes	M11630	AZ63A	50	1	1	1	No	3
M11810[n]	AZ81A	505	790 (421)	882 (472)	1115 (602)	5/32 (13.0)	2	1	2	Yes	M11810[n]	AZ81A	505	1	2	2	No	1
M11914[m]	AZ91C	504	785 (418)	875 (468)	1105 (596)	5/32 (13.0)	2	1	2	Yes	M11914[m]	AZ91C	504	1	2	2	No	1
M11920[m]	AZ92A	500	770 (410)	830 (443)	1100 (593)	5/32 (13.0)	2	1	2	Yes	M11920[m]	AZ92A	500	1	2	2	No	2
M12390[p]	EZ33A	506	—	1010 (543)	1189 (643)	3/16 (15.5)	1	2	1	Yes	M12390[p]	EZ33A	506	1	1	1	No	1
M13310	HK31A	507	—	1092 (589)	1204 (651)	7/32 (18.0)	1	2	1	Yes	M13310	HK31A	507	1	1	1	—[k]	1
M13320[n]	HZ32A		—	1026 (552)	1198 (648)	3/16 (15.5)	1	2	1	Yes	M13320[n]	HZ32A		1	—[k]	2	—[k]	2
M13010[n]	K1A		—	1205 (652)		3/16 (15.5)	2	2	2	No	M13010[n]	K1A		1	3	2	—[k]	1
M18220[p]	QE22A		—	1020 (549)	1190 (643)	5/32 (13.0)	2	2	2	Yes	M18220[p]	QE22A		1	2	1	—[k]	1
M16410[p]	ZE41A		—	950 (510)	1184 (640)	3/16 (15.5)	—[k]	2	—[k]	Yes	M16410[p]	ZE41A		1	1	1	No	1
M16630[p]	ZE63A		—	510 (266)	950 (510)	3/16 (15.5)	1	2	1	Yes	M16630[p]	ZE63A		1	—[k]	1	No	1
M16620	ZH62A	508	—		1169 (632)	5/32 (13.0)	2	2	2	Yes	M16620	ZH62A	508	1	1	1	No	—[k]
M16510	ZK51A	509	—	1020 (549)	1185 (641)	5/32 (13.0)	3	2	3	Yes	M16510	ZK51A	509	1	2	2	No	3
M16610	ZK61A	513	—	985 (529)	1175 (635)	5/32 (13.0)	3	2	3	Yes	M16610	ZK61A	513	1	2	1	No	3

[a] As measured on metal solidified under normal casting conditions.
[b] Allowance for average castings. Shrinkage requirements will vary with intricacy of design and dimensions. (1 in/ft × 8.333 = % Shrinkage.)
[c] Rating of 1 indicates best of group; 3 indicates poorest of group.
[d] Ability of liquid alloy to flow readily in mold and fill thin sections.
[e] Based on radiographic evidence.
[f] Composite rating based on ease of cutting, chip characteristics, quality of finish, and tool life. Ratings, in the case of heat-treatable alloys, based on -T6 type temper. Other tempers, particularly the annealed temper, may have lower ratings.
[g] Ability of casting to take and hold an electroplate applied by present standard methods.
[h] Ability of castings to be cleaned in standard pickle solutions and to be conditioned for best paint adhesion.
[i] Refers to suitability of alloy to withstand brazing temperature without excessive distortion or melting.
[j] Based on ability of material to be fusion welded with filler rod of same alloy.
[k] Inexperience with these alloys under wide production conditions makes it undesirable to supply ratings at this time.
[m] Properties applicable for permanent mold and investment castings.
[n] Properties applicable for permanent mold castings also.
[p] Properties applicable for investment castings also.

TABLE 3—SIMILAR SPECIFICATIONS OF MAGNESIUM CASTING ALLOYS

Alloy Designation			Form	ASTM	AMS	Federal	Military
UNS	ASTM and SAE	Old SAE					
M10100	AM100A	502	Sand Cast Permanent Mold Investment	B80 B199 B403	— 4483 4455	— QQ-M-55 —	— — —
M11630	AZ63A	50	Sand Cast	B80	4420 4422 4424	QQ-M-56	—
M11810	AZ81A	505	Sand Cast Permanent Mold Investment	B80 B199 B403	— — —	QQ-M-56 QQ-M-55 —	— — —
M11910	AZ91A	501	Die Cast	B94	4490	QQ-M-38	—
M11912	AZ91B	501A	Die Cast	B94	—	—	—
M11914	AZ91C	504	Sand Cast Permanent Mold Investment	B80 B199 B403	4437 — —	QQ-M-56 QQ-M-55 —	Mil-M-46062 — —
M11920	AZ92A	500	Sand Cast Permanent Mold Investment	B80 B199 B403	4434 4484 4453	QQ-M-56 QQ-M-55 —	Mil-M-46062 Mil-M-46062 —
M12390	EZ33A	506	Sand Cast Permanent Mold Investment	B80 B199 B403	4442 — —	QQ-M-56 QQ-M-55 —	— — —
M13310	HK31A	507	Sand Cast Permanent Mold Investment	B80 B199 B403	4445 — —	QQ-M-56 QQ-M-55 —	Mil-M-46062 Mil-M-46062 —
M13320	HZ32A	—	Sand Cast	B80	4447	QQ-M-56 QQ-M-55	—
M13010	K1A	—	Sand Cast Investment	B80 B403	— —	— —	Mil-M-45207 —
M18220	QE22A	—	Sand Cast Permanent Mold Investment	B80 B199 B403	4418 — —	QQ-M-56 QQ-M-55 —	Mil-M-46062 — —
M16410	ZE41A	—	Sand Cast	B80	4439	QQ-M-56	—
M16510	ZK51A	509	Sand Cast	B80	4443	QQ-M-56	Mil-M-46062
M16610	ZK61A	513	Sand Cast Investment	B80 B403	4444 —	QQ-M-56 —	Mil-M-46042 —
M16630	ZE63A	—	Sand Cast Investment	B80 —	— —	— —	— —
M16620	ZH62A	508	Sand Cast	B80	4438	QQ-M-56	Mil-M-46062
M10600	AM60A	—	Die Cast	—	—	—	—
M10410	AS41A	—	Die Cast	—	—	—	—

6. *Investment Mold Castings*—As with permanent mold castings, any of the alloys listed in Table 4 may be used for investment castings. The complexity and quality requirements of investment castings has limited the application of most alloys except M11914, M10100, and, to a lesser extent, M11920. However, alloys such as M18220, M12390, and M13010 are frequently used.

Specifications applicable to sand and permanent mold castings (Table 4) are used commonly for investment castings, including composition and minimum properties limitations as called out in Tables 3 and 4. AMS specifications exist for investment castings in M11920-T6 (4453A) and M10100-T6 (4455A). An ASTM specification exists for investment castings in M13010, M12390, M10100, M11810, M11920, M11914, M18220, M16610, and M13310 (ASTM B 403).

Magnesium investment castings are used widely in applications requiring moderate to high degrees of configuration complexity, including coring; and minimum weight, section thicknesses of 0.060 in (1.52 mm) being normal and 0.040 in (1.02 mm) possible in some cases over smaller areas. Tool costs are usually high relative to sand and permanent mold tools; hence, volume considerations are intermediate as between sand and die casting and roughly equivalent to permanent mold or shell molding. The consideration of design change flexibility is in the same relationship as original tool cost.

7. Die Castings

7.1 Introduction—The die casting process offers many advantages as a method of fabricating magnesium alloys, including low cost in quantity production, decrease in amount of machining, excellent surface finish, dimensional accuracy, and metal saving, by virtue of being able to cast thin sections.

While most magnesium die castings are still produced on conventional cold-chamber die-casting machines, the use of hot-chamber machines for magnesium die castings is growing rapidly. With the exception of the metal melting equipment, both the machines and dies used in the cold-chamber process are practically interchangeable with those used for aluminum die casting.

The melting of magnesium is done in a non-oxidizing atmosphere, usually with a protective flux. Casting temperatures range from 1150–1250°F (621–677°C). When automatic metering of the magnesium is used, the metal is usually protected with a layer of molten flux. Some installations are using protective gas atmospheres. When hand ladling, the metal is protected with either flux, sulfur dioxide, or an atmosphere of air-SF_6. Metal in hot-chamber die-casting machines is generally protected with an air-SF_6 atmosphere. Metal injection pressures lie between 2000 and 15 000 psi (14 and 103 MPa).

In amenability to intricate coring, magnesium die castings rank between zinc, which is the best, and aluminum. Required draft is greater than for zinc and less than for aluminum.

Magnesium castings do not have a tendency to solder or adhere to the die. Consequently, in contrast with aluminum, die coating solutions are not necessary and the need for die lubrication is decreased.

Molten magnesium does not react with iron or steel and can, therefore, be transferred in the molten state through steel pipes. This makes magnesium adaptable to automatic ladling and metal handling devices.

Due to the low heat content of magnesium, a part made with equivalent machines and dies may be cast at rates comparable to those obtained with zinc alloys, and at higher rates than for aluminum. The magnesium cools faster, permitting earlier removal from the die.

7.2 General Data—Alloys M11910 and M11912—Magnesium alloy die castings have been used more extensively on automobiles than magnesium in any other form. Magnesium has been accepted as a competitive material in such applications as steering column parts such as shrouds, brackets, collars, and signal switches; instrument and transmission components; convertible top mechanism; generator end plates; clutch housings; fuel pump body and parts; oil pumps; and crankcases of air-cooled engines.

M11910 and M11912 are equivalent for most applications, with M11910 having greater purity and somewhat better corrosion resistance upon exposure to salt water. M11912 is more readily available and cheaper. Both alloys are dimensionally stable and will withstand relatively high stresses without cold flow. In many die casting applications magnesium can be substituted for other materials using the same section thickness as the metal replaced.

Alloy M10600—Magnesium alloy M10600 approaches alloys M11910 and M11912 in castability. Because of its excellent ductility, this alloy is used in applications where impact resistance is important.

Alloy M10410—Magnesium die casting M10410 has excellent resistance to creep at temperature and is used for castings which operate under a combination of high temperature (up to 350°F (177°C)) and stress.

7.3 Mechanical Properties—The typical properties obtained with magnesium die-casting alloys on separately cast-to-shape test bars similar to Fig. 13 in ASTM B 557-74 are shown in Table 1.

TABLE 4—COMPOSITION OF MAGNESIUM CASTING ALLOYS

Alloy Designation			Elements, wt. %									
UNS	ASTM and SAE	Old SAE	Al	Mn, min	Zn	Th	Rare Earths	Zr	Cu, max	Ni, max	Si, max	Total other Elements, max
M10600	AM60A	—	5.5–6.5	0.13	0.22	—	—	—	0.35	0.03	0.50	—
M10100	AM100A	502	9.3–10.7	0.10	0.30 max	—	—	—	0.10	0.01	0.30	0.30
M10410	AS41A	—	3.7–4.8	0.22–0.48	0.10 max	—	—	—	0.04	0.01	0.60–1.4	0.30
M11630	AZ63A	50	5.3–6.7	0.15	2.5–3.5	—	—	—	0.25	0.01	0.30	0.30
M11810	AZ81A	505	7.0–8.1	0.13	0.40–1.0	—	—	—	0.10	0.01	0.30	0.30
M11910	AZ91A	501	8.3–9.7	0.13	0.35–1.0	—	—	—	0.10	0.03	0.50	0.30
M11912	AZ91B	501A	8.3–9.7	0.13	0.35–1.0	—	—	—	0.35	0.03	0.50	0.30
M11914	AZ91C	504	8.1–9.3	0.13	0.4–1.0	—	—	—	0.10	0.01	0.30	0.30
M11920	AZ92A	500	8.3–9.7	0.10	1.6–2.4	—	—	—	0.25	0.01	0.30	0.30
M12390	EZ33A	506	—	—	2.0–3.1	—	2.5–4.0	0.50–1.0	0.10	0.01	—	0.30
M13310	HK31A	507	—	—	0.30 max	2.5–4.0	—	0.40–1.0	0.10	0.01	—	0.30
M13320	HZ32A	—	—	—	1.7–2.5	2.5–4.0	0.10 max	0.50–1.0	0.10	0.01	—	0.30
M13010	K1A	—	—	—	—	—	—	0.40–1.0	—	—	—	0.30
M18220	QE22A[a]	—	—	—	—	—	1.8–2.5[b]	0.40–1.0	0.10	0.01	—	0.30
M16410	ZE41A	—	—	0.15 max	3.5–5.0	—	0.75–1.75	0.40–1.0	0.10	0.01	—	0.30
M16630	ZE63A	—	—	—	5.5–6.0	—	2.1–3.0	0.40–1.0	0.10	0.01	—	0.30
M16620	ZH62A	508	—	—	5.2–6.2	1.4–2.2	—	0.50–1.0	0.10	0.01	—	0.30
M16510	ZK51A	509	—	—	3.6–5.5	—	—	0.50–1.0	0.10	0.01	—	0.30
M16610	ZK61A	513	—	—	5.5–6.5	—	—	0.6–1.0	0.10	0.01	—	0.30

[a] Silver content in M18220 shall be 2.0–3.0.
[b] Rare earth elements in M18220 are in the form of didymium; in alloys M12390, M16410, and M16630, in the form of mischmetal.

TABLE 5—MINIMUM MECHANICAL PROPERTIES OF SEPARATELY CAST TEST BARS MAGNESIUM SAND CASTING ALLOYS[b]

Alloy or Temper Designation				Ultimate Tensile Strength		Yield Strength 0.2% Offset,		Elongation in 2 in (50.8 mm), %
UNS	ASTM and SAE		Temper	psi	MPa	psi	MPa	
M10100-F	AM100A-F		As-cast	20 000	138	—[a]	—[a]	—[a]
		-T4	Solution heat treated	34 000	234	—[a]	—[a]	6
		-T6	Solution heat treated and artificially aged	35 000	241	17 000	117	—[a]
M11630-F	AZ63A-F		As-cast	26 000	179	11 000	76	4
		-T4	Solution heat treated	34 000	234	11 000	76	7
		-T5	Artificially aged only	26 000	179	12 000	83	2
		-T6	Solution heat treated and artificially aged	34 000	234	16 000	110	3
M11810	AZ81A	-T4	Solution heat treated	34 000	234	11 000	76	7
M11914-F	AZ91C-F		As-cast	23 000	159	11 000	76	—[a]
		-T4	Solution heat treated	34 000	234	11 000	76	7
		-T5	Artificially aged only	23 000	159	12 000	83	2
		-T6	Solution heat treated and artificially aged	34 000	234	16 000	110	3
M11920-F	AZ92A-F		As-cast	23 000	159	11 000	76	—[a]
		-T4	Solution heat treated	34 000	234	11 000	76	6
		-T5	Artificially aged only	23 000	159	12 000	83	—[a]
		-T6	Solution heat treated and artificially aged	34 000	234	18 000	124	1
M12330	EZ33A	-T5	Artificially aged only	20 000	138	14 000	97	2
M13310	HK31A	-T6	Solution heat treated and artificially aged	27 000	186	13 000	90	4
M13320	HZ32A	-T5	Artificially aged only	27 000	186	13 000	90	4
M18010-F	K1A-F	-T6	As-cast	24 000	165	6 000	41	14
M18220	QE22A	-T5	Solution heat treated and artificially aged	35 000	241	25 000	172	2
M16410	ZE41A	-T6	Artificially aged only	29 000	200	19 500	134	2.5
M16630	ZE63A	-T5	Solution heat treated and artificially aged	40 000	276	27 000	186	5
M16620	ZH62A	-T5	Artificially aged only	35 000	241	22 000	152	5
M16510	ZK51A	-T6	Artificially aged only	34 000	234	20 000	138	5
M16610	ZK61A		Solution heat treated and artificially aged	40 000	276	26 000	179	5

[a] Not required.
[b] Alloy suitable for permanent mold and/or investment castings should meet these properties.

TABLE 6—MINIMUM MECHANICAL PROPERTIES OF TEST SPECIMENS FROM DESIGNATED AREAS OF HIGH STRENGTH CASTINGS OF MAGNESIUM ALLOYS
(According to Specifications in Mil-M-46062)

Alloy Designation			Temper	Class[a]	Guaranteed Minimum Properties in Designated Areas		
UNS	ASTM and SAE	Old SAE			Ultimate Tensile Strength, psi (MPa)	Yield Strength 0.2% Offset, psi (MPa)	Elongation in 2 in (50.8 mm), %
M11814	AZ91C	504	-T6	1	35 000 (241)	18 000 (124)	4
				2	29 000 (200)	16 000 (110)	3
				3	27 000 (186)	14 000 (97)	2
				X	17 000 (117)	12 000 (83)	0.75
M11920	AZ92A	500	-T6	1	40 000 (276)	25 000 (172)	3
				2	34 000 (234)	20 000 (138)	1
				3	30 000 (207)	18 000 (124)	0.75
				X	17 000 (117)	13 500 (93)	0.25
M13310	HK31A	507	-T6	1	33 000 (228)	16 000 (110)	6
				2	29 000 (200)	14 000 (97)	3
				3	25 000 (172)	12 000 (83)	1
				X	19 000 (131)	10 500 (72)	1
M18220	QE22A	—	-T6	1	40 000 (276)	28 000 (193)	4
				2	37 000 (255)	26 000 (179)	2
				3	33 000 (228)	23 000 (159)	2
				X	28 000 (193)	20 000 (138)	2
M16220	ZH62A	508	-T5	1	38 000 (262)	23 000 (159)	5
				2	34 000 (234)	21 000 (145)	3
				3	31 500 (217)	19 000 (131)	2
				X	28 500 (197)	17 500 (121)	1.25
M16510	ZK51A	509	-T5	1	36 000 (248)	21 000 (145)	6
				2	32 000 (221)	19 000 (131)	4
				3	29 000 (200)	17 000 (117)	3
				X	24 000 (165)	14 000 (97)	1.25
M16610	ZK61A	513	-T6	1	42 000 (290)	29 000 (200)	6
				2	37 000 (255)	26 000 (179)	4
				3	34 000 (234)	23 000 (159)	2
				X	30 000 (207)	21 000 (145)	1.25

[a] Stress levels of various sections of the castings should be carefully considered before specifying the class of mechanical properties for any particular casting section. Since a uniform stress level is seldom required in casting design, it would be advantageous from the design and foundry aspect to have higher properties in local designated areas with the remainder of the casting having lower properties. Three classes (1–3) of mechanical properties are therefore incorporated in the specification for various stress levels. In addition, minimum properties are given for test specimens taken from castings in unspecified areas (X) of castings.

MAGNESIUM WROUGHT ALLOYS—SAE J466c   SAE Standard

Report of Nonferrous Metals Division approved January 1940 and completely revised by Nonferrous Metals Committee April 1979.

1. Purpose—This SAE Standard covers the most common magnesium alloys used in wrought forms, and lists chemical composition and minimum mechanical properties for the various forms. A general indication of the usage of the various materials is also provided.

2. Introduction—Magnesium wrought alloys are produced and fabricated by all the common production methods such as rolling, extrusion, and forging. Forms available are sheet, plate, wire, rod, bar, shapes, tubes, forgings, and impact extrusions. Magnesium alloys can be formed by bending, drawing, spinning, and pressing. The work is generally done hot except for simple operations. When done hot, magnesium alloys have exceptional workability. The temperature used varies from 300–750°F (149–399°C), depending on operation, alloy, and condition. All of the wrought alloys can be joined by adhesive bonding, spot welding, riveting, and bolting. Most of them are readily fusion welded and some do not require stress relief after welding. As with the cast alloys, all wrought alloys machine readily.

The temper designations used for wrought magnesium are similar to those used for aluminum alloys. Temper designations are covered by ASTM B296-67 (1972), Recommended Practice for Temper Designations of Magnesium Alloys, Cast and Wrought. Mechanical properties are obtained by standard ASTM procedures. The tensile and compressive yield strength is taken at an offset of 0.2% from the initial modulus line.

Table 1 lists similar ASTM, AMS, Military, and Federal specifications covering the SAE wrought alloys in this SAE Standard.

3. Sheet and Plate
3.1 Introduction—Magnesium alloy sheet is rolled to a thickness of 0.006–0.249 in (0.15–6.32 mm). Plate is 0.250 in (6.35 mm) or over in thickness. Dimensional tolerances used are the same as for aluminum alloys and are given in the current issue of ASTM B90-70.

Magnesium sheet and plate is flattened thermally and can be obtained commercially, with smaller flatness tolerances than for most other metals. One grade of specially flattened plate is used extensively as tooling plate. The annealed condition (-0) is used for maximum formability and ductility. The cold rolled and partially annealed condition (-H24) has better strength and less ductility than the -0 temper. Tensile properties of the sheet and alloys covered by this standard are given in Table 3 and the chemical composition limits are given in Table 2.

3.2 General Data—Alloy M11311 is the most commonly used of the sheet alloys and is available in either the annealed (-0) or cold rolled and partially annealed (-H24 and -H26) conditions. M11311 alloy can be formed and welded readily. It has found widespread use. Applications most familiar in the automotive field would be its use in truck bodies, ramps, and dockboards and the various places such as patterns, jigs, and fixtures in which tooling plate has been used.

M16100 is not as strong as M11311. It has good formability and excellent weldability and does not require stress relieving after welding. It is used in place of M11311 primarily in tanks and large structures where stress relieving

TABLE 1—SIMILAR SPECIFICATIONS OF MAGNESIUM WROUGHT ALLOYS

Alloy Designation			Form	ASTM	AMS	Military or Federal
UNS	ASTM and SAE	Old SAE				
M11311	AZ31B	510	Sheet and plate Bar, rod, shapes Tube Forgings	B90 B107 B107 B91	4375, 4376, 4377 — — —	QQ-M-44 QQ-M-31 WW-T-825 QQ-M-40
M11610	AZ61A	520	Bar, rod, shapes Tube Wire (welding rod) Forgings	B107 B107 — B91	4350 4350 — 4358[a]	QQ-M-31 WW-T-825 Mil-R-6944 QQ-M-40
M11800	AZ80A	523	Bar, rod, shapes Forgings	B107 B91	— 4360[a]	QQ-M-31 QQ-M-40
M14141	LA141A		Sheet and plate	B90	—	—
M13310	HK31A	507	Sheet and plate	B90	4384, 4385	Mil-M-26075
M13210	HM21A		Sheet and plate Forgings	B90 —	4383, 4390 4363	Mil-M-8917 QQ-M-40
M13312	HM31A		Bar, rod, shapes	—	4388, 4389	Mil-M-8916
M15100	M1A	522	Bar, rod, shapes Forgings	B107 —	— —	QQ-M-31 QQ-M-40
M16100	ZE10A	534	Sheet and plate	B90	—	Mil-M-46037
M16400	ZK40A		Bar, rod, shapes	B107	—	—
M16600	ZK60A	524	Bar, rod, shapes Tube Forgings	B107 B107 B91	4352 4352 4362	QQ-M-31 WW-T-825 QQ-M-40

[a] Noncurrent specifications.

would be a problem. It is available in either the annealed (-0) or cold rolled and partially annealed (-H24) conditions.

M13310 was developed primarily for elevated temperature use in the 300–700°F (149–371°C) range. It is more costly than M11311 and M16100. M13310 alloy has excellent weldability and good formability. It is available in either the annealed (-0) or cold rolled and partially annealed (-H24) conditions. It has been used primarily in aircraft and missiles. M13210 was also developed for elevated temperature use in aircraft, missiles, and electronics.

M14141, containing 14% Li, is the only magnesium alloy with a body-centered cubic rather than hexagonal-close-packed crystal structure. It was developed as a very ductile and highly formable alloy.

4. Extrusions

4.1 Introduction—Magnesium alloys in general are extruded to size without subsequent drawing operations. Some sizing or *shaving* has been used to get better tolerances than can be obtained by extrusion. Wire, rod, bar, tubes, and special shaped sections are produced as extrusions. Dimensional tolerances on the various forms are given in the current issue of ASTM B107-70.

Magnesium alloys produced as extrusions are available in the as-extruded (-F) condition. In some alloys an increase in strength is obtained by artificial aging to the extruded and aged (-T5) condition. Minimum mechanical properties of the SAE extrusion alloys are shown in Table 4. The chemical composition of the SAE alloys used for extrusions is given in Table 2.

4.2 General Data—M11311, M11610, and M11800 contain aluminum and zinc as the principal alloying elements. M11311-F has moderate strength, good ductility, and good weldability. It is used where maximum strength is not a requirement. M11610-F has slightly better strength than M11311 but less than M11800. It has been supplanted to a large degree by the higher strength alloys, although still widely used as welding wire. M11800 has the highest strength of these three alloys. Low ductility has caused it to be replaced largely by M16600. Alloy M11800 is not as weldable as M11311 and M11610, which have excellent welding characteristics. M11311 has been used in truck bodies, ramps, and docks, and with tooling plate in making jigs and fixtures.

M15100 has excellent weldability and good corrosion resistance. It is a low strength alloy with good ductility. It has been replaced for most applications by M11311.

M16600 combines high strength with good ductility and toughness. However, it has limited weldability. M16600 costs more than M11311, M16610, M11800, and M15100. It has been used primarily in military applications and aircraft. M13312 alloy was developed for use at elevated temperatures in the range of 300–800°F (149–427°C). M16400 possesses high yield strength and has better extrusion characteristics than M16610.

5. Forgings

5.1 Introduction—Magnesium alloys are available as both hammer forgings and press forgings. The stronger alloys are too tender at hot working temperature to stand the shock of hammer forging and must be worked slowly under hydraulic presses. They may be forged sometimes to advantage by first pressing to shape and finishing on the hammer.

Die equipment built for aluminum alloy forgings can, in many cases, be used without change for producing magnesium alloy hammer forgings. This also applies to small die forgings made by pressing. Large press forgings comparable in size to an aircraft radial motor crankcase require special equipment and can usually be supplied only as oversize die forgings. Compared with an aluminum forging, magnesium press forgings frequently require an extra blocking die.

Forgings subject to shock, vibration, or repeated stresses must be carefully designed and carefully machined to avoid notches, sharp corners, tool marks, and other stress raisers. Minimum machining radius is 0.040 in (1.02 mm), and all sharp corners and feather edges must be broken. Magnesium alloy forgings may have marked directional properties, especially with regard to yield strength in tension and in compression. For this reason it is advisable for user to consult with manufacturer on design of forgings.

5.2 General Data—M11311, M11610, M11800, M13210, and M16600 alloys are used principally as press forgings. M11311 and M11610 are used where moderately high strength and good ductility are desired. Alloys M11800 and M16600 are used where greater strength is required. M13210 alloy was developed for use at temperatures of 300–800°F (149–427°C). The chemical composition of the SAE alloys used for forgings are given in Table 2.

5.3 Mechanical Properties—The properties for forgings are those obtained from tensile test specimens taken with the longitudinal axis of the specimen parallel to the direction of maximum flow of the metal or from separately forged coupons. Minimum properties of magnesium alloy forgings are given in Table 5.

TABLE 2—COMPOSITION OF WROUGHT MAGNESIUM ALLOYS

Alloy Designation			Elements, weight %												
UNS	ASTM and SAE	Old SAE	Al	Mn min	Zn	Zr	Rare Earths	Th	Ca, max	Cu, max	Fe, max	Na, max	Ni, max	Si, max	Total Other Elements, max
M11311	AZ31B	510	2.5–3.5	0.20	0.6–1.4	—	—	—	0.04	0.05	0.005	—	0.005	0.10	0.30
M11610	AZ61A	520	5.8–7.2	0.15	0.40–1.5	—	—	—	—	0.05	0.005	—	0.005	0.10	0.30
M11800	AZ80A	523	7.8–9.2	0.12	0.20–0.8	—	—	—	—	0.05	0.005	—	0.005	0.10	0.30
M13310	HK31A	507	—	—	0.30 max	0.40–1.0	—	2.5–4.0	—	0.10	—	—	0.01	—	0.30
M13210	HM21A		—	0.45–1.1	—	—	—	1.5–2.5	—	—	—	—	—	—	0.30
M13312	HM31A		—	1.2	—	—	—	2.5–3.5	—	—	—	—	—	—	0.30
M14141[a]	LA141A		1.0–1.5	0.15	—	—	—	—	—	0.04	0.005	0.005	0.005	0.10	0.30
M15100	M1A	522	—	1.2	—	—	—	—	0.30	0.05	—	—	0.01	0.10	0.30
M16100	ZE10A	534	—	—	1.0–1.5	—	0.12–0.22[b]	—	—	—	—	—	—	—	0.30
M16400	ZK40A		—	—	3.5–4.5	0.45 min	—	—	—	—	—	—	—	—	0.30
M16600	ZK60A	524	—	—	4.8–6.2	0.45 min	—	—	—	—	—	—	—	—	0.30

[a] LA141 contains 13–15% Li.
[b] Rare earth elements are in the form of mischmetal.

Table 3—MINIMUM MECHANICAL PROPERTIES OF MAGNESIUM ALLOY SHEET AND PLATE

Alloy Designation			Temper	Thickness in (mm)	Tensile Strength psi (MPa)	Yield Strength 0.2% Offset, psi (MPa)	Elongation in 2 in (50.8 mm), %
UNS	ASTM and SAE	Old SAE					
M11311	AZ31B	510	-O	0.016–0.250 (0.41–12.70) 0.251–2.000 (12.73–50.80) 2.001–3.000 (50.83–76.20)	32 000 (221)[a] 32 000 (221)[a] 32 000 (221)[a]	— — —	12 10 9
			-H24	0.016–0.249 (0.41–6.32) –0.374 (6.35–9.50) –0.500 (9.52–12.70) 0.501–1.000 (12.73–25.40) 1.001–2.000 (25.43–50.80) 2.001–3.000 (50.83–76.20)	39 000 (269) 38 000 (262) 37 000 (255) 36 000 (248) 34 000 (234) 34 000 (234)	29 000 (200) 26 000 (179) 24 000 (165) 22 900 (152) 20 000 (138) 18 000 (124)	6 8 8 8 8 8
			-H26	0.250–0.374 (6.35–9.50) –0.500 (9.52–12.70) 0.501–0.750 (12.73–19.05) 0.751–1.000 (19.08–25.40) 1.001–1.500 (25.43–38.10) 1.501–2.000 (38.13–50.80)	39 000 (269) 38 000 (262) 37 000 (255) 37 000 (255) 35 000 (241) 35 000 (241)	27 000 (186) 26 000 (179) 25 000 (172) 23 000 (159) 22 000 (152) 21 000 (145)	6 6 6 6 6 6
M13310	HK31A	507	-O	0.016–0.250 (0.41–6.35) 0.251–0.500 (6.38–12.70) 0.501–1.000 (12.73–25.40) 1.001–3.000 (25.43–76.20)	30 000 (207)[b] 30 000 (207) 30 000 (207) 29 000 (200)	— 16 000 (110) 15 000 (103) 14 000 (97)	12 12 12 12
			-H24	0.016–0.125 (0.41–3.18) 0.126–0.250 (3.20–6.35) 0.251–1.000 (6.38–25.40) 1.001–3.000 (25.43–76.20)	34 000 (234) 34 000 (234) 34 000 (234) 33 000 (228)	26 000 (179) 24 000 (165) 23 000 (159) 23 000 (159)	4 4 4 4
M13210	HM21A		-T8	0.016–0.250 (0.41–6.35) 0.251–0.500 (6.38–12.70) 0.501–3.000 (12.73–76.20)	33 000 (228) 32 000 (221) 30 000 (207)	18 000 (124) 21 000 (145) 21 000 (145)	6 6 6
M14141	LA141A		-T7	0.010–0.090 (0.25–2.29) 0.091–0.250 (2.31–6.35) 0.251–2.000 (6.38–50.80)	19 000 (131) 19 000 (131) 18 000 (124)	15 000 (103) 14 000 (97) 13 000 (90)	10 10 10
M16100	ZE10A	534	-O	0.016–0.060 (0.41–1.52) 0.061–0.250 (1.55–6.35) 0.251–0.500 (6.38–12.70)	30 000 (207) 30 000 (207) 29 000 (200)	18 000 (124) 15 000 (103) 12 000 (83)	15 15 12
			-H24	0.016–0.125 (0.41–3.18) 0.126–0.188 (3.20–4.78) 0.189–0.250 (4.80–6.35)	36 000 (248) 34 000 (234) 31 000 (214)	25 000 (172) 22 000 (152) 20 000 (138)	4 4 4

[a] Maximum tensile strength shall be 40 000 psi (276 MPa).
[b] Maximum tensile strength shall be 38 000 psi (262 MPa).

TABLE 4—MINIMUM MECHANICAL PROPERTIES OF MAGNESIUM ALLOY EXTRUSIONS

Alloy Designation				Form	Dia or Thickness, in (mm)	Cross Sectional Area, in² (cm²)	Minimum Properties		
UNS	ASTM and SAE	Old SAE	Temper				Tensile Strength, psi (MPa)	Yield Strength 0.2% Offset, psi (MPa)	Elongation in 2 in (50.8 mm), %
M11311	AZ31B	510	-F	Bars, rods, shapes	0.249 (6.32) and under 0.250–1.499 (6.35–38.07) 1.500–2.499 (38.10–63.47) 2.500–4.999 (63.50–126.97)	All All All All	35 000 (241) 35 000 (241) 34 000 (234) 32 000 (221)	21 000 (145) 22 000 (152) 22 000 (152) 20 000 (138)	7 7 7 7
				Hollow shapes	All	All	32 000 (221)	16 000 (110)	8
				Tubes	0.028–0.250 (0.71–6.35) 0.251–2.499 (6.38–63.47)	6.000 (38.71) and under 6.000 (38.71) and under	32 000 (221) 32 000 (221)	16 000 (110) 16 000 (110)	8 4
M11610	AZ61A	520	-F	Bars, rods, shapes	0.249 (6.32) and under 0.250–2.499 (6.35–63.47) 2.500–4.999 (63.50–126.97)	All All All	38 000 (262) 39 000 (269) 40 000 (276)	21 000 (145) 24 000 (165) 22 000 (152)	8 9 7
				Hollow shapes	All	All	36 000 (248)	16 000 (110)	7
				Tubes	0.028–0.750 (0.71–19.05)	6.000 (38.71) and under	36 000 (248)	16 000 (110)	7
M11800	AZ80A	523	-F	Bars, rods, shapes	0.249 (6.32) and under 0.250–1.499 (6.35–38.07) 1.500–2.499 (38.10–63.47) 2.500–4.999 (63.50–126.97)	All All All All	43 000 (296) 43 000 (296) 43 000 (296) 42 000 (290)	28 000 (193) 28 000 (193) 28 000 (193) 27 000 (186)	9 8 6 4
			-T5	Bars, rods, shapes	0.249 (6.32) and under 0.250–2.499 (6.35–63.47) 2.500–4.999 (63.50–126.97)	All All All	47 000 (324) 48 000 (331) 45 000 (310)	30 000 (207) 33 000 (228) 30 000 (207)	4 4 2
M13312	HM31A		-T5	Bars, rods, shapes	Under 1.000 (25.40) 1.000–3.999 (25.40–101.57)	All All	37 000 (255) 37 000 (255)	26 000 (179) 26 000 (179)	4 4
M15100	M1A	522	-F	Bars, rods, shapes	0.249 (6.32) and under 0.250–1.499 (6.35–38.07) 1.500–4.999 (38.10–126.97)	All All All	30 000 (207) 32 000 (221) 29 000 (200)	—[a] —[a] —[a]	2 3 2
				Hollow shapes	All	All	28 000 (193)	—[a]	2
				Tubes	0.028–0.750 (0.71–19.05)	6.000 (38.71) and under	28 000 (193)	—[a]	2
M16400	ZK40A		-T5	Bars, rods, shapes, and wires	All	4.999 (32.25) and under	40 000 (276)	37 000 (255)	4
				Hollow shapes	All	All	40 000 (276)	37 000 (255)	4
				Tubes	0.062–0.500 (1.57–12.70)	3.000 (19.35) and under	40 000 (276)	36 000 (248)	4
M16600	ZK60A	523	-F	Bars, rods, shapes	All	4.999 (32.25) and under	43 000 (296)	31 000 (214)	4
				Hollow shapes	All	All	40 000 (276)	28 000 (193)	5
				Tubes	0.028–0.750 (0.71–19.05)	3.000 (19.35) and under	40 000 (276)	28 000 (193)	5
		524	-T5	Bars, rods, shapes	All	4.999 (32.25) and under	45 000 (310)	36 000 (248)	4
				Hollow shapes	All	All	46 000 (317)	38 000 (262)	4
				Tubes	0.028–0.250 (0.71–6.35)	3.000 (19.35) and under	46 000 (317)	38 000 (262)	4

[a] Not required.

TABLE 5—MINIMUM MECHANICAL PROPERTIES OF MAGNESIUM ALLOY FORGINGS

Alloy Designation			Temper	Tensile Strength psi (MPa)	Yield Strength 0.2% Offset, psi (MPa)	Elongation in 2 in (50.8 mm), %
UNS	ASTM and SAE	Old SAE				
M11311	AZ31B	510	-F	34 000 (234)	19 000 (131)	6
M11610	AZ61A	520	-F	38 000 (262)	22 000 (152)	6
M11800	AZ80A	523	-F	42 000 (290)	26 000 (179)	5
			-T5	42 000 (290)	28 000 (193)	2
M13210	HM21A		-F	34 000 (234)	25 000 (172)	3
			-T5[a]	33 000 (228)	25 000 (172)	3
M16600	ZK60A	524	-T5[b]	42 000 (290)	26 000 (179)	7
			-T6[c]	42 000 (290)	32 000 (221)	4

[a] For forgings 4 in (102 mm) or less in thickness.
[b] For forgings 3 in (76 mm) or less in thickness.
[c] For forgings 2 in (51 mm) or less in thickness.

SPECIAL PURPOSE ALLOYS ("SUPERALLOYS")—SAE J467b

SAE Information Report

Report of Nonferrous Metals Committee approved January 1956 and last revised July 1968. Editorial change October 1968. Prepared in cooperation with Aerospace Materials Division of SAE Aerospace Council.

The data given in Tables 1–4 are typical values only and are not intended for design parameters. Mechanical properties of the special purpose alloys depend greatly upon processing variables and heat treatment. It is recommended that design data be obtained by actual testing or by consultation with the producers of the alloys.

TABLE 1 — SIMILAR MATERIAL SPECIFICATION DESIGNATIONS

Material Commercial Designation	AISI No.	ASTM No.	SAE No.	Military	Aerospace Material Specifications			
					Castings	Bars and Forgings	Sheet and Plate	Tubing	Wire
Martensitic Low Alloy Steels										
"17-22-A"	—	—	—	—	—	6304	—	—	—
"17-22-A" S	—	—	—	—	—	6302	6385	—	6458
"17-22-A" V	—	—	—	—	—	6303	6436	—	—
Chromoloy	—	—	—	—	—	—	—	—	—
D6A	—	—	—	—	—	6431	6438	—	—
300M	—	—	—	—	—	6416	—	—	—
UCX2	—	—	—	—	—	—	—	—	—
Martensitic Secondary Alloy Steels										
H11	—	—	J438	—	—	6485	6437	—	—
H12	—	—	J438	—	—	—	—	—	—
H13	—	—	J438	—	—	—	—	—	—
M2	—	—	J438	—	—	—	—	—	—
M10	—	—	—	—	—	—	—	—	—
M50	—	—	—	—	—	6490	—	—	—
Martensitic Chromium Steels										
410	—	A176	51410	QQ-S-763	5350	5612	5504	—	5776
	—	A276	60410	MIL-S-16993	5351	5613	5505	5591	5821
Greek Ascoloy	—	—	—	—	5354	5616	5508	—	5817
422	—	—	—	—	—	5655	—	—	—
422M	—	—	—	—	—	—	—	—	—
422M (Cast)	—	—	—	—	—	—	—	—	—
440C	—	—	51440C	QQ-S-763	5352	5630	—	—	—
14Cr-4Mo	—	—	—	—	—	—	—	—	—
Lapelloy	—	—	—	—	—	—	—	—	—
Lapelloy C	—	—	—	—	—	—	—	—	—
H-46	—	—	—	—	—	—	—	—	—
Semi Austenitic Precipitation and Transformation Hardening Steels										
AM-350	—	—	—	QQ-S-763	—	5745	5546	—	5774
	—	—	—	MIL-S-8840	—	5548	5554	5775
AM-355	—	A461	—	—	5359	—	5547	—	5780
	—	—	—	—	5368	5743	5549	—	5781
Stainless W	—	—	—	—	—	—	—	—	—
14-4PH	—	—	—	—	5340	—	—	—	5727
17-4PH	—	A461	—	—	5355	—	—	—	—
	—	—	—	MIL-S-853	5398	5643	—	—	5825
17-7PH	—	A461	—	—	—	—	5528	—	—
	—	—	—	—	—	5644	5529	5568	5673
PH15-7Mo	—	A461	—	—	—	—	—	—	5812
	—	—	—	—	—	5657	5520	—	5813
Austenitic Nickel-Chromium-Iron Steels										
302	302	—	30302	—	—	5636	5515	—	—
	—	—	60302	QQ-S-763	5358	5637	5516	—	5688
304	304	—	30304	QQ-S-763	—	—	—	5560	—
	—	—	60304	MIL-T-8506	—	5639	5513	5565	5697
304L	304L	—	30304L	—	5370	—	—	—	—
	—	—	60304L	QQ-S-763	5371	5647	5511	—	—
309S	309S	—	—	QQ-S-763	—	5650	5523	5574	—
310	310	—	30310	—	5365	—	—	5572	5694
	—	—	60310	QQ-S-766	5366	5651	5521	5577	5695
314	314	—	30314	—	—	5652	5522	—	—
316	316	—	30316	—	5360	—	—	—	5690
	—	—	60316	QQ-S-763	5361	5648	5524	5573	5691
321	321	A269	—	—	—	—	—	5570	—
	—	A271	30321	QQ-S-763	—	5645	5510	5576	5689
347	347	—	30347	—	5362	—	—	5571	5680
	—	—	60347	QQ-S-763	5363	5646	5512	5575	5681

(Table continued on next page)

TABLE 1 — SIMILAR MATERIAL SPECIFICATION DESIGNATIONS (CONTINUED)

Material Commercial Designation	AISI No.	ASTM No.	SAE No.	Military	Aerospace Material Specifications				
					Castings	Bars and Forgings	Sheet and Plate	Tubing	Wire
Austenitic Iron Base Alloys									
A286	—	—	—	—	—	5731–32	—	—	—
	—	—	—	—	—	5734–35	—	—	5804
	—	—	—	—	—	5736–37	5525	—	5805
V-57	—	—	—	—	—	—	—	—	—
Discaloy	—	—	—	—	—	5733	—	—	—
N-155	—	—	—	—	—	5768	5531	—	—
	—	—	—	—	5376	5769	5532	5585	5794
D-979	—	—	—	—	—	5746	—	—	—
W-545	—	—	—	—	—	5741	5543	—	—
S590	—	—	—	—	—	5770	5533	—	—
RA330	—	—	—	—	—	5716	5592	—	—
Unitemp 212	—	—	—	—	—	—	—	—	—
CRM-6D	—	—	—	—	—	—	—	—	—
CRM-15D	—	—	—	—	—	—	—	—	—
16-25-6	—	—	—	MIL-S-16538	—	5727	—	—	—
						5728			
19-9DL	—	—	—	—	—	5720	5526	—	—
						5722	5527		
19-9DX	—	—	—	—	—	5723	5538	—	—
						5724	5539		
17-14CuMo	—	—	—	—	—	—	—	—	—
G-192	—	—	—	—	—	—	—	—	—
AF-71	—	—	—	—	—	—	—	—	—
Incoloy 800	—	—	—	—	—	—	—	—	—
Incoloy 801	—	—	—	—	—	—	5552	—	—
Incoloy 805	—	—	—	—	—	—	—	—	—
Incoloy 810	—	—	—	—	—	—	—	—	—
Cobalt Base Alloys									
L-605	—	—	—	—	—	5759	5537	—	5797
S-816	—	—	—	—	—	5765	5534	—	—
HS-31 (X-40)	—	—	—	—	5382	—	—	—	—
HS-21	—	—	—	—	5385	—	—	—	—
Stellite 6	—	—	—	—	5373	—	—	—	—
	—	A399	—	MIL-R-17131	5387	—	—	—	5788
Haynes No. 151	—	—	—	—	—	—	—	—	—
WI-52	—	—	—	—	—	—	—	—	—
V-36	—	—	—	—	—	—	—	—	—
MAR-M 302	—	—	—	—	—	—	—	—	—
MAR-M 322	—	—	—	—	—	—	—	—	—
UMCo-50	—	—	—	—	—	—	—	—	—
UMCo-51	—	—	—	—	—	—	—	—	—
Nivco-10	—	—	—	—	—	—	—	—	—
Elgiloy	—	—	—	—	—	—	—	—	—
MAR-M 509	—	—	—	—	—	—	—	—	—
Austenitic Nickel Base Alloys									
Hastelloy X	—	—	—	—	5390	5754	5536	—	5799
Incoloy 901	—	—	—	—	—	5660	—	—	—
Incoloy 901 Mod.	—	—	—	—	—	5661	—	—	—
Rene 41	—	—	—	—	—	5712	—	—	—
	—	—	—	—	—	5713	5545	—	5800
Udimet 500	—	—	—	—	—	5751	—	—	—
	—	—	—	—	5384	5753	—	—	—
Waspaloy	—	—	—	—	—	5544	—	5586	5828
	—	—	—	—	—	5704	—	—	—
	—	—	—	—	—	5706	—	—	—
	—	—	—	—	—	5707	—	—	—
	—	—	—	—	—	5708	—	—	—
	—	—	—	—	—	5709	—	—	—
R-235	—	—	—	—	—	—	—	—	—
Udimet 700	—	—	—	—	—	—	—	—	—
Inconel X-750	—	—	—	MIL-N-7786	—	5667	—	—	5698
	—	—	—	MIL-N-8550	—	5668	5542	5582	5699
M-252	—	—	—	—	—	5756	—	—	—
	—	—	—	—	—	5757	5551	—	—
Refractaloy 26	—	—	—	—	—	—	—	—	—
Astroloy	—	—	—	—	—	—	—	—	—
GMR-235	—	—	—	—	—	—	—	—	—
GMR-235D	—	—	—	—	—	—	—	—	—

(Table continued on next page)

TABLE 1 — SIMILAR MATERIAL SPECIFICATION DESIGNATIONS (CONTINUED)

Material Commercial Designation	AISI No.	ASTM No.	SAE No.	Military	Aerospace Material Specifications				
					Castings	Bars and Forgings	Sheet and Plate	Tubing	Wire
Austenitic Iron Base Alloys (Continued)										
Hastelloy B	—	B333	—	—	—	—	—	—	—
	—	B335	—	MIL-R-5031	5396	—	—	—	—
Hastelloy C	—	B334	—	—	5388	—	—	—	—
	—	B336	—	MIL-N-18088	5389	5750	5530	—	—
Hastelloy F	—	—	—	—	—	—	—	—	—
Hastelloy N	—	—	—	—	—	—	—	—	—
Hastelloy W	—	—	—	—	—	5755	—	—	5786
Inconel 600	—	—	—	MIL-N-6840	—	—	—	—	—
	—	—	—	MIL-N-6710	—	5665	5540	5580	—
Inconel 604	—	—	—	—	—	—	—	—	—
Inconel 610	—	—	—	—	—	—	—	—	—
Inconel 700	—	—	—	—	—	—	—	—	—
Inconel 702	—	—	—	—	—	—	5550	—	—
Inconel 705	—	—	—	—	—	—	—	—	—
Alloy 713C	—	—	—	—	5391	—	—	—	—
Inconel 718	—	—	—	—	—	—	5596	—	—
Inconel 722	—	—	—	—	—	—	5541	—	—
AF 1753	—	—	—	—	—	—	—	—	—
IN100	—	—	—	—	5397	—	—	—	—
Nimonic 75	—	—	—	—	—	—	—	—	—
Nimonic 80A	—	—	—	—	—	—	—	—	—
Nimonic 90	—	—	—	—	—	—	—	—	—
Nimonic 105	—	—	—	—	—	—	—	—	—
Nimonic 115	—	—	—	—	—	—	—	—	—
MAR-M 200	—	—	—	—	—	—	—	—	—
RA-333	—	—	—	—	—	5717	5593	—	—
Titanium Base Alloys — Commercially Pure										
A40	—	—	—	—	—	—	—	4941	—
	—	—	—	MIL-T-9046	—	—	4902	4942	4951
A55	—	—	—	MIL-T-9046	—	—	4900	—	—
A70	—	—	—	MIL-T-9046	—	—	—	—	—
	—	—	—	MIL-T-9047	—	4921	4901	—	—
Titanium Base Alloys — Alpha Alloy Grades										
5Al-2.5 Sn	—	—	—	MIL-T-9046	—	4926	—	—	—
	—	—	—	MIL-T-9047	—	4966	4910	—	4953
5Al-5 Sn-5 Zr	—	—	—	—	—	4968	—	—	—
7Al-12 Zr	—	—	—	—	—	—	—	—	—
7Al-2 Cb-1 Ta	—	—	—	—	—	—	—	—	—
8Al-1 Mo,1 V	—	—	—	—	—	4972	4915	—	4955
	—	—	—	—	—	4973	4916	—	—
Ti-679	—	—	—	—	—	—	—	—	—
Titanium Base Alloys — Alpha-Beta Alloy Grades										
Ti-155A	—	—	—	—	—	4929	—	—	—
8Mn	—	—	—	MIL-T-009046	—	—	4908	—	—
2.5Al-16V	—	—	—	—	—	—	—	—	—
3Al-2.5V	—	—	—	—	—	—	—	—	—
4Al-4Mn	—	—	—	MIL-T-9047	—	—	—	—	—
	—	—	—	MIL-T-12117	—	4925	—	—	—
4Al-3Mo-1V	—	—	—	—	—	—	4912	—	—
	—	—	—	MIL-T-8884	—	—	4913	—	—
5Al-1.25 Fe-2.75 Cr	—	—	—	—	—	—	—	—	—
6Al-4V	—	—	—	MIL-T-9046	—	4928	—	—	—
	—	—	—	MIL-T-9047	—	4935	4911	—	4954
6Al-6V-2 Sn	—	—	—	MIL-T-46035	—	—	—	—	—
	—	—	—	MIL-T-46038	—	4971	4918	—	—
7Al-4 Mo	—	—	—	—	—	4970	—	—	—
2Cr-2 Fe-2 Mo	—	—	—	MIL-T-9047	—	4923	—	—	—
Titanium Base Alloys — Beta Alloy Grades										
1Al-8V-5 Fe	—	—	—	—	—	—	—	—	—
3Al-13V-11 Cr	—	—	—	—	—	—	4917	—	—

TABLE 2 — NOMINAL CHEMICAL COMPOSITIONS, %

Material Commercial Designation	C	Mn	Si	Cr	Ni	Co	Mo	W	Cb	Ti	Al	B	Fe	V	Zr	Cu	N	Other
Martensitic Low Alloy Steels																		
"17-22-A"	0.45	0.55	0.65	1.25	—	—	0.55	—	—	—	—	—	Bal	0.30	—	—	—	—
"17-22-A" S	0.30	0.55	0.65	1.25	—	—	0.50	—	—	—	—	—	Bal	0.25	—	—	—	—
"17-22-A" V	0.28	0.75	0.65	1.25	—	—	0.50	—	—	—	—	—	Bal	0.85	—	—	—	—
Chromoloy	0.20	0.50	0.75	1.00	—	—	1.00	—	—	—	—	—	Bal	0.10	—	—	—	—
D6A	0.47	0.75	0.22	1.05	0.55	—	1.00	—	—	—	—	—	Bal	0.10	—	—	—	—
300M	0.40	0.75	1.60	0.85	1.85	—	0.40	—	—	—	—	—	Bal	0.08	—	—	—	—
UCX2	0.39	0.70	1.00	1.10	—	1.00	0.25	—	—	—	—	—	Bal	0.15	—	—	—	—
Martensitic Secondary Alloy Steels																		
H11	0.35	0.30	1.00	5.10	—	—	1.50	—	—	—	—	—	Bal	0.40	—	—	—	—
H12	0.35	0.35	1.05	5.10	—	—	1.35	1.25	—	—	—	—	Bal	0.30	—	—	—	—
H13	0.35	0.30	1.00	5.10	—	—	1.50	—	—	—	—	—	Bal	1.00	—	—	—	—
M2	0.84	0.30	0.30	4.20	—	—	5.00	6.15	—	—	—	—	Bal	1.90	—	—	—	—
M10	0.87	0.20	0.30	4.00	—	—	8.25	—	—	—	—	—	Bal	1.90	—	—	—	—
M50	0.81	0.30	0.20	4.08	—	—	4.25	—	—	—	—	—	Bal	1.00	—	—	—	—
Martensitic Chromium Steels																		
410	0.12	0.50	0.35	12.25	0.40	—	0.30	—	—	—	—	—	Bal	—	—	—	—	—
Greek Ascoloy	0.15	0.40	0.30	13.0	2.00	—	0.15	3.00	—	—	—	—	Bal	—	—	0.15	—	—
422	0.22	0.75	0.40	12.5	0.75	—	1.00	1.00	—	—	—	—	Bal	0.22	—	—	—	—
422M	0.85	0.84	0.25	12.0	0.20	—	2.25	1.70	—	—	—	—	Bal	0.50	—	—	—	—
422M (Cast)	0.26	1.00	0.40	13.0	—	—	2.50	1.50	—	—	—	—	Bal	0.50	—	—	—	—
440C	1.10	0.50	0.40	17.5	—	—	0.50	—	—	—	—	—	Bal	—	—	—	—	—
14Cr-4Mo	1.05	0.50	0.30	14.5	—	—	4.00	—	—	—	—	—	Bal	0.12	—	—	—	—
Lapelloy	0.30	1.00	0.25	12.0	0.30	—	2.75	—	—	—	—	—	Bal	0.25	—	—	—	—
Lapelloy C	0.22	0.80	0.25	11.5	0.20	—	2.75	—	—	—	—	—	Bal	—	—	2.00	0.08	—
H-46	0.17	0.65	0.40	12.0	0.45	—	0.65	—	0.40	—	—	—	Bal	0.30	—	—	0.08	—
Semi Austenitic Precipitation and Transformation Hardening Steels																		
AM-350	0.10	1.00	0.40	16.5	4.25	—	2.75	—	—	—	—	—	Bal	—	—	—	0.10	—
AM-355	0.15	1.00	0.40	15.5	4.25	—	2.75	—	—	—	—	—	Bal	—	—	—	0.10	—
Stainless W	0.06	0.55	0.60	17.0	7.00	—	—	—	—	0.80	0.20	—	Bal	—	—	—	0.02	—
14-4PH	0.03	0.35	0.75	14.1	4.25	—	2.38	—	0.25	—	—	—	Bal	—	—	3.25	0.02	—
17-4PH	0.04	0.28	0.60	16.0	4.25	—	—	—	0.27	—	—	—	Bal	—	—	3.30	—	—
17-7PH	0.07	0.50	0.30	17.0	7.10	—	—	—	—	—	1.17	—	Bal	—	—	—	—	—
PH 15-7 Mo	0.07	0.50	0.30	15.0	7.00	—	2.20	—	—	—	1.17	—	Bal	—	—	—	—	—
Austenitic Nickel-Chromium-Iron Steels																		
302	0.08	1.00	0.50	18.0	9.0	—	—	—	—	—	—	—	Bal	—	—	—	—	—
304	0.04	1.00	0.50	19.0	10.0	—	—	—	—	—	—	—	Bal	—	—	—	—	—
304L	0.02	1.00	0.40	19.0	10.0	—	—	—	—	—	—	—	Bal	—	—	—	—	—
309S	0.04	1.00	0.50	23.0	13.5	—	0.25	—	—	—	—	—	Bal	—	—	0.25	—	—
310	0.12	1.00	0.40	25.0	20.5	—	—	—	—	—	—	—	Bal	—	—	—	—	—
314	0.12	1.00	2.25	24.5	20.5	—	—	—	—	—	—	—	Bal	—	—	—	—	—
316	0.05	1.00	0.40	17.0	12.5	—	2.50	—	—	—	—	—	Bal	—	—	—	—	—
321	0.04	1.00	0.40	18.5	11.0	—	—	—	—	0.40	—	—	Bal	—	—	—	—	—
347	0.05	1.00	0.40	18.5	11.0	—	—	—	0.70	—	—	—	Bal	—	—	—	—	—
Austenitic Iron Base Alloys																		
A286	0.05	1.40	0.40	15.0	26.0	—	1.30	—	—	2.15	0.20	0.004	Bal	0.30	—	—	—	—
V-57	0.05	0.20	0.35	14.75	27.25	—	1.30	—	—	3.00	0.20	0.01	Bal	0.30	—	—	—	—
Discaloy	0.08	0.90	0.80	13.5	26.0	—	2.75	—	—	1.75	0.07	0.005	Bal	—	—	—	—	—
N-155	0.15	1.50	0.50	21.0	20.0	20.0	3.00	2.50	1.00	—	—	—	Bal	—	—	—	0.13	—
D-979	0.06	0.25	0.20	14.90	44.30	—	4.05	3.65	—	3.00	1.05	0.01	Bal	—	—	—	—	—
W-545	0.03	1.65	0.80	13.5	26.0	—	1.75	—	—	3.00	0.15	0.02	Bal	—	—	—	—	—
S590	0.43	1.25	0.40	21.0	20.0	20.0	4.00	4.00	4.00	—	—	—	Bal	—	—	—	—	—
RA330	0.06	1.00	1.25	19.0	35.0	—	—	—	—	—	—	—	Bal	—	—	—	—	—
Unitemp 212	0.08	0.05	0.15	16.0	25.0	—	—	—	0.50	4.00	0.15	0.06	Bal	—	0.05	—	—	—
CRM-6D	1.00	5.00	0.50	20.0	5.00	—	1.00	1.00	1.00	—	—	—	Bal	—	—	—	—	—
CRM-15D	1.00	5.00	0.50	20.0	5.00	—	2.00	2.00	2.00	—	—	—	Bal	—	—	—	0.20	—
16-25-6	0.50	1.75	—	16.0	25.0	—	6.00	—	—	—	—	—	Bal	—	—	—	0.15	—
19-9DL	0.32	1.15	0.55	18.5	9.00	—	1.40	1.35	0.40	0.25	—	—	Bal	—	—	—	—	—
19-9DX	0.32	1.15	0.55	18.5	9.00	—	1.60	1.35	—	0.55	—	—	Bal	—	—	—	—	—
17-14-CuMo	0.12	0.75	0.50	15.9	14.1	—	2.50	—	0.45	0.25	—	—	Bal	—	—	3.00	—	—
G-192	0.60	8.50	0.50	22.0	—	—	—	—	—	—	—	—	Bal	—	—	—	0.35	—
AF71	0.30	18.0	0.30	12.5	—	—	3.00	—	—	—	—	0.20	Bal	0.90	—	—	0.20	—

(Table continued on next page)

TABLE 2 — NOMINAL CHEMICAL COMPOSITIONS, % (CONTINUED)

Material Commercial Designation	C	Mn	Si	Cr	Ni	Co	Mo	W	Cb	Ti	Al	B	Fe	V	Zr	Cu	N	Other
Austenitic Iron Base Alloys (Continued)																		
Incoloy 800	0.04	0.75	0.35	20.5	32.0	—	—	—	—	—	—	—	Bal	—	—	0.30	—	—
Incoloy 801	0.04	0.75	0.35	20.5	32.0	—	—	—	—	1.10	—	—	Bal	—	—	0.15	—	—
Incoloy 805	0.12	0.60	0.50	7.50	36.0	—	0.50	—	—	—	—	—	Bal	—	—	0.10	—	—
Incoloy 810	0.25	0.90	0.80	21.0	32.0	—	—	—	—	—	—	—	Bal	—	—	0.50	—	—
Cobalt Base Alloys																		
L-605	0.10	1.50	0.60	20.0	10.0	Bal	—	15.0	—	—	—	—	1.60	—	—	—	—	—
S-816	0.37	1.50	0.55	20.0	20.0	Bal	4.0	4.0	4.0	—	—	—	3.40	—	—	—	—	—
HS-31 (X-40)	0.50	0.75	0.75	25.5	10.5	Bal	—	7.5	—	—	—	—	1.50	—	—	—	—	—
HS-21	0.25	—	—	27.0	2.5	Bal	5.5	—	—	—	—	0.007	1.75	—	—	—	—	—
Stellite 6	1.15	0.45	0.55	29.0	1.5	Bal	0.75	4.5	—	—	—	—	1.50	—	—	—	—	—
Haynes No. 151	0.47	0.50	0.50	20.0	—	Bal	—	12.8	—	0.15	—	0.05	—	—	—	—	—	—
WI-52	0.45	0.25	0.25	21.0	0.50	Bal	—	11.0	2.0	—	—	—	2.00	—	—	—	—	—
V-36	0.29	0.60	0.50	25.0	20.0	Bal	4.0	2.0	2.0	—	—	—	2.40	—	—	—	—	—
MAR-M 302	0.85	0.10	0.20	21.5	—	Bal	—	10.0	—	—	—	0.005	0.75	—	0.20	—	—	Ta, 9.0
MAR-M 322	1.00	0.10	0.10	21.5	—	Bal	—	9.0	—	0.75	—	—	0.75	—	2.25	—	—	Ta, 4.5
UMCo-50	0.08	0.65	0.75	28.0	—	50.0	—	—	—	—	—	—	20.5	—	—	—	—	—
UMCo-51	0.32	0.75	0.75	28.0	—	50.0	—	2.1	—	—	—	—	18.0	—	—	—	—	—
Nivco-10	0.03	0.35	0.20	—	22.5	Bal	—	—	—	1.75	0.22	—	0.50	—	—	0.62	—	—
Elgiloy	0.15	2.00	—	20.0	15.00	Bal	7.00	—	—	—	—	—	16.00	—	—	—	—	Be, 0.04
MAR-M 509	0.60	0.05	0.05	23.0	10.0	Bal	—	7.0	—	0.20	—	0.005	—	—	0.50	—	—	Ta, 3.50
Austenitic Nickel Base Alloys																		
Hastelloy X	0.10	0.65	0.60	22.0	Bal	1.50	9.00	0.60	—	—	—	—	18.5	—	—	—	—	—
Incoloy 901	0.05	0.24	0.12	12.5	Bal	—	6.00	—	—	2.70	0.15	0.015	34.0	—	—	—	—	—
Incoloy 901 Mod.	0.05	0.09	0.08	12.5	Bal	—	5.80	—	—	2.90	—	0.015	34.0	—	—	—	—	—
Rene 41	0.09	0.25	0.25	19.0	Bal	11.0	10.0	—	—	3.10	1.50	0.005	1.80	—	—	—	—	—
Udimet 500	0.10	0.10	0.10	17.50	Bal	18.45	4.25	—	—	3.00	3.00	0.005	0.50	—	—	0.06	—	—
Waspaloy	0.07	0.10	0.10	19.75	Bal	13.50	4.45	—	—	3.00	1.40	0.005	0.75	—	—	0.04	—	—
R-235	0.12	0.10	0.30	15.00	Bal	1.15	5.50	—	—	2.50	2.00	—	10.00	—	—	—	—	—
Udimet 700	0.07	—	—	15.00	Bal	18.50	5.25	—	—	3.50	4.25	0.03	0.50	—	—	—	—	—
Inconel X-750	0.04	0.70	0.30	15.00	Bal	—	—	—	0.85	2.50	0.80	—	6.75	—	—	—	—	—
M-252	0.15	0.50	0.50	20.00	Bal	10.00	10.00	—	—	2.60	1.00	0.005	—	—	—	—	—	—
Refractaloy 26	0.03	0.80	1.00	18.00	Bal	20.00	3.20	—	—	2.75	0.20	—	16.00	—	—	—	—	—
Astroloy	0.06	—	—	15.0	Bal	15.0	5.25	—	—	3.50	4.40	0.03	—	—	—	—	—	—
GMR-235	0.15	0.13	0.30	15.5	Bal	—	5.25	—	—	2.00	3.00	0.06	10.0	—	—	—	—	—
GMR-235D	0.15	0.05	0.15	15.5	Bal	—	5.00	—	—	2.50	3.50	0.05	4.50	—	—	—	—	—
Hastelloy B	0.10	0.80	0.70	0.60	Bal	1.25	28.0	—	—	—	—	—	5.50	0.30	—	—	—	—
Hastelloy C	0.07	0.80	0.70	16.0	Bal	1.25	17.0	4.0	—	—	—	—	5.75	0.30	—	—	—	—
Hastelloy F	0.02	1.50	0.50	22.0	Bal	1.25	6.5	0.50	2.10	—	—	—	21.0	—	—	—	—	—
Hastelloy N	0.06	0.40	0.25	7.0	Bal	0.25	16.5	0.20	—	—	—	0.01	3.0	—	—	—	0.10	—
Hastelloy W	0.06	0.50	0.50	5.0	Bal	1.25	24.5	—	—	—	—	—	5.5	0.6	—	—	—	—
Inconel 600	0.04	0.20	0.20	15.8	Bal	—	—	—	—	—	—	—	7.20	—	—	0.10	—	—
Inconel 604	0.04	0.20	0.20	15.8	Bal	—	—	—	2.00	—	—	—	7.20	—	—	0.10	—	—
Inconel 610	0.20	0.90	2.00	15.5	Bal	—	—	—	1.00	—	—	—	9.00	—	—	0.50	—	—
Inconel 700	0.12	0.10	0.30	15.0	46.0	28.5	3.75	—	—	2.20	3.00	—	0.70	—	—	0.05	—	—
Inconel 702	0.04	0.05	0.20	15.6	Bal	—	—	—	—	0.70	3.40	—	0.35	—	—	0.10	—	—
Inconel 705	0.30	0.90	5.50	15.5	Bal	—	—	—	—	—	—	—	8.00	—	—	0.50	—	—
Alloy 713C	0.12	0.10	0.30	12.5	Bal	—	4.50	—	2.00	0.60	6.00	0.012	1.00	—	0.10	—	—	—
Alloy 713LC	0.05	0.10	0.30	12.00	Bal	—	4.50	—	2.00	0.60	5.90	0.01	0.30	—	0.10	—	—	—
Inconel 718	0.04	0.20	0.20	18.6	Bal	—	3.10	—	5.00	0.90	0.40	—	18.50	—	—	—	—	—
Inconel 722	0.04	0.55	0.20	15.0	Bal	—	—	—	—	2.40	0.60	—	6.50	—	—	—	—	—
AF 1753	0.24	0.05	0.10	16.25	Bal	7.20	1.60	8.40	—	3.20	1.90	0.008	9.50	—	—	0.06	—	—
IN 100	0.18	—	—	10.0	Bal	15.0	3.00	—	—	4.75	5.50	0.015	—	—	0.06	—	—	—
Nimonic 75	0.10	0.45	0.50	19.5	Bal	—	—	—	—	0.40	0.20	—	2.40	—	—	0.05	—	—
Nimonic 80A	0.06	0.10	0.70	19.5	Bal	1.1	—	—	—	2.50	1.30	—	2.40	—	—	—	—	—
Nimonic 90	0.07	0.50	0.75	19.5	Bal	18.0	—	—	—	2.40	1.40	—	2.50	—	—	0.05	—	—
Nimonic 105	0.13	0.10	0.25	15.0	Bal	20.0	5.0	—	—	1.20	4.50	—	0.40	—	—	0.25	—	—
Nimonic 115	0.15	—	—	15.0	Bal	15.0	3.50	—	—	4.00	5.00	—	—	—	—	—	—	—
MAR-M200	0.15	—	—	9.0	Bal	10.0	—	12.5	1.00	2.00	5.00	0.015	—	—	0.05	—	—	—
RA-333	0.04	1.00	1.15	25.5	Bal	3.25	3.25	3.25	—	—	—	—	17.0	—	—	0.10	—	—
Titanium Base Alloys— Commercially Pure																		

Material Commercial Designation	C	Mn	O	Cr	H	Sn	Mo	Ta	Cb	Ti	Al	B	Fe	V	Zr	Cu	N	Other
A40	0.04	—	0.08	—	0.006	—	—	—	—	Bal	—	—	0.25	—	—	—	0.02	—
A55	0.04	—	0.08	—	0.007	—	—	—	—	Bal	—	—	0.25	—	—	—	0.03	—
A70	0.05	—	0.09	—	0.005	—	—	—	—	Bal	—	—	0.35	—	—	—	0.03	—

(Table continued on next page)

TABLE 2 — NOMINAL CHEMICAL COMPOSITIONS, % (CONTINUED)

Material Commercial Designation	C	Mn	Si	Cr	Ni	Co	Mo	W	Cb	Ti	Al	B	Fe	V	Zr	Cu	N	Other
Titanium Base Alloys — Alpha Alloy Grades																		
5Al-2.5Sn	0.15	—	0.09	—	0.009	2.50	—	—	—	Bal	5.00	—	—	—	—	—	0.03	
5Al-5Sn-5Zr	0.02	—	0.06	—	0.007	4.80	—	—	—	Bal	5.00	—	0.07	—	5.20	—	0.02	
7Al-12Zr	0.02	—	0.04	—	0.005	—	—	—	—	Bal	7.00	—	0.07	—	12.0	—	0.015	
7Al-2Cb-1Ta	0.04	—	0.08	—	0.009	—	—	1.00	2.00	Bal	7.00	—	—	—	—	—	—	
8Al-1Mo-1V	0.04	—	0.07	—	0.008	—	1.00	—	—	Bal	8.00	—	0.15	1.00	—	—	0.02	
Ti-679	0.02	—	0.07	—	0.004	11.0	1.00	—	—	Bal	2.25	—	—	—	5.00	—	0.02	
Titanium Base Alloys — Alpha-Beta Alloy Grades																		
Ti-155A	0.04	—	0.06	1.40	0.005	—	1.20	—	—	Bal	5.50	—	1.40	—	—	—	0.02	
8Mn	0.10	8.00	0.10	—	0.008	—	—	—	—	Bal	—	—	—	—	—	—	0.04	
2.5Al-16V	0.04	—	0.07	—	0.007	—	—	—	—	Bal	2.50	—	—	16.0	—	—	0.01	
3Al-2.5V	0.02	—	0.06	—	0.007	—	—	—	—	Bal	3.00	—	—	2.50	—	—	0.01	
4Al-4Mn	0.05	4.00	0.11	—	0.012	—	—	—	—	Bal	4.00	—	—	—	—	—	0.02	
4Al-3Mo-1V	0.04	—	0.10	—	0.005	—	3.00	—	—	Bal	4.25	—	0.15	1.00	—	—	0.03	
5Al-1.25 Fe-2.75 Cr	0.08	—	0.08	2.75	0.006	—	—	—	—	Bal	5.00	—	1.25	—	—	—	0.02	
6Al-4V	0.023	—	0.097	—	0.008	—	—	—	—	Bal	6.18	—	0.22	3.81	—	—	0.026	
6Al-6V-2Sn	0.02	—	0.10	—	0.006	2.00	—	—	—	Bal	5.50	—	0.70	5.50	—	0.70	0.02	
7Al-4Mo	0.05	—	0.10	—	0.006	—	4.00	—	—	Bal	6.90	—	0.15	—	—	—	0.02	
2Cr-2 Fe-2Mo	0.05	—	0.10	2.25	0.005	—	2.25	—	—	Bal	—	—	2.25	—	—	—	0.05	
Titanium Base Alloys — Beta Alloy Grades																		
1Al-8V-5Fe	0.05	—	0.09	—	0.012	—	—	—	—	Bal	1.00	—	5.00	8.00	—	—	0.07	
3Al-13V-11Cr	0.02	—	0.10	11.0	0.007	—	—	—	—	Bal	3.00	—	—	13.5	—	—	0.03	

TABLE 3 — AVERAGE PHYSICAL PROPERTIES

Material Commercial Designation	Density at 70 F lb/in³	Specific Gravity g/cc	Melting Range, F	Thermal Conductivity		Electrical Resistivity		Coefficient of Linear Thermal Expansion in./in./F × 10⁻⁶										Modulus of Elasticity in Tension (E) 10⁶ psi				Modulus of Rigidity (G) 10⁶ psi				Poisson's Ratio at 70 F
				F	Btu/ft²/hr/F/in.	F	Microhm, in.	70-200	70-400	70-600	70-800	70-1000	70-1200	70-1400	70-1600	70-1800	70-2000	F	E	F	E	F	G	F	G	
Martensitic Low Alloy Steels																										
"17-22-A"	0.283	—	—	1000	227	—	—	—	—	—	7.66	7.80	7.93	—	—	—	—	70	30.8	1000	25.8	70	11.9	1000	9.8	0.296
"17-22-A" S	0.283	7.84	2700-2750	1200	200	—	—	6.80	6.88	7.32	7.57	7.77	7.95	—	—	—	—	70	29.5	1000	20.0	70	11.9	1000	9.5	0.296
"17-22-A" V	—	—	—	—	—	—	—	—	6.67	7.04	7.19	7.39	7.50	—	—	—	—	70	31.0	1000	25.2	70	12.1	1000	9.6	0.296
Chromoloy	0.285	—	—	—	—	—	—	7.0	7.15	7.30	7.55	7.90	8.30	—	—	—	—	70	31.7	1000	25.3	70	12.3	1000	9.6	0.280
D6A	0.284	7.87	—	—	—	—	—	7.31	—	—	—	—	—	—	—	—	—	80	29.65	1000	23.2	—	—	—	—	0.269
300M	0.283	—	—	70	260	—	—	—	—	—	—	—	—	—	—	—	—	80	29.5	—	—	—	—	—	—	
UCX2	0.276	7.68	—	—	—	—	—	—	—	5.68	—	—	—	—	—	—	—	80	29.4	—	—	—	—	—	—	
Martensitic Secondary Alloy Steels																										
H11	0.282	7.77	2500-2600	890	196	—	—	6.1	6.4	—	6.68	6.90	7.11	—	—	—	—	70	30.5	1000	22.7	70	12.0	1000	9.0	0.27
H12	0.282	—	—	—	—	—	—	6.1	6.5	—	7.0	—	7.1	—	—	—	—	70	30.0	—	—	—	—	—	—	
H13	0.282	7.76	—	890	196	—	—	5.8	—	—	6.9	7.0	7.4	—	—	—	—	80	30.0	1000	23.0	—	—	—	—	
M2	0.293	8.16	2500-2600	—	—	—	—	5.69	6.09	6.42	6.67	6.97	—	—	—	—	—	70	29.5	—	—	—	—	—	—	
M10	0.285	7.88	2500-2600	—	—	—	—	—	—	—	—	6.95	—	—	—	—	—	70	29.5	—	—	—	—	—	—	
M50	0.286	7.87	2500-2600	—	—	—	—	6.23	6.58	6.83	7.05	7.38	—	—	—	—	—	70	29.5	1000	23.5	—	—	—	—	
Martensitic Chromium Steels																										
410	0.280	7.78	2700-2790	932	199	750	34.6	5.5		5	6		6.4	6.5	—	—	—	—	70	29.8	900	23.8	70	12.0	900	10.0	0.24
Greek Ascoloy	0.286	7.87	2600-2700	—	—	77	31.0	5.83	6.12	6.20	6.37	6.52	6.63	—	—	—	—	70	29.0	1000	21.5	—	—	—	—	
422	0.280	7.78	2600-2700	800	190	—	—	5.90	6.10	6.20	6.30	6.50	6.70	—	—	—	—	70	29.8	1000	21.5	—	—	—	—	0.23
422M																										
422M (Cast)																										
440C	0.277	7.68	2450-2550	212	168	—	—	—	—	—	—	5.60	—	—	—	—	—	70	29.0	—	—	—	—	—	—	
14Cr-4Mo	0.280	7.78	2450-2550	—	—	—	—	—	—	—	6.67	—	—	—	—	—	—	70	29.0	—	—	—	—	—	—	
Lapelloy	0.285	7.85	2700-2750	1000	198	—	—	5.90	6.10	6.20	6.50	6.70	—	—	—	—	—	70	30.0	1000	23.1	—	—	—	—	
Lapelloy C	0.285	7.85	2650-2750	—	—	—	—	5.90	—	—	6.30	6.70	—	—	—	—	—	80	30.0	—	—	—	—	—	—	
H-46	0.280	7.75	2650-2750	—	—	—	—	—	—	—	6.72	—	—	—	—	—	—	80	31.3	—	—	—	—	—	—	

(Table continued on next page)

TABLE 3 — AVERAGE PHYSICAL PROPERTIES (CONTINUED)

Material Commercial Designation	Density at 70 F lb/in³	Specific Gravity g/cc	Melting Range F	Thermal Conductivity		Electrical Resistivity		Coefficient of Linear Thermal Expansion in./in./F × 10⁻⁶										Modulus of Elasticity in Tension (E) 10⁶ psi				Modulus of Rigidity (G) 10⁶ psi				Poisson's Ratio at 70 F
				F	Btu/ft²/hr/F/in.	F	Microhm, in.	70-200	70-400	70-600	70-800	70-1000	70-1200	70-1400	70-1600	70-1800	70-2000	F	E	F	E	F	G	F	G	

Semi Austenitic Precipitation and Transformation Hardening Steels

							cm																			
AM-350	0.282	—	2500-2550	800	141	80	78.8	6.3	—	6.8	—	7.2	—	6.72	—	—	—	70	29.4	800	24.3	80	11.3	800	9.3	
AM-355	0.281	—	2500-2550	800	139	80	75.6	6.4	—	6.8	—	7.2	—	6.5	—	—	—	80	29.3	800	24.6	80	11.4	800	9.4	
Stainless W	0.280	7.65	—	752	170	70	85	—	—	—	—	—	—	—	—	—	—	77	28.2	—	—	70	11.3	—	—	
14-4PH																										
17-4PH	0.282	7.80	2660-2720	900	157	70	77	6.0	6.1	6.3	6.5	—	—	—	—	—	—	70	28.5	600	26.0	70	11.2	600	10.2	0.272
17-7PH	0.276	7.65	—	900	146	70	83	5.7	6.6	6.8	6.9	—	—	—	—	—	—	70	29.0	—	—	70	11.0	—	—	
PH15-7Mo	0.277	7.68	—	1000	150	70	83	5.0	5.4	5.6	5.9	6.1	—	—	—	—	—	70	29.0	—	—	70	11.0	—	—	

Austenitic Nickel-Chromium-Iron Steels

302	0.29	—	2550-2590	—	—	1200	109	—	—	—	—	—	—	—	—	—	—	70	27.9	1200	21.8	70	12.5	1200	8.2	
304	0.287	7.94	2550-2650	—	—	1200	114	8.8	9.2	9.5	9.8	10.0	10.4	—	—	—	—	70	27.9	1200	21.1	70	12.5	1200	8.3	
304L	0.287	7.94	2550-2650	—	—	68	72	—	—	9.9	—	10.2	10.4	—	—	—	—	70	28.0	—	—	70	12.5	—	—	
309S	0.285	7.88	2550-2650	—	—	1200	116	—	—	9.3	—	9.6	10.0	—	—	11.5	—	70	29.0	1200	21.8	—	—	—	—	
310	0.288	7.98	2550-2650	—	—	1200	124	8.4	8.8	9.0	9.3	9.5	—	—	—	—	—	70	28.2	1200	21.8	—	—	1200	7.9	
314	0.279	7.72	—	932	—	68	77	—	—	—	—	—	—	—	—	—	—	70	29.0	—	—	—	—	—	—	
316	0.286	7.91	2500-2550	—	—	1200	110	8.9	9.3	9.6	9.8	10.0	10.3	—	—	—	—	70	28.1	1200	21.5	—	—	1200	8.1	
321	0.285	7.89	2550-2600	1000	150	1200	117	—	—	9.5	—	10.3	10.7	—	—	11.2	—	70	28.0	1200	21.2	—	—	1200	7.9	
347	0.290	8.02	2500-2600	—	—	1200	111	9.0	9.4	9.7	9.9	10.1	—	—	—	11.1	—	70	28.2	1200	21.4	—	—	1200	8.1	

Austenitic Iron Base Alloys

							in.																			
A286	0.286	7.94	2500-2600	1200	172	1200	46.8	9.17	9.35	9.47	9.64	9.78	9.88	10.32	—	—	—	70	29.1	1500	18.7	70	11.0	1500	6.8	0.306
V-57	0.287	7.96	—	—	—	—	—	9.0	9.2	—	—	9.9	—	10.8	—	—	—	80	28.5	1200	22.2	—	—	—	—	0.29
Discaloy	0.288	7.97	2515-2665	1200	158	—	—	8.5	8.7	9.1	9.4	9.5	9.6	9.8	—	—	—	75	28.4	1200	21.0	—	—	—	—	0.298
N-155	0.298	8.20	2350-2475	1200	144	—	—	7.75	8.50	8.52	8.65	9.13	9.46	9.72	9.90	10.10	10.26	70	29.3	1500	20.8	—	—	—	—	
D-979	0.295	8.17	—	1472	151	—	—	—	—	—	—	—	—	—	—	—	—	75	30.0	1500	21.9	75	11.6	1500	8.2	0.29
W-545	0.285	—	2455-2530	1400	141	—	—	—	—	9.67	—	10.1	—	11.0	—	—	—	80	28.4	1400	17.5	80	11.5	1400	6.65	0.23
S590	0.301	8.34	2400-2500	1292	155	—	—	8.0	—	8.3	—	—	—	9.2	—	—	—	80	31.1	—	—	80	11.9	—	—	
RA330	0.286	7.86	2500-2600	500	130	—	—	—	—	—	—	—	8.9	—	9.3	—	9.8	80	29.0	—	—	—	—	—	—	
Unitemp 212	0.286	—	2430-2530	1292	170	—	—	—	8.8	9.0	9.2	9.4	9.6	9.8	10.3	—	—	80	29.0	1400	20.5	70	10.4	—	—	
CRM-6D	0.284	7.86	2400-2500	800	170	—	—	7.5	8.1	8.7	9.3	9.5	9.7	9.85	9.9	10.1	10.2	70	28.7	1200	18.0	70	11.5	—	—	0.26
CRM-15D	0.286	7.92	2400-2500	—	—	—	—	—	—	—	—	—	—	—	—	—	—	—	—	—	—	—	—	—	—	
16-25-6	0.291	8.06	2550-2650	1100	180	—	—	—	—	9.28	9.29	9.36	9.52	10.8	11.7	—	—	70	29.1	1100	22.8	70	11.0	1100	8.2	0.295
19-9 DL	0.286	7.93	2525-2625	1200	147	—	—	8.50	9.11	9.31	9.59	9.78	9.97	—	—	—	—	70	29.5	1200	22.1	70	11.4	1200	8.2	0.286
19-9 DX	0.287	7.94	2500-2600	—	—	—	—	8.52	9.11	9.31	9.59	9.78	9.97	—	—	—	—	70	29.5	—	—	70	11.4	—	—	
17-14 CuMo	0.287	8.01	—	600	128	—	—	—	—	—	—	9.69	—	—	—	—	—	70	28.0	900	23.0	—	—	—	—	
G-192	0.279	—	2400-2600	—	—	—	—	8.6	8.9	9.3	9.7	10.1	10.3	10.5	10.7	—	—	70	29.9	1500	19.6	70	11.6	1500	7.34	0.29
AF71	0.281	—	2400-2500	—	—	—	—	5.6	7.4	—	—	—	—	—	—	—	—	70	28.4	1500	19.0	70	11.1	1500	7.14	0.27
Incoloy 800	0.290	8.02	2475-2525	212	97	—	—	8.0	8.4	8.7	9.0	9.3	9.6	9.8	10.0	10.2	—	70	28.2	—	—	—	—	—	—	
Incoloy 801	0.288	7.97	—	—	—	—	—	—	—	—	—	—	—	—	—	—	—	70	29.0	1350	17.0	—	—	—	—	
Incoloy 805	—	—	—	—	—	—	—	—	—	—	—	—	—	—	—	—	—	—	—	—	—	—	—	—	—	
Incoloy 810	—	—	—	—	—	—	—	—	—	—	—	—	—	—	—	—	—	—	—	—	—	—	—	—	—	

Cobalt Base Alloys

L-605	0.330	9.13	2425-2570	1200	153	75	34.9	6.83	7.19	7.59	7.77	8.02	8.24	8.61	9.06	9.41	9.84	70	34.2	1500	25.6	—	—	—	—	0.294
S-816	0.313	8.66	2350-2450	1200	161	—	—	7.38	7.45	7.65	7.91	8.11	8.40	8.75	9.00	9.25	—	70	35.2	1500	26.2	80	13.6	1500	9.6	0.294
HS-31 (X-40)	0.311	8.60	2445-2545	1200	158	70	38.2	—	—	7.84	8.08	8.39	8.75	9.19	—	—	—	70	32.8	1600	22.8	—	—	—	—	
HS-21	0.300	8.30	2465	1112	142	70	34.4	—	—	7.83	7.96	8.18	8.38	—	—	—	—	—	—	—	—	—	—	—	—	
Stellite 6	0.303	8.38	2310-2460	72	102.7	72	35.8	7.44	7.72	8.03	8.18	8.32	8.65	8.93	9.30	9.62	—	70	30.4	—	—	—	—	—	—	
Haynes No. 151																										
WI-52	0.321	8.87	2400-2450	1200	195	—	—	—	7.5	7.6	7.8	8.0	8.3	8.6	9.0	9.2	9.7	—	—	—	—	—	—	—	—	
V-36	0.303	8.41	2350-2450	—	—	—	—	—	—	—	—	—	—	9.1	—	—	—	70	32.4	1500	23.8	70	12.4	1500	8.84	0.30
MAR-M 302	0.333	9.21	2400-2450	1200	155	—	—	—	6.9	7.2	7.4	7.6	7.8	8.0	8.3	8.7	9.2	70	36.2	1800	24.2	—	—	—	—	
MAR-M 322	0.322	8.91	2425-2475	—	—	—	—	—	—	—	—	—	—	—	—	—	—	—	—	—	—	—	—	—	—	
UMCo-50	0.291	8.05	2515-2540	70	63	70	32.5	—	—	—	—	—	—	—	—	9.33	—	70	31.5	—	—	—	—	—	—	
UMCo-51	0.288	7.79	2420-2450	—	—	70	40.1	—	—	—	—	—	—	—	—	6.68	—	70	27.6	—	—	—	—	—	—	
Nivco-10	0.312	8.60	—	1472	198	70	—	—	—	—	—	7.36	—	—	—	—	—	70	30.5	1200	25.3	—	—	—	—	0.32
Elgiloy	0.300	8.3	—	392	112	70	39.27	—	—	—	—	8.43	—	—	—	—	—	70	29.5	—	—	—	—	—	—	
MAR-M 509	0.320	8.86	2350-2550	—	—	—	—	—	—	—	—	—	—	—	—	—	—	—	—	—	—	—	—	—	—	

(Table continued on next page)

TABLE 3 — AVERAGE PHYSICAL PROPERTIES (CONTINUED)

Material Commercial Designation	Density at 70 F lb/in³	Specific Gravity g/cc	Melting Range F	Thermal Conductivity		Electrical Resistivity		Coefficient of Linear Thermal Expansion in./in./F × 10⁻⁶										Modulus of Elasticity in Tension (E) 10⁶ psi				Modulus of Rigidity (G) 10⁶ psi				Poisson's Ratio at 70 F
				F	Btu/ft²/hr/F/in.	F	Microhm, in.	70-200	70-400	70-600	70-800	70-1000	70-1200	70-1400	70-1600	70-1800	70-2000	F	E	F	E	F	G	F	G	
Austenitic Nickel Base Alloys																										
							cm																			
Hastelloy X	0.297	8.23	2300-2400	1500	173	R.T.	—	7.70	7.82	7.90	8.15	8.39	8.56	8.81	9.02	9.20	—	70	28.6	1800	18.5	—	—	—	—	0.320
Incoloy 901	0.297	8.23	2245-2580	1400	142	—	—	7.75	7.85	8.02	8.27	8.50	8.79	9.15	—	—	—	75	29.9	1200	22.1	—	—	—	—	—
Incoloy 901 Mod	—	—	—	—	—	—	—	—	—	—	—	—	—	—	—	—	—	—	—	—	—	—	—	—	—	—
Rene 41	0.298	8.26	2400-2500	1600	160	—	—	6.7	6.8	7.0	7.2	7.5	7.8	8.2	8.70	9.3	9.9	80	31.9	1600	23.6	80	12.1	1550	8.8	0.31
Udimet 500	0.290	8.02	2350-2450	1800	177.1	1800	135.9	6.75	7.15	7.40	7.60	7.80	8.05	8.50	8.95	9.85	—	72	32.1	1800	21.0	—	—	—	—	—
Waspaloy	0.296	8.19	2425-2475	1800	182	—	—	6.8	7.1	7.3	7.6	7.8	8.0	8.5	8.9	9.8	10.4	70	29.0	1600	20.2	—	—	—	—	—
R-235	0.296	8.19	2460-2530	1800	173.1	—	133	6.70	7.17	7.51	7.72	7.98	8.13	8.44	8.92	9.58	10.18	70	30.5	1600	18.8	—	—	—	—	0.330
Udimet 700	0.285	7.91	2230-2450	1800	240	1800	146.1	—	7.50	7.52	7.60	7.74	7.98	8.35	8.95	9.65	10.30	70	32.4	1800	22.1	—	—	—	—	—
Inconel X-750	0.298	8.25	2540-2600	1600	164	—	—	6.96	7.14	7.46	7.76	8.10	8.41	8.84	9.33	9.75	—	80	31.0	1800	20.0	80	11.0	—	—	0.29
M-252	0.298	8.26	2400-2500	1500	149	—	—	6.7	6.8	7.0	7.2	7.5	7.8	8.2	8.8	9.3	—	80	29.8	1600	21.0	80	12.0	1600	8.3	0.314
Refractaloy 26	0.296	8.19	2450-2500	1200	177	—	—	7.8	7.9	8.0	8.1	8.2	8.4	—	—	—	—	70	30.6	1600	19.5	—	—	—	—	—
Astroloy	0.290	—	—	—	—	—	—	—	—	—	—	—	—	—	—	—	—	—	—	—	—	—	—	—	—	—
GMR-235	—	—	—	800	106	—	—	—	—	—	7.6	—	8.1	8.6	9.1	—	—	—	—	—	—	—	—	—	—	—
GMR-235D	0.291	8.05	—	—	—	—	—	—	—	—	—	—	—	—	—	—	—	70	28.7	1800	17.0	—	—	—	—	—
Hastelloy B	0.334	9.24	2400-2460	1112	114	—	—	—	—	6.41	6.57	6.66	6.73	—	7.78	—	—	80	28.5	—	—	—	—	—	—	—
Hastelloy C	0.323	8.94	2310-2380	1112	118	—	—	—	—	7.02	7.35	7.44	7.73	—	8.20	—	—	80	28.5	—	—	—	—	—	—	—
Hastelloy F	0.295	8.16	2300-2400	—	—	—	112	8.1	8.3	8.7	8.8	8.9	9.2	9.5	9.8	10.2	—	80	29.0	—	—	—	—	—	—	0.305
Hastelloy N	0.317	8.79	2470-2555	1112	140.1	—	—	—	6.45	6.76	7.09	7.43	7.81	8.16	8.51	8.85	—	80	31.3	1500	25.8	—	—	—	—	—
Hastelloy W	0.298	8.26	2540-2600	—	—	—	—	—	—	—	—	—	—	—	—	—	—	80	31.0	—	—	—	—	—	—	—
Inconel 600	0.301	8.33	2500-2600	1600	200	—	—	7.4	7.7	7.9	8.1	8.4	8.6	8.9	9.1	9.3	—	80	31.4	1600	23.1	80	11.0	—	—	—
Inconel 604	0.305	8.45	—	—	—	—	—	6.75	—	—	7.75	—	8.25	—	—	—	—	80	31.0	1350	24.5	—	—	—	—	—
Inconel 610	—	—	—	—	—	—	—	—	—	—	—	—	—	—	—	—	—	—	—	—	—	—	—	—	—	—
Inconel 700	0.295	8.16	2450-2600	1600	126	—	—	6.84	7.21	7.48	7.79	8.02	8.32	8.68	9.27	—	—	70	32.5	1600	23.6	—	—	—	—	—
Inconel 702	0.304	8.41	—	1200	235	—	—	6.7	7.5	7.8	8.0	8.3	8.7	9.1	9.4	10.0	—	70	31.5	1800	20.8	—	—	—	—	—
Inconel 705	0.292	8.06	2500-2580	—	—	—	—	—	—	—	—	—	—	—	9.20	—	—	80	25.0	—	—	—	—	—	—	—
Alloy 713C	0.286	7.91	2300-2350	1600	218	1600	158	5.92	6.61	7.00	7.26	7.52	7.81	8.17	8.63	9.13	9.48	70	29.9	1600	22.6	—	—	—	—	—
Inconel 718	0.296	8.19	2200-2450	—	—	—	—	7.1	7.5	7.7	7.9	8.0	8.4	8.9	—	—	—	70	29.0	1400	23.3	—	—	—	—	0.293
Inconel 722	0.298	8.26	2540-2600	—	—	—	—	6.5	7.3	7.6	7.7	7.9	8.2	8.6	9.1	9.3	9.5	70	31.0	1600	22.6	—	—	—	—	—
AF 1753	0.305	8.45	2525-2575	—	—	—	132.0	—	—	6.7	7.0	7.3	7.5	7.8	8.2	8.5	—	70	31.0	1800	20.4	—	—	—	—	—
IN 100	0.280	7.75	2305-2435	—	—	—	—	—	7.2	7.5	7.5	7.7	8.0	8.3	8.8	9.3	10.1	70	30.8	1600	23.4	—	—	—	—	—
Nimonic 75	0.301	8.33	2530-2600	1472	180	1472	113	—	—	—	—	—	—	—	—	—	—	70	27.0	—	—	—	—	—	—	—
Nimonic 80A	0.295	8.16	2480-2540	1400	142	—	—	7.0	7.2	7.4	7.6	7.7	7.9	8.2	8.6	—	—	70	31.2	1600	22.7	—	—	—	—	—
Nimonic 90	0.296	8.19	2480-2540	1400	144	1472	124	6.4	7.0	7.2	7.5	7.7	8.1	8.5	9.0	—	—	70	32.1	1600	22.9	—	—	—	—	—
Nimonic 105	0.289	8.00	2440-2520	1400	129	1472	142	6.8	7.2	7.5	7.8	8.0	8.3	8.8	9.5	10.6	—	70	32.9	1600	24.7	—	—	—	—	—
Nimonic 115	0.284	7.85	—	—	—	—	—	—	7.05	7.15	7.40	7.70	8.00	8.45	9.00	9.75	—	70	32.4	1600	23.8	—	—	—	—	—
MAR-M 200	0.308	8.53	2400-2450	2000	198	—	—	—	6.6	6.9	7.1	7.3	7.5	7.8	8.2	8.8	9.8	80	31.6	1800	22.9	—	—	—	—	—
RA-333	0.298	8.26	—	—	—	—	—	—	—	—	—	—	—	—	—	8.85	—	75	31.3	1600	16.6	—	—	—	—	—
Titanium Base Alloys— Commercially Pure																										
A40	0.163	4.54	2950-3100	—	—	70	56	4.8	—	5.2	—	5.5	5.6	—	—	—	—	70	14.9	—	—	70	6.5	—	—	—
A55	0.163	4.54	2950-3100	1000	128.4	1000	132	4.8	—	5.3	—	5.5	5.6	—	—	—	—	70	15.0	—	—	70	6.5	—	—	—
A70	0.163	4.54	2985-3085	1000	126.0	1000	150	4.8	4.9	5.0	5.1	5.3	5.4	5.7	—	—	—	70	15.1	1000	10.0	70	6.5	—	—	0.340
Titanium Base Alloys — Alpha Alloy Grades																										
5Al-2.5Sn	0.162	—	2820-3000	800	86.4	800	183	5.2	5.2	5.3	5.3	5.4	5.5	5.6	5.7	—	—	70	16.0	—	—	70	7.0	—	—	—
5Al-5Sn-5Zr	0.166	—	2950-3050	—	—	—	—	—	—	—	—	—	—	—	—	—	—	70	16.0	—	—	—	—	—	—	—
7Al-12Zr	0.165	—	2950-3050	—	—	—	—	—	—	—	—	—	—	—	—	—	—	70	16.5	—	—	—	—	—	—	—
7Al-2Cb-1Ta	0.160	—	3065-3115	—	—	—	—	—	—	—	—	5.0	—	—	—	—	—	70	17.7	—	—	—	—	—	—	—
8Al-1Mo-1V	0.156	4.37	—	—	—	800	201	4.7	4.9	5.0	—	5.6	5.7	—	—	—	—	70	18.5	—	—	—	—	—	—	—
Ti-679	0.174	—	—	800	81.6	800	185	5.0	—	5.7	—	5.8	—	—	—	—	—	70	—	—	—	—	—	—	—	—
Titanium Base Alloys — Alpha-Beta Alloy Grades																										
Ti-155A	0.163	—	—	800	83.6	800	180	—	—	—	5.7	—	—	—	—	—	—	70	16.5	1000	12.3	80	6.3	1000	4.5	0.327
8Mn	0.171	—	2730-2970	800	108	800	140	4.8	5.1	5.4	5.7	6.0	6.5	—	—	—	—	70	16.4	—	—	70	7.0	—	—	—
2.5Al-16V	0.166	—	3050-3150	—	—	—	—	—	—	—	4.9	5.0	—	—	—	—	—	70	15.0	—	—	—	—	—	—	—
3Al-2.5V	0.162	—	3050-3150	—	—	—	—	5.3	—	5.5	—	5.5	—	—	—	—	—	70	15.5	—	—	—	—	—	—	—
4Al-4Mn	0.163	—	2920-3050	800	88.8	800	172	4.9	4.9	5.1	5.3	5.4	5.6	—	—	—	—	70	16.4	—	—	70	7.3	—	—	—
4Al-3Mo-1V	0.163	—	2950-3050	800	81.6	—	—	5.0	5.1	5.3	5.4	5.5	—	—	—	—	—	70	16.5	—	—	70	7.0	—	—	—
5Al-1.25 Fe-2.75 Cr	0.162	—	—	—	—	—	—	5.2	—	5.3	—	5.5	—	—	—	—	—	70	16.5	—	—	—	—	—	—	—
6Al-4V	0.160	4.424	2950-3050	800	81.6	800	187	4.8	5.0	5.1	5.2	5.3	6.1	—	—	—	—	70	16.5	—	—	70	6.2	—	—	—
6Al-6V-2Sn	0.164	—	3050-3150	—	—	—	—	5.0	5.05	5.20	5.25	5.30	—	—	—	—	—	70	16.5	—	—	—	—	—	—	—
7Al-4Mo	0.162	—	2950-3050	800	81.6	800	183	4.92	5.09	5.21	5.38	5.55	5.80	6.15	—	—	—	70	16.2	800	13.2	70	6.5	—	—	—
2Cr-2Fe-2Mo	—	—	—	—	—	—	—	—	—	—	—	—	—	—	—	—	—	—	—	—	—	—	—	—	—	—
Titanium Base Alloys— Beta Alloy Grades																										
1Al-8V-5Fe	0.168	—	—	—	—	—	—	5.2	—	5.6	—	5.9	—	—	—	—	—	70	14.2	—	—	70	—	—	—	—
3Al-13V-11Cr	0.175	—	—	800	98.4	—	—	5.2	—	5.6	—	5.9	—	—	—	—	—	70	14.7	1000	11.6	70	6.2	—	—	0.304

TABLE 4 — TYPICAL MECHANICAL PROPERTIES

Material Commercial Designation	Material Condition	Ultimate Tensile Strength				Yield Strength at 0.2% offset				Elongation in 2 in. or 4D				Reduction of Area				Charpy Impact, ft-lb	Hardness No.	Stress Rupture Strengths, 1000 psi							
																				100	1000	100	1000	100	1000	100	1000
		F	Ksi	F	Ksi	F	Ksi	F	Ksi	F	%	F	%	F	%	F	%			800 F		1000 F		1200 F		1400 F	

Martensitic Low Alloy Steels

"17-22-A"	Oil quench (1200 F)	80	169	1000	86	80	161	1000	74	80	13	1000	28	80	46	1000	82	58	Bhn 341	84	72	49	36	—	—	—	—
"17-22-A" S	Norm (1200 F)	80	153	1000	108	80	134	1000	92	80	18	1000	21	80	53	1000	70	25	Bhn 320	—	—	75	60	14	6	—	—
"17-22-A" V	Norm (1200 F)	80	160	1000	100	80	145	1000	92	80	17	1000	21	80	52	1000	66	18	Bhn 341	—	—	67	49	25	14	—	—
Chromoloy	Air cooled, (1200 F)	70	138	1000	110	70	117	1000	85	70	7	1000	—	70	45	1000	—	—	Rc 30	—	—	75	51	—	—	—	—
D6A	Oil quench (600 F)	80	267	—	—	80	247	—	—	80	10	—	—	80	41	—	—	13	Rc 53								
300M	Oil quench (600 F)	80	289	—	—	80	245	—	—	80	9	—	—	80	34	—	—	22	Bhn 525								
UCX2	Oil quench (600 F)	80	272	—	—	80	235	—	—	80	6	—	—	80		—	—		Rc 51								

Martensitic Secondary Alloy Steels

H11	Air cooled, (1050 F)	70	262	1000	180	70	215	1000	141	70	10	1000	12	70	35	1000	41	—	Rc 52	205	190	100	47	—	—	—	—
H12	Air cooled, (1050 F)	80	205	1000	137	80	185	1000	120	80	12	1000	25	80	42	1000	64	15	Rc 44								
H13	Air cooled, (1100 F)	70	215	1000	143	70	184	1000	103	70	13	1000	19	70	45	1000	65	18	Rc 45								
M2	—																										
M10	—																										
M50	Air cooled, (1025 F)	70	411	1000	309	70	338	1000	250	70	2	1000	6	70	2	1000	12		Rc 64								

Martensitic Chromium Steels

410	Oil quench, (1000 F)	70	157	1000	101	70	145	1000	93	70	13	1000	16	70	69	1000	77	25	Bhn 300	60	55	32	26	8	6	3	
Greek Ascoloy	Oil quench, (1050 F)	70	160	1000	95	70	135	1000	84	70	16	1000	17	70	45	1000	64	19	Rc 35	—	—	52	41	19	11	—	
422	Oil quench, (1200 F)	70	149	1000	96	70	125	1000	82	70	18	1000	25	70	52	1000	67	19	Rc 43	—	—	63	57	25	18	—	
422M	—																					86	72	26	16	—	
422M (Cast)	—																							34	20		
440C	Oil quench, (600 F)	70	285	1000	122	70	275			70	2	—	—	70	10				Bhn 580								
14Cr-4Mo	—																										
Lapelloy	Oil quench, (1275 F)	70	155	1000	95	70	140	1000	85	70	15	1000	15	70	35	1000	45	—	Rc 35	—	—	65	55	25	15	—	
Lapelloy C	Oil quench, (1200 F)	80	155	1000	98	80	120	1000	95	80	17	1000	19	80	44	1000	60	30	Rc 33	—	—	70	55	24	13	—	
H-46	Air cooled, (1200 F)	80	150	1000	99	80	128	1000	88	80	20	1000	25	80	56	1000	61	—	Bhn 302	—	—	75	65	35	25	—	

Semi Austenitic Precipitation and Transformation Hardening Steels

AM-350	SCT[a], (850 F)	70	203	1000	106	70	170	1000	85	70	13	1000	16	—	—	—	—	14	—	183	181	—	—	—	—	—	—
AM-350	SCT, (850 F)	70	169	800	129	70	147	800	104	70	15	800	8	—	—	—	—			130	127						
AM-355	SCT, (850 F)	70	216	1000	144	70	182	1000	97	70	19	1000	16	70	39	1000	57	17	Rc 48	186	180	70	57	—			
AM-355	SCT, (1000 F)	70	186	1000	115	70	171	1000	96	70	19	1000	19	70	57	1000	65	45	Rc 38	134	132	73	61	—			
Stainless W	Solution + 1000 F	70	192	1000	94	70	187	1000	54	70	13	1000	22	70	53	1000	75	13	Rc 44	—	—	31	—	12	—		
14-4PH	—																										
17-4PH	H900	70	203	900	149	70	186	900	132	70	11	900	10	70	50	900	30	20	Rc 44	140	128						
17-7PH	RH 950 Sheet	70	230	900	133	70	217	900	114	70	6	900	15	70	30				Rc 48	113	92						
17-7PH	TH 1050 Sheet	70	193	900	124	70	182	900	100	70	10	900	10	—	—				Rc 42	110	90						
PH15-7Mo	RH 950	70	240	1000	130	70	225	1000	105	70	6	1000	14	—	—	—	—		Rc 48	174	171						

Austenitic Nickel-Chromium-Iron Steels

302	Annealed	70	92	1200	44	70	38	1200	11	70	68	1200	40	70	78	1200	62	—	Rb 85	—	—	35	—	14	—	6	
304	Annealed	70	85	1200	45	70	30	1200	14	70	60	1200	14	—	—	70	70	1200	—	Rb 80	—	—	35	—	14	—	6	
304L	Annealed	70	80	1000	52	70	30	1000	14	70	60	1000	12	70	60	1000	45		70	77	1000	67	Rb 80	—	—	32	26	14	11	5	
309S	Annealed	70	90	1400	36	70	40	1400	26	70	50	1400	40	70	65	1400	42	—	Rb 83	—	—	—	—	—	20	—	7
310	Annealed	70	92	1300	50	70	40	1300	22	70	47	1300	36	70	73	1300	52	80	Rb 89	—	—	32	—	17	—	6	
314		70	100	—	—	70	50			70	45			70	60				Rb 89					22		14	
316	Annealed	70	85	1300	46	70	38	1300	19	70	60	1300	42	70	77	1300	58	80	Rb 80	—	—	—	—	—	24	—	11
321	Annealed	70	85	1300	37	70	33	1300	16	70	58	1300	56	70	75	1300	78	70	Rb 80	—	—	—	—	17	—	5	
347	Annealed	70	91	1300	40	70	39	1300	24	70	50	1300	51	70	71	1300	74	48	Bhn 160	—	—	32	26	19	—	7	

Austenitic Iron Base Alloys

A286	1800, 1325 F	70	145	1200	103	70	95	1200	88	70	24	1200	13	70	45	1200	14	64	Rc 26	99	88	61	46	25	—	—	—
A286	1650, 1325 F	70	157	1200	109	70	102	1200	90	70	25	1200	18	70	46	1200	25	—	Bhn 302	—	—	59	47				
V-57	1800, 1350 F	70	175	1300	120	80	125	1300	—	80	21	1300	—	80	35	1300	—	—	Bhn 331	112	102	77	64	35	—	—	
Discaloy	1850, 1350, 1200 F	70	145	1200	104	70	106	1200	91	70	19	1200	19	70	23	1200	24	—	Bhn 293	90	72	52	43	18	8	—	
N-155	2150, 1400 F	70	119	1400	60	70	57	1400	36	70	43	1400	31	70	50	1400	41	43	Bhn 210	—	—	52	38	24	20	12	8
D-979	1900, 1550, 1300 F	70	204	1500	74	70	145	1500	63	70	15	1500	23	70	23	1500	57	8	—	—	—	88	70	43	30	—	
W-545	1875, 1415, 1350 F	70	181	1400	91	70	132	1400	82	70	19	1400	28	70	30	1400	56	30	—	120	105	82	66	40	23	—	
S590	2250, 1400 F	70	145	1200	95	70	82	1200	71	70	20	1200	22	70	25	1200	24	14	—	90	74	48	38	—	—	12	9
RA330	Annealed	70	89	1600	28	—	—	—	—	70	43	1600	21	70	68	1600	24	—	Rb 80	—	—	—	—	88	46	28	4
Unitemp 212	1850, 1325 F	80	187	1400	102	80	130	1400	97	80	23	1400	16	80	40				Rc 39	—	—	100	88	46	28	—	
CRM-6D	Aged	70	110	1500	51	70	78	1500	40	70	2	1500	12	70	3	1500	17		Rc 35	90	—	61	52	36	29	20	16
CRM-15D	Aged	70	115	1500	58	70	90	1500	46	70	11	1500	11	70	3	1500	16		Rc 37	93	—	67	54	28	20	15	
16-25-6	Hot cold worked	70	162	1200	106	70	143	1200	93	70	15	1200	13	70	34	1200	28		Bhn 326	57	42	36	20	18	11	—	

[a] Subcritical Transformation.

(Table continued on next page)

TABLE 4 — TYPICAL MECHANICAL PROPERTIES (CONTINUED)

Material Commercial Designation	Material Condition	Ultimate Tensile Strength		Yield Strength at 0.2% offset		Elongation in 2 in. or 4D		Reduction of Area		Charpy Impact, ft-lb	Hardness No.	Stress Rupture Strengths, 1000 psi							
												100	1000	100	1000	100	1000	100	1000
		F	Ksi	F	Ksi	F	%	F	%			1000 F		1200 F		1400 F		1600 F	

Austenitic Iron Base Alloys (Continued)

Material	Condition	F	Ksi	F	Ksi	F	%	F	%	Charpy	Hardness	100/1000F	1000/1000F	100/1200F	1000/1200F	100/1400F	1000/1400F	100/1600F	1000/1600F
19-9 DL	Stress relieved	70 118 1200	75	70 69 1200	37	70 56 1200	34	70 55 1200	34	46	Bhn 215	64	56	44	37	20	10	—	—
19-9 DX	Stress relieved	70 118 1200	75	70 69 1200	37	70 55 1200	33	70 54 1200	33	46	Bhn 216	64	56	48	39	—	—	—	—
17-14CuMo	2250, 1350 F	75 86 1000	72	75 42 1000	28	75 45 1000	32	75 63 1000	45	26	Rb 80	—	—	43	37	—	—	—	—
G-192	2150, 1400 F	70 136 1500	41	70 86 1500	32	70 5 1500	24	70 4 1500	28	—	—	—	—	42	—	—	—	—	—
AF71	2050, 1325 F	70 151 1500	48	70 106 1500	43	70 25 1500	29	70 35 1500	58	—	—	—	—	69	54	—	—	—	—
Incoloy 800	Annealed	80 87 1400	35	80 47 1400	27	80 42 1400	70	80 69 1400	64	207	—	62	49	32	23	12	8	5	3
Incoloy 801		80 90 1400	43	80 40 1400	27	80 36 1400	16	—	—	—	—	—	—	33	26	—	—	—	—
Incoloy 805		—	—	—	—	—	—	—	—	—	—	—	—	—	—	—	—	—	—
Incoloy 810		—	—	—	—	—	—	—	—	—	—	—	—	—	—	—	—	—	—

Cobalt Base Alloys

Temperature headers for stress rupture: 1200 F, 1400 F, 1600 F, 1800 F

Material	Condition	F	Ksi	F	Ksi	F	%	F	%	Charpy	Hardness	100/1200F	1000/1200F	100/1400F	1000/1400F	100/1600F	1000/1600F	100/1800F	1000/1800F
L-605	Solution at 2250 F	70 160 1600	40	70 86 1600	37	70 47 1600	25	—	—	—	Bhn 230	70	54	35	27	15	10	7	3
S-816	Solution, precipitate at 1400 F	70 140 1600	51	70 55 1600	35	70 35 1600	17	70 29 1600	20	—	Rc 31	65	50	36	28	16	9	—	—
HS-31 (X-40)	As cast	70 108 1500	63	70 76 1500	—	70 9 1500	15	70 11 1500	18	6	Bhn 228	56	51	37	33	20	16	11	10
HS-21	As cast	70 101 1600	55	70 82 1600	—	70 8 1600	23	70 9 1600	48	10	Bhn 237	52	42	24	15	16	11	—	—
Stellite 6	As cast	70 115 1500	70	70 96 1500	—	70 3 1500	—	70 3 1500	12	9	Rc 41	—	—	—	—	—	—	—	—
Haynes No. 151	As cast	70 106 1500	64	70 74 1500	43	70 8 1500	11	70 — 1500	12	—	Rc 33	73	68	—	—	27	24	—	—
WI-52	As cast	70 120 1400	94	70 89 1400	49	70 2 1400	10	70 — 1400	12	—	—	—	—	—	—	25	22	13	10
V-36	Water quench (2250 F) Air cooled (1400 F)	70 146 1500	61	70 83 1500	47	70 15 1500	18	70 — 1500	2	—	—	64	43	32	24	17	12	—	—
MAR-M 302	As cast	70 136 1600	67	70 100 1600	45	70 2 1600	10	70 — 1600	2	—	Rc 37	—	—	—	—	27	—	14	—
MAR-M 322	As cast	70 121 1600	80	70 91 1600	50	70 6 1600	12	70 4 1600	12	—	Rc 35	—	—	—	—	32	—	20	—
UMCo-50	As cast	70 78 1650	18	70 46 1650	15	70 8 1650	8	70 — 1650	9	27	Dpn 250	—	—	—	—	—	—	—	—
UMCo-51	As cast	70 91 900	30	70 72 900	26	70 2 900	14	70 — 900	—	—	Dpn 280	—	—	—	—	—	—	—	—
Nivco-10	Bar	70 165 1200	105	70 110 1200	75	70 25 1200	20	70 28 1200	35	31	—	51	38	—	—	—	—	—	—
Elgiloy	Strip	70 368 —	—	70 280 —	—	70 — —	—	70 — —	—	—	Rc 58	—	—	—	—	—	—	—	—
MAR-M 509	As cast	70 112 1600	68	70 83 1600	45	70 3 1600	7	70 6 1600	13	—	Rc 32	—	—	—	—	29	—	17	—

Austenitic Nickel Base Alloys

Material	Condition	F	Ksi	F	Ksi	F	%	F	%	Charpy	Hardness	100/1200F	1000/1200F	100/1400F	1000/1400F	100/1600F	1000/1600F	100/1800F	1000/1800F
Hastelloy X	Solution at 2175 F	70 115 1800	23	70 52 1800	21	70 52 1800	66	70 64 1800	50 15	88	Rb 90	42	31	21	15	9	6	4	3
Incoloy 901	2000, 1450, 1325 F	70 175 1300	130	70 125 1300	114	70 15 1300	11	70 19 1300	21	—	—	78	61	37	—	—	—	—	—
Incoloy 901	2000, 1450, 1325 F	70 175 1300	129	70 130 1300	111	70 14 1300	13	70 17 1300		—	—	90	76	44	—	—	—	—	—
Rene 41	1950, 1400 F	70 206 1600	90	70 154 1600	80	70 14 1600	19	70 —	—	—	Rc 31	110	100	64	40	23	14	10	—
Udimet 500	1975, 1550, 1400 F	70 175 1800	40	70 115 1800	35	70 16 1800	22	70 16 1800	40	—	—	135	110	65	47	30	18	12	—
Waspaloy	Solution, stabilize, precipitate	70 185 1600	76	70 115 1600	75	70 25 1600	34	70 20 1600	54	—	Rc 36	110	86	60	42	25	16	6	—
R-235	Solution, precipitate	70 169 1600	76	70 116 1600	58	70 21 1600	14	—	—	—	—	52	40	39	29	22	15	8	5
Udimet 700	Solution, stabilize, precipitate	70 204 1800	52	70 140 1800	44	70 17 1800	28	70 20 1800	28	—	—	—	102	79	62	42	28	16	7
Inconel X-750	2100, 1550, 1300 F	80 162 1500	52	80 92 1500	44	80 24 1500	22	80 30 1500	34	37	Rc 33	80	68	40	30	13	7	3	2
M-252	1950, 1400 F	70 176 1500	91	70 110 1500	84	70 25 1500	24	70 — 1500	—	—	Bhn 363	100	79	52	38	23	13	—	—
Refractaloy 26	Solution, precipitate	70 154 1500	71	70 96 1500	68	70 19 1500	29	70 20 1500	35	—	—	77	63	40	3	—	—	—	—
Astroloy	—	—	—	—	—	—	—	—	—	—	—	—	—	—	37	—	—	—	—
GMR-235	—	80 103 1500	105	80 93 1500	78	80 3 1500	3	—	—	—	—	—	—	26	18	11	—	—	—
GMR-235D	—	80 112 1500	121	80 103 1500	81	80 3 1500	3	—	—	—	—	110	100	77	60	32	23	17	12
Hastelloy B	As cast	80 85 1500	58	80 53 —	—	80 15 1500	19	80 15 1500	13	13	Rb 93	51	40	—	—	13	9	—	—
Hastelloy C	As cast	80 89 1500	56	80 52 —	—	80 11 1500	18	80 12 1500	15	—	—	49	42	—	—	—	—	—	—
Hastelloy F	Solution at 2125 F	80 102 1400	54	80 45 1400	29	80 46 1400	47	—	—	—	—	42	36	—	—	9	—	—	—
Hastelloy N	Solution at 2150 F	80 115 1500	56	80 45 1500	29	80 50 1500	24	—	—	85	—	42	29	—	—	8	3	—	—
Hastelloy W	2000, 1300 F	80 158 1500	59	80 82 1500	46	80 26 1500	24	—	—	—	—	74	54	—	—	7	5	3	—
Inconel 600	Annealed	80 90 1400	27	80 37 1400	17	80 47 1400	46	80 64 1400	60	—	—	23	14	12	8	5	3	3	2
Inconel 604	Annealed 2050 F	80 102 —	—	80 33 —	—	80 48 —	—	80 62 —	—	—	—	37	27	16	11	—	—	—	—
Inconel 610	—	—	—	—	—	—	—	—	—	—	—	—	—	—	—	—	—	—	—
Inconel 700	Solution, aged 1600 F	80 171 1500	107	80 104 1500	75	80 25 1500	6	80 27 1500	8	—	—	100	87	—	—	27	17	6	3
Inconel 702	2000, 1350 F	80 148 1400	72	80 84 1400	62	80 35 1400	4	—	—	—	—	56	43	24	15	7	4	3	2
Inconel 705	—	—	—	—	—	—	—	—	—	—	—	—	—	—	—	—	—	—	—
Alloy 713C	As cast	70 123 1600	105	70 106 1600	72	70 8 1600	14	70 11 1600	20	—	Rc 38	—	—	83	65	42	28	21	13
Inconel 718	1800, 1325, 1150 F	80 208 1400	107	80 172 1400	138	80 21 1400	25	—	—	—	—	105	86	44	25	—	—	—	—
Inconel 722	1975, 1300 F	80 158 1500	59	80 82 1500	46	80 26 1500	24	—	—	—	—	74	54	37	22	7	5	3	—
AF 1753	2150, 1400 F	70 194 1500	113	70 127 1500	108	70 20 1500	13	70 22 1500	19	—	—	115	98	65	51	32	22	10	6
IN 100	As cast	70 144 1800	80	70 124 1800	66	70 8 1800	6	70 — 1800	8	—	—	—	—	89	72	55	37	25	15
Nimonic 75	Annealed	70 116 1400	33	70 49 1400	22	70 44 1400	47	70 62 1400	35	—	—	—	—	—	—	—	—	—	—
Nimonic 80A	Solution, aged 1290 F	70 145 1400	87	70 90 1400	73	70 39 1400	17	70 38 1400	19	—	—	76	61	37	23	11	—	—	—
Nimonic 90	Solution, aged 1290 F	70 179 1400	95	70 117 1400	78	70 33 1400	12	70 — 1400	—	—	—	79	66	45	30	15	9	—	—
Nimonic 105	2100, 1925, 1550 F	70 140 1400	118	70 115 1400	94	70 7 1400	17	—	—	—	—	107	89	66	50	28	19	8	4
Nimonic 115	2175, 2010 F	70 180 1800	67	70 125 1800	35	70 27 1800	23	—	—	—	—	—	—	79	61	38	27	16	9
MAR-M 200	As cast	80 135 1600	123	80 120 1600	110	80 7 1600	4	—	—	—	—	—	—	94	84	58	43	26	18
RA-333	Annealed	70 108 1600	28	70 51 1600	24	70 43 1600	30	—	—	—	—	—	—	18	—	7	—	—	—

(Table continued on next page)

TABLE 4 — TYPICAL MECHANICAL PROPERTIES (CONTINUED)

Material Commercial Designation	Material Condition	Ultimate Tensile Strength		Yield Strength at 0.2% offset		Elongation in 2 in. or 4D		Reduction of Area		Charpy Impact, ft-lb	Hardness No.	Stress Rupture Strengths, 1000 psi							
												100	1000	100	1000	100	1000	100	1000
		F	Ksi	F	Ksi	F	%	F	%	F		1200 F		1400 F		1600 F		1800 F	

Titanium Base Alloys — Commercially Pure

												400 F		600 F		800 F		1000 F	
A40	Annealed	75	60	600	28	75	45	600	13	75	28	600	45	75	50	—	—	32	Bhn 200	—	—	—	—	—	—	—	—
A55	Annealed	75	75	600	33	75	60	600	19	75	25	600	33	75	45	600	73	28	Bhn 225	—	42	—	36	20	10	—	4
A70	Annealed	75	96	600	43	75	78	600	27	75	20	600	28	75	49	600	57	13	Bhn 265	—	44	—	32	—	16	—	5

Titanium Base Alloys — Alpha Alloy Grades

5Al-2.5Sn	Annealed	75	125	600	82	75	117	600	65	75	18	600	19	75	40	600	45	19	Rc 36	84	78	70	65	64	58	32	20
5Al-5Sn-5Zr	Annealed 1650 F	75	125	600	94	75	120	600	74	75	18	600	20	75	26	600	40	—	—	—	—	—	—	—	—	60	—
7Al-12Zr	Annealed 1650 F	75	135	600	109	75	130	600	86	75	15	600	21	75	24	600	40	—	—	—	—	—	—	—	—	—	—
7Al-2Cb-1Ta	Annealed 1650 F	75	126	600	100	75	120	600	81	75	17	600	25	75	28	600	30	—	Rc 36	—	—	—	—	—	—	—	32
8Al-1Mo-1V	Duplex annealed 1650 F	75	145	600	110	75	138	600	84	75	15	600	20	75	28	600	38	24	—	—	—	—	—	—	—	—	—
Ti-679	Aged	70	154	600	118	70	134	600	96	70	13	600	13	—	—	—	—	—	—	—	—	—	—	—	—	—	—

Titanium Base Alloys — Alpha-Beta Alloy Grades

Ti-155A	Annealed	75	155	600	80	75	140	600	69	75	15	600	15	75	35	600	48	11	—	—	—	—	100	—	91	—	26
8Mn	Annealed	75	137	600	98	75	125	600	75	75	15	600	13	75	32	—	—	—	—	—	—	—	—	90	—	—	—
2.5Al-16V	Solution + aged	75	180	600	145	75	165	600	127	75	6	600	5	—	—	—	—	—	—	—	—	—	—	—	—	—	—
3Al-2.5V	Annealed	75	100	600	70	75	85	600	50	75	20	600	25	—	—	—	—	—	—	—	—	—	—	—	—	—	—
4Al-4Mn	Annealed	75	148	600	110	75	135	600	90	75	15	600	17	75	25	1000	65	16	Rc 35	—	—	—	110	—	95	—	46
4Al-3Mo-1V	Annealed	75	140	1000	65	75	120	1000	55	75	15	1000	35	—	—	—	—	—	Rc 35	—	—	—	—	—	—	—	—
5Al-1.25 Fe-2.75 Cr	Annealed	75	155	600	122	75	145	600	102	75	15	600	20	75	25	1000	38	—	Rc 35	—	—	—	—	100	—	75	—	13
6Al-4V	Annealed	75	138	600	105	75	128	600	95	75	12	600	11	75	37	1000	68	18	Rc 32	—	95	98	78	68	50	—	10
6Al-6V-2Sn	Annealed	75	165	600	132	75	150	600	117	75	15	600	20	75	42	—	—	15	—	—	—	—	—	—	—	—	—
7Al-4Mo	Annealed	75	160	600	127	75	150	600	108	75	16	600	18	75	22	1000	50	18.0	Rc 38	—	—	—	—	—	—	—	—
2Cr-2Fe-2Mo	Annealed	75	137	800	75	75	125	800	55	75	18	800	30	—	—	—	—	13	—	—	—	—	—	—	—	—	—

Titanium Base Alloys — Beta Alloy Grades

| 1Al-8V-5Fe | Annealed | 75 | 177 | 600 | 128 | 75 | 170 | 600 | 115 | 75 | 8 | 600 | 19 | — | — | — | — | 8 | Rc 34 | — | — | — | — | — | — | — | — |
| 3Al-13V-11 Cr | Solution | 75 | 185 | 800 | 160 | 75 | 175 | 800 | 120 | 70 | 8 | 800 | 12 | — | — | — | — | 8 | Rc 34 | — | — | — | — | — | — | — | — |

ZINC ALLOY INGOT AND DIE CASTING COMPOSITIONS—SAE J468b

SAE Standard

Report of Nonferrous Metals Division approved June 1934 and last revised by Nonferrous Metals Committee July 1965. Reaffirmed without change January 1971. Editorial change March 1972.

Similar Specifications—SAE 903 ingot is similar to ASTM B 240-64, Alloy AG40A (XXIII); and SAE 903 die casting is similar to ASTM B 86-71, Alloy AG40A (XXIII). SAE 925 ingot is similar to ASTM B 240-64, Alloy AC41A (XXV); and SAE 925 die casting is similar to ASTM B 86-71, Alloy AC41A (XXV).

TABLE 1—ZINC ALLOY INGOT AND DIE CASTING COMPOSITIONS

	Composition[a,b], %						
	Al	Cu	Mg	Fe	Pb	Cd	Sn
SAE 903							
Ingot	3.9–4.3	0.10	0.025–0.05	0.075	0.004	0.003	0.002
Castings	3.5–4.3	0.25[c]	0.020–0.05[d]	0.10	0.005	0.004	0.003
SAE 925							
Ingot	3.9–4.3	0.75–1.25	0.03–0.06	0.075	0.004	0.003	0.002
Castings	3.5–4.3	0.75–1.25	0.03–0.08	0.10	0.005	0.004	0.003

[a] Percentages given are maximum except where indicated as a range. Zinc is remainder.
[b] Zinc alloy die castings may contain nickel, chromium, silicon, and manganese in amounts up to their solubility (0.02, 0.02, 0.035, and about 0.5%, respectively) at the freezing temperature. No harmful effects have ever been noted due to the presence of these elements in these concentrations and, therefore, analyses are not required for these elements.
[c] For the majority of commercial applications, a copper content in the range of 0.25 to 0.75% will not adversely affect the serviceability of die castings and should not serve as a basis for rejection.
[d] Magnesium may be as low as 0.005% provided that at least 0.005% nickel is present, and lead, cadmium, and tin do not exceed 0.0030, 0.0020, and 0.0010%, respectively.

ZINC DIE CASTING ALLOYS—SAE J469a

SAE Information Report

Report of Nonferrous Metals Division approved June 1934 and last revised by Nonferrous Metals Committee July 1965. Reaffirmed with editorial change January 1971.

Because of the drastic chilling involved in die casting and the fact that the solid solubilities of both aluminum and copper in zinc change with temperature, these alloys are subject to some aging changes, one of which is a dimensional change. Both of the alloys undergo a slight shrinkage after casting, which at room temperature is about two-thirds complete in five weeks. It is possible to accelerate this shrinkage by a stabilizing anneal, after which no further changes occur. The recommended stabilizing anneal is 3 to 6 hr at 100 C (212 F), or 5 to 10 hr at 85 C (185 F), or 10 to 20 hr at 70 C (158 F). The time in each case is measured from the time at which the castings reach the annealing temperature. The parts may be air cooled after annealing. Such a treatment will cause a shrinkage (0.0004 in. per in.) of about two-thirds of the total, and the remaining shrinkage will occur at room temperature during the subsequent few weeks. Stabilizing results in a decrease in dimensions of about 0.0005 in. per in. from the original size of the casting. Stabilizing is, of course, unnecessary if the machine or fitting operations can be delayed until the castings have aged five weeks at room temperature.

When exposed to stagnant moisture or condensation with limited access to oxygen, a nonuniform type of corrosion may occur on zinc die castings, which often results in the formation of a bulky film of white corrosion products. This may hinder the operation of such parts as automobile lock cylinders, fuel pumps, and carburetors, and in severe cases result in rather rapid loss of zinc. Various types of chromate films are available to satisfactorily overcome this condition.

The same electroplating or enameling procedure is used with both alloys. Organic finishes are quite variable in their ability to adhere well to zinc surfaces. The phosphate type of chemical pretreatment has received widest commercial utilization, and most zinc die castings which are to be finished with lacquers or enamels are phosphate pretreated. In general, a much wider selection of finishes can be used on pretreated die castings.

The relative merits of the two SAE alloys may be outlined as follows:

SAE 903—When the shrinkage referred to above has been removed by normal aging or by a stabilizing anneal, the dynamic and dimensional properties of this alloy are permanent at service temperatures up to 100 C (212 F). Castings stabilized at elevated or at room temperature prior to final machining, assembling, or other adjusting of dimensions will permanently maintain such dimensions within a tolerance of ±0.00025 in. per in. in the absence of excessive moisture. When exposed to high humidity and temperature (as wet steam at atmospheric pressure or humid tropical climates), any change of either properties or dimensions will be only that resulting from surface corrosion analogous to that occurring in any other materials which are not totally resistant to oxidation.

SAE 925—This alloy is somewhat stronger and harder than SAE 903. At room temperature, it is equal to SAE 903 in permanence of dimensions and impact strength. At elevated temperatures it is subject to slight growth in dimensions and some loss of impact strength.

TABLE 1—TYPICAL PHYSICAL AND MECHANICAL PROPERTIES IN THE AS-CAST STATE AT ROOM TEMPERATURE

	SAE 903	SAE 925[a]
Tensile strength, psi[c]	41,000	47,000
Impact strength[b], ft-lb (1/4 x 1/4 in. bar)	43	48
Elongation in 2 in., %	10	7
Transverse deflection, in.	0.27	0.16
Compression strength, psi	60,000	87,000
Shearing strength, psi	31,000	38,000
Electrical conductivity, mhos per cm at 20 C	157,000	153,000
Thermal conductivity, cal per sec per sq cm per C at 18 C	0.27	0.26
Thermal expansion, in. per in. per C	0.0000274	0.0000274
Thermal expansion, per F	0.0000152	0.0000152
Specific gravity	6.6	6.7
Specific heat, cal per gm per C	0.10	0.10
Brinell hardness	82	91

[a] Die castings of SAE Alloy 925 shall not be used in applications where they will be subjected to prolonged temperature, above 200 F.

[b] Impact strength drops rapidly below 32 F to approximately 4 ft-lb at −4 F and 2 ft-lb at −40 F.

[c] Tensile properties are determined from test specimens cast in a die and conforming to chemical composition specified. Test bars machined from castings do not provide a reliable measure of the strength properties of the casting and this method should not be used to determine conformance to data shown in Table 1.

WROUGHT NICKEL AND NICKEL-RELATED ALLOYS—SAE J470c

SAE Information Report

Report of Nonferrous Metals Committee approved January 1946 and last revised July 1976.

Scope—This Report presents general information on over 50 alloys in which nickel either predominates or is a significant alloying element. It covers primarily wrought materials, and is not necessarily all inclusive. Values given are in most cases average or nominal, and if more precise values are required the producer(s) should be contacted. This report does not cover the so-called "superalloys," or the iron base stainless steels. Refer to SAE J467, Special Purpose Alloys, and SAE J405, Chemical Compositions of SAE Wrought Stainless Steels, respectively, for data on these alloys.

TABLE 1

Key No.*	Alloy Groups and Alloys	Commercial Designations**	Characteristics and Applications
1	NICKEL		
1A	Nickel UNS N02200	NICKEL 200[1a] HARDER[12a] 200	Commercially pure, malleable nickel, containing about 99.40% nickel and including a few tenths of a percent of cobalt, which is counted as nickel because its effect upon the significant properties of the alloy is not detrimental. Applications—Electron tube cathodes and grilles, hot caustic handling equipment, catalysts, printed circuits.
1B	Age Hardenable Nickel	DURANICKEL[1a] Alloy 301[1]	Age hardenable, high nickel alloy with high strength and hardness as well as the general corrosion characteristics of nickel. Alloy has good spring properties. Applications—Extrusion press parts, molds used in glass industry, clips, diaphragms and springs.
1C	High Purity Nickel UNS N02270	Nickel 270[1]	A high purity product containing about 99.98% nickel and maximum of 0.001% cobalt. Due to the low level of impurities, the alloy exhibits good thermal conductivity. Applications—Cathode shanks, fluorescent lamps, hydrogen-thyratron components, plates (anodes) and passive cathodes, heat exchangers and heat shields.
2	NICKEL-BERYLLIUM		
2A	2 Be—97 Ni—0.5 Ti	BERYLCO[9a] Nickel 440	Age hardenable alloy possessing high strength, extreme hardness and good ductility. Used up to 420°C (800°F). Good impact and fatigue properties. Applications—Heat resistant springs and switches, diaphragms, bellows, retainer clips, feather valves, contact springs, electrical shunts.
3	NICKEL-MANGANESE		
3A	95 Ni—4 Mn—1 Si	Alloy 667[2] R63 alloy[5]	Highly resistant to attack by the corrosive elements of internal combustion engine fuels—particularly sulphur and lead compounds. Applications—General purpose spark plug electrode.
4	NICKEL-MANGANESE-ALUMINUM-SILICON		
4A	95 Ni—2 Mn—2 Al—1 Si	NIAL[3a] T-2[5] ALUMEL[2a]	Applications—Negative leg of ANSI Type K thermocouples; used with 90 Ni—20 Cr alloy as positive element.
5	NICKEL-COPPER		
5A	70 Ni—30 Cu UNS N04400	MONEL[1a] alloy 400 CUNEL[11a] HARPER[12a] 400 D-H[5a] 400 alloy	These alloys have high strength and hardness, good resistance to corrosion. The age hardenable alloy contains approximately 2.75% aluminum and is non-magnetic at temperatures down to −100°C (−150°F). Its mechanical properties, particularly in large sections, are comparable with those of heat treated alloy steels. The free-machining alloy is suitable for use in automatic screw machines, its free machining characteristics being achieved by a sulphur content of approximately 0.035%. Adjustments in carbon and titanium contents of the age hardenable grade result in an alloy with improved machinability. Applications—Heat exchanger tubing, transmission oil cooler, marine engine components. The age hardenable grade is used for fasteners, pump and propeller shafts, and valve stems.
5B	Age Hardenable 70 Ni—30 Cu—2.75 Al UNS N05500	MONEL[1a] alloy K 500	
5C	Free Machining 70 Ni—30 Cu UNS N04405	MONEL[1a] alloy R-405	
5D	Age hardenable-Free Machining 70 Ni—30 Cu—2.75 Al UNS N05502	MONEL[1a] alloy 502	
6	COPPER-NICKEL		
6A	55 Cu—45 Ni	ADVANCE[5a] CUPRON[3a]	These alloys are of the "Constantan" type and are used extensively with iron or copper as thermoelectric elements for temperature measurement and control. They are used for electrical resistance purposes at temperatures up to 500°C (930°F) and for thermocouple purposes up to 760°C (1400°F). Their temperature coefficient of electrical resistivity is very small in the temperature range 20–100°C (68–212°F), or higher. Applications—Type T, J and E thermocouples, wire wound resistors, rheostats, low temperature heaters.
6B	77 Cu—23 Ni	MIDOHM[5a] 180 alloy[3]	These alloys are widely used in instruments and controls where resistivity and temperature coefficient must be held within very close limits. This is accomplished by careful control in melting and fabricating. Applications—Radio and automotive resistors, high current edge-wound resistors, resistor leads, voltage control relays and rheostats.
6C	89 Cu—11 Ni	90 alloy[3,5]	
6D	93 Cu—6 Ni	L OHM[5a] 60 alloy[3]	
6E	97.5 Cu—2.5 Ni	30 alloy[3,5]	
7	COPPER-NICKEL-IRON		
7A	60 Cu—20 Ni—20 Fe	Cunife I[13,14,15]	A permanent magnet alloy used in speedometers and small synchronous motors. It is ductile and easily formed after heat treatment. Applications—Permanent magnets which require ductility
8	COPPER-NICKEL-MANGANESE		
8A	83 Cu—13 Mn—4 Ni	Manganin 130 manganin[3]	These alloys are extremely stable with respect to electrical resistance change with time, and have very low temperature coefficients of electrical resistance over certain temperature ranges. Consequently they find wide use as windings for precision and standard resistors. Shunt manganin is designed to carry high currents which cause it to heat-up in service. Therefore the composition of shunt manganin is adjusted so that the temperature range, over which it possesses a low temperature coefficient of resistance, is higher than that at which the other manganins exhibit this property. Applications—Precision resistors, standard resistors and shunts.
8B	86 Cu—10 Mn—4 Ni	Shunt manganin Manganin (shunt)[3]	
9	NICKEL-IRON		
9A	70 Ni—30 Fe	BALCO[3a] HYTEMCO[5a]	Alloy has a high temperature coefficient of resistance along with moderate resistivity useful in various electrical instruments. Applications—Ballast resistors, voltage regulators, resistive thermometers, temperature compensators, low temperature heaters and ballistic devices in instruments and controls.

TABLE 1 (continued)

Key No.*	Alloy Groups and Alloys	Commercial Designations**	Characteristics and Applications
	NICKEL-IRON (cont.)		
9B	50 to 51.5 Ni—48.5 to 50 Fe	152 alloy[5] NIRON[3a] 52 Glass sealing 52[7] UNISEAL[6a] 52	Alloy has expansion characteristics for certain glass to metal seals, and has high magnetic permeability for high field strengths. Applications—Reed switches, mercury switches, contact rectifiers, amplifier coils. Glass to metal seals for matching to lead sealing glasses.
9C	47 to 50 Ni—50 to 53 Fe	49 PERMALLOY[3a] SIMALLOY[4a] magnetic 50 UNIMAG[6a] 50 Low expansion 49[7]	A medium high initial permeability alloy and medium high magnetic saturation alloy employed in manufacture of laminated magnetic cores and solid magnetic core configurations. Applications—Magnetic applications in communications industry, sensitive control devices and relays.
9D	46 Ni—54 Fe	146 alloy[5] Glass sealing 46 gas free[7] NIROMET 46[3a]	Expansion properties and inflection temperature between 50 Ni—50 Fe alloys and 42 Ni—58 Fe alloys. Applications—Terminals on vitreous enamelled resistors.
9E	42 Ni—58 Fe	NIROMET[3a] 42 SIMALLOY[4a] glass seal 42 UNISEAL[6a] 42 142 alloy[5] Glass sealing 42 gas free[7]	Alloy has low expansion, matching thermal expansion of some common glasses. It is used for sealing in glass and other controls at temperatures above those for which Invar is suitable. Applications—Headlights, lamps, audio transformers, coils, relays.
9F	36 Ni—64 Fe	Free cut Invar 36[7] NIRON[3a] 36 SIMALLOY[4a] glass seal 36 UNISPAN[6a] 36 NILBAR[5a] CARPENTER[7a] invar UNISPAN[6a] LR 35	Generally known as Invar and has the lowest coefficient of thermal expansion of any known alloy (up to 150C). Used extensively for thermostats and precision instrument parts for aeronautical use, struts in aluminum pistons, and other applications for low thermal expansion. Applications—Thermostat and precision instrument components. Base metal on which to solder silicon chips, bimetal component.
9G	32.5 Ni—67.5 Fe	SIMALLOY[4a] compensator #1 Temperature compensator 32[7]	These two alloys are generally known as magnetic compensator alloys. (The magnetism and permeability of these alloys change gradually and predictably with changes in temperature.) They are used extensively in automotive speedometer applications. Both alloys are representative of a family of alloys in this composition range. Applications—Instrumentation components.
9H	30 Ni—70 Fe	SIMALLOY[4a] compensator #4 Temperature compensator 30[7]	
10	NICKEL-CHROMIUM		
10A	90 Ni—10 Cr	CHROMEL[2a] T-1[5] TOPHEL[3a]	This alloy is the positive element in standard ANSI Type K thermocouples. The negative element is basically nickel with an approximate total of 5% of manganese, aluminum, and silicon. Applications—Thermocouple Type K.
10B	80 Ni—20 Cr	NICHROME[5a] V CHROMEL[2a] A TOPHET[3a] A PYROMET[7a] 80–20	A commercially iron-free, non-magnetic alloy developed especially to give maximum life as electrical heating elements which are expected to standup under the most adverse conditions up to surface temperatures of 1175°C (2150°F) in air. Its exceptional resistance to oxidation at elevated temperatures, high electrical resistivity, high tensile strength at temperature, and resistance to chemical corrosion has given it a wide variety of applications. Applications—Heating elements, wire wound resistors, high temperature conveyor belts, thermocouple wire, furnace components and thermocouple tubes.
10C	78 Ni—20 Cr—1 Cb—1 Si	242 alloy[5] TOPHET[3a] A + Cb	A special grade of 80 Ni—20 Cr. Inhibits "green rot" attack in high temperature mechanical or structural applications in reducing or marginal atmospheres.
10D	74 Ni—20 Cr—3 Al—2 Cu—1 Si	EVANOHM[3a] K	These alloys have low temperature coefficient of resistance up to 150°C and high resistivity. Applications—Precision wound resistors and potentiometers.
10E	72 Ni—20 Cr—3 Al—5 Mn	EVANOHM[3a] S	
10F	70 Ni—30 Cr	TOPHET[3a] 30	Excellent high temperature oxidation resistance. High resistivity and low temperature coefficient of resistance. Applications—Heating elements, wire wound power resistors, high temperature conveyor belts, thermostats.
10G	50 Ni—50 Cr	INCONEL[1a] 671 50 Nickel—50 Chromium[1]	Excellent elevated temperature liquid phase corrosion resistance; especially suitable in highly sulfidizing, high temperature environments. Applications—High temperature and exhaust components, high temperature baffles, supports in high sulphur atmospheres.
11	NICKEL-CHROMIUM-IRON		
11A	76 Ni—15 Cr—9 Fe UNS N06600	SIMALLOY[4a] 600 INCONEL[1a] alloy 600 NIREX[5a] PYROMET[7a] 600 HARPER[12a] 600	Good resistance to a great variety of corrosive media, to high temperature oxidation and scaling, and to intercrystalline attack at elevated temperatures. A restricted chemistry modification of this alloy is used for spark plug electrodes. Applications—Heat treating, nitriding and carburizing fixtures. High temperature belts, screens, pickling baskets, retorts, radiant tubes, exhaust control afterburners.
11B	Age Hardenable 73 Ni—15.5 Cr—8 Fe— 2.5 Ti—0.95 Cb + Ta UNS N07750	UNITEMP[6a] 750 SIMALLOY[4a] 750 INCONEL[1a] alloy X750[1a]	Maximum strength and resistance to oxidation at temperatures of 650–815°C (1200–1500°F) for gas turbine and heat engine components. Desirable spring characteristics at temperatures up to 540°C (1000°F). Applications—Diesel exhaust valves, high temperature springs, gas turbine parts, bolts, nuclear reactors.
11C	61 Ni—21.5 Cr—5 Fe—9 Mo— 3.6 Cb + Ta UNS N06625	INCONEL[1a] alloy 625 SIMALLOY[4a] 625	High strength and toughness from cryogenic temperatures to 1090°C (2000°F). High fatigue strength. Good oxidation resistance and resistance to many corrosive media. Virtually immune to chloride stress-corrosion cracking. Good fabrication properties. Applications—Ducting and combustion systems, thrust reversers, fuel nozzles, afterburners, spray bars.
11D	60.5 Ni—14 Cr—24 Fe—1.4 Al UNS N06601	INCONEL[1a] alloy 601	Excellent resistance to oxidizing, carburizing, and sulphur-containing environments. Resistance to oxidation and scaling up to temperatures as high as 1260°C (2300°F). Applications—Heat treating baskets and fixtures, radiant furnace tubes, strand-annealing tubes, thermocouple protection tubes, and furnace muffles and retorts. Thermal reactors for controlling automotive emissions.
11E	60 Ni—16 Cr—23 Fe—1 Si UNS N06004	NICHROME[5a] TOPHET[3a] C CHROMEL[2a] C	Maximum life as electrical heating elements up to 1065°C (1950°F) in air, non-magnetic, resists chemical corrosion, high electrical resistivity and low thermal coefficient of electrical resistance. Applications—Heating elements, rheostats, potentiometers.

TABLE 1 (continued)

Key No.*	Alloy Groups and Alloys	Commercial Designations**	Characteristics and Applications
	NICKEL-CHROMIUM-IRON (cont.)		
11F	42 Ni—21.5 Cr—32 Fe—3 Mo UNS N08825	INCOLOY[1a] alloy 825	For use in aggressively corrosive environments. Resistant to chloride-ion stress-corrosion cracking. Resistant to reducing acids, as well as to sulphuric acid and phosphoric acid solutions and to sea water. Applications—Phosphoric acid evaporators, pickling tank heaters, hooks and equipment, propellor shafts and tank trucks.
11G thru 11J	32 to 37 Ni—18 to 21 Cr—42 Fe—(Si, Cb)		This family of alloys is one of the most versatile and widely used groups of alloys made. There are numerous modifications around the basic 35-20 analysis to produce alloys for specific, highly demanding applications. Included are chemical corrosion resistance, high temperature strength and oxidation resistance and electrical resistance applications. Some of the specific properties of each alloy are listed next to the names below.
11G		NICHROME[5a] I CHROME[2a] I	Electrical resistance, high temperature strength, oxidation resistance.
11H	UNS N08330	CHROMEL[2a] D CHROMAX[5a] 525 RA 330[8a] SIMALLOY[4a] TOPHET[3a] D	High temperature strength, oxidation resistance and carburization resistance
11I	UNS N08800	INCOLOY[1a] alloy 800 HARPER[12a] 800	High temperature, strength, oxidation resistance, carburization resistance and chemical corrosion resistance.
11J		CHROMAX[5a] 520	Higher strength modification of UNS N08330 alloy for mechanical applications.
12	NICKEL-IRON-COBALT		
12A	38 Ni—41 Fe—15 Co—3 Cb—1.6 Ti—0.8 Al—(.008 B)	INCOLOY[1a] alloy 903 PYROMET[7a] CTX-1	Precipitation-hardenable alloy which has a constant, low coefficient of thermal expansion, a constant modulus of elasticity, and high strength. Applications—Rocket engine thrust chambers, steam turbine bolts, springs, gage blocks and ordnance hardware.
12B	29 Ni—54 Fe—17 Co	THERLO[5a] KOVAR[7a] LOCKINVAR[4a] RODAR[3a]	Sealing alloy for hard (borosilicate) glass-to-metal alloy seals.
13	NICKEL-MOLYBDENUM-IRON		
13A	65 Ni—28 Mo—5.5 Fe UNS N10001	HASTELLOY[10a] alloy B	Chromium-free alloy used for handling hydrochloric acid. Gas turbine applications, bolting, shafting, high stresses up to 760°C (1400°F) in oxidizing atmosphere, and higher temperatures in reducing atmospheres. Applications—Components for hydrochloric acid service. Gas turbine bolting and shafting.
13B	69 Ni—28 Mo—2 Fe	HASTELLOY[10a] alloy B-2 UNILOY[6a] LR-HB	This alloy is a lower carbon, more ductile version of UNS N10001, above. The lower carbon reduces formation of grain boundry carbide formation during welding. Other properties are very similar.
14	NICKEL-CHROMIUM-MOLYBDENUM-(COBALT-TUNGSTEN-COPPER)		
14A	64 Ni—16 Cr—16 Mo—3 Fe	HASTELLOY[10a] alloy C-4	Excellent high temperature strength and oxidation resistance. Outstanding corrosion resistance in certain environments. Good thermal shock properties up to 980°C (1800°F). Applications—Combustion cups for diesel engines, turbine blade jet engine components. Fixtures in nitric acid and organic acid salts service.
14B	58 Ni—16 Cr—6 Mo—6 Fe—3.5 W UNS N10002	CARPENTER[7a] alloy C UNILOY[6a] C	One of the most universally corrosion resistant alloys available, with excellent high-temperature properties. Resistant to oxidizing and reducing atmospheres up to 1090°C (2000°F). Is particularly useful where parts are either highly stressed or subject to repeated thermal shock at temperatures from 870–980°C (1600 to 1800°F). Exceptional resistance to strong oxidizing agents such as ferric chloride and cupric chloride.
14C	57 Ni—16 Cr—16 Mo—6 Fe—4 W UNS N10276	HASTELLOY[10a] alloy C 276 UNITEMP[6a] C 276	A modified version of UNS N10002 with improved fabricability. Resists formation of grain boundry precipitates in weld heat-affected zone. Excellent resistance to pitting, stress-corrosion cracking and to oxidizing environments up to 1040°C (1900°F).
14D	45 Ni—22 Cr—9 Mo—1 Co—0.8 W UNS N06002	SIMALLOY[4a] HX UNITEMP[6a] HX HASTELLOY[10a] alloy X PYROMET[7a] 680	Excellent high temperature properties. Suitable for sheet metal and bar components in jet engines, valve parts, furnace parts and heat treat containers and fixtures. Noted for excellent carburization, oxidation resistance and strength properties at temperatures up to 1150°C (2100°F). Good resistance to stress corrosion cracking. Applications—Combustion cans, heat treat fixtures and components.
14E	45 Ni—25 Cr—3 Mo—3 W—3 Co—1.25 Si UNS N06333	RA 333[8a]	
14F	43 Ni—22 Cr—7 Mo—1 W—20 Fe—2 Cu	HASTELLOY[10a] G	Excellent resistance to hot sulfuric and phosphoric acids. Resists corrosive effects of both oxidizing and reducing agents and both acid and alkaline solutions. Resists stress corrosion cracking and formation of grain boundary precipitates.
14G	35 Ni—20 Cr—2.5 Mo—3.5 Cu—37 Fe—1 Cb + Ta UNS N08020	CARPENTER[7a] 20 Cb 3	Austenitic stainless steel with superior resistance to 10–40% sulfuric acid, and many other corrosive media. Weldable. Applications—Mixing tanks, heat exchanges, process piping, pump shafts and rods.
15	COPPER-NICKEL-ZINC (NICKEL SILVERS)		
15A	72 Cu—18 Ni—10 Zn UNS C73500	CDA 735	These alloys, known as nickel-silvers, find application as car keys (especially, the leaded versions), electrically conductive springs, and switches. They have excellent spring characteristics, mechanical properties, are corrosion resistant, machinable and formable, along with a relatively high electrical conductivity.
15B	55 Cu—18 Ni—27 Zn UNS C77000	CDA 770	
15C	60 Cu—12 Ni—28 Zn UNS C76200	CDA 762	

* See same Key Number in each table for complete information on each alloy. (Key is used to avoid duplication of data in each table.)
** Superscript numerals (1–15) denote producer as shown in Table 6; superscript letter (a) denotes trademark.

TABLE 2—NOMINAL CHEMICAL COMPOSITIONS

Key* No.	Percent by Weight							
	Ni**	Cu	Cr	Fe	Mn	Si	C	Other
1A	99.4	0.1	—	0.15	0.18	0.18	0.05	0.005 S
1B	93.7	0.13	—	0.35	0.3	0.5	0.17	4.4 Al, 0.005 S, 0.63 Ti
1C	99.98	<0.001	<0.001	0.003	<0.001	<0.001	0.01	<0.001 S, Ti, Co, Mg
2A	97.5	—	—	—	—	—	—	1.95 Be, 0.5 Ti
3A	95	—	—	—	4	1	—	—
4A	94	—	—	0.25	2.5	1.0	—	2.0 Al
5A	67	30	—	1.4	1	0.25	0.15	0.012 S
5B	66	29	—	1	0.75	0.25	0.15	0.005 S, 2.75 Al, 0.60 Ti
5C	67	30	—	1.4	1	0.25	0.15	0.035 S
5D	66.5	28	—	1	0.75	0.25	0.05	0.005 S, 3 Al, 0.25 Ti
6A	43	55	—	0.25	0.5–1.0	—	0.05	—
6B	23	77	—	—	—	—	—	—
6C	11	89	—	—	—	—	—	—
6D	6	93	—	—	—	—	—	—
6E	2.5	97.5	—	—	—	—	—	—
7A	20	60	—	20	—	—	—	—
8A	4	83	—	—	1	—	—	—
8B	4	86	—	—	10	—	—	—
9A	70	—	—	29	1	0.05	0.05	—
9B	50–51.5	—	—	47.5–50	0.05	0.35	0.02	—
9C	47–50	—	—	50–53	0.5	0.40	0.05	—
9D	46	—	—	54	0.1	0.05	0.05	—
9E	42	—	—	58	0.1	0.05	0.05	—
9F	36	—	—	64	0.1	0.05	0.05	—
9G	32.5	—	—	67.5	0.7	0.3	0.1	—
9H	30	—	—	70	0.75	0.2	0.1	—
10A	90	—	10	—	—	—	—	—
10B	77	—	20	0.5	0.2–2	1.25–1.4	0.06–0.1	—
10C	78	—	20	—	—	1	—	1 Cb
10D	74	2	20	—	—	1	—	3 Al
10E	72	—	20	—	5	1	—	3 Al
10F	70	—	30	—	—	—	—	—
10G	51	—	48	—	—	—	0.5	0.35 Ti
11A	76	0.5	15.5	9	1	0.5	0.15	—
11B	73	0.5	15.5	8	1	0.5	0.08	0.7 Al, 2.5 Ti, 0.9 Co + Ta
11C	68	—	21.5	5	0.5	0.5	0.1	0.4 Al, 0.4 Ti, 1.0 Co, 9.0 Mo
11D	60.5	1	23	13	1	0.5	0.1	1.3 Al
11E	60	—	16	23	—	1.0	—	—
11F	42	2.25	21.5	—	1	—	0.05	0.2 Al, 0.9 Ti, 3.0 Mo
11G	36	—	20	42	—	2	—	—
11H	35	—	19	45	—	1.25	—	—
11I	32.5	0.75	21	—	1.5	1	0.1	0.4 Al, 0.45 Ti
11J	35	—	21	42	—	2	—	1.0 Cb
12A	38	0.50 max	0.20 max	41	0.20 max	0.20 max	0.03	0.8 Al, 1.6 Ti, 15 Co, 3 Cb, (0.008 B)
12B	29	—	—	54	—	—	—	17 Co
13A	65	—	—	5.5	1 max	1 max	0.05 max	28 Mo
13B	69	—	—	2–5 max	1 max	0.1 max	0.02 max	28 Mo
14A	64	—	16	3 max	1 max	0.08 max	0.015 max	16 Mo, 0.7 Ti
14B	57	—	16	6	1 max	1 max	0.08 max	16 Mo, 4 W
14C	57	—	16	6	1 max	0.05 max	0.02 max	16 Mo, 4 W
14D	45	—	22	18	1 max	1 max	0.10	9 Mo, 1 Co, 0.8 W
14E	45	—	25	20	2 max	1.25	0.08 max	3 Mo, 3 W, 3 Co
14F	43	2	22	20	2 max	1 max	0.05 max	7 Mo, 1 W, 2 Cb + Ta
14G	35	3.5	20	37	2 max	1 max	0.06 max	2.5 Mo, 1 Cb + Ta
15A	18	72	—	—	0.25	—	—	10 Zn
15B	18	55	—	—	0.25	—	—	27 Zn
15C	18	57	—	—	0.25	—	—	28 Zn

* See Key Number in each table for complete information on each alloy.
** Includes a small amount of cobalt which is counted as nickel.

TABLE 3—AVERAGE PHYSICAL CONSTANTS*

Key No.	Density at 20°C g/cm³ (lb/in³)	Melting Range, Solidus-Liquidus °C (°F)	Thermal Conductivity 0–100°C (32–212°F) W/m·K (Btu/hr/ft²/F/in)	Coefficient of Linear Thermal Expansion °C⁻¹ × 10⁻⁶ (°F⁻¹ × 10⁻⁶) 0–100°C (32–212°F)	Coefficient of Linear Thermal Expansion 0–1000°C (32–1832°F)	Specific Heat 0–100°C (32–212°F) J/kg K (Btu/lb·°F)	Electrical Resistivity at R.T. Ω·mm²/m (ohm circular-mil/ft)	Temperature Coefficient of Electrical Resistance 20–100°C Per °C	Temperature Coefficient of Electrical Resistance 20–1000°C Per °C	Magnetic Properties Condition at R.T.	Curie Temp. °C (°F)	Max Operating Temp. in Air (Sulfur Free) °C (°F)
** →	×27.68	—	×0.1441	×1.8	×1.8	×4186.8	×0.16624	—	—	—	—	—
1A	8.88 (0.321)	1435–1446 (2615–2635)	60.52 (420)	13.0 (7.2)	—	544 (0.13)	9.5 (57)	0.00432	—	Ferromagnetic	360 (680)	1038 (1900)
1B	8.75 (0.316)	1435–1446 (2615–2635)	60.52 (420)	13.0 (7.2)	—	544 (0.13)	15.7 (94.5)	0.0036	—	Ferromagnetic	290–299 (555–570)	—
1C	8.88 (0.321)	1455 (2650)	79 (548)	13.3 (7.4)	—	460 (0.11)	7.5 (45)	—	—	Ferromagnetic	353 (667)	—
2A	8.36 (0.302)	1220–1370 (2240–2500)	32 (220)	14.5 (8.0)¹	—	473 (0.113)	23.8 (143)	—	—	Ferromagnetic	—	—
3A	8.40 (0.3035)	1416–1445 (2550–2600)	28.5 (198)	13.2 (7.33)	16.0 (8.89)	—	22 (130)	—	0.00135	Strongly Magnetic	—	—
4A	8.60 (0.3107)	1380–1410 (2525–2575)	29.7 (206)	12.0 (6.66)	—	523 (0.125)	29–32 (177–191)	0.00188	—	Strongly Magnetic	—	1260 (2300)
5A	8.83 (0.319)	1300–1350 (2370–2460)	25.94 (180)	14.04 (7.8)	14.22 (8.9)	544.3 (0.13)	48.2 (290)	0.00198	—	Slightly Ferromagnetic	43–60 (110–140)	540 (1000)
5B	8.47 (0.306)	1315–1350 (2400–2460)	18.73 (130)	14.04 (7.8)	14.22 (8.9)	544.3 (0.13)	58.1 (350)	0.00198	—	Ferromagnetic	−101 (−150)	—
5C	8.83 (0.319)	1300–1350 (2370–2460)	24.94 (180)	14.04 (7.8)	—	544.3 (0.13)	48.2 (290)	0.00198	—	Slightly Ferromagnetic	43–60 (110–140)	—
5D	8.44 (0.305)	1315–1350 (2400–2460)	—	13.68 (7.6)	—	418.7 (0.10)	61.5 (370)	—	—	Paramagnetic	<−101 (<−150)	—
6A	8.88 (0.321)	—	22.9 (159)	14.6 (8.1)	18.7 (10.4)	394 (0.094)	49 (294)	0.00002	—	Paramagnetic	—	760 (1400)
6B	8.88 (0.321)	1130–1210 (2065–2210)	33.4 (232)	15.8 (8.8)	17.5 (9.7)	385 (0.092)	30 (180)	0.0003	—	Paramagnetic	—	540 (1000)
6C	8.88 (0.321)	1105–1150 (2020–2100)	38.4 (267)	16.0 (8.9)	—	385 (0.092)	15 (90)	0.0004	—	Paramagnetic	—	430 (800)
6D	8.88 (0.321)	1030–1110 (1995–2030)	37.6 (261)	16.4 (9.1)	18.0 (10.0)	385 (0.092)	10 (60)	0.0005	—	Paramagnetic	—	320 (600)
6E	8.88 (0.321)	1083–1100 (1980–2010)	37.7 (262)	16.6 (9.2)	—	385 (0.092)	5 (30)	0.0013	—	Paramagnetic	—	320 (600)
7A	8.61 (0.311)	1146–1330 (2095–2425)	—	—	—	—	18 (108)	—	—	Ferromagnetic	355 (670)	260 (500)
8A	8.41 (0.304)	1020– (1870–)	20.5 (142)	18.72 (10.4)²	—	—	48.2 (290)	0.0000144²	—	Nonmagnetic	—	—
8B	8.48 (0.31)	1020– (1870–)	20.5 (142)	18.72 (10.4)	—	—	38.3 (230)	0.0000144³	—	Nonmagnetic	—	—
9A	8.44 (0.305)	1432 (2610)	9.2 (64)	12.42 (6.9)	14.94 (8.3)	523 (0.125)	20 (120)	0.0045	0.0054⁴	Ferromagnetic	610 (1130)	594 (1100)
9B	8.30 (0.300)	1424– (2596–)	13 (90)	9.36 (5.2)	12.96 (7.2)	481 (0.115)	43 (260)	0.00306	0.0018	Ferromagnetic	500 (932)	524 (975)
9C	8.25 (0.298)	1427– (2600–)	13 (90)	9.36 (5.2)	—	500 (0.12)	48 (290)	—	0.0036⁵	Ferromagnetic	450–500 (840–930)	—
9D	8.17 (0.295)	1435–1441 (2615–2625)	14.1 (98)	13.5⁶ (7.5)	15.48 (8.6)⁶	490 (0.117)	46 (275)	0.0032	—	Ferromagnetic	490 (914)	600 (1112)
9E	8.11 (0.293)	1441–1452 (2625–2645)	10.7 (74.5)	4.86 (2.7)	12.96 (7.2)	498 (0.119)	67 (400)	0.00216	0.000936	Ferromagnetic	380 (716)	374 (705)
9F	8.05 (0.291)	1446–1460 (2635–2660)	10.5 (73)	1.44 (0.8)	14.4 (8)	515 (0.123)	80 (484)	0.00135	—	Ferromagnetic	280 (536)	200 (390)
9G	8.11 (0.293)	—	11.5 (80)	10.8 (6.0)	—	502 (0.12)	80 (480)	0.00126	—	Ferromagnetic	—	—
9H	8.19 (0.296)	—	11.5 (80)	7.70 (4.28)	—	502 (0.12)	80 (480)	0.00126	—	Ferromagnetic	—	—
10A	8.72 (0.315)	1430– (2610–)	19.16 (133)	23.4 (13)	—	452 (0.107)	70.6 (425)	0.00036	0.000324⁷	Paramagnetic	—	1149 (2100)
10B	8.41 (0.304)	1400– (2552–)	15 (104)	11.9–13.7 (6.6–7.6)	17.1–17.6 (9.5–9.8)	435–452 (0.104–0.107)	108 (650)	0.000056	0.0000014–0.0000067	Paramagnetic	—	1177 (2150)
10C	8.41 (0.304)	1400– (2552–)	15 (104)	—	17 (9.4)	435–452 (0.104–0.107)	111 (670)	—	0.00011⁸	Paramagnetic	—	1150 (2100)
10D	8.41 (0.304)	1400– (2552–)	15 (104)	12.6 (7.0)	—	—	133 (800)	—	±0.000005	Nonmagnetic	—	315 (600)
10E	7.14 (0.258)	1350–1380 (2460–2510)	14.6 (101)	13 (7.2)	—	—	137 (825)	—	<±0.000005	Nonmagnetic	—	315 (600)

TABLE 3— AVERAGE PHYSICAL CONSTANTS* (continued)

Key No.	Density at 20°C g/cm³ (lb/in³)	Melting Range, Solidus-Liquidus °C (°F)	Thermal Conductivity 0-100°C (32-212°F) W/m·K (Btu/hr/ft²/F/in)	Coefficient of Linear Thermal Expansion °C⁻¹ × 10⁻⁶ (°F⁻¹ × 10⁻⁶) 0-100°C (32-212°F)	Coefficient of Linear Thermal Expansion 0-1000°C (32-1832°F)	Specific Heat 0-100°C (32-212°F) J/kg K (Btu/lb·°F)	Electrical Resistivity at R.T. Ω·mm²/m (ohm circular-mil/ft)	Temperature Coefficient of Electrical Resistance 20-100°C Per °C	Temperature Coefficient of Electrical Resistance 20-1000°C Per °C	Magnetic Properties Condition at R.T.	Curie Temp. °C (°F)	Max Operating Temp. in Air (Sulfur Free) °C (°F)
** →	×27.68	—	×0.1441	×1.8	×1.8	×4186.8	×0.16624	—	—	—	—	—
10F	8.11 (0.293)	1377– (2510–)	15.85 (110)	12.2 (6.8)	—	452 (0.107)	118 (710)	0.00009	—	Paramagnetic	—	1260 (2300)
10G	7.89 (0.285)	1308–1318 (2386–2404)	16.43 (114)	10.0 (5.54)	13.8 (7.66)	456 (0.109)	93 (556)	—	—	—	—	—
11A	8.50 (0.307)	1343–1427 (2540–2600)	14.98 (104)	11.5 (6.4)	12.4 (6.9)	460 (0.11)	103 (620)	—	—	Paramagnetic	−40 (−40)	1094 (2000)
11B	8.30 (0.3)	1343–1427 (2540–2600)	14.7 (102)	13.7 (7.6)	17.1 (9.5)	460 (0.11)	122 (734)	—	—	Paramagnetic	−174 (−280)	815 (1500)
11C	8.44 (0.305)	1288–1349 (2350–2460)	9.80 (68)	12.8 (7.1)	15.7 (8.7)	410 (0.098)	129 (776)	—	—	Paramagnetic	<−196 (<−320)	—
11D	8.05 (0.291)	1301–1368 (2374–2494)	12.54 (87)	13.7 (7.6)	16.7 (9.3)	448 (0.107)	119 (717)	—	—	Paramagnetic	<−196 (<−320)	—
11E	8.41 (0.304)	1350–1375 (2460–2500)	13.3 (92)	—	17 (9.4)	460 (0.11)	108 (650)	0.000133	0.00009	Paramagnetic	—	1010 (1850)
11F	8.14 (0.294)	1371–1399 (2500–2550)	—	14.0 (7.8)	17.3 (9.6)	—	113 (678)	—	—	Paramagnetic	<−196 (<−320)	—
11G	7.94 (0.287)	1380– (2515–)	13.0 (90)	14.0 (7.8)	18 (10)	502 (0.12)	108 (650)	0.0004	0.00025	Paramagnetic	—	994 (1800)
11H	8.0 (0.289)	1399–1427 (2550–2600)	13.15 (91.2)	14.9 (8.3)	18.0 (10)	460 (0.11)	102 (612)	0.00035	0.00023	Paramagnetic	—	954 (1750)
11I	7.94 (0.287)	1368–1380 (2475–2525)	11.53 (80)	14.2 (7.9)	18.2 (10.1)	502 (0.12)	100 (600)	—	—	Paramagnetic	−115 (−175)	—
11J	7.94 (0.287)	1382– (2520–)	13.0 (90)	14.0 (7.8)	18 (10)	502 (0.12)	106 (640)	0.0004	0.00025	Paramagnetic	—	994 (1800)
12A	8.14 (0.294)	1318–1393 (2405–2539)	17.1 (119)	7.9 (4.4)	—	435 (0.104)	61.0 (367)	—	—	Ferromagnetic	460 (860)	—
12B	8.36 (0.302)	1450– (2640–)	16.4 (114)	9.2 (5.1–5.5)⁹	12 (6.7)	461 (0.11)	49 (294)	0.0033	—	Ferromagnetic	435 (815)	—
13A	9.25 (0.334)	1302–1368 (2375–2495)	11.3 (78.5)	10.0 (5.6)	14.6 (8.1)	381 (0.091)	135 (812)	—	—	Paramagnetic	—	760 (1400)
13B	9.22 (0.333)	1302–1365 (2375–2490)	11.7 (81)	10.26 (5.7)	—	381 (0.091)	130 (785)	—	—	—	—	—
14A	8.64 (0.312)	—	10.66 (74.0)	10.8 (6.0)	15.7 (8.7)	419 (0.100)	125 (752)	—	—	Paramagnetic	—	1038 (1900)
14B	8.94 (0.323)	1210–1305 (2318–2381)	11.3 (78)	11.3 (6.3)	15.3 (8.5)	385 (0.092)	130 (779)	—	—	Paramagnetic	—	1038 (1900)
14C	8.88 (0.321)	1324–1371 (2415–2500)	10.23 (71.0)	11.2 (6.2)	15.8 (8.8)	427 (0.102)	130 (782)	—	—	Paramagnetic	—	1038 (1900)
14D	8.23 (0.297)	1260–1355 (2300–2470)	11.6 (80.5)	13.8 (7.7)	14.9 (8.4)	486 (0.116)	118 (712)	—	—	Paramagnetic	—	1204 (2200)
14E	8.25 (0.298)	1327–1352 (2420–2465)	11.0 (76.2)	13.9 (7.7)	17.5 (9.7)	460 (0.11)	114 (687)	0.00027	0.000127	Paramagnetic	—	1204 (2200)
14F	8.30 (0.300)	1260–1343 (2300–2450)	10.66 (74.0)	13.5 (7.5)	—	460 (0.110)	—	—	—	—	—	—
14G	8.05 (0.291)	—	—	14.96 (8.31)	16.74 (9.43)	502 (0.12)	104 (625)	—	—	—	—	—
15A	8.80 (0.318)	—	38.6 (268)	14.9 (8.3)	16.4 (9.1)¹⁰	—	21.6 (130)	—	—	Paramagnetic	—	—
15B	8.69 (0.314)	—	29.4 (204)	16.7 (9.3)	16.7 (9.3)	—	31.4 (189)	0.000093	—	Paramagnetic	—	—
15C	8.69 (0.314)	—	41.8 (290)	15.8 (8.8)	16.0 (8.9)	—	19.9 (120)	—	—	Paramagnetic	—	—

*Consult producer for more specific values.
**Metric values shown were computed by use of these conversion factors.
[1] At 20–550°C
[2] At 59–95°F
[3] At 104–140°F
[4] At 0–500°C
[5] At −20–500°C
[6] At 30–350°C
[7] At 25–870°C
[8] At 20–500°C
[9] At 30–450°C
[10] At 68–572°F

TABLE 4—RANGE OF MECHANICAL PROPERTIES*

Key No.	Available Forms**	Yield Strength 0.2% Offset 10^6 Pa (10^3 psi)	Tensile Strength 10^6 Pa (10^3 psi)	Elongation in 50 mm (2 in) %	Reduction of Area %	Hardness Brinell 3000 kg	Hardness Rockwell	Modulus of Elasticity 10^9 Pa (10^6 psi)	Endurance Limit 10^8 Cycles 10^6 Pa (10^3 psi)	Impact at R.T. Standard Izod J (ft-lb)	Impact at R.T. Standard Charpy J (ft-lb)
*** →		×6.9	×6.9	—	—	—	—	×6.9	×6.9	×1.356	×1.356
1A	A,B,C,D,E,F,G	103–1070 (15–155)	415–1140 (60–165)	50–2	75–50	90–230	B40–100	207 (30)	160–250 (23–36)	163 (120)	301–264 (222–195)
1B	A,C,E	207–1030 (30–150)	620–1030 (90–150)	50–2	65–15	140–380	B75–C46	207 (30)	360–407 (52–59)	163–34 (120–25)	325–49 (240–36)
1C	A,C,D,E,F	110–620 (16–90)	345–655 (50–95)	50–4	—	80–210	B35–95	207 (30)	—	—	—
2A	D,E	310–1588 (45–230)	724–1863 (105–270)	40–2	—	—	B70–C51	186–207 (27–30)	655–966 (95–140)	—	149–81 (110–60)
3A	C,E	—	—	—	—	—	—	—	—	—	—
4A	A,C,E	193 (28)	587–1173 (85–170)	45	—	—	—	—	—	—	—
5A	A,B,C,D,E,F,G	172–1104 (25–160)	483–1173 (70–170)	50–2	75–50	110–250	B60–C23	179.4 (26.0)	207–345 (30–50)	163–102 (120–75)	298–203 (220–150)
5B	A,B,C,D,E,F	276–1207 (40–175)	620–1380 (90–200)	45–2	70–25	140–320	B75–C40	179.4 (26.0)	283–407 (41–59)	163–35 (120–26)	230–57 (170–42)
5C	A,C	172–897 (25–130)	483–966 (70–140)	50–4	70–50	110–230	B60–100	179.4 (26.0)	193–283 (28–41)	163–130 (120–96)	266–190 (196–140)
5D	A,C,D,E,F	255–648 (37–94)	586–973 (85–141)	47–25	—	135–255	B74–C25	179.4 (26.0)	—	—	—
6A	A,B,C,E	207–828 (30–120)	414–931 (60–135)	45–0.5	70–25	—	B50–93	—	—	—	—
6B	A,B,C,E	138–483 (20–70)	276–552 (40–80)	55–2	75–50	—	B50–85	—	—	—	—
6C	A,B,C,E	138–483 (20–70)	276–552 (40–80)	55–2	75–50	—	B50–85	—	—	—	—
6D	A,B,C,E	138–483 (20–70)	276–552 (40–80)	55–2	75–50	—	B50–85	—	—	—	—
6E	A,B,C,E	138–483 (20–70)	276–552 (40–80)	55–2	75–50	—	B50–85	—	—	—	—
7A	C,E	—	—	—	—	—	—	—	—	—	—
8A	C	— (25–40)	276–621 (40–90)	15–30	—	—	—	—	—	—	—
8B	E	— (25–40)	345–690 (50–100)	15–30	—	—	—	—	—	—	—
9A	A,B,C,E	241–965 (35–140)	483–1035 (70–150)	35–0.5	60–30	—	B58–93	—	—	—	—
9B	A,B,C,D,E	241–965 (35–140)	483–1035 (70–150)	35–0.5	60–30	—	B58–100	165 (24)	—	—	—
9C	A,B,C,D,E	241–965 (35–140)	483–1035 (70–150)	35–0.5	60–30	—	B58–100	165 (24)	—	—	—
9D	A,C,E	207–965 (30–140)	552–1035 (80–150)	35–0.5	70–40	—	B58–93	—	—	—	—
9E	A,B,C,D,E	241–965 (35–140)	483–1035 (70–150)	35–0.5	65–30	—	B58–93	145 (21)	—	—	—
9F	A,B,C,D,E	241–965 (35–140)	483–1035 (70–150)	40–0.5	65–30	—	B58–98	141 (20.5)	—	—	—
9G	E	276 (40)	483 (70)	35	—	—	B75 Ann.	152 (22)	—	—	—
9H	E	276 (40)	483 (70)	35	—	—	B75 Ann.	152 (22)	—	—	—
10A	C	—	655–1138 (95–165)	—	—	—	—	—	—	—	—
10B	A,B,C,D,E,F,G	345–1311 (50–190)	690–1380 (100–200)	35–0.5	70–40	150–320	B85–C30	214 (31)	—	—	—
10C	A,C,E	345–1311 (50–190)	690–1380 (100–200)	35–0.5	70–40	150–320	B80–85	214 (31)	—	—	—
10D	A,C,E	517–655 (75–95)	966–1104 (140–160)	35–15	—	—	—	—	—	—	—
10E	A,C,E	517–655 (75–95)	966–1104 (140–160)	35–15	—	—	—	—	—	—	—

TABLE 4—RANGE OF MECHANICAL PROPERTIES* (continued)

Key No.	Available Forms**	Yield Strength 0.2% Offset 10^6 Pa (10^3 psi)	Tensile Strength 10^6 Pa (10^3 psi)	Elongation in 50 mm (2 in) %	Reduction of Area %	Hardness Brinell 3000 kg	Hardness Rockwell	Modulus of Elasticity 10^9 Pa (10^6 psi)	Endurance limit 10^8 Cycles 10^6 Pa (10^3 psi)	Impact at R.T. Standard Izod J (ft-lb)	Impact at R.T. Standard Charpy J (ft-lb)
***		×6.9	×6.9	—	—	—	—	×6.9	×6.9	×1.356	×1.356
10F	A,C,E	414–586 (60–85)	828–1035 (120–150)	30–10	—	—	—	166 (24)	—	—	—
10G	A,D,G	497 (72)	1035 (150)	19	25	—	—	—	—	—	—
11A	A,B,C,D,E,F,G	172–1207 (25–175)	552–1276 (80–185)	50–2	70–40	120–190	B65–C30	214 (31)	283–414 (41–60)	163–95 (120–70)	312–205 (230–151)
11B	A,B,C,D,E,F,G	345–1656 (50–240)	759–1897 (110–275)	45–2	60–30	200–500	B93–C47	214 (31)	310–448 (45–65)	54 (40)	—
11C	A,B,C,D,E,F	290–759 (42–110)	724–1104 (105–160)	65–30	—	110–240	—	207 (30)	—	—	—
11D	A,B,C,D,E,F	193–345 (28–50)	586–759 (85–110)	58–45	—	115–160	—	206 (29.9)	—	—	—
11E	A,C,E	311–1346 (45–195)	656–1380 (95–200)	40–20	50–20	83	B85–C30	214 (31)	—	—	—
11F	A,B,C,D,E,F	241–448 (35–65)	586–724 (85–105)	50–30	—	120–180	—	193 (28.0)	—	—	—
11G	A,C,E	207–1000 (30–145)	483–1035 (70–150)	35–0.5	—	—	B80	—	—	—	—
11H	A,C,D,E	290 (42)	586 (85)	45	65	—	B75–85	197 (28.5)	—	—	>325 (>240)
11I	A,B,C,D,E,F	241–862 (35–125)	517–1035 (75–150)	60–10	—	120–130	—	197 (28.5)	—	—	—
11J	A,C,E	207–1000 (30–145)	483–1035 (70–150)	35–0.5	—	—	B80	—	—	—	—
12A	A,C,D,E,F	362–1242 (52.5–180)	652–1476 (94.5–214)	52–13	—	—	—	152 (22)	—	—	—
12B	A,C,D,E	276–966 (40–140)	517–1035 (75–150)	30–0.5	—	—	B82–100	—	—	—	—
13A	A,B,C,D,E,F	462 (67)	925 (134)	51	—	—	B96	215 (31.1)	—	81 (60)	—
13B	A,B,C,D,E,F	504 (73)	987 (143)	51	—	—	—	217 (31.4)	—	—	240 (177)
14A	A,B,C,D,E,F	414 (60)	766 (111)	52	—	—	B90	212 (30.8)	—	—	—
14B	A,B,C,D,E,F	462–828 (67–120)	759–1000 (110–115)	24–60	—	—	B90–105	206 (29.8)	—	—	28–31 (21–23)
14C	A,B,C,D,E,F	462–828 (67–120)	759–1000 (110–145)	24–60	—	—	B90–105	206 (29.8)	—	—	28–31 (21–23)
14D	A,B,C,D,E,F	380 (55)	780 (113)	50	65	192	B88	196 (28.5)	—	121 (89)	119–73 (88–54)
14E	A,B,C,D,E	345 (50)	690 (100)	50	57	160–202	B82–92	193 (28)	—	—	150 (110)
14F	A,B,C,D,E,F	317 (46)	704 (102)	61	52	—	B84	—	—	—	—
14G	A,B,C,D,E,F	310 (45)	628 (91)	45	67	160	B84–90	193 (28)	—	—	—
15A	A,B,C,D,E,F,G	179–497 (26–72)	373–545 (54–79)	43	—	—	B35–90	—	—	—	—
15B	A,B,C,D,E,F,G	186–586 (27–85)	414–690 (60–100)	40	—	—	—	124 (18)	—	—	—
15C	A,B,C,D,E,F,G	152–566 (22–82)	380–662 (55–96)	40	—	—	—	—	—	—	—

*Consult producers for more specific values.
** A—Rods and bars over 1/2 in dia.
   B—Forgings
   C—Wire 1/2 in dia. max.
   D—Sheet
   E—Strip
   F—Tubing
   G—Castings
*** Metric values shown were computed by use of these conversion factors.

TABLE 5—AVAILABLE SPECIFICATIONS

Key No.	ASTM	Federal	Military	AMS	ASME	Other
1A	B160, B161, B162, B163, B366	—	—	—	SB160, SB161, SB162, SB163	—
1B	—	—	—	—	—	—
1C	F239	—	—	—	—	—
2A	—	—	—	—	—	—
3A	—	—	—	—	—	—
4A	E230	—	MIL-W-5846	—	—	ANSI C96.1
5A	B127, B163, B164, B165, B366, B564	QQ-N-281	MIL-N-894, MIL-N-24106, MIL-T-842, MIL-T-1368, MIL-T-23520	4544, 4575, 4674, 4675, 4730, 4731, 7322	SB127, SB163, SB164, SB165	—
5B	B164	QQ-N-286	MIL-N-894, —	4674, 4676, 7234	SB164	—
5C	—	QQ-N-286	MIL-F-23999, MIL-N-17506, MIL-W-4471	4676	—	—
5D	—	QQ-N-286	—	—	—	—
6A	B267	QQ-R-175	MIL-W-5845, MIL-W-5908	—	—	—
6B	B267	—	—	—	—	—
6C	B267	—	—	—	—	—
6D	B267	—	—	—	—	—
6E	B267	—	—	—	—	—
7A	—	—	—	—	—	—
8A	B267	—	—	—	—	—
8B	B267	—	—	—	—	—
9A	B267	—	—	—	—	—
9B	F30	—	MIL-I-23011	—	—	—
9C	F30	—	—	—	—	—
9D	F30	—	MIL-I-23011	7717, 7718, 7719	—	—
9E	F30	—	MIL-I-22011	—	—	—
9F	—	—	MIL-S-16598	—	—	—
9G	—	—	—	—	—	—
9H	—	—	—	—	—	—
10A	E120	—	—	—	—	—
10B	B344, B267	QQ-R-175	MIL-W-16970, MIL-W-14593	5676, 5677	—	—
10C	—	—	—	—	—	—
10D	B267	—	—	—	—	—
10E	—	—	—	—	—	—
10F	—	—	—	—	—	—
10G	—	—	—	—	—	—
11A	B163, B166, B167, B168	QQ-W-390	MIL-N-6710, MIL-T-7840, MIL-N-6840, MIL-N-23228, MIL-N-23229	5665, 5580, 5540, 5687, 7232	SB163, SB166, SB167, SB168	—
11B	—	—	MIL-N-24114, MIL-N-8550, MIL-N-7786	5542, 5598, 5667, 5668, 5669, 5670, 5671, 5698, 5699	—	Air Research EMS 56 9A, GE B50T1232, GE B50YP44, Westinghse 15125
11C	B443, B444, B446	—	—	5599, 5666, 5837	SB443, SB444, SB446, CC1409-1, CC1422	—
11D	—	—	—	—	—	—
11E	B344	QQ-R-175	—	—	—	—
11F	B163, B423, B424, B425	—	—	—	SB163, SB423, SB424, SB425	—
11G	B344	—	—	—	—	—
11H	—	—	—	—	—	—
11I	B163, B407, B408, B409	—	—	—	SB163, SB407, SB408, SB409, CC1325-5	—
12A	—	—	—	—	—	—
12B	F15	—	MIL-I-23011	7726, 7727, 7728	—	—
13A	B333	—	MIL-N-18008	—	Case 1323	—
13B	B333, B335	—	—	5752	Case 1642	—
14A	—	—	—	—	Case 1641	—
14B	A296, A494, A567, B334, B336, B366	—	—	5388, 5389, 5530, 5750	SFA 5.14, SFA 15.11, SB334, SB336	—
14C	B574, B575	—	—	5388, 5389, 5530, 5750	Case 1410	—
14D	B434, A567, B366, B572, B435	—	—	5754, 5536	SB434, SB435, Case 1321	GE B50T, F24-54
14E	—	—	—	5593, 5717	—	—
14F	—	—	—	—	Case 1472	—
14G	B462, B463, B464, B468, B471, B472, B473, B474, B475	—	—	—	SB462, SB463, SB464	—
15A	B122, B121	QQ-C-585	—	—	—	—
15B	B122, B151, B206	QQ-C-585, QQ-C-586, QQ-C-W321	—	—	—	—
15C	B122, B151, B206	QQ-C-585	—	—	—	—

TABLE 6—PRODUCERS

Superscript Numeral*	Name and Address
1	Huntington Alloys Inc., Huntington, WV 25720
2	Hoskins Manufacturing Co., 4445 Lawton Ave., Detroit, MI 48208
3	Wilber B. Driver Co., 1875 McCarter Highway, Newark, NJ 07104
4	Simonds Steel, Wallace Murray Corp., Lockport, NY 14094
5	Driver-Harris, Harrison, NJ 07029
6	Cyclops Corp., 650 Washington Rd., Pittsburgh, PA 15228
7	Carpenter Technology, Reading, PA 19603
8	Rolled Alloys, Inc., 5311 Concord Ave., Detroit, MI 48211
9	Kawecki Berylco Industries, Inc., 220 East 42nd St., New York, NY 10017
10	Cabot Corp., Stellite Div., 1020 West Park Ave., Kokomo, IN 46901
11	Wolverine Tube Div., UOP, Box 2202, 2100 Market St., N.E., Decatur, AL 35601
12	H. M. Harper Co., Morton Grove, IL 60053
13	Indiana General, 407 Elm St., Valparaiso, IN 46383
14	Arnold Engineering Co., Railroad Ave. and West, Box G, Marengo, IL 60152
15	Colt Industries, Crucible Magnetic Div., Box 100, Elizabethtown, KY 42701

*As shown in Table 1.

ELECTROPLATING OF NICKEL AND CHROMIUM ON METAL PARTS—AUTOMOTIVE ORNAMENTATION AND HARDWARE—SAE J207

SAE Standard

Report of Nonferrous Metals Committee approved August 1970.

1. Scope—This standard covers requirements for several types and grades of electrodeposited nickel/chromium coatings on ferrous or copper alloy basis metals and copper/nickel/chromium on zinc or aluminum alloys for the finishing and corrosion protection of decorative ornamentation and hardware of motor vehicles and marine controls and fittings. Four grades of coatings are provided to correlate with the service conditions under which each is expected to provide satisfactory performance, namely: very severe, severe, moderate, and mild. Definitions and typical examples of these service conditions are provided in Appendix A.[1] Information contained in this document generally conforms to the information contained in ASTM B 456, Specification for Electrodeposited Coatings of Nickel plus Chromium.

2. Manufacture

2.1 Only those parts shall be plated that are free from visible surface defects, such as scratches, porosity, nonconducting inclusions, roll and die marks, cold shuts, cracks, etc., which may adversely affect the appearance and the performance of coatings. In order to minimize problems of this sort, the specifications covering the basis material or the item to be plated should be appropriately specified.

2.2 When required, the basis metal shall be subjected to such polishing, buffing, or finishing operations as are necessary to yield deposits with the desired final appearance. (See paragraph 4.)

2.3 Proper preparatory procedures and thorough cleaning of the basis metal surface are essential in order to assure satisfactory adhesion and corrosion performance of the coating. Accordingly, it is suggested that the following ASTM documents on the preparation of various basis metals for electroplating be followed where appropriate:

ASTM B 183, Preparation of Low Carbon Steel for Electroplating
ASTM B 242, Preparation of High Carbon Steel for Electroplating
ASTM B 253-68, Preparation of and Electroplating on Aluminum Alloys by the Zincate Process
ASTM B 252, Preparation of Zinc Die Castings for Electroplating
ASTM B 281, Preparation of Copper and Copper Alloys for Electroplating
ASTM B 320, Preparation of Iron Castings for Electroplating

3. Significant Surfaces—Significant surfaces are defined as those normally visible—directly or by reflection—which are essential to the appearance or serviceability of the article when assembled in normal position, or which can be the source of corrosion products that deface visible surfaces on the assembled article. When necessary, the significant surface shall be the subject of agreement between purchaser and manufacturer and shall be indicated on the drawings of the parts, or by the provision of suitably marked samples.[2]

4. Appearance

4.1 The significant surface of the plated article shall be free from clearly visible plating defects, such as blisters, pits, roughness, cracks, or unplated areas, and shall not be stained or discolored. On articles where a visible contact mark is unavoidable, its position shall be the subject of agreement between the manufacturer and the purchaser.

4.2 The plated article shall be clean and free from damage. The purchaser shall specify the appearance required, for example, bright, dull or satin. Alternatively, samples showing the required finish or range of finish shall be supplied or approved by the purchaser.

5. Manner of Specifying Requirements

5.1 Coating Classification Number or Service Condition Number—When ordering articles to be plated in accordance with this standard, the purchaser shall state, in addition to SAE J207, either the classification number

---

[1] It is recognized that uses exist in which either thicker or thinner coatings than those covered by these specifications may be required. In such cases, the particular thickness desired by the purchaser (minimum, maximum, or range of thickness permissible) should be the subject of agreement between the purchaser and the plater.

[2] When significant surfaces are involved on which the specified thickness of deposit cannot readily be controlled, such as threads, holes, deep recesses, bases of angles, and similar areas, the purchaser and the manufacturer should recognize the necessity for either thicker deposits on the more accessible surfaces or for special racking. Special racks may involve the use of conforming, auxiliary electrodes, or nonconducting shields.

of the particular coating required (see paragraph 5.4) or the basis metal and the service condition number denoting the severity of the conditions it is required to withstand (see paragraph 5.2). If the service condition number is quoted but not the classification number, the manufacturer is free to supply any of the classes of coating corresponding to the service condition number; but when requested to do so, he shall inform the purchaser of the classification number of the coating supplied.

5.2 Service Condition Number—The service condition number indicates the severity of the service conditions in accordance with the following scale:

SC 4—very severe service
SC 3—severe service
SC 2—moderate service
SC 1—mild service

Typical service conditions for which the various service condition numbers are appropriate are given in Appendix A.

5.3 Coating Classification Number—The coating classification number comprises:

(a) The chemical designation for the basis metal (or for the principle metal if an alloy). The following chemical symbols are used:

FE—for steel (or iron)
ZN—for zinc alloy
CU—for copper or copper alloy
AL—for aluminum

(b) The chemical designation for nickel (NI).
(c) A number indicating the minimum thickness of the nickel coating, in micrometers[3].
(d) A letter designating the type of nickel deposit.
(e) The chemical designation for chromium (CR).
(f) A letter (or letters) designating the type of chromium deposit.

5.3.1 TYPE OF NICKEL AND DEPOSIT THICKNESS

5.3.1.1 *Type of Nickel*—The type of nickel[4] is designated by the following symbols:

B—For nickel deposited in the fully bright condition.

P—For dull or semi-bright nickel requiring polishing to give full brightness, containing less than 0.005% sulfur[5], and having an elongation not less than 8% when tested by the method given in Appendix C.

D—For a double-layer or triple-layer nickel coating, of which the bottom layer contains less than 0.005% sulfur[5], and has an elongation not less than 8% when tested by the method given in Appendix C, and the top layer contains more than 0.04% sulfur[5]. The thickness of the bottom layer in double-layer coatings shall be not less than 75% of the total nickel thickness, and in triple-layer coatings shall be not less than 50% of the total nickel thickness, the thickness of the top layer in either case being not less than 10% of the total nickel thickness. If there are three layers, the intermediate layer shall contain more sulfur than the top layer and shall not exceed 10% of the total nickel thickness.

5.3.1.2 *Thickness of Nickel Deposits*—The number following the chemical designation NI indicates, in micrometers[3], the minimum thickness of the nickel deposit, measured in accordance with ASTM A 219, Methods of Test for Local Thickness of Electrodeposited Coatings, at points on the significant surface.

5.3.2 TYPE OF CHROMIUM AND DEPOSIT THICKNESS—The thickness of the chromium deposit shall be measured by the method given in ASTM A 219, at points on the significant surface.

The type of chromium and thickness of deposit is designated by the following symbols placed after the chemical designation CR (numerals are not used in this case to specify thickness as in the case of nickel):

R—For regular (that is, conventional) chromium, having a minimum thickness of 0.25 μm (0.01 mil), except in the case of SC 1 where the minimum thickness is 0.12 μm (0.005 mil).

MC—For microcracked chromium, having more than 300 cracks/cm (750 cracks/in.) (method for measurement is given in Appendix D) in any direction over the whole of the significant surface, and having a minimum thickness of 0.8 μm (0.03 mil), unless it can be demonstrated by the plater that equally good performance can be obtained with a lower thickness.

MP—For microporous chromium, having a minimum thickness of 0.25 μm (0.01 mil) and containing a minimum of 10,000 pores per cm^2 (64,500 per in.^{2}).

5.3.3 EXAMPLE OF COMPLETE CLASSIFICATION NUMBER—A coating on steel comprising 40 μm (1.6 mils) minimum bright nickel plus 0.8 μm (0.03 mil)

---

[3]The approximately equivalent coating thicknesses in mils are given in Tables 1–4. (1 μm = approximately 0.04 mil; 1 mil = 0.001 in. = 25 μm.)

[4]ASTM A 219, section on microscopic examination of polished and etched.

[5]The sulfur contents are specified in order to indicate the type of nickel plating solution that is to be used. Although no simple method exists for determining the sulfur content of a nickel deposit on a coated article, the x-ray fluorescence technique can be used. A chemical determination is possible on a specially prepared test specimen.

minimum microcracked chromium has the classification number: FE/N140B/CRMC.

5.4 Coatings Appropriate to Each Service Condition Number—Tables 1–3 show, for the various basis metals, the coating classification numbers appropriate for each service condition number.

5.5 Adhesion—The coating shall be sufficiently adherent to the basis metal, and the separate layers of multilayer coatings shall be sufficiently adherent to each other, to pass the tests described in Appendix B. The particular test or tests to be used shall be subject to agreement between the purchaser and the manufacturer.

5.6 Corrosion Resistance or Corrosion Protection of Coatings

5.6.1 Coated articles shall be subjected to one of the corrosion tests for the stated time shown in Tables 1–4 to be appropriate for the particular service condition number. The particular test to be used in any instance shall be specified by the purchaser or shall be the subject of agreement between the

TABLE 1—NICKEL/CHROMIUM COATINGS ON STEEL[a]

Service Condition No.	Classification No.	Equiv. Nickel Thickness, mils (approx)[b]	Corrosion Test Duration, h		
			Cass ASTM B 368	Corrodkote ASTM B 380	Acetic-Salt ASTM B 287
SC 4	FE/NI40D/CRR	1.6[d]	22	20	144
	FE/NI30D/CRMC	1.2	22	20	144
	FE/NI30D/CRMP	1.2	22	20	144
SC 3	FE/NI30D/CRR	1.2[d]	16	16	96
	FE/NI25D/CRMC	1.0	16	16	96
	FE/NI25D/CRMP	1.0	16	16	96
	FE/NI40P/CRR	1.6	16	16	96
	FE/NI30P/CRMC	1.2	16	16	96
	FE/NI30P/CRMP	1.2	16	16	96
SC 2[c]	FE/NI20B/CRR	0.8	—	—	24
	FE/NI15B/CRMC	0.6	—	—	24
	FE/NI15B/CRMP	0.6	—	—	24
SC 1[c]	FE/NI10B/CRR	0.4	—	—	8

[a]When agreed by the purchaser and the manufacturer, copper may be used as an undercoat for the nickel but is not substitutable for any part of the nickel thickness specified.

[b]1 mil = 0.001 in = 25.4 μm.

[c]P or D nickel may be substituted for B nickel and MC or MP chromium may be substituted for R chromium in service condition No. 1; P or D nickel may be substituted for B nickel in service condition No. 2.

[d]Copper can contribute to the protection of the basis metal. This protection is enhanced when the chromium deposit is microcracked or microporous and the final nickel layer is an active sulfur-containing nickel deposit.

In double nickel systems, buffing one of the nickel deposits may be beneficial to the corrosion resistance.

TABLE 2—COPPER/NICKEL/CHROMIUM COATINGS ON ZINC ALLOY[a]

Service Condition No.	Classification No.	Equiv. Nickel Thickness, mils (approx)[b]	Corrosion Test Duration, h		
			Cass ASTM B 368	Corrodkote ASTM B 380	Acetic-Salt ASTM B 287
SC 4	ZN/NI40D/CRR	1.6	22	20	144
	ZN/NI30D/CRMC	1.2	22	20	144
	ZN/NI30D/CRMP	1.2	22	20	144
SC 3	ZN/NI30D/CRR	1.2	16	16	96
	ZN/NI25D/CRMC	1.0	16	16	96
	ZN/NI25D/CRMP	1.0	16	16	96
	ZN/NI40P/CRR	1.6	16	16	96
	ZN/NI30P/CRMC	1.2	16	16	96
	ZN/NI30P/CRMP	1.2	16	16	96
SC 2[c]	ZN/NI20B/CRR	0.8	4	4	24
	ZN/NI15B/CRMC	0.6	4	4	24
	ZN/NI15B/CRMP	0.6	4	4	24
SC 1[c]	ZN/NI10B/CRR	0.4	—	—	8

[a]All these coatings shall be applied over an undercoat of copper or yellow brass having a minimum thickness on significant surfaces of 5 μm (0.2 mil) as measured in accordance with ASTM A 219, Methods of Test for Local Thickness of Electrodeposited Coatings.

[b]1 mil = 0.001 in = 25.4 μm.

[c]P or D nickel may be substituted for B nickel and MC or MP chromium may be substituted for R chromium in service condition No. 1; P or D nickel may be substituted for B nickel in service condition No. 2 or 1.

TABLE 3—NICKEL/CHROMIUM COATINGS ON COPPER OR COPPER ALLOYS[a]

Service Condition No.	Classification No.	Equiv. Nickel Thickness, mils (approx)[b]	Corrosion Test Duration, h		
			Cass ASTM B 368	Corrodkote ASTM B 380	Acetic-Salt ASTM B 287
SC 4	CU/NI30D/CRR	1.2	22	20	144
	CU/NI25D/CRMC	1.0	22	20	144
	CU/NI25D/CRMP	1.0	22	20	144
SC 3	CU/NI25D/CRR	1.0	16	16	96
	CU/NI20D/CRMC	0.8	16	16	96
	CU/NI20D/CRMP	0.8	16	16	96
	CU/NI30B/CRR	1.2	16	16	96
	CU/NI25B/CRMC	1.0	16	16	96
	CU/NI25B/CRMP	1.0	16	16	96
	CU/NI25P/CRR	1.0	16	16	96
	CU/NI20P/CRMC	0.8	16	16	96
	CU/NI20P/CRMP	0.8	16	16	96
SC 2[c]	CU/NI15B/CRR	0.6	—	—	24
	CU/NI10B/CRMC	0.4	—	—	24
	CU/NI10B/CRMP	0.4	—	—	24
SC 1[c]	CU/NI5B/CRR	0.2	—	—	8

[a] All these coatings may be applied directly to the basis metal or over a copper strike.
[b] 1 mil = 0.001 in = 25.4 μm.
[c] P or D nickel may be substituted for B nickel in service condition Nos. 2 and 1, and MC or MP chromium may be substituted for R chromium in service condition No. 1.

TABLE 4—COPPER/NICKEL/CHROMIUM COATINGS ON ALUMINUM[a]

Service Condition No.	Classification No.	Equiv. Nickel Thickness, mils (approx)[b]	Corrosion Test Duration, h		
			Cass ASTM B 368	Corrodkote ASTM B 380	Acetic-Salt ASTM B 287
SC 4	AL/NI40D/CRR	1.6	22	20	144
	AL/NI30D/CRMC	1.2	22	20	144
	AL/NI30D/CRMP	1.2	22	20	144
SC 3	AL/NI30D/CRR	1.2	16	16	96
	AL/NI25D/CRMC	1.0	16	16	96
	AL/NI25D/CRMP	1.0	16	16	96
	AL/NI40P/CRR	1.6	16	16	96
	AL/NI30P/CRMC	1.2	16	16	96
	AL/NI30P/CRMP	1.2	16	16	96
SC 2[c]	AL/NI20B/CRR	0.8	—	—	24
	AL/NI15B/CRMC	0.6	—	—	24
	AL/NI15B/CRMP	0.6	—	—	24
SC 1[c]	AL/NI10B/CRR	0.4	—	—	8

[a] All these coatings shall be applied over an undercoat of copper having a minimum thickness on significant surfaces of 5 μm (0.2 mil) as measured in accordance with ASTM A 219, Methods of Test for Local Thickness of Electrodeposited Coatings.
[b] 1 mil = 0.001 in = 25.4 μm.
[c] P or D nickel may be substituted for B nickel and MC or MP chromium may be substituted for R chromium in service condition No. 1; P or D nickel may be substituted for B nickel in service condition Nos. 2 or 1.

purchaser and the manufacturer. The tests are described in detail in the referenced ASTM documents.[6]

5.6.2 After subjecting the article to the treatment described in the relevant test method, it shall be examined for evidence of corrosion of the basis metal or blistering of the coating. Any evidence of basis metal corrosion or blistering of the coating shall be cause for rejection, unless otherwise agreed between the purchaser and the manufacturer.[7]

5.6.3 Surface deterioration of the coating itself is expected to occur during the testing of some types of coatings. The extent to which such surface deterioration will be tolerated shall be subject to agreement between purchaser and manufacturer.[7]

6. Sampling

6.1 Since test methods may be destructive and since 100% inspection is expensive and usually unnecessary, it is recommended that the purchaser select suitable sampling plans for the acceptance testing of lots of coated items. General information on sampling procedures is given in ASTM E 105 and E 122. Standard sampling plans have been published by several sources. In order that the manufacturer know the quality he is expected to meet, the plans selected should be made a part of the purchase contract.

APPENDIX A
DEFINITIONS OF SERVICE CONDITIONS FOR WHICH THE VARIOUS SERVICE CONDITION NUMBERS ARE APPROPRIATE

Service Condition No. SC 4 (Very Severe)—Service conditions which include exposure to very severe, heavy corrosive environments such as those found in an area where there is heavy industry accompanied by snow and below-freezing temperatures, and conditions where parts are subjected to continued exposure in a salt water environment.

Service Condition No. SC 3 (Severe)—Exposure which is likely to include severe industrial or seacoast environments and areas where frequent wetting by rain or dew is experienced.

Severe exposure is defined as that which is likely to include occasional or frequent wetting by rain, dew, or snow in an industrial or seacoast environment.

Service Condition No. SC 2 (Moderate)—Moderate exposure is defined as that which is likely to include normally dry sheltered locations, but with coating subject to occasional condensation of moisture, wear, or abrasion.

Service Condition No. SC 1 (Mild)—Exposure to normally warm, dry interior atmospheres.

APPENDIX B
TESTS FOR ADHESION OF COATINGS[8]

1. Bend Test for Adhesion—The plated part shall be repeatedly flexed or deformed in some manner until fracture occurs. The ability to peel the coating or to separate the different layers of the coating, at other than the immediate area of the fracture, is evidence of failure to conform to the adhesion requirement.

2. File Test for Adhesion—Saw a piece off the plated article, hold it in a vise, and apply a coarse file (12–20 cuts per inch) to the cut edge in such a manner as to attempt to raise the deposit. Peeling of the coating away from the cut edge or separation of the different layers of the coating is evidence of failure to conform to the adhesion requirement.

3. Quenching Test for Adhesion[9]—Heat the plated article for 1 h in an oven maintained at the following temperatures, within ±10 C:

Basis Metal	Temperature, C (F)
Steel	350 (662)
Zinc alloy	150 (302)
Copper or copper alloy	250 (482)
Aluminum	250 (482)

Quench the articles in water at room temperature.

Any lifting or blistering of the coating is evidence of failure to conform to the adhesion requirement.

APPENDIX C
DUCTILITY TEST[10]

1. Preparation of Test Piece—Prepare a plated test strip, 150 mm long, 10 mm wide, and 1 mm thick (approximately 6 in. x 0.4 in. x 0.040 in.), by the following method.

Polish a sheet of the appropriate basis metal, similar to that of the articles being plated, except that the sheet may be of soft brass if the basis metal is zinc alloy. (Use a sheet that is sufficiently large to allow the test strip to be cut from it after trimming off a border at least 25 mm (1.0 in.) wide all around.) Place the sheet on one side with nickel to a thickness of 25 μm (1.0 mil) under the same conditions and in the same bath as the corresponding articles.

---

[6] The corrosion tests indicated in Tables 1–4 are a means of controlling the continuity and quality of the coatings and the duration of the tests does not necessarily have a fixed relationship with the service life of the finished article.

[7] It is to be understood that occasional, widely scattered, small corrosion defects may be observed after the testing period. In general, "acceptable resistance" shall mean that such defects are not significantly defacing or otherwise deleterious to the function of the plated part.

[8] There is no single satisfactory test for evaluating the adhesion of electrodeposited coatings. Those given above are widely used; however, other tests may prove more applicable in specific cases. If so, such tests should be made a part of the purchase order. A review of methods of measuring adhesion is given in the 50th Technical Proceedings of the American Electroplaters' Society (1963).

[9] CAUTION: This test may have an adverse effect on the mechanical properties of the article tested.

[10] This test is used to check that the type of nickel deposit complies with the appropriate definition given in paragraph 5.4.1.1, and should not be used to assess the acceptability of a plated article.

Cut the test strip from the plated sheet with a flat shear. Round or chamfer the longer edges of the test strip, at least on the plated side, by careful filing or grinding.

2. Procedure—Bend the test strip with the plated side in tension, by steadily applied pressure, through 180 deg over a mandrel of diameter 11.5 mm (0.45 in.) until the two ends of the test strip are parallel. Ensure that contact between the test strip and the mandrel is maintained during bending.

3. Assessment—The plating is deemed to comply with the minimum requirement of an elongation of 8% provided that after testing there are no cracks passing completely across the convex surface. Small cracks at the edges do not signify failure.

APPENDIX D
DETERMINATION OF CHROMIUM DISCONTINUITY

Equipment

    Metallurgical microscope (B & L 31-20-6637 or equivalent)
        Eyepiece 10X (B & L 31-15-09 or equivalent)
        Objective 5X (B & L 42-33-51 or equivalent)
        Objective 10X (B & L 42-33-53 or equivalent)
        Howard disc (B & L 31-16-15 or equivalent)
        Cross-line disc (B & L 31-16-30 or equivalent)
    Stage micrometer (B & L 31-16-89)
    Soft bristle brush
    Hot alkaline cleaner (4 oz/gal trisodium phosphate plus 2 oz/gal sodium hydroxide)
    Stainless steel beaker, 4000 ml capacity
    Copper plating solution (28–32 oz/gal copper sulfate ($CuSO_4 \cdot 5H_2O$) plus 6–8 fluid oz/gal sulfuric acid (H_2SO_4))
    Plastic plating tank—Capacity depending on parts to be tested
    Copper anodes
    Low voltage rectifier (E. H. Sargent Co., Cat. No. S-30968 or equivalent)
    Platers tape (3M Co., Pressure Sensitive Tape No. 470, ¾ in. width)
    Acid dip (5% by volume sulfuric acid in water)
    Hot plate

Procedure	Explanation	Precautions
A. Preparation of Solutions		
1. Alkaline Cleaner		
(a) Into the stainless steel beaker place 1 gal of water, 4 oz trisodium phosphate, and 2 oz sodium hydroxide.	(a) This is the alkaline cleaner.	(a) Wear rubber gloves to avoid caustic burns.
(b) Stir the solution until the chemicals are dissolved and then heat to 150–160 F.	(b) This is the operating temperature.	(b) Wear suitable safety equipment.
2. Copper Plating Solution		
(a) Fill the plastic plating tank about 3/4 full with water.	(a) Measure the amount of water.	
(b) For each gallon of water in the tank, add 28–32 oz of copper sulfate and stir to dissolve.		
(c) For each gallon of water in the tank, add 6–8 oz of sulfuric acid and stir to mix.	(c) This is the copper plating solution. This solution is used at room temperature.	(c) Wear suitable safety equipment.
(d) Insert the copper anodes in the bath and connect them to the positive terminal of the rectifier.		

Procedure	Explanation	Precautions
B. Plating of Parts		
1. Connect a copper wire to the part long enough to immerse in the plating tank.	1. The wire should be masked with platers tape where it does not make electrical contacts.	
2. If the part has been cut, tape the edges with platers tape.	2. Cut edges will interfere with copper plating.	
3. Rinse the parts, if necessary, with an organic solvent.	3. To remove any grease or oil.	
4. Dip the brush in the hot alkaline cleaner and gently wash the part, avoiding scrubbing.	4. If you scratch the test part, the scratches will plate.	4. Do not dip the part in the hot cleaner; this may crack chromium that otherwise would not crack.
5. Rinse in running water.	5. To remove alkaline cleaner.	
6. Dip the part for 5 s in the 5% sulfuric acid solution.		
7. Rinse in running water.	7. To remove acid.	
8. Inspect the part for water breaks. If the part shows no water break, proceed to the next step. If the part shows water breaks, repeat steps 4–7.	8. To insure a clean part.	8. Do not touch the part with hands.
9. Connect a copper wire to the negative terminal of the rectifier and to the racking wire (step B-1).		
10. Immerse the part into the plating solution and turn on the rectifier.		10. Do not let the part touch the anodes as this will cause a short circuit.
11. Adjust the rectifier so that the voltage reading is about 0.3 V and plate for 3–15 min.	11. The time for plating will vary depending on many factors, such as the type of chromium, chromium coverage, etc. The parts should be plated until a copper deposit is barely visible to the naked eye.	
12. Shut off the rectifier, remove the part from the plating bath, and rinse in running warm water. Allow it to dry.	12. This the end of the plating cycle.	12. Handle the part by the edges only, as the copper deposit can be wiped off.
C. Calibration of Microscope		
1. Insert the Howard disc into a 10X eyepiece.		
2. Insert the eyepiece and 5X objective into the microscope.	2. This will give a 50X magnification used for measuring microporous chromium.	
3. Place the stage micrometer on the stage of the microscope and focus the microscope on it.		
4. Rotate the eyepiece so that the grid can be measured with the stage micrometer.		
5. Determine the length of the grid to the nearest 0.001 in.	5. This is L.	
6. Rotate the eyepiece 90 deg and determine the width of the grid to the nearest 0.001 in.	6. This is W.	
7. Calculate the area of the grid.	7. L × W = Area	
8. Calculate the factor for the microscope. This will be used when measuring microporous chromium.	8. Factor = $\dfrac{1}{\text{Area, in}^2}$	
9. Insert the cross-line disc in a 10X eyepiece.		
10. Insert the eyepiece and the 10X objective into the microscope.	10. This will give 100X magnifications used for measuring microcracked chromium.	
11. Place the stage micrometer on the stage of the microscope and focus the microscope on it.		
12. Rotate the eyepiece until one of the lines in the cross is at right angles to the lines of the stage micrometer.		
13. Determine the length of the line within the field of view to the nearest 0.001 in.		
14. Calculate the factor for this microscope setup. This will be used when measuring microcracked chromium.	14. Factor = $\dfrac{1}{\text{Length, in}}$	
D. Measuring Chromium Pattern		
1. Determine what type of chromium is on the part.	1. When viewed under the microscope microporous chromium will show numerous black spots, while microcracked chromium will show numerous fine cracks.	
2. Set the microscope up for the particular type of chromium.	2. For microporous chromium use the 5X objective and the 10X eyepiece with the Howard disc. For microcracked chromium, use the 10X objective and the 10X eyepiece with the cross-line disc.	2. Be sure to use the correct setup for each type of chromium.
3. Place the treated test part under the microscope and focus the microscope on the part.	3. The test part has been plated with copper as per section B of this procedure.	3. Handle the part carefully as the copper can be wiped off.
4. If the part is microporous chromium, count the number of spots in the grid and multiply this by the factor determined in C-8.	4. This will give the pore count per square inch.	
5. If the part is microcracked chromium, count the number of cracks which cross one of the lines in the cross-line eyepiece and multiply this by the factor determined in step C-14.	5. This will give the cracks per inch.	
6. Repeat this procedure (steps D-3, D-4, or D-5) at three locations and average the three values.		
E. Alternate Methods of Measuring Chromium Pattern		
1. Photographic—Using a Platers Microscope		
(a) Insert a 10X eyepiece and a 5X objective into the microscope.	(a) This will give a 50X magnification for use with microporous chromium.	
(b) Attach the camera to the microscope.		

(continued on next page)

Procedure	Explanation	Precautions	Procedure	Precautions	Explanation
(c) Place the stage micrometer on the stage of the microscope and focus the microscope on it.			(k) If the discontinuities in the chromium are microcracks use a 10X eyepiece and a 10X objective and proceed with the calibration as in steps E-1-b,c,d, and e.	(k) The magnification determined by this method is also the factor for microcracked chromium	
(d) Photograph the slide and develop it.			(l) Place the test part under the microscope and focus on it.		
(e) Measure with an accurate scale the separation between each 0.01 in marking of the stage micrometer on the photograph and determine the magnification.	(e) Magnification = $\dfrac{100}{\text{Measurements from photograph, in}}$		(m) Photograph this area and develop the photograph.		
			(n) Draw a straight line across the photograph and count the number of cracks which intersect the line per linear inch.		
(f) Determine factor for this setup.	(f) Factor = Magnification $\times$ magnification		(o) Multiply the number of cracks by the factor determined in k.	(o) This is the cracks per linear inch.	
(g) Place test part prepared as in Section B under the microscope and focus on H.			2. Using a Metallograph		
			(a) If a photograph is taken, proceed as in step E-1.		
(h) Photograph this area and develop the photograph.			(b) If the image is projected on a ground glass screen and not photographed, proceed as in step E-1, treating the image on the ground glass screen as the photograph.		
(i) Count the number of spots in an area 1.0 in $\times$ 1.0 in. Repeat this three times on different areas of the photograph.	(i) A linen tester will be of great assistance for this measurement.				
(j) Average the number of spots and multiply by the factor determined in step E-1-f.	(j) This will give the pores per square inch.				

SOLDERS—SAE J473a

SAE Standard

Report of Nonferrous Metals Division approved June 1911 and last revised by Nonferrous Metals Committee June 1962.

The choice of the type and grade of solder for any specific purpose will depend on the materials to be joined and the method of applying. Those with higher amounts of tin usually wet and bond more readily and have a narrower semi-molten range than lower amounts of tin.

For strictly economic reasons, it is recommended that the grade of solder metal be selected that contains least amount of tin required to give suitable flowing and adhesive qualities for application.

All the lead-tin solders, with or without antimony, are usually suitable for joining steel and copper base alloys. For galvanized steel or zinc, only Class A solders should be used. Class B solders, containing antimony usually as a substitute for some of the tin or to increase strength and hardness of the filler metal, form intermetallic antimony-zinc compounds, causing the joint to become embrittled. Lead-tin solders are not recommended for joining aluminum, magnesium, or stainless steel.

Permissible impurity levels are shown:

MAX IMPURITIES, %

Bismuth	0.25	Zinc	0.005
Copper	0.08	Aluminum	0.005
Iron	0.02	Other elements, total	0.08

In dipping solders, 0.5% max copper is permissible because of pickup in bath.

Compositions, temperatures, and similar specifications of these SAE solders are shown in Table 1.

TABLE 1—COMPOSITIONS, TEMPERATURES, AND SIMILAR SPECIFICATIONS

SAE No.	Sn	Pb	Sb	Temperature, F Solidus	Temperature, F Liquidus	Similar ASTM Grades in Specification B 32-58T
1A	45.0, −1.0	Remainder	0.4 max	360	440	Alloy 45B
1B	43.0, +0.5	Remainder	1.5–2.00	365	435	
2A	40.0, −1.0	Remainder	0.4 max	360	455	Alloy 40B
2B	38.0, +0.5	Remainder	1.5–2.00	365	450	
3A	30.0, −1.0	Remainder	0.5 max	360	490	Alloy 30B
3B	30.0, −1.0	Remainder	0.75–1.25	365	485	
4A	25.0, −1.0	Remainder	0.4 max	360	510	Alloy 25B
4B	25.0, −1.0	Remainder	1.25–1.75	365	500	
5A	20.0, −1.0	Remainder	0.4 max	360	535	Alloy 20B
5B	20.0, −1.0	Remainder	1.25–1.75	365	510	
6A	15.0, −1.0	Remainder	0.4 max	435	555	Alloy 15B
6B	15.0, −1.0	Remainder	As specified[a]	435–445	530–555	Alloy 50B
7A	51.0, −2.0	Remainder	0.4 max	360	420	
8A	35.0, −1.0	Remainder	0.4 max	360	475	Alloy 35B
9B	2.75, −0.25	Remainder	4.90–5.40[b]	465	555	

[a] Maximum, 2.75%.
[b] Also contains 0.40–0.60 arsenic; this solder should be used only with previously tinned base metal. Pure tin or higher tin-lead alloys may be used.

ELECTROPLATING AND RELATED FINISHES—SAE J474b

SAE Information Report

Report of Nonferrous Metals Division approved January 1930 and last revised by Nonferrous Metals Committee June 1972.

Electroplating is a process whereby an object is coated with one or more relatively thin, tightly adherent layer of one or more metals. It is accomplished by placing the object to be coated on a plating rack or a fixture, or in a basket or in a rotating container in such a manner that a suitable current may flow through it, and then immersing it in a series of solutions and rinses in planned sequence. The advantage to be gained by electroplating may be considerable; broadly speaking, the process is used when it is desired to endow the basis material (selected for cost, material conservation, and physical property reasons) with surface properties it does not possess.

It should be noted that although electroplating is the most widely used process for applying metals to a substrate, they may also be applied by spraying, vacuum deposition, cladding, hot dipping, chemical reduction, mechanical plating, etc. The purpose for applying an electroplate and the metals used for various applications follow.

Decorative-Protective Coatings—This type of coating has as its prime purpose the maintenance of an acceptable appearance on a product exposed to various service conditions involving wear and/or corrosion. Typical examples are door handles, bumpers, nameplates, and other bright finished hardware. For this application the copper/nickel/chromium, the nickel/chromium, or other combinations of these metals are most frequently used. However, zinc, brass, tin, cadmium, gold, silver, and rhodium may also be used where a unique appearance and/or a specific protective quality is desired. See SAE J207.

Protective Coatings—Protective coatings can be classified as either sacrificial or barrier type. Both cadmium and zinc are well known as sacrificial coatings, being more active chemically than the substrate and offering protection by being preferentially attacked. Since it is relatively inexpensive and readily applied in a plating barrel or tank, or mechanically applied, zinc is often preferred for coating ferrous parts. However, due to the lesser amount of corrosion products that may form under similar corrosive conditions, cadmium is preferred over zinc in applications where the buildup of corrosion products would have a detrimental effect, such as restricted movement of closely fitted parts or the prevention of current flow in electrical components. In addition, cadmium is more readily solderable.

Tin and its alloys are examples of barrier-type coatings. These coatings protect by serving as an inactive barrier between the substrate and the environment. Such coatings must be thick enough to be free of discontinuities, otherwise corrosion will take place at any void in the coating. Applications of protective coatings include screws, nuts, bolts, and other fasteners; components of mechanical assemblies, and control mechanisms.

Engineering Coatings—Functional enhancement of a component by the use of an electroplated coating is recognized. Among the many applications, the more important usages include:

1. Abrasion and scratch protection as provided by a coating that is harder than the basis metal. Coatings of chromium, nickel (both electrodeposited and electroless), tin-nickel, or iron are examples of hard coatings.

2. Use of soft electrodeposits, such as silver, lead, tin-lead, indium, or lead-indium which adjust to minor imperfections in mating surfaces, are suitable as bearing surfaces.

3. Use of electrodeposits such as chromium, iron, or nickel for rebuilding undersized parts.

4. The use of various electrodeposits in specialized fields such as in coating of conductors, plating of plastics and ceramics, in both the electrical and electronic fields.

Electroforming—Electroforming is the production or reproduction of articles by electrodeposition. Typical applications are the production of printing plates, phonograph matrices, patterns, molds, dies, and paint masks made of electrodeposited copper, iron, nickel, and other metals.

NOTE: It should be stressed that this writing is only a very brief introduction to the subject of electroplating and related finishes. Detailed information on the subject may be found in one or more of the publications appearing in the bibliography and in the specifications listing.

BIBLIOGRAPHY

1. Allen Gray, *Modern Electroplating*. The Electrochemical Society, Inc. Vols. 1 and 2. New York: John Wiley & Sons.
2. Herbert H. Uhlig, *Corrosion Handbook*. The Electrochemical Society, Inc. New York: John Wiley & Sons.
3. William Blum and George B. Hogaboom, *Principles of Electroplating and Electroforming*. New York: McGraw-Hill.
4. A. K. Graham and H. L. Pinkerton, *Electroplating Engineers Handbook* (Second Edition). New York: Reinhold Publishing Co.
5. *Metal Finishing Guide Book and Directory*. Westwood, N.J.: Metal and Plastics Publications.
6. Lester F. Spencer, "Electroplated Coatings—Selection Factors." *Metal Finishing*, Vol. 69, No. 9 (September 1971) and No. 10 (October 1971).
7. J. B. Mohler, "Primer on Electrodeposited Coatings." *Materials Engineering*, Vol. 75, No. 1 (January 1972).

ASTM Specifications[1]

Decorative-Protective Coatings

  B 456 Electrodeposited Coatings of Nickel Plus Chromium
  B 253 Preparation of and Electroplating on Aluminum Alloys by the Zincate Process
  B 254 Preparation of and Electroplating on Stainless Steel

Protective Coatings

  A 165 Electrodeposited Coatings of Cadmium on Steel
  B 545 Electrodeposited Coatings of Tin
  A 164 Electrodeposited Coatings of Zinc on Steel

Engineering Coatings

  B 488 Electrodeposited Coatings of Gold for Engineering Uses
  B 200 Electrodeposited Coatings of Lead on Steel
  B 177 Chromium Plating on Steel for Engineering Use

Electroforming

  B 503 Use of Copper and Nickel Electroplating Solutions for Electroforming

---

[1] These and other specifications related to electroplating may be found in ASTM Book of Standards, Part 7.

NONMETALLIC MATERIALS

NONMETALLIC MATERIALS

12 Nonmetallic Materials

† Classification System for Rubber Materials for Automotive Applications—SAE J200 APR80 .... 12.01
Recommended Guidelines for Fatigue Testing of Elastomeric Materials and Components—SAE J1183 .... 12.29
Test for Dynamic Properties of Elastomeric Isolators—SAE J1085a 12.34
§ Latex Foam Rubbers—SAE J17 OCT79 .... 12.37
† Sponge- and Expanded Cellular-Rubber Products—SAE J18 DEC79 .... 12.40
Latex Dipped Goods and Coatings for Automotive Applications—SAE J19 ..... 12.45
Synthetic Resin Plastic Sealers, Nondrying Type—SAE J250 .... 12.45
Methods of Tests for Automotive-Type Sealers, Adhesives, and Deadeners—SAE J243 12.46
Sound Deadeners and Underbody Coatings—SAE J671 .... 12.56
Test Method for Measuring Weight of Organic Trim Materials—SAE J860 ..... 12.58
Test Method for Determining Dimensional Stability of Automotive Textile Materials—SAE J883 .... 12.58
Method of Testing Resistance to Crocking of Organic Trim Materials—SAE J861 ... 12.58
Test Method for Measuring Thickness of Automotive Textiles and Plastics—SAE J882 12.59
Test Method for Determining Cold Cracking of Flexible Plastic Materials—SAE J323 12.59
Nonmetallic Trim Materials—Test Method for Determining the Staining Resistance to Hydrogen Sulfide Gas—SAE J322 .... 12.60
Test Method for Determining Window Fogging Resistance of Interior Trim Materials—SAE J275 .... 12.61
† Test for Chip Resistance of Surface Coatings—SAE J400 JUN80 .... 12.63
Florida Exposure of Automotive Finishes—SAE J951 FEB80 .... 12.66
Glossary of Fiberboard Terminology—SAE J947b 12.67
Fiberboard Test Procedure—SAE J315b 12.68
Test Method for Wicking of Automotive Fabrics and Fibrous Materials—SAE J913a 12.69
Method of Testing Resistance to Scuffing of Trim Materials—SAE J365a 12.69
Test Method for Determining Blocking Resistance and Associated Characteristics of Automotive Trim Materials—SAE J912a 12.71
Fiberboard Crease Bending Test—SAE J119a 12.72
Definitions of Acoustical Terms—SAE J1184 .... 12.72
Flammability of Automotive Interior Materials—Horizontal Test Method—SAE J369a 12.73
Test Method for Determining Stiffness (Modulus of Bending) of Fiberboards—SAE J949a 12.74
Test Method for Determining Resistance to Snagging and Abrasion of Automotive Bodycloth—SAE J948a 12.75
Test Method for Determining Visual Color Match to Master Specimen for Fabrics, Vinyls, Coated Fiberboards, and Other Automotive Trim Materials—SAE J361a 12.75
Test Method of Stretch and Set of Textiles and Plastics—SAE J855a 12.76
Felts—Wool and Part Wool—SAE J314b 12.78
Nonmetallic Automotive Gasket Materials—SAE J90b 12.80
Rubber Rings for Automotive Applications—SAE J120a 12.87
Engine Coolants—SAE J814c 12.93
Engine Coolant Concentrate—Ethylene-Glycol Type—SAE J1034 .... 12.97
Coolant System Hoses—SAE J20e 12.98
Fuel and Oil Hoses—SAE J30d 12.102

* New
† Technical revision
§ Editorial change

Power Steering Pressure Hose—High Volumetric Expansion Type—SAE J188 ..... 12.107
Power Steering Return Hose—Low Pressure—SAE J189 ..... 12.108
Power Steering Pressure Hose—Low Volumetric Expansion Type—SAE J191 ..... 12.110
Power Steering Pressure Hose—Wire Braid—SAE J190 ..... 12.111
Windshield Wiper Hose—SAE J50a 12.112
Automotive Air Conditioning Hose—SAE J51b 12.112
Emission Control Hose—SAE J1010 ..... 12.114
Windshield Washer Tubing—SAE J1037 ..... 12.115

13 Fuels and Lubricants

Effective Dates of Revisions—SAE J301 ..... 13.01
† Engine Oil Performance and Engine Service Classification—SAE J183 FEB80 ..... 13.01
Engine Oil Viscosity Classification—SAE J300d 13.05
Engine Oil Tests—SAE J304c 13.06
Physical and Chemical Properties of Engine Oils—SAE J357a 13.08
Axle and Manual Transmission Lubricants—SAE J308c 13.12
§ Axle and Manual Transmission Lubricant Viscosity Classification—SAE J306 OCT79 ..... 13.13
§ Automotive Lubricating Greases—SAE J310 MAR80 ..... 13.14
Fluid for Passenger Car Type Automatic Transmissions—SAE J311b 13.17
* Powershift Transmission Fluid Classification—SAE J1285 FEB80 ..... 13.19
Central Fluid Systems—SAE J71a 13.20
† Automotive Gasolines—SAE J312 JUN80 ..... 13.22
† Diesel Fuels—SAE J313 JUN80 ..... 13.28
* Alternative Automotive Fuels—SAE J1297 APR80 ..... 13.32
The Engine Oil Performance and Engine Service Classification Maintenance Procedure—SAE J1146 ..... 13.34

12 Nonmetallic Materials

CLASSIFICATION SYSTEM FOR RUBBER MATERIALS FOR AUTOMOTIVE APPLICATIONS—
SAE J200 APR80

SAE Recommended Practice

Report of the Nonmetallic Materials Committee, approved May 1962, last revised April 1980.

This classification system was prepared jointly by the Society of Automotive Engineers and the American Society for Testing and Materials and bears the designation SAE J200/ANSI/ASTM D 2000. This system supersedes and replaces SAE J14—ASTM D 735 and is to be used as a source of material quality "line call-out" specifications on procurement documents and drawings. As SAE J14 and ASTM D 735 are no longer published, Tables 1-6 from these documents have been extracted and are published as part of the Appendix to SAE J200/ANSI/ASTM D 2000 as a reference only for existing procurement documents and drawings where they have been employed.

This classification system is based on physical properties of desired rubber products rather than material compositions produced from selected polymers and is, therefore, much more flexible than the SAE J14—ASTM D 735 system.

1. Scope

1.1 This classification system tabulates the properties of vulcanized rubber materials (natural rubber, reclaimed rubber, synthetic rubbers, alone or in combination) that are intended for, but not limited to, use in rubber products for automotive applications.

NOTE 1—This classification system may serve many of the needs of other industries in much the same manner as SAE numbered steels. It must be remembered, however, that this system is subject to revision when required by automotive needs. It is recommended that the latest revision always be used.

1.2 This classification system is based on the premise that the properties of all rubber products can be arranged into characteristic material designations. These designations are determined by *types*, based on resistance to heat aging, and *classes*, based on resistance to swelling oil. Basic levels are thus established which, together with values describing additional requirements, permit complete description of the quality of all rubber materials.

1.3 In all cases where provisions of this classification system would conflict with those of the detailed specifications for a particular product, the latter shall take precedence.

NOTE 2—When the rubber product is to be used for purposes where the requirements are too specific to be completely prescribed by this classification system, it is necessary for the purchaser to consult the supplier in advance to establish the appropriate properties, test methods, and specification test limits.

2. Purpose

2.1 The purpose of this classification system is to provide guidance to the engineer in the selection of practical, commercially available rubber materials, and further to provide a method for specifying these materials by the use of a simple *line call-out* designation.

2.2 This classification system was developed to permit the addition of descriptive values for future rubber materials without complete reorganization of the classification system and to facilitate the incorporation of future new methods of test to keep pace with changing industry requirements.

3. Type and Class

3.1 The prefix letter M shall be used to indicate that this classification system is based on SI units.

NOTE 3—Call-outs not prefixed by the letter M refer to an earlier classification system based on U. S. customary units. This was published in editions prior to 1979.

3.2 Rubber materials shall be designated on the basis of *type* (heat resistance) and *class* (oil resistance). Type and class are each indicated by letter designations as shown in Tables 1 and 2 and illustrated in paragraph 8.1.

3.3 *Type* is based on changes in tensile strength of not more than ±30%, elongation of not more than −50%, and hardness of not more than ±15 points after heat aging for 70 h at an appropriate temperature. The temperatures at which these materials shall be tested for determining type are listed in Table 1.

3.4 *Class* is based on the resistance of the material to swelling in ASTM Oil No. 3 after 70 h immersion at a temperature determined from Table 1, except that a maximum temperature of 150°C (the upper limit of oil stability) shall be used. Limits of swelling for each class are shown in Table 2.

NOTE 4—The selection of type based on heat resistance is understood to be indicative of the inherent heat resistance that can be normally expected from commercial compositions. Likewise, choice of class is based on the range of volume swell normally expected from such commercial compositions as established by type. *The fact that a type and class of material is listed in Table 6, under Basic Requirements, indicates that materials that meet these requirements for heat and oil resistance are commercially available.*

3.5 The letter designations shall always be followed by a three-digit number to specify the hardness and the tensile strength—for example, 505. The first digit indicates durometer hardness, for example, 5 for 50 ± 5, 6 for 60 ± 5. The next two digits indicate the minimum tensile strength—for example, 05 for 5 MPa, 14 for 14 MPa. *Correlation of available materials for desired hardness and tensile strength is obtained through the elongation values in Table 6. See paragraph 6.2.*

4. Grade Numbers, Suffix Letters, and Numbers

4.1 **Grade Numbers**—Since the basic requirements do not always sufficiently describe all the necessary qualities, provision is made for deviation or adding requirements through a system of prefix grade numbers. Grade No. 1 indicates that only the basic requirements are compulsory and no suffix requirements are permitted. Grades other than No. 1 are used for expressing deviations or additional requirements and are listed as *Available Suffix Grade Numbers* in the last column under Basic Requirements in Table 6. A grade number is written as a material prefix number preceeding the letters for type and class (see paragraph 8.1).

4.2 **Suffix Letters**—The suffix letters that may be used together with their meaning, appear in Table 3.

4.3 **Suffix Numbers**—Each suffix letter should preferably be followed by two suffix numbers (see Note 6 in paragraph 7.1). *The first suffix number always indicates the method of test;* time of test is part of the method and is taken from the listings in Table 4. *The second suffix number, if used, always indicates the temperature of test* and is taken from Table 5. Where three-digit numbers are required, a dash, (-), is used for separation, for example: A1-10; B4-10; F1-11.

5. Composition and Manufacture

5.1 This classification is predicated upon materials, furnished under a specification based thereon, being manufactured from natural rubber, reclaimed rubber, synthetic rubber, alone or in combination, together with added compounding materials of such nature and quantity as to produce vulcanizates that comply with the specified requirements. All materials and workmanship shall be in accordance with good commercial practice, and the resulting product shall be free of porous areas, weak sections, bubbles, foreign matter, or other defects affecting serviceability.

5.2 **Color**—With the exception of FC, FE, FK, and GE materials, the values in the material tables are based on black compounds and comparable values may not be available in color.

6. Basic Requirements

6.1 The basic requirements for physical properties specified in Table 6[1] are based on values obtained from standard laboratory test specimens prepared and tested in accordance with the applicable ASTM methods of test.

Test results from specimens prepared from finished products may not duplicate values obtained from standard test specimens.

NOTE 5—When standard test specimens can be cut from finished parts in accordance with ASTM D 3183. Standard Recommended Practice for Rubber-Preparation of Pieces for Test from Other than Standard Vulcanized Sheets,[2] a deviation to the extent of 10% (on tensile strength and elongation values only) is permissible when agreed upon by the purchaser and the supplier. This deviation is permissible only because of the recognized effects of knitting, grain, and buffing on the material when test specimens are prepared from finished parts and tested for tensile strength and elongation. This deviation is intended to apply to goods purchased by the Government. For all other uses, when differences due to the method of processing or to the difficulty in obtaining suitable test specimens from the finished part arise, the purchaser and the supplier may agree on acceptable deviations. This can be done by comparing results of standard test specimens with those obtained on actual parts.

6.2 The available materials are listed in the appropriate material section of the table, giving each hardness and tensile strength grade with its appropriate elongation value. Also, there is a repetition of the values for the basic heat and oil aging requirements for the material resulting from the assignment of type and class. Compression set values are basic requirements to ensure proper vulcanization.

7. Suffix Requirements

7.1 Suffix requirements shall be specified *only as needed* to define qualities necessary to meet service requirements. These suffix requirements are set forth for the various grade numbers. Suffix letters and suffix numbers describing these suffix requirements may be used singly or in combination, *but not all suffix values available for a given material need be specified.*

NOTE 6—Examples of the use of suffix letters and numbers would be A14 and EO34. Suffix A (Table 3) stands for heat resistance. Suffix 1 (Table 4) specifies that the test be run according to ASTM Method D 573[2] for 70 h, and Suffix 4 (Table 5) indicates the temperature of test as 100°C. Similarly, Suffix EO34 indicates resistance in ASTM Oil No. 3 in accordance with ASTM Method D 471[2] for 70 h at 100°C.

7.2 Basic requirements are always in effect, unless superseded by specific suffix requirements in the *line call-out*.

8. Line Call-Outs

8.1 A *line call-out, which is a specification,* shall contain: The document designations, the prefix letter M, the grade number, the material designation (type and class), and the hardness and tensile strength, followed by the appropriate suffix requirements. Following is an example of a *line call-out:*

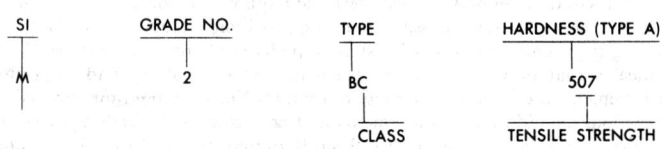

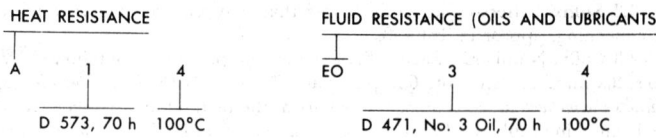

In this example, the basic requirements for heat resistance and oil resistance are superseded for suffix requirements. However, the basic requirements of 80% for compression set, which is not included as a suffix requirement, is not superseded and therefore shall be met as specified in Table 6.

9. Methods of Test

9.1 The applicable methods of test are listed in Table 4.

10. Sampling and Inspection

10.1 A lot, unless otherwise specified, shall consist of all products of the same material submitted for inspection at the same time.

10.2 When proof of conformance with a specification based on this classification system is required, the supplier shall, upon request of the purchaser at the time of ordering, furnish a sufficient number of samples to perform the required tests. Test specimens shall be prepared as prescribed in paragraph 6.1. The samples shall be warranted to have equivalent cure and to be from the same run or batch of compound used in the lot.

TABLE 1—BASIC REQUIREMENTS FOR ESTABLISHING TYPE BY TEMPERATURE

Type	Test Temperature, °C
A	70
B	100
C	125
D	150
E	175
F	200
G	225
H	250
J	275

TABLE 2—BASIC REQUIREMENTS FOR ESTABLISHING CLASS BY VOLUME SWELL

Class	Volume Swell, max, %
A	no requirement
B	140
C	120
D	100
E	80
F	60
G	40
H	30
J	20
K	10

TABLE 3—MEANING OF SUFFIX LETTERS

Suffix Letter	Test Required
A	Heat Resistance
B	Compression Set
C	Ozone or Weather Resistance
D	Compression-Deflection Resistance
EA	Fluid Resistance (Aqueous)
EF	Fluid Resistance (Fuels)
EO	Fluid Resistance (Oils and Lubricants)
F	Low Temperature Resistance
G	Tear Resistance
H	Flex Resistance
J	Abrasion Resistance
K	Adhesion
M	Flammability Resistance
N	Impact Resistance
P	Staining Resistance
R	Resilience
Z	Any special requirement which shall be specified in detail

---

[1] Tensile strength values shown as (psi) in Table 6 are for information purposes only.
[2] Annual Book of ASTM Standards, Part 37.

TABLE 4—ASTM METHODS OF TEST[a]

Requirement or Suffix Letter / Basic Requirements and First Suffix No.	Basic	1	2	3	4	5	6	7	8
Durometer Hardness (Type A)	D 2240	—	—	—	—	—	—	—	—
Tensile Strength, Elongation	D 412 die C	—	—	—	—	—	—	—	—
Suffix A, Heat Resistance	—	D 573, 70 h	D 865, 70 h	D 865, 168 h	—	—	—	—	—
Suffix B, Compression Set	—	D 395, 22 h, Method B, solid	D 395, 70 h, Method B, solid	D 395, 22 h, Method B, plied	D 395, 70 h, Method B, plied	—	—	—	—
Suffix C, Ozone or Weather Resistance	—	D 1171,[c] ozone exposure, Method A	D 1171,[f] weather exposure	D 1171,[g] ozone exposure, Method B	—	—	—	—	—
Suffix D, Compression-Deflection Resistance	—	D 575, Method A	D 575, Method B	—	—	—	—	—	—
Suffix EA, Fluid Resistance (Aqueous)	—	D 471, water 70 h[d]	D 471, water-ethylene glycol, 70 h[e]	—	—	—	—	—	—
Suffix EF, Fluid Resistance (Fuels)	—	D 471, Reference Fuel A, 70 h	D 471, Reference Fuel B, 70 h	D 471, Reference Fuel C, 70 h	—	—	—	—	—
Suffix EO, Fluid Resistance (Oils and Lubricants)	—	D 471, ASTM Oil No. 1,[h] 70 h	D 471, ASTM Oil No. 2,[h] 70 h	D 471, ASTM Oil No. 3,[h] 70 h	D 471, ASTM Oil No. 1, 168 h	D 471, ASTM Oil No. 2, 168 h	D 471, ASTM Oil No. 3, 168 h	D 471, Service Fluid 101, 70 h	D 471, Fluid as designated in Table 6, 70 h
Suffix F, Low-Temperature Resistance	—	D 2137, Method A, paragraph 9.3.2 3 min	D 1053, 5 min T_2 or T_5 or T_{10} or T_{50} or T_{100}	D 2137, Method A, paragraph 9.3.2, 22 h	D 1329, 38.1 mm die, 50% elongation, 10 minutes, paragraph 7.5, retraction 10% min	D 1329, 38.1 mm 50% elongation, 10 minutes, paragraph 7.5, retraction 50% min	—	—	—
Suffix G, Tear Resistance	—	D 624, die B	D 624, die C	—	—	—	—	—	—
Suffix H, Flex Resistance	—	D 430, Method A	D 430, Method B	D 430, Method C	—	—	—	—	—
Suffix J, Abrasion Resistance[b]	—	—	—	—	—	—	—	—	—
Suffix K, Adhesion	—	D 429, Method A	D 429, Method B	[i]	—	—	—	—	—
Suffix M, Flammability Resistance[b]	—	—	—	—	—	—	—	—	—
Suffix N, Impact Resistance[b]	—	—	—	—	—	—	—	—	—
Suffix P, Staining Resistance	—	D 925, Method A	D 925, Method B, Control Panel	—	—	—	—	—	—
Suffix R, Resilience	—	D 945	—	—	—	—	—	—	—
Suffix Z, Special Requirement[b]	—	—	—	—	—	—	—	—	—

[a] The designations refer to the following methods of the American Society for Testing and Materials.
D 394, Test for Abrasion Resistance of Rubber Compounds, discontinued, see 1969 Annual Book of ASTM Standards, Part 28.
D 395, Tests for Rubber Property-Compression Set.
D 412, Tests for Rubber Properties in Tension.
D 429, Tests for Rubber Property-Adhesion to Rigid Substrates.
D 430, Tests for Rubber Deterioration-Dynamic Fatigue.
D 471, Test for Rubber Property-Effect of Liquids.
D 573, Test for Rubber Deterioration in an Air Oven.
D 575, Tests for Rubber Properties in Compression.
D 624, Test For Rubber Property-Tear Resistance.
D 865, Test for Rubber Deterioration by Heating in a Test Tube.
D 925, Tests for Rubber Property-Staining of Surfaces (Contact, Migration, and Diffusion).
D 945, Tests for Rubber Properties in Compression or Shear (Mechanical Oscillograph).
D 1053, Measuring Rubber Property-Stiffening at Low Temperature Using a Torsional Wire Apparatus.
D 1171, Test for Rubber Deterioration-Surface Ozone Cracking Outdoors or Chamber (Triangular Specimens).
D 1329, Evaluating Rubber Property-Retraction at Low Temperatures (TR Test).
D 2137, Test for Rubber Property-Brittleness Point of Flexible Polymers and Coated Fabrics.
D 2240, Test for Rubber Property-Durometer Hardness.
[b] Test Method to be specified.
[c] Use ozone chamber exposure method of ASTM Method D 1171, Method A.
[d] Distilled water shall be used. Volume increase by water displacement method, except alcohol dip omitted. When determining changes in tensile strength, elongation, and hardness, test tube to be 3/4 full after specimens are immersed. Determination to be made after 30 min. Cool in distilled water; acetone dip to be omitted.
[e] Equal parts by volume of distilled water and reagent grade ethylene glycol. Volume increase by displacement method, except alcohol dip omitted. When determining changes in tensile strength, elongation, and hardness, test tube to be 3/4 full after specimens are immersed. Determination to be made after 30 min. Cool in distilled water; acetone dip to be omitted.
[f] ASTM Method D 1171, Weather Test, is 6 weeks duration. Test area and time of year to be agreed upon by the purchaser and the manufacturer.
[g] Use ozone chamber exposure method of ASTM Method D 1171, Method B.
[h] ASTM Oil No. 1 is available from: MZF Assoc., 11200 Homedale Street, Los Angeles, CA 90049.
ASTM Oils No. 2 and 3 are available from: R. E. Carroll, P.O. Box 139, Trenton, NJ 08601.
ASTM Service Fluid 101 is available as Anderol 774 from Tenneco Chemicals, Inc., Intermediates Division, P.O. Box 2, Turner Place, Piscataway, NJ 08854.
ASTM Service Fluid 102 consists of 95 mass percent ASTM No. 1 Oil + 5 mass percent Anglamol 99. Anglamol 99 is available from Lubrizol Corp., P.O. Box 17100, Cleveland, OH 44117.
[i] Bond made after vulcanization.

TABLE 5—SUFFIX NUMBERS TO INDICATE TEMPERATURE OF TEST

Applicable Suffix Requirements	Second Suffix No.	Test Temperature °C[a]
A, B, C, EA, EF, EO, G, K	11	275
	10	250
	9	225
	8	200
	7	175
	6	150
	5	125
	4	100
	3	70
	2	38
	1	23
	0	[b]
F	1	23
	2	0
	3	−10
	4	−18
	5	−25
	6	−35
	7	−40
	8	−50
	9	−55
	10	−65
	11	−75
	12	−80

[a] These test temperatures are based on the ASTM Recommended Practice D 1349, Rubber Standard Temperatures and Atmospheres for Testing and Conditioning. Annual Book of ASTM Standards, Parts 37 and 38.
[b] Ambient temperature in the case of outdoor testing.

TABLE 6—BASIC AND SUPPLEMENTARY (SUFFIX) REQUIREMENTS FOR CLASSIFICATION OF RUBBER MATERIALS

AA Materials

Basic Requirements

Durometer Hardness, ±5 points	Tensile Strength, min MPa	Tensile Strength, min (psi)	Ultimate Elongation, min, %	Heat Aged, ASTM D 573, 70 h at 70°C	Oil Immersion, ASTM D 471, No. 3 Oil, 70 h at 70°C	Compression Set, ASTM D 395, Method B, Solid, max, %, 22 h at 70°C	Available Suffix Grade Numbers
30	7	1015	400				2, 4
30	10	1450	400				2, 4
30	14	2031	400				2, 4
40	7	1015	400				2, 4
40	10	1450	400				2, 4
40	14	2031	400				2, 4
40	17	2466	500				2, 4
40	21	3046	600				2, 4
50	3	435	250				2
50	6	870	250				2
50	7	1015	400				2, 3
50	8	1160	400				2, 3
50	10	1450	400				2, 3, 4, 5
50	14	2031	400				2, 3, 4, 5
50	17	2466	400				2, 3, 4, 5
50	21	3046	500				2, 3, 4, 5
60	3	435	250	Change in durometer hardness, ±15 points Change in tensile strength, ±30% Change in ultimate elongation −50% max	No requirement	Compression set, 50% max	2
60	6	870	250				2
60	7	1015	300				2, 3
60	8	1160	300				2, 3
60	10	1450	350				2, 3, 4, 5
60	14	2031	400				2, 3, 4, 5
60	17	2466	400				2, 3, 4, 5
60	21	3046	400				2, 3, 4, 5
60	24	3481	500				2, 3, 4, 5
70	3	435	150				2
70	6	870	150				2
70	7	1015	200				2, 3
70	8	1160	200				2, 3
70	10	1450	250				2, 3, 4, 5
70	14	2031	300				2, 3, 4, 5
70	17	2466	300				2, 3, 4, 5
70	21	3046	350				2, 3, 4, 5
80	3	435	100				2
80	7	1015	100				2
80	10	1450	150				2
80	14	2031	200				2
80	17	2466	200				2
90	3	435	75				2
90	7	1015	100				2
90	10	1450	125				2

φAA Materials

Suffix Requirements		Grade 1	Grade 2	Grade 3	Grade 4	Grade 5	Grade 6	Grade 7	Grade 8
A13	Heat resistance, ASTM D 573, 70 h at 70°C:	Basic Requirements Only							
	Change in hardness, max, points		±15		+10	+10			
	Change in tensile strength, max, %		±30		−25	−25			
	Change in ultimate elongation, max, %		−50		−25	−25			
B13	Compression set, ASTM D 395, Method B, 22 h at 70°C, max, %			25	25	25			
B33	Compression set, ASTM D 395, Method B, 22 h at 70°C, max, %			35	35	35			
C12	Resistance to ozone, ASTM D 1171, quality retention rating, min, %		85	a	85	a			
C20	Resistance to outdoor aging, ASTM D 1171, quality retention rating, min, %		85	85	85	85			
EA14	Fluid resistance, ASTM D 471, water, 70 h at 100°C, volume change, max, %		10	10	10	10			
F17	Low-temperature resistance, ASTM D 2137, Method A, paragraph 9.3.2, nonbrittle after 3 min at −40°C		pass	pass	pass	pass			
G21	Tear resistance, ASTM D 624, Die C:								
	Under 7.0 MPa tensile strength, min, kN/m			22	22	22			
	Over 7.0 MPa tensile strength, min, kN/m			26	26	26			
K11	Adhesion, ASTM D 429, Method A, min, MPa		1.4	2.8	1.4	2.8			
K21	Adhesion, ASTM D 429, Method B, min, MPa		7	7	7	9			
P2	Staining resistance, ASTM D 925, Method B, Control Panel, Nonstaining		pass	pass	pass	pass			
Z	(Special Requirements) Any special requirements should be specified in detail, including test methods.								

[a] The requirement is applicable and materials are available having these characteristics, but values have not yet been established.

AK Materials

Basic Requirements

Durometer Hardness, ±5 points	Tensile Strength, min		Ultimate Elongation, min, %	Heat Aged, ASTM D 573, 70 h at 70°C	Oil Immersion, ASTM D 471, No. 3 Oil, 70 h at 70°C	Compression Set, ASTM D 395, Method B, Solid, max, %, 22 h at 70°C	Available Suffix Grade Numbers
	MPa	(psi)					
40	3	435	400	Change in durometer hardness, ±15 points Change in tensile strength, ±30% Change in ultimate elongation, −50% max	Volume change, +10% max	Compression set, 50% max	2
50	3	435	400				2
60	5	725	300				2
70	7	1015	250				2
80	7	1015	150				3
90	7	1015	100				3

Suffix Requirements		Grade 1	Grade 2	Grade 3	Grade 4	Grade 5	Grade 6	Grade 7	Grade 8
A14	Heat resistance, ASTM D 573, 70 h at 100°C:	Basic Requirements Only							
	Change in durometer hardness, max, points		+15	+15					
	Change in tensile strength, max, %		−15	−15					
	Change in ultimate elongation, max, %		−40	−40					
B33	Compression set, ASTM D 395, Method B, 22 h at 70°C, max, %		50	50					
EO14	Fluid resistance, ASTM D 471, No. 1 Oil, 70 h at 100°C:								
	Change in durometer hardness, max, points		a	a					
	Change in tensile strength, max, %		a	a					
	Change in ultimate elongation, max, %		a	a					
	Change in volume, %		−3 to +5	−3 to +5					
EO34	Fluid resistance, ASTM D 471, No. 3 Oil, 70 h at 100°C:								
	Change in durometer hardness, points		−5 to +10	−5 to +10					
	Change in tensile strength, max, %		−30	−30					
	Change in ultimate elongation, max, %		−50	−50					
	Change in volume, %		a	a					
F17	Low temperature resistance, ASTM D 2137, Method A, paragraph 9.3.2, nonbrittle, 3 min at −40°C		pass						
Z	(Special Requirements) Any special requirements should be specified in detail, including test methods.								

[a] The requirement is applicable, and materials are available having these characteristics, but values have not yet been established.

BA Materials

Basic Requirements

Durometer Hardness, ±5 points	Tensile Strength, min MPa	Tensile Strength, min (psi)	Ultimate Elongation, min, %	Heat Aged, ASTM D 573, 70 h at 100°C	Oil Immersion, ASTM D 471, No. 3 Oil, 70 h at 100°C	Compression Set, ASTM D 395, Method B, Solid, max, %, 22 h at 70°C	Available Suffix Grade Numbers
30	7	1015	400				2
30	10	1450	400				2, 3, 4, 5
30	14	2031	400				2, 3, 4, 5
40	3	435	300				2, 8
40	7	1015	300				2, 8
40	10	1450	400				2, 3, 4, 5, 6
40	14	2031	400				2, 3, 4, 5
50	7	1015	300				2, 8
50	10	1450	400				2, 3, 4, 5, 6
50	14	2031	400				2, 3, 4, 5
50	17	2466	400				2, 3, 4, 5
60	3	435	250	Change in durometer hardness, ±15 points	No requirement	Compression set, 50% max	8
60	6	870	250	Change in tensile strength, ±30%			8
60	7	1015	300	Change in ultimate elongation, −50% max			2, 8
60	10	1450	350				2, 3, 4, 5, 6
60	14	2031	400				2, 3, 4, 5, 6
60	17	2466	400				2, 3, 4, 5, 6
70	3	435	150				8
70	6	870	150				8
70	7	1015	200				2, 8
70	8	1160	200				8
70	10	1450	250				2, 3, 4, 5, 6
70	14	2031	300				2, 3, 4, 5
70	17	2466	300				2, 3, 4, 5
80	7	1015	100				2, 7
80	10	1450	150				2, 4
80	14	2031	200				2, 4
90	3	435	75				7
90	7	1015	100				2, 7
90	10	1450	125				2, 4

Suffix Requirements		Grade 1	Grade 2	Grade 3	Grade 4	Grade 5	Grade 6	Grade 7	Grade 8
A14	Heat resistance ASTM D 573, 70 h at 100°C: Change in hardness, max, points / Change in tensile strength, max, % / Change in ultimate elongation, max, %	Basic Requirements Only		+10 / −25 / −25	+10 / −25 / −25				
B13	Compression set, ASTM D 395, Method B, 22 h at 70°C, max, %			25		25		25	
C12	Resistance to ozone, ASTM D 1171, quality retention rating, min, %		100	100	100	100	100	100	100
F17	Low-temperature resistance, ASTM D 2137, Method A, paragraph 9.3.2, nonbrittle after 3 min at −40°C		pass	pass	pass	pass			
F19	Low-temperature resistance, ASTM D 2137, Method A, paragraph 9.3.2, nonbrittle after 3 min at −55°C					pass			
K11	Adhesion, ASTM D 429, Method A, min, MPa			1.4	1.4	1.4	1.4		
K21	Adhesion, ASTM D 429, Method B, min, kN/m			7.0 [a]	7.0 [a]	7.0 [a]			
K31	Adhesion, bond made after vulcanization								
Z	(Special Requirements) Any special requirements should be specified in detail, including test methods.								

[a] Materials must be free from surface conditions and compound constituents that are or may become deleterious to cement adhesion.

BC Materials

Basic Requirements

Durometer Hardness, ±5 points	Tensile Strength, min MPa	Tensile Strength, min (psi)	Ultimate Elongation, min, %	Heat Aged, ASTM D 573, 70 h at 100°C	Oil Immersion, ASTM D 471, No. 3 Oil, 70 h at 100°C	Compression Set, ASTM D 395, Method B, Solid, max, %, 22 h at 100°C	Available Suffix Grade Numbers
30	3	435	300				2, 5
30	7	1015	400				2, 5
30	10	1450	500				2, 5
30	14	2031	500				2
40	3	435	300				2
40	7	1015	400				2, 5
40	10	1450	500				2, 5
40	14	2031	500				2, 5
40	17	2466	500				2
50	3	435	300				2, 5
50	7	1015	300				2, 5
50	10	1450	350				2, 5, 6
50	14	2031	400				2, 5, 6
50	17	2466	450				2, 6
50	21	3046	500				2, 6
50	24	3481	500				2, 6
60	3	435	300	Change in durometer hardness, ±15 points	Volume change, +120% max	Compression set, 80% max	3, 5
60	7	1015	300	Change in tensile strength, ±30%			3, 5
60	10	1450	350	Change in ultimate elongation, −50% max			3, 5, 6
60	14	2031	350				3, 5, 6
60	17	2466	400				3, 6
60	21	3046	400				3, 6
60	24	3481	400				3, 6
70	3	435	200				3, 5
70	7	1015	200				3, 5
70	10	1450	250				3, 5, 6
70	14	2031	300				3, 5, 6
70	17	2466	300				3, 6
70	21	3046	300				3, 6
80	3	435	100				4
80	7	1015	100				4
80	10	1450	100				4
80	14	2031	150				4
90	3	435	50				4
90	7	1015	100				4
90	10	1450	150				4
90	14	2031	150				4

Suffix Requirements		Grade 1	Grade 2	Grade 3	Grade 4	Grade 5	Grade 6	Grade 7	Grade 8
A14	Heat resistance, ASTM D 573, 70 h at 100°C:	Basic Requirements Only							
	Change in hardness, max, points		+15	+15	+15	+15	+15		
	Change in tensile strength, max, %		−15	−15	−15	−15	−15		
	Change in ultimate elongation, max, %		−40	−40	−40	−40	−40		
B14	Compression set, ASTM D 395, Method B, 22 h at 100°C, max, %		35	35	35	35	35		
C12	Resistance to ozone, ASTM D 1171, quality retention rating, min, %		100	100	100	100	100		
C20	Resistance to outdoor aging, ASTM D 1171		a	a	a	a	a		
EO14	Fluid resistance, ASTM D 471, No. 1 Oil, 70 h at 100°C:								
	Change in hardness, points		±10	±10	±10	±10	±10		
	Change in tensile strength, max, %		−30	−30	−30	−30	−30		
	Change in ultimate elongation, max, %		−30	−30	−30	−30	−30		
	Change in volume, %		−10 to +15	−10 to +15	−10 to +15	−10 to +15	−10 to +15		
EO34	Fluid resistance, ASTM D 471, No. 3 Oil, 70 h at 100°C:								
	Change in tensile strength, max, %		−70	−60	−45	−60	−60		
	Change in ultimate elongation, max, %		−55	−50	−30	−60	−50		
	Change in volume, max, %		+120	+100	+80	+100	+100		
F17	Low-temperature resistance, ASTM D 2137, Method A, paragraph 9.3.2, nonbrittle after 3 min at −40°C		pass	pass	pass		pass		
F19	Low-temperature resistance, ASTM D 2137, Method A, paragraph 9.3.2, nonbrittle after 3 min at −55°C					pass			
G21	Tear resistance, ASTM D 624, Die C:								
	Under 7.0 MPa tensile strength, min, kN/m		22	22	22				
	7.0–10 MPa tensile strength, min, kN/m		26	26	26				
	10 MPa tensile strength and over, min, kN/m		26	26	26	26	26		
K11	Adhesion, ASTM D 429, Method A, min, MPa		1.4	1.4	1.4	1.4	2.8		
P2	Staining resistance, ASTM D 925, Method B, Control Panel, Nonstaining		pass	pass	pass				
Z	(Special Requirements) Any special requirements should be specified in detail, including test methods.								

a The requirement is applicable, and materials are available having these characteristics, but values have not yet been established.

BE Materials

Basic Requirements

Durometer Hardness, ±5 points	Tensile Strength, min MPa	Tensile Strength, min (psi)	Ultimate Elongation, min, %	Heat Aged, ASTM D 573, 70 h at 100°C	Oil Immersion, ASTM D 471, No. 3 Oil, 70 h at 100°C	Compression Set, ASTM D 395, Method B, Solid, max, %, 22 h at 100°C	Available Suffix Grade Numbers
40	3	435	500			40	2
40	7	1015	500			40	2
50	3	435	350			40	2
50	6	870	350			40	2
50	7	1015	400			40	2
50	10	1450	400			40	2, 3
50	14	2031	400			40	2
60	3	435	300			40	2
60	6	870	300	Change in durometer hardness, ±15 points		40	2
60	7	1015	350	Change in tensile strength, ±30%	Volume change, +80% max	40	2
60	10	1450	350	Change in ultimate elongation, −50% max		40	2, 3
60	14	2031	350			40	2
70	3	435	200			50	2
70	6	870	200			50	2
70	7	1015	200			50	2
70	10	1450	250			50	2, 3
70	14	2031	250			50	2
70	17	2466	250			50	2
80	7	1015	100			50	2
80	10	1450	100			50	2
80	14	2031	150			50	2
80	17	2466	150			50	2
90	7	1015	100			50	2
90	10	1450	100			50	2
90	14	2031	150			50	2

Suffix Requirements		Grade 1	Grade 2	Grade 3	Grade 4	Grade 5	Grade 6	Grade 7	Grade 8
A14	Heat resistance, ASTM D 573, 70 h at 100°C:								
	Change in hardness, max, points		+15	+15					
	Change in tensile strength, max, %		−15	−15					
	Change in ultimate elongation, max, %		−40	−40					
B14	Compression set, ASTM D 395, Method B 22 h at 100°C, max, %		25	25					
C12	Resistance to ozone, ASTM D 1171, quality retention rating, min, %	Basic Requirements Only	100	100					
C20	Resistance to outdoor aging, ASTM D 1171		a	a					
EO14	Fluid resistance, ASTM D 471, No. 1 Oil, 70 h at 100°C:								
	Change in hardness, points		±10	±10					
	Change in tensile strength, max, %		−30	−30					
	Change in ultimate elongation, max, %		−30	−30					
	Change in volume, %		−10 to +15	−10 to +15					
EO34	Fluid resistance, ASTM D 471, No. 3 Oil, 70 h at 100°C:								
	Change in tensile strength, max, %		−50	−50					
	Change in ultimate elongation, max, %		−40	−40					
F17	Low-temperature resistance, ASTM D 2137, Method A, paragraph 9.3.2, nonbrittle after 3 min at −40°C		pass						
F19	Low-temperature resistance, ASTM D 2137, Method A, paragraph 9.3.2, nonbrittle after 3 min at −55°C			pass					
G21	Tear resistance, ASTM D 624, Die C: 10 MPa tensile strength and over, min, kN/m			26					
K11	Adhesion, ASTM D 429, Method A, min, MPa			1.4					
Z	(Special Requirements) Any special requirements should be specified in detail including test methods.								

[a] The requirement is applicable, and materials are available having those characteristics, but values have not yet been established.

BF Materials

Basic Requirements

Durometer Hardness, ±5 points	Tensile Strength, min		Ultimate Elongation, min, %	Heat Aged, ASTM D 573, 70 h at 100°C	Oil Immersion, ASTM D 471, No. 3 Oil, 70 h at 100°C	Compression Set, ASTM D 395, Method B, Solid, max, %, 22 h at 100°C	Available Suffix Grade Numbers
	MPa	(psi)					
60	3	435	200				2
60	6	870	200				2
60	7	1015	250				2
60	8	1160	250				2
60	10	1450	300				2
60	14	2031	350				2
60	17	2466	350				2
70	3	435	150	Change in durometer hardness, ±15 points	Volume change, +60% max	Compression set, 50% max	2
70	6	870	150	Change in tensile strength, ±30%			2
70	7	1015	200	Change in ultimate elongation, −50% max			2
70	8	1160	200				2
70	10	1450	250				2
70	14	2031	250				2
70	17	2466	300				2
80	3	435	100				2
80	7	1015	100				2
80	10	1450	125				2
80	14	2031	125				2

	Suffix Requirements	Grade 1	Grade 2	Grade 3	Grade 4	Grade 5	Grade 6	Grade 7	Grade 8
B14	Compression set, ASTM D 395, Method B, 22 h at 100°C, max, %		25						
B34	Compression set, ASTM D 395, Method B, 22 h at 100°C, max, %		25						
EO14	Fluid resistance, ASTM D 471, No. 1 Oil, 70 h at 100°C:	Basic Requirements Only							
	Change in hardness, points		±10						
	Change in tensile strength, max, %		−25						
	Change in ultimate elongation, max, %		−45						
	Change in volume, %		−10 to +10						
EO34	Fluid resistance, ASTM D 471, No. 3 Oil, 70 h at 100°C:								
	Change in hardness max, points		−20						
	Change in tensile strength, max, %		−45						
	Change in ultimate elongation, max, %		−45						
	Change in volume, %		0 to +60						
F19	Low-temperature resistance, ASTM D 2137, Method A, paragraph 9.3.2, nonbrittle after 3 min at −55°C		pass						
K11	Adhesion		[a]						
Z	(Special Requirements) Any special requirements should be specified in detail, including test methods.								

[a] Materials are available which can be bonded to metal during vulcanization. Because of the wide variety of compounds in use, combined with manifold end-use service requirements, no values are shown. Test method ASTM D 429 and requirements should be agreed upon by supplier and user.

BG Materials

Basic Requirements

Durometer Hardness, ±5 points	Tensile Strength, min MPa	(psi)	Ultimate Elongation, min, %	Heat Aged, ASTM D 573, 70 h at 100°C	Oil Immersion, ASTM D 471, No. 3 oil, 70 h at 100°C	Compression Set, ASTM D 395, Method B, Solid, max, %, 22 h at 100°C	Available Suffix Grade Numbers
40	7	1015	450				2, 5
40	10	1450	450				2, 5
50	3	435	300				2, 5
50	6	870	300				2
50	7	1015	350				2, 5
50	8	1160	350				2
50	10	1450	300				2, 3, 4, 5
50	14	2031	350				2, 3, 4, 5
50	21	3046	400				3, 4
60	3	435	200				2, 5
60	6	870	200				2
60	7	1015	250				2, 5
60	8	1160	250				2
60	10	1450	300				2, 5
60	14	2031	300				2, 3, 4, 5
60	17	2466	350	Change in durometer hardness, ±15 points Change in tensile strength, ±30% Change in ultimate elongation, −50% max	Volume change, +40% max	Compression set, 50% max	2
60	21	3046	350				3, 4
60	28	4061	400				3, 4
70	3	435	150				2, 5
70	6	870	150				2
70	7	1015	200				2, 5
70	8	1160	200				2
70	10	1450	250				2, 5
70	14	2031	250				2, 3, 4, 5
70	17	2466	300				2, 3
70	21	3046	350				3, 4
70	28	4061	400				3, 4
80	3	435	100				6, 7
80	7	1015	100				6, 7
80	10	1450	125				6, 7
80	14	2031	125				3, 4, 6, 7
80	21	3046	300				3, 4
80	28	4061	350				3, 4
90	3	435	50				6, 7
90	7	1015	100				6, 7
90	10	1450	100				6, 7

BG Materials

Suffix	Requirements	Grade 1	Grade 2	Grade 3	Grade 4	Grade 5	Grade 6	Grade 7	Grade 8
A14	Heat resistance, ASTM D 573, 70 h at 100°C: Change in hardness, max, points; Change in tensile strength, max, %; Change in ultimate elongation, max, %				±5; ±15; −15	±15; −20; −40	±15; −20; −40		
B14	Compression set, ASTM D 395, Method B, 22 h at 100°C max, %		25	50	50	25	25	25	
B34	Compression set, ASTM D 395, Method B, 22 h at 100°C max, %		25			25	25		
C12	Resistance to ozone, ASTM D 1171			a	a				
C20	Resistance to outdoor aging, ASTM D 1171			a	a				
EA14	Fluid resistance, ASTM D 471, water, 70 h at 100°C: Change in hardness, points; Change in volume, %	Basic Requirements Only	±10; ±15					±10; ±15	
EF11	Fluid resistance, ASTM D 471, Reference Fuel A, 70 h at room temperature: Change in hardness, points; Change in tensile strength, max, %; Change in ultimate elongation, max, %; Change in volume, %		±10; −25; −25; −5 to +10					±10; −25; −25; −5 to +10	
EF21	Fluid resistance, ASTM D 471, Reference Fuel B, 70 h at room temperature: Change in hardness, points; Change in tensile strength, max, %; Change in ultimate elongation, max, %; Change in volume, %		0 to −30; −60; −60; 0 to +40					0 to −30; −60; −60; 0 to +40	
EO14	Fluid resistance, ASTM D 471, No. 1 Oil, 70 h at 100°C: Change in hardness, max, points; Change in tensile strength, max, %; Change in ultimate elongation, max, %; Change in volume, %		−5 to +10; −25; −45; −10 to +5	−7 to +5; −20; −40; −5 to +10	−7 to +5; −20; −40; −5 to +5	−5 to +15; −25; −45; −10 to +5	−5 to +15; −25; −45; −10 to +5	−5 to +15; −25; −45; −10 to +5	
EO34	Fluid resistance, ASTM D 471, No. 3 Oil, 70 h at 100°C: Change in hardness, points; Change in tensile strength, max, %; Change in ultimate elongation, max, %; Change in volume, %		−10 to +5; −45; −45; 0 to +25	−10 to +5; −35; −40; +16 to +35	−10 to +5; −35; −40; 0 to +6	0 to −15; −45; −45; 0 to +35	0 to −20; −45; −45; 0 to +35	−10 to +5; −45; −45; 0 to +25	
F16	Low-temperature resistance, ASTM D 2137, Method A, paragraph 9.3.2, nonbrittle after 3 min at −35°C							pass	
F17	Low-temperature resistance, ASTM D 2137, Method A, paragraph 9.3.2, nonbrittle after 3 min at −40°C		pass				pass		
F19	Low-temperature resistance, ASTM D 2137, Method A, paragraph 9.3.2, nonbrittle after 3 min at −55°C			pass	pass	pass			
K11	Adhesion		b	b	b	b	b	b	
P2	Staining resistance, ASTM D 925, Method B, Control Panel, Nonstaining			pass	pass				
Z	(Special Requirements) Any special requirements should be specified in detail, including test methods.								

[a] The requirement is applicable, and materials are available having these characteristics but values have not yet been established.

[b] Materials are available which can be bonded to metal during vulcanization. Because of the wide variety of compounds in use, combined with manifold end-use service requirements, no values are shown. Test method ASTM D 429 and requirements should be agreed upon by supplier and user.

BK Materials

Basic Requirements

Durometer Hardness, ±5 points	Tensile Strength, min MPa	Tensile Strength, min (psi)	Ultimate Elongation, min, %	Heat Aged, ASTM D 573, 70 h at 100°C	Oil Immersion, ASTM D 471 No. 3 Oil, 70 h at 100°C	Compression Set, ASTM D 395, Method B, Solid, max, %, 22 h at 100°C	Available Suffix Grade Numbers
60	3	435	200				4
60	6	870	200				4
60	7	1015	250				4
60	8	1160	250				4
60	10	1450	300				4
60	14	2031	350				4
60	17	2466	350				4
70	3	435	150				4
70	6	870	150	Change in durometer hardness, ±15 points			4
70	7	1015	200	Change in tensile strength, ±30%	Volume change, +10% max	Compression set, 50% max	4
70	8	1160	200	Change in ultimate elongation, −50% max			4
70	10	1450	250				4
70	14	2031	250				4
70	17	2466	300				4
80	3	435	100				4
80	7	1015	100				4
80	10	1450	125				4
80	14	2031	125				4
90	3	435	50				4
90	7	1015	100				4
90	10	1450	100				4

Suffix Requirements		Grade 1	Grade 2	Grade 3	Grade 4	Grade 5	Grade 6	Grade 7	Grade 8
A24	Heat resistance, ASTM D 865, 70 h 100°C:								
	Change in hardness, points				±10				
	Change in tensile strength, max, %				−20				
	Change in ultimate elongation, max, %				−30				
B14	Compression set, ASTM D 395, Method B, 22 h at 100°C, max, %	Basic Requirements Only			25				
B34	Compression set, ASTM D 395, Method B, 22 h at 100°C, max, %				25				
EF11	Fluid resistance, ASTM D 471, Reference Fuel A, 70 h at room temperature:								
	Change in hardness, points				±5				
	Change in tensile strength, max, %				−20				
	Change in ultimate elongation, max, %				−20				
	Change in volume, %				±5				
EF21	Fluid resistance, ASTM D 471, Reference Fuel B, 70 h at room temperature:								
	Change in hardness, points				0 to −20				
	Change in tensile strength, max, %				−50				
	Change in ultimate elongation, max, %				−50				
	Change in volume, %				0 to +25				
EO14	Fluid resistance, ASTM D 471, No. 1 Oil, 70 h at 100°C:								
	Change in hardness, points				±5				
	Change in tensile strength, max, %				−20				
	Change in ultimate elongation, max, %				−20				
	Change in volume, %				−10 to 0				
EO34	Fluid resistance, ASTM D 471, No. 3 Oil, 70 h at 100°C:								
	Change in hardness, points				−10 to +5				
	Change in tensile strength, max, %				−20				
	Change in ultimate elongation, max, %				−30				
	Change in volume, %				0 to +5				
K11	Adhesion				[a]				
Z	(Special Requirements) Any special requirement should be specified in detail, including test methods.								

[a] Materials are available which can be bonded to metal during vulcanization. Because of the wide variety of compounds in use, combined with manifold end-use service requirements, no values are shown. Test Method ASTM D 429 and requirements should be agreed upon by supplier and user.

CA Materials

Basic Requirements

Durometer Hardness, ±5 points	Tensile Strength, min MPa	Tensile Strength, min (psi)	Ultimate Elongation, min, %	Heat Aged, ASTM D 573, 70 h at 125°C	Oil Immersion, ASTM D 471, No. 3 Oil, 70 h at 125°C	Compression Set, ASTM D 395, Method B, Plied, max, %, 22 h at 100°C	Available Suffix Grade Numbers
30	7	1015	500				2
30	10	1450	500				2
40	7	1015	400				2
40	10	1450	400				2
40	14	2031	400				2
50	7	1015	300	Change in durometer hardness, ±15 points			3
50	10	1450	300	Change in tensile strength, ±30%			4
50	14	2031	350	Change in ultimate elongation, −50% max	No requirements	Compression set, 60% max	4
50	17	2466	350				4
60	7	1015	250				3
60	10	1450	250				4
60	14	2031	250				4
70	7	1015	200				3
70	10	1450	200				4, 5
70	14	2031	200				4, 5
80	7	1015	150				6
80	10	1450	150				7, 8
80	14	2031	150				7, 8
90	7	1015	100				6
90	10	1450	100				7, 8

Suffix Requirements

	Suffix Requirements	Grade 1	Grade 2	Grade 3	Grade 4	Grade 5	Grade 6	Grade 7	Grade 8
A25	Heat resistance, ASTM D 865, 70 h at 125°C:								
	Change in hardness, max, points		+10	+10	+10	+10	+10	+10	+10
	Change in tensile strength, max, %		−20	−20	−20	−20	−20	−20	−20
	Change in ultimate elongation, max, %		−40	−40	−40	−40	−40	−40	−40
B35	Compression set, ASTM D 395, Method B, 22 h at 125°C, max, %	Basic Requirements Only	70	70	70	50	70	70	50
B44	Compression set, ASTM D 395, Method B, 70 h at 100°C max, %		35	50					
C32	Resistance to ozone, ASTM D 1171, Method B		pass	pass	pass	pass	pass	pass	pass
EA14	Fluid resistance, ASTM D 471, water, 70 h at 100°C, volume change, %		±5	±5	±5	±5	±5	±5	±5
F17	Low-temperature resistance, ASTM D 2137, Method A, paragraph 9.3.2 nonbrittle after 3 min at −40°C		pass	pass	pass	pass	pass	pass	pass
F18	Low-temperature resistance, ASTM D 2137, Method A, paragraph 9.3.2, nonbrittle after 3 min at −50°C		pass	pass	pass	pass		pass	
F19	Low-temperature resistance, ASTM D 2137, Method A, paragraph 9.3.2, nonbrittle after 3 min at −55°C					pass			
G11	Tear resistance, ASTM D 624, Die B, min, kN/m		17	26	26	26	26	26	26
G21	Tear resistance, ASTM D 624, Die C, min, kN/m		17	26	26	26	26	26	26
K11	Adhesion, ASTM D 429, Method A, min, MPa			1.4	2.8	2.8	1.4	2.8	2.8
P2	Staining resistance, ASTM D 925, Method B, Control Panel, Nonstaining		pass	pass	pass	pass	pass	pass	pass
R11	Resilience in compression, ASTM D 945, min, %		70	50	60				
Z	(Special Requirements) Any special requirements should be specified in detail, including test methods.								

CE Materials

Basic Requirements

Durometer Hardness, ±5 points	Tensile Strength, min		Ultimate Elongation, min, %	Heat Aged, ASTM D 573, 70 h at 125°C	Oil Immersion, ASTM D 471, No. 3 Oil, 70 h at 125°C	Compression Set, ASTM D 395, Method B, Solid, max, %, 22 h at 70°C	Available Suffix Grade Numbers
	MPa	(psi)					
50	14	2031	400				2, 3
60	10	1450	350				2, 3
60	14	2031	400				2, 3
60	17	2466	400	Change in durometer hardness, ±15 points	Volume change +80% max	Compression set, 80% max	2, 3
70	7	1015	200	Change in tensile strength, ±30%			2, 3
70	10	1450	250	Change in ultimate elongation, −50% max			2, 3
70	14	2031	300				2, 3
70	17	2466	300				2, 3
80	7	1015	200				2, 3
80	10	1450	250				2, 3
80	14	2031	250				2, 3

Suffix Requirements	Grade 1	Grade 2	Grade 3	Grade 4	Grade 5	Grade 6	Grade 7	Grade 8
A16 Heat resistance, ASTM D 573, 70 h at 150°C:	—Basic Requirements Only—							
Change in hardness, points		±20						
Change in tensile strength, %		±30						
Change in ultimate elongation, max, %		−60						
B15 Compression set, ASTM D 395, Method B, 22 h at 125°C, max, %		60	80					
C12 Resistance to ozone, ASTM D 1171		a	a					
C20 Resistance to outdoor aging, ASTM D 1171		a	a					
F19 Low-temperature resistance, ASTM D 2137, Method A, paragraph 9.3.2, nonbrittle after 3 min at −55°C		pass	pass					
P2 Staining resistance, ASTM D 925, Method B, Control Panel, Nonstaining		pass	pass					
Z (Special Requirements) Any special requirements should be specified in detail, including test methods.								

[a] The requirement is applicable, and materials are available having these characteristics, but values have not yet been established.

CH Materials

Basic Requirements

Durometer Hardness, ±5 points	Tensile Strength, min MPa	Tensile Strength, min (psi)	Ultimate Elongation, min, %	Heat Aged, ASTM D 865, 70 h at 125°C	Oil Immersion, ASTM D 471, No. 3 Oil, 70 h at 125°C	Compression Set, ASTM D 395, Method B, Solid, max, %, 22 h at 100°C	Available Suffix Grade Numbers
60	3	435	200	Change in durometer hardness, ±15 points; Change in tensile strength, ±30%; Change in ultimate elongation, −50% max	Volume change, +30% max	Compression set, 50% max	2, 3
60	6	870	200				2, 3
60	7	1015	250				2, 3
60	8	1160	250				2, 3
60	10	1450	300				2, 3, 5, 6
60	14	2031	350				2, 3
60	17	2466	350				2, 3
70	3	435	150				2, 3
70	6	870	150				2, 3
70	7	1015	200				2, 3
70	8	1160	200				2, 3
70	10	1450	250				2, 3
70	14	2031	250				2, 3, 5, 6
70	17	2466	300				2, 3
80	3	435	100				3, 4
80	7	1015	100				3, 4
80	10	1450	125				3, 4
80	14	2031	125				3, 4, 5, 6
90	3	435	50				3, 4
90	7	1015	100				3, 4
90	10	1450	100				3, 4, 5, 6

	Suffix Requirements	Grade 1	Grade 2	Grade 3	Grade 4	Grade 5	Grade 6	Grade 7	Grade 8
A25	Heat resistance, ASTM D 865, 70 h at 125°C:	Basic Requirements Only							
	Change in hardness, points		0 to +15	0 to +15	0 to +15	0 to +10	0 to +10		
	Change in tensile strength, max, %		−25	−25	−25	−10	−20		
	Change in ultimate elongation, max, %		−50	−50	−50	−40	−30		
B14	Compression set, ASTM D 395, Method B, 22 h at 100°C, max, %		25	25	25	30	25		
B34	Compression set, ASTM D 395, Method B, 22 h at 100°C, max, %		25	25		30	25		
C12	Resistance to ozone ASTM D 1171, quality retention rating, min, %					100	100		
C20	Resistance to outdoor aging, ASTM D 1171					a	a		
EF31	Fluid resistance, ASTM D 471, Reference Fuel C, 70 h at 23°C:								
	Change in hardness, points		0 to −30		0 to −30	0 to −20	0 to −20		
	Change in tensile strength, max, %		−60		−60	−50	−50		
	Change in ultimate elongation, max, %		−60		−60	−60	−50		
	Change in volume, %		0 to +50		0 to +50	0 to +40	0 to +40		
EO15	Fluid resistance, ASTM D 471, No. 1 Oil, 70 h at 125°C:								
	Change in hardness, points		0 to +10		0 to +10				
	Change in tensile strength, max, %		−20		−20				
	Change in ultimate elongation, max, %		−35		−35				
	Change in volume, %		−15 to +5		−15 to +5				
EO16	Fluid resistance, ASTM D 471, No. 1 Oil, 70 h at 150°C:								
	Change in hardness, points			0 to +10					
	Change in tensile strength, max, %			−20					
	Change in ultimate elongation, max, %			−40					
	Change in volume, %			−15 to +5					
EO35	Fluid resistance, ASTM D 471, No. 3 Oil, 70 h at 125°C:								
	Change in hardness, points		±10		±10				
	Change in tensile strength, max, %		−15		−15				
	Change in ultimate elongation, max, %		−30		−30				
	Change in volume, %		0 to +25		0 to +25				
EO36	Fluid resistance, ASTM D 471, No. 3 Oil, 70 h at 150°C:								
	Change in hardness, points			±10		−5 to +10	−5 to +10		
	Change in tensile strength, max, %			−35		−10	−15		
	Change in ultimate elongation, max, %			−35		−50	−40		
	Change in volume, %			0 to +25		0 to +10	0 to +15		
F14	Low-temperature resistance, ASTM D 2137, Method A, paragraph 9.3.2, nonbrittle after 3 min at −18°C					pass			
F16	Low-temperature resistance, ASTM D 2137, Method A, paragraph 9.3.2, nonbrittle after 3 min at −35°C				pass				
F17	Low-temperature resistance, ASTM D 2137, Method A, paragraph 9.3.2, nonbrittle after 3 min at −40°C		pass				pass		
K11	Adhesion		b	b	b	b	b		
Z	(Special Requirements) Any special requirements should be specified in detail, including test methods.								

[a] The requirement is applicable, and materials are available having these characteristics, but values have not yet been established.

[b] Materials are available which can be bonded to metal during vulcanization. Because of the wide variety of compounds in use, combined with manifold end-use service requirements, no values are shown. Test method ASTM D 429 and requirements should be agreed upon by supplier and user.

DA Materials

Basic Requirements

Durometer Hardness, ±5 points	Tensile Strength, min		Ultimate Elongation, min, %	Heat Aged, ASTM D 573, 70 h at 150°C	Oil Immersion, ASTM D 471, No. 3 Oil, 70 h at 150°C	Compression Set, ASTM D 395, Method B, Plied, max, % 22 h at 150°C	Available Suffix Grade Numbers
	MPa	(psi)					
50	7	1015	300				2
50	10	1450	300				2
50	14	2031	350				2
60	7	1015	250	Change in durometer hardness, ±15 points			2, 3
60	10	1450	250	Change in tensile strength, ±30%			2, 3
60	14	2031	300	Change in ultimate elongation, −50% max	No requirement	Compression set, 50% max	2, 3
70	7	1015	200				2, 3
70	10	1450	200				2, 3
70	14	2031	200				2, 3
80	7	1015	150				2, 3
80	10	1450	150				2, 3
80	14	2031	150				2, 3

	Suffix Requirements	Grade 1	Grade 2	Grade 3	Grade 4	Grade 5	Grade 6	Grade 7	Grade 8
A26	Heat resistance, ASTM D 865, 70 h at 150°C:								
	Change in hardness, max, points		+10	+10					
	Change in tensile strength, max, %		−20	−20					
	Change in ultimate elongation, max, %		−20	−20					
B36	Compression set, ASTM D 395, Method B, 22 h at 150°C, max, %	Basic Requirements Only	40	25					
C32	Resistance to ozone, ASTM D 1171, exposure Method B		pass	pass					
EA14	Fluid resistance, ASTM D 471, water, 70 h at 100°C, volume change, max, %		±5	±5					
F19	Low-temperature resistance, ASTM D 2137, Method A, paragraph 9.3.2, nonbrittle after 3 min at −55°C		pass	pass					
G11	Tear resistance, ASTM D 624, Die B, min, kN/m		17	17					
G21	Tear resistance, ASTM D 624, Die C, min, kN/m		17	17					
K11	Adhesion, ASTM D 429, Method A, min, MPa			1.4					
P2	Staining resistance, ASTM D 925, Method B, Control Panel, Nonstaining		pass	pass					
R11	Resilience in compression, D 945, min, %		60	60					
Z	(Special Requirements) Any special requirements should be specified in detail, including test methods.								

DF Materials

Basic Requirements

Durometer Hardness, ±5 points	Tensile Strength, min		Ultimate Elongation, min, %	Heat Aged, ASTM D 865, 70 h at 150°C	Oil Immersion, ASTM D 471, No. 3 Oil, 70 h at 150°C	Compression Set, ASTM D 395, Method B, Solid max, %, 22 h at 150°C	Available Suffix Grade Numbers
	MPa	(psi)					
40	6	870	225			80	2
50	7	1015	225			80	2
60	8	1160	175	Change in durometer hardness, ±15 points		80	2
70	6	870	100	Change in tensile strength, ±30%	Volume change, +60% max	90	5
70	8	1160	150	Change in ultimate elongation, −50% max		80	2
80	6	870	100			90	5
80	8	1160	150			80	3
90	7	1015	125			85	4

Suffix Requirements		Grade 1	Grade 2	Grade 3	Grade 4	Grade 5	Grade 6	Grade 7	Grade 8
A26	Heat resistance, ASTM D 865, 70 h at 150°C:								
	Change in hardness, max, points		+10	+10	+10	+10			
	Change in tensile strength, max, %		−25	−25	−25	−25			
	Change in ultimate elongation, max, %		−30	−30	−30	−30			
B16	Compression set, ASTM D 395, Method B, 22 h at 150°C, max, %		50	60	75	80			
B36	Compression set, ASTM D 395, Method B, 22 h at 150°C, max, %		75	80	85				
C12	Resistance to ozone, ASTM D 1171	Basic Requirements Only	a	a	a	a			
C20	Resistance to weather aging, ASTM D 1171		a	a	a	a			
EO16	Fluid resistance, ASTM D 471, No. 1 Oil, 70 h at 150°C:								
	Change in hardness, points		−8 to +15	−8 to +10	−8 to +10	−8 to +10			
	Change in tensile strength, max, %		−20	−20	−20	−30			
	Change in ultimate elongation, max, %		−30	−30	−30	−50			
	Change in volume, %		−5 to +10	−5 to +10	−5 to +10	−5 to +10			
EO36	Fluid resistance, ASTM D 471, No. 3 Oil, 70 h at 150°C:								
	Change in hardness, max, points		−30	−30	−30	−30			
	Change in tensile strength, max, %		−60	−60	−60	−60			
	Change in ultimate elongation, max, %		−40	−30	−30	−50			
	Change in volume, max, %		+50	+50	+50	+50			
F14	Low-temperature resistance, ASTM D 2137, Method A, paragraph 9.3.2, nonbrittle after 3 min at −18°C			pass	pass	pass			
F15	Low-temperature resistance, ASTM D 2137, Method A, paragraph 9.3.2, nonbrittle after 3 min at −25°C		pass						
K11	Adhesion, ASTM D 429, Method A, min, MPa		1.4	1.4	1.4	1.4			
Z	(Special Requirements) Any special requirements should be specified in detail, including test methods.								

[a] The requirement is applicable, and materials are available having these characteristics, but values have not yet been established.

DH Materials

Basic Requirements

Durometer Hardness, ±5 points	Tensile Strength, min		Ultimate Elongation, min, %	Heat Aged, ASTM D 865, 70 h at 150°C	Oil Immersion, ASTM D 471, No. 3 Oil, 70 h at 150°C	Compression Set, ASTM D 395, Method B, Solid, max, %, 22 h at 150°C	Available Suffix Grade Numbers
	MPa	(psi)					
40	7	1015	300			60	2
50	8	1160	250			60	2
60	8	1160	200			60	2
60	10	1450	200			60	2
70	6	870	100	Change in durometer hardness, ±15 points		75	5
70	8	1160	200	Change in tensile strength, ±30%	Volume change, +30% max	60	3
70	10	1450	200	Change in ultimate elongation, −50%		60	3
80	6	870	100			75	5
80	8	1160	175			60	3
80	10	1450	175			60	3
90	10	1450	100			60	4

Suffix Requirements

		Grade 1	Grade 2	Grade 3	Grade 4	Grade 5	Grade 6	Grade 7	Grade 8
A26	Heat resistance, ASTM D 865, 70 h at 150°C:								
	Change in hardness, max, points		+10	+10		+10			
	Change in tensile strength, max, %		−25	−25		−25			
	Change in ultimate elongation, max, %		−30	−30		−30			
B16	Compression set, ASTM D 395, Method B, 22 h at 150°C, max, %		30	30		60			
B36	Compression set, ASTM D 395, Method B, 22 h at 150°C, max, %	Basic Requirements Only	50	50					
C12	Resistance to ozone, ASTM D 1171		a	a		a			
C20	Resistance to outdoor aging, ASTM D 1171		a	a		a			
EO16	Fluid resistance, ASTM D 471, No. 1 Oil, 70 h at 150°C:								
	Change in hardness, points		−5 to +10	−5 to +10		−5 to +10			
	Change in tensile strength, max, %		−20	−20		−20			
	Change in ultimate elongation, max, %		−30	−30		−40			
	Change in volume, %		±5	±5		±5			
EO36	Fluid resistance, ASTM D 471, No. 3 Oil, 70 h at 150°C:								
	Change in hardness, max, points		−15	−15		−15			
	Change in tensile strength, max, %		−40	−30		−40			
	Change in ultimate elongation, max, %		−40	−30		−40			
	Change in volume, max, %		+25	+25		+25			
F13	Low-temperature resistance, ASTM D 2137, Method A, paragraph 9.3.2, nonbrittle after 3 min at −10°C			pass		pass			
F14	Low-temperature resistance, ASTM D 2137, Method A, paragraph 9.3.2, nonbrittle after 3 min at −18°C		pass						
K11	Adhesion, ASTM D 429, Method A, min, MPa		1.4	1.4		1.4			
Z	(Special Requirements) Any special requirements should be specified in detail, including test methods.								

[a] The requirement is applicable, and materials are available having these characteristics, but values have not yet been established.

FC Materials

Basic Requirements

Durometer Hardness, ±5 points	Tensile Strength, min MPa	Tensile Strength, min (psi)	Ultimate Elongation, min, %	Heat Aged, ASTM D 573, 70 h at 200°C	Oil Immersion, ASTM D 471, No. 3 Oil, 70 h at 150°C	Compression Set, ASTM D 395, Method B, Plied, max, %, 22 h at 175°C	Available Suffix Grade Numbers
30	3	435	350			60	2
30	5	725	400			60	2
40	7	1015	400	Change in durometer hardness, ±15 points		60	3
50	7	1015	400	Change in tensile strength, ±30%	Volume change, +120% max	60	3
50	8	1160	500	Change in ultimate elongation, −50% max		80	4
60	7	1015	300			60	3
60	8	1160	400			80	4
70	7	1015	200			60	3

Suffix Requirements

Suffix Requirements		Grade 1	Grade 2	Grade 3	Grade 4	Grade 5	Grade 6	Grade 7	Grade 8
A19	Heat resistance, ASTM D 573, 70 h at 225°C:								
	Change in hardness, max, points		+10	+10	+15				
	Change in tensile strength, max, %		−40	−40	−50				
	Change in ultimate elongation, max, %		−40	−40	−50				
B37	Compression set, ASTM D 395, Method B, 22 h at 175°C, max, %		40	45	60				
C12	Resistance to ozone, ASTM D 1171	Basic Requirements Only	a	a	a				
C20	Resistance to outdoor aging, ASTM D 1171		a	a	a				
EA14	Fluid resistance, ASTM D 471, water, 70 h at 100°C:								
	Change in hardness, points		±5	±5	±5				
	Change in volume, %		±5	±5	±5				
EO16	Fluid resistance, ASTM D 471, No. 1 Oil, 70 h at 150°C:								
	Change in hardness, points		0 to −10	0 to −15	0 to −15				
	Change in tensile strength, max, %		−50	−50	−50				
	Change in ultimate elongation, max, %		−30	−50	−50				
	Change in volume, %		0 to +20	0 to +20	0 to +20				
F1-11	Low-temperature resistance, ASTM D 2137, Method A, paragraph 9.3.2, nonbrittle after 3 min at −75°C		pass	pass	pass				
G11	Tear resistance, ASTM D 624, Die B:								
	Under 7.0 MPa tensile strength, min, kN/m		5						
	7.0–10 MPa tensile strength, min, kN/m			17	26				
Z	(Special Requirements) Any special requirements should be specified in detail, including test methods.								

[a] The requirement is applicable, and materials are available having these characteristics, but values have not yet been established.

FE Materials

Basic Requirements

Durometer Hardness, ±5 points	Tensile Strength, min MPa	Tensile Strength, min (psi)	Ultimate Elongation, min, %	Heat Aged, ASTM D 573, 70 h at 200°C	Oil Immersion, ASTM D 471, No. 3 Oil, 70 h at 150°C	Compression Set, ASTM D 395, Method B, Solid, max, %, 22 h at 175°C	Available Suffix Grade Numbers
30	3	435	400	Change in durometer hardness, ±15 points	Volume change, +80% max	60	2
30	7	1015	500	Change in tensile strength, ±30%		60	5
40	8	1160	500	Change in ultimate elongation, −50% max		60	3
50	8	1160	500			80	4

	Suffix Requirements	Grade 1	Grade 2	Grade 3	Grade 4	Grade 5	Grade 6	Grade 7	Grade 8
A19	Heat resistance, ASTM D 573, 70 h at 225°C:	Basic Requirements Only							
	Change in hardness, max, points		+10	+10	+15	+10			
	Change in tensile strength, max, %		−60	−40	−40	−50			
	Change in ultimate elongation, max, %		−60	−60	−60	−50			
B37	Compression set, ASTM D 395, Method B, 22 h at 175°C, max, %		45	50	65	35			
C12	Resistance to ozone, ASTM D 1171		a	a	a				
C20	Resistance to outdoor aging, ASTM D 1171		a	a	a				
EA14	Fluid resistance, ASTM D 471, water, 70 h at 100°C:								
	Change in hardness, points		±5	±5	±5	±5			
	Change in volume, %		±5	±5	±5	±5			
EO16	Fluid resistance, ASTM D 471, No. 1 Oil, 70 h at 150°C:								
	Change in hardness, points		0 to −10	0 to −10	0 to −10	0 to −10			
	Change in tensile strength, max, %		−50	−50	−50	−40			
	Change in ultimate elongation, max, %		−50	−50	−50	−40			
	Change in volume, %		0 to +20	0 to +20	0 to +20	0 to +20			
EO36	Fluid resistance, ASTM D 471, No. 3 Oil, 70 h at 150°C:								
	Change in hardness, points			a	−40				
	Change in volume, max, %		+80	+80	+65				
F19	Low-temperature resistance, ASTM D 2137, Method A, paragraph 9.3.2, nonbrittle after 3 min at −55°C		pass	pass	pass				
G11	Tear resistance, ASTM D 624, Die B:								
	Under 7.0 MPa tensile strength, min, kN/m		9						
	7.0–10 MPa, tensile strength, min, kN/m			22	26	25			
K11 & K21	Adhesion, ASTM D 429, Methods A and B		a	a	a				
K31	Adhesion, bond made after vulcanization		b	b	b	b			
P2	Staining resistance, ASTM D 925, Method B, Control Panel, Nonstaining		pass	pass	pass				
Z	(Special Requirements) Any special requirements should be specified in detail, including test methods.								

[a] The requirement is applicable, and materials are available having these characteristics, but values have not yet been established.

[b] Materials must be free from surface conditions and compound constituents that are or may become deleterious to adhesion.

FK Materials

Basic Requirements

Durometer Hardness, ±5 points	Tensile Strength, min		Ultimate Elongation, min, %	Heat Aged, ASTM D 573, 70 h at 200°C	Oil Immersion, ASTM D 471, No. 3 Oil, 70 h at 150°C	Compression Set, ASTM D 395, Method B, Plied, max, %, 22 h at 175°C	Available Suffix Grade Numbers
	MPa	(psi)					
60	6	870	150	Change in durometer hardness, ±15 points Change in tensile strength, ±30% Change in ultimate elongation, −50% max	Volume change, +10% max	50	2

Suffix Requirements		Grade 1	Grade 2	Grade 3	Grade 4	Grade 5	Grade 6	Grade 7	Grade 8
A19	Heat resistance, ASTM D 573, 70 h at 225°C: Change in hardness, max, points Change in tensile strength, max, % Change in ultimate elongation, max, %	Basic Requirements Only	+15 −45 −45						
C12	Resistance to ozone, ASTM D 1171		a						
C20	Resistance to outdoor aging, ASTM D 1171		a						
EF31	Fluid resistance, ASTM D 471, Reference Fuel C, 70 h at room temperature: Change in hardness, points Change in tensile strength, max, % Change in ultimate elongation, max, % Change in volume, %		0 to −15 −60 −50 0 to +25						
EO36	Fluid resistance, ASTM D 471, No. 3 Oil, 70 h at 150°C: Change in hardness, points Change in tensile strength, max, % Change in ultimate elongation, max, % Change in volume, %		0 to −10 −35 −30 0 to +10						
F19	Low-temperature resistance, ASTM D 2137, Method A, paragraph 9.3.2, nonbrittle after 3 min at −55°C		pass						
Z	(Special Requirements) Any special requirements should be specified in detail, including test methods.								

[a] The requirement is applicable, and materials are available having these characteristics but values have not yet been established.

GE Materials

Basic Requirements

Durometer Hardness, ±5 points	Tensile Strength, min MPa	Tensile Strength, min (psi)	Ultimate Elongation, min, %	Heat Aged, ASTM D 573, 70 h at 225°C	Oil Immersion, ASTM D 471, No. 3 Oil, 70 h at 150°C	Compression Set, ASTM D 395, Method B, Plied, max, %, 22 h at 175°C	Available Suffix Grade Numbers
30	3	435	300			50	2
30	5	725	400			50	2
30	6	870	400			50	8
40	3	435	200			50	2
40	5	725	300			50	2
40	6	870	300			50	8
50	3	435	200	Change in durometer hardness, ±15 points		50	3
50	5	725	250	Change in tensile strength, ±30%	Volume change, +80% max	70	4, 5
50	6	870	250	Change in ultimate elongation, −50% max		50	5
50	8	1160	400			60	9
60	3	435	100			50	3
60	5	725	200			70	4, 5
60	6	870	200			50	5
70	3	435	60			50	6
70	5	725	150			50	7
70	6	870	150			50	5
80	3	435	50			50	6
80	5	725	100			50	7
80	6	870	100			50	5

	Suffix Requirements	Grade 1	Grade 2	Grade 3	Grade 4	Grade 5	Grade 6	Grade 7	Grade 8	Grade 9
A19	Heat resistance, ASTM D 573, 70 h at 225°C:									
	Change in hardness, max, points		+10	+10	+10	+10	+10	+10	+10	+10
	Change in tensile strength, %		−25	−25	−30	−25	−25	−25	−25	−30
	Change in ultimate elongation, max, %		−30	−30	−30	−30	−30	−30	−25	−30
B37	Compression set, ASTM D 395, Method B, 22 h at 175°C, max, %	Basic Requirements Only	25	30	50	25	30	30	25	40
C12	Resistance to ozone, ASTM D 1171		a	a	a	a	a	a	a	a
C20	Resistance to outdoor aging, ASTM D 1171		a	a	a	a	a	a	a	a
EA14	Fluid resistance, ASTM D 471, water, 70 h at 100°C:									
	Change in hardness, max, points		±5	±5	±5	±5	±5	±5	±5	±5
	Change in volume, %		±5	±5	±5	±5	±5	±5	±5	±5
EO16	Fluid resistance, ASTM D 471, No. 1 Oil, 70 h at 150°C:									
	Change in hardness, points		0 to −10	0 to −15	0 to −15	0 to −15	0 to −15	0 to −15	0 to −10	0 to −10
	Change in tensile strength, max, %		−30	−20	−20	−20	−20	−20	−30	−30
	Change in ultimate elongation, max, %		−30	−20	−20	−20	−20	−20	−20	−30
	Change in volume, %		0 to +15	0 to +10	0 to +15	0 to +10	0 to +10	0 to +15	0 to +15	0 to +10
EO36	Fluid resistance, ASTM D 471, No. 3 Oil, 70 h at 150°C:									
	Change in hardness points, max		−30	−35	−30	−40	−40		a	−30
	Change in volume, max, %		+60	+60	+60	+60	+60	+60	+60	+60
F19	Low-temperature resistance, ASTM D 2137, Method A, paragraph 9.3.2, nonbrittle after 3 min at −55°C		pass	pass	pass	pass	pass	pass	pass	pass
G11	Tear resistance, ASTM D 624, Die B:									
	Under 7.0 MPa tensile strength, min, kN/m		5	6	9	9	5	9	9	
	7.0–10 MPa tensile strength, min, kN/m									25
K11 & K21	Adhesion, ASTM D 429, Methods A and B		a	a	a	a	a	a	a	a
K31	Adhesion, bond made after vulcanization		b	b	b	b	b	b	b	b
P2	Staining resistance, ASTM D 925, Method B, Control Panel, Nonstaining		pass	pass	pass	pass	pass	pass	pass	pass
Z	(Special Requirements) Any special requirements should be specified in detail, including test methods.									

[a] The requirement is applicable, and materials are available having these characteristics, but values have not yet been established.

[b] Materials must be free from surface conditions and compound constituents that are or may become deleterious to adhesion.

HK Materials

Basic Requirements

Durometer Hardness, ±5 points	Tensile Strength, min MPa	Tensile Strength, min (psi)	Ultimate Elongation, min, %	Heat Aged, ASTM D 573, 70 h at 250°C	Oil Immersion, ASTM D 471, No. 3 Oil, 70 h at 150°C	Compression Set, ASTM D 395, Method B, Plied, max, %, 22 h at 175°C	Available Suffix Grade Numbers
60	7	1015	200				2, 4, 6
60	10	1450	200				2, 4, 6
60	14	2031	200				2, 4, 6
70	7	1015	175	Change in durometer hardness, ±15 points			2, 4, 6
70	10	1450	175	Change in tensile strength, ±30%	Volume change, +10% max	Compression set, 35% max	2, 4, 6
70	14	2031	175	Change in ultimate elongation, −50% max			2, 4, 6
80	7	1015	150				2, 4, 6
80	10	1450	150				2, 4, 6
80	14	2031	150				2, 4, 6
90	7	1015	100				3, 5, 7
90	10	1450	100				3, 5, 7
90	14	2031	100				3, 5, 7

Suffix Requirements

	Suffix Requirements	Grade 1	Grade 2	Grade 3	Grade 4	Grade 5	Grade 6	Grade 7	Grade 8
A1-10	Heat resistance, ASTM D 573, 70 h at 250°C:								
	Change in hardness, max, points		+10	+10			+10	+10	
	Change in tensile strength, max, %		−25	−25			−25	−25	
	Change in ultimate elongation, max, %		−25	−25			−25	−25	
A1-11	Heat resistance, ASTM D 573, 70 h at 275°C:								
	Change in hardness, max, points				+10	+10	−5 to +10	−5 to +10	
	Change in tensile strength, max, %				−40	−40	−40	−40	
	Change in ultimate elongation, max, %				−20	−20	−20	−20	
B31	Compression set, ASTM D 395, Method B, 22 h at 23°C, max, %						15	20	
B37	Compression set, ASTM D 395, Method B, 22 h at 175°C, max, %		50	30					
B38	Compression set, ASTM D 395, Method B, 22 h at 200°C, max, %	Basic Requirements Only	50	50	50	50	15	20	
C12	Resistance to ozone, ASTM D 1171		pass	pass	pass	pass	pass	pass	
C20	Resistance to outdoor aging, ASTM D 1171		pass	pass	pass	pass	pass	pass	
EF31	Fluid resistance, ASTM D 471, Reference Fuel C, 70 h at room temperature:								
	Change in hardness, points		±5	±5	±5	±5	±5	±5	
	Change in tensile strength, max, %		−25	−25	−25	−25	−25	−25	
	Change in ultimate elongation, max, %		−20	−20	−20	−20	−20	−20	
	Change in volume, %		0 to +10	0 to +10	0 to +10	0 to +10	0 to +10	0 to +10	
EO78	Fluid resistance, ASTM D 471, Service Fluid No. 101, 70 h at 200°C:								
	Change in hardness, points		−15 to +5	−15 to +5	−15 to +5	−15 to +5			
	Change in tensile strength, max, %		−40	−40	−40	−40			
	Change in ultimate elongation, max, %		−20	−20	−20	−20			
	Change in volume, %		0 to +15	0 to +15	0 to +15	0 to +15			
EO88	Fluid resistance, ASTM D 471, SAE Fluid 2, Stauffer 7700, 70 h at 200°C:								
	Change in hardness, points						−15 to +5	−15 to +5	
	Change in tensile strength, max, %						−40	−40	
	Change in ultimate elongation, max, %						−20	−20	
	Change in volume, max, %						+25	+25	
F15	Low-temperature resistance, ASTM D 2137, Method A, paragraph 9.3.2, nonbrittle after 3 min at −25°C		pass			pass	pass		
F17	Low-temperature resistance, ASTM D 2137, Method A, paragraph 9.3.2, nonbrittle after 3 min at −40°C					pass			
Z	(Special Requirements) Any special requirements should be specified in detail, including test methods.								

APPENDIX

This Appendix is intended to assist in the conversion from earlier systems to the present classification system. It is not to be considered as part of the system.

TABLE A-1—EQUIVALENTS FOR CONVERSION

SAE J14—ASTM D 735 Material Designation (Type and Class)	SAE J200/ANSI/ ASTM D 2000 Material Designation (Type and Class)	Type of Polymer Most Often Used[a]
R	AA	NR, SBR, IIR, BIIR, CIIR, EPM, EPDM, BR, IR, Reclaim Rubber
SA	AK	T
R	BA	SBR, IIR, BIIR, CIIR, EPM, EPDM
SC	BC	CR
—	BE	CR
—	BF	NBR
SB	BG	NBR, AU, EU
SA	BK	T, NBR
—	CA	EPM, EPDM
—	CE	CSM
—	CH	NBR, CO, ECO
—	DA	EPM, EPDM
—	DF	ACM
TB	DH	ACM
TA	FC	PVMQ
—	FE	MQ
—	FK	FVMQ
TA	GE	VMQ
—	HK	FKM

[a] ASTM D 1418—Standard Recommended Practice for Rubber and Rubber Latices—Nomenclature.

Tables 1-6 from SAE J14 and ASTM D 735 are extracted to aid in the interpretation of existing specifications where SAE J14 or ASTM D 735 was the source document. These tables are not intended, nor are they now suitable, for the preparation of new specifications for rubber materials for automotive applications.

TABLE 1—PHYSICAL REQUIREMENTS OF TYPE R COMPOUNDS, NONOIL RESISTANT,[a] SAE 10R1 (FROM SAE J14 AND ASTM D 735)

	Basic Requirements							Requirements Added by Suffix Letter[b]		
								D	B	R
Grade No.	Durometer Hardness No.	Min Tensile Strength, psi	Min Ultimate Elongation, %	Heat Aged 70 h at 158°F (70°C)			Max Compression Set after 22 h at 158°F (70°C), %	Load at 20% Deformation, psi	Max Compression Set after 22 h at 158°F (70°C), %	Min Yerzley Resilience at 20% Deformation, %
				Max Change in Tensile Strength, %	Max Change in Ultimate Elongation, %	Max Change in Durometer Hardness				
R310	30 ± 5	1000	400	−25	−35	+10	50	—	25	—
R315	30 ± 5	1500	500	−25	−25	+10	50	70 ± 10	25	—
R320	30 ± 5	2000	600	−25	−25	+10	50	70 ± 10	25	—
R325	30 ± 5	2500	600	−25	−25	+10	50	70 ± 10	35	—
R410	40 ± 5	1000	400	−25	−35	+10	50	—	25	—
R415[c]	40 ± 5	1500	500	−25	−25	+7	50	100 ± 15	25	70
R420[c]	40 ± 5	2000	500	−25	−25	+7	50	100 ± 15	25	75
R425	40 ± 5	2500	500	−25	−25	+7	50	100 ± 15	25	80
R430	40 ± 5	3000	600	−25	−25	+7	50	100 ± 15	35	80
R505	50 ± 5	500	300	−25	−35	+10	50	—	—	—
R508	50 ± 5	800	350	−25	−35	+10	50	—	—	—
R510[c]	50 ± 5	1000	400	−25	−35	+10	50	—	25	—
R512	50 ± 5	1200	400	−25	−35	+10	50	—	25	—
R515[c]	50 ± 5	1500	400	−25	−25	+7	50	140 ± 20	25	65
R520[c]	50 ± 5	2000	500	−25	−25	+7	50	140 ± 20	25	65
R525[c]	50 ± 5	2500	500	−25	−25	+7	50	140 ± 20	25	75
R530	50 ± 5	3000	600	−25	−25	+7	50	140 ± 20	35	75
R535	50 ± 5	3500	600	−25	−25	+7	50	140 ± 20	35	75
R605	60 ± 5	500	300	−25	−35	+10	50	—	—	—
R608	60 ± 5	800	300	−25	−35	+10	50	—	—	—
R610[c]	60 ± 5	1000	300	−25	−35	+10	50	—	25	—
R612	60 ± 5	1200	300	−25	−35	+10	50	—	25	—
R615[c]	60 ± 5	1500	350	−25	−25	+7	50	195 ± 30	25	60
R620[c]	60 ± 5	2000	400	−25	−25	+7	50	195 ± 30	25	60
R625[c]	60 ± 5	2500	450	−25	−25	+7	50	195 ± 30	25	70
R630	60 ± 5	3000	500	−25	−25	+7	50	195 ± 30	35	70
R635	60 ± 5	3500	550	−25	−25	+7	50	195 ± 30	35	70
R705	70 ± 5	500	150	−25	−35	+10	50	—	—	—
R708	70 ± 5	800	150	−25	−35	+10	50	—	—	—
R710[c]	70 ± 5	1000	200	−25	−35	+10	50	—	25	—
R712	70 ± 5	1200	200	−25	−35	+10	50	—	25	—
R715[c]	70 ± 5	1500	250	−25	−25	+7	50	300 ± 70	25	50
R720[c]	70 ± 5	2000	300	−25	−25	+7	50	300 ± 70	25	50
R725	70 ± 5	2500	300	−25	−25	+7	50	300 ± 70	25	60
R730	70 ± 5	3000	400	−25	−25	+7	50	300 ± 70	35	60
R805	80 ± 5	500	100	−25	−35	+10	50	—	—	—
R810	80 ± 5	1000	100	−25	−35	+10	50	—	—	—
R815	80 ± 5	1500	150	−25	−25	+7	50	475 ± 100	—	—
R820	80 ± 5	2000	200	−25	−25	+7	50	475 ± 100	—	—
R825	80 ± 5	2500	200	−25	−25	+7	50	475 ± 100	—	—
R905	90 ± 5	500	75	−25	−35	+10	50	—	—	—
R910	90 ± 5	1000	100	−25	−35	+10	50	—	—	—
R915	90 ± 5	1500	125	−25	−25	+7	50	—	—	—

[a] See section, Methods of Testing.
[b] See section on Types, Classes, and Grades of Compounds.
[c] In order to keep to a minimum the number of grades required, it is suggested that, wherever practical, these grades be used.

TABLE 2—PHYSICAL REQUIREMENTS OF TYPE S COMPOUNDS, CLASS SA, OIL RESISTANT,[a] SAE 10R2 (FROM SAE J14 AND ASTM D 735)

	Basic Requirements							Requirements Added by Suffix Letter[b]						
				Oil Aged 70 h at 212°F (100°C)	Heat Aged 70 h at 212°F (100°C)			Max Compression Set after 22 h at 158°F (70°C), %	E1				E3		
									70 h at 212°F (100°C), ASTM Oil No. 1				70 h at 212°F (100°C), ASTM Oil No. 3		
Grade No.	Durometer Hardness No.	Min Tensile Strength, psi	Min Ultimate Elongation, %	ASTM Oil No. 3 Volume Change, % (Limits)	Max Change in Tensile Strength, %	Max Change in Ultimate Elongation, %	Max Change in Durometer Hardness		Max Change in Tensile Strength, %	Max Change in Ultimate Elongation, %	Change in Durometer Hardness (Limits)	Volume Change, % (Limits)	Max Change in Tensile Strength, %	Max Change in Ultimate Elongation, %	Change in Durometer Hardness (Limits)
SA405	40 ± 5	500	400	0 to +10	−15	−40	+15	65	−20	−50	−5 to +10	−3 to +5	−20	−50	−5 to +10
SA505	50 ± 5	500	400	0 to +10	−15	−40	+15	60	−20	−50	−5 to +10	−3 to +5	−20	−50	−5 to +10
SA607	60 ± 5	700	300	0 to +10	−15	−40	+10	50	−30	−50	−5 to +10	−3 to +5	−30	−50	−5 to +10
SA710	70 ± 5	1000	250	0 to +10	−15	−40	+10	50	−30	−50	−5 to +10	−3 to +5	−30	−50	−5 to +10
SA810	80 ± 5	1000	150	0 to +10	−15	−40	+10	50	−30	−50	−5 to +10	−3 to +5	−30	−50	−5 to +10
SA910	90 ± 5	1000	100	0 to +10	−15	−40	+10	50	−30	−50	−5 to +10	−3 to +5	−30	−50	−5 to +10

[a] See section, Methods of Testing.
[b] See section, Types, Classes, and Grades of Compounds.

TABLE 3—PHYSICAL REQUIREMENTS OF TYPE S COMPOUNDS, CLASS SB, OIL RESISTANT,[a] SAE 10R3 (FROM SAE J14 AND ASTM D 735)

	Basic Requirements							Requirements Added by Suffix Letter[b]							
								B	E1				E3			
				Oil Immersion 70 h at 212°F (100°C)	Heat Aged 70 h at 21°F (100°C)			Max Compression Set after 22 h at 212°F (100°C), %	Max Compression Set after 22 h at 212°F (100°C), %	70 h at 212°F (100°C) in ASTM Oil No. 1				70 h at 212°F (100°C) in ASTM Oil No. 3			
Grade No.	Durometer Hardness No.	Min Tensile Strength, psi	Min Ultimate Elongation, %	ASTM Oil No. 3 Volume Change, % (Limits)	Max Change in Tensile Strength, M	Max Change in Ultimate Elongation, %	Max Change in Durometer Hardness			Max Change in Tensile Strength, %	Max Change in Ultimate Elongation, %	Volume Change, % (Limits)	Change in Durometer Hardness (Limits)	Max Change in Tensile Strength, %	Max Change in Ultimate Elongation, %	Volume Change, % (Limits)	Change in Durometer Hardness (Limits)
SB410	40 ± 5	1000	450	−5 to +40	−25	−50	+20	60	30	−25	−50	−15 to +5	−5 to +15	−40	−50	0 to +30	−15 to +5
SB415	40 ± 5	1500	450	0 to +40	−25	−50	+20	60	30	−30	−50	−10 to +5	−5 to +10	−45	−50	0 to +35	−15 to +5
SB505	50 ± 5	500	300	0 to +40	−25	−50	+15	55	25	−25	−45	−10 to +5	−5 to +10	−45	−50	0 to +25	−15 to +5
SB508	50 ± 5	800	300	0 to +40	−25	−50	+15	55	25	−25	−45	−10 to +5	−5 to +10	−45	−50	0 to +25	−15 to +5
SB510	50 ± 5	1000	350	0 to +40	−25	−50	+15	55	25	−25	−45	−10 to +5	−5 to +10	−45	−50	0 to +25	−15 to +5
SB512	50 ± 5	1200	350	0 to +40	−25	−50	+15	55	25	−25	−45	−10 to +5	−5 to +10	−45	−50	0 to +25	−15 to +5
SB515[c]	50 ± 5	1500	400	0 to +60	−25	−50	+15	55	25	−25	−45	−10 to +5	−5 to +10	−45	−50	0 to +30	−15 to +5
SB520	50 ± 5	2000	400	0 to +60	−25	−50	+15	55	25	−30	−45	−10 to +5	−5 to +10	−50	−55	0 to +30	−15 to +5
SB605	60 ± 5	500	200	0 to +40	−20	−50	+15	55	25	−20	−40	−10 to +5	−5 to +10	−35	−40	0 to +25	−10 to +5
SB608	60 ± 5	800	200	0 to +40	−20	−50	+15	55	25	−20	−40	−10 to +5	−5 to +10	−35	−40	0 to +25	−10 to +5
SB610[c]	60 ± 5	1000	250	0 to +40	−20	−50	+15	55	25	−20	−40	−10 to +5	−5 to +10	−35	−40	0 to +25	−10 to +5
SB612	60 ± 5	1200	250	0 to +40	−20	−50	+15	55	25	−20	−40	−10 to +5	−5 to +10	−35	−40	0 to +25	−10 to +5
SB615[c]	60 ± 5	1500	300	0 to +50	−20	−50	+15	55	25	−20	−40	−10 to +5	−5 to +10	−40	−45	0 to +25	−10 to +5
SB620	60 ± 5	2000	350	0 to +50	−20	−50	+15	55	25	−20	−40	−10 to +5	−5 to +10	−40	−45	0 to +30	−10 to +5
SB625	60 ± 5	2500	350	0 to +50	−20	−50	+15	55	25	−20	−40	−10 to +5	−5 to +10	−45	−50	0 to +30	−10 to +5
SB705	70 ± 5	500	150	0 to +35	−20	−50	+15	55	25	−20	−40	−10 to +5	−5 to +10	−35	−35	0 to +25	−10 to +5
SB708	70 ± 5	800	150	0 to +35	−20	−50	+15	55	25	−20	−40	−10 to +5	−5 to +10	−35	−35	0 to +25	−10 to +5
SB710	70 ± 5	1000	200	0 to +35	−20	−50	+15	55	25	−20	−40	−10 to +5	−5 to +10	−35	−35	0 to +25	−10 to +5
SB712	70 ± 5	1200	200	0 to +35	−20	−50	+15	55	25	−20	−40	−10 to +5	−5 to +10	−35	−35	0 to +25	−10 to +5
SB715[c]	70 ± 5	1500	250	0 to +40	−20	−50	+15	55	25	−20	−40	−10 to +5	−5 to +10	−35	−35	0 to +25	−10 to +5
SB720	70 ± 5	2000	250	0 to +40	−20	−50	+15	55	25	−20	−40	−10 to +5	−5 to +10	−40	−40	0 to +30	−10 to +5
SB725	70 ± 5	2500	300	0 to +40	−20	−50	+15	55	25	−20	−40	−10 to +5	−5 to +10	−40	−40	0 to +30	−10 to +5
SB805	80 ± 5	500	100	0 to +30	−20	−50	+15	55	25	−20	−40	−10 to +5	−5 to +10	−35	−35	0 to +20	−10 to +5
SB810	80 ± 5	1000	100	0 to +30	−20	−50	+15	55	25	−20	−40	−10 to +5	−5 to +10	−35	−35	0 to +20	−10 to +5
SB815	80 ± 5	1500	125	0 to +30	−20	50	+15	55	25	−20	−40	−10 to +5	−5 to +10	−35	−35	0 to +20	−10 to +5
SB820	80 ± 5	2000	125	0 to +30	−20	−50	+15	55	25	−20	−40	−10 to +5	−5 to +10	−35	−35	0 to +20	−10 to +5
SB905	90 ± 5	500	50	0 to +25	−20	−50	+10	55	25	−20	−40	−10 to +5	−5 to +10	−35	−35	0 to +20	−10 to +5
SB910	90 ± 5	1000	100	0 to +25	−20	−50	+10	55	25	−20	−40	−10 to +5	−5 to +10	−35	−35	0 to +20	−10 to +5
SB915	90 ± 5	1500	100	0 to +25	−20	−50	+10	55	25	−20	−40	−10 to +5	−5 to +10	−35	−35	0 to +20	−10 to +5

Grades meeting either or both Suffix E1 and E3 requirements and meeting the requirements of Suffix F2 may NOT be commercially available.

[a] See section, Methods of Testing.
[b] See section, Types, Classes, and Grades of Compounds.
[c] In order to keep to a minimum the number of grades required, it is suggested that, wherever practical, these grades be used.

TABLE 4—PHYSICAL REQUIREMENTS OF TYPE S COMPOUNDS, CLASS SC, OIL RESISTANT,[a] SAE 10R4 (FROM SAE J14 AND ASTM D 735)

Grade No.	Durometer Hardness No.	Min Tensile Strength, psi	Min Ultimate Elongation, %	Oil Immersion 70 h at 212°F (100°C) ASTM Oil No. 3 Volume Change, % (Limits)	Heat Aged 70 h at 212°F (100°C) Max Change in Tensile Strength, %	Heat Aged 70 h at 212°F (100°C) Max Change in Ultimate Elongation, %	Heat Aged 70 h at 212°F (100°C) Max Change in Durometer Hardness	Max Compression Set after 22 h at 212°F (100°C), %	B Max Compression Set after 22 h at 212°F (100°C), %	E1 70 h at 212°F (100°C) in ASTM Oil No.1 Max Change in Tensile Strength, %	E1 Max Change in Ultimate Elongation, %	E1 Change in Durometer Hardness (Limits)	E1 Volume Change, % (Limits)	E3 70 h at 212°F (100°C) in ASTM Oil No.3 Max Change in Tensile Strength, %	E3 Max Change in Ultimate Elongation, %	R Min Yerzley Resilience at 20% Deformation, %
SC305	30 ± 5	500	300	+50 to +120	−15	−40	+20	80	35	−20	−30	±10	−10 to +15	−80	−60	—
SC310	30 ± 5	1000	400	+50 to +120	−15	−40	+20	80	35	−20	−30	±10	−10 to +15	−80	−60	—
SC315	30 ± 5	1500	500	+50 to +120	−15	−40	+20	80	35	−30	−30	±10	−10 to +15	−80	−60	70
SC320	30 ± 5	2000	500	+50 to +140	−15	−40	+20	80	35	−30	−30	±10	−10 to +15	−80	−60	70
SC405	40 ± 5	500	300	+50 to +120	−15	−40	+20	80	35	−20	−30	±10	−10 to +15	−75	−55	—
SC410	40 ± 5	1000	400	+50 to +120	−15	−40	+20	80	35	−20	−30	±10	−10 to +15	−75	−55	60
SC415[c]	40 ± 5	1500	500	+50 to +120	−15	−40	+15	80	35	−20	−30	±10	−10 to +15	−75	−55	65
SC420[c]	40 ± 5	2000	500	+50 to +140	−15	−40	+15	80	35	−30	−30	±10	−10 to +15	−80	−55	70
SC425	40 ± 5	2500	500	+50 to +140	−15	−40	+15	80	35	−30	−30	±10	−10 to +15	−80	−55	70
SC505	50 ± 5	500	300	+40 to +100	−15	−40	+20	80	35	−30	−30	±10	−10 to +15	−70	−55	—
SC508	50 ± 5	800	300	+40 to +100	−15	−40	+20	80	35	−30	−30	±10	−10 to +15	−70	−55	—
SC510	50 ± 5	1000	300	+40 to +100	−15	−40	+20	80	35	−30	−30	±10	−10 to +15	−70	−55	—
SC512	50 ± 5	1200	300	+40 to +100	−15	−40	+20	80	35	−30	−30	±10	−10 to +15	−70	−55	—
SC515[c]	50 ± 5	1500	350	+40 to +110	−15	−40	+15	80	35	−30	−30	±10	−10 to +15	−70	−55	60
SC520[c]	50 ± 5	2000	400	+40 to +120	−15	−40	+15	80	35	−30	−30	±10	−10 to +15	−70	−55	65
SC525	50 ± 5	2500	450	+40 to +130	−15	−40	+15	80	35	−40	−30	±10	−10 to +15	−70	−55	70
SC530	50 ± 5	3000	500	+40 to +140	−15	−40	+15	80	35	−40	−30	±10	−10 to +15	−80	−55	70
SC535	50 ± 5	3500	500	+40 to +150	−15	−40	+15	80	35	−40	−30	±10	−10 to +15	−80	−55	70
SC605	60 ± 5	500	300	+40 to +100	−15	−40	+20	80	35	−30	−30	±10	−10 to +15	−65	−55	—
SC608	60 ± 5	800	300	+40 to +100	−15	−40	+20	80	35	−30	−30	±10	−10 to +15	−65	−55	—
SC610[c]	60 ± 5	1000	300	+40 to +100	−15	−40	+15	80	35	−30	−30	±10	−10 to +15	−65	−55	—
SC612	60 ± 5	1200	300	+40 to +100	−15	−40	+15	80	35	−30	−30	±10	−10 to +15	−65	−55	—
SC615[c]	60 ± 5	1500	350	+40 to +110	−15	−40	+15	80	35	−30	−30	±10	−5 to +10	−65	−55	55
SC620[c]	60 ± 5	2000	350	+40 to +110	−15	−40	+15	80	35	−30	−30	±10	−5 to +10	−65	−55	60
SC625	60 ± 5	2500	400	+40 to +120	−15	−40	+15	80	35	−30	−30	±10	−5 to +10	−70	−55	65
SC630	60 ± 5	3000	400	+40 to +140	−15	−40	+15	80	35	−40	−30	±10	−5 to +10	−80	−55	65
SC635	60 ± 5	3500	400	+40 to +150	−15	−40	+15	80	35	−40	−30	±10	−5 to +10	−80	−55	65
SC705	70 ± 5	500	200	+30 to +100	−15	−40	+15	80	35	−15	−30	−5 to +10	−5 to +10	−65	−50	—
SC708	70 ± 5	800	200	+30 to +100	−15	−40	+15	80	35	−15	−30	−5 to +10	−5 to +10	−65	−50	—
SC710[c]	70 ± 5	1000	200	+30 to +100	−15	−40	+15	80	35	−15	−30	−5 to +10	−5 to +10	−65	−50	—
SC712	70 ± 5	1200	200	+30 to +100	−15	−40	+15	80	35	−15	−30	−5 to +10	−5 to +10	−65	−50	—
SC715[c]	70 ± 5	1500	250	+30 to +100	−15	−40	+15	80	35	−15	−30	−5 to +10	−5 to +10	−65	−50	50
SC720	70 ± 5	2000	300	+30 to +100	−15	−40	+15	80	35	−15	−30	−5 to +10	−5 to +10	−65	−50	55
SC725	70 ± 5	2500	300	+30 to +120	−15	−40	+15	80	35	−20	−30	−5 to +10	−5 to +10	−70	−50	60
SC805	80 ± 5	500	100	+20 to +80	−15	−40	+15	80	35	−15	−30	−5 to +10	−5 to +10	−45	−30	—
SC810	80 ± 5	1000	100	+20 to +80	−15	−40	+15	80	35	−15	−30	−5 to +10	−5 to +10	−45	−30	—
SC815	80 ± 5	1500	100	+20 to +80	−15	−40	+15	80	35	−15	−30	−5 to +10	−5 to +10	−45	−30	—
SC820	80 ± 5	2000	150	+20 to +80	−15	−40	+15	80	35	−15	−30	−5 to +10	−5 to +10	−45	−30	—
SC905	90 ± 5	500	50	+10 to +70	−15	−40	+10	80	35	−15	−30	±5	−5 to +10	−40	−20	—
SC910	90 ± 5	1000	100	+10 to +70	−15	−40	+10	80	35	−15	−30	±5	−5 to +10	−40	−20	—
SC915	90 ± 5	1500	150	+10 to +70	−15	−40	+10	80	35	−15	−30	±5	−5 to +10	−40	−20	—
SC920	90 ± 5	2000	150	+10 to +70	−15	−40	+10	80	35	−15	−30	±5	−5 to +10	−40	−20	—

[a] See section, Methods of Testing.
[b] See section, Types, Classes, and Grades of Compounds.
[c] In order to keep to a minimum the number of grades required, it is suggested that, wherever practical, these grades be used.

TABLE 5—PHYSICAL REQUIREMENTS OF TYPE T COMPOUNDS, CLASS TA, HEAT RESISTANT[a] SAE 10R5 (FROM SAE J14 AND ASTM D 735)

	Basic Requirements							Requirements Added by Suffix Letter[b]								
				Heat Aged[c] 70 h at 437°F (225°C)			B[d]	E1				E3		G[e]	L	
								70 h at 302°F (150°C), ASTM Oil No. 1				70 h at 302°F (150°C), ASTM Oil No. 3			Water Absorption after 70 h in Boiling Water	
Grade No.	Durometer Hardness No.	Min Tensile Strength, psi	Min Ultimate Elongation, %	Max Change in Durometer Hardness	Max Change in Tensile Strength, %	Max Change in Ultimate Elongation, %	Max[d] Compression Set, %	Max Change in Tensile Strength, %	Max Change in Ultimate Elongation, %	Max[d] Change in Durometer Hardness	Volume Change, % (Limits)	Max Change in Durometer Hardness	Max Volume Change, %	Min Tear Strength lb/in	Max Change in Durometer Hardness	Max Volume Change %
TA505	50 ± 5	500	200	+10	−25	−30	35	−20	−20	−15	0 to +10	−30	+60	—	±5	+5
TA507	50 ± 5	700	250	+10	−30	−30	50	−20	−20	−15	0 to +15	−30	+60	50	±5	+5
TA510	50 ± 5	1000	400	+10	−40	−40	40	−50	−50	−15	0 to +20	—	—	100	±5	±5
TA512	50 ± 5	1200	450	+15	−50	−50	60	−50	−50	−15	0 to +20	—	—	150	±5	±5
TA605	60 ± 5	500	100	+10	−25	−30	25	−20	−20	−15	0 to +10	−35	+60	—	±5	+5
TA607	60 ± 5	700	200	+10	−30	−30	40	−20	−20	−15	0 to +15	−35	+60	50	±5	+5
TA610	60 ± 5	1000	300	+10	−40	−40	40	−50	−50	−15	0 to +15	—	—	100	±5	±5
TA612	60 ± 5	1200	400	+15	−50	−50	60	−50	−50	−15	0 to +15	—	—	150	±5	±5
TA705	70 ± 5	500	75	+10	−25	−30	25	−20	−20	−15	0 to +15	−40	+60	—	±5	+5
TA707	70 ± 5	700	150	+10	−30	−30	40	−20	−20	−15	0 to +15	−40	+60	50	±5	+5
TA710	70 ± 5	1000	200	+10	−40	−40	50	−50	−50	−15	0 to +15	—	—	100	±5	±5
TA805	80 ± 5	500	50	+10	−25	−30	30	−20	−20	−15	0 to +15	−45	+60	—	±5	+5
TA807	80 ± 5	700	150	+10	−30	−40	40	−20	−20	−15	0 to +15	−45	+60	50	±5	+5

NOTE: TA510, 512, 610, 612, and 710 grades are not resistant to ASTM Oil No. 3 at elevated temperatures.
[a] See section, Methods of Testing.
[b] See section 2.
[c] Heat aging tests shall be according to ASTM D 573, except that a temperature of 437 ± 5°F (225 ± 2.8°C) shall be used. Experience has shown that most commercial laboratory ovens vary widely in temperature between different sections of the oven. It is recommended that thermometers or thermocouples be placed at several points immediately adjacent to the test specimen to indicate accurately the test temperature.
[d] For either 22 h at 347°F or 70 h at 302°F, 22 h at 347°F perferred.
[e] Die B.

TABLE 6—PHYSICAL REQUIREMENTS OF TYPE T COMPOUNDS, CLASS TB, HEAT RESISTANT,[a] SAE 10R6 (FROM SAE J14 AND ASTM D 735)

	Basic Requirements							Requirements Added by Suffix Letter[b]											
				Heat Aged 70 h at 302°F (150°C)			Max Compression Set after 22 h at 302°F (150°C), %[c]	E1				E3				E4			
								70 h at 302°F (150°C) ASTM Oil No. 1				70 h at 302°F (150°C) ASTM Oil No. 3				70 h at 302°F (150°C) Hydrocarbon Test Fluid			
Grade No.	Durometer Hardness No.	Min Tensile Strength, psi	Min Ultimate Elongation, %	Change in Durometer Hardness (Limits)	Max Change in Tensile Strength, %	Max Change in Ultimate Elongation, %		Max Change in Tensile Strength, %	Max Change in Ultimate Elongation, %	Change in Durometer Hardness (Limits)	Volume Change, % (Limits)	Max Change in Tensile Strength, %	Max Change in Ultimate Elongation, %	Max Change in Durometer Hardness	Max Volume Change, %	Max Change in Tensile Strength, %	Max Change in Ultimate Elongation, %	Change in Durometer Hardness (Limits)	Volume Change, % (Limits)
TB614	60 ± 5	1400	200	0 to +10	−30	−30	40	−20	−30	−7 to +10	−5 to +5	−30	−30	−20	+25	−20	−10	−5 to +15	−3 to +5
TB715	70 ± 5	1500	175	0 to +10	−30	−30	40	−20	−30	−7 to +10	−5 to +5	−30	−30	−20	+25	−20	−10	−5 to +15	−3 to +5
TB815	80 ± 5	1500	150	0 to +10	−30	−30	40	−20	−30	−7 to +10	−5 to +5	−30	−30	−20	+25	−20	−10	−5 to +15	−3 to +5

[a] See section, Methods of Testing.
[b] See section, Types, Classes, and Grades of Compounds.
[c] Compression set—ASTM D 395-55T, paragraph 5 (b).

RECOMMENDED GUIDELINES FOR FATIGUE TESTING OF ELASTOMERIC MATERIALS AND COMPONENTS—SAE J1183

SAE Recommended Practice

Report of Nonmetallic Materials Committee approved January 1978.

1. Scope—The purpose of this document is to review factors that influence the behavior of elastomers under conditions of dynamic stress and to provide guidance concerning laboratory procedures for defining and specifying the fatigue characteristics of elastomeric materials and fabricated elastomeric components.

2. Introduction—These guidelines describe:

2.1 The manner in which elastomeric materials and components undergo changes due to stresses or strains in a fatigue environment that ultimately culminate in failure.

2.2 Factors to be considered in selecting from available test methods or in developing a test method to meet specific requirements.

2.3 A set of definitions and terminology to allow interchange of information on a common basis.

2.4 Important considerations in the evaluation and reporting of test information.

3. Elastomeric Behavior—An elastomer is a viscoelastic material. It acts as though it is composed of an elastic component and a viscous component. The elastic component controls stress versus strain behavior. Because of an elastomer's visco-elastic nature, the dynamic response and mechanical behavior are dependent upon stress or strain history, rate of loading, frequency and amplitude of strain, and specimen temperature.

The viscous component determines internal energy loss, or hysteresis. The lost energy is converted into heat and since elastomers are poor heat conductors this can result in a considerable temperature rise.

In addition to being viscoelastic, elastomers have a lower elastic modulus and lower strength than most metals and plastics. Although softer and weaker than structural metals and plastics, elastomers are like these materials from an energy per unit volume standpoint. For elastomers, metals, and plastics, a power form of a fatigue correlation exists:

$$N_f W^b = C$$

Where N_f = cycles to failure
W = energy input ($\sim \frac{1}{2}$ stress $\times$ strain)
b and C = constants for specific materials

The application of this fatigue law to elastomers is discussed in Appendix A. References 1, 4, 7, 8, and 9 listed in the Bibliography of this document provide comprehensive information on general elastomeric behavior.

4. Failure Criteria—Failure should be defined in such a manner that it can be accurately detected and the time of its occurrence accurately determined. Since different types of elastomeric components may fail in different ways, it is necessary that the definition of failure and the means used to detect it apply equally well to all materials scheduled for evaluation.

4.1 Commonly used failure criteria:

4.1.1 Complete rupture of the specimen, i.e., total separation in tensile specimens, bond failure, or metal-to-metal contact between opposing mounting surfaces in bushings or compression specimens.

4.1.2 Time until appearance of visible cracks of a specified size or growth of a crack, or rupture to a specified point. A cut may be used to initiate the crack before the start of the test.

4.1.3 A specified level of change in physical properties such as hardness, spring rate, spring rate under dynamic conditions, or damping.

4.1.4 A specified change in displacement due to creep, set, or abrasive wear.

4.1.5 Chemical changes as evidenced by porosity.

4.1.6 Failure to function as intended.

4.2 Since different failure criteria will rank various elastomers differently, it is important that the definition of failure has relevance to the type of failure that occurs in the intended application. If an elastomer material is replaced in service by another having a tendency to fail in a different manner, fatigue test criteria used in quality control may have to be changed to assure applicability.

5. Specimen History—Prior to fatigue testing it is important to know if the component has experienced any mechanical preflexing or temperature changes and to know the timing of these influences. This is necessary for consistency in comparing test results.

5.1 Mechanical Preflexing—It is well known that elastomers undergoing load deflection tests will progressively soften for the first 3-10 cycles until a steady state condition is reached. Depending on the elastomer, this change ranges from about 3-15%. It is also known that most of the stiffness on the first cycle can be regained if a period of about 8 h or more elapses between load deflection tests. Consequently, if some specific change in spring rate of the specimen is the failure definition, preflexing influences must be considered to establish the *initial* stiffness. For shock type applications the first cycle data might be the basis. For steady state vibration applications the third (or more) cycle data might be required. For preflexing to be effective the load or deflection must be at least that at which the stiffness is to be determined.

5.2 Temperature Changes—Because elastomers are viscoelastic materials, it is necessary to know the temperature of the specimen and its temperature history prior to testing. Temperature influences all the failure definitions previously mentioned. One common situation concerns periodic evaluation of some physical property during fatigue testing. The specimen will heat up during testing due to internal heat generation. Consequently, when the periodic test is run, the specimen must be allowed to cool down or the *initial* property must be run at this elevated temperature. This is especially important when running low ambient temperature tests.

Another influence is the temperature history. A specimen stored in a very cold or hot atmosphere prior to testing influences the first cycle data in mechanical preflexing as well as the time to wait before regaining the first cycle data after preflexing.

5.3 Aging—Oil assembled components such as silentblocs require sufficient time for the elastomer to absorb the assembly oil. One week is recommended. Oven aging for 3 h at 70°C is sometimes used as a substitute for natural aging.

6. Test Parameters

6.1 Mechanical.

6.1.1 Direction of deformation.

6.1.2 Types of deformation.

A. Static.

1. Applying a specified constant load.
2. Applying a fixed displacement.
3. Loading to a specified initial displacement and maintaining the load.
4. Displacing until a specified initial load is reached and fixing this displacement.

B. Dynamic.

1. Applying a specified dynamic load.
2. Applying a specified dynamic displacement.
3. Dynamically loading to a specified initial displacement and maintaining the load.
4. Dynamically displacing until a specified initial dynamic load is reached and fixing this displacement.

6.1.3 Magnitude of load or displacement (see Section 9.2).

6.1.4 Frequency of dynamic deformation.

6.1.5 Dynamic load or deflection waveform, i.e.: Sinusoidal, square wave, continuous or intermittent; positive, negative, or positive to negative reversal.

6.1.6 Compensation for specimen changes during test such as adjusting displacement to maintain a specified load as specimen stress relaxation or wear takes place.

6.1.7 In applying load(s) or deflection(s) the following shall be considered: Constant load application will favor hard specimens and penalize soft specimens. Fixed displacement will favor soft specimens, especially those having high set, and penalize hard specimens.

6.1.8 Methods 3 and 4 (Section 6.1.2 (A) and (B)) have advantages over 1 and 2 in that they tend to minimize differences due to hardness and are, therefore, useful for material comparisons, but they represent conditions rarely found in elastomeric component applications.

6.2 Application of Parameters

6.2.1 Methods of applying static and dynamic deformation must be studied carefully to ensure that only the intended parameter is applied to the specimen. Most methods have inherent characteristics resulting from mass, friction, geometry, compliance, misalignment, and nonlinearity which may affect the parameter being applied. Through design, many of these undesirable effects can be reduced to an acceptable level.

6.2.2 Proper instrumentation is a good aid to accuracy in that error can be seen in the parameter measurement and, if recognized as such, and the source of error identified, be corrected. In measuring displacement, direct specimen deflection measurement is recommended rather than that of a test machine component attached to the specimen. In measuring loads, a load measuring device located between the specimen and the loading mechanism is recommended.

6.2.3 Instrumentation alone is not assurance of equivalent data between machines of different design used to run the same test. Machine geometry can affect specimen restraint, and if significantly different, can dramatically influence test results even though the measured parameters are identical.

6.2.4 In many tests, the specimen will roll, shift, bulge or otherwise react when the major parameters are applied. These reactions are often of high force magnitude. Improper attempts to completely restrict motions of this type often cause bending and friction in machine components which may adversely affect test repeatability.

6.2.5 It is recommended that the influencing machine geometry that affords the best compromise of the following be chosen:

6.2.5.1 Restrains specimen similar to the intended application.

6.2.5.2 Has no inherent adverse effect on test repeatability.

6.2.5.3 Has minimum effect on data as a result of wear, changes in friction, minor misadjustment, or other slight loss of precision.

6.2.6 Test fixtures, poorly designed or improperly specified, may constitute a test variable. The following should be considered in the specifying of test fixtures:

6.2.6.1 Fixture stiffness.

6.2.6.2 Cleanliness and finish of surfaces in contact with specimen.

6.2.6.3 Heat sink effects.

6.2.7 To assure that a test can be duplicated at a future date, the following information should be recorded:

6.2.7.1 Test parameters used.

6.2.7.2 Load and/or deflection versus time photographs or charts.

6.2.7.3 Detailed and dimensioned sketches of the influencing machine components and their location.

6.2.7.4 Prints of fixtures used.

6.2.7.5 Test setup procedure.

6.2.7.6 Data obtained.

6.3 Part Temperature

6.3.1 IMPORTANCE OF PART TEMPERATURE—Elastomers are functional over a rather narrow temperature range compared to other materials such as metals. Further, each compound of a given elastomer has its own temperature range where it is functional. Within that functional range will lie a band of temperatures at which maximum fatigue life is obtained. It is not unusual for fatigue life to change by a factor of two or more over a 20°C change in temperature near the boundaries of that band. Therefore, select a temperature that is representative of service conditions, and control part temperature during the test.

6.3.2 DEFINITION OF PART TEMPERATURE—Since rubber is a poor heat conductor, thick parts will usually have large temperature gradients. Measurements should, therefore, be made by placing the temperature sensing element as close to the area of heat generation as possible. The location chosen and the type of temperature measurement should be carefully defined and consistently adhered to.

6.3.3 PART TEMPERATURE CONTROL—Part temperature is a function of ambient temperature, hysteresis of the specimen, energy input, and external friction.

Ambient temperature control is necessary. First, it is recommended that the part and associated fixturing be allowed to reach equilibrium with the environment before starting the test. Guidelines for achieving this are given in the Appendix to SAE J1085[19] for elevated temperature testing. For elevated temperature testing, it is suggested that the part be enclosed in an air circulating heat chamber. At moderate temperatures, circulation of air over the specimen is commonly used to control part temperature. It should be recognized, however, that in some situations this may lower specimen surface temperature but have a relatively small effect on temperature within the specimen. Air cooling exaggerates the heat sink effect of fixturing in contact with the specimen so care must be taken to assure consistency in fixture contact area, shape, and mass. In cases where correlation between test facilities is necessary, air cooling may be undesirable as another source of variability.

The heating due to the combination of hysteresis and energy input should not cause the part to exceed the desired test temperature. High hysteresis elastomers will experience a relatively high temperature rise above ambient as compared to low hysteresis elastomers when tested under high frequency and/or amplitude. This sometimes makes it necessary to adjust test conditions when elastomers of different hysteresis levels are tested. In most cases, it is desirable to design the test in such a way that a significant portion of the testing takes place after the part temperature has stabilized.

Sometimes elastomer hysteresis is falsely blamed for high specimen temperature when the source of heat is actually friction due to slip between elastomer and metal components and/or test fixtures. When this is the case, and the elastomer has low hysteresis, reducing the test amplitude and/or load and increasing frequency will sometimes reduce temperature without adding significantly to test time.

6.4 Other Parameters

6.4.1 OZONE CONCENTRATION—Some elastomers are inherently ozone resistant so that ozone has little effect on their fatigue life. Other elastomers are not ozone resistant and must be chemically protected to prevent ozone cracking in stressed areas. Ozone cracking results in shortened flex life, particularly so for specimens with a high ratio of exposed surface to mass. Ozone crack rate increases with stress level and temperature.

It is desirable to avoid uncontrolled and excessive ozone concentrations as can be found in close proximity to electrical discharges or some motors. In critical situations, ozone concentration should be measured and reported in test conditions. ASTM D 1149, Standard Method of Test for Accelerated Ozone Cracking of Vulcanized Rubber[16] describes ozone concentration measurement.

The antiozonants used in many elastomer compositions must migrate to the surface of the specimen before they become fully effective. Testing of recently molded specimens should not be conducted before protective agents have migrated to the surface. Usually, 24 h is the minimum time for migration.

6.4.2 OXIDATION—The reaction of oxygen (oxidation) with many elastomers can initiate crack formation as well as result in hardening or softening. At temperatures higher than room temperature, the effect of oxygen is accelerated. Elastomers may be protected by chemicals introduced during mixing or by coatings. However, protection is seldom complete. Test specimens should not be stored for long time periods at elevated temperatures unless this is a necessary and controlled part of the test requirement.

6.4.3 DELETERIOUS FLUIDS AND GASES—No elastomer is resistant to all fluids and gases. Oils, oil vapor, and solvents can seriously degrade non-resistant elastomers. Water, steam, coolants, acids, and alkalies in fluid or vapor form can reduce specimen fatigue life. The atmosphere surrounding the test specimen should be free of deleterious fluids and gases unless they are a necessary and controlled part of the test requirement.

7. Property Measurement

Since the properties of different elastomeric specimens in a fatigue environment change differently, it is desirable to measure as many of these changes as possible. The instrumentation required will depend on the nature and purpose of the test, i.e., a materials evaluation would call for more detailed data than a quality control test. In all cases, however, the instrumentation must be adequate to observe both:

7.1 Changes corresponding to those that adversely affect performance in the intended application and which, therefore, qualify as criteria for failure.

7.2 Changes which can affect the severity of the test, obscure the point of failure or affect the mode of failure, thereby giving misleading results. Stress relaxation, set, and excessive heat buildup due to accelerated test conditions are examples of such changes.

Table A shows changes that can be anticipated and examples of the types of instrumentation that can be used to detect them.

8. Test Apparatus

This document is intended to apply to all elastomer and elastomeric component fatigue testing apparatus. Typical commercially available testers are:

8.1 Chrysler "Diving Board".

8.2 De Mattia Flexing Machine (ASTM D 430 and D 813).[16]

8.3 E. I. duPont Flexing Machine (ASTM D 430).[16]

8.4 Firestone Flexometer (ASTM D 623).[16]

8.5 Goodrich Flexometer (ASTM D 623).[16]

8.6 Monsanto Flex to Failure Tester.[2][3]

8.7 Roelig Machine

8.8 Ross Flexing Machine (ASTM D 1052).[16]

8.9 St. Joe Flexometer.

8.10 Sonntag Low Frequency Fatigue Testing Machine.

Other applicable test machines may be proprietary or especially constructed to evaluate a specific component.

9. Degree of Test Acceleration

9.1 Elastomeric material and component fatigue tests are accelerated to various degrees depending on the type and/or purpose of the test. Most tests fall into one of the following categories:

9.1.1 ENGINEERING EVALUATION TESTS—The purpose of evaluation testing is to rank and/or optimize material or component design performance under test conditions simulating the intended application as closely as practical.

9.1.2 QUALITY TEST—The purpose of quality testing is to measure the fatigue life of a specimen against a standard that is based on tests run on known quality specimens. The test conditions used may or may not simulate the type or direction of deformation found in the intended application, and are usually highly accelerated.

9.1.3 COMPARISON TESTS—This type of test is performed to compare fatigue performance of materials or components.

The initial comparison testing or screening may be performed under accelerated quality test conditions and final evaluation under conditions simulating the intended application.

9.2 Effects of Acceleration—Acceleration can introduce obvious or subtle factors that affect the test by changing the point of failure initiation, final location of failure, propagation, and the major cause of failure. This can be very misleading when materials for end use are chosen based on the results of such a test.

Table B describes examples of acceleration methods and possible effects they may have on the test and/or specimen.

TABLE A

Change in Specimen	Method of Observation	Notes
Abrasive wear.	Weight change.	May be dry or tacky depending on polymer type and formulation.
Amplitude of vibration under fixed force input.	LVDT[a], velocity transducer (integrated), accelerometer (integrated twice), leaf spring with strain gauge, optical methods, or micro switches.	If not fixed or controlled, amplitude usually increases during test due to the combined effect of temp. rise, chemical degradation, tearing, abrasion, etc. In some configurations amplitude can decrease due to overall movement relative to constraints.
Bond failure (to metal or fabric).	Visual.	Type and percentage of failure may be indicated using terminology of ASTM D429.
Cracks or tearing-initiation and rate of growth.	Visual and optical. May also be inferred from changes in deflection, damping or elastic rate.	Possibility of internal failure must be considered with thick specimens, in which case sectioning is required.
Deflection (mid-point) or drift.	LVDT[a], leaf spring with strain gauge or micrometer head (if member maintaining fixed load is different than member applying oscillating load) optical methods, micro-switches.	If constant force is maintained by a dead weight or servo system, deflection usually increases due to changes in the material, tearing, abrasion, etc. Temperature rise may result in decreased or increased deflection.
Distortion.	Visual.	Buckling, bending, etc. can lead to typical failure modes.
Dynamic properties- elastic rate and viscous damping.	Analysis of force and amplitude signals (magnitude and phase angle).	Increases or decreases in either property can occur due to chemical changes or changes in physical dimensions due to set.
Force—static and dynamic in displacement controlled test equipment.	Load cell—strain gauge or piezoelectric type.	Load cell must be placed so as to avoid the effects of the weight of surrounding machine elements and extraneous inertial forces.

Property Undergoing Change	Instrumentation	Notes
Permanent set.	Direct measurement after a specified period of recovery.	Method of measurement must be carefully defined. Usually not applicable to badly cracked or degraded specimens.
Porosity (internal).	Visual examination of sectioned specimen, comparison with standard specimens.	Indicative of chemical degradation due to internal heat build-up.
Temperature.	Thermocouple, thermistor in, on or adjacent to test specimen. Infrared pyrometer for surface temperature.	Sample temperature is normally non-uniform throughout the part due to the internal viscosity and poor heat transfer characteristics of elastomers.

[a] Linear variable differential transformer.

TABLE B

Method of Acceleration	Possible Effects on Test & Specimen
Increase static load or displacement.	Increase or decrease in cycles to failure.
	Failure by splitting and tearing (tensile failure) rather than by abrasive wear or fatigue cracking.
	Increased bulge area (compression).
	Decreased cross-sectional area (tension).
	Increased creep or slip.
	More data scatter (hardness sensitivity).
Increase dynamic load or displacement.	Decrease in cycles to failure.
	Increase in temp. due to hysteresis.
	Increase in temp. due to slip between specimen and fixturing.
	Decrease in modulus.
	Tensile failure rather than abrasive wear or fatigue cracking.
	Increase bulge area (compression).
	Decreased cross-sectional area (tension).
	More data scatter (hardness sensitivity).
Increase frequency of dynamic load or displacement.	Increased heat generation per unit time.
	Change from mechanical to chemical failure.
	Change in load or displacement waveform.
	Change in dynamic response of specimen.
	Increase or decrease in cycles to failure.
Increase ambient temperature.	Increased specimen temperature.
	Decrease in modulus.
	Change in cycles to failure.
	Change in dynamic response of specimen.
	Change in mode of failure.

10. Experiment Design—A designed experiment can obtain more information for less material and process cost than can be obtained by traditional methods. References 12, 13, and 14 listed in the Bibliography of this document cover experiment design in detail.

11. Reporting Data—Reporting test results in a clear, concise manner is every bit as important as assuring that the test conducted was valid and accurate. Also, hardware is often disassembled after a test is completed; the test report is needed to assure that the information of interest is not lost.

Following is a suggested outline of the minimum information that should be presented in a fatigue test report.

11.1 Summary—Present only the major important findings with some background information so that the report's contents can be rapidly digested and analyzed.

11.2 Material Specification and Properties—The minimum information presented should include the designation and/or specification, form of product, condition, chemical composition and note of any special treatment applied.

Also, to be included, a presentation of the mechanical properties of the material in the test component, designation of the test method used to procure those properties and identification of the location from which the samples were taken.

11.3 Component Dimensions—Present a drawing(s) or sketch(s) showing test section details, grip section, orientation with respect to force application and geometry of any induced notches.

11.4 Specimen Preparation—Report any observed deterioration of the specimen during storage since fabrication and changes in shape, dimensions or mechanical properties. Also desirable would be the environment in which the specimens were stored and any protection applied.

11.5 Information on Test Procedures—Included should be information on the test machine, its functional characteristics (electrohydraulic, pneumatic, etc.), frequency of load application, forcing function, method of calibration and load monitoring procedures. Further information would encompass the type of test (axial, torsional, etc.), failure criteria, number of cycles to run out and the statistical techniques used to design the test program and accommodate expected or unexpected deviations. Also desirable would be to spell out the procedure for mounting the specimen in the machine, grip details and precautions taken to ensure that unknown stresses induced by vibration, friction and eccentricity are negligible. Ambient conditions to include temperature and humidity average values and ranges together with controls applied should be reported. Special items of interest such as ozone level, deleterious substance presence, and so on, should also appear.

To complete the section, presentation of the reason for test termination for each specimen and a description of the failure and its location would be desirable.

11.6 Original Data—It is desirable in a test report to include copies of the original data sheets and test logs if these are not excessively voluminous. Proper planning before hand as to the data to be recorded and the method of recording to include log sheets and other forms will allow controlled compilation. All of the information expected to be utilized in the project write-up will be obtained from this single source so sufficient time must be spent in planning.

A primary item often overlooked is provision for and encouragement of the use of a remarks section in the log sheets. Orientation of the test technician to the log will often provide explanation for events, which otherwise would go unnoted, unnoticed, unexplained, or unknown. Be certain in this regard that the test personnel understand the use of the forms provided and the information that is of importance to the test.

As a minimum, the log sheets would be the primary original data. These should provide a chronological record of the test and its typical and atypical events. Intermittent readings of load, deflections, temperature and other important parameters as well as notes of interest should be input here. In conjunction with the sheets, it is desirable to have strip or X-Y charts providing direct, intermittent read-out of the essential parameters to back up and/or expand the information on the log sheets.

11.7 Presentation of Results—The most straightforward methods of presenting fatigue data are the tabulation and S-N_f curve. When used, the tabular form should include specimen identification, test sequence, stresses applied, cycles to end of test, cause of termination, results of post test examination and identification of station and machine used for each test.

On the S-N_f curve, the dependent variable fatigue life (N_f) in cycles is plotted on the abscissa, a logarithmic scale. The independent variable, maximum stress (S), is plotted on the ordinate and may be an arithmetic or logarithmic scale. If the data curve is fitted by regression analysis the stress-life relation equation and concomitant statistical measures of dispersion should be presented.

As discussed in Appendix A, the straight line obtained by plotting strain versus N_f (number of cycles) on log-log paper may be a more usable form of S-N_f data representation.

Photographs of failures, together with an explanation, provide a permanent record and valuable supplement to S-N_f curves.

12. Description of Terms—The following terms and definitions are applicable to this document:

12.1 Aging—The irreversible change of material properties after exposure to an environment for an interval of time.[16]

12.2 Ambient Temperature—The temperature of the environment surrounding the test specimen.[19]

12.3 Compound—An intimate admixture of a polymer with all the materials necessary for the finished article.[16]

12.4 Compression—Reduction of dimension from an external force.

12.5 Creep—The time-dependent part of a strain resulting from stress.[16]

12.6 Elastomer—Macromolecular material that returns rapidly to approximately the initial dimensions and shape after substantial deformation by a weak stress and release of the stress.[16]

12.7 Equilibrium Temperature—Stable temperature at which heat loss equals heat input.

12.8 Failure—When a material or component ceases to fulfill the design specified responses essential to the successful operation as a sub unit of a system. A rubber part may fail from tearing, cracking, rupture, hardening, softening, heat or chemical degradation, creep, set, or a combination thereof.

12.9 Fatigue—The process of progressive localized permanent structural changes occurring in a material or component subject to conditions which produce fluctuating stresses and strains at some point or points and which may culminate in loss of load bearing ability, cracks or complete fracture after a sufficient number of fluctuations.[11]

12.10 Fatigue Life—The number of cycles of stress or strain of a specified character that a given specimen sustains before failure of a specified nature occurs.[11]

12.11 Frequency—The number of complete cycles, whole periods, of forced vibrations per unit of time caused and maintained by a periodic excitation, usually sinusoidal.[19]

12.12 Hysteresis—The percent energy lost per deformation cycle.[18]

12.13 Maximum Stress—S_{max}—The stress having the highest algebraic value in the stress cycle, tensile stress being considered positive, and compressive stress negative. In this definition the nominal stress is used most commonly.[11]

12.14 Mean Stress (or Steady Component of Stress)—S_m—The algebraic average of the maximum and minimum stresses in one cycle, that is,

$$S_m = \frac{S_{max} + S_{min}}{2} \qquad (11)$$

12.15 Minimum Stress—S_{min}—The stress having the lowest algebraic value in the cycle, tensile stress being considered positive and compressive stress negative.[11]

12.16 Modulus of Elasticity—Ratio of stress to the strain produced by that stress. $E = \frac{Stress}{Strain}$ property of material.[18]

12.17 Nominal Stress—S—The stress at a point calculated on the net cross-section by simple elastic theory, without taking into account the effect on the stress produced by geometric discontinuities such as holes, grooves, fillets, etc.[11]

12.18 Permanent Set—The residual deformation of a specimen or component after removal of the external load.

12.19 Polymer—A macromolecular material formed by the chemical combination of monomers having either the same or different chemical composition.[16]

12.20 Preload—An external static load producing a strain in a test specimen. Preload is imposed prior to forced vibration testing. Preload is usually expressed in pounds of load instead of inches of deflection.[19]

12.21 Resilience—The ratio of energy output to energy input in a rapid (or instantaneous) full recovery of a deformed specimen.[16]

12.22 Resonant Frequency—The frequency at which maximum amplitude occurs for a given input force in a forced vibration system.

12.23 S-N_f Diagram—A plot of stress against the number of cycles to failure. The stress can be S_{max}, S_{min}, or S_a. The diagram indicates the S-N_f relationship and a specified probability of survival. For N_f a log scale is almost always used. For S a log scale is used most often but a linear scale is sometimes used.[11]

12.24 Shear—Force which causes two contiguous parts of the same body to slide relative to each other in a direction parallel to their plane of contact.[17]

12.25 Specimen Temperature—The temperature obtained by placing or locating a temperature sensing device in or on the specimen. In most cases, temperature gradients that develop within flexing rubber specimens make it necessary to define the precise points and techniques used to measure temperature.

12.26 Spring Rate—Ratio of force to the deflection produced by that force. Spring rate = $\dfrac{\text{force}}{\text{deflection}}$ property of the particular elastic body under consideration.[18]

12.27 Strain—Change (in length) per unit length in a linear dimension of a part or specimen.[20]

12.28 Stress (Uniaxial)—Load on a specimen divided by the area through which it acts.

12.29 Stress Amplitude—(or Variable Component of Stress), S_a—One-half the range of stress, that is,

$$S_a = \frac{S}{2} = \frac{S_{max} - S_{min}}{2} \tag{11}$$

12.30 Stress Relaxation—The decrease in stress after a given time at a constant strain.

12.31 Tension—Increase in dimension from an external force.

12.32 Elongation—Extension produced by a tensile stress.[16]

12.33 Ozone—O_3—An allotropic form of oxygen. It is a gas with a characteristic odor and is a powerful oxidizing agent.

12.34 Torsion—A twisting action resulting in shear stresses and strains.

12.35 Damping—Decreasing the time of vibrations in the motion of a body subject to influences which cause vibration.

12.36 Bushing—A cylindrical bearing or guide.[20]

12.37 Silentbloc—A type of bushing consisting of a thin wall, elastomeric cylinder compressed between concentric metal sleeves.

12.38 Friction—The resistance to relative motion between two bodies in contact.[17]

13. Bibliography

13.1 The following ASTM Recommended Practices are applicable to this document:

ASTM D 430, Dynamic Testing for Ply Separation and Cracking of Rubber Products. (Scott Flexing Machine, De Mattia Flexing Machine and E. I. du Pont de Nemours and Company Flexing Machine)

ASTM D 623, Compression Fatigue of Vulcanized Rubber, tests for. (Goodrich Flexometer and Firestone Flexometer)

ASTM D 813, Crack Growth of Rubber, test for. (De Mattia Flexing Machine)

ASTM D 1052, Measurement of Cut Growth of Rubber by the Use of the Ross Flexing Machine, test for.

ASTM D 1349, Standard Test Temperature for Rubber and Rubberlike Materials.

ASTM D 1566, Terms Relating to Rubber and Rubberlike Materials.

ASTM E 4, Standard Method of Verification of Testing Machines.

ASTM E 74, Standard Methods of Verification of Calibration Devices for Verifying Testing Machines.

13.2 The following references are applicable to this document:

1. P. W. Allen, P. B. Lindley, and A. R. Payne, "Use of Rubber in Engineering." London: Maclaren and Sons, Ltd., 1967, pp. 60–71.
2. Anonymous, "Fatigue Failure and Its Reduction in Natural Rubber." Akron: Monsanto Technical Bulletin O/RC-7.
3. Anonymous, "Fatigue to Failure Tester." Akron: Monsanto Literature with attachments.
4. A. B. Davey and A. R. Payne, "Rubber in Engineering Practice." New York: Palmerton Publishers, 1964.
5. T. A. Knurek and R. P. Salisbury, "Carbon Black Effect on Engine Mount Compounds." Rubber World (August 1964), pp. 45–57.
6. G. J. Lake, "Corrosive Aspects of Fatigue." Rubber Age (August 1972), pp. 30–42.
7. McPherson and Klemin, "Engineering Uses of Rubber." New York: Rheinhold, 1956, pp. 132–134, 139, 165–167, 170.
8. A. R. Payne and J. R. Scott, "Engineering Design with Rubber." New York: Interscience Publishers, Inc., 1960, pp. 104–106.
9. J. R. Scott, "Physical Testing of Rubber." New York: Palmerton Publishing Co., 1965, pp. 129–143.
10. F. L. Yost, "Fatigue Characteristics of Rubber." ASME Transactions, Vol. 65, pp. 881–888.
11. SAE Fatigue Design Handbook, Vol. 4, 1968.
12. Charles R. Hicks, "Fundamental Concepts in the Design of Experiments." New York: Holt, Rinehart and Winston, 1964.
13. Bernard Ostle, "Statistics in Research." Iowa State University Press, 1964.
14. United States Department of Commerce, "Experimental Statistics." Washington, D.C.: U.S. Government Printing Office (1963).
15. "ASTM General Test Methods." Annual Book of ASTM Standards, Part 41, 1976.
16. "ASTM Rubber Products, Industrial Specifications and Related Test Methods; Carbon Black; Gaskets; Tires." Annual Book of ASTM Standards, Part 37, 1976.
17. "Webster's New Collegiate Dictionary." Springfield: G & C Merriam Co.
18. "Handbook of Molded and Extruded Rubber." Akron: The Goodyear Tire and Rubber Co., Third Edition, 1969.
19. SAE J1085, Test for Dynamic Properties of Elastomeric Isolators.
20. "Properties and Selection of Metals." ASM Metals Handbook, Vol. 1.

APPENDIX A
FATIGUE CORRELATION FOR ELASTOMERS

1. $N_f W^b = C$

   Where N_f = cycles to failure
   W = energy input (½ stress × strain)
   C = constant for specific material
   b = ~2 for natural rubber
   b = ~4 for styrene butadiene rubber

   Assuming linearity for small strains:

2. $W = \frac{1}{2}$ stress × strain
   Stress = G strain

   Where G = shear modulus of material

3. Therefore, $W = \frac{1}{2} G (\text{strain})^2$

   if $b = 2$ (for natural rubber) and substituting (3) into (1):

4. $N_f [\frac{1}{4} G^2 (\text{strain})^4] = C$

   Rearranging:

5. $N_f = \left[\dfrac{K^1}{\text{strain}}\right]^4 \quad$ Where $K^1 = \sqrt[4]{\dfrac{4C}{G^2}}$

This shows the fatigue life of natural rubber to be dependent on the 4th power of strain and implies that plotting strain versus N_f (cycles to failure) on log-log paper is a straight line. This applies directly to fatigue tests where the elastomer is stressed in tension only as with the Monsanto Flex to Failure Tester.

The constants b and c could be determined for a component, such as a mount, with two tests at two different strain levels.

APPENDIX B
APPLICATION OF TEST ACCELERATION METHODS

1. Increase Dynamic Load or Displacement—This method illustrates use of the S-N_f curve and is valid if two conditions are met:

   A. The type of failure is known and is not changed by acceleration of the test.

   B. A plot of strain versus life (N_f) in cycles on log-log paper is a straight line.

The method requires two points on the straight line plot—either two levels of acceleration or the actual load (or displacement) on the specimen in the intended application and one level of acceleration. This is illustrated schematically as follows:

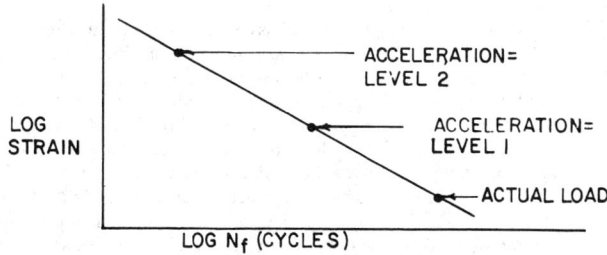

2. Increase Dynamic Load or Displacement During Test—This method is intended to predict life at the load (or displacement) on the specimen in the intended application by increasing the load or displacement. The test is started at the actual load (or displacement) on the specimen but prior to failure the load (or displacement) is increased to a higher value. The test is continued at this higher value until failure occurs.

An additional test is conducted at the higher value only. This test is conducted to failure. A plot of percent time at the higher load (or displacement) value versus percent time at the actual load is made. From this plot, the life of the specimen may be predicted by extrapolating to the 100% time at the actual load (or displacement) axis. The method is illustrated schematically as follows:

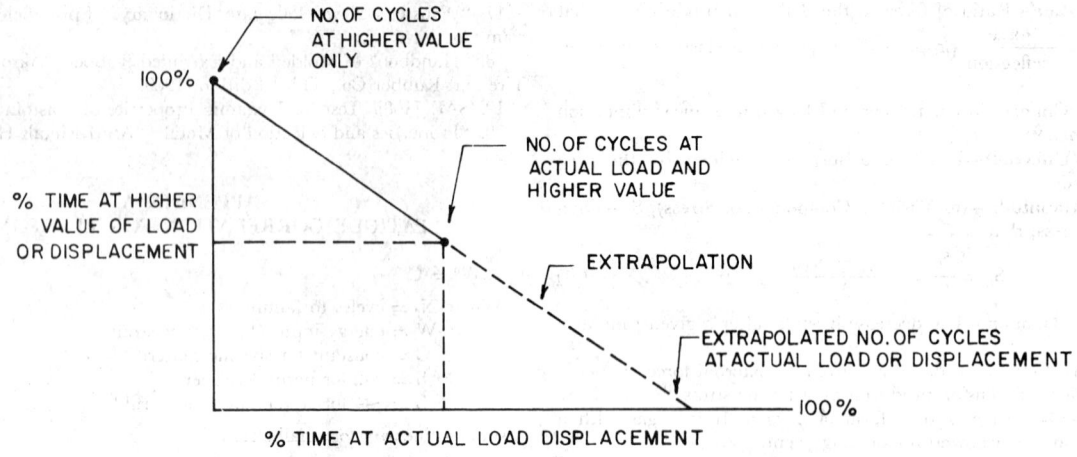

TEST FOR DYNAMIC PROPERTIES OF ELASTOMERIC ISOLATORS—SAE J1085a

SAE Recommended Practice

Report of Nonmetallic Materials Committee approved September 1974 and last revised August 1978.

1. **Scope**—These methods cover testing procedures for defining and specifying the dynamic characteristics of elastomers and fabricated elastomeric isolators used in vehicle components.

2. **Summary**—These methods describe procedures for measuring the dynamic characteristics of automotive elastomeric mountings using forced vibration testing machines. These characteristics are the elastic spring rate, damping coefficient, and loss tangent. Either fabricated mountings or elastomer specimens may be tested. Since measured dynamic properties are highly dependent upon test conditions, emphasis has been placed on the definition of suitable conditions.

3. **Description of Terms**—These terms are in common use throughout the automotive industry:

   3.1 Test Temperature

   3.1.1 AMBIENT TEMPERATURE—The temperature of the environment surrounding the test specimen. Unless otherwise specified, it is assumed that the sample is at the ambient temperature before being subjected to dynamic flexing.

   3.1.2 PART TEMPERATURE—The temperature obtained by locating a temperature-sensing device in or on the specimen. In most cases, temperature gradients that develop within flexing rubber specimens make it necessary to define the precise points and techniques used to measure temperature.

   3.2 Frequency (f)—The number of complete cycles, whole periods, of forced vibrations per unit of time caused and maintained by a periodic excitation, usually sinusoidal.

   3.3 Preload—An external static load producing a strain in a test specimen. Preload is imposed prior to forced vibration testing. Preload is usually expressed in newtons (pounds) of force instead of metres (inches) of deflection.

   3.4 Double Amplitude Displacement (DA)—The peak-to-peak amplitude of the elastomer specimen measured in the direction of the applied vibrational force. Two times the single peak value in either the plus or minus direction may not be equivalent to the peak-to-peak value.

   3.5 Complex Spring Rate (K^*)—The effective spring rate of a part under sinusoidal dynamic stress. It is equal in magnitude to the peak-to-peak force across the sample divided by the double amplitude. The complex spring rate can be visualized as being the vector sum of an elastic component and a viscous damping component.

   3.6 Dynamic Spring Rate (K)—The proportionality factor between the component of the applied force vector that is in phase with the displacement and the displacement vector. The dynamic spring rate is equal to the elastic component of the complex spring rate.

   3.7 Damping Coefficient (C)—The proportionality factor between the component of the applied force vector which is in phase with velocity and the velocity vector.

   3.8 Loss Rate ($C\omega$)—The proportionality factor between the magnitude of the component of the applied force vector that is in phase with the velocity and the magnitude of the displacement vector, where:

   $$\omega = 2\pi f$$

   NOTE: The magnitudes of the complex spring rate, elastic spring rate and loss rate are related by the equation:

   $$K^* \equiv \text{complex spring rate}$$
   $$|K^*| = \sqrt{(K)^2 + (C\omega)^2}$$

   The last equation is sometimes written:

   $$(K^*)^2 = (K')^2 + (K'')^2$$

   3.9 Loss Tangent (loss factor, tan δ)—The tangent of the phase angle between the applied force and the resulting displacement:

   $$\tan \delta = \frac{C\omega}{K}$$

4. **General Testing Methods**

   4.1 Outline of Procedure

   4.1.1 Virgin specimens must be allowed to age between manufacturing and testing. Typically, a minimum of 24 h is suggested. Elastomeric specimens that have undergone some permanent deformation (such as that due to an assembly operation) may require additional time to permit relaxation of any internal stresses that may exist.

   4.1.2 Parts that have been kept at temperatures other than the test temperature (for example, during shipment, storage, or environmental testing) must be conditioned at the test temperature long enough to achieve uniform temperature stabilization throughout. Minimum conditioning time depends upon many factors, including temperature difference, specimen size and shape, and airflow around the specimen. Guidance for determining the required conditioning time is given in the Appendix.

   4.1.3 All test equipment instrumentation should be fully stabilized per manufacturer's instructions. At least $\frac{1}{2}$ h is required. Greater stability is obtained by leaving electronic equipment on permanently. It is recommended that the start and conclusion of every test session or operator change, a quick check of calibration be performed with a control specimen.

   4.1.4 Insert the specimen.
   4.1.5 Apply the preload.
   4.1.6 Apply and maintain the dynamic conditions of test for 5 min.
   4.1.7 Data should then be recorded within 1 min.
   4.1.8 Remove the specimen if no further measurements are to be made.

   4.2 Preferred Test Conditions

   4.2.1 Where a single measurement is to be made on a specimen, the following reference test conditions are suggested in the interests of standardization. They take into account the requirements of precision of equipment, minimization of heat buildup, avoidance of regions in which elastomers are most sensitive to changes in test conditions, and relevance to most practical applications.

   PRELOAD—Selected to correspond to that existing in the intended application. Sufficient preload should be applied to prevent any separation of sample-to-machine interfaces unless all interfaces are securely attached. The preload should be chosen so that any sharp changes in the slope of the load-deflection curve are avoided.

DOUBLE AMPLITUDE—0.50 mm (0.020 in).
FREQUENCY—15 Hz.
AMBIENT TEMPERATURE—23 ±2°C (73.4 ±3.6°F).

4.2.2 Additional or alternate test conditions should follow the guidelines in ASTM D 2231, Recommended Practice for Forced Vibration Testing of Vulcanizates, and ASTM D 1349, Recommended Practice for Standard Test Temperatures for Rubber and Rubber-like Materials. The following ambient temperatures are suggested for testing elastomers used in automotive applications:

°C	°F
− 40	− 40
− 10	+ 14
+ 23	+ 73.4
+100	+212
+150	+302

φ 4.2.3 ALTERNATIVE COLD TEST PROCEDURE
φ 4.2.3.1 *Discussion*—When a specimen has been temperature stabilized at a test temperature such as −40°C (−40°F), any test data obtained in the first *few* thousand cycles will be transient data. Each unit of energy input will change the specimen's dynamic properties. The following procedure is suggested so data can be obtained in a repeatable manner when specimen response is changing.

φ 4.2.3.2 *Pre Test Preparation*
φ 4.2.3.2.1 Prepare the test machine to record test data continuously during test so that information pertaining to any particular cycle can be determined. Include the following:

1. Test cycles count.
2. Dynamic spring rate.
3. Damping coefficient.
4. Test chamber ambient temperature—Temperature sensor will be located to best sense the temperature of the test chamber ambient to which the sample is subjected.
5. Energy input ($\int_0^t F(t)V(t)dt$).
6. Specimen temperature—Is assumed the same as chamber ambient when correctly stabilized at start of test.

φ 4.2.3.2.2 Prior to Test—Weigh the specimen (include all specimen elements that are molded together or fastened together).
φ 4.2.3.2.3 Insert the specimen.
φ 4.2.3.2.4 Condition the specimen at test temperature.
φ 4.2.3.2.5 Preload—There are two ways to soak the specimen at ambient temperature with or without preload. Measured dynamic characteristics are influenced by preload history. The desired condition should be specified in test requests.

φ 4.2.3.2.5.1 *Soak Period with Preload*—The preload will be maintained until the entire soak period and test is complete.

Start of Preload:
Load Control—No stabilization required.
Displacement Control—Preload stabilization period will be required.

φ 4.2.3.2.5.2 *Soak Period without Preload*—The preload is added following the soak period and maintained until the test is complete.

φ 4.2.3.3 *Test Machine Control During Test*—Equipment used to perform extended environmental tests should have the required capability. When equipment does not have the required capability this method makes no suggested compromises. Machine control must be directed at measuring transient response and obtaining repeatable data.

φ 4.2.3.4 *Data Reduction*—Should include the following information pertaining to the cycle of interest:

φ 4.2.3.4.1 Number of cycles.
φ 4.2.3.4.2 Spring rate.
φ 4.2.3.4.3 Damping coefficient.
φ 4.2.3.4.4 Total energy input.
φ 4.2.3.4.5 Total energy input per unit weight of specimen. When reading the dynamic spring rate or damping, use the average for the specified cycle such as; cycle 100 equals the end of 99 to the beginning of 101.

4.3 Specimens
4.3.1 Standard compression specimens used for comparing elastomer properties or standardizing test machines should be chosen based on the following considerations:

4.3.1.1 The size of the specimens shall be chosen to suit the load capacities of the test machine but should be no less than 12.7 mm (0.50 in) nor more than 50.8 mm (2.0 in) high; the recommended height is 25.4 mm (1.0 in).

4.3.1.2 The preferred shape factor for comparing elastomer properties is 0.5 where:

$$\text{Shape factor} = \frac{\text{area of one loaded face}}{\text{area free to bulge}}$$

4.3.1.3 The preferred shape is a right circular cylinder with faces parallel within 0.001 mm/mm or 0.001 in./in.

4.3.1.4 *Test Interface*—For best reproducibility, the sample mentioned in paragraph 4.3.1.2 should have metal plates bonded to both faces during vulcanization.

4.3.1.5 *Optional Test Interface*—The test machine will be equipped with loading plates top and bottom of sufficient area to support the loaded specimen. The specimen will be held in place with 300 grit sandpaper, securely bonded to both loading plates. The sandpaper prevents specimen lateral movement and aids in bulge control. The two plates exciting the specimen shall be parallel within 0.001 mm/mm (0.001 in./in.) of platen length in the neutral position. Plate parallelism will be within tolerances on orthogonal lines.

4.3.2 Standard shear specimens shall comply with ASTM D 2231 for general configuration. Dimensions may be adjusted to provide required spring rates. Supporting fixtures should be sufficiently rigid to maintain parallelism of all plates.

4.3.3 The following considerations shall apply when fabricated mountings or bushings are tested:

4.3.3.1 Supporting fixtures shall be designed to restrain lateral movement of the top or bottom surfaces of the mounting as a result of forces applied in the test direction.

4.3.3.2. It shall be carefully determined that any lateral forces which may develop as a result of forces applied in the test direction do not influence the test readings.

NOTE—A correction factor will be required unless the overall spring rate of the fixtures and the machine elements that are included in the measurement is sufficiently high. With an infinitely rigid specimen installed, the machine and fixture spring rate should be at least 100 times greater than the nominal spring rate of the test specimen. If this degree of rigidity cannot be achieved, the correction factor shall be calculated and applied.

4.3.4 Standard specimens to be tested must be clearly marked for identification.

4.3.5 Standard specimens used for standardizing test machines shall be aged no less than one month and shall be accepted only after repetitive testing indicates that dynamic properties have stabilized.

4.4 Specific Procedures
4.4.1 FORCED NONRESONANT SYSTEMS—A forced nonresonant system is comprised of a drive mechanism which forces the specimen through a desired sinusoidal load, displacement, or energy. The desired frequency and amplitude of the test are not affected by the specimen's dynamic response; therefore, test conditions may be quite easily changed.

4.4.1.1 *Theory*—In a forced nonresonant system, the sample is excited with a sinusoidal oscillation which is either force or displacement controlled. The forcing medium causing this sinusoidal oscillation can be an electromechanical, electrohydraulic system, or a pure mechanical system.

This method assumes that the existing force or displacement and the response of the specimen can be considered to be sinusoidal. If this is not the case, special methods of analysis are required.

The transmitted force is measured by a load cell in contact with the sample, preferably on the stationary side to minimize errors due to acceleration of the mass of the fixture. The component of this force which is in phase with velocity and the component that is in phase with the displacement are usually determined electronically. From this information, the values of C and K are usually determined. The vector phase relationships are illustrated in Fig. 1.

4.4.1.2 *Apparatus*—The basic elements of a forced nonresonant system are the drive system, the control system, a loading frame, transducers, and the instrumentation for readout.

4.4.1.2.1 Drive System—The system should be capable of providing sinusoidal dynamic operation, with minimum harmonic distortion, in the same direction as the applied force.

4.4.1.2.2 Control—A means of precise control over the input drive unit is required for repeatable test results. Controls for mean input, dynamic input, and frequency should be independently selectable for the desired test condition.

4.4.1.2.3 Transducers and Instrumentation—Common transducer signals used in forced nonresonant systems for obtaining data are load, displacement, and/or velocity. Each transducer should be calibrated to the following minimum accuracies.

Load: ±0.5% of full-scale for each calibrated range.
Displacement: ±0.5% of full-scale for each calibrated range.
Velocity: ±1.0% of full-scale for each calibrated range.

Readout instrumentation for load, displacement, and velocity transducers should provide a sufficient number of ranges so that it will not be necessary to use outputs less than 20% of range. Scales in the ratio of 1, 2, 5, 10 are suggested.

Precautions listed in ASTM D 2231, paragraph 4.2 should be observed in selecting transducers, electronics, and techniques of calibration.

4.4.1.2.4 *Preferred Location of Transducers*—The load cell shall be mounted on the stationary side of the sample being tested.

The displacement and/or velocity transducer should be located directly across the sample and should not include a compliant structure within its measurement path. It should be located parallel to and as close as possible to the centerline of the exciting force.

4.4.1.2.5 *Fixture Mass*—The mass of the fixture located on the stationary platen shall be minimized to reduce errors due to mass-inertia accelerations. The fixture located on the moving platen shall be rigid to eliminate any possibility of structural resonances near operating frequencies.

5. Report—The report shall include the following:

5.1 Type of testing machine used.
5.1.1 Test specimen(s) identification.
5.1.2 Type of specimen loading, for example, compression or shear. For specimens of complex configuration, full description of fixtures used, with diagrams if necessary.
5.1.3 Date of test.
5.1.4 Preload.
5.1.5 Frequency used at each test point in hertz.
5.1.6 Double amplitude displacement used at each test point.
5.1.7 Ambient temperature.
5.1.8 Specimen internal temperature (optional). If internal temperature is used, the following additional information is required:
5.1.8.1 Internal temperature before flexing.
5.1.8.2 Ambient temperature.
5.1.8.3 Exact location of the temperature measuring transducer.
5.1.8.4 Time from start of flexing until temperature and dynamic property readings are taken.

5.2 The method for computing C and K from the measured variable shall be described.
5.2.1 Dynamic (elastic) spring rate, K.
5.2.2 Damping coefficient, C.
5.2.3 Loss tangent, C_ω/K.
5.2.4 For all test machines, as applicable:
5.2.4.1 Range scale settings.
5.2.4.2 Mode of test control, that is, stroke or load.
5.2.4.3 All observed and recorded data on which calculations are based.

6. Precision or Reproducibility—Precision as defined in ASTM E 177–68T is a function of the operator, compound, and maintenance of constant test conditions. The test conditions that can influence "level" are: preload, frequency, double amplitude displacement, temperature, and others not yet well defined.

7. References—The following ASTM Standards:

D 832, Recommended Practice for Rubber Conditioning for Low Temperature Testing.

D 1053, Test for Rubber Property—Stiffening at Low Temperature Using A Torsional Wire Apparatus.

D 1229, Test for Rubber Property—Compression Set at Low Temperature.

D 1329, Test for Rubber Property—Retraction at Low Temperatures (TR Test).

D 1349, Recommended Practice for Rubber—Standard Temperatures and Atmospheres for Testing and Conditioning.

D 1566, Definition of Terms Relating to Rubber.

D 2231, Recommended Practice for Rubber Properties in Forced Vibration.

E 4, Verification of Testing Machines.

E 74, Calibration of Force-Measuring Instruments for Verifying the Load Indication of Testing Machines.

8. Bibliography

8.1 B. M. Hillberry and A. F. Hegerich, "The Measurement of the Dynamic Properties of Elastomers and Elastomeric Mounts." SAE SP-375, ASTM STP-535. Proceedings of Symposium presented at SAE International Automotive Engineering Congress, Detroit, January 1973.

8.2 D. Hands, "Simple Methods for Heat Flow Calculations." RAPRA Technical Review No. 60, Class No. 96, July 1971.

8.3 Marion D. Thomson, "Cooling Rubber Slabs." RAPRA Bulletin, May 1972.

8.4 S. D. Gehman, "Heat Transfer in Processing and Use of Rubber." Rubber Chem. & Tech., 1967, pp. 36–99.

APPENDIX

The following information is presented for guidance in determining the amount of time required for rubber parts to substantially reach equilibrium with the ambient test temperature.

Fig. A-1 shows the time required for the center of a rubber part to reach 90% of the desired temperature change under the following conditions:

1. Unrestricted free convection.
2. Thermal conductivity of the elastomer = 0.173 W/m °C (0.1 Btu/h-ft² °F/ft).
3. Thermal diffusivity of the elastomer = 0.00024 m²/H (0.0026 ft²/H)—a conservative value for most elastomer compositions.
4. Film coefficients = 13.3 W/m² °C (2.35 Btu/h-ft² °F) and ∞.

The curves apply to shapes approximating spheres, cylinders whose radii are less than $\frac{1}{4}$ that of their lengths or slabs whose total thickness is less than $\frac{1}{4}$ that of their length and width.

For example, for a 1.27 cm (0.5 in.) radius sphere in still air, 35 min are required to reach at least 90% of the desired temperature change throughout; for a 2.54 cm (1 in.) thick slab, 2 h are required. Times for actual specimens may be estimated by relating them to these basic shapes or by calculations using the information in Refs. 8.2 and 8.4. Ref. 8.4 also discusses the effect of metal plates in, or attached to, elastomeric shapes.

Extra time should be allowed if circulation around each part is restricted or if more precise part temperature control is required, as, for example, at low temperatures where dynamic response is most sensitive to temperature. Conditioning time may be shortened by forced convection, immersion in water baths, or other means suggested in Ref. 8.3. The curves for film coefficient equal to ∞ show the minimum times that can be achieved by such methods.

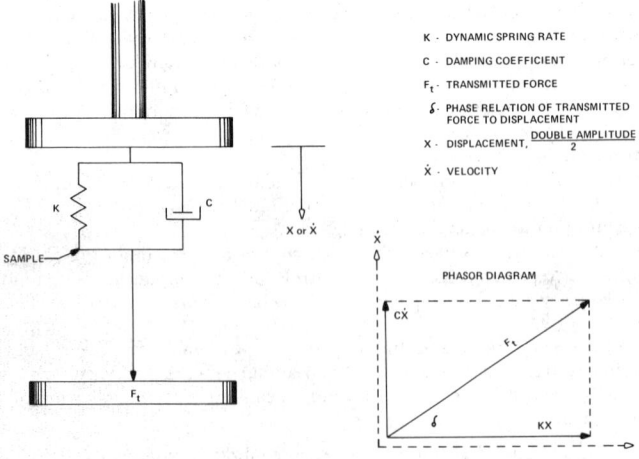

FIG. 1

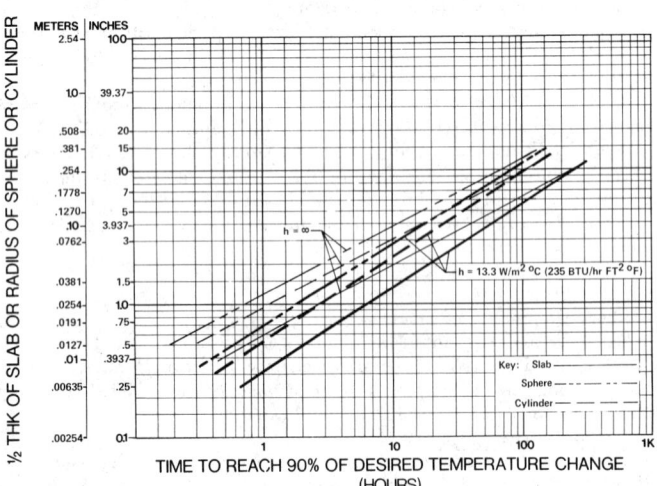

FIG. A-1—TIME TO REACH 90% OF DESIRED TEMPERATURE CHANGE FOR VARIOUS SIZES OF SPHERES, CYLINDERS, AND SLABS; AND TWO VALUES OF FILM COEFFICIENT (h)

LATEX FOAM RUBBERS—SAE J17 OCT79

SAE Standard

Report of the Nonmetallic Materials Committee, approved January 1952, completely revised July 1979, editorial change October 1979. Conforms substantially with ASTM D 1055.

[This SAE Standard was formulated by the SAE-ASTM Technical Committee on Automotive Rubber.]

1. Scope—These specifications and methods of testing apply to cellular-rubber products known as latex foam rubbers but do not apply to sponge and expanded rubbers. The base material used in their manufacture may be natural rubber, reclaimed rubber, synthetic rubber, or rubberlike materials, alone or in combination. In case of conflict between the provisions of these general specifications and those of detailed specifications or methods of test for a particular product, the latter shall take precedence. Reference to methods for testing cellular-rubber products should specifically state the particular test or tests desired.

2. Description of Terms

  2.1 Flexible Cellular Rubber—A cellular organic polymeric material that will not rupture when a specimen 200 x 25 x 25 mm (8 x 1 x 1 in) is bent around a 25 mm (1 in) diameter mandrel at a uniform rate of 1 lap in 5 s at a temperature between 18 and 29°C (65 and 85°F). In the case of latex foam rubbers, these cells are open and interconnecting.

  2.2 Cellular Rubbers—Cellular-rubber products all contain cells or small hollow receptacles. In the case of latex foam rubbers, these cells are open and interconnecting.

  2.3 Rubber—The term rubber is used to include both natural and synthetic types.

  2.4 Skin—The smooth surface of the latex foam rubber product, formed by contact with mold or cover plates, is defined as a natural skin.

3. Manufacture

  3.1 Latex Foam Rubbers—The structure of latex foam rubbers consist of a network of open or interconnecting cells. Latex foam rubbers are made from rubber latices or liquid rubbers. They are manufactured in sheet, strip, molded, or specific shapes. Latex foam rubbers shall have a vulcanized cellular structure with a porous surface. The cells shall be interconnecting and of a uniform character. Latex foam rubbers may be either cored or solid. Size, shape, and distribution of coring shall be at the producer's option but subject to the approval of the purchaser.

4. Grades of Latex Foam Rubbers—Latex foam rubbers shall have their grade numbers designated by two letters which identify the kind of latex foam rubber as follows:

  RC—Latex foam rubbers, cored and
  RU—Latex foam rubbers, uncored

Digits following the letters are used to indicate the degree of firmness, the softer grades being identified with the lower numbers and the firmer grades with the higher numbers (see Table 1).

Suffix letters may be added singly or in combination after any grade numbers to indicate additional requirements beyond those specified in Table 1 as basic requirements.

The significance of the approved suffix letters are shown below. The test methods and values must be arranged by agreement between the purchaser and supplier.

  C   Weather Resistance.
  D   Load Deflection.
  E   Oil Resistance. Note that there are no requirements for oil resistance in these specifications.
  F1   Low Temperature at −40°C (−40°F). Required with values as specified in Table 1.
  F2   Low Temperature at −55°C (−67°F).
  G   Tear Resistance.
  H   Flexing Resistance. Test required with values as specified in Table 1.
  J   Abrasion Resistance.
  K1   Adhesion to Metal—Bond made during vulcanization.
  K2[1] Adhesion—Cemented bond made after vulcanization.
  L   Water Resistance.
  M   Flammability Resistance.
  N   Impact Resistance.
  P   Staining Test Required.
  R   Resilience.
  Z   Optional Requirements.

Example—Grade RC20 F1H denotes soft, cored latex foam rubber made from natural, reclaim synthetic, or a blend with a load deflection value of 89 ± 18 N (20 ± 4 lb) and requiring, in addition to the basic tests, a low temperature test at −40°C (−40°F) and a flexing test.

5. Material and Workmanship—Latex foam rubbers furnished under these

---

[1] Suffix K2 denotes that the finished vulcanized part will be adhered to a rigid surface sometime after vulcanization and that all surface imperfections and/or the use of materials which might be on or bloom to the surface and be detrimental to obtaining good bonds must be avoided.

TABLE 1—PHYSICAL REQUIREMENT OF LATEX FOAM RUBBERS

Grade Number	Basic Requirements				Requirements Added by Suffix Letters			
	Indentation of 325 cm² (50 in²) (0.03 m²), 25% Deflection[a] (Limits)		Air Oven Aged 22 h at 100°C (212°F) Change from Original Load Deflection or Indentation Value (Limits), %	Constant Deflection Compression Set, 22 h at 70°C (158°F), 50% Deflections, max, %		Suffix F	Suffix H	
	Newtons	Pounds		C_h[b]	C_d[b]	Low Temperature Test, Change from Original Deflection, max, %	Flexing Test Compression Set, max, %	
							C_h[b]	C_d[b]
Latex Foam Rubbers (Cored)								
RC 5	22 ± 13	5 ± 3	±20	10	20	75	5	10
RC 10	44 ± 13	10 ± 3	±20	10	20	75	5	10
RC 15	67 ± 18	15 ± 4	±20	10	20	75	5	10
RC 20	89 ± 18	20 ± 4	±20	10	20	75	5	10
RC 25	111 ± 22	25 ± 5	±20	10	20	75	5	10
RC 30	133 ± 27	30 ± 6	±20	10	20	75	5	10
RC 40	178 ± 31	40 ± 7	±20	10	20	75	5	10
RC 50	222 ± 36	50 ± 8	±20	10	20	75	5	10
RC 60	267 ± 40	60 ± 9	±20	10	20	75	5	10
RC 70	311 ± 53	70 ± 12	±20	10	20	75	5	10
RC 90	400 ± 62	90 ± 14	±20	10	20	75	5	10
Latex Foam Rubbers (Uncored)								
RU 11	49 ± 18	11 ± 4	±20	10	20	75	5	10
RU 20	89 ± 22	20 ± 5	±20	10	20	75	5	10
RU 35	156 ± 44	35 ± 10	±20	10	20	75	5	10
RU 55	245 ± 44	55 ± 10	±20	10	20	75	5	10
RU 80	356 ± 67	80 ± 15	±20	10	20	75	5	10
RU 150	667 ± 245	150 ± 55	±20	10	20	75	5	10

[a] Rubber Manufacturers Association buyers' specification designation.
[b] As defined in section on compression set.

specifications shall be manufactured from natural rubber, synthetic rubber, or rubberlike materials together with added compounding ingredients of such nature and quality that the finished product complies with the specification requirements. In permitting choice in use of those materials by the producer it is not intended to imply that the different rubber materials are equivalent in respect to all physical properties. Any special characteristics other than those prescribed in these specifications which may be desired for specific applications shall be specified in the product specifications as they may influence the choice of the type of rubber materials or other ingredients used. All materials and workmanship shall be in accordance with good commercial practice and the resulting cellular rubber shall be free from defects affecting serviceability.

Because of manufacturing conditions, material may have to be altered or repaired. This repaired or altered material will be acceptable under these specifications provided the material used in such repairs or alterations shall be of the same composition and quality as the original product and provided such alterations do not affect the serviceability, size, and shape beyond tolerances as provided herein.

6. Color—Unless otherwise specified, the color of latex foam rubbers shall be optional with the manufacturer.

7. Physical Properties—The various grades of latex foam rubber shall conform to the requirements as to physical properties prescribed in Table 1, together with any additional requirements indicated. When subjected to the static fatigue test, the latex foam specimen shall show no cracking at the folded edge.

8. Methods of Testing—Unless specifically stated otherwise, all tests shall be made in accordance with the methods specified for the following:

8.1 Basic Tests
8.1.1 Accelerated Aging Tests.
8.1.2 Compression Set Under Constant Deflection.
8.1.3 Indentation Test.
8.2 Suffix Tests
8.2.1 H—Flexing Test.
8.2.2 F—Low Temperature Test.

9. Tolerances on Dimensions—Tolerances on dimensions of latex-foam-rubber products are given in the Appendix, Tables 2 and 3. These tolerances are published as information for guidance only and shall not be considered as a part of these specifications.

10. Packaging and Marketing—The material shall be properly and adequately packaged. Each package or container shall be legibly marked with the name of the material, name or trademark of the manufacturer, and any required purchaser's designations.

11. Inspection and Rejection—All tests and inspection shall be made at the place of manufacture prior to shipment, unless otherwise specified. The manufacturer shall afford the inspector all reasonable facilities for tests and inspection.

The purchaser may make the tests and inspection to govern acceptance or rejection of the material at his own laboratory or elsewhere.

All samples for testing, provided as specified in Section 13, Sampling, shall be visually inspected to determine compliance with the material, workmanship, and color requirements.

Any material which fails in one or more of the test requirements may be retested. For this purpose, two additional tests shall be made for the requirement in which failure occurred. Failure of either of the retests shall be cause for final rejection.

Rejected material shall be disposed of as directed by the manufacturer.

12. General Methods—Except as otherwise specified in the methods of testing latex foam rubbers given in ASTM D 1055, the following methods of test of the American Society for Testing and Materials, applicable in general to vulcanized rubber, shall be complied with as required and hereby made a part of these test methods.

12.1 Aging Test—See ASTM D 572, ASTM D 573, and ASTM D 454.
12.2 Compression Set, Suffix B—See ASTM D 395.
12.3 Low Temperature Test, Suffix F1 and F2—See method described in ASTM D 1055.
12.4 Compression—Deflection—See ASTM D 575, ASTM D 1055, and the Indentation Test in this SAE Recommended Practice. In case of conflict between provisions of the above methods and the procedures herein specifically described for latex foam rubbers, the latter shall take precedence. In case of conflict between the procedure herein described for latex foam rubbers and the methods of a particular specification or for a particular latex-foam-rubber product, the latter shall take precedence.

13. Sampling—When possible, the completed manufactured product shall be used for the tests specified. Representative samples of the lot being examined shall be selected at random as required.

When it is necessary or advisable to obtain test specimens from the article, as in those cases where the entire sample is not required or adaptable for testing, the method of cutting and the exact position from which specimens are to be taken shall be specified. The apparent density and the state of cure may vary in different parts of the finished product, more especially if the article is of complicated shape or of varying thickness, and these factors affect the physical properties of the specimens. Also, the apparent density is affected by the number of cut surfaces as opposed to the number of skin-covered surfaces on the test specimen.

When the finished product does not lend itself to testing or to the taking of test specimens because of complicated shape or other reasons, manufacturer and purchaser shall agree on the preparation of a suitable test specimen. When differences due to the difficulty in obtaining suitable test specimens from the finished part arise, manufacturer and purchaser may agree on acceptable deviations. This can be done by comparing results of standard test specimens and those obtained on actual parts.

14. Measurement of Test Specimens—Test specimens are to be measured in accordance with ASTM D 1055.

15. Accelerated Aging Tests

15.1 Test Specimens—The test specimen used in any of the aging tests shall be that required by the latex-foam-rubber methods for the particular determination which is to be employed for measuring the effect of the aging exposure.

15.2 Procedure—Either the oxygen-pressure-chamber aging test as described in ASTM D 572, the air-oven aging test as described in ASTM D 573, or the air-pressure heat test as described in ASTM D 454, respectively, may be used for latex foam rubbers as specified, except that in the air-pressure heat test, an air pressure of 415 ± 15 kPa (60 ± 2 psig) shall be used in place of the 550 ± 15 kPa (80 ± 2 psig) prescribed in ASTM D 454. Deterioration may be expressed as a percent change of compression-deflection values, or the results may be determined by visual observation. No relation between accelerated aging tests and natural aging is given or implied.

16. Compression Set Under Constant Deflection (Calculation Based on Amount of Deflection)

16.1 Test Specimens—The specimen for this test shall have parallel top and bottom surfaces. A cylinder 29 mm (1.129 in) in diameter shall be suitable for slab or uncored stock. Cored stock specimens may be round or rectangular. The minimum dimension on the top and bottom surfaces must be greater than the height of the sample and the surface shall have a minimum area of 0.01 m^2 (16 in^2). The thickness of the test specimen may vary, but shall not be less than 19 mm (0.75 in) for slab or uncored stock. The thickness shall be measured and stated in the report.

16.2 Procedure—The apparatus and procedure shall be the same as prescribed in Method B of ASTM D 395, except as follows: Test specimens shall be compressed 50% of their original thickness. The load shall be released at the end of the test period and the thickness measured after 30 min rest at room temperature. Thickness measurements shall be made as described in Section 14, Measurement of Test Specimens. The temperature of the test shall be $70 \pm 2°C$ ($158 \pm 3.6°F$). The time of the test shall be as specified. Chromium-plated metal plates are not required. Aluminum plates, or any stiff plates that are clean and smooth and that will not deflect measurably under load necessary for deflection of the specimen, may be used.

16.3 Calculations—Calculate the constant deflection compression set, expressed as a percentage of the original height as follows:

$$C_h = \frac{t_0 - t_1}{t_0} \times 100$$

where: C_h = compression set expressed as a percentage of the original height
t_0 = original height of test specimen
t_1 = height of test specimen 30 min + 10 or −0 min after removal from the apparatus

Calculate the constant deflection compression set, expressed as a percentage of the original deflection as follows:

$$C_d = \frac{t_0 - t_1}{t_0 - t_s} \times 100$$

where: C_d = compression expressed as a percentage of the original deflection
t_0 = original height of test specimen
t_s = height of spacer bar used
t_1 = height of test specimen 30 min + 10 or −0 min after removal from apparatus

17. Indentation Test

17.1 Scope—This test consists of measuring the load necessary to produce a 25% indentation in the latex-foam-rubber product.

17.2 Apparatus—An apparatus having a flat circular indentor foot 0.03 m^2 (50 in^2) in area, connected to a force measuring device by means of a ball-and-socket joint, and mounted in such a manner that the product or specimen can be deflected at a rate of 0.2–10 mm/s (0.5–25.0 in/min) shall be used for this test. A maximum radius of 2 mm (0.07874 in) is allowable on the edge of the indentor foot. The apparatus shall be arranged to support the specimen on a level horizontal plate which is perforated with 6 mm (0.25 in) holes on 20 mm (0.75 in) centers to allow for rapid escape of air during the test.

NOTE: When testing products with parallel top and bottom surfaces, the ball-and-socket joint is not required.

17.3 Test Specimens—The test specimen shall consist of the entire product sample or a suitable portion of it, except that in no case shall the surface for identation have dimensions less than 300 x 300 mm (12 x 12 in). The full thickness of the product shall be used.

17.4 Procedure—The procedure for identation test should be made in accordance with ASTM D 1055. In cases of dispute, the compression readings shall be performed at a temperature of 23 ± 1.1°C (73.4 ± 2°F) and in an atmosphere having a relative humidity of 50 ± 2%. The product shall be conditioned undeflected and undistorted at this temperature and humidity for at least 12 h before being tested. Ordinarily only one test will be made, but in case of dispute the result shall be expressed as the average of a minimum of three tests.

18. Flexing Test (Suffix H)

18.1 Scope—The flexing test consists of subjecting the test specimen to repeated compression and noting the effect on the cellular structure.

18.2 Test Specimens—The test specimen shall consist of the entire product sample or a suitable portion of it as agreed upon by manufacturer and purchaser. The full thickness of the product shall be used.

18.3 Procedure—Flexing test shall be made in accordance with ASTM D 1055.

19. Low-Temperature Test (Suffix F1, −40°C (−40°F); Suffix F2, −55°C (−67°F))

19.1 Apparatus—The apparatus shall consist of two parallel plates at least 38 mm (1.5 in) in diameter, one of which is movable and the other one stationary, a means of applying a load and a means of accurately measuring the distance between the parallel plates.

19.2 Test Specimens—Cylinders 29 mm (1.129 in) in diameter shall be used for this test. The minimum thickness shall be 19 mm (0.75 in). They shall be dried in a desiccator for not less than 16 h before testing. The thickness shall be measured and recorded.

19.3 Procedure—The compression deflection of the specimen shall first be measured at room temperature and the load in N/m² (psi) necessary to obtain a 25% deflection recorded. The specimen shall then be placed in the cold box for 5 h at the specified temperature, at the end of which time the previously determined load shall be applied as rapidly as possible while the specimens are still in the cold box and the deflection recorded 30 s later.

19.4 Calculation—The percent change in deflection shall be calculated as follows:

$$C = \frac{D - E}{D} \times 100$$

where: C = % change in deflection
D = deflection at room temperature
E = deflection at temperature of test

20. Static Fatigue Test

20.1 Procedure

20.1.1 SLAB STOCK—Bend a specimen of latex foam 100 x 230 mm (4 x 9 in) parallel to the 100 mm (4 in) dimension to an angle of 180 deg between two compression plates and place in a Geer oven at 70 ± 2°C (158 ± 3.6°F) for 22 h. The opening between the two plates should be equal to twice the thickness of the unfolded specimen. The folded edge of the specimen should not extend beyond the edge of the compression plates.

20.1.2 CORED STOCK—Test the specimen of cored stock as above except that the specimen shall be 100 mm (4 in) wide and the length shall be approximately three times the thickness. Fold out the skin side, if present for testing.

APPENDIX

Tolerances on Dimensions of Latex-Foam-Rubber Products—The tolerances on dimensions of latex-foam-rubber products shown in Tables 2 and 3 are for guidance only and shall not be considered as part of these specifications.

TABLE 2—TOLERANCES ON DIMENSIONS OF LATEX FOAM RUBBER PRODUCTS FOR GENERAL APPLICATIONS

Type	Dimension	Tolerance		Dimension	Tolerance	
		Plus	Minus		Plus	Minus
	Thickness, mm			Thickness, in		
Cored	0 to 76, incl	3	2	0 to 3, incl	1/8	1/16
	76 to 127, incl	5	3	3 to 5, incl	3/16	1/8
	127 and over	6	5	5 and over	1/4	3/16
Uncored	Up to and including 12.7	2	2	Up to and including 1/2	1/16	1/16
	From 12.7 to 25.4, incl	3	2	From 1/2 to 1, incl	1/8	1/16
	Over 25.4	3	5	Over 1	1/8	3/16
	Length and Width, mm			Length and Width, in		
Cored	0 to 152, incl	5	2	0 to 6, incl	3/16	1/16
	152 to 305, incl	10	3	6 to 12, incl	3/8	1/8
	305 to 610, incl	13	6	12 to 24, incl	1/2	1/4
	610 to 914, incl	16	10	24 to 36, incl	5/8	3/8
	914 to 1219, incl	19	13	36 to 48, incl	3/4	1/2
	1219 to 1524, incl	22	16	48 to 60, incl	7/8	5/8
	1524 to 1829, incl	25	19	60 to 72, incl	1	3/4
	1829 and over	29	22	72 and over	1-1/8	7/8
Uncored	0 to 152, incl	8	2	0 to 6, incl	5/16	1/16
	152 to 305, incl	13	3	6 to 12, incl	1/2	1/8
	305 to 610, incl	18	6	12 to 24, incl	11/16	1/4
	610 to 914, incl	22	10	24 to 36, incl	7/8	3/8
	914 to 1219, incl	29	13	36 to 48, incl	1-1/16	1/2
	1219 to 1524, incl	35	16	48 to 60, incl	1-1/4	5/8
	1524 to 1829, incl	38	19	60 to 72, incl	1-3/8	3/4
	1829 and over	41	22	72 and over	1-1/2	7/8

TABLE 3—TOLERANCES FOR SPECIAL APPLICATIONS OF LATEX FOAM RUBBERS, SUCH AS AUTOMOTIVE TOPPER PADS, SPRING COVERINGS, ETC.

Type	Dimension	Tolerance		Dimension	Tolerance	
		Plus	Minus		Plus	Minus
	Thickness, mm			Thickness, in		
Cored	0 to 76, incl	5	2	0 to 3, incl	3/16	1/16
	76 to 127, incl	6	3	3 to 5, incl	1/4	1/8
	127 and over	8	5	5 and over	5/16	3/16
Uncored	Up to and including 12.7	2	2	Up to and including 1/2	1/16	1/16
	From 12.7 to 25.4, incl	3	2	From 1/2 to 1, incl	1/8	1/16
	Over 25.4	3	3	Over 1	1/8	1/8
	Length and Width, mm			Length and Width, in		
Cored and Uncored	0 to 152, incl	8	2	0 to 6, incl	5/16	1/16
	152 to 305, incl	13	3	6 to 12, incl	1/2	1/8
	305 to 610, incl	18	6	12 to 24, incl	11/16	1/4
	610 to 914, incl	22	10	24 to 36, incl	7/8	3/8
	914 to 1219, incl	29	13	36 to 48, incl	1-1/16	1/2
	1219 to 1524, incl	35	16	48 to 60, incl	1-3/8	5/8
	1524 to 1829, incl	38	19	60 to 72, incl	1-1/2	3/4
	1829 and over	41	22	72 and over	1-5/8	7/8

SPONGE- AND EXPANDED CELLULAR-RUBBER PRODUCTS—SAE J18 DEC79

SAE Recommended Practice

Report of the Nonmetallic Materials Committee, approved January 1952, completely revised December 1979. Conforms substantially with ASTM D 1056.

1. Scope—These specifications and methods of testing apply to cellular-rubber products known as sponge rubbers and expanded rubbers, but do not apply to latex foam rubbers. The base material used in their manufacture may be natural rubber, synthetic rubber, and rubberlike materials, alone or in combination. Ebonite cellular rubbers are not included.

Extruded or molded shapes or sizes too small for cutting standard test specimens are difficult to classify or test by these methods and will usually require special testing procedures.

In cases of conflict between the provisions of this general specification and those of a detailed specification, the latter shall take precedence. References to these methods for testing cellular-rubber products should specifically state the particular test or tests desired.

2. Description of Terms

2.1 Flexible—A flexible cellular organic polymeric material will not rupture when a specimen 200 x 25 x 25 mm (8 x 1 x 1 in) is bent around a 25 mm (1 in) diameter mandrel at a uniform rate of one lap in 5 s at a temperature between 18 and 29°C (65 and 85°F).

2.2 Cellular Rubbers—A generic term for materials containing many cells (either open, closed, or both) dispersed throughout the mass.

2.3 Rubber—A material that is capable of recovering from large deformation quickly and forcibly and can be, or already is, modified to a state in which it is essentially insoluble (but can swell) in boiling solvent such as benzene, methyl ethyl ketone, and ethanol-toluene azeotrope. A rubber in its modified state, free of diluents, retracts within 1 min to less than 1.5 times its original length after being stretched at room temperature (18 to 29°C or 65 to 85°F) to twice its length and held 1 min before release.

2.4 Skin—A relatively dense layer at the surface of a cellular material. Normally, this skin is formed by contact with the mold or cover plates during manufacture. Molded open-cell (sponge) parts usually have skin on all surfaces except when cut to length from longer strips. Parts made by cutting from open-cell (sponge) sheets usually have skin on two faces and open cells at the cut edges. Closed-cell (expanded) rubber sheets are frequently split from thicker pieces and consequently do not have the skin faces. On some products, it is desirable to add a solid rubber skin coating. The use to which a cellular rubber product is put determines the thickness of the added skin required. Products subject to abrasion or those which must withstand absorption of water or transmission of gases will ordinarily require an applied skin coating. Expanded (closed cell) rubber does not usually require an added skin because it is somewhat more abrasion resistant than open cell sponges and does not absorb water or transmit gases because of its closed cell structure. In all cases where a skin is applied, there should be good adhesion between it and the cellular rubber.

2.5 Sponge Rubber—Cellular rubber consisting predominantly of open interconnecting cells made from a solid rubber compound.

2.6 Expanded Rubber—Cellular rubbers having closed cells made from a solid rubber compound.

3. Manufacture

3.1 Sponge Rubbers—Sponge rubbers are made by incorporating into the compound an inflating agent, such as sodium bicarbonate that gives off a gas which expands the mass during the vulcanization process. Sponge rubbers are manufactured in sheet, strip, molded, or special shapes. Unless otherwise specified, sheet and strip sponge rubber shall have a natural skin on both the top and bottom surfaces. Fabric surface impressions are ordinarily not objectionable. The coarseness of the impressions shall be agreed upon by the parties concerned.

3.2 Expanded Rubbers—Closed-cell rubbers are made by incoporating gas-forming ingredients in the rubber compound or by subjecting the compound to high-pressure gas, such as nitrogen. Expanded rubbers are manufactured in sheet, strip, molded, and special shapes by molding or extruding. Unless otherwise specified, the presence of skin on the top or bottom surfaces of sheet and strip expanded rubber shall be optional. Extruded shapes have skin on all surfaces except cut ends.

4. Types of Cellular Rubbers—These specifications cover three types of cellular rubbers designated by the prefix letters R, S, and T as follows:

4.1 Type R—Cellular rubbers made from natural rubber, reclaimed rubber, synthetic rubber, or rubberlike materials, alone or in combination, where specific resistance to the action of petroleum-base oils is not required.

4.2 Type S—Cellular rubbers made from synthetic rubbers or rubberlike materials, alone or in combination, having specific requirements for resistance to the action of petroleum-base oils or other organic fluids.

4.3 Type T—Cellular rubber made from synthetic rubber or rubberlike materials, alone or in combination, for resistance to extreme temperatures.

5. Classes, Grades, and Suffixes of Cellular Rubbers

5.1 Classes—Type S rubbers are divided into two classes designated by the letters B and C added to the prefix S as follows:

5.1.1 CLASS SB—Cellular rubbers made from synthetic rubber or rubberlike materials having oil resistance with low swell.

5.1.2 CLASS SC—Cellular rubbers made from synthetic rubber or rubberlike materials having oil resistance with medium swell.

Type R, S, and T rubbers are divided into two classes designated by the letter O for open cell sponge rubbers and E for closed cell (expanded) rubbers.

5.2 Grades—Each type of cellular rubber has a number of grades. Grades are designated by numbers following prefix letters.

Grade numbers shall consist of two digits, the first of which identifies the kind of cellular rubber as follows:

> 1—Sponge rubbers
> 4—Expanded rubbers

The second digit is used to indicate the degree of firmness of the cellular rubbers, the softer grades being identified with the lower numbers and the firmer grades being identified with the higher numbers (see Tables 1, 2, and 4).

5.3 Suffix Letters—Suffix letters may be added singly or in combination after any grade number to indicate additional requirements beyond those specified in Tables 1, 2, and 4 as basic requirement. The significance of the approved suffix letters is shown in Table 3. The test methods and values must be arranged between the purchaser and supplier.

6. Material and Workmanship—Cellular rubbers furnished under these specifications shall be manufactured from natural rubber, synthetic rubber, or rubberlike materials together with added compounding ingredients of such nature and quality that a finished product complies with the specification requirements. In permitting choice in use of those materials by the producer, it does not imply that the different rubber materials are equivalent in respect to all physical properties. Any special characteristics other than those prescribed in these specifications which may be desired for specific applications shall be specified in the product specifications as they may influence the choice of the type of rubber material or other ingredients used. All materials and workmanship shall be in accordance with good commercial practice, and the resulting cellular rubbers shall be free from defects affecting serviceability.

6.1 Color—Unless otherwise specified, the color of cellular rubbers shall be black.

6.2 Physical Properties—The various grades of cellular rubber shall conform to the requirements as to physical properties prescribed in Tables 1, 2, and 4 together with any additional requirements indicated by suffix letters in the grade designations as described in Section 5, Classes, Grades, and Suffixes of Cellular Rubbers.

6.3 Methods of Testing—Unless specifically stated otherwise, all tests shall be made in accordance with the methods specified in Section 7, General Methods.

6.4 Tolerances on Dimensions—Tolerances on dimensions of cellular rubber products are given in the Appendix (Table 5). These tolerances are published as information for guidance only and shall not be considered as part of these specifications.

6.5 Packaging and Marking—The material shall be properly and adequately packaged. Each package or container shall be legibly marked with the name of the material, name or trademark of the manufacturer, and any required purchaser's designations.

6.6 Inspection and Rejection—All tests and inspections shall be made at the place of manufacture prior to shipment, unless otherwise specified. The manufacturer shall afford the inspector all reasonable facilities for tests and inspections.

The purchaser may make the tests and inspection to govern acceptance or rejection of the material at his own laboratory or elsewhere. All samples for testing, provided as specified in paragraph 7.6, Sampling, shall be visually inspected to determine compliance with material, workmanship, and color requirements.

Any material which fails in one or more of the test requirements may be retested. For this purpose, two additional tests shall be made for the requirement in which failure occurred. Failure of either of the retests shall be cause for final rejection.

Rejected material shall be disposed of as directed by the manufacturer.

7. General Methods—Except as otherwise specified in the methods of testing cellular rubbers given in ASTM D 1056, the following methods of test of the American Society for Testing and Materials, applicable in general to vulcan-

TABLE 1—PHYSICAL REQUIREMENTS OF CELLULAR RUBBERS, CLASS O, OPEN-CELL SPONGE

Grade No.	Basic Requirements					Replacements Added by Suffix Letters		
	Compression Deflection, 25% Deflection (Limits)		Oil-Aged 22 h at 70°C (158°F), Change in Volume in ASTM Oil No. 3 (Limits), %	Oven-Aged 7 d at 70°C (158°F), Change from Original Deflection Values (Limits), %	Compression Set, 22 h at 70°C (158°F), −50% Deflection, max, %	Suffix B	Suffix F1	Suffix F2
						Compression Set, 22 h at 70°C (158°F), 50% Deflection, max, %	Low-Temperature Test at −40°C (−40°F), Change from Original Deflection Values, max, %	Low-Temperature Test at −55°C (−67°F), Change from Original Deflection Values, max, %
	kPa	psi						
Type R, Non-Oil Resistant									
RO 10	3.5–14	0.5–2	—	±20[a]	15	—	25	25
RO 11	14–35	2–5	—	±20	15	—	25	25
RO 12	35–63	5–9	—	±20	15	—	25	25
RO 13	63–91	9–13	—	±20	15	—	25	25
RO 14	91–119	13–17	—	±20	15	—	25	25
RO 15	119–168	17–24	—	±20	15	—	25	25
Type S, Class SB, Oil-Resistant, Low Swell									
SBO 10	3.5–14	0.5–2	−25 to +10	±20[a]	40	—	50	—
SBO 11	14–35	2–5	−25 to +10	±20	40	—	50	—
SBO 12	35–63	5–9	−25 to +10	±20	40	—	50	—
SBO 13	63–91	9–13	−25 to +10	±20	40	—	50	—
SBO 14	91–119	13–17	−25 to +10	±20	40	—	50	—
SBO 15	119–168	17–24	−25 to +10	±20	40	—	50	—
Type S, Class SC, Oil-Resistant, Medium Swell									
SCO 10	3.5–14	0.5–2	+10 to +60	±20[a]	50	25	50	—
SCO 11	14–35	2–5	+10 to +60	±20	50	25	50	—
SCO 12	35–63	5–9	+10 to +60	±20	50	25	50	—
SCO 13	63–91	9–13	+10 to +60	±20	50	25	50	—
SCO 14	91–119	13–17	+10 to +60	±20	50	25	50	—
SCO 15	119–168	17–24	+10 to +60	±20	50	25	50	—

[a] If this grade after aging still falls within the compression-deflection requirement of 7 + 7, −3.5 kPa (1 + 1, −1/2 psi), it shall be considered acceptable even though the change from the original load deflection is greater than ±20%.

TABLE 2—PHYSICAL REQUIREMENTS OF CELLULAR RUBBERS, CLASS E, CLOSED-CELL, EXPANDED

Grade No.	Basic Requirements					Requirement Added by Suffix Letters
	Compression Deflection, 25% Deflection (Limits)		Fluid Immersion, 7 d at 23°C (73.4°F), Change in Weight in ASTM Reference Fuel B, max, %[a]	Oven-Aged 7 d at 70°C (158°F) Change from Original Deflection Values (Limits), %	Water Absorption, max, weight %[b]	Suffix B
						Compression Set, 22 h at Room Temperature, 50% Deflection, After 24 h Recovery at Room Temperature, max, %
	kPa	psi				
Type R, Non-Oil Resistant						
RE 41	14–35	2–5	—	±30	5	25
RE 42	35–63	5–9	—	±30	5	25
RE 43	63–91	9–13	—	±30	5	25
RE 44	91–119	13–17	—	±30	5	25
RE 45	119–168	17–24	—	±30	5	25
Type S, Class SB, Oil-Resistant, Low Swell						
SBE 41	14–35	2–5	50	±30	5	25
SBE 42	35–63	5–9	50	±30	5	25
SBE 43	63–91	9–13	50	±30	5	25
SBE 44	91–119	13–17	50	±30	5	25
SBE 45	119–168	17–24	50	±30	5	25
Type S, Class SC, Oil-Resistant, Medium Swell						
SCE 41	14–35	2–5	—	±30	5	25
SCE 42	35–63	5–9	—	±30	5	25
SCE 43	63–91	9–13	—	±30	5	25
SCE 44	91–119	13–17	—	±30	5	25
SCE 45	119–168	17–24	—	±30	5	25

[a] This test of weight change in Reference Fuel B is used in place of the usual oil resistance test of volume change in No. 3 oil for the following reason. Oil or solvent immersion of flexible closed-cellular materials usually causes loss of gas by diffusion through the softened cell walls, which results in some shrinkage of the test sample. This shrinkage counteracts the swell which would normally occur, thus invalidating test data based on volume change. Reference Fuel B is used because it produces a wider and more consistent differentiation among the R, SB, and SC grades than does the No. 3 oil.

[b] For cellular materials with densities of 160 kg/m³ (10 lb/ft³) or less, the value of water absorption allowed is 10% max by weight. For densities of more than 160 kg/m³ (10 lb/ft³), the value of water absorption is 5% max by weight.

TABLE 3—ASTM METHODS OF TEST[a]

Basic Requirements and Suffix Number Requirement or Suffix Letter	Basic	1	2	3	4
Compression Deflection	D 1056 Sections 17–20				
Heat Resistance	D 1056 Sections 15–16 Change in compression deflection after aging 7 days at 70°C (158°F)				
Oil Resistance (SBO and SCO Rubbers Only)	D 1056 Sections 24–25 22 h at 70°C (158°F)				
Compression Set (RO, SBO, and SCO Rubbers Only)	D 1056 Sections 21–23 22 h at 70°C (158°F) 50% deflection 30 min recovery at RT				
Water Absorption (RE and TE Rubbers Only)	D 1056 Sections 30–32				
Suffix A, Heat Resistance		D 1056 Sections 15–16 Change in compression deflection after aging 22 h at 100 ± 1°C (212°F)	D 1056 Sections 15–16 Change in compression deflection after aging 22 h at 125 ± 1°C (257°F)	D 1056 Sections 15–16 Change in compression deflection after aging 22 h at 150 ± 1°C (302°F)	D 1056 Sections 15–16 Change in compression deflection after aging 22 h at 175 ± 1°C (347°F)
Suffix B, Compression Set		D 1056 Sections 21–23 22 h at 70°C (158°F), 50% deflection, 30 min recovery at RT	D 1056 Sections 21–23 22 h at RT, 50% deflection, 24 h recovery at RT		
Suffix C, Ozone or Weather Resistance		D 1171[c] Ozone exposure Method A	D 1171[c] Outdoor exposure	D 1171[c] Ozone exposure Method B	
Suffix D, Load Deflection[b]					
Suffix E, Fluid Resistance		D 1056[d] Sections 33–36 150% max	D 1056[d] Sections 33–36 50% max		
Suffix F, Low Temperature Resistance		D 1056 Sections 26–29 5 h at −40°C (−40°F)	D 1056 Sections 26–29 5 h at −55°C (−67°F)	D 1056 Sections 26–29 5 h at −75°C (−103°F)	
Suffix G, Tear Resistance[b]					
Suffix H, Flex Resistance[b]					
Suffix J, Abrasion Resistance[b]					
Suffix K, Adhesion Capability[b]					
Suffix L, Water Absorption[b]					
Suffix M, Flammability Resistance[b]					
Suffix N, Impact Resistance[b]					
Suffix P, Staining Resistance[b]					
Suffix R, Resilience[b]					
Suffix Z, Special Requirements[b]					

[a] The designations refer to the following methods of the American Society for Testing and Materials.*
  D 1056, Specification for Sponge and Expanded Cellular Rubber Products, Sections 12–36.
  D 1171, Test for Weather Resistance Exposure of Automotive Rubber Compounds.
[b] Test method and values to be arranged between the purchaser and the supplier.
[c] Ratings to be arranged between the purchaser and the supplier.
[d] See Table 2 for materials having densities of 160 kg/m³ (10 lb/ft³) or less.

NOTE: Example—Grade RO11 C1F1 denotes soft sponge rubber containing natural, reclaimed, synthetic, or blends of these rubbers with a compression deflection value of 14–35 kPa (2–5 psi), having no specific solvent or oil resistance, and requiring in addition to the basic tests, a weather resistance test run in accordance with ASTM D 1171, Test for Weather Resistance Exposure of Automotive Rubber Compounds.* Ozone Chamber Exposure Method A, and a low temperature test at −40°C (−40°F). Examples of specification conversion are given in Table 5.

* 1978 Annual Book of ASTM Standards Part 37.

TABLE 4—PHYSICAL REQUIREMENTS OF TYPE T—EXTREME TEMPERATURE RESISTANT CELLULAR RUBBER −75 to +175°C (−103 to +347°F)

	Basic Requirements					Requirements Added by Suffix Letters	
	Compression Deflection at 25% Deflection	Compression Deflection After Heat Aging	Compression Deflection at Low Temperature	Compression Set Under Constant Deflection (50%)	Water Absorption	Compression Deflection After Heat Aging (Suffix A4)	Compression Deflection at Low Temperature (Suffix F3)
	23 ± 3°C (73 ± 5°F)	22 h/1.50 ± 2°C (302 ± 3.6°F)	5 h/−55 ± 2°C (−67 ± 3.6°F)	22 h/100 ± 1°C (212 ± 1.8°F)	3 min/Room Temperature	22 h/175 ± 2°C (347 ± 3.6°F)	5 h/−75 ± 2°C (−103 ± 3.6°F)
Grade Numbers	(limits) kPa (psi)	Change from Original Compression Deflection, (Limits), %	Change from Original Compression Deflection, Max %	Max %	Water Max, Mass %	Change from Original Compression Deflection, (Limits), %	Change from Original Compression Deflection, Max %
TO 11	14–35 (2–5)	±5	5	50	—	±25	25
TO 12	35–63 (5–9)	±5	5	30	—	±25	25
TO 13	63–105 (9–15)	±5	5	30	—	±25	25
TO 14	105–147 (15–21)	±5	5	30	—	±25	25
TO 15	147–203 (21–29)	±5	5	30	—	±25	25
TE 41	14–35 (2–5)	±5	5	80	5	—	—
TE 42	35–63 (5–9)	±5	5	60	5	—	—
TE 43	63–105 (9–15)	±5	5	60	5	—	—
TE 44	105–147 (15–21)	±5	5	60	5	—	—
TE 45	147–203 (21–29)	±5	5	60	5	—	—

ized rubber, shall be complied with as required and hereby made a part of these test methods.

7.1 General Physical Requirements—See ASTM D 1056.

7.2 Aging Test—See ASTM D 572, ASTM D 573, and ASTM D 1056.

7.3 Compression Set, Suffix B—See ASTM D 395.

7.4 Low Temperature Test, Suffix F—See method described in ASTM D 1056.

7.5 Fluid Immersion—See ASTM D 471. In case of conflict between provisions of the above methods and the procedure herein specifically described for cellular rubbers, the latter shall take precedence.

7.6 Sampling—When possible, the completed manufactured product shall be used for the tests specified. Representative samples of the lot being examined shall be selected at random as required.

When it is necessary or advisable to obtain test specimens from the article, as in those cases where the entire sample is not required or adaptable for testing, the method of cutting and the exact position from which specimens are to be taken shall be specified. The apparent density and the state of cure may vary in different parts of the finished product, more especially if the article is of complicated shape or of varying thickness, and these factors affect the physical properties of the specimens. Also, the apparent density is affected by the number of cut surfaces as opposed to the number of skin-covered surfaces on the test specimen.

When the finished product does not lend itself to testing or to the taking of test specimens because of complicated shape, small size, metal or fabric inserts, solid covers, adhesion to metal, or other reasons, standard test slabs shall be prepared. When differences due to the difficulty in obtaining suitable test specimens from the finished part arise, manufacturer and purchaser may agree on acceptable deviations. This can be done by comparing results of standard test specimens and those obtained on actual parts.

7.7 Standard Test Specimens and Slabs—See ASTM D 1056.

7.8 Measurement of Test Specimens—This shall be according to ASTM D 1056.

TABLE 5—TOLERANCES ON DIMENSIONS OF CELLULAR RUBBER PRODUCTS FOR GENERAL APPLICATIONS

Form	Thickness				Length and Width			
	Dimensions		Tolerance		Dimensions		Tolerance	
	mm	in	mm	in	mm	in	mm	in
Sponge Rubbers								
Sheet and strip	3.18 and under Over 3.18 to 12.7, incl Over 12.7	1/8 and under Over 1/8 to 1/2, incl Over 1/2	0.40 0.79 1.19	1/64 1/32 3/64	152 and under Over 152 to 457, incl Over 457	6 and under Over 6 to 18, incl Over 18	1.59 3.18 5%	1/16 1/8 5%
Molded or special shapes	6.35 and under Over 6.35 to 76.2, incl	1/4 and under Over 1/4 to 3, incl	0.79 1.59	1/32 1/16	6.35 and under Over 6.35 to 76.2, incl Over 76.2 to 457, incl Over 457	1/4 and under Over 1/4 to 3, incl Over 3 to 18, incl Over 18	0.79 1.59 3.18 5%	1/32 1/16 1/8 5%

Form	Thickness				Length and Width			
	Dimensions		Tolerance		Dimensions		Tolerance	
	mm	in	mm	in	mm	in	mm	in
Expanded Rubbers								
Sheet and strip	3.18 to 12.7, incl Over 12.7	1/8 to 1/2, incl Over 1/2	1.59 2.38	1/16 3/32	152 and under Over 152 to 305, incl Over 305	6 and under Over 6 to 12, incl Over 12	6.35 9.53 3%	1/4 3/8 3%
Molded or special shapes	3.18 to 12.7, incl Over 12.7 to 38.1, incl Over 38.1 to 76.2, incl	1/8 to 1/2, incl Over 1/2 to 1-1/2, incl Over 1-1/2 to 3, incl	1.59 2.38 3.18	1/16 3/32 1/8	152 and under Over 152 to 305, incl Over 305	6 and under Over 6 to 12, incl Over 12	6.35 9.53 3%	1/4 3/8 3%

8. Test Procedures

8.1 Accelerated Aging Tests

8.1.1 PROCEDURE—The air oven aging test as described in ASTM D 573, Standard Test Method for Rubber Deterioration in an Air Oven, shall be used for cellular rubbers, except that the sample size shall be appropriate for compression-deflection testing. Deterioration shall be expressed as a percentage change of compression-deflection values. No relation between accelerated aging tests and natural aging is given or implied.

8.2 Compression-Deflection Tests

8.2.1 APPARATUS—Any compression machine which meets the following requirements will be satisfactory: The machine shall be capable of compressing the specimen at a rate of 12–50 mm/min (0.5–2 in/min) gently without impact. The machine may be motor or hand driven. It shall be equipped with a gage to measure the deflection caused by the increase in load. The rate of compression of the sample is specified rather than the rate of the compressing platform of the machine. This is an important consideration when scales are used, since sponges of various compression-deflection characteristics will require different times to compress 25% due to the travel of the scale platform under varying loads.

The deflection shall be read on a dial gage graduated in 0.025 mm (or 0.001 in). No gage is necessary if the machine automatically compresses the sample 25%.

8.2.2 TEST SPECIMENS—Standard test specimens shall be used in this test. They shall be cut so that opposite edges are parallel, either from the finished product in a manner agreed upon by the parties concerned or, as shown in ASTM D 1056, from standard test slabs or from commercial flat sheets. The thickness of the test specimen may vary, but shall be measured and stated in the report. The minimum thickness shall be 6 mm ($\frac{1}{4}$ in). Thin samples may be piled-up to obtain this thickness or a standard test slab may be used if agreed upon by the manufacturer and purchaser.[1]

The compression-deflection test is ordinarily made on sponge rubbers and expanded rubbers. The indentation test is usually preferred on latex foam rubbers.

8.2.3 PROCEDURE—See ASTM D 1056.

8.2.4 REPORT—The unit load required, expressed in kilopascals (or pounds per square inch) shall be reported as the result of the compression-deflection test.

8.3 Compression Set Under Constant Deflection—Calculations are based on amount of deflection.

8.3.1 TEST SPECIMENS—Standard test specimens shall be used for this test. They shall be cut so that opposite edges are parallel either from the finished product in a manner agreed upon by the parties concerned, or as shown in ASTM D 1056, from standard test slabs or from commercial flat sheets. The thickness of the test specimens may vary, but shall be measured and stated in the report. The minimum thickness shall be 6 mm ($\frac{1}{4}$ in). Thin samples may be piled-up to obtain this thickness, or a standard test slab may be used if agreed upon by the manufacturer and purchaser.

8.3.2 PROCEDURE—The apparatus and procedure shall be the same as that prescribed in Method B of ASTM D 395, except as follows: For open-cell (sponge) rubbers, compress test specimens to 50% of original thickness. Release the load at the end of the test period and measure the thickness after 30 min at room temperature.

For closed cell (expanded) rubbers, compress test specimens to 50% of their original thickness. Release the load at the end of the test period and measure the thickness after 24 h at room temperature.

In both cases (open-cell sponge and closed-cell expanded rubbers) measure the thickness as described in ASTM D 1056. The temperature of the test for sponge rubbers (open cell) shall be 70°C (158°F). The temperature of the test for expanded rubber (closed cell) shall be 23°C (73°F). The time of the test shall be specified. Chromium plated plates are not required. Aluminum plates or any stiff plates that are clean and smooth, and that will not deflect measurably under load necessary for compression of the specimens may be used.

8.3.3 CALCULATIONS—The percent compression set shall be calculated as follows:

$$\text{Compression Set}, \% = \frac{t_0 - t_1}{t_0 - t_s} \times 100$$

where: t_0 = original thickness
t_1 = thickness of specimen after test
t_s = thickness of spacers used

8.4 Oil Immersion Test (Suffix E)

8.4.1 TEST SPECIMENS—Standard test specimens approximately 12.5 mm ($\frac{1}{2}$ in) in thickness shall be used for this test. The diameter and thickness shall be measured before and after immersion in the specified petroleum-base oil for 22 h at 70°C (158°F) and the percent change in volume calculated. Three specimens shall be run on each test and the average of the three values reported.

8.4.2 PROCEDURE—Follow the procedure for ASTM D 471, Standard Test Method for Rubber Property—Effects of Liquids, Using Petroleum Base Oil No. 3.

8.5 Fluid Immersion Test, Closed Cell (Suffix E)—See ASTM D 1056.

8.6 Low-Temperature Test—Suffix F1 −40°C (−40°F); Suffix F2 −55°C (−67°F); Suffix F3 −75°C (−103°F).

8.6.1 TEST SPECIMENS—Standard test specimens shall be used for this test. The thickness shall be measured and stated in the report. The minimum thickness shall be 6 mm ($\frac{1}{4}$ in). Plied up samples are not satisfactory. The specimen shall be dried in a desiccator for 16 h before testing.

8.6.2 PROCEDURE—The compression-deflection of the specimen shall be measured at room temperature and the load in kilopascals (pounds per square inch) necessary to obtain a 25% deflection recorded. The specimen shall then be placed in the cold box for 5 h at the specified temperature, at the end of which time the previously determined load shall be applied as rapidly as possible while the specimens are still in the cold box, and the deflection recorded 30 s later.

8.6.3 CALCULATION—The percentage change in deflection shall be recorded as follows:

$$\% = \frac{D - E}{D} \times 100$$

where: D = deflection at room temperature
E = deflection at temperature of test

8.7 Water Absorption Test

8.7.1 SCOPE—The water absorption test is applicable to expanded rubbers (closed cell). It should not be used on sponge rubbers or latex foam rubbers (open-cell type) unless they are completely encased in an added skin.

8.7.2 TEST SPECIMENS—Test specimens shall measure 2500 mm² (4 in²) in area by thickness of the product furnished. Specimens should be round, although square samples may be used.

Some products do not lend themselves to flat samples, profile extrusions, for example. When testing such parts, a specimen approximately 50 mm (2 in) long may be used.

In cases of dispute between customer and supplier, 12.5 mm ($\frac{1}{2}$ in) thick by 2500 mm² (4 in²) specimens shall be used.

8.7.3 PROCEDURE—Submerge specimens in distilled water at room temperature, 18–35°C (65–95°F) at least 50 mm (2 in) below the surface of the water and subject to a vacuum of 84 kPa (25 in Hg) for 5 min. Release the vacuum and allow the specimens to remain submerged for 3 min at atmospheric pressure. Remove the specimens, blot dry, and calculate the percentage change in weight.

APPENDIX

A1. Tolerances of Dimensions of Cellular Rubber Products—The tolerances of dimensions of cellular rubber products shown in Table 5 are for guidance only, and shall not be considered as part of these specifications.

---

[1] Using the same compound, thin sections under 6 mm ($\frac{1}{4}$ in) do not blow in the same manner as those over 6 mm ($\frac{1}{4}$ in). The thinner sections are usually higher in compression and density.

LATEX DIPPED GOODS AND COATINGS FOR AUTOMOTIVE APPLICATIONS—SAE J19

SAE Standard

Report of Nonmetallic Materials Committee approved September 1960. Conforms with ASTM D 1764.

1. Scope

(a) These specifications cover dipped goods and coatings made from compounded latex. Products manufactured from this material include boots, coated clips, coated sponge parts, and coated fabrics for automotive applications.

(b) The compounds listed in the following tables are grouped in classifications based primarily on physical properties, which are prescribed in the tables. These values, together with any additional requirements, indicated by suffix letters in the grade designations as described in Section 2, define the properties of the compounds after vulcanization. These values apply to test specimens obtained from standard laboratory dipped films prepared in accordance with procedures described in the applicable ASTM methods. TEST RESULTS FROM FINISHED PRODUCTS MAY NOT DUPLICATE THE VALUES OBTAINED FROM STANDARD TEST FILMS. When differences due to the difficulty in obtaining suitable test specimens from the finished part arise, the purchaser and the supplier may agree on acceptable deviations. This can be done by comparing results obtained on standard test films with those obtained on actual parts.

2. Types, Classes, and Grades of Compounds

(a) *Types*—These specifications cover two types of compounds designated by the prefix letters LR and LS as follows:

TYPE LR—Compounds made from natural rubber, synthetic rubber, or rubberlike materials, alone or in combination, for services where specific resistance to the action of petroleum base fluids is not required.

TYPE LS—Compounds made from synthetic rubber or rubberlike materials for services where specific resistance to the action of petroleum base fluids is required.

(b) *Classes*—Type LR compounds are of one class only, Type LS compounds are divided into two classes, LSB[1] and LSC.

CLASS LSB—Compounds made from synthetic rubber or rubberlike materials having low volume swell in low aniline point oils.

CLASS LSC—Compounds made from synthetic rubber or rubberlike materials having medium volume swell in low aniline point oils.

(c) *Grades*—Each of compounds may have a number of different grades, each having different physical properties. The grade shall be designated by a grade number following the prefix letters and, when necessary, by suffix letters after the grade number.

Suffix Letters may be added singly or in combination after any grade number to indicate additional requirements beyond those specified in the tables as basic requirements for that particular grade.

If no value for the suffix letter requirement is specified in the table, or when no method of test is provided, agreement as to the required value and method of test must be arranged between the purchaser and the supplier.

3. Materials and Workmanship

All materials and workmanship shall be in accordance with good commercial practice, and the resulting product shall be free of bubbles, voids, foreign matter, or other defects affecting serviceability.

4. Color

Unless otherwise specified, these compounds shall be black and free from objectionable bloom.

5. Methods of Testing

The properties enumerated in these specifications shall be determined in accordance with the following methods of the American Society for Testing Materials, except as modified in accordance with certain provisions stated herein. All exposure periods and temperatures prescribed in the following tables shall be given precedence over those specified in the ASTM methods.

(a) **Standard Test Film Preparation**—Standard test films shall be prepared using procedures (coagulation, leach, dry, and cure) identical to the preparation of production parts except thickness which shall be 0.025–0.030 in.

(b) **Durometer Hardness**—ASTM D 676, Tentative Method of Test for Indentation of Rubber by Means of a Durometer, except plied up thickness of test sample to be between 0.175 and 0.250 in.

(c) **Tensile Strength, Elongation and Tension Set**—ASTM D 412, Tentative Method of Tension Testing of Vulcanized Rubber, using Die C. Film thickness to be 0.025–0.030 in.

(d) **Immersion, Including Suffix E2**—ASTM D 471, Tentative Method of Test for Change in Properties of Elastomeric Vulcanizates Resulting from Immersion in Liquids.

(e) **Heat Aging**—ASTM D 573, Method of Test for Accelerated Aging of Vulcanized Rubber by the Oven Method.

(f) **Low Temperature Brittleness**—ASTM D 746, Tentative Method of Test for Brittleness Temperature of Plastics and Elastomers by Impact, except sample thickness to be 0.025–0.030 in.

(g) **Nonstaining, Suffix P**—ASTM D 925, Standard Method of Test for Contact and Migration Stain of Vulcanized Rubber in Contact with Organic Finishes.

6. Sampling and Inspection

A lot, unless otherwise specified, shall consist of all products of the same composition and same grade submitted for inspection at the same time.

TABLE 1—PHYSICAL REQUIREMENTS OF TYPE LR COMPOUNDS, NONOIL RESISTANT

GRADE NUMBER	LR 420
DUROMETER HARDNESS	45 ± 5
TENSILE STRENGTH, MIN PSI	2000
ULTIMATE ELONGATION, MIN %	500
HEAT AGED 70 HOURS AT 158 F	
Change in Durometer Hardness, Max points	±5
Tensile Strength, Min psi	1500
Ultimate Elongation, Min %	400
PERMANENT SET AT 400% ELONGATION, MAX %	10

TABLE 2—PHYSICAL REQUIREMENTS OF TYPE LS COMPOUNDS CLASS LSC, OIL RESISTANT

GRADE NUMBER	LSC 515
BASIC REQUIREMENTS	
Durometer Hardness	55 ± 5
Tensile Strength, Min psi	1500
Ultimate Elongation, Min %	400
Oil Immersion 22 Hr at 212 F. ASTM Oil No. 2, Volume Change, Max %	80
Heat Aged 70 Hr at 212 F	
Change in Durometer Hardness, Max points	+10
Tensile Strength, Min psi	1200
Ultimate Elongation, Min %	300
Permanent Set at 300% Elongation, Max %	20
ADDED REQUIREMENTS	
Suffix E2, 22 Hr at 212 F in ASTM Oil No. 2, Volume Change, Max %	50

[1] No tables have been established for this class; however, this will be done at a later date when values become available.

Suffix Letters	Test Required
C1	Ozone Resistance
E2	Oil Resistance—ASTM Oil No. 2
F1	Low Temperature Brittleness at −40 F
G	Tear Resistance
H	Flex Resistance
K1	Adhesion to metal—Bond made during vulcanization
K2	Adhesion—Cemented bond made after vulcanization
P	Nonstaining
Z	Special Requirements

SYNTHETIC RESIN PLASTIC SEALERS, NONDRYING TYPE—SAE J250

SAE Recommended Practice

Report of Nonmetallic Materials Committee approved May 1959.

Scope—The material desired under this recommended practice is a synthetic resin plastic sealer of the nondrying, nonbleeding, and noncorrosive type that may be extruded to the specified size and used as a medium for producing a water tight seal between 2 pressed steel sections or between rubber and steel.

1. Color

1.1 Gray
The color of the sealer must be light and shall not smear in order to facilitate clean up in assembly.

1.2 Black

2. Physical Properties

2.1 Consistency

Requirements No. 1—Cone penetration, 7–10 mm.

Method of Test—ASTM D 217 using 150 g cone plus 150 g additional weight.

Requirement No. 2—Needle penetration, 18–22 mm.

Method Test—ASTM D 1321 using total weight including needle of 100 g.

2.2 Solids—99.0% minimum.

Method of Test—ASTM D 553—Weight 3 g sample and bake for 3 hr at 215 F.

2.3 Cold Resistance

Requirement No. 1—Pliable at −20 F.

Method of Test—A sample of the material applied to a thin polyethylene or polyester film is exposed to a temperature of −20 F for 4 hr and immediately upon removal from this temperature bent across a 2½ in. diameter mandrel. Cracks shall not be evident.

Requirement No. 2—No evidence of cracking, embrittlement, or loss of adhesion shall be present.

Method of Test—Apply 3/16 x 6 in. bead of sealer to a 4 x 12 in. enameled or lacquered panel. A piece of waxed Kraft paper is placed on top of the beads and rolled with 3 single passes of a 3 lb roller. The panels are then exposed to −20 F for 4 hr and slammed at 90 deg in a cold slam fixture, see Fig. 1.

2.4 Bleeding—No discoloration or migratory staining of the enameled or lacquered finish will be tolerated when the material is subjected to the following test:

Method of Test—Apply a 3/16 x 6 in. bead of the material to a steel panel which has been primed, painted, and baked according to the users specification and subject to a baking temperature of 280 F for 1 hr.

2.5 Stability

Method of Test No. 1—A sample of the material shall be exposed for 24 hr to a temperature of 158 F and after exposure the surface must remain tacky with no appreciable hardening. Bead must be readily removed from backing without stretching over 15%. As received the sealer shall not stretch over 10%.

Method of Test No. 2—A sample of the material after being exposed for 2 weeks to a temperature of 158 F shall not leach its oils or show cracks when bent across a 1 in. diameter mandrel.

2.6 Water Resistance—The water resistance of the material shall conform to the following:

Method of Test—A sample of the material is held under water, at room temperature, and kneaded with the fingers for 1 minute. During the kneading the material shall not disintegrate nor become short showing loss of cohesion. The sealer must be unaffected by water or a 5% sodium chloride solution when immersed for a period of 24 hr at room temperature.

2.7 Solvent Resistance (When Specified)—The solvent resistance of the material shall conform to the following:

Method of Test—A 3/16 x 2 in. bead shall be immersed in gasoline for a period of 70 hr at room temperature. There shall be no deterioration, evidence of solubility, or loss of adhesion to the metal.

2.8 Sag Characteristics—The material shall not sag, blister, nor pull away from a horizontal metal surface when tested under the following conditions:

Method of Test—Apply a ½ in. bead of sealer to the angle joint of the fixture shown in Fig. 2. (Fixture is to be made from 036–042 gage standard body steel and phosphate coated.) Brush or trowel the bead against the sides of the angle so that the material takes a concave shape. Thir may be done with the angle pointing up. Immediately invert the panel and block so that one leg is vertical. Place in the inverted position in an oven and heat for ½ hr at 400 F.

2.9 Aged Adhesion—The adhesion shall be greater than the cohesive strength of the material when an attempt is made to strip or peel a bead of the sealer from a metal panel after being subjected to the following test:

Method of Test—Apply a 3/16 x 6 in. bead of the material to a 4 x 12 in. steel panel which has been primed, painted, and baked according to the users specifications. A piece of waxed Kraft paper is placed on top of the bead and rolled with 3 single passes of a 3 lb roller. The panel shall then be aged for 1 hr at 280 F followed by 1 week at 158 F and a cold cycle of −20 F for 6 hr.

3. Packaging—Beads shall be supplied as specified by the user and must be packaged in such a manner as to prevent the beads from sticking together.

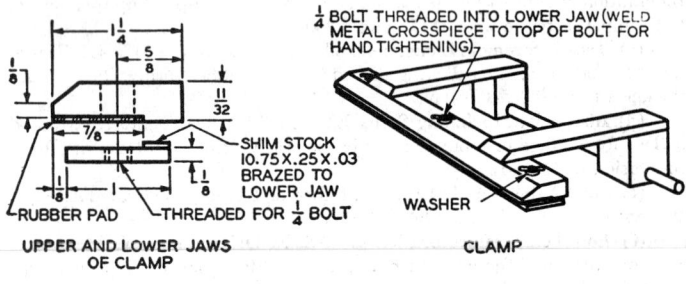

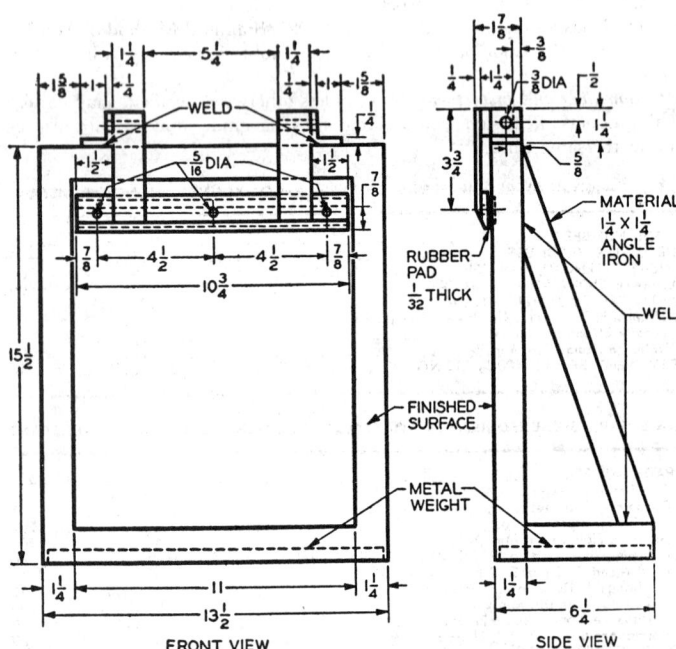

FIG. 1—DETAILED VIEW OF SLAMMING FIXTURE FOR COLD ADHESION TEST

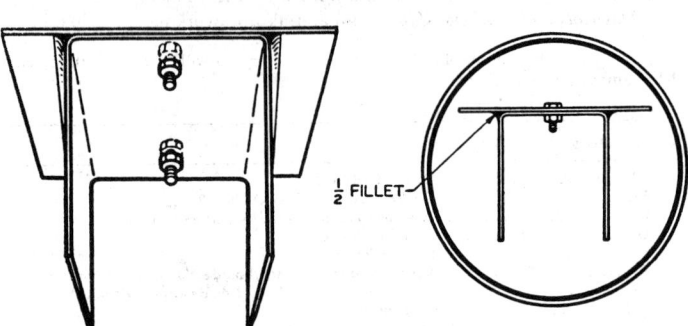

FIG. 2—FIXTURE FOR SAG CHARACTERISTICS TEST

METHODS OF TESTS FOR AUTOMOTIVE-TYPE SEALERS, ADHESIVES, AND DEADENERS—SAE J243

SAE Recommended Practice

Report of Nonmetallic Materials Committee approved October 1971.

Scope—This SAE Recommended Practice contains a series of test methods for use in measuring characteristics of automotive-type sealers, adhesives, and deadeners. The test methods which are contained in this document are as follows:

ADS-1 —Methods of Determining Viscosity
ADS-2 —Low Temperature Tests
ADS-3 —Weld-Through Tests
ADS-4 —Enamel, Lacquer, and Fabric Staining Test

ADS-5 —Wash-Off Resistance Test
ADS-7 —Solids Test
ADS-8 —Flash Point Test
ADS-9 —Sag and Bridging Tests
ADS-10 —Flow Test

The intent of this document is to provide a series of test methods which can be used in testing the various qualities of sealers, adhesives, and deadener material. In later revisions of this document, attempts will be made to reduce the number of tests now presented. The specific temperatures and times at which some of these tests are to be conducted are not dictated in these test procedures, but they will be found in the material standards which govern each type of material to be tested.

ADS-1—METHODS OF DETERMINING VISCOSITY

1. Methods of Conditioning Test Materials Prior to Checking Viscosity.

1.1 For Viscosity Unagitated

1.1.1 CONDITIONING METHOD A—Check the submitted sample as received. Test in the original container or transfer to the test vessel with minimum handling.

1.2 For Viscosity Agitated

1.2.1 CONDITIONING METHOD A—Material shall be subject to the specified number of cycles in a standard mechanical greaseworker as outlined in ASTM D217, Test for Cone Penetration of Lubricating Grease.

1.2.2 CONDITIONING METHOD B—Pass the sample once through the sealer cup using specified pressure and orifice.

1.2.3 CONDITIONING METHOD C—Stir a pint sample to 50 stirs with a 1 x 6 in. steel-bladed spatula.

1.3 For Viscosity Aged—Conditioning (agitated or unagitated as specified):

(a) Condition the sample in a sealed ½ pt can for 72 h in an oven at specified temperature.

(b) Remove from the oven, condition the sample to 77 ±2 F, and determine viscosity.

2. Viscosity Tests

2.1 Viscosity, Pressure Flow Method

2.1.1 APPLICATION—This procedure is used to determine the viscosity of adhesives, sealers, and deadeners. The time required for a specified weight of the material to pass through a specified orifice under a given pressure indicates the viscosity of the material.

2.1.2 EQUIPMENT REQUIRED

2.1.2.1 Castor-Severs Rheometer or Pressure Flowmeter—The pressure flowmeter required for this test is detailed in Fig. 1. The flowmeter is not available commercially but must be fabricated. For example:

	Sealer Cup Orifices							
	A	B	C	D	E	F	G	H
Diameter of Orifice, in	0.052	0.063	0.073	0.104	0.104	0.125	0.200	0.250
Lengths of Orifice, in	0.531	2.00	0.531	0.531	0.750	2.00	0.750	2.00

2.1.2.2 Ring stand and clamps for supporting pressure flowmeter.

2.1.2.3 Pressure gage, 100 lb maximum air gage, calibrated in 2 lb increments.

2.1.2.4 Pressure relief valve—This is an air cock which opens or closes at a single turn.

2.1.2.5 Shutoff valve—Same type as pressure relief valve.

2.1.2.6 Pressure regulator and extractor—This unit may be of any suitable type which will remove oil and water from the air and which will control the pressure of the air delivered to the pressure flowmeter. The regulator and extractor unit is assembled between the shutoff valve and an adequate air source.

2.1.2.7 Pipe cleaners suitable for cleaning orifice of flowmeter.

2.1.2.8 Stopwatch or other timing device calibrated in seconds.

2.1.2.9 Balance, double beam type or equivalent, sensitivity to 0.01 g.

2.1.2.10 Connections 0.25 in. pipe and fittings with standard pipe threads as are necessary for assembling equipment, as shown in Fig. 1.

2.1.2.11 Mechanical convection oven capable of maintaining a temperature of ±2 F.

2.1.3 PROCEDURE

2.1.3.1 Fill the clean and dry sealer cup equipped with specified orifice with the test material, allowing room for the plunger disc, and assemble the apparatus. Care should be taken to avoid air entrapment.

NOTE: Test material and equipment shall be maintained at a temperature of 77 ±2 F during the test.

FIG. 1—CASTOR-SEVERS RHEOMETER OR PRESSURE FLOWMETER (OR KEIL RHEOMETER)

2.1.3.2 Adjust the air line pressure to the flowmeter as designated by the material standard and bleed until free of air. This should be done while the test material is passing through the pressure flowmeter or Severs Rheometer.

2.1.3.3 Close air line valve, place a paper on the balance pan under the flowmeter and bring balance to equilibrium. Add specified weight.

2.1.3.4 Open the air line valve and start the timer when the material touches the paper on the weighing pan.

2.1.3.5 When the specified weight of the sealer has accumulated on the balance pan, stop the timer, close the air line valve, and open the pressure relief valve.

2.1.3.6 Report the viscosity of the material as the number of seconds required for a specified amount of the material to pass through the orifice at the specified pressure. (Note: Take the average of three readings.)

2.2 Brookfield Method

2.2.1 APPLICATION—This procedure is to determine the viscosity of adhesives, deadeners, and thin body sealers. The viscosity is indicated by the resistance produced upon a spindle rotating at a definite speed while immersed in the material under test.

2.2.2 EQUIPMENT—Commercially available Brookfield Viscometer[1] (Fig. 2).

2.2.3 PROCEDURE—Test material and equipment shall be maintained at a temperature of 77 ±2 F during the test.

2.2.3.1 Insert the specified spindle in a pint of test material, keeping the fluid's level below the immersion groove cut in the spindle shaft.

2.2.3.2 Attach spindle to the lower shaft.

2.2.3.3 Lower viscometer so that the groove cut in spindle shaft is flush with the fluid's level.

2.2.3.4 Level the viscometer and set viscometer speed at specified rpm.

2.2.3.5 Depress the clutch and turn on the viscometer motor. Release clutch and allow dial to rotate for 1 min. Take reading at this position. If the pointer has not stabilized at a fixed position after 1 min, the reading shall not be taken until the pointer has stabilized. The time shall then be recorded.

2.2.3.6 Using conversion table, convert to centipoise.

2.2.3.7 When reporting viscosity, the spindle, rpm, and viscometer and model number shall be indicated. Average of three readings.

2.3 MacMichael Method—Used for measuring the viscosity of both Newtonian and non-Newtonian liquids such as sealers, adhesives, and deadeners. The viscosity is given in degrees MacMichael (M).

2.3.1 EQUIPMENT REQUIRED—A commercially available Fisher-MacMichael Viscometer[2] with sample cups, plungers, and different gage wires (Fig. 3).

2.3.2 PROCEDURE

2.3.2.1 Insert specified wire in hollow spindle.

2.3.2.2 Attach specified plunger to the spindle.

2.3.2.3 Suspend spindle assembly from the pointer assembly support.

---

[1] Available from Brookfield Engineering Laboratories, 240 Cushing St., Stoughton, Mass.

[2] Available from Fischer Scientific Co., 1458 N. Lamon Ave., Chicago, Illinois 60651.

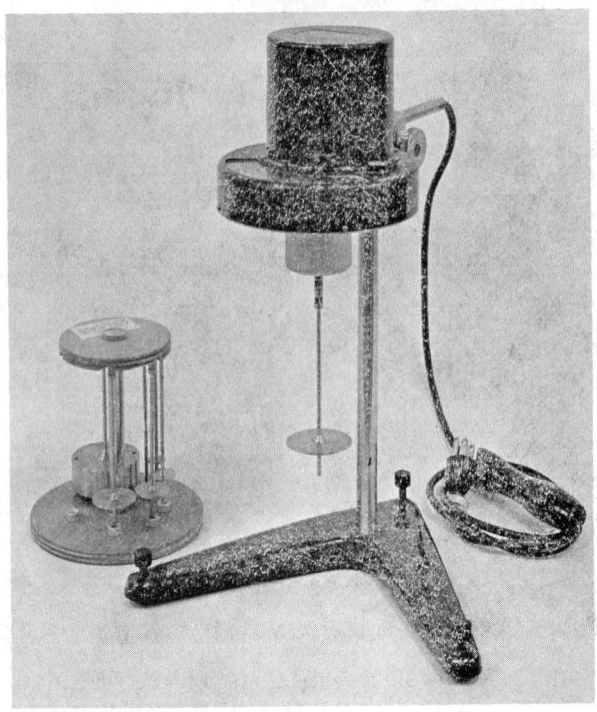

FIG. 2—BROOKFIELD VISCOMETER

2.3.2.4 Level apparatus.
2.3.2.5 Zero the dial.
2.3.2.6 Lift spindle assembly off support.
2.3.2.7 Adjust hot plate rotation speed to rotational speed specified.
2.3.2.8 Adjust the sample temperature to temperature specified.
2.3.2.9 Fill specified clean sample cup (sample depth specified) and place on hot plate.
2.3.2.10 If test temperature is greater than 77 F, adjust thermostat to desired temperature.
2.3.2.11 Replace spindle.
2.3.2.12 Cover sample cup.
2.3.2.13 Check that sample is at desired temperature.
2.3.2.14 Take reading by turning on the rotate switch and read degrees M from the dial at point spindle becomes stationary or at specified time.
2.3.2.15 When reporting degrees M viscosity, the temperature of material, wire gage, plunger, hot plate rotational speed, sample cup size, and sample depth should be indicated.

2.4 Ford Cup Method—Particularly suited for measuring the viscosity of relatively thin adhesives, sealers, and deadeners. The viscosity is given in seconds and is the amount of time it takes for specific amounts of fluid material to pass through a known size orifice.

2.4.1 EQUIPMENT
2.4.1.1 Commercially available Ford Cups[3] (Fig. 4).
2.4.1.2 Ring stand and ring for holding Ford Cup.
2.4.1.3 Timing device for measuring seconds.

[3] Available from Ford Viscosimeter Corp., 7730 W. Fort St., Detroit, Michigan 48209.

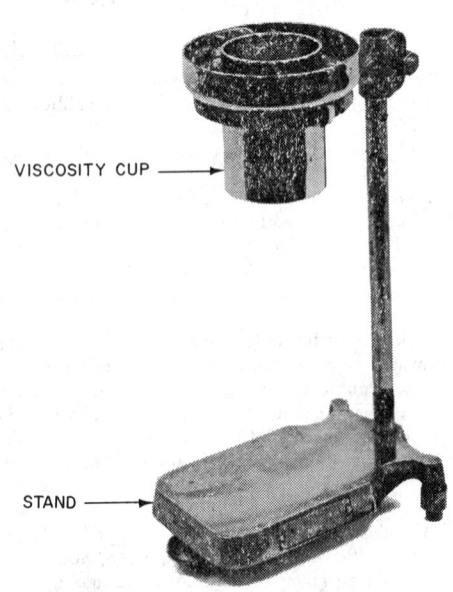

FIG. 3—FISHER-MacMICHAEL VISCOMETER

FIG. 4—FORD CUP

2.4.2 PROCEDURE—Test material and equipment shall be at a temperature of 77 ±2 F during the test.

2.4.2.1 Holding finger over aperture, fill the specified clean cup and orifice with the material being tested.

2.4.2.2 Simultaneously remove finger from aperture and start timing device.

2.4.2.3 When a break in the flow of material through the open aperture occurs or a specified amount of material has flowed, stop the timing device.

2.4.2.4 When reporting Ford viscosity, indicate the Ford Cup used.

2.5 Penetrometer Method—This procedure is used to determine the viscosity of heavy bodied sealers and deadeners. Viscosity is a measure of depth of penetration of a cone or needle into a standard body of material.

2.5.1 EQUIPMENT REQUIRED

2.5.1.1 Commercially available universal penetrometer as described in ASTM D 217 or ASTM D 5, Test for Penetration of Bituminous Materials (Fig. 5).

2.5.1.2 Weights to place on loading bar.

2.5.1.3 Penetrating instrument (cone—ASTM D 217, needle—ASTM D 5).

2.5.1.4 Stopwatch.

2.5.1.5 Sample cup as specified.

2.5.2 PROCEDURE—Test material and equipment shall be at 77 ±2 F during the test.

2.5.2.1 Level penetrometer.

2.5.2.2 Insert specified penetrating instrument into chuck.

2.5.2.3 Set dial reading to zero.

2.5.2.4 Add weights as required to loading bar to achieve specified load. (Load is the total weight of rod and penetrating instrument.)

2.5.2.5 Fill clean sample cup level full with test material, smooth surface, and place it in position centered under the penetrating instrument.

2.5.2.6 Adjust height so as to bring the point of the penetrating instrument exactly into contact with the smooth surface of the sample.

2.5.2.7 Release test rod by pushing the clutch trigger down and holding it down during specified time of the test.

2.5.2.8 At the end of the specified time, lock the test rod by releasing the clutch trigger.

2.5.2.9 Push down the depth gage rod as far as it will go and read the depth of penetration in tenths of millimetres.

2.5.2.10 When reporting penetrometer viscosity, indicate the load, penetration time, and cup size (average of three samples).

2.6 Gardner Mobilometer Method—This method is used for determining the viscosity of adhesives and "thin" bodied sealers. Viscosity is expressed as the time in seconds for a standard plunger assembly (disc, piston rod, weight pan), loaded or unloaded, to fall through 10 cm of the test product.

FIG. 5—PENETROMETER

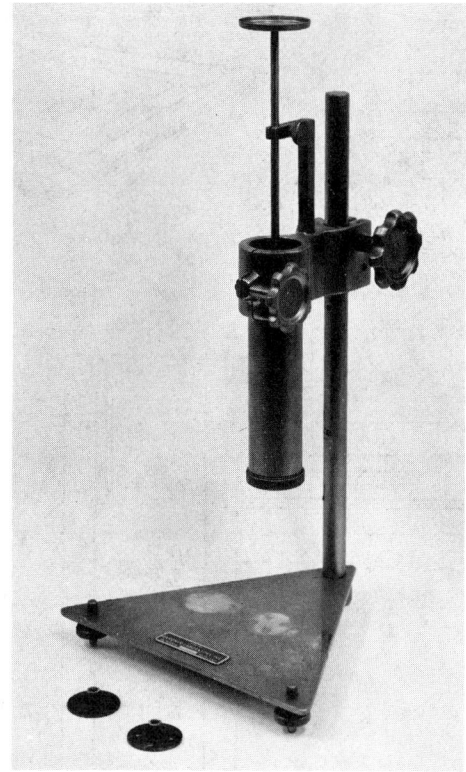

FIG. 6—REGULAR GARDNER MOBILOMETER

2.6.1 EQUIPMENT

2.6.1.1 Commercially available regular Gardner Mobilometer[4] with 51-hole disc, solid disc, and 4-hole disc (Fig. 6).

2.6.1.2 Weights—Various gram amounts in 50 g increments.

2.6.1.3 Timing device for measuring seconds.

2.6.2 PROCEDURE—Test material and equipment shall be maintained at a temperature of 77 ±2 F during the test.

2.6.2.1 Fill clean cylinder to a depth of 20 cm with material to be tested.

2.6.2.2 Level instrument.

2.6.2.3 Attach collar bracket cylinder so top of collar bracket is 4 in. down from top of cylinder.

2.6.2.4 Place piston rod through piston guide and attach specified disc to piston rod.

2.6.2.5 Lower disc into material until lowest mark on the piston rod is flush with the top of the piston guide. (If there is only two marks on piston rod, lower the piston rod into the material until the bottom mark is ½ in. above the piston guide top.)

2.6.2.6 If a load is called for, add weights to weight pan to give specified load. By definition, load shall be considered zero if there is no weights added to the weight pan.

2.6.2.7 Release piston rod and start timer when second lowest mark reaches top of piston guide. (If there is only two marks on piston rod, start timer when bottom mark reaches top of piston guide.)

2.6.2.8 Stop timer when the last mark reaches the top of the piston guide and report viscosity.

2.6.2.9 When reporting Gardner viscosity, indicate the disc used and the load applied.

ADS-2—LOW TEMPERATURE TESTS

1. Method A (Impact Test)—This procedure is used to determine the adhesion properties of sealers when subjected to an impact at a low temperature.

1.1 Equipment Required

1.1.1 Slam fixture capable of delivering a uniform impact to the test panel. See Fig. 7.

1.1.2 Cold box capable of maintaining a temperature of −40 ±2 F.

1.1.3 Circulating air oven capable of maintaining temperatures up to 400 ±2 F.

1.1.4 Analytical balance accurate to 1 mg.

1.1.5 Metal panels—12 x 12 x 0.036 in. cold rolled, low carbon body stock steel, primed or painted as specified.

[4]Gardner Laboratory, Inc., P. O. Box 5728, Bethesda, Maryland 20014.

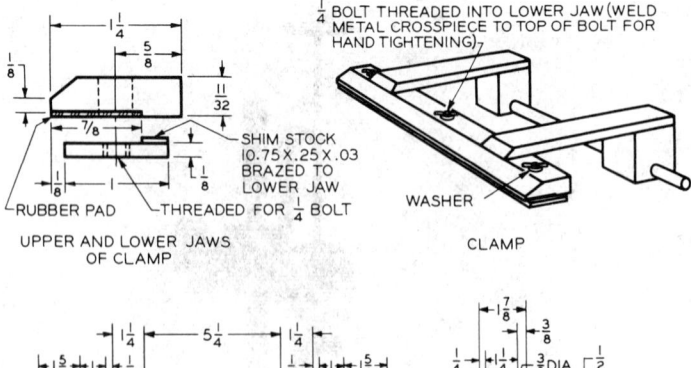

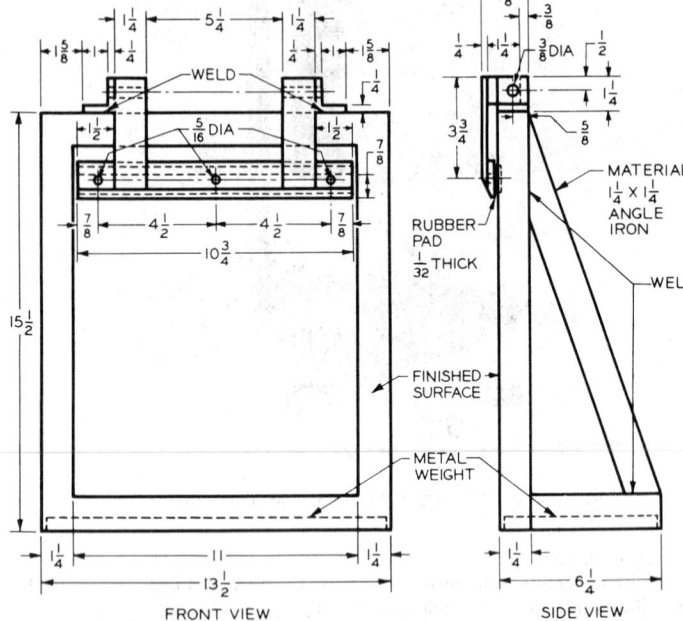

FIG. 7—DETAILED VIEW OF SLAMMING FIXTURE FOR COLD ADHESION TEST

1.1.6 Draw down fixture with opening made to produce the bead size specified.

1.2 General Procedure

1.2.1 Samples shall be mixed or otherwise treated as specified before applying.

1.2.2 Weigh panel and prepare as specified and record weight.

1.2.3 A size bead or ribbon of metal as specified shall be applied to the panel.

1.2.4 The applied bead or ribbon of material shall be conditioned, that is, air dried or baked for the time and temperature specified.

1.2.5 Weigh test panel and cured material and record weight.

1.2.6 Place the panels and specified slam fixture in the cold box for the time and temperature specified.

1.2.7 Insert the test panel in the slam fixture with sealer facing out and in the vertical position. Raise panel to a horizontal position. Release panel and allow it to slam against the test fixture. The number of slams to be a specified.

1.2.8 Inspect for loss of adhesion and remove loose material; calculate weight of material loss and record results as percent loss of adhesion.

2. Method B (Bend Test)—This procedure is used to determine the adhesion properties of seals when bent around a mandrel.

2.1 Equipment Required

2.1.1 Steel mandrel of size diameter as specified.

2.1.2 Aluminum foil 0.001–0.003 in. thick.

2.1.3 Cold box capable of maintaining a temperature of -40 ± 2 F.

2.1.4 Circulating air oven capable of maintaining temperatures up to 400 ± 2 F.

2.1.5 Draw down fixture with opening made to produce the bead size specified.

2.2 General Procedure

2.2.1 Samples shall be mixed or treated as specified.

2.2.2 Aluminum foil shall be prepared as specified.

2.2.3 A size bead or ribbon of material as specified shall be applied to the aluminum foil.

2.2.4 The applied bead or ribbon of material shall be conditioned, that is, air dried or baked for the time and temperature specified.

2.2.5 Place the aluminum foil and the mandrel in the cold box for the time and temperature specified.

2.2.6 While at the test temperature, wrap the aluminum foil 180 deg around the mandrel.

2.2.7 Inspect for cracking and loss of adhesion. Record number and size of cracks and loss of adhesion.

ADS-3—WELD-THROUGH-TESTS

1. Scope—These tests are used to determine acceptability of weld-through sealers.

2. Method A

2.1 Equipment and Supplies

2.1.1 Single point spot welder with low inertia head, transformer tap setting of 4–5 V.

2.1.2 Two WA 2510 spot welding electrodes with $\frac{1}{4}$ in. diameter face and a 45 deg truncated cone.

2.1.3 A galvanometer-type oscillograph for recording sine wave of the secondary current.

2.1.4 A mechanical convection oven capable of maintaining 130 ± 2 F.

2.1.5 Test coupons 1 x 3 x 0.036 in. cold rolled, low carbon, open hearth steel, free from burrs or ragged edges that might provide a shunt path for welding current.

2.2 Welding Schedule

2.2.1 Electrode force: 550 lb.

2.2.2 Weld time: 9 cycles.

2.2.3 Secondary amperes: 11,000 A.

2.2.4 Secondary current time: Full sine wave starting at 0 deg point on first half cycle.

2.3 Procedure—Weld-through characteristics shall be tested in two groups. Each group shall be tested using 25 sets of test coupons with sealer. Bare test coupons shall be welded at the start and finish of the test.

2.3.1 Prepare 50 coupon assemblies with 1 x 0.093 ± 0.015 in. of sealer, as shown in Fig. 8.

2.3.2 Age 25 of the assemblies for 30 d at 72 ± 5 F (group 1).

2.3.3 Condition 25 assemblies for 72 h at 130 ± 2 F in a mechanical convection oven (group 2).

2.3.4 Verify the above specified weld schedule on a set of bare coupons.

2.3.5 Make 25 successive welds on the oven-aged coupons (group 2), one weld for each set of coupons through a bead of sealer. Exercise care to assure that the weld is directly through the sealer material. After each weld, examine the sine wave trace. Acceptable materials must show 90% of full welding current on or before the second half cycle of all welds. The height of the sine wave obtained in welding bare steel of the same thickness shall indicate full welding current.

2.3.6 Repeat the welding test on the 25 sets of coupons which were aged for 30 d (group 1), and examine the sine wave trace. Acceptable materials must show 90% of full welding current on or before the second half cycle of all welds.

2.3.7 During the above welding tests, the sealer shall show no tendency to ignite. There shall be no fouling of the spot welder points if the sealer is in direct contact.

2.3.8 Repeat the verification of the specified weld schedule on a set of bare coupons.

2.3.9 All welds must tear the metal when the "sealer-prepared" and "bare" welded panels are pulled in a tensile shear testing machine.

2.3.10 Measure the diameter of the weld buttons on the destructed shear strength panels after separating them with a chisel.

2.3.11 The material shall flow back around the weld to form a complete seal. Test method: Drill the spot welds from five panels of each of the two groups tested and visually observe for flowback.

3. Method B

3.1 Equipment and Supplies

3.1.1 A press type stationary spot welder.

3.1.2 Two water-cooled electrodes with $\frac{5}{8}$ in. diameter shank having a 45 deg truncated cone with a $\frac{1}{4}$ in. diameter welding face.

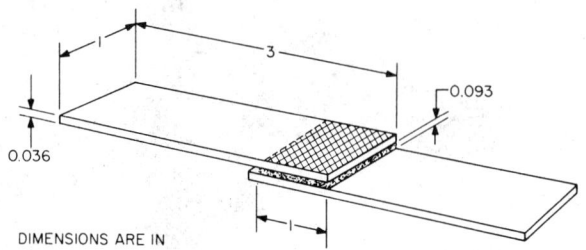

FIG. 8—COUPON ASSEMBLIES WITH SEALER

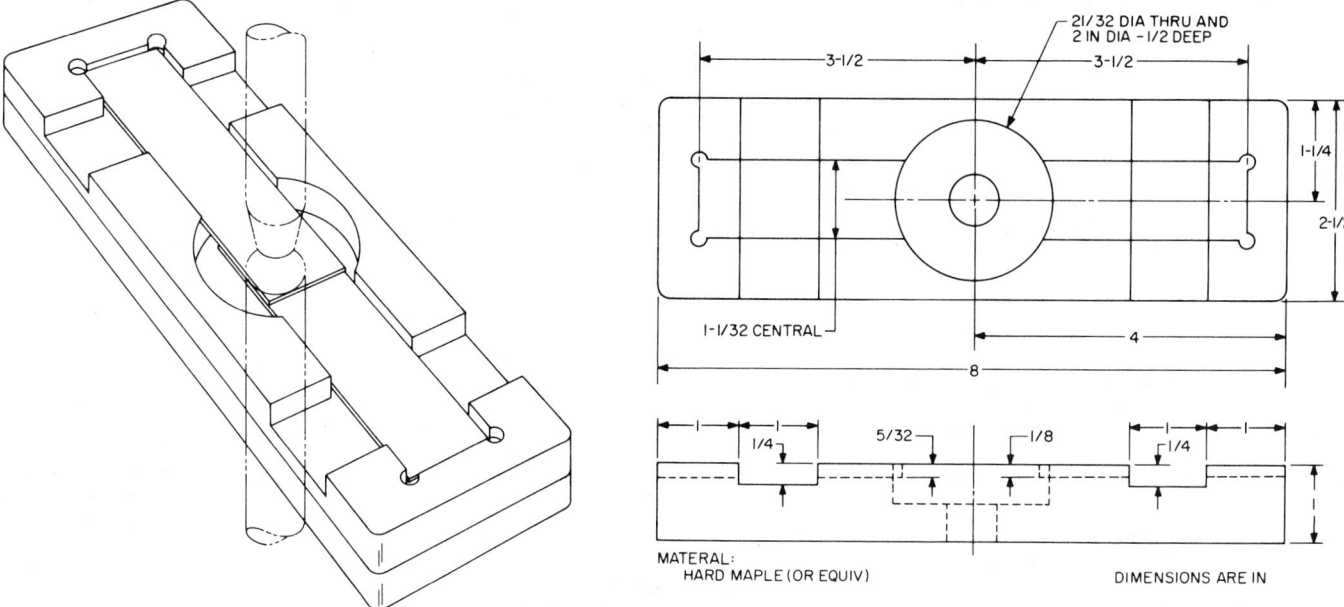

FIG. 9—LOCATING FIXTURE

3.1.3 Test coupons of flat, cold rolled steel free from edge burrs and rust. Size 1 x 4 x 0.035 in. and 4 x 24 x 0.035 in.
3.1.4 Suitable tensile test machine.
3.1.5 Locating fixture (Figs. 9A and 9B).
3.1.6 Notched spreader bar to give $\frac{1}{32}$ in. effective film thickness.

3.2 Welding Schedule
3.2.1 Electrode force: 500 lb.
3.2.2 Weld time: 5 cycles.
3.2.3 Weld current: 9500–10,500 A.

3.3 Procedure
3.3.1 Flat weld 1 x 4 x 0.035 in. coupons without sealer to determine if proper weld is being obtained. Pull in tensile tester to check tensile shear.
3.3.2 WELDABILITY AT VARIOUS DRYING TIMES—Condition sealer sample to 77 ±2 F before testing.
Using 1 x 4 in. cold rolled steel test panels (Fig. 10), apply the spot weld sealer with the notched spreader bar to only one coupon of the weld sample. Material will spread under pressure to form uniform coating of approximately $\frac{1}{32}$ in. thickness. Weld samples using the following drying times after application:
 (a) 0 h.
 (b) 24 h—If not weldable at this drying time, recheck to find limit of drying time relative to weldability.
 (c) 48 h.
 (d) 72 h.
 (e) 96 h.
Weld coupons so that the applied material is in the 1 in. lap joint and with a single spot weld in the center of the lap area (Fig. 10). Use the locating jig for locating coupons to insure accurate alignment. (See Figs. 9A and 9B.) Three welding samples are required for each test.
Weld samples are to be pulled on a tensile test machine. Samples should yield a plug-type failure and a minimum tensile strength of 650 lb.
3.3.3 EFFECT OF MATERIAL UPON WELDABILITY AND ELECTRODE LIFE—Use same equipment as listed in paragraph 3.1.
 3.3.3.1 *Application of Materials*—Use same procedure as outlined in paragraph 3.3.2.
Weld a strip sample 4 x 24 in. immediately after application so that the material, applied to only one strip of the sample, contacts the movable electrode (Fig. 11) and note:
 (a) Flashing.
 (b) Effect on electrode life.
 (c) Amount of electrode pickup.
 (d) Effect of electrode pickup on weldability.
Welds should be at a $\frac{1}{2}$ in. (approximately) spot spacing. Speed of operation: 100 spots/min, repetitive welds to a minimum of 800 welds.
Chisel test the welded strip and examine the sample for weld quality.
Note burn-out condition of material around weld nugget.
Do not redress the electrodes during this test. This test is to be continued until additional welding is considered impractical because of one of the four factors listed above.
Additional welding is considered impractical when any one of these factors exists to such a degree that:
 (a) Flashing is an operator hazard.
 (b) The electrode sticks to the work.
 (c) An insulating coating forms on the electrode face prohibiting the flow of welding current.
 (d) The tensile strength of the welds falls below the 650 lb minimum.
Several trials should be made in order to permit a more accurate evaluation of results.

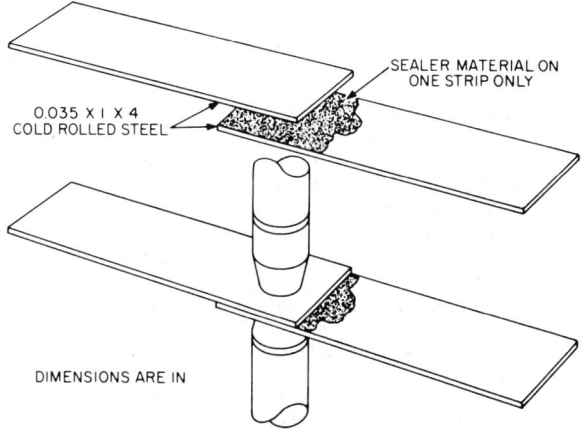

FIG. 10—WELDABILITY TEST

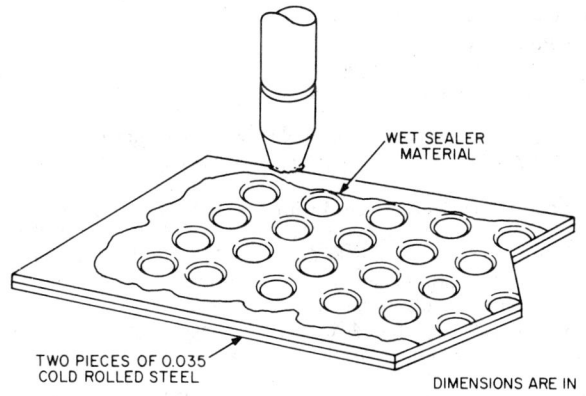

FIG. 11—TIP FOULING TEST

```
WELD TEST DATA SHEET FOR PRIMERS, SEALERS, AND COATED STEELS
MANUFACTURER'S NAME _____ NO. _____
TYPE OF MATERIAL _____
FISHER SPECIFICATION NO. _____ DATE _____
DEPARTMENTAL IDENTIFYING NO. _____

CURRENT:   9500 - 10,500 A          ELECTRODE FORCE - 500 LB
WELD TIME: 5 CYCLES                 ELECTRODE TIP DIA - 1/4 IN - FLAT
STOCK SIZE: 20 GAGE (0.035)                          (UPPER AND LOWER)
```

TEST 1 — WELDABILITY AT VARIOUS DRYING TIMES WITH MATERIAL APPLIED BETWEEN WELD JOINT.

SAMPLE NUMBER	TYPE OF FAILURE PLUG	TYPE OF FAILURE SHEAR	SHEAR STRENGTH, LB	PLUG DIA	DRYING TIME, H	REMARKS
1-0					0	
2-0					0	
3-0					0	
1-24					24	
2-24					24	
3-24					24	
1-48					48	
2-48					48	
3-48					48	
1-72					72	
2-72					72	
3-72					72	
1-96					96	
2-96					96	
3-96					96	
COMPARISON SAMPLES - UNCOATED METAL						
1	X				--	
2	X				--	
3	X				--	

TEST 2 — EFFECT OF MATERIAL UPON WELDABILITY AND ELECTRODE LIFE WHEN APPLIED TO THE OUTER SURFACE OF A SAMPLE IN CONTACT WITH MOVABLE ELECTRODE. SPEED OF OPERATION - 100-200 SPOTS/MIN

TRIAL NUMBER	EFFECT ON FLASHING	EFFECT ON ELECTRODE	NUMBER OF CONTINUOUS SPOT WELDS	REMARKS
1				
2				
3				
4				
5				

COMMENTS:

RECOMMENDED ☐   TEST PERFORMED BY: _____
NOT RECOMMENDED ☐   DATA SHEET NO. _____

FIG. 12—WELD TEST DATA SHEET FOR PRIMERS, SEALERS, AND COATED STEELS

3.3.4 REPORTED RESULTS—Fig. 12 shows a suggested report form.

4. Method C

4.1 Equipment and Supplies

4.1.1 150 KVA gun type spot welder.

4.1.2 Two water-cooled $\frac{5}{8}$ in. diameter electrodes with 45 deg truncated cone and $\frac{1}{4}$ in. diameter welding tip.

4.1.3 Panels, 1.5 x 36 x 0.035 in. and 4 x 12 x 0.035 in. clean, flat, cold rolled steel free from all edge burrs.

4.1.4 Spreader bar to coat film 1.25 in. wide and $\frac{1}{32}$ in. thick.

4.2 Welding Schedule

4.2.1 Electrode force: 600 lb.

4.2.2 Weld time: 8 cycles.

4.2.3 Weld current: 9500–12,000 A.

4.3 Procedure

4.3.1 WELD-THROUGH PERFORMANCE

4.3.1.1 Make a test weld through two pieces of 0.035 in. thick steel with no sealer.

4.3.1.2 Mix sealer sample thoroughly prior to application.

4.3.1.3 Apply $\frac{1}{8}$ in. diameter x 8 in. long bead along center of 4 x 12 in. panel.

4.3.1.4 Within 1 h cover bead with a second 4 x 12 in. panel.

4.3.1.5 Immediately spot weld the assembly directly through the sealer.

4.3.1.6 Weld nugget should be 0.16 in. in diameter. Make a sharp 45 deg angle bend at the weld. If failure occurs, a hole at least equal to the diameter of the weld must be pulled from one of the sheets.

4.3.1.7 Repeat steps 4.3.1.4–4.3.1.6 after aging prepared panel samples 1 week at room temperature (78 $\pm$2 F).

4.3.2 ELECTRODE FOULING

4.3.2.1 Use the spreader bar to apply a film 1.25 in. wide and $\frac{1}{32}$ in. thick the entire length of two 1.5 x 36 in. strips.

4.3.2.2 Place a second 1.5 x 36 in. strip in back of each of the coated strips leaving the sealer exposed.

4.3.2.3 Within 1 h after coating, weld the strips together, bringing one electrode in direct contact with the sealer.

4.3.2.4 Place 2 rows of welds on the strips, 50 in each row on the constructions, for a total of 200 welds. Do not adjust or dress the electrodes during this test.

4.3.2.5 Weld at the rate of 60 welds/min with a 30 s cooling period between each group of 25 welds.

4.3.2.6 During this test observe any electrode sticking, corrosion, or excessive degraded sealer building up on the electrode face. This buildup will act as an insulator, resulting in extreme deterioration or complete stoppage of the weld.

4.3.2.7 Welds should pass the test outlined in paragraph 4.3.1.6.

ADS-4—ENAMEL, LACQUER, AND FABRIC STAINING TEST

1. Application—The methods A and B outlined in this section are the two general procedures used for determining the staining effects of sealers and adhesives materials on or under painted finishes or fabrics. The equipment suggested in paragraph 2.2 are examples of that most generally used by the industry. The procedure and equipment used for this test are predicated on the application and are subject to agreement by the supplier and user.

2. Equipment and Materials Required

2.1 Phosphatized, primed, enameled, lacquered panels or plastics or specified fabrics.

2.2 Groven Fluorescent (F-20) UV Cabinet, S-1 Sunlamp Cabinet, Standard Weather-Ometer, or Fade-Ometer.

3. Procedure

3.1 Method A—Adhesive or Sealant Applied Prior to Paint Bake Cycle

3.1.1 Apply a specified amount of the material under test onto the specified test sample or panel.

3.1.2 Condition test assembly at specified time and temperature before exposure.

3.1.3 Paint and bake the assembly according to the user's regular paint operation using a light-colored, currently released production enamel or lacquer.

3.1.4 Expose the test panel assembly or fabric in a Groven Fluorescent UV Cabinet, S-1 Sunlamp Cabinet, Standard Weather-Ometer, or Fade-Ometer at a specified distance from the light source and at a specified temperature and cycle for a length of time, as indicated on the engineering drawing and/or material specification.

3.1.5 Examine for contact and/or migration stains.

3.2 Method B—Adhesive or Sealant Applied After Paint Bake Cycle

3.2.1 Paint and bake test assembly according to the user's regular paint operation using a light-colored, currently released production enamel or lacquer.

3.2.2 Apply a specified amount of the material under test onto the specified test sample or panel.

3.2.3 Condition test assembly at specified time and temperature before exposure.

3.2.4 Expose the test panel assembly or fabric in a Groven Fluorescent UV Cabinet, S-1 Sunlamp Cabinet, Standard Weather-Ometer, or Fade-Ometer at a specified distance from the light source and at a specified temperature and cycle for a length of time, as indicated on the engineering drawing and/or material specification.

3.2.5 Examine for contact and/or migration stains.

ADS-5—WASH-OFF RESISTANCE TEST

1. Application—This procedure is used to determine the resistance of automotive sealer, deadeners, and adhesives to wash-off during rinsing and phosphatizing operations.

1.2 Equipment Required—See Fig. 13.

1.2.1 Cold rolled, low carbon steel body stock panels 12 x 12 in., 20 gage.

1.2.2 Light paraffinic oil having the following properties:
Viscosity SUS at 100 F: 70–100 s.
Flash COC: 300 F min.
Pour point: 30 F min.

1.2.3 Nozzle—Spraying System No. $\frac{1}{2}$ GG-25 full jet or as specified on material standard.

1.2.4 Water supply capable of maintaining 160 F and a pressure of 20 psi.

1.2.5 Burette graduated in 0.1 cc.

1.2.6 Rubber hose $\frac{1}{2}$ in. ID of adequate lengths to connect water supply outlet with nozzle.

1.2.7 Pressure gage with minimum dial diameter of 3.5 in., graduated in 1 lb increments and having a 60 psi range.

1.2.8 Metal bar 24 x 2.5 x 0.187 in. with opening of 0.750 x 22.0 in. cut out of center. This is used for a guide for the nozzle during testing.

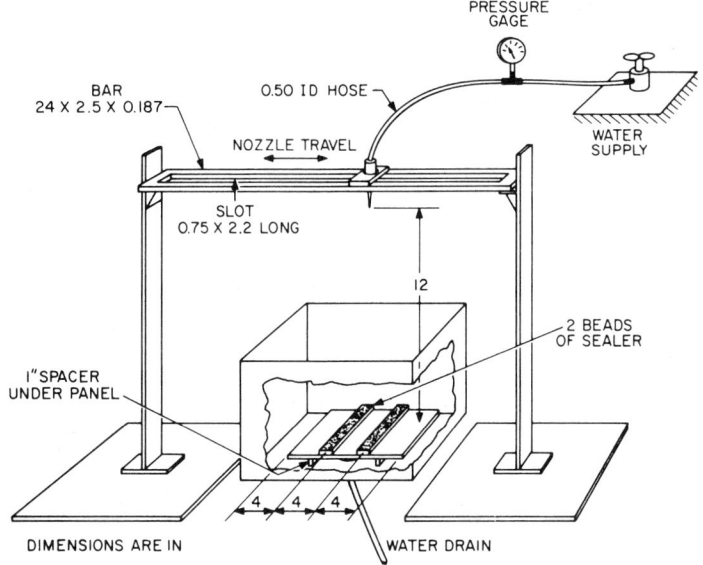

FIG. 13—WASH-OFF RESISTANCE TEST

1.3 Procedure: A Horizontal Wash

1.3.1 Clean the panels by washing with aliphatic hydrocarbon solvent having a boiling range between 200 and 300 F. (Solvents commonly known as VM&P meet this requirement.) Dry the panels with a lint-free cloth.

1.3.2 From a burette apply 0.5 cc of oil to a clean test panel and distribute evenly over the surface by rubbing with two finger tips.

1.3.3 Apply two equally spaced beads or ribbons of material to the panel, using the size of bead or ribbon specified in the material standard.

1.3.4 Air dry for the period of time specified on the material standards.

1.3.5 Refer to Fig. 13. Assemble the hose, nozzle, and gage and connect to the water supply. Adjust the water temperature to 135 ±5 F and the throughput for 1 gal per 20 s or as specified on material standard. Record the pressure required for the desired throughput. Subsequent tests using this equipment can be run at this pressure setting without rechecking volume of flow.

1.3.6 Place the panel horizontal with a minimum clearance of 1.0 in. between the panel and the bottom of the sink or reservoir to allow the spray water to drain out without accumulating over the panel and giving erroneous test results.

1.3.7 Position the nozzle guide directly above the center of the panel so that the nozzle tip is 12.0 in. above the test panel and so that the long axis of the guide is at a 90 deg angle with the long axis of the ribbon or bead of material.

1.3.8 Place the nozzle on top of the guide so that the spray of water is through the cutout section of the guide and move the nozzle back and forth across the guide in such a manner that the spray of water crosses the panel each 2 s.

1.3.9 Visually examine the material for wash-off and displacement. Record the results.

ADS-7—SOLIDS TEST

1. Method A (Fixed Time)—This procedure is used for determining the solid content of adhesives, sealers, and deadeners containing organic solvents.

1.1 Equipment Required

1.1.1 Low form weighing bottle, Fisher Model 3-420 or equivalent.
1.1.2 Circulating air oven capable of maintaining temperatures between 70–150 C ±2% (158–302 F ±2%).
1.1.3 Desiccator, with drying agent and tray.
1.1.4 Analytical balance, accurate to 1 mg.
1.1.5 Spatula, square-tipped blade.

1.2 Procedure

1.2.1 Weigh bottle and cover to nearest milligram. Record weight.
1.2.2 Mix sample thoroughly using square-tipped spatula. Care should be taken to avoid entrapment of air and/or loss of solids.
1.2.3 Transfer 5–10 g (or other specified weight) of the material into the tared bottle, cover and weigh to nearest milligram. Record weight.
1.2.4 Place cover and uncovered bottle containing the sample in the oven at 215 ±5 F for 3 h unless otherwise specified.
1.2.5 Remove bottle containing the residue and cover from the oven and immediately place in the desiccator. Allow to cool to room temperature for 3 h minimum.
1.2.6 Weigh the covered bottle with the residue to nearest milligram. Record weight.
1.2.7 Subtract the weight of bottle and cover as determined in step 1.2.1 from the weight recorded in steps 1.2.3 and 1.2.6.
1.2.8 Calculate percent solids, and record the results.

$$\frac{\text{Weight (step 1.2.6)} - \text{Weight (1.2.1)}}{\text{Weight (step 1.2.3)} - \text{Weight (step 1.2.1)}} \times 100 = \% \text{ solids}$$

2. Method B (Solids to Constant Weight)—This procedure is used to determine the solid content of adhesives, sealers, and deadeners containing organic solvents. Reference: ASTM D 553, Standard Method of Test for Viscosity and Total Solids Content of Rubber Cements.

2.1 Equipment Required

2.1.1 Low form weighing bottle, Fisher Model 3-402 or equivalent.
2.1.2 Circulating air oven capable of maintaining temperatures between 70–150 C ±2% (158–302 F ±2%).
2.1.3 Desiccator, with drying agent and tray.
2.1.4 Analytical balance, accurate to 1 mg.
2.1.5 Spatula, square-tipped blade.

2.2 Procedure

2.2.1 Weigh bottle and cover to nearest milligram. Record weight.
2.2.2 Mix sample thoroughly using square-tipped spatula.
2.2.3 Transfer 5–10 g or other specified weights of the material into the tared bottle, cover and weigh to nearest milligram. Record weight.
2.2.4 Place cover and uncovered bottle containing the sample in the oven at 215 ±5 F for 3 h unless otherwise specified.
2.2.5 Remove bottle containing the residue and cover from the oven and immediately place in the desiccator. Allow to cool to room temperature for 3 h minimum at 73–78 F.
2.2.6 Weigh the covered bottle with the residue to nearest milligram. Record weight.
2.2.7 Repeat steps 2.2.4, 2.2.5, and 2.2.6 until a constant weight is obtained.
2.2.8 Subtract the weight of bottle and cover as determined in step 2.2.1 from the weights recorded in steps 2.2.3 and 2.2.7.
2.2.9 Calculate percent solids:

$$\frac{\text{Weight (step 2.2.7)} - \text{Weight (step 2.2.1)}}{\text{Weight (step 2.2.3)} - \text{Weight (step 2.2.1)}} \times 100 = \% \text{ solids}$$

3. Method C (ASTM D 1582)—This method is suitable for determining the nonvolatile content phenol, resorcinol, and melamine adhesives, with or without hardener and containing high boiling and low boiling volatile organic solvents and water, or both.

Procedure reference: ASTM D 1582, Standard Method of Test of Non-Volatile Content of Phenol, Resorcinol and Melamine Adhesives.

ADS-8—FLASH POINT TEST

1. Method A (Pensky Marten's Closed Cup Method)—This procedure is used to determine the flash point of fuel oils as well as viscous materials and suspensions of solids.

Reference: ASTM D 93, Standard Method of Test for Flash Point by Pensky-Martin's Closed Tester. (See note in paragraph 7 of this test for use with materials having solids in suspension.)

2. Method B (Tag Open Cup Method)—This method covers procedures for the determination of flash points of liquids having flash points between 0 and 235 F. This method, when applied to paints and resin solutions which tend to skin over or which are very viscous, gives less reproducible results than when applied to solvents.

Reference: ASTM D 1310, Standard Method of Test for Flash Point of Volatile Flammable Materials by Tag Open Cup Apparatus.

ADS-9—SAG AND BRIDGING TESTS

1. Application—These tests are used to determine the ability of the material to remain in position and bridge gaps.

1.1 Equipment Required

1.1.1 Circulating air oven capable of maintaining temperatures up to 400 ±5 F.
1.1.2 50 ml burette graduated in 1 ml.
1.1.3 Light paraffine oil having the following properties:
   Viscosity SUS at 100 F: 70–100 s
   Flash COC: 300 F minimum
   Pour point: 30 F minimum
1.1.4 Metal panels 12 x 12 in., 20 gage cold rolled, low carbon steel.
   1.1.4.1 Phosphatized panels[5] as specified.
   1.1.4.2 Oiled panels. Clean the panels by first wiping with VM&P

---

[5] Phosphatized panels are commercially available from: Q-Panel Co., 15610 Industrial Parkway, Cleveland, Ohio; or Parker Rust Proof Co., Division of Hooker Chemical Corp., 2177 E. Milwaukee Avenue, Detroit, Michigan 48211.

naphtha and then dry with a lint-free cloth. Apply 0.5 cc of oil from the burette and distribute it evenly on the surface by rubbing with two finger tips.

1.1.4.3 Painted panels to be prepared as specified by customer.

1.1.5 Draw down fixture. These are made in a "U" configuration as shown on Fig. 14 (flow type), part A with the opeining made to give the bead size specified.

1.2 General Procedure—Samples shall be mixed or otherwise treated as specified by the customer before applying.

2. Sag Tests

2.1 Method A (Inverted Brake Test)

2.1.1 Apply a prescribed size bead or ribbon of material to a panel treated as specified.

2.1.2 Invert panel and air dry at room temperature (material on under side of panel) for 15 min or as specified. Examine at the end of the exposure period and record the results.

2.1.3 Bake in the inverted position for specified time and temperature.

2.1.4 Cool to room temperature and examine and record the results.

2.2 Method B (Horizontal and Vertical Bake Test)

2.2.1 Apply two prescribed size ribbons or beads of material, equally spaced on two panels. The panels are to be of the size and surface treatment as specified.

2.2.2 Mark the position of the ribbons or beads.

2.2.3 Place one panel in a vertical position, with the ribbon(s) or bead(s) horizontal. Place one panel in a vertical position with the ribbon(s) or bead(s) vertical.

2.2.4 Air dry at 73–78 F for 15 min or as specified.

2.2.5 Heat cure panels at specified time and temperature, in above position and protect from direct air flow if specified.

2.2.6 Remove from oven and measure the amount of sag or slump by measuring from the bottom of the curved material to the marked bottom edge of the original bead or ribbon. Record the results.

3. Bridging Tests

3.1 Equipment—See Fig. 14.

3.2 Method C (Coach Joint Test)

3.2.1 Clamp the two steel panels together to form a coach joint.

3.2.2 Drill three $\frac{1}{4}$ in. diameter aligning holes in the flanges.

3.2.3 Assemble the panels to form a coach joint, using the three spacers to give the opening specified. See Fig. 14.

3.2.4 Apply a uniform ribbon or bead of material over the gap of the assembled coach joint. The ribbon or bead shall be of the size specified.

3.2.5 Keep the assembly and material horizontal for the specified time at room temperature.

3.2.6 Measure the depth to which the material has slumped and record.

The measurement from the top of the material bead or ribbon at the time of application.

3.2.7 Place the assembly, in a horizontal position, in the oven for the specified time(s) and temperature(s).

3.2.8 Cool to room temperature.

3.2.9 Measure the depth the sealer has slumped and record.

3.3 Method D (Metal Perforation)

3.3.1 Extrude a $\frac{1}{2}$ in. diameter bead of material lengthwise over the centerline of the holes of the specified bridging test fixture, as shown in Figs. 15 and 16.

3.3.2 Air dry assembly at the specified temperature and time.

3.3.3 Heat cure the assembly at specified time and temperature.

3.3.4 Measure the length of the material that flows through the holes and record as sag, in fixture 1 of Fig. 15 (nonflow).

3.3.5 Record largest hole size that material bridges in fixture 2 of Fig. 16 (controlled flow).

ADS-10—FLOW TEST

1. Application—These tests are used to determine the ability of the material to flow a controlled amount.

1.1 Equipment

1.1.1 Circulating air oven capable of maintaining any temperature up to 400 ±5 F.

1.1.2 Fixtures as specified in methods A, B, and C.

1.2 General Procedures—Samples shall be mixed or otherwise tested as specified.

2. Method A (45 deg Flow Channel)

2.1 Block the $\frac{1}{4}$ in. groove in the flow rate fixture (Fig. 17) 6 in. from the top edge.

2.2 Fill the 6 in. upper portion of the groove with material.

2.3 Remove the block and air dry the fixture for the specified time at the specified temperature.

2.4 Heat cure the fixture at the specified time(s) and at the specified temperature(s).

2.5 After heat exposure, measure the amount of flow from the original position and record.

3. Method B (45 deg Flow Channel)

3.1 Measure 4 in. from the top of a flow channel (Fig. 18) and apply a strip of masking tape down the sides and across the bottom of the channel at this point. Also apply a strip of masking tape across the end and down the side $\frac{1}{2}$ in. at the top of the channel.

3.2 Place quantity of the material into the 4 in. area in the manner specified.

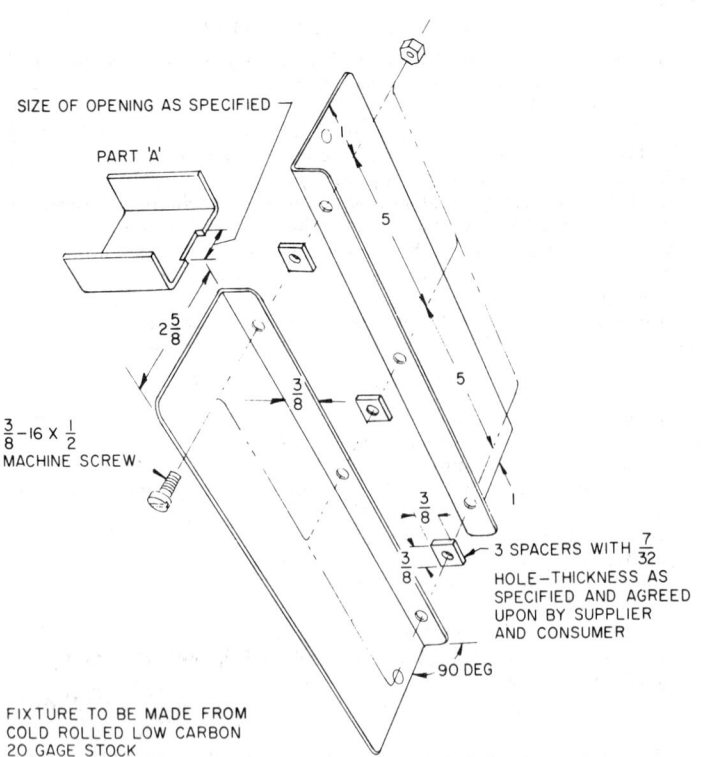

FIG. 14—DRAW DOWN FIXTURE (FLOW-TYPE MATERIALS)

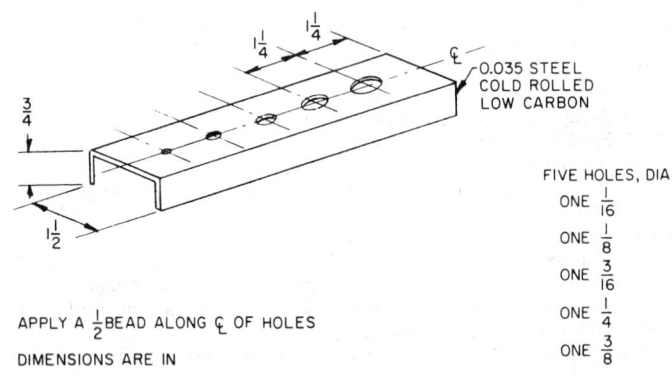

FIG. 15—BRIDGING TEST FIXTURE 1, (NONFLOW)

FIG. 16—BRIDGING TEST FIXTURE 2 (CONTROLLED FLOW)

3.3 Immediately knife the material flush with the legs of the flow channel, remove the tape that was placed at the 4 in. mark, and place the flow channel on the fixture at a 45 deg angle. (See Fig. 18).

3.4 Allow the fixture to remain at room temperature for time specified.

3.5 Place the fixture, without disturbing the flow channel, in a circulating air type oven at the specified temperature(s) and for the specified time(s).

3.6 Remove the fixture from the oven, allow it to return to room temperature, and measure the total distance of material flow. Record the results.

4. Method C (Boeing Flowmeter Test)

4.1 Position the fixture as shown in Fig. 19 with the opening on top.

4.2 Fill the opening with material and knife off flush, and air dry in a horizontal position at room temperature for time specified.

4.3 Place the fixture in an upright position.

4.4 Push the plunger forward at the specified time, forcing the material to be discharged from the opening, and air dry at room temperature for the time specified.

4.5 Measure the amount of flow on the calibrated face of the instrument and record as room temperature flow.

4.6 Place the fixture and material in an oven in an upright position for the specified heat cure time and temperature.

4.7 Measure the amount of flow shown by the material on the calibrated face of the instrument.

4.8 Report both room temperature flow and oven flow.

NOTE: Fixtures as shown in Fig. 19 are not readily available commercially. They may be made by local machine shop. The diameter and depth of the opening may be varied as mutually agreed upon between buyer and seller, should the opening shown not be suitable for the material being tested.

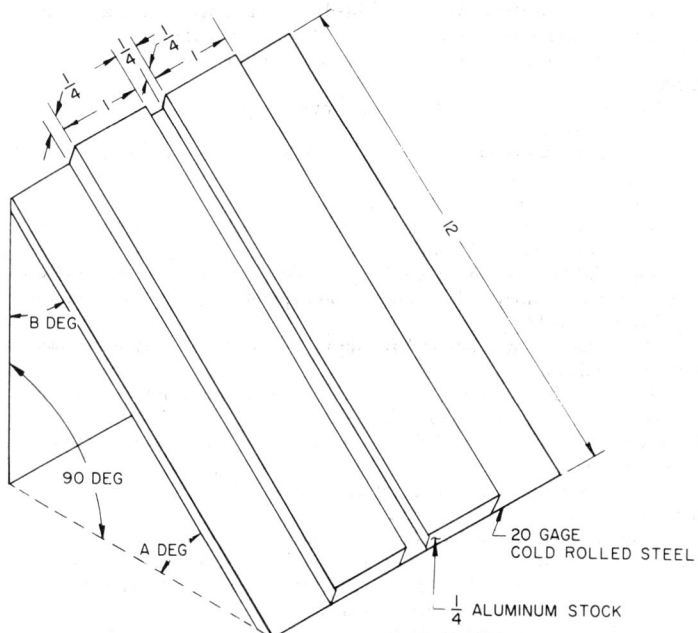

FIG. 17—FLOW TEST FIXTURE

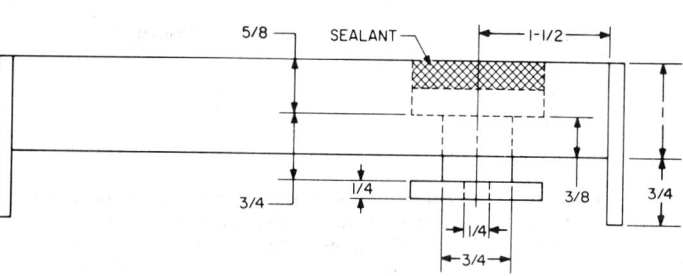

JIG PLACED ON TABLE, FRONT FACE UP

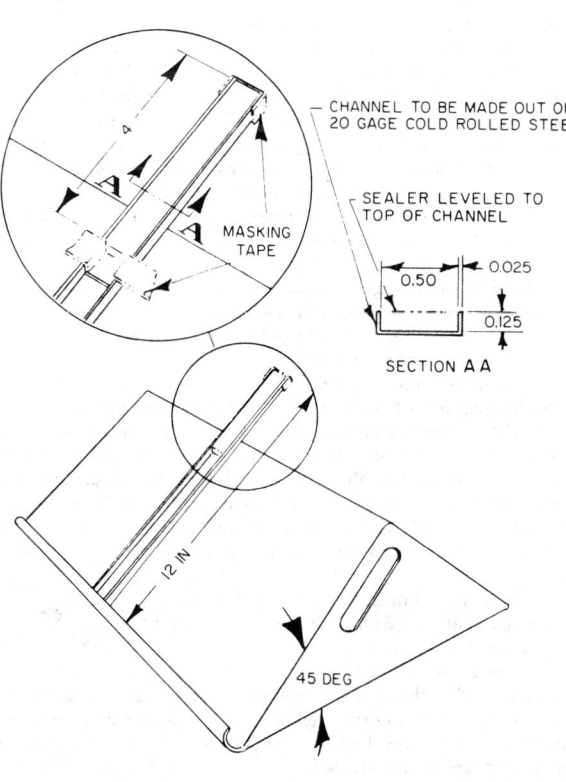

FIG. 18—FLOW CHANNEL

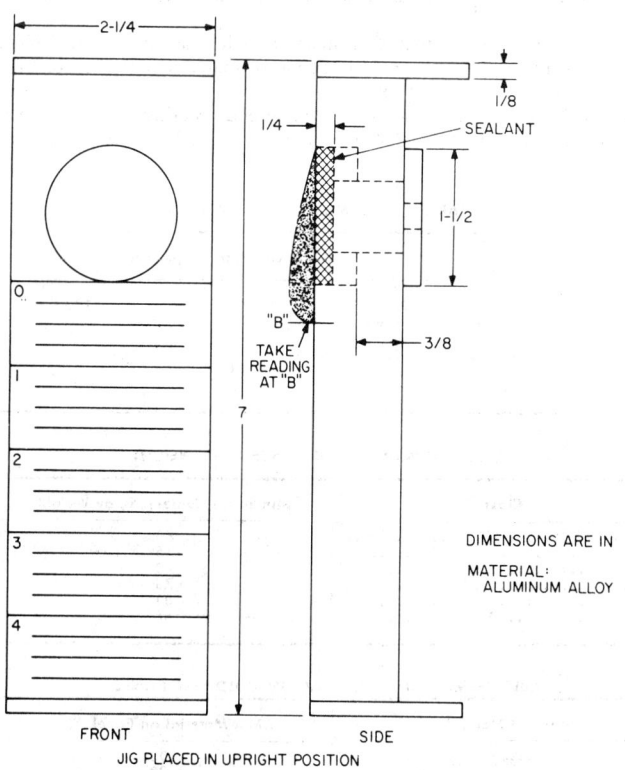

FIG. 19—BOEING FLOW TEST FIXTURE

SOUND DEADENERS AND UNDERBODY COATINGS—SAE J671

SAE Standard

Report of Passenger Car Body Engineering and Nonmetallic Materials Committee approved November 1951 and last revised by Body Engineering and Nonmetallic Materials Committees May 1958.

Scope—Description of Material—The materials classified under this specification are:

1. Mastic sound deadeners used to reduce the sound emanating from metal panels.
2. Mastic underbody coatings used to give protection and some sound deadening to motor vehicle underbodies, fenders, and other parts.

Scope—Numbering System—The prefix "D" is used to indicate a cutback sound deadener, and the prefix "U" is used to indicate a cutback underbody coating. Should a water emulsion be desired, the prefix should be followed by the letter "E". Should a solvent water emulsion be desired, the prefix should be followed by the letter "F".

The materials are further divided into Types, based on the decay rate (decibels per sound at 70 F), and Classes, based on the percentage of solids contained in the material.

Types and classes are as follows:

Type	Nominal Decay Rate, db per sec
5	5
10	10
15	15
20	20

Class	Nominal Solids Content, %
60	60
65	65
70	70
75	75
85	85
90	90

Scope—Example of the Use of Numbers—A cutback underbody coating having a decay rate of 5 db per sec and a solids content of 65% would be designated by the number U-565. An emulsion body deadener of medium decay rate and maximum solids content would be numbered DE-1090. Any combination of type and class could be used to suit the needs of the application.

Physical Properties—Test Specifications—For testing procedures, see section, Method of Test.

Sound Deadening—The material shall have a decay rate as shown in Table 1.

Solids—The solids content of the material shall be as specified in Table 2.

Cold Adhesion—The average retention of material on three cold test panels shall be as shown in Table 3.

Flash Point—The flash point of an underbody coating shall be at least 100 F; that of a sound deadener shall be as agreed upon by the purchaser and the manufacturer.

TABLE 1—MINIMUM DECAY RATE

Type	Min Decay Rate, db per sec		
	70 F	0 F	100 F
5	5	2	2
10	10	4	4
15	15	4	4
20	20	5	5

TABLE 2—MINIMUM SOLIDS CONTENT BY WEIGHT

Class	Min Solids Content, % by Weight
60	60
65	65
70	70
75	75
85	83
90	88

TABLE 3—MINIMUM MATERIAL ON COLD-TEST PANEL

Angle of Slam, deg	Min Material on Panel, %
70	100
50	80
60	50

Abrasion Resistance (When Specified)—The average retention on three abrasion test panels shall be not less than 95% of the material by weight. Each panel must retain at least 90% of the material, and there shall be no exposed metal surfaces.

Sagging—The material shall not sag more than $\frac{1}{4}$ in. when tested on an inverted 45-deg panel.

Sprayability—Flow Rate—The material shall have a flow rate as follows:
Underbody coatings . . . . 45 sec per qt, max
Sound deadeners . . . . 60 sec per qt, max

Sprayability—Stability—Emulsion materials shall not break under usual working pressures or conditions.

Sprayability—Sprayback and Fogging—The average of three sprayback and three fogging tapes shall contain no more sprayback and fogging than the standard tapes held by the consumer.

Consistency (Penetrometer Viscosity)—This shall be as agreed between consumer and supplier.

Settling—The material shall not settle nor separate within a reasonable period of transit and storage to such an extent as to cause difficulty in its use. For the purpose of this specification, the minimum time period shall be 30 days at room temperature.

Toxic Properties—The material shall not contain dangerous amounts of toxic ingredients.

Special Note: Although this specification defines certain properties necessary in a satisfactory product, it is not to be construed that compliance with this specification relieves the vendor of the responsibility of supplying material commercially suitable for the use specified.

Physical Properties—Method of Test

Decay Rate (Sound Deadening)—A 20 x 20 x $\frac{1}{4}$ in. cold rolled steel panel with a decay rate at room temperature of not more than 0.5 db per sec and a natural frequency of 145 to 165 cps shall be sprayed with a uniform coating of the material to the test weight of $\frac{1}{2}$ (0.48 to 0.52) psf dry weight. After air drying at room temperature for a minimum of 12 hr, the panels containing sound deadening materials shall be baked 3 hr $\pm$ 15 min at 275 $\pm$ 5 F. Those containing underbody coatings shall be baked 24 hr $\pm$ 15 min at 160 $\pm$ 5 F.

The decay rate shall be determined as follows:

Support the panel at one or more nodal points for the fundamental natural frequency (the nodal pattern is a square connecting the midpoints of the edges of the panel). Excite vibration of the panel at its fundamental frequency and measure the rate of decay of free vibration. This rate of decay expressed in decibels per second is the decay rate. (NOTE: The number of decibels corresponding to the ratio between any two vibration amplitudes is equal to twenty times the logarithm to the base ten of that ratio.)

Solids—The solid content shall be determined by placing 3 to 5 g of the material in a weighted container (an ointment can, approximately 2 in. in diameter) and drying 3 hr at 221 $\pm$ 5 F (105 $\pm$ 3 C).

Cold Adhesion—Thoroughly clean three 12 x 12 x 0.040 in. cold rolled steel test panels by immersing them in a solution of

30% by volume concentrated phosphoric acid
30% by volume cellosolve
40% by volume water

and washing thoroughly with a clean rag. Remove the panel from the solution and wipe dry with another clean rag. These panels must have a surface finish of not less than 35 nor more than 60 Mu in., and each panel may be used only once for cold adhesion testing. Proceed according to the following schedule:

1. Spray a 9 x 9 in. area on each of three panels with a uniform coating to a test weight of 0.48 to 0.52 psf dry weight of material.
2. Air dry at room temperature for a minimum of 12 hr.
3. Bake
(a) Sound deadener—3 hr $\pm$ 15 min at 275 $\pm$ 5 F.
(b) Underbody coating—24 hr $\pm$ 15 min at 160 $\pm$ 5 F.
4. Cool to room temperature.
5. Cool for 3 hr at −10 $\pm$ 2 F.
6. Slam each panel in the cold adhesion test fixture, Fig. 1, beginning at an angle of 10 deg and increasing the angle of slam 10 deg each successive slam until the 90-deg position is reached. The amount of material remaining on the panel is estimated between each slam.

Flash Point—The flash point shall be determined by the ASTM Pensky-Martens closed tester, ASTM D 93.

Abrasion Resistance—Clean three 5 x 12 x 0.040 in. cold rolled steel panels as specified for the Cold Adhesion Test. Mask each panel to give an exposed area 4 x 12 in. and spray to the test weight of 0.48 to 0.52 psf dry weight of material. Air dry at room temperature 12 hr minimum, and bake 24 hr ± 15 min at 160 ± 5 F.

Insert each panel in the abrasion tester, Fig. 2, and subject to 10 cycles of abrasion with 100 lb (a total of 1000 lb) of No. 780 iron shot at 80 psi air pressure. Allow the panel to cool to room temperature between cycles. The amount of material retained on each panel shall be determined by weighing.

Sagging—Thoroughly clean two 12 x 12 x 0.040 in. cold rolled steel test panels as specified for the Cold Adhesion Test. Spray a 9 x 9 in. area on each panel with a uniform coating of the material to the test weight of 0.65 to 0.75 psf dry weight. The panels shall then be supported at an inverted 45-deg angle, air dried for 15 min, and baked for 30 min at 275 ± 5 F. For undercoating materials, the test weight is to be 0.48 to 0.52 psf dry weight, and the baking requirements are not necessary.

Sprayability—Flow Rate—The flow rate of a body deadener shall be determined by placing the material in a 2-gal bottom outlet pressure tank with 15 ft of $\frac{3}{4}$-in. ID fluid hose with a suitable deadener spray gun and $\frac{1}{2}$-in. round nozzle. The flow rate shall be measured with 60 psi pressure on the tank and no atomizing pressure at a temperature of 75 ± 5 F.

The flow rate of an underbody coating shall be determined in a like manner except that a $\frac{1}{4}$-in. round nozzle shall be used.

Sprayability—Sprayback and Fogging—This is a comparative method for determining the sprayback and fogging characteristics of a material. The particles which are blown back during the spraying are caught on the adhesive side of transparent cellulose tapes. The tapes can be mounted on a sheet of white paper for observation. The tapes with the adhesive side facing the panel are designated as sprayback tapes, and those facing away from the panel are fogging tapes.

A 12 x 12 in. test panel shall be mounted in the sprayback and fogging test booth as shown in Fig. 3. With the spraygun and test tapes located as shown, the panel shall be sprayed with approximately $\frac{1}{2}$ lb of material. The tapes should then be removed and attached to white cards.

Consistency—For inspection purposes the consistency may be determined by ASTM D 217 penetration cone.

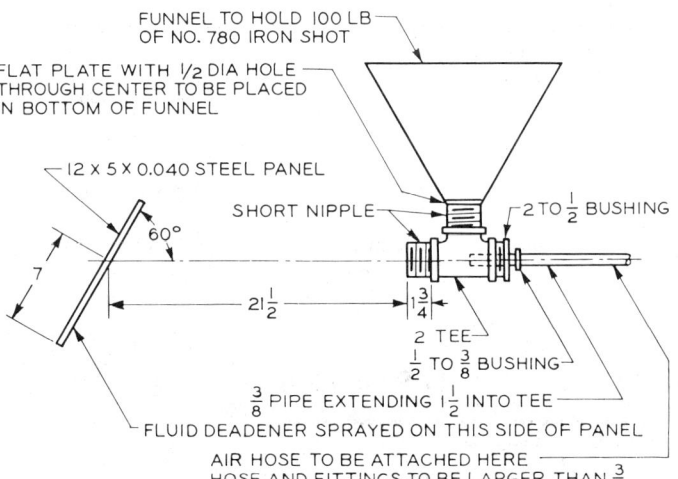

FIG. 2—ABRASION TEST FIXTURE

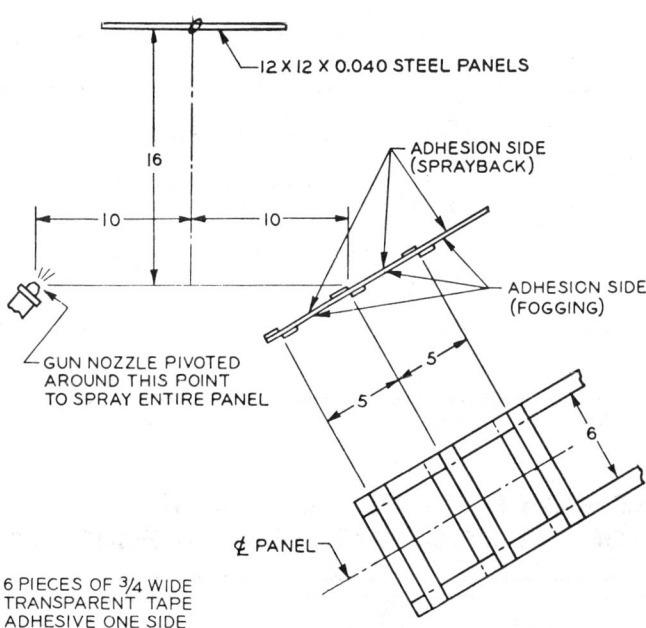

FIG. 3—RELATIVE POSITIONS OF PANEL, SPRAYGUN, AND TAPES FOR SPRAYBACK AND FOGGING TESTS—PLAIN VIEW

TEST METHOD FOR MEASURING WEIGHT OF ORGANIC TRIM MATERIALS—SAE J860

SAE Standard

Report of Nonmetallic Materials Committee approved June 1963. Editorial change May 1978.

Scope—This procedure is used to determine the weight, in ounce per square yard, of materials used for trimming automobile interiors.

Procedure

1. Condition all test specimens in a standard atmosphere of $70 \pm 2°F$ ($21 \pm 1°C$) and $65 \pm 2\%$ relative humidity for 24 h.

2. Cut three samples 4 in x 4 in (102 mm x 102 mm) (or as specified) not closer than one tenth the width of the material from each selvage and from the center. In materials other than roll goods, cut three samples from representative areas.

3. Weigh all three test specimens accurately.

4. Calculate the weight per square yard (or square metre) as follows:

$$\frac{1296 \times A}{B \times 28.35} = oz/yd^2 \quad \text{or} \quad \frac{A \times 10^6}{B(mm^2)} = g/m^2$$

where A = Average weight in grams of the three test specimens
B = Area of specimen, in^2 (or mm^2).

TEST METHOD FOR DETERMINING DIMENSIONAL STABILITY OF AUTOMOTIVE TEXTILE MATERIALS—SAE J883

SAE Recommended Practice

Report of Nonmetallic Materials Committee approved April 1965. Editorial change May 1978.

1. Scope—This test method can be used to determine the dimensional stability of textile materials and vinyl coated fabrics when subjected to conditions which cause changes in the moisture content of the materials.

2. Test Specimens—A test specimen 12 x 12 in (305 x 305 mm) shall be cut from the material to be tested with one direction parallel to the warp yarns (machine direction) and the other direction parallel to the filling yarns (across machine direction).

3. Conditioning—The test specimen shall be conditioned for a minimum of 24 h at $70 \pm 2°F$ ($21 \pm 1°C$) and $65 \pm 2\%$ relative humidity.

4. Procedure

4.1 Mark off accurately a 10 x 10 in (254 x 254 mm) square concentric with the square outline of the specimen. This can be done with indelible ink, indelible pencil, or other suitable method, on which ever side of the material is more markable. Also mark an arrow to indicate the warp (machine direction) of the specimen.

4.2 Place the specimen face side up flat without wrinkles on a 4 mesh screen surface measuring a minimum of 13 x 13 in (330 x 330 mm). Position a similar screen over the specimen using spacers at the corners of the two screens, so that the top screen is not in contact with the top surface of the test specimen. To test more than one specimen at one time, use additional spacers and screens as required.

4.3 Immerse the specimen(s) and screens in a pan or tank of clean tap water containing 1 ml of alkylarylsulfonate synthetic detergent[1] per 21 ml of water at $70-80°F$ ($21-27°C$) for 1 h.

4.4 Remove the specimen(s) and screens from the water and allow to drip dry in an atmosphere having a temperature of $70 \pm 2°F$ ($21 \pm 1°C$) and a relative humidity of $65 \pm 2\%$ for 30 min. If more than one specimen is being tested, separate the screens so that no specimen will drip on any other specimen.

4.5 Lay the specimen(s) flat on a table top. Measure the original 10 in (254 mm) square with a scale calibrated in hundredths of an inch (or ¼ mm). Make three measurements in both the warp and filling directions. The measurements shall be made along the centerlines of the square and along lines parallel to and 2 in (51 mm) in from each side.

4.6 Average the three measurements in each direction and substitute in the following formulas:

$$S_{w1} = \frac{W_1 - 10(254)}{10(254)} \times 100 \qquad S_{f1} = \frac{F_1 - 10(254)}{10(254)} \times 100$$

S_{w1} = Warp Stability $\qquad S_{f1}$ = Filling Stability
W_1 = Warp Measurement Wet $\qquad F_1$ = Filling Measurement Wet

A plus result indicates expansion and a minus result indicates shrinkage.

4.7 Replace the test specimen(s) on the screens and place the specimen(s) and screens in an air circulating oven maintained at $175-180°F$ ($79-82°C$) for 24 h.

4.8 Remove the specimen and screen from the oven and allow to cool in the standard atmosphere described in paragraph 4.4 for 10 min. After cooling, place the specimen on a flat table top and remeasure as described in paragraph 4.5.

4.9 Average the three measurements in each direction and substitute in the following formulas:

$$S_{w2} = \frac{W_2 - 10(254)}{10(254)} \times 100 \qquad S_{f2} = \frac{F_2 - 10(254)}{10(254)} \times 100$$

S_{w2} = Warp Stability $\qquad S_{f2}$ = Filling Stability
W_2 = Warp Measurement after Drying $\qquad F_2$ = Filling Measurement after Drying

A plus result indicates expansion and a minus result indicates shrinkage.

[1] Alkylarylsulfonate type of synthetic detergent is available under various trade names from manufacturers of detergents.

METHOD OF TESTING RESISTANCE TO CROCKING OF ORGANIC TRIM MATERIALS—SAE J861

SAE Standard

Report of Nonmetallic Materials Committee approved June 1963. Reaffirmed with editorial change January 1971. Editorial change May 1978.

1. Scope—This test can be used to determine the resistance to color rub-off (crocking) of organic trim materials such as fabrics, vinyl coated fabrics, leather, coated fiberboard, and carpet.

2. Materials and Equipment Required

Crockmeter—Official AATCC Apparatus available from Atlas Electrical Devices, 4114 N. Ravenswood Ave., Chicago, Illinois 60613.

Crock Cloth—80 x 80 (31.5 x 31.5/cm) count white cotton print cloth, completely desized and bleached, available from Testfabrics, Inc. 55 Van Dam St., New York, New York 10013. The crock cloth is cut into approximately 2 x 2 in (51 x 51 mm) squares for this test.

AATCC Color Transference Chart—Available from the American Association of Textile Chemists and Colorists, P.O. Box 12215, Research Triangle Park, North Carolina 27709.

Distilled Water—pH of 6.5–7.5.

3. Test Specimens—A minimum of two test specimens are required, one for a dry crock test and one for a wet crock test. A multicolored and/or multipatterned material may require additional specimens for a complete evaluation. Specimens should be as flat as possible and at least 2 x 5 in (51 x 127 mm) in size.

4. Procedure

A. Dry Crock Test

1. Condition the test specimen(s) and the crock cloth(s) for a minimum of 16 h at $70 \pm 2°F$ ($21 \pm 1°C$) and $65 \pm 2\%$ relative humidity.

2. Place the test specimen with the long dimension in the direction of rubbing on the base of the crockmeter so that it is flat and uniformly in contact with the abrasive cloth mount.

3. Fasten the crock cloth over the end of the peg with the spiral wire clip provided for this purpose. The weave of the crock cloth should be at an approximate 45 deg (0.8 rad) angle to the direction of rubbing.

4. Lower the peg into contact with the test specimen. Turn the crank at a rate of one turn per second so that the peg moves back and forth 20 times (10 cycles) on the test specimen.

5. Remove the crock cloth from the peg and evaluate the color transfer onto the cloth using the AATCC Color Transference Chart. Attach the crock cloth to the face of the test specimen and submit it with the report.

B. Wet Crock Test

1. The procedure for the wet test is exactly the same as for the dry crock test except that the crock cloth is wetted before testing as follows:

2. Immerse the crock cloth in distilled water at 70 ± 2°F (21 ± 1°C) long enough to wet the cloth thoroughly throughout. Drip dry or wring the cloth out until the moisture content reaches 65 ± 2% based on the original conditioned weight of the cloth. Proceed with the test immediately when the desired moisture content is reached.

TEST METHOD FOR MEASURING THICKNESS OF AUTOMOTIVE TEXTILES AND PLASTICS—SAE J882

SAE Recommended Practice

Report of Nonmetallic Materials Committee approved April 1965. Editorial change May 1978.

1. Scope—This test is designed to measure the thickness of relatively flat textiles, plastics, and similar materials.

2. Apparatus Required

2.1 A dead weight type of dial micrometer capable of accurately measuring to 0.001 in (0.025 mm). The pressure foot shall have a diameter of:

3.00 in (76.2 mm) for testing tufted floor covering.
1.13 in (28.7 mm) for testing felts and other nonwovens.
0.250 in (6.35 mm) for testing bodycloth.[1]
0.250 in (6.35 mm) for testing unsupported vinyl film and coated fabrics.

2.2 The pressure foot and connected parts shall be weighted to apply to a total load of:

12.0 oz (340 g) for testing tufted floor coverings.
10.0 oz (283 g) for testing felt and other nonwovens.
3.0 oz (85 g) for testing bodycloth.[1]
3.0 oz (85 g) for testing unsupported vinyl film and coated fabrics.

[1] When measuring the overall thickness of three dimensional materials, the thickness shall be determined by calculating the average of three readings taken on the high points of the material.

3. Test Specimens—Cut three test specimens approximately 4 x 4 in (102 x 102 mm) from the center and two opposite sides of the roll. Unless otherwise specified, samples shall be taken no nearer the selvage edge than one-tenth the width of the material, or nearer than 12 in (305 mm) from either end of the roll. If the material does not come in rolls, the sample should be taken from the most representative areas.

4. Procedure—Condition the samples at 70 ± 2°F (21 ± 1°C) and 65 ± 2% relative humidity for 24 h. Raise the pressure foot of the thickness gage and place the sample on the base. Gently lower the pressure foot until it contacts the sample. Completely release the pressure foot and after 10 s read the thickness of the dial micrometer. Repeat in two additional areas and record the average thickness of all (3) test specimens.

TEST METHOD FOR DETERMINING COLD CRACKING OF FLEXIBLE PLASTIC MATERIALS—SAE J323

SAE Recommended Practice

Report of Nonmetallic Materials Committee approved August 1968. Editorial change May 1978.

Scope—This method of test is applicable for determining the cold characteristics of vinyl coated fabrics and other automotive plastic materials, as applicable. It consists of three different methods for determining low temperature properties of materials depending on type of material and end use.

Method A, Mandrel Test

Apparatus and Materials

MANDREL—Steel mandrel ¼ in (152 mm) in diameter and 6 in (6.35 mm) long attached to a suitable stand. Other diameters may be specified depending on the thickness and rigidity of the material to be tested.

OVEN—Air circulating oven capable of maintaining a temperature of 180 ± 5°F (82 ± 2°C).

COLD BOX—A cold box capable of maintaining a temperature of −30°F (−34°C) and large enough to permit bending the test specimen while it remains in the box.

GLOVES—Heavy cloth gloves to prevent heat transfer when handling samples.

Procedure—Cut 2 x 8 in (51 x 203 mm) samples in the machine and across machine direction and condition in the oven at 180°F (82°C) for 24 h or as specified. (The dimensions of the sample may vary for extruded or molded parts.) Remove samples from the oven and condition at room temperature to maintain equilibrium. Place samples, gloves, and mandrel with stand in the cold box at −30°F (−34°C) for 4 h or as specified. Put on gloves and grasp each end of the sample and bend, finish side out, 180 deg (3.1 rad) around the mandrel in approximately ½ sec with a uniform motion. Remove samples from cold box and examine visually for evidence of cracks.

Method B, Impact Test

Apparatus and Materials

IMPACT TESTER—Impact tester capable of applying an 8 ft-lb (10.8 J) impact with a spherical ball head having a radius of ¹⁵⁄₁₆ in (23.81 mm). (See Fig. 1.)

BASE—A 4 x 4 x ¾ in (102 x 102 x 19 mm) thick wood base with a 3½ x 3½ in (89 x 89 mm) square marked off in the center for stapling sample and a 3 x 3 in (76 x 76 mm) square marked off for positioning urethane foam pad.[1]

OVEN—Air circulating oven capable of maintaining a temperature of 180 ± 5°F (82 ± 2°C).

Procedure—Cut a 4 x 4 in (102 x 102 mm) sample and age in the oven at 180°F (82°C) for 7 days (or as specified). Place a 3 x 3 x ¾ in (76 x 76 x 19 mm) thick urethane pad in the center of the wood base and attach the aged sample to the base by stapling ¼ in (6 mm) from the edge of the sample as indicated on Fig. 2 (staples should be ⁵⁄₁₆ in (7.9 mm) minimum length). Place the sample in the cold chamber at −20°F (−29°C) for a minimum of 12 h (or as specified). While still at −20°F (−29°C), the center of the test material shall be impacted with a force of 8 ft-lb (10.8 J). The height of the impactor shall be as specified between the user and supplier. Remove the sample from the cold box and examine for cracks.

[1] Unless otherwise specified, the foam base shall be either a urethane foam or a latex foam material.

The urethane material shall have the following load deflection characteristics: a load deflection at 25% of RT shall be 0.440–0.600 psi (3.034–4.137 kPa) and a load deflection of 8–18% at −20°F (−29°C) when tested with a load of 0.375 ± 0.01 psi (2.59 ± 0.07 kPa).

The latex material shall have the following load deflection characteristics: a load of 25% at RT shall be 0.360–0.450 psi (2.482–3.103 kPa) and a deflection of 37–47% at −20°F (−29°C) when tested with a load of 0.375 ± 0.01 psi (2.59 ± 0.07 kPa).

The sample used for checking the −20°F (−29°C) cold deflection property shall be a 4 x 4 in (102 x 102 mm) and preconditioned for a minimum of 12 h at −20°F (−29°C) prior to testing.

NOTE: The wood and urethane foam base described above was established primarily for vinyl coated fabrics. Other plastic materials may require modifications in the base depending on the flexibility of the material to be tested.

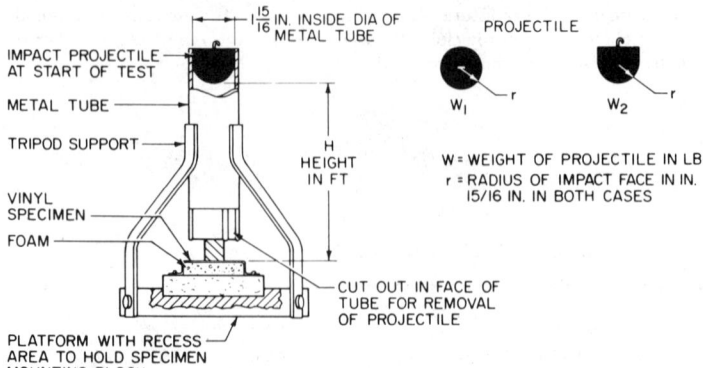

FIG. 1—IMPACT APPARATUS

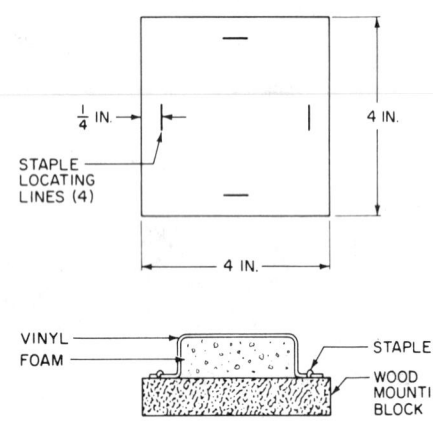

FIG. 2

Method C, Dynamic Flex Test

Apparatus

COLD BOX—A cold box capable of maintaining a temperature of $-20°F$ ($-29°C$) and large enough to hold flexing equipment.

OVEN—Air circulating oven capable of maintaining a temperature of $180 \pm 5°F$ ($82 \pm 2°C$).

FLEXING EQUIPMENT—Dynamic flex equipment with reciprocating motion (Fig. 3).

Procedure—Cut four specimens 3 x 2 in (76 x 51 mm) with two specimens having the long dimension in the warp direction and the remaining two having the long dimension in the filling direction. Condition two specimens (warp and filling directions) for 7 days at $180 \pm 5°F$ ($82 \pm 2°C$) in an air circulating oven followed by 4 h at $-20°F$ ($-29°C$). Condition the remaining two samples at $-20°F$ ($-29°C$) for 4 h. Clamp the two ends of the unaged samples in the cold flex apparatus with the vinyl side facing out. Flex at $-20°F$ ($-29°C$) for 700 cycles at 90 cycles/min. Remove and examine for cracks. Flex the two aged samples for 600 cycles. Remove and examine for cracks.

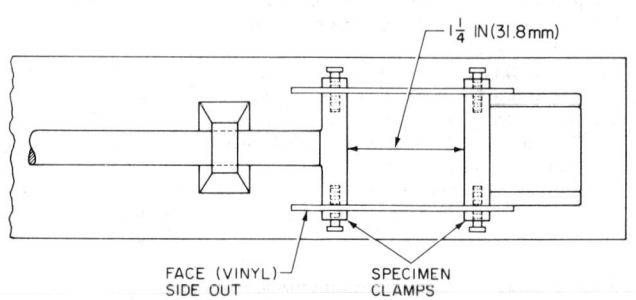

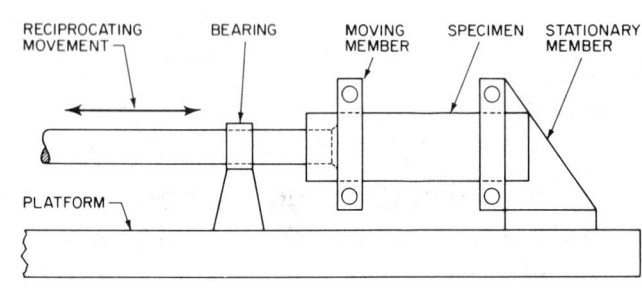

FIG. 3—BASIC DETAILS OF COLD FOLD TEST APPARATUS

NONMETALLIC TRIM MATERIALS— TEST METHOD FOR DETERMINING THE STAINING RESISTANCE TO HYDROGEN SULFIDE GAS—SAE J322

SAE Recommended Practice

Report of Nonmetallic Materials Committee approved December 1967. Reaffirmed without change May 1978.

1. Scope—This procedure is designed to reveal discoloration which may occur when nonmetallic materials used for trimming automobiles are exposed for a limited time to an atmosphere containing hydrogen sulfide.

2. Materials and Equipment Required
  2.1 Hydrogen sulfide cylinder with valve.
  2.2 Two test tubes approximately 38 x 200 mm.
  2.3 Two two-hole stoppers to fit above.
  2.4 Miscellaneous glass tubing, cotton, plastic tubing, and pinch clamp as per Fig. 1.

3. Test Specimen
  3.1 Cut a test specimen which can be conveniently inserted into the lower portion of tube B. The test specimen must be of such a size and shape to allow for free passage of gas on all sides.

4. Procedure
  4.1 Assemble apparatus per Fig. 1 and place under fume hood.
  4.2 Insert test specimen into tube B.
  4.3 Add approximately 50 ml of tap water to tube A. Insert cotton packing per Fig. 1. Stopper tubes A and B.

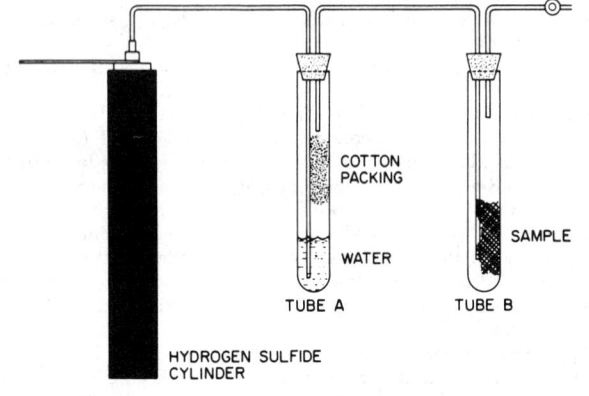

FIG. 1

4.4 Remove pinch clamp from exhaust hose of tube B and cautiously open the valve on the hydrogen sulfide cylinder. Adjust the valve so hydrogen sulfide gas bubbles through the system at a uniform and controlled rate.

4.5 Allow the system to purge for 1-1/2 minutes, then close the valve and immediately replace the pinch clamp on the exhaust hose of tube B.

4.6 Allow the system to remain in the closed position for an additional 2 minutes.

4.7 Remove the stopper of tube B and take out the test specimen.

4.8 Compare the test specimen with an unexposed sample of the same material and report any discoloration which has occurred.

TEST METHOD FOR DETERMINING WINDOW FOGGING RESISTANCE OF INTERIOR TRIM MATERIALS—SAE J275

SAE Recommended Practice

Report of Nonmetallic Materials Committee approved November 1971. Reaffirmed without change May 1978.

1. Scope

1.1 To simulate the formation of a condensate on the inside surface of clear glass when interior trim materials are exposed to an external source of radiant energy.

1.2 To provide a method of measuring the effect of condensate formation on light transmittance.

2. Materials and Equipment Required

2.1 Heating Unit (Fig. 1)

2.1.1 500W infrared lamp (nonreflective type) (General Electric "Industrial" or equivalent).

2.1.2 Parabolic reflector 10 in. (254 mm) deep with 10 in. (254 mm) ID bell.

2.1.3 Cylindrical draft shield of 20 gage galvanized steel metal 15 in. (381 mm) in height and 23 in. (584.2 mm) ID lined with 0.001 in. (0.0254 mm) gage aluminum foil.

2.1.4 Circular piece of 0.001 in. (0.0254 mm) gage aluminum foil 18 in. (457.2 mm) in diameter with a 10 in. (254 mm) diameter concentric circle marked on surface. This piece of foil to lie flat on an 18 in. (457.2 mm) diameter turntable capable of revolving at 3 rpm. A ring with a thickness of 1/4 in. (6.35 mm) and 1-5/8 in. (41.28 mm) ID to be fastened to the foil at the center point.

2.1.5 Thermocouple and lamp controller to maintain specified temperature on surface of reference specimen (West Instrument Corp. Model JP Controller or equivalent).

2.2 Condensing Unit (Figs. 2 and 3)

2.2.1 Six watch glasses, plain, ground edge, 75 mm in diameter. Lip of one watch glass to be notched to accommodate thermocouple lead to reference cell.

2.2.2 Six octagonal gaskets of 1/32 in. (0.79 mm) silicone rubber, 3-5/8 in. (92.08 mm) cross corner with 2-1/4 in. (57.15 mm) diameter hole in center (DuPont No. SR-5550 silicone rubber or equivalent).

2.2.3 Six clear glass cover plates, 1/8 in. (3.18 mm) thick, octagonal in shape, 3-5/8 in. (92.08 mm) cross corner.

2.2.4 Circular reference specimen, 2-1/2 in. (63.5 mm) in diameter. This reference specimen should be black in color and of similar composition to materials to be tested.

2.2.5 Circular test specimens, 2-1/2 in. (63.5 mm) diameter.

2.2.6 Twelve No. 1 pad clips.

2.3 Measuring Assembly (Figs. 4 and 5)

2.3.1 Foot-candle meter (Weston's Model 614 or equivalent).

2.3.2 Light-tight box (Fig. 4).

2.3.3 Microscope illuminator with variable transformer (American Optical's Model 353 or equivalent).

2.3.4 Constant voltage transformer (Iola Electric Co. Model 302554 or equivalent).

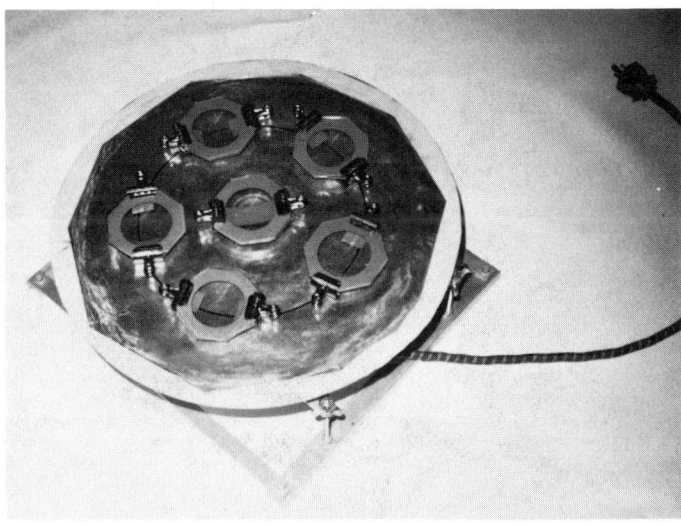

FIG. 2—ARRANGEMENT OF CONDENSING UNITS ON ALUMINUM FOIL

FIG. 1—HEATING UNIT

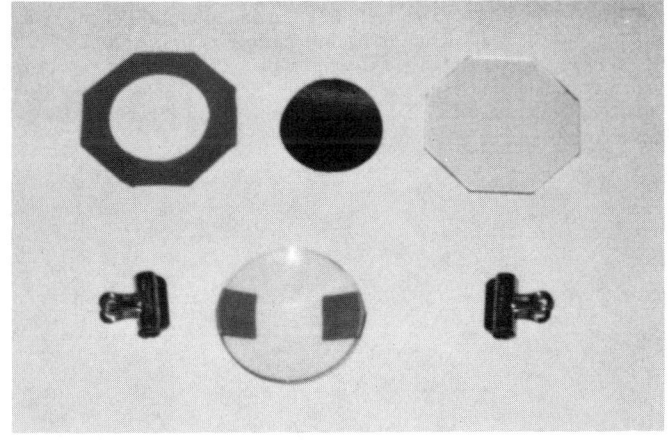

FIG. 3—COMPONENTS OF SINGLE CONDENSING UNIT

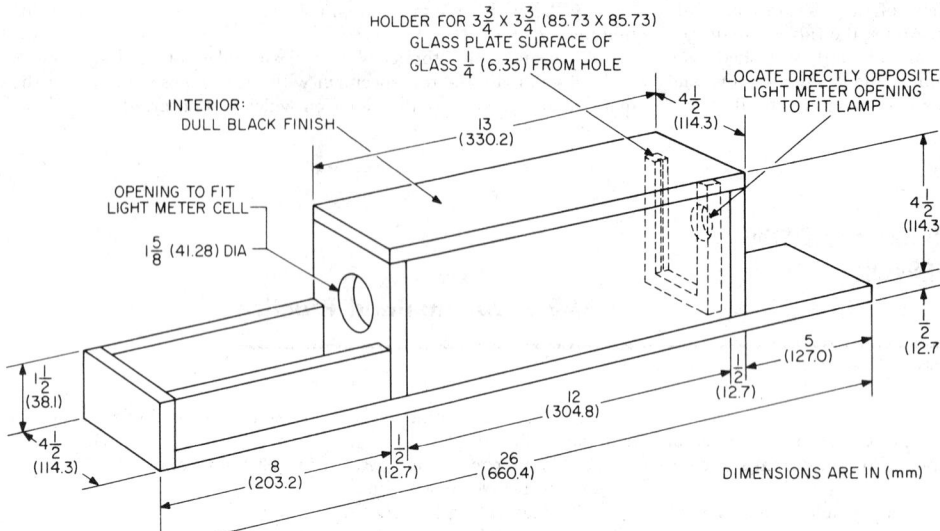

FIG. 4—LIGHT-TIGHT BOX

FIG. 5—MEASURING ASSEMBLY

3. Procedure

3.1 Place a circular test specimen into the concavity of a clean, dry, watch glass with the finished surface of the specimen facing upward. The convex surface of the watch glass will be down and against the circular piece of aluminum foil. A maximum of five specimens may be tested.

3.1.1 Place a gasket and then a clean glass plate on the ground edge of the watch glass and clip the condensing unit together with two pad clips. A strip of masking tape attached to the convex side of the watch glass will prevent slippage of the clips.

3.1.2 Prepare a reference condensing unit in the same manner as above, except introduce a thermocouple lead through the notched watch glass. Establish intimate contact between the black surface of the reference specimen and the thermocouple lead. The lead must remain in contact with the reference specimen throughout the test. Center this reference condensing unit upon the raised ring described in paragraph 2.1.4 so that the thermocouple wire will restrain it from revolving with the turntable. If the thermocouple wire is not stiff enough to restrain the reference unit, a fine wire between the reference unit clips and the draft shield will secure the unit in the central position.

3.1.3 Position the condensing units containing the test specimens on the circular piece of aluminum foil so that they are equally spaced and the test specimens are tangent to the inside of the 10 in. (254 mm) circle. A small disc of double-faced masking tape between the bottom of the watch glass and the foil will prevent dislocation of the condensing units due to vibration in the turntable.

3.1.4 Surround the condensing unit arrangement on the turntable with the cylindrical draft shield. Center the lamp and reflector above the reference condensing unit. The bottom of the lamp shall be 10 in. (254 mm) from the plane formed by the top of the glass plate of the reference unit.

3.2 Connect the thermocouple to the lamp controller, start the turntable, and expose the samples at the specified temperature and time. Timing shall start when reference specimen has reached the desired temperature.

3.3 Disassemble the condensing units containing the test specimens and place the cover plates with the condensate upward on a table top at room temperature for approximately 30 min.

3.4 Standardize the measuring unit by inserting a clean glass plate and adjusting the light beam to its highest spot concentration. Then adjust the light intensity by means of the variable transformer to give a reading a 100 on the foot-candle meter.

3.4.1 Remove the clean glass plate and replace it with the cover plates from step 3.3. The side of the plate with the condensate should be inserted in the slot next to the illuminator lamp. Readings should be made within a few hours of the completion of the specified exposure time.

3.4.2 Record the light transmittance reading as measured on the foot-candle meter. Four readings should be made on each glass plate by rotating the plate 90 deg after each reading. Report the average of the four readings as the light transmittance of the sample.

TEST FOR CHIP RESISTANCE OF SURFACE COATINGS—SAE J400 JUN80

SAE Recommended Practice

Report of the Nonmetallic Materials Committee, approved July 1968, completely revised June 1980.

1. Scope—This SAE Recommended Practice covers a laboratory procedure for testing and evaluating the resistance of surface coatings to chipping by gravel impact. The test is designed to reproduce the effect of gravel striking exposed paint or coated surfaces of an automobile and has been correlated with actual field results. The specific intent of the test is to evaluate organic surface coatings or systems on flat test panels; however, it may be possible to extend this type of testing to finished parts or other types of materials such as anodized aluminum or plated plastics if the results are interpreted with respect to the limitations and intent implied by the original testing procedure and rating system.

2. Summary of Method—The test consists of projecting a standardized road gravel by means of a controlled air blast onto a suitable test panel. The testing apparatus is contained in a box on wheels, called a gravelometer, designed to contain road gravel, a test panel holder, and a gravel projecting mechanism. The projecting mechanism, located in front of the test panel, consists of an air nozzle in the base of an inverted pipe tee. The stem of the pipe tee points upward and is connected to a funnel into which the gravel is poured. The gravel, falling into the air blast, is projected toward and impacts upon the test panel which is usually held perpendicular to the impinging gravel. All testing is conducted under controlled temperature conditions, generally room temperature or 0°F. After gravel impact, masking tape is applied to remove any loose paint chips remaining on the panel, and the degree of chipping is determined by visual comparison with the SAE Chipping Rating Standards[1] or by counting the number and sizes of all chips.

3. Equipment and Materials

3.1 Gravelometer—A gravel projecting test apparatus which is constructed according to the design specifications shown in Fig. 1.[2]

3.2 Gravel—The gravel for this test shall be water-worn road gravel,[3] not crushed limestone or rock, which will pass through 5/8 in (15.86 mm) space screen when graded but be retained on 3/8 in (9.53 mm) space screen. It is important to note that mesh screen is not a substitute for space screen. The gravelometer has 3/8 in (9.53 mm) space screen in the bottom to separate fractured pieces of rock and dust smaller than 3/8 in (9.53 mm) so that the retained gravel on this screen may be reused; however, even this reusable gravel will tend to blunt or fragment after repeated impacts, so a periodic gravel change after reasonable service use is mandatory.

3.3 Masking Tape—Four inches (10 cm) wide, Minnesota Mining and Manufacturing Co., No. 202-2, or equivalent.

3.4 Temperature Conditioning Equipment—Gravelometer tests are usually run at room temperature or a lower temperature, generally 0°F (−17.8°C), which shall be mutually agreed upon by supplier and purchaser. When tests are to be conducted at temperatures lower than room temperature, one of the following must be used:

3.4.1 PREFERRED METHOD—A cold room (or freezer of sufficient size) in which the gravelometer and test panels both are maintained at the specified temperature of testing.

3.4.2 A freezer in which the test panels are cooled to 10°F (5.6°C) below the test temperature before they are individually transferred and tested immediately in a gravelometer at room temperature located nearby.

3.5 Tally Counter—A mechanical counter such as a Veeder Root Tally, Model No. SC-149000, four counters wide by one tier high, or an equivalent type, which is used to tally the number of chips of four specified sizes on the tested panel.

3.6 Transparent Grid—A chip counting aid constructed of transparent plastic approximately 1/8 x 5 x 5 in (3.176 mm x 12.7 cm x 12.7 cm), on which a 4 x 4 in (10.16 x 10.16 cm) grid of 1 in (2.54 cm) squares has been etched or scribed.

3.7 Chipping Rating Standards—A set of photographic transparencies, each depicting one size of chip and the fewest number of chips in each rating category. See Fig. 2 for a representation of these transparencies.

3.8 Test Panels—The test panels should be flat and their size should be 4 x 12 in (10.16 x 30.48 cm) in order to fit into the panel holder of the gravelometer. The test panel material, the panel's thickness or gage, and preliminary surface treatments (such as phosphatizing or anodizing) should be the same for all tests in any one series and as representative as possible of the actual part. Any deviations in these three parameters may produce misleading test results. Specification of these parameters shall be by mutual agreement between supplier and purchaser. A typical test panel specification would be: 4 x 12 in (10.16 x 30.48 cm) 20 gage, 0.0359 in (0.091 cm) phosphated, cold rolled steel.

4. Procedure

4.1 Paint or process the test panels as specified for the systems under test. It should be noted that gravelometer results will be dependent upon the nature of the coating's formulation, the method and degree of drying or curing of the various coats, and the film thickness involved. Uniformity of film

---

[1] Available from Society of Automotive Engineers, Inc., 400 Commonwealth Drive, Warrendale, PA 15096—Identified as EA-400.

[2] A similar type of commercial apparatus meeting these specifications can be obtained from: The Q-Panel Co., 15610 Industrial Parkway, Cleveland, OH.

[3] Gravel meeting these specifications can be obtained from: The Q-Panel Co., 15610 Industrial Parkway, Cleveland, OH.

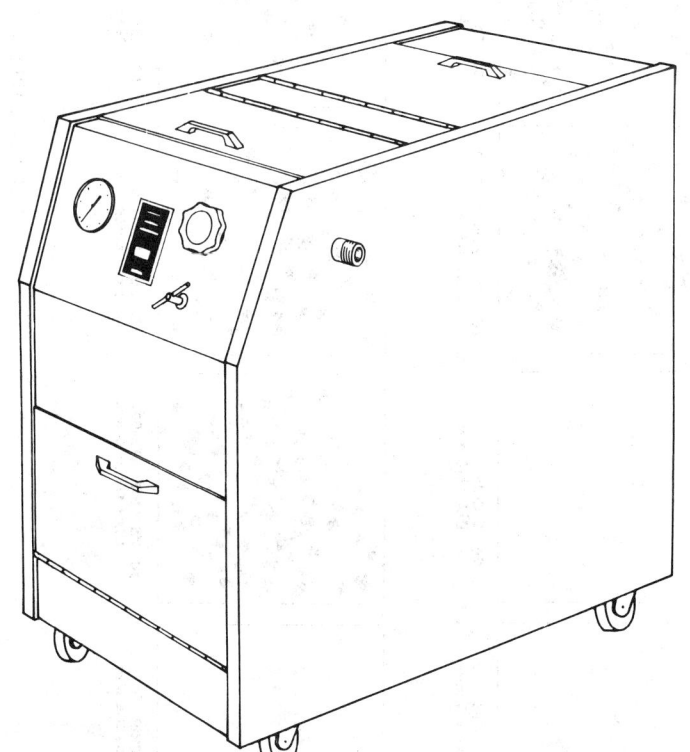
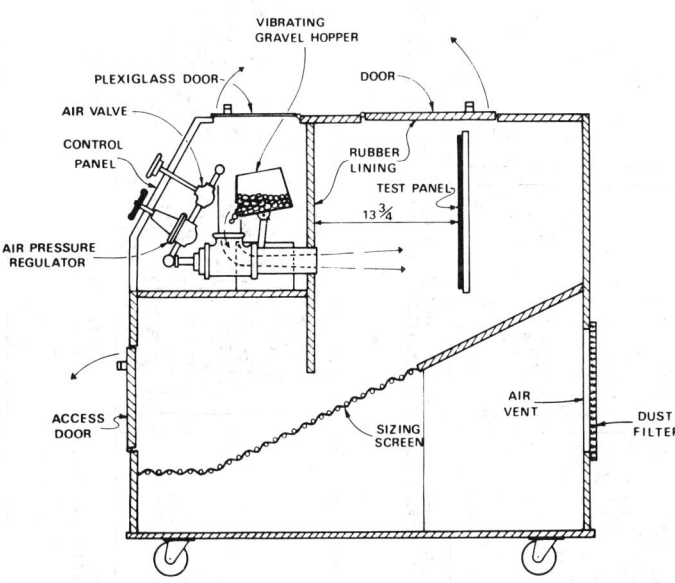

FIG. 1—GRAVELOMETER

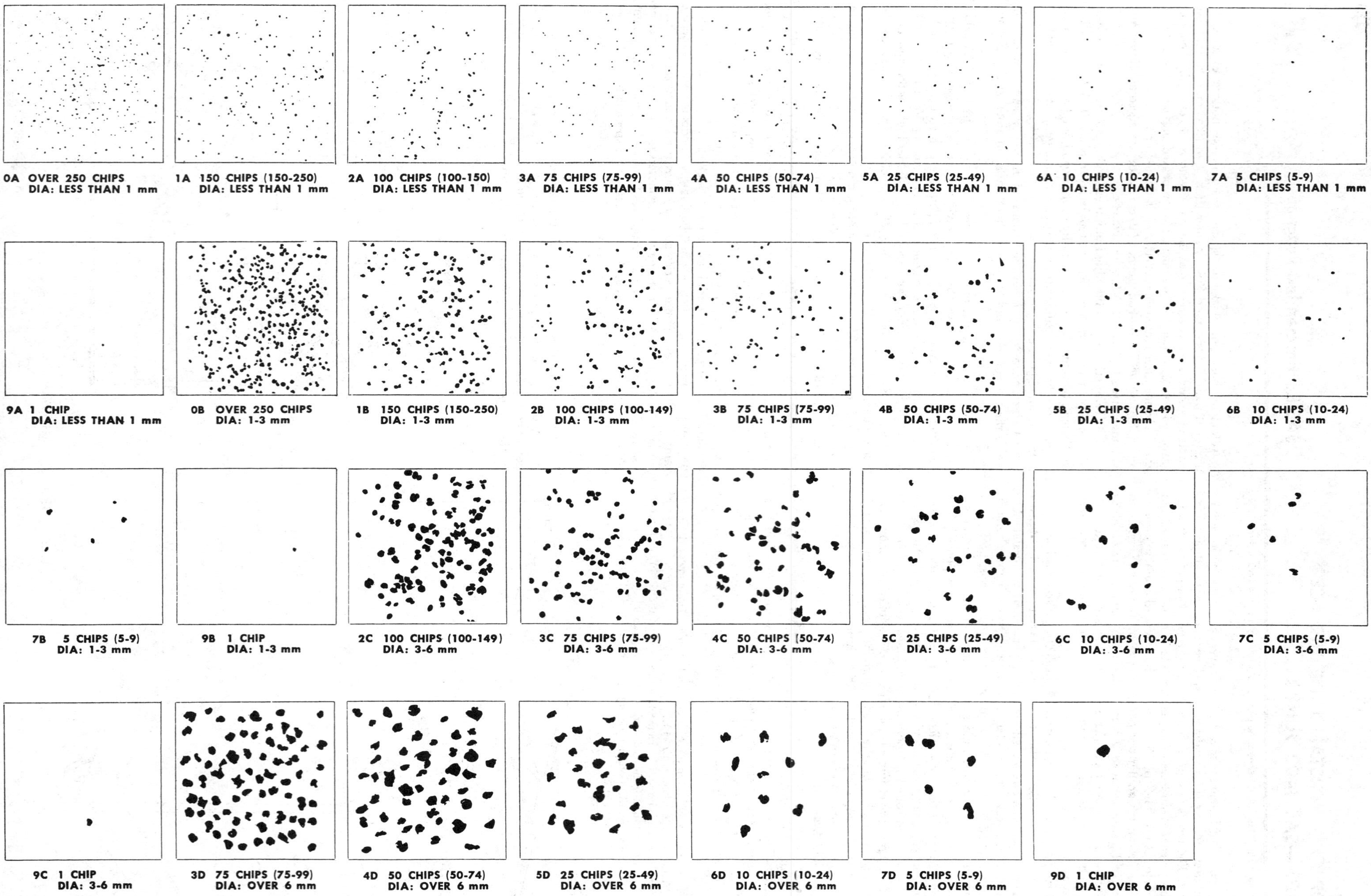

FIG. 2—CHIPPING RATING STANDARDS (REPRESENTATION ONLY)

thickness is extremely important, and each component of the system should be controlled to ±10% of the coating's thickness of ±0.1 mil, whichever is greater.

4.2 Age prepare test panels a minimum of 72 h at room temperature before initiation of gravelometer testing.

4.3 Condition test panels for a minimum of 1 h at the specified test temperature, mutually agreed upon by supplier and purchaser, in accordance with the appropriate method specified in paragraph 3.4. Make certain the test panels are separated and have free access to the conditioning environment so that optimum heat transfer is facilitated.

4.4 Adjust air pressure on the gravelometer to 70 ± 3 psi (482.6 ± 20.68 kPa) with the air valve open. Keep lid to gravel chamber on the gravelometer closed during this operation as a safety precaution.

4.5 After the air pressure is adjusted, shut off air valve, open the lid to the gravel chamber, and collect 1 pt (0.47317 L) of graded gravel (approximately 250–300 stones) in a suitable container. Gravel should be collected by scraping across the screen to allow fines to fall through. (No more than two or three pints of gravel shall be allowed to collect on the screen.)

4.6 Place one test panel conditioned at the desired test temperature in the panel holder with the coated side facing the gravel projecting mechanism, close lid to gravel chamber only. Panel holder should have heavy aluminum back up plate to provide for rigid support.[4] Open the air valve and feed 1 pt (0.47317 L) of gravel slowly over a period of 5–10 s into the funnel.[5] Do not dump the gravel into the funnel because an instantaneous charge such as this may partially choke the pipe tee and lead to inconsistent results due to low particle velocity.

4.7 Shut off air valve, open lid to gravel chamber, and remove the tested panel. If necessary, allow panels to return to room temperature and dry to remove any condensed moisture.

4.8 Cover the tested panel with a 4 in (10 cm) strip of masking tape, press down firmly, and then pull off slowly to remove chips of paint that have not been completely separated from the panel.

4.9 Determine the degree of chipping by one of the following methods of the Gravelometer Rating System.

5. Gravelometer Rating System

5.1 Methods Available

5.1.1 There are two methods available for determining the degree of chipping of the tested panel. In Method I, the exact number of chips in each size range is tabulated for the specified test area, while Method II utilizes a visual comparison of the tested panel with the SAE Chipping Rating Standards[1] which depict various degrees of chipping severity and are arranged sequentially from best to worst according to chipping size and frequency. Method I is the most precise and should be used where definitive accuracy is required or as the referee method in case differences arise between laboratories; however, it is more time-consuming than the visual comparison method. Method II is much faster and, while more of an approximation than the first method, can be used for many routine laboratory evaluations where the accuracy of Method I is not required. Method II also lends itself to field survey work where the chipped areas can be rated by direct comparison with the Chipping Rating Standards.

5.1.2 With both methods, the chipped area to be evaluated on the tested panel should be the 10 x 10 cm (4 x 4 in) square that exhibits the worst degree of chipping.

5.2 Basic Structure of Rating System
Generally, the basic structure of the chip rating system consists of one or more number-letter combinations in which rating numbers 10–0 indicate the number of chips of each size and rating letters A–D designate the sizes of the corresponding chips. A point of failure notation may also be included in the rating if more descriptive refinement is desired.

5.2.1 NUMBER OF CHIPS—A whole rating number selected from the range of 10–0 in Table 1 is used to indicate the number of chips of each size in the 10 x 10 cm test area.

5.2.2 SIZE OF CHIPS—The size of the chips is specified by a rating letter selected from A–D in Table 2. Due to the irregular nature of chipping, the size cannot always be measured exactly so it has to be approximated. Two general types of chipping occur: glassy fracturing or flaking of the film from impact of the gravel and knife-like cut-throughs or gouges caused by sharp edges on the gravel. With true fracturing the longest average dimension of the chip is used for size measurements; with cut-throughs or gouges where little material is actually lost, the most representative shorter dimensions or axis are used.

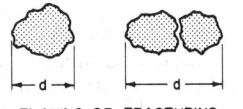
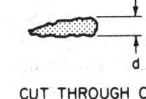

FIG. 3—CHIP SIZE MEASUREMENT APPROXIMATION

Examples of chip size determination, expressed by dimension (d), are shown in Fig. 3.

5.2.3 POINT OF FAILURE—The coating layer at which the most predominant chipping failure occurs is designated as the point of failure. The notations in Table 3 can be used to designate this information if desired. In cases where multiple layers of different primers and/or sealers, multiple topcoats, or other more complex film/substrate combinations are encountered, the point of failure notation can be expanded accordingly in the sense of the descriptive terms of Table 3, once suitable agreement between vendor and purchaser has been established.

5.3 Details of Method I and Method II

5.3.1 METHOD I—EXACT COUNTING PROCEDURE

5.3.1.1 Counting can be facilitated by the use of a transparent overlay onto which has been etched a grid of 1 in (2.5 cm) squares. The grid is placed over the area to be rated as a guide to remembering the areas that have been counted.

5.3.1.2 The mechanical tally is used to record the number of chips, using one counter for each of the four sizes.

5.3.1.3 The operator examines the area within each 1 x 1 in (2.5 x 2.5 cm) square, decides on the size of each chip as it is encountered, and registers it on the appropriate counter. After all 16 squares have been rated, the number of chips of each size can be read on the tally. A 4 x 4 in (10 x 10 cm) area can be rated in this manner in about 90 s.

5.3.1.4 The actual number of chips counted for each size is then converted into the number-letter combinations utilizing Tables 1 and 2. The number-letter rating is then arranged with the most numerous size first, followed by the next more numerous, etc. For example, for a panel on which there are 20 chips less than 1 mm (A size), 40 chips of 1–3 mm (B size), and 3 chips of 3–6 mm (C size) with primer-topcoat failure, the number-letter rating would be 5B-6A-8C (P/T). This rating can be condensed by converting the total number of chips on the panel to the corresponding number category, which is then followed by the size designations in the same order. In this example, with a total of 63 chips, the rating would be summarized as 4 BAC (P/T).

5.3.2 METHOD II—VISUAL COMPARISON PROCEDURE

5.3.2.1 Chipping Rating Standards[1] are utilized. These have been prepared so that chips of only one size are shown in each illustration. The number of chips illustrated in each standard is the fewest number of chips in each rating number category; for example, the No. 5 standards all show 25 chips, the No. 3 standards show 75 chips. All of the No. 8 and No. 10

TABLE 1—NUMBER CATEGORIES FOR CHIP RATING

Rating Number	Number of Chips	Rating Number	Number of Chips
10	0	4	50–74
9	1	3	75–99
8	2–4	2	100–149
7	5–9	1	150–250
6	10–24	0	>250
5	25–49		

TABLE 2—SIZE CATEGORIES FOR CHIP RATING

Rating Letter	Size of Chips (see "d," Fig. 2)
A	<1 mm (<approximately 0.03 in)
B	1–3 mm (approximately 0.03–0.12 in)
C	3–6 mm (approximately 0.12–0.25 in)
D	>6 mm (>approximately 0.25 in)

TABLE 3—POINT OF FAILURE NOTATION

Notation	Level of Failure	Failure Type
(S/P)	Substrate to primer	Adhesional
(S/T)	Substrate to topcoat	Adhesional
(P)	Prime	Cohesional
(P/T)	Primer to topcoat	Adhesional
(T)	Topcoat	Cohesional

---

[4] Back up plate shall be 7 gage, 0.1793 x 4 x 12 in (0.455 x 10.16 x 30.48 cm).

[5] When the test panel has to be cooled to a temperature 10°F lower than the specified test temperature in a freezer located next to a gravelometer at room temperature, the total elapsed time between removal of the panel from the freezer and completion of the gravelometer run shall be no more than 15 s. The 10°F lower conditioning temperature is to provide a working margin for the panel temperature rise toward the specified testing temperature during the transferring and testing interval.

categories and the lower number D size categories have not been included in order to keep the photographs to a manageable number.

5.3.2.2 Visually compare the area to be rated with the standards. Since each standard exhibits only one chip size and actual chipping seldom occurs in only one size, one or more standards should be superimposed until that combination of standards which more nearly resembles the panel is obtained. Record the standards that were used to achieve the match with the panel under examination.

5.3.2.3 As with Method I, the most numerous chips should be listed first, the next most numerous second, etc. Again the number-letter ratings may be summarized to give a condensed single number rating based on the total number of chips of all sizes followed by the letter ratings to indicate the relative number of chips of each size. For example, a panel requiring the superimposition of a 6A standard, a 5B standard, and an 8C standard would be described as 5B-6A-8C (P/T) and summarized as 4 BAC (P/T).

6. *Precision*—Because of the possibility of slight variations in the number, size, type, and distribution of gravel in each test sequence, some variation in the raw counts of chips in the various size categories will be reflected in the data. However, when these counts are converted into the condensed rating of Method I or the rating that can be obtained by Method II, if the results differ by greater than one number-letter rating, they should be considered suspect.

7. *Reporting of Results*—Reports of the gravelometer tests shall include the number-letter rating and all applicable test conditions that deviated from the standard as outlined. In addition, reports should include the material type, thickness, and any preliminary surface treatment of the test panel together with the type of surface coating(s), baking or pertinent processing schedules, and the film thicknesses of finishing system being evaluated.

FLORIDA EXPOSURE OF AUTOMOTIVE FINISHES—SAE J951 FEB80

SAE Information Report

Report of the Nonmetallic Materials Committee, approved June 1966, reaffirmed without change February 1980.

Purpose—The purpose of this SAE Information Report is to compare the results of Florida exposure at 45 deg from the horizontal, facing south, with those at 5 deg from the horizontal, facing south, using various types and colors of automotive finishes.

Summary—Fifteen different automotive finishes were used: five nonoxidizing type alkyd enamels, five thermoplastic acrylics, and five thermosetting acrylics. Each group of five included the following colors: white, black, red, blue metallic, and silver metallic.

Several of the finishes selected were not of the maximum durability type currently in use, to insure that differences in gloss retention on exposure would show up.

Sixteen panels of each finish were prepared under constant conditions and the eight sets of 30 panels were exposed (15 at 45 deg and 15 at 5 deg) at six southern Florida exposure sites within a radius of approximately 40 miles of Miami. The total number of panels exposed was 240.

Part of each panel was washed each month using the same washing procedure at each exposure site.

After 12 months exposure all the panels were returned to one location and 60 and 20 deg glossmeter readings were made on both unwashed and washed areas. Readings on all panels were made by one operator using the same glossmeter; the total number of readings was 960.

The panels were then reexposed for an additional 6 months, for a total of 18 months exposure. Gloss readings were again made in the same manner.

One set of panels was returned for additional exposure and was exposed for 32 months.

Conclusions—

1. Exposure at 5 deg south resulted in greater loss of gloss, on the average, than at 45 deg.
2. Exposure at 5 deg caused faster failure of weaker finishes than did exposure at 45 deg.
3. Finishes with very good gloss retention showed little difference in loss of gloss when exposed at 5 and 45 deg within the duration of this study.
4. While some differences in severity of exposure were noted between the six exposure sites employed, differences were not large and the preceding conclusions hold true regardless of site.
5. The preceding conclusions are supported by gloss retention data from both washed and unwashed areas of the panels.

Recommendation—To most quickly evaluate automotive finishes under southern Florida exposure conditions, it is recommended that panels be exposed at 5 deg from the horizontal, facing south.

Discussion—Table 1 shows typical examples selected from the data to illustrate the conclusions.

Exposure Sites—Six different exposure sites were used. At site 1, three sets of 30 panels were exposed; at each of sites 2-6, one set of panels was exposed.

Glossmeter Readings—Although only 60 deg glossmeter readings are shown in the preceding examples, the conclusions reached are supported by both 60 and 20 deg glossmeter readings. The 60 deg glossmeter is generally felt to be more suitable for exposure work as the 20 deg glossmeter is more sensitive and 20 deg readings can be affected by panel surface conditions other than gloss.

TABLE 2 — AVERAGE OF WASHED 60 DEG READINGS FOR EACH SET

Site No.	Panel Set	12 Months		18 Months	
		45 deg	5 deg	45 deg	5 deg
1	1	76.0	72.7	59.8	55.7
1	2	79.7	78.7	61.9	55.3
1	3	79.7	77.8	58.3	51.6
2	4	82.8	78.3	60.6	55.0
3	5	73.9	65.9	53.9	48.6
4	6	77.1	76.1	—a	—a
5	7	77.9	75.8	52.4	47.4
6	8	82.9	80.2	59.9	53.8
Average of all sets		78.8	75.6	58.1	52.5

a Data not available.

TABLE 1 — TYPICAL RESULTS OF STUDY

		60 deg Gloss			
		Washed		Unwashed	
		45 deg	5 deg	45 deg	5 deg
Finish with Poorer Gloss Retention					
Silver metallic alkyd enamel	Original	94	92	94	92
	6 months	89	78	78	67
	12 months	37	18	34	13
	18 months	12	7	9	5
	24 months	10	8	7	6
	30 months	8	6	7	5
	32 months	8	7	6	5
Finish with Good Gloss Retention					
Blue metallic alkyd enamel	Original	91	91	91	91
	6 months	87	84	76	75
	12 months	76	73	65	57
	18 months	48	38	32	19
	24 months	21	10	12	6
	30 months	14	11	6	5
	32 months	13	15	6	5

		60 deg Gloss			
		Washed		Unwashed	
		45 deg	5 deg	45 deg	5 deg
Finish with Very Good Gloss Retention					
Blue metallic thermosetting acrylic	Original	90	89	90	89
	6 months	87	84	77	69
	12 months	85	83	73	66
	18 months	84	80	66	54
	24 months	80	75	56	46
	30 months	75	69	52	39
	32 months	75	71	46	37
Black thermosetting acrylic	Original	99	98	99	98
	6 months	99	99	95	89
	12 months	99	99	86	72
	18 months	100	100	68	52
	24 months	100	100	60	44
	30 months	100	100	53	36
	32 months	100	100	50	32

GLOSSARY OF FIBERBOARD TERMINOLOGY—SAE J947b

SAE Information Report

Report of Nonmetallic Materials Committee approved June 1966 and last revised August 1972. Reaffirmed without change July 1978.

1. Fiberboard—

Description—A broad general term for fibrous structures produced on any of the several types of fiber forming machines. The primary composition of these boards is normally refined cellulosic or matted wood fibers which may or may not be supplemented by the use of synthetic materials or chemical additives. The manufacture of fiberboards normally involves the forming of a wet web of suspended fibers, which is subsequently pressed, dried, and often calendered or laminated to develop desired end use properties.

Physical Properties—Except for the characteristic fibrous structure, the physical properties may vary over a wide range. The term fiberboard is normally limited to thicknesses of 0.009 in. (0.23 mm) or above.

Applications—The normal uses for this material include nearly all automotive applications where fibrous board structures are specified.

2. Panel Board—

Description—A fiberboard constructed in such a manner that its inherent rigidity lends its usage to applications which require a minimum of attachment supports.

Physical Properties—A panel board may be either single ply or laminated, but will normally require thicknesses of 0.060 in. (1.5 mm) or more to satisfy functional requirements. Normally these boards are characterized by high stiffness and high strength, and usually require water resistance. Various applications may require the surface to be painted, grained or coated for dielectric bonding, and other decorative treatments.

Applications—Typical applications include door and rear quarter panels and trunk liners.

3. Chipboard—

Description—A general term describing a type of fiberboard produced primarily from mixed grades of waste paper and most often produced on a cylinder machine. The final product may be sold as either a single ply or laminated board.

Physical Properties—This material is usually characterized by low density and gray color and is used where strength and quality are not required. The final product may be modified by the addition of nonfibrous components to impart water resistance or other special properties. The normal range of thickness is from about 0.009 in. to 0.045 in. (0.23–1.1 mm) for single ply and 0.050 in. to over 0.200 in. (1.3–5.1 mm) for laminated constructions.

Applications—Used in applications where appearance and ultimate strength are not important. Typical uses include visor cores, trim panel subfoundations, and some gasket applications.

4. Kraft Paper—

Description—Kraft is the generic name for paper of high strength and is derived from the Swedish word Kraft or sulfate pulp from which the paper is made. The sulfate pulping process involves the cooking of wood fibers in an alkaline medium to produce strong, hard cellulosic fiber which is normally converted to paper on a Fourdrinier paper machine.

Physical Properties—The term kraft paper is normally restricted to materials 0.009 in. (0.23 mm) or under in caliper and usually less than 26 lbm/ft^2 (1.3 kg/m^2). The paper is characterized by a reddish brown color in the unbleached state, but may be bleached to a very high brightness white for some applications. This material normally has high strength and is relatively dense.

Applications—Uses include wire wrapping, braided insulators, liner for laminated fiberboards and water shields.

5. Subfoundation Board—

Description—A fiberboard or liner used as a subfoundation in combination with a foundation board. It is used as a carrier for subsequent trim or product applications.

Physical Properties—These boards are often chosen for their flexibility, and as a result caliper usually ranges 0.010–0.030 in. (0.25–0.76 mm). In some cases the boards may be coated to facilitate dielectric bonding. They are usually characterized by medium strength, good plybond, good dimensional stability and are usually treated for water resistance.

Applications—Typical automotive uses are in conjunction with door panels and rear quarter panels.

6. Forming Board—

Description—A board suitable for forming into random three dimensional shapes through the use of heat and pressure applied in a matched set of dies. This board may be a one ply or a laminated combination of the basic board types.

Physical Properties—The primary composition of these boards is normally refined cellulosic fibers which may or may not be supplemented by the addition of some synthetic fibres. Usually these boards contain various amounts of thermoplastic or thermosetting resins to facilitate formability and to enhance stability and rigidity of the formed part. Normal thicknesses range from about 0.070 to 0.120 in. (1.8–3.1 mm).

Applications—Used in parts requiring three dimensional shapes with rounded corners, such as formed arm rests, heater ducts, firewall components, package trays, etc.

7. Bending Board—

Description—A paperboard, either single ply or laminated, the components of which are comprised primarily of refined cellulosic fibres.

Physical Properties—This material is constructed in such a way that the liner or liners are capable of accepting a suitable score, and can later be bent on this score to varying degrees with little or no fracture of the surface fibres. This requirement is usually satisfied by using relatively long, strong fibres on the bending surfaces. In some cases the visual requirements of the bent scores are obtained by covering the surface with an extensible coating or by laminating a pliable film to the surface, prior to scoring and bending. Thickness is usually confined to the 0.016–0.100 in. (0.5–2.5 mm) range.

Applications—Typical applications include scored glove boxes, package trays, visors, trunk liners, etc.

8. Laminated Board—

Description—A general term describing a board comprised of two or more single plies of board, paper, or other sheet materials in any combination, which is firmly adhered to by the use of an adhesive substance between the plies. The adhesion and cohesion of the entire finished structure is such that it will function as a single unit.

Physical Properties—Except for the multiply structure, the physical characteristics of laminated boards vary over a wide range of properties. Because of the general nature of the term there are few typical physical properties.

Applications—Typical uses include head liners, trunk liners, glove boxes, door panels, etc.

9. Wet Machine Board (Homogeneous)—

Description—This material is produced on a one cylinder wet machine. It is manufactured by the building up on a roll of a number of wet plies of paper stock (refined cellulose fibres) from a continuous web. The wet plies adhere mechanically to one another in the wet state and, when the desired thickness of board has been reached, the wet stock (approximately 40% solids) is removed from the make roll as a sheet. It is then pressed, dried, and calendered to desired finished thickness. The pressing and drying operations develop strong fibre to fibre chemical and mechanical bonds within the plies and between the ply interfaces.

Physical Properties—This board is characterized by high density, stiffness and strength. This material is commonly produced in calipers ranging from a minimum of 0.050 or 0.060 in. (1.3 or 1.5 mm) up to a thickness of 0.500–1.0 in. (1.3–25 mm) for various applications. This material frequently contains nonfibrous components such as resins or asphalt to develop water resistance, formability or other special properties.

Applications—Typical uses include tacking strips, dash insulators, etc.

10. Creasing—Method of scoring without cutting. See definition for scoring, paragraph 13.

11. Bending—The folding movement applied to fiberboard, usually along impressed or cut score lines. See definition for scoring, paragraph 13.

12. Foundation Board—

Description—A fiberboard, usually a hardboard or a laminated kraft paper board, that is used as a structural foundation or a supporting member in a trim panel assembly.

Physical Properties—Boards selected for foundation applications generally require a high degree of strength, rigidity, and dimensional stability; hence, most foundation boards are specified in thicknesses of 0.08 in. or greater. Various applications may require this board to be coated to facilitate dielectric bonding or to be painted, embossed, or perforated for decorative purposes.

Applications—Typical applications include door and rear quarter panels, package tray panels, and headlining applications.

13. Scoring—The method by which fiberboard may be depressed, or partially cut, in basically linear configurations in any direction which will later facilitate bending along with the depressed or cut scores into various three-dimensional shapes.

14. Sizing—Sizing is a broad term referring to the resistance of fiberboard to the penetration of liquids and to the process and chemicals for developing this resistance. Surface sizing (surface application) and beater sizing (internal application) are the methods of applying the sizing materials.

FIBERBOARD TEST PROCEDURE—SAE J315b

SAE Standard

Report of Passenger Car Body Engineering Committee and Nonmetallic Materials Committee approved August 1951 and last revised by Nonmetallic Materials Committee August 1972. Reaffirmed without change July 1978.

1. Scope—This SAE Standard provides test methods for determining the critical characteristics of basic or finished fiberboard products. Where applicable, methods of test developed by SAE and ASTM have been referenced.

2. Fiberboard Terminology—See SAE J947.

3. Recommendations—Fiberboard fabrication and finishing techniques, such as crease bending, scoring, forming, perforating, and the application of barrier coatings or paints, will modify the characteristics of the producer's basic material. Consequently, it is recommended that separate but related specifications be established for (1) the properties of the basic product and (2) the finished processed material.

4. Conditioning—Test for material classification and for arbitration purposes shall be made on material conditioned to a constant weight in a controlled atmosphere of 70 ±2 F (21 ±1 C) and 50 ±5% relative humidity. Quality control tests can be conducted on unconditioned specimens unless otherwise specified by the user.

5. Thickness—Thickness shall be measured by a micrometer having two plane, parallel faces the smaller of which should be circular and 0.25–0.33 in.^{2} (161–212 mm^2) in area. When the specimen is clamped between the faces, it should be under a steady pressure of 7.0–9.0 psi (48.23–62.0 kPa). The graduations of the dial face should be such as to permit estimating the thickness to at least 0.0005 in. (0.013 mm).

The sample should be comprised of at least three representative specimens, each of which should be tested in four separate places. The test should be made by placing the specimen between the jaws of the micrometer and lowering the pressure foot gently upon the surface of the specimen, taking care that the edge of the foot is at least 0.25 in. (6.3 mm) from the edge of the specimen.

The average thickness should be reported in decimals of an inch (millimeter) to the nearest 0.0005 in. (0.013 mm) and may be supplemented by maximum and minimum readings.

Fundamental technique and apparatus used shall be similar to ASTM D 645.

NOTE: Specimens cut for dimensional stability tests are satisfactory for these measurements.

6. Weight—The weight shall be determined by weighing 1 x 1 ft (305 x 305 mm) of material to the nearest 0.10 g. Dimensions shall be measured accurately to the nearest 0.01 in. (0.25 mm). Three representative specimens shall be weighed and the average computed and reported in pounds per 1000 ft^2 (kilograms per 93 m^2).

7. Density—Density in pounds per cubic feet (kilograms per cubic meter) shall be computed using data obtained from the average thickness and weight report.

8. Bursting Strength—The bursting strength shall be determined using the conventional power-driven hydraulic type machine. The average value, to the nearest 5 lb (22.25 N), obtained by making five bursts on each side of three specimens is to be reported. Fundamental technique and apparatus used shall conform to ASTM D 774, Method of Test for Bursting Strength of Paper, or ASTM D 2529, Methods of Test for Bursting Strength of Paperboard or Linerboard.

9. Moisture Content—The moisture content shall be determined by observing the loss in weight of a 4 x 4 in. (100 x 100 mm) specimen (the test specimen may be delaminated to facilitate moisture removal) upon drying in an air circulating oven maintained at 215 ±5 F (102 ±3 C) until a constant weight is obtained. The weight loss shall be expressed as percent moisture on the basis of the initial weight of the specimen. For reference purposes, see ASTM D 644, Method of Test for Moisture in Paper.

Paperboard and Fiberboard Containers—In cases where appreciable volatile other than water is known to exist, the Dean and Stark apparatus may be used. See ASTM D 95, Method of Test for Water in Petroleum Products and Other Bituminous Materials.

10. Water Absorption—The percent of water absorption shall be determined by observing the gain in weight of each of three 4 x 4 in. (100 x 100 mm) specimens upon immersion in water. The test specimens shall be cut with a paper cutter or band saw to prevent delamination of the edges. The specimens shall be weighed to the nearest 0.01 g and then submerged horizontally under 1 in. (25 mm) of water maintained at 70 ±2 F (21 ±1 C) and at a pH of 7.0 ±0.5. After periods of 2.5 and 24 h, the samples are removed from the water and the surplus water blotted off. The specimens shall be immediately reweighed to the nearest 0.01 g. The weight of absorbed water shall be calculated and the water absorption expressed as percent by weight based on the initial weight. The average value for each time period is reported.

11. Thickness Swell—The thickness shall be determined to the nearest 0.001 in. (0.025 mm) by averaging four readings taken at the center of each side of the water absorption specimen and 1 in. (25 mm) from the edge. The caliper reading shall be taken using the same apparatus as described in paragraph 5. The specimen shall be soaked and treated in the same manner as established in paragraph 10. Immediately following the tests, the specimen shall be recalipered in the same location and manner, and the average reading established for each soaked specimen. The following formula shall be used when calculating the percent of swelling:

$$S = \left(\frac{T_2}{T_1} - 1.0\right)100$$

S = swelling, %.
T_1 = average thickness before soaking, in. (mm).
T_2 = average thickness after soaking, in. (mm).

12. Surface Water Absorption—Refer to ASTM D 2045, Test Methods for Water Absorption on Non-bibulous Paper and Paper Boards (Cobb Test).

13. Dimensional Stability—The linear expansion and contraction shall be determined in the following manner:

(a) Cut three 12 x 12 in. (305 x 305 mm) test specimens from three different samples of fiberboard which are representative of a shipment.

(b) Inscribe a 10 x 10 in. (254 x 254 mm) on one side of each test specimen.

(c) Follow Method A and/or Method B, as required by the material specification, followed by Method C. Method C may also be used individually as a drying test.

Method A—Hang the test specimens in a vertical position in a humidity cabinet maintained at a temperature of 100 ±2 F (38 ±1 C) and a relative humidity of 98 ±2% for a period of 24 h. On highly water resistant board, the exposure period may be continued to 7 d.

NOTE: The test specimens shall be protected from condensation water droplets by a slanted rustproof metal shield.

Method B—Place each test specimen between two 12 x 12 in. (305 x 305 mm) fine mesh stainless steel screens. Then immerse the specimens horizontally in a tank 1 in. (25 mm) below the surface of water maintained at 70 ±2 F (21 ±1 C) for periods of 2.5 and 24 h. On highly water resistant boards, the immersion may be continued to 48 h.

Method C—Place the three test specimens in an air circulating oven maintained at 190 ±5 F (88 ±3 C) for 24 h.

(d) At the end of the specified exposure period, the test specimen shall be removed and the gage lines measured to the nearest 0.01 in. (0.25 mm), both with machine direction and across machine direction. Calculate and report the average percent expansion or contraction of the three specimens.

14. Spew Test—Two test methods are used to evaluate the tendency of colored extractable materials to stain automotive trim when such trim is cleaned.

Method A, Solvent Extractable Discoloration—An area of the fiberboard is thoroughly wet with cleaners naphtha Hi-Flash VM and P, distillation range 240–300 F (115–149 C) and sandwiched between sheets of white absorbent (filter) paper. The sandwich is then pressed together with a load of 10 psi (6.89 kPa) and maintained for 30 s.

The white paper is examined for color stain and rated as positive or negative or in accord with mutually acceptable standards between producer and user.

Method B, Aqueous Extractable Discoloration—Method A of the spew test is repeated using water containing a 1% by weight alkyl aryl sulfonate or similar surface tension depressant.

The resultant discoloration, if present, is rated positive, and the intensity may be rated in accordance with standards mutually accepted by producer and user.

15. Odor in Fiberboard—Odors are evaluated only by means of the human nose. No instrumentation has been devised to measure intensity or enable a classification of odor.

Objectionable odors are best determined by obtaining a consensus of opinion by a panel of people.

16. Stiffness of Fiberboard—For test procedure, refer to SAE J949.

17. Wicking—For test procedure, refer to SAE J913.

18. Blocking Resistance—For test procedure, refer to SAE J912.

19. Burning Rate—For test procedure, refer to SAE J369.

20. Color Matching—For test procedure, refer to SAE J361.

21. Additional Test Methods Applicable to Coated Boards—In many instances, fiberboards are grained, painted, and/or printed for decorative interior automotive trim applications. The test methods for fade, scuff, and wear generally rely on visual appearance rather than numerical values for determination of the acceptability of the decorated surfaces. Consequently, it is recommended that an "appearance master sample" be established by the consumer as a control for each color, pattern, grain, and finish desired.

22. Resistance to Fading, Cracking, Crazing, or Discoloring of Coated Fiberboard

Method A—The coated fiberboard shall be exposed to ultraviolet light produced by a carbon arc in a suitable machine such as a Fadeometer[1] (or equivalent). The time of exposure shall be 150 standard fade hours (SFH[2]) unless otherwise specified. The machine used shall be controlled to maintain a black panel temperature of 145–160 F (63–71 C), as measured by a black panel thermometer exposed in the specimen mounting rack. The rack must be kept fully loaded at all times to maintain the specified temperature properly even if dummy specimens must be used. After exposure for the specified number of hours, the specimen shall be allowed to cool to room temperature and shall then be examined closely for fading, discoloring, crazing, tackiness, cracking, or other detrimental change.

Method B—The coated fiberboard shall be exposed to continuous ultraviolet light produced by a carbon arc, and to a direct water spray 9 min out of each hour, in a suitable machine such as a Weatherometer[1] (or equivalent). This test is recommended in addition to Method A because some pigments and/or dyes are more susceptible to fading under high humidity conditions. Paint and fiberboard surface crazing or cracking failures also are more likely to be predicted by this test, especially on formed fiberboard parts. The time of exposure shall be 100 SFH unless otherwise specified. The machine used shall be controlled to maintain a black panel temperature of 145–160 F (63–71 C), during the period in which no water is spraying. The temperature shall be measured with a black panel thermometer exposed in the specimen mounting rack. The rack shall be kept fully loaded at all times to maintain the specified temperature properly even if dummy specimens must be used. After exposure for the specified number of hours, the specimen shall be allowed to cool to room temperature and shall then be examined closely for fading, discoloring, crazing, cracking, or other detrimental change.

23. Scuff Resistance—For detailed test information, see SAE J365.

24. Abrasion Resistance—Visual acceptability of the abrasion resistance of coated fiberboards shall be determined on a Model 174 Taber Abrader[3] (or equivalent) employing CS 10 wheels, with a specified load for a specified number of cycles. Products of abrasion shall be continuously removed during test with vacuum attachment. For further procedural information for running test, see SAE J948.

25. Topcoat Adhesion—This procedure is used to determine the topcoat adhesion of coated fiberboard by means of a tape pull test. A 6 in. (150 mm) length of masking tape[4] is applied to the specimen in such a way that approximately 4 in. (100 mm) of the tape is in contact with the surface of the specimen. The tape should not contact specimen edges or openings. Rub the tape with an object (paper clip) until a color change (whitening) of the tape indicates that intimate contact has been established between the tape and the specimen. Grip the loose end of the tape and bend it back upon itself at an angle of approximately 180 deg. With a quick pull, peel the tape from the specimen.

---

[1] Equipment which meets the requirements of this test may be obtained from Atlas Electrical Devices Co., Chicago, Illinois 60610. Maintenance of the equipment shall be in accordance with the manufacturer's instructions.

[2] The equipment shall be calibrated to check the performance and to predict the number of clock hours to produce a specific number of standard fade hours (SFH). Standardized light sensitive papers and a booklet of standardized fade strips may be obtained from the National Bureau of Standards, Textile Section, Washington, D. C. 20025. A letter circular LC 1004, which describes the papers and explains their use, can be obtained from the Bureau.

[3] Equipment which meets the requirements of this test can be obtained from the Taber Instrument Crop., North Tonawanda, New York, U. S. A. The equipment and abrasive wheels shall be maintained and operated in accordance with the manufacturer's instructions. The wheels shall be refaced every 1000 cycles or after each test, whichever occurs first.

[4] Available from Minnesota Mining & Manufacturing Co., Scotch Brand 1 in. masking tape, No. 271. The maximum shelf life of this tape is one year when stored at 70 F (21 C).

TEST METHOD FOR WICKING OF AUTOMOTIVE FABRICS AND FIBROUS MATERIALS—SAE J913a

SAE Standard

Report of Nonmetallic Materials Committee approved July 1965 and last revised August 1972. Reaffirmed without change May 1978.

1. Scope—This test method is applicable for determining the wicking characteristics of seat fabrics, convertible tops, headlining, fiber padding, and other automotive textile materials.

2. Apparatus and Materials Required—

2.1 Indelible pencil.

2.2 Solution:

(A) 0.1 g fluorescein dye (fluorescein sodium salt, "Uramine") dissolved in 1000 ml of distilled water (pH 7 max).

(B) 50% fluroescein dye as in (A); 50% alkyl aryl sulfonate solution (10 g of "Nacconal 40F"-40% alkyl aryl sulfonate dissolved in 1000 ml of distilled water—pH 7 max).

(C) Distilled water.

2.3 Apparatus for suspending samples in pan.

2.4 A Black Light (Hanovia Lamp No. 16180 or equivalent).

3. Test Specimens—Cut strips of material to be tested in 8 in. (203 mm) long and 2 in. (51 mm) wide in both the machine direction and cross machine direction, and condition at 70 ± 2 F (21 ± 1 C) and 50 ± 5% relative humidity for 24 h.

4. Procedure—Draw a line with an indelible pencil 2 in. (51 mm) from end of samples to be immersed. Prepare a suitable container with a 3 in. (76 mm) minimum depth of solutions (A), (B), or (C) as specified and immerse samples to a point where upper meniscus just touches line marked with indelible pencil. Allow samples to remain for a specified period in solution while maintaining solution to within 1/16 in. (2 mm). Tests shall be run in a controlled atmosphere of 70 ± 2 F (21 ± 1 C) and 50 ± 5% relative humidity.

At end of the specified period, remove samples from test solution and examine under "Black Light." Travel of fluorescein dye above marked line indicates degree of wicking. Samples tested in solution (C) shall be examined for wicking, migration, and discoloration under normal light.

METHOD OF TESTING RESISTANCE TO SCUFFING OF TRIM MATERIALS—SAE J365a

SAE Standard

Report of Nonmetallic Materials Committee approved October 1968 and last revised August 1972. Reaffirmed without change May 1978.

1. Scope—This test can be used to determine the resistance to scuffing of test specimens such as fiberboards, fabrics, vinyl coated fabrics, leathers, and similar trim materials.

2. Materials and Equipment Required

Abraser—Taber Model 174, or equivalent. Equipment which meets the requirements of this test can be obtained from the Taber Instrument Corp., North Tonawanda, New York.

Specimen Holder—Catalog No. E-100-125[1], 4 1/4 in. (108 mm) OD.

Hold-Down Ring—Catalog No. E-100-101[1], 4 1/4 in. (108 mm) OD.

Rubber Pad—Catalog No. S-19[1].

Clamp Plate—2 1/8 in. (54 mm) OD for fabrics, leather, coated fabrics, and similar flexible materials. 1 1/4 in. (32 mm) OD for carpets and other floor covering materials.

Scuff Fixture—The special scuff fixture head, weight, and other components are shown in Fig. 1 and are assembled as shown in Fig. 2. The scuff fixture is attached to the abrader as shown in Fig. 3. The scuff head is held at

---

[1] Taber catalog numbers.

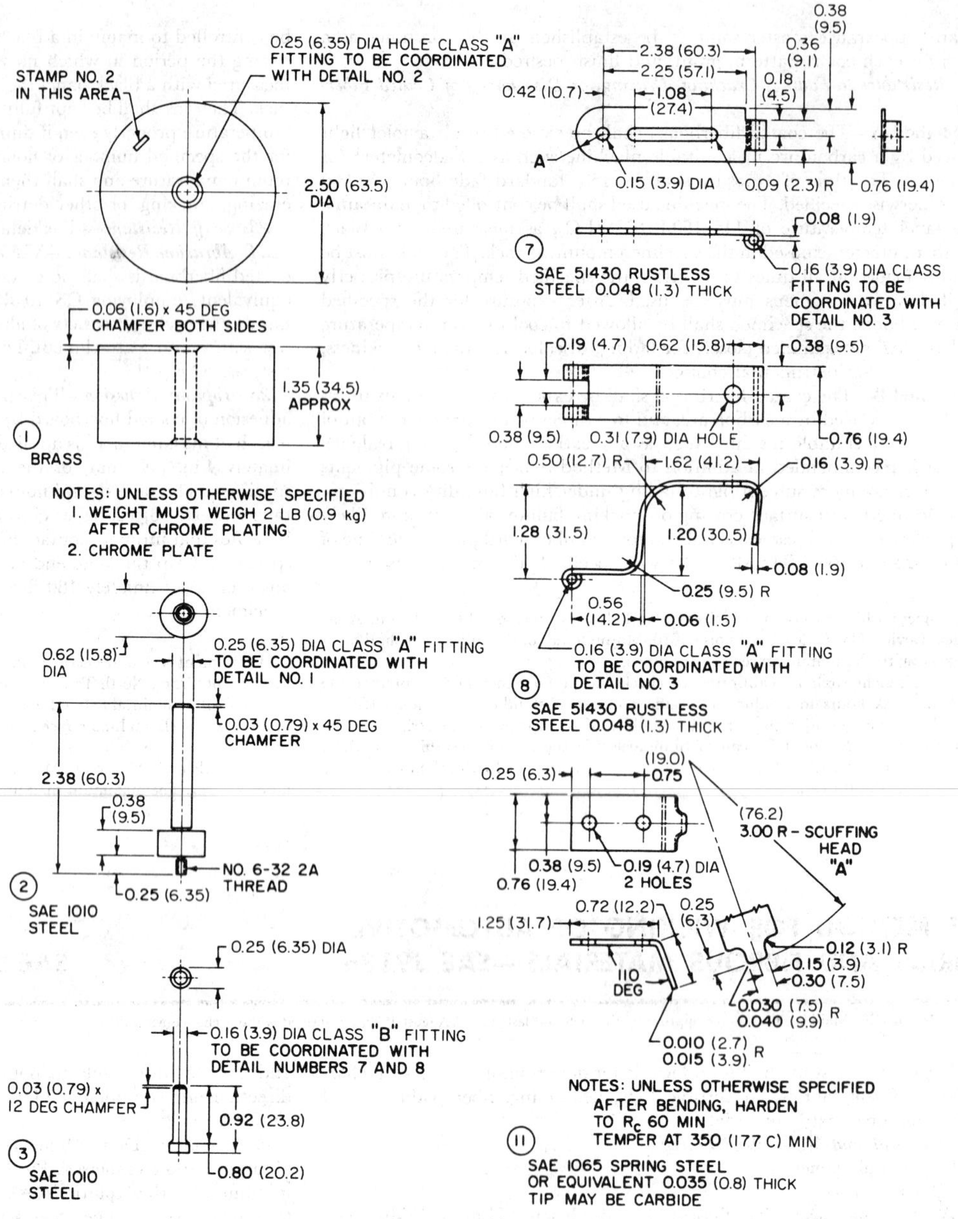

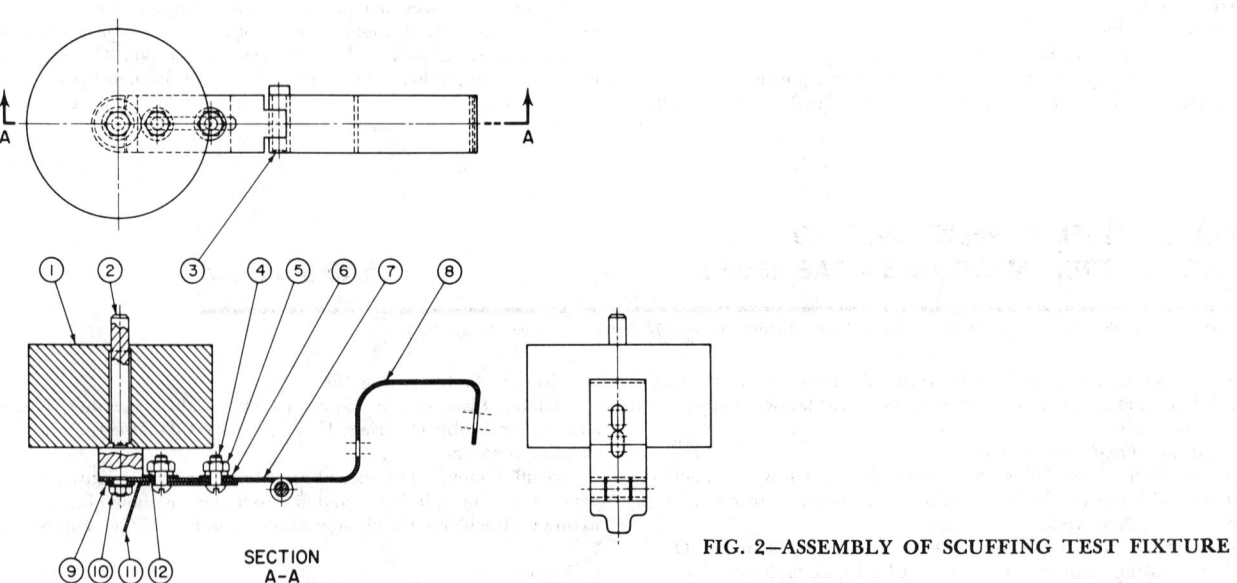

FIG. 1—COMPONENTS OF SCUFFING ASSEMBLY

*ITEMS 4, 5, 6, 9, 10 AND 12 MAY BE ANY CONVENIENT SIZE
DIMENSIONS ARE IN (mm)

FIG. 2—ASSEMBLY OF SCUFFING TEST FIXTURE

FIG. 3—ATTACHMENT OF SCUFFING TEST FIXTURE TO ABRADER

a 110 deg angle. The vertical centerline of the scuff head is 1¼ in. (32 mm) from the specimen holder center pin. The tip is centered under the 2 lb (0.9 kg) weight and in a horizontal alignment with the center pin as shown in Fig. 3.

NOTE: The attachment bracket Fig. 1 (detail 8) may be modified for other abrader models, provided the scuff head position is maintained and test results correlate.

The scuff head tip shown in Fig. 1 (detail 11) must be frequently checked for dimensions and reground or replaced if found to deviate from the specified tolerances.

3. Procedure

Conditioning—The test specimens shall be conditioned for a minimum of 24 h at 70 ±2 F (21 ±1 C) and 50 ±5% relative humidity for this test. Unless otherwise specified, the test shall be conducted under the same controlled conditions since a change in relative humidity and temperature can affect the test results.

Test Samples for Textiles, Coated Fabrics, Leather, or Similar Flexible Materials

1. Cut a ¼ in. (6.4 mm) hole in the center of a 5 5/32 in. (131 mm) diameter specimen.
2. Place the specimen on the rubber pad of the specimen holder.
3. Place a 2⅛ in. (54 mm) OD clamp plate over the material and tighten down securely with the clamping nut.
4. Press the hold-down ring over the test specimen so that the material is drawn taut over the specimen holder with no wrinkles or bulges.
5. Tighten the adjusting screw of the hold-down ring just enough to hold the test specimen but not so tight as to cause wrinkling or bulging.
6. Place the assembled test specimen on the abrading machine and lower the scuff fixture onto the test specimen as shown in Fig. 3.
7. Scuff for the number of cycles indicated by the engineering specification.

Test Samples for Fiberboard, Rubber Floor Mats, Carpets, and Other Semirigid Materials

1. Cut a ¼ in. (6.4 mm) hole in the center of a 4 3/16 in. (106 mm) diameter test specimen.
2. Place the test specimen on the rubber pad on the specimen holder and tighten down securely with a 1¼ in. (32 mm) clamp plate and nut.
3. Place the hold-down ring over the test specimen, press the ring with the fingers, and tighten it. The specimen, when properly secured, shall be free of wrinkles and bulges.
4. Place the assembled test specimen on the abrading machine and lower the scuff fixture onto the test specimen as shown in Fig. 3.
5. Scuff the test specimen for the number of cycles indicated by the engineering specification.

4. Reporting
Observe and report scuff resistance by comparing the test specimen to an approved master scuff specimen established by the consumer.

TEST METHOD FOR DETERMINING BLOCKING RESISTANCE AND ASSOCIATED CHARACTERISTICS OF AUTOMOTIVE TRIM MATERIALS—SAE J912a

SAE Standard

Report of Nonmetallic Materials Committee approved June 1965 and last revised August 1972. Reaffirmed without change May 1978.

1. Scope—This test method is designed to indicate the degree of surface tackiness, color transfer, loss of embossment, and surface marring when two trim materials are placed face to face under specific conditions of time, temperature, and pressure. These specific conditions are not dictated in this test procedure but will be found in the material standards which govern each type of trim material to be tested.

2. Materials and Equipment Required—

2.1 Weights or compression fixture capable of exerting uniform pressure over a 2 x 2 in. (51 x 51 mm) area.

2.2 Air circulating oven.

2.3 Small capacity tensile machine capable of determining increments of 0.1 lb (0.4 N) over a 0–22 lb (0–88 N) range. Front and back jaws shall have a minimum width of 2 in. (51 mm).

3. Test Specimens—Cut one 2 x 3 in. (51 x 76 mm) test specimen from each trim material to be tested. If only one material is to be tested, cut two test specimens.

4. Procedure—

4.1 Place two test specimens face to face and align all edges.

4.2 Apply specified pressure to only upper ⅔ area of sandwich. This will allow an end flap of 1 x 2 in. (25 x 51 mm) on each specimen which will not be under pressure. See Fig. 1.

4.3 Place assembly into an oven which has been heated to specified temperature.

4.4 After specified time has elasped, remove assembly from oven, release pressure, and allow a cooling period at room temperature for at least ½ h.

4.5 Position specimens in tensile machine so that one flap is in upper jaw and other flap is in lower jaw. Disengage pawls. Separate specimens at a rate of 2 in./min (50 mm/min) and record average reading to nearest 0.1 lb (0.4 N).

4.6 Record also, observations concerning surface appearance of specimens as listed under paragraph 1.

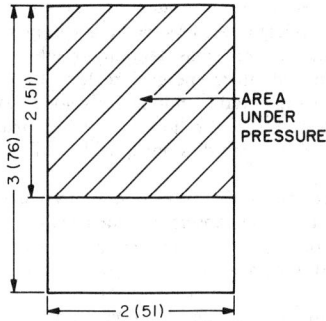

FIG. 1

DIMENSIONS ARE IN (mm)

FIBERBOARD CREASE BENDING TEST—SAE J119a — SAE Standard

Report of Nonmetallic Materials Committee approved September 1969 and last revised August 1972. Reaffirmed without change July 1978.

1. *Scope*—This test method is designed to determine the suitability of a painted or unpainted fiberboard for application involving creasing and bending. The specific purpose of the test is to determine whether a given material, properly creased, can be bent along the impressed crease without objectionable failure on the surface of the bend.

2. *Equipment Required*
   2.1 Press[1], with adequate tonnage to crease the test specimen to the desired configuration.
   2.2 Matched male and female die sections, selected for the caliper of sample to be tested and for the desired end result (see paragraph 4).

3. *Test Specimen*
   3.1 Size—Minimum of 8 x 8 in. (203 x 203 mm).
   3.2 Condition Prior to Testing—See SAE J315 (paragraph 4) unless otherwise specified.

4. *Creasing Rule (Male Die Section)*
   4.1 The thickness of the creasing rule is designated commercially by the printer's point system (1 point equals approximately 0.014 in. (0.36 mm)).
   4.2 Commercially available in either flat or round face and in the following thicknesses (flat face rule should be used unless otherwise specified):

Thickness	Approx. in	Approx. mm	Thickness	Approx. in	Approx. mm
2 points	0.028	0.71	10 points	0.140	3.50
3 points	0.042	1.07	12 points	0.166	4.29
4 points	0.056	1.42	1/8 in (3 mm)	0.125	3.17
6 points	0.083	2.12	3/16 in (5 mm)	0.188	4.78
8 points	0.112	2.85	1/4 in (6 mm)	0.250	6.35

   4.3 For the normal range of bending fiberboards, the following may serve as a guide in the selection of creasing rule:

Material in	Material mm	Creasing Rule	Material in	Material mm	Creasing Rule
0.030	0.76	3–6 points	0.070	1.78	6–12 points
0.040	1.02	3–6 points	0.080	2.03	8 points–3/16 in (5 mm)
0.050	1.26	4–8 points	0.090	2.28	12 points–1/4 in (6 mm)
0.060	1.52	6–8 points	0.100	2.54	12 points–1/4 in (6 mm)

   4.4 Penetration of the creasing rule should be such that it clears the bottom of the female die by the caliper of the test material plus 0.005–0.008 in. (0.13–0.20 mm).

[1] Suitable press can be obtained from Box Board Research and Development Association, 350 S. Burdick Mall, Kalamazoo, Michigan.

5. *Female Die Section*
   5.1 The female die opening may be determined by adding the creasing rule thickness, in inches (millimetres), to twice the caliper of the test specimen.
   5.2 Depth should be equivalent to the caliper of test specimen unless otherwise specified.
   5.3 Both female die section and creasing rule should extend beyond the edges of test specimen.

6. *Procedure*
   6.1 Matched male and female die sections should be determined by the caliper of the test specimen (see paragraph 4.3) and mounted in press (see paragraph 2.1). Stops should be provided so that penetration of male creasing rule is as specified in paragraph 4.4.
   6.2 Insert test specimen in die and crease with one stroke of press. Using separate specimens, crease both with the grain and across the grain and on one or both sides as specified.
   6.3 Where specified more than one crease may be impressed into the test specimen simultaneously, but the creases should not be less than 3 in. (76 mm) on centers nor less than 1 in. (25 mm) from the edges of the test specimen parallel with the creasing rule. If creases less than 3 in. (76 mm) on centers or less than 1 in. (25 mm) from the edge parallel with the creasing rule are used, special consideration should be given in specifying the end results.

7. *Evaluation*
   7.1 The recommended practice is to bend sides of specimen away from male side of die. In actual practice, the design of the part may require bending both ways.
   7.2 The specimen should be bent on a supporting form over an edge conforming to the specified bending angle. An angle of 60 deg (Fig. 1) is recommended for material comparison tests unless otherwise specified.
   7.3 The test specimen is evaluated visually after bending. In general fracturing of the fiberboard surface and/or the paint coating is subject to rejection. A master sample showing the minimum acceptable appearance after testing may be used for comparison at the discretion of the user.

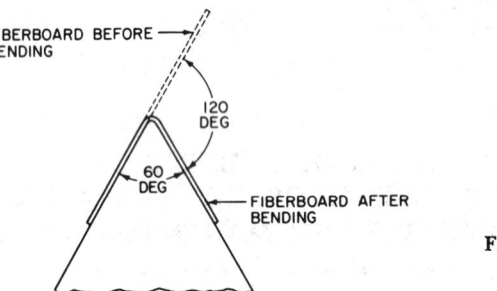

FIG. 1

DEFINITIONS OF ACOUSTICAL TERMS—SAE J1184 — SAE Information Report

Report of Nonmetallic Materials Committee approved June 1978.

1. *Scope*—This information report provides definitions of acoustical terms relating to sound insulation materials. Appropriate methods of test are being developed by SAE and where applicable, ASTM methods will be referenced.

2. *Deadening*—Deadening, also referred to as damping or vibration damping, is the process of converting vibratory energy into heat by internal or interface friction. An example of a material with inherently good deadening is lead, while one with poor deadening is cold rolled steel. When struck with a hammer, lead produces a dull thud which dies out quickly while cold rolled steel rings for a long time.

3. *Absorption*—Sound absorption is the process of converting incident sound energy into heat energy. The sound absorption coefficient is the ratio of the absorbed sound energy to the incident sound energy.

4. *Transmission Loss*—Sound transmission loss of a partition, in a specified frequency range, is the ratio, expressed in decibels, of the airborne sound power incident on the partition to the sound power transmitted by the partition and radiated on the other side.

5. *Decibel (dB)*—The decibel is a logarithmic unit denoting the ratio between two like quantities that are proportional to power. The ratio is expressed in decibels by multiplying its common logarithm by 10.

6. *Sound Pressure Level*—A unit expressed in decibels equal to 20 times the logarithm of the ratio of the measured sound pressure to the reference pressure. The reference pressure is usually 0.0002 microbars (2×10^{-5} N/m^2) = 2×10^{-5} Pa.

7. *Hertz*—Hertz is the accepted unit for frequency. One hertz = 1 cycle per second.

8. *Decay Rate*—Decay rate is the rate at which vibratory or acoustic energy dies out, usually expressed in decibels per second. In the Geiger Plate test for acoustic deadening materials, decay rate is used as the criterion of deadening efficiency of a material.

9. *Rayl*—The rayl is the magnitude of a specific acoustic resistance, reactance, or impedance for which a sound pressure of 1 microbar produces a linear velocity of 1 centimeter per second (dyne-s/m^3). When expressed in Newton-s/m^3 it is called the mks rayl, (from *American Institute of Physics Handbook*, Section 3a-2).

FLAMMABILITY OF AUTOMOTIVE INTERIOR MATERIALS— HORIZONTAL TEST METHOD—SAE J369a

SAE Recommended Practice

Report of Nonmetallic Materials Committee approved March 1969 and last revised June 1972. Reaffirmed without change May 1978.

1. Scope—This method of test is intended for use in the measurement of the burning rate of automotive interior materials as specified by the applicable standard.

2. Apparatus Required

2.1 Burner—A Tirrill, Bunsen, or equivalent, burner with a gas flow regulating valve and 0.375 in. (9.53 mm) inside diameter tube, so positioned in the cabinet that the center of the end of the specimen shall be directly above the tip of the flame when the specimen is in place.

2.2 Burner Fuel—The gas supplied to the burner shall have a flame temperature equivalent to that of natural gas. Recommend 900–1100 BTU/ft^3 (3.36 x 10^7–4.13 x 10^7 J/m^3), if so required.

2.3 Specimen Holder—Consisting of two identical U-shaped metal frames made from chrome or nickel-plated steel, or other metal that will not corrode.

Dimensions for these frames are shown in Fig. 1. Lines shall be engraved or scribed on both surfaces of each frame located as shown in Fig. 1.

2.4 Specimen Holder Support—The specimen holder shall be supported horizontally so that the top of the burner is 0.75 in. (19.1 mm) below the top surface of the lower specimen frame.

2.5 Metal Cabinet—The cabinet (Fig. 2) for protecting the specimen from drafts shall be fabricated from noncorroding metal and shall be 15 in. (381 mm) long, 8 in. (203 mm) wide, and 14 in. (356 mm) high. It shall have a removable top and a glass observation window in front. For ventilation, the base shall have five 0.75 in. (19.1 mm) diameter holes equally spaced along each side of the cabinet. In addition, there shall be a 0.50 in. (12.7 mm) ventilating clearance running around the perimeter of the cabinet just below the top. At one end of the cabinet there shall be a door to permit insertion of the specimen holder and the specimen. A small hole may be drilled in the cabinet to accommodate the tubing which connects the gas line to the burner. The cabinet shall have 0.375 in. (9.53 mm) risers to permit the circulation of air.

2.6 Combing Device—A comb 4 in. (102 mm) wide with 7–8 smooth round teeth per inch (25.4 mm).

2.7 Timing Device—A stop watch which will indicate time to 0.1 of a second.

3. Test Specimens Size—In all instances, the largest possible specimen size is to be cut from the material up to the standard specimen size of 4 in. (102 mm) x 14 in. (356 mm) x thickness. The maximum thickness of any specimen shall be 0.50 in. (12.7 mm). If any material to be tested exceeds this, it shall be cut down to a thickness of 0.50 in. (12.7 mm) and shall include the primary surface of the part.

Where the maximum available width of the specimen is 2 in. (50.8 mm) or less so that the sides of the specimen cannot be held in the two matching U-shaped frames, it is to be supported by the use of 0.010 in. (0.254 mm) wires of heat resistance composition spanning the top surface of the bottom U-shaped frame at 1.0 in. (25.4 mm) intervals, as shown in Fig. 3. The U-shaped wire frame shall also be used for a specimen that softens and bends at the flaming end.

3.1 Selection and Direction—Shall be as specified in the applicable standard.

3.2 Surface Preparation—When materials to be tested contain either a napped or a tufted-type surface, this test specimen shall be placed on a hard, flat surface and combed twice against the nap prior to testing.

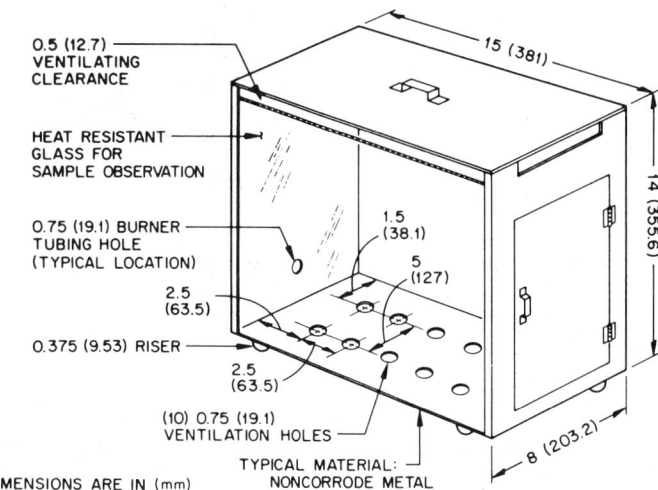

FIG. 2—HORIZONTAL FLAMMABILITY CABINET

4. Test Procedure

4.1 Prior to testing, each specimen is conditioned for 24 h at a temperature of 70 F (21.1 C) and 50% relative humidity or as otherwise specified.

4.2 For convenience, samples may be stored up to 1 h in closed polyethylene bags after conditioning and prior to testing.

4.3 Tests should be conducted with the metal cabinet in a draft-free fume hood to prevent fumes from spreading throughout the room.

4.4 Close the air intake ports on the burner and adjust the gas flow to produce a flame 1.5 in. (38.1 mm) in height.

4.5 Place the mounted specimen in a horizontal position in the center of the cabinet.

4.6 Position the burner so that the center of the barrel will be directly below the center of the open end of the mounted test specimen.

4.7 Expose the specimen to the flame for 15 s, then extinguish the burner flame or remove the burner from the specimen.

4.8 From the time of initial burner flame contact with the specimen, observe for any rapid burning or flame front progression across the top or bottom surface of the material. Begin timing (without reference to the 15 s burner flame application), when the leading flame front reaches the first scribed line 1.5 in. (38.1 mm) from the open end of the U-shaped frame. If the leading flame front progresses more than 2 in. (50.8 mm) beyond the first scribed line at such a rapid rate that it cannot be measured with any degree of accuracy, the material shall be reported as "rapid burning."

4.9 Stop timing when the flame is either extinguished or has burned the additional 10 in. (254 mm) to the second engraved line on the specimen holder.

4.10 Record time in seconds required for the flame to travel the 10 in. (254 mm) between scribed lines on the specimen holder. Or, record time in seconds and burned length beyond the 1.5 in. (38.1 mm) scribed line if the flame is extinguished before traveling the full 10 in. (254 mm).

FIG. 1—SPECIMEN HOLDER, CONSISTING OF TWO IDENTICAL U-SHAPED FRAMES

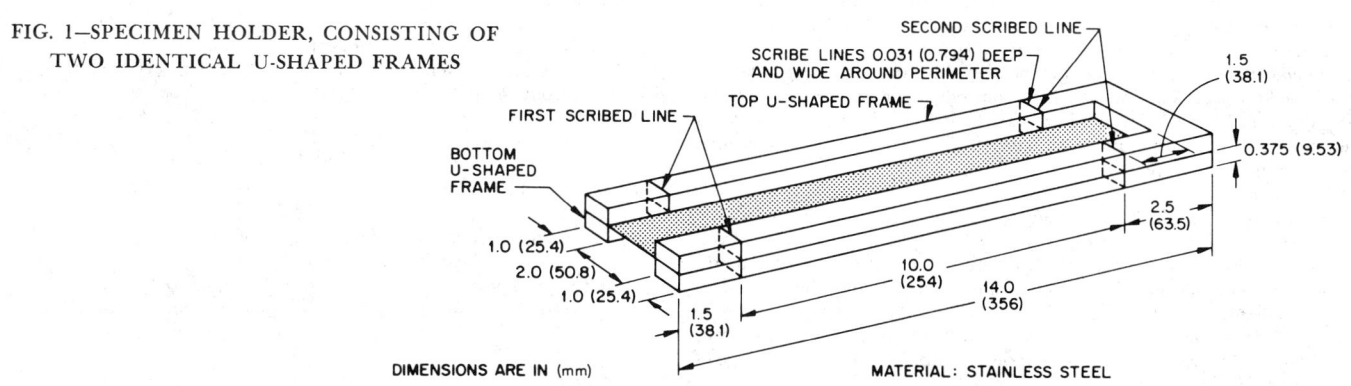

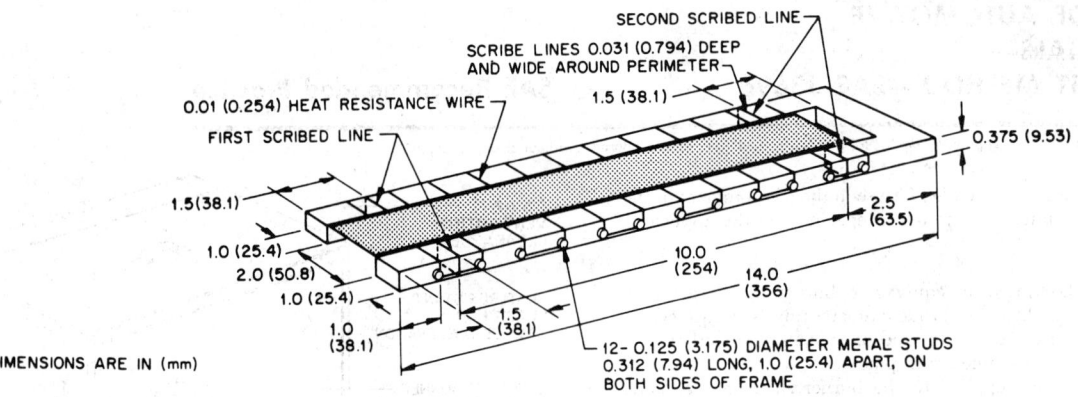

FIG. 3—BOTTOM U-SHAPED WIRED FRAME (USED FOR SPECIMENS THAT ARE 2 IN (50.8 mm) OR LESS IN WIDTH OR THAT BURN NONUNIFORMLY. USED IN CONJUNCTION WITH EITHER FRAME SHOWN IN FIG. 1

4.11 Use the following definitions to report complete flammability results as illustrated in Fig. 4.

5. Definitions

5.1 Does not Ignite (DNI)—The material does not support combustion during or following the 15 s ignition period and does not transmit a flame front across either surface to the first scribed line. (No calculation required.)

Report results as: DNI.

5.2 Self-Extinguishing (SE)—The material ignites on either surface, but the flame extinguishes itself before reaching the first scribed line. (No calculation required.)

Report results as: SE.

5.3 Self-Extinguishing/No Burn Rate (SE/NBR)—The material stops burning before it has burned for 60 s from the start of timing, and has not burned more than 2 in. (50.8 mm) from the point where timing was started. (No calculation required.)

Report results as: SE/NBR.

5.4 Self-Extinguishing with a Burn Rate in inches (millimeters)/minute (SE/B)—When the leading flame front on either surface progresses beyond the first scribed line, but extinguishes itself before reaching the second scribed line, time and measure its progress to the furthest point where the burning stops and calculate and report the burn rate only if the burner distance exceeds 2 in. (50.8 mm) or the burn time is 60 s or greater.

Report results as: SE/B. Calculate burn rate.

5.5 Burn Rate in inches (millimeters)/minute (B)—The material burns the full 10 in. (254 mm).

Report results as: B. Calculate burn rate.

5.6 Rapid Burning (RB)—The material transmits a flame across either surface more than 2 in. (50.8 mm) beyond the first scribed line at a rate too fast to measure accurately; and, therefore, no calculation is required. Examples of materials in this category are extremely thin films which burn rapidly, or napped surfaces which "flash".

Report results as: RB.

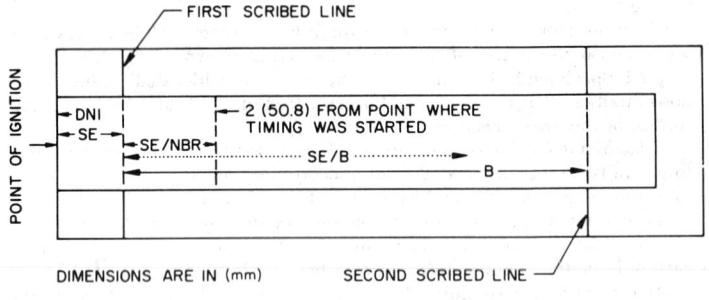

FIG. 4

6. Calculating and Recording—Calculate and record the burn rate for the conditions SE/B (paragraph 5.4) and B (paragraph 5.5) from the formula:

$$B = 60 \times \frac{D}{T}$$

where: B = burn rate, in./min (mm/min) (rounded to one decimal)
D = length the flame travels starting from the first scribed line, in (mm) (rounded to one decimal)
T = time starting from the first scribed line for the flame to travel D, s

TEST METHOD FOR DETERMINING STIFFNESS (MODULUS OF BENDING) OF FIBERBOARDS—SAE J949a

SAE Standard

Report of Nonmetallic Materials Committee approved March 1966 and last revised August 1972. Reaffirmed without change July 1978.

1. Scope—This SAE Standard presents a method of determining the stiffness of fiberboards.

2. Test Specimens—Cut three specimens each 3 x 12 in. (76 x 305 mm), with the long dimension in the machine direction of the fiberboard. Cut three additional specimens each 3 x 12 in. (76 x 305 mm), with the long dimension in the across-machine direction of the fiberboard.

3. Conditioning—Unless otherwise specified, the test specimens shall be conditioned to a constant weight in a controlled atmosphere of 70 ± 2 F (21 ± 1 C) and $50 \pm 5\%$ relative humidity. This test can also be conducted after soaking the specimens in water under specified conditions to determine relative stiffness when wet.

4. Procedure—

4.1 Measure the thickness of each specimen with a micrometer to the nearest 0.001 in. (0.025 mm) at the point of load application (center of specimen) and at the support points (3 in. (76 mm) on each side of the center measured in the long direction of the specimen). Determine the average of these three readings and record.

4.2 Measure the width of the specimen to the nearest 0.01 in. (0.25 mm) in the same locations as given in paragraph 4.1 and determine the average of the three readings and record.

4.3 Support the flat surface of the specimen on parallel supports 6 in. (152 mm) apart and apply the load on a bearing parallel to the end supports. The bearing and supports shall be rounded to a radius of $\frac{3}{8}$ in. (10 mm) and shall have a minimum length of 3 in. (76 mm).

4.4 Apply the load at the rate of 0.5 in./min (13 mm/min) until the specimen is deflected $\frac{1}{4}$ in. (6.3 mm) at the midspan.

4.5 Calculate and report the average stiffness for each direction at the fiberboard from the following formula:

$$E = \frac{PL^3}{4\,b\,d^3\,Y}$$

where: E = stiffness = modulus of bending, psi (Pa)
P = load, lb (N)
L = length of span, in (mm) = 6 in. (152 mm)
b = width of specimen, in. (mm)
d = thickness of specimen, in. (mm)
Y = deflection of specimen at midspan = $\frac{1}{4}$ in. (6.3 mm)

TEST METHOD FOR DETERMINING RESISTANCE TO SNAGGING AND ABRASION OF AUTOMOTIVE BODYCLOTH—SAE J948a

SAE Standard

Report of Nonmetallic Materials Committee approved June 1966 and last revised August 1972. Editorial change May 1978.

Scope—This method of test is applicable for determining the resistance to snagging and abrasion of automotive fabrics.

Apparatus Required—
Taber Abraser Model No. 174 or equivalent, complete with vacuum accessory.
H-18 wheels (snagging).
CS-10 wheels (abrasion).
Diamond wheel dresser.
S-11 abrasive paper.
Camel's hair brush.

Test Specimens—Test specimens are prepared by folding a $4\frac{1}{4}$ x $4\frac{1}{4}$ in (108 x 108 mm) sample once in each direction and then clipping the folded point to produce a small central hole to fit over the turntable clamping screw. Specimens are then conditioned at $70 \pm 2°F$ ($21 \pm 1°C$) and $50 \pm 5\%$ relative humidity for 24 h.

Unless otherwise specified samples shall be taken no nearer the selvage edge than $\frac{1}{10}$ the width of the material or no nearer than 12 in (305 mm) from either end of the roll.

Procedure—
1. Mount the refacing disc holder on the Taber Abraser and fasten to the disc holder a piece of S-11 abrasive paper.
2. Adjust test instrument for a 1000 g wheel load. Loosen the knurled cap nuts and install the new set of wheels on their respective flanged holders as indicated by the printing on the side of the wheel. The one marked right side fits on the right hand mounting with printed side out; the same with the left. The nut is then replaced and moderately tightened. Check the wheels for alignment. H-18 wheels shall be used when testing snagging resistance and CS-10 wheels shall be used when testing abrasion resistance.
3. Reface abrasive wheels 30 cycles by running them against the S-11 abrasive paper disc mounted on the refacing disc holder. Remove any rough edges on the wheels by manually sanding lightly with the abrasive paper.
4. The wheels must be refaced before each test run to remove abraded materials from the wheels that collected in the prior test.
5. If the wheels are worn out of round, crowned or excessively clogged with abraded material, they should be dressed against a diamond point until the condition is corrected. In cases where arbitration is necessary, new wheels shall be used.
6. Dust the refaced abrasive wheels with a small camel's hair brush and remove the refacing disc holder.
7. With specimen turntable removed from the abraser, place test specimen on the turntable. Adjust the clamping ring to a tight fit over the specimen and holder and press the hold-down ring over the circumference of the holder to pull the test material taut.
8. Remove any wrinkles in the test specimen by adjusting the fabric edges which extend below the clamping ring. Then, tighten the adjusting screw of the ring. Place the washer over the turntable screw and tighten the nut. Trim off the excess test specimen which extends beyond the lower edge of the clamping ring.
9. Lower the abrasive wheels carefully from their upright position to the surface of the test specimen. Set the counter mechanism at zero.
10. Position the vacuum nozzle along the diameter of the turntable $\frac{1}{32}$ to $\frac{1}{16}$ in (0.8–1.6 mm) above the surface of the test specimen and set the vacuum dial in the range of 60–70.
11. Turn on the vacuum and start the Taber Abraser.
12. Run the specimen the number of cycles specified and remove for evaluation.

TEST METHOD FOR DETERMINING VISUAL COLOR MATCH TO MASTER SPECIMEN FOR FABRICS, VINYLS, COATED FIBERBOARDS, AND OTHER AUTOMOTIVE TRIM MATERIALS—SAE J361a

SAE Standard

Report of Nonmetallic Materials Committee approved September 1968 and last revised August 1972. Editorial change May 1978.

1. Scope—This SAE Standard presents a method of matching the color of a test specimen to that of an approved appearance master specimen.

2. Test Specimen—The test specimen shall be approximately letter size ($8\frac{1}{2}$ x 11 in. (215 x 280 mm)) or as close as possible[1] and shall be cut from an area of the sample having the same pattern and finish as the appearance master specimen.

3. Appearance Master Specimen—An appearance master specimen or specimens shall be established for use as a control for each color, pattern, and finish of material. The master specimen shall be approximately letter size ($8\frac{1}{2}$ x 11 in. (215 x 280 mm)) or as close as possible[1], and shall be stored in a dark, chemically inert (except for normal oxygen content of air) container, such as a filing cabinet, at a maximum temperature of 75 F (24 C). Specimens containing rubber, asphalt, or other materials which are potential stainers, shall be isolated to themselves. It is recommended that whenever possible, a numerical color rating be established for the appearance master specimen by use of suitable color measuring equipment such as a colorimeter. The appearance master specimen shall be checked against the numerical color rating at least every six months. The appearance master specimen shall be replaced if necessary with a new specimen having the same numerical color rating as the original appearance master specimen.

4. Color Matching Booth—The enclosed area for color matching shall be free from extraneous light. The background on which the specimens are examined and all portions of the color matching booth shall be finished in a neutral gray tone (Munsell N7/). The color matching booth shall be provided with an overhead simulated north sky daylight light source capable of providing light at a color temperature of 7500 $\pm$300 K and an illumination of a minimum of 100 ft-c (1081 x) measured at the center of the viewing table. The 1000 W incandescent light bulbs used shall be replaced at or before 750 h ($\frac{3}{4}$ of their rated life). The booth shall also have an overhead simulated horizon sunlight light source using 300 W flood lamps capable of providing light at a color temperature of 2300 $\pm$100 K. Each bulb must be replaced immediately after burning out. The critical dimensions and the relative positions of the examining table and the light sources are shown in Fig. 1.[2]

5. Procedure

5.1 Secure the appearance master specimen and the test specimen side by side to the center of the examining table which is set initially at an angle of +30 deg (0.525 rad) (above the horizontal). The grain or machine direction of each specimen shall be running fore and aft, relative to the examining table.

5.2 Compare the specimens under the simulated north sky daylight starting at the +30 deg (0.525 rad) (above the horizontal) and continue comparing the specimens while rotating them away from the observer through the horizontal until an angle of −50 deg (−0.875 rad) (below the horizontal) is reached.

5.3 Return the specimens to the original angle of +30 deg (0.525 rad) (above the horizontal) and turn both specimens at right angles to their original position on the examining table (the grain or machine direction will now be running side to side, relative to the examining table).

5.4 Repeat step 5.2.

5.5 Repeat steps 5.1, 5.2, 5.3, and 5.4 under the simulated horizon sunlight.

5.6 All comparisons of highly textured three-dimensional materials should be made at a series of angles of view.

5.7 The specimens should match under both light sources and all viewing angles to be considered a satisfactory color match.

---

[1] A specimen at least 8 in. (203 mm) long shall be used for welts, straps, and similar parts.

[2] A light source which meets the requirements of this test can be obtained from: The Macbeth Daylighting Corp., P. O. Box 950, Newburgh, New York 12553.

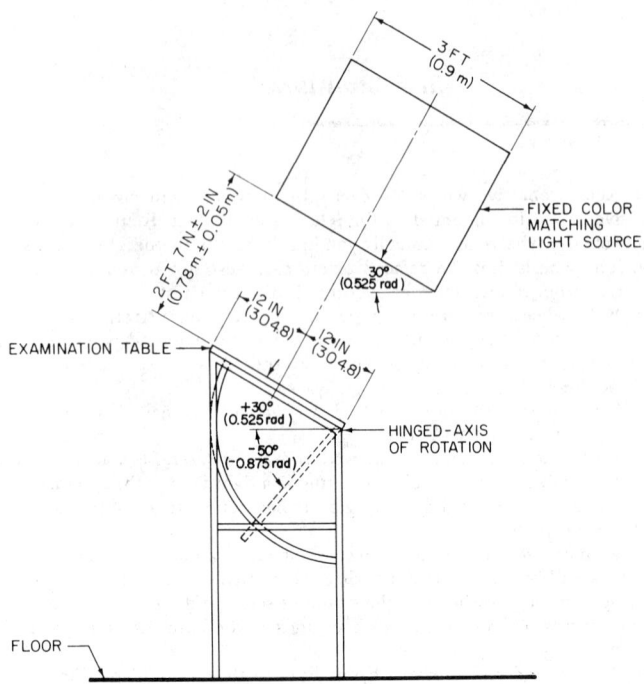

FIG. 1—COLOR MATCHING EQUIPMENT

5.8 Specimens which match under one type of light, but not under the other light source, are called metamers and are not perfect matches. (This type of color match is not usually considered satisfactory unless a deviation is allowable in the material specification.)

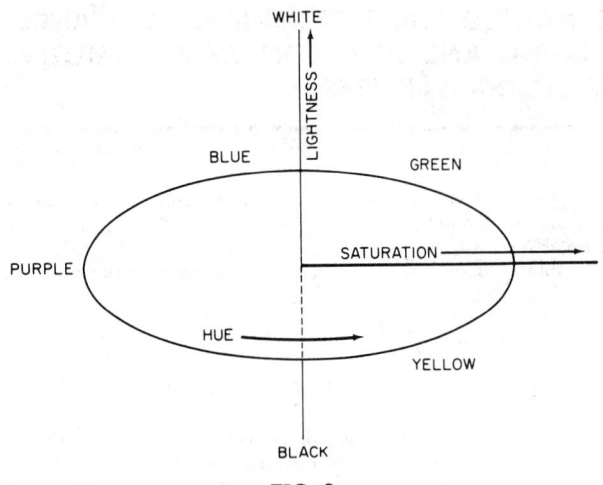

FIG. 2

5.9 Report the color departures of the test specimen from the appearance master specimen in terms of lightness, saturation, and hue. Also define the condition under which each departure occurs, such as: north sky daylight, −50 deg, grain or machine direction running side to side. Also report the name or names of the observers.

6. Definitions—See Fig. 2.

6.1 Hue—The attribute of color perception by means of which an object is judged to be red, yellow, green, blue, purple, or intermediate between some adjacent pair of these.

6.2 Lightness—The attribute of color perception by means of which an object is judged to reflect diffusely more or less light than another object.

6.3 Saturation—The attribute of color perception that expresses the degree of departure from gray of the same lightness.

TEST METHOD OF STRETCH AND SET OF TEXTILES AND PLASTICS—SAE J855a

SAE Standard

Report of Nonmetallic Materials Committee approved June 1963 and last revised June 1964. Editorial change May 1978.

Scope—This method of test applies to the measurement of elastic and recovery properties of materials after being subjected to a low static load.

Apparatus Required

Fixtures—Either of the fixtures shown in Figs. 1 and 3 is satisfactory for this test. The fixture shown with the vise grip clamps is preferred because of its speed and ease of operation.

Weight—A weight is attached to the clamp on the suspended end of the test specimen. The clamp and the weight together shall total 27 lb (12.25 kg) unless otherwise specified.

Test Specimens—At 70 ± 2°F (21 ± 1°C) and 65 ± 2% relative humidity cut three test specimens 3 x 9 in (76 x 229 mm) (or as specified) each in both the warp and fill directions. The warp stretch and set is measured on the samples with the long dimension parallel to the warp yarns and the fill stretch and set is measured on the samples with the long dimension parallel to the filling yarns. In nondirectional materials such as unsupported or nonwoven backed vinyl, one set of samples should have the long dimension cut in the machine direction of the roll, while the other set should be cut in the opposite or cross-machine direction. For woven fabrics such as bodycloth, sidewall, or headlining, the samples should be cut slightly oversized and the yarns unraveled on each side until the 3 in (76 mm) dimension is attained.

Unless otherwise specified, samples shall be taken no nearer the selvage edge than one tenth the width of the material or nearer than 12 in (305 mm) from either end of the roll.

Procedure—Mark two sharp clear lines 3 in (76 mm) apart (or as specified) across the center portion of the sample. Fasten the clamps firmly at each end of the test specimen. One clamp is attached to the supporting fixture and the weight is attached to the clamp on the other end. The weight is lowered carefully to prevent undue stress on the sample, and the assembly is suspended vertically for 10 min. Measure the length of the section between the bench marks to the closest $1/64$ in (0.40 mm) and record the results at L_2. Remove the clamps and weight and allow the sample to recover in a horizontal position for 10 min. Again measure the length of the section between bench marks and record as L_3.

Calculations and Report—Report the data as percent stretch and percent set calculated as follows:

$$\% \text{ stretch} = \frac{L_2 - L_1}{L_2(L_1)} \times 100$$

$$\% \text{ set} = \frac{L_3 - L_1}{L_1} \times 100$$

where L_1 = Original length between bench marks
L_2 = Measured length after the weight is applied for 10 min
L_3 = Measured length after the 10 min recovery period

Report the average of the results for the three warp (or machine) direction samples as percentage warp stretch and set. Report the average of the results for the three filling (or cross-machine) direction samples as percentage fill stretch and set.

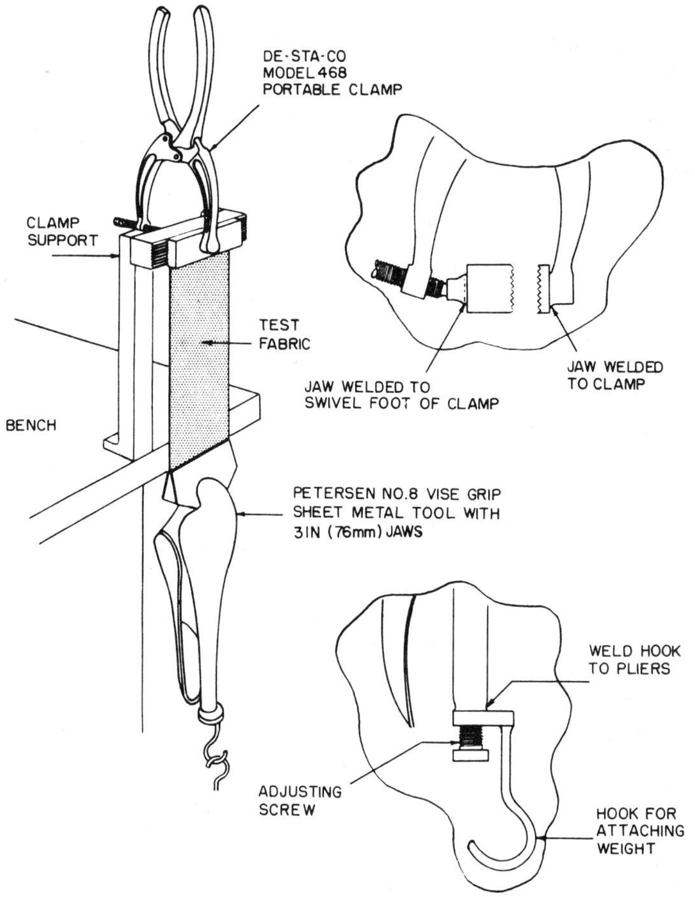

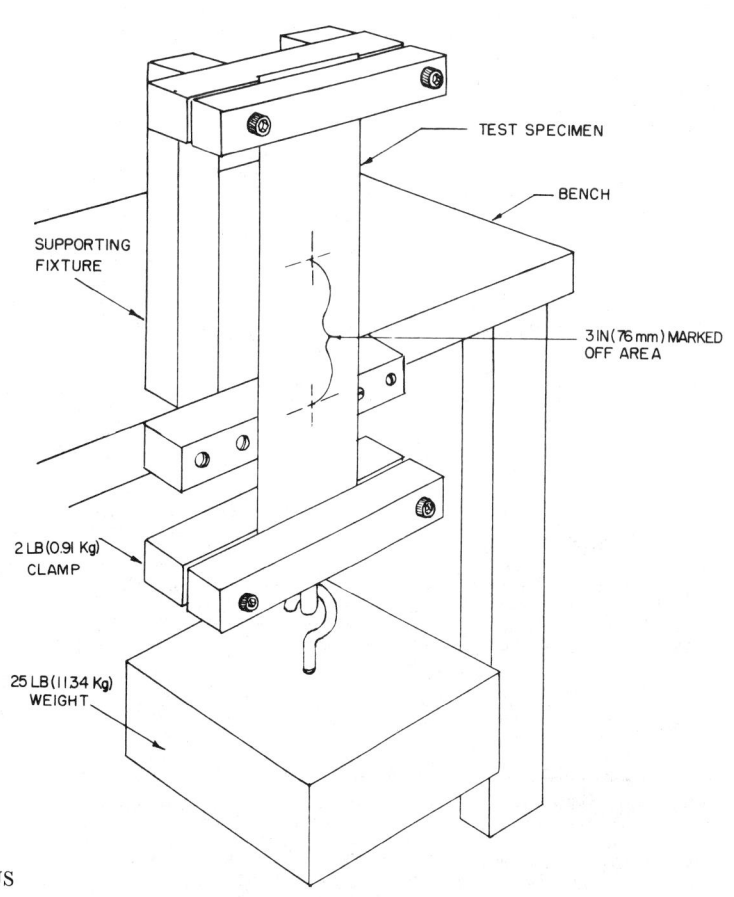

FIG. 1—CLAMP AND FIXTURE ASSEMBLY FOR STRETCH AND SET OF TEXTILES

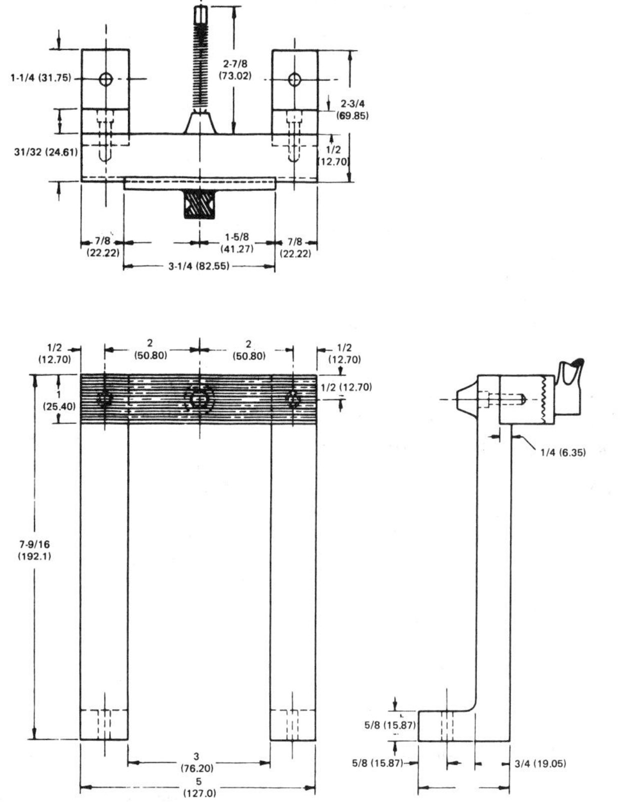

FIG. 2—DETAIL OF CLAMP SUPPORT

FIG. 3—STRETCH AND SET APPARATUS

FELTS—WOOL AND PART WOOL—SAE J314b

SAE Standard

Report of Parts and Fittings Division approved July 1923 and last revised by Nonmetallic Materials Committee January 1976.

1. *Scope*—This SAE Standard covers types and qualities of felts required for general automotive uses. It was developed with the cooperation of the Standardization Committee of the Felt Association, Inc., and in accordance with the ASTM tests indicated in the standard.

The commercial trade designations of the more commonly used grades of automotive felts are given along with complete specifications and tolerances for thickness, mass, wool content, chemical and physical requirements, color, and width.

General information, recommended uses, etc., are published in an Appendix as a guide in the selection of felts for particular uses, but the requirements for each application should be taken into consideration in making final selections.

2. *Chemical and Mechanical Properties*

   2.1 The chemical and mechanical requirements for the several grades of automotive felts given in Table 1 include actual wool content (chemical basis), methyl chloroform soluble (percentage of residual oil and grease), water soluble (sizing and nonfibrous impurities), ash (the amount of residual inorganic matter), tensile strength, and splitting resistance.

   2.2 All tests shall be made in accordance with ASTM D461, Standard Methods of Testing Wool Felt. If it is desired to detect the presence of, and identify, fibers other than wool, such as other animal fibers, vegetable, and synthetic fibers, the felt shall be tested as described in ASTM D276, Identification of Fibers in Textiles.

3. *Thickness, Width and Mass*

   3.1 Thickness and mass requirements are given in Table 2.

   3.2 Felt shall be furnished in standard width as shown in Table 1, unless otherwise specified.

   3.3 The thickness tolerances given in Table 2 vary, depending on the density, thickness, and grade or quality of the felt, and are expressed as the permissible minimum and maximum thickness for each grade and thickness rather than as a percentage variation from the nominal thickness.

   3.4 Any density or mass determinations shall be based on the thickness of the felt as ordered and no correction shall be made for variations in the thickness of the felt as received. For example, SAE F-1, back-check felt, in 12.7 mm ($\frac{1}{2}$ in) thickness may, according to Table 2, vary in mass from 4.12 to 4.55 kg/m^2 (7.60 to 8.40 lb/yd^2), while the thickness may, according to Table 1, vary from 12.22 to 13.18 mm (0.481 to 0.519 in). The combination of mass and thickness tolerances control the degree of felting or matting of the fibers, in other words, the hardness or the density, and conversely, the resiliency of the finished felt. Therefore, to maintain the normal density for each grade or type of felt, no correction in mass is permitted to compensate for a variation from the nominal thickness as specified.

   NOTE TO USERS: The mass or density requirements for the several grades of automotive felts given in Table 2 are expressed as the mass in kilogram per square metre (pounds per square yard) for each commercial thickness. This is the established standard unit of mass employed in the felt industry. Density may also be expressed as the mass in grams per cubic centimeter (ounces per cubic inch), specific gravity as compared with water, percentage specific gravity (specific gravity × 100), or as surface density in kilograms per square metre (pounds per square yard) of nominal 25.4 mm (1 in) thickness. The mass or density of cut parts may be expressed as the mass per one hundred (100) parts based on the nominal mass of the felt in the thickness specified.

4. *Other Requirements*

   4.1 Color requirements are given in Table 1.

   4.2 Special sizing, adhesives, and impregnating materials used to impart specific properties may alter the chemical and physical requirements specified in Table 1. The specific properties and methods of test for special products shall be agreed upon by supplier and purchaser.

   4.3 When specified by purchaser for ball and roller bearing oil retaining washers felt shall be sheared on both sides to give a smooth surface free from "surface fuzz" or "flock."

   4.4 The quality, appearance, and oil absorption characteristics may be specified by the purchaser to be in accord with approved samples.

APPENDIX
(For guidance only; not a part of the specification)

A1. *General*

   A1.1 Felt is a fabric built up by the interlocking of fibers by a suitable combination of mechanical work, chemical action, moisture, and heat, without stitching, weaving, or knitting. Felt may consist of one or more classes of fibers, wool, reprocessed wool, and reused wool, with or without admixture with animal, vegetable, and synthetic fibers.

   A1.2 Felt as defined here is commonly referred to as wool felt and does not include needle loomed, woven, synthetically bonded, stitched, quilted, paper, or other materials of felt-like appearance which are products of entirely different constructions and properties.

   A1.3 Clip wools or noils, which are the short fiber combings resulting from the preparation of wool for spinning, as well as reprocessed and reused wools are used in the manufacture of automotive felts. The best grades of wool are white and are used without admixture with other fibers in the highest grade felts.

   A1.4 Varying amounts of cotton, rayon, and other fibers may be used as a filler to reduce the cost of the felt or to impart certain desired characteristics to the finished material. Traces of cotton are found in most commercial "all-wool" felts.

   A1.5 Raw wool contains "wool fat" and "wool perspiration" in addition to mechanically adhering impurities and foreign matter. The foreign matter

TABLE 1—STANDARD MECHANICAL ROLL FELT SPECIFICATIONS

SAE No.	Min Actual Wool Content,[a] %	Max Methyl Chloroform Soluble, %	Max Water Soluble, %	Max Combined Methyl Chloroform and Water, %	Max Ash, %	Min Tensile Strength kPa	Min Tensile Strength psi	Min Splitting Resistance[b] N 5 cm width	Min Splitting Resistance[b] lb 2 in width	Trade Designation	Color	Standard Width cm	Standard Width in
F-1	95	2.5	2.5	3.0	1.5	3450	500	142	32	Back check	White	152	60
F-2	90	2.5	2.5	4.0	2.0	3450	500	125	28	Back check	Any color except gray or black	152	60
F-3	85	2.5	3.0	4.5	2.5	2760	400	98	22	Back check	Grey	152	60
F-5	95	2.5	2.5	3.0	2.0	2760	400	80	18	Extra firm pad	White	152	60
F-6	87	2.5	2.5	4.5	2.5	1900	275	71	16	Extra firm pad	Grey	152 or 183	60 or 72
F-7	80	3.0	4.0	7.0	3.0	1730	250	53	12	Extra firm pad	Grey	183	72
F-10	95	2.5	2.5	3.0	2.5	1550	225	36	8	Firm pad	White	183	72
F-11	87	3.0	2.5	4.5	3.0	1380	200	27	6	Firm pad	Grey	183	72
F-12	85	4.0	2.5	6.5	3.5	690	100	13	3	Firm pad	Grey	183	72
F-13	75	4.0	4.0	8.0	3.5	518	75	9	2	Firm pad	Grey	183	72
F-15	55	4.0	5.0	9.0	4.0	518	75	9	2	Firm pad	Grey	183	72
F-26	45	8.0	6.0	14.0	5.0	—	—	—	—	Soft pad	Grey	183	72
F-50	95	2.5	2.5	3.0	1.5	3450	500	—	—	Ball bearing felt	White	152 or 183	60 or 72
F-51	92	2.5	2.5	4.5	2.5	2070	300	—	—	Ball bearing felt	Grey	152 or 183	60 or 72
F-55	75	4.0	4.0	8.0	3.0	1380	200	—	—	Lining	Grey or Black	152 or 183	60 or 72

[a] The actual wool content indicates the percent of wool by chemical analysis and is exclusive of traces of other fibers and impurities present in the wool used in fabricating the several grades of felt. For example, SAE F-1, fabricated from 100% wool, may contain incidental traces of cotton and other fibers, residual wool fats, and oils or soaps used in processing which may reduce the actual wool fiber content on analysis to a minimum of 95%.

[b] Splitting resistance is not applicable to felts where the thickness is less than 4.75 mm 3/16 in). For materials less than 4.75 mm (3/16 in) in thickness, breaking strength only is recommended as an indicative test.

TABLE 2—STANDARD MECHANICAL ROLL FELT SPECIFICATIONS

SAE No.	Thickness, mm		Thickness, in		Mass, kg/m²		Mass (Weight) lb/yd²	
	Nominal	Limits	Nominal	Limits	Nominal	Limits	Nominal	Limits
F-1[a]	3.2	2.87–3.48	1/8	0.113–0.137	1.08	1.03–1.14	2.0	1.90–2.10
	4.8	4.45–5.11	3/16	0.175–0.201	1.63	1.54–1.71	3.0	2.85–3.15
	6.4	5.99–6.71	1/4	0.236–0.264	2.17	2.06–2.28	4.0	3.80–4.20
	8.0	7.57–8.33	5/16	0.298–0.328	2.71	2.57–2.85	5.0	4.75–5.25
	9.5	9.12–9.93	3/8	0.359–0.391	3.25	3.09–3.41	6.0	5.70–6.30
	12.7	12.22–13.18	1/2	0.481–0.519	4.34	4.12–4.55	8.0	7.60–8.40
	15.9	15.32–16.43	5/8	0.603–0.647	5.42	5.15–5.69	10.0	9.50–10.50
	19.1	18.42–19.69	3/4	0.725–0.775	6.50	6.18–6.83	12.0	11.40–12.60
	22.2	21.51–22.94	7/8	0.847–0.903	7.59	7.21–7.97	14.0	13.30–14.70
	25.4	24.61–26.19	1	0.969–1.031	8.67	8.24–9.11	16.0	15.20–16.80
F-2	3.2	2.87–3.48	1/8	0.113–0.137	1.08	1.03–1.14	2.0	1.90–2.10
	4.8	4.45–5.11	3/16	0.175–0.201	1.63	1.54–1.71	3.0	2.85–3.15
	6.4	5.99–6.71	1/4	0.236–0.264	2.17	2.06–2.28	4.0	3.80–4.20
	8.0	7.57–8.33	5/16	0.298–0.328	2.71	2.57–2.85	5.0	4.75–5.25
	9.5	9.12–9.93	3/8	0.359–0.391	3.25	3.09–3.41	6.0	5.70–6.30
	12.7	12.22–13.18	1/2	0.481–0.519	4.34	4.12–4.55	8.0	7.60–8.40
	15.9	15.32–16.43	5/8	0.603–0.647	5.42	5.15–5.69	10.0	9.50–10.50
	19.1	18.42–19.69	3/4	0.725–0.775	6.50	6.18–6.83	12.0	11.40–12.60
	22.2	21.51–22.94	7/8	0.847–0.903	7.59	7.21–7.97	14.0	13.30–14.70
	25.4	24.61–26.19	1	0.969–1.031	8.67	8.24–9.11	16.0	15.20–16.80
F-3[a]	3.2	2.87–3.48	1/8	0.113–0.137	1.07	0.98–1.14	1.97	1.80–2.10
	4.8	4.45–5.11	3/16	0.175–0.201	1.59	1.47–1.71	2.93	2.71–3.15
	6.4	5.99–6.71	1/4	0.236–0.264	2.11	1.96–2.27	3.90	3.61–4.19
	8.0	7.57–8.33	5/16	0.298–0.328	2.64	2.44–2.84	4.87	4.50–5.24
	9.5	9.12–9.93	3/8	0.359–0.391	3.17	2.93–3.41	5.85	5.41–6.29
	12.7	12.22–13.18	1/2	0.481–0.519	4.23	3.91–4.55	7.80	7.21–8.39
	15.9	15.32–16.43	5/8	0.603–0.647	5.28	4.88–5.69	9.75	9.01–10.49
	19.1	18.42–19.69	3/4	0.725–0.775	6.34	5.86–6.82	11.70	10.81–12.59
	22.2	21.51–22.94	7/8	0.847–0.903	7.40	6.83–7.96	13.65	12.61–14.69
	25.4	24.61–26.19	1	0.969–1.031	8.46	7.81–9.10	15.60	14.41–16.79
F-5	3.2	2.82–3.53	1/8	0.111–0.139	0.83	0.79–0.87	1.53	1.45–1.61
	4.8	4.37–5.18	3/16	0.172–0.204	1.24	1.18–1.31	2.29	2.17–2.41
	6.4	5.89–6.81	1/4	0.232–0.268	1.66	1.57–1.75	3.06	2.90–3.22
	8.0	7.44–8.46	5/16	0.293–0.333	2.07	1.96–2.18	3.82	3.62–4.02
	9.5	8.97–10.08	3/8	0.353–0.397	2.49	2.36–2.62	4.59	4.35–4.83
	12.7	12.04–13.36	1/2	0.474–0.526	3.32	3.14–3.49	6.12	5.80–6.44
	15.9	15.11–16.64	5/8	0.595–0.655	4.15	3.93–4.36	7.65	7.25–8.05
	19.1	18.19–19.91	3/4	0.716–0.784	4.98	4.72–5.24	9.18	8.70–9.66
	22.2	21.26–23.19	7/8	0.837–0.913	5.80	5.50–6.11	10.71	10.15–11.27
	25.4	24.33–26.47	1	0.958–1.042	6.63	6.29–6.98	12.24	11.60–12.88
F-6	3.2	2.82–3.53	1/8	0.111–0.139	0.83	0.79–0.87	1.53	1.45–1.61
	4.8	4.37–5.18	3/16	0.172–0.204	1.24	1.18–1.31	2.29	2.17–2.41
	6.4	5.89–6.81	1/4	0.232–0.268	1.66	1.57–1.75	3.06	2.90–3.22
	8.0	7.44–8.46	5/16	0.293–0.333	2.07	1.96–2.18	3.82	3.62–4.02
	9.5	8.97–10.08	3/8	0.353–0.397	2.49	2.36–2.62	4.59	4.35–4.83
	12.7	12.04–13.36	1/2	0.474–0.526	3.32	3.14–3.49	6.12	5.80–6.44
	15.9	15.11–16.64	5/8	0.595–0.655	4.15	3.93–4.36	7.65	7.25–8.05
	19.1	18.19–19.91	3/4	0.716–0.784	4.98	4.72–5.24	9.18	8.70–9.66
	22.2	21.26–23.19	7/8	0.837–0.913	5.80	5.50–6.11	10.71	10.15–11.27
	25.4	24.33–26.47	1	0.958–1.042	6.63	6.29–6.98	12.24	11.60–12.88
F-7[a]	3.2	2.82–3.53	1/8	0.111–0.139	0.83	0.79–0.87	1.53	1.45–1.61
	4.8	4.37–5.18	3/16	0.172–0.204	1.24	1.18–1.31	2.29	2.17–2.41
	6.4	5.89–6.81	1/4	0.232–0.268	1.66	1.57–1.75	3.06	2.90–3.22
	8.0	7.44–8.46	5/16	0.293–0.333	2.07	1.96–2.18	3.82	3.62–4.02
	9.5	8.97–10.08	3/8	0.353–0.397	2.49	2.36–2.62	4.59	4.35–4.83
	12.7	12.04–13.36	1/2	0.474–0.526	3.32	3.14–3.49	6.12	5.80–6.44
	15.9	15.11–16.64	5/8	0.595–0.655	4.15	3.93–4.36	7.65	7.25–8.05
	19.1	18.19–19.91	3/4	0.716–0.784	4.98	4.72–5.24	9.18	8.70–9.66
	22.2	21.26–23.19	7/8	0.837–0.913	5.80	5.50–6.11	10.71	10.15–11.27
	25.4	24.33–26.47	1	0.958–1.042	6.63	6.29–6.98	12.24	11.60–12.88
F-10	3.2	2.67–3.68	1/8	0.105–0.145	0.57	0.53–0.62	1.06	0.98–1.14
	4.8	4.19–5.36	3/16	0.165–0.211	0.86	0.80–0.93	1.59	1.47–1.71
	6.4	5.69–7.01	1/4	0.224–0.276	1.15	1.06–1.24	2.12	1.96–2.28
	8.0	7.21–8.69	5/16	0.284–0.342	1.44	1.33–1.54	2.65	2.45–2.85
	9.5	8.71–10.34	3/8	0.343–0.407	1.72	1.59–1.85	3.18	2.94–3.42
	12.7	11.73–13.67	1/2	0.462–0.538	2.30	2.12–2.47	4.24	3.92–4.56
	15.9	14.76–16.99	5/8	0.581–0.669	2.87	2.66–3.09	5.30	4.90–5.70
	19.1	17.78–20.32	3/4	0.700–0.800	3.45	3.19–3.71	6.36	5.88–6.84
	22.2	20.80–23.65	7/8	0.819–0.931	4.02	3.72–4.33	7.42	6.86–7.98
	25.4	23.83–26.97	1	0.938–1.062	4.60	4.25–4.94	8.48	7.84–9.12
F-11	3.2	2.67–3.68	1/8	0.105–0.145	0.57	0.53–0.62	1.06	0.98–1.14
	4.8	4.19–5.36	3/16	0.165–0.211	0.86	0.80–0.93	1.59	1.47–1.71
	6.4	5.69–7.01	1/4	0.224–0.276	1.15	1.06–1.24	2.12	1.96–2.28
	8.0	7.21–8.69	5/16	0.284–0.342	1.44	1.33–1.54	2.65	2.45–2.85
	9.5	8.71–10.34	3/8	0.343–0.407	1.72	1.59–1.85	3.18	2.94–3.42
	12.7	11.73–13.67	1/2	0.462–0.538	2.30	2.12–2.47	4.24	3.92–4.56
	15.9	14.76–16.99	5/8	0.581–0.669	2.87	2.66–3.09	5.30	4.90–5.70
	19.1	17.78–20.32	3/4	0.700–0.800	3.45	3.19–3.71	6.36	5.88–6.84
	22.2	20.80–23.65	7/8	0.819–0.931	4.02	3.72–4.33	7.42	6.86–7.98
	25.4	23.83–26.97	1	0.938–1.062	4.60	4.25–4.94	8.48	7.84–9.12
F-12	3.2	2.67–3.68	1/8	0.105–0.145	0.57	0.53–0.62	1.06	0.98–1.14
	4.8	4.19–5.36	3/16	0.165–0.211	0.86	0.80–0.93	1.59	1.47–1.71
	6.4	5.69–7.01	1/4	0.224–0.276	1.15	1.06–1.24	2.12	1.96–2.28
	8.0	7.21–8.69	5/16	0.284–0.342	1.44	1.33–1.54	2.65	2.45–2.85
	9.5	8.71–10.34	3/8	0.343–0.407	1.72	1.59–1.85	3.18	2.94–3.42
	12.7	11.73–13.67	1/2	0.462–0.538	2.30	2.12–2.47	4.24	3.92–4.56
	15.9	14.76–16.99	5/8	0.581–0.669	2.87	2.66–3.09	5.30	4.90–5.70
	19.1	17.78–20.32	3/4	0.700–0.800	3.45	3.19–3.71	6.36	5.88–6.84
	22.2	20.80–23.65	7/8	0.819–0.931	4.02	3.72–4.33	7.42	6.86–7.98
	25.4	23.83–26.97	1	0.938–1.062	4.60	4.25–4.94	8.48	7.84–9.12

[a] For thicknesses less than 3.2 mm (1/8 in) for SAE F-1, see SAE F-50; F-3, see SAE F-51; and F-7, see SAE F-55.

TABLE 2—STANDARD MECHANICAL ROLL FELT SPECIFICATIONS (continued)

SAE No.	Thickness, mm Nominal	Thickness, mm Limits	Thickness, in Nominal	Thickness, in Limits	Mass, kg/m² Nominal	Mass, kg/m² Limits	Mass (Weight) lb/yd² Nominal	Mass (Weight) lb/yd² Limits
F-13	3.2	2.67-3.68	1/8	0.105-0.145	0.57	0.53-0.62	1.06	0.98-1.14
	4.8	4.19-5.36	3/16	0.165-0.211	0.86	0.80-0.93	1.59	1.47-1.71
	6.4	5.69-7.01	1/4	0.224-0.276	1.15	1.06-1.24	2.12	1.96-2.28
	8.0	7.21-8.69	5/16	0.284-0.342	1.44	1.33-1.54	2.65	2.45-2.85
	9.5	8.71-10.34	3/8	0.343-0.407	1.72	1.59-1.85	3.18	2.94-3.42
	12.7	11.73-13.67	1/2	0.462-0.538	2.30	2.12-2.47	4.24	3.92-4.56
	15.9	14.76-16.99	5/8	0.581-0.669	2.87	2.66-3.09	5.30	4.90-5.70
	19.1	17.78-20.32	3/4	0.700-0.800	3.45	3.19-3.71	6.36	5.88-6.84
	22.2	20.80-23.65	7/8	0.819-0.931	4.02	3.72-4.33	7.42	6.86-7.98
	25.4	23.83-26.97	1	0.938-1.062	4.06	4.25-4.94	8.48	7.84-9.12
F-15	3.2	2.67-3.68	1/8	0.105-0.145	0.57	0.53-0.62	1.06	0.98-1.14
	4.8	4.19-5.36	3/16	0.165-0.211	0.86	0.80-0.93	1.59	1.47-1.71
	6.4	5.69-7.01	1/4	0.224-0.276	1.15	1.06-1.24	2.12	1.96-2.28
	8.0	7.21-8.69	5/16	0.284-0.342	1.44	1.33-1.54	2.65	2.45-2.85
	9.5	8.71-10.34	3/8	0.343-0.407	1.72	1.59-1.85	3.18	2.94-3.42
	12.7	11.73-13.67	1/2	0.462-0.538	2.30	2.12-2.47	4.24	3.92-4.56
	15.9	14.76-16.99	5/8	0.581-0.669	2.87	2.66-3.09	5.30	4.90-5.70
	19.1	17.78-20.32	3/4	0.700-0.800	3.45	3.19-3.71	6.36	5.88-6.84
	22.2	20.80-23.65	7/8	0.819-0.931	4.02	3.72-4.33	7.42	6.86-7.98
	25.4	23.83-26.97	1	0.938-1.062	4.60	4.25-4.94	8.48	7.84-9.12
F-26	3.2	2.16-4.19	1/8	0.085-0.165	0.49	0.44-0.54	0.90	0.81-0.99
	6.4	4.93-7.77	1/4	0.194-0.306	0.98	0.88-1.07	1.80	1.62-1.98
	9.5	7.70-11.35	3/8	0.303-0.447	1.46	1.32-1.61	2.70	2.43-2.97
	12.7	10.46-14.94	1/2	0.412-0.588	1.95	1.76-2.15	3.60	3.24-3.96
	19.1	16.00-22.10	3/4	0.630-0.870	2.93	2.63-3.22	5.40	4.86-5.94
	25.4	21.54-29.26	1	0.848-1.152	3.90	3.71-4.29	7.20	6.84-7.92
F-50	1.2	1.02-1.37	3/64	0.040-0.054	0.41	0.39-0.43	0.750	0.712-0.788
	1.6	1.42-1.78	1/16	0.056-0.070	0.53	0.51-0.55	0.975	0.937-1.013
	2.0	1.80-2.16	5/64	0.071-0.085	0.65	0.63-0.67	1.200	1.162-1.238
	2.4	2.21-2.57	3/32	0.087-0.101	0.77	0.75-0.79	1.425	1.387-1.463
F-51	1.2	1.02-1.37	3/64	0.040-0.054	0.41	0.39-0.43	0.750	0.712-0.788
	1.6	1.42-1.78	1/16	0.056-0.070	0.53	0.51-0.55	0.975	0.937-1.013
	2.0	1.80-2.16	5/64	0.071-0.085	0.65	0.63-0.67	1.200	1.162-1.238
	2.4	2.21-2.57	3/32	0.087-0.101	0.77	0.75-0.79	1.425	1.387-1.463
F-55	1.6	1.42-1.78	1/16	0.056-0.070	0.41	0.39-0.43	0.750	0.712-0.788
	2.4	2.21-2.57	3/32	0.087-0.101	0.61	0.59-0.63	1.125	1.087-1.163

and some of the wool fat is removed in the scouring operation. Oils and soaps are added in the fabricating process to obtain the necessary degree of felting. Sizing or filler may be used in some of the lower grade felts and in special applications to stiffen or strengthen the finished material. Adhesives and impregnating materials may be used in special purpose felts to impart specific properties.

A1.6 Methyl chloroform soluble, water soluble, and ash determinations indicate the cleanliness of the fiber and the amount of fats, oils, and sizing materials present in the finished product.

A2. *Terminology*—The terms wool, reprocessed wool, and reused wool are defined essentially in accordance with the Wool Products Labelling Act, 1939, as follows:

A2.1 *Wool*—The term "wool" means the fiber from the fleece of the sheep or lamb, or hair of the angora or cashmere goat (and may include the so-called specialty fibers from the hair of the camel, alpaca, llama, and vicuna) which has never been reclaimed from any woven or felted wool product.

A2.2 *Reprocessed Wool*—The term "reprocessed wool" means the resulting fiber when wool has been woven or felted into a wool product which, without ever having been utilized in any way by the ultimate consumer, subsequently has been made into a fibrous state.

A2.3 *Reused Wool*—The term "reused wool" means the resulting fiber when wool or reprocessed wool has been spun, woven, knitted, or felted into a wool product which, after having been used in any way by the ultimate consumer, subsequently has been made into a fibrous state.

A3. *Recommended Uses*

A3.1 SAE F-1 is suitable for oil retention in installations where the felt is not compressed, for feeding low viscosity or light oil, and where unusual strength and hardness are required. Washers, bushings, wicks, door bumpers, polishing blocks, and parts where wear and resistance to abrasion are required, are typical uses.

A3.2 SAE F-2 and F-3 are recommended for vibration mountings and the same general purposes as SAE F-1 and where a felt of slightly lower quality is satisfactory. SAE F-5, F-6, and F-7 are recommended for dust shields, wipers, grease retainer washers, wicks, vibration mountings, and in uses where a resilient felt is required.

A3.3 SAE F-10, F-11, and F-12 are recommended for grease and oil retention where the felt is confined and compressed in assembly. Also recommended for dust shields under less severe operating conditions where F-5, F-6, and F-7 are not required.

A3.4 SAE F-13 and F-15 are recommended for sound deadening, chassis strips, spacers, dust shields, pedal pads, dash liners, and for mechanical purposes where abrasion and wear are not important factors.

A3.5 SAE F-26 is suitable for packing or padding when held in place between other materials as in shipping and packaging. This grade should not be used for mechanical purposes.

A3.6 SAE F-1 to F-26 inclusive can be obtained in thicknesses up to 1 in. The weight and thickness tolerance shall be agreed upon by supplier and purchaser.

A3.7 SAE F-50 is recommended for ball and roller bearing oil retainer washers and small dust excluding washers; also for mechanical purposes where an accurate, thin, smooth, high grade felt is required.

A3.8 SAE F-51 is recommended for the same general uses as F-50 but in installations where tolerances and length of life are not as important; also for thin cut parts such as gaskets and liners.

A3.9 SAE F-55 is recommended for anti-squeak strips and for lining when cemented to fiber board or metal panels.

NONMETALLIC AUTOMOTIVE GASKET MATERIALS—SAE J90b

SAE Recommended Practice

Report of Nonmetallic Materials Committee approved January 1952 and last revised August 1970. Editorial change July 1974.

[This report has been developed by the Joint SAE-ASTM Committee on Gaskets. This document supersedes the information shown in SAE J90a, and conforms with the information shown in ASTM D 104-68T. The information shown as Appendix B is SAE J90a. This document is obsolete and will be removed within three years. SAE J90a is, however, being retained for this three year period as Appendix B in order to relate the writing from the old gasket classification system to the new gasket classification system.]

1. *Scope*

1.1 The classification system provides a means for specifying or describing pertinent properties of commercial nonmetallic gasket materials. Materials composed of asbestos, cork cellulose, and other organic or inorganic materials in combination with various binders or impregnants are included. Materials normally classified as rubber compounds are not included, since they are covered in SAE J200—ASTM D 2000, classification System for Elastomeric Materials for Automotive Applications. Gasket coatings are not covered, since details thereof are intended to be given on engineering drawings or in separate specifications.

1.2 Since all of the properties that contribute to gasket performance are not included, use of the writing as a basis for selecting materials is limited.

2. Purpose

2.1 This classification is intended to encourage uniformity in reporting properties, to provide a common language for communications between suppliers and consumers, to guide engineers and designers in the test methods commonly used for commercially available materials, and to be versatile enough to cover new materials and test methods as they are introduced.

2.2 It is based on the principle that nonmetallic gasket materials should be described, insofar as is possible, in terms of specific physical and mechanical characteristics, and that an infinite number of such descriptions can be formulated by use of one or more standard statements based on standard tests. Thus, users of gasket materials can, by selecting different combinations of statements, specify different combinations of properties desired in various parts. Suppliers, likewise, can report properties available in their respective products.

3. Numbering System

3.1 To permit "line call-out" of the descriptions mentioned in paragraph 2.2, this classification system establishes letter or number symbols or both for various performance levels of each property or characteristic. (See Table 1.)

3.2 In specifying or describing gasket materials, each "line call-out" shall include the number of this system (minus date symbol) followed by the letter "F" and six numerals; for example: SAE J90a (F125400). Since each numeral of the call-out represents a characteristic (as shown in Table 1), six numerals are always required. The numeral "0" is used when the description of any characteristic is not desired. The numeral "9" is used when the description of any characteristic (or test related thereto) is specified by some supplement to this classification system, such as notes on engineering drawings.

3.3 To further specify or describe gasket materials, each "line call-out" may include one or more suffix letter-numeral symbols, as listed in Table 2; for example: SAE J90a (F125400-B2M4). Various levels of definition may be established by increasing or decreasing the number of letter-numeral symbols used in the "line call-out."

3.4 For convenience, gasket materials are referred to by Type 1, Type 2, or Type 3, according to the principal fibrous or particulate reinforcement or other material from which the gasket material is made and by Class 1, 2, or 3, according to the manufacturing method or the common trade designation. Type numbers correspond with the first numeral, and class numbers correspond with the second numeral of the basic six-digit line call-out, as shown in Table 1.

NOTE: While this "cell-type" format provides the means for close characterization and specification of each property and combinations of properties for a broad range of materials, it is subject to possible misapplications, since impossible property combinations can be coded if the user is not familiar with available commercial materials. Table A1 in Appendix A of this classification indicates properties, characteristics, and test methods that are normally considered applicable to each type of material.

4. Physical and Mechanical Requirements

4.1 Gasket materials identified by this classification shall have the characteristics or properties indicated by the first six numerals of the line call-out, within the limits shown in Table 1, and by additional letter-numeral symbols shown in Table 2.

5. Thickness Requirements

Gasket materials identified by this classification system shall conform to the thickness tolerances specified in Table 3.

6. Sampling

6.1 Specimens shall be selected from finished gaskets or sheets of suitable size, whichever is the more practicable. If sheets are used, they shall, where applicable, be cut squarely with the grain of the stock, and the grain direction shall be noted by an arrow.

6.2 For qualification purposes, thickness shall be 0.03 in. (0.8 mm); except where the compressibility load of 100 or 400 psi (0.689 or 2.758 mN/m^2) is specified, the thickness shall be 0.06–0.25 in. (1.5–6.4 mm). When thicknesses other than those shown above are to be tested, the specification limits shall be agreed to in writing between the purchaser and the supplier.

6.3 Sufficient specimens shall be selected to provide a minimum of three determinations for each test specified. The average of the determinations shall be considered as the result.

7. Conditioning

7.1 Prior to all tests, specimens shall be conditioned as follows:

7.1.1 When the first numeral of line call-out is "1" (Type 1 materials), specimens shall be conditioned in an oven at 212 ±3.6 F (100 ±2 C) for 1 hr and allowed to cool to 70–85 F (21–30 C) in a desiccator containing anhydrous calcium chloride; *except* when second numeral of line call-out is "3" (Class 3 materials), the specimens shall be conditioned in an oven for 4 hr at 212 ±3.6 F (100 ±2 C).

TABLE 1—BASIC PHYSICAL AND MECHANICAL CHARACTERISTICS

Basic Six-Digit Number	Basic Characteristic
First Numeral	**"Type" of material** (the principal fibrous or particulate reinforcement material from which the gasket is made) shall conform to the first numeral of the basic six-digit number, as follows: 0 = not specified 1 = asbestos or other inorganic fibers (Type 1) 2 = cork (Type 2) 3 = cellulose or other organic fibers (Type 3) 9 = as specified[a]
Second Numeral	**Class of material** (method of manufacture or common trade designation) shall conform to the second numeral of the basic six-digit number, as follows: When **first** numeral is "1," second numeral: 0 = not specified 1 = compressed asbestos (Class 1) 2 = beater addition asbestos (Class 2) 3 = asbestos paper and millboard (Class 3) 9 = as specified[a] When **first** numeral is "2," second numeral: 0 = not specified 1 = cork composition (Class 1) 2 = cork and elastomeric (Class 2) 3 = cork and cellular rubber (Class 3) 9 = as specified[a] When **first** numeral is "3," second numeral: 0 = not specified 1 = untreated fiber—tag, chipboard, vulcanized fiber, etc. (Class 1) 2 = protein treated (Class 2) 3 = elastomeric treated (Class 3) 9 = as specified[a]
Third Numeral	**Compressibility characteristics**, determined in accordance with paragraph 8.2, shall conform to the percent indicated by the third numeral of the basic six-digit number. (Example: 4 = 15–25%) 0 = not specified 5 = 20–30% 1 = 0–10% 6 = 25–40% 2 = 5–15% 7 = 30–50% 3 = 10–20% 8 = 40–60% 4 = 15–25% 9 = as specified[a]
Fourth Numeral	**Thickness increase when immersed in ASTM No. 3 Oil**, determined in accordance with paragraph 8.3.1, shall conform to the percent indicated by the fourth numeral of the basic six-digit number. (Example: 4 = 15–30%) 0 = not specified 5 = 20–40% 1 = 0–15% 6 = 30–50% 2 = 5–20% 7 = 40–60% 3 = 10–25% 8 = 50–70% 4 = 15–30% 9 = as specified[a]
Fifth Numeral	**Weight increase when immersed in ASTM No. 3 Oil**, determined in accordance with paragraph 8.3.3, shall conform to the percent indicated by the fifth numeral of the basic six-digit number. (Example: 4 = 30%, max) 0 = not specified 5 = 40%, max 1 = 10%, max 6 = 60%, max 2 = 15%, max 7 = 80%, max 3 = 20%, max 8 = 100%, max 4 = 30%, max 9 = as specified[a]
Sixth Numeral	**Weight increase when immersed in water**, determined in accordance with paragraph 8.3.3, shall conform to the percent indicated by the sixth numeral of the basic six-digit number. See left and below. (Example: 4 = 30%, max) 0 = not specified 5 = 40%, max 1 = 10%, max 6 = 60%, max 2 = 15%, max 7 = 80%, max 3 = 20%, max 8 = 100%, max 4 = 30%, max 9 = as specified[a]

[a] On engineering drawings or other supplement to this classification system.

7.1.2 When the first numeral of line call-out is "2" (Type 2 materials), specimens shall be conditioned at least 46 hr in a controlled humidity room or in a closed chamber with gentle mechanical circulation of the air at 70–85 F (21–30 C) and 50–55% relative humidity.

NOTE: If a mechanical means of maintaining 50–55% relative humidity is not available, a tray containing a saturated solution of reagent grade magnesium nitrate, $Mg(NO_3)_2 \cdot 6H_2O$, shall be placed in the chamber to provide the required relative humidity.

7.1.3 When the first numeral of line call-out is "3" (Type 3 materials), specimens shall be preconditioned for 4 hr at 70–85 F (21–30 C) in a closed chamber containing anhydrous calcium chloride as a desiccant. The air in the chamber shall be circulated by gentle mechanical agitation. Specimens shall then be transferred immediately to a controlled humidity room or closed chamber with gentle mechanical circulation of the air and conditioned for at least 20 hr at 70–85 F (20–30 C) and 50–55% relative humidity.

7.1.4 When the first numeral of a line call-out is "0" or "9" specimens shall be conditioned as in paragraph 7.1.3, unless otherwise specified in supplements to this classification.

7.2 In all cases where testing is conducted outside the area of specified humidity, specimens shall be removed from the chamber one at a time just prior to testing.

8. Methods of Test

8.1 Thickness

8.1.1 Measure the specimens with a dial-type micrometer actuated by a dead-weight load. The dial shall be graduated in 0.001 in. (0.02 mm) or smaller units, and readings shall be estimated to the nearest 0.0001 in. (0.002 mm). The presser foot shall be 0.252 ±0.005 in. (6.40 ±0.13 mm) in diameter. The anvil shall have a diameter not less than that of the presser foot. The pressure on the sample shall be as specified in Table 4.

8.1.2 Take the reading by lowering the presser foot gently until it is in contact with the specimen. Take a sufficient number of readings, depending on the size of the specimen, to provide a reliable average value.

8.2 Compressibility—Test specimens in accordance with ASTM F 36, Test for Compressibility and Recovery of Gasket Materials, using the procedure which is applicable to the material described by the first two numerals of the basic six-digit number, as given in Table 5.

TABLE 2—SUPPLEMENTARY PHYSICAL AND MECHANICAL CHARACTERISTICS

Suffix Symbol	Supplementary Characteristics
A9	**Sealability** characteristics shall be determined in accordance with paragraph 8.4. External load, internal pressure, other details of test, and results shall be as specified on engineering drawing or other supplement to this classification.
B1 through B9	**Creep relaxation** characteristics shall be determined in accordance with paragraph 8.5. Loss of stress at end of 24 hr shall not exceed the amount indicated by the numeral of the B symbol. B1 = 10% B5 = 30% B2 = 15% B6 = 40% B3 = 20% B7 = 50% B4 = 25% B8 = 60% B9 = as specified[a]
D00 through D99	**Adhesion and corrosion** characteristics shall be determined in accordance with paragraph 8.6. **Adhesion** shall not exceed the standard rating number indicated by the **first** numeral of the two digit number of the D symbol. **Corrosion** shall not exceed the standard rating number indicated by the **second** numeral of the D symbol. Adhesion Rating (first numeral) Corrosion Rating (second numeral) D0 = not specified D0 = not specified D1 = 1 D1 = 1 D2 = 2 D2 = 2 D3 = 3 D3 = 3 D4 = 4 D4 = 4 D5 = 5 D5 = 5 D9 = as specified[a] D6 = 6 D9 = as specified[a]
M1 through M9	**Tensile strength** characteristics shall be determined in accordance with paragraph 8.7. Results in psi (or mN/m^2) shall be no less than the value indicated by the numeral of the M symbol. M1 = 100 (0.689) M5 = 1500 (10.342) M2 = 250 (1.724) M6 = 2000 (13.790) M3 = 500 (3.447) M7 = 3000 (20.684) M4 = 1000 (6.895) M8 = 4000 (27.579) M9 = as specified[a]
R	**Flotation** characteristics shall be determined in accordance with paragraph 8.8. There shall be no evidence of disintegration at conclusion of test.
S9	**Volume change** characteristics, when immersed in ASTM No. 1 Oil, ASTM No. 3 Oil, and ASTM Reference Fuel A, shall be determined in accordance with paragraph 8.3.2. Results shall be as specified on engineering drawing or other supplement to this classification.
T	**Flexibility** characteristics shall be determined in accordance with paragraph 8.9. There shall be no evidence of cracks, breaks, or separation at conclusion of test.
Z	**Other characteristics** shall be as specified on engineering drawing or other supplement to this classification.

[a] On engineering drawing or other supplement to this classification.

TABLE 3—THICKNESS TOLERANCES

Type and Class of Material (First Two Numerals of Basic Six-Digit Number)	Thickness Specified, in. (mm)	Applicable Tolerance[a], in. (mm)
11 and 12	0.016 (0.41) and under	+0.005 (+0.13) −0.002 (−0.05)
	Over 0.016 (0.41) and under 0.062 (1.57)	±0.005 (±0.13)
	0.062 (1.57) and over	±0.008 (±0.20)
13	Up to 0.125 (3.18)	±0.005 (±0.13)
	0.125–0.500 (3.18–12.70)	±0.010 (±0.25)
21	All thicknesses	±10%, or ±0.010 (±0.25) whichever is the greater
22	Under 0.062 (1.57)	±0.010 (±0.25)
	0.062 (1.57) and over	±0.015 (±0.38)
23	0.062 (1.57) and over	±0.015 (±0.38)
31, 32, and 33 (also 00 and 99)[b]	0.016 (0.41) and under	±0.0035 (±0.089)
	Over 0.016–0.062 (0.41–1.57)	±0.005 (±0.13)
	Over 0.062–0.094 (1.57–2.39)	±0.008 (±0.20)
	Over 0.094 (2.39)	±0.016 (±0.41)

[a] Tolerances listed are permissible variations applicable to a given lot of sheets or gaskets. Where other thickness tolerances are necessary due to the gasket application, tolerances applicable to individual sheet or gasket may be agreed to in writing between the purchaser and the supplier.

[b] Unless otherwise specified on engineering drawing or other supplement to this classification.

TABLE 4—THICKNESS MEASUREMENT STRESSES AND FORCES

Type of Material of First Numeral of Six-Digit Number	Pressure on Sample, psi (kN/m^2)	Total Force on Presser Foot, oz (N) (Reference)
1	11.5 ± 1.0 (80.3 ± 6.9)	9.0 (2.50)
2	5.1 ± 1.0 (35 ± 6.9)	4.0 (1.11)
3	8.0 ± 1.0 (55 ± 6.9)	6.3 (1.75)
0 and 9[a]	8.0 ± 1.0 (55 ± 6.9)	6.3 (1.75)

[a] Unless otherwise specified on engineering drawing or other supplement to this classification.

TABLE 5—COMPRESSIBILITY TEST METHODS

First Two Numerals of Six-Digit Number	Procedure, ASTM F 36	Pressure, psi (mN/m^2)
11 and 12	A	5000 (34.474)
13	H	1000 (6.895)
21 and 23	F	100 (0.689)
22	B	400 (2.758)
31, 32, and 33	G	1000 (6.895)
00 and 99	G[a]	1000 (6.895)

[a] Unless otherwise specified on engineering drawing or other supplement to this classification.

8.3 Immersion in Liquids—Test specimens in accordance with the procedures outlined below for the type of material described by the first numeral of the basic six-digit number:

8.3.1 TYPE 1 MATERIALS

8.3.1.1 For ASTM No. 3 Oil immersion tests, maintain bath at 302 ±3.6 F (150 ±2 C) and immerse specimens for 5 hr. For ASTM Reference Fuel B immersion tests, maintain bath at 70–85 F (21–30 C) and immerse specimens for 5 hr.

8.3.1.2 Use apparatus as specified in Section 6 of ASTM D 471, Test for Change in Properties of Elastomeric Vulcanizates Resulting from Immersion in Liquids. Use fresh test fluid for each determination. Test specimens immersed in fuel within 30 sec after removal from the test medium. Remove specimens immersed in hot oil and immediately immerse in a cool, clean portion of the test medium for 30–60 min, then dip quickly in acetone and blot lightly with filter paper. Thickness and weight change specimens shall be 1 x 2 in. (approximately 25 x 50 mm) in size. Measure thickness before and after immersion in accordance with the procedure described in paragraph 8.1, in the same areas of the specimen. Calculate the change as a percentage of the original thickness. Weigh the specimens in a sealed, tared container, and calculate the change as a percentage of the original weight.

8.3.2 TYPE 2 MATERIALS

8.3.2.1 For ASTM No. 1 and No. 3 Oil immersion tests, maintain the bath at 212 ±3.6 F (100 ±2C) and immerse the specimens for 70 hr. For ASTM Reference Fuel A immersion tests, maintain the bath at 70–85 F (21–30 C) and immerse the specimens for 22 hr.

8.3.2.2 Test specimens for volume change in accordance with ASTM D 471. For materials having a specific gravity of less than 1.00, use the following procedure when using a Jolly Balance:

(1) Level and zero the Jolly Balance which shall be properly shielded from drafts.

(2) Attach a small metal sinker (5 g is usually sufficient) to the weighing hook, so that it is totally immersed in water.

(3) Weigh the specimen in air and record the scale reading, SR_1.

(4) Then weigh the specimen in water and record the scale reading, SR_2.

(5) The original volume, V_1, then equals $(SR_1 - SR_2)$.

(6) After removing the specimen from the test medium, repeat steps 3, 4, and 5. (Caution: Use the same sinker throughout.) This gives the final volume, V_2. The distilled water used in the test shall be changed frequently.

Calculate the change in volume, V, as follows:

$$V = (V_2 - V_1)/V_1 \times 100$$

where: V = change in volume, %
V_2 = volume after removal from liquid
V_1 = original volume

Immerse specimens for flexibility as described in ASTM D 471. Prepare and test the specimens in accordance with the procedure described in paragraph 8.9 of this classification.

8.3.3 TYPE 3 MATERIALS

8.3.3.1 For ASTM No. 3 Oil, ASTM Reference Fuel B, and water immersion tests, maintain the baths at 70–85 F (21–30 C) and immerse the specimens for 22 hr.

8.3.3.2 Cut the specimens to a minimum width of 1 in. (25 mm) or to a minimum diameter of 1.129 in. (28.08 mm) so that the area will be 1–4 in.^{2} (640–2580 mm^2) prior to conditioning. After conditioning, transfer a specimen as rapidly as possible to a tared weighing bottle of suitable dimensions and weigh to the nearest 1 mg. Remove and measure the specimen for thickness in accordance with the procedure described in paragraph 8.1. Immerse the specimen in the test medium for the specified length of time. Use a container of convenient size to provide a minimum of 10 ml of test fluid per specimen. Light wire screens may be used, if necessary, to keep the specimens immersed or separated from each other. Upon removal, blot the specimen as rapidly as possible with sheets of rapid qualitative filter paper, or paper of similar texture, to remove excess liquid from the surface. Take care to remove all of the surface excess but not to exert any squeezing action upon the specimen. Blot specimens over 0.03 in. (0.8 mm) in thickness on the edges. Place the specimen immediately in the tared weighing bottle and reweigh to the nearest 1 mg. Calculate the change in weight as a percentage of the initial weight of the specimen. Remove the specimen from the weighing bottle and measure for thickness in the same areas as the initial set of measurements. Average both sets of measurements to the nearest 0.0001 in. (0.002 mm), and calculate the change as a percentage of the initial thickness.

8.3.4 OTHER TYPES OF MATERIALS (as indicated by 0 or 9 first numeral of basic six-digit number)—Use the same apparatus and general procedure outlined in paragraph 8.3.3, unless otherwise specified in the engineering drawing or other supplement to this classification.

8.4 Sealability—Test specimens in accordance with ASTM Method F 37, Test for Sealability of Gasket Materials.

8.5 Creep Relaxation—Test specimens in accordance with Method B of ASTM F 38, Test for Stress Relaxation of a Gasket material.

8.6 Corrosion and Adhesion—Test specimens in accordance with ASTM F 64, Test for Corrosive and Adhesive Effects of Gasket Materials on Metal Surfaces, time and temperature as follows:

8.6.1 TYPE 1 MATERIALS—Fourteen days at 70–80 F (21–27 C) or 70 hr at 250 ±3.6 F (121 ±2 C).

8.6.2 TYPE 2 MATERIALS—Fourteen days at 70–80 F (21–27 C).

8.6.3 TYPE 3 MATERIALS—Fourteen days at 70–80 F (21–27 C) or 70 hr at 250 ±3.6 F (121 ±2 C).

8.6.4 OTHER TYPES OF MATERIALS—Same as Type 3, unless otherwise specified on engineering drawing or other supplement to this classification.

8.7 Tensile Strength—Test for specimens in accordance with the following, using the procedure applicable to the type of material described by first numeral of the basic six-digit number, as indicated:

8.7.1 TYPE 1 MATERIALS—Use ASTM F 39, Testing Compressed Asbestos Sheet Packing.

8.7.2 TYPE 2 MATERIALS

8.7.2.1 Use specimens 2 in. (approximately 50 mm) wide. Place in the test apparatus with a 2 in. (approximately 50 mm) distance between the jaws and a 1 in. (approximately 25 mm) minimum amount of material gripped by each jaw.

8.7.2.2 Use a tension testing apparatus of the spring-actuated type, pendulum-type, or equivalent. Drive the power-actuated jaw at the rate of 12 ±1 in. (approximately 300 ±25 mm)/min. Equip the jaws with a load indicator that will indicate the point of maximum load after rupture of the specimen. The normally used range of the machine should be 15–85% of the full range of the load indicator scale in use. Prepare the specimens and place in the testing machine as follows, after measuring for thickness as described in paragraph 7.1.

8.7.23 In all cases, cut the specimens cleanly and align carefully with the direction of travel of the jaws.

8.7.2.4 Calculate the tensile strength by dividing the breaking load by the original cross-sectional area of the specimen, and express in pounds per square inch (or meganewtons per square meter).

8.7.3 TYPE 3 MATERIALS

8.7.3.1 Use specimens 1 in. (approximately 25 mm) wide by 6 in. (approximately 150 mm) long. (The lengthwise direction of the specimen shall be perpendicular to the grain direction of the material.) Place in the test apparatus with a 4 in. (approximately 100 mm) distance between the jaws. A specimen 0.5 in. (approximately 13 mm) wide may be used where necessary to fall within the range of the load indicator.

8.7.3.2 Use the apparatus and general procedure as outlined in paragraphs 8.7.2.2, 8.7.2.3, and 8.7.2.4.

8.7.4 OTHER TYPES OF MATERIALS (first two numerals of the basic six-digit number, 00 or 99)—Use specimens and general procedures outlined in paragraph 8.7.3, unless otherwise specified on the engineering drawing or other supplement to this classification.

8.8 Flotation—Float specimens 1 in.^{2} (645 mm^2) or greater in area in boiling hydrochloric acid having a concentration of 35% vol for $\frac{1}{2}$ hr. Use a reflux condenser to maintain the concentration. Examine the specimens at the completion of the test period for evidence of disintegration.

8.9 Flexibility—Test the specimens in accordance with the following, using the procedure applicable to the type of material described by the first two numerals of the basic six-digit number, as indicated:

8.9.1 FIRST TWO NUMERALS 21—Use specimens 0.5 in. (approximately 13 mm) wide, 6 in. (approximately 150 mm) long, and 0.1875 in. (4.76 mm) maximum thickness. Condition specimens for 24 ±1 hr at 70–85 F (21–30 C) at 50–55% relative humidity. Immediately after conditioning slowly bend the specimens 180 deg around a mandrel, the diameter of which is five times the thickness of the specimens. Inspect the specimens for cracks, breaks, or separations.

8.9.2 FIRST TWO NUMERALS 22 OR 23—Conduct the test as outlined in paragraph 8.9.1; also using same size specimens, age the specimens in the oven at 212 ±3.6 F (100 ±2 C) for 70 hr. Remove from the oven and follow the same conditioning and procedure as paragraph 8.9.1, except use a mandrel having a diameter of 16 times the thickness of the specimens.

APPENDIX A—APPLICABLE TEST METHODS

Table A1 indicates properties, characteristics, and test methods that are normally considered applicable to each type of material. It is not intended to limit the use of numeral-symbols as provided in ASTM F 104 where experience indicates that the related properties, characteristics, or test methods, or both, are applicable.

APPENDIX B—NONMETALLIC GASKETS FOR GENERAL AUTOMOTIVE PURPOSES—SAE J90a

Report of Nonmetallic Materials Committee approved January 1952 and last revised April 1963. Reaffirmed without change May 1968. This section of this SAE Recommended Practice on Method of Test for Compressibility and Recovery of Gasket Materials Conforms with ASTM F 36, Tentative Method of Test for Compressibility and Recovery of Gasket Materials. The remainder of this SAE Recommended Practice conforms with ASTM D 1170, Tentative Specifications for Nonmetallic Gasket Materials for General Automotive and Aeronautical Purposes.

NOTE: THESE NUMBERS ARE NOT TO BE SPECIFIED ON NEW DRAWINGS OR MATERIAL SPECIFICATIONS.

[This SAE Recommended Practice was formulated by the SAE—ASTM Technical Committee on Automotive Rubber.]

Scope—These specifications for SAE J90 are intended to define the basic properties of commercial nonmetallic gasketing materials commonly used in automotive applications. These include materials composed of asbestos or other inorganic fibers, cork, or cellulose or other organic fibers, in combination with various binders or impregnants. Rubber compounds without fibrous or cork reinforcement are not included since they are covered in SAE Standard, Specifications for Elastomer Compounds for Automotive Applications—SAE J14, and in ASTM D 735-61T. Although the test methods and values are designed to describe the basic properties of the material in each category, they do not define all of the properties which govern gasket performance. Caution should, therefore, be exercised in using these specifications as a basis for the selection of materials.

Numbering System

Types—The various categories of material are grouped into three types, according to whether the principal fibrous or particulate reinforcement is asbestos or other inorganic fibers (Type 1), cork (Type 2), or cellulose or other organic fibers (Type 3).

Identification—Each category is identified by a number consisting of four digits and a suffix letter. The significance of these is shown in Table for System of Identification.

Coated Gaskets—Gaskets from materials conforming to these specifications may be coated on one or more sides. Because of the many possible coating materials and thicknesses, it is suggested that coating specifications be given on the blueprints.

Detailed Requirements—Gasket materials shall conform to the requirements as prescribed in the tables as follows:

Type 1—Table 1
Type 2—Table 2
Type 3—Table 3

Methods of Test

Sampling—Specimens shall be selected from finished gaskets or sheets of suitable size, whichever is the most practicable. If sheets are used, they shall, where applicable, be cut squarely with the grain of the stock and the grain direction noted by an arrow.

For qualification purposes, thicknesses shall be used as follows:

Type 1—$1/32$ in.
Type 2—$1/16$ to $1/4$ in.
Type 3—$1/32$ in.

When thicknesses other than those shown above are to be tested, the specification limits shall be agreed to, in writing, between purchaser and supplier.

Sufficient specimens shall be selected to provide a minimum of three determinations for each test described in subsequent paragraphs. The average of the determinations shall be considered as the result.

Conditioning—Prior to all tests, specimens shall be conditioned as follows:

Type 1—Specimens shall be conditioned in an oven at 212 ± 2 F for 1 hr and allowed to cool to 70-85 F in a desiccator containing anhydrous calcium chloride, except that P1300 series shall be conditioned in an oven for 4 hr at 212 ± 2 F.

Type 2—Specimens shall be conditioned at least 46 hr in a controlled humidity room or in a closed chamber with gentle mechanical circulation of the air at 70-85 F and 50-55% relative humidity.

Type 3—Specimens shall be preconditioned for 4 hr at 70-85 F in a closed chamber containing anhydrous calcium chloride as a desiccant. The air in the chamber shall be circulated by gentle mechanical agitation. Specimens shall then be transferred immediately to a controlled humidity room or closed chamber with gentle mechanical circulation of the air and conditioned for at least 20 hr at 70-85 F and 50-55% relative humidity.

If a mechanical means of maintaining 50-55% relative humidity is not

TABLE A1—TYPICAL TYPES OF MATERIALS

Properties, Characteristics, and Test Methods	Type 1, Asbestos or Other Inorganic Fibers			Type 2, Cork			Type 3, Cellulose or Other Organic Fibers		
	Compressed Asbestos	Beater Addition Asbestos	Asbestos Paper and Millboard	Cork Composition	Cork and Elastomeric	Cork and Cellular Rubber	Untreated Fiber	Protein Treated	Elastomeric Treated
Compressibility:									
5000 psi load (ASTM F 36, Procedure A)	X	X	—	—	—	—	—	—	—
100 psi load (ASTM F 36, Procedure F)	—	—	—	X	—	X	—	—	—
1000 psi load (ASTM F 36, Procedure H)	—	—	X	—	—	—	—	—	—
(ASTM F 36, Procedure G)	—	—	—	—	—	—	X	X	X
400 psi load (ASTM F 36, Procedure B)	—	—	—	—	X	—	—	—	—
Tensile strength	X	X	X	X	X	X	X	X	X
Resistance to exposure in ASTM No. 3 Oil:									
Volume change, 70 hr at 212 F	—	—	—	—	X	X	—	—	—
Weight increase, 22 hr at 70-85 F	—	—	—	—	—	—	X	X	X
Thickness increase: 22 hr at 70-85 F	—	—	—	—	—	—	X	X	X
5 hr at 300 F	X	X	—	—	—	—	—	—	—
Resistance to exposure in ASTM Fuel B:									
Weight increase: 22 hr at 70-85 F	—	—	—	—	—	—	X	X	X
5 hr at 70-85 F	X	X	—	—	—	—	—	—	—
Thickness change: 22 hr at 70-85 F	—	—	—	—	—	—	X	X	X
5 hr at 70-85 F	X	X	—	—	—	—	—	—	—
Resistance to exposure in ASTM No. 1 Oil:									
Volume change, 70 hr at 212 F	—	—	—	—	X	X	X	—	—
Resistance to exposure in ASTM Fuel A:									
Volume change, 22 hr at 70-85 F	—	—	—	—	X	X	X	—	—
Resistance to exposure in distilled water:									
Weight increase, 22 hr at 70-85 F	—	—	—	—	—	—	X	X	X
Thickness change, 22 hr at 70-85 F	—	—	—	—	—	—	X	X	X
Sealability	X	X	X	X	X	X	X	X	X
Stress relaxation	X	X	—	—	—	—	—	—	—
Adhesion and corrosion	X	X	X	X	X	X	X	X	X
Flotation	—	—	—	X	—	—	—	—	—
Flexibility	—	—	—	X	X	X	—	—	—

Note: "X" indicates that the test conditions shown in first column have been used to characterize the type of material named in column heading. "Dash" (—) indicates that the test method is either "not applicable" to the material named or has not been commonly used in characterizing the material.

TABLE FOR SYSTEM OF IDENTIFICATION

	Type 1	Type 2	Type 3
First digit (principal fibrous or particulate material)	1. Asbestos or other inorganic fibers	2. Cork	3. Cellulose or other organic fibers
Second digit (trade designation)	1. Compressed asbestos sheet 2. Asbestos beater sheet 3. Asbestos paper and millboard	1. Cork composition 2. Cork and rubber 3. Cork and cellular rubber	0. Tag 1. Chipboard 2. Vulcanized fiber 3. Cellulose fiber 4. Fiber and filler compositions
Third digit (binder or treatment other than sizing)	(Same for all three types) 0. None 1. Protein (glue-glycerine or equivalent) 2. Resin 3. Rubber, Type S, Class SA (polysulfide or equivalent) 4. Rubber, Type S, Class SB (acrylonitrile or equivalent) 5. Rubber, Type S, Class SC (chloroprene or equivalent) 6. Rubber, Type R (natural, reclaim, styrene or equivalent)		
Fourth digit (Compressibility Index ASTM D 1147-61T Procedure G, total load—1000 psi)	(Same for all three types) 0. 0–5% 5. 46–55% 1. 6–15% 6. 56–65% 2. 16–25% 7. 66–75% 3. 26–35% 8. 76–85% 4. 36–45% 9. 86–95% For identification purposes only. May not agree with compressibility in tables where other loads are employed.		
Suffix letter	Used to distinguish grades of material within one four digit category which differ sufficiently to justify separate tabular values. If only one grade of material is listed in the table, the letter "A" is used.		

EXAMPLE:
```
          ┌─ Letter indicating a gasket material included in SAE J90
          │ ┌─ Cellulose or other organic fibers
          │ │ ┌─ Rope and/or chemical wood
          │ │ │ ┌─ Binder or treatment, rubber, Type S, Class SC
          │ │ │ │ ┌─ Compressibility index is 26–35%
          │ │ │ │ │ ┌─ Grade
          P 3 3 5 3 – A
```

available, a tray containing a saturated solution of reagent grade magnesium nitrate Mg(NO$_3$)$_2$ 6H$_2$O shall be placed in the chamber to provide the required relative humidity. In all cases where testing is conducted outside the area of specified humidity, specimens shall be removed from the chamber one at a time as needed.

Compressibility and Recovery—Specimens shall be tested according to procedures outlined in ASTM D 1147-61T, Method of Test for Compressibility and Recovery of Gasket Materials, as follows:

Type 1—P1100 and P1200 series—Procedure A
       P1300 series   —Procedure H
Type 2—P2100 and P2300 series—Procedure F
       P2200 series   —Procedure B
Type 3   —Procedure G

Thickness—Specimens shall be measured with a dial type micrometer actuated by a dead weight load. The dial shall be graduated in 0.001 in. or smaller units and readings shall be estimated to the nearest 0.0001 in. The anvil shall have a diameter not less than that of the presser foot.

Loads and presser foot diameters shall be as follows:

Type	Total Load on Presser Foot, oz. (Reference)	Load on Sample, psi	Presser Foot Dia, in.
1	9.0	11.5±1.0	0.252±0.005
2	4.0	5.1±1.0	0.252±0.005
3	5.3	8.0±1.0	0.252±0.005

Readings shall be taken by lowering the presser foot gently until it is in contact with the specimen. A sufficient number of readings shall be taken, depending on the size of the specimen, to provide a reliable average value.

Tensile Strength—An acceptable tension testing machine, spring actuated type, pendulum type, or equivalent shall be used. The power actuated jaw shall be driven at the rate of 12 ± 1 in. per minute. It shall be equipped with a load indicator which will indicate the point of maximum load after rupture of the specimen. The normally used range of the machine should be 15–85% of the full range of the load indicator scale in use.

Specimens shall be prepared and placed in the testing machine as follows, after being measured for thickness according to paragraph on Thickness:

Type 1—Paragraph 6(c). ASTM D 733-59, Methods of Testing Compressed Asbestos Sheet Packing, except that only the direction perpendicular to the grain shall be tested.

Type 2—Specimens shall be 2 in. in width. They shall be placed in the machine with a 2 in. distance between the jaws and shall be gripped for at least 1 in. in each jaw.

Type 3—Specimens shall be 1 in. in width by 6 in. long. The lengthwise direction of the specimen shall be perpendicular to the grain direction of the material. They shall be placed in the machine with a 4 in. distance between the jaws. Specimens of 1/2 in. in width may be used where necessary to fall within the range of the load indicator.

In all cases, specimens shall have cleanly cut edges and shall be carefully aligned with the direction of travel of the jaws.

Tensile strength shall be calculated by dividing the breaking load in pounds by the original cross sectional area of the specimen in square inches and shall be expressed in pounds per square inch.

Immersion in Liquids

Type 1—Apparatus used shall be as specified in Paragraph 6, ASTM D 471-62T, Method of Test for Change in Properties of Elastomeric Vulcanizates Resulting from Immersion in Liquids. Fresh test fluid shall be used for each determination. Specimens immersed in fuel shall be tested within 30 sec after removal from the test medium. Specimens immersed in hot oil shall be removed and immediately immersed in a cool, clean portion of the test medium for 30 to 60 minutes, than dipped quickly in acetone and blotted lightly with filter paper.

TABLE B1—TYPE 1—ASBESTOS OR OTHER INORGANIC FIBERS

Identification No.[a]	Former "G" No. (For Reference Only)	Original Properties — Compressibility Total Load, psi	Compressibility, %	Recovery, min, %	Tensile Strength, min, psi	Ignition Loss, max, %	After Aging 5 Hrs at 300 F in ASTM Oil No. 3 — Compressibility, max, %	Loss in Tensile Strength, max, %	Thickness Increase, %	After Aging 5 Hrs at 70–85 F in ASTM Reference Fuel B — Weight Increase, max, %	Thickness Increase, %
P1141A	G1122-1	5000	7–17	40	2000	—	20	30	0–13	20	0–15
P1151A	G1123-1	5000	7–17	40	2000	—	30	50	15–30	30	10–25
P1161A	G1111-1	5000	7–17	40	2000	—	—	70	20–50	40	15–35
P1161B		5000	7–17	40	2000	—	—	80	40–70	50	25–45
P1162A	G1111-2	5000	15–25	30	1600	—	—	70	20–50	40	15–35
P1241C		5000	13–23	35	1000	—	30	35	5–20	30	0–15
P1242C		5000	30–40	30	1700	—	45	15	0–20	50	0–15
P1242D		5000	20–30	35	2000	—	35	20	0–20	40	0–15
P1243A	G1422-2	5000	35–50	15	500	—	55	25	0–5	50	0–5
P1251A	G1423-1	5000	10–20	40	2000	—	35	40	10–20	35	0–15
P1252A	G1423-2	5000	20–30	35	1000	—	—	60	20–35	50	5–20
P1252D		5000	30–40	35	1200	—	45	30	10–25	50	5–20
P1252E		5000	20–30	35	1200	—	40	40	10–25	45	5–20
P1253A	G1423-3	5000	35–50	20	1000	—	—	50	0–15	55	0–10
P1261A		5000	15–30	30	1200	—	—	60	10–25	60	5–20
P1262B		5000	25–40	35	1000	—	—	80	10–40	70	0–30
P1301A	G4131	1000	6–15	40	200	20	—	—	—	—	—
P1302A	G4111	1000	16–25	30	175	20	—	—	—	—	—

[a] These thickness tolerances are permissible variations applicable to given lot of sheets of gaskets. Where special thickness tolerances are necessary due to application, the tolerance on individual sheet or gasket shall be agreed to in writing, between the supplier and customer.

Series	Thickness, in.	Tolerance, in.	Series	Thickness, in.	Tolerance, in
P1100 and P1200	1/64 and under	+0.005 −0.002	P1301A P1302A	— Up to 1/8 1/8 to 1/2	±10% ±0.005 ±0.010
	Over 1/64 and under 1/16	±0.005			
	1/16 and over	±0.008			

TABLE B2—TYPE 2—CORK

Identification No.[a,b]	Former "G" No. (For Reference Only)	Original Properties — Compressibility — Total Load, psi	Compressibility, %	Recovery, min, %	Tensile Strength, min, psi	Density min, lb per cu. ft	Flotation Tests — 3 Hr in Boiling Water	1/2 Hr in Boiling 35% HCL	2 Hr at 212 F in ASTM Oil No. 1	Flexibility Factor, F	After Oven Aging 70 Hr at 212 F — Flexibility Factor, F	After Aging 70 Hr at 212 F in ASTM Oil No. 1 — Flexibility Factor, F	After Aging 70 Hr at 212 F in ASTM Oil No. 1 — Volume Change, %	After Aging 70 Hr at 212 F in ASTM Oil No. 3 — Volume Change, %	After Aging 22 Hr at 70–85 F in ASTM Reference Fuel A — Volume Change, %
Cork Composition															
P2116A	G2114	100	10–25	60	175	24	N	—	N	5	—	—	—	—	—
P2117A	G2113	100	15–30	65	150	20	N	—	N	5	—	—	—	—	—
P2117B	G2112	100	20–40	75	100	17	N	—	N	5	—	—	—	—	—
P2118A	G2111	100	30–50	80	75	14	N	—	N	5	—	—	—	—	—
P2126A	G2214	100	10–25	60	175	24	N	N	N	5	—	—	—	—	—
P2127A	G2213	100	15–30	65	150	20	N	N	N	5	—	—	—	—	—
P2127B	G2212	100	20–40	75	100	17	N	N	N	5	—	—	—	—	—
P2128A	G2211	100	30–50	80	75	14	N	N	N	5	—	—	—	—	—
Cork and Rubber															
P2236A	G1221-3	400	25–45	75	200	—	—	—	—	5	16	16	−5 to +5	0 to +10	−5 to +5
P2243A	G1222-2	400	15–25	75	250	—	—	—	—	5	16	16	−5 to +10	−2 to +15	−2 to +10
P2245A	G1222-3	400	25–35	75	250	—	—	—	—	5	16	16	−5 to +10	−2 to +15	−2 to +10
P2245B	—	400	40–55	70	150	—	—	—	—	5	16	16	−15 to +15	0 to +25	−5 to +15
P2246A	G1222-4	400	35–45	75	200	—	—	—	—	5	16	16	−5 to +10	−2 to +15	−2 to +10
P2254A	G1223-2	400	15–25	75	250	—	—	—	—	5	16	16	−2 to +20	+15 to +50	0 to +15
P2255A	G1223-3	400	25–35	75	250	—	—	—	—	5	16	16	−2 to +20	+15 to +50	0 to +15
P2255B	—	400	40–55	75	125	—	—	—	—	5	16	16	−10 to +5	+15 to +50	0 to +35
P2256A	G1223-4	400	35–45	75	220	—	—	—	—	5	16	16	−2 to +20	+15 to +50	0 to +15
P2265A	G1211-3	400	25–45	75	150	—	—	—	—	5	16	—	—	—	—
P2268A	G1211-5	400	40–60	75	75	—	—	—	—	5	16	—	—	—	—
Cork and Cellular Rubber															
P2347A	—	100	35–50	75	100	—	—	—	—	5	16	—	−20 to −5	−10 to +5	−10 to +5
P2357A	—	100	35–50	75	75	—	—	—	—	5	16	—	−10 to +10	15 to 50	0 to 25
P2367A	—	100	35–50	75	100	—	—	—	—	5	16	—	—	—	—

N = No disintegration

[a] Thickness Tolerances:
P2100 Series—±10% or ±0.010 in. whichever is the greater.
P2200 Series—Under 1/16 in. ±0.010 in., 1/16 in. and over, ±0.015 in.
P2300 Series—1/16 in. (minimum thickness) and over ±0.015 in.

[b] Grain size may be specified for certain applications. If so, the following definitions will usually apply:
Fine—Pass a No. 20 sieve and retained on a No. 40 seive.
Medium—Pass a No. 10 sieve and retained on a No. 20 sieve.
Coarse—Pass a No. 5 sieve and retained on a No. 10 sieve.
(Sieve sizes are as specified in ASTM E 11—39, Specification for Sieves for Testing Purposes (Wire Cloth Sieves, Round-Hole and Square-Hole Screens, or Sieves), Table 1.

TABLE B3—TYPE 3—CELLULOSE OR OTHER ORGANIC FIBERS

Identification No.[c]	Former "G" No. (For Reference Only)	Total Load, psi	Compressibility, %	Recovery min, %	Tensile Strength min, psi	ASTM Reference Fuel B — Thickness Increase, max, %	ASTM Reference Fuel B — Weight Increase, max, %	ASTM Oil, No. 3 — Thickness Increase, max, %	ASTM Oil, No. 3 — Weight Increase, max, %	Distilled Water — Thickness Increase, max, %	Distilled Water — Weight Increase, max, %
P3002A	G3111	1000	10–25	50	1500	—	—	—	—	—	—
P3102A	G3141	1000	20–30	45	750	—	—	—	—	—	—
P3200A	G3261-3262	1000	0–10	70	6000	—	—	—	—	—	—
P3301A	G3151	1000	5–15	60	3000	—	—	—	—	—	—
P3302C	—	1000	15–30	40	1500	—	—	—	—	—	—
P3313B	G3212	1000	25–40	40	2000	5	15	5	15	30	90
P3341A	—	1000	5–15	50	4000	5	10	5	15	90	80
P3341D	—	1000	6–16	55	3000	12	35	5	35	40	40
P3342C	—	1000	16–26	45	3500	20	35	5	20	20	30
P3342F	—	1000	17–27	40	1800	30	80	30	80	50	80
P3342G	—	1000	20–30	35	1500	10	50	10	50	35	70
P3345A	G3232-6	1000	40–60	20	500	5	80	10	80	25	65
P3353A	—	1000	25–40	25	1500	12	80	12	90	30	75
P3353B	—	1000	20–35	40	2000	15	50	15	55	40	50
P3354A	—	1000	30–50	30	800	15	80	15	90	25	65
P3361A[a]	G3233-5A	1000	5–15	40	2000	10	25	10	20	30	25
P3362A[b]	G3234-2	1000	15–30	40	2000	15	45	15	60	40	50
P3365A	G3234-3	1000	40–60	20	500	5	95	5	120	20	70
P3413A	G3234-6B	1000	25–45	40	1500	5	15	5	15	15	85
P3415A	G3221	1000	40–55	40	1000	5	30	5	30	30	100
P3421A	G3222	1000	10–20	50	1500	20	35	5	25	15	30
P3423A	G3223	1000	25–35	50	1000	5	50	5	40	5	20
P3441A	—	1000	5–15	45	2200	20	30	20	30	30	30
P3442A	—	1000	10–20	35	2500	15	30	10	25	40	60
P3443B	—	1000	25–35	55	1100	25	65	10	50	15	25
P3443C	—	1000	25–35	40	1200	10	65	10	70	35	50
P3444A	G3242-3	1000	30–45	25	800	10	75	10	70	20	55
P3464A	G3243-3	1000	30–45	30	700	30	100	30	105	20	70

[a] Formerly P3561A.
[b] Formerly P3562A.
[c] These thickness tolerances are permissible variations applicable to given lot of sheets of gaskets. Where special thickness tolerances are necessary due to application, the tolerance on individual sheet or gasket shall be agreed to in writing between the supplier and customer.

Thickness, in.	Tolerance, in.	Thickness, in.	Tolerance, in.
1/64 and under	±0.0035	Over 1/16 to 3/32	±0.008
Over 1/64 to 1/16	±0.005	Over 3/32	±0.016

Tensile strength shall be determined according to paragraph on Tensile Strength and shall be based on the cross sectional area before immersion.

Compressibility shall be determined according to paragraph on Compressibility and Recovery and shall be used on the thickness after immersion.

Thickness and weight change specimens shall be 1 x 2 in. in size. Thickness before and after immersion shall be measured according to paragraph on Thickness and determined in the same areas of the specimen. The change shall be calculated as a percentage of the original thickness. Weight measurements before and after immersion shall be made to the nearest milligram in a sealed, tared container and the change calculated as a percentage of the original weight.

Type 2—Specimens for volume change shall be tested in accordance with ASTM D 471–62T, Method of Test for Change in Properties of Elastomeric Vulcanizates Resulting from Immersion in Liquids.

For materials having a specific gravity of less than 1.00, the following procedure shall be used if a Jolly Balance is employed:

(a) The Jolly Balance, properly shielded from drafts, is leveled and zeroed.
(b) A small metal sinker (5 g is usually sufficient) is attached to the weighing hook so that it is totally immersed in water.
(c) The specimen is then weighed in air and the scale reading SR_1 is recorded.
(d) The specimen is then weighed in water and the scale reading SR_2 is recorded.
(e) V_1 then equal $(SR_1 - SR_2)$.
(f) After removing the specimen from the test medium, steps c,d, and e are repeated. (Caution: Use same sinker throughout.) This gives V_2. The distilled water used in the test must be changed frequently.

$$\frac{V_2 - V_1}{V_1} \times 100 = \% V$$

where: V_1 is original volume.
V_2 is volume after removal from liquid.
%V is per cent volume change.

Specimens for flexibility shall be immersed according to ASTM D 471–62T, Method of Test for Change in Properties of Elastomeric Vulcanizates Resulting from Immersion in Liquids. The specimens shall be prepared and tested according to paragraph on Flexibility (Type 2 only).

Type 3—Specimens shall be cut with a minimum width of 1 in. or a minimum diameter of 1.129 in. and ranging from 1 to 4 sq in. in area prior to conditioning. After conditioning, a specimen shall be transferred as rapidly as possible to a tared weighing bottle of suitable dimensions and weighed to the nearest milligram. The specimen shall then be removed and measured for thickness according to paragraph on Thickness. The specimen shall then be immersed in the test medium for the specified length of time. A container of convenient size shall be used so as to provide a minimum of 10 ml of test fluid per sample. Light wire screens may be used, if necessary, to keep specimens immersed or separated from each other.

Upon removal, the specimen shall be blotted as rapidly as possible with sheets of Whatman No. 1 filter paper, or paper of similar texture, to remove excess liquid from the surface. Care must be exercised to remove all of the surface excess but not to exert any squeezing action upon the specimen. Specimens over $\frac{1}{32}$ in. in thickness shall also be blotted on the edges.

The specimen shall then be placed immediately in the tared weighing bottle and reweighed to the nearest milligram. The change in weight shall be calculated as a percentage of the initial weight of the specimen. The specimen shall then be removed from the weighing bottle and measured for thickness in the same areas as the initial set of measurements according to the paragraph on Thickness. Both sets of measurements shall be averaged to the nearest 0.0001 in. and the change calculated as a percentage of the initial thickness.

Loss on Ignition (Type 1 Only)—A specimen weighing from 5 to 10 g shall be disintegrated, conditioned as in paragraph on Conditioning, cooled in a desiccator and weighed. It shall be ignited in a crucible at 1500 F for not less than 1 hr, cooled in a desiccator and reweighed. The loss on ignition shall be calculated as a percentage of the original weight.

Oven Aging (Type 2 Only)—Specimens shall be aged in accordance with ASTM D 573–53, Standard Method of Test for Accelerated Aging of Vulcanized Rubber by the Oven Method.

Flexibility shall be determined in accordance with paragraph on flexibility (Type 2 only) after conditioning the oven aged specimens for 24 ±1 hr at 70–85 F at 50–55% relative humidity.

Flexibility (Type 2 Only)—Specimens shall be $\frac{1}{2}$ in. wide, 6 in. long, and $\frac{3}{16}$ in. maximum thickness. The material shall not crack, break, or separate when bent slowly 180 deg around a mandrel of appropriate diameter. The diameter of the mandrel is determined by multiplying the thickness of the material by its flexibility factor "F."

Density (Type 2 Only)—Specimens shall have an area of at least 2 sq in. Thickness shall be measured according to paragraph on Thickness. Length and width shall be measured to an accuracy of 0.01 in. Weight shall be determined to an accuracy of ±1%. Density shall be calculated by dividing the weight of the specimen by its volume and shall be expressed in pounds per cubic foot.

Flotation (Type 2 Only)—Specimens of not less than 1 sq in. in area shall be floated in liquids under the conditions specified and shall be examined for evidence of disintegration upon conclusion of the test.

RUBBER RINGS FOR AUTOMOTIVE APPLICATIONS—SAE J120a

SAE Recommended Practice

Report of Nonmetallic Materials Committee approved January 1954 and last revised November 1962. Editorial change July 1968.

120R1—RUBBER O-RINGS FOR AUTOMOTIVE APPLICATION

1. Scope

(a) This SAE Recommended Practice covers the dimensional and material requirements of rubber O-rings for use in automotive sealing applications. Included are the recommended gland dimensions for standard size O-rings.

(b) The material and dimensional requirements established by this specification are based upon the following:
(1) Static and dynamic applications.
(2) Compounds of nominal 70 durometer hardness.
(3) Fluid pressures of 1500 psi max (increased compound hardness, back-up rings, or reduced clearances, or both, promote sealing to much greater pressures).
(4) Rotary seal applications are sometimes possible, but special design considerations are necessary and special compound recommendations can be obtained from seal suppliers.

2. Material Classifications—Two classes of material are covered by this specification. O-rings shall be classified in accordance with their intended use as follows:
Class I —Oil resistant service
Class II—Gasoline resistant service

Applications requiring resistance to special fluids can usually be satisfied by numerous compounds available. Suppliers should be contacted for their recommendations whenever possible. (See tables for compound designations in SAE J200.)

3. Basis of Purchase
Material and sizes required are specified by listing the SAE specification number, service classification, size number, and identification, when desired.

EXAMPLE: SAE Specification No. 120R1
Class II
Size 128

4. Physical Requirements

(a) Quality control and source approval tests shall be conducted on the actual seal ring when the section diameter is 0.100 in. or larger, and the inside diameter is 1 in. or larger. Smaller section parts do not provide reliable data. If, by necessity, smaller sizes are tested, then special specification limits and test procedures shall be based on actual part performance as agreed between the purchaser and the supplier.

(b) Test specimens shall conform to the requirements specified in Table 1.

5. Dimensions and Tolerances
O-rings shall conform to the dimensions and tolerances prescribed in Table 2. For dimensional standardization purposes, O-rings shall be numbered as follows:

Number	Section Dia, in	Number	Section Dia, in
006 to 045	0.070 ±0.003	325 to 349	0.210 ±0.005
110 to 163	0.103 ±0.003	425 to 460	0.275 ±0.006
210 to 281	0.139 ±0.004		

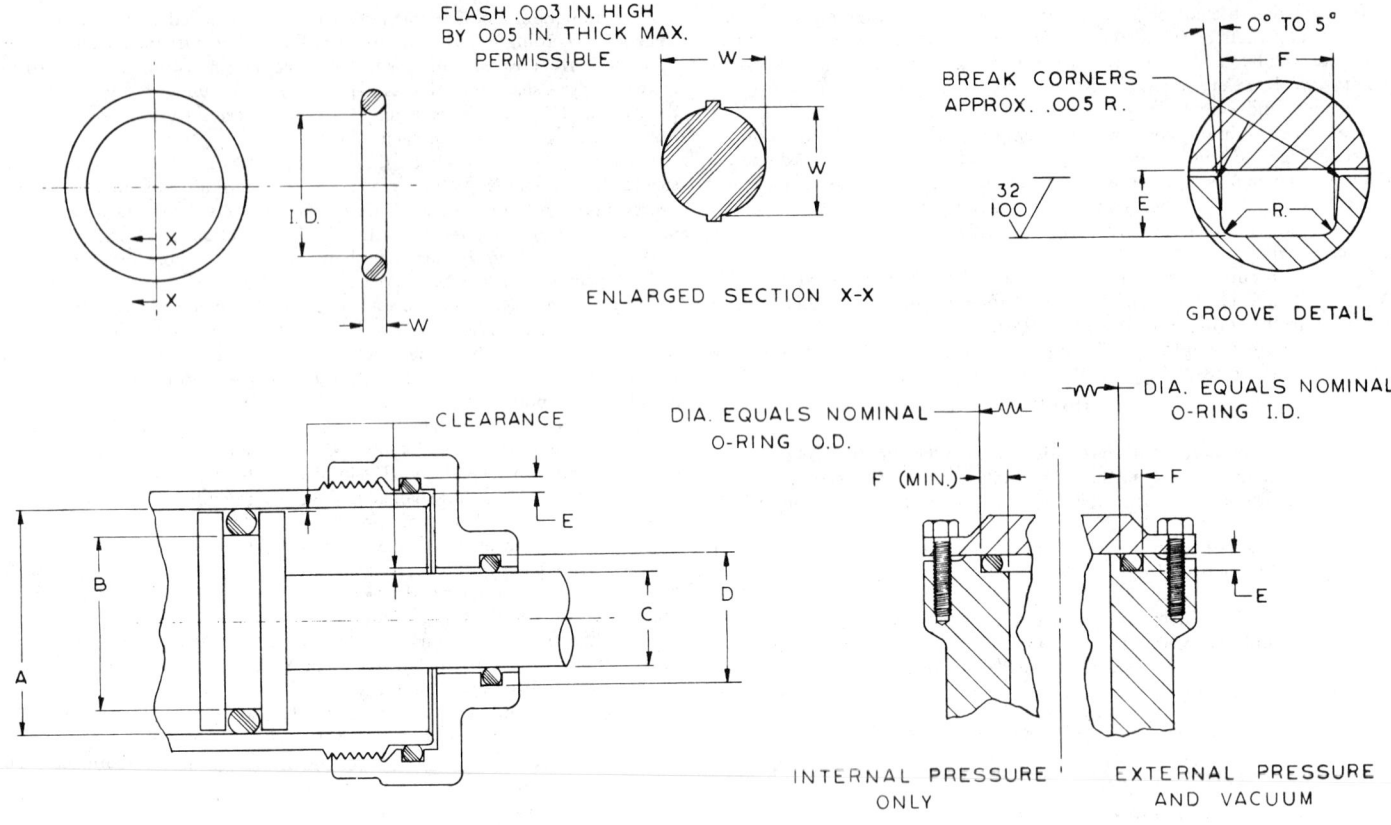

MAXIMUM RECOMMENDED CLEARANCE, IN (INCLUDING PISTON OR ROD TO BORE CONCENTRICITY)

psi	70 Hardness	80 Hardness	90 Hardness	psi	70 Hardness	80 Hardness	90 Hardness
0	0.010	0.010	0.010	1500	0.003	0.005	0.008
250	0.010	0.010	0.010	2000	—	0.004	0.005
500	0.008	0.010	0.010	3000	—	0.003	0.004
1000	0.005	0.008	0.010	5000	—	—	0.003

NOTE: Data in bold face indicates basis for establishment of Table 2.

FIG. 1—O-RING DESIGN EXAMPLES

TABLE 1—PHYSICAL REQUIREMENTS

	Class I Oil Service		Class II Fuel Service			Class I Oil Service		Class II Fuel Service	
	ASTM Specimen	Actual Part	ASTM Specimen	Actual Part		ASTM Specimen	Actual Part	ASTM Specimen	Actual Part
(a) Original Physical Properties					Elongation, max, % change	−40	−40	−25	−25
Hardness points	70 ±5	—	70 ±5	—	Volume change, max, %	±5	±5	±3	±3
Tensile strength, min, psi	1500	1200	1500	1200	Dry-out volume change, max, % (4 h at 158 F)	—	—	−6	−6
Ultimate elongation, min, %	200	150	200	150					
Ultimate elongation, max, %	400	400	400	400	(e) Fluid Aging, 70 h at __F				
Modulus at 100 elongation, min, psi	450	400	450	400	Type of ASTM Fluid	302 No. 3 Oil	302 No. 3 Oil	Room Temp Fuel B	Room Temp Fuel B
(b) Heat Aging, 70 h at __F	257	257	212	212	Hardness, points change	0 to −15	—	−20	−20
Hardness, points change	+10	—	+10	—	Tensile strength, max, % change	−30	−30	−50	−50
Tensile strength, max, % change	−15	−15	−15	−15	Elongation, max, % change	−40	−40	−50	−50
Elongation, max, % change	−50	−50	−50	−50	Volume change, max, %	+20	+20	+25	+25
(c) Compression Set, 70 h at __F	257	257	212	212	Dry-out volume change, max, % (4 h at 158 F)	—	—	−10	−10
ASTM specimen, max, %	35	—	25	—					
0.200 to 0.300 in, max, %	—	45	—	35	(f) Fluid Aging, 70 h at __F				
0.100 to 0.200 in, max, %	—	65	—	55	Type of ASTM Fluid			Room Temp Fuel C	Room Temp Fuel C
(d) Fluid Aging, 70 h at __F	302	302	Room Temp	Room Temp	Volume change, max, %			+35	+35
Type of ASTM Fluid	No. 1 Oil	No. 1 Oil	Fuel A	Fuel A	(g) Low Temperature Aging				
Hardness, points change	−5 to +8	—	0 to −8	—	Brittleness at −20 F	No cracks	—	No cracks	—
Tensile strength, max, % change	−15	−15	−15	−15	Fig. 8 bend, 5 h at −40 F	—	No cracks	—	No cracks

NOTE: The limits in this table for actual parts are for parts 0.100 in cross section and larger. If smaller cross section parts are tested, special limits shall be agreed upon between the purchaser and the supplier.

6. Methods of Test—The properties enumerated in these specifications shall be determined in accordance with the following methods of the American Society for Testing and Materials:

(a) *Hardness*—The hardness reading shall be taken before and after aging by means of a durometer in accordance with ASTM D 676, Tentative Method of Test for Indentation of Rubber by Means of a Durometer. Take hardness readings on sheet stock only.

(b) *Tensile Strength, Elongation, and Modulus*—The tensile strength, elongation, and modulus of sheet stock shall be taken in accordance with ASTM D 412, Tentative Method of Tension Testing of Vulcanized Rubber, and for seals in accordance with ASTM D 1414, Tentative Method for Tension Testing of Rubber O-Rings. For testing the molded parts, pull the entire ring, using ring spool tester attachments on the testing machine. An exception is made in paragraph (d) on Fluid Aging.

(c) *Heat Aging*—Heat aging tests shall be conducted in accordance with ASTM D 573, Method of Test for Accelerated Aging of Vulcanized Rubber by the Oven Method. If ASTM D 865, Method of Heat Aging of Vulcanized Natural or Synthetic Rubber by Test Tube Method, is used, it shall be referenced in the report.

(d) *Fluid Aging*—The fluid immersion tests shall be conducted in accordance with ASTM D 471, Tentative Method of Test for Change in Properties of Elastomeric Vulcanizates Resulting from Immersion in Liquids. When oil aging is to be performed on rings of a diameter greater than the width of the aging container, cut segments from the rings. This is to eliminate stress cracking which may occur where a large ring is bent, as in a test tube. When segments are used for aged data, use them for original property determinations. The dry-out volume after 4 h at 158 F is based on the original unaged volume of the sample.

(e) *Compression Set*—The compression set shall be conducted in accordance with Method B of ASTM D 395, Methods of Test for Compression Set of Vulcanized Rubber. Express compression set as a percentage of the original deflection. The specimen may be either $\frac{1}{2}$ in. disc-molded or plied up from sheet stock, or O-rings 0.100 in. in nominal cross section or larger, or both.

(f) *Low Temperature Aging*—Low temperature evaluation shall be made on sheet stock by ASTM D 746, Tentative Method of Test for Brittleness Temperature of Plastics and Elastomers by Impact. Molded parts shall be subjected to a 180 deg twist forming a figure eight while at the required temperature. This is to be accomplished by the use of suitable hooks or tongs, preconditioned at the testing temperature.

7. Identification—When requested by the purchaser, O-rings shall be marked with a nonpermanent colored stripe approximately $\frac{1}{16}$ in. wide and approximately 45 deg of the seal circumference of $\frac{1}{2}$ in., whichever is smaller. The color of the stripe shall be as follows:

Oil resistant service—yellow
Fuel resistant service—red

120R2-RECTANGULAR SECTION RUBBER SEAL RINGS FOR AUTOMOTIVE APPLICATIONS

1. Scope

(a) This SAE Recommended Practice covers the dimensional and material requirements of rectangular section rubber seal rings for use in automotive sealing applications.

(b) The materials and groove dimensions covered in these specifications are recommended for static applications only in which the fluid pressures do not exceed 1500 psi. For higher pressures, either extended service tests should be performed or the suppliers should be contacted for recommendations. Rectangular section seals should never be used in rotary or oscillating applications and are not generally intended for use as dynamic seals, although they have performed satisfactorily in some dynamic applications. It is advised that extensive service tests be conducted before rectangular section rings are released for dynamic uses.

2. Basis of Purchase—To specify the rectangular section seal desired, each engineering drawing shall list the SAE specification number, material classification, and size number covering the proper size ring. The letter "R" in the size number will designate a rectangular section seal ring.

EXAMPLE: SAE Specification No. 120R2
Class I
Size R128

3. Detail Requirements

(a) Refer to 120R1 of SAE J120 for material identification, physical properties, and test methods. It is intended that the same material shall be used for both O-rings and rectangular section seal rings for each material classification.

(b) For applications requiring resistance to special fluids, numerous compounds are available. For these applications the suppliers should be contacted for their compound recommendations. (See tables in SAE J200 for designation).

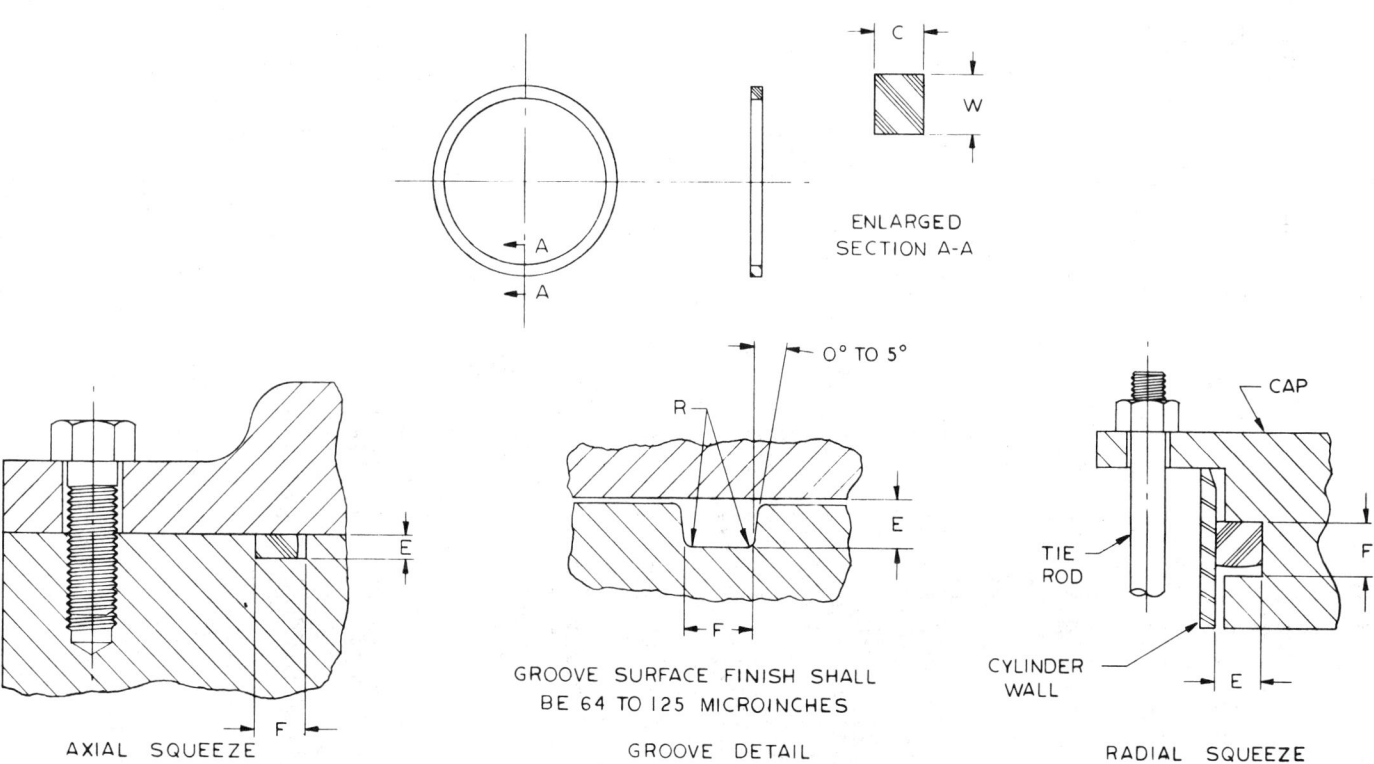

FIG. 2—TYPICAL DESIGN EXAMPLES FOR RECTANGULAR SECTION SEAL RINGS

TABLE 2—DIMENSIONS AND TOLERANCES, IN

Size No.	Actual Size ID	Actual Size OD (Ref)	Dynamic Seals A Cylinder Bore Dia	Dynamic Seals B Piston Groove Dia	Dynamic Seals C Rod Dia	Dynamic Seals D Rod Gland Groove Dia
Tolerance	±0.005	—	+0.002 −0.000	+0.000 −0.002	+0.000 −0.002	+0.002 −0.000
006	0.114	0.254	0.250	0.140	0.125	0.235
007	0.145	0.285	0.281	0.171	0.156	0.266
008	0.176	0.316	0.312	0.202	0.188	0.298
009	0.208	0.348	0.343	0.233	0.219	0.329
010	0.239	0.379	0.375	0.265	0.250	0.360
011	0.301	0.441	0.437	0.327	0.312	0.422
012	0.364	0.504	0.500	0.390	0.375	0.485
013	0.426	0.566	0.563	0.453	0.437	0.547
014	0.489	0.629	0.625	0.515	0.500	0.610
015	0.551	0.691	0.688	0.578	0.562	0.672
016	0.614	0.754	0.750	0.640	0.625	0.735
017	0.676	0.816	0.812	0.702	0.688	0.798
018	0.739	0.879	0.875	0.765	0.750	0.860
Tolerance	±0.006	—	+0.002 −0.000	+0.000 −0.002	+0.000 −0.002	+0.002 −0.000
019	0.801	0.941	0.937	0.827	0.812	0.922
020	0.864	1.004	1.000	0.890	0.875	0.985
021	0.926	1.066	1.062	0.952	0.937	1.047
022	0.989	1.129	1.125	1.015	1.000	1.110
023	1.051	1.191	1.188	1.078	1.062	1.172
024	1.114	1.254	1.250	1.140	1.125	1.235
025	1.176	1.316	1.312	1.202	1.188	1.298
026	1.239	1.379	1.375	1.265	1.250	1.360
027	1.301	1.441	1.437	1.327	1.312	1.422
028	1.364	1.504	1.500	1.390	1.375	1.485
Tolerance	±0.010	—	+0.002 −0.000	+0.000 −0.002	+0.000 −0.002	+0.002 −0.000
029	1.489	1.629	1.625	1.515	1.500	1.610
030	1.614	1.754	1.750	1.640	1.625	1.735
031	1.739	1.879	1.875	1.765	1.750	1.860
032	1.864	2.004	2.000	1.890	1.875	1.985
033	1.989	2.129	2.125	2.015	2.000	2.110
034	2.114	2.254	2.250	2.140	2.125	2.235
035	2.239	2.379	2.375	2.265	2.250	2.360
036	2.364	2.504	2.500	2.390	2.375	2.485
037	2.489	2.629	2.625	2.515	2.500	2.610
038	2.614	2.754	2.750	2.640	2.625	2.735
Tolerance	±0.015	—	+0.002 −0.000	+0.000 −0.002	+0.000 −0.002	+0.002 −0.000
039	2.739	2.879	2.875	2.765	2.750	2.860
040	2.864	3.004	3.000	2.890	2.875	2.985
041	2.989	3.129	3.125	3.015	3.000	3.110
042	3.239	3.379	3.375	3.265	3.250	3.360
043	3.489	3.629	3.625	3.515	3.500	3.610
044	3.739	3.879	3.875	3.765	3.750	3.860
045	3.989	4.129	4.125	4.015	4.000	4.110

W—Actual seal width 0.070 ±0.003 in
E—Groove width, in:
Dynamic—0.055–0.057
Static:
Radial—0.049–0.055
Axial—0.045–0.050
F—Groove length, in—0.090–0.100
R—Radius, in—0.005–0.015
Cross-sectional squeeze, in, min:
Dynamic—0.010
Static:
Radial—0.012
Axial—0.017
Piston groove diameter shall be concentric with piston OD within 0.002 total indicator reading and rod gland groove diameter shall be concentric with rod bore diameter within 0.002 total indicator reading.
NOTE: Sizes 013 to 045 are not generally used for dynamic seals and are not ordinarily recommended for such service. In special cases where space or other factors dictate the desirability of small cross-sections, these sizes should be incorporated into a design only after a careful engineering analysis and thorough testing.

Size No.	Actual Size ID	Actual Size OD (Ref)	Dynamic Seals A Cylinder Bore Dia	Dynamic Seals B Piston Groove Dia	Dynamic Seals C Rod Dia	Dynamic Seals D Rod Gland Groove Dia
Tolerance	±0.005	—	+0.003 −0.000	+0.000 −0.003	+0.000 −0.003	+0.003 −0.000
110	0.362	0.568	0.562	0.388	0.375	0.549
111	0.424	0.630	0.625	0.451	0.437	0.611
112	0.487	0.693	0.688	0.514	0.500	0.674
113	0.549	0.755	0.750	0.576	0.562	0.736
114	0.612	0.818	0.812	0.638	0.625	0.799
115	0.674	0.880	0.875	0.701	0.688	0.862
116	0.737	0.943	0.937	0.763	0.750	0.924
Tolerance	±0.006	—	+0.003 −0.000	+0.000 −0.003	+0.000 −0.003	+0.003 −0.000
117	0.799	1.005	1.000	0.826	0.812	0.986
118	0.862	1.068	1.062	0.888	0.875	1.049
119	0.924	1.130	1.125	0.951	0.937	1.111
120	0.987	1.193	1.188	1.014	1.000	1.174
121	1.049	1.255	1.250	1.076	1.062	1.236
122	1.112	1.318	1.312	1.138	1.125	1.299
123	1.174	1.380	1.375	1.201	1.188	1.362
124	1.237	1.443	1.437	1.263	1.250	1.424
125	1.299	1.505	1.500	1.326	1.312	1.486
126	1.362	1.568	1.562	1.389	1.375	1.549
127	1.424	1.630	1.625	1.451	1.437	1.611
128	1.487	1.693	1.688	1.514	1.500	1.674
Tolerance	±0.010	—	+0.003 −0.000	+0.000 −0.003	+0.000 −0.003	+0.003 −0.000
129	1.549	1.755	1.750	1.576	1.562	1.736
130	1.612	1.818	1.812	1.638	1.625	1.799
131	1.674	1.880	1.875	1.701	1.688	1.862
132	1.737	1.943	1.937	1.763	1.750	1.924
133	1.799	2.005	2.000	1.826	1.812	1.986
134	1.862	2.068	2.062	1.888	1.875	2.049
135	1.925	2.131	2.125	1.951	1.937	2.111
136	1.987	2.193	2.188	2.014	2.000	2.174
137	2.050	2.256	2.250	2.076	2.062	2.236
138	2.112	2.318	2.312	2.138	2.125	2.299
139	2.175	2.381	2.375	2.201	2.188	2.362
140	2.237	2.443	2.437	2.263	2.250	2.424
141	2.300	2.506	2.500	2.326	2.312	2.486
142	2.362	2.568	2.563	2.389	2.375	2.549
143	2.425	2.631	2.625	2.451	2.437	2.611
144	2.487	2.693	2.688	2.514	2.500	2.674
145	2.550	2.756	2.750	2.576	2.562	2.736
146	2.612	2.818	2.812	2.638	2.625	2.799
Tolerance	±0.015	—	+0.003 −0.000	+0.000 −0.003	+0.000 −0.003	+0.003 −0.000
147	2.675	2.881	2.875	2.701	2.688	2.862
148	2.737	2.943	2.937	2.763	2.750	2.924
149	2.800	3.006	3.000	2.826	2.812	2.986
150	2.862	3.068	3.062	2.888	2.875	3.049
151	2.987	3.193	3.188	3.014	3.000	3.174
152	3.237	3.442	3.437	3.263	3.250	3.424
153	3.487	3.693	3.688	3.514	3.500	3.674
154	3.737	3.943	3.937	3.763	3.750	3.924
155	3.987	4.193	4.188	4.014	4.000	4.174
156	4.237	4.443	4.437	4.263	4.250	4.424
157	4.487	4.693	4.687	4.514	4.500	4.674
158	4.737	4.943	4.937	4.763	4.750	4.924
159	4.987	5.193	5.188	5.014	5.000	5.174
Tolerance	±0.023	—	+0.003 −0.000	+0.000 −0.003	+0.000 −0.003	+0.003 −0.000
160	5.237	5.442	5.437	5.263	5.250	5.424
161	5.487	5.693	5.688	5.514	5.500	5.674
162	5.737	5.943	5.937	5.763	5.750	5.924
163	5.987	6.193	6.188	6.014	6.000	6.174

W—Actual seal width 0.103 ±0.003 in
E—Groove width, in:
Dynamic—0.087–0.090
Static:
Radial—0.080–0.086
Axial—0.072–0.080
F—Groove length, in—0.140–0.150
R—Radius, in—0.005–0.020
Cross-sectional squeeze, in, min:
Dynamic—0.010
Static:
Radial—0.014
Axial—0.020
Piston groove diameter shall be concentric with piston OD within 0.003 total indicator reading and rod gland groove diameter shall be concentric with rod bore diameter within 0.003 total indicator reading.
NOTE: Sizes 117 to 163 are not generally used for dynamic seals and are not ordinarily recommended for such service. In special cases where space or other factors dictate the desirability of small cross-sections, these sizes should be incorporated into a design only after a careful engineering analysis and thorough testing.

Size No.	Actual Size ID	Actual Size OD (Ref)	Dynamic Seals A Cylinder Bore Dia	Dynamic Seals B Piston Groove Dia	Dynamic Seals C Rod Dia	Dynamic Seals D Rod Gland Groove Dia
Tolerance	±0.006	—	+0.004 −0.000	+0.000 −0.004	+0.000 −0.004	+0.004 −0.000
210	0.734	1.012	1.000	0.762	0.750	0.988
211	0.796	1.074	1.062	0.824	0.812	1.050
212	0.859	1.137	1.125	0.887	0.870	1.113
213	0.921	1.199	1.188	0.950	0.937	1.175
214	0.984	1.262	1.250	1.012	1.000	1.238
215	1.046	1.324	1.312	1.075	1.062	1.300
216	1.109	1.387	1.375	1.137	1.125	1.363
217	1.171	1.449	1.437	1.199	1.188	1.426
218	1.234	1.512	1.500	1.262	1.250	1.488
219	1.296	1.574	1.562	1.324	1.312	1.550
220	1.359	1.637	1.625	1.387	1.375	1.613
221	1.421	1.699	1.688	1.450	1.437	1.675
222	1.484	1.762	1.750	1.512	1.500	1.738
Tolerance	±0.010	—	+0.004 −0.000	+0.000 −0.004	+0.000 −0.004	+0.004 −0.000
223	1.609	1.887	1.875	1.637	1.625	1.863
224	1.734	2.012	2.000	1.762	1.750	1.988
225	1.859	2.137	2.125	1.887	1.875	2.113
226	1.984	2.262	2.250	2.012	2.000	2.238
227	2.109	2.287	2.375	2.137	2.125	2.363
228	2.234	2.512	2.500	2.262	2.250	2.488
229	2.359	2.637	2.625	2.387	2.375	2.613
230	2.484	2.762	2.750	2.512	2.500	2.738
231	2.609	2.887	2.875	2.637	2.625	2.863
Tolerance	±0.015	—	+0.004 −0.000	+0.000 −0.004	+0.000 −0.004	+0.004 −0.000
232	2.734	3.012	3.000	2.762	2.750	2.988
233	2.859	3.137	3.125	2.887	2.875	3.113
234	2.984	3.262	3.250	3.012	3.000	3.238
235	3.109	3.387	3.375	3.137	3.125	3.363
236	3.234	3.512	3.500	3.262	3.250	3.488
237	3.359	3.637	3.625	3.387	3.375	3.613
238	3.484	3.762	3.750	3.512	3.500	3.738
239	3.609	3.887	3.875	3.637	3.625	3.863
240	3.734	4.012	4.000	3.762	3.750	3.988
241	3.859	4.137	4.125	3.887	3.875	4.113
242	3.984	4.262	4.250	4.012	4.000	4.238
243	4.109	4.387	4.375	4.137	4.125	4.363
244	4.234	4.512	4.500	4.262	4.250	4.488
245	4.359	4.637	4.625	4.387	4.375	4.613
246	4.484	4.762	4.750	4.512	4.500	4.738
247	4.609	4.887	4.875	4.637	4.625	4.863
248	4.734	5.012	5.000	4.762	4.750	4.988
249	4.859	5.137	5.125	4.887	4.875	5.113
250	4.984	5.262	5.250	5.012	5.000	5.238
Tolerance	±0.023	—	+0.004 −0.000	+0.000 −0.004	+0.000 −0.004	+0.004 −0.000
251	5.109	5.387	5.375	5.137	5.125	5.363
252	5.234	5.512	5.500	5.262	5.250	5.488
253	5.359	5.637	5.625	5.385	5.375	5.613
254	5.484	5.762	5.750	5.512	5.500	5.730
255	5.609	5.887	5.875	5.637	5.625	5.863
256	5.734	6.012	6.000	5.762	5.750	5.988
257	5.859	6.137	6.125	5.887	5.875	6.113
258	5.984	6.262	6.250	2.012	6.000	6.238
259	6.234	6.512	6.500	6.262	6.250	6.488
260	6.484	6.762	6.750	6.512	6.500	6.738
261	6.734	7.012	7.000	6.762	6.750	6.988
262	6.984	7.262	7.250	7.012	7.000	7.238
Tolerance	±0.030	—	+0.004 −0.000	+0.000 −0.004	+0.000 −0.004	+0.004 −0.000
263	7.234	7.512	7.500	7.262	7.250	7.488
264	7.484	7.762	7.750	7.512	7.500	7.738
265	7.734	8.012	8.000	7.762	7.750	7.988
266	7.984	8.262	8.250	8.012	8.000	8.238
267	8.234	8.512	8.500	8.262	8.250	8.488
268	8.484	8.762	8.750	8.512	8.500	8.738

W—Actual seal width 0.139 ±0.004 in
E—Groove width, in:
Dynamic—0.119–0.123
Static:
Radial—0.112–0.118
Axial—0.100–0.110
F—Groove length, in—0.180–0.190
R—Radius, in—0.005–0.030 in
Cross-sectional squeeze, in, min:
Dynamic—0.012
Static:
Radial—0.017
Axial—0.025
Piston groove diameter shall be concentric with piston OD within 0.004 total indicator reading and rod gland groove diameter shall be concentric with rod bore diameter within 0.004 total indicator reading.
NOTE: Sizes 223 to 281 are not generally used for dynamic seals and are not ordinarily recommended for such service. In special cases where space or other factors dictate the desirability of small cross-sections, these sizes should be incorporated into a design only after a careful engineering analysis and thorough testing.

(Table 2 continued on next page)

TABLE 2 — DIMENSIONS AND TOLERANCES, IN (CONTINUED)

Size No.	Actual Size ID	Actual Size OD (Ref)	Dynamic Seals A Cylinder Bore Dia	Dynamic Seals B Piston Groove Dia	Dynamic Seals C Rod Dia	Dynamic Seals D Rod Gland Groove Dia	
Tolerance	±0.030	—	+0.004 / −0.000	+0.000 / −0.004	+0.000 / −0.004	+0.004 / −0.000	
269	8.734	9.012	9.000	8.762	8.750	8.988	
270	8.984	9.262	9.250	9.012	9.000	9.238	
271	9.234	9.512	9.500	9.262	9.250	9.488	
272	9.484	9.762	9.750	9.512	9.500	9.738	
273	9.734	10.012	10.000	9.762	9.750	9.988	
274	9.984	10.262	10.250	10.012	10.000	10.238	
Tolerance	±0.040	—	+0.004 / −0.000	+0.000 / −0.004	+0.000 / −0.004	+0.004 / −0.000	
275	10.484	10.762	10.750	10.512	10.500	10.738	
276	10.984	11.262	11.250	11.012	11.000	11.238	
277	11.484	11.762	11.750	11.512	11.500	11.738	
278	11.984	12.262	12.250	12.012	12.000	12.238	
Tolerance	±0.050	—	+0.004 / −0.000	+0.000 / −0.004	+0.000 / −0.004	+0.004 / −0.000	
279	12.984	13.262	13.250	13.012	13.000	13.238	
280	13.984	14.262	14.250	14.012	14.000	14.238	
Tolerance	±0.060	—	+0.004 / −0.000	+0.000 / −0.004	+0.000 / −0.004	+0.004 / −0.000	
281	14.984	15.262	15.250	15.012	15.000	15.238	
Tolerance	±0.010	—	+0.005 / −0.000	+0.000 / −0.005	+0.000 / −0.005	+0.005 / −0.000	W—Actual seal width 0.210 ±0.005 in E—Groove width, in: Dynamic— 0.183–0.188 Static: Radial— 0.176–0.184 Axial— 0.165–0.175 F—Groove length, in— 0.280–0.290 R—Radius, in— 0.005–0.050 Cross-sectional squeeze, in, min: Dynamic—0.017 Static: Radial—0.021 Axial—0.030 Piston groove diameter shall be concentric with piston OD within 0.006 total indicator reading and rod gland groove diameter shall be concentric with rod bore diameter within 0.006 total indicator reading.
325	1.475	1.895	1.875	1.509	1.500	1.866	
326	1.600	2.020	2.000	1.634	1.625	1.991	
327	1.725	2.145	2.125	1.759	1.750	2.116	
328	1.850	2.270	2.250	1.884	1.875	2.241	
329	1.975	2.395	2.375	2.009	2.000	2.366	
330	2.100	2.520	2.500	2.134	2.125	2.491	
331	2.225	2.645	2.625	2.259	2.250	2.616	
332	2.350	2.770	2.750	2.384	2.375	2.741	
333	2.475	2.895	2.875	2.509	2.500	2.866	
334	2.600	3.020	3.000	2.634	2.625	2.991	
335	2.725	3.145	3.125	2.759	2.750	3.116	
336	2.850	3.270	3.250	2.884	2.875	3.241	
337	2.975	3.395	3.375	3.009	3.000	3.366	
338	3.100	3.520	3.500	3.134	3.125	3.491	
339	3.225	3.645	3.625	3.259	3.250	3.616	
340	3.350	3.770	3.750	3.384	3.375	3.741	
341	3.475	3.985	3.875	3.509	3.500	3.866	
342	3.600	4.020	4.000	3.634	3.625	3.991	
343	3.725	4.145	4.125	3.759	3.750	4.116	
344	3.850	4.270	4.250	3.884	3.875	4.241	
345	3.975	4.395	4.375	4.009	4.000	4.366	
346	4.100	4.520	4.500	4.134	4.125	4.491	
347	4.225	4.645	4.625	4.259	4.250	4.616	
348	4.350	4.770	4.750	4.384	4.375	4.741	
349	4.475	4.895	4.875	4.509	4.500	4.866	
Tolerance	±0.015	—	+0.006 / −0.000	+0.000 / −0.006	+0.000 / −0.006	+0.006 / −0.000	W—Actual seal width 0.275 ±0.006 in E—Groove width, in: Dynamic— 0.234–0.240 Static: Radial— 0.225–0.235 Axial— 0.220–0.230 F—Groove length, in— 0.370–0.380 R—Radius, in— 0.005–0.060 Cross-sectional squeeze, in, min: Dynamic—0.029 Static: Radial—0.034 Axial—0.039 Piston groove diameter shall be concentric with piston OD within 0.008 total indicator reading and rod gland groove diameter shall be concentric with rod bore diameter within 0.008 total indicator reading.
425	4.475	5.025	5.000	4.532	4.500	4.968	
426	4.600	5.150	5.125	4.657	4.625	5.093	
427	4.725	5.275	5.250	4.782	4.750	5.218	
428	4.850	5.400	5.375	4.907	4.875	5.343	
429	4.975	5.525	5.500	5.032	5.000	5.468	
Tolerance	±0.023	—	+0.006 / −0.000	+0.000 / −0.006	+0.000 / −0.006	+0.006 / −0.000	
430	5.100	5.650	5.625	5.157	5.125	5.593	
431	5.225	5.775	5.750	5.282	5.250	5.718	
432	5.350	5.900	5.875	5.407	5.375	5.843	
433	5.475	6.025	6.000	5.532	5.500	5.968	
434	5.600	6.150	6.125	5.657	5.625	6.093	
435	5.725	6.275	6.250	5.782	5.750	6.218	
436	5.850	6.400	6.375	5.907	5.875	6.343	
437	5.975	6.525	6.500	6.032	6.000	6.468	
438	6.225	6.775	6.750	6.282	6.250	6.718	
439	6.475	7.025	7.000	6.532	6.500	6.986	
Tolerance	±0.015	—	+0.006 / −0.000	+0.000 / −0.006	+0.000 / −0.006	+0.006 / −0.000	
440	6.725	7.275	7.250	6.782	6.750	7.218	
441	6.975	7.525	7.500	7.032	7.000	7.468	
442	7.225	7.775	7.750	7.282	7.250	7.718	
443	7.475	8.025	8.000	7.532	7.500	7.968	
444	7.725	8.275	8.250	7.782	7.750	8.218	
445	7.975	8.525	8.500	8.032	8.000	8.468	
446	8.475	9.025	9.000	8.532	8.500	8.968	
447	8.975	9.525	9.500	9.032	9.000	9.468	
448	9.475	10.025	10.000	9.532	9.500	9.968	
449	9.975	10.525	10.500	10.032	10.000	10.468	
450	10.475	11.025	11.000	10.532	10.500	10.968	
451	10.975	11.525	11.500	11.032	11.000	11.468	
452	11.475	12.025	12.000	11.532	11.500	11.968	
453	11.975	12.525	12.500	12.032	12.000	12.468	
454	12.475	13.025	13.000	12.532	12.500	12.968	
455	12.975	13.525	13.500	13.032	13.000	13.468	
456	13.475	14.025	14.000	13.532	13.500	13.968	
457	13.975	14.525	14.500	14.032	14.000	14.468	
458	14.475	15.025	15.000	14.532	14.500	14.968	
459	14.975	15.525	15.500	15.032	15.000	15.468	
460	15.475	16.025	16.000	15.532	15.500	15.968	

TABLE 3 — GROOVES FOR RECTANGULAR SECTION SEAL RINGS

Size No.	Nominal ID	Nominal OD	W	Actual ID	Actual OD (Ref)	See Fig. 2, for identification of dimensions W, C, E, F, and R
Tolerance, ±0.005						
R006	1/8	1/4	1/16	0.114	0.246	W—Actual seal width 0.066 ±0.004 in C—Actual seal length 0.066 ±0.003 in E—Groove width 0.055 to 0.057 in F—Groove length 0.090 to 0.100 in R—Radius 0.005 to 0.015 in Cross-sectional squeeze, 0.005 in, min
R007	5/32	9/32	1/16	0.145	0.277	
R008	3/16	5/16	1/16	0.176	0.308	
R009	7/32	11/32	1/16	0.208	0.340	
R010	1/4	3/8	1/16	0.239	0.371	
R011	5/16	7/16	1/16	0.301	0.433	
R012	3/8	1/2	1/16	0.364	0.496	
R013	7/16	9/16	1/16	0.426	0.558	
R014	1/2	5/8	1/16	0.489	0.621	
R015	9/16	11/16	1/16	0.551	0.683	
R016	5/8	3/4	1/16	0.614	0.746	
R017	11/16	13/16	1/16	0.676	0.808	
R018	3/4	7/8	1/16	0.739	0.871	
Tolerance, ±0.006						
R019	13/16	15/16	1/16	0.801	0.933	
R020	7/8	1	1/16	0.864	0.996	
R021	15/16	1-1/16	1/16	0.926	1.058	
R022	1	1-1/8	1/16	0.989	1.121	
R023	1-1/16	1-3/16	1/16	1.051	1.183	
R024	1-1/8	1-1/4	1/16	1.114	1.246	
R025	1-3/16	1-5/16	1/16	1.176	1.308	
R026	1-1/4	1-3/8	1/16	1.239	1.371	
R027	1-5/16	1-7/16	1/16	1.301	1.433	
R028	1-3/8	1-1/2	1/16	1.364	1.496	
Tolerance, ±0.010						
R029	1-1/4	1-5/8	1/16	1.489	1.621	
R030	1-5/8	1-3/4	1/16	1.614	1.746	
R031	1-3/4	1-7/8	1/16	1.739	1.871	
R032	1-7/8	2	1/16	1.864	1.996	
R033	2	2-1/8	1/16	1.989	2.121	
R034	2-1/8	2-1/4	1/16	2.114	2.246	
R035	2-1/4	2-3/8	1/16	2.239	2.371	
R036	2-3/8	2-1/2	1/16	2.364	2.496	
R037	2-1/2	2-5/8	1/16	2.489	2.621	
R038	2-5/8	2-3/4	1/16	2.614	2.746	
Tolerance, ±0.015						
R039	2-3/4	2-7/8	1/16	2.739	2.871	
R040	2-7/8	3	1/16	2.864	2.996	
R041	3	3-1/8	1/16	2.989	3.121	
R042	3-1/4	3-3/8	1/16	3.239	3.371	
R043	3-1/2	3-5/8	1/16	3.489	3.621	
R044	3-3/4	3-7/8	1/16	3.739	3.871	
R045	4	4-1/8	1/16	3.989	4.121	
Tolerance, ±0.005						
R110	3/8	9/16	3/32	0.362	0.560	W—Actual seal width 0.099 ±0.004 in C—Actual seal length 0.099 ±0.003 in E—Groove width 0.087 to 0.090 in F—Groove length 0.140 to 0.150 in R—Radius 0.005 to 0.015 in Cross-sectional squeeze, 0.005 in, min
R111	7/16	5/8	3/32	0.424	0.622	
R112	1/2	11/16	3/32	0.487	0.685	
R113	9/16	3/4	3/32	0.549	0.747	
R114	5/8	13/16	3/32	0.612	0.810	
R115	11/16	7/8	3/32	0.674	0.872	
R116	3/4	15/16	3/32	0.737	0.935	
Tolerance, ±0.006						
R117	13/16	1	3/32	0.799	0.997	
R118	7/8	1-1/16	3/32	0.862	1.060	
R119	15/16	1-1/8	3/32	0.924	1.122	
R120	1	1-3/16	3/32	0.987	1.185	
R121	1-1/16	1-1/4	3/32	1.049	1.247	
R122	1-1/8	1-5/16	3/32	1.112	1.310	
R123	1-3/16	1-3/8	3/32	1.174	1.372	
R124	1-1/4	1-7/16	3/32	1.237	1.435	
R125	1-5/16	1-1/2	3/32	1.299	1.497	
R126	1-3/8	1-9/16	3/32	1.362	1.560	
R127	1-7/16	1-5/8	3/32	1.424	1.622	
R128	1-1/2	1-11/16	3/32	1.487	1.685	
Tolerance, ±0.010						
R129	1-9/16	1-3/4	3/32	1.549	1.747	
R130	1-5/8	1-13/16	3/32	1.612	1.820	
R131	1-11/16	1-7/8	3/32	1.674	1.872	
R132	1-3/4	1-15/16	3/32	1.737	1.935	
R133	1-13/16	2	3/32	1.799	1.997	
R134	1-7/8	2-1/16	3/32	1.862	2.060	
R135	1-15/16	2-1/8	3/32	1.925	2.123	
R136	2	2-3/16	3/32	1.987	2.185	
R137	2-1/16	2-1/4	3/32	2.050	2.248	
R138	2-1/8	2-5/16	3/32	2.112	2.310	
R139	2-3/16	2-3/8	3/32	2.175	2.373	

(Table 3 continued on next page)

TABLE 3 — GROOVES FOR RECTANGULAR SECTION SEAL RINGS (CONTINUED)

Size No.	Nominal ID	Nominal OD	W	Actual ID	Actual OD (Ref)
Tolerance, ±0.010					
R140	2-1/4	2-7/16	3/32	2.237	2.435
R141	2-5/16	2-1/2	3/32	2.300	2.498
R142	2-3/8	2-9/16	3/32	2.362	2.560
R143	2-7/16	2-5/8	3/32	2.425	2.623
R144	2-1/2	2-11/16	3/32	2.487	2.685
R145	2-9/16	2-3/4	3/32	2.550	2.748
R146	2-5/8	2-13/16	3/32	2.612	2.810
Tolerance, ±0.015					
R147	2-11/16	2-7/8	3/32	2.675	2.873
R148	2-3/4	2-15/16	3/32	2.737	2.935
R149	2-13/16	3	3/32	2.800	2.998
R150	2-7/8	3-1/16	3/32	2.862	3.060
R151	3	3-3/16	3/32	2.987	3.185
R152	3-1/4	3-7/16	3/32	3.237	3.435
R153	3-1/2	3-11/16	3/32	3.487	3.685
R154	3-3/4	3-15/16	3/32	3.737	3.935
R155	4	4-3/16	3/32	3.987	4.185
R156	4-1/4	4-7/16	3/32	4.237	4.435
R157	4-1/2	4-11/16	3/32	4.487	4.685
R158	4-3/4	4-15/16	3/32	4.737	4.935
R159	5	5-3/16	3/32	4.987	5.185
Tolerance, ±0.023					
R160	5-1/4	5-7/16	3/32	5.237	5.435
R161	5-1/2	5-11/16	3/32	5.487	5.685
R162	5-3/4	5-15/16	3/32	5.737	5.935
R163	6	6-3/16	3/32	5.987	6.185
Tolerance, ±0.006					
R210	3/4	1	1/8	0.734	1.002
R211	13/16	1-1/16	1/8	0.796	1.064
R212	7/8	1-1/8	1/8	0.859	1.127
R213	15/16	1-3/16	1/8	0.921	1.189
R214	1	1-1/4	1/8	0.984	1.252
R215	1-1/16	1-5/16	1/8	1.046	1.314
R216	1-1/8	1-3/8	1/8	1.109	1.377
R217	1-3/16	1-7/16	1/8	1.171	1.439
R218	1-1/4	1-1/2	1/8	1.234	1.502
R219	1-5/16	1-9/16	1/8	1.296	1.564
R220	1-3/8	1-5/8	1/8	1.359	1.627
R221	1-7/16	1-11/16	1/8	1.421	1.689
R222	1-1/2	1-3/4	1/8	1.484	1.752
Tolerance, ±0.010					
R223	1-5/8	1-7/8	1/8	1.609	1.877
R224	1-3/4	2	1/8	1.734	2.002
R225	1-7/8	2-1/8	1/8	1.859	2.127
R226	2	2-1/4	1/8	1.984	2.252
R227	2-1/8	2-3/8	1/8	2.109	2.377
R228	2-1/4	2-1/2	1/8	2.234	2.502
R229	2-3/8	2-5/8	1/8	2.359	2.627
R230	2-1/2	2-3/4	1/8	2.484	2.752
R231	2-5/8	2-7/8	1/8	2.609	2.877
Tolerance, ±0.015					
R232	2-3/4	3	1/8	2.734	3.002
R233	2-7/8	3-1/8	1/8	2.859	3.127
R234	3	3-1/4	1/8	2.984	3.252
R235	3-1/8	3-3/8	1/8	3.109	3.337
R236	3-1/4	3-1/2	1/8	3.234	3.502
R237	3-3/8	3-5/8	1/8	3.359	3.627
R238	3-1/2	3-3/4	1/8	3.484	3.752
R239	3-5/8	3-7/8	1/8	3.609	3.877
R240	3-3/4	4	1/8	3.734	4.002
R241	3-7/8	4-1/8	1/8	3.859	4.127
R242	4	4-1/4	1/8	3.984	4.252
R243	4-1/8	4-3/8	1/8	4.109	4.377
R244	4-1/4	4-1/2	1/8	4.234	4.502
R245	4-3/8	4-5/8	1/8	4.359	4.627
R246	4-1/2	4-3/4	1/8	4.484	4.752
R247	4-5/8	4-7/8	1/8	4.609	4.877
R248	4-3/4	5	1/8	4.734	5.002
R249	4-7/8	5-1/8	1/8	4.859	5.127
R250	5	5-1/4	1/8	4.984	5.252
Tolerance, ±0.023					
R251	5-1/8	5-3/8	1/8	5.109	5.377
R252	5-1/4	5-1/2	1/8	5.234	5.502
R253	5-3/8	5-5/8	1/8	5.359	5.627
R254	5-1/2	5-3/4	1/8	5.484	5.752
R255	5-5/8	5-7/8	1/8	5.609	5.877
R256	5-3/4	6	1/8	5.734	6.002
R257	5-7/8	6-1/8	1/8	5.859	6.127
R258	6	6-1/4	1/8	5.984	6.252
R259	6-1/4	6-1/2	1/8	6.234	6.502
R260	6-1/2	6-3/4	1/8	6.484	6.752
R261	6-3/4	7	1/8	6.734	7.002
R262	7	7-1/4	1/8	6.984	7.252
R263	7-1/4	7-1/2	1/8	7.234	7.502
R264	7-1/2	7-3/4	1/8	7.484	7.752
R265	7-3/4	8	1/8	7.734	8.002
R266	8	8-1/4	1/8	7.984	8.252
R267	8-1/4	8-1/2	1/8	8.234	8.502
R268	8-1/2	8-3/4	1/8	8.484	8.752
R269	8-3/4	9	1/8	8.734	9.002
R270	9	9-1/4	1/8	8.984	9.252
R271	9-1/4	9-1/2	1/8	9.234	9.502
R272	9-1/2	9-3/4	1/8	9.484	9.752
R273	9-3/4	10	1/8	9.734	10.002
R274	10	10-1/4	1/8	9.984	10.252
Tolerance, ±0.040					
R275	10-1/2	10-3/4	1/8	10.484	10.752
R276	11	11-1/4	1/8	10.984	11.252
R277	11-1/2	11-3/4	1/8	11.484	11.752
R278	12	12-1/4	1/8	11.984	12.252
Tolerance, ±0.050					
R279	13	13-1/4	1/8	12.984	13.252
R280	14	14-1/4	1/8	13.984	14.252
Tolerance, ±0.060					
R281	15	15-1/4	1/8	14.984	15.252
Tolerance, ±0.010					
R325	1-1/2	1-7/8	3/16	1.475	1.881
R326	1-5/8	2	3/16	1.600	2.006
R327	1-3/4	2-1/8	3/16	1.725	2.131
R328	1-7/8	2-1/4	3/16	1.850	2.256
R329	2	2-3/8	3/16	1.975	2.381
R330	2-1/8	2-1/2	3/16	2.100	2.506
R331	2-1/4	2-5/8	3/16	2.225	2.631
R332	2-3/8	2-3/4	3/16	2.350	2.756
R333	2-1/2	2-7/8	3/16	2.475	2.881
R334	2-5/8	3	3/16	2.600	3.006
Tolerance, ±0.015					
R335	2-3/4	3-1/8	3/16	2.725	3.131
R336	2-7/8	3-1/4	3/16	2.850	3.256
R337	3	3-3/8	3/16	2.975	3.381
R338	3-1/8	3-1/2	3/16	3.100	3.506
R339	3-1/4	3-5/8	3/16	3.225	3.631
R340	3-3/8	3-3/4	3/16	3.350	3.756
R341	3-1/2	3-7/8	3/16	3.475	3.881
R342	3-5/8	4	3/16	3.600	4.006
R343	3-3/4	4-1/8	3/16	3.725	4.131
R344	3-7/8	4-1/4	3/16	3.850	4.256
R345	4	4-3/8	3/16	3.975	4.381
R346	4-1/8	4-1/2	3/16	4.100	4.506
R347	4-1/4	4-5/8	3/16	4.225	4.631
R348	4-3/8	4-3/4	3/16	4.350	4.756
R349	4-1/2	4-7/8	3/16	4.475	4.881
Tolerance, ±0.015					
R425	4-1/2	5	1/4	4.475	5.005
R426	4-5/8	5-1/8	1/4	4.600	5.130
R427	4-3/4	5-1/4	1/4	4.725	5.255
R428	4-7/8	5-3/8	1/4	4.850	5.380
R429	5	5-1/2	1/4	4.975	5.505
Tolerance, ±0.023					
R430	5-1/8	5-5/8	1/4	5.100	5.630
R431	5-1/4	5-3/4	1/4	5.225	5.755
R432	5-3/8	5-7/8	1/4	5.350	5.880
R433	5-1/2	6	1/4	5.475	6.005

See Fig. 2, for identification of dimensions W, C, E, F, and R

W — Actual seal width
   0.134 ±0.004 in
C — Actual seal length
   0.134 ±0.004 in
E — Groove width
   0.119 to 0.123 in
F — Groove length
   0.180 to 0.190 in
R — Radius 0.005 to 0.030 in
Cross-sectional squeeze
   0.007 in, min

W — Actual seal width
   0.203 ±0.005 in
C — Actual seal length
   0.203 ±0.005 in
E — Groove width
   0.183 to 0.188 in
F — Groove length
   0.280 to 0.290 in
R — Radius 0.005 to 0.050 in
Cross-sectional squeeze
   0.010, in, min

W — Actual seal width
   0.265 ±0.005 in
C — Actual seal length
   0.265 ±0.005 in
E — Groove width
   0.234 to 0.240 in
F — Groove length
   0.370 to 0.380 in
R — Radius 0.005 to 0.060 in
Cross-sectional squeeze
   0.020 in, min

(Table 3 continued on next page)

TABLE 3—GROOVES FOR RECTANGULAR SECTION SEAL RINGS (CONTINUED)

Size No.	Size, in. Nominal ID	Nominal OD	W	Actual ID	Actual OD (Ref)		Size No.	Size, in. Nominal ID	Nominal OD	W	Actual ID	Actual OD (Ref)
						See Fig. 2, for identification of dimensions W, C, E, F, and R						
\multicolumn{6}{c}{Tolerance, ±0.023}		\multicolumn{6}{c}{Tolerance, ±0.030}
R434	5-5/8	6-1/8	1/4	5.600	6.130		R445	8	8-1/2	1/4	7.975	8.505
R435	5-3/4	6-1/4	1/4	5.725	6.255		R446	8-1/2	9	1/4	8.475	9.005
R436	5-7/8	6-3/8	1/4	5.850	6.380		R447	9	9-1/2	1/4	8.975	9.505
R437	6	6-1/2	1/4	5.975	6.505		R448	9-1/2	10	1/4	9.475	10.005
R438	6-1/4	6-3/4	1/4	6.225	6.755		R449	10	10-1/2	1/4	9.975	10.505
R439	6-1/2	7	1/4	6.475	7.005		R450	10-1/2	11	1/4	10.475	11.005
R440	6-3/4	7-1/4	1/4	6.725	7.255		R451	11	11-1/2	1/4	10.975	11.505
R441	7	7-1/2	1/4	6.975	7.505		R452	11-1/2	12	1/4	11.475	12.005
\multicolumn{6}{c}{Tolerance, ±0.030}		R453	12	12-1/2	1/4	11.975	12.505
R442	7-1/4	7-3/4	1/4	7.225	7.755		R454	12-1/2	13	1/4	12.475	13.005
R443	7-1/2	8	1/4	7.475	8.005		R455	13	13-1/2	1/4	12.975	13.505
R444	7-3/4	8-1/4	1/4	7.725	8.255		R456	13-1/2	14	1/4	13.475	14.005
							R457	14	14-1/2	1/4	13.975	14.505
							R458	14-1/2	15	1/4	14.475	15.005
							R459	15	15-1/2	1/4	14.975	15.505
							R460	15-1/2	16	1/4	15.475	16.005

ENGINE COOLANTS—SAE J814c

SAE Information Report

Report of the Nonmetallic Materials Committee approved March 1962 and completely revised October 1978.

1. Scope

1.1 This report is intended as a source of information concerning the basic properties of engine coolants which are satisfactory for use in internal combustion engines to provide corrosion protection, lower the freezing point, and raise the boiling point. For additional information on engine coolants see SAE J1034a, Engine Coolant Concentrate—Ethylene Glycol Type.

1.2 The values which are presented describe desirable basic properties. The results from laboratory corrosion tests are not conclusive, and it should be recognized that the final selection of satisfactory coolants can be accomplished only after a series of performance tests in vehicles.

1.3 The report also describes in general terms the maintenance procedures which should be applied to insure a properly functioning cooling system. It is not intended to cover maintenance of engine cooling systems and components which are discussed in detail in SAE HS 40, Maintenance of Automotive Engine Cooling Systems.

2. Types of Coolants

2.1 *Water*—Water has been the most commonly used constituent of engine coolants for internal combustion engines because it has the ability to transfer heat and can be readily obtained. Some properties of water, such as its boiling point and freezing point, limit its usefulness when used alone as a coolant. The natural corrosive action of water on metals is definitely undesirable. Some natural water impurities, such as sulfates, chlorides, and bicarbonates, can increase corrosion. Others, such as calcium and magnesium carbonate, reduce heat transfer by the formation of scale, particularly at hot spots. They can also contribute to radiator clogging if excessive additions of hard water are made to replenish coolant losses.

When water freezes, it forms solid ice and expands approximately 9% in volume. If water is allowed to freeze inside the cooling system, the resultant extreme pressure can cause serious damage. In order to prevent coolant freeze damage, an antifreeze compound must be added to the water. Because of the inherent properties of water, the increasing heat loads placed on the cooling system, and the design factors of modern engines, water alone or water with inhibitors is not recommended as an engine coolant.

2.2 *Antifreeze Compounds—Coolant Concentrates*

2.2.1 Water containing the proper amount of antifreeze will not cause freeze-cracking damage from expansion and can be circulated freely in the cooling system at temperatures lower than the freezing point of water. There are many requirements for an acceptable antifreeze. The most essential of these are:

1. The ability to lower the freezing point of water to the lowest winter operating temperatures likely to be encountered.
2. The ability to protect the cooling system metals from corrosion or deposits.
3. A minimum of undesirable effects on engine cooling and heat transfer.
4. No deleterious effect on rubber.
5. Chemical stability.
6. Low cost.
7. Little or no odor.
8. Minimum effect on automobile finishes.
9. An acceptable viscosity at low temperatures.
10. Low coefficient of expansion.
11. Usability for at least one year of service.
12. Easily checked freezing point.

In addition, low toxicity, suitable boiling point characteristics, low foaming and operating losses, and nonflammability are desirable. No one chemical meets each of these requirements to the fullest. However, there are materials which represent satisfactory compromises.

2.2.2 ETHYLENE GLYCOL BASE COOLANTS—The most commonly used antifreeze material is ethylene glycol. A typical formulated glycol base antifreeze coolant will contain 85% min ethylene glycol, corrosion inhibitors, up to 5% total water, a dye, and an antifoam agent. Occasionally, up to 10% of other glycols, such as diethylene glycol and propylene glycol are also used.

When compared to water alone, solutions of glycol base coolant in water have higher boiling points, lower freezing points, and slightly lower heat transfer characteristics. See Table 1.

Automobile manufacturers fill the cooling systems of cars at the factory year-round and across the country with approximately a 50% concentration of ethylene glycol water base antifreeze coolant. The higher boiling points of ethylene glycol water solutions have been found to be beneficial for hot weather operation and in high altitude areas. Glycol base coolant solutions should be used with thermostats having opening temperatures of 180°F (82°C) or higher. When a glycol base coolant is used, it is recommended that approximately a 50% concentration be maintained year-round to provide adequate corrosion protection and a sufficiently high boiling point. Concentrations over 70% result in a loss in freezing protection and an increase in viscosity at low temperature and therefore should be avoided.

2.2.3 ALCOHOL BASE COOLANTS—Alcohol base coolants constitute an extremely small part of the coolant market. Methyl alcohol (methanol) is gener-

TABLE 1

Vol % Antifreeze	Freezing Point °F	Freezing Point °C	Boiling Point[a] °F	Boiling Point[a] °C
40	−12	−24	222	106
50	−34	−37	226	108
60	−62	−52	232	111
70	−84	−64	238	114

[a] At 760 mm pressure (atmospheric). A higher boiling point is obtained by using a radiator pressure cap which permits the development of pressure within the cooling system.

Note: The above data represent industry standards that necessitate a minimum glycol content in the concentrated antifreeze.

ally used as the base in these coolants. Although alcohol water solutions have lower freezing points than water, their boiling points are also lower than that of water. Because late model cars are designed to use glycol base coolants, and the engine overheating warning devices are keyed to the boiling point of this coolant, methanol base products are not recommended by vehicle manufacturers.

2.2.4 OTHER ANTIFREEZE COOLANTS—Various other chemicals are used as antifreeze coolants in special applications.

2.2.4.1 An ethylene glycol, water and glycol ether combination is sometimes used for arctic service. This type of coolant is suitable for service down to $-90°F$ ($-68°C$).

2.2.4.2 Propylene glycol is not as effective a freeze depressant as ethylene glycol, and because its specific gravity is very close to that of water, it is very difficult to obtain a satisfactory field check for concentration by hydrometer.

2.2.4.3 Inhibited glycol ether, specifically methoxy propanol, is available for use as a coolant for heavy-duty equipment. It is not as effective a freeze depressant as ethylene glycol, nor can the concentration be checked with a standard antifreeze hydrometer. It does not have the same sludge forming properties as ethylene glycol coolant solutions in the event of accidental leakage into the engine crankcase, but as with any aqueous coolant, chronic leakage can adversely affect engine service life and lubrication efficiency.

2.2.4.4 All specialty coolant products should be used with strict adherence to the engine manufacturer's and coolant manufacturer's recommendations.

2.2.5 Do not mix different types of coolants.

3. *Properties*—Properties which are considered in testing and evaluating the concentrated antifreze coolant and its solutions are described in the following paragraphs. The ASTM test methods for measuring these properties are listed below:

D 1119, Ash Content of Engine Antifreezes, Antirusts, and Coolants.

D 1120, Standard Method of Test for Boiling Point of Engine Coolants.

D 1121, Reserve Alkalinity of Engine Antifreezes, Antirusts, and Coolants.

D 1122, Specific Gravity of Engine Antifreezes by Hydrometer.

D 1177, Standard Method of Test for Freezing Point of Aqueous Engine Coolant Solution.

D 1287, pH of Engine Antifreezes, Antirusts, and Coolants.

D 1881, Foaming Tendencies of Engine Coolants in Glassware.

D 1882, Effect of Antifreeze and Cooling System Chemical Solutions on Organic Finishes for Automotive Vehicles.

3.1 Ash Content—Ash content of an antifreeze coolant is the residue which remains after ignition. For most formulations, the ash content will be less than 5.0% by weight. While the ash results from the inhibitors used, it is not a measure of inhibitor concentration.

3.2 Equilibrium Boiling Point—Equilibrium boiling point indicates the temperatures at which the coolant begins to boil in a cooling system under equilibrium conditions at atmospheric pressure. The boiling point of a coolant is an important property, especially when high opening temperature thermostats are used or when the cooling system is operated at high ambient temperatures or under high heat load conditions. See Appendix, Table A-2.

3.3 Reserve Alkalinity—Reserve alkalinity is defined as the number of milliliters, to the nearest 0.1 mL of 0.100 N hydrochloric acid required for the titration to a pH of 5.5 of a 10 mL sample of undiluted antifreeze, antirust, or coolant additive.

Mildly alkaline pH solutions are generally less corrosive than strongly acid or strongly alkaline solutions. Neither pH nor reserve alkalinity is a sufficient criterion to indicate the efficiency of corrosion inhibitors. Different inhibitor systems have individual optimum pH ranges. Many of the very efficient corrosion inhibitors have very little effect on the reserve alkalinity.

The reserve alkalinity is most useful as part of a qualification test to indicate that: (1) A submitted lot is similar to a previously qualified sample, or (2) as a measure of the degradation of a coolant in service or under test. In the latter case, the rate of change of the reserve alkalinity is more important than the absolute value. The measurement gives an indication of the capacity of the system to neutralize acids which may be present or form in the system.

3.4 Specific Gravity—Specific gravity is the ratio of the weight of a given volume of liquid at 60°F (15.5°C) to the weight of an equal volume of gas-free distilled water at 60°F (15.5°C). Its measurement offers a convenient means of identifying a liquid or of determining the degree of dilution and hence the freezing point. While the specific gravities for pure methanol and ethylene glycol are 0.793 and 1.115 at 68/68°F (20/20°C), respectively, the specific gravities of the concentrated antifreeze, using either as a base, will be somewhat different depending upon the type and the amount of inhibitors used, on the amount of water present, and, in the case of ethylene glycol, on the amount of other glycols present.

3.5 Freezing Point—Freezing point is defined as the temperature at which ice crystallization begins in the absence of supercooling, or the maximum temperature reached immediately after initial ice crystal formation in the case of supercooling. The ASTM method of freezing point determination should be used when freezing point accuracy is desired, as in determining the limiting values for specifications. Hydrometers or refractometers are used in the field to determine the freezing point of coolants. (The field testers will give erroneous results for ethylene glycol based coolants if the content of other glycols, such as diethylene glycol or propylene glycol is too high.) Because of the limited accuracy of hydrometer-thermometer testers, the results should be viewed as an approximate freezing point rather than a precise measurement. (ASTM D 1124 describes the hydrometer-thermometer field tester for engine coolants). Freezing points can be determined more accurately by the refractometer type field tester. (ASTM D 3321, Use of the Refractometer for Determining the Freezing Point of Aqueous Engine Coolants, describes the use of a refractometer for determining freezing points.) See Appendix, Table A-1.

3.6 pH—pH is a measurement of the hydrogen ion concentration and indicates whether a coolant is acid, neutral, or alkaline. pH measurements are sometimes used for production quality control, but they are not significant from the standpoint of predicting service life. The pH of a used coolant is not a dependable indication of either existing effectiveness or remaining life of the solution.

3.7 Foaming Tendency—Foaming tendency of a coolant is measured by the amount of foam generated during aeration under controlled conditions and the time required for this foam to subside. (If the coolant foams excessively and the system is open, coolant losses may take place through the overflow tube of the radiator.) Foaming tendency of the coolant may be tested in the laboratory according to the ASTM method. To provide satisfactory performance, the foam volume in this test is normally less than 150 mL. The time required for the foam to subside sufficiently for a portion of the liquid surface to be seen is normally 5 s or less.

3.8 Organic Finishes—Organic finishes shall not be adversely affected by the coolant. This is of particular importance to the vehicle manufacturer during assembly operations and to the car owner during coolant installation in the event of accidental spillage. Most coolant concentrates contain a dye as a positive visual identification that an antifreeze coolant is being used. It has been customary to identify methanol base concentrates with a violet dye. Many other colors are used for glycol base antifreeze coolants. Green through blue green is recommended in SAE J1034.

3.9 Effect on Nonmetallics—The coolant solution should not accelerate failure of the radiator hose, gaskets, and nonmetallic coatings on metallic gaskets. Many immersion tests, as well as other tests, have been proposed for this evaluation. Each supplier and consumer has his favorite procedure. The final evaluation, of course, is service experience. Preliminary indications are obtained when the coolant is tested for corrosion inhibition in simulated service and engine dynamometer tests.

3.10 Storage Stability—Storage stability of the concentrate cannot be determined conclusively by accelerated tests. However, it is evident that the packaged concentrate must be stable for at least two years under many different climatic conditions.

4. Corrosion Inhibition

4.1 Corrosion—If corrosion of the cooling system of an internal combustion engine is allowed to proceed without interruption, it not only shortens the life of metallic components but it also effectively decreases the operational characteristics of the coolant—that is, the transfer of heat from the engine to the air where it can be dissipated. For these reasons effective corrosion inhibition must be provided and maintained in the cooling system.

4.2 Corrosion Testing—Because of the elapsed time involved in the field testing of inhibitors and inhibited antifreeze coolants, accelerated tests are used to determine the quality of these products. The results from accelerated tests can be indicative of quality only if the tests incorporate many of the factors which affect corrosion in the cooling system. Some of the more important factors to be included are:

1. Coolant.
2. Flow.
3. Aeration.
4. Temperature.
5. Water quality.
6. Galvanic couples.
7. Corrosion products.
8. Hot spots.

As a test incorporates more of these factors, it more nearly simulates service performance, but it usually requires more labor and it becomes more costly. The generally accepted order of evaluation tests is:

1. Screening test (glassware corrosion test).
2. Simulated service test.
3. Dynamometer test.
4. Field service test.

Special tests are required to evaluate the performance of coolants with regard to specific forms of corrosive attack. These stepwise procedures are used to avoid the unnecessary expenditure of time and money on obviously poor materials, and to ensure that better materials will meet service requirements by the use of more rigorous test conditions. Vehicle service tests are desirable as the final evaluation method because of the difficulty in reproducing all service variables in accelerated tests.

4.2.1 GLASSWARE TESTS—A glassware test procedure can be used to evaluate all types of antifreeze coolants and inhibitors. The advantages of this type of test are its simplicity and brevity of operation (usually about two weeks). For these reasons, it is easy to screen a large number of samples with minimum of effort.

Weighed specimens of metals common to the engine cooling system are immersed in a heated, aerated test solution for the entire period of the test. Corrosion products are removed from each metal at the end of the test, and the metal weight losses determined. Corrosion inhibition is evaluated on the basis of these metal weight loss values. Weight losses in a properly inhibited solution should be only a few milligrams at most. It is a common practice to establish weight loss limits based upon the performace of well inhibited solutions. A material is presumed to fail the test if the weight loss of any metal is above the limit.

A typical glassware procedure for the evaluation of coolants is that found in ASTM D 1384, Corrosion Test for Engine Coolants in Glassware.

4.2.2 SIMULATED SERVICE TESTS—This test concept involves the circulation of coolant, at a preselected operating temperature, in a rig simulating an engine cooling system. Several automobile parts are generally used in the construction of equipment including a coolant pump, coolant outlet, radiator, and radiator hoses. By the use of a large reservoir, or an engine block, a volume of coolant equivalent to that in a cooling system can be used. Coolant flow rates within the normal operating range are achieved by driving the coolant pump with an electric motor. This test simulates the engine cooling system more closely than the glassware test and is more discriminating in coolant performance evaluation. Corrosion inhibition is measured from weight losses of metal specimens placed in the system and by visual examination of the components. If standard radiator hoses are used, coolant effects on the hoses can be observed. By sampling the coolant at intervals during the test, coolant concentration can be controlled and solution properties can be monitored. Experience with this test method has shown that it is a useful development tool, but for some specimen metals, repeatability between tests may not be consistent and reproducibility among different laboratories may vary widely. The use of components of different materials, such as an aluminum reservoir versus a cast iron one or an aluminum radiator versus a brass one, can significantly affect certain metal specimen weight losses. In view of these circumstances, it is desirable to gain experience by running multiple tests with coolants of known service performance.

Although there are many variations in procedure, a recommended test procedure has been developed under the sponsorship of ASTM Committee D-15. This method is listed as ASTM D 2570, Simulated Service Corrosion Testing of Engine Coolants.

4.2.3. DYNAMOMETER TESTS—Inhibited engine coolants can be evaluated by an engine dynamometer test procedure. The advantage of this method over either the glassware test or the simulated service test is that test conditions such as load, speed, coolant temperature, and heat transfer are obtained and varied through the operation of a standard automobile engine. The performance of the engine is monitored throughout the test to ensure the reproduction of the proper service conditions. This type of test is more significant if the radiator, pump, and engine combination is the same as that normally used in a single vehicle, since the flow rate can affect the continued performance of the coolant.

Corrosion protection is determined by the measurement of weight losses of metal specimens in contact with the coolant, and by the inspection of parts at the conclusion of the test. Inhibitor stability is determined from changes in pH, reserve alkalinity, and solution appearance. Quantitative analysis may be obtained for the inhibitors known to be present. Although this test is expensive to run, it provides more meaningful results than other laboratory tests.

ASTM D 2758, Method of Testing Engine Coolants by Engine Dynamometer, prescribes a 700 h test in a standard passenger car engine.

4.2.4. FIELD TESTS—The final test of any coolant is its performance in cars in the field. Closely controlled tests can be made under the supervision of technical personnel. General field tests can be conducted in a large number of cars to provide a broad service pattern. Metal specimens may be installed in the coolant stream to determine corrosion rates. The coolant can be sampled periodically to determine its chemical and physical properties. The condition of the cooling system and components should be examined visually at the end of the test. Information should be obtained from the participants concerning their observations regarding cooling system performance. If at all possible, vehicles from different areas of the country should be utilized for the tests, which will involve a range of antifreeze coolant concentrations with a variety of local waters.

ASTM D 2847 is a standard recommended practice for testing engine coolants in vehicle service. Metal corrosion specimens, mounted in special capsules, are installed in the coolant flow of the test vehicles. The test duration in terms of time or mileage is consistent with the recommended service life of the coolant under test.

4.2.5 CAVITATION TESTS—Cavitation erosion corrosion of water pumps constructed to aluminum can be a serious problem because of the rate at which damage may occur. This type of corrosion is usually the result of pump design and cooling system characteristics. At certain coolant velocities, low and high pressure areas develop that cause the formation of vapor bubbles which implode on the surface of the metal and cause segments of metal to be removed. The composition of the coolant has been found to have a contributing effect on cavitation erosion, and it is often necessary to evaluate coolant formulations under cavitating conditions. Many different methods have been used to induce cavitation for laboratory studies, but it is preferable to use an operating pump under controlled conditions.

ASTM D 2966, Method of Test for Cavitation-Erosion Characteristics of Aluminum in Engine Coolants Using Ultrasonic Energy, is intended as a screening test. Aluminum specimens are totally immersed in a 15% by volume antifreeze coolant solution for 20 h at $180 \pm 3°F$ ($82 \pm 2°C$) in an ultrasonic tank. The capability of the coolant in preventing cavitation-erosion damage is determined by comparing the average weight loss incurred by the specimens in the test solution with that of specimens in a reference coolant solution.

Final evidence of satisfactory cavitation-erosion prevention should be confirmed by ASTM D 2809, Test for Cavitation Erosion Corrosion Characteristics of Aluminum Pumps with Engine Coolants. In this procedure, test coolant is pumpd through a pressurized, simulated automotive cooling system at a temperature of $235°F$ ($112°C$) for 100h. The pump is driven by an electric motor at a high speed and cavitation occurs. After the test, the pump is examined and rated by a numerical system.

4.2.6 ALUMINUM CORROSION TRANSPORT TESTS—Aluminum corrosion transport deposition may occur with engines containing aluminum cylinder heads or other aluminum heat rejecting (from metal to coolant) surfaces. While the metal loss from aluminum surfaces may be relatively small and have no effect on the strength or durability of the aluminum components, the volume of corrosion products deposited in coolant passages of radiator or heater cores will reduce heat transfer and may result in plugged passages. The presence of certain corrosion inhibitors in the antifreeze formulation can affect this type of corrosion.

Several test methods have been used to evaluate engine coolants for aluminum corrosion transport deposition including car tests. The accelerated laboratory tests involve circulating the test coolant in an aparatus so that the coolant is heated by an aluminum part and cooled with a copper, brass, or aluminum part. Performance is evaluated by the amount of corrosion (weight loss) from the heating aluminum part and/or the increase in weight of the cooling part. If the cooling part is a radiator or heater core, evaluation is based on loss in heat transfer or visually observing the quantity of deposits in the water passages.

5. Maintenance of Engine Coolants—Satisfactory performance of the coolants previously discussed depends upon the use of the proper coolant, periodic changes, coolant volume, coolant concentrate, cleanliness, and tightness of the engine cooling system.

5.1 Coolant Volume—It is important that the engine coolant be maintained at the level specified by the vehicle manufacturer. The proper level for most modern cars is shown in the coolant recovery tank or container which collects the overflow coolant when the coolant expands and from which the coolant is drawn back into the radiator when the coolant cools. The level marks are usually labeled *cold* and *hot* or *full cold* and *full hot* because the proper level is associated with the temperature of the coolant. In cars that do not have a coolant recovery system the level may be shown by a mark on the tank of the radiator. Periodic inspection of the coolant level in the recovery tank should be made to ensure that the coolant is at the proper level.

If overheating should occur even though the coolant level in the recovery tank indicates sufficent coolant, the level in the radiator should also be checked after the coolant has been cooled to ensure that the radiator is full. (Air can be trapped in the radiator, particularly after the system has been drained and filled, and the air may not be released through the recovery tank.) When it is necessary to add coolant to either the radiator or the recovery tank, add a 50–50 mixture of an ethylene glycol coolant concentrate and water or a coolant of the same type as that already in the system. Do not add water alone or an alcohol water mixture because neither one meets all the requirements for a satisfactory coolant; that is, adequate corrosion inhibition, proper freezing and boiling protection, and satisfactory heat transfer characteristics.

Loss of coolant may usually be attributed to one or more of the following:

5.1.1 OVERFLOWING, AFTER-BOILING, AND OVERHEATING—Overflowing is the loss of coolant during normal driving. The loss occurs because the system has been filled above the manufacturer's recommended level when the system was cold. As the coolant temperature increases, the coolant expands and is forced out of the coolant recovery tank, or radiator in the case of older model cars.

Proper coolant level minimizes the possibility for loss of coolant through overflow.

After-boiling is the boiling of the coolant in the engine block after the engine is stopped when the vehicle has been driven at high speeds or has been under a heavy load, such as pulling a trailer or climbing a hill. The residual heat in the engine causes the coolant temperature to rise above its boiling point, because the coolant is no longer being circulated and cooled.

Overheating may be defined as a condition where the coolant temperature exceeds the normal operating range as indicated by the coolant warning light or temperature gauge. When this occurs, it is best to reduce the load on the cooling system by shifting the transmission to neutral and turning off the air conditioner. It also helps to reduce coolant temperature by increasing the engine idle speed for 2 or 3 min to increase fan cooling and coolant pump output. Other factors can contribute such as thermostat, radiator cap, or fan belt malfunction, radiator tube plugging, or the accumulation of corrosion deposits or scale in the radiator that reduce heat transfer. In addition, the use of coolant solutions with lower boiling points than that provided by the recommended ethylene glycol water mixture can also be a factor in after-boiling or overheating. After-boiling or overheating can be minimized by ensuring that components are operating properly and that a satisfactory coolant is used and maintained.

5.1.2 COOLANT LEAKAGE—Leaks usually occur because of loose fitting parts or through cracks or pin holes caused by corrosion or deterioration of materials. Susceptible locations are hose connections, gasketed parts such as the cylinder head or thermostat, core hole plugs, pump seals, and radiator or heater core assemblies. When a leak occurs, the source should be identified and the leakage rate determined before attempting to stop the leak. The leak often can be corrected by tightening a clamp or bolt, but in many cases it may be necessary to replace the component part. Minor leaks may be sealed through the use of commercial stop-leak materials, but one is cautioned that this is never a substitute for mechanical repairs when the leak is sufficiently bad. Furthermore, overuse of stop-leak products contributes to the solids in the cooling system, which can affect heat transfer and tube plugging.

5.1.3 EXHAUST GAS LEAKAGE, AIR LEAKAGE, AND FOAMING—Air can be drawn into the coolant at loose hose connections or through the water pump. Exhaust gas leakage into the coolant occurs between the combustion chamber and the water jacket, usually because of a loose gasket or a cracked casting. Gases that enter the system become entrained in the coolant, occupying space and increasing the volume of the coolant. Gases can also rise to high points in the system as foam. Antifoam agents are usually added to the coolant concentrate to reduce the formation of foam, but these agents may be dissipated rather rapidly.

The increased volume occupied by the gas-liquid mixture can cause coolant losses during thermal expansion. In addition, the entrained gases can reduce the transfer of heat from the casting to the coolant and subsequently to the outside air through the radiator. Any of these effects can contribute to overheating.

Continued aeration of the system accelerates depletion of coolant inhibitors and can contribute to increased corrosion and/or erosion in the system. Air leakage to the coolant can be minimized by maintaining a proper coolant level and by good maintenance to ensure satisfactory operation of parts and tight joints.

Exhaust gases are normally acidic and leakage into the cooling system contributes not only to foaming and overheating but to inhibitor depletion, accelerated corrosion, and a more rapid breakdown of ethylene glycol. Any evidence of leakage requires immediate mechanical repair.

5.2 Antifreeze Coolant Concentration—Essentially all liquid cooled engines are designed for aqueous systems. Since water alone is inadequate (see paragraphs 2.1 and 2.2), an antifreeze coolant concentrate is usually added. Most automobile manufacturers fill the cooling system with approximately a 50% ethylene glycol base coolant. This concentration will supply more freeze protection than is required for some areas. A concentration of approximately 50% is recommended to ensure adequate concentration of corrosion inhibitors. The maximum freeze point depression with ethylene glycol is obtained with 68% concentrate. To achieve higher boiling points for use at higher ambient temperatures, under heavy loads, or at higher altitudes, a maximum of 70% ethylene glycol base coolant can be used. When replenishing the cooling system, the proper proportions of water and glycol should be added to maintain the desired ratio.

Should the coolant freeze, the automobile should not be operated until normal circulation is restored by a suitable thawing operation. The coolant can form a slush which will clog the radiator, resulting in overheating, loss of coolant and consequent damage to the engine such as cylinder head cracking at the combustion chamber.

If the coolant becomes overheated, the entire coolant solution must be cooled down before opening the radiator cap. If the cooling system is opened, the pressure will be reduced to atmospheric, coolant above its boiling point will flash to a vapor and force out the liquid coolant. The risk of personal injury is very great.

5.3 Coolant Replacement—The vehicle manufacturer's and/or coolant manufacturer's directions for periodic changes should be followed. Periodic replacement of coolant solutions is required because the solutions may become corrosive due to chemical reaction, contamination, or inhibitor depletion.

5.4 Cleaning—A properly inhibited, normally functioning cooling system should not require cleaning. Cooling systems should be cleaned only when indicated by the appearance of rust or other sediment in the solution or persistent overheating. The type of cleaner to be used is indicated by the condition of the system and the metal components.

Conventional cooling systems, with brass, copper, and cast iron used in major components, can be treated with available flush-type cleaners for oily deposits and acid-type cleaners or chelators for rust deposits. Cooling systems utilizing aluminum components should use only those cleaners specified as safe for aluminum. In general these cleaners are neither strong acids nor strong alkalies.

Residual cleaning components (including neutralizers) should be flushed from the system because some cleaners, if left in the cooling system, may attack components and shorten the service life of newly installed antifreeze coolants.

Clogged radiators will not respond to chemical treatment and will require repair or replacement.

APPENDIX

TABLE A-1—FREEZING POINTS OF SOLUTIONS OF COOLANT CONCENTRATE[a]

Cooling System Capacity, qt	Concentrate Required, qt														
	3	4	5	6	7	8	9	10	11	12	13	14	15	16	17
6	−34	−85													
7	−17	−54													
8		−34	−69												
9		−21	−50	−85											
10		−12	−34	−62	−84										
11			−23	−47	−75										
12			−15	−34	−57	−85									
13				−25	−45	−66	−86								
14				−17	−34	−54	−78								
15				−12	−26	−43	−62	−85							
16					−19	−34	−52	−69							
17					−14	−27	−42	−58	−79						
18						−21	−34	−50	−65	−85					
19	Bold numbers show the 50%				−16	−28	−42	−56	−74	−89				
20	solution of concentrate				−12	−22	−34	−48	−62	−80	−84			
21	and water.					−17	−28	−41	−54	−68	−85			
22						−14	−23	−34	−47	−59	−75	−92		
23							−19	−29	−40	−52	−64	−80	−84	
24							−15	−24	−34	−46	−57	−69	−85	
25							−12	−20	−29	−40	−50	−62	−75	−92

[a]Temperatures are shown in degrees Fahrenheit. To convert to degrees Celsius, $t_C = 5/9 (t_F − 32)$.

Note: 1 gal = 3.785 l.

TABLE A-2—BOILING POINTS OF SOLUTIONS OF COOLANT CONCENTRATIONS

Vol % Concentrate	Boiling Point			
	At Atmospheric Pressure (760 mm)		Using 14 lb (62N) Pressure Cap in Good Condition	
	°F	°C	°F	°C
40	222	105.5	259	126.1
50	226	107.8	263	128.3
60	232	111.1	268	131.1
70	238	114.4	274	134.4

ENGINE COOLANT CONCENTRATE—ETHYLENE-GLYCOL TYPE—SAE J1034

SAE Recommended Practice

Report of Nonmetallic Materials Committee approved June 1973.

Scope—This standard covers glycol-type compounds which, when added to engine cooling systems at concentrations of 40–70% by volume of coolant concentrate in water, provide corrosion protection, lower the freezing point, and raise the boiling point of the coolant. Such compounds are intended for a minimum of 1 year (approximately 12,000 miles) service in a properly maintained cooling system. (Reference: SAE HS-40, Maintenance of Automotive-Engine Cooling Systems.) Coolants meeting this standard do not require the use of supplementary materials. For additional information on engine coolants, see SAE J814.

Material—The base material shall be essentially ethylene-glycol. Other glycols such as propylene or diethylene-glycol may be incorporated if the physical and chemical properties mentioned below are met. Solutions of the coolant concentrate, when installed in a properly maintained cooling system, shall not adversely affect the normal fluid flow.

Freezing Points and Boiling Points—The engine coolant concentrate shall have a freezing point of no higher than 0°F (−18°C) and a boiling point of no lower than 300°F (149°C)[1].

A 50% solution of engine coolant concentrate in water shall have a freezing point no higher than −34°F (−37°C) and a boiling point no lower than 226°F (108°C)[1]. (See Appendix.)

ASTM D 1120, Standard Method of Test for Boiling Point of Engine Antifreezes, and D 1177, Standard Method of Test for Freezing Point of Aqueous Engine Antifreeze Solution, shall be used for these determinations.

Specific Gravity and Refractive Index—The coolant concentrate shall have a gravity of 1.115–1.145 at 60/60°F or 1.113–1.143 at 20/20°C when tested according to ASTM D 1122, Standard Method of Test for Specific Gravity of Engine Antifreezes by the Hydrometer. A mixture of one part concentrate to one part water shall have a refractive index of 1.3840–1.3855 at 20°C. These requirements permit checking the freezing protection of aqueous solutions with commercially available antifreeze testers of either the hydrometer or refractometer type.

Ash Content—The ash content of the concentrated product shall not exceed 5.0% by weight when determined by ASTM D 1119, Standard Method of Test for Ash Content of Engine Antifreezes and Antirusts.

pH Value—The pH of a solution of one part of concentrate to one part of distilled water by volume, as determined with a pH meter following the procedures outlined in ASTM D 1287, Standard Method of Test for pH of Engine Antifreezes and Antirusts, shall fall within the range of 7.5–11.0.

Reserve Alkalinity—The reserve alkalinity of the concentrated product shall not be less than 10 when determined by ASTM D 1121, Standard Method of Test for Reserve Alkalinity of Engine Antifreezes and Antirusts. NOTE: Reserve alkalinity in itself is not an adequate indication of the quality of an antifreeze product. It should not be used to measure a product's ability to prevent corrosion.

Compatibility with Cooling System Nonmetals—Solutions of the coolant concentrate as normally used in cooling systems shall not have deleterious effects on the nonmetallic components, as determined from examination of the nonmetallic components used in conjunction with the test outlined in the paragraph on corrosion inhibition. The hoses used in the test shall conform to SAE J20. After test, the tube of the coolant hose must meet the physical requirements of the coolant immersion test of SAE J20 (J20R4, Class D-1).

[1] At 760 mm pressure (atmospheric).

APPENDIX

TABLE A-1—FREEZING POINTS OF SOLUTIONS OF COOLANT CONCENTRATE[a]

Cooling System Capacity, qt	Concentrate Required, qt														
	3	4	5	6	7	8	9	10	11	12	13	14	15	16	17
6	−34	−85													
7	−17	−54													
8		**−34**	−69												
9		−21	−50	−85											
10		−12	**−34**	−62	−84										
11			−23	−47	−75										
12			−15	**−34**	−57	−85									
13				−25	−45	−66	−86								
14				−17	**−34**	−54	−78								
15				−12	−26	−43	−62	−85							
16					−19	**−34**	−52	−69							
17					−14	−27	−42	−58	−79						
18						−21	**−34**	−50	−65	−85					
19		Bold numbers show the 50%				−16	−28	−42	−56	−74	−89				
20		solution of concentrate				−12	−22	**−34**	−48	−62	−80	−84			
21		and water.					−17	−28	−41	−54	−68	−85			
22							−14	−23	**−34**	−47	−59	−75	−92		
23								−19	−29	−40	−52	−64	−80	−84	
24								−15	−24	**−34**	−46	−57	−69	−85	
25								−12	−20	−29	−40	−50	−62	−75	−92

[a] Temperatures are shown in degrees Fahrenheit. To convert to degress Celsius, $t_C = 5/9\,(t_F − 32)$.

Note: 1 gal = 3.785 l.

Foaming—The coolant concentrate when tested in accordance with ASTM D 1881, Standard Method of Test for Foaming Tendencies of Engine Antifreezes in Glassware, shall not have an increase in volume greater than 150 ml and the foam shall break within 5 s after aeration is stopped.

Dye—The coolant concentrate shall be distinctively colored (preferably green through blue green) with a stable dye.

Effect on Car Finishes—The coolant shall have no effect on standard original finishes used on automotive vehicles when evaluated by ASTM D 1882, Standard Method of Test for Effect of Antifreeze and Cooling System Chemical Solutions on Organic Finishes for Automotive Vehicles.

Storage Stability—The utility of the coolant concentrate as packaged shall not be adversely effected by storage for a minimum period of 1 year.

Corrosion Inhibition—The coolant concentrate shall be suitably inhibited to mitigate corrosion of the cooling system metals in a properly maintained system. The performance is indicated by the following weight loss limits per test specimen when tested according to ASTM D 2570, Standard Method for Simulated Service Corrosion Testing of Engine Coolants:

Copper	20 mg max
Solder	60 mg max
Brass	20 mg max
Steel	20 mg max
Cast iron	20 mg max
Cast aluminum	60 mg max

Cavitation and Erosion—There shall be no pitting, cavitation damage, or erosion of the water pump which will produce a rating below 8 when tested according to ASTM D 2809, Standard Method of Test for Cavitation Erosion-Corrosion Characteristics of Aluminum Automotive Water Pumps with Coolants.

TABLE A-2—BOILING POINTS OF SOLUTIONS OF COOLANT CONCENTRATIONS

Vol % Concentrate	Boiling Point			
	At Atmospheric Pressure (760 mm)		Using 14 lb (62N) Pressure Cap in Good Condition	
	°F	°C	°F	°C
40	222	105.5	259	126.1
50	226	107.8	263	128.3
60	232	111.1	268	131.1
70	238	114.4	274	134.4

COOLANT SYSTEM HOSES—SAE J20e   SAE Standard

Report of Nonmetallic Materials Division approved January 1944 and last revised by Nonmetallic Materials Committee January 1974.

[This specification was originally formulated in 1944 under the joint SAE-ASTM Technical Committee on Automotive Rubber to meet the critical shortage of rubber. It can be used as a guide in the selection of suitable materials for use in military and commercial vehicles and passenger cars. For the purposes of simplification, this SAE standard is divided into six parts as follows:
Part I (SAE 20R1)—Heavy-Duty Type—for service such as combat vehicles and heavy-duty trucks. This type is available in two wall thicknesses as indicated.
Part II (SAE 20R2)—Flexible Heavy-Duty Type—for the same service as Part I with helical wire built into the wall of the hose.
Part III (SAE 20R3)—Heater Hose—for standard service.
Part IV (SAE 20R4)—Normal Service Type—Knitted or Braided Hose.
Part V (SAE 20R5)—Normal Service Type—Flexible Wire Reinforced.
Part VI—Physical Test Procedures.

Compounds based on five different synthetic rubber grades are specified and designated Class A, high-temperature resistant; Class B, high oil swell resistant; Class C, medium oil swell resistant; Class D-1, low oil resistant, low compression set; Class D-2, low oil resistant, medium compression set. Compounds normally available for each type hose are shown in Table 3.

In accordance with customary procedure in adopting recommendations of this Technical Committee, these product specifications for coolant system hose have been approved by the SAE, and all test methods and requirements are specified in accordance with ASTM standards except as noted. While the specifications have been accepted as satisfactory for current requirements and practice, it should be borne in mind that they are subject to frequent change to meet new developments or requirements in the manufacture and use of coolant system hose.]

General Requirements—The following requirements are minimal product standards:

Initial Sample Approval—Acceptance testing for initial sample approval shall be based upon testing in accordance with the complete specification. The testing may be conducted in the customer's laboratory or in the supplier's laboratory with certified test results provided to the customer.

Dimensional and Visual Inspection—Dimensional and visual inspection shall be conducted by the vendor.

Quality Assurance Sampling and Testing—Production hose of each basic size, construction, and material will be sampled weekly and tested for original physical properties, including tests that are performed on the tube and cover and on the whole hose. The first occurrence of production in each month will be sampled and tested in accordance with the complete specification.

Upon failure of any characteristic, two samples will be selected at random for retesting from the same code date. In the event of failure on either of the two resamples, production of the same code date shall be isolated and suitable disposition instituted.

Measurement of both hose and parts thereof shall be made in accordance with ASTM D 380 unless otherwise specified.

All applicable requirements in this section on General Requirements and those in Part VI, Physical Test Procedures, shall be part of SAE 20R1, 20R2, 20R3, 20R4, and 20R5.

Workmanship and Finish—Workmanship and finish shall be in accordance with precision or commercial practice covering this class of work.

Packing, Marking, and Shipping—Details of packing, marking, and shipping are subject to individual arrangements between the consumer and producer.

Length Tolerance—Straight Hose—Unless otherwise specified, the length tolerances shall be as follows:

Length		Precision Tolerance		Commercial Tolerance	
in	m	in	mm	in	mm
0–12	0 –0.30	±0.12	± 3.18	+0.38, −0.12	+ 9.65, − 3.05
12–24	0.30–0.61	±0.19	± 4.76	+0.38, −0.19	+ 9.65, − 4.76
24–36	0.61–0.91	±0.25	± 6.35	+0.50, −0.25	+12.70, − 6.35
36–48	0.91–1.22	±0.38	± 9.65	+0.50, −0.38	+12.70, − 9.65
48–72	1.22–1.83	±0.50	±12.70	+0.75, −0.50	+19.05, −12.70
Over 72	1.83	±1%	±1%	±2%	±2%

Tolerances on arm lengths of SAE 20R4 hose measured from end to intersection of nearest centerline are: precision: +0.19, −0.12 in. (+4.76, −3.05 mm); commercial: +0.38, −0.12 in. (+9.65, −3.05 mm).

On curved SAE 20R3 hose, the tolerances on arm lengths, measured from end to intersection of nearest centerline, each end shall be as follows:

Arm Length		Precision Tolerance		Commercial Tolerance	
in	m	in	mm	in	mm
0–12	0 –0.30	+0.19, −0.12	+4.76, −3.18	+0.38, −0.12	+ 9.65, − 3.05
12–24	0.30–0.61	±0.19	± 4.76	+0.38, −0.19	+ 9.65, − 4.76
24–36	0.61–0.91	±0.25	± 6.35	+0.50, −0.25	+12.70, − 6.35
36–48	0.91–1.22	±0.38	± 9.65	+0.50, −0.38	+12.70, − 9.65
48–72	1.22–1.83	±0.50	±12.70	+0.75, −0.50	+19.05, −12.70
Over 72	1.83	±1%	±1%	±2%	±2%

General Layout Tolerances, Curved Hose—Dimensions locating bend intersections are to establish the theoretical centerline of hose. Actual outside contour of hose must be held within 0.19 in. (4.76 mm) of all planes with respect to theoretical outside contour of hose.

For hose check, hose ends should first be placed in nominal position before checking (hose may have to be flexed to correct for any distortion caused by handling or during shipment). SAE end length tolerances shall apply per type hose being checked.

When the ID of one end of the hose is enlarged, the wall gage of the enlarged end normally changes. Allowable change in wall gage permitted is +0.06, −0.02 in. (+1.59, −0.53 mm) from nominal. Enlarged ends should be considered arm lengths for tolerance purposes. Tolerances apply to all arm and body lengths in addition to contour tolerances. Dimensions covering more than one arm or body length are reference only and have no tolerance.

The wall gage within bends of a curved hose may differ from the wall gage of the straight portion. This difference shall not be more than 33% from nominal.

Squareness of Ends on Curved or Formed Parts—The tolerance on squareness of ends of curved or formed parts will be a maximum of +0.25 in. (+6.25 mm) on all hose in sizes up through 2 in. ID, and +0.38 in. (+9.65 mm) on all sizes over 2 in. ID. End squareness will be measured from a point projected from the short side to the inside of the long side by a line perpendicular to the wall of the hose, to the end of the hose at its longest point. End squareness measurements will be considered superimposed upon the arm length tolerance. The midpoint of deviation from squareness must fall within the arm length tolerance.

Finish on Connections—Users of coolant hose should take every precaution to obtain connections with as smooth a finish as possible.

Clamps—Recommended hose clamps are listed in SAE Standard J536.

PART I—HEAVY-DUTY TYPE (SAE 20R1)

Scope—This type of hose is primarily for severe, critical, and heavy-duty service of which the diesel-locomotive application is a typical example. The hose is intended to withstand the effects of pressure systems, various present types of coolants, cold temperatures down to −40°F (−40°C), and unskilled application to connections not particularly designed for ease of assembly with the hose.

TABLE 1—SAE 20R1, STANDARD WALL

Nominal Size, in	Burst		Tolerance				Vacuum Resistance	
			ID		Wall			
	psi	MPa	±in	±mm	in	mm	in Hg	kPa
3/8	475	3.29	0.03	0.80	0.17–0.22	4.32–5.59	10	33.8
1/2	425	2.93	0.03	0.80	0.17–0.22	4.32–5.59	10	33.8
5/8	375	2.59	0.03	0.80	0.17–0.22	4.32–5.59	8	27.0
3/4	325	2.24	0.03	0.80	0.17–0.22	4.32–5.59	8	27.0
1	300	2.06	0.03	0.80	0.17–0.22	4.32–5.59	7	23.6
1-1/8	300	2.06	0.03	0.80	0.17–0.22	4.32–5.59	6	20.3
1-1/4	275	1.90	0.03	0.80	0.17–0.22	4.32–5.59	5	16.9
1-1/2	250	1.72	0.03	0.80	0.17–0.22	4.32–5.59	3	10.1
1-3/4	225	1.55	0.03	0.80	0.17–0.22	4.32–5.59	1	3.4
2	200	1.38	0.06	1.59	0.17–0.22	4.32–5.59	—	—
2-1/4	175	1.21	0.06	1.59	0.17–0.22	4.32–5.59	—	—
2-1/2	150	1.03	0.06	1.59	0.17–0.22	4.32–5.59	—	—
2-3/4	125	0.86	0.06	1.59	0.17–0.22	4.32–5.59	—	—
3	100	0.69	0.06	1.59	0.17–0.22	4.32–5.59	—	—
3-1/2	75	0.52	0.06	1.59	0.17–0.22	4.32–5.59	—	—
4	50	0.34	0.06	1.59	0.17–0.22	4.32–5.59	—	—

TABLE 2—SAE 20R1, HEAVY WALL

Nominal Size, in	Burst		Tolerance				Vacuum Resistance	
			ID		Wall			
	psi	MPa	±in	±mm	in	mm	in Hg	kPa
1-1/4	500	3.45	0.03	0.80	0.23–0.28	5.84–7.11	10	33.8
1-1/2	450	3.10	0.03	0.80	0.23–0.28	5.84–7.11	10	33.8
1-3/4	400	2.76	0.03	0.80	0.23–0.28	5.84–7.11	5	16.9
2	350	2.41	0.06	1.59	0.23–0.28	5.84–7.11	3	10.1
2-1/4	350	2.41	0.06	1.59	0.23–0.28	5.84–7.11	1	3.4
2-1/2	300	2.06	0.06	1.59	0.23–0.28	5.84–7.11	—	—
2-3/4	250	1.72	0.06	1.59	0.23–0.28	5.84–7.11	—	—
3	250	1.72	0.06	1.59	0.23–0.28	5.84–7.11	—	—
3-1/2	200	1.38	0.06	1.59	0.23–0.28	5.84–7.11	—	—
4	150	1.03	0.06	1.59	0.23–0.28	5.84–7.11	—	—

When desired, hose with one class of material in the tube and another in the cover may be obtained. In such cases, the physical properties specified for respective parts shall apply. Adhesion requirements shall be based on the class of material for which the lower value is specified.

Construction

Tube—Minimum thickness shall be 0.06 in. (1.59 mm).

Reinforcement—The reinforcement shall be multiple plies of woven or cord fabric, or ply or plies of braided or knitted yarns and shall be such that the hose meets the minimum burst and vacuum requirements as given in Tables 1 and 2.

Cover—Minimum thickness shall be 0.03 in. (0.80 mm), measured as the thickness from the reinforcement out.

Dimensions—Dimensions and tolerances shall be as specified in Table 1 for light wall and Table 2 for heavy wall. These dimensions shall be measured at a section not including a lap.

Physical Tests—Tests shall be performed on sections and parts from the finished hose. The preparation of test samples shall be in accordance with ASTM D 380. See Table 3 and SAE J20, Part VI.

PART II—HEAVY-DUTY WIRE INSERTED TYPE (SAE 20R2)

Scope—This type of hose is primarily for severe, critical, and heavy-duty type of service of which diesel engine applications are examples. The hose is intended to withstand high vacuum with some forced curvature, the effects of pressure systems, various present types of coolants, cold temperatures down to −40°F (−40°C), and unskilled application to connections not particularly designed for ease of assembly with the hose. This is similar to Part I hose except that it embodies wire helix or helices built into the wall of the hose without soft ends or "cuffs."

Construction

Tube—Minimum thickness shall be 0.06 in. (1.59 mm).

Reinforcement—The reinforcement shall be multiple plies of woven or cord fabric, or ply or plies of braided or knitted yarns, including wire helix or helices, such that the minimum vacuum and burst requirements as given in Table 4 are met.

Cover—Minimum thickness shall be 0.03 in. (0.80 mm) measured as the thickness from the reinforcement out.

Dimensions—Dimensions and tolerances shall be as specified in Table 4.

Physical Tests—Tests shall be performed on sections and parts from the finished hose. The preparation of test samples shall be in accordance with ASTM D 380. See Table 3, and SAE J20, Part VI.

PART III—HEATER HOSE (SAE 20R3)

Scope—This type of hose is used in connecting hot water heaters in the coolant circulating systems of ground vehicles.

Construction

Tube—Minimum thickness shall be 0.06 in. (1.59 mm).

Reinforcement—The reinforcement shall be multiple plies of wrapped fabric, or ply or plies of braided or knitted yarn, and shall be such as to meet the minimum burst requirements in Table 5.

Cover—Minimum thickness shall be 0.03 in. (0.80 mm).

The dimensions and tolerances and burst and vacuum requirements shall be as specified in Table 5.

Physical Tests—Details for laboratory tests are shown under Part VI, Physical Test Procedures. Tests are intended to be performed on sections and parts from the finished hose. The preparation of test samples is in accordance with ASTM D 380. See Table 3, and SAE J20, Part VI.

PART IV—NORMAL SERVICE TYPE—KNITTED OR BRAIDED HOSE (SAE 20R4)

Scope—This part covers hose for coolant circulating systems of automotive type engines, commonly known as radiator hose. When resistance to vacuum collapse is required, an inserted wire helix should be specified.

Construction

Tube—Minimum thickness shall be 0.06 in. (1.59 mm).

Reinforcement—The reinforcement shall consist of one or more plies of knitted or braided yarn and shall be such that the hose will meet the minimum burst requirements in Table 6.

Cover—Minimum thickness shall be 0.03 in. (0.80 mm).

Dimensions—Dimensions, tolerances, burst, and vacuum resistance shall be as specified in Table 6.

Physical Tests—Tests shall be performed on sections and parts from the finished hose. The preparation of test samples is to be in accordance with ASTM D 380. See Table 3, and SAE J20, Part VI.

PART V—NORMAL SERVICE TYPE—FLEXIBLE, WIRE REINFORCED HOSE (SAE 20R5)

Scope—This part covers a wire reinforced hose for coolant circulating systems of automotive type engines, commonly known as universal type hose.

TABLE 3—COOLANT SYSTEM HOSE

(Previous SAE Designation)	Class A (New)	Class B (SB)	Class C (SC)	Class D-1 (New)	Class D-2 (R)
Tube					
Original properties					
Durometer, points	55 to 75	55 to 75	55 to 75	55 to 75	55 to 75
Tensile, min, psi (MPa)	800(5.52)	1250(8.62)	1000(6.90)	1000(6.90)	700(4.83)
Elongation, %	200	250	200	300	150
Oven aging conditions and change limits					
Hours	70	70	70	70	70
Temperature, °F(°C)	350(176.7)	212(100)	212(100)	250(121.1)	250(121.1)
Durometer, points	+10	+15	+20	+15	+15
Tensile, max, %	−10	−15	−20	−20	−20
Elongation, max, %	−35	−50	−50	−50	−50
Oil immersion change limits ASTM No. 3 Oil					
Volume, max, %	0 to +40	−5 to +25	+80	+150	+120
Tensile, max, %	−40	−20	−50	—	—
Coolant immersion change limits					
Volume, %	0 to +40	0 to +20	0 to +20	−5 to +20	−5 to +20
Durometer, points	−5 to +10	−10 to +10	−10 to +10	−10 to +10	−10 to +10
Tensile, max, %	−30	−20	−20	−10	−10
Elongation, max, %	−25	−40	−40	−25	−25
Cover					
Original properties					
Durometer, points	55 to 75	55 to 70	55 to 70	55 to 70	55 to 70
Tensile, min, psi (MPa)	800(5.52)	1000(6.80)	1000(6.80)	1000(6.80)	700(6.80)
Elongation, %	200	250	200	300	150
Oven aging change limits (conditions as above)					
Tensile, max, %	−10	−20	−20	−20	−20
Elongation, max, %	−35	−50	−50	−50	−50
Oil immersion change limits ASTM No. 3 Oil					
Volume, max, %	0 to +40	+80	+80	+150	+120
Tensile, max, %	−40	−50	−50	—	—
Adhesion (original)					
Tube to ply, lb/in (N/m)	8(1400)	8(1400)	8(1400)	10(1750)	8(1400)
Ply to ply, lb/in (N/m)	8(1400)	8(1400)	8(1400)	10(1750)	8(1400)
Compression set, temp °F(°C)	250(121.1)	212(100)	212(100)	250(121.1)	250(121.1)
(70 h), %	40	50	75	75	85
Normally Available on					
Part I, Standard wall	X	X	X	X	X
Part I, Heavy wall	—	X	—	—	—
Part II, Wire inserted	—	X	—	—	—
Part III	—	X	X	X	X
Part IV	—	—	—	X	—
Part V	—	—	X	—	X

Construction

Tube—Minimum thickness shall be 0.06 in. (1.59 mm) measured on the straight ends of the hose.

Reinforcement—One of the following constructions shall be used at the manufacturer's option. The hose must meet the minimum burst requirements listed in Table 7.

1. Tube and cover are integral and of the same compound: a ply of fabric, cord, or fabric and cord, a helical high carbon steel wire imbedded in the convoluted section of the hose running out into the plain ends, with the ends reinforced with a ply of fabric.

2. A tube, a helical high carbon steel wire imbedded in the convoluted section of the hose running out into the plain ends, a ply of frictioned or skimmed fabric, and the plain ends reinforced with a ply of fabric.

3. A tube, a helical high carbon steel wire imbedded in the convoluted section of the hose running out into the plain ends, with external rubber coating, the plain ends to be reinforced with a fabric.

4. A tube, a helical high carbon steel wire imbedded in the convoluted section of the hose running out into the plain ends, a ply of fabric, with external rubber cover, the plain ends to be reinforced with a ply of fabric.

Dimensions—Dimensions, tolerances, vacuum resistance and burst shall be as specified in Table 7.

TABLE 4—WIRE INSERTED HOSE (SAE 20R2)

Nominal Size, in	Burst		Tolerance				Vacuum Resistance	
			ID		Wall (exclusive of wire gauge)			
	psi	MPa	±in	±mm	in	mm	in Hg	kPa
1	300	2.06	0.03	0.80	0.17–0.25	4.32–6.35	25	84.4
1-1/4	275	1.90	0.03	0.80	0.17–0.25	4.32–6.35	25	84.4
1-1/2	250	1.72	0.03	0.80	0.17–0.25	4.32–6.35	25	84.4
1-3/4	225	1.55	0.03	0.80	0.17–0.25	4.32–6.35	25	84.4
2	200	1.38	0.06	1.59	0.17–0.25	4.32–6.35	25	84.4
2-1/4	175	1.21	0.06	1.59	0.17–0.25	4.32–6.35	25	84.4
2-1/2	150	1.03	0.06	1.59	0.17–0.25	4.32–6.35	25	84.4
2-3/4	125	0.86	0.06	1.59	0.17–0.25	4.32–6.35	25	84.4
3	100	0.69	0.06	1.59	0.17–0.25	4.32–6.35	25	84.4
3-1/2	75	0.52	0.06	1.59	0.17–0.25	4.32–6.35	25	84.4
4	50	0.34	0.06	1.59	0.17–0.25	4.32–6.35	25	84.4

TABLE 5—NORMAL SERVICE HEATER HOSE (SAE 20R3)

Size, in	Burst		Tolerance				Vacuum Resistance	
			ID		OD			
	psi	MPa	±in	±mm	in	mm	in Hg	kPa
3/8	250	1.72	0.03	0.80	0.69 ±0.03	17.53 ±0.80	10	33.8
1/2	250	1.72	0.03	0.80	0.81 ±0.03	20.57 ±0.80	10	33.8
5/8	250	1.72	0.03	0.80	0.94 ±0.03	23.88 ±0.80	8	27.0
3/4	200	1.38	0.03	0.80	1.06 ±0.03	26.92 ±0.80	7	23.6
1	175	1.21	0.06	1.59	1.34 ±0.06	34.04 ±1.59	6	20.3

TABLE 6—KNITTED OR BRAIDED HOSE (SAE 20R4)

Size, in	Burst		Tolerance			
			ID		Wall	
	psi	MPa	±in	±mm	in	mm
3/8	180	1.24	0.03	0.80	0.17–0.22	4.32–5.59
1/2	170	1.17	0.03	0.80	0.17–0.22	4.32–5.59
5/8	160	1.10	0.03	0.80	0.17–0.22	4.32–5.59
3/4	150	1.03	0.03	0.80	0.17–0.22	4.32–5.59
1	140	0.97	0.03	0.80	0.17–0.22	4.32–5.59
1-1/4	130	0.90	0.03	0.80	0.17–0.22	4.32–5.59
1-1/2	120	0.83	0.03	0.80	0.17–0.22	4.32–5.59
1-3/4	110	0.76	0.03	0.80	0.17–0.22	4.32–5.59
2	100	0.69	0.03	0.80	0.17–0.22	4.32–5.59
2-1/4	90	0.62	0.03	0.80	0.17–0.25	4.32–6.35
2-1/2	80	0.55	0.03	0.80	0.17–0.25	4.32–6.35

TABLE 7—FLEXIBLE WIRE REINFORCED HOSE (SAE 20R5)

Size, in	Burst		Tolerance			
			ID		Wall	
	psi	MPa	in	mm	in	mm
1	140	0.97	+0.03, −0.06	+0.80, −1.59	0.14–0.19	3.56–4.83
1-1/4	130	0.90	+0.03, −0.06	+0.80, −1.59	0.14–0.19	3.56–4.83
1-1/2	120	0.83	+0.03, −0.06	+0.08, −1.59	0.14–0.19	3.56–4.83
1-3/4	110	0.76	+0.03, −0.06	+0.80, −1.59	0.14–0.19	3.56–4.83
2	100	0.69	+0.03, −0.06	+0.08, −1.59	0.14–0.19	3.56–4.83
2-1/4	90	0.62	+0.03, −0.06	+0.80, −1.59	0.14–0.19	3.56–4.83
2-1/2	80	0.55	+0.03, −0.06	+0.80, −1.59	0.14–0.19	3.56–4.83

Physical Tests—See Table 3. The test samples will be taken from the straight ends of the hose. The preparation of test samples is to be in accordance with ASTM D 380. Tests will be run only on the cover when applicable. Due to construction and application of this type hose, it is recognized that ozone deterioration is accelerated and resistance requirements should be agreed upon between the supplier and purchaser.

PART VI—PHYSICAL TEST PROCEDURES

Specifications for Immersion Liquids
Oil—ASTM Oil No. 3. See ASTM D 471.
  Ethylene Glycol—This shall be what is termed "refined grade of ethylene glycol" and shall conform to the following requirements:
  Acidity: Not more than 0.01% as acetic acid.
  Specific Gravity: 1.1150–1.1158 at 20°C (68°F).
  Color: Water white.
  Boiling Range at 760 mm Hg: Above 210°C (410°F), none; below 190°C (374°F), below 195°C (383°F), not more than 5%; below 202°C (396°F), not less than 95%.
  Water: Not more than 0.5% by weight.
  Ash: Not more than 0.005 g/100 cm^3.
  Odor: Mild, nonresidual.
  Average Weight: 9.28 lb (20.42 kg) per gal at 20°C (68°F).
Durometer Hardness—Hardness shall be measured with an instrument according to ASTM D 2240, Tentative Method of Test for Indentation of Rubber by Means of the Durometers, except that the reading shall be made immediately. Tests shall be made on not less than 1/4 in. (6.35 mm) thickness to be made up by laying together smoothly buffed layers of the compound.
Tensile Strength and Elongation—Test according to ASTM D 380. The maximum percent change in elongation shall be computed as a percentage of the actual original percent elongation determined for the particular sample.

Example: If the original requirement is 250% minimum elongation, and the maximum percent change permitted in elongation after oven aging is −60%, and if the actual original elongation of the material is found to be 350%, then the minimum elongation permitted after the aging test is 140%.
  Oven Aging—Shall conform to ASTM D 573, except that test specimens shall be prepared in accordance with ASTM D 380.
  Oil Immersion—Shall conform to ASTM D 471, except that the test specimens shall be in accordance with ASTM D 380 in the cases of the volume change test and the strength deterioration test.
  Three tensile test samples and two volume change samples shall be placed in a glass test tube having an outside diameter of 38 mm and an overall length of 300 mm fitted tightly with a cork stopper and a reflux condenser. The tube shall be three-quarters full of test fluid. The test shall be run for 70 h at 100°C (212°F).
  The tensile strength of the specimens after removal from the liquid (ASTM D 471) shall be calculated on the original unaged cross-sectional area and not according to the formula given in ASTM D 471. Measurement and calculation of volume change shall be in accordance with ASTM D 471.
  Coolant Immersion—Percent swell, tensile, and durometer changes shall be observed after 70 h immersion in the following mixture maintained at the boiling point under a water-cooled reflux condenser:
  1/2 by volume, distilled water
  1/2 by volume, ethylene glycol (see above)
  All other details of test shall be the same as under Section 5. Test specimens shall be in accordance with ASTM D 380, in the cases of the swelling test and the strength deterioration test. Measurements of tensile, durometer, and volume change shall be made in accordance with ASTM D 471.
  Adhesion—Test according to ASTM D 413. Wrapped fabric hose is to be separated one ply from the adjacent part. Braided or knitted constructions may be separated as two or more plies from the adjacent part. In SAE 20R4, the adhesion values apply to "rubber-to-rubber" contacts only.
  Cold Flexibility—The following procedure shall be used:
  1. Time of exposure to low temperature is to be 5 h.
  2. Temperature is −40°C (−40°F).
  3. Medium shall be air.
  4. For hose 1 in. ID and below, specimen shall consist of complete hose of length sufficient to perform bend test described as follows: The hose shall be placed in a cold box for 5 h at −40°C (−40°F). The hose shall then be flexed in the cold chamber through 180 deg from the centerline to a diameter of 10 times the maximum outside diameter of the hose. This flexing shall be within 4 s. The hose shall not fracture and shall not show any cracks, checks, or breaks in the tube or cover. Cracking of the tube may be determined by application of a proof pressure of 50 psi (0.34 MPa).
  For hose over 1 in. ID, specimens are to be 1 in. (25.4 mm) wide section of the complete hose. The specimen shall be placed in a cold box for 5 h at −40°F (−40°C). The specimen is then compressed 50% of its original inside diameter between parallel plates within 4 s. The specimen shall not crack, check, or break. The testing fixture shall be in the cold box during the entire test.
  Burst—This test shall be performed on a straight length for SAE 20R1, 20R2, and 20R3 hose in accordance with ASTM D 380, except that the length tested shall be 10 in. (254 mm) with 8 in. (203 mm) of free length.
  For SAE 20R4 hose, the same reference applies, except that the test shall be performed on the individual curved hose with one end free and unrestrained and the rate of application of pressure shall be not less than 300 psi (2.06 MPa) nor more than 1000 psi (6.89 MPa) per minute.
  Ozone Test and Requirements—The following test and requirements apply:
  For Hoses 1 in. ID and Above—A specimen 1/2 in. (12.7 mm) wide is taken lengthwise from the hose. Place bench marks on the cover of the test specimen while it is flat on a work surface. The specimen is to be mounted so as to give 15% elongation with cover out.
  Fasten the ends of the specimen with tape or twine. The test specimens, while still elongated, shall be conditioned for 24 h in air at room temperature, and then while still on the mandrel shall be placed in an exposure chamber containing air mixed with ozone in the proportion of 50 ±5 parts of ozone per 100 million parts of air by volume. The ambient air temperature in the chamber during the test shall be 40 ±1°C (104 ±2°F). The specimen shall be exposed to the ozone and air mixture for a period of 70 h. To determine conformance to this requirement, the cover of the specimen when examined under 7X magnification, ignoring the area covered by the tape or twine or immediate adjacent area, shall meet a "0" rating. The ozone cabinet shall be in accordance with ASTM D 1149.
  For Hoses Under 1 in. ID—A specimen of hose having a length equal to the full circumference of the test mandrel plus approximately 10 in. (254 mm) shall be bent around the mandrel, the diameter of which shall be eight times the nominal outside diameter of the hose being tested, and bound with a tape or twine where the ends cross one another. If collapse of the hose occurs when bent around the mandrel, provision should be made to support the hose

internally. Three specimens shall be conditioned for 24 h in air at room temperature, and then, while still on the mandrel, shall be placed in an exposure chamber containing air mixed with ozone in the proportion of 50 ±5 parts of ozone per 100 million parts of air by volume. The ambient air temperature in the chamber during the test shall be 40 ±1°C (104 ±2°F). The specimen shall be exposed to the ozone and air mixture for a period of 70 h. To determine conformance to this requirement, the cover of the specimen when examined under 7X magnification, ignoring the area covered by the tape or twine or immediate adjacent area, shall meet a "0" rating. The ozone cabinet shall be in accordance with ASTM D 1149.

Kink Test—This test applies to SAE 20R3 hose only. See Part III.

Test Procedure—Condition specimen length of hose at room temperature for at least 2 h. Measure the OD at the approximate center of the specimen length. Insert one end of the hose into one hole of the specified test fixture, carefully bend the hose (in direction of natural curvature) and insert the remaining hose end into the remaining test fixture hole. Do not overbend or bend hose with sharp motion to prevent excessive kinking or collapse. Within 30 s measure the OD at the point of greatest collapse percent. This test applies to lengths of straight hose only which equal or exceed specimen lengths.

When a sufficient length of hose is available, it is permissible and suggested that a length in excess of the specimen length be used in an effort to minimize the human handling variable and overbending. The second or "remaining" hose end can then be inserted entirely through the hole in the test fixture and adjusted to the correct specimen length.

Test Fixture—Shall consist of a 1 in. (25.4 mm) thick flat plate drilled with holes not to exceed the hose OD by more than 1/16 in. (1.6 mm) and separated by the specified center distances.

Test Data—See Table 8.

Compression Set—Test to be performed per ASTM D 395, Method B.

Vacuum Collapse Test—Hose is to be subjected to the specified inches of mercury vacuum. The entire hose shall be tested when practical. No diameter shall decrease by more than 20% during application of vacuum for 15 and not over 30 s.

TABLE 8—KINK TEST DATA FOR SAE 20R3 HOSE

Hose ID, in	Specimen Length		Center Distance		Collapse Allowed, %
	in	cm	in	cm	
3/8	10	25.40	3	7.62	25
1/2	10	25.40	4	10.16	25
5/8	12	30.48	6	15.24	25
3/4	18	45.72	7	17.78	25
1	24	60.96	10	25.40	25

FUEL AND OIL HOSES—SAE J30d

SAE Standard

Report of Nonmetallic Materials Committee approved January 1946 and last revised February 1977.

[This SAE Standard was formulated by SAE-ASTM Technical Committee on Automotive Rubber.]

COUPLED AND UNCOUPLED SYNTHETIC RUBBER TUBE AND COVER (SAE 30R1)

1. Scope—This specification covers hose which may be supplied either coupled or uncoupled for use with gasoline, diesel fuel, lubricating oil, or the vapor present in either the fuel system or in the crankcase of internal combustion engines in mobile, stationary, and marine applications. The hose may be furnished in long lengths, specific cut lengths, or as a part preformed to a specific configuration.

2. Hose Construction—The construction of this hose embodies a smooth bore tube of suitable synthetic rubber material, reinforced with one ply of braided, knit, spiral, or woven fabric, and finished with a suitable oil- and ozone-resisting synthetic rubber cover.

3. Dimensions[1]—Dimensions and tolerances for inside diameter and outside diameter are shown in Table 1.

Other tolerances applicable are as follows:

3.1 When hose is supplied in specific cut or long lengths the length tolerance shall be as follows:

Length		Precision		Commercial	
in	m	in	mm	in	mm
0 to 12	0 to 0.30	±0.12	± 3.18	+0.38 −0.12	+ 9.6 − 3.0
Over 12 to 24	0.30 to 0.61	±0.19	± 4.76	+0.38 −0.19	+ 9.6 − 4.7
Over 24 to 36	0.61 to 0.91	±0.25	± 6.35	+0.50 −0.25	+12.7 − 6.3
Over 36 to 48	0.91 to 1.22	±0.38	± 9.65	+0.50 −0.38	+12.7 − 9.6
Over 48 to 72	1.22 to 1.83	±0.50	±12.70	+0.75 −0.50	+19.0 −12.7
Over 72	1.83	±1%		±2%	

Commercial tolerances shall apply unless otherwise specified.

The ends of hose shall be square within 0.12 in. (3.05 mm) for sizes up to and including 3/4 in., 0.25 in. (6.35 mm) for sizes 1 through 2 in., and 0.38 in. (9.65 mm) for sizes over 2 in. End squareness shall be measured between a point on the outside of the long side which is on a circumferential line emanating from the end of the short side and at 90 deg to the axis and a point at the end of the long side. End squareness measurement will be considered equally applicable with length tolerance. The midpoint of the measurement of the deviation from squareness taken on the long side shall fall within the length tolerance.

3.2 When hose is supplied as a preformed item, the tolerances shall be as follows:

3.2.1 Squareness of Ends—The tolerance on squareness of ends on preformed parts shall be a maximum of 0.25 in. (6.35 mm) on all sizes through 2 in. (50.8 mm) ID and 0.38 in. (9.65 mm) for sizes over 2 in. (50.8 mm) ID. End squareness shall be measured between a point on the outside of the long side which is on a circumferential line emanating from the end of the short side and at 90 deg to the axis and a point at the end of the long side. Squareness measurements will be considered equally applicable with length tolerance. The midpoint of the measurement of the deviation from squareness taken on the long side shall fall within the arm length of the tolerance.

3.2.2 Arm Lengths—Measured from end to intersection of nearest centerline. Each end shall be as shown in paragraph 3.1. These tolerances apply also to the length of an expanded end.

3.2.3 General Layout—Dimensions locating bend intersections are to establish the theoretical centerline of the hose. Actual outside contour of the hose must be held within ±0.19 in. (4.76 mm) in all planes with respect to the

TABLE 1—DIMENSIONS AND TOLERANCES FOR SAE 30R1

Size, in	Inside Diameter				Outside Diameter[a]			
	in		mm		in		mm	
5/32	0.156	±0.016	3.97	±0.40	0.360	±0.023	9.13	±0.58
3/16	0.188	±0.016	4.76	±0.40	0.406	±0.023	10.32	±0.58
7/32	0.219	±0.016	5.56	±0.40	0.438	±0.023	11.11	±0.58
1/4	0.250	±0.016	6.35	±0.40	0.500	±0.023	12.70	±0.58
9/32	0.281	±0.016	7.14	±0.40	0.531	±0.023	13.49	±0.58
5/16	0.312	±0.016	7.94	±0.40	0.562	±0.023	14.29	±0.58
11/32	0.344	±0.016	8.73	±0.40	0.594	±0.023	15.08	±0.58
3/8	0.375	±0.016	9.53	±0.40	0.625	±0.023	15.88	±0.58
7/16	0.438	±0.023	11.11	±0.58	0.719	±0.031	18.26	±0.79
1/2	0.500	±0.023	12.70	±0.58	0.781	±0.031	19.84	±0.79
5/8	0.625	±0.031	15.88	±0.79	0.938	±0.031	23.81	±0.79
3/4	0.750	±0.031	19.05	±0.79	1.125	±0.031	28.58	±0.79
1	1.000	±0.062	25.40	±1.59	1.375	±0.062	34.93	±1.59
1-1/4	1.25	±0.03	31.75	±0.76	Not specified[b]			
1-1/2	1.50	±0.03	38.10	±0.76	Not specified[b]			
1-3/4	1.75	±0.03	44.45	±0.76	Not specified[b]			
2	2.00	±0.03	50.80	±0.76	Not specified[b]			
2-1/4	2.25	±0.03	57.35	±0.76	Not specified[b]			
2-1/2	2.50	±0.03	63.50	±0.76	Not specified[b]			

[a]Concentricity based on total indicator reading between the inside bore of the hose and the outer surface of the hose shall not exceed the values given below:
  Sizes 1/4 in and under: 0.030 in (0.762 mm)
  Sizes over 1/4 up to 1/2: 0.040 in (1.016 mm)
  Sizes over 1/2: 0.050 in (1.270 mm).

[b]The wall gage for sizes 1-1/4 through 2 in shall be between 0.17 and 0.22 (4.32 and 5.89 mm), and for sizes 2-1/4 to 2-1/2 in between 0.17 and 0.25 (4.32 and 6.35 mm).

[1]Consideration will be given to the development of a separate standard applicable to tolerances which would be used on all automotive hoses, particularly tolerances on cut length and preformed hoses. Adoption of such a standard would give cause to remove the great amount of detail from this standard as well as others. However, its inclusion in this revision is felt mandatory to satisfy present needs.

theoretical outside contour. To check contour, hose ends should first be placed in nominal position (hose may have to be flexed to correct any distortion caused by handling after vulcanization in the producing plant or in shipment) in a checking fixture from which contour deviation can be measured. Allowance shall be provided in the end mounting area of the fixture for the arm length tolerances which are applicable.

When the ID of an end of the hose is enlarged, the wall gage of the enlarged end normally changes. Allowable change should be +0.03, −0.02 in. (+0.76, −0.51 mm).

The wall gage within bends of a preformed hose may differ from the gage in straight portions. This difference shall not exceed 33%.

4. Retests and Rejection—Any hose or assembly that fails in one or more tests shall be resampled and retested. Twice the number of specimens shall be selected from the lot in question for any retests, and failure of any of the retested samples shall be cause for rejection.

5. Tests—Procedures described by ASTM D 380, Methods of Testing Rubber Hose, are to be followed wherever applicable.

5.1 Qualification Tests—For the qualification tests, one 25 ft (7.6 m) length of bulk hose, 10 preformed parts, or 10 assemblies of each size to be qualified shall be furnished. In order to qualify under this specification, hose and hose assemblies must meet the requirements of the following tests: (1) burst, (2) vacuum collapse, (3) cold flexibility, (4) tensile strength and elongation, tube and cover, (5) dry heat resistance, (6) fuel resistance, (7) oil resistance, (8) waxy hydrocarbon extractibles, (9) ozone resistance, (10) adhesion, (11) visual inspection.

5.2 Frequency of Testing for Inspection—On uncoupled hose, tests shall be conducted on samples representing each lot of 500–10,000 ft (152–3048 m). Where a lot is 500 ft (152 m) or less, no tests shall be conducted, but materials and workmanship shall be the same on such lots as on hose previously qualified under this specification.

On preformed parts or coupled assemblies in lot sizes of 100–10,000 parts, not less than two parts or assemblies shall be selected and subjected to inspection tests, except the visual test which shall apply to 100% of lots of assemblies.

5.3 Inspection Test—On uncoupled and preformed hose, these inspection tests shall apply: (1) burst, (2) vacuum collapse, (3) cold flexibility, (10) adhesion, (11) visual inspection; and for coupled assemblies, these shall apply: (1) burst, (11) visual inspection.

6. Test Requirements

ϕ **6.1 Burst Test**—Minimum burst for sizes through 1 in (25.4 mm) shall be 175 psi (1.21 MPa) and for sizes over 1 in (25.4 mm) shall be 80 psi (0.55 MPa).

ϕ **6.2 Vacuum Collapse Test**—Sizes less than ½ in (12.7 mm) diameter shall be subjected to 24 in Hg (81 kPa) vacuum. Sizes ½ (12.7 mm) through 1 in (25.4 mm) shall be subjected to 10 in Hg (34 kPa) vacuum. This requirement shall not apply to sizes larger than 1 in (25.4 mm).

During the vacuum test described, a 3 ft (915 mm) length of hose or a hose assembly shall be held in a straight line, and no diameter shall decrease by more than 20% during application of vacuum for 15 and not over 30 s.

The vacuum collapse test on preformed parts should be performed on the finished parts on a straight section from as near midpoint as is possible.

6.3 Cold Flexibility—For straight hose ¾ in. (19.05 mm) ID and under, the whole hose shall be used for this test.

Two samples shall be used, one to be unaged and the other immersed in ASTM Oil No. 3 for 70 h at 100 ±1°C (212 ±2°F). The aged and unaged samples are conditioned at −40 ±1°C (−40 ±2°F) for 5 h and then flexed in the cold chamber through 180 deg from the centerline to a diameter of 10 times the maximum OD of the hose. The flexing shall be within 4 s and the hose must not fracture or show any cracks, checks, or breaks in the tube or cover. Proof pressure of 100 psi (690 kPa) may be applied to determine tube damage.

For straight hose over ¾ in. (19.05 mm) ID and all preformed hose, prepare six specimens 4 x 0.25 in. (100 x 6.4 mm) of tube and cover thickness. One set of three shall be unaged and the other aged in ASTM Oil No. 3 for 70 h at 100 ±1°C (212 ±2°F). Both sets shall be conditioned for 5 h at −40 ±1°C (−40 ±2°F) in an unrestrained loop position between two jaws 2 in. (50 mm) wide and 2½ in. (63 mm) apart. After conditioning and while still in the cold chamber, the jaws shall be rapidly brought together until they are 1 in. (25.4 mm) apart. The specimens shall not fracture nor show any cracks, checks, or breaks.

6.4 Tensile Strength and Elongation
  Original tensile strength of cover: 1000 psi (6.89 MPa) min
  Original tensile strength of tube: 1000 psi (6.89 MPa) min
  Original elongation of cover and tube: 200% min

6.5 Dry Heat Resistance—After oven aging for 70 h at a temperature of 100 ±1°C (212 ±2°F), the reductions in tensile strength and elongation of specimens taken from the tube and cover shall not exceed the following values:
  Original tensile strength: −20%
  Original elongation: −50%

6.6 Fuel Resistance—After 48 h immersion at room temperature in ASTM Reference Fuel B, the reductions in tensile strength and elongation of specimens taken from the tube shall not exceed the following values:
  Original tensile strength: −30%
  Original elongation: −30%

In addition, the volume change of specimens taken from the tube shall not exceed the following values:
  Tube, volume change: −5 to +25%

6.7 Oil Resistance—After 70 h immersion at a temperature of 100 ±1°C (212 ±2°F) in ASTM Oil No. 3, the reductions in tensile strength and elongation of specimens taken from the tube shall not exceed the following values:
  Original tensile strength: −40%
  Original elongation: −40%

In addition the volume change of specimens taken from the tube and cover shall not exceed the following values:
  Tube, volume change: −5 to +25%
  Cover, volume change: 0 to +100%

ϕ **6.8 Waxy Hydrocarbon Extractables**—This test is restricted to hose sizes under ½ in ID. A length of hose 18–24 in (457–610 mm) shall be selected, suitably clamped to the proper size plug which is inserted 1 in (25.4 mm) into one end. The free length (actual −1 in (−25.4 mm)) and the ID to the nearest 0.01 and 0.001 in (0.25 and 0.025 mm), respectively, shall be measured and recorded. The hose shall be filled with ASTM Fuel B. The other end shall be suitably sealed and allowed to stand 24 h at temperature of 24 ± 2°C (75 ± 4°F). At the conclusion, the remaining fluid shall be poured from the hose into a tared beaker. Rinse the inside of the test hose free length with an amount of fresh ASTM Fuel B approximately equal to the volume of the original filling, and add to original extraction.

Remove solvent by evaporation on a steam bath at 66 ±1°C (150 ±2°F) to 82 ±1°C (180 ±2°F). Cool to room temperature and take up residue with 10 ml methanol. Filter this solution on a tared "gooch" crucible, rinsing beaker with not more than 5 ml of methanol. Dry with suction air used for filtering.

Weigh and calculate weight of extractables per square inch (square centimetre) of inside surface area.

Extractables shall not exceed 5.0 mg/in.^{2} (0.775 mg/cm^2).

6.9 Ozone Resistance—Test procedure and apparatus shall be in accordance with ASTM D 1149, Test for Accelerated Ozone Cracking of Vulcanized Rubber, where applicable.

For straight hose 1 in. (25.4 mm) ID and under, a specimen of hose of sufficient length shall be bent around a mandrel with OD eight times the nominal OD of the sample. The two ends shall be tied at their crossing with enameled copper or aluminum wire. After mounting, the specimen shall be allowed to rest in an ozone-free atmosphere for 24 h at room temperature. The mounted specimen shall be placed in a test chamber with ozone concentration of 50 ±5 pphm at a temperature of 40 ±1°C (104 ±2°F).

After 70 h of exposure, the specimen shall be removed and allowed to cool to room temperature and then inspected. The specimen shall be visually inspected under 7X magnification and must meet a "O" rating except for the area immediately adjacent to the wire, which shall be ignored.

From hose over 1 in. (25.4 mm) and preformed parts, prepare a specimen by cutting a strip of whole hose ½ in. wide x 4 in. long (12.7 x 100 mm) and tie the specimen (cover out) around a ½ in. (12.7 mm) diameter mandrel.

Condition in same manner as for whole hose and apply same requirements. This test applies to the cover only and cracks in the exposed tube or cut edges of the cover shall be ignored.

6.10 Adhesion Test—The minimum load required to separate the tube from the reinforcing ply and the cover from the reinforcing ply shall be 6 lb (27 N).

6.11 Visual Inspection—All assemblies shall be inspected to see that the correct fittings are properly applied.

COUPLED AND UNCOUPLED SYNTHETIC RUBBER TUBE AND COVER (SAE 30R2)

1. Scope—This specification covers three types of fuel and oil hose, coupled and uncoupled, for use with gasoline oil and diesel fuel.

2. Hose Construction

2.1 Type 1—The construction of this hose embodies a smooth bore tube of suitable synthetic rubber material, reinforced with one ply of braided, knit, spiral, or woven fabric, and finished with a suitable oil- and ozone-resisting synthetic rubber cover.

2.2 Type 2—The construction of this hose embodies a smooth bore tube of suitable synthetic rubber material, reinforced with two braided plies or multiples of woven fabric, and finished with a suitable oil- and ozone-resisting synthetic rubber cover.

TABLE 2—DIMENSIONS AND TOLERANCES FOR SAE 30R2

Size, in	Inside Diameter		Outside Diameter[a]							
			Types 1 and 3				Type 2			
	in	mm	in		mm		in		mm	
			Min	Max	Min	Max	Min	Max	Min	Max
1/8	0.125 ±0.010	3.18 ±0.25	0.328	0.375	8.33	9.53	0.438	0.500	11.13	12.70
3/16	0.188 ±0.016	4.78 ±0.40	0.391	0.438	9.93	11.13	0.500	0.562	12.70	14.27
1/4	0.250 ±0.016	6.35 ±0.40	0.453	0.500	11.51	12.70	0.562	0.625	14.27	15.88
5/16	0.312 ±0.016	7.92 ±0.40	0.516	0.562	13.11	14.27	0.625	0.688	15.88	17.48
3/8	0.375 ±0.016	9.53 ±0.40	0.578	0.625	14.68	15.88	0.688	0.750	17.48	19.05
7/16	0.438 ±0.023	11.13 ±0.58	—	—	—	—	0.750	0.812	19.05	20.62
1/2	0.500 ±0.023	12.70 ±0.58	0.719	0.781	18.26	19.84	0.812	0.875	20.62	22.23
5/8	0.625 ±0.023	15.88 ±0.58	0.844	0.906	21.44	23.01	0.938	1.000	23.83	25.40
3/4	0.750 ±0.023	19.05 ±0.58	1.062	1.125	26.97	28.58	1.062	1.125	26.97	28.58
7/8	0.875 ±0.031	22.23 ±0.79	—	—	—	—	1.188	1.250	30.18	31.75
1	1.000 ±0.031	25.40 ±0.79	1.281	1.375	32.54	34.93	1.281	1.375	32.54	34.93
1-1/8	1.125 ±0.031	28.58 ±0.79	—	—	—	—	1.500	1.625	38.10	41.28
1-1/4	1.250 ±0.039	31.75 ±0.99	—	—	—	—	1.625	1.750	41.28	44.45
1-3/8	1.375 ±0.039	34.93 ±0.99	—	—	—	—	1.750	1.875	44.45	47.63
1-1/2	1.500 ±0.039	38.10 ±0.99	—	—	—	—	1.875	2.000	47.63	50.80
1-5/8	1.625 ±0.039	41.28 ±0.99	—	—	—	—	2.000	2.125	50.80	53.98
1-3/4	1.750 ±0.039	44.45 ±0.99	—	—	—	—	2.125	2.250	53.98	57.15
2	2.000 ±0.039	50.80 ±0.99	—	—	—	—	2.375	2.500	60.33	63.50

[a]Concentricity based on total indicator reading between the inside bore of the hose and the outer surface of the hose shall not exceed the values given below:
Sizes 1/4 in and under: 0.030 in (0.76 mm)
Sizes over 1/4 up to 7/8 in: 0.040 in (1.02 mm)
Sizes over 7/8 in: 0.050 in (1.27 mm)

2.3 Type 3—The construction of this hose embodies a smooth bore tube of suitable synthetic rubber material, a single braided ply of textile reinforcement, and finished with a suitable oil- and ozone-resisting synthetic rubber cover.

3. Dimensions—Dimensions and tolerances are shown in Table 2.

4. Retests and Rejection—Any hose or assembly that fails in one or more tests shall be resampled and retested. Twice the number of specimens shall be selected from the lot in question for any retests, and failure of any of the retested samples shall be cause for rejection.

5. Tests

5.1 Qualification Tests—For qualification tests, one 25 ft (7.6 m) length of bulk hose or 10 assemblies of each size to be qualified shall be furnished. In order to qualify under this specification, hose and hose assemblies must meet the requirements of the following tests: (1) change-in-length followed by (2) burst, (3) vacuum collapse, (4) cold flexibility, (5) tensile strength and elongation, tube and cover, (6) dry heat resistance, (7) fuel resistance, (8) oil resistance, (9) ozone resistance, (10) adhesion.

In addition to the above, hose assemblies shall be subjected to qualification tests as follows: (11) proof, (12) tensile test of assembly, (13) leakage, (14) corrosion, (15) visual inspection.

5.2 Frequency of Testing for Inspection—On uncoupled hose, tests shall be conducted on samples representing each lot of 500–10,000 ft (152–3048 m). Where a lot is 500 ft (152 m) or less, no tests shall be conducted, but materials and workmanship shall be the same on such lots as on hose previously qualified under this specification.

On coupled hose lots from 100–10,000 pieces, not less than two assemblies shall be subjected to all inspection tests except the visual test, which shall apply to 100% of the assemblies.

5.3 Inspection Tests—On coupled hose, these inspection tests shall apply: (1) change-in-length, (2) burst, (3) vacuum collapse, (4) cold flexibility, (10) adhesion.

On coupled hose assemblies, these tests shall apply: (2) burst, (11) proof, (12) tension test of assembly (unaged), (15) visual inspection (100%).

6. Test Requirements

6.1 Change-in-Length Tests—Tests for change in length shall be conducted in accordance with ASTM D 380, except that the original measurement shall be at 0 psi (0 Pa) pressure. The change in length shall be determined at the pressures specified in Table 3. Requirements are as follows:

6.1.1 Type 1—All sizes ±5% maximum.

6.1.2 Type 2—Up to 1/2 in. (12.7 mm) ID hose, 0 to −8% change in length; 1/2 in. (12.7 mm) ID hose and larger, 0 to −6% change in length.

6.1.3 Type 3—All sizes, ±5% maximum.

6.2 Burst Test—The minimum bursting strength shall be as specified in Table 3.

6.3 Vacuum Collapse Test

6.3.1 Types 1 and 3—Sizes less than 1/2 in. (12.7 mm) ID shall be subjected to 20 in. Hg (67.5 kPa) vacuum; 1/2 in. (12.7 mm) ID hose shall be subjected to 10 in. Hg (34 kPa) vacuum. Sizes greater than 1/2 in. (12.7 mm) ID shall be excluded from the vacuum test.

6.3.2 Type 2—Sizes less than 5/8 in (15.88 mm) ID shall be subjected to 20 in Hg (67.5 kPa) vacuum. Hose 5/8 in (15.88 mm) through 1 in (15.8 mm-25.4 mm) ID shall be subjected to 10 in Hg (34 kPa) vacuum; sizes greater than 1 in (25.4 mm) ID shall be excluded from the vacuum test.

During the vacuum test described, a 3 ft (915 mm) length of hose or a hose assembly shall be held in a straight line, and no diameter shall decrease by more than 20% during application of vacuum for 15 and not over 30 s.

6.4 Cold Flexibility—Hose 3/4 in. (19.05 mm) ID and under, with or without couplings, shall be used for this test. Two samples shall be used for this test. One sample shall be unaged and the other sample shall be immersed in ASTM Oil No. 3 for 70 h at 100 ±1°C (212 ±2°F). The aged and unaged samples shall then be subjected to a temperature of −40 ±1°C (−40 ±2°F) for a period of 5 h, after which the hose shall be flexed in the cold chamber through 180 deg from the centerline to a diameter of 10 times the maximum OD of the hose. This flexing shall be within 4 s. The hose shall not fracture and shall not show any cracks, checks, or breaks in the tube or cover. Cracking of the tube may be determined by application of the proof pressure specified in Table 3.

TABLE 3—BURST AND CHANGE-IN-LENGTH TESTS FOR SAE 30R2

Size, in	Burst Test, min						Change-in-Length Test			
	Type 1		Type 2		Type 3		Types 1 and 2		Type 3	
	psi	MPa	psi	MPa	psi	MPa	psi	MPa	psi	MPa
1/8	700	4.82	700	4.82	—	—	115	0.79	—	—
3/16	700	4.82	700	4.82	2000	13.80	115	0.79	500	3.45
1/4	700	4.82	700	4.82	1600	11.04	115	0.79	400	2.76
5/16	700	4.82	700	4.82	1600	11.04	115	0.79	400	2.76
3/8	700	4.82	700	4.82	1600	11.04	115	0.79	400	2.76
7/16	700	4.82	700	4.82	—	—	115	0.79	—	—
1/2	700	4.82	700	4.82	1600	11.04	115	0.79	400	2.76
5/8	500	3.45	500	3.45	1400	9.66	85	0.59	350	2.42
3/4	500	3.45	500	3.45	1200	8.27	85	0.59	300	2.07
7/8	—	—	500	3.45	—	—	85	0.59	—	—
1	500	3.45	500	3.45	—	—	85	0.59	—	—
1-1/8	—	—	400	2.76	—	—	65	0.45	—	—
1-1/4	—	—	400	2.76	—	—	65	0.45	—	—
1-3/8	—	—	400	2.76	—	—	65	0.45	—	—
1-1/2	—	—	400	2.76	—	—	65	0.45	—	—
1-5/8	—	—	250	1.73	—	—	40	0.28	—	—
1 3/4	—	—	250	1.73	—	—	40	0.28	—	—
2	—	—	250	1.73	—	—	40	0.28	—	—

Hose over 3/4 in. (19.05 mm) ID shall have specimens 4 x 0.25 in. (100 x 6.4 mm) tube and cover thickness. The thickness shall be 0.10 in. (2.54 mm) maximum. Cut from the tube and cover if necessary. One set of samples shall be unaged. The other set of samples shall be completely immersed in ASTM Oil No. 3 for 70 h at 100 ±1°C (212 ±2°F). The unaged and aged specimens shall then be subjected to a temperature of −40 ±1°C (−40 ±2°F) for a period of 5 h in an unrestrained loop position between two jaws 2 in. (50 mm) and 2½ in. (63 mm) apart. At the end of 5 h and while still in the cold chamber, the jaws shall be rapidly brought together until they are 1 in. (25.4 mm) apart. The specimen shall not fracture and shall not show any cracks, checks, or breaks.

6.5 Tensile Strength and Elongation

Original tensile strength of cover: 1000 psi (6.89 MPa) min
Original tensile strength of tube: 1200 psi (8.27 MPa) min
Original elongation of tube and cover: 200% min

6.6 Dry Heat Resistance—After oven aging for 70 h at a temperature of 100 ±1°C (212 ±2°F), the reductions in tensile strength and elongation of specimens taken from the tube and cover shall not exceed the following values:

Original tensile strength: −20%
Original elongation: −50%

6.7 Fuel Resistance—After 48 h immersion at room temperature in ASTM Reference Fuel B, the reductions in tensile strength and elongation of specimens taken from the tube shall not exceed the following values:

Original tensile strength: −30%
Original elongation: −30%

In addition, the volume change of specimens taken from the tube shall not exceed the following values:

Tube, volume change: −5 to +25%

6.8 Oil Resistance—After 70 h immersion at a temperature of 100 ±1°C (212 ±2°F) in ASTM Oil No. 3, the reduction in tensile strength and elongation of specimens taken from the tube shall not exceed the following values:

Original tensile strength: −40%
Original elongation: −40%

In addition, the volume change of specimens taken from the tube and cover shall not exceed the following values:

Tube, volume change: −5 to +25%
Cover, volume change: 0 to +100%

6.9 Ozone Resistance—Test procedure shall be in accordance with ASTM D 1149 where applicable.

For hose 1 in. (25.4 mm) ID and under, a specimen of hose of sufficient length shall be bent around a mandrel with an outside diameter equal to eight times the nominal OD of the sample. The two ends shall be tied at their crossing with enameled copper or aluminum wire. After mounting, the specimen shall be allowed to rest in an ozone-free atmosphere for 24 h at room temperature. The mounted specimen shall be placed in a test chamber with ozone concentration of 50 pphm at a temperature of 40 ±1°C (104 ±2°F). After 70 h of exposure, the specimen shall be removed and allowed to cool to room temperature and then inspected visually under 7X magnification. It must meet a rating of "0" except for the area immediately adjacent to the wire, which shall be ignored.

For hose over 1 in. (25.4 mm) ID, prepare a specimen by cutting a strip of the whole hose ½ wide by 4 in. long (12.7 x 100 mm) and tie to a specimen (cover out) around a ½ in. (12.7 mm) diameter mandrel. Condition in the same manner as specified above for the whole hose and apply the same conditions and requirements. This test applies to the cover only and cracks in the exposed tube or cut edges of the cover shall be ignored.

6.10 Adhesion Test

6.10.1 TYPES 1 AND 3—The minimum load required to separate tube from reinforcing ply and cover from reinforcing ply shall be 6 lb (27 N).

6.10.2 TYPE 2—The minimum load required to separate tube from ply, cover from ply, and ply from ply shall be 12 lb (53.4 N).

6.11 Proof Test—Before shipment by the vendor, a suitable number of assemblies from each lot shall be proof tested at 50% of the minimum burst pressure specified in Table 3 for a period of not less than 30, or more than 60 s, to ensure an acceptable quality level.

6.12 Tensile Test of Assembly—The hose complete with fittings shall be dry-air aged at 100 ±1°C (212 ±2°F) for 70 h and then permitted to rest at room temperature for 2 h. The end fittings of the assembly shall be clamped in the jaws of a tension testing machine so that a straight pull may be applied. The jaws of the test machine shall separate at a rate not greater than 1 in. (25.4 mm)/min. The hose assembly shall withstand, after the aging test, a minimum pull of 100 lb (444 N) on sizes up to, and including, the ¼ in. (6.35 mm). All sizes over ¼ in. (6.35 mm) ID shall withstand a minimum pull of 150 lb (667 N). For inspection tests, unaged samples may be used.

6.13 Leakage Test

6.13.1 TYPES 1 AND 2—Coupled hose shall show no leakage under a hydrostatic pressure of 70% of the minimum burst specified in Table 3, following an aging period of 70 h at 100 ±1°C (212 ±2°F) oven temperature. The pressure shall be held for a period of not less than 5, or more than 7, min.

6.13.2 TYPE 3—Coupled hose shall show no leakage under a hydrostatic pressure of 50% of the minimum burst specified in Table 3, following an aging period of 70 h at 100 ±1°C (212 ±2°F) oven temperature. The pressure shall be held for a period of not less than 5, or more than 7, min.

6.14 Corrosion Test—The assembly shall be tested in accordance with ASTM B 117, Method of Salt Spray (Fog) Testing. The period shall be 48 h. There shall be no evidence of corrosion or other deterioration at the expiration of this test.

6.15 Visual Inspection—All assemblies shall be inspected to see that the correct fittings are properly applied.

LIGHTWEIGHT BRAIDED REINFORCED LACQUER, CEMENT, OR RUBBER COVERED HOSE (SAE 30R3)

1. Scope—This specification covers one type and one style of fuel and oil hose, coupled and uncoupled, for use with gasoline, oil, or diesel fuel.

2. Hose Construction—The construction of this hose embodies a smooth bore tube of suitable synthetic rubber material, reinforced with one braided ply of cotton or other suitable material and finished with a gasoline-, oil-, and water-resistant flexible coating of lacquer, cement, or synthetic rubber.

3. Dimensions—Dimensions and tolerances are shown in Table 4.

4. Retests and Rejection—Any hose or assembly that fails in one or more tests shall be resampled and retested. Twice the number of specimens shall be selected from the lot in question for any retests, and failure of any of the retested samples shall be cause for rejection.

5. Tests—Procedures described in ASTM D 380, Methods of Testing Rubber Hose, shall be followed wherever applicable.

5.1 Qualification Tests—For qualification tests, one 25 ft (7.6 m) length of bulk hose or 10 assemblies of each size to be qualified shall be furnished. In order to qualify under this specification, hose and hose assemblies must meet the requirements of the following tests: (1) change-in-length followed by (2) burst, (3) vacuum collapse, (4) cold flexibility, (5) tensile strength and elongation of tube, (6) dry heat resistance, (7) fuel resistance, (8) oil resistance, (9) ozone resistance.

In addition to the above, hose assemblies shall be subjected to qualification tests as follows: (10) proof, (11) tensile test of assembly, (12) leakage, (13) corrosion, (14) visual inspection.

5.2 Frequency of Testing for Inspection—On uncoupled hose, tests shall be conducted on samples representing each lot of 500–10,000 ft (152–3048 m). Where a lot is 500 ft (152 m) or less, no tests shall be conducted, but materials and workmanship shall be the same on such lots as on hose previously qualified under this specification.

On coupled hose lots of 100–10,000 pieces, not less than two assemblies shall be subjected to all inspection tests except the visual test, which shall apply to 100% of the assemblies.

5.3 Inspection Tests—On uncoupled hose, these inspection tests shall apply: (1) change-in-length, (3) vacuum collapse, (4) cold flexibility.

On coupled hose assemblies, these tests shall apply: (2) burst, (10) proof, (11) tension test of assembly (unaged), (14) visual inspection (100%).

6. Test Requirements

6.1 Change-in-Length Tests—Tests for change in length shall be conducted in accordance with ASTM D 380, except that the original measurement shall be made at 0 psi (0 Pa) pressure. The change in length shall not exceed ±5% at the change-in-length pressures specified in Table 5.

6.2 Burst Test—The minimum bursting strength shall be as specified in Table 5.

6.3 Vacuum Collapse Test—Two assemblies shall be subjected to a vacuum of 20 in. Hg (67.5 kPa), and the reduction in OD shall not exceed 20%

TABLE 4—DIMENSIONS AND TOLERANCES FOR SAE 30R3

Size, in	Inside Diameter				Outside Diameter[a]			
					Min		Max	
	in		mm		in	mm	in	mm
3/16	0.188	±0.016	4.76	±0.40	0.344	8.73	0.375	9.53
1/4	0.250	±0.016	6.35	±0.40	0.406	10.32	0.438	11.13
5/16	0.312	±0.016	7.94	±0.40	0.500	12.70	0.531	13.49
3/8	0.375	±0.016	9.53	±0.40	0.578	14.68	0.625	15.88

[a]Concentricity based on total indicator reading between the inside bore of the hose and the outer surface of the hose shall not exceed the values given below:
Size 1/4 in and under: 0.030 in (0.762 mm)
Size over 1/4 in: 0.040 in (1.016 mm)

TABLE 5—BURST, HYDROSTATIC, AND CHANGE-IN-LENGTH TESTS FOR SAE 30R3

Size, in	Burst Test, min		Hydrostatic Proof		Change-in-Length	
	psi	MPa	psi	MPa	psi	MPa
3/16	2000	13.80	1000	6.89	500	3.45
1/4	1600	11.04	800	5.51	400	2.76
5/16	1200	8.27	600	4.14	300	2.07
3/8	900	6.20	450	3.10	225	1.53

of the original diameter during application of vacuum for 15, and not over 30, s. Tests shall be conducted with hose in a straight position.

6.4 Cold Flexibility—Two samples shall be used for this test. One sample shall be unaged and the other sample shall be immersed in ASTM Oil No. 3 for 70 h at 100 ±1°C (212 ±2°F). The aged and unaged samples shall then be subjected to a temperature of −40 ±1°C (−40 ±2°F) for a period of 5 h, after which the hose shall be flexed in the cold chamber through 180 deg from the centerline to a diameter of 10 times the maximum OD of the hose. This flexing shall be within 4 s. The hose shall not fracture and shall not show any cracks, checks, or breaks in the tube or cover. Cracking of the tube may be determined by application of the proof pressure specified in Table 5.

6.5 Tensile Strength and Elongation

Original tensile strength of tube: 1200 psi (8.27 MPa) min
Original elongation of tube: 200% min

6.6 Dry Heat Resistance—After oven aging for 70 h at a temperature of 100 ±1°C (212 ±2°F), the reductions in tensile strength and elongation of specimens taken from the tube shall not exceed the following values:

Original tensile strength: −20%
Original elongation: −50%

6.7 Fuel Resistance—After 48 h immersion at room temperature in ASTM Reference Fuel B, the reductions in tensile strength and elongation of specimens taken from the tube shall not exceed the following values:

Original tensile strength: −30%
Original elongation: −30%

In addition, the volume change of specimens taken from the tube shall not exceed the following values:

Tube, volume change: −5 to +25%

6.8 Oil Resistance—After 70 h immersion at a temperature of 100 ±1°C (212 ±2°F) in ASTM Oil No. 3, the reductions in tensile strength and elongation of specimens taken from the tube shall not exceed the following values:

Original tensile strength: −40%
Original elongation: −40%

In addition, the volume change of specimens taken from the tube shall not exceed the following values:

Tube, volume change: −5 to +25%

6.9 Ozone Resistance—Test procedure shall be in accordance with ASTM D 1149 where applicable.

A specimen of hose of sufficient length shall be bent around a mandrel with an OD equal to eight times the nominal OD of the sample. The two ends shall be tied at their crossing with enameled copper or aluminum wire. After mounting, the specimen shall be allowed to rest in an ozone-free atmosphere for 24 h at room temperature. The mounted specimen shall be placed in a test chamber with ozone concentration of 50 pphm at a temperature of 40 ±1°C (104 ±2°F).

After 70 h of exposure, the specimen shall be removed and allowed to cool to room temperature and then be inspected visually under 7X magnification. It must meet a rating of "0" except for the area immediately adjacent to the wire, which shall be ignored.

6.10 Proof Test—Before shipment by the vendor, a suitable number of assemblies from each lot shall be tested at the hydrostatic proof pressure listed in Table 5 for a period of not less than 30, or more than 60 s, to ensure an acceptable quality level.

6.11 Tensile Test of Assembly—The hose complete with fittings shall be dry aged at 100 ±1°C (212 ±2°F) for 70 h and then permitted to rest at room temperature for 2 h. The end fittings of the assembly shall be clamped in the jaws of a tension testing machine so that a straight pull may be applied. The jaws of the test machine shall separate at a rate not greater than 1 in. (25.4 mm)/min. The hose assembly shall withstand, after the aging test, a minimum pull of 100 lb (444 N) on the 3/16 in. (4.76 mm) size, 125 lb (556 N) on the 1/4 and 5/16 in. (6.35 and 7.94 mm) sizes, and 150 lb (667 N) on the 3/8 in. (9.53 mm) size. For inspection tests, unaged samples may be used.

6.12 Leakage Test—Coupled hose filled with ASTM Oil No. 3 shall show no leakage under the hydrostatic proof test pressure specified in Table 5 following an aging period of 70 h at 100 ±1°C (212 ±2°F). The pressure shall be held for a period of not less than 5, or more than 7, min.

6.13 Corrosion Test—The assembly shall be tested in accordance with ASTM B 117, Method of Salt Spray (Fog) Testing. The period shall be 48 h. There shall be no evidence of corrosion or other deterioration at the expiration of this test.

6.14 Visual Inspection—All assemblies shall be inspected to see that the correct fittings are properly applied.

WIRE INSERTED SYNTHETIC RUBBER TUBE AND COVER (SAE 30R5)

1. Scope—This specification covers a wire inserted hose for fuel and oil filler and vent use in mobile, stationary, or marine applications. The hose is furnished uncoupled in specific lengths and is secured in application by the use of suitable clamps. The hose is particularly useful in applications where it must be installed in a curved configuration and where resistance to collapse is desirable.

2. Hose Construction—The hose will consist of a fuel- and oil-resistant tube, a helical high-carbon steel wire embedded in the convoluted section of this hose and running out into the plain[2] ends and an ozone- and oil-resistant cover. A ply of fabric or cord may be applied between the tube or cover and the helical wire. A ply of fabric must be used to reinforce the ends.

3. Dimensions and Tolerances—The applicable dimensions and tolerances are shown in Table 6.

4. Physical Tests—The minimum burst shall be 90 psi (0.621 MPa) for all sizes.

Test samples for other physical properties shall be taken from the plain ends and shall be conducted per ASTM D 380 unless otherwise agreed between vendor and purchaser.

The physical properties of the tube and cover shall be as shown in Table 7.

5. Low-Temperature Flexibility—Specimens of the plain end shall be cut 1 in. (25.4 mm) wide. They shall be exposed in a cold box for 5 h at −40 ±1°C (−40 ±2°F). Without removing it from the cold box, the specimen shall be compressed to 50% of its original ID between parallel plates in 8–12 s. After removal and allowing it to come to room temperature, it shall be carefully examined visually. Any evidence of crack or breaking shall be cause for rejection.

6. Vacuum Collapse—The hose shall be subjected to 20 in. Hg (67.5 kPa) vacuum. During the test, the entire hose shall be mounted on suitable nipples. The OD shall not collapse more than 20% in the body section and there shall be no evidence of separation of the tube from the body wire when examined after application of the vacuum for 30–60 s.

7. Ozone Test—Due to the construction and application of this type of hose,

[2] Synonyms may be "straight ends" or "soft ends."

TABLE 6—DIMENSIONS AND TOLERANCES FOR SAE 30R5

ID		Tolerance		Wall		Length		Tolerance	
in	mm	in	mm	in	mm	in	mm	in	mm
3/4	19.05	+0.03 −0.06	+0.76 −1.52	0.12/0.22	3.05/5.89	0 thru 12	0–305	±0.25	± 6.35
1	25.40	+0.03 −0.06	+0.76 −1.52	0.12/0.22	3.05/5.89	Over 12 thru 24	305–610	±0.50	±12.70
1-1/4	31.75	+0.03 −0.09	+0.76 −2.28	0.12/0.22	3.05/5.89	Over 24 thru 36	610–915	±0.75	±19.05
1-1/2	38.10	+0.03 −0.09	+0.76 −2.28	0.12/0.22	3.05/5.89	Over 36	915	±2%	±2%
1-3/4	44.45	+0.03 −0.09	+0.76 −2.28	0.12/0.22	3.05/5.89	—	—	—	—
2	50.80	+0.03 −0.09	+0.76 −2.28	0.12/0.22	3.05/5.89	—	—	—	—
2-1/4	57.35	+0.03 −0.09	+0.76 −2.28	0.12/0.22	3.05/5.89	—	—	—	—
2-1/2	63.50	+0.03 −0.09	+0.76 −2.28	0.12/0.22	3.05/5.89	—	—	—	—

The minimum tube gage for all sizes shall be 0.062 in (1.57 mm).

TABLE 7—PHYSICAL PROPERTIES OF 30R5 TUBE AND COVER

Property	Tube	Cover
Original tensile, min	1200 psi (8.27 MPa)	1000 psi (6.89 MPa)
Original elongation, min, %	200	200
Tensile after oven aging 100 h at 100 ±1°C (212 ±2°F) (ASTM D 573), min, % of original	80	80
Elongation after oven aging 100 h at 100 ±1°C (212 ±2°F) (ASTM D 573), min, % of original	70	80
Tensile after immersion in Fuel B, 48 h (ASTM D 471), min, % of original	70	Not required
Elongation after immersion in Fuel B, 48 h (ASTM D 471), min, % of original	70	Not required
Volume change after immersion in Fuel B, 48 h (ASTM D 471), %	−5 to +25	Not required
Tensile after immersion in ASTM No. 3 Oil, 70 h at 100 ±1°C (212±2°F), min, % of original	60	Not required
Elongation after immersion in ASTM No. 3 Oil, 70 h at 100 ±1°C (212±2°F), min, % of original	60	Not required
Volume change after immersion in ASTM No. 3 Oil, 70 h at 100±1°C (212±2°F), %	−5 to +25	0 to +100

ozone deterioration of the cover is particularly significant. The test procedure shall be in accordance with ASTM D 1149 where applicable. The entire hose or a section removed therefrom shall be used as agreed between supplier and user and the specimen shall be mounted as agreed. After mounting, the specimen shall be allowed to rest in an ozone-free atmosphere for 24 h at room temperature. This shall then be placed in a test chamber with ozone concentration of 50 pphm at a temperature of 40 ±1°C (104 ±2°F).

After 70 h exposure, the specimen shall be removed and allowed to cool to room temperature and then shall be inspected visually under 7X magnification. It must meet a rating of "0" except for areas which have been agreed can be ignored.

8. Qualification and Inspection Testing—For qualifications, 10 lengths of each size to be qualified shall be furnished and shall be subjected to all tests as described above. For inspection testing, not less than two samples shall be selected from each lot which shall not exceed 10,000 pieces and shall be subjected to the burst, vacuum collapse, and low-temperature flexibility tests described above. Any hose that fails in one or more tests shall be resampled and retested. Twice the number of specimens shall be selected from the lot in question for any retest. Failure of any of the retest samples shall be cause for rejection.

POWER STEERING PRESSURE HOSE—HIGH VOLUMETRIC EXPANSION TYPE—SAE J188

SAE Standard

Report of Nonmetallic Materials Committee approved August 1970.

The specifications in this SAE Standard originated in the SAE-ASTM Technical Committee on Automotive Rubber (other than tires). They represent the correlation of the best information available from research investigation and production experience on the minimum constructional and performance characteristics essential for new power steering hose assemblies used as original or replacement equipment. The standard applies to passenger cars. It may prove useful to truck manufacturers, but it is not to be presented as present practices.

They also represent the minimum quality recognized by original equipment manufacturers and hose suppliers as essential for satisfactory and safe operation by the hose itself and other coacting parts of the power steering system. The original equipment manufacturer may, at his option, add or alter tests through OEM specifications.

Scope—This specification covers two types of hose fabricated from fabric braid and synthetic rubber, assembled with end fittings for use in automotive power steering applications as flexible connections within the temperature range of −40 to +121 C (−40 to +250 F) average, +149 C (+300 F) maximum peaks.

These hoses are intended for use in applications where reduction in amplitude of pump pressure pulsations is required.

Type 1 hose shall be suitable for 1500 psi maximum working pressure.

Type 2 hose shall be suitable for 1300 psi maximum working pressure.

Hose Construction—The construction of this hose embodies a smooth bore inner tube of suitable synthetic rubber material, reinforced with two plies of braided fabric and covered with a synthetic rubber outer cover.

Dimensions

	Type 1	Type 2
ID	0.365/0.405	0.359/0.398
OD	0.750/0.813	0.750/0.813

Concentricity based upon full indicator reading between the inside bore and the outer surface of the hose shall not exceed 0.030 in.

Test Procedures—Procedures described in ASTM D 380, Methods of Testing Rubber Hose, shall be followed wherever applicable.

Qualification Tests—To qualify hose under this specification, all of the requirements shown under Test Requirements must be met.

Inspection Tests—Production shipments or lots of qualified hose shall be tested in accordance with Table 1 and shall conform to the applicable test requirements, but the user may test hose or hose assemblies from any or all such production shipments or lots to all the test requirements. Fourteen sample hose assemblies, selected at random, as listed in Table 1, are required to conduct a complete test. In the event of a failure, the test or tests which have failed shall be retested using twice the number of samples indicated in Table 1. Failure of any of the retested samples shall be cause for rejection of the entire lot.

Frequency of Testing for Inspection—All inspection tests except Impulse shall be performed on either a bulk hose lot or a coupled hose lot basis or tests may be split between a bulk hose lot and a coupled hose lot.

A coupled hose lot shall not exceed 10,000 hose assemblies and a bulk hose lot shall not exceed 20,000 ft of bulk hose. The lot size for Impulse testing shall not exceed 100,000 ft of bulk hose.

TABLE 1—INSPECTION TESTS

Test	Samples Required
1. Volumetric Expansion (paragraph 7) followed by Length Change (paragraph 11) followed by Bursting Strength (paragraph 5)	3
2. Tensile (paragraph 2)	3
3. Adhesion (paragraph 4)	1
4. Low Temperature Flexibility (paragraph 3)	1
5. Impulse (paragraph 1)	6

Test Requirements

1. Impulse Test

TEST CONDITIONS

Oil Temperature: 135 ±2 C (275 ±5 F).
Ambient Temperature: 104 ±11 C (220 ±20 F).
Cycle Rate: 30–40 per minute.
Cycle Data: Pressure rise time, 0.20 ±0.10 sec. High pressure hold time, 0.65 ±0.20 sec. Pressure drop time, 0.20 ±0.10 sec.
Pressure Variation:
  Type 1 0–100 to 1500 psi
  Type 2 0–100 to 1300 psi

HYDRAULIC FLUID AND TEST FIXTURE—As specified by the original equipment manufacturer.

CYCLE LIFE—Samples submitted to this test shall exceed 100,000 cycles for inspection acceptance and 225,000 cycles for qualification testing, without failure.

2. Tensile Test—When tested in accordance with ASTM D 571, Testing Automotive Hydraulic Brake Hose, hose assemblies shall withstand a minimum tensile load of 1200 lb without the fittings pulling off or rupture of the hose.

3. Low Temperature Flexibility—Samples shall be subjected to a temperature of −40 ±1 C (−40 ±2 F) for a period of 24 hr, after which the hose shall be flexed in the cold chamber through 180 deg from the centerline around a mandrel whose diameter is eight times the nominal hose OD. Flexing shall be accomplished within 4 sec. Hose shall not fracture or show any cracks, checks, or breaks in the tube or cover.

4. Adhesion Test—When tested in accordance with ASTM D 413, Tests for Adhesion of Vulcanized Rubber (Friction Test), a pull of not less than 10 lb shall be required to separate a 1 in. wide ring section of the bond between any adjacent layers of the hose.

5. Bursting Strength—Samples shall meet the following minimum bursting strength requirements:
  Type 1—6000 psi
  Type 2—5200 psi

6. Ozone Resistance—Test procedure and apparatus shall be in accordance with ASTM D 1149, Test for Accelerated Ozone Cracking of Vulcanized Rubber, where applicable. A specimen of the hose shall be bent around a mandrel having an outside diameter seven times the nominal outside diameter of the hose under test. The two ends of the hose shall be tied where they cross one another with enameled copper or aluminum wire. After mounting, the specimens shall be permitted to rest in a relatively ozone-free atmosphere for 24 hr at room temperature. The mounted specimen shall be placed in a suitable ozone test chamber that is maintained at an ozone concentration of 50 ±5 parts ozone per 100 million parts of air (by volume) and a chamber ambient temperature of 38 ±1 C (100 ±2 F).

After 72 hr of exposure, specimens shall be removed from the chamber and permitted to cool to room temperature and then, while still on the mandrel, shall be visually inspected for signs of cracking under 7X magnification. There shall be no evidence of cracking of the cover. The area immediately adjacent to the wire shall be ignored in making the visual inspection.

7. Volumetric Expansion—Samples shall be installed on the test fixture in a manner which allows the hose to lie straight but not under tension when test pressure is applied. Samples shall be preconditioned at 1300 ±20 psi for 30 ±5 sec and allowed to recover at 0 psig for 2 min ±5 sec prior to the test. Test pressure shall then be applied and held for 2 min ±5 sec. At the end of this period, the inlet valve shall be closed and the test fluid allowed to rise in the burrette for a period of 1 min ±5 sec. The outlet valve shall then be closed and the reading taken. The average of three volumetric expansion readings per sample shall not exceed the values listed below:

	Volumetric Expansion, cc/ft at 0–1300 psi	
	Max	Min
Type 1	8	3
Type 2	17	8

8. Tensile Strength and Elongation
  Original tensile strength of cover: 1000 psi min
  Original tensile strength of tube: 1000 psi min
  Original elongation of cover: 175% min
  Original elongation of tube: 150% min

9. Dry Heat Resistance—After oven aging for 70 hr at a temperature of 100 C (212 F), the reduction in tensile strength and elongation of specimens taken from tube and cover material shall not exceed the following values:
  Original tensile strength: 20%
  Original elongation: 50%

10. Oil Resistance—After 70 hr immersion in ASTM No. 3 oil at a temperature of 121 C (250 F), the reductions in tensile strength and elongation of specimens taken from the tube material shall not exceed the following values:
  Original tensile strength: 65%
  Original elongation: 50%

In addition, the volume change of specimens taken from the tube and cover materials shall not exceed 0 to +100%.

11. Length Change—Type 1 hose shall not change in length more than +2%, −5% when pressure is increased from 0 to 1500 psig.

Type 2 hose shall not change in length more than +0%, −8% when pressure is increased from 0 to 1300 psig.

12. Identification Marking—Hose shall be identified with the SAE number, type, size of inside diameter in fractions, date code in days of the year and last digit of the year (for example, 1707 represents the 170th day of 1967), and hose manufacturer's and/or coupling manufacturer's code marking. This marking shall appear on the outer cover of the hose at intervals of not greater than 10 in.

Additional identification may be added as agreed upon by user and supplier.

13. 100% Proof Pressure Test—Each hose assembly shall be proof pressure tested using air, oil, or water as the pressure medium. Test pressure shall be 1500 psi when tested with air and 2000 psi when tested with oil or water. Test pressure shall be held for not less than 30 nor more than 60 sec. Care should be taken when testing with air due to its explosive nature at high pressure.

POWER STEERING RETURN HOSE— LOW PRESSURE—SAE J189

SAE Standard

Report of Nonmetallic Materials Committee approved August 1970.

The specifications in this SAE standard originated in the SAE-ASTM Technical Committee on Automotive Rubber (other than tires). They represent the correlation of the best information available from research investigation and production experience on the minimum constructional and performance characteristics essential for new power steering hose assemblies used as original or replacement equipment. This standard applies to passenger cars. It may prove useful to truck manufacturers, but it is not to be presented as present practices.

They also represent the minimum quality recognized by original equipment manufacturers and hose suppliers as essential for satisfactory and safe operation by the hose itself and other coacting parts of the power steering system. The original equipment manufacturer may, at his option, add or alter tests through OEM specifications.

Scope—This specification covers hose fabricated from fabric braid and synthetic rubber, assembled with end fittings or user applied clamps for use in automotive power steering applications as flexible connections within the temperature range of −40 C (−40 F) to +121 C (+250 F) average, 149 C (300 F) maximum peaks. Hose assemblies shall be suitable for 250 psi maximum working pressure with end fittings and 100 psi maximum working pressure with user applied clamps.

Hose Construction—The construction of this hose embodies a smooth bore inner tube of suitable synthetic rubber material, reinforced with one ply of braided fabric and covered with a synthetic rubber outer cover.

Dimensions—Suggested hose dimensions are given in Table 1, but it is not the intent of this specification to exclude hose with different dimensions that comply with all other requirements of this specification.

In addition, concentricity based upon full indicator reading between the inside bore and the outer surface of the hose shall not exceed 0.030 in.

Test Procedures—Procedures described in ASTM D 380, Methods of Testing Rubber Hose, shall be followed wherever applicable.

Qualification Tests—To qualify hose under this specification all of the requirements shown under Test Requirements must be met.

Inspection Tests—Production shipments or lots of qualified hose shall be tested in accordance with Table 2 and shall conform to the applicable test requirements, but the user may test hose or hose assemblies from any or all such production shipments or lots to all the test requirements. Fourteen sample hose assemblies, selected at random, as listed in Table 2, are required to conduct a complete test. In the event of a failure, the test or tests which have failed shall be retested using twice the number of samples indicated in Table 2. Failure of any of the retested samples shall be cause for rejection of the entire lot.

Frequency of Testing for Inspection—All inspection tests except Impulse shall be performed on either a bulk hose lot or a coupled hose lot basis or tests may be split between a bulk hose lot and a coupled hose lot.

A coupled hose lot shall not exceed 10,000 hose assemblies and a bulk hose lot shall not exceed 20,000 ft of bulk hose. The lot size for Impulse testing shall not exceed 100,000 ft of bulk hose.

Test Requirements

1. Impulse Test—(Not applicable to hose assembled with user applied clamps.)

TEST CONDITIONS

Oil Temperature: 135 ±2 C (275 ±5 F).
Ambient Temperature: 93 C (200 F) max.
Cycle Rate: 30–40 per minute.
Cycle Data: Pressure rise time, 0.20 ±0.10 sec. High pressure hold time, 0.65 ±0.20 sec. Pressure drop time, 0.20 ±0.10 sec.
Pressure Variation: 0–25 maximum recommended working pressure.

HYDRAULIC FLUID AND TEST FIXTURE—As specified by the original equipment manufacturer.

CYCLE LIFE—Samples submitted to this test shall exceed 100,000 cycles for inspection acceptance and 225,000 cycles for qualification testing, without failure.

2. Tensile Test—When tested in accordance with ASTM D 571, Testing Automotive Hydraulic Brake Hose, end fittings shall withstand a minimum tensile load as shown in Table 1 without the fittings pulling off or rupture of the hose.

3. Low Temperature Flexibility—Samples shall be subjected to a temperature of −40 ±1 C (−40 ±2 F) for a period of 24 hr, after which the hose shall be flexed in the cold chamber through 180 deg from the centerline around a mandrel having a diameter of eight times the nominal OD of the hose. Flexing shall be accomplished within 4 sec. Hose shall not fracture or show any cracks, checks, or breaks in the tube or cover.

4. Adhesion Test—When tested in accordance with ASTM D 413, Tests for Adhesion of Vulcanized Rubber (Friction Test), a pull of not less than 8 lb shall be required to separate a 1 in. wide ring section of the bond between any adjacent layers of the hose.

TABLE 2—INSPECTION TESTS

Test	Sample Required
1. Length Change (paragraph 10) followed by Burst Strength (paragraph 5)	3
2. Tensile (paragraph 2)	3
3. Low Temperature Flexibility (paragraph 3)	1
4. Adhesion (paragraph 4)	1
5. Impulse (paragraph 1)	6

5. Bursting Strength—Samples shall meet the minimum bursting strength requirements shown in Table 1.

6. Ozone Resistance—Test procedure and apparatus shall be in accordance with ASTM D 1149, Test for Accelerated Ozone-Cracking of Vulcanized Rubber, where applicable. A specimen of the hose shall be bent around a mandrel having an outside diameter seven times the nominal outside diameter of the hose under test. The ends of the hose shall be tied where they cross one another with enameled copper or aluminum wire. After mounting, the specimens shall be permitted to rest in a relatively ozone-free atmosphere for 24 hr at room temperature. The mounted specimens shall be placed in a suitable ozone test chamber that is maintained at an ozone concentration of 50 ±5 parts ozone per 100 million parts of air (by volume) and a chamber ambient temperature of 38 ±1 C (100 ±2 F). After 72 hr of exposure, specimens shall be removed from the chamber and permitted to cool to room temperature and then, while still on the mandrel, shall be visually inspected for signs of cracking under 7X magnification. There shall be no evidence of cracking of the cover. The area immediately adjacent to the wire shall be ignored in making the visual inspection.

7. Tensile Strength and Elongation

Original tensile strength of cover: 1000 psi min
Original tensile strength of tube: 1000 psi min
Original elongation of cover: 175% min
Original elongation of tube: 150% min

8. Dry Heat Resistance—After oven aging for 70 hr at a temperature of 100 C (212 F), the reductions in tensile strength and elongation of specimens taken from tube and cover material shall not exceed the following values:

Original tensile strength: 20%
Original elongation: 50%

9. Oil Resistance—After 70 hr immersion in ASTM No. 3 oil at a temperature of 121 C (250 F), the reductions in tensile strength and elongation of specimens taken from the tube material shall not exceed the following values:

Original tensile strength: 65%
Original elongation: 50%

In addition, the volume change of specimens taken from the tube and cover shall not exceed 0 to +100%.

10. Length Change—Hose shall not change more than +0%, −10% when the pressure is increased from 0 psi to the maximum recommended working pressure shown in Table 1.

11. Identification Marking—Hose shall be identified with the SAE number, size of inside diameter in fractions, date code in days of the year and last digit of the year (for example, 1707 represents the 170th day of 1967), and hose manufacturer's and/or coupling manufacturer's code marking. This marking shall appear on the outer cover of the hose at intervals of not greater than 10 in. Additional identification may be added as agreed upon by user and supplier.

12. 100% Proof Pressure Test—Hose or hose assemblies shall be proof pressure tested at the recommended working pressure using air, oil, or water as the pressure medium. Pressure shall be held for not less than 30 nor more than 60 sec. Care should be taken when testing with air due to its explosive nature at high pressures.

TABLE 1

Nominal ID, in.	Nominal OD, in.	ID Tolerance, in.	OD Tolerance, in.	Recommended Working Pressure, max, psi		Tensile Load, min, lb		Burst Strength, min, psi
				With End Fittings	With User Applied Clamps	With End Fittings	With User Applied Clamps	
3/8	21/32	0.390/0.344	0.688/0.625	250	100	450	Not applicable	1000

POWER STEERING PRESSURE HOSE—LOW VOLUMETRIC EXPANSION TYPE—SAE J191

SAE Standard

Report of Nonmetallic Materials Committee approved August 1970.

The specifications in this SAE Standard originated in the SAE-ASTM Technical Committee on Automotive Rubber (other than tires). They represent the correlation of the best information available from research investigation and production experience on the minimum constructional and performance characteristics essential for new power steering assemblies used as original or replacement equipment. This standard applies to passenger cars. It may prove useful to truck manufacturers, but it is not to be presented as present practices.

They also represent the minimum quality recognized by original equipment manufacturers and hose suppliers as essential for satisfactory and safe operation by the hose itself and other coacting parts of the power steering system. The original equipment manufacturer may, at his option, add or alter tests through OEM specifications.

Scope—The specification covers hose fabricated from fabric braid and synthetic rubber, assembled with end fittings for use in automotive power steering applications at pressures as indicated in Table 1, as flexible connections within the temperature range of −40 C (−40 F) to 121 C (250 F) average, 149 C (300 F) maximum peaks.

These hoses are intended for use in applications where reduction in amplitude of pump pressure pulsations is not required.

Hose Construction—This construction of this hose embodies a smooth bore inner tube of suitable synthetic rubber material, reinforced with two plies of braided fabric and covered with a synthetic rubber outer cover.

Dimensions—Hose must be within the tolerances shown in Table 1. In addition, the concentricity based upon full indicator reading between the inside bore and the outer surface of the hose shall not exceed 0.030 in.

Test Procedure—Procedures described in ASTM D 380, Methods of Testing Rubber Hose, shall be followed wherever applicable.

Qualification Tests—To qualify hose under this specification, all of the requirements shown under Test Requirements must be met.

Inspection Tests—Production shipments or lots of qualified hose shall be tested in accordance with Table 2 and shall conform to the applicable test requirements, but the user may test hose or hose assemblies from any or all such production shipments or lots to all the test requirements. Fourteen sample hose assemblies, selected at random, as listed in Table 2 are required to conduct a complete test. In the event of a failure, the test or tests which have failed shall be retested using twice the number of samples indicated in Table 2. Failure of any of the retested samples shall be cause for rejection of the entire lot.

Frequency of Testing for Inspection—All inspection tests except Impulse shall be performed on either a bulk hose lot or a coupled hose lot basis or tests may be split between a bulk hose lot and a coupled hose lot.

A coupled hose lot shall not exceed 10,000 hose assemblies and a bulk hose lot shall not exceed 20,000 ft of bulk hose. The lot size for Impulse testing shall not exceed 100,000 ft of bulk hose.

Test Requirements

1. Impulse Test

TEST CONDITIONS:
Oil Temperature: 135 ±2 C (275 ±5 F).
Ambient Temperature: 104 ±11 C (220 ±20 F).
Cycle Rate: 30–40 per minute.
Cycle Data: Pressure rise time, 0.20 ±0.10 sec. High pressure hold time, 0.65 ±0.20 sec. Pressure drop time, 0.20 ±0.10 sec.
Pressure Variation: 0–100 psi to maximum working pressure listed in Table 1.

HYDRAULIC FLUID AND TEST FIXTURE: As specified by the original equipment manufacturer.

Cycle Life: Samples submitted to this test shall exceed 100,000 cycles for inspection acceptance and 225,000 cycles for qualification testing, without failure.

2. Tensile Test—When tested in accordance with ASTM D 571, Testing Automotive Hydraulic Brake Hose, hose assemblies shall withstand a minimum tensile load as specified in Table 1 without the fittings pulling off or rupture of the hose.

3. Low Temperature Flexibility—Samples shall be subjected to a temperature of −40 ±1 C (−40 ±2 F) for a period of 24 hr, after which the hose shall be flexed in the cold chamber through 180 deg from the centerline around a mandrel whose diameter is eight times the nominal hose OD. Flexing shall be accomplished within 4 sec. Hose shall not fracture or show any cracks, checks, or breaks in the tube or cover.

4. Adhesion Test—When tested in accordance with ASTM D 413, Tests for Adhesion of Vulcanized Rubber (Friction Test), a pull of not less than 10 lb shall be required to separate a 1 in. wide ring section of the bond between any adjacent layers of the hose.

5. Bursting Strength—Samples shall meet the minimum bursting strength requirements shown in Table 1.

6. Ozone Resistance—Test procedure and apparatus shall be in accordance with ASTM D 1149, Test for Accelerated Ozone Cracking of Vulcanized Rubber, where applicable. A specimen of the hose shall be bent around a mandrel having an outside diameter seven times the nominal outside diameter of the hose under test. The two ends of the hose shall be tied where they cross one another with enameled copper or aluminum wire. After mounting, the specimens shall be permitted to rest in a relatively ozone-free atmosphere for 24 hr at room temperature. The mounted specimen shall be placed in a suitable ozone test chamber that is maintained at an ozone concentration of 50 ±5 parts ozone per 100 million parts of air (by volume) and a chamber ambient temperature of 38 ±1 C (100 ±2 F).

After 72 hr of exposure, specimens shall be removed from the chamber and permitted to cool to room temperature and then, while still on the mandrel, shall be visually inspected for signs of cracking under 7X magnification. There shall be no evidence of cracking of the cover. The area immediately adjacent to the wire shall be ignored in making the visual inspection.

7. Length Change—Hose shall not change in length more than +2%, −4% when pressure is increased from 0 psi to the maximum working pressure shown in Table 1.

8. Tensile Strength and Elongation
Original tensile strength of cover: 1000 psi min
Original tensile strength of tube: 1000 psi min
Original elongation of cover: 175% min
Original elongation of tube: 150% min

9. Dry Heat Resistance—After oven aging for 70 hr at a temperature of 100 C (212 F), the reduction in tensile strength and elongation of specimens taken from tube and cover material shall not exceed the following values:
Original tensile strength: 20%
Original elongation: 50%

10. Oil Resistance—After 70 hr immersion in ASTM No. 3 oil at a temperature of 121 C (250 F), the reduction in tensile strength and elongation of specimens taken from the tube material shall not exceed the following values:
Original tensile strength: 65%
Original elongation: 50%

TABLE 2—INSPECTION TESTS

Test	Samples Required
1. Length Change (paragraph 7) followed by Bursting Strength (paragraph 5)	3
2. Tensile (paragraph 2)	3
3. Adhesion (paragraph 4)	1
4. Low Temperature Flexibility (paragraph 3)	1
5. Impulse (paragraph 1)	6

TABLE 1

Nominal ID, in.	Nominal OD, in.	ID Tolerance, in.	OD Tolerance, in.	Working Pressure, max, psi	Proof Pressure, psi	Burst Strength, min, psi	Tensile Strength, min, lb
1/4	9/16	0.275/0.240	0.592/0.552	1300	2000	6000	450
3/8	3/4	0.400/0.365	0.781/0.719	1200	2000	4800	1500
1/2	15/16	0.530/0.490	0.969/0.906	1000	2000	4000	1500

In addition, the volume change of specimens taken from the tube and cover material shall not exceed 0 to +100%.

11. Identification Marking—Hose shall be identified with the SAE number, type, size of inside diameter in fractions, date code in days of the year and last digit of the year (for example, 1707 represents the 170th day of 1967), and hose manufacturer's and/or coupling manufacturer's code marking. This marking shall appear on the outer cover of the hose at intervals of not greater than 10 in. Additional identification may be added as agreed upon by user and supplier.

12. 100% Proof Pressure Test—Each hose assembly shall be proof pressure tested using air, oil, or water as the pressure medium. Hose shall be tested at the maximum working pressure when air is used or at the proof pressure shown in Table 1, when tested with oil or water. Care should be taken when testing with air due to its explosive nature at high pressure.

POWER STEERING PRESSURE HOSE—WIRE BRAID—SAE J190

SAE Standard

Report of Nonmetallic Materials Committee approved August 1970.

The specifications in this SAE Standard originated in the SAE-ASTM Technical Committee on Automotive Rubber (other than tires). They represent the correlation of the best information available from research investigation and production experience on the minimum constructional and performance characteristics essential for new power steering assemblies used as original or replacement equipment. This standard applies to passenger cars. It may prove useful to truck manufacturers, but it is not to be presented as present practices.

They also represent the minimum quality recognized by original equipment manufacturers and hose suppliers as essential for satisfactory and safe operation by the hose itself and other coacting parts of the power steering system. The original equipment manufacturer may, at his option, add or alter tests through OEM specifications.

Scope—This specification covers hose fabricated from wire braid and synthetic rubber, assembled with end fittings for use in automotive applications up to 1500 psi maximum pressure, as flexible connections within the temperature range of −40 to +121 C (−40 to +250 F) average, 149 C (300 F) maximum peaks.

Hose Construction—The construction of this hose embodies a smooth bore inner tube of suitable synthetic rubber material, reinforced with one ply of wire braid and covered with a synthetic rubber outer cover.

Dimensions—Hose must be within the tolerances shown in Table 1. In addition, the concentricity, based upon full indicator reading, between the inside bore and the outer surface of the hose shall not exceed 0.030 in.

Test Procedures—Procedures described in ASTM D 380, Methods of Testing Rubber Hose, shall be followed wherever applicable.

Qualification Tests—To qualify hose under this specification, all of the requirements shown under Test Requirements must be met.

Inspection Tests—Production shipments or lots of qualified hose shall be tested in accordance with Table 2 and shall conform to the applicable test requirements, but the user may test hose or hose assemblies from any or all such production shipments or lots to all the test requirements. Fourteen sample hose assemblies, selected at random, as listed in Table 2 are required to conduct a complete test. In the event of a failure, the test or tests which have failed shall be retested using twice the number of samples indicated in Table 2. Failure of any of the retested samples shall be cause for rejection of the entire lot.

Frequency of Testing for Inspection—All inspection tests except Impulse shall be performed on either a bulk hose lot or a coupled hose lot basis or tests may be split between a bulk hose lot and a coupled hose lot.

A coupled hose lot shall not exceed 10,000 hose assemblies and a bulk hose lot shall not exceed 20,000 ft of bulk hose. The lot size for Impulse testing shall not exceed 100,000 ft of bulk hose.

Test Requirements

1. Impulse Test

TEST CONDITIONS

Oil Temperature: 135 ±2 C (275 ±5 F).
Ambient Temperature: 104 ±11 C (220 ±20 F).
Cycle Rate: 30–40 per minute.
Cycle data: Pressure rise time, 0.20 ±0.10 sec. High pressure hold time, 0.65 ±0.20 sec. Pressure drop time, 0.20 ±0.10 sec.
Pressure Variation: 0–100 psi to maximum working pressure listed in Table 1.

HYDRAULIC FLUID AND TEST FIXTURE—As specified by the original equipment manufacturer.

CYCLE LIFE—Samples submitted to this test shall exceed 100,000 cycles for inspection acceptance and 225,000 cycles for qualification testing, without failure.

2. Tensile Test—When tested in accordance with ASTM D 571, Testing Automotive Hydraulic Brake Hose, hose assemblies shall withstand a minimum tensile load as specified in Table 1 without the fittings pulling off or rupture of the hose.

3. Low Temperature Flexibility—Hose and/or hose assemblies shall be subjected to −40 ±1 C (−40 ±2 F) for 24 hr. After this time and while still at −40 ±1 C (−40 ±2 F), the samples shall be flexed over a mandrel having a diameter equal to twice the minimum bend radius specified in Table 1 in 4 sec or less. Hose shall be bent through 180 deg over the mandrel. After flexing, the sample shall be allowed to warm to room temperature and be visually examined for cover cracks and subjected to the proof test. There shall be no cover cracks or leakage.

4. Adhesion Test—When tested in accordance with ASTM D 413, Tests for Adhesion of Vulcanized Rubber (Friction Test), a pull of not less than 10 lb shall be required to separate a 1 in. wide ring section of the bond between the cover and the reinforcement.

5. Bursting Strength—Samples shall meet the minimum bursting strength requirements shown in Table 1.

6. Ozone Resistance—Test procedure and apparatus shall be in accordance with ASTM D 1149, Test for Accelerated Ozone Cracking of Vulcanized Rubber, where applicable. A specimen of the hose shall be bent around a mandrel having an outside diameter seven times the nominal outside diameter of the hose under test. The two ends of the hose shall be tied where they cross one another with enameled copper or aluminum wire. After mounting, the specimens shall be permitted to rest in a relatively ozone-free atmosphere for 24 hr at room temperature. The mounted specimen shall be placed in a suitable ozone test chamber that is maintained at an ozone concentration of 50 ±5 parts ozone per 100 million parts of air (by volume) and a chamber ambient temperature of 38 ±1 C (100 ±2 F).

After 72 hr of exposure, specimens shall be removed from the chamber and permitted to cool to room temperature and then, while still on the mandrel, shall be visually inspected for signs of cracking under 7X magnification. There shall be no evidence of cracking of the cover.

TABLE 2—INSPECTION TESTS

Test	Samples Required
1. Length Change (paragraph 7) followed by Bursting Strength (paragraph 5)	3
2. Tensile (paragraph 2)	3
3. Adhesion (paragraph 4)	1
4. Low Temperature Flexibility (paragraph 3)	1
5. Impulse (paragraph 1)	6

TABLE 1

Nominal ID, in.	Nominal OD, in.	ID Tolerance, in.	OD Tolerance, in.	Wire OD, in.	Working Pressure, max, psi	Proof Pressure, psi	Burst Strength, min, psi	Tensile Strength, min, lb	Bend Radius, min, in.
3/8	25/32	0.398/0.367	0.812/0.750	0.617/0.571	1500	2000	9000	1000	5
1/2	29/32	0.531/0.485	0.938/0.875	0.750/0.688	1500	2000	8000	1200	7

7. Length Change—Hose shall not change in length more than +2%, −4% when pressure is increased from 0 psig to maximum working pressure shown in Table 1.

8. Tensile Strength and Elongation

Original tensile strength of cover: 1000 psi min
Original tensile strength of tube: 1000 psi min
Original elongation of cover: 175% min
Original elongation of tube: 100% min

9. Dry Heat Resistance—After oven aging for 70 hr at a temperature of 100 C (212 F), the reduction in tensile strength and elongation of specimens taken from the tube and cover material shall not exceed the following values:

Original tensile strength: 20%
Original elongation: 50%

10. Oil Resistance—After 70 hr immersion in ASTM No. 3 oil at a temperature of 121 C (250 F), the reductions in tensile strength and elongation of specimens taken from the tube material shall not exceed the following values:

Original tensile strength: 65%
Original elongation: 50%

In addition, the volume change of specimens taken from the tube material shall not exceed 0 to +100%.

11. Identification Marking—Hose shall be identified with the SAE number, size of inside diameter in fractions, date code in days of the year and last digit of the year (for example, 1707 represents the 170th day of 1967), and hose manufacturer's and/or coupling manufacturer's code marking. This marking shall appear on the outer cover of the hose at intervals of not greater than 10 in. Additional identification may be added as agreed upon by user and supplier.

12. 100% Proof Pressure Test—Each hose assembly shall be proof pressure tested using air, oil, or water as the pressure medium. Hose shall be tested at the maximum working pressure when air is used or at the proof pressure shown in Table 1 when tested with oil or water. Care should be taken when testing with air due to its explosive nature at high pressures.

WINDSHIELD WIPER HOSE—SAE J50a
SAE Standard

Report of Nonmetallic Materials Division approved January 1945 and last revised by Nonmetallic Materials Committee July 1964. Editorial change August 1973.

[This SAE Standard was formulated by SAE—ASTM Technical Committee on Automotive Rubber.]

Construction—The hose shall consist of an inner tube of flexible material, reinforced with one or more plies of wrapped fabric or braided cord, or a combination of both, and a cover of flexible material. The tube and cover shall be smooth and uniform in thickness in accordance with good manufacturing practice.

The hose shall be so manufactured as to comply with the test requirements set forth below. Where not specifically designated, the materials and construction shall be suitable for the purpose intended.

Sizes and Tolerances—The inside diameter shall be $\frac{5}{32}$ and $\frac{7}{32}$ in. (3.96 and 5.56 mm); outside diameter shall be $\frac{11}{32}$ and $\frac{13}{32}$ in. (8.74 and 10.31 mm). The inside diameter shall be subject to a tolerance of $\pm\frac{1}{64}$ in. (± 0.40 mm); and the outside diameter to a tolerance of $+\frac{3}{64}, -\frac{1}{64}$ in. (+1.2, −0.4 mm).

Length—Unless otherwise specified, the hose shall be furnished in coils of nominal 50 ft (15 m) lengths or in "reel lengths" of nominal 500 ft (152 m). The reels may consist of a maximum of four lengths, but the minimum length shall be 25 ft (7.6 m).

When "cut lengths" are specified on the contract or order, the length shall not vary by more than $+\frac{1}{16}, -\frac{1}{32}$ in. (+1.6, −0.8 mm) on lengths up to 3 in. (76 mm); $\pm\frac{1}{16}$ in. (± 1.60 mm) on lengths 3–18 in. (76–457 mm); $\pm\frac{1}{4}$ in. (± 6.35 mm) on lengths 18–100 in. (457–2540 mm); and $\pm 1\%$ on lengths over 100 in. (2540 mm).

Methods of Test—Measurements and tests shall be made in accordance with ASTM D 380, Methods of Test of Hose, unless otherwise specified in this SAE Standard.

Test Requirements

Burst Test—The hose shall not burst at a pressure less than 250 psi (1.72 MPa).

Bend Test—No breaking of reinforcement or other components shall occur when the hose is bent at a 90 deg angle while installed on the proper size tubing ($\frac{3}{16}$ in. (4.78 mm) for $\frac{5}{32}$ in. (3.96 mm) size and $\frac{1}{4}$ in. (6.35 mm) for $\frac{7}{32}$ in. (5.56 mm) size).

Vacuum Collapse—The hose shall be subjected to a vacuum of 24 in. Hg (81 kPa) while being curved to a radius equal to five times the maximum OD of the hose. After 5 min, the OD shall not have been reduced by more than 30%. During this test, the inner tube shall not collapse nor pull away from the reinforcement. Tube collapse may be determined by sectioning the hose lengthwise and examining.

Cold Flexibility—Samples of hose shall be subjected to a temperature of $-20°F$ ($-28.8°C$) for a period of 5 h, after which the hose shall be flexed in the cold chamber through 180 deg from the centerline in each direction to a diameter 10 times the maximum OD of the hose at each extreme of the cycle for five cycles. The rate of cycling shall be approximately one cycle in 4 s. The hose shall not fracture nor show any cracks, checks, or breaks in the tube or cover.

Tensile Test—The ends of a sample of hose shall be fastened in the jaws of a testing machine. The jaws shall then be separated at the rate of 1 in. (25.4 mm)/min until rupture of the hose occurs. Failure shall not occur at a tension of less than 60 lb (267 N) for the $\frac{5}{32}$ in. (3.96 mm) size and 80 lb (359 N) for the $\frac{7}{32}$ in. (5.56 mm) size.

Aging Test—After being subjected to dry air aging in accordance with ASTM D 573, Method of Test for Accelerated Aging of Vulcanized Rubber by the Oven Method, for 70 h at 158 $\pm 2°F$ (70 $\pm 1°C$), the hose shall meet the requirements for vacuum collapse and cold flexibility outlined previously.

Ozone Resistance—A specimen of hose having a length equal to the full circumference of the test mandrel plus approximately 10 in. (254 mm) shall be bent around the mandrel, the diameter of which shall be eight times the nominal OD of the hose being tested and bound with a tape or twine where the ends cross one another. If collapse of the hose occurs when bent around the mandrel, provision should be made to support the hose internally. Three specimens shall be conditioned for 70–72 h in air at room temperature and then, while still on the mandrel, shall be placed in an exposure chamber containing air mixed with ozone in the proportion of 50 ± 5 parts of ozone per 100 million parts of air by volume. The ambient temperature in the chamber during the test shall be 100 $\pm 2°F$ (37.7 $\pm 1°C$). The specimen shall be exposed to the ozone and air mixture for a period of 70 h. To determine conformance to this requirement, the cover of the specimen shall not show any cracks when examined under 2X magnification, ignoring areas immediately adjacent to or within the area covered by the tape or twine. The ozone cabinet shall be in accordance with ASTM D 1149, Method of Test for Accelerated Ozone Cracking of Vulcanized Rubber, and the rating per ASTM D 1171, Methods of Test for Fluoride Ion in Industrial Water and Industrial Waste Water.

Sampling—From each lot of 25,000 ft (7620 m) or less of hose, representative samples shall be taken to determine conformance with this specification.

Samples of suitable length for conducting the tests shall be used but shall not exceed 12 in. (305 mm), except in the case of the burst test which shall be conducted on a sample not exceeding 18 in. (457 mm). A separate sample shall be used for each test.

For methods of inspection, retest, and rejection, refer to ASTM D 380.

AUTOMOTIVE AIR CONDITIONING HOSE—SAE J51b
SAE Standard

Report of Nonmetallic Materials Committee approved September 1960 and last revised June 1968. Editorial change December 1971.

Scope—This specification covers reinforced synthetic rubber hose and reinforced thermoplastic hose, or hose assemblies, intended for conducting liquid and gaseous dichlorodifluoromethane (refrigerant-12) in automotive air conditioning systems.

Manufacture

Type A—Textile Reinforced—Hose shall be mandrel built having a seamless oil resistant synthetic rubber tube designed to minimize permeation of dichlorodifluoromethane and contamination of the system and serviceable over a temperature range of −20 to +250 F. The reinforcement shall consist of one or more plies of braided yarn firmly adhered to the tube and cover. The outer cover shall be an abrasion, oil, and weather resistant synthetic rubber compound. The cover shall be pin-pricked or perforated.

Type B—Wire Reinforced—Hose shall be mandrel built having a seamless, oil resistant synthetic rubber tube designed to minimize permeation of dichlorodifluoromethane and contamination of the system and serviceable over a temperature range of −20 to +250 F. The reinforcement shall consist of a single ply of steel wire braid firmly adhered to the rubber tube. There shall be a layer of rubber over the wire braid to prevent corrosion of the wire. The cover shall consist of a synthetic rubber cement-impregnated ply of braided cotton yarn.

Type C—Reinforced thermoplastic hose shall have a thermoplastic tube designed to minimize permeation of dichlorodifluoromethane and contamination of the system and serviceable over a temperature range of −20 to +250 F. The reinforcement shall consist of one or more plies of suitable material. Outer cover shall be abrasion, oil, and weather resistant thermoplastic and may be perforated.

Hose Identification—The hose shall be identified with the SAE number, type, size of inside diameter in fractions, and hose manufacturer's code marking. This marking shall appear on the outer cover of the hose at intervals not greater than 15 in.

Sizes—The suggested hose dimensions are given in Table 1, but it is not the intent of this specification to exclude hose with different dimensions that comply with all other requirements of this specification.

Sampling—Hose assemblies required for qualification and inspection tests are shown in the following chart. Assemblies required for inspection tests are marked "a."

Type of Test	Number of Assemblies	Free Length of Hose, in.
Permeability	12	18
Aging	1	12 to 30
Cold Test	1	18
Vacuum	1	24 to 36
Length Change [a]	1	24 to 36
Burst	1	24 to 36
Tensile [a]	1	12
Extraction	1	18
Cleanliness [a]	1	24

[a] Required for inspection tests.

If uncoupled hose is to be used, a 35 ft length shall be submitted for qualification. For inspection tests a representative sample of hose approximately 7 ft in length shall be selected from each lot to be tested. If a single length of 7 ft is not available, several sections, each of sufficient length to provide the required test specimens may be taken.

Retests and Rejections—Any hose which fails in one or more tests may be resampled and the tests shall be repeated with twice the number of samples selected from the hose for the test that failed to meet the requirements. Failure of any of the retested samples shall be cause for final rejection.

Test Requirements—All measurements and tests necessary for determining the conformity of the hose with these specifications shall be made in accordance with the latest ASTM D 1680, Methods of Testing Automotive Air Conditioning Hose. To qualify hose under this specification, all the requirements shown under qualification tests and inspection tests must be met. Production shipments or lots of the qualified hose shall meet the requirements shown under inspection tests, but the user may in addition, if he so desires, test hose from any or all such production shipments or lots to the requirements under the qualification tests.

Qualification Tests

Permeability Tests—The combined hose and coupling assembly shall not permit effusion of dichlorodifluoromethane at a rate greater than that listed in Table 2 when tested at the specified temperature.

These tests can be accomplished by sealing liquid dichlorodifluoromethane within the hose and calculating effusion from accurate weight loss measurements at the temperature and time listed in Table 2. Since the volume of saturated liquid dichlorodifluoromethane shows a substantial increase with an increase in temperature, care must be taken not to overfill test assemblies or they may burst due to liquid expansion. (An 18 in. hose assembly of 1/2 in. inside diameter hose should be loaded with 35 ±6 gm of dichlorodifluoromethane for a proper test.)

The permeation rate is proportional to the temperature. Fig. 1 may be used to estimate the maximum permeation rate at temperatures other than 212 F.

Aging Test—The hose shall show no cracks, charring, or disintegration externally or internally when tested as specified after aging at 168 h at 250 ±4 F. The mandrel used shall have a diameter eight times the nominal OD for Types A and B and shall have a diameter twice the minimum bend radius shown in Table 3 for Type C.

Cold Test—The hose shall show no evidence of cracking or breaking when tested as specified. The mandrel used for Types A and B shall have a diameter eight times the nominal OD of the hose and for Type C shall have a diameter twice the minimum bend radius shown in Table 3.

Vacuum Test—The collapse of the outside diameter of the hose shall not exceed 20% of the original outside diameter when subjected to the specified vacuum.

Extraction Test—After being subjected to the Extraction Test the extractables shall not exceed 76 mg per sq in. of inside tube surface, and any extractables shall be oily or soft-greasy in nature and not harmful to the refrigerant system.

TABLE 2—PERMEABILITY TEST REQUIREMENTS

Temperature, F	Test Time, Days	Reference Pressure, psi	Load/cu in. of Hose Volume, gm	Max Loss of Refrigerant gm/ft/72 hr		Max Loss of Refrigerant lb/ft/yr of 1/2 in. ID Hose[a]	
				Type A and B	Type C	Type A and B	Type C
212	4	470	10 ± 2	9.3	3.73	2.5	1.00

[a] The tests are normally conducted on 1/2 in. ID hose. The maximum rate of effusion for other sizes shall vary in proportion with the ID, for example, the maximum rate of effusion for the 1-in. size shall be double that for the 1/2 in. size. The tube compound for other sizes shall be identical with that furnished in the 1/2-in. size.

TABLE 1—AUTOMOTIVE AIR CONDITIONING HOSE DIMENSIONS, IN. [a, b]

ID Size		ID Tolerance			OD Tolerance		
Nom	Dim	Type A	Type B	Type C	Type A	Type B	Type C
3/16	(0.187)	—	—	−0.015 −0.005	—	—	0.328 max
1/4	(0.250)	+0.025 −0.005	+0.025 −0.005	+0.015 −0.010	0.563 ±0.031	0.625 ±0.023	0.450 max
5/16	(0.312)	+0.025 −0.005	+0.025 −0.005	+0.015 −0.012	0.750 ±0.031	0.719 ±0.023	0.530 max
3/8	(0.375)	+0.025 −0.005	+0.025 −0.005	±0.015	0.875 ±0.031	0.781 ±0.023	0.600 max
13/32	(0.406)	+0.030 −0.005	+0.030 −0.005	±0.017	0.906 ±0.031	0.812 ±0.023	0.635 max
1/2	(0.500)	+0.035 −0.010	+0.035 −0.010	±0.020	1.000 ±0.031	0.906 ±0.031	0.740 max
5/8	(0.625)	+0.035 −0.010	+0.035 −0.010	±0.025	1.125 ±0.031	1.031 ±0.031	0.920 max
3/4	(0.750)	+0.035 −0.010	+0.035 −0.010	+0.035 −0.015	1.312 ±0.031	1.156 ±0.031	1.040 max
7/8	(0.875)	+0.035 −0.010	+0.035 −0.010	—	1.438 ±0.031	1.312 ±0.031	—
1	(1.000)	+0.040 −0.015	+0.040 −0.015	—	1.563 ±0.047	1.438 ±0.031	—
1-1/8	(1.125)	+0.047 −0.015	+0.047 −0.015	—	1.688 ±0.047	1.563 ±0.047	—
1-1/4	(1.250)	+0.047 −0.015	+0.047 −0.015	—	1.844 ±0.062	1.688 ±0.047	—

[a] Concentricity based on total indicator reading between inside bore of hose and outer surface of the hose shall not exceed the following values:

Types A and B		Type C	
Sizes 1/4 in. and under	0.030 in.	Sizes 1/4 in. and under	0.020 in.
Sizes over 1/4 to 7/8 in.	0.040 in.	Sizes over 1/4 to 1/2 in.	0.025 in.
Sizes over 7/8 in.	0.050 in.	Sizes over 1/2 in.	0.030 in.

[b] Fitting compatibility—fittings for thermoplastic hose may not necessarily be interchangeable. Therefore, it is recommended that fittings for hose be properly matched. Fittings and/or hose manufacturers' recommendations should be followed.

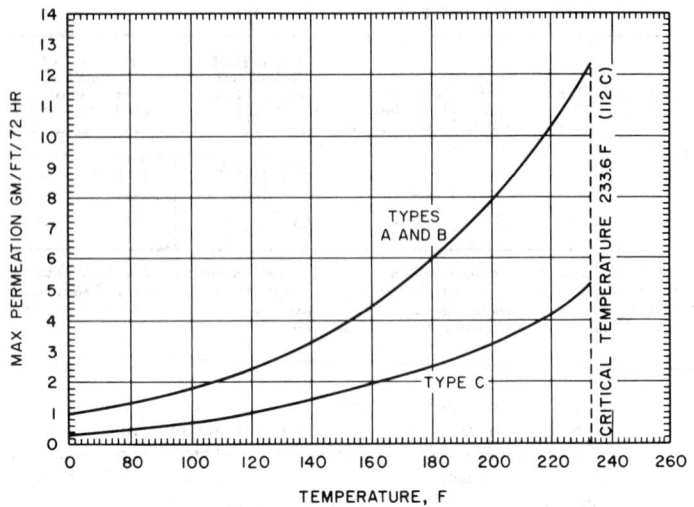

FIG. 1

Change in Volume—A specimen prepared from the inner tube of the hose shall show a volume change between −5 and +20% when measured within 5 minutes after removal from ASTM Oil No. 3 in which it has been immersed for 70 hr at a temperature of 212 ±2 F.

Inspection Tests

Length Change—For Types A and B, the hose shall not contract in length more than 4% nor elongate more than 2% when tested under a pressure of 350 psi. For Type C, hose length change shall not exceed ±3% when tested at 450 psi.

Bursting Strength—The minimum bursting strength for Type A hose shall be 1750 psi, for Type B 2500 psi, and for Type C 2500 psi.

Proof Test—All hose shall satisfactorily withstand a proof test with hydraulic pressure equivalent to 50% of the specified minimum burst pressure for a period not less than 30 sec nor more than 5 minutes.

Cleanliness Test—The bore of all hose shall be clean and dry. When subjected to this test there shall not be more than 25 mg of foreign material per square foot of tube surface.

Tensile Test of Assembly—When coupled hose assemblies are furnished the minimum load required to separate the hose from the coupling or break the hose itself shall not be less than specified in Table 4.

TABLE 3 — MINIMUM BEND RADIUS FOR TYPE C HOSE

Nom	Size Dim	Min Bend Radius	Nom	Size Dim	Min Bend Radius
3/16	(0.187)	2.0	13/32	(0.406)	4.5
1/4	(0.250)	3.0	1/2	(0.500)	5.0
5/16	(0.312)	3.5	5/8	(0.625)	6.5
3/8	(0.375)	4.0	3/4	(0.750)	8.0

TABLE 4 — TENSILE STRENGTH OF ASSEMBLY

Size, in.	Type A	Type B	Type C	Size, in.	Type A	Type B	Type C
1/4	250	400	250	3/4	550	725	550
5/16	350	600	350	7/8	550	725	550
3/8	450	725	450	1	550	725	550
13/32	500	725	500	1-1/8	550	725	550
1/2	550	725	550	1-1/4	550	725	550
5/8	550	725	550				

EMISSION CONTROL HOSE—SAE J1010

SAE Standard

Report of Nonmetallic Materials Committee approved February 1973.

1. Scope—This specification covers hose suitable for use with emission control systems requiring the injection of compressed air. Two types are covered: Type A for general purpose use where high air and underhood temperatures are encountered, and Type B for use where moderately high air and environmental temperatures are encountered along with the occasional presence of fuel and/or oil vapors and their condensates. The hose may be furnished in long lengths, specific cut lengths, or as parts preformed to specific configurations.

2. Hose Construction—The hose shall consist of a heat resistant tube, a reinforcement of braided, knit, spiral yarn cord, or of woven fabric, and a heat and ozone resistant cover.

3. Dimensions and Tolerances

3.1 The ID and OD shall be as shown below:

Nominal Size[a]	ID and Tolerance		OD and Tolerance	
in	in	mm	in	mm
15/32	0.47 ±0.03	11.91 ±0.79	0.81 ±0.03	20.64 ±0.79
19/32	0.59 ±0.03	15.08 ±0.79	0.94 ±0.03	23.81 ±0.79
23/32	0.72 ±0.03	18.26 ±0.79	1.12 ±0.03	28.58 ±0.79
15/16	0.94 ±0.06	23.81 ±1.59	1.38 ±0.06	34.93 ±1.59

[a] Hose with other nominal dimensions may be specified. It shall meet all the other requirements of this standard.

3.2 Length tolerance shall be as follows:[1]

Length		Precision Tolerance		Commercial Tolerance	
in	m	in	mm	in	mm
0 to 12 incl	0 to 0.30	±0.12	± 3.18	+ 0.38 −0.12	+ 9.65 −3.05
Over 12 to 24 incl	0.30 to 0.61	±0.19	± 4.76	+0.38 −0.19	+ 9.65 −4.76
Over 24 to 36 incl	0.61 to 0.91	±0.25	± 6.35	+0.50 −0.25	+12.70 −6.35
Over 36 to 48 incl	0.91 to 1.22	±0.38	± 9.65	+0.50 −0.38	+12.70 −9.65
Over 48 to 72 incl	1.22 to 1.83	±0.50	±12.70	+0.75 −0.50	+19.05 −9.65
Over 72	1.83	±1%		±2%	

Commercial tolerances shall apply unless otherwise specified.

3.3 **Squareness of Ends**—The ends of hose shall be square within 0.12 in. (3.05 mm) for sizes up to and including 23/32 in. (18.26 mm) and 0.25 in. (6.35 mm) for 15/16 in. (23.81 mm) size and all preformed parts. End squareness shall be measured between a point on the outside of the long side which is on a circumferential line emanating from the end of the short side and at 90 deg to the axis and a point at the end of the long side. End squareness measurement will be considered equally applicable with length tolerance. The mid-

[1] These length tolerances also apply to arm lengths of preformed parts which are measured from end to intersection of nearest centerline, and expanded end lengths.

point of the measurement of the deviation from squareness taken on the long side shall fall within the length tolerance.

3.4 Dimensions of Preformed Parts—General layout dimensions locating bend intersections are to establish the theoretical centerline of the hose. Actual outside contour of the hose should be held within ±0.19 in. (4.76 mm) in all planes with respect to the theoretical outside contour. To check contour, hose ends should first be placed in nominal position (hose may have to be flexed to correct for any distortion caused by handing after vulcanization in the producing plant or in shipment) in a checking fixture from which contour deviation can be measured. Allowance shall be provided in the end mounting area of the fixture for the arm length tolerances which are applicable.

When the ID of an end is enlarged, the wall gage or the enlarged end normally changes. Allowable change shall be +0.03, −0.02 in. (+0.76, −0.51 mm).

The wall gage within bends of a preformed hose may differ from the gage in straight portions. This difference shall not exceed 33%.

4. Physical Tests—Tests shall be performed on sections and parts removed from the finished hose and shall be carried out in accordance with the provision of ASTM D 380, except where otherwise specified. Sampling and inspection are not a part of this specification; however, suggested procedures for quality assurance are listed at the end of the specification.

5. Qualification Tests—The length of hose or the number of cut lengths or preformed parts which shall be supplied for qualification testing shall be as prescribed by the buyer. All tests shall be conducted as the basis for qualifications.

6. Test Requirements

6.1 Burst—The minimum burst for all sizes shall be 175 psi (1.21 MPa).

6.2 Adhesion—The tube-to-reinforcement and cover-to-reinforcement adhesion shall be a minimum of 8 lb (3.63 kg) per inch (25.4 mm) of width.

6.3 Original properties of compounds:

Types A & B	Tube	Cover
Tensile	(8.27 MPa) 1200 psi min	(6.89 MPa) 1000 psi min
Elongation	250% min	250% min
Hardness	60–75 duro	60–75 duro

6.4 Dry Heat Resistance of Tube and Cover Compounds (using ASTM Method D 865)

Type A—After aging 168 h at 150 ±1°C (302 ±2°F)
  Change in tensile strength from original: ±30%
  Change in elongation from original: −50% max
  Change in durometer hardness from original: ±15 points

Type B—After aging 70 h at 125 ±1°C (257 ±2°F)
  Change in tensile strength from original: ±30%
  Change in elongation from original: −50% max
  Change in durometer hardness: ±15 points

6.5 Fuel Resistance, Type B Tube—After immersion in Fuel C at room temperature for 120 h.
  Volume Change: +45% max

6.6 Oil Resistance, Type B Tube—After immersion in ASTM No. 3 Oil for 70 h at 125 ±1°C (257 ±2°F).
  Volume Change: +80% max

6.7 Ozone Resistance, Types A and B—Test procedure and apparatus shall be in accordance with ASTM D 1149, Test for Accelerated Ozone Cracking of Vulcanized Rubber, where applicable.

A specimen of hose of sufficient length shall be bent around a mandrel with an outside diameter eight times the nominal OD of the sample. The two ends shall be tied at their crossing with enameled copper or aluminum wire. After mounting, the specimen shall be allowed to rest in an ozone-free atmosphere for 24 h at room temperature. After rest, the mounted specimens shall be placed in a test chamber with ozone concentration of 50 ±5 pphm at a temperature of 40 ±1°C (104 ±2°F).

After 70 h of exposure, the specimens shall be removed and allowed to cool to room temperature and then visually inspected under 7X magnification. The cover must meet a rating of "0," except for the area immediately adjacent to the wire which shall be ignored.

6.7.1 Ozone Testing of Preformed Parts—From preformed parts, prepare a specimen by cutting a strip of whole hose ½ in. wide by 4 in. long (12.7 x 101.6 mm) and tie the specimen (cover out) around a ½ in. (12.7 mm) diameter mandrel.

Condition same as for whole hose and apply same requirements. This test applies to the cover only and cracks in the exposed tube or cut edges of the cover shall be ignored.

6.8 Cold Flexibility—A sample length equal to 10 times the nominal OD plus 12 in. (305 mm) shall be placed in a low-temperature cabinet in a straight position for 5 h at −40 ±1°C (−40 ±2°F). At the end of the exposed period, without removing the sample from the cabinet, it shall be gripped on the ends and flexed[2] 180 deg over a mandrel having a diameter equal to 10 times the nominal hose OD within 4–8 s. The sample shall be removed from the cabinet, allowed to come to room temperature, and then it shall be subjected to a hydrostatic proof pressure test of 100 psi (0.69 MPa) for a period of 30–60 s without evidence of leaks or other failure. The outside shall be examined for any cracks. Evidence of cracks or other failure shall be considered a cause for rejection.

7. Suggested Quality Assurance Procedures

7.1 Sampling for Inspection and Test

7.1.1 A lot shall be all hose of one type and size vulcanized on any shift.

7.1.2 Dimensional and visual inspection shall be conducted by vendor on all production lots in accordance with MIL-STD-105, latest revision Tables IIA, with AQL's to apply as follows:
  Critical characteristics 0.25
  Major characteristics 2.5
  Minor characteristics 6.5

Parts failing to meet above AQL's may be 100% reinspected and resubmitted.

The classification of defects shall be agreed upon between vendor and purchaser.

7.1.3 Quality assurance sampling and testing will be conducted by sampling every lot of each size and type and testing for burst, adhesion, and original properties of the tube and cover compounds. In the event of a failure of any characteristic on any lot, two additional samples shall be selected at random and tested. Failure of either shall be cause for rejection. Conformance of both shall be reason to accept.

The sample from the first lot of each size and tube in any month shall be tested to all test requirements. In the event of failure of any characteristic, two additional samples shall be taken at random and tested. If both resamples pass all tests, no further action need be taken. In the event of failure, that lot shall be rejected and all lots produced since the previous monthly sample shall be suspect and notification must be given to all customers in accordance with accepted practice. All inventory shall be sampled and tested. All lots produced following a lot on which there was failure shall be tested for all requirements until 10 consecutive lots meeting all requirements are produced, at which time it will be permissible to resume the first lot in the month practice.

[2] Alternate procedure to be applied to Type B hose only: It may occur that Type B hose is too stiff to bend in the manner and time prescribed above. In this instance, samples of the tube and cover are to be prepared and tested in accordance with paragraph 24 of ASTM D 380 5 h at −40 ±1°C (−40 ±2°F).

WINDSHIELD WASHER TUBING—SAE J1037

SAE Standard

Report of Nonmetallic Materials Committee approved September 1973.

1. Scope—This SAE Standard covers two types of nonreinforced, extruded, flexible tubing:
  Type 1—Oil resistant elastomer
  Type 2—Non-oil resistant elastomer

2. Application—The tubing is generally used for all windshield washer lines requiring good resistance to the engine compartment environment.

3. Requirements

3.1 Tubing shall be free of visible wax bloom or other contaminants which may enter and deleteriously affect the windshield washer system.

3.2 The primary elastomer shall have an infrared curve that corresponds to the curve of the tubing sample upon which original product approval was based.

3.3 Except where noted, the following tests shall be conducted on full sections of tubing:

3.3.1 Hardness, Durometer A (ASTM D 2240)—70 ±5 points.

3.3.2 Tensile Strength—Slabs if available (ASTM D 412, or optional test method): 1000 psi min (6.9 MPa min).

3.3.3 Elongation—Slabs if available: 200% min.

3.3.3.1 *Optional Test Method*—ASTM D 412, except use die D if the geometry of the tubing permits. Or tubing is to be fastened in the tensile tester by means of knots tied in both ends with two washers located between the knots and the washers fastened in the jaws. The tests shall be run until failure occurs. If the tubing fails in the knots or within 1 in. (25 mm) of the knots, then the tests should be rerun until failure occurs in the section between the knots. To obtain the cross-sectional area of tubing, an optical comparator or similar instrument may be used.

3.3.4 BURST PRESSURE (ASTM D 380)

ID		Burst Pressure, min	
in	mm	psi	MPa
0.203–0.235	5.16–5.97	150	1.03
0.140–0.172	3.56–4.37	100	0.69
0.090–0.110	2.29–2.79	100	0.69

3.3.5 TENSION TEST

ID		Tension	
in	mm	psi	MPa
0.203–0.235	5.16–5.97	150	1.03
0.140–0.172	3.56–4.37	85	0.59
0.090–0.110	2.29–2.79	40	0.28

Ten inch (254 mm) lengths of tubing shall be fastened to the jaws on a tensile testing machine. The jaw separation rate shall be 20 ±1 in. (508 ±25 mm)/min until rupture of the tube occurs within the section bounded by 1 in. (25 mm) from the jaws.

3.3.6 VACUUM COLLAPSE TEST—30%.

The collapse of the OD of the tubing under internal vacuum of 24 in. (610 mm) of mercury for 5 min. The test shall be made with the tubing curved to a radius equal to five times the maximum OD.

3.3.7 WAX BLOOM—There shall be no visible evidence of wax or any other contaminants exuding from the inside or outside diameter of the tubing.

3.3.7.1 *Test Method*—Condition a 6 in. (150 mm) section of tubing for 45 min at −40°F (−40°C). Remove the specimen from the cold chamber and permit recovery to room temperature for 1 h. The tubing shall then be twisted 360 deg for 10 successive cycles, after which the center 2 in. (50 mm) section shall be compressed 10 successive cycles by finger pressure or utilization of a compression device to full closure of the ID. The tubing shall then be sectioned longitudinally and examined for evidence of wax bloom or other contaminants.

3.3.8 TEAR TEST—The tubing shall not tear when expanded to a minimum internal diameter of two times the nominal ID by forcing the tubing over a 30 deg tapered, clean, metal case which has been lubricated with a silicone parting agent. The metal case shall have a maximum finish of rms 20.

3.3.9 COMPRESSION SET—70% max. (ASTM D 395, Method B, 22 h at 212°F (100°C)).

The inner diameter of a 2 in. (50 mm) section of tubing shall be coated with talc to prevent heat fusion. The test specimen shall be compressed 50% of the original OD. The compression set shall be calculated on the basis of the original deflection.

3.3.10 COLD RESISTANCE—Shall not show fractures, cracks, checks, or breaks.

3.3.10.1 *Test Method*—The tubing shall be subjected to a temperature of −40°F (−40°C) for a period of 5 h, after which the tubing shall be flexed in the cold chamber through 180 deg from the centerline in each direction to a diameter 10 times the maximum OD of the tube at each extreme of cycle for five cycles. The rate of cycling shall be approximately one cycle in 4 s.

3.3.11 OZONE RESISTANCE (ASTM D 1171 rating)—0.

3.3.11.1 *Test Method*—Elongate the specimen around a wooden or aluminum mandrel of random selection to attain approximately 25% elongation. Condition the specimen on the mandrel 24 h at room temperature in an ozone-free atmosphere. Then hang the specimen(s) in an ozone box with an ozone concentration of 50 ±5 pphm for 70 h at 104 ±2°F (40 ±1°C). Examine for rating using 7X power glass.

3.3.12 HEAT AGING—After being subjected to dry air aging in accordance with ASTM D 573 method of test for accelerated aging of vulcanized rubber by the oven method for 70 h at 212°F (100°C), the tubing must then meet the following test requirements:

Vacuum collapse: 30% max
Cold resistance: Shall not crack, etc.
Hardness change, Durometer: +15 points max
Tensile change: −15% max
Elongation change: −40% max
Tear resistance: Must not tear
Ozone resistance, ASTM rating: 0

3.3.13 STAIN TEST (Water Solution)—When tested as follows, slight staining of the paint is permitted so long as it can be removed by employing the usual cleaning materials.

3.3.13.1 *Test Method*—Use 200 ml of a 50% aqueous solution of approved windshield washer solution and reflux small pieces cut from a 6 in. (150 mm) length of tubing for 4 h. Cool the resulting liquid to room temperature and cover about 1 sq in. (645 sq mm) of surface on a freshly prepared white paint panel and convertible and vinyl top materials of your production. Expose to ultraviolet (S-1 Sunlamp, ASTM D 925) radiation for 24 h. Cool and examine.

3.3.14 OIL AGING REQUIRED FOR TYPE 1 ONLY (ASTM OIL NO. 1)—After having been totally immersed in ASTM Oil No. 1 for 70 h with an oil temperature of 212 ±5°F (100°C), the tubing shall meet the requirements of:

Vacuum collapse test (paragraph 3.3.6): 30% max
Cold resistance test (paragraph 3.3.10): Shall not crack

3.3.15 OIL AGING REQUIRED FOR TYPE 1 ONLY (ASTM OIL NO. 3)—Immerse sample for 70 h in ASTM Oil No. 3 at 212 ±5°F (100°C) (ASTM D 471):

Hardness change, Durometer: 0 to −30 points
Tensile change: −65% max
Elongation change: −55% max
Volume change: +70% max

13 Fuels and Lubricants

EFFECTIVE DATES OF REVISIONS—SAE J301

SAE Information Report

Report of Lubricants Division approved January 1935 and last revised by Fuels and Lubricants Technical Committee January 1949. Reaffirmed with editorial change May 1964. Reaffirmed without change May 1973.

When a new classification is recommended by the Fuels and Lubricants Technical Committee, it shall be effective in use immediately upon approval by the Society. When an existing classification is modified or cancelled, field application immediately following adoption of such changes is optional but shall be fully effective eighteen months after such adoption.

ENGINE OIL PERFORMANCE AND ENGINE SERVICE CLASSIFICATION—SAE J183 FEB80

SAE Recommended Practice

Report of the Fuels and Lubricants Technical Committee, approved June 1970, last revised February 1980.

ϕ Automotive engine oils are classified by SAE J300 in terms of viscosity only. In order to permit the recommendation of oils by classes which would include factors other than viscosity, the American Petroleum Institute adopted in 1947 a system which divided crankcase oils into three classes depending on the properties of the oil and the operating conditions under which it was intended to be used. In this system crankcase oils were classified as: Regular Type, Premium Type, or Heavy Duty Type. Generally, the Regular Type oils were straight mineral oils, Premium Type contained oxidation inhibitors, and Heavy Duty Type contained oxidation inhibitors plus detergent-dispersant additives.

These early classifications did not recognize that diesel and gasoline engines
ϕ might have different engine oil requirements or that the requirements for either type of engine are influenced significantly by the characteristics of the fuel burned and operating conditions, especially cold weather "start and stop" operation. Consequently, the API developed a new classification system based on the severity of engine service. This system was developed in 1952 and revised in 1955 and again in 1960.

The API Engine Service Classification System described and classified, in general terms, the service conditions under which engines were operated. It included three classifications for gasoline engines (ML, MM, and MS) and three for diesel engines (DG, DM, and DS). Detail regarding these classifications was given in SAE J303, Internal Combustion Engine Service Classifications, which last appeared in the 1971 SAE Handbook.

Because the API Engine Service Classification lacked precise technical definitions of quality, gasoline and diesel engine oils were described using combinations of the API Service Classification, and individual company and military specifications. Supplementary quality definitions were found necessary and these supplemental definitions were inncorporated into individual company specifications. This practice encouraged the development of special lubricants acceptable to only one equipment manufacturer. Also, the performance level indicated by each category changed periodically and thus it became necessary to include supplementary definitions in communications regarding engine oil.

It became apparent that more effective means must be found to communicate engine oil performance and engine service classification information between the automotive equipment manufacturer, the petroleum industry, and the customer. Accordingly, in 1969 and 1970 the API, ASTM, and SAE cooperated in establishing the present classification as a joint effort to provide these means. By this classification, engine oils can be more precisely defined and selected according to their performance characteristics than heretofore, and they can be more easily related to the type of usage for which each is intended.

The Engine Oil Performance and Engine Service Classification is an "open-ended" system; that is, new categories can be added as required without changing or deleting existing categories. New categories will be promulgated only when clear need is shown to exist after investigation in depth by SAE in cooperation with API and ASTM. (For more detail the interested reader is referred to the latest revision of API Publication 1509 and ASTM Research Report D2:1002.) Petitions for addition (or change) may be initiated by individuals or organizations through API, ASTM, or SAE.

The primary responsibility of the cooperating organizations in establishment and administration of the classification is:

SAE—evaluation of categories suggested and *promulgation* of the categories to be included.

ASTM—establishment of test methods and performance limits, that is, *development* of a technical language describing the categories to engine builders and oil formulators.

API—identification of the categories and description of their service use, that is, *elucidation* of the technical language to consumers.

Tables 1 and 2, which summarize the Engine Oil Performance and Engine Service Classification, were prepared in cooperation with API and ASTM. Lubricants meeting more than one category may be so designated. The ϕ
lubricants covered by this classification are designated SA, SB, SC, SD, SE, SF, CA, CB, CC, and CD and apply to gasoline and diesel powered passenger ϕ cars, trucks, and off-highway equipment.

The categories shown in Tables 1 and 2 include all engine oils which have ϕ been, or currently are marketed in substantial volume for use in passenger cars, gasoline and diesel powered trucks, and gasoline and diesel powered off-highway equipment. Some of the categories are not currently recom- ϕ mended by manufacturers for use in their equipment. It should be noted that some individual recommendations may also include compositional or proprietary considerations in a choice product. Should any doubt arise in the appli- ϕ cation of this category, the engine builder and oil supplier should be consulted.

This document will be reviewed in its entirety on an annual basis to insure conformance with the requirements of the automotive and petroleum industries and the consumer.

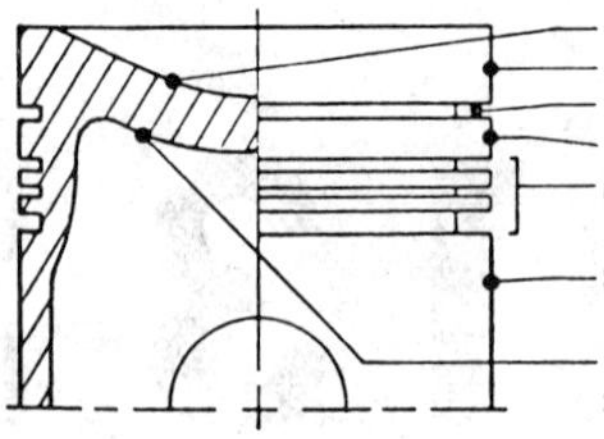

FIG. 1—CRC PISTON ZONE DEFINITIONS

- CROWN—PISTON TOP SURFACE
- LAND NO. 1 (CROWNLAND)—SIDE AREA ABOVE TOP RING
- GROOVE NO. 1—TOP COMPRESSION RING GROOVE
- LAND NO. 2—SIDE BETWEEN GROOVES NO. 1 AND 2
- OTHER GROOVES AND LANDS—NO. 2, 3, ETC. DOWN SIDE
- SKIRT—SIDE BELOW BOTTOM GROOVE INCLUDING ANY RELIEVED AREAS OR SKIRT RING GROOVES
- UNDERCROWN—UNDERSIDE TOP TO FIRST VERTICAL SIDE SURFACE

φ TABLE 1—DESIGNATION, IDENTIFICATION AND DESCRIPTIONS OF CATEGORIES

Letter Designation	API Engine Service Description	ASTM Engine Oil Description	Letter Designation	API Engine Service Description	ASTM Engine Oil Description
SA	**Formerly for Utility Gasoline and Diesel Engine Service** Service typical of older engines operated under such mild conditions that the protection afforded by compounded oils is not required. This category has no performance requirements and oils in this category should not be used in any engine unless specifically recommended by the equipment manufacturer.	Oil without additive except that it may contain pour and or foam depressants.	SF	**1980 Gasoline Engine Warranty Maintenance Service** Service typical of gasoline engines in passenger cars and some trucks beginning with the 1980 model operating under engine manufacturers' recommended maintenance procedures. Oils developed for this service provide increased oxidation stability and improved anti-wear performance relative to oils which meet the minimum requirements for API Service Category SE. These oils also provide protection against engine deposits, rust, and corrosion. Oils meeting API Service Category SF may be used where API Service Categories SE, SD, or SC are recommended.	Oil meeting the 1980 warranty requirements of the automobile manufacturers. Intended primarily for use in gasoline engine passenger cars. Provides protection against sludge, varnish, rust, wear, and high-temperature thickening.
SB	**Minimum Duty Gasoline Engine Service** Service typical of older gasoline engines operated under such mild conditions that only minimum protection afforded by compounding is desired. Oils designed for this service have been used since the 1930s and provide only antiscuff capability and resistance to oil oxidation and bearing corrosion. They should not be used in any engine unless specifically recommended by the equipment manufacturer.	Provides some antioxidant and antiscuff capabilities.	CA for Diesel Engine Service	**Light Duty Diesel Engine Service** Service typical of diesel engines operated in mild to moderate duty with high-quality fuels and occasionally has included gasoline engines in mild service. Oils designed for this service provide protection from bearing corrosion and from ring belt deposits in some naturally aspirated diesel engines when using fuels of such quality that they impose no unusual requirements for wear and deposit protection. They were widely used in the late 1940s and 1950s but should not be used in any engine unless specifically recommended by the equipment manufacturer.	Oil meeting the requirements of MIL-L-2104A. For use in gasoline and naturally aspirated diesel engines operated on low sulfur fuel. The MIL-L-2104A Specification was issued in 1954.
SC	**1964 Gasoline Engine Warranty Service** Service typical of gasoline engines in 1964–1967 models of passenger cars and trucks operating under engine manufacturers' warranties in effect during those model years. Oils designed for this service provide control of high and low temperature deposits, wear, rust, and corrosion in gasoline engines.	Oil meeting the 1964–1967 requirements of the automobile manufacturers. Intended primarily for use in passenger cars. Provides low temperature antisludge and antirust performance.	CB for Diesel Engine Service	**Moderate Duty Diesel Engine Service** Service typical of diesel engines operated in mild to moderate duty, but with lower quality fuels which necessitate more protection from wear and deposits. Occasionally has included gasoline engines in mild service. Oils designed for this service were introduced in 1949. Such oils provide necessary protection from bearing corrosion and from high temperature deposits in normally aspirated diesel engines with higher sulfur fuels.	Oil for use in gasoline and naturally aspirated diesel engines. Includes MIL-L-2104A oils where the diesel engine test was run using high sulfur fuel.
SD	**1968 Gasoline Engine Warranty Maintenance Service** Service typical of gasoline engines in 1968 through 1970 models of passenger cars and some trucks operating under engine manufacturers' warranties in effect during those model years. Also may apply to certain 1971 and or later models, as specified (or recommended) in the owners' manuals. Oils designed for this service provide more protection against high and low temperature engine deposits, wear, rust, and corrosion in gasoline engines than oils which are satisfactory for API Engine Service Category SC and may be used when API Engine Service Category SC is recommended.	Oil meeting the 1968–1971 requirements of the automobile manufacturers. Intended primarily for use in passenger cars. Provides low temperature antisludge and antirust performance.	CC for Diesel Engine Service	**Moderate Duty Diesel and Gasoline Engine Service** Service typical of lightly supercharged diesel engines operated in moderate to severe duty and has included certain heavy duty, gasoline engines. Oils designed for this service were introduced in 1961 and used in many trucks and in industrial and construction equipment and farm tractors. These oils provide protection from high temperature deposits in lightly supercharged diesels and also from rust, corrosion, and low temperature deposits in gasoline engines.	Oil meeting requirements of MIL-L-2104B. Provides low temperature antisludge, antirust, and lightly supercharged diesel engine performance. The MIL-L-2104B specification was issued in 1964.
SE	**1972 Gasoline Engine Warranty Maintenance Service** Service typical of gasoline engines in passenger cars and some trucks beginning with 1972 and certain 1971 models operating under engine manufacturers' warranties. Oils designed for this service provide more protection against oil oxidation, high temperature engine deposits, rust, and corrosion in gasoline engines than oils which are satisfactory for API Engine Service Categories SD or SC and may be used when either of these categories are recommended.	Oil meeting the 1972–1979 requirements of the automobile manufacturers. Intended primarily for use in passenger cars. Provides high temperature antioxidation, low temperature antisludge, and antirust performance.	CD for Diesel Engine Service	**Severe Duty Diesel Engine Service** Service typical of supercharged diesel engines in high speed, high output duty requiring highly effective control of wear and deposits. Oils designed for this service were introduced in 1955, and provide protection from bearing corrosion and from high temperature deposits in supercharged diesel engines when using fuels of a wide quality range.	Oil meeting Caterpillar Tractor Co. certification requirements for Superior Lubricants (Series 3) for Caterpillar diesel engines. Provides moderately supercharged diesel engine performance. The certification of Series 3 oil was established by Caterpillar Tractor Co. in 1955. The related MIL-L-45199 specification was issued in 1958.

◆ TABLE 2—TEST TECHNIQUES AND PRIMARY PERFORMANCE CRITERIA

Letter Designation	Test Techniques[a]	Primary Performance Criteria[a]		
SA	None	None		
SB	L-4[e] or L-38[b]	Bearing weight loss, mg, max	**L-4** 500	**L-38** 500
SC	Sequence IV[e]	Cam scuffing Lifter scuff rating, max	None 2	
SC	Sequences IIA[e] and IIIA[e]	Cam and lifter scuffing Avg cam plus lifter wear, in (mm) max Avg rust rating, min Avg sludge rating, min Avg varnish rating, min	None 0.0025 (0.064) 8.2 9.5 9.7	
SC	Sequence IV[e]	Cam scuffing Lifter scuff rating, max	None 2	
SC	Sequence V[e]	Total engine sludge rating, min Avg piston skirt varnish rating, min Total engine varnish rating, min Avg intake valve tip wear, in (mm) max Ring sticking Oil ring clogging, %, max Oil screen plugging, %, max	40 7.0 35 0.0020 (0.051) None 20 20	
SC	L-38[c]	Bearing weight loss, mg, max	50	
SC	L-1 (0.95% min sulfur fuel)[e,f]	Groove No. 1 (top) carbon fill, % vol, max Groove No. 2 and below	25 Essentially clean	
SD	Sequences IIB[e] and IIIB[e]	Cam and lifter scuffing Avg cam plus lifter wear, in (mm) max Avg rust rating, min Avg sludge rating, min Avg varnish rating, min	None 0.0030 (0.076) 8.8 9.6 9.6	
SD	Sequence IV[e]	Cam scuffing Lifter scuff rating, max	None 1	
SD	Sequence VB[e]	Total engine sludge rating, min Avg piston skirt varnish rating, min Total engine varnish rating, min Avg intake valve tip wear, in (mm) max Oil ring clogging, %, max Oil screen plugging, %, max	42.5 8.0 37.5 0.0015 (0.038) 5 5	
SD	L-38[c]	Bearing weight loss, mg, max	40	
SD	L-1 (0.95% min sulfur fuel)[c,e,f] or 1H[c,e,f]	Groove No. 1 (top) carbon fill, % vol, max Groove No. 2 lacquer coverage, % area, max Groove No. 2 and below Land No. 3 and below	**L-1** 25 — Essentially clean —	**1H** 30 50 — Essentially clean
SD	Falcon[c,e]	Avg engine rust rating, min	9.0	
SE	Sequence IIB[e], IIC[e], or IID	Avg engine rust rating, min Number stuck lifters	**IIB** 8.9 None	**IIC** 8.4 None **IID** 8.5 None

SE	Sequence IIIC[e] or IIID	Viscosity increase at 100°F (37.78°C) and 40 test h, %, max
		Viscosity increase at 40°C and 40 test h, %, max
		Avg engine ratings at 64 test h
		Avg sludge rating, min (CRC manual No. 12)
		Avg piston skirt varnish rating, min (CRC manual No. 9)
		Avg oil ring land deposit rating, min (CRC manual No. 9)
		Ring sticking
		Lifter sticking
		Scuffing and wear at 64 test h
		Cam or lifter scuffing
		Cam plus lifter wear, in (mm)
		  Average
		  Maximum
			IIIC
			400
			—
			
			9.2
			9.3
			6.0
			None
			None
			
			None
			
			0.0010 (0.025)
			0.0020 (0.051)
				IIID
				400
				375
				
				9.2
				9.1
				4.0
				None
				None
				
				None
				
				0.0040 (0.102)[d]
				0.0100 (0.254)[d]

TABLE 2—TEST TECHNIQUES AND PRIMARY PERFORMANCE CRITERIA (continued)

Letter Designation	Test Techniques[a]	Primary Performance Criteria[a]			

SE (continued)	Sequence VC[e] or V-D[e]	Avg engine sludge rating, min (CRC manual No. 12)
		Avg piston skirt varnish rating, min (CRC manual No. 9)
		Avg engine varnish rating, min
		Oil screen clogging, %, max
		Oil ring clogging, %, max
		Compression ring sticking
		Cam wear, in (mm)
		  Average, max
		  Maximum, max
			VC
			8.7
			7.9
			8.0
			5
			5
			None
			
			—
			—
					V-D
					9.2
					6.4
					6.3
					10.0
					10.0
					None
					
					Rate and Report[g]
					Rate and Report[g]	

	L-38	Bearing weight loss, mg, max	40			
SF	Sequence IID	Avg engine rust rating, min Number stuck lifters	8.5 None			

	Sequence IIID	Viscosity increase at 40°C (64 test h)
		Avg sludge rating, min (CRC manual No. 12)
		Avg piston skirt varnish rating, min (CRC manual No. 9)
		Avg oil ring land deposit rating, min (CRC manual No. 9)
		Ring sticking
		Lifter sticking
		Scuffing and wear
		  Cam and lifter scuffing
		  Cam plus lifter wear, in (mm)
		    Average, max
		    Maximum, max
			375
			9.2
			9.2
			4.8
			None
			None
			
			None
			
			0.0040 (0.102)
			0.0080 (0.203)			

	Sequence V-D	Avg engine sludge rating, min (CRC manual No. 12)
		Avg piston skirt varnish rating, min (CRC manual No. 9)
		Avg engine varnish rating, min
		Oil ring clogging, %, max
		Oil screen clogging, %, max
		Compression ring sticking
		Cam wear, in (mm)
		  Average, max
		  Maximum, max
			9.4
			6.7
			6.6
			10.0
			7.5
			None
			
			0.0010 (0.025)
			0.0025 (0.064)			

	L-38	Bearing weight loss, mg, max	40			
CA	L-4[e] or L-38[b]	Bearing weight loss, mg, max Piston skirt varnish rating, min	L-4 120–135 9.0		L-38 50 9.0	
	L-1 (0.35% min sulfur fuel)[e,f]	Groove No. 1 (top) carbon fill, % vol, max Groove No. 2 and below	25 Essentially clean			
CB	L-4[e] or L-38[b]	Same as CA				
	L-1 (0.95% min sulfur fuel)[e,f]	Same as CA, except Groove No. 1 (top) carbon fill, % vol, max	30			
CC	L-38	Bearing weight loss, mg, max Piston skirt varnish rating, min	50 9.0			
	LTD[e] or Modified LTD[b,e]	Piston skirt varnish rating, min Total engine varnish rating, min Total engine sludge rating, min Oil ring plugging, %, max Oil screen clogging, %, max	LTD 7.5 — 35 25 25			Modified LTD 7.5 42 42 10 10
	Sequence IIA[e], IIB[b,e], IIC[e], or IID	Avg engine rust rating, min	IIA 8.2	IIB 8.2	IIC 7.6	IID 7.7
	IH[e,f] or IH2[f]	Groove No. 1 (top) carbon fill, % vol, max Groove No. 2 lacquer coverage, % area, max Land No. 3 and below Weighted total demerits, max Ring side clearance Loss, in (mm) max	IH 30 50 Essentially clean — —			IH2 45 — — 140 0.0005 (0.013)

φ TABLE 2—TEST TECHNIQUES AND PRIMARY PERFORMANCE CRITERIA (continued)

Letter Designation	Test Techniques[a]	Primary Performance Criteria[a]		
CD	ID[f]	Groove No. 1 (top) carbon fill, % vol, max Groove No. 2 and below	75 Essentially clean	
	IG[e,f] or IG2[f]		IG	IG2
		Groove No. 1 (top) carbon fill, % vol, max	60	80
		Land No. 2 carbon and lacquer coverage, % area, max	50	—
		Groove No. 2 carbon and lacquer coverage, % area, max	30	—
		Land No. 3 and below	Essentially clean	—
		Weighted total demerits, max	—	300
		Ring side clearance Loss, in (mm) max	—	0.0005 (0.013)
	L-38	Bearing weight loss, mg, max Piston skirt varnish rating, min	50 9.0	

[a]Detail regarding many of these test techniques, including a description of their objectives or
φ significance, may be found in SAE J304, ASTM STP 315 and 509, and Federal Test Method Standard 791. The dimensionless numbers listed under Primary Performance Criteria refer to arbitrary scales as follows:

(1) When a maximum rating is quoted as 2, the reference is to a scale of 1–6 with 1 being perfect.

(2) When minimum ratings are quoted at about 4.0–9.7 the reference is to a scale where 10.0 is "clean."

(3) When minimum ratings are quoted at about 35.0–45.0, the reference is to a scale where 50.0 is "clean."

[b]Test conditions or performance requirements changed since originally promulgated.
[c]This test technique has also been used in this evaluation.
[d]Because of the change in test techniques, no meaningful correlation can be established between IIIC and IIID wear data. These recommended values are intended to assure adequate wear protection.
[e]This test is obsolete; engine parts, and/or test fuel, and/or reference oils are no longer generally available and the test is no longer being monitored by the test developer or ASTM.
[f]Refer to Fig. 1 for nomenclature of piston zones.
[g]Because Sequence VC Test contained no value train wear measurement requirement, no φ meaningful relationship could be established with VC Tests. However, for informational guidance in development of SE-quality oils, cam wear values of 0.0020 in (0.051 mm) average and 0.0040 in (0.102 mm) max should be considered.

ENGINE OIL VISCOSITY CLASSIFICATION—SAE J300d

SAE Recommended Practice

Report of Miscellaneous Division approved June 1911 and last revised by Fuels and Lubricants Technical Committee July 1977.

The SAE viscosity grades in Table 1 constitute a classification for engine lubricating oils in terms of viscosity only. Other oil characteristics are not considered. This SAE Recommended Practice is intended for use by engine manufacturers in defining the engine oil viscosity requirements of their engines, and by oil marketers in labeling their products.

The classification is based on viscosities determined at 100°C and −18°C (212°F and −0.4°F). In Table 1, viscosity grades without the letter W are based on 100°C viscosities. Viscosity grades with the letter W are based on −18°C viscosities. A multiviscosity graded oil is one whose −18°C viscosity is within the prescribed range of one of the W grade classifications and whose 100°C viscosity is within the prescribed range of one of the non-W grade classifications.

The oil viscosity at 100°C is measured according to ASTM D 445, Method of Test for Kinematic Viscosity of Transparent and Opaque Liquids, and the results reported in centistokes. Viscosities so measured are useful as a guide in selecting the proper viscosity oil for use under normal engine operating temperatures. In the summer, the oil selection should consider the highest ambient temperatures expected.

The oil viscosity at −18°C is measured according to ASTM D 2602, Method of Test for Apparent Viscosity of Motor Oils at Low Temperature Using the Cold Cranking Simulator, and the results reported in centipoises. Such measurements are intended to ensure that a given oil will permit satisfactory engine cranking under low ambient temperature conditions. The selection of a winter engine oil should consider the lowest anticipated ambient temperature.

It is recognized that certain low-temperature, viscosity-related phenomena are not measured by the test methods of this recommended practice. The reader is referred to the Appendix for a review of progress toward a quantitative definition of these phenomena.

Some engine oils are prediluted, usually to assist in mixing with fuel when used in two-stroke-cycle engines. In the case of such oils, SAE Viscosity Grades

φ TABLE 1—SAE VISCOSITY GRADES FOR ENGINE OILS

SAE Viscosity Grade	Viscosity Range		
	Centipoises (cP) at −18°C (ASTM D 2602)	Centistokes (cSt) at 100°C (ASTM D 445)	
	Max	Min	Max
5W	1 250	3.8	—
10W	2 500	4.1	—
20W[a]	10 000	5.6	—
20	—	5.6	less than 9.3
30	—	9.3	less than 12.5
40	—	12.5	less than 16.3
50	—	16.3	less than 21.9

Note: 1 cP = 1 mPa·s; 1 cSt = 1 mm²/s

[a]SAE 15W may be used to identify SAE 20W oils which have a maximum viscosity at −18°C of 5 000 cP.

by which the oils are classified shall be determined by the viscosity of the undiluted oils. Wherever the SAE Viscosity Grades are used on prediluted oils, the containers should show that the SAE grade applies to the undiluted oil.

APPENDIX

This appendix outlines current low temperature fluidity technology. It is for information only, and is not intended to be used for classification purposes at this time. SAE Fuels and Lubricants Subcommittee 2—Engine Oils, and ASTM Research and Development Division VII on Flow Properties, have been studying the low temperature fluidity of engine oils since 1969. An extensive review of these investigations was included in the May, 1973 SAE/ASTM Symposium on Viscometry and Its Applications to Automotive Lubricants.[1]

Although, in some cases, oil flow to critical engine parts may be reduced to unacceptable levels because of restrictions downstream from the oil pump, work to date has concentrated on studying oil flow from the oil pan to the pump, defined here as pumpability. This work indicates that, at the temperatures and soak conditions involved, oils must be sufficiently fluid (1) to flow to the oil screen under the existing hydrostatic head and (2) to flow through the restrictions imposed by the oil screen and pump inlet tube. This flow, further, must be at a rate sufficient to satisfy pump demand under service conditions.

Fig. 1 illustrates the best current thinking on a relationship between limiting viscosity and shear rate, both of which influence oil flow in engines. The shear rates of importance in pumpability seem to be below $100\ s^{-1}$. For reference purposes the viscosity/shear relationship pertinent to cold cranking[2] is also shown. Oils having viscosities no greater than those represented by the shaded areas should provide adequate pumpability at the existing use temperature and soak conditions. The temperatures of importance are those ambient temperatures at which engine oils are used in automotive equipment. Minimum use temperatures for the various W classifications are being developed.

Several viscometric procedures have been used to evaluate oil pumpability at the shear rates indicated (see Fig. 1 and SP-382). These and other instruments are being evaluated by ASTM to establish those which exhibit the best correlation with engine pumpability experience. The viscometer selection will be critical with instruments subjecting oils to shear rates less than $100\ s^{-1}$ being most applicable.

The ASTM program consists of several parts. The engine tests have been completed and the data have been published in ASTM report DS 57 (Low-Temperature Pumpability Characteristics of Engine Oils in Full-Scale Engines). These data are being used as the basis for selection of a suitable bench test method(s). As additional information is accumulated this appendix will be revised. A recommended practice incorporating low temperature oil pumpability requirements should be available in 1978 or 1979.

[1] Symposium papers are included in SP-382, available from SAE Headquarters.
[2] Cranking limits in Fig. A-1 based on data at 0°F (−17.8°C) only.

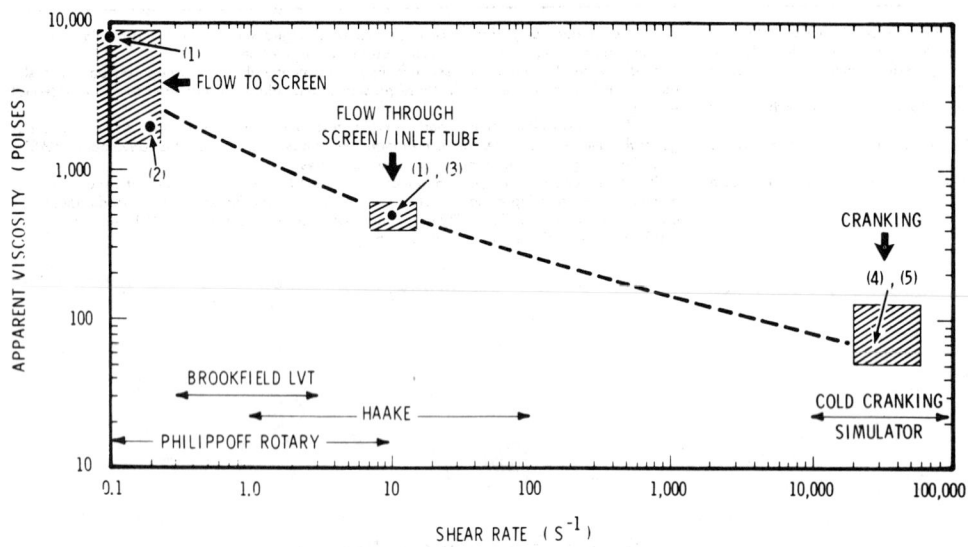

1. F. F. Tao and W. E. Waddey, "Low Shear Viscometry and Cold Flow Mechanism-Engine Oils." Paper 730481, SAE SP-382.
2. M. L. McMillan and C. K. Murphy, "The Relationship of Low-Temperature Rheology to Engine Oil Pumpability." Paper 730478, SAE SP-382.
3. R. M. Stewart and M. F. Smith, Jr., "Proposed Laboratory Methods for Predicting the Low Temperature Pumpability Properties of Crankcase Oils." Paper 730479, SAE SP-382.
4. "Determination of Low-Temperature Cranking Characteristics of Engine Oils at −20°F using CRC L-49-663 Research Test Technique." CRC Report 405, January 1967.
5. M. L. Haviland, "Engine and Transmission Lubricant Viscosity Effects on Low Temperature Cranking and Starting." SAE Transactions, Vol. 78 (1969), paper 690768.

FIG. A-1—ESTIMATED MAXIMUM OIL VISCOSITIES AT LOW TEMPERATURES (NUMBERS IN PARENTHESES DESIGNATE REFERENCES LISTED AT SIDE OF ILLUSTRATION

ENGINE OIL TESTS—SAE J304c

SAE Information Report

Report of Lubricants Division approved January 1942 and last revised by Fuels and Lubricants Technical Committee July 1977.

Engine tests are used to determine many performance characteristics of gasoline and diesel internal combustion engine oils. Such tests are included in the requirements for oils in SAE J183, and for United States military specification MIL-L-2104C and MIL-L-46152 oils.

Diesel Engine Tests—Many tests for piston ring sticking, piston ring and cylinder wear, and general deposit accumulation have been developed using single-cylinder compression ignition (diesel) engines. The more generally used tests are the L-1, modified L-1, 1-D, 1-G, and 1-H tests. They evaluate the effectiveness of base stocks and additives in minimizing ring and cylinder wear, ring sticking, and piston deposits. Test conditions are given in Table 1. Also shown in Table 1 are the related military specifications and SAE J183 performance categories, as well as the corresponding Method Number in U. S. Federal Test Method Standard 791 (Ref. 1).

Gasoline Engine Tests—Lubricant performance is evaluated in gasoline engine tests in terms of rust and corrosion, sludge, varnish, ring zone deposits and condition, wear, bearing corrosion, PCV valve clogging, and oil consumption. For those tests listed in SAE J183, Table 2 gives the test conditions, procedure references, evaluation factors, and the performance categories and military specifications in which the tests are included.

References

1. Federal Test Method Standard 791, "Lubricants, Liquid Fuels and Related Products; Methods of Testing." Available from General Services Administration, Business Service Center, Region 3, Seventh and D Streets, SW, Washington, D.C. 20025.
2. Copies of the Caterpillar Engine Lubricant Test Manual are supplied to diesel lubricants test engine owners by the Basic Engine Engineering, Caterpillar Tractor Co., Peoria, Illinois 61602.
3. ASTM Special Technical Publication 509, "Single Cylinder Engine Tests for Evaluating Performance of Crankcase Lubricants." Available from American Society for Testing and Materials, 1916 Race Street, Philadelphia, Pennsylvania 19103.
4. ASTM Special Technical Publication 315E, F, and G, "Multicylinder Test Sequences for Evaluating Automotive Engine Oils." Available from American Society for Testing and Materials, 1916 Race Street, Philadelphia, Pennsylvania 19103.
5. Available from Ford Motor Co., Engine and Foundry Division, Product Engineering Office, 20000 Rotunda Drive, P.O. Box 2053, Dearborn, Michigan, 48121.

TABLE 1—DIESEL ENGINE TESTS[a] (REF. 2)

	L-1	Modified L-1	1-D	1-G	1-H
Engine displacement, in³ (cm³)	208 (3400)	208 (3400)	208 (3400)	134 (2200)	134 (2200)
Test duration, h	480	480	480	480	480
Oil change period, h	120	120	120	120	120
Engine speed, rpm	1000	1000	1200	1800	1800
Fuel rate, Btu/min (kW)	2950 (51.8)	2950 (51.8)	5600 (98.4)	5850 (102.8)	4950 (87.0)
Load, bmep, psi (kPa)	~75 (520)	~75 (520)	~135 (925)	~141 (972)	~108 (745)
Air-to-engine temperature, °F (°C)	Room (not over 100°F) (38)	Room (not over 100°F) (38)	200 (93)	255 (124)	170 (77)
Air-to-engine pressure, in Hg (kPa), Abs	Atmospheric	Atmospheric	44.5 (150.3)	53 (179)	40 (135)
Water outlet temperature, °F (°C)	177.5 (80.83)	177.5 (80.83)	200 (93.3)	190 (87.8)	160 (71.1)
Oil-to-bearing temperature, °F (°C)	147.5 (64.17)	147.5 (64.17)	175 (79.4)	205 (96.1)	180 (82.2)
Fuel sulfur, wt %	0.35 min	1.0 ± 0.05	1.0 ± 0.05	0.35 min[b]	0.35 min[b]
Test included in: Military specifications			MIL-L-2104C	MIL-L-2104C	MIL-L-46152
Oil category	CA	CB	CD	CD	CC
Federal Test Method	FTMS 791A—FTM 332	FTMS 791A—FTM 345	FTMS 791B—FTM 340.3	FTMS 791B—FTM 341.4	FTMS 791B—FTM 346.2

[a] See Appendix for conversion factors.
[b] Sulfur must be of "natural origin"

TABLE 2—GASOLINE ENGINE TESTS[a,b]

Test	Engine Type	Displacement in³	Displacement cm³	Time Total, h	Time Each Phase	Load bhp	Load kW	Fuel Flow lb/h	Fuel Flow kg/h	Speed, rpm	Coolant °F	Coolant °C	Oil °F	Oil °C
L-4	L6	216	3540	36		30	22.4		2.04/2.27	3150	200	93.3	280	137.8
L-38	Single	42.5	700	40				4.5–5.0	2.04/2.27	3150	200	93.3	290	143.3
L-38 (Revised)	Single	42.5	700	40				4.5–5.0	2.04/2.27	3150	200	93.3	290	143.3
LTD	Single	42.5	700	180	3 h 1 h			4.7 4.7	2.13 2.13	1800 1800	120 200	48.9 93.3		
Modified LTD	Single	42.5	700	180	3 h 1 h			4.7 4.7	2.13 2.13	1800 1800	120 200	48.9 93.3		
Sequence I	V8	394	60	5	10 min	2	1.5			2500	95	35.0	120	48.9
Sequence II	V8	394	60	30	3 h	25	18.6			1500	95	35.0	120	48.9
Sequence III	V8	394	60	40		85	63.4			3400	200	93.3	265	129.4
Sequence IIA	V8	394	60	22	20 h 2 h	25 25	18.6 18.6			1500 1500	95 120	35.0 48.9	120 120	48.9 48.9
Sequence IIIA	V8	394	60	40		85	63.4			3400	200	93.3	265	129.4
Sequence IIB	V8	425	60	24	20 h 2 h 2 h	25 25 100	18.6 18.6 74.6			1500 1500 3600	105 120 200	40.6 48.9 93.3	120 120 275	48.9 48.9 135.0
Sequence IIC	V8	425	60	32	28 h 2 h 2 h	25 25 100	18.6 18.6 74.6			1500 1500 3600	110 120 200	43.3 48.9 93.3	120 120 260	48.9 48.9 126.7
Sequence IIIB	V8	425	6960	56	7 h 7 h	100 100	74.6 74.6			3600 3600	150 200	65.6 93.3	200 275	93.3 135.0
Sequence IIIC	V8	425	6960	64		100	74.6			3000	245	118.3	300	148.9
Sequence IV	V8	361	5920	24	2 h 2 h	None None	None None			3600 0	180 55	82.2 12.8	220 Not controlled	104.4
Sequence V	V8	368	6030	192	45 min 2 h 75 min	None 105 105	None 78.3 78.3			500 2500 2500	115 125 170	46.1 51.7 76.7	120 175 205	48.9 79.4 96.1
Sequence VB	V8	289	4740	192	45 min 2 h 75 min	None 86.6 86.6	None 64.58 64.58			500 2500 2500	115 125 171	46.1 51.7 77.2	120 175 201	48.9 79.4 93.9
Sequence VC	V8	302	4950	192	2 h 75 min 45 min	86.6 86.6 2	64.58 64.58 1.5			2500 2500 500	135 170 115	57.2 76.7 46.1	175 200 120	79.4 93.3 48.9
Falcon	L6	170	2790	55	45 min 2 h	None 30.9	None 23.04			500 2500	115 125	46.1 51.7	120 175	48.9 79.4

TABLE 2—GASOLINE ENGINE TESTS[a,b] (continued)

Test	Air-Fuel Ratio	Procedure Reference (See Refs. 1, 3–5)	Evaluation					Oil Categories Defined
			Rust and Corrosion	Sludge	Varnish	Wear	Other	
L-4	14.5	FTMS 791A-FTM 3402		X	X		Bearing corrosion	CA, CB, SB
L-38	14.0	FTMS 791B-FTM 3405.1		X	X		Bearing corrosion	CA, CB, CC, SB, SC, SD
L-38 (Revised)	14.0	FTMS 791B-FTM 3405.2		X	X		Bearing corrosion	CD, SE MIL-L-2104C, MIL-L-46152
LTD	14.5 14.5	FTMS 791B-FTM 348.1		X	X			CC
Modified LTD	15.25 15.25	FTMS 791B-FTM 348.2		X	X			CC
Sequence I	14.0							MS (obsolete)
Sequence II	14.0	ASTM STP 315, 315A	X					MS (obsolete)
Sequence III	15.0			X	X			MS (obsolete)
Sequence IIA	12.0 12.0	ASTM STP 315B, 315C	X					CC, SC
Sequence IIIA	16.5			X	X	X		
Sequence IIB	13.0 13.0 16.5	ASTM STP 315D, 315E	X					CC, SD, SE MIL-L-2104C, MIL-L-46152
Sequence IIC	13.0 13.0 16.5	ASTM STP 315F	X					SE MIL-L-2104C, MIL-L-46152
Sequence IIIB	16.5 16.5	ASTM STP 315D		X	X	X		SD
Sequence IIIC	16.5	ASTM STP 315E, F		X	X	X	Thickening	SE, MIL-L-46152
Sequence IV	Not controlled	ASTM STP 315, 315A, B, C, D				X		SB, SC, SD
Sequence V	9.5 15.5 15.5	ASTM STP 315, 315A, B		X	X			SC
Sequence VB	9.5 15.5 15.5	ASTM STP 315C, D		X	X		PCV valve clogging	SD
Sequence VC		ASTM STP 315E, F		X	X		PCV valve clogging	SE MIL-L-2104C, MIL-L-46152
Falcon	Max. Vac. 15.5	Ford Motor Co. FLTM BJ 11-2	X					SD

[a] See Appendix for conversion factors
[b] Leaded fuels are used in all tests, either commercial gasolines or special reference fuels.

PHYSICAL AND CHEMICAL PROPERTIES OF ENGINE OILS—SAE J357a

SAE Information Report

Report of Fuel and Lubricants Technical Committee approved August 1969 and completely revised November 1977.

1. Introduction—This SAE Information Report discusses a number of the physical and chemical properties of new and used engine oils. Where appropriate, standardized methods of test for these properties are listed. This report was prepared to provide those concerned with the design and maintenance of internal combustion engines with information relative to the terms used to describe engine lubricants.

The lubricants used in modern engines must be multifunctional. In addition to the basic function of lubricating (reducing friction and wear between moving parts), the oil must also provide a seal between cylinder walls and pistons, it must carry heat away from critical areas, and it must help keep the engine clean by dispersion of contaminants. Protection against rust and corrosion on internal engine parts is also an important function of the lubricant. In performing these functions, it is subjected to extreme environmental conditions. Engine component temperatures can range from below zero during shutdown periods to 200°C or higher while the engine is in operation. An oxidizing atmosphere is usually present in the engine. A wide variety of contaminants is introduced into the engine oil system. Some are inert; others either are chemically active, or they catalyze undesirable chemical reactions.

2. General Description of Engine Oil Components—Fully formulated lubricants consist of (1) base stocks and (2) the additives that are necessary to produce the required performance in the finished product. Base stocks are of two general types; petroleum and synthetic. The additives used to enhance performance consist of many types of chemicals. These engine oil components will be described in the following sections.

2.1 Petroleum Base Stocks—Crude petroleum oils have as principal components three basic types of hydrocarbon molecules; that is, paraffinic, naphthenic, and aromatic. The type or types of molecules that predominate are a basis for classifying crude oils as follows:

Crude Oil Type	Predominant Hydrocarbon
Paraffinic	Paraffinic
Naphthenic	Naphthenic
Asphaltic	Naphthenic/Aromatic
Mixed Base	Paraffinic/Naphthenic/Aromatic

Crude oils as they come from the ground can be a mixture of the following: gaseous products, gasoline, diesel fuel, lubricating oil stocks, asphalt, etc. The various classes of products are separated initially through vacuum and/or atmospheric distillation. Precipitation of the heaviest, most viscous fractions using a solvent such as propane can also be practiced. The lubricating oil fractions resulting therefrom provide a series of base stocks of various volatilities and viscosities that are referred to as neutral fractions and bright stock. Generally speaking, the viscosities for the neutral fractions range from about 13–130 cSt at 40°C. The bright stock fraction will have a viscosity of from 500–1100 cSt at 40°C. These fractions generally require further refining to make them suitable for engine oil applications.

Where wax is present in the lubricant fraction, a process such as methyl-ethyl ketone solvent dewaxing or propane solvent dewaxing is used to remove some waxy materials which crystallize and/or congeal at low temperatures and thereby impede low temperature flow. Also found in the as-distilled base stock fractions will be unstable components such as nitrogen and sulfur-containing compounds, metal-containing compounds, and aromatic hydrocarbons of varying structures. Many of these compounds can have an adverse effect on the efficiency of the additives used in engine oil applications, and they should be removed through extraction processes using solvents such as phenol or furfural, or modified by hydrotreating or hydrocracking.

Normally, wax is removed after solvent refining (fufural, phenol). Thus, in a simplified description, the refining process consists of (1) fractionation of the crude oil into neutrals and bright stocks suitable for blending to various viscosities, (2) removal of unstable components to enhance base oil stability and additive response and, (3) removal of some waxy hydrocarbons to improve low temperature fluidity characteristics.

The physical and chemical properties of the finished base stocks will not be solely a function of crude source but will also be dependent upon the processes employed and the degree of refining severity employed. The engine oil compounder blends the various components to achieve the viscometric properties required and adds additive agents to achieve the performance levels that are desired for a given application.

2.2 Synthetic Base Stocks—A number of chemical compounds also have been found to be suitable as base stocks for engine oils. These base stocks are manufactured by organic reactions such as alkylation, condensation, esterification, polymerization, etc. The starting material(s) may be one or more relatively pure organic compounds, generally of a simple composition, which are usually obtained by processing fractions from petroleum, vegetable, and animal sources.

Classes of chemical compounds which might be used as a source of synthetic base stocks are shown in the following table. Those marked with an asterisk are of greatest current interest for technical and economic reasons.

(a) Synthetic hydrocarbons such as: Alkylated Aromatics*
Polyalphaolefins*
Polybutenes*

(b) Organic Esters such as: Dibasic Acid Esters*
Polyol Esters*
Polyesters

(c) Others such as: Halogenated Hydrocarbons
Phosphate Esters
Polyglycols*
Polyphenyl Esters
Silicate Esters
Silicones

A synthetic lubricant base stock may consist of any of the above or any mixture of the above which are compatible with each other.

The additive agents necessary in both petroleum base stock and synthetic base stock engine oils are also synthesized materials. However, even though these materials are synthesized they should be referred to as additives and not included in the base stock description.

Some synthetic base stocks are compatible with petroleum base stocks and the two types may be blended together to obtain desired physical and chemical properties. Such combinations should be identified as blends of synthetic and petroleum base stocks.

2.3 Additive Agents—A lubricant additive agent is defined as a material designed to enhance the performance properties of the base oil. These additive agents are used at concentration levels ranging from several parts per million to greater than 10 volume percent. Generally these are materials which have been chemically synthesized to supply the desired performance features, and they frequently contain an oil-solubilizing hydrocarbon as part of the molecule. Some additive agents are naturally occurring materials which have undergone only minor modifications to obtain the desired property. The various types of additives are classified according to their primary functions as follows: antifoam agent, antirust agent, antiwear agent, corrosion inhibitor, detergent, dispersant, extreme pressure agent, friction reducer, oxidation inhibitor, pour point depressant, and viscosity index improver. Some additives possess multifunctional properties.

The unique performance features contributed by additives are required to satisfy the lubrication needs of today's engines, oil change intervals, and service conditions. Some additives may either enhance or interfere with the function of another. Therefore, the finished engine oil is formulated to achieve a balanced performance through combined properties attributable to both the base oil and the additives.

3. Physical and Chemical Properties—A tabulation of the physical and chemical properties of an oil can assist the user and the oil refiner in defining a consistently uniform product. However, these properties cannot be used to establish oil performance which is related to the requirements of the engine as defined by design parameters and the service in which it is used. Some of the performance characteristics of engine oils are discussed in SAE J304 and SAE J183.

While the physical and chemical properties of an oil do not generally define oil performance, these individual properties are meaningful and are related to the oil's ability to fulfill its function as a lubricant. The following sections discuss these properties and their lubrication function.

3.1 Tests Pertinent to Both New and Used Oils

3.1.1 VISCOSITY AND VISCOSITY INDEX—The viscosity of an oil is a measure of its resistance to flow. The viscosity of an oil decreases with increasing temperature and vice versa. Newtonian oils do not experience a viscosity change with change in shear rate at a given temperature; whereas non-Newtonian oils, such as engine oils compounded with viscosity index improvers or oils with some wax particles present, do show a decrease in viscosity when the shear rate is increased. Since some of the viscosity decrease encountered at high shear rate will be regained when the shear stress is removed, this decrease is referred to as a temporary shear loss. That viscosity which is not regained is referred to as a permanent shear loss. Therefore, non-Newtonian oil may exhibit several different *temporary* or apparent viscosities at various locations within an engine due to differences in shear rates. The magnitude of the temporary and permanent loss of viscosity depends upon the nature of the viscosity index improver and the conditions under which it is used. Furthermore, the oil viscosity also varies with pressure. Oil viscosity will change in service through shearing, dilution by fuel fractions, oxidation, volatilization, contamination by combustion by-products, etc. To be considered pertinent, the viscosity of an engine oil must be determined using procedures which have been shown to give results which correlate with those found in engines.

Low shear rate measurement of viscosity is usually made using ASTM Method D445 and reported in centistokes. The test temperatures most commonly used are 40°C and 100°C. Several special methods are in use to measure oil viscosity at high shear rates. Results obtained from ASTM Method D2602, using a high shear rate instrument known as the Cold Cranking Simulator, have been correlated with the low temperature cranking performance of engine oils. It is the method adopted for classifying oils according to viscosity at −18°C in SAE J300.

The viscosity-temperature relationship of an oil depends primarily on the nature or composition of the oil. A minimum change in viscosity with temperature is desirable. ASTM Method D341, Standard Viscosity-Temperature Charts, provides a series of six charts so constructed that for any given oil, the viscosity-temperature values can be approximated by a straight line over the temperature range in which the oil is a Newtonian liquid. ASTM Method D2270, Viscosity Index, provides a means for calculating an empirical number that yields a relative evaluation of the viscosity-temperature characteristics of the fluid in question when compared to two reference oils having the same viscosity at 100°C. The higher the viscosity index, the smaller will be the change in viscosity with change in temperature.

The relationship of viscosity and temperature defined by Viscosity Index is not affected by shear rate for Newtonian oils. However, the physical properties of non-Newtonian oils (such as those containing polymers) are not fully described by the Viscosity Index derived from low shear rate viscosity measurements only. The viscosity-temperature relationship of polymer thickened oils is a function of the shear rate applied in measuring viscosity. At high shear rates, the improvement in viscosity-temperature relationship imparted by the polymer may be reduced relative to the data found at low shear rates.

The viscosity of an oil at low temperature is important in cold starting due to the effect on cranking speed as well as the power required to accelerate the engine to sustained operating speed. Viscosity is also a factor in the time required to develop the necessary oil pressure and lubrication throughout the engine after cold starting. A relatively low viscosity oil, defined by low temperature measurement techniques, is desirable to aid in both cold starting ability and lubrication. The viscosity, as measured at low temperature with the Cold Cranking Simulator, correlates with engine cold cranking performance. However, low viscosity at low temperature as measured by the Cold Cranking Simulator, a high shear rate device, does not assure prompt development of engine oil pressure and flow from the pump to remote areas of the engine. Low shear rate viscosity is critical in determining flow to the pump inlet and throughout the oil system.

The viscosity of engine oils decreases as temperature increases. An oil with proper high temperature viscosity characteristics must be selected for the high temperature operation associated with continuous running of an engine. Increasing the high temperature viscosity of the oil in use will generally reduce oil consumption, leakage, blowby, and wear. Alternatively, both the oil pressure and friction associated with oil film shearing are increased with the use of higher viscosity oils.

The Fuels and Lubricants Technical Committee is currently evaluating new methods of test for viscosity which may lead to a revision of the SAE Engine Oil Viscosity Classification System.

3.1.2 CLOUD POINT AND POUR POINT—The cloud point of an oil is defined as the temperature at which a cloud or haze appears at the bottom of a test jar when testing a moisture-free oil by ASTM Method D2500. The haze indicates the presence of some insoluble fractions, such as wax, at the temperature noted. In most applications this haze will have little practical significance.

The pour point of an oil is defined as the lowest temperature at which it can be poured when tested by ASTM Method D97. The pour point has some relationship to the rate at which oil will be supplied to the suction side of the oil pump. However, more precise and correlatable viscometric methods are being developed which will predict the ability of an oil flow to the oil pump and throughout the system at low temperature. In actual practice, the oil in the crankcase will be a mixture of oil and small amounts of fuel fractions, the composition depending on several factors. (See paragraph on Fuel Dilution.)

Some oils display an increase in pour point when exposed to a repeated temperature variation. Federal Test Method, Standard No. 791a, Method 203 of January 15, 1969 describes a procedure for evaluating the tendency of the pour point to increase as follows: "This method is used for determining the pour stability of blends of winter grade (regular, heavy duty, and diluted heavy duty) motor oil, and certain types of hydraulic fluids. It consists of subjecting the sample to specified temperature changes for six days, and determining the lowest temperature at which the oil will remain fluid."

The cloud and pour points of engine oils do not have a direct relationship with the cranking or starting of engines.

3.1.3 FLASH AND FIRE POINT—The flash point of a petroleum product is the temperature to which the product must be heated under specific conditions to give off sufficient vapor to form a mixture with air that can be ignited momentarily by a specified flame.

Fire point is the temperature to which the product must be heated under prescribed conditions to burn continuously when the mixture of vapor and air is ignited by a specified flame.

Flash and fire points are significant from the viewpoint of safety and should be related to the temperatures to which petroleum products will be subjected in storage, transportation, and use. Normally, engine oils will present no hazards in this respect. For engine oil, relatively low flash and fire points are indications of oil volatility, and thus may be related to oil consumption at high temperature. The minimum flash point that can be tolerated must be determined in each application. Flash point is also used to indicate contamination by a volatile product such as gasoline. ASTM Method D56, Flash Point by Tag Closed Tester, ASTM Method D93, Flash Point by Pensky-Martens Closed Tester, ASTM Method D92, Flash and Fire Points by Cleveland Open Cup, and ASTM Method D1310, Flash Point by Tag Open Cup Tester, are all methods of obtaining the above type of information. ASTM Method D92 is the preferred method for engine oils.

3.1.4 DISTILLATION DATA—The volatility characteristics of engine oils can be defined by distillation procedures outlined in the ASTM Methods. ASTM Method D1160 is a reduced pressure (vacuum) distillation method. This is needed with engine oils since the components generally used in such product require distillation at reduced pressure (hence lower temperature) to prevent cracking. ASTM Method D2887, which gives boiling range distribution data by gas chromatography, is gaining acceptance and is increasingly used instead of ASTM Method D1160. Correlations between performance characteristics, such as oil consumption and the volatility characteristics of the oil in use must be developed with actual engine test data.

3.1.5 ALKALINITY-ACIDITY—The alkalinity or acidity characteristics of petroleum products are defined by the neutralization number, the acid number, or the base number obtained by one of several standardized methods. Methods currently used include ASTM Method D664, Neutralization Value by Potentiometric Titration, ASTM Method D974, Neutralization Value by Color-Indicator Titration, and ASTM Method D2896, Total Base Number of Petroleum Products by Potentiometric Perchloric Acid Titration. By ASTM Method D664, engine oils may have both acidic and alkaline characteristics, depending upon the nature of the additives used. Certain salts which are commonly used as engine oil additives, such as zinc dithiophosphates, will undergo exchange reactions with the standard titrant KOH and thereby produce false acid values. Changes in alkalinity or acidity with use give some indication of the nature of the changes taking place in the engine oil. For instance, a reduction in alkalinity can be ascribed to depletion of additive components. An increase in acidity may be ascribed to oxidation and/or contamination by products of combustion. No general relationship between bearing corrosion and acid or base number is known. None of the aforementioned methods is intended to be used to predict performance of an oil under service conditions. They can be used to follow general deterioration rates of oils in service.

3.1.6 CARBON RESIDUE—The base stock components of engine oils are mixtures of many compounds which differ widely in their physical and chemical properties. Some vaporize at atmospheric pressure without leaving an appreciable residue. When destructively distilled, the non-volatile compounds may leave a carbonaceous material known as carbon residue. Two ASTM Methods are used for evaluating base stocks in this respect. These are ASTM Method D189, Conradson Carbon Residue, and ASTM Method D524, Ramsbottom Carbon Residue of Petroleum Products. Engine oils which contain ash-forming constituents, such as the additives commonly used in formulating oils, may have misleadingly high carbon residues by either method. Carbon residue has little value as a guide for predicting deposit-forming tendencies in engines.

3.1.7 ASH CONTENT—The amount of ash formed from burning engine oils may be obtained by ASTM Method D482, Ash from Petroleum Products, but, ASTM Method D874, Sulfated Ash from Lubricating Oils and Additives, is the method currently and most commonly used. The ash produced from burning new engine oils is principally related to the quantity of ash-producing additive contained therein. The ash produced by used oils will be a function, not only of the amount of ash-producing additive agents in the original oil, but also of the amount of contaminants such as lead compounds present in the engine oil if the engine is operated on a leaded fuel. High values can also result from other contaminants such as dirt, iron oxide, wear metals, and corrosion products. The ash from an oil will contribute to deposits on combustion chamber surfaces, spark plugs, and exhaust valves. However, the mechanism for the buildup of deposit in these areas is very complex and depends upon many variables in addition to the ash content of the oil.

3.1.8 COMPATIBILITY—Engine oils are expected to be homogeneous and completely miscible with all types of engine oils with which they might be mixed in service. When oils are mixed in any proportion, there should be no evidence of separation either of the components or of the oils when the mixed oils are heated to a temperature as high as 225°C and cooled to a temperature as low as the pour point of the mixture. The homogeneity and miscibility test currently used to evaluate automotive engine oils is Standard No. 791B, Federal Test Method 3470.

3.1.9 FOAMING—Antifoam quality is a performance characteristic which can be and has been evaluated by specially developed engine tests. A bench test for determining this quality is ASTM Method D892, Foaming Characteristics of Lubricating Oils. As with any bench test, the degree of correlation with actual service should be determined before applying ASTM D892 results to specifications, etc.

3.1.10 GRAVITY, COLOR, ODOR—The gravity (density) and color of engine oils may frequently be used in specifying engine oils. ASTM Method D287, API Gravity of Crude Petroleum Products (Hydrometer Method), ASTM Method D1298, Density, Specific Gravity or API Gravity of Crude Petroleum and Liquid Petroleum Products by Hydrometer Method, and ASTM Method D941, Density and Specific Gravity by Lipkin Bicapillary Pycnometer, may be used to determine the gravity and density characteristics of oils. The color of engine oils may be specified by using ASTM Method D156, Saybolt Color of Petroleum Products or ASTM Method D1500, ASTM Color of Petroleum Products (ASTM Color Scale). These factors are generally associated with the quality control of manufactured products rather than with performance characteristics.

It is expected that engine oils will not produce offensive odors due to the nature of the base oils or the additive agents with which the oil is compounded. Offensive odors should not be generated during either use or prolonged storage of an engine oil. There are no standardized odor tests available.

3.1.11 ELEMENTAL ANALYSIS—Elemental analysis of engine oils is often used as a means of quality control. Instrumental analytical techniques such as emission spectrography, atomic absorption spectrometry, x-ray diffraction, etc. are useful in this respect. Similar analysis of used oils will provide information relative to the changes in the chemical content of the engine oil. These data can also give a measure of contamination by materials such as products of combustion, particularly with engines using leaded gasoline. They can also provide information relative to the extent of wear in the engine. Concentrations of the following elements are commonly determined.

1. Additive elements such as barium, calcium, zinc, phosphorus, sulfur, and magnesium.
2. Contaminants such as lead, silica, etc.
3. Wear metals such as iron, copper, lead, tin, aluminum, etc.

3.1.12 HYDROCARBON STRUCTURE—Infrared spectrophotometry techniques are valuable in determining the hydrocarbon structures found in lubricant stocks and additives. Changes in these structures can be determined by analysis of used oil samples. It is also possible to identify the presence of hydrocarbon contaminants, water, glycol, and similar materials in used oil samples.

3.2 Tests Pertinent to Used Oils

3.2.1 USED OIL PROPERTIES—The analysis of used engine oils may be of value in establishing the condition of the oil with respect to its useful life, and may be helpful in estimating the condition of the engine. To be of most value, used oil analyses must be compared with similar analyses of the new oil. The conditions of usage must also be considered in evaluating used oil analyses.

3.2.2 INSOLUBLE CONTENT—Insoluble materials found in both new and used engine oils may be determined using ASTM Method D91, Precipitation Number of Lubricating Oils, or the more frequently used ASTM Method D893, Insolubles in Used Lubricating Oils. Use of these methods permits an evaluation of the contaminant content and buildup of insoluble materials through oxidation, etc. However, the results must be judged with care, because minor changes in the analytical procedure can produce different results. For example, the age and purity of the coagulant solutions specified in ASTM Method D893 can affect the results obtained.

With modern, highly dispersant oils, insoluble determinations become exceedingly difficult. Because of the extremely small particle size generally found in these oils, coagulants, super centrifuges, ultrafine filters, and other procedural and equipment refinements may be required in order to make accurate determinations. ASTM Committee RDD-VI-B is investigating means of accurately and reproducibly determining the insoluble content of highly dispersant oils.

3.2.3 COOLANT (MOISTURE) CONTENT—Small quantities of water will frequently be found in used engine oil as contamination from products of combustion, leakage from the cooling system, or condensation from atmospheric moisture. ASTM Method D95, Water in Petroleum Products and Bituminous Materials by Distillation, defines a procedure for determination of the water content of used oil. For a qualitative determination, a commonly used simple test is to heat a drop of oil on aluminum foil. A snapping or crackling sound indicates free or suspended water in the oil. Cooling system leakage can be suspected when water is found in the oil on cooldown after operation for several hours under high temperature conditions such as interstate highway driving. The presence of glycol can be a more definite indication of leakage. It is detected best by distillation of the aqueous material, followed by chemical analyses or infrared spectrophotometry on the distillate. A less complicated procedure that is adaptable to field kit use which gives positive, trace, or negative results is ASTM Method D2982, Detecting Glycol-Base Antifreeze in Used Lubricating Oils. Some additives commonly used in formulating crankcase oils contain glycol at a level which will give a positive result in this test. If the new oil gives a positive result, the test in its simple form will be inadequate for detecting coolant glycol in used oils and the oil supplier should be consulted for advice.

3.2.4 FUEL DILUTION—Engines in good mechanical condition and operated at normal temperatures will usually show a small amount of fuel dilution in the used engine oil. Low operating temperatures, rich mixtures of fuel and air, and low ambient temperatures will promote fuel dilution, particularly if the engine is in poor mechanical condition or crankcase ventilation is inadequate. High dilution reduces oil viscosity and pour point. The presence of such dilution can cause accelerated wear and promote the formation of sludge, varnish, and rust. The presence of a high dilution content may indicate a need for engine maintenance. Dilution may be determined through the use of ASTM D322, Dilution of Gasoline-Engine Crankcase Oils. This method is useful only with gasoline engines since the diesel fuel oil distillation range in many cases overlaps that of the engine oil distillation range. Recently developed procedures applicable to both diesel and gasoline engine oils are ASTM Method D3524, Diesel Fuel Diluent in Used Diesel Engine Oil by Gas Chromatography, and ASTM Method D3525, Gasoline Diluent in Used Gasoline Engine Oils by Gas Chromatography.

4. Performance Characteristics—In the operation of an internal combustion engine, engine oils are expected to lubricate, cool, seal, maintain cleanliness, and protect against wear and corrosion. An oil's ability to perform these functions depends on the combined effectiveness of its base stocks and additives. Although the physical and chemical tests described in the preceding sections can be used for quality control to insure manufacturing uniformity, they are not effective for accurately evaluating performance characteristics. Only actual performance evaluations in special laboratory engines and in field tests will define the capabilities of an engine oil.

A variety of laboratory diesel and gasoline engine tests which have become industry standards for evaluating engine oils are described in SAE J304. SAE J183 classifies oils according to performance criteria based on results from these engine tests. Combustion chamber deposit control and oil consumption characteristics of engine oils are not completely defined by the test procedures outlined in SAE J304. Where these properties are of interest, specific tests must be developed using the equipment and conditions most relevant to a given situation.

Although the laboratory engine tests are necessary and valuable aids to engine oil development and evaluation, even they have limitations. In many instances, the final proof-of-performance is established by field tests of the oil in actual vehicle service. No industry standardized field test methods are currently available. Rather, a given oil is evaluated in the types of service typical of the use for which it was designed.

5. Conclusions—The lubrication requirements for modern engines are extremely complex. Current engine oils are the result of extensive research and development aimed at meeting these requirements. It is not the objective of this information report to treat the subject in detail. Rather, the purpose of this document is to define very briefly the terms frequently encountered in discussions of engine oils and engine oil performance for those technical people not directly associated with lubricants and lubricant development. For more detailed information on these matters, the reader is referred to the technical services offered by lubricant suppliers, the lubricants groups of engine manufacturers, and pertinent literature available through the Society of Automotive Engineers, American Society of Lubrication Engineers, American Society of Mechanical Engineers, American Petroleum Institute, American Society for Testing and Materials, etc. Information directly related to this report may be found in the following publications:

1. "Significance of ASTM Tests for Petroleum Products." ASTM Special Technical Publication 7-C, 1977.
2. "The Physical Properties of Lubricants." American Society of Lubrication Engineers, 1951.
3. C. W. Georgi, "Motor Oils and Engine Lubrication." New York: Reinhold Publishing Corp., 1950.
4. William A. Gruse, "Motor Oils, Performance and Evaluation." New York: Reinhold Publishing Corp., 1967.
5. A. Schilling, "Automobile Engine Lubrication." Broseley, England: Scientific Publications (G.B.) Ltd., 1972.
6. R. C. Gunderson and A. W. Hart, "Synthetic Lubricants." New York: Reinhold Publishing Corp., 1962.
7. ASTM STP 315F, "ASTM Multicylinder Test Sequences for Evaluating Automotive Engine Oils."
8. ASTM STP 509, "ASTM Single Cylinder Engine Tests for Evaluating the Performance of Crankcase Lubricants."

AXLE AND MANUAL TRANSMISSION LUBRICANTS—SAE J308c

SAE Information Report

Report of Lubricants Division approved February 1924 and last revised by Fuels and Lubricants Technical Committee April 1976.

This SAE Information Report was prepared by the SAE Fuels and Lubricants Technical Committee for two purposes: (1) to assist the users of automotive equipment in the selection of differential and manual transmission lubricants for field use, and (2) to promote a uniform practice for use by marketers of lubricants and by equipment builders in identifying and recommending these lubricants by type.

In both differentials and transmissions, gears and bearings of different designs are employed under a variety of service conditions. Therefore, the selection of a lubricant involves careful consideration of the performance characteristics required.

1. Load-Carrying Capacity—One of the most important gear lubricant performance characteristics is load-carrying capacity. Some gears are operated under such loads and speeds that the very low capacity of straight mineral oil is adequate. However, most gears require lubricants of greater load-carrying capacity, which is provided through the use of additives.

Gear lubricants compounded to achieve increased load-carrying capacity are referred to as "extreme-pressure" (EP) lubricants. However, when this term is applied to gear lubricant it means only that the load-carrying capacity of the lubricant is greater than that of straight mineral oil, with no distinction as to how much greater it may be. Therefore, to differentiate among EP lubricants of various load-carrying capacities, it is necessary to classify them further. The Coordinating Research Council (CRC) and the American Petroleum Institute (API) have devised ways to aid in this classification.

CRC has available reference gear oils (RGO) that can be used to rate the load-carrying capacity of a gear lubricant. The rating, or comparison, can be done in either automotive equipment on the road or in full-scale laboratory equipment. The reference oils use a numerical rating scale that ranges from RGO-100, very low load-carrying capacity, to RGO-115, very high load-carrying capacity. RGO-100 is a solvent refined straight mineral oil and RGO-115 is the same mineral oil containing 15% (wt.) of a particular EP additive. The complete designation shows the percent additive followed by the SAE viscosity grade, such as RGO-110-90 or RGO-112-80.

The following designations are quoted from API Publication 1560, Lubricant Service Designations for Automotive Manual Transmissions[1] and Axles, February 1976.

"**API-GL-1** Designates the type of service characteristic of automotive spiral-bevel and worm-gear axles and some manually operated transmissions operating under such mild conditions of low unit pressures and sliding velocities, that straight mineral oil can be used satisfactorily. Oxidation and rust inhibitors, defoamers, and pour depressants may be utilized to improve the characteristics of lubricants for this service. Frictional modifiers and extreme pressure agents shall not be utilized."

Lubricants suitable for this type of service are, therefore, considered to be "straight mineral gear oils." In antiscoring protection, these lubricants are comparable to CRC RGO-100.

Due to speeds and loads involved, straight mineral oil is not a satisfactory lubricant for most 4-speed and some 3-speed passenger car manual transmissions.[1] For some truck and tractor manual transmissions, straight mineral oil is suitable.

"**API-GL-2** Designates the type of service characteristic of automotive type worm-gear axles operating under such conditions of load, temperature, and sliding velocities, that lubricants satisfactory for API-GL-1 service will not suffice."

Products suited for this type of service contain antiwear or very mild extreme-pressure agents which provide protection for worm gears.

There are relatively very few differentials in use that are equipped with worm gears. The GL-2 designation is included in this list for those worm gears used in a service that has been found to require a lubricant other than straight mineral oil.

"**API-GL-3** Designates the type of service characteristic of manual transmissions and spiral-bevel axles operating under moderately severe conditions of speed and load. These service conditions require a lubricant having load carrying capacities greater than those which will satisfy API-GL-1 service, but below the requirements of lubricants satisfying API-GL-4 service."

Lubricants designated for this service typically contain additives which react with tooth surfaces at the temperatures resulting from high speed or load. Due to the rate of reactivity or the relatively low concentration of the additives, products designated for GL-3 are not formulated to provide adequate protection for hypoid gears. The scoring resistance of such oils is comparable to that provided by CRC reference gear oils below RGO-104.

"**API-GL-4** This classification is still used commercially to describe lubricants, but the equipment required for the anti-scoring test procedures to verify lubricant performance is no longer available.

Designates the type of service characteristic of gears, particularly hypoid[2] in passenger cars and other automotive type equipment operated under high-speed, low-torque, and low-speed, high-torque conditions.

Lubricants suitable for this service are those which provide antiscore protection equal to or better than defined by CRC Reference Gear Oil RGO-105 and have been subjected to the test procedures and provide the performance levels described in ASTM STP-512 dated April 1972.[3]"

"**API-GL-5** Designates the type of service characteristic of gears, particularly hypoid[2] in passenger cars and other automotive equipment operated under high-speed, shock-load; high-speed, low-torque; and low-speed, high-torque conditions.

Lubricants suitable for this service are those which provide anti-score protection equal to or better than defined by CRC Reference Gear Oil RGO-110 and have been subjected to the test procedures and provide the performance levels described in ASTM STP-512 dated April 1972.[3]"

"**API-GL-6** This is an obsolete classification. The equipment required for the test procedures to verify lubricant performance is no longer available.

The type of service designated by API-GL-6 is characteristic of gears, specifically high offset hypoid[2] gears (above 2 in offset and approaching 25% of ring gear diameter) in passenger cars and other automotive equipment operated under high-speed, high-performance conditions.

Lubricants suitable for this service are those which provide anti-score protection equal to or greater than Reference Gear Oil L-1000[4] and have been subjected to the test procedures and provide the performance levels described in ASTM STP-512 dated April 1972.[3]"

In addition to load-carrying capacity, the equipment manufacturer, lubricant supplier, and the vehicle operator should consider the following factors.

2. Viscosity and Viscosity Index—Viscosity determines ease of gear shifting and influences channeling at low temperatures. It has some relation to load-carrying capacity, leakage, and gear noise.

Viscosity, which is a measure of resistance to flow, increases with decreasing temperature. At a given temperature, Newtonian oils will not experience a viscosity change with a change in shear rate. A non-Newtonian oil may exhibit several different apparent viscosities in different parts of a differential due to variations in shear rate, and a "permanent" viscosity loss due to high shear rate conditions. The method of determining viscosity must give values which have been shown to correlate with performance.

Two viscosity methods are pertinent to gear oils. ASTM D 445 measures kinematic viscosity based on the time for oil to flow by gravity through the capillary tube. The common unit is centistokes (cSt) and the corresponding numerically equal Systems International (SI) unit is mm^2/s. The most common test temperatures are 100° and 210°F (38° and 99°C). ASTM D 2983 Brookfield method has significance for measuring apparent viscosity at low temperatures (for example at $+10°F [-12°C]$ or lower) because most oils are non-Newtonian at these temperatures. The common unit is centipoise (cP); the Pascal second (Pa·s) is the SI unit.

The viscosity-temperature relationship of an oil depends primarily on the nature or composition of the oil. A minimum change in viscosity with temperature is desirable. ASTM D 341, Standard Viscosity-Temperature Charts, provides a series of six charts so constructed that for any given petroleum oil, the viscosity-temperature values can be approximated by a straight line over the applicable temperature range in which the oil is a Newtonian liquid. At low temperatures there are serious limitations with these charts. To be accurate, only interpolated values should be taken from these charts. Measure-

---

NOTE: Lubricants suitable for more than one service classification may be so designated.

[1] Automatic or semi-automatic transmissions, fluid couplings, torque converters, and tractor hydraulic systems usually require special lubricants. For the proper lubricant to be used, consult the manufacturer or lubricant supplier.

[2] Limited slip differentials generally have special lubrication requirements. The lubricant supplier should be consulted regarding the suitability of his lubricant for such differentials. Information helpful in evaluating lubricants for this type of service may be found in ASTM STP-512 dated April 1972.

[3] The complete publication is titled "Laboratory Performance Tests for Automotive Gear Lubricants Intended for API-GL-4, GL-5, and GL-6 Services."

[4] Reference Gear Oil L-1000 is available for a fee from Southwest Research Institute, P. O. Drawer 28510, San Antonio, Texas 78284.

ments of the viscosity of a gear lubricant should be made at or near the temperature at which it is expected to perform. This is especially true at low temperatures by ASTM D 2983 since viscosity-related problems may occur with regard to flow to bearings in the differential and shifting ability in transmissions. For the high temperature operation associated with continuous, heavy load operation of the gear box, an oil with proper high temperature viscosity characteristics should be selected. Increasing the viscosity will reduce leakage, wear, and noise. It will also increase fluid friction which may contribute to excessive temperatures in mechanisms such as transmissions with a large number of gear ratios or in differentials operated at relatively high speeds.

Viscosity index is an empirical number indicating the effect of change of temperature on the viscosity of an oil. The higher the viscosity index, the smaller will be the change in viscosity with change in temperature. (See ASTM D 2270, Viscosity Index.)

The relationship of viscosity and temperature defined by viscosity index is not affected by shear rate for Newtonian oils. However, non-Newtonian oils (such as those containing polymers or waxy oils at low temperatures) do not have their physical properties fully described by a viscosity index derived from low shear rate measurements only. The viscosity index of non-Newtonian oils becomes a function of the shear rate at the time of measurement, and tends to become smaller with increase in shear rate.

In general, for passenger car service, one viscosity grade or a multi-viscosity lubricant is recommended for year-round service; for trucks, buses, and other heavy duty equipment, two viscosity grades or a multi-viscosity lubricant may be recommended. However, for extremely low or high operating temperatures and for extremes in climatic conditions, even lower or higher SAE viscosity grades may be necessary. The six SAE viscosity grades for transmission and rear axle gear lubricants, and their limits, are detailed in Table 1 of SAE J306.

3. Channeling—It is important that under low temperature conditions, the lubricant should not "set" or "solidify." Lubricants should flow readily between contacting surfaces. This is particularly true in the case of highly stressed gears and axle front pinion bearings where the absence of lubricant even for a few seconds may result in rapid failure.

The pour point of a gear lubricant is not a reliable index in determining whether the lubricant will flow satisfactorily and not channel at low temperatures.

4. Stability and Oxidation Resistance—The major factors affecting these characteristics are operating temperatures and mileage in service, as well as time and temperature during storage. Abnormally high or low temperatures during storage may cause formation or separation of materials that will alter the original properties of the lubricant. Ideally, storage temperature should be as close to room temperature as possible.

For passenger car axles and transmissions in mild service, temperatures of the lubricant may not be high enough to make resistance to oxidation of paramount importance. However, for passenger cars in severe conditions of service such as pulling trailers or for trucks or buses in service where high temperatures occur, oxidation resistance is an important factor. Accordingly, oils with a high degree of oxidation resistance should be used in this type of service.

5. Foaming—Excessive foaming may interfere with proper lubrication of gear and bearing surfaces and is extremely objectionable. It can cause leakage and deplete the supply of lubricant in the housing. Defoamers are used to minimize this problem.

6. Chemical Activity or Corrosion—In order to obtain gear lubricants with adequate load-carrying capacity, chemical agents are usually employed as compounding ingredients. Such agents may introduce corrosion problems. However, blackening or darkening of brightly polished copper or other metal surfaces in contact with the lubricant does not necessarily indicate that the lubricant will cause harmful corrosion in service. Corrosion may be minimized by choice of a lubricant containing the proper combination of chemical additives, particularly those stable in the presence of water.

7. Seal Compatibility—While the primary function of a gear lubricant is to protect gears and bearings, consideration must be given to the effect of a lubricant on elastomers or other seal materials used in the design of the component. Simple immersion tests such as those described in ASTM D 471 may be used to establish the relative compatibility of the lubricant and the seal material.

8. Mixing Gear Lubricants—When most gear lubricants of different types and viscosities are mixed, the load-carrying properties of the mixtures will be intermediate to the load-carrying properties of the individual lubricants. For example, mixing a gear lubricant intended for API Service GL-1 with a gear lubricant intended for API Service GL-4 will increase the load-carrying capacity of the former but will decrease that of the latter. The extent of this change in load-carrying capacity depends in part on the quality of the individual lubricants and on the relative amount of each present in the mixture. Other properties of the lubricants may be affected in a similar manner. The mixing of different types of gear lubricants is not recommended.

9. Solid Matter—In general, lubricants containing any solid materials are undesirable for ball or roller bearings in transmissions and rear axles. They may cause abrasion and wear.

AXLE AND MANUAL TRANSMISSION LUBRICANT VISCOSITY CLASSIFICATION—SAE J306 OCT79

SAE Recommended Practice

Report of the Lubricants Division, approved February 1924, last revised by the Fuels and Lubricants Committee November 1977, editorial change October 1979.

This SAE Recommended Practice is intended for equipment manufacturers in defining and recommending axle and manual transmission lubricants, for oil marketers in labeling such lubricants with respect to viscosity, and for users in following their owner's manual recommendations. The SAE viscosity grades shown in Table 1 constitute a classification for axle and transmission lubricants in terms of viscosity only; the change in viscosity with use, or other gear lubricant qualities, are not considered.

Axle and transmission lubricant SAE viscosity grades should not be confused with engine oil SAE viscosity grades. (Compare Table 1 in this report with Table 1 in SAE J300.) A gear lubricant and an engine oil having the same viscosity will have widely different SAE viscosity grade designations as defined in the two viscosity classifications. For instance, an SAE 80W gear lubricant can have the same viscosity characteristics as an SAE 20W-20 engine oil; and an SAE 90 gear lubricant viscosity can be similar to that of an SAE 40 or SAE 50 engine oil.

This classification is based on the lubricant viscosity measured at both high and low temperatures. The high-temperature values are determined according to ASTM D 445, Method of Test for Viscosity of Transparent and Opaque Liquids, with the results reported in centistokes (cSt).[b] The low-temperature values are determined according to ASTM D 2983, Method of Test for

Apparent Viscosity at Low Temperature Using the Brookfield Viscometer, and these results are reported in centipoises (cP).[a] These two viscosity units are related as follows:

$$\frac{cP}{\text{Density, kg/dm}^3} = cSt$$

Density is measured at the test temperature.

This relationship is valid for Newtonian fluids; it is an approximation for non-Newtonian fluids.

A multiviscosity graded lubricant, such as SAE 80W-90, meets both the low and high-temperature requirements shown in Table 1. That is, it conforms to the SAE 80W requirement at low temperature and is in the range provided for SAE 90 at high temperature.

The selection of an axle or transmission lubricant should be based on the lowest and highest service temperatures. The multiviscosity graded lubricants may be satisfactory at both temperature extremes. The 150 000 cP viscosity value used for the definition of low-temperature properties is based on a series of tests in a specific rear axle design. These tests have shown that pinion bearing failure has occurred at viscosities higher than 150 000 cP and the Brookfield method was shown to give adequate precision at this viscosity level. However, it should be pointed out that other axle designs may tolerate higher viscosities or fail at lower viscosities. The Brookfield low-temperature viscosity curves for several gear lubricants, made with conventional petroleum base stocks, are shown in a viscosity-temperature chart in Fig. 1. It must be recognized that some gear lubricants can have viscosity-temperature relationships different than those shown in this chart.

Other applications may require considerably different Brookfield viscosity limits. For example, experience has indicated that, for satisfactory ease of shifting, many manual transmissions require a lubricant viscosity not exceeding 20 000 cP at the shifting temperature.

In recommending axle and transmission lubricants by SAE viscosity grade, the following temperatures are suggested as a uniform practice: $-40°C$, $-26°C$, and $-12°C$.

NOTE: [a] and [b] refer to footnotes in Table 1.

TABLE 1—AXLE AND MANUAL TRANSMISSION LUBRICANT VISCOSITY CLASSIFICATION

SAE Viscosity Grade	Maximum Temperature for Viscosity of 150 000 cP[a] °C	Viscosity at 100°C[b] cSt	
		Minimum	Maximum
75W	−40	4.1	—
80W	−26	7.0	—
85W	−12	11.0	—
90	—	13.5	<24.0
140	—	24.0	<41.0
250	—	41.0	—

[a] Centipoise (cP) is the customary absolute viscosity unit and is numerically equal to the corresponding SI unit of millipascal-second (mPa·s).

[b] Centistokes (cSt) is the customary kinematic viscosity unit and is numerically equal to the corresponding SI unit of square millimetre per second (mm²/s).

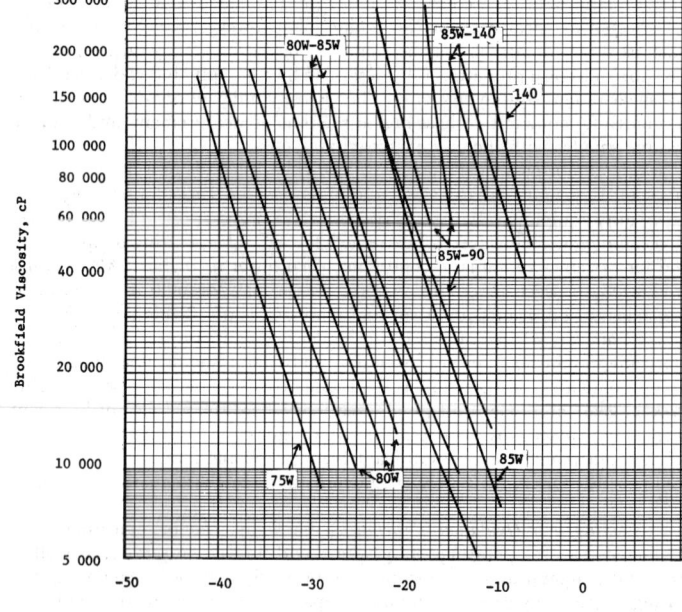

FIG. 1—BROOKFIELD VISCOSITY VERSUS TEMPERATURE FOR TYPICAL GEAR LUBRICANTS (SAE VISCOSITY GRADES INDICATED)

AUTOMOTIVE LUBRICATING GREASES— SAE J310 MAR80

SAE Information Report

Report of the Fuels and Lubricants Technical Committee, approved September 1951, last revised April 1967, reaffirmed with editorial change February 1976, editorial change March 1980.

This SAE Information Report was prepared by Subcommittee 4, Lubricating Greases, of the SAE Fuels and Lubricants Technical Committee to assist those concerned with the design of automotive components, and with the selection and marketing of greases for the lubrication of certain of those components on passenger cars, trucks, and buses. It is expected that the information contained will be helpful in the standardization of terms related to the types of greases and the means of designation.

Definition of Lubricating Grease—A lubricating grease is a solid to semifluid mixture of a fluid lubricant and a thickening agent. Additives to impart special properties or performance characteristics may be incorporated. The fluid component may be a mineral (petroleum) oil or a synthetic fluid; the thickener may be a metallic soap or soaps or a nonsoap substance such as an organophilic modified clay, a urea compound, carbon black, or other material. The soaps commonly used are lithium, calcium (lime), sodium, aluminum, and barium, or certain combinations of these with other materials, such as calcium-lead. The viscosity of the fluid, the ratio of fluid to thickener, and the chemical nature of the thickener may vary widely. The properties of the finished grease are influenced by the processes of its manufacture as well as by the materials used.

Basic Performance Requirements—Greases are most often used instead of fluids where a lubricant is required to maintain its original position in a mechanism, especially where opportunities for frequent relubrication may be limited or economically unjustifiable. This requirement may be due to the physical configuration of the mechanism, the type of motion, the type of sealing, or to the need for the lubricant to perform all or part of any sealing function in the prevention of lubricant loss or the entrance of contaminants.

Because of their essentially solid nature, greases do not perform the cooling and cleaning functions associated with the use of a fluid lubricant. With these exceptions, greases are expected to accomplish all other functions of fluid lubricants.

A satisfactory grease for a given application is expected to:
1. Provide adequate lubrication to reduce friction and to prevent harmful wear of bearing components.
2. Protect against corrosion.
3. Act as a seal to prevent entry of dirt and water.
4. Resist leakage, dripping, or undesirable throw-off from the lubricated surfaces.
5. Resist objectionable change in structure or consistency with mechanical working (in the bearing) during prolonged service.
6. Not stiffen excessively to cause undue resistance to motion in cold weather.
7. Have suitable physical characteristics for the method of application.
8. Be compatible with elastomer seals and other materials of construction in the lubricated portion of the mechanism.
9. Tolerate some degree of contamination, such as moisture, without loss of significant characteristics.

Properties of Greases

Consistency—A measure of relative hardness. This property is commonly expressed in terms of the ASTM penetration or NLGI consistency number. The ASTM penetration is a numerical statement of the actual penetration of the grease sample, in tenths of a millimeter, by a standard test cone under stated conditions. The higher the penetration value, the softer the grease. The National Lubricating Grease Institute classifies greases according to their ASTM penetration as shown in Table 1.

The consistency of a grease is an important factor in its ability to lubricate, seal, and remain in place, and to the methods and ease by which it can be dispensed and applied. Most automotive greases are in the NLGI No. 1, 2, or 3 range, that is, ranging from soft to medium consistency.

Texture and Structure—The appearance and feel of greases. A grease may be described as smooth, buttery, fibrous, long- or short-fibered, stringy, tacky, etc. These characteristics are influenced by the viscosity of the fluid, type of thickener, proportion of each of these components, presence of certain additives, and process of manufacture. There are no standard test methods for quantitative definitions of these properties. Texture and structure are factors in the adhesiveness and ease of handling of a grease.

Structural Stability—The ability of a grease to retain its as-manufactured consistency and texture despite age, temperature, mechanical working, and other influences, or its ability to return to its original state when a transient influence is removed.

Mechanical Stability—The resistance of a grease to permanent changes in consistency due to the continuous application of shearing forces.

The stability of a grease is important to its ability to provide adequate lubrication and sealing and to remain properly in place during use.

Apparent Viscosity—The ratio of shear stress to rate of shear at a stated temperature and shear rate. Grease is by nature a plastic material. Therefore, the usual concept of viscosity valid for simple fluids (that is, internal resistance to flow) is not entirely applicable. The ratio of shear stress to shear rate varies as the shear rate changes. The apparent viscosity of most greases decreases with increase of either temperature or shear rate. Apparent viscosity greatly influences the ease of handling and dispensing a grease.

Dropping Point—The temperature at which the grease generally passes from a plastic solid to a liquid state, and flows through an orifice under standard test conditions. The dropping point is incorrectly regarded by some as establishing the maximum temperature for acceptable use. Performance at high temperature also depends on other factors such as duration of exposure, evaporation resistance, and design of the lubricated mechanism.

Oxidation Resistance—The resistance to chemical deterioration in storage and in service caused by exposure to air. It depends basically on the stability of the individual grease components, and can be improved by use of antioxidants. High oxidation resistance is important wherever long storage or service life is required or where *high temperatures* prevail even for short periods.

Protection Against Friction and Wear—A protection greatly influenced by the viscosity and type of the fluid component and by grease structural and consistency characteristics. This performance characteristic can be altered by use of additives.

Protection Against Corrosion—A protection depending on grease composition, the ability to form and maintain a seal against the entrance of corrosive (and other undesirable) materials, and the reaction to water. Some greases are *water resistant* or *waterproof,* meaning that they resist the washing effect of water and do not absorb it to any extent. Others can absorb varying amounts of water without appreciable damage to their structure or consistency, and may in service provide better protection against corrosion than waterproof greases, since the latter may permit the accumulation of free water in bearings. Rust protection may be improved by the use of suitable additives.

Bleeding or Oil Separation—The separation of liquid lubricant from a grease. Slight bleeding is regarded as desirable by some as indicative of good lubricating ability in rolling element bearings. *ed.*

Color—A superficial grease property without performance significance.

Of the above properties, oxidation resistance, protection against friction and wear, protection against corrosion, and structural stability are probably of most importance in automotive service as far as actual performance in bearings is concerned.

There is, of course, the problem of getting grease to the bearings to be lubricated. Certain terms, by no means of strict, rigid interpretation, are used to describe the factors involved: feedability, pumpability, and dispensability.

Feedability, or Slumpability—The ability to flow to the suction of the grease-dispensing equipment or mechanism to be lubricated.

Pumpability—The ability to flow through the grease-dispensing lines at a satisfactory rate, without the necessity of using excessively high pressures.

Dispensability—The ease with which a grease may be transferred from its container to the point of application. For practical purposes, it is a combination of feedability and pumpability.

Grease Testing—Many of the above grease properties are determined by tests which have been standardized or otherwise accorded industry recognition. These, in conjunction with simulated performance tests, permit some approximate judgment for the proper selection of greases for a given application. They are, however, not considered to be replacements for, or equivalent to, long-time service tests.

Table 2 shows some of the more important tests, both standard and otherwise, identified as to sponsor, title, and purpose.

Designation of Greases by Fields of Use—Greases are commonly classified and designated according to chemical composition, such as calcium-soap grease; by broad type of usage, such as antifriction bearing grease or multipurpose grease; by specific properties, such as high temperature grease; by special additives, such as extreme-pressure grease or graphite grease; and by specific applications, such as automotive-wheel-bearing grease. SAE recognizes the following designations for greases used in servicing passenger cars, trucks, and buses according to their specific applications.

Wheel-Bearing Grease—This term designates lubricating greases of such composition, structure, and consistency as to be suitable for longtime use in antifriction wheel bearings.

NOTE: Generally, these greases have high resistance to the deteriorating effects of temperature and the separating effects of centrifugal action. They should have good antirust properties. They should not exhibit oil-soap separation or excessive softening which could result in leakage that could lead to braking failure.

Universal Joint Grease—This term designates lubricating greases of such composition, structure, and consistency as to be suitable for the lubrication of those types of automotive universal joints requiring grease lubrication.

NOTE: Some designs of universal joints require lubricants other than the usual universal joint type. Manufacturers' recommendations or lubrication charts should be consulted.

Chassis Grease—This term designates lubricating greases of proper consistency to be applied periodically at intervals in accordance with equipment manufacturers' recommendations, with grease guns through grease fittings, into the various parts of automotive chassis requiring grease for lubricants.

NOTE: A grease with a relatively high apparent viscosity at high shear rates may be required for heavy-duty service. This characteristic is a function of the viscosity of the oil component as well as of grease consistency. The oil viscosity, of course, is a function of operating temperature; however, grease viscosity is influenced also by the nature and amount of thickener.

ELI (Extended-Lubrication-Interval) Chassis Grease—This term designates lubricating greases of proper composition, structure, consistency, life, and antiwear and anticorrosion abilities to permit their use in suspension, driveline, and steering systems employing sealed joints, prepacked during manufacture or assembly and normally not requiring relubrication for comparatively long intervals.

NOTE: The design of the sealed joint naturally has an important role in the ability of ELI greases to achieve their objective. Seal life is especially critical,

TABLE 1—NLGI[a] CONSISTENCY NUMBERS

NLGI Consistency No.	ASTM Worked (60 Strokes) Penetration at 25°C (77°F) tenths of a millimetre	NLGI Consistency No.	ASTM Worked (60 Strokes) Penetration at 25°C (77°F) tenths of a millimetre
000	445 to 475	3	220 to 250
00	400 to 430	4	175 to 205
0	355 to 385	5	130 to 160
1	310 to 340	6	85 to 115
2	265 to 295		

[a] National Lubricating Grease Institute, 4635 Wyandotte St., Kansas City, Missouri 64112

TABLE 2—GREASE TESTS

Test Designation	Test Purpose	Test Designation	Test Purpose	Test Designation	Test Purpose
ASTM D 128, Analysis of Lubricating Grease[a]	Determination of nominal chemical composition, such as soap, unsaponifiable matter (mineral oil), water, free alkalinity, free fatty acid, glycerine, and insolubles. NOTE: This procedure has a supplementary method useful for greases containing nonsoap thickeners or synthetic fluids.	ASTM D 1403, Cone Penetration of Lubricating Grease Using One-Quarter and One-Half Scale Cone Equipment[a]	Essentially same as ASTM D 217, using small grease samples, but reserved to greases of NLGI Nos. 0 to 4 consistency.	ASTM D 2509, Measurement of Extreme-Pressure Properties of Lubricating Grease (Timken Method)	Determination of load carrying ability of lubricating greases by Timken Extreme-Pressure Tester. In this device, a rectangular steel test block is forced against a rotating steel ring. Scar width and surface conditions are noted. Method differentiates between lubricants of various extreme-pressure levels; not a replacement for actual service tests.
ASTM D 217, Cone Penetration of Lubricating Grease[a]	Measurement of relative hardness by unworked and worked penetration; test satisfactory up to penetrations of 400 (tenths of millimeter) at 25°C (77°F), or to 475 with alternate cone.	ASTM D 1741, Functional Life of Ball-Bearing Greases	Endurance life of grease lubricated 306 ball bearings at 3600 rpm; evaluation valid up to 125°C (257°F) operating temperature; is primarily a screening test and does not replace long-time service tests.		
ASTM D 566, Dropping Point of Lubricating Grease[a]	Establishment of temperature at which grease generally passes from plastic to liquid state; not regarded as indicative of service suitability; limited to dropping points up to 260°C (500°F). (In this test, some nonsoap-thickened greases may release oil before the grease flows which is defined as their dropping point.)	ASTM D 1742, Oil Separation from Lubricating Grease During Storage[a]	Determination of tendency of oil constituent to separate from parent grease while in containers; suitable for NLGI No. 1 or harder greases; results are indicative of oil separation in containers, but not of oil separation under dynamic service conditions.	ASTM D 2595, Evaporation Loss of Lubricating Grease Over Wide-Temperature Range	Evaluation of weight loss by evaporation at temperatures between 93 and 316°C (200 and 600°F).
		ASTM D 1743, Corrosion Preventive Properties of Lubricating Greases[a]	Determination of surface damage due to corrosion, such as pitting, etching, rusting, or black stains on raceways and rollers of tapered roller bearings which have been run-in and stored for a prescribed period at a definite temperature and 100% relative humidity.	ASTM D 2596, Measurement of Extreme-Pressure Properties of Lubricating Grease (Four-Ball Method)	Load-carrying properties up to extremely high pressures: (a) load-wear index (formerly mean-Hertz load) and (b) weld point by Four-Ball EP Tester.
ASTM D 942, Oxidation Stability of Lubricating Greases by the Oxygen Bomb Method[a]	Determination of resistance to oxidation under static conditions of thin grease films (such as coatings on machine parts); not indicative of storage stability in containers nor service under actual conditions of use in bearings.			ASTM D 3232, Measurement of Flow Properties of Lubricating Greases at High Temperatures	Measurement of the flow properties of lubricating greases under high-temperature low-shear conditions.
		ASTM D 1831, Roll Stability of Lubricating Grease[a]	Determination of changes in consistency after working in tester for 2 h at room temperature. Although test significance has not been determined, changes in worked penetration of a grease after rolling are believed to be an indication of its shear stability.	ASTM D 3428, Torque Stability Wear and Brine Sensitivity Evaluation of Ball Joint Greases	Evaluation of grease performance by two procedures in tension-type automotive ball joints as determined by noise level, wear and torque stability; used as a screening test to aid in selection of greases for the lubrication of automotive chassis ball joints.
ASTM D 972, Evaporation Loss of Lubricating Oils and Greases[a]	Evaluation of weight loss by evaporation at temperatures up to 149°C (300°F).				
ASTM D 1092, Apparent Viscosity of Lubricating Greases[a]	Determination of apparent viscosity in temperature range of −54 to 38°C (−65 to 100°F); results relatable to ease of handling and dispensing.	ASTM D 2265, Dropping Point of Lubricating Grease of Wide-Temperature Range	See remarks under ASTM D 566; test is also valid for high temperature greases (up to 330°C (625°F)).	ASTM D 3527, Life Performance of Automotive Wheel Bearing Grease	Evaluation of the high-temperature life performance of wheel bearing grease.
		ASTM D 2266, Wear Preventive Characteristics of Lubricating Grease (Four-Ball Method)	Determination of wear preventive characteristics of grease when a rotating loaded steel ball slides against three similar stationary steel balls; measured by wear-scar diameters on stationary balls after completion of test; not indicative of results in actual service, and cannot distinguish between extreme-pressure (EP) and nonextreme-pressure (non-EP) greases.	Ford Ball-Joint Grease Test	Evaluation of grease performance in tension and compression type automotive ball joints as determined by general surface condition, rust, joint and seal wear, and noise level under simulated service conditions.
ASTM D 1263, Leakage Tendencies of Automotive Wheel-Bearing Greases[a]	Evaluation of leakage tendencies from an unsealed wheel bearing assembly, run for 6 h at 104°C (220°F); permits screening candidate greases; not a replacement for long-time service tests.				
ASTM D 1264, Water Washout Characteristics of Lubricating Greases[a]	Evaluation of resistance to water washout from rotating bearings at 38°C (100°F) and at 80°C (175°F) under prescribed conditions; not a replacement for actual service tests; not suitable for fibrous greases.			NLGI Method, Matching Lubricating Grease Flow Properties with Lubricating Grease Dispensing Pump Delivery Behavior at Low Temperatures[b]	Determination of ability of a pump to dispense a grease at a stated temperature, and therefore primarily a pump test; can supply comparative data on two or more greases using same pumping equipment and test temperatures.

[a] Also approved as American National Standard by American National Standards Institute, Inc.

[b] NLGI Spokesman, May 1960, page 47.

as only seals in good condition can effectively bar the entrance of water, dirt, and other contaminants, and minimize loss of grease by leakage.

Multipurpose Grease—This term designates lubricating greases of such composition, structure, and consistency to meet the performance requirements for chassis grease (more than 3200 km (2000 mile) service life), wheel bearing grease, universal joint grease, and other automotive uses of a miscellaneous nature, such as fifth-wheel service.

NOTE: Some ELI chassis lubricants are satisfactory as multipurpose greases. The grease manufacturer should be consulted as to the multipurpose qualities of his product.

"Extreme-Pressure or EP"—This term is not a designation by usage, but is applied to greases with high load-carrying capacity, determined usually by the Timken and Four-Ball Machines. In some cases, the EP property results from a surface-active additive that imparts antiwear or antiseize properties beyond the capabilities of the usual fluid-thickener or other finely dispersed inert solids in the grease. Extreme-pressure or wear-reducing properties may be incorporated in any of the usage types, most frequently those designated as ELI or multipurpose.

Greases for Other Vehicle Needs—Automotive equipment may require special greases not as yet designated by SAE. Examples of such applications are speedometer cables and brake adjustors.

Grease Application—Automotive greases are applied by hand packing, by hand and power operated pressure guns, and by hand and power operated central systems fitted to individual vehicles. In wheel bearing lubrication, a bearing packing device is often used, as more effective, faster, and less wasteful of grease than hand packing. Mixing of different types of greases in wheel bearings should be avoided since it might result in excessive thinning and leakage.

The prime consideration in using and applying greases is that of cleanliness: of containers and dispensing and pumping equipment and in the removal of surface grease and dirt accumulations from application points such as plugs and grease gun fittings.

Excessive dispensing pressures and pumping rates are to be avoided. They tend to cause seal rupture and deformation and are wasteful of lubricant.

Automotive servicing literature is voluminous on the subject of grease lubrication. Important sources are vehicle manufacturers' service bulletins, oil company bulletins and lubrication charts, and trade organization manuals. Among the latter are three publications issued by National Lubricating Grease Institute: "Recommended Practice for Lubricating Passenger Car Wheel Bearings," "Recommended Practices for Lubricating Passenger Car Ball Joint Front Suspensions," and "Recommended Practice for Grease Lubricated Truck Wheel Bearings."

Grease Properties as Related to Types of Service—Service requirements determine the relative importance of the above grease properties for each kind of application and set the level of performance needed. Table 3 is a rough summary of the grease properties of primary importance in the several fields of automotive use previously discussed. Certain properties as, for example, texture or structure, consistency, and apparent viscosity are not included in the summary, since it is assumed they will be appropriate to the purposes of the individual grease types.

TABLE 3—RELATIVE IMPORTANCE OF LUBRICATING GREASE PROPERTIES FOR AUTOMOTIVE USES SHOWN[a]

Property	Wheel Bearings	Universal Joints	Chassis	ELI Chassis	Multipurpose Applications
Structural Stability (inc. Mechanical Stability)	H	M	L	H	H
High Dropping Point (High-Temp. Service)	H	M	L	M	H
Oxidation Resistance	H	M	L	H	H
Protection Against Friction and Wear	M	H	M	H	H
Protection Against Corrosion	M	M	L	H	M
Protection Against Washout	M	M	M	H	M

[a] H = Highest, M = Moderate, L = Least.

FLUID FOR PASSENGER CAR TYPE AUTOMATIC TRANSMISSIONS—SAE J311b

SAE Information Report

Report of Fuels and Lubricants Technical Committee approved January 1952 and last revised May 1971.

The information which follows has been compiled by Subcommittee 5, Fluids, of the SAE Fuels and Lubricants Technical Committee. It details some of the equipment and procedures used to measure critical characteristics of hydraulic fluids used in automatic shift transmissions current at the time of this revision. These transmissions characteristically use a torque converter and planetary gear type drive.

The fluid used in passenger car, light truck, and school bus automatic transmissions must perform in five distinct functions. These are:

1. Hydrodynamic energy transmission medium for use in the torque converter.
2. Hydrostatic energy transmission medium for use in the hydraulic control-logic circuits and for servomechanisms.
3. Lubricating medium for shaft bearings, thrust bearings, and involute or spur gear load surfaces.
4. Sliding-friction energy transmission medium for use with lubricated bands and clutches.
5. Heat transfer medium with liquid- or air-cooled systems for maintenance of a suitable automatic transmission temperature range.

The functional capability of a fluid is determined by measurement of physical and mechanical characteristics that relate to performance of the fluid in an automatic transmission.

Reference—For more detailed information on hydrodynamic transmissions and fluids, refer to "Design Practices—Passenger Car Automatic Transmissions," SAE Advances in Engineering, Vols. 1 and 2, and various other references indicated in this report.

Viscosity—Typical automatic transmission fluids are non-Newtonian. Low shear viscosities are specified.

Viscosities chosen vary with the individual manufacturer, but are generally within the range of 7.0–8.5 cST at 210 F (99 C). Viscosity units used are generally confined to kinematic (centistokes) or absolute (centipoises).

The viscosities specified are functional compromises dictated by conflicting requirements of transmission subsystems. As fluid viscosity is increased, hydraulic control system and pump leakage is decreased, but converter efficiency also is decreased. Viscosities are also important in the design of bearings, gears, and clutch and band friction surfaces.

Detailed discussion of fundamental principles and measuring equipment is available in current literature.

Viscosity at low temperature should be low enough to assure good cranking, starting, and running performance in cold weather. Typical low temperature viscosity requirements are 1400 cp maximum at 0 F, 4000 cp maximum at −10 F, 23,000 cp maximum at −20 F, 55,000 cp maximum at −40 F. Good shear stability is required in order to retain designed viscometric properties. ASTM D 445, Method of Test for Kinematic Viscosity, is the standard laboratory method for measuring kinematic viscosity. Low temperature viscosities can be determined by the CRC-L45, Technique for Determining the Low Temperature Fluidity of Automatic Transmission Fluids. Shear stability may be determined by the method outlined in SAE J72 or field tests in actual equipment.

References

1. M. L. Haviland, "Engine and Transmission Lubricant Viscosity Effects on Low Temperature Cranking and Starting." SAE Transactions, Vol. 78 (1969), paper 690768.
2. R. A. Kobe and J. C. Wagner, "The Chrysler TorqueFlit and Automatic Transmission Fluid." Paper 680036 presented at SAE Automotive Engineering Congress, Detroit, January 1968.
3. R. L. Stambaugh and A. L. Preuss, "Laboratory Methods for Predicting

Viscosity Loss of Polymer Thickened Hydraulic Fluids." Fuels and Lubricants, 1968 Papers, SAE Activity Proceedings, AP-1, paper 680438, p. 74.

Density or Specific Gravity, Specific Heat, and Thermal Coefficient of Expansion—Petroleum-derived base fluids having viscosity characteristics in the range of current automatic transmission fluids do not vary appreciably in density (or specific gravity), specific heat, and thermal expansion coefficients at a given temperature. These properties must be considered in transmission design, but purposeful manipulation of these properties is not likely unless mixed with other than mineral oil base fluids.

Typical values for automatic transmission fluids are:

Specific gravity at	60 F (16 C)	0.88
	210 F (100 C)	0.821
	300 F (150 C)	0.789
Specific heat at	60 F (16 C)	0.46 Btu/lb-F
	210 F (100 C)	0.52
	300 F (150 C)	0.509
Coefficient of expansion at 60 F (16 C)		0.0004 in^3/in^{3}F (0.00072 cm^3/cm^3–C)

Specific gravity, specific heat, and coefficient of expansion values may be calculated using procedures detailed in numerous textbooks. One such test is "Petroleum Refinery Engineering," by W. L. Nelson. New York: McGraw-Hill Book Co., pp. 136–138 (specific heat); pp. 257–258 (coefficient of expansion); and pp. 161–168 (specific gravity).

Oxidation and Thermal Stability—Fluids used in automatic transmissions must be capable of operating at temperatures in excess of 300 F (150 C) in services such as trailer towing, and under conditions encountered in hot-weather high-density urban traffic. Introduction of air, through normal transmission breathing, results in severe oxidizing conditions which changes many new-fluid characteristics.

Some effects which oxidation can produce are:

1. Alteration of frictional characteristics which result in excessive clutch and band slippage, producing high localized clutch temperatures, which in turn make oxidizing conditions more severe.
2. Acids or peroxides formed in fluid oxidation which may be corrosive of bushing and thrust washer materials, and detrimental to elastomeric seal materials and the composition clutch plates.
3. Viscosity increases great enough to degrade transmission operation.
4. Sludge which can plug hydraulic controls and fluid passages.
5. Varnish formation which can lead to control valve or governor sticking and ultimate transmission failure.
6. Oxidation products which can reduce antifoamant effectiveness.

Full-scale transmission tests, such as the Powerglide Oxidation Test, the Mercomatic Oxidation Test, and the Chrysler Transmission Fluid Shift Cycle Test, are used for evaluating oxidation resistance of fluids. Laboratory tests such as the Ford Aluminum Beaker Test and the Chrysler Bench Oxidation Test are also used as screening tests.

References

1. R. L. Anderson and N. A. Hunstad, "Dexron Automatic Transmission Fluid." Fuels and Lubricants, 1968 Papers, SAE Activity Proceedings, AP-1, paper 680038, p. 29.
2. W. D. Ross and B. A. Pearson, "ATF-Type F Keeps Pace with Fill-for-Life Requirements." Fuels and Lubricants, 1968 Papers, SAE Activity Proceedings, AP-1, paper 680037, p. 10.
3. B. A. Pearson and J. L. Thompson, "Ford Aluminum Beaker Test: A New Tool for the Study of ATF Oxidation." Paper 670023 presented at SAE Automotive Engineering Congress, Detroit, January 1967.

Friction Characteristic—Fluid friction characteristics are important in automatic transmissions that utilize lubricated clutches to change gear ratios. Extensive performance and durability testing is performed in actual transmissions and bench friction test apparatus. No single fluid satisfies the optimum friction requirements of all existing transmission types. Some modification of mineral oil fluid-friction levels can be effected by use of additive substances at sliding velocities above 200 ft/min (1 m/s). Extensive modification of friction values below 200 ft/min (1 m/s) is possible with additive treatment and is responsible for the two main automatic transmission fluid types now in regular use, that is, ATF Type F and Dexron.[1] In one type, the coefficient of friction increases with decreasing sliding speed, while in the other type the coefficient of friction decreases with a decreasing sliding speed.

Evolution of friction materials and consequent changes in reaction member surfaces, both clutch and band, has emphasized the friction controlling role of automatic transmission fluid. Matching fluid-friction properly with clutch and band materials is a fundamental design consideration in all currently produced automatic transmissions.

References

1. L. E. Coleman, "Development of Type F Automatic Transmission Fluids." Paper 680039 presented at SAE Automotive Engineering Congress, Detroit, January 1968.
2. M. L. Haviland and J. J. Rodgers, "2,000,000 Miles of Fluid Evaluation in City Bus Automatic Transmissions." SAE Transactions, Vol. 76, paper 670185.
3. G. R. Smith, W. D. Ross, P. L. Silbert, and W. B. Herndon, "Putting Automatic Transmission Clutch Friction Researchers on Speaking Terms." Paper 670051 presented at SAE Automotive Engineering Congress, Detroit, January 1967.
4. N. A. Nann and F. H. Pinchbeck, "Trailoring Automatic Transmission Fluid Shift Quality in the Laboratory." Paper 650466 presented at SAE Mid-Year Meeting, Chicago, May 1965.

Antifoam Characteristics—Suppression of the foaming tendency of fluids in an automatic transmission is essential to proper operation. Foaming of the transmission fluid can produce erratic pump, converter, and hydraulic control response; and frequently results in fluid loss through the breather or filler tube. Measurement of foaming tendency and foam stability is used for evaluating fluid suitability. Techniques for evaluating fluid foaming tendencies include transmission tests, bench tests such as ASTM D 892 (IP-146), Method of Test for Foaming characteristics of Lubricating Oils, and the Dexron[1] Foam Test.

Fluid/Seal Compatibility—Compatibility of the automatic transmission fluid with elastomeric seal materials must be established during fluid formulation. Accepted design procedure involves use of a reference elastomer to determine seal swell, shrink, and hardening tendencies in a candidate fluid. Seal materials must be selected to meet transmission performance requirement with the established fluid formulation. Bench test procedures such as ASTM D 471, Method of Test for Change in Properties of Elastomeric Vulcanizates Resulting from Immersion in Liquids, and ASTM D 2240, Method of Test for Indentation of Rubber and Plastics by Means of a Durometer, are of value for screening purposes. Bench test devices for seal assemblies are also useful for compatibility studies; see SAE J110.

Score and Wear Resistance—Current automatic transmission fluids contain additive substances which inhibit score and wear of rubbing surfaces. The additives must be compatible with the variety of materials used in the transmission. Many bench test devices exist for evaluation of score and wear, but no single test is acceptable to all and full-scale transmission tests are mandatory.

Corrosion Properties—Automatic transmission fluids generally contain corrosion inhibitors to assure protection of transmission components. There is no universally accepted bench test procedure for evaluating moisture and/or chemical corrosion resistance of transmission fluids, although the following tests are useful: ASTM D 1748, Method of Test for Rust Protection by Metal Preservatives in the Humidity Cabinet, ASTM D 665 (IP-135), Method of Test for Rust-Preventing Characteristics of Steam-Turbine Oil in the Presence of Water, ASTM D 130 (IP-154), Method of Test of Copper Strip Corrosion by Petroleum Products, or ASTM D 1275, Method of Test for Corrosive Sulfur in Electrical Insulating Oils.

---

[1] Registered Trademark.

POWERSHIFT TRANSMISSION FLUID CLASSIFICATION—SAE J1285 FEB80

SAE Recommended Practice

Report of the Fuels and Lubricants Technical Committee, approved February 1980.

This SAE Recommended Practice was prepared by the SAE Fuels and Lubricants Technical Committee:
 (a) to assist the designers and users of heavy-duty transmissions in the selection of powershift transmission fluids for field use and
 (b) to promote a uniform practice for use by marketers of lubricants and equipment builders in identifying and recommending these fluids by type.

This classification is designed for fluids used in heavy-duty truck, bus, earthmoving, and marine transmissions or steering clutches. The fluids must perform the following five functions:
 1. Transmit hydrodynamic energy in a torque converter.
 2. Transmit hydrostatic energy in hydraulic circuits.
 3. Lubricate bearings, bushings, gears, and moving parts.
 4. Provide proper frictional properties in lubricated bands and clutches.
 5. Provide heat transfer medium for liquid- or air-cooled systems to maintain suitable operational temperature range.

SIGNIFICANCE AND METHODS OF MEASURING POWERSHIFT TRANSMISSION FLUID PROPERTIES

1. Viscosity—Viscosity controls the efficiency of the hydraulic control and torque converter systems. Viscosity grades and recommended temperature ranges are chosen by the individual manufacturer. Both Newtonian and non-Newtonian fluids are used. The viscosity grades are classified by SAE J300d. All other properties of the fluids are covered by this classification.

Powershift transmission fluids must flow readily at low temperatures to oil screens and through inlet tubes. This property is evaluated by the Brookfield Viscometer test, ASTM D 2983, and by the Pour Point test, ASTM D 97.

2. Foaming Characteristics—Suppression of the foaming tendency of fluids in a powershift transmission is essential to proper operation. Foaming of the transmission fluid can produce erratic pump, converter, and hydraulic control response; and frequently results in fluid loss through the breather or filler tube. Measurement of foaming tendency and foam stability is used for evaluating fluid suitability. The technique for measuring this property in a powershift transmission fluid is the ASTM D 892 Foaming Characteristics test.

3. Fluid/Seal Compatibility—Compatibility of the powershift transmission fluid with elastomeric seal materials must be established during fluid formulation. Accepted design procedure involves use of a reference elastomer to determine seal swell, shrink, and hardening tendencies in a candidate fluid. Seal materials must be selected to meet transmission performance requirement with the established fluid formulation. Bench test procedures such as ASTM D 471, Method of Test for Change in Properties of Elastomeric Vulcanizates Resulting from Immersion in Liquids, and ASTM D 2240, Method of Test for Indentation of Rubber and Plastics by Means of a Durometer, are of value for screening purposes.

4. Rust Protection—The bench test used to evaluate the rust-preventative properties of powershift transmission fluids is ASTM D 1748, Method of Test for Rust Protection by Metal Preservatives in the Humidity Cabinet.

5. Wear Resistance—Powershift transmission fluids must inhibit scoring and wear of rubbing surfaces in the transmission. There is no transmission test available which correlates with wear in field service. A power steering pump test is used for evaluation of the wear performance of powershift transmission fluids.

6. Oxidation Stability—Fluids used in powershift transmissions must be capable of operating at temperatures up to 150°C. Introduction of air, through normal transmission breathing, results in severe oxidizing conditions which change many new fluid characteristics. Some effects which oxidation can produce are:
 (a) Alteration of frictional characteristics which result in excessive clutch and band slippage, producing high localized clutch temperatures, which in turn make oxidizing conditions more severe.
 (b) Acids or peroxides formed in fluid oxidation which may be corrosive to bushing and thrust washer materials, and detrimental to elastomeric seal materials and the composition clutch plates.
 (c) Viscosity increases great enough to degrade transmission operation.
 (d) Sludge which can plug hydraulic controls and fluid passages.
 (e) Varnish formation which can lead to control valve or governor sticking and ultimate transmission failure.
 (f) Oxidation products which can reduce antifoamant effectiveness.

7. Friction Retention—Matching fluid-friction properly with clutch and band materials is a fundamental design consideration in all currently produced powershift transmissions.

Fluid friction characteristics are important in automatic transmissions that utilize lubricated clutches to change gear ratios. Extensive performance and durability testing is performed in actual transmissions and bench friction test apparatus. No single friction test can evaluate the requirements of different clutch plate friction materials. The performance of powershift transmission fluids with bronze faced friction plates is evaluated using a 15 000 cycle test procedure and the performance with graphite faced plates is evaluated with a second 5500 cycle procedure. Both procedures are conducted in the SAE No. 2 friction machine.

References

1. C. R. Potter, and R. H. Schaefer, "Development of Type C-3 Torque Fluid for Heavy-Duty Power Shift Transmission." Paper 770513 presented at SAE Earthmoving Industry Conference, Peoria, April 1977.

2. J. A. McLain, "Oil Friction Retention Measured by Caterpillar Oil Test No. TO-2." SAE Transactions, Vol. 86 (1977), Paper 770512.

TABLE 1—TEST TECHNIQUES AND PERFORMANCE CRITERIA

Test Technique	Limits	Test Method
1. Brookfield Viscosity at −18°C, cSt	2500 max for 5W and 10W oils report result for other SAE grades	ASTM D 2983
2. Pour Point, °C	−30 max for SAE 5W or 10W oils −25 max for SAE 15W oils −15 max for SAE 30 or 40 oils	ASTM D 97
3. Foaming Characteristics Sequence 1, cm³ Sequence 2, cm³ Sequence 3, cm³	 25/0 max 50/0 max 25/0 max	ASTM D 892
4. Seal Compatibility		a
(a) Total Immersion Test[b] (Nitrile Rubber)		
Volume Change, %	0 to +5 for SAE 5W and 10W oils −1 to +5 for other SAE grades	
Hardness Change, Points	0 to ±5	
(b) Dip Cycle Test[b] (Polyacrylate)		
Volume Change, %	0 to +10	
Hardness Change, Points	0 to +5	
(c) Tip Cycle Test[b] (Silicone)		
Volume Change, %	0 to +5	
Hardness Change, Points	0 to −10	
5. Rust Resistance	No rust ("no more than 3 random spots 1 mm or less in diameter at least 6 mm away from edge of panel")	a
Other Properties		
6. Wear Test	Pump Cam Ring shall show grinding pattern for 360 deg and be free from scuffing, scoring, or chatter marks	a
7. Oxidation Test	—No significant varnish or sludge on transmission parts. —No blackening or flaking of copper containing parts. —Oil shall not gain more than 15% in viscosity at 100°C.	a
8. Friction Retention Tests		
(a) Test A—(5500 cycles)	Slip Time—0.85 s max Dynamic Torque—102 N·m min Decrease in Dynamic Torque—40 N·m max	a
(b) Test B—(15 000 cycles)	Stopping Time Increase—15% max Wear—Bronze Disc 0.25 mm max Steel Plate 0.1 mm max	c

[a] Detroit Diesel Allison Division Transmission Engineering Specifications Serial No. TES 122.
[b] Nominal values which are adjusted for each elastomer batch.
[c] Caterpillar Engineering Specification No. TO-2.

CENTRAL FLUID SYSTEMS—SAE J71a

SAE Recommended Practice

Report of Fuels and Lubricants Technical Committee and Nonmetallic Materials Committee approved June 1960 and last revised April 1976.

Specifications for "central system fluids" were developed in 1960 and identified as SAE J71, R1 for petroleum base fluids and R2 for synthetic fluids. Associated with these specifications was a series of test procedures identified as SAE J72. These documents were intended to facilitate the development, specification, and use of central hydraulic systems which combine actuating power for steering, engine starting, brakes, seat adjusting, windshield wipers, window regulators, hydropneumatic suspension, and/or automatic transmissions. Since the anticipated central systems did not materialize in the United States in a period of over 15 years, both documents were deleted from the handbook at the end of 1976. Anyone having need for useful information on central system fluids or related test procedures may refer to J71 and J72, in copies of the handbook published in 1961-1976.

TABLE 1—SPECIFICATIONS FOR PETROLEUM BASE CENTRAL SYSTEM FLUIDS (SAE 71R1)
The latest ASTM test method shall be used unless otherwise noted.

Tests	Test Limits
VISCOSITY—KINEMATIC	2000 cSt max at −40 F (before and after shear). 5.5 cSt min at 210 F (after shear).
	The −40 F viscosity of the fluid shall be determined by actual test (not extrapolation) using Low Temperature (−40 F) Viscosity—Brookfield Procedure[a].
	The 210 F viscosity of the fluid shall be determined by procedure outlined by ASTM D 445.
SHEAR TEST	The shear stability of the fluid shall be determined by Shear Test for Fluids—Pump[a].
FLASH POINT	225 F min. The flash point shall be determined by the ASTM D 92 method.
INITIAL BOILING POINT	400 F min. The initial boiling point shall be determined by ASTM D 158 Method.
COLD TEST (a)	−50 F min (6 days at −50 F). The cold test (a) shall be determined by SAE 70R3 brake fluid test 8.2[b].
COLD TEST (b)	−70 F min (6 hours at −70 F). The cold test (b) shall be determined by SAE 70R3 brake test 8.2[b].
FOAMING	100 ml foam volume max at end of 5-minute blowing period. No foam at end of 4-minute settling period. The foam tests shall be determined by ASTM D 892 method except that the settling period shall be 4 minutes instead of 10 minutes.
ANTI-WEAR	Pump delivery at 700 rpm and 600 psi discharge shall not decrease more than 0.2 gpm during 100 hr as indicated by measurements on Standard Reference Fluid at start and end of test as determined by Wear and Pump Delivery Test[a].
	Pump parts, by visual inspection, shall show no signs of excessive wear. Parts should be burnished and show no signs of galling.
OXIDATION STABILITY	Rating 80 min. The oxidation stability shall be determined by Oxidation Test—Automatic Transmission[a].
CORROSION RESISTANCE	No visible rust. The corrosion resistance shall be determined by ASTM D 665 turbine oil test with a steel test specimen using procedure "A" for distilled water.
SEAL COMPATIBILITY (Rubber Swelling)	The increase in the base diameter of nitrile rubber cups after 70 hr exposure to the fluid at 250 ± 5 F shall not be less than 0.005 in. nor more than 0.055 in. The surface shall not be tacky or show any sloughing as may be indicated by carbon black on the surface.
	This test shall be similar to SAE 70R3 brake fluid test 7.3.2[b].
(Lubrication)	Pass SAE 70R3 brake fluid test 7.5[b].
	This test shall be determined by SAE 70R3 brake fluid test procedure "B", section 8.12[b] except that nitrile rubber cups shall be used both in wheel cylinders and master cylinder. Steel tubing 3/16 in. dia shall be substituted for copper tubing and steel fittings used.

Part	Test Procedure B Part Number[c]
Wheel Cylinder	NXA-18210
Master Cylinder	NXA-18209
Primary Cup	NXA-18212
Secondary Cup	NXA-18213
Residual Check Valve Seat	NXA-18214
Residual Check Valve Seal	NXA-18215
Wheel Cylinder Cups	NXA-18211

WATER TOLERANCE	Water tolerance is not practical for petroleum base fluids. Provision must be made in the system for the elimination of water or prevention of its entrance.
COMPATIBILITY	No liquid stratification or precipitation shall be evident.
	The test shall be the same as SAE 70R3 brake fluid test 8.11[b]. Only petroleum base central system fluids meeting all other SAE 71R1 specifications are to be used. Compatibility between petroleum base and synthetic fluids is not required.

[a] See SAE Recommended Practice, Tests for Central System Fluids—SAE J72.
[b] See SAE Standard, Hydraulic Brake Fluid—SAE J70.
[c] These parts or exact equivalent should be used. The test parts can be ordered from Bendix Products Division, Automotive Sales Dept., 401 Bendix Drive, South Bend, Indiana.

TABLE 2—SPECIFICATIONS FOR SYNTHETIC BASE CENTRAL SYSTEM FLUID (SAE 71R2)
The latest ASTM test method shall be used unless otherwise noted.

Tests	Test Limits
VISCOSITY—KINEMATIC	1800 cSt max at −40 F (before and after shear). 5.5 cSt min at 210 F (after shear) for oils containing VI improver; or 4.5 cSt at 210 F (after shear) for oils not containing VI improver. The −40 F viscosity of the fluid shall be determined by actual test (not extrapolation) by ASTM D 445 method. The 210 F viscosity of the fluid shall be determined by procedure outlined by ASTM D 445.
SHEAR TEST	The shear stability of the fluid shall be determined by Shear Test for Fluids—Pump[a].
FLASH POINT	205 F min. The flash point shall be determined by the ASTM D 92 method.
BOILING POINT	400 F min. The boiling point shall be determined by method used for SAE 70R3 brake fluid test 8.3[b].
COLD TEST (a)	−50 F min (6 days at −50 F). The cold test (a) shall be determined by SAE 70R3 brake fluid test 8.2[b].
COLD TEST (b)	−70 F min (6 hours at −70 F). The cold test (b) shall be determined by SAE 70R3 brake test 8.2[b].
FOAMING	100 ml foam value max at end of 5-minute blowing period. No foam at end of 4-minute settling period. The foam tests shall be determined by ASTM D 892 method except that the settling period shall be 4-minutes instead of 10-minutes.
ANTI-WEAR	Pump delivery at 700 rpm and 600 psi discharge shall not decrease more than 0.2 gpm during 100 hr as indicated by measurements on Standard Reference Fluid at start and end of test as determined by Wear and Pump Delivery Test[a]. Pump parts, by visual inspection, shall show no signs of excessive wear. Parts should be burnished and show no signs of galling.
OXIDATION STABILITY	Rating 80 min. The oxidation stability shall be determined by Oxidation Test—Automatic Transmission[a].
CORROSION RESISTANCE	The fluid when tested under conditions outlined for SAE 70R3 brake fluid 8.8[b] shall not cause corrosion exceeding the following limits:

Test Strip	Permissible Loss mg per sq cm of surface, max
Tin Coated Iron cut from 1.19 lb per base box min. sheet (Type 1, Grade 1 of Federal Spec. QQ. 1-706A)	0.2
Steel, SAE 1010 (cold-rolled)	0.2
Aluminum, AA2024	0.1
Cast Iron, SAE 111, or strips from housing of wheel brake cylinders (smooth machined surface)	0.2
Brass, SAE 70A	0.5
Copper, SAE 71	0.5

The fluid mixture shall show no gelling at room temperature and shall not contain more than a medium[c] precipitate. Discoloration or slight generalized etching of the test strips shall not be cause for rejection of the fluid.

SEAL COMPATIBILITY (Rubber Swelling)	The increase in the base diameter of both natural[d] and SBR[e] rubber cups after 70 hr at 250 ± 5 F exposure to the fluid shall not be less than 0.005 in nor more than 0.055 in. The surface shall not be tacky or show any sloughing as may be indicated by carbon black on the surface. This test shall be similar to SAE 70R3 brake fluid test 7.3.2.
(Lubrication)	Pass SAE 70R3 brake fluid test 7.5[b]. This test shall be determined by SAE 70R3 brake fluid test procedure "B", section 8.12 except that 3/16 in dia steel tubing shall be substituted for copper tubing and steel fittings used.

Part	Test Procedure B Part Number
Master Cylinder Primary Cup	FD-2398-(AA) (Compound W-6730-B)[e]
Master Cylinder Secondary Cup	FD-2399-(AA) (Compound W-6730-B)[e]
Compensator Check Valve	FC-14424
Wheel Cylinder Cups	1674080-AA (Compound W-6100)[e]

WATER TOLERANCE COMPATIBILITY	Pass SAE brake fluid 70R3 water tolerance test 8.6[b]. No liquid stratification or precipitation shall be evident. The test shall be determined by SAE 70R3 brake fluid test 8.11[b]. Only synthetic central system fluids meeting all other SAE 70R3 specifications are to be used. Compatibility between synthetic and petroleum base fluids is not required.

[a] See SAE Recommended Practice, Tests for Central System Fluids—SAE J72.
[b] See SAE Standard, Hydraulic Brake Fluid—SAE J70.
[c] The following are definitions for types of precipitates:
  A trace precipitate shall be flocculent and remain in suspension. A light precipitate shall be flocculent, may settle, but possess no crystalline formations.
  A medium precipitate shall be primarily flocculent material with a few small crystals.
  A heavy precipitate shall be large in volume and contain both flocculent and crystalline material.
[d] Goodrich Compound 15 JM 800 or exact equivalent. This material must be used to assure uniformity of testing and may be procured from Chemical Specialties Manufacturers Association, Inc., 50 East 41 St., New York 17, N.Y.
[e] Goodrich Compound 15 JM 581 or exact equivalent. This material must be used to assure uniformity of testing and may be procured from Chemical Specialties Manufacturers Association, Inc., 50 East 41 St., New York 17, N.Y.

AUTOMOTIVE GASOLINES—SAE J312 JUN80

SAE Information Report

Report of the Lubricants Division, approved January 1931, completely revised by the Fuels and Lubricants Technical Committee June 1980.

Automotive gasolines are used to fuel internal combustion spark-ignition engines. While gasolines discussed herein are used primarily in passenger car and highway truck service, they are also used extensively in off-highway utility vehicles and farm machinery, two-stroke and four-stroke cycle marine engines, and other spark-ignition engines employed in a variety of different service applications.

Automotive gasolines are essentially blends of hydrocarbons derived from petroleum; in addition, they may contain selected additives that impart specific features to the finished gasoline. The hydrocarbons may be derived from fractional distillation of crude oil as well as from complex processes that increase either the amount or quality of the gasoline obtained. These components vary, from individual compounds such as normal butane to wide-boiling-range products that contain hundreds of different hydrocarbons. The properties of commercial gasolines are influenced by the refinery practices employed and the nature of the crude oils from which they are produced. Finished gasolines encompass a boiling range of about 30–225°C (85–437°F).

Gasolines are blended to satisfy diverse automobile requirements. Antiknock quality, distillation characteristics, vapor pressure, sulfur content, oxidation stability, and anticorrosion behavior are balanced to provide satisfactory vehicle performance. Additives may be used to provide or enhance specific performance features.

This information report summarizes the significance of the more important physical and chemical characteristics of automotive gasolines, and describes pertinent test methods for defining or evaluating these properties. Information on properties of automotive gasolines currently marketed in service stations throughout the United States can be found in semi-annual reports issued by the Bartlesville Energy Technology Center (BETC) of the United States Department of Energy (DOE). Prior to the winter of 1974–1975, these reports were issued by the U.S. Department of the Interior, Bureau of Mines.

Historically, automotive gasolines have been classified on the basis of antiknock quality.

1. Antiknock Quality—The antiknock quality of an automotive gasoline is of prime importance. If antiknock quality is too low, knock occurs. Knock is a high pitch metallic rapping noise. In addition to being audibly annoying, in severe cases knock may lead to uncontrolled preignition which can cause burning of piston crowns and other engine damage. There is also evidence that knock increases the rate of engine wear. The potential durability, power, and fuel economy of a given engine is realized only when the gasoline antiknock quality is adequate. However, there is no advantage in using a gasoline having antiknock quality higher than the engine requires.

Knock depends on complex physical and chemical phenomena highly interrelated with engine design and engine operating conditions. It has not been possible to characterize completely the antiknock performance of gasoline with any single measurement. The antiknock performance of a gasoline is intimately related to the engine in which it is used and to the engine operating conditions. Furthermore, this relationship varies from one engine design to another and may even be different among engines of the same design due to normal production variations.

The antiknock quality of a gasoline is measured by several methods. These employ single-cylinder laboratory engines and more realistic, but much less precise, multicylinder engines in cars. The American Society for Testing and Materials (ASTM) has standardized the following two single-cylinder methods: ASTM D 2699, Test for Knock Characteristics of Motor Fuels by the Research Method, and ASTM D 2700, Test for Knock Characteristics of Motor and Aviation-Type Fuels by the Motor Method.

Both of these test procedures employ a variable-compression-ratio engine. The Motor method operates at a higher speed and inlet mixture temperature than does the Research method. They relate the knocking characteristics of a test gasoline to standard fuels which are blends of two pure hydrocarbons—isooctane (2,2,4-trimethylpentane) and n-heptane. These blends are called primary reference fuels. By definition, the octane number of isooctane is 100 and the octane number of n-heptane is zero. At octane levels below 100, the octane number of a given gasoline is the percentage by volume of isooctane in a blend with n-heptane that knocks with the same intensity at the same compression ratio as the gasoline when compared by one of the standardized engine test methods. The octane number of a gasoline greater than 100 is based upon the milliliters of tetraethyllead required to be added to isooctane to produce knock with the same intensity as the gasoline. The number of milliliters of tetraethyllead in isooctane is converted to octane numbers greater than 100 by use of tables published by ASTM.

The octane number of a given blend of either isooctane and n-heptane or tetraethyllead in isooctane is, by definition, the same for the Research and Motor methods. However, the Research and Motor octane numbers will rarely be the same for commercial gasolines. Therefore, when considering the octane number of a given gasoline, it is necessary to know the engine test method. Commercial gasoline typically will have an octane number approximately eight units lower by the Motor method than by the Research method, illustrating the important effect of engine operating conditions on antiknock performance. The numerical difference between the Research and Motor octane numbers is called *sensitivity*. Research octane number is, in general, the better indicator of antiknock quality for engines operating at full throttle and low engine speed. Motor octane number is the better indicator at full throttle, high engine speed and part throttle, low and high engine speed.

The antiknock performance of gasolines in cars, or road octane number, may be determined directly by using primary reference fuels and manually controlled ignition timing to vary the knocking tendency of the engine. The most commonly used test methods are the Coordinating Research Council, Inc. (CRC) Modified Borderline (F-27) and Modified Uniontown (F-28) research techniques. When using these procedures, the knocking tendencies of test gasolines and reference fuels are usually compared at the lowest audible level of knock. Road octane number of a gasoline is defined as being equal to the octane number of the primary reference fuel blend that produces the same knock intensity while operating under the specified test conditions.

For most automotive engines and operating conditions, the road octane number of a gasoline will be between its Research and Motor octane numbers. The relationship between Research and Motor octane numbers used to predict the road antiknock quality of gasolines is dependent upon the vehicles and operating conditions. The correlations between laboratory octane numbers and road octane numbers that have been developed usually employ a combination of Research and Motor octane numbers.

The average of Research (RON) and Motor (MON), that is (RON + MON)/2, is considered to be a reasonable guide to a gasoline's road octane performance, and ASTM and Federal gasoline specifications contain (RON + MON)/2 minimums for the various gasoline grades.

The tendency of gasolines to knock can be markedly decreased by the addition of very small quantities of chemical additives known as antiknock compounds. The lead alkyls (tetraethyllead, tetramethyllead, and physical and chemically reacted mixtures of these two materials) are the most commonly used antiknock compounds. These compounds are added to gasoline in the form of fluids that also contain an identifying dye and organic halides. The organic halides aid in scavenging the combustion products of the lead alkyls from the combustion chamber during the exhaust cycle.

2. Grades of Gasoline—Historically, leaded automotive gasolines have been classified into two grades—*premium* and *regular*—on the basis of octane number. *Super-premium, intermediate, sub-regular,* and blends of high and low octane grades have also been marketed. In high-altitude areas, octane levels have been somewhat lower than at sea level, because car octane requirements decrease with altitude.

With the exception of one premium gasoline marketed on the east coast and southern areas of the United States, all automotive gasolines until 1970 contained lead antiknock compounds to increase antiknock quality. Because lead antiknock compounds were found to be detrimental to the performance of catalytic emission control systems then under development, U.S. passenger car manufacturers in 1971 began to build engines designed to operate satisfactorily on gasolines of nominal 91 Research octane number. Some of these engines were designed to operate on unleaded fuel while others required leaded fuel or the occasional use of leaded fuel. The 91 Research octane level was chosen in the belief that unleaded gasoline at this level could be made available in quantities required using then current refinery processing equipment. Accordingly, during 1970 unleaded and low-lead gasolines were introduced to supplement the conventional gasolines already available.

Beginning with the 1975 model year, numerous new car models were equipped with catalytic exhaust treatment devices as one means of compliance with the 1975 legal restrictions in the United States on automobile emissions. The need for gasolines that do not adversely affect such catalytic devices has lead to the large scale availability and growing use of unleaded gasolines. Phosphorus also causes adverse effects on catalytic devices and consequently the phosphorus content of unleaded gasolines is restricted.

At present, ASTM categorizes gasolines into six levels on the basis of Antiknock Index, (RON + MON)/2 and lead content, see Table 1. Since road octane requirements decrease with increases in altitudes, reductions in the Antiknock Index are permitted in high altitude areas as shown in Fig. 1.

Leaded gasolines are defined as those to which lead antiknock compounds have been added in amounts that do not exceed 4.23 g Pb per U.S. gallon (1.12 g Pb/L) as outlined in U.S. Public Health Service Publication No. 712 (1959).

TABLE 1—GASOLINE ANTIKNOCK INDEXES AND THEIR APPLICATION

Leaded Gasoline (For Vehicles Which Can or Must Use Leaded Gasoline)

Antiknock Index (RON + MON)/2, Min[a]	Application
87	For vehicles with low antiknock requirements.
89	Meets antiknock requirements of most 1970 and prior model vehicles which were designed to operate on leaded gasoline and most 1971 and later model vehicles that can use leaded gasoline.
93	For 1970 and most prior model vehicles with high antiknock requirements and for later model vehicles with high antiknock requirements that can use leaded gasoline.

Unleaded Gasoline (For Vehicles Which Can or Must Use Unleaded Gasoline)

Antiknock Index (RON + MON)/2, Min[a]	Application
85	For vehicles with low antiknock requirements.
87[b]	Meets antiknock requirements of most 1971 and later model vehicles.
90	For most 1971 and later model vehicles with high antiknock requirements.

[a] Reductions for altitude are allowed in accordance with Fig. 1.
[b] In addition, Motor octane number must not be less than 82.0.

Unleaded gasolines are defined as those to which no lead antiknock has been added deliberately. However, a small amount of lead pick-up can occur in gasoline distribution. Beginning July 1, 1974 the Environmental Protection Agency has required the widespread availability of unleaded gasoline containing no more than 0.05 g Pb per U.S. gallon (0.013 g Pb/L) and no more than 0.005 g of phosphorus per U.S. gallon (0.0013 g P/L) and these limits are also contained in Federal Specification VV-G-1690 B and in ASTM D 439.

Some gasolines may contain manganese in the form of Methyl Cyclopentadienyl Manganese Tricarbonyl (MMT) for octane improvement. The Environmental Protection Agency banned the use of MMT for unleaded gasoline in the United States effective October 27, 1978 because of adverse effects on exhaust hydrocarbon emissions.

3. Volatility—The volatility characteristics of a gasoline are of prime importance. Gasoline is used in a great variety of engines with large variations in operating conditions and under a wide range of atmospheric temperatures and barometric pressures. These variations impose many limitations on the volatility of gasoline if it is to give satisfactory performance.

Gasolines that do not varporize readily may cause hard starting of cold engines and poor vehicle driveability during warm-up and acceleration as well as unequal distribution of fuel to the individual cylinders which may cause knock. Conversely, gasoline that vaporizes too readily in pumps, fuel lines, and carburetors will cause decreased liquid fuel flow to the engine, resulting in rough engine operation or stoppage. Fuels that vaporize too readily also may, if certain atmospheric conditions exist, cause ice formation in the carburetor throat, resulting in rough idle and stalling. Exhaust emissions and fuel economy are to a lesser degree related to fuel volatility.

3.1 Properties Related to Volatility—Since gasoline is a mixture of many hydrocarbons, it does not have a single boiling temperature. Therefore, one way to characterize automotive gasoline volatility is a distillation curve. This curve is a series of temperatures at which various percentages of the gasoline have evaporated. (See ASTM D 86, Distillation of Petroleum Products.) Another measure of volatility is vapor pressure which is the force per unit area exerted on the walls of a closed container by the varporized portion of the liquid contained therein. For hydrocarbon mixtures such as gasoline, the vapor pressure depends on the ratio of vapor to liquid volume in the container and on the temperature. Vapor pressure of gasoline when measured at 37.8°C (100°F) in a bomb having a 4:1 ratio of air to liquid is known as Reid Vapor Pressure (Rvp). (See ASTM D 323, Vapor Pressure of Petroleum Products (Reid Method).) Gasoline vaporization tendency may also be expressed in terms of vapor/liquid ratio (V/L) at temperatures approximating those found in critical parts of the fuel system. (See ASTM D 2533, Vapor/Liquid Ratio of Gasoline.)

ASTM D 439, Standard Specification for Automotive Gasolines and Federal Specification VVG-1690 B define five volatility classes in terms of limits

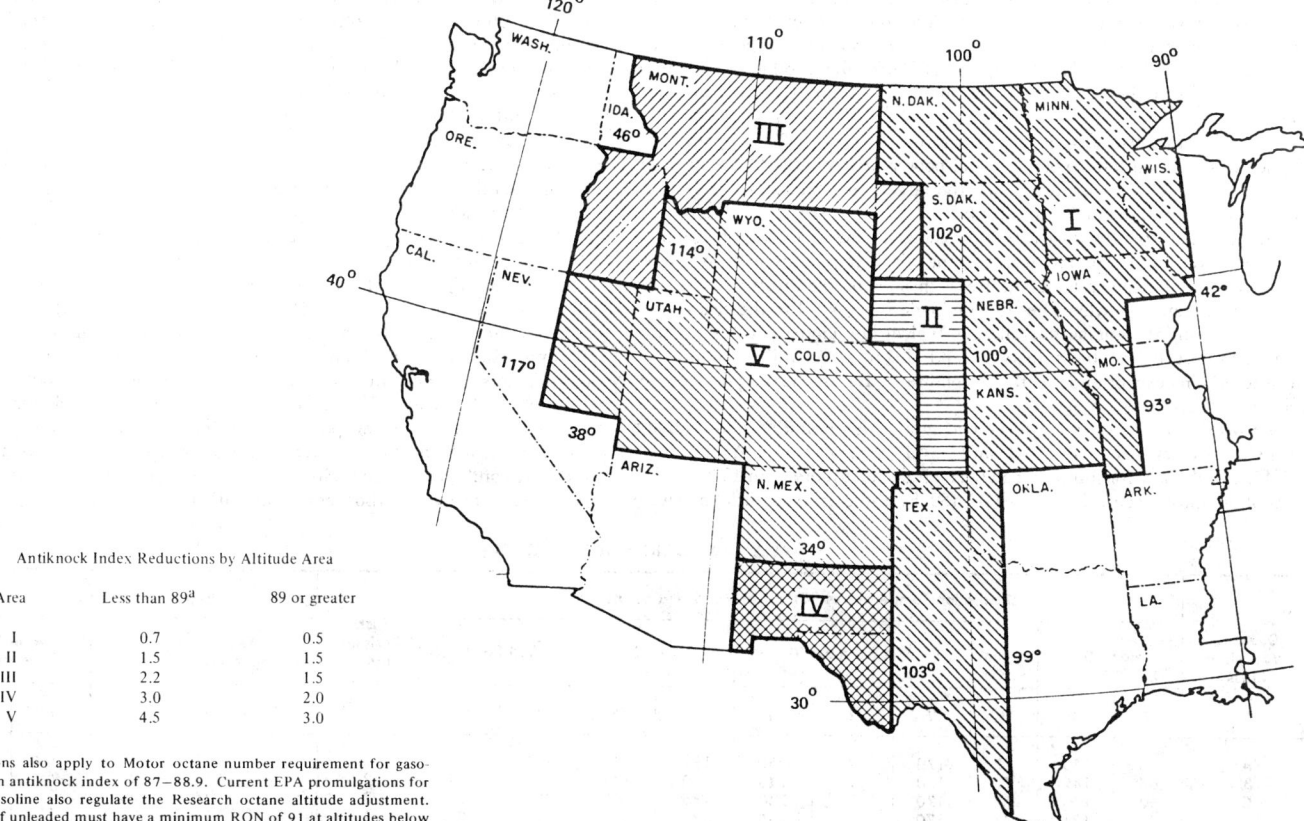

Antiknock Index Reductions by Altitude Area

Area	Less than 89[a]	89 or greater
I	0.7	0.5
II	1.5	1.5
III	2.2	1.5
IV	3.0	2.0
V	4.5	3.0

[a] Reductions also apply to Motor octane number requirement for gasolines with an antiknock index of 87–88.9. Current EPA promulgations for unleaded gasoline also regulate the Research octane altitude adjustment. One grade of unleaded must have a minimum RON of 91 at altitudes below 600 m (2000 ft), a minimum RON of 90 at altitudes between 600 m (2000 ft) and 900 m (3000 ft), a minimum RON of 89 at altitudes between 900 M (3000 ft) and 1200 m (4000 ft), and a minimum RON of 88 at altitudes over 1200 m (4000 ft).

FIG. 1—ANTIKNOCK INDEX REDUCTIONS FOR ALTITUDE

on maximum and minimum vaporization tendency as shown in Table 2. Shown in Table 3 are the geographic and seasonal recommendations for use of gasoline meeting the specifications of each volatility class within the United States.

For sea level areas outside of the United States, ASTM has suggested the following air temperatures for guidance in selecting the appropriate volatility class:

Volatility Class	10 Percentile Minimum Daily Temperatures, °C (°F)	90 Percentile Maximum Daily Temperatures, °C (°F)
A	>16 (60)	≥43 (110)
B	>10 (50)	<43 (110)
C	>4 (40)	<36 (97)
D	>−7 (20)	<29 (85)
E	≤−7 (20)	<21 (69)

For areas above sea level outside of the United States, the 90 percentile maximum daily temperature for the area should be increased by 4.4°C per 1000 m (2.4°F per 1000 ft) of altitude for comparing it to the suggested sea level temperature. This correction compensates for changes in fuel volatility caused by changes in barometric pressure due to altitude.

These ASTM volatility specifications have been established on the basis of broad experience and cooperation between gasoline suppliers and manufacturers of automotive vehicles and equipment. Gasolines meeting this specification have usually provided satisfactory performance in typical passenger car service. Obviously certain equipment or operating conditions may require or permit variations from these limits.

3.2 Driveability—Vehicle driveability is influenced by fuel volatility, ambient temperatures, and engine calibrations such as carburetor and choke settings, ignition timing, and exhaust gas recirculation. The main symptoms (malfunctions) of poor driveability are stalling, stumble, hesitation, surge, and stretch. Stalling is the inability of the engine to continue running. Stumble is a short, sharp reduction in acceleration after the vehicle is in motion. Hesitation is a temporary delay in acceleration when the throttle is opened. Surging is a cycling of engine power output. Stretch is characterized by an abnormal lack of power during an acceleration.

3.2.1 COLD START DRIVEABILITY—Any or all of the above malfunctions can occur after a cold start at any ambient temperature. Generally, the frequency of occurrences and the severity of malfunctions increase with decreasing ambient temperatures. Tests conducted by the Coordinating Research Council (CRC) have shown that cold start and cool weather driveability are affected by changes in the 10%, 50%, and 90% evaporated temperatures of a gasoline. (See ASTM D 86, Distillation of Petroleum Products.) A simple expression that predicts relative cool weather driveability performance fairly well is:

0.5 (10% temperature) + 50% temperature + 0.5 (90% temperature)

Higher values mean lower volatility and poorer driveability. This expression shows that the 50% temperature has the largest effect but that the entire boiling range can influence driveability.

3.2.2 CARBURETOR ICING—Carburetor icing is caused by vaporization of gasoline lowering the temperature of the air/fuel mixture, which causes moisture in the air to condense and freeze into ice on the throttle blade. The ice blocks the air flow when the throttle blade is returned to the closed, idle position causing the engine to stall. The typical late model engine has sufficient heat directed to the intake air and the intake manifold to prevent carburetor icing except under extreme conditions, but the heating devices can become inoperative with age.

The worst conditions for carburetor icing are high humidity and ambient temperatures between about −1 and 13°C (30 and 55°F). Above 13°C (55°F), the chances for forming ice are small because more than 13°C (23°F) of cooling must occur. Below −1°C (30°F) there is very little moisture available in the air, even at 100% relative humidity, to form ice. Fuels that are more volatile cause more carburetor icing because they evaporate more completely in the carburetor, causing larger temperature decreases. Carburetor icing correlates with mid-range distillation temperature.

The Department of the Army completed a study in 1973 concerning the meteorological conditions that contribute to carburetor icing. Isopleth maps were proposed to identify geographical regions where carburetor icing is likely to occur.

Two types of gasoline additives are used to control carburetor icing. The freezing point depressant reduces ice formation by acting as an antifreeze. Surfactant additives are also effective. They function by reducing adhesion between ice and metal surfaces.

3.2.3 VAPOR LOCK—At high ambient temperatures, vapor lock and other driveability problems can occur. Vapor lock is defined as the inability of the fuel system to supply the required quantity of gasoline to the engine because of the formation of excessive vapor in the system. The effect of vapor lock may range from a condition of slightly lean air/fuel ratio to engine stoppage from lack of fuel.

Considerable test work has been conducted on vapor lock over a period of years. A vapor lock test was developed by CRC in 1958 and further refined in later test programs. The CRC procedure was developed from and is similar to the procedure used by automobile manufacturers for equipment development.

Expressions which relate fuel volatility to vapor lock are of prime importance for specifying fuel properties for control of warm engine driveability. A large number of volatility expressions have been developed over the years and in 1962 and 1964 CRC conducted studies to evaluate seventeen of these. One of these, the temperature at vapor to liquid ratio (V/L) of 20, is used in ASTM D 439 as a control for vapor lock.

Vehicle design factors affecting vapor lock are use of vapor return lines, fuel tank and line placement, and fuel pump location and capacity.

3.2.4 HOT START AND DRIVEAWAY—Hot starting characteristics of automobiles are defined as the ability or inability of a hot engine to start after an engine-off soak period. Hot driveaway problems of stalling, idle roughness, hesitation, and stumble can occur because of enrichment caused by fuel vaporization in the carburetor or because of leaning due to vapor lock. Different techniques have been developed to evaluate two types of performance parameters: (a) hot start and hot driveaway and (b) vapor lock. The fuel property that has the greatest influence on hot start and driveaway is the temperature at V/L 20 but the limiting temperature for V/L 20 may be different for the two parameters and for individual vehicles.

3.2.5 DRIVEABILITY TEST TECHNIQUES—Many driveability procedures are used throughout the automotive and petroleum industries for vehicle and fuel testing. One set of procedures, those of the CRC, was developed cooperatively and is commonly used by both industries to determine the effects of fuel parameters on vehicle driveability. Three procedures are used by CRC; (a) the Cold Start and Warm-Up Driveability Procedure for testing at low and intermediate ambient temperatures, generally below 26.7°C (80°F); (b) Hot Start and Driveaway Procedure, and (c) the Vapor Lock Procedure used for testing at ambient temperatures above 26.7°C (80°F).

3.3 Emissions

3.3.1 EVAPORATIVE EMISSIONS—The amount of gasoline which will evaporate into the atmosphere from a fuel tank depends on a variety of factors. Tank configuration, vehicle operating temperature, and the volume of vapor in the tank all contribute. The Reid Vapor Pressure (Rvp) is a good fuel characteristic for predicting the amount of fuel which will evaporate.

Evaporative emissions from the carburetor vary widely with design. If design factors are constant, the percent evaporated at 71°C (160°F) is a good predictor of relative loss. Evaporative control systems are being used to control fuel tank and carburetor evaporative emissions from late model vehicles.

3.3.2 EXHAUST EMISSIONS—Exhaust emission studies related to volatility have produced conflicting results. However, a majority of these reports suggest a *slight decrease* in hydrocarbon emissions with increasing volatility. The com-

TABLE 2—VOLATILITY REQUIREMENTS

Gasoline Volatility Class	Distillation Temperatures °C(°F), at Vol % Evaporated										Dist. Residue max. Vol %	Vapor/Liquid Ratio[a]				
	max 10[a]		min 50[a]		max 50[a]		max 90[a]		End Point			Test Temperature		V/L, max	Reid Vapor Pressure, max	
	°C	°F	°C	°F	°C	°F	°C	°F	°C	°F		°C	°F		kPa	(lb)
A	70	158	77	170	121	250	190	374	225	437	2	60	140	20	62	(9.0)
B	65	149	77	170	118	245	190	374	225	437	2	56	133	20	69	(10.0)
C	60	140	77	170	116	240	185	365	225	437	2	51	124	20	79	(11.5)
D	55	131	77	170	113	235	185	365	225	437	2	47	116	20	93	(13.5)
E	50	122	77	170	110	230	185	365	225	437	2	41	105	20	103	(15.0)

[a] At 760 mm Hg pressure (101.3 kPa).

TABLE 3—SCHEDULE OF SEASONAL AND GEOGRAPHICAL VOLATILITY CLASSES

This schedule, subject to agreement between buyer and seller, represents the time and place of use of the gasoline. Shipments intended for future use may anticipate this schedule. Where alternative classes are permitted, the option shall be exercised by the seller.

State	Jan.	Feb.	March	April	May	June	July	Aug.	Sept.	Oct.	Nov.	Dec.
Alabama	D	D	D/C	C	C/B	B	B	B	B/C	C	C/D	D
Alaska	E	E	E	E/D	D	D	D	D	D/E	E	E	E
Arizona	D/C	C	C/B	B	B/A	A	A	A	A	A/B	B/C	C/D
Arkansas	D	D	D/C	C	C/B	B	B	B	B/C	C	C/D	D
California:[a]												
North Coast	D	D	D	D/C	C	C	C	C	C	C	C/D	D
South Coast	D	D	D/C	C	C	C/B	B	B	B	B/C	C	C/D
Southeast	D/C	C	C/B	B	B/A	A	A	A	A	A/B	B/C	C/D
Interior	D	D	D/C	C	C/B	B	B	B	B	B/C	C/D	D
Colorado	D	D	D/C	C/B	B	B/A	A	A	A/B	B/C	C/D	D
Connecticut	E	E	E/D	D	D/C	C	C	C	C	C/D	D/E	E
Delaware	E	E/D	D	D/C	C	C	C	C	C	C	C/D	D/E
District of Columbia	E	E/D	D	D/C	C	C	B/C	B/C	C	C	D	D/E
Florida	D/C	C	C	C	C	C	C	C	C	C	C	C/D
Georgia	D	D	D/C	C	C/B	B	B	B	B/C	C	C/D	D
Hawaii	C	C	C	C	C	C	C	C	C	C	C	C
Idaho	E	E/D	D	D/C	C/B	B	B	B	B	B/C	C/D	D/E
Illinois:												
N 40° Latitude	E	E	E/D	D/C	C	C	C	C	C	C/D	D	D/E
S 40° Latitude	E	E/D	D	D/C	C	C/B	B	B	B/C	C	D	D/E
Indiana	E	E	E/D	D/C	C	C	C	C	C	C/D	D	D/E
Iowa	E	E	E/D	D/C	C	C/B	B	B	B/C	C	D	D/E
Kansas	E/D	D	D/C	C	C/B	B	B	B	B	B/C	C/D	D
Kentucky	E	E/D	D/C	C	C	C	C/B	B	B	C	C	D/E
Louisiana	D	D	D/C	C	C/B	B	B	B	B/C	C	C/D	D
Maine	E	E	E/D	D	D/C	C	C	C	C	C/D	D/E	E
Maryland	E	E/D	D	D/C	C	C	B/C	B/C	C	C	D	D/E
Massachusetts	E	E	E/D	D	D/C	C	C	C	C	C/D	D/E	E
Michigan	E	E	E/D	D	D/C	C	C	C	C	C/D	D/E	E
Minnesota	E	E	E/D	D/C	C	C	C	C	C	C/D	D/E	E
Mississippi	D	D	D/C	C	C/B	B	B	B	B/C	C	C/D	D
Missouri	E	E/D	D/C	C	C/B	B	B	B	B/C	C	C/D	D/E
Montana	E	E/D	D	D/C	C	C/B	B	B	B/C	C	C/D	D/E
Nebraska	E	E/D	D/C	C	C	B	B	B	B	B/C	C/D	D/E
Nevada:												
N 38° Latitude	E/D	D	D/C	C	C/B	B/A	A	A	A/B	B/C	C/D	D/E
S 38° Latitude	D	D/C	C/B	B	B/A	A	A	A	A	A/B	B/C	C/D
New Hampshire	E	E	E/D	D	D/C	C	C	C	C	C/D	D/E	E
New Jersey	E	E	E/D	D/C	C	C	C	C	C	C/D	D/E	E
New Mexico:												
N 34° Latitude	D	D/C	C	C/B	B/A	A	A	A/B	B	B/C	C/D	D
S 34° Latitude	D/C	C	C/B	B	B/A	A	A	A	A/B	B/C	C	C/D
New York	E	E	E/D	D/C	C	C	C	C	C	C	D/E	E
North Carolina	E/D	D	D/C	C	C/B	B	B	B	B/C	C	C/D	D/E
North Dakota	E	E	E/D	D/C	C	C/B	B	B	B/C	C	D/E	E
Ohio	E	E	E/D	D/C	C	C	C	C	C	C/D	D	D/E
Oklahoma	D	D	D/C	C	C	C/B	B	B	B	B/C	C/D	D
Oregon:												
E 122° Longitude	E	E/D	D	D/C	C/B	B	B	B	B	B/C	C/D	D/E
W 122° Longitude	E	E	E	D	D/C	C	C	C	C	C	D/E	E
Pennsylvania	E	E	E/D	D/C	C	C	C	C	C	C	D	D/E
Rhode Island	E	E	E/D	D	D/C	C	C	C	C	C/D	D/E	E
South Carolina	D	D	D/C	C	C/B	B	B	B	B/C	C	C	D
South Dakota	E	E/D	D	D/C	C	C/B	B	B	B	B/C	C/D	D/E
Tennessee	E/D	D	D/C	C	C	C/B	B	B	B/C	C	C/D	D/E
Texas:												
E 99° Longitude	D/C	C	C	C/B	B	B	B	B	B	B/C	C	C/D
W 99° Longitude	D/C	C	C/B	B	B/A	A	A	A	A/B	B/C	C	C/D
Utah	E	E/D	D/C	C/B	B/A	A	A	A	A/B	B/C	C/D	D/E
Vermont	E	E	E/D	D	D/C	C	C	C	C	C/D	D/E	E
Virginia	E	E/D	D/C	C	C	C	B/C	B/C	C	C	C/D	D/E
Washington:												
E 122° Longitude	E	E	E/D	D/C	C	C/B	B	B	B/C	C/D	D/E	E
W 122° Longitude	E	E	E	D	D/C	C	C	C	C	C/D	D/E	E
West Virginia	E	E/D	D	D/C	C	C	C	C	C	C	C/D	D/E
Wisconsin	E	E	E/D	D	D/C	C	C	C	C	C/D	D/E	E
Wyoming	E	E/D	D/C	C	C/B	B/A	A	A/B	B	B/C	C/D	D/E

[a] Details of State Climatological Division by county as indicated:
California, North Coast—Alameda, Contra Costa, Del Norte, Humbolt, Lake, Marin, Mendocino, Monterey, Napa, San Benito, San Francisco, San Mateo, Santa Clara, Santa Cruz, Solano, Sonoma, Trinity.
California, Interior—Lassen, Modoc, Plumas, Sierra, Siskiyou, Alpine, Amador, Butte, Calaveras, Colusa, El Dorado, Fresno, Glenn, Kern (except that portion lying east of the Los Angeles County Aqueduct), Kings, Madera, Mariposa, Merced, Placer, Sacramento, San Joaquin, Shasta, Stanislaus, Sutter, Tehama, Tulare, Tuolumne, Yolo, Yuba, Nevada.
California, South Coast—Orange, San Diego, San Luis Obispo, Santa Barbara, Ventura, Los Angeles (except that portion north of the San Gabriel Mountain range and east of the Los Angeles County Aqueduct).
California, Southeast—Imperial, Riverside, San Bernardino, Los Angeles (that portion north of the San Gabriel Mountain range and east of the Los Angeles County Aqueduct), Mono, Inya, Kern (that portion lying east of the Los Angeles County Aqueduct).

parative characteristic in a majority of these studies was the 50% distillation temperature.

3.4 Fuel Economy—Fuel economy can be influenced by specific gravity and energy content of the fuel. Where fuel volatility changes result in change of these properties, fuel economy can be affected. However, within practical limits, the effects of volatility on fuel economy are very small.

4. Cleanliness—In addition to providing acceptable antiknock quality and volatility characteristics, automotive gasolines must also provide for satisfactory engine cleanliness, both physical and chemical. The following properties have a direct bearing on the overall performance of a gasoline.

4.1 Workmanship and Contamination—A finished gasoline is essentially a blend of petroleum fractions visually free of undissolved water, sediment, and suspended matter. It is clear and bright when observed at 21°C (70°F). Physical contamination may occur during distribution of the fuel to the eventual power source. Control of such contamination is a matter requiring constant vigilance. Solid and liquid contamination can lead to restriction of fuel metering points, improper seating of inlet valves, corrosion, fuel line freezing, gel formation, and filter plugging.

4.2 Rust and Corrosion—Filter plugging and engine wear problems are reduced by minimizing rust and corrosion in fuel distribution and in vehicle fuel systems when engines are idle. Modifications of ASTM D 665, Rust-Preventing Characteristics of Steam-Turbine Oil in the Presence of Water, such as NACE Standard TM-01-72, are used to measure rust protection of gasolines.

4.3 Sulfur Content and Copper Strip Corrosion—Crude petroleum contains sulfur compounds, some of which are removed during refining. ASTM D 439 limits sulfur content of unleaded gasolines to a maximum of 0.10 mass % and leaded gasolines to a maximum of 0.15 mass %. Currently, average sulfur content of gasolines distributed in the United States is about 0.03 mass % with the maximum reported values approaching the specification limits.

Sulfur oxides formed during combustion may be converted to acids that promote rusting and corrosion of engine parts and piston ring and cylinder wall wear. Sulfur oxides formed in the exhaust are undesirable atmospheric pollutants; however, the contribution of automotive exhaust to total sulfur oxide emissions is negligible. The fate of these sulfur compounds upon passage through the catalytic exhaust treatment devices and their possible contribution to sulfates, sulfuric acid, and hydrogen sulfide in automotive exhaust and the atmosphere is receiving careful study. Sulfur content is usually determined by ASTM D 1266, Sulfur in Petroleum Products (Lamp Method). Presence of corrosive sulfur compounds can be detected by ASTM D 130, Corrosion by Petroleum Products, Copper Strip Test.

4.4 Existent Gum and Stability—During storage, gasolines may oxidize slowly in presence of air, forming undesirable oxidation products called gum. The gum is usually soluble in the gasoline but may appear as a sticky residue on evaporation. These residues may deposit on carburetor surfaces, intake manifolds, pistons, and intake valves, stems, and guides. ASTM D 439 limits gasoline to a maximum of 5 mg of gum per 100 mL of gasoline. ASTM D 381, Existent Gum in Fuels by Jet Evaporation, is a test to determine the amount of existing gum.

Automotive gasolines have a negligible gum content when manufactured, but may form gum during extended storage. ASTM D 525, Oxidation Stability of Gasoline (Induction Period Method), is a test to indicate the tendency of a gasoline to form gum in storage. ASTM has set a minimum limit of 240 min for Oxidation Stability by this method. It should be recognized, however, that the method's correlation may vary markedly under different storage conditions and with different gasoline blends. Most automotive gasolines contain special chemicals (antioxidants) to inhibit oxidation and gum formation. Some gasolines also contain metal deactivators for this purpose. Commercial gasoline is not designed for abnormally severe conditions of storage. Gasoline purchased for severe bulk storage conditions or for prolonged storage in vehicle fuel systems generally has additional amounts of antioxidant and metal deactivator added.

5. Other Properties—Many chemical and physical properties of gasoline, not directly controlled by specifications, are important considerations in combustion, fuel system design and calibration, and safe handling in storage and distribution systems. These properties are particularly important when comparing the performance of alternative fuels or alternative fuel components with gasoline.

5.1 Hydrocarbon Composition—The stoichiometry of gasoline combustion is determined by the concentrations of hydrogen and carbon in gasoline, which in turn depend upon the types of hydrocarbons present. The main hydrocarbon constituents are saturates, olefins, and aromatics; these are identified using ASTM D 1319, Hydrocarbon Types in Petroleum Products by Fluorescent Indicator Absorption. The concentrations of these hydrocarbons in commercial gasolines typically cover the range of 15–40 vol% for aromatics, 0–20% for olefins, and 50–80% for saturates. Non-standardized analytical procedures are available which permit identification of specific compounds and the ultimate determination of carbon and hydrogen content. Stoichiometric air-fuel ratios of commercial gasolines typically range from 14.3 for gasolines with high concentrations of aromatics to 14.8 for low aromatic fuels.

5.2 Heating Value—The heating value of gasoline is a measure of its energy content and has a small effect on fuel economy. The lower heating value of gasoline is about 44 MJ/kg (18 900 Btu/lb) and can be measured by ASTM D 240, Heat of Combustion of Liquid Hydrocarbon Fuels by Bomb Calorimeter or by ASTM D 2332, Heat of Combustion of Hydrocarbon Fuels by Bomb Calorimeter (High Precision Method). Volumetric heating value of gasolines generally varies directly with specific gravity which is more convenient to measure than is heat of combustion. The relationship between specific gravity and fuel economy is quantified in SAE J1082a, Fuel Economy Measurements—Road Test Procedure, by the correction factor 1.0 + 0.8 (0.737 − specific gravity). Typical extremes in specific gravity (or heating value) of commercial gasolines represent a fuel economy difference of about 3%.

5.3 Gravity—*Gravity* is a term used to denote the density of gasolines, and is useful as a means of identification. Two methods of expressing gravity are commonly used. Specific gravity is the ratio of the mass of a given volume of gasoline at a temperature of 15.6°C (60°F) to the mass of an equal volume of water at the same temperature. Typically, automotive gasolines have specific gravities between 0.72 and 0.75 and API gravities between 65 and 57. API gravity is based on an arbitrary hydrometer scale and is related to specific gravity as follows:

$$\text{Degrees API} = \frac{141.5}{\text{Sp. Gr. } (60°F/60°F)} - 131.5$$

Gravity of gasoline is determined using ASTM D 1298, Density, Specific Gravity, or API Gravity of Crude Petroleum and Liquid Petroleum Products by Hydrometer Method.

It should be noted that in the change to the SI metric system of measurement, API gravity and specific gravity are to be eliminated from usage. API gravity is to be replaced by absolute density (kg/m^3) at 15°C and 101.325 kPa. Specific gravity is to be replaced by relative density at 15°C and 101.325 kPa, where the reference fluids for liquids and gases are water and air respectively.

5.4 Viscosity and Surface Tension—Viscosity and surface tension of gasolines affect calibration of carburetors and metering orifices. Viscosity is measured by ASTM D 445, Viscosity of Transparent and Opaque Liquids (Kinematic and Dynamic Viscosities). Surface tension is measured by non-standard techniques. Both viscosity and surface tension vary with ambient temperature. Viscosity variation due to temperature is greater than the variability caused by fuel compositions. Typical values for viscosity and surface tension at ambient and low temperatures are shown in the following table:

	Temperature	
	20°C (68°F)	−20°C (−4°F)
Viscosity, mm^2/s (cSt)	(0.5–0.6)	(0.8–1.0)
Surface Tension, N/m	20 × 10^{-3}	24 × 10^{-3}
dynes/cm	20	24

5.5 Fire Safety—Fire hazards of gasolines are related to the typical properties shown in the following table:

Flash Point, °C (°F)	∼ −43 (−45)
Autoignition Temperature, °C (°F)	∼260 (500)
Flammability Limits, vol % in air	1.4–7.6
Vapor Pressure at 21°C (70°F), kPa (psi)	28–55 (4–8)
Concentration in saturated air at 20°C (68°F), vol %	25–50

Flash point can be measured by use of ASTM D 56, Flash Point by Tag Closed Tester and autoignition temperature by ASTM D 2155, Autoignition Temperature of Liquid Petroleum Products. The flash point of automotive gasoline is so low that a flammable mixture should always be presumed to exist over the liquid. However, gasoline does not normally present an explosive hazard in covered storage tanks, because the saturated vapor above the liquid is too rich to ignite. Explosive conditions may occur at extremely low temperatures or during transfer operations.

6. Additives—Gasoline additives are used to enhance or provide various performance features related to the satisfactory operation of engines, as well as to minimize gasoline handling and storage problems. These compounds complement refinery processing in attaining the desired level of product quality.

The amount and variety of additives used in gasolines have grown rapidly. For many years, antiknocks were almost the only additives used; however, the list now also includes combustion chamber deposit modifiers, antioxidants, metal deactivators, corrosion or rust inhibitors, carburetor anti-icing additives, fuel line antifreeze agents, gasoline detergents, gasoline dispersants, and identifying dyes.

Gasoline additives currently available are shown by class and function in Table 4.

7. Special Test Fuels—Special fuels are used in test programs within the automotive and petroleum industries to insure engine-fuel compatibility. Most special fuels can be classified as follows.

7.1 Antiknock Requirement Testing—In addition to primary reference fuels, full-boiling range reference fuels varying in Research and Motor octane numbers are used to determine antiknock requirements of vehicles. *Full-boiling range* means that the fuel is similar to commercial fuels, consisting of many hydrocarbons with different boiling points. The Coordinating Research Council (CRC) sponsors preparation of these reference fuels each year.

7.2 Driveability Testing—Gasolines with specific volatility specifications are used in controlled tests for evaluating vapor lock, evaporative control systems, tank filling, cold starting, and warm-up. The CRC sponsors preparation of volatility test fuels for specific programs.

7.3 Emission Test Fuel—The federal government specifies gasolines for emission testing; see the Federal Register, June 28, 1977.

7.4 Engine Oil Qualification—Gasolines are specified for engine oil qualification tests according to ASTM engine oil test sequences described in ASTM Special Technical Publication (STP) 315G.

7.5 Fuel System Materials—Rubber-swelling test fluids for evaluating

TABLE 4—COMMERCIAL GASOLINE ADDITIVES—FUNCTION AND TYPE

Class or Function	Common Additive Type
1. **Antiknock Compounds**—To improve Research, Motor, and road octane quality.	Lead alkyls, such as tetraethyllead, tetramethyllead, and physical or reacted mixtures. Organo-manganese compounds, such as methyl cyclopentadienyl manganese tricarbonyl (MMT).
2. **Combustion Deposit Modifiers**—To minimize surface ignition, rumble, preignition, and spark plug fouling.	Organic or organo-metallic compounds usually containing phosphorus.
3. **Antioxidants**—To minimize oxidation and gum formation in gasoline and to improve handling and storage characteristics.	Phenylene diamine, phenol, and aminophenol compounds.
4. **Metal Deactivators**—To deactivate copper ions that are powerful oxidation catalysts.	N,N'-disalicylidene-1,2-propane diamine.
5. **Corrosion of Rust Inhibitors**—To minimize corrosion and rusting of fuel system and storage handling facilities.	Derivatives of carboxylic, sulfonic, or phosphoric acids, many of which have surface-active properties.
6. **Carburetor Anti-Icing Additives**—To minimize engine stalling due to ice accumulation on the throttle.	Derivatives of carboxylic, sulfonic, or phosphoric acid that have surface-active properties. Freeze point depressants such as alcohols and glycols.
7. **Gasoline Detergents**—To remove and/or minimize the accumulation of deposits in the throttle section of the carburetor, which adversely affect the metering characteristics.	Amines and derivatives of carboxylic, sulfonic, or phosphoric acid having surface-active properties, some of which are polymeric.
8. **Gasoline Dispersants**—To extend PCV valve life, reduce engine sludge, and remove and/or minimize the accumulation of deposits in the carburetor, intake manifold, intake ports, and underside of the intake valves.	Amines and low-molecular-weight synthetic polymers. Specific fractions of special oils.
9. **Dyes**—For various identification purposes.	Oil-soluble solid and liquid dyes.

NOTE: Some materials may also be marketed as multifunctional or multipurpose additives, performing more than one of the functions described in items 5, 6, 7, and 8.

the compatibility of rubber or rubber-like fuel system components with gasolines are specified in ASTM D 471.

7.6 Flow Testing of Carburetors and Injectors—Special fluids are used in flow test stands. These fluids have carefully controlled density and viscosity, but have a narrower boiling range than commercial fuels.

8. Bibliography—Additional information on automotive gasolines may be obtained from the following references:

1. R. V. Kerley and K. W. Thurston, "Knocking Behavior of Fuels and Engines." SAE Transactions, Vol. 64 (1956), p. 554.
2. W. E. Bettoney, J. G. Burt, and H. J. Scheule, "Factors Affecting the Performance of Additives in Gasoline." Sixth World Petroleum Conference Proceedings, Section VI, Paper 22, June 1963, pp. 111–130.
3. J. R. Lodwick, "Chemical Additives in Petroleum Fuels: Some Uses and Action Mechanisms." Journal Institute of Petroleum, Vol. 50, No. 491, November 1964, pp. 297–318.
4. K. M. Elliot, W. D. Myers, and J. G. Porter, Jr., "Automotive Fuel Quality Trends." Proceedings American Petroleum Institute, Section III—Refining, Vol. 44 (1964), pp. 222–232.
5. M. M. Roensch and R. L. Courtney, "Gasoline Engines Take a New Look at Fuels." Proceedings American Petroleum Institute, Section III—Refining, Vol. 44 (1964), pp. 200–207.
6. M. R. Barusch and J. H. MacPherson, "Engine Fuel Additives." Advances in Petroleum Chemistry and Refining, Vol. 10 (1965), Chap. 10, pp. 457–538.
7. Bonner and Moore Associates, Inc. (Houston, Texas), "U.S. Motor Gasoline Economics." Vols. 1 and 2, June and November 1967.
8. K. T. Dishart and W. C. Harris, "The Effect of Gasoline Hydrocarbon Composition on Automotive Exhaust Emissions." Paper presented at the Midyear Meeting of American Petroleum Institute, Division of Refining, Philadelphia, May 1968.
9. W. G. Agnew, "Future Emission-Controlled Spark-Ignition Engines and Their Fuels." Proceedings American Petroleum Institute, Division of Refining, Vol. 49 (1969), pp. 242–280.
10. T. J. Sheahan, C. J. Dorer, Jr., and C. O. Miller, "Detergent-Dispersant Fuel Performance and Handling." Paper 690516 presented at SAE Midyear Meeting, Chicago, May 1969.
11. R. K. Stone and B. H. Eccleston, "Vehicle Emissions vs Fuel Composition." (API-Bureau of Mines), Part I presented at the Midyear Meeting of American Petroleum Institute, Division of Refining, Philadelphia, May 1968. Part II presented at the Midyear Meeting of American Petroleum Institute, Division of Refining, Chicago, May 1969.
12. B. Dimitriades, B. H. Eccleston, and R. W. Hurn, "An Evaluation of the Fuel Through Direct Measurement of Photochemical Reactivity of Emissions." Paper presented at the 62nd Annual Air Pollution Control Association Meeting, New York, June 1969.
13. "Public Health Aspects of Increasing Tetraethyl Lead Content in Motor Fuel." Public Health Service Publication No. 712, Washington, DC, 1959.
14. "Gasoline Modification—Its Potential as an Air Pollution Control Measure in Los Angeles County." A Joint Report by the California Air Resources Board, Los Angeles County Air Pollution Control District, and Western Oil and Gas Association, November 1969.
15. "CRC Revised Road Rating Techniques." Report No. 436, Coordinating Research Council, Inc., July 1970.
16. "Automotive Fuels and Air Pollution." Report of the Panel on Automotive Fuels and Air Pollution to the Commerce Technical Advisory Board, March 1971.
17. "Mathematical Expressions Relating Evaporative Emissions from Motor Vehicles to Gasoline Volatility." Report by Scott Research Laboratories, Inc., prepared for the American Petroleum Institute, March 1971.
18. "Exhaust Emission Standards and Test Procedures." Federal Register, Washington, DC, Vol. 36, No. 128, July 2, 1971.
19. L. C. Lichty, "Internal Combustion Engines." New York and London: McGraw-Hill Book Co., Inc., 1951.
20. "Significance of ASTM Test for Petroleum Products." American Society for Testing and Materials, STP 7C, January 1977.
21. "Symposium on Current Research on Motor Gasoline Which May Affect Future Specifications." American Society for Testing and Materials, STP No. 298, 1961.
22. W. A. Gruse, "Motor Fuels, Performance and Testing." New York: Reinhold Publishing Corp., 1967.
23. "Test Methods for Rating Motor, Diesel, Aviation Fuels." Annual ASTM Standards, Part 47. See most recent edition.
24. "Annual ASTM Standards." Parts 23 and 24. See most recent edition.
25. P. Polss, "What Additives Do for Gasolines." Hydrocarbon Processing, February 1973.
26. P. J. Clarke, "The Effect of Gasoline Volatility on Emissions and Driveability." Paper 710136 presented at SAE Automotive Engineering Congress, Detroit, January 1971.
27. B. S. Bailey, E. J. Forster, W. E. Morris, "Road Rating Trends of U.S. Motor Cars—A Review of Recent CRC Programs." Paper 730012 presented at SAE Automotive Engineering Congress, Detroit, January 1973.
28. W. E. Morris, "Engine—Octane Quality Relationships—State of the Art." Paper presented at the 37th Midyear Meeting of API, Div. of Refining, New York, May 1972.
29. L. A. McReynolds, "Automotive Fuels to Meet the Requirements of Future Exhaust Emission Control Systems—A Petroleum Man's View." Paper AM-73-26 presented at NPRA Annual Meeting, San Antonio, April 1973.
30. "Automobile Fuel Economy." Motor Vehicles Manufacturers Association, September 1973.
31. J. P. Doner, "A Predictive Study of the Occurrence of Meteorological Conditions Contributing to Automotive Carburetor Icing." CCL Report No. 3006, September 1973, NITS.
32. "1958 CRC Vapor Lock Tests." CRC Report No. 343, available through SAE Publications Department.
33. "CRC Vapor Lock Technique—Its Development and Application." SAE Transactions (1963), p. 122.
34. "1962 CRC Vapor Lock Tests." CRC Report No. 378, available through Coordinating Research Council, Inc.
35. "1964 CRC Vapor Lock Tests." CRC Report No. 386, available through Coordinating Research Council, Inc.
36. "Evaluation of Expressions for Fuel Volatility." CRC Report No. 403, available through Coordinating Research Council, Inc.
37. "Driveability Performance of 1977 Passenger Cars at Intermediate Ambient Temperatures Paso Robles." CRC Report No. 499, September 1978, available through Coordinating Research Council, Inc.
38. "Driveability Performance of 1975 Passenger Cars at High Temperatures." CRC Report No. 490, November 1976, available through Coordinating Research Council, Inc.
39. R. W. Hurn, "Evaporative Emissions—Which Volatility Factors Count." American Society for Testing and Materials, STP 487, Effect of Automotive Emission Requirements on Gasoline Characteristics, 1971, available from American Society for Testing and Materials.

40. M. W. Jackson and R. L. Everett, "Effect of Fuel Composition on Amount and Reactivity of Evaporative Emissions." SAE Publication No. 690088, 1969, available through SAE Publications Department.

41. P. J. Clarke, "The Effect of Gasoline Volatility on Exhaust Emissions." SAE Publication No. 720932, 1972, available through SAE Publications Department.

42. R. P. Doelling, A. F. Garber, and M. P. Walsh, "Effect of Gasoline Characteristics on Automotive Exhaust Emissions." American Society for Testing and Materials, STP 487, Effect of Automotive Emission Requirements on Gasoline Characteristics, 1971, available through American Society for Testing and Materials.

43. R. E. Burtner, "Gasoline Characteristics and Air Conservation—A Review of API Engine Fuel Studies." American Society for Testing and Materials, STP 487, Effect of Automotive Emission Requirements on Gasoline Characteristics, 1971, available through American Society for Testing and Materials.

44. B. H. Eccleston and R. W. Hurn, "Effect of Fuel Front-End and Mid-Range Volatility on Automobile Emissions." Bureau of Mines Report of Investigation 7707, 1972, available through the U.S. Department of Energy.

DIESEL FUELS—SAE J313 JUN80

SAE Information Report

Report of the Fuels and Lubricants Technical Committee, approved January 1955, last revised June 1980.

Automotive and railroad diesel fuels, in general, are derived from petroleum refinery products which are commonly referred to as middle distillates. Middle distillates represent products which have a higher boiling range than gasoline and are obtained from fractional distillation of the crude oil or from streams from other refining processes. Finished diesel fuels represent blends of middle distillates.

The properties of commercial distillate diesel fuels depend on the refinery practices employed and the nature of the crude oils from which they are derived. Thus they may differ both with and within the region in which they are manufactured. Such fuels generally boil over a range between 163 and 371°C (325-700°F). Their makeup can represent various combinations of volatility, ignition quality, viscosity, sulfur level, gravity, and other characteristics. Additives may be used to impart special properties to the finished diesel fuel.

Classification of Diesel Fuel Oils, ASTM D 975[1]—ASTM D 975, Standard Specification for Diesel Fuel Oils, which includes 1-D, 2-D, and 4-D grades of fuel, is intended as a statement of permissible limits of significant fuel properties used for classifying the wide variety of commercially available diesel fuels, in accordance with their service application. The classification chart contained in ASTM D 975 is shown herein as Table 1.

National Survey of Diesel Fuels—Since diesel fuels can vary in characteristics, a national diesel fuel survey is conducted annually to provide an indication of the properties of fuels produced throughout the United States. This survey was started in 1950 and was continued until 1975 by the U.S. Department of Interior, Bureau of Mines. It is now issued by the Bartlesville Energy Technology Center (BETC) of the Department of Energy (DOE). Laboratory inspection data on diesel fuels supplied and marketed in 14 regions of the country are furnished by the manufacturers. Prior to 1957 these data were classified in the DOE reports according to ASTM grade number (1-D, 2-D,

[1] References to ASTM Standards in this report are all to be construed as the most recent ASTM publications.

etc.), but since that date they have been classified according to fuel type as follows:

Type C-B—Diesel fuel oils for city-bus and similar operations.
Type T-T—Fuels for diesel engines in trucks, tractors, and similar service.
Type R-R—Fuel for railroad diesel service.
Type S-M—Heavy distillate and residual fuels for stationary and marine diesel engines.

Fuels that meet requirements of ASTM grade No. 1-D and 2-D are so identified in the DOE reports and are generally satisfactory for use in all automotive diesel engines including those used in light duty applications such as passenger cars and light trucks.

Significance and Methods of Measurement of Diesel Fuel Properties

1. **Hydrocarbon Composition**—The hydrocarbon composition of diesel fuel determines many other properties: the ignition quality, heating value, volatility, gravity, oxidation stability, etc. Thus, it directly affects the power and economy, wear, deposit formation, starting, and smoke performance of diesel engines.

Since diesel fuels are complex mixtures of many individual hydrocarbons, the measurement of their hydrocarbon composition is not simple. A number of procedures are used, and the reader is referred to current ASTM publications for suggested methods.

2. **Ignition Quality**—Engine performance factors which may be influenced by ignition quality of the fuel are: cold starting, warmup, combustion roughness, acceleration, deposits under idle and light load operation, and exhaust smoke density. In each of these cases, other fuel factors as well as engine factors can also affect performance. The ignition quality requirement of an engine depends on design, size, mechanical condition, operating conditions, atmospheric temperature, and altitude. Increase in ignition quality over the level required does not materially improve engine performance.

Ignition quality is measured by ASTM D 613, Ignition Quality of Diesel Fuels by the Cetane Method. The ASTM cetane number of a diesel fuel may

φ TABLE 1—DETAILED REQUIREMENTS FOR DIESEL FUEL OILS[a,b,i]

Grade of Diesel Fuel Oil	Flash Point, °C (°F)	Cloud Point, °C (°F)	Water and Sediment, volume %	Carbon Residue on 10 percent Residuum, %	Ash, weight %	Distillation Temperatures, C(°F) 90% Point		Viscosity				Sulfur,[d] weight %	Copper Strip Corrosion	Cetane Number[e]
								Kinematic cSt or mm²/s[g] at 40°C		Saybolt, SUS at 100°F				
	Min	Max	Max	Max	Max	Min	Max	Min	Max	Min	Max	Max	Max	Min
No. 1-D A volatile distillate fuel oil for engines in service requiring frequent speed and load changes.	38 (100)	b	0.05	0.15	0.01	—	288 (550)	1.3	2.4	—	34.4	0.50	No. 3	40[f]
No. 2-D A distillate fuel oil of lower volatility for engines in industrial and heavy mobile service.	52 (125)	b	0.05	0.35	0.01	282[c] (540)	338 (640)	1.9	4.1	32.6	40.1	0.50	No. 3	40[f]
No. 4-D A fuel oil for low and medium speed engines.	55 (130)	b	0.50	—	0.10	—	—	5.5	24.0	45.0	125.0	2.0	—	30[f]

[a] To meet special operating conditions, modifications of individual limiting requirements may be agreed upon between purchaser, seller, and manufacturer.

[b] It is unrealistic to specify low-temperature properties that will ensure satisfactory operation on a broad basis. Satisfactory operation should be achieved in most cases if the cloud point (or wax appearance point) is specified at 6°C (10°F) above the tenth percentile minimum ambient temperature for the area in which the fuel will be used. The tenth percentile minimum ambient temperatures for the months of October, November, December, January, February, and March are given in ASTM D 975 in maps for the 48 contiguous states and Alaska. This guidance is of a general nature; some equipment designs, use of flow improver additives, fuel properties, and/or operations may allow higher or require lower cloud point fuels. Appropriate low temperature operability properties should be agreed on between the fuel supplier and purchaser for the intended use and expected ambient temperatures.

φ [c] When cloud point less than −12°C (10°F) is specified, the minimum viscosity shall be 1.7 cSt and the 90% point shall be waived.

[d] In countries outside the U.S.A., other sulfur limits may apply.

[e] Where cetane number by Method D 613, is not available, ASTM Method D 976, Calculated Cetane Index of Distillate Fuels may be used as an approximation. Where there is disagreement, Method D 613 shall be the referee method.

[f] Low-atmospheric temperatures as well as engine operation at high altitudes may require use of fuels with higher cetane ratings.

[g] Millimeter squared per second (official SI unit) (Note: 1 cST = 1 mm²/s).

[h] The values in SI units are to be regarded as the standard. The values in U. S. customary units are for information only.

[i] Nothing in this specification shall preclude observance of federal, state, or local regulations which may be more restrictive.

be defined as the percentage by volume of normal cetane in a blend with heptamethylnonane required to match the ignition quality of the test fuel. The method requires the use of a standard test engine equipped with accepted instrumentation and operated under prescribed conditions.

When a test engine is not available for determining cetane number, or when the quantity of sample is too small for an engine rating, this property may be approximated by the ASTM calculated cetane index formula provided in ASTM D 976, Calculated Cetane Index of Distillate Fuels. A nomograph from ASTM D 976 based on this formula is shown in Fig. 1. The method for determining the ASTM calculated cetane index from the API gravity and the mid-boiling point is indicated by the illustrative example on the nomograph.

The ASTM calculated cetane index is particularly applicable to straight-run fuels, catalytically cracked stocks, and blends of the two. Its application is limited as follows:

(a) It is not applicable to fuels containing additives for raising cetane number.

(b) It is not applicable to pure hydrocarbons, synthetic fuels, alkylates, or coal-tar products.

(c) Substantial inaccuracies in correlation may occur if used for crude oils, residuals, or products having a distillation end point below 260°C (500°F).

The ASTM calculated cetane index is not recommended as a substitute for actual measured cetane number in the preparation of fuel specifications. However, because good correlations can be established between cetane index ϕ and cetane number for a given crude oil and processing scheme, cetane index is often used for refinery control of diesel fuel ignition quality.

3. **Heating Value**—An important property of a diesel fuel is its heat of combustion, which is a measure of the energy available from the fuel. A knowledge of this value is essential when considering the thermal efficiency of equipment for producing power.

The gross heat of combustion at constant volume is determined by the bomb calorimeter, which measures the amount of heat actually released by burning a known quantity of fuel. ASTM D 240, Heat of Combustion of Liquid Hydrocarbon Fuels by Bomb Calorimeter, is applicable to diesel fuels. In the determination of the heating value of fuels by the bomb calorimeter method, the water vapor formed is condensed, which results in the gross or higher heating value for the fuel. The net or lower heating value of fuel is determined by subtracting the latent heat of condensation of the water vapor formed in the calorimeter process from the reported gross or higher heating value. It is customary to use the lower heating value for calculating thermal efficiency.

Heating values may be expressed as J/kg, J/L, Btu/lb, Btu/gal, cal/g or cal/L. Since diesel engine fuel consumption is expressed on a kg/kW·h or lb/bhp·h basis and since diesel fuel is purchased on a liter or gallon basis, both weight and volume bases are of interest to the engine manufacturer and user.

When bomb calorimeter heating values are not available, it is customary to use empirical relations which result in heating values sufficiently accurate for many purposes. Two sources of such empirical data are the Proposed Method for Estimation of Net and Gross Heat of Combustion of Burner and Diesel Fuels, 1969 ASTM Book of Standards, Part 17, Appendix VI, D-2—1968, p. 1120, and the U.S. Bureau of Standards Miscellaneous Publication No. 97, 1929, "Thermal Properties of Petroleum Products," by C. S. Cragoe. In both of these cases average heating values are tabulated in terms of API gravity and sulfur content of the fuel.

Two nomographs developed from empirical formulas and theoretical considerations are shown in Figs. 2, 3A, and 3B. Fig. 2 permits the estimation of either gross or net heat of combustion in Btu/lb from the aniline point (ASTM D 611, a measure of aromatic and naphthenic constituents) and API ϕ gravity of a fuel. The gross heating value of a fuel in Btu/gal can be estimated from the API gravity and the boiling point temperature for 50% recovered by the ASTM D 86 distillation. A nomograph for determining Btu/gal is shown in Fig. 3A. A similar nomograph in SI metric units is shown in Fig. 3B.

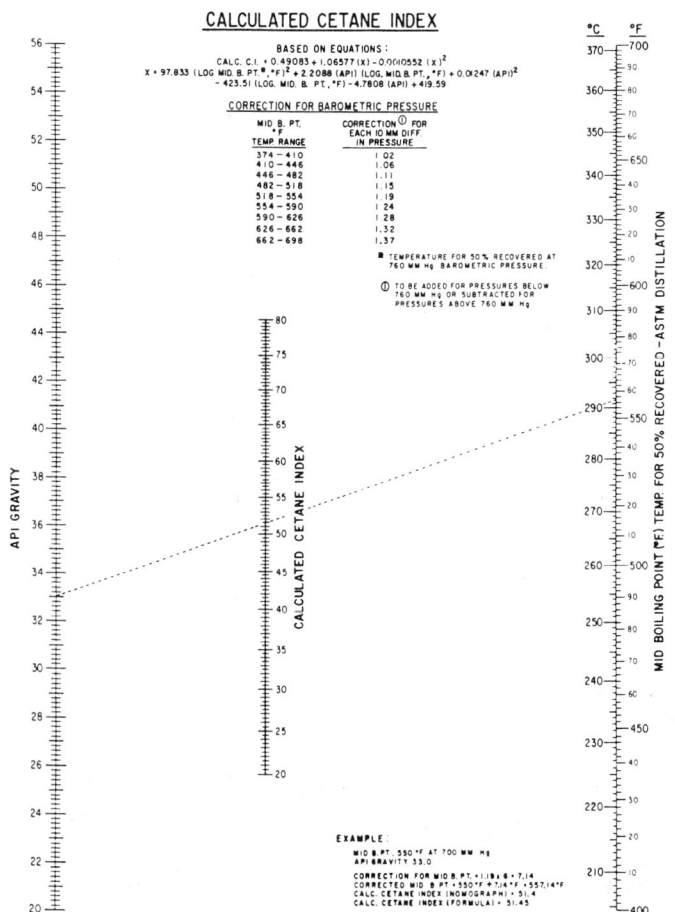

NOTE.—The Calculated Cetane Index formula represents a useful tool for estimating cetane number. Due to inherent limitations in its application, Index values may not be a valid substitute for ASTM Cetane Numbers as determined in a test engine.

From ASTM D 976, Calculated Cetane Index of Distillate Fuels. Copies of the Nomograph for Calculated Cetane Index are available from the American Society for Testing and Materials, 1916 Race St., Philadelphia, Pennsylvania 19103.

ϕ FIG. 1—NOMOGRAPH FOR CALCULATED CETANE INDEX (ECS-1 METER BASIS—METHOD D 613)

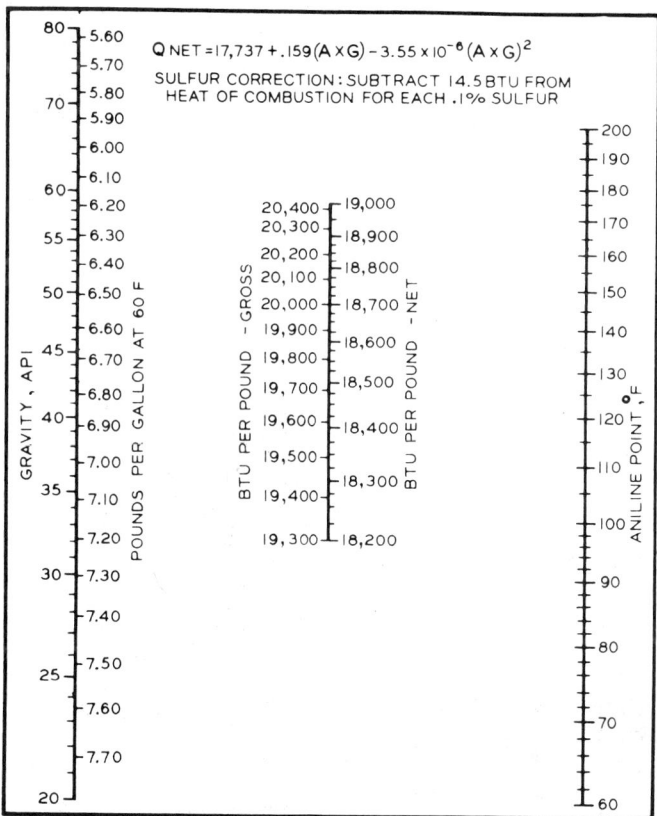

FIG. 2—RELATION BETWEEN HEAT OF COMBUSTION AND ANILINE POINT AND GRAVITY OF DISTILLATE PETROLEUM FUELS

ϕ NOTE.—A = Aniline Point; G = API Gravity

From Exxon Research and Engineering Co., Linden, New Jersey.

Heating values determined from the various empirical relations may not agree precisely, nor will they necessarily indicate the true difference in heating value between two specific fuels. However, as average values, they do offer useful estimates for many purposes.

4. Volatility—Power and economy of diesel engines are insensitive to volatility directly, although less volatile fuels normally have a higher heating value (see Figs. 3A and 3B) and thus indirectly affect these performance factors. On the other hand, starting and warmup are favored by high front end volatility, and deposition, wear, and exhaust smoke in some engines are increased by high 90% and end points.

The volatility of No. 1-D and No. 2-D diesel fuels is normally measured by ASTM D 86, Test for Distillation of Petroleum Products.

5. Gravity/Density—Knowledge of fuel gravity/density, along with volatility, provides useful information about the composition of the fuel, which in turn relates to heating value (Figs. 2, 3A, and 3B), power and economy, deposition, wear, and exhaust smoke. It should be noted from Figs. 2 and 3 that, for fixed aniline point and mid-boiling point, decreased API gravity (increased pound per gallon) increases the heating value per gallon, but decreases the heating value per pound. On the other hand, it should be understood that gravity is not the sole determining factor of heating value.

Gravity and density are normally measured by ASTM D 287, API Gravity of Crude Petroleum and Petroleum Products (Hydrometer Method) or by ASTM D 1298a, Density, Specific Gravity or API Gravity of Crude Petroleum and Liquid Petroleum Products (Hydrometer Method).

It should be noted that in the change to the SI metric system of measurement, API gravity and specific gravity are to be eliminated from usage. API gravity is to be replaced by absolute density (kg/m³) at 15°C and 101.325 kPa. Specific gravity is to be replaced by relative density at 15°C and 101.325 kPa where the reference fluids for liquids and gases are water and air respectively.

6. Viscosity—Diesel fuel viscosity can affect the performance of fuel injection systems. Low viscosity can result in excessive wear in some injection pumps and power loss due to pump and injector leakage. High viscosity can result in excessive pump resistance or filter damage. Fuel spray characteristics are also influenced by viscosity.

Diesel fuel viscosity is measured by ASTM D 445, Viscosity of Transparent and Opaque Liquids (Kinematic and Dynamic Viscosities).

7. Pour Point and Cloud Point—Pour point and cloud point are important in relation to the lowest temperature at which the fuel is sufficiently fluid to be pumped or transferred. The cloud point is the temperature at which some component of the fuel first becomes visibly insoluble under specified conditions of test. It may indicate a tendency toward filter plugging with some fuel system designs and is expressed in increments of 1°C (2°F). Pour point is the lowest temperature, expressed as a multiple of 3°C (or 5°F) at which the oil is observed to flow when cooled and examined under prescribed conditions. The use of flow improver additives can increase the difference between cloud and pour point to 10°C (18°F) or more.

The applicable measurement methods are ASTM D 97, Pour Point, and ASTM D 2500, Cloud Point of Petroleum Oils.

While it is not a U.S. specification, another test used in Europe to predict low-temperature performance is the Cold Filter Plugging Point of Distillate Fuels (CFPP), IP Test Method No. 309/76. This test measures the temperature at which fuel will no longer flow through a fine wire mesh having a nominal aperture width of 45 μm (BS410: 1974, Close Tolerance).

8. Flash Point—The flash point is the lowest fuel temperature at which application of a test flame causes the vapor above the sample to ignite momentarily under prescribed conditions. It is not directly related to engine performance, but is important in connection with legal requirements and safety precautions involved in fuel handling and storage. An abnormally low flash point may denote contamination with lighter products such as gasoline.

The usual method of measurement is ASTM D 93, Flash Point by the Pensky-Martens Closed Tester.

9. Sulfur Content—The effect of sulfur content on engine wear and deposits varies with engine and operating conditions. Higher sulfur contents than those stipulated in ASTM D 975 should be avoided because of these effects. In some engines, higher sulfur fuels may be used if a crankcase lubricant formulated to combat the effects of high sulfur is used along with an appropriate shortening of the oil drain intervals. Sulfur content of diesel fuels may also be limited by regulations directed at exhaust emission control.

Sulfur content may be measured by one of several methods: ASTM D 129, Sulfur in Petroleum Products by the Bomb Method (prescribed in ASTM D 975); ASTM D 1552, Sulfur in Petroleum Products (High Temperature Method); ASTM D 1266, Sulfur in Petroleum Products and Liquified Petro-

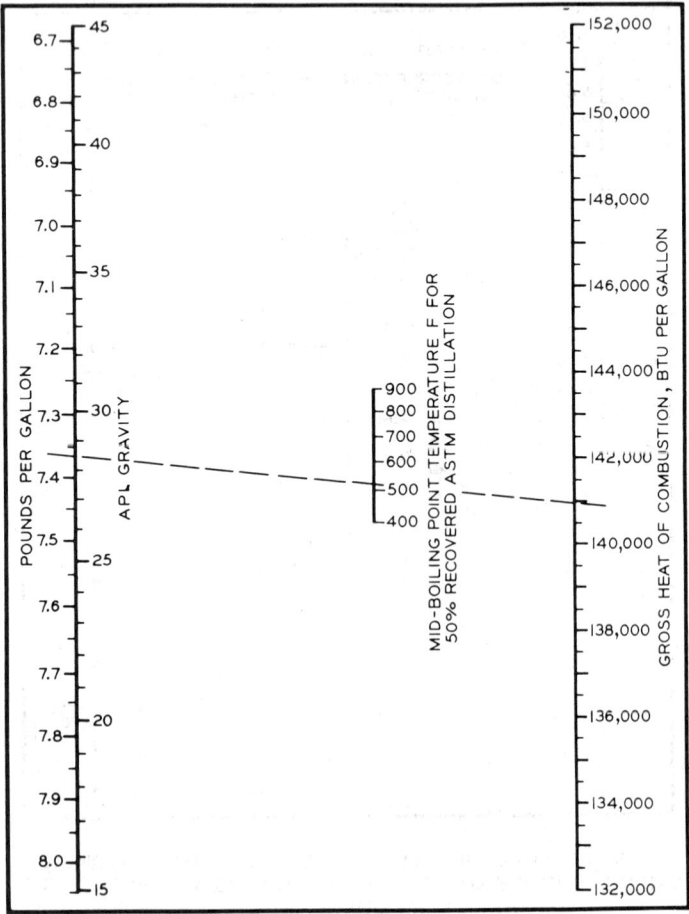

FIG. 3A—GROSS HEATING VALUE FOR DIESEL FUELS

From "Diesel Engine Fuel Specifications," compiled by Ethyl Corporation, Detroit, Michigan, August, 1967.

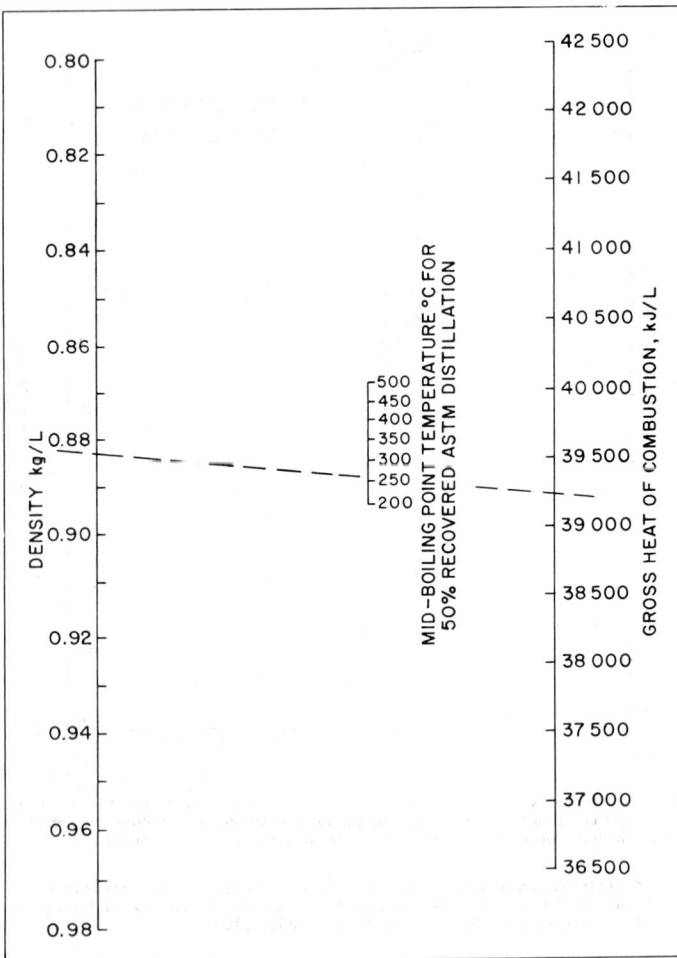

FIG. 3B—GROSS HEATING VALUE FOR DIESEL FUELS

leum (LP) Gases (Lamp Method); and ASTM D 2622, Sulfur in Petroleum Products (X-Ray Spectrographic Method).

10. Carbon Residue—Carbon residue may be related to the formation of deposits in diesel engine injection systems and combustion chambers.

Carbon residue is measured by ASTM D 524, Ramsbottom Carbon Residue of Petroleum Products.

11. Ash—Ash forming materials may be present in diesel fuel either as suspended solids or soluble metallic compounds. Suspended solids and certain dissolved organometallic compounds containing sodium, vanadium, etc., may contribute to injector, fuel pump, and ring wear and also to rapid deterioration of engine parts exposed to high temperatures such as turbochargers and valves. Soluble metallic compounds may also contribute to engine deposits.

ASTM D 482, Ash from Petroleum Products, is commonly used to measure ash.

12. Copper-Strip Corrosion—Corrosion of copper, brass, or bronze parts of the fuel system can result from certain sulfur compounds in diesel fuel. ASTM D 130, Copper Corrosion by Petroleum Products, Copper Strip Test, is an indication of possible difficulties of this type.

13. Water and Sediment Content—Contamination of diesel fuel by water and sediment sometimes occurs during distribution of diesel fuels from the refinery to the users and during the use in diesel equipment. Control of such contamination is a matter requiring constant vigilance. In engines such contamination can lead to filter plugging and injection system wear, and may promote corrosion. ASTM D 1796, Water and Sediment in Fuel Oils by Centrifuge, indicates the amount of these contaminants present in a sample.

14. Oxidation Stability—The products of oxidation of diesel fuel in storage can result in deposits, filter plugging, and lacquering of fuel pump and injector parts.

ASTM D 2274, Stability of Distillate Fuel Oil (Accelerated Method), is available as a measure of this property.

15. Rust Protection—Minimizing rust in fuel distribution and diesel equipment fuel handling systems when engines are idle minimizes filter plugging and wear problems. ASTM D 665, Rust-Preventing Characteristics of Steam-Turbine Oil in the Presence of Water, is a possible measure of rust protection of diesel fuels, although it has not been clearly correlated with field experience.

Relationships Among Properties—Certain properties of distillate diesel fuels, such as volatility, viscosity, gravity, ignition quality, cloud and pour points, and heating value, exhibit interrelationships. Low volatility, straight run fuels from the same crude source usually have lower API gravity and higher viscosity, cetane number, cloud and pour points, and heating value on a Btu/gal or J/L basis than fuels of higher volatility. Further, since diesel fuels are composed of complex mixtures of hydrocarbon compounds, their influence is likewise reflected in the interrelationship of the properties mentioned above. It has been established, through extensive experimental investigation, that certain characteristics of a fuel can be estimated with reasonable accuracy from two or more measured characteristics of the fuel, such as volatility and API gravity.

A useful chart showing the approximate interrelationship between the boiling point temperature for 50% recovered by the ASTM distillation, viscosity at 100°F, API gravity, and calculated cetane index is given in Fig. 4. This chart should not be used for specification purposes.

Diesel Fuel Additives—Commercial diesel fuels may contain a variety of additives to enhance or impart certain desirable properties. Among those which may be found in current fuels are ignition quality improvers, oxidation inhibitors, biocides, rust preventives, metal deactivators, flow improvers, demulsifiers, smoke suppressants, detergent-dispersants, conductivity improvers, and dyes.

Diesel fuel additives are shown by class and function in Table 2. As with any system in which a variety of additives may be used, care should be taken to avoid incompatibilities among additives and unanticipated interactions which may produce undesirable fuel effects.

Additional Pertinent Information—Additional information will be found in the following references:

1. ASTM Standards on Petroleum Products, Parts 23, 24, and 25 by ASTM Committee D-2.
2. "Significance of Tests of Petroleum Products." ASTM Special Technical Publication No. 7C, January 1977.
3. "Significance of Cetane Number of Diesel Fuels." Coordinating Research Council Report 291, August 1955.
4. "Front-End Volatility of Diesel Fuels." Coordinating Research Council Report 276, October 1952.
5. "A Survey of Available Information on the Deposit-Forming Characteristics of Diesel Fuels and Engines." Coordinating Research Council Report 288, March 1955.
6. R. W. Seniff and F. A. Robbins, "Full Scale Field Service Tests of Railroad Diesel Fuels," *SAE* Transactions, Vol. 65 (1957), pp. 46–69.
7. "Investigation of the Effect of Fuel Sulfur Content on the GMC 278A Submarine Diesel Engine Under Snorkel Conditions." Coordinating Research Council Report 298, October 1955.
8. L. Eltinge, D. S. Gray, and H. R. Taliaferro, "How Fuel Composition Affects Deposits and Wear." Paper 403 presented at SAE Fuels and Lubricants Meeting, Tulsa, November 1954.
9. ASTM Symposium on Diesel Fuel Oils, June 29, 1966, Atlantic City, NJ.

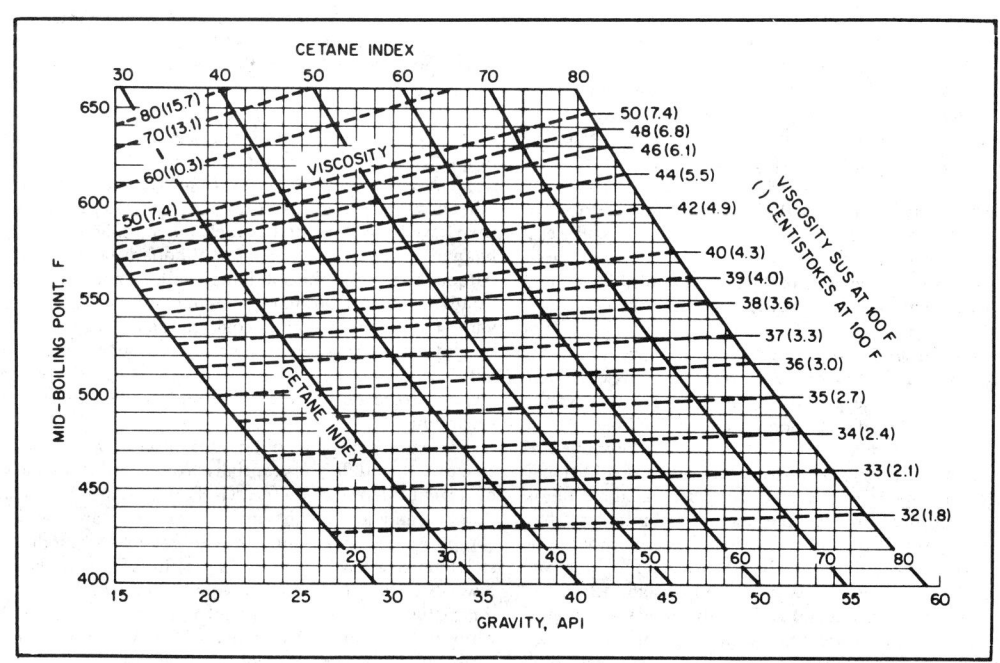

FIG. 4—RELATED PROPERTIES OF DISTILLATE DIESEL FUELS

Adapted from A. J. Blackwood and G. H. Cloud, "Characteristics of Diesel Fuels Influencing Power and Economy." SAE Transactions, Vol. 46, (1940), pp. 49–53.

⊕ TABLE 2—COMMERCIAL DIESEL FUEL ADDITIVES—FUNCTION AND TYPE

Class or Function	Common Additive Type
1. **Ignition Quality Improvers**—Raise Cetane Number thereby promoting faster starts and less white smoke.	Alkyl nitrates.
2. **Oxidation Inhibitors**—Minimize oxidation and gum and precipitate formation, improve storage life.	Alkyl amines and amine-containing complex materials.
3. **Biocides**—Inhibit the growth of bacteria and fungi which feed on hydrocarbons, help prevent filter-clogging caused by these organisms.	Boron compounds, ethers of ethylene glycol, quaternary amine compounds.
4. **Rust Preventives**—Minimize rust formation in fuels systems and storage facilities.	Organic acids and amine salts. A widely-used type is based on dimerized linoleic acid.
5. **Metal Deactivator**—Deactivates copper ions which are powerful oxidation catalysts.	N,N′-disalicylidene-1,2-propane diamine.
6. **Flow Improvers**—Improve low-temperature handling properties of distillate fuels by interfering with formation of wax crystals in the fuel.	Generally consist of polymeric materials such as polyolefins, polyacrylates, polymethacrylates, modified polystyrenes, and ethylene-vinyl acetate copolymers; are often polymeric materials which are absorbed on surface of crystals thereby preventing further growth.
7. **Demulsifiers and Dehazers**—Improve the separation of water from distillate fuels and prevent haze.	Surface-active materials which increase the rate of water/oil separation. Usually quite complex mixtures.
8. **Smoke Suppressants**—Minimize exhaust smoke by catalyzing more complete combustion of carbonaceous materials or by helping to maintain fuel spray patterns.	Catalyst types are generally overbased barium compounds. Maintenance of spray patterns is helped by detergents.
9. **Detergent-Dispersants**—Promote engine fuel system cleanliness, help prevent nozzle deposit formation and injector sticking, interfere with precipitate agglomeration, thus maintaining optimum filtration characteristics.	These are usually surface-active agents. They are often polymeric materials containing amines and other functional groups.
10. **Conductivity Improvers**—Improve dissipation of electrostatic charge.	Amine salts, metallic salts, and polymeric compounds.
11. **Dyes**—Various identification purposes including tax status.	Oil-soluble solid and liquid dyes.

NOTE: Some materials may also be marketed as multifunctional or multipurpose additives, performing more than one of the functions.

10. R. P. Lindeman, D. K. Lawrence, and T. O. Wagner, "Classification of Diesel Fuels." Paper 680467 presented at SAE Mid-Year Meeting, Detroit, May 1968.

11. T. J. Fallon, "Cold Flow Properties of Diesel Fuels." Paper 680537 presented at SAE West Coast Meeting, San Francisco, August 1968.

12. Manual on Hydrocarbon Analysis, ASTM Special Technical Publication 332-A.

13. "Effects of Fuels and Fuel System Designs on Low Temperature Operability of Diesel Vehicles." Coordinating Research Council Report 463, (June 1973).

ALTERNATIVE AUTOMOTIVE FUELS—SAE J1297 APR80

SAE Information Report

Report of the Fuels and Lubricants Technical Committee, approved April 1980.

This report provides information on certain fuels that are being used or have been suggested as alternatives to motor gasoline (SAE J312, Automotive Gasoline) or automotive diesel fuel (SAE J313, Diesel Fuels) for use in spark-ignition or compression-ignition engines. Some of these fuels are derived from petroleum while others are from non-petroleum sources.

Other than gasoline, the only fuel widely used now in spark-ignition engines is propane, sometimes referred to as LPG (liquefied petroleum gas). Other possible alternative fuels include compressed or liquefied methane, or alcohols and ethers, either blended into gasoline or used undiluted. Alcohols are sometimes used in racing engines as the sole fuel, and blends of alcohols in gasoline presently are being offered commercially on a limited basis in some areas. Alcohols are also used in gasoline as octane-improving agents. Except when used at low concentrations as blending agents in gasoline, all these alternative fuels would require engine and/or vehicle modifications, and might also require changes in engine oil characteristics and in distribution and storage facilities.

Only diesel fuel is used in compression-ignition engines at present, and the only practical alternative is a wide-boiling range petroleum fuel. Use of such fuel would likely require some engine and/or vehicle changes but would probably not require any changes in distribution or storage facilities.

Petroleum Gases—Of the petroleum gases, only methane and propane are practical automotive fuels. Methane is available as natural gas and additional methane an be made from non-petroleum sources such as coal. Propane is available in commercial quantities, although the amount is limited to that produced along with natural gas and crude petroleum and as a by-product of refinery reforming and craking processes. Ethane is fully utilized for chemicals manufacture, and butane is already used in gasoline and in some LPG.

Methane and propane have both advantages and disadvantages as automotive fuels relative to gasoline. They have good octane quality, are clean burning, easy to meter, and generally produce lower vehicle exhaust emissions, (Refs. 1, 2) but are more difficult to store than gasoline. Methane can be stored as a compressed gas, or as a liquid if cooled to very low temperatures, but is seldom used in automotive engines because of storage problems. Propane can be stored as a liquid at pressures of approximately 1.4 MPa (200 psi) at normal ambient temperature; however, special equipment is needed to assure good pressure reduction and gasification. Specifications for propane as a motor fuel are given in ASTM D 1835, Standard Specification for Liquefied Petroleum (LP) Gases. (Ref. 3.)

The present supply of methane and/or propane is inadequate for large-scale automotive use. Additional methane could be made from coal or other non-petroleum sources with new facilities but at a cost substantially higher than that of price-regulated natural gas. Ability to increase the domestic supply of propane is very limited, and imports would be necessary.

Alcohols and Other Oxygenated Compounds—Alcohols such as methanol and ethanol have occasionally been blended into motor fuel in the past. Recently they have been suggested as fuels for spark-ignition engines because they can be made from *renewable* or more abundant non-petroleum resources. (Ref. 4.) Although most of the methanol and ethanol manufactured in the United States now is made from petroleum gases, methanol can be synthesized from a variety of materials such as coal, municipal wastes, and biomass. Ethanol can be synthesized by direct fermentation of sugars or from starches and cellulose with chemical or enzyme pretreatment. As a matter of national policy, Brazil has embarked on a program to replace gasoline with ethanol derived from sugar cane.

The various alcohols differ substantially in chemical and physical properties, and they differ substantially from gasoline, as shown for several alcohols of current interest in Table 1. (Refs. 5, 6, 7, 8.) Most alcohols have high octane number and low cetane number, so they are not well-suited as fuels for compression-ignition engines. Compared to gasoline, they all have lower stoichiometric air/fuel (A/F) ratio, higher latent heat of vaporization, and lower Btu content per unit of volume. These differences are most pronounced for the lighter alcohols, methanol and ethanol. Because the stoichiometric A/F ratios differ so much, a vehicle calibrated for gasoline cannot operate satisfactorily on any of these alcohols alone without modification; correspondingly, a vehicle calibrated for alcohol will not operate satisfactorily on gasoline. However, most vehicles can operate on fuel blends containing up to about 2% oxygen by weight without serious problems. Limited tests in laboratory engines suggest that the lighter alcohols burn with slightly greater thermal

TABLE 1—PROPERTIES OF GASOLINE AND OXYGENATED COMPOUNDS

Compound	Chemical Formula	Molecular Weight	Oxygen Content Mass %	Net Heat of Combustion, Btu/Gal[a]	Stoichiometric A/F Ratio, Mass Air/Mass Fuel	Latent Heat of Vaporization, Btu/Gal[a]	Octane Number[b]		Relative U.S. Wholesale Price[c] (Unleaded Regular Gasoline = 1.00)	
							Research	Motor	Volumetric Basis	Energy Basis
Methanol	CH_3OH	32	50	57 000	6.4	3300	112	91	0.70	1.45
Ethanol	C_2H_5OH	46	35	76 000	9.0	2600	111	92	2.20	3.35
Isopropyl Alcohol (IPA)	C_3H_7OH	60	27	86 000	10.3	2100	106	99	1.80	2.40
Tertiary Butyl Alcohol (TBA)	C_4H_9OH	74	22	93 000	11.1	1700	113	110	2.50	3.10
Methyl Tertiary Butyl Ether (MTBE)	$C_4H_9OCH_3$	88	18	94 000	11.7	900	117	102	1.65	2.00
Gasoline (Typical)[d]	C_8H_{15}	111	0	115 000	14.6	800	91 97	82 87	1.00[e] 1.07[f]	1.00[e] 1.07[f]

[a] 1 Btu/gal = 279 J/L.
[b] Octane numbers shown are for pure compounds. Blending octane numbers of the oxygenated compounds in gasoline at low concentration (0–10%) may be substantially different.
[c] Approximate prices obtained from Refs. 7, 8, and 18 (Summer, 1979).
[d] Mixture of C_4 to C_{12} hydrocarbons.
[e] For unleaded regular gasoline.
[f] For unleaded premium gasoline.

efficiency than gasoline. (Refs. 9, 10.) However, because of their relatively low Btu content per unit of volume, use of these alcohols results in higher volumetric fuel consumption. One well-controlled test with 15% methanol in gasoline showed 5% higher fuel consumption than with gasoline. (Ref. 11.) When tested at the same equivalence ratio, vehicle exhaust emissions with alcohol fuels are about the same as with gasoline, (Refs. 12, 13) although emissions of unburned fuel and aldehydes may be higher. (Refs. 10, 11, 12.)

Because of the high latent heat of vaporization and low vapor pressure of alcohols, cars operated on neat alcohols do not start as well at low temperatures, nor do they provide as good driveability as cars operated on gasoline (Ref. 12); although alcohol-gasoline blends generally present no starting difficulties, driveability may still be a problem (Ref. 14.) Because of the lower volumetric energy content of alcohols, more storage volume would be required with alcohols than with gasoline, both in the distribution system and on the vehicle. Methanol, for example, would require about twice the volume.

Use of pure methanol as a motor fuel would present several problems. Saturated vapor in a liquid methanol tank is flammable, so increased fire hazards might be expected. Methanol is also corrosive to some metals such as terneplate (tin-lead coated steel used for fuel tanks), copper, and brass. (Ref. 5.) It is also incompatible with certain elastomers and sealants. Ingestion, inhalation, or skin absorption of methanol can cause blindness and ultimately death. (Refs. 15, 16.) Ethanol and higher alcohols are mild narcotics and irritants. Most of these properties would be retained in proportion to their concentration in blends of alcohols and gasoline used as motor fuel.

All alcohols have strong affinity for water, and water could cause serious problems if it contaminated alcohol-gasoline blends. The low molecular weight alcohols have limited solubility in gasoline compared to their solubility in water. Solubility of alcohols in gasoline decreases with decreasing temperature and generally increases with decreasing gasoline paraffin content. (Ref. 5.)

If the alcohol-gasoline blend is saturated with water, a reduction in ambient temperature may cause the alcohol and gasoline to form two layers. Water concentrations above the saturation limit will cause immediate separation. Methanol gasoline blends are most prone to phase separation when contaminated with water, but ethanol-gasoline blends suffer from the same problem. Distribution facilities would have to be kept free of water if blends of gasoline and the lighter alcohols were utilized. This would be difficult with the distribution facilities presently employed by the petroleum industry. Refinery, pipeline, and terminal tanks now used are not designed to fully exclude rainwater, and many ships and barges are ballasted with water that can contact the fuel being transported. Condensation in storage tanks can cause additional water contamination; this problem may never be fully eliminated.

Of the alcohols added as octane-blending agents, only tertiary butyl alcohol (TBA) is used on a commercial basis at present. It is used at concentrations up to 7% in special refining situations, but the volume dispensed represents less than 0.1% of total motor gasoline. Chemical and physical properties of another oxygenated compound, methyl tertiary butyl ether (MTBE), are also shown in Table 1. (Refs. 17, 18.) Because of its high octane number, some U.S. and European refiners are using MTBE as an octane-improving agent.

Use of 10% ethanol in gasoline (Gasohol) is being promoted as a way to utilize surplus farm commodities and to extend the motor fuel supply through use of renewable resources. Gasohol is available in some service stations now, but the volume of fuel dispensed is small.

Wide-Boiling Range Fuels—Wide-boiling range fuels have been used or proposed as a way to reduce refinery process energy and cost and to increase the supply of petroleum fuels. (Ref. 19.) Such products do not differ fundamentally from current petroleum-based automotive fuels but have much broader distillation limits, encompassing all or parts of both gasoline and diesel fuel. The only fuel of this type now used commercially is JP-4, described by Specification MIL-T-5624, or its civilian counterpart, Jet B, described in ASTM D 1655, Standard Specification for Aviation Turbine Fuels. (Ref. 20.) JP-4 was originally developed as a *maximum-volume* fuel for military use under wartime conditions. Typically, JP-4 is simply distilled from crude oil with few of the major compositional changes required for gasoline. To produce JP-4 in large quantities, some cracking of heavy crude fractions would be necessary, as is now done for gasoline and diesel fuel. The boiling range of JP-4 covers the heavier part of motor gasoline and the lighter part of No. 1 and No. 2 diesel fuels. JP-4 and similar wide-boiling range fuels have low octane number, perhaps 40 Motor ON versus 80–90 Motor ON for gasoline. While it is used primarily in military jet aircraft, JP-4 and fuels of even broader boiling range have been considered for land-based vehicles powered by gas turbines or stratified-charge engines. Even heavier broad-boiling range fuels could be used in diesel engines, although none is used at the present time.

Hydrogen—Hydrogen has been studied as an alternative fuel largely because of its clean-burning characteristics and other possible performance advantages. (Refs. 21, 22, 23, 24, 25.) The use of hydrogen would require revolutionary changes in fuel manufacturing, distribution, and handling systems. At present, application of hydrogen to vehicles as motor fuel is in the research stage, and problems in its use are not clearly defined. Safety and methods for storage of hydrogen fuel will require additional investigation.

Hydrogen has almost three times the heat content of gasoline by weight, but in liquid form has only one-quarter the heat content of gasoline by volume. Hydrogen can be handled as a compressed gas, as a cryogenic liquid, or as a chemically unstable hydride. Although they are being promoted, none of these systems is now practical or economical for vehicle use. High cost of hydrogen, storage problems, and safety considerations represent substantial deterrents to its use as a vehicle fuel.

Fuel from Synthetic Crudes—Synthetic crudes can be prepared from coal, shale and tar sands, and liquid fuels similar to gasoline and diesel fuel can be refined from the crude by currently available processes. However, compared to petroleum, these synthetic crudes are of low quality and must be extensively upgraded before they can be refined into useful products. Characteristics of the synthetic crudes and the products made from them are highly dependent on how they are produced and upgraded. Intense research is underway on processes for production and upgrading, and it is impossible at present to define the methods that will be used ultimately. In any event, upgrading processes and subsequent refining can, at some cost, provide products of any quality desired, including that of present-day gasoline and diesel fuel. Although production of tar sands liquids in Canada is already practical on commercial scale, large quantities of synthetic crudes are not likely to be available in the United States, or even Canada, for many years.

Other Possible Alternatives—Other materials such as ammonia, acetylene, powdered coal, and hydrazine have been suggested as possible alternative automotive fuels (Ref. 2), but their wide-scale use is unlikely.

References

1. A. E. Felt and R. V. Kerley, "Anti-Knock Compounds Applied to LP-Gas." Presented at 65th Annual Meeting of American Society for Testing and Materials, New York, NY, June 23, 1962.
2. N. E. Gallopoulos, "Alternative Fuels for Reciprocating Internal Combustion Engines." Reprinted from "Alternative Hydrocarbon Fuels:

Combustion and Chemical Kinetics." Edited by Craig T. Bowman and Jorgen Birkeland, Vol. 62, 1978, of *Progress in Astronautics and Aeronautics.*

3. American Society for Testing and Materials, 1979 Annual Book of Standards, Part 24, Philadelphia, PA.

4. W. M. Scott, "Alternative Fuels for Automotive Diesel Engines." *Future Automotive Fuels,* Plenum Press, New York, 1977, pp. 263–290.

5. "Alcohols—A Technical Assessment of Their Application as Fuels." Publication No. 4261, July 1976, American Petroleum Institute, Washington, DC.

6. T. O. Wagner, et al., "Practicality of Alcohols as Motor Fuel." Paper 790429 presented at SAE Congress and Exposition, Detroit, February 26–March 2, 1979.

7. "Chemical Marketing Reporter." New York, Schnell Publishing Co., August 20, 1979.

8. "The Oil Daily." New York, The Oil Daily, Inc., August 24, 1979.

9. J. A. Harrington and R. M. Pilot, "Combustion and Emission Characteristics of Methanol." Paper 750420 presented at SAE Automotive Engineering Congress and Exposition, Detroit, February 24-28, 1975.

10. D. L. Hilden and F. B. Parks, "A Single-Cylinder Engine Study of Methanol Fuel—Emphasis on Organic Emissions." Paper 760378 presented at SAE Automotive Engineering Congress and Exposition, Detroit, February 23–27, 1976.

11. A. Koenig, W. Lee, and W. E. Bernhardt, "Technical and Economic Aspects of Methanol as an Automotive Fuel." Paper 760545 presented at SAE Fuels and Lubricants/Powerplant Meeting, St. Louis, June 7–10, 1976.

12. N. D. Brinkman, "Vehicle Evaluation of Neat Methanol—Compromises Among Exhaust Emissions, Fuel Economy, and Driveability." *Proceedings of the Fourth International Symposium on Automotive Propulsion Systems,* NATO/CCMS Vol. 2, 1977, Session 4.

13. N. D. Brinkman, N. E. Gallopoulos, and M. W. Jackson, "Exhaust Emissions, Fuel Economy and Driveability of Vehicles Fueled with Alcohol-Gasoline Blends." SAE Transactions, Vol 84 (1975), Paper 750120.

14. A. W. Crowley, et al., "Methanol-Gasoline Blends-Performance in Laboratory Tests and in Vehicles." Paper 750419 presented at SAE Automotive Engineering Congress and Exposition, Detroit, February 24–28, 1975.

15. H. S. Posner, "Biohazards of Methanol in Proposed New Uses." *Journal of Toxicology and Environmental Health* 1: 153-171, 1975.

16. N. I. Sax, *Dangerous Properties of Industrial Materials,* Third Edition, Van Nostrand Reinhold Co., New York, 1968.

17. R. T. Johnson and B. Y. Taniguchi, "Methyl Tertiary-Butyl Ether-Evaluation as a High Octane Blending Component for Unleaded Gasoline." Presented at Meeting of American Chemical Society, Miami, September 1978.

18. "Chemical and Engineering News." The American Chemical Society, Washington, DC, June 25, 1979.

19. W. T. Tierney, E. M. Johnson, and N. R. Crawford, "Energy Conservation, Optimization of the Vehicle-Fuel-Refinery System." Paper 750673 presented at SAE Fuels and Lubricants Meeting, Houston, June 3–5, 1975.

20. American Society for Testing and Materials, 1979 Annual Book of Standards, Part 23, Philadelphia, PA.

21. F. J. Salzano and C. Braun, "Hydrogen Energy Assessment." March 1977, National Center for Analysis of Energy Systems, Brookhaven National Laboratory, Upton, NY.

22. W. J. D. Escher, "Hydrogen-Fueled Internal Combustion Engine. A Technical Survey of Contemporary U.S. Projects." TEC-75/605, September 1975, Escher Technology Associates, St. Johns, MI.

23. R. B. Cole, "The Performance of Hydrogen-Fueled Reciprocating Engines." *Proceedings of the Fourth International Symposium on Automotive Propulsion Systems,* NATO/CCMS, Vol. 2, 1977, Session 6.

24. R. E. Billings, "Hydrogen's Potential as a Vehicular Fuel for Transportation." Tenth Intersociety Energy Conversion Engineering Conference, Newark, DE, August 18–22, 1975, IEEE Catalogue No. 75 CHO 983-7 TAB.

25. F. B. Simpson, J. H. Lofthouse, and D. R. Swope, "Modification Techniques and Performance Characteristics of Hydrogen-Powered IC Engines—State of the Art, 1975." *Proceedings of the First World Hydrogen Energy Conference,* The University of Miami, Vol. 3, 1976, pp. 6C/75–6C/95.

THE ENGINE OIL PERFORMANCE AND ENGINE SERVICE CLASSIFICATION MAINTENANCE PROCEDURE—SAE J1146

SAE Information Report

Report of Fuels and Lubricants Technical Committee approved June 1976. Editorial change February 1977.

The Engine Oil Performance and Engine Service Classification is designed to keep abreast of changing requirements by redefining existing, or adding new oil categories. To accomplish such action expeditiously requires close coordination among the API, ASTM, and SAE.

Although it is neither possible nor desirable to develop rigid operating rules, the following guidelines are recommended:

1. Any individual, company, or society can request changes in, or additions to, the oil categories. Such requests shall be referred to SAE (Fuels and Lubricants Technical Committee).

2. SAE shall inform both API (Marketing Department—Fuels and Lubricants Committee) and ASTM (Committee D-2, Technical Division B on Automotive Lubricants) of the request and ask for a member from each to serve on an SAE Task Force to study the request. These API and ASTM representatives shall reflect the viewpoints of their respective societies.

3. The SAE Task Force shall consider whether the request is consistent with the overall classification objectives which include:

(a) retention of flexibility,

(b) avoidance of unnecessary and unsound changes or proliferation,

(c) discouragement of the use of obsolete categories.

4. The SAE Task Force shall recommend acceptance or rejection of the request with the concurrence of both the API and ASTM representatives. The recommendation of the Task Force shall be forwarded to API and ASTM.

(a) Criteria for justifying establishment of a new category shall include the following:

1. A reasonable existing or potential market for oil of the proposed category.

2. A service, engine, or requirement not covered by existing categories.

3. A significant difference in performance capability (either increased or decreased) of newly-developed lubricants compared to that of previously available lubricants.

(b) Revision or redefinition of existing classifications is to be discouraged. However, changes such as updating existing test techniques may be desirable, provided that they do not result in significant performance differences.

(c) Revised or new categories must be describable by suitable tests; if suitable techniques are not available, they must be developed. New tests may be suggested by any individual or company, or they may be developed by groups such as CRC.

5. SAE shall letter ballot the Task Force recommendation for either a new or revised category. A consensus negative ballot, as determined by normal SAE policy, would reject the requested revision. An affirmative ballot would initiate a request from SAE: (1) to ASTM to select the test techniques and to develop performance criteria; and (2) to API to develop the user language.

6. ASTM shall appoint a Task Force, that includes at least one representative from API and one from SAE, to select test techniques and to develop performance criteria.

(a) Criteria considered in selecting test techniques shall include:

1. Correlation with service.
2. Precision.
3. An adequate supply of spare parts.
4. Availability of reference oils and test fuel(s).

(b) The development of performance criteria shall include such factors as:

1. Correlation of laboratory and field data.
2. Performance of reference oils.

COORDINATION GUIDE

This guide indicates the coordination required among API, ASTM, and SAE in developing new or revising existing oil categories. It should also be of value in indicating the minimum time required to effect such a change.

Date	SAE	ASTM	API
May–June 19XX	Receive expression of approximate requirements. (Consult with ASTM and API). Appoint Task Force that includes an ASTM and an API representative, to determine if revision is needed.		
October 19XX	Letter ballot Task Force recommendation.		
November 19XX	Request: 1) ASTM to establish test techniques and performance criteria, and 2) API to develop user language.	Establish test techniques and performance criteria.	Develop user language.
May 19XX + 1		Letter ballot test technique and performance criteria recommendations.	
June 19XX + 1			Letter ballot user language recommendations.
August 19XX + 1	Issue final report for publication in SAE Handbook.		

(c) Selection of test techniques and performance criteria shall consider the availability of laboratory and vehicle tests data.

1. Ideally, data should be available from more than one engine make, oil type, and oil performance level.

2. If an inadequate or marginal data bank exists, requests for further data will be made to all concerned and particularly to the group(s) requesting the new category.

7. ASTM shall letter ballot the test techniques and performance criteria. A consensus negative ballot, as determined by normal ASTM policy, would result in reconvening the ASTM Task Force to resolve the problem. If the problem cannot be resolved, ASTM will advise SAE and request further direction. The results of an affirmative ballot would be sent to both API and SAE.

8. API shall appoint a Task Force, that includes at least one representative from ASTM and one from SAE, to develop user language.

9. API shall letter ballot the user language. A consensus negative ballot, as determined by normal API policy, would result in reconvening the API Task Force for further study and development of a user language for reballoting. The results of an affirmative ballot would be sent to ASTM and SAE. If either society is opposed to the user language, API shall be notified and representatives of the three societies will convene to resolve the differences.

10. The SAE Task Force shall issue a final report for inclusion in the SAE Handbook.

A coordination guide for suggested category changes is attached.

NOTE: The above procedures are also recommended for use in establishing categories for other lubricants such as hydraulic/transmission fluids, etc.

THREADS, FASTENERS, AND COMMON PARTS

THREADS, FASTENERS, AND COMMON PARTS

14 Threads
- Screw Threads—SAE J475a 14.01
- Dryseal Pipe Threads—SAE J476a 14.01

15 Fasteners
- Hex Bolts—SAE J105 .... 15.01
- Hi-Head Finished Hex Bolts—SAE J871a 15.04
- Round Head Bolts—SAE J481 .... 15.04
- Slotted and Recessed Head Screws—SAE J478a 15.07
- Slotted Headless Setscrews—SAE J479a 15.40
- Square Head Setscrews—SAE J102 .... 15.41
- Flanged 12-Point Screws—SAE J58 .... 15.42
- Lag Screws—SAE J103 .... 15.46
- Square and Hex Nuts—SAE J104 .... 15.47
- Hexagon High Nuts—SAE J482a 15.53
- Alignment of Nut Slots—SAE J484 .... 15.54
- Crown (Blind, Acorn) Nuts—SAE J483a 15.54
- Spring Nuts—SAE J891a 15.55
- Push-On Spring Nuts—SAE J892a 15.63
- Steel Stamped Nuts of One Pitch Thread Design—SAE J1053 .... 15.64
- Plain Washers—SAE J488 .... 15.68
- Lock Washers—SAE J489b 15.68
- Conical Spring Washers—SAE J773b 15.73
- Nut and Conical Spring Washer Assemblies—SAE J238 .... 15.74
- Torque-Tension Test Procedure for Steel Threaded Fasteners—SAE J174 15.76
- Cotter Pins—SAE J487a 15.77
- Holes in Bolt and Capscrew Shanks and Slots in Nuts for Cotter Pins—SAE J485 . 15.79
- Straight Pins (Solid)—SAE J495 .... 15.79
- Grooved Straight Pins—SAE J494 .... 15.80
- Spring Type Straight Pins—SAE J496 .... 15.82
- Unhardened Ground Dowel Pins—SAE J497 .... 15.84
- Rod Ends and Clevis Pins—SAE J493 .... 15.84
- Rivets and Riveting—SAE J492 .... 15.85
- Blind Rivets—Break Mandrel Type—SAE J1200 .... 15.95
- Bolt and Capscrew Sizes for Use in Construction and Industrial Machinery—SAE J370 15.101
- Hose Clamps—SAE J536b 15.101

16 Ball Studs and Joints
- Ball Studs and Ball Stud Socket Assemblies—SAE J491b 16.01
- Ball Stud and Socket Assembly Test Procedure—SAE J193 SEP79 .... 16.06
- Ball Joints—SAE J490c 16.08
- † Spherical Rod Ends—SAE J1120 SEP79 .... 16.12
- * Metric Spherical Rod Ends—SAE J1259 APR80 .... 16.15

17 Splines
- Involute Splines and Inspections—SAE J498c 17.01
- Parallel Side Splines for Soft Broached Holes in Fittings—SAE J499a 17.01

* New
† Technical revision
§ Editorial change

Serrated Shaft Ends—SAE J500 . . . . 17.03
Shaft Ends—SAE J501 . . . . 17.04
Woodruff Keys—SAE J502 . . . . 17.05
Woodruff Key Slots and Keyways—SAE J503 . . . . 17.06

18 V-Belts
V-Belts and Pulleys—SAE J636c 18.01
Automotive V-Belt Drives—SAE J637b 18.04

19 Springs
Leaf Springs for Motor Vehicle Suspension—Metric Bar Sizes—SAE J1123a 19.01
Leaf Springs for Motor Vehicle Suspension—SAE J510c 19.01
Pneumatic Spring Terminology—SAE J511a 19.08
Helical Compression and Extension Spring Terminology—SAE J1121 . . . . 19.09
Helical Springs: Specification Check Lists—SAE J1122 . . . . 19.14
Rated Spring Capacity—SAE J274 . . . . 19.15

20 Speedometers
Factors Affecting Automotive Odometer-Speedometer Accuracy—SAE J862b 20.01
Speedometers and Tachometers—Automotive—SAE J678e 20.03
Speedometer Test Procedure—SAE J1059 . . . . 20.07
† Electric Tachometer Specification—On Road—SAE J196 JUN80 . . . . 20.08
Electric Tachometer Specification—Off Road—SAE J197 . . . . 20.10
Automatic Vehicle Speed Control—Motor Vehicles—SAE J195 . . . . 20.12

21 Tubing and Fittings
§ Coding System for Identification of Tube, Pipe, and Hose Fittings—SAE J846 MAY80 21.01
† Automotive Tube Fittings—SAE J512 NOV79 . . . . 21.04
§ Spherical and Flanged Sleeve (Compression) Tube Fittings—SAE J246 JUN80 . . . . 21.18
Refrigeration Tube Fittings—SAE J513f 21.26
§ Hydraulic Tube Fittings—SAE J514 APR 80 . . . . 21.46
§ Hydraulic Hose Fittings—SAE J516 SEP79 . . . . 21.72
Hose Push-On Fittings—SAE J1231   21.89
† Hydraulic Hose—SAE J517 JUN80 . . . . 21.92
* Selection, Installation, and Maintenance of Hose and Hose Assemblies—SAE J1273
  SEP79 . . . . 21.104
§ Tests and Procedures for SAE 100R Series Hydraulic Hose and Hose Assemblies—SAE
  J343 JUN80 . . . . 21.105
Flex-Impulse Test Procedure for Hydraulic Hose Assemblies—SAE J1405 . . . . 21.106
Tests and Procedures for High-Temperature Transmission Oil Hose, Lubricating Oil Hose
  and Hose Assemblies—SAE J1019 . . . . 21.107
Hydraulic Flanged Tube, Pipe, and Hose Connections,
  4-Bolt Split Flange Type—SAE J518c 21.107
Hydraulic "O" Ring—SAE J515a 21.112
Formed Tube Ends for Hose Connections—SAE J962b 21.112
Flares for Tubing—SAE J533b 21.113
† Seamless Low Carbon Steel Tubing Annealed for Bending and Flaring—SAE J524 JAN80 21.115
† Welded and Cold Drawn Low Carbon Steel Tubing Annealed for Bending and Flaring—
  SAE J525 JAN80 . . . . 21.115
† Welded Low Carbon Steel Tubing—SAE J526 JAN80 . . . . 21.116
† Brazed Double Wall Low Carbon Steel Tubing—SAE J527 JAN80 . . . . 21.117
† Welded Flash Controlled Low Carbon Steel Tubing Normalized for Bending, Double
  Flaring, and Beading—SAE J356 JAN80 . . . . 21.118
Pressure Ratings for Hydraulic Tubing and Fittings—SAE J1065 . . . . 21.119
† Seamless Copper Tube—SAE J528 JAN80 . . . . 21.121
Metallic Air Brake System Tubing and Pipe—SAE J1149 . . . . 21.122
Nonmetallic Air Brake System Tubing—SAE J844d 21.124
Performance Requirements for SAE J844d Nonmetallic Tubing and Fitting Assemblies
  Used in Automotive Air Brake Systems—SAE J1131 . . . . 21.126
Fuel Injection Tubing—SAE J529b 21.127
Automotive Pipe Fittings—SAE J530a 21.128
Automotive Pipe, Filler, and Drain Plugs—SAE J531a 21.134
Automotive Straight Thread Filler and Drain Plugs—SAE J532a 21.138
Lubrication Fittings—SAE J534c 21.143

14 Threads

SCREW THREADS—SAE J475a — SAE Standard

SAE Regular (NF Series) approved June 1911; SAE Fine (NEF Series) approved January 1915; SAE Special Pitch Series approved June 1926; Coarse (NC Series) approved January 1935; 8, 12, and 16 (National Series) approved January 1935. Last revised by Screw Threads Committee June 1964. Conforms in general to American Standard Unified Screw Threads, (including American National), ASA B1.1.

This report is available as ANSI B1.1 from American National Standards Institute, 1430 Broadway, New York, New York 10018.

DRYSEAL PIPE THREADS—SAE J476a — SAE Standard

Report of Miscellaneous Division approved March 1921 and last revised by Screw Threads Committee June 1961.
Values in Table 1 conform to those in Table 9, Limits on Crest and Root of proposed American Standard, Dryseal Pipe Threads, ASA B2.2

OUTLINE OF STANDARD

General Information
    Table 1—Limits on Crest and Root Truncation
    Table 2—Basic Dimensions (NPTF)
        Dryseal American Standard Taper Pipe Thread
    Table 3—Basic Dimensions (PTF—SAE Short External)
        Dryseal SAE Short External Taper Pipe Thread
    Table 4—Basic Dimensions (PTF—SAE Short Internal)
        Dryseal SAE Short Internal Taper Pipe Thread
    Table 5—Pipe Thread Limits (NPSF)
        Dryseal American Standard Fuel Internal Straight Pipe Thread
    Table 6—Pipe Thread Limit (NPSI)
        Dryseal American Intermediate Internal Straight Pipe Thread
Appendix A—Supplementary Thread Information
    Terminology
    Formulas
    Table A1—Blank Dimensions for External Pipe Thread (NPTF) and (PTF—SAE Short) Dryseal American Standard Taper Pipe Thread and Dryseal SAE Short External Taper Pipe Thread
    Table A2—Drilled Hole Specifications for Straight and Taper Pipe Threads
Appendix B—Chaser and Tap Information
    General Information
    Table B1—Tap and Chaser Teeth Crest and Root Flats
    Table B2—Taper Taps for Dryseal American Standard Fuel Internal Taper Pipe Thread (NPTF) Ground Thread Limits
    Table B3—Taper Taps for Dryseal SAE Short Internal Taper Pipe Thread, Ground Thread Limits
    Table B4—Straight Taps for Dryseal American Standard Fuel Internal Straight Pipe Thread (NPSF) Ground Thread Limits
    Table B5—Straight Taps for Dryseal American Intermediate Internal Straight Pipe Thread (NPSI) Ground Thread Limits
Appendix C—Dryseal Pipe Thread Gaging
    General Information
    Positional Gaging
    Step-Limit Gaging
    Turns Engagement Gaging
    Table C1—Basic Dimensions (L_1 and L_2) Basic-Notch Ring Gages
        Dryseal American Standard Taper Pipe Thread
    Table C2—Basic Dimensions (L_1) Basic-Notch Plug Gages
        Dryseal American Standard Taper Pipe Thread
    Table C3—Basic Dimensions (L_3) Basic-Notch Length Plug Gages
        Dryseal American Standard Taper Pipe Thread
    Table C4—Basic Dimensions (L_1) Step-Limit Thin-Ring Gages
        Dryseal American Standard Taper Pipe Thread
    Table C5—Basic Dimensions (L_2) Step-Limit Full-Ring Gages
        Dryseal American Standard Taper Pipe Thread
    Table C6—Basic Dimensions (L_1) Step-Limit Plug Gages
        Dryseal American Standard Taper Pipe Thread
    Table C7—Basic Dimensions (L_3) Length Step-Limit Plug Gages
        Dryseal American Standard Taper Pipe Thread
    Table C8—Basic Dimensions (L_1 Short) Step-Limit Thin-Ring Gages Dryseal SAE Short Taper Pipe Thread
    Table C9—Basic Dimensions (L_2 Short) Step-Limit Full-Ring Gages Dryseal SAE Short Taper Pipe Thread
    Table C10—Basic Dimensions (L_1 Short) Step-Limit Plug Gages
        Dryseal SAE Short Taper Pipe Thread and Dryseal American Standard Fuel Internal Straight Pipe Thread
    Table C11—Basic Dimensions (L_3 Short) Length Step Limit Plug Gages
        Dryseal SAE Short Taper Pipe Thread
    Table C12—Basic Dimensions (L_1) Step-Limit Plug Gages
        Dryseal American Intermediate Internal Straight Pipe Threads
Appendix D—Special and Fine Dryseal Pipe Threads
    General Information
    Dryseal Special Short Taper Pipe Thread
    Dryseal Special Extra Short Taper Pipe Thread
    Fine Thread Series
    Special Thread Series
    Formulae for Diameter and Length of Thread
    Designations
    Table D1—Basic Dimensions of Dryseal Taper Pipe Thread, Fine, F-PTF
    Table D2—Basic Dimensions of Dryseal Taper Pipe Thread, Special, SPL-PTF, for Thin Wall Nominal Size OD Tubing

GENERAL INFORMATION

Introduction—The Dryseal American Standard Taper Pipe Thread, the Dryseal American Fuel Internal Straight Pipe Thread and the Dryseal American Intermediate Internal Straight Pipe Thread covered by this standard conform with the American Standard ASA-B2.2. The Dryseal SAE-Short Taper Pipe Thread in this standard conforms with the Dryseal American Standard Taper Pipe Thread except for the length of thread, which is shortened for increased clearance and economy of material.

The significant feature of the Dryseal thread is controlled truncation at the crest and root to assure metal to metal contact coincident with or prior to flank contact. Contact at the crest and root prevents spiral leakage and insures pressure-tight joints without the use of a lubricant or sealer.

Lubricants, if not functionally objectionable, may be used to minimize the possibility of galling in assembly.

TRUNCATION—DRYSEAL AMERICAN STANDARD EXTERNAL AND INTERNAL PIPE THREADS
For Pressure-Tight Joints without Lubricant or Sealer

Thread Form—The angle between the flanks of the thread is 60 deg when measured on an axial plane and the line bisecting this angle is perpendicular to the axis of both the taper and straight threads.

Diametral taper of tapered threads is 0.75 in. ±0.06 in. per 12.00 in. of length.

Although the crests and roots of the Dryseal threads are theoretically flat, they may be rounded provided their contour is within the limits specified in Table 1.

Thread Series Symbols—The identification symbols which have been adopted for designating the various Dryseal Pipe Thread Series are as follows:
NPTF for Dryseal American Standard Taper Pipe Thread.
PTF—SAE for Dryseal SAE Short Taper Pipe Thread.
NPSF for Dryseal American Fuel Internal Straight Pipe Thread.
NPSI for Dryseal American Intermediate Internal Straight Pipe Thread.
Where: N stands for American Standard [formerly American (National) Standard].
P stands for Pipe
T stands for Taper
F stands for Fuel
S stands for Straight
I stands for Intermediate

Thread Designation—Dryseal pipe threads are designated by specifying in sequence the nominal size, number of threads per inch, form (Dryseal), and symbol of the thread series.
EXAMPLE: 1/8—27 DRYSEAL NPTF
1/8—27 DRYSEAL PTF—SAE SHORT
1/8—27 DRYSEAL NPSF
1/8—27 DRYSEAL NPSI

Straight Pipe Threads—An assembly with straight internal pipe threads and taper external pipe threads is frequently more advantageous than an all taper thread assembly, particularly in automotive and other allied industries where economy and rapid production are paramount considerations. Dryseal threads are not used on assemblies in which both components have straight pipe threads.

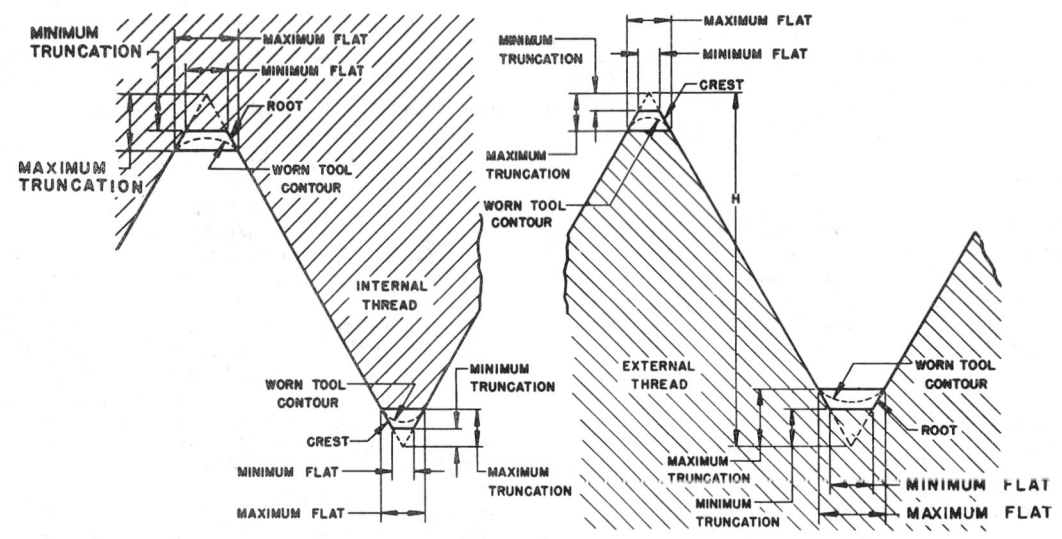

TABLE 1—LIMITS ON CREST AND ROOT TRUNCATION

Threads per in.		Depth of Sharp-V Thread, H	Truncation				Equivalent Width of Flat[a]			
			Min		Max		Min		Max	
		in.	Formula	in.	Formula	in.	Formula	in.	Formula	in.
27	Crest	0.03208	0.047p	0.0017	0.094p	0.0035	0.054p	0.0020	0.108p	0.0040
	Root		0.094p	0.0035	0.140p	0.0052	0.108p	0.0040	0.162p	0.0060
18	Crest	0.04811	0.047p	0.0026	0.078p	0.0043	0.054p	0.0030	0.090p	0.0050
	Root		0.078p	0.0043	0.109p	0.0061	0.090p	0.0050	0.126p	0.0070
14	Crest	0.06186	0.036p	0.0026	0.060p	0.0043	0.042p	0.0030	0.070p	0.0050
	Root		0.060p	0.0043	0.085p	0.0061	0.070p	0.0050	0.098p	0.0070
11-1/2	Crest	0.07531	0.040p	0.0035	0.060p	0.0052	0.046p	0.0040	0.069p	0.0060
	Root		0.060p	0.0052	0.090p	0.0078	0.069p	0.0060	0.103p	0.0090
8	Crest	0.10825	0.042p	0.0052	0.055p	0.0069	0.048p	0.0060	0.064p	0.0080[b]
	Root		0.055p	0.0069	0.076p	0.0095	0.064p	0.0080	0.088p	0.0110

[a] The major diameter of plug gages and minor diameter of ring gages used for gaging dryseal threads shall be truncated an amount sufficient to produce a flat width as shown in Appendix C, Tables C1-1 to C12-1 inclusive.

[b] There is reason to doubt the correctness of the 8 threads per in. flat widths on account of the volume of metal to be displaced.

DRYSEAL AMERICAN STANDARD TAPER PIPE THREAD (NPTF)

This series applies to both the external and internal threads of full length and is suitable for pipe joints in practically every type of service. These threads are generally conceded to be superior for strength and seal. Use of the tapered internal thread in hard or brittle materials having thin sections will minimize trouble from fracture. Dimensional data for (NPTF) threads is given in Table 2. See Appendix D for limitations of assembly of NPTF threads with other series Dryseal pipe threads.

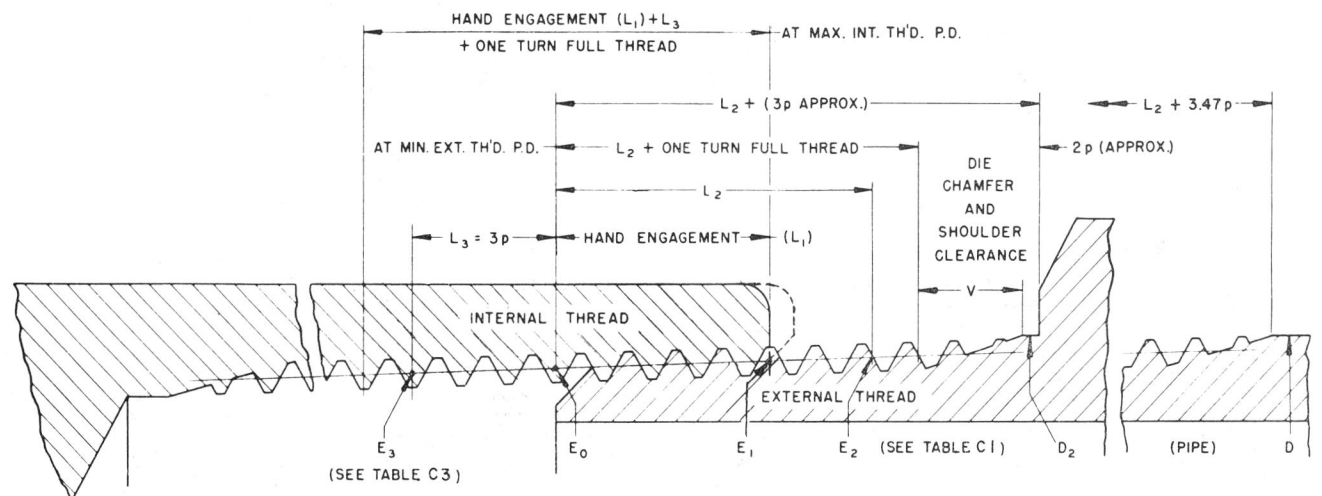

TABLE 2—BASIC DIMENSIONS OF DRYSEAL AMERICAN STANDARD TAPER PIPE THREAD[a]

NPTF Size	Pitch, p	Pitch Diameter at End of External Thread, E_0	Pitch Diameter at End of Internal Thread, E_1	Hand Engagement, L_1		Length of Full Thread,[b] L_2		Vanish Threads V Plus Full Thread Tolerance Plus Shoulder Clearance $(V + 1p + 1/2p)$		Shoulder Length $L_2 + (3p$ Approx)	External Thread for Draw $(L_2 - L_1)$		Length of Internal Full Thread,[c] $(L_1 + L_3)$		OD of Fitting, D_2	OD of Pipe, D
	in.	in.	in.	in.	Thread	in.	Thread	in.	Thread	in.	in.	Thread	in.	Thread	in.	in.
1	2	3	4	5	6	7	8	9	10	11	12	13	14	15	16	17
1/16—27	0.03704	0.27118	0.28118	0.160	4.32	0.2611	7.05	0.1139	3.075	0.3750	0.1011	2.73	0.2711	7.32	0.315	0.3125
1/8—27	0.03704	0.36351	0.37360	0.1615	4.36	0.2639	7.12	0.1112	3.072	0.3750	0.1024	2.76	0.2726	7.36	0.407	0.405
1-4—18	0.05556	0.47739	0.49163	0.2278	4.10	0.4018	7.23	0.1607	2.892	0.5625	0.1740	3.13	0.3945	7.10	0.546	0.540
3/8—18	0.05556	0.61201	0.62701	0.240	4.32	0.4078	7.34	0.1547	2.791	0.5625	0.1678	3.02	0.4067	7.32	0.681	0.675
1/2—14	0.07143	0.75843	0.77843	0.320	4.48	0.5337	7.47	0.2163	3.028	0.7500	0.2137	2.99	0.5343	7.48	0.850	0.840
3/4—14	0.07143	0.96768	0.98887	0.339	4.75	0.5457	7.64	0.2043	2.860	0.7500	0.2067	2.89	0.5533	7.75	1.060	1.050
1—11-1/2	0.08696	1.21363	1.23863	0.400	4.60	0.6828	7.85	0.2547	2.929	0.9375	0.2828	3.25	0.6609	7.60	1.327	1.315
1-1/4—11-1/2	0.08696	1.55713	1.58338	0.420	4.83	0.7068	8.13	0.2620	3.013	0.9688	0.2868	3.30	0.6809	7.83	1.672	1.660
1-1/2—11-1/2	0.08696	1.79609	1.82234	0.420	4.83	0.7235	8.32	0.2765	3.180	1.0000	0.3035	3.49	0.6809	7.83	1.912	1.900
2—11-1/2	0.08696	2.26902	2.29627	0.436	5.01	0.7565	8.70	0.2747	3.159	1.0312	0.3205	3.69	0.6969	8.01	2.387	2.375
2-1/2—8	0.12500	2.71953	2.76216	0.682	5.46	1.1375	9.10	0.3781	3.025	1.5156	0.4555	3.64	1.0570	8.46	2.893	2.875
3—8	0.12500	3.34062	3.38850	0.766	6.13	1.2000	9.60	0.3781	3.025	1.5781	0.4340	3.47	1.1410	9.13	3.518	3.500

[a] See general specifications preceding tables.
For gaging methods, gages, cut thread blanks, taps, drilled hole sizes, hole depths, and full thread lengths, see Appendixes A, B, and C.

[b] External thread tabulated full thread lengths include chamfers not exceeding one and one-half pitches (threads) length.

[c] Internal thread tabulated full thread lengths do not include countersink beyond the intersection of the pitch line and the chamfer cone (gaging reference point).

DRYSEAL SAE SHORT EXTERNAL TAPER PIPE THREAD (PTF—SAE SHORT EXTERNAL)

For assembly with Dryseal American Intermediate Internal Straight (Table 6) or Dryseal American Standard Taper (Table 2) Pipe Threads

External threads of this series conform in all respects with the NPTF threads except that the full thread length has been shortened by eliminating one thread from the small end. These threads are primarily intended for assembly with NPSI internal threads but may also be used with NPTF internal threads. They are not designed for and at extreme tolerance limits may not assemble with PTF—SAE Short or NPSF internal threads. Dimensional data for PTF—SAE Short External Threads is given in Table 3. See Appendix D for limitations of assembly of PTF—SAE Short external threads with other series Dryseal pipe threads.

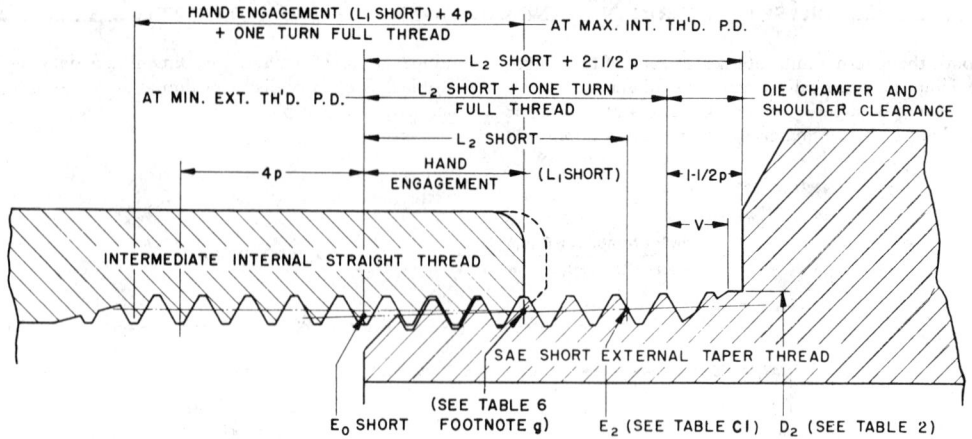

TABLE 3—BASIC DIMENSIONS OF DRYSEAL SAE SHORT EXTERNAL TAPER PIPE THREAD [a]

PTF—SAE Short Size	Pitch, p	Pitch Diameter at End of External Thread, E_0 Short	L_1		Hand Engagement, L_1 Short		Length of Full Thread, L_2 Short [b]		Vanish Threads V Plus Full Thread Tolerance Plus Shoulder Clearance $(V + 1p + 1/2 p)$		Min Shoulder Length (L_2 Short $+ 2$-$1/2$p)		External Thread for Draw (L_2 Short— L_1 Short)		Length of Internal Full Thread, [c] (L_1 Short + 4p)	
	in.	in.	in.	Thread	in.	Thread	in.	Thread	in.	Thread	in.	in.	Thread	in.	Thread	
1	2	3	4	5	6	7	8	9	10	11	12	13	14	15	16	
1/16—27	0.03704	0.27349	0.160	4.32	0.1230	3.32	0.2241	6.05	0.0926	2.50	0.3167	0.1011	2.73	0.2711	7.32
1/8—27	0.03704	0.36582	0.1615	4.36	0.1244	3.36	0.2268	6.12	0.0926	2.50	0.3194	0.1024	2.76	0.2726	7.36
1/4—18	0.05556	0.48086	0.2278	4.10	0.1722	3.10	0.3462	6.23	0.1389	2.50	0.4851	0.1740	3.13	0.3945	7.10
3/8—18	0.05556	0.61548	0.240	4.32	0.1844	3.32	0.3522	6.34	0.1389	2.50	0.4911	0.1678	3.02	0.4067	7.32
1/2—14	0.07143	0.76289	0.320	4.48	0.2486	3.48	0.4623	6.47	0.1786	2.50	0.6409	0.2137	2.99	0.5343	7.48
3/4—14	0.07143	0.97214	0.339	4.75	0.2676	3.75	0.4743	6.64	0.1786	2.50	0.6528	0.2067	2.89	0.5533	7.75
1—11-1/2	0.08696	1.21906	0.400	4.60	0.3130	3.60	0.5958	6.85	0.2174	2.50	0.8132	0.2828	3.25	0.6609	7.60
1-1/4—11-1/2	0.08696	1.56256	0.420	4.83	0.3330	3.83	0.6198	7.13	0.2174	2.50	0.8372	0.2868	3.30	0.6809	7.83
1-1/2—11-1/2	0.08696	1.80152	0.420	4.83	0.3330	3.83	0.6365	7.32	0.2174	2.50	0.8539	0.3035	3.49	0.6809	7.83
2—11-1/2	0.08696	2.27445	0.436	5.01	0.3490	4.01	0.6695	7.70	0.2174	2.50	0.8869	0.3205	3.69	0.6969	8.01
2-1/2—8	0.12500	2.72734	0.682	5.46	0.5570	4.46	1.0125	8.10	0.3125	2.50	1.3250	0.4555	3.64	1.0570	8.46
—8	0.12500	3.34844	0.766	6.13	0.6410	5.13	1.0750	8.60	0.3125	2.50	1.3875	0.4340	3.47	1.1410	9.13

[a] See general specifications preceding tables. For gaging methods, gages, cut thread blanks, taps, drilled hole sizes, hole depths, and full thread lengths, see Appendixes A, B, and C.

[b] External thread tabulated full thread lengths include chamfers not exceeding one and one-half pitches (threads) length.

[c] Internal thread tabulated full thread lengths do not include countersink beyond the intersection of the pitch line and the chamfer cone (gaging reference point).

DRYSEAL SAE SHORT INTERNAL TAPER PIPE THREAD (PTF—SAE SHORT INTERNAL)

For assembly with American Standard External Taper Pipe Thread (Table 2)

Internal Threads of this series conform in all respects with the NPTF threads except that the full thread length has been shortened by eliminating on thread from the large end. These threads are primarily intended for assembly with NPTF external threads. They are not designed for and at extreme tolerance limits may not assemble with PTF—SAE Short external threads. Dimensional data for PTF—SAE Short Internal Threads is given in Table 4. See Appendix D for limitations of assembly of PTF—SAE Short internal threads with other series Dryseal pipe threads.

Trouble-free assemblies and pressure-tight joints without the use of lubricant or sealer can best be assured where both components are threaded with NPTF (full length) threads. This should be considered before specifying PTF—SAE Short External or Internal Thread.

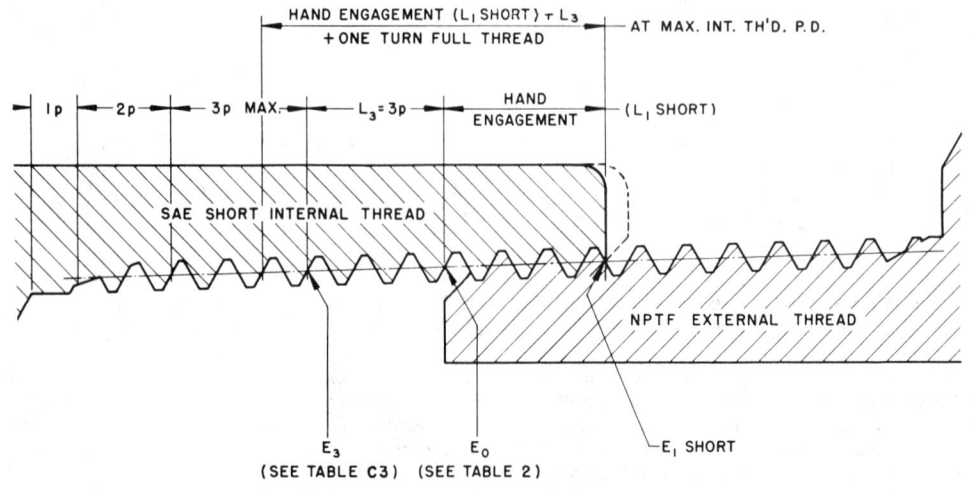

TABLE 4—BASIC DIMENSIONS OF DRYSEAL SAE SHORT INTERNAL TAPER PIPE THREAD [a]

PTF—SAE Short Size	Pitch p	Pitch Diameter at End of Internal Thread E_1 Short	L_1		Hand Engagement [b], L_1 Short		Length of Internal Full Thread [b] (L_1 Short + L_3)		Hole Depth for SAE Short Tap (Table B3)
	in.	in.	in.	Thread	in.	Thread	in.	Thread	in.
1	2	3	4	5	6	7	8	9	10
1/16—27	0.03704	0.27887	0.160	4.32	0.1230	3.32	0.2341	6.32	0.4564
1/8—27	0.03704	0.37129	0.1615	4.36	0.1244	3.36	0.2356	6.36	0.4578
1/4—18	0.05556	0.48815	0.2278	4.10	0.1722	3.10	0.3389	6.10	0.6722
3/8—18	0.05556	0.62354	0.240	4.32	0.1844	3.32	0.3511	6.32	0.6844
1/2—14	0.07143	0.77397	0.320	4.48	0.2486	3.48	0.4629	6.48	0.8915
3/4—14	0.07143	0.98441	0.339	4.75	0.2676	3.75	0.4819	6.75	0.9105
1—11-1/2	0.08696	1.23320	0.400	4.60	0.3130	3.60	0.5739	6.60	1.0956
1-1/4—11-1/2	0.08696	1.57795	0.420	4.83	0.3330	3.83	0.5939	6.83	1.1156
1-1/2—11-1/2	0.08696	1.81691	0.420	4.83	0.3330	3.83	0.5939	6.83	1.1156
2—11-1/2	0.08696	2.29084	0.436	5.01	0.3490	4.01	0.6099	7.01	1.1316
2-1/2—8	0.12500	2.75435	0.682	5.46	0.5570	4.46	0.9320	7.46	1.6820
3—8	0.12500	3.38069	0.766	6.13	0.6410	5.13	1.0160	8.13	1.7660

[a] See general specification preceding tables. For gaging methods, gages, taps, drilled hole sizes, hole depths, and full thread lengths, see Appendixes A, B, and C.

[b] Internal thread tabulated full thread lengths do not include countersink beyond the intersection of the pitch line and the chamfer cone (gaging reference point).

DRYSEAL AMERICAN STANDARD FUEL INTERNAL STRAIGHT PIPE THREAD (NPSF)

For assembly with Dryseal American Standard External Taper Pipe Thread (Table 2)

Threads of this series are straight (cylindrical) instead of tapered. They are generally used in soft or ductile materials which will adjust at assembly to the taper of external threads but may also be used in hard or brittle materials where the section is heavy. These threads are primarily intended for assembly with full length NPTF external taper threads. Dimensional data for NPSF threads is given in Table 5. See Appendix D for limitations of assembly of NPSF internal threads with other series Dryseal pipe threads.

TABLE 5—DRYSEAL AMERICAN STANDARD FUEL INTERNAL STRAIGHT PIPE THREAD LIMITS [a]

NPSF Size	Pitch Diameter [b]		Minor Diameter [c]	Desired Min Length of Full Thread [g]	
	Max [d,e]	Min [e,f]	Min	in.	Thread
1	2	3	4	5	6
1/16—27	0.2803	0.2768	0.2482	0.31	8.44
1/8—27	0.3727	0.3692	0.3406	0.31	8.44
1/4—18	0.4904	0.4852	0.4422	0.47	8.44
3/8—18	0.6257	0.6205	0.5776	0.50	9.00
1/2—14	0.7767	0.7700	0.7133	0.66	9.19
3/4—14	0.9872	0.9805	0.9238	0.66	9.19
1—11-1/2	1.2365	1.2284	1.1600	0.78	8.98

Notes for Table 5

[a] See general specifications preceding tables. For gaging methods, gages, taps, drilled hole sizes, hole depths, and full thread lengths, see Appendixes A, B, and C.

[b] The pitch diameter of the tapped hole as indicated by the taper plug gage is slightly larger than the values given due to the gage having to enter approximately 3/8 turn to engage first full thread.

[c] As the Dryseal American Standard pipe thread form is maintained, the major and minor diameters of the internal thread vary with the pitch diameter.

[d] Column 2 is the same as the E_1 pitch diameter of thread at large end of internal thread (Table 2) minus (small) 3/8 thread taper.

[e] Taps specified in Table B4 produce tapped holes to the above limits in cast iron, steel, and brass. In zinc and similar soft metals, they produce tapped holes approximately 0.001 smaller. Plug-gage turns engagement should be reduced accordingly.

[f] Column 3 is Column 2 reduced by 1-1/2 turns.

[g] Internal thread tabulated full thread lengths do not include countersink beyond the intersection of the pitch line and the chamfer cone (gaging reference point).

DRYSEAL AMERICAN INTERMEDIATE INTERNAL STRAIGHT PIPE THREAD (NPSI)

For assembly with Dryseal SAE Short External Taper (Table 3) or American Standard Taper Pipe Thread (Table 2)

Threads of this series are straight (cylindrical) instead of tapered. They are generally used in hard or brittle materials where the section is heavy and where there is little expansion at assembly with the external taper threads. These threads are primarily intended for assembly with PTF—SAE Short External Taper Threads, but will also assemble with full length NPTF External Taper Threads. Dimensional data for NPSI threads is given in Table 6. See Appendix D for limitations of assembly of NPSI internal threads with other series Dryseal pipe threads.

TABLE 6—DRYSEAL AMERICAN INTERMEDIATE INTERNAL STRAIGHT PIPE THREAD LIMITS [a]

NPSI Size	Pitch Diameter [b]		Minor Diameter [c]	Desired Min Length of Full Thread [g]	
	Max [d,e]	Min [e,f]	Min	in.	Thread
1	2	3	4	5	6
1/16—27	0.2826	0.2791	0.2505	0.31	8.44
1/8—27	0.3750	0.3715	0.3429	0.31	8.44
1/4—18	0.4938	0.4886	0.4457	0.47	8.44
3/8—18	0.6292	0.6240	0.5811	0.50	9.00
1/2—14	0.7812	0.7745	0.7180	0.66	9.19
3/4—14	0.9917	0.9850	0.9283	0.66	9.19
1—11-1/2	1.2420	1.2338	1.1655	0.78	8.98

Notes for Table 6

[a] See general specifications preceding tables. For gaging methods, gages, taps, drilled hole sizes, hole depths, and full thread lengths see Appendixes A, B, and C.

[b] The pitch diameter of the tapped hole as indicated by the taper plug gage is slightly larger than the values given due to the gage having to enter approximately 3/8 turn to engage first full thread.

[c] As the Dryseal American Standard pipe thread form is maintained, the major and minor diameters of the internal thread vary with the pitch diameter.

[d] Column 2 is the E_1 pitch diameter of thread at large end of internal thread (Table 2) plus (large) 5/8 thread taper.

[e] Taps specified in Table B5 produce tapped holes to the above limits in cast iron, steel, and brass. In zinc and similar soft metals, they produce tapped holes approximately 0.001 smaller. Plug-gage turns engagement should be reduced accordingly.

[f] Column 3 is Column 2 reduced by 1-1/2 turns.

[g] Internal thread tabulated full thread lengths do not include countersink beyond the intersection of the pitch line and the chamfer cone (gaging reference point).

APPENDIX A—SUPPLEMENTARY THREAD INFORMATION

Terminology—For definitions of terms relating to size of parts, geometrical elements, or dimensions of threads see SAE Standards Screw Threads, Appendix A—Terminology.

DRYSEAL AMERICAN STANDARD AND SAE SHORT EXTERNAL TAPER PIPE THREAD BLANKS, CUT THREADS

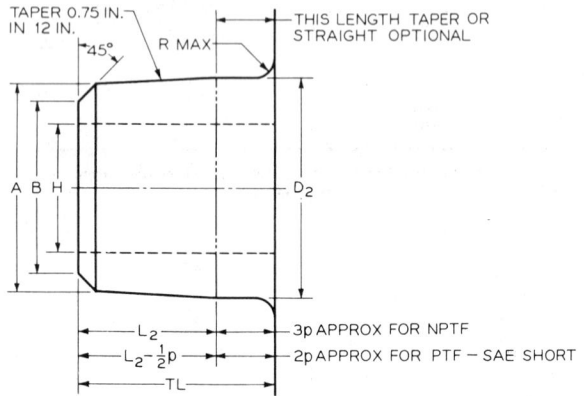

Formulas for Diameter and Length of Thread

Basic diameter and length of thread for different sizes given in Tables 2, 3, and 4, are based on the following formulas:

Basic pitch diameter of thread at small end of NPTF External Thread.

$$E_o = D - (0.05D + 1.1)p$$

Basic pitch diameter of thread at small end of PTF—SAE Short External Thread.

$$E_o \text{ Short} = D - (0.05D + 1.037)p$$

Basic pitch diameter of thread at large end of NPTF Internal Thread.

$$E_1 = E_o + (0.0625 \times L_1 \text{ Basic})$$

Basic pitch diameter of thread at large end of PTF—SAE Short Internal Thread.

$$E_1 \text{ Short} = E_o + (0.0625 \times L_1 \text{ Short})$$

Basic length of NPTF external full and effective length thread.

$$L_2 = (0.8D + 6.8)p$$

Basic length of PTF—SAE Short external full and effective length thread.

$$L_2 \text{Short} = (0.8D + 5.8)p$$

Basic length of NPTF internal full and effective length thread = L_1 Basic + L_3

Basic length of PTF—SAE Short internal full and effective length thread = L_1 Short + L_3

where D = outside diameter of pipe
P = pitch of thread in inches
NPSG (for oil and grease cup) is Dryseal American Standard Pipe Thread Form—use NPSF tap drill sizes.

The drilled hole sizes given above for Dryseal straight and taper internal pipe threads are the diameters produced by drills which are closest to the minimum minor diameters as shown in Table A2.

They represent the diameters of the holes which would be cut with a twist drill correctly ground when drilling a material without tearing or flow of metal. This is approximately the condition obtained when a correctly sharpened twist drill is cutting a hole in SAE 1112 or 1113 steel, or SAE 72 brass. When Dryseal taps are used, these holes produce an acceptable pipe thread with the required thread height.

When flat drills are used, the width of the cutting edge may have to be adjusted to produce a hole of the required diameter.

When hard metals and other similar materials are to be drilled and tapped, it may be found necessary to use a drill of slightly smaller diameter to produce a hole of a size that will make it possible for the tap to cut an acceptable pipe thread with the required thread height.

When soft metals and other similar materials are to be drilled and tapped, it may be found necessary to use a drill of slightly larger diameter to produce a hole of a size that will allow for a flow of the metal or material without loading the tap or tearing the material and make it possible for the tap to produce an acceptable pipe thread with the required thread height.

TABLE A1—DIMENSIONS OF DRYSEAL AMERICAN STANDARD EXTERNAL TAPER PIPE THREAD BLANKS (CUT THREADS)

Size	OD at Large End NPTF at L_2 Length D_2 PTF-SAE Short at $L_2 - \frac{1}{2}p$ Length (Basic Thread One Turn Large with Max Truncation) +0.003 −0.000	OD at Small End, A		Chamfer Dia[b], B (Minor Dia[b] at Small End)	Min Length from Small End to Shoulder, TL		Corner Radius, R Max	Recommended Hole Size[a], H
		NPTF (Basic Thread Two Turns Large with Max Truncation) +0.003 −0.000	PTF-SAE Short (Basic Thread 2-1/2 Turns Large with Max Truncation) +0.003 −0.000		NPTF $L_2 +$ (3 p Approx)	PTF-SAE Short L_2 Short + (2-1/2 p Approx)		
1/16—27	0.315	0.301	0.302	0.23	0.38	0.3167	0.03	0.12
1/8—27	0.407	0.393	0.394	0.32	0.38	0.3194	0.03	0.19
1/4—18	0.546	0.523	0.525	0.42 +0.00	0.56	0.4851	0.06	0.28
3/8—18	0.681	0.658	0.660	0.55	0.56	0.4911	0.06	0.41
1/2—14	0.850	0.820	0.822	0.68 −0.02	0.75	0.6409	0.08	0.56
3/4—14	1.060	1.029	1.031	0.89	0.75	0.6528	0.08	0.72
1 —11-1/2	1.327	1.289	1.292	1.12	0.94	0.8132	0.09	0.94
1-1/4 —11-1/2	1.672	1.633	1.636	1.46 +0.00	0.97	0.8372	0.09	1.25
1-1/2 —11-1/2	1.912	1.872	1.875	1.70	1.00	0.8539	0.09	1.47
2 —11-1/2	2.387	2.345	2.348	2.17 −0.03	1.03	0.8869	0.09	1.94
2-1/2—8	2.893	2.829	2.833	2.59	1.52	1.3250	0.12	2.31
3 —8	3.518	3.450	3.454	3.21	1.58	1.3875	0.12	2.91

[a] The hole sizes recommended represent a desirable maximum, strength of wall being considered. However, as considerations other than wall strength frequently control the hole size in specific applications, the recommendations should not be construed as a requirement of this SAE Standard.

[b] External pipe threads shall be chamfered from a diameter (rounded to a two-place decimal) obtained by subtracting 0.016 in. for sizes below 1 in. and 0.025 in. for larger sizes from the minimum minor diameter at small end to produce a length of chamfered or partial thread equivalent to 1 to 1-1/2 times the pitch (rounded to a three-place decimal).

PIPE-THREAD DRILLED HOLE SIZES FOR DRYSEAL AMERICAN STANDARD INTERNAL PIPE THREAD

It should be understood that this table of drilled hole sizes is intended to help only the occasional user of drills in the application of this SAE Standard. When internal pipe threads are produced in larger quantities in a particular type of material and with specially designed machinery, it may be found to be more advantageous to use a drilled hole size not given in the table, even one requiring a nonstandard diameter drill size.

TABLE A2—PIPE-THREAD DRILLED HOLE SIZES

Size	Straight Pipe Thread						Taper Pipe Thread							Countersink
	Fuel (NPSF)		Intermediate (NPSI)		Desired Length of Full Thread	Hole Depth for Plug End Tap, Tables B4 and B5	NPTF (Not Reamed)				NPTF[d] (Taper Reamed)		Desired Length of Full Thread	Hole Depth for Standard Tap, Table B2	90 Deg x Dia[d]
							2 FF Thread[a]		4 FF Thread[a]						
	Minor Dia[c]	Drilled Hole Size	Minor Dia[c]	Drilled Hole Size			Minor Dia 2 Thread Small from Large End	Drilled Hole Size	Minor Dia 4 Thread Small from Large End	Drilled Hole Size	Desired Minor Dia at Small End	Drilled Hole Size			
	Min	+0.003 −0.001	Min	+0.003 −0.001	Min		Min	+0.003 −0.001	Min	+0.003 −0.001	Min	+0.003 −0.001	Min		
1/16—27	0.2482	0.2500	0.2505	0.2500	0.31	0.47	0.2480	0.2460	0.2434	0.2420	0.2356	0.2344	0.31	0.56	0.33
1/8 —27	0.3406	0.3437	0.3429	0.3437	0.31	0.47	0.3403	0.3390	0.3357	0.3320	0.3279	0.3281	0.31	0.56	0.42
1/4 —18	0.4422	0.4440	0.4457	0.4440	0.47	0.72	0.4417	0.4375	0.4348	0.4300	0.4241	0.4219	0.47	0.81	0.55
3/8 —18	0.5776	0.5781	0.5811	0.5781	0.50	0.72	0.5771	0.5781	0.5702	0.5700	0.5587	0.5625	0.50	0.81	0.69
1/2 —14	0.7133	0.7187	0.7180	0.7187	0.66	0.94	0.7127	0.7031	0.7038	0.6960	0.6873	0.6875	0.66	1.06	0.85
3/4 —14	0.9238	—	0.9283	—	0.66	0.94	0.9232	0.9219	0.9143	0.9062	0.8976	0.8906	0.66	1.06	1.06
1 —11-1/2	1.1600	—	1.1655	—	0.78	1.16	1.1593	1.1562	1.1484	1.1406	1.1290	1.1250	0.78	1.25	1.34
1-1/4 —11-1/2	—	—	—	—	—	—	1.5041	1.5000	1.4932	1.4844	1.4725	1.4687	0.81	1.31	1.58
1-1/2 —11-1/2	—	—	—	—	—	—	1.7430	1.7344	1.7321	1.7188	1.7115	1.7031	0.81	1.31	1.92
2 —11-1/2	—	—	—	—	—	—	2.2170	2.2187	2.2061	2.2031	2.1844	2.1875	0.81	1.31	2.39
2-1/2 —8	—	—	—	—	—	—	2.6488	2.6406	2.6336	2.6250	2.5983	2.5937	1.25	1.84	2.89
3 —8	—	—	—	—	—	—	3.2751	3.2656	3.2595	3.2500	3.2194	3.2187	1.34	1.91	3.52

Countersink column tolerances:
- 1/16—27 through 3/8—18: +0.02 / −0.00
- 1/2—14 through 1-1/4—11-1/2: +0.02 / −0.00
- 1-1/2—11-1/2 through 3—8: +0.03 / −0.00

[a] NPTF (not reamed) drilled hole sizes are recommended for taper tapping without reaming. [NPTF (2 FF thread)] minimum minor diameter two threads small from large end and closest drilled hole sizes are recommended only for low pressure use. [NPTF (4 FF thread)] minimum minor diameter four threads small from large end and closest drilled hole sizes are recommended for all pressures. Thread lengths so produced are designated "Effective Thread" on drawings.

[b] NPTF (taper reamed) drilled hole sizes are recommended for taper reaming before tapping. They also are used without taper reaming by taper drilling or allowing the tap to act as a reamer. Thread lengths so produced are designated "Full or Complete Thread" on drawings.

[c] Minimum minor diameter for internal straight pipe threads is based upon minimum pitch diameter and minimum truncation and will vary with the pitch diameter.

[d] Internal pipe threads shall be countersunk 90 deg included angle to a diameter (rounded to a two-place decimal) obtained by adding 0.016 in. for sizes below 1 in. and 0.025 in. for larger sizes to the maximum major diameter at large end.

APPENDIX B—DRYSEAL PIPE THREAD TAPS AND CHASERS

General Information—While production taps will usually be purchased to specification, occasions may arise requiring adaptations of taps, dies, or chasers at hand.

American Standard Taper Taps, Dies, or Chasers (NPT) may be adapted for producing the Dryseal American Standard Taper Pipe Threads (NPTF) by truncating the outside diameter of taps and the inside diameter of dies or chasers the amount necessary to obtain flats shown in Table B1 for producing the limits on the product specified in Table 1. The pitch diameter of taps and dies or chasers so modified will remain standard. American Standard Coupling Straight Pipe Taps (NPSC) used for tapping the American Standard Coupling Straight Pipe Thread (NPSC), with the exception of one size only, may be adapted for tapping the Dryseal American Intermediate Straight Pipe Thread (NPSI) by truncating the outside diameter the amount necessary to obtain flats shown in Table B3 for producing the limits on the product shown in Table 1. The exception is the 1/4-18 size which has a minimum pitch diameter under that required. With the exception of one size only, taps designed to other standards cannot be adapted for tapping the Dryseal American Fuel Straight Pipe Thread (NPSF). The exception is the 1/8-27 American Standard Grease Fitting Tap (NPSG) which, if made in conformity with Tap Manufacturers' Standards of 1939 to 1941 issue, may be used without change for tapping of the 1/8-27 Dryseal American Fuel Straight Pipe Thread (NPSF).

Chamfer—2 to 3 threads.

Lead Tolerance—A maximum lead error of ±0.0005 in. in 1-in. of thread is permitted.

Angle Tolerance—Error in half angle of ±30 min is permitted.

Taper Tolerance—A maximum taper error of ±1/32 in. per ft is permitted.

Marking—In addition to regular markings, Dryseal American Standard Taper Taps will be marked NPTF.

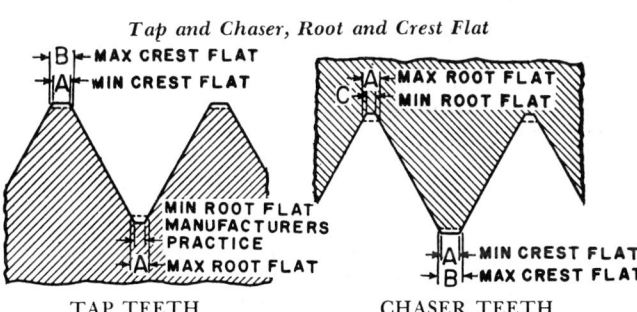

Tap and Chaser, Root and Crest Flat

TABLE B1—WIDTH OF FLATS

Threads per in.	A	B	C
27	0.004	0.005	0.002
18	0.005	0.006	0.003
14	0.005	0.006	0.003
11-1/2	0.006	0.008	0.004
8	0.008	0.010	0.006[a]

[a] There is reason to doubt the correctness of the 8 threads per in. flat widths because of the volume of metal to be displaced.

TABLE B2—DRYSEAL TAPER PIPE TAPS FOR DRYSEAL AMERICAN STANDARD INTERNAL TAPER PIPE THREAD, GROUND THREAD LIMITS

PTF Size	Gage[a] Measure		Major Dia Flat		Minor Dia Flat[b]
	Min	Max	Min	Max	Max
1/16—27	0.250	0.375	0.004	0.005	0.004
1/8 —27	0.250	0.375	0.004	0.005	0.004
1/4 —18	0.397	0.521	0.005	0.006	0.005
3/8 —18	0.392	0.516	0.005	0.006	0.005
1/2 —14	0.517	0.641	0.005	0.006	0.005
3/4 —14	0.503	0.627	0.005	0.006	0.005
1 —11-1/2	0.584	0.772	0.006	0.008	0.006
1-1/4 —11-1/2	0.592	0.780	0.006	0.008	0.006
1-1/2 —11-1/2	0.605	0.793	0.006	0.008	0.006
2 —11-1/2	0.573	0.761	0.006	0.008	0.006
2-1/2 —8[c]	0.831	1.019	0.008	0.010	0.008
3 —8[c]	0.831	1.019	0.008	0.010	0.008

[a] Distance small end of tap projects beyond face of American Standard Thin-Ring Gage.
[b] Minor diameter as specified or sharper.
[c] There is reason to doubt the correctness of the 8 threads per in. flat widths because of the volume of metal to be displaced.

TABLE B3—DRYSEAL TAPER PIPE TAPS FOR DRYSEAL SAE SHORT INTERNAL TAPER PIPE THREAD, GROUND THREAD LIMITS

PTF—SAE Short Size	Gage[a] Measure		Major Dia Flat		Minor Dia Flat[b]
	Min (6p)	Max (7p)	Min	Max	Max
1/16—27	0.222	0.259	0.004	0.005	0.004
1/8 —27	0.222	0.259	0.004	0.005	0.004
1/4 —18	0.333	0.389	0.005	0.006	0.005
3/8 —18	0.333	0.389	0.005	0.006	0.005
1/2 —14	0.429	0.500	0.005	0.006	0.005
3/4 —14	0.429	0.500	0.005	0.006	0.005

[a] Distance small end of tap projects beyond face of American Standard Thin-Ring Gage.
[b] Minor diameter as specified or sharper.

Chamfer—1½ to 2 threads.
Marking—In addition to regular markings, Dryseal SAE Short Taper Taps will be marked PTF—SAE Short.
Tolerances—Same as for Dryseal American Standard Taper Taps.

TABLE B4—DRYSEAL STRAIGHT PIPE TAPS FOR DRYSEAL AMERICAN FUEL INTERNAL STRAIGHT PIPE THREAD, GROUND THREAD LIMITS

NPSF Size	Pitch Diameter			Major Dia Flat[a]		Minor Dia Flat[b]	Major Dia		
	Basic	Min	Max	Min	Max	Max	Basic	Min	Max
1/16—27	0.2812	0.2772	0.2777	0.004	0.005	0.004	0.3108	0.3008	0.3018
1/8 —27	0.3748	0.3696	0.3701	0.004	0.005	0.004	0.4044	0.3932	0.3942
1/4 —18	0.4899	0.4859	0.4864	0.005	0.006	0.005	0.5343	0.5239	0.5249
3/8 —18	0.6270	0.6213	0.6218	0.005	0.006	0.005	0.6714	0.6593	0.6603
1/2 —14	0.7784	0.7712	0.7717	0.005	0.006	0.005	0.8356	0.8230	0.8240
3/4 —14	0.9889	0.9817	0.9822	0.005	0.006	0.005	1.0460	1.0335	1.0345
1 —11-1/2	1.2386	1.2295	1.2305	0.006	0.008	0.006	1.3082	1.2933	1.2943

[a] For reference only. Major-diameter flats specified may be slightly larger or smaller with extreme combinations of pitch diameter, major diameter, and half angle.
[b] Minor diameter as specified or sharper.

Chamfer—Plug end 3 to 5 threads, intermediate end 2 to 3 threads, bottom end 1½ to 2 threads.
Lead Tolerance—A maximum lead error of ±0.0005 in. in 1 in. of thread is permitted.
Angle Tolerance—Error in half angle of ±30 min permitted.
Marking—In addition to regular markings, Dryseal American Fuel Straight Pipe Taps will be marked NPSF.
Maximum Major Diameter equals minimum pitch diameter plus sharp-V thread height minus twice tool crest minimum truncation.
Minimum Major Diameter equals maximum major diameter minus tolerance.
Maximum Pitch Diameter equals the E_1 pitch diameter at large end of internal thread (Table 2) minus (small) 1½ threads taper.
Minimum Pitch Diameter equals maximum diameter minus tolerance.

TABLE B5—DRYSEAL STRAIGHT PIPE TAPS FOR DRYSEAL AMERICAN INTERMEDIATE INTERNAL STRAIGHT PIPE THREAD, GROUND THREAD LIMITS

NPSI Size	Pitch Diameter			Major Dia Flat[a]		Minor Dia Flat[b]	Major Dia		
	Basic	Min	Max	Min	Max	Max	Basic	Min	Max
1/16—27	0.2812	0.2795	0.2800	0.004	0.005	0.004	0.3108	0.3031	0.3041
1/8 —27	0.3748	0.3719	0.3724	0.004	0.005	0.004	0.4044	0.3955	0.3965
1/4 —18	0.4899	0.4894	0.4899	0.005	0.006	0.005	0.5343	0.5274	0.5284
3/8 —18	0.6270	0.6248	0.6253	0.005	0.006	0.005	0.6714	0.6628	0.6638
1/2 —14	0.7784	0.7757	0.7762	0.005	0.006	0.005	0.8356	0.8275	0.8285
3/4 —14	0.9889	0.9862	0.9867	0.005	0.006	0.005	1.0460	1.0380	1.0390
1 —11-1/2	1.2386	1.2349	1.2359	0.006	0.008	0.006	1.3082	1.2987	1.2997

[a] For reference only. Major-diameter flats specified may be slightly larger or smaller with extreme combinations of pitch diameter, major diameter, and half angle.
[b] Minor diameter as specified or sharper.

Chamfer—Plug end 3 to 5 threads, intermediate end 2 to 3 threads, bottom end 1½ to 2 threads.
Lead Tolerance—A maximum lead error of ±0.0005 in. in 1 in. of thread is permitted.
Angle Tolerance—Error in half angle of ±30 min permitted.
Marking—In addition to regular markings, American Intermediate Straight Pipe Taps will be marked NPSI.
Maximum Major Diameter equals minimum pitch diameter plus sharp-V thread height minus twice tool crest minimum truncation.
Minimum Major Diameter equals maximum major diameter minus tolerance.
Maximum Pitch Diameter equals the E_1 pitch diameter at large end of internal thread (Table 2) minus (small) ½ thread taper.
Minimum Pitch Diameter equals maximum pitch diameter minus tolerance.

APPENDIX C—DRYSEAL PIPE-THREAD GAGING

General Information—There are three accepted methods of checking Dryseal pipe threads with threaded plug and ring gages. The methods separately described in the following sections are:
Section I—Position Method of Gaging with Basic Notch Gages.
Section II—Limit Method of Gaging with Step-Limit Gages.
Section III—Turns-Engagement Method of Gaging with Basic-Notch or Step-Limit Gages.

All methods of gaging external Dryseal threads involve the use of two ring thread gages, the (L_1) thin-ring thread gage for checking the pitch diameter over the hand engagement or (L_1) thread length and the (L_2) full-thread ring gage for checking the pitch diameter over the full thread length to insure adequate threads for wrench tightening.

All methods of gaging internal Dryseal threads involve the use of two plug thread gages, the (L_1) plug thread gage for checking the pitch diameter over the hand engagement or (L_1) thread length and the (L_3) plug thread gage for checking pitch diameter of the thread beyond the hand engagement length.

As indicated in the separate descriptions of the various gaging methods, coordination of the two ring thread gages for external threads and coordination of the two plug thread gages for internal threads control and check thread taper and length. The gages cannot be correlated, however, for external threads of minimum pitch diameter or internal threads of maximum pitch diameter unless the length of the threads is one thread longer than basic full thread length.

Working gages shall not be used where worn beyond the basic dimensions by more than ½ turn (thread). Proper allowance shall be made for any variation from basic when using a gage.

The threads of tools and the threads of a percentage of the product or casts in the case of internal threads should be projected as a check on thread form and truncation. Although projection is strongly recommended, the truncation at major diameter of internal thread and minor diameter of external thread may be checked respectively with special plug and ring gages with thread angle reduced to clear the flank of the threads; and the truncation at minor diameter of internal taper thread and major diameter of external taper thread may be checked respectively with plain taper plug gages and plain taper ring gages. Internal straight thread truncation at minor diameter may be checked with plain plug gages.

Section I—Position Method of Gaging with Basic Notch Gages

The position method of gaging Dryseal threads with plug thread and ring thread gages is a visual check of the position of the gages in relation to the product. It involves estimating the position of a notch or step on the thread gages in relation to the gaging point of the product within the allowable tolerance.

While the method is the same as that used for years past in checking conventional pipe threads without the Dryseal feature, the gages are different with respect to truncation of threads, the crests of the threads at the minor diameter of the ring gages and the major diameter of the plug gages being truncated to a greater extent to clear the increased truncation of the product thread. Another distinction is that the Dryseal (L_2) ring is counterbored larger than the thread diameter at the small end a distance equal to the (L_1) thread length minus one pitch. Conventional rings and plugs, however, may be converted to Dryseal by grinding the crests to conform with the width of flats specified for Dryseal gages, and grinding a counterbore in the (L_2) ring gage.

The gages are turned or screwed hand-tight into or onto the threaded product, the position of the gage notch in relation to the product reference point being noted to determine whether the standoff exceeds the allowable tolerance. Allowance must be made for excessive chamfer at the small end of external threads and the large end of internal threads, the product reference point in the first instance being the beginning of the first thread on the chamfer, and in the second instance being the intersecton of the pitch diameter cone and the chamfer cone.

External Threads—Dryseal American Standard External Taper Pipe Threads (NPTF) are gaged with the NPTF (L_1) basic-notch Dryseal ring thread gages (Table C1) and the NPTF (L_2) basic-notch Dryseal ring thread gages (Table C1). Threads are within the allowable tolerance when the product reference point is flush with the gage reference point within a tolerance of plus (small) one turn, minus (large) one turn. As a check on taper, the (L_1) and (L_2) ring thread gages shall gage the same within $\frac{1}{2}$ turn.

Dryseal SAE Short External Taper Pipe Threads PTF—SAE Short, which are one thread shorter at the small end than standard full thread length, are gaged with the NPTF (L_1) basic-notch Dryseal ring thread gages (Table C1) and the NPTF (L_2) basic-notch Dryseal ring thread gages (Table C1). Threads are within the allowable tolerance when the product reference point is flush with the gage reference point within a tolerance of plus zero, minus (large) $1\frac{1}{2}$ turns. As a check on taper, the (L_1) and (L_2) ring thread gages shall gage the same within $\frac{1}{2}$ turn.

Internal Threads—Dryseal American Standard Internal Taper Pipe Threads (NPTF) are gaged with the NPTF (L_1) basic-notch Dryseal plug thread gages (Table C2) and the NPTF (L_3) basic-notch Dryseal plug thread gages (Table C3). Threads are within the allowable tolerance when the product reference point is flush with the gage notch within the following tolerances:

Plus (large) 1 turn, minus (small) 1 turn.

As a check on taper, the (L_1) and the (L_3) plug thread gages shall gage the same with relation to their respective notches within $\frac{1}{2}$ turn.

Dryseal SAE Short Internal Taper Pipe Threads PTF—SAE Short, which are one thread shorter at the large end than standard full thread length, are gaged with the NPTF (L_1) basic-notch Dryseal plug thread gages (Table C2) and the NPTF (L_3) basic-notch Dryseal plug thread gages (Table C3). Threads are within the allowable tolerance when the product reference point is flush with the gage notch within the following tolerances:

Plus (large) 0 turns, minus (small) $1\frac{1}{2}$ turns.

Dryseal American (National) Standard Fuel Internal Straight Pipe Threads (NPSF) are gaged with the NPTF (L_1) basic-notch Dryseal plug thread gages (Table C2). Threads are within the allowable tolerance when the product reference point is flush with the gage notch within the following tolerances:

Plus (large) 0 turns, minus (small) $1\frac{1}{2}$ turns.

As depth gages without regard to gage notches, any of the (L_3) Dryseal plug thread gages may be used to check the full thread length of internal straight pipe threads.

Dryseal American (National) Standard Intermediate Internal Straight Pipe Threads (NPSI) are gaged with the NPTF (L_1) basic-notch Dryseal plug thread gages (Table C2). Threads are within the allowable tolerance when the product reference point is flush with the gage notch within the following tolerances:

Plus (large) 1 turn, minus (small) $\frac{1}{2}$ turn.

As depth gages without regard to gage notches, any of the (L_3) Dryseal plug thread gages may be used to check the full thread length of internal straight pipe threads.

Section II—Limit Method of Gaging with Step-Limit Gages

The limit-gage or step-limit method of checking threaded product with plug thread and ring thread gages is a visual check of the position of the gages in relation to the product. Plug and ring gages with maximum and minimum limit notches are provided for the different thread types: NPTF, PTF—SAE Short, NPSF, and NPSI. The location of the limit notches on the $\frac{1}{8}$- and $\frac{1}{4}$-in. plugs eliminates the necessity for gaging correction.

The gages are turned or screwed hand-tight into or onto the threaded product, the position of the product reference point in relation to the limit notches on the gage being noted. Allowance must be made for excessive chamfer at the small end of external threads and the large end of internal threads, the product reference point in the first instance being the beginning of the first thread on the chamfer, and in the second instance being the intersection of the pitch diameter cone and the chamfer cone.

External Threads—Dryseal American Standard External Taper Pipe Threads (NPTF) are gaged with the NPTF (L_1) step-limit Dryseal ring thread gages (Table C4) and the NPTF (L_2) step-limit Dryseal ring thread gages (Table C5). Threads are within the allowable tolerance when the product reference point is on or between the limit notches. As a check on taper, the (L_1) and the (L_2) Dryseal ring thread gages shall gage the same in relation to their respective notches within $\frac{1}{2}$ turn.

Dryseal SAE Short External Taper Pipe Threads PTF—SAE Short, which are one thread shorter at the small end than standard full thread length, are gaged with the PTF—SAE (L_1short) step-limit Dryseal ring thread gages (Table C8) and the PTF—SAE (L_2 Short) step-limit ring thread gages (Table C9). Threads are within the allowable tolerance when the product reference point is on or between the limit notches. As a check on taper, the (L_1 Short) and the (L_2 Short) Dryseal ring thread gages shall gage the same with relation to their respective notches within $\frac{1}{2}$ turn.

Internal Threads—Dryseal American Standard Internal Taper Pipe Threads (NPTF) are gaged with the NPTF (L_1) step-limit Dryseal plug thread gages (Table C6) and the NPTF (L_3) step-limit Dryseal plug thread gages (Table C7). Threads are within the allowable tolerance when the product reference point is on or between the limit notches. As a check on taper, the (L_1) and (L_3) Dryseal plug thread gages shall gage the same with relation to their respective notches within $\frac{1}{2}$ turn.

Dryseal SAE Short Internal Taper Pipe Threads PTF—SAE Short, which are one thread shorter at the large end than standard full thread length, are gaged with the PTF—SAE (L_1 Short) step-limit Dryseal plug thread gages (Table C10) and the PTF—SAE (L_3 Short) step-limit Dryseal plug thread gages (Table C11). Threads are within the allowable tolerance when the product reference point is on or between the limit notches. As a check on taper, the (L_1 Short) and (L_3 Short) Dryseal plug thread gages shall gage the same with relation to their respective notches within $\frac{1}{2}$ turn.

Dryseal American Standard Fuel Internal Straight Pipe Threads (NPSF) are gaged with the NPSF (L_1 Short) step-limit Dryseal plug thread gages (Table C10). Threads are within the allowable tolerance when the product reference point is on or between the limit notches. As depth gages without regard to limit notches, any of the (L_3) Dryseal plug thread gages may be used to check the full thread length of internal straight pipe threads.

Dryseal American Standard Intermediate Internal Straight Pipe Threads (NPSI) are gaged with the NPSI (L_1) step-limit Dryseal plug thread gages (Table C12). Threads are within the allowable tolerance when the product reference point is on or between the limit notches. As depth gages without regard to limit notches, any of the (L_3) Dryseal plug thread gages may be used to check the full thread length of internal straight pipe threads.

Section III—Turns-Engagement Method of Gaging with Basic-Notch or Step-Limit Gages

The turns-engagement method of checking threaded product with plug thread and ring thread gages is a tactual check of the position of the gages in relation to the product. In checking by this method, either the basic-notch or the step-limit gages may be used. The gages are turned or screwed into or onto the threaded product and the turns to remove the gages are counted. This method compensates for gage chamfer and eliminates the variable of product chamfer.

Basic Turns Engagement of Ring Gages—The basic turns engagement of the (L_1) ring thread gages (Tables C1, C4, and C8) with Dryseal external taper pipe threads is the product of the (L_1) thread length of the ring gage used and the threads per inch, minus one turn to compensate for chamfer of the external threads and chamfer of the ring gages. (See accompanying tabulation of basic turns engagement.)

The basic turns engagement of the (L_2) ring thread gages (Tables C1, C5, and C9) with Dryseal external taper pipe threads is the product of the (L_2) thread length and the threads per inch, minus $1\frac{1}{4}$ turns to compensate for chamfer of the external threads and the chamfer and taper of the ring gages. (See accompanying tabulation of basic turns engagement.)

External Threads—Dryseal American Standard External Taper Pipe Threads (NPTF) by the turns-engagement method may be gaged with any combination of (L_1) and (L_2) Dryseal ring thread gages (Tables C1, C4, C5, C8, and C9). Nominal turns engagement equals basic turns engagement. The tolerance is plus (small) 1 turn, minus (large) 1 turn. As a check on taper, the difference in turns engagement with the (L_1) and the (L_2) Dryseal ring thread gages shall be within $\frac{1}{2}$ turn of the difference between the basic turns engagement of the ring gages.

Dryseal SAE Short External Taper Pipe Threads PTF—SAE Short, which are one thread shorter at the small end than standard full thread length, may be gaged by the turns-engagement method with any combination of (L_1) and (L_2) Dryseal ring thread gages (Tables C1, C4, C5, C8, and C9). Nominal turns engagement is one turn less than basic turns engagement. The tolerance is plus (small) 1 turn, minus (large) $\frac{1}{2}$ turn. As a check on taper, the difference in turns engagement with the (L_1) and the (L_2) Dryseal ring thread gages shall be within $\frac{1}{2}$ turn of the difference between the basic turns engagement of the ring gages.

Basic Turns Engagement of Plug Gages—The basic turns engagement of the (L_1) Dryseal plug thread gages (Tables C2, C6, C10, and C12) and Dryseal internal pipe threads is the product of the (L_1) thread length (Table 2) and the threads per inch, minus $\frac{1}{2}$ turn to compensate for chamfer on plug gages. (See accompanying tabulation of basic turns engagement.)

The basic turns engagement of the (L_3) Dryseal plug thread gages (Tables C3, C7, and C11) with Dryseal internal pipe threads is the (L_1) thread length (Table 2) plus three threads multiplied by the threads per inch, minus $\frac{3}{4}$ turn to compensate for chamfer and taper on plug gages. (See accompanying tabulation of basic turns engagement.)

Internal Threads—Dryseal American Standard Internal Taper Pipe Threads (NPTF) are gaged with any combination of (L_1) and (L_3) Dryseal plug thread gages (Tables C2, C3, C6, C7, C10, C11, and C12). The nominal turns engagement equals basic turns engagement. The tolerance is plus (large) 1 turn, minus (small) 1 turn. As a check on taper, the difference in turns engagement of the (L_1) and the (L_3) Dryseal plug thread gages shall not be less than $2\frac{1}{4}$ turns nor more than $3\frac{1}{4}$ turns.

Dryseal SAE Short Internal Taper Pipe Threads PTF—SAE Short, which are one thread shorter at the large end than standard full thread length, are gaged with any combination of (L_1) and (L_3) Dryseal plug thread gages (Tables C2, C3, C6, C7, C10, C11, and C12). The nominal turns engagement is one turn less than basic turns engagement. The tolerance is plus (large) 1 turn, minus (small) $\frac{1}{2}$ turn. As a check on taper, the difference in turns engagement of the (L_1) and the (L_3) Dryseal plug thread gages shall not be less than $2\frac{1}{4}$ turns nor more than $3\frac{1}{4}$ turns.

Dryseal American Standard Fuel Internal Straight Pipe Threads (NPSF) are gaged with any of the (L_1) Dryseal plug thread gages (Tables C2, C6, C10, and C12). The nominal turns engagement is one turn less than basic turns engagement. The tolerance is plus (large) 1 turn, minus (small) $\frac{1}{2}$ turn. As depth gages without regard to limit notches, any of the (L_3) Dryseal plug thread gages may be used to check the full thread length of internal straight pipe threads.

Dryseal American Standard Intermediate Internal Straight Pipe Threads (NPSI) are gaged with any of the (L_1) Dryseal plug thread gages (Tables C2, C6, C10, and C12). The nominal turns engagement equals basic turns engagement. The tolerance is plus (large) 1 turn, minus (small) $\frac{1}{2}$ turn. As depth gages without regard to limit notches, any of the (L_3) Dryseal plug thread gages may be used to check the full thread length of internal straight pipe threads.

BASIC TURNS ENGAGEMENT[a]

Size	Thread Gage				
	L_1 Rings		All L_2 Rings	All L_1 Plugs	All L_3 Plugs
	Basic-Notch, Table C1	Step-Limit, Tables C4 and C8			
1/16—27	3.32	3.32	5.80	3.82	6.57
1/8 —27	3.36	3.36	5.87	3.86	6.61
1/4 —18	3.10	3.10	5.98	3.60	6.35
3/8 —18	3.32	3.32	6.09	3.82	6.57
1/2 —14	3.48	3.48	6.22	3.98	6.73
3/4 —14	3.75	3.75	6.39	4.25	7.00
1 —11-1/2	3.60	3.60	6.60	4.10	6.85
1-1/4 —11-1/2	3.83	3.83	6.88	4.33	7.08
1-1/2 —11-1/2	3.83	3.83	7.07	4.33	7.08
2 —11-1/2	4.01	4.01	7.45	4.51	7.26
2-1/2 —8	4.46	4.46	7.85	4.96	7.71
3 —8	5.13	5.13	8.35	5.63	8.38

[a] Derivation of nominal turns engagement and tolerance for the different thread types, NPTF, PTF—SAE Short, NPSF, and NPSI, is explained in accompanying text.

DRYSEAL AMERICAN TAPER PIPE THREAD (L_1 AND L_2) RING GAGES

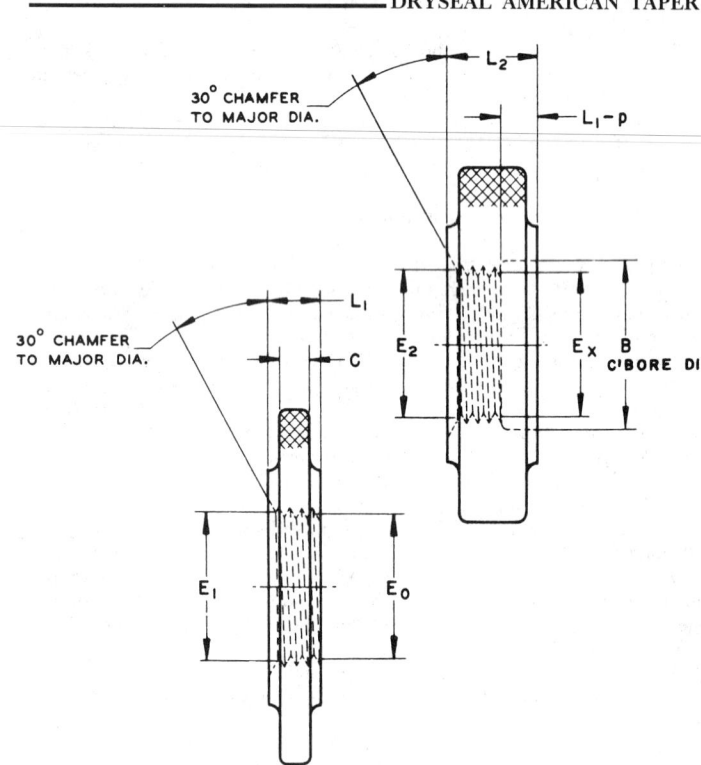

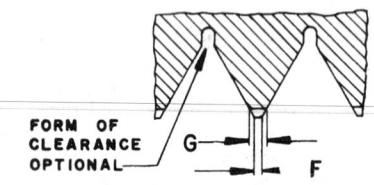

TABLE C1-1—THREAD FLATS

Threads per in.	F	G
27	0.0086	0.0107
18	0.0128	0.0160
14	0.0165	0.0206
11-1/2	0.0201	0.0251
8	0.0289	0.0361

Marking—In addition to the regular markings, Dryseal American Standard Taper Thread Thin-Ring Gages will be marked NPTF (L_1) and Full-Ring Gages will be marked NPTF (L_2) on the entering side of gage.

Thread Form—The threads in all particulars excepting truncation shall conform to American Standard Taper Pipe Thread practice. Crests of Threads at the minor diameter of ring gages and major diameter of plug gages shall be truncated 0.20p minimum to 0.25p maximum producing the minimum and maximum widths of flats specified in Table C1-1.

All other thread dimensions shall be within tolerances specified for the Dryseal American Standard Pipe Thread Working Plug Gages (ASA B2.2). Other gage details shall conform to American Gage Design Standards published in Commercial Standard CS8.

TABLE C1—BASIC DIMENSIONS OF DRYSEAL AMERICAN TAPER PIPE THREAD (L_1 AND L_2) BASIC-NOTCH RING GAGES

Size	(L_2) Basic-Notch Full-Ring Gages							(L_1) Basic-Notch Thin-Ring Gages				
	L_2	Pitch Dia, E_2	Minor Dia at Large End[a]	Pitch Dia at $L_1 - p$, E_x	Minor Dia at $L_1 - p$[a]	$L_1 - p$	B	L_1	Pitch Dia, E_1	Minor Dia at Large End[a]	Pitch Dia, E_0	Minor Dia at Small End[a]
1/16—27	0.26113	0.28750	0.27024	0.27886	0.26160	0.12296	0.38	0.1600	0.28118	0.26392	0.27118	0.25392
1/8 —27	0.26385	0.38000	0.36274	0.37129	0.35403	0.12446	0.47	0.1615	0.37360	0.35634	0.36351	0.34625
1/4 —18	0.40178	0.50250	0.47661	0.48816	0.46227	0.17224	0.59	0.2278	0.49163	0.46574	0.47739	0.45150
3/8 —18	0.40778	0.63750	0.61161	0.62354	0.59765	0.18444	0.72	0.2400	0.62701	0.60112	0.61201	0.58612
1/2 —14	0.53371	0.79179	0.75850	0.77396	0.74067	0.24857	0.88	0.3200	0.77843	0.74514	0.75843	0.72514
3/4 —14	0.54571	1.00179	0.96850	0.98440	0.95111	0.26757	1.09	0.3390	0.98887	0.95558	0.96768	0.93439
1 —11-1/2	0.68278	1.25630	1.21577	1.23320	1.19267	0.31304	1.34	0.4000	1.23863	1.19810	1.21363	1.17310
1-1/4 —11-1/2	0.70678	1.60130	1.56077	1.57794	1.53741	0.33304	1.69	0.4200	1.58338	1.54285	1.55713	1.51660
1-1/2 —11-1/2	0.72348	1.84130	1.80077	1.81690	1.77637	0.33304	1.94	0.4200	1.82234	1.78181	1.79609	1.75556
2 —11-1/2	0.75652	2.31630	2.27577	2.29084	2.25031	0.34904	2.50	0.4360	2.29627	2.25574	2.26902	2.22849
2-1/2 —8	1.13750	2.79062	2.73237	2.75434	2.69609	0.55700	2.94	0.6820	2.76216	2.70391	2.71953	2.66128
3 —8	1.20000	3.41562	3.35737	3.38068	3.32243	0.54100	3.56	0.7660	3.38850	3.33025	3.34062	3.28237

[a] Minor diameter is based on crest minimum truncation of 0.20 p.

14.11

DRYSEAL AMERICAN TAPER PIPE THREAD (L_1) PLUG GAGES
Taper lock design, range 1/8 to 3 in., inclusive

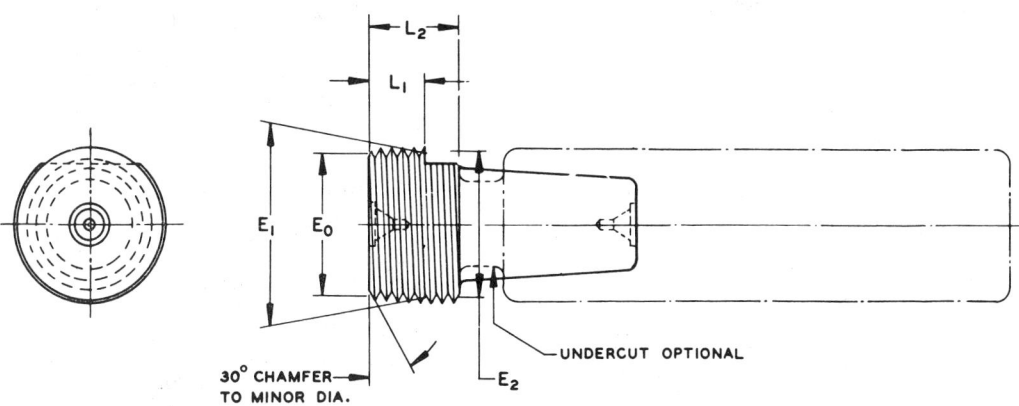

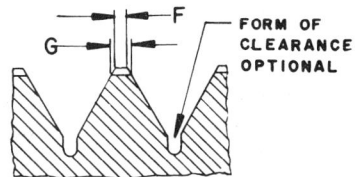

TABLE C2-1—THREAD FLATS

Threads per in.	F	G
27	0.0086	0.0107
18	0.0128	0.0160
14	0.0165	0.0206
11-1/2	0.0201	0.0251
8	0.0289	0.0361

Marking—In addition to the regular markings, Dryseal American Standard Taper Pipe Thread (L_1) Plug Gages will be marked NPTF (L_1).

Thread Form—The threads in all particulars excepting truncation shall conform to American Standard Taper Pipe Thread practice. Crests of threads at the minor diameter of ring gages and major diameter of plug gages shall be truncated 0.20p minimum to 0.25p maximum producing the minimum and maximum widths of flats specified in Table C2-1.

All other thread dimensions shall be within tolerances specified for the Dryseal American Standard Pipe Thread Working Plug Gages (ASA B2.2). Other gage details shall conform to American Gage Design Standards published in Commercial Standard CS8.

TABLE C2—BASIC DIMENSIONS OF DRYSEAL AMERICAN TAPER PIPE THREAD (L_1) BASIC-NOTCH PLUG GAGES

Size	L_1	L_2	Small End		Gaging Notch		Large End	
			Pitch Dia, E_0	Major Dia[a]	Pitch Dia, E_1	Major Dia[a]	Pitch Dia, E_2	Major Dia[a]
1/16—27	0.1600	0.26113	0.27118	0.28844	0.28118	0.29844	0.28750	0.30476
1/8—27	0.1615	0.26385	0.36351	0.38077	0.37360	0.39086	0.38000	0.39726
1/4—18	0.2278	0.40178	0.47739	0.50328	0.49163	0.51752	0.50250	0.52839
3/8—18	0.2400	0.40778	0.61201	0.63790	0.62701	0.65290	0.63750	0.66339
1/2—14	0.3200	0.53371	0.75843	0.79170	0.77843	0.81170	0.79179	0.82506
3/4—14	0.3390	0.54571	0.96768	1.00095	0.98887	1.02214	1.00179	1.03506
1 —11-1/2	0.4000	0.68278	1.21363	1.25416	1.23863	1.27916	1.25630	1.29683
1-1/4—11-1/2	0.4200	0.70678	1.55713	1.59766	1.58338	1.62391	1.60130	1.64183
1-1/2—11-1/2	0.4200	0.72348	1.79609	1.83662	1.82234	1.86287	1.84130	1.88183
2 —11-1/2	0.4360	0.75652	2.26902	2.30955	2.29627	2.33680	2.31630	2.35683
2-1/2—8	0.6820	1.13750	2.71953	2.77778	2.76216	2.82041	2.79062	2.84887
3 —8	0.7660	1.20000	3.34062	3.39887	3.38850	3.44675	3.41562	3.47387

[a] Major diameter is based upon crest minimum truncation of 0.20 p.

DRYSEAL AMERICAN TAPER PIPE THREAD (L_3) LENGTH PLUG GAGES

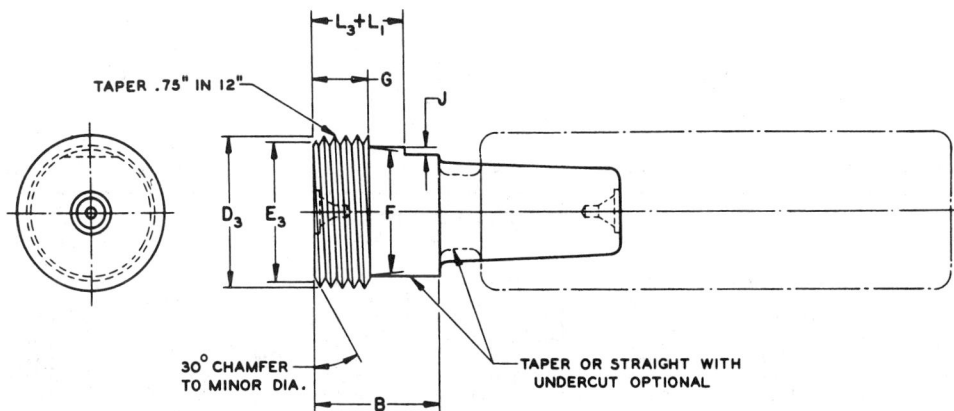

Marking—In addition to the regular markings, Dryseal American Standard Taper Pipe Thread (L_3) Plug Gages will be marked NPTF (L_3).

Thread Form—The threads in all particulars excepting truncation shall conform to American Standard Taper Pipe Thread practice. Crests of threads at major diameter shall be truncated 0.20p minimum to 0.25p maximum, producing the minimum and maximum widths of flat specified in Table C3-1.

All other thread dimensions shall be within tolerances specified for the Dryseal American Standard Pipe Thread Working Plug Gages (ASA B2.2). Other gage details shall conform to American Gage Design Standards published in Commercial Standard CS8.

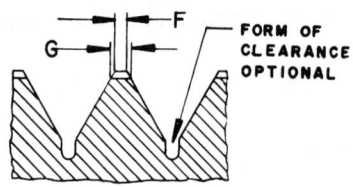

FORM OF CLEARANCE OPTIONAL

TABLE C3-1—THREAD FLATS

Threads per in.	F	G
27	0.0086	0.0107
18	0.0128	0.0160
14	0.0165	0.0206
11-1/2	0.0201	0.0251
8	0.0289	0.0361

TABLE C3—BASIC DIMENSIONS OF DRYSEAL AMERICAN TAPER PIPE THREAD (L₂) BASIC-NOTCH LENGTH PLUG GAGES

Size	Small End Pitch Dia, E₁	Small End Major Dia,ᵃ D₂	Relief Dia, F [E₂+(0.0625 × 4p)—Sharp-V Thread Height—0.020 to 0.025 below Sharp Root] +0.005 −0.000	Four Threads, G (L₂+p)	L₁ Plus 3 Threads (L₁+L₃)	Blank Length, B	Notch Depth, J +0.005 −0.000
1/16—27	0.2642	0.2815	0.216	0.1482	0.2711	0.38	0.030
1/8 —27	0.3566	0.3738	0.309	0.1482	0.2726	0.41	0.030
1/4 —18	0.4670	0.4928	0.409	0.2222	0.3945	0.50	0.030
3/8 —18	0.6016	0.6275	0.542	0.2222	0.4067	0.56	0.030
1/2 —14	0.7451	0.7783	0.676	0.2857	0.5343	0.69	0.040
3/4 —14	0.9543	0.9876	0.886	0.2857	0.5533	0.72	0.040
1 —11-1/2	1.1973	1.2379	1.118	0.3478	0.6609	0.88	0.050
1-1/4 —11-1/2	1.5408	1.5814	1.462	0.3478	0.6809	0.88	0.050
1-1/2 —11-1/2	1.7798	1.8203	1.701	0.3478	0.6809	0.88	0.050
2 —11-1/2	2.2527	2.2932	2.174	0.3478	0.6969	0.88	0.050
2-1/2 —8	2.6961	2.7543	2.590	0.5000	1.0570	1.50	0.050
3 —8	3.3172	3.3754	3.214	0.5000	1.1410	1.50	0.050

ᵃ Major diameter is based upon crest minimum truncation of 0.20 p.

DRYSEAL AMERICAN STANDARD TAPER PIPE THREAD (L₁) STEP-LIMIT THIN-RING GAGES

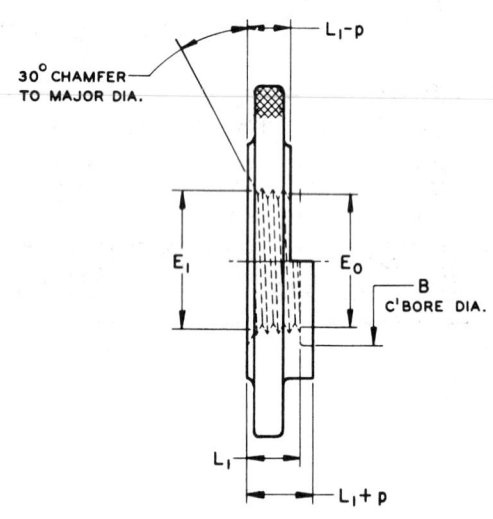

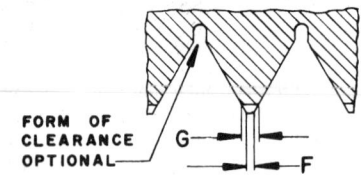

FORM OF CLEARANCE OPTIONAL

TABLE C4-1—THREAD FLATS

Threads per in.	F	G
27	0.0086	0.0107
18	0.0128	0.0160
14	0.0165	0.0206
11-1/2	0.0201	0.0251
8	0.0289	0.0361

Marking—In addition to the regular markings, Dryseal American Standard Taper Thread Ring Gages will be marked NPTF (L₁) on the entering side of gage.

Thread Form—The threads in all particulars excepting truncation shall conform to American Standard Taper Pipe Thread practice. Crests of threads at the minor diameter shall be truncated 0.20p minimum to 0.25p maximum, producing the minimum and maximum widths of flat specified in Table C4-1.

All other thread dimensions shall be within tolerances specified for the Dryseal American Standard Pipe Working Gages (ASA B2.2). Other gage details shall conform to American Gage Design Standard Published in Commercial CS8.

TABLE C4—BASIC DIMENSIONS OF DRYSEAL AMERICAN STANDARD TAPER PIPE THREAD (L₁) STEP-LIMIT THIN-RING GAGES

Size	(L₁) Step-Limit Thin-Ring Gages							
	L₁	Max Pitch Dia Gaging Step L₁−p	Min Pitch Dia Gaging Step L₁+p	Pitch Dia, E₁	Minor Dia at Large Endᵃ	Pitch Dia at Small End Counterbore E₀	Minor Dia at Small End Counterboreᵃ	B
1/16—27	0.1600	0.12296	0.19704	0.28118	0.26392	0.27118	0.25392	0.38
1/8 —27	0.1615	0.12446	0.19854	0.37360	0.35634	0.36351	0.34625	0.47
1/4 —18	0.2278	0.17224	0.28336	0.49163	0.46574	0.47739	0.45150	0.59
3/8 —18	0.2400	0.18444	0.29556	0.62701	0.60112	0.61201	0.58612	0.72
1/2 —14	0.3200	0.24857	0.39143	0.77843	0.74514	0.75843	0.72514	0.88
3/4 —14	0.3390	0.26757	0.41043	0.98887	0.95558	0.96768	0.93439	1.09
1 —11-1/2	0.4000	0.31304	0.48696	1.23863	1.19810	1.21363	1.17310	1.34
1-1/4 —11-1/2	0.4200	0.33304	0.50696	1.58338	1.54285	1.55713	1.51660	1.69
1-1/2 —11-1/2	0.4200	0.33304	0.50696	1.82234	1.78181	1.79609	1.75556	1.94
2 —11-1/2	0.4360	0.34904	0.52296	2.29627	2.25574	2.26902	2.22849	2.50
2-1/2 —8	0.6820	0.55700	0.80700	2.76216	2.70391	2.71953	2.66128	2.94
3 —8	0.7660	0.64100	0.89100	3.38850	3.33025	3.34062	3.28237	3.56

ᵃ Minor diameter is based on crest minimum truncation of 0.20 p.

DRYSEAL AMERICAN STANDARD TAPER PIPE THREAD (L₂) STEP-LIMIT FULL-RING GAGES

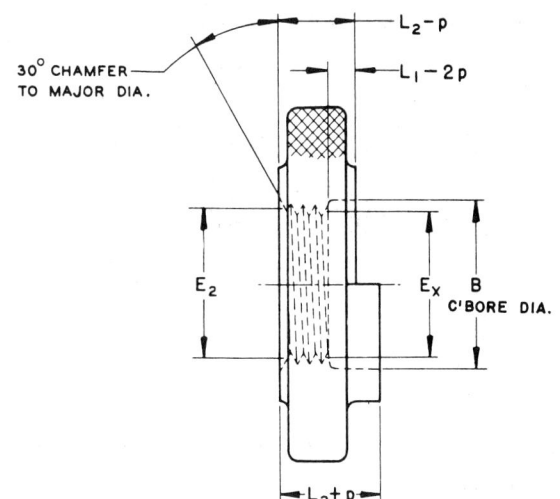

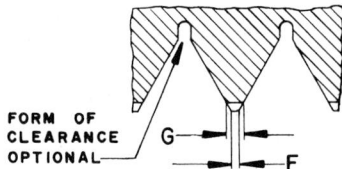

TABLE C5-1—THREAD FLATS

Threads per in.	F	G
27	0.0086	0.0107
18	0.0128	0.0160
14	0.0165	0.0206
11-1/2	0.0201	0.0251
8	0.0289	0.0361

Marking—In addition to the regular markings, Dryseal American Standard Taper Pipe Thread Ring Gages will be marked NPTF (L₂) on the entering side of gage.

Thread Form—The threads in all particulars excepting truncation shall conform to American Standard Taper Pipe Thread practice. Crests of threads at the minor diameter shall be truncated 0.20p minimum to 0.25p maximum, producing the minimum and maximum widths of flat specified in Table C5-1.

All other thread dimensions shall be within tolerances specified for the Dryseal American Standard Pipe Working Gages (ASA B2.2). Other gage details shall conform to American Gage Design Standard published in Commercial Standard CS8.

TABLE C5—BASIC DIMENSIONS OF DRYSEAL AMERICAN STANDARD TAPER PIPE THREAD (L₂) STEP-LIMIT FULL-RING GAGES

Size	(L₂) Step-Limit Full-Ring Gages								
	L₂	Max Pitch Dia Gaging Step L₂−p	Min Pitch Dia Gaging Step L₂+p	Pitch Dia, E₂	Minor Dia at Large End[a]	Pitch Dia at L₁ from Min Pitch Dia Gaging Step (Eₓ)	Minor Dia at Small End Counterbore[a]	L₁−2p	B
1/16—27	0.26113	0.22409	0.29817	0.28750	0.27024	0.27886	0.26160	0.08592	0.38
1/8—27	0.26385	0.22681	0.30089	0.38000	0.36274	0.37129	0.35403	0.08742	0.47
1/4—18	0.40178	0.34622	0.45734	0.50250	0.47661	0.48816	0.46227	0.11668	0.59
3/8—18	0.40778	0.35222	0.46334	0.63750	0.61161	0.62354	0.59765	0.12888	0.72
1/2—14	0.53371	0.46228	0.60514	0.79179	0.75850	0.77396	0.74067	0.17714	0.88
3/4—14	0.54571	0.47428	0.61714	1.00179	0.96850	0.98440	0.95111	0.19614	1.09
1—11-1/2	0.68278	0.59582	0.76974	1.25630	1.21577	1.23320	1.19267	0.22608	1.34
1-1/4—11-1/2	0.70678	0.61982	0.79374	1.60130	1.56077	1.57794	1.53741	0.24608	1.69
1-1/2—11-1/2	0.72348	0.63652	0.81044	1.84130	1.80077	1.81690	1.77637	0.24608	1.94
2—11-1/2	0.75652	0.66956	0.84348	2.31630	2.27577	2.29084	2.25031	0.26208	2.50
2-1/2—8	1.13750	1.01250	1.26250	2.79062	2.73237	2.75434	2.69609	0.43200	2.94
3—8	1.20000	1.07500	1.32500	3.41562	3.35737	3.38068	3.32243	0.51600	3.56

[a] Minor diameter is based on crest minimum truncation of 0.20 p.

DRYSEAL AMERICAN STANDARD TAPER PIPE THREAD (L₁) STEP-LIMIT PLUG GAGES

Taper lock design, range 1/8 to 3 in., inclusive

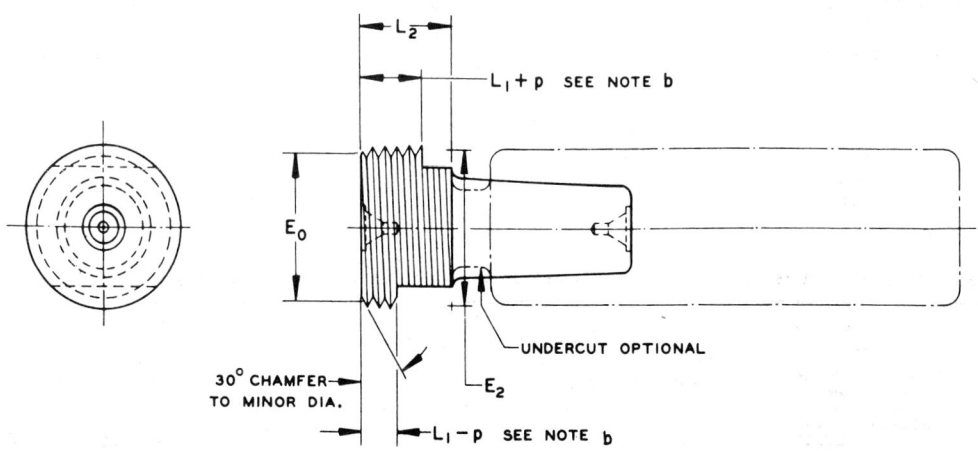

Marking—In addition to the regular markings, Dryseal American Standard Taper Pipe Thread (L₁) Plug Gages will be marked NPTF (L₁).

Thread Form—The threads in all particulars excepting truncation shall conform to American Standard Taper Pipe Thread practice. Crests of threads at major diameter shall be truncated 0.20p minimum to 0.25p maximum, producing the minimum and maximum widths of flat specified in Table C6-1.

All other thread dimensions shall be within tolerances specified for the Dryseal American Standard Pipe Thread Working Plug Gages (ASA B2.2). Other gage details shall conform to American Gage Design Standards published in Commercial Standards CS8.

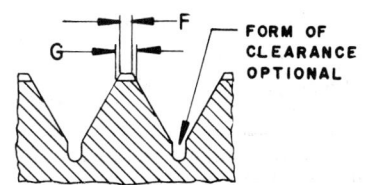

TABLE C6-1—THREAD FLATS

Threads per in.	F	G
27	0.0086	0.0107
18	0.0128	0.0160
14	0.0165	0.0206
11-1/2	0.0201	0.0251
8	0.0289	0.0361

TABLE C6—BASIC DIMENSIONS OF DRYSEAL AMERICAN STANDARD TAPER PIPE THREAD (L_1) STEP-LIMIT PLUG GAGES

Size	L_1	L_2	Small End		Min Pitch Dia Gaging Step[b]		Max Pitch Dia Gaging Step[b]		Large End	
			Pitch Dia, E_0	Major Dia[a]	L_1-p	Pitch Dia	L_1+p	Pitch Dia	Pitch Dia, E_2	Major Dia[a]
1/16—27	0.1600	0.26113	0.27118	0.28844	0.12296	0.27887	0.19704	0.28350	0.28750	0.30476
1/8 —27	0.1615	0.26385	0.36351	0.38077	0.12446	0.37129	0.19854	0.37592	0.38000	0.39726
1/4 —18	0.2278	0.40178	0.47739	0.50328	0.17224	0.48816	0.28336	0.49510	0.50250	0.52839
3/8 —18	0.2400	0.40778	0.61201	0.63790	0.18444	0.62354	0.29556	0.63048	0.63750	0.66339
1/2 —14	0.3200	0.53371	0.75843	0.79172	0.24857	0.77397	0.39143	0.78289	0.79179	0.82508
3/4 —14	0.3390	0.54571	0.96768	1.00097	0.26757	0.98441	0.41043	0.99333	1.00179	1.03508
1 —11-1/2	0.4000	0.68278	1.21363	1.25416	0.31304	1.23320	0.48696	1.24407	1.25630	1.29683
1-1/4 —11-1/2	0.4200	0.70678	1.55713	1.59766	0.33304	1.57795	0.50696	1.58882	1.60130	1.64183
1-1/2 —11-1/2	0.4200	0.72346	1.79609	1.83662	0.33304	1.81691	0.50696	1.82778	1.84130	1.88183
2 —11-1/2	0.4360	0.75652	2.26902	2.30955	0.34904	2.29084	0.52296	2.30171	2.31630	2.35683
2-1/2 —8	0.6820	1.13750	2.71953	2.77778	0.55700	2.75435	0.80700	2.76997	2.79062	2.84887
3 —8	0.7660	1.20000	3.34062	3.39887	0.64100	3.38069	0.89100	3.39631	3.41562	3.47387

[a] Major diameter is based on crest minimum truncation of 0.20 p.
[b] Maximum and minimum pitch-diameter steps are gaging limits. Notch formulas on drawing apply to all sizes.

DRYSEAL AMERICAN STANDARD TAPER THREAD (L_3) LENGTH STEP-LIMIT PLUG GAGES

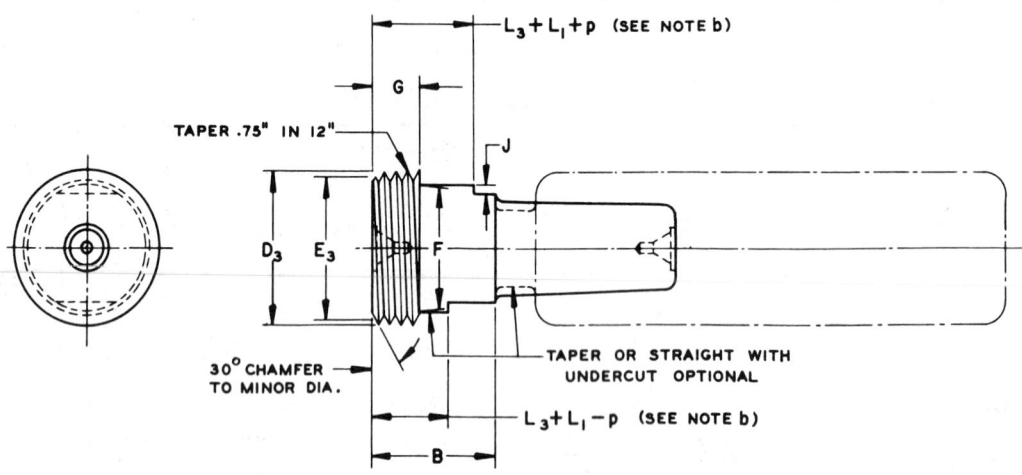

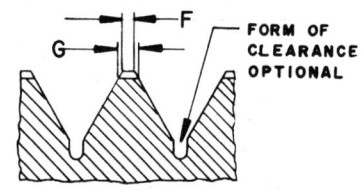

TABLE C7-1—THREAD FLATS

Threads per in.	F	G
27	0.0086	0.0107
18	0.0128	0.0160
14	0.0165	0.0206
11-1/2	0.0201	0.0251
8	0.0289	0.0361

Marking—In addition to the regular markings, Dryseal American Standard Taper Pipe Thread (L_3) Plug Gages will be marked PTF (L_3).

Thread Form—The threads in all particulars excepting truncation shall conform to American Standard Taper Pipe Thread practice. Crests of threads at major diameter shall be truncated 0.20p minimum and 0.25p maximum, producing the minimum and maximum widths of flat specified in Table C7-1.

All other thread dimensions shall be within tolerances specified for the Dryseal American Standard Pipe Thread Working Gages (ASA B2.2). Other gage details shall conform to American Gage Design Standards published in Commercial Standard CS8.

TABLE C7—BASIC DIMENSIONS OF DRYSEAL AMERICAN STANDARD TAPER THREAD (L_3) LENGTH STEP-LIMIT PLUG GAGES

Size	Small End		Relief Dia, F [$E_3+(0.0625\times 4p)$—Sharp-V Thread Height—0.020 to 0.025 below Sharp Root] +0.005 −0.000	Four Threads, G (L_3+p)	Pitch Dia Gaging Step[b] Plus 3 Threads		Blank Length B	Notch Depth, J +0.005 −0.000
	Pitch Dia, E_3	Major Dia,[a] D_3			(L_3+L_1-p) Min	(L_3+L_1+p) Max		
1/16—27	0.2642	0.2815	0.216	0.1482	0.2341	0.3082	0.38	0.030
1/8 —27	0.3566	0.3738	0.309	0.1482	0.2356	0.3097	0.41	0.030
1/4 —18	0.4670	0.4928	0.409	0.2222	0.3389	0.4500	0.50	0.030
3/8 —18	0.6016	0.6275	0.542	0.2222	0.3511	0.4622	0.56	0.030
1/2 —14	0.7451	0.7783	0.676	0.2857	0.4628	0.6057	0.69	0.040
3/4 —14	0.9543	0.9876	0.886	0.2857	0.4818	0.6247	0.72	0.040
1 —11-1/2	1.1973	1.2379	1.118	0.3478	0.5739	0.7478	0.88	0.050
1-1/4 —11-1/2	1.5408	1.5814	1.462	0.3478	0.5939	0.7678	0.88	0.050
1-1/2 —11-1/2	1.7798	1.8203	1.701	0.3478	0.5939	0.7678	0.88	0.050
2 —11-1/2	2.2527	2.2932	2.174	0.3478	0.6099	0.7838	0.88	0.050
2 1/2 —8	2.6961	2.7543	2.590	0.5000	0.9320	1.1820	1.50	0.050
3 —8	3.3172	3.3754	3.214	0.5000	1.0160	1.2660	1.50	0.050

[a] Major diameter is based upon crest minimum truncation of 0.20 p.
[b] Maximum and minimum pitch-diameter steps are gaging limits. Notch formulas on drawing apply to all sizes.

DRYSEAL SAE SHORT TAPER PIPE THREAD (L_1 SHORT) STEP-LIMIT THIN-RING GAGES

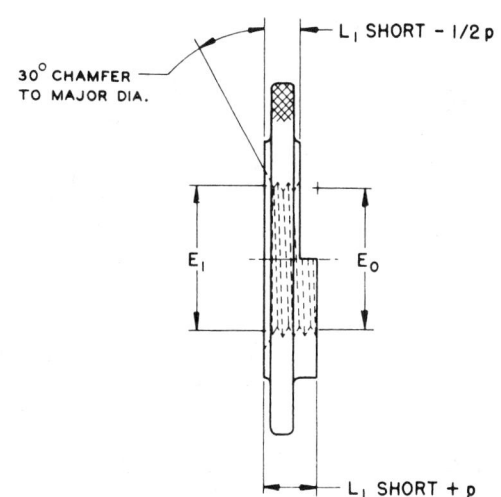

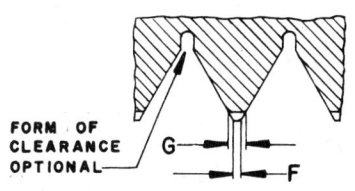

TABLE C8-1—THREAD FLATS

Threads per in.	F	G
27	0.0086	0.0107
18	0.0128	0.0160
14	0.0165	0.0206
11-1/2	0.0201	0.0251
8	0.0289	0.0361

TABLE C8—BASIC DIMENSIONS OF DRYSEAL SAE SHORT TAPER PIPE THREAD (L_1 SHORT) STEP-LIMIT THIN-RING GAGES

Size	(L_1 Short) Step-Limit Thin-Ring Gages						
	L_1 Short	Max Pitch Dia Gaging Step L_1 Short − 1/2p	Min Pitch Dia Gaging Step L_1 Short + p	Pitch Dia, E_1	Minor Dia at Large End [a]	Pitch Dia at Min Pitch Dia Gaging Step, E_0	Minor Dia at Small End [a]
1/16—27	0.12296	0.10444	0.16000	0.28118	0.26392	0.27118	0.25392
1/8 —27	0.12446	0.10594	0.16150	0.37360	0.35634	0.36351	0.34625
1/4 —18	0.17224	0.14446	0.22780	0.49163	0.46574	0.47739	0.45150
3/8 —18	0.18444	0.15666	0.24000	0.62701	0.60112	0.61201	0.58712
1/2 —14	0.24857	0.21286	0.32000	0.77843	0.74514	0.75843	0.72514
3/4 —14	0.26757	0.23186	0.33900	0.98887	0.95558	0.96768	0.93439
1 —11-1/2	0.31304	0.26956	0.40000	1.23863	1.19810	1.21363	1.17310
1-1/4 —11-1/2	0.33304	0.28956	0.42000	1.58338	1.54285	1.55713	1.51660
1-1/2 —11-1/2	0.33304	0.28956	0.42000	1.82234	1.78181	1.79609	1.75556
2 —11-1/2	0.34904	0.30556	0.43600	2.29627	2.25574	2.26902	2.22849
2-1/2 —8	0.55700	0.49450	0.68200	2.76216	2.70391	2.71953	2.66128
3 —8	0.64100	0.57850	0.76600	3.38850	3.33025	3.34062	3.28237

[a] Minor diameter is based on crest minimum truncation of 0.20 p.

Marking—In addition to the regular markings, Dryseal SAE Short Taper Pipe Thread Ring Gages will be marked PTF—SAE Short (L_1 Short) on the entering side of gage.

Thread Form—The threads in all particulars excepting truncation shall conform to American Standard Taper Pipe Thread practice. Crests of threads at the minor diameter shall be truncated 0.20p minimum to 0.25p maximum, producing the minimum and maximum widths of flat specified in Table C8-1.

All other thread dimensions shall be within tolerances specified for the Dryseal American Standard Pipe Working Gages (ASA B2.2). Other gage details shall conform to American Gage Design Standard published in Commercial Standard CS8.

DRYSEAL SAE SHORT TAPER PIPE THREAD (L_2 SHORT) STEP-LIMIT FULL-RING GAGES

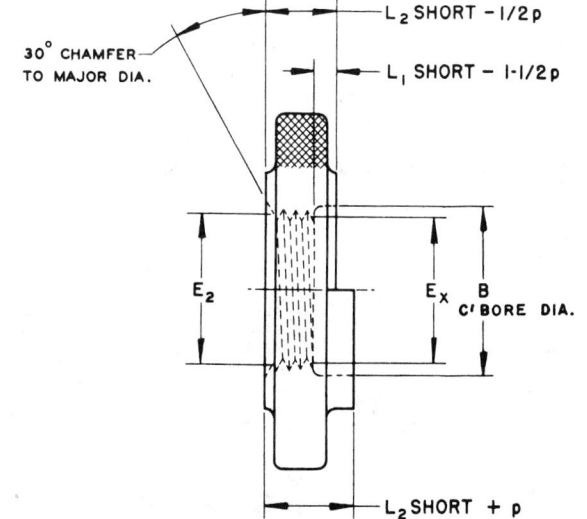

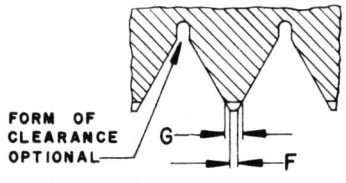

TABLE C9-1—THREAD FLATS

Threads per in.	F	G
27	0.0086	0.0107
18	0.0128	0.0160
14	0.0165	0.0206
11-1/2	0.0201	0.0251
8	0.0289	0.0361

Marking—In addition to the regular markings, Dryseal SAE Short Taper Pipe Thread Ring Gages will be marked PTF—SAE Short (L_2 Short) on the entering side of gage.

Thread Form—The threads in all particulars excepting truncation shall conform to American Standard Taper Pipe Thread practice. Crests of threads at the minor diameter shall be truncated 0.20p minimum to 0.25p maximum, producing the minimum and maximum widths of flat specified in Table C9-1.

All other thread dimensions shall be within tolerance specified for the Dryseal American Standard Pipe Working Gages (ASA B2.2). Other gage details shall conform to American Gage Design Standard published in Commercial Standard CS8.

TABLE C9—BASIC DIMENSIONS OF DRYSEAL SAE SHORT TAPER PIPE THREAD (L_2 SHORT) STEP-LIMIT FULL-RING GAGES

Size	L_2 Short	Max Pitch Dia Gaging Step L_2 Short − 1/2p	Min Pitch Dia Gaging Step L_2 Short + p	Pitch Dia, E_2	Minor Dia at Large End [a]	Pitch Dia at L_1 Short − 1-1/2p from Min Pitch Dia Gaging Step, E_x	Minor Dia at Small End Counterbore [a]	L_1 Short − 1-1/2p	B
1/16 —27	0.2241	0.20557	0.26113	0.28750	0.27024	0.27886	0.26160	0.06740	0.38
1/8 —27	0.2268	0.20829	0.26385	0.38000	0.36274	0.37129	0.35403	0.06890	0.47
1/4 —18	0.3462	0.31845	0.40178	0.50250	0.47661	0.48816	0.46227	0.08891	0.59
3/8 —18	0.3522	0.32445	0.40778	0.63750	0.61161	0.62354	0.59765	0.10111	0.72
1/2 —14	0.4623	0.42657	0.53371	0.79179	0.75850	0.77396	0.74067	0.14143	0.88
3/4 —14	0.4743	0.43857	0.54571	1.00179	0.96850	0.98440	0.95111	0.16043	1.09
1 —11-1/2	0.5958	0.55235	0.68278	1.25630	1.21577	1.23320	1.19267	0.18260	1.34
1-1/4 —11-1/2	0.6198	0.57635	0.70678	1.60130	1.56077	1.57794	1.53741	0.20260	1.69
1-1/2 —11-1/2	0.6365	0.59305	0.72348	1.84130	1.80077	1.81690	1.77637	0.20260	1.94
2 —11-1/2	0.6695	0.62609	0.75652	2.31630	2.27577	2.29084	2.25031	0.21860	2.50
2-1/2 —8	1.0125	0.95000	1.13750	2.79062	2.73237	2.75434	2.69609	0.36950	2.94
3 —8	1.0750	1.01250	1.20000	3.41562	3.35737	3.38068	3.32243	0.45350	3.56

[a] Minor diameter is based on crest minimum truncation of 0.20 p.

DRYSEAL SAE SHORT TAPER PIPE THREAD AND DRYSEAL AMERICAN STANDARD FUEL INTERNAL STRAIGHT PIPE THREAD (L_1 SHORT) STEP-LIMIT PLUG GAGES

Taper lock design, range 1/8 to 2 in., inclusive

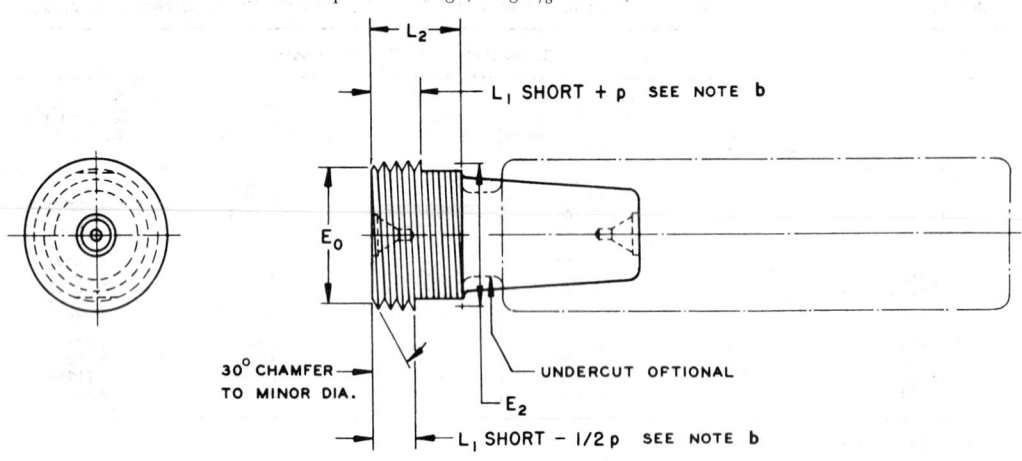

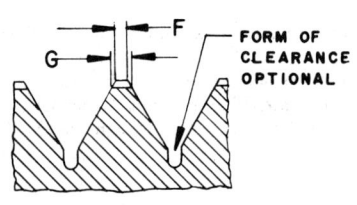

TABLE C10-1—THREAD FLATS

Threads per in.	F	G
27	0.0086	0.0107
18	0.0128	0.0160
14	0.0165	0.0206
11-1/2	0.0201	0.0251
8	0.0289	0.0361

Marking—In addition to the regular markings, Dryseal SAE Short Taper Pipe Thread L_1 Short Plug Gages will be marked PTF—SAE Short (L_1 Short). Dryseal American Standard Fuel Internal Straight Pipe Thread Taper Plug Gages will be marked NPSF (L_1 short).

Thread Form—The threads in all particulars excepting truncation shall conform to American Standard Taper Pipe thread practice. Crests of threads at major diameter shall be truncated 0.20p minimum to 0.25p maximum, producing the minimum and maximum widths of flat specified in Table C10-1.

All other thread dimensions shall be within tolerances specified for the Dryseal American Standard Pipe Thread Working Plug Gages (ASA B2.2). Other gage details shall conform to American Gage Design Standards published in Commercial Standard CS8.

TABLE C10—BASIC DIMENSIONS OF DRYSEAL SAE SHORT TAPER PIPE THREAD AND DRYSEAL AMERICAN STANDARD FUEL INTERNAL STRAIGHT PIPE THREAD (L_1 SHORT) STEP-LIMIT PLUG GAGES

Size	L_1 Short	L_2	Small End Pitch Dia, E_0	Small End Major Dia [a]	Min Pitch Dia Gaging Step [b] L_1 Short − 1/2p	Min Pitch Dia Gaging Step [b] Pitch Dia	Max Pitch Dia Gaging Step [b] L_1 Short + p	Max Pitch Dia Gaging Step [b] Pitch Dia	Large End Pitch Dia, E_2	Large End Major Dia [a]
1/16 —27	0.12296	0.26113	0.27118	0.28844	0.10444	0.27771	0.16000	0.28118	0.28750	0.30476
1/8 —27	0.12446	0.26385	0.36351	0.38077	0.10594	0.37013	0.16150	0.37360	0.38000	0.39726
1/4 —18	0.17224	0.40178	0.47739	0.50328	0.14446	0.48642	0.22780	0.49163	0.50250	0.52839
3/8 —18	0.18444	0.40778	0.61201	0.63790	0.15666	0.62180	0.24000	0.62701	0.63750	0.66339
1/2 —14	0.24857	0.53371	0.75843	0.79170	0.21286	0.77174	0.32000	0.77843	0.79179	0.82506
3/4 —14	0.26757	0.54571	0.96768	1.00095	0.23186	0.98218	0.33900	0.98887	1.00179	1.03506
1 —11-1/2	0.31304	0.68278	1.21363	1.25416	0.26956	1.23048	0.40000	1.23863	1.25630	1.29683
1-1/4 —11-1/2 [c]	0.33304	0.70678	1.55713	1.59766	0.28956	1.57523	0.42000	1.58338	1.60130	1.64183
1-1/2 —11-1/2 [c]	0.33304	0.72348	1.79609	1.83662	0.28956	1.81419	0.42000	1.82234	1.84130	1.88183
2 —11-1/2 [c]	0.34904	0.75652	2.26902	2.30955	0.30556	2.28812	0.43600	2.29627	2.31630	2.35683
2-1/2 —8 [c]	0.55700	1.13750	2.71953	2.77778	0.49450	2.75044	0.68200	2.76216	2.79062	2.84887
3 —8 [c]	0.64100	1.20000	3.34062	3.39887	0.57850	3.37678	0.76600	3.38850	3.41562	3.47387

[a] Major diameter is based on crest minimum truncation of 0.20 p.
[b] Maximum and minimum pitch-diameter steps are gaging limits. Notch formulas on drawing apply to all sizes.
[c] For reference only above 1 — 11-1/2 NPSF.

DRYSEAL SAE SHORT TAPER PIPE THREAD (L_3 SHORT) LENGTH STEP-LIMIT PLUG GAGES

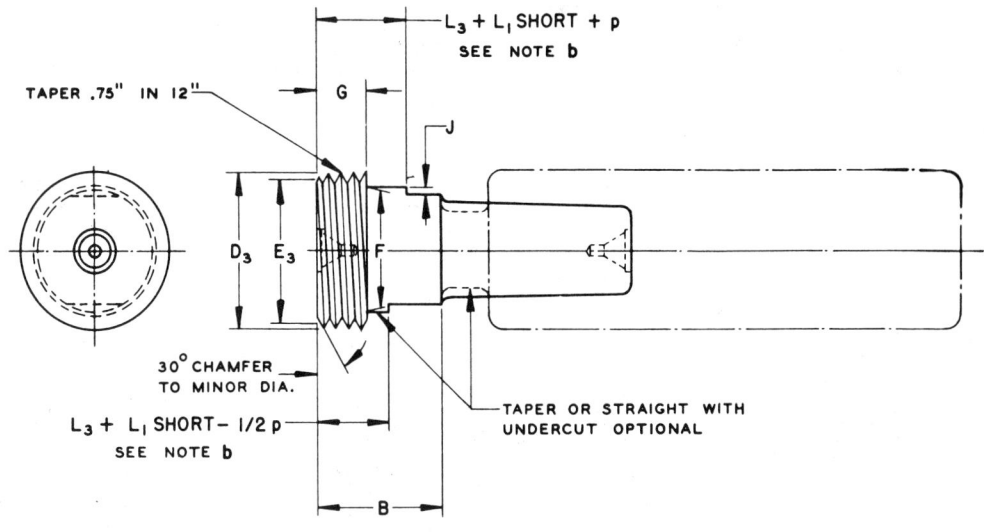

TABLE C11-1—THREAD FLATS

Threads per in.	F	G
27	0.0086	0.0107
18	0.0128	0.0160
14	0.0165	0.0206
11-1/2	0.0201	0.0251
8	0.0289	0.0361

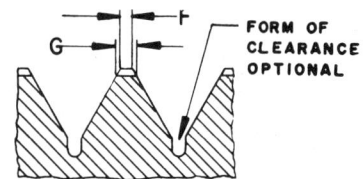

Marking—In addition to the regular markings, Dryseal SAE Short Taper Pipe Thread (L_3) Plug Gages will be marked PTF—SAE Short (L_3 Short).

Thread Form—The threads in all particulars excepting truncation shall conform to American Standard Taper Pipe Thread practice. Crests of threads at major diameter shall be truncated 0.20p minimum to 0.25p maximum, producing the minimum and maximum widths of flat specified in Table C11-1.

All other thread dimensions shall be within tolerances specified for the Dryseal American Standard Pipe Thread Working Plug Gages (ASA B2.2). Other gage details shall conform to American Gage Design Standards published in Commercial Standard CS8.

TABLE C11—BASIC DIMENSIONS OF DRYSEAL SAE SHORT TAPER PIPE THREAD (L_3 SHORT) LENGTH STEP-LIMIT PLUG GAGES

Size	Small End		Relief Dia, F [E_3 + (0.0625 × 4p) — Sharp-V Thread Height — 0.020 to 0.025 below Sharp Root] +0.005 −0.000	Four Threads, (G) ($L_3 + p$)	Pitch Dia Gaging Step[b] Plus 3 Threads		Black Length, (B)	Notch Depth, (J) +0.005 −0.000
	Pitch Dia, E_3	Major Dia,[a] D_3			($L_3 + L_1$ Short − 1/2 p) Min	($L_3 + L_1$ Short + p) Max		
1/16—27	0.2642	0.2815	0.216	0.1482	0.2156	0.2711	0.38	0.030
1/8 —27	0.3566	0.3738	0.309	0.1482	0.2171	0.2726	0.41	0.030
1/4 —18	0.4670	0.4928	0.409	0.2222	0.3111	0.3945	0.50	0.030
3/8 —18	0.6016	0.6275	0.542	0.2222	0.3233	0.4067	0.56	0.030
1/2 —14	0.7451	0.7783	0.676	0.2857	0.4271	0.5343	0.69	0.040
3/4 —14	0.9543	0.9876	0.886	0.2857	0.4462	0.5533	0.72	0.040
1 —11-1/2	1.1973	1.2379	1.118	0.3478	0.5304	0.6609	0.88	0.050
1-1/4 —11-1/2	1.5408	1.5814	1.462	0.3478	0.5504	0.6809	0.88	0.050
1-1/2 —11-1/2	1.7798	1.8203	1.701	0.3478	0.5504	0.6809	0.88	0.050
2 —11-1/2	2.2527	2.2932	2.174	0.3478	0.5644	0.6969	0.88	0.050
2-1/2 —8	2.6961	2.7543	2.590	0.5000	0.8695	1.0570	1.50	0.050
3 —8	3.3172	3.3754	3.214	0.5000	0.9535	1.1410	1.50	0.050

[a] Major diameter is based upon crest minimum truncation of 0.20 p.

[b] Maximum and minimum pitch-diameter steps are gaging limits. Notch formulas on drawing apply to all sizes.

DRYSEAL AMERICAN INTERMEDIATE INTERNAL STRAIGHT PIPE THREAD (L_1) STEP-LIMIT PLUG GAGES

Taper lock design, range 1/8 to 1 in., inclusive

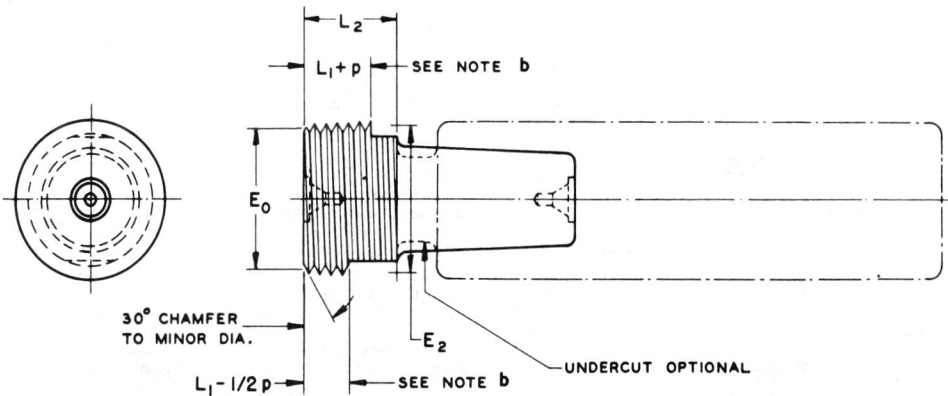

Marking—In addition to the regular markings, Dryseal American Intermediate Internal Straight Pipe Thread Taper Plug Gages will be marked NPSI (L_1).

Thread Form—The threads in all particulars excepting truncation shall conform to American Standard Taper Pipe Thread Practice. Crests of threads at major diameter shall be truncated 0.20p minimum to 0.25p maximum,

TABLE C12-1—THREAD FLATS

Threads per in.	F	G
27	0.0086	0.0107
18	0.0128	0.0160
14	0.0165	0.0206
11-1/2	0.0201	0.0251
8	0.0289	0.0361

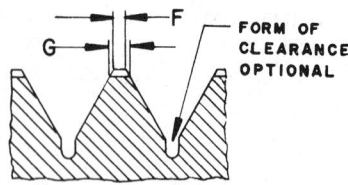

producing the minimum and maximum widths of flat specified in Table C12-1.

All other thread dimensions shall be within tolerances specified for the Dryseal American Standard Pipe Thread Working Plug Gages (ASA B2.2). Other gage details shall conform to American Gage Design Standards published in Commercial Standard CS8.

TABLE C12—BASIC DIMENSIONS OF DRYSEAL AMERICAN INTERMEDIATE INTERNAL STRAIGHT PIPE THREAD (L_1) STEP-LIMIT PLUG GAGES

Size	L_1	L_2	Small End		Min Pitch Dia Gaging Step[b]		Max Pitch Dia Gaging Step[b]		Large End	
			Pitch Dia, E_0	Major Dia[a]	$L_1 - 1/2p$	Pitch Dia	$L_1 + p$	Pitch Dia	Pitch Dia, E_2	Major Dia[a]
1/16 —27	0.1600	0.26113	0.27118	0.28844	0.14148	0.28002	0.19704	0.28350	0.28750	0.30476
1/8 —27	0.1615	0.26385	0.36351	0.38077	0.14298	0.37245	0.19854	0.37592	0.38000	0.39726
1/4 —18	0.2278	0.40178	0.47739	0.50328	0.20002	0.48989	0.28336	0.49510	0.50250	0.52839
3/8 —18	0.2400	0.40778	0.61201	0.63790	0.21222	0.62527	0.29556	0.63048	0.63750	0.66339
1/2 —14	0.3200	0.53371	0.75843	0.79170	0.28428	0.77620	0.39143	0.78289	0.79179	0.82506
3/4 —14	0.3390	0.54571	0.96768	1.00095	0.30328	0.98664	0.41043	0.99333	1.00179	1.03506
1 —11-1/2	0.4000	0.68278	1.21363	1.25416	0.35652	1.23592	0.48696	1.24406	1.25630	1.29683
1-1/4 —11-1/2[c]	0.1200	0.70678	1.55713	1.59766	0.37652	1.58066	0.50696	1.58882	1.60130	1.64183
1-1/2 —11-1/2[c]	0.1200	0.72348	1.79609	1.83662	0.37652	1.81962	0.50696	1.82778	1.84130	1.88183
2 —11-1/2[c]	0.4360	0.75652	2.26902	2.30955	0.39252	2.29355	0.52296	2.30170	2.31630	2.35683
2-1/2 —8[c]	0.6820	1.13750	2.71953	2.77778	0.61950	2.75825	0.80700	2.76997	2.79062	2.84887
3 —8[c]	0.7660	1.20000	3.34062	3.39887	0.70350	3.38459	0.89100	3.39631	3.41562	3.47387

[a] Major diameter is based on crest minimum truncation of 0.20 p.
[b] Maximum and minimum pitch-diameter steps are gaging limits. Notch formulas on drawing apply to all sizes.
[c] For reference only

APPENDIX D—SPECIAL SHORT, SPECIAL EXTRA SHORT, FINE, AND SPECIAL DIAMETER PITCH COMBINATION DRYSEAL PIPE THREADS

General Information—The SAE Dryseal Pipe Thread Series are based on thread length. Full thread lengths and clearances for Dryseal Standard and SAE Short Series are shown in Tables 2, 3, and 4 of the standard and the differences between them are described in the text under the series headings. These full thread lengths and clearances should be used in design applications wherever possible.

Design limitations, economy of material, permanent installation or other limiting conditions may not permit the use of either of the full thread lengths and shoulder lengths in the preceding tables for the above thread series. To meet these conditions two special thread series have been established as shown in Fig. 1. The deviations from standard practice are described below.

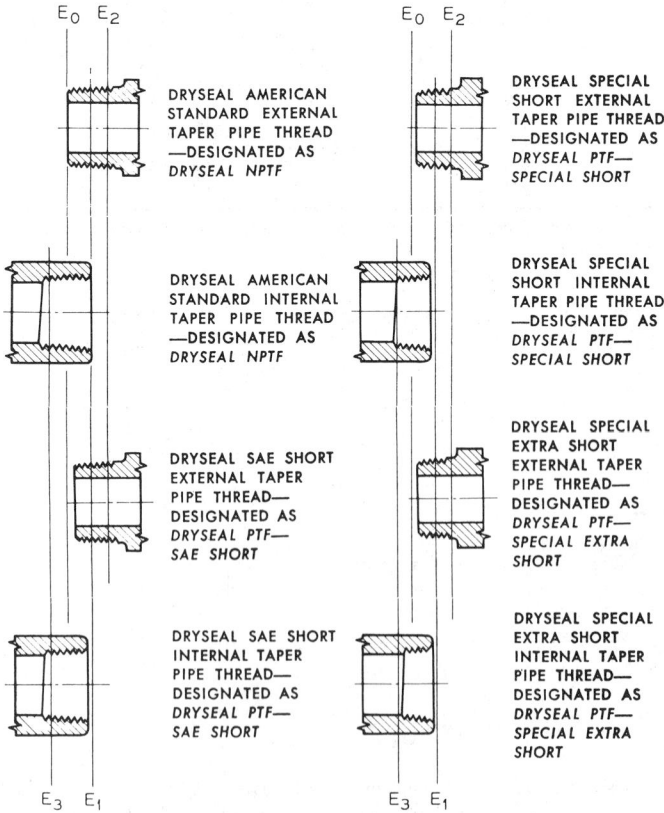

FIG. 1—THREAD LENGTH AND DESIGNATION

Dryseal Special Short Taper Pipe Thread (PTF—SPL Short)—Threads of this series conform in all respects to the PTF—SAE Short threads except that the full thread length has been further shortened by eliminating one thread at the large end of external threads or eliminating one thread at the small end of internal threads. Gaging is the same as for PTF—SAE Short except the L_2 ring thread gage for external thread length and taper or the L_3 plug thread gage for internal thread length and taper cannot be used. Tolerance must be altered and co-ordinated as described in paragraph on Limitation of Assembly. The designation of this series thread is for example:

$\frac{1}{8}$—27 DRYSEAL PTF—SPL Short

Dryseal Special Extra Short Taper Pipe Thread (PTF—SPL Extra Short)—Threads of this series conform in all respects to the PTF—SAE Short threads except that the full thread length has been further shortened by eliminating two threads at the large end of external threads or eliminating two threads at the small end of internal threads. Gaging is the same as for PTF—SAE Short except the L_2 ring thread gage for external thread length and taper or the L_2 plug thread gage for internal thread length and taper cannot be used. Tolerance must be altered and co-ordinated as described in paragraph on Limitation of Assembly. The designation of this series thread is for example:

$\frac{1}{8}$—27 DRYSEAL PTF—SPL Extra Short

Limitation of Assembly—Standard combinations and applications of the various series Dryseal Pipe Threads are given in the preceding thread descriptions. However, where special combinations are used, additional considerations as outlined below must be observed. These should be designated with the suffix "SPL" and gaging tolerance should be specified.

PTF—SPL Short External	May[a]	PTF—SAE Short Internal
PTF—SPL Extra Short External	Assemble With	NPSF Internal
		PTF—SPL Short Internal
PTF—SPL Short Internal	May[a]	PTF—SPL Extra Short Internal
PTF—SPL Extra Short Internal	Assemble With	PTF—SAE Short External

[a] Only when the external thread or the internal thread or both are held closer than the standard tolerance, the external toward the minimum and the internal toward the maximum pitch diameter to provide a minimum of one turn hand engagement. At extreme tolerance limits the shortened full thread lengths reduce hand engagement and threads may not start

PTF—SPL Short External	May[a]	NPTF or NPSI Internal
PTF—SPL Extra Short External	Assemble With	
PTF—SPL Short Internal	May[a]	NPTF External
PTF—SPL Extra Short Internal	Assemble With	

[a] Only when both the internal thread and the external thread are held closer than the standard tolerance, the internal toward the minimum and the external toward the maximum pitch diameter to provide a minimum of two turns or wrench make up and sealing. At extreme tolerance limits the shortened full thread lengths reduce wrench make up and threads may not seal.

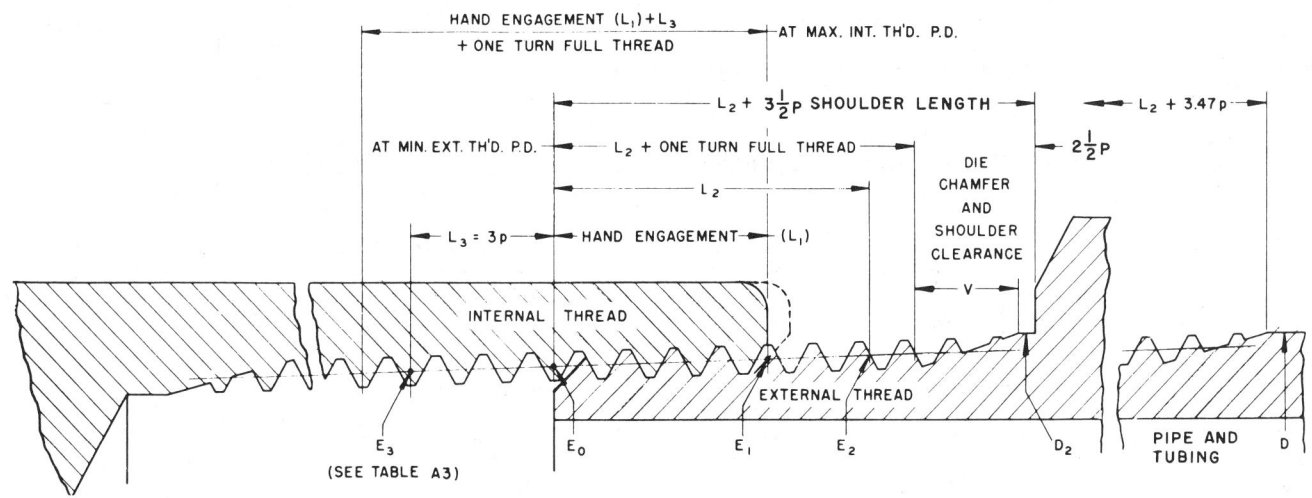

TABLE D1—BASIC DIMENSIONS OF DRYSEAL TAPER PIPE THREAD, FINE, F-PTF

F-PTF Size (Fine)	Pitch, p	Pitch Dia at Small End of External Thread, E_0	Pitch Dia at Large End of Internal Thread, E_1	Pitch Dia at Large End of External Thread, E_2	Pitch Dia at Small End of Internal Thread, E_3	Hand Engagement, L_1		Length of Full Thread,a,b Internal (L_1+L_3) and External (L_2)		Vanish Threads- V Plus Full Thread Tolerance Plus Shoulder Clearance, ($V+1p+1/2p$)		Shoulder Length, L_2+3-1/2p	Thread for Draw		OD of Fitting, D_2	OD of Pipe, D
	in.	in.	in.	in.	in.	in.	Thread	In.	Thread	in.	Thread	in.	in.	Thread	in.	in.
1	2	3	4	5	6	7	8	9	10	11	12	13	14	15	16	17
1/4—27	0.03704	0.49826	0.50807	0.51501	0.49132	0.157	4.23	0.268	7.23	0.1296	3.5	0.3975	0.1111	3.0	0.546	0.540
3/8—27	0.03704	0.63301	0.64307	0.65001	0.62607	0.161	4.34	0.272	7.34	0.1296	3.5	0.4015	0.1111	3.0	0.681	0.675
1/2—18	0.05556	0.77655	0.79205	0.80249	0.76613	0.248	4.47	0.415	7.47	0.1944	3.5	0.6096	0.1667	3.0	0.850	0.840
3/4—18	0.05556	0.98597	1.00210	1.01247	0.97555	0.258	4.64	0.424	7.64	0.1944	3.5	0.6189	0.1667	3.0	1.060	1.050
1 —14	0.07143	1.23173	1.25342	1.26679	1.21834	0.347	4.85	0.561	7.85	0.2500	3.5	0.8109	0.2143	3.0	1.327	1.315
1-1/4—14	0.07143	1.57550	1.59837	1.61181	1.56211	0.366	5.13	0.581	8.13	0.2500	3.5	0.8306	0.2143	3.0	1.672	1.660
1-1/2—14	0.07143	1.81464	1.83839	1.85176	1.80125	0.380	5.32	0.594	8.32	0.2500	3.5	0.8443	0.2143	3.0	1.912	1.900
2 —14	0.07143	2.28794	2.31338	2.32675	2.27455	0.407	5.70	0.621	8.70	0.2500	3.5	0.8714	0.2143	3.0	2.387	2.375

a External thread tabulated full thread lengths include chamfers not exceeding one and one-half pitches (threads) length.

b Internal thread tabulated full thread lengths do not include countersink beyond the intersection of the pitch line and the chamfer cone (gaging reference point).

TABLE D2—BASIC DIMENSIONS OF DRYSEAL TAPER PIPE THREAD, SPECIAL, SPL-PTF, FOR THIN WALL NOMINAL SIZE OD TUBING

Tubing Dia,c D	Threads per in.	Pitch, p	Pitch Dia at Small End of External Thread, E_0	Pitch Dia at Large End of Internal Thread, E_1	Pitch Dia at Large End of External Thread, E_2	Pitch Dia at Small End of Internal Thread, E_3	Hand Engagement, L_1		Length of Full Thread,a,b Internal (L_1+L_3) and External (L_2)		Thread for Draw	
in.		in.	in.	in.	in.	in.	in.	Thread	in.	Thread	in.	Thread
1	2	3	4	5	6	7	8	9	10	11	12	13
1/2	27	0.03704	0.45833	0.46806	0.47500	0.45139	0.1556	4.2	0.2667	7.2	0.1111	3.0
5/8	27	0.03704	0.58310	0.59306	0.60000	0.57616	0.1593	4.3	0.2704	7.3	0.1111	3.0
3/4	27	0.03704	0.70787	0.71806	0.72500	0.70093	0.1630	4.4	0.2741	7.4	0.1111	3.0
7/8	27	0.03704	0.83264	0.84306	0.85000	0.82570	0.1667	4.5	0.2778	7.5	0.1111	3.0
1	27	0.03704	0.95740	0.96805	0.97500	0.95046	0.1704	4.6	0.2815	7.6	0.1111	3.0

a External thread tabulated full thread lengths include chamfers not exceeding one and one-half pitches (threads) length.

b Internal thread tabulated full thread lengths do not include countersink beyond the intersection of the pitch line and the chamfer cone (gaging reference point).

c This denotes nominal outside diameter of tubing and should not be confused with nominal pipe diameter and thread designations.

Fine Thread Series—The need for finer pitches for nominal pipe sizes has brought into use applications of 27 threads per in. to 1/4 and 3/8 in. pipe sizes. There may be other needs which require finer pitches for larger pipe sizes. It is recommended that the existing threads per in. be applied to the next size larger pipe size for a fine thread series such as shown in Table D1. This series applies to external and internal threads of full length and is suitable for applications where threads finer than NPTF are required.

Special Thread Series—Other applications of diameter-pitch combinations have also come into use where taper pipe threads are applied to nominal size thin wall tubing such as shown in Table D2. This series applies to external and internal threads of full length and is applicable to thin wall nominal outside diameter tubing. The pitch is uniform at 27 threads per in. Dimensions of other combinations of diameter and pitch, in addition to those listed in Table D2, may be developed by the use of formulae.

Formulae for Diameter and Length of Thread—Basic diameter and length of thread for sizes of Dryseal Taper Pipe Thread Fine (F-PTF), and Dryseal Taper Pipe Thread Special (SPL—PTF) given in Tables D1 and D2 are based on the following formulae:

D = outside diameter of pipe or tubing (in.)
p = pitch of thread (in.)
Diametral taper = 0.75 in. per 12.00 in. of length
Basic pitch diameter at small end of external thread
  $E_0 = D - (0.05D + 1.1)p$
Basic pitch diameter at large end of internal thread
  $E_1 = E_0 + 0.0625 L_1 = D - 0.8625\ p$
Basic pitch diameter at large end of external thread
  $E_2 = E_0 + 0.0625 L_2 = D - 0.675\ p$
Basic pitch diameter at small end of internal thread
  $E_3 = E_0 - 0.0625 L_3 = D - (0.05D + 1.2875)\ p$
Basic length of thread for hand engagement
  $L_1 = (0.8D + 3.8)\ p$
Basic length of full and effective thread
  $L_2 = (0.8D + 6.8)\ p$
Basic length of internal thread from end of hand engagement (E_0) to small end of internal thread (E_3)
  $L_3 = 3p$

Tolerance shall be equal to plus or minus the taper of 1 thread on the diameter

Designations—The designation for a fine thread series pipe thread should include letter F and omit N, for example: $\frac{1}{4}$—27 Dryseal F-PTF. The designation for a special thread series pipe thread should include abbreviation SPL for special and omit letter N. Also the outside diameter of tubing should be given, for example: $\frac{1}{2}$—27 Dryseal SPL-PTF, OD 0.500.

APPENDIX E—SUPERSEDED GAGE DIMENSIONS AND GAGING PRACTICE FOR $\frac{1}{8}$ AND $\frac{1}{4}$ SIZE DRYSEAL PIPE THREADS

In this standard, the L_1 dimensions for the $\frac{1}{8}$—27 and $\frac{1}{4}$—18 sizes have been revised to correct for a disproportionate number of threads for hand engagement.

In the previous issue of this standard, the values of L_1 hand engagements in the tables of basic dimensions for the product were corrected, but the values in the tables of basic dimensions for gages were left unaltered since users were able to apply existing gages by modifying gaging practices and this allowed gage manufacturers an opportunity to reduce existing inventories. In this issue of the standard, the L_1 hand engagement dimensions affecting gages in Tables C1, C2 and C3 have been revised to agree with the product dimensions for future gage procurement.

Therefore, it should be noted that where basic-notch thread gages having superseded dimensions (see Table E1) are being used for gaging the $\frac{1}{8}$—27 and $\frac{1}{4}$—18 sizes, the formerly observed deviations from specified gaging practice should be applied as follows:

Internal threads gaged by the Position Method should be $\frac{1}{2}$ turn smaller for the $\frac{1}{8}$—27 size and $\frac{1}{2}$ turn larger than the $\frac{1}{4}$—18 size than the specified tolerances given in Appendix C.

External threads gaged by the Turns Engagement Method should be $\frac{1}{2}$ greater for the $\frac{1}{8}$—27 size and $\frac{1}{2}$ turn less for the $\frac{1}{4}$—18 size than the basic turns specified in Appendix C.

Table E1 lists the dimensions derived from the superseded L_1 dimensions of 0.1800 in. for the $\frac{1}{8}$—27 size and 0.2000 in. for the $\frac{1}{4}$—18.

TABLE E1—BASIC DIMENSIONS OF SUPERSEDED BASIC-NOTCH GAGES

Size	L_2 Ring Gage			L_1 Plug and Ring Gages		L_1 Ring Gage	L_1 Plug Gage	L_3 Plug Gage
	Pitch Dia at $L_1 - p$ (E_2)	Minor Dia at $L_1 - p$	$L_1 - p$	L_1	Pitch Dia (E_1)	Minor Dia at Large End	Major Dia at Gaging Notch	3 Threads Plus L_1 ($L_3 + L_1$)
1/8—27	0.37244	0.35518	0.14296	0.1800	0.37476	0.35750	0.39202	0.2911
1/4—18	0.48642	0.46053	0.14444	0.2000	0.48989	0.46400	0.51578	0.3667

15 Fasteners

HEX BOLTS—SAE J105

SAE Standard

Report of Screw Threads Division approved June 1911, revised by Screw Threads Committee January 1957 and last revised by Fasteners Committee June 1969. Conforms in general, for the products included, with USA Standard-Square and Hex Bolts and Screws, USAS B18.2.1.

Scope—Included herein are complete general and dimensional data for the various types of square and hex bolts recognized as SAE Standard and commonly used in automotive and other ground based motor vehicles and industrial equipment. Also included are appendixes covering wrench openings for bolts and thread runout sleeve gages.

The inclusion of dimensional data in this standard is not intended to imply that all of the products described are stock production sizes. Users are requested to consult with manufacturers concerning lists of stock production sizes.

Dimensions—All dimensions in this standard are in inches unless otherwise stated.

Options—Options, where specified, shall be at the discretion of the manufacturers unless otherwise agreed by manufacturer and user.

CONTENTS

General Specifications
Dimensions
  Finished Hexagon Bolts..................................... Table 1
  Heavy Finished Hexagon Bolts............................ Table 2
  Thread Runout Sleeve Gages .... Appendix A
  Wrench Openings for Bolts................................ Appendix B

GENERAL SPECIFICATIONS

General—The following specifications cover all of the bolt categories included in this standard. Specifications which are peculiar to a particular product or products are so indicated by references to the applicable dimensional tables.

Heads

Head Height—The head height specified in the dimensional tables is the overall distance, measured parallel to the axis of bolt, from the top of the head to the bearing surface and includes the thickness of the washer face where provided.

Head Chamfer—The top of head shall be flat with chamfered corners. The diameter of chamfer circle shall be equal to the maximum width across flats within a tolerance of −15%.

Head Taper—No transverse section through the head between 25 and 75% of the actual head height as measured from the bearing surface shall be less than the minimum width across flats. The maximum width across flats shall not be exceeded except for milled-from-bar nonferrous products where the maximum (basic) width may conform with the commercial tolerances of the bar stock material.

Head Concentricity—The axis of the head shall be concentric with the axis of the bolt body (determined by one diameter length of body under the head) within a tolerance equal to 3% (6% FIR) of the maximum width across flats.

Bearing Surface—The bearing surface shall be flat and washer faced and shall be at a right angle to the axis of body within a tolerance of 2 deg for 1 in. size or smaller, and within 1 deg for larger sizes. The bearing surface for finished hex bolts in Table 1 may have chamfered corners. The diameter of bearing surface shall be equal to 95% of maximum width across flats within a tolerance of ±5%.

Length

Measurement—The length of bolt shall be measured, parallel to the axis, from the bearing surface of the head to the extreme end of the bolt, including the point.

Tolerance on Length—Tolerance on length shall be as tabulated below:

Nominal Bolt Length	Nominal Bolt Size				
	1/4 to 3/8	7/16 and 1/2	9/16 to 3/4	7/8 and 1	1-1/8 to 1-1/2
	Tolerance on Length				
Up to 1, incl	−0.03	−0.03	−0.03		
Over 1 to 2-1/2, incl	−0.04	−0.06	−0.08	−0.10	−0.12
Over 2-1/2 to 4, incl	−0.06	−0.08	−0.10	−0.14	−0.16
Over 4 to 6, incl	−0.10	−0.10	−0.10	−0.16	−0.18
Over 6	−0.18	−0.18	−0.18	−0.20	−0.22

Threads

Form and Tolerance—Threads on all products shall be Unified Standard, Class 2A. For external threads with additive finish, the maximum diameters of Class 2A may be exceeded by the amount of the allowance; that is, the Class 2A maximum diameters shall apply to an unplated part or to a part before plating, whereas the basic diameters (Class 2A maximum diameters plus the allowance) shall apply to a part after plating.

Series—Threads shall be coarse (UNC), fine (UNF), or 8-thread (8UN) series.

Runout—the total runout (eccentricity and angularity) of thread in relation to the body shall be such that for sizes up to and including $\frac{3}{4}$ in., the product shall screw at least two full threads into a tapped hole counterbored to provide 0.031 in. diametral clearance over the maximum body diameter, to a depth equal to the length of the product less one diameter. The starting thread of the tapped hole shall be countersunk to the diameter of the counterbore. The tapped hole shall have a Class 2B maximum pitch diameter. The inspection fixture shall be hardened. For products over $\frac{3}{4}$ in., the diametral clearance of the counterbored hole shall be 0.062 in. Suitable gages are shown in Appendix A.

Thread Length

Measurement—For purposes of this standard, thread length shall be the distance, measured parallel to the axis of bolt, from the extreme end of the bolt to and including the last complete (full form) thread.

Long Bolts—The minimum thread length on all products shall be equal to twice the basic bolt diameter plus 0.25 in. for length of 6 in., or shorter, and plus 0.50 in. for longer lengths. A tolerance of +0.19 in. or $2\frac{1}{2}$ threads, whichever is greater, shall apply.

Short Bolts—Products too short to accommodate the minimum thread length the distance from the bearing surface to the first complete thread, measured with a thread ring gage, shall not exceed the length of $2\frac{1}{2}$ threads on sizes 1 in. or smaller, and $3\frac{1}{2}$ threads on larger sizes.

Points—Finished hex bolts shall have the point chamfered from approximately 0.016 in. below the minor diameter of the thread to a length of $\frac{1}{2}$ to

Finished Hex Bolts

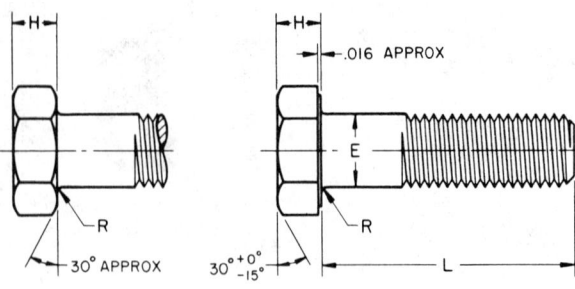
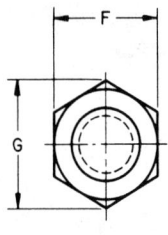

TABLE 1—DIMENSIONS OF FINISHED HEX BOLTS

Nominal Size or Basic Bolt Diameter	Body Dia, E		Width Across Flats, F			Width Across Corners, G		Height, H			Radius of Fillet, R	
	Max	Min	Basic	Max	Min	Max	Min	Basic	Max	Min	Max	Min
1/4	0.2500	0.2450	7/16	0.4375	0.428	0.505	0.488	5/32	0.163	0.150	0.025	0.015
5/16	0.3125	0.3065	1/2	0.5000	0.489	0.577	0.557	13/64	0.211	0.195	0.025	0.015
3/8	0.3750	0.3690	9/16	0.5625	0.551	0.650	0.628	15/64	0.243	0.226	0.025	0.015
7/16	0.4375	0.4305	5/8	0.6250	0.612	0.722	0.698	9/32	0.291	0.272	0.025	0.015
1/2	0.5000	0.4930	3/4	0.7500	0.736	0.866	0.840	5/16	0.323	0.302	0.025	0.015
9/16	0.5625	0.5545	13/16	0.8125	0.798	0.938	0.910	23/64	0.371	0.348	0.045	0.020
5/8	0.6250	0.6170	15/16	0.9375	0.922	1.083	1.051	25/64	0.403	0.378	0.045	0.020
3/4	0.7500	0.7410	1-1/8	1.1250	1.100	1.299	1.254	15/32	0.483	0.455	0.045	0.020
7/8	0.8750	0.8660	1-5/16	1.3125	1.285	1.516	1.465	35/64	0.563	0.531	0.065	0.040
1	1.0000	0.9900	1-1/2	1.5000	1.469	1.732	1.675	39/64	0.627	0.591	0.095	0.060
1-1/8	1.1250	1.1140	1-11/16	1.6875	1.631	1.949	1.859	11/16	0.718	0.658	0.095	0.060
1-1/4	1.2500	1.2390	1-7/8	1.8750	1.812	2.165	2.066	25/32	0.813	0.749	0.095	0.060
1-3/8	1.3750	1.3630	2-1/16	2.0625	1.994	2.382	2.273	27/32	0.878	0.810	0.095	0.060
1-1/2	1.5000	1.4880	2-1/4	2.2500	2.175	2.598	2.480	15/16	0.974	0.902	0.095	0.060

[a] Where specifying nominal size in decimals, zeros in the fourth decimal place shall be omitted. "Finished" as used in title refers to the quality of manufacture and the degree of tolerance and does not indicate that surfaces are completely machined.

Sizes to and including 1-1/2 in. are unified dimensionally with British and Canadian Standards Additional requirements given in the General Specifications shall apply.

Heavy Finished Hex Bolts

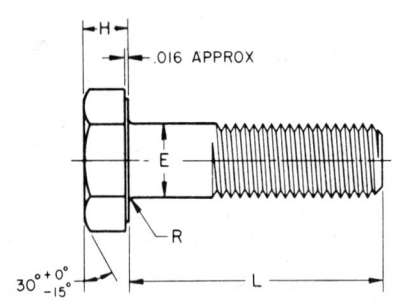
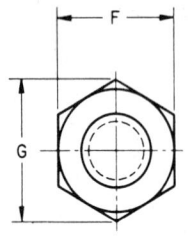

TABLE 2—DIMENSIONS OF HEAVY FINISHED HEX BOLTS

Nominal Size or Basic Bolt Diameter	Body Dia, E[b]		Width Across Flats, F			Width Across Corners, G		Height, H			Radius of Fillet, R	
	Max	Min	Basic	Max	Min	Max	Min	Basic	Max	Min	Max	Min
1/2	0.5000	0.4930	7/8	0.8750	0.850	1.010	0.969	5/16	0.323	0.302	0.031	0.009
5/8	0.6250	0.6170	1-1/16	1.0625	1.031	1.227	1.175	25/64	0.403	0.378	0.062	0.021
3/4	0.7500	0.7410	1-1/4	1.2500	1.212	1.443	1.383	15/32	0.483	0.455	0.062	0.021
7/8	0.8750	0.8660	1-7/16	1.4375	1.394	1.660	1.589	35/64	0.563	0.531	0.062	0.031
1	1.0000	0.9900	1-5/8	1.6250	1.575	1.876	1.796	39/64	0.627	0.591	0.093	0.062
1-1/8	1.1250	1.1140	1-13/16	1.8125	1.756	2.093	2.002	11/16	0.718	0.658	0.093	0.062
1-1/4	1.2500	1.2390	2	2.0000	1.938	2.309	2.209	25/32	0.813	0.749	0.093	0.062
1-3/8	1.3750	1.3630	2-3/16	2.1875	2.119	2.526	2.416	27/32	0.878	0.810	0.093	0.062
1-1/2	1.5000	1.4880	2-3/8	2.3750	2.300	2.742	2.622	15/16	0.974	0.902	0.093	0.062

[a] Where specifying nominal sizes in decimals, zeros in the fourth decimal place shall be omitted.
[b] Products within the hot forged size range may have a reasonable swell for a distance of up to two bolt diameters measured from underside of head. Maximum body diameter at the swell shall not exceed 1.03 times basic bolt diameter for sizes up to and including 3/4 in., and 1.02 times basic bolt diameter for sizes over 3/4 in.

"Finished" as used in title refers to the quality of manufacture and the degree of tolerance and does not indicate that surfaces are completely machined.

Sizes to and including 2 in. are unified dimensionally, except for thickness and minimum body diameter, with British and Canadian Standards.

Additional requirements given in the General Specifications shall apply.

1½ threads. Finished heavy hex bolts shall have chamfered or round point, the length of which shall not exceed 1½ threads.

Materials
Steel—Suitable properties for steel bolts are covered in SAE J429. Products shall normally conform to SAE Grades 2, 5, or 8 as specified by user.

Other Materials—Where specified, products may be made from brass, corrosion resisting steel, aluminum alloy, or other materials as agreed upon by the manufacturer and user.

Finish
Plain—Unless otherwise specified, bolts shall be supplied plain (unplated or uncoated), as processed.

Plated—Where plating is specified, the thickness or quality of plating shall be measured or tested on the side of the bolt head.

Defects—Bolts shall be free from burrs, seams, laps, loose scale, and any other defects that affect serviceability.

APPENDIX A—THREAD RUNOUT SLEEVE GAGE

Gages capable of checking the thread eccentricity and bow of shank on bolts are illustrated below.

The construction at right permits use of various length sleeves to accommodate different lengths of product. Ring gage A is centered on the sleeve B by means of positioning plug E and is secured in position by means of attachment screws C. The ring gage is also set to Class 2B maximum pitch diameter by positioning plug E.

Diameter D, of counterbore or hole in sleeve, equals the nominal diameter of the bolt or screw plus the run-out allowance. The sleeve length or counterbore depth L should be such that entering face of gage extends beyond the last thread of the product to be inspected, but should not exceed 3 in.

Failure of the product to enter the threads of gage or interference between the sides of hole and the product while engaging threads of gage indicates excessive thread runout.

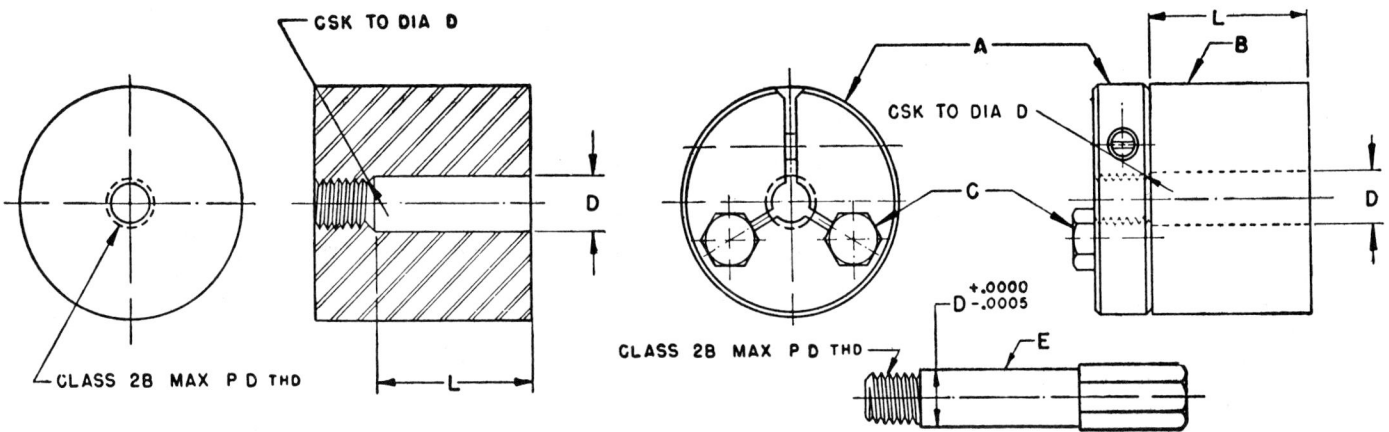

OPTIONAL CONSTRUCTIONS

A —Ring gage set to Class 2B maximum pitch diameter.
B —Sleeve.
C —Attachment screws.
D —Diameter of hole in sleeve equals maximum bolt diameter plus clearance allowance.
E —Positioning and setting plug for adjusting ring gage to Class 2B maximum pitch diameter and centering sleeve.
L —Length of sleeve equals length of bolt minus one diameter of bolt. L shall not exceed 3 in.

To insure adequate service life, gages shall be hardened.

APPENDIX B—WRENCH OPENINGS FOR HEX BOLTS

Nominal Size of Wrench[a] also Basic (Maximum) Width Across Flats of Bolt and Screw Heads	Allowance[b] between Bolt or Screw Head and Jaws of Wrench	Wrench Openings[c]			Finished Hex and Hi-Head Finished Hex Bolts	Heavy Finished Hex Bolt
		Min	Tol	Max		
7/16	0.4375	0.440	0.006	0.446	1/4	—
1/2	0.5000	0.504	0.006	0.510	5/16	—
9/16	0.5625	0.566	0.007	0.573	3/8	—
5/8	0.6250	0.629	0.007	0.636	7/16	—
3/4	0.7500	0.755	0.008	0.763	1/2	—
13/16	0.8125	0.818	0.008	0.826	9/16	—
7/8	0.8750	0.880	0.008	0.888	—	1/2
15/16	0.9375	0.944	0.009	0.953	5/8	—
1-1/16	1.0625	1.068	0.009	1.077	—	5/8
1-1/8	1.1250	1.132	0.010	1.142	3/4	—
1-1/4	1.2500	1.257	0.010	1.267	—	3/4
1-5/16	1.3125	1.320	0.011	1.331	7/8	—
1-7/16	1.4375	1.446	0.011	1.457	—	7/8
1-1/2	1.5000	1.508	0.012	1.520	1	—
1-5/8	1.6250	1.634	0.012	1.646	—	1
1-11/16	1.6875	1.696	0.012	1.708	1-1/8	—
1-13/16	1.8125	1.822	0.013	1.835	—	1-1/8
1-7/8	1.8750	1.885	0.013	1.898	1-1/4	—
2	2.0000	2.011	0.014	2.025	—	1-1/4
2-1/16	2.0625	2.074	0.014	2.088	1-3/8	—
2-3/16	2.1875	2.200	0.015	2.215	—	1-3/8
2-1/4	2.2500	2.262	0.015	2.277	1-1/2	—
2-3/8	2.3750	2.388	0.016	2.404	—	1-1/2

[a] Wrenches shall be marked with the "nominal size of wrench" which is equal to the basic (maximum) width across flats of the corresponding bolt or screw head.

[b] Allowance (minimum clearance) between maximum width across flats of bolt or screw head and jaws of wrench equals (0.005W + 0.001).

[c] Tolerance on wrench opening equals plus (0.005W + 0.004 from minimum). W equals nominal size of wrench.

HI-HEAD FINISHED HEX BOLTS—SAE J871a

SAE Standard

Report of Fasteners Committee approved June 1963 and last revised June 1969.

GENERAL SPECIFICATIONS

Scope—Included herein are the detailed general and dimensional specifications applicable to hi-head finished hex bolts. All general specifications and dimensions not shown shall conform to those applicable to finished hex bolts in Table 4 of SAE J104. Hi-head finished hex bolts are primarily intended for use in the heavy construction and industrial equipment industry. The increase in head height over that of the finished hex bolt assures good wrenchability where frequent servicing is necessary, where high torquing is a requirement, or where wear on the bolt head is a problem.

Materials

Steel—Suitable properties for steel bolts are covered in SAE J429. Hi-head bolts shall normally conform to SAE Grade 5, or higher grades, as specified by user.

TABLE 1—DIMENSIONS OF HI-HEAD FINISHED HEX BOLTS[a]

Nominal Size[b] or Basic Bolt Dia		Height, H			Nominal Size[b] or Basic Bolt Dia		Height, H		
		Basic	Max	Min			Basic	Max	Min
1/4	0.2500	3/16	0.194	0.181	9/16	0.5625	27/64	0.433	0.410
5/16	0.3125	15/64	0.242	0.227	5/8	0.6250	15/32	0.481	0.456
3/8	0.3750	9/32	0.290	0.273	3/4	0.7500	9/16	0.577	0.548
7/16	0.4375	21/64	0.338	0.319	7/8	0.8750	21/32	0.672	0.640
1/2	0.5000	3/8	0.386	0.364	1	1.0000	11/16	0.706	0.670

[a] All other dimensions are identical to those of finished hex bolts given in Table 4 of SAE J104. "Finished" as used in the title refers to quality of manufacture and the degree of tolerance and does not indicate that surfaces are completely machined.

[b] Where specifying nominal size in decimals, zeros in the fourth decimal place shall be omitted.

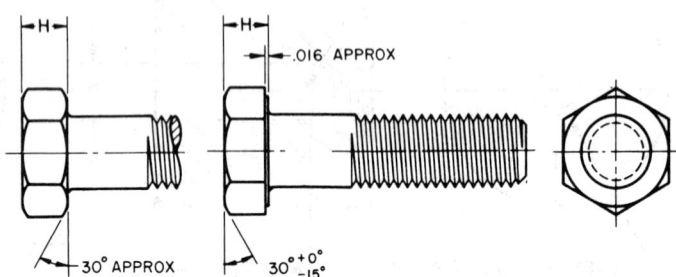

ROUND HEAD BOLTS—SAE J481

SAE Standard

Report of Screw Threads Division approved February 1929 and last revised by Screw Threads Committee January 1953. Reaffirmed without change by Fasteners Committee January 1961. Conforms with American Standard for Round Head Bolts, ASA B18.5-1952, except that ASA Tables 8, 9, and 11 are not included in this SAE Standard, and this SAE Standard covers only bolts with undersize body diameters.

Threads—Threads shall be National Coarse, Class 2A. For an interim period, Class 2 will be acceptable.

Thread Lengths—All threads are measured from end of bolt to last complete thread. Minimum thread length is two diameters plus 1/4 in. for bolts up to and including 6 in. long, and two diameters plus 1/2 in. for bolts over 6 in. long. When bolts are too short for formula thread length, thread will extend as close to neck as practicable.

Length of Bolt—Length, L, is measured from the largest diameter of head to end of bolt.

Length Tolerance—Tolerances for 6-in. length and less are plus or minus 1/32 in. for sizes 1/4 to 3/8 in.; plus or minus 1/16 in. for sizes 7/16 and 1/2 in.; plus or minus 1/8 in. for sizes 5/8 to 1 1/4 in.; and plus or minus 1/4 in. for sizes 1 3/8 to 1 1/2 in. Tolerances for lengths over 6 in. are plus or minus 1/16 in. for sizes 1/4 to 3/8 in.; plus or minus 3/32 in. for sizes 7/16 and 1/2 in.; plus or minus 3/16 in. for sizes 5/8 to 1 1/4 in.; and plus or minus 1/4 in. for sizes 1 3/8 to 1 1/2 in.

Points—Bolts may be sheared end with first thread incomplete, or they may be round end, or chamfered 45 deg to the minor diameter.

Fillets—Maximum radius of fillet under bolt head for sizes No. 10 to 1/2 in. diameter inclusive is 1/32 in. and for sizes 5/8 to 1 in. diameter inclusive is 1/16 in. These maximum fillets may be somewhat increased for nonferrous and corrosion resisting materials.

Body Diameter—Round head bolts are made in two body styles: (a) A full size body, with a maximum diameter somewhat greater than the nominal diameter (not an SAE Standard except for ribbed neck bolts, Table 3); and (b) an undersize body, with a minimum diameter approximating the pitch diameter of the thread and a maximum diameter never exceeding nominal (SAE Standard for all types except ribbed neck bolts). The body diameter of either style may be exceeded by a reasonable swelling or fin under the head, or under corners of the square necks, to the extent that serviceability is not affected.

Square Neck—The corners of the square neck need not be filled out for the total depth of square.

Heads—Because the heads of these products are not machined or trimmed, circumference may be somewhat irregular and edge may be rounded or flat.

Material—Suitable material for mild steel bolts is covered in SAE Standard, Mechanical and Chemical Requirements for Threaded Fasteners—SAE J429. Other materials will be as agreed upon by manufacturer and user.

Measurements—All dimensions are given in inches.

Defects—Bolts shall be free from burrs, seams, laps, loose scale, irregular surfaces, and so forth, that affect serviceability.

Round Head Square Neck Carriage Bolt

TABLE 1—DIMENSIONS OF ROUND HEAD SQUARE NECK BOLT[a]

Size		Body Dia		Dia of Head, A			Height of Head, H		Depth of Square, P		Width of Square, B	
D	Threads per in.	Max	Min	Basic		Max	Min	Max	Min	Max	Min	Max
No. 10	24	0.190	—	0.438	7/16	0.438	0.469	3/32	0.094	0.114	0.094	0.125	0.185	0.199
1/4	20	0.250	—	0.563	9/16	0.563	0.594	1/8	0.125	0.145	0.125	0.156	0.245	0.260
5/16	18	0.312	—	0.688	11/16	0.688	0.719	5/32	0.156	0.176	0.156	0.187	0.307	0.324
3/8	16	0.375	—	0.782	13/16	0.813	0.844	3/16	0.188	0.208	0.188	0.219	0.368	0.388
7/16	14	0.438	—	0.907	15/16	0.938	0.969	7/32	0.219	0.239	0.219	0.250	0.431	0.452
1/2	13	0.500	—	1.032	1-1/16	1.063	1.094	1/4	0.250	0.270	0.250	0.281	0.492	0.515
5/8	11	0.625	—	1.219	1-5/16	1.313	1.344	5/16	0.313	0.344	0.313	0.344	0.616	0.642
3/4	10	0.750	—	1.469	1-9/16	1.563	1.594	3/8	0.375	0.406	0.375	0.406	0.741	0.768
7/8	9	0.875	—	1.719	1-13/16	1.813	1.844	7/16	0.438	0.469	0.438	0.469	0.865	0.895
1	8	1.000	—	1.969	2-1/16	2.063	2.094	1/2	0.500	0.531	0.500	0.531	0.990	1.022

[a] In ordering, the maximum body diameter indicated in the table should be specified.

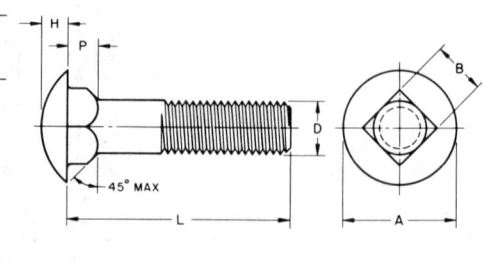

Round Head Short Square Neck Bolt

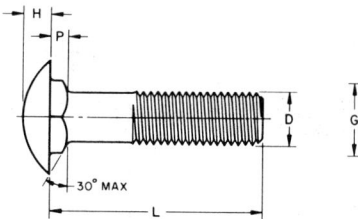
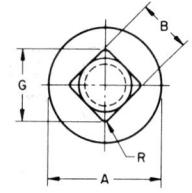

TABLE 2—DIMENSIONS OF ROUND HEAD SHORT SQUARE NECK BOLT[a]

Size		Body Dia	Dia of Head, A		Height of Head, H		Depth of Square, P		Width of Square, B		Width Across Corners, G	Radius Corners of Square, R
D	Threads per in.	Max	Min	Max	Min	Max	Min	Max	Min	Max	Max[b]	Max
1/4	20	0.250	0.563	0.594	0.125	0.145	0.093	0.124	0.245	0.260	0.368	1/32
5/16	18	0.312	0.688	0.719	0.156	0.176	0.093	0.124	0.307	0.324	0.458	1/32
3/8	16	0.375	0.782	0.844	0.188	0.208	0.125	0.156	0.368	0.388	0.549	3/64
7/16	14	0.438	0.907	0.969	0.219	0.239	0.125	0.156	0.431	0.452	0.639	3/64
1/2	13	0.500	1.032	1.094	0.250	0.270	0.125	0.156	0.492	0.515	0.728	3/64
5/8	11	0.625	1.219	1.344	0.313	0.344	0.187	0.218	0.616	0.642	0.908	5/64
3/4	10	0.750	1.469	1.594	0.375	0.406	0.187	0.218	0.741	0.768	1.086	5/64

[a] In ordering, the maximum body diameter indicated in the table should be specified.   [b] Based on theoretical sharp corners of maximum square.

Round Head Ribbed Neck Carriage Bolt

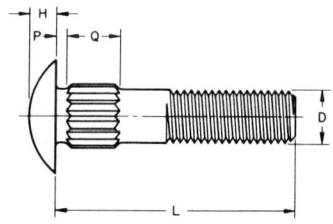
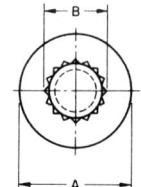

TABLE 3—DIMENSIONS OF ROUND HEAD RIBBED NECK[a] CARRIAGE BOLT[b]

Size		Body Dia	Dia of Head, A			Height of Head, H			Dia of Ribbed Neck, B	Distance of Ribs Below Head, P			Length of Ribs, q			Approximate Number of Ribs
D	Threads per in.	Max	Min	Basic	Max	Min		Max	Min	For L=7/8 or Less	For L=1 or More	Plus or Minus	For L=7/8 or Less	For L=1 and 1-1/8	For L=1-1/4 or More	
No. 10	24	0.199	0.438	7/16	0.438	0.469	3/32	0.094	0.114	0.210	0.031	0.063	0.031	0.188	0.313	0.500	9
1/4	20	0.260	0.563	9/16	0.563	0.594	1/8	0.125	0.145	0.274	0.031	0.063	0.031	0.188	0.313	0.500	10
5/16	18	0.324	0.688	11/16	0.688	0.719	5/32	0.156	0.176	0.340	0.031	0.063	0.031	0.188	0.313	0.500	12
3/8	16	0.388	0.782	13/16	0.813	0.844	3/16	0.188	0.208	0.405	0.031	0.063	0.031	0.188	0.313	0.500	12
7/16	14	0.452	0.907	15/16	0.938	0.969	7/32	0.219	0.239	0.470	0.031	0.063	0.031	0.188	0.313	0.500	14
1/2	13	0.515	1.032	1-1/16	1.063	1.094	1/4	0.250	0.270	0.534	0.031	0.063	0.031	0.188	0.313	0.500	16
5/8	11	0.642	1.210	1-5/16	1.313	1.344	5/16	0.313	0.344	0.660	0.094	0.094	0.031	0.188	0.313	0.500	19
3/4	10	0.768	1.469	1-9/16	1.563	1.594	3/8	0.375	0.406	0.785	0.094	0.094	0.031	0.188	0.313	0.500	22

[a] Included angle of ribs shall be approximately 90 deg.   [b] Only full size body bolts are furnished.

Round Head Fin Neck Carriage Bolt

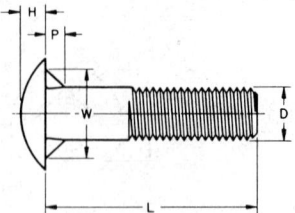 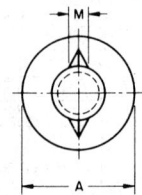

TABLE 4—DIMENSIONS OF ROUND HEAD FIN NECK CARRIAGE BOLT[a]

Size		Body Dia		Dia of Head, A			Height of Head, H			Depth of Fins, P			Distance Across Fins, W			Thickness of Fins, M		
D	Threads per in.	Max	Min	Basic	Min	Max	Basic	Min	Max	Basic	Min	Max	Basic	Min	Max	Basic	Min	Max
No. 10	24	0.190	0.438	7/16	0.438	0.469	3/32	0.094	0.114	5/64	0.078	0.088	3/8	0.375	0.395	5/64	0.078	0.098
1/4	20	0.250	0.563	9/16	0.563	0.594	1/8	0.125	0.145	3/32	0.094	0.104	7/16	0.438	0.458	3/32	0.094	0.114
5/16	18	0.312	0.688	11/16	0.688	0.719	5/32	0.156	0.176	1/8	0.125	0.135	17/32	0.531	0.551	1/8	0.125	0.145
3/8	16	0.375	0.782	13/16	0.813	0.844	3/16	0.188	0.208	9/64	0.141	0.151	5/8	0.625	0.645	9/64	0.141	0.161
7/16	14	0.438	0.907	15/16	0.938	0.969	7/32	0.219	0.239	11/64	0.172	0.182	23/32	0.719	0.739	11/64	0.172	0.192
1/2	13	0.500	1.032	1-1/16	1.063	1.094	1/4	0.250	0.270	3/16	0.188	0.198	13/16	0.813	0.833	3/16	0.188	0.208

[a] In ordering, the maximum body diameter indicated in the table should be specified.

114-Deg Countersunk Square Neck Carriage Bolt

TABLE 5—DIMENSIONS OF 114-DEG COUNTERSUNK SQUARE NECK CARRIAGE BOLT[a]

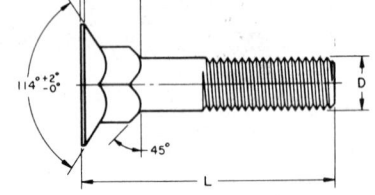 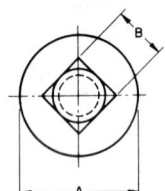

Size		Body Dia	Dia of Head, A			Feed Thickness, F	Depth of Square and Countersink, P			Width of Square, B	
D	Threads per in.	Max	Min	Max		F	Basic	Min	Max	Min	Max
No. 10	24	0.190	1/2	0.500	0.520	0.016	7/32	0.219	0.250	0.185	0.199
1/4	20	0.250	5/8	0.625	0.645	0.016	9/32	0.281	0.312	0.245	0.260
5/16	18	0.312	3/4	0.750	0.770	0.031	11/32	0.344	0.375	0.307	0.324
3/8	16	0.375	7/8	0.875	0.895	0.031	13/32	0.406	0.437	0.368	0.388
7/16	14	0.438	1	1.000	1.020	0.031	15/32	0.469	0.500	0.431	0.452
1/2	13	0.500	1-1/8	1.125	1.145	0.031	17/32	0.531	0.562	0.492	0.515
5/8	11	0.625	1-3/8	1.375	1.400	0.031	21/32	0.656	0.687	0.616	0.642
3/4	10	0.750	1-5/8	1.625	1.650	0.047	25/32	0.781	0.812	0.741	0.768

[a] In ordering, the maximum body diameter indicated in the table should be specified.

Round (Button) Head Bolt

TABLE 6—DIMENSIONS OF BUTTON HEAD BOLT[a]

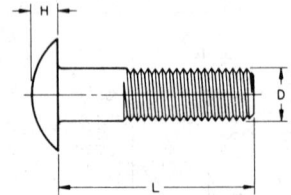 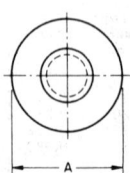

Size		Body Dia	Dia of Head, A				Height of Head, H		
D	Threads per in.	Max	Min	Basic		Max	Basic	Min	Max
No. 10	24	0.190	0.438	7/16	0.438	0.469	3/32	0.094	0.114
1/4	20	0.250	0.563	9/16	0.563	0.594	1/8	0.125	0.145
5/16	18	0.312	0.688	11/16	0.688	0.719	5/32	0.156	0.176
3/8	16	0.375	0.782	13/16	0.813	0.844	3/16	0.188	0.208
7/16	14	0.438	0.907	15/16	0.938	0.969	7/32	0.219	0.239
1/2	13	0.500	1.032	1-1/16	1.063	1.094	1/4	0.250	0.270
5/8	11	0.625	1.219	1-5/16	1.313	1.344	5/16	0.313	0.344
3/4	10	0.750	1.469	1-9/16	1.563	1.594	3/8	0.375	0.406
7/8	9	0.875	1.719	1-13/16	1.813	1.844	7/16	0.438	0.469
1	8	1.000	1.969	2-1/16	2.063	2.094	1/2	0.500	0.531

[a] In ordering, the maximum body diameter indicated in the table should be specified.

Step Bolt

TABLE 7—DIMENSIONS OF STEP BOLT[a]

Size		Body Dia		Dia of Head, A		Height of Head, H		Depth of Square, P		Width of Square, B	
D	Threads per in.	Max	Min	Max	Min	Max	Min	Max	Min	Max	Min
No. 10	24	0.190		0.625	0.656	0.094	0.114	0.094	0.125	0.185	0.199
1/4	20	0.250		0.813	0.844	0.125	0.145	0.125	0.156	0.245	0.260
5/16	18	0.312		1.000	1.031	0.156	0.176	0.156	0.187	0.307	0.324
3/8	16	0.375		1.188	1.219	0.188	0.208	0.188	0.219	0.368	0.388
7/16	14	0.438		1.375	1.406	0.219	0.239	0.219	0.250	0.431	0.452
1/2	13	0.500		1.563	1.594	0.250	0.270	0.250	0.281	0.492	0.515

[a] In ordering, the maximum body diameter indicated in the table should be specified.

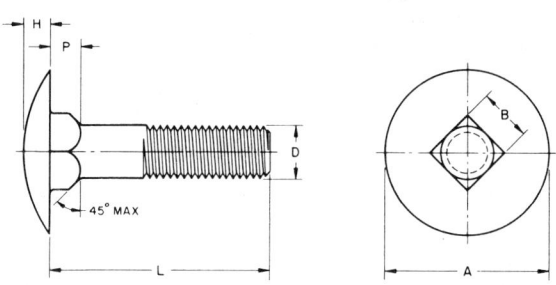

Countersunk Bolt

TABLE 8—DIMENSIONS OF COUNTERSUNK BOLT[a]

Size		Body Dia	Dia of Head[b], A			Depth of Head, H[c]	Width of Slot		Depth of Slot	
D	Threads per In.	Max	Max Sharp[d]	Min Sharp[e]	Absolute Min of Rounded or Flat Edged Irregular Shaped Head[f]		Max	Min	Max	Min
1/4	20	0.250	0.493	0.477	0.445	0.140	0.075	0.064	0.069	0.046
5/16	18	0.312	0.618	0.598	0.558	0.176	0.075	0.064	0.084	0.072
3/8	16	0.375	0.740	0.715	0.668	0.210	0.084	0.072	0.094	0.081
7/16	14	0.438	0.803	0.778	0.726	0.210	0.094	0.081	0.103	0.069
1/2	13	0.500	0.935	0.905	0.845	0.250	0.106	0.091	0.103	0.069
5/8	11	0.625	1.169	1.132	1.066	0.313	0.133	0.116	0.137	0.092
3/4	10	0.750	1.402	1.357	1.285	0.375	0.149	0.131	0.171	0.115
7/8	9	0.875	1.637	1.584	1.511	0.438	0.167	0.147	0.206	0.139
1	8	1.000	1.869	1.810	1.735	0.500	0.188	0.166	0.240	0.162
1-1/8	7	1.125	2.104	2.037	1.962	0.563	—	—	—	—
1-1/4	7	1.250	2.337	2.262	2.187	0.625	—	—	—	—
1-3/8	6	1.375	2.571	2.489	2.414	0.688	—	—	—	—
1-1/2	6	1.500	2.804	2.715	2.640	0.750	—	—	—	—

[a] In ordering, the maximum body diameter indicated in the table should be specified.
[b] Head is unslotted unless otherwise specified.
[c] Depth of head, H, is given for construction purposes only. Variations in this dimension are controlled by the diameters A and D and the included angle of the head.
[d] Calculated on nominal head height, nominal diameter of bolt, and 82 deg included angle, and extended to a sharp corner.
[e] Calculated on nominal head height, nominal diameter of bolt, and 78 deg included angle, and extended to a sharp corner.
[f] See section on Heads in text at beginning of this SAE Standard.

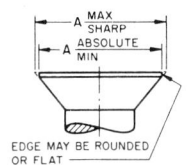
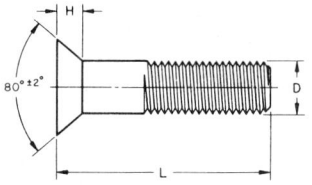
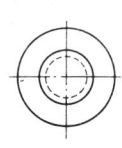

SLOTTED AND RECESSED HEAD SCREWS—SAE J478a

SAE Standard

Report of Screw Threads Committee approved October 1947, revised June 1960 and last revised by Fasteners Committee June 1962. Editorial change June 1964. Conforms to American Standards Association, ASA B18.6.2-1956, Hexagon and Slotted Head Cap Screws—Square Head Set Screws—Slotted Headless Set Screws; proposed revision to ASA B18.6.4-1959, Slotted and Recessed Head Tapping Screws and Metallic Drive Screws; proposed revision to ASA B18.6.1, Slotted and Recessed Head Wood Screws; and proposed ASA B18.6.3 Slotted and Recessed Head Machine Screws and Machine Screw Nuts.

Scope—Included herein are complete general and dimensional data for the types of slotted and recessed head machine, tapping, and wood screws, and slotted capscrews recognized as SAE Standard. Also included are performance data for tapping screws and appendixes which provide instructions for protrusion gaging flat heads, penetration gaging of recessed heads, across corners gaging of hexagon heads, recommended hole sizes for tapping screws, and a cross reference of tapping screw type designations.

The inclusion of dimensional data in this standard is not intended to imply that all of the products described are stock production sizes. Consumer interests are requested to consult with manufacturers concerning lists of stock production sizes.

CONTENTS—TAPPING, MACHINE, AND WOOD SCREWS; AND CAPSCREWS

General Specifications for Tapping, Machine, and Wood Screws
Tapping Screw Tables
  Hole Sizes for Drive Test
    Inspection 1
  Torsional Strengths for
    Referee Testing 2
  Flat Head 3, 3A, 3B
  Undercut Flat Head 4, 4A, 4B
  Flat Trim Head 5, 5A, 5B
  Oval Head 7, 7A, 7B
  Undercut Oval Head 8, 8A, 8B
  Oval Trim Head 9, 9A, 9B
  Pan Head 10, 10A, 10B
  Fillister Head 11, 11A, 11B
  Truss Head 12, 12A, 12B
  Round Head 13, 13A, 13B
  Hexagon Head 15
  Hexagon Washer Head 16
  Type A Screw Threads
    and Points 19
  Types B, BF, FG, BP, and
    BT Screw Threads and
    Points 20
  Types C, D, F, G, and T
    Screw Threads and Points ... 21

Recommended Hole Sizes
  (See Appendix E) 30, 31, 32, 33,
                                34, 35, 36, 37,
                                38, 39, 40, 41
                                and 42

Machine Screw Tables
  Flat Head 3, 3A, 3B
  Undercut Flat Head 4, 4A, 4B
  Flat Trim Head 5, 5A, 5B
  100 Deg Flat Head 6, 6A, 6B
  Oval Head 7, 7A, 7B
  Undercut Oval Head 8, 8A, 8B
  Oval Trim Head 9, 9A, 9B
  Pan Head 10, 10A, 10B
  Fillister Head 11, 11A, 11B
  Truss Head 12, 12A, 12B
  Round Head 13, 13A, 13B
  Binding Head 14, 14A, 14B
  Hexagon Head 15
  Hexagon Washer Head 16
  Drilled Fillister Head 17
  Threads and Header Points ... 18
Wood Screw Tables
  Flat Head 3, 3C, 3D
  Oval Head 7, 7C, 7D

Round Head 13, 13C, 13D
Threads and Points 22
General Specifications for Capscrews
Capscrew Tables
  Flat Head 23
  Fillister Head 24
  Round Head 25
Appendix A
  Protrusion Gaging
Appendix B
  Penetration Gaging of Cross
    Recess Screws and Tables ... 26, 27, 28 and 29
Appendix C
  Across Corners Gaging of Hexagon Heads
Appendix D
  Cross Reference of Tapping Screw Point Designations
Appendix E
  Recommended Hole Sizes
    for Tapping Screws and
    Tables 30, 31, 32, 33,
                                 34, 35, 36, 37
                                 38, 39, 40,
                                 and 42

MACHINE, TAPPING, AND WOOD SCREWS—GENERAL SPECIFICATIONS

Heads

Head Height—The head height indicated in the dimensional tables represents a metal to metal measurement. In other words, on heads having rounded top surfaces, the truncation of the rounded surface due to recess or slot is not considered part of the head height.

On countersunk type heads, the head height is a reference dimension measured parallel to the axis of the screw from the largest diameter of the bearing surface of the head to the point of intersection of the bearing surface of the head and basic major diameter of the screw. This point of intersection is not necessarily the same as the actual junction of head and shank.

Depth of Recess—The recess depth shown in the tables is measured parallel to the axis of the screw from the intersection of the maximum diameter of the recess with the rounded head surface to the bottom of the recess.

Penetration Gaging—Penetration gaging values are shown in the tables, and the method is explained in Appendix B.

Depth of Slots—The depth of slots is measured parallel to axis of screw from the highest part of the head to the intersection of the bottom of the slot with the head surface.

Bearing Surface—The bearing surface of perpendicular bearing surface type machine and tapping screw heads shall be at right angles to the axis of the body within 2 deg.

Eccentricity—Eccentricity is defined as one-half of the full indicator reading.

Eccentricity of Head—Machine and tapping screw heads shall not be eccentric with the axis of the screw by more than 3% of the maximum head diameter.

Eccentricity of Slots—Slots in machine and tapping screw heads shall not be eccentric with the axis of the screw by more than 6% of the basic screw diameter or 0.010 in., whichever is greater. Slots in wood screw heads shall not be eccentric with the axis of the screw by more than 6% of the basic screw diameter or 0.015 in., whichever is greater.

Eccentricity of Recesses—Recesses in machine, tapping, and wood screw heads shall not be eccentric with the axis of the screw by more than 6% of the basic screw diameter or 0.015 in., whichever is greater.

Underhead Fillets—The radius of the fillet under perpendicular bearing surface head and undercut countersunk head machine and tapping screws shall be not greater than 15% of the basic screw diameter. The radius of fillet under countersunk type heads shall be not greater than 40% of the basic screw diameter.

The maximum radius of the fillet under the head of perpendicular bearing surface head wood screws shall be 0.016 in. for sizes No. 0 through No. 4; 0.031 in. for sizes No. 5 through No. 12; and 0.046 for sizes larger than No. 12. The maximum radius of fillet under countersunk head wood screws shall be 0.031 for sizes No. 0 through No. 4; 0.062 in. for sizes No. 5 through No. 12; and 0.093 in. for sizes larger than No. 12.

Length

Measurement—The length of screws shall be measured from the largest diameter of the bearing surface of the head to the extreme point parallel with the screw axis.

Tolerances on Length—Tolerances on length for the various screw types shall be tabulated as follows:

Screw Type	Nominal Screw Length, in.	Tolerance on Length, in.
Machine Screws	Up to 1, incl Over 1 to 2, incl Over 2	−0.031 −0.062 −0.094
Tapping Screws Types A and BP	Up to 1, incl Over 1	±0.031 ±0.047
Tapping Screws Types B, BF, BG, BT, C, D, F, G, and T	Up to 3/4, incl Over 3/4 to 1-1/2, incl Over 1-1/2	−0.031 −0.047 −0.062
Wood Screws	Up to 5/8, incl Over 5/8 to 1-1/2, incl Over 1-1/2 to 2-3/4, incl Over 2-3/4 to 5, incl	−0.031 −0.047 −0.062 −0.094

Threads and Points

Machine Screws—Threads shall be UNC or UNF, Class 2A as indicated in Table 18. For threads with additive finish, the maximum diameters of Class 2A may be exceeded by the amount of the allowance; that is, the 2A maximum diameters apply to an unplated part or to a part before plating, whereas the basic diameters (the 2A maximum diameters plus the allowance) apply to a part after plating. Unless otherwise specified, machine screws shall have plain sheared ends. When specified, header points are obtainable as shown in Table 18. Other points of longer lengths will require machining.

Thread Forming Tapping Screws—For application in the more ductile materials where large internal stresses are permissible or desirable to increase resistance to loosening, tapping screws shall be of the following types:

Type A—Spaced thread screw with gimlet point primarily for use in materials such as light sheet metal, resin impregnated plywood, and asbestos compositions. See Table 19.

Type B—Spaced thread screw with pitches generally somewhat finer than Type A with a blunt point for use in materials such as light and heavy sheet metal, nonferrous castings, plastics, resin impregnated plywood, and asbestos compositions. See Table 20.

Type BP—Spaced thread screw, the same as Type B but having a cone point, used for piercing fabrics or in assemblies where holes are misaligned. See Table 20.

Type C—Screws having machine screw diameter pitch combinations with threads approximating Unified form and with blunt tapered points. Used where the use of a machine screw thread is preferable to the use of spaced thread types of thread forming screws. Also useful when chips from thread cutting screws are objectionable. It should be recognized that in specific applications, this type of screw may require extreme driving torques due to long thread engagement or use in hard materials. See Table 21.

Thread Cutting Tapping Screws—For application in materials where disruptive internal stresses are not desirable, or where excessive driving torques are encountered with thread forming screws. Tapping screws shall be of the following types:

Types D, F, G, and T—Screws with threads approximating machine screw threads, with blunt point, and with tapered entering threads having one or more cutting edges and chip cavities. The tapered threads of the Type F may be complete or incomplete at the producer's option; all other types have incomplete tapered threads. These screws can be used in materials such as aluminum, zinc and lead die castings, steel sheets and shapes, cast iron, brass, and plastics. See Table 21.

Types BF, BG, and BT—Spaced thread screws as in Type B with blunt points, with one or more cutting edges and chip cavities, for use in plastics, die castings, metal clad and resin impregnated plywoods, asbestos, and other similar compositions. See Table 20.

Wood Screws—Wood screws shall be furnished with gimlet points and thread style shall conform with that shown in the illustration above Table 22.

Length of Thread

Machine Screws—Screws up to and including 2 in. nominal length shall have full form threads extending to within 2 threads of the bearing surface of the head or closer if practicable. Screws over 2 in. nominal length shall have a minimum full thread length of 1.75 in.

Trim Head Machine Screws—Screws up to and including 3 diameters nominal length shall have full form threads extending to within 1 thread of the head or closer if practicable. Screws over 3 diameters, up to and including 1½ in. nominal length, shall have full form threads extending to within 2 threads of the head or closer if practicable. Screws over 1½ in. long shall have a minimum full thread length of 1.25 in.

Tapping Screws, Types A, B, BF, BG, BP, and BT—Screws up to and including 8 diameters nominal length shall be threaded as close for the head as is practicable in manufacturing process. For nominal lengths greater than 8 diameters, unless otherwise specified thread length shall be optional with the producer, but shall not be less than 6 diameters.

Trim Head Tapping Screws, Type A—For sizes up to and including No. 12 body diameter, the thread on screws up to and including 1 in. nominal length shall extend as close to the head as is practicable in the manufacturing process; on nominal lengths greater than 1 in., threads shall extend to within 0.500 in. of the head with a tolerance of plus 2 pitches. For sizes larger than No. 12 body diameter, the threads on screws having nominal lengths up to and including 8 diameters, shall extend as close to the head as is practicable in the manufacturing process; and on nominal lengths greater than 8 diameters, threads shall extend to within 0.750 in. of the head with a tolerance of plus 2 pitches.

Tapping Screws, Types C, D, F, G, and T—Screws up to and including 3 diameters nominal length shall be threaded to within 1 full thread of the head or closer if practicable. For nominal lengths over 3 diameters, up to and including 8 diameters, threads shall extend to within 2 full threads of the head or closer if practicable. For nominal lengths greater than 8 diameters, unless otherwise specified, thread length shall be optional with producer but shall be not less than 6 diameters.

Trim Head Tapping Screws, Types B, BF, BG, BP, BT, C, D, F, G, and T—Screws up to and including 3 diameters nominal length shall be threaded to within 1 full thread of the head or closer if practicable. For nominal lengths greater than 3 diameters, up to and including 1½ in., the thread shall extend to within 2 full threads of the head or closer if practicable. For nominal lengths greater than 1½ in., minimum full thread length shall be 1.25 in.

Wood Screws—The length of thread shall be equal to approximately two-thirds of the length of the screw.

Diameter of Unthreaded Body

Machine Screws—The diameter of the unthreaded body shall be not less than the minimum pitch diameter of the thread nor greater than the basic major diameter of the thread.

Trim Head Machine Screws—The diameter of the unthreaded body shall be not less than the minimum pitch diameter of the thread nor greater than the basic major diameter of the thread. Nominal lengths over 1½ in. shall have a 0.062 in. minimum length shoulder under the head as specified in Tables 5 and 9.

Tapping Screws, Types A, B, BF, BG, BP, and BT—The diameter of the unthreaded body shall be not less than the minimum minor diameter nor greater than the maximum major diameter of the thread.

Trim Head Tapping Screws, Types A, B, BF, BG, BP, and BT—For screws which are not threaded to the head, the diameter of the unthreaded body shall be not less than the minimum minor diameter nor greater than the maximum major diameter of the thread except for a 0.062 in. minimum length full body diameter shoulder under the head as specified in Tables 5 and 9. At manufacturer's option, the full body diameter may extend entire length from the head to the thread.

Tapping Screws, Types C, D, F, G, and T—The diameter of the unthreaded body shall be not less than the Class 2A thread minimum pitch diameter nor greater than maximum major diameter of the thread.

Trim Head Tapping Screws, Types C, D, F, G, and T—The diameter of the unthreaded body shall be not less than the minimum pitch diameter of Class 2A nor greater than the maximum major diameter of the thread. Nominal lengths over 1½ in. shall have a 0.062 in. minimum length shoulder under the head as specified in Tables 5 and 9.

Wood Screws—Unthreaded body detail shown in Table 22.

Materials

Machine and Wood Screws—Steel, brass, or stainless steel, as specified by the purchaser. Unless otherwise specified, no physical requirements apply.

Tapping Screws—Steel used in the manufacture of tapping screws shall be of high quality and suitable to meet the performance requirements of the specifications. Tapping screws may also be made of corrosion resistant steel, brass, monel, and aluminum alloys; however, the performance requirements do not apply to screws of these materials.

TABLE 1—STANDARD HOLE SIZES FOR DRIVE TEST INSPECTION OF TAPPING SCREWS

Screw Size	Test Plate Thickness					Test Hole Size[a]										
	Types A, B, BP, and C			Types D, F, G, and T		Type A		Types B and BP[b]		Type C				Types D, F, G, and T			
										Coarse Thread		Fine Thread		Coarse Thread		Fine Thread	
	Gage	Max	Min	Max	Min	Drill Size	Hole Dia	Drill Size	Hole Dia	Drill Size	Hole Dia	Drill Size	Hole Dia	Drill Size	Hole Dia	Drill Size	Hole Dia
2	18	0.0500	0.0460	0.0800	0.0760	48	0.0760	48	0.0760	48	0.0760	48	0.0760	49	0.0730	—	—
3	18	0.0500	0.0460	0.0960	0.0920	46	0.0810	46	0.0810	44	0.0860	43	0.0890	46	0.0810	—	—
4	18	0.0500	0.0460	0.1110	0.1070	44	0.0860	44	0.0860	41	0.0960	40	0.0980	41	0.0960	—	—
5	18	0.0500	0.0460	0.1110	0.1070	36	0.1065	36	0.1065	35	0.1100	35	0.1100	37	0.1040	—	—
6	14	0.0770	0.0730	0.1425	0.1385	32	0.1160	32	0.1160	31	0.1200	1/8	0.1250	31	0.1200	—	—
7	14	0.0770	0.0730	—	—	30	0.1285	30	0.1285	—	—	—	—	—	—	—	—
8	14	0.0770	0.0730	0.1420	0.1380	29	0.1360	29	0.1360	27	0.1440	26	0.1470	26	0.1470	—	—
10	1/8	0.1270	0.1230	0.1905	0.1845	21	0.1590	21	0.1590	19	0.1660	11/64	0.1719	17	0.1730	16	0.1770
12	1/8	0.1270	0.1230	0.1905	0.1845	3/16	0.1875	3/16	0.1875	11	0.1910	10	0.1935	8	0.1990	—	—
14	1/8	0.1270	0.1230	—	—	5.5mm	0.2165	—	—	—	—	—	—	—	—	—	—
1/4	3/16	0.1905	0.1845	0.2530	0.2470	—	—	5.5mm	0.2165	7/32	0.2188	1	0.2280	1	0.2280	A	0.2340
16	3/16	0.1905	0.1845	—	—	B	0.2380	—	—	—	—	—	—	—	—	—	—
5/16	3/16	0.1905	0.1845	0.3155	0.3095	—	—	I	0.2720	J	0.2770	L	0.2900	L	0.2900	M	0.2950
18	3/16	0.1905	0.1845	—	—	G	0.2610	—	—	—	—	—	—	—	—	—	—
3/8	3/16	0.1905	0.1845	0.3780	0.3720	—	—	21/64	0.3281	R	0.3390	11/32	0.3438	T	0.3580	T	0.3580
20	3/16	0.1905	0.1845	—	—	L	0.2900	—	—	—	—	—	—	—	—	—	—
24	3/16	0.1905	0.1845	—	—	11/32	0.3438	—	—	—	—	—	—	—	—	—	—
7/16	3/16	0.1905	0.1845	—	—	—	—	13/32	0.4062	—	—	—	—	—	—	—	—
1/2	3/16	0.1905	0.1845	—	—	—	—	15/32	0.4688	—	—	—	—	—	—	—	—

[a] Inches unless otherwise specified.
[b] Drive test data not applicable to Types BF, BG, and BT tapping screws.

Tapping Screw Performance and Testing

Hardness—Tapping screws made of carbon steel shall be case hardened to meet the performance requirements in the test specifications. Drive test requirements apply to Types A, B, BP, C, D, F, G, and T tapping screws only. Referee test requirements shall apply to all types of carbon steel case hardened tapping screws.

Test Plates—Test plates shall be low carbon cold rolled steel, having 80-107 Rockwell B (RB) or equivalent hardness. Test holes shall be drilled and redrilled or reamed to ±0.001 in. of nominal. See Table 1 for test hole sizes and material gages or thickness.

Drive Test—Thread forming tapping screws shall, without deforming their own threads, form a mating thread in the test plate until the tapered threads of the point are completely through the test plate. Thread cutting tapping screws shall, without deforming their own threads, cut or form a mating thread until the tapered threads of the point are completely through the test plate. If the screws fail while driving, then additional screws from the same lot shall be subjected to the Referee Test.

Referee Test—Screws shall be clamped by suitable means, making certain that the clamped portion of threads are not damaged, that at least two full threads project above clamping device, and that at least two full threads exclusive of point or end slot are held within the clamping device. By means of a suitably calibrated torque measuring device, torque shall be applied to the screw until failure occurs. The torque required to cause failure shall equal or exceed the minimum values given in Table 2.

TABLE 2—TORSIONAL STRENGTHS FOR REFEREE TESTING OF TAPPING SCREWS

Screw Size	Min Torsional Strength, in.-lbs					
	Type A	Types B, BF, BG, BP, and BT	Type C		Types D, F, G, and T	
			Coarse Thread	Fine Thread	Coarse Thread	Fine Thread
2	4	4	5	6	5	6
3	9	9	9	10	9	10
4	12	13	13	15	13	15
5	18	18	18	20	18	20
6	24	24	23	27	23	27
7	30	30	—	—	—	—
8	39	39	42	47	42	47
10	48	56	56	74	56	74
12	83	88	93	108	93	108
14	125	—	—	—	—	—
1/4	—	142	140	179	140	179
16	152	—	—	—	—	—
5/16	—	290	306	370	306	370
18	196	—	—	—	—	—
3/8	—	590	560	710	560	710
20	250	—	—	—	—	—
24	492	—	—	—	—	—

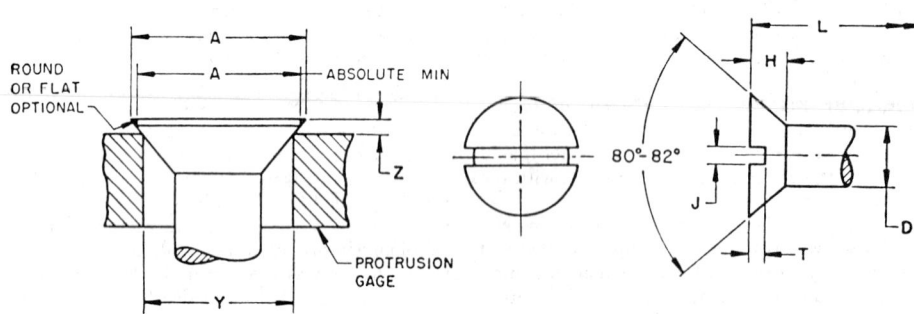

TABLE 3—DIMENSIONS OF FLAT HEAD MACHINE, TAPPING, AND WOOD SCREWS, IN.[a]

Nominal Size	Applicable to Screw Types[f] Code Letters	D Basic Dia of Screw	A Dia of Head			H[b] Height of Head Ref	J Width of Slot		T Depth of Slot		Y[c] Gage Dia	Z[c] Protrusion of Head Above Gage		L[c] Screws These Lengths and Shorter are Undercut[d]	
			Max, Sharp	Min, Sharp	Abs, Min		Min	Max	Min	Max		Max	Min	Machine	Tapping
0	MABW	0.060	0.119	0.105	0.099	0.035	0.016	0.023	0.010	0.015	0.078	0.026	0.016	1/8	1/8[e]
1	MABW	0.073	0.146	0.130	0.123	0.043	0.019	0.026	0.012	0.019	0.101	0.028	0.016	1/8	5/32[e]
2	MTW	0.086	0.172	0.156	0.147	0.051	0.023	0.031	0.015	0.023	0.124	0.029	0.017	1/8	3/16
3	MTW	0.099	0.199	0.181	0.171	0.059	0.027	0.035	0.017	0.027	0.148	0.031	0.018	1/8	7/32
4	MTW	0.112	0.225	0.207	0.195	0.067	0.031	0.039	0.020	0.030	0.172	0.032	0.019	3/16	1/4
5	MTW	0.125	0.252	0.232	0.220	0.075	0.035	0.043	0.022	0.034	0.196	0.034	0.020	3/16	1/4
6	MTW	0.138	0.279	0.257	0.244	0.083	0.039	0.048	0.024	0.038	0.220	0.036	0.021	3/16	5/16
7	ABW	0.151	0.305	0.283	0.268	0.091	0.039	0.048	0.027	0.041	0.243	0.037	0.022	—	3/8
8	MTW	0.164	0.332	0.308	0.292	0.100	0.045	0.054	0.029	0.045	0.267	0.039	0.023	1/4	7/16
9	W	0.177	0.358	0.334	0.316	0.108	0.045	0.054	0.032	0.049	—	—	—	—	—
10	MTW	0.190	0.385	0.359	0.340	0.116	0.050	0.060	0.034	0.053	0.313	0.042	0.025	5/16	1/2
12	MTW	0.216	0.438	0.410	0.389	0.132	0.056	0.067	0.039	0.060	0.362	0.045	0.027	3/8	9/16
14	AW	0.242	0.491	0.461	0.437	0.148	0.064	0.075	0.044	0.068	0.410	0.049	0.029	—	5/8
16	AW	0.268	0.544	0.512	0.485	0.164	0.064	0.075	0.049	0.075	0.457	0.052	0.031	—	3/4
18	AW	0.294	0.597	0.563	0.534	0.180	0.072	0.084	0.054	0.083	0.505	0.055	0.033	—	13/16
20	AW	0.320	0.650	0.614	0.582	0.196	0.072	0.084	0.059	0.090	0.553	0.058	0.035	—	13/16
24	AW	0.372	0.756	0.716	0.679	0.228	0.081	0.094	0.069	0.105	0.648	0.065	0.039	—	1
1/4	MBCD	0.250	0.507	0.477	0.452	0.153	0.064	0.075	0.046	0.070	0.424	0.050	0.029	7/16	5/8
5/16	MBCD	0.3125	0.635	0.600	0.568	0.191	0.072	0.084	0.058	0.088	0.539	0.057	0.034	1/2	5/8
3/8	MBCD	0.375	0.762	0.722	0.685	0.230	0.081	0.094	0.070	0.106	0.653	0.065	0.039	9/16	5/8
7/16	MB	0.4375	0.812	0.767	0.723	0.223	0.081	0.094	0.066	0.103	0.690	0.073	0.044	5/8	3/4
1/2	MB	0.500	0.875	0.831	0.775	0.223	0.091	0.106	0.065	0.103	0.739	0.081	0.049	3/4	3/4
9/16	M	0.5625	1.000	0.950	0.889	0.260	0.102	0.118	0.077	0.120	0.851	0.089	0.053	13/16	—
5/8	M	0.6250	1.125	1.069	1.002	0.298	0.116	0.133	0.088	0.137	0.962	0.097	0.058	15/16	—
3/4	M	0.7500	1.375	1.306	1.230	0.372	0.131	0.149	0.111	0.171	1.186	0.112	0.067	1-1/8	—

[a] Additional specifications and details pertinent to bodies and threads are given in the General Specifications.
[b] These values are calculated to maximum formula.
[c] Protrusion gaging and undercut heads not applicable to wood screws.
[d] See Table 4, Dimensions of Undercut Flat Head Machine and Tapping Screws.
[e] 3/16 for Type A tapping screws.
[f] Application code:
M — Machine Screws.
T — Tapping Screws, all types.
A — Tapping Screws, Type A.
B — Tapping Screws, Types B, BF, BG, BP, and BT.
C — Tapping Screws, Type C.
D — Tapping Screws, Types D, F, G, and T.
W — Wood Screws.

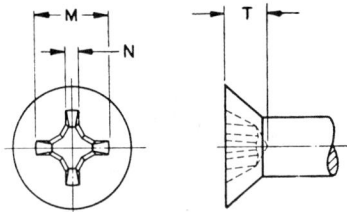

This type of recess has a large center opening, tapered wings, and blunt bottom, with all edges relieved or rounded.

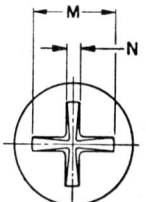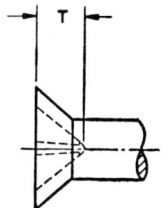

This type of recess consists of two intersecting slots with parallel sides converging to a slightly truncated apex at bottom of recess.

TABLE 3A—TYPE I CROSS RECESS DIMENSIONS OF FLAT HEAD MACHINE AND TAPPING SCREWS, IN.

Nominal Size	M Dia of Recess		T Depth of Recess		N Width of Recess	Driver Size	Penetration Gaging Depth[a]	
	Min	Max	Min	Max	Min		Min	Max
0	0.056	0.069	0.027	0.043	0.014	0	0.020	0.036
1	0.064	0.077	0.035	0.051	0.015	0	0.028	0.044
2	0.089	0.102	0.047	0.063	0.017	1	0.040	0.056
3	0.094	0.107	0.052	0.068	0.018	1	0.045	0.061
4	0.115	0.128	0.073	0.089	0.018	1	0.066	0.082
5	0.141	0.154	0.063	0.086	0.027	2	0.052	0.075
6	0.161	0.174	0.083	0.106	0.029	2	0.072	0.095
7	0.169	0.182	0.091	0.114	0.030	2	0.080	0.103
8	0.176	0.189	0.098	0.121	0.030	2	0.087	0.110
10	0.191	0.204	0.113	0.136	0.032	2	0.102	0.125
12	0.255	0.268	0.133	0.156	0.035	3	0.116	0.139
14	0.270	0.283	0.148	0.171	0.036	3	0.131	0.154
16	0.290	0.303	0.168	0.191	0.039	3	0.151	0.174
18	0.352	0.365	0.194	0.216	0.061	4	0.174	0.196
20	0.365	0.378	0.208	0.230	0.062	4	0.188	0.210
24	0.380	0.393	0.223	0.245	0.065	4	0.203	0.225
1/4	0.270	0.283	0.148	0.171	0.036	3	0.131	0.154
5/6	0.352	0.365	0.194	0.216	0.061	4	0.174	0.196
3/8	0.380	0.393	0.223	0.245	0.065	4	0.203	0.225
7/16	0.396	0.409	0.239	0.261	0.068	4	0.219	0.241
1/2	0.411	0.424	0.254	0.276	0.069	4	0.234	0.256
9/16	0.431	0.454	0.278	0.300	0.073	4	0.258	0.280
5/8	0.473	0.496	0.316	0.342	0.070	5	0.283	0.309
3/4	0.537	0.560	0.380	0.406	0.077	5	0.347	0.373

[a] See Appendix B.

TABLE 3B—TYPE II CROSS RECESS DIMENSIONS OF FLAT HEAD MACHINE AND TAPPING SCREWS, IN.

Nominal Size	M Dia of Recess		T Depth of Recess		N Width of Recess	Driver Size	Penetration Gaging Depth[a]	
	Min	Max	Min	Max	Min		Min	Max
0	0.073	0.083	0.031	0.042	0.021		—[b]	—[b]
1	0.091	0.102	0.042	0.054	0.024		—[b]	—[b]
2	0.109	0.120	0.054	0.066	0.027		0.029	0.040
3	0.127	0.139	0.066	0.079	0.030		0.041	0.053
4	0.145	0.157	0.075	0.088	0.032		0.052	0.064
5	0.162	0.176	0.087	0.101	0.035		0.064	0.077
6	0.180	0.195	0.098	0.113	0.038		0.075	0.089
7	0.198	0.213	0.110	0.125	0.040		0.087	0.101
8	0.216	0.232	0.117	0.132	0.043	Point	0.099	0.113
10	0.251	0.269	0.140	0.156	0.048	same	0.122	0.137
12	0.287	0.307	0.163	0.181	0.054	on all	0.145	0.162
14	0.323	0.344	0.178	0.197	0.059	drivers	0.168	0.186
16	0.358	0.381	0.201	0.221	0.064		0.191	0.210
18	0.394	0.418	0.225	0.245	0.070		0.215	0.234
20	0.430	0.455	0.248	0.269	0.075		0.238	0.258
24	0.501	0.529	0.294	0.318	0.086		0.284	0.307
1/4	0.334	0.355	0.186	0.204	0.061		0.176	0.193
5/16	0.420	0.444	0.242	0.262	0.074		0.232	0.251
3/8	0.495	0.523	0.291	0.314	0.086		0.281	0.303
7/16	0.540	0.568	0.320	0.343	0.092		0.310	0.332
1/2	0.578	0.608	0.345	0.370	0.098		0.335	0.359
9/16	0.623	0.656	0.374	0.400	0.104		0.364	0.389
5/8	0.623	0.656	0.374	0.400	0.104		0.364	0.389
3/4	0.623	0.656	0.374	0.400	0.104		0.364	0.389

[a] See Appendix B. [b] Not practical to gage.

TABLE 3C—TYPE I CROSS RECESS DIMENSIONS OF FLAT HEAD WOOD SCREWS, IN.

Nominal Size	M Dia of Recess		T Depth of Recess		N Width of Recess	Driver Size	Penetration Gauging Depth[a]	
	Min	Max	Min	Max	Min		Min	Max
0	0.056	0.069	0.027	0.043	0.014	0	0.020	0.036
1	0.064	0.077	0.035	0.051	0.015	0	0.028	0.044
2	0.089	0.102	0.047	0.063	0.017	1	0.040	0.056
3	0.094	0.107	0.052	0.068	0.018	1	0.045	0.061
4	0.115	0.128	0.073	0.089	0.018	1	0.066	0.082
5	0.141	0.154	0.063	0.086	0.027	2	0.052	0.075
6	0.161	0.174	0.083	0.106	0.029	2	0.072	0.095
7	0.176	0.189	0.098	0.121	0.030	2	0.087	0.110
8	0.191	0.204	0.113	0.136	0.032	2	0.102	0.125
9	0.201	0.214	0.123	0.146	0.033	2	0.112	0.135
10	0.245	0.258	0.123	0.146	0.034	3	0.106	0.129
12	0.270	0.283	0.148	0.171	0.036	3	0.131	0.154
14	0.290	0.303	0.168	0.191	0.039	3	0.151	0.174
16	0.314	0.327	0.193	0.216	0.045	3	0.176	0.199
18	0.365	0.378	0.208	0.230	0.062	4	0.188	0.210
20	0.380	0.393	0.223	0.245	0.065	4	0.203	0.225
24	0.411	0.424	0.254	0.276	0.069	4	0.234	0.256

[a] See Appendix B.

TABLE 3D—TYPE II CROSS RECESS DIMENSIONS OF FLAT HEAD WOOD SCREWS, IN.

Nominal Size	M Dia of Recess		T Depth of Recess		N Width of Recess	Driver Size	Penetration Gauging Depth[a]	
	Min	Max	Min	Max	Min		Min	Max
0	0.073	0.083	0.031	0.042	0.021		—[b]	—[b]
1	0.091	0.102	0.042	0.054	0.024		—[b]	—[b]
2	0.109	0.120	0.054	0.066	0.027		0.029	0.040
3	0.127	0.139	0.066	0.079	0.030		0.041	0.053
4	0.145	0.157	0.075	0.088	0.032		0.052	0.064
5	0.162	0.176	0.087	0.101	0.035		0.064	0.077
6	0.180	0.195	0.098	0.113	0.038		0.075	0.089
7	0.198	0.213	0.110	0.125	0.040		0.087	0.101
8	0.216	0.232	0.117	0.132	0.043	Point	0.099	0.113
9	0.234	0.251	0.128	0.145	0.046	same	0.110	0.126
10	0.251	0.269	0.140	0.156	0.048	on all	0.122	0.137
12	0.287	0.307	0.163	0.181	0.054	drivers	0.145	0.162
14	0.323	0.344	0.178	0.197	0.059		0.168	0.186
16	0.358	0.381	0.201	0.221	0.064		0.191	0.210
18	0.394	0.418	0.225	0.245	0.070		0.215	0.234
20	0.430	0.455	0.248	0.269	0.075		0.238	0.258
24	0.501	0.529	0.294	0.318	0.086		0.284	0.307

[a] See Appendix B. [b] Not practical to gage.

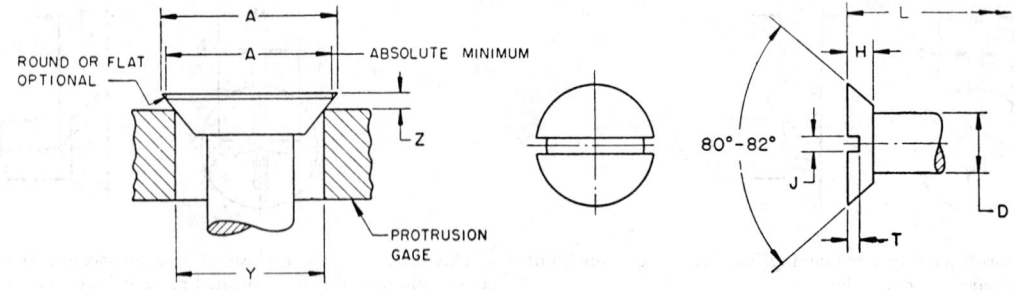

TABLE 4—DIMENSIONS OF UNDERCUT FLAT HEAD MACHINE AND TAPPING SCREWS, IN.[a]

Nominal Size	Applicable to Screw Types[c] Code Letters	D Basic Dia of Screw	L Screws These Lengths and Shorter are Undercut Machine	L ... Tapping	A Dia of Head Max Sharp	A Dia of Head Min Sharp	A Dia of Head Abs Min	H Height of Head Max	H Height of Head Min	J Width of Slot Min	J Width of Slot Max	T Depth of Slot Min	T Depth of Slot Max	Y Gage Dia	Z Protrusion of Head Above Gage Max	Z Protrusion of Head Above Gage Min
0	MAB	0.060	1/8	1/8[b]	0.119	0.105	0.099	0.025	0.018	0.016	0.023	0.007	0.011	—	—	—
1	MAB	0.073	1/8	5/32[b]	0.146	0.130	0.123	0.031	0.023	0.019	0.026	0.009	0.014	—	—	—
2	MT	0.086	1/8	3/16	0.172	0.156	0.147	0.036	0.028	0.023	0.031	0.011	0.016	0.124	0.029	0.017
3	MT	0.099	1/8	7/32	0.199	0.181	0.171	0.042	0.033	0.027	0.035	0.012	0.019	0.148	0.031	0.018
4	MT	0.112	3/16	1/4	0.225	0.207	0.195	0.047	0.038	0.031	0.039	0.014	0.022	0.172	0.032	0.019
5	MT	0.125	3/16	1/4	0.252	0.232	0.220	0.053	0.043	0.035	0.043	0.016	0.024	0.196	0.034	0.020
6	MT	0.138	3/16	5/16	0.279	0.257	0.244	0.059	0.048	0.039	0.048	0.017	0.027	0.220	0.036	0.021
7	AB	0.151	—	3/8	0.305	0.283	0.268	0.064	0.053	0.039	0.048	0.019	0.030	0.243	0.037	0.022
8	MT	0.164	1/4	7/16	0.332	0.308	0.292	0.070	0.058	0.045	0.054	0.021	0.032	0.267	0.039	0.023
10	MT	0.190	5/16	1/2	0.385	0.359	0.340	0.081	0.068	0.050	0.060	0.024	0.037	0.313	0.042	0.025
12	MT	0.216	3/8	9/16	0.438	0.410	0.389	0.092	0.078	0.056	0.067	0.028	0.043	0.362	0.045	0.027
14	A	0.242	—	5/8	0.491	0.461	0.437	0.104	0.088	0.064	0.075	0.031	0.048	0.410	0.049	0.029
16	A	0.268	—	3/4	0.544	0.512	0.485	0.115	0.098	0.064	0.075	0.035	0.053	0.457	0.052	0.031
18	A	0.294	—	13/16	0.597	0.563	0.534	0.126	0.108	0.072	0.084	0.038	0.058	0.505	0.055	0.033
20	A	0.320	—	13/16	0.650	0.614	0.582	0.137	0.118	0.072	0.084	0.042	0.064	0.553	0.058	0.035
24	A	0.372	—	1	0.756	0.716	0.679	0.160	0.139	0.081	0.094	0.049	0.074	0.648	0.065	0.039
1/4	MBCD	0.250	7/16	5/8	0.507	0.477	0.452	0.107	0.092	0.064	0.075	0.032	0.050	0.424	0.050	0.029
5/16	MBCD	0.3125	1/2	5/8	0.635	0.600	0.568	0.134	0.116	0.072	0.084	0.041	0.062	0.539	0.057	0.034
3/8	MBCD	0.375	9/16	5/8	0.762	0.722	0.685	0.161	0.140	0.081	0.094	0.049	0.075	0.653	0.065	0.039
7/16	MB	0.4375	5/8	3/4	0.812	0.767	0.723	0.156	0.133	0.081	0.094	0.045	0.072	0.690	0.073	0.044
1/2	MB	0.500	3/4	3/4	0.875	0.831	0.775	0.156	0.130	0.091	0.106	0.046	0.072	0.739	0.081	0.049
9/16	M	0.5625	13/16	—	1.000	0.950	0.889	0.182	0.153	0.102	0.118	0.054	0.084	0.851	0.089	0.053
5/8	M	0.6250	15/16	—	1.125	1.069	1.002	0.208	0.176	0.116	0.133	0.062	0.096	0.962	0.097	0.058
3/4	M	0.7500	1-1/8	—	1.375	1.306	1.230	0.260	0.223	0.131	0.149	0.078	0.120	1.186	0.112	0.067

[a] Additional specifications and details pertinent to bodies and threads are given in the General Specifications.
[b] 3/16 for Type A tapping screws.
[c] Application code:
  M—Machine Screws.

T —Tapping Screws, all types.
A —Tapping Screws, Type A.
B —Tapping Screws, Types B, BF, BG, BP, and BT.
C —Tapping Screws, Type C.
D —Tapping Screws, Types D, F, G, and T.

TABLE 4A—TYPE I CROSS RECESS DIMENSIONS OF UNDERCUT FLAT HEAD MACHINE AND TAPPING SCREWS, IN.

Nominal Size	M Dia of Recess Min	M Dia of Recess Max	T Depth of Recess Min	T Depth of Recess Max	N Width of Recess Min	Driver Size	Penetration Gaging Depth[a] Min	Penetration Gaging Depth[a] Max
0	0.056	0.069	0.027	0.043	0.014	0	0.020	0.036
1	0.064	0.077	0.035	0.051	0.015	0	0.028	0.044
2	0.082	0.095	0.040	0.056	0.017	1	0.033	0.049
3	0.089	0.102	0.047	0.063	0.018	1	0.040	0.056
4	0.103	0.116	0.062	0.078	0.018	1	0.055	0.071
5	0.115	0.128	0.073	0.089	0.018	2	0.066	0.082
6	0.133	0.146	0.055	0.078	0.025	2	0.044	0.067
7	0.141	0.154	0.063	0.086	0.027	2	0.052	0.075
8	0.161	0.174	0.083	0.106	0.029	2	0.072	0.095
10	0.176	0.189	0.098	0.121	0.030	2	0.087	0.110
12	0.220	0.233	0.098	0.121	0.030	3	0.081	0.104
14	0.237	0.250	0.113	0.136	0.032	3	0.096	0.119
16	0.255	0.268	0.133	0.156	0.035	3	0.116	0.139
18	0.304	0.317	0.146	0.168	0.053	4	0.126	0.148
20	0.322	0.335	0.167	0.186	0.055	4	0.144	0.166
24	0.352	0.365	0.194	0.216	0.061	4	0.174	0.196
1/4	0.237	0.250	0.113	0.136	0.032	3	0.096	0.119
5/16	0.304	0.317	0.146	0.168	0.053	4	0.126	0.148
3/8	0.352	0.365	0.194	0.216	0.061	4	0.174	0.196
7/16	0.380	0.393	0.223	0.245	0.065	4	0.203	0.225
1/2	0.396	0.409	0.242	0.261	0.068	4	0.219	0.241

[a] See Appendix B.

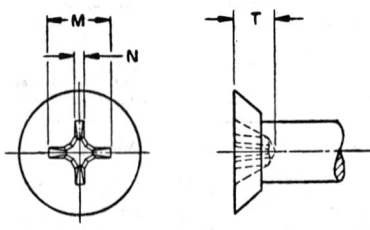

This type of recess has a large center opening, tapered wings, and blunt bottom, with all edges relieved or rounded.

TABLE 4B—TYPE II CROSS RECESS DIMENSIONS OF UNDERCUT FLAT HEAD MACHINE AND TAPPING SCREWS, IN.

Nominal Size	M Dia of Recess		T Depth of Recess		N Width of Recess	Driver Size	Penetration Gaging Depth[a]	
	Min	Max	Min	Max	Min		Min	Max
0	0.062	0.072	0.023	0.035	0.020		—[b]	—[b]
1	0.077	0.087	0.033	0.045	0.022		—[b]	—[b]
2	0.095	0.105	0.044	0.056	0.025		0.020	0.030
3	0.106	0.116	0.052	0.064	0.026		0.027	0.038
4	0.123	0.135	0.063	0.076	0.029		0.038	0.050
5	0.141	0.153	0.073	0.086	0.032		0.050	0.062
6	0.155	0.167	0.082	0.095	0.034		0.059	0.071
7	0.171	0.185	0.092	0.107	0.036	Point	0.069	0.083
8	0.190	0.204	0.105	0.119	0.039	same	0.082	0.095
10	0.228	0.244	0.125	0.140	0.045	on all	0.107	0.121
12	0.251	0.269	0.140	0.156	0.048	drivers	0.122	0.137
14	0.285	0.305	0.162	0.180	0.053		0.144	0.161
16	0.314	0.336	0.173	0.192	0.058		0.163	0.181
18	0.341	0.365	0.190	0.210	0.062		0.180	0.200
20	0.380	0.404	0.216	0.236	0.068		0.206	0.225
24	0.439	0.467	0.254	0.277	0.077		0.244	0.266
1/4	0.294	0.314	0.160	0.178	0.054		0.150	0.167
5/16	0.369	0.393	0.208	0.229	0.066		0.198	0.218
3/8	0.439	0.467	0.254	0.277	0.077		0.244	0.266
7/16	0.483	0.513	0.283	0.307	0.083		0.273	0.296
1/2	0.533	0.563	0.315	0.340	0.090		0.305	0.329

[a] See Appendix B. [b] Not practical to gage.

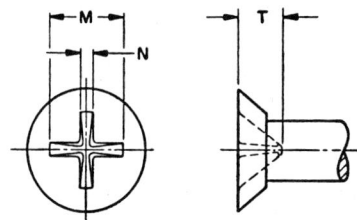

This type of recess consists of two intersecting slots with parallel sides converging to a slightly truncated apex at bottom of recess.

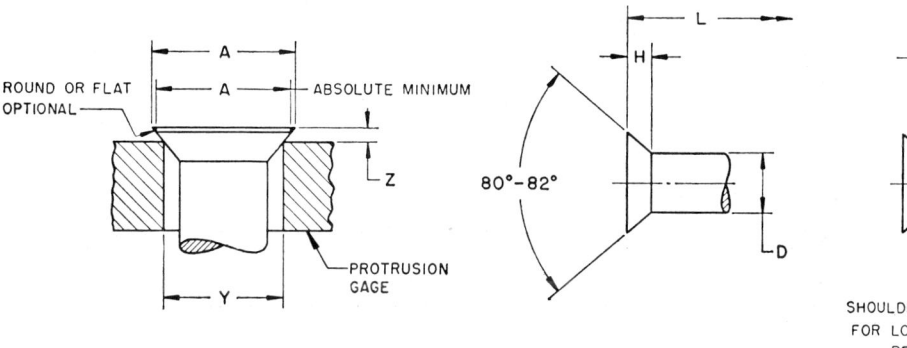

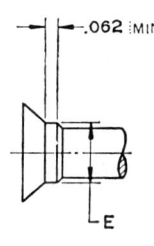

SHOULDER CONSTRUCTION FOR LONG SCREWS WITH REDUCED BODY (SEE GENERAL SPECIFICATIONS)

TABLE 5—DIMENSIONS OF FLAT TRIM HEAD MACHINE AND TAPPING SCREWS, IN.[a,b]

Nominal Size		Applicable to Screw Types[d]	D Basic Dia of Screw	E Dia of Body or Shoulder				A Dia of Head			H[c] Height of Head	Y Gage Dia	Z Protrusion of Head Above Gage	
				Type A Tapping		Machine and Other Types of Tapping								
Body	Head	Code Letters		Max	Min	Max	Min	Max Sharp	Min Sharp	Abs Min	Ref		Max	Min
4	3	MT	0.112	0.116	0.105	0.112	0.106	0.199	0.181	0.171	0.052	0.148	0.031	0.018
5	4	MT	0.125	0.129	0.118	0.125	0.119	0.225	0.207	0.195	0.060	0.172	0.032	0.019
6	4	MT	0.138	0.142	0.131	0.138	0.131	0.225	0.207	0.195	0.052	0.172	0.032	0.019
6	5	MT	0.138	0.142	0.131	0.138	0.131	0.252	0.232	0.220	0.068	0.196	0.034	0.020
8	5	MT	0.164	0.168	0.157	0.164	0.157	0.252	0.232	0.220	0.052	0.196	0.034	0.020
8	6	MT	0.164	0.168	0.157	0.164	0.157	0.279	0.257	0.244	0.069	0.220	0.036	0.021
10	8	MT	0.190	0.194	0.183	0.190	0.181	0.332	0.308	0.292	0.085	0.267	0.039	0.023
12	8	MT	0.216	0.220	0.209	0.216	0.207	0.332	0.308	0.292	0.069	0.267	0.039	0.023
12	10	MT	0.216	0.220	0.209	0.216	0.207	0.385	0.359	0.340	0.101	0.313	0.042	0.025
14	10	A	0.242	0.246	0.235	—	—	0.385	0.359	0.340	0.080	0.313	0.042	0.025
14	12	A	0.242	0.246	0.235	—	—	0.438	0.410	0.389	0.112	0.362	0.045	0.027
1/4	10	MBCD	0.250	—	—	0.250	0.240	0.385	0.359	0.340	0.080	0.313	0.042	0.025
1/4	12	MBCD	0.250	—	—	0.250	0.240	0.438	0.410	0.389	0.112	0.362	0.045	0.027
5/16	12	MBCD	0.3125	—	—	0.312	0.302	0.438	0.410	0.389	0.075	0.362	0.045	0.027
5/16	1/4	MBCD	0.3125	—	—	0.312	0.302	0.507	0.477	0.452	0.116	0.424	0.049	0.029
3/8	5/16	MBCD	0.375	—	—	0.375	0.364	0.635	0.600	0.568	0.155	0.539	0.057	0.034

[a] Additional specifications and details pertinent to bodies and threads are given in the General Specifications.
[b] Flat trim head machine and tapping screws are furnished in cross recess head only, see Tables 5A and 5B for recess dimensions.
[c] These values are calculated to maximum formula.
[d] Application code:

M — Machine Screws.
T — Tapping Screws, all types.
A — Tapping Screws, Type A.
B — Tapping Screws, Types B, BF, BG, BP, and BT.
C — Tapping Screws, Type C.
D — Tapping Screws, Types D, F, G, and T.

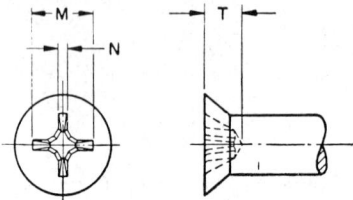

This type of recess has a large center opening, tapered wings, and blunt bottom, with all edges relieved or rounded.

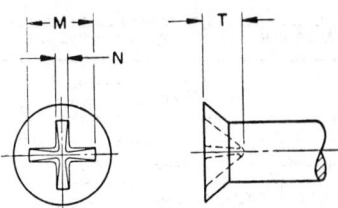

This type of recess consists of two intersecting slots with parallel sides converging to a slightly truncated apex at bottom of recess.

TABLE 5A—TYPE I CROSS RECESS DIMENSIONS OF FLAT TRIM HEAD MACHINE AND TAPPING SCREWS, IN.

Nominal Size		M Dia of Recess		T Depth of Recess		N Width of Recess	Driver Size	Penetration Gaging Depth[a]	
Body	Head	Min	Max	Min	Max	Min		Min	Max
4	3	0.094	0.107	0.052	0.068	0.018	1	0.045	0.061
5	4	0.115	0.128	0.073	0.089	0.018	1	0.066	0.082
6	4	0.115	0.128	0.073	0.089	0.018	1	0.066	0.082
6	5	0.141	0.154	0.063	0.086	0.027	2	0.052	0.075
8	5	0.151	0.164	0.073	0.096	0.029	2	0.062	0.085
8	6	0.169	0.182	0.091	0.114	0.030	2	0.080	0.103
10	8	0.176	0.189	0.098	0.121	0.031	2	0.087	0.110
12	8	0.185	0.198	0.108	0.131	0.032	2	0.097	0.120
12	10	0.191	0.204	0.113	0.136	0.032	2	0.102	0.125
14	10	0.191	0.204	0.113	0.136	0.032	2	0.102	0.125
14	12	0.255	0.268	0.133	0.156	0.035	3	0.116	0.139
1/4	10	0.191	0.204	0.113	0.136	0.032	2	0.102	0.125
1/4	12	0.255	0.268	0.133	0.156	0.035	3	0.116	0.139
5/16	12	0.255	0.268	0.133	0.156	0.035	3	0.116	0.139
5/16	1/4	0.270	0.283	0.148	0.171	0.036	3	0.131	0.154
3/8	5/16	0.352	0.365	0.194	0.216	0.061	4	0.174	0.196

[a] See Appendix B.

TABLE 5B—TYPE II CROSS RECESS DIMENSIONS OF FLAT TRIM HEAD MACHINE AND TAPPING SCREWS, IN.

Nominal Size		M Dia of Recess		T Depth of Recess		N Width of Recess	Driver Size	Penetration Gaging Depth[a]	
Body	Head	Min	Max	Min	Max	Min		Min	Max
4	3	0.127	0.139	0.066	0.079	0.030		0.041	0.053
5	4	0.145	0.157	0.075	0.088	0.032		0.052	0.064
6	4	0.145	0.157	0.075	0.088	0.032		0.052	0.064
6	5	0.162	0.176	0.087	0.101	0.035		0.064	0.077
8	5	0.162	0.176	0.087	0.101	0.035		0.064	0.077
8	6	0.180	0.195	0.098	0.113	0.038		0.075	0.089
10	8	0.216	0.232	0.117	0.132	0.043	Point same on all drivers	0.099	0.113
12	8	0.216	0.232	0.117	0.132	0.043		0.099	0.113
12	10	0.251	0.269	0.140	0.156	0.048		0.122	0.137
14	10	0.251	0.269	0.140	0.156	0.048		0.122	0.137
14	12	0.287	0.307	0.163	0.181	0.054		0.145	0.162
1/4	10	0.251	0.269	0.140	0.156	0.048		0.122	0.137
1/4	12	0.287	0.307	0.163	0.181	0.054		0.145	0.162
5/16	12	0.287	0.307	0.163	0.181	0.054		0.145	0.162
5/16	1/4	0.334	0.355	0.186	0.204	0.061		0.176	0.193
3/8	5/16	0.420	0.444	0.242	0.262	0.074		0.232	0.251

[a] See Appendix B.

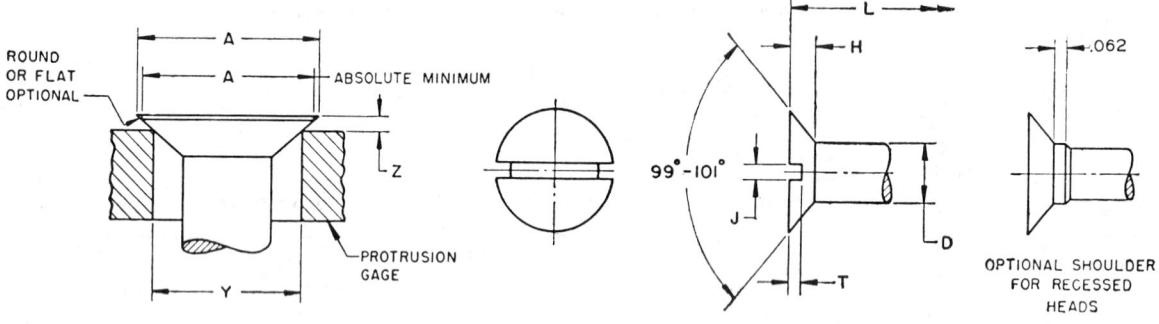

TABLE 6—DIMENSIONS OF 100 DEG FLAT HEAD MACHINE SCREWS, IN.[a]

Nominal Size	D Basic Dia of Screw	A Dia of Head			H[b] Height of Head	J Width of Slot		T Depth of Slot		Y Gage Dia	Z Protrusion of Head Above Gage	
		Max Sharp	Min Sharp	Abs Min	Ref	Min	Max	Min	Max		Max	Min
4	0.112	0.225	0.207	0.191	0.049	0.031	0.039	0.017	0.024	0.167	0.025	0.016
6	0.138	0.279	0.257	0.238	0.060	0.039	0.048	0.022	0.030	0.214	0.028	0.017
8	0.164	0.332	0.308	0.285	0.072	0.045	0.054	0.027	0.036	0.261	0.031	0.019
10	0.190	0.385	0.359	0.333	0.083	0.050	0.060	0.031	0.042	0.307	0.034	0.021
1/4	0.250	0.507	0.477	0.442	0.110	0.064	0.075	0.042	0.055	0.415	0.040	0.025
5/16	0.3125	0.635	0.600	0.556	0.138	0.072	0.084	0.053	0.069	0.526	0.047	0.030
3/8	0.375	0.762	0.722	0.670	0.165	0.081	0.094	0.065	0.083	0.638	0.053	0.034

[a] Additional specifications and details pertinent to bodies and threads are given in the General Specifications.

[b] These values calculated to maximum formula.

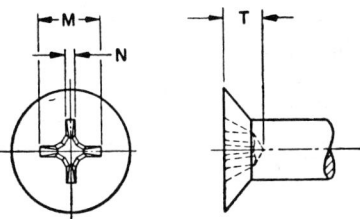
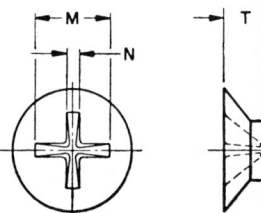

This type of recess has a large center opening, tapered wings, and blunt bottom, with all edges relieved or rounded.

This type of recess consists of two intersecting slots with parallel sides converging to a slightly truncated apex at bottom of recess.

TABLE 6A—TYPE I CROSS RECESS DIMENSIONS OF 100 DEG FLAT MACHINE SCREWS, IN.

Nominal Size	M Dia of Recess		T Depth of Recess		N Width of Recess	Driver Size	Penetration Gaging Depth[a]	
	Min	Max	Min	Max	Min		Min	Max
4	0.103	0.116	0.062	0.078	0.018	1	0.055	0.071
6	0.141	0.154	0.063	0.086	0.027	2	0.052	0.075
8	0.156	0.169	0.078	0.101	0.028	2	0.067	0.090
10	0.171	0.184	0.093	0.116	0.030	2	0.082	0.105
1/4	0.234	0.247	0.112	0.135	0.033	3	0.095	0.118
5/16	0.304	0.317	0.146	0.168	0.053	4	0.126	0.148
3/8	0.329	0.342	0.171	0.193	0.056	4	0.151	0.173

[a] See Appendix B.

TABLE 6B—TYPE II CROSS RECESS DIMENSIONS OF 100 DEG FLAT HEAD MACHINE SCREWS, IN.

Nominal Size	M Dia of Recess		T Depth of Recess		N Width of Recess	Driver Size	Penetration Gaging Depth[a]	
	Min	Max	Min	Max	Min		Min	Max
4	0.126	0.136	0.050	0.089	0.029		0.036	0.051
6	0.152	0.162	0.065	0.106	0.033		0.051	0.068
8	0.179	0.191	0.081	0.124	0.037	Point	0.067	0.086
10	0.213	0.225	0.097	0.147	0.042	same on all	0.087	0.109
1/4	0.282	0.294	0.136	0.192	0.053	drivers	0.132	0.154
5/16	0.349	0.361	0.179	0.235	0.063		0.175	0.197
3/8	0.426	0.440	0.230	0.287	0.075		0.226	0.249

[a] See Appendix B.

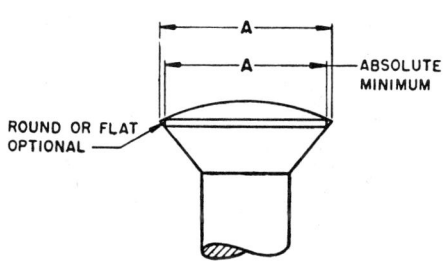
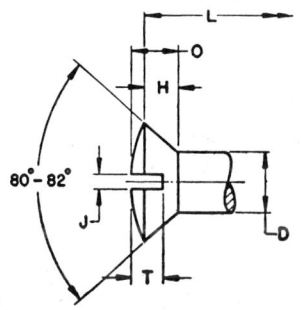

TABLE 7—DIMENSIONS OF OVAL HEAD MACHINE, TAPPING, AND WOOD SCREWS, IN.[a]

Nominal Size	Applicable to Screw Types[e] Code Letters	D Basic Dia of Screw	A Dia of Head			H[b] Side Height of Head	O Total Height of Head	J Width of Slot		T Depth of Slot		L[c] Screws These Lengths and Shorter are Undercut	
			Max Sharp	Min Sharp	Abs Min	Ref	Max	Min	Min	Max	Min	Max	Machine	Tapping
0	MABW	0.060	0.119	0.105	0.099	0.035	0.056	0.041	0.016	0.023	0.025	0.030	1/8	1/8[d]
1	MABW	0.073	0.146	0.130	0.123	0.043	0.068	0.052	0.019	0.026	0.031	0.038	1/8	5/32[d]
2	MTW	0.086	0.172	0.156	0.147	0.051	0.080	0.063	0.023	0.031	0.037	0.045	1/8	3/16
3	MTW	0.099	0.199	0.181	0.171	0.059	0.092	0.073	0.027	0.035	0.043	0.052	1/8	7/32
4	MTW	0.112	0.225	0.207	0.195	0.067	0.104	0.084	0.031	0.039	0.049	0.059	3/16	1/4
5	MTW	0.125	0.252	0.232	0.220	0.075	0.116	0.095	0.035	0.043	0.055	0.067	3/16	1/4
6	MTW	0.138	0.279	0.257	0.244	0.083	0.128	0.105	0.039	0.048	0.060	0.074	3/16	5/16
7	ABW	0.151	0.305	0.283	0.268	0.091	0.140	0.116	0.039	0.048	0.066	0.081	—	3/8
8	MTW	0.164	0.332	0.308	0.292	0.100	0.152	0.126	0.045	0.054	0.072	0.088	1/4	7/16
9	W	0.177	0.358	0.334	0.316	0.108	0.164	0.137	0.045	0.054	0.078	0.095	—	—
10	MTW	0.190	0.385	0.359	0.340	0.116	0.176	0.148	0.050	0.060	0.084	0.103	5/16	1/2
12	MTW	0.216	0.438	0.410	0.389	0.132	0.200	0.169	0.056	0.067	0.096	0.117	3/8	9/16
14	AW	0.242	0.491	0.461	0.437	0.148	0.224	0.190	0.064	0.075	0.108	0.132	—	5/8
16	AW	0.268	0.544	0.512	0.485	0.164	0.248	0.212	0.064	0.075	0.120	0.146	—	3/4
18	AW	0.294	0.597	0.563	0.534	0.180	0.272	0.233	0.072	0.084	0.132	0.160	—	13/16
20	AW	0.320	0.650	0.614	0.582	0.196	0.296	0.254	0.072	0.084	0.144	0.175	—	13/16
24	AW	0.372	0.756	0.716	0.679	0.228	0.344	0.297	0.081	0.094	0.168	0.204	—	1
1/4	MBCD	0.250	0.507	0.477	0.452	0.153	0.232	0.197	0.064	0.075	0.112	0.136	7/16	5/8
5/16	MBCD	0.3125	0.635	0.600	0.568	0.191	0.290	0.249	0.072	0.084	0.141	0.171	1/2	5/8
3/8	MBCD	0.375	0.762	0.722	0.685	0.230	0.347	0.300	0.081	0.094	0.170	0.206	9/16	5/8
7/16	MB	0.4375	0.812	0.767	0.723	0.223	0.345	0.295	0.081	0.094	0.174	0.210	5/8	3/4
1/2	MB	0.500	0.875	0.831	0.775	0.223	0.354	0.299	0.091	0.106	0.176	0.216	3/4	3/4
9/16	M	0.5625	1.000	0.950	0.889	0.260	0.410	0.350	0.102	0.118	0.207	0.250	13/16	—
5/8	M	0.6250	1.125	1.069	1.002	0.298	0.467	0.399	0.116	0.133	0.235	0.285	15/16	—
3/4	M	0.7500	1.375	1.306	1.230	0.372	0.578	0.497	0.131	0.149	0.293	0.353	1-1/8	—

[a] Additional specifications and details pertinent to bodies and threads are given in the General Specifications.
[b] These values calculated to maximum formula.
[c] Undercut heads not applicable to wood screws. See Table 8, Dimensions of Undercut Oval Head Machine and Tapping Screws.
[d] 3/16 for Type A tapping screws.
[e] Application code:

M — Machine Screws.
T — Tapping Screws, all types.
A — Tapping Screws, Type A.
B — Tapping Screws, Types B, BF, BG, BP, and BT.
C — Tapping Screws, Type C.
D — Tapping Screws, Types D, F, G, and T.
W — Wood Screws.

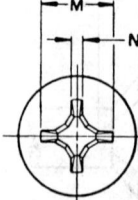 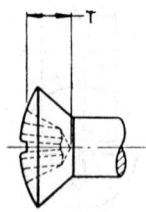 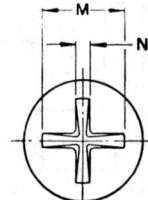 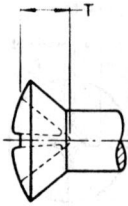

This type of recess has a large center opening, tapered wings, and blunt bottom, with all edges relieved or rounded.

This type of recess consists of two intersecting slots with parallel sides converging to a slightly truncated apex at bottom of recess.

TABLE 7A—TYPE I CROSS RECESS DIMENSIONS OF OVAL HEAD MACHINE AND TAPPING SCREWS, IN.

Nominal Size	M Dia of Recess		T Depth of Recess		N Width of Recess	Driver Size	Penetration Gaging Depth[a]	
	Min	Max	Min	Max	Min		Min	Max
0	0.061	0.074	0.027	0.045	0.014	0	0.020	0.038
1	0.064	0.077	0.030	0.048	0.015	0	0.023	0.041
2	0.099	0.112	0.052	0.069	0.018	1	0.045	0.062
3	0.111	0.124	0.064	0.081	0.019	1	0.057	0.074
4	0.123	0.136	0.077	0.094	0.019	1	0.070	0.087
5	0.145	0.158	0.061	0.085	0.028	2	0.050	0.074
6	0.165	0.178	0.080	0.105	0.030	2	0.069	0.094
7	0.170	0.183	0.086	0.111	0.030	2	0.075	0.100
8	0.179	0.192	0.095	0.119	0.031	2	0.084	0.108
10	0.196	0.209	0.113	0.137	0.033	2	0.102	0.126
12	0.257	0.270	0.128	0.152	0.038	3	0.111	0.135
14	0.275	0.288	0.140	0.165	0.039	3	0.123	0.148
16	0.319	0.332	0.189	0.214	0.046	3	0.172	0.197
18	0.368	0.381	0.202	0.226	0.064	4	0.182	0.206
20	0.387	0.400	0.221	0.245	0.066	4	0.201	0.225
24	0.423	0.436	0.258	0.282	0.072	4	0.238	0.262
1/4	0.277	0.290	0.148	0.173	0.040	3	0.131	0.156
5/16	0.377	0.390	0.214	0.238	0.065	4	0.194	0.218
3/8	0.397	0.410	0.233	0.257	0.068	4	0.213	0.237
7/16	0.409	0.422	0.245	0.269	0.070	4	0.225	0.249
1/2	0.424	0.437	0.259	0.283	0.071	4	0.239	0.263
9/16	0.450	0.473	0.292	0.316	0.075	4	0.272	0.296
5/8	0.568	0.591	0.324	0.356	0.081	5	0.291	0.323
3/4	0.630	0.653	0.390	0.421	0.088	5	0.357	0.388

[a] See Appendix B.

TABLE 7B—TYPE II CROSS RECESS DIMENSIONS OF OVAL HEAD MACHINE AND TAPPING SCREWS, IN.

Nominal Size	M Dia of Recess		T Depth of Recess		N Width of Recess	Driver Size	Penetration Gaging Depth[a]	
	Min	Max	Min	Max	Min		Min	Max
0	0.073	0.083	0.031	0.042	0.021		—[b]	—[b]
1	0.091	0.102	0.042	0.054	0.024		—[b]	—[b]
2	0.109	0.120	0.054	0.066	0.027		0.029	0.040
3	0.127	0.139	0.066	0.079	0.030		0.041	0.053
4	0.145	0.157	0.075	0.088	0.032		0.052	0.064
5	0.162	0.176	0.087	0.101	0.035		0.064	0.077
6	0.180	0.195	0.098	0.113	0.038		0.075	0.089
7	0.198	0.213	0.110	0.125	0.040	Point	0.087	0.101
8	0.216	0.232	0.117	0.132	0.043	same	0.099	0.113
10	0.251	0.269	0.140	0.156	0.048	on all	0.122	0.137
12	0.287	0.307	0.163	0.181	0.054	drivers	0.145	0.162
14	0.323	0.344	0.178	0.197	0.059		0.168	0.186
16	0.358	0.381	0.201	0.221	0.064		0.191	0.210
18	0.394	0.418	0.225	0.245	0.070		0.215	0.234
20	0.430	0.455	0.248	0.269	0.075		0.238	0.258
24	0.501	0.529	0.294	0.318	0.086		0.284	0.307
1/4	0.334	0.355	0.186	0.204	0.061		0.176	0.193
5/16	0.420	0.444	0.242	0.262	0.074		0.232	0.251
3/8	0.495	0.523	0.291	0.314	0.086		0.281	0.303
7/16	0.540	0.568	0.320	0.343	0.092		0.310	0.332
1/2	0.578	0.608	0.345	0.370	0.098		0.335	0.359
9/16	0.623	0.656	0.374	0.400	0.104		0.364	0.389
5/8	0.623	0.656	0.374	0.400	0.104		0.364	0.389
3/4	0.623	0.656	0.374	0.400	0.104		0.364	0.389

[a] See Appendix B. [b] Not practical to gage.

TABLE 7C—TYPE I CROSS RECESS DIMENSIONS OF OVAL HEAD WOOD SCREWS, IN.

Nominal Size	M Dia of Recess		T Depth of Recess		N Width of Recess	Driver Size	Penetration Gauging Depth[a]	
	Min	Max	Min	Max	Min		Min	Max
0	0.061	0.074	0.027	0.045	0.014	0	0.020	0.038
1	0.064	0.077	0.030	0.048	0.015	0	0.023	0.041
2	0.099	0.112	0.052	0.069	0.018	1	0.045	0.062
3	0.111	0.124	0.064	0.081	0.019	1	0.057	0.074
4	0.123	0.136	0.077	0.094	0.019	1	0.070	0.087
5	0.145	0.158	0.061	0.085	0.028	2	0.050	0.074
6	0.165	0.178	0.080	0.105	0.030	2	0.069	0.094
7	0.176	0.189	0.092	0.115	0.031	2	0.081	0.104
8	0.192	0.205	0.106	0.131	0.033	2	0.095	0.120
9	0.203	0.216	0.119	0.144	0.034	2	0.108	0.133
10	0.248	0.261	0.118	0.142	0.037	3	0.101	0.125
12	0.270	0.283	0.140	0.165	0.040	3	0.123	0.148
14	0.292	0.305	0.163	0.188	0.042	3	0.146	0.171
16	0.319	0.332	0.189	0.214	0.046	3	0.172	0.197
18	0.368	0.381	0.202	0.226	0.064	4	0.182	0.206
20	0.387	0.400	0.221	0.245	0.066	4	0.201	0.225
24	0.423	0.436	0.258	0.282	0.072	4	0.238	0.262

[a] See Appendix B.

TABLE 7D—TYPE II CROSS RECESS DIMENSIONS OF OVAL HEAD WOOD SCREWS, IN.

Nominal Size	M Dia or Recess		T Depth of Recess		N Width of Recess	Driver Size	Penetration Gauging Depth[a]	
	Min	Max	Min	Max	Min		Min	Max
0	0.073	0.083	0.031	0.042	0.021		—[b]	—[b]
1	0.091	0.102	0.042	0.054	0.024		—[b]	—[b]
2	0.109	0.120	0.054	0.066	0.027		0.029	0.040
3	0.127	0.139	0.066	0.079	0.030		0.041	0.053
4	0.145	0.157	0.075	0.088	0.032		0.052	0.064
5	0.162	0.176	0.087	0.101	0.035		0.064	0.077
6	0.180	0.195	0.098	0.113	0.038		0.075	0.089
7	0.198	0.213	0.110	0.125	0.040	Point	0.087	0.101
8	0.216	0.232	0.117	0.132	0.043	same	0.099	0.113
9	0.234	0.251	0.128	0.145	0.046	on all	0.110	0.126
10	0.251	0.269	0.140	0.156	0.048	drivers	0.122	0.137
12	0.287	0.307	0.163	0.181	0.054		0.145	0.162
14	0.323	0.344	0.178	0.197	0.059		0.168	0.186
16	0.358	0.381	0.201	0.221	0.064		0.191	0.210
18	0.394	0.418	0.225	0.245	0.070		0.215	0.234
20	0.430	0.455	0.248	0.269	0.075		0.238	0.258
24	0.501	0.529	0.294	0.318	0.086		0.284	0.307

[a] See Appendix B. [b] Not practical to gage.

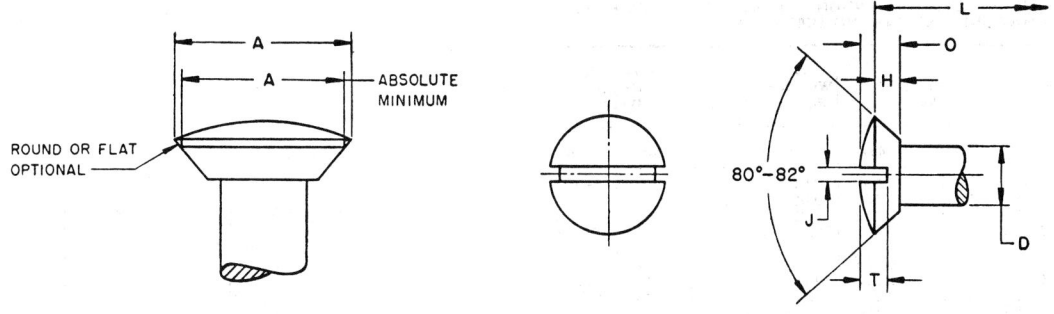

TABLE 8—DIMENSIONS OF UNDERCUT OVAL HEAD MACHINE AND TAPPING SCREWS, IN.[a]

Nominal Size	Applicable to Screw Types[d] Code Letters	D Basic Dia of Screw	L Screws These Lengths and Under are Undercut		A Dia of Head			H[b] Side Height of Head	O Total Height of Head		J Width of Slot		T Depth of Slot	
			Machine	Tapping	Max Sharp	Min Sharp	Abs Min	Ref	Max	Min	Min	Max	Min	Max
0	MAB	0.060	1/8	1/8[c]	0.119	0.105	0.099	0.025	0.046	0.033	0.016	0.023	0.022	0.028
1	MAB	0.073	1/8	5/32[c]	0.146	0.130	0.123	0.031	0.056	0.042	0.019	0.026	0.027	0.034
2	MT	0.086	1/8	3/16	0.172	0.156	0.147	0.036	0.065	0.050	0.023	0.031	0.033	0.040
3	MT	0.099	1/8	7/32	0.199	0.181	0.171	0.042	0.075	0.059	0.027	0.035	0.038	0.047
4	MT	0.112	3/16	1/4	0.225	0.207	0.195	0.047	0.084	0.067	0.031	0.039	0.043	0.053
5	MT	0.125	3/16	1/4	0.252	0.232	0.220	0.053	0.094	0.076	0.035	0.043	0.048	0.059
6	MT	0.138	3/16	5/16	0.279	0.257	0.244	0.059	0.104	0.084	0.039	0.048	0.053	0.065
7	AB	0.151	—	3/8	0.305	0.283	0.268	0.064	0.113	0.093	0.039	0.048	0.059	0.071
8	MT	0.164	1/4	7/16	0.332	0.308	0.292	0.070	0.123	0.101	0.045	0.054	0.064	0.078
10	MT	0.190	5/16	1/2	0.385	0.359	0.340	0.081	0.142	0.118	0.050	0.060	0.074	0.090
12	MT	0.216	3/8	9/16	0.438	0.410	0.389	0.092	0.161	0.135	0.056	0.067	0.085	0.103
14	A	0.242	—	5/8	0.491	0.461	0.437	0.104	0.180	0.152	0.064	0.075	0.095	0.115
16	A	0.268	—	3/4	0.544	0.512	0.485	0.115	0.199	0.169	0.064	0.075	0.107	0.128
18	A	0.294	—	13/16	0.597	0.563	0.534	0.126	0.218	0.186	0.072	0.084	0.116	0.140
20	A	0.320	—	13/16	0.650	0.614	0.582	0.137	0.238	0.203	0.072	0.084	0.127	0.153
24	A	0.372	—	1	0.756	0.716	0.679	0.160	0.276	0.237	0.081	0.094	0.148	0.178
1/4	MBCD	0.250	7/16	5/8	0.507	0.477	0.452	0.107	0.186	0.158	0.064	0.075	0.098	0.119
5/16	MBCD	0.3125	1/2	5/8	0.635	0.600	0.568	0.134	0.232	0.198	0.072	0.084	0.124	0.149
3/8	MBCD	0.375	9/16	5/8	0.762	0.722	0.685	0.161	0.278	0.239	0.081	0.094	0.149	0.179
7/16	MB	0.4375	5/8	3/4	0.812	0.767	0.723	0.156	0.279	0.239	0.081	0.094	0.154	0.184
1/2	MB	0.500	3/4	3/4	0.875	0.831	0.775	0.156	0.288	0.244	0.091	0.106	0.169	0.204
9/16	M	0.5625	13/16	—	1.000	0.950	0.889	0.182	0.332	0.283	0.102	0.118	0.194	0.233
5/8	M	0.6250	15/16	—	1.125	1.069	1.002	0.208	0.377	0.323	0.116	0.133	0.220	0.263
3/4	M	0.7500	1-1/8	—	1.375	1.306	1.230	0.260	0.467	0.402	0.131	0.149	0.270	0.322

[a] Additional specifications and details pertinent to bodies and threads are given in the General Specifications.
[b] These values calculated to maximum formula.
[c] 3/16 for Type A tapping screws.
[d] Application code:

M — Machine Screws.
T — Tapping Screws, all types.
A — Tapping Screws, Type A.
B — Tapping Screws, Types B, BF, BG, BP, and BT.
C — Tapping Screws, Type C.
D — Tapping Screws, Types D, F, G, and T.

TABLE 8A—TYPE I CROSS RECESS DIMENSIONS OF UNDERCUT OVAL HEAD MACHINE AND TAPPING SCREWS, IN.

Nominal Size	M Dia of Recess		T Depth of Recess		N Width of Recess	Driver Size	Penetration Gaging Depth[a]	
	Min	Max	Min	Max	Min		Min	Max
0	0.061	0.074	0.027	0.045	0.014	0	0.020	0.038
1	0.064	0.077	0.030	0.048	0.015	0	0.023	0.041
2	0.099	0.112	0.052	0.069	0.018	1	0.045	0.062
3	0.111	0.124	0.064	0.081	0.019	1	0.057	0.074
4	0.123	0.136	0.077	0.094	0.019	1	0.070	0.087
5	0.145	0.158	0.061	0.085	0.028	2	0.050	0.074
6	0.165	0.178	0.080	0.105	0.030	2	0.069	0.094
7	0.170	0.183	0.086	0.111	0.030	2	0.075	0.100
8	0.179	0.192	0.095	0.119	0.031	2	0.084	0.108
10	0.196	0.209	0.113	0.137	0.033	2	0.102	0.126
12	0.257	0.270	0.128	0.152	0.038	3	0.111	0.135
14	0.275	0.288	0.140	0.165	0.039	3	0.123	0.148
16	0.294	0.307	0.166	0.190	0.040	3	0.149	0.173
18	0.352	0.365	0.185	0.209	0.061	4	0.165	0.189
20	0.377	0.390	0.214	0.238	0.064	4	0.194	0.218
24	0.387	0.400	0.221	0.245	0.066	4	0.201	0.225
1/4	0.277	0.290	0.148	0.173	0.040	3	0.131	0.156
5/16	0.368	0.381	0.202	0.226	0.064	4	0.182	0.206
3/8	0.387	0.400	0.221	0.245	0.066	4	0.201	0.225
7/16	0.397	0.410	0.233	0.257	0.068	4	0.213	0.237
1/2	0.409	0.422	0.245	0.269	0.070	4	0.225	0.249

[a] See Appendix B.

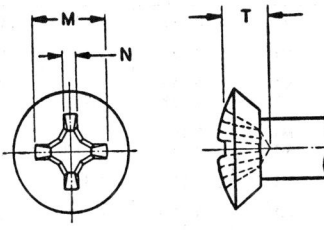

This type of recess has a large center opening, tapered wings, and blunt bottom, with all edges relieved or rounded.

15.18

TABLE 8B—TYPE II CROSS RECESS DIMENSIONS OF UNDERCUT OVAL HEAD MACHINE AND TAPPING SCREWS, IN.

Nominal Size	M Dia of Recess		T Depth of Recess		N Width of Recess	Driver Size	Penetration Gaging Depth[a]	
	Min	Max	Min	Max	Min		Min	Max
0	0.062	0.072	0.023	0.035	0.020		—[b]	—[b]
1	0.077	0.087	0.033	0.045	0.022		—[b]	—[b]
2	0.095	0.105	0.044	0.056	0.025		0.020	0.030
3	0.106	0.116	0.052	0.064	0.026		0.027	0.038
4	0.123	0.135	0.063	0.076	0.029		0.038	0.050
5	0.141	0.153	0.073	0.086	0.032		0.050	0.062
6	0.155	0.167	0.082	0.095	0.034		0.059	0.071
7	0.171	0.185	0.092	0.107	0.036		0.069	0.083
8	0.190	0.204	0.105	0.119	0.039	Point	0.082	0.095
10	0.228	0.244	0.125	0.140	0.045		0.107	0.121
12	0.251	0.269	0.140	0.156	0.048	same	0.122	0.137
14	0.285	0.305	0.162	0.180	0.053		0.144	0.161
16	0.314	0.336	0.173	0.192	0.058	on all	0.163	0.181
18	0.341	0.365	0.190	0.210	0.062		0.180	0.200
20	0.380	0.404	0.216	0.236	0.068	drivers	0.206	0.225
24	0.439	0.467	0.254	0.277	0.077		0.244	0.266
1/4	0.294	0.314	0.160	0.178	0.054		0.150	0.167
5/16	0.369	0.393	0.208	0.229	0.066		0.198	0.218
3/8	0.439	0.467	0.254	0.277	0.077		0.244	0.266
7/16	0.483	0.513	0.283	0.307	0.083		0.273	0.296
1/2	0.533	0.563	0.315	0.340	0.090		0.305	0.329

[a] See Appendix B.   [b] Not practical to gage.

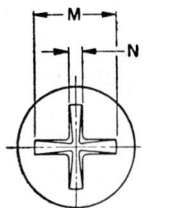
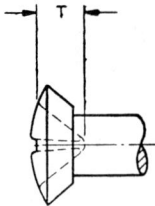

This type of recess consists of two intersecting slots with parallel sides converging to a slightly truncated apex at bottom of recess.

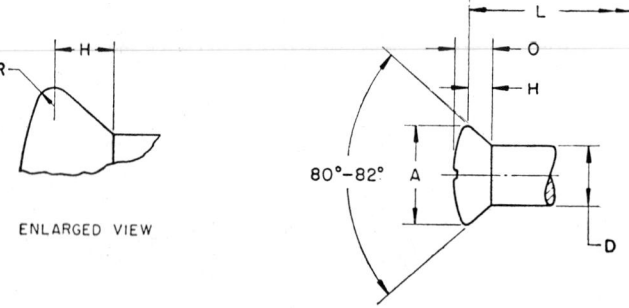
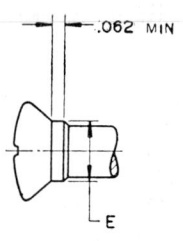

SHOULDER CONSTRUCTION FOR LONG SCREWS WITH REDUCED BODY (SEE GENERAL SPECIFICATIONS)

TABLE 9—DIMENSIONS OF OVAL TRIM HEAD MACHINE AND TAPPING SCREWS, IN.[a, b]

Nominal Size		Applicable to Screw Types[d]	D Basic Dia of Screw	E Dia of Body or Shoulder				A Dia of Head		H[c] Side Height of Head	O Total Height of Head		R Radius of Head
				Type A Tapping		Machine and Other Types of Tapping							
Body	Head	Code Letters		Max	Min	Max	Min	Max	Min	Ref	Max	Min	Nominal
4	3	MT	0.112	0.116	0.105	0.112	0.106	0.199	0.181	0.057	0.086	0.074	0.014
5	4	MT	0.125	0.129	0.118	0.125	0.119	0.225	0.207	0.066	0.099	0.086	0.017
6	4	MT	0.138	0.142	0.131	0.138	0.131	0.225	0.207	0.058	0.091	0.079	0.017
6	5	MT	0.138	0.142	0.131	0.138	0.131	0.252	0.232	0.075	0.112	0.098	0.019
8	5	MT	0.164	0.168	0.157	0.164	0.157	0.252	0.232	0.060	0.096	0.083	0.019
8	6	MT	0.164	0.168	0.157	0.164	0.157	0.279	0.257	0.076	0.117	0.102	0.021
10	8	MT	0.190	0.194	0.183	0.190	0.181	0.332	0.308	0.094	0.141	0.124	0.025
12	8	MT	0.216	0.220	0.209	0.216	0.207	0.332	0.308	0.078	0.125	0.109	0.025
12	10	MT	0.216	0.220	0.209	0.216	0.207	0.385	0.359	0.111	0.166	0.148	0.029
14	10	A	0.242	0.246	0.235	—	—	0.385	0.359	0.091	0.146	0.129	0.029
14	12	A	0.242	0.246	0.235	—	—	0.438	0.410	0.124	0.187	0.167	0.033
1/4	10	MBCD	0.250	—	—	0.250	0.240	0.385	0.359	0.091	0.146	0.129	0.029
1/4	12	MBCD	0.250	—	—	0.250	0.240	0.438	0.410	0.124	0.187	0.167	0.033
5/16	12	MBCD	0.3125	—	—	0.312	0.302	0.438	0.410	0.087	0.150	0.131	0.033
5/16	1/4	MBCD	0.3125	—	—	0.312	0.302	0.507	0.477	0.130	0.202	0.181	0.038
3/8	5/16	MBCD	0.375	—	—	0.375	0.364	0.635	0.600	0.173	0.265	0.240	0.048

[a] Additional specifications and details pertinent to bodies and threads are given in the General Specifications.
[b] Oval trim head machine and tapping screws are furnished in cross recess head only. See Tables 9A and 9B for recess dimensions.
[c] These are maximum values taken from layouts of these heads.
[d] Application code:

M — Machine Screws.
T — Tapping Screws, all types.
A — Tapping Screws, Type A.
B — Tapping Screws, Types B, BF, BG, BP, and BT.
C — Tapping Screws, Type C.
D — Tapping Screws, Types D, F, G, and T.

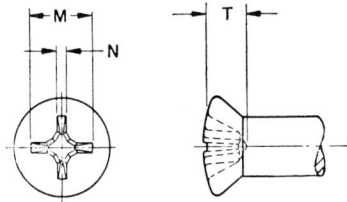

This type of recess has a large center opening, tapered wings, and blunt bottom, with all edges relieved or rounded.

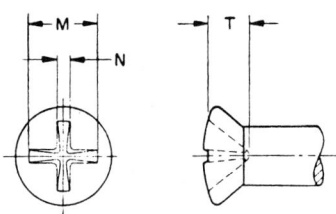

This type of recess consists of two intersecting slots with parallel sides converging to a slightly truncated apex at bottom of recess.

TABLE 9A—TYPE I CROSS RECESS DIMENSIONS OF OVAL TRIM HEAD MACHINE AND TAPPING SCREWS, IN.

Nominal Size		M Dia of Recess		T Depth of Recess		N Width of Recess	Driver Size	Penetration Gaging Depth[a]	
Body	Head	Min	Max	Min	Max	Min		Min	Max
Short Screws[b]									
4	3	0.111	0.124	0.064	0.081	0.019	1	0.057	0.074
5	4	0.123	0.136	0.077	0.094	0.019	1	0.070	0.087
6	4	0.123	0.136	0.077	0.094	0.019	1	0.070	0.087
6	5	0.145	0.158	0.061	0.085	0.028	2	0.050	0.074
8	5	0.145	0.158	0.061	0.085	0.028	2	0.050	0.074
8	6	0.165	0.178	0.080	0.105	0.030	2	0.069	0.094
10	8	0.179	0.192	0.095	0.119	0.031	2	0.084	0.108
12	8	0.179	0.192	0.095	0.119	0.031	2	0.084	0.108
12	10	0.196	0.209	0.113	0.137	0.033	2	0.102	0.126
14	10	0.196	0.209	0.113	0.137	0.033	2	0.102	0.126
14	12	0.257	0.270	0.128	0.152	0.038	3	0.111	0.135
1/4	10	0.196	0.209	0.113	0.137	0.033	2	0.102	0.126
1/4	12	0.257	0.270	0.128	0.152	0.038	3	0.111	0.135
5/16	12	0.257	0.270	0.128	0.152	0.038	3	0.111	0.135
5/16	1/4	0.277	0.290	0.148	0.173	0.040	3	0.131	0.156
3/8	5/16	0.377	0.390	0.214	0.238	0.065	4	0.194	0.218
Long Screws[c]									
4	3	0.111	0.124	0.064	0.081	0.019	1	0.057	0.074
5	4	0.123	0.136	0.077	0.094	0.019	1	0.070	0.087
6	4	0.123	0.136	0.077	0.094	0.019	1	0.070	0.087
6	5	0.156	0.169	0.072	0.096	0.028	2	0.061	0.085
8	5	0.156	0.169	0.072	0.096	0.028	2	0.061	0.085
8	6	0.181	0.194	0.096	0.121	0.031	2	0.085	0.110
10	8	0.194	0.207	0.110	0.135	0.031	2	0.099	0.124
12	8	0.194	0.207	0.110	0.135	0.031	2	0.099	0.124
12	10	0.207	0.220	0.123	0.148	0.033	2	0.112	0.137
14	10	0.207	0.220	0.123	0.148	0.033	2	0.112	0.137
14	12	0.257	0.270	0.128	0.152	0.038	3	0.111	0.135
1/4	10	0.207	0.220	0.123	0.148	0.033	2	0.112	0.137
1/4	12	0.257	0.270	0.128	0.152	0.038	3	0.111	0.135
5/16	12	0.257	0.270	0.128	0.152	0.038	3	0.111	0.135
5/16	1/4	0.277	0.290	0.148	0.173	0.040	3	0.131	0.156
3/8	5/16	0.377	0.390	0.214	0.238	0.065	4	0.194	0.218

[a] See Appendix B.
[b] Short oval trim head screws are defined as follows—Machine and Types: B, BF, BG, BP, BT, C, D, F, G, and T tapping screws having lengths of 1-1/2 in. or shorter. Type A tapping screws in sizes through No. 12 having lengths of 1 in. or shorter and in sizes over No. 12 having lengths equal to or less than 8 times the diameter.
[c] Long oval trim head screws are screws having longer lengths than those described in footnote [b].

TABLE 9B—TYPE II CROSS RECESS DIMENSIONS OF OVAL TRIM HEAD MACHINE AND TAPPING SCREWS, IN.

Nominal Size		M Dia of Recess		T Depth of Recess		N Width of Recess	Driver Size	Penetration Gaging Depth[a]	
Body	Head	Min	Max	Min	Max	Min		Min	Max
All Length Screws									
4	3	0.127	0.139	0.066	0.079	0.030		0.041	0.053
5	4	0.145	0.157	0.075	0.088	0.032		0.052	0.064
6	4	0.145	0.157	0.075	0.088	0.032		0.052	0.064
6	5	0.162	0.176	0.087	0.101	0.035		0.064	0.077
8	5	0.162	0.176	0.087	0.101	0.035		0.064	0.077
8	6	0.180	0.195	0.098	0.113	0.038	Point same on all drivers	0.075	0.089
10	8	0.216	0.232	0.117	0.132	0.043		0.099	0.113
12	8	0.216	0.232	0.117	0.132	0.043		0.099	0.113
12	10	0.251	0.269	0.140	0.156	0.048		0.122	0.137
14	10	0.251	0.269	0.140	0.156	0.048		0.122	0.137
14	12	0.287	0.307	0.163	0.181	0.054		0.145	0.162
1/4	10	0.251	0.269	0.140	0.156	0.048		0.122	0.137
1/4	12	0.287	0.307	0.163	0.181	0.054		0.145	0.162
5/16	12	0.287	0.307	0.163	0.181	0.054		0.145	0.162
5/16	1/4	0.334	0.355	0.186	0.204	0.061		0.176	0.193
3/8	5/16	0.420	0.444	0.242	0.262	0.074		0.232	0.251

[a] See Appendix B.

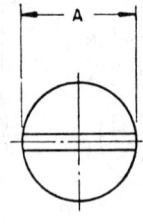 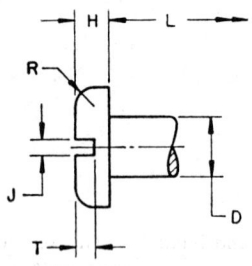 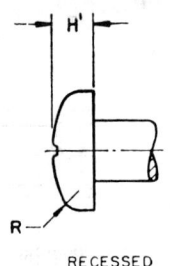

RECESSED

TABLE 10—DIMENSIONS OF PAN HEAD MACHINE AND TAPPING SCREWS, IN.[a]

Nominal Size	Applicable to Screw Types[b] Code Letters	D Basic Dia of Screw	A Dia of Head		H Height of Head (Slotted)		H' Height of Head (Recessed)		J Width of Slot		T Depth of Slot		R Radius of Head (Slotted)	R' Radius of Head (Recessed)
			Max	Min	Max	Min	Max	Min	Min	Max	Min	Max	Max	Min
0	MAB	0.060	0.116	0.104	0.039	0.031	0.044	0.036	0.016	0.023	0.014	0.022	0.020	0.005
1	MAB	0.073	0.142	0.130	0.046	0.038	0.053	0.044	0.019	0.026	0.018	0.027	0.025	0.005
2	MT	0.086	0.167	0.155	0.053	0.045	0.062	0.053	0.023	0.031	0.022	0.031	0.035	0.010
3	MT	0.099	0.193	0.180	0.060	0.051	0.071	0.062	0.027	0.035	0.026	0.036	0.037	0.010
4	MT	0.112	0.219	0.205	0.068	0.058	0.080	0.070	0.031	0.039	0.030	0.040	0.042	0.010
5	MT	0.125	0.245	0.231	0.075	0.065	0.089	0.079	0.035	0.043	0.034	0.045	0.044	0.015
6	MT	0.138	0.270	0.256	0.082	0.072	0.097	0.087	0.039	0.048	0.037	0.050	0.046	0.015
7	AB	0.151	0.296	0.281	0.089	0.079	0.106	0.096	0.039	0.048	0.041	0.054	0.049	0.015
8	MT	0.164	0.322	0.306	0.096	0.085	0.115	0.105	0.045	0.054	0.045	0.058	0.052	0.015
10	MT	0.190	0.373	0.357	0.110	0.099	0.133	0.122	0.050	0.060	0.053	0.068	0.061	0.020
12	MT	0.216	0.425	0.407	0.125	0.112	0.151	0.139	0.056	0.067	0.061	0.077	0.078	0.025
14	A	0.242	0.476	0.457	0.139	0.126	0.169	0.156	0.064	0.075	0.068	0.085	0.087	0.035
16	A	0.268	0.528	0.508	0.153	0.139	0.187	0.173	0.064	0.075	0.074	0.093	0.094	0.035
18	A	0.294	0.579	0.558	0.168	0.153	0.205	0.191	0.072	0.084	0.080	0.100	0.099	0.035
20	A	0.320	0.431	0.608	0.182	0.166	0.223	0.208	0.072	0.084	0.087	0.108	0.121	0.040
24	A	0.372	0.734	0.709	0.211	0.193	0.259	0.242	0.081	0.094	0.100	0.123	0.143	0.040
1/4	MBCD	0.250	0.492	0.473	0.144	0.130	0.175	0.162	0.064	0.075	0.070	0.087	0.087	0.035
5/16	MBCD	0.3125	0.615	0.594	0.178	0.162	0.218	0.203	0.072	0.084	0.085	0.106	0.099	0.040
3/8	MBCD	0.375	0.740	0.716	0.212	0.195	0.261	0.244	0.081	0.094	0.100	0.124	0.143	0.040

[a] Additional specifications and details pertinent to bodies and threads are given in the General Specifications.
[b] Application code:
 M—Machine Screws.
 T—Tapping Screws, all types.

A—Tapping Screws, Type A.
B—Tapping Screws, Types B, BF, BG, BP, and BT.
C—Tapping Screws, Type C.
D—Tapping Screws, Types D, F, G, and T.

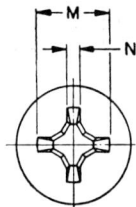 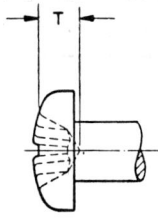

This type of recess has a large center opening, tapered wings, and blunt bottom, with all edges relieved or rounded.

TABLE 10A—TYPE I CROSS RECESS DIMENSIONS OF PAN HEAD MACHINE AND TAPPING SCREWS, IN.

Nominal Size	M Dia of Recess		T Depth of Recess		N Width of Recess	Driver Size	Penetration Gaging Depth[a]	
	Min	Max	Min	Max	Min		Min	Max
0	0.054	0.067	0.021	0.039	0.013	0	0.014	0.032
1	0.061	0.074	0.025	0.045	0.014	0	0.022	0.040
2	0.091	0.104	0.041	0.059	0.017	1	0.034	0.052
3	0.099	0.112	0.050	0.068	0.019	1	0.043	0.061
4	0.109	0.122	0.060	0.078	0.019	1	0.053	0.071
5	0.145	0.158	0.057	0.083	0.028	2	0.046	0.072
6	0.153	0.166	0.066	0.091	0.028	2	0.055	0.080
7	0.163	0.176	0.075	0.100	0.029	2	0.064	0.089
8	0.169	0.182	0.082	0.108	0.030	2	0.071	0.097
10	0.186	0.199	0.100	0.124	0.031	2	0.089	0.113
12	0.246	0.259	0.115	0.141	0.034	3	0.098	0.124
14	0.268	0.281	0.135	0.161	0.036	3	0.118	0.144
16	0.279	0.292	0.149	0.175	0.037	3	0.132	0.158
18	0.322	0.335	0.154	0.178	0.056	4	0.134	0.158
20	0.337	0.350	0.169	0.193	0.059	4	0.149	0.173
24	0.380	0.393	0.210	0.233	0.065	4	0.190	0.213
1/4	0.268	0.281	0.135	0.161	0.036	3	0.118	0.144
5/16	0.337	0.350	0.169	0.193	0.059	4	0.149	0.173
3/8	0.376	0.389	0.210	0.233	0.065	4	0.190	0.213

[a] See Appendix B.

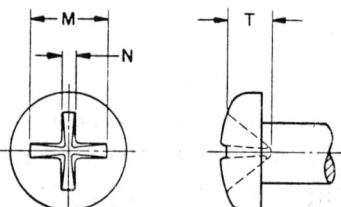

This type of recess consists of two intersecting slots with parallel sides converging to a slightly truncated apex at bottom of recess.

TABLE 10B—TYPE II CROSS RECESS DIMENSIONS OF PAN HEAD MACHINE AND TAPPING SCREWS, IN.

Nominal Size	M Dia of Recess		T Depth of Recess		N Width of Recess	Driver Size	Penetration Gaging Depth[a]	
	Min	Max	Min	Max	Min		Min	Max
0	0.068	0.075	0.027	0.037	0.021		—[b]	—[b]
1	0.080	0.088	0.035	0.045	0.023		—[b]	—[b]
2	0.101	0.109	0.049	0.059	0.026		0.024	0.033
3	0.117	0.125	0.059	0.069	0.028		0.034	0.043
4	0.133	0.142	0.068	0.079	0.031		0.045	0.055
5	0.150	0.159	0.079	0.090	0.033		0.056	0.066
6	0.166	0.175	0.089	0.100	0.035		0.066	0.076
7	0.183	0.192	0.100	0.111	0.038	Point	0.077	0.087
8	0.199	0.209	0.111	0.122	0.040	same	0.088	0.098
10	0.232	0.242	0.127	0.139	0.045	on all	0.109	0.120
12	0.265	0.276	0.149	0.161	0.050	drivers	0.131	0.142
14	0.297	0.309	0.162	0.174	0.055		0.152	0.163
16	0.330	0.343	0.183	0.197	0.060		0.173	0.186
18	0.363	0.376	0.205	0.218	0.064		0.195	0.207
20	0.391	0.405	0.223	0.237	0.069		0.213	0.226
24	0.461	0.477	0.268	0.284	0.081		0.258	0.273
1/4	0.307	0.320	0.168	0.182	0.057		0.158	0.171
5/16	0.386	0.400	0.220	0.234	0.069		0.210	0.223
3/8	0.465	0.481	0.271	0.286	0.081		0.261	0.275

[a] See Appendix B. [b] Not practical to gage.

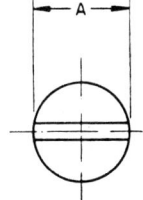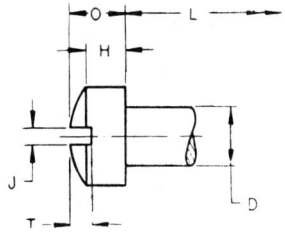

TABLE 11—DIMENSIONS OF FILLISTER HEAD MACHINE AND TAPPING SCREWS, IN.[a,b]

Nominal Size	Applicable to Screw Types[c] Code Letters	D Basic Dia of Screw	A Dia of Head Max	A Dia of Head Min	H Side Height of Head Max	H Side Height of Head Min	O Total Height of Head Max	O Total Height of Head Min	J Width of Slot Min	J Width of Slot Max	T Depth of Slot Min	T Depth of Slot Max
0	MAB	0.060	0.096	0.083	0.045	0.037	0.059	0.043	0.016	0.023	0.015	0.025
1	MAB	0.073	0.118	0.104	0.053	0.045	0.071	0.055	0.019	0.026	0.020	0.031
2	MT	0.086	0.140	0.124	0.062	0.053	0.083	0.066	0.023	0.031	0.025	0.037
3	MT	0.099	0.161	0.145	0.070	0.061	0.095	0.077	0.027	0.035	0.030	0.043
4	MT	0.112	0.183	0.166	0.079	0.069	0.107	0.088	0.031	0.039	0.035	0.048
5	MT	0.125	0.205	0.187	0.088	0.078	0.120	0.100	0.035	0.043	0.040	0.054
6	MT	0.138	0.226	0.208	0.096	0.086	0.132	0.111	0.039	0.048	0.045	0.060
7	AB	0.151	0.248	0.229	0.105	0.094	0.144	0.122	0.039	0.048	0.049	0.065
8	MT	0.164	0.270	0.250	0.113	0.102	0.156	0.133	0.045	0.054	0.054	0.071
10	MT	0.190	0.313	0.292	0.130	0.118	0.180	0.156	0.050	0.060	0.064	0.083
12	MT	0.216	0.357	0.334	0.148	0.134	0.205	0.178	0.056	0.067	0.074	0.094
14	A	0.242	0.400	0.376	0.165	0.151	0.230	0.201	0.064	0.075	0.084	0.105
1/4	MBCD	0.250	0.414	0.389	0.170	0.155	0.237	0.207	0.064	0.075	0.087	0.109
5/16	MBCD	0.3125	0.518	0.490	0.211	0.194	0.295	0.262	0.072	0.084	0.110	0.137
3/8	MBCD	0.375	0.622	0.590	0.253	0.233	0.355	0.315	0.081	0.094	0.133	0.164
7/16	MB	0.4375	0.625	0.589	0.265	0.242	0.368	0.321	0.081	0.094	0.135	0.170
1/2	MB	0.500	0.750	0.710	0.297	0.273	0.412	0.362	0.091	0.106	0.151	0.190
9/16	M	0.5625	0.812	0.768	0.336	0.308	0.466	0.410	0.102	0.118	0.172	0.214
5/8	M	0.6250	0.875	0.827	0.375	0.345	0.521	0.461	0.116	0.133	0.193	0.240
3/4	M	0.7500	1.000	0.945	0.441	0.406	0.612	0.542	0.131	0.149	0.226	0.281

[a] Additional specifications and details pertinent to bodies and threads are given in the General Specifications.
[b] A slight rounding of the edges at periphery of head shall be permissible provided the diameter of the bearing circle is not less than 90% of the minimum head diameter.
[c] Application code:
A—Tapping Screws, Type A.
B—Tapping Screws, Types B, BF, BG, BP, and BT.
C—Tapping Screws, Type C.
D—Tapping Screws, Types D, F, G, and T.
M—Machine Screws.
T—Tapping Screws, all types.

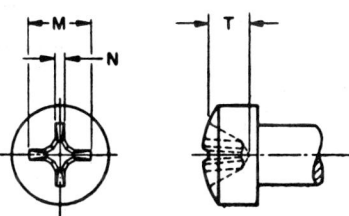

This type of recess has a large center opening, tapered wings, and blunt bottom, with all edges relieved or rounded.

TABLE 11A—TYPE I CROSS RECESS DIMENSIONS OF FILLISTER HEAD MACHINE AND TAPPING SCREWS, IN.

Nominal Size	M Dia of Recess Min	M Dia of Recess Max	T Depth of Recess Min	T Depth of Recess Max	N Width of Recess Min	Driver Size	Penetration Gaging Depth[a] Min	Penetration Gaging Depth[a] Max
0	0.054	0.067	0.021	0.039	0.013	0	0.014	0.032
1	0.061	0.074	0.025	0.045	0.014	0	0.022	0.040
2	0.091	0.104	0.041	0.059	0.017	1	0.034	0.052
3	0.099	0.112	0.050	0.068	0.019	1	0.043	0.061
4	0.109	0.122	0.060	0.078	0.019	1	0.053	0.071
5	0.135	0.148	0.042	0.067	0.027	2	0.031	0.056
6	0.153	0.166	0.066	0.091	0.028	2	0.055	0.080
7	0.163	0.176	0.075	0.100	0.029	2	0.064	0.089
8	0.169	0.182	0.082	0.108	0.030	2	0.071	0.097
10	0.186	0.199	0.100	0.124	0.031	2	0.089	0.113
12	0.246	0.259	0.115	0.141	0.034	3	0.098	0.124
14	0.268	0.281	0.135	0.161	0.036	3	0.118	0.144
1/4	0.268	0.281	0.135	0.161	0.036	3	0.118	0.144
5/16	0.309	0.322	0.177	0.203	0.042	3	0.160	0.186
3/8	0.376	0.389	0.210	0.233	0.065	4	0.190	0.213
7/16	0.400	0.413	0.234	0.259	0.068	4	0.214	0.239
1/2	0.422	0.435	0.255	0.280	0.071	4	0.235	0.260
9/16	0.436	0.459	0.279	0.304	0.075	4	0.259	0.284
5/8	0.547	0.570	0.306	0.341	0.079	5	0.273	0.308
3/4	0.594	0.617	0.353	0.388	0.084	5	0.320	0.355

[a] See Appendix B.

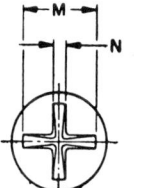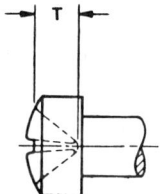

This type of recess consists of two intersecting slots with parallel sides converging to a slightly truncated apex at bottom of recess.

TABLE 11B—TYPE II CROSS RECESS DIMENSIONS OF FILLISTER HEAD MACHINE AND TAPPING SCREWS, IN.

Nominal Size	M Dia of Recess Min	M Dia of Recess Max	T Depth of Recess Min	T Depth of Recess Max	N Width of Recess Min	Driver Size	Penetration Gaging Depth[a] Min	Penetration Gaging Depth[a] Max
0	0.062	0.072	0.023	0.035	0.020		—[b]	—[b]
1	0.078	0.088	0.034	0.045	0.022		—[b]	—[b]
2	0.093	0.105	0.044	0.056	0.024		0.019	0.030
3	0.109	0.121	0.054	0.067	0.027		0.029	0.041
4	0.124	0.137	0.064	0.077	0.029		0.039	0.051
5	0.140	0.154	0.072	0.086	0.031		0.049	0.062
6	0.156	0.170	0.083	0.097	0.034		0.060	0.073
7	0.171	0.186	0.092	0.107	0.036	Point	0.069	0.083
8	0.187	0.202	0.103	0.118	0.039	same	0.080	0.094
10	0.219	0.235	0.119	0.134	0.043	on all	0.101	0.115
12	0.250	0.268	0.139	0.156	0.048	drivers	0.121	0.137
14	0.282	0.300	0.160	0.176	0.053		0.142	0.157
1/4	0.290	0.308	0.165	0.182	0.054		0.147	0.163
5/16	0.367	0.388	0.207	0.226	0.066		0.197	0.215
3/8	0.442	0.466	0.256	0.277	0.077		0.246	0.266
7/16	0.442	0.466	0.256	0.277	0.077		0.246	0.266
1/2	0.532	0.562	0.315	0.339	0.091		0.305	0.328
9/16	0.576	0.609	0.343	0.370	0.097		0.333	0.359
5/8	0.576	0.609	0.343	0.370	0.097		0.333	0.359
3/4	0.576	0.609	0.343	0.370	0.097		0.333	0.359

[a] See appendix B. [b] Not practical to gage.

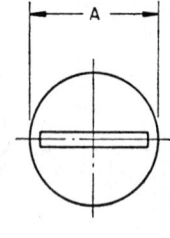
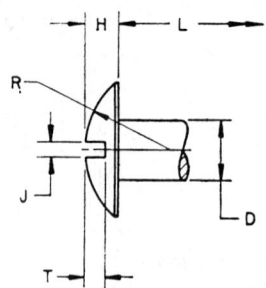
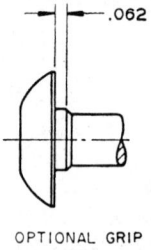

OPTIONAL GRIP FOR RECESSED HEAD MACHINE SCREWS

TABLE 12—DIMENSIONS OF TRUSS HEAD MACHINE AND TAPPING SCREWS, IN.[a]

Nominal Size	Applicable to Screw Types[b] Code Letters	D Basic Dia of Screw	A Diameter of Head		H Height of Head		J Width of Slot		T Depth of Slot		R Radius of Head
			Max	Min	Max	Min	Min	Max	Min	Max	Max
0	MAB	0.060	0.131	0.119	0.037	0.029	0.016	0.023	0.014	0.022	0.087
1	MAB	0.073	0.164	0.149	0.045	0.037	0.019	0.026	0.018	0.027	0.107
2	MT	0.086	0.194	0.180	0.053	0.044	0.023	0.031	0.022	0.031	0.129
3	MT	0.099	0.226	0.211	0.061	0.051	0.027	0.035	0.026	0.036	0.151
4	MT	0.112	0.257	0.241	0.069	0.059	0.031	0.039	0.030	0.040	0.169
5	MT	0.125	0.289	0.272	0.078	0.066	0.035	0.043	0.034	0.045	0.191
6	MT	0.138	0.321	0.303	0.086	0.074	0.039	0.048	0.037	0.050	0.211
7	AB	0.151	0.352	0.333	0.094	0.081	0.039	0.048	0.041	0.054	0.231
8	MT	0.164	0.384	0.364	0.102	0.088	0.045	0.054	0.045	0.058	0.254
10	MT	0.190	0.448	0.425	0.118	0.103	0.050	0.060	0.053	0.068	0.283
12	MT	0.216	0.511	0.487	0.134	0.118	0.056	0.067	0.061	0.077	0.336
14	A	0.242	0.557	0.530	0.146	0.129	0.064	0.075	0.068	0.085	0.375
16	A	0.268	0.609	0.580	0.159	0.141	0.064	0.075	0.074	0.093	0.410
18	A	0.294	0.661	0.630	0.173	0.153	0.072	0.084	0.080	0.100	0.446
20	A	0.320	0.713	0.680	0.186	0.165	0.072	0.084	0.087	0.108	0.484
24	A	0.372	0.817	0.780	0.213	0.190	0.081	0.094	0.100	0.123	0.557
1/4	MBCD	0.250	0.573	0.546	0.150	0.133	0.064	0.075	0.070	0.087	0.375
5/16	MBCD	0.3125	0.698	0.666	0.183	0.162	0.072	0.084	0.085	0.106	0.457
3/8	MBCD	0.375	0.823	0.787	0.215	0.191	0.081	0.094	0.100	0.124	0.538
7/16	MB	0.4375	0.948	0.907	0.248	0.221	0.081	0.094	0.116	0.142	0.619
1/2	MB	0.500	1.073	1.028	0.280	0.250	0.091	0.106	0.131	0.161	0.701
9/16	M	0.5625	1.198	1.149	0.312	0.279	0.102	0.118	0.146	0.179	0.783
5/8	M	0.6250	1.323	1.269	0.345	0.309	0.116	0.133	0.162	0.196	0.863
3/4	M	0.7500	1.573	1.511	0.410	0.368	0.131	0.149	0.182	0.234	1.024

[a] Additional specifications and details pertinent to bodies and threads are given in the General Specifications.
[b] Application code:
M — Machine Screws.
T — Tapping Screws, all types.
A — Tapping Screws, Type A.
B — Tapping Screws, Types B, BF, BG, BP, and BT.
C — Tapping Screws, Type C.
D — Tapping Screws, Types D, F, G and T.

TABLE 12A—TYPE I CROSS RECESS DIMENSIONS OF TRUSS HEAD MACHINE AND TAPPING SCREWS, IN.

Nominal Size	M Dia of Recess		T Depth of Recess		N Width of Recess	Driver Size	Penetration Gaging Depth[a]	
	Min	Max	Min	Max	Min		Min	Max
0	0.050	0.063	0.019	0.037	0.013	0	0.012	0.030
1	0.058	0.071	0.027	0.045	0.014	0	0.020	0.038
2	0.091	0.104	0.041	0.059	0.018	1	0.034	0.052
3	0.097	0.110	0.049	0.066	0.018	1	0.042	0.059
4	0.099	0.112	0.051	0.069	0.018	1	0.044	0.062
5	0.115	0.128	0.067	0.085	0.019	1	0.060	0.078
6	0.145	0.158	0.059	0.084	0.027	2	0.048	0.073
7	0.152	0.165	0.066	0.091	0.028	2	0.055	0.080
8	0.160	0.173	0.074	0.099	0.029	2	0.063	0.088
10	0.175	0.188	0.090	0.115	0.030	2	0.079	0.104
12	0.235	0.248	0.103	0.128	0.032	3	0.086	0.111
14	0.250	0.263	0.118	0.143	0.033	3	0.101	0.126
16	0.254	0.267	0.126	0.151	0.034	3	0.109	0.134
18	0.310	0.323	0.141	0.165	0.054	4	0.121	0.145
20	0.343	0.356	0.171	0.197	0.059	4	0.151	0.177
24	0.370	0.383	0.202	0.226	0.063	4	0.182	0.206
1/4	0.250	0.263	0.118	0.143	0.033	3	0.101	0.126
5/16	0.339	0.352	0.168	0.193	0.059	4	0.148	0.173
3/8	0.370	0.383	0.202	0.226	0.063	4	0.182	0.206
7/16	0.401	0.414	0.232	0.257	0.068	4	0.212	0.237
1/2	0.431	0.444	0.263	0.288	0.072	4	0.243	0.268
9/16	0.458	0.481	0.278	0.302	0.074	4	0.258	0.282
5/8	0.536	0.559	0.289	0.322	0.077	5	0.256	0.289
3/4	0.597	0.620	0.352	0.384	0.085	5	0.319	0.351

[a] See Appendix B.

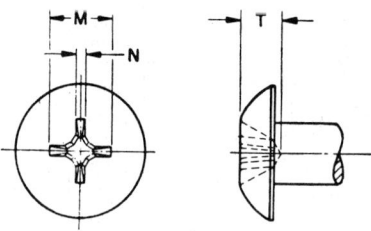

This type of recess has a large center opening, tapered wings, and blunt bottom, with all edges relieved or rounded.

TABLE 12B—TYPE II CROSS RECESS DIMENSIONS OF TRUSS HEAD MACHINE AND TAPPING SCREWS, IN.

Nominal Size	M Dia of Recess		T Depth of Recess		N Width of Recess	Driver Size	Penetration Gaging Depth[a]	
	Min	Max	Min	Max	Min		Min	Max
0	0.067	0.073	0.027	0.036	0.021		—[b]	—[b]
1	0.083	0.091	0.037	0.047	0.023		—[b]	—[b]
2	0.101	0.109	0.049	0.059	0.026		0.024	0.033
3	0.118	0.127	0.060	0.071	0.028		0.035	0.045
4	0.135	0.144	0.069	0.080	0.031		0.046	0.056
5	0.152	0.162	0.080	0.092	0.033		0.057	0.068
6	0.170	0.180	0.091	0.103	0.035		0.068	0.079
7	0.186	0.197	0.102	0.114	0.038		0.079	0.090
8	0.204	0.215	0.109	0.121	0.041	Point	0.091	0.102
10	0.238	0.251	0.131	0.145	0.046	same	0.113	0.126
12	0.273	0.286	0.154	0.167	0.051	on all	0.136	0.148
14	0.297	0.312	0.162	0.176	0.055	drivers	0.152	0.165
16	0.325	0.341	0.180	0.195	0.059		0.170	0.184
18	0.348	0.365	0.195	0.211	0.062		0.185	0.200
20	0.381	0.399	0.216	0.233	0.068		0.206	0.222
24	0.436	0.456	0.252	0.270	0.076		0.242	0.259
1/4	0.306	0.321	0.167	0.182	0.057		0.157	0.171
5/16	0.373	0.391	0.211	0.228	0.067		0.201	0.217
3/8	0.436	0.456	0.252	0.270	0.076		0.242	0.259
7/16	0.500	0.523	0.294	0.314	0.086		0.284	0.303
1/2	0.576	0.601	0.343	0.365	0.097		0.333	0.354
9/16	0.576	0.601	0.343	0.365	0.097		0.333	0.354
5/8	0.576	0.601	0.343	0.365	0.097		0.333	0.354
3/4	0.576	0.601	0.343	0.365	0.097		0.333	0.354

[a] See Appendix B. [b] Not practical to gage.

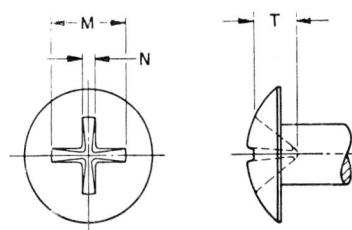

This type of recess consists of two intersecting slots with parallel sides converging to a slightly truncated apex at bottom of recess.

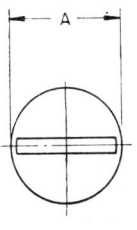 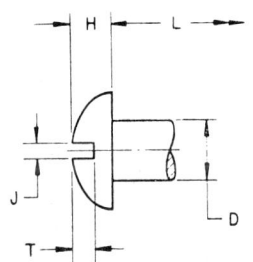

NOTE: ROUND HEAD MACHINE AND TAPPING SCREWS ARE NOT RECOMMENDED FOR NEW DESIGN. SUBSTITUTION OF PAN HEAD IS FAVORED.

TABLE 13—DIMENSIONS OF ROUND HEAD MACHINE, TAPPING, AND WOOD SCREWS, IN[a]

Nominal Size	Applicable to Screw Types[b] Code Letters	D Basic Dia of Screw	A Dia of Head		H Height of Head		J Width of Slot		T Depth of Slot	
			Max	Min	Max	Min	Min	Max	Min	Max
0	MABW	0.060	0.113	0.099	0.053	0.043	0.016	0.023	0.029	0.039
1	MABW	0.073	0.138	0.122	0.061	0.051	0.019	0.026	0.033	0.044
2	MTW	0.086	0.162	0.146	0.069	0.059	0.023	0.031	0.037	0.048
3	MTW	0.099	0.187	0.169	0.078	0.067	0.027	0.035	0.040	0.053
4	MTW	0.112	0.211	0.193	0.086	0.075	0.031	0.039	0.044	0.058
5	MTW	0.125	0.236	0.217	0.095	0.083	0.035	0.043	0.047	0.063
6	MTW	0.138	0.260	0.240	0.103	0.091	0.039	0.048	0.051	0.068
7	ABW	0.151	0.285	0.264	0.111	0.099	0.039	0.048	0.055	0.072
8	MTW	0.164	0.309	0.287	0.120	0.107	0.045	0.054	0.058	0.077
9	W	0.177	0.334	0.311	0.128	0.115	0.045	0.054	0.062	0.082
10	MTW	0.190	0.359	0.334	0.137	0.123	0.050	0.060	0.065	0.087
12	MTW	0.216	0.408	0.382	0.153	0.139	0.056	0.067	0.073	0.096
14	AW	0.242	0.457	0.429	0.170	0.155	0.064	0.075	0.080	0.106
16	AW	0.268	0.506	0.476	0.187	0.171	0.064	0.075	0.087	0.115
18	AW	0.294	0.555	0.523	0.204	0.187	0.072	0.084	0.094	0.125
20	AW	0.320	0.604	0.570	0.220	0.203	0.072	0.084	0.101	0.134
24	AW	0.372	0.702	0.664	0.254	0.235	0.081	0.094	0.116	0.154
1/4	MBCD	0.250	0.472	0.443	0.175	0.160	0.064	0.075	0.082	0.109
5/16	MBCD	0.3125	0.590	0.557	0.216	0.198	0.072	0.084	0.099	0.132
3/8	MBCD	0.375	0.708	0.670	0.256	0.237	0.081	0.094	0.117	0.155
7/16	MB	0.4375	0.750	0.707	0.328	0.307	0.081	0.094	0.148	0.196
1/2	MB	0.500	0.813	0.766	0.355	0.332	0.091	0.106	0.159	0.211
9/16	M	0.5625	0.938	0.887	0.410	0.385	0.102	0.118	0.183	0.242
5/8	M	0.6250	1.000	0.944	0.438	0.411	0.116	0.133	0.195	0.258
3/4	M	0.7500	1.250	1.185	0.547	0.516	0.131	0.149	0.242	0.320

[a] Additional specifications and details pertinent to bodies and threads are given in the General Specifications.
[b] Application code:
  M — Machine Screws.
  T — Tapping Screws, all types.

A — Tapping Screws, Type A.
B — Tapping Screws, Types B, BF, BG, BP, and BT
C — Tapping Screws, Type C.
D — Tapping Screws, Types D, F, G, and T.
W — Wood Screws.

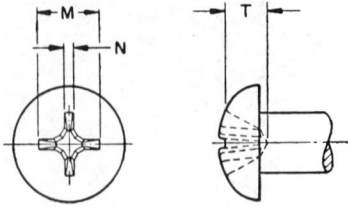

This type of recess has a large center opening, tapered wings, and blunt bottom, with all edges relieved or rounded.

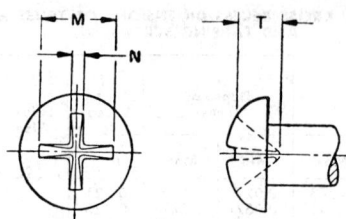

This type of recess consists of two intersecting slots with parallel sides converging to a slightly truncated apex at bottom of recess.

TABLE 13A—TYPE I CROSS RECESS DIMENSIONS OF ROUND HEAD MACHINE AND TAPPING SCREWS, IN.

Nominal Size	M Dia of Recess		T Depth of Recess		N Width of Recess	Driver Size	Penetration Gaging Depth[a]	
	Min	Max	Min	Max	Min		Min	Max
0	0.060	0.073	0.022	0.042	0.014	0	0.015	0.035
1	0.069	0.082	0.033	0.052	0.015	0	0.026	0.045
2	0.087	0.100	0.034	0.053	0.017	1	0.027	0.046
3	0.096	0.109	0.042	0.062	0.018	1	0.035	0.055
4	0.105	0.118	0.053	0.072	0.019	1	0.046	0.065
5	0.141	0.154	0.046	0.074	0.027	2	0.035	0.063
6	0.149	0.162	0.056	0.084	0.027	2	0.045	0.073
7	0.157	0.170	0.066	0.092	0.028	2	0.055	0.081
8	0.165	0.178	0.075	0.101	0.030	2	0.064	0.090
10	0.182	0.195	0.093	0.119	0.031	2	0.082	0.108
12	0.236	0.249	0.099	0.125	0.032	3	0.082	0.108
14	0.252	0.265	0.116	0.142	0.034	3	0.099	0.125
16	0.268	0.281	0.136	0.159	0.038	3	0.119	0.142
18	0.316	0.329	0.141	0.167	0.055	4	0.121	0.147
20	0.331	0.344	0.157	0.183	0.057	4	0.137	0.163
24	0.374	0.387	0.202	0.228	0.064	4	0.182	0.208
1/4	0.255	0.268	0.121	0.147	0.034	3	0.104	0.130
5/16	0.295	0.308	0.161	0.187	0.040	3	0.144	0.170
3/8	0.374	0.387	0.202	0.228	0.064	4	0.182	0.208
7/16	0.389	0.402	0.216	0.241	0.066	4	0.196	0.221
1/2	0.403	0.416	0.231	0.256	0.068	4	0.211	0.236
9/16	0.436	0.459	0.265	0.292	0.075	4	0.245	0.272
5/8	0.531	0.554	0.277	0.318	0.077	5	0.244	0.285
3/4	0.631	0.654	0.379	0.418	0.088	5	0.346	0.385

[a] See Appendix B

TABLE 13B—TYPE II CROSS RECESS DIMENSIONS OF ROUND HEAD MACHINE AND TAPPING SCREWS, IN.

Nominal Size	M Dia of Recess		T Depth of Recess		N Width of Recess	Driver Size	Penetration Gaging Depth[a]	
	Min	Max	Min	Max	Min		Min	Max
0	0.064	0.073	0.025	0.036	0.020		—[b]	—[b]
1	0.079	0.090	0.034	0.047	0.022		—[b]	—[b]
2	0.095	0.105	0.045	0.056	0.025		0.020	0.030
3	0.110	0.122	0.055	0.067	0.027		0.030	0.041
4	0.125	0.137	0.064	0.077	0.029		0.039	0.051
5	0.141	0.153	0.073	0.086	0.032		0.050	0.062
6	0.156	0.169	0.083	0.096	0.034		0.060	0.072
7	0.172	0.185	0.093	0.107	0.036		0.070	0.083
8	0.187	0.201	0.103	0.117	0.039		0.080	0.093
10	0.217	0.233	0.117	0.133	0.043	Point	0.099	0.114
12	0.248	0.265	0.138	0.154	0.048	same	0.120	0.135
14	0.279	0.297	0.158	0.175	0.052	on all	0.140	0.156
16	0.309	0.329	0.169	0.187	0.057	drivers	0.159	0.176
18	0.340	0.361	0.190	0.208	0.062		0.180	0.197
20	0.370	0.393	0.209	0.229	0.066		0.199	0.218
24	0.432	0.456	0.249	0.270	0.076		0.239	0.259
1/4	0.288	0.307	0.164	0.181	0.054		0.146	0.162
5/16	0.362	0.383	0.204	0.223	0.065		0.194	0.212
3/8	0.435	0.460	0.251	0.273	0.076		0.241	0.262
7/16	0.460	0.487	0.268	0.290	0.080		0.258	0.279
1/2	0.498	0.528	0.292	0.317	0.086		0.282	0.306
9/16	0.577	0.610	0.344	0.370	0.098		0.334	0.359
5/8	0.614	0.650	0.368	0.397	0.103		0.358	0.386
3/4	0.614	0.650	0.368	0.397	0.103		0.358	0.386

[a] See appendix B [b] Not practical to gage

TABLE 13C—TYPE I CROSS RECESS DIMENSIONS OF ROUND HEAD WOOD SCREWS, IN.

Nominal Size	M Dia of Recess		T Depth of Recess		N Width of Recess	Driver Size	Penetration Gauging Depth[a]	
	Min	Max	Min	Max	Min		Min	Max
0	0.060	0.073	0.022	0.042	0.014	0	0.015	0.035
1	0.069	0.082	0.033	0.052	0.015	0	0.026	0.045
2	0.101	0.114	0.042	0.064	0.018	1	0.035	0.057
3	0.109	0.122	0.052	0.073	0.019	1	0.045	0.066
4	0.117	0.130	0.063	0.083	0.019	1	0.056	0.076
5	0.141	0.154	0.046	0.074	0.027	2	0.035	0.063
6	0.149	0.162	0.056	0.084	0.027	2	0.045	0.073
7	0.157	0.170	0.066	0.092	0.028	2	0.055	0.081
8	0.165	0.178	0.075	0.101	0.030	2	0.064	0.090
9	0.173	0.186	0.084	0.110	0.030	2	0.073	0.099
10	0.182	0.195	0.093	0.119	0.031	2	0.082	0.108
12	0.236	0.249	0.099	0.125	0.032	3	0.082	0.108
14	0.252	0.265	0.116	0.142	0.034	3	0.099	0.125
16	0.268	0.281	0.136	0.159	0.038	3	0.119	0.142
18	0.326	0.339	0.151	0.176	0.057	4	0.131	0.156
20	0.343	0.356	0.168	0.194	0.059	4	0.148	0.174
24	0.374	0.387	0.202	0.228	0.064	4	0.182	0.208

[a] See Appendix B

TABLE 13D—TYPE II CROSS RECESS DIMENSIONS OF ROUND HEAD WOOD SCREWS, IN.

Nominal Size	M Dia of Recess		T Depth of Recess		N Width of Recess	Driver Size	Penetration Gauging Depth[a]	
	Min	Max	Min	Max	Min		Min	Max
0	0.069	0.079	0.028	0.039	0.021		—[b]	—[b]
1	0.085	0.097	0.038	0.051	0.023		—[b]	—[b]
2	0.102	0.114	0.049	0.062	0.026		0.024	0.036
3	0.118	0.131	0.060	0.073	0.028		0.035	0.047
4	0.135	0.148	0.069	0.082	0.031		0.046	0.058
5	0.152	0.165	0.080	0.094	0.033		0.057	0.070
6	0.168	0.182	0.090	0.105	0.036		0.067	0.081
7	0.185	0.199	0.102	0.116	0.038	Point	0.079	0.092
8	0.201	0.216	0.107	0.122	0.041	same	0.089	0.103
9	0.218	0.234	0.118	0.133	0.043	on all	0.100	0.114
10	0.234	0.251	0.128	0.145	0.046	drivers	0.110	0.126
12	0.267	0.286	0.150	0.167	0.051		0.132	0.148
14	0.300	0.320	0.163	0.182	0.056		0.153	0.171
16	0.333	0.354	0.185	0.204	0.061		0.175	0.193
18	0.366	0.388	0.206	0.226	0.066		0.196	0.215
20	0.399	0.423	0.228	0.249	0.071		0.218	0.238
24	0.465	0.491	0.271	0.293	0.081		0.261	0.282

[a] See Appendix B [b] Not practical to gage.

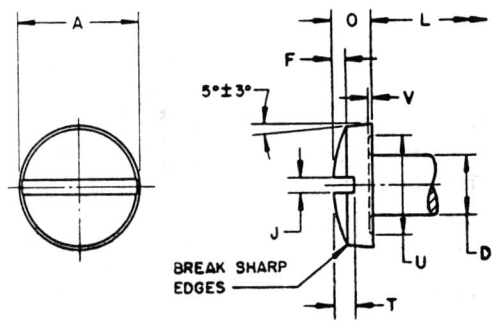

TABLE 14—DIMENSIONS OF BINDING HEAD MACHINE SCREWS, IN.[a]

Nominal Size	D Basic Dia of Screw	A Dia of Head		O Total Height of Head		F Height of Crown		J Width of Slot		T Depth of Slot		U Dia of Undercut[b]		V Depth of Undercut[b]	
		Max	Min	Max	Min	Max	Min	Min	Max	Min	Max	Min	Max	Min	Max
2	0.086	0.181	0.171	0.050	0.041	0.018	0.013	0.023	0.031	0.024	0.030	0.124	0.141	0.005	0.010
3	0.099	0.208	0.197	0.059	0.048	0.022	0.016	0.027	0.035	0.029	0.036	0.143	0.162	0.006	0.011
4	0.112	0.235	0.223	0.068	0.056	0.025	0.018	0.031	0.039	0.034	0.042	0.161	0.184	0.007	0.012
5	0.125	0.263	0.249	0.078	0.064	0.029	0.021	0.035	0.043	0.039	0.048	0.180	0.205	0.009	0.014
6	0.138	0.290	0.275	0.087	0.071	0.032	0.024	0.039	0.048	0.044	0.053	0.199	0.226	0.010	0.015
8	0.164	0.344	0.326	0.105	0.087	0.039	0.029	0.045	0.054	0.054	0.065	0.236	0.269	0.012	0.017
10	0.190	0.399	0.378	0.123	0.102	0.045	0.034	0.050	0.060	0.064	0.077	0.274	0.312	0.015	0.020
12	0.216	0.454	0.430	0.141	0.117	0.052	0.039	0.056	0.067	0.074	0.089	0.311	0.354	0.018	0.023
1/4	0.250	0.513	0.488	0.165	0.138	0.061	0.046	0.064	0.075	0.088	0.105	0.360	0.410	0.021	0.026
5/16	0.3125	0.641	0.609	0.209	0.174	0.077	0.059	0.072	0.084	0.112	0.134	0.450	0.513	0.027	0.032
3/8	0.375	0.769	0.731	0.253	0.211	0.094	0.071	0.081	0.094	0.136	0.163	0.540	0.615	0.034	0.039

[a] Additional specifications and details pertinent to bodies and threads are given in the General Specifications.
[b] The use of undercut is optional. Where undercut is not used, maximum radius of underhead fillet shall be 0.15 D

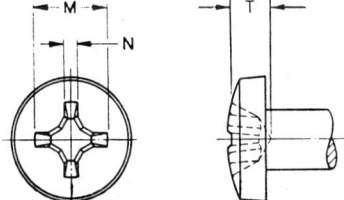

This type of recess has a large center opening, tapered wings, and blunt bottom, with all edges relieved or rounded.

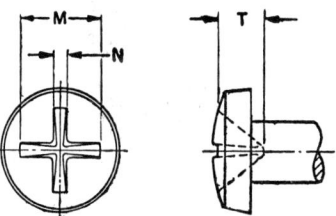

This type of recess consists of two intersecting slots with parallel sides converging to a slightly truncated apex at bottom of recess.

TABLE 14A—TYPE I CROSS RECESS DIMENSIONS OF BINDING HEAD MACHINE SCREWS, IN.

Nominal Size	M Dia of Recess		T Depth of Recess		N Width of Recess	Driver Size	Penetration Gaging Depth[a]	
	Min	Max	Min	Max	Min		Min	Max
2	0.087	0.100	0.041	0.058	0.017	1	0.034	0.051
3	0.097	0.110	0.051	0.068	0.017	1	0.044	0.061
4	0.105	0.118	0.059	0.077	0.017	1	0.052	0.070
5	0.135	0.148	0.051	0.075	0.018	2	0.040	0.064
6	0.147	0.160	0.064	0.088	0.026	2	0.053	0.077
8	0.173	0.186	0.090	0.114	0.028	2	0.079	0.103
10	0.192	0.205	0.109	0.134	0.029	2	0.098	0.123
12	0.254	0.267	0.126	0.151	0.032	3	0.109	0.134
1/4	0.268	0.281	0.140	0.164	0.046	3	0.123	0.147
5/16	0.337	0.350	0.172	0.195	0.068	4	0.152	0.175
3/8	0.387	0.400	0.223	0.247	0.076	4	0.203	0.227

[a] See Appendix B.

TABLE 14B—TYPE II CROSS RECESS DIMENSIONS OF BINDING HEAD MACHINE SCREWS, IN.

Nominal Size	M Dia of Recess		T Depth of Recess		N Width of Recess	Driver Size	Penetration Gaging Depth[a]	
	Min	Max	Min	Max	Min		Min	Max
2	0.109	0.116	0.054	0.064	0.027		0.029	0.038
3	0.127	0.134	0.065	0.075	0.029		0.040	0.049
4	0.144	0.152	0.075	0.085	0.032		0.052	0.061
5	0.162	0.171	0.087	0.097	0.035	Point	0.064	0.073
6	0.180	0.189	0.098	0.109	0.038	same	0.075	0.085
8	0.212	0.224	0.114	0.127	0.042		0.096	0.108
10	0.235	0.248	0.129	0.143	0.046	on all	0.111	0.124
12	0.279	0.295	0.158	0.173	0.052	drivers	0.140	0.154
1/4	0.318	0.334	0.175	0.191	0.058		0.165	0.180
5/16	0.401	0.422	0.229	0.248	0.071		0.219	0.237
3/8	0.465	0.490	0.271	0.292	0.081		0.261	0.281

[a] See appendix B.

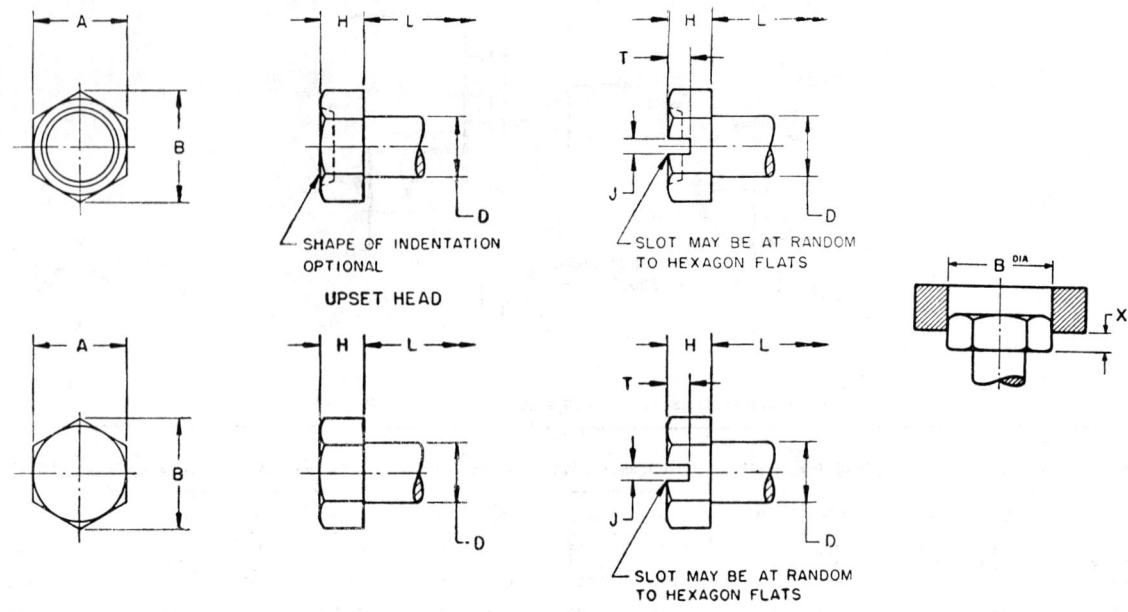

UPSET HEAD

TRIMMED HEAD

TABLE 15—DIMENSIONS OF HEXAGON HEAD MACHINE AND TAPPING SCREWS, IN[a]

Nominal Size	Applicable to Screw Types[f] Code Letters	D Basic Dia of Screw	A[b] Width Across Flats Standard Trimmed or Upset Head		B[b,e] Width Across Corners	A[b] Optional Upset Head[c] Width Across Flats		B[b,e] Width Across Corners	H Height of Head		J[d] Width of Slat		T[d] Depth of Slot		X[e] Protrusion Beyond Gaging Ring
			Max	Min	Min	Max	Min	Min	Max	Min	Min	Max	Min	Max	Min
2	MAB	0.086	0.125	0.120	0.134	—	—	—	0.050	0.040	—	—	—	—	0.024
3	MABD	0.099	0.187	0.181	0.202	—	—	—	0.055	0.044	—	—	—	—	0.026
4	MT	0.112	0.187	0.181	0.202	0.219	0.213	0.238	0.060	0.049	0.031	0.039	0.025	0.036	0.029
5	MT	0.125	0.187	0.181	0.202	0.250	0.244	0.272	0.070	0.058	0.035	0.043	0.030	0.042	0.035
6	MT	0.138	0.250	0.244	0.272	—	—	—	0.093	0.080	0.039	0.048	0.033	0.046	0.048
7	AB	0.151	0.250	0.244	0.272	—	—	—	0.093	0.080	0.039	0.048	0.040	0.054	0.048
8	MT	0.164	0.250	0.244	0.272	0.312	0.305	0.340	0.110	0.096	0.045	0.054	0.052	0.066	0.058
10	MT	0.190	0.312	0.305	0.340	—	—	—	0.120	0.105	0.050	0.060	0.057	0.072	0.063
12	MT	0.216	0.312	0.305	0.340	0.375	0.367	0.409	0.155	0.139	0.056	0.067	0.077	0.093	0.083
14	A	0.242	0.375	0.367	0.409	—	—	—	0.190	0.172	0.064	0.075	0.083	0.101	0.103
20	A	0.320	0.500	0.489	0.545	—	—	—	0.230	0.208	0.072	0.084	0.100	0.122	0.125
24	A	0.372	0.562	0.551	0.614	—	—	—	0.295	0.270	0.081	0.094	0.131	0.156	0.162
1/4	MBCD	0.250	0.375	0.367	0.409	0.437	0.428	0.477	0.190	0.172	0.064	0.075	0.083	0.101	0.103
5/16	MBCD	0.3125	0.500	0.489	0.545	—	—	—	0.230	0.208	0.072	0.084	0.100	0.122	0.125
3/8	MBCD	0.375	0.562	0.551	0.614	—	—	—	0.295	0.270	0.081	0.094	0.131	0.156	0.162

[a] Additional specifications and details pertinent to bodies and threads are given in the General Specifications.

[b] Dimensions across flats and corners shall be measured at the point of maximum metal. Taper of sides of hexagon (angle between one side and the axis) shall not exceed 2 deg or 0.004 in., whichever is greater, the specified width across flats being the largest dimension. A slight rounding of all edges and corners of the hexagon surfaces of upset type heads shall be permissible provided the diameter of bearing circle is not less than 90% of the minimum across flats dimension and the side flat width is not less than 0.43 times the maximum width across flats.

[c] Applicable only to screw and washer assemblies or other applications requiring large bearing area.

[d] Unless otherwise specified, hexagon head screws are not slotted.

[e] The rounding due to lack of fill on all six corners of head shall be reasonably uniform and width across corners of the head shall be such that when a sharp ring having an inside diameter equal to the minimum width across corners is placed on the top and bottom of the head, the head shall protrude by an amount equal to, or greater than, the value tabulated. See Appendix C, Across Corners Gaging of Hexagon Heads.

[f] Application code:
M—Machine Screws.
T—Tapping Screws, all types.
A—Tapping Screws, Type A.
B—Tapping Screws, Types B, BF, BG, BP, and BT.
C—Tapping Screws, Type C.
D—Tapping Screws, Types D, F, G, and T.

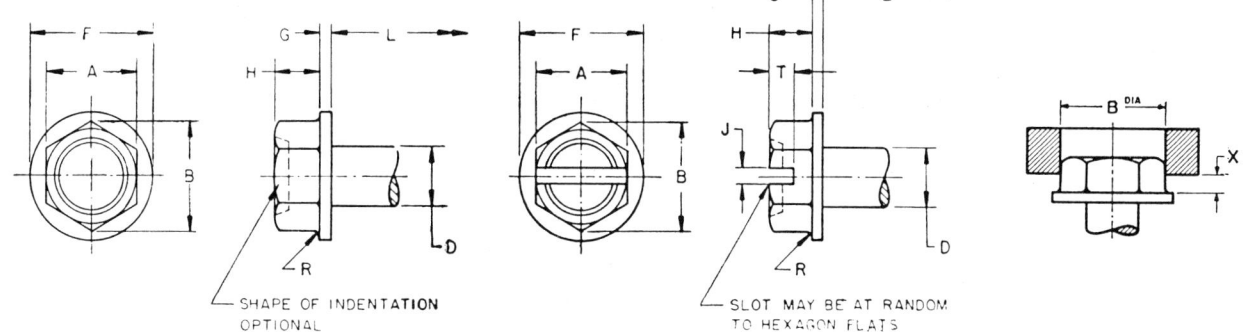

TABLE 16—DIMENSIONS OF HEXAGON WASHER HEAD MACHINE AND TAPPING SCREWS, IN.[a]

Nominal Size	Applicable to Screw Types[e] Code Letters	D Basic Dia of Screw	A[b] Width Across Flats Max	A[b] Width Across Flats Min	B[b,d] Width Across Corners Min	F Dia of Washer Max	F Dia of Washer Min	G Thickness of Washer Max	G Thickness of Washer Min	H Height of Head Max	H Height of Head Min	J[c] Width of Slot Min	J[c] Width of Slot Max	T[c] Depth of Slot Min	T[c] Depth of Slot Max	R Radius of Fillet Max	X[d] Protrusion Beyond Gaging Ring Min
2	MABD	0.086	0.125	0.120	0.134	0.166	0.154	0.016	0.010	0.050	0.040	—	—	—	—	0.013	0.024
3	MABD	0.099	0.125	0.120	0.134	0.177	0.163	0.016	0.010	0.055	0.044	—	—	—	—	0.015	0.026
4	MT	0.112	0.187	0.181	0.202	0.243	0.225	0.019	0.011	0.060	0.049	0.031	0.039	0.025	0.042	0.017	0.029
5	MT	0.125	0.187	0.181	0.202	0.260	0.240	0.025	0.015	0.070	0.058	0.035	0.043	0.030	0.049	0.019	0.035
6	MT	0.138	0.250	0.244	0.272	0.328	0.302	0.025	0.015	0.093	0.080	0.039	0.048	0.033	0.053	0.021	0.048
7	AB	0.151	0.250	0.244	0.272	0.328	0.302	0.029	0.017	0.093	0.080	0.039	0.048	0.040	0.062	0.021	0.048
8	MT	0.164	0.250	0.244	0.272	0.348	0.322	0.031	0.019	0.110	0.096	0.045	0.054	0.052	0.074	0.025	0.058
10	MT	0.190	0.312	0.305	0.340	0.414	0.384	0.031	0.019	0.120	0.105	0.050	0.060	0.057	0.080	0.028	0.063
12	MT	0.216	0.312	0.305	0.340	0.432	0.398	0.039	0.022	0.155	0.139	0.056	0.067	0.077	0.103	0.032	0.083
14	A	0.242	0.375	0.367	0.409	0.520	0.480	0.050	0.030	0.190	0.172	0.064	0.075	0.083	0.111	0.036	0.103
20	A	0.320	0.500	0.489	0.545	0.676	0.624	0.055	0.035	0.230	0.208	0.072	0.084	0.100	0.134	0.048	0.125
24	A	0.372	0.562	0.551	0.614	0.780	0.720	0.063	0.037	0.295	0.270	0.081	0.094	0.131	0.168	0.056	0.162
1/4	MBCD	0.250	0.375	0.367	0.409	0.520	0.480	0.050	0.030	0.190	0.172	0.064	0.075	0.083	0.111	0.038	0.103
5/16	MBCD	0.3125	0.500	0.489	0.545	0.676	0.624	0.055	0.035	0.230	0.208	0.072	0.084	0.100	0.134	0.047	0.125
3/8	MBCD	0.375	0.562	0.551	0.614	0.780	0.720	0.063	0.037	0.295	0.270	0.081	0.094	0.131	0.168	0.056	0.162

[a] Additional specifications and details pertinent to bodies and threads are given in the General Specifications.
[b] Dimensions across flats and corners shall be measured at the point of maximum metal. Taper of sides of hexagon (angle between one side and the axis) shall not exceed 2 deg or 0.004 in., whichever is greater, the specified width across flats being the largest dimension. A slight rounding of all edges and corners of the hexagon surfaces is permissible provided that the side flat width is not less than 0.43 times the maximum width across flats.
[c] Unless otherwise specified, hexagon washer head screws are not slotted.
[d] The rounding due to lack of fill on all six corners of head shall be reasonably uniform and width across corners of the head shall be such that when a sharp ring having an inside diameter equal to the minimum width across corners is placed on the top and bottom of the head, the head shall protrude by an amount equal to, or greater than, the value tabulated. See Appendix C, Across Corners Gaging of Hexagon Heads
[e] Application code:
M—Machine Screws.
T—Tapping Screws, all types.
A—Tapping Screws, Type A.
B—Tapping Screws, Types B, BF, BG, BP, and BT.
C—Tapping Screws, Type C.
D—Tapping Screws, Types D, F, G, and T.

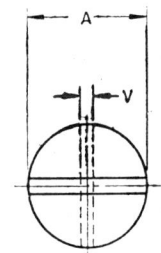
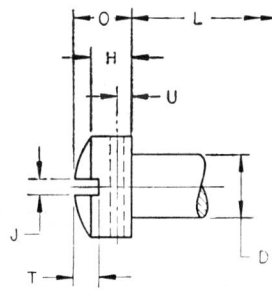

TABLE 17—DIMENSIONS OF DRILLED FILLISTER HEAD MACHINE SCREWS, IN.[a,c]

Nominal Size	D Basic Dia of Screw	A Dia of Head Max	A Dia of Head Min	H Side Height of Head Max	H Side Height of Head Min	O Total Height of Head Max	O Total Height of Head Min	J Width of Slot Min	J Width of Slot Max	T Depth of Slot Min	T Depth of Slot Max	U Centerline of Hole	V[b] Dia of Hole
2	0.086	0.140	0.124	0.062	0.055	0.083	0.070	0.023	0.031	0.022	0.030	0.026	0.031
3	0.099	0.161	0.145	0.070	0.064	0.095	0.082	0.027	0.035	0.028	0.034	0.030	0.037
4	0.112	0.183	0.166	0.079	0.072	0.107	0.094	0.031	0.039	0.030	0.038	0.035	0.037
5	0.125	0.205	0.187	0.088	0.081	0.120	0.106	0.035	0.043	0.033	0.042	0.038	0.046
6	0.138	0.226	0.208	0.096	0.089	0.132	0.118	0.039	0.048	0.035	0.045	0.043	0.046
8	0.164	0.270	0.250	0.113	0.106	0.156	0.141	0.045	0.054	0.054	0.065	0.043	0.046
10	0.190	0.313	0.292	0.130	0.123	0.180	0.165	0.050	0.060	0.064	0.075	0.043	0.046
12	0.216	0.357	0.334	0.148	0.139	0.205	0.188	0.056	0.067	0.074	0.087	0.053	0.046
1/4	0.250	0.414	0.389	0.170	0.161	0.237	0.219	0.064	0.075	0.087	0.102	0.062	0.062
5/16	0.3125	0.518	0.490	0.211	0.201	0.295	0.276	0.072	0.084	0.110	0.130	0.078	0.070
3/8	0.375	0.622	0.590	0.253	0.242	0.355	0.333	0.081	0.094	0.134	0.154	0.094	0.070

[a] Additional specifications and details pertinent to bodies and threads are given in the General Specifications.
[b] Drilled hole to be approximately at right angles to the slot and may be permitted to break through the bottom of the slot. Edges of hole shall be free from burrs.
[c] A slight rounding of the edges at periphery of head shall be permissible provided the diameter of the bearing circle is not less than 90% of the minimum head diameter.

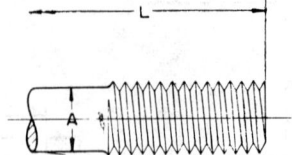
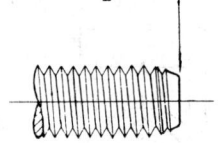
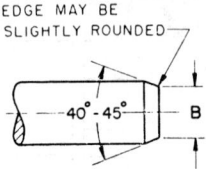

PLAIN POINT   HEADER POINT

VIEW OF BLANK BEFORE ROLLING THREAD

TABLE 18—DIMENSIONS OF MACHINE SCREW THREADS AND HEADER POINTS, IN.[a]

Nominal Size	Threads per in.	Series Designation	A Body Dia		Class 2A—Thread Detail					B Point Dia on Blank		L Determinant Length for Header Points[d]
					Major Dia		Pitch Dia					
			Max[b]	Min	Max	Min	Max	Min	Basic[c]	Max	Min	Max
0	80	UNF	0.0600	0.0496	0.0595	0.0563	0.0514	0.0496	0.0519	—	—	—
1	64	UNC	0.0730	0.0603	0.0724	0.0686	0.0623	0.0603	0.0629	—	—	—
1	72	UNF	0.0730	0.0615	0.0724	0.0689	0.0634	0.0615	0.0640	—	—	—
2	56	UNC	0.0860	0.0717	0.0854	0.0813	0.0738	0.0717	0.0744	—	—	—
2	64	UNF	0.0860	0.0733	0.0854	0.0816	0.0753	0.0733	0.0759	—	—	—
3	48	UNC	0.0990	0.0825	0.0983	0.0938	0.0848	0.0825	0.0855	—	—	—
3	56	UNF	0.0990	0.0845	0.0983	0.0942	0.0867	0.0845	0.0874	—	—	—
4	40	UNC	0.1120	0.0925	0.1112	0.1061	0.0950	0.0925	0.0958	0.074	0.065	1/2
4	48	UNF	0.1120	0.0954	0.1113	0.1068	0.0978	0.0954	0.0985	0.079	0.070	1/2
5	40	UNC	0.1250	0.1054	0.1242	0.1191	0.1080	0.1054	0.1088	0.086	0.076	1/2
5	44	UNF	0.1250	0.1070	0.1243	0.1195	0.1095	0.1070	0.1102	0.088	0.079	1/2
6	32	UNC	0.1380	0.1141	0.1372	0.1312	0.1169	0.1141	0.1177	0.090	0.080	3/4
6	40	UNF	0.1380	0.1184	0.1372	0.1321	0.1210	0.1184	0.1218	0.098	0.087	3/4
8	32	UNC	0.1640	0.1399	0.1631	0.1571	0.1428	0.1399	0.1437	0.114	0.102	1
8	36	UNF	0.1640	0.1424	0.1632	0.1577	0.1452	0.1424	0.1460	0.118	0.106	1
10	24	UNC	0.1900	0.1586	0.1890	0.1818	0.1619	0.1586	0.1629	0.125	0.112	1-1/4
10	32	UNF	0.1900	0.1658	0.1891	0.1831	0.1688	0.1658	0.1697	0.138	0.124	1-1/4
12	24	UNC	0.2160	0.1845	0.2150	0.2078	0.1879	0.1845	0.1889	0.149	0.134	1-3/8
12	28	UNF	0.2160	0.1886	0.2150	0.2085	0.1918	0.1886	0.1928	0.156	0.141	1-3/8
1/4	20	UNC	0.2500	0.2127	0.2489	0.2408	0.2164	0.2127	0.2175	0.170	0.153	1-1/2
1/4	28	UNF	0.2500	0.2225	0.2490	0.2425	0.2258	0.2225	0.2268	0.187	0.169	1-1/2
5/16	18	UNC	0.3125	0.2712	0.3113	0.3026	0.2752	0.2712	0.2764	0.221	0.200	1-1/2
5/16	24	UNF	0.3125	0.2806	0.3114	0.3042	0.2843	0.2806	0.2854	0.237	0.215	1-1/2
3/8	16	UNC	0.3750	0.3287	0.3737	0.3643	0.3331	0.3287	0.3344	0.270	0.244	1-1/2
3/8	24	UNF	0.3750	0.3430	0.3739	0.3667	0.3468	0.3430	0.3479	0.295	0.267	1-1/2
7/16	14	UNC	0.4375	0.3850	0.4361	0.4258	0.3897	0.3850	0.3911	0.316	0.287	1-1/2
7/16	20	UNF	0.4375	0.3995	0.4362	0.4281	0.4037	0.3995	0.4050	0.342	0.310	1-1/2
1/2	13	UNC	0.5000	0.4435	0.4985	0.4876	0.4485	0.4435	0.4500	0.367	0.333	1-1/2
1/2	20	UNF	0.5000	0.4619	0.4987	0.4906	0.4662	0.4619	0.4675	0.399	0.362	1-1/2
9/16	12	UNC	0.5625	0.5016	0.5609	0.5495	0.5068	0.5016	0.5084	—	—	—
9/16	18	UNF	0.5625	0.5205	0.5611	0.5524	0.5250	0.5205	0.5264	—	—	—
5/8	11	UNC	0.6250	0.5589	0.6234	0.6113	0.5644	0.5589	0.5660	—	—	—
5/8	18	UNF	0.6250	0.5828	0.6236	0.6149	0.5875	0.5828	0.5889	—	—	—
3/4	10	UNC	0.7500	0.6773	0.7482	0.7353	0.6832	0.6773	0.6850	—	—	—
3/4	16	UNF	0.7500	0.7029	0.7485	0.7391	0.7079	0.7029	0.7094	—	—	—

[a] Additional specifications and details pertinent to thread length and points are given in General Specifications.
[b] Also basic major diameter of screw and maximum major diameter of screw after plating. See Section, Threads in General Specifications.
[c] Maximum after plating. See Section, Threads in General Specifications.
[d] Screws this length and shorter can be pointed in the heading operation. Conical points on longer screws will require machining.

TABLE 19—DIMENSIONS OF TYPE A TAPPING SCREW THREADS AND POINTS, IN.[a, c]

Nominal Size	Threads per in.		D Major Dia			E Minor Dia		F Flat on Crest of Thread	L Determinant Length for Threads per in.[d]		L Min Practical Length	
	Screws Longer Than Determinant Length	Screws Equal to or Less Than Determinant Length	Basic	Max	Min	Max	Min	Max	90 Deg Heads	Csk Heads[b]	90 Deg Heads	Csk Heads
0	40	48	0.060	0.060	0.057	0.042	0.039	0.004	1/8	3/16	3/32	7/64
1	32	42	0.073	0.075	0.072	0.051	0.048	0.004	1/8	3/16	1/8	9/64
2	32	32	0.086	0.088	0.084	0.061	0.056	0.004	5/32	3/16	9/64	11/64
3	28	28	0.099	0.101	0.097	0.076	0.071	0.004	3/16	7/32	11/64	3/16
4	24	24	0.112	0.114	0.110	0.083	0.078	0.004	3/16	1/4	3/16	7/32
5	20	20	0.125	0.130	0.126	0.095	0.090	0.004	3/16	1/4	3/16	1/4
6	18	20	0.138	0.141	0.136	0.102	0.096	0.004	1/4	5/16	7/32	17/64
7	16	19	0.151	0.158	0.152	0.114	0.108	0.004	5/16	3/8	17/64	5/16
8	15	18	0.164	0.168	0.162	0.123	0.116	0.004	3/8	7/16	9/32	21/64
10	12	16	0.190	0.194	0.188	0.133	0.126	0.006	3/8	1/2	21/64	3/8
12	11	14	0.216	0.221	0.215	0.162	0.155	0.006	7/16	9/16	3/8	13/32
14	10	14	0.242	0.254	0.248	0.185	0.178	0.006	1/2	5/8	13/32	15/32
16	10	12	0.268	0.280	0.274	0.197	0.189	0.006	9/16	3/4	31/64	1/2
18	9	12	0.294	0.306	0.300	0.217	0.209	0.006	5/8	13/16	17/32	19/32
20	9	12	0.320	0.333	0.327	0.234	0.226	0.006	11/16	13/16	9/16	5/8
24	9	12	0.372	0.390	0.383	0.291	0.282	0.006	3/4	1	5/8	3/4

[a] Additional specifications and details pertinent to thread length and points are given in General Specifications.
[b] Except for trim head styles, countersunk head screws these lengths and shorter shall be furnished with undercut heads.
[c] See Appendix D, Cross Reference of Tapping Screw Type Designations.
[d] Except for trim head styles, screws this length and shorter shall have threads per inch shown in column 3. Screws having greater lengths shall have threads per inch shown in column 2.

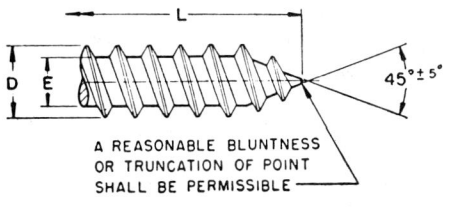

TYPE A

A REASONABLE BLUNTNESS OR TRUNCATION OF POINT SHALL BE PERMISSIBLE

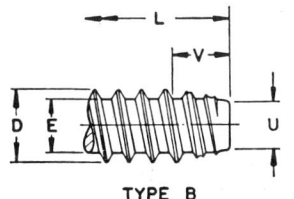

TYPE B — **TYPE BP**

THREAD FORMING POINTS

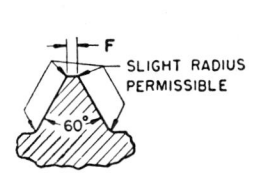

THREAD DETAIL

SLIGHT RADIUS PERMISSIBLE

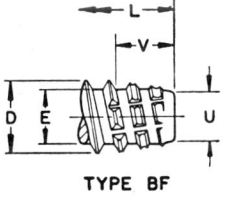 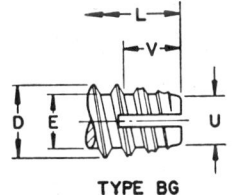 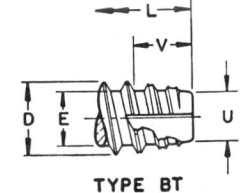

TYPE BF — **TYPE BG** — **TYPE BT**

THREAD CUTTING POINTS

TABLE 20—DIMENSIONS OF TYPES B, BF, BG, BP, AND BT TAPPING SCREW THREADS AND POINTS, IN.[a, i]

Nominal Size	Threads per in.	D Major Dia			E Minor Dia		F Flat on Crest of Thread	U[b] Point Dia		V[c], Taper Length				L[g] Determinant Length for Taper[j]	
										Screws Longer Than L		Screws Equal to or Less Than L			
		Basic	Max	Min	Max	Min	Max	Max	Min	Min[e]	Max[d]	Min[f]	Max[e]	90 Deg Heads	Csk Heads[h]
0	48	0.060	0.060	0.057	0.036	0.033	0.004	0.031	0.027	0.042	0.052	0.031	0.042	5/64	1/8
1	42	0.073	0.075	0.072	0.049	0.046	0.004	0.044	0.040	0.048	0.060	0.036	0.048	5/64	5/32
2	32	0.086	0.088	0.084	0.064	0.060	0.004	0.058	0.054	0.062	0.078	0.047	0.062	7/64	3/16
3	28	0.099	0.101	0.097	0.075	0.071	0.004	0.068	0.063	0.071	0.089	0.054	0.071	9/64	7/32
4	24	0.112	0.114	0.110	0.086	0.082	0.004	0.079	0.074	0.083	0.104	0.063	0.083	3/16	1/4
5	20	0.125	0.130	0.126	0.094	0.090	0.004	0.087	0.082	0.100	0.125	0.075	0.100	3/16	1/4
6	20	0.138	0.139	0.135	0.104	0.099	0.004	0.095	0.089	0.100	0.125	0.075	0.100	1/4	5/16
7	19	0.151	0.154	0.149	0.115	0.109	0.004	0.105	0.099	0.105	0.132	0.079	0.105	5/16	3/8
8	18	0.164	0.166	0.161	0.122	0.116	0.004	0.112	0.106	0.111	0.139	0.083	0.111	5/16	7/16
10	16	0.190	0.189	0.183	0.141	0.135	0.006	0.130	0.123	0.125	0.156	0.094	0.125	3/8	1/2
12	14	0.216	0.215	0.209	0.164	0.157	0.006	0.152	0.145	0.143	0.179	0.107	0.143	7/16	9/16
1/4	14	0.250	0.246	0.240	0.192	0.185	0.006	0.179	0.171	0.143	0.179	0.107	0.143	1/2	5/8
5/16	12	0.3125	0.315	0.308	0.244	0.236	0.006	0.230	0.222	0.167	0.208	0.125	0.167	1/2	5/8
3/8	12	0.375	0.380	0.371	0.309	0.299	0.006	0.293	0.285	0.167	0.208	0.125	0.167	1/2	5/8
7/16	10	0.4375	0.440	0.431	0.359	0.349	0.006	0.343	0.335	0.200	0.250	0.150	0.200	5/8	3/4
1/2	10	0.500	0.504	0.495	0.423	0.413	0.006	0.407	0.399	0.200	0.250	0.150	0.200	5/8	3/4

[a] Additional specifications and details pertinent to thread length and points are given in General Specifications.
[b] Point diameters specified are values before roll threading.
[c] Tapered threads shall have unfinished crests.
[d] Values are equal to 2-1/2 times the pitch distance rounded off to three decimal places.
[e] Values are equal to 2 times the pitch distance rounded off to three decimal places.
[f] Values are equal to 1-1/2 times the pitch distance rounded off to three decimal places.
[g] Screws this length or shorter are impractical to manufacture in BP style.
[h] Except for trim head styles, countersunk head screws these lengths and shorter shall be furnished with undercut heads.
[i] See Appendix D, Cross Reference of Tapping Screw Type Designations.
[j] Screws this length and shorter shall have taper lengths shown in columns 13 and 14. Screws having greater lengths shall have taper lengths shown in columns 11 and 12.

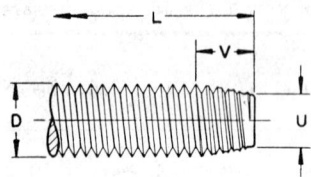

TYPE C
THREAD FORMING POINT

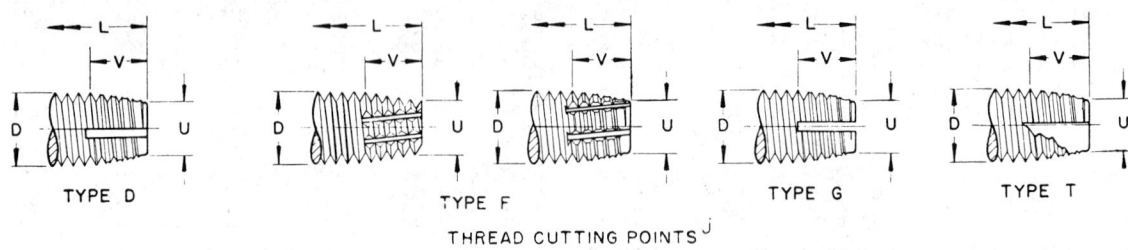

TYPE D   TYPE F   TYPE G   TYPE T

THREAD CUTTING POINTS[j]

TABLE 21—DIMENSIONS OF TYPES C, D, F, G, AND T TAPPING SCREW THREADS AND POINTS, IN.[a,k]

Nominal Size	Threads per in.	D Major Dia			U Point Dia		V^d, Taper Length				L Determinant Length for Taper[l]	
							Screws Longer Than L		Screws Equal to or Less Than L			
		Basic	Max	Min	Max[b]	Min[c]	Min[f]	Max[e]	Min[g]	Max[i]	90 Deg Heads[h]	Csk Heads[i]
2	56	0.0860	0.0860	0.0820	0.067	0.061	0.062	0.080	0.045	0.062	9/64	3/16
2	64	0.0860	0.0860	0.0822	0.070	0.064	0.055	0.070	0.039	0.055	1/8	11/64
3	48	0.0990	0.0990	0.0946	0.077	0.070	0.073	0.094	0.052	0.073	11/64	7/32
3	56	0.0990	0.0990	0.0950	0.080	0.074	0.062	0.080	0.045	0.062	9/64	3/16
4	40	0.1120	0.1120	0.1072	0.086	0.077	0.088	0.112	0.062	0.088	13/64	1/4
4	48	0.1120	0.1120	0.1076	0.090	0.083	0.073	0.094	0.052	0.073	11/64	7/32
5	40	0.1250	0.1250	0.1202	0.099	0.090	0.088	0.112	0.062	0.088	13/64	9/32
5	44	0.1250	0.1250	0.1204	0.101	0.093	0.080	0.102	0.057	0.080	3/16	1/4
6	32	0.1380	0.1380	0.1326	0.106	0.095	0.109	0.141	0.078	0.109	1/4	5/16
6	40	0.1380	0.1380	0.1332	0.112	0.103	0.088	0.112	0.062	0.088	13/64	17/64
8	32	0.1640	0.1640	0.1586	0.132	0.121	0.109	0.141	0.078	0.109	1/4	21/64
8	36	0.1640	0.1640	0.1590	0.135	0.125	0.097	0.125	0.069	0.097	15/64	19/64
10	24	0.1900	0.1900	0.1834	0.147	0.133	0.146	0.188	0.104	0.146	11/32	27/64
10	32	0.1900	0.1900	0.1846	0.158	0.147	0.109	0.141	0.078	0.109	1/4	11/32
12	24	0.2160	0.2160	0.2094	0.173	0.159	0.146	0.188	0.104	0.146	11/32	7/16
12	28	0.2160	0.2160	0.2098	0.179	0.167	0.125	0.161	0.089	0.125	19/64	25/64
1/4	20	0.2500	0.2500	0.2428	0.198	0.181	0.175	0.225	0.125	0.175	13/32	33/64
1/4	28	0.2500	0.2500	0.2438	0.213	0.201	0.125	0.161	0.089	0.125	19/64	13/32
5/16	18	0.3125	0.3125	0.3043	0.255	0.236	0.194	0.250	0.139	0.194	29/64	19/32
5/16	24	0.3125	0.3125	0.3059	0.269	0.255	0.146	0.188	0.104	0.146	11/32	15/32
3/8	16	0.3750	0.3750	0.3660	0.310	0.289	0.219	0.281	0.156	0.219	1/2	47/64
3/8	24	0.3750	0.3750	0.3684	0.332	0.318	0.146	0.188	0.104	0.146	11/32	1/2

[a] Additional specifications and details pertinent to thread length and points are given in General Specifications.
[b] Values are equal to basic minor diameter plus 20% of double thread depth.
[c] Values are equal to basic minor diameter less the Class 2 pitch diameter tolerance.
[d] Tapered threads shall have unfinished crests except as noted in General Specifications.
[e] Values are equal to 4-1/2 times the pitch distance rounded off to three decimal places.
[f] Values are equal to 3-1/2 times the pitch distance rounded off to three decimal places.
[g] Values are equal to 2-1/2 times the pitch distance rounded off to three decimal places.
[h] Values are equal to 8 times the pitch distance rounded upward to the nearest 1/64 in.
[i] Values are equal to 8 times the pitch distance plus the maximum head side height rounded upward to the nearest 1/64 in.
[j] See drawing—Thread Cutting Points. Details of taper and flute design shall be optional with manufacturer provided they meet the performance requirements and the flutes extend through the first full thread.
[k] See Appendix D, Cross Reference of Tapping Screw Type Designations.
[l] Screws this length and shorter shall have taper lengths shown in columns 10 and 11. Screws having greater lengths shall have taper lengths shown in columns 8 and 9.

TABLE 22—DIMENSIONS OF WOOD SCREW THREADS AND POINTS[a]

Nominal Size	Threads per in.[b]	D, Screw Dia[c]		
	Nominal	Basic	Max	Min
0	32	0.060	0.064	0.053
1	28	0.073	0.077	0.066
2	26	0.086	0.090	0.079
3	24	0.099	0.103	0.092
4	22	0.112	0.116	0.105
5	20	0.125	0.129	0.118
6	18	0.138	0.142	0.131
7	16	0.151	0.155	0.144
8	15	0.164	0.168	0.157
9	14	0.177	0.181	0.170
10	13	0.190	0.194	0.183
12	11	0.216	0.220	0.209
14	10	0.242	0.246	0.235
16	9	0.268	0.272	0.261
18	8	0.294	0.298	0.287
20	8	0.320	0.324	0.313
24	7	0.372	0.376	0.365

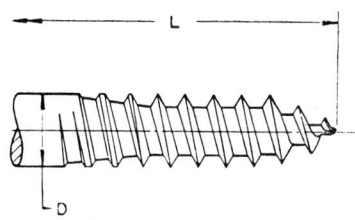

[a] Additional specifications and details pertinent to thread length and points are given in General Specifications.
[b] The maximum permissible variation in number of threads per inch shall be ±10% of nominal specified.
[c] Except for the tapered portion of point, the major diameter of the thread shall be not less than the minimum screw diameter.

CAPSCREWS—GENERAL DATA

Defects—Slotted capscrews shall be free from burrs, seams, laps, loose scale, irregular surfaces and so forth, that affect serviceability.

Bearing Surface—The bearing surface of capscrew heads except flat heads shall be within 2 deg of the right angles to the body.

Eccentricity of Heads—The heads of flat, fillister, and round-head capscrews shall not be eccentric with the body beyond a total indicator reading amounting to 2% of the maximum head diameter, or 0.010 in., whichever is greater.

Eccentricity of Slots—Slots shall not be off center from axis of the screw more than 6% of the nominal diameter of the screw, or 0.010 in., whichever is greater.

Total Runout (Eccentricity and Angularity) of Thread—For sizes up to and including 3/4 in., slotted head capscrews shall screw at least two threads into a tapped hole counterbored for 1/32 in. diametral clearance over the maximum body diameter to a depth equal to the length of the screw less one screw diameter. The starting thread of the tapped hole shall be countersunk to the diameter of the counterbore. The tapped hole shall have class 2B maximum pitch diameter, and the inspection fixture shall be hardened. For screws over 3/4 in., the diametral clearance of the counterbored hole shall be 1/16 in. A suggested gage is shown in the appendix.

Depth of Slots—Slot depths are measured from the highest point of the head to the intersection of the bottom of the slot with the head surface.

Length of Capscrews—The length of capscrews shall be measured from the largest diameter of the bearing surface of the head to the extreme point, in a line parallel to the axis of the screw.

Tolerance in Length—The length of capscrews shall not vary from that specified by more than the following: Up to 1 in. in length, minus 1/32 in.; over 1 to 2 in. inclusive, minus 1/16 in.; over 2 in., minus 3/32 in.

Length of Thread—The minimum length of thread on capscrews shall be equal to 2D plus 1/4 in. A tolerance of plus 2½ pitch minus 0 will be allowed. When capscrews are too short for the specified minimum thread length, the complete threads shall extend to within two and one half threads of the head.

Points—Points shall be flat and chamfered from approximately 1/64 in. below the minor diameter, the length of point to be from 1/2 to 1½ threads.

Material—Suitable material for steel capscrews is covered in SAE Standard, Mechanical and Chemical Requirements for Threaded Fasteners—SAE J429.

Thread Series and Tolerances—The threads on capscrews shall be either the Unified and American Coarse or Fine Series, Class 2A.

Finish—Slotted head capscrew heads shall be machined. Slotted head capscrews are usually not heat treated.

All table dimensions are in inches.

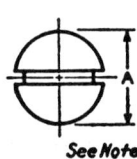 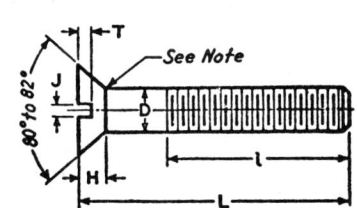

NOTE—Refer to Capscrews—General Data.

The maximum and minimum head diameters, A, are extended to the theoretical sharp corners.

The radius of the fillet at the base of the head shall not exceed 0.4 maximum body diameter.

TABLE 23—DIMENSIONS OF FLAT HEAD CAPSCREWS, IN.

Nominal Size	D Body Dia		A Head Dia			G Gaging Dia	H Height of Head	J Width of Slot		T Depth of Slot	
	Max	Min	Max	Min	Absolute Min with Flat		Average	Max	Min	Max	Min
1/4	0.250	0.245	0.500	0.477	0.452	0.4245	0.140	0.075	0.064	0.068	0.045
5/16	0.3125	0.307	0.625	0.598	0.567	0.5376	0.177	0.084	0.072	0.086	0.057
3/8	0.375	0.369	0.750	0.720	0.682	0.6507	0.210	0.094	0.081	0.103	0.068
7/16	0.4375	0.431	0.8125	0.780	0.736	0.7229	0.210	0.094	0.081	0.103	0.068
1/2	0.500	0.493	0.875	0.841	0.791	0.7560	0.210	0.106	0.091	0.103	0.068
9/16	0.5625	0.555	1.000	0.962	0.906	0.8691	0.244	0.118	0.102	0.120	0.080
5/8	0.625	0.617	1.125	1.083	1.020	0.9822	0.281	0.133	0.116	0.137	0.091
3/4	0.750	0.742	1.375	1.326	1.251	1.2085	0.352	0.149	0.131	0.171	0.115
7/8	0.875	0.866	1.625	1.568	1.480	1.4347	0.423	0.167	0.147	0.206	0.138
1	1.000	0.990	1.875	1.811	1.711	1.6610	0.494	0.188	0.166	0.240	0.162
1-1/8	1.125	1.114	2.062	1.992	1.880	1.8262	0.529	0.196	0.178	0.257	0.173
1-1/4	1.250	1.239	2.312	2.235	2.110	2.0525	0.600	0.211	0.193	0.291	0.197
1-3/8	1.375	1.363	2.562	2.477	2.340	2.2787	0.665	0.226	0.208	0.326	0.220
1-1/2	1.500	1.488	2.812	2.720	2.570	2.5050	0.742	0.258	0.240	0.360	0.244

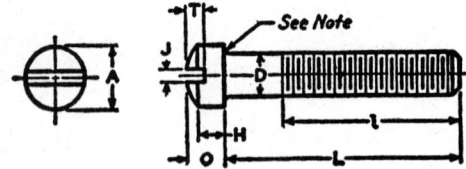

NOTE—Refer to Capscrews—General Data.
Radius of the fillet at the base of the head for sizes:
¼ to ⅜ in. inclusive is 0.016 min. and 0.031 max.
⁷⁄₁₆ to ⁹⁄₁₆ in. inclusive is 0.016 min. to 0.047 max.
⅝ to 1 in. inclusive is 0.031 min. and 0.062 max.

TABLE 24—DIMENSIONS OF FILLISTER HEAD CAPSCREWS, IN.

Nominal Size	D Body Dia		A Head Dia		H Height of Head		O Total Height of Head		J Width of Slot		T Depth of Slot	
	Max	Min	Max	Min	Max	Min	Max	Min	Max	Min	Max	Min
1/4	0.250	0.245	0.375	0.363	0.172	0.157	0.216	0.194	0.075	0.064	0.097	0.077
5/16	0.3125	0.307	0.437	0.424	0.203	0.186	0.253	0.230	0.084	0.072	0.115	0.090
3/8	0.375	0.369	0.562	0.547	0.250	0.229	0.314	0.284	0.094	0.081	0.142	0.112
7/16	0.4375	0.431	0.625	0.608	0.297	0.274	0.368	0.336	0.094	0.081	0.168	0.133
1/2	0.500	0.493	0.750	0.731	0.328	0.301	0.413	0.376	0.106	0.091	0.193	0.153
9/16	0.5625	0.555	0.812	0.792	0.375	0.346	0.467	0.427	0.118	0.102	0.213	0.168
5/8	0.625	0.617	0.875	0.853	0.422	0.391	0.521	0.478	0.133	0.116	0.239	0.189
3/4	0.750	0.742	1.000	0.976	0.500	0.466	0.612	0.566	0.149	0.131	0.283	0.223
7/8	0.875	0.866	1.125	1.098	0.594	0.556	0.720	0.668	0.167	0.147	0.334	0.264
1	1.000	0.990	1.312	1.282	0.656	0.612	0.803	0.743	0.188	0.166	0.371	0.291

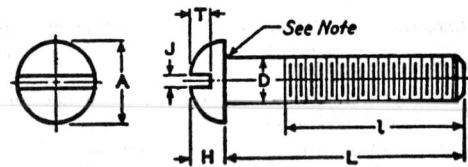

NOTE—Refer to Capscrews—General Data.
Radius of the fillet at the base of the head for sizes:
¼ to ⅜ in. inclusive is 0.016 min. and 0.031 max.
⁷⁄₁₆ to ⁹⁄₁₆ in. inclusive is 0.016 min. to 0.047 max.
⅝ to 1 in. inclusive is 0.031 min. and 0.062 max.

TABLE 25—DIMENSIONS OF ROUND HEAD CAPSCREWS, IN.

Nominal Size	D Body Dia		A Head Dia		H Height of Head		J Width of Slot		T Depth of Slot	
	Max	Min	Max	Min	Max	Min	Max	Min	Max	Min
1/4	0.250	0.245	0.437	0.418	0.191	0.175	0.075	0.064	0.117	0.097
5/16	0.3125	0.307	0.562	0.540	0.245	0.226	0.084	0.072	0.151	0.126
3/8	0.375	0.369	0.625	0.603	0.273	0.252	0.094	0.081	0.168	0.138
7/16	0.4375	0.431	0.750	0.725	0.328	0.302	0.094	0.081	0.202	0.167
1/2	0.500	0.493	0.812	0.786	0.354	0.327	0.106	0.091	0.218	0.178
9/16	0.5625	0.555	0.937	0.909	0.409	0.378	0.118	0.102	0.252	0.207
5/8	0.625	0.617	1.000	0.970	0.437	0.405	0.133	0.116	0.270	0.220
3/4	0.750	0.742	1.250	1.215	0.546	0.507	0.149	0.131	0.338	0.278

APPENDIX A—PROTRUSION GAGING

Suitability of flat head machine and tapping screws except for No. 0 and No. 1 undercut heads for application in countersinks designed to the principal dimensions of the screws, may be tested by use of a protrusion gage as shown in the figure below.

The gaging dimensions and the gage diameters are specified in the dimensional tables for flat head and undercut flat head screws. The protrusion limits shown in the tables apply when the gaging diameter is exactly as indicated with the gaging edge of a sharpness obtained by lapping the hole and the top surface to the gage. Any variation in the gaging diameter will require recalculation of protrusion values by the original formulas:

[1]Maximum protrusion

$$= \frac{\text{max sharp dia} - \text{gage hole dia}}{2} \times \tan\left(90 \text{ deg} - \frac{\text{min head angle}}{2}\right)$$

[1]Minimum protrusion

$$= \frac{\text{min sharp dia} - \text{gage hole dia}}{2} \times \tan\left(90 \text{ deg} - \frac{\text{max head angle}}{2}\right)$$

or correction in accordance with the following formula:

$$F' = F\left(\frac{A - G'}{A - G}\right)$$

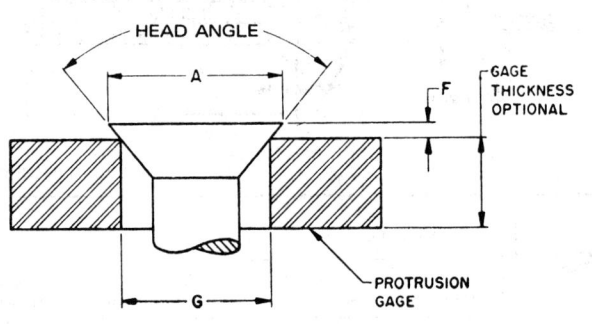

where: F = tabulated protrusion value;
F' = corrected protrusion value;
A = head diameter (maximum or minimum for maximum or minimum protrusion respectively);
G = tabulated gage diameter
G' = measured gage diameter

[1]Protrusion values shown in tables were calculated with these formulas and rounded to the nearest 0.001 in. upward for the maximum and downward for the minimum.

To insure adequate service life, the protrusion gage should be made of tool steel having hardness of not less than 60 RC.

APPENDIX B—PENETRATION GAGING OF CROSS RECESS SCREWS

Penetration gaging is a test to determine suitability of the recesses in the heads of screws and may be used to indicate deficiencies in the dimensions of the recesses specified in the dimensional tables. However, the penetration gaging test in itself shall not be cause for rejection of the screws should the screws meet all other dimensional requirements.

Specified herein are dimensions of gage points to be used for penetration gaging Type I and Type II Recesses. In all cases, the gage points approach as nearly as possible perfect driver form. Also specified are gage heads and bushings which adapt the gage points to standard dial gages. Penetration gaging values for the various styles of recessed heads are included in the dimensional tables for the heads.

Penetration is gaged relative to a reference plane defined by the intersection of the edge of the recess wings with the top surface of the screw head. This plane is the same as the top surface of a flat head screw but is somewhat below the topmost portion of heads which have rounded top surfaces. Knife edges or tapered ridges on the gage head are used to establish the reference plane. A reverse reading dial gage is used to indicate the penetration of the gage point into the recess. The gage may be zeroed on any flat surface.

TYPE I RECESS PENETRATION GAGES

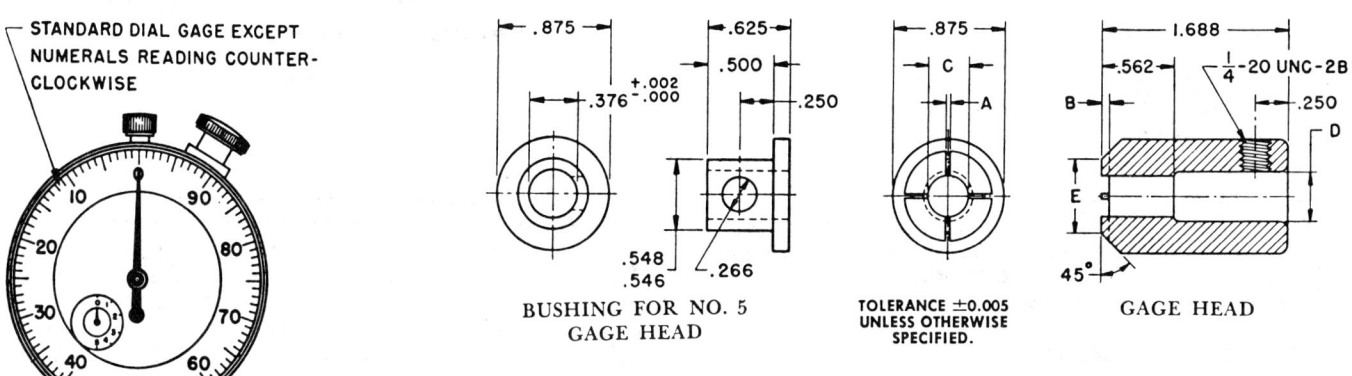

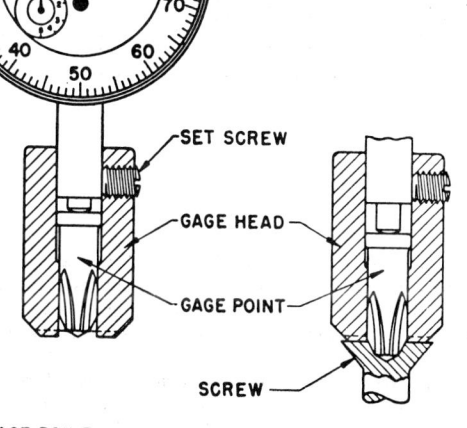

TABLE 26—DIMENSIONS OF GAGE HEADS, IN.

Size of Recess Gage	A ±0.002	B ±0.003	C ±0.0002	D +0.002 −0.000	E ±0.005
No. 0	0.008	0.015	0.0460	0.376	0.562
No. 1	0.012	0.020	0.0880	0.376	0.562
No. 2	0.018	0.031	0.1420	0.376	0.562
No. 3	0.022	0.037	0.2100	0.376	0.562
No. 4	0.031	0.062	0.3130	0.376	0.562
No. 5	0.041	0.094	0.5010	0.549	0.750[a]

[a] Minimum dimension, 45° chamfer may be omitted.

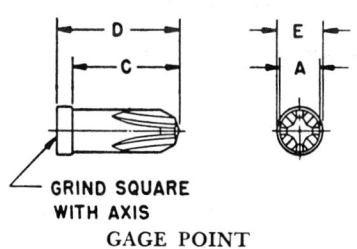

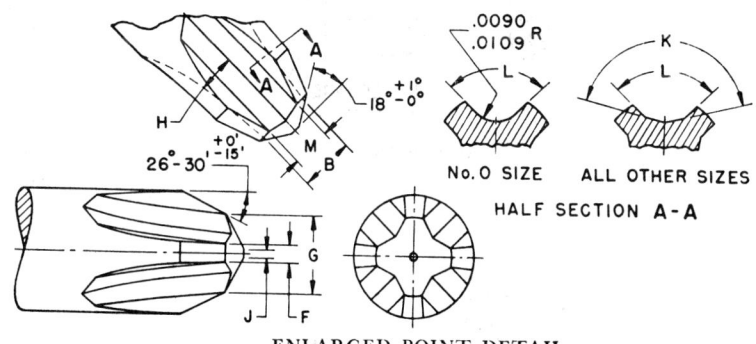

TABLE 27—DIMENSIONS OF GAGE POINTS[a]

Size of Recess Gage	A Point Dia. ±0.002	B Point Width +0.0000 −0.0010	C Length ±0.005	D Length ±0.005	E Dia. ±0.005	F Wing Thickness Max.	F Wing Thickness Min.	G Point Width +0.0010 −0.0000	H Milling Angle +0' −15'	J Flat on End Max.	J Flat on End Min.	K Base Flute Angle +15' −0'	L Side Flute Angle +15' −0'	M Flute Width at Bottom +0.0000 −0.0010
No. 0	0.0450	0.0240	0.656	0.781	0.094	0.012	0.010	0.0320	7°	0.015	0.010	[b]	92°	0.0151[c]
No. 1	0.0870	0.0394	0.688	0.812	0.156	0.020	0.018	0.0500	7°	0.020	0.015	138°	92°	0.0202
No. 2	0.1410	0.0606	0.750	0.875	0.219	0.025	0.023	0.0900	5°45'	0.020	0.015	140°	92°	0.0434
No. 3	0.2090	0.0983	0.781	0.906	0.250	0.031	0.029	0.1500	5°45'	0.020	0.015	146°	92°	0.0826
No. 4	0.3120	0.1407	0.844	0.969	0.359	0.044	0.042	0.2000	7°	0.020	0.015	153°	92°	0.1078
No. 5	0.5000	0.2310	1.031	1.156	0.531	0.063	0.061	0.3110	7°	0.025	0.020	162°46'	92°	0.1730

[a] Inches unless otherwise specified.
[b] Base of flute on size No. 0 is 0.0090 to 0.0109 radius.
[c] Tolerance on size No. 0 is +0.0000, −0.0026.

TYPE II RECESS PENETRATION GAGES

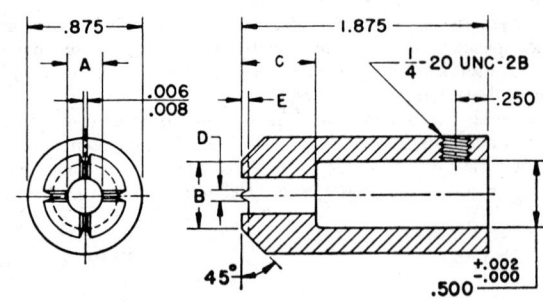

BUSHING

GAGE HEAD

TOLERANCE ±0.005 UNLESS OTHERWISE SPECIFIED

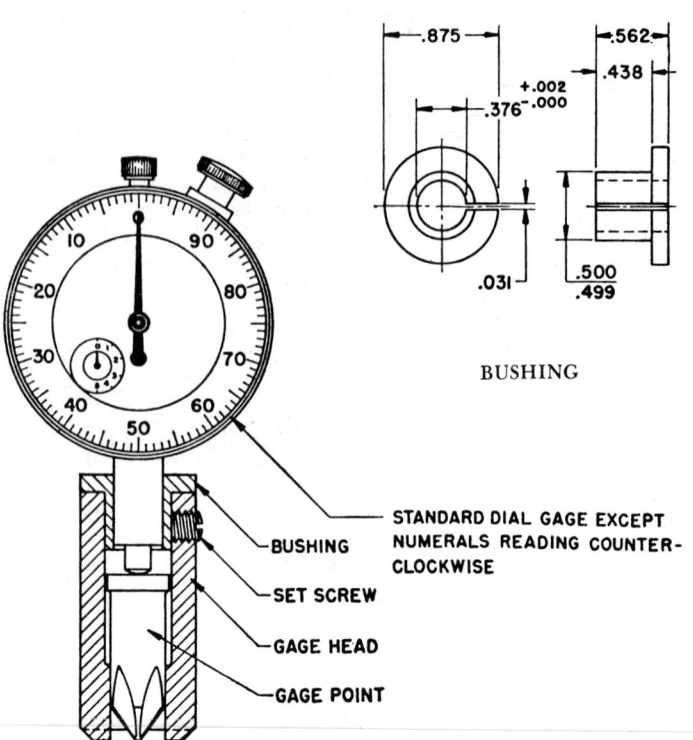

GAGE POINT IN ZERO POSITION
PENETRATION GAGE

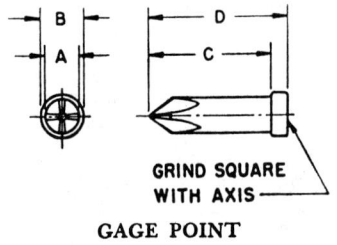

GRIND SQUARE WITH AXIS

GAGE POINT

TABLE 28—DIMENSIONS OF GAGE HEADS, IN.

Gage Dia	A ±0.0001	B ±0.005	C ±0.005	D +0.002 −0.000	E +0.002 −0.000
0.093	0.0930	0.344	0.375	0.030	0.024
0.141	0.1410	0.375	0.438	0.046	0.036
0.246	0.2460	0.500	0.500	0.062	0.051
0.436	0.4360	0.688	0.562	0.094	0.078

Penetrations which are too deep indicate the possibility of a thin section between head and shank of screw, a weakness which might result in twisting off screw heads during tightening of the screws. Use of screws having shallow penetrations might result in production problems such as reaming of recesses or excessive wear on driver bits.

Applicability of Gage Diameters to Recess Diameters and Screw Sizes —Although these gages may be used interchangeably showing identical readings on those sizes of screws where dimension "B" of gage head is greater than the recess diameter, the following recommendations may be applied.

Use 0.093 Gage for recess diameters up to 0.150.[1]
Use 0.141 Gage for recess diameters of 0.150 and up to 0.270.
Use 0.246 Gage for recess diameters of 0.270 and up to 0.460.
Use 0.436 Gage for recess diameters of 0.460 and up to 0.670.

[1] It is not practical to gage screw sizes No. 0 and No. 1 having recess diameters of less than 0.102 maximum.

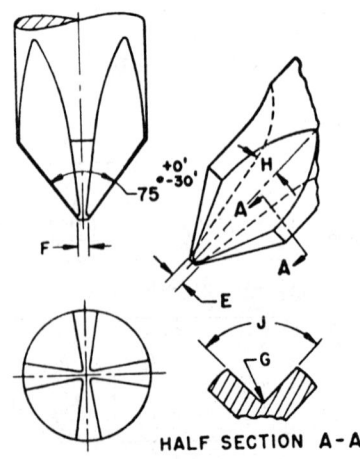

HALF SECTION A-A
ENLARGED POINT DETAIL

TABLE 29—DIMENSIONS OF GAGE POINTS[a]

Gage Dia	A Point Dia ±0.0001	B Dia ±0.005	C Length ±0.005	D Length ±0.005	E Point Width +0.001 −0.000	F Flat on End ±0.002	G Radius +0.000 −0.001	H Milling Angle +5′ −0′	J Side Flute Angle +15′ −0′
0.093	0.0926	0.188	0.750	1.875	0.027	0.062	0.005	8° 45′	90°
0.141	0.1406	0.250	0.875	1.000	0.027	0.062	0.005	8° 45′	90°
0.246	0.2456	0.312	0.938	1.062	0.027	0.062	0.005	8° 45′	90°
0.436	0.4356	0.469	1.125	1.250	0.027	0.062	0.005	8° 45′	90°

[a] Inches unless otherwise specified.

APPENDIX C—ACROSS CORNERS GAGING OF HEXAGON HEADS

Acceptance of width across corners of hexagon head and hexagon washer head screws may be determined by the use of gaging rings as described below.

When the gaging ring is placed on the top of a hex or hex washer head screw, and also the bottom of a hex head screw, at right angles to the axis of the screw; the head (hex portion of washer head) must protrude beyond the ring by an amount equal to 60% of the minimum head height, H. For convenience, the minimum protrusion values are given in the dimensional tables for hex and hex washer head screws.

The gaging ring shall have an inside diameter equal to the tabulated minimum width across corners, within a tolerance of plus 0.0003 in. The gaging edges of the ring shall be sharp and opposite faces shall be parallel. To insure adequate service life, the ring should be made of tool steel and have a hardness of not less than 60 Rockwell C.

A typical gaging fixture is shown below with an explanation of its application; however, any equivalent means may be used.

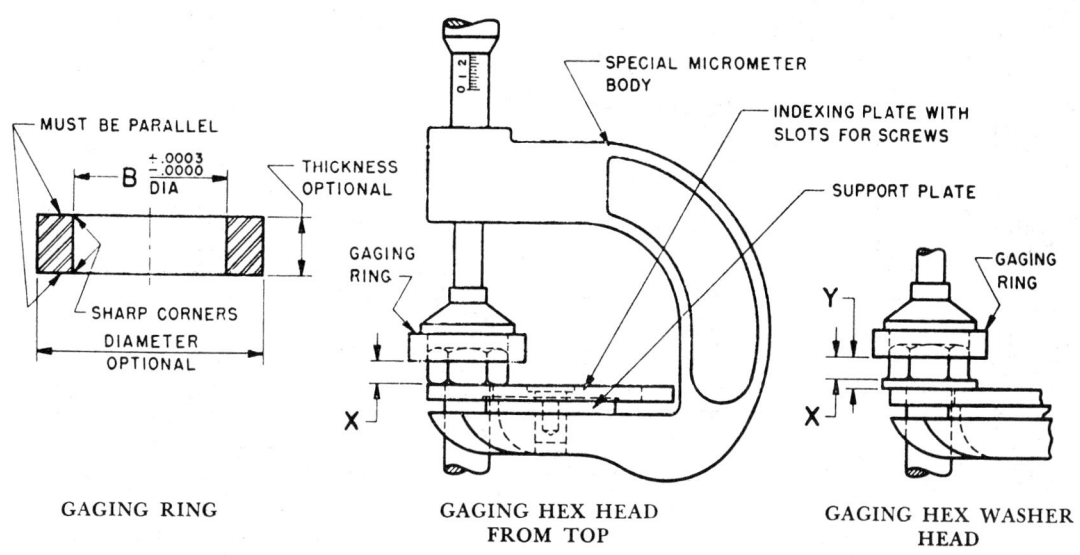

GAGING RING **GAGING HEX HEAD FROM TOP** **GAGING HEX WASHER HEAD**

TYPICAL GAGING FIXTURE

To check hex head screws from the top, an initial reading shall be taken with the gaging ring placed on the indexing plate. Then, with the screw placed in the fixture, the gaging ring shall be placed on top of the screw head and a second reading taken. The difference between the two readings is equal to the protrusion, X, of the head beyond the gaging ring.

To check hex washer head screws, the gaging procedure shall be exactly the same as that for checking hex head screws from the top. However, in this case, the difference, Y, between the two readings includes the washer thickness and it is necessary to deduct the actual (measured) thickness of the washer portion from the difference, Y, to obtain the protrusion, X, of the hex beyond the gaging ring.

Gaging the bottom of the head on hex head screws may be accomplished in the same manner as gaging the top, except the ring is placed below the head. The same protrusion values shall apply.

APPENDIX D—CROSS REFERENCE OF TAPPING SCREW TYPE DESIGNATIONS

Type	SAE and ASA	Manufacturer	Federal
	A	A	A
	B	B or Z	B
	BP	BP or ZP	BP
	C	C	C
	D	1	CS Alternate 1
	F	F	CF
	F	F	CF
	G	G	CS Alternate 2
	T	23	CG
	BF	FZ	BF
	BG	H	BG
	BT	25	BG

APPENDIX E—RECOMMENDED HOLE SIZES FOR TAPPING SCREWS

The hole sizes tabulated herein represent recommendations based on practical experience of screw producers and users, and are not to be misconstrued as being a part of the tapping screw specifications. (The performance requirements are given in the General Specifications.)

For certain conditions of hardness and material composition the facility of assembly and strength of the fastening might be improved by a slight variation from the recommended hole size. Only by experiment in specific applications can the user determine the advisability of such deviation.

TABLE 30—RECOMMENDED HOLES FOR TYPE A SCREWS IN SHEET METAL
Steel, Stainless Steel, Monel Metal, Brass, Aluminum Alloy

Screw Size	Metal Thickness	Pierced or Extruded Hole	Drilled or Clean Punched Hole	
		Hole Dia	Hole Dia[a]	Drill Size
4	0.015	—	0.086	44
	0.018	—	0.086	44
	0.024	0.098	0.094	42
	0.030	0.098	0.094	42
	0.036	0.098	0.098	40
6	0.015	—	0.104	37
	0.018	—	0.104	37
	0.024	0.111	0.104	37
	0.030	0.111	0.104	37
	0.036	0.111	0.106	36
7	0.015	—	0.116	32
	0.018	—	0.116	32
	0.024	0.120	0.116	32
	0.030	0.120	0.116	32
	0.036	0.120	0.116	32
	0.048	0.120	0.120	31
8	0.018	—	0.125	1/8
	0.024	0.136	0.125	1/8
	0.030	0.136	0.125	1/8
	0.036	0.136	0.125	1/8
	0.048	0.136	0.128	30
10	0.018	—	0.136	29
	0.024	0.157	0.136	29
	0.030	0.157	0.136	29
	0.036	0.157	0.136	29
	0.048	0.157	0.149	25
12	0.024	—	0.161	20
	0.030	0.185	0.161	20
	0.036	0.185	0.161	20
	0.048	0.185	0.161	20
14	0.024	—	0.185	13
	0.030	0.209	0.189	12
	0.036	0.209	0.191	11
	0.048	0.209	0.196	9

[a] Decimals shown are standard drill sizes to nearest thousandth.

TABLE 31—RECOMMENDED HOLES FOR TYPE A SCREWS IN RESIN IMPREGNATED PLYWOOD
Compreg, Pregwood, and Similar Material

Screw Size	Hole Dia[c]	Drill Size	Min[a] Material Thickness	Penetration in Blind Holes[b]	
				Max	Min
4	0.098	40	3/16	3/4	1/4
6	0.110	35	3/16	3/4	1/4
7	0.128	30	1/4	3/4	5/16
8	0.140	28	1/4	3/4	5/16
10	0.170	18	5/16	1	3/8
12	0.189	12	5/16	1	3/8
14	0.228	1	7/16	1	1/2

[a] Based on full thread engagement through the entire material, with the tapered portion of the screw projecting completely.
[b] Total length of thread in engagement.
[c] Decimals shown are standard drill sizes to nearest thousandth.

TABLE 32—RECOMMENDED HOLES FOR TYPE A SCREWS IN ASBESTOS COMPOSITIONS
Transite, Ebony Asbestos, and Similar Material

Screw Size	Hole Dia[c]	Drill Size	Min[a] Material Thickness	Penetration in Blind Holes[b]	
				Max	Min
4	0.094	42	3/16	3/4	1/4
6	0.106	36	3/16	3/4	1/4
7	0.125	1/8	1/4	3/4	5/16
8	0.136	29	1/4	3/4	5/16
10	0.161	20	5/16	1	3/8
12	0.185	13	5/16	1	3/8
14	0.213	3	7/16	1	1/2

[a] Based on full thread engagement through the entire material, with the tapered portion of the screw projecting completely.
[b] Total length of thread in engagement.
[c] Decimals shown are standard drill sizes to nearest thousandth.

TABLE 33—RECOMMENDED HOLES FOR TYPES B AND BP SCREWS IN SHEET METAL

Screw Size	Metal Thickness	Steel, Stainless Steel, Monel Metal, Brass			Aluminum Alloy		
		Pierced or Extruded Hole	Drilled or Clean Punched Hole		Pierced or Extruded Hole	Drilled or Clean Punched Hole	
		Hole Dia	Hole Dia[a]	Drill Size	Hole Dia	Hole Dia[a]	Drill Size
2	0.015	—	0.064	52	—	—	—
	0.018	—	0.064	52	—	—	—
	0.024	—	0.067	51	—	0.064	52
	0.030	—	0.070	50	—	0.064	52
	0.036	—	0.073	49	—	0.064	52
	0.048	—	0.073	49	—	0.067	51
	0.060	—	0.076	48	—	0.070	50
4	0.015	0.086	0.086	44	—	—	—
	0.018	0.086	0.086	44	—	—	—
	0.024	0.098	0.089	43	0.086	—	—
	0.030	0.098	0.094	42	0.086	0.086	44
	0.036	0.098	0.094	42	0.086	0.086	44
	0.048	—	0.096	41	0.086	0.086	44
	0.060	—	0.100	39	—	0.089	43
	0.075	—	0.102	38	—	0.089	43
	0.105	—	—	—	—	0.094	42
6	0.015	0.111	0.104	37	—	—	—
	0.018	0.111	0.104	37	—	—	—
	0.024	0.111	0.106	36	0.111	—	—
	0.030	0.111	0.106	36	0.111	0.104	37
	0.036	0.111	0.110	35	0.111	0.104	37
	0.048	—	0.111	34	—	0.104	37
	0.060	—	0.116	32	—	0.106	36
	0.075	—	0.120	31	—	0.110	35
	0.105	—	0.128	30	—	0.111	34
	0.128–0.250	—	—	—	—	0.120	31
7	0.018	0.120	0.116	32	—	—	—
	0.024	0.120	0.116	32	0.120	—	—
	0.030	0.120	0.116	32	0.120	0.113	33
	0.036	0.120	0.116	32	0.120	0.113	33
	0.048	0.120	0.120	31	0.120	0.116	32
	0.060	—	0.128	30	—	0.120	31
	0.075	—	0.136	29	—	0.128	30
	0.105	—	0.140	28	—	0.136	29
	0.128–0.250	—	—	—	—	0.136	29
8	0.018	0.136	—	—	0.136	—	—
	0.024	0.136	0.125	48	0.136	—	—
	0.030	0.136	0.125	48	0.136	0.116	32
	0.036	0.136	0.125	48	0.136	0.120	31
	0.048	0.136	0.128	30	0.136	0.128	30
	0.060	—	0.136	29	—	0.136	29
	0.075	—	0.140	28	—	0.140	28
	0.105	—	0.150	25	—	0.147	26
	0.125	—	0.150	25	—	0.147	26
	0.135	—	0.152	24	—	0.149	25
	0.162–0.375	—	—	—	—	0.152	24
10	0.018	0.157	—	—	0.157	—	—
	0.024	0.157	0.144	27	0.157	—	—
	0.030	0.157	0.144	27	0.157	—	—
	0.036	0.157	0.147	26	0.157	0.144	27
	0.048	0.157	0.152	24	0.157	0.144	27
	0.060	—	0.152	24	—	0.144	27
	0.075	—	0.157	22	—	0.147	26
	0.105	—	0.161	20	—	0.147	26
	0.125	—	0.169	18	—	0.154	23
	0.135	—	0.169	18	—	0.154	23
	0.164	—	0.173	17	—	0.159	21
	0.200–0.375	—	—	—	—	0.166	19
12	0.024	0.185	0.166	19	—	—	—
	0.030	0.185	0.166	19	—	—	—
	0.036	0.185	0.166	19	—	—	—
	0.048	0.185	0.170	18	—	0.161	20
	0.060	—	0.177	16	—	0.166	19
	0.075	—	0.182	14	—	0.173	17
	0.105	—	0.185	13	—	0.180	15
	0.125	—	0.196	9	—	0.182	14
	0.135	—	0.196	9	—	0.182	14
	0.164	—	0.201	7	—	0.189	12
	0.200–0.375	—	—	—	—	0.196	9
1/4	0.030	0.209	0.194	10	—	—	—
	0.036	0.209	0.194	10	—	—	—
	0.048	0.209	0.194	10	—	—	—
	0.060	—	0.199	8	—	0.199	8
	0.075	—	0.204	6	—	0.201	7
	0.105	—	0.209	4	—	0.204	6
	0.125	—	0.228	1	—	0.209	4
	0.135	—	0.228	1	—	0.209	4
	0.164	—	0.234	15/64	—	0.213	3
	0.187	—	0.234	15/64	—	0.213	3
	0.194	—	0.234	15/64	—	0.221	2
	0.200–0.375	—	—	—	—	0.228	1

[a] Decimals shown are standard drill sizes to nearest thousandth.

TABLE 34—RECOMMENDED HOLES FOR TYPES B AND BP SCREWS IN RESIN IMPREGNATED PLYWOOD

	Compreg, Pregwood, and Similar Material				
Screw Size	Hole Dia[c]	Drill Size	Min Material Thickness[a]	Penetration in Blind Holes[b]	
				Max	Min
2	0.073	49	1/8	1/2	3/16
4	0.100	39	3/16	5/8	1/4
6	0.125	1/8	3/16	5/8	1/4
7	0.136	29	3/16	3/4	1/4
8	0.144	27	3/16	3/4	1/4
10	0.173	17	1/4	1	5/16
12	0.194	10	5/16	1	3/8
1/4	0.228	1	5/16	1	3/8

[a] Based on full thread engagement through the entire material, with the tapered portion of the screw projecting completely.
[b] Total length of thread in engagement.
[c] Decimals shown are standard drill sizes to nearest thousandth.

TABLE 35—RECOMMENDED HOLES FOR TYPES B AND BP SCREWS IN ASBESTOS COMPOSITIONS

	Transite, Ebony Asbestos, and Similar Material				
Screw Size	Hole Dia[c]	Drill Size	Min Material Thickness[a]	Penetration in Blind Holes[b]	
				Max	Min
2	0.076	48	1/8	1/2	3/16
4	0.101	38	3/16	5/8	1/4
6	0.120	31	3/16	5/8	1/4
7	0.136	29	1/4	3/4	5/16
8	0.147	26	5/16	3/4	3/8
10	0.166	19	5/16	1	3/8
12	0.196	9	5/16	1	3/8
1/4	0.228	1	7/16	1	1/2

[a] Based on full thread engagement through the entire material with the tapered portion of the screw projecting completely.
[b] Total length of thread in engagement.
[c] Decimals shown are standard drill sizes to nearest thousandth.

TABLE 36—RECOMMENDED HOLES FOR TYPES B AND BP SCREWS IN CAST METALS

	Aluminum, Magnesium, Zinc, Brass, Bronze		
Screw Size	Hole Dia[b]	Drill Size	Penetration in Blind Holes[a]
			Min
2	0.078	47	1/8
4	0.104	37	3/16
6	0.128	30	1/4
7	0.144	27	1/4
8	0.152	24	1/4
10	0.177	16	1/4
12	0.199	8	9/32
1/4	0.234	15/64	5/16

[a] Total length of thread in engagement.
[b] Decimals shown are standard drill sizes to nearest thousandth.

TABLE 37—RECOMMENDED HOLES FOR TYPES B AND BP SCREWS IN PLASTICS

	Phenol Formaldehyde		Cellulose Acetate, Cellulose Nitrate Acrylic Resin, Styrene Resin		
Screw Size	Hole Dia[a]	Drill Size	Hole Dia[a]	Drill Size	Penetration in Blind Holes Min
2	0.078	47	0.078	47	3/16
4	0.100	39	0.094	42	1/4
6	0.128	30	0.120	31	1/4
7	0.136	29	0.128	30	1/4
8	0.150	25	0.144	27	5/16
10	0.177	16	0.170	18	5/16
12	0.199	8	0.191	11	3/8
1/4	0.234	15/64	0.221	2	3/8

[a] Decimals shown are standard drill sizes to nearest thousandth.

TABLE 38—RECOMMENDED HOLES FOR TYPES BF, BG, AND BT SCREWS IN DIE CAST METALS

Zinc and Aluminum

Screw Size	Metal Thickness	Hole Dia[a]	Drill Size	Screw Size	Metal Thickness	Hole Dia[a]	Drill Size
2	0.060	0.073	49		0.125	0.166	19
	0.083	0.073	49		0.140	0.166	19
	0.109	0.076	48	10	0.188	0.166	19
	0.125	0.076	48		0.250	0.170	18
	0.140	0.076	48		0.312	0.172	11/64
					0.375	0.172	11/64
3	0.060	0.086	44				
	0.083	0.086	44		0.125	0.191	11
	0.109	0.086	44		0.140	0.191	11
	0.125	0.086	44	12	0.188	0.191	11
	0.140	0.089	43		0.250	0.196	9
	0.188	0.089	43		0.312	0.196	9
					0.375	0.196	9
4	0.109	0.098	40				
	0.125	0.100	39		0.125	0.221	2
	0.140	0.100	39		0.140	0.221	2
	0.188	0.100	39	1/4	0.188	0.221	2
	0.250	0.102	38		0.250	0.228	1
					0.312	0.228	1
					0.375	0.228	1
5	0.109	0.111	34				
	0.125	0.111	34		0.125	0.281	K
	0.140	0.113	33		0.140	0.281	K
	0.188	0.113	33		0.188	0.281	K
	0.250	0.116	32	5/16	0.250	0.281	K
					0.312	0.290	L
					0.375	0.290	L
6	0.125	0.120	31				
	0.140	0.120	31		0.125	0.344	11/32
	0.188	0.120	31		0.140	0.344	11/32
	0.250	0.125	1/8		0.188	0.344	11/32
	0.312	0.125	1/8	3/8	0.250	0.344	11/32
					0.312	0.348	S
					0.375	0.348	S
8	0.125	0.149	25				
	0.140	0.149	25				
	0.188	0.149	25				
	0.260	0.152	24				
	0.312	0.152	24				

[a] Decimals shown are standard drill sizes to nearest thousandth.

TABLE 39—RECOMMENDED HOLES FOR TYPES BF, BG, AND BT SCREWS IN PLASTICS

Screw Size	Phenol Formaldehyde		Cellulose Acetate, Cellulose Nitrate Acrylic Resin, Styrene Resin			
	Hole Dia[b]	Drill Size	Hole Dia[b]	Drill Size	Penetration in Blind Holes[a]	
					Max	Min
2	0.078	5/64	0.076	48	1/4	3/32
3	0.089	43	0.089	43	5/16	1/8
4	0.104	37	0.100	39	5/16	1/8
5	0.116	32	0.113	33	3/8	3/16
6	0.125	1/8	0.120	31	3/8	3/16
8	0.147	26	0.144	27	1/2	1/4
10	0.170	18	0.166	19	5/8	5/16
12	0.194	10	0.189	12	5/8	3/8
1/4	0.228	1	0.221	2	3/4	3/8

[a] Total length of thread in engagement.
[b] Decimals shown are standard drill sizes to nearest thousandth.

TABLE 40—RECOMMENDED HOLES FOR TYPE C SCREWS IN SHEET METAL

Screw Size	Threads per in.	Metal Thickness	Steel	
			Hole Dia[a]	Drill Size
4	40	0.037	0.094	42
		0.048	0.094	42
		0.062	0.096	41
		0.075	0.100	39
		0.105	0.102	38
		0.134	0.102	38
6	32	0.037	0.113	33
		0.048	0.116	32
		0.062	0.116	32
		0.075	0.122	3.1 MM
		0.105	0.125	1/8
		0.134	0.125	1/8
8	32	0.037	0.136	29
		0.048	0.144	27
		0.062	0.144	27
		0.075	0.147	26
		0.105	0.150	25
		0.134	0.150	25
10	24	0.037	0.154	23
		0.048	0.161	20
		0.062	0.166	19
		0.075	0.170	18
		0.105	0.173	17
		0.134	0.177	16
	32	0.037	0.170	18
		0.048	0.170	18
		0.062	0.170	18
		0.075	0.173	17
		0.105	0.177	16
		0.134	0.177	16
12	24	0.037	0.189	12
		0.048	0.194	10
		0.062	0.194	10
		0.075	0.199	8
		0.105	0.199	8
		0.134	0.199	8
1/4	20	0.037	0.221	2
		0.048	0.221	2
		0.062	0.228	1
		0.075	0.234	A
		0.105	0.234	A
		0.134	0.236	6 MM
	28	0.037	0.224	5.7 MM
		0.048	0.228	1
		0.062	0.232	5.9 MM
		0.075	0.234	A
		0.105	0.238	B
		0.134	0.238	B
5/16	18	0.037	0.290	L
		0.048	0.290	L
		0.062	0.290	L
		0.075	0.295	M
		0.105	0.295	M
		0.134	0.295	M

[a] Decimals shown are standard drill sizes to nearest thousandth.

TABLE 41—RECOMMENDED HOLES FOR TYPES D, F, G, AND T SCREWS IN SHEET METAL AND CASTINGS

Screw Size	Threads per in.	Metal Thickness	Sheet Metal				Castings			
			Steel		Aluminum		Iron		Zinc and Aluminum	
			Hole Dia[a]	Drill Size	Hole Dia[a]	Drill Size	Hole Dia[a]	Drill Size	Hole Dia[a]	Drill Size
2	56	0.050	0.073	49	0.070	50	0.076	48	0.073	49
		0.060	0.073	49	0.073	49	0.076	48	0.073	49
		0.083	0.073	49	0.073	49	0.076	48	0.076	48
		0.109	0.073	49	0.073	49	0.078	5/64	0.076	48
		0.125	0.076	48	0.073	49	0.078	5/64	0.076	48
		0.140	0.076	48	0.073	49	0.078	5/64	0.076	48
3	48	0.050	0.081	46	0.078	5/64	0.089	43	0.082	45
		0.060	0.081	46	0.081	46	0.089	43	0.082	45
		0.083	0.082	45	0.082	45	0.089	43	0.082	45
		0.109	0.086	44	0.082	45	0.089	43	0.086	44
		0.125	0.086	44	0.082	45	0.089	43	0.089	43
		0.140	0.086	44	0.086	44	0.094	42	0.089	43
		0.187	0.089	43	0.086	44	0.094	42	0.089	43
4	40	0.050	0.089	43	0.089	43	0.100	39	0.096	41
		0.060	0.089	43	0.089	43	0.100	39	0.096	41
		0.083	0.094	42	0.089	43	0.102	38	0.096	41
		0.109	0.096	41	0.094	42	0.102	38	0.096	41
		0.125	0.098	40	0.094	42	0.102	38	0.100	39
		0.140	0.098	40	0.094	3/32	0.102	38	0.100	39
		0.187	0.102	38	0.098	40	0.104	37	0.100	39
5	40	0.050	0.106	—	0.102	38	0.111	34	0.106	—
		0.060	0.106	—	0.102	38	0.111	34	0.106	—
		0.083	0.106	—	0.104	37	0.113	33	0.106	—
		0.109	0.106	36	0.104	37	0.113	33	0.110	35
		0.125	0.109	7/64	0.106	36	0.116	32	0.110	35
		0.140	0.110	35	0.106	36	0.116	32	0.110	35
		0.187	0.116	32	0.110	35	0.116	32	0.111	34
		0.250	0.116	32	0.113	33	0.116	32	0.113	33

(Table continued on next page)

TABLE 41—RECOMMENDED HOLES FOR TYPES D, F, G, AND T SCREWS IN SHEET METAL AND CASTINGS (CONTINUED)

Screw Size	Threads per in.	Metal Thickness	Sheet Metal Steel Hole Dia[a]	Drill Size	Sheet Metal Aluminum Hole Dia[a]	Drill Size	Castings Iron Hole Dia[a]	Drill Size	Castings Zinc and Aluminum Hole Dia[a]	Drill Size
6	32	0.050	0.110	35	0.109	7/64	0.120	31	0.116	32
		0.060	0.113	33	0.109	7/64	0.120	31	0.120	31
		0.083	0.116	32	0.111	34	0.125	1/8	0.120	31
		0.109	0.116	32	0.113	33	0.125	1/8	0.120	31
		0.125	0.116	32	0.116	32	0.125	1/8	0.120	31
		0.140	0.120	31	0.116	32	0.125	1/8	0.120	31
		0.187	0.125	1/8	0.120	31	0.128	30	0.120	31
		0.250	0.125	1/8	0.125	1/8	0.128	30	0.120	31
8	32	0.050	0.136	29	0.136	29	0.147	26	0.144	27
		0.060	0.140	28	0.136	29	0.150	25	0.144	27
		0.083	0.140	28	0.136	29	0.150	25	0.144	27
		0.109	0.144	27	0.140	28	0.150	25	0.144	27
		0.125	0.144	27	0.140	28	0.150	25	0.147	26
		0.140	0.147	26	0.144	27	0.150	25	0.147	26
		0.187	0.150	25	0.147	26	0.154	23	0.147	26
		0.250	0.150	25	0.150	25	0.154	23	0.150	25
		0.312	0.150	25	0.150	25	0.154	23	0.150	25
10	24	0.050	0.152	24	0.150	25	0.170	18	0.161	20
		0.060	0.154	23	0.152	24	0.170	18	0.166	19
		0.083	0.161	20	0.154	23	0.172	11/64	0.166	19
		0.109	0.161	20	0.157	22	0.173	17	0.166	19
		0.125	0.166	19	0.159	21	0.173	17	0.166	19
		0.140	0.170	18	0.161	20	0.173	17	0.166	19
		0.187	0.173	17	0.166	19	0.177	16	01.70	18
		0.250	0.173	17	0.172	11/64	0.177	16	0.170	18
		0.312	0.173	17	0.173	17	0.177	16	0.172	11/64
		0.375	0.173	17	0.173	17	0.177	16	0.172	11/64
10	32	0.050	0.159	21	0.161	20	0.173	17	0.170	18
		0.060	0.166	19	0.161	20	0.173	17	0.170	18
		0.083	0.166	19	0.161	20	0.177	16	0.172	11/64
		0.109	0.170	18	0.166	19	0.177	16	0.172	11/64
		0.125	0.170	18	0.166	19	0.177	16	0.172	11/64
		0.140	0.170	18	0.166	19	0.177	16	0.172	11/64
		0.187	0.177	16	0.172	—	0.180	15	0.172	11/64
		0.250	0.177	16	0.177	16	0.180	15	0.173	17
		0.312	0.177	16	0.177	16	0.180	15	0.173	17
		0.375	0.177	16	0.177	16	0.180	15	0.177	16
12	24	0.060	0.180	15	0.177	16	0.196	9	0.189	12
		0.083	0.182	14	0.180	15	0.199	8	0.191	11
		0.109	0.188	3/16	0.182	14	0.199	8	0.191	11
		0.125	0.191	11	0.185	13	0.199	8	0.191	11
		0.140	0.191	11	0.188	3/16	0.199	8	0.194	10
		0.187	0.199	8	0.191	11	0.203	13/64	0.194	10
		0.250	0.199	8	0.199	8	0.204	6	0.196	9
		0.312	0.199	8	0.199	8	0.204	6	0.196	9
		0.375	0.199	8	0.199	8	0.204	6	0.199	8
		0.500	0.199	8	0.199	8	0.204	6	0.199	8
1/4	20	0.083	0.213	3	0.206	5	0.228	1	0.219	7/32
		0.109	0.219	7/32	0.209	4	0.228	1	0.219	7/32
		0.125	0.221	2	0.213	3	0.228	1	0.221	2
		0.140	0.221	2	0.213	3	0.228	1	0.221	2
		0.187	0.228	1	0.221	2	0.234	15/64	0.221	2
		0.250	0.228	1	0.228	1	0.234	15/64	0.228	1
		0.312	0.228	1	0.228	1	0.234	15/64	0.228	1
		0.375	0.228	1	0.228	1	0.234	15/64	0.228	1
		0.500	0.228	1	0.228	1	0.234	15/64	0.228	1
1/4	28	0.083	0.221	2	0.219	7/32	0.234	A	0.228	1
		0.109	0.228	1	0.221	2	0.234	15/64	0.228	1
		0.125	0.228	1	0.221	2	0.234	15/64	0.228	1
		0.140	0.234	A	0.221	2	0.234	15/64	0.228	1
		0.187	0.234	15/64	0.228	1	0.238	B	0.228	1
		0.250	0.234	15/64	0.234	15/64	0.238	B	0.234	A
		0.312	0.234	15/64	0.234	15/64	0.238	B	0.234	A
		0.375	0.234	15/64	0.234	15/64	0.238	B	0.234	15/64
		0.500	0.234	15/64	0.234	15/64	0.238	B	0.234	15/64
5/16	18	0.109	0.277	J	0.266	H	0.290	L	0.277	J
		0.125	0.277	J	0.272	I	0.290	L	0.281	K
		0.140	0.281	—	0.272	I	0.290	L	0.281	K
		0.187	0.290	L	0.281	K	0.295	M	0.281	9/32
		0.250	0.290	L	0.290	L	0.295	M	0.281	9/32
		0.312	0.290	L	0.290	L	0.295	M	0.290	L
		0.375	0.290	L	0.290	L	0.295	M	0.290	L
		0.500	0.290	L	0.290	L	0.295	M	0.290	L
5/16	24	0.109	0.290	L	0.281	K	0.295	M	0.290	L
		0.125	0.290	L	0.281	9/32	0.295	M	0.290	L
		0.140	0.290	L	0.281	9/32	0.295	M	0.290	L
		0.187	0.295	M	0.290	L	0.302	N	0.290	L
		0.250	0.295	M	0.295	M	0.302	N	0.290	L
		0.312	0.295	M	0.295	M	0.302	N	0.295	M
		0.375	0.295	M	0.295	M	0.302	N	0.295	M
		0.500	0.295	M	0.295	M	0.302	N	0.295	M
3/8	16	0.125	0.339	R	0.328	21/64	0.348	S	0.339	R
		0.140	0.339	R	0.332	Q	0.348	S	0.339	R
		0.187	0.348	S	0.339	R	0.348	S	0.339	R
		0.250	0.358	T	0.348	S	0.348	S	0.344	11/32
		0.312	0.358	T	0.348	S	0.348	S	0.344	11/32
		0.375	0.358	T	0.348	S	0.348	S	0.348	S
		0.500	0.358	T	0.348	S	0.348	S	0.348	S
3/8	24	0.125	0.348	S	0.344	11/32	0.358	T	0.348	S
		0.140	0.348	S	0.344	11/32	0.358	T	0.348	S
		0.187	0.358	T	0.348	S	0.358	T	0.348	S
		0.250	0.358	T	0.358	T	0.358	T	0.358	T
		0.312	0.358	T	0.358	T	0.358	T	0.358	T
		0.375	0.358	T	0.358	T	0.358	T	0.358	T
		0.500	0.358	T	0.358	T	0.358	T	0.358	T

[a] Decimals shown are standard drill sizes to nearest thousandth.

TABLE 42—RECOMMENDED HOLES FOR TYPES D, F, G, AND T SCREWS IN PLASTICS

Screw Size	Threads per in.	Phenol Formaldehyde				Cellulose Acetate, Cellulose Nitrate, Acrylic Resins, Styrene Resins			
		Hole Dia[b]	Drill Size	Depth of Penetration[a]		Hole Dia[b]	Drill Size	Depth of Penetration[a]	
				Max	Min			Max	Min
2	56	0.078	5/64	3/8	7/32	0.076	48	3/8	7/32
3	48	0.089	43	3/8	7/32	0.086	44	3/8	7/32
4	40	0.098	40	5/16	1/4	0.093	42	5/16	1/4
5	40	0.113	33	7/16	1/4	0.110	35	7/16	1/4
6	32	0.116	32	5/16	1/4	0.116	32	5/16	1/4
8	32	0.144	27	1/2	5/16	0.144	27	1/2	5/16
10	24	0.166	19	1/2	3/8	0.166	19	1/2	3/8
	32	0.161	20	1/2	3/8	0.161	20	1/2	3/8
1/4	20	0.228	1	5/8	3/8	0.228	1	1	3/8

[a] Total length of thread in engagement.

[b] Decimals shown are standard drill sizes to nearest thousandth.

SLOTTED HEADLESS SETSCREWS—SAE J479a

SAE Standard

Report of Screw Threads Committee approved October 1947 and last revised by Fasteners Committee July 1971. Conforms with American National Standard for Slotted Headless Setscrews, ANSI B18.6.2.

GENERAL SPECIFICATIONS

Scope—Included herein are complete general and dimensional data for slotted headless set screws.

The inclusion of dimensional data in this standard is not intended to imply that all of the products described are stock production sizes. Consumers should consult with manufacturers concerning availability of products.

Dimensions—All dimensions in this standard are in inches unless stated otherwise.

Headless Ends

End Configuration—The slotted end of screws may be crowned as depicted and dimensioned in Table 1 or be flat, at the option of manufacturer.

Depth of Slot—The depth of slot shall be measured, parallel to the axis of screw, from the end of the screw to the intersection of the bottom of the slot with the thread major diameter.

Eccentricity of Slot—The slot shall not be eccentric with the axis of screw by an amount equivalent to more than 6% of the basic screw diameter or 0.010 in., whichever is greater. (Eccentricity is defined as one-half of the full or total indicator reading.)

Length

Measurement—The length of headless set screws shall be measured overall, parallel to the axis of screw.

Tolerance on Length—The tolerance on length for slotted headless set screws shall be as tabulated below:

Nominal Screw Length, in	Tolerance on Length, in
Up to 1, incl.	−0.03
Over 1 to 2, incl.	−0.06
Over 2	−0.09

Threads—The threads on slotted headless set screws shall be Unified Standard, Class 2A; UNC and UNF Series or UNRC and UNRF Series, at option of manufacturer. For threads with additive finish, the maximum diameters of Class 2A may be exceeded by the amount of the allowance; that is, the Class 2A maximum diameters shall apply to an unplated or uncoated part or to a part before plating or coating, whereas the basic diameters (Class 2A maximum diameters plus the allowance) shall apply to a part after plating or coating. The minimum major diameter of plated or coated screws may approach but shall not be less than the Class 2A minimum limit.

Inasmuch as standard thread gages provide only for lengths of engagement up to $1\frac{1}{2}$ times the basic screw diameter, changes in the pitch diameter of either or both the external and mating internal thread may be required for applications involving longer lengths of engagement.

Thread Lengths—Slotted headless set screws shall have complete (full form) threads extending over that portion of the screw length which is not affected by the point or the crown on headless end. Threads through angular or crowned portions of length shall have fully formed roots with partial crests.

Points

Point Types—Unless otherwise specified, slotted headless set screws shall be supplied with cup points. Where so specified by purchaser, screws shall have cone, dog, half-dog, flat, or oval points conforming with specifications in Table 1.

Point Angles—The external point angles specified shall apply only to those portions of the point which lie below the thread root diameter, it being recognized that angles within the thread profile may vary due to manufacturing practices.

Dog Points—Dog points are not supplied on screws where the length of usable (effective) thread is less than the basic screw diameter. Half-dog points should be specified for such screw lengths.

The permissible eccentricity of the axis of dog and half-dog points with respect to the axis of thread shall be equivalent to 3% of the basic screw diameter for screw sizes up to and including 6 (0.1380) and 0.005 in. for larger screw sizes. (Total runout shall not exceed twice the permissible eccentricity.)

Materials

Steel—Slotted headless set screws shall normally be made from carbon steel and hardened in accordance with manufacturer's practices.

Other Materials—Where specified, screws may be made from corrosion resistant steel, or nonferrous materials, as mutually agreed upon by the manufacturer and purchaser.

Finish—Unless otherwise specified, set screws shall be supplied with a natural (as processed) finish, unplated or uncoated.

Workmanship—Slotted headless set screws shall be free from burrs, seams, laps, loose scale, and any other defects affecting their serviceability.

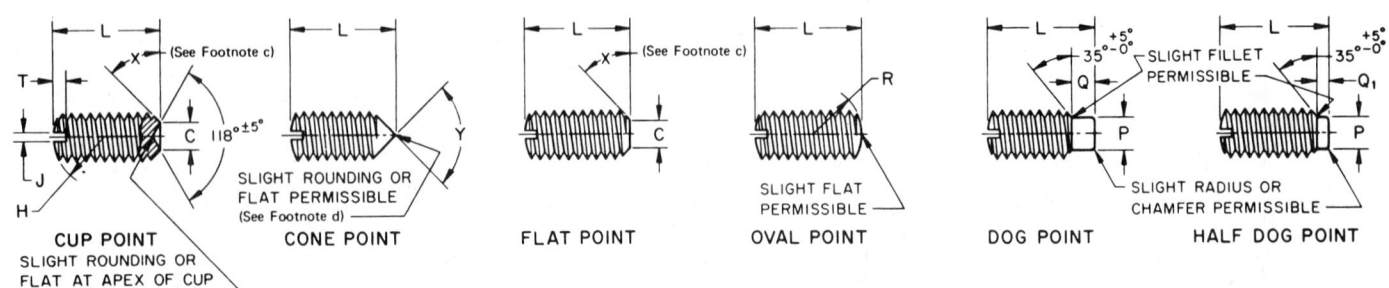

TABLE 1—DIMENSIONS OF SLOTTED HEADLESS SETSCREWS, IN

Nominal Size[a] or Basic Screw Dia		H End Crown Radius[b] Basic	J Slot Width		T Slot Depth		C Cup and Flat Point Dia		Full Dog and Half Dog Point						R Oval Point Radius[b] Basic	Y Cone Point Angle 90 ±2 deg for These Nominal Lengths or Longer; 118 ±2 deg for Shorter Screws
									P Dia		Q Length of Full Dog		Q₁ Length of Half Dog			
			Max	Min	Max	Min	Max	Min	Max	Min	Max	Min	Max	Min		
0	0.0600	0.060	0.014	0.010	0.020	0.016	0.033	0.027	0.040	0.037	0.032	0.028	0.017	0.013	0.045	5/64
1	0.0730	0.073	0.016	0.012	0.020	0.016	0.040	0.033	0.049	0.045	0.040	0.036	0.021	0.017	0.055	3/32
2	0.0860	0.086	0.018	0.014	0.025	0.019	0.047	0.039	0.057	0.053	0.046	0.042	0.024	0.020	0.064	7/64
3	0.0990	0.099	0.020	0.016	0.028	0.022	0.054	0.045	0.066	0.062	0.052	0.048	0.027	0.023	0.074	1/8
4	0.1120	0.112	0.024	0.018	0.031	0.025	0.061	0.051	0.075	0.070	0.058	0.054	0.030	0.026	0.084	5/32
5	0.1250	0.125	0.026	0.020	0.036	0.026	0.067	0.057	0.083	0.078	0.063	0.057	0.033	0.027	0.094	3/16
6	0.1380	0.138	0.028	0.022	0.040	0.030	0.074	0.064	0.092	0.087	0.073	0.067	0.038	0.032	0.104	3/16
8	0.1640	0.164	0.032	0.026	0.046	0.036	0.087	0.076	0.109	0.103	0.083	0.077	0.043	0.037	0.123	1/4
10	0.1900	0.190	0.035	0.029	0.053	0.043	0.102	0.088	0.127	0.120	0.095	0.085	0.050	0.040	0.142	1/4
12	0.2160	0.216	0.042	0.035	0.061	0.051	0.115	0.101	0.144	0.137	0.115	0.105	0.060	0.050	0.162	5/16
1/4	0.2500	0.250	0.049	0.041	0.068	0.058	0.132	0.118	0.156	0.149	0.130	0.120	0.068	0.058	0.188	5/16
5/16	0.3125	0.312	0.055	0.047	0.083	0.073	0.172	0.156	0.203	0.195	0.161	0.151	0.083	0.073	0.234	3/8
3/8	0.3750	0.375	0.068	0.060	0.099	0.089	0.212	0.194	0.250	0.241	0.193	0.183	0.099	0.089	0.281	7/16
7/16	0.4375	0.438	0.076	0.068	0.114	0.104	0.252	0.232	0.297	0.287	0.224	0.214	0.114	0.104	0.328	1/2
1/2	0.5000	0.500	0.086	0.076	0.130	0.120	0.291	0.270	0.344	0.334	0.255	0.245	0.130	0.120	0.375	9/16
9/16	0.5625	0.562	0.096	0.086	0.146	0.136	0.332	0.309	0.391	0.379	0.287	0.275	0.146	0.134	0.422	5/8
5/8	0.6250	0.625	0.107	0.097	0.161	0.151	0.371	0.347	0.469	0.456	0.321	0.305	0.164	0.148	0.469	3/4
3/4	0.7500	0.750	0.134	0.124	0.193	0.183	0.450	0.425	0.562	0.549	0.383	0.367	0.196	0.180	0.562	7/8

[a] Where specifying nominal size in decimals, zeros preceding decimal and in fourth decimal place shall be omitted.
[b] Tolerance on radius for nominal sizes up to and including 5 (0.125 in) shall be +0.015, −0.000; and for larger sizes +0.031, −0.000. Slotted ends on screws may be flat at option of manufacturer.
[c] Point angle X shall be 45, +5, −0 deg for screws of nominal lengths equal to or greater than those listed under Column Y; and 30 deg minimum for screws of shorter nominal lengths.
[d] The extent of rounding or flat at apex of cone point shall not exceed an amount equivalent to 10% of the basic screw diameter.
For dimensions and tolerances not specified above, see General Specifications.

SQUARE HEAD SETSCREWS—SAE J102

SAE Standard

Report of Screw Threads Division approved June 1911, revised by Screw Threads Committee January 1957, and last revised by Fasteners Committee June 1969. Conforms with proposed revision of USA Standard, USAS B18.6.2.

Scope—Included herein are complete general and dimensional data for square head setscrews.

The inclusion of dimensional data in this standard is not intended to imply that all of the products described are stock production sizes. Users are requested to consult with manufacturers concerning lists of stock production sizes.

Dimensions—All dimensions in this standard are in inches unless otherwise stated.

Options—Options, where specified, shall be at the discretion of the manufacturers, unless otherwise agreed by manufacturer and user.

Heads

Head Height—The head height specified in the dimensional tables is the overall distance, measured parallel to the axis of screw, from the top of the head to the intersection of the side of the head with the under surface.

Head Taper—No transverse section through the head between 25 and 75% of the actual head height shall be less than the minimum width across flats. The maximum width across flats shall not be exceeded except for milled-from-bar nonferrous screws where the maximum (basic) width may conform with the commercial tolerances of the bar stock material.

Head Concentricity—The axis of the head shall be concentric with the axis of the screw (determined by one diameter length of shank under the head) within a tolerance equal to 3% (6% FIR) of the basic screw diameter.

Length

Measurement—The length of screw shall be measured, parallel to the axis of the screw, from the intersection of the side of the head with the undersurface to the extreme point.

Tolerance on Length—The tolerance on length shall be as tabulated below:

	Nominal Screw Size	
Nominal Screw Length	5/8 and Smaller	Over 5/8
	Tolerance on Length	
Up to 1, incl	−0.03	−0.06
Over 1 to 2, incl	−0.06	−0.12
Over 2	−0.09	−0.18

Threads

Form and Tolerance—Threads shall be Unified Standard Class 2A. For external threads with additive finish, the maximum diameters of Class 2A may be exceeded by the amount of the allowance; that is, the Class 2A maximum diameters shall apply to an unplated part or to a part before plating, whereas the basic diameters (Class 2A maximum diameters plus the allowance) shall apply to a part after plating.

Long Engagement—Since standard gages provide only for lengths of engagement up to 1½ diameters, changes in pitch diameter of either or both the external and internal thread may be required for applications involving longer lengths of engagement.

Series—Threads shall be coarse (UNC), fine (UNF), or 8-thread (8UN) series. Sizes ¼ in. and larger are normally stocked in coarse thread series only.

Thread Length—Screws shall have complete (full form) threads extending over that portion of the screw length which is not affected by the point, into the neck relief, to the conical underside of head or to within one thread (as measured with a ring thread gage) from the flat underside of the head. Threads through angular or crowned portions of points shall have fully formed roots with partial crests.

Points

Point Types—Unless otherwise specified, square head setscrews shall be supplied with cup points. Where so specified by the user, screws shall have cone, dog, half dog, flat, or oval points conforming to the dimensions tabulated.

Point Angles—The external point angles specified shall apply only to those portions of the angles which lie below the thread root diameter, it being recognized the angle within the thread profile may be varied due to manufacturing processes.

Oval Point Radius—The tolerance on radius on oval points shall be +0.031 in.

Dog Points—Dog points are not supplied on screws where the usable (effective) length of thread is less than the basic screw diameter. Half dog points shall apply.

The axis of dog and half dog points shall be concentric with the axis of the screw within a tolerance equal to 3% (6% FIR) of the basic screw diameter or 0.005 in. (0.010 in. FIR), whichever is the smaller.

Materials

Steel—Square head setscrews shall be made from alloy or carbon steel, suitably hardened.

Other Materials—Where specified, screws may be made from brass, corrosion resisting steel, or other materials as agreed upon by the manufacturer and user.

Finish—Unless otherwise specified, setscrews shall be supplied with natural, as-processed finish (unplated or uncoated).

Defects—Square head setscrews shall be free from burrs, seams, laps, loose scale, and any other defects that affect serviceability.

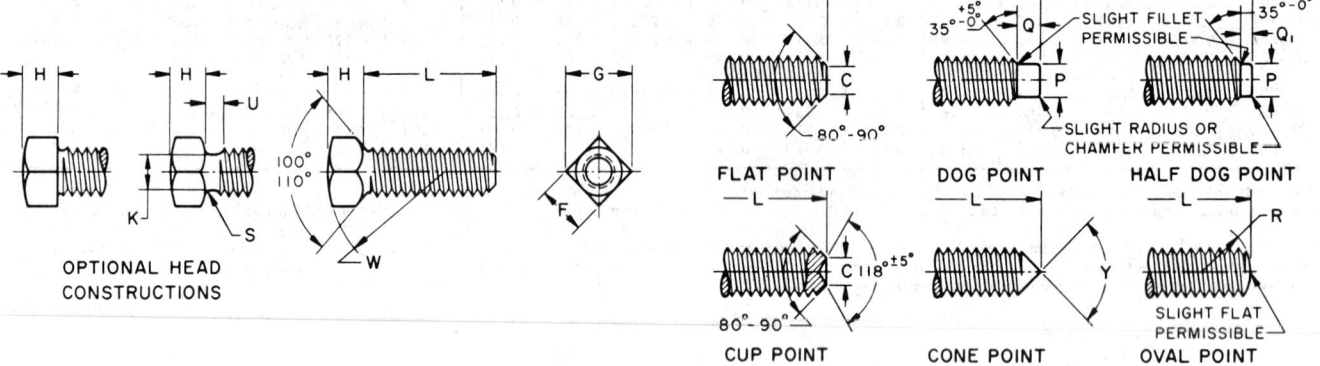

TABLE 1—DIMENSIONS OF SQUARE HEAD SETSCREWS

Nominal Size[a] or Basic Screw Diameter	Width Across Flats, F		Width Across Corners, G		Height of Head, H		Dia of Neck Relief, K		Radius of Head, W	Radius of Neck Relief, S		Width of Neck Relief, U		Cup and Flat Point Dia, C		Oval Point Radius, R	Full Dog and Half Dog Point					
																	Dia, P		Length Full Dog, Q		Length Half Dog, Q₁	
	Max	Min	Max	Min	Max	Min	Max	Min	Nom	Max	Min	Max	Min	Max	Min	Nom	Max	Min	Min	Max	Min	Max
No. 10	0.1900	0.1875	0.180	0.265	0.247	0.148	0.134	0.145	0.140	0.484	0.027	0.083	0.102	0.088	0.141	0.127	0.120	0.085	0.095	0.040	0.050		
No. 12	0.2160	0.2160	0.208	0.305	0.292	0.163	0.147	0.162	0.156	0.547	0.029	0.091	0.115	0.101	0.156	0.144	0.137	0.105	0.115	0.050	0.060		
1/4	0.2500	0.2500	0.241	0.354	0.331	0.196	0.178	0.185	0.170	0.625	0.032	0.100	0.132	0.118	0.188	0.156	0.149	0.120	0.130	0.058	0.068		
5/16	0.3125	0.3125	0.302	0.442	0.415	0.245	0.224	0.240	0.225	0.781	0.036	0.111	0.172	0.156	0.234	0.203	0.195	0.151	0.161	0.083	0.073		
3/8	0.3750	0.3750	0.362	0.530	0.497	0.293	0.270	0.294	0.279	0.938	0.041	0.125	0.212	0.194	0.281	0.250	0.241	0.183	0.193	0.089	0.099		
7/16	0.4375	0.4375	0.423	0.619	0.581	0.341	0.315	0.345	0.330	1.094	0.046	0.143	0.252	0.232	0.328	0.297	0.287	0.214	0.224	0.104	0.114		
1/2	0.5000	0.5000	0.484	0.707	0.665	0.389	0.361	0.400	0.385	1.250	0.050	0.154	0.291	0.270	0.375	0.344	0.334	0.245	0.255	0.120	0.130		
9/16	0.5625	0.5625	0.545	0.795	0.748	0.437	0.407	0.454	0.439	1.406	0.054	0.167	0.332	0.309	0.422	0.391	0.379	0.275	0.287	0.134	0.146		
5/8	0.6250	0.6250	0.606	0.884	0.833	0.485	0.452	0.507	0.492	1.562	0.059	0.182	0.371	0.347	0.469	0.469	0.456	0.305	0.321	0.148	0.164		
3/4	0.7500	0.7500	0.729	1.060	1.001	0.582	0.544	0.620	0.605	1.875	0.065	0.200	0.450	0.425	0.563	0.563	0.549	0.367	0.383	0.180	0.196		
7/8	0.8750	0.8750	0.852	1.237	1.170	0.678	0.635	0.731	0.716	2.188	0.072	0.222	0.530	0.502	0.656	0.656	0.642	0.430	0.446	0.211	0.227		
1	1.0000	1.0000	0.974	1.414	1.337	0.774	0.726	0.838	0.823	2.500	0.081	0.250	0.609	0.579	0.750	0.750	0.734	0.490	0.510	0.240	0.260		
1-1/8	1.1250	1.1250	1.096	1.591	1.505	0.870	0.817	0.939	0.914	2.812	0.092	0.283	0.689	0.655	0.844	0.844	0.826	0.562	0.572	0.271	0.291		
1-1/4	1.2500	1.2500	1.219	1.768	1.674	0.966	0.908	1.064	1.039	3.125	0.092	0.283	0.767	0.733	0.938	0.938	0.920	0.615	0.635	0.303	0.323		
1-3/8	1.3750	1.3750	1.342	1.944	1.843	1.063	1.000	1.159	1.134	3.438	0.109	0.333	0.848	0.808	1.031	1.031	1.011	0.678	0.698	0.334	0.354		
1-1/2	1.5000	1.5000	1.464	2.121	2.010	1.159	1.091	1.284	1.259	3.750	0.109	0.333	0.926	0.886	1.125	1.125	1.105	0.740	0.760	0.365	0.385		

[a] Where specifying nominal size in decimals, zeros in the fourth decimal place shall be omitted.
Where nominal screw length is equal to or less than basic screw diameter, cone point angle Y = 118 ± 2 deg. For longer lengths, Y = 90 ± 2 deg.
Additional requirements given in the General Specifications shall apply.

FLANGED 12-POINT SCREWS—SAE J58

SAE Standard

Report of Fasteners Committee approved September 1972.

GENERAL SPECIFICATIONS

Scope—Included herein are the detailed general and dimensional specifications applicable to flanged 12-point screws recognized as SAE Standard and intended for general use in automotive and other ground based vehicles and industrial equipment. Also included is an appendix covering runout sleeve gages and gaging.

The inclusion of dimensional data in this standard is not intended to imply that all of the products described are stock production sizes. Consumers should consult manufacturers concerning availability of product.

Dimensions—All dimensions in this standard are in inches unless otherwise stated.

Options—Options, where specified, shall be at the discretion of the manufacturer unless otherwise agreed by manufacturer and purchaser.

Heads

Head Height—The head height shall be measured, parallel to the axis of screw, from the top of the head to the bearing surface of the flange.

Top of Head—The top of head may be full form or indented at the option of the manufacturer. If full form, the top of head shall be chamfered or rounded with the diameter of chamfer circle or start of rounding being equal to the specified maximum width across flats within a tolerance of −15%. If the top of head is indented, the periphery may be rounded.

Corner Fill—The rounding due to lack of fill at all 12 corners of the head shall be reasonably uniform and the width across corners of the head shall be such that when a sharp ring, having an inside diameter equal to the T dimension and a thickness within the limits for A specified in Table 1, is placed on the top of the head, normal to the screw axis, the head may enter but shall not protrude through the gage.

Wrenching Height—The wrenching height shall be measured, parallel to the axis of screw, from the intersection of the top contour of flange with any corner of head to the top of the head.

Bearing Surface—The outer periphery may be rounded or chamfered to the extent permitted by X, as measured on the bearing face. The plane of the bearing face shall be perpendicular to the axis of screw within the runout limit specified in Table 1. Measurement of runout shall be made as close to the periphery of bearing face as possible while the screw is held in a collet or other gripping device at a distance equal to one screw diameter from the underside of the head.

Underhead Fillets—For all lengths of screws, the form of the fillet shall be

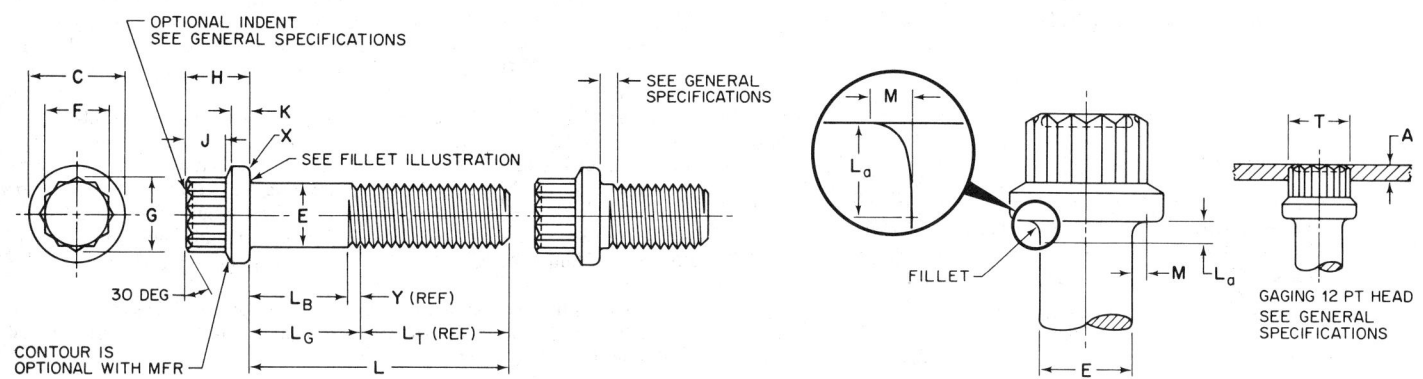

TABLE 1—DIMENSIONS OF FLANGED 12-POINT SCREWS AND HEAD GAGING RINGS

Nominal Size or Basic Screw Dia		E Body Dia Min (Max Equal to Basic Screw Dia)	C Flange Dia		F Width Across Flats		G Width Across Corners	H Head Height	J Wrenching Height	K Flange Thickness	Runout of Bearing Surface FIR
			Max	Min	Max	Min	Min	Max	Min	Min	Max
1/4	0.2500	0.2435	0.375	0.365	0.252	0.244	0.278	0.260	0.15	0.058	0.007
5/16	0.3125	0.3053	0.469	0.457	0.315	0.306	0.348	0.312	0.18	0.074	0.008
3/8	0.3750	0.3678	0.562	0.550	0.377	0.368	0.420	0.375	0.21	0.095	0.010
7/16	0.4375	0.4294	0.656	0.642	0.438	0.429	0.489	0.438	0.26	0.109	0.011
1/2	0.5000	0.4919	0.750	0.735	0.502	0.493	0.562	0.500	0.29	0.129	0.013
9/16	0.5625	0.5538	0.844	0.828	0.564	0.555	0.633	0.563	0.33	0.145	0.015
5/8	0.6250	0.6163	0.938	0.921	0.627	0.618	0.705	0.625	0.36	0.166	0.016
3/4	0.7500	0.7406	1.125	1.107	0.752	0.743	0.847	0.750	0.44	0.200	0.020
7/8	0.8750	0.8647	1.312	1.293	0.877	0.866	0.987	0.875	0.51	0.234	0.023
1	1.0000	0.9886	1.500	1.479	1.003	0.991	1.130	1.000	0.60	0.268	0.026
1-1/8	1.1250	1.1086	1.688	1.665	1.128	1.115	1.271	1.125	0.66	0.310	0.029
1-1/4	1.2500	1.2336	1.875	1.852	1.253	1.240	1.414	1.250	0.73	0.350	0.033
1-3/8	1.3750	1.3568	2.062	2.038	1.378	1.365	1.556	1.375	0.80	0.392	0.036
1-1/2	1.5000	1.4818	2.250	2.224	1.503	1.489	1.697	1.500	0.87	0.433	0.039

Nominal Size[a] or Basic Screw Dia		M Fillet Extension		L_a Fillet Length		Bearing Surface Juncture Radius	X Chamfer or Radius	A Gaging Ring Thickness		T Gaging Ring Dia		L_T Thread Length	Y Transition Thread Length
		Max	Min	Max	Min	Max	Max	Max	Min	Max	Min	Basic	Max
1/4	0.2500	0.014	0.009	0.087	0.007	0.020	0.020	0.0525	0.0522	0.2783	0.2780	1.000	0.25
5/16	0.3125	0.017	0.012	0.087	0.009	0.020	0.020	0.0600	0.0597	0.3483	0.3480	1.125	0.28
3/8	0.3750	0.020	0.015	0.087	0.012	0.020	0.020	0.0711	0.0708	0.4203	0.4200	1.250	0.31
7/16	0.4375	0.023	0.018	0.087	0.014	0.030	0.030	0.0840	0.0837	0.4893	0.4890	1.375	0.36
1/2	0.5000	0.026	0.020	0.087	0.016	0.030	0.030	0.0948	0.0945	0.5623	0.5620	1.500	0.38
9/16	0.5625	0.029	0.022	0.157	0.018	0.030	0.030	0.1071	0.1068	0.6333	0.6330	1.625	0.42
5/8	0.6250	0.032	0.024	0.157	0.021	0.040	0.040	0.1179	0.1176	0.7053	0.7050	1.750	0.46
3/4	0.7500	0.039	0.030	0.157	0.025	0.040	0.040	0.1416	0.1413	0.8473	0.8470	2.000	0.50
7/8	0.8750	0.044	0.034	0.227	0.031	0.040	0.040	0.1656	0.1653	0.9873	0.9870	2.250	0.56
1	1.0000	0.050	0.040	0.332	0.034	0.040	0.040	0.1893	0.1890	1.1303	1.1300	2.500	0.62
1-1/8	1.1250	0.055	0.045	0.332	0.039	0.050	0.050	0.2109	0.2106	1.2713	1.2710	2.750	0.71
1-1/4	1.2500	0.060	0.050	0.332	0.044	0.050	0.050	0.2331	0.2328	1.4143	1.4140	3.000	0.71
1-3/8	1.3750	0.065	0.055	0.332	0.048	0.050	0.050	0.2544	0.2541	1.5563	1.5560	3.250	0.83
1-1/2	1.5000	0.070	0.060	0.332	0.052	0.050	0.050	0.2763	0.2760	1.6973	1.6970	3.500	0.83

[a] Where specifying nominal size in decimals, zeros preceding decimal and in fourth decimal place shall be omitted.

Additional requirements given in the General Specifications shall apply.

optional provided: it is tangent to the shank of the screw at a distance no greater than L_a from the underside of the head; it is tangent to the bearing surface within the limits of the basic screw diameter plus M max, and E min plus M min; and it is a smooth continuous curve having a bearing surface juncture radius no less than that specified in Table 1.

For reduced diameter body screws threaded full length, the minimum fillet extension, M min, may be reduced by an amount equivalent to one-half the difference between the basic screw diameter and the specified minimum pitch diameter of the thread.

Length

Measurement—The length of screw shall be measured, parallel to the axis of screw, from the bearing surface of the head to the extreme end of the shank.

Tolerance on Length—The tolerance on length shall be as tabulated below:

Nominal Screw Size	0 thru 3/8	7/16 thru 3/4	7/8 thru 1-1/2
Nominal Screw Length, in		**Tolerance on Length**	
Up to 1, incl	−0.03	−0.03	−0.05
Over 1 to 2-1/2, incl	−0.04	−0.06	−0.10
Over 2-1/2 to 6, incl	−0.06	−0.08	−0.14
Over 6	−0.12	−0.12	−0.20

Body Diameter—On screws threaded full length, the diameter of the screw under the head shall not be less than the specified minimum pitch diameter of the thread.

Threads—Threads, when produced by roll threading, shall be Unified coarse or fine thread series UNRC or UNRF, Class 2A or Class 3A as specified by purchaser. Threads produced by other methods shall preferably be UNRC or UNRF series, but at the option of the manufacturer may be UNC or UNF series, Class 2A or Class 3A as specified by purchaser.

For threads with additive finish, the maximum diameters of Class 2A may be exceeded by the amount of the allowance, that is, the Class 2A maximum diameters apply to an unplated or uncoated part or to a part before plating or coating, whereas the basic diameters (Class 2A maximum diameters plus the allowance) apply to a part after plating or coating. The maximum diameters of Class 3A threads apply to screws with or without additive finish.

Length of Thread—The length of thread on screws shall be controlled by the grip gaging length L_G max and the body length L_B min as set forth in the following:

Grip Gaging Length—L_G max is the distance, measured parallel to the axis of screw, from the underside of head to the face of a noncounterbored or nonchamfered standard GO thread ring gage assembled by hand as far as the thread will permit. It represents the minimum design grip length and shall be used as the criterion for inspection and for determining thread availability when selecting screw lengths even though usable threads may extend beyond this point. Values for L_G max applicable to common nominal screw lengths are shown in Table 2.

For screws having nominal lengths which fall between those tabulated in Table 2, the L_G max value shown for the next shorter tabulated nominal length and respective screw size shall apply.

TABLE 2—MAXIMUM GRIP GAGING LENGTHS (L_G MAX) FOR FLANGED 12-POINT SCREWS

Nominal Length[a]	Nominal Size													
	1/4	5/16	3/8	7/16	1/2	9/16	5/8	3/4	7/8	1	1-1/8	1-1/4	1-3/8	1-1/2
1-1/2	0.500	—	—	—	—	—	—	—	—	—	—	—	—	—
1-3/4	0.500	0.625	0.500	—	—	—	—	—	—	—	—	—	—	—
2	1.000	0.625	0.500	0.625	—	—	—	—	—	—	—	—	—	—
2-1/4	1.000	1.125	1.000	0.625	0.750	0.750	—	—	—	—	—	—	—	—
2-1/2	1.500	1.125	1.000	1.125	0.750	0.750	0.750	—	—	—	—	—	—	—
2-3/4	1.500	1.625	1.500	1.125	0.750	0.750	0.750	—	—	—	—	—	—	—
3	2.000	1.625	1.500	1.625	1.500	1.500	0.750	1.000	—	—	—	—	—	—
3-1/4	2.000	2.125	2.000	1.625	1.500	1.500	1.500	1.000	1.000	—	—	—	—	—
3-1/2	2.500	2.125	2.000	2.125	1.500	1.500	1.500	1.000	1.000	1.000	—	—	—	—
3-3/4	2.500	2.625	2.500	2.125	2.250	2.250	1.500	1.000	1.000	1.000	1.000	—	—	—
4	3.000	2.625	2.500	2.625	2.250	2.250	2.250	2.000	1.000	1.000	1.000	1.000	—	—
4-1/4	3.000	3.125	3.000	2.625	2.250	2.250	2.250	2.000	2.000	1.000	1.000	1.000	1.000	—
4-1/2	3.500	3.125	3.000	3.125	3.000	3.000	2.250	2.000	2.000	2.000	1.000	1.000	1.000	1.000
4-3/4	3.500	3.625	3.500	3.125	3.000	3.000	3.000	2.000	2.000	2.000	1.000	1.000	1.000	1.000
5	4.000	3.625	3.500	3.625	3.000	3.000	3.000	3.000	2.000	2.000	2.000	2.000	1.000	1.000
5-1/4	—	4.125	4.000	3.625	3.750	3.750	3.000	3.000	3.000	2.000	2.000	2.000	2.000	1.000
5-1/2	—	4.125	4.000	4.125	3.750	3.750	3.750	3.000	3.000	3.000	2.000	2.000	2.000	2.000
5-3/4	—	4.625	4.500	4.125	3.750	3.750	3.750	3.000	3.000	3.000	2.000	2.000	2.000	2.000
6	—	4.625	4.500	4.625	4.500	4.500	3.750	4.000	3.000	3.000	3.000	2.000	2.000	2.000
6-1/4	—	5.125	5.000	4.625	4.500	4.500	4.500	4.000	4.000	3.000	3.000	3.000	3.000	2.000
6-1/2	—	5.125	5.000	5.125	4.500	4.500	4.500	4.000	4.000	4.000	3.000	3.000	3.000	3.000
6-3/4	—	—	5.500	5.125	5.250	5.250	4.500	4.000	4.000	4.000	4.000	3.000	3.000	3.000
7	—	—	5.500	5.625	5.250	5.250	5.250	5.000	4.000	4.000	4.000	4.000	3.000	3.000
7-1/4	—	—	6.000	5.625	5.250	5.250	5.250	5.000	5.000	4.000	4.000	4.000	4.000	3.000
7-1/2	—	—	—	6.125	6.000	6.000	5.250	5.000	5.000	5.000	4.000	4.000	4.000	4.000
7-3/4	—	—	—	6.125	6.000	6.000	6.000	5.000	5.000	5.000	5.000	4.000	4.000	4.000
8	—	—	—	6.625	6.000	6.750	6.000	6.000	5.000	5.000	5.000	5.000	4.000	4.000
8-1/2	—	—	—	7.125	7.000	6.750	6.750	6.000	6.000	6.000	5.000	5.000	5.000	4.000
9	—	—	—	7.625	7.000	7.750	6.750	7.000	6.000	6.000	6.000	5.000	5.000	5.000
9-1/2	—	—	—	—	8.000	7.750	7.750	7.000	7.000	6.000	6.000	6.000	5.000	5.000
10	—	—	—	—	8.000	9.25	7.750	8.000	7.000	7.000	6.000	6.000	6.000	5.000
11	—	—	—	—	—	10.25	9.25	9.000	8.000	8.000	7.000	7.000	6.000	6.000
12	—	—	—	—	—	—	10.25	10.000	9.000	9.000	8.000	7.000	7.000	6.000
13	—	—	—	—	—	—	—	11.000	10.000	10.000	9.000	8.000	7.000	7.000
14	—	—	—	—	—	—	—	12.000	11.000	11.000	10.000	9.000	8.000	7.000
15	—	—	—	—	—	—	—	13.000	12.000	12.000	11.000	10.000	9.000	8.000
16	—	—	—	—	—	—	—	—	13.000	13.000	12.000	11.000	10.000	9.000
17	—	—	—	—	—	—	—	—	14.000	14.000	13.000	12.000	11.000	10.000
18	—	—	—	—	—	—	—	—	15.000	15.000	14.000	13.000	12.000	11.000
19	—	—	—	—	—	—	—	—	—	16.000	15.000	14.000	13.000	12.000
20	—	—	—	—	—	—	—	—	—	17.000	16.000	15.000	14.000	13.000

[a] For nominal screw lengths falling between tabulated lengths, and nominal lengths shorter or longer than tabulated lengths, see Thread Length in the General Specifications.

For screws having nominal lengths longer than those specified in Table 2 for the respective screw size, the L_G max value shall be determined by subtracting the L_T value in Table 1 from the nominal screw length.

For screws having shorter nominal lengths than those specified in Table 2 for the respective screw size, the complete (full form) thread, as measured with a thread ring gage, shall extend to within two threads (pitches) of the bearing face for sizes up to and including $\frac{5}{8}$ in., and as close to the head as practicable for larger sizes.

Body Length—L_B min is the distance, measured parallel to the axis of screw, from the underside of the head to the last scratch of thread or the top of extrusion angle. The minimum body length for any screw length shall be equal to the maximum grip length minus the transition thread length (L_B min = L_G max − Y max). It shall be used as a criterion for inspection.

Basic Thread Length—L_T is a reference dimension, intended for calculation purposes only, which represents the distance from the extreme end of the screw to the last complete (full form) thread.

Transition Thread Length—Y max is a reference dimension, intended for calculation purposes only, which represents the length of incomplete threads and the tolerance on grip length.

Point—The point shall be flat and chamfered from a diameter approximately 0.016 in. below the minor diameter of the thread to produce a length of point equivalent to $\frac{1}{2}$ to $1\frac{1}{2}$ threads (pitches).

Incomplete Thread Diameter—The major diameter of incomplete thread shall not exceed the actual major diameter of the full form thread.

Thread Runout—The total runout between thread, body, and head of screws shall be such that screw may be assembled for a minimum of two full turns into the threaded hole of a concentricity sleeve gage as shown in the Appendix.

Materials

Steel—Suitable properties for steel screws are covered in SAE J429, Grades 2, 5, and 8, as specified by purchaser.

Other Materials—Where specified, screws may be made from brass, corrosion resisting steel, or other materials as agreed upon by the manufacturer and purchaser.

Finish—Unless otherwise specified, screws shall be supplied plain (unplated or uncoated), as processed.

Workmanship—Screws shall be free from burrs, seams, laps, loose scale, and any defects that affect serviceability.

APPENDIX—THREAD RUNOUT SLEEVE GAGE

A gage capable of checking the head and thread eccentricity along with the bow of shank on screws is illustrated in Fig. A-1.

The gage construction incorporates three primary components: ring gage A, shank sleeve B_2, and head sleeve B_1. Shank sleeve length can be varied to accommodate different lengths of product and also diameters D_2 and D_1 can be varied depending upon the screw diameter and head size, respectively. Ring gage A and head sleeve B_1 are centered on the sleeve B_2 by means of the positioning plug shown below the gage proper and are secured in position by means of attachment screws C. The ring gage A is also set to Class 2B or 3B maximum pitch diameter as applicable.

Diameter D_1, of the head sleeve, equals the maximum flange diameter of the screw plus 0.031 in allowance. Diameter D_2, of the shank sleeve, equals the max screw diameter plus the runout allowance given below as a function of product length. The shank sleeve length L_2 should be such that entering face of gage A extends beyond the last thread of the product to be inspected.

Failure of the screw to enter the threads of gage or interference between the sides of hole D_1 or D_2 and the screw while engaging threads of the gage indicates excessive runout.

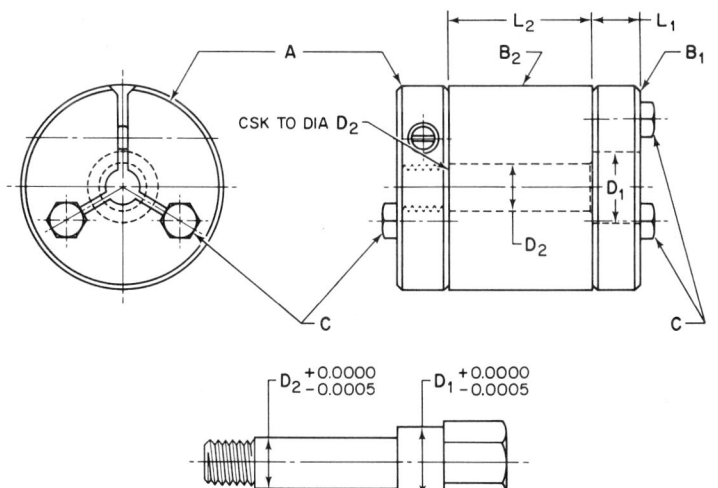

LEGEND
A — THREAD RING GAGE SET TO 2B OR 3B MAX PITCH DIAMETER, AS APPLICABLE
B_1 — HEAD SLEEVE
B_2 — SHANK SLEEVE
C — ATTACHMENT SCREWS
D_1 — DIAMETER OF HOLE IN HEAD SLEEVE = MAX FLANGE DIAMETER OF SCREW PLUS 0.031 IN ALLOWANCE
D_2 — DIAMETER OF HOLE IN SHANK SLEEVE = MAX SCREW DIAMETER PLUS 0.016 IN FOR LENGTHS UP TO 2 IN, INCL; 0.031 IN FOR LENGTHS OVER 2 TO 6 IN, INCL; AND 0.062 IN FOR LENGTHS OVER 6 IN
L_1 — LENGTH OF HEAD SLEEVE = HEAD HEIGHT OF SCREW
L_2 — LENGTH OF SHANK SLEEVE = LENGTH OF SCREW MINUS ONE SCREW DIAMETER

FIG. A-1

LAG SCREWS—SAE J103

SAE Standard

Report of Screw Threads Division approved June 1911 and last revised by Fasteners Committee June 1969. Conforms with USA Standard, USAS B18.2.1.

GENERAL SPECIFICATIONS

Scope—Included herein are complete general and dimensional data for lag screws.

The inclusion of dimensional data in this standard is not intended to imply that all of the products described are stock production sizes. Users are requested to consult with manufacturers concerning lists of stock production sizes.

Dimensions—All dimensions in this standard are in inches unless otherwise stated.

Options—Options, where specified, shall be at the discretion of the manufacturers, unless otherwise agreed by manufacturer and user.

Heads

Head Height—The head height specified in the dimensional tables is the overall distance, measured parallel to the axis of screw, from the top of the head to the bearing surface.

Head Chamfer—The top of head shall be flat with chamfered corners. The diameter of chamfer circle shall be equal to the maximum width across flats within a tolerance of −15%.

Head Taper—No transverse section through the head between 25 and 75% of the actual head height, as measured from the bearing surface, shall be less than the minimum width across flats. The maximum width across flats shall not be exceeded except for milled-from-bar nonferrous screws where the maximum (basic) width may conform with the commercial tolerances of the bar stock material.

Head Concentricity—The axis of the head shall be concentric with the axis of the screw body (determined by one diameter length of body under the head) within a tolerance equal to 3% (6% FIR) of the maximum width across flats.

Bearing Surface—The bearing surface shall be at a right angle to the axis of the body within 3 deg for 1 in. size or smaller, and within 2 deg for larger sizes.

Length

Measurement—The length of the screw shall be measured, parallel to the axis of the screw, from the bearing surface of the head to the extreme end of the screw, including the point.

Tolerance on Length—The tolerance on length of lag screws shall be ±0.125 in. for sizes up to and including ½ in. and lengths up to and including 6 in. For sizes over ½ in. and lengths over 6 in., the length tolerance shall be ±0.250 in.

Threads

$$\text{Pitch} = \frac{1}{\text{No. of threads per inch}}$$

Flat at root = pitch × 0.4305
Depth of single thread = pitch × 0.385

Thread Length

Measurement—For purposes of this standard, thread length shall be the distance, measured parallel to the axis of the screw, from the extreme end of the screw to and including the last complete (full form) thread.

Long Screws—The minimum thread length shall be ½ of the length of screw +0.50 in. or 6.00 in., whichever is shorter.

Short Screws—Screws too short to accommodate the minimum thread length shall be threaded as close to the head or shoulder as practical.

Points—Lag screws shall have gimlet or cone points as illustrated.

Materials

Steel—Suitable material for steel lag screws shall be low carbon steel.

Other Materials—Other materials shall be as agreed upon by the manufacturer and user.

Defects—Lag screws shall be free from burrs, seams, laps, loose scale, and any other defects that affect serviceability.

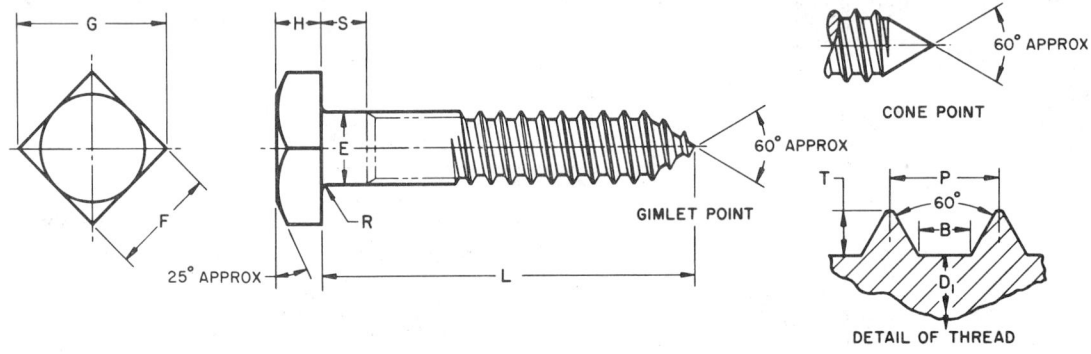

TABLE 1 DIMENSIONS OF LAG SCREWS

Nominal Size[a] or Basic Screw Diameter	Body or Shoulder Dia, E		Width Across Flats, F			Width Across Corners, G		Height, H			Length of Shoulder, S	Radius of Fillet, R	Threads per in.	Thread Dimensions			
	Max[b]	Min[c]	Basic	Max	Min	Max	Min	Basic	Max	Min	Min	Max		Pitch, P	Flat at Roof, S	Depth of Thread, T	Roof Dia, D_1
No. 10	0.1900	0.199	0.178	9/32	0.2810	0.271	0.398	0.372	1/8	0.140	0.110	0.094	0.031	11	0.091	0.039	0.035	0.120
1/4	0.2500	0.260	0.237	3/8	0.3750	0.362	0.530	0.498	11/64	0.188	0.156	0.094	0.031	10	0.100	0.043	0.039	0.173
5/16	0.3125	0.324	0.298	1/2	0.5000	0.484	0.707	0.665	13/64	0.220	0.186	0.125	0.031	9	0.111	0.048	0.043	0.227
3/8	0.3750	0.388	0.360	9/16	0.5625	0.544	0.795	0.747	1/4	0.268	0.232	0.125	0.031	7	0.143	0.062	0.055	0.265
7/16	0.4375	0.452	0.421	5/8	0.6250	0.603	0.884	0.828	19/64	0.316	0.278	0.156	0.031	7	0.143	0.062	0.055	0.328
1/2	0.5000	0.515	0.482	3/4	0.7500	0.725	1.061	0.995	21/64	0.348	0.308	0.156	0.031	6	0.167	0.072	0.064	0.371
5/8	0.6250	0.642	0.605	15/16	0.9375	0.906	1.326	1.244	27/64	0.444	0.400	0.312	0.062	5	0.200	0.086	0.077	0.471
3/4	0.7500	0.768	0.729	1-1/8	1.1250	1.088	1.591	1.494	1/2	0.524	0.476	0.375	0.062	4-1/2	0.222	0.096	0.085	0.579
7/8	0.8750	0.895	0.852	1-5/16	1.3125	1.269	1.856	1.742	19/32	0.620	0.568	0.375	0.062	4	0.250	0.108	0.096	0.683
1	1.0000	1.022	0.976	1-1/2	1.5000	1.450	2.121	1.991	21/32	0.684	0.628	0.625	0.093	3-1/2	0.286	0.123	0.110	0.780
1-1/8	1.1250	1.149	1.098	1-11/16	1.6875	1.631	2.386	2.239	3/4	0.780	0.720	0.625	0.093	3-1/4	0.308	0.133	0.119	0.887
1-1/4	1.2500	1.277	1.223	1-7/8	1.8750	1.812	2.652	2.489	27/32	0.876	0.812	0.625	0.093	3-1/4	0.308	0.133	0.119	1.012

[a] Where spec'fying nominal size in decimals, zeros in the fourth decimal place shall be omitted.
[b] There may be a reasonable swell or fin under the head, or die seam on the body, not to exceed the basic screw diameter by 0.030 in. for sizes up to 1/2 in., 0.050 in. for sizes 5/8 and 3/4 in., and 0.063 in. for sizes over 3/4 in. to 1-1/4 in.
[c] Screws in sizes 1/2 in. by 6 in. long, and smaller, are normally stocked in "reduced diameter body" and the body diameter is reduced to the blank diameter before threading. A shoulder of full body diameter under the head shall be provided.

Additional requirements given in the General Specifications shall apply.
For wrench openings applicable to Lag Screws, see Appendix B, SAE J105.

SQUARE AND HEX NUTS—SAE J104

SAE Standard

Report of Screw Threads Division approved June 1911, revised by Screw Threads Committee January 1957, and last revised by Fastener Committee June 1969. Conforms in general, for the products included, with USA Standards, USAS B18.2.2 and USAS B18.6.3.

Scope—Included herein are complete general and dimensional data for the various types of square and hex nuts recognized as SAE Standard and commonly used in automotive and other ground based motor vehicles and industrial equipment. Also included are appendixes covering wrench openings for nuts.

The inclusion of dimensional data in this standard is not intended to imply that all of the products described are stock production sizes. Users are requested to consult with manufacturers concerning lists of stock production sizes.

Dimensions—All dimensions in this standard are in inches unless otherwise stated.

Options—Options, where specified, shall be at the discretion of the manufacturers unless otherwise agreed by manufacturer and user.

CONTENTS

General specifications
Dimensions

Square Machine Screw Nuts	Table 1
Square Nuts	Table 2
Heavy Square Nuts	Table 3
Hex Machine Screw Nuts	Table 4
Hex Flat and Hex Flat Jam Nuts	Table 5
Hex, Hex Jam and Hex Slotted Nuts	Table 6
Hex Thick and Hex Thick Slotted Nuts	Table 7
Hex Castle Nuts	Table 8
Heavy Hex Flat and Flat Jam Nuts	Table 9
Heavy Hex, Hex Jam and Hex Slotted Nuts	Table 10
Wrench Openings for Nuts	Appendix A

GENERAL SPECIFICATIONS

General—The following specifications cover all of the nut categories included in this standard. Specifications which are peculiar to a particular product or products are so indicated by references to the applicable dimensional tables.

Nut Thickness—The nut thickness specified in the dimensional tables is the overall distance measured parallel to the axis of nut, from the top of the nut to the bearing surface and includes the thickness of the washer face where provided.

Taper of Sides—No transverse section through the nut between 25 and 75% of the actual nut thickness as measured from the bearing face shall be less than the minimum width across flats. The maximum width across flats shall not be exceeded except for milled-from-bar nonferrous nuts where the maximum (basic) width may conform with the commercial tolerances of the bar stock material.

Construction

Square Nuts—The square machine screw nuts in Table 1 shall have both top and bottom flat without chamfer.

The square nuts in Table 2 and heavy square nuts in Table 3 shall have the bearing surface flat without chamfer and a flat top, either chamfered or washer crowned. The diameter of the chamfer circle or washer crown shall be equal to the maximum width across flats within a tolerance of −15%.

Hex Nuts—The hex machine screw nuts in Table 4 shall normally have the bearing surface flat without chamfer and a flat chamfered top, but for special purposes, where so specified, may be double chamfered. The diameter of chamfer circle shall be equal to the maximum width across flats within a tolerance of −15%.

The hex flat and heavy hex flat nut products in Tables 5 and 9, respectively, shall have the bearing surface flat without chamfer and the top flat and chamfered. The diameter of the chamfer circle shall be equal to the maximum width across flats within a tolerance of −15%. The length of chamfer at the hex corners shall be 5–15% of the basic thread diameter. The surface of the chamfer may be slightly convex or rounded.

The hex nut products in Tables 6 and 7, for sizes up to and including $5/8$ in., and the heavy hex nut products in Table 10, for sizes up to and including $7/16$ in., shall be double chamfered. Hex castle nuts in Table 8, for sizes up to and including $5/8$ in., shall have a flat bearing face with chamfered corners. Larger sizes of these nuts shall have a washer faced bearing surface with, except castle nuts, a flat chamfered top or be double chamfered. The diameter of the chamfer circle on double chamfered nuts or chamfered castle nuts and the diameter of washer face shall be equal to the maximum width across flats within a tolerance of −5%. The diameter of a chamfer circle on top of washer faced nuts shall be equal to the maximum width across flats within a tolerance of −15%. The length of chamfer at the hex corners shall be 5–15% of the basic thread diameter. The surface of the chamfer may be slightly convex or rounded.

Rounding at Corners

Hex Nuts—Except for machine screw nuts, a rounding or lack of fill at the junction of hex corners with chamfer shall be permissible, provided the minimum width across corners is reached and maintained beyond a distance equal to 17.5% of the basic thread diameter from the chamfered face and the junction of top of hexagon faces with the castle fillet on castle nuts.

Angularity of Bearing Surface

Machine Screw Nuts—The bearing surface of machine screw nuts in Tables 1 and 4 shall be at a right angle to the axis of the tapped hole within a tolerance of 4 deg.

Square Nut Products—The bearing surface of square and heavy square nuts in Tables 2 and 3, respectively, shall be at a right angle to the axis of the tapped hole within a tolerance of 3 deg for 1 in. nuts or smaller, and 2 deg for larger sizes.

Hex Flat Nut Products—The bearing surface of hex flat nut products in Table 5 and heavy hex flat nut products in Table 9 shall be at right angle to the axis of the tapped hole within a tolerance of 2 deg.

Hex Nut Products—The bearing surface of the hex nut products in Tables 6–8 and heavy hex nut products in Table 10 shall be at a right angle to the axis of tapped hole within 2 deg for $5/8$ in. nuts or smaller, and 1 deg for larger sizes.

Total Runout—The maximum total runout of the bearing surface shall equal the tangent of the angular deviation times the distance across flats.

Threads

Form and Tolerance—Threads on all products shall be Unified Standard, Class 2B.

Series—Threads for products covered in Tables 1–5 and 9 shall be coarse thread series (UNC). Threads for products in Tables 6, 7, and 10 shall be coarse (UNC), fine (UNF), or 8-thread (8UN) series. Threads for hex castle nuts in Table 8 shall be limited to coarse (UNC) or fine (UNF) thread series, with fine thread being furnished unless otherwise specified.

Concentricity—On other than machine screw nuts (where no control is established), the axis of the tapped hole shall be concentric with the axis of nut body within a tolerance equal to 5% (10% TIR) of the maximum width across flats for all square nut products and within a tolerance equal to 3% (6% TIR) for all hex nut products.

Countersink—For all hex nut products in Tables 6–8 and 10, the tapped holes shall be countersunk on the bearing face or faces. The maximum countersink diameter shall be the thread basic (nominal) major diameter plus 0.025 in. for $3/8$ size or smaller, and 1.06 times the basic major diameter for larger sizes. No part of the threaded portion shall extend beyond the bearing surface.

Slots—Contour of bottom of slots in slotted nuts and castle nuts shall be at manufacturer's option.

Materials

Steel—Suitable properties for steel nuts are covered in SAE J995.

Other Materials—Other materials shall be as agreed upon by the manufacturer and user.

Finish

Plain—Unless otherwise specified, nuts shall be supplied plain (unplated or uncoated), as processed.

Plated—Where plating is specified, the thickness or quality of plating shall be measured or tested on the side of the nut.

Defects—Nuts shall be free from burrs, seams, laps, loose scale, and any other defects that affect serviceability.

Square Machine Screw Nuts

TABLE 1—DIMENSIONS OF SQUARE MACHINE SCREW NUTS

Nominal Size[a] or Basic Major Dia of Thread		Width Across Flats, F			Width Across Corners, G		Thickness, H		
		Basic	Max	Min	Max	Min	Basic	Max	Min
No. 0	0.0600	5/32	0.156	0.150	0.221	0.206	3/64	0.050	0.043
No. 1	0.0730	5/32	0.156	0.150	0.221	0.206	3/64	0.050	0.043
No. 2	0.0860	3/16	0.187	0.180	0.265	0.247	1/16	0.066	0.057
No. 3	0.0990	3/16	0.187	0.180	0.265	0.247	1/16	0.066	0.057
No. 4	0.1120	1/4	0.250	0.241	0.354	0.331	3/32	0.098	0.087
No. 5	0.1250	5/16	0.312	0.302	0.442	0.415	7/64	0.114	0.102
No. 6	0.1380	5/16	0.312	0.302	0.442	0.415	7/64	0.114	0.102
No. 8	0.1640	11/32	0.344	0.332	0.486	0.456	1/8	0.130	0.117
No. 10	0.1900	3/8	0.375	0.362	0.530	0.497	1/8	0.130	0.117
No. 12	0.2160	7/16	0.437	0.423	0.619	0.581	5/32	0.161	0.148
1/4	0.2500	7/16	0.437	0.423	0.619	0.581	3/16	0.193	0.178
5/16	0.3125	9/16	0.562	0.545	0.795	0.748	7/32	0.225	0.208
3/8	0.3750	5/8	0.625	0.607	0.884	0.833	1/4	0.257	0.239

[a] Where specifying nominal size in decimals, zeros in the fourth decimal place shall be omitted. Additional requirements given in the General Specifications shall apply.

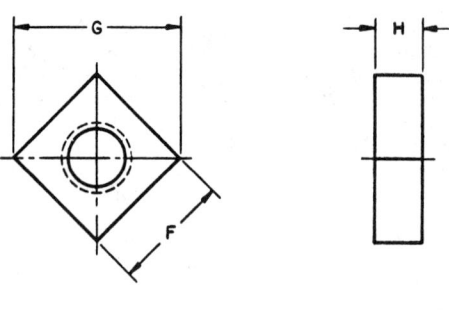

Square Nuts

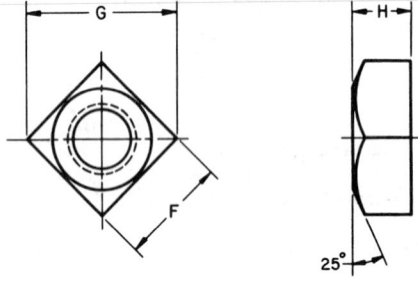

TABLE 2—DIMENSIONS OF SQUARE NUTS

Nominal Size[a] or Basic Major Dia of Thread		Width Across Flats, F			Width Across Corners, G		Thickness, H		
		Basic	Max	Min	Max	Min	Basic	Max	Min
1/4	0.2500	7/16	0.4375	0.425	0.619	0.584	7/32	0.235	0.203
5/16	0.3125	9/16	0.5625	0.547	0.795	0.751	17/64	0.283	0.249
3/8	0.3750	5/8	0.6250	0.606	0.884	0.832	21/64	0.346	0.310
7/16	0.4375	3/4	0.7500	0.728	1.061	1.000	3/8	0.394	0.356
1/2	0.5000	13/16	0.8125	0.788	1.149	1.082	7/16	0.458	0.418
5/8	0.6250	1	1.0000	0.969	1.414	1.330	35/64	0.569	0.525
3/4	0.7500	1-1/8	1.1250	1.088	1.591	1.494	21/32	0.680	0.632
7/8	0.8750	1-5/16	1.3125	1.269	1.856	1.742	49/64	0.792	0.740
1	1.0000	1-1/2	1.5000	1.450	2.121	1.991	7/8	0.903	0.847
1-1/8	1.1250	1-11/16	1.6875	1.631	2.386	2.239	1	1.030	0.970
1-1/4	1.2500	1-7/8	1.8750	1.812	2.652	2.489	1-3/32	1.126	1.062
1-3/8	1.3750	2-1/6	2.0625	1.994	2.917	2.738	1-13/64	1.237	1.169
1-1/2	1.5000	2-1/4	2.2500	2.175	3.182	2.986	1-5/16	1.348	1.276

[a] Where specifying nominal size in decimals, zeros in the fourth decimal place shall be omitted. All sizes are unified dimensionally with British and Canadian Standards. Additional requirements given in the General Specifications shall apply.

Heavy Square Nuts

TABLE 3—DIMENSIONS OF HEAVY SQUARE NUTS

Nominal Size[a] or Basic Major Dia of Thread		Width Across Flats, F			Width Across Corners, G		Thickness, H		
		Basic	Max	Min	Max	Min	Basic	Max	Min
1/4	0.2500	1/2	0.5000	0.488	0.707	0.670	1/4	0.266	0.218
5/16	0.3125	9/16	0.5625	0.546	0.795	0.750	5/16	0.330	0.280
3/8	0.3750	11/16	0.6875	0.669	0.973	0.919	3/8	0.393	0.341
7/16	0.4375	3/4	0.7500	0.728	1.060	1.000	7/16	0.456	0.403
1/2	0.5000	7/8	0.8750	0.850	1.237	1.167	1/2	0.520	0.464
5/8	0.6250	1-1/16	1.0625	1.031	1.503	1.416	5/8	0.647	0.587
3/4	0.7500	1-1/4	1.2500	1.212	1.768	1.665	3/4	0.774	0.710
7/8	0.8750	1-7/16	1.4375	1.394	2.033	1.914	7/8	0.901	0.833
1	1.0000	1-5/8	1.6250	1.575	2.298	2.162	1	1.028	0.956
1-1/8	1.1250	1-13/16	1.8125	1.756	2.563	2.411	1-1/8	1.155	1.079
1-1/4	1.2500	2	2.0000	1.938	2.828	2.661	1-1/4	1.282	1.187
1-3/8	1.3750	2-3/16	2.1875	2.119	3.094	2.909	1-3/8	1.409	1.310
1-1/2	1.5000	2-3/8	2.3750	2.300	3.359	3.158	1-1/2	1.536	1.433

[a] Where specifying nominal size in decimals, zeros in the fourth decimal place shall be omitted. Additional requirements given in the General Specifications shall apply.

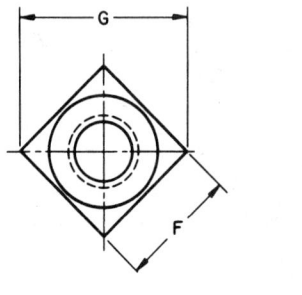

Hex Machine Screw Nuts

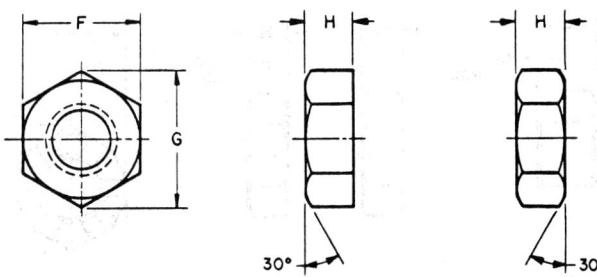

TABLE 4—DIMENSIONS OF HEX MACHINE SCREW NUTS

Nominal Size[a] or Basic Major Dia of Thread		Width Across Flats, F			Width Across Corners, G		Thickness, H		
		Basic	Max	Min	Max	Min	Basic	Max	Min
No. 0	0.0600	5/32	0.156	0.150	0.180	0.171	3/64	0.050	0.043
No. 1	0.0730	5/32	0.156	0.150	0.180	0.171	3/64	0.050	0.043
No. 2	0.0860	3/16	0.187	0.180	0.217	0.205	1/16	0.066	0.057
No. 3	0.0990	3/16	0.187	0.180	0.217	0.205	1/16	0.066	0.057
No. 4	0.1120	1/4	0.250	0.241	0.289	0.275	3/32	0.098	0.087
No. 5	0.1250	5/16	0.312	0.302	0.361	0.344	7/64	0.114	0.102
No. 6	0.1380	5/16	0.312	0.302	0.361	0.344	7/64	0.114	0.102
No. 8	0.1640	11/32	0.344	0.332	0.397	0.378	1/8	0.130	0.117
No. 10	0.1900	3/8	0.375	0.362	0.433	0.413	1/8	0.130	0.117
No. 12	0.2160	7/16	0.437	0.423	0.505	0.482	5/32	0.161	0.148
1/4	0.2500	7/16	0.437	0.423	0.505	0.482	3/16	0.193	0.178
5/16	0.3125	9/16	0.562	0.545	0.650	0.621	7/32	0.225	0.208
3/8	0.3750	5/8	0.625	0.607	0.722	0.692	1/4	0.257	0.239

[a] Where specifying nominal size in decimals, zeros in the fourth decimal place shall be omitted.   Additional requirements given in the General Specifications shall apply.

Hex Flat and Hex Flat Jam Nuts

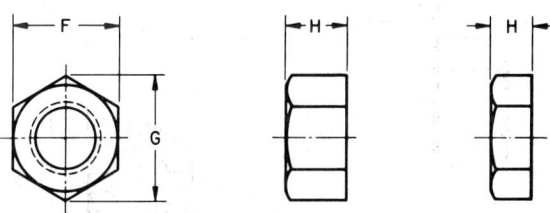

TABLE 5—DIMENSIONS OF HEX FLAT AND HEX FLAT JAM NUTS[a]

Nominal Size[b] or Basic Major Dia of Thread		Width Across Flats, F			Width Across Corners, G		Thickness, Flat Nuts, H			Thickness, Flat Jam Nuts, H_1		
		Basic	Max	Min	Max	Min	Basic	Max	Min	Basic	Max	Min
1-1/8	1.1250	1-11/16	1.6875	1.631	1.949	1.859	1	1.030	0.970	5/8	0.655	0.595
1-1/4	1.2500	1-7/8	1.8750	1.812	2.165	2.066	1-3/32	1.126	1.062	3/4	0.782	0.718
1-3/8	1.3750	2-1/16	2.0625	1.994	2.382	2.273	1-13/64	1.237	1.169	13/16	0.846	0.778
1-1/2	1.5000	2-1/4	2.2500	2.175	2.598	2.480	1-5/16	1.348	1.276	7/8	0.911	0.839

[a] For sizes 1/4 in. through 1 in., see hex nuts and hex jam nuts, Table 6.
[b] Where specifying nominal size in decimals, zeros in the fourth decimal place shall be omitted.

All sizes are unified dimensionally with British and Canadian Standards.
Additional requirements given in the General Specifications shall apply.

Hex, Hex Jam and Hex Slotted Nuts

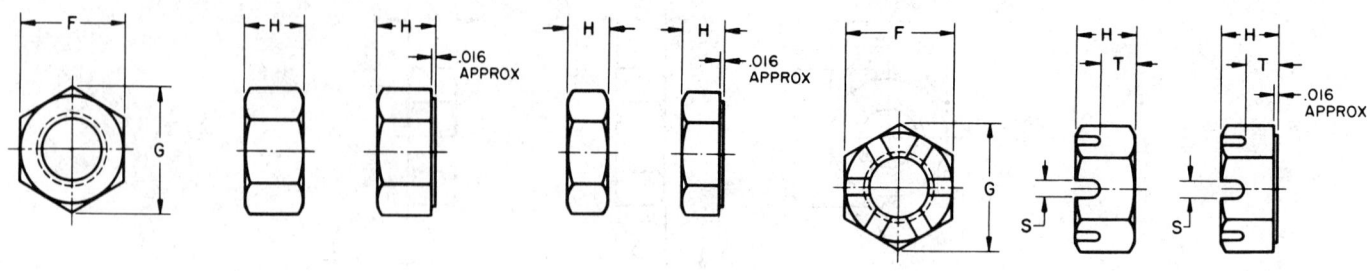

TABLE 6—DIMENSIONS OF HEX, HEX JAM AND HEX SLOTTED NUTS

Nominal Size[a] or Basic Major Dia of Thread		Width Across Flats, F			Width Across Corners, G		Thickness, Nuts, H			Thickness, Jam Nuts, H₁			Slot Width, S		Unslotted Thickness, T	
		Basic	Max	Min	Max	Min	Basic	Max	Min	Basic	Max	Min	Min	Max	Max	Min
1/4	0.2500	7/16	0.4375	0.428	0.505	0.488	7/32	0.226	0.212	5/32	0.163	0.150	0.07	0.10	0.14	0.12
5/16	0.3125	1/2	0.5000	0.489	0.577	0.557	17/64	0.273	0.258	3/16	0.195	0.180	0.09	0.12	0.18	0.16
3/8	0.3750	9/16	0.5625	0.551	0.650	0.628	21/64	0.337	0.320	7/32	0.227	0.210	0.12	0.15	0.21	0.19
7/16	0.4375	11/16	0.6875	0.675	0.794	0.768	3/8	0.385	0.365	1/4	0.260	0.240	0.12	0.15	0.23	0.21
1/2	0.5000	3/4	0.7500	0.736	0.866	0.840	7/16	0.448	0.427	5/16	0.323	0.302	0.15	0.18	0.29	0.27
9/16	0.5625	7/8	0.8750	0.861	1.010	0.982	31/64	0.496	0.473	5/16	0.324	0.301	0.15	0.18	0.31	0.29
5/8	0.6250	15/16	0.9375	0.922	1.083	1.051	35/64	0.559	0.535	3/8	0.387	0.363	0.18	0.24	0.34	0.32
3/4	0.7500	1-1/8	1.1250	1.088	1.299	1.240	41/64	0.665	0.617	27/64	0.446	0.398	0.18	0.24	0.40	0.38
7/8	0.8750	1-5/16	1.3125	1.269	1.516	1.447	3/4	0.776	0.724	31/64	0.510	0.458	0.18	0.24	0.52	0.49
1	1.0000	1-1/2	1.5000	1.450	1.732	1.653	55/64	0.887	0.831	35/64	0.575	0.519	0.24	0.30	0.59	0.56
1-1/8	1.1250	1-11/16	1.6875	1.631	1.949	1.859	31/32	0.999	0.939	39/64	0.639	0.579	0.24	0.33	0.64	0.61
1-1/4	1.2500	1-7/8	1.8750	1.812	2.165	2.066	1-1/16	1.094	1.030	23/32	0.751	0.687	0.31	0.40	0.70	0.67
1-3/8	1.3750	2-1/16	2.0625	1.994	2.382	2.273	1-11/64	1.206	1.138	25/32	0.815	0.747	0.31	0.40	0.82	0.78
1-1/2	1.5000	2-1/4	2.2500	2.175	2.598	2.480	1-9/32	1.317	1.245	27/32	0.880	0.808	0.37	0.46	0.86	0.82

[a] Where specifying nominal size in decimals, zeros in the fourth decimal place shall be omitted. All sizes are unified dimensionally, except for slot dimensions, with British and Canadian Standards.

Additional requirements given in the General Specifications shall apply.

Hex Thick and Hex Thick Slotted Nuts

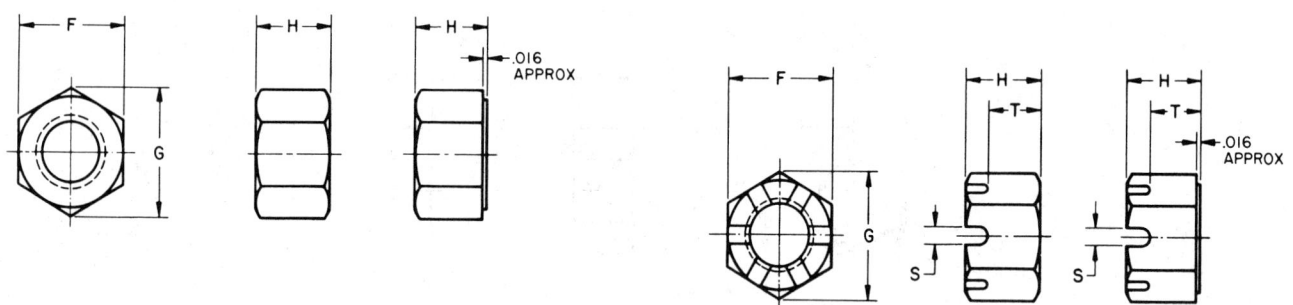

TABLE 7—DIMENSIONS OF HEX THICK AND HEX THICK SLOTTED NUTS

Nominal Size[a] or Basic Major Dia of Thread		Width Across Flats, F			Width Across Corners, G		Thickness, H			Slot Width, S		Unslotted Thickness, T	
		Basic	Max	Min	Max	Min	Basic	Max	Min	Min	Max	Max	Min
1/4	0.2500	7/16	0.4375	0.428	0.505	0.488	9/32	0.288	0.274	0.07	0.10	0.20	0.18
5/16	0.3125	1/2	0.5000	0.489	0.577	0.557	21/64	0.336	0.320	0.09	0.12	0.24	0.22
3/8	0.3750	9/16	0.5625	0.551	0.650	0.628	13/32	0.415	0.398	0.12	0.15	0.29	0.27
7/16	0.4375	11/16	0.6875	0.675	0.794	0.768	29/64	0.463	0.444	0.12	0.15	0.31	0.29
1/2	0.5000	3/4	0.7500	0.736	0.866	0.840	9/16	0.573	0.552	0.15	0.18	0.42	0.40
9/16	0.5625	7/8	0.8750	0.861	1.010	0.982	39/64	0.621	0.598	0.15	0.18	0.43	0.41
5/8	0.6250	15/16	0.9375	0.922	1.083	1.051	23/32	0.731	0.706	0.18	0.24	0.51	0.49
3/4	0.7500	1-1/8	1.1250	1.088	1.299	1.240	13/16	0.827	0.798	0.18	0.24	0.57	0.55
7/8	0.8750	1-5/16	1.3125	1.269	1.516	1.447	29/32	0.922	0.890	0.18	0.24	0.67	0.64
1	1.0000	1-1/2	1.5000	1.450	1.732	1.653	1	1.018	0.982	0.24	0.30	0.73	0.70
1-1/8	1.1250	1-11/16	1.6875	1.631	1.949	1.859	1-5/32	1.176	1.136	0.24	0.33	0.83	0.80
1-1/4	1.2500	1-7/8	1.8750	1.812	2.165	2.066	1-1/4	1.272	1.228	0.31	0.40	0.89	0.86
1-3/8	1.3750	2-1/16	2.0625	1.994	2.382	2.273	1-3/8	1.399	1.351	0.31	0.40	1.02	0.98
1-1/2	1.5000	2-1/4	2.2500	2.175	2.598	2.480	1-1/2	1.526	1.474	0.37	0.46	1.08	1.04

[a] Where specifying nominal size in decimals, zeros in the fourth decimal place shall be omitted. All sizes of Hex Thick Slotted Nuts are unified dimensionally, except for slot dimensions, with British and Canadian Standards.

Additional requirements given in the General Specifications shall apply.

Hex Castle Nuts

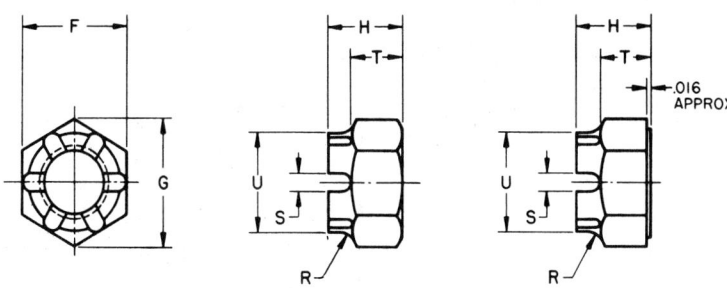

TABLE 8—DIMENSIONS OF HEX CASTLE NUTS

Nominal Size[a] or Basic Major Dia of Thread		Width Across Flats, F			Width Across Corners, G		Thickness, H			Slot Width, S		Unslotted Thickness and Height of Flats, T[b]		Radius of Fillet, R ±0.010	Dia of Cylindrical Part, U[c] Min
		Basic	Max	Min	Max	Min	Basic	Max	Min	Min	Max	Max	Min		
1/4	0.2500	7/16	0.4375	0.428	0.505	0.488	9/32	0.288	0.274	0.07	0.10	0.20	0.18	0.094	0.371
5/16	0.3125	1/2	0.5000	0.489	0.577	0.557	21/64	0.336	0.320	0.09	0.12	0.24	0.22	0.094	0.425
3/8	0.3750	9/16	0.5625	0.551	0.650	0.628	13/32	0.415	0.398	0.12	0.15	0.29	0.27	0.094	0.478
7/16	0.4375	11/16	0.6875	0.675	0.794	0.768	29/64	0.463	0.444	0.12	0.15	0.31	0.29	0.094	0.582
1/2	0.5000	3/4	0.7500	0.736	0.866	0.840	9/16	0.573	0.552	0.15	0.18	0.42	0.40	0.125	0.637
9/16	0.5625	7/8	0.8750	0.861	1.010	0.982	39/64	0.621	0.598	0.15	0.18	0.43	0.41	0.156	0.744
5/8	0.6250	15/16	0.9375	0.922	1.083	1.051	23/32	0.731	0.706	0.18	0.24	0.51	0.49	0.156	0.797
3/4	0.7500	1-1/8	1.1250	1.088	1.299	1.240	13/16	0.827	0.798	0.18	0.24	0.57	0.55	0.188	0.941
7/8	0.8750	1-5/16	1.3125	1.269	1.516	1.447	29/32	0.922	0.890	0.18	0.24	0.67	0.64	0.188	1.097
1	1.0000	1-1/2	1.5000	1.450	1.732	1.653	1	1.018	0.982	0.24	0.30	0.73	0.70	0.188	1.254
1-1/8	1.1250	1-11/16	1.6875	1.631	1.949	1.859	1-5/32	1.176	1.136	0.24	0.33	0.83	0.80	0.250	1.411
1-1/4	1.2500	1-7/8	1.8750	1.812	2.165	2.066	1-1/4	1.272	1.228	0.31	0.40	0.89	0.86	0.250	1.570
1-3/8	1.3750	2-1/16	2.0625	1.994	2.382	2.273	1-3/8	1.399	1.351	0.31	0.40	1.02	0.98	0.250	1.726
1-1/2	1.5000	2-1/4	2.2500	2.175	2.598	2.480	1-1/2	1.526	1.474	0.37	0.46	1.08	1.04	0.250	1.881

[a] Where specifying nominal size in decimals, zeros in the fourth decimal place shall be omitted.
[b] Height of the hexagon is measured from bearing surface to the top of arc.
[c] Maximum diameter of cylindrical portion shall not exceed maximum width across flats. Additional requirements given in the General Specifications shall apply.

Heavy Hex Flat and Hex Flat Jam Nuts

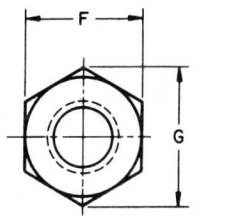  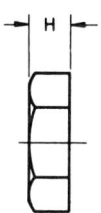

TABLE 9—DIMENSIONS OF HEAVY HEX FLAT AND HEX FLAT JAM NUTS[a]

Nominal Size[b] or Basic Major Dia of Thread		Width Across Flats, F			Width Across Corners, G		Thickness, Heavy Flat Nuts, H			Thickness, Heavy Flat Jam Nuts, H₁		
		Basic	Max	Min	Max	Min	Basic	Max	Min	Basic	Max	Min
1-1/8	1.1250	1-13/16	1.8125	1.756	2.093	2.002	1-1/8	1.155	1.079	5/8	0.655	0.579
1-1/4	1.2500	2	2.0000	1.938	2.309	2.209	1-1/4	1.282	1.187	3/4	0.782	0.687
1-3/8	1.3750	2-3/16	2.1875	2.119	2.526	2.416	1-3/8	1.409	1.310	13/16	0.846	0.747
1-1/2	1.5000	2-3/8	2.3750	2.300	2.742	2.622	1-1/2	1.536	1.433	7/8	0.911	0.808
1-3/4	1.7500	2-3/4	2.7500	2.662	3.175	3.035	1-3/4	1.790	1.679	1	1.040	0.929
2	2.0000	3-1/8	3.1250	3.025	3.608	3.449	2	2.044	1.925	1-1/8	1.169	1.050
2-1/4	2.2500	3-1/2	3.5000	3.388	4.041	3.862	2-1/4	2.298	2.155	1-1/4	1.298	1.155
2-1/2	2.5000	3-7/8	3.8750	3.750	4.474	4.275	2-1/2	2.552	2.401	1-1/2	1.552	1.401
2-3/4	2.7500	4-1/4	4.2500	4.112	4.907	4.688	2-3/4	2.806	2.647	1-5/8	1.681	1.522
3	3.0000	4-5/8	4.6250	4.475	5.340	5.102	3	3.060	2.893	1-3/4	1.810	1.643
3-1/4	3.2500	5	5.0000	4.838	5.774	5.515	3-1/4	3.314	3.124	1-7/8	1.939	1.748
3-1/2	3.5000	5-3/8	5.3750	5.200	6.207	5.928	3-1/2	3.568	3.370	2	2.068	1.870
3-3/4	3.7500	5-3/4	5.7500	5.562	6.640	6.341	3-3/4	3.822	3.616	2-1/8	2.197	1.990
4	4.0000	6-1/8	6.1250	5.925	7.073	6.755	4	4.076	3.862	2-1/4	2.326	2.112

[a] For sizes 1/4 in. through 1 in., see heavy hex nuts and heavy hex jam nuts in Table 10.
[b] Where specifying nominal size in decimals, zeros in the fourth decimal place shall be omitted.

Sizes 1-1/8 through 1-1/4 in. are unified dimensionally with British and Canadian Standards. Additional requirements given in the General Specifications shall apply.

Heavy Hex, Hex Jam and Hex Slotted Nuts

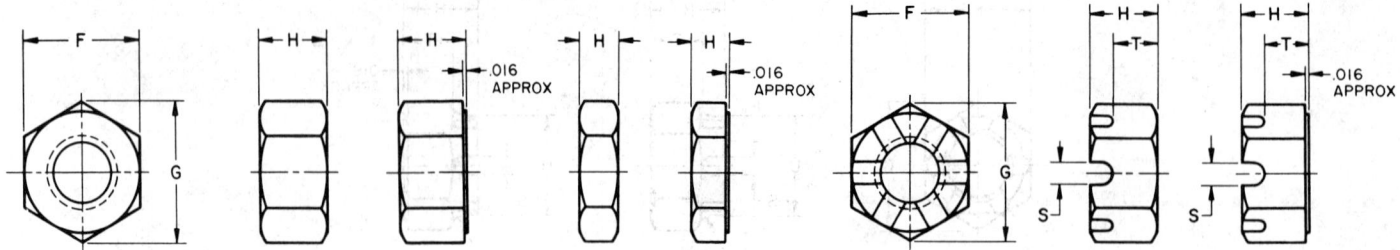

TABLE 10—DIMENSIONS OF HEAVY HEX, HEX JAM AND HEX SLOTTED NUTS

Nominal Size or Basic Major Dia of Thread		Width Across Flats, F			Width Across Corners, G		Thickness, Heavy Nuts, H			Thickness, Heavy Jam Nuts, H_1			Slot Width, S		Unslotted Thickness, T	
		Basic	Max	Min	Max	Min	Basic	Max	Min	Basic	Max	Min	Min	Max	Max	Min
1/4	0.2500	1/2	0.5000	0.488	0.577	0.556	15/64	0.250	0.218	11/64	0.188	0.156	0.07	0.10	0.15	0.13
5/16	0.3125	9/16	0.5625	0.546	0.650	0.622	19/64	0.314	0.280	13/64	0.220	0.186	0.09	0.12	0.21	0.19
3/8	0.3750	11/16	0.6875	0.669	0.794	0.763	23/64	0.377	0.341	15/64	0.252	0.216	0.12	0.15	0.24	0.22
7/16	0.4375	3/4	0.7500	0.728	0.866	0.830	27/64	0.441	0.403	17/64	0.285	0.247	0.12	0.15	0.28	0.26
1/2	0.5000	7/8	0.8750	0.850	1.010	0.969	31/64	0.504	0.464	19/64	0.317	0.277	0.15	0.18	0.34	0.32
9/16	0.5625	15/16	0.9375	0.909	1.083	1.037	35/64	0.568	0.526	21/64	0.349	0.307	0.15	0.18	0.37	0.35
5/8	0.6250	1-1/16	1.0625	1.031	1.227	1.175	39/64	0.631	0.587	23/64	0.381	0.337	0.18	0.24	0.40	0.38
3/4	0.7500	1-1/4	1.2500	1.212	1.443	1.382	47/64	0.758	0.710	27/64	0.446	0.398	0.18	0.24	0.49	0.47
7/8	0.8750	1-7/16	1.4375	1.394	1.660	1.589	55/64	0.885	0.833	31/64	0.510	0.458	0.18	0.24	0.62	0.59
1	1.0000	1-5/8	1.6250	1.575	1.876	1.796	63/64	1.012	0.956	35/64	0.575	0.519	0.24	0.30	0.72	0.69
1-1/8	1.1250	1-13/16	1.8125	1.756	2.093	2.002	1-7/64	1.139	1.079	39/64	0.639	0.579	0.24	0.33	0.78	0.75
1-1/4	1.2500	2	2.0000	1.938	2.309	2.209	1-7/32	1.251	1.187	23/32	0.751	0.687	0.31	0.40	0.86	0.83
1-3/8	1.3750	2-3/16	2.1875	2.119	2.526	2.416	1-11/32	1.378	1.310	25/32	0.815	0.747	0.31	0.40	0.99	0.95
1-1/2	1.5000	2-3/8	2.3750	2.300	2.742	2.622	1-15/32	1.505	1.433	27/32	0.880	0.808	0.37	0.46	1.05	1.01
1-5/8	1.6250	2-9/16	2.5625	2.481	2.959	2.828	1-19/32	1.632	1.556	29/32	0.944	0.868	—	—	—	—
1-3/4	1.7500	2-3/4	2.7500	2.662	3.175	3.035	1-23/32	1.759	1.679	31/32	1.009	0.929	0.43	0.52	1.24	1.20
1-7/8	1.8750	2-15/16	2.9375	2.844	3.392	3.242	1-27/32	1.886	1.802	1-1/32	1.073	0.989	—	—	—	—
2	2.0000	3-1/8	3.1250	3.025	3.608	3.449	1-31/32	2.013	1.925	1-3/32	1.138	1.050	0.43	0.52	1.43	1.38
2-1/4	2.2500	3-1/2	3.5000	3.388	4.041	3.862	2-13/64	2.251	2.155	1-13/64	1.251	1.155	0.43	0.52	1.67	1.62
2-1/2	2.5000	3-7/8	3.8750	3.750	4.474	4.275	2-29/64	2.505	2.401	1-29/64	1.505	1.401	0.55	0.64	1.79	1.74
2-3/4	2.7500	4-1/4	4.2500	4.112	4.907	4.688	2-45/64	2.759	2.647	1-37/64	1.634	1.522	0.55	0.64	2.05	1.99
3	3.0000	4-5/8	4.6250	4.475	5.340	5.102	2-61/64	3.013	2.893	1-45/64	1.763	1.643	0.62	0.71	2.23	2.17
3-1/4	3.2500	5	5.0000	4.838	5.774	5.515	3-3/16	3.252	3.124	1-13/16	1.876	1.748	0.62	0.71	2.47	2.41
3-1/2	3.5000	5-3/8	5.3750	5.200	6.207	5.928	3-7/16	3.506	3.370	1-15/16	2.006	1.870	0.62	0.71	2.72	2.65
3-3/4	3.7500	5-3/4	5.7500	5.562	6.640	6.341	3-11/16	3.760	3.616	2-1/16	2.134	1.990	0.62	0.71	2.97	2.90
4	4.0000	6-1/8	6.1250	5.925	7.073	6.755	3-15/16	4.014	3.862	2-3/16	2.264	2.112	0.62	0.71	3.22	3.15

[a] Where specifying nominal size in decimals, zeros in the fourth decimal place shall be omitted.
Sizes 1/2 through 2 in., except 9/16, 1-5/8 and 1-7/8 in., are unified dimensionally, except for slot dimensions, with British and Canadian Standards.
Additional requirements given in the General Specifications shall apply.

APPENDIX A—WRENCH OPENINGS FOR NUTS

Nominal Size of Wrench[a] also Basic (Maximum) Width Across Flats of Nuts		Allowance[b] between Nut Flats and Jaws of Wrench	Wrench Openings[c]			Square Nut	Hex Flat / Hex Flat Jam / Hex / Hex Jam / Hex Slotted / Hex Thick / Hex Thick Slotted / Hex Castle	Heavy Square / Heavy Hex Flat / Heavy Hex Flat Jam / Heavy Hex / Heavy Hex Jam / Heavy Hex Slotted
			Min	Tol	Max			
7/16	0.4375	0.003	0.440	0.006	0.446	1/4	1/4	—
1/2	0.5000	0.004	0.504	0.006	0.510	—	5/16	1/4
9/16	0.5625	0.004	0.566	0.007	0.573	5/16	3/8	5/16
5/8	0.6250	0.004	0.629	0.007	0.636	3/8	—	—
11/16	0.6875	0.004	0.692	0.007	0.699	—	7/16	3/8
3/4	0.7500	0.005	0.755	0.008	0.763	7/16	1/2	7/16
13/16	0.8125	0.005	0.818	0.008	0.826	1/2	—	—
7/8	0.8750	0.005	0.880	0.008	0.888	—	9/16	1/2
15/16	0.9375	0.006	0.944	0.009	0.953	—	5/8	9/16
1	1.0000	0.006	1.006	0.009	1.015	5/8	—	—
1-1/16	1.0625	0.006	1.068	0.009	1.077	—	—	5/8
1-1/8	1.1250	0.007	1.132	0.010	1.142	3/4	3/4	—
1-1/4	1.2500	0.007	1.257	0.010	1.267	—	—	3/4

(Table continued on next page)

APPENDIX A—WRENCH OPENINGS FOR NUTS (continued)

Nominal Size of Wrench[a] also Basic (Maximum) Width Across Flats of Nuts		Allowance[b] between Nut Flats and Jaws of Wrench	Wrench Openings[c]			Square Nut	Hex Flat / Hex Flat Jam / Hex / Hex Jam / Hex Slotted / Hex Thick / Hex Thick Slotted / Hex Castle	Heavy Square / Heavy Hex Flat / Heavy Hex Flat Jam / Heavy Hex / Heavy Hex Jam / Heavy Hex Slotted
			Min	Tol	Max			
1-5/16	1.3125	0.008	1.320	0.011	1.331	7/8	7/8	—
1-3/8	1.3750	0.008	1.383	0.011	1.394	—	—	—
1-7/16	1.4375	0.008	1.446	0.011	1.457	—	—	7/8
1-1/2	1.5000	0.008	1.508	0.012	1.520	1	1	—
1-5/8	1.6250	0.009	1.634	0.012	1.646	—	—	1
1-11/16	1.6875	0.009	1.696	0.012	1.708	1-1/8	1-1/8	—
1-13/16	1.8125	0.010	1.822	0.013	1.835	—	—	1-1/8
1-7/8	1.8750	0.010	1.885	0.013	1.898	1-1/4	1-1/4	—
2	2.0000	0.011	2.011	0.014	2.025	—	—	1-1/4
2-1/16	2.0625	0.011	2.074	0.014	2.088	1-3/8	1-3/8	—
2-3/16	2.1875	0.012	2.200	0.015	2.215	—	—	1-3/8
2-1/4	2.2500	0.012	2.262	0.015	2.277	1-1/2	1-1/2	—
2-3/8	2.3750	0.013	2.388	0.016	2.404	—	—	1-1/2
2-7/16	2.4375	0.013	2.450	0.016	2.466	—	—	—
2-9/16	2.5625	0.014	2.576	0.017	2.593	—	—	1-5/8
2-5/8	2.6250	0.014	2.639	0.017	2.656	—	—	—
2-3/4	2.7500	0.014	2.766	0.017	2.783	—	—	1-3/4
2-13/16	2.8125	0.015	2.827	0.018	2.845	—	—	—
2-15/16	2.9375	0.016	2.954	0.019	2.973	—	—	1-7/8
3	3.0000	0.016	3.016	0.019	3.035	—	—	—
3-1/8	3.1250	0.017	3.142	0.020	3.162	—	—	2
3-3/8	3.3750	0.018	3.393	0.021	3.414	—	—	—
3-1/2	3.5000	0.019	3.518	0.022	3.540	—	—	2-1/4
3-3/4	3.7500	0.020	3.770	0.023	3.793	—	—	—
3-7/8	3.8750	0.020	3.895	0.023	3.918	—	—	2-1/2
4-1/8	4.1250	0.022	4.147	0.025	4.172	—	—	—
4-1/4	4.2500	0.022	4.272	0.025	4.297	—	—	2-3/4
4-1/2	4.5000	0.024	4.524	0.026	4.550	—	—	—
4-5/8	4.6250	0.024	4.649	0.027	4.676	—	—	3
4-7/8	4.8750	0.025	4.900	0.028	4.928	—	—	—
5	5.0000	0.026	5.026	0.029	5.055	—	—	3-1/4
5-1/4	5.2500	0.027	5.277	0.030	5.307	—	—	3-1/2
5-3/8	5.3750	0.028	5.403	0.031	5.434	—	—	3-1/2
5-5/8	5.6250	0.029	5.654	0.032	5.686	—	—	3-3/4
5-3/4	5.7500	0.030	5.780	0.033	5.813	—	—	3-3/4
6	6.0000	0.031	6.031	0.034	6.065	—	—	4
6-1/8	6.1250	0.032	6.157	0.035	6.192	—	—	4

[a] Wrenches shall be marked with the "nominal size of wrench" which is equal to the basic (maximum) width across flats of the corresponding nut.
[b] Allowance (minimum clearance) between maximum width across flats of the nut and jaws of wrench equals (0.005W + 0.001).
[c] Tolerance on wrench opening equals plus (0.005W + 0.004 from minimum). W equals nominal size of wrench.

HEXAGON HIGH NUTS—SAE J482a

SAE Standard

Report of Screw Threads Division approved July 1924, revised by Screw Threads Committee May 1959, and last revised by Fasteners Committee June 1966.

GENERAL SPECIFICATIONS

Scope—Included herein are the detailed general and dimensional specifications applicable to high hex nuts. All general specifications not shown here shall conform with those applicable to hex thick nuts and hex thick slotted nuts appearing in SAE J104, Table 7. High hex nuts are primarily intended for use in automotive and other ground based vehicles and industrial equipment where a long length of hexagon is required for wrenching purposes.

Threads—
Form and Tolerance—Threads shall conform to Unified Standard, Class 2B.

Series—Threads shall be coarse (UNC) or fine (UNF) thread series.
Countersink—Tapped holes shall be countersunk (unless counterbore option is specified) on the bearing face or faces. The maximum countersink diameter shall be the thread basic (nominal) diameter plus 0.025 in. for 3/8 size or smaller, and 1.06 times the basic major diameter for larger sizes. No part of the threaded portion shall extend beyond the bearing surface.

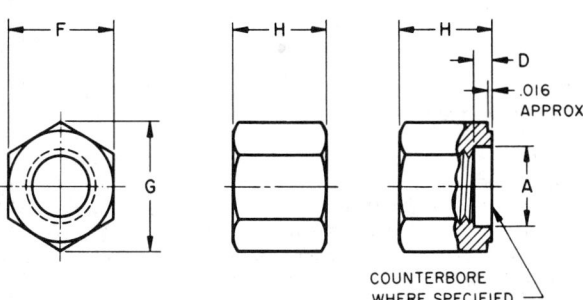

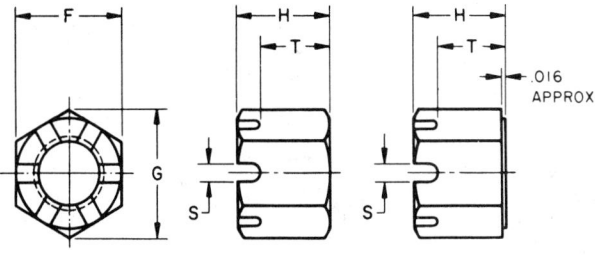

TABLE 1—DIMENSIONS OF HEX HIGH AND HEX SLOTTED HIGH NUTS

Nominal Size[a] or Basic Major Dia of Thread		Width Across Flats F			Width Across Corners G		Thickness H			Slot Width, S		Unslotted Thickness, T		Counterbore (Optional)	
		Basic	Max	Min	Max	Min	Basic	Max	Min	Min	Max	Max	Min	Dia A	Depth, D
1/4	0.2500	7/16	0.4375	0.428	0.505	0.488	3/8	0.382	0.368	0.07	0.10	0.29	0.27	0.266	0.062
5/16	0.3125	1/2	0.5000	0.489	0.577	0.557	29/64	0.461	0.445	0.09	0.12	0.37	0.35	0.328	0.078
3/8	0.3750	9/16	0.5625	0.551	0.650	0.628	1/2	0.509	0.491	0.12	0.15	0.38	0.36	0.391	0.094
7/16	0.4375	11/16	0.6875	0.675	0.794	0.768	39/64	0.619	0.599	0.12	0.15	0.46	0.44	0.453	0.109
1/2	0.5000	3/4	0.7500	0.736	0.866	0.840	21/32	0.667	0.645	0.15	0.18	0.51	0.49	0.516	0.125
9/16	0.5625	7/8	0.8750	0.861	1.010	0.982	49/64	0.778	0.754	0.15	0.18	0.59	0.57	0.594	0.141
5/8	0.6250	15/16	0.9375	0.922	1.083	1.051	27/32	0.857	0.831	0.18	0.24	0.63	0.61	0.656	0.156
3/4	0.7500	1-1/8	1.1250	1.088	1.299	1.240	1	1.015	0.985	0.18	0.24	0.76	0.73	0.781	0.188
7/8	0.8750	1-5/16	1.3125	1.269	1.516	1.447	1-5/32	1.172	1.140	0.18	0.24	0.92	0.89	0.906	0.219
1	1.0000	1-1/2	1.5000	1.450	1.732	1.653	1-5/16	1.330	1.294	0.24	0.30	1.05	1.01	1.031	0.250
1-1/8	1.1250	1-11/16	1.6875	1.631	1.949	1.859	1-1/2	1.520	1.480	0.24	0.33	1.18	1.14	1.156	0.281
1-1/4	1.2500	1-7/8	1.8750	1.812	2.165	2.066	1-11/16	1.710	1.666	0.31	0.40	1.34	1.29	1.281	0.312

[a] Where specifying nominal size in decimals, zeros in the fourth decimal place shall be omitted.

ALIGNMENT OF NUT SLOTS—SAE J484

SAE Information Report

Report of Screw Threads Division approved February 1925 and last revised by Screw Threads Committee January 1956. Editorial change June 1964. Reaffirmed without change by Fasteners Committee June 1969.

This method of gaging alignment of nut slots allows equal variations for location of the cotter pin hole in the bolt and location of slots in the nut. To inspect the nut, the slotted gage is inserted through the nut hole from the bearing surface of the nut. Alignment of slots is considered satisfactory if the gage pin can be slipped into the gage and nut slots without interference.

Some approximations may be necessary because of differences in the nut minor diameter which result from the method of producing the hole and threads in the nut. Nut manufacturers may reduce errors by basing the diameter D of the plug on the minimum drilled or punched hole. Users may reduce errors by basing the diameter D of the plug on the minimum minor diameter of the thread. The diameter B of the slotted gage pilot is purposely undersize to clear burrs which may result from blanking or slotting the nut.

The gaging pin diameter is based on the mean of the nut slot width and the nominal cotter pin diameter—the pin diameter specified in the table being the drill rod size next larger than this mean figure. The pin should preferably be tempered and should fit the gage slot with minimum functional clearance.

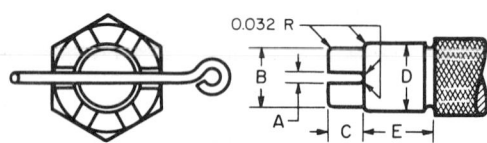

FIG. 1—GAGE DIMENSIONS

TABLE 1—GAGE DIMENSIONS FOR ALIGNMENT OF NUT SLOTS

Nut		Cotter Pin	Gage Pin		Gage					
Nominal Size	Slot Width	Nominal Size	No.	Dia	A	B	C	Minor, Dia D[a]		E
								Fine Threads	Coarse Threads	
1/4	0.078	1/16	49	0.073	0.073	0.182	0.156	0.211	0.196	0.250
5/16	0.094	5/64	44	0.086	0.086	0.234	0.156	0.267	0.252	0.297
3/8	0.125	3/32	34	0.110	0.110	0.297	0.188	0.330	0.307	0.344
7/16	0.125	3/32	34	0.110	0.110	0.344	0.188	0.383	0.360	0.391
1/2	0.156	1/8	28	0.141[b]	0.141[b]	0.406	0.250	0.446	0.417	0.438

[a] Minimum minor diameter of Class 2B threads. If the nuts are not threaded in accordance with SAE J475, the diameter shall be the same as the diameter of the Go-plug gage for the bore.
[b] This dimension or 9/64.

CROWN (BLIND, ACORN) NUTS—SAE J483a

SAE Recommended Practice

Report of Screw Threads Division approved January 1933, revised by Screw Threads Committee May 1959, and last revised by Fasteners Committee June 1966.

GENERAL SPECIFICATIONS

Scope—Included herein are complete general and dimensional data for the high and low types of crown nuts recognized as SAE Standard. These nuts are primarily intended for application in automotive and other ground based motor vehicles and industrial equipment to provide an ornamental or protective closure over end of bolts, studs, or screws.

Dimensions—All dimensions in this standard are in inches unless otherwise stated.

Options—Options, where specified, shall be at the discretion of the manufacturer unless otherwise agreed by manufacturer and user.

Construction—Nuts may be either solid or of two piece construction. The bearing surface may be flat with chamfered corners or washer faced. The diameter of chamfer circle or washer face shall be equal to the maximum width across flats within a tolerance of −5%. The length of chamfer at hexagon corners shall be from 5 to 15% of the basic thread diameter. The surface of chamfer may be slightly convex or rounded.

Rounding at Corners—A rounding or lack of fill at junction of hex corners with chamfer shall be permissible, provided the minimum width across corners is reached and maintained beyond a distance equal to 17.5% of the basic thread diameter from the chamfered face and the junction of hexagon faces with crown fillet.

Taper of Sides—No transverse section through hexagon portion of nut between 25 and 75% of the actual hexagon thickness, as measured from the bearing face, shall be less than the minimum width across flats. The maximum width across flats shall not be exceeded except for milled-from-bar nonferrous nuts where the maximum (basic) width may conform with the commercial tolerances of the bar stock material.

Angularity of Bearing Surface—The bearing surface shall be at right angle to the axis of the tapped hole within 2 deg for 1 in. size or smaller, and within 1 deg for larger sizes. Therefore, the maximum total runout of bearing face shall equal the tangent of the angular deviation times the distance across flats.

Threads—

Form and Tolerance—Threads shall conform to Unified Standard, Class 2B.

Series—Threads shall be coarse (UNC) or fine (UNF) thread series.

Countersink—The tapped hole shall be countersunk on the bearing face. The maximum countersink diameter shall be the thread basic (nominal) major diameter plus 0.025 in. for ⅜ size or smaller, and 1.06 times the basic major diameter for larger sizes. No part of the threaded portion shall extend beyond the bearing surface.

Materials—

Steel—Suitable properties for steel nuts are covered in SAE J429.

Other Materials—Other materials shall be as agreed upon by the manufacturer and user.

Finish—

Plain—Unless otherwise specified, nuts shall be supplied plain (unplated or uncoated), as processed.

Plated—Where plating is specified, the thickness or quality of plating shall be measured or tested on the side of the nut.

*Defects—*Nuts shall be free from burrs, seams, laps, loose scale, and any other defects that affect serviceability.

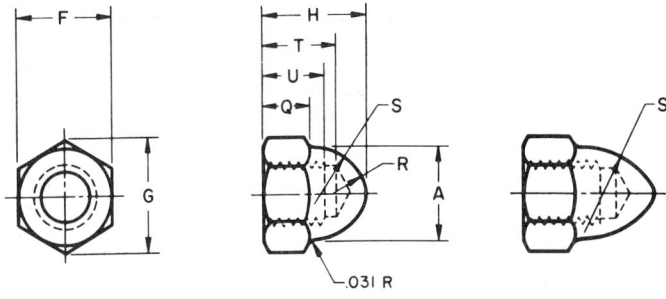

FIG. 1—DIMENSIONS OF NUTS

TABLE 1—DIMENSIONS OF HIGH AND LOW CROWN NUTS

Nominal Size or Basic Major Dia of Thread		Width Across Flats F		Width Across Corners G		Body Dia A	High Crown						Low Crown					
							Over-all Height H	Hexa-gon Height Q	Nose Radius R	Body Radius S	Drill Depth T	Full Thread U	Over-all Height H	Hexa-gon Height Q	Nose Radius R	Body Radius S	Drill Depth T	Full Thread U
		Max (Basic)	Min	Max	Min						Max	Min					Max	Min
No. 6	0.1380	5/16	0.3125	0.302	0.361	0.344	0.30	0.42	0.17	0.05	0.25	0.28	0.19	0.34	0.16	0.08	0.17	0.25	0.16
No. 8	0.1640	5/16	0.3125	0.302	0.361	0.344	0.30	0.42	0.17	0.05	0.25	0.28	0.19	0.34	0.16	0.08	0.17	0.25	0.16
No. 10	0.1900	3/8	0.3750	0.362	0.433	0.413	0.36	0.52	0.20	0.06	0.30	0.34	0.25	0.41	0.19	0.09	0.22	0.28	0.19
No. 12	0.2160	3/8	0.3750	0.362	0.433	0.413	0.36	0.52	0.20	0.06	0.30	0.38	0.28	0.41	0.19	0.09	0.22	0.31	0.22
1/4	0.2500	7/16	0.4375	0.428	0.505	0.488	0.41	0.59	0.23	0.06	0.34	0.41	0.31	0.47	0.22	0.11	0.25	0.34	0.25
5/16	0.3125	1/2	0.5000	0.489	0.577	0.557	0.47	0.69	0.28	0.08	0.41	0.47	0.38	0.53	0.25	0.12	0.28	0.41	0.31
3/8	0.3750	9/16	0.5625	0.551	0.650	0.628	0.53	0.78	0.31	0.09	0.44	0.56	0.47	0.62	0.28	0.14	0.33	0.45	0.38
7/16	0.4375	5/8	0.6250	0.612	0.722	0.698	0.59	0.88	0.34	0.09	0.50	0.62	0.53	0.69	0.31	0.16	0.36	0.52	0.44
1/2	0.5000	3/4	0.7500	0.736	0.866	0.840	0.72	1.03	0.42	0.12	0.59	0.75	0.62	0.81	0.38	0.19	0.42	0.59	0.50
9/16	0.5625	7/8	0.8750	0.861	1.010	0.982	0.84	1.19	0.48	0.16	0.69	0.81	0.69	0.94	0.44	0.22	0.50	0.69	0.56
5/8	0.6250	15/16	0.9375	0.922	1.083	1.051	0.91	1.28	0.53	0.16	0.75	0.91	0.78	1.00	0.47	0.23	0.53	0.75	0.62
3/4	0.7500	1-1/16	1.0625	1.045	1.227	1.191	1.03	1.45	0.59	0.17	0.84	1.06	0.94	1.16	0.53	0.27	0.59	0.88	0.75
7/8	0.8750	1-1/4	1.2500	1.231	1.443	1.403	1.22	1.72	0.70	0.20	0.98	1.22	1.09	1.36	0.62	0.31	0.70	1.00	0.88
1	1.0000	1-7/16	1.4375	1.417	1.660	1.615	1.41	1.97	0.81	0.23	1.14	1.38	1.25	1.55	0.72	0.36	0.81	1.12	1.00
1-1/8	1.1250	1-5/8	1.6250	1.602	1.876	1.826	1.59	2.22	0.92	0.27	1.28	1.59	1.41	1.75	0.81	0.41	0.92	1.31	1.12
1-1/4	1.2500	1-13/16	1.8125	1.788	2.093	2.038	1.78	2.47	1.03	0.28	1.44	1.75	1.56	1.95	0.91	0.45	1.03	1.44	1.25

[a] Where specifying nominal size in decimals, zeros in the fourth decimal place shall be omitted.

SPRING NUTS—SAE J891a

SAE Standard

Report of Fasteners Committee approved August 1964 and last revised June 1970.

GENERAL SPECIFICATIONS

*Scope—*Included herein are complete general and dimensional specifications for the types of spring nuts recognized as SAE Standard. These nuts are intended for general use where the engagement of a single thread on the mating screw is considered adequate for the application.

It should be noted that spring nuts having other configurations, dimensions, provisions for ground, etc., are available and manufacturers should be consulted.

*Dimensional Tolerance—*Tolerance on dimensions in the tables shall be plus and minus 0.010 in. unless otherwise specified.

*Boss Detail—*The detail of boss shall be such as to assemble readily and function satisfactorily with the specified screw and meet the performance requirements of this specification except as indicated otherwise.

Both the Type P and Type T bosses are designed to function with either Types A, AB, or B tapping screws in sizes 6-18 or 6-20, 8-15 or 8-18, and 10-12 or 10-16. The Type P boss also is designed to function with either Types A, AB, or B tapping screws in sizes 14-10 or ¼-14. The Type T bosses for the 14-10 Type A and the ¼-14 Types B and AB tapping screws have not been compromised and separate nuts must be specified for the respective thread sizes. Type P and Type T bosses are also made for use with 6-32, 8-32, 10-24, and ¼-20 sizes of machine screws.

The sides of the Type P boss (Fig. 1) shall be formed to provide an opening conforming to the helix of the mating thread. The opening shall be round and equal to, or slightly larger than, the minor diameter of the mating thread.

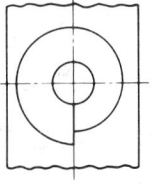

FIG. 1—TYPE P BOSS

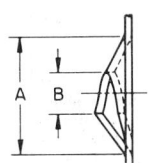

FIG. 2—TYPE T BOSS

The prongs of the Type T boss (Fig. 2) shall be formed to provide a circular opening conforming to the helix of the mating thread and, at the opening, the prongs shall be normal to the axis of the mating thread. The round portion of the opening shall be equal to, or slightly larger than, the minor diameter of the mating thread.

The size and formation of the helical opening (see Table 1) shall be such as to permit ready assembly of the specified screw or gage when inserted from the base of the boss at 90 deg to the plane of the nut, or component thereof which contains the boss. For Unified threads, basic GO thread plug gages shall be used to check assembly. For Type A and Type B pitch tapping screw threads, special gages conforming to the maximum limits of the screws may be used in place of the specified screws to check assembly.

Retaining Extrusion Detail—The size and configuration of the extrusion in the lower leg of "J" shape and "U" shape spring nuts shall be such that nuts will meet the performance requirements of this specification. The size and relative location of the hole and extrusion to the boss shall be such that when nut is assembled onto a test panel having minimum hole size, located at maximum edge distance, the extrusion will snap into the hole and permit the specified screw of maximum size (or special threaded plug gage, Fig. 3) to be assembled into the boss normal to the base of boss with interference at the extrusion or the sides of either hole. The screw or gage is to be entered into the boss until the head of the screw or shoulder on gage lightly contacts the bottom of the lower leg. The extrusion shall have a uniform shape and blend evenly from the specified height at point X into the upper surface of the lower leg at points Y and Y' as shown. The critical edges of the extrusion shall be free from burrs which would cause interference as spring nut is assembled to panel.

Material—Spring steel: SAE 1050, or higher carbon; suitably processed to meet the performance requirements of this specification.

Hardness—Hardness shall be as follows:

Material Thickness	Rockwell Scale	Dial Reading	Conversion to Rockwell C Scale
Up to 0.016	15N	80.4–85.5	40–50
0.017–0.024	30N	59.5–68.5	40–50
0.025–0.039	45N	43.1–55.0	40–50
0.040 and over	C	40–50	40–50

Finish—Spring nuts are normally supplied with rustproof finish as specified by the purchaser. Nuts subjected to corrosion preventative treatment which might induce hydrogen embrittlement shall be baked or otherwise treated to obviate such embrittlement.

Workmanship—Spring nuts shall be free from cracks, burrs, splits, loose scale, or any other defects that might affect their serviceability.

Performance—Spring nuts shall perform in accordance with the requirements specified in Table 2 except as indicated otherwise.

Assembly Detail—The recommended design data pertaining to assembly of "J" shape and "U" shape spring nuts for guidance of users is presented in Tables 4 and 5. The proper method of assembling these nuts to panels is described as follows:

"J" shape nuts are assembled to panel by placing nut against the edge of the panel as shown opposite and rocking onto panel in the direction indicated by the arrow.

"U" shape nuts are assembled to panel by placing nut over edge of the panel as shown opposite and pushing onto the panel in the direction indicated by the arrow.

TABLE 1—DETAIL OF BOSSES[a]

Screw		Type P Boss		Type T Boss		
Type	Size	A Base Dia, Ref.	B Hole Dia, +0.010 −0.000	C Width of Shear Basic	D[b] End of Slit to Edge, Min	E Width of Blank, Min
Machine	6-32	0.26	0.104	0.157	0.050	0.312
	8-32	0.23	0.130	0.184	0.050	0.406
	10-24	0.28	0.149	0.210	0.050	0.375
	1/4-20	0.37	0.203	0.270	0.090	0.500
Type A Tapping	6-18	0.25	0.105	0.157	0.050	0.312
	8-15	0.28	0.123	0.184	0.060	0.406
	10-12	0.28	0.142	0.210	0.080	0.500
	14-10	0.38	0.193	0.270	0.090	0.562
Types B and AB Tapping	6-20	0.25	0.105	0.157	0.050	0.312
	8-18	0.28	0.123	0.184	0.060	0.406
	10-16	0.28	0.142	0.210	0.080	0.500
	1/4-14	0.38	0.193	0.270	0.090	0.562

[a]The above dimensions are design criteria and are not to be applied to part drawings.
[b]The tabulated values are applicable to standard spring nuts only. This factor shall be sufficient to meet the performance requirements for torque, tensile strength and vibration as set forth in Table 2.

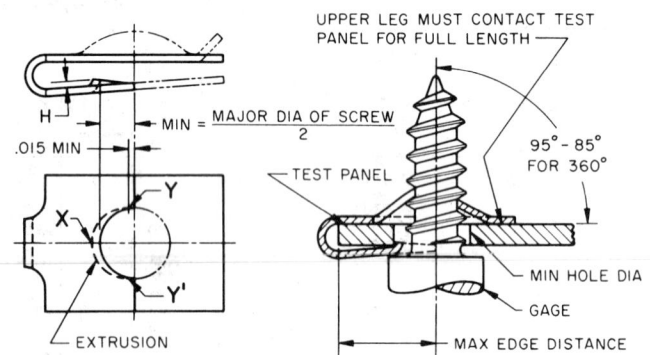

FIG. 3—EXTRUSION AND GAGING DETAIL

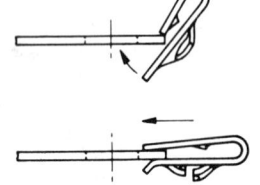

TABLE 2—PERFORMANCE REQUIREMENTS FOR SPRING NUTS

Nut Design		Wide-Range				Regular					Wide-Range and Regular			
Screw		Recommended Installation Torque, Max, lb-in.	Clamp-Load at Recommended Installation Torque, Min, lb	Destructive Torque, Min, lb-in.	Ultimate Tensile Strength, Min, lb	Recommended Installation Torque, Max, lb-in.	Destructive Torque, Min, lb-in.	Test Plate Separation, in.	Tensile Yield Strength, lb	Ultimate Tensile Strength, Min, lb	Back-Off Torque[b], Min, oz-in.			Vibration Test Load, lb
								Type T	Type P	Type T	Type P	Type T at 5 deg	Type T at 40 deg	
Type	Size													
Machine	6-32[a]	—	—	—	—	6	8	0.004	89	155	156	33	26	16.5	25
	8-32	—	—	—	—	8	10	0.005	134	200	189	45	24.8	20.6	38
	10-24	—	—	—	—	14	17	0.006	221	315	274	78	48	40	50
	1/4-20	—	—	—	—	35	45	0.008	350	570	550	195	256	216	85
Type A Tapping	6-18[a]	—	—	—	—	12	17	0.004	312	410	425	48	42.4	13.8	70
	8-15	—	—	—	—	20	25	0.005	420	570	534	64	120.6	54	95
	10-12	—	—	—	—	35	44	0.006	525	700	672	140	113.6	8	135
	14-10	—	—	—	—	60	80	0.008	891	900	1158	240	432	150.4	185
Types B and AB Tapping	6-20[a]	12	240	17	450	12	17	0.004	312	500	425	57	42.4	13.8	70
	8-18	20	400	25	500	20	25	0.005	420	620	534	96	120.6	54	95
	10-16	35	550	44	750	35	44	0.006	525	1000	672	168	113.6	8	135
	1/4-14	60	750	80	900	60	80	0.008	891	1130	1158	288	432	150.4	185

[a]These requirements shall not apply to Type P, Style I, Flat Spring Nuts shown in Table 6 due to the limitations on Loss design imposed by the narrow widths and long lengths.
[b]Back-off torque values shown for Type P shall apply to wide-range design nuts incorporating Type P boss design and those shown for Type T shall apply to the other wide-range design nut construction. Back-off torque shall not be less than the value specified after nuts have been subjected to the vibration test.

TESTS AND TEST FIXTURES FOR EVALUATING SPRING NUT PERFORMANCE

Spring nuts shall be subjected to the following tests to determine conformance with the performance requirements specified in Table 2, except as indicated otherwise.

Test Plates and Screws for Tests—To assure uniformity of test results, the test plates and screws used for the tests shall conform to the following specifications.

Test plates shall have boundary dimensions and hole sizes as depicted in Fig. 4 and specified in Table 3. The thickness of test plates shall be equal to the mean of the specified panel range within a tolerance of plus and minus 0.001 in. The holes in test plates shall be located at the maximum edge distance specified for the particular spring nut within a tolerance of plus and minus 0.001 in. Test plates and panels shall have a minimum hardness of Rockwell C50-54.

The screws used for test purposes shall conform to the specifications in SAE J478 for the respective sizes and types. They shall be Hexagon Head Slotted style, 1 in. long, with a 72 hr phosphate finish.

Torque Tests—Spring nut samples shall be assembled with a test screw onto a test plate (see Table 3) and tightened to the recommended installation torque. For wide-range design spring nuts, this test shall be performed with a device capable of measuring the clamp load developed and, when assembly is tightened to the recommended installation torque, the clamp load obtained shall not be less than the value tabulated. The back-off torques, readings being taken when the screw first moves, shall not be less than the values specified.

Upon disassembly, the boss shall return to a position that will accept reentry of the test screw or "GO" thread gage.

The spring nut, when reassembled and tightened on the test plate, shall not strip the threads on the screw nor fail the nut boss at less than the ultimate torque specified.

Tensile Tests—When the spring nut on a test plate is assembled to suitable back-up plates at the recommended installation torque and pulled in a tensile testing machine, the spring nut shall meet the minimum yield and ultimate strengths specified. The tensile yield strength shall be considered reached when the back-up plates have separated by the amount specified. The ultimate strength shall be considered reached when the boss or the thread on the screw is destroyed. In performing tensile test, care should be taken to assure there is no interference between the screw and the holes in the plates. A typical tensile test fixture is illustrated in Fig. 5.

Vibration Test—After tightening the spring nut on a test plate to the recommended installation torque, and subjecting the assembly to 20 hr vibration, the back-off torque shall not be less than the values specified. The vibration test shall be conducted while the nut is subject to the vibration test load specified. The vibration amplitude shall be 0.030 in. in a direction parallel to the axis of the screw and the frequency shall be 3600 rpm, or cpm.

The vibration test panel shall be an assembly consisting of a test plate, a panel, a screw, a spring nut and a means for applying the vibration test load. The vibration panel assembly shall be rigidly fastened to the source of vibration and the fastening shall not interfere with the assembly being tested. A typical vibration test fixture is depicted in Fig. 6.

Preassembly and Retention Test—The "J" shape and "U" shape spring nuts shall preassemble onto test panels of thickness equal to the two extremes of the panel ranges specified, having minimum holes, located at the maximum edge distance. The extrusion in the lower leg shall snap into the hole and, when nuts are so assembled, a pull force of 3 lb minimum, applied parallel to the upper leg in line with the axis of the nut, shall be required to remove the nut from the panel.

TABLE 3—DIMENSIONS OF TEST PLATES (FIG. 4)

Screw Type	Size	A Edge to Center, ±0.001	B Plate Thickness, ±0.001	C Hole Dia, ±0.001
Machine	6-32			0.250
	8-32			0.281
	10-24			0.312
	1/4-20			0.375
Type A Tapping	6-18	Equals maximum edge distance specified in assembly data or on part drawing	Equals mean of panel range specified in dimensional tables	0.250
	8-15			0.281
	10-12			0.312
	14-10			0.375
Types B and AB Tapping	6-20			0.250
	8-18			0.281
	10-16			0.312
	1/4-14			0.375

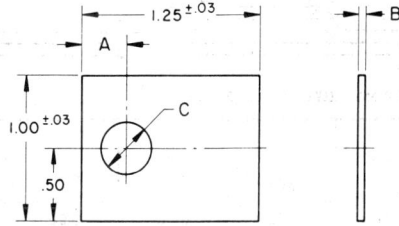

FIG. 4—TEST PLATES

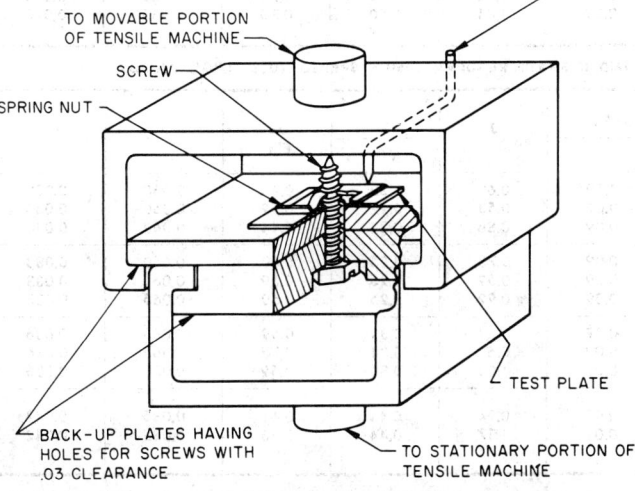

FIG. 5—TYPICAL TENSILE TEST FIXTURE

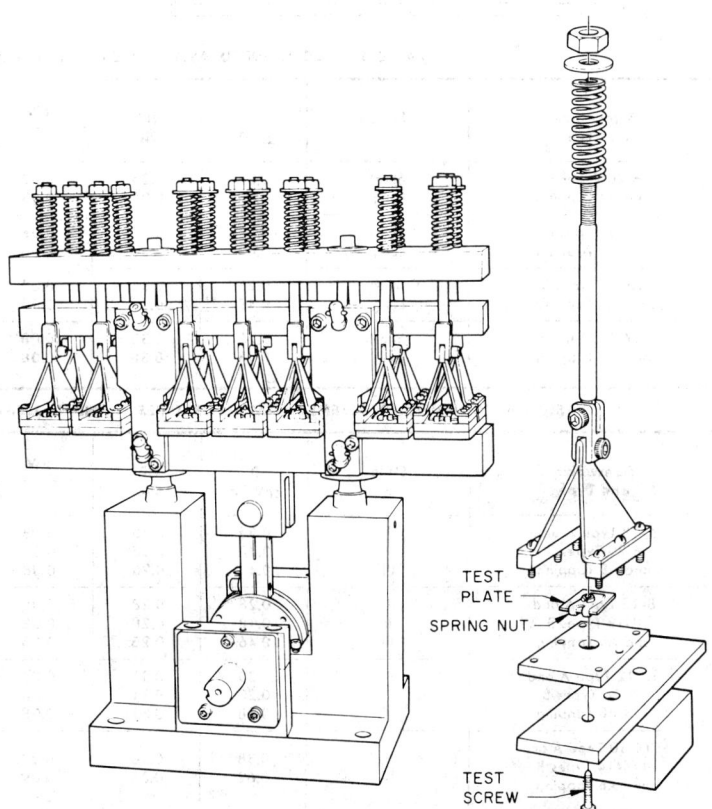

FIG. 6—TYPICAL VIBRATION TEST FIXTURE

J SHAPE AND U SHAPE ASSEMBLY DATA

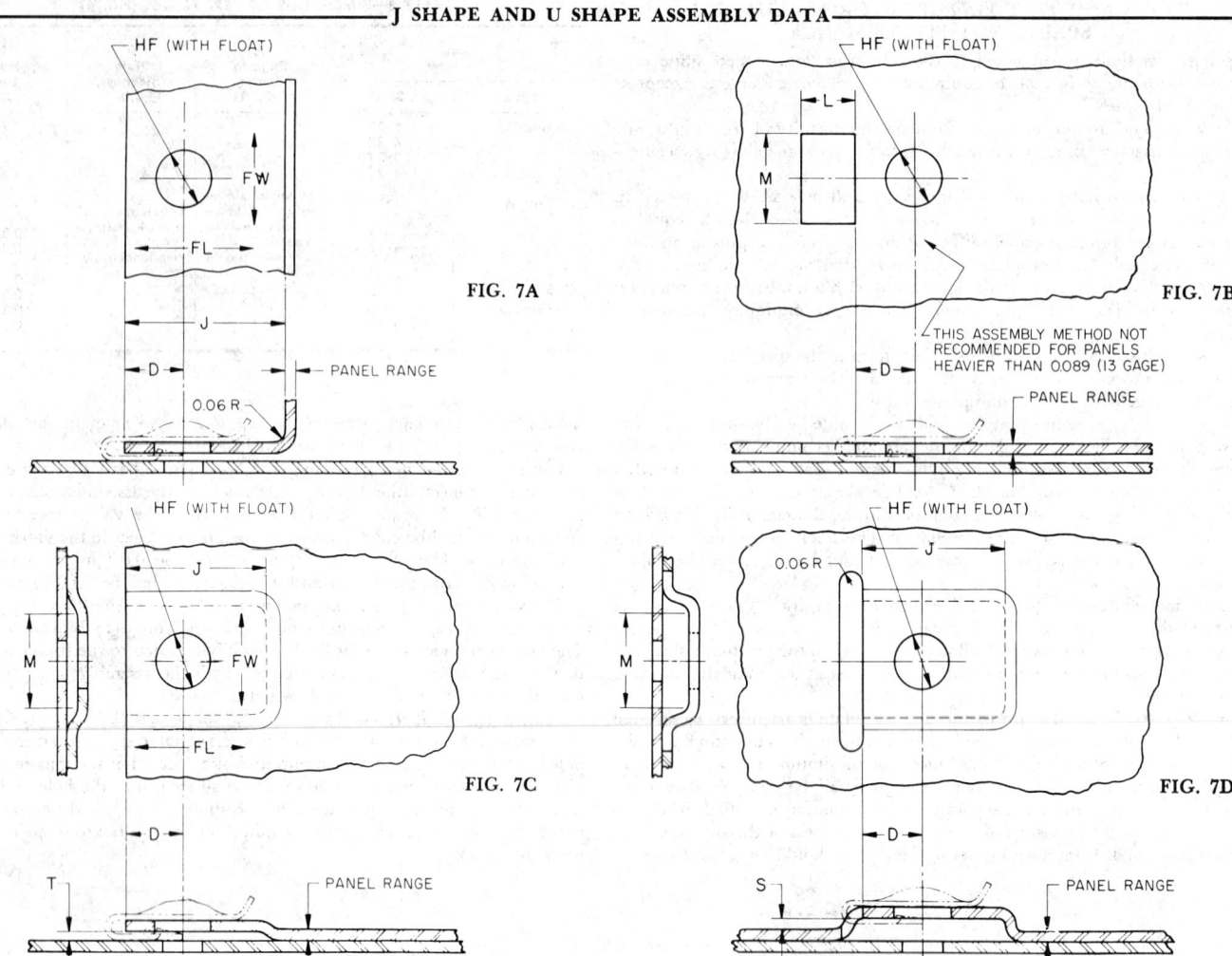

FIG. 7A   FIG. 7B   FIG. 7C   FIG. 7D

TABLE 4—RECOMMENDED ASSEMBLY DATA FOR WIDE-RANGE DESIGN SPRING NUTS[a] (FIGS. 7A-7D)

Screw Size and Type	Style	D ±0.02	HF Dia	FW	FL	J	L	M	S	T
				Float			Min	Flat		
6-20 Types B and AB Tapping	Short	0.25	0.25	0.08	0.09	0.78	0.25	0.63	0.065	0.027
	Long	0.48	0.25	0.08	0.09	1.01	0.25	0.63	0.065	0.027
8-18 Types B and AB Tapping	Short	0.25	0.28	0.08	0.09	0.78	0.34	0.63	0.065	0.030
	Long	0.48	0.28	0.08	0.09	1.01	0.34	0.63	0.065	0.030
10-16 Types B and AB Tapping	Short	0.25	0.31	0.08	0.09	0.78	0.43	0.73	0.065	0.035
	Long	0.48	0.31	0.08	0.09	1.01	0.43	0.73	0.065	0.035
1/4-14 Types B and AB Tapping	Short	0.25	0.38	0.08	0.09	0.78	0.50	0.73	0.065	0.040
	Long	0.48	0.38	0.08	0.09	1.01	0.50	0.73	0.065	0.040

TABLE 5—RECOMMENDED ASSEMBLY DATA FOR TYPES P AND T, J SHAPE AND U SHAPE REGULAR DESIGN SPRING NUTS[a] (FIGS. 7A-7D)

Screw Size and Type	Style	D ±0.02	HF Dia	FW	FL	J	L	M	S	T
				Float			Min	Flat		
6-18 Type A and 6-20 Types B and AB Tapping	I	0.25	0.25	0.08	0.09	0.69	0.22	0.44	0.060	0.030
	II	0.17	0.25	0.08	0.09	0.53	0.22	0.59	0.060	0.030
	III	0.42	0.25	0.08	0.09	0.88	0.22	0.44	0.060	0.030
8-15 Type A and 8-18 Types B and AB Tapping	I	0.28	0.28	0.08	0.09	0.73	0.25	0.50	0.060	0.033
	II	0.18	0.28	0.08	0.09	0.59	0.25	0.59	0.065	0.033
	III	0.46	0.28	0.08	0.09	0.92	0.25	0.50	0.065	0.033
10-12 Type A and 10-16 Types B and AB Tapping	I	0.30	0.31	0.08	0.09	0.81	0.31	0.59	0.065	0.036
	II	0.20	0.31	0.08	0.09	0.62	0.31	0.72	0.065	0.036
	III	0.48	0.31	0.08	0.09	1.00	0.31	0.59	0.070	0.036
14-10 Type A or 1/4-14 Types B and AB Tapping	I	0.38	0.38	0.09	0.09	0.94	0.44	0.66	0.085	0.042
	III	0.62	0.38	0.09	0.09	1.17	0.44	0.66	0.085	0.042

[a] These data are intended for design guidance only and are not to be considered a mandatory part of the standard. The dimensions specified have been selected to accommodate the optional constructions of wide-range design spring nuts and both Type P and Type T regular design spring nuts, respectively, covered by the standard. Tolerances on the nuts and tolerances entailed in the manufacturing processes used to emboss and punch the various assembly features were not considered in the derivation of these dimensions.

TYPE P AND TYPE T FLAT

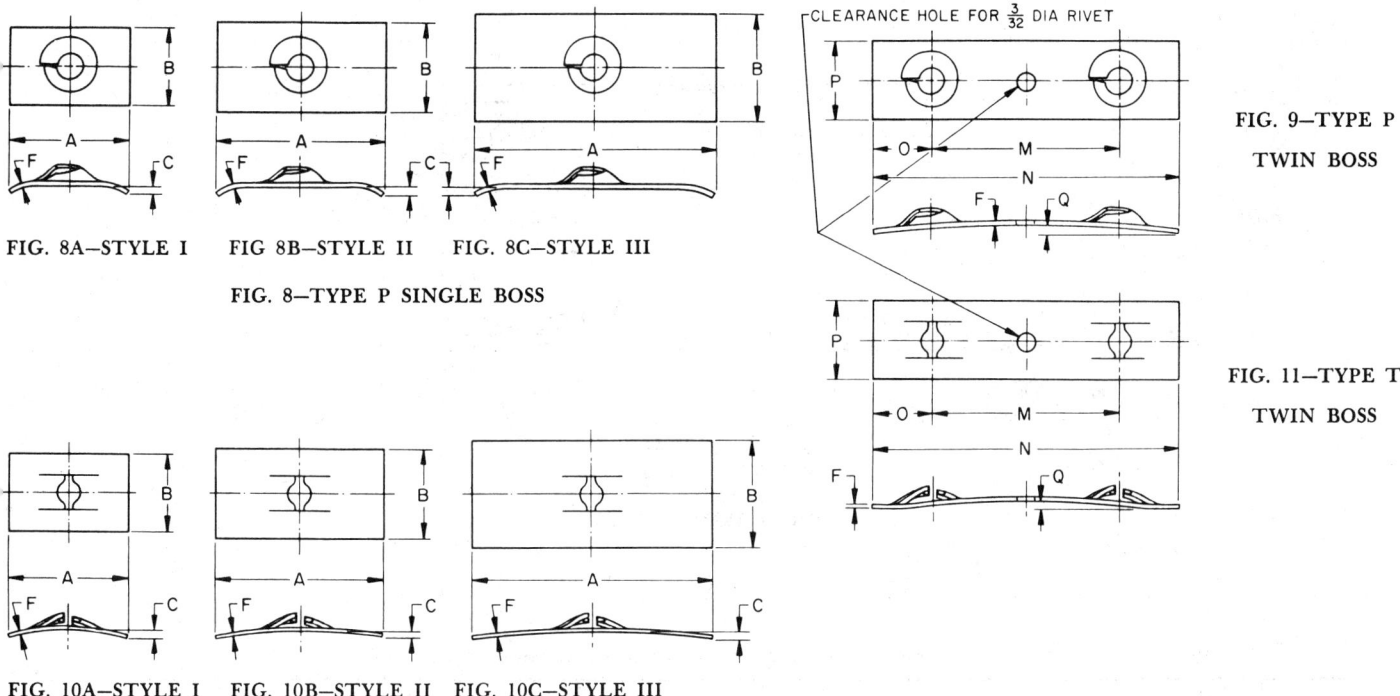

FIG. 8A—STYLE I   FIG. 8B—STYLE II   FIG. 8C—STYLE III

FIG. 8—TYPE P SINGLE BOSS

FIG. 9—TYPE P TWIN BOSS

FIG. 10A—STYLE I   FIG. 10B—STYLE II   FIG. 10C—STYLE III

FIG. 10—TYPE T SINGLE BOSS

FIG. 11—TYPE T TWIN BOSS

TABLE 6—DIMENSIONS OF TYPE P AND TYPE T FLAT SPRING NUTS (FIGS. 8–11)

Screw Size and Type[a]		Single Boss				F Stock Thickness		Twin Boss					
	Style	A Nut Length ±0.02	B Nut Width	C Arch Height				M Boss Center to Center	N Nut Length ±0.02	O End to Center	P Nut Width	Q Arch Height	
				Max	Min	Max	Min					Max	Min
6-32 Machine	I[b]	0.44	0.281	0.040	0.020	0.018	0.015	0.500	1.12	0.312	0.375	0.045	0.005
	II	0.88	0.375	0.060	0.030			0.625	1.25	0.312	0.375	0.045	0.005
	III	1.38	0.500	0.050	0.010			0.750	1.38	0.312	0.375	0.045	0.005
								0.875	1.50	0.312	0.375	0.065	0.005
								1.000	1.62	0.312	0.375	0.065	0.005
6-18 Type A and 6-20 Types B and AB Tapping	I[b]	0.50	0.312	0.040	0.020	0.026	0.019	0.500	1.12	0.312	0.375	0.045	0.005
	II	0.88	0.438	0.060	0.030			0.625	1.25	0.312	0.375	0.045	0.005
	III	1.38	0.531	0.060	0.020			0.750	1.38	0.312	0.375	0.045	0.005
								0.875	1.50	0.312	0.375	0.065	0.005
								1.000	1.62	0.312	0.375	0.065	0.005
8-32 Machine	I	0.50	0.312	0.050	0.030	0.018	0.015	0.500	1.12	0.312	0.375	0.045	0.005
	II	0.88	0.438	0.070	0.040			0.625	1.25	0.312	0.375	0.045	0.005
	III	1.38	0.531	0.055	0.015			0.750	1.38	0.312	0.375	0.045	0.005
								0.875	1.50	0.312	0.375	0.065	0.005
								1.000	1.62	0.312	0.375	0.065	0.005
8-15 Type A and 8-18 Types B and AB Tapping	I	0.62	0.406	0.055	0.025	0.029	0.024	0.500	1.12	0.312	0.375	0.045	0.005
	II	0.88	0.469	0.065	0.035			0.625	1.25	0.312	0.375	0.045	0.005
	III	1.25	0.562	0.050	0.010			0.750	1.38	0.312	0.375	0.045	0.005
								0.875	1.50	0.312	0.375	0.065	0.005
								1.000	1.62	0.312	0.375	0.065	0.005
10-24 Machine	I	0.62	0.375	0.065	0.035	0.023	0.019	0.500	1.12	0.312	0.500	0.045	0.005
	II	1.00	0.469	0.065	0.035			0.625	1.25	0.312	0.500	0.045	0.005
	III	1.38	0.562	0.080	0.040			0.750	1.38	0.312	0.500	0.045	0.005
								0.875	1.50	0.312	0.500	0.065	0.005
								1.000	1.62	0.312	0.500	0.065	0.005
10-12 Type A and 10-16 Types B and AB Tapping	I	0.75	0.500	0.065	0.035	0.033	0.028	0.500	1.12	0.312	0.500	0.045	0.005
	II	1.12	0.594	0.050	0.020			0.625	1.25	0.312	0.500	0.045	0.005
	III	1.38	0.688	0.080	0.040			0.750	1.38	0.312	0.500	0.045	0.005
								0.875	1.50	0.312	0.500	0.065	0.005
								1.000	1.62	0.312	0.500	0.065	0.005
1/4-20 Machine	I	0.75	0.500	0.075	0.045	0.027	0.024	0.750	1.50	0.375	0.562	0.065	0.005
	II	1.12	0.594	0.050	0.020			0.875	1.62	0.375	0.562	0.065	0.005
	III	1.38	0.688	0.085	0.045			1.000	1.75	0.375	0.562	0.065	0.005
14-10 Type A or 1/4-14 Types B and AB Tapping	I	0.88	0.562	0.075	0.045	0.039	0.028	0.750	1.50	0.375	0.562	0.065	0.005
	II	1.06	0.625	0.050	0.020			0.875	1.62	0.375	0.562	0.065	0.005
	III	1.38	0.688	0.085	0.045			1.000	1.75	0.375	0.562	0.065	0.005

[a] See Boss Detail under General Specifications for applicability of types and sizes.
[b] The Type P nuts in this style will not meet the performance requirements in Table 2 due to the limitations on boss design imposed by the narrow width and the long length.

WIDE-RANGE DESIGN, U SHAPE

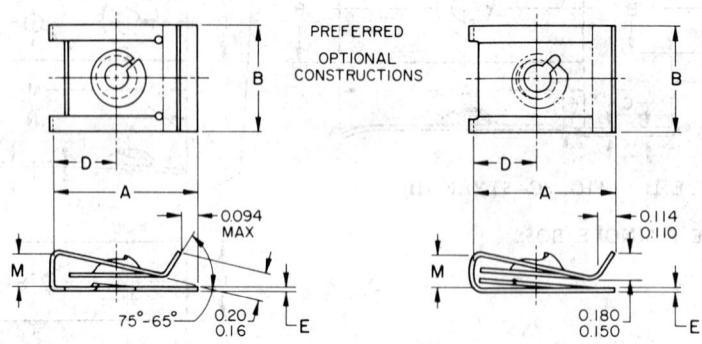

FIG. 12—U SHAPE
SHORT THROAT STYLE
AND
LONG THROAT STYLE

TABLE 7—DIMENSIONS OF U SHAPE WIDE-RANGE DESIGN SPRING NUTS (FIG. 12)

Screw Size and Type	Panel Thickness Range	Style	A Leg Length		B Nut Width		D Throat Depth		E Stock Thickness		M Width at Fold	
			Max	Min	Max	Min	Max	Min	Max	Min	Max	Min
6-20 Types B and AB Tapping	0.025-0.144	Short	0.785	0.755	0.550	0.520	0.360	0.340	0.027	0.018	0.190	0.150
	0.025-0.144	Long	1.015	0.985	0.550	0.520	0.600	0.570	0.027	0.018	0.190	0.150
8-18 Types B and AB Tapping	0.025-0.144	Short	0.785	0.755	0.550	0.520	0.360	0.340	0.030	0.023	0.190	0.150
	0.025-0.144	Long	1.015	0.985	0.550	0.520	0.600	0.570	0.030	0.023	0.190	0.150
10-16 Types B and AB Tapping	0.025-0.144	Short	0.785	0.755	0.650	0.620	0.360	0.340	0.035	0.028	0.190	0.150
	0.025-0.144	Long	1.015	0.985	0.650	0.620	0.600	0.570	0.035	0.028	0.190	0.150
1/4-14 Types B and AB Tapping	0.025-0.144	Short	0.785	0.755	0.650	0.620	0.360	0.340	0.039	0.033	0.190	0.150
	0.025-0.144	Long	1.015	0.985	0.650	0.620	0.600	0.570	0.039	0.033	0.190	0.150

REGULAR DESIGN, TYPE P, J SHAPE, AND U SHAPE

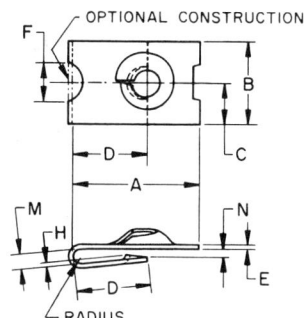

FIG. 13—J SHAPE
STYLE I—STANDARD,
STYLE II—SHORT
AND
STYLE III—LONG
THROAT

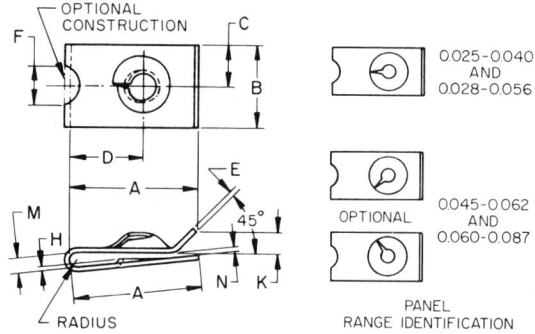

FIG. 14—U SHAPE
STYLE I—STANDARD,
STYLE II—SHORT
AND
STYLE III—LONG
THROAT

TABLE 8—DIMENSIONS OF TYPE P, J SHAPE AND U SHAPE REGULAR DESIGN SPRING NUTS (FIGS. 13 AND 14)

Screw Size and Type[a]	Panel Thickness Range	Style	A Leg Length, ±0.02 J Shape	A Leg Length, ±0.02 U Shape	B Nut Width	C Edge to Center	D Throat Depth, ±0.015	E Stock Thickness Max	E Stock Thickness Min	F[b] Hole Dia	H Height, +0.010 −0.000	K Tang Height, Approx U Shape	M Width at Fold	N Gap Opening, Min J Shape	N Gap Opening, Max U Shape
6-18 Type A and 6-20 Types B and AB Tapping	0.025-0.040 0.045-0.062	I	0.53	0.59			0.330						0.06 0.08	0.02 0.04	0.025 0.041
	0.025-0.040 0.045-0.062	II	0.47	0.53	0.34	0.17	0.270	0.021	0.019	0.218	0.025	0.09	0.06 0.08	0.02 0.04	0.025 0.041
	0.025-0.040 0.045-0.062	III	0.70	0.76			0.520						0.06 0.08	0.02 0.04	0.025 0.041
8-15 Type A and 8-18 Types B and AB Tapping	0.025-0.040 0.045-0.062	I	0.61	0.67			0.390						0.06 0.08	0.02 0.04	0.025 0.041
	0.025-0.040 0.045-0.062	II	0.51	0.57	0.40	0.20	0.280	0.026	0.024	0.218	0.025	0.09	0.06 0.08	0.02 0.04	0.025 0.041
	0.025-0.040 0.045-0.062	III	0.81	0.88			0.580						0.06 0.08	0.02 0.04	0.025 0.041
10-12 Type A and 10-16 Types B and AB Tapping	0.025-0.040 0.045-0.062	I	0.64	0.70			0.390						0.06 0.08	0.02 0.04	0.025 0.041
	0.025-0.040 0.045-0.062	II	0.61	0.64	0.50	0.25	0.330	0.032	0.028	0.218	0.025	0.09	0.06 0.08	0.02 0.04	0.025 0.041
	0.025-0.040 0.045-0.062	III	0.83	0.89			0.580						0.06 0.08	0.02 0.04	0.025 0.041
14-10 Type A or 1/4-14 Types B and AB Tapping	0.028-0.056 0.060-0.087	I	0.80	0.86	0.56	0.28	0.470	0.032	0.028	0.218	0.025	0.08	0.08 0.10	0.04 0.05	0.035 0.041
	0.028-0.056 0.060-0.087	III	1.09	1.16			0.720						0.08 0.10	0.04 0.05	0.035 0.041

[a]See Boss Detail under General Specifications for applicability of types and sizes. Type P spring nuts of similar proportions are also available in respective machine screw sizes 6-32, 8-32, 10-24, and 1/4-20; and manufacturers should be consulted for dimensions.
[b]Diameter of hole punched in blank before forming.

REGULAR DESIGN, TYPE T, J SHAPE, AND U SHAPE

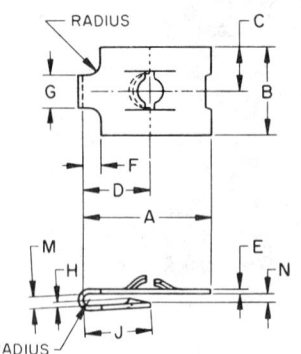
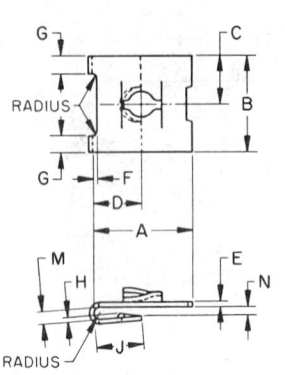
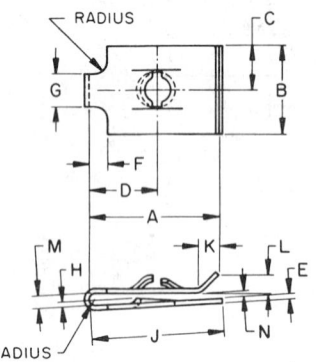
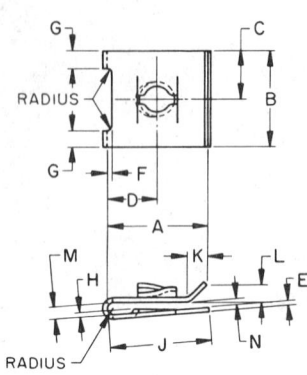

PANEL RANGE IDENTIFICATION FOR J SHAPE NUTS — 0.025–0.040 AND 0.028–0.056 ; 0.045–0.062 AND 0.060–0.087

PANEL RANGE IDENTIFICATION FOR U SHAPE NUTS — 0.025–0.040 AND 0.028–0.056 ; 0.045–0.062 AND 0.060–0.087

FIG. 15A— STYLE I—STANDARD THROAT AND STYLE III—LONG THROAT

FIG. 15B— STYLE II—SHORT THROAT

FIG. 16A— STYLE I—STANDARD THROAT AND STYLE III—LONG THROAT

FIG. 16B— STYLE II—SHORT THROAT

FIG. 15—J SHAPE

FIG. 16—U SHAPE

TABLE 9—DIMENSIONS OF TYPE T, J SHAPE AND U SHAPE REGULAR DESIGN SPRING NUTS (FIGS. 15A–16B)

Screw Size and Type[a]	Panel Thickness Range	Style	A Leg Length, ±0.02	B Nut Width	C Edge to Center	D Throat Depth, ±0.015	E Stock Thickness Max	E Stock Thickness Min	F Notch Depth, ±0.015	G Width, ±0.02	H Height, ±0.005	J Leg Length, ±0.02 J Shape	K Tang Length, ±0.015 U Shape	L Tang Height, ±0.015 U Shape	M Width at Fold	N Gap Opening, Min J Shape	N Gap Opening, Max U Shape
6-18 Type A and 6-20 Types B and AB Tapping	0.025–0.040 0.045–0.062	I	0.64	0.312	0.156	0.340			0.085	0.11		0.35			0.06 0.08	0.02 0.04	0.025 0.041
	0.025–0.040 0.045–0.062	II	0.48	0.500	0.250	0.265	0.027	0.023	0.025	0.09	0.020	0.26	0.070	0.080	0.06 0.08	0.02 0.04	0.025 0.041
	0.025–0.040 0.045–0.062	III	0.81	0.312	0.156	0.510			0.245	0.14		0.51			0.06 0.08	0.02 0.04	0.025 0.041
8-15 Type A and 8-18 Types B and AB Tapping	0.025–0.040 0.045–0.062	I	0.68	0.406	0.203	0.370			0.095	0.18		0.37			0.06 0.08	0.02 0.04	0.025 0.041
	0.025–0.040 0.045–0.062	II	0.52	0.500	0.250	0.260	0.030	0.026	0.020	0.09	0.025	0.27	0.090	0.090	0.06 0.08	0.02 0.04	0.025 0.041
	0.025–0.040 0.045–0.062	III	0.86	0.406	0.203	0.550			0.275	0.18		0.56			0.06 0.08	0.02 0.04	0.025 0.041
10-12 Type A and 10-16 Types B and AB Tapping	0.025–0.040 0.045–0.062	I	0.76	0.500	0.250	0.420			0.110	0.12		0.42			0.06 0.08	0.02 0.04	0.025 0.041
	0.025–0.040 0.045–0.062	II	0.57	0.625	0.312	0.305	0.033	0.029	0.020	0.13	0.025	0.31	0.090	0.090	0.06 0.08	0.02 0.04	0.025 0.041
	0.025–0.040 0.045–0.062	III	0.96	0.500	0.250	0.610			0.255	0.17		0.61			0.06 0.08	0.02 0.04	0.025 0.041
14-10 Type A or 1/4-14 Types B and AB Tapping	0.028–0.056 0.060–0.087	I	0.92	0.562	0.281	0.495	0.042	0.035	0.065	0.18	0.030	0.49	0.110	0.100	0.09 0.12	0.04 0.05	0.035 0.041
	0.028–0.056 0.060–0.087	III	1.16	0.562	0.281	0.745	0.039	0.035	0.310	0.25		0.75			0.09 0.12	0.04 0.05	0.035 0.041

[a] See Boss Detail under General Specifications for applicability of types and sizes. Type T spring nuts of similar proportions are also available in respective machine screw sizes 6-32, 8-32, 10-24, and 1/4-20; and manufacturers should be consulted for dimensions.

PUSH-ON SPRING NUTS—SAE J892a

SAE Standard

Report of Fasteners Committee approved August 1964 and last revised June 1970.

GENERAL SPECIFICATIONS

Scope—It should be noted that push-on spring nuts having other configurations are available and manufacturers should be consulted.

Dimensional Tolerance—Tolerance on dimensions in the tables shall be plus and minus 0.010 in. unless otherwise specified. Tolerance on the thickness of material shall be plus and minus 0.001 in.

Boss—Size and formation of boss and other detail shall be such as to assemble readily and function satisfactorily with the specified stud.

Material—Spring steel suitably processed to meet the hardness requirements of this specification.

Hardness—Hardness shall be as follows:

Material Thickness	Rockwell Scale	Dial Reading	Conversion to Rockwell C Scale
Up to 0.016	15N	82.5–85.5	44–50
0.017–0.024	30N	63.0–68.5	44–50

Finish—Spring nuts are normally supplied with rustproof finish as specified by purchaser. Nuts subjected to corrosion preventative treatment which might induce hydrogen embrittlement shall be baked or otherwise treated to obviate such embrittlement.

Workmanship—Spring nuts shall be free from cracks, burrs, splits, loose scale, or any other defects which might affect their serviceability.

Application and Design—Where nut is to function only as a locking means, Style I is recommended. Where greater area of load distribution is a requirement, i.e.; nut is to function also as spanner washer, Style II is recommended. The Light Series are for use on plastic studs; the Medium Series are for use on soft metal studs, and the Heavy Series are for use on hardened metal or chromium plated studs.

Assembly Considerations—Since performance of push-on spring nuts is dependent upon the studs to which they are applied, it is essential that stud diameters and plating recommendations as set forth in Fig. 1 and Table 1 be adhered to as closely as possible. The actual stud length is determined by adding the thickness of mating panel or panels "T," through which the stud protrudes, to the factors tabulated under "C" (the minimum stud protrusion

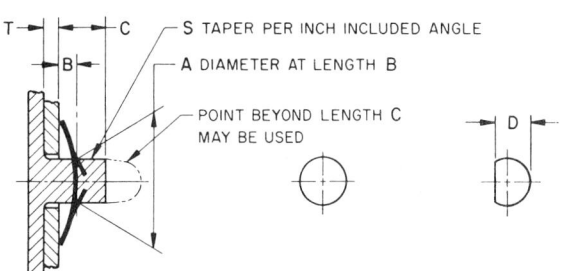

FIG. 1A—ROUND STUD FIG. 1B—"D" SHAPED STUD

FIG. 1—RECOMMENDED STUD DESIGN

TABLE 1—DIMENSIONS OF RECOMMENDED STUDS (FIGS. 1A and 1B)

Nominal Stud Size	A[a] Stud Diameter		B Straight Length, Min	C Stud Length, Min	D Stud Width		S Taper, Max
	Max	Min			Max	Min	
1/16	0.065	0.059	T + 0.24	T + 0.36	0.054	0.044	0.052
3/32	0.097	0.091	T + 0.24	T + 0.36	0.079	0.069	0.052
1/8	0.128	0.122	T + 0.24	T + 0.36	0.105	0.095	0.052
5/32	0.159	0.153	T + 0.24	T + 0.36	0.130	0.120	0.040
3/16	0.191	0.185	T + 0.24	T + 0.36	0.155	0.145	0.040
7/32	0.222	0.216	T + 0.24	T + 0.36	0.180	0.170	0.040
1/4	0.253	0.247	T + 0.24	T + 0.36	0.205	0.195	0.040
5/16	0.315	0.309	T + 0.24	T + 0.36	0.255	0.245	0.033
3/8	0.378	0.372	T + 0.24	T + 0.36	0.305	0.295	0.033

[a]Diameter limits include thickness of plating on studs. Chromium or nickel plating is permissible only on studs to be used with Heavy Series Nuts and then it is recommended that plating thickness along length of die cast studs be held to within 0.0015 in. wherever possible and that in no case should the plating thickness exceed 0.003 in.

TABLE 2—DIMENSIONS OF PUSH-ON SPRING NUTS (FIGS. 2A–2C)

Nominal Stud Size	Style	Series[a]	A Nut Length	B Nut Width	C Arch Height Max	C Arch Height Min	F Stock Thickness
1/16	I	Light	0.38	0.22	0.025	0.005	0.012
		Medium					0.014
		Heavy					0.017
	II	Light	0.56	0.34	0.040	0.010	0.012
		Medium					0.014
		Heavy					0.017
3/32	I	Light	0.45	0.23	0.040	0.010	0.012
		Medium					0.014
		Heavy					0.017
	II	Light	0.70	0.38	0.050	0.020	0.012
		Medium					0.014
		Heavy					0.017
1/8	I	Light	0.58	0.31	0.040	0.010	0.012
		Medium					0.014
		Heavy					0.017
	II	Light	0.45	0.50	0.080	0.050	0.012
		Medium					0.014
		Heavy					0.017
5/32	I	Light	0.56	0.38	0.040	0.010	0.012
		Medium					0.014
		Heavy					0.017
5/32	II	Light	0.88	0.56	0.075	0.045	0.012
		Medium					0.014
		Heavy					0.017
3/16	I	Light	0.62	0.38	0.060	0.030	0.012
		Medium					0.017
		Heavy					0.020
	II	Light	0.98	0.56	0.080	0.050	0.012
		Medium					0.017
		Heavy					0.020
7/32	I	Light	0.62	0.44	0.050	0.020	0.012
		Medium					0.017
1/4	I	Light	0.62	0.44	0.050	0.020	0.012
		Medium					0.017
		Heavy					0.020
	II	Light	0.98	0.62	0.095	0.065	0.012
		Medium					0.017
		Heavy					0.020
5/16	I	Light	0.69	0.50	0.060	0.030	0.014
		Medium					0.020
3/8	I	Light	0.75	0.56	0.060	0.030	0.014
		Medium					0.020

[a]See General Specifications, Application and Design.

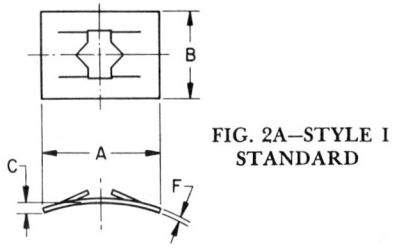 FIG. 2A—STYLE I STANDARD

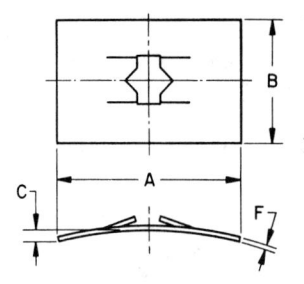 FIG. 2B—STYLE II SPANNER

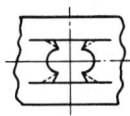 FIG. 2C—HEAVY SERIES BOSS DESIGN

SEE TABLE 2, NOTE a

FIG. 2—PUSH-ON SPRING NUTS

required for normal installation). It may be necessary to increase this factor to provide adequate stud protrusion where uncompressed materials or mismatch of trim contours are encountered. It should be noted by users desiring to standardize on stud designs that the studs applicable to self-threading stamped nuts may be utilized for push-on spring nuts where economics justify and the additional stud protrusion is not objectionable.

Heavy Series Nuts are used on round studs only. All other nuts in this standard may be used on either round or "D" shaped studs. Nuts are used on round studs in applications where the assembly is permanent and on "D" shaped studs where disassembly is a consideration.

STEEL STAMPED NUTS OF ONE PITCH THREAD DESIGN—SAE J1053

SAE Standard

Report of Fasteners Committee approved August 1973.

1. General Specifications

1.1 Scope—Included herein are general, dimensional, and performance specifications for those types, styles, and sizes of stamped nuts of one pitch thread design recognized as SAE Standard. These nuts are intended for general use where the engagement of a single thread on the mating screw or unthreaded stud is considered adequate for the application.

It should be noted that stamped nuts having other sizes and configurations are available and manufacturers should be consulted.

1.2 Dimensional Tolerance—Tolerance on dimensions shown in the tables shall be ±0.010 in. unless otherwise specified.

1.3 Miscellaneous Dimensions—Taper on the sides of hexagon portions of nuts (angle between one side and the axis of nut) shall not exceed 1 deg, the maximum limit specified being the largest dimension. Tolerance on steel stock thickness shall be ±0.001 in. for thicknesses up to and including 0.028 in., and ±0.0015 in. for thicknesses exceeding 0.028 in.

1.4 Thread Embossments

1.4.1 Formed Thread Embossment—Detail of the thread engaging portion of formed thread type nuts shall be such as to permit nut to assemble readily with the specified screw and not strip or deform at the minimum torques shown in Table 2. The edges around the opening shall be spirally formed to conform to the helix of the mating thread and, as indicated on illustrations, the top or top and bottom corners on edges of holes shall be swaged to provide flats for bearing on flanks of the mating thread.

1.4.2 Self-Threading Embossment—The configuration of self-threading embossment may vary with manufacturer; however, the detail and formation of embossment shall be such as to enable the nuts to cut and/or form threads on cast or wire studs, conforming to the recommended stud designs contained in paragraph 3.2, at or below the maximum driving torques shown in Table 3.

1.5 Material—Nuts shall be fabricated from carbon spring steel suitably processed to meet the performance requirements of this standard.

1.6 Finish—Stamped nuts are normally supplied with finishes as specified by the purchaser. Nuts processed with supplemental finishes shall be suitably treated to obviate hydrogen embrittlement.

1.7 Workmanship—Stamped nuts shall be free from cracks, burrs, splits, loose scale, or any defects that might affect their serviceability.

2. Test Procedures and Performance Requirements

2.1 Formed Thread Embossment

2.1.1 Ultimate Torque Test—Insert hardened steel (Rockwell C53 min) unplated or uncoated test socket head cap screws of the respective size and 1.00 in. length, Class 3A thread, as-received with light coating of oil, into holes in the test fixture. The test fixture is to consist of a hardened steel (Rockwell C58-62) bar, 1.00 x 0.25 x 18.00 in. or equivalent, having 12 equally spaced test holes of the diameter given in Table 1 for respective size.

Hand assemble the test nuts to the test screws. In turn, hold each nut and tighten the test screw to the torque value shown in Table 2 for the respective size. The test shall be performed with a device capable of measuring the clamp load developed, and the load attained shall not be less than the minimum tension values specified in Table 2. After initial breakaway, the nuts must disassemble, by hand, from the test screws.

2.1.2 Embrittlement Test—Insert hardened steel (Rockwell C53 min) unplated or uncoated test socket head cap screws of the respective size and 1.00 in. length, Class 3A thread, as-received with light coating of oil, into holes in the test fixture described in paragraph 2.1.1.

Hand assemble new test nuts, from the same lot, to test screws. In turn, hold each nut and tighten test screw to the minimum torque value shown in

TABLE 1—TEST BAR HOLE SIZES, IN

Nominal Screw and Nut Size	Hole Diameter		Nominal Screw and Nut Size	Hole Diameter	
	Max	Min		Max	Min
6-32	0.149	0.144	1/4-20	0.262	0.257
8-32	0.178	0.173	5/16-18	0.328	0.323
10-24	0.204	0.199	3/8-16	0.391	0.386

TABLE 2—ULTIMATE TORQUE SPECIFICATIONS

Nominal Nut Size, in	Torque, lb-in	Tension, lb	Nominal Nut Size, in	Torque, lb-in	Tension, lb
	Min	Min		Min	Min
6-32	8	120	1/4-20	27	340
8-32	12	150	5/16-18	32	450
10-24	17	220	3/8-16	40	480

TABLE 3—TORQUE AND RELATED TENSION SPECIFICATIONS

Nominal Stud or Test Rod Dia, in	Test Rod Dia, in		Driving Torque, lb-in	Nut Flange Dia, in	Test Torque, lb-in	Tension, lb
	Max	Min	Max	Basic		Min
1/8	0.126	0.123	8	0.437	34	180
3/16	0.189	0.186	26	0.562	68	280
1/4	0.251	0.248	35	0.687	90	350

Table 2 for respective size. After 48 h in this state, the test nuts shall be examined. No cracks are permitted.

2.1.3 SCREW THREAD DAMAGE APPRAISAL TEST—Insert 12 nonheat treated, unplated or uncoated, steel screws of respective size and 1.00 in. length, into holes in the test fixture described in paragraph 2.1.1.

Hand assemble new test nuts, from the same lot, to test screws. In turn, hold each nut and tighten test screw to the torque value shown in Table 2 for the respective size. Remove nuts from screws and screws from test bar and examine threads on screws for visible damage. Continue test by assembling, with the fingers, untested nuts from same lot onto tested screws. The new nuts must pass over the area on the screw where the previously tested nut engaged the threads.

2.2 Self-Threading Embossment

2.2.1 STARTING EASE TEST—The test nut must start onto the chamfered end (0.030 x 45 deg) of an unplated or uncoated cold rolled steel (Rockwell 30T 78–81) rod of the diameter specified in Table 3, within one revolution of nut when applied with an appropriate socket affixed to a screwdriver handle.

2.2.2 ULTIMATE TORQUE TEST—Insert a test rod (see paragraph 2.2.1) into a suitable holding device exposing the chamfered end to a height equivalent to the nut height plus 0.125 in. or, for closed end nuts, equivalent to the wrenching height. Place an unplated or uncoated soft steel (Rockwell 30T 78–82) flat test plate on the exposed test rod. The test plate shall have a minimum thickness of 0.030 in., an inside diameter 0.031 in. larger than the diameter of test rod, and shall be at least 1.00 in. square. A new test plate shall be used for each torque test. The test rod and assembled plate must be retained in a suitable clamping device to prevent rotation of the rod and plate and tilting of the plate. The test shall be performed with a device capable of measuring the clamp load developed.

Assemble test nut on the test rod with a suitable torque indicating device. The maximum driving torque shall be recorded and this shall not exceed the maximum driving torque values shown in Table 3 for the respective size. At the torque test values specified in Table 3, the minimum tension values indicated shall be achieved.

2.2.3 EMBRITTLEMENT TEST—Assemble new test nut from the same lot to test rod using test torques shown in Table 3 for respective size. After 48 h, inspect the assembled nut for cracks. No cracks are permitted.

3. Design Criteria

3.1 Formed Thread Embossment—To insure proper starting of formed thread type stamped nuts, the length of the mating externally threaded component shall be such that it will protrude beyond the embossment in nut a minimum distance equivalent to two pitches (threads), exclusive of the length of any chamfer or point provision, under limit stack conditions. Recommended minimum protrusion lengths beyond panels with no allowance for pointing are presented in Fig. 1 and Table 4 for respective nut types.

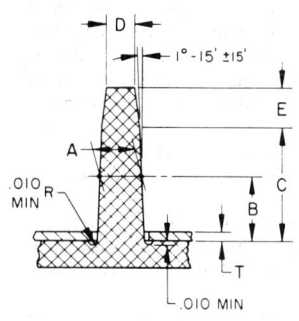

FIG. 2—DIE CAST STUD

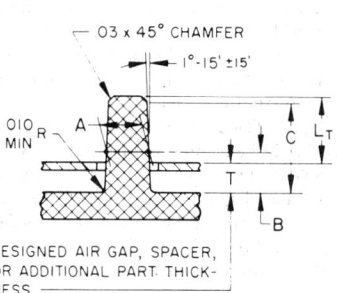

FIG. 3—SHORT DIE CAST STUD

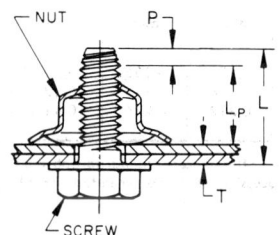

FIG. 1A—FACETED FLANGE TYPE NUTS

FIG. 1B—ACORN OR REGULAR TYPE NUTS

WHERE:
L = Lp + T + P
L = MINIMUM LENGTH OF SCREW OR STUD
Lp = MINIMUM PROTRUSION OF FULL FORM THREAD LENGTH BEYOND PANEL (SEE TABLE 4 FOR RESPECTIVE NUT TYPES)
LT = MAXIMUM PROTRUSION OF MATING PART BEYOND PANEL ALLOWABLE FOR ACORN TYPE NUTS (SEE TABLE 4)
P = LENGTH OF POINT ON SCREW OR STUD
T = MAXIMUM THICKNESS OF PANEL OR PANELS TO BE ASSEMBLED, INCLUDING ALLOWANCE, IF NECESSARY, TO ACCOMMODATE MISMATCH OF SURFACES, ETC.

FIG. 1

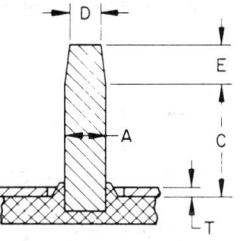

FIG. 4—WIRE OR ROD STUD

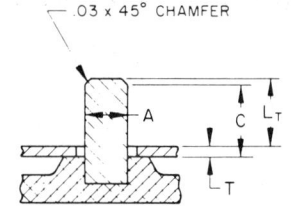

FIG. 5—SHORT WIRE OR ROD STUD

TABLE 4—PROTRUSION LENGTHS FOR FORMED THREAD TYPE STAMPED NUTS, IN

Nominal Thread Size	Lp		LT
	Protrusion of Threaded Length on Mating Part Beyond Panel		Total Protrusion of Mating Part Beyond Panel
	Faceted Flange Type	Acorn or Regular Types	Acorn Types
	Min[a]	Min[b]	Max[c]
6-32	0.29	0.13	0.21
8-32	0.29	0.13	0.24
10-24	0.33	0.16	0.25
1/4-20	0.40	0.19	0.28
5/16-18	0.44	0.22	0.36
3/8-16	—	0.23	0.34

[a]Values shown are applicable to nuts shown in Table 7. For sealer styles, add height of uncompressed sealer.
[b]Values shown are applicable to nuts shown in Tables 8 and 9, respectively.
[c]Values shown apply to nuts shown in Table 8.

TABLE 5—RECOMMENDED DIMENSIONS OF STUDS FOR USE WITH SELF-THREADING TYPES OF STAMPED NUTS, IN

Nominal Stud Size	A[a] Stud Dia		B[b,c] Length		C[c] Length		D[a] Point Dia	E[d] Point Length		LT Stud Protrusion Acorn Type
			Faceted Flange Type	Acorn and Regular Types	Faceted Flange Type	Acorn and Regular Types				
	Max	Min	Min	Min	Min	Min	±0.005	Max	Min	Max
1/8[e]	0.128	0.122	T+0.24	T+0.08	T+0.45	T+0.12	0.073	0.175	0.145	0.26
3/16	0.191	0.185	T+0.24	T+0.11	T+0.45	T+0.16	0.130	0.190	0.150	0.36
1/4	0.253	0.247	T+0.27	T+0.11	T+0.45	T+0.16	0.180	0.260	0.220	0.38

[a]Maximum 0.003 in plating per side.
[b]Point on shank of die cast studs where A diameter must be within the specified limits.
[c]The T dimension in illustrations represents the distance from base of part to the bearing face of nut in the installed position. The factors to be added represent the minimum length required for normal installation of sealer styles of nuts. Where the factors specified would create an interference condition or otherwise be objectionable, it may be necessary to reduce the factor to that which is required for the respective nut size, type and style.
[d]On studs for acorn type nuts, it may be necessary to shorten point or apply the chamfer specified for short studs in order to keep protrusion of stud beyond panel within maximum permissible.
[e]Due to susceptibility to breakage in handling and processing, it is recommended use of 1/8 in size die cast studs be avoided wherever possible.

3.2 Self-Threading Embossment—To assure proper function and performance of self-threading types of stamped nuts and to provide flexibility for changing nut designs, it is essential that studs and clearance holes in mating panels be designed in conformance with the recommendations set forth in the following.

3.2.2 STUD DESIGN—Studs which are integral features of die cast components should comply as closely as possible with the recommendations presented in Figs. 2 and 3 and Table 5. Studs fabricated from wire or rod shall be in accord with recommendations shown in Figs. 4 and 5 and Table 5. Consideration should also be given to the recommendations for fillets, plating and alignment which follow:

3.2.2.1 *Fillets*—The fillet at the junction of stud with die casting base shall have as generous a radius as the design will permit, but not less than 0.010 in. Where the panel is to fit tight against the die casting, an annular relief groove should be provided in the die casting at base of stud to accommodate the fillet (see Fig. 2) and the fillet radius should be made larger wherever the design will permit.

3.2.2.2 *Angularity*—It is preferable that the axis of the stud be kept perpendicular to the base of the part or as nearly so as possible. However, where design conditions or parting lines on die castings dictate the axis of stud must deviate from square with base, the departure from perpendicular shall not exceed 20 deg. Similarly, where sufficient driver clearance cannot be provided in line with the stud axis, the angular deviation from axis should in no case exceed 15 deg in order to insure the socket will have adequate engagement with the nut for assembly.

3.2.2.3 *Stud Location*—On drawings for parts entailing multiple studs, the studs should be located in accordance with the dimensioning and tolerancing practices set forth in the SAE Drawing Standards.

3.3 Panel Clearance Holes—The clearance holes in mating panels for stamped nuts should be designed in conformance with Fig. 6 and Table 6. A selection of three hole sizes for each stud size is provided to best satisfy varying design conditions as explained in the following:

(a) Preferred hole sizes listed under "X" are recommended and should be used for all attachments requiring normal provisions for clearance and adjustment.

(b) Maximum clearance holes tabulated under "X_1" should be used only in applications where maximum adjustment capability is a requirement. These holes provide maximum clearance while assuring that the hole can effectively be sealed with sealer styles of the faceted flange type nuts contained herein.

(c) Minimum clearance holes shown under "X_2" may necessarily have to be used where the width of the part being fastened is at or approaches the minimum "Z" dimension. These holes provide adequate clearance for studs under limit stack conditions while insuring that the fastened part will cover the hole. It follows, therefore, that the "Z" dimension shall be the design criterion for the width of portions of parts adjacent to studs.

3.3.1 PANEL HOLE LOCATION—On drawings for panels, multiple holes shall be located in a manner which is compatible with that used to position studs on the mating part.

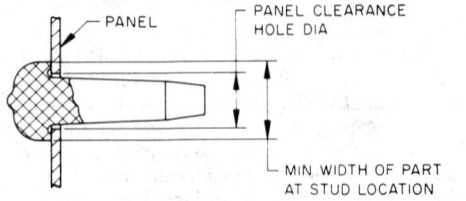

FIG. 6

TABLE 6—PANEL CLEARANCE HOLES, IN

Nominal Stud Dia	X		X_1		X_2		Z
	Clearance Hole Diameter[a]						Part Width
	Preferred		Maximum Clearance		Minimum Clearance		
	Max	Min	Max	Min	Max	Min	Min
1/8[b]	0.188	0.172	0.219	0.203	0.171	0.155	0.24
3/16	0.250	0.234	0.281	0.265	0.234	0.219	0.32
1/4	0.344	0.328	0.406	0.390	0.312	0.296	0.41

[a] For recommendations on application of the three choices offered, refer to paragraph 3.3.
[b] Due to susceptibility to breakage in handling and processing, it is recommended use of 1/8 in size die cast studs be avoided wherever possible.

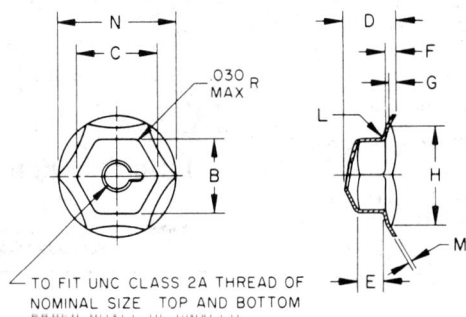

FIG. 7

TABLE 7—DIMENSIONS OF FORMED THREAD FACETED FLANGE TYPE STAMPED NUTS, IN[a]

Nominal Size[b] or Basic Thread Dia	Threads per in	B Hexagon Across Flats		C Hexagon Across Corners		D Overall Height		E Height of Flat	F Depth to Radius		G Dish Depth		H Dish Diameter		L Fillet Radius		M Stock Thickness	N Flange Diameter	
		Max	Min	Max	Min	Max	Min	Min	Max	Min	Max	Min	Max	Min	Max	Min	Basic	Max	Min
6 0.1380	32	0.312	0.306	0.360	0.348	0.218	0.198	0.067	0.017	0.007	—	—	0.401	0.387	0.033	0.027	0.013	0.442	0.432
8 0.1640	32	0.343	0.337	0.396	0.382	0.225	0.205	0.072	0.018	0.008	—	—	0.429	0.415	0.035	0.029	0.014	0.474	0.464
10 0.1900	24	0.375	0.369	0.433	0.418	0.246	0.226	0.076	0.019	0.009	—	—	0.457	0.443	0.037	0.031	0.018	0.505	0.495
1/4 0.2500	20	0.437	0.430	0.505	0.488	0.302	0.282	0.090	0.047	0.035	0.043	0.033	0.572	0.552	0.039	0.033	0.021	0.692	0.682
5/16 0.3125	18	0.500	0.492	0.577	0.557	0.330	0.310	0.100	0.057	0.045	0.050	0.040	0.674	0.650	0.041	0.035	0.023	0.817	0.807

[a] Sealer styles are also available, consult nut manufacturers.
[b] Where specifying nominal size in decimals, zeros preceding decimal and in fourth decimal place shall be omitted.

For recommended assembly data refer to Design Criteria in paragraph 3.
Additional requirements given in General Specifications shall apply.

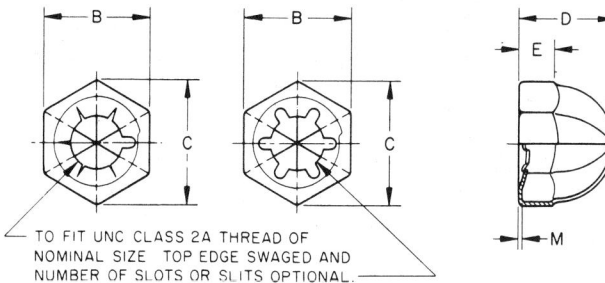

FIG. 8

TABLE 8—DIMENSIONS OF FORMED THREAD ACORN TYPE STAMPED NUTS, IN

Nominal Size[a] or Basic Thread Dia		Threads per in	B Hexagon Across Flats		C Hexagon Across Corners		D Overall Height	E Height at Corner of Hexagon	M Stock Thickness
			Max	Min	Max	Min	±0.010	Min	Basic
6	0.1380	32	0.312	0.306	0.360	0.348	0.261	0.084	0.013
8	0.1640	32	0.343	0.337	0.397	0.383	0.297	0.097	0.013
10	0.1900	24	0.375	0.368	0.433	0.418	0.324	0.110	0.017
1/4	0.2500	20	0.437	0.429	0.505	0.488	0.380	0.122	0.021
5/16	0.3125	18	0.562	0.553	0.650	0.627	0.484	0.157	0.024
3/8	0.3750	16	0.562	0.553	0.650	0.627	0.474	0.157	0.020

[a]Where specifying nominal size in decimals, zeros preceding decimal and in fourth decimal place shall be omitted.
For recommended assembly data, refer to Design Criteria in paragraph 3.
Additional requirements given in General Specifications shall apply.

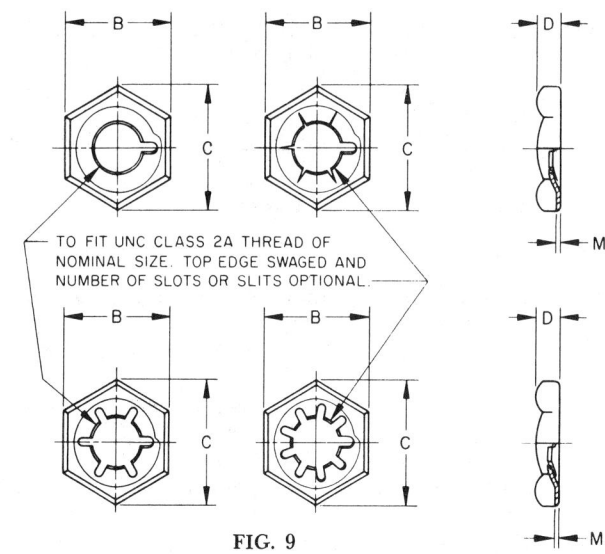

FIG. 9

TABLE 9—DIMENSIONS OF FORMED THREAD REGULAR TYPE STAMPED NUTS, IN

Nominal Size[a] or Basic Thread Dia		Threads per in	B Hexagon Across Flats		C Hexagon Across Corners		D Overall Height		M Stock Thickness
			Max	Min	Max	Min	Max	Min	Basic
6	0.1380	32	0.312	0.305	0.361	0.348	0.102	0.082	0.013
8	0.1640	32	0.343	0.336	0.397	0.383	0.109	0.089	0.013
10	0.1900	24	0.375	0.368	0.433	0.418	0.115	0.095	0.017
1/4	0.2500	20	0.437	0.429	0.505	0.488	0.133	0.113	0.021
5/16	0.3125	18	0.562	0.553	0.650	0.627	0.155	0.135	0.024
3/8	0.3750	16	0.625	0.615	0.722	0.697	0.166	0.146	0.027

[a]Where specifying nominal size in decimals, zeros preceding decimal and in fourth decimal place shall be omitted.
For recommended assembly data, refer to Design Criteria in paragraph 3.
Additional requirements given in General Specifications shall apply.

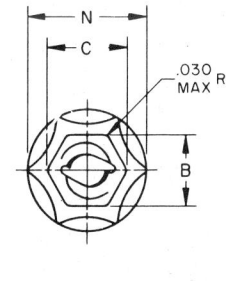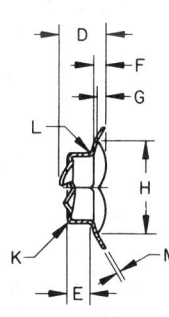

FIG. 10

TABLE 10—DIMENSIONS OF SELF-THREADING FACETED FLANGE TYPE STAMPED NUTS, IN[a]

Nominal Size[b] or Basic Stud Dia		B Hexagon Across Flats		C Hexagon Across Corners		D Overall Height		E Height of Flat	F Depth to Radius		G Dish Depth		H Dish Diameter		K Corner Radius		L Fillet Radius		M Stock Thickness	N Flange Diameter	
		Max	Min	Max	Min	Max	Min	Min	Max	Min	Max	Min	Max	Min	Max	Min	Max	Min	Basic	Max	Min
1/8	0.125	0.312	0.304	0.360	0.348	0.199	0.179	0.067	0.017	0.007	—	—	0.401	0.387	0.035	0.033	0.027		0.020	0.442	0.432
3/16	0.188	0.375	0.366	0.433	0.418	0.239	0.219	0.078	0.037	0.025	0.036	0.026	0.468	0.448	0.037	0.037	0.031		0.020	0.567	0.557
1/4	0.250	0.437	0.428	0.505	0.488	0.273	0.253	0.090	0.047	0.035	0.043	0.033	0.572	0.552	0.043	0.039	0.033		0.021	0.692	0.682

[a]Sealer styles are also available, consult nut manufacturers.
[b]Where specifying nominal size in decimals, zeros preceding decimal shall be omitted.
For recommended assembly data refer to Design Criteria in paragraph 3.
Additional requirements given in General Specifications shall apply.

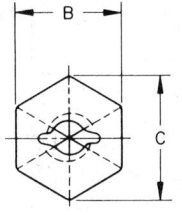 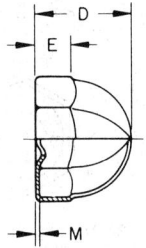 FIG. 11

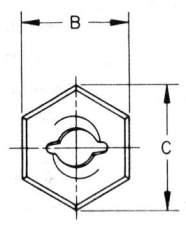 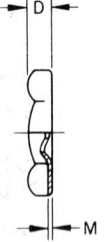 FIG. 12

TABLE 11—DIMENSIONS OF SELF-THREADING ACORN TYPE STAMPED NUTS, IN[a]

Nominal Size[b] or Basic Stud Dia		B Hexagon Across Flats		C Hexagon Across Corners		D Overall Height	E Height at Corner of Hexagon	M Stock Thickness
		Max	Min	Max	Min	±0.010	Min	Basic
1/8	0.125	0.375	0.368	0.433	0.418	0.324	0.119	0.017
3/16	0.188	0.500	0.491	0.577	0.557	0.437	0.147	0.020
1/4	0.250	0.562	0.553	0.650	0.628	0.484	0.166	0.024

[a]Sealer Styles are also available, consult nut manufacturers.
[b]Where specifying nominal size in decimals, zeros preceding decimal shall be omitted.
For recommended assembly data, refer to Design Criteria in paragraph 3.
Additional requirements given in General Specifications shall apply.

TABLE 12—DIMENSIONS OF SELF-THREADING REGULAR TYPE STAMPED NUTS, IN[a]

Nominal Size[b] or Basic Stud Dia		B Hexagon Across Flats		C Hexagon Across Corners		D Overall Height		M Stock Thickness
		Max	Min	Max	Min	Max	Min	Basic
1/8	0.125	0.312	0.306	0.360	0.348	0.110	0.090	0.017
3/16	0.188	0.500	0.492	0.577	0.557	0.139	0.119	0.019
1/4	0.250	0.500	0.492	0.577	0.557	0.150	0.130	0.026

[a]Sealer styles are also available, consult nut manufacturers.
[b]Where specifying nominal size in decimals, zeros preceding decimal shall be omitted.
For recommended assembly data refer to Design Criteria in paragraph 3.
Additional requirements given in General Specifications shall apply.

PLAIN WASHERS—SAE J488   SAE Standard

Report of Parts and Fittings Division approved August 1922 and last revised by Parts and Fittings Committee January 1953. Conforms to ASA B27.2, American Standard Plain Washers, except for the manner of expressing thickness.

TABLE 1—DIMENSIONS OF PLAIN WASHERS, IN.

Inside Dia[a]	Outside Dia[b]	Thickness[c] Max	Thickness[c] Min	Inside Dia[a]	Outside Dia[b]	Thickness[c] Max	Thickness[c] Min	Inside Dia[a]	Outside Dia[b]	Thickness[c] Max	Thickness[c] Min	Inside Dia[a]	Outside Dia[b]	Thickness[c] Max	Thickness[c] Min
5/64	3/16	0.025	0.016	5/16	7/8	0.080	0.051	5/8	2-1/8	0.160	0.108	1-3/8	3	0.192	0.136
3/32	7/32	0.025	0.016	11/32	11/16	0.080	0.051	21/32	1-5/16	0.121	0.074	1-7/16	3	0.213	0.153
3/32	1/4	0.025	0.016	3/8	3/4	0.080	0.051	11/16	1-1/2	0.160	0.108	1-1/2	3-1/4	0.213	0.153
1/8	1/4	0.028	0.017	3/8	7/8	0.104	0.064	11/16	1-3/4	0.160	0.108	1-9/16	3-1/4	0.213	0.153
1/8	5/16	0.040	0.025	3/8	1-1/8	0.080	0.051	11/16	2-3/8	0.192	0.136	1-5/8	3-1/2	0.213	0.153
5/32	5/16	0.048	0.027	13/32	13/16	0.080	0.051	13/16	1-1/2	0.160	0.108	1-11/16	3-1/2	0.213	0.153
5/32	3/8	0.065	0.036	7/16	7/8	0.104	0.064	13/16	1-3/4	0.177	0.122	1-3/4	3-3/4	0.213	0.153
11/64	13/32	0.065	0.036	7/16	1	0.104	0.064	13/16	2	0.177	0.122	1-13/16	3-3/4	0.213	0.153
3/16	3/8	0.065	0.036	7/16	1-3/8	0.104	0.064	13/16	2-7/8	0.192	0.136	1-7/8	4	0.213	0.153
3/16	7/16	0.065	0.036	15/32	59/64	0.080	0.051	15/16	1-3/4	0.160	0.108	1-15/16	4	0.213	0.153
13/64	15/32	0.065	0.036	1/2	1-1/8	0.104	0.064	15/16	2	0.192	0.136	2	4-1/4	0.213	0.153
7/32	7/16	0.065	0.036	1/2	1-1/4	0.104	0.064	15/16	2-1/4	0.192	0.136	2-1/16	4-1/4	0.213	0.153
7/32	1/2	0.065	0.036	1/2	1-5/8	0.104	0.064	15/16	3-3/8	0.213	0.153	2-1/8	4-1/2	0.213	0.153
15/64	17/32	0.065	0.036	17/32	1-1/16	0.121	0.074	1-1/16	2	0.160	0.108	2-3/8	4-3/4	0.248	0.193
1/4	1/2	0.065	0.036	9/16	1-1/4	0.132	0.086	1-1/16	2-1/4	0.192	0.136	2-5/8	5	0.280	0.210
1/4[d]	9/16	0.065	0.036	9/16	1-3/8	0.132	0.086	1-1/16	2-1/2	0.192	0.136	2-7/8	5-1/4	0.310	0.228
1/4[d]	9/16	0.080	0.051	9/16	1-7/8	0.132	0.086	1-1/16	3-7/8	0.280	0.210	3-1/8	5-1/2	0.327	0.249
17/64	5/8	0.065	0.036	19/32	1-3/16	0.121	0.074	1-3/16	2-1/2	0.192	0.136				
9/32	5/8	0.080	0.051	5/8	1-3/8	0.132	0.086	1-1/4	2-3/4	0.192	0.136				
5/16	3/4	0.080	0.051	5/8	1-1/2	0.132	0.086	1-5/16	2-3/4	0.192	0.136				

[a] Tolerance is ± 0.005 on ID to and including 7/32 ID, and ± 0.010 on ID greater than 7/32, except for two 1/4 x 9/16 sizes, whose tolerances are given in footnote d.
[b] Tolerance is ± 0.010 on OD for all sizes.
[c] Nominal thicknesses of washers are Birmingham gage sizes. The limits specified represent a tolerance of ± one gage, or the spread from the minimum of one gage minus to the maximum of one gage plus.
[d] Tolerance is ± 0.005 on ID.

LOCK WASHERS—SAE J489b   SAE Standard

Report of Parts and Fittings Division approved August 1922, revised by Parts and Fittings Committee January 1956 and last revised by Fasteners Committee February 1974. Conforms to American National Standard, ANSI B18.21.1—1972, Lock Washers.

1. Introduction

1.1 Scope—This SAE Standard covers the complete dimensional and general specifications, including physical properties and methods of testing, for those types of lock washers intended for use in automotive and general industrial applications.

1.1.1 It should be noted that the word "lock" appearing in the names of the products in this standard is a generic term historically associated with their identification and is not intended to imply an indefinite permanency of fixity in attachments where they are used.

1.1.2 The inclusion of dimensional data in this standard is not intended to imply that all of the products described are stock production items. Consumers should consult with manufacturers concerning the availability of products.

1.2 Washer Types and Application—The two varieties of lock washers covered by this standard and their applications are described in the following:

1.2.1 HELICAL SPRING LOCK WASHERS—This standard includes helical spring lock washers of carbon steel, corrosion resistant steel, aluminum-zinc alloy, phosphor-bronze, silicon-bronze, and K-Monel in the series indicated in the respective detailed specifications. Helical spring lock washers are intended for general industrial applications and provide hardened bearing surfaces for the threaded fasteners used.

1.2.2 TOOTH LOCK WASHERS—This standard includes tooth lock washers of carbon steel having internal teeth, external teeth, and both internal and external teeth in two constructions designated Type A and Type B. Countersunk external tooth lock washers are provided for use with countersunk head screws, and heavy series internal tooth lock washers are available in nominal sizes over $\frac{1}{4}$ in. The internal-external tooth lock washers have a selection of outside diameters to satisfy various design requirements. Tooth lock washers are intended for general industrial applications and serve to increase the friction between fasteners and the component parts of an assembly. Where used on painted surfaces, the tooth lock washer provides electrical grounding by cutting through most finishes when put under tension.

1.3 Dimensions—All dimensions in this standard are given in inches, unless otherwise stated.

1.4 Responsibility for Modifications—The washer manufacturer shall not be held responsible for malfunctions of product determined to be due to plating or other modifications when such plating or modification is done by the purchaser or his designee.

2. Helical Spring Lock Washers

2.1 General Specifications—These General Specifications apply and are common to helical spring lock washers of the various materials covered by this standard. The additional requirements peculiar to washers of a specific material are presented in the detailed specifications under the respective materials which follow.

2.1.1 DIMENSIONS—The dimensions of helical spring lock washers shall be as specified in Table 2. Selection should be made from the regular, heavy, extra duty, or hi collar series to suit design requirements. It should be noted, however, that washers in certain materials are not produced in all series. For series in which washers are available, reference should be made to detailed specifications for respective materials.

2.1.2 WASHER SECTION—The section of finished washers shall be slightly trapezoidal in shape with the thickness at the inner periphery greater than the thickness at the outer periphery by an amount varying from a minimum of 0.0005 in. to a maximum of 0.001 in. per 0.0156 in. of the minimum section width. The minimum thickness specified in the table of dimensions represents the nominal mean thickness of the trapezoid. Reduced to formulas, the increase in thickness from the outer periphery to inner periphery is t_i minus t_o or 0.032 W (minimum) to 0.064 W (maximum). The tolerance on the nominal mean thickness of the trapezoid shall be plus and equal to the following:

Size 12 and smaller	0.006
Size 1/4 to 13/16	0.010
Size 7/8 to 1-1/2	0.020

2.1.2.1 *Corners*—The corners at the outer and inner peripheries of washer shall be rounded sufficiently to avoid checks. However, the radius at corners shall not exceed the equivalent of 15% of the minimum width of washer section and need not necessarily be tangent to the face of the washer section.

2.1.3 COILING—Washers shall be coiled so that free height is approximately equivalent to twice the thickness of the washer section. The gap and relationship of the severed ends shall be such as to prevent the washers tangling and ensure that washers will compress flat. Washers which require extra operations, such as flattening to reduce free height, etc., shall be classified as specials.

2.1.4 FINISHES—Unless otherwise specified, washers shall be supplied with a natural (as processed) finish, unplated or uncoated. Where corrosion preventative treatment is required, washers shall be plated or coated as agreed upon between the manufacturer and the purchaser. However, where carbon steel washers are plated or coated and subject to hydrogen embrittlement, they shall be suitably treated subsequent to the plating or coating operation to obviate such embrittlement and subjected to an embrittlement test.

2.1.5 WORKMANSHIP—The flat faces of washers and the inner and outer peripheries shall be smooth and free from knurling, serrations, die marks, deep scratches, etc., although slight feed roll marks shall be permissible. Washers shall also be free from burrs, rust, pit marks, loose scale, and defects which might affect their serviceability.

2.1.6 DESIGNATION—Nominal washer sizes are intended for use with comparable nominal screw or bolt sizes. Helical spring lock washers shall be designated by the following data, in the sequence shown: product name, nominal size (number, fraction, or decimal equivalent[1]), series, material, and protective finish, if required. For example:

Helical spring lock washer, $\frac{1}{4}$ regular, phosphor bronze
Helical spring lock washer, .375 extra duty, steel, phosphate coated
Helical spring lock washer, $\frac{1}{2}$ hi-collar, steel

2.1.7 TESTS—Helical spring lock washers shall meet the requirements of the following tests:

2.1.7.1 *Temper Test*—After compressing to flat and releasing, the free height of a washer shall be at least two-thirds of the original free height.

2.1.7.2 *Twist Test*—The washer shall be gripped in vise jaws having sharp edges. The ends of the washers shall be free and an axis passing through the gap shall be parallel to and slightly above the top of the vise. A 90 deg segment of the free end of the washer shall be gripped in wrench jaws as shown in Fig. 1. Edges of the wrench jaws shall be sharp and in a plane parallel to the vise. Movement of the wrench in a direction that increases the free height of the spring lock washer shall twist carbon steel washers through an angle of 90 deg and corrosion resistant steel and nonferrous washers through an angle of 45 deg, without sign of fracture. When the washer fractures at a greater degree of twist, the structure at the point of fracture shall show a fine grain, and the washer up to the instant of fracture shall deliver a tough, springy reaction.

2.2 Carbon Steel Helical Spring Lock Washers—The following additional requirements peculiar to carbon steel helical spring lock washers supplement the requirements set forth in the General Specifications applying to all helical spring lock washers.

2.2.1 DIMENSIONS—Carbon steel washers are available in the four series specified in Table 2. Namely, regular, heavy, extra-duty, and hi-collar series.

2.2.1.1 When carbon steel helical spring lock washers are to be hot-dipped galvanized for use with hot-dipped galvanized bolts or screws, they shall be coiled to limits 0.020 in. in excess of those specified in Table 2 for minimum inside diameter and maximum outside diameter. Galvanizing on washers under $\frac{1}{4}$ in. nominal size is not recommended.

2.2.2 MATERIAL AND HARDNESS—Washers shall be made from carbon steel lock washer wire meeting the limits for decarburization shown in Table 1. They shall be heat treated to a hardness of Rockwell C45-51, or equivalent. The washer shall be prepared for checking the material hardness by grinding the sides flat and parallel sufficiently to assure the removal of a decarburized or plated surface. During the grinding operation, care shall be exercised to prevent the temperature of the washer from exceeding 250°F. Galvanized washers shall have a hardness of Rockwell C45-51, or equivalent, before galvanizing.

2.2.3 EMBRITTLEMENT TEST FOR PLATED WASHERS—Plated carbon steel washers shall not break after having been compressed flat between hardened and ground steel plates for a minimum period of 24 h.

[1]Where specifying nominal size in decimals, zeros preceeding the decimal shall be omitted.

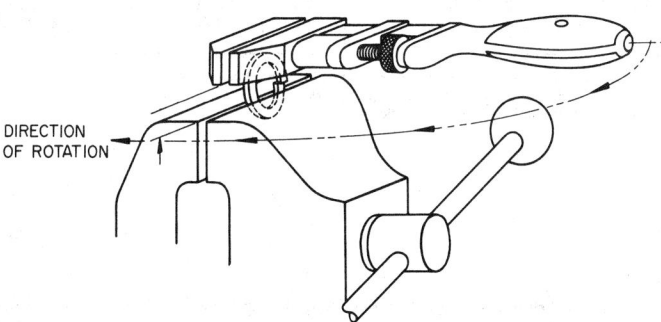

FIG. 1—WASHER TWIST TEST

TABLE 1—LIMITS FOR DECARBURIZATION

Diameter of Round Wire, or Sections of Equivalent Area	Max Depth Free Ferrite	Max Total Affected Depth (Free Ferrite Plus Partial Decarburization)
Up to 0.140, incl	0.002	0.006
Over 0.140 to 0.250, incl	0.003	0.008
Over 0.250 to 0.375, incl	0.004	0.010
Over 0.375 to 0.500, incl	0.006	0.015

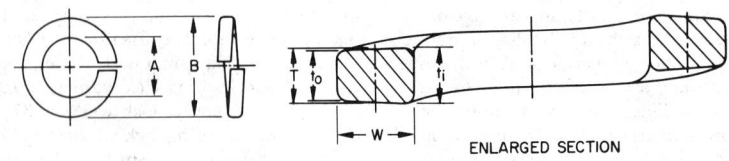

ENLARGED SECTION

TABLE 2—DIMENSIONS OF REGULAR, HEAVY, EXTRA-DUTY AND HI-COLLAR HELICAL SPRING LOCK WASHERS

Nominal Washer Size	A Inside Dia		Regular[a]			Heavy[b]			Extra-Duty[c]			Hi-Collar[d]		
			B Outside Dia[e]	T Mean Section Thickness $\left(\dfrac{t_i + t_o}{2}\right)$	W Section Width	B Outside Dia[e]	T Mean Section Thickness $\left(\dfrac{t_i + t_o}{2}\right)$	W Section Width	B Outside Dia[e]	T Mean Section Thickness $\left(\dfrac{t_i + t_o}{2}\right)$	W Section Width	B Outside Dia[e]	T Mean Section Thickness $\left(\dfrac{t_i + t_o}{2}\right)$	W Section Width
	Max	Min	Max	Min	Min	Max	Min	Min	Max	Min	Min	Max	Min	Min
No. 2	0.094	0.088	0.172	0.020	0.035	0.182	0.025	0.040	0.208	0.027	0.053	—	—	—
No. 3	0.107	0.101	0.195	0.025	0.040	0.209	0.031	0.047	0.239	0.034	0.062	—	—	—
No. 4	0.120	0.114	0.209	0.025	0.040	0.223	0.031	0.047	0.253	0.034	0.062	0.173	0.022	0.022
No. 5	0.133	0.127	0.236	0.031	0.047	0.252	0.040	0.055	0.300	0.045	0.079	0.202	0.030	0.030
No. 6	0.148	0.141	0.250	0.031	0.047	0.266	0.040	0.055	0.314	0.045	0.079	0.216	0.030	0.030
No. 8	0.174	0.167	0.293	0.040	0.055	0.307	0.047	0.062	0.375	0.057	0.096	0.267	0.047	0.042
No. 10	0.200	0.193	0.334	0.047	0.062	0.350	0.056	0.070	0.434	0.068	0.112	0.294	0.047	0.042
No. 12	0.227	0.220	0.377	0.056	0.070	0.391	0.063	0.077	0.497	0.080	0.130	—	—	—
1/4	0.262	0.254	0.489	0.062	0.109	0.491	0.077	0.110	0.535	0.084	0.132	0.365	0.078	0.047
5/16	0.326	0.317	0.586	0.078	0.125	0.596	0.097	0.130	0.622	0.108	0.143	0.460	0.093	0.062
3/8	0.390	0.380	0.683	0.094	0.141	0.691	0.115	0.145	0.741	0.123	0.170	0.553	0.125	0.076
7/16	0.455	0.443	0.779	0.109	0.156	0.787	0.133	0.160	0.839	0.143	0.186	0.647	0.140	0.090
1/2	0.518	0.506	0.873	0.125	0.171	0.883	0.151	0.176	0.939	0.162	0.204	0.737	0.172	0.103
9/16	0.582	0.570	0.971	0.141	0.188	0.981	0.170	0.193	1.041	0.182	0.223	—	—	—
5/8	0.650	0.635	1.079	0.156	0.203	1.093	0.189	0.210	1.157	0.202	0.242	0.923	0.203	0.125
11/16	0.713	0.698	1.176	0.172	0.219	1.192	0.207	0.227	1.258	0.221	0.260	—	—	—
3/4	0.775	0.760	1.271	0.188	0.234	1.291	0.226	0.244	1.361	0.241	0.279	1.111	0.218	0.154
13/16	0.843	0.824	1.367	0.203	0.250	1.391	0.246	0.262	1.463	0.261	0.298	—	—	—
7/8	0.905	0.887	1.464	0.219	0.266	1.494	0.266	0.281	1.576	0.285	0.322	1.296	0.234	0.182
15/16	0.970	0.950	1.560	0.234	0.281	1.594	0.284	0.298	1.688	0.308	0.345	—	—	—
1	1.042	1.017	1.661	0.250	0.297	1.705	0.306	0.319	1.799	0.330	0.366	1.483	0.250	0.208
1-1/16	1.107	1.080	1.756	0.266	0.312	1.808	0.326	0.338	1.910	0.352	0.389	—	—	—
1-1/8	1.172	1.144	1.853	0.281	0.328	1.909	0.345	0.356	2.019	0.375	0.411	1.669	0.313	0.236
1-3/16	1.237	1.208	1.950	0.297	0.344	2.008	0.364	0.373	2.124	0.396	0.431	—	—	—
1-1/4	1.302	1.271	2.045	0.312	0.359	2.113	0.384	0.393	2.231	0.417	0.452	1.799	0.313	0.236
1-5/16	1.366	1.334	2.141	0.328	0.375	2.211	0.403	0.410	2.335	0.438	0.472	—	—	—
1-3/8	1.432	1.398	2.239	0.344	0.391	2.311	0.422	0.427	2.439	0.458	0.491	2.041	0.375	0.292
1-7/16	1.497	1.462	2.334	0.359	0.406	2.406	0.440	0.442	2.540	0.478	0.509	—	—	—
1-1/2	1.561	1.525	2.430	0.375	0.422	2.502	0.458	0.458	2.638	0.496	0.526	2.170	0.375	0.292

Additional requirements given in General Specifications shall apply.
[a] Formerly designated medium helical spring lock washers.
[b] Not recommended for new applications.
[c] Formerly designated extra-heavy helical spring lock washers.
[d] For use with 1960 series socket head cap screws.
[e] The maximum outside diameters specified allow for commercial tolerances on cold drawn wire.

2.3 Corrosion Resistant Steel Helical Spring Lock Washers—The following additional requirements peculiar to corrosion resistant steel helical spring lock washers supplement the requirements set forth in the General Specifications applying to all helical spring lock washers.

2.3.1 DIMENSIONS—Corrosion resistant steel helical spring lock washers are available in the regular series specified in Table 2.

2.3.2 MATERIAL AND HARDNESS—Washers shall be made from SAE 30302 or SAE 30305 series corrosion resistant steels. They shall be fabricated and cold worked to a hardness of Rockwell C35-43, or equivalent.

2.4 Aluminum-Zinc Alloy Helical Spring Lock Washers—The following additional requirements peculiar to aluminum-zinc alloy helical spring lock washers supplement the requirements set forth in the General Specifications applying to all helical spring lock washers.

2.4.1 DIMENSIONS—Aluminum-zinc alloy helical spring lock washers are available in the regular series given in Table 2.

2.4.2 MATERIAL AND HARDNESS—Washers shall be made from aluminum-zinc alloy physical properties of which are such that after fabrication and heat treatment the washers will meet the requirements specified in paragraph 2.4.2.1.

2.4.2.1 The washers shall receive a solution and precipitation heat treatment resulting in a hardness of Rockwell B75-97, or equivalent; or a minimum tensile strength of 77,000 psi.

2.4.2.2 Tensile test pieces of keystone wire heat treated with the washers they represent shall be furnished upon agreement between the purchaser and the manufacturer.

2.5 Phosphor-Bronze Helical Spring Lock Washers—The following requirements peculiar to phosphor-bronze helical spring lock washers supplement the requirements set forth in the General Specifications applying to all helical spring lock washers.

2.5.1 DIMENSIONS—Phosphor-bronze helical spring lock washers are available in the regular series specified in Table 2.

2.5.2 MATERIAL AND HARDNESS—Washers shall be made from phosphor-bronze conforming to SAE CA510. They shall be fabricated and cold worked to a minimum hardness of Rockwell B90, or equivalent.

2.6 Silicon-Bronze Helical Spring Lock Washers—The following requirements peculiar to silicon-bronze helical spring lock washers supplement the requirements set forth in the General Specifications applying to all helical spring lock washers.

2.6.1 DIMENSIONS—Silicon-bronze helical spring lock washers are available in the regular series specified in Table 2.

2.6.2 MATERIAL AND HARDNESS—Washers shall be made from silicon-bronze conforming to SAE CA651 or CA655. They shall be fabricated and cold worked to a minimum hardness of Rockwell B90, or equivalent.

2.7 K-Monel Helical Spring Lock Washers—The following requirements peculiar to K-Monel helical spring lock washers supplement the requirements set forth in the General Specifications applying to all helical spring lock washers.

2.7.1 DIMENSIONS—K-Monel helical spring lock washers are available in the regular series specified in Table 2.

2.7.2 MATERIAL AND HARDNESS—Washers shall be made from K-Monel conforming to Federal Specification QQ-N-286. The washers shall be manufactured from cold-worked material and subsequently age hardened to a hardness of Rockwell C33–40, or equivalent, or a minimum tensile strength of 155,000 psi.

3. Tooth Lock Washers

3.1 General Specifications—The following General Specifications shall apply to all tooth lock washers covered by this standard.

3.1.1 DIMENSIONS—Dimensions of internal tooth lock washers, heavy internal tooth lock washers, external tooth lock washers, countersunk external tooth lock washers, and internal-external tooth lock washers shall be as specified in Tables 3–7, respectively.

3.1.2 MANUFACTURING DETAIL—The number of teeth, the length of the teeth, the width of the rim, and the thickness of the washer over the teeth (free height) shall be optional with the manufacturer, with the provision, however, that the projection of the teeth on both sides of the washer shall be uniform within a tolerance equal to one-half of the projection on one side.

3.1.3 MATERIAL AND HARDNESS—Washers shall be made from carbon steel, fabricated and heat treated to a hardness of Rockwell C40–50, or equivalent. It is recommended, however, that the lighter, more sensitive depth reading Rockwell A scale be used when testing washers of thin section. Decarburization shall be removed before testing hardness.

3.1.4 FINISHES—Unless otherwise specified, washers shall be supplied with a natural (as processed) finish, unplated or uncoated. Where corrosion preventative treatment is required, washers shall be plated or coated as agreed upon between the manufacturer and the purchaser. However, where carbon steel washers are plated or coated and subject to hydrogen embrittlement, they shall be suitably treated subsequent to the plating or coating operation to obviate such embrittlement and subjected to an embrittlement test.

3.1.5 WORKMANSHIP—Washers shall be symmetrical in shape and free from rust, loose scale, and defects that might affect their serviceability.

3.1.6 DESIGNATION—Nominal washer sizes are intended for use with comparable screw or bolt sizes. Tooth lock washers shall be designated by the following data, in the sequence shown: Product name, nominal size (number, fraction, or decimal equivalent[1]), maximum washer outside diameter (internal-external tooth washers only), type, material, and protective finish, if required. For example:

Internal tooth lock washer, 1/4, Type A, steel
External tooth lock washer, .562, Type B, steel, phosphate coated
Internal-external tooth lock washer, No. 12, (.900 O.D.), Type A, steel, cadmium plated

3.1.7 TESTS—Tooth lock washers shall meet the requirements of the following tests:

3.1.7.1 *Temper Test*—Washers after being compressed to a height equal to the actual material thickness plus 0.005 in. and then released, shall have a free height greater than the compressed height. Compression shall be accomplished between parallel flat surfaces for flat varieties of tooth washers and between mating countersunk holes and cones for countersunk tooth washers.

3.1.7.2 *Embrittlement Test*—Plated washers shall be clamped to a height equal to the actual material thickness plus 0.005 in. for a minimum period of 24 h without breaking. The clamping shall be accomplished between parallel flat surfaces for flat varieties of tooth washers, and between mating countersunk holes and cones for countersunk tooth washers.

3.1.7.3 *Twist Test*—The rim of the lock washer shall be cut or severed with a cold chisel or cutting pliers and the severed ends shall be gripped by pliers or vise and pliers. Separation of the ends in the form of a helix, to a distance equal to the inside diameter of the washer, shall not result in fracture. When the ends are separated a greater distance and the washer does fracture, the structure at the point of fracture shall show a fine grain, and the washer up to the instant of fracture shall deliver a tough, springy reaction.

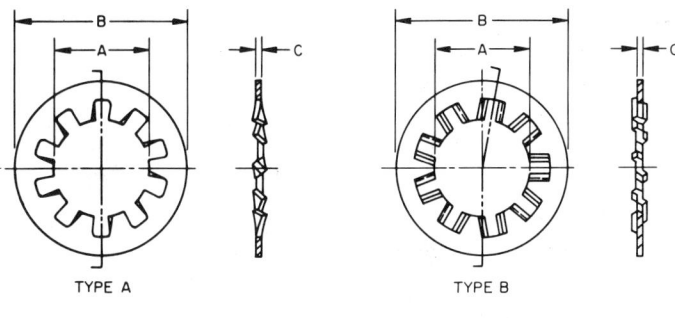

TYPE A TYPE B

TABLE 3—DIMENSIONS OF INTERNAL TOOTH LOCK WASHERS

Nominal Washer Size		A Inside Dia		B Outside Dia		C Thickness	
		Max	Min	Max	Min	Max	Min
No. 2	0.086	0.095	0.089	0.200	0.175	0.015	0.010
No. 3	0.099	0.109	0.102	0.232	0.215	0.019	0.012
No. 4	0.112	0.123	0.115	0.270	0.255	0.019	0.015
No. 5	0.125	0.136	0.129	0.280	0.245	0.021	0.017
No. 6	0.138	0.150	0.141	0.295	0.275	0.021	0.017
No. 8	0.164	0.176	0.168	0.340	0.325	0.023	0.018
No. 10	0.190	0.204	0.195	0.381	0.365	0.025	0.020
No. 12	0.216	0.231	0.221	0.410	0.394	0.025	0.020
1/4	0.250	0.267	0.256	0.478	0.460	0.028	0.023
5/16	0.312	0.332	0.320	0.610	0.594	0.034	0.028
3/8	0.375	0.398	0.384	0.692	0.670	0.040	0.032
7/16	0.438	0.464	0.448	0.789	0.740	0.040	0.032
1/2	0.500	0.530	0.512	0.900	0.867	0.045	0.037
9/16	0.562	0.596	0.576	0.985	0.957	0.045	0.037
5/8	0.625	0.663	0.640	1.071	1.045	0.050	0.042
11/16	0.688	0.728	0.704	1.166	1.130	0.050	0.042
3/4	0.750	0.795	0.769	1.245	1.220	0.055	0.047
13/16	0.812	0.861	0.832	1.315	1.290	0.055	0.047
7/8	0.875	0.927	0.894	1.410	1.364	0.060	0.052
1	1.000	1.060	1.019	1.637	1.590	0.067	0.059
1-1/8	1.125	1.192	1.144	1.830	1.799	0.067	0.059
1-1/4	1.250	1.325	1.275	1.975	1.921	0.067	0.059

Additional requirements given in General Specifications shall apply.

TABLE 4—DIMENSIONS OF HEAVY INTERNAL TOOTH LOCK WASHERS

Nominal Washer Size		A Inside Dia		B Outside Dia		C Thickness	
		Max	Min	Max	Min	Max	Min
1/4	0.250	0.267	0.256	0.536	0.500	0.045	0.035
5/16	0.312	0.332	0.320	0.607	0.590	0.050	0.040
3/8	0.375	0.398	0.384	0.748	0.700	0.050	0.042
7/16	0.438	0.464	0.448	0.858	0.800	0.067	0.050
1/2	0.500	0.530	0.512	0.924	0.880	0.067	0.055
9/16	0.562	0.596	0.576	1.034	0.990	0.067	0.055
5/8	0.625	0.663	0.640	1.135	1.100	0.067	0.059
3/4	0.750	0.795	0.768	1.265	1.240	0.084	0.070
7/8	0.875	0.927	0.894	1.447	1.400	0.084	0.075

Additional requirements given in General Specifications shall apply.

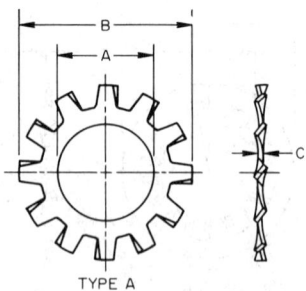
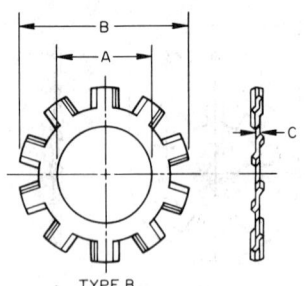

TYPE A   TYPE B

TABLE 5—DIMENSIONS OF EXTERNAL TOOTH LOCK WASHERS

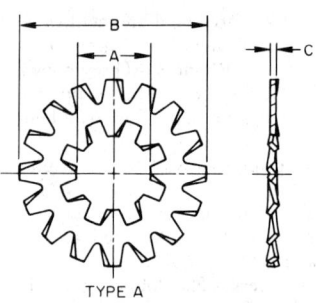
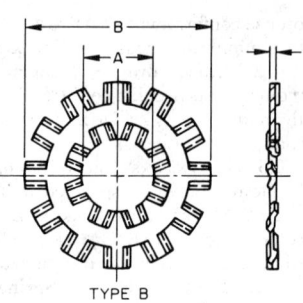

TYPE A   TYPE B

TABLE 7—DIMENSIONS OF INTERNAL-EXTERNAL TOOTH LOCK WASHERS

Nominal Washer Size		A Inside Dia		B Outside Dia		C Thickness	
		Max	Min	Max	Min	Max	Min
No. 3	0.099	0.109	0.102	0.235	0.220	0.015	0.013
No. 4	0.112	0.123	0.115	0.260	0.245	0.019	0.014
No. 5	0.125	0.136	0.129	0.285	0.270	0.019	0.014
No. 6	0.138	0.150	0.141	0.320	0.305	0.022	0.016
No. 8	0.164	0.176	0.168	0.381	0.365	0.023	0.018
No. 10	0.190	0.204	0.195	0.410	0.395	0.025	0.020
No. 12	0.216	0.231	0.221	0.475	0.460	0.028	0.023
1/4	0.250	0.267	0.256	0.510	0.494	0.028	0.023
5/16	0.312	0.332	0.320	0.610	0.588	0.034	0.028
3/8	0.375	0.398	0.384	0.694	0.670	0.040	0.032
7/16	0.438	0.464	0.448	0.760	0.740	0.040	0.032
1/2	0.500	0.530	0.513	0.900	0.880	0.045	0.037
9/16	0.562	0.596	0.576	0.985	0.960	0.045	0.037
5/8	0.625	0.663	0.641	1.070	1.045	0.050	0.042
11/16	0.688	0.728	0.704	1.155	1.130	0.050	0.042
3/4	0.750	0.795	0.768	1.260	1.220	0.055	0.047
13/16	0.812	0.861	0.833	1.315	1.290	0.055	0.047
7/8	0.875	0.927	0.897	1.410	1.380	0.060	0.052
1	1.000	1.060	1.025	1.620	1.590	0.067	0.059

Additional requirements given in General Specifications shall apply.

TABLE 7—DIMENSIONS OF INTERNAL-EXTERNAL TOOTH LOCK WASHERS

Nominal Washer Size		A Inside Dia		B Outside Dia		C Thickness	
		Max	Min	Max	Min	Max	Min
No. 4	0.112	0.123	0.115	0.475	0.460	0.021	0.016
				0.510	0.495	0.021	0.017
				0.610	0.580	0.021	0.017
No. 6	0.138	0.150	0.141	0.510	0.495	0.028	0.023
				0.610	0.580	0.028	0.023
				0.690	0.670	0.028	0.023
No. 8	0.164	0.176	0.168	0.610	0.580	0.034	0.028
				0.690	0.670	0.034	0.028
				0.760	0.740	0.034	0.028
No. 10	0.190	0.204	0.195	0.610	0.580	0.034	0.028
				0.690	0.670	0.040	0.032
				0.760	0.740	0.040	0.032
				0.900	0.880	0.040	0.032
No. 12	0.216	0.231	0.221	0.690	0.670	0.040	0.032
				0.760	0.725	0.040	0.032
				0.900	0.880	0.040	0.032
				0.985	0.965	0.045	0.037
1/4	0.250	0.267	0.256	0.760	0.725	0.040	0.032
				0.900	0.880	0.040	0.032
				0.985	0.965	0.045	0.037
				1.070	1.045	0.045	0.037
5/16	0.312	0.332	0.320	0.900	0.865	0.040	0.032
				0.985	0.965	0.045	0.037
				1.070	1.045	0.050	0.042
				1.155	1.130	0.050	0.042
3/8	0.375	0.398	0.384	0.985	0.965	0.045	0.037
				1.070	1.045	0.050	0.042
				1.155	1.130	0.050	0.042
				1.260	1.220	0.050	0.042
7/16	0.438	0.464	0.448	1.070	1.045	0.050	0.042
				1.155	1.130	0.050	0.042
				1.260	1.220	0.055	0.047
				1.315	1.290	0.055	0.047
1/2	0.500	0.530	0.512	1.260	1.220	0.055	0.047
				1.315	1.290	0.055	0.047
				1.410	1.380	0.060	0.052
				1.620	1.590	0.067	0.059
9/16	0.562	0.596	0.576	1.315	1.290	0.055	0.047
				1.430	1.380	0.060	0.052
				1.620	1.590	0.067	0.059
				1.830	1.797	0.067	0.059
5/8	0.625	0.663	0.640	1.410	1.380	0.060	0.052
				1.620	1.590	0.067	0.059
				1.830	1.797	0.067	0.059
				1.975	1.935	0.067	0.059

Additional requirements given in General Specifications shall apply.

TABLE 6—DIMENSIONS OF COUNTERSUNK EXTERNAL TOOTH LOCK WASHERS

Nominal Washer Size		A Inside Dia		B Outside Dia[a]	C Thickness		D Length	
		Max	Min	Approx	Max	Min	Max	Min
No. 4	0.112	0.123	0.113	0.213	0.019	0.015	0.065	0.050
No. 6	0.138	0.150	0.140	0.289	0.021	0.017	0.092	0.082
No. 8	0.164	0.177	0.167	0.322	0.021	0.017	0.105	0.088
No. 10	0.190	0.205	0.195	0.354	0.025	0.020	0.099	0.083
No. 12	0.216	0.231	0.220	0.421	0.025	0.020	0.128	0.118
1/4	0.250	0.267	0.255	0.454	0.025	0.020	0.128	0.113
No. 16	0.268	0.287	0.273	0.505	0.028	0.023	0.147	0.137
5/16	0.312	0.333	0.318	0.599	0.028	0.023	0.192	0.165
3/8	0.375	0.398	0.383	0.765	0.034	0.028	0.255	0.242
7/16	0.438	0.463	0.448	0.867	0.045	0.037	0.270	0.260
1/2	0.500	0.529	0.512	0.976	0.045	0.037	0.304	0.294

Additional requirements given in General Specifications shall apply.
[a] For reference purposes only, not subject to inspection.

CONICAL SPRING WASHERS—SAE J773b

SAE Standard

Report of Fasteners Committee approved June 1961 and last revised February 1976.

1. **Scope**—This SAE Standard covers dimensional, material, and general specifications and methods of test for two types of general purpose conical spring washers, designated type L and type H, for use as loose washers over screws and bolts, and also for use as pre-assembled washers in screw and washer assemblies.

    1.1 Both the type L and type H washers are available in three washer series (narrow, regular and wide), having varied proportions designed to fulfill specific application requirements for load distribution.

    1.2 Where so specified by the user, washers shall be supplied with peripheral teeth.

    1.3 All sizes and types of washers specified in this standard are not necessarily stock production items. Users should consult with manufacturers concerning availability.

2. **Designation**—Washers shall be specified or designated as shown in the following example:

    Washer, Conical, ½, SAE Type L, Wide

3. **Use and Application**—Type L washers are intended for use with screws and bolts equivalent to SAE Grade 1 and 2. Type H washers are intended for use with SAE Grade 5 or equivalent bolts or screws (SAE J429).

    3.1 The desired installed position of this washer is as near flat as possible. The flattening will occur at a load equal to approximately 27 500 bolt psi for the Type L washer and 60 000 bolt psi for the Type H washer. The spring return will vary due to the compromises in washer diameter, thickness, and tolerances, which have been made to maintain this standard in a commercial category (see Paragraph 7.1, Recovery Test).

    3.2 The relatively high supporting load and spring return makes this washer very effective where bolt tension may be subject to loss due to such factors as compensating for wear, thermal expansion, or compression set.

    3.3 When used to span over-size clearance holes, it is recommended that (1) if the full periphery is supported, at least 70% of the washer annular area be bearing or (2) if the periphery is partially supported, as over a slot, the slot should be no wider than 1½ times the I.D. Narrow series should always be fully supported. Insufficient bearing will reduce spring return.

    3.4 Washers with peripheral teeth are used for non-slip or positive electrical grounding purposes.

4. **Dimensions**—Dimensions of Type L and Type H conical spring washers are specified in Table 1.

    4.1 **Manufacturing Detail**—Washers shall be symmetrical in shape. The radial section of the washer shall be flat to convex upward with flat preferred

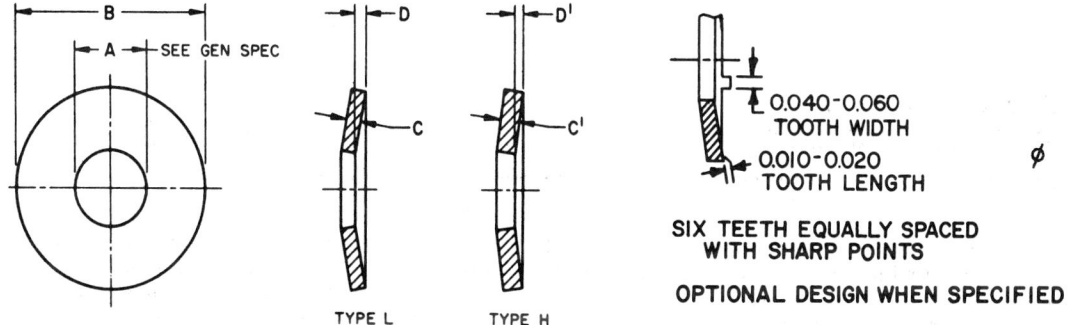

TABLE 1—DIMENSIONS OF CONICAL SPRING WASHERS, in

Nominal Screw or Bolt Size	Washer Series	A[a] ID		B OD		Type L				Type H				
						C Thickness			D Crown Height		C^1 Thickness			D^1 Crown Height	
		Min	Max	Max	Min	Nom	Max	Min	Min	Max	Nom	Max	Min	Min	Max
6	Narrow Regular Wide	0.151	0.156	0.320 0.446 0.570	0.307 0.433 0.557	0.025 0.030 0.030	0.029 0.034 0.034	0.023 0.028 0.028	0.010 0.014 0.021	0.016 0.020 0.031	0.035 0.040 0.040	0.040 0.046 0.046	0.033 0.037 0.037	0.015 0.015 0.019	0.025 0.025 0.029
8	Narrow Regular Wide	0.183	0.188	0.383 0.508 0.640	0.370 0.495 0.620	0.035 0.035 0.035	0.040 0.040 0.040	0.033 0.033 0.033	0.010 0.020 0.027	0.016 0.030 0.037	0.040 0.045 0.045	0.046 0.050 0.050	0.037 0.042 0.042	0.015 0.016 0.030	0.025 0.026 0.040
10	Narrow Regular Wide	0.203	0.208	0.446 0.570 0.765	0.433 0.557 0.743	0.035 0.040 0.040	0.040 0.046 0.046	0.033 0.037 0.037	0.010 0.017 0.026	0.016 0.027 0.036	0.050 0.055 0.055	0.056 0.060 0.060	0.047 0.052 0.052	0.015 0.016 0.024	0.025 0.026 0.034
12	Narrow Regular Wide	0.230	0.240	0.446 0.640 0.890	0.433 0.620 0.868	0.040 0.040 0.045	0.046 0.046 0.050	0.037 0.037 0.042	0.011 0.023 0.034	0.017 0.033 0.044	0.055 0.055 0.064	0.060 0.060 0.071	0.052 0.052 0.059	0.015 0.016 0.023	0.025 0.026 0.033
1/4	Narrow Regular Wide	0.271	0.281	0.515 0.765 1.015	0.495 0.743 0.993	0.045 0.050 0.055	0.050 0.056 0.060	0.042 0.047 0.052	0.014 0.023 0.030	0.024 0.033 0.040	0.064 0.079 0.079	0.071 0.087 0.087	0.059 0.074 0.074	0.015 0.022 0.029	0.025 0.032 0.039
5/16	Narrow Regular Wide	0.334	0.344	0.640 0.890 1.140	0.620 0.868 1.118	0.055 0.064 0.064	0.060 0.071 0.071	0.052 0.059 0.059	0.016 0.031 0.034	0.026 0.041 0.044	0.079 0.095 0.095	0.087 0.103 0.103	0.074 0.090 0.090	0.016 0.019 0.030	0.026 0.029 0.040
3/8	Narrow Regular Wide	0.396	0.406	0.765 1.015 1.265	0.743 0.993 1.243	0.071 0.071 0.079	0.079 0.079 0.087	0.066 0.066 0.074	0.015 0.033 0.037	0.025 0.043 0.047	0.095 0.118 0.118	0.103 0.126 0.126	0.090 0.112 0.112	0.015 0.023 0.035	0.025 0.033 0.045
7/16	Narrow Regular Wide	0.470	0.480	0.890 1.140 1.530	0.868 1.118 1.493	0.079 0.095 0.095	0.087 0.103 0.103	0.074 0.090 0.090	0.018 0.031 0.049	0.028 0.041 0.059	0.128 0.128 0.132	0.136 0.136 0.140	0.122 0.122 0.126	0.016 0.028 0.039	0.026 0.038 0.049
1/2	Narrow Regular Wide	0.530	0.540	1.015 1.265 1.780	0.993 1.243 1.743	0.100 0.111 0.111	0.108 0.120 0.120	0.094 0.106 0.106	0.021 0.033 0.052	0.031 0.043 0.062	0.142 0.142 0.152	0.150 0.150 0.160	0.136 0.136 0.146	0.020 0.027 0.042	0.030 0.037 0.052

[a] Not applicable to washers assembled with screw blanks. See General Specifications.

φ (see Fig. 1). Unless otherwise specified by the user, the direction of blanking the outside diameter should permit the sharper edge to be on the underside of the washer. Washers shall be free from sharp edges, burrs, cracks, checks, embrittlement, loose scale, and all other defects that might affect their serviceability.

4.2 Assembly Detail—The inside diameters of washers for pre-assembly on unthreaded screw blanks shall be optional, but shall be such that the washer will be retained on the screw after thread rolling, but shall not bind on the screw shank before and during tightening of the assembly.

5. Material and Hardness—Washers shall be made from SAE 1050 to 1065 carbon steel, fabricated and heat treated to a Rockwell hardness of 44-48 C scale (Rockwell hardness of C46-50 if austempering is used) or equivalent for loose washers, and 40-48 C scale or equivalent for pre-assembled washers, heat treated as an integral part of heat treated bolt or screw and washer assemblies. Washer hardness shall be checked by grinding or filing a flat spot on the top conical surface of the washer to rest on the anvil, with reading to be taken on the undisturbed inner face of the washer. If washer hardness, as obtained above, is not within specification, washers may be qualified by checking hardness on a cut-out section of the washer on which both sides have been ground. However, an excessive decarburized surface, especially on the lighter gage material, may be grounds for rejection if the performance of the washer is affected.

6. Finish—Electroplated washers or screw or bolt and washer assemblies shall be baked at 400°F as soon as practicable after plating, in order to relieve hydrogen or acid embrittlement. If washers so treated fail to meet the prescribed tests, the baking time and/or the temperature shall be increased, but not to approach annealing temperature.

7. Tests

φ **7.1 Recovery Test**—The washers shall retain at least one-third their original crown height after flattening between two hardened plates and release. (Note: Conical washers which have a higher angle of elevation than covered by this standard are not expected to have the same percentage of recovery.)

7.2 Embrittlement Test—As a constant quality control check, a minimum of 12 pieces shall be taken from each batch after plating or final finishing operations and subjected to a load test sufficient to flatten washers for a minimum period of 24 h. Upon examination after testing, washers shall not exhibit cracks or fractures.

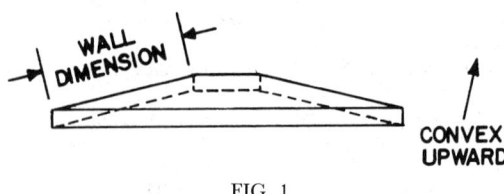

FIG. 1

NUT AND CONICAL SPRING WASHER ASSEMBLIES—SAE J238

SAE Standard

Report of Fasteners Committee approved August 1973.

1. Scope—This SAE Standard covers general, dimensional data, and methods of test for two types of general purpose nut and conical spring washer assemblies, designated Type LN and Type HN, intended for mass production and other operations where speed and convenience are paramount factors.

1.1 Both the Type LN and Type HN assemblies are available in three washer series (narrow, regular, and wide), having varied proportions designed to fulfill specific purposes of distributing the load over various areas, as shown in Table 1.

1.2 Where so specified by user, assemblies shall be supplied with toothed washers for nonslip or positive electrical grounding purposes. Toothed washers shall have six teeth, of proportions depicted in Fig. 1, equally spaced on the outer periphery. Teeth shall have sharp edges.

1.3 The inclusion of dimensional data in this standard is not intended to imply that all of the products described are stock production items. Users should consult with manufacturers concerning availability.

2. Designation—Nut and conical spring washer assemblies shall be specified or designated as shown in the following examples: $\frac{1}{4}$-20 nut and conical spring washer assembly, Type LN, wide; No. 10-24 nut and toothed conical spring washer assembly, Type HN, regular. (Unless otherwise specified, threads will be furnished as Class UNC 2B.)

3. Identification—Assemblies for No. 10 and $\frac{1}{4}$ in. nominal sizes are available in Types LN and HN. To identify the HN type in these sizes, parts should be finished in accordance with paragraph 7.

4. Use and Application—Type LN assemblies are intended for use with mating fasteners equivalent to SAE Grades 1 and 2, and Type HN assemblies are for use with mating fasteners equivalent to SAE Grade 5. (See SAE J429.)

4.1 In the installed position, it is desirable to have the washer compressed flat. Such flattening is designed to occur at a load in the bolt equivalent to approximately 27,500 psi for the Type LN assemblies and 60,000 psi for the Type HN assemblies.

4.2 The relatively high load supporting and spring return characteristics of the washer components make these assemblies very effective in applications where bolt tension may be subject to loss due to such factors as brinnelling, thermal set of parts, compression set of gaskets, etc.

5. Dimensions—All dimensions in this standard are in inches unless otherwise specified. Dimensions for both Type LN and Type HN assemblies are given in Table 1.

5.1 Nut Manufacturing Detail—The nut thickness specified in Table 1 is the overall distance, measured parallel to the axis of nut, from the top of nut to the surface which bears against top of washer. No transverse section through the nut between 25 and 75% of the actual nut thickness, as measured from the top of the nut, shall be less than the minimum width across flats. The maximum width across flats shall not be exceeded. Tops of nuts shall be flat. Corners on top and bottom of hexagon portion of nuts shall be chamfered to a diameter equal to the maximum width across flats within a tolerance of −15%. The length of chamfer at hexagon corners shall be 5-15% of the basic

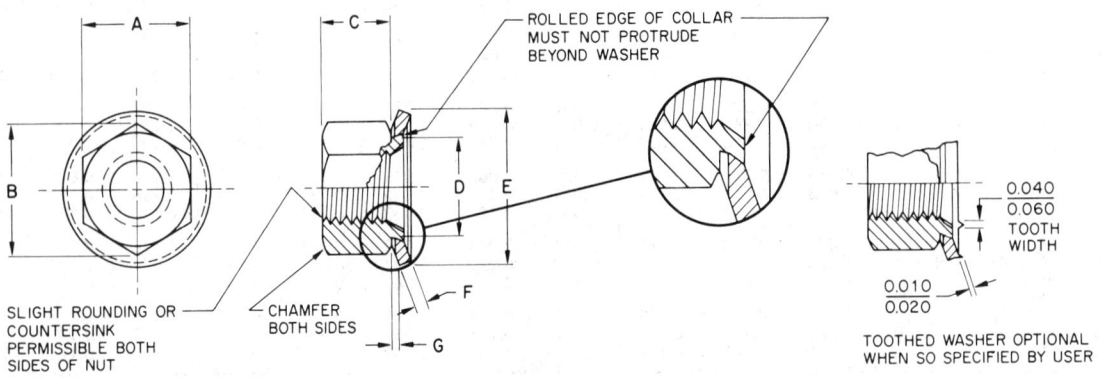

FIG. 1

TABLE 1—DIMENSIONS OF NUT AND CONICAL SPRING WASHER ASSEMBLIES

Nom Size	Basic Major Dia of Thread	Thds per in	Types LN and HN						Type LN						Type HN					
			Washer Series	Washer OD	Nut					Nut		Washer				Nut		Washer			
				E	A		B	D		C		F		G		C		F		G	
				±0.010	Max	Min	Min	Max	Min	Max	Min	Max	Min	Max	Min	Max	Min	Max	Min	Max	Min
No. 8	0.1540	32	Narrow Regular Wide	0.375 0.500 0.625	0.343	0.332	0.378	0.236	0.232	0.130	0.117	0.040 0.040 0.040	0.033 0.033 0.033	0.025 0.025 0.035	0.015 0.015 0.025	—	—	—	—	—	—
No. 10	0.1900	24	Narrow Regular Wide	0.438 0.562 0.750	0.375	0.365	0.413	0.274	0.270	0.130	0.117	0.040 0.040 0.046	0.033 0.033 0.037	0.025 0.025 0.030	0.015 0.015 0.020	0.207	0.187	0.043 0.051 0.056	0.037 0.042 0.047	0.025 0.025 0.030	0.015 0.015 0.020
1/4	0.2500	20	Narrow Regular Wide	0.625 0.750 1.000	0.437	0.428	0.488	0.332	0.328	0.193	0.178	0.051 0.056 0.065	0.042 0.047 0.055	0.025 0.025 0.030	0.015 0.015 0.025	0.226	0.212	0.065 0.079 0.087	0.055 0.066 0.074	0.025 0.025 0.030	0.015 0.015 0.020
5/16	0.3125	18	Narrow Regular Wide	0.750 1.000 1.125	0.500	0.489	0.557	0.405	0.400	—	—	—	—	—	—	0.273	0.258	0.079 0.103 0.103	0.066 0.090 0.090	0.025 0.030 0.032	0.015 0.020 0.022
3/8	0.3750	15	Narrow Regular Wide	1.000 1.125 1.250	0.562	0.551	0.628	0.470	0.465	—	—	—	—	—	—	0.337	0.320	0.103 0.120 0.120	0.090 0.106 0.106	0.025 0.032 0.035	0.015 0.022 0.025
7/16	0.4375	14	Narrow Regular Wide	1.125 1.250 1.500	0.687	0.675	0.768	0.550	0.545	—	—	—	—	—	—	0.385	0.365	0.126 0.136 0.136	0.112 0.122 0.122	0.027 0.036 0.036	0.017 0.026 0.026
1/2	0.5000	13	Narrow Regular Wide	1.250 1.500 1.750	0.750	0.736	0.840	0.610	0.605	—	—	—	—	—	—	0.448	0.427	0.140 0.150 0.150	0.126 0.136 0.136	0.027 0.035 0.035	0.017 0.025 0.025

thread diameter. The surface of chamfer may be slightly convex or rounded. A rounding or lack of fill at the junction of hexagon corners with chamfer shall be permissible provided the minimum width across corners is reached and maintained beyond a distance equal to 17.5% of the basic thread diameter from the chamfered faces.

5.1.1 TAPER OF SIDES OF HEX—Nut (angle between one side and the axis) shall not exceed 2 deg, the specified width across flats being the largest dimension.

5.2 Washer Manufacturing Detail—The washers shall be symmetrical in shape and shall be tumbled (except toothed washers) or otherwise processed to remove sharp edge at top inner periphery prior to assembly to nuts.

5.2.1 A diametral section through the washer shall show the surface element to be straight, subject to the following tolerances (see Fig. 2):

Wall Dimension	Tolerance (convex upward only), in
Up to 1/4	0.010
Over 1/4 to 1/2	0.015
Over 1/2	0.020

5.3 Assembly Detail—The size and shape of the hole in washers and the collar on the nuts shall be such that washers after assembly to nuts—by spinning, swaging, or staking of collar—will be firmly retained on the nuts and yet be free to rotate at a torque not to exceed 5 lb-in. The length of the collar on the nuts shall be such as to be wholly contained within the thickness of the washer after the assembly operation. No protusion of the collar beyond the washer in the retention area shall be permissible.

5.3.1 COLLAR CRACKS—Collar cracks may occur due to the application of pressure to the collar lip during assembly of the washer. Providing these cracks are limited to the contour of the collar, such cracks shall be permissible discontinuities and not considered cause for rejection of otherwise acceptable assemblies.

6. Material—Nut and washer components of assemblies shall be made from materials specified below:

6.1 Nuts shall be manufactured in accordance with SAE J995 (latest issue). Type LN shall be Grade 2 and Type HN shall be Grade 5.

6.2 Washers shall be made from SAE 1050 to 1065 carbon steel, fabricated and heat treated to a hardness of Rockwell C44–48 (or equivalent) and shall be capable of meeting the embrittlement tests set forth in paragraph 8.2. When the austempering process is used, washers shall be heat treated to a Rockwell C 46–50.

When heat treatment takes place after assembly of the washer and nut, a hardness range of Rockwell C 40–48 is permitted. Washer hardness shall be checked by grinding or filing a flat spot on the top side of the washer to rest on the anvil with the reading to be taken on the undisturbed inner face of the washer. If washer hardness, as thus obtained, is not within specification, washers may be qualified by checking hardness on a cutout section of the washer on which both sides have been ground flat and parallel. Excessive decarburization which adversely affects the performance of the washer may be grounds for rejection of the assembly.

7. Finish—Finish shall be as specified by purchaser. Where assemblies are to be used for electrical ground, cadmium or zinc plating is recommended. To identify the No. 10 and 1/4 in. nominal sizes Type HN when used for electrical grounding, surface treatment with yellow dichromate solution is recommended. Where electrical grounding is not a consideration, it is recommended that the No. 10 and 1/4 in. nominal sizes Type HN be phosphate coated.

7.1 Assemblies shall be free from hydrogen embrittlement or acid embrittlement. It is recommended that electroplated assemblies be baked at approximately 400°F for 3 h as soon as practicable after plating. If assemblies so treated fail to meet the test described in paragraph 8, the baking time and/or the baking temperature shall be increased.

8. Tests

8.1 Recovery Test—Conical washers shall not remain flat after deflection and release. The washers covered by this standard shall retain at least one-third the original minimum crown height after flattening between two hardened plates and release. (Note: Conical washers which have a higher angle of elevation than covered by this standard are not expected to have the same percentage of recovery.)

8.2 Embrittlement Test—As a constant quality control check, a minimum of 12 assemblies shall be taken from each batch after plating or final finishing operations and subjected to a load test sufficient to flatten washers for a minimum period of 24 h. Upon examination after testing, washers shall not exhibit any signs of cracks or fractures.

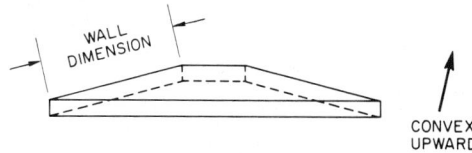

FIG. 2

TORQUE-TENSION TEST PROCEDURE FOR STEEL THREADED FASTENERS—SAE J174

SAE Recommended Practice

Report of Fasteners Committee approved June 1970. Editorial change April 1971.

Introduction—On some applications of threaded fasteners, it is desirable to control the amount of developed tension when a specified range of torque has been applied or the torque required to develop a specified range of tension. Accurate torque-tension relationships can be achieved only by uniquely defining and controlling the many related test parameters.

Scope—This test procedure is intended to provide a standard method for checking torque-tension relationships of nonprevailing torque type threaded steel fasteners $\frac{1}{4}$ through 1 in. nominal diameters.

Test Material

Test Bolt—Bolts conforming to SAE J429, Grade 8, requirements shall be used to evaluate nuts. Threads shall gage to the same class of fit as the nut. Threads on all bolts shall be produced by rolling.

Bolts shall be free from burrs, loose scale, fins, and contamination. The finish shall be zinc phosphate and oil (dry to the touch), meeting a 72 hr salt spray life when tested in accordance with ASTM D 117.

Test Washer—Washer shall conform to the dimensional, metallurgical, mechanical, and finish requirements given in Table 1. Optionally, clipped washers or multihole plates or strips may be used providing they conform to the above requirements.

Test Nut—Nuts conforming to SAE J995, Grade 8, requirements shall be used to evaluate bolts. Threads shall gage to the same class of fit as the bolt.

Nuts shall be free from burrs, loose scale, and contamination. The finish shall be zinc phosphate and oil (dry to the touch), meeting a 72 hr salt spray life when tested in accordance with ASTM B 117.

NOTE: Lubricant shall neither be removed nor added to test material.

Test Equipment

Tension Measuring Device—The tension measuring device shall be capable of measuring the axial tension induced in the bolt as it is tightened. The device shall be accurate within ±5% of the test load.

Torque Measuring Device—The torque measuring device shall have an accuracy within ±1% of a given torque reading.

Test Socket—A hexagon socket is preferred, features shall be provided in the socket to prevent the socket from contacting either the test washer or the threaded end of the bolt.

Test Spacer (If Required)—The test spacer (used only for testing bolts) shall be placed under the nut. The spacer must be hardened to Rockwell C52 minimum and the faces shall be parallel to each other and perpendicular to the axis within 0.0005 in./in. The spacer hole diameter shall be equivalent to Table 1, dimension A, and minimum spacer wall thickness shall be equivalent to one-half the bolt diameter. A feature of preventing the nut and spacer from rotating shall be provided.

Test Method

Testing Bolt—The bolt, as received, shall be inserted in the tension measuring device with the test washer placed under the bolt head. The test nut, and spacer if required, shall be assembled onto the bolt by turning the bolt head until the bolt is seated against the hardened washer. The test shall be such that a minimum of two threads protrude through the nut.

The bolt shall then be continuously and uniformly tightened at a speed not to exceed 30 rpm, with a torque measuring device or equivalent means, until either the torque or the tension value, as required, is developed, at which time both torque and tension readings shall be recorded. NOTE: The nut must not have engaged incomplete bolt threads.

During all tests, the test washer shall be prevented from turning and contacting bolt shank. A new bolt, nut, and washer shall be used for each test.

Testing Nut—To test a nut, the nut and bolt exchange positions and the above procedure shall apply.

TABLE 1—TEST WASHERS

Nominal Fastener Size	Washer Dimensions				
	Inside Dia A[a]	Outside Dia B	Width D	Thickness C	
				Max	Min
1/4	0.281	0.750	0.656	0.080	0.073
5/16	0.344	0.875	0.776	0.080	0.073
3/8	0.406	1.000	0.892	0.080	0.073
7/16	0.469	1.125	1.018	0.080	0.073
1/2	0.531	1.312	1.152	0.121	0.114
9/16	0.625	1.500	1.274	0.121	0.114
5/8	0.688	1.625	1.422	0.121	0.114
11/16	0.750	1.687	1.500	0.121	0.114
3/4	0.812	1.750	1.678	0.160	0.153
7/8	0.969	1.875	1.916	0.160	0.153
1	1.025	2.000	2.184	0.160	0.153

[a] The washer ID is intended for use with finished hex bolts and all nuts. To accommodate other bolts with larger under head fillet radii, washer hole diameter shall be increased proportionally to allow bearing surface of bolt head to seat.

NOTES:
1. All dimensions are in inches.
2. Square washers are preferred. Use of round washers is acceptable during a transition period to exclusive use of square washers.
3. Material shall be carbon steel with a chemical composition of C, 0.48–0.60%; Mn, 0.60–1.50%; P, 0.035% max; and S, 0.045% max; quenched and tempered, with a surface hardness of Rockwell 15N 85–88, and a core hardness of Rockwell A73–78.
4. Washers shall be electrodeposited zinc plated to a coating thickness of 0.0002–0.0004 in. and shall be subjected to no additional surface treatment. As soon as practicable following plating, washers shall be baked for 1 hr at 375 ± 25 F. Plating thickness shall be checked in accordance with ASTM A 219 (Microscopic Test).
5. Washers shall be free from burrs and sharp edges.

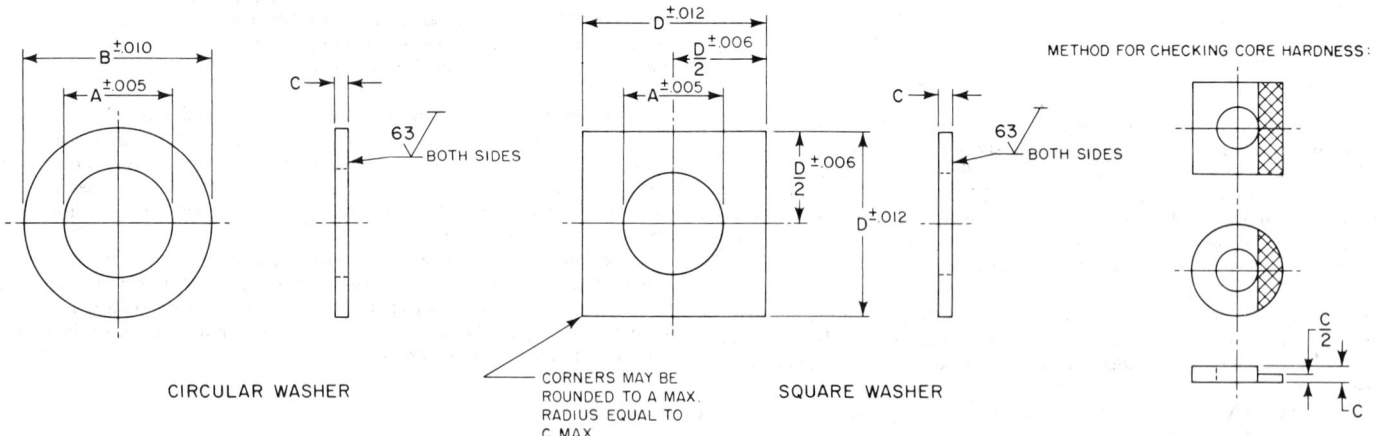

CIRCULAR WASHER — CORNERS MAY BE ROUNDED TO A MAX. RADIUS EQUAL TO C MAX. — SQUARE WASHER — METHOD FOR CHECKING CORE HARDNESS

COTTER PINS—SAE J487a

SAE Standard

Report of Miscellaneous Division approved August 1915 and last revised by Fasteners Committee August 1972.

1. Introductory Notes

1.1 Scope—This standard contains complete dimensional and general data applicable to the two types of cotter (split) pins most commonly used (extended prong square cut and hammer lock). These cotter pins are intended for use in clevis pin, pinned bolt and slotted nut assemblies, and other types of free-fitting pinned assemblies in general industrial applications.

1.2 Dimensions—Table 1 gives all dimensions in inches.

1.3 Other Types—Not shown but manufactured are additional types of cotter pins: standard square cut, mitre end, extended mitre end, bevel point, and chisel point.

1.4 Availability—The inclusion of dimensional data in this standard is not intended to imply that all of the sizes and lengths are stock items. Users are advised to consult with suppliers as to type, size, and length availability.

2. General Data for Cotter Pins

2.1 Heads

2.1.1 Head Design—A degree of leeway shall be permissible in the design of the head provided, however, the specified minimum outside diameter is maintained.

2.2 Length

2.2.1 Measurement—The length of pin, L, shall be measured, parallel to the axis of pin, from the plane of contact of a ring gage (see Fig. 1) with the head of pin to the end of prong or pin as depicted in illustrations for the respective point types. The ring gage shall have a hole of diameter equal to the specified recommended hole size within a tolerance of ±0.001 in. The permissible break or rounding of the gaging edges of gaging holes shall not exceed 0.005 in. Pin shall be inserted into the ring gage with finger pressure (force not to exceed 8 oz).

Where pins having point types other than those illustrated herein are gaged, the length, L, shall be measured from the plane of contact of gage with head to the end of the shortest prong.

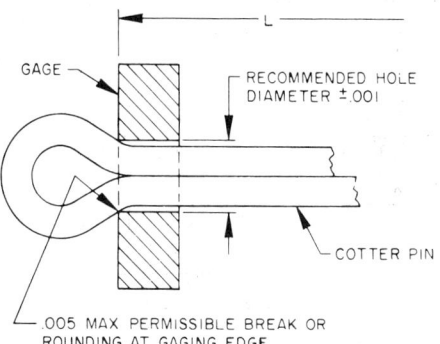

FIG. 1—COTTER PIN LENGTH GAGE

2.2.2 Tolerance on Length—The tolerance on length of cotter pins shall be as tabulated below:

Nominal Pin Length	Tolerance on Length
Up to 1 in	±0.03
1 in and longer	±0.06

2.2.3 Preferred Lengths—Table 2 depicts the preferred sizes and lengths of pins that are normally available. Other sizes and lengths are produced, as required by the purchaser.

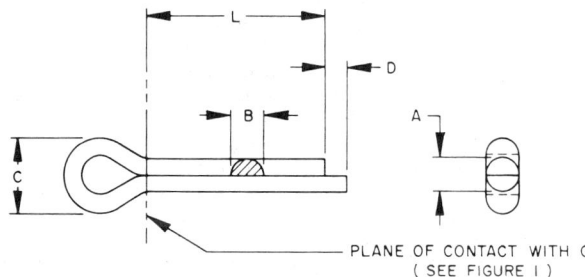

EXTENDED PRONG SQUARE CUT TYPE

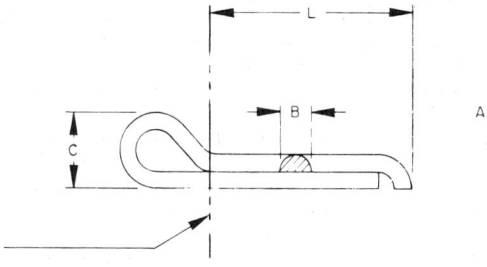

HAMMER LOCK TYPE

TABLE 1—DIMENSIONS OF COTTER PINS, IN

Nominal Size[a] or Basic Pin Diameter	A Total Shank Diameter		B Wire Width		C Head Diameter	D Extended Prong Length	Recommended Hole Size
	Max	Min	Max	Min	Min	Min	
1/32 0.031	0.032	0.028	0.032	0.022	0.06	0.01	0.047
3/64 0.047	0.048	0.044	0.048	0.035	0.09	0.02	0.062
1/16 0.062	0.060	0.056	0.060	0.044	0.12	0.03	0.078
5/64 0.078	0.076	0.072	0.076	0.057	0.16	0.04	0.094
3/32 0.094	0.090	0.086	0.090	0.069	0.19	0.04	0.109
7/64 0.109	0.104	0.100	0.104	0.080	0.22	0.05	0.125
1/8 0.125	0.120	0.116	0.120	0.093	0.25	0.06	0.141
9/64 0.141	0.134	0.130	0.134	0.104	0.28	0.06	0.156
5/32 0.156	0.150	0.146	0.150	0.116	0.31	0.07	0.172
3/16 0.188	0.176	0.172	0.176	0.137	0.38	0.09	0.203
7/32 0.219	0.207	0.202	0.207	0.161	0.44	0.10	0.234
1/4 0.250	0.225	0.220	0.225	0.176	0.50	0.11	0.266
5/16 0.312	0.280	0.275	0.280	0.220	0.62	0.14	0.312
3/8 0.375	0.335	0.329	0.335	0.263	0.75	0.16	0.375
7/16 0.438	0.406	0.400	0.406	0.320	0.88	0.20	0.438
1/2 0.500	0.473	0.467	0.473	0.373	1.00	0.23	0.500
5/8 0.625	0.598	0.590	0.598	0.472	1.25	0.30	0.625
3/4 0.750	0.723	0.715	0.723	0.572	1.50	0.36	0.750

[a] Where specifying nominal size in decimals, zeros preceding decimal shall be omitted.

TABLE 2—PREFERRED SIZES AND LENGTHS

Nominal Pin Length, in	Nominal Pin Size, in											
	1/32	3/64	1/16	3/32	1/8	5/32	3/16	1/4	5/16	3/8	1/2	5/8
1/4	X	X O										
3/8	X		X O									
1/2	X	X O	X O	X O	X O							
3/4	X	X O	X O	X O	X O	X O	X O					
1	X	X	X O	X O	X O	X O	X O	X O	X O			
1-1/4			X O	X O	X O	X O	X O	X O	X O	X		
1-1/2			X O	X O	X O	X O	X O	X O	X O	X O		
1-3/4			X O	X O	X O	X O	X O	X O	X O	X	X	
2			X O	X O	X O	X O	X O	X O	X O	X O	X	
2-1/4			O	X	X	X O	X O	X O	X O	X	X	
2-1/2				X	X	X O	X O	X O	X O	X O	X	
2-3/4					X	X O	X O	X O	X O	X	X	X
3					X	X O	X O	X O	X O	X	X	X
3-1/2							X O	X O	X O	X O	X	X
4							X O	X O	X O	X O	X	X
5									X	X	X	X
6										X	X	X

X—Extended prong   O—Hammer lock

2.3 Points

2.3.1 The preferred point types shall be the extended prong square cut or hammer lock designs illustrated, as specified by purchaser. Variations of the extended prong design and other types of points are also available, subject to mutual agreement between the purchaser and manufacturer.

2.3.2 The ends of pins shall not be open and any gaps occurring between the prongs along the shank portion of pins beyond the point shall not exceed 0.015 in. The misalignment of prongs over entire length of shank shall not exceed 0.015 in.

2.4 Material

2.4.1 STEEL—Unless otherwise specified, pins shall be made from one of the following steels: SAE 1005 through SAE 1012.

2.4.2 OTHER MATERIALS—When so specified by purchaser, pins may also be made from SAE 30302 or 30304 corrosion resistant steel, SAE CA260 brass, or monel alloy as agreed upon between manufacturer and purchaser.

2.4.3 DUCTILITY—Each prong of the cotter pin shall be capable of withstanding being bent back upon itself (180 deg) once with no visible indication of fracture occurring at the point of bend.

2.4.4 WIRE SECTION—Cotter pins are manufactured from approximately half-round wire and it is desirable that the flat side of the wire have a small degree of rounding at edges rather than sharp corners.

2.5 Finish—Cotter pins shall normally be furnished with a natural (as processed) finish, unplated or uncoated. Other finishes, where required, shall be subject to agreement between the manufacturer and purchaser.

2.6 Workmanship—Cotter pins shall be free from burrs, loose scale, sharp edges, and all other defects affecting their serviceability.

2.7 Designation—Cotter pins shall be designated by the following date, in the sequence shown: Product name (noun first), nominal size (fractional or decimal equivalent), pin length, point type, material, and protective finish, if required. See examples below:

Pin, Cotter, $\frac{1}{8} \times 1\frac{1}{4}$, Extended Prong Type, Steel, Zinc Plated.

Pin, Cotter, 0.250×1.50, Hammer Lock Type, Corrosion Resistant Steel.

HOLES IN BOLT AND CAPSCREW SHANKS AND SLOTS IN NUTS FOR COTTER PINS—SAE J485

SAE Recommended Practice

Report of Screw Threads Division approved July 1926 and last revised by Screw Threads Committee January 1956. Reaffirmed without change by Fasteners Committee January 1962.

TABLE 1—DIMENSIONS FOR COTTER PIN HOLES AND SLOTS

Bolt, Screw and Nut, Nominal Size, in.	Hole in Bolt or Screw Shank		Slot in Nut		Cotter Pin			Bolt, Screw and Nut, Nominal Size, in.	Hole in Bolt or Screw Shank		Slot in Nut		Cotter Pin		
	Dia in.	Distance, Extreme Point of Bolt or Screw to Hole Center[a], in.	Width, in.	Depth, in.	Nominal Size, in.	Min	Max		Dia in.	Distance, Extreme Point of Bolt or Screw to Hole Center[a], in.	Width, in.	Depth, in.	Nominal Size, in.	Min	Max
1/4	5/64	7/64	0.078	0.094	1/16	0.056	0.060	1-1/4	15/64	13/32	0.312	0.375	7/32	0.202	0.207
5/16	3/32	7/64	0.094	0.094	5/64	0.072	0.076	1-3/8	15/64	7/16	0.312	0.375	7/32	0.202	0.207
3/8	7/64	9/64	0.125	0.125	3/32	0.086	0.090	1-1/2	17/64	31/64	0.375	0.438	1/4	0.220	0.225
7/16	7/64	11/64	0.125	0.156	3/32	0.086	0.090	1-5/8	17/64	31/64	0.375	0.438	1/4	0.220	0.225
1/2	9/64	11/64	0.156	0.156	1/8	0.116	0.120	1-3/4	5/16	35/64	0.438	0.500	5/16	0.275	0.280
9/16	9/64	13/64	0.156	0.188	1/8	0.116	0.120	1-7/8	5/16	35/64	0.438	0.562	5/16	0.275	0.280
5/8	11/64	15/64	0.188	0.219	5/32	0.146	0.150	2	5/16	41/64	0.438	0.562	5/16	0.275	0.280
3/4	11/64	17/64	0.188	0.250	5/32	0.146	0.150	2-1/4	5/16	41/64	0.438	0.562	5/16	0.275	0.280
7/8	11/64	9/32	0.188	0.250	5/32	0.146	0.150	2-1/2	3/8	3/4	0.562	0.688	3/8	0.329	0.335
1	13/64	5/16	0.250	0.281	3/16	0.172	0.176	2-3/4	3/8	3/4	0.562	0.688	3/8	0.329	0.335
1-1/8	13/64	25/64	0.250	0.344	3/16	0.172	0.176	3	1/2	3/4	0.625	0.750	1/2	0.467	0.473

[a] This dimension is suggested to determine the distance from the cotter pin hole to under face of bolt or screw head.

STRAIGHT PINS (SOLID)—SAE J495

SAE Standard

Report of Parts and Fittings Committee approved January 1957. Editorial change June 1964.

TABLE 1—DIMENSIONS OF STRAIGHT PINS, IN.[a]

A, Pin Dia			B Chamfer
Nominal	Max	Min	
0.062	0.0625	0.0605	0.015
0.094	0.0937	0.0917	0.015
0.109	0.1094	0.1074	0.015
0.125	0.1250	0.1230	0.015
0.156	0.1562	0.1542	0.015
0.188	0.1875	0.1855	0.015
0.219	0.2187	0.2167	0.015
0.250	0.2500	0.2480	0.015
0.312	0.3125	0.3095	0.030
0.375	0.3750	0.3720	0.030
0.438	0.4375	0.4345	0.030
0.500	0.500	0.4970	0.030

[a] These pins must be straight and free from burrs or any other defects that will affect their serviceability.

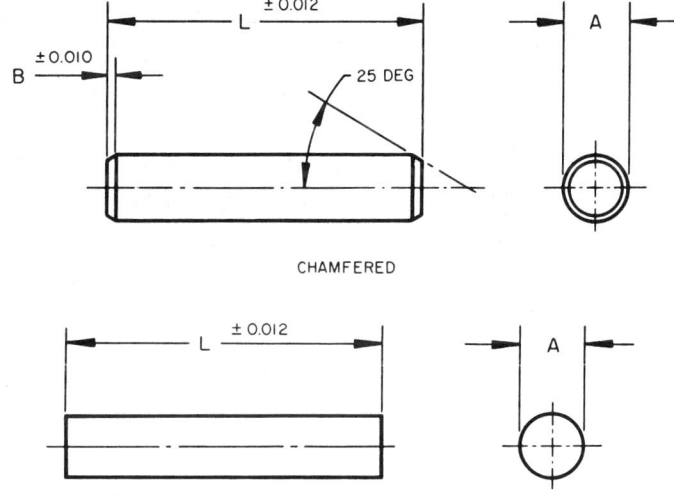

GROOVED STRAIGHT PINS—SAE J494

SAE Standard

Report of Parts and Fittings Committee approved May 1955 and last revised May 1959. Editorial change June 1962. Reaffirmed without change June 1964.

GENERAL DATA

Material—Cold drawn SAE 1112 or 1113 steel, alloy steel, stainless steel or copper alloy as specified by purchaser.

Finishes—Unless otherwise specified, steel pins shall have a flash plate of cadmium or zinc for protection of pins in transit or storage.

Defects—Grooved Straight Pins must be free from burrs and all other defects that might affect their use and serviceability.

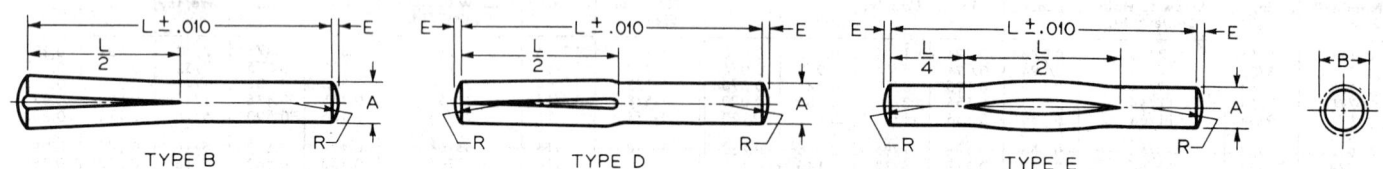

TYPE B   TYPE D   TYPE E

TABLE 1—TYPES B, D, AND E GROOVED STRAIGHT PINS, IN.

	Nominal Size	3/64	1/16	5/64	3/32	7/64	1/8	5/32	3/16	7/32	1/4	5/16	3/8	7/16	1/2
A	Diameter, max	0.0469	0.0625	0.0781	0.0938	0.1094	0.1250	0.1563	0.1875	0.2188	0.2500	0.3125	0.3750	0.4375	0.5000
	Diameter, min	0.0459	0.0615	0.0771	0.0928	0.1084	0.1230	0.1543	0.1855	0.2168	0.2480	0.3105	0.3730	0.4355	0.4980
	Recommended Hole, max	0.0478	0.0640	0.0798	0.0956	0.1113	0.1271	0.1587	0.1903	0.2219	0.2534	0.3166	0.3797	0.4428	0.5060
	Recommended Hole, min	0.0465	0.0625	0.0781	0.0938	0.1094	0.1250	0.1563	0.1875	0.2188	0.2500	0.3125	0.3750	0.4375	0.5000
	Crown Height, ±0.005	0.0000	0.0065	0.0087	0.0091	0.0110	0.0130	0.0170	0.0180	0.0220	0.0260	0.0340	0.0390	0.0470	0.0520
R	Radius at Nom Crown Height, ±0.010	—	0.0781	0.0938	0.125	0.1406	0.1562	0.1875	0.2500	0.2812	0.3125	0.3750	0.4688	0.5312	0.6250
	Length	\multicolumn{14}{c}{**B Diameter, Max and Min Limits (Measured with Ring Gages)**}
	1/4	0.052 / 0.050	0.069 / 0.067	0.085 / 0.083	0.102 / 0.100	0.118 / 0.116	0.136 / 0.132	—	—	—	—	—	—	—	—
	3/8	0.052 / 0.050	0.069 / 0.067	0.085 / 0.083	0.102 / 0.100	0.118 / 0.116	0.136 / 0.132	0.168 / 0.164	0.200 / 0.196	—	—	—	—	—	—
	1/2	0.052 / 0.050	0.069 / 0.067	0.085 / 0.083	0.102 / 0.100	0.118 / 0.116	0.136 / 0.132	0.168 / 0.164	0.200 / 0.196	0.232 / 0.228	0.265 / 0.261	—	—	—	—
	5/8	0.052 / 0.050	0.069 / 0.067	0.085 / 0.083	0.102 / 0.100	0.118 / 0.116	0.136 / 0.132	0.168 / 0.164	0.200 / 0.196	0.232 / 0.228	0.265 / 0.261	0.331 / 0.327	—	—	—
	3/4	—	0.069 / 0.067	0.085 / 0.083	0.102 / 0.100	0.118 / 0.116	0.136 / 0.132	0.168 / 0.164	0.200 / 0.196	0.232 / 0.228	0.265 / 0.261	0.331 / 0.327	0.396 / 0.392	—	—
	7/8	—	0.069 / 0.067	0.085 / 0.083	0.102 / 0.100	0.118 / 0.116	0.136 / 0.132	0.168 / 0.164	0.200 / 0.196	0.232 / 0.228	0.265 / 0.261	0.331 / 0.327	0.396 / 0.392	0.461 / 0.457	—
	1	—	0.069 / 0.067	0.085 / 0.083	0.102 / 0.100	0.118 / 0.116	0.136 / 0.132	0.168 / 0.164	0.200 / 0.196	0.232 / 0.228	0.265 / 0.261	0.331 / 0.327	0.396 / 0.392	0.461 / 0.457	0.527 / 0.523
	1-1/4	—	—	—	0.102 / 0.100	0.118 / 0.116	0.136 / 0.132	0.168 / 0.164	0.200 / 0.196	0.232 / 0.228	0.265 / 0.261	0.331 / 0.327	0.396 / 0.392	0.461 / 0.457	0.527 / 0.523
	1-1/2	—	—	—	—	0.136 / 0.132	0.168 / 0.164	0.200 / 0.196	0.232 / 0.228	0.265 / 0.261	0.331 / 0.327	0.396 / 0.392	0.461 / 0.457	0.527 / 0.523	
	1-3/4	—	—	—	—	—	0.167 / 0.163	0.200 / 0.196	0.232 / 0.228	0.265 / 0.261	0.331 / 0.327	0.396 / 0.392	0.461 / 0.457	0.527 / 0.523	
	2	—	—	—	—	—	0.167 / 0.163	0.200 / 0.196	0.232 / 0.228	0.265 / 0.261	0.331 / 0.327	0.396 / 0.392	0.461 / 0.457	0.527 / 0.523	
	2-1/4	—	—	—	—	—	—	0.199 / 0.195	0.232 / 0.228	0.265 / 0.261	0.331 / 0.327	0.396 / 0.392	0.461 / 0.457	0.527 / 0.523	
	2-1/2	—	—	—	—	—	—	—	0.232 / 0.228	0.265 / 0.261	0.331 / 0.327	0.396 / 0.392	0.461 / 0.457	0.527 / 0.523	
	2-3/4	—	—	—	—	—	—	—	0.231 / 0.227	0.264 / 0.260	0.331 / 0.327	0.396 / 0.392	0.461 / 0.457	0.527 / 0.523	
	3	—	—	—	—	—	—	—	0.231 / 0.227	0.264 / 0.260	0.331 / 0.327	0.396 / 0.392	0.461 / 0.457	0.527 / 0.523	
	3-1/4	—	—	—	—	—	—	—	—	0.264 / 0.260	0.330 / 0.326	0.395 / 0.391	0.461 / 0.457	0.527 / 0.523	
	3-1/2	—	—	—	—	—	—	—	—	—	0.330 / 0.326	0.395 / 0.391	0.461 / 0.457	0.527 / 0.523	
	3-3/4	—	—	—	—	—	—	—	—	—	—	0.395 / 0.391	0.460 / 0.456	0.527 / 0.523	
	4	—	—	—	—	—	—	—	—	—	—	0.395 / 0.391	0.460 / 0.456	0.527 / 0.523	
	4-1/4	—	—	—	—	—	—	—	—	—	—	0.395 / 0.391	0.460 / 0.456	0.526 / 0.522	
	4-1/2	—	—	—	—	—	—	—	—	—	—	—	0.460 / 0.456	0.526 / 0.522	

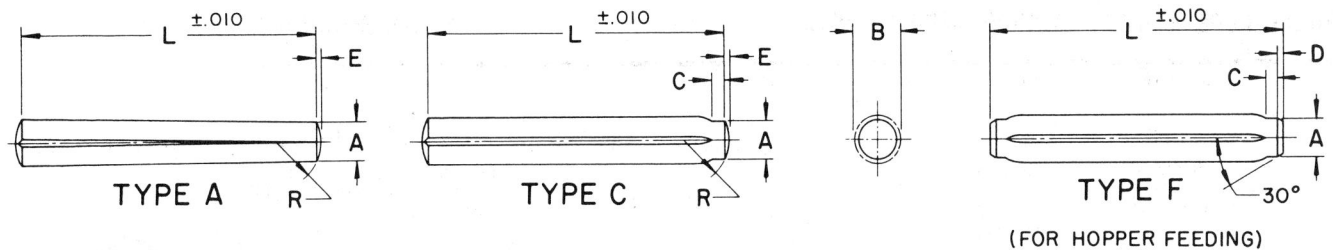

TABLE 2—TYPES A, C, AND F GROOVED STRAIGHT PINS, IN.

	Nominal Size	3/64	1/16	5/64	3/32	7/64	1/8	5/32	3/16	7/32	1/4	5/16	3/8	7/16	1/2
A	Diameter, max Diameter, min	0.0469 0.0459	0.0625 0.0615	0.0781 0.0771	0.0938 0.0928	0.1094 0.1084	0.1250 0.1230	0.1563 0.1543	0.1875 0.1855	0.2188 0.2168	0.2500 0.2480	0.3125 0.3105	0.3750 0.3730	0.4375 0.4355	0.5000 0.4980
	Recommended Hole, max Recommended Hole, min	0.0478 0.0465	0.0640 0.0625	0.0798 0.0781	0.0956 0.0938	0.1113 0.1094	0.1271 0.1250	0.1587 0.1563	0.1903 0.1875	0.2219 0.2188	0.2534 0.2500	0.3166 0.3125	0.3797 0.3750	0.4428 0.4375	0.5060 0.5000
E	Crown Height, ±0.005	0.0000	0.0065	0.0087	0.0091	0.0110	0.0130	0.0170	0.0180	0.0220	0.0260	0.0340	0.0390	0.0470	0.0520
R	Radius at Nom Crown Height, ±0.010	—	0.0781	0.0938	0.1250	0.1406	0.1562	0.1875	0.2500	0.2812	0.3125	0.3750	0.4688	0.5312	0.6250
C	Pilot Length	—	0.0312	0.0312	0.0312	0.0312	0.0312	0.0625	0.0625	0.0625	0.0625	0.0938	0.0938	0.0938	0.0938
D[a]	Chamfer Length (Type F Only)	—	0.0156	0.0156	0.0156	0.0156	0.0156	0.0312	0.0312	0.0312	0.0312	0.0469	0.0469	0.0469	0.0469
Length		B Diameter, Max and Min Limits (Measured with Ring Gages)													
1/8		0.052 0.050	—	—	—	—	—	—	—	—	—	—	—	—	—
3/16		0.052 0.050	—	—	—	—	—	—	—	—	—	—	—	—	—
1/4		0.052 0.050	0.069 0.067	0.085 0.083	0.102 0.100	0.118 0.116	0.136 0.132	—	—	—	—	—	—	—	—
3/8		0.052 0.050	0.069 0.067	0.085 0.083	0.102 0.100	0.118 0.116	0.136 0.132	0.168 0.164	0.200 0.196	—	—	—	—	—	—
1/2		0.052 0.050	0.069 0.067	0.085 0.083	0.102 0.100	0.118 0.116	0.136 0.132	0.168 0.164	0.200 0.196	0.232 0.228	0.265 0.261	—	—	—	—
5/8		0.052 0.050	0.069 0.067	0.085 0.083	0.102 0.100	0.118 0.116	0.136 0.132	0.168 0.164	0.200 0.196	0.232 0.228	0.265 0.261	0.331 0.327	—	—	—
3/4		—	0.069 0.067	0.085 0.083	0.102 0.100	0.117 0.115	0.136 0.132	0.168 0.164	0.200 0.196	0.232 0.228	0.265 0.261	0.331 0.327	0.396 0.392	—	—
7/8		—	0.069 0.067	0.085 0.083	0.102 0.100	0.117 0.115	0.135 0.131	0.167 0.163	0.200 0.196	0.232 0.228	0.265 0.261	0.331 0.327	0.396 0.392	0.461 0.457	—
1		—	0.069 0.067	0.085 0.083	0.102 0.100	0.116 0.114	0.135 0.131	0.167 0.163	0.200 0.196	0.232 0.228	0.265 0.261	0.331 0.327	0.396 0.392	0.461 0.457	0.527 0.523
1-1/4		—	—	—	0.102 0.100	0.116 0.114	0.134 0.130	0.166 0.162	0.199 0.195	0.232 0.228	0.265 0.261	0.331 0.327	0.396 0.392	0.461 0.457	0.527 0.523
1-1/2		—	—	—	—	—	0.134 0.130	0.166 0.162	0.199 0.195	0.231 0.227	0.264 0.260	0.331 0.327	0.396 0.392	0.461 0.457	0.527 0.523
1-3/4		—	—	—	—	—	—	0.165 0.161	0.199 0.195	0.231 0.227	0.264 0.260	0.330 0.326	0.395 0.391	0.461 0.457	0.527 0.523
2		—	—	—	—	—	—	0.165 0.161	0.198 0.194	0.231 0.227	0.264 0.260	0.330 0.326	0.395 0.391	0.460 0.456	0.527 0.523
2-1/4		—	—	—	—	—	—	—	0.198 0.194	0.231 0.227	0.264 0.260	0.330 0.326	0.395 0.391	0.460 0.456	0.526 0.522
2-1/2		—	—	—	—	—	—	—	—	0.230 0.226	0.263 0.259	0.329 0.325	0.395 0.391	0.460 0.456	0.526 0.522
2-3/4		—	—	—	—	—	—	—	—	0.230 0.226	0.263 0.259	0.329 0.325	0.395 0.391	0.460 0.456	0.526 0.522
3		—	—	—	—	—	—	—	—	0.229 0.225	0.262 0.258	0.329 0.325	0.394 0.390	0.459 0.455	0.525 0.521
3-1/4		—	—	—	—	—	—	—	—	—	0.262 0.258	0.328 0.324	0.394 0.390	0.459 0.455	0.525 0.521
3-1/2		—	—	—	—	—	—	—	—	—	—	0.328 0.324	0.393 0.389	0.458 0.454	0.524 0.520
3-3/4		—	—	—	—	—	—	—	—	—	—	—	0.393 0.389	0.458 0.454	0.524 0.520
4		—	—	—	—	—	—	—	—	—	—	—	0.392 0.388	0.457 0.453	0.523 0.519
4-1/4		—	—	—	—	—	—	—	—	—	—	—	0.392 0.388	0.457 0.453	0.523 0.519
4-1/2		—	—	—	—	—	—	—	—	—	—	—	—	0.456 0.452	0.522 0.518

[a] On agreement between user and supplier a suitable radius may be substituted optionally for the chamfers on the ends of Type F pins for the 1/4 in. size and below.

SPRING TYPE STRAIGHT PINS—SAE J496

SAE Standard

Report of Parts and Fittings Committee approved January 1957. Editorial change November 1972.

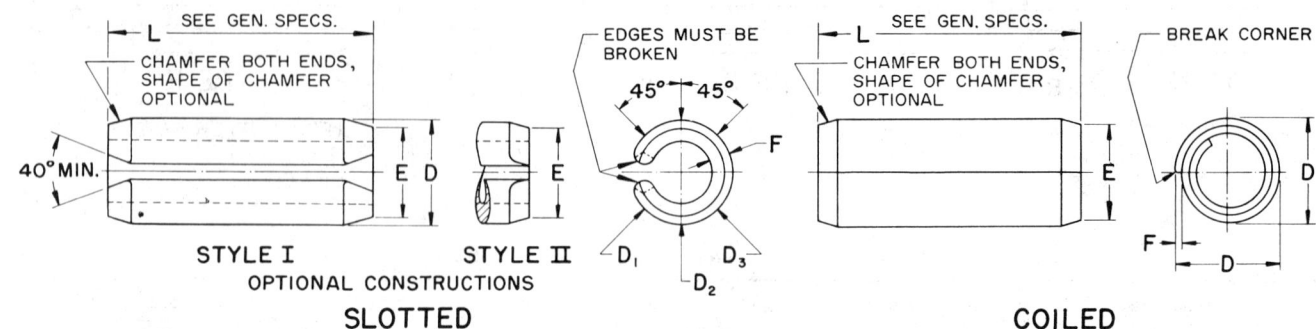

TABLE 1—PIN DIMENSIONS

Nominal Pin Size	D[a]								E	F						Double Shear Strength (lb)[e]			Recommended Hole Size			
	Slotted		Coiled						Slotted and Coiled	Slotted			Coiled			Slotted and Coiled			Slotted		Coiled	
	Series A[b] and B[b]		Series A		Series B		Series C		Series A, B, and C	Series A Nominal	Series B		Series A Nominal	Series B Nominal	Series C Nominal	Series A	Series B	Series C	Series A and B		Series A, B, and C	
											Style I Nominal	Style II Nominal										
	Max	Min[c]	Max	Min	Max	Min	Max	Min	Max							Min	Min	Min	Max	Min	Max	Min
1/32	—	—	—	—	0.035	0.033	—	—	0.029[d]	—	—	—	—	0.003	—	75	—	—	—	—	0.0325	0.0310
3/64	—	—	—	—	0.052	0.049	—	—	0.045[d]	—	—	—	—	0.004	—	170	—	—	—	—	0.0485	0.0470
0.052	—	—	—	—	0.057	0.054	—	—	0.050[d]	—	—	—	—	0.004	—	230	—	—	—	—	0.0535	0.0520
1/16	0.069	0.066	0.070	0.066	0.071	0.067	0.072	0.067	0.059	0.012	—	0.006	0.007	0.005	0.003	425	300	160	0.065	0.062	0.065	0.061
5/64	0.086	0.083	0.086	0.082	0.087	0.083	0.088	0.083	0.075	0.018	—	0.008	0.007	0.005	0.003	650	480	260	0.081	0.078	0.081	0.077
3/32	0.103	0.099	0.103	0.098	0.104	0.099	0.105	0.099	0.091	0.022	0.012	0.012	0.007	0.005	0.005	1000	690	370	0.097	0.094	0.097	0.093
7/64	0.118	0.113	0.118	0.113	0.119	0.114	0.120	0.114	0.106	0.022	—	0.018	0.010	0.007	0.005	1410	940	510	0.112	0.109	0.112	0.108
1/8	0.135	0.131	0.136	0.130	0.137	0.131	0.138	0.131	0.122	0.028	0.012	0.018	0.014	0.010	0.007	1840	1000	660	0.129	0.125	0.129	0.124
9/64	0.149	0.145	0.151	0.145	0.152	0.146	0.153	0.146	0.136	0.028	—	0.022	0.014	0.010	0.007	2200	1550	830	0.141	0.140	0.144	0.139
5/32	0.167	0.162	0.168	0.161	0.170	0.163	0.171	0.163	0.152	0.032	0.018	0.022	0.017	0.011	0.007	2880	1750	1040	0.160	0.156	0.160	0.155
3/16	0.199	0.194	0.202	0.194	0.204	0.196	0.206	0.196	0.182	0.040	0.022	0.028	0.020	0.015	0.010	4140	2500	1500	0.192	0.187	0.192	0.185
7/32	0.232	0.226	0.235	0.226	0.238	0.229	0.240	0.229	0.214	0.048	0.028	0.032	0.024	0.017	0.011	5640	3760	2040	0.224	0.219	0.224	0.217
1/4	0.264	0.258	0.268	0.258	0.270	0.260	0.272	0.260	0.245	0.048	0.028	0.032	0.028	0.020	0.015	7360	4600	2660	0.256	0.250	0.256	0.248
5/16	0.328	0.321	0.340	0.327	0.341	0.327	0.342	0.327	0.306	0.062	—	0.040	0.032	0.024	0.017	11500	7670	4160	0.318	0.312	0.318	0.308
3/8	0.392	0.385	0.407	0.391	0.408	0.391	0.409	0.391	0.368	0.077	—	0.048	0.040	0.028	0.020	16580	11040	6000	0.382	0.375	0.382	0.368
7/16	0.456	0.448	0.475	0.457	0.476	0.457	0.478	0.457	0.430	0.077	—	0.048	0.047	0.036	0.024	20000	15020	8160	0.445	0.437	0.445	0.429
1/2	0.521	0.513	0.542	0.522	0.543	0.522	0.545	0.522	0.490	0.094	—	0.062	0.055	0.040	0.028	25800	19600	10640	0.510	0.500	0.510	0.490

[a] Maximum D shall be checked by a "GO" ring gage.
[b] Series designation applies to stock thickness, A being heaviest.
[c] Minimum D shall be the average of the D_1, D_2, and D_3 diameters.
[d] Series B coiled.
[e] Applies to pins made from SAE 1070 to 1095 steel and SAE 51410 or AISI 420 corrosion resistant steel. SAE 30302 stainless steel has a minimum shear strength equal to 85% of values shown for coiled pins.

GENERAL SPECIFICATIONS

Length and Availability—Table 3, Practical Length Increments and Ranges, indicates spring pin sizes. Information on availability of individual lengths in the various types, weights, and materials may be obtained from suppliers.

The tolerance on length for coiled type spring pins shall be ±0.010 in. for sizes up to and including 5/16 in.; and ±0.015 in. for sizes larger than 5/16 in.

The tolerance on length for slotted type spring pins shall be in accordance with the following tabulation:

Length Range	3/16 to 1 Incl	Over 1 to 2 Incl	Over 2 to 3 Incl	Over 3 to 4 Incl	Over 4 in.
Tolerance on Length	±0.015	±0.020	±0.025	±0.030	±0.035

Surface Treatment—Where corrosion preventive treatment applied to carbon steel spring pins is such that it might produce hydrogen embrittlement, the spring pins shall be baked or treated in such a manner as to obviate such embrittlement.

Material and Hardness—Hardness shall be tested in the following manner: For slotted pins the readings shall be taken near the center of a longitudinal flat ground or cut on the pin at right angles to the slot. Coiled pins shall be ground or cut in half along the longitudinal axis and the hardness readings shall be taken on the inside surface of the outer half coil. Table 2 designates materials and the proper Rockwell scale to be used for the various wall thickness ranges.

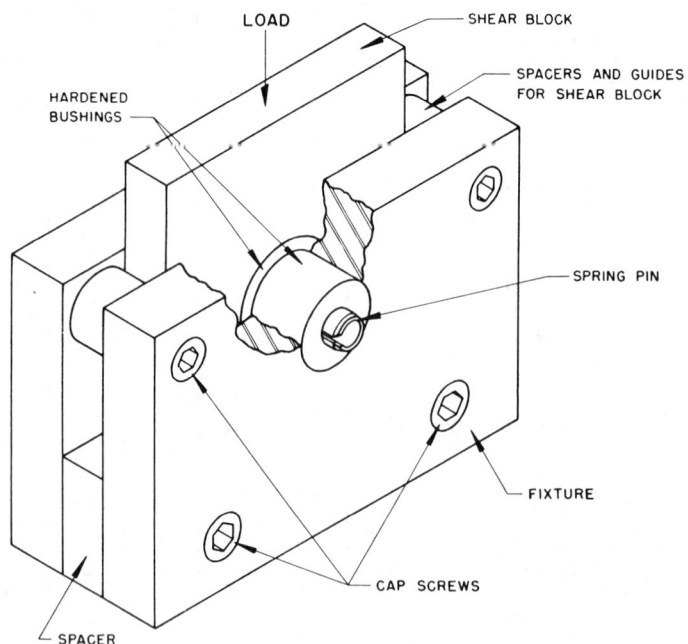

TYPICAL SPRING PIN SHEAR TEST FIXTURE

Defects—Spring pins shall be free from burrs, loose scale, seams, notches, sharp edges, or other defects which might affect their serviceability.

Performance—Spring pins shall withstand the minimum double shear loads specified in the dimensional tables when tested in accordance with the following procedure.

The shear test shall be performed in a fixture in which the pin support members and the member for applying the load shall have holes of a diameter conforming to the recommended hole size for the pins being tested and shall have a Rockwell hardness of C 58 or equivalent. The clearance between the supporting members and the loading member shall not exceed 0.005 in. and the shear plane shall be at least one diameter away from each end of the pin. Pins shall be located so that the slot is approximately at right angles to the line of application of the load. The speed of load application shall not exceed $\frac{1}{2}$ in. per min. Pins too short to be tested in double shear shall be tested by shearing two pins simultaneously in single shear. Spring pins which have been tested for shear strength shall show a ductile shear without longitudinal cracks.

TABLE 2—PIN HARDNESS

Wall Thickness Range	Over 0.001 to 0.010	Over 0.010 to 0.025	Over 0.025 to 0.050	Over 0.050 to 0.094
Rockwell Scale	Dph[a]-Tukon	15N	A	C
Pin Type / Material	Hardness Reading			
Slotted — SAE 1070–1095 Steel	458–562	83.6–87	73.6–78	46–53
Slotted — AISI 420 Corrosion Resistant Steel	413–545	82–86.6	72–77	43–52
Coiled — SAE 1070 Steel	393–515	80.4–85.5	70.4–75.9	40–50
Coiled — SAE 51410 or AISI 420 Corrosion Resistant Steel	393–515	80.4–85.5	70.4–75.9	40–50
Coiled — SAE 30302 Stainless Steel	—	—	—	—

[a] Diamond pyramidal hardness.

TABLE 3—PRACTICAL LENGTH INCREMENTS AND RANGES

(Shaded cells in the original table indicate available length/diameter combinations for Nominal Diameters: 1/32[a], 3/64[a], 0.052[a], 1/16, 5/64, 3/32, 7/64, 1/8, 9/64, 5/32, 3/16, 7/32, 1/4, 5/16, 3/8, 7/16, 1/2; and Lengths: 1/8, 3/16, 1/4, 5/16, 3/8, 7/16, 1/2, 9/16, 5/8, 11/16, 3/4, 13/16, 7/8, 15/16, 1, 1-1/8, 1-1/4, 1-3/8, 1-1/2, 1-5/8, 1-3/4, 1-7/8, 2, 2-1/4, 2-1/2, 2-3/4, 3, 3-1/4, 3-1/2, 3-3/4, 4.)

[a] Coiled type only.

UNHARDENED GROUND DOWEL PINS—SAE J497

SAE Standard

Report of Parts and Fittings Committee approved January 1957. Editorial change June 1964.

TABLE 1—DIMENSIONS OF UNHARDENED GROUND DOWEL PINS, IN.[a]

Nominal	Diameter, A		Chamfer, B
	Max	Min	
0.062	0.0600	0.0595	0.015
0.094	0.0912	0.0907	0.015
0.109	0.1068	0.1063	0.015
0.125	0.1223	0.1218	0.015
0.156	0.1535	0.1530	0.015
0.188	0.1847	0.1842	0.015
0.219	0.2159	0.2154	0.015
0.250	0.2470	0.2465	0.015
0.312	0.3094	0.3089	0.030
0.375	0.3717	0.3712	0.030
0.438	0.4341	0.4336	0.030
0.500	0.4964	0.4959	0.030
0.625	0.6211	0.6206	0.045
0.750	0.7458	0.7453	0.045
0.875	0.8705	0.8700	0.060
1.000	0.9952	0.9947	0.060

[a] Maximum diameters are graduated from 0.0005 on 0.062 in. pins to 0.0028 on 1.000 in. pins under the minimum commercial bar stock sizes.

ROD ENDS AND CLEVIS PINS—SAE J493

SAE Standard

Report of Miscellaneous Division approved January 1915 and last revised by Parts and Fittings Committee May 1959. Editorial change June 1961.

GENERAL SPECIFICATIONS FOR CLEVIS PINS

Material—SAE 1010 or SAE 1111 steel or equivalent.

Heat Treatment—Clevis pins shall be supplied either soft or cyanide hardened as specified.

Defects—Clevis pins must be free from burrs, loose scale, sharp edges, and all other defects that might affect their serviceability.

Tolerances—General tolerances for all dimensions are ±0.010 unless otherwise specified.

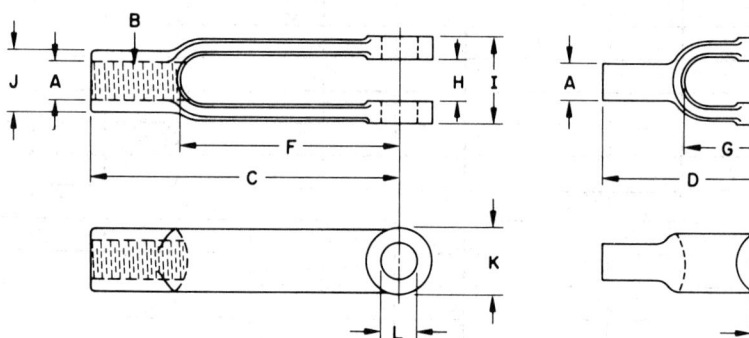
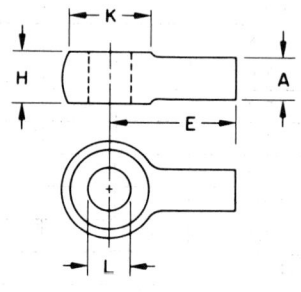

FIG. 1—ADJUSTABLE AND PLAIN YOKES   FIG. 2—EYE

TABLE 1—ADJUSTABLE AND PLAIN YOKE AND EYE, FIGS. 1 AND 2

Series	A	Threads per in., B	C	D	E	F	G	H Fork ±0.010 Eye +0.000 −0.010	I ±0.010	J	K	L Nominal	L Tolerance Plus 0.001; Minus
Light	No. 10	32	1-9/16	1-1/4	1-1/4	1	7/16	3/16	7/16	5/16	3/8	3/16	0.001
	1/4	28	2	1-3/4	1-1/4	1-1/4	5/8	9/32	5/8	7/16	1/2	1/4	0.001
	5/16	24	2-1/4	2	1-3/8	1-7/16	3/4	11/32	3/4	1/2	19/32	5/16	0.001
	3/8	24	2-1/2	2-1/8	1-1/2	1-5/8	7/16	7/16	7/8	5/8	11/16	3/8	0.001
	7/16	20	2-7/8	2-1/4	1-1/2	1-7/8	1	1/2	1	23/32	13/16	7/16	0.001
	1/2	20	3	2-1/2	1-3/4	1-7/8	1-1/8	9/16	1-1/8	13/16	15/16	1/2	0.002
Heavy	1/2	20	4-3/16	2-1/2	1-3/4	3-1/4	1-1/8	9/16	1-1/8	13/16	15/16	1/2	0.002
	5/8	18	4-15/16	2-7/8	2	3-11/16	1-7/16	11/16	1-5/16	1-1/16	1-3/16	5/8	0.002
	3/4	16	6-1/16	3-5/8	2-3/8	4-9/16	1-11/16	13/16	1-1/2	1-1/4	1-7/8	3/4	0.002
	7/8	14	7-1/8	4	2-3/4	5-1/4	2	15/16	1-3/4	1-7/16	1-11/16	7/8	0.002
	1	14	8	4-1/2	3	6	2-1/2	1-1/16	2-1/8	1-5/8	1-15/16	1	0.002

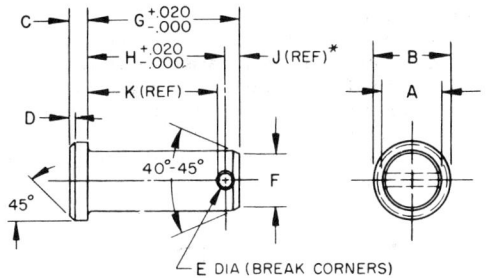

FIG. 3—PIN

* THE "J" DIMENSION (DISTANCE FROM CENTERLINE OF HOLE TO END OF PIN) IS FOR CALCULATION ONLY. ON DETAIL DRAWINGS OF CLEVIS PINS, HOLE LOCATION WILL BE SHOWN AS THE DISTANCE FROM THE UNDERSIDE OF THE HEAD TO THE CENTERLINE OF THE HOLE

TABLE 2—CLEVIS PINS, FIG. 3

Nominal Size	A Body Dia		B Head Dia	C Head Height	D Head Chamfer	E Hole Dia		F Chamfer Dia		Length (H+J) Nom G[a]	H Under Head to Center of Hole	J Center of Hole to End	K Under Head to Edge of Nom Hole
	Max	Min				Min	Max	Max	Min			Ref	Ref
3/16	0.186	0.181	0.312	0.062	0.016	0.073	0.088	0.152	0.147	0.578	0.484	0.094	0.445
1/4	0.248	0.243	0.375	0.094	0.031	0.073	0.088	0.214	0.209	0.766	0.672	0.094	0.633
5/16	0.311	0.306	0.438	0.094	0.031	0.104	0.119	0.265	0.259	0.938	0.812	0.125	0.757
3/8	0.373	0.368	0.500	0.125	0.031	0.104	0.119	0.327	0.321	1.062	0.938	0.125	0.883
7/16	0.436	0.431	0.562	0.156	0.047	0.104	0.119	0.390	0.384	1.188	1.062	0.125	1.008
1/2	0.496	0.491	0.625	0.156	0.047	0.136	0.151	0.439	0.431	1.359	1.203	0.156	1.133
5/8	0.621	0.616	0.812	0.203	0.062	0.136	0.151	0.564	0.556	1.609	1.453	0.156	1.383
3/4	0.746	0.741	0.938	0.250	0.078	0.167	0.182	0.678	0.668	1.906	1.719	0.188	1.633
7/8	0.871	0.866	1.031	0.312	0.094	0.167	0.182	0.803	0.793	2.156	1.969	0.188	1.883
1	0.996	0.991	1.188	0.344	0.109	0.167	0.182	0.928	0.918	2.406	2.219	0.188	2.133

[a] Tabulated lengths intended for use with standard clevises without spacers, where other lengths are required it is recommended that 1/16 in. increments be used.

RIVETS AND RIVETING—SAE J492

SAE Standard

Report of Parts and Fittings Division approved February 1928 and last revised by Fasteners Committee June 1961. Editorial change May 1968.

GENERAL SPECIFICATIONS FOR SMALL SOLID RIVETS

General—This small solid rivet standard covers the complete general and dimensional data for flat head, pan head, button head, truss head, countersunk head, copper's, tinner's and belt rivets. Design and assembly data are given in the Appendix—Rivet Selection and Design Considerations.

The inclusion of dimensional data in this standard is not intended to imply that all of the products described are stock production sizes. Consumers are requested to consult with manufacturers concerning stock production sizes.

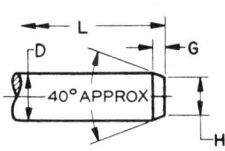

TABLE 1—DIMENSIONS OF STANDARD POINTS FOR SMALL SOLID RIVETS

Nominal Size or Basic Shank Dia		G Point Length	H Point Dia	L Rivet Length
		Ref	Approx[a]	Max
1/16	0.062	0.015	0.051	9/16
3/32	0.094	0.023	0.077	3/4
1/8	0.125	0.031	0.102	3/4
5/32	0.156	0.039	0.127	1
3/16	0.188	0.047	0.154	1
7/32	0.219	0.055	0.179	1-3/8
1/4	0.250	0.062	0.204	1-3/8
9/32	0.281	0.070	0.230	1-1/2
5/16	0.312	0.078	0.255	1-1/2
11/32	0.344	0.086	0.281	1-5/8
3/8	0.375	0.094	0.307	1-5/8
13/32	0.406	0.102	0.332	2
7/16	0.438	0.110	0.358	3

[a] No standard tolerances are contemplated.

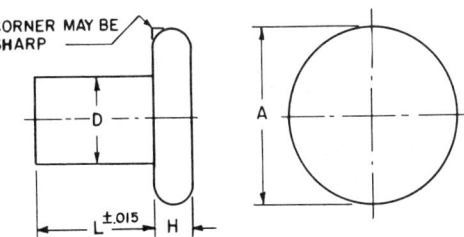

TABLE 2—FLAT HEAD RIVETS

Nominal Size or Basic Shank Dia		D Dia of Shank		A Dia of Head		H Height of Head	
		Max	Min	Max	Min	Max	Min
1/16	0.062	0.065	0.059	0.140	0.120	0.027	0.017
3/32	0.094	0.096	0.090	0.200	0.180	0.038	0.026
1/0	0.125	0.127	0.121	0.260	0.240	0.048	0.036
5/32	0.156	0.158	0.152	0.323	0.301	0.059	0.045
3/16	0.188	0.191	0.182	0.307	0.361	0.069	0.055
7/32	0.219	0.222	0.213	0.453	0.427	0.080	0.065
1/4	0.250	0.253	0.244	0.515	0.485	0.091	0.075
9/32	0.281	0.285	0.273	0.579	0.545	0.103	0.085
5/16	0.312	0.316	0.304	0.641	0.607	0.113	0.095
11/32	0.344	0.348	0.336	0.705	0.667	0.124	0.104
3/8	0.375	0.380	0.365	0.769	0.731	0.135	0.115
13/32	0.406	0.411	0.396	0.834	0.790	0.146	0.124
7/16	0.438	0.443	0.428	0.896	0.852	0.157	0.135

For dimensions and tolerances not specified above, see General Specifications.

Tolerances—The tolerances given on the dimensional tables are those for rivets made by the normal cold heading process. The tolerance for rivets made by the hot heading or forging process shall be as agreed upon between the purchaser and supplier.

Heads—Because the heads of these rivets are not machined or trimmed, the circumference may be slightly irregular and the edges may be rounded or flat.

Underhead Fillets—Rivets, other than countersunk type, shall be furnished with a definite fillet under the head but radius of fillet shall not exceed 10% of maximum shank diameter or 0.03 in., whichever is the smaller.

Material—Rivets shall be steel, copper, brass, aluminum, or other metals as specified by purchaser.

Suitable material for steel small solid rivets is covered by SAE Recommended Practice, Mechanical and Chemical Requirements for Non-threaded Fasteners—SAE J430.

Requirements of rivets made from other materials shall be as agreed upon between the purchaser and supplier.

Points—Unless otherwise specified, rivets shall have plain sheared ends. Ends shall be at right angles, within 2 deg, to the axis of the rivet and the end shall be reasonably flat, sufficient for the purpose of driving that end satisfactorily. When so specified, rivets with standard upset points are obtainable on lengths up to the maximum lengths shown in Table 1.

Workmanship—Rivets shall be free from surface seams, loose scale, and all other defects that might affect their serviceability.

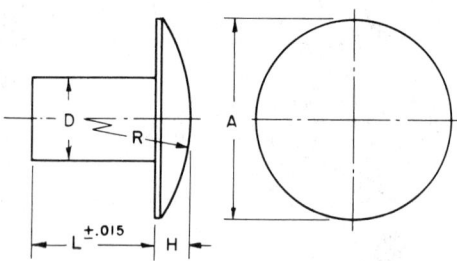

TABLE 5 — TRUSS HEAD OR WAGON BOX RIVETS

Nominal Size or Basic Shank Dia	D Dia of Shank		A Dia of Head		H Height of Head		R Radius of Head
	Max	Min	Max	Min	Max	Min	Approx
3/32 0.094	0.096	0.090	0.226	0.206	0.038	0.026	0.239
1/8 0.125	0.127	0.121	0.297	0.277	0.048	0.036	0.314
5/32 0.156	0.158	0.152	0.368	0.348	0.059	0.045	0.392
3/16 0.188	0.191	0.182	0.442	0.422	0.069	0.055	0.470
7/32 0.219	0.222	0.213	0.515	0.495	0.080	0.066	0.555
1/4 0.250	0.253	0.244	0.590	0.560	0.091	0.075	0.628
9/32 0.281	0.285	0.273	0.661	0.631	0.103	0.085	0.706
5/16 0.312	0.316	0.304	0.732	0.702	0.113	0.095	0.784
11/32 0.344	0.348	0.336	0.806	0.776	0.124	0.104	0.862
3/8 0.375	0.380	0.365	0.878	0.848	0.135	0.115	0.942
13/32 0.406	0.411	0.396	0.949	0.919	0.145	0.123	1.028
7/16 0.438	0.443	0.428	1.020	0.990	0.157	0.135	1.098

For dimensions and tolerances not specified above, see General Specifications.

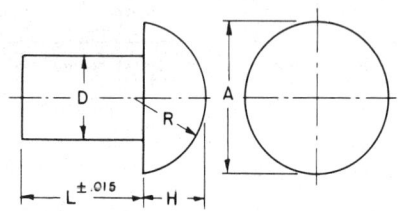

TABLE 3 — BUTTON HEAD RIVETS

Nominal Size or Basic Shank Dia	D Dia of Shank		A Dia of Head		H Height of Head		R Rad of Head
	Max	Min	Max	Min	Max	Min	Approx
1/16 0.062	0.065	0.059	0.122	0.102	0.052	0.042	0.055
3/32 0.094	0.096	0.090	0.182	0.162	0.077	0.065	0.084
1/8 0.125	0.127	0.121	0.235	0.215	0.100	0.088	0.111
5/32 0.156	0.158	0.152	0.290	0.268	0.124	0.110	0.138
3/16 0.188	0.191	0.182	0.348	0.322	0.147	0.133	0.166
7/32 0.219	0.222	0.213	0.405	0.379	0.172	0.158	0.195
1/4 0.250	0.253	0.244	0.460	0.430	0.196	0.180	0.221
9/32 0.281	0.285	0.273	0.518	0.484	0.220	0.202	0.249
5/16 0.312	0.316	0.304	0.572	0.538	0.243	0.225	0.276
11/32 0.344	0.348	0.336	0.630	0.592	0.267	0.247	0.304
3/8 0.375	0.380	0.365	0.684	0.646	0.291	0.271	0.332
13/32 0.406	0.411	0.396	0.743	0.699	0.316	0.294	0.358
7/16 0.438	0.443	0.428	0.798	0.754	0.339	0.317	0.387

For dimensions and tolerances not specified above, see General Specifications.

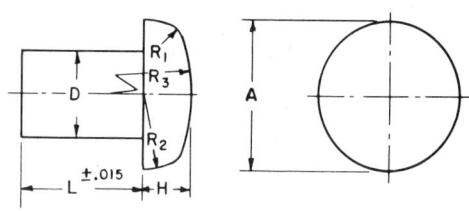

TABLE 4 — PAN HEAD RIVETS

Nominal Size or Basic Shank Dia	D Dia of Shank		A Dia of Head		H Height of Head		Radii of Head		
	Max	Min	Max	Min	Max	Min	R_1	R_2	R
							Approx		
1/16 0.062	0.065	0.059	0.118	0.098	0.040	0.030	0.019	0.052	0.217
3/32 0.094	0.096	0.090	0.173	0.153	0.060	0.048	0.030	0.080	0.326
1/8 0.125	0.127	0.121	0.225	0.205	0.078	0.066	0.039	0.106	0.429
5/32 0.156	0.158	0.152	0.279	0.257	0.096	0.082	0.049	0.133	0.535
3/16 0.188	0.191	0.182	0.334	0.308	0.114	0.100	0.059	0.159	0.641
7/32 0.219	0.222	0.213	0.391	0.365	0.133	0.119	0.069	0.186	0.754
1/4 0.250	0.253	0.244	0.444	0.414	0.151	0.135	0.079	0.213	0.858
9/32 0.281	0.285	0.273	0.499	0.465	0.170	0.152	0.088	0.239	0.963
5/16 0.312	0.316	0.304	0.552	0.518	0.187	0.169	0.098	0.266	1.070
11/32 0.344	0.348	0.336	0.608	0.570	0.206	0.186	0.108	0.292	1.176
3/8 0.375	0.380	0.365	0.663	0.625	0.225	0.205	0.118	0.319	1.286
13/32 0.406	0.411	0.396	0.719	0.675	0.243	0.221	0.127	0.345	1.392
7/16 0.438	0.443	0.428	0.772	0.728	0.261	0.239	0.137	0.372	1.500

For dimensions and tolerances not specified above, see General Specifications.

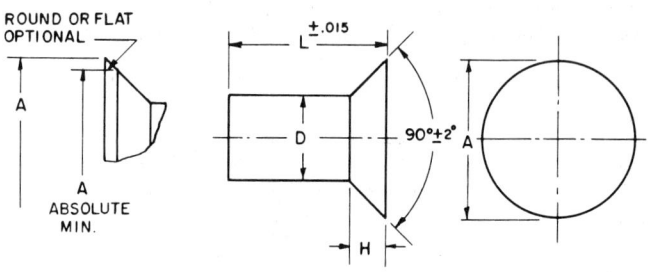

TABLE 6 — COUNTERSUNK HEAD RIVETS

Nominal Size or Basic Shank Dia	D Dia of Shank		A Dia of Head		H[a] Height of Head
	Max	Min	Max Sharp	Abs Min	
1/16 0.062	0.065	0.059	0.118	0.110	0.027
3/32 0.094	0.096	0.090	0.176	0.163	0.040
1/8 0.125	0.127	0.121	0.235	0.217	0.053
5/32 0.156	0.158	0.152	0.293	0.272	0.066
3/16 0.188	0.191	0.182	0.351	0.326	0.079
7/32 0.219	0.222	0.213	0.413	0.384	0.094
1/4 0.250	0.253	0.244	0.469	0.437	0.106
9/32 0.281	0.285	0.273	0.528	0.491	0.119
5/16 0.312	0.316	0.304	0.588	0.547	0.133
11/32 0.344	0.348	0.336	0.646	0.602	0.146
3/8 0.375	0.380	0.365	0.704	0.656	0.159
13/32 0.406	0.411	0.396	0.763	0.710	0.172
7/16 0.438	0.443	0.428	0.823	0.765	0.186

[a] Height of head, H, is given for construction purposes only. Variations in this dimension are controlled by the diameters A and D and the included angle of the head.

For dimensions and tolerances not specified above, see General Specifications.

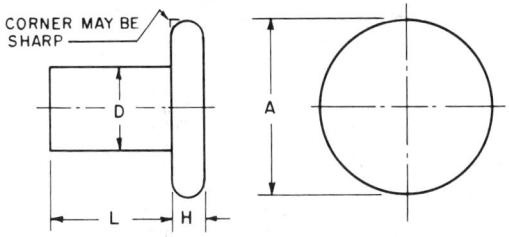

TABLE 7—TINNERS' RIVETS

Nominal Size[a]	D Dia of Shank		A Dia of Head		H Height of Head		Length	
	Max	Min	Max	Min	Max	Min	Max	Min
6 oz	0.081	0.075	0.213	0.193	0.028	0.016	0.135	0.115
8 oz	0.091	0.085	0.225	0.205	0.036	0.024	0.166	0.146
10 oz	0.097	0.091	0.250	0.230	0.037	0.025	0.182	0.162
12 oz	0.107	0.101	0.265	0.245	0.037	0.025	0.198	0.178
14 oz	0.111	0.105	0.275	0.255	0.038	0.026	0.198	0.178
1 lb	0.113	0.107	0.285	0.265	0.040	0.028	0.213	0.193
1-1/4 lb	0.122	0.116	0.295	0.275	0.045	0.033	0.229	0.209
1-1/2 lb	0.132	0.126	0.316	0.294	0.046	0.034	0.244	0.224
1-3/4 lb	0.136	0.130	0.331	0.309	0.049	0.035	0.260	0.240
2 lb	0.146	0.140	0.341	0.319	0.050	0.036	0.276	0.256
2-1/2 lb	0.150	0.144	0.311	0.289	0.069	0.055	0.291	0.271
3 lb	0.163	0.154	0.329	0.303	0.073	0.059	0.323	0.303
3-1/2 lb	0.168	0.159	0.348	0.322	0.074	0.060	0.338	0.318
4 lb	0.179	0.170	0.368	0.342	0.076	0.062	0.354	0.334
5 lb	0.190	0.181	0.388	0.362	0.084	0.070	0.385	0.365
6 lb	0.206	0.197	0.419	0.393	0.090	0.076	0.401	0.381
7 lb	0.223	0.214	0.431	0.405	0.094	0.080	0.416	0.396
8 lb	0.227	0.218	0.475	0.445	0.101	0.085	0.448	0.428
9 lb	0.241	0.232	0.490	0.460	0.103	0.087	0.463	0.443
10 lb	0.241	0.232	0.505	0.475	0.104	0.088	0.479	0.459
12 lb	0.263	0.251	0.532	0.498	0.108	0.090	0.510	0.490
14 lb	0.288	0.276	0.577	0.543	0.113	0.095	0.525	0.505
16 lb	0.304	0.292	0.597	0.563	0.128	0.110	0.541	0.521
18 lb	0.347	0.335	0.706	0.668	0.156	0.136	0.603	0.583

[a] Nominal size refers to the approximate weight of 1,000 rivets.
For dimensions and tolerances not specified above, see General Specifications.

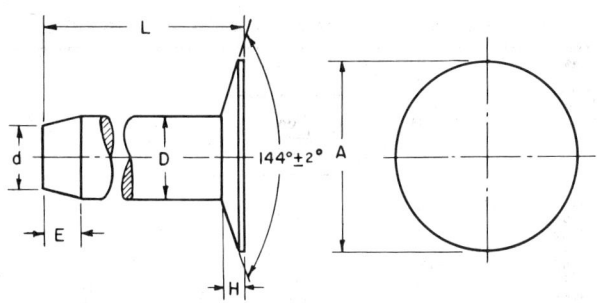

TABLE 8—COOPERS' RIVETS

Nominal Size[a]	D Dia of Shank		A Dia of Head		H Height of Head		Point		L Length	
							d Dia	E Length		
	Max	Min	Max	Min	Max	Min	Approx	Approx	Max	Min
1 lb	0.111	0.105	0.291	0.271	0.045	0.031	Not Pointed		0.249	0.219
1-1/4	0.122	0.116	0.324	0.302	0.050	0.036	Not Pointed		0.285	0.255
1-1/2	0.132	0.126	0.324	0.302	0.050	0.036	Not Pointed		0.285	0.255
1-3/4	0.136	0.130	0.324	0.302	0.052	0.034	Not Pointed		0.318	0.284
2	0.142	0.136	0.355	0.333	0.056	0.038	Not Pointed		0.322	0.288
3	0.158	0.152	0.386	0.364	0.058	0.040	0.123	0.062	0.387	0.353
4	0.168	0.159	0.388	0.362	0.058	0.040	0.130	0.062	0.418	0.388
5	0.183	0.174	0.419	0.393	0.063	0.045	0.144	0.062	0.454	0.420
6	0.206	0.197	0.482	0.456	0.073	0.051	0.160	0.094	0.498	0.457
7	0.223	0.214	0.513	0.487	0.076	0.054	0.175	0.094	0.561	0.523
8	0.241	0.232	0.546	0.516	0.081	0.059	0.182	0.094	0.597	0.559
9	0.248	0.239	0.578	0.548	0.085	0.063	0.197	0.094	0.601	0.563
10	0.253	0.244	0.578	0.548	0.085	0.063	0.197	0.094	0.632	0.594
12	0.263	0.251	0.580	0.546	0.086	0.060	0.214	0.094	0.633	0.575
14	0.275	0.263	0.611	0.577	0.091	0.065	0.223	0.094	0.670	0.612
16	0.285	0.273	0.611	0.577	0.089	0.063	0.223	0.094	0.699	0.641
18	0.285	0.273	0.642	0.608	0.108	0.082	0.230	0.125	0.749	0.691
20	0.316	0.304	0.705	0.671	0.128	0.102	0.250	0.125	0.769	0.711
3/8	0.380	0.365	0.800	0.762	0.136	0.106	0.312	0.125	0.840	0.778

[a] Nominal size refers to the approximate weight of 1,000 rivets.
For dimensions and tolerances not specified above, see General Specifications.

TABLE 9—BELT RIVETS

Nominal Size[a]	D Dia of Shank		A Dia of Head		H Height of Head		Point	
							d Dia	E Length
	Max	Min	Max	Min	Max	Min	Approx	Approx
14	0.085	0.079	0.260	0.240	0.042	0.030	0.065	0.078
13	0.097	0.091	0.322	0.302	0.051	0.039	0.073	0.078
12	0.111	0.105	0.353	0.333	0.054	0.040	0.083	0.078
11	0.122	0.116	0.383	0.363	0.059	0.045	0.097	0.078
10	0.136	0.130	0.417	0.395	0.065	0.047	0.109	0.094
9	0.150	0.144	0.448	0.426	0.069	0.051	0.122	0.094
8	0.167	0.161	0.481	0.455	0.072	0.054	0.135	0.094
7	0.183	0.174	0.513	0.487	0.075	0.056	0.151	0.125
6	0.206	0.197	0.606	0.580	0.090	0.068	0.165	0.125
5	0.223	0.214	0.700	0.674	0.105	0.083	0.185	0.125
4	0.241	0.232	0.921	0.893	0.138	0.116	0.204	0.141

[a] Nominal size refers to the Stubs iron wire gage number of the stock used in the shank of the rivet.
For dimensions and tolerances not specified above, see General Specifications.

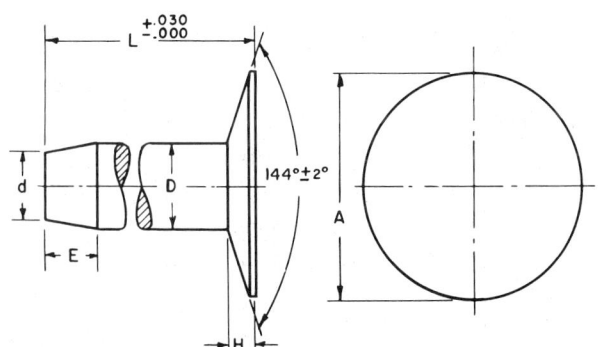

GENERAL SPECIFICATIONS FOR TUBULAR RIVETS

General—This tubular rivet standard covers the complete general and dimensional data for oval head, truss head, flat head, 90- and 120-deg countersunk head semitubular rivets and oval head, truss head and countersunk head full tubular rivets. Design and assembly data are given in the Appendix—Rivet Selection and Design Considerations.

The inclusion of dimensional data in this standard is not intended to imply that all of the products described are stock production sizes. Consumers are requested to consult with manufacturers concerning stock production sizes.

Heads—The bearing surface of flat, oval, and truss head rivets shall be at right angles to the axis of the body within 2 deg. Heads of all tubular rivets shall not be eccentric with the shank beyond a tolerance of 3% of the maximum head diameter. Because the heads are not machined or trimmed, the circumference may be slightly irregular and the edges rounded or flat.

Underhead Fillets—Rivets, other than countersunk type, shall be furnished with a definite fillet under the head but radius of fillet shall not exceed 10% of maximum shank diameter.

Material—Tubular rivets shall be low carbon steel, or brass, standard with manufacturer; or stainless steel, aluminum, copper, or other metals as agreed upon between the purchaser and supplier.

Length—Length of rivets shall be measured as indicated in the illustrations for each head style. Tubular rivets are available in length increments specified in Table 10.

Tolerance on length of tubular rivets shall be as specified in Table 11.

Workmanship—Tubular rivet end irregularities shall not be such that usability of the rivet is impaired. Rivets shall be free from surface seams, splits, and all other defects that might affect their serviceability.

TABLE 10—LENGTH INCREMENTS AND MINIMUM LENGTHS

Nominal Size	Length Increments	Min Lengths		
		Right Angle Heads	90 deg Csk Heads	120 deg Csk Heads
1/16	1/64	1/16	—	—
5/64	1/64	5/64	—	—
3/32	1/64	5/64[a]	1/8[a]	7/64[a]
7/64	1/64	3/32	—	—
1/8	1/64	7/64[a]	11/64[a]	5/32[a]
9/64	1/32	1/8[a]	3/16[a]	3/16[a]
3/16	1/32	5/32[a]	1/4[a]	1/4[a]
7/32	1/16	3/16[a]	5/16[a]	9/32[a]
1/4	1/16	7/32[a]	11/32[a]	5/16[a]
5/16	1/16	1/4[a]	—	—

[a] Hole depth to point of apex shall not exceed shank length for straight hole rivets of these lengths.

TABLE 11—TOLERANCES ON LENGTH

Nominal Size	Rivet Length		
	To and Including 4 x Dia	Over 4 Dia to and Including 8 x Dia	Over 8 x Dia
	Tolerance		
1/16	±0.007	±0.008	±0.010
5/64	±0.007	±0.008	±0.010
3/32	±0.007	±0.008	±0.010
7/64	±0.007	±0.008	±0.010
1/8	±0.007	±0.010	±0.015
9/64	±0.010	±0.012	±0.015
3/16	±0.010	±0.012	±0.015
7/32	±0.010	±0.015	±0.020
1/4	±0.010	±0.015	±0.020
5/16	±0.010	±0.015	±0.020

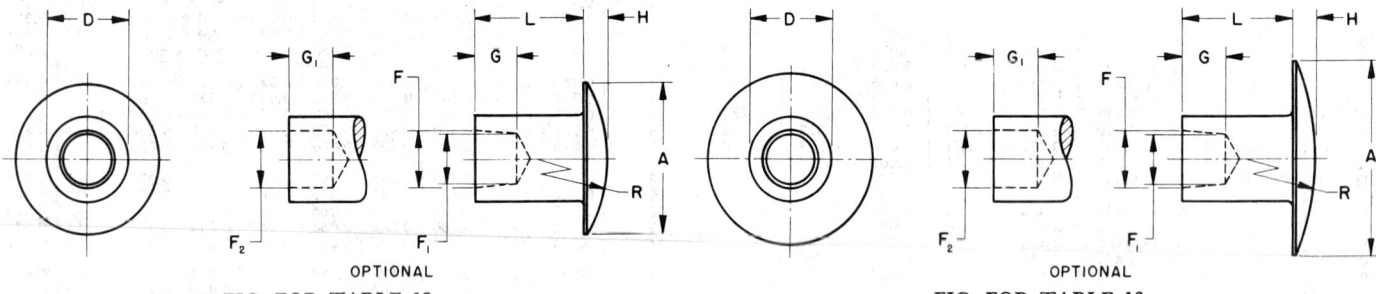

FIG. FOR TABLE 12   FIG. FOR TABLE 13

TABLE 12—DIMENSIONS OF OVAL HEAD SEMITUBULAR RIVETS

Nominal Size	D Dia of Shank		A Dia of Head		H Height of Head		Tapered Hole				Straight Hole				R Head Radius (Ref)
							Dia			G Depth	F₂ Dia		G₁ Depth		
							F		F₁						
	Max	Min	Max	Min	Max	Min	Max	Min	Min	Min	Max	Min	Max	Min	Min
1/16	0.061	0.058	0.114	0.104	0.020	0.014	0.046	0.042	0.036	0.042	0.044	0.039	0.057	0.042	0.084
5/64	0.075	0.072	0.133	0.123	0.023	0.017	0.053	0.047	0.040	0.053	0.051	0.045	0.068	0.053	0.101
3/32	0.089	0.085	0.152	0.142	0.026	0.020	0.069	0.065	0.057	0.057	0.068	0.062	0.072	0.057	0.120
7/64	0.099	0.095	0.192	0.182	0.032	0.026	0.076	0.072	0.063	0.065	0.076	0.070	0.088	0.073	0.158
1/8	0.123	0.118	0.223	0.213	0.038	0.030	0.095	0.091	0.079	0.082	0.090	0.084	0.104	0.089	0.183
9/64	0.146	0.141	0.239	0.229	0.045	0.035	0.112	0.106	0.091	0.104	0.107	0.100	0.135	0.120	0.182
3/16	0.188	0.182	0.318	0.306	0.065	0.055	0.145	0.139	0.120	0.135	0.141	0.134	0.166	0.151	0.232
7/32	0.217	0.210	0.444	0.430	0.075	0.061	0.166	0.158	0.136	0.151	0.163	0.155	0.198	0.183	0.381
1/4	0.252	0.244	0.507	0.493	0.085	0.071	0.191	0.181	0.155	0.183	0.184	0.176	0.229	0.214	0.439
5/16	0.310	0.302	0.570	0.554	0.100	0.086	0.235	0.225	0.201	0.214	0.219	0.211	0.260	0.245	0.473

For dimensions and tolerances not specified above, see General Specifications.

TABLE 13—DIMENSIONS OF TRUSS HEAD SEMITUBULAR RIVETS

Nominal Size	D Dia of Shank		A Dia of Head		H Height of Head		Tapered Hole				Straight Hole				R Head Radius (Ref)
							Dia			G Depth	F₂ Dia		G₁ Depth		
							F		F₁						
	Max	Min	Max	Min	Max	Min	Max	Min	Min	Min	Max	Min	Max	Min	Min
1/16	0.061	0.058	0.130	0.120	0.020	0.014	0.046	0.042	0.036	0.042	0.044	0.039	0.057	0.042	0.110
3/32	0.089	0.085	0.192	0.182	0.026	0.020	0.069	0.065	0.057	0.057	0.068	0.062	0.072	0.057	0.189
1/8	0.123	0.118	0.286	0.276	0.038	0.030	0.095	0.091	0.079	0.082	0.090	0.084	0.104	0.089	0.300
9/64	0.146	0.141	0.318	0.306	0.045	0.035	0.112	0.106	0.091	0.104	0.107	0.100	0.135	0.120	0.313
3/16	0.188	0.182	0.381	0.369	0.065	0.055	0.145	0.139	0.120	0.135	0.141	0.134	0.166	0.151	0.324

For dimensions and tolerances not specified above, see General Specifications.

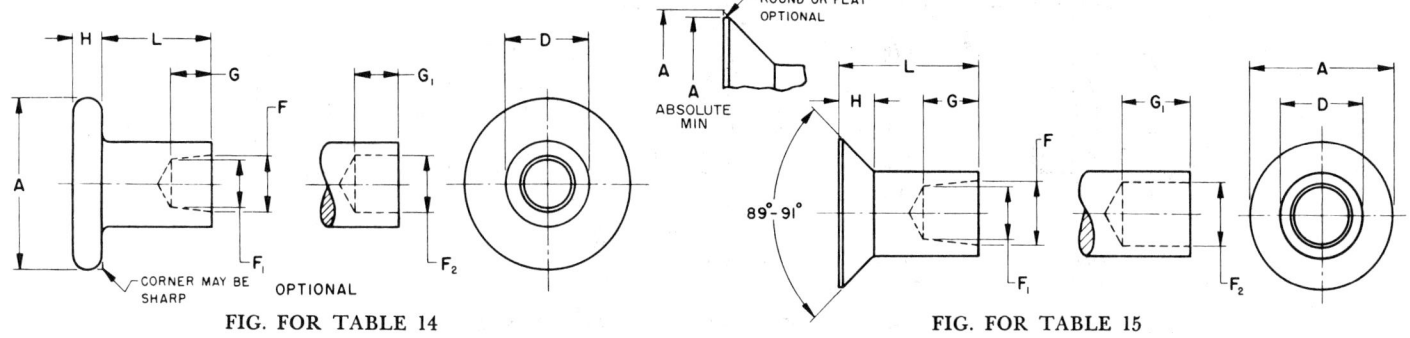

FIG. FOR TABLE 14

FIG. FOR TABLE 15

TABLE 14—DIMENSIONS OF FLAT HEAD SEMITUBULAR RIVETS

Nominal Size	D Dia of Shank		A Dia of Head		H Height of Head		Tapered Hole				Straight Hole			
							Dia			G Depth	F₂ Dia		G₁ Depth	
							F		F₁					
	Max	Min	Max	Min	Max	Min	Max	Min	Min	Min	Max	Min	Max	Min
1/16	0.061	0.058	0.114	0.104	0.027	0.023	0.046	0.042	0.036	0.042	0.044	0.039	0.057	0.042
3/32	0.089	0.085	0.161	0.151	0.034	0.028	0.069	0.065	0.057	0.057	0.068	0.062	0.072	0.057
1/8	0.123	0.118	0.223	0.213	0.041	0.034	0.095	0.091	0.079	0.082	0.090	0.084	0.104	0.089
9/64	0.146	0.141	0.317	0.307	0.052	0.042	0.112	0.106	0.091	0.104	0.107	0.100	0.135	0.120
3/16	0.188	0.182	0.381	0.369	0.067	0.057	0.145	0.139	0.120	0.135	0.141	0.134	0.166	0.151
1/4	0.252	0.244	0.507	0.493	0.090	0.076	0.191	0.181	0.155	0.183	0.184	0.176	0.229	0.214

For dimensions and tolerances not specified above, see General Specifications.

TABLE 15—DIMENSIONS OF 90 DEGREE COUNTERSUNK HEAD SEMITUBULAR RIVETS

Nominal Size	D Dia of Shank		A Dia of Head		H Height of Head	Tapered Hole				Straight Hole			
						Dia			G Depth	F₂ Dia		G₁ Depth	
						F		F₁					
	Max	Min	Max Sharp	Abs Min	Ref[a]	Max	Min	Min	Min	Max	Min	Max	Min
3/32	0.089	0.085	0.176	0.163	0.045	0.069	0.065	0.057	0.057	0.068	0.062	0.072	0.057
1/8	0.123	0.118	0.235	0.217	0.057	0.095	0.091	0.079	0.082	0.090	0.084	0.104	0.089
9/64	0.146	0.141	0.270	0.250	0.060	0.112	0.106	0.091	0.104	0.107	0.100	0.135	0.120
3/16	0.188	0.182	0.351	0.326	0.083	0.145	0.139	0.120	0.135	0.141	0.134	0.166	0.151
7/32	0.217	0.210	0.413	0.384	0.100	0.166	0.158	0.136	0.151	0.163	0.155	0.198	0.183
1/4	0.252	0.244	0.469	0.437	0.112	0.191	0.181	0.155	0.183	0.184	0.176	0.229	0.214

[a] Height of head, H, is given for reference purposes only. Variations in this dimension are controlled by diameters A and D and included angle of the head.

For dimensions and tolerances not specified above, see General Specifications.

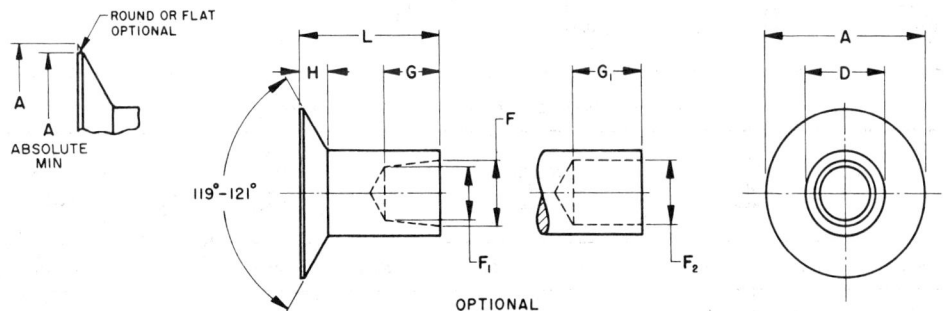

TABLE 16—DIMENSIONS OF 120 DEGREE COUNTERSUNK HEAD SEMITUBULAR RIVETS

Nominal Size	D Dia of Shank		A Dia of Head		H Height of Head	Tapered Hole				Straight Hole			
						Dia			G Depth	F₂ Dia		G₁ Depth	
						F		F₁					
	Max	Min	Max Sharp	Abs Min	Ref[a]	Max	Min	Min	Min	Max	Min	Max	Min
3/32	0.089	0.085	0.223	0.203	0.041	0.069	0.065	0.057	0.057	0.068	0.062	0.072	0.057
1/8	0.123	0.118	0.271	0.245	0.045	0.095	0.091	0.079	0.082	0.090	0.084	0.104	0.089
9/64	0.146	0.141	0.337	0.307	0.057	0.112	0.106	0.091	0.104	0.107	0.100	0.135	0.120
3/16	0.188	0.182	0.404	0.369	0.065	0.145	0.139	0.120	0.135	0.141	0.134	0.166	0.151
7/32	0.217	0.210	0.472	0.430	0.077	0.166	0.158	0.136	0.151	0.163	0.155	0.198	0.183
1/4	0.252	0.244	0.540	0.493	0.087	0.191	0.181	0.155	0.183	0.184	0.176	0.299	0.214

[a] Height of head, H, is given for reference purposes only. Variations in this dimension are controlled by diameters A and D and included angle of the head.

For dimensions and tolerances not specified above, see General Specifications.

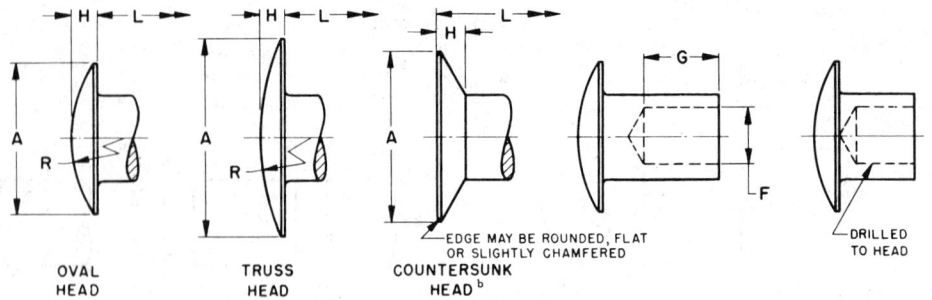

TABLE 17—DIMENSIONS OF FULL TUBULAR RIVETS

Head Style	Nominal Size	D Dia of Shank		A Dia of Head		H Height of Head		F Dia of Hole		G Depth of Hole	R Head Radius (Ref)
		Max	Min	Max	Min	Max	Min	Max	Min	Min[a]	Min
Oval	9/64	0.146	0.141	0.239	0.229	0.045	0.035	0.107	0.100	0.375	0.182
	9/64	0.146	0.141	0.318	0.306	0.045	0.035	0.107	0.100	0.375	0.313
Truss	3/16	0.188	0.182	0.381	0.369	0.065	0.055	0.141	0.134	0.375	0.324
Countersunk[b]	9/64	0.146	0.141	0.317	0.307	0.050	0.040	0.107	0.100	0.375	—
	3/16	0.188	0.182	0.381	0.369	0.060	0.048	0.141	0.134	0.375	—

[a] Full tubular rivets having length of 3/8 or shorter shall be drilled to head.
[b] Angle of head not specified since it is assumed this type of rivet would generally be used in soft materials and therefore form its own countersink.

For dimensions and tolerances not specified above, see General Specifications.

GENERAL SPECIFICATIONS FOR SPLIT RIVETS

General—This standard covers the complete general and dimensional data for oval head and countersunk head split rivets. Design and assembly data are given in the Appendix—Rivet Selection and Design Considerations.

Heads—The bearing surface of oval head split rivets shall be at right angles to the axis of the body within 2 deg. Because the heads are not machined or trimmed, the circumference may be slightly irregular and the edges may be rounded or flat.

Material—Split rivets shall be low carbon steel, brass, or other metals as agreed upon between the purchaser and supplier.

Workmanship—Rivets shall be free from surface seams, and all other defects that might affect their serviceability.

Length—Rivet length shall be measured as indicated in the illustrations for each head style. Tolerance on length shall be as specified in Table 18.

TABLE 18—TOLERANCES ON LENGTH

Nominal Size	Rivet Length		
	To and Including 4 x Dia	Over 4 to and Including 8 x Dia	Over 8 x Dia
	Tolerance		
3/32	±0.007	±0.008	±0.010
1/8	±0.007	±0.010	±0.015
9/64	±0.010	±0.012	±0.015
3/16	±0.010	±0.012	±0.015

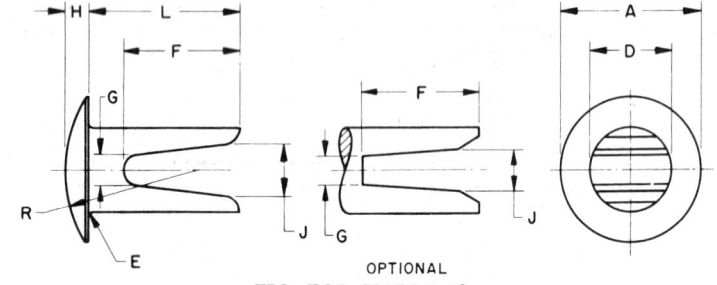

OPTIONAL
FIG. FOR TABLE 19

TABLE 19—DIMENSIONS OF OVAL HEAD SPLIT RIVETS

Nominal Size	D Dia of Shank		A Dia of Head		H Height of Head		E Radius of Fillet	R Radius of Head	L[a] Length of Rivet	F Depth of Slot ±0.015	Width of Slot	
	Max	Min	Max	Min	Max	Min	Max	Ref			G ±0.005	J ±0.005
3/32	0.092	0.086	0.151	0.141	0.027	0.019	0.010	0.130	3/16	0.156	0.030	0.037
									1/4	0.219	0.030	0.039
									5/16	0.250	0.030	0.039
									3/8 and over	0.312	0.030	0.039
1/8	0.122	0.114	0.223	0.213	0.037	0.027	0.014	0.210	3/16	0.156	0.040	0.047
									1/4	0.219	0.040	0.052
									5/16	0.266	0.040	0.057
									3/8 and over	0.312	0.040	0.057
9/64	0.152	0.144	0.317	0.307	0.049	0.037	0.018	0.304	3/16	0.156	0.050	0.060
									1/4	0.219	0.050	0.073
									5/16	0.281	0.050	0.078
									3/8	0.328	0.050	0.081
									7/16	0.344	0.050	0.083
									1/2 and over	0.391	0.052	0.077
3/16	0.195	0.185	0.350	0.338	0.064	0.050	0.022	0.300	1/4	0.219	0.065	0.120
									5/16	0.281	0.065	0.125
									3/8	0.312	0.065	0.127
									7/16	0.375	0.065	0.130
									1/2 and over	0.437	0.068	0.133

[a] Lengths over those tabulated shall be in increments of 1/16 in.

For dimensions and tolerances not specified above, see General Specifications.

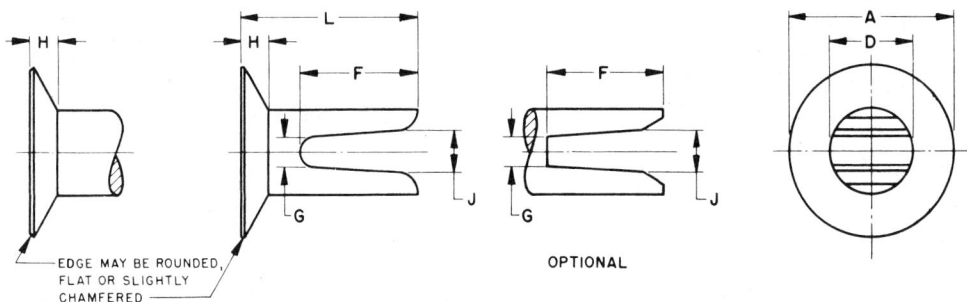

TABLE 20—DIMENSIONS OF COUNTERSUNK[a] HEAD SPLIT RIVETS

Nominal Size	D Dia of Shank		A Dia of Head		H Height of Head		L[b] Length of Rivet	F Depth of Slot ±0.016	Width of Slot	
									G ±0.005	J ±0.005
	Max	Min	Max	Min	Max	Min				
1/8	0.122	0.114	0.223	0.213	0.036	0.026	1/4 5/16 3/8 7/16 and over	0.156 0.219 0.281 0.312	0.040 0.040 0.040 0.040	0.047 0.052 0.057 0.057
9/64	0.152	0.144	0.317	0.307	0.053	0.043	1/4 5/16 3/8 7/16 1/2 9/16 and over	0.156 0.219 0.281 0.328 0.344 0.391	0.050 0.050 0.050 0.050 0.050 0.052	0.060 0.073 0.078 0.081 0.083 0.077
			0.380	0.370	0.062	0.052	1/4 5/16 3/8 7/16 1/2 9/16 and over	0.156 0.219 0.281 0.328 0.344 0.391	0.050 0.050 0.050 0.050 0.050 0.052	0.060 0.073 0.078 0.081 0.083 0.077
3/16	0.195	0.185	0.443	0.431	0.061	0.051	5/16 3/8 7/16 1/2 9/16 and over	0.219 0.281 0.312 0.375 0.437	0.065 0.065 0.065 0.065 0.068	0.120 0.125 0.127 0.130 0.133

[a] Lengths over those tabulated shall be in increments of 1/16 in.
[b] Angle of head not specified since it is assumed this type of rivet would generally be used in soft materials and therefore form its own countersink.

For dimensions and tolerances not specified above, see General Specifications.

GENERAL SPECIFICATIONS FOR RIVET CAPS

General—This standard covers the complete general and dimensional data for rivet caps used with full tubular and split rivets where appearance is a consideration.

Materials—Rivet caps shall be brass or steel, standard with manufacturer.

Workmanship—Rivet caps shall be free from all defects that might affect their serviceability.

TABLE 21—DIMENSIONS OF RIVET CAPS

Style	D Dia of Hole		A Outside Dia		H Height	
	Max	Min	Max	Min	Max	Min
1[a]	0.233 0.233 0.233	0.203 0.203 0.203	0.288 0.311 0.358	0.258 0.299 0.346	0.098 0.098 0.098	0.068 0.068 0.068
2[b]	0.233 0.281	0.203 0.251	0.350 0.442	0.320 0.412	0.098 0.129	0.068 0.099

[a] Style 1 rivet caps are designed for use with split rivets.
[b] Style 2 rivet caps are designed for use with full tubular rivets.

GENERAL SPECIFICATIONS FOR EYELETS

General—This standard covers the complete general and dimensional data for rolled flange eyelets. Design and assembly data are given in the Appendix—Rivet Selection and Design Considerations.

Flanges—Flanges of eyelets shall not be eccentric with the shank by more than 0.0075 in.

Material—Eyelets shall be brass, steel, or aluminum, standard with manufacturer.

Length—Length of eyelets shall be measured as indicated in the illustration. They are available in length increments of $\frac{1}{32}$ in. between the limits specified in Table 22.

Workmanship—Eyelet end irregularities shall not be such that usability of the eyelet is impaired. Eyelets shall be free from surface seams, splits, and all other defects that might affect their serviceability.

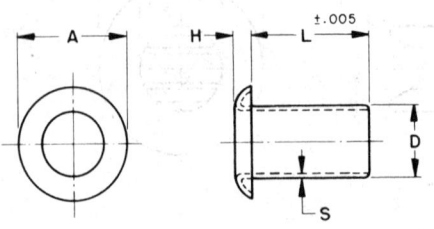

TABLE 22—DIMENSIONS OF EYELETS

Nominal Size	D Dia of Shank		A Dia of Flange		H Height of Flange	S[a] Material Thickness	L Available Lengths	
	Max	Min	Max	Min	Max		Max	Min
1/16	0.061	0.057	0.110	0.100	0.025	0.007	7/32	1/16
3/32	0.091	0.087	0.155	0.145	0.030	0.009	5/16	3/32
1/8	0.123	0.119	0.205	0.195	0.035	0.0095	11/32	3/32
5/32	0.154	0.150	0.250	0.240	0.040	0.010	11/32	3/32
3/16	0.185	0.181	0.295	0.285	0.045	0.0105	7/16	3/32
7/32	0.217	0.213	0.345	0.335	0.050	0.011	3/8	3/32
1/4	0.248	0.244	0.390	0.380	0.055	0.011	13/32	3/32

[a] Thicknesses tabulated are those from which eyelets are fabricated; therefore, thickness at shank may be slightly less than specified values.

APPENDIX—RIVET SELECTION AND DESIGN CONSIDERATIONS

General—This appendix is a guide intended to aid the user in the proper selection and application of rivets as a fastening means. It consists of general information on the advantages of riveting, various methods of riveting, selection of rivets and design considerations.

Advantages of Riveting—Riveting as a means of fastening is popular because of its simplicity, dependability, and low cost. Where the parts to be assembled do not normally need to be disassembled and, in the case of tubular, semitubular and split rivets, the tensile and fatigue strength of the joints made are not critical, riveting has many advantages. Some of the more outstanding of these are:

1. Metallic rivets are almost universally made by cold heading in high speed headers, and this makes a rivet a very economical fastener.
2. Investment in assembly equipment is low.
3. Maintenance costs of assembly equipment are low.
4. Rate of assembly is high and due to its simplicity, riveting lends itself to automation.
5. A minimum of skill is required to perform the operation.
6. Metallic or nonmetallic materials, or combinations thereof may be joined.
7. Rivets can be produced in a great variety of metals, ranging from low carbon steel to precious metals such as silver or gold.
8. Rivets may be used, not only as fasteners, but as functional components, such as pivots, electrical contacts, spacers, or supports.
9. Riveting normally requires no supplementary parts such as plain washers, lock washers, nuts, or safety wiring, nor are additional operations required such as assembly of nuts or locking devices as in the case of threaded fasteners.
10. Except for tubular, semitubular and split rivets, the rivet, when driven, usually fills the hole and prevents shifting of the parts joined.

Methods of Riveting—Riveting operations are performed by a number of methods, some of which are applicable only to particular types of rivets. The most commonly used methods are as follows:

Impact—This method employs a header die which strikes repetitive blows thus forming a head while the preformed end of the rivet is backed up with a tool called a buck or bucking bar. The header die may also be rotated while striking the repetitive blows. In machine riveting the buck is usually a part of the holding fixture. The method is applicable to solid rivets driven either hot or cold. Hot riveting is usually confined to large rivets used for structural purposes, while cold riveting is the method generally used for industrial applications on manufactured products. During the riveting operation the rivet material is displaced outward and downward into contact with the sides of the hole in which it is being assembled. The remainder of the material at rivet end forms the head. Upsetting of the shank can be controlled by using the proper impact force. See Fig. 1.

Squeeze—As its name implies, this method consists of applying steady pressure with a formed header die while the preformed end of rivet is backed up with a buck which may be made a part of the holding fixture. This method is applicable to solid rivets driven hot or cold. As in the case of impact riveting, the rivet material is displaced outward and downward into contact with the sides of the hole in which it is being assembled. The remainder of the material forms the head. See Fig. 2.

Clinch—This method of riveting involves forming the hollow end of tubular rivets and eyelets or prongs of split rivets back against the material being fastened and, depending on the shape and extent of the forming, is referred to as roll clinching, star or corrugated clinching, or scored clinching.

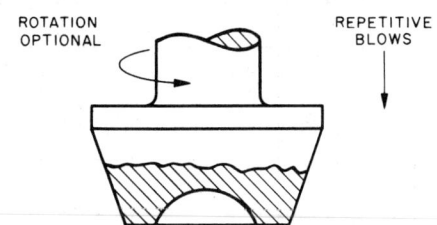

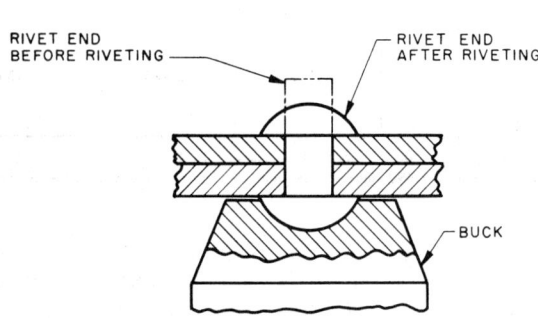

FIG. 1—IMPACT RIVETING

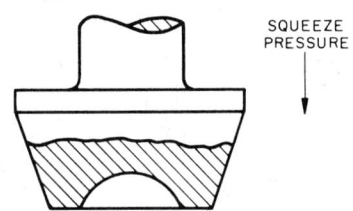

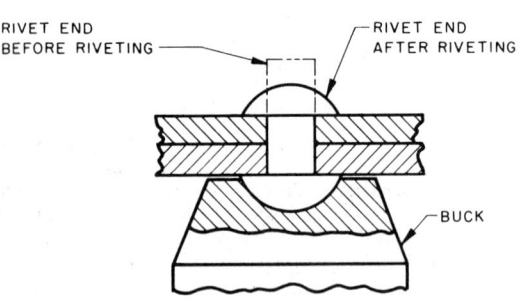

FIG. 2—SQUEEZE RIVETING

See Fig. 3. Roll clinching is accomplished by applying pressure with a formed header die, commonly called anvil, which turns or rolls the tubular shank or prongs of the rivet outward and over to bring it into contact with the part being assembled and is the method generally used to rivet semitubular and full tubular rivets and eyelets when used in metals or other hard materials. Star or corrugated clinching is accomplished by applying pressure with a formed header die which first splits or splays the tubular shank of the rivet and then turns or rolls the splayed portions outward and over to bring it into contact with the parts being assembled and is the method generally used to rivet full tubular rivets or eyelets when used in soft or resilient materials. When the splayed portions are actually turned back into the material being fastened, the method is often referred to as scored clinching. Where a finished appearance on both sides of the assembly is desirable, tubular and split rivets may be clinched into rivet caps designed for the purpose.

Shear—This method of riveting is accomplished by the use of a circular shear tool resembling a hollow punch. The method is applicable to solid rivets and the operating is performed cold. With the rivet properly bucked the tool having a hollow portion smaller than the rivet shank shears an annulus of material from the shank and with squeeze pressure upsets or displaces it into a flat annular head formed around the stub portion of the shank left by the hollow in the tool. The annular head is in contact with the part being assembled. The shearing action terminates flush with the top of the head thus leaving the head integral with the shank. See Fig. 4.

Staking—Staking consists of deforming the material of assembled rivets in such a way as to prevent their loosening or becoming disassembled under operating conditions. It does not include the forming of a head. It is done with a sharp tool at one or more points which forces the metal at these points tightly against the mating part. Where rivets are used in soft or thin materials and where light riveting is sufficient, the end of the rivet may be staked or slightly peened over the hole in a plain washer, commonly referred to as a riveting burr, to provide more bearing area on the staked side. See Fig. 5.

Rivet Selection—Requirement Considerations—With the wide variety of rivet types available, no fixed rule can be established to cover the selection of a type best suited for a given application. Generally, however, solid rivets are indicated for maximum strength while semitubular are preferred where cost is a prime factor and tensile or fatigue strength is not as critical. Full tubular rivets can, in some cases, be used with materials such as plastic, leather, canvas fabric, and wood in which the rivet under pressure pierces its own hole. The deep hole allows the slugs of pierced material to compress inside the rivet thereby exposing the required rivet material for clinching. Split rivets are also used extensively in soft materials such as those mentioned herein. The prongs pierce their own holes and are then clinched to effect the assembly. Split rivets may be used in the self-piercing and fastening of light gage metal as well. Semitubular rivets may also be used in the self-piercing and fastening of light metal wherever the appearance of the clinch is not important. Self-piercing riveting is economical and lends itself to high speed assembly operations.

Strength—A rivet is primarily strong only in shear. When set it is not stressed in tension. Thus, the designer must select the rivet size and material which will provide the necessary shear resistance needed in the application.

Diameter—The shear strength of a rivet is a direct function of the diameter so it is important to select a diameter which will provide the necessary shear strength.

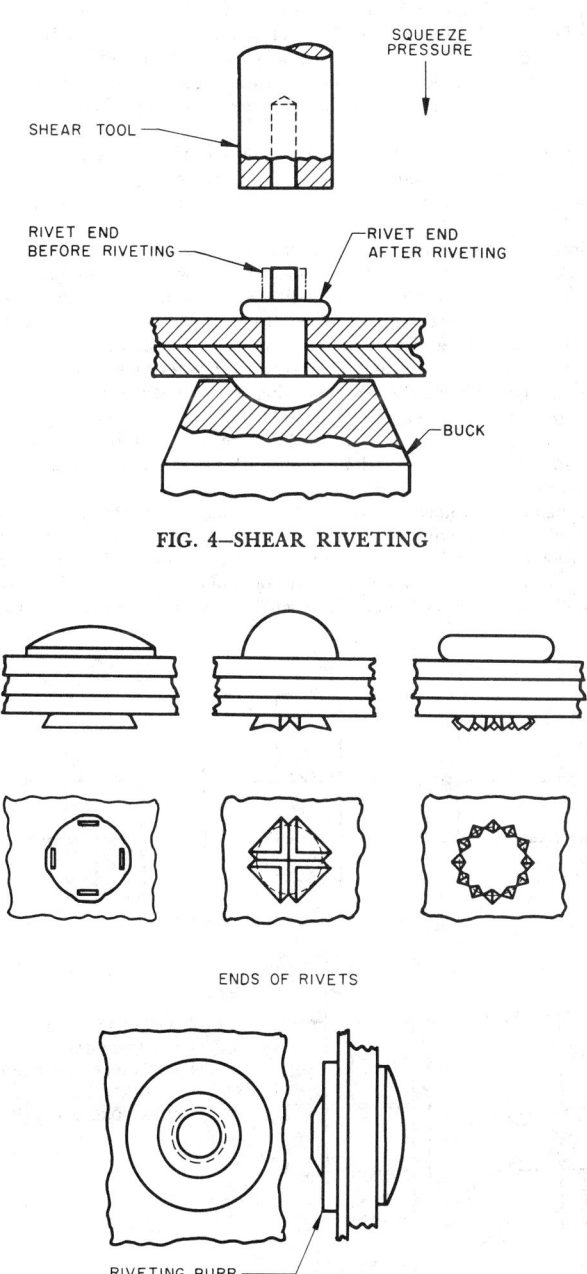

FIG. 4—SHEAR RIVETING

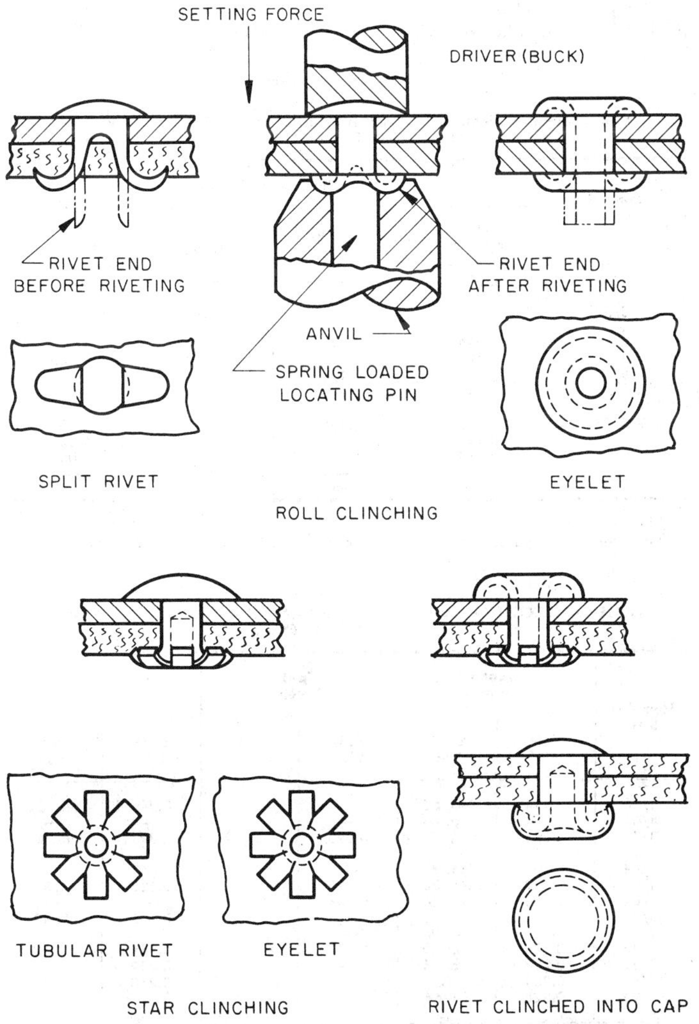

FIG. 5—STAKING

FIG. 3—CLINCH RIVETING

Head Design—The type of head specified will, of necessity, be dictated by the requirements of the application such as clearance, appearance, bearing area, and so forth. Round, truss, oval, flat, pan, and similar head styles with flat bearing surfaces provide good holding power at minimum cost. The use of flat head rivets where appearance is not a consideration minimizes tooling and production problems. Countersunk head rivets should be employed only where a flush surface is required since the countersinking or dimpling of parts to be fastened increases cost and production time on the assembly line.

Length—The length of rivet is affected by conditions such as the total compressed thickness of the members to be joined, the kind of rivet being used, the method of riveting being employed, the head style being formed, and the clearance hole into which the rivet is being assembled. The length of rivet required to provide optimum assembly conditions for a particular application can best be determined by experiment.

The following recommendations are often used to determine the length of various types of rivets for general applications and as a starting point in specific applications. The approximate length of solid rivets, when impact or squeeze riveted, required to form the head and fill the clearance space in the hole should be in excess of the thickness of the material to be riveted by an amount equal to approximately 0.75 to 1.00 times the rivet diameter for forming countersunk heads and from 1.3 to 1.7 times the rivet diameter for forming round or pan heads. See Fig. 6.

The approximate lengths for tubular rivets, split rivets, and eyelets should be determined by adding the total compressed thickness of the work to be assembled to the appropriate clinch allowance specified in Table 23. If the length so determined does not conform to the length increments shown in the specifications for the particular fastener, the next longer length should be used. See Fig. 6.

Design Considerations—After the design of rivet to be used has been determined which includes diameter, type of head, material, and other factors, the designer must then establish other related features of the design of the assembly.

Spacing—Where more than one rivet is indicated, the spacing between rivets must be such that there is sufficient room for the driving tools. Also a minimum pitch of 3 times the rivet diameter should be provided. For thin sheets it is recommended that the pitch be not greater than 24 times the thickness of the sheet. For functional strength consideration the strength afforded by the portion of metal between rivet holes should be determined and compared with the shear and bearing strength of the rivets.

Edge Distance—Failure of the metal between the rivet hole and the edge of the sheet, where solid rivets are used, can be prevented by maintaining an edge distance of $1\frac{1}{2}$ times the hole diameter for hot driven rivets and 2 times the hole diameter for cold driven rivets.

Clinching of tubular rivets exerts little radial force on the sides of the hole compared to the driving of solid rivets. The edge distance, where tubular rivets are used, can, therefore, be less than the values given herein. It can, in most applications, be dictated by the strength of the material and the load to be applied on the riveted joint. The small amount of radial force need be considered only where fastening very brittle materials such as ceramics and some plastics.

Accessibility—When using standard rivets, it is necessary to have both the preformed and driven head ends of the rivet accessible so that both the forming die and the buck may be properly used. Sufficient space should be provided to permit the use of power or manually operated rivet sets and bucks.

Hole Size—Holes should be held as close to the rivet shank diameter as possible and still permit easy and rapid assembly. Possible misalignment of holes must be considered in establishing hole sizes. Holes that are too large may result in buckling of the rivet shank or other detrimental effect when the rivet is being driven. The most suitable hole size for a given application can best be determined by experiment.

A general rule often applied to determine the hole size for solid rivets is to provide a clearance of from 0.003 to 0.008 over the maximum shank diameter of the rivet. The clearance can be increased to 0.015 where necessary for rivets $\frac{1}{4}$ in. and under and to 0.030 for rivets over $\frac{1}{4}$ in. in size. These increases, however, often result in poor riveting especially in applications requiring long grip lengths.

A general rule often applied to determine the preformed hole size for tubular rivets is to allow approximately 7% of the maximum shank diameter of the rivet for clearance. Recommended hole size values for these fasteners are given in Table 24. See Fig. 6.

Countersinking—Where the rivet heads on one or both sides of a riveted assembly must be flush with the surface, rivets with countersunk heads are used and the hole is countersunk to conform with the size and contour of the heads.

Dimpling—Dimpling involves the deformation of a sheet surface by pressure to form a countersunk recess on the one side and a corresponding projecting cone on the other. In the case of 2 or more sheets to be joined by riveting the dimpling is done on each sheet. When assembled the nesting of the dimples and projecting cones provides a large shear area and the rivet merely serves as a compression anchor to keep the dimples in contact. A relatively thin sheet may also be dimpled to match a countersunk recess in a thick sheet or manufactured part. Dimpling may be produced by a die set or by pressure exerted on the rivet head.

TABLE 23—RECOMMENDED CLINCH ALLOWANCES FOR TUBULAR AND SPLIT RIVETS AND EYELETS

Nominal Size	Clinch Allowances			
	Semitubular Rivets[a]	Full Tubular Rivets[b]	Split Rivets	Eyelets[a]
1/16	0.034	—	—	0.043
5/64	0.041	—	—	—
3/32	0.048	—	0.078	0.048
7/64	0.059	—	—	—
1/8	0.074	—	0.094	0.048
9/64	0.088	0.125	0.125	—
5/32	—	—	—	0.053
3/16	0.122	0.188	0.141	0.053
7/32	0.141	—	—	0.058
1/4	0.162	—	—	0.058
5/16	0.202	—	—	—

[a] For rolled clinch.
[b] For star or corrugated clinch, where roll clinch is desired, use semitubular values

TABLE 24—RECOMMENDED WORK HOLE DIAMETERS FOR TUBULAR AND SPLIT RIVETS AND EYELETS

Nominal Size	Tubular Rivets		Split Rivets		Eyelets	
	Hole Dia[a]	Drill Size	Hole Dia[a]	Drill Size	Hole Dia	Drill Size
1/16	0.064	No. 52	—	—	0.063	No. 52
5/64	0.081	No. 46	—	—	—	—
3/32	0.094	3/32	0.093	No. 42	0.093	No. 42
7/64	0.104	No. 37	—	—	—	—
1/8	0.129	No. 30	0.128	No. 30	0.125	1/8
9/64	0.152	No. 24	0.154	No. 23	—	—
5/32	—	—	—	—	0.156	5/32
3/16	0.196	No. 9	0.199	No. 8	0.188	No. 12
7/32	0.228	No. 1	—	—	0.219	7/32
1/4	0.261	G	—	—	—	—
5/16	0.328	21/64	—	—	0.250	1/4

[a] Applicable to full tubular and split rivets only when one of the parts to be assembled is prepunched or drilled.

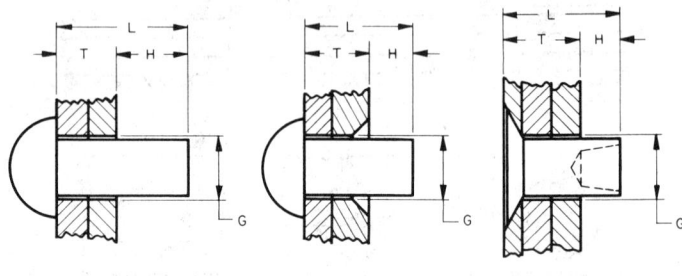
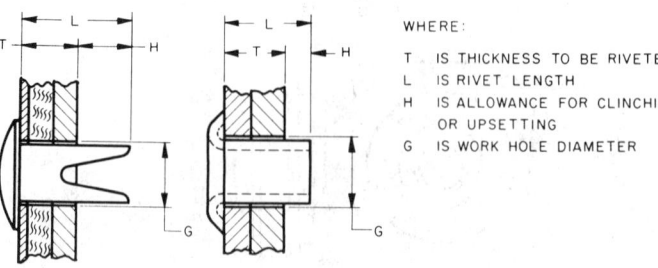

WHERE:
T IS THICKNESS TO BE RIVETED
L IS RIVET LENGTH
H IS ALLOWANCE FOR CLINCHING OR UPSETTING
G IS WORK HOLE DIAMETER

FIG. 6—TYPES OF RIVETS

BLIND RIVETS—BREAK MANDREL TYPE—SAE J1200 — SAE Standard

Report of Fasteners Committee approved July 1977.

1. General Specifications

1.1 Scope—This standard establishes the dimensional, mechanical, and performance requirements of inch and metric break mandrel blind rivets suitable for use in joining the component parts of an assembly.

1.2 Definitions

1.2.1 BLIND RIVET—A blind rivet is a blind fastener which has a self-contained mechanical, chemical, or other feature which permits the formation of an upset on the blind end of the rivet and expansion of the rivet shank during rivet setting to join the component parts of an assembly.

1.2.2 BREAK MANDREL BLIND RIVET—Break mandrel blind rivets are pull mandrel-type blind rivets, where during the setting operation the mandrel is pulled into or against the rivet body and breaks at or near the junction of the mandrel shank and its upset end.

1.3 Designations—These rivets are designated by styles and grades as described below in addition to size, length, and finish.

1.3.1 RIVET STYLES—The two basic styles of break mandrel blind rivets are designated as protruding head and flush head. Protruding head rivets are available in two styles designated as regular head and large head. Flush head rivets are available in the 120 deg countersunk head.

1.3.2 RIVET GRADES—The material combination of break mandrel blind rivets are designated as grades, with each material combination representing a different combination of rivet body material and mandrel material as given in Table 1.

TABLE 1—GRADES OF BREAK MANDREL BLIND RIVETS

Grade Designation	Rivet Body Material	Mandrel Material
10	Aluminum Alloy 5050	Aluminum Alloy 7178 or 2024
11	Aluminum Alloy 5052	Aluminum Alloy 7178 or 2024
16	Aluminum Alloy 5154	Carbon Steel
18	Aluminum Alloy 5052	Carbon Steel
19	Aluminum Alloy 5056	Carbon Steel
20	Copper Alloy No. 110	Carbon Steel
30	Low Carbon Steel	Carbon Steel
40	Nickel-Copper Alloy (Monel)	Carbon Steel
50	Stainless Steel (300 Series)	Carbon Steel
51	Stainless Steel (300 Series)	Stainless Steel (300 Series)

1.4 Dimensions and Tolerances—The design of break mandrel type blind rivets shall be in accordance with the practice of the manufacturer providing the dimensions shown in Tables 6A, 6B, 7A, and 7B are maintained and rivets meet the mechanical and performance requirements of this standard. Tolerance on dimensions in tables, not designated otherwise, shall be ±0.010 in (0.25 mm).

1.5 Materials—Rivet bodies and mandrels shall be made of the material specified for the grade in Table 1. When the specific material analysis is not given, the analysis shall be selected by the manufacturer and shall be such to assure that rivets meet the mechanical and performance requirements specified under paragraph 2.

1.6 Finishes—Grade 30 rivet bodies are either zinc or cadmium plated with a minimum plating thickness of 0.00015 in (0.004 mm). Rivet bodies of all other grades are furnished plain (bare metal), unless otherwise specified. Because mandrels are discarded following rivet setting, mandrels of all materials may be furnished plain or with a protective coating at the option of the manufacturer, unless otherwise specified.

2. Mechanical and Performance Requirements

2.1 Shear Strength—Rivets, except those described in paragraph 2.2.1, shall have ultimate shear loads not less than the minimum ultimate shear loads specified for the applicable size and grade given in Tables 2A and 2B, when tested in accordance with paragraph 3.1.

2.2 Tensile Strength—Rivets, except those described in paragraph 2.2.1, shall have ultimate tensile loads not less than the minimum ultimate tensile loads specified for the applicable size and grade given in Tables 3A and 3B when tested in accordance with paragraph 3.2.

2.2.1 Grade 20 rivet is not subject to either shear or tensile testing. For all other grades, protruding head rivets with specified maximum grip lengths shorter than 1.0 times the nominal rivet diameter, and flush head rivets with specified maximum grip lengths shorter than 1.5 times the nominal rivet diameter shall not be subject to either shear or tensile testing.

2.3 Mandrel Break Load—While the rivet is being set, the axially applied load necessary to break the mandrel shall be within the limits specified for the applicable rivet size and grade in Tables 4A and 4B when tested in accordance with paragraph 3.3.

2.4 Mandrel Retention—The mandrel shall be retained within the rivet body such that a force in excess of 2 lb (8.9 N) is required to reduce the mandrel protrusion to its specified minimum.

TABLE 2A—ULTIMATE SHEAR LOADS OF BREAK MANDREL BLIND RIVETS

Nominal Rivet Size or Basic Shank Dia		Ultimate Shear Load[a] (Force) Min, lb				
		Grades 10, 11, & 18	Grades 16 & 19	Grade 30	Grade 40	Grades 50 & 51
3/32	0.0938	70	90	130	200	230
1/8	0.1250	120	170	260	350	420
5/32	0.1562	190	260	370	550	650
3/16	0.1875	260	380	540	800	950
1/4	0.2500	460	700	1000	1400	1700

[a] Grade 20 rivet is not subject to shear testing.

TABLE 2B—ULTIMATE SHEAR LOADS OF BREAK MANDREL BLIND RIVETS

Nominal Rivet Size mm	Ultimate Shear Load[a] (Force) Min/N				
	Grades 10, 11, & 18	Grades 16 & 19	Grade 30	Grade 40	Grades 50 & 51
2.4	310	400	580	890	1020
3.2	530	760	1160	1560	1870
4.0	850	1160	1650	2450	2890
4.8	1160	1690	2400	3560	4230
6.3	2050	3110	4450	6230	7560

[a] Grade 20 rivet is not subject to shear testing.

TABLE 3A—ULTIMATE TENSILE LOADS OF BREAK MANDREL BLIND RIVETS

Nominal Rivet Size or Basic Shank Dia		Ultimate Tensile Load[a] (Force) Min, lb				
		Grades 10, 11, & 18	Grades 16 & 19	Grade 30	Grade 40	Grades 50 & 51
3/32	0.0938	80	120	170	250	280
1/8	0.1250	150	220	310	450	530
5/32	0.1562	230	350	470	700	820
3/16	0.1875	320	500	680	1000	1200
1/4	0.2500	560	920	1240	1850	2100

[a] Grade 20 rivet is not subject to tensile testing.

TABLE 3B—ULTIMATE TENSILE LOADS OF BREAK MANDREL BLIND RIVETS

Nominal Rivet Size mm	Ultimate Tensile Load[a] (Force) Min/N				
	Grades 10, 11, & 18	Grades 16 & 19	Grade 30	Grade 40	Grades 50 & 51
2.4	360	530	760	1110	1250
3.2	670	980	1380	2000	2360
4.0	1020	1560	2090	3110	3650
4.8	1420	2220	3020	4450	5340
6.3	2490	4090	5520	8230	9340

[a] Grade 20 rivet is not subject to tensile testing.

TABLE 4A—MANDREL BREAK LOADS OF BREAK MANDREL BLIND RIVETS

Nominal Rivet Size or Basic Shank Dia	Limit	Mandrel Break Load[a] lb						
		Grades 10 & 11	Grades 16, 18, & 19	Grade 20	Grade 30	Grade 40	Grade 50	Grade 51
3/32 0.0938	Min	140	175	175	260	300	300	300
	Max	240	275	275	360	450	500	500
1/8 0.1250	Min	250	400	400	600	650	650	650
	Max	400	600	600	800	850	950	950
5/32 0.1562	Min	425	600	600	750	950	1150	1150
	Max	600	850	850	1000	1200	1450	1450
3/16 0.1875	Min	625	750	750	1150	1450	1400	1400
	Max	825	1050	1050	1450	1750	1900	1900
1/4 0.2500	Min	1100	1450	1450	1950	2500	3000	3000
	Max	1400	1850	1850	2350	2900	3600	3600

[a] Mandrel break load is defined as the load in pounds necessary to break the mandrel when setting break mandrel types of pull mandrel blind rivets.

TABLE 4B—MANDREL BREAK LOADS OF BREAK MANDREL BLIND RIVETS

Nominal Rivet Size or Basic Shank Dia mm	Limit	Mandrel Break Load[a] N						
		Grades 10 & 11	Grades 16, 18, & 19	Grade 20	Grade 30	Grade 40	Grade 50	Grade 51
2.4	Min	620	780	780	1160	1330	1330	1330
	Max	1070	1220	1220	1600	2000	2220	2220
3.2	Min	1110	1780	1780	2670	2890	2890	2890
	Max	1780	2670	2670	3560	3780	4230	4230
4.0	Min	1890	2670	2670	3340	4230	5120	5120
	Max	2670	3780	3780	4450	5340	6450	6450
4.8	Min	2780	3340	3340	5120	6450	6230	6230
	Max	3670	4670	4670	6450	7780	8450	8450
6.3	Min	4890	6450	6450	8670	11 100	13 300	13 300
	Max	6230	8230	8230	10 500	12 900	16 000	16 000

[a] Mandrel break load is defined as the load in Newtons necessary to break the mandrel when setting break mandrel types of pull mandrel blind rivets.

2.5 Blind Head Formation—The axially applied load necessary to upset the end of the rivet body, that is, form the blind side head, shall not exceed 80% of the actual mandrel break load, when tested in accordance with paragraph 3.3.

3. Test Methods

3.1 Shear Test—The test shall be comprised of loading a single lap joint assembled with one rivet so that the direction of applied load induces transverse shear against the rivet body. The test specimen shall be mounted in a tensile testing machine capable of applying load at a controllable rate. The grips shall be self-aligning and care shall be taken when mounting the specimen to assure that the load will be transmitted in a straight line through the test rivet.

The specimen shall be loaded at a speed of testing as determined with a free running cross head not less than 0.3 in (7.6 mm) nor greater than 0.5 in (13.0 mm)/min. Loading shall be continued until failure of the rivet occurs.

The maximum load in pounds or Newtons applied to the specimen coincident with or prior to rivet failure shall be recorded as the ultimate shear strength of the rivet.

The test specimen shall be comprised of two plates, of equal nominal thickness, axially aligned and assembled into a single lap joint with the test rivet, as shown in Fig. 1 and Tables 5A and 5B. The design of test plates may be modified to include holes for shear testing two or more rivets using the same plates. Such holes shall be located on the longitudinal centerline of the plate, and center distances between adjacent holes shall be at least 4 times the diameter of the larger test hole. Ends of plates may be drilled for pin-type mounting in testing machine. Plates shall be alloy steel, quenched and tempered to a hardness of Rockwell C46-50.

The test rivet shall be set with a setting tool standard for that type of rivet and in accordance with the setting procedures recommended by the rivet manufacturer.

TABLE 5A—DIMENSIONS OF TEST PLATES, IN (FIGS. 1, 2, AND 3)

Nominal Rivet Size or Basic Shank Dia	G Shear and Tensile Test Plate Hole Dia		G_1 Break Mandrel Test Restraining Plate Hole Dia	S End to Center Length	T Shear and Tensile Test Plate Thickness Protruding Head Styles	T_1 Shear and Tensile Test Plate Thickness Flush Head Styles
	Max	Min	Basic[a]	Min[b]	Min[c]	Min[d]
3/32 0.0938	0.100	0.098	0.067	0.375	0.047	0.056
1/8 0.1250	0.132	0.130	0.086	0.500	0.062	0.093
5/32 0.1562	0.164	0.162	0.105	0.625	0.078	0.117
3/16 0.1875	0.196	0.194	0.125	0.750	0.094	0.141
1/4 0.2500	0.260	0.258	0.161	1.000	0.125	0.188

[a] Values shown are equal to nominal mandrel diameter plus 0.010 in.
[b] Values shown are equal to 4 times basic shank diameter of rivet.
[c] Minimum values shown are equal to 0.50 times basic shank diameter of rivet. Maximum thickness shall not exceed 0.50 times maximum grip length specified for applicable rivet in Table 8A.
[d] Minimum values shown are equal to 0.75 times basic shank diameter of rivet. Maximum thickness shall not exceed 0.50 times maximum grip length specified for applicable rivet in Table 8A.
[e] The protrusion diameter of the mandrel (W diameter), including the point burr, shall be less than basic G_1 plate hole diameter.

TABLE 5B—DIMENSIONS OF TEST PLATES, MM (FIGS. 1, 2, AND 3)

Nominal Rivet Size or Basic Shank Dia	G Shear and Tensile Test Plate Hole Dia		G_1 Break Mandrel Test Restraining Plate Hole Dia	S End To Center Length	T Shear and Tensile Test Plate Thickness Protruding Head Styles	T_2 Shear and Tensile Test Plate Thickness Flush Head Styles
	Max	Min	Basic[a]	Min[b]	Min[c]	Min[d]
2.4	2.54	2.49	1.70	9.6	1.2	1.8
3.2	3.35	3.30	2.18	12.8	1.6	2.4
4.0	4.16	4.11	2.66	16.0	2.0	3.0
4.8	4.98	4.93	3.15	19.2	2.4	3.6
6.3	6.60	6.55	4.09	25.2	3.2	4.7

[a] Values shown are equal to nominal mandrel diameter plus 0.25 mm.
[b] Values shown are equal to 4 times basic shank diameter of rivet.
[c] Minimum values shown are equal to 0.50 times basic shank diameter of rivet. Maximum thickness shall not exceed 0.50 times maximum grip length specified for applicable rivet in Table 8B.
[d] Minimum values shown are equal to 0.75 times basic shank diameter of rivet. Maximum thickness shall not exceed 0.50 times maximum grip length specified for applicable rivet in Table 8B.
[e] The protrusion diameter of the mandrel (W diameter), including the point burr, shall be less than basic G_1 plate hole diameter.

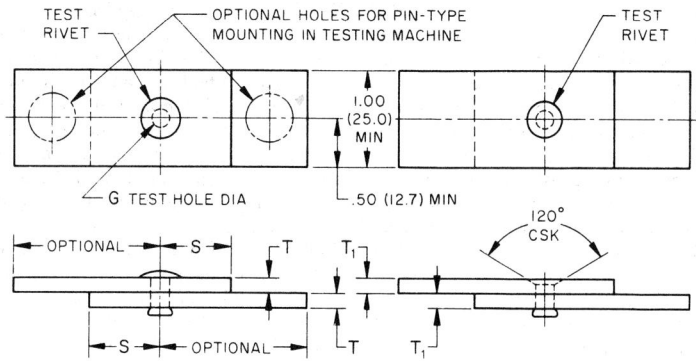

FIG. 1—TEST SPECIMENS FOR SHEAR TESTING BREAK MANDREL BLIND RIVETS

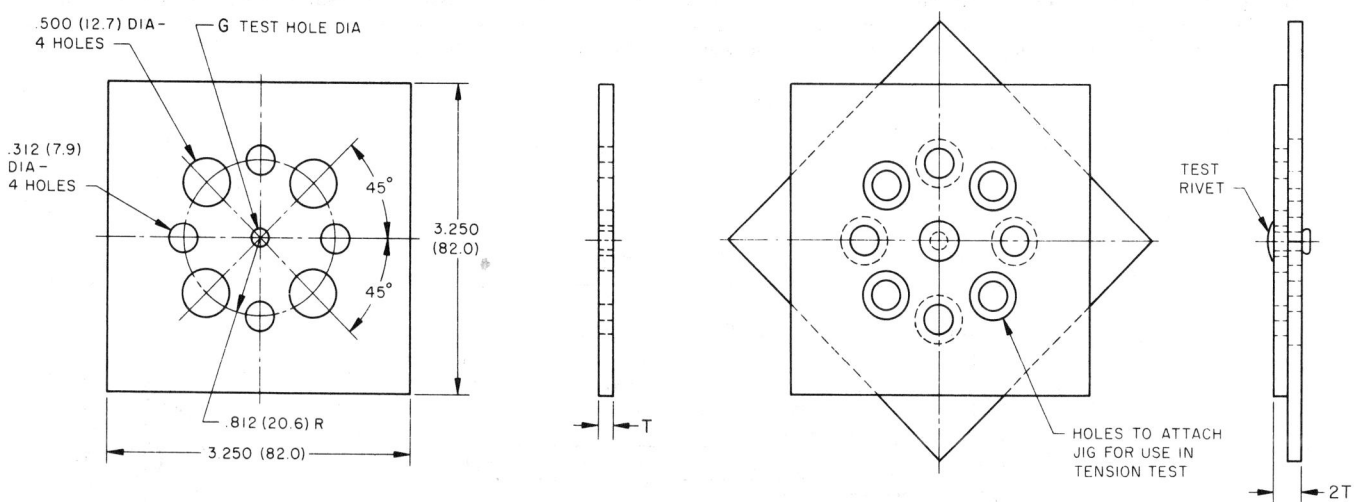

FIG. 2A—DETAIL OF PLATE USED FOR TENSION TESTS

FIG. 2B—ASSEMBLY OF TENSION TEST PLATES BEFORE ATTACHING TO JIG

FIG. 2—TEST FIXTURE FOR TENSILE TESTING BREAK MANDREL BLIND RIVETS

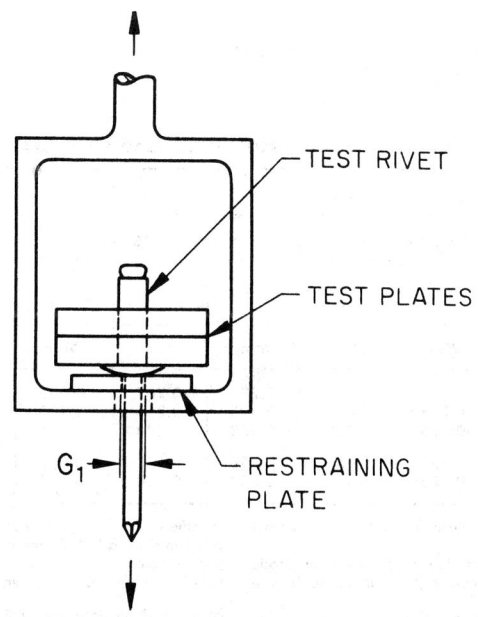

FIG. 3—TEST FIXTURE FOR TESTING MANDREL BREAK LOADS AND BLIND HEAD FORMATION

3.2 Tensile Test—The test shall be comprised of separating two plates of a joint assembled with one blind rivet. The test rivet shall be installed in a test fixture, as depicted in Fig. 2 and Tables 5A and 5B, or another comparable arrangement if an alternate test fixture is used, and the fixture placed between the compression heads of a testing machine. For referee purposes the test fixture shown in Fig. 2 shall be used. Care shall be exercised to locate the fixture at the center of the piston when hydraulic testing machines are used. Load shall be applied to the joint at a speed of testing, as determined with a free running cross head, not less than 0.3 in (7.6 mm) nor greater than 0.5 in (13.0 mm)/min. Loading shall be continued to failure with failure occuring when the rivet body fractures or is pulled through one of the plates. The maximum load in pounds (Newtons) applied to the joint coincident with or prior to rivet failure shall be recorded as the ultimate strength of the rivet.

The test specimen shall be comprised of two plates of equal nominal thickness, aligned and assembled into a joint with the test rivet. The plates shall be of alloy steel, quenched, and tempered to a hardness of Rockwell C46-50.

The test rivet shall be set with a setting tool which is standard for that type of rivet and in accordance with the setting procedures recommended by the rivet manufacturer.

3.3 Mandrel Break Load and Blind Head Formation Test—The test rivet shall be installed in a test plate(s), and the assembly mounted in the fixture of a tensile testing machine. A suggested test fixture is illustrated in Fig. 3. Load shall be applied axially to the mandrel. The load at which it is visually observed that the rivet body end is upset or otherwise deformed to form a head on the blind side, shall be recorded as the blind head formation load. (Note: The blind head formation load is a load applied to the mandrel sufficient to pull the mandrel head into the rivet body and initiate an expansion of the length of rivet body projecting beyond the blind side surface of the joined parts. When the formation of the blind side upset occurs there will normally be a period of tensile machine cross head travel with little or no increase in applied load.) Loading shall be continued until the mandrel breaks, and the maximum load occuring coincident with or prior to failure shall be recorded as the mandrel break load.

The test plate(s) may be of any material capable of supporting the test load without permanent deformation. Thickness of test plate(s) shall be as close as practicable to the maximum of the grip range of the test rivet as specified in Tables 8A and 8B. The hole in the test plate(s) shall conform to the recommended hole size given for the rivet size in Tables 8A and 8B.

The restraining plate shall be alloy steel, quenched, and tempered to a hardness of Rockwell C42-46. The hole in the plate shall conform to G diameter as specified in Tables 5A and 5B.

4. Inspection—Break mandrel blind rivets shall be inspected to determine conformance with dimensional, mechanical, and performance requirements. Inspection procedures shall be as specified by the purchaser on the purchase order or engineering drawings.

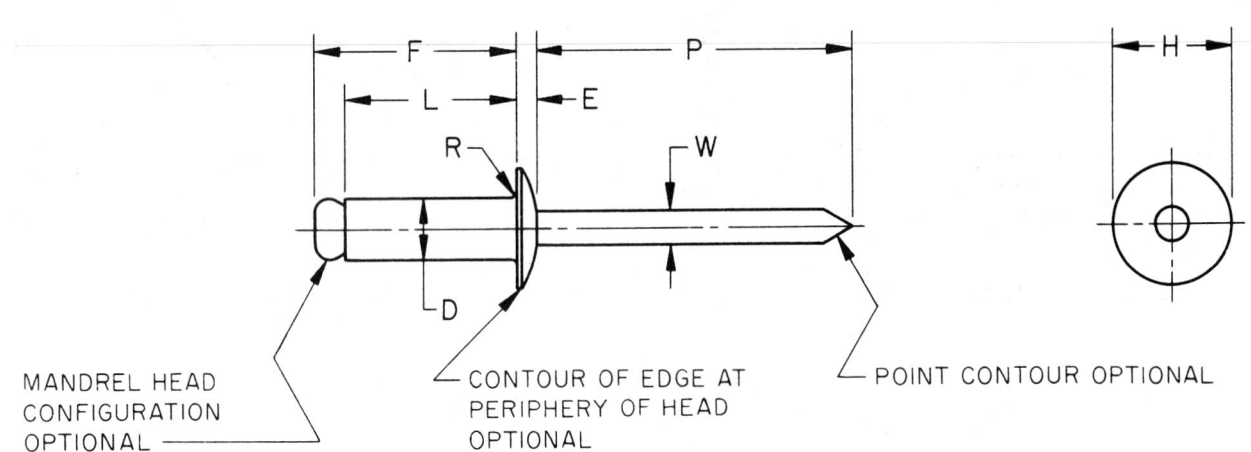

TABLE 6A—DIMENSIONS OF REGULAR AND LARGE PROTRUDING HEAD STYLE BREAK MANDREL BLIND RIVETS, IN

Nominal Rivet Size[a] or Basic Shank Dia		D Rivet Shank Dia		Style 1—Regular Head				Style 2—Large Head				R Fillet Radius[b]	W Mandrel Dia	P Mandrel Protrusion	F Blind Side Protrusion[c]
				H Head Dia		E Head Height		H Head Dia		E Head Height					
		Max	Min	Max	Min	Max		Max	Min	Max		Max	Nom	Min	Max
3/32	0.0938	0.096	0.090	0.198	0.178	0.032		0.293	0.269	0.040		0.015	0.057	1.00	L + 0.100
1/8	0.1250	0.128	0.122	0.262	0.238	0.040		0.390	0.360	0.065		0.020	0.076	1.00	L + 0.120
5/32	0.1562	0.159	0.153	0.328	0.296	0.050		0.488	0.448	0.075		0.020	0.095	1.06	L + 0.140
3/16	0.1875	0.191	0.183	0.394	0.356	0.060		0.650	0.600	0.092		0.025	0.114	1.06	L + 0.160
1/4	0.2500	0.255	0.246	0.525	0.475	0.080		0.780	0.720	0.107		0.030	0.151	1.25	L + 0.180

[a] Where specifying nominal size in decimals, zeros preceding decimal and in fourth decimal place shall be omitted.

[b] The junction of head and shank shall have a fillet with a max radius as shown. For Grade 40, 50, and 51 rivets, the max fillet radius for 3/16 in rivets shall be 0.035 in and for 1/4 in rivets shall be 0.060 in.

[c] When computing the blind side protrusion (F), the max length of rivet (L), as given in Table 8A for the applicable grip shall be used. Minimum blind side clearance may be calculated by subtracting the actual grip (G), (that is, total thickness of the material to be joined), from the specified blind side protrusion (F). (Example: To join two plates, each 0.100 in thick, with a 5/32 in rivet, a 0.425 length rivet would be used. Minimum blind side clearance necessary to permit proper rivet setting would be L + 0.140 in − G, which is 0.425 in + 0.140 in − 0.200 in, and equals 0.365 in.)

For application data see Table 8A.
Additional requirements given in General Specifications shall apply.

TABLE 6B—DIMENSIONS OF REGULAR AND LARGE PROTRUDING HEAD STYLE BREAK MANDREL BLIND RIVETS, MM

Nominal Rivet Size or Basic Shank Dia	D Rivet Shank Dia		Style 1—Regular Head			Style 2—Large Head			R Fillet Radius[a]	W Mandrel Dia	P Mandrel Protrusion	F Blind Side Protrusion[b]
			Head Dia		Head Height	Head Dia		Head Height				
	Max	Min	Max	Min	Max	Max	Min	Max	Max	Nom	Min	Max
2.4	2.44	2.29	5.03	4.52	0.81	7.44	6.83	1.02	0.4	1.45	25.0	L + 2.5
3.2	3.25	3.10	6.65	6.05	1.02	9.91	9.14	1.65	0.5	1.93	25.0	L + 3.0
4.0	4.04	3.89	8.33	7.52	1.27	12.40	11.38	1.90	0.5	2.41	27.0	L + 3.5
4.8	4.85	4.65	10.01	9.04	1.52	16.51	15.24	2.34	0.7	2.90	27.0	L + 4.0
6.3	6.48	6.25	13.33	12.07	2.03	19.81	18.29	2.72	0.8	3.84	31.0	L + 4.5

[a] The junction of head and shank shall have a fillet with a max radius as shown. For Grade 40, 50, and 51 rivets, the max fillet radius for 4.8 mm rivets shall be 0.9 mm and for 6.3 mm rivets shall be 1.5 mm.

[b] When computing the blind side protrusion (F), the max length of rivet (L) as given in Table 8B for the applicable grip shall be used. Minimum blind side clearance may be calculated by subtracting the actual grip (G), (that is, total thickness of the material to be joined), from the specified blind side protrusion (F). (Example: To join two plates, each 2.5 mm thick, with a 4.0 mm rivet, a 10.8 mm length rivet would be used. Minimum blind side clearance necessary to permit proper rivet setting would be L + 3.5 mm − G, which is 10.8 mm + 3.5 mm − 5.0 mm and equals 9.3 mm).

For application data see Table 8B.
Additional requirements given in General Specifications shall apply.

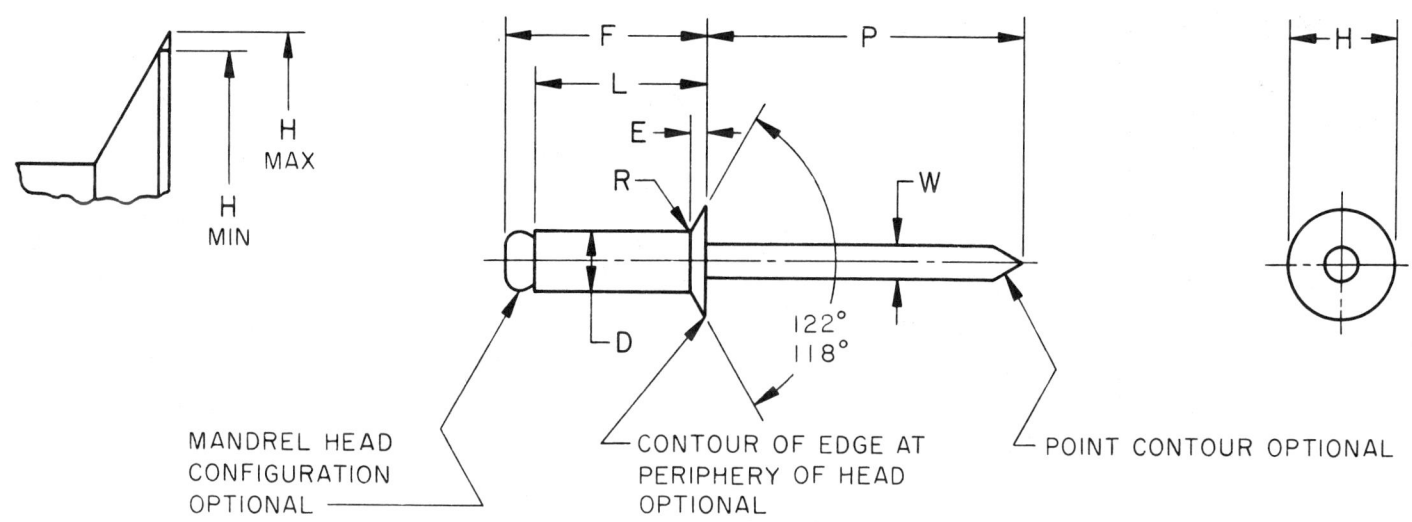

TABLE 7A—DIMENSIONS OF 120 DEG COUNTERSUNK FLUSH HEAD STYLE BREAK MANDREL BLIND RIVETS, IN

Nominal Rivet Size[a] or Basic Shank Dia		D Rivet Shank Dia		H Head Dia[b]		E Head Height[c]	R Fillet Radius	W Mandrel Dia	P Mandrel Protrusion	F Blind Side Protrusion[d]
		Max	Min	Max	Min	Ref	Max	Nom	Min	Max
3/32	0.0938	0.096	0.090	0.187	0.161	0.027	0.020	0.057	1.00	L + 0.100
1/8	0.1250	0.128	0.122	0.233	0.207	0.031	0.025	0.076	1.00	L + 0.120
5/32	0.1562	0.159	0.153	0.294	0.268	0.040	0.030	0.095	1.06	L + 0.140
3/16	0.1875	0.191	0.183	0.361	0.335	0.050	0.035	0.114	1.06	L + 0.160

[a] Where specifying nominal size in decimals, zeros preceding decimal and in fourth decimal place shall be omitted.

[b] Max head diameter is calculated on nominal rivet diameter and nominal head angle extended to sharp corner. Min head diameter is absolute.

[c] Head height is given for reference purposes only. Variations in this dimension are controlled by the diameters (H) and (D) and the included angle of the head.

[d] When computing the blind side protrusion (F), the max length of rivet (L), as given in Table 8A for the applicable grip shall be used. Minimum blind side clearance may be calculated by subtracting the actual grip (G), (such as, total thickness of the material to be joined), from the specified blind side protrusion (F). (Example: To join two plates, each 0.187 in thick, with a 3/16 in rivet, a 0.575 length rivet would be used. Minimum blind side clearance necessary to permit proper rivet setting would be L + 0.160 in − G, which is 0.575 in + 0.160 in − 0.374 in which equals 0.361 in).

For application data see Table 8A.
Additional requirements given in General Specifications shall apply.

TABLE 7B—DIMENSIONS OF 120 DEG COUNTERSUNK FLUSH HEAD STYLE BREAK MANDREL BLIND RIVETS, MM

Nominal Rivet Size or Basic Shank Dia	D Rivet Shank Dia		H Head Dia[a]		E Head Height[b]	R Fillet Radius	W Mandrel Dia	P Mandrel Protrusion	F Blind Side Protrusion[c]
	Max	Min	Max	Min	Ref	Max	Nom	Min	Max
2.4	2.44	2.29	4.75	4.09	0.69	0.5	1.45	25.0	L + 2.5
3.2	3.25	3.10	5.92	5.26	0.79	0.7	1.93	25.0	L + 3.0
4.0	4.04	3.89	7.47	6.81	1.02	0.8	2.41	27.0	L + 3.5
4.8	4.85	4.65	9.17	8.51	1.27	0.9	2.90	27.0	L + 4.0

[a] Max head diameter is calculated on nominal rivet diameter and nominal head angle extended to sharp corner. Min head diameter is absolute.
[b] Head height is given for reference purposes only. Variations in this dimension are controlled by the diameters (H) and (D) and the included angle of the head.
[c] When computing the blind side protrusion (F), the max length of rivet (L), as given in Table 8B for the applicable grip shall be used. Minimum blind side clearance may be calculated by subtracting the actual grip (G), (that is, total thickness of the material to be joined), from the specified blind side protrusion (F). (Example: To join two plates, each 4.7 mm thick, with a 4.8 mm rivet, a 14.6 mm length rivet would be used. Minimum blind side clearance necessary to permit proper rivet setting would be L + 4.0 mm − G, which is 14.6 mm + 4.0 mm − 9.4 mm which equals 9.2 mm.)

For application data see Table 8B.
Additional requirements given in General Specifications shall apply.

TABLE 8A—APPLICATION DATA FOR PROTRUDING HEAD AND FLUSH HEAD STYLE BREAK MANDREL BLIND RIVETS, IN

Nominal Rivet Size or Basic Shank Dia		Recommended Hole Size			Grip Range		L Rivet Length[b]
		Drill Size[a]	Hole Dia		For Protruding Style Heads	For Flush Style Heads	Max
			Max	Min			
3/32	0.0938	No. 41	0.100	0.097	0.020–0.125	0.079–0.125	0.250
					0.126–0.250	0.126–0.250	0.375
					0.251–0.375	—	0.500
1/8	0.1250	No. 30	0.133	0.129	0.020–0.062	—	0.212
					0.063–0.125	0.092–0.125	0.275
					0.126–0.187	0.126–0.187	0.337
					0.188–0.250	0.188–0.250	0.400
					0.251–0.312	0.251–0.312	0.462
					0.313–0.375	0.313–0.375	0.525
					0.376–0.500	0.376–0.500	0.650
					0.501–0.625	—	0.775
5/32	0.1562	No. 20	0.164	0.160	0.020–0.125	—	0.300
					0.126–0.187	0.120–0.187	0.362
					0.188–0.250	0.188–0.250	0.425
					0.251–0.375	0.251–0.375	0.550
					0.376–0.500	0.376–0.500	0.675
					0.501–0.625	—	0.800
3/16	0.1875	No. 11	0.196	0.192	0.020–0.125	—	0.325
					0.126–0.187	0.151–0.187	0.387
					0.188–0.250	0.188–0.250	0.450
					0.251–0.375	0.251–0.375	0.575
					0.376–0.500	0.376–0.500	0.700
					0.501–0.625	0.501–0.625	0.825
					0.626–0.750	—	0.950
					0.751–0.875	—	1.075
					0.876–1.000	—	1.200
					1.001–1.125	—	1.325
1/4	0.2500	F	0.261	0.257	0.020–0.125	—	0.375
					0.126–0.250	—	0.500
					0.251–0.375	—	0.625
					0.376–0.500	—	0.750
					0.501–0.625	—	0.875
					0.626–0.750	—	1.000
					0.751–0.875	—	1.125
					0.876–1.000	—	1.250
					1.001–1.125	—	1.375
					1.126–1.250	—	1.500

[a] Recommended drill sizes are those which normally produce holes within the specified hole size limits.
[b] Where blind side clearances permit and it is economically feasible, rivets of the next longer length than those recommended for a given grip may be substituted to limit the number of different inventory items.

TABLE 8B—APPLICATION DATA FOR PROTRUDING HEAD AND FLUSH HEAD STYLE BREAK MANDREL BLIND RIVETS, MM

Nominal Rivet Size or Basic Shank Dia	Recommended Hole Size			Grip Range		L Rivet Length[b]
	Drill Size[a]	Hole Dia		For Protruding Style Heads	For Flush Style Heads	Max
		Max	Min			
2.4	2.5	2.54	2.46	0.5–3.2	2.0–3.2	6.4
				Over 3.2–6.4	Over 3.2–6.4	9.5
				Over 6.4–9.5	—	12.7
3.2	3.3	3.38	3.28	0.5–1.6	—	5.4
				Over 1.6–3.2	2.3–3.2	7.0
				Over 3.2–4.8	Over 3.2–4.8	8.6
				Over 4.8–6.4	Over 4.8–6.4	10.2
				Over 6.4–7.9	Over 6.4–7.9	11.7
				Over 7.9–9.5	Over 7.9–9.5	13.4
				Over 9.5–12.7	Over 9.5–12.7	16.5
				Over 12.7–15.9	—	19.7
4.0	4.1	4.16	4.06	0.5–3.2	—	7.6
				Over 3.2–4.8	3.0–4.8	9.2
				Over 4.8–6.4	Over 4.8–6.4	10.8
				Over 6.4–9.5	Over 6.4–9.5	14.0
				Over 9.5–12.7	Over 9.5–12.7	17.2
				Over 12.7–15.9	—	20.3
4.8	4.9	4.98	4.88	0.5–3.2	—	8.3
				Over 3.2–4.8	3.8–4.8	9.8
				Over 4.8–6.4	Over 4.8–6.4	11.5
				Over 6.4–9.5	Over 6.4–9.5	14.6
				Over 9.5–12.7	Over 9.5–12.7	17.8
				Over 12.7–15.9	Over 12.7–15.9	21.0
				Over 15.9–19.1	—	24.2
				Over 19.1–22.2	—	27.3
				Over 22.2–25.4	—	30.5
				Over 25.4–28.6	—	33.7
6.3	6.5	6.63	6.53	0.5–3.2	—	9.5
				Over 3.2–6.4	—	12.7
				Over 6.4–9.5	—	15.9
				Over 9.5–12.7	—	19.1
				Over 12.7–15.9	—	22.2
				Over 15.9–19.1	—	25.4
				Over 19.1–22.2	—	28.6
				Over 22.2–25.4	—	31.8
				Over 25.4–28.6	—	34.9
				Over 28.6–31.8	—	38.1

[a] Recommended drill sizes are those which normally produce holes within the specified hole size limits.
[b] Where blind side clearances permit and it is economically feasible, rivets of the next longer length than those recommended for a given grip may be substituted to limit the number of different inventory items.

15.101

BOLT AND CAPSCREW SIZES FOR USE IN CONSTRUCTION AND INDUSTRIAL MACHINERY—SAE J370

SAE Recommended Practice

Report of Fasteners Committee and Construction and Industrial Machinery Committee approved November 1968.

Scope—This SAE Recommended Practice covers the use of preferred bolt and capscrew sizes, wherever applicable, on construction and industrial machinery to ease serviceability.

Purpose—Designers, in their efforts to determine the proper fastener for heavy equipment, sometimes overlook the practical aspects of serviceability. When a particular component requires frequent service, it is much faster and easier for the serviceman to work with common size, easily distinguished bolts and capscrews which require a minimum of tools.

Below are listed eight diameters of recommended bolts and capscrews, some of which can be found in any hardware store, require only standard tools, are easy to distinguish, and at the same time cover proof loads from 2300 lb on up at convenient intervals.

While some applications may require other sizes and/or fine threaded fasteners for greater clamping forces and/or area restrictions, the convenience of UNC threads in the sizes indicated is preferred wherever possible.

Threads—Unified Coarse Thread Series (UNC).

Diameters—$\frac{3}{8}$, $\frac{1}{2}$, $\frac{5}{8}$, $\frac{7}{8}$, 1, $1\frac{1}{8}$, $1\frac{1}{4}$.

Diameters above $1\frac{1}{4}$ in. to increase in $\frac{1}{4}$ in. increments.

Diameter deviations below $\frac{1}{2}$ in. decrease in $\frac{1}{16}$ in. increments.

Reference

Bolts—SAE J102, J103, J104, J105 (formerly SAE J477)

Bolts—SAE J871

Capscrews—SAE J478

Fasteners—SAE J429

HOSE CLAMPS—SAE J536b

SAE Standard

Report of Parts and Fittings Committee approved May 1958, revised May 1959, and last revised by Fasteners Committee March 1971.

GENERAL SPECIFICATIONS

General—Types A, B, C, and D clamps shall be supplied in open position with the nut retained on the screw by staking or other means agreeable to user. Where so specified by purchaser, Types B and D clamps shall have provision to retain nut in base leg when axial pressure is applied to screw. Types A, B, C, D, and F clamps shall open over or close tight upon round mandrels of the sizes indicated in the respective open and closed diameter columns. All clamps shall be free from burrs, seams, laps, loose scale, or any other defects that may affect their serviceability.

Material

Types A, B, C, and D—SAE 1010 steel or equivalent. Band on Type C clamps in size numbers 13 through 21, 22S, 23, 24S, 25 and 26S is corrosion resistant steel.

Type E—SAE 1065-1080 heat treated to a minimum Rockwell hardness of 53 C to meet the performance and ductility requirements specified hereinafter.

Type F—Bands—corrosion resistant steel, standard with manufacturer. Saddle housings—corrosion resistant steel or carbon steel, standard with manufacturer as required by user. Screws—corrosion resistant steel or low carbon steel, standard with manufacturer as required by user.

Finish—Carbon steel components of clamps are normally supplied with rust proof finish as specified by purchaser. It is recommended that a reasonable latitude be allowed in the inspection of finish on parts fabricated from precoated steel and the over-lapping areas on clamps treated after assembly of component parts.

Threads

Types A, B, C, and D—Unified Standard Class 2A external and Class 2B internal threads shall apply.

Type F—Modified buttress thread standard with manufacturer.

Screws—Shall conform to the section on Machine Screws in SAE J478, except for special head and point details specified herein or unspecified detail specifically left to manufacturers option.

Type A—10-24 hexagon washer head slotted or round head slotted machine screws, unless otherwise specified.

Type B—10-24 hexagon washer head slotted, 10-24 fillister head slotted, 10-24 fillister washer head slotted, 12-24 fillister head slotted or 12-24 round head cross recess screws, as specified.

Type C—6-32 Hexagon head slotted machine screws; and 10-24 or 12-24 upset hexagon head slotted machine screws; or 10-24 or 12-24 fillister head slotted machine screws; with flat or pilot point; as specified.

Type D—10-24 hexagon washer head slotted or fillister washer head slotted machine screws.

Type F—Conform to Styles 1, 2, 3, or 4 (Table 6) shown with unspecified details standard with manufacturer.

Nuts shall be square or rectangular as indicated, of a size to suit clamp design, except Type A which may have a flat trunnion nut standard with manufacturer.

Identification—Type C clamps shall be marked with SAE size number or fractional equivalent thereof. Type E clamps shall be identified by SAE size number die stamped on top of the ring or on one tang. Also, Type E clamps shall be color coded as noted in Table 5. Type F clamps shall be identified by size number stamped on the band. At manufacturers option, manufacturers' marks may appear adjacent to size identifications.

Type B Clamps—Normally manufactured with one slot in sizes up to and including No. 40; two slots in sizes No. 42 through No. 96 and three slots in sizes No. 100 and larger. Widths of slots and tongues shall not be greater than 40% of band width and not less than 30% of band width. Slots shall be centered in the band width.

Type E Clamps—**Performance and Acceptance**—Acceptability of clamps will be determined by the following tests and inspections:

EXPANSION AND PERMANENT SET—Expansion and permanent set of clamps shall be inspected by subjecting the clamps to the following tests and inspections in sequence:

1. Expand clamp to fit diameter A of gage. When so expanded, the span across the outside edges of the tips of the tangs measured in a plane perpendicular to the axis of the gage shall be between the tabulated limits for dimension X. NOTE: Care should be taken to avoid overexpansion during this operation.

2. Clamp shall be fitted respectively to gage diameters B and C. When clamps are so fitted, a wire of Z diameter shall not pass between the gage and the clamp when inserted in a direction parallel to the axis of the gage.

BRITTLENESS—Clamps subjected to corrosion preventative treatment which might produce hydrogen embrittlement shall be baked or otherwise treated to obviate such embrittlement and shall be capable of being expanded on a nominal diameter plug for a continuous 24 hr period without signs of breaking or cracking.

DUCTILITY—Ductility of clamps shall be inspected by subjecting the clamps to the following tests:

1. The clamp shall be gripped in a vise in a manner such that the gripping edge of the vise will coincide with the clamp axis which bisects the angle between tangs as illustrated in Fig. 1. Clamp shall be expanded by moving the free tang as shown in Fig. 1 to a point where the free tang will pass around the stationary tang. There shall be no evidence of fracture during or after this test.

2. When clamp is expanded by movement of the free tang beyond the stationary tang to where the clamp fractures, the structure at the point of fracture shall show a fine grain and the clamp up to the instant of fracture shall deliver a tough springy reaction.

ASSEMBLY TOOLS—Settings of assembly tools should be governed by tabulated dimension W (max) at maximum opening of tool and dimension X (min) at minimum closure of tool.

Type F Clamps

Band Ductility—When subjected to 180 deg bend around a 0.188 in. diameter mandrel and then restraightened, the band shall at no time show breaking, cracks, or other indications of failure.

Durability—Screw threads and slots in the band shall show no evidence of deformation or excessive wear when clamps are tightened once on a steel mandrel to an applied screw torque of 50 lb-in. for carbon steel screws and 60 lb-in. for stainless steel screws.

TYPE A HOSE CLAMPS

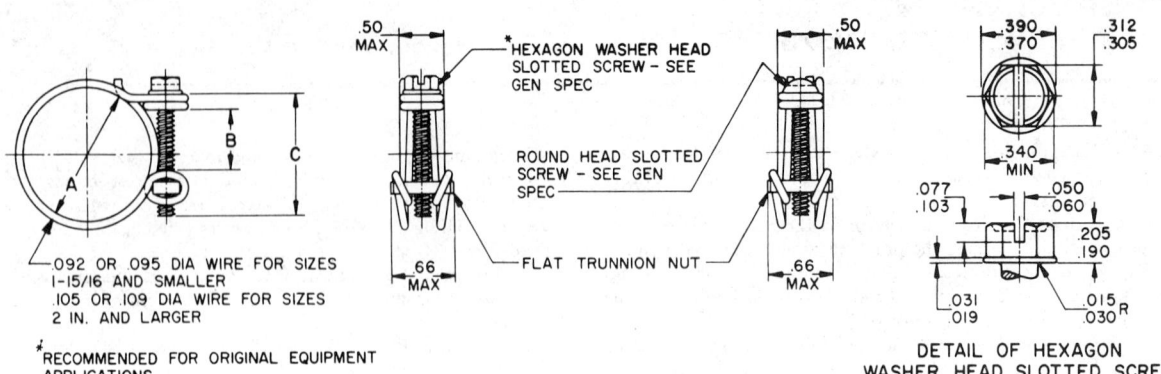

TABLE 1—DIMENSIONS OF TYPE A HOSE CLAMPS

SAE Size No.	A Dia Nom	A Dia Open	A Dia Closed	B[a] Gap	C Screw Length Min	SAE Size No.	A Dia Nom	A Dia Open	A Dia Closed	B[a] Gap	C Screw Length Min	SAE Size No.	A Dia Nom	A Dia Open	A Dia Closed	B[a] Gap	C Screw Length Min	SAE Size No.	A Dia Nom	A Dia Open	A Dia Closed	B[a] Gap	C Screw Length Min
16	0.44	0.50	0.44	0.31	0.88	54	1.62	1.69	1.50	0.75	1.25	90	2.75	2.81	2.59	0.94	1.50	126	3.88	3.94	3.69	1.12	1.75
18	0.50	0.56	0.48	0.38	0.88	56	1.69	1.75	1.56	0.81	1.38	92	2.81	2.88	2.66	0.94	1.50	128	3.94	4.00	3.75	1.12	1.75
20	0.56	0.62	0.55	0.38	0.88	58	1.75	1.81	1.62	0.81	1.38	94	2.88	2.94	2.72	0.94	1.50	130	4.00	4.06	3.81	1.12	1.75
22	0.62	0.69	0.58	0.50	0.88	60	1.81	1.88	1.69	0.81	1.38	96	2.94	3.00	2.78	0.94	1.50	132	4.06	4.12	3.88	1.12	1.75
24	0.69	0.75	0.64	0.50	1.12	62	1.88	1.94	1.75	0.81	1.38	98	3.00	3.06	2.84	0.94	1.50	134	4.12	4.19	3.94	1.12	1.75
26	0.75	0.81	0.69	0.50	1.12	64	1.94	2.00	1.81	0.81	1.38	100	3.06	3.12	2.91	0.94	1.50	136	4.19	4.25	4.00	1.12	1.75
28	0.81	0.88	0.75	0.50	1.12	66	2.00	2.06	1.88	0.81	1.38	102	3.12	3.19	2.97	0.94	1.50	138	4.25	4.31	4.06	1.12	1.75
30	0.88	0.94	0.81	0.50	1.12	68	2.06	2.12	1.94	0.81	1.38	104	3.19	3.25	3.03	0.94	1.50	140	4.31	4.38	4.12	1.12	1.75
32	0.94	1.00	0.88	0.56	1.12	70	2.12	2.19	2.00	0.81	1.50	106	3.25	3.31	3.09	0.94	1.50	142	4.38	4.44	4.19	1.12	1.75
34	1.00	1.06	0.94	0.56	1.12	72	2.19	2.25	2.06	0.81	1.50	108	3.31	3.38	3.16	0.94	1.50	144	4.44	4.50	4.25	1.12	1.75
36	1.06	1.12	0.95	0.56	1.12	74	2.25	2.31	2.12	0.81	1.50	110	3.38	3.44	3.22	0.94	1.50	146	4.50	4.56	4.31	1.12	1.75
38	1.12	1.19	1.06	0.56	1.12	76	2.31	2.38	2.19	0.81	1.50	112	3.44	3.50	3.25	1.12	1.75	148	4.56	4.62	4.38	1.12	1.75
40	1.19	1.25	1.09	0.62	1.12	78	2.38	2.44	2.25	0.88	1.50	114	3.50	3.56	3.31	1.12	1.75	150	4.62	4.69	4.44	1.12	1.75
42	1.25	1.31	1.16	0.62	1.25	80	2.44	2.50	2.31	0.88	1.50	116	3.56	3.62	3.38	1.12	1.75	152	4.69	4.75	4.50	1.12	1.75
44	1.31	1.38	1.19	0.62	1.25	82	2.50	2.56	2.34	0.94	1.50	118	3.62	3.69	3.44	1.12	1.75	154	4.75	4.81	4.56	1.12	1.75
46	1.38	1.44	1.25	0.75	1.25	84	2.56	2.62	2.41	0.94	1.50	120	3.69	3.75	3.50	1.12	1.75	156	4.81	4.88	4.62	1.12	1.75
48	1.44	1.50	1.31	0.75	1.25	86	2.62	2.69	2.47	0.94	1.50	122	3.75	3.81	3.56	1.12	1.75	158	4.88	4.94	4.69	1.12	1.75
50	1.50	1.56	1.38	0.75	1.25	88	2.69	2.75	2.53	0.94	1.50	124	3.81	3.88	3.62	1.12	1.75	160	4.94	5.00	4.75	1.12	1.75
52	1.56	1.62	1.44	0.75	1.25																		

[a] Reference dimension. When gap is at value tabulated, clamp diameter shall approximate the open diameter.

TYPE B HOSE CLAMPS

TABLE 2—DIMENSIONS OF TYPE B HOSE CLAMPS

SAE Size No.	A Dia Nom	A Dia Open	A Dia Closed	B[a] Gap	C Band Width ±0.01	D Screw Length Min	SAE Size No.	A Dia Nom	A Dia Open	A Dia Closed	B[a] Gap	C Band Width ±0.01	D Screw Length Min
18	0.50	0.58	0.44	0.38	0.50[b]	1.00	58	1.75	1.83	1.64	0.50	0.62[c]	1.12
20	0.56	0.64	0.48	0.38	0.50[b]	1.00	60	1.81	1.89	1.70	0.50	0.62[c]	1.12
22	0.62	0.70	0.55	0.38	0.50[b]	1.00	62	1.88	1.95	1.77	0.50	0.62[c]	1.12
24	0.69	0.77	0.61	0.38	0.50[b]	1.00	64	1.94	2.02	1.83	0.50	0.62[c]	1.12
26	0.75	0.83	0.67	0.38	0.50[b]	1.00	67	2.03	2.11	1.92	0.50	0.62[c]	1.12
28	0.81	0.89	0.73	0.38	0.50[b]	1.00							
30	0.88	0.95	0.80	0.38	0.50[b]	1.00	70	2.12	2.20	2.02	0.50	0.62[c]	1.12
32	0.94	1.02	0.86	0.38	0.50[b]	1.00	72	2.19	2.27	2.08	0.50	0.62[c]	1.12
35	1.03	1.11	0.95	0.38	0.50[b]	1.00	75	2.28	2.36	2.17	0.50	0.62[c]	1.12
36	1.06	1.14	0.98	0.38	0.50[b]	1.00	79	2.38	2.48	2.27	0.50	0.62[c]	1.25
38	1.12	1.20	1.02	0.38	0.50[b]	1.12	83	2.50	2.61	2.39	0.50	0.62[c]	1.25
40	1.19	1.27	1.08	0.50	0.50[b]	1.12	88	2.62	2.75	2.52	0.50	0.62[c]	1.25
42	1.25	1.33	1.14	0.50	0.62[c]	1.12	92	2.75	2.88	2.64	0.50	0.62[c]	1.25
44	1.31	1.39	1.20	0.50	0.62[c]	1.12	96	2.88	3.00	2.77	0.50	0.62[c]	1.25
46	1.38	1.45	1.27	0.50	0.62[c]	1.12	100	3.00	3.12	2.89	0.50	0.62	1.25
48	1.44	1.52	1.33	0.50	0.62[c]	1.12	104	3.12	3.25	3.02	0.50	0.62	1.25
50	1.50	1.58	1.39	0.50	0.62[c]	1.12	108	3.25	3.38	3.14	0.50	0.62	1.25
52	1.56	1.64	1.45	0.50	0.62[c]	1.12	112	3.38	3.50	3.27	0.50	0.62	1.25
54	1.62	1.70	1.52	0.50	0.62[c]	1.12	122	3.56	3.81	3.42	0.62	0.75	1.38
56	1.69	1.77	1.58	0.50	0.62[c]	1.12							

[a] Reference dimension. When gap is at value tabulated, clamp diameter shall approximate the nominal diameter.
[b] 0.62 in. width optional with user.
[c] 0.50 in. width optional with user.

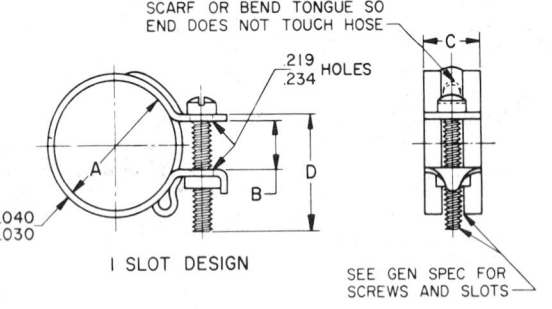

1 SLOT DESIGN

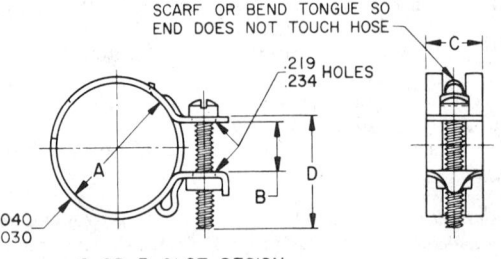

2 OR 3 SLOT DESIGN

TYPE C HOSE CLAMPS

TABLE 3—DIMENSIONS OF TYPE C HOSE CLAMPS

SAE Size No.[a]	A Diameter Open	A Diameter Closed	B Bridge Stock Thickness ±0.002	C[b] Bridge Width Max	D Band Width ±0.010	E[c] Band Thickness ±0.001	F Screw Size and Length	G[b] Height Over Screw Max
13	0.40	0.34	0.035	0.41	0.281	0.010	6-32 x 0.50	0.64
14	0.43	0.37	0.035	0.41	0.281	0.010	6-32 x 0.50	0.64
15	0.46	0.40	0.035	0.41	0.281	0.010	6-32 x 0.50	0.64
16	0.50	0.37	0.035	0.41	0.281	0.010	6-32 x 0.50	0.64
17	0.53	0.40	0.035	0.41	0.281	0.010	6-32 x 0.50	0.64
18	0.56	0.43	0.035	0.41	0.281	0.010	6-32 x 0.50	0.64
19	0.59	0.46	0.035	0.41	0.281	0.010	6-32 x 0.50	0.64
20	0.62	0.50	0.035	0.41	0.281	0.010	6-32 x 0.50	0.64
21	0.65	0.53	0.035	0.41	0.281	0.010	6-32 x 0.50	0.64
22	0.69	0.38	0.050	0.64	0.438	0.017	10-24 x 0.88	1.13
22N	0.69	0.56	0.035	0.41	0.281	0.010	6-32 x 0.50	0.64
23	0.71	0.59	0.035	0.41	0.281	0.010	6-32 x 0.50	0.64
24	0.75	0.44	0.050	0.64	0.438	0.017	10-24 x 0.88	1.13
24N	0.75	0.62	0.035	0.41	0.281	0.010	6-32 x 0.50	0.64
25	0.78	0.66	0.035	0.41	0.281	0.010	6-32 x 0.50	0.64
26	0.81	0.50	0.050	0.64	0.438	0.017	10-24 x 0.88	1.13
26N	0.81	0.69	0.035	0.41	0.281	0.010	6-32 x 0.50	0.64
28	0.88	0.56	0.050	0.64	0.438	0.017	10-24 x 0.88	1.13
30	0.94	0.62	0.050	0.72	0.505	0.017	12-24 x 0.88	1.13
30N	0.94	0.62	0.050	0.64	0.438	0.017	10-24 x 0.88	1.13
32	1.00	0.69	0.050	0.72	0.505	0.017	12-24 x 0.88	1.13
32N	1.00	0.69	0.050	0.64	0.438	0.017	10-24 x 0.88	1.13
34	1.06	0.75	0.050	0.72	0.505	0.020	12-24 x 0.88	1.13
34N	1.06	0.75	0.050	0.64	0.438	0.017	10-24 x 0.88	1.13
36	1.12	0.81	0.050	0.72	0.505	0.020	12-24 x 0.88	1.13
36N	1.12	0.81	0.050	0.64	0.438	0.017	10-24 x 0.88	1.13
38	1.19	0.88	0.050	0.72	0.505	0.020	12-24 x 0.88	1.13
38N	1.19	0.88	0.050	0.64	0.438	0.017	10-24 x 0.88	1.13
40	1.25	0.94	0.062	0.72	0.505	0.020	12-24 x 0.88	1.13
40N	1.25	0.94	0.050	0.64	0.438	0.017	10-24 x 0.88	1.13
42	1.31	1.00	0.062	0.72	0.505	0.020	12-24 x 0.88	1.13
42N	1.31	1.00	0.050	0.64	0.438	0.017	10-24 x 0.88	1.13
44	1.38	1.06	0.062	0.72	0.505	0.020	12-24 x 0.88	1.13
44N	1.38	1.06	0.050	0.64	0.438	0.017	10-24 x 0.88	1.13
46	1.44	1.12	0.062	0.72	0.505	0.020	12-24 x 0.88	1.13
46N	1.44	1.12	0.050	0.64	0.438	0.017	10-24 x 0.88	1.13
48	1.50	1.19	0.062	0.72	0.505	0.020	12-24 x 0.88	1.13
48N	1.50	1.19	0.050	0.64	0.438	0.017	10-24 x 0.88	1.13
50	1.56	1.25	0.062	0.72	0.505	0.020	12-24 x 0.88	1.13
52	1.62	1.31	0.062	0.72	0.505	0.020	12-24 x 0.88	1.13
54	1.69	1.38	0.062	0.72	0.505	0.020	12-24 x 0.88	1.13
56	1.75	1.44	0.062	0.72	0.505	0.020	12-24 x 0.88	1.13
58	1.81	1.50	0.062	0.72	0.505	0.020	12-24 x 0.88	1.13
60	1.88	1.56	0.062	0.72	0.505	0.020	12-24 x 0.88	1.13
62	1.94	1.62	0.062	0.72	0.505	0.020	12-24 x 0.88	1.13
64	2.00	1.69	0.062	0.72	0.505	0.020	12-24 x 0.88	1.13
66	2.06	1.69	0.062	0.72	0.505	0.020	12-24 x 1.00	1.25
68	2.12	1.75	0.062	0.72	0.505	0.020	12-24 x 1.00	1.25
70	2.19	1.81	0.062	0.72	0.505	0.020	12-24 x 1.00	1.25
72	2.25	1.88	0.062	0.72	0.505	0.020	12-24 x 1.00	1.25
74	2.31	1.94	0.062	0.72	0.505	0.020	12-24 x 1.00	1.25
76	2.38	2.00	0.062	0.72	0.505	0.020	12-24 x 1.00	1.25
78	2.44	2.06	0.062	0.72	0.505	0.020	12-24 x 1.00	1.25
80	2.50	2.12	0.062	0.72	0.505	0.020	12-24 x 1.00	1.25
82	2.56	2.19	0.062	0.72	0.505	0.020	12-24 x 1.00	1.25
84	2.62	2.25	0.062	0.72	0.505	0.020	12-24 x 1.00	1.25
86	2.69	2.31	0.062	0.72	0.505	0.020	12-24 x 1.00	1.25
88	2.75	2.38	0.062	0.72	0.505	0.020	12-24 x 1.00	1.25
90	2.81	2.44	0.062	0.72	0.505	0.020	12-24 x 1.00	1.25
92	2.88	2.50	0.062	0.72	0.505	0.020	12-24 x 1.00	1.25
94	2.94	2.56	0.062	0.72	0.505	0.020	12-24 x 1.00	1.25
96	3.00	2.62	0.062	0.72	0.505	0.020	12-24 x 1.00	1.25
100	3.12	2.75	0.062	0.72	0.505	0.020	12-24 x 1.00	1.25
104	3.25	2.88	0.062	0.72	0.505	0.020	12-24 x 1.00	1.25
110	3.44	3.06	0.062	0.72	0.505	0.020	12-24 x 1.00	1.25
114	3.56	3.19	0.062	0.72	0.505	0.020	12-24 x 1.00	1.25
118	3.69	3.31	0.062	0.72	0.505	0.020	12-24 x 1.00	1.25
138	4.31	3.94	0.062	0.72	0.505	0.020	12-24 x 1.00	1.25

[a] The N suffix applied to SAE size numbers designates the smaller series clamp design where sizes overlap in two clamp designs.
[b] Reference dimension for clearance purposes only.
[c] For size numbers 30–138, clamps having 0.020 tabulated band thickness are also available with 0.018–0.016 and 0.027–0.025 band thickness where so specified by user.

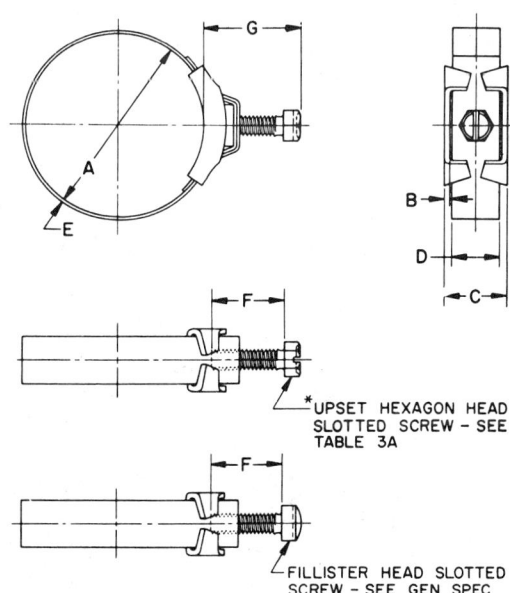

SIZES 13 THRU 21, 22S, 23, 24S, 25 AND 26S

SIZES 22, 24, 26 AND 28 THRU 138

STYLE 1 — FLAT POINT SCREW

STYLE 2 — PILOT POINT SCREW

BRIDGE DETAILS

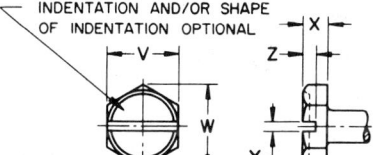

*RECOMMENDED FOR ORIGINAL EQUIPMENT APPLICATION

TABLE 3A—DIMENSIONS OF HEXAGON SCREW HEADS

Screw Size	V Across Flats Max	V Across Flats Min	W Across Corners Min	X Head Height Max	X Head Height Min	Y Slot Width Max	Y Slot Width Min	Z Slot Depth Max	Z Slot Depth Min
6	0.250	0.244	0.272	0.080	0.067	0.048	0.039	0.046	0.033
10	0.375	0.367	0.409	0.145	0.120	0.060	0.050	0.072	0.057
12	0.375	0.367	0.409	0.155	0.139	0.067	0.056	0.077	0.093

Torque required to draw band through bridge on free clamp shall not exceed 4 in.-lb for sizes having 6-32 screws, 8 in.-lb for sizes having 10-24 screws, and 10 in.-lb for sizes having 12-24 screws.

It is recommended that Type C Clamps not be tightened beyond maximum torques of 9 in.-lb for sizes having 6-32 screws, 22 in.-lb for sizes having 10-24 screws, and 30 in.-lb for sizes having 12-24 screws.

TYPE D HOSE CLAMPS

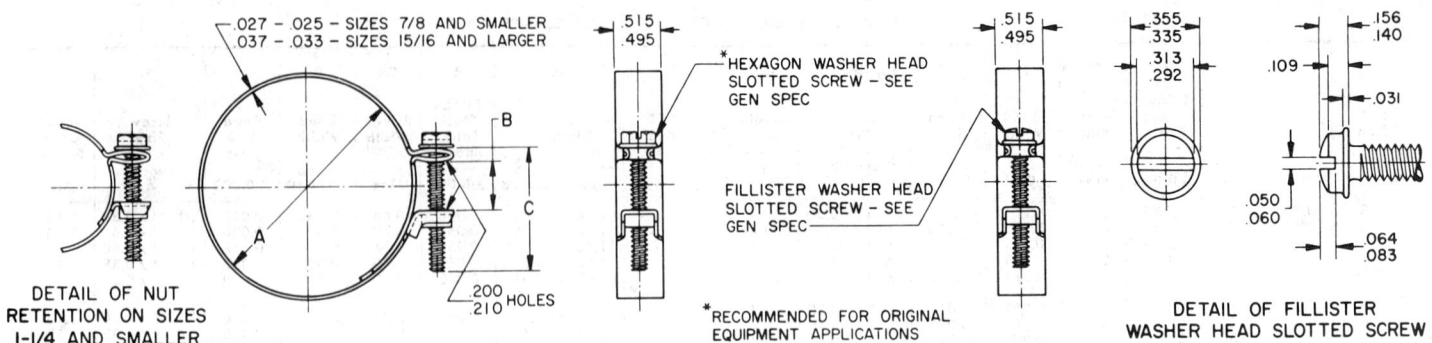

TABLE 4—DIMENSIONS OF TYPE D HOSE CLAMPS

SAE Size No.	A, Dia			B[a], Gap	C, Screw Length	SAE Size No.	A, Dia			B[a], Gap	C, Screw Length	SAE Size No.	A, Dia			B[a], Gap	C, Screw Length
	Nom	Open	Closed		Min		Nom	Open	Closed		Min		Nom	Open	Closed		Min
23	0.62	0.72	0.53	0.38	1.12	83	2.50	2.59	2.34	0.62	1.38	143	4.38	4.47	4.16	0.75	1.50
25	0.69	0.78	0.59	0.38	1.12	85	2.56	2.66	2.41	0.62	1.38	145	4.44	4.53	4.22	0.75	1.50
27	0.75	0.84	0.66	0.38	1.12	87	2.62	2.72	2.47	0.62	1.38	147	4.50	4.59	4.28	0.75	1.50
29	0.81	0.91	0.72	0.38	1.12	89	2.69	2.78	2.53	0.62	1.38	149	4.56	4.66	4.34	0.75	1.50
31	0.88	0.97	0.78	0.38	1.12	91	2.75	2.84	2.59	0.62	1.38	151	4.62	4.72	4.41	0.75	1.50
33	0.94	1.03	0.84	0.38	1.12	93	2.81	2.91	2.66	0.62	1.38	153	4.69	4.78	4.47	0.75	1.50
35	1.00	1.09	0.91	0.38	1.12	95	2.88	2.97	2.72	0.62	1.38	155	4.75	4.84	4.53	0.75	1.50
37	1.06	1.16	0.97	0.38	1.12	97	2.94	3.03	2.78	0.62	1.38	157	4.81	4.91	4.59	0.75	1.50
39	1.12	1.22	1.03	0.38	1.12	99	3.00	3.09	2.84	0.62	1.38	159	4.88	4.97	4.66	0.75	1.50
41	1.19	1.28	1.06	0.50	1.25	101	3.06	3.16	2.91	0.62	1.38	161	4.94	5.03	4.72	0.75	1.50
43	1.25	1.34	1.12	0.50	1.25	103	3.12	3.22	2.97	0.62	1.38	163	5.00	5.09	4.78	0.75	1.50
45	1.31	1.41	1.19	0.50	1.25	105	3.19	3.28	3.03	0.62	1.38	165	5.06	5.16	4.84	0.75	1.50
47	1.38	1.47	1.25	0.50	1.25	107	3.25	3.34	3.09	0.62	1.38	167	5.12	5.22	4.91	0.75	1.50
49	1.44	1.53	1.31	0.50	1.25	109	3.31	3.41	3.16	0.62	1.38	169	5.19	5.28	4.97	0.75	1.50
51	1.50	1.59	1.38	0.50	1.25	111	3.38	3.47	3.22	0.62	1.38	171	5.25	5.34	5.03	0.75	1.50
53	1.56	1.66	1.44	0.50	1.25	113	3.44	3.53	3.28	0.62	1.38	173	5.31	5.41	5.09	0.75	1.50
55	1.62	1.72	1.50	0.50	1.25	115	3.50	3.59	3.34	0.62	1.38	175	5.38	5.47	5.16	0.75	1.50
57	1.69	1.78	1.56	0.50	1.25	117	3.56	3.66	3.34	0.75	1.50	177	5.44	5.53	5.22	0.75	1.50
59	1.75	1.84	1.62	0.50	1.25	119	3.62	3.72	3.41	0.75	1.50	179	5.50	5.59	5.28	0.75	1.50
61	1.81	1.91	1.69	0.50	1.25	121	3.69	3.78	3.47	0.75	1.50	181	5.56	5.66	5.34	0.75	1.50
63	1.88	1.97	1.75	0.50	1.25	123	3.75	3.84	3.53	0.75	1.50	183	5.62	5.72	5.41	0.75	1.50
65	1.94	2.03	1.81	0.50	1.25	125	3.81	3.91	3.59	0.75	1.50	185	5.69	5.78	5.47	0.75	1.50
67	2.00	2.09	1.88	0.50	1.25	127	3.88	3.97	3.66	0.75	1.50	187	5.75	5.84	5.53	0.75	1.50
69	2.06	2.16	1.94	0.50	1.25	129	3.94	4.03	3.72	0.75	1.50	189	5.81	5.91	5.59	0.75	1.50
71	2.12	2.22	2.00	0.50	1.25	131	4.00	4.09	3.78	0.75	1.50	191	5.88	5.97	5.66	0.75	1.50
73	2.19	2.28	2.06	0.50	1.25	133	4.06	4.16	3.84	0.75	1.50	193	5.94	6.03	5.72	0.75	1.50
75	2.25	2.34	2.12	0.50	1.25	135	4.12	4.22	3.91	0.75	1.50	195	6.00	6.09	5.78	0.75	1.50
77	2.31	2.41	2.19	0.50	1.25	137	4.19	4.28	3.97	0.75	1.50						
79	2.38	2.47	2.22	0.62	1.38	139	4.25	4.34	4.03	0.75	1.50						
81	2.44	2.53	2.28	0.62	1.38	141	4.31	4.41	4.09	0.75	1.50						

[a]Reference dimension. When gap is at value tabulated, clamp diameter shall approximate the nominal diameter.

TYPE E HOSE CLAMPS

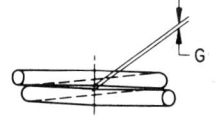

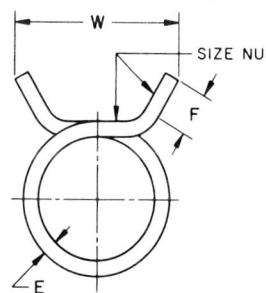

TABLE 5—DIMENSIONS OF TYPE E HOSE CLAMPS

SAE Size No.	Clamp Dia Range[a] A Max	B Nom	C Min	D NOT GO Gage Dia	E Wire Dia[d] Max	Min	F Length of Tang Max	Min	G Clearance at Overlap Max	W Width Over Tangs Free Max	X Expanded Max	Min	Y Overall Height Ref	X Gaging Clearance Max
6	0.380	0.375	0.370	0.350	0.083	0.081	0.38	0.34	0.015	0.88	0.435	0.415	1.06	0.004
7[b]	0.442	0.438	0.432	0.405	0.088	0.086	0.38	0.34	0.015	0.94	0.435	0.415	1.12	0.004
8[c]	0.510	0.500	0.490	0.462	0.093	0.091	0.38	0.34	0.025	1.00	0.435	0.415	1.19	0.005
9	0.573	0.562	0.551	0.520	0.108	0.106	0.38	0.34	0.025	1.06	0.435	0.415	1.38	0.006
10[b]	0.640	0.625	0.610	0.580	0.108	0.106	0.38	0.34	0.025	1.06	0.445	0.425	1.38	0.006
11[c]	0.703	0.688	0.671	0.635	0.113	0.111	0.38	0.34	0.025	1.12	0.445	0.425	1.50	0.006
12	0.770	0.750	0.730	0.690	0.113	0.111	0.38	0.34	0.031	1.19	0.445	0.425	1.50	0.008
13[b]	0.832	0.812	0.792	0.740	0.118	0.116	0.38	0.34	0.031	1.25	0.445	0.425	1.50	0.008
14[c]	0.900	0.875	0.850	0.800	0.123	0.121	0.38	0.34	0.031	1.25	0.445	0.425	1.62	0.008
15	0.968	0.938	0.906	0.855	0.123	0.121	0.38	0.34	0.062	1.25	0.460	0.440	1.69	0.008
16[b]	1.031	1.000	0.969	0.915	0.133	0.131	0.38	0.34	0.062	1.31	0.460	0.440	1.75	0.008
17[c]	1.090	1.062	1.034	0.960	0.143	0.141	0.41	0.34	0.062	1.50	0.520	0.500	1.88	0.010
18	1.150	1.125	1.100	1.030	0.153	0.151	0.41	0.34	0.062	1.62	0.545	0.525	2.00	0.010
19[b]	1.218	1.188	1.156	1.095	0.153	0.151	0.41	0.34	0.062	1.62	0.545	0.525	2.02	0.010
20[c]	1.280	1.250	1.219	1.145	0.153	0.151	0.41	0.34	0.062	1.75	0.545	0.525	2.00	0.010
21	1.344	1.312	1.281	1.210	0.163	0.161	0.41	0.34	0.062	1.75	0.570	0.540	2.31	0.010
22[b]	1.406	1.375	1.344	1.250	0.163	0.161	0.41	0.34	0.062	1.88	0.570	0.540	2.31	0.010
24	1.531	1.500	1.469	1.350	0.163	0.161	0.44	0.38	0.062	1.88	0.570	0.540	2.40	0.010
26	1.672	1.625	1.578	1.455	0.174	0.170	0.44	0.38	0.062	2.00	0.610	0.580	2.69	0.010
28	1.797	1.750	1.703	1.550	0.174	0.170	0.44	0.38	0.062	2.12	0.610	0.580	2.75	0.010
30	1.937	1.875	1.812	1.675	0.179	0.175	0.44	0.38	0.093	2.25	0.610	0.580	2.88	0.010
31	2.000	1.938	1.875	1.720	0.179	0.175	0.44	0.38	0.093	2.25	0.610	0.580	3.00	0.010
32	2.061	2.000	1.939	1.750	0.179	0.175	0.44	0.38	0.093	2.31	0.610	0.580	3.00	0.010
34	2.187	2.125	2.062	1.860	0.184	0.180	0.44	0.38	0.093	2.31	0.635	0.600	3.19	0.010
35	2.250	2.188	2.125	1.925	0.184	0.180	0.44	0.38	0.093	2.31	0.635	0.600	3.25	0.010
36	2.312	2.250	2.187	2.000	0.184	0.180	0.44	0.38	0.093	2.38	0.635	0.600	3.25	0.010
38	2.437	2.375	2.312	2.100	0.194	0.190	0.44	0.38	0.093	2.38	0.635	0.600	3.44	0.010
40	2.561	2.500	2.439	2.187	0.194	0.190	0.44	0.38	0.093	2.38	0.635	0.600	3.62	0.010
42	2.688	2.625	2.562	2.320	0.204	0.200	0.44	0.38	0.093	2.38	0.655	0.620	3.75	0.010

[a] To be used for corresponding gage diameter. Gage diameter tolerance +0.001, −0.000.
[b] These sizes shall be furnished with greenish hue.
[c] These sizes shall be furnished with reddish hue.
[d] Wire diameters shown are before forming.

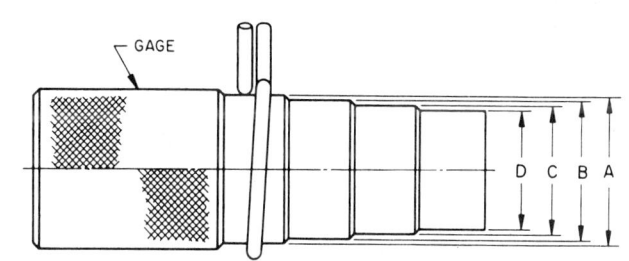

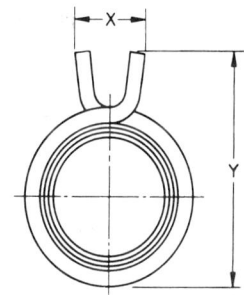

TYPE F HOSE CLAMPS

TABLE 6—DIMENSIONS OF TYPE F HOSE CLAMPS

SAE Size No.[a]	A, Dia Open	Closed	SAE Size No.[a]	A, Dia Open	Closed	SAE Size No.[a]	A, Dia Open	Closed	SAE Size No.[a]	A, Dia Open	Closed
6	0.78	0.44	S 20	1.94	1.25	44	3.25	2.31	S 60	4.88	3.63
8	0.91	0.50	24	2.00	1.06	48	3.50	2.56	72	5.00	4.06
S 8	1.00	0.50	28	2.25	1.31	S 50	3.75	2.56	S 65	5.50	4.25
10	1.06	0.56	S 30	2.25	1.44	52	3.75	2.81	80	5.50	4.62
12	1.25	0.69	32	2.50	1.56	56	4.00	3.06	S 70	6.00	4.75
S 10	1.38	0.50	36	2.75	1.81	60	4.25	3.31	88	6.00	5.12
16	1.50	0.81	S 40	2.94	2.12	S 55	4.38	3.19			
20	1.75	0.81	40	3.00	2.06	64	4.50	3.56			

[a] The S prefix applied to SAE size numbers designates special size clamp. Additional sizes are available from manufacturers.

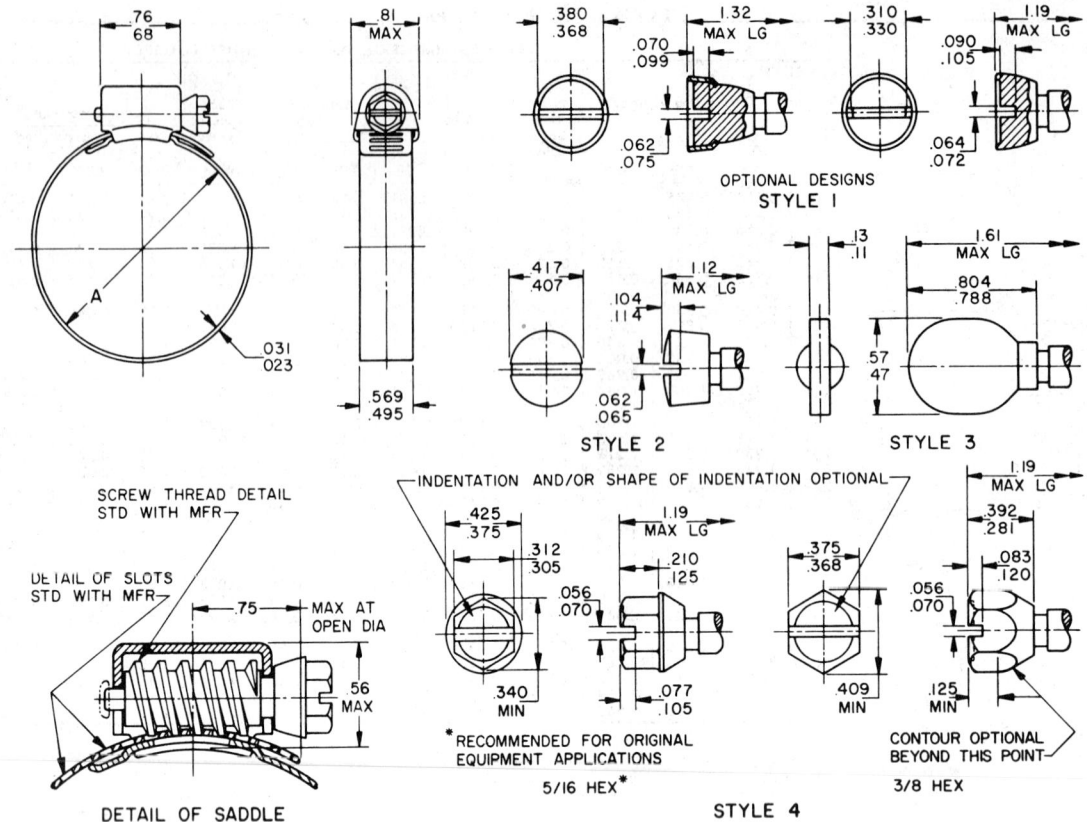

16 Ball Studs and Joints

BALL STUDS AND BALL STUD SOCKET ASSEMBLIES—SAE J491b

SAE Recommended Practice

Report of Parts and Fittings Division approved August 1922 and revised by Parts and Fittings Committee January 1951 and last revised by Ball Stud and Tie Rod Socket Committee February 1977.

General Specification

Purpose—This SAE Recommended Practice has been established for the purpose of providing design criteria and suggested dimensional proportions which may be used for ball studs and ball stud socket assemblies as used on steering systems or control mechanisms of passenger vehicles, trucks, and off-road equipment.

The recommended practice does not cover all applications. It is intended to provide assistance in obtaining functional satisfaction and interchangeability.

The inclusion of dimensional data in this report is not intended to imply that all the products described are stock production sizes. Consumers are requested to consult with manufacturers concerning stock production parts.

Terminology

Master Gage—A taper gage that serves as a standard or base, designed to specific dimensions within blueprint specifications.

Blueing—A nondrying light paste with a pigment or die (such as "prussian blue") that colors a contacting surface. Blueing must be distributed evenly with minimum thickness on the taper surface of the master gage.

Blueing in a Master Gage—Blueing as applied to a master gage and gage forced onto a taper by hand pressure with a slight twisting motion followed by a rocking motion of the stud in the gage. (The gage should not rock.) Gage is removed for visual check of contact surface of taper.

Ball Studs

Selection of Size—Tensile and compressive forces and functional load requirements of the tie rod must be considered in selecting the proper ball stud size.

The proper ball stud size for a specific application can be reasonably well estimated by considering the stud as a cantilever beam supported at the junction with the mating boss, and loaded radially through the ball end by a force, or forces, the magnitude and direction of which have been previously determined. The type of loading, ball stud and bearing material, heat treatment, and safety factor requirements will influence the size of unit chosen for a specific application.

Design requirements which cannot be satisfied by the tabulated dimensions may be safely fulfilled by deviating wherever necessary provided due consideration is given to the functional stresses.

Materials—Plain carbon and alloy steels are widely used for ball stud fabrication. The principal requirements for either type of steel are: case hardenability to provide a wear resistant skin, good machinability or formability, and good quenching and drawing properties to provide a tough core without surface cracks or brittleness under impact loading. Some of the more popular standard materials used for ball studs lie in the same category as SAE 1019, 4615, 8115, 8615, 8620, and 8640 steels.

Processing—Processing of ball studs is usually dictated by the size, volume of production, and equipment available. Ball studs used on passenger car and light truck steering systems are usually cold formed. Larger ball studs of relatively low production volume such as used on heavy duty trucks and off-road equipment are often fabricated as machined parts. Hot upset forging methods have also been used. The forming or machining is followed by total or selective case hardening. Depending on the application and method of fabrication, it may be desirable to control the surface roughness.

Attachment—The ball stud locking taper is usually used in conjunction with screw thread and nut attachment to allow for repair or replacement. The

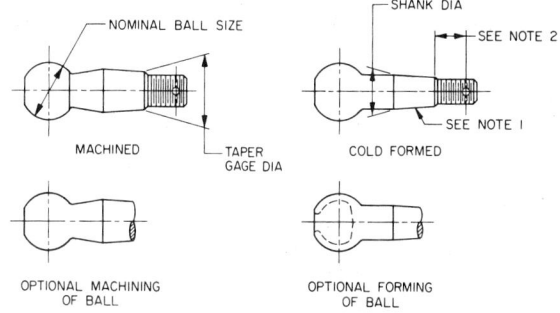

NOTES:
1. TAPER 1.5 IN./FT ON DIAMETER UNLESS OTHER SPECIFIED. MUST SHOW 60% MINIMUM AREA OF CONTACT WHEN BLUED IN A MASTER GAGE (SEE GENERAL SPECIFICATIONS).
2. COTTER PIN HOLE DIAMETER SHOULD CONFORM TO SAE J485. THE HOLE LOCATION IS DETERMINED BY THE ATTACHMENT EYE AND NUT SIZE. THE HOLE SHOULD BE LOCATED TO PROVIDE A MINIMUM OF 50% ENGAGEMENT OF THE COTTER PIN WITH NUT SLOT AFTER ALLOWING FOR DRAW-IN AT SPECIFIED NUT TORQUE.

COTTER PIN HOLE MAY BE OMITTED AND OTHER THAN SLOTTED NUT USED WHEN APPLICATION PERMITS.

FIG. 1—"FULL BALL" BALL STUDS

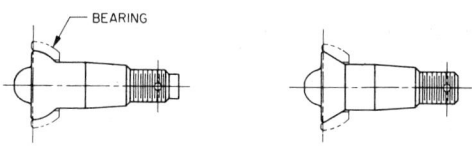

FIG. 2—"HALF BALL" BALL STUDS FOR SEPARATE SPHERICAL BEARINGS

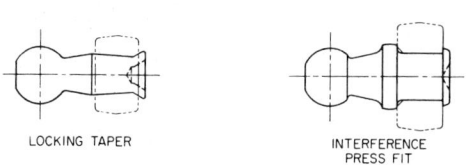

FIG. 3—PERMANENTLY ATTACHED STUDS—RIVETED, WELDED, SPUN, OR STAKED

16.01

taper (usually 1.5 in. taper per foot) is designed to attach into mating parts made of steel; however, by proper design the ball studs can be adapted to mate with other materials for which different tapers may be desirable.

The nut selection should be determined by design requirements.

The specified nut torque or seating force may draw the stud locking taper into the mating arm significantly depending on several variables such as amount of taper, size of taper, material, heat treat, surface, and operating conditions. When using a slotted nut, the nut should be tightened to specification and then tightened further if necessary to align a slot with the cotter hole.

The amount of stud taper draw-in should be determined experimentally for each specific application.

The boss thickness and hole gage diameter at the nut face of the boss must be so related to the stud gage diameter as to provide sufficient unexposed threads to allow for draw-in at specified nut torque (reference Fig. 5). To obtain the full stud cantilever strength, the face of the mounting boss should correspond nominally with the large end of the stud taper.

In addition to the screw thread and nut attachment method, permanent attachment means may be used such as riveting (upsetting), welding, spinning, or staking. These methods may be used in conjunction with a locking taper or an interference press fit, in which case the stud is usually a straight shank end with a shoulder for locating purposes.

The permanently attached studs are used in conjunction with tubular ("horizontal type") sockets which can be assembled onto the stud. The screw thread and nut tapered studs are usually preassembled into ball stud sockets, but can also be used with horizontal type sockets.

Sockets

Selection of Size—The socket size is generally dependent upon the ball stud size necessary for a specific application; however, the tensile and compressive forces and functional load requirements of the tie rod must also be considered in selecting the proper socket size.

Materials—Tie rod sockets are often made of SAE 1030, 1038, 1040, 1041, and 1541 steels; however, a number of standard and special steels may be used to provide the desired mechanical properties for each application. Tubular

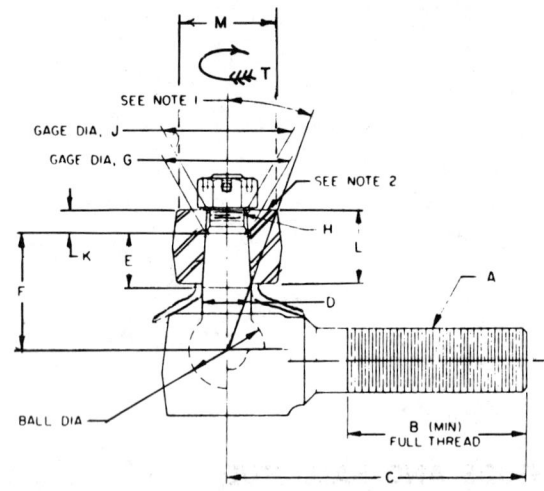

FIG. 5—TYPICAL BALL STUD SOCKET ASSEMBLY WITH EXTERNAL THREADED STEM

NOTES:
1. 0 - ANGULAR TRAVEL DETERMINED AS REQUIRED FOR SPECIFIC APPLICATION.
2. NUT SURFACE ON BOSS MUST BE SUFFICIENTLY SQUARE WITH TAPERED HOLE TO PREVENT EXCESSIVE STRESS ON STUD THREADED END.

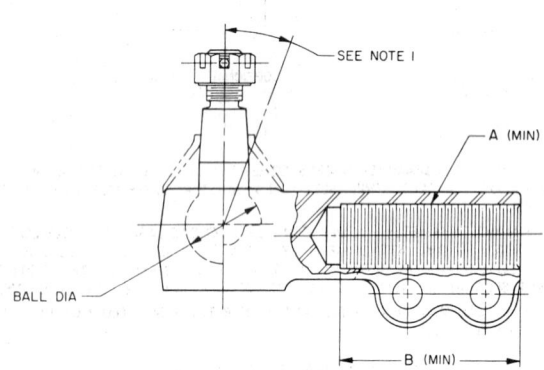

FIG. 4—TYPICAL BALL STUD SOCKET ASSEMBLY WITH INTERNAL THREADED STEM

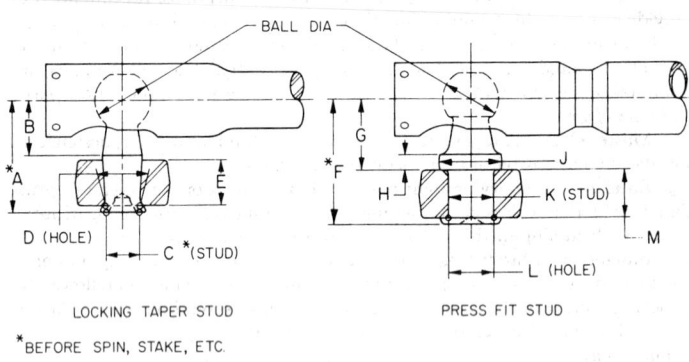

FIG. 6—TYPICAL TUBULAR ("HORIZONTAL TYPE") SOCKET ASSEMBLY FOR RIVETED, SPUN, STAKED, WELDED, OR THREADED TAPER STUD ATTACHMENT. (FOR THREADED TAPER STUDS, USE TABLE 1 DIMENSIONS)

("horizontal type") sockets may also be made from seamless or welded tubing, frequently in SAE 1010, 1020, and 1025 steels.

Processing and Attachment—Tie rod sockets are usually machined forgings with external threaded stems for attaching to mating components. In heavy duty or off-road applications, the socket is often forged with integral clamp ears and threaded with internal threads for attaching to mating components. Sockets may also be forged into both ends of an integral link. In applications of sockets with threaded stems for attachment to tubular intermediates or turnbuckles, a right-hand thread is usually paired with a left-hand thread to provide fast assembly and adjustment.

Threads—Unified Class 2A of the size indicated in Table 1. Ends of threaded features should be chamfered from approximately 0.02 in. below the minor diameter, the length of chamfer to be $\frac{1}{2}$ to $1\frac{1}{2}$ threads.

In applications of sockets with threaded stems, the length of engagement of stem threads with the attaching rod or sleeve depends on design and may vary, but should be approximately $2\frac{1}{2}$ times the thread diameter. Under extremes of adjustment, where the application permits, engagement may be as low as $1\frac{1}{2}$ times the thread diameter.

Lubrication—Lubrication fittings, if required, may be placed at any convenient location on the periphery or face of the socket provided the location does not create forging processing, assembly, or functional complications.

TABLE 1A—BALL STUD SOCKET ASSEMBLY DIMENSIONS[a] (DIMENSIONS IN UNITED STATES CUSTOMARY UNITS)

Nominal Ball Dia	Socket Thread			Ball Stud						Attachment Arm			Stud Nut
	Size A	Length B	Socket Length C	Shank Dia D	Taper Length E	Gage Dia Location F	Gage Dia G	Thread Size H	Gage Dia J	Draw-In Clearance K[b]	Thickness L	Boss Dia M[c]	Tightening Torque T[d]
in	in	in	in	in	in	in	in	in	in	in	in	in	in
5/8	1/2-20	1.74	2.62	0.469	0.41	1.06	0.418	3/8-24	0.402	0.128	0.50	1.25	15-30
3/4	9/16-18	1.94	2.94	0.547	0.45	1.20	0.490	7/16-20	0.473	0.136	0.56	1.38	30-45
7/8	11/16-18	2.06	3.12	0.625	0.52	1.36	0.560	1/2-20	0.543	0.136	0.62	1.50	35-55
1	11/16-18	2.21	3.38	0.703	0.72	1.62	0.613	9/16-18	0.590	0.184	0.88	1.75	55-80
1 1/8	7/8-18	2.75	3.88	0.781	0.84	1.88	0.675	5/8-18	0.652	0.184	1.00	1.88	80-110
1 1/4	1-16	3.19	4.25	0.875	0.94	2.03	0.758	5/8-18	0.731	0.216	1.12	2.00	80-110
1 1/2	1 1/8-12	3.75	4.88	1.031	1.12	2.44	0.890	3/4-16	0.863	0.216	1.31	2.25	100-140
1 3/4	1 1/4-12		5.75	1.250	1.28	2.81	1.074	7/8-14	1.043	0.248	1.50	2.75	120-170
2				1.350	1.50	3.22	1.166	1-12	1.131	0.280	1.75	3.00	140-220
2 1/4				1.510	1.78	3.72	1.285	1 1/8-12	1.250	0.280	2.00	3.25	180-270
2 1/2				1.700	2.06	4.34	1.441	1 1/4-12	1.406	0.280	2.25	3.50	230-320

[a]These dimensions may be varied as required for specific applications. See General Specifications.
[b]Before tightening nut.
[c]"The boss diameter or "hoop size" was determined by using the recommended nut tightening torque to determine "hoop size" at recommended thickness "L" with the stress level below the yield strength of medium carbon steel forgings.
[d]Ranges of tightening torque for 1.5 in./ft taper in medium carbon or alloy steel forgings. For other materials or tapers, these torque values must be adjusted. The torque values recommended are empirical values determined by combined experience of SAE Ball Stud and Tie Rod Socket Committee members.

TABLE 1B—BALL STUD SOCKET ASSEMBLY DIMENSIONS[a] (DIMENSIONS IN SI UNITS)

Nominal Ball Dia[e]	Socket Thread			Ball Stud						Attachment Arm			Stud Nut
	Size A[e]	Length B	Socket Length C	Shank Dia D	Taper Length E	Gage Dia Location F	Gage Dia G	Thread Size H[e]	Gage Dia J	Draw-In Clearance K[b]	Thickness L	Boss Dia M[c]	Tightening Torque T[d]
in	in	mm	mm	mm	mm	mm	mm	in	mm	mm	mm	mm	N·m
5/8	1/2-20	44.2	66.6	11.91	10.4	26.9	10.62	3/8-24	10.21	3.3	12.7	31.8	20-40
3/4	9/16-18	49.3	74.7	13.89	11.4	30.5	12.45	7/16-20	12.01	3.5	14.2	35.0	40-60
7/8	11/16-18	52.3	79.3	15.87	13.2	34.5	14.22	1/2-20	13.79	3.5	15.8	38.1	50-75
1	11/16-18	56.1	85.9	17.86	18.3	41.2	15.57	9/16-18	14.99	4.7	22.4	44.5	75-110
1 1/8	7/8-18	69.9	98.6	19.84	21.3	47.8	17.15	5/8-18	16.56	4.7	25.4	47.7	110-150
1 1/4	1-16	81.0	108.0	22.23	23.9	51.6	19.25	5/8-18	18.57	5.5	28.5	50.8	110-150
1 1/2	1 1/8-12	95.3	124.0	26.19	28.5	62.0	22.61	3/4-16	21.92	5.5	33.3	57.2	140-190
1 3/4	1 1/4-12		146.1	31.75	32.5	71.4	27.28	7/8-14	26.49	6.3	38.1	69.9	160-230
2				34.29	38.1	81.8	29.62	1-12	28.73	7.1	44.5	76.2	190-300
2 1/4				38.35	45.2	94.5	32.64	1 1/8-12	31.75	7.1	50.8	82.6	240-370
2 1/2				43.18	52.3	110.2	36.60	1 1/4-12	35.71	7.1	57.2	88.9	310-430

[a] These dimensions may be varied as required for specific applications. See General Specifications.
[b] Before tightening nut.
[c] The boss diameter or "hoop size" was determined by using the recommended nut tightening torque to determine "hoop size" at recommended thickness "L" with the stress level below the yield strength of medium carbon steel forgings.
[d] Ranges of tightening torque for one unit per eight units of length taper in medium carbon steel forgings. For other materials or tapers, these torque values must be adjusted. The torque values recommended are empirical values determined by combined experience of SAE Ball Stud and Tie Rod Socket Committee members.
[e] No direct SI conversion for nominal sizes that are not measurements.

TABLE 2A—TYPICAL DIMENSIONS FOR TUBULAR SOCKET STUDS (DIMENSIONS IN UNITED STATES CUSTOMARY UNITS)

Nominal Ball Dia	Locking Taper Studs					Press Fit Studs						
	A	B	C	D	E	F	G	H	J	K	L	M
in	in	in	in	in	in	in	in	in	in	in	in	in
5/8	1.31	0.66	0.387	0.402	0.50							
3/4	1.47	0.75	0.457	0.437	0.56	1.49	0.87	0.18	0.75	0.598	0.592	0.50
7/8	1.62	0.84	0.527	0.543	0.62	1.78	1.03	0.28	0.88	0.693	0.637	0.62
1	1.84	0.91	0.585	0.601	0.78	2.06	1.12	0.28	1.00	0.770	0.764	0.81
1 1/8	2.14	1.03	0.642	0.660	0.94	2.15	1.21	0.28	1.12	0.839	0.833	0.81
1 1/4	2.34	1.09	0.719	0.739	1.06							
1 1/2	2.72	1.31	0.855	0.879	1.19							
1 3/4	3.16	1.53	1.031	1.059	1.38							
2	3.50	1.72	1.131	1.162	1.50							
2 1/4	4.02	1.94	1.248	1.281	1.75							
2 1/2	4.69	2.28	1.399	1.437	2.00							

TABLE 2B—TYPICAL DIMENSIONS FOR TUBULAR SOCKET STUDS (DIMENSIONS IN SI UNITS)

Nominal Ball Dia	Locking Taper Studs					Press Fit Studs						
	A	B	C	D	E	F	G	H	J	K	L	M
in	mm	mm	mm	mm	mm	mm	mm	mm	mm	mm	mm	mm
5/8	33.3	16.8	9.83	10.21	12.7							
3/4	37.3	19.1	11.61	11.10	14.2	37.9	22.1	4.6	19.1	15.19	15.04	12.7
7/8	41.2	21.3	13.39	13.79	15.8	45.2	26.2	7.1	22.4	17.60	17.45	15.8
1	46.7	23.1	14.86	15.27	19.8	52.3	28.5	7.1	25.4	19.56	19.41	20.6
1 1/8	54.5	26.2	16.31	16.76	23.9	54.6	30.7	7.1	28.5	21.31	21.16	20.6
1 1/4	59.4	27.7	18.26	18.77	26.9							
1 1/2	69.1	33.3	21.72	22.33	30.2							
1 3/4	80.3	38.9	26.19	26.90	35.1							
2	88.9	43.7	28.73	29.51	38.1							
2 1/4	102.1	49.3	31.70	32.54	44.5							
2 1/2	119.1	57.9	35.53	36.50	50.8							

BALL STUD AND SOCKET ASSEMBLY TEST PROCEDURE—SAE J193 SEP79

SAE Recommended Practice

Report of the Ball Stud and Tie Rod Socket Committee, approved August 1970, completely revised April 1979.

1. Scope—The purpose of this test procedure is to provide a uniform method of testing ball stud and socket assemblies to determine their functional characteristics. This procedure is an extension of the dimensional recommendations for ball studs as used in integral socket assemblies. All tests, except ball stud yield, may be run using complete integral assemblies representing the application.

2. Objective—To provide adequate testing format to ensure that the parts will meet functional requirements of the individual application.

3. Test Procedures—The test procedures cover the following characteristics:

3.1 Ball stud to socket rotating and oscillating torque.
3.2 Ball stud to socket axial end movement.
3.3 Ball stud to socket cam out strength.
3.4 Ball stud and socket assembly fatigue and wear test.
3.5 Ball stud yield load.

4. Objectives and Test Procedures

4.1 Ball Stud to Socket Rotating and Oscillating Torque

4.1.1 OBJECTIVE—To ensure desired rotating and oscillating torque is obtained.

4.1.2 PROCEDURE—The assembly should be clamped at an area away from the socket to prevent addition of external clamping pressures which may affect torque readings.

4.1.2.1 *Breakaway Torque*—Assemblies should be filled with specified application lubricant when it is required.

For some designs and applications it is necessary to store the assembly (with lubricant) for 48 h without movement prior to test to ascertain the cold flow characteristics of the materials and congelation effect of the selected lubricant on breakaway torque.

The torque is read with a torque device with gradual application of a rotating force.

Breakaway torque values may be varied to suit the application.

4.1.2.2 *Rotating or Oscillating Torque*—Assemblies should be filled with specified application lubricant when it is required.

Rotate stud a minimum of five complete revolutions to minimize congelation and other factors prior to recording torque.

The torque is read with a torque device while the stud is being revolved or oscillated at approximately 5 r/min.

Rotating and oscillating torque values may be varied to suit the application.

4.2 Ball Stud to Socket Axial End Movement

4.2.1 OBJECTIVE—To determine end movement measurement.

4.2.2 PROCEDURE

4.2.2.1 *Spring Loaded Type*—For axial movement the following is commonly used. The stud should be set perpendicular to the socket. Socket should be supported on the bottom of assembly. A force is applied to the stud (less nut) and the axial movement of the stud is noted and recorded. (Fig. 1A depicts typical fixture.)

NOTE: Ensure that the top of the stud is flat at the contact point of force (grind if necessary).

4.2.2.2 *All Other Types of Socket Assemblies*—With the shank of the socket assembly clamped to prevent squeezing of socket and the stud, pull upward. After the movement of the stud is noted and recorded, the operation is repeated with a force pushing downward. Fig. 1B depicts a typical fixture.

4.3 Ball Stud to Socket Cam Out Strength

4.3.1 OBJECTIVE—To determine retention of the ball stud in the socket at angular positions and to determine the angle of separation.

4.3.2 PROCEDURE—The ball stud and socket assemblies should be mounted in a tensile test machine with the test specimen stud held in a fixture which permits unrestricted angular travel. Fig. 2 depicts a typical fixture.

A tensile load is applied to the assembly parallel to the normal load direction when the test stud is in full angular travel. The test is repeated with a

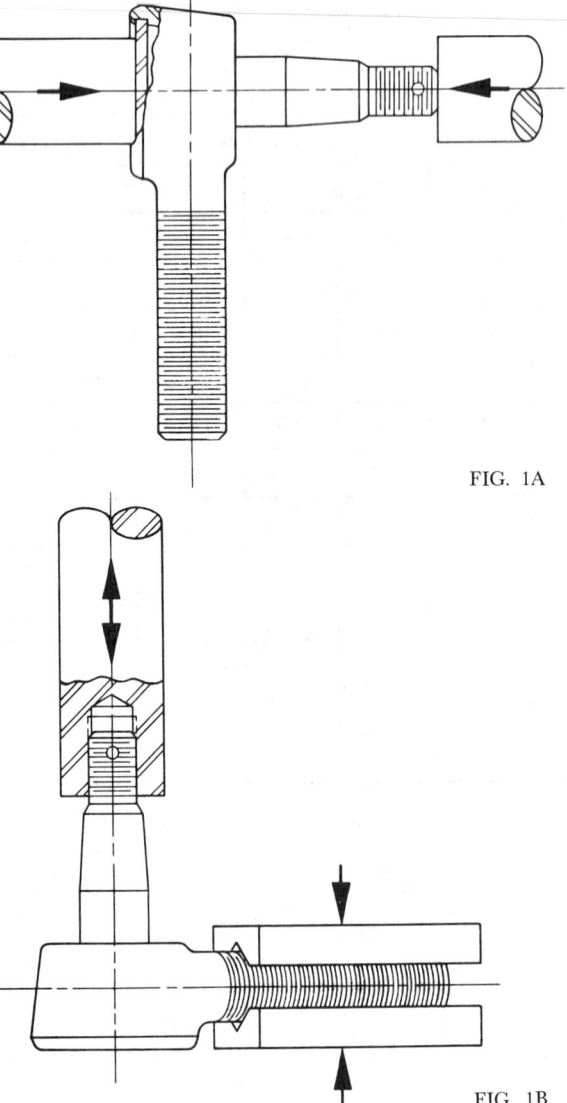

FIG. 1A

FIG. 1B

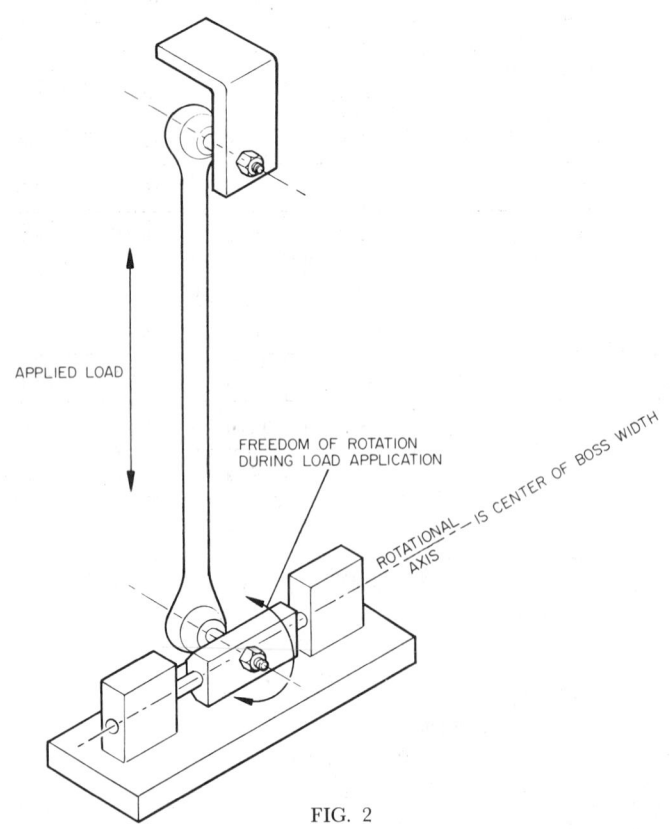

FIG. 2

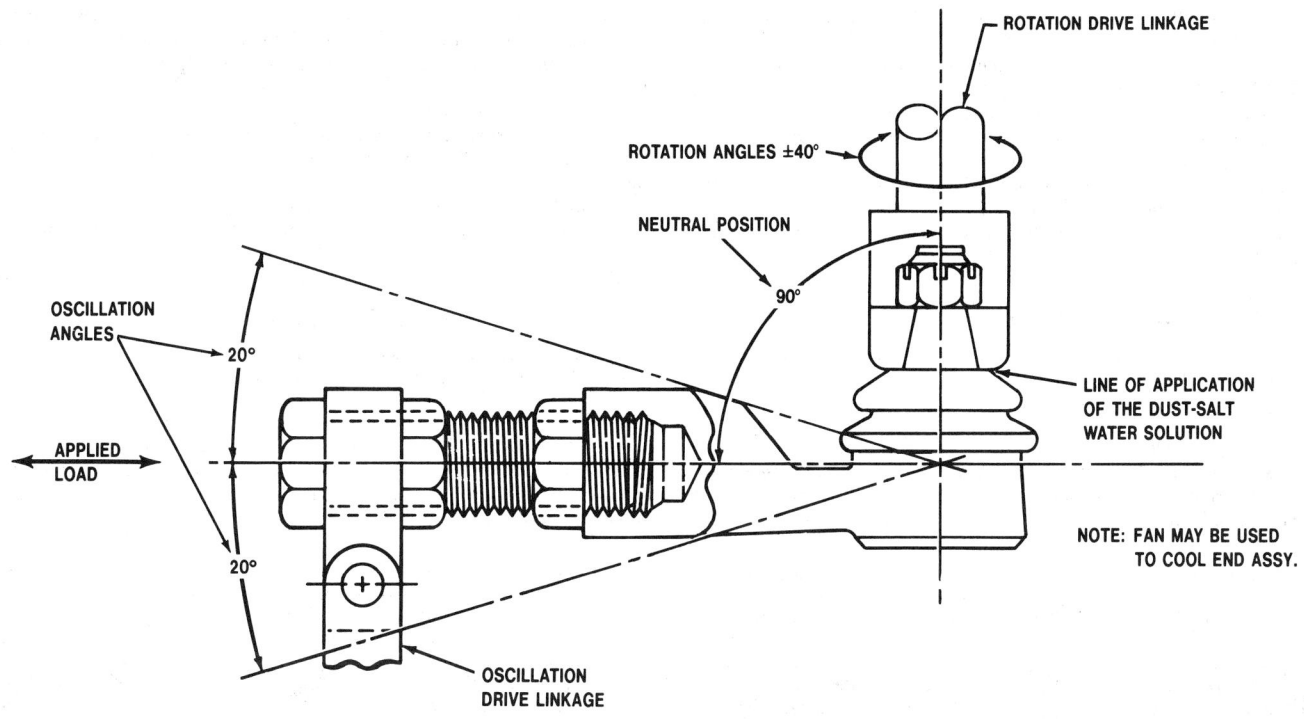

FIG. 3

new sample, using a compression load. The maximum load and angle induced prior to separating the stud from the socket is recorded.

4.4 Ball Stud and Socket Assembly Fatigue and Wear Test

4.4.1 OBJECTIVE—To determine fatigue and wear characteristics of ball stud and socket assemblies.

4.4.2 PROCEDURE—Use socket assemblies which have been tested according to paragraphs 4.1 and 4.2 and found acceptable. Socket assemblies should be filled with recommended lubricant when required in the application. Socket assemblies should be installed, with seals when required, in a fixture by placing the taper shank in the mating tapered hole with the retaining nut torqued to design specification. For each type to be tested, the ball stud and socket assemblies may be cut and threaded to suit the test machine. Securely clamp the link in a manner to achieve the required motions. The following are typical motions which may be used: Refer to Fig. 3.

1. Angular Oscillation—±20 deg in a plane parallel to the link centerline at 60 c/min.
2. Angular Rotation—±40 deg measured about the ball stud shank centerline at 32 c/min.
3. Load—Alternating designated tension and compression load at 60 c/min.

Load application angle may be varied to suit the application. The socket assemblies are then tested to required angles, frequencies, and load applications concurrently and completed in two phases.

NOTE: When actual use dictates, other application loads, angles, and frequencies may be substituted for above.

4.4.2.1 *Phase I Test: Peak Load*—To correlate the cycle life with the maximum operating load to which the assembly will be subjected in its actual application.

The cycle life varies for each type of application and environment; therefore, a program loading procedure for the specific application is required to establish load and cycle life required for this test.

In the absence of complete program loading data, and to provide a basis for standardized testing of the assembly, 7500 cycles is a reasonable cycle life for this test.

4.4.2.2 *Phase II Test: Endurance Load*—To correlate the cycle life of the assembly for the average load to which the assembly will be subjected in its application and environment, with life in actual use, and to establish the load which provides for extended fatigue and wear life.

The cycle life varies for each type of application and environment; therefore a program loading procedure for the specific application is required to establish load and cycle life required for this test.

In the absence of complete program loading data, and to provide a basis for standardized testing of the assembly, 250 000 cycles is a reasonable cycle life for this test.

4.4.2.3 During the Phase I and Phase II tests, artificial cooling may be used where deemed necessary to prevent heat build-up which would not be experienced in the application.

If the application of the ball stud and socket assembly includes environmental contamination, contaminants should be provided to correlate with these conditions in the test. This procedure will determine seal durability and effectiveness.

Typical solutions commonly used are:

1. Ozone atmosphere.
2. Saline and dust—4 L of water, 50% saturated solution of common salt at 21–24°C, and 0.15 kg of SAE air cleaner test dust fine grade. See SAE J726b (November, 1976).
3. Steam.

4.4.3 Conditions to be examined at completion of test to determine adequacy for the application:

4.4.3.1 The rotating and oscillating torque condition (see paragraph 4.1).
4.4.3.2 The end movement (see paragraph 4.2).
4.4.3.3 The ball stud should be examined for any surface cracks, determined by a dye check or approved equivalent.

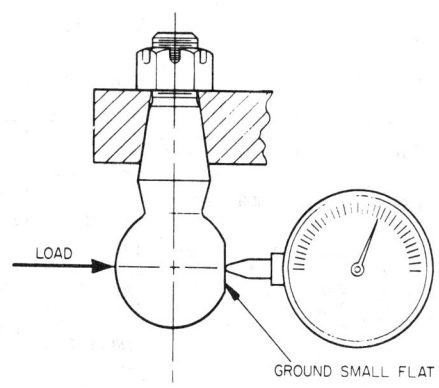

FIG. 4

4.4.3.4 The ball stud should be examined for any local yielding, determined by a dimension check.

4.4.3.5 The internal components should be examined for damage.

4.5 Ball Stud Yield Load

4.5.1 OBJECTIVE—To determine at what load condition the ball stud will take a permanent set without fracture.

4.5.2 PROCEDURE—Fig. 4 depicts fixture.

4.5.2.1 Grind a small flat on the head of the stud for accuracy of reading, to receive the dial indicator or other measuring device.

4.5.2.2 Install ball stud in the fixture with mating taper hole in such a manner that a load can be applied to the stud at right angles to the stud centerline and opposite the flat ground on stud. Lock stud in fixture (with hardened and ground eye) by torquing the retaining nut to design specification.

4.5.2.3 Preload stud.

4.5.2.4 Set dial indicator or other measuring device to zero.

4.5.2.5 Take deflection and set readings in desired increments to permanent set range.

4.5.2.6 Stud yield load is equal to load required to permanently set stud without surface cracks or failure and should be used to select stud application.

BALL JOINTS—SAE J490c   SAE Standard

Report of Miscellaneous Division approved March 1920, revised by Parts and Fittings Committee November 1948, and last revised by Ball Joints Committee August 1973. Editorial change May 1974.

1. General Specifications

1.1 Scope—This SAE Standard covers the general and dimensional data for various types of ball joints commonly used on control linkages in automotive, marine, and construction and industrial equipment applications.

1.1.1 Inasmuch as the load carrying and wear capabilities of ball joints vary considerably with their design and fabrication, it is suggested that the manufacturers be consulted in regard to these features and for recommendations relating to application of the different types and styles available.

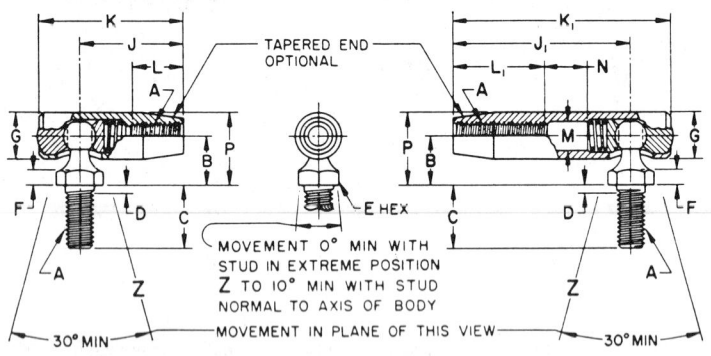
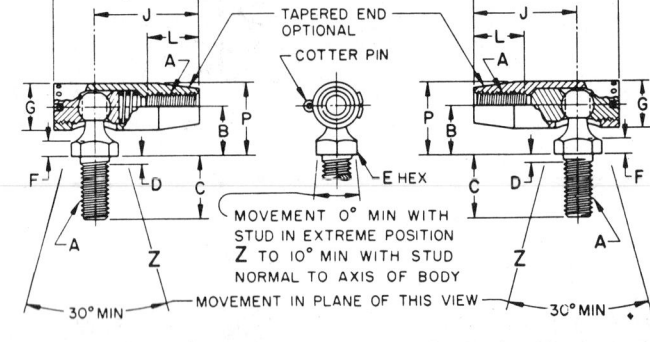

FIG. 1—TYPE A   FIG. 2—TYPE AL   FIG. 3—TYPE B   FIG. 4—TYPE C

CRIMPED END PLUG WITH SPRING CONSTRUCTION   THREADED END PLUG WITH SPRING CONSTRUCTION   THREADED END PLUG WITHOUT SPRING CONSTRUCTION

TABLE 1—DIMENSIONS OF TYPES A, B, C AND AL BALL JOINTS (FIGS. 1–4)

Nominal Ball Joint Size and Thread Diameter, A, in	Threads per in	B ±0.02		C ±0.02		D Max		E Hex		F Min		G Dia		J ±0.03		J_1 ±0.03	
		in ±0.5	mm ±0.5	in ±0.5	mm ±0.5	in	mm	in	mm	in	mm	in	mm	in ±0.8	mm ±0.8	in ±0.8	mm ±0.8
No. 10 0.190	32	0.44	11.2	0.44	11.2	0.06	1.5	0.312	7.92	0.12	3.0	0.38	9.7	0.88	22.3	1.47	37.3
No. 12 0.216	32	0.44	11.2	0.44	11.2	0.06	1.5	0.312	7.92	0.12	3.0	0.38	9.7	0.88	22.3	1.47	37.3
1/4 0.250	28	0.47	11.9	0.56	14.2	0.06	1.5	0.375	9.52	0.12	3.0	0.44	11.2	0.97	24.6	1.81	46.0
5/16 0.3125	24	0.53	13.5	0.69	17.5	0.09	2.3	0.438	11.12	0.16	4.1	0.50	12.7	1.12	28.4	1.94	49.3
3/8 0.375	24	0.69	17.5	0.88	22.3	0.09	2.3	0.500	12.70	0.19	4.8	0.62	15.8	1.38	35.0	—	—
7/16 0.4375	20	0.88	22.3	1.12	28.4	0.12	3.0	0.625	15.88	0.25	6.4	0.75	19.0	1.94	49.3	—	—
1/2 0.500	20	0.88	22.3	1.12	28.4	0.12	3.0	0.625	15.88	0.25	6.4	0.75	19.0	1.94	49.3	—	—

Nominal Ball Joint Size and Thread Diameter, A, in	K ±0.03		K_1 ±0.03		L Min Full Thread		L_1 Min Full Thread		M Dia +0.01/−0.00	+0.3/−0.0	N[a] (Ref)		P[a] Max (Ref)		Stud Ball Diameter (Ref)[a]			
															Max		Min	
	in ±0.8	mm ±0.8	in ±0.8	mm ±0.8	in	mm	in	mm	in	mm	in	mm	in	mm	in	mm	in	mm
No. 10 0.190	1.25	31.8	1.81	46.0	0.44	11.2	0.56	14.2	0.20	5.1	0.50	12.7	0.65	16.5	0.255	6.48	0.250	6.35
No. 12 0.216	1.25	31.8	1.81	46.0	0.44	11.2	0.56	14.2	0.23	5.8	0.50	12.7	0.65	16.5	0.255	6.48	0.250	6.35
1/4 0.250	1.38	35.0	2.25	57.2	0.50	12.7	0.88	22.3	0.27	6.9	0.50	12.7	0.72	18.3	0.305	7.75	0.300	7.62
5/16 0.3125	1.56	39.6	2.38	60.5	0.56	14.2	1.00	25.4	0.33	8.4	0.50	12.7	0.81	20.6	0.350	8.89	0.345	8.76
3/8 0.375	1.94	49.3	—	—	0.75	19.0	—	—	—	—	—	—	1.03	26.2	0.424	10.77	0.419	10.64
7/16 0.4375	2.62	66.5	—	—	1.00	25.4	—	—	—	—	—	—	1.28	32.5	0.555	14.10	0.550	13.97
1/2 0.500	2.62	66.5	—	—	1.00	25.4	—	—	—	—	—	—	1.28	32.5	0.555	14.10	0.550	13.97

[a] These dimensions are given for design purposes only and are not intended for inspection.

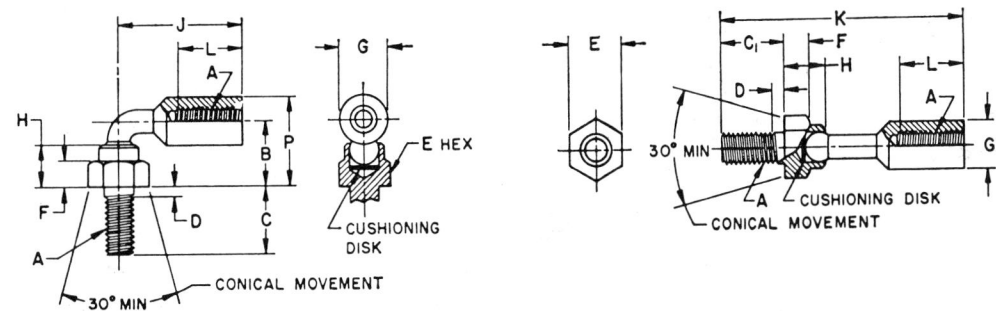

*TYPES D AND DS ARE NOT RECOMMENDED FOR APPLICATIONS INVOLVING TENSION OR SEVERE VIBRATION.

FIG. 5—TYPE D*   FIG. 6—TYPE DS*

CUSHIONED TWO PIECE CONSTRUCTION

TABLE 2—DIMENSIONS OF TYPES D AND DS BALL JOINTS (FIGS. 5 AND 6)

Nominal Ball Joint Size and Thread Diameter, A, in	Thds per in	B ±0.03 in	B ±0.8 mm	C ±0.02 in	C ±0.5 mm	C_1 ±0.02 in	C_1 ±0.5 mm	D Max in	D Max mm	E Hex in	E Hex mm	F Min in	F Min mm	G Dia in	G Dia mm	H ±0.03 in	H ±0.8 mm	J ±0.03 in	J ±0.8 mm	K ±0.03 in	K ±0.8 mm	L Min Full Thread in	L Min Full Thread mm	P^a Max (Ref) in	P^a Max (Ref) mm
No. 10 0.190	32	0.53	13.5	0.44	11.2	0.56	14.2	0.06	1.5	0.375	9.52	0.19	4.8	0.28	7.1	0.33	8.4	1.03	26.2	2.03	51.6	0.50	12.7	0.70	17.8
No. 10 0.190	32	0.53	13.5	0.44	11.2	—	—	0.06	1.5	0.375	9.52	0.19	4.8	0.28	7.1	0.33	8.4	0.78	19.8	—	—	0.38	9.7	0.70	17.8
No. 12 0.216	24	0.53	13.5	0.56	14.2	0.56	14.2	0.06	1.5	0.375	9.52	0.19	4.8	0.28	7.1	0.33	8.4	1.03	26.2	2.03	51.6	0.50	12.7	0.70	17.8
No. 12 0.216	32	0.53	13.5	0.56	14.2	0.56	14.2	0.06	1.5	0.375	9.52	0.19	4.8	0.28	7.1	0.33	8.4	1.03	26.2	2.03	51.6	0.50	12.7	0.70	17.8
1/4 0.250	28	0.56	14.2	0.56	14.2	0.56	14.2	0.06	1.5	0.438	11.12	0.19	4.8	0.31	7.9	0.35	8.9	1.06	26.9	2.09	53.1	0.56	14.2	0.75	19.0
5/16 0.3125	24	0.69	17.5	0.69	17.5	0.69	17.5	0.09	2.3	0.562	14.28	0.28	7.1	0.44	11.2	0.45	11.4	1.31	33.3	2.59	65.8	0.69	17.5	0.94	23.9

^{a}These dimensions are given for design purposes only and are not intended for inspection.

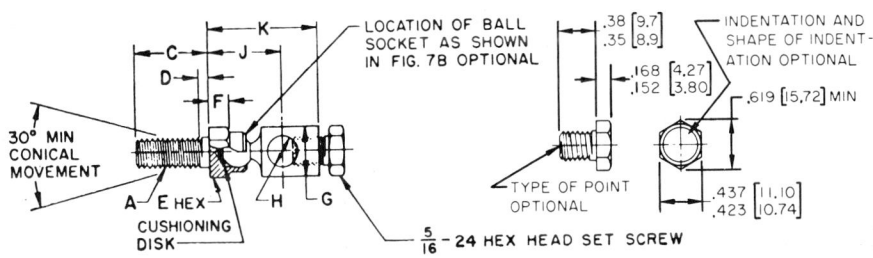

FIG. 7A—STYLE 1   FIG. 7B—STYLE 2

FIG. 7—TYPE DC

TABLE 3—DIMENSIONS OF TYPE DC BALL JOINTS (FIGS. 7A AND 7B)

Nominal Ball Joint Size and Thread Diameter, A, in	Thds per in	C ±0.02 in	C ±0.5 mm	D Max in	D Max mm	E Hex in	E Hex mm	F Min in	F Min mm	G Dia in	G Dia mm	H Dia ±0.005 in	H Dia ±0.13 mm	J ±0.03 in	J ±0.8 mm	K ±0.03 in	K ±0.8 mm	L ±0.02 in	L ±0.5 mm	M Dia in	M Dia mm	N ±0.01 in	N ±0.3 mm	P ±0.005 in	P ±0.13 mm
STYLE 1																									
No. 10 0.190	32	0.31	7.9	0.06	1.5	0.438	11.12	0.19	4.8	0.50	12.7	0.328	8.33	0.75	19.0	1.12	28.4	—	—	—	—	—	—	—	—
1/4 0.250	20	0.44	11.2	0.09	2.3	0.438	11.12	0.19	4.8	0.50	12.7	0.328	8.33	0.75	19.0	1.12	28.4	—	—	—	—	—	—	—	—
1/4 0.250	20	0.56	14.2	0.09	2.3	0.438	11.12	0.19	4.8	0.50	12.7	0.328	8.33	0.75	19.0	1.12	28.4	—	—	—	—	—	—	—	—
1/4 0.250	28	0.44	11.2	0.06	1.5	0.438	11.12	0.19	4.8	0.50	12.7	0.328	8.33	0.75	19.0	1.12	28.4	—	—	—	—	—	—	—	—
1/4 0.250	28	0.56	14.2	0.06	1.5	0.438	11.12	0.19	4.8	0.50	12.7	0.328	8.33	0.75	19.0	1.12	28.4	—	—	—	—	—	—	—	—
5/16 0.3125	24	0.62	15.8	0.09	2.3	0.438	11.12	0.19	4.8	0.50	12.7	0.328	8.33	0.75	19.0	1.12	28.4	—	—	—	—	—	—	—	—
5/16 0.3125	24	0.75	19.0	0.09	2.3	0.438	11.12	0.19	4.8	0.50	12.7	0.328	8.33	0.75	19.0	1.12	28.4	—	—	—	—	—	—	—	—
3/8 0.375	24	0.62	15.8	0.09	2.3	0.438	11.12	0.19	4.8	0.50	12.7	0.328	8.33	0.75	19.0	1.12	28.4	—	—	—	—	—	—	—	—
STYLE 2																									
No. 10 0.190	32	0.50	12.7	0.06	1.5	0.375	9.52	0.09	2.3	0.44	11.2	0.197	5.00	0.62	15.8	0.78	19.8	0.40	10.2	0.56	14.2	0.12	3.0	0.250	6.35
1/4 0.250	20	0.44	11.2	0.09	2.3	0.438	11.12	0.09	2.3	0.50	12.7	0.328	8.33	0.78	19.8	1.02	25.9	0.34	8.6	0.62	15.8	0.12	3.0	0.250	6.35
1/4 0.250	28	0.44	11.2	0.06	1.5	0.438	11.12	0.09	2.3	0.50	12.7	0.328	8.33	0.78	19.8	1.02	25.9	0.34	8.6	0.62	15.8	0.12	3.0	0.250	6.35
5/16 0.3125	24	0.62	15.8	0.09	2.3	0.438	11.12	0.11	2.8	0.56	14.2	0.380	9.65	0.75	19.0	1.03	26.2	0.53	13.5	0.75	19.0	0.19	4.8	0.344	8.74

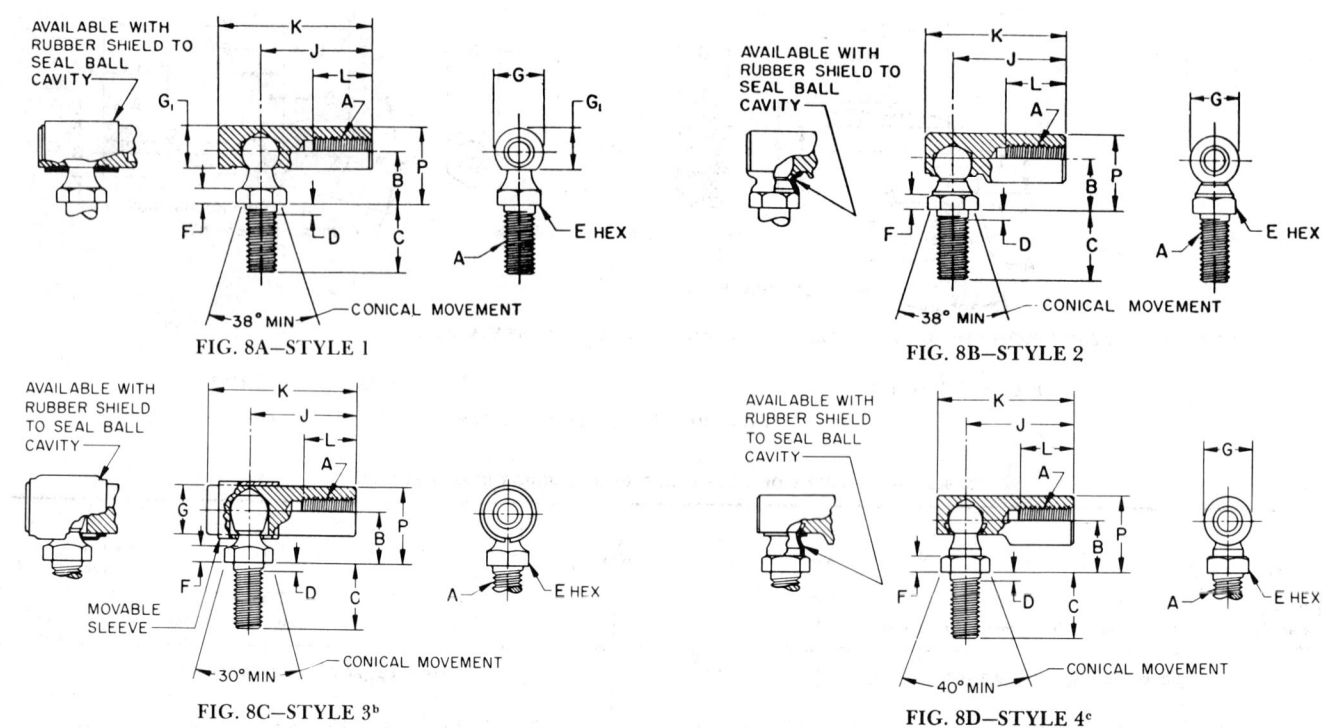

FIG. 8A—STYLE 1
FIG. 8B—STYLE 2
FIG. 8C—STYLE 3[b]
FIG. 8D—STYLE 4[c]

FIG. 8—TYPE G NONDETACHABLE CONSTRUCTION WITHOUT SPRING

TABLE 4—DIMENSIONS OF TYPE G BALL JOINTS (FIGS. 8A–8D)

Nominal Ball Joint Size and Thread Diameter, A, in		Thds per in	B ±0.02		C ±0.02		D Max		E Hex		F Min		G		G₁ Min		J ±0.02		K ±0.02		L Min Full Thread		P[a] Max (Ref)	
			in	mm ±0.5	in	mm ±0.5	in	mm	in	mm	in	mm	in	mm	in	mm ±0.5	in	mm ±0.5	in	mm ±0.5	in	mm	in	mm
STYLE 1																								
No. 10	0.190	32	0.44	11.2	0.44	11.2	0.06	1.5	0.312	7.92	0.12	3.0	0.38	9.7	0.31	7.9	0.88	22.3	1.16	29.5	0.47	11.9	0.65	16.5
1/4	0.250	28	0.47	11.9	0.56	14.2	0.06	1.5	0.375	9.52	0.12	3.0	0.44	11.2	0.38	9.7	0.97	24.6	1.31	33.3	0.53	13.5	0.72	18.3
5/16	0.3125	24	0.53	13.5	0.69	17.5	0.09	2.3	0.438	11.12	0.16	4.1	0.50	12.7	0.44	11.2	1.12	28.4	1.56	39.6	0.59	15.0	0.81	20.6
3/8	0.375	24	0.69	17.5	0.88	22.3	0.09	2.3	0.500	12.70	0.19	4.8	0.62	15.8	0.56	14.2	1.38	35.0	1.81	46.0	0.81	20.6	1.03	26.2
7/16	0.4375	20	0.88	22.3	1.12	28.4	0.12	3.0	0.625	15.88	0.25	6.4	0.75	19.0	0.69	17.5	1.94	49.3	2.50	63.5	1.12	28.4	1.28	32.5
1/2	0.500	20	0.88	22.3	1.12	28.4	0.12	3.0	0.625	15.88	0.25	6.4	0.75	19.0	0.69	17.5	1.94	49.3	2.50	63.5	1.12	28.4	1.28	32.5
STYLE 2																								
No. 10	0.190	32	0.44	11.2	0.44	11.2	0.06	1.5	0.312	7.92	0.12	3.0	0.38	9.7	—	—	0.88	22.3	1.06	26.9	0.47	11.9	0.65	16.5
1/4	0.250	28	0.47	11.9	0.56	14.2	0.06	1.5	0.375	9.52	0.12	3.0	0.44	11.2	—	—	0.97	24.6	1.22	31.0	0.50	12.7	0.72	18.3
5/16	0.3125	24	0.53	13.5	0.69	17.5	0.09	2.3	0.438	11.12	0.16	4.1	0.50	12.7	—	—	1.12	28.4	1.41	35.8	0.56	14.2	0.81	20.6
3/8	0.375	24	0.69	17.5	0.88	22.3	0.09	2.3	0.500	12.70	0.19	4.8	0.62	15.8	—	—	1.38	35.0	1.69	42.9	0.75	19.0	1.03	26.2
7/16	0.4375	20	0.88	22.3	1.12	28.4	0.12	3.0	0.625	15.88	0.25	6.4	0.75	19.0	—	—	1.94	49.3	2.38	60.5	1.00	25.4	1.28	32.5
1/2	0.500	20	0.88	22.3	1.12	28.4	0.12	3.0	0.625	15.88	0.25	6.4	0.75	19.0	—	—	1.94	49.3	2.38	60.5	1.00	25.4	1.28	32.5
5/8	0.625	18	1.00	25.4	1.12	28.4	0.12	3.0	0.750	19.05	0.31	7.9	0.88	22.3	—	—	2.06	52.3	2.58	65.5	1.00	25.4	1.47	37.3
3/4	0.750	16	1.06	26.9	1.12	28.4	0.12	3.0	0.875	22.22	0.31	7.9	1.00	25.4	—	—	2.12	53.8	3.00	76.2	1.12	28.4	1.59	40.4
STYLE 3[b]																								
No. 10	0.190	32	0.44	11.2	0.44	11.2	0.06	1.5	0.312	7.92	0.12	3.0	0.38	9.7	—	—	0.88	22.3	1.16	29.5	0.47	11.9	0.65	16.5
1/4	0.250	28	0.47	11.9	0.56	14.2	0.06	1.5	0.375	9.52	0.12	3.0	0.44	11.2	—	—	0.97	24.6	1.31	33.3	0.53	13.5	0.72	18.3
5/16	0.3125	24	0.53	13.5	0.69	17.5	0.09	2.3	0.438	11.12	0.16	4.1	0.50	12.7	—	—	1.12	28.4	1.56	39.6	0.59	15.0	0.81	20.6
3/8	0.375	24	0.69	17.5	0.88	22.3	0.09	2.3	0.500	12.70	0.19	4.8	0.62	15.8	—	—	1.38	35.0	1.81	46.0	0.81	20.6	1.03	26.2
1/2	0.500	20	0.88	22.3	1.12	28.4	0.12	3.0	0.625	15.88	0.28	7.1	0.75	19.0	—	—	1.94	49.3	2.62	66.5	1.12	28.4	1.28	32.5
5/8	0.625	18	1.06	26.9	1.12	28.4	0.12	3.0	0.875	22.22	0.31	7.9	1.00	25.4	—	—	2.12	53.8	3.00	76.2	1.12	28.4	1.59	40.4
3/4	0.750	16	1.06	26.9	1.12	28.4	0.12	3.0	0.875	22.22	0.31	7.9	1.00	25.4	—	—	2.12	53.8	3.00	76.2	1.12	28.4	1.59	40.4
STYLE 4[c]																								
No. 10	0.190	32	0.47	11.9	0.44	11.2	0.06	1.5	0.375	9.52	0.12	3.0	0.44	11.2	—	—	0.97	24.6	1.22	31.0	0.44	11.2	0.72	18.3
1/4	0.250	28	0.47	11.9	0.56	14.2	0.09	2.3	0.375	9.52	0.12	3.0	0.44	11.2	—	—	0.97	24.6	1.22	31.0	0.50	12.7	0.72	18.3
5/16	0.3125	24	0.53	13.5	0.69	17.5	0.09	2.3	0.438	11.12	0.16	4.1	0.50	12.7	—	—	1.12	28.4	1.41	35.8	0.56	14.2	0.81	20.6
3/8	0.375	24	0.69	17.5	0.88	22.3	0.09	2.3	0.500	12.70	0.19	4.8	0.62	15.8	—	—	1.38	35.0	1.69	42.9	0.75	19.0	1.03	26.2
7/16	0.4375	20	0.88	22.3	1.12	28.4	0.12	3.0	0.625	15.88	0.25	6.4	0.75	19.0	—	—	1.94	49.3	2.38	60.5	1.00	25.4	1.28	32.5
1/2	0.500	20	0.88	22.3	1.12	28.4	0.12	3.0	0.625	15.88	0.25	6.4	0.75	19.0	—	—	1.94	49.3	2.38	60.5	1.00	25.4	1.28	32.5
5/8	0.625	18	1.00	25.4	1.12	28.4	0.12	3.0	0.750	19.05	0.31	7.9	0.88	22.3	—	—	2.06	52.3	2.58	65.5	1.00	25.4	1.47	37.3

[a] These dimensions are given for design purposes only and are not intended for inspection.
[b] Type G Style 3 ball joints are furnished with ball studs and ball cavities (ball stud only on 5/8 and 3/4 in sizes) hardened to assure longer wear.
[c] Type G, Style 4 ball joints in all sizes are furnished with both ball studs and ball sockets hardened to assure longer wear.

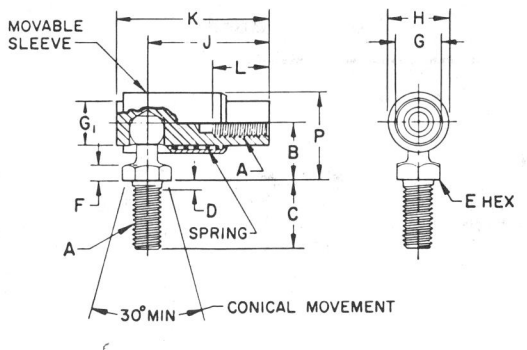

FIG. 9A—STYLE 1

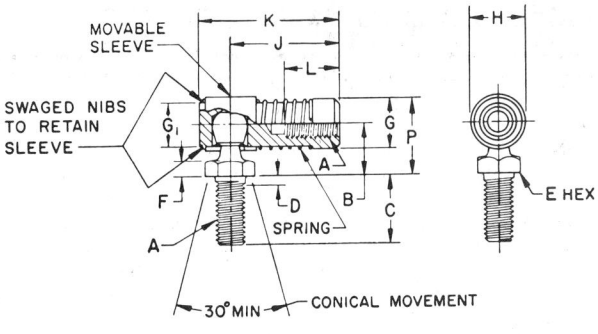

FIG. 9B—STYLE 2

FIG. 9—TYPE S DETACHABLE CONSTRUCTION

TABLE 5—DIMENSIONS OF TYPE S BALL JOINTS (FIGS. 9A AND 9B)

Nominal Ball Joint Size and Thread Diameter, A, in		Thds per in	B		C		D Max		E Hex		F Min		G Dia		G₁ Dia		H Dia		J		K		L Min Full Thread		Pa Max (Ref)	
			in	mm	in	mm	in	mm	in	mm	in	mm	in	mm	in	mm	in	mm	in	mm	in	mm	in	mm	in	mm
			±0.02	±0.5	±0.02	±0.5							±0.010	±0.25	±0.010	±0.25	±0.01	±0.3	±0.02	±0.5	±0.02	±0.5				
STYLE 1																										
No. 10	0.190	32	0.47	11.9	0.44	11.2	0.06	1.5	0.312	7.92	0.12	3.0	0.312	7.92	0.312	7.92	0.312	7.92	0.44	11.2	0.91	23.1	1.09	27.7	0.44	11.2	0.72	18.3
No. 12	0.216	32	0.47	11.9	0.44	11.2	0.06	1.5	0.312	7.92	0.12	3.0	0.312	7.92	0.312	7.92	0.312	7.92	0.44	11.2	0.91	23.1	1.09	27.7	0.44	11.2	0.72	18.3
1/4	0.250	28	0.47	11.9	0.56	14.2	0.06	1.5	0.312	7.92	0.12	3.0	0.312	7.92	0.312	7.92	0.312	7.92	0.44	11.2	0.91	23.1	1.09	27.7	0.50	12.7	0.72	18.3
5/16	0.3125	24	0.59	15.0	0.69	17.5	0.09	2.3	0.438	11.12	0.16	4.1	0.438	11.12	0.438	11.12	0.62	15.7	1.25	31.8	1.56	39.6	0.56	14.2	0.93	23.6
3/8	0.375	24	0.78	19.8	0.88	22.3	0.09	2.3	0.500	12.70	0.19	4.8	0.562	14.28	0.562	14.27	0.75	19.0	1.56	39.6	1.94	49.3	0.75	19.0	1.19	30.2
7/16	0.4375	20	0.97	24.6	1.12	28.4	0.12	3.0	0.625	15.88	0.25	6.4	0.750	12.70	0.750	19.05	1.00	25.4	2.03	51.6	2.53	64.3	1.00	25.4	1.50	38.1
1/2	0.500	20	0.97	24.6	1.12	28.4	0.12	3.0	0.625	15.88	0.25	6.4	0.750	12.70	0.750	19.05	1.00	25.4	2.03	51.6	2.53	64.3	1.00	25.4	1.50	38.1
5/8	0.625	18	1.12	28.4	1.12	28.4	0.12	3.0	0.750	19.05	0.31	7.9	0.875	22.22	0.875	22.22	1.12	28.4	2.31	58.7	2.88	73.2	1.00	25.4	1.71	43.4
STYLE 2																										
1/4	0.250	28	0.47	11.9	0.56	14.2	0.06	1.5	0.375	9.52	0.12	3.0	0.562	14.28	0.438	11.12	0.53	13.5	0.97	24.6	1.25	31.8	0.53	13.5	0.78	19.8
5/16	0.3125	24	0.53	13.5	0.69	17.5	0.09	2.3	0.438	11.12	0.16	4.1	0.625	15.88	0.500	12.70	0.59	15.0	1.12	28.4	1.45	36.8	0.59	15.0	0.87	22.1
3/8	0.375	24	0.69	17.5	0.88	22.3	0.09	2.3	0.500	12.70	0.19	4.8	0.750	19.05	0.625	15.88	0.75	19.0	1.38	35.0	1.75	44.4	0.81	20.6	1.09	27.7
1/2	0.500	20	0.88	22.3	1.12	28.4	0.12	3.0	0.625	15.88	0.28	7.1	0.938	23.82	0.750	19.05	0.89	22.6	1.94	49.3	2.38	60.5	1.12	28.4	1.39	35.3

aThese dimensions are given for design purposes only and are not intended for inspection.

1.1.2 The inclusion of dimensional data in this standard is not intended to imply that all the products described are stock production sizes. Consumers are requested to consult with manufacturers concerning availability of stock production parts.

1.2 Dimensions and Tolerances—Except for nominal sizes and thread designations which are inch values only, dimensions and tolerances are given in both U. S. customary and SI units, as designated. Tabulated dimensions shall apply to the finished parts, plated or otherwise processed, as specified by the user. Limits on hexagon or round bar shapes shall be within the commercial tolerance of the bar stock material from which the components are produced.

1.3 Threads—Unified Standard Class 2A external threads and Class 2B internal threads shall apply to plain finish (unplated) parts. For externally threaded components with additive finish, the maximum diameters of Class 2A may be exceeded by the amount of the allowance; that is, the basic diameters (Class 2A maximum diameters plus the allowance) apply to an externally threaded part after plating. For internally threaded components with additive finish, the Class 2B diameters apply after plating. See ANSI B 1.1.

1.3.1 External threads shall be chamfered to a diameter 0.01 in. (0.3 mm) less than the minor diameter to produce a length of chamfered or partial thread equivalent to 3/4 to 1 1/4 times the pitch (rounded to a three-place decimal).

1.3.2 Internal threads shall be countersunk 90 deg included angle to a diameter 0.01 in. (0.3 mm) greater than the major diameter of the thread (rounded to a two-place decimal).

1.4 Material

1.4.1 BALL JOINTS—Ball joints are normally made from low carbon free machining steel. The ball stud and mating plug components of Types A, AL, B, and C and the ball sockets on Type G, Styles 3 and 4, ball joints shall be case hardened unless otherwise specified. For special application, ball joints can be produced from alloy steel, corrosion resistant steel, brass, bronze, or other materials.

1.4.2 CUSHIONING DISCS—Cushioning discs shall be Neoprene, Buna N rubber, or equivalent material.

1.5 Finishes—Unless otherwise specified, carbon steel ball joints shall be furnished with cadmium or zinc protective finish and shall meet the requirements of 32 h salt spray test in accordance with ASTM B 117, Method of Salt Spray (Fog) Testing. At manufacturer's option, a subsequent chromate treatment may be used. Plated, hardened carbon steel components of ball joints (subject to hydrogen embrittlement) shall be baked or otherwise processed to obviate such embrittlement.

1.6 Lubrication—Unless otherwise specified by user, ball joints shall be supplied with ball sockets suitably lubricated in accordance with manufacturers practice.

1.7 Dust Covers—Where so specified by the user, Type G ball joints shall be supplied with an oil resistant rubber shield of such construction as to prevent dirt and dust from entering the ball cavity. However, shields for Style 3 are available in sizes 5/8 and 3/4 only.

1.8 Workmanship—Ball joints must be free from burrs, loose scale, sharp edges, and any other defects which might affect their serviceability.

SPHERICAL ROD ENDS—SAE J1120 SEP79

SAE Standard

Report of the Ball Joint Committee, approved July 1975, last revised September 1979.

1. General Specifications

1.1 Scope—This SAE Standard covers the general and dimensional data for industrial quality spherical rod ends commonly used on control linkages in automotive, marine, construction, and industrial equipment applications.

The rod ends described are available from several manufacturers within the range of the interchangeable specifications. The sliding contact spherical self-aligning bearing members (ball and socket) are available in a variety of materials in types shown. The load capacities and wear capabilities vary considerably with the design and fabrication. It is suggested that the manufacturers be consulted for recommendations for the type and design appropriate to particular applications.

1.2 Sizes—Spherical rod end sizes are normally specified by a number indicating the ball bore size in sixteenths of an inch (size 5 = $\frac{5}{16}$ bore). The housing threads (external or internal) used for mounting, as well as the stud thread if required, are equal in size to the nominal ball bore. Sizes larger than those listed are available in both standard and special configurations.

1.3 Threads—Unified Standard fine thread series (UNF) Class 2A external threads and Class 2B internal threads shall apply to plain finish (unplated) parts. For externally threaded components with additive finish, the maximum diameters of Class 2A may be exceeded by the amount of the allowance; that is, the basic diameters (Class 2A maximum diameters plus the allowance) apply to an externally threaded part after plating. For internally threaded components with additive finish, the Class 2B diameters apply after ϕ plating. See SAE J475 (ANSI B1.1-1974).

Housing threads left or right hand may be specified as required. Standard studs are threaded right hand.

External and internal threads must be chamfered to insure a clean start according to good industrial practice. Roll formed internal and external threads are preferred.

1.4 Material—Spherical rod end housing members are normally made from low carbon steel turned, forged, headed, or press stamped blanks.

Race and ball materials vary according to manufacturers preference for bearing materials.

For special applications spherical rod ends can be produced from alloy steel, corrosion resistant steel, brass, bronze or other materials. The charted combinations illustrate the preferred materials in each category available as standard.

Spherical rod ends are available with ball and race material options listed below:

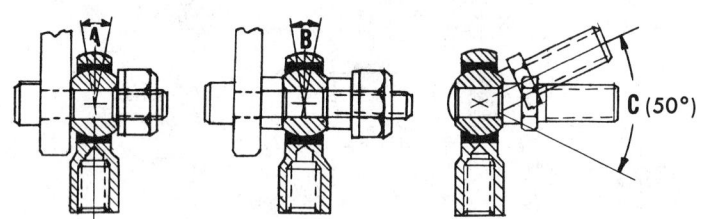

Rod End Size	Min A	Min B
3	10°	34°
4	14°	34°
5	12°	28°
6	10°	30°
7	14°	32°
8	10°	32°
10	14°	30°
12	14°	25°

FIG. 1—A—HOUSING STRIKES YOKE OR LEVER
B—WASHER OR SHOULDERED SHAFT WITH DIA "O" STRIKES RACE ID
C—STUD STRIKES RACE ID

Studs (Fig. 6) which may be secured in the bore of any of the ball variations are normally made from turned low carbon steel or headed blanks. Studs with greater strength to resist bending are also available as standard, employing high tensile bar stock or heat treatment during fabrication.

Ball studs which combine ball and stud as a single part are mild steel case hardened.

1.5 Angle of Misalignment—If a spherical rod end is mounted between the legs of a fork or clevis, the total misalignment angle will be limited by the diameter of the housing head as it contacts the legs. This angle varies from 18 deg to 34 deg in race type spherical rod ends and from 12 deg to 30 deg in raceless construction. Specific information for a given size and type should be requested from the manufacturer if this is a critical element of the application. See illustration, Fig. 1A.

If a spherical rod end is mounted on a shouldered shaft or with washers having a diameter equal to ball dimension "O" the shaft cone angle will vary from 25 deg to 34 deg. See illustration, Fig. 1B.

The use of a stud for mounting increases the limit of total misalignment to a minimum of 50 deg. See illustration, Fig. 1C.

1.6 Finishes—Unless otherwise specified, low carbon steel housings, races and studs shall be furnished with cadmium or zinc protective finish and shall meet the requirements of 32 h Salt Spray (Fog) Testing in accordance with ASTM B-117. At manufacturers option, a subsequent chromate treatment may be used. Black oxide treatment for studs may also be employed.

ϕ MATERIAL OPTIONS

Rod End	Housing	Race	Ball
Type A (Fig. 2)	Mild Steel, Alloy Steel, Stainless Steel, Hardened Steel, Aluminum Bronze, Brass	Sintered Phosphor Bronze	Hardened Sintered Nickel Steel, Oil Impregnated Case Hardened Steel, Tin Nickel Plated
		Wrought Bronze, Brass	Hardened Sintered Steel
		Mild Steel, Cad Plated	Hardened 52100 Steel, Chrome Plated Hardened Sintered Steel
		Hardened Steel	Hardened Sintered Nickel Steel, Oil Impregnated Sintered Bronze, Oil Impregnated Hardened 52100
Type B (Fig. 3)		Nylon Reinforced, Delrin, TFE Lined	Case Hardened Steel, Cad or Tin Nickel Plated Hardened Sintered Nickel Steel, Oil Impregnated Hardened 52100
Type C (Fig. 4)		None	Hardened 52100 Hardened Sintered Iron, Oil Impregnated Case Hardened Steel, Tin Nickel Plated
Type D (Fig. 5)		None	Mild Steel—Case Hardened, Cad Plated

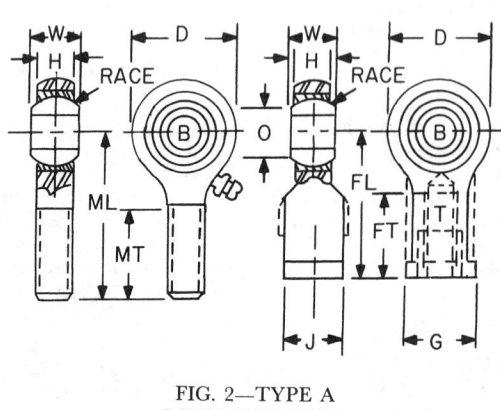

FIG. 2—TYPE A METALLIC RACE

Hardened steel races shall be black oxide treated and oiled. Non sintered balls and ball studs shall be plated according to manufacturer's preference for corrosion protection appropriate to their use as bearing elements.

1.7 Lubrication—Unless otherwise specified by the user spherical rod ends shall be supplied with ball sockets suitably lubricated in accordance with manufacturer's practice, including vacuum impregnation of self lubricating sintered bearing elements.

Grease fittings for supplemental lubrication are provided upon request for most types. Standard location is shown. Special locations at 12 o'clock and 3 o'clock positions are also available.

1.8 Workmanship—Industrial quality spherical rod ends must be free from burrs, loose scale, sharp edges, and any other defects.

1.9 Ball Bore Chamfer—Ball bores are chamfered at both faces to break the edge 0.005 in (0.13 mm) or up to a maximum of 0.03 in (0.8 mm) according to manufacturer's preference and method of fabrication. The user is cautioned against seating bolt heads against the ball face during mounting because bolt fillets under the head may distort or crack the ball. This is especially true of hex bolts and screws meeting ANSI B18.2.1-1972 specifications. The use of a washer or other suitable alternate is recommended.

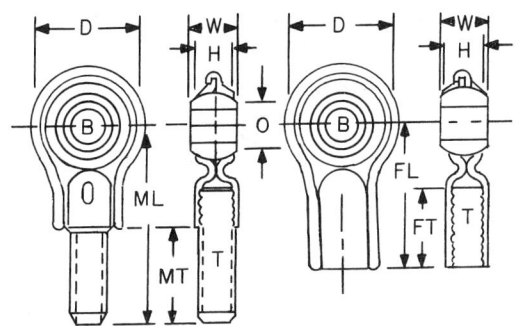

FIG. 5—TYPE D
RACELESS, STAMPED HOUSING

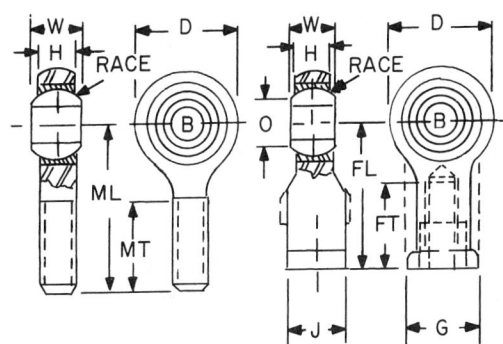

FIG. 3—TYPE B
MOLDED RACE

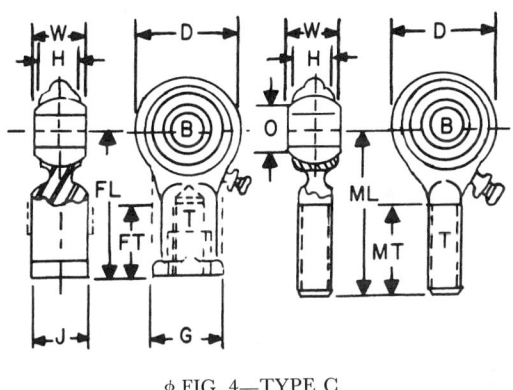

ɸ FIG. 4—TYPE C
RACELESS

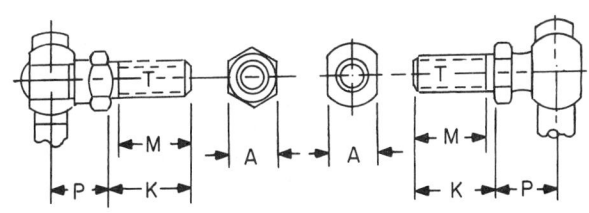

Rod End Size	A ±0.010	0.25	K ±0.03	0.8	M ±0.03	0.8	P ±0.04	±1.0	T Nominal Thread Size
	in	mm	in	mm	in	mm	in	mm	
3	0.312	7.92	0.50	12.7	0.44	11.2	0.48	12.2	10-32
4	0.375	9.52	0.56	14.2	0.50	12.7	0.48	12.2	1/4-28
5	0.437	11.10	0.69	17.5	0.59	15.0	0.54	13.7	5/16-24
6	0.500	12.70	0.89	22.6	0.81	20.6	0.65	16.5	3/8-24
7	0.625	15.88	1.09	27.7	0.97	24.6	0.84	21.3	7/16-20
8	0.625	15.88	1.12	28.4	1.00	25.4	0.88	22.4	1/2-20
10	0.750	19.05	1.50	38.1	1.38	35.1	1.00	25.4	5/8-18
12	1.000	25.40	1.81	46.0	1.63	41.4	1.19	30.2	3/4-16

FIG. 6—STUD ASSEMBLED TO BALLS HAVING ANY STANDARD MATERIAL OPTIONS. ONE PIECE BALL STUD, LOW CARBON STEEL, CASE HARDENED

ɸ TABLE 1—DIMENSIONS FOR TYPE A ROD ENDS (FIG. 2)

Rod End Size	B +0.0025 −0.0005	+0.064 −0.013	D Max		G Ref		H Ref		J ±0.015	±0.38	T Nominal Thread Size		W ±0.005	±0.13	FL +0.06 −0.03	+1.5 −0.8	FT ±0.06	±1.5	ML +0.06 −0.03	+1.5 −0.8	MT ±0.06	±1.5	Ball Dia Ref		O Ref	
	in	mm	in	mm	in	mm	in	mm	in	mm	in	mm	in	mm	in	mm	in	mm	in	mm	in	mm	in	mm	in	mm
3	0.1900	4.826	0.76	19.3	0.41	10.4	0.25	6.4	0.312	7.92	10-32		0.312	7.92	1.06	26.9	0.56	14.2	1.25	31.8	0.75	19.0	0.44	11.2	0.31	7.9
4	0.2500	6.350	0.89	22.6	0.47	11.9	0.28	7.1	0.375	9.52	1/4-28		0.375	9.52	1.31	33.3	0.75	19.0	1.56	39.6	1.00	25.4	0.51	13.0	0.35	8.9
5	0.3125	7.938	1.01	25.7	0.50	12.7	0.34	8.6	0.438	11.12	5/16-24		0.438	11.12	1.38	35.1	0.75	19.0	1.88	47.8	1.25	31.8	0.62	15.7	0.45	11.4
6	0.3750	9.525	1.11	28.2	0.69	17.5	0.41	10.4	0.562	14.27	3/8-24		0.500	12.70	1.62	41.1	0.94	23.9	1.94	49.3	1.25	31.8	0.72	18.3	0.52	13.2
7	0.4375	11.112	1.20	30.5	0.75	19.0	0.44	11.2	0.625	15.88	7/16-20		0.562	14.27	1.81	46.0	1.06	26.9	2.12	53.8	1.38	35.1	0.81	20.6	0.59	15.0
8	0.5000	12.700	1.39	35.3	0.88	22.4	0.50	12.7	0.750	19.05	1/2-20		0.625	15.88	2.12	53.8	1.19	30.2	2.44	62.0	1.50	38.1	0.94	23.9	0.70	17.8
10	0.6250	15.875	1.57	39.9	1.00	25.4	0.56	14.2	0.875	22.22	5/8-18		0.750	19.05	2.50	63.5	1.50	38.1	2.62	66.5	1.62	41.1	1.12	28.4	0.81	20.6
12	0.7500	19.050	1.82	46.2	1.12	28.4	0.69	17.5	1.000	25.40	3/4-16		0.875	22.22	2.88	73.2	1.75	44.4	2.88	73.2	1.75	44.4	1.32	33.5	1.02	25.9

ɸ TABLE 2—DIMENSIONS FOR TYPE B ROD ENDS (FIG. 3)

Rod End Size	B +0.0025 −0.0005		B +0.064 −0.013		D Max		G Ref		H Ref		J ±0.015		T ±0.38		W Nominal Thread Size		W ±0.005		W ±0.13		FL +0.06 −0.03		FL +1.5 −0.8		FT ±0.06		FT ±1.5		ML +0.06 −0.03		ML +1.5 −0.8		MT ±0.06		MT ±1.5		Ball Dia Ref		O Ref	
	in	mm	in	mm	in	mm	in	mm	in	mm	in	mm	in	mm			in	mm	in	mm	in	mm	in	mm	in	mm	in	mm	in	mm	in	mm	in	mm
3	0.1900	4.826	0.76	19.3	0.41	10.4	0.25	6.4	0.312	7.92	10–32		0.312	7.92	1.06	26.9	0.56	14.2	1.25	31.8	0.75	19.0	0.44	11.2	0.31	7.9
4	0.2500	6.350	0.89	22.6	0.47	11.9	0.28	7.1	0.375	9.52	1/4–28		0.375	9.52	1.31	33.3	0.75	19.0	1.56	39.6	1.00	25.4	0.51	13.0	0.35	8.9
5	0.3125	7.938	1.01	25.7	0.50	12.7	0.34	8.6	0.438	11.12	5/16–24		0.438	11.12	1.38	35.1	0.75	19.0	1.88	47.8	1.25	31.8	0.62	15.7	0.45	11.4
6	0.3750	9.525	1.11	28.2	0.69	17.5	0.41	10.4	0.562	14.27	3/8–24		0.500	12.70	1.62	41.1	0.94	23.9	1.94	49.3	1.25	31.8	0.72	18.3	0.52	13.2
7	0.4375	11.112	1.20	30.5	0.75	19.0	0.44	11.2	0.625	15.88	7/16–20		0.562	14.27	1.81	46.0	1.06	26.9	2.12	53.8	1.38	35.1	0.81	20.6	0.59	15.0
8	0.5000	12.700	1.39	35.3	0.88	22.4	0.50	12.7	0.750	19.05	1/2–20		0.625	15.88	2.12	53.8	1.19	30.2	2.44	62.0	1.50	38.1	0.94	23.9	0.70	17.8
10	0.6250	15.875	1.51	38.4	1.00	25.4	0.56	14.2	0.875	22.22	5/8–18		0.750	19.05	2.50	63.5	1.50	38.1	2.62	66.5	1.62	41.1	1.12	28.4	0.81	20.6

ɸ TABLE 3—DIMENSIONS FOR TYPE C ROD ENDS (FIG. 4)

Rod End Size	B +0.0025 −0.0005		B +0.064 −0.013		D Max		G Ref		H Ref		J ±0.015		T ±0.38		W Nominal Thread Size		W ±0.005		W ±0.13		FL +0.06 −0.03		FL +1.5 −0.8		FT ±0.06		FT ±1.5		ML +0.06 −0.03		ML +1.5 −0.8		MT ±0.06		MT ±1.5		Ball Dia Ref		O Ref	
	in	mm	in	mm	in	mm	in	mm	in	mm	in	mm	in	mm			in	mm	in	mm	in	mm	in	mm	in	mm	in	mm	in	mm	in	mm	in	mm
3	0.1900	4.826	0.62	15.7	0.41	10.4	0.25	6.4	0.312	7.92	10–32		0.312	7.92	1.06	26.9	0.50	12.7	1.25	31.8	0.75	19.0	0.45	11.4	0.35	8.9
4	0.2500	6.350	0.76	19.3	0.47	11.9	0.28	7.1	0.375	9.52	1/4–28		0.375	9.52	1.31	33.3	0.69	17.5	1.56	39.6	1.00	25.4	0.53	13.5	0.42	10.7
5	0.3125	7.938	0.88	22.4	0.50	12.7	0.34	8.6	0.438	11.12	5/16–24		0.438	11.12	1.38	35.1	0.69	17.5	1.88	47.8	1.25	31.8	0.64	16.3	0.49	12.4
6	0.3750	9.525	1.01	25.7	0.69	17.5	0.41	10.4	0.562	14.27	3/8–24		0.500	12.70	1.62	41.1	0.81	20.6	1.94	49.3	1.25	31.8	0.72	18.3	0.51	13.2
7	0.4375	11.112	1.12	28.4	0.75	19.0	0.44	11.2	0.625	15.88	7/16–20		0.562	14.27	1.81	46.0	0.94	23.9	2.12	53.8	1.38	35.1	0.81	20.6	0.58	14.7
8	0.5000	12.700	1.31	33.3	0.88	22.4	0.50	12.7	0.750	19.05	1/2–20		0.625	15.88	2.12	53.8	1.06	26.9	2.44	62.0	1.50	38.1	0.96	24.4	0.79	20.1
10	0.6250	15.875	1.50	38.1	1.00	25.4	0.56	14.2	0.875	22.22	5/8–18		0.750	19.05	2.50	63.5	1.38	35.1	2.62	66.5	1.62	41.1	1.16	29.5	0.92	23.4
12	0.7500	19.050	1.75	44.4	1.12	28.4	0.69	17.5	1.000	25.40	3/4–16		0.875	22.22	2.88	73.2	1.56	39.6	2.88	73.2	1.75	44.4	1.34	34.0	1.06	26.9

TABLE 4—DIMENSIONS FOR TYPE D ROD ENDS (FIG. 5)

Rod End Size	B +0.0025 −0.0005		B +0.064 −0.013		D Max		H Ref		T Nominal Thread Size		W ±0.005		W ±0.13		FL ±0.06		FL ±1.5		FT ±0.06		FT ±1.5		ML ±0.09		ML ±2.3		MT ±0.06		MT ±1.5		Ball Dia Ref		O Ref	
	in	mm	in	mm	in	mm	in	mm			in	mm	in	mm	in	mm	in	mm	in	mm	in	mm	in	mm	in	mm	in	mm
3	0.1900	4.826	0.78	18.8	0.25	6.4	10–32		0.312	7.92	1.06	26.9	0.56	14.2	1.50	38.1	0.75	19.0	0.44	11.2	0.31	7.9
4	0.2500	6.350	0.88	22.4	0.29	7.1	1/4–28		0.375	9.52	1.31	33.3	0.75	19.0	1.86	47.2	1.00	25.4	0.52	13.2	0.35	8.9
5	0.3125	7.938	1.05	26.7	0.31	7.9	5/16–24		0.438	11.12	1.38	35.1	0.75	19.0	2.25	57.2	1.25	31.8	0.62	15.7	0.45	11.4
6	0.3750	9.525	1.16	29.5	0.41	10.4	3/8–24		0.500	12.70	1.62	41.1	0.94	23.9	2.39	60.7	1.25	31.8	0.72	18.3	0.52	13.2
7	0.4375	11.112	1.37	34.8	0.44	11.2	7/16–20		0.562	14.27	1.91	48.5	1.06	26.9	2.74	69.6	1.38	35.1	0.81	20.6	0.59	15.0
8	0.5000	12.700	1.51	38.4	0.50	12.7	1/2–20		0.625	15.88	2.12	53.8	1.19	30.2	3.04	77.2	1.50	38.1	0.94	23.9	0.70	17.8

METRIC SPHERICAL ROD ENDS—SAE J1259 APR80

SAE Standard

16.15

Report of the Ball Joint Spherical Rod End Committee, approved April 1980.

1. General Specifications

1.1 Scope—This SAE Standard covers the general and dimensional data for industrial quality spherical rod ends commonly used on control linkages in metric automotive, marine, construction, and industrial equipment applications.

The rod ends described are available from several manufacturers within the range of the interchangeable specifications. The sliding contact spherical self-aligning bearing members (ball and socket) are available in a variety of materials in types shown. The load capacities and wear capabilities vary considerably with the design and fabrication. It is suggested that the manufacturers be consulted for recommendations for the type and design appropriate to particular applications.

1.2 Dimensions—All dimensions are in millimeters. See SAE J1120 for the U.S. Customary unit specification for spherical rod ends.

1.3 Sizes—Spherical rod end sizes are normally specified by a number indicating the ball bore in millimeters (size 5 = 5 mm). The housing threads (external or internal) used for mounting, as well as the stud thread if required, are equal in size to the nominal ball bore. Sizes larger than those listed are available in both standard and special configurations.

1.4 Threads—Thread form, diameter, and associated pitches are in accordance with ISO 965/II and ANSI B1.13, tolerance class 6g external and 6H internal.

Threads shall be right hand unless otherwise specified. Threads must be chamfered to insure a clean start according to good industrial practice.

1.5 Material—Spherical rod end housing members are normally made from low carbon steel turned, forged, or headed.

Race and ball materials vary according to manufacturer's preference for bearing materials.

For special applications spherical rod ends can be produced from alloy steel, corrosion resistant steel, brass, bronze, or other materials. The charted combinations illustrate the preferred materials in each category available as standard.

Spherial rod ends are available with the ball and race material options listed below:

1.6 Angle of Misalignment—If a spherical rod end is mounted between the legs of a fork or clevis, the total misalignment angle will be limited by that portion of the housing head which contacts the legs. This angle varies from 12–18 deg. Specific information for a given size and type should be requested from the manufacturer if this is a critical element of the application. See illustration, Fig. 1A.

If a spherical rod end is mounted on a shouldered shaft or with washers having a diameter equal to ball dimension "O" the shaft cone angle will vary from 24–30 deg. See illustration, Fig. 1B.

The use of a stud for mounting increases the limit of total misalignment to a minimum of 44 deg. See illustration, Fig. 1C.

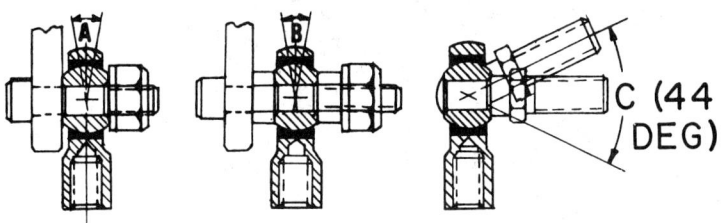

Rod End Size	Min A Deg	Min B Deg
5	13	24
6	12	24
8	14	26
10	14	26
12	14	26
14	18	30
16	17	30
20	17	28

FIG. 1—A—HOUSING STRIKES YOKE OR LEVER
B—WASHER OR SHOULDERED SHAFT WITH DIA "O" STRIKES RACE ID
C—STUD STRIKES RACE ID

MATERIAL OPTIONS

Rod End	Housing	Race	Ball
Type A (Fig. 2)	Mild Steel, Alloy Steel, Stainless Steel, Hardened Steel, Aluminum Bronze, Brass	Sintered Phosphor Bronze	Hardened Sintered Nickel Steel, Oil Impregnated Case Hardened Steel, Tin Nickel Plated
		Wrought Bronze, Brass	Hardened Sintered Steel
		Mild Steel, Cad Plated	Hardened 52100 Steel, Chrome Plated Hardened Sintered Steel
		Hardened Steel	Hardened Sintered Nickel Steel, Oil Impregnated Sintered Bronze, Oil Impregnated Hardened 52100
Type B (Fig. 3)		Nylon Reinforced, Delrin, TFE Lined	Case Hardened Steel, Cad or Tin Nickel Plated Hardened Sintered Nickel Steel, Oil Impregnated Hardened 52100
Type C (Fig. 4)		None	Hardened 52100 Hardened Sintered Iron, Oil Impregnated Case Hardened Steel, Tin Nickel Plated

Studs (Fig. 5) which may be secured in the bore of any of the ball variations are normally made from turned low carbon steel or headed blanks. Studs with greater strength to resist bending are also available by agreement between user and manufacturer.

Ball studs which combine ball and stud as a single part are mild steel case hardened.

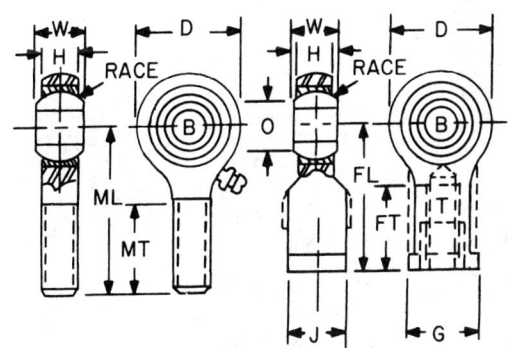

FIG. 2—TYPE A—METALLIC RACE

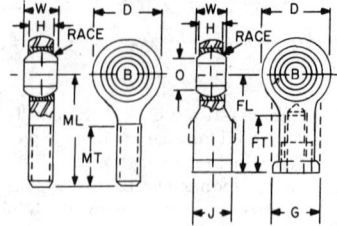

FIG. 3—TYPE B—MOLDED RACE

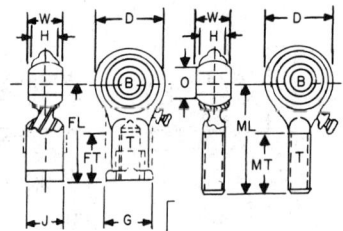

FIG. 4—TYPE C—RACELESS

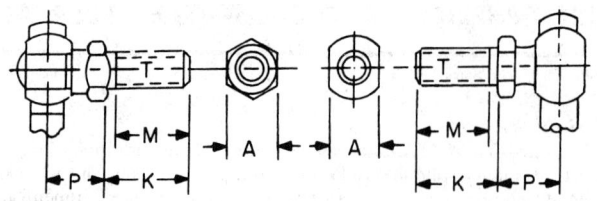

Rod End Size	A Ref	K ±0.25	M Min	P ±1	T Nominal Thread Size
5	8	13.0	10.0	9.0	M5 x 0.80
6	10	14.0	11.0	10.0	M6 x 1.00
8	12	17.5	14.0	12.0	M8 x 1.25
10	14	23.0	19.5	16.5	M10 x 1.50
12	16	28.5	24.5	19.5	M12 x 1.75
14	20	33.0	29.0	20.5	M14 x 2.00
16	22	38.0	34.0	24.0	M16 x 2.00
20	25	46.0	40.0	28.5	M20 x 1.50

FIG. 5—STUDDED ASSEMBLIES

1.7 Finishes—Unless otherwise specified, low carbon steel housing, races, and studs shall be furnished with cadmium or zinc protective finish and shall meet the requirements of 32 h Salt Spray (Fog) Testing in accordance with ASTM B 117. At manufacturer's option, a subsequent chromate treatment may be used. Black oxide treatment for studs may also be employed.

Non-sintered balls and ball studs shall be plated according to manufacturer's preference for corrosion protection appropriate to their use as bearing elements.

1.8 Lubrication—Unless otherwise specified by the user, spherical rod ends shall be supplied with ball sockets suitably lubricated in accordance with manufacturer's practice, including vacuum impregnation of self-lubrication sintered bearing elements.

Grease fittings for supplemental lubrication are provided upon request for most types. Standard location is shown. Special locations at 12 o'clock and 3 o'clock position are also available.

1.9 Workmanship—Industrial quality spherical rod ends must be free from burrs, loose scale, sharp edges, and any other defects.

1.10 Ball Bore Chamfer—Ball bores are chamfered at both faces to break the edge 0.13 mm or up to a maximum of 0.8 mm according to manufacturer's preference and method of fabrication. The user is cautioned against seating bolt heads against the ball face during mounting because bolt fillets under the head may distort or crack the ball. The use of a washer or other suitable alternate is recommended.

TABLE 1—DIMENSIONS FOR ROD ENDS—TYPE A (FIG. 2), TYPE B (FIG. 3), AND TYPE C (FIG. 4)

Rod End Size	B +0.07 −0.00	D Max	G ±0.25	H ±0.15	J Ref	T Nominal Thread Size	W ±0.15	FL Min	FT Min	ML Min	MT Min	Ball Dia Ref	O Ref
5	5	18	11	6.00	9.0	M5 x 0.80	8	26	9	32	19	11.1	7.7
6	6	22	13	6.75	10.0	M6 x 1.00	9	29	11	35	21	12.7	8.9
8	8	26	16	9.00	12.5	M8 x 1.25	12	35	15	41	24	15.8	10.4
10	10	30	19	10.50	15.0	M10 x 1.50	14	42	19	47	28	19.1	12.9
12	12	34	22	12.00	17.5	M12 x 1.75	16	49	21	54	32	22.2	15.4
14	14	38	25	13.50	20.0	M14 x 2.00	19	56	24	59	35	25.4	16.8
16	16	42	27	15.00	22.0	M16 x 2.00	21	63	27	65	39	28.6	19.3
20	20	50	34	18.00	27.5	M20 x 1.50	25	76	32	77	46	34.9	24.3

17 Splines

INVOLUTE SPLINES AND INSPECTIONS—SAE J498c — SAE Standard

This report has been published separately as ANSI B92.1, available from Society of Automotive Engineers, Inc., 400 Commonwealth Drive, Warrendale, Penna. 15096.

PARALLEL SIDE SPLINES FOR SOFT BROACHED HOLES IN FITTINGS—SAE J499a — SAE Standard

Report of Broaches Division approved January 1914, revised by Shaft Fittings Division March 1920, and reviewed January 1936. Last revised by ANSI B92 Committee—Involute Splines and Inspection—October 1975.

This Information Report along with SAE J500 and J501 is generally understood to be technically obsolete for the design of new applications. However, it is listed for those existing applications where it may be required. For the design of new applications, consult ANSI B92.1-1970—Involute Splines and Inspections Standard.

[The dimensions, given in inches, apply only to soft broached holes. The shaft dimensions depend upon the shape and material of the parts, their heat treatment, and methods of machining to give the required fit. The method and amount of "breaking" sharp corners and edges also depend upon the conditions and requirements of each application.

The formula for theoretical torque capacity (pressure on sides of spline) in inch-pounds per inch of bearing length (L) and at 1000 psi pressure is:

T = Torque = 1000 x No. of splines x mean radius x h x L

The tolerances allowed are for good construction and may be readily maintained by usual broaching methods. The tolerances selected for the large and small diameters will depend upon whether the fit between the mating parts, as finally made, is on the large or the small diameter. The other diameter, being designed for clearance may have a wider manufacturing tolerance. If the final fit between the parts is on only the sides of the spline, wider tolerances may be permitted on both the large and small diameters.]

Radii on corners of splines are not to exceed 0.015 in.

Splines shall not be more than 0.006 in per ft out of parallel with respect to the axis of the shaft.

No allowance is made for radii on corners or for clearance. Dimensions are intended to apply to only the soft broached hole. Allowance must be made for machining.

For values of D, W, d, h, and T for four-, six-, ten-, and sixteen-spline fittings, see Tables 2, 3, 4, and 5, respectively.

TABLE 1—W, h, AND d, IN TERMS OF LARGE DIAMETER, D

No. of Splines	W For All Fits	A Permanent Fit		B To Slide when Not under Load		C To Slide under Load	
		h	d	h	d	h	d
4	0.241[a]	0.075	0.850	0.125	0.750	—	—
6	0.250	0.050	0.900	0.075	0.850	0.100	0.800
10	0.156	0.045	0.910	0.070	0.860	0.095	0.810
16	0.098	0.045	0.910	0.070	0.860	0.095	0.810

[a] Four splines, for fits A and B only.

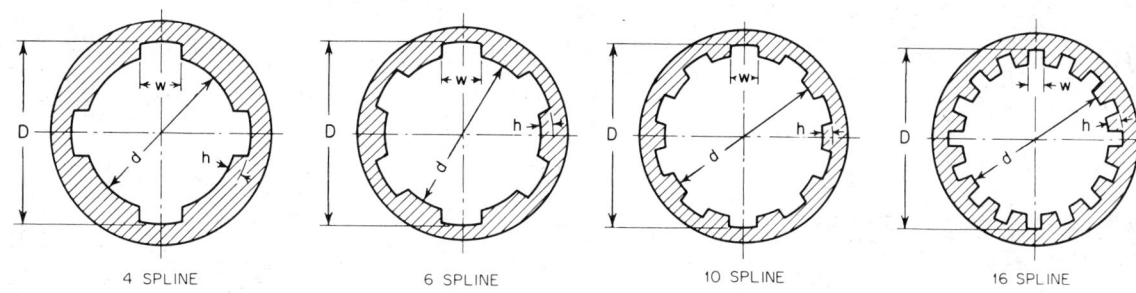

FIG. 1—DIMENSIONS FOR 4, 6, 10, AND 16 SPLINE FITTINGS (SEE TABLE 1)

TABLE 2—FOUR SPLINE FITTINGS

Nominal Dia	For All Fits D		W		4A, Permanent Fit d		h		T	4B, To Slide when Not under Load d		h		T
	Min	Max	Min	Max	Min	Max	Min	Max		Min	Max	Min	Max	
3/4	0.749	0.750	0.179	0.181	0.636	0.637	0.055	0.056	78	0.561	0.562	0.093	0.094	123
7/8	0.874	0.875	0.209	0.211	0.743	0.744	0.065	0.066	107	0.655	0.656	0.108	0.109	167
1	0.999	1.000	0.239	0.241	0.849	0.850	0.074	0.075	139	0.749	0.750	0.124	0.125	219
1-1/8	1.124	1.125	0.269	0.271	0.955	0.956	0.083	0.084	175	0.843	0.844	0.140	0.141	277
1-1/4	1.249	1.250	0.299	0.301	1.061	1.062	0.093	0.094	217	0.936	0.937	0.155	0.156	341
1-3/8	1.374	1.375	0.329	0.331	1.168	1.169	0.102	0.103	262	1.030	1.031	0.171	0.172	414
1-1/2	1.499	1.500	0.359	0.361	1.274	1.275	0.111	0.112	311	1.124	1.125	0.186	0.187	491
1-5/8	1.624	1.625	0.389	0.391	1.380	1.381	0.121	0.122	367	1.218	1.219	0.202	0.203	577
1-3/4	1.749	1.750	0.420	0.422	1.486	1.487	0.130	0.131	424	1.311	1.312	0.218	0.219	670
2	1.998	2.000	0.479	0.482	1.698	1.700	0.148	0.150	555	1.498	1.500	0.248	0.250	875
2-1/4	2.248	2.250	0.539	0.542	1.910	1.912	0.167	0.169	703	1.685	1.687	0.279	0.281	1106
2-1/2	2.498	2.500	0.599	0.602	2.123	2.125	0.185	0.187	865	1.873	1.875	0.310	0.312	1365
3	2.998	3.000	0.720	0.723	2.548	2.550	0.223	0.225	1249	2.248	2.250	0.373	0.375	1969

TABLE 3—SIX SPLINE FITTINGS

Nominal Dia	For All Fits D		W		6A, Permanent Fit d		T	6B, To Slide when Not under Load d		T	6C, To Slide when under Load c		T
	Min	Max	Min	Max	Min	Max		Min	Max		Min	Max	
3/4	0.749	0.750	0.186	0.188	0.674	0.675	80	0.637	0.638	117	0.599	0.600	152
7/8	0.874	0.875	0.217	0.219	0.787	0.788	109	0.743	0.744	159	0.699	0.700	207
1	0.999	1.000	0.248	0.250	0.899	0.900	143	0.849	0.850	208	0.799	0.800	270
1-1/8	1.124	1.125	0.279	0.281	1.012	1.013	180	0.955	0.956	263	0.899	0.900	342
1-1/4	1.249	1.250	0.311	0.313	1.124	1.125	223	1.062	1.063	325	0.999	1.000	421
1-3/8	1.374	1.375	0.342	0.344	1.237	1.238	269	1.168	1.169	393	1.099	1.100	510
1-1/2	1.499	1.500	0.373	0.375	1.349	1.350	321	1.274	1.275	468	1.199	1.200	608
1-5/8	1.624	1.625	0.404	0.406	1.462	1.463	376	1.380	1.381	550	1.299	1.300	713
1-3/4	1.749	1.750	0.436	0.438	1.574	1.575	436	1.487	1.488	637	1.399	1.400	827
2	1.998	2.000	0.497	0.500	1.798	1.800	570	1.698	1.700	833	1.598	1.600	1080
2-1/4	2.248	2.250	0.560	0.563	2.023	2.025	721	1.911	1.913	1052	1.798	1.800	1367
2-1/2	2.498	2.500	0.622	0.625	2.248	2.250	891	2.123	2.125	1300	1.998	2.000	1688
3	2.998	3.000	0.747	0.750	2.698	2.700	1283	2.548	2.550	1873	2.398	2.400	2430

TABLE 4—TEN SPLINE FITTINGS

Nominal Dia	For All Fits D		W		10A, Permanent Fit d		T	10B, To Slide when Not under Load d		T	10C, To Slide when under Load d		T
	Min	Max	Min	Max	Min	Max		Min	Max		Min	Max	
3/4	0.749	0.750	0.115	0.117	0.682	0.683	120	0.644	0.645	183	0.607	0.608	241
7/8	0.874	0.875	0.135	0.137	0.795	0.796	165	0.752	0.753	248	0.708	0.709	329
1	0.999	1.000	0.154	0.156	0.909	0.910	215	0.859	0.860	326	0.809	0.810	430
1-1/8	1.124	1.125	0.174	0.176	1.023	1.024	271	0.967	0.968	412	0.910	0.911	545
1-1/4	1.249	1.250	0.193	0.195	1.137	1.138	336	1.074	1.075	508	1.012	1.013	672
1-3/8	1.374	1.375	0.213	0.215	1.250	1.251	406	1.182	1.183	614	1.113	1.114	813
1-1/2	1.499	1.500	0.232	0.234	1.364	1.365	483	1.289	1.290	732	1.214	1.215	967
1-5/8	1.624	1.625	0.252	0.254	1.478	1.479	566	1.397	1.398	860	1.315	1.316	1135
1-3/4	1.749	1.750	0.271	0.273	1.592	1.593	658	1.504	1.505	997	1.417	1.418	1316
2	1.998	2.000	0.309	0.312	1.818	1.820	860	1.718	1.720	1302	1.618	1.620	1720
2-1/4	2.248	2.250	0.348	0.351	2.046	2.048	1088	1.933	1.935	1647	1.821	1.823	2176
2-1/2	2.498	2.500	0.387	0.390	2.273	2.275	1343	2.148	2.150	2034	2.023	2.025	2688
3	2.998	3.000	0.465	0.468	2.728	2.730	1934	2.578	2.580	2929	2.428	2.430	3869
3-1/2	3.497	3.500	0.543	0.546	3.183	3.185	2632	3.007	3.010	3987	2.832	2.835	5266
4	3.997	4.000	0.621	0.624	3.637	3.640	3438	3.437	3.440	5208	3.237	3.240	6878
4-1/2	4.497	4.500	0.699	0.702	4.092	4.095	4351	3.867	3.870	6591	3.642	3.645	8705
5	4.997	5.000	0.777	0.780	4.547	4.550	5371	4.297	4.300	8137	4.047	4.050	10746
5-1/2	5.497	5.500	0.855	0.858	5.002	5.005	6500	4.727	4.730	9846	4.452	4.455	13003
6	5.997	6.000	0.933	0.936	5.457	5.460	7735	5.157	5.160	11718	4.857	4.860	15475

TABLE 5—SIXTEEN SPLINE FITTINGS

Nominal Dia	For All Fits D		W		16A, Permanent Fit d		T	16B, To Slide when Not under Load d		T	16C, To Slide when under Load d		T
	Min	Max	Min	Max	Min	Max		Min	Max		Min	Max	
2	1.997	2.000	0.193	0.196	1.817	1.820	1375	1.717	1.720	2083	1.617	1.620	2751
2-1/2	2.497	2.500	0.242	0.245	2.273	2.275	2149	2.147	2.150	3255	2.022	2.025	4299
3	2.997	3.000	0.291	0.294	2.727	2.730	3094	2.577	2.580	4687	2.427	2.430	6190
3-1/2	3.497	3.500	0.340	0.343	3.182	3.185	4212	3.007	3.010	6378	2.832	2.835	8426
4	3.997	4.000	0.389	0.392	3.637	3.640	5501	3.437	3.440	8333	3.237	3.240	11005
4-1/2	4.497	4.500	0.438	0.441	4.092	4.095	6962	3.867	3.870	10546	3.642	3.645	13928
5	4.997	5.000	0.487	0.490	4.547	4.550	8595	4.297	4.300	13020	4.047	4.050	17195
5-1/2	5.497	5.500	0.536	0.539	5.002	5.005	10395	4.727	4.730	15754	4.452	4.455	20806
6	5.997	6.000	0.585	0.588	5.457	5.460	12377	5.157	5.160	18749	4.857	4.860	24760

SERRATED SHAFT ENDS—SAE J500

SAE Recommended Practice

Report of Parts and Fittings Division approved 1922 and last revised by Parts and Fittings Committee June 1955.

This Recommended Practice is intended for service only. Use SAE Standard, Involute Splines, Serrations, and Inspection—SAE J498 for new applications.

STRAIGHT SHAFTS

N = Number of serrations.

b = Included angle of the space in the hole, and the tooth on the shaft.

The pitch diameter (PD) and hole are basic.

The pitch line is midway between the inner and outer sharp points.

The minimum hole with maximum shaft as measured across wires in Table 1, produce basic (no clearance) fit.

The wire diameter in Table 1 is the diameter that will bear on the pitch line.

Tolerance for diameter across wires = −0.001 on tooth thickness of hole and shaft sizes from 1/8 to 1-3/4 in., and −0.0015 on sizes from 2 to 3 in., inclusive.

Tooth thickness on the shaft may be varied from the tolerance given, to secure desired fit.

Wc = Constant to be added to measurement across wires for shaft, subtracted for hole, for each 0.001-in. increase in wire diameter used over wire size in Table 1.

When serrations are hobbed, the sides of teeth are involute. This departure from flat sides is slight and is ignored.

FORMULAS

1. Diameter over sharp points

   (OD) = 1.0476479 PD for 36 serrations.
   = 1.0349592 PD for 48 serrations.

2. Diameter under sharp points

   (RD) = 0.9523521 PD for 36 serrations.
   = 0.9650408 PD for 48 serrations.

3. Diameter of wire that will bear on pitch line of hole

   (W_h) = 0.05309792 PD for 36 serrations.
   = 0.04133332 PD for 48 serrations.

   Diameter of wire that will bear on pitch line of shaft

   (W_s) = 0.06585005 PD for 36 serrations.
   = 0.0485955 PD for 48 serrations.

4. Measurement across wires

   for hole = 0.9119441 PD for 36 serrations.
   = 0.9309375 PD for 48 serrations.

   Measurement across wires

   for shaft = 1.1113285 PD for 36 serrations.
   = 1.0823601 PD for 48 serrations.

5. Tolerance constant for measurement across wires per 0.001-in. tooth

   thickness for hole $= 0.001 \dfrac{\sin 45}{\sin\left(45 - \dfrac{180}{N}\right)}$

   for shaft $= 0.001 \dfrac{\sin\left(45 - \dfrac{180}{N}\right)}{\sin 45}$

6. Wire diameter constant for use of other diameter wire than that which bears on pitch line, (W_c)

   $= 0.001 + \dfrac{0.001}{\sin\left(45 - \dfrac{180}{N}\right)}$ for hole.

   $= 0.001 + \dfrac{0.001}{\cos 45}$ for shaft.

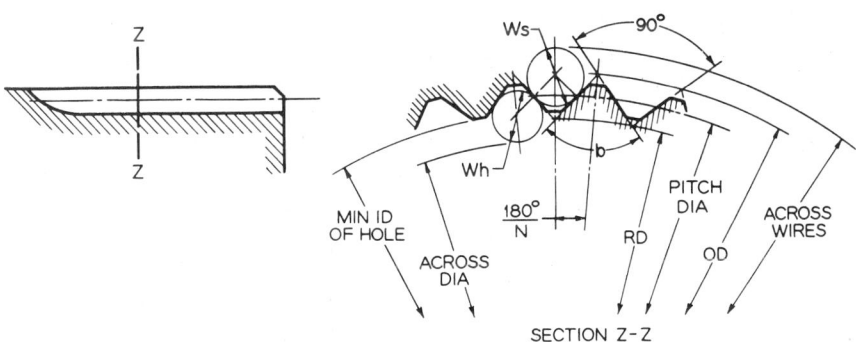

TABLE 1—DIMENSIONS OF HOLES AND SHAFTS, IN.

Hole and Shaft						Hole							Shaft					
Nominal Dia	Pitch Dia	N	b, Deg	Theoretical Dia of Points		Large Dia, min	Small Dia		Dia Across Wires		Wire Size, Wh	Wc	Root Dia, max	Outside Dia		Dia Across Wires		Wire Size, Ws	Wc
				OD	RD		Max	Min	Min	Max				Max	Min	Max	Min		
1/8	0.122	36	80	0.1278	0.1162	0.125	0.118	0.117	0.1113	0.1124	0.0065	0.0026	0.116	0.124	0.123	0.1356	0.1347	0.0080	0.0024
3/16	0.182	36	80	0.1907	0.1733	0.187	0.176	0.175	0.1660	0.1671	0.0097	0.0026	0.174	0.186	0.185	0.2023	0.2014	0.0120	0.0024
1/4	0.243	36	80	0.2546	0.2314	0.250	0.235	0.234	0.2216	0.2227	0.0129	0.0026	0.233	0.249	0.248	0.2701	0.2692	0.0160	0.0024
5/16	0.303	36	80	0.3174	0.2886	0.312	0.293	0.292	0.2763	0.2774	0.0161	0.0026	0.291	0.311	0.310	0.3367	0.3358	0.0200	0.0024
3/8	0.363	36	80	0.3803	0.3457	0.375	0.352	0.351	0.3310	0.3321	0.0193	0.0026	0.350	0.374	0.373	0.4034	0.4025	0.0239	0.0024
1/2	0.485	36	80	0.5081	0.4619	0.500	0.469	0.468	0.4423	0.4434	0.0258	0.0026	0.467	0.499	0.498	0.5390	0.5381	0.0319	0.0024
5/8	0.605	36	80	0.6338	0.5762	0.625	0.584	0.583	0.5517	0.5528	0.0321	0.0026	0.582	0.624	0.623	0.6724	0.6715	0.0398	0.0024
3/4	0.733	48	82-1/2	0.7586	0.7074	0.750	0.716	0.714	0.6824	0.6835	0.0303	0.0025	0.713	0.749	0.747	0.7934	0.7925	0.0356	0.0024
7/8	0.855	48	82-1/2	0.8849	0.8251	0.875	0.835	0.833	0.7960	0.7971	0.0353	0.0025	0.832	0.874	0.872	0.9254	0.9245	0.0415	0.0024
1	0.977	48	82-1/2	1.0112	0.9428	1.000	0.954	0.952	0.9095	0.9106	0.0404	0.0025	0.951	0.999	0.997	1.0575	1.0566	0.0475	0.0024
1-1/8	1.098	48	82-1/2	1.1364	1.0596	1.125	1.071	1.069	1.0222	1.0233	0.0454	0.0025	1.068	1.124	1.122	1.1884	1.1875	0.0534	0.0024
1-1/4	1.220	48	82-1/2	1.2626	1.1773	1.250	1.190	1.188	1.1357	1.1368	0.0504	0.0025	1.187	1.249	1.247	1.3205	1.3196	0.0593	0.0024
1-3/8	1.342	48	82-1/2	1.3889	1.2951	1.375	1.309	1.307	1.2493	1.2504	0.0555	0.0025	1.306	1.374	1.372	1.4525	1.4516	0.0652	0.0024
1-1/2	1.464	48	82-1/2	1.5152	1.4128	1.500	1.428	1.426	1.3629	1.3640	0.0605	0.0025	1.425	1.499	1.497	1.5846	1.5837	0.0711	0.0024
1-3/4	1.708	48	82-1/2	1.7677	1.6483	1.750	1.666	1.664	1.5900	1.5911	0.0706	0.0025	1.663	1.749	1.747	1.8487	1.8478	0.0830	0.0024
2	1.952	48	82-1/2	2.0202	1.8838	2.000	1.904	1.902	1.8172	1.8188	0.0807	0.0025	1.901	1.999	1.997	2.1128	2.1114	0.0949	0.0024
2-1/4	2.196	48	82-1/2	2.2728	2.1192	2.250	2.142	2.140	2.0443	2.0459	0.0908	0.0025	2.139	2.249	2.247	2.3769	2.3755	0.1067	0.0024
2-1/2	2.440	48	82-1/2	2.5253	2.3547	2.500	2.380	2.378	2.2715	2.2731	0.1009	0.0025	2.377	2.499	2.497	2.6410	2.6396	0.1180	0.0024
2-3/4	2.684	48	82-1/2	2.7778	2.5902	2.750	2.618	2.616	2.4986	2.5002	0.1109	0.0025	2.615	2.749	2.747	2.9051	2.9037	0.1304	0.0024
3	2.928	48	82-1/2	3.0304	2.8256	3.000	2.856	2.854	2.7258	2.7274	0.1210	0.0025	2.853	2.999	2.997	3.1692	3.1678	0.1423	0.0024

SHAFT ENDS—SAE J501

SAE Standard

Report of Broaches Division approved June 1914 and last revised by Parts and Fittings Committee May 1948.

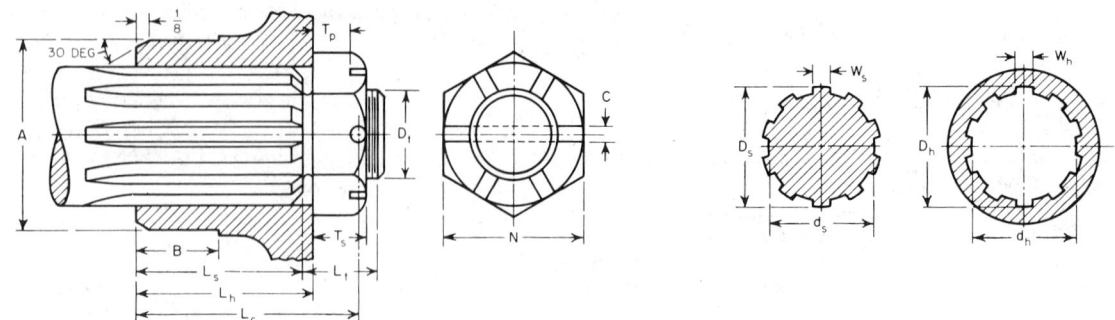

FIG. 1—PERMANENT FIT SPLINE SHAFT ENDS (SEE TABLE 1)

TABLE 1—PERMANENT FIT SPLINE SHAFT ENDS FOR UNIVERSAL JOINTS AND SIMILAR APPLICATIONS

Nominal Shaft Dia	10-Spline Shaft[a]			10-Spline Hole[a]			Hub Dimensions									C	A[b]	B
	D_s +0.000 −0.001	W_s +0.000 −0.0015	d_s +0.000 −0.010	D_h	W_h +0.000 −0.0015	d_h +0.010 −0.000	L_c	L_s	L_h	L_t	D_t	Threads per in.	T_s	T_p	N			
3/4	0.749	0.1170	0.632	0.751 / 0.749	0.1170	0.682	1-11/32	15/16	1	17/32	1/2	28	7/16	1/4	13/16	1/8	1-1/8	5/8
7/8	0.874	0.1370	0.745	0.876 / 0.874	0.1370	0.795	1-11/16	1-1/8	1-1/4	11/16	5/8	24	1/2	5/16	15/16	5/32	1-1/4	3/4
1	0.999	0.1560	0.859	1.001 / 0.999	0.1560	0.909	1-15/16	1-3/8	1-1/2	11/16	3/4	20	1/2	5/16	1-1/16	5/32	1-3/8	7/8
1-1/8	1.124	0.1760	0.973	1.126 / 1.124	0.1760	1.023	1-15/16	1-3/8	1-1/2	11/16	7/8	20	1/2	5/16	1-1/4	5/32	1-1/2	7/8
1-1/4	1.249	0.1950	1.087	1.251 / 1.249	0.1950	1.137	1-15/16	1-3/8	1-1/2	11/16	1	20	1/2	5/16	1-7/16	5/32	1-3/4	7/8
1-3/8	1.374	0.2150	1.200	1.376 / 1.374	0.2150	1.250	2-7/16	1-7/8	2	11/16	1	20	1/2	5/16	1-7/16	5/32	2	1
1-1/2	1.499	0.2340	1.304	1.501 / 1.499	0.2340	1.364	2-7/16	1-7/8	2	13/16	1-1/4	18	5/8	7/16	1-13/16	5/32	2-1/4	1
1-5/8	1.624	0.2540	1.347	1.627 / 1.624	0.2540	1.397	2-13/16	2-1/8	2-1/4	13/16	1-1/4	18	5/8	7/16	1-13/16	5/32	2-3/8	1-1/4
1-3/4	1.749	0.2730	1.454	1.752 / 1.749	0.2730	1.504	2-13/16	2-1/8	2-1/4	13/16	1-1/4	18	5/8	7/16	2-3/16	5/32	2-1/2	1-1/4
2	1.999	0.3120	1.668	2.002 / 1.999	0.3120	1.718	3-9/16	2-7/8	3	13/16	1-1/4	18	5/8	7/16	2-3/16	5/32	2-3/4	1-1/2
2-1/4	2.249	0.3510	1.883	2.252 / 2.249	0.3510	1.933	3-9/16	2-7/8	3	13/16	1-1/2	18	5/8	7/16	2-3/8	5/32	3	1-1/2
2-1/2	2.499	0.3900	2.098	2.502 / 2.499	0.3900	2.148	4-9/32	3-3/8	3-1/2	1-1/4	2	16	1	5/8	3-1/8	7/32	3-1/2	1-3/4
3	2.999	0.4680	2.528	3.002 / 2.999	0.4680	2.578	4-25/32	3-7/8	4	1-1/4	2	16	1	5/8	3-1/8	7/32	4	2

[a] SAE Standard, Involute Splines, Serrations, and Inspection—SAE J498 optional.
[b] Tolerance for ground finish, nominal +0.003, −0.002; and when specified, the maximum eccentricity with respect to the hole shall be 0.002 (indicator reading 0.004). Tolerance for lathe finish, nominal +1/32, −0.

TABLE 2—TAPER SHAFT END

Nominal Shaft Dia	D_s Shaft Dia		D_h Hole Dia		L_c	L_s	L_h	L_t	D_t	Threads per in.	T_s	T_p	Nut Width (Flats)	C	W		H +0.004 −0.000	Square Key		A[a]	B
	Max	Min	Max	Min											Max	Min		Max	Min		
1/4	0.250	0.249	0.248	0.247	9/16	5/16	3/8	5/16	#10	40	7/32	9/64	5/16	5/64	0.0625	0.0615	0.033	0.0635	0.0625	1/2	3/16
3/8	0.375	0.374	0.373	0.372	47/64	7/16	1/2	23/64	5/16	32	17/64	3/16	1/2	5/64	0.0937	0.0927	0.049	0.0947	0.0937	11/16	1/4
1/2	0.500	0.499	0.498	0.497	63/64	11/16	3/4	23/64	5/16	32	17/64	3/16	1/2	5/64	0.1250	0.1240	0.065	0.1260	0.1250	7/8	3/8
5/8	0.625	0.624	0.623	0.622	1-3/32	11/16	3/4	17/32	1/2	28	7/16	1/4	3/4	1/8	0.1562	0.1552	0.080	0.1572	0.1562	1-1/16	3/8
3/4	0.750	0.749	0.748	0.747	1-11/32	15/16	1	17/32	1/2	28	7/16	1/4	3/4	1/8	0.1875	0.1865	0.096	0.1885	0.1875	1-1/4	5/8
7/8	0.875	0.874	0.873	0.872	1-11/16	1-1/8	1-1/4	11/16	5/8	24	1/2	5/16	15/16	5/32	0.2500	0.2490	0.127	0.2510	0.2500	1-1/2	3/4
1	1.001	0.999	0.997	0.995	1-15/16	1-3/8	1-1/2	11/16	3/4	20	1/2	5/16	1-1/16	5/32	0.2500	0.2490	0.127	0.2510	0.2500	1-3/4	7/8
1-1/8	1.126	1.124	1.122	1.120	1-15/16	1-3/8	1-1/2	11/16	7/8	20	1/2	5/16	1-1/4	5/32	0.3125	0.3115	0.158	0.3135	0.3125	2	7/8
1-1/4	1.251	1.249	1.247	1.245	1-15/16	1-3/8	1-1/2	11/16	1	20	1/2	5/16	1-7/16	5/32	0.3125	0.3115	0.158	0.3135	0.3125	2-1/8	7/8
1-3/8	1.376	1.374	1.372	1.370	2-7/16	1-7/8	2	11/16	1	20	1/2	5/16	1-7/16	5/32	0.3750	0.3740	0.190	0.3760	0.3750	2-1/4	1
1-1/2	1.501	1.499	1.497	1.495	2-7/16	1-7/8	2	11/16	1	20	1/2	5/16	1-7/16	5/32	0.3750	0.3740	0.190	0.3760	0.3750	2-1/2	1
1-5/8	1.626	1.624	1.622	1.620	2-13/16	2-1/8	2-1/4	13/16	1-1/4	18	5/8	7/16	2-3/16	5/32	0.4375	0.4365	0.221	0.4385	0.4375	2-3/4	1-1/4
1-3/4	1.751	1.749	1.747	1.745	2-13/16	2-1/8	2-1/4	13/16	1-1/4	18	5/8	7/16	2-3/16	5/32	0.4375	0.4365	0.221	0.4385	0.4375	3	1-1/4
1-7/8	1.876	1.874	1.872	1.870	3-1/16	2-3/8	2-1/2	13/16	1-1/4	18	5/8	7/16	2-3/16	5/32	0.4375	0.4365	0.221	0.4385	0.4375	3-1/8	1-1/4
2	2.001	1.999	1.997	1.995	3-9/16	2-7/8	3	13/16	1-1/4	18	5/8	7/16	2-3/16	5/32	0.5000	0.4990	0.252	0.5010	0.5000	3-1/4	1-1/2
2-1/4	2.252	2.248	2.245	2.242	3-9/16	2-7/8	3	13/16	1-1/2	18	5/8	7/16	2-3/8	5/32	0.5625	0.5610	0.283	0.5640	0.5625	3-1/2	1-1/2
2-1/2	2.502	2.498	2.495	2.492	4-9/32	3-3/8	3-1/2	1-1/4	2	16	1	5/8	3-1/8	7/32	0.6250	0.6235	0.315	0.6265	0.6250	4	1-3/4
2-3/4	2.752	2.748	2.745	2.742	4-9/32	3-3/8	3-1/2	1-1/4	2	16	1	5/8	3-1/8	7/32	0.6875	0.6860	0.346	0.6890	0.6875	4-3/8	1-3/4
3	3.002	2.998	2.995	2.992	4-25/32	3-7/8	4	1-1/4	2	16	1	5/8	3-1/8	7/32	0.7500	0.7485	0.377	0.7515	0.7500	4-3/4	2
3-1/4	3.252	3.248	3.245	3.242	5-1/32	4-1/8	4-1/4	1-1/4	2	16	1	5/8	3-1/8	7/32	0.7500	0.7485	0.377	0.7515	0.7500	5	2-1/8
3-1/2	3.502	3.498	3.495	3.492	5-7/16	4-3/8	4-1/2	1-3/8	2-1/2	16	1-1/8	3/4	3-7/8	9/32	0.8750	0.8735	0.440	0.8765	0.8750	5-1/2	2-1/4
4	4.002	3.998	3.995	3.992	6-7/16	5-3/8	5-1/2	1-3/8	2-1/2	16	1-1/8	3/4	3-7/8	9/32	1.0000	0.9985	0.502	1.0015	1.0000	6-1/4	2-3/4

[a] Tolerance for ground finish, nominal +0.003, −0.002; and when specified, the maximum eccentricity with respect to the hole shall be 0.002 (indicator reading 0.004). Tolerance for lathe finish, nominal +1/32, −0.

(Continued on next page)

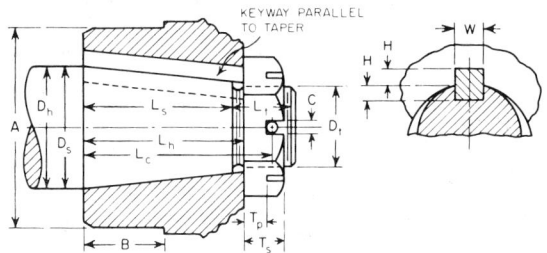

Taper per foot = 1.500 ± 0.002 in. Dimension H is measured normal to the key and at the large end of the taper.

C = cotter pin hole or slot. The centerline of the cotter pin hole shall be 90 deg from the position of the keyway, as shown in Fig. 2.

FIG. 2—TAPER SHAFT END (SEE TABLE 2)

WOODRUFF KEYS—SAE J502

SAE Standard

Report of Parts and Fittings Division approved February 1928 and last revised by Parts and Fittings Committee January 1956. Editorial change September 1972.

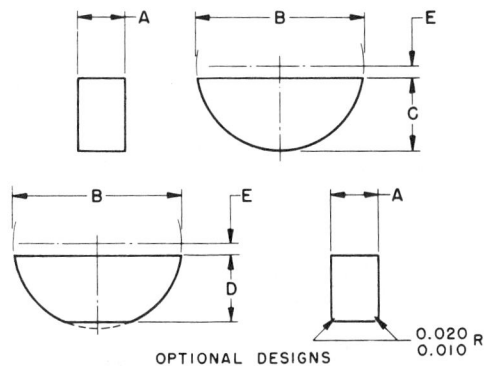

TABLE 1A—KEY DIMENSIONS

Part No.	SAE Nominal Size	Width, A +0.001 −0.000	Dia, B +0.000 −0.010	Heights C +0.000 −0.005	D +0.000 −0.006	E Nominal	Key Area at Shear Line	Approximate Weight, Lb per 1000
201	1/16 x 1/4	0.0625	0.250	0.109	—	1/64	0.0145	0.6
206	1/16 x 5/16	0.0625	0.312	0.140	—	1/64	0.0184	0.7
207	3/32 x 5/16	0.0938	0.312	0.140	—	1/64	0.0264	0.9
211	1/16 x 3/8	0.0625	0.375	0.172	—	1/64	0.0225	0.9
212	3/32 x 3/8	0.0938	0.375	0.172	—	1/64	0.0328	1.3
213	1/8 x 3/8	0.1250	0.375	0.172	—	1/64	0.0420	1.5
1	1/16 x 1/2	0.0625	0.500	0.203	0.194	3/64	0.0296	1.3
2	3/32 x 1/2	0.0938	0.500	0.203	0.194	3/64	0.0434	1.9
3	1/8 x 1/2	0.1250	0.500	0.203	0.194	3/64	0.0512	2.5
4	3/32 x 5/8	0.0938	0.625	0.250	0.240	1/16	0.0523	3.0
5	1/8 x 5/8	0.1250	0.625	0.250	0.240	1/16	0.0716	3.9
6	5/32 x 5/8	0.1563	0.625	0.250	0.240	1/16	0.0871	4.9
61	3/16 x 5/8	0.1875	0.625	0.250	0.240	1/16	0.0105	5.8
7	1/8 x 3/4	0.1250	0.750	0.313	0.303	1/16	0.0884	6.1
8	5/32 x 3/4	0.1563	0.750	0.313	0.303	1/16	0.1086	7.5
9	3/16 x 3/4	0.1875	0.750	0.313	0.303	1/16	0.1279	9.0
91	1/4 x 3/4	0.2500	0.750	0.313	0.303	1/16	0.1623	12.0
10	5/32 x 7/8	0.1563	0.875	0.375	0.365	1/16	0.1294	11.0
11	3/16 x 7/8	0.1875	0.875	0.375	0.365	1/16	0.1531	13.0
12	7/32 x 7/8	0.2188	0.875	0.375	0.365	1/16	0.1813	14.9
A	1/4 x 7/8	0.2500	0.875	0.375	0.365	1/16	0.1976	17.0
13	3/16 x 1	0.1875	1.000	0.438	0.428	1/16	0.1781	17.0
14	7/32 x 1	0.2188	1.000	0.438	0.428	1/16	0.2100	20.1
15	1/4 x 1	0.2500	1.000	0.438	0.428	1/16	0.2320	23.0
B	5/16 x 1	0.3125	1.000	0.438	0.428	1/16	0.2811	29.0
16	3/16 x 1-1/8	0.1875	1.125	0.484	0.475	5/64	0.2007	22.0
17	7/32 x 1-1/8	0.2188	1.125	0.484	0.475	5/64	0.2320	25.0
18	1/4 x 1-1/8	0.2500	1.125	0.484	0.475	5/64	0.2622	29.0
C	5/16 x 1-1/8	0.3125	1.125	0.484	0.475	5/64	0.3193	36.0
19	3/16 x 1-1/4	0.1875	1.250	0.547	0.537	5/64	0.2284	27.1
20	7/32 x 1-1/4	0.2188	1.250	0.547	0.537	5/64	0.2608	31.8
21	1/4 x 1-1/4	0.2500	1.250	0.547	0.537	5/64	0.2955	36.0
D	5/16 x 1-1/4	0.3125	1.250	0.547	0.537	5/64	0.3621	45.0
E	3/8 x 1-1/4	0.3750	1.250	0.547	0.537	5/64	0.4243	54.0
22	1/4 x 1-3/8	0.2500	1.375	0.594	0.584	3/32	0.3259	43.0
23	5/16 x 1-3/8	0.3125	1.375	0.594	0.584	3/32	0.4003	54.0
F	3/8 x 1-3/8	0.3750	1.375	0.594	0.584	3/32	0.4705	65.0
24	1/4 x 1-1/2	0.2500	1.500	0.641	0.631	7/64	0.3562	50.0
25	5/16 x 1-1/2	0.3125	1.500	0.641	0.631	7/64	0.4384	63.0
G	3/8 x 1-1/2	0.3750	1.500	0.641	0.631	7/64	0.5166	75.0

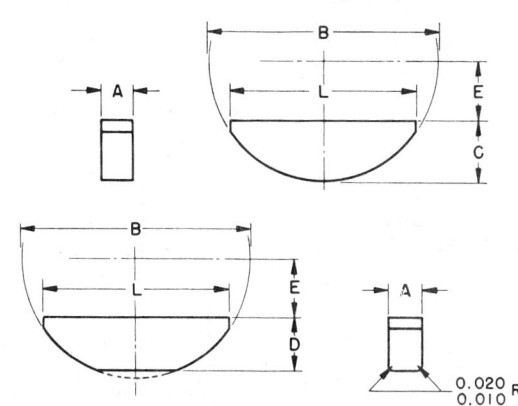

TABLE 1B—KEY DIMENSIONS

Part No.	SAE Nominal Size	Width, A +0.001 −0.000	Dia, B +0.000 −0.010	Heights C +0.000 −0.005	D +0.000 −0.006	E Nominal	Length, L +0.000 −0.010	Key Area at Shear Line	Approximate Weight, Lb per 1000
126	3/16 x 2-1/8	0.1875	2.125	0.406	0.396	21/32	1.380	0.2578	23.4
127	1/4 x 2-1/8	0.2500	2.125	0.406	0.396	21/32	1.380	0.3437	31.2
128	5/16 x 2-1/8	0.3125	2.125	0.406	0.396	21/32	1.380	0.4296	39.3
129	3/8 x 2-1/8	0.3750	2.125	0.406	0.396	21/32	1.380	0.4833	47.2
26	3/16 x 2-1/8	0.1875	2.125	0.531	0.521	17/32	1.723	0.3222	36.3
27	1/4 x 2-1/8	0.2500	2.125	0.531	0.521	17/32	1.723	0.4178	48.2
28	5/16 x 2-1/8	0.3125	2.125	0.531	0.521	17/32	1.723	0.5062	60.1
29	3/8 x 2-1/8	0.3750	2.125	0.531	0.521	17/32	1.723	0.5868	72.3
Rx	1/4 x 2-3/4	0.2500	2.750	0.594	0.584	25/32	2.000	0.5000	64.8
Sx	5/16 x 2-3/4	0.3125	2.750	0.594	0.584	25/32	2.000	0.6286	80.8
Tx	3/8 x 2-3/4	0.3750	2.750	0.594	0.584	25/32	2.000	0.6943	96.6
Ux	7/16 x 2-3/4	0.4375	2.750	0.594	0.584	25/32	2.000	0.8253	112.9
Vx	1/2 x 2-3/4	0.5000	2.750	0.594	0.584	25/32	2.000	0.9094	129.3
R	1/4 x 2-3/4	0.2500	2.750	0.750	0.740	5/8	2.317	0.5718	91.6
S	5/16 x 2-3/4	0.3125	2.750	0.750	0.740	5/8	2.317	0.7071	114.2
T	3/8 x 2-3/4	0.3750	2.750	0.750	0.740	5/8	2.317	0.8319	136.6
U	7/16 x 2-3/4	0.4375	2.750	0.750	0.740	5/8	2.317	0.9499	159.2
V	1/2 x 2-3/4	0.5000	2.750	0.750	0.740	5/8	2.317	1.0606	191.8
30	3/8 x 3-1/2	0.3750	3.500	0.938	0.927	13/16	2.880	1.0781	216.0
31	7/16 x 3-1/2	0.4375	3.500	0.938	0.927	13/16	2.880	1.2371	252.0
32	1/2 x 3-1/2	0.5000	3.500	0.938	0.927	13/16	2.880	1.3905	288.0
33	9/16 x 3-1/2	0.5625	3.500	0.938	0.927	13/16	2.880	1.5368	325.0
34	5/8 x 3-1/2	0.6250	3.500	0.938	0.927	13/16	2.880	1.6755	359.0
35	11/16 x 3-1/2	0.6875	3.500	0.938	0.927	13/16	2.880	1.8062	399.0
36	3/4 x 3-1/2	0.7500	3.500	0.938	0.927	13/16	2.880	1.9281	435.0

Material—Keys are to be carbon steel or alloy heat treated steel as specified. Carbon steel keys are to be 0.30 carbon minimum, with hardness of Rockwell B 90 minimum. Alloy steel keys are to be SAE 2330 or 8630 steel, heat treated to a hardness of 40–50 RC; or other alloy steels having equal physical properties at the same hardness. Alloy heat treated keys are to be marked with depressions on the top to distinguish them from carbon steel keys.

Dimensions—Dimensions are to be as given in Tables 1A, 1B, and 2.

Width A (Shown in Illustrations Accompanying Tables)—Values shown were set with the maximum key slot width as that figure which will receive a key with the greatest amount of looseness consistent with assuring the key's sticking in the slot. Minimum key slot width is that figure permitting the largest shaft distortion acceptable when assembling maximum key in minimum key slot.

Dimensions A, B, C, and E—These dimensions are to be taken at the side intersection.

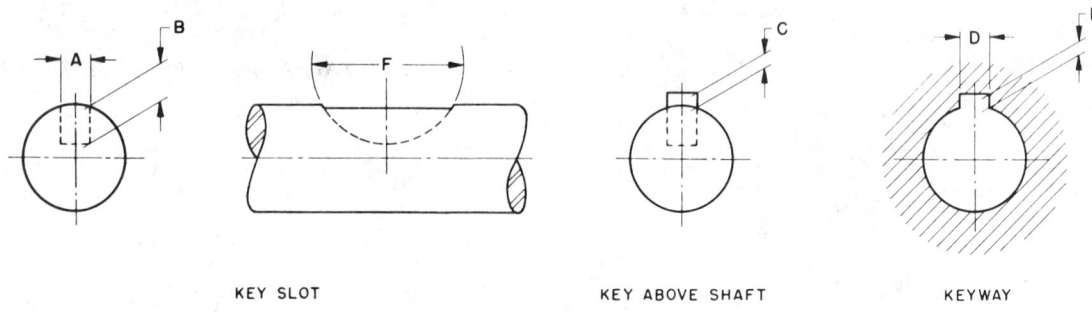

KEY SLOT   KEY ABOVE SHAFT   KEYWAY

TABLE 2—KEY SLOT AND KEYWAY DIMENSIONS

Part No.	SAE Nominal Size	Key Slot Width, A Min	Key Slot Width, A Max	Key Slot Depth, B +0.005 −0.000	Dia, F Min	Dia, F Max	Key above Shaft Height, C ±0.005	Keyway Width, D +0.002 −0.000	Keyway Depth, E +0.005 −0.000	Mfrs Part No.	SAE Nominal Size	Key Slot Width, A Min	Key Slot Width, A Max	Key Slot Depth, B +0.005 −0.000	Dia, F Min	Dia, F Max	Key above Shaft Height, C ±0.005	Keyway Width, D +0.002 −0.000	Keyway Depth, E +0.005 −0.000
201	1/16 x 1/4	0.0615	0.0630	0.0728	0.250	0.268	0.0312	0.0635	0.0372	E	3/8 x 1-1/4	0.3735	0.3755	0.3545	1.250	1.273	0.1875	0.3760	0.1935
206	1/16 x 5/16	0.0615	0.0630	0.1038	0.312	0.330	0.0312	0.0635	0.0372	22	1/4 x 1-3/8	0.2487	0.2505	0.4640	1.375	1.398	0.1250	0.2510	0.1310
207	3/32 x 5/16	0.0928	0.0943	0.0882	0.312	0.330	0.0469	0.0948	0.0529	23	5/16 x 1-3/8	0.3111	0.3130	0.4328	1.375	1.398	0.1562	0.3135	0.1622
211	1/16 x 3/8	0.0615	0.0630	0.1358	0.375	0.393	0.0312	0.0635	0.0372	F	3/8 x 1-3/8	0.3735	0.3755	0.4015	1.375	1.398	0.1875	0.3760	0.1935
212	3/32 x 3/8	0.0928	0.0943	0.1202	0.375	0.393	0.0469	0.0948	0.0529	24	1/4 x 1-1/2	0.2487	0.2505	0.5110	1.500	1.523	0.1250	0.2510	0.1310
213	1/8 x 3/8	0.1240	0.1255	0.1045	0.375	0.393	0.0625	0.1260	0.0685	25	5/16 x 1-1/2	0.3111	0.3130	0.4798	1.500	1.523	0.1562	0.3135	0.1622
1	1/16 x 1/2	0.0615	0.0630	0.1668	0.500	0.518	0.0312	0.0635	0.0372	G	3/8 x 1-1/2	0.3735	0.3755	0.4485	1.500	1.523	0.1875	0.3760	0.1935
2	3/32 x 1/2	0.0928	0.0943	0.1511	0.500	0.518	0.0469	0.0948	0.0529										
3	1/8 x 1/2	0.1240	0.1255	0.1355	0.500	0.518	0.0625	0.1260	0.0685	126	3/16 x 2-1/8	0.1863	0.1880	0.3073	2.125	2.160	0.0937	0.1885	0.0997
4	3/32 x 5/8	0.0928	0.0943	0.1981	0.625	0.643	0.0469	0.0948	0.0529	127	1/4 x 2-1/8	0.2487	0.2505	0.2760	2.125	2.160	0.1250	0.2510	0.1310
5	1/8 x 5/8	0.1240	0.1255	0.1825	0.625	0.643	0.0625	0.1260	0.0685	128	5/16 x 2-1/8	0.3111	0.3130	0.2448	2.125	2.160	0.1562	0.3135	0.1622
6	5/32 x 5/8	0.1553	0.1568	0.1669	0.625	0.643	0.0781	0.1573	0.0841	129	3/8 x 2-1/8	0.3735	0.3755	0.2135	2.125	2.160	0.1875	0.3760	0.1935
61	3/16 x 5/8	0.1863	0.1880	0.1513	0.625	0.643	0.0937	0.1885	0.0997	26	3/16 x 2-1/8	0.1863	0.1880	0.4323	2.125	2.160	0.0937	0.1885	0.0997
7	1/8 x 3/4	0.1240	0.1255	0.2455	0.750	0.768	0.0625	0.1260	0.0685	27	1/4 x 2-1/8	0.2487	0.2505	0.4010	2.125	2.160	0.1250	0.2510	0.1310
8	5/32 x 3/4	0.1553	0.1568	0.2299	0.750	0.768	0.0781	0.1573	0.0841	28	5/16 x 2-1/8	0.3111	0.3130	0.3698	2.125	2.160	0.1562	0.3135	0.1622
9	3/16 x 3/4	0.1863	0.1880	0.2143	0.750	0.768	0.0937	0.1885	0.0997	29	3/8 x 2-1/8	0.3735	0.3755	0.3385	2.125	2.160	0.1875	0.3760	0.1935
91	1/4 x 3/4	0.2487	0.2505	0.1830	0.750	0.768	0.1250	0.2510	0.1310										
10	5/32 x 7/8	0.1553	0.1568	0.2919	0.875	0.895	0.0781	0.1573	0.0841	Rx	1/4 x 2-3/4	0.2487	0.2505	0.4640	2.750	2.785	0.1250	0.2510	0.1310
										Sx	5/16 x 2-3/4	0.3111	0.3130	0.4328	2.750	2.785	0.1562	0.3135	0.1622
11	3/16 x 7/8	0.1863	0.1880	0.2763	0.875	0.895	0.0937	0.1885	0.0997	Tx	3/8 x 2-3/4	0.3735	0.3755	0.4015	2.750	2.785	0.1875	0.3760	0.1935
12	7/32 x 7/8	0.2175	0.2193	0.2607	0.875	0.895	0.1093	0.2198	0.1153	Ux	7/16 x 2-3/4	0.4360	0.4380	0.3703	2.750	2.785	0.2187	0.4385	0.2247
A	1/4 x 7/8	0.2487	0.2505	0.2450	0.875	0.895	0.1250	0.2510	0.1310	Vx	1/2 x 2-3/4	0.4985	0.5005	0.3390	2.750	2.785	0.2500	0.5010	0.2560
13	3/16 x 1	0.1863	0.1880	0.3393	1.000	1.020	0.0937	0.1885	0.0997										
14	7/32 x 1	0.2175	0.2193	0.3237	1.000	1.020	0.1093	0.2198	0.1153	R	1/4 x 2-3/4	0.2487	0.2505	0.6200	2.750	2.785	0.1250	0.2510	0.1310
15	1/4 x 1	0.2487	0.2505	0.3080	1.000	1.020	0.1250	0.2510	0.1310	S	5/16 x 2-3/4	0.3111	0.3130	0.5888	2.750	2.785	0.1562	0.3135	0.1622
B	5/16 x 1	0.3111	0.3130	0.2768	1.000	1.020	0.1562	0.3135	0.1622	T	3/8 x 2-3/4	0.3735	0.3755	0.5575	2.750	2.785	0.1875	0.3760	0.1935
16	3/16 x 1-1/8	0.1863	0.1880	0.3853	1.125	1.145	0.0937	0.1885	0.0997	U	7/16 x 2-3/4	0.4360	0.4380	0.5263	2.750	2.785	0.2187	0.4385	0.2247
17	7/32 x 1-1/8	0.2175	0.2193	0.3697	1.125	1.145	0.1093	0.2198	0.1153	V	1/2 x 2-3/4	0.4985	0.5005	0.4950	2.750	2.785	0.2500	0.5010	0.2560
18	1/4 x 1-1/8	0.2487	0.2505	0.3540	1.125	1.145	0.1250	0.2510	0.1310	30	3/8 x 3-1/2	0.3735	0.3755	0.7455	3.500	3.535	0.1875	0.3760	0.1935
C	5/16 x 1-1/8	0.3111	0.3130	0.3228	1.125	1.145	0.1562	0.3135	0.1622	31	7/16 x 3-1/2	0.4360	0.4380	0.7143	3.500	3.535	0.2187	0.4385	0.2247
19	3/16 x 1-1/4	0.1863	0.1880	0.4483	1.250	1.273	0.0937	0.1885	0.0997	32	1/2 x 3-1/2	0.4985	0.5005	0.6830	3.500	3.535	0.2500	0.5010	0.2560
										33	9/16 x 3-1/2	0.5610	0.5630	0.6518	3.500	3.535	0.2812	0.5635	0.2872
20	7/32 x 1-1/4	0.2175	0.2193	0.4327	1.250	1.273	0.1093	0.2198	0.1153	34	5/8 x 3-1/2	0.6235	0.6255	0.6205	3.500	3.535	0.3125	0.6260	0.3185
21	1/4 x 1-1/4	0.2487	0.2505	0.4170	1.250	1.273	0.1250	0.2510	0.1310	35	11/16 x 3-1/2	0.6860	0.6880	0.5893	3.500	3.535	0.3437	0.6885	0.3497
D	5/16 x 1-1/4	0.3111	0.3130	0.3858	1.250	1.273	0.1562	0.3135	0.1622	36	3/4 x 3-1/2	0.7485	0.7505	0.5580	3.500	3.535	0.3750	0.7510	0.3810

WOODRUFF KEY SLOTS AND KEYWAYS—SAE J503

SAE Information Report

Report of Parts and Fittings Division approved February 1928 and last revised by Parts and Fittings Committee January 1949. Editorial change May 1959.

Shaft Diameter, L—Decimal equivalents are given to four places in Table 1. All figures are calculated from this basic dimension. Any change in the shaft diameter from basic will necessarily change all other figures; and, in this case, should accurate dimensions be required, the formula given below should be used.

Versed Sine, G—The versed sines specified are determined from the following formula:

$$G = \frac{L}{2} - \sqrt{\frac{L^2 - A^2}{4}}$$

where A is the minimum width of the key.

Bottom of Key Slot to Opposite Side of Shaft, H—Obtain by subtracting the versed sine G, and depth of key slot, B, from the shaft diameter, L

Top of Key to Opposite Side of Shaft, J—Obtain by subtracting the versed sine, G, from the shaft diameter, L, and then adding to this figure the height of key above shaft, C.

Bottom of Keyway to Opposite Side of Bore, K—Obtain by subtracting the versed sine, G, from the shaft diameter, L, and then adding to this figure the depth of keyway, E.

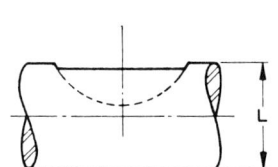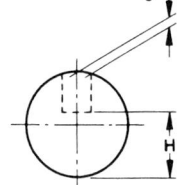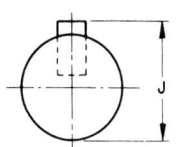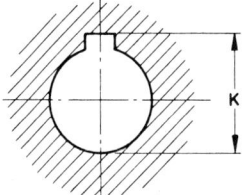

TABLE 1—VERSED SINE DIMENSION, G[a]

L Shaft Dia	Key Width														
	1/16	3/32	1/8	5/32	3/16	7/32	1/4	5/16	3/8	7/16	1/2	9/16	5/8	11/16	3/4
0.3125	0.0032	—	—	—	—	—	—	—	—	—	—	—	—	—	—
0.3437	0.0029	0.0065	—	—	—	—	—	—	—	—	—	—	—	—	—
0.3750	0.0026	0.0060	0.0107	—	—	—	—	—	—	—	—	—	—	—	—
0.4060	0.0024	0.0055	0.0099	—	—	—	—	—	—	—	—	—	—	—	—
0.4375	0.0022	0.0051	0.0091	—	—	—	—	—	—	—	—	—	—	—	—
0.4687	0.0021	0.0047	0.0085	0.0134	—	—	—	—	—	—	—	—	—	—	—
0.5000	0.0020	0.0044	0.0079	0.0125	—	—	—	—	—	—	—	—	—	—	—
0.5625	—	0.0039	0.0070	0.0111	0.0161	—	—	—	—	—	—	—	—	—	—
0.6250	—	0.0035	0.0063	0.0099	0.0144	0.0198	—	—	—	—	—	—	—	—	—
0.6875	—	0.0032	0.0057	0.0090	0.0130	0.0179	0.0235	—	—	—	—	—	—	—	—
0.7500	—	0.0029	0.0052	0.0082	0.0119	0.0163	0.0214	0.0341	—	—	—	—	—	—	—
0.8125	—	0.0027	0.0048	0.0076	0.0110	0.0150	0.0197	0.0312	—	—	—	—	—	—	—
0.8750	—	0.0025	0.0045	0.0070	0.0102	0.0139	0.0182	0.0288	—	—	—	—	—	—	—
0.9375	—	—	0.0042	0.0066	0.0095	0.0129	0.0170	0.0268	0.0391	—	—	—	—	—	—
1.0000	—	—	0.0039	0.0061	0.0089	0.0121	0.0159	0.0250	0.0365	—	—	—	—	—	—
1.0625	—	—	0.0037	0.0058	0.0083	0.0114	0.0149	0.0235	0.0342	—	—	—	—	—	—
1.1250	—	—	0.0035	0.0055	0.0079	0.0107	0.0141	0.0221	0.0322	0.0443	—	—	—	—	—
1.1875	—	—	0.0033	0.0052	0.0074	0.0102	0.0133	0.0209	0.0304	0.0418	—	—	—	—	—
1.2500	—	—	0.0031	0.0049	0.0071	0.0097	0.0126	0.0198	0.0288	0.0395	—	—	—	—	—
1.3750	—	—	—	0.0045	0.0064	0.0088	0.0115	0.0180	0.0261	0.0357	0.0471	—	—	—	—
1.5000	—	—	—	0.0041	0.0059	0.0080	0.0105	0.0165	0.0238	0.0326	0.0429	—	—	—	—
1.6250	—	—	—	0.0038	0.0054	0.0074	0.0097	0.0152	0.0219	0.0300	0.0394	0.0502	—	—	—
1.7500	—	—	—	—	0.0050	0.0069	0.0090	0.0141	0.0203	0.0278	0.0365	0.0464	—	—	—
1.8750	—	—	—	—	0.0047	0.0064	0.0084	0.0131	0.0189	0.0259	0.0340	0.0432	0.0536	—	—
2.0000	—	—	—	—	0.0044	0.0060	0.0078	0.0123	0.0177	0.0242	0.0318	0.0404	0.0501	—	—
2.1250	—	—	—	—	—	0.0056	0.0074	0.0116	0.0167	0.0228	0.0298	0.0379	0.0470	0.0572	0.0684
2.2500	—	—	—	—	—	—	0.0070	0.0109	0.0157	0.0215	0.0281	0.0357	0.0443	0.0538	0.0643
2.3750	—	—	—	—	—	—	0.0103	0.0149	0.0203	0.0266	0.0338	0.0419	0.0509	0.0608	
2.5000	—	—	—	—	—	—	—	0.0141	0.0193	0.0253	0.0321	0.0397	0.0482	0.0576	
2.6250	—	—	—	—	—	—	—	0.0135	0.0184	0.0240	0.0305	0.0377	0.0457	0.0547	
2.7500	—	—	—	—	—	—	—	—	0.0175	0.0229	0.0291	0.0360	0.0437	0.0521	
2.8750	—	—	—	—	—	—	—	—	0.0168	0.0219	0.0278	0.0344	0.0417	0.0498	
3.0000	—	—	—	—	—	—	—	—	—	0.0210	0.0266	0.0329	0.0399	0.0476	

[a] Listed for the different shaft sizes and keyway widths for reference in checking dimensions H, J, and K.

18 V-Belts

ɸ V-BELTS AND PULLEYS—SAE J636c   SAE Standard

Report of Miscellaneous Division approved August 1915 and completely revised by V-Belt Committee July 1977. Editorial change January 1978.

1. *Preface to 1977 Revision*—This revision includes the addition of three sections to the standard to conform with existing use practices—0.250 (6A), 0.315 (8A), and 0.440 (11A).

2. *Preface to 1970 Revision*—In this standard the dimensioning and measuring practices for V-belts and pulleys have been unified. No change has been made in 0.380 (10A) and 0.500 (13A) sizes. The ⅝ size (28 deg groove) has been dropped. On the other sizes ¹¹⁄₁₆ (15A), ¾ (17A), ⅞ (20A) and 1 in (23A) the new dimensions in Table 1 are designed to permit the same actual grooves as in the 1953 SAE Standard. The measuring of belts of SAE sizes is done in new pulleys having the following effects on belt measurements as compared with measurements in the previously specified notched pulleys:

Center Distances—The new measuring pulleys give 0.289 in (7.34 mm), 0.822 in (20.88 mm), and 0.181 in (4.60 mm) shorter center distances for ¹¹⁄₁₆ (15A), ⅞ (20A), and 1 in (23A) sizes respectively, and 0.035 in (0.89 mm) longer center distances on the ¾ (17A) size.

Ride-Out—Due to lowering the base for ride-out measurement, the new pulleys give approximately ³⁄₆₄ in (1.2 mm) more ride-out than shown by the notched pulleys for ¹¹⁄₁₆ (15A), ¾ (17A), ⅞ (20A), and 1 in (23A) sizes.

Belt Length Factor—With the notched pulleys the belt length factor was based on the corresponding outside diameter for each size, plus ⅛ in (3.2 mm) to allow for the average ride-out of ¹⁄₁₆ in (1.6 mm). With present measuring pulleys, the belt length factor is based on the effective diameter without allowance for ride-out.

3. *Scope*—This specification covers standard dimensions, tolerances, and methods of measurement of V-belts and pulleys for automotive V-belt drives.

4. *V-Belt Types*—Automotive V-belts are produced in a variety of constructions in a basic trapezoidal shape. The inside circumference of the V-belt can be a plain straight line or corrugated by means of cogs or notches for the purpose of increasing the belt(s) flexibility for use with pulleys in the lower proposed diameter. Belts are to be dimensioned in such a way that they are functional in pulleys dimensioned as described in subsequent sections.

5. *Pulleys*—Pulleys are to conform to requirements of Fig. 1 and Table 1.

6. *V-Belt Measurement*—Belt length and SAE size are defined by using effective length and ride-out as measured in standard pulleys. These are determined by use of a measuring fixture comprised of two pulleys of equal diameter, a method of applying force, and a means of measuring the center distance between the two pulleys. One of the two pulleys is fixed in position while the other is movable along a graduated scale. The fixture is shown

TABLE 1A—V-BELT PULLEY DIMENSIONS, IN

SAE Size	Recommended[a] Min Effective Dia	A Groove Angle (deg) ±0.5	W Effective Groove Width	D Groove Depth Min	d Ball or Rod Dia (±0.0005)	2K 2x Ball Extension	2X[b]	S Groove[c] Spacing (±0.015)
0.250	2.25	36	0.248	0.276	0.2188	0.164	0.04	0.315
0.315	2.25	36	0.315	0.354	0.2812	0.222	0.05	0.413
0.380	2.40	36	0.380	0.433	0.3125	0.154	0.06	0.541
0.440	2.75	36	0.441	0.512	0.3750	0.231	0.07	0.591
0.500	3.00	36	0.500	0.551	0.4375	0.314	0.08	0.661
11/16 (0.600)	3.00	34	0.597	0.551	0.500	0.258	0.00	0.778
	Over 4.00	36				0.280		
	Over 6.00	38				0.302		
3/4 (0.660)	3.00	34	0.660	0.630	0.5625	0.328	0.02	0.841
	Over 4.00	36				0.352		
	Over 6.00	38				0.374		
7/8 (0.790)	3.50	34	0.785	0.709	0.6875	0.472	0.04	0.966
	Over 4.50	36				0.496		
	Over 6.00	38				0.520		
1 (0.910)	4.00	34	0.910	0.827	0.8125	0.616	0.06	1.091
	Over 6.00	36				0.642		
	Over 8.00	38				0.666		

[a] Pulley effective diameters below those recommended should be used with caution, because power transmission and belt life may be reduced.
[b] 2X is to be subtracted from the effective diameter to obtain "pitch diameter" for speed ratio calculation.
[c] These values are intended for adjacent grooves of the same effective width (W). Choice of pulley manufacture or belt design parameter may justify variance from these values. The S dimension shall be the same on all multiple groove pulleys in a drive using matched belts.

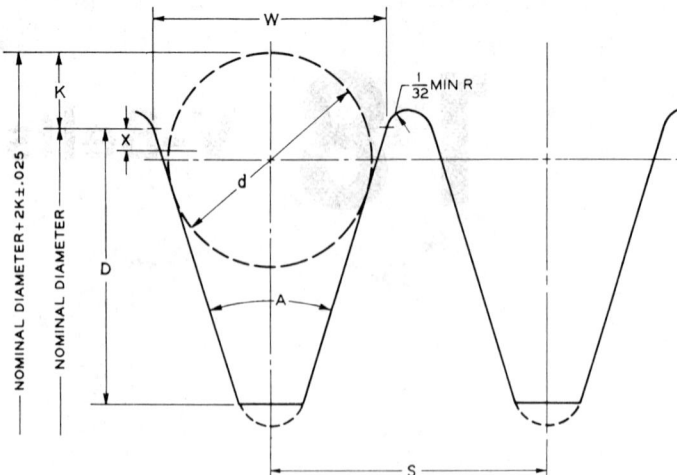

Notes:
1. The sides of the groove are to be 125 μin (3.2 μm) A. A. maximum.
2. Radial run-out is not to exceed 0.015 in (0.38 mm) full indicator movement (FIM). Axial run-out is not to exceed 0.015 in (0.38 mm) FIM. Run-out in the two directions is measured separately with a ball mounted under spring pressure to follow the groove as the pulley is rotated. Diameter, load, and overhang conditions may require or permit variations in the above specified run-out limits.
3. Bottom corner radii optional, but if used, it shall be below the depth, D.
4. In pulleys for use with belts in multiple on common centers, the diameters over the ball gages are not to vary from groove to groove in the same pulley more than 0.002 in/in (0.05 mm/25 mm) of diameter, with top limit of 0.012 in (0.30 mm) for diameters 6 in (152 mm) and above.
5. Centerline of groove is to be 90 ± 2 deg with pulley axis.
6. The X dimension is radial. 2X is to be subtracted from the effective diameter to obtain "pitch diameter" for speed ratio calculation.

FIG. 1—V-BELT PULLEY DIMENSIONS

TABLE 1B—V-BELT PULLEY DIMENSIONS, MM

SAE Size	Recommended[a] Min Effective Dia	A Groove Angle (deg) ±0.5	W Effective Groove Width	D Groove Depth Min	d Ball or Rod Dia (±0.013)	2K 2x Ball[d] Extension	2X	S Groove[c] Spacing (±0.38)
6A	57	36	6.3	7	5.558	4.16	1.0	8.00
8A	57	36	8.0	9	7.142	5.63	1.3	10.49
10A	61	36	9.7	11	7.938	3.77	1.5	13.74
11A	70	36	11.2	13	9.525	5.88	1.8	15.01
13A	76	36	12.7	14	11.113	7.99	2.0	16.79
15A	76	34				6.42		
	Over 102	36	15.2	14	12.700	7.02	0	19.76
	Over 152	38				7.56		
17A	76	34				8.21		
	Over 102	36	16.8	15	14.288	8.82	0.5	21.36
	Over 152	38				9.38		
20A	89	34				11.77		
	Over 114	36	20.0	18	17.463	12.42	1.0	24.54
	Over 152	38				13.02		
23A	102	34				15.67		
	Over 152	36	23.1	21	20.638	16.33	1.5	27.71
	Over 203	38				16.94		

[a] Pulley effective diameters below those recommended should be used with caution, because power transmission and belt life may be reduced.
[b] 2X is to be subtracted from the effective diameter to obtain "pitch diameter" for speed ratio calculation.
[c] These values are intended for adjacent grooves of the same effective width (W). Choice of pulley manufacture or belt design parameter may justify variance from these values. The S dimension shall be the same on all multiple groove pulleys in a drive using matched belts.
[d] 2K dimensions are calculated in millimetres.

schematically in Fig. 2. Specifications for measuring pulley dimensions are given in Table 2 and Fig. 3.

NOTE: The outside diameter and the effective diameter on the measuring pulley are one and the same.

6.1 Length—To measure the length, the belt is placed on the measuring fixture at the force shown in Table 3, and rotated around the pulleys at least two revolutions of the belt to seat the belt properly in the pulley grooves and to divide the total force equally between the two strands of the belt. The midpoint of the center distance travel of the movable pulley defines the center distance and will be measured through one revolution of the belt minimum after the two seating revolutions. The belt effective length is equal to two times the center distance plus the effective pulley circumference. Standard belt center distance tolerances are shown in Table 4.

6.2 Ride-Out—The ride-out standard and ride-out tolerance are shown in Table 3. The ride-out of a belt section is determined by measuring from a straight edge across the top of the belt to the rim of the measuring pulley, as shown in Fig. 4.

6.3 Matched Belt Sets—For V-belts used in sets of two or more for a general application, the difference in center distance between the belts cannot exceed the values shown in Table 5.

7. Standard Lengths—Standard lengths up to and including 80 in (2032 mm) are to be in 1/2 in (12.7 mm) increments. Standard lengths over 80 in (2032 mm) up to and including 100 in (2540 mm) are to be 1 in (25.4 mm) increments without fractions.

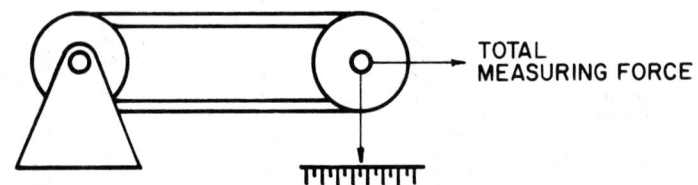

FIG. 2—DIAGRAM OF A FIXTURE FOR MEASURING V-BELTS

18.03

TABLE 2A—MEASURING PULLEY DIMENSIONS, IN

SAE Size US	d_1 Effective Dia (±0.002)	Effective Pulley Circumference	A Groove Angle (deg) ±0.15	W Effective Groove Width	D Groove Depth Min	d Ball or Rod Dia (±0.0005)	d_2 Dia Over Balls or Rods (±0.002)
0.250	3.820	12.000	36	0.248	0.276	0.2188	3.984
0.315	3.820	12.000	36	0.315	0.354	0.2812	4.042
0.380	3.820	12.000	36	0.380	0.433	0.3125	3.974
0.440	3.820	12.000	36	0.441	0.512	0.3750	4.051
0.500	3.820	12.000	36	0.500	0.551	0.4375	4.134
11/16 (0.600)	3.820	12.000	34	0.597	0.551	0.5000	4.078
3/4 (0.660)	3.820	12.000	34	0.660	0.630	0.5625	4.148
7/8 (0.790)	4.775	15.000	34	0.785	0.709	0.6875	5.247
1 (0.910)	4.775	15.000	34	0.910	0.827	0.8125	5.391

TABLE 2B—MEASURING PULLEY DIMENSIONS, MM

SAE Size	d_1 Effective Dia (±0.05)	Effective Pulley Circumference	A Groove Angle (deg) ±0.15	W Effective Groove Width	D Groove Depth Min	d Ball or Rod Dia (±0.013)	d_2 Dia[a] Over Balls or Rods (±0.05)
6A	97.03	304.8	36	6.3	7	5.558	101.18
8A	97.03	304.8	36	8.0	9	7.142	102.66
10A	97.03	304.8	36	9.7	11	7.938	100.80
11A	97.03	304.8	36	11.2	13	9.525	102.91
13A	97.08	304.8	36	12.7	14	11.113	105.02
15A	97.03	304.8	34	15.2	14	12.700	103.45
17A	97.03	304.8	34	16.8	16	14.288	105.24
20A	121.29	381.0	34	20.0	18	17.463	133.06
23A	121.29	381.0	34	23.1	21	20.638	136.96

[a] d_2 dimensions are calculated in millimetres.

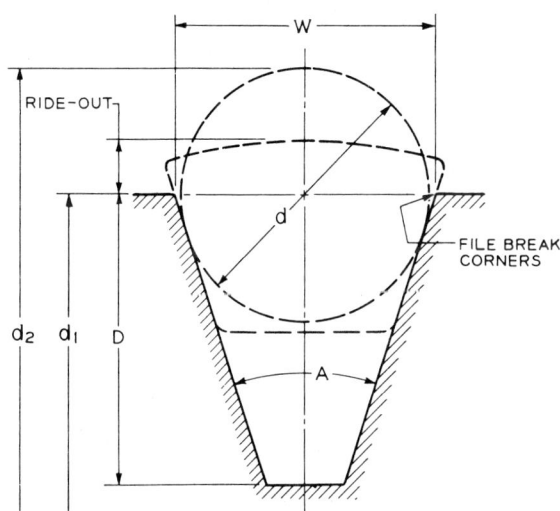

NOTE: The outside diameter and the effective diameter on the measuring pulley are one and the same.

FIG. 3—MEASURING PULLEY DIMENSIONS

TABLE 3A—MEASURING CONDITIONS AND RIDE-OUT, U.S. CUSTOMARY UNITS

SAE Size in	Total Measuring Force lb	Ride-[a] Out in	Ride-[a] Out Tolerance in
0.250	50	0.031	±0.031
0.315	50	0.031	±0.031
0.380	60	0.060	±0.045
0.440	60	0.040	±0.045
0.500	60	0.060	±0.045
11/16 (0.600)	60	0.090	±0.045
3/4 (0.660)	80	0.090	±0.045
7/8 (0.790)	100	0.090	±0.045
1 (0.910)	120	0.090	±0.045

[a] The belt ride-out, as measured along the circumference of the belt, must fall within the specified tolerance at all points with the exception of measurements at points of dimension variations inherent to the manufacturing process or product such as material splices, belt identifications, etc.

TABLE 3B—MEASURING CONDITIONS AND RIDE-OUT, SI UNITS

SAE Size Metric	Total Measuring Force N	Ride-[a] Out mm	Ride-[a] Out Tolerance mm
6A	222	0.8	±0.8
8A	222	0.8	±0.8
10A	267	1.5	±1.1
11A	267	1.0	±1.1
13A	267	1.5	±1.1
15A	267	2.3	±1.1
17A	356	2.3	±1.1
20A	445	2.3	±1.1
23A	534	2.3	±1.1

[a] The belt ride-out, as measured along the circumference of the belt, must fall within the specified tolerance at all points with the exception of measurements at points of dimension variations inherent to the manufacturing process or product such as material splices, belt identifications, etc.

TABLE 4A—STANDARD BELT CENTER DISTANCE TOLERANCES, IN

Belt Length	Tolerance on Center Distance
50 and less	±0.12
Over 50–60, incl.	±0.16
Over 60–80, incl.	±0.19
Over 80–100, incl.	±0.22

TABLE 4B—STANDARD BELT CENTER DISTANCE TOLERANCES, MM

Belt Length	Tolerance on Center Distance
1270 and less	±3.0
Over 1270–1524, incl.	±4.1
Over 1524–2032, incl.	±4.8
Over 2032–2540, incl.	±5.6

TABLE 5A—MAXIMUM CENTER DISTANCE DIFFERENCE FOR BELTS IN A SET, IN

SAE Size	
0.250	0.03
0.315	0.03
0.380	0.04
0.440	0.04
0.500	0.04
11/16 (0.600)	0.06
3/4 (0.660)	0.06
7/8 (0.790)	0.06
1 (0.910)	0.06

TABLE 5B—MAXIMUM CENTER DISTANCE DIFFERENCE FOR BELTS IN A SET, MM

SAE Size	
6A	0.8
8A	0.8
10A	1.0
11A	1.0
13A	1.0
15A	1.5
17A	1.5
20A	1.5
23A	1.5

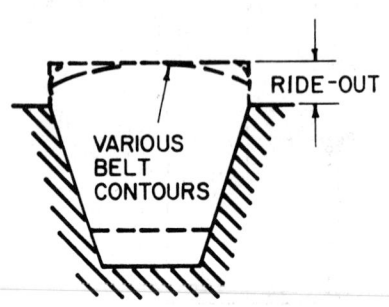

FIG. 4—MEASURING BELT RIDE

AUTOMOTIVE V-BELT DRIVES—SAE J637b

SAE Recommended Practice

Report of Engine Committee approved January 1954 and last revised by V-Belt Committee April 1974. Editorial change June 1977.

1. Introduction—Selection and specification of belts has been a major problem due to the lack of a recognized industry standard for classifying V-belts according to performance and quality level.

From the very beginning of the use of V-belts on automotive drives, the automotive manufacturers and the V-belt manufacturers have employed laboratory tests on the product for such purposes as product development, source approval, and quality verification. This standard is the result of the combined effort of the users and suppliers.

2. Scope—The following information is intended as a guide to be used for evaluating belt construction, source approval, or quality audit. This recommendation has been prepared from existing literature, including standards, specifications, and data supplied by both producers and users.

These recommendations cover drive layout details and V-belt testing methods, including test layout, pulley diameters, torque loads and guidance for interpreting test data.

3. General Drive Layout Considerations

3.1 Power Transmission—When the engine is used to drive an external unit equipped with industrial type pulleys and belts, it is recommended that the power takeoff pulley on the engine be grooved according to the appropriate industrial standard. There are four such standards. Three of these standards are published by the RMA-MTPA (Rubber Manufacturers Association—Mechanical Power Transmission Association) and include Classical Multi-V-Belt (A, B, C, D, and E belt sections), Narrow Multi-V-Belts (3V, 5V, and 8V), and Single V-Belts (2L, 3L, 4L, and 5L). The fourth is published as an

TABLE 1A—MINIMUM ALLOWANCE FOR BELT INSTALLATION, IN
(IN TERMS OF BELT LENGTH, NOT CENTER DISTANCE)

Belt Length	SAE Belt Size											
	0.380		0.500		11/16 (0.597)		3/4 (0.660)		7/8 (0.785)		1 (0.910)	
	Belt Drives											
	Single	Multiple	Single	Multiple	Single	Multiple	Single	Multiple	Single	Multiple	Single	Multiple
50 and less	11/16	15/16	3/4	1.0	7/8	1-3/16	7/8	1-3/16	15/16	1-5/16	1.0	1-3/8
Over 50 to 60, incl	3/4	1.0	13/16	1-1/16	15/16	1-1/4	15/16	1-1/4	1.0	1-3/8	1-1/16	1-7/16
Over 60 to 80, incl	13/16	1-1/16	7/8	1-1/8	1.0	1-5/16	1.0	1-5/16	1-1/16	1-7/16	1-1/8	1-1/2
Over 80 to 100, incl	7/8	1-1/8	15/16	1-3/16	1-1/16	1-3/8	1-1/16	1-3/8	1-1/8	1-1/2	1-3/16	1-9/16

TABLE 1B—MINIMUM ALLOWANCE FOR BELT INSTALLATION, MM
(IN TERMS OF BELT LENGTH, NOT CENTER DISTANCE)

Belt Length	SAE Belt Size											
	10		13		15		17		20		23	
	Belt Drives											
	Single	Multiple	Single	Multiple	Single	Multiple	Single	Multiple	Single	Multiple	Single	Multiple
1270 and less	18	24	19	26	22	30	22	30	24	33	26	35
Over 1270 to 1520, incl	19	26	21	27	24	32	24	32	26	35	27	37
Over 1520 to 2030, incl	21	27	22	29	26	33	26	33	27	37	29	38
Over 2030 to 2540, incl	22	29	24	30	27	35	27	35	29	38	30	40

American Society of Agricultural Engineers standard, V-Belt Drives for Farm Machines.

The grooves in these four standards differ from each other in the reference dimensions. They are not interchangeable with SAE grooves which were standardized for engine accessory and other engine compartment drives.

3.2 Belt Speed—It is recommended that the pulley diameters be as large as possible without exceeding 7000 ft/min (35.6 m/s) belt speed.

3.3 Pulley Sizes—Pulleys should never be smaller than in Tables 3 and 4.

3.4 Allowance for Belt Installation—The recommended minimum allowance for belt installation is given in Table 1 in terms of belt length (not center distance).

To allow for belt installation and take-up, one pulley should be adjustable from its initial position with the mean length belt at operating tension. Table 1 gives the absolute minimum allowance for easy installation of the belt without prying or otherwise forcing it over the sides of the grooves. This allowance includes the minus length tolerance from Table 3 in SAE J636.

3.5 Allowance for Belt Take-Up—The recommended minimum allowance for belt take-up is given in Table 2 in terms of belt length (not center distance).

Table 2 gives the minimum allowance for take-up consisting of the plus tolerance from Table 3 in SAE J636, and an allowance of 1½% of belt length for belt growth and wear in service. Because of greater belt wear, an additional take-up allowance of at least ½% of belt length is generally advisable on heavy-duty drives such as those of large trucks, buses, earthmoving equipment, and the like.

3.6 Pulley Misalignment—The recommended maximum misalignment between pulleys is $\frac{1}{16}$ in. per foot of span length (1.6 mm per 300 mm span length) or approximately $\frac{1}{3}$ of 1 deg.

4. V-Belt Fatigue Test Method—The belt shall be mounted on a test layout as shown in Fig. 1 with pulley diameters and speeds as given in Tables 3 and 4. The power (kilowatts) to be absorbed at the driven pulley shall be compatible with the tension pulley diameter and belt length as shown in Tables 3 and 4.

The driver pulley speed (rpm) shall be used in the torque load calculation and the torque load shall be kept constant without compensation for loss of driven pulley rpm resulting from belt slippage.

$$\text{Torque, lb} \cdot \text{in} = \frac{\text{Specified power} \times 63025}{\text{Driver rpm}}$$

$$\text{Torque, N} \cdot \text{m} = \frac{\text{Specified kilowatts} \times 9550}{\text{Driver rpm}}$$

Measurable parasitic loads due to bearing losses, lubricants, etc., shall be deducted from the specified power (kilowatts) in the above calculation.

The tension shall be applied by weights equal in number of pounds to 10 times the number of units of the specified power (in number of kilograms to 60 times the number of units of the specified kilowatts).

4.1 The test procedure shall be as follows:

4.1.1 Condition the belt by running 5 min under the prescribed test details but without the dynamometer load. Maintain a constant tension during this period by operating with the tension pulley center position unlocked.

4.1.2 Stop the machine, allow to stand for a minimum of 10 min, and lock the tension pulley center position midway of the limits of travel during belt rotation.

4.1.3 Restart with the dynamometer load and run until the slip reaches 8% or until the belt will no longer transmit the load uniformly because of breakage or rough running.

4.1.4 Whenever the slip reaches 8%, stop the machine, allow to stand for a minimum of 20 min, unlock the tension pulley center, restore the initial tension, relock, and restart the machine.

4.1.5 Record the number of hours run and the number of resets (exclusive of the 5 min run-in).

4.1.6 The ambient temperature shall be 80–90°F (27–32°C). An increase in internal belt temperature will reduce belt life. Internal belt temperature is dependent upon ambient temperature as well as other test conditions.

5. Test Performance Guidelines—The test life which a belt must attain shall be according to agreement between user and manufacturer. However, typical curves of average test life versus belt length are shown in Fig. 2. The typical curves of Fig. 2 are constructed with the belt life varying as the 2.75 power of belt length for the test conditions given in Tables 3 and 4. The acceptable number of retentions after the initial 5 min run-in shall be according to agreement between the manufacturer and user.

TABLE 2A—MINIMUM ALLOWANCE FOR BELT TAKE-UP, IN
(IN TERMS OF BELT LENGTH, NOT CENTER DISTANCE)

Belt Length	All SAE Belt Sizes, Take-Up Allowance
50 and less	1.0
Over 50 to 60, incl	1-1/4
Over 60 to 80, incl	1-5/8
Over 80 to 100, incl	1-15/16

TABLE 2B—MINIMUM ALLOWANCE FOR BELT TAKE-UP, MM
(IN TERMS OF BELT LENGTH, NOT CENTER DISTANCE)

Belt Length	All SAE Belt Sizes, Take-Up Allowance
1270 and less	25
Over 1270 to 1520, incl	32
Over 1520 to 2030, incl	41
Over 2030 to 2540, incl	49

TABLE 3A—TEST CONDITIONS[a]
PLAIN SECTION BELTS, IN

SAE Belt Size	Standard Groove Width	Diameter Where Specified Groove Width Occurs (without width tolerance)		Driver Pulley Speed, rpm ±2%	Load, hp	Length Range	
		Driver and Driven Pulleys ±0.010	Tension Pulley ±0.010			Total	Preferred
0.380	0.380	4.750	2.500	4900	10.0 / 11.0 / 12.5	Under 40 / 40-55 / 55 and over	36-40 / 45-50 / 55-60
0.470		See Appendix					
0.500	0.500	5.000	3.000	4700	11.5 / 13.0 / 14.0	Under 40 / 40-55 / 55 and over	36-40 / 45-50 / 55-60
11/16 (0.597)	0.597	5.000	3.500	4700	—[b]		
3/4 (0.660)	0.660	5.000	3.625	4700	—[b]		
7/8 (0.785)	0.785	6.000	4.000	3900	—[b]		
1 (0.910)	0.910	7.000	4.625	3350	—[b]		

[a] Groove details as given in SAE J636, Table 1, Fig. 1.
[b] The horsepower to be absorbed at the driven pulley is to be according to agreement between user and manufacturer.

TABLE 3B—TEST CONDITIONS[a]
PLAIN SECTION BELTS, MM

SAE Belt Size	Standard Groove Width	Diameter Where Specified Groove Width Occurs (without width tolerance)		Driver Pulley Speed, rpm ±2%	Load, kW	Length Range	
		Driver and Driven Pulleys ±0.25	Tension Pulley ±0.25			Total	Preferred
10	9.65	120.5	63.5	4900	7.5 / 8.2 / 9.4	Under 1020 / 1020-1400 / 1400 and over	920-1020 / 1140-1270 / 1400-1520
12		See Appendix					
13	12.70	127.0	76.0	4700	8.6 / 9.8 / 10.4	Under 1020 / 1020-1400 / 1400 and over	920-1020 / 1140-1270 / 1400-1520
15	15.16	127.0	89.0	4700	—[b]		
17	16.76	127.0	92.0	4700	—[b]		
20	19.94	152.5	101.5	3900	—[b]		
23	23.11	178.0	117.5	3350	—[b]		

[a] Groove details as given in SAE J636, Table 1, Fig. 1.
[b] The horsepower to be absorbed at the driven pulley is to be according to agreement between user and manufacturer.

TABLE 4A—TEST CONDITIONS[a]
COG OR NOTCHED BELTS, IN

SAE Belt Size	Standard Groove Width	Diameter Where Specified Groove Width Occurs (without width tolerance)		Driver Pulley Speed, rpm ±2%	Load, hp	Length Range	
		Driver and Driven Pulleys ±0.010	Tension Pulley ±0.010			Total	Preferred
0.380	0.380	4.750	2.250	4900	10.0 / 11.0 / 12.5	Under 50 / 40-55 / 55 and over	36-40 / 45-50 / 55-60
0.470		See Appendix					
0.500	0.500	5.000	2.750	4700	11.5 / 13.0 / 14.0	Under 40 / 40-55 / 55 and over	36-40 / 45-50 / 55-60
11/16 (0.597)	0.597	5.000	3.250	4700	—[b]		
3/4 (0.660)	0.660	5.000	3.375	4700	—[b]		
7/8 (0.785)	0.785	6.000	3.750	3900	—[b]		
1 (0.910)	0.910	7.000	4.375	3350	—[b]		

[a] Groove details as given in SAE J636, Table 1, Fig. 1.
[b] The horsepower to be absorbed at the driven pulley is to be according to agreement between user and manufacturer.

TABLE 4B—TEST CONDITIONS[a]
COG OR NOTCHED BELTS, MM

SAE Belt Size	Standard Groove Width	Diameter Where Specified Groove Width Occurs (without width tolerance)		Driver Pulley Speed, rpm ±2%	Load, kW	Length Range	
		Driver and Driven Pulleys ±0.25	Tension Pulley ±0.25			Total	Preferred
10	9.65	120.5	57.0	4900	7.5 / 8.2 / 9.4	Under 1020 / 1020-1400 / 1400 and over	920-1020 / 1140-1270 / 1400-1520
12		See Appendix					
13	12.70	127.0	70.0	4700	8.6 / 9.8 / 10.4	Under 1020 / 1020-1400 / 1400 and over	920-1040 / 1140-1270 / 1400-1520
15	15.16	127.0	82.5	4700	—[b]		
17	16.76	127.0	85.5	4700	—[b]		
20	19.94	152.5	95.0	3900	—[b]		
23	23.11	178.0	111.0	3350	—[b]		

[a] Groove details as given in SAE J636, Table 1, Fig. 1.
[b] The horsepower to be absorbed at the driven pulley is to be according to agreement between user and manufacturer.

The belt manufacturer's test data on belts of a certain construction specification shall be considered valid for evaluation of all belts of the same construction specification regardless of the intended user. Belts shall be considered to be of the same construction specification when they are the same with respect to the manufacturer's cross-section dimensions, material specifications, and method of manufacture.

In evaluating for part source approval and for production quality surveillance, test data for the entire length group containing a part in question shall be considered pertinent. The design of some test machines may not accommodate the shortest lengths shown in Fig. 2. In such cases, test data on some longer belt(s) of the same construction specification and within the length group 28–40 in. (710–1020 mm) shall be used. Similarly, test data on belt(s) within the length group 56–68 in. (1420–1730 mm) shall be used for lengths beyond 68 in. (1730 mm).

Whether testing is performed for part source approval or for production quality surveillance, a realistic statistical guide to acceptability would be "not more than 10% of test lives shall be permitted to fall below 50% of the specified average life."

For part source approval, test data of the immediately preceding 3 month period shall be considered pertinent. When such data are not sufficient for the statistical evaluation, the manufacturer may have the option of submitting data for source approval on a "sample" of the part under consideration or on samples of the same length group and construction specification. Because the data would be limited in this situation, a guide to approval could be to permit no test result to be below 50% of the specified average life.

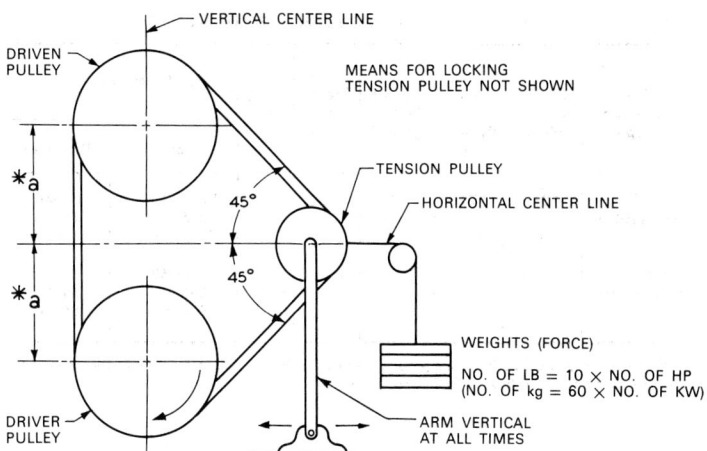

*Dimension a is adjusted for various length belts to maintain tension pulley midway vertically between driver and driven pulleys.
**45 deg is specified for initial test configuration and may change slightly with resets as test progresses.

FIG. 1—V-BELT FATIGUE TEST

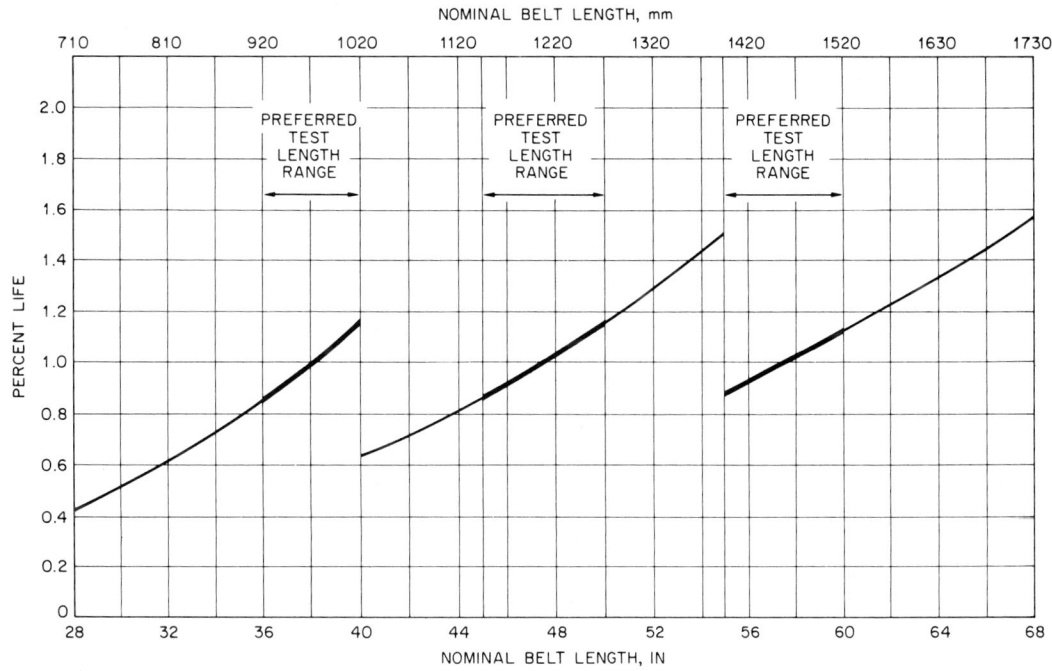

FIG. 2—TYPICAL LIFE-LENGTH CURVES (FOR TEST CONDITIONS IN TABLE 1)

APPENDIX
STANDARDIZED LABORATORY TESTING
OF
NONSTANDARDIZED AUTOMOTIVE V-BELTS

The following information is supplementary to the test conditions for the six standard SAE top width belts shown in Tables 3 and 4.

"High ride" belts have been tried in a number of different standard SAE top width sheave grooves. At least one of these "high ride" belts has had sufficient usage to be recognized and test conditions specified for it. The 0.470 in. (12 mm) nominal top width belt originated as a "high ride" belt used in standard 0.380 in. (9.65 mm) top width sheaves.

In addition to the 0.470 in. (12 mm) belt in the 0.380 in. (9.65 mm) standard groove, there has been some use of this belt in nonstandard sheave grooves of 0.470 and 0.440 in. (11.94 and 11.18 mm) top width. If a new standard groove is adopted, it would most likely be the 0.440 in. (11.18 mm) top width which is the midpoint between existing standards.

Considerations for the "high ride" belt in the standard 0.380 in. (9.65 mm) groove are:

1. Improved belt life since the belt diameters are effectively larger slightly increasing belt velocity and decreasing bending stress.

2. Decreased tension decay resulting from No. 1 above, and a slight increase in belt tensile width.

3. Change in speed ratio with a reduced driven speed when used on such accessories as alternator drives.

4. Questions on belt stability.

Decisions associated with the use of standard 0.380 or 0.500 in. (10 or 13 mm) belts in their standard top width grooves, or the 0.470 in. (12 mm) nominal top width belt in either a 0.470 or 0.440 in. (11.94 or 11.18 mm) top width groove, should be made on the basis of drive design data or experience.

The test conditions shown in Table 5 are applicable to the 0.470 in. (12 mm) top width belt as a "high ride" belt in 0.380 in. (9.65 mm) standard top width sheaves.

TABLE 5A—TEST CONDITIONS, 0.470 IN BELT, IN

Belt Size and Type	Standard Groove Width	Driver and Driven Pulleys ±0.010	Tension Pulley ±0.010	Driver Speed, rpm ±2%	Load, hp	Length Range	
						Total	Preferred
0.470[a] Plain section	0.380	4.750	2.500	4900	11.0 12.5 13.5	Under 40 40–55 55 and over	36–40 45–50 55–60
0.470[a] Notched and cog	0.380	4.750	2.250	4900	11.0 12.5 13.5	Under 40 40–55 55 and over	36–40 45–50 55–60

[a] Ride-out in SAE 0.380 in pulley grooves shall be 1/8 ± 1/32 in consistent with the measuring requirements of Table 2 of SAE J636.

TABLE 5B—TEST CONDITIONS, 12 MM BELT, MM

Belt Size and Type	Standard Groove Width	Driver and Driven Pulleys ±0.25	Tension Pulley ±0.25	Driver Speed, rpm ±2%	Load, kW	Length Range	
						Total	Preferred
12 Plain section	9.65	120.5	63.5	4900	8.2 9.4 10.1	Under 1020 1020–1400 1400 and over	920–1020 1140–1270 1400–1520
12 Notched and cog	9.65	120.5	57.0	4900	8.2 9.4 10.1	Under 1020 1020–1400 1400 and over	920–1020 1140–1270 1400–1520

19 Springs

LEAF SPRINGS FOR MOTOR VEHICLE SUSPENSION—METRIC BAR SIZES—SAE J1123a

SAE Standard

Report of Spring Committee approved November 1975 and last revised January 1977.

This SAE Standard is limited for the present to the presentation of metric bar sizes and tolerances. They are not identical with those in SAE J510, but the sizes as well as the tolerances follow an analogous pattern. The Spring Committee is now engaged in writing a new MANUAL ON DESIGN AND APPLICATION OF LEAF SPRINGS IN SI (METRIC) UNITS (which will eventually replace SAE J788), and as this work progresses the new Standard will be expanded and will eventually replace SAE J510.

As with all leaf spring bars for automotive springs adopted as SAE standard since 1938, the metric bars shall be of flat rolled steel having two flat surfaces and two rounded (convex) edges. They are subject to the tolerances shown in Table 1. These cross section tolerances permit the two flat surfaces to be slightly concave. When that occurs, the radii of the arcs of the two concave surfaces shall be of approximately equal length.

The rounding of the convex edges shall be an arc with a radius of curvature that may vary from 65% to 85% of the thickness of the bar.

Bars shall be substantially straight and free from physical characteristics known as "kinks" or "twists" which render them unsatisfactory for spring manufacturing purposes.

Distortions due to a bar being bent about either major axis of section shall be measured with the bar against a flat checking surface so as to make contact with this surface near both bar ends. Gaps between the bar and the checking surface shall not exceed 4.0 mm/1 m of bar length out of contact with the checking surface when this bar length is greater than 1 m. Also, a gap between the bar and a straight edge 1 m long applied along any portion of the surface or edge of the bar shall not exceed 4.0 mm.

It is recommended that all leaf spring bars which have been cold straightened be identified by the steel mill so that the spring manufacturer can use them selectively.

The bars which are generally provided in alloy steel shall be specified and rolled in the following mm widths and thicknesses:

TABLE 1—CROSS SECTION TOLERANCES, mm

Width	Width Tolerance	Tolerance In Thickness (±)[a] And In Flatness (−)[b]			Maximum Difference In Thickness[c]		
		For Thickness			For Thickness		
	Minus 0.00	5.00 to 9.50	10.00 to 21.20	22.40 to 37.50	5.00 to 9.50	10.00 to 21.20	22.40 to 37.50
40.0	+0.75	0.13	0.15	—	0.05	0.05	—
45.0	+0.75	0.13	0.15	—	0.05	0.05	—
50.0	+0.75	0.13	0.15	—	0.05	0.05	—
56.0	+0.75	0.13	0.15	—	0.05	0.05	—
63.0	+0.75	0.13	0.15	—	0.05	0.05	—
75.0	+1.15	0.15	0.20	0.30	0.08	0.10	0.15
90.0	+1.15	0.15	0.20	0.30	0.08	0.10	0.15
100.0	+1.15	0.15	0.20	0.30	0.08	0.10	0.15
125.0	+1.65	0.18	0.25	0.40	0.10	0.13	0.20
150.0	+2.30	—	0.30	0.50	—	0.15	0.25

[a] Thickness measurements shall be taken at the edge of the bar where the flat surfaces intersect the rounded edge.
[b] This tolerance represents the maximum amount by which the thickness at the center of the bar may be less than the thickness at the edges. Thickness at the center may never exceed the thickness at the edges.
[c] Maximum difference in thickness between the two edges of each bar.

Widths		Thicknesses					
40.0	75.0	5.00	7.10	10.00	14.00	20.00	28.00
45.0	90.0	5.30	7.50	10.60	15.00	21.20	30.00
50.0	100.0	5.60	8.00	11.20	16.00	22.40	31.50
56.0	125.0	6.00	8.50	11.80	17.00	23.60	33.50
63.0	150.0	6.30	9.00	12.50	18.00	25.00	35.50
		6.70	9.50	13.20	19.00	26.50	37.50

It should be noted that all the widths and thicknesses are "Preferred Numbers" in accordance with American National Standard ANSI Z17.1-1973.

LEAF SPRINGS FOR MOTOR VEHICLE SUSPENSION—SAE J510c

SAE Standard

Report of Springs Division approved August 1951 and last revised by Spring Committee August 1973. Editorial change March 1974.

This SAE Standard is limited to concise specifications promoting an adequate understanding between spring maker and spring user on all practical requirements in the finished spring. THE BASIC CONCEPTS FOR THE SPRING DESIGN AND FOR MANY OF THE DETAILS HAVE BEEN FULLY DEALT WITH IN HS-J788A, SAE INFORMATION REPORT, MANUAL ON DESIGN AND APPLICATION OF LEAF SPRINGS, which is available from SAE Headquarters, will be cited in many instances for a more thorough explanation or discussion.

Bar Sizes and Tolerances—Round edge flat spring steel has been adopted as the SAE standard.

The bars shall be of flat rolled steel having two flat surfaces and two rounded (convex) edges. They are subject to the tolerances shown in Table 1. These cross section tolerances permit the two flat surfaces to be slightly concave. When that occurs, the radii of the arcs of the two concave surfaces shall be of approximately equal length.

TABLE 1—CROSS SECTION TOLERANCES

Nominal Width				Tolerance in Width		For Thickness		Tolerance in Thickness[a]		Tolerance in Flat Surfaces[b]		Max Difference[c] in Thickness	
Over		To and Including		in	mm			±in	±mm	in	mm	in	mm
in	mm	in	mm	−0.00	−0.0	in	mm						
0.00	0.0	2.50	63.5	+0.030	+0.76	0.375 or under	9.52	0.005	0.13	−0.005	−0.13	0.002	0.05
						Over 0.375 to 0.875, incl	9.52–22.22	0.006	0.15	−0.006	−0.15	0.002	0.05
2.50	63.5	4.00	101.6	+0.045	+1.14	0.375 or under	9.52	0.006	0.15	−0.006	−0.15	0.003	0.08
						Over 0.375 to 0.875, incl	9.52–22.22	0.008	0.20	−0.008	−0.20	0.004	0.10
						Over 0.875 to 1.500, incl	22.22–38.10	0.012	0.30	−0.012	−0.30	0.006	0.15
4.00	101.6	5.00	127.0	+0.065	+1.65	0.375 or under	9.52	0.007	0.18	−0.007	−0.18	0.004	0.10
						Over 0.375 to 0.875, incl	9.52–22.22	0.010	0.25	−0.010	−0.25	0.005	0.13
						Over 0.875 to 1.500, incl	22.22–38.10	0.016	0.41	−0.016	−0.41	0.008	0.20
5.00	127.0	6.00	152.4	+0.090	+2.29	Over 0.375 to 0.875, incl	9.52–22.22	0.012	0.30	−0.012	−0.30	0.006	0.15
						Over 0.875 to 1.500, incl	22.22–38.10	0.020	0.51	−0.020	−0.51	0.010	0.25

[a] Thickness measurements shall be taken at the edge of the bar where the flat surfaces intersect the rounded edge.

[b] This tolerance represents the maximum amount by which the thickness at the center of the bar may be less than the thickness at the edges. Thickness at the center may never exceed the thickness at the edges.

[c] Maximum difference in thickness between the two edges of each bar.

The rounding of the convex edges shall approximate a circular arc with a radius equal to at least three-quarters the thickness of the bar.

Bars shall be substantially straight and free from physical characteristics known as "kinks" or "twists" which render them unsatisfactory for spring manufacturing purposes.

Distortions due to a bar being bent about either major axis of section shall be measured with the bar against a flat checking surface so as to make contact with this surface near both bar ends. Gaps between the bar and the checking surface shall not exceed 0.05 in./ft (4.2 mm/m) of bar length out of contact with the checking surface when this bar length is greater than 3 ft (0.9 m). Also, a gap between the bar and a straight edge 3 ft (0.9 m) long applied along any portion of the surface or edge of the bar shall not exceed 0.15 in. (3.8 mm).

It is recommended that all leaf spring bars which have been cold straightened be identified by the steel mill so that the spring manufacturer can use them selectively.

Leaf spring bars are generally available in the following widths:

in	mm	in	mm	in	mm
1.75	44.4	2.50	63.5	4.00	101.6
2.00	50.8	3.00	76.2	5.00	127.0
2.25	57.2	3.50	88.9	6.00	152.4

Spring drawings shall specify steel of the following nominal thicknesses, to which all bars shall be rolled:

in	mm	in	mm	in	mm	in	mm	in	mm
0.194	4.93	0.276	7.01	0.401	10.19	0.625	15.88	0.999	25.37
0.204	5.18	0.291	7.39	0.423	10.74	0.662	16.81	1.061	26.95
0.214	5.44	0.307	7.80	0.447	11.35	0.702	17.83	1.127	28.63
0.225	5.72	0.323	8.20	0.473	12.01	0.744	18.90	1.197	30.40
0.237	6.02	0.341	8.66	0.499	12.67	0.788	20.02	1.273	32.33
0.249	6.32	0.360	9.14	0.527	13.39	0.836	21.23	1.354	34.39
0.262	6.65	0.380	9.65	0.558	14.17	0.887	22.53	1.440	36.58
				0.590	14.99	0.941	23.90		

Definitions, Dimensions, and Tolerances

Leaf Spring—A spring of full elliptic, semi-elliptic, or quarter-elliptic shape with one or more leaves. The term "multi-leaf" has generally applied to springs of constant width and with stepped leaves, each of constant thickness except where leaf ends may be tapered in thickness. More recently, the term has been extended to include an assembly of stacked "single" leaves, each of which is characterized by tapering either in width or in thickness or by a combination of both. Examples of multi-leaf springs are shown in Figs. 1, 3, 4, 5, 6, and 8; Fig. 7 shows a single leaf spring.

The leaves of a multi-leaf spring are usually held together with a center bolt and prevented from lateral shifting by alignment clips. Prior to assembly, the leaves are formed (cambered) and heat treated by heating, quenching, and tempering to the required hardness. Quench dies or fixtures are used to maintain the required camber within tolerances.

Datum Line—Reference line used with many of the subsequently defined terms. In Fig. 1 (where the springs are shown inverted as in a machine for load and rate checking), it is shown as the line X-X. On springs with eyes, the datum line passes through the centers of the eyes. On other springs it passes through the points where the load is applied near the ends of the spring. These points must be indicated on the drawing. When load and rate are checked, the spring ends shall be free to move in the direction of the datum line.

Loaded Length—Distance between spring eye centers when the spring is deflected to the specified load position. On springs without eyes, it is the distance between the lines where load is applied under the specified conditions. Tolerance, ±0.12 in. (±3.0 mm).

Loaded Fixed End Length—Distance from the center of the fixed eye to the projection on the datum line of the point where the centerline of the center bolt intersects the spring surface in contact with the spring seat. Tolerance, ±0.06 in. (±1.5 mm).

Straight Length—Distance between spring eye centers when the main leaf is flat, measured parallel to the flat leaf. Tolerance, ±0.12 in. (±3.0 mm).

Seat Length—Length of spring that is in physical engagement with the spring seat when installed on a vehicle at design height. It is always greater than the inactive or clamp length.

Clamp Length—Length of spring rendered inactive by the clamp located on the side opposite the spring seat. It is always less than the length in physical engagement with the clamp.

Seat Angle—Angle between the tangent to the center of the spring seat and the seat angle baseline which is a line drawn through the terminal points of the active spring length at each eye, taken along the tension surface of the main leaf (see Fig. 1). When both ends of the spring have eyes of identical configuration and diameter (or have plain ends without eyes), the seat angle is the angle between the tangent to the center of the spring seat and the datum line.

The seat angle is considered either positive or negative, depending upon the direction in which the tangent to the center of the spring seat is disposed from the seat angle base line.

With the fixed end of the spring shown to the left of the drawing and the load applied to the shortest leaf from above (see Fig. 1: spring upside down from normal vehicle position, as in the spring tester), the seat angle is considered positive when that tangent is disposed counterclockwise; it is considered negative when that tangent is disposed clockwise.

Consequently, with the fixed end of the spring shown to the left of the drawing and the load applied to the shortest leaf from below (spring in normal vehicle position), the seat angle is considered positive when that tangent is disposed clockwise; it is considered negative when that tangent is disposed counterclockwise.

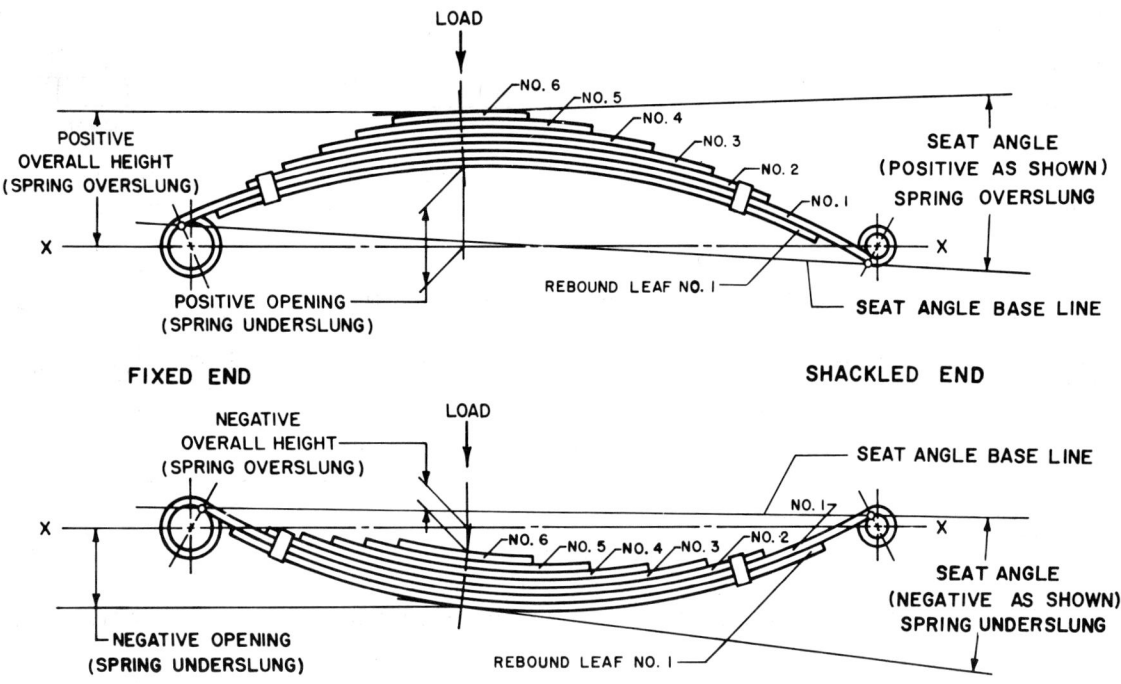

FIG. 1—MEASUREMENT OF OPENING, OVERALL HEIGHT, AND SEAT ANGLE

Finished Width—Width to which the spring leaves are ground or milled to give the edges a flat bearing surface. If the edges of the spring ends are finished, the required length of the finished edge must also be indicated. Where the design calls for edge finishing of spring leaves, the finished dimension will usually equal the nominal width of the leaf (which equals the minimum limit of the rolled bar) with a minus tolerance of 0.010 in. (0.25 mm) for widths up to and including 2.00 in. (50.8 mm), 0.015 in. (0.38 mm) for widths to 2.50 in. (63.5 mm), and 0.020 in. (0.51 mm) for widths exceeding 2.50 in. (63.5 mm). Where required, a finished dimension substantially less than the nominal width may be obtained by machining or trimming with tolerances to suit the installation.

Assembled Spring Width—Where more than one leaf constitutes a

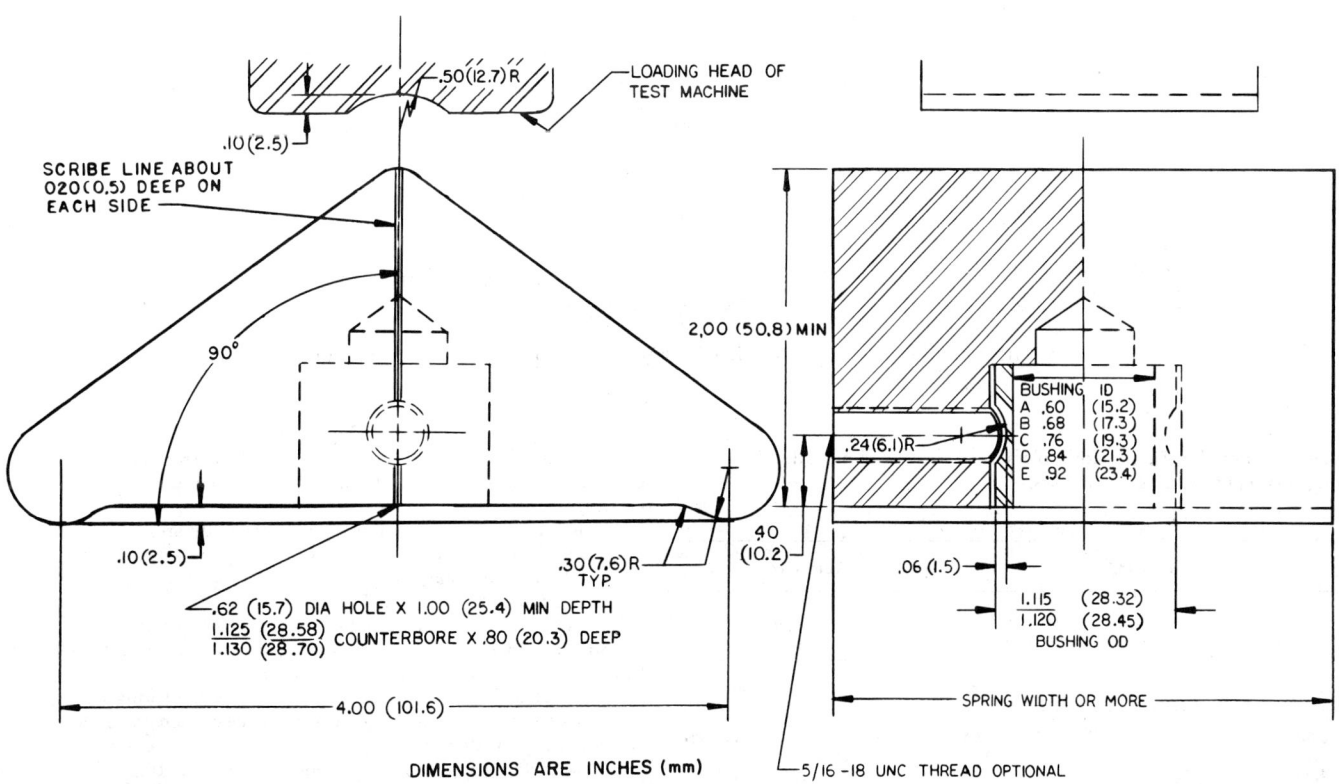

FIG. 2—SPRING LOADING BLOCK

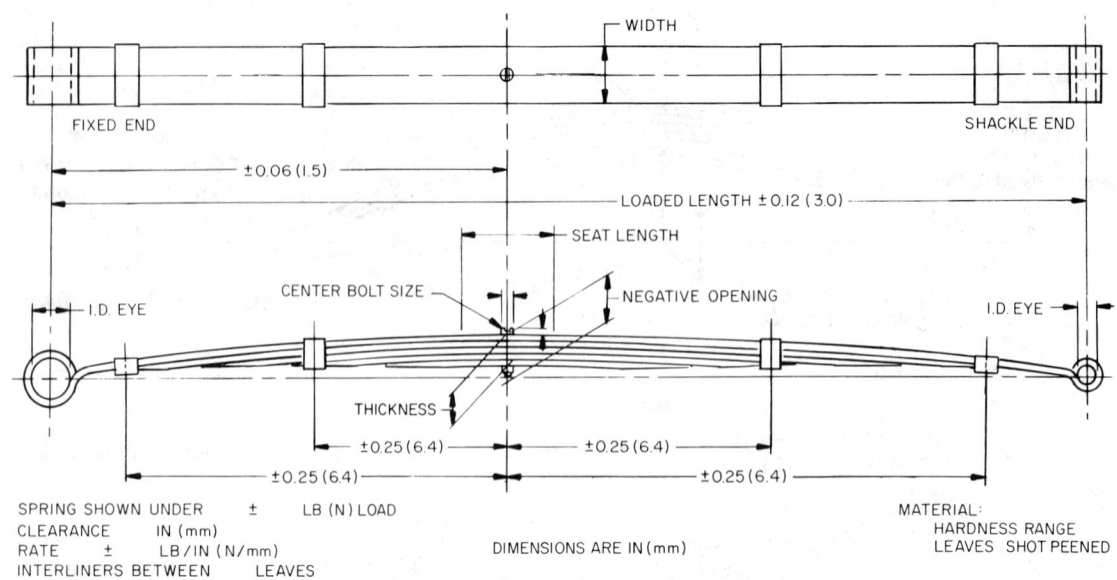

FIG. 3—MINIMUM SPECIFICATION REQUIREMENTS FOR UNDERSLUNG SPRINGS WITH NEGATIVE OPENING

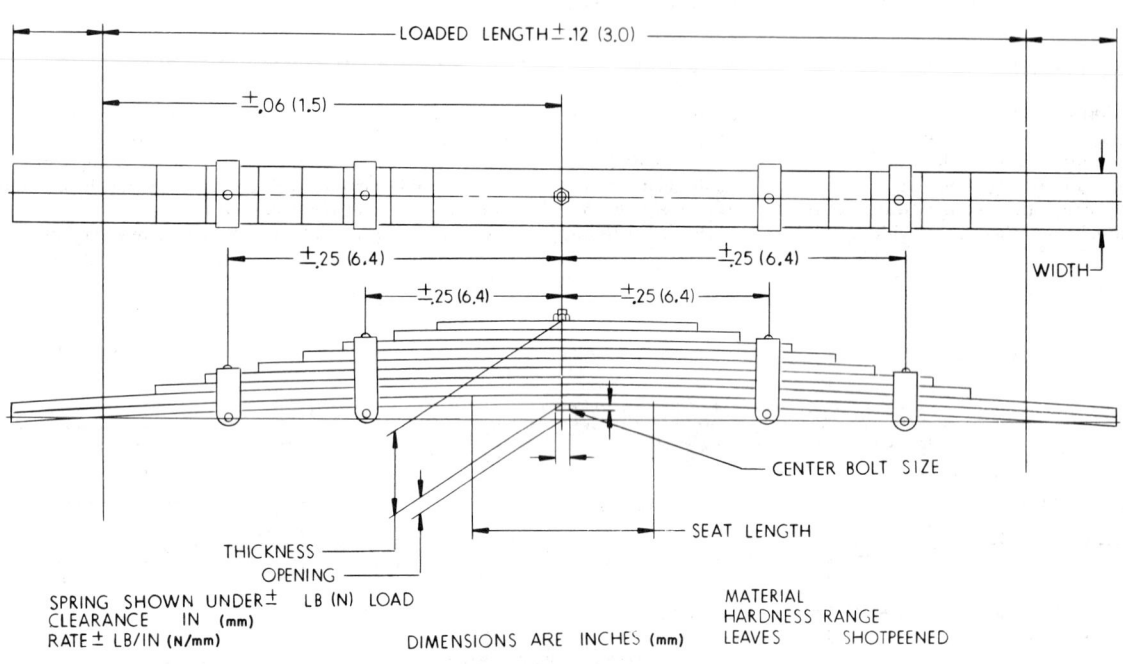

FIG. 4—MINIMUM SPECIFICATION REQUIREMENTS FOR SPRINGS WITH PLAIN ENDS

spring assembly, the overall width tolerance of the assembly made within the spring seat length shall be as follows:

Leaf Width				Tolerance	
Over		To and including		−0.000	
in	mm	in	mm	in	mm
0.00	0.0	2.50	63.5	+0.100	+2.5
2.50	63.5	4.00	101.6	+0.120	+3.0
4.00	101.6	5.00	127.0	+0.145	+3.7
5.00	127.0	6.00	152.4	+0.175	+4.4

Thickness—Aggregate of the nominal thicknesses of all leaves of the spring including any liners and spacer plates which are part of the spring at the seat.

Leaf Ends—The leaf ends used most generally are:
Square as sheared
Trimmed to a shape
Taper rolled
Taper rolled; trimmed or forged to a shape or both.

Surface Finish—Condition of the surface of the spring leaves after the steel has been heat treated and prior to coating.

"As Heat Treated" Finish—the surface of the spring leaves is in the condition as taken from the heat treating furnace where generally the leaves have a finish of oxide coating.

"Shot Peened" Finish—the tension surface of the spring leaves has been exposed to the shot peening operation where the oxide coating and scale are removed and a matte luster finish is produced.

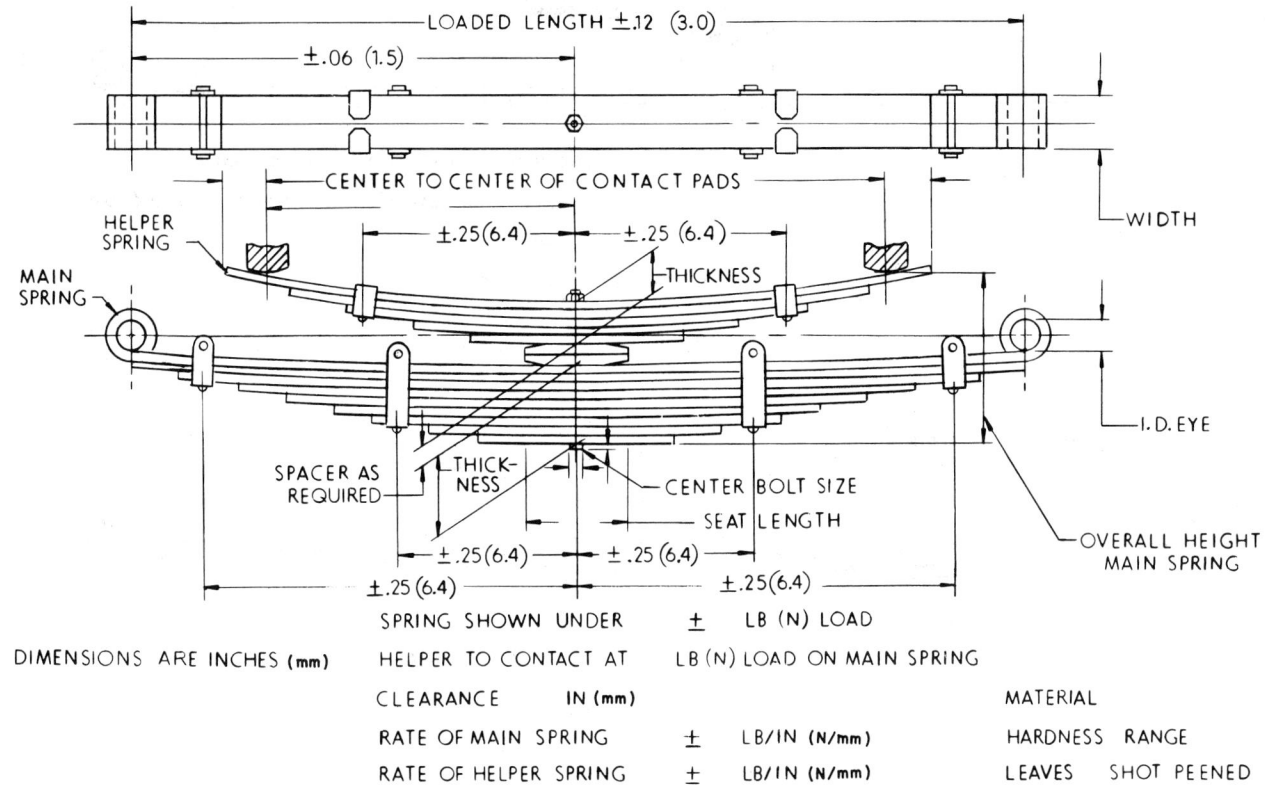

FIG. 5—MINIMUM SPECIFICATION REQUIREMENTS FOR OVERSLUNG COMMERCIAL VEHICLE SPRINGS

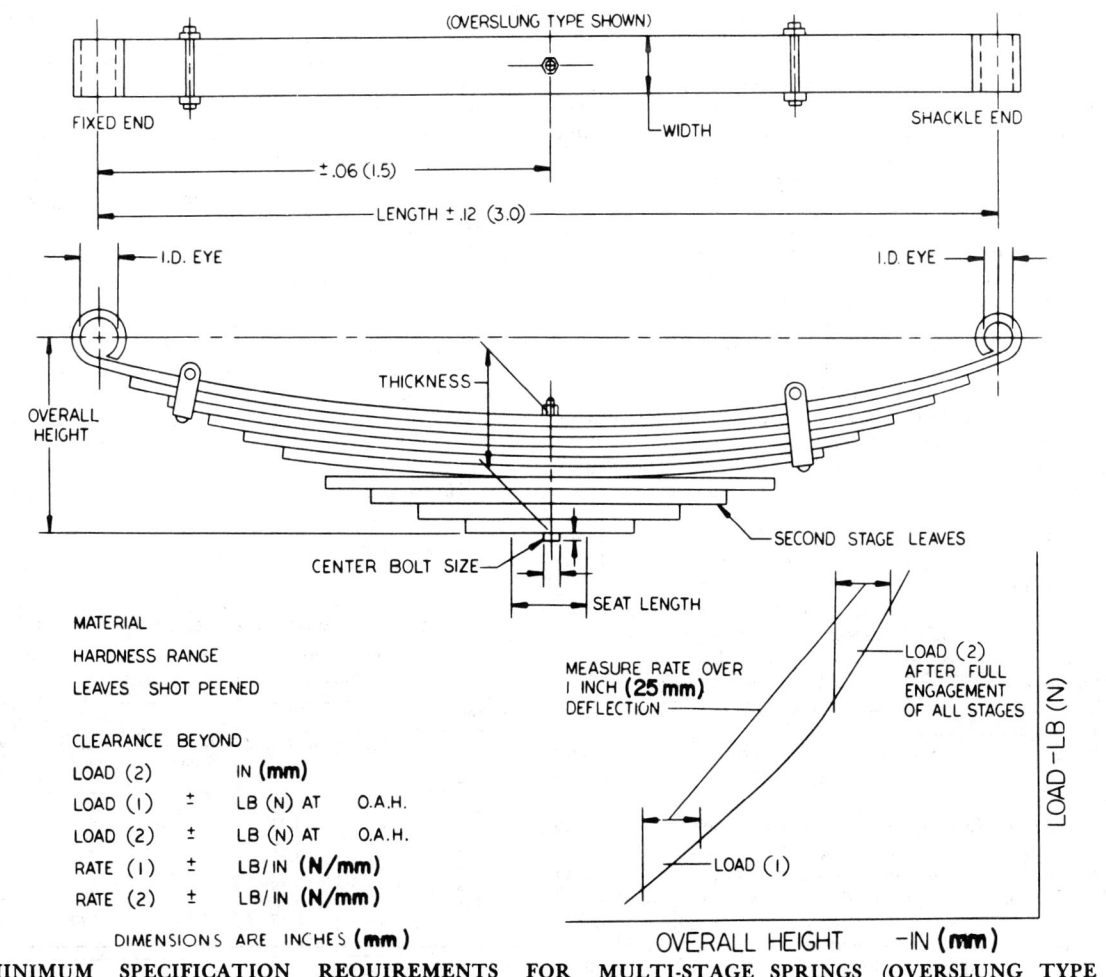

FIG. 6—MINIMUM SPECIFICATION REQUIREMENTS FOR MULTI-STAGE SPRINGS (OVERSLUNG TYPE SHOWN)

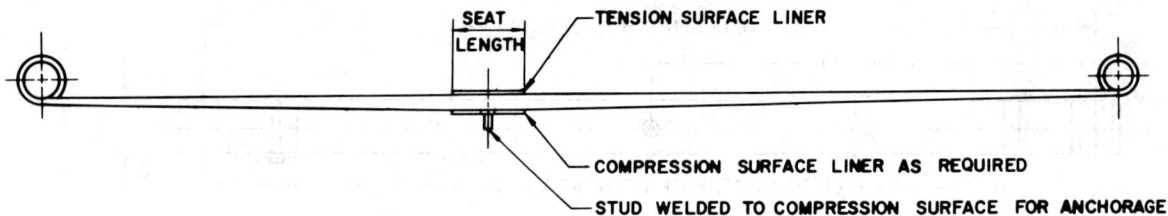

FIG. 7—SINGLE LEAF SPRINGS

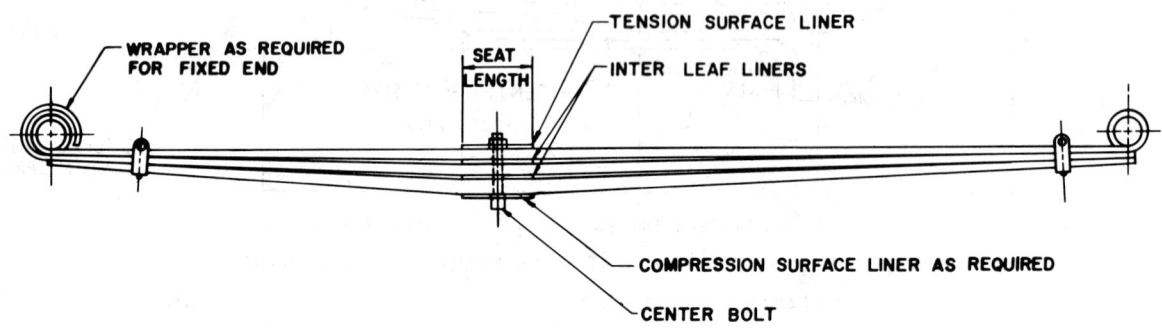

FIG. 8—STACKED SINGLE LEAF SPRINGS

"Ground or Polished Leaf Ends"—the bearing areas of leaves are ground or polished to produce a smooth surface for reduced friction. The distance or length to be ground or polished should be specified.

Protective Coating—Material added to surface of spring leaves or exposed areas of assembled springs. For additional information, see HS-J788a.

Leaf Numbers—See Fig. 1. Leaves are designated by numbers, starting with the main leaf which is No. 1. The adjoining leaf is No. 2, and so on. If rebound leaves are used, the rebound leaf adjoining the main leaf is rebound leaf No. 1, the next one rebound leaf No. 2, and so on. (Rebound leaves are on that side of the main leaf on which load is applied to the spring ends; away from the side on which load is applied to the spring center.) Helper springs are considered as separate units.

Opening and Overall Height—See Fig. 1. Distance from the datum line to the point where the center bolt centerline or cup center intersects the surface of the spring that is in contact with the spring seat.

If the surface in contact with the seat is on the main leaf or a rebound leaf (as on underslung springs), this distance is called "opening."

If the surface in contact with the seat is on the shortest leaf (as on overslung springs), this distance is called "overall height."

"Opening" and "overall height" may be positive or negative (see Fig. 1). They are fixed reference dimensions. Tolerance is expressed in terms of load. See paragraph on load.

Clearance—Difference in opening, or overall height, between the design load position and the extreme position (of maximum stress) to which the spring can be deflected on the vehicle.

Load—The force exerted by the spring at the specified opening or overall height. The total tolerance on load at the specified overall height or opening is usually expressed as a load range (lb, N) which is equivalent to a deflection (in., mm) at the nominal rate (lb/in., N/mm). This deflection may be as small as 0.25 in. (6.4 mm) for a passenger car spring and as large as 0.50 in. (12.7 mm) for a heavy truck spring.

Rate—Half the difference between the loads measured 1 in. above and 1 in. below the specified load position. Tolerance, ±5% on light springs and ±7% on heavy springs.

Load and Rate Checking—Load and rate are the terms usually employed to describe the basic characteristics of a leaf spring and, as specified on the spring drawing, refer to quantities measured on the spring without center clamp and without shackles. They are, therefore, not the same as those of the installed spring.

When the load is measured, the spring ends are free to move in the direction of the datum line; the ends are usually mounted on carriages with rollers.

The spring shall be supported on its ends, and the load shall be applied to the shortest leaf from above. It shall be transmitted from the testing machine head through a standard SAE loading block, shown in Fig. 2. The loading block shall be centered over the center bolt with the legs of the V resting on the spring. It is understood that the load specified on the spring drawing does not include either the weight of the spring or the weight of the loading block.

Just before the spring is checked for load or rate, it shall undergo a preloading operation. During the initial preloading by the spring maker, the spring shall be deflected at least to the position defined under the paragraph on Clearance. During any subsequent preloading, the spring shall be deflected only to and not beyond this "clearance position" in order to remove any temporary recovery from the set incurred during the initial preloading. After the spring has been preloaded, it shall be released to the free position before the load is applied for load and rate checking. For additional information on preloading, see HS-J788a.

Load and rate shall be measured in terms of the forces exerted by the spring during compression of the spring (compression loads) and not during release of the spring (release loads). The compression load in any position shall be read only after the spring has been thoroughly rapped in that position with a plastic or soft metal hammer.

Specification Requirements—Minimum specification requirements are given in Figs. 3-6.

TABLE 2—RECOMMENDED CENTER BOLT AND NUT DIMENSIONS, IN

Bolt Diameter	Threads[a]		Head Size		Nut Size	
	Per Inch	Minimum Length	Nominal Diameter	Nominal Height	Nominal Width Across Flats	Nominal Thickness
5/16	24	1.00	1/2	1/4	1/2	17/64
3/8	24	1.00	9/16	5/16	9/16	21/64
7/16	20	1.25	5/8	3/8	11/16	3/8
1/2	20	1.25	3/4	7/16	3/4	7/16
5/8	18	1.50	7/8	9/16	15/16	35/64
3/4	16	1.75	1	5/8	1-1/8	41/64
7/8	14	2.00	1-1/8	11/16	1-5/16	3/4
1	12	2.25	1-5/16	25/32	1-1/2	55/64

[a] Threads are Unified Standard Fine, Class 2A.

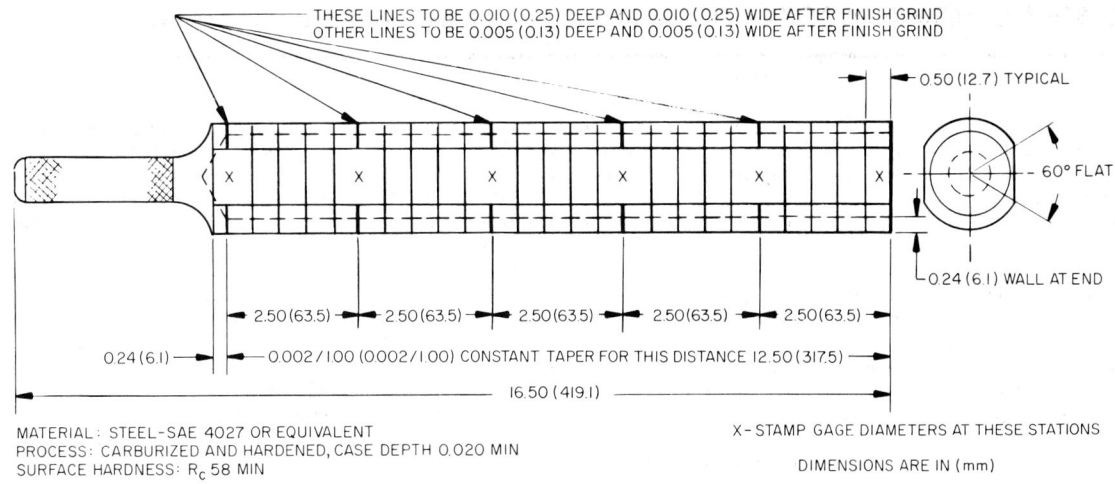

FIG. 9—GAGE-LEAF SPRING EYE PLUG

Spring Eyes and Bushings—For some types of currently used spring eyes, spring ends, bushings, and shackle constructions, see HS-J788a.

For eyes with specified inside diameter, the size and roundness of the eye should be checked by means of a round plug gage from which two opposite segments of 60 deg have been removed. The gage shall have a taper on diameter per unit of length of 0.002:1 (see Fig. 9). The gage shall be inserted into the eye three times from each side at angular positions differing by about 60 deg. The eye is acceptable only if the gage reading on the side of the eye from which the gage is inserted is within the specified diametral limits at each of the six checks.

Also, the eye should be checked with a round plug, GO/NO GO gage, to determine if the eye is cone shaped or tapered. The GO diameter must pass completely through the eye and the NO GO diameter must not enter the eye from either side.

Total tolerance—for 1 in (25.4 mm) or less diameter: 0.010 in. (0.25 mm).
—for larger than 1 in. (25.4 mm) diameter: 1% of nominal diameter (example: 0.015 in. (0.38 mm) for 1.50 in. (38.1 mm) ID eye); where bushing retention is critical, the 1% tolerance may be reduced to 0.75%.

For a bushing where the ID may have been affected by pressing into the spring eye, it should be checked with a round plug gage. Total tolerance: 0.005 in. (0.13 mm) unless otherwise specified.

Eyes of the main leaf in the assembled spring, measured in the unloaded condition, shall be parallel to the surface at the spring seat, and square with a tangent to either edge of the main leaf at the spring seat, within ±1 deg.

Alignment Clips—Most automotive leaf springs are fitted with clips of some form which serve primarily to prevent sidewise spread and vertical separation of the leaves.

Clips employed for passenger car springs show a great variety in design, but commercial vehicle springs are generally equipped with either bolt type or clinch type clips. See HS-J788a. Dimensions must be chosen to suit the individual service requirement.

Center Bolt—The center bolt is required to hold the spring leaves together, and the center bolt head is used as a locating dowel during installation on the vehicle. For underslung springs, the head should be adjacent to the main leaf; and for overslung springs, the head should be adjacent to the short leaf. The center bolt should not be depended upon to prevent the shifting of leaves due to driving and braking forces.

In most cases, center bolts are highly stressed in the handling of the springs and in service. Therefore, it is necessary to use heat treated bolts of high mechanical properties. See Table 2.

Cup Center—Cup centers are often used in heavy duty springs which may not safely depend alone on center clamps and bolts to prevent a shifting of the spring on the axle seat due to driving and braking forces.

When the main leaf is assembled adjacent to the axle seat as in underslung springs, the main leaf only is cupped (away from the No. 2 leaf); and when the short leaf is mounted above the axle seat as in overslung springs, all the leaves must be cupped toward the short leaf.

This method of cupping locks in the main leaf which is subject to horizontal forces to the axle seat and eliminates the shear from the clamp and center bolts.

There are several types of cup centers in general use, two of which are shown in Fig. 10. The cup dimensions are listed according to center bolt diameter; however, the diameter of the cup should not exceed one-half of the leaf width, and the depth of the cup should not exceed one-half of the leaf thickness.

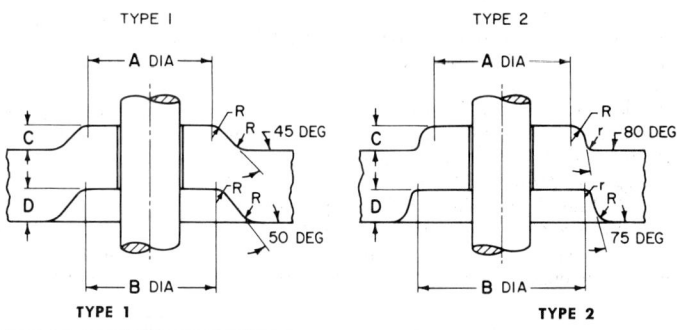

5/16, 3/8, 7/16, 1/2		5/8, 3/4		Center Bolt Dia			5/16, 3/8, 7/16, 1/2		5/8, 3/4	
				Dimension	Tolerance					
in	mm	in	mm		+0.00 in	+0.0 mm	in	mm	in	mm
0.84	21.3	1.24	31.5	Dia A	−0.02	−0.5	1.06	26.9	1.30	33.0
0.88	22.4	1.30	33.0	Dia B	−0.02	−0.5	1.16	29.5	1.42	36.1
0.14	3.6	0.20	5.1	Height C	−0.02	−0.5	0.18	4.6	0.24	6.1
0.18	4.6	0.24	6.1	Depth D	−0.02	−0.5	0.22	5.6	0.28	7.1
0.10	2.5	0.12	3.0	Radius R	−0.01	−0.3	0.10	2.5	0.12	3.0
—	—	—	—	Radius r	−0.01	−0.3	0.04	1.0	0.06	1.5

FIG. 10—CUP CENTERS

PNEUMATIC SPRING TERMINOLOGY—SAE J511a

SAE Information Report

Report of Spring Committee approved March 1960 and last revised October 1970.

Scope—This pneumatic spring terminology has been developed to assist engineers and designers in the preparation of specifications and descriptive material relating to pneumatic springs and their components. It does not include gas supply or control systems.

1. Pneumatic Spring—A spring which utilizes the elasticity of a confined gas as the energy medium.

2. General Terms

2.1 Flexible Member—The flexible portion of the pneumatic spring.

2.1.1 BEAD—That portion of the flexible member adjacent to any attachment part which provides an anchor and a gas seal.

NOTE: The bead can be classified as either mechanically fastened, which produces a seal through a positive clamping medium, or self-sealing, which produces a seal through gas pressure and/or bead displacement.

2.1.2 REINFORCEMENT—A structure of cord built into the flexible member to control its shape and strengthen its wall structure against internal gas pressure.

2.1.2.1 *Cord Angle*—The acute angle between a plane through the axial centerline of the flexible member and the centerline of any cord. This angle can pertain to the as-molded shape of the flexible member and will vary according to position of measurement and cross-sectional shape. It also can pertain to inflated shape and will vary according to position of measurement, cross-sectional shape, and inflation pressure.

NOTE: The cord angle is a determining factor of the inflated shape of the flexible member, and may affect the load-deflection characteristics of the assembly. Since it does not totally govern the load-deflection characteristics, cord angle is not usually specified.

2.1.3 COVER—The external layer of elastic substance which protects the reinforcement against abrasion, weathering, or other undesirable effects.

2.1.4 LINER—The internal layer of elastic substance which affords resistance to gas permeability and protects the reinforcement against aging or the effects of a harmful environment.

2.2 Piston (Internal Support)—The portion of the pneumatic spring which supports the smaller diameter of the flexible member and controls the inward movement of the flexible member during the working stroke, thereby affecting the shape of the load-deflection curve.

2.3 External Support—(See Fig. 1, A, B, C, D.) A component of some pneumatic springs which controls the outside configuration of the flexible member, thereby affecting the shape of the load-deflection curve. The external support may be either fixed (A, B) or floating (C) in relation to one of the beads.

NOTE: Some pneumatic springs do not employ an external support, but rely on the self-restraining construction of the flexible member (D) to perform the functions of the external support.

3. Types of Pneumatic Springs

3.1 Piston Type—This type uses a piston which is attached to the inner bead of a reversible flexible member. See Fig. 1.

3.1.1 REVERSIBLE DIAPHRAGM—In this type, the piston bead usually passes through the opposite bead of the flexible member. See Fig. 1, A and B.

3.1.2 REVERSIBLE SLEEVE—In this type, the piston bead travels within the flexible member and does not pass through the opposite bead. See Fig. 1, C and D.

3.2 Bellows Type—This type utilizes a nonreversible flexible member and relies upon its self-restraining characteristics to affect the load-deflection curve. See Fig. 2.

NOTE: The flexible member (round or oblong in section) may consist of one or more convolutions. A girdle ring is usually used between the convolutions of the round section multiconvolution bellows type pneumatic spring.

3.3 Piston and Cylinder Type—This type uses a piston and cylinder, but does not require a flexible member. A gas tight sliding seal is provided between the piston and cylinder.

3.4 Bladder Type—This type utilizes no integral reinforcement. It relies on being contained with a restrictive structure, such as a coil spring, for its support.

3.5 Hydropneumatic Type—This type contains both liquid and gas. Spring characteristics are provided by the confined gas, while damping may be provided by forcing the liquid through a restriction.

4. Pneumatic Spring Characteristics

4.1 Spring Rate—The change in load per unit of deflection.

NOTE: Fig. 3 illustrates a typical load-deflection curve of a pneumatic spring which has a variable effective area versus spring deflection. (Spring supports and cord construction control the degree of variation of the effective area.) With such a pneumatic spring, the rate varies throughout the spring travel. Pneumatic spring rate also varies with the gas compression process, that is, adiabatic, isothermal, or polytropic. However, it is usually specified as the adiabatic rate at the design position.

4.1.1 ADIABATIC RATE—That rate which results when there is no heat transfer to or from the gas during spring deflection. It is usually approached during rapid spring deflection when there is insufficient time for heat transfer.

4.1.2 ISOTHERMAL RATE—That rate which results when the spring deflects at a constant gas temperature. Isothermal rate is approached when the spring is deflected very slowly to allow time for the transfer of the heat.

4.1.3 POLYTROPIC RATE—That rate which results when there is limited heat transfer to or from the gas during spring deflection. Polytropic rate results during spring deflections which produce neither adiabatic nor isothermal rate.

4.2 Working Volume—The confined gas volume of the pneumatic spring. It is usually specified at design position.

4.3 Design Position—The selected position of the pneumatic spring which satisfies the vehicle requirements. It is usually specified by a dimension between reference points on the fixed and movable parts of the pneumatic spring.

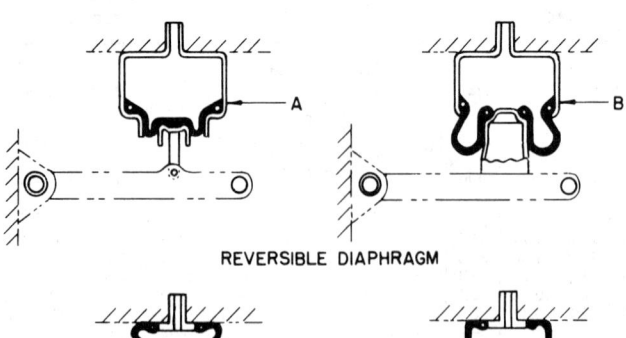

REVERSIBLE DIAPHRAGM

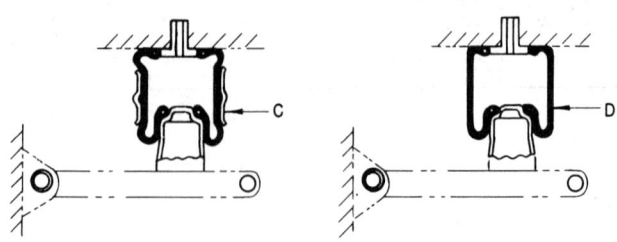

REVERSIBLE SLEEVE

FIG. 1—PISTON TYPE PNEUMATIC SPRINGS

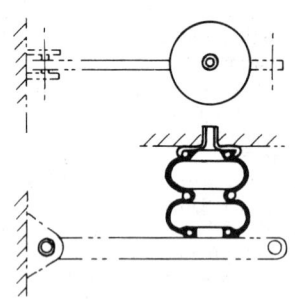

TWO CONVOLUTION CIRCULAR SECTION

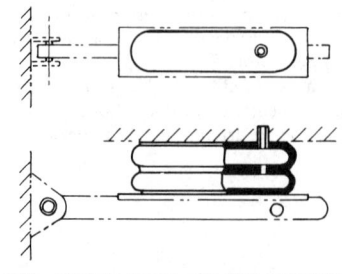

TWO CONVOLUTION OBLONG SECTION

FIG. 2—BELLOWS TYPE PNEUMATIC SPRINGS

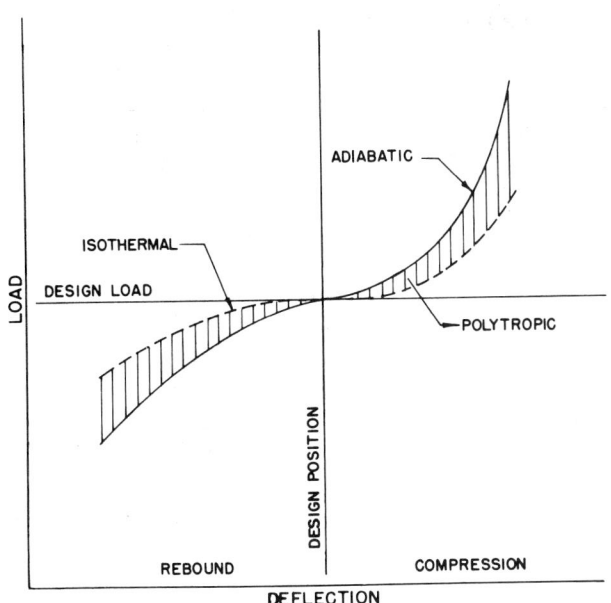

FIG. 3—PNEUMATIC SPRING LOAD-DEFLECTION CURVE

4.4 Total Spring Travel—The distance between the extremes of the spring position measured at the spring axis. It is designated as the total of the compression and rebound deflections from the design position.

4.5 Design Load—The pneumatic spring load at design position.

4.6 Design Pressure—The internal gas pressure required to support the design load at the design position.

4.6.1 PRESSURE LIMITS—The minimum and maximum permissible pressures at the design position which provide satisfactory pneumatic spring operation.

4.7 Effective Area—A nominal area found by dividing the load of the pneumatic spring by its gas pressure at any given spring position.

5. Color Coding To Identify Pneumatic Springs—Pneumatic springs may be color coded for specific properties and operational environments by placing permanent color markings at least 0.25 in. in diameter on a visible section of the flexible member. The recommended guide for normal capabilities is shown in Table 1.

TABLE 1—IDENTIFICATION CODE

Color	Usage Characteristic	General Temperature Range, F
Yellow	Oil resistant	−20 to +150
Red	High temperature	−20 to +180
Green	Low temperature	−65 to +150
No color	General service	−20 to +150

NOTE: These color codes may be used in combination. Other temperature ranges and usage characteristics are available with special materials.

HELICAL COMPRESSION AND EXTENSION SPRING TERMINOLOGY—SAE J1121

SAE Recommended Practice

Report of Spring Committee approved November 1975. SAE J507, J508, and J509 have been discontinued and replaced by this report.

1. Scope—The following recommended practice has been developed to assist engineers and designers in the preparation of specifications for the major types of helical compression and extension springs. It is restricted to a concise presentation of items which will promote an adequate understanding between spring manufacturer and spring user of the major practical requirements in the finished spring. Closer tolerances are obtainable where greater accuracy is required and the increased cost is justified.

For the basic concepts underlying the spring design and for many of the details see the SAE Information Report MANUAL ON DESIGN AND APPLICATION OF HELICAL AND SPIRAL SPRINGS, SAE J795, which is available from SAE Headquarters in Warrendale, PA 15096. A uniform method for specifying design information is shown in the TYPICAL DESIGN CHECK LISTS FOR HELICAL SPRINGS, SAE J1122.

Two types of helical springs are considered:
  Hot-coiled compression springs for general automotive use as well as for motor vehicle suspensions.
  Cold-wound compression and extension springs for general automotive use.

This recommended practice uses SI (metric) units in accordance with the provisions of SAE J916, dated February 1974.

2. Hot-Coiled Springs

2.1 Materials and Heat Treatment—Round spring steel bars are available in carbon and alloy analyses. The bars are generally used in the "as rolled" condition (either commercial hot rolled or precision hot rolled), but they may be centerless ground before coiling.

The heat treatment necessary to develop the required physical properties of the material may be accomplished by direct quench immediately after coiling, or by allowing the coiled spring to cool to a temperature below the critical, then reheating to the required temperature and quenching; the quench is followed by tempering to produce the specified hardness.

Table 1 lists available materials. Their hardenability limitations dictate maximum bar size. For tensile and torsional properties see MANUAL, SAE J795, Chapter IV.

2.2 Shot Peening—Shot peening is used to increase the fatigue life of springs. It consists of subjecting the spring to a stream of metallic shot moving at high velocity. The peening action of the shot reduces the effect of surface defects and sets up beneficial stresses in a thin surface layer. It also results in cold working this layer. To be effective, the peening must reach the area of highest stress which for helical compression springs is the inside diameter of the coil.

The fatigue life of hot-coiled springs is greatly impaired when the bar surface is afflicted by such flaws as impurities, cracks, seams, or decarburization; but it can be increased by the peening operation in the order of 4 to 1. Even the much better fatigue life attainable in hot-coiled springs with nearly perfect bar surface will be improved by peening in the order of more than 2 to 1. For further details see MANUAL, SAE J795, Chapters I and IV, also SHOT PEENING MANUAL, SAE J808.

2.3 Presetting—Presetting (also called scragging, cold setting, or bulldozing) is an operation during the manufacturing process in which the spring is compressed beyond the yield point of the heat treated material. In preparation for this, the spring is coiled to a free length in excess of the designated free length. The yielding in the surface layers of the bar which occurs during presetting produces beneficial residual stresses, thus increasing the elastic limit and thereby reducing the chances of settling in subsequent service. The yielding causes the spring to take a permanent set, thus bringing it down to the designated free length. See also Preset Length, paragraph 2.6.3.

2.4 Bar Diameter and Length—Round bars can be hot rolled to any desired size between 9.00 and 75.00 mm. Table 2 shows the tolerances for commercial hot rolled bars. Bars may be precision hot rolled with 50% of the tolerances in Table 2, or they may be centerless ground with 25% of the tolerances in Table 2.

The bars are commonly purchased in the exact length required to produce one spring. The length tolerance is +13.0 mm for bars 3.0 m and under, and +19.0 mm for bars over 3.0 m long.

2.5 Coil Diameter—The coil diameter can be expressed in terms of the mean coil diameter (D) which is used in the rate and stress formulae. However, coil diameter tolerances should be specified on either the inside diameter

TABLE 1—MATERIALS FOR HOT-COILED COMPRESSION SPRINGS

Materials	Specification	Max. Bar Size, mm
Carbon Steels	SAE 1085	12
	SAE 1095	12
Carbon Boron Steel	SAE 15B62 or 15B62H	20
Alloy Steels	SAE 5160 or 5160H	20
	SAE 9260 or 9260H	20
	SAE 51B60 or 51B60H	50
	SAE 4161 or 4161H	75

TABLE 2—TOLERANCES FOR COMMERCIAL HOT ROLLED ROUND BARS, MM

Specified Bar Size	Size Tolerance, Plus or Minus	Out of Round
9.00 to 12.00 incl.	0.15	0.23
Over 12.00 to 16.00 incl.	0.17	0.25
Over 16.00 to 22.00 incl.	0.20	0.30
Over 22.00 to 28.00 incl.	0.25	0.38
Over 28.00 to 35.00 incl.	0.30	0.45
Over 35.00 to 45.00 incl.	0.35	0.53
Over 45.00 to 55.00 incl.	0.40	0.60
Over 55.00 to 65.00 incl.	0.45	0.68
Over 65.00 to 75.00 incl.	0.55	0.83

Note: Bars over 50 mm are normally supplied with a positive tolerance only which is twice that listed for the plus or minus tolerance.

(ID) or the outside diameter (OD) of the coils, depending upon the importance of the respective dimensions to the user. Tolerances are shown in Table 3, based on coil diameter and spring length.

For motor vehicle suspension springs it is customary to specify the ID in order to facilitate the coiling of a family of springs on a single arbor.

2.6 Spring Lengths—Spring lengths are to be measured after preloading (see Preload Length, paragraph 2.6.4), as the distance parallel to the spring axis between the end surfaces, or else between two reference points specified on the spring drawing.

2.6.1 FREE LENGTH—Free length is the length when no external load is applied. When load is specified, free length is used as a reference dimension only. When load is not specified, free length tolerance equals $\pm[1.5 \text{ mm} + 4\%$ of free-to-solid deflection].

2.6.2 SOLID LENGTH (see also Number of Coils, paragraph 2.7)—Solid length is the length when the spring is compressed with an applied load sufficient to bring all coils in contact; for practical purposes, this applied load is taken to equal approximately 150% of the load beyond which no appreciable deflection takes place.

2.6.3 PRESET LENGTH—In the presetting operation (see Presetting, paragraph 2.3) the spring is usually compressed solid. However, if the stress at solid length is so high that the spring would be excessively distorted, the presetting operation may only be carried to a specified preset length. If more than one preset compression is desired, this must be specified on the drawing. See also MANUAL, SAE J795, Chapters I and IV.

2.6.4 PRELOAD LENGTH—Preloading is the operation of deflecting the spring to the preload length in order to remove temporary recovery of free length before the spring is checked for load and rate.

If the spring was preset during the manufacturing process to the solid length, the preloading may also be carried to the solid length; but it may be restricted to a preload length slightly greater than the solid length, provided the maximum deflection during subsequent service will not go below the preload length.

If the spring was preset to a specified preset length greater than the solid length, the preloading should be restricted to a preload length greater than the preset length.

However, the preload length must not exceed the minimum spring length possible in the mechanism for which the spring is designed. In suspensions this is called the "length at metal-to-metal position". The metal-to-metal contact will occur in the suspension mechanism when rubber bumpers are disregarded. The spring deflection from the specified loaded length to the metal-to-metal position is called "clearance".

2.6.5 LOADED LENGTH—Loaded length is the length while the load is being measured; it is a fixed dimension, with the tolerance applied to the load.

TABLE 3—COIL DIAMETER TOLERANCES

For Specified or Computed Outside Diameter—mm	Inside or Outside Diameter Tolerance, ±mm				
	For Free Spring Length, mm				
	250 and Under	Over 250 to 450 incl.	Over 450 to 650 incl.	Over 650 to 850 incl.	Over 850 to 1050 incl.
75.0 to 110.0 incl.	0.8	1.3	2.5	3.6	4.6
Over 110.0 to 150.0 incl.	1.3	2.5	3.6	4.6	5.6
Over 150.0 to 200.0 incl.	2.5	3.6	4.6	5.6	6.6
Over 200.0 to 300.0 incl.	3.6	4.6	5.6	6.6	6.6

TABLE 4—TOTAL COILS AND NOMINAL SOLID LENGTH

End Configuration	Total Coils (N_t)	Nominal Solid Length
Both ends taper rolled	$N + 2$	$1.01 \, d \, (N_t - 1) + 2t$
Both ends with tangent tail	$N + 1.33$	$1.01 \, d \, (N_t + 1)$
Both ends with pigtail	$N + 1.50$	$1.01 \, d \, (N_t - 1.25)$
Taper rolled plus tangent tail	$N + 1.67$	$1.01 \, d \, N_t + t$
Taper rolled plus pigtail	$N + 1.75$	$1.01 \, d \, (N_t - 1) + t$
Tangent tail plus pigtail	$N + 1.42$	$1.01 \, d \, N_t$

where d = bar diameter
t = tip thickness of taper rolled bar
1.01 = factor used to compensate for the cosine effect of the coil helix angle

The bracketed term in the solid length formula for springs with two pigtails may vary between $(N_t - 0.90)$ and $(N_t - 1.60)$, depending on the pigtail details.

2.7 Number of Coils—Total number of coils (N_t) are counted tip to tip, active number of coils (N) are specified as the number of working coils at free length. With increasing load, N may progressively decrease due to the "bottoming out" effect. If no appreciable bottoming out occurs, the relationships between N and N_t are as shown in Table 4 which also gives the formulae for nominal solid length.

Since nominal solid length may be exceeded somewhat by actual solid length due to manufacturing variations, a frequent practice is to specify nominal solid length together with a maximum solid length, as shown in Table 5.

2.8 Spring Ends—Four types of ends are used (Fig. 1):

1. A flat end formed from a tapered bar end. The bar end is usually tapered for a length equal to $\tfrac{2}{3}$ coil and to a tip thickness of approximately $\tfrac{1}{3}$ of the bar diameter. When the spring is coiled, the tip shall be in approximate contact with the adjacent coil and shall not protrude beyond the outside diameter by more than 20% of the bar diameter.

When stipulated, the bearing surface of the spring end shall be ground perpendicular to the axis of the spring helix in order to produce a firm bearing. The actual ground bearing surface shall not be shorter than two-thirds of the mean coil circumference, nor narrower than half the width of the hot tapered surface of the bar. However, this grinding is usually not required if the tapering and coiling operations are performed adequately.

2. An untapered end coil formed substantially smaller than the central coils of the spring and in such a fashion as to have an outboard bearing surface perpendicular to the axis of the spring helix, the so-called "pigtail" end.

3. An untapered end coil formed as a helix having a pitch substantially equal to the bar diameter. To facilitate coiling, a straight end portion about 25 mm long is permitted to project tangent to the helix of this end construction, the so-called "tangent tail" end. The use of this type of end requires a spring seat formed at the same pitch of helix as that of the spring end.

4. An untapered end coil formed perpendicular to the axis of the spring helix for a circumference of at least 220 deg, the so-called "flat tangent tail" end. To facilitate coiling, a straight end portion about 25 mm long is permitted to project tangent to the outer circumference.

Springs can be specified to have any combination of the four types of ends. The combination of two tangent tail ends may involve a complex arrangement for indexing the spring seats, unless the design of every spring is adjusted to an identical number of total coils. Springs for general automotive use generally have two flat tapered ends. Spring ends and seats are usually so formed as to render approximately two-thirds to one coil inactive at each end.

2.9 Squareness of Ends—Unless otherwise specified, the tapered ends of any spring having an outside diameter to bar diameter ratio of 4 or more, and a free length to outside diameter of 4 or less, shall not deviate more than 3 deg from the perpendicular to the spring axis, as determined by standing the spring on its end and measuring the angular deviation of the outer helix from a perpendicular to the plate on which the spring is standing. In the case of a

TABLE 5—NOMINAL AND MAXIMUM SOLID LENGTH

Nominal Solid Length, mm	Maximum Solid Length may exceed Nominal Solid Length by, mm
Up to 175 incl.	1.5
Over 175 to 250 incl.	2.3
Over 250 to 325 incl.	3.0
Over 325 to 400 incl.	4.0
Over 400 to 480 incl.	4.8
Over 480 to 560 incl.	5.6
Over 560 to 650 incl.	6.5

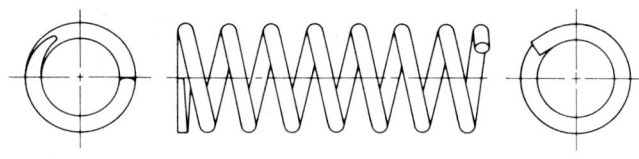

FIG. 1—HOT-COILED COMPRESSION SPRINGS WITH THREE TYPES OF ENDS

tangent tail end, the spring must stand on a seat with matching helical ramp. Tolerances for springs outside these limits are subject to special agreement.

2.10 Load—Load is the force in newtons (N) measured on the load testing machine required to deflect the spring to the specified loaded length. It is to be measured during compression of the spring (compression load) and not during release of the spring (release load), unless otherwise specified.

With loaded length fixed, the usual tolerance for motor vehicle suspension springs is expressed in terms of load equivalent to a deflection of ±5 mm at the nominal rate. Where the demand for greater accuracy warrants the cost of additional presetting or other operations, the load tolerance may be specified as low as ±1.50 mm at the nominal rate.

In the springs for general automotive use, the load tolerance (with loaded length fixed) typically equals ±(1.50 mm + 3% of free-to-solid deflection)· nominal rate. This tolerance is limited to springs where the free length does not exceed 900 mm, does not exceed six times the free-to-solid deflection, and is not less than 0.8 times the OD.

2.11 Rate—Rate is the change of load per unit length of spring deflection (N/mm).

In the springs for motor vehicle suspension the rate is expressed in terms of the load increase per 25 mm deflection (N/25 mm). It is therefore determined as one-half the difference between the loads measured 25 mm above and 25 mm below the specified loaded length. Tolerance is ±3% with centerless ground or with precision rolled bars, and ±4% when commercial hot rolled bars are used.

In the springs for general automotive use the rate is determined between 20 and 60% of the total deflection unless otherwise defined. Typical tolerance is ±5%. In non-critical applications, this may be increased to ±10%.

2.12 Direction of Coiling—For most applications the direction of coiling is unimportant; however, right hand coiling is preferred because most spring manufacturers are so equipped. When direction of coiling is important, as in the case of concentrically nested springs, it must be specified for each component spring, maintaining opposite directions for adjacent springs. For tangent tail springs the direction of coiling must conform with the installation conditions.

2.13 Uniformity of Pitch—The pitch of coils in a compression spring must be sufficiently uniform so that when the spring is compressed, unsupported laterally, to a length representing a deflection of 80% of the nominal free-to-solid deflection, none of the coils must be in contact with one another, excluding the inactive end coils. This requirement does not apply when the design of the spring calls for variable pitch, or when it is such that the spring cannot be compressed to solid length without lateral support.

2.14 Concentricity of Coils—At free length the center of all coils must be concentric with the spring axis within 1.5 mm. This axis is the straight line connecting the centers of the end coils.

3. Cold-Wound Springs

3.1 Material—Round wire sizes and tolerances may be found in the individual wire specifications, such as:

Music Wire	SAE J178
Carbon Steel	
Spring Wire—Oil Tempered	SAE J316
—Hard Drawn	SAE J113
—Special Quality High Tensile Hard Drawn	SAE J271
—Valve Spring Quality Oil Tempered	SAE J351
—Valve Spring Quality Hard Drawn	SAE J172
Chromium Vanadium Wire—	ASTM A231
—Valve Spring Quality	SAE J132
Chromium Silicon Alloy Steel Wire	SAE J157
18-8 Stainless Steel Wire	SAE J230
17-7 Stainless Steel Wire	SAE J217
Phosphor-Bronze Wire, SAE CA510	SAE J463
Beryllium-Copper Wire, SAE CA172	SAE J463
Nickel-Silver Wire	ASTM B206
Silicon-Bronze Wire	ASTM B99
Brass Wire, SAE CA260	SAE J463

3.2 Shot Peening—Shot peening is used to increase the fatigue life of springs. It consists of subjecting the spring to a stream of metallic shot moving at high velocity. The peening action of the shot reduces the effect of surface defects and sets up beneficial stresses in a thin surface layer. It also results in cold working this layer. To be effective, the peening must reach the area of highest stress which for helical compression and extension springs is the inside diameter of the coil.

Even when the wire surface is virtually flawless, the fatigue life of the cold-wound spring can be increased by peening in the order of more than 2 to 1. See MANUAL, SAE J795, Chapter I, also SHOT PEENING MANUAL, SAE J808.

3.3 Presetting—Presetting (also called cold setting) is an operation during the manufacturing of helical compression springs in which the spring is compressed beyond the yield point of the material. In preparation for this, the spring is coiled to a free length in excess of the designated free length. The yielding in the surface layers of the wire which occurs during presetting produces beneficial residual stresses, thus increasing the elastic limit and thereby reducing the chances of settling in subsequent service. The yielding causes the spring to take a permanent set, thus bringing it down to the designated free length. See also Preset Length, paragraph 3.5.3.

3.4 Coil Diameter—Coil diameter tolerances can be specified on either the inside diameter (ID) or the outside diameter (OD) of the coils, depending upon the importance of the respective dimensions to the user. Tolerances are functions of the "Spring Index", which is the ratio of mean coil diameter (D) to wire diameter (d). They are to be considered as manufacturing tolerances and do not take into account the effects of changes in diameter due to applied loads.

3.5 Spring Lengths—Spring lengths of compression springs are overall dimensions measured parallel to the axis of the spring.

Spring lengths of extension springs are measured inside to inside of the hooks (overall length minus two wire diameters).

3.5.1 FREE LENGTH—Free length is the length under no load. When load is specified, free length is used as a reference dimension only. When load is not specified, free length is specified for control and inspection purposes by using Fig. 2 for compression springs and Fig. 3 for extension springs.

The tolerances in Fig. 2 are based on the number of active coils (N), the free length (L_o), and the spring index (D/d). With these parameters known, the N/L_o value is established on the abscissa, and the tolerance is found by multiplying the corresponding ordinate value by L_o. Round off the index to the nearest whole number and interpolate when this is an odd number. The

TABLE 6—COIL DIAMETER TOLERANCES

Wire Size, mm		Inside or Outside Diameter Tolerance, ±mm						
		for D/d Ratio						
Over	Incl.	4	6	8	10	12	14	16
	0.60		0.1	0.1	0.2	0.2	0.2	0.3
0.60	0.90	0.1	0.1	0.1	0.2	0.2	0.3	0.3
0.90	1.40	0.1	0.1	0.2	0.3	0.3	0.4	0.4
1.40	2.00	0.1	0.2	0.2	0.3	0.4	0.5	0.5
2.00	2.50	0.1	0.2	0.3	0.4	0.5	0.5	0.6
2.50	3.00	0.2	0.2	0.3	0.4	0.5	0.6	0.7
3.00	3.50	0.2	0.3	0.4	0.5	0.6	0.7	0.8
3.50	4.00	0.2	0.3	0.4	0.5	0.7	0.8	0.9
4.00	4.50	0.2	0.3	0.4	0.5	0.7	0.8	0.9
4.50	5.00	0.2	0.3	0.5	0.6	0.8	0.9	1.0
5.00	6.00	0.3	0.4	0.5	0.7	0.9	1.0	1.2
6.00	7.00	0.3	0.4	0.6	0.8	1.0	1.2	1.4
7.00	8.00	0.3	0.4	0.6	0.8	1.0	1.3	1.6
8.00	9.00	0.4	0.5	0.7	0.9	1.1	1.4	1.8
9.00	10.00	0.4	0.6	0.8	1.0	1.3	1.6	2.0
10.00	11.00	0.5	0.6	0.9	1.2	1.5	1.9	2.3
11.00	13.00	0.5	0.7	1.0	1.4	1.8	2.2	2.7

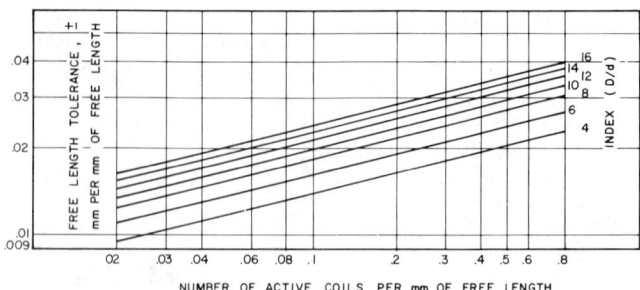

FIG. 2—FREE LENGTH TOLERANCE FOR COMPRESSION SPRINGS

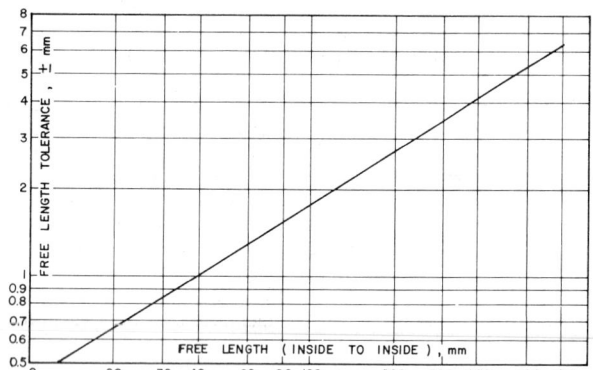

FIG. 3—FREE LENGTH TOLERANCE FOR EXTENSION SPRINGS

tolerances shown in Fig. 1 are for springs with ends closed and ground. For springs with the ends closed but not ground, multiply by 1.7.

The tolerances in Fig. 3 may also be expressed by the formula:

$$\text{Free length tolerance} = \pm 0.105 \cdot (\text{free length})^{0.615}$$

3.5.2 SOLID LENGTH (see also Number of Coils, paragraph 3.6)—In compression springs this is the length with all active coils closed, to be specified as a maximum dimension allowing the manufacturer any tolerance required by the variations in wire size, total coils, and the amount of grind at the ends; platings and coatings increase the wire diameter and must be considered.

For springs with ground ends, the maximum solid length is the total number of coils times the wire diameter; for springs with ends not ground the solid length is the total number of coils plus one, times the wire diameter.

3.5.3 PRESET LENGTH—After the compression spring has been coiled to a free length in excess of the designated free length, it is compressed solid or to a specified preset length; this produces yielding, which results in bringing the spring to the designated free length. If more than one preset compression is desired, it must be specified on the drawing. See also MANUAL, SAE J795, Chapter I.

3.5.4 LOADED LENGTH—This is the length while the load is being measured. It is a fixed reference dimension, with the tolerance applied to the load.

3.5.5 MAXIMUM EXTENDED LENGTH—Extension springs normally do not have a definite stop to their deflection, therefore the drawing specifications should include a statement of the maximum extended length which must be attained without encountering permanent set.

3.6 **Number of Coils**—In compression springs it is often necessary to vary the number of coils in order to meet the requirements on load, rate, free length, and solid length. Therefore, the number of coils should be specified as an approximate figure. For reference only, the tolerance for the number of coils is given in Table 7 compression springs and in Table 8 for extension springs. It is expressed in degrees as a function of the number of active coils.

TABLE 7—NUMBER OF COILS TOLERANCE OF COMPRESSION SPRINGS

Active Coils	Tolerance, ±deg
3–10	90
For each additional 10 coils, add	30

TABLE 8—NUMBER OF COILS TOLERANCE OF EXTENSION SPRINGS

Active Coils	Tolerance, ±deg	
	Close Wound	Open Wound
3	30	90
4–10	45	90
For each additional 10 coils, add	15	30

In extension springs, either the number of coils in the body of the spring or the length over the coils may be specified, but only as an approximate figure. In computing the length over coils it should be recognized that there is always one more wire diameter in the length than the number of coils in a close-wound spring.

3.7 **Spring Ends**—In compression springs there are four typical end configurations (Fig. 4):
 1. Plain end (with the end coil having the same pitch as all other coils);
 2. Plain end ground (the end surface being ground perpendicular to the spring axis);
 3. Closed end (with the tip of the wire contacting the adjacent coil);
 4. Closed and ground end (the closed end being ground perpendicular to the spring axis).

The unground ends may be used for reasons of economy, but they give eccentric loading with some increase in maximum spring wire stress and space required. The plain ends similarly produce eccentric loading and additionally present a handling problem due to springs tangling together.

In extension springs many types of hooks, loops, eyes, etc. are used (see MANUAL, SAE J795, Fig. 14). Details such as hook opening restraint of the loop within the body diameter should be specified on the drawing. The position of hooks relative to each other can be in line, at right angles, or at any other angular position as required. If this relative position is important, the spring drawing should emphasize the importance by a statement as well as by pictorial representation. Sharp bends in forming the end hooks should be avoided because they produce stress concentrations.

3.8 **Squareness of Ends**—In compression springs with closed and ground ends the squareness of the ends, as measured in the unloaded position, is to be maintained within a limit of 3 deg with the axis of the spring.

3.9 **Load**—Load is the force in newtons (N) measured on the load testing machine required to deflect the spring to the specified loaded length.

For compression springs the load is to be measured during compression of the spring (compression load) unless otherwise specified. Tolerances are shown in Fig. 5 as functions of the nominal free length tolerance (Fig. 2) and the deflection from free length to loaded length. Round off the percent load tolerance values to the next larger whole number. Interpolate when this is an odd number and when it is between 8 and 20 percent.

For extension springs the load is to be measured during extension of the spring. Tolerances are computed as the product of the appropriate tolerance factor in Fig. 6 and the appropriate multiplying factor in Fig. 7.

Cold coiled extension springs may be wound with tension between the coils so that a load must be applied to separate them, the so-called initial tension in the spring.

3.10 **Rate**—Rate is the change of load per unit length of spring deflection (N/mm). The rate is to be determined between 20 and 60% of the total deflection. Tolerances depending on the number of active coils are given in Table 9.

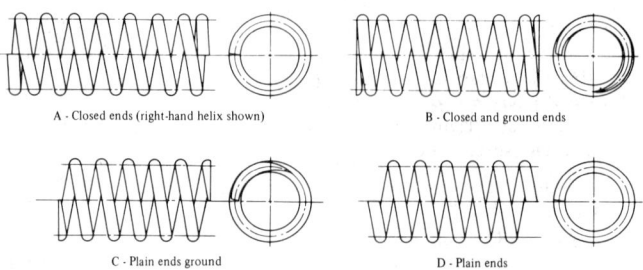

FIG. 4—COLD-WOUND COMPRESSION SPRINGS WITH FOUR TYPES OF ENDS

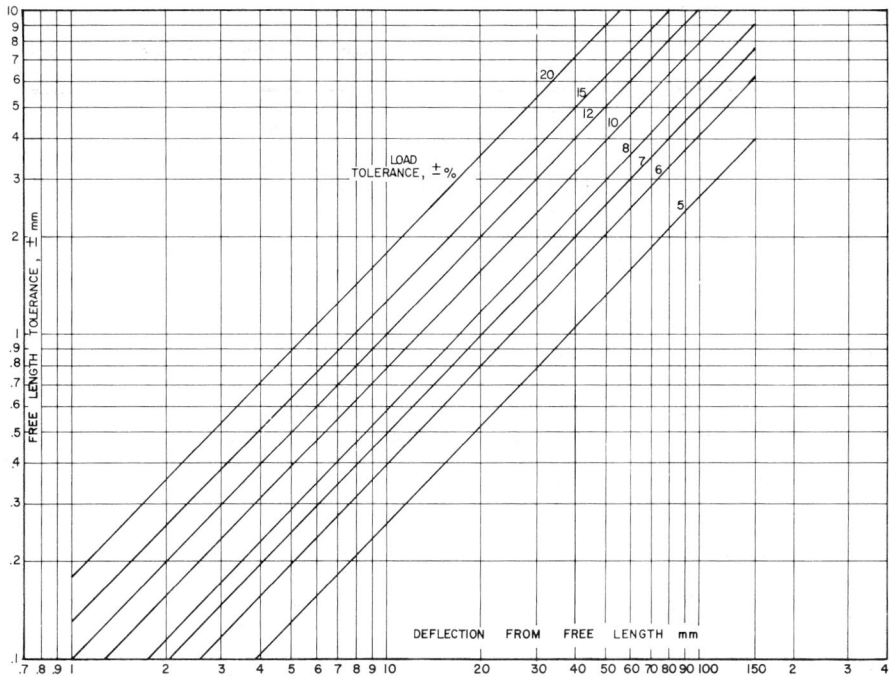

FIG. 5—LOAD TOLERANCE FOR COMPRESSION SPRINGS

3.11 Direction of Coiling—For most applications the direction of coiling is unimportant; however, right hand coiling is preferred because most spring manufacturers are so equipped.

3.12 Uniformity of Pitch—The pitch of coils in a compression spring must be sufficiently uniform so that when the spring is compressed, unsupported laterally, to a length representing a deflection of 80% of the nominal free-to-solid deflection, none of the coils must be in contact with one another, excluding the inactive end coils. This requirement does not apply when the design of the spring calls for variable pitch, or when it is such that the spring cannot be compressed to solid length without lateral support.

TABLE 9—RATE TOLERANCE

Number of Active Coils	Rate Tolerance, ±%
3 or less	10
Over 3 to 9 incl.	8
Over 9 to 15 incl.	6
Over 15	5

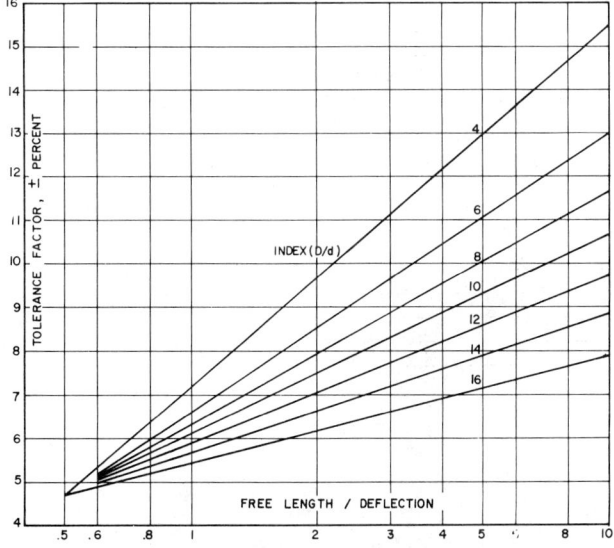
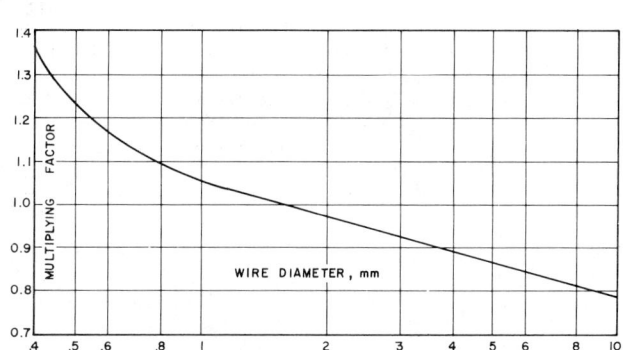

FIGS. 6 AND 7—LOAD TOLERANCE FOR EXTENSION SPRINGS

HELICAL SPRINGS: SPECIFICATION CHECK LISTS—SAE J1122

SAE Recommended Practice

Report of Spring Committee approved February 1976. SAE J653 has been discontinued and replaced by this report.

Scope—The following Recommended Practice furnishes sample forms for helical compression, extension, and torsion springs to provide a uniform method for specifying design information. It is not necessary to fill in all the data, but sufficient information must be supplied to fully describe the part and to satisfy the requirements of its application. For detailed information see Design and Application of Helical and Spiral Springs—SAE J795a, also Helical Compression and Extension Spring Terminology—SAE J1121. For springs to be designed in the metric system with SI units, the terms shown in parentheses are to be used.

A—HELICAL COMPRESSION SPRINGS   Application _____

OD _____ in (mm) to work inside _____ in (mm) Dia Hole
ID _____ in (mm) to work over _____ in (mm) Dia Rod
Note: Specify only those diameters that are necessary for assembly and operation
Wire Dia _____ in (mm)   Total Coils _____   Active Coils _____
Free Length _____ in (mm) approx
Direction of Coil Winding: Right Hand, Left Hand, or Optional _____
Type of Ends _____ Square with Axis within _____ deg
Max Solid Length _____ in (mm)
Load _____ lb (N) ± _____ lb (N) at _____ in (mm) length
Load _____ lb (N) ± _____ lb (N) at _____ in (mm) length
Rate _____ lb/in (N/mm) Ref
After being compressed to a length of _____ in (mm) for _____ hours at a temperature of _____ °F (°C) the spring must not show a load loss in excess of _____ lb (N) at a length of _____ in (mm)
Spring Index = Mean Coil Diameter/Wire Diameter (D/d) _____
Wahl Stress Correction Factor K_W _____
Stress at _____ in (mm) length: _____ psi (MPa) Corrected or Uncorrected _____
Stress at _____ in (mm) length: _____ psi (MPa) Corrected or Uncorrected _____
Material _____
Hardness or Tensile Strength _____
Surface Treatment/Finish _____
Identification _____
Remarks _____

B—HELICAL EXTENSION SPRINGS   Application _____

OD _____ in (mm)   Wire Dia _____ in (mm)
Number of Coils _____
Direction of Coil Winding: Right Hand, Left Hand, or Optional _____
Free Length inside Hooks _____ in (mm) approx
Type of Ends (use sketch if necessary) _____
Load _____ lb (N) ± _____ lb (N) at _____ in (mm) length
Load _____ lb (N) ± _____ lb (N) at _____ in (mm) length
Initial Tension _____ lb (N) Ref
Rate _____ lb/in (N/mm) Ref
Stress at _____ in (mm) length: _____ psi (MPa) Corrected or Uncorrected _____
Stress at _____ in (mm) length: _____ psi (MPa) Corrected or Uncorrected _____
Max Extended Length without set _____ in (mm)
Stress at Max Extended Length _____ psi (MPa) Corrected or Uncorrected _____
Material _____
Hardness or Tensile Strength _____
Surface Treatment/Finish _____
Identification _____
Remarks _____

C—HELICAL TORSION SPRINGS   Application _____

OD _____ in (mm)
ID _____ in (mm) to work over _____ in (mm) Dia Shaft
Note: Specify only those diameters that are necessary for assembly and operation
Wire Dia _____ in (mm)   Number of Coils _____
Max Free Length _____ in (mm)
Direction of Coil Winding: Right Hand or Left Hand _____
Type of Ends (use sketch if necessary) _____
Moment _____ lb in (N mm) ± _____ lb in (N mm) at _____ deg between ends
Moment _____ lb in (N mm) ± _____ lb in (N mm) at _____ deg between ends
Rate _____ lb in/deg (N mm/deg) approx
Stress at _____ deg _____ psi (MPa)
Stress at _____ deg _____ psi (MPa)
Max Wound Position _____ deg
Stress at Max Wound Position _____ psi (MPa)
Material _____
Hardness or Tensile Strength _____
Surface Treatment/Finish _____
Identification _____
Remarks _____

RATED SPRING CAPACITY—SAE J274

SAE Recommended Practice

Report of Spring Committee approved September 1972.

1. Scope—The rated spring capacity definition has been developed to assist engineers and designers in the preparation of specifications and of descriptive material and values relating thereto.

2. Purpose—The following definition of rated spring capacity is applicable to all types of suspensions for vehicles used predominantly on the highway. This capacity provides a basis for comparison of spring load-carrying abilities in a particular suspension application. This definition is intended to clarify a commonly used term which has heretofore been used indiscriminately.

3. Definition—Rated spring capacity is a load rating assigned to each spring installation and vehicle application which will provide adequate spring durability and vehicle stability under all intended load conditions. The value of the load rating must equal or exceed that portion of the maximum allowable weight at the ground which relates direction to the spring. The load rating is therefore based upon the total of sprung and unsprung weights of the loaded vehicle.

4. Related Terms

4.1 Spring—Includes all types of suspension springs (such as leaf, coil, torsion bar, rubber, air bags, etc.).

4.2 Load Rating—Is expressed in units of weight, determined vertically with the vehicle on a horizontal plane.

4.3 Spring Installation—Any spring as used in a particular suspension.

4.4 Vehicle Application—The usage of the vehicle as intended by the vehicle manufacturer.

4.5 Adequate Spring Durability—The endurance life characteristics regarded as sufficient by the vehicle manufacturer to satisfy customer requirements.

4.6 Adequate Vehicle Stability—The ride and handling characteristics of the vehicle regarded by the vehicle manufacturer as sufficient for safe operation.

4.7 Intended Load Conditions—The various payloads and payload distribution applied to the vehicle within the prescribed limits of gross vehicle weight, vehicle full rated load, and component capacities as established by the vehicle manufacturer.

4.8 Sprung Weight and Unsprung Weight—Defined in SAE J670, Vehicle Dynamics Terminology.[1]

4.9 Loaded Vehicle—A vehicle which satisfies the conditions described in paragraph 4.7.

4.10 Maximum Allowable Weight at the Ground—The vehicle full rated load or gvw acting at the ground.

4.11 Related Directly to the Spring—The load at the ground, which is transmitted through the suspension components to the spring and includes that portion of the unsprung weight.

The rated spring capacity does not indicate spring payload capability, but rather the total of payload and vehicle weight. The assignment of a rated spring capacity value is the responsibility of the vehicle manufacturer.

[1] Available as HS J670 from Society of Automotive Engineers, Inc., 400 Commonwealth Drive, Warrendale, Penna. 15096.

20 Speedometers

FACTORS AFFECTING AUTOMOTIVE ODOMETER-SPEEDOMETER ACCURACY—SAE J862b

SAE Information Report

Report of Speedometer and Tachometer Committee approved June 1963, and last revised April 1969.

1. Scope—This report is concerned with factors which affect accuracy of mileage indication and speed indication of automotive type odometer-speedometers. It is the intent to supply information regarding all items which affect the instrument.

2. Mileage Indication—Distance traveled is indicated by a numbered set of wheels, called the odometer, normally viewed through a slot in the dial of the speedometer. The wheels incorporate internal gear teeth which engage a pinion interposed between each set of wheels. The odometer can then be said to be a set of gears with numerals on their outer surface. The odometer is driven by a system of reduction gearing within the speedometer instrument. This reduction gearing is, in turn, driven by the speedometer cable core. SAE J678 specifies that 1000 revolutions of the speedometer cable core shall cause a 1 mile indication on the odometer where speedometers are driven from the transmission. In front wheel driven speedometers, the nominal number of wheel revolutions per mile shall cause a 1 mile indication on the odometer. Because of the positive gear drive mechanism, *no inherent error of mileage indication exists in the speedometer head.*

3. Factors Affecting Odometer Accuracy

3.1 Overall Assembly in Vehicle—The ideal of achieving the exact nominal value of speedometer cable core revolutions in one mile of vehicle travel can seldom be realized. This becomes apparent when consideration is given to the overall design problem.

3.1.1 The speedometer cable core is driven by a gear called the takeoff pinion gear which is driven by the worm drive gear connected to the output shaft from the transmission which, in turn, connects with the vehicle drive shaft (Fig. 1). The drive shaft drives the rear wheels through the differential. The distance traveled is dependent on the number of tire revolutions in a mile. By experimentation, a nominal figure of tire revolutions per mile is determined for the vehicle. Knowing the differential ratio, it is possible to calculate the necessary ratio in the transmission and the takeoff pinion gear for the speedometer cable core to achieve 1000 rpm. In the case of front wheel drives, it is then necessary to calculate the proper gearing within the speedometer head itself to achieve nominal conditions.

3.1.2 The exact ratio frequently results in a fraction which must, of course, be rounded to a whole number of teeth for the takeoff pinion gear. SAE J678 recommends that the error in rounding the odometer gearing ratio be such that the odometer accuracy will be within the limits of −1% to +3.75% at 45 mph. Thus, from this factor alone an indication of 99 to 103.75 miles may be indicated for 100 actual miles traveled.

3.1.3 Because of different axle ratios used, it is necessary in any one line of

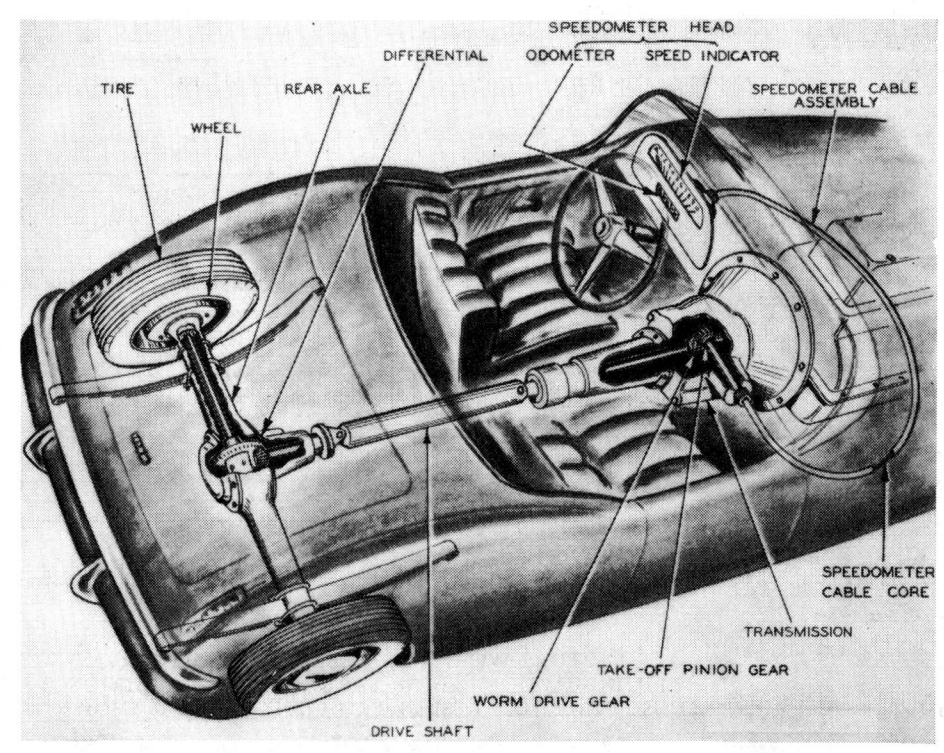

FIG. 1—OVERALL ASSEMBLY FOR VEHICLE

automobiles to have a variety of takeoff pinion gears with different numbers of teeth. The number of teeth in the worm drive gear is not readily subject to change, since this gear is assembled within the transmission and is usually uniform for any transmission model.

3.2 Tires and Load—Tires are elastic members subject to variations from nominal size caused by manufacturing tolerances, temperature, inflation pressure, wear, and loading. A tire will change in size due to aging, after it is placed on a rim and inflated. These size variations, plus differences in construction, material, and in the type of tread on tires from the same or different manufacturers, can result in a different number of tire revolutions per mile. It is obvious that these variations from the nominal originally selected can directly affect mileage indication.

3.3 Speed—An average tire experiences a 3% change in revolutions per mile from a 30 mph speed to a 90 mph speed due to a change in rolling radius caused by centrifugal force.

3.4 Analysis and Summary—Fig. 2 is a chart which demonstrates the magnitudes of error which might occur in odometer readings. The average individual effect will be less than the maximum indicated by the chart since some of the conditions tend to compensate for others. For instance, tire wear and aging growth are compensating factors. Tire wear has the effect of increasing odometer indication and the tire aging growth will decrease the indication. When reading the chart, however, it should be appreciated that the errors may be additive.

3.5 Corrective Measures

3.5.1 In the foregoing, it has been shown that there are factors present which cannot be economically reduced or controlled which will cause mileage indication errors. Some of the factors, however, can, in some degree, be controlled by proper tire inflation and replacement of worn tires.

3.5.2 Inadvertent installation of an improper pinion for a particular axle ratio will, of course, result in considerable errors in odometer reading. Such a condition is, however, easily remedied by installation of the correct takeoff pinion gear.

3.5.3 A vehicle operator, especially one who modifies a standard vehicle, can determine his percentage of odometer error by driving an accurately known distance at approximately nominal operating conditions of speed, load, temperature, and proper tire inflation. The reading of the odometer shall be compared to the known distance traveled. A reading greater than the distance traveled indicates a plus error (under registration), conversely a reading less than the distance traveled, indicates a negative error (over registration). For example, if the odometer indicates 5.2 miles as compared to a nominal 5.0 miles distance traveled, an error of $2/10$ divided by five or +4% exists. A grossly plus error can be compensated for by using a takeoff pinion gear with a greater number of teeth or a large minus error may be corrected with a takeoff pinion gear with less teeth.

4. Speed Indication—Speed indication in an automotive speedometer is commonly accomplished through use of a principle known as the eddy current drive. The speedometer cable core drives a magnet shaft of the speedometer to which a permanent magnet is affixed. This magnet is located inside an aluminum or copper speed cup. The speed cup is attached to the same spindle on which the speedometer pointer is affixed. Also affixed to this spindle is a

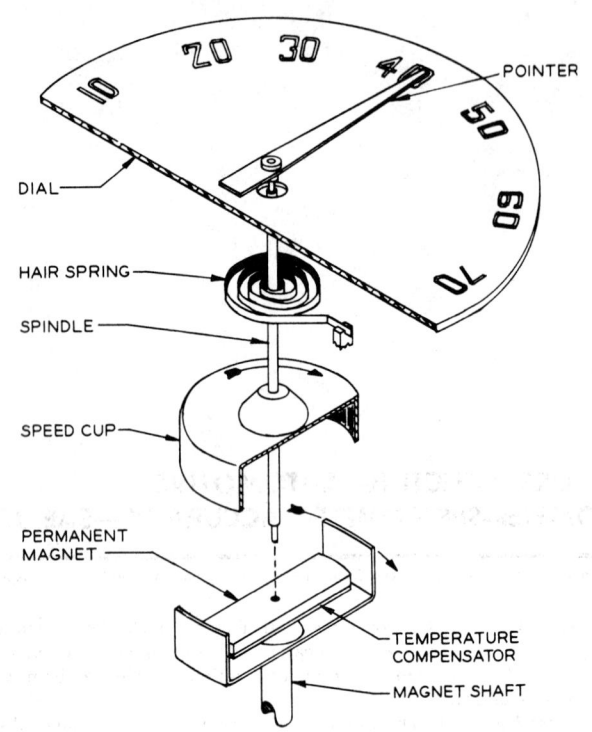

FIG. 3—EDDY CURRENT DRIVE

hair-spring. A force applied to the speed cup results in a controlled reaction of the speedometer pointer. As the magnet rotates inside the speed cup, a force proportional to the speed of rotation is developed, thus providing measurement of speed indicated on the dial (Fig. 3).

5. Speedometer Calibration

The speedometer is calibrated at room temperature, 70 F by the instrument manufacturer. SAE J678 recommends that at a cable speed of 167 rpm, the instrument shall read from 8-12 mph; at 500 rpm, the instrument shall read 30-33 mph; at 1000 rpm cable speed, the instrument shall read 60-63 mph and at 1500 rpm, a reading of 90-94 mph is required.

6. Factors Affecting Speedometer Accuracy

6.1 Drive Errors—Indication of speed is subject to the same errors as mileage indication because the same speedometer cable core drives both the odometer and speed indicator. Some of the error may be compensated by calibration of the speed indicator. Note should be taken that individual errors due to takeoff pinion, tire size, tire wear, tire pressure, the effect of temperature on tire pressure, and load in the vehicle may be additive or subtractive to the speed indication.

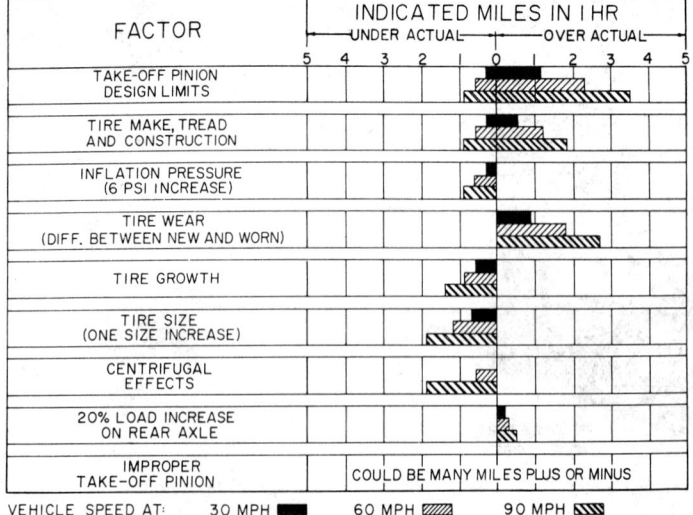

FIG. 2—FACTORS WHICH AFFECT ODOMETER MILEAGE (CROSS SECTION OF ALL U.S. MAKES)

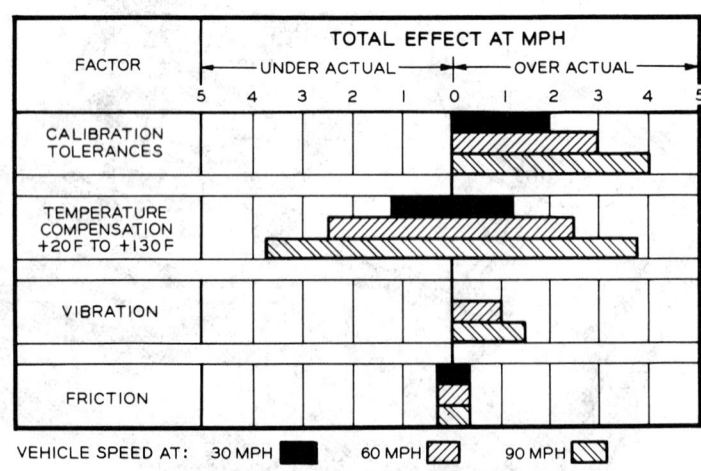

FIG. 4—FACTORS WITHIN A SPEEDOMETER WHICH AFFECT SPEED INDICATION. THESE ERRORS ARE IN ADDITION TO THOSE THAT MAY RESULT FROM THE FACTORS SHOWN IN FIG. 2

6.2 Temperature—The speed indication will vary with changes in temperature. Temperature affects the reaction between the speed cup and the magnet. To counteract this effect, an element called a temperature compensator is incorporated in all speedometers. Due to variations in the materials which cannot be perfectly controlled, the temperature compensation effect is not perfect. SAE J678 allows a ±2½ mph change in reading at 60 mph between temperatures of 20 and 130 F. This change is approximately a 4% change, thus the effect at 30 mph would be 1.25 mph.

6.3 Vibration—Another condition which affects speed indication is vibration. At 60 mph an effect of 0.5 to 1 mph may be noted and this may increase to 1.5 mph at 90 mph.

6.4 Friction—A minor error which affects speed indication is frictional lag. This is a condition in which the speedometer will not read exactly the same under acceleration and deceleration. Investigations indicate that the error from frictional lag is less than 1 mph at 60 mph.

6.5 Summary—Fig. 4 illustrates, in composite form, the various factors affecting speedometer indication.

SPEEDOMETERS AND TACHOMETERS— AUTOMOTIVE—SAE J678e

SAE Recommended Practice

Report of Parts and Fittings Division approved January 1939 and last revised by Speedometer and Tachometer Committee June 1975.

1. Scope—This SAE Recommended Practice applies to speedometers, odometers, and speedometer drives typical of passenger vehicles, buses and trucks used for personal or commercial purposes. The method of determining wheel revolutions per mile (paragraph 2.1) and overall system design variation (paragraph 2.3.2) are applicable to passenger cars only. Comparable recommendations for trucks and buses are under development. The data on tachometers is applicable to vehicular use as previously described and also to stationary and marine engines and special vehicles.

2. Speedometer

2.1 Wheel Revolutions per Mile[1]—The nominal number of vehicle wheel revolutions per mile is to be determined by the vehicle manufacturer and the information is to be used as a basis for design calculations of gearing and speedometer calibrations. Vehicle wheel revolutions shall be determined at 45 mph. Tire inflation for measuring wheel revolutions is to be in accordance with the vehicle manufacturers' recommended pressure with tires at ambient test temperatures. This test is to be run immediately after a 5 mile test run at 45 mph to stabilize tire pressures and with the vehicle at curb weight plus driver and one passenger.

2.2 Types of Drive—The practice for cable driven speedometers is to drive the system from either the transmission or the front wheel of the vehicle.

2.2.1 TRANSMISSION DRIVE—The design of a transmission drive for a speedometer requires that the vehicle manufacturer determine the nominal number of vehicle wheel revolutions per mile. This information is then used as a basis for calculating the speedometer drive ratio. The number of teeth on the worm drive gear in the transmission and the number of teeth on the take-off pinion gear (drive ratio) are selected to provide a proper odometer drive and speed indication as defined in paragraphs 2.3.1, 2.3.2 and 2.4.1. It is urged that those unfamiliar with conditions which affect vehicle wheel revolutions consult SAE J862 so that they may properly control conditions when determining the nominal value.

2.2.2 FRONT WHEEL DRIVE—This type of drive also requires that the nominal number of vehicle wheel revolutions be determined. The information is used to provide proper odometer drive gearing and for calibration purposes to achieve a proper speed indication.

2.3 Mileage Indication (Odometer)—

2.3.1 ALLOWABLE VARIATION WITHIN THE INSTRUMENT—The odometer shall indicate 1 mile for every 1000 or 1001 revolutions of the flexible shaft if driven from the transmission. The odometer of front wheel drive units shall indicate 1 mile when the flexible shaft is rotated a specified number of revolutions as determined by the vehicle manufacturer.

[1]See also SAE J966.

2.3.2 OVERALL SYSTEM DESIGN VARIATION—The vehicle manufacturer shall specify odometer drive ratios that will produce a 1 mile indication within −1% to +3.75% for each actual mile of travel at 45 mph.[2] The design limits thus derived should not however be construed as absolute under operating conditions. Factors which cause variations from nominal wheel revolutions under operating conditions are covered in SAE J862. It is recommended that SAE J862 be studied to determine probable effects under service conditions.

2.4 Speed Indication—

2.4.1 ALLOWABLE VARIATION WITHIN THE INSTRUMENT—The speedometer head speed indication calibration shall be within the limits shown in Table 1 when driven at the specified rpm at a temperature of 75 F. Spacing of the graduations on the speedometer dial may be adjusted to compensate for different wheel revolutions per mile at various speeds. It should be noted that the variations in speedometer reading on the road may lie outside the limits of Table 1 by the factors described in Table 2 of SAE J862. All calibration of speedometers in production shall be made with the instrument in approximately the same angular position that it will have when mounted in the vehicle.

[2]This accuracy is required by many state and local regulations. Vehicles which are introduced into the rental or leasing market should have accuracies within the 3.75% error allowed in this SAE Recommended Practice.

TABLE 1—SPEEDOMETER HEAD CALIBRATION, MPH[a]

For shaft input calculated to produce exactly	10	30	60	90
Speed indication shall read within	8–12	30–33	60–63	90–94

[a]When full range of dial is 100 mph or less, calibration at 1500 rpm may not apply.

TABLE 2—DIMENSIONS FOR HEAVY DUTY FLEXIBLE SHAFTS ILLUSTRATED IN FIGS. 4 AND 5

SAE Size	A Flexible Shaft Lower Tip	B Hole in Take-Off	C Tip Key Height
5/32	0.155 dia 0.150	0.161 No. 20 Drill	0.210 0.190
3/16	0.188 dia 0.183	0.191 No. 11 Drill	0.250 0.225
13/64	0.206 dia 0.200	0.213 No. 3 Drill	0.250 0.235

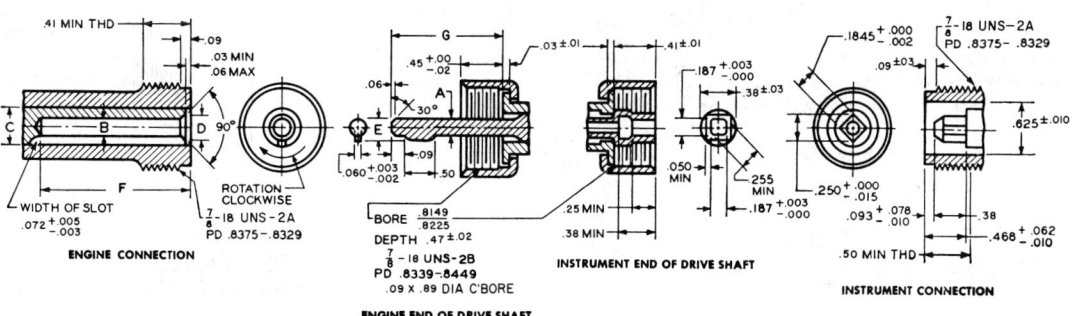

FIG. 1—SAE REGULAR DRIVE FOR SPEEDOMETERS AND TACHOMETERS

Variations of indication at the design rpm for 60 mph in mile speedometer readings, due to ambient temperature changes from 20 to 130 F, shall not exceed ±2½ mph from indicated mph at 75 F. For the kilometer speedometers, this variation shall not exceed ±4 kmph.

2.5 Identification—Identification should appear on all speedometers. It should consist of a distinct stamping of model number or part number on the instrument and the date of its manufacture. All kilometer speedometers should be identified by stamping "Kilo" on the back of the speedometer case, unless identification of sufficient nature appears on the face dial or other suitable means of identification is applied, such as marking the speedometer head with a red dot. Drive ratio information shall appear on front wheel drive instruments in an area readily visible when the instrument is removed.

3. Speedometer and Tachometer Drive—Flexible shafts for driving mechanical speedometers and tachometers shall consist of a flexible casing and a flexible cable capable of transmitting motion from a suitable take-off to operate the instrument. Recommended take-off and instrument fittings are shown in the following illustrations.

In routing of flexible shafts, bends of less than 6 in. R should be avoided. Figs. 1–5 and Table 2 give dimensions for SAE regular, light, square, and heavy duty drive for speedometers and tachometers.

3.1 Miscellaneous Drive Ends of Flexible Shafts—Specific detail and dimensions to be determined between user and supplier.

3.1.1 PLUG TYPE LOWER FERRULE—This type of drive end (shown in Fig. 6) is widely used for speedometer drives in the automotive industry. No specific dimensional standard is recommended because dimensions vary depending on design of transmission with which it is used. The end of the cable has a standard 0.101–0.104 square in. passenger car service.

3.1.2 FRONT WHEEL DRIVE (See Fig. 7)—Some speedometer drives are taken from the vehicle front wheel. This involves spindle machined to accept the flexible shaft drive end and providing a watertight joint. The cable terminates in a standard 0.101–0.104 square in. passenger car service. No specific dimensional standard is recommended because of the many variations in front wheel suspensions, steering mechanism, and spindle designs which will affect the flexible shaft design.

4. Tachometers

4.1 Mechanical Eddy Current Tachometers—

4.1.1 GENERAL—Illumination, waterproofness, and corrosion resistance are not within the scope of this recommended practice.

4.1.1.1. *Tachometer Drive Connections*—SAE regular optional and heavy. Clockwise and counter clockwise rotation.

4.1.1.2. *Tachometer Dials*—Recommended dial ranges in rpm are: 0–2500, 0–4000, 0–6000, and 0–8000, for a minimum of 270 deg full deflection, clockwise or counterclockwise rotation. Light graduations, numerals, and pointers on a dark background and with graduations outside numerals for highest accuracy in scale readings are also recommended.

4.1.1.3. *Tachometer Drive Ratio*—Tachometer to indicate two times flexible shaft speed and to be driven ½ times engine speed.

4.1.1.4. *Tachometer with Hour Meter*—Hour meter to indicate, as closely as practical, engine hours at a specific speed on an hour meter recording up to 9999.99 hr before repeating from zero. (See Fig. 8.)

4.1.1.5. *Tachometer Mounting*—Tachometer case to be provided with studs for easy mounting by suitable U-clamp or similar means. The mounting position to be with tachometer faced backward 5–45 deg from a vertical plane. See Fig. 8 for general dimensions of tachometer housing.

4.1.1.6. *Tachometer Calibration*—Recommended calibration limits are as follows:

0–2500 scale ±50 rpm at 500, 1500, 2000 indicated rpm.
0–4000 scale ±50 rpm at 500, 2000, 3000 indicated rpm.
0–6000 scale ±50 rpm at 1000, ±75 rpm at 3000, 5000 indicated rpm.
0–8000 scale ±50 rpm at 1000, ±100 rpm at 4000, 6000 indicated rpm.

4.2 Mechanical Centrifugal Tachometers—Illustrations of these speedometers and tachometers and their principal features and mounting dimensions are shown in Fig. 9.

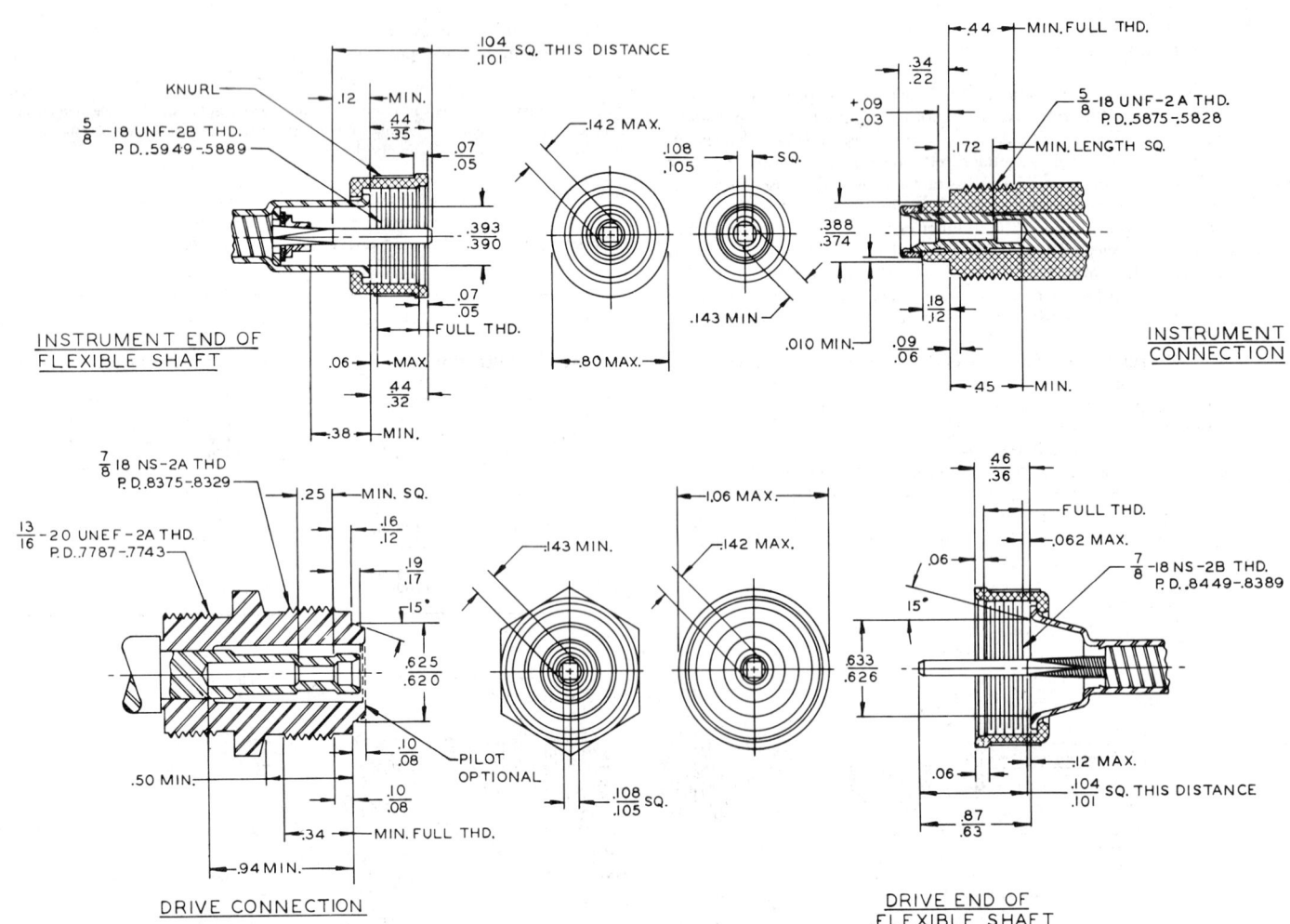

FIG. 2—LIGHT DUTY DRIVE (FOR USE WITH EDDY CURRENT TYPE INSTRUMENTS)

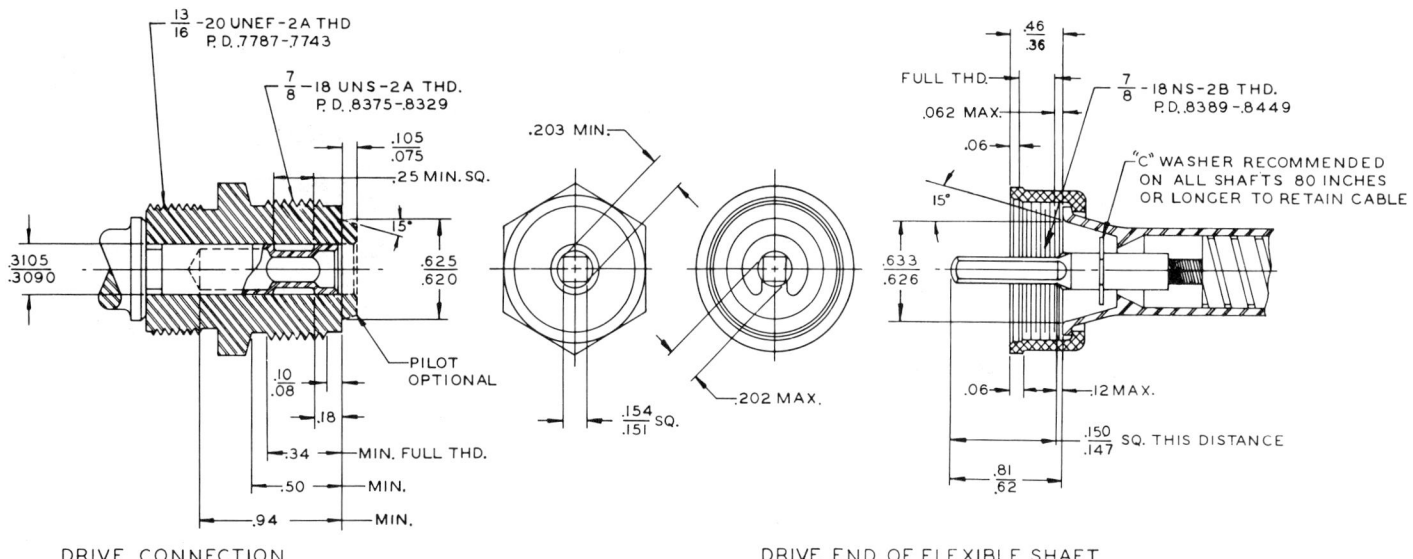

DRIVE CONNECTION

DRIVE END OF FLEXIBLE SHAFT

FIG. 3—SQUARE DRIVE

DRIVE CONNECTION

DRIVE END OF FLEXIBLE SHAFT

FOR LETTERED DIMENSIONS SEE TABLE "2"

INSTRUMENT END OF FLEXIBLE SHAFT

INSTRUMENT CONNECTION

FIG. 4—HEAVY DUTY DRIVE (FOR USE WITH CENTRIFUGAL TYPE INSTRUMENTS)

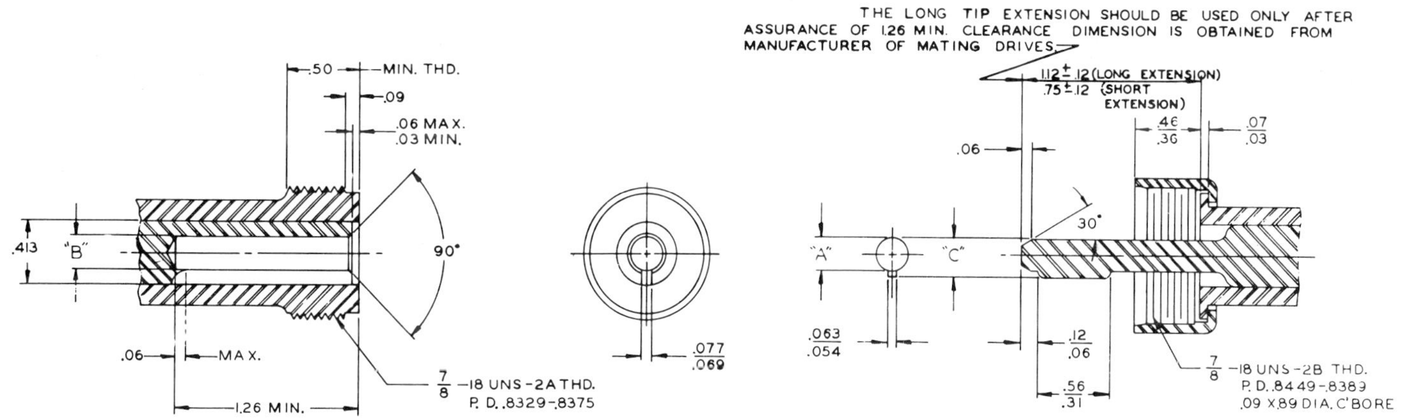

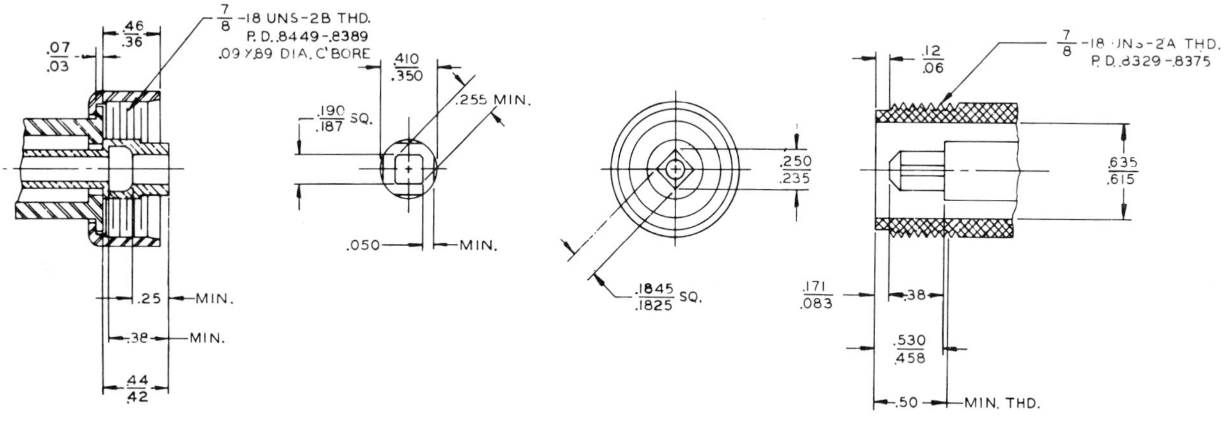

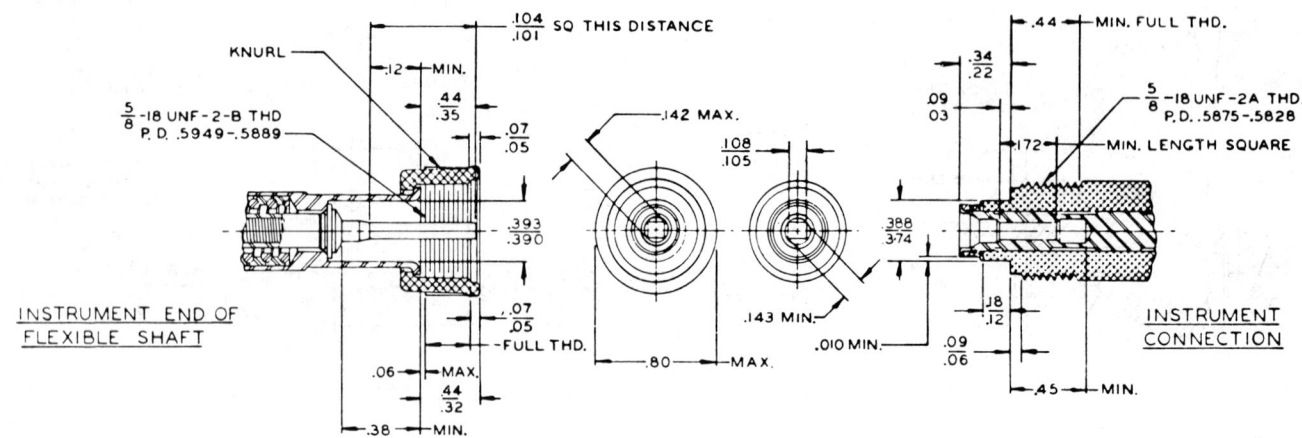

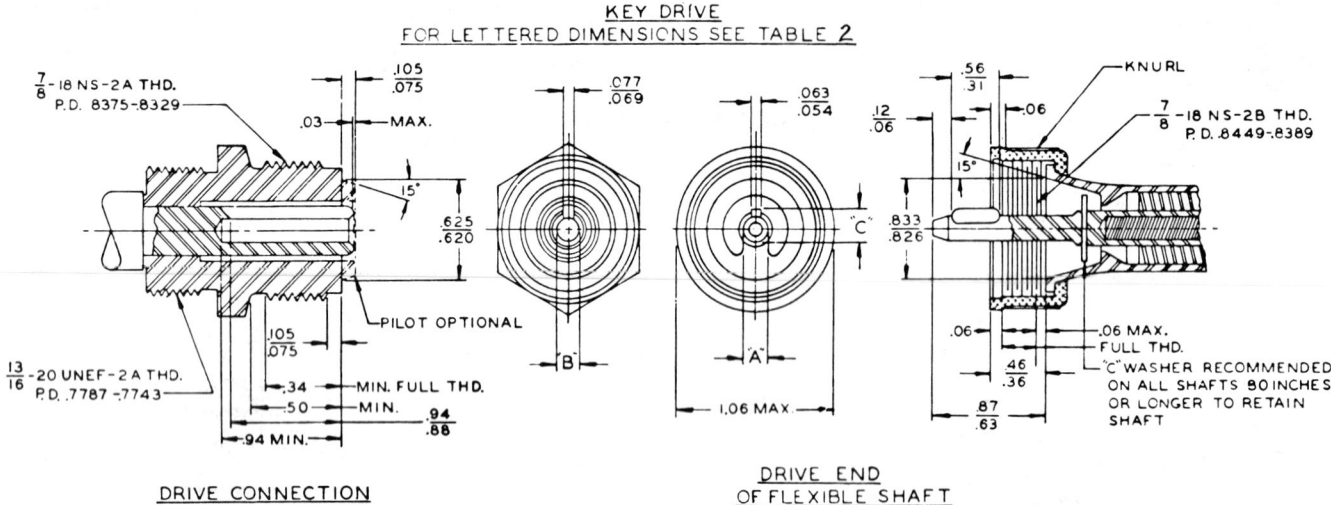

FIG. 5—HEAVY DUTY DRIVE (FOR USE WITH EDDY CURRENT TYPE INSTRUMENTS AND KEY DRIVE OR SQUARE DRIVE)

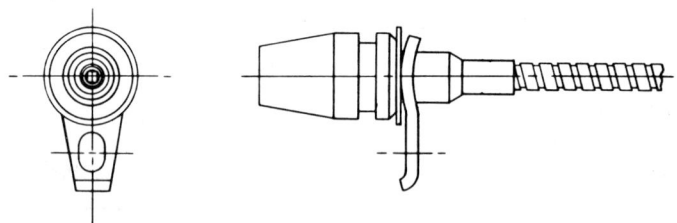

FIG. 6—PLUG TYPE LOWER FERRULE

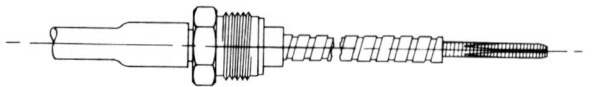

FIG. 7—FRONT WHEEL DRIVE

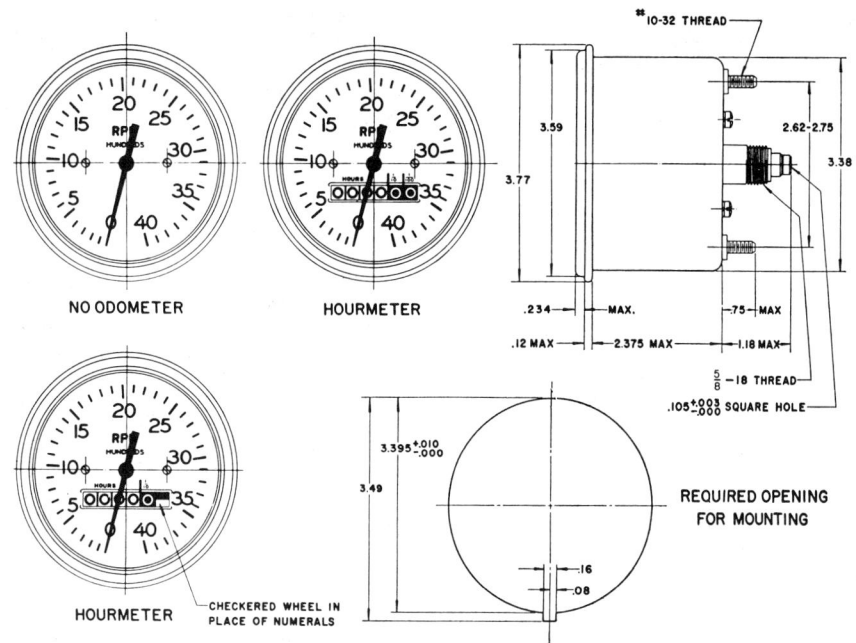

FIG. 8—MECHANICAL EDDY CURRENT TACHOMETER

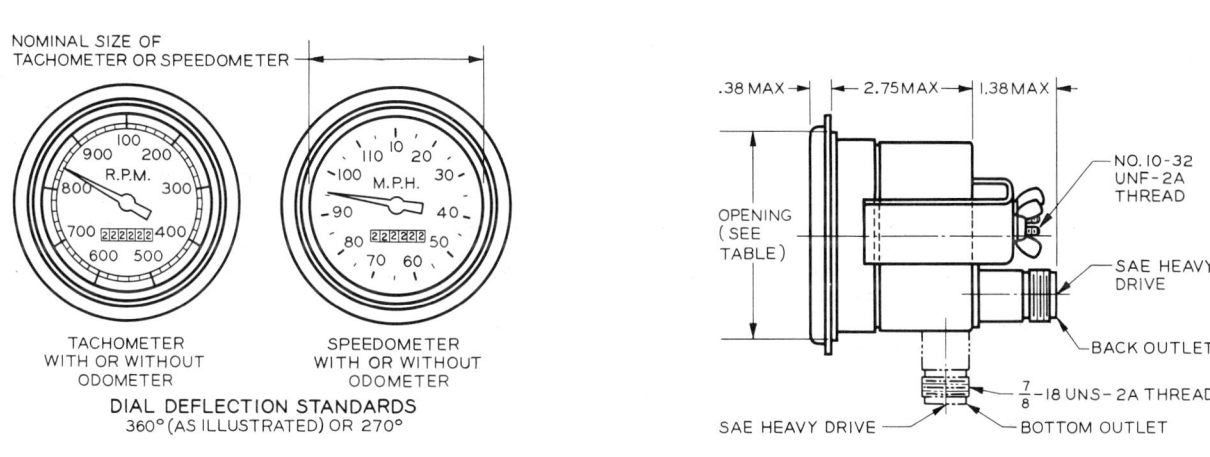

FIG. 9—MECHANICAL CENTRIFUGAL TACHOMETERS AND SPEEDOMETERS

SPEEDOMETER TEST PROCEDURE—SAE J1059

SAE Recommended Practice

Report of Speedometer and Tachometer Committee approved November 1973.

1. Scope—This SAE Recommended Practice provides a test procedure for eddy current speedometers, including the odometer if an integral portion of the speedometer, for passenger car service.

2. Performance Tests—All performance tests shall be made with the dial tilted at its design angle but shall not be less than 5 deg backward. Reference temperature for all performance tests is 75 ±5°F (24 ±2.8°C).

2.1 The calibration shall be as in Table 1 of SAE J678.

2.1.1 The temperature compensation shall be as noted in paragraph 2.4 of SAE J678.

2.1.2 The speedometer shall smoothly break away from the design rest position in a manner agreed to by the customer and manufacturer.

2.1.3 The indicator shall always return to its rest position when the drive becomes immobile. This must be accomplished throughout the range of specified temperature and without external vibration. The condition may also be tested at zero drive speed by releasing the indicator from the 5 mph (8 km/h) position. The indicator must return to its design rest position.

2.1.4 The total backlash (hysteresis) in an instrument with a live bearing indicator staff system shall not exceed 1.5 mph (2.4 km/h), or 3 mph

(4.8 km/h) for a stationary bearing pointer staff system, on both accelerating and decelerating without external vibration being applied to the instrument. This condition shall be checked at 500 rpm and the checkpoint shall be approached at the rate of 1 mph (1.6 km/h) from 25 mph (40 km/h) ascending or 35 mph (56 km/h) descending. A 2 s time interval must be allowed for dissipation of the damping effect prior to observing the readings.

2.1.5 The balance of the speed cup and indicator assemblies shall be such that no more than a total of 6 mph (10 km/h) change of indication occurs when the speedometer is driven at 500 rpm and the instrument is rotated 360 deg about the indicator axis and in the design mounting angle.

2.1.6 The rotation of the internal parts of the speedometer shall not result in unusual indicator flutter or waver or in erratic deflections of the indicator. This condition shall be checked at random speeds and be observed from a 2 ft (0.6 m) distance perpendicular to and at 45 deg angle to the dial. During this test, the speedometer shall be driven by means which exclude excitement caused by or transmitted through the speedometer cable.

2.1.7 The speedometer shall be so damped that when being driven at 500 rpm and the indicator is physically displaced to 70 mph (110 km/h), and released, the indicator shall reverse direction not more than four times; and if it does not reverse direction, it shall return to the original reading within 1.5 s.

2.1.8 The torque required to rotate the magnet shaft and odometer shall not exceed 0.00085 N·m for single odometer units nor more than 0.00150 N·m with total and trip odometers. During these tests, all odometer numerals shall be in operation. The test shall be conducted between a drive speed of 3 and 1000 rpm at a temperature of 75 $\pm$5°F (24 $\pm$2.8°C).

3. *Vibration Tests*

3.1 Test speedometers shall be vibrated for 3 h. For 1 h in each direction along three mutually perpendicular axis with a total excursion of 0.020 in. (0.5 mm) and a frequency varying 16–50 c/s, the frequency shall be cycled from 16 to 50 to 16 over a 2 min period. The mounting to be at design angle but no less than 5 deg backward.

3.2 After completion of the vibration test, the performance deviation listed in paragraph 6 will be permitted.

4. *Laboratory Endurance Tests*

4.1 Endurance life tests for the speedometer shall be 50,000 miles (80,000 km) or a duration test for an equivalent number of driveshaft revolutions, with speed and temperature cycling as follows:

4.1.1 SPEED CYCLING—The speed shall be cycled from 167 rpm reverse to 1500 rpm forward to 167 rpm reverse every 2 min.

4.1.2 TEMPERATURE CYCLING—Elevate the test chamber to 120 $\pm$5°F (49 $\pm$2.8°C) each day for three consecutive days, 6 h per day, and speed cycle as per paragraph 4.1.1. The speedometers are operated at room temperature and 1500 rpm for the remainder of the 24 h period.

4.1.3 Elevate the test chamber to 170–180°F (80 $\pm$2.8°C) for 6 h one day each week. The speedometers are not operated during this heat cycle. The speedometers are then continued operating at room temperature and 1500 rpm for the remaining hours of the day.

4.1.4 Reduce the test chamber to 0 $\pm$5°F ($-$17.8 $\pm$2.8°C) for 6 h one day each week. During the first hour, the speedometer shall not be operated. During the next 5 h, the speedometer shall be operated according to the speed cycling test of paragraph 4.1.1, except that the maximum speed shall be 1000 rpm. The test sample is then to be operated at room temperature and 1500 rpm for the remainder of the day. Throughout the cold test and any subsequent testing, the speedometer shall not seize or exhibit an appreciable increase in noise level when tapped by a wooden drafting pencil held loosely in the fingers.

4.1.5 Test speedometers shall be run at room temperature and 1500 rpm constant speed for two days to complete the weekly cycle.

4.2 After completing the life test (paragraph 4.1), the performance shall be as specified by paragraph 6.

5. *Vehicle Testing*—Test speedometers shall be installed in test vehicles and subjected to a 25,000 mile (40,250 km) (or equivalent driveshaft revolutions) general endurance road test whereby a great variety of road conditions are encountered. After completing the vehicle test, the performance deviation permissible in paragraph 6 will be permitted.

6. *Performance Checks after Vibration, Endurance, or Vehicle Tests*—After completion of the endurance, vibration, or vehicle test, the performance of the instruments shall be checked against the readings taken during the initial performance check. Deviations are allowed as follows:

6.1 The calibration shall not deviate more than 3% of the test speed from the limits of Table 1 of SAE J678.

6.2 The temperature compensation shall be as stated in paragraph 2 of SAE J678.

6.3 The indicator must have a positive smooth breakaway movement from its design rest position, as agreed to by the manufacturer and customer.

6.4 The indicator must return to design rest position.

6.5 The indicator backlash (hysteresis) shall not exceed 2.5 mph (4.0 km/h) for live bearing units or 4.0 mph (6.5 km/h) for a stationary bearing unit.

6.6 The rotation of the internal parts of the speedometer shall not result in unusual indicator flutter or waver or in erratic deflections of the indicator. This condition shall be checked at random speeds and be observed from a 2 ft (0.6 m) distance perpendicular to and at a 45 deg angle to the dial. During this test, the speedometer shall be driven by means which exclude excitement caused by or transmitted through the speedometer cable.

6.7 The damping shall be within the original specification, except five reversals of direction shall be allowed.

6.8 The drive torque shall be no more than 0.00100 N·m for single odometer units or 0.00175 N·m for double odometer units. Tests shall be as described in paragraph 2.1.9.

6.9 The balance of the indicator assembly may change no more than 2 mph (3 km/h), as compared to the reading obtained as per paragraph 2.1.6.

ELECTRIC TACHOMETER SPECIFICATION— ON ROAD—SAE J196 JUN80

SAE Recommended Practice

Report of the Speedometer and Tachometer Committee, approved February 1972, completely revised June 1980.

1. *Scope*—This SAE Recommended Practice establishes minimum requirements for electric tachometer systems with and without hourmeter or revolution counter, for general on-road (passenger car, multi-purpose passenger vehicle, truck, and bus) applications.

2. *Electric Tachometer System*—A typical electric tachometer system for engines using a Kettering ignition system or the newer electronic ignition systems, consists of an indicating unit that obtains a signal proportional to engine speed from the ignition system.

If the tachometer is intended for use on a diesel engine, a sender may be used to supply a signal proportional to engine speed. A signal may also be obtained from an AC tap on the alternator if the alternator is so equipped. If a sending unit is used it will often be one of the following types: permanent magnet generator, magnetic switch, or magnetic sensor. The sender may be mounted on the engine outlet provided for mechanical tachometer cables, or it may be mounted so as to sense the number of teeth on the flywheel ring gear or some other location where a rotating element with teeth, slots, holes, or bosses may be sensed.

The indicating unit may contain an hourmeter. The hourmeter in an electric tachometer may be a true time indicator rather than an indication proportional to the number of engine revolutions. The latter indication is usually found in mechanical tachometers.

3. *Factors Affecting Tachometer and Hourmeter Accuracy*—Changes in ambient temperature and voltage may affect the tachometer and/or the hourmeter indication.

4. *Tachometer and True Hourmeter Indication (Allowable System Variation)*

4.1 *Tachometers Driven by Signal from Ignition System or Alternator AC Tap*—The tachometer indication shall be within $\pm$2% of full scale with nominal voltage applied at a temperature of 75 $\pm$ 5°F (24 $\pm$ 3°C) when the tachometer is driven with a signal from an ignition system or from an alternator AC tap, as applicable. If a calibrator is used it must supply a signal having the same characteristics as that supplied by an ignition system or an alternator AC tap. Calibration of tachometers shall be made with the instrument in approximately the same angular position that it will have when mounted in the vehicle. See Environmental Conditions for allowable variation within the instrument due to changes in ambient temperatures and voltage.

4.2 *Sender Driven Units*—The tachometer indication shall be within $\pm$2% of full scale with nominal voltage applied at a temperature of 75 $\pm$ 5°F (24 $\pm$ 3°C), when the tachometer is driven with a signal from a sender either rotated or excited in a fashion simulating actual operation. If a calibrator is used, it must supply a signal having the same characteristics as the sender. Calibration of tachometers shall be made with the instrument in approxi- mately the same angular position that it will have when mounted in the vehicle. See Environmental Conditions for allowable variation within the instrument due to changes in ambient temperatures and voltage.

4.3 True Hourmeter—The hourmeter indication shall be within ±2% with nominal voltage applied at a temperature of 75 ± 5°F (24 ± 3°C).

4.4 Hourmeter Proportional to Number of Engine Revolutions—The time indication shall be within ±0.3% with nominal voltage applied, nominal input rpm required to indicate one hour, and at a temperature of 75 ± 5°F (24 ± 3°C).

5. Effects of Environmental Conditions

5.1 Temperature (Allowable System Variation)

5.1.1 TACHOMETER INDICATION—With nominal voltage applied, the tachometer indication shall not vary more than ±2% of full scale from the reading determined in Section 4, while the indicating unit is operating over the range of 20–130°F (−7–54°C) and the sender (if required) is operating over the range of −40–250°F (−40–121°C). No permanent damage shall result from operating the indicating unit in a range of −40–180°F (−40–82°C).

5.1.2 TRUE HOURMETER—With nominal voltage applied, the time indication shall not vary more than ±1% from a reading obtained at 75 ± 5°F (24 ± 3°C) while the instrument is operating over the range of 20–130°F (−7–54°C). No permanent damage shall result from operating the instrument in a range of −40–180°F (−40–82°C).

5.1.3 HOURMETER PROPORTIONAL TO NUMBER OF ENGINE REVOLUTIONS—With nominal voltage applied and nominal input rpm required to indicate 1 h, the time indication shall not vary more than ±0.3% from a reading obtained in Section 4, while the instrument is operating over the range of 20–130°F (−7–54°C). No permanent damage shall result from operating the instrument in a range of −40–180°F (−40–82°C).

5.2 Temperature Extremes (Sender Only)—It will be necessary to evaluate the specific application to specify the allowable temperature extremes.

5.3 Storage Temperature (Indication Unit Only)

5.3.1 TACHOMETER—A 4 h exposure of the indicating unit to a temperature of −40–185°F (−40–85°C) shall result in no more than ±1% of full scale permanent calibration change from the reading obtained in Section 4. The rate of temperature change during this test shall not exceed 3.6°F (2°C) per minute.

5.3.2 HOURMETERS—A 4 h exposure of the indicating unit to a temperature of −40–185°F (−40–85°C) shall result in no more than ±1% permanent calibration change from the reading obtained in Section 4. The rate of temperature change during this test shall not exceed 3.6°F (2°C) per minute.

5.4 Voltage Variation (Indicating Unit)

5.4.1 TACHOMETER—The indication shall not change more than ±1% of full scale from the reading obtained in Section 4, within the following voltage ranges.

12 Volt System	24 Volt System
12–16 VDC	24–32 VDC

Twelve and twenty-four volt tachometers shall not change more than ±3% of full scale, from the reading obtained in Section 4, at 11 and 22 V respectively.

5.4.2 TRUE HOURMETER—The indication shall not change more than ±1% from the reading obtained in Section 4, within the following voltage ranges.

12 Volt System	24 Volt System
12–16 VDC	24–32 VDC

Twelve and twenty-four volt hourmeters shall not change more than ±3% of full scale, from the reading obtained in Section 4, at 11 and 22 V respectively.

5.4.3 HOURMETER PROPORTIONAL TO NUMBER OF ENGINE REVOLUTIONS—At 75 ± 5°F (24 ± 3°C) and with nominal input rpm required to indicate 1 h, the time indication shall not vary more than ±0.3% from a reading obtained in Section 4, when operating within the voltage ranges given in paragraph 5.4.2.

5.5 Abnormal Voltage Conditions—Tachometer and True Hourmeter

5.5.1 TRANSIENT VOLTAGE PROTECTION—The indicating unit shall be capable of withstanding supply voltage transients without permanent damage and shall remain within the calibration specification of Section 4 at the conclusion of this test. The instrument shall be connected and operated for a total of 1 h with a means provided to impress upon the nominal battery voltage a repetitive rectangular voltage pulse of plus and minus six times nominal battery voltage with a duration of 300 microseconds and 1% duty cycle with a current of no more than 1.0 A.

For applications with transient voltages having a magnitude, duration, or duty cycle exceeding the above requirements, contact the instrument manufacturer for recommendations.

5.5.2 OVERVOLTAGE AND REVERSE POLARITY—Provisions for protection against booster starts with double battery voltage and/or reversed polarity must be negotiated between the user and the manufacturer.

5.6 Moisture Resistance

5.6.1 HUMIDITY (INDICATING UNIT)—Indicating unit shall withstand exposure to 95% relative humidity at 100°F (38°C) for 48 h.

5.6.2 SALT SPRAY (SENDER UNIT)—Sender units shall be corrosion resistant and shall withstand a salt spray (fog) test of 48 h duration with 5% salt solution (Reference ASTM B117-73).

5.7 Vibration Test (Indicating Unit)—The indicating unit shall be capable of withstanding without mechanical or electrical failure, 3 h of vibration, 1 h in each direction along the three mutually perpendicular axes. One of said axes is to be parallel to the indicator shaft. The vibration test shall be run at a double amplitude of 0.030 in (0.76 mm) with the frequency varying from 10–30–10 Hz at intervals of 1 min. After completion of test, the calibration shall remain within tolerances as specified in Sections 4 and 5.

5.8 Vibration Test (Sender Only)

5.8.1 ENGINE MOUNTED—The sender shall be capable of withstanding 6 h of vibration without mechanical or electrical failure, 2 h in each direction along the three mutually perpendicular axes. One of said axes is to be parallel to the input shaft. The vibration test shall be run at a double amplitude of 0.020 in (0.51 mm) with the frequency varying from 10–120–10 Hz at intervals of 1 min.

5.9 Shock Test (Indicating Unit Only)—The indicating unit shall be capable of withstanding without mechanical or electrical failure, the following series of shocks and still maintain the calibration tolerances specified in Sections 4 and 5.

The indicating unit shall be subjected to one shock in each direction along each of three mutually perpendicular axes. One of said axes is to be parallel to the indicator shaft. Each shock shall have an amplitude of 23–27 g, half sine of 9–13 ms duration.

5.10 Shock Test (All Senders)—The sender shall be capable of withstanding, without mechanical or electrical failure, 6 shocks of 44–55 g, half sine of 9–13 ms duration in each direction along each of three mutually

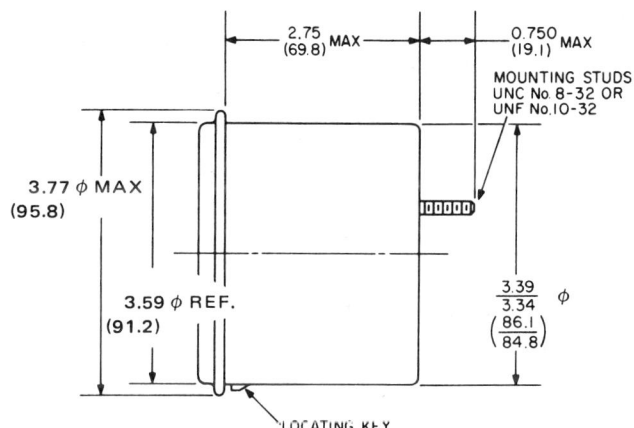

DIMENSIONS ARE INCHES (MILLIMETERS)

FIG. 1—ENVELOPE AND MOUNTING STUDS

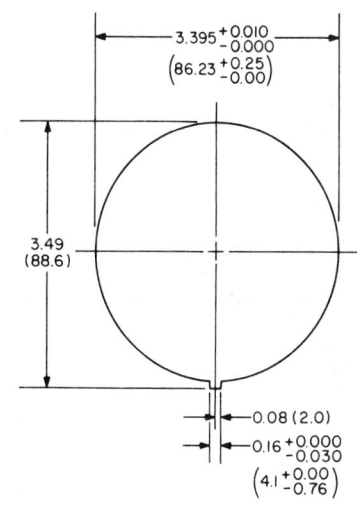

DIMENSIONS ARE INCHES (MILLIMETERS)

FIG. 2—MOUNTING CUTOUT DETAIL

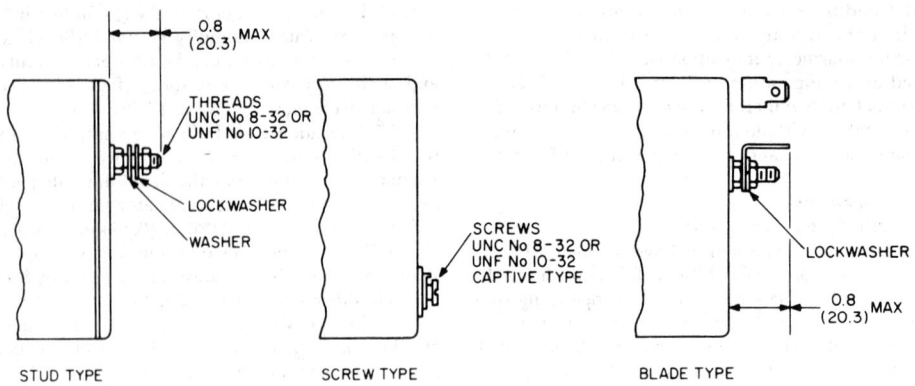

DIMENSIONS ARE INCHES (MILLIMETERS)

FIG. 3—TERMINALS

perpendicular axes. One of said axes is to be perpendicular to the mounting plane.

5.11 Design Detail Recommendations (Indicating Unit Only)

5.11.1 When analog displays are used, the display shall be accomplished by a pointer or other indicator traversing in a clockwise or left to right direction as applicable, to register increasing revolutions per minute over a suitable scale on the indicating unit dial.

5.11.2 Graduations shall be designed for the best practical legibility and accuracy of reading.

5.11.3 Unless otherwise specified: pointers and dial printing shall be white, dial background shall be low gloss black, and visible portions of the indicating unit should exhibit low reflectivity; the time or revolution indicator shall have white numerals on a low gloss black background except for the tenths indicator, which shall have black numerals on a white background.

5.11.4 The indicating unit case shall be provided with studs for mounting by suitable U-clamps or similar means.

5.11.5 Typical envelope, mounting studs, and terminal designations are displayed in Figs. 1 and 2.

5.12 Identification

5.12.1 INDICATING UNIT

5.12.1.1 To be legibly indicated on outside of case:
(a) Manufacturer's or user's part number
(b) Manufacturer's or user's serial number and/or date of manufacture

5.12.1.2 To be printed on dial and/or indicated on case:
(a) Manufacturer's or user's name or trademark

5.12.1.3 Electrical connections shall be clearly identified for proper wiring of instrument into circuit.

5.12.2 SENDER—Sender identification is to be as agreed between manufacturer and user.

ELECTRIC TACHOMETER SPECIFICATION—OFF ROAD—SAE J197

SAE Recommended Practice

Report of Speedometer and Tachometer Committee approved April 1973.

1. Scope—To specify minimum requirements for electric tachometers for general off-road applications and to provide dimensional envelopes for determining space allowances. Two general classes are provided to correspond with different applications, Class 1 being less severe than Class 2.

2. Identification

2.1 To be legibly stamped on outside of case:
2.1.1 Manufacturer's part number.
2.1.2 Manufacturer's serial number and/or date of manufacture.

2.2 To be printed on dial and/or stamped on case: manufacturer's name or trademark.

2.3 Electrical connections shall be clearly identified for proper wiring of instrument into circuit.

3. Calibration—For instruments mounted with dials in any plane from vertical to 35 deg back from vertical, calibration accuracy shall be ±3% of full-scale reading at 75°F (24°C) (instrument to be tapped lightly when checking calibration). For other instrument mounting angles, effect on calibration should be determined by contact with manufacturer. Recommended calibration points shall be at 20 and 80% of full-scale dial reading.

4. Environmental Conditions

4.1 Temperature

4.1.1 Instrument indication shall not vary in excess of ±2% of the full-scale reading of the values determined at a temperature of 75°F (24°C) while operating through a range of +20–130°F (−7 to +54°C). No permanent damage shall result from operating in a range of −40 to +160°F (−40 to +71°C).

4.1.2 STORAGE—Varying the temperature of the instrument from −40°F (−40°C) for 2 h, allow to recover to room temperature before checking then for 2 h at +180°F (+82°C) exposure. Again allow to recover to room temperature. The check shall result in not more than ±1% of full-scale permanent calibration change from readings determined at 75°F (24°C) before exposure (paragraph 3) when checked at 75°F (24°C) after exposure.

4.2 Humidity—Instrument shall not have its function impaired by exposure to 95% relative humidity at 100°F (38°C) for 48 h.

4.3 Vibration Test—The electric tachometer and/or tachometer and electric sender shall be capable of withstanding without mechanical or electrical failure 6 h of vibration, 2 h in each direction along the three mutually perpendicular axes. One of said axes shall be parallel to the indicator shaft. The vibration test shall be run at a double amplitude of 0.060 in. (1.52 mm) with the frequency variation of 10–55–10 Hz for Class 1 and 10–80–10 Hz for Class 2 at intervals of 1 min. After test, the calibration shall remain within the tolerances specified in paragraph 3. The sample shall be mounted in accordance with the manufacturer's recommendations.

4.4 Shock Test—The electric tachometer and/or tachometer and electric sender shall be capable of withstanding without mechanical or electrical failure the following series of shocks and still maintain the calibration tolerances specified in paragraph 3.

The instrument shall be mounted in a vertical mounted position and subjected to 25 shocks with a vertical acceleration of 23–27 g for Class 1, 44–55 g for Class 2, half sine characteristic of 9–13 ms duration. The instrument shall then be rotated 90 deg counterclockwise in the plane of mounting and subjected to 25 shocks of the same magnitude and direction as previously. The instrument shall then be mounted face upward and subjected to 25 shocks of the same magnitude and direction as previously. The sample shall be mounted in accordance with the manufacturer's recommendations.

4.5 Voltage Variations—The electric tachometer calibration shall not change more than ±1% of full-scale rpm at a half-scale dial graduation due to a voltage change of a nominal 12 V system from 12 to 16 V d-c and a nominal 24 V system from 24 to 32 V d-c.

4.6 Transient Protection—The instrument shall be capable of withstanding supply voltage transients without permanent damage and shall remain within the calibration specification of paragraph 3 at the conclusion of this test.

The instrument shall be connected and operated for a total of 1 h with a means provided to impress upon the nominal battery voltage a single or repetitive rectangular voltage pulse as follows, limiting current flow to 0.5–1 A:

(a) Plus six times nominal battery voltage of 300 μs duration and 1% duty cycle.

(b) Minus four times nominal battery voltage of 300 μs duration and 1% duty cycle.

Any transient voltage that has a magnitude, duration, or duty cycle exceeding the above requirement shall be considered destructive and paragraph 5.8 shall be followed.

4.7 Secondary Losses—The effect of the tachometer on the ignition system should not reduce the available secondary voltage by more than 4%. Testing is to be done with the exact ignition system to be used in actual practice. The distributor is to be run with the coil input voltage held constant at 14.0 V and the coil secondary open circuited. The exact test procedure for measurements shall be established by the supplier and consumer.

5. Detail Recommendations

5.1 Indication shall be accomplished by a pointer traversing in a clockwise or left-to-right direction as applicable to register increasing rpm over a suitable scale on the instrument dial.

5.2 Graduations shall be designed for the best practical legibility and accuracy of reading.

5.3 Unless otherwise specified, pointers and dial printing shall be white, dial background shall be low gloss black, and visible portions of the instrument should exhibit low reflectivity.

5.4 All exposed surfaces shall be corrosion resistant for limited exposure. (NOTE: If instruments are required for extreme exposure installations, contact manufacturer for recommendations.)

5.5 Instruments shall be moisture and dust resistant. (See note in paragraph 5.4.)

5.6 Sending units shall be corrosion resistant to withstand external exposure. Engine mounted signal source sending units shall be excluded from this specification until clarification of the mounting environment is investigated.

5.7 Tachometer mounting—The tachometer case shall be provided with studs for mounting by suitable U-clamp or similar means.

5.8 For instrument requirements not covered by this specification, contact the manufacturer for recommendations.

5.9 Typical envelope, mounting studs, and terminal designations are displayed in Figs. 1 and 2.

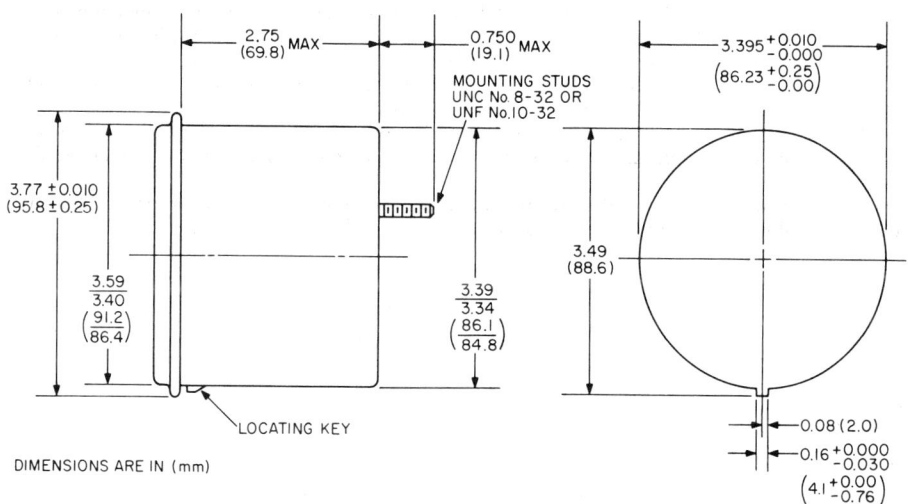

FIG. 1—ENVELOPE AND MOUNTING STUDS

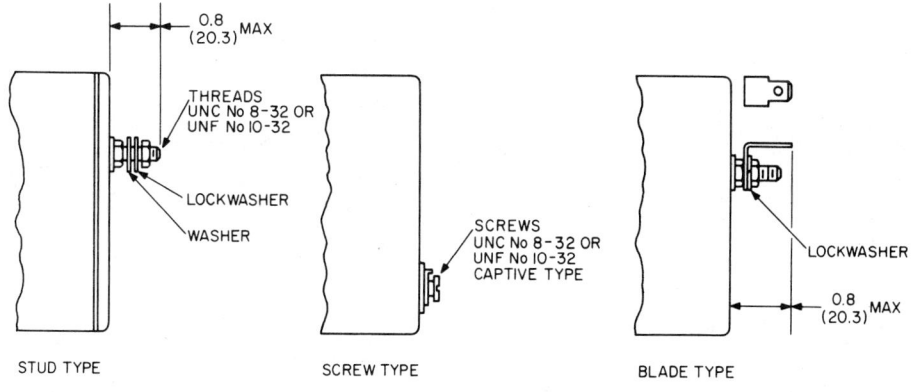

FIG. 2—TERMINALS

AUTOMATIC VEHICLE SPEED CONTROL—MOTOR VEHICLES—SAE J195

SAE Recommended Practice

Report of Automotive Safety Committee approved October 1970.

1. *Purpose*—The purpose of this SAE Recommended Practice is to provide a series of engineering guidelines for the design of an automatic vehicle speed control, and to define the minimum control performance which a device must provide in order to be classified an automatic vehicle speed control.

2. *Definition of Automatic Vehicle Speed Control*—An automatic vehicle speed control is a device capable of maintaining selected vehicle speeds in the presence of changing road load conditions.

3. *Scope*—This SAE Recommended Practice is intended to apply only to the design of an automatic vehicle speed control. It is not intended to encourage or discourage the installation of automatic vehicle speed controls on any class of vehicles, nor is it intended to influence the requirements of engine speed governors.

4. *Design Recommendations*

   4.1 The speed control shall require a deliberate action of the driver to cause activation and reactivation. Systems reactivated solely by the operation of the accelerator pedal shall include a signal to the driver to indicate reactivation.

   4.2 The speed control shall be deactivated upon application of the service brakes (depressing the clutch pedal on manual clutch equipped vehicles shall also deactivate the speed control) and shall not reactivate without the deliberate action of the driver.

   4.3 When deactivated, the speed control shall have no effect on vehicle operation.

   4.4 If a speed signal source other than the drivetrain or a driving wheel is used, suitable precautions shall be provided to prevent runaway when a driving wheel loses traction.

   4.5 An alternate hand-operated deactivation control within the reach of the driver, in addition to the brake, clutch (if so equipped), and ignition key, shall be provided.

   4.6 The system shall be capable of deactivation or capable of being made inoperative by a control within the reach of the driver under the following conditions:

   4.6.1 Failure of any power source to the device.
   4.6.2 Failure of speed signal to the device.
   4.6.3 Short circuit of electrical leads of the device.
   4.6.4 Failure of other vehicle components upon which the device is dependent for function.

   4.7 The device shall not be operable below 20 mph (32 km/hr).

   4.8 The speed control linkage shall be designed and installed to prevent inadvertent interference with normal accelerator control operation under all operating and environmental conditions consistent with the environmental capabilities of the vehicle upon which it is installed.

5. *Performance Requirements*—The automatic vehicle speed control shall regulate the output power of the engine to provide a stable and essentially constant vehicle speed. The following test defines the minimum performance requirements of an automatic vehicle speed control.

   5.1 Test Conditions—The performance evaluation shall be performed under the following conditions:

   5.1.1 Ambient temperature of $+30$–80 F (-1 to $+27$ C).
   5.1.2 Altitude not to exceed 2500 ft (760 m) above sea level.
   5.1.3 Component wind velocity in the direction of travel of the vehicle not to exceed 10 mph (16 km/hr).
   5.1.4 Selected vehicle speed of 40–70 mph (64–113 km/hr).
   5.1.5 Test road shall be hard surfaced and shall include a minimum of one $\frac{1}{8}$ mile (200 m) (minimum) grade of -2% and one $\frac{1}{8}$ mile (200 m) (minimum) grade of $+2\%$, and shall be at least 5 miles (8 km) long.

   5.2 Test Vehicle—The vehicle used for performance evaluation shall be capable of a level road 40–70 mph (64–113 km/hr) acceleration time of not over 16 sec. The vehicle qualification test shall be run in the highest available transmission gear or range. Downshifting is prohibited in meeting this requirement.

   5.3 Performance Limits—Maximum vehicle speed variation shall not exceed ± 3 mph (5 km/hr).

21 Tubing and Fittings

CODING SYSTEM FOR IDENTIFICATION OF TUBE, PIPE, AND HOSE FITTINGS—SAE J846 MAY80

SAE Recommended Practice

Report of the Tube, Pipe, Hose, and Lubrication Fittings Committee, approved January 1963, last revised by Fluid Conductors and Connectors Technical Committee May 1978, editorial change May 1980.

NOTE: It should be noted that the code numbers assigned to the applicable standards covered by the coding system could possibly change. Therefore, it is recommended that for the purpose of transmitting technical or engineering information relating to the various tube, pipe, and hose fittings, the applicable code numbers and standard number and revision letter be specified for proper identification.

Scope—This coding system is intended to provide a convenient means of identifying the various tube, pipe, and hose fittings and of transmitting technical or engineering information relating to them wherever drawings or other pictorial media may not be readily available. The code has been kept flexible to permit expansion to cover new fitting categories or styles and, if the need develops, the inclusion of materials. The system is also compatible with automatic data processing equipment.

It is not intended that this code should supersede established systems or means of identification. However, because the SAE code for automotive flare fittings shown in SAE J512 is also applicable to corresponding refrigeration fittings in SAE J513, both an SAE code and the existing code ANSI B70.1, Refrigeration Flare Fittings, are included throughout SAE J513. Therefore, it should be the prerogative of the user to apply that code which best satisfies his requirements.

GENERAL SPECIFICATIONS

Code—The code shall consist of two groups of numbers and one group of letters: the first group of numbers symbolizing the size identification, the second group of numbers symbolizing the fitting and hose identification, and the third group of letters symbolizing the material and assembly as delineated below.

1. Size Identification—The fitting size shall be identified by a series of dash numbers, each representing the size of the respective fitting ends. The size of the tube end shall precede the size of the pipe, hose, or other ends of the fitting. The dash shall be given in the sequence defined below.

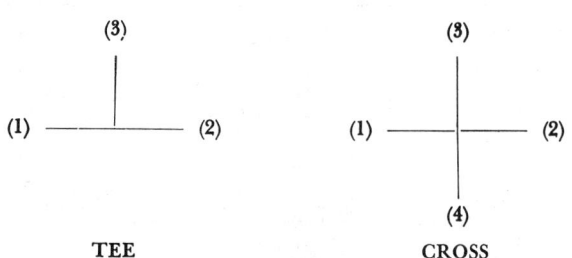

TEE   CROSS

When special size combinations of tube to tube fitting ends are specified, the largest tube size shall precede the smaller tube size for unions and union elbows. For tees, the (1) shall be the larger tube size of (1) and (2). For crosses the (1) shall be the largest tube size of (1) and (2) and (3) shall be equal to or larger than (4).

The dash size symbol applicable to all tube ends and straight thread O-ring boss ends shall consist of the number of sixteenth inch increments contained in the outside diameter of the tubing (nominal tube OD) they are designed to be used with, as listed below:

Nominal Tube OD, in	Dash Size Symbol	Nominal Tube OD, in	Dash Size Symbol	Nominal Tube OD, in	Dash Size Symbol
1/8	-2	7/16	-7	7/8	-14
3/16	-3	1/2	-8	1	-16
1/4	-4	9/16	-9	1-1/4	-20
5/16	-5	5/8	-10	1-1/2	-24
3/8	-6	3/4	-12	2	-32

The dash size symbol for pipe thread ends shall be the number of sixteenth inch increments contained in the nominal pipe thread size as listed below:

Nominal Pipe Thread Size, in	Dash Size Symbol	Nominal Pipe Thread Size, in	Dash Size Symbol	Nominal Pipe Thread Size, in	Dash Size Symbol
1/16	-1	3/8	-6	1	-16
1/8	-2	1/2	-8	1-1/4	-20
1/4	-4	3/4	-12	1-1/2	-24
				2	-32

The dash size symbol for hose shall be the number of sixteenth inch increments contained in the inside diameter of the hose (nominal hose ID), except in the case of SAE 100R5 hose where it is equivalent to the number of sixteenth inch increments in the outside diameter of tubing having approximately the same inside diameter as the hose. See tabulation below for respective hose types.

Nominal Hose ID, in	Dash Size Symbols For Hose Type						
	SAE 100R1	SAE 100R2	SAE 100R3	SAE 100R4	SAE 100R5	SAE 100R6	SAE 100R7
3/16	−3	−3	−3	—	−4	−3	−3
1/4	−4	−4	−4	—	−5	−4	−4
5/16	−5	−5	−5	—	−6	−5	−5
3/8	−6	−6	−6	—	—	−6	−6
13/32	−6.5	—	—	—	−8	—	—
1/2	−8	−8	−8	—	−10	−8	−8
5/8	−10	−10	—	—	−12	−10	—
3/4	−12	−12	−12	−12	—	—	−12
7/8	−14	−14	—	—	−16	—	—
1	−16	−16	−16	−16	—	—	−16
1-1/8	—	—	—	—	−20	—	—
1-1/4	−20	−20	−20	−20	—	—	—
1-3/8	—	—	—	—	−24	—	—
1-1/2	−24	−24	—	−24	—	—	—
1-13/16	—	—	—	—	−32	—	—
2	−32	−32	—	−32	—	—	—
2-1/2	—	—	—	—	−40	—	—
3	—	—	—	—	−48	—	—
3-1/2	—	—	—	—	−56	—	—
4	—	—	—	—	−64	—	—

The dash size symbol for 4-bolt split flange O-ring connections shall be the number of sixteenth inch increments contained in the nominal flange size as listed below:

Nominal Flange Size, in	Dash Size Symbol	Nominal Flange Size, in	Dash Size Symbol	Nominal Flange Size, in	Dash Size Symbol
1/2	-8	1-1/2	-24	3-1/2	-56
3/4	-12	2	-32	4	-64
1	-16	2-1/2	-40	5	-80
1-1/4	-20	3	-48		

The dash size symbols for straight thread pipe plugs and filler and drain plugs shall be the number of sixteenth inch increments contained in the nominal straight thread size except in the case of metric thread size where the thread size shall be designated as the dash size listed below:

Nominal Straight Thread Size in	Dash Size Symbol	Nominal Metric Thread Size (mm)	Dash Size Symbol
5/16	-5	M10 x 1	-M10
3/8	-6	M14 x 1.25	-M14
1/2	-8	M18 x 1.5	-M18
5/8	-10		
3/4	-12		
7/8	-14		
1	-16		
1-1/4	-20		
1-1/2	-24		
1-3/4	-28		
2	-32		

2. Fitting Identification—The fitting identification shall consist of a six-digit number made up of three groups of two digits each symbolizing in sequence: (a) the fitting type, (b) the fitting shape, and (c) the fitting connecting ends. (For convenient reference, the fitting identificaton codes applicable to fittings appearing in SAE J246, SAE J512, SAE J513, SAE J514, SAE J516, SAE J518, SAE J530, SAE J531, and SAE J532 are shown in brackets adjacent to the respective figure numbers.) The identification symbols for each of the three groups shall be as follows:

(a) FITTING TYPE IDENTIFICATION—The two-digit symbols applicable to the various fitting types and styles shall be as tabulated below:

Fitting Type Symbol	Fitting Type and Styles
01	45 deg flared, automotive
02	
03	
04	Inverted flared, automotive
05	
06	Tapered sleeve, automotive
07	37 deg flared, hydraulic
08	Flareless, hydraulic
09	SAE O-ring boss, hydraulic
10	Flanged sleeve, automotive, nylon tube
11	4-bolt split flange O-ring, hydraulic
12	Spherical sleeve, automotive, copper tube
13	Pipe, automotive
14	Pipe, hydraulic
15	Male pipe, hose, permanently attached
16	Male pipe, hose, field attachable
17	Male pipe, hose, field attachable segment clamp
18	SAE O-ring boss, hose, permanently attached
19	SAE O-ring boss, hose, field attachable
20	SAE O-ring boss, hose, field attachable segment clamp
21	37 deg male flared, hose, permanently attached
22	37 deg male flared, hose, field attachable
23	37 deg male flared, hose, field attachable segment clamp
24	37 deg female flared, hose, permanently attached
25	37 deg female flared, hose, field attachable (2 piece)
26	37 deg female flared, hose, field attachable segment clamp
27	37 deg female flared, hose, field attachable (3 piece)
28	45 deg male flared, hose, permanently attached
29	45 deg male flared, hose, field attachable
30	45 deg female flared, hose, permanently attached
31	45 deg female flared, hose, field attachable (2 piece)
32	45 deg female flared, hose, field attachable (3 piece)
33	Male flareless, hose, permanently attached
34	Male flareless, hose, field attachable
35	Male flareless, hose, field attachable segment clamp
36	Female flareless, hose, permanently attached
37	Female flareless, hose, field attachable
38	Female flareless, hose, field attachable segment clamp
39	Split flange, hose, permanently attached, standard pressure series
40	Split flange, hose, field attachable, standard pressure series
41	Split flange, hose, field attachable segment clamp, standard pressure series
42	Straight thread filler and drain plug
43	Push-on hose
45	Capillary, refrigeration
49	Split flange, hose, permanently attached, high pressure series
50	Split flange, hose, field attachable, high pressure series
51	Split flange, hose, field attachable, segment clamp, high pressure series

(b) FITTING SHAPE IDENTIFICATION—The two-digit symbols applicable to the various shapes of fittings shall be as tabulated below:

Fitting Shape Symbol	Fitting Shape
01	Straight
02	90 deg elbow
03	45 deg elbow
04	Tee
05	Cross
06	Straight bulkhead union
07	90 deg bulkhead elbow union
08	45 deg bulkhead elbow union
09	Bulkhead tee
10	22-1/2 deg elbow
11	30 deg elbow
12	60 deg elbow
13	67-1/2 deg elbow
14	90 deg elbow short drop
15	90 deg elbow long drop
16	90 deg elbow extra long drop

(c) FITTING CONNECTING END IDENTIFICATION—The two-digit symbols applicable to the various threaded, hose, connecting ends or combinations thereof for the fittings covered shall be as tabulated below:

Fitting Connecting End Symbol	Connecting Ends, Hose, and Combinations
01	Tube, all ends
02	Tube to external pipe
03	Tube to internal pipe
04	Tube to solder connection
05	Internal flare to external flare
06	Internal flare to external pipe
07	Internal flare union
08	Swivel flare union internal
09	Plug
10	Short nut
11	Long nut
12	Cap
13	Flare gasket
14	Flare seal bonnet
15	Tube sleeve
16	Reducing nut
17	Lock nut, small hex
18	Lock nut, large hex
19	Connector, large hex
20	Tube to SAE O-ring boss
21	Tube to SAE straight swivel connection

(Table continued on next page)

(continued)

Fitting Connecting End Symbol	Connecting Ends, Hose, and Combinations
22	Connector, long (tube to SAE O-ring boss)
23	Reducer seat to tube
24	External pipe on run
25	External pipe on branch
26	Internal pipe on run
27	Internal pipe on branch
28	Straight thread on run
29	Straight thread on branch
30	Swivel straight pipe to external pipe
31	Swivel straight pipe to internal pipe
32	Tube to SAE swivel connection on run
33	Tube to SAE swivel connection on branch
34	Tube to internal pipe on run and external pipe on branch
35	Tube to external pipe on run and internal pipe on branch
36	Straight thread to SAE O-ring boss
37	External pipe to external pipe
38	Internal pipe to internal pipe
39	Internal pipe to external pipe
40	Internal pipe to external pipe (bushing)
41	Seat insert
42	100R1 hose, Type A
43	100R2 hose, Types A and B
44	100R3 hose
45	100R4 hose
46	100R5 hose
47	100R6 hose
48	100R7 hose
49	100R8 hose
50	100R9 hose
51	100R10 hose
52	100R11 hose
53	100R1 hose, Type AT
54	100R2 hose, Types AT and BT
55	
56	
57	Swivel straight pipe to SAE O-ring boss
58	Tube to bulkhead on run
59	Tube to bulkhead on branch
60	Push-on hose to external pipe
61	Split flange to flanged head, standard pressure series
62	Split flange to flanged head, high pressure series
63	Tube to pipe fusible
64	Plug, pipe, fusible
65	Tube to straight pipe drum adapter
66	Nut, short, refrigeration
67	Nut, long, refrigeration
68	Nut, short, reducing, refrigeration
69	Nut, long, reducing, refrigeration

3. Material and Assembly Identification—The material and assembly identification shall consist of two letters symbolizing the material and assembly of multiple tube fitting components supplied assembled together rather than as separate pieces. For convenient reference, this code is applicable to SAE J246, SAE J512, and SAE J514. The identification symbols shall be as tabulated below:

Material and Assembly Symbol	Material and Assembly
BA	Brass assembly
CA	Carbon steel assembly
SA	Stainless steel assembly

Application of Code—The identification code shall be applied to the various fittings as depicted in the examples below.

1. Examples of Code Applied to Tube Fittings

The 45 deg flared tube connector for 1/8 in. tube OD, shown in Fig. 1A and Table 2 of SAE J512, would be coded as follows:

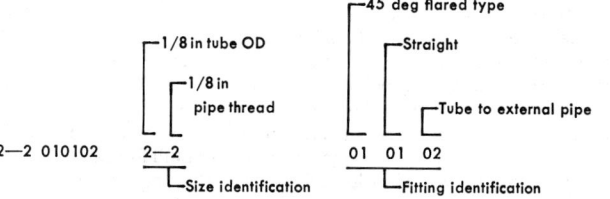

The hydraulic flareless tube tee for 3/8 in. tube OD, shown in Fig. 22D and Table 7 of SAE J514, would be coded as follows:

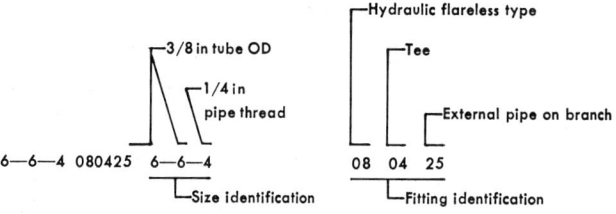

2. Examples of Code Applied to Hose Fittings—The male 45 deg flared type permanently attached style hose fitting for 1/4 in. tube OD (−4 thread size) and 1/4 in. ID SAE 100R1 hydraulic hose, shown in Fig. 8A and Table 5A of SAE J516, would be coded as follows:

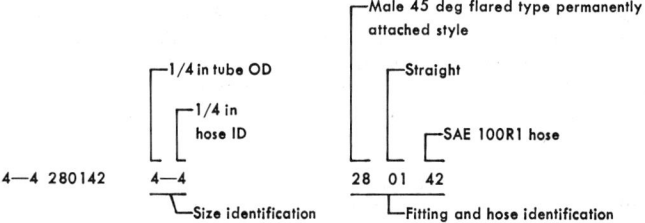

The 4-bolt split flange type field attachable screw style 45 deg angle hose fitting for 1/2 in. flange size and 3/4 in. ID SAE 100R2 hydraulic hose, shown in Fig. 17B and Table 9B of SAE J516, would be coded as follows:

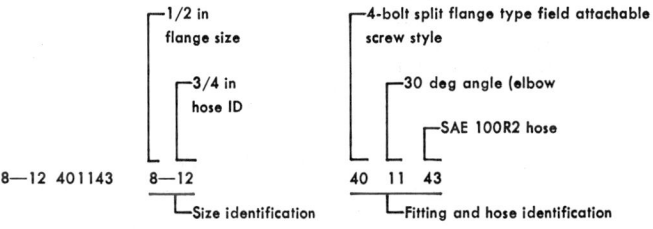

3. Example of Code Applied to 4-Bolt Split Flange Connections

The 4-bolt split flange connection for the 1 1/4 in. flange size and split flange to flanged head, standard pressure series, shown in Fig. 1 and Table 1 of SAE J518, would be coded as follows:

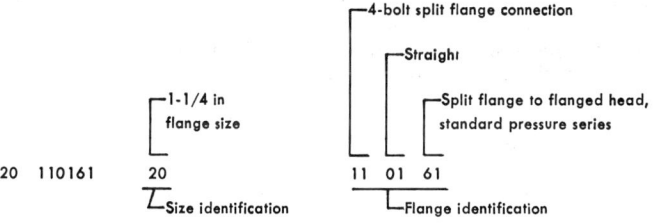

4. Example of Code Applied to Assembled Tapered Sleeve, Automotive, Union—The tapered sleeve, automotive union assembly for 1/4 in. tube OD, the union shown in Fig. 12B and Table 8, the sleeve shown in Fig. 15 and Table 9, and the nut shown in Fig. 16 and Table 10 of SAE J512 would be coded as follows:

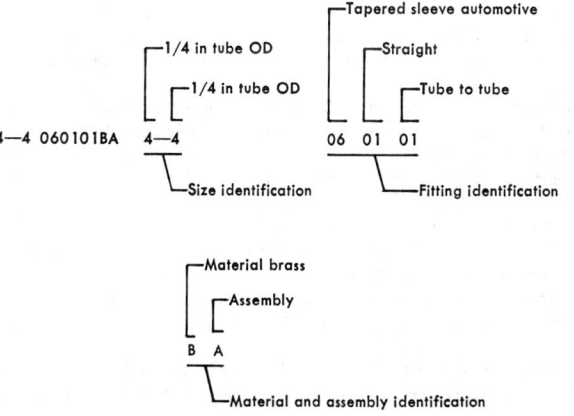

AUTOMOTIVE TUBE FITTINGS—SAE J512 NOV79   SAE Standard

Report of the Carburetor Fittings Division, approved June 1912, last revised by the Fluid Conductors and Connectors Technical Committee November 1979.

GENERAL SPECIFICATIONS

Scope—This standard covers complete general and dimensional specifications for the various types of tube fittings intended for general application in the automotive, appliance, and allied fields.

Flare type fittings shall be as specified in Figs. 1–4 and Tables 1–3.

NOTE: For sizes 3/16–3/8 and 1/2–3/4 the flare type fittings depicted in Figs. 1A–3C are identical with the corresponding refrigeration tube fittings specified in SAE J513 (October, 1977). Special size combination fittings 3/16–3/8 and 1/2–3/4 shall be as specified in J513 (October, 1977).

Inverted flared type fittings shall be as specified in Figs. 5–11 and Tables 1, 4–7. Gages and gaging procedures pertaining to inverted flared tube fittings are given in Appendix A.

NOTE: The seat dimensions specified in Table 4 are predicated on practical threading limitations in steel fittings and use of these fittings with double flared tubing. Therefore, wherever purchasers contemplate using these fittings with single flared tubing, it is recommended fitting manufacturers be consulted even though it has been common practice to provide slightly deeper threads in brass fittings.

Tapered sleeve compression type fittings intended for general use with annealed copper alloy tubings shall be as specified in Figs. 12–17, and Tables 1, 8–10. To assure satisfactory performance, spherical sleeve compression type fitting components (SAE J246 (July, 1977)) should not be intermixed with tapered sleeve compression type fitting components when assembling connections in areas where both types are available.

Dimensions of single and double 45 deg flares on tubing to be used in conjunction with flared and inverted flared fittings are given in Fig. 2 and Table 1 of SAE J533 (January, 1972).

The following general specifications supplement the dimensional data for all types of fittings contained in Tables 1–11 with respect to all unspecified detail.

Size Designations—Fitting sizes are designated by the corresponding outside diameter of the tubing for the various types of tube ends and by the corresponding standard nominal pipe size for pipe thread ends.

Dimensions and Tolerances—Except for nominal sizes and thread specifications, dimensions and tolerances are given in both U. S. customary and SI units as designated. Tabulated dimensions shall apply to the finished parts, plated or otherwise processed, as specified by the purchaser. Unless otherwise specified, the maximum and minimum across flats dimensions shall be within the commercial tolerance of bar or extruded stock from which the fittings are produced. The minimum across corners dimensions of hexagons shall be 1.092 times the nominal width across flats, but shall not result in a side flat width less than 0.43 times the nominal width across flats.

Unless otherwise specified, tolerance on hole diameters designated drill in the dimensional tables shall be as tabulated below:

Drill Size Range		Tolerance on Hole Diameter			
		Plus		Minus	
in	mm	in	mm	in	mm
0.0135 thru 0.1850	0.343 thru 4.699	0.003	0.08	0.002	0.05
0.1875 thru 0.2480	4.762 thru 6.299	0.004	0.10	0.002	0.05
0.2500 thru 0.7500	6.350 thru 19.050	0.006	0.15	0.003	0.08
0.7579 thru 1.0000	19.251 thru 25.400	0.007	0.18	0.004	0.10

Tolerance on all dimensions not otherwise limited shall be ±0.010 in (0.25 mm). Angular tolerance on axis of ends on elbows, tees, and crosses shall be ±2.50 deg for sizes up to and including 3/8 in, and ±1.50 deg for sizes larger than 3/8 in.

Integral internal seats in inverted flared tube fittings shall be concentric with straight thread pitch diameters within 0.005 in (0.13 mm) full indicator reading (FIR). Unless otherwise specified, fitting seats shall be concentric with straight thread pitch diameters within 0.010 in (0.25 mm) full indicator reading (FIR).

Where so illustrated and not otherwise specified, hexagon corners shall be chamfered 30 ± 5 deg to a diameter equal to the nominal width across flats, with a tolerance of −0.016 in (0.41 mm); or where design permits, corners may be chamfered to the diameter of the abutting surface provided the length of chamfer does not exceed that produced by the 30 deg chamfer previously described.

Passages—Where passages in straight fittings are machined from opposite ends, the offset at the meeting point shall not exceed 0.015 in (0.38 mm). The cross sectional area at the junction of passages in angle fittings shall not be less than that of the smallest passage. Where the passage is specified as a maximum, or as tap drill diameter or less, the minimum shall be no less than the minimum diameter of the smallest passage in the fitting.

Wall Thickness—Unless otherwise designated, the wall thickness at any point on fittings shall not be less than the thickness established by the specified dimensions, tolerances, and eccentricities for inner and outer surfaces.

Contour—Details of contour shall be optional with manufacturer provided the tabulated dimensions are maintained and serviceability of the fittings is not impaired. Wrench flats on elbows and tees shall be optional. Where extruded or forged shapes are reduced to conserve material, the wall thickness, unless otherwise specified, shall not be less than the respective minimum values tabulated below:

Nom Tube OD, in	Wall Thickness, Min			
	Extruded Shape[a]		Forged Shape	
	in	mm	in	mm
1/8	0.04	1.0	0.060	1.52
3/16	0.04	1.0	0.070	1.78
1/4	0.04	1.0	0.075	1.90
5/16	0.05	1.3	0.075	1.90
3/8	0.05	1.3	0.090	2.29
7/16	0.06	1.5	0.090	2.29
1/2	0.06	1.5	0.090	2.29
9/16	0.06	1.5	0.090	2.29
5/8	0.08	2.0	0.100	2.54
3/4	0.08	2.0	0.100	2.54
7/8	0.08	2.0	0.120	3.05
1	0.08	2.0	0.120	3.05

[a]Applies to reduction to one plane of shape only.

Straight Threads—Unified Standard Class 2A external threads and Class 2B internal threads with minor diameters, where specified, modified to Class 3B limits, shall apply to plain finish (unplated) fittings of all types. For externally threaded parts with additive finish, the maximum diameters of Class 2A may be exceeded by the amount of the allowance; that is, the basic diameters (Class 2A maximum diameters plus the allowance) shall apply to an externally threaded part after plating. For internally threaded parts with additive finish, the Class 2B diameters and modified minor diameters shall apply after plating.

The pitch diameter tolerance shall be the same as the corresponding diameter-pitch combination and the class of the Unified coarse and fine thread series or for special diameter-pitch combinations shall be based on diameter, pitch, and a length of engagement of 9 times the pitch. See ANSI B1.1, Screw Threads.

Where external threads are produced by roll threading and the body is not undercut, the unthreaded portion of body adjacent to the shoulder may be reduced to the minimum pitch diameter.

External threads shall be chamfered to the diameter of abutting surfaces, or to the diameters specified, to produce a length of chamfered or partial thread equivalent to 3/4 to 1 1/4 times the pitch (rounded to a three-place decimal). Internal threads shall be countersunk 90 deg included angle to the diameters specified to the dimensional tables.

Thread Eccentricity Tolerances—The various thread elements of Class 2A external and Class 2B, modified, internal threads on tube fittings shall be concentric within the following limitations:

External Thread (Screw)

1. Where screw pitch diameter is maximum and screw major diameter is maximum, these two thread elements must be concentric. However, if the screw major diameter is out-of-round, undersize, these two thread elements may be eccentric at the point of out-of-roundness, a full indicator reading amount equal to the screw major diameter tolerance.

2. Where screw pitch diameter is minimum and screw major diameter is maximum, these two thread elements may be eccentric a full indicator reading amount equal to the screw pitch diameter tolerance.

3. Where screw pitch diameter is maximum and screw major diameter is minimum, these two thread elements may be eccentric a full indicator reading amount equal to the screw major diameter tolerance.

TABLE 1—STRAIGHT THREAD SPECIFICATION DATA, IN

Nominal Size	Series Designation	External Thread Pitch Diameter		Internal Thread Pitch Diameter		Internal Thread Minor Diameter			Nominal Size	Series Designation	External Thread Pitch Diameter		Internal Thread Pitch Diameter		Internal Thread Minor Diameter		
		Max	Min	Max	Min[a]	Max[b]	Max[c]	Min[c]			Max	Min	Max	Min[a]	Max[b]	Max[c]	Min[c]
5/16-24	UNF	0.2843	0.2806	0.2902	0.2854	0.2754	0.277	0.267	11/16-18	UNS	0.6500	0.6455	0.6573	0.6514	0.6335	0.640	0.627
5/16-28	UN	0.2883	0.2849	0.2937	0.2893	0.2807	0.282	0.274	11/16-20	UN	0.6537	0.6494	0.6606	0.6550	0.6412	0.645	0.633 φ
3/8-24	UNF	0.3468	0.3430	0.3528	0.3479	0.3372	0.340	0.330	3/4-16	UNF	0.7079	0.7029	0.7159	0.7094	0.6908	0.696	0.682
7/16-20	UNF	0.4037	0.3995	0.4104	0.4050	0.3916	0.395	0.383	3/4-18	UNS	0.7125	0.7079	0.7199	0.7139	0.6980	0.703	0.690
7/16-24	UNS	0.4093	0.4055	0.4153	0.4104	0.3994	0.402	0.392	13/16-18	UNS	0.7750	0.7704	0.7824	0.7764	0.7605	0.765	0.752
1/2-20	UNF	0.4662	0.4619	0.4731	0.4675	0.4537	0.457	0.446	7/8-14	UNF	0.8270	0.8216	0.8356	0.8286	0.8068	0.814	0.798
1/2-24	UNS	0.4717	0.4678	0.4780	0.4729	0.4619	0.465	0.455	7/8-18	UNS	0.8375	0.8239	0.8449	0.8389	0.8230	0.828	0.815
φ 9/16-24	UNEF	0.5342	0.5303	0.5354	0.5405	0.5244	0.527	0.517	1-18	UNS	0.9625	0.9578	0.9701	0.9639	0.9480	0.953	0.940
5/8-18	UNF	0.5875	0.5828	0.5949	0.5889	0.5730	0.578	0.565	1-1/16-14	UNS	1.0145	1.0092	1.0230	1.0161	0.9940	1.001	0.985
5/8-24	UNEF	0.5967	0.5927	0.6031	0.5979	0.5869	0.590	0.580	1-1/16-16	UN	1.0204	1.0154	1.0284	1.0219	1.0033	1.009	0.995
11/16-16	UN	0.6455	0.6407	0.6531	0.6469	0.6284	0.634	0.620	1-1/4-12	UNF	1.1941	1.1879	1.2039	1.1959	1.1698	1.178	1.160
									1-3/8-12	UNF	1.3127	1.3190	1.3291	1.3209	1.2948	1.303	1.285

[a] These values are also the basic pitch diameter.
[b] Class 3B maximum minor diameter limits shall apply where so designated in respective dimensional tables.
[c] Class 2B minor diameter limits shall apply unless otherwise designated.

4. Where screw pitch diameter is minimum and screw major diameter is minimum, these two thread elements may be eccentric a full indicator reading amount equal to the sum of the screw pitch diameter tolerance and the screw major diameter tolerance.

Internal Thread (Nut)

1. Where nut pitch diameter is minimum and nut minor diameter is minimum, these two thread elements must be concentric. However, if the nut minor diameter is out-of-round, oversize, the two thread elements may be eccentric at the point of out-of-roundness, a full indicator reading amount equal to the nut minor diameter tolerance.

2. Where nut pitch diameter is maximum and nut minor diameter is minimum, these two thread elements may be eccentric a full indicator reading amount equal to the nut pitch diameter tolerance.

3. Where nut pitch diameter is minimum and nut minor diameter is

═══════ **FLARED TYPE** ═══════

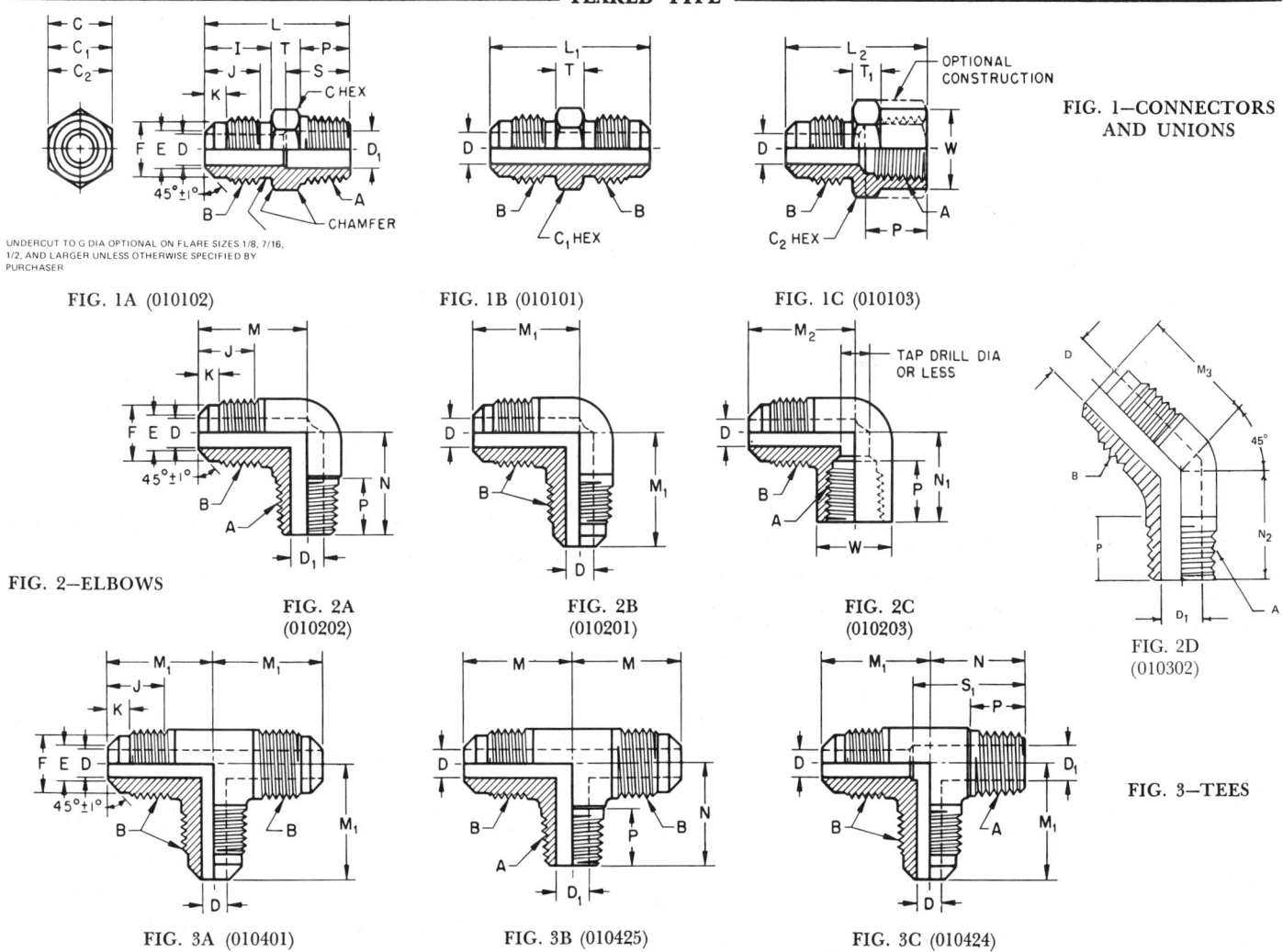

FIG. 1A (010102) FIG. 1B (010101) FIG. 1C (010103) FIG. 1—CONNECTORS AND UNIONS

UNDERCUT TO G DIA OPTIONAL ON FLARE SIZES 1/8, 7/16, 1/2, AND LARGER UNLESS OTHERWISE SPECIFIED BY PURCHASER

FIG. 2—ELBOWS
FIG. 2A (010202) FIG. 2B (010201) FIG. 2C (010203) FIG. 2D (010302)

FIG. 3A (010401) FIG. 3B (010425) FIG. 3C (010424) FIG. 3—TEES

NOTES: UNSPECIFIED DETAIL WITH RESPECT TO DIMENSIONS, TOLERANCES, CONTOURS, MATERIAL, WORKMANSHIP, ETC., MUST CONFORM TO GENERAL SPECIFICATIONS FOR AUTOMOTIVE TUBE FITTINGS. THE DIMENSIONAL DESIGNATIONS ON THE FIRST FIGURE IN EACH GROUP SHALL APPLY TO ALL OTHER FIGURES IN THAT GROUP EXCEPT AS SHOWN φ OTHERWISE. CODES SHOWN IN BRACKETS ADJACENT TO FIGURE NUMBERS REPRESENT RESPECTIVE FITTING IDENTIFICATION IN ACCORDANCE WITH SAE J846 (MAY, 1978).

TABLE 2—DIMENSIONS OF CONNECTORS, UNIONS, ELBOWS, AND TEES (FIGS. 1A–3C)

Nom Tube OD, in	A Dryseal Taper Thread NPTF[a], in	B, Nom Thread Size, in Class 2A, Ext.	C Nom, in	C_1 Nom, in	C_2 Nom, in	D[e] Dia Drill in	mm	D_1[e] Dia Drill in	mm	E Dia in	mm	F Dia in	mm	G[d] Dia in +0.000 −0.010	mm +0.00 −0.25
1/8	1/8	5/16-24	7/16	5/16	9/16	0.078	1.98	0.219	5.56	0.109	2.77	0.234	5.94	0.250	6.35
3/16	1/8	3/8 -24	7/16	3/8	9/16	0.125	3.18	0.219	5.56	0.156	3.96	0.297	7.54	—	—
1/4	1/8	7/16-20	7/16	7/16	9/16	0.188	4.78	0.219	5.56	0.219	5.56	0.344	8.74	—	—
5/16	1/8	1/2 -20	1/2	1/2	9/16	0.219	5.56	0.219	5.56	0.250	6.35	0.406	10.31	—	—
3/8	1/4	5/8 -18	5/8	5/8	11/16	0.281	7.14	0.312	7.92	0.312	7.92	0.531	13.49	—	—
7/16	1/4	11/16-16	11/16	11/16	11/16	0.312	7.92	0.312	7.92	0.344	8.74	0.578	14.63	0.596	15.14
1/2	3/8	3/4 -16	3/4	3/4	13/16	0.406	10.31	0.406	10.31	0.438	11.13	0.641	16.28	0.659	16.74
5/8	1/2	7/8 -14	7/8	7/8	1	0.500	12.70	0.562	14.27	0.531	13.49	0.750	19.05	0.770	19.56
3/4	1/2	1-1/16-14	1-1/16	1-1/16	1-1/16	0.625	15.88	0.562	14.27	0.719	18.26	0.938	23.83	0.958	24.33
7/8	3/4	1-1/4 -12	1-1/4	1-1/4	1-1/4	0.750	19.05	0.750	19.05	0.797	20.24	1.125	28.58	1.128	28.65
1	1	1-3/8 -12	1-3/8	1-3/8	1-1/2	0.875	22.22	0.938	23.82	0.938	23.83	1.250	31.75	1.253	31.83

Nom Tube OD, in	I in	mm	J[d] Full Thread Min in	mm	K in	mm	L[b] ±0.03 in	±0.8 mm	L_1 ±0.03 in	±0.8 mm	L_2[b,c] ±0.03 in	±0.8 mm	M ±0.03 in	±0.8 mm	M_1 ±0.03 in	±0.8 mm	M_2 ±0.03 in	±0.8 mm	ϕM_3 ±0.03 in	±0.8 mm
1/8	0.38	9.7	0.31	7.9	0.12	3.0	0.92	23.4	0.92	23.4	0.91	23.1	0.62	15.7	0.62	15.7	0.75	19.0	0.59	15.0
3/16	0.44	11.2	0.38	9.7	0.12	3.0	1.00	25.4	1.06	26.9	0.97	24.6	0.75	19.0	0.75	19.0	0.81	20.6	0.62	15.7
1/4	0.50	12.7	0.41	10.4	0.16	4.1	1.06	26.9	1.19	30.2	1.03	26.2	0.81	20.6	0.88	22.4	0.88	22.4	0.67	17.0
5/16	0.56	14.2	0.47	11.9	0.19	4.8	1.16	29.5	1.34	34.0	1.06	26.9	0.91	23.1	0.91	23.1	0.94	23.9	0.78	19.8
3/8	0.62	15.7	0.54	13.7	0.22	5.6	1.44	36.6	1.50	38.1	1.31	33.3	1.00	25.4	1.06	26.9	1.09	27.7	0.89	22.6
7/16	0.69	17.5	0.56	14.2	0.25	6.4	1.53	38.9	1.66	42.2	1.41	35.8	1.12	28.4	1.12	28.4	1.12	28.4	0.97	24.6
1/2	0.75	19.0	0.66	16.8	0.25	6.4	1.62	41.1	1.81	46.0	1.50	38.1	1.22	31.0	1.22	31.0	1.28	32.5	1.06	26.9
5/8	0.88	22.4	0.76	19.3	0.28	7.1	2.00	50.8	2.12	53.8	1.81	46.0	1.41	35.8	1.41	35.8	1.50	38.1	1.23	31.2
3/4	1.00	25.4	0.90	22.9	0.28	7.1	2.19	55.6	2.44	62.0	1.91	48.5	1.62	41.1	1.66	42.2	1.62	41.1	1.41	35.8
7/8	1.12	28.4	0.95	24.1	0.38	9.7	2.37	60.2	2.74	69.6	2.25	57.2	1.75	44.5	1.75	44.5	1.88	47.8	1.62	41.1
1	1.12	28.4	0.97	24.6	0.38	9.7	2.62	66.5	2.80	71.1	2.44	62.0	1.94	49.3	1.94	49.3	2.06	52.3	1.69	42.9

Nom Tube OD, in	N[b] ±0.03 in	±0.8 mm	N_1[b,c] ±0.03 in	±0.8 mm	ϕN_2[b] ±0.03 in	±0.8 mm	P[b,c] in	mm	S[b,e] Max in	mm	S_1[b,e] Min in	mm	T[f] Ref in	mm	T_1 Min in	mm	W[g] Dia in +0.00 −0.02	mm +0.0 −0.5
1/8	0.69	17.5	0.42	10.7	0.52	13.2	0.38	9.7	0.46	11.7	0.77	19.6	0.15	3.8	0.21	5.3	0.56	14.2
3/16	0.75	19.0	0.44	11.2	0.52	13.2	0.38	9.7	0.48	12.2	0.85	21.6	0.18	4.6	0.21	5.3	0.56	14.2
1/4	0.78	19.8	0.47	11.9	0.64	16.3	0.38	9.7	0.48	12.2	0.92	23.4	0.18	4.6	0.24	6.1	0.56	14.2
5/16	0.78	19.8	0.47	11.9	0.64	16.3	0.38	9.7	—	—	—	—	0.21	5.3	0.24	6.1	0.56	14.2
3/8	1.06	26.9	0.69	17.5	0.86	21.8	0.56	14.2	0.69	17.5	1.24	31.5	0.24	6.1	0.30	7.6	0.69	17.5
7/16	1.06	26.9	0.72	18.3	0.86	21.8	0.56	14.2	—	—	—	—	0.27	6.9	0.30	7.6	0.69	17.5
1/2	1.12	28.4	0.75	19.0	0.95	24.1	0.56	14.2	—	—	—	—	0.30	7.6	0.37	9.4	0.81	20.6
5/8	1.38	35.1	1.00	25.4	1.17	29.7	0.75	19.0	0.94	23.9	1.67	42.4	0.37	9.4	0.43	10.9	1.00	25.4
3/4	1.50	38.1	1.06	26.9	1.20	30.5	0.75	19.0	1.22	31.0	2.02	51.3	0.43	10.9	0.49	12.4	1.06	26.9
7/8	1.69	42.9	1.12	28.4	1.27	32.2	0.75	19.0	—	—	—	—	0.49	12.4	0.52	13.2	1.25	31.8
1	1.94	49.3	1.38	35.1	1.48	37.6	0.94	23.9	1.22	31.0	2.42	61.5	0.55	14.0	0.58	14.7	1.50	38.1

[a] Dryseal American Standard Taper Pipe Thread. See General Specifications.
[b] Where SAE Short Pipe Thread is authorized by purchaser, dimensions L, L_2, N, N_1, P, S, and ϕS_1 are reduced in accordance with reduction of pipe thread length. See SAE J476 (June, 1961), Tables 3 and 4.
[c] Tap drill depths given require use of bottoming taps to produce standard full thread length. For increased tap clearance, see Internal Taper Pipe Threads in the General Specifications.
[d] Where thread relief undercut is used, the last thread shall be chamfered 1/2 to 1 pitch long from G diameter and dimension 1 may be reduced by an amount equal to 1/2 pitch.
[e] At manufacturers' option, through passages in fittings shown in Figs. 1A and 3C may conform with the smaller diameter specified or be counterbored to the larger diameter from the appropriate end for depths S or S_1, respectively.
[f] Minimum design thickness, not subject to inspection.
[g] Basic dimensions shown shall apply as minimum diameter or across flats for bosses. The −0.02 in (−0.5 mm) tolerance shall apply only to chamber diameter on full hexagon versions of fitting in Fig. 1C.

maximum, these two thread elements may be eccentric a full indicator reading amount equal to the nut minor diameter tolerance.

4. Where nut pitch diameter is maximum and nut minor diameter is maximum, these two thread elements may be eccentric a full indicator reading amount equal to the sum of the nut pitch diameter tolerance and the nut minor diameter tolerance.

Pipe Threads—Pipe threads, unless there is specific authorization to the contrary, shall conform to the Dryseal American Standard Taper Pipe Thread (NPTF). At purchaser's option, the pipe thread on automotive tube fittings may be shortened in conformity with the SAE Short Dryseal Taper Pipe Thread (PTF-SAE Short). Specifications for pipe threads are given in detail in ϕ SAE J476 (June, 1961).

ϕ The length of full form external thread shall not be shorter than L_2, plus
ϕ one pitch (thread) for Dryseal NPTF and L_2, for Dryseal PTF-SAE Short, except that where thread is cut through into a relieved body or undercut on the fitting, the minimum full thread length may be reduced by one pitch (thread).

Where external pipe threads are produced by roll threading, the diameter of the unthreaded portion of shank adjacent to shoulder may be reduced to the E, basic diameter.

The tube fitting dimensions tabulated herein are based on length of the Dryseal American Standard Taper Pipe Thread (NPTF), it being the consensus of manufacturers and users that trouble-free assembly cannot be assured unless a full length is used. However, the tap drill depths and overall lengths specified in the tables for fittings with internal taper pipe threads are not consistent with the tap drill depths and overall thread lengths of the Dryseal American Standard Taper Pipe Threads (NPTF) given in Table A2, Appendix A, of SAE J476 (June, 1961). The full length Dryseal American ϕ Standard Taper Pipe Taps specified in Table B2 of SAE J476 (June, 1961), ϕ cannot be used, as the tap drill depths and overall lengths of the fittings have been reduced to the minimum required by bottoming taps to produce standard full length thread. The deviations described above are peculiar to these tube fittings and as special tooling is required, caution should be exercised in specifying such deviations for any other products.

External pipe threads shall be chamfered from the diameters shown in Table 11 to produce the specified length of chamfered or partial thread. Internal pipe threads shall be countersunk 90 deg, included angle, to the diameters shown in Table 11.

Material and Manufacture—All types of automotive tube fittings shall be

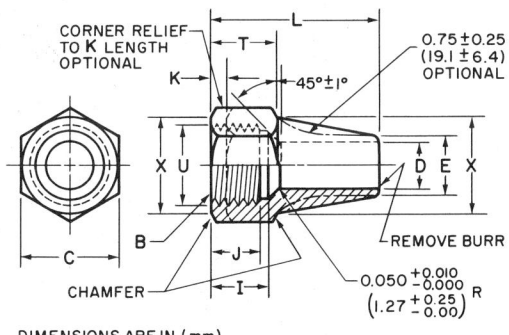

FIG. 4—FLARED TYPE NUTS
(010110) Short Nut
(010111) Long Nut

TABLE 3—DIMENSIONS OF NUTS (FIG. 4)

Nom Tube OD, in	B, Nom Thread Size, in Class 2B, int	C Nom, in	D Dia +0.005 −0.000 in	D Dia +0.13 −0.00 mm	E Dia in	E Dia mm	I in	I mm	J Full Thread Min in	J Full Thread Min mm	K in	K mm
1/8	5/16-24	3/8	0.130	3.30	0.25	6.4	0.22	5.6	0.16	4.1	0.09	2.3
3/16	3/8 -24	7/16	0.192	4.88	0.31	7.9	0.28	7.1	0.22	5.6	0.09	2.3
1/4	7/16-20	9/16	0.255	6.48	0.38	9.7	0.34	8.6	0.28	7.1	0.12	3.0
5/16	1/2 -20	5/8	0.317	8.05	0.44	11.2	0.38	9.7	0.31	7.9	0.12	3.0
3/8	5/8 -18	3/4	0.380	9.65	0.50	12.7	0.44	11.2	0.38	9.7	0.12	3.0
7/16	11/16-16	13/16	0.442	11.23	0.56	14.2	0.47	11.9	0.41	10.4	0.12	3.0
1/2	3/4 -16	7/8	0.505	12.83	0.62	15.7	0.53	13.5	0.47	11.9	0.12	3.0
5/8	7/8 -14	1-1/16	0.630	16.00	0.75	19.1	0.66	16.8	0.59	15.0	0.12	3.0
3/4	1-1/16-14	1-1/4	0.755	19.18	0.88	22.4	0.78	19.8	0.69	17.5	0.12	3.0
7/8	1-1/4 -12	1-1/2	0.880	22.35	1.06	26.9	0.78	19.8	0.69	17.5	0.12	3.0
1	1-3/8 -12	1-5/8	1.005	25.53	1.19	30.2	0.81	20.6	0.69	17.5	0.12	3.0

Nom Tube OD, in	L Long in	L Long mm	L Short in	L Short mm	T in	T mm	U Dia Min in	U Dia Min mm	U Dia Max in	U Dia Max mm	X Dia in	X Dia mm
1/8	0.75	19.1	0.50	12.7	0.22	5.6	0.33	8.4	0.35	8.9	0.38	9.7
3/16	0.81	20.6	0.62	15.7	0.28	7.1	0.39	9.9	0.41	10.4	0.44	11.2
1/4	0.94	23.9	0.75	19.1	0.38	9.7	0.45	11.4	0.47	11.9	0.56	14.2
5/16	1.12	28.4	0.88	22.4	0.44	11.2	0.51	13.0	0.53	13.5	0.62	15.7
3/8	1.31	33.3	1.00	25.4	0.50	12.7	0.64	16.3	0.67	17.0	0.75	19.1
7/16	1.50	38.1	1.06	26.9	0.56	14.2	0.71	18.0	0.74	18.8	0.81	20.6
1/2	1.62	41.1	1.12	28.4	0.62	15.7	0.77	19.6	0.80	20.3	0.88	22.4
5/8	1.88	47.8	1.31	33.3	0.75	19.1	0.90	22.9	0.93	23.6	1.06	26.9
3/4	2.19	55.6	1.50	38.1	0.88	22.4	1.08	27.4	1.11	28.2	1.25	31.8
7/8	2.31	58.7	1.62	41.1	1.00	25.4	1.27	32.3	1.30	33.0	1.50	38.1
1	2.50	63.5	1.81	46.0	1.00	25.4	1.39	35.3	1.42	36.1	1.62	41.1

21.08

―――――――INVERTED FLARED TYPE―――――――

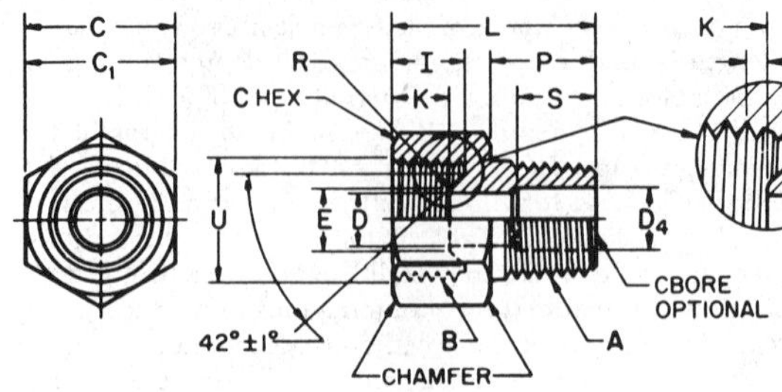

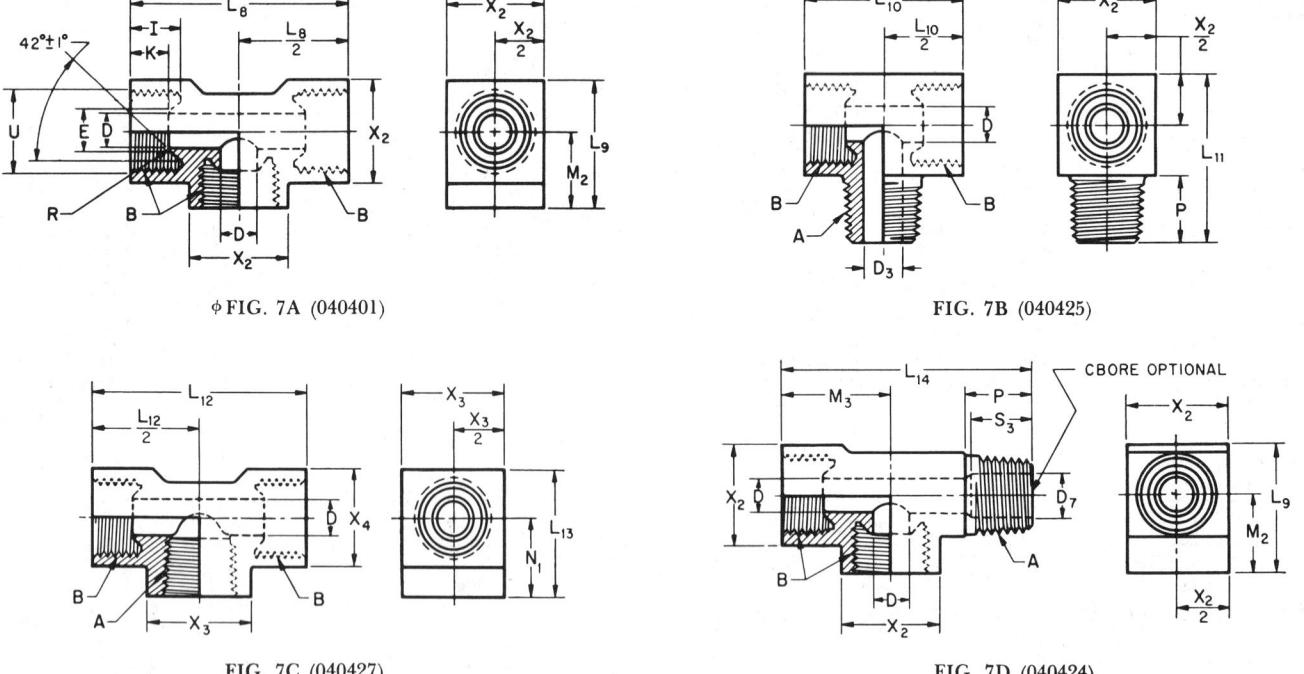

φ FIG. 5A (040102) 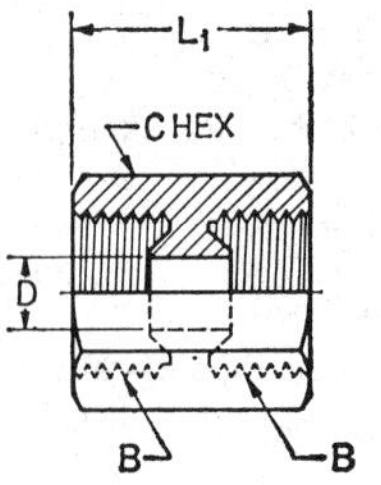   FIG. 5B (040101) 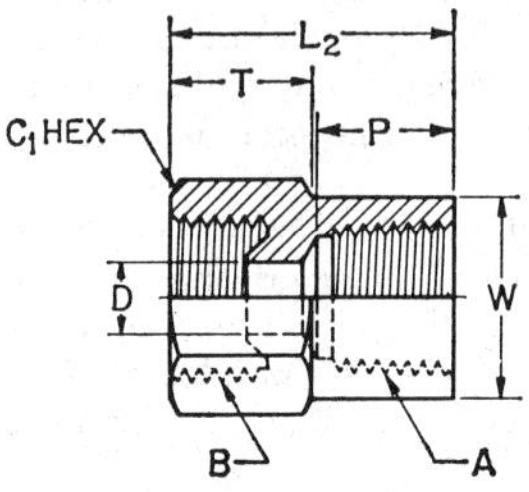   FIG. 5C (040103)

FIG. 5—CONNECTORS AND UNIONS

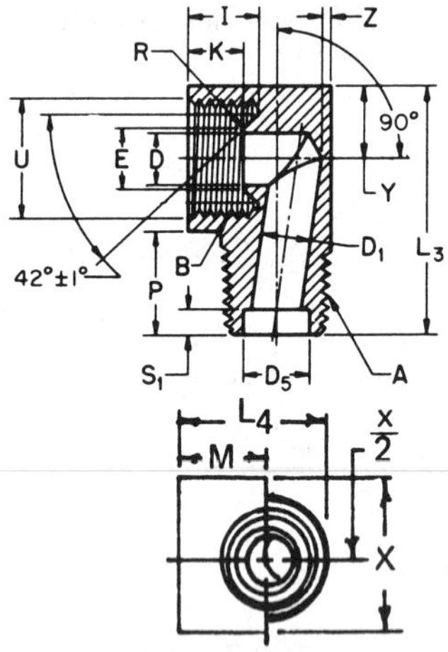

φ FIG. 6A (040202) 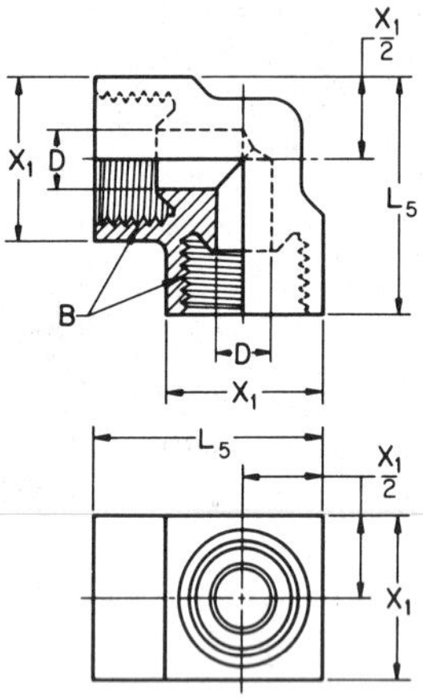   FIG. 6B (040201) 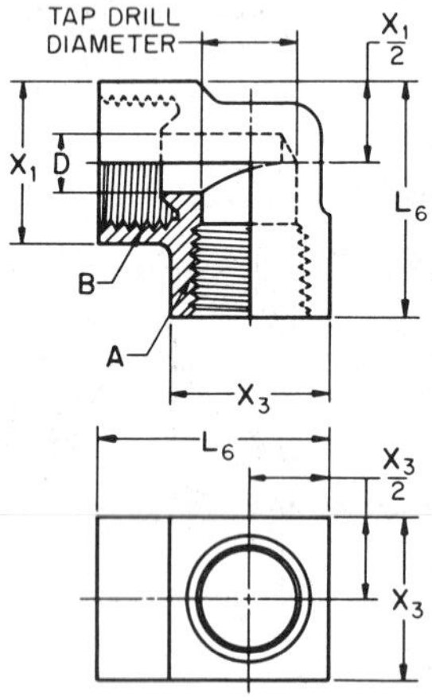   FIG. 6C (040203) 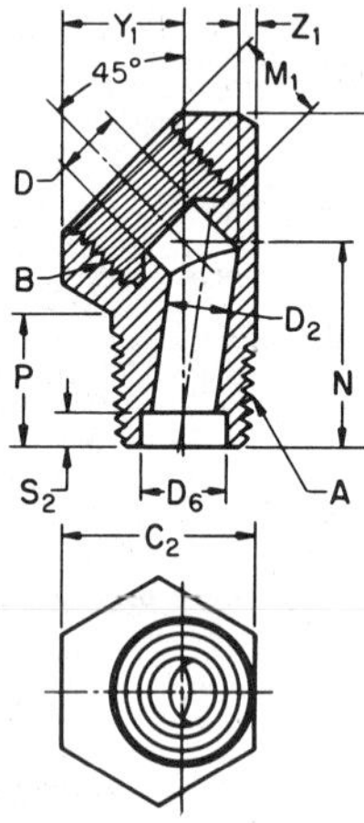   FIG. 6D (040302)

FIG. 6— ELBOWS

φ FIG. 7A (040401)   FIG. 7B (040425)

FIG. 7C (040427)   FIG. 7D (040424)

FIG. 7—TEES

NOTES: UNSPECIFIED DETAIL WITH RESPECT TO DIMENSIONS, TOLERANCES, CONTOURS, MATERIAL, WORKMANSHIP, ETC., MUST CONFORM TO GENERAL SPECIFICATIONS FOR AUTOMOTIVE TUBE FITTINGS. THE DIMENSIONAL DESIGNATIONS ON THE FIRST FIGURE IN EACH GROUP SHALL APPLY TO ALL OTHER FIGURES IN THAT GROUP EXCEPT AS SHOWN
φ OTHERWISE. CODES SHOWN IN BRACKETS ADJACENT TO FIGURE NUMBERS REPRESENT RESPECTIVE FITTING IDENTIFICATION IN ACCORDANCE WITH SAE J846 (MAY, 1978).

φ TABLE 4—DIMENSIONS OF CONNECTORS, UNIONS, ELBOWS, AND TEES (FIGS. 5A–7D)

Nom Tube OD, in	A Dryseal Taper Thread NPTF[a], in	B, Nom Thread Size, in Class 2B, Int[d]	C Nom, in		C_1 Nom, in		C_2 Nom, in		D Dia Drill		D_1 Dia Drill		D_2 Dia Drill		D_3 Dia Drill		D_4 Dia Drill	
			Brass	Steel	Brass	Steel	Brass	Steel	in	mm	in	mm	in	mm	in	mm	in	mm
1/8	1/8	5/16–28	13/32	7/16	1/2	1/2	1/2	1/2	0.078	1.98	0.116	2.95	0.125	3.17	0.219	5.56	0.219	5.56
3/16	1/8	3/8–24	15/32	1/2	1/2	1/2	17/32	9/16	0.125	3.17	0.125	3.17	0.156	3.96	0.219	5.56	0.219	5.56
1/4	1/8	7/16–24	17/32	9/16	17/32	9/16	9/16	9/16	0.188	4.77	0.177	4.50	0.188	4.77	0.219	5.56	0.219	5.56
5/16	1/8	1/2–20	19/32	5/8	19/32	5/8	5/8	5/8	0.219	5.56	0.219	5.56	0.203	5.16	0.219	5.56	0.219	5.56
5/16	1/4	1/2–20	19/32	5/8	11/16	11/16	5/8	5/8	0.219	5.56	0.219	5.56	0.203	5.16	0.312	7.92	0.312	7.92
3/8	1/4	5/8–18	3/4	3/4	3/4	3/4	25/32	13/16	0.281	7.14	0.281	7.14	0.281	7.14	0.344	8.74	0.344	8.74
7/16	1/4	11/16–18	13/16	13/16	13/16	13/16	13/16	13/16	0.344	8.74	0.328	8.33	0.312	7.92	0.344	8.74	0.344	8.74
1/2	3/8	3/4–18	29/32	15/16	29/32	15/16	7/8	7/8	0.406	10.31	0.375	9.52	0.375	9.52	0.406	10.31	0.406	10.31
5/8	1/2	7/8–18	1-1/16	1-1/16	1-1/16	1-1/16	1-1/16	1-1/16	0.531	13.49	0.500	12.70	0.500	12.70	0.562	14.27	0.594	15.09
3/4	3/4	1-1/16–16	1-1/4	1-1/4	1-1/4	1-1/4	1-5/16	1-5/16	0.625	15.87	0.625	15.87	0.625	15.87	0.750	19.05	0.750	19.05

Nom Tube OD, in	D_5 Dia Drill		D_6 Dia Drill		D_7 Dia Drill		E Dia				I Seat Depth	
							Min		Max		+0.010 / −0.005	+0.25 / −0.13
	in	mm	in	mm	in	mm	in	mm	in	mm	in	mm
1/8	0.250	6.35	0.250	6.35	0.188	4.77	0.104	2.64	0.108	2.74	0.250	6.35
3/16	0.250	6.35	0.250	6.35	0.188	4.77	0.151	3.84	0.155	3.94	0.266	6.76
1/4	0.250	6.35	0.250	6.35	—	—	0.214	5.44	0.218	5.54	0.266	6.76
5/16	0.250	6.35	—	—	—	—	0.276	7.01	0.280	7.11	0.297	7.54
5/16	0.312	7.92	0.281	7.14	—	—	0.276	7.01	0.280	7.11	0.297	7.54
3/8	—	—	0.344	8.74	—	—	0.342	8.69	0.346	8.79	0.344	8.74
7/16	—	—	0.344	8.74	—	—	0.405	10.29	0.409	10.39	0.375	9.53
1/2	0.469	11.91	0.469	11.91	—	—	0.467	11.86	0.471	11.96	0.391	9.93
5/8	0.594	15.09	0.562	14.27	—	—	0.592	15.04	0.596	15.14	0.406	10.31
3/4	0.750	19.05	0.688	17.47	—	—	0.703	17.86	0.708	17.98	0.469	11.91

Nom Tube OD, in	J[e] End Full Thread Max		K Seat Depth		L[b]		L_1		L_2[b,c]		L_3[b]		L_4		L_5	
			+0.010 / −0.005	+0.25 / −0.13	+0.03 / −0.00	+0.8 / −0.0	+0.03 / −0.00	+0.8 / −0.0	+0.03 / −0.00	+0.8 / −0.0	+0.03 / −0.00	+0.8 / −0.0			+0.03 / −0.00	+0.8 / −0.0
	in	mm	in	mm	in	mm	in	mm	in	mm	in	mm	in	mm	in	mm
1/8	0.011	0.28	0.187	4.75	0.62	15.7	0.59	15.0	0.72	18.3	0.78	19.8	0.47	11.9	0.70	17.8
3/16	0.013	0.33	0.203	5.16	0.69	17.5	0.62	15.7	0.75	19.1	0.84	21.3	0.47	11.9	0.70	17.8
1/4	0.026	0.66	0.203	5.16	0.73	18.5	0.62	15.7	0.75	19.1	0.91	23.1	0.55	14.0	0.77	19.6
5/16	0.033	0.84	0.234	5.94	0.78	19.8	0.70	17.8	0.78	19.8	0.97	24.6	0.67	17.0	0.86	21.8
5/16	0.033	0.84	0.234	5.94	0.97	24.6	—	—	1.00	25.4	1.15	29.2	0.75	19.1	—	—
3/8	0.023	0.58	0.266	6.76	1.02	25.9	0.80	20.3	1.03	26.2	1.31	33.3	0.81	20.6	1.03	26.2
7/16	0.029	0.74	0.296	7.52	1.05	26.7	0.88	22.4	1.06	26.9	1.41	35.8	0.88	22.4	1.14	29.0
1/2	0.029	0.74	0.312	7.93	1.06	26.9	0.91	23.1	1.09	27.7	1.47	37.3	0.94	23.9	1.25	31.8
5/8	0.041	1.04	0.328	8.33	1.31	33.3	0.97	24.6	1.31	33.3	1.81	46.0	1.11	28.2	1.47	37.3
3/4	0.016	0.41	0.359	9.12	1.38	35.1	1.12	28.4	1.50	38.1	2.00	50.8	1.28	32.5	1.72	43.7

Nom Tube OD, in	L_6[b,c]		L_7[b]		L_8		L_9		L_{10}		L_{11}[b]		L_{12}		L_{13}		L_{14}[b]	
	+0.03 / −0.00	+0.8 / −0.0			+0.03 / −0.00	+0.8 / −0.0			+0.03 / −0.00	+0.8 / −0.0			+0.03 / −0.00	+0.8 / −0.0				
	in	mm	in	mm	in	mm	in	mm	in	mm	in	mm	in	mm	in	mm	in	mm
1/8	—	—	0.81	20.6	0.94	23.9	0.53	13.5	0.78	19.8	0.78	19.8	1.09	27.7	0.62	15.7	1.16	29.5
3/16	—	—	0.88	22.4	1.09	27.7	0.62	15.7	0.81	20.6	0.84	21.3	1.09	27.7	0.62	15.7	1.25	31.8
1/4	0.79	20.1	0.94	23.9	1.12	28.4	0.69	17.5	0.84	21.3	0.91	23.1	1.12	28.4	0.69	17.5	1.31	33.3
5/16	0.88	22.4	1.00	25.4	1.25	31.8	0.75	19.1	0.94	23.9	0.97	24.6	1.25	31.8	0.75	19.1	1.47	37.3
5/16	0.97	24.6	1.16	29.5	—	—	0.75	19.1	1.06	26.9	1.16	29.5	1.38	35.1	0.88	22.4	1.53	38.9
3/8	1.03	26.2	1.34	34.0	1.47	37.3	0.94	23.9	1.16	29.5	1.31	33.3	1.47	37.3	0.94	23.9	1.83	46.5
7/16	1.16	29.5	1.41	35.8	1.62	41.1	1.03	26.2	1.22	31.0	1.38	35.1	1.62	41.1	1.03	26.2	1.88	47.8
1/2	1.25	31.8	1.44	36.6	1.75	44.5	1.12	28.4	1.38	35.1	1.47	37.3	1.75	44.5	1.12	28.4	2.00	50.8
5/8	1.47	37.3	1.75	44.5	1.94	49.3	1.28	32.5	1.56	39.6	1.81	46.0	1.94	49.3	1.28	32.5	2.38	60.5
3/4	1.72	43.7	2.00	50.8	2.25	57.2	1.50	38.1	1.81	46.0	2.00	50.8	2.25	57.2	1.50	38.1	2.62	66.5

(Table continued on next page)

TABLE 4—DIMENSIONS OF CONNECTORS, UNIONS, ELBOWS, AND TEES (FIGS. 5A–7D) (CONTINUED)

Nom Tube OD, in	M in	M mm	M_1 in	M_1 mm	M_2 in	M_2 mm	M_3 in	M_3 mm	N^b in	N^b mm	N_1 in	N_1 mm	$P^{b,c}$ in	$P^{b,c}$ mm	R Radius Max in	R Radius Max mm
1/8	0.27	6.9	0.22	5.6	0.33	8.4	0.47	11.9	0.52	13.2	0.39	9.9	0.38	9.7	0.010	0.25
3/16	0.27	6.9	0.25	6.4	0.39	9.9	0.53	13.5	0.55	14.0	0.39	9.9	0.38	9.7	0.010	0.25
1/4	0.33	8.4	0.27	6.9	0.42	10.7	0.56	14.2	0.58	14.7	0.42	10.7	0.38	9.7	0.036	0.91
5/16	0.47	11.9	0.34	8.6	0.45	11.4	0.62	15.7	0.56	14.2	0.45	11.4	0.38	9.7	0.036	0.91
5/16	0.45	11.4	0.23	5.8	0.45	11.4	0.62	15.7	0.83	21.1	0.50	12.7	0.56	14.2	0.036	0.91
3/8	0.53	13.5	0.38	9.7	0.56	14.2	0.75	19.1	0.84	21.3	0.56	14.2	0.56	14.2	0.036	0.91
7/16	0.59	15.0	0.41	10.4	0.62	15.7	0.81	20.6	0.86	21.8	0.62	15.7	0.56	14.2	0.036	0.91
1/2	0.59	15.0	0.38	9.6	0.67	17.0	0.88	22.4	0.91	23.1	0.67	17.0	0.56	14.2	0.036	0.91
5/8	0.67	17.0	0.45	11.4	0.75	19.1	0.97	24.6	1.09	27.7	0.75	19.1	0.75	19.1	0.036	0.91
3/4	0.75	19.1	0.50	12.7	0.88	22.4	1.12	28.4	1.22	31.0	0.88	22.4	0.75	19.1	0.036	0.91

Nom Tube OD, in	S^b in	S^b mm	S_1^b in	S_1^b mm	S_2^b in	S_2^b mm	S_3^b in	S_3^b mm	T in	T mm	U Dia Min in	U Dia Min mm	U Dia Max in	U Dia Max mm	W Dia +0.00/−0.01 in	W Dia +0.00/−0.25 mm
1/8	0.19	4.8	0.250	6.35	0.093	2.36	0.38	9.7	0.34	8.6	0.32	8.1	0.34	8.6	0.50	12.7
3/16	0.23	5.8	0.093	2.36	0.093	2.36	0.38	9.7	0.38	9.7	0.39	9.9	0.41	10.4	0.50	12.7
1/4	0.28	7.1	0.093	2.36	0.093	2.36	—	—	0.38	9.7	0.45	11.4	0.47	11.9	0.53	13.5
5/16	—	—	0.093	2.36	—	—	—	—	0.41	10.4	0.51	13.0	0.53	13.5	0.59	15.0
5/16	0.50	12.7	0.500	12.70	0.170	4.32	—	—	0.44	11.2	0.51	13.0	0.53	13.5	0.69	17.5
3/8	0.41	10.4	—	—	0.093	2.36	—	—	0.47	11.9	0.64	16.3	0.67	17.0	0.75	19.1
7/16	—	—	—	—	0.093	2.36	—	—	0.50	12.7	0.70	17.8	0.73	18.5	0.81	20.6
1/2	—	—	0.156	3.96	0.093	2.36	—	—	0.53	13.5	0.77	19.6	0.80	20.3	0.91	23.1
5/8	0.53	13.5	0.219	5.56	0.188	4.77	—	—	0.62	15.7	0.89	22.6	0.92	23.4	1.06	26.9
3/4	0.53	13.5	0.312	7.95	0.250	6.35	—	—	0.69	17.5	1.08	27.4	1.11	28.2	1.25	31.8

Nom Tube OD, in	X in	X mm	X_1 in	X_1 mm	X_2 in	X_2 mm	X_3 in	X_3 mm	X_4 in	X_4 mm	Y +0.010/−0.000 in	Y +0.25/−0.00 mm	Y_1^f in	Y_1^f mm	Z Min in	Z Min mm	Z_1 Min in	Z_1 Min mm
1/8	0.44	11.2	0.41	10.4	0.41	10.4	0.50	12.7	0.47	11.9	0.205	5.21	0.30	7.6	0.03	0.8	0.08	2.0
3/16	0.47	11.9	0.47	11.9	0.47	11.9	0.50	12.7	0.47	11.9	0.230	5.84	0.33	8.4	0.03	0.8	0.06	1.5
1/4	0.53	13.5	0.53	13.5	0.53	13.5	0.53	13.5	0.53	13.5	0.260	6.60	0.36	9.1	0.03	0.8	0.05	1.3
5/16	0.59	15.0	0.59	15.0	0.59	15.0	0.59	15.0	0.59	15.0	0.290	7.37	0.42	10.7	0.05	1.3	0.09	2.3
5/16	0.59	15.0	0.59	15.0	0.59	15.0	0.69	17.5	0.59	15.0	0.290	7.37	0.31	7.9	—	—	0.06	1.5
3/8	0.75	19.1	0.72	18.3	0.75	19.1	0.75	19.1	0.75	19.1	0.370	9.40	0.50	12.7	0.03	0.8	0.05	1.3
7/16	0.84	21.3	0.78	19.8	0.81	20.6	0.81	20.6	0.81	20.6	0.415	10.54	0.53	13.5	0.05	1.3	0.03	0.8
1/2	0.91	23.1	0.88	22.4	0.91	23.1	0.91	23.1	0.91	23.1	0.450	11.43	0.53	13.5	0.03	0.8	0.06	1.5
5/8	1.06	26.9	1.06	26.9	1.06	26.9	1.06	26.9	1.06	26.9	0.525	13.34	0.64	16.3	0.05	1.3	0.06	1.5
3/4	1.25	31.8	1.25	31.8	1.25	31.8	1.25	31.8	1.25	31.8	0.620	15.75	0.78	19.8	0.05	1.3	0.08	2.0

[a] Dryseal American Standard Taper Pipe Thread. See General Specifications.
[b] Where SAE Short Pipe Thread is authorized by purchaser, dimensions L, L_2, L_3, L_5, L_7, L_{11}, L_{14}, N, P, S, S_1, S_2, and S_3 are reduced in accordance with reduction of pipe thread length. See φ SAE J476 (June, 1961), Tables 3 and 4.
[c] Tap drill depths given require use of bottoming taps to produce standard full thread length. For increased tap clearance, see Internal Taper Pipe Threads in General Specifications.
[d] Class 3B minor diameter limits apply to copper alloy fittings and Class 2B minor diameter limits apply to steel fittings.
[e] End full thread is measured from face of cone seat to last full form thread at major diameter, see Appendix A. A minimum of 3/4 partial thread beyond the last full form thread is required.
[f] For steel parts, the Y_1 dimension shall be 0.35 in (8.9 mm) for 3/16 in tube size and 0.52 in (13.2 mm) for 3/8 in tube size.

made from brass or steel as specified by the purchaser and, at manufacturer's option and in accordance with his process of manufacture, may be milled from the bar or forged. Brass shall be UNS C36000 (half-hard), UNS C34500, or UNS C35000 for bar or extruded stock and UNS C37700 for forgings.

Finish—Unless otherwise specified by the purchaser, steel fittings shall be furnished cadmium or zinc plated to a thickness of 0.0002 in (0.005 mm) minimum. These parts must meet the requirements of a 32 h salt spray test in accordance with ASTM B 117. At manufacturer's option, plated fittings may be given a subsequent chromate treatment.

Workmanship—Workmanship shall conform to the best commercial practice to produce high quality fittings. Fittings shall be free from all hanging burrs, loose scale, and slivers which might become dislodged in usage and all other defects which might affect their serviceability. All sealing surfaces must be smooth except that annular tool marks up to 100 μin (2.5 μm) maximum shall be permissible.

Assembly Considerations—Where it is not objectionable from a function or production standpoint, the use of a compatible lubricant or sealant may be desirable in assembling Dryseal pipe threads on automotive tube fittings to minimize galling and effect a pressure-tight seal.

21.11

TABLE 4A—DIMENSIONS OF SPECIAL SIZE CONNECTORS (FIG. 5A) AND ELBOWS (FIG. 6A)

Nom Tube OD, in	A, Dryseal Taper Thread NPTF[a] in	B, Nom Thread Size, in Class 2B, Int[d]	C Nom, in	D Dia Drill		D_1 Dia Drill		D_4 Dia Drill		D_5 Dia Drill		L^b +0.03 −0.00		L^b +0.8 −0.0		L_3^b +0.03 +0.00		L_3^b +0.8 −0.0		L_4	
				in	mm	in	mm	in	mm	in	mm	in	mm	in	mm	in	mm	in	mm	in	mm
1/4	1/4	7/16-24	9/16	0.188	4.78	0.188	4.78	0.312	7.92	0.344	8.74	0.88	22.4	1.09	27.7	0.56	14.2
3/8	1/8	5/8-18	3/4	0.281	7.14	0.219	5.56	0.219	5.56	—	—	0.88	22.4	1.12	28.4	0.73	18.5
3/8	3/8	5/8-18	3/4	0.281	7.14	0.312	7.92	0.406	10.31	0.437	11.10	1.00	25.4	1.31	33.3	0.84	21.3
1/2	1/4	3/4-18	29/32	0.406	10.31	0.281	7.14	0.344	8.74	—	—	1.06	26.9	1.46	37.1	0.88	22.4
1/2	1/2	3/4-18	29/32	0.406	10.31	0.406	10.31	0.562	14.27	0.500	12.70	1.25	31.8	1.66	42.2	1.09	27.7
5/8	3/8	7/8-18	1-1/16	0.531	13.49	0.437	11.10	0.406	10.31	0.469	11.91	1.12	28.4	1.62	41.1	1.11	28.2
3/4	1/2	1-1/16-16	1-1/4	0.625	15.82	0.531	13.49	0.562	14.27	—	—	1.38	35.1	2.00	50.8	1.30	33.0

Nom Tube OD, in	A, Dryseal Taper Thread NPTF[a] in	B, Nom Thread Size, in Class 2B, Int[d]	M		$P^{b,c}$		$S^{b,g}$		S_1^b		X		Y +0.010 −0.000		Y +0.25 −0.00		Z Min	
			in	mm	in	mm	in	mm	in	mm	in	mm	in	mm	in	mm	in	mm
1/4	1/4	7/16-24	0.28	7.1	0.56	14.2	0.44	11.2	0.240	6.10	0.56	14.2	0.275	6.98	0.03	0.8
3/8	1/8	5/8-18	0.53	13.5	0.38	9.7	0.38	9.7	—	—	0.75	19.0	0.370	9.40	—	—
3/8	3/8	5/8-18	0.50	12.7	0.56	14.2	0.41	10.4	0.437	11.10	0.75	19.0	0.370	9.40	0.05	1.3
1/2	1/4	3/4-18	0.59	15.0	0.56	14.2	0.44	11.2	—	—	0.91	23.1	0.450	11.43	—	—
1/2	1/2	3/4-18	0.66	16.8	0.75	19.0	0.62	15.7	0.625	15.88	−0.91	23.1	0.450	11.43	—	—
5/8	3/8	7/8-18	0.75	19.0	0.56	14.2	0.44	11.2	0.156	3.96	1.06	26.9	0.525	13.34	0.05	1.3
3/4	1/2	1-1/16-16	0.85	21.6	0.75	19.0	0.47	11.9	—	—	1.25	31.8	0.620	15.75	—	—

[g] Measured from end containing the largest passage.

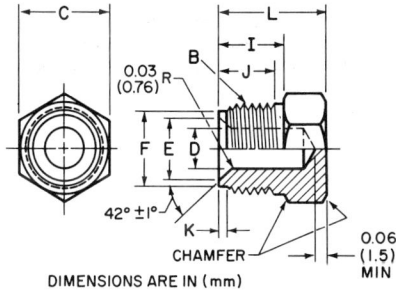

FIG. 8—INVERTED FLARE PLUGS (040109)

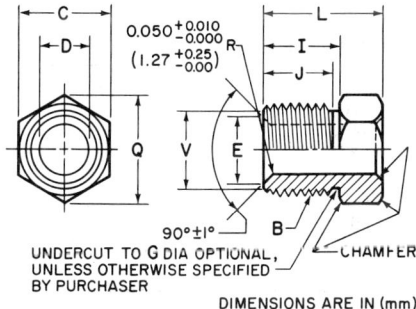

FIG. 9—INVERTED FLARE NUTS (040110)

TABLE 5—DIMENSIONS OF PLUGS (FIG. 8)

Nom Tube OD, in	B, Nom Thread Size, in Class 2A Ext.	C Hex Nom, in	D Dia		E Dia		F Dia		I		J Full Thread Min		K		L	
			in ±0.005	mm ±0.13	in +0.004 −0.000	mm +0.10 −0.00	in +0.000 −0.005	mm +0.00 −0.03	in	mm	in	mm	in	mm	in	mm
3/16	3/8 -24	3/8	0.188	4.77	0.233	5.92	0.290	7.37	0.34	8.6	0.28	7.1	0.04	1.0	0.53	13.5
1/4	7/16-24	7/16	0.188	4.77	0.296	7.52	0.355	9.02	0.34	8.6	0.28	7.1	0.04	1.0	0.54	13.7
5/16	1/2 -20	1/2	0.250	6.35	0.358	9.09	0.417	10.59	0.38	9.7	0.31	7.9	0.06	1.5	0.59	15.0
3/8	5/8 -18	5/8	0.312	7.92	0.446	11.33	0.507	12.89	0.41	10.4	0.33	8.4	0.06	1.5	0.66	16.8
7/16	11/16-18	11/16	0.375	9.52	0.510	12.95	0.580	14.73	0.50	12.7	0.39	9.9	0.06	1.5	0.75	19.1
1/2	3/4 -18	3/4	0.437	11.10	0.571	14.50	0.640	16.26	0.50	12.7	0.41	10.4	0.06	1.5	0.81	20.6
5/8	7/8 -18	7/8	0.562	14.27	0.696	17.68	0.771	19.58	0.50	12.7	0.43	10.9	0.06	1.5	0.88	22.4
3/4	1-1/16-16	1-1/16	0.656	16.66	0.865	21.97	0.937	23.80	0.61	15.5	0.50	12.7	0.08	2.0	0.95	24.1

TABLE 6—DIMENSIONS OF NUTS (FIG. 9)

Nom Tube OD, in	B, Nom Thread Size, in Class 2A, Ext.	C[b] in	C[b] mm	D[c] Dia in	D[c] Dia mm	E Dia in	E Dia mm	G Dia in	G Dia mm	I in	I mm	J Full Thread Min in	J Full Thread Min mm	L in	L mm	Q Min in	Q Min mm	V in	V mm
		+0.005 −0.000	+0.13 −0.00	+0.005 −0.000	+0.13 −0.00	+0.004 −0.000	+0.10 −0.00	+0.000 −0.005	+0.00 −0.13									+0.000 −0.015	+0.00 −0.38
1/8	5/16–28	0.307	7.80	0.130	3.30	0.221	5.61	0.259	6.58	0.32	8.1	0.26	6.6	0.52	13.2	0.347	8.81	0.261	6.63
3/16	3/8 –24	0.370	9.40	0.194	4.93	0.268	6.81	0.313	7.95	0.36	9.1	0.30	7.6	0.56	14.2	0.418	10.62	0.316	8.03
1/4	7/16–24	0.433	11.00	0.257	6.53	0.331	8.41	0.375	9.53	0.36	9.1	0.30	7.6	0.56	14.2	0.488	12.40	0.378	9.60
5/16	1/2 –20	0.495	12.57	0.319	8.10	0.393	9.98	0.427	10.85	0.40	10.2	0.32	8.1	0.62	15.7	0.555	14.10	0.430	10.92
3/8	5/8 –18	0.620	15.75	0.382	9.70	0.487	12.37	0.543	13.79	0.40	10.2	0.32	8.1	0.66	16.8	0.701	17.81	0.548	13.92
7/16	11/16–18	0.683	17.35	0.443	11.25	0.550	13.97	0.605	15.37	0.44	11.2	0.34	8.6	0.68	17.3	0.772	19.61	0.610	15.49
1/2	3/4 –18	0.745	18.92	0.506	12.85	0.612	15.54	0.668	16.97	0.46	11.7	0.36	9.1	0.74	18.8	0.842	21.39	0.673	17.09
5/8	7/8 –18	0.870	22.10	0.631	16.03	0.737	18.72	0.793	20.14	0.48	12.2	0.38	9.7	0.80	20.3	0.973	24.71	0.798	20.27
3/4	1-1/16–16	1.057	26.85	0.757	19.23	0.906	23.01	0.971	24.66	0.54	13.7	0.44	11.2	0.88	22.4	1.193	30.30	0.976	24.79

[a] On 3/16, 1/4, 5/16, and 3/8 in size nuts, the maximum hole diameter limits may be exceeded at the hexagon end and the hole slightly tapered to a depth not exceeding 0.10 in (2.5 mm) below the hexagon end. A slight flash at this point shall be not be objectionable provided the tabulated hole diameter limits are maintained. A definite rounding at the edge of the hole and a slight rounding of hexagon corners shall also be permissible provided the nuts meet the Wrenching Test requirements in Table 6A.

[b] Tabulated values apply only to cold formed nuts and nuts made from steel bar stock. Limits for across flats on nuts made from brass bar stock shall conform with commercial tolerances on the bar stock.

[c] Hole diameter D and 90 deg seat shall be concentric with thread pitch diameter within 0.007 in (0.18 mm) full indicator reading (FIR).

TABLE 6A—WRENCHING TEST REQUIREMENTS

Nom Tube OD, in	Torque Requirements for Steel Nuts lb-in	Torque Requirements for Steel Nuts N·m	Wrench Opening Max in	Wrench Opening Max mm
1/8	60	6.8	0.322	8.18
3/16	120	13.6	0.384	9.75
1/4	150	16.9	0.446	11.33
5/16	180	20.3	0.510	12.95
3/8	210	23.7	0.636	16.15
7/16	300	33.9	0.699	17.76
1/2	400	45.2	0.763	19.38
5/8	500	56.5	0.888	22.56
3/4	650	73.4	1.077	27.36

Wrenching Test—Steel nuts when assembled without tubing into mating brass fittings which are held securely by a suitable means, such as in a vise, shall be capable of being tightened by means of a standard open end wrench, having an opening as tabulated to the minimum torque values specified without failure (rounding) of the hexagon corner.

NOTE: The tabulated torque requirements should not in any case be misconstrued as being installation torques. They are intended solely for determining the adequacy of the hexagon corners. (See Table 6A)

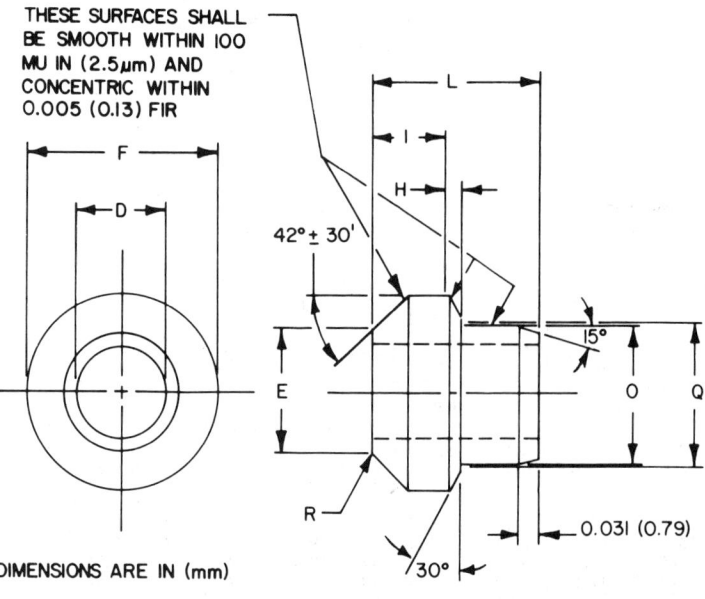

φ FIG. 10—SEAT INSERT (040141)

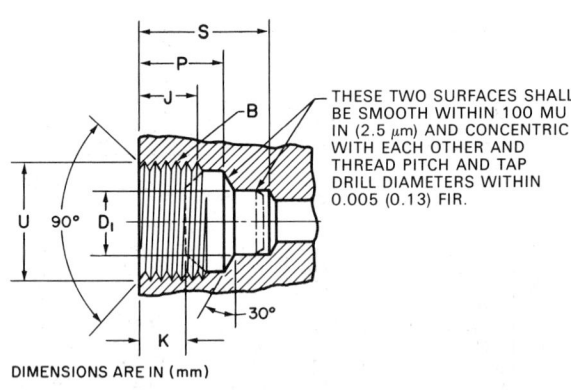

FIG. 11—DETAIL OF SEAT ASSEMBLY

TABLE 7A—DIMENSIONS OF TYPE A INVERTED FLARED TUBE SEAT INSERTS AND SEAT DETAIL[a] (FIGS. 10 AND 11)

Nom Tube OD, in	B, Nom Thread Size, in Class 2B, Int.	D Dia Min in	D Dia Min mm	D Dia Max in	D Dia Max mm	D_1 Dia +0.003 −0.002 in	D_1 Dia +0.08 −0.05 mm	E Dia +0.000 −0.004 in	E Dia +0.00 −0.10 mm	F Dia +0.000 −0.005 in	F Dia +0.00 −0.13 mm
1/8	5/16–28	0.075	1.91	0.081	2.06	0.161	4.09	0.108	2.74	0.264	6.71
3/16	3/8 –24	0.122	3.10	0.128	3.25	0.213	5.41	0.155	3.94	0.321	8.15
1/4	7/16–24	0.184	4.67	0.190	4.83	0.272	6.91	0.218	5.54	0.383	9.73
5/16	1/2 –20	0.215	5.46	0.221	5.61	0.302	7.67	0.280	7.11	0.435	11.05
3/8	5/8 –18	0.278	7.06	0.284	7.21	0.359	9.12	0.346	8.79	0.554	14.10
7/16	11/16–18	0.340	8.64	0.346	8.79	0.422	10.72	0.409	10.39	0.616	16.65
1/2	3/4 –18	0.403	10.24	0.409	10.39	0.484	12.29	0.471	11.96	0.679	17.25
5/8	7/8 –18	0.528	13.41	0.534	13.56	0.625	15.88	0.596	15.14	0.804	29.42
3/4	1-1/16–16	0.621	15.77	0.629	15.98	0.750	19.05	0.708	17.98	0.983	24.97

Nom Tube OD, in	H Min in	H Min mm	H Max in	H Max mm	I +0.000 −0.004 in	I +0.00 −0.10 mm	J Full Thread Min in	J Full Thread Min mm	K Ref in	K Ref mm	L in	L mm	O Dia +0.000 −0.003 in	O Dia +0.00 −0.08 mm
1/8	0.015	0.38	0.020	0.51	0.146	3.71	0.214	5.44	0.187	4.75	0.28	7.1	0.161	4.09
3/16	0.015	0.38	0.020	0.51	0.151	3.84	0.225	5.72	0.203	5.16	0.30	7.6	0.213	5.41
1/4	0.015	0.38	0.020	0.51	0.151	3.84	0.225	5.72	0.203	5.16	0.32	8.1	0.272	6.91
5/16	0.025	0.64	0.030	0.76	0.166	4.22	0.258	6.55	0.234	5.94	0.33	8.4	0.302	7.67
3/8	0.030	0.76	0.035	0.89	0.197	5.00	0.289	7.34	0.266	6.76	0.38	9.6	0.359	9.12
7/16	0.030	0.76	0.035	0.89	0.197	5.00	0.320	8.13	0.296	7.52	0.39	9.9	0.422	10.72
1/2	0.035	0.89	0.045	1.14	0.216	5.49	0.336	8.53	0.312	7.93	0.40	10.2	0.484	12.29
5/8	0.035	0.89	0.045	1.14	0.236	5.99	0.351	8.92	0.328	8.33	0.46	11.7	0.625	15.88
3/4	0.045	1.14	0.055	1.40	0.285	7.24	0.414	10.52	0.359	9.12	0.54	13.7	0.750	19.05

Nom Tube OD, in	P ±0.005 in	P ±0.13 mm	Q Dia +0.0000 −0.0025 in	Q Dia +0.000 −0.064 mm	R Radius in	R Radius mm	S Min in	S Min mm	U Dia Min in	U Dia Min mm	U Dia Max in	U Dia Max mm
1/8	0.325	8.26	0.1670	4.242	0.006	0.15	0.50	12.7	0.32	8.1	0.34	8.6
3/16	0.348	8.84	0.2190	5.563	0.006	0.15	0.54	13.7	0.39	9.9	0.41	10.4
1/4	0.348	8.84	0.2780	7.061	0.031	0.79	0.56	14.2	0.45	11.4	0.47	11.9
5/16	0.392	9.96	0.3080	7.823	0.031	0.79	0.59	15.0	0.51	13.0	0.53	13.5
3/8	0.455	11.56	0.3650	9.271	0.031	0.79	0.68	17.3	0.64	16.3	0.67	17.0
7/16	0.484	12.29	0.4280	10.871	0.031	0.79	0.72	18.3	0.70	17.8	0.73	18.5
1/2	0.519	13.18	0.4900	12.446	0.031	0.79	0.75	19.0	0.77	19.6	0.80	20.3
5/8	0.555	14.10	0.6310	16.027	0.031	0.79	0.82	20.8	0.89	22.6	0.92	23.4
3/4	0.635	16.13	0.7560	19.202	0.031	0.79	0.93	23.6	1.08	27.4	1.11	28.2

[a] Type A seat inserts are intended for general purpose applications in cast or malleable iron and steel. Where standard inserts are assembled into other materials, modifications of seat may be necessary to assure proper installation.

TAPERED SLEEVE (COMPRESSION) TYPE

TABLE 7B—DIMENSIONS OF TYPE B INVERTED FLARED TUBE SEAT INSERTS AND SEAT DETAIL[a] (FIGS. 10 and 11)

Nom Tube OD, in	B, Nom Thread Size, in Class 2B, Int.	D Dia		D_1 Dia		E[b] Dia		F Dia			
								Max		Min	
		in	mm	in	mm	in	mm	in	mm	in	mm
		+0.007 −0.000	+0.18 −0.00	±0.001	±0.03	+0.000 −0.004	+0.00 −0.10				
3/16	3/8 –24	0.108	2.74	0.189	4.80	0.129	3.28	0.317	8.05	0.307	7.80
1/4	7/16–24	0.108	2.74	0.189	4.80	0.191	4.85	0.380	9.65	0.360	9.14

Nom Tube OD, in	H Dia		I		J Full Thread Min		K				L	
							Max		Min			
	in	mm	in	mm	in	mm	in	mm	in	mm	in	mm
	+0.008 −0.000	+0.20 −0.00	+0.000 −0.004	+0.00 −0.00								
3/16	0.017	0.43	0.185	4.70	0.334	8.48	0.244	6.20	0.224	5.69	0.34	8.6
1/4	0.033	0.84	0.215	5.46	0.390	9.91	0.260	6.60	0.230	5.84	0.39	9.9

Nom Tube OD, in	O Dia		P		Q Dia		R Radius		S Min		U Dia Max	
	in	mm	in	mm	in	mm	in	mm	in	mm	in	mm
	+0.000 −0.004	+0.00 −0.10			+0.000 −0.002	+0.00 −0.05	±0.001	±0.03				
3/16	0.191	4.85	0.406	10.31	0.195	4.95	0.031	0.79	0.615	15.62	0.437	11.10
1/4	0.191	4.85	0.470	11.94	0.195	4.95	0.031	0.79	0.680	17.27	0.500	12.70

[a] Type B seat inserts are used extensively in cast or malleable iron and steel components of hydraulic brake systems. Where standard inserts are assembled into other materials, modifications of seat may be necessary to assure proper installation.

[b] On Type B inserts, E diameter defines center of R radius.

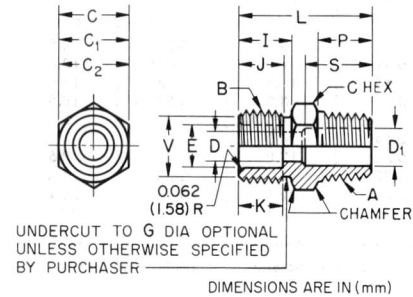

FIG. 12A (060102)

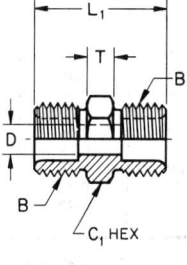

FIG. 12B (060101)

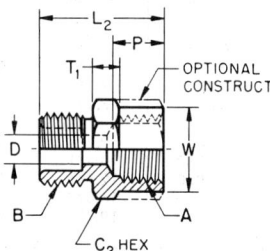

FIG. 12C (060103)

FIG. 12—CONNECTORS AND UNIONS

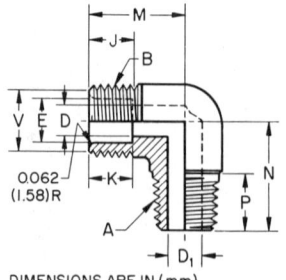

FIG. 13A (060202)

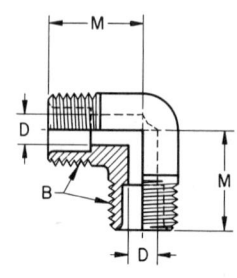

FIG. 13B (060201)

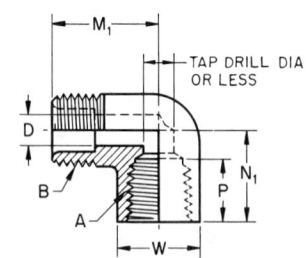

FIG. 13C (060203)

FIG. 13—ELBOWS

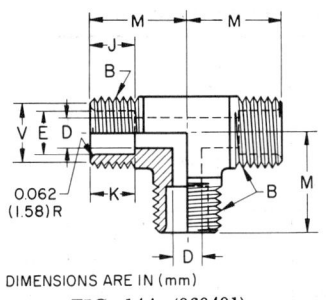
FIG. 14A (060401)

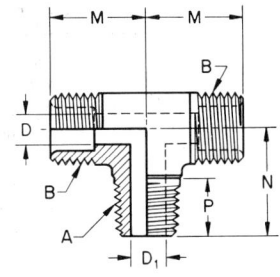
FIG. 14B (060425)

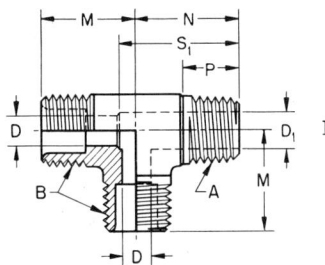
FIG. 14C (060424)

FIG. 14—TEES

DIMENSIONS ARE IN (mm)

NOTES: UNSPECIFIED DETAIL WITH RESPECT TO DIMENSIONS, TOLERANCES, CONTOURS, MATERIAL, WORKMANSHIP, ETC., MUST CONFORM TO GENERAL SPECIFICATIONS FOR AUTOMOTIVE TUBE FITTINGS. THE DIMENSIONAL DESIGNATIONS ON THE FIRST FIGURE IN EACH GROUP SHALL APPLY TO ALL OTHER FIGURES IN THAT GROUP EXCEPT AS SHOWN OTHERWISE. CODES SHOWN IN BRACKETS ADJACENT TO FIGURE NUMBERS REPRESENT RESPECTIVE FITTING IDENTIFICATION IN ACCORDANCE WITH SAE J846 (MAY, 1978).

TABLE 8—DIMENSIONS OF CONNECTORS, UNIONS, ELBOWS, AND TEES (FIGS. 12A–14C)

Nom Tube OD, in	A Dryseal Taper Thread NPTF[a], in	B, Nom Thread Size, in Class 2A, Ext.	C Nom, in	C_1 Nom, in	C_2 Nom, in	D^e Dia Drill in	D^e Dia Drill mm	D_1^e Dia Drill in	D_1^e Dia Drill mm	E Dia ±0.002 in	E Dia ±0.05 mm	G^d Dia +0.000 −0.010 in	G^d Dia +0.00 −0.25 mm	I in	I mm
1/8	1/8	5/16-24	7/16	5/16	9/16	0.094	2.39	0.219	5.56	0.134	3.40	0.250	6.35	0.25	6.4
3/16	1/8	3/8 -24	7/16	3/8	9/16	0.125	3.17	0.219	5.56	0.195	4.95	0.313	7.95	0.28	7.1
1/4	1/8	7/16-24	7/16	7/16	9/16	0.188	4.77	0.219	5.56	0.260	6.60	0.364	9.25	0.31	7.9
5/16	1/8	1/2 -24	1/2	1/2	9/16	0.250	6.35	0.234	5.94	0.323	8.20	0.438	11.13	0.34	8.6
5/16	1/4	1/2 -24	9/16	—	—	0.250	6.35	0.312	7.92	0.323	8.20	0.438	11.13	0.34	8.6
3/8	1/4	9/16-24	9/16	9/16	11/16	0.312	7.92	0.344	8.74	0.387	9.83	0.500	12.70	0.38	9.7
7/16	1/4	5/8 -24	5/8	5/8	11/16	0.312	7.92	0.344	8.74	0.451	11.46	0.563	14.30	0.41	10.4
1/2	3/8	11/16-20	11/16	11/16	13/16	0.406	10.31	0.406	10.31	0.516	13.11	0.614	15.60	0.44	11.2
5/8	1/2	13/16-18	7/8	13/16	1	0.500	12.70	0.562	14.27	0.644	16.36	0.730	18.54	0.50	12.7
3/4	1/2	1 -18	1	1	1	0.562	14.27	0.562	14.27	0.772	19.61	0.918	23.32	0.56	14.2

Nom Tube OD, in	J Full Thread Min in	J Full Thread Min mm	K in	K mm	L^b ±0.03 in	L^b ±0.8 mm	L_1 ±0.03 in	L_1 ±0.8 mm	$L_2^{b,c}$ in	$L_2^{b,c}$ mm	M ±0.03 in	M ±0.8 mm	M_1 ±0.03 in	M_1 ±0.8 mm	N^b ±0.03 in	N^b ±0.8 mm
1/8	0.19	4.8	0.19	4.8	0.79	20.1	0.66	16.8	0.75	19.1	0.62	15.7	0.69	17.5	0.69	17.5
3/16	0.22	5.6	0.22	5.6	0.85	21.6	0.75	19.1	0.78	19.8	0.62	15.7	0.69	17.5	0.69	17.5
1/4	0.25	6.4	0.25	6.4	0.88	22.4	0.81	20.6	0.78	19.8	0.62	15.7	0.69	17.5	0.75	19.1
5/16	0.28	7.1	0.28	7.1	0.91	23.1	0.87	22.1	0.81	20.6	0.62	15.7	0.69	17.5	0.75	19.1
5/16	0.28	7.1	0.28	7.1	1.09	27.7	—	—	—	—	0.69	17.5	—	—	0.84	21.3
3/8	0.31	7.9	0.31	7.9	1.16	29.5	0.98	24.9	1.06	26.9	0.75	19.1	0.81	20.6	0.94	23.9
7/16	0.34	8.6	0.34	8.6	1.19	30.2	1.04	26.4	1.06	26.9	0.84	21.3	0.91	23.1	1.00	25.4
1/2	0.38	9.7	0.38	9.7	1.22	31.0	1.10	27.9	1.12	28.4	0.94	23.9	1.00	25.4	1.12	28.4
5/8	0.44	11.2	0.38	9.7	1.50	38.1	1.25	31.8	1.38	35.1	1.06	26.9	1.06	26.9	1.31	33.3
3/4	0.50	12.7	0.44	11.2	1.62	41.1	1.43	36.3	1.50	38.1	1.19	30.2	1.19	30.2	1.50	38.1

Nom Tube OD, in	$N_1^{b,c}$ ±0.03 in	$N_1^{b,c}$ ±0.8 mm	$P^{b,c}$ in	$P^{b,c}$ mm	$S^{b,e}$ Max in	$S^{b,e}$ Max mm	$S_1^{b,e}$ Min in	$S_1^{b,e}$ Min mm	T^f Ref in	T^f Ref mm	T_1 Min in	T_1 Min mm	V Dia +0.00 −0.02 in	V Dia +0.0 −0.5 mm	W^g +0.00 −0.02 in	W^g +0.0 −0.5 mm
1/8	0.56	14.2	0.38	9.7	0.46	11.7	0.78	19.8	0.15	3.8	0.21	5.3	0.25	6.3	0.56	14.2
3/16	0.56	14.2	0.38	9.7	0.48	12.2	0.79	20.1	0.18	4.6	0.21	5.3	0.31	7.9	0.56	14.2
1/4	0.56	14.2	0.38	9.7	0.48	12.2	0.89	22.6	0.18	4.6	0.24	6.1	0.38	9.7	0.56	14.2
5/16	0.56	14.2	0.38	9.7	0.42	10.7	0.79	20.1	0.18	4.6	0.24	6.1	0.44	11.2	0.56	14.2
5/16	—	—	0.56	14.2	0.66	16.8	1.01	25.6	0.18	4.6	—	—	0.44	11.2	—	—
3/8	0.75	19.1	0.56	14.2	0.67	17.0	1.14	29.0	0.21	5.3	0.30	7.6	0.50	12.7	0.69	17.5
7/16	0.75	19.1	0.56	14.2	0.67	17.0	1.20	30.5	0.21	5.3	0.30	7.6	0.56	14.2	0.69	17.5
1/2	0.88	22.4	0.56	14.2	—	—	—	—	0.21	5.3	0.37	9.4	0.62	15.7	0.81	20.6
5/8	1.00	25.4	0.75	19.1	0.88	22.4	1.60	40.6	0.24	6.1	0.43	10.9	0.73	18.5	1.00	25.4
3/4	1.00	25.4	0.75	19.1	—	—	—	—	0.30	7.6	0.43	10.9	0.92	23.4	1.00	25.4

[a] Dryseal American Standard Taper Pipe Thread. See General Specifications.
[b] Where SAE Short Pipe Thread is authorized by purchaser, dimensions L, L_2, N, N_1, P, S, and S_1 are reduced in accordance with reduction of pipe thread length. See SAE J476 (June, 1961), Tables 3 and 4.
[c] Tap drill depths given require use of bottoming taps to produce standard full thread length. For increased tap clearance, see Internal Taper Pipe Threads in General Specifications.
[d] Where thread relief undercut is used, the last thread shall be chamfered 1/2 to 1 pitch long from G diameter.
[e] At manufacturer's option, through passages in fittings shown in Figs. 12A and 14C may conform with smaller diameter specified or the appropriate end may be counterbored to larger diameter for depths defined by S or S_1, respectively.
[f] Minimum design thickness, not subject to inspection.
[g] Basic dimensions shown shall apply as minimum diameter or across flats for bosses. The —0.02 in (—0.5 mm) tolerance shall apply only to chamfer diameter on full hexagon version of fitting in Fig. 12C.

FIG. 15—TAPERED SLEEVES (060115)

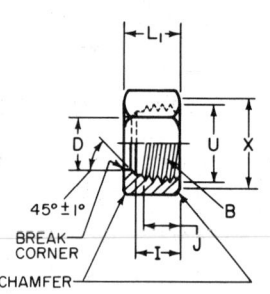

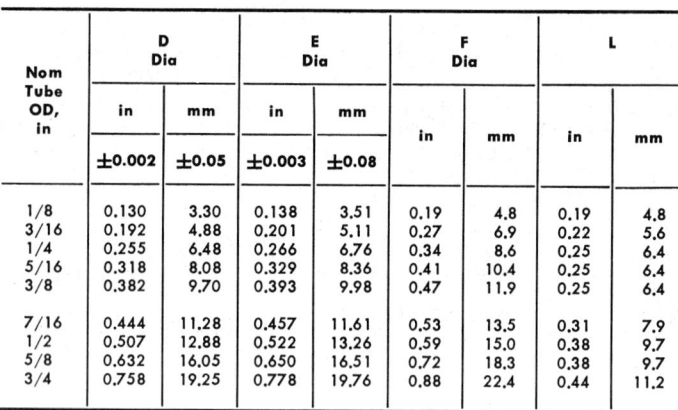

TABLE 9—DIMENSIONS OF TAPERED SLEEVES (FIG. 15)

Nom Tube OD, in	D Dia		E Dia		F Dia		L	
	in	mm	in	mm	in	mm	in	mm
	±0.002	±0.05	±0.003	±0.08				
1/8	0.130	3.30	0.138	3.51	0.19	4.8	0.19	4.8
3/16	0.192	4.88	0.201	5.11	0.27	6.9	0.22	5.6
1/4	0.255	6.48	0.266	6.76	0.34	8.6	0.25	6.4
5/16	0.318	8.08	0.329	8.36	0.41	10.4	0.25	6.4
3/8	0.382	9.70	0.393	9.98	0.47	11.9	0.25	6.4
7/16	0.444	11.28	0.457	11.61	0.53	13.5	0.31	7.9
1/2	0.507	12.88	0.522	13.26	0.59	15.0	0.38	9.7
5/8	0.632	16.05	0.650	16.51	0.72	18.3	0.38	9.7
3/4	0.758	19.25	0.778	19.76	0.88	22.4	0.44	11.2

FIG. 16—SHORT NUTS (060110)

FIG. 17—LONG NUTS (060111)

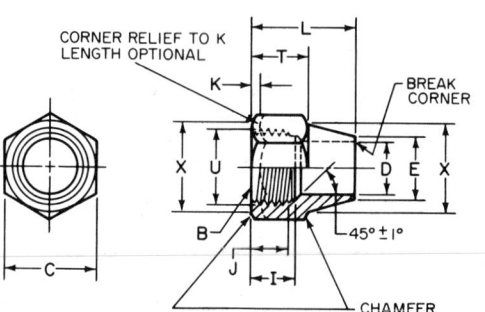

φ TABLE 10—DIMENSIONS OF SHORT AND LONG NUTS (FIGS. 16 AND 17)

Nom Tube OD, in	B Straight Thread Nom Size, in Class 2B, Int.	C Hex, in	D Dia		E Dia		I		J Full Thread Min		K		L		L₁		T		U Dia				X Dia	
		Nom	±.002	±0.05	in	mm	+0.000 −0.010	+0.00 −0.25	in	mm	in	mm	in	mm	in	mm	in	mm	Min		Max		in	mm
			in	mm			in	mm											in	mm	in	mm		
1/8	5/16-24	3/8	0.130	3.30	0.25	6.4	0.25	6.4	0.19	4.8	0.09	2.3	0.50	12.7	0.38	9.7	0.25	6.4	0.33	8.4	0.35	8.9	0.38	9.7
3/16	3/8 -24	7/16	0.192	4.88	0.31	7.9	0.28	7.1	0.22	5.6	0.09	2.3	0.62	15.7	0.41	10.4	0.28	7.1	0.39	9.9	0.41	10.4	0.44	11.2
1/4	7/16-24	1/2	0.255	6.48	0.38	9.7	0.31	7.9	0.25	6.4	0.12	3.0	0.75	19.1	0.44	11.2	0.38	9.7	0.45	11.4	0.47	11.9	0.50	12.7
5/16	1/2 -24	9/16	0.318	8.08	0.44	11.2	0.31	7.9	0.25	6.4	0.12	3.0	0.84	21.3	0.44	11.2	0.44	11.2	0.51	13.0	0.53	13.5	0.56	14.2
3/8	9/16-24	5/8	0.382	9.70	0.50	12.7	0.34	8.6	0.28	7.1	0.12	3.0	0.97	24.6	0.47	11.9	0.50	12.7	0.58	14.7	0.60	15.2	0.62	15.7
7/16	5/8 -24	11/16	0.444	11.28	0.56	14.2	0.37	9.4	0.31	7.9	0.12	3.0	1.03	26.2	0.50	12.7	0.56	14.2	0.64	16.3	0.66	16.8	0.69	17.5
1/2	11/16-20	13/16	0.507	12.88	0.62	15.7	0.49	12.4	0.41	10.4	0.12	3.0	1.06	26.9	0.62	15.7	0.62	15.7	0.70	17.8	0.73	18.5	0.81	20.6
5/8	13/16-18	15/16	0.632	16.05	0.75	19.1	0.50	12.7	0.41	10.4	0.12	3.0	1.19	30.2	0.62	15.7	0.72	18.3	0.83	21.1	0.86	21.8	0.94	23.9
3/4	1-18	1-3/16	0.758	19.25	0.88	22.4	0.53	13.5	0.44	11.2	0.12	3.0	1.38	35.1	0.69	17.5	0.75	19.1	1.02	25.9	1.05	26.7	1.19	30.2

TABLE 11

Nominal Pipe Thread Size, in	External Thread								Internal Thread			
	Chamfer Diameter[a]				Length of Chamfered or Partial Thread				Countersink Diameter[a]			
	Max		Min		Min		Max		Min		Max	
	in	mm	in	mm	in	mm	in	mm	in	mm	in	mm
1/8	0.32	8.1	0.30	7.6	0.037	0.94	0.055	1.40	0.42	10.7	0.44	11.2
1/4	0.42	10.7	0.40	10.2	0.056	1.42	0.084	2.13	0.55	14.0	0.57	14.5
3/8	0.55	14.0	0.53	13.5	0.056	1.42	0.084	2.13	0.69	17.5	0.71	18.0
1/2	0.68	17.3	0.66	16.8	0.071	1.80	0.107	2.72	0.85	21.6	0.87	22.1
3/4	0.89	22.6	0.87	22.1	0.071	1.80	0.107	2.72	1.06	26.9	1.08	27.4
1	1.12	28.4	1.09	27.7	0.087	2.21	0.130	3.30	1.34	34.0	1.37	34.8

φ [a]Tabulated diameters conform with Appendix A, SAE J476 (June, 1961).

APPENDIX A—GAGES AND GAGING FOR INVERTED FLARED TYPE FITTINGS

General—The information contained herein is intended to provide and promote the use of uniform gaging practices for determining conformance of inverted flared type tube fittings with the specifications given in the standard.

Gaging of Cone Seats

Depth of Face—Proper location of the face of the cone seat relative to the face of the fitting shall be determined by use of the step limit gage depicted in Fig. A-1. One-half of the top surface of body of this gage shall be machined and/or ground 0.015 in (0.38 mm) below the balance of surface to provide a low limit step. The gage pin shall have a minimum diameter equal to "E max" (see Table 4) and shall be of a length equivalent to the sum of the height of the high-limit side of gage body plus "K min" (see Table 4). The opposite faces of gage body and ends of gage pin shall be flat, smooth, parallel, and square with the axis of pin. The fit between gage pin and gage body shall be such that the pin alignment is maintained yet free to move, of its own weight, within the gage body. All gage components should be made from steel suitably hardened to assure adequate service life.

When this gage is placed on the face of fitting, in line with the axis of seat and with gage pin contacting face of seat, the top of gage pin must not protrude above the high limit nor be below the low limit top surfaces of the gage body for fitting to be acceptable.

Gaging of Internal Threads

Size and Form—Conformance of internal threads with the dimensions specified shall be determined in accordance with standard thread gaging procedures.

Full Form Thread Depth—Suitability of the relationship between the last full form thread to the face of the cone seat shall be determined by use of the gage depicted in Fig. A-2. This gage shall have external threads, conforming in all respects with the corresponding GO thread plug gage for the respective thread size, of a minimum length equal to "K min" (see Table 4) plus one pitch (thread) and shall have the partial thread at starting end removed. Use of a thread relief undercut beyond a length equal to "K min" shall be permissible. The gage pin hole at starting end shall be suitably counterbored to clear the fitting cone seat to a depth equivalent to $1\frac{1}{2}$ pitches (threads). The length of the gage pin shall be equivalent to the overall length of the body of gage plus the difference between "J max" (see Table 4) and $\frac{1}{2}$ pitch (thread). All other features of gage body, gage pin, fit and material shall be as specified for the gage in Fig. A-1

When this gage is assembled by hand into the fitting as far as the thread will permit, with pin contacting face of cone seat, the top of the gage pin must be flush with or protrude above the top surface of the gage body for fitting to be acceptable.

Concentricity Gage and Gaging

Cone Seat Concentricity—Conformance with specified limitations on concentricity of cone seats with respect to thread pitch cylinder shall be checked by the use of functional gages of the type described herein and depicted in Fig. A-3, or equivalent means.

The gage consists of a body or frame providing a means for mounting a dial indicator gage in such a manner that it can be actuated through a rotating pin on the opposite end of which is a machined stylus designed to ride on the conical surface of the seat. The gage is centered on the threads of the fitting by means of an interchangeable male threaded gage insert. The threads on the insert shall conform in all respects with the corresponding GO thread plug gage for the respective thread size and the length of full form thread shall be equivalent, to *K min* (see Table 3) plus one pitch (thread). The entering end of gage insert shall have a pilot of length equal to one pitch (thread) and diameter which will clear the minor diameter of the seat threads. The first thread beyond the pilot shall be chamfered or the partial portion shall be removed. Use of a thread relief undercut beyond a length equivalent to *K min* shall be permissible. The fit between the stylus extension pin and the hole through gage body and threaded gage insert shall be such that pin will move freely yet be retained in alignment. Fitting manufacturers may be consulted with regard to details of existing gages.

The fitting to be inspected shall be gaged in accordance with the following procedure:

1. The threaded end of gage shall be assembled by hand into the seat on the fitting as far as the thread will permit.

2. The indicator gage shall be zeroed and the stylus then rotated slowly through a complete 360 deg revolution by twisting the knurled collar through the finger notches in gage body.

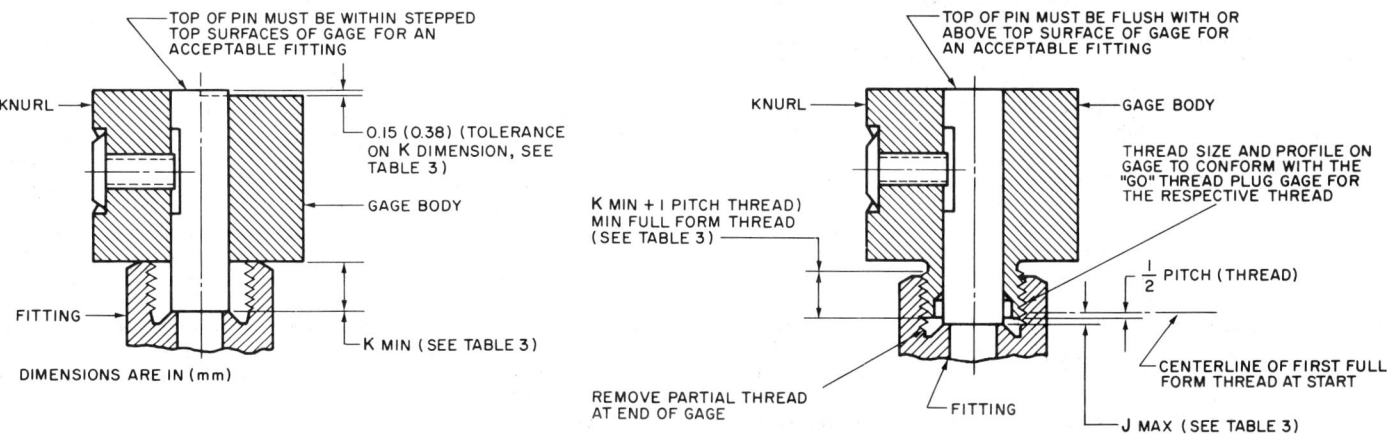

FIG. A-1—FLUSH PIN GAGE FOR CHECKING LOCATION OF FACE OF CONE SEAT

FIG. A-2—FLUSH PIN GAGE FOR CHECKING RELATION OF END OF FULL THREAD WITH FACE OF CONE SEAT

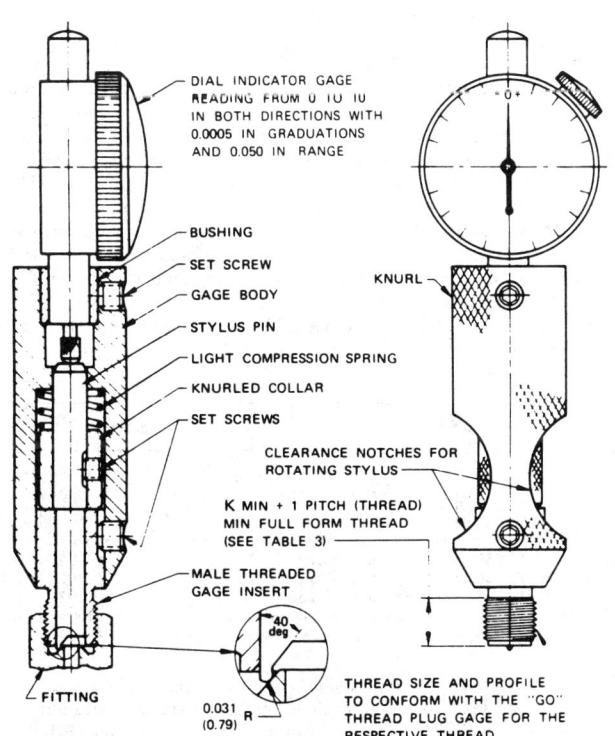

FIG. A-3—TYPICAL CONE SEAT CONCENTRICITY GAGE

3. The total runout indicated throughout the rotation of stylus shall be read and recorded. It shall not exceed 0.0056 (0.142 mm) for the fitting to be acceptable. (The measured deviation is in a plane perpendicular to the plane of the cone seat. The concentricity variation therefore may be determined from the tabulation below, or by multiplying the dial indicator reading by the tangent of 42 deg, or 0.90040.)

Dial Reading		Concentricity		Dial Reading		Concentricity	
in	mm	in	mm	in	mm	in	mm
3 Pl.		4 Pl.		3 Pl.		4 Pl.	
0.0005	0.013	0.00045	0.0114	0.0035	0.089	0.00315	0.0800
0.0010	0.025	0.00090	0.0229	0.0040	0.102	0.00360	0.0914
0.0015	0.038	0.00135	0.0343	0.0045	0.114	0.00405	0.1029
0.0020	0.051	0.00180	0.0457	0.0050	0.127	0.00450	0.1143
0.0025	0.064	0.00225	0.0572	0.0055	0.140	0.00495	0.1257
0.0030	0.076	0.00270	0.0686	0.0060	0.152	0.00540	0.1372

SPHERICAL AND FLANGED SLEEVE (COMPRESSION) TUBE FITTINGS—SAE J246 JUN80

SAE Standard

Report of the Tube, Pipe, Hose, and Lubrication Fittings Committee, approved May 1971, last revised July 1977, editorial change June 1980.

GENERAL SPECIFICATIONS

Scope—This standard covers complete general and dimensional specifications for tube fittings of the spherical and flanged sleeve compression types for use in the piping of air brake systems on automotive vehicles. The spherical sleeve compression type Figs. 1A–5 and Tables 3–5 is intended for use with annealed copper alloy tubing per SAE J1149 (July, 1976), Type 1. The flanged sleeve compression type Figs. 6A–11 and Tables 6–8 is intended for use with nylon tubing per SAE J844 (July, 1976). It is not intended to restrict or preclude other designs of a tube fitting for use with SAE J844 (July, 1976), air brake tubing. Performance requirements for SAE J844 (July, 1976), Nonmetallic Tubing and Fittings Assemblies, are covered in J1131 (January, 1976).

CAUTION: To assure satisfactory performance, tapered sleeve compression type fitting components (SAE J512 (July, 1976)) should not be intermixed with the spherical or flanged sleeve components, nor should the spherical sleeve compression type components be intermixed with the flanged sleeve compression type components when assembling connection in areas where the three types are available.

The following general specifications supplement the dimensional data contained in Tables 1–8 with respect to all unspecified details.

Identification—At manufacturer's option, or where so specified by the purchaser, the fittings, except sleeves, may be permanently and legibly marked *air brake*. The nut shall be permanently and legibly marked according to the current U. S. Department of Transportation FMVSS 106 Regulation (NHTSA). The location of such markings shall be optional with manufacturer.

Size Designations—Fitting sizes are designated by the corresponding nominal outside diameter of the tubing for the various sizes of tube ends and by the corresponding standard nominal pipe size for pipe thread ends.

Dimensions and Tolerances—Except for nominal size and thread specifications, dimensions and tolerances are given in both U. S. customary and SI units as designated. Tabulated dimensions shall apply to the finished parts. The maximum and minimum across flat dimensions shall be within the commercial tolerance of bar or extruded stock from which the fittings are produced. The minimum across corners dimensions of hexagons shall be 1.092 times the nominal width across flats, but shall not result in a side flat width less than 0.43 times the nominal width across flats.

Unless otherwise specified, tolerance on hole diameters designated drill in the dimensional tables shall be as tabulated below:

Drill Size Range		Tolerance			
		Plus		Minus	
in	mm	in	mm	in	mm
0.0135 thru 0.1850	0.343 thru 4.699	0.003	0.08	0.002	0.05
0.1875 thru 0.2480	4.762 thru 6.299	0.004	0.10	0.002	0.05
0.2500 thru 0.7500	6.350 thru 19.050	0.006	0.15	0.003	0.08
0.7579 thru 1.0000	19.251 thru 25.400	0.007	0.18	0.004	0.10

Tolerance on all dimensions not otherwise specified shall be ±0.010 in. (0.25 mm). Tube seat diameters E shall be concentric with straight thread pitch diameters within 0.010 in. (0.25 mm) full indicator reading (FIR). Large seat diameters F shall be concentric with tube seat diameters E within 0.005 in. (0.13 mm) full indicator reading (FIR). The surface of tube stop at base of seat shall be flat and perpendicular to the axis of thread. Angular tolerance on axis of ends of elbows and tees shall be ±2.50 deg for sizes up to and including 3/8 in., and ±1.50 deg for sizes larger than 3/8 in.

Where so illustrated, hexagon corners shall be chamfered 30 ±5 deg to a diameter equal to the nominal width across flats, with a tolerance of −0.016 in. (−0.41 mm); or, where design permits, corners may be chamfered to the diameter of the abutting surface, provided the length of chamfer does not exceed that produced by the 30 deg chamfer previously described.

Passages—Where passages in straight fittings are machined from opposite ends, the offset at the meeting point shall not exceed 0.015 in. (0.38 mm). The cross-sectional area at the junction of passages in angle fittings shall not be less than that of the smallest passage. At manufacturer's option, all passages in a particular fitting may conform with the smallest diameter specified for that fitting. Where the passage is specified as tap drill diameter or less, the minimum shall be no less than the minimum diameter of the smallest passage in the fitting.

Wall Thickness—Unless otherwise designated, the wall thickness at any point on fittings shall not be less than the thickness established by the specified dimensions, tolerances, and eccentricities for inner and outer surfaces.

Contour—Details of contour shall be optional with manufacturer, provided the tabulated dimensions are maintained and serviceability of the fittings is not impaired. Wrench flats on elbows and tees shall be optional. Where extruded and forged shapes are reduced to conserve material, the wall thickness, unless otherwise specified, shall not be less than the respective minimum values tabulated below:

Nominal Tube OD, in	Wall Thickness, Min			
	Extruded Shapes[a]		Forged Shapes	
	in	mm	in	mm
1/4	0.04	1.0	0.075	1.90
5/16	0.05	1.3	0.075	1.90
3/8	0.05	1.3	0.085	2.16
1/2	0.06	1.5	0.090	2.29
5/8	0.08	2.0	0.100	2.54
3/4	0.08	2.0	0.100	2.54
1	0.08	2.0	0.120	3.05

[a] Applies to reduction to one plane of shape only.

Straight Threads—Unified standard Class 2A external and Class 2B internal threads shall apply to plain finish (unplated) fittings of all types. For internally threaded parts with additive finish, the Class 2B diameters shall apply after plating.

The pitch diameter tolerance shall be the same as the corresponding diameter-pitch combination and class of the Unified 18 and 20 thread series or for special diameter-pitch combinations shall be based on diameter, pitch, and a length of engagement of 9 times the pitch. See ANSI B1.1.

For convenient reference, the data generally required to specify threads are given in Table 1. (Inasmuch as threads are normally produced and gaged with

SPHERICAL SLEEVE (COPPER TUBE) TYPE

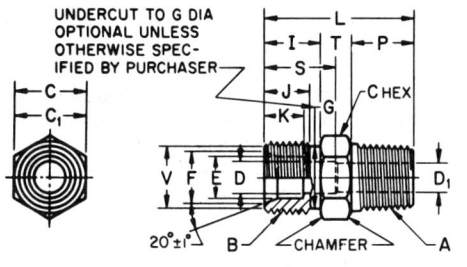

FIG. 1A (120102)

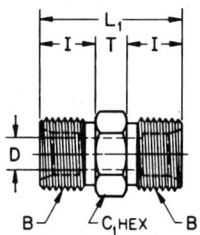

FIG. 1B (120101)

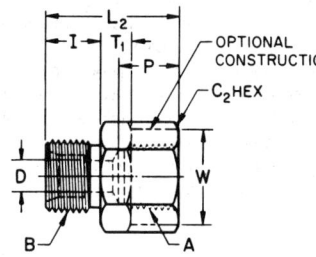

FIG. 1C (120103)

FIG. 1—CONNECTORS AND UNIONS

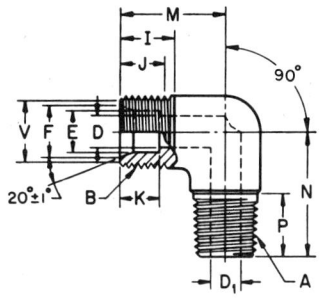

FIG. 2A (120202)

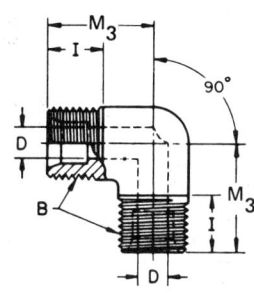

φ FIG. 2B (120201)

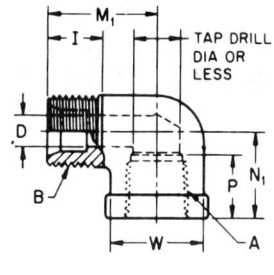

φ FIG. 2C (120203)

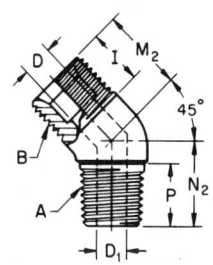

FIG. 2D (120302)

FIG. 2—ELBOWS

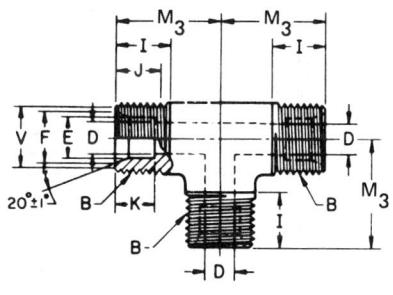

φ FIG. 3A (120401)

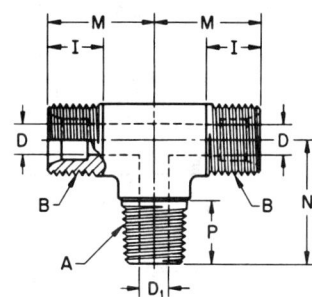

FIG. 3B (120425)

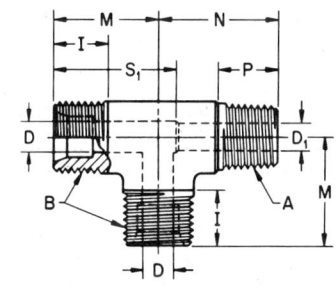

FIG. 3C (120424)

FIG. 3—TEES

NOTES: UNSPECIFIED DETAIL WITH RESPECT TO DIMENSIONS, TOLERANCES, CONTOURS, MATERIAL, WORKMANSHIP, ETC., MUST CONFORM TO GENERAL SPECIFICATIONS FOR SPHERICAL AND FLANGED SLEEVE (COMPRESSION) TUBE FITTINGS. THE DIMENSIONAL DESIGNATIONS ON THE FIRST FIGURE IN EACH GROUP SHALL APPLY TO ALL OTHER FIGURES IN THAT GROUP EXCEPT AS SHOWN OTHERWISE. CODES SHOWN IN BRACKETS ADJACENT TO FIGURE NUMBERS REPRESENT RESPECTIVE FITTING IDENTIFI-
ed. CATION IN ACCORDANCE WITH SAE J846 (MAY, 1978).

equipment made to the inch system of measurement, conversion of size designations and dimensions to SI units is considered unnecessary.)

Where external threads are produced by roll threading and the body is not undercut, the unthreaded portion of body adjacent to the shoulder may be reduced to the minimum pitch diameter.

External threads shall be chamfered to the diameters specified to produce a length of chamfered or partial thread equivalent to $3/4$ to $1\,1/4$ times the pitch (rounded to a three-place decimal).

Internal threads shall be countersunk 90 deg included angle to the diameters specified.

TABLE 1—STRAIGHT THREAD SPECIFICATION DATA, IN

Nominal Size	Series Desig-nation	External Thread Pitch Diameter		Internal Thread Pitch Diameter		Internal Thread Minor Diameter	
		Max	Min	Max	Min[a]	Max	Min
7/16–24	UNS	0.4093	0.4055	0.4153	0.4104	0.402	0.392
1/2 –24	UNS	0.4717	0.4678	0.4780	0.4729	0.465	0.455
17/32–24	UNS	0.5030	0.4991	0.5092	0.5041	0.496	0.486
11/16–20	UN	0.6537	0.6494	0.6606	0.6550	0.645	0.633
13/16–18	UNS	0.7750	0.7704	0.7824	0.7764	0.765	0.752
1 –18	UNS	0.9625	0.9578	0.9701	0.9639	0.953	0.940
1-1/4 –16	UN	1.2079	1.2028	1.2160	1.2094	1.196	1.182

[a] These values are also the basic pitch diameter.

TABLE 2

Nom Pipe Thread Size, in	External Thread Chamfer Diameter[a]				External Thread Length of Chamfered or Partial Thread				Internal Thread Countersink Diameter[a]			
	Max		Min		Min		Max		Min		Max	
	in	mm	in	mm	in	mm	in	mm	in	mm	in	mm
1/8	0.32	8.1	0.30	7.6	0.037	0.94	0.056	1.42	0.42	10.7	0.44	11.2
1/4	0.42	10.7	0.40	10.2	0.056	1.42	0.083	2.11	0.55	14.0	0.57	14.5
3/8	0.55	14.0	0.53	13.5	0.056	1.42	0.083	2.11	0.69	17.5	0.71	18.0
1/2	0.68	17.3	0.66	16.8	0.071	1.80	0.107	2.72	0.85	21.6	0.87	22.1
3/4	0.89	22.6	0.87	22.1	0.071	1.80	0.107	2.72	1.06	26.9	1.08	27.4
1	1.12	28.4	1.09	27.7	0.087	2.21	0.130	3.30	1.34	34.0	1.37	34.8

ed. [a] Tabulated diameters conform with Appendix A, SAE J476 (June, 1961).

Thread Eccentricity Tolerances—The various thread elements of Class 2A external and Class 2B internal straight threads on tube fittings shall be concentric within the limitations specified in the General Specifications of ed. SAE J512 (July, 1976).

Pipe Threads—Taper pipe threads, unless there is specific authorization to the contrary, shall conform to the Dryseal American Standard Taper Pipe Thread (NPTF). Specifications for pipe threads are given in detail in SAE ed. J476 (June, 1971).

The length of full form external thread shall not be shorter than L_2 plus one pitch (thread), except that where thread is cut through into a relieved body or undercut on the fitting, the minimum full threaded length may be reduced by one pitch (thread).

Where external pipe threads are produced by roll threading, the diameter of the unthreaded portion of shank adjacent to shoulder may be reduced to the E_2 basic pitch diameter.

The tube fitting dimensions tabulated herein are based on length of the Dryseal American Standard Taper Pipe Thread (NPTF), it being the consensus of manufacturers and users that trouble-free assembly cannot be assured unless a full length thread is used. However, the tap drill depths and overall lengths specified in the tables for fittings with internal taper pipe threads are

TABLE 3—DIMENSIONS OF CONNECTORS, UNIONS, ELBOWS, AND TEES (FIGS. 1A–3C)

Nom Tube OD, in	A[a] Dryseal Taper Thread in	B, Nom Thread Size, in Class 2A Ext.	C Nom, in	C_1 Nom, in	C_2 Nom, in	D Dia Drill in	D mm	D_1 Dia Drill in	D_1 mm	E Dia in +0.004 −0.000	E mm +0.10 −0.00	F Dia in +0.004 −0.000	F mm +0.10 −0.00	G[b] Dia in +0.000 −0.010	G mm +0.00 −0.25
1/4	1/8	7/16-24	7/16	7/16	9/16	0.188	4.78	0.188	4.78	0.254	6.45	0.311	7.90	0.392	9.96
5/16	1/8	1/2 -24	1/2	1/2	9/16	0.250	6.35	0.188	4.78	0.318	8.08	0.378	9.60	0.454	9.96
3/8	1/4	17/32-24	9/16	9/16	11/16	0.312	7.92	0.312	7.92	0.382	9.70	0.448	11.38	0.486	12.34
1/2	3/8	11/16-20	11/16	11/16	7/8	0.406	10.31	0.406	10.31	0.507	12.88	0.577	14.66	0.633	16.08
5/8	1/2	13/16-18	7/8	13/16	1-1/16	0.531	13.49	0.531	13.49	0.632	16.05	0.709	18.01	0.752	19.10
3/4	1/2	1 -18	1	1	1-1/16	0.656	16.66	0.531	13.49	0.758	19.25	0.850	21.59	0.939	23.85
3/4	3/4	1 -18	1-1/16	1-1/4	1-1/4	0.656	16.66	0.750	19.05	0.758	19.25	0.850	21.59	0.939	23.85
1	1	1-1/4 -16	1-3/8	1-1/4	1-5/8	0.875	22.22	0.875	22.22	1.008	25.60	1.122	28.50	1.182	30.02

Nom Tube OD, in	I[c] Min in	I Min mm	J Full Thread Min in	J mm	K in	K mm	L in ±0.03	L mm ±0.8	L_1 in ±0.03	L_1 mm ±0.8	L_2 in ±0.03	L_2 mm ±0.8	M in ±0.03	M mm ±0.8	M_1 in ±0.03	M_1 mm ±0.8
1/4	0.33	8.4	0.27	6.9	0.25	6.4	0.90	22.9	0.85	21.6	0.85	21.6	0.63	16.0	0.70	17.8
5/16	0.34	8.6	0.28	7.1	0.28	7.1	0.94	23.9	0.90	22.9	0.86	21.8	0.67	17.0	0.70	17.8
3/8	0.44	11.2	0.38	9.7	0.31	7.9	1.22	31.0	1.10	27.9	1.19	30.2	0.80	20.3	0.90	22.9
1/2	0.53	13.5	0.44	11.2	0.44	11.2	1.34	34.0	1.31	33.3	1.28	32.5	0.94	23.9	1.04	26.4
5/8	0.59	15.0	0.50	12.7	0.44	11.2	1.59	40.4	1.43	36.3	1.51	38.4	1.10	27.9	1.20	30.5
3/4	0.66	16.8	0.53	13.5	0.56	14.2	1.69	42.9	1.60	40.6	1.58	40.1	1.20	30.5	1.23	31.2
3/4	0.66	16.8	0.53	13.5	0.56	14.2	1.72	43.7	—	—	1.63	41.4	1.28	32.5	1.36	34.5
1	0.72	18.3	0.59	15.0	0.78	19.8	2.00	50.8	1.78	45.2	1.91	48.5	1.44	36.6	1.59	40.4

Nom Tube OD, in	M_2 in ±0.03	M_2 mm ±0.8	M_3 in ±0.03	M_3 mm ±0.8	N in ±0.03	N mm ±0.8	N_1 in ±0.03	N_1 mm ±0.8	N_2 in ±0.03	N_2 mm ±0.8	P in	P mm	S Max in	S Max mm
1/4	0.50	12.7	0.63	16.0	0.67	17.0	0.54	13.7	0.64	16.3	0.38	9.7	—	—
5/16	0.54	13.7	0.67	17.0	0.70	17.8	0.57	14.5	0.64	16.3	0.38	9.7	0.45	11.4
3/8	0.72	18.3	0.80	20.3	0.93	23.6	0.78	19.8	0.86	21.8	0.56	14.2	—	—
1/2	0.85	21.6	0.94	23.9	1.00	25.4	0.83	21.1	0.95	24.1	0.56	14.2	—	—
5/8	0.94	23.9	1.10	27.9	1.25	31.8	1.08	27.4	1.17	29.7	0.75	19.1	—	—
3/4	1.12	28.4	1.25	31.8	1.34	34.0	1.14	29.0	1.20	30.5	0.75	19.1	0.80	20.3
3/4	1.20	30.5	—	—	1.34	34.0	1.14	29.0	1.20	30.5	0.75	19.1	0.86	21.8
1	1.27	32.3	1.44	36.6	1.66	42.2	1.44	36.6	1.48	37.6	0.94	23.9	—	—

Nom Tube OD, in	S_1 Min in	S_1 Min mm	T[d] Ref in	T Ref mm	T_1 Min in	T_1 Min mm	V Dia Max in	V Dia Max mm	V Dia Min in	V Dia Min mm	W[e] Forged or Bar Min in +0.00 −0.02	W mm +0.0 −0.5
1/4	—	—	0.18	4.6	0.18	4.6	0.38	9.7	0.36	9.1	0.56	14.2
5/16	0.78	19.8	0.21	5.3	0.21	5.3	0.44	11.2	0.42	10.7	0.56	14.2
3/8	—	—	0.21	5.3	0.21	5.3	0.48	12.2	0.46	11.7	0.69	17.5
1/2	—	—	0.24	6.1	0.24	6.1	0.62	15.7	0.61	15.5	0.81	20.6
5/8	—	—	0.24	6.1	0.24	6.1	0.73	18.5	0.72	18.3	1.00	25.4
3/4	1.50	38.1	0.27	6.9	0.27	6.9	0.92	23.4	0.90	22.9	1.00	25.4
3/4	1.72	43.7	0.30	7.6	0.30	7.6	0.92	23.4	0.90	22.9	1.25	31.8
1	—	—	0.33	8.4	0.33	8.4	1.16	29.5	1.14	29.0	1.50	38.1

[a] Dryseal American Standard Taper Pipe Thread, except as noted in General Specifications.
[b] Where thread relief undercut is used, last thread shall be chamfered 1/2 to 1 pitch long from G diameter. Thread marks on surface of undercut shall be permissible.
[c] For elbows and tees, the 1 dimensions shown shall apply to turned or finished length. Where body is relieved beyond thread, length of turned boss 1 may be reduced to the minimum full thread length J.
[d] Minimum design thickness, not subject to inspection.
[e] Basic dimensions shown shall apply as min dia or across flats for bosses. The −0.02 in (−0.5 mm) tolerance shall apply only to chamfer diameter on full hexagon version in Fig. 1C.

TABLE 3A—DIMENSIONS OF SPECIAL SIZE FITTINGS,[b]
CONNECTORS, ELBOWS, AND TEES (FIGS. 1A, 1C, 2A, 2C, 2D, 3B, AND 3C)

Nom Tube OD, in	A[a] Dryseal Taper Thread, in	B, Nom Thread Size, in Class 2A Ext.	C Nom, in	C₂ Nom, in	L in ±0.03	L mm ±0.8	L₂ in ±0.03	L₂ mm ±0.8	M in ±0.03	M mm ±0.8	M₁ in ±0.03	M₁ mm ±0.8	M₂ in ±0.03	M₂ mm ±0.8	N in ±0.03	N mm ±0.8	N₁ in ±0.03	N₁ mm ±0.8	N₂ in ±0.03	N₂ mm ±0.8	W in +0.00 −0.02	W mm +0.0 −0.5
1/4	1/4	7/16-24	9/16	11/16	1.11	28.2	1.08	27.4	0.69	17.5	0.79	20.1	0.61	15.5	0.88	22.4	0.72	18.3	0.86	21.8	0.69	17.5
1/4	3/8	7/16-24	11/16	7/8	1.14	29.0	1.08	27.4	0.74	18.8	0.84	21.3	0.65	16.5	0.87	22.1	0.72	18.3	0.95	24.1	0.81	20.6
1/4	1/2	7/16-24	7/8	1-1/16	1.33	33.8	1.25	31.8	0.84	21.3	0.94	23.9	0.68	17.3	1.06	26.9	0.91	23.1	1.17	29.7	1.00	25.4
5/16	1/4	1/2-24	9/16	11/16	1.12	28.4	1.09	27.7	0.70	17.8	0.80	20.3	0.62	15.7	0.91	23.1	0.75	19.0	0.86	21.8	0.69	17.5
5/16	3/8	1/2-24	11/16	7/8	1.15	29.2	1.09	27.7	0.75	19.0	0.85	26.6	0.66	16.8	0.91	23.1	0.75	19.0	0.95	24.1	0.81	20.6
5/16	1/2	1/2-24	7/8	1-1/16	1.34	34.0	1.26	32.0	0.85	21.6	0.95	24.1	0.69	17.5	1.09	27.7	0.94	23.9	1.17	29.7	1.00	25.4
3/8	1/8	17/32-24	9/16	9/16	1.04	26.4	1.01	25.7	0.73	18.5	0.84	21.3	0.72	18.3	0.75	19.0	0.60	15.2	0.68	17.3	0.56	14.2
3/8	3/8	17/32-24	11/16	7/8	1.25	31.8	1.19	30.2	0.85	21.6	0.95	24.1	0.76	19.3	0.92	23.4	0.78	19.8	0.95	24.1	0.81	20.6
3/8	1/2	17/32-24	7/8	1-1/16	1.44	36.6	1.36	34.5	0.95	24.1	1.05	26.7	0.79	20.1	1.11	28.2	0.97	24.6	1.17	29.7	1.00	25.4
1/2	1/8	11/16-20	11/16	11/16	1.16	29.5	1.10	27.9	0.80	20.3	0.91	23.1	0.85	21.6	0.82	20.8	0.65	16.5	0.77	19.6	0.56	14.2
1/2	1/4	11/16-20	11/16	11/16	1.34	34.0	1.28	32.5	0.87	22.1	0.97	24.6	0.85	21.6	1.00	25.4	0.83	21.1	0.95	24.1	0.69	17.5
1/2	1/2	11/16-20	7/8	1-1/16	1.53	38.9	1.45	36.8	1.04	26.4	1.14	29.0	0.88	22.4	1.19	30.2	1.02	25.9	1.17	29.7	1.00	25.4
1/2	3/4	11/16-20	1-1/16	1-1/4	1.59	40.4	1.50	38.1	1.15	29.2	1.23	31.2	1.07	27.2	1.18	30.0	1.01	25.7	1.20	30.5	1.25	31.8
1/2	1	11/16-20	1-3/8	1-5/8	1.81	46.0	1.72	43.7	1.25	31.8	1.40	35.6	1.08	27.4	1.38	35.1	1.21	30.7	1.48	37.6	1.50	38.1
5/8	1/4	13/16-18	13/16	13/16	1.40	35.6	1.32	33.5	0.95	24.1	1.04	26.4	0.94	23.9	1.06	26.9	0.89	22.6	0.98	24.8	0.69	17.5
5/8	3/8	13/16-18	13/16	7/8	1.40	35.6	1.32	33.5	1.01	25.7	1.10	27.9	0.94	23.9	1.06	26.9	0.89	22.6	0.98	24.8	0.81	20.6
5/8	3/4	13/16-18	1-1/16	1-1/4	1.65	41.9	1.56	39.6	1.21	30.7	1.29	32.8	1.13	28.7	1.25	31.8	1.08	27.4	1.20	30.5	1.25	31.8
5/8	1	13/16-18	1-3/8	1-5/8	1.87	47.5	1.78	45.2	1.31	33.3	1.46	37.1	1.14	29.0	1.44	36.6	1.27	32.3	1.48	37.6	1.50	38.1
3/4	1/4	1-18	1	1	1.50	38.1	1.39	35.3	1.05	26.7	1.07	27.2	1.12	28.4	1.15	29.2	0.95	24.1	1.01	25.7	0.69	17.5
3/4	3/8	1-18	1	1	1.50	38.1	1.39	35.3	1.10	27.9	1.13	28.7	1.12	28.4	1.15	29.2	0.95	24.1	1.01	25.7	0.81	20.6
3/4	1	1-18	1-3/8	1-5/8	1.94	49.3	1.85	47.0	1.38	35.1	1.53	38.9	1.21	30.7	1.53	38.9	1.33	33.8	1.48	37.6	1.50	38.1
1	1/2	1-1/4-16	1-1/4	1-1/4	1.81	46.0	1.72	43.7	1.20	30.5	1.34	34.0	1.27	32.3	1.47	37.3	1.25	31.8	1.29	32.8	1.00	25.4
1	3/4	1-1/4-16	1-1/4	1-1/4	1.81	46.0	1.72	43.7	1.31	33.3	1.46	37.1	1.27	32.3	1.47	37.3	1.25	31.8	1.29	32.8	1.25	31.8

[a] Dryseal American Standard Taper Pipe Thread, except as noted in General Specifications.
[b] Dimensions of the sizes listed are derived through the use of the appendix "Table for Calculating Dimensions of Special Sizes".

not consistent with the tap drill depths and overall thread lengths of the Dryseal American Standard Taper Pipe Threads (NPTF) given in Table A2, Appendix A of SAE J476. The full length Dryseal American Standard Taper Pipe Taps specified in Table B2 of SAE J476 cannot be used, as the tap drill depths and overall lengths of the fittings have been reduced to the minimum required by bottoming taps to produce standard full thread length. The deviations described above are peculiar to these tube fittings and as special tooling is required, caution should be exercised in specifying such deviations for any other products.

External pipe threads shall be chamfered from the diameters tabulated in Table 2 to produce the specified length of chamfered or partial thread. Internal pipe threads shall be countersunk 90 deg, included angle, to the diameters tabulated in Table 2.

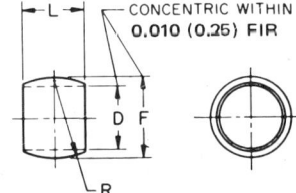

FIG. 4—SPHERICAL SLEEVES (120115)

NOTE: DIMENSIONS ARE IN (mm)

Material and Manufacture—Fittings shall be made from brass and, at manufacturer's option and in accordance with his process of manufacture, may be milled from the bar and forged. Brass shall be UNS C36000 (half-hard), UNS C34500, or UNS C35000 for bar or extruded stock and UNS C37700 for forgings. Sleeves shown in Fig. 4 shall be made from UNS C36000 brass and annealed to a maximum hardness of Rockwell F70. Sleeves shown in Fig. 10 shall be made from UNS C36000 brass. Tube supports shown in Fig. 9 shall be made from UNS C26000 or 300 series stainless steel, according to manufacturer's process standard. Nuts may be made from steel when so specified by the purchaser.

Finish—Steel nuts shall be furnished cadmium plated or zinc plated to a thickness of 0.0002 in. (0.005 mm) minimum followed by a chromate treatment. Plated nuts must meet the requirements of 32 h salt spray test in accordance with ASTM B 117.

Workmanship—Workmanship shall conform to the best commercial practice to produce high quality fittings. Fittings shall be free from all machining fluids, chips, hanging burrs, loose scale, and slivers which might become dislodged in usage and all other defects which might affect their serviceability. All seating surfaces for the spherical sleeve must be smooth except that annular tool marks up to 100 μin. (2.5 μm) maximum shall be permissible.

NOTES: UNSPECIFIED DETAIL WITH RESPECT TO DIMENSIONS, TOLERANCES, CONTOUR, MATERIAL, WORKMANSHIP, ETC., MUST CONFORM TO GENERAL SPECIFICATIONS FOR SPHERICAL AND FLANGED SLEEVE (COMPRESSION) TUBE FITTINGS. CODES SHOWN IN BRACKETS ADJACENT TO FIGURE NUMBERS REPRESENT RESPECTIVE FITTING IDENTIFICATION IN ACCORDANCE WITH SAE J846 (MAY, 1978).

TABLE 4—DIMENSIONS OF SPHERICAL SLEEVES (FIG. 4)

Nominal Tube OD, in	D Dia in +0.004 −0.000	D Dia mm +0.10 −0.00	F Dia in +0.004 −0.000	F Dia mm +0.10 −0.00	L in ±0.005	L mm ±0.13	R Radius in +0.004 −0.000	R Radius mm +0.10 −0.00
1/4	0.253	6.43	0.320	8.13	0.250	6.35	0.312	7.92
5/16	0.316	8.03	0.386	9.80	0.281	7.14	0.358	9.09
3/8	0.380	9.65	0.459	11.66	0.313	7.95	0.404	10.26
1/2	0.505	12.83	0.592	15.04	0.375	9.52	0.498	12.65
5/8	0.630	16.00	0.732	18.59	0.438	11.13	0.593	15.06
3/4	0.756	19.20	0.872	22.15	0.500	12.70	0.686	17.42
1	1.006	25.55	1.150	29.21	0.626	15.90	0.873	22.17

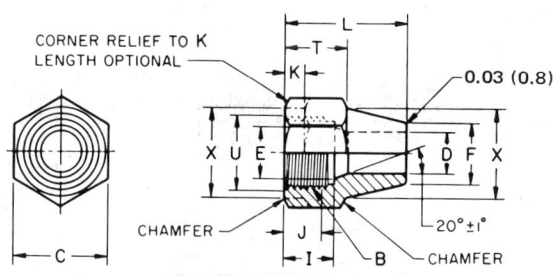

NOTE: DIMENSIONS ARE IN (mm)

FIG. 5—SPHERICAL SLEEVE NUTS (120111)

TABLE 5—DIMENSIONS OF SPHERICAL SLEEVE NUTS (FIG. 5)

Nom Tube OD, in	B, Nom Thread Size, in Class 2B Int	C Nom, in	D Dia		E		F Dia		I	
			in +0.004 −0.000	mm +0.10 −0.00	in +0.004 −0.000	mm +0.10 −0.00	in	mm	in	mm
1/4	7/16-24	9/16	0.254	6.45	0.311	7.90	0.38	9.7	0.31	7.9
5/16	1/2 -24	5/8	0.318	8.08	0.378	9.60	0.44	11.2	0.31	7.9
3/8	17/32-24	5/8	0.382	9.70	0.448	11.38	0.47	11.9	0.41	10.4
1/2	11/16-20	13/16	0.507	12.88	0.582	14.78	0.63	16.0	0.50	12.7
5/8	13/16-18	15/16	0.632	16.05	0.714	18.14	0.77	19.6	0.56	14.2
3/4	1 -18	1-1/8	0.758	19.25	0.850	21.59	0.92	23.4	0.62	15.7
1	1-1/4 -16	1-3/8	1.008	25.60	1.122	28.50	1.19	30.2	0.69	17.5

Nom Tube OD, in	J Full Thread Min		K		L		T		U Dia		X Dia Min	
	in	mm	in	mm	in	mm	in	mm	in +0.02 −0.00	mm +0.5 −0.0	in	mm
1/4	0.25	6.4	0.13	3.3	0.75	19.0	0.38	9.7	0.45	11.4	0.53	13.5
5/16	0.25	6.4	0.13	3.3	0.88	22.4	0.50	12.7	0.51	13.0	0.59	15.0
3/8	0.35	8.9	0.13	3.3	1.13	28.7	0.53	13.5	0.54	13.7	0.61	15.5
1/2	0.42	10.7	0.13	3.3	1.25	31.8	0.63	16.0	0.70	17.8	0.78	19.8
5/8	0.48	12.2	0.19	4.8	1.38	35.1	0.72	18.3	0.83	21.1	0.91	23.1
3/4	0.54	13.7	0.19	4.8	1.56	39.6	0.75	19.0	1.02	25.9	1.09	27.7
1	0.60	15.2	0.19	4.8	1.69	42.9	0.94	23.9	1.27	32.3	1.34	34.0

FLANGED SLEEVE TYPE

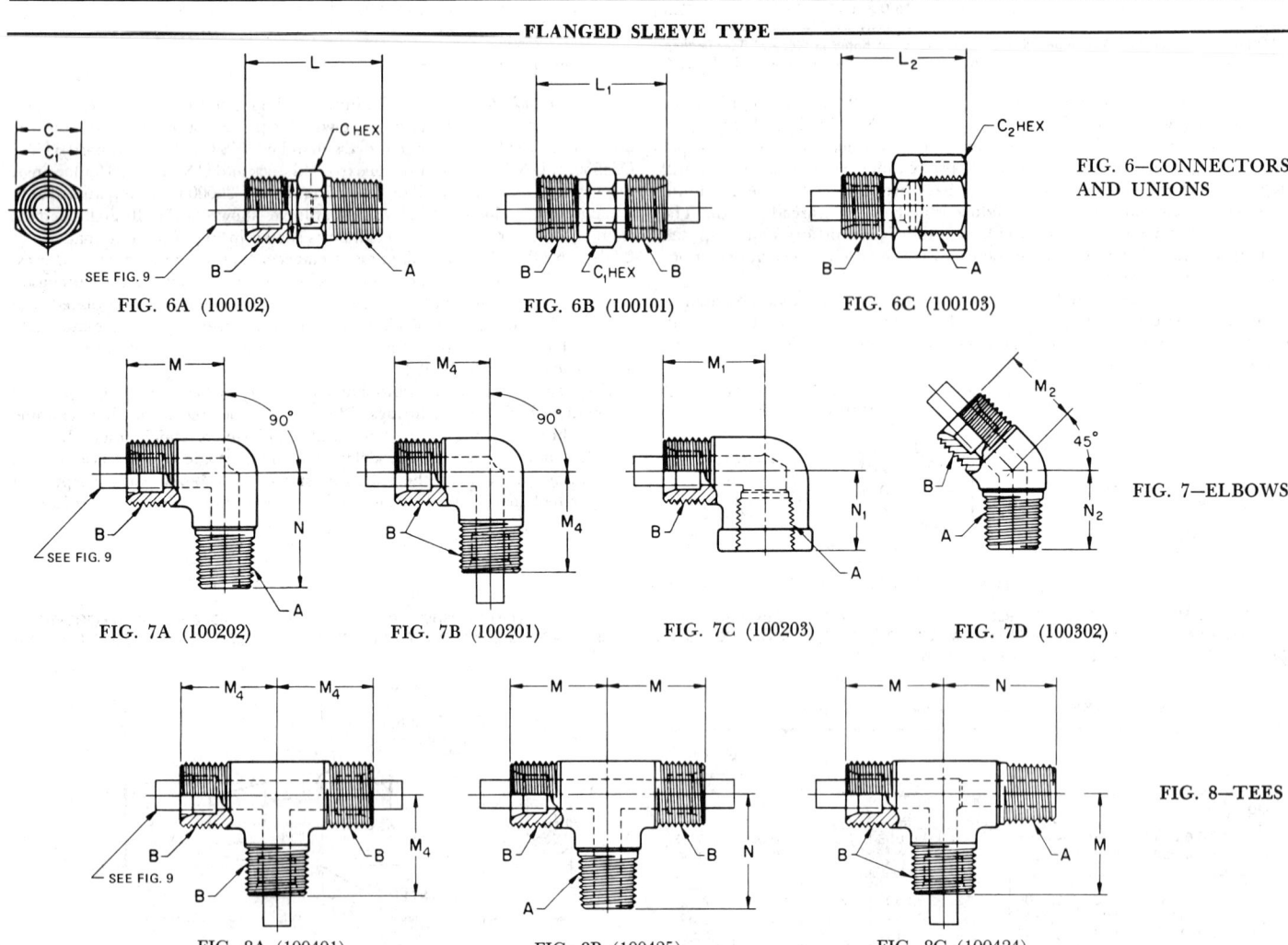

FIG. 6—CONNECTORS AND UNIONS
FIG. 6A (100102)   FIG. 6B (100101)   FIG. 6C (100103)

FIG. 7—ELBOWS
FIG. 7A (100202)   FIG. 7B (100201)   FIG. 7C (100203)   FIG. 7D (100302)

FIG. 8—TEES
FIG. 8A (100401)   FIG. 8B (100425)   FIG. 8C (100424)

ed.   NOTES: FLANGED SLEEVE TYPE FITTINGS HAVING TUBE SUPPORT FEATURES (FIGS. 6A–8C) ARE INTENDED ONLY FOR USE WITH NYLON TUBING PER SAE J844 (JULY, 1976) IN CONJUNCTION WITH FLANGED SLEEVE IN FIG. 10 AND FLANGED SLEEVE NUT IN FIG. 11.
UNSPECIFIED DETAIL WITH RESPECT TO DIMENSIONS, TOLERANCES, CONTOURS, MATERIAL, WORKMANSHIP, ETC., MUST CONFORM TO GENERAL SPECIFICATIONS FOR SPHERICAL AND FLANGED SLEEVE (COMPRESSION) TUBE FITTINGS. ALL DIMENSIONS FOR FIGS. 6A–8C SHALL CORRESPOND WITH RESPECTIVE FIGS. 1A–3C AND TABLE 3 EXCEPT FOR ADDITION OF THE TUBE SUPPORT SHOWN IN FIG. 9 AND TABLE 6. CODES SHOWN IN BRACKETS ADJACENT TO FIGURE NUMBERS REPRESENT RESPECTIVE FITTING IDENTIFICATION IN ACCORDANCE
ed.   WITH SAE J846 (MAY, 1978).

FIG. 9—TUBE SUPPORT

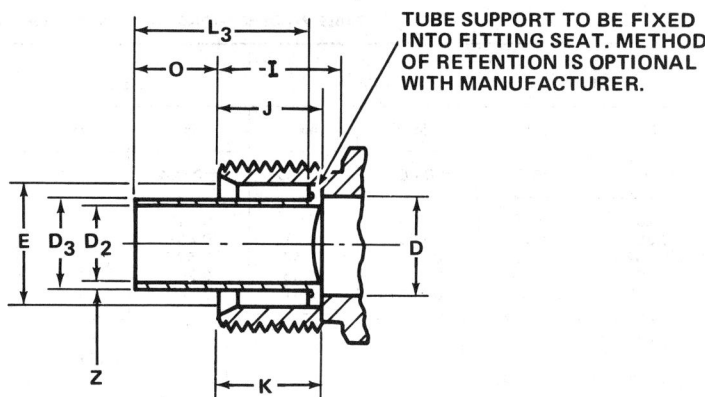

TUBE SUPPORT TO BE FIXED INTO FITTING SEAT. METHOD OF RETENTION IS OPTIONAL WITH MANUFACTURER.

TABLE 6—DIMENSIONS OF TUBE SUPPORT (FIG. 9a)

Nom Tube OD, in	D_2 Dia Min		D_3 Dia				L_3 Min		O				Z			
			Min		Max				Min		Max		Min		Max	
	in	mm	in	mm	in	mm	in	mm	in	mm	in	mm	in	mm	in	mm
1/4	0.123	3.12	0.161	4.09	0.165	4.19	0.45	11.4	0.24	6.1	0.30	7.6	0.012	0.30	0.017	0.43
3/8	0.205	5.21	0.243	6.17	0.247	6.27	0.57	14.5	0.29	7.4	0.35	8.9	0.013	0.33	0.017	0.43
1/2	0.326	8.28	0.368	9.35	0.372	9.45	0.75	19.0	0.34	8.6	0.40	10.2	0.013	0.33	0.019	0.48
5/8	0.384	9.75	0.432	10.97	0.436	11.07	0.79	20.1	0.40	10.2	0.46	11.7	0.016	0.41	0.022	0.56
3/4	0.501	12.73	0.557	14.15	0.561	14.25	0.97	24.6	0.46	11.7	0.52	13.2	0.016	0.41	0.026	0.66

[a] For dimensions depicted on Fig. 9 but not shown in Table 6, refer to Table 3.

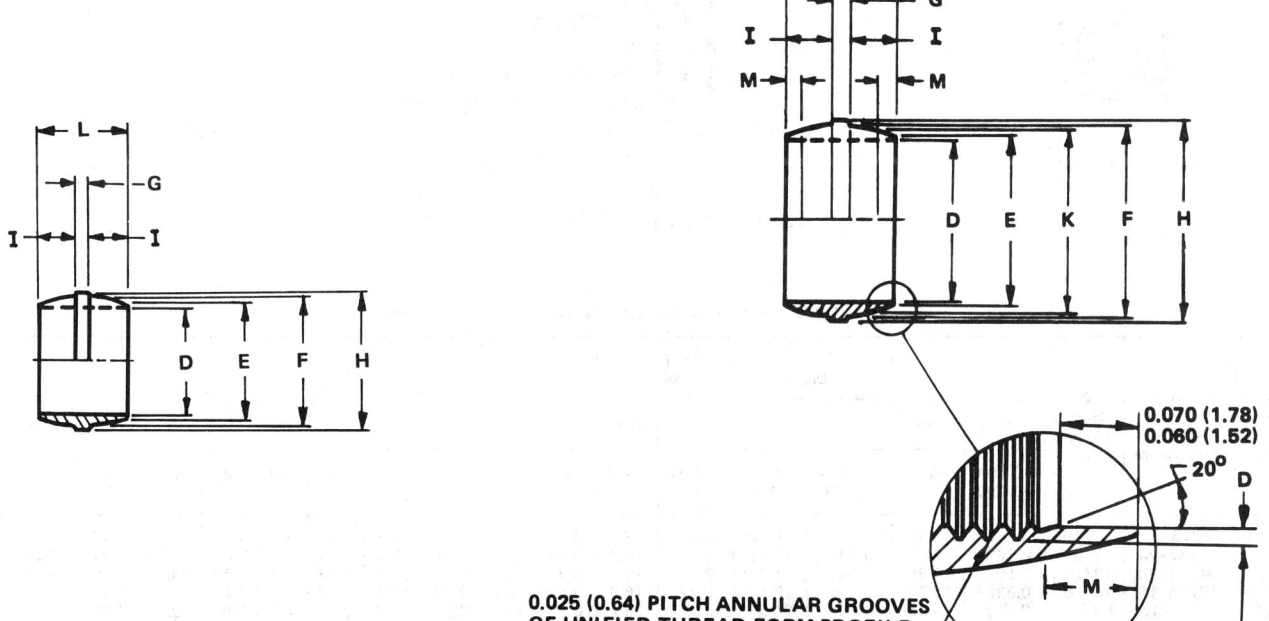

FIG. 10A—FLANGED SLEEVE, 1/4, 3/8 AND 1/2 IN SIZES (100115)

FIG. 10B—FLANGED SLEEVE, 5/8 AND 3/4 IN SIZES (100115)

NOTES: UNSPECIFIED DETAIL WITH RESPECT TO DIMENSIONS, TOLERANCES, CONTOUR, MATERIAL, WORKMANSHIP, ETC., MUST CONFORM TO GENERAL SPECIFICATIONS FOR SPHERICAL AND FLANGED SLEEVE (COMPRESSION) TUBE FITTINGS. CODES SHOWN IN BRACKETS ADJACENT TO FIGURE NUMBERS REPRESENT RESPECTIVE IDENTIFICATION IN ACCORDANCE WITH SAE J846 (MAY, 1978).

TABLE 7—DIMENSIONS OF FLANGED SLEEVES (FIGS. 10A-10B)

Nom Tube OD, in	D Dia		E Dia		F Dia		G Ref		H Dia			
	in	mm	in	mm	in	mm	in	mm	Max		Min	
	±0.002	±0.05	±0.002	±0.05	±0.002	±0.05			in	mm	in	mm
1/4	0.256	6.50	0.276	7.01	0.319	8.10	0.050	1.27	0.364	9.25	0.354	8.99
3/4	0.384	9.75	0.406	10.31	0.458	11.63	0.060	1.52	0.481	12.22	0.477	12.12
1/2	0.509	12.93	0.538	13.67	0.595	15.11	0.060	1.52	0.628	15.95	0.620	15.75
5/8	0.634	16.10	0.644	16.36	0.721	18.31	0.080	2.03	0.748	19.00	0.744	18.90
3/4	0.760	19.30	0.772	19.61	0.864	21.95	0.090	2.29	0.927	23.55	0.917	23.90

Nom Tube OD, in	I				K Dia		L		M			
	Max		Min		in	mm	in	mm	in	mm		
	in	mm	in	mm	±0.002	±0.05	±0.005	±0.13	±0.005	±0.13		
1/4	0.125	3.18	0.120	3.05	—	—	0.295	7.49	—	—		
3/8	0.170	4.32	0.160	4.06	—	—	0.390	9.91	—	—		
1/2	0.190	4.83	0.180	4.57	—	—	0.430	10.92	—	—		
5/8	0.210	5.33	0.200	5.08	0.684	17.37	0.490	12.45	0.055	1.40		
3/4	0.230	5.84	0.220	5.59	0.827	21.01	0.540	13.72	0.075	1.90		

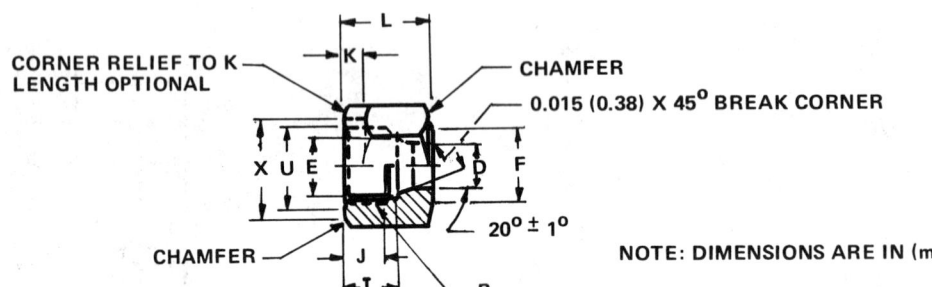

FIG. 11—FLANGED SLEEVE NUTS (100110)

TABLE 8—DIMENSIONS OF FLANGED SLEEVE NUTS (FIG. 11)

Nom Tube OD, in	B, Nom Thread Size, in Class 2B Int	C Hex Nom, in	D Dia		E Dia		F Dia		I		J Full Thd Min		K		L		U Dia		X Dia Min	
			in	mm	in	mm	in	mm	in	mm	in	mm	in	mm	in	mm	in	mm	in	mm
			+0.004 -0.000	+0.10 -0.00	+0.004 -0.000	+0.10 -0.00	±0.02	±0.5									+0.02 -0.00	+0.5 -0.0		
1/4	7/16-24	9/16	0.254	6.45	0.311	7.90	0.34	8.6	0.31	7.9	0.25	6.4	0.13	3.3	0.45	11.4	0.45	11.4	0.53	13.5
3/8	17/32-24	5/8	0.382	9.70	0.448	11.38	0.47	11.9	0.44	11.2	0.38	9.7	0.13	3.3	0.63	16.0	0.54	13.7	0.61	15.5
1/2	11/16-20	13/16	0.507	12.88	0.582	14.78	0.61	15.5	0.51	13.0	0.44	11.2	0.13	3.3	0.72	18.3	0.70	17.8	0.78	19.8
5/8	13/16-18	15/16	0.632	16.05	0.714	18.14	0.74	18.8	0.58	14.7	0.50	12.7	0.19	4.8	0.77	19.6	0.83	21.1	0.91	23.1
3/4	1-18	1-1/8	0.758	19.25	0.850	21.59	0.88	22.4	0.62	15.7	0.54	13.7	0.19	4.8	0.81	20.6	1.02	25.9	1.09	27.7

APPENDIX
TABLES FOR CALCULATING DIMENSIONS ON SPECIAL SIZES

The tables in this Appendix present various factors to be used in determining the dimensions applicable to special size combination fittings not contained in SAE J246 (July, 1977). (See Table 3A for calculated sizes.)

For any special size fitting, be it a connector, 45 or 90 deg elbow, tee, or cross, one end is always standard. Considering this end to be the largest on the fitting, it may then be used as the basis for establishing the stock size and the length (either overall or end to center) by deducting factors equivalent to the reduction in machining requirements from the appropriate standard lengths.

The factors applicable to the various end configurations and size reductions tabulated below were determined on the following basis:

Those pertaining to lengths were derived by maintaining the standard hexagon thickness for straight fittings and the standard centerline to machining start for shaped fittings.

Tables A1 and A2 were derived by substracting the standard machining length required for the smaller end from that required for the larger end.

Table A3 factors are equal to one-half the difference in tube end thread diameters.

Table A4 factors are equal to one-half the difference in pipe end thread dimensions.

Table A5 factors are equal to one-half the difference in the tube end D drill diameter.

Table A6 factors are equal to one-half the difference in the pipe end W diameter or width.

Straight, Tube Size Reduced

$L(L1, L2)$ special size = $L(L1, L2)$ std size − factor from Table A2

EXAMPLE: For a straight connector (Figs. 1A and 1C) with 3/8 in. tube and 3/8 in. NPTF, the overall length would be determined as follows:

1.34 in. (34.0 mm) = L overall length for 1/2 tube to 3/8 NPTF from Table 1
−0.09 in. (2.3 mm) = factor from Table A2 for 3/8 machining on 1/2 size fitting 1.25 in. (31.8 mm) = overall length

The proper hex size will be the larger of the values for the tube size and external pipe size shown in Tables A7 and A8, respectively. The 3/8 in. tube to 3/8 in. NPTF fitting's hex size will (from Table A8) be 11/16 in.

Straight, Pipe Size Reduced

$L(L1, L2)$ special size = $L(L1, L2)$ std size − factor from Table A1

Elbows with External Pipe Threads, Tube Size Reduced

M special size = M std size − factor from Table A2
N special size = N std size − factor from Table A3

EXAMPLE: For elbows and tees (Figs. 2A, 3B, 3C) with 1/4 in. tube and 1/4 in. NPTF, the end to center length M would be derived as follows:
0.80 in. (20.3 mm) = M dimension for 3/8 tube to 1/4 NPTF from Table 1 − 0.11 in. (2.8 mm) = factor from Table A2 for 1/4 machined on 3/8 size fitting 0.69 in. (17.5 mm) = end-to-center length

TABLE A1—LENGTH FACTORS
Standard Pipe Size

Reduced Pipe Size, in	1/4		3/8		1/2		3/4		1	
	in	mm	in	mm	in	mm	in	mm	in	mm
1/8	0.18	4.6	0.18	4.6	0.38	9.7	0.38	9.7	0.56	14.2
1/4	—	—	0.00	0.0	0.19	4.8	0.19	4.8	0.38	9.7
3/8	—	—	—	—	0.19	4.8	0.19	4.8	0.38	9.7
1/2	—	—	—	—	—	—	0.00	0.0	0.19	4.8
3/4	—	—	—	—	—	—	—	—	0.19	4.8

TABLE A2—LENGTH FACTORS
Standard Machining Size

Reduced Tube Size, in	5/16		3/8		1/2		5/8		3/4		1	
	in	mm	in	mm	in	mm	in	mm	in	mm	in	mm
1/4	0.01	0.3	0.11	2.8	0.20	5.1	0.26	6.6	0.33	8.4	0.39	9.9
5/16	—	—	0.10	2.5	0.19	4.8	0.25	6.4	0.32	8.1	0.38	9.7
3/8	—	—	—	—	.09	2.3	0.15	3.8	0.22	5.6	0.28	7.1
1/2	—	—	—	—	—	—	0.06	1.5	0.13	3.3	0.19	4.8
5/8	—	—	—	—	—	—	—	—	0.07	1.8	0.13	3.3
3/4	—	—	—	—	—	—	—	—	—	—	0.06	1.5

TABLE A3—LENGTH FACTORS
Standard Machining Size

Reduced Tube Size, in	5/16		3/8		1/2		5/8		3/4		1	
	in	mm	in	mm	in	mm	in	mm	in	mm	in	mm
1/4	0.03	0.8	0.05	1.3	0.13	3.3	0.19	4.8	0.28	7.1	1.41	10.4
5/16	—	—	0.02	0.5	0.09	2.3	0.16	4.1	0.25	6.4	0.38	9.7
3/8	—	—	—	—	0.08	2.0	0.14	3.6	0.23	5.8	0.36	9.1
1/2	—	—	—	—	—	—	0.06	1.5	0.16	4.1	0.28	7.1
5/8	—	—	—	—	—	—	—	—	0.09	2.3	0.22	5.6
3/4	—	—	—	—	—	—	—	—	—	—	0.13	3.3

TABLE A4—LENGTH FACTORS
Standard Pipe Sizes

Reduced Pipe Size, in	1/4		3/8		1/2		3/4		1	
	in	mm	in	mm	in	mm	in	mm	in	mm
1/8	0.07	1.8	0.14	3.6	0.22	5.6	0.33	8.4	0.46	11.7
1/4	—	—	0.07	1.8	0.15	3.8	0.28	7.1	0.39	9.9
3/8	—	—	—	—	0.09	2.3	0.19	4.8	0.32	8.1
1/2	—	—	—	—	—	—	0.11	2.8	0.24	6.1
3/4	—	—	—	—	—	—	—	—	0.13	3.3

TABLE A5—LENGTH FACTORS
Standard Machining Size

Reduced Tube OD, in	5/16		3/8		1/2		5/8		3/4		1	
	in	mm	in	mm	in	mm	in	mm	in	mm	in	mm
1/4	0.03	0.8	0.06	1.5	0.11	2.8	0.17	4.3	0.23	5.8	0.34	8.6
5/16	—	—	0.03	0.8	0.08	2.0	0.14	3.6	0.20	5.1	0.31	7.9
3/8	—	—	—	—	0.05	1.3	0.11	2.8	0.17	4.3	0.28	7.1
1/2	—	—	—	—	—	—	0.06	1.5	0.13	3.3	0.23	5.8
5/8	—	—	—	—	—	—	—	—	0.06	1.5	0.17	4.3
3/4	—	—	—	—	—	—	—	—	—	—	0.11	2.8

TABLE A6—LENGTH FACTORS
Standard Pipe Size

Reduced Pipe Size, in	1/4		3/8		1/2		3/4		1	
	in	mm	in	mm	in	mm	in	mm	in	mm
1/8	0.06	1.5	0.13	3.3	0.22	5.6	0.35	8.9	0.47	11.9
1/4	—	—	0.07	1.8	0.16	4.1	0.29	7.4	0.41	10.4
3/8	—	—	—	—	0.10	2.5	0.22	5.6	0.35	8.9
1/2	—	—	—	—	—	—	0.13	3.3	0.25	6.4
3/4	—	—	—	—	—	—	—	—	0.13	3.3

TABLE A7—MINIMUM STOCK SIZE FOR TUBE ENDS

Nom Tube OD, in	Hexagon Width, Min in	Width over Wrench Pads, Min	
		in	mm
1/4	7/16	0.44	11.2
5/16	1/2	0.50	12.7
3/8	9/16	0.53	13.5
1/2	11/16	0.69	17.5
5/8	13/16	0.81	20.6
3/4	1	1.00	25.4
1	1-1/4	1.25	31.8

TABLE A8—MINIMUM STOCK SIZE FOR EXTERNAL PIPE ENDS

Nom Pipe Size, in	Hexagon Width, Min in	Width over Wrench Pads, Min	
		in	mm
1/8	7/16	0.41	10.4
1/4	9/16	0.55	14.0
3/8	11/16	0.69	17.5
1/2	7/8	0.85	21.6
3/4	1-1/16	1.06	26.9
1	1-3/8	1.33	33.8

TABLE A9—MINIMUM STOCK SIZE FOR INTERNAL PIPE ENDS

Nom Pipe Size, in	Hexagon Width, Min in	Width over Wrench Pads, Min	
		in	mm
1/8	9/16	0.56	14.2
1/4	11/16	0.69	17.5
3/8	7/8	0.81	20.6
1/2	1-1/16	1.00	25.4
3/4	1-1/4	1.25	31.8
1	1-5/8	1.50	38.1

The end to center length N would be derived as follows:
0.93 in. (23.6 mm) = N dimension for $\frac{3}{8}$ tube to $\frac{1}{4}$ NPTF from Table 1 − 0.05 in. (1.3 mm) = factor from Table A3 for $\frac{1}{4}$ machined on $\frac{3}{8}$ size fitting 0.88 in. (22.3 mm) = end-to-center length

The proper wrench pad width will be the larger value for the tube size and external pipe size shown in Tables A7 and A8, respectively. The wrench pad width for $\frac{1}{4}$ in. tube to $\frac{1}{4}$ in. NPTF fitting will (from Table A8) be 0.55 in. (14.0 mm).

Elbows with External Pipe Thread, Pipe Size Reduced

M special size = M std size − factor from Table A4
N special size = N std size − factor from Table A1

Elbows, All Tube Ends

M_4 reduced end = M_4 std size − factor from Table A2
M_4 end not reduced = M_4 std size − factor from Table A3

Tees with External Pipe Thread, Tube Size Reduced—Figure as two elbows. Use larger of the two figures obtained for the branch centerline-to-end dimension.

Tees with External Pipe Thread, Pipe Size Reduced—Figure as two elbows. Use larger of the two figures obtained for the branch centerline-to-end dimension.

Tees, All Tube Ends—Figure as two elbows. Use larger of the two figures obtained for the branch centerline-to-end dimension.

Elbow with Internal Threads, Tube Size Reduced

M_1 special size = M_1 std size − factor from Table A2
N_1 special size = N_1 std size − factor from Table A5

Elbows with Internal Pipe Threads, Pipe Size Reduced

M_1 special size = M_1 std size − factor from Table A6
N_1 special size = N_1 std size − factor from Table A1

45 deg Elbow, Tube Size Reduced

M_2 special size = M_2 std size − factor from Table A2
N_2 special size = N_2 std size

45 deg Elbow, Pipe Size Reduced

M_2 special size = M_2 std size
N_2 special size = N_2 std size − factor from Table A1

REFRIGERATION TUBE FITTINGS—SAE J513f

SAE Standard

Report of Parts and Fittings Division approved January 1936 and last revised by Tube, Pipe, Hose, and Lubrication Fittings Committee December 1976. Conforms in general to ANSI B70-1974, Refrigeration Flare Type Fittings. Editorial change October 1977.

GENERAL SPECIFICATIONS

Scope—This standard covers complete general and dimensional specifications for refrigeration tube fittings of the flare type specified in Figs. 1–42 and Tables 1–16. These fittings are intended for general use with flared annealed copper tubing in refrigeration applications.

Dimensions of single and double 45 deg flares on tubing to be used in conjunction with these fittings are given in Fig. 2 and Table 1 of SAE J533.

The following general specifications supplement the dimensional data contained in Tables 1–16 with respect to all unspecified detail.

Pressure Ratings and Service Limitations—Fittings covered by this standard are satisfactory for operating pressures up to 500 psi (3450 kPa) and are suitable for use in systems conducting most fluorinated hydrocarbon refrigerants. Fitting manufacturers should be consulted for recommendations.

Size Designations—Fitting sizes throughout the dimensional tables are designated by the corresponding outside diameter of the tubing for flared type or solder type tube ends and by the corresponding standard nominal pipe size for pipe thread ends.

Dimensions and Tolerances—Except for nominal sizes and thread specifications, dimensions and tolerances are given in both U. S. customary and SI units as designated. Tabulated dimensions shall apply to the finished parts, plated or otherwise processed, as specified by the purchasers. Unless otherwise specified, the maximum and minimum across flat dimensions shall be within the commercial tolerance of bar or extruded stock from which the fittings are produced. The minimum across corners dimensions of hexagons shall be 1.092 times the nominal width across flats, but shall not result in a size flat width less than 0.43 times the nominal width across flats.

Unless otherwise specified, tolerance on hole diameters designated drill in the dimensional tables shall be as tabulated below:

Drill Size Range		Tolerance			
		Plus		Minus	
in	mm	in	mm	in	mm
0.0135 thru 0.1850	0.343 thru 4.699	0.003	0.08	0.002	0.05
0.1875 thru 0.2480	4.762 thru 6.299	0.004	0.10	0.002	0.05
0.2500 thru 0.7500	6.350 thru 19.050	0.006	0.15	0.003	0.08
0.7579 thru 1.0000	19.25 thru 25.400	0.007	0.18	0.004	0.10

Tolerance on all dimensions not otherwise limited shall be ±0.010 in (0.25 mm). Fitting seats shall be concentric with the straight thread pitch diameters within 0.010 in (0.25 mm) full indicator reading (FIR). Angular tolerance on axis of ends on elbows, tees, and crosses shall be ±2.50 deg for sizes up to and including 3/8 in, and ±1.50 deg for sizes larger than 3/8 in.

Where so illustrated and not otherwise specified, hexagon corners shall be chamfered 30 ± 5 deg, to a diameter equal to the nominal width across flats, with a tolerance of −0.016 in (0.41 mm); or where design permits, corners may be chamfered to the diameter of the abutting surface provided the length of chamfer does not exceed that produced by the 30 deg chamfer previously described.

Passages—Where passages in straight fittings are machined from opposite ends, the offset at the meeting point shall not exceed 0.015 in (0.38 mm). The cross sectional area at the junction of passages in angle fittings shall be not less than that of the smallest passage. Where the passage is specified as a maximum or as tap drill diameter or less, the minimum shall be no less than the minimum diameter of the smallest passage in the fitting.

Wall Thickness—Unless otherwise designated, the wall thickness at any point on fittings shall not be less than the thickness established by the specified dimensions, tolerances, and eccentricities for inner and outer surfaces.

Contour—Details of contour shall be optional with the manufacturer, providing the tabulated dimensions are maintained and serviceability of the fittings is not impaired. Wrench flats on elbows and tees shall be optional. Where extruded or forged shapes are reduced to conserve material, the wall thickness, unless otherwise specified, shall not be less than the respective minimum values tabulated below:

Nominal Tube OD, in	Wall Thickness, min			
	Extruded Shapes[a]		Forged Shapes	
	in	mm	in	mm
3/16	0.04	1.0	0.060	1.52
1/4	0.04	1.0	0.075	1.90
5/16	0.05	1.3	0.075	1.90
3/8	0.05	1.3	0.085	2.16
1/2	0.06	1.5	0.090	2.29
5/8	0.08	2.0	0.100	2.54
3/4	0.08	2.0	0.100	2.54

[a]Applies to reduction in one plane of shape only.

Straight Threads—Unified Standard Class 2A external and Class 2B internal threads with minor diameters, where specified, modified to Class 3B limits shall apply to plain finish (unplated) fittings of all types. For externally threaded parts with additive finish, the maximum diameters of Class 2A may be exceeded by the amount of the allowance, that is, the basic diameters (Class 2A maximum diameters plus the allowance) shall apply after plating. For internally threaded parts with additive finish, the Class 2B diameters and modified minor diameters shall apply after plating.

The pitch diameter tolerance shall be the same as the corresponding diameter-pitch combination and class of the Unified fine thread series or for special diameter-pitch combinations shall be based on diameter, pitch, and a length of engagement of 9 times the pitch. See ANSI B1.1, Screw Threads.

For convenient reference, the data generally required to specify threads are given in Table 1. (Inasmuch as threads are normally produced and gaged with equipment made to the inch system of measurement, conversion of size designations and dimensions to SI units is considered unnecessary.)

Where external threads are produced by roll threading and the body is not undercut, the unthreaded portion of body adjacent to the shoulder may be reduced to the minimum pitch diameter.

External threads shall be chamfered to the diameter of abutting surfaces, or to the diameters specified, to produce a length of chamfered or partial thread equivalent to 3/4 to 1 1/4 times the pitch (rounded to a three-place decimal).

Internal threads shall be countersunk 90 deg included angle to the diameters specified in the dimensional tables.

Where external threads are produced by roll threading and the body is not undercut, the unthreaded portion of body adjacent to the shoulder may be reduced to the minimum pitch diameter.

External threads shall be chamfered from the diameter of abutting surface to produce a length of chamfered or partial thread equivalent to 3/4 to 1 1/4 times the pitch (rounded to a three-place decimal).

Internal threads shall be countersunk 90 deg included angle to the diameters specified in the dimensional tables.

Thread Eccentricity Tolerances—The various thread elements of Class 2A external and Class 2B, modified, internal threads on tube fittings shall be concentric within the limitations specified in SAE J512.

Pipe Threads—Taper pipe threads, unless there is specific authorization to the contrary, shall conform to the Dryseal American Standard Taper Pipe Thread (NPTF). Specifications for pipe threads are given in detail in SAE J476.

The length of full form external thread shall not be shorter than L_2 plus one pitch (thread), except that where thread is cut through into a relieved body or undercut on the fitting, the minimum full threaded length may be reduced by one pitch (thread).

Where external pipe threads are produced by roll threading, the diameter of the unthreaded portion of shank adjacent to shoulder may be reduced to the E_2 basic pitch diameter.

The tube fitting dimensions tabulated herein are based on length of the Dryseal American Standard Taper Pipe Thread (NPTF), it being the consensus of manufacturers and users that trouble-free assembly cannot be assured unless a full length thread is used. However, the tap drill depths and overall lengths specified in the tables for fittings with internal taper pipe threads are not consistent with the tap drill depths and overall thread lengths of the Dryseal American Standard Taper Pipe Threads (NPTF) given in Table A2, Appendix A of SAE J476. The full length Dryseal American Standard Taper Pipe Taps specified in Table B2 of SAE J476 cannot be used, as the tap drill depths and overall lengths of the fittings have been reduced to the minimum required by bottoming taps to produce standard full thread length. The deviations described above are peculiar to these tube fittings and as special tooling is required, caution should be exercised in specifying such deviations for any other products.

Straight pipe threads, where specified, shall conform to American Standard Straight Pipe Threads for Mechanical Joints (NPSM) in ANSI B2.1, Pipe Threads.

External pipe threads shall be chamfered from the diameters tabulated in Table 2 to produce the specified length of chamfered or partial thread. Internal pipe threads shall be countersunk 90 deg included angle to the diameters tabulated in Table 2.

Material and Manufacture—Fittings shall be made from SAE CA360 brass (half-hard), CA345, or CA350 brass bar or extruded shapes or from SAE CA377 brass forgings in accordance with the manufacturer's processes. Nuts may be made from SAE CA377 brass forging, or steel as specified by the purchaser. Seal bonnets and gaskets shall be made from copper conforming to SAE CA102, CA110, or CA122. As specified by purchaser, fusible metal alloys shall be supplied for temperature ranges 158–165, 203–219, or 275–290°F (70–74, 95–104, 135–143°C).

Finish—As specified by purchaser, steel nuts shall be furnished plain, cadmium or zinc plated to a thickness of 0.0002 in (0.005 mm) minimum followed by a chromate treatment, or with a phosphate coating (oil finished). Plated or coated nuts must meet the requirements of 32 or 16 h salt spray test, respectively, in accordance with ASTM B 117.

Workmanship—Workmanship shall conform to the best commercial practice to produce high quality fittings. Fittings shall be free from all hanging burrs, loose scale, and slivers which might become dislodged in usage and all other defects which might affect their serviceability. All sealing surfaces must be smooth except that annular tool marks up to 100 μin. (2.5 μm) maximum shall be permissible.

Assembly Considerations—Use of a compatible lubricant or sealant is desirable in assembling Dryseal pipe threads on refrigeration tube fittings to minimize galling and effect a pressure-tight seal.

TABLE 1—STRAIGHT THREAD SPECIFICATION DATA, IN

Nominal Size	Series Designation	External Thread		Internal Thread			
		Pitch Dia		Pitch Dia		Minor Dia[b]	
		Max	Min	Max	Min[a]	Max	Min
5/16–24	UNF	0.2843	0.2806	0.2902	0.2854	0.2754	0.2670
3/8–24	UNF	0.3468	0.3430	0.3528	0.3479	0.3372	0.3300
7/16–20	UNF	0.4037	0.3995	0.4104	0.4050	0.3916	0.3830
1/2–20	UNF	0.4462	0.4619	0.4731	0.4675	0.4537	0.4460
5/8–18	UNF	0.5875	0.5828	0.5949	0.5889	0.5730	0.5650
3/4–16	UNF	0.7079	0.7029	0.7159	0.7094	0.6908	0.6820
7/8–14	UNF	0.8270	0.8216	0.8356	0.8286	0.8068	0.7980
1-1/16–14	UNS	1.0145	1.0092	1.0230	1.0161	0.9940	0.9850

[a]These values are also the basic pitch diameter.
[b]Class 3B minor diameter limits.

◊ TABLE 2

Nominal Pipe Thread Size, in	External Thread								Internal Thread			
	Chamfer Diameter[a]				Length of Chamfered or Partial Thread				Countersink Diameter[a]			
	Max		Min		Min		Max		Min		Max	
	in	mm	in	mm	in	mm	in	mm	in	mm	in	mm
1/8	0.32	8.1	0.30	7.6	0.037	0.90	0.056	1.42	0.42	10.7	0.44	11.2
1/4	0.42	10.7	0.40	10.2	0.056	1.42	0.083	2.11	0.55	14.0	0.57	14.5
3/8	0.55	14.0	0.53	13.5	0.056	1.42	0.083	2.11	0.69	17.5	0.71	18.0
1/2	0.68	17.3	0.66	16.8	0.071	1.80	0.107	2.72	0.85	21.6	0.87	22.1
3/4	0.89	22.6	0.87	22.1	0.071	1.80	0.107	2.72	1.06	26.9	1.08	27.4

[a] Tabulated diameters conform with Appendix A, SAE J476.

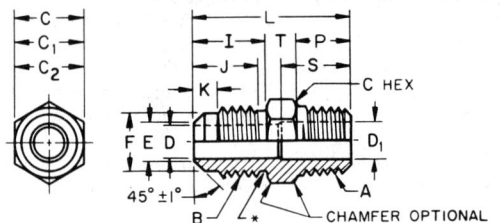

*UNDERCUT TO G DIA OPTIONAL ON FLARE SIZES 1/2 AND LARGER UNLESS OTHERWISE SPECIFIED BY PURCHASER

FIG. 1—CONNECTOR (HALF UNION) (010102) (U1)

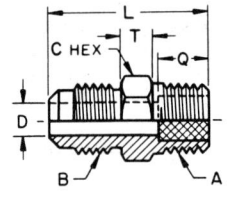

FIG. 2—FUSIBLE CONNECTOR (HALF UNION) (010163) (FU)

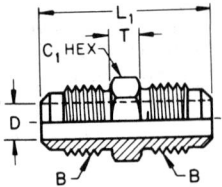

FIG. 3—UNION (010101) (U2)

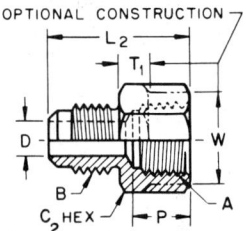

FIG. 4—INTERNAL PIPE THREAD CONNECTOR (HALF UNION) (010103) (U3)

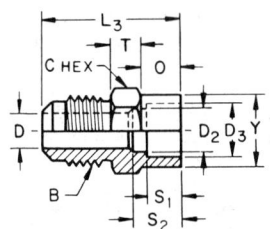

FIG. 5—SOLDER CONNECTOR (HALF UNION) (010104) (US3)

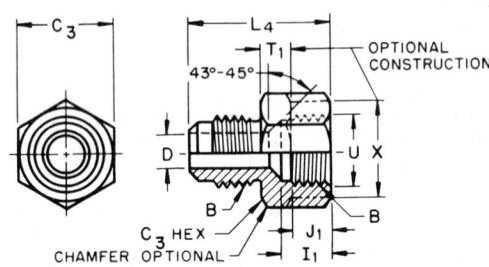

FIG. 6—INTERNAL FLARE TO EXTERNAL FLARE ADAPTOR (010105) (UR3)

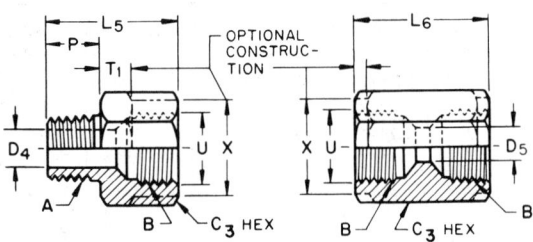

FIG. 7—INTERNAL FLARE TO EXTERNAL PIPE ADAPTOR (010106) (U5)

FIG. 8—INTERNAL FLARE UNION (010107) (U4)

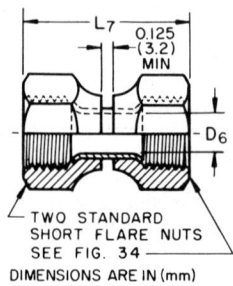

FIG. 9—INTERNAL FLARE SWIVEL UNION (010108) (US4)

21.29

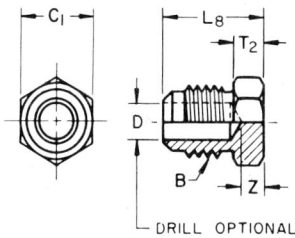

FIG. 10—PLUG
(010109) (P2)

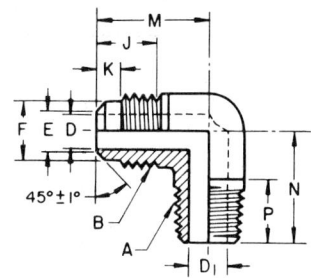

FIG. 11—90 DEG ELBOW
(010202) (E1)

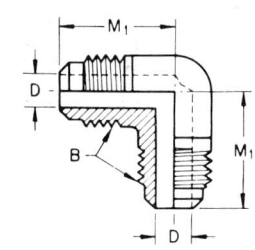

FIG. 12—90 DEG
ELBOW UNION
(010201) (E2)

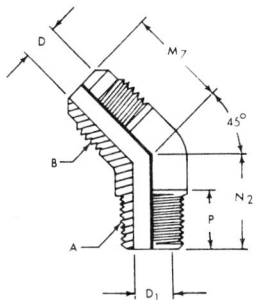

ϕFIG. 13—45 DEG ELBOW
(010302) (E5)

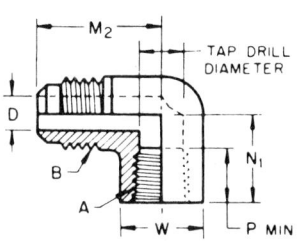

ϕFIG. 14—90 DEG
INTERNAL PIPE
THREAD ELBOW
(010203) (E3)

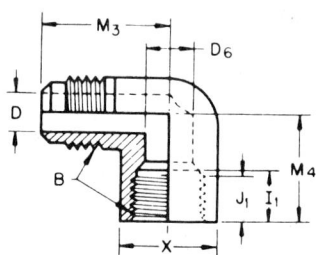

ϕFIG. 15—INTERNAL FLARE
TO EXTERNAL FLARE
90 DEG ELBOW
(010205) (E4)

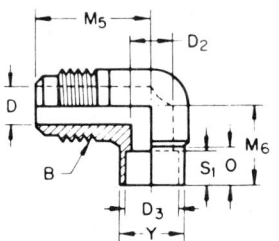

ϕFIG. 16—90 DEG
SOLDER ELBOW
(010204) (ES)

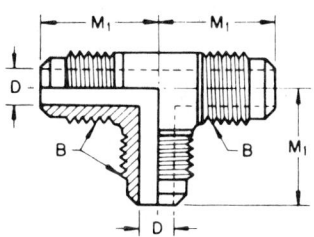

ϕFIG. 17—THREE
WAY TEE
(010401) (T2)

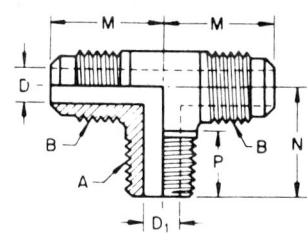

ϕFIG. 18—TWO WAY TEE
(010425) (T1)

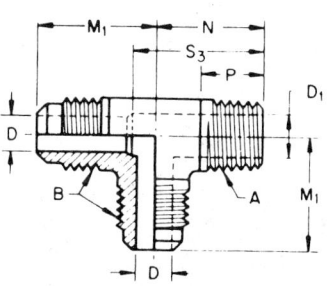

ϕFIG. 19—RIGHT ANGLE
TWO WAY TEE
(010424) (T3)

NOTE: UNSPECIFIED DETAIL WITH RESPECT TO DIMENSIONS, TOLERANCES, CONTOUR, MATERIAL, WORKMANSHIP, ETC., MUST CONFORM TO GENERAL SPECIFICATIONS FOR REFRIGERATION TUBE FITTINGS. THE DIMENSIONAL DESIGNATIONS IN FIGS. 1, 6 AND 11 AND THE FIRST FIGURE IN EACH GROUP SHALL APPLY TO CORRESPONDING FEATURES OF OTHER FIGURES ON THIS PAGE UNLESS SHOWN OTHERWISE. THE ILLUSTRATIONS ON THIS PAGE APPLY TO TABLE 3. CODES SHOWN IN BRACKETS ADJACENT TO FIGURE NUMBERS REPRESENT RESPECTIVE FITTING IDENTIFICATION IN ACCORDANCE WITH SAE J846 (FIRST NUMBER) AND ANSI B70.1 (SECOND NUMBER).

TABLE 3—DIMENSIONS OF CONNECTORS, UNIONS, ADAPTORS, ELBOWS, TEES, AND CROSSES (FIGS. 1–20)

Nom Tube OD, in	A Dryseal Pipe Thread NPTF[b]	B Straight Thread Nominal Size	C Hex, in Nom	C_1 Hex, in Nom	C_2 Hex, in Nom	C_3 Hex, in Nom	D[e] Drill in	D[e] Drill mm	D_1[e] Drill in	D_1[e] Drill mm	D_2 Drill in	D_2 Drill mm	D_3[h] Dia in ±0.0010	D_3[h] Dia mm ±0.025
3/16	1/8	3/8 –24	7/16	3/8	9/16	1/2	0.125	3.18	0.219	5.56	0.156	3.96	0.1915	4.864
1/4	1/8	7/16–20	7/16	7/16	9/16	5/8	0.188	4.78	0.219	5.56	0.188	4.78	0.2540	6.452
5/16	1/8	1/2 –20	1/2	1/2	9/16	1-1/16	0.219	5.56	0.219	5.56	0.250	6.35	0.3165	8.039
3/8	1/4	5/8 –18	5/8	5/8	11/16	13/16	0.281	7.14	0.312	7.92	0.312	7.92	0.3790	9.627
1/2	3/8	3/4 –16	3/4	3/4	13/16	15/16	0.406	10.31	0.406	10.31	0.438	11.13	0.5040	12.802
5/8	1/2	7/8 –14	7/8	7/8	1	1-1/16	0.500	12.70	0.562	14.27	0.547	13.89	0.6290	15.977
3/4	1/2	1-1/16-14	1-1/16	1-1/16	1-1/16	1-5/16	0.625	15.88	0.562	14.27	0.688	17.48	0.7540	19.152

Nom Tube OD, in	D_4 Drill in	D_4 Drill mm	D_5 Drill in	D_5 Drill mm	D_6 Tube ID in	D_6 Tube ID mm	E Dia in	E Dia mm	F Dia in	F Dia mm	G[c] Dia in +0.000 −0.010	G[c] Dia mm +0.00 −0.25	I in	I mm	I_1 in	I_1 mm
3/16	0.188	4.78	0.188	4.78	0.117	2.97	0.156	3.96	0.297	7.54	—	—	0.44	11.2	0.28	7.1
1/4	0.219	5.56	0.250	6.35	0.180	4.57	0.219	5.56	0.344	8.74	—	—	0.50	12.7	0.34	8.6
5/16	0.219	5.56	0.312	7.92	0.242	6.15	0.250	6.35	0.406	10.31	—	—	0.56	14.2	0.38	9.7
3/8	0.344	8.74	0.375	9.52	0.305	7.75	0.312	7.92	0.531	13.49	—	—	0.62	15.7	0.44	11.2
1/2	0.406	10.31	0.500	12.70	0.430	10.92	0.438	11.13	0.641	16.28	0.659	16.74	0.75	19.0	0.53	13.5
5/8	0.562	14.27	0.625	15.88	0.555	14.10	0.531	13.49	0.750	19.05	0.770	19.56	0.88	22.4	0.66	16.8
3/4	0.562	14.27	0.750	19.05	0.680	17.27	0.719	18.26	0.938	23.83	0.958	24.33	1.00	25.4	0.78	19.8

Nom Tube OD, in	J[c] Full Thread Min in	J[c] Full Thread Min mm	J_1 Full Thread Min in	J_1 Full Thread Min mm	K in	K mm	L in ±0.03	L mm ±0.8	L_1 in ±0.03	L_1 mm ±0.8	L_2 in ±0.03	L_2 mm ±0.8	L_3 in ±0.03	L_3 mm ±0.8	L_4 in ±0.03	L_4 mm ±0.8
3/16	0.38	9.7	0.22	5.6	0.12	3.0	1.00	25.4	1.06	26.9	0.97	24.6	0.94	23.9	0.94	23.9
1/4	0.41	10.4	0.27	6.9	0.16	4.1	1.06	26.9	1.19	30.2	1.03	26.2	1.00	25.4	1.06	26.9
5/16	0.47	11.9	0.30	7.6	0.19	4.8	1.16	29.5	1.34	34.0	1.06	26.9	1.09	27.7	1.12	28.4
3/8	0.54	13.7	0.34	8.6	0.22	5.6	1.44	36.6	1.50	38.1	1.31	33.3	1.19	30.2	1.31	33.3
1/2	0.66	16.8	0.44	11.2	0.25	6.4	1.62	41.1	1.81	46.0	1.50	38.1	1.44	36.6	1.56	39.6
5/8	0.76	19.3	0.55	14.0	0.28	7.1	2.00	50.8	2.12	53.8	1.81	46.0	1.75	44.4	1.81	46.0
3/4	0.90	22.9	0.67	17.0	0.28	7.1	2.19	55.6	2.44	62.0	1.91	48.5	2.06	52.3	2.06	52.3

Nom Tube OD, in	L_5 in ±0.03	L_5 mm ±0.8	L_6 in ±0.03	L_6 mm ±0.8	L_7 Min in	L_7 Min mm	L_8 in ±0.03	L_8 mm ±0.8	M in ±0.03	M mm ±0.8	M_1 in ±0.03	M_1 mm ±0.8	M_2 in ±0.03	M_2 mm ±0.8	M_3 in ±0.03	M_3 mm ±0.8
3/16	0.81	20.6	0.88	22.4	1.31	33.3	0.59	15.0	0.75	19.0	0.75	19.0	0.81	20.6	—	—
1/4	0.91	23.1	1.00	25.4	1.31	33.3	0.69	17.5	0.81	20.6	0.88	22.4	0.88	22.4	0.94	23.9
5/16	0.94	23.9	1.06	26.9	1.38	35.1	0.78	19.8	0.91	23.1	0.91	23.1	0.94	23.9	—	—
3/8	1.28	32.5	1.25	31.8	1.50	38.1	0.88	22.4	1.00	25.4	1.06	26.9	1.09	27.7	1.16	29.5
1/2	1.33	35.1	1.44	36.6	1.75	44.4	1.06	26.9	1.22	31.0	1.22	31.0	1.28	32.5	1.34	34.0
5/8	1.66	42.2	1.69	42.9	2.00	50.8	1.19	30.2	1.41	35.8	1.41	35.8	1.50	38.1	—	—
3/4	1.88	47.8	2.00	50.8	2.38	60.5	1.31	33.3	1.62	41.1	1.66	42.2	1.62	41.1	—	—

Nom Tube OD, in	M_4 in ±0.03	M_4 mm ±0.8	M_5 in ±0.03	M_5 mm ±0.8	M_6 in ±0.03	M_6 mm ±0.8	M_7 in ±0.03	M_7 mm ±0.8	N in ±0.03	N mm ±0.8	N_1 in ±0.03	N_1 mm ±0.8	N_2 in ±0.03	N_2 mm ±0.8	O in	O mm	P in	P mm	Q[d] in	Q[d] mm
3/16	—	—	0.72	18.3	0.59	15.0	0.62	15.7	0.75	19.0	0.44	11.2	0.52	13.2	0.31	7.9	0.38	9.7	—	—
1/4	0.78	19.8	0.81	20.6	0.62	15.7	0.67	17.0	0.78	19.8	0.47	11.9	0.64	16.3	0.31	7.9	0.38	9.7	—	—
5/16	—	—	0.91	23.1	0.66	16.8	0.78	19.8	0.78	19.8	0.47	11.9	0.64	16.3	0.31	7.9	0.38	9.7	—	—
3/8	0.97	24.6	1.03	26.2	0.72	18.3	0.89	22.6	1.06	26.9	0.69	17.5	0.86	21.8	0.31	7.9	0.56	14.2	0.69	17.5
1/2	1.12	28.4	1.22	31.0	0.84	21.3	1.06	26.9	1.12	28.4	0.75	19.0	0.95	24.1	0.38	9.7	0.56	14.2	0.94	23.9
5/8	—	—	1.41	35.8	1.03	26.2	1.23	31.2	1.38	35.1	1.00	25.4	1.17	29.7	0.50	12.7	0.75	19.0	—	—
3/4	—	—	1.62	41.1	1.25	31.8	1.41	35.8	1.50	38.1	1.06	26.9	1.20	30.5	0.62	15.7	0.75	19.0	—	—

(Table 3 continued on next page)

TABLE 3—DIMENSIONS OF CONNECTORS, UNIONS, ADAPTORS, ELBOWS, TEES, AND CROSSES (FIGS. 1–20)[a] (continued)

Nom Tube OD, in	S^e Max		S_1		S_2^e Max		S_3^e Min		T^f Ref		T_1^f Min		T_2^f Ref	
	in	mm	in	mm	in	mm	in	mm	in	mm	in	mm	in	mm
3/16	0.48	12.2	0.31	7.9	0.41	10.4	0.85	21.6	0.18	4.6	0.21	5.3	0.15	3.8
1/4	0.48	12.2	0.31	7.9	—	—	0.92	23.4	0.18	4.6	0.24	6.1	0.18	4.6
5/16	—	—	0.31	7.9	0.42	10.7	—	—	0.21	5.3	0.24	6.1	0.21	5.3
3/8	0.69	17.5	0.31	7.9	0.44	11.2	1.24	31.5	0.24	6.1	0.30	7.6	0.24	6.1
1/2	—	—	0.38	9.7	0.54	13.7	—	—	0.30	7.6	0.37	9.4	0.30	7.6
5/8	0.94	23.9	0.50	12.7	0.69	17.5	1.67	42.4	0.37	9.4	0.43	10.9	0.30	7.6
3/4	1.22	31.0	0.62	15.7	0.84	21.3	2.02	51.3	0.43	10.9	0.49	12.4	0.30	7.6

Nom Tube OD, in	U Dia				W^g Dia		X^g Dia		Y Dia Min		Z Min	
	Min		Max		in	mm	in	mm				
	in	mm	in	mm	+0.00 −0.02	+0.00 −0.5	+0.00 −0.02	+0.00 −0.5	in	mm	in	mm
3/16	0.39	9.9	0.41	10.4	0.56	14.2	0.50	12.7	0.28	7.1	0.06	1.5
1/4	0.45	11.4	0.47	11.9	0.56	14.2	0.62	15.7	0.34	8.6	0.05	1.3
5/16	0.51	13.0	0.53	13.5	0.56	14.2	0.69	17.5	0.40	10.2	0.06	1.5
3/8	0.64	16.3	0.67	17.0	0.69	17.5	0.81	20.6	0.48	12.2	0.06	1.5
1/2	0.77	19.6	0.80	20.3	0.81	20.6	0.94	23.9	0.60	15.2	0.08	2.0
5/8	0.90	22.9	0.93	23.6	1.00	25.4	1.06	26.9	0.74	18.8	0.10	2.5
3/4	1.08	27.4	1.11	28.2	1.06	26.9	1.31	33.3	0.86	21.8	0.10	2.5

[a] For reducing sizes of Unions, Internal Flare to External Flare Adaptors, and 90 Deg Elbow Unions, see Table 4; for reducing sizes of Solder Connectors and 90 Deg Solder Elbows, see Table 5; for reducing sizes of Connectors, Internal Pipe Thread Connectors, Internal Flare to External Pipe Adapters, 90 Deg Elbow, 45 Deg Elbow, and Internal Pipe Thread 90 Deg Elbow, see Table 6; for reducing sizes of Tees, see Tables 7 and 8.
[b] Dryseal American Standard Taper Pipe Thread.
[c] Where thread relief undercut is used, last thread shall be chamfered 1/2 to 1 pitch long from G diameter and dimension J may be reduced by an amount equal to 1/2 pitch.
[d] Available with three types of fusible alloys as specified in general specifications.
[e] At manufacturer's option through passages in fittings shown in Figs. 1, 5, and 19 may conform with the smaller diameter specified or be counterbored to the larger diameter from the appropriate end for depths S, S_2 or S_3, respectively.
[f] Minimum design thickness, not subject to inspection.
[g] Basic dimensions shown shall apply as minimum diameter for bosses or across flats. The −0.02 in (0.5 mm) tolerance shall apply only to chamfer diameters on full hexagon versions of fittings shown in Figs. 4, 6–8.
[h] ID of solder cup shall not be out of round by more than 0.0003 in (0.08 mm).

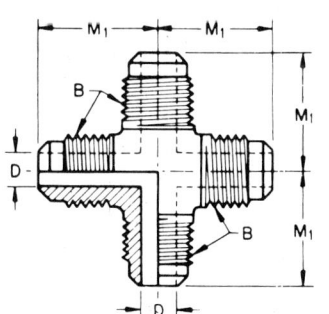

φFIG. 20—CROSS
(010501) (C1)

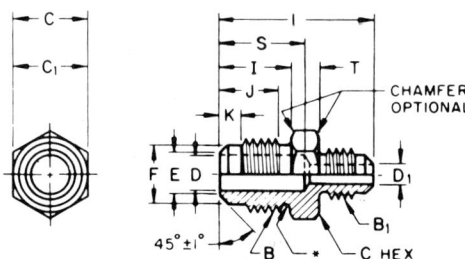

*UNDERCUT TO G DIA OPTIONAL ON FLARE SIZES 1/2 AND LARGER UNLESS OTHERWISE SPECIFIED BY PURCHASER (SEE FOOTNOTE c)

φFIG. 21—REDUCING UNION
(010101) (UR2)

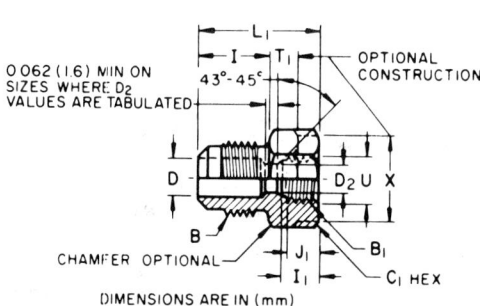

φFIG. 22—INTERNAL FLARE TO EXTERNAL FLARE REDUCING ADAPTOR
(010105) (UR3)

NOTE: UNSPECIFIED DETAIL WITH RESPECT TO DIMENSIONS, TOLERANCES, CONTOUR, MATERIAL, WORKMANSHIP, ETC., MUST CONFORM TO GENERAL SPECIFICATIONS FOR REFRIGERATION TUBE FITTINGS. THE ILLUSTRATIONS ON THIS PAGE APPLY TO TABLE 4. CODES SHOWN IN BRACKETS ADJACENT TO FIGURE NUMBERS REPRESENT RESPECTIVE FITTING IDENTIFICATION IN ACCORDANCE WITH SAE J846 (FIRST NUMBER) AND ANSI B70.1 (SECOND NUMBER).

TABLE 4—DIMENSIONS OF REDUCING UNIONS, REDUCING ADAPTORS, AND REDUCING ELBOW UNIONS (FIGS. 21–23)[a]

B[c] Nom Tube OD in	B$_1$[c] Nom Tube OD in	C, in Nom	C$_1$, in Nom	D[d] Drill in	D[d] Drill mm	D$_1$[d] Drill in	D$_1$[d] Drill mm	D$_2$ Drill in	D$_2$ Drill mm	L in ±0.03	L mm ±0.8	L$_1$ in ±0.03	L$_1$ mm ±0.8
3/16	1/4	7/16	5/8	0.125	3.18	0.188	4.78	—	—	1.12	28.4	1.03	26.2
3/16	5/16	1/2	11/16	0.125	3.18	0.219	5.56	—	—	1.22	31.0	1.06	26.9
3/16	3/8	5/8	13/16	0.125	3.18	0.281	7.14	—	—	1.31	33.3	1.19	30.2
3/16	1/2	3/4	15/16	0.125	3.18	0.406	10.31	—	—	1.50	38.1	1.34	34.0
3/16	5/8	7/8	1-1/16	0.125	3.18	0.500	12.70	—	—	1.69	42.9	1.53	38.9
3/16	3/4	1-1/16	1-5/16	0.125	3.18	0.625	15.88	—	—	1.88	47.8	1.75	44.4

B[c] Nom Tube OD in	B$_1$[c] Nom Tube OD in	M in ±0.03	M mm ±0.8	M$_1$ in ±0.03	M$_1$ mm ±0.8	S[d] Max in	S[d] Max mm	T[e] Ref in	T[e] Ref mm	T$_1$ Min in	T$_1$ Min mm	X[f] Dia in +0.00/−0.02	X[f] Dia mm +0.0/−0.5
3/16	1/4	0.75	19.0	0.88	22.4	0.60	15.2	0.18	4.6	0.24	6.1	0.62	15.7
3/16	5/16	0.78	19.8	0.91	23.1	0.67	17.0	0.21	5.3	0.24	6.1	0.69	17.5
3/16	3/8	0.84	21.3	1.06	26.9	0.75	19.0	0.24	6.1	0.30	7.6	0.81	20.6
3/16	1/2	0.91	23.1	1.22	31.0	0.91	23.1	0.30	7.6	0.37	9.4	0.94	23.9
3/16	5/8	0.97	24.6	1.41	35.8	1.07	27.2	0.37	9.4	0.43	10.9	1.06	26.9
3/16	3/4	1.06	26.9	1.66	42.2	1.22	31.0	0.43	10.9	0.49	12.4	1.31	33.3

B[c] Nom Tube OD in	B$_1$[c] Nom Tube OD in	C, in Nom	C$_1$, in Nom	D[d] Drill in	D[d] Drill mm	D$_1$[d] Drill in	D$_1$[d] Drill mm	D$_2$ Drill in	D$_2$ Drill mm	L in ±0.03	L mm ±0.8	L$_1$ in ±0.03	L$_1$ mm ±0.8
1/4	3/16	7/16	1/2	0.188	4.78	0.125	3.18	—	—	1.12	28.4	0.97	24.6
1/4	5/16	1/2	11/16	0.188	4.78	0.219	5.56	—	—	1.28	32.5	1.12	28.4
1/4	3/8	5/8	13/16	0.188	4.78	0.281	7.14	—	—	1.38	35.1	1.22	31.0
1/4	1/2	3/4	15/16	0.188	4.78	0.406	10.31	—	—	1.56	39.6	1.38	35.1
1/4	5/8	7/8	1-1/16	0.188	4.78	0.500	12.70	—	—	1.75	44.4	1.56	39.6
1/4	3/4	1-1/16	1-5/16	0.188	4.78	0.625	15.88	—	—	1.94	49.3	1.69	42.9

B[c] Nom Tube OD in	B$_1$[c] Nom Tube OD in	M in ±0.03	M mm ±0.8	M$_1$ in ±0.03	M$_1$ mm ±0.8	S[d] Max in	S[d] Max mm	T[e] Ref in	T[e] Ref mm	T$_1$ Min in	T$_1$ Min mm	X[f] Dia in +0.00/−0.02	X[f] Dia mm +0.0/−0.5
1/4	3/16	0.88	22.4	0.75	19.0	0.60	15.2	0.18	4.6	0.21	5.3	0.50	12.7
1/4	5/16	0.84	21.3	0.91	23.1	0.67	17.0	0.21	5.3	0.24	6.1	0.69	17.5
1/4	3/8	0.91	23.1	1.06	26.9	0.75	19.0	0.24	6.1	0.30	7.6	0.81	20.6
1/4	1/2	0.97	24.6	1.22	31.0	0.91	23.1	0.30	7.6	0.37	9.4	0.94	23.9
1/4	5/8	1.03	26.2	1.41	35.8	1.07	27.2	0.37	9.4	0.43	10.9	1.06	26.2
1/4	3/4	1.12	28.4	1.66	42.2	1.22	31.0	0.43	10.9	0.49	12.4	1.31	33.3

B[c] Nom Tube OD in	B$_1$[c] Nom Tube OD in	C, in Nom	C$_1$, in Nom	D[d] Drill in	D[d] Drill mm	D$_1$[d] Drill in	D$_1$[d] Drill mm	D$_2$ Drill in	D$_2$ Drill mm	L in ±0.03	L mm ±0.8	L$_1$ in ±0.03	L$_1$ mm ±0.8
5/16	3/16	1/2	1/2	0.219	5.56	0.125	3.18	0.188	4.78	1.22	31.0	1.00	25.4
5/16	1/4	1/2	5/8	0.219	5.56	0.188	4.78	—	—	1.28	32.5	1.09	27.7
5/16	3/8	5/8	13/16	0.219	5.56	0.281	7.14	—	—	1.44	36.6	1.25	31.8
5/16	1/2	3/4	15/16	0.219	5.56	0.406	10.31	—	—	1.62	41.1	1.41	35.8
5/16	5/8	7/8	1-1/16	0.219	5.56	0.500	12.70	—	—	1.81	46.0	1.59	40.4
5/16	3/4	1-1/16	1-5/16	0.219	5.56	0.625	15.88	—	—	2.00	50.8	1.81	46.0

(Table continued on next page)

TABLE 4—DIMENSIONS OF REDUCING UNIONS, REDUCING ADAPTORS, AND REDUCING ELBOW UNIONS (FIGS. 21–23)[a] (CONTINUED)

B[c] Nom Tube OD in	B_1[c] Nom Tube OD in	M in ±0.03	M mm ±0.8	M_1 in ±0.03	M_1 mm ±0.8	S[d] Max in	S[d] Max mm	T[e] Ref in	T[e] Ref mm	T_1 Min in	T_1 Min mm	X[f] Dia in +0.00 −0.02	X[f] Dia mm +0.0 −0.5
5/16	3/16	0.91	23.1	0.78	19.8	0.67	17.0	0.21	5.3	0.21	5.3	—	—
5/16	1/4	0.91	23.1	0.84	21.3	0.67	17.0	0.21	5.3	0.24	6.1	0.50	12.7
5/16	3/8	0.97	24.6	1.06	26.9	0.75	19.0	0.24	6.1	0.30	7.6	0.62	15.7
5/16	1/2	1.03	26.2	1.22	31.0	0.91	23.1	0.30	7.6	0.37	9.4	0.81	20.6
5/16	5/8	1.09	27.7	1.41	35.8	1.07	27.2	0.37	9.4	0.43	10.9	0.94	23.9
5/16	3/4	1.19	30.2	1.66	42.2	1.22	31.0	0.43	10.9	0.49	12.4	1.06	26.9
												1.31	33.3

B[c] Nom Tube OD in	B_1[c] Nom Tube OD in	C, in Nom	C_1, in Nom	D[d] Drill in	D[d] Drill mm	D_1[d] Drill in	D_1[d] Drill mm	D_2 Drill in	D_2 Drill mm	L in ±0.03	L mm ±0.8	L_1 in ±0.03	L_1 mm ±0.8
3/8	3/16	5/8	5/8	0.281	7.14	0.125	3.18	0.188	4.78	1.31	33.3	1.03	26.2
3/8	1/4	5/8	5/8	0.281	7.14	0.188	4.78	0.250	6.35	1.38	35.1	1.12	28.4
3/8	5/16	5/8	11/16	0.281	7.14	0.219	5.56	—	—	1.44	36.6	1.19	30.2
3/8	1/2	3/4	15/16	0.281	7.14	0.406	10.31	—	—	1.69	42.9	1.44	36.6
3/8	5/8	7/8	1-1/16	0.281	7.14	0.500	12.70	—	—	1.88	47.8	1.62	41.1
3/8	3/4	1-1/16	1-5/16	0.281	7.14	0.625	15.89	—	—	2.06	52.3	1.84	46.7

B[c] Nom Tube OD in	B_1[c] Nom Tube OD in	M in ±0.03	M mm ±0.8	M_1 in ±0.03	M_1 mm ±0.8	S[d] Max in	S[d] Max mm	T[e] Ref in	T[e] Ref mm	T_1 Min in	T_1 Min mm	X[f] Dia in +0.00 −0.02	X[f] Dia mm +0.0 −0.5
3/8	3/16	1.06	26.9	0.84	21.3	0.75	19.0	0.24	6.1	0.24	6.1	0.62	15.7
3/8	1/4	1.06	26.9	0.91	23.1	0.75	19.0	0.24	6.1	0.24	6.1	0.62	15.7
3/8	5/16	1.06	26.9	0.97	24.6	0.75	19.0	0.24	6.1	0.24	6.1	0.69	17.5
3/8	1/2	1.09	27.7	1.22	31.0	0.91	23.1	0.30	7.6	0.37	9.4	0.94	23.9
3/8	5/8	1.16	29.5	1.41	35.8	1.07	27.2	0.37	9.4	0.43	10.9	1.06	26.9
3/8	3/4	1.25	31.8	1.66	42.2	1.22	31.0	0.43	10.9	0.49	12.4	1.31	33.3

B[c] Nom Tube OD in	B_1[c] Nom Tube OD in	C, in Nom	C_1, in Nom	D[d] Drill in	D[d] Drill mm	D_1[d] Drill in	D_1[d] Drill mm	D_2 Drill in	D_2 Drill mm	L in ±0.03	L mm ±0.8	L_1 in ±0.03	L_1 mm ±0.8
1/2	3/16	3/4	3/4	0.406	10.31	0.125	3.18	0.188	4.78	1.50	38.1	1.16	29.5
1/2	1/4	3/4	3/4	0.406	10.31	0.188	4.78	0.250	6.35	1.56	39.6	1.25	31.8
1/2	5/16	3/4	3/4	0.406	10.31	0.219	5.56	0.312	7.92	1.62	41.1	1.28	32.5
1/2	3/8	3/4	13/16	0.406	10.31	0.281	7.14	0.375	9.52	1.69	42.9	1.41	35.8
1/2	5/8	7/8	1-1/16	0.406	10.31	0.500	12.70	—	—	2.00	50.8	1.69	42.9
1/2	3/4	1-1/16	1-5/16	0.406	10.31	0.625	15.88	—	—	2.19	55.6	1.91	48.5

B[c] Nom Tube OD in	B_1[c] Nom Tube OD in	M in ±0.03	M mm ±0.8	M_1 in ±0.03	M_1 mm ±0.8	S[d] Max in	S[d] Max mm	T[e] Ref in	T[e] Ref mm	T_1 Min in	T_1 Min mm	X[f] Dia in +0.00 −0.02	X[f] Dia mm +0.0 −0.5
1/2	3/16	1.22	31.0	0.91	23.1	0.91	23.1	0.30	7.6	0.30	7.6	0.75	19.0
1/2	1/4	1.22	31.0	0.97	24.6	0.91	23.1	0.30	7.6	0.30	7.6	0.75	19.0
1/2	5/16	1.22	31.0	1.03	26.2	0.91	23.1	0.30	7.6	0.30	7.6	0.75	19.0
1/2	3/8	1.22	31.0	1.09	27.7	0.91	23.1	0.30	7.6	0.30	7.6	0.81	20.6
1/2	5/8	1.28	32.5	1.41	35.8	1.07	27.2	0.37	9.4	0.43	10.9	1.06	26.9
1/2	3/4	1.38	35.1	1.66	42.2	1.22	31.0	0.43	10.9	0.49	12.4	1.31	33.3

(Table continued on next page)

TABLE 4—DIMENSIONS OF REDUCING UNIONS, REDUCING ADAPTORS, AND REDUCING ELBOW UNIONS (FIGS. 21-23)[a] (CONTINUED)

B[c] Nom Tube OD in	B_1[c] Nom Tube OD in	C, in Nom	C_1, in Nom	D[d] Drill in	D[d] Drill mm	D_1[d] Drill in	D_1[d] Drill mm	D_2 Drill in	D_2 Drill mm	L in ±0.03	L mm ±0.8	L_1 in ±0.03	L_1 mm ±0.8
5/8	3/16	7/8	7/8	0.500	12.70	0.125	3.18	0.188	4.78	1.69	42.9	1.25	31.8
5/8	1/4	7/8	7/8	0.500	12.70	0.188	4.78	0.250	6.35	1.75	44.4	1.31	33.3
5/8	5/16	7/8	7/8	0.500	12.70	0.219	5.56	0.312	7.92	1.81	46.0	1.38	35.1
5/8	3/8	7/8	7/8	0.500	12.70	0.281	7.14	0.375	9.52	1.88	47.8	1.47	37.3
5/8	1/2	7/8	15/16	0.500	12.70	0.406	10.31	—	—	2.00	50.8	1.62	41.1
5/8	3/4	1-1/16	1-5/16	0.500	12.70	0.625	15.88	—	—	2.31	58.7	1.97	50.0

B[c] Nom Tube OD in	B_1[c] Nom Tube OD in	M in ±0.03	M mm ±0.8	M_1 in ±0.03	M_1 mm ±0.8	S[d] Max in	S[d] Max mm	T[e] Ref in	T[e] Ref mm	T_1 Min in	T_1 Min mm	X[f] Dia in +0.00/−0.02	X[f] Dia mm +0.0/−0.5
5/8	3/16	1.41	35.8	0.97	24.6	1.07	27.2	0.37	9.4	0.37	9.4	0.88	22.4
5/8	1/4	1.41	35.8	1.03	26.2	1.07	27.2	0.37	9.4	0.37	9.4	0.88	22.4
5/8	5/16	1.41	35.8	1.09	27.7	1.07	27.2	0.37	9.4	0.37	9.4	0.88	22.4
5/8	3/8	1.41	35.8	1.16	29.5	1.07	27.2	0.37	9.4	0.37	9.4	0.88	22.4
5/8	1/2	1.41	35.8	1.28	32.5	1.07	27.2	0.37	9.4	0.37	9.4	0.94	23.9
5/8	3/4	1.50	38.1	1.66	42.2	1.22	31.0	0.43	10.9	0.49	12.4	1.31	33.3

B[c] Nom Tube OD in	B_1[c] Nom Tube OD in	C, in Nom	C_1, in Nom	D[d] Drill in	D[d] Drill mm	D_1[d] Drill in	D_1[d] Drill mm	D_2 Drill in	D_2 Drill mm	L in ±0.03	L mm ±0.8	L_1 in ±0.03	L_1 mm ±0.8
3/4	3/16	1-1/16	1-1/16	0.625	15.88	0.125	3.18	0.188	4.78	1.88	47.8	1.44	36.6
3/4	1/4	1-1/16	1-1/16	0.625	15.88	0.188	4.78	0.250	6.35	1.94	49.3	1.44	36.6
3/4	5/16	1-1/16	1-1/16	0.625	15.88	0.219	5.56	0.312	7.92	2.00	50.8	1.44	36.6
3/4	3/8	1-1/16	1-1/16	0.625	15.88	0.281	7.14	0.375	9.52	2.06	52.3	1.53	38.9
3/4	1/2	1-1/16	1-1/16	0.625	15.88	0.406	10.31	0.500	12.70	2.19	55.6	1.69	42.9
3/4	5/8	1-1/16	1-1/16	0.625	15.88	0.500	12.70	—	—	2.31	58.7	1.88	47.8

B[c] Nom Tube OD in	B_1[c] Nom Tube OD in	M in ±0.03	M mm ±0.8	M_1 in ±0.03	M_1 mm ±0.8	S[d] Max in	S[d] Max mm	T[e] Ref in	T[e] Ref mm	T_1 Min in	T_1 Min mm	X[f] Dia in +0.00/−0.02	X[f] Dia mm +0.0/−0.5
3/4	3/16	1.66	42.2	1.06	26.9	1.22	31.0	0.43	10.9	0.43	10.9	1.06	26.9
3/4	1/4	1.66	42.2	1.12	28.4	1.22	31.0	0.43	10.9	0.43	10.9	1.06	26.9
3/4	5/16	1.66	42.2	1.19	30.2	1.22	31.0	0.43	10.9	0.43	10.9	1.06	26.9
3/4	3/8	1.66	42.2	1.25	31.8	1.22	31.0	0.43	10.9	0.43	10.9	1.06	26.9
3/4	1/2	1.66	42.2	1.38	35.1	1.22	31.0	0.43	10.9	0.43	10.9	1.06	26.9
3/4	5/8	1.66	42.2	1.50	38.1	1.22	31.0	0.43	10.9	0.43	10.9	1.06	26.9

[a] For flare dimensions shown on Figs. 21–23 but not covered in Table 4, see corresponding dimensions for the specified Tube OD in Table 3.
[b] In these sizes the reducing unions and reducing elbows are the reverses of sizes already specified in Table.
[c] Where thread relief undercut is used last thread shall be chamfered 1/2 to 1 pitch long from G diameter and dimension J may be reduced by an amount equal to 1/2 pitch.
[d] At manufacturer's option through passages in fittings shown in Fig. 21 may conform with the smaller diameter specified or be counterbored to the larger diameter from the appropriate end for depth S.
[e] Minimum design thickness, not subject to inspection.
[f] Basic dimensions shown shall apply as minimum for bosses. The −0.02 in (0.51 mm) tolerance shall apply only to chamfer diameter on full hexagon version of fittings in Fig. 22.

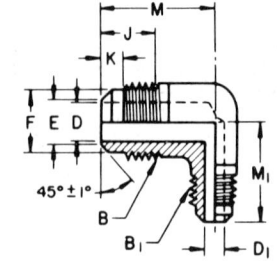

FIG. 23—90 DEG REDUCING ELBOW UNION (010201) (ER2)

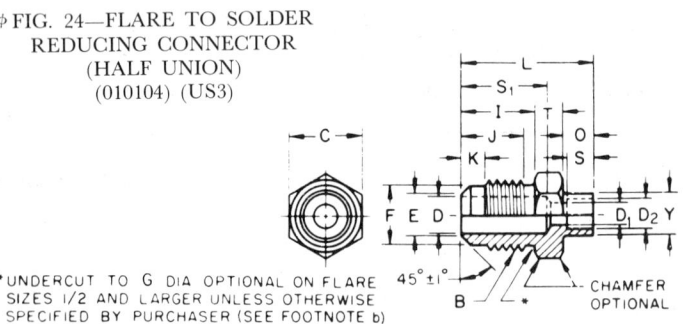

FIG. 24—FLARE TO SOLDER REDUCING CONNECTOR (HALF UNION) (010104) (US3)

*UNDERCUT TO G DIA OPTIONAL ON FLARE SIZES 1/2 AND LARGER UNLESS OTHERWISE SPECIFIED BY PURCHASER (SEE FOOTNOTE b)

NOTE: UNSPECIFIED DETAIL WITH RESPECT TO DIMENSIONS, TOLERANCES, CONTOUR, MATERIAL, WORKMANSHIP, ETC., MUST CONFORM TO GENERAL SPECIFICATIONS FOR REFRIGERATION TUBE FITTINGS. THE ILLUSTRATIONS ON THIS PAGE APPLY TO TABLE 4. CODES SHOWN IN BRACKETS ADJACENT TO FIGURE NUMBERS REPRESENT RESPECTIVE FITTING IDENTIFICATION IN ACCORDANCE WITH SAE J846 (FIRST NUMBER) AND ANSI B70.1 (SECOND NUMBER).

TABLE 5—DIMENSIONS OF REDUCING SOLDER CONNECTORS AND REDUCING SOLDER ELBOWS (FIGS. 24 AND 25)[a]

B Nom Tube OD, in	Solder Tube OD, in	C Hex, in Nom	D[c] Drill in	D[c] Drill mm	D_1[c] Drill in	D_1[c] Drill mm	D_2[e] Dia in ±0.0010	D_2[e] Dia mm ±0.025	L in ±0.03	L mm ±0.8	M in ±0.03	M mm ±0.8
3/16	1/8	3/8	0.125	3.18	0.094	2.39	0.1290	3.277	0.91	23.1	0.72	18.3
3/16	1/4	7/16	0.125	3.18	0.188	4.78	0.2540	6.452	0.94	23.9	0.72	18.3
3/16	5/16	7/16	0.125	3.18	0.250	6.35	0.3165	8.039	0.94	23.9	0.75	19.0
3/16	3/8	1/2	0.125	3.18	0.312	7.92	0.3790	9.627	0.94	23.9	0.78	19.8
3/16	1/2	5/8	0.125	3.18	0.438	11.13	0.5040	12.802	1.03	26.2	0.84	21.3
3/16	5/8	3/4	0.125	3.18	0.547	13.89	0.6290	15.977	1.19	30.2	0.91	23.1
3/16	3/4	7/8	0.125	3.18	0.688	17.48	0.7540	19.152	1.38	35.1	0.97	24.6
3/16	7/8	1	0.125	3.18	0.781	19.84	0.8790	22.327	1.56	39.6	1.06	26.9

B Nom Tube OD, in	Solder Tube OD, in	M_1 in ±0.03	M_1 mm ±0.8	O in	O mm	S in	S mm	S_1[c] Max in	S_1[c] Max mm	T[d] Ref in	T[d] Ref mm	Y Dia Min in	Y Dia Min mm
3/16	1/8	0.59	15.0	0.31	7.9	0.31	7.9	0.52	13.2	0.15	3.8	0.22	5.6
3/16	1/4	0.59	15.0	0.31	7.9	0.31	7.9	0.41	10.4	0.18	4.6	0.34	8.6
3/16	5/16	0.62	15.7	0.31	7.9	0.31	7.9	0.41	10.4	0.18	4.6	0.40	10.2
3/16	3/8	0.66	16.8	0.31	7.9	0.31	7.9	0.41	10.4	0.18	4.6	0.48	12.2
3/16	1/2	0.78	19.8	0.38	9.7	0.38	9.7	0.49	12.4	0.21	5.3	0.60	15.2
3/16	5/8	0.97	24.6	0.50	12.7	0.50	12.7	0.63	16.0	0.24	6.1	0.74	18.8
3/16	3/4	1.16	29.5	0.62	15.7	0.62	15.7	0.78	19.8	0.30	7.6	0.86	21.8
3/16	7/8	1.38	35.1	0.75	19.0	0.75	19.0	0.94	23.9	0.37	9.4	0.98	24.9

B Nom Tube OD, in	Solder Tube OD, in	C Hex, in Nom	D[c] Drill in	D[c] Drill mm	D_1[c] Drill in	D_1[c] Drill mm	D_2[e] Dia in ±0.0010	D_2[e] Dia mm ±0.025	L in ±0.03	L mm ±0.8	M in ±0.03	M mm ±0.8
1/4	1/8	7/16	0.188	4.78	0.094	2.39	0.1290	3.277	1.00	25.4	0.81	20.6
1/4	3/16	7/16	0.188	4.78	0.188	4.78	0.1915	4.864	1.00	25.4	0.81	20.6
1/4	5/16	7/16	0.188	4.78	0.250	6.35	0.3165	8.039	1.00	25.4	0.81	20.6
1/4	3/8	1/2	0.188	4.78	0.312	7.92	0.3790	9.627	1.00	25.4	0.84	21.3
1/4	1/2	5/8	0.188	4.78	0.438	11.13	0.5040	12.802	1.09	27.7	0.91	23.1
1/4	5/8	3/4	0.188	4.78	0.547	13.89	0.6290	15.977	1.25	31.8	0.97	24.6
1/4	3/4	7/8	0.188	4.78	0.688	17.48	0.7540	19.152	1.44	36.6	1.03	26.2
1/4	7/8	1	0.188	4.78	0.781	19.84	0.8790	22.327	1.62	41.1	1.12	28.4

B Nom Tube OD, in	Solder Tube OD, in	M_1 in ±0.03	M_1 mm ±0.8	O in	O mm	S in	S mm	S_1[c] Max in	S_1[c] Max mm	T[d] Ref in	T[d] Ref mm	Y Dia Min in	Y Dia Min mm
1/4	1/8	0.62	15.7	0.31	7.9	0.31	7.9	0.60	15.2	0.18	4.6	0.22	5.6
1/4	3/16	0.62	15.7	0.31	7.9	0.31	7.9	0.60	15.2	0.18	4.6	0.28	7.1
1/4	5/16	0.62	15.7	0.31	7.9	0.31	7.9	0.41	10.4	0.18	4.6	0.40	10.2
1/4	3/8	0.66	16.8	0.31	7.9	0.31	7.9	0.41	10.4	0.18	4.6	0.48	12.2
1/4	1/2	0.78	19.8	0.38	9.7	0.38	9.7	0.49	12.4	0.21	5.3	0.60	15.2
1/4	5/8	0.97	24.6	0.50	12.7	0.50	12.7	0.63	16.0	0.24	6.1	0.74	18.8
1/4	3/4	1.16	29.5	0.62	15.7	0.62	15.7	0.78	19.8	0.30	7.6	0.86	21.8
1/4	7/8	1.38	35.1	0.75	19.0	0.75	19.0	0.94	23.9	0.37	9.4	0.98	24.9

B Nom Tube OD, in	Solder Tube OD, in	C Hex, in Nom	D[c] Drill in	D[c] Drill mm	D_1[c] Drill in	D_1[c] Drill mm	D_2[e] Dia in ±0.0010	D_2[e] Dia mm ±0.025	L in ±0.03	L mm ±0.8	M in ±0.03	M mm ±0.8
5/16	1/8	1/2	0.219	5.56	0.094	2.39	0.1290	3.277	1.09	27.7	0.91	23.1
5/16	3/16	1/2	0.219	5.56	0.156	3.96	0.1915	4.864	1.09	27.7	0.91	23.1
5/16	1/4	1/2	0.219	5.56	0.188	4.78	0.2540	6.452	1.09	27.7	0.91	23.1
5/16	3/8	1/2	0.219	5.56	0.312	7.92	0.3790	9.627	1.09	27.7	0.91	23.1
5/16	1/2	5/8	0.219	5.56	0.438	11.13	0.5040	12.802	1.16	29.5	0.97	24.6
5/16	5/8	3/4	0.219	5.56	0.547	13.89	0.6290	15.977	1.31	33.3	1.03	26.2
5/16	3/4	7/8	0.219	5.56	0.688	17.48	0.7540	19.152	1.50	38.1	1.09	27.7
5/16	7/8	1	0.219	5.56	0.781	19.84	0.8790	22.327	1.69	42.9	1.19	30.2

(Table 5 continued on next page)

TABLE 5—DIMENSIONS OF REDUCING SOLDER CONNECTORS AND REDUCING SOLDER ELBOWS (FIGS. 24 AND 25)[a] (CONTINUED)

B Nom Tube OD, in	Solder Tube OD, in	M_1		O		S		S_1[c] Max		T[d] Ref		Y Dia Min	
		in ±0.03	mm ±0.8	in	mm	in	mm	in	mm	in	mm	in	mm
5/16	1/8	0.66	16.8	0.31	7.9	0.31	7.9	0.67	17.0	0.21	5.3	0.22	5.6
5/16	3/16	0.66	16.8	0.31	7.9	0.31	7.9	0.67	17.0	0.21	5.3	0.28	7.1
5/16	1/4	0.66	16.8	0.31	7.9	0.31	7.9	0.67	17.0	0.21	5.3	0.34	8.6
5/16	3/8	0.66	16.8	0.31	7.9	0.31	7.9	0.42	10.7	0.21	5.3	0.48	12.2
5/16	1/2	0.78	19.8	0.38	9.7	0.38	9.7	0.49	12.4	0.21	5.3	0.60	15.2
5/16	5/8	0.97	24.6	0.50	12.7	0.50	12.7	0.63	16.0	0.24	6.1	0.74	18.8
5/16	3/4	1.16	29.5	0.62	15.7	0.62	15.7	0.78	19.8	0.30	7.6	0.86	21.8
5/16	7/8	1.38	35.1	0.75	19.0	0.75	19.0	0.94	23.9	0.37	9.4	0.98	24.9

B Nom Tube OD, in	Solder Tube OD, in	C Hex, in Nom	D[c] Drill		D_1[c] Drill		D_2[e] Dia		L		M	
			in	mm	in	mm	in ±0.0010	mm ±0.025	in ±0.03	mm ±0.8	in ±0.03	mm ±0.8
3/8	1/8	5/8	0.281	7.14	0.094	2.39	0.1290	3.277	1.19	30.2	1.03	26.2
3/8	3/16	5/8	0.281	7.14	0.156	3.96	0.1915	4.864	1.19	30.2	1.03	26.2
3/8	1/4	5/8	0.281	7.14	0.188	4.78	0.2540	6.452	1.19	30.2	1.03	26.2
3/8	5/16	5/8	0.281	7.14	0.250	6.35	0.3165	8.039	1.19	30.2	1.03	26.2
3/8	1/2	5/8	0.281	7.14	0.438	11.13	0.5040	12.802	1.25	31.8	1.03	26.2
3/8	5/8	3/4	0.281	7.14	0.547	13.89	0.6290	15.977	1.38	35.1	1.09	27.7
3/8	3/4	7/8	0.281	7.14	0.688	17.48	0.7540	19.152	1.56	39.6	1.16	29.5
3/8	7/8	1	0.281	7.14	0.781	19.84	0.8790	22.327	1.75	44.4	1.25	31.8

B Nom Tube OD, in	Solder Tube OD, in	M_1		O		S		S_1[c] Max		T[d] Ref		Y Dia Min	
		in ±0.03	mm ±0.8	in	mm	in	mm	in	mm	in	mm	in	mm
3/8	1/8	0.72	18.3	0.31	7.9	0.31	7.9	0.75	19.0	0.24	6.1	0.22	5.6
3/8	3/16	0.72	18.3	0.31	7.9	0.31	7.9	0.75	19.0	0.24	6.1	0.28	7.1
3/8	1/4	0.72	18.3	0.31	7.9	0.31	7.9	0.75	19.0	0.24	6.1	0.34	8.6
3/8	5/16	0.72	18.3	0.31	7.9	0.31	7.9	0.75	19.0	0.24	6.1	0.40	10.2
3/8	1/2	0.78	19.8	0.38	9.7	0.38	9.7	0.51	13.0	0.24	6.1	0.60	15.2
3/8	5/8	0.97	24.6	0.50	12.7	0.50	12.7	0.63	16.0	0.24	6.1	0.74	18.8
3/8	3/4	1.16	29.5	0.62	15.7	0.62	15.7	0.78	19.8	0.30	7.6	0.86	21.8
3/8	7/8	1.25	31.8	0.75	19.0	0.75	19.0	0.94	23.9	0.37	9.4	0.98	24.9

B Nom Tube OD, in	Solder Tube OD, in	C Hex, in Nom	D[c] Drill		D_1[c] Drill		D_2[e] Dia		L		M	
			in	mm	in	mm	in ±0.0010	mm ±0.025	in ±0.03	mm ±0.8	in ±0.03	mm ±0.8
1/2	1/8	3/4	0.406	10.31	0.094	2.39	0.1290	3.277	1.38	35.1	1.22	31.0
1/2	3/16	3/4	0.406	10.31	0.156	3.96	0.1915	4.864	1.38	35.1	1.22	31.0
1/2	1/4	3/4	0.406	10.31	0.188	4.78	0.2540	6.452	1.38	35.1	1.22	31.0
1/2	5/16	3/4	0.406	10.31	0.250	6.35	0.3165	8.039	1.38	35.1	1.22	31.0
1/2	3/8	3/4	0.406	10.31	0.312	7.92	0.3790	9.627	1.38	35.1	1.22	31.0
1/2	5/8	3/4	0.406	10.31	0.547	13.89	0.6290	15.977	1.56	39.6	1.22	31.0
1/2	3/4	7/8	0.406	10.31	0.688	17.48	0.7540	19.152	1.69	42.9	1.28	32.5
1/2	7/8	1	0.406	10.31	0.781	19.84	0.8790	22.327	1.88	47.8	1.38	35.1

B Nom Tube OD, in	Solder Tube OD, in	M_1		O		S		S_1[c] Max		T[d] Ref		Y Dia Min	
		in ±0.03	mm ±0.8	in	mm	in	mm	in	mm	in	mm	in	mm
1/2	1/8	0.78	19.8	0.31	7.9	0.31	7.9	0.91	23.1	0.30	7.6	0.22	5.6
1/2	3/16	0.78	19.8	0.31	7.9	0.31	7.9	0.91	23.1	0.30	7.6	0.28	7.1
1/2	1/4	0.78	19.8	0.31	7.9	0.31	7.9	0.91	23.1	0.30	7.6	0.34	8.6
1/2	5/16	0.78	19.8	0.31	7.9	0.31	7.9	0.91	23.1	0.30	7.6	0.40	10.2
1/2	3/8	0.78	19.8	0.31	7.9	0.31	7.9	0.91	23.1	0.30	7.6	0.48	12.2
1/2	5/8	0.97	24.6	0.50	12.7	0.50	12.7	0.66	16.8	0.30	7.6	0.74	18.8
1/2	3/4	1.16	29.5	0.62	15.7	0.62	15.7	0.78	19.8	0.30	7.6	0.86	21.8
1/2	7/8	1.38	35.1	0.75	19.0	0.75	19.0	0.94	23.9	0.37	9.4	0.98	24.9

(Table 5 continued on next page)

TABLE 5—DIMENSIONS OF REDUCING SOLDER CONNECTORS AND REDUCING SOLDER ELBOWS (FIGS. 24 AND 25)[a] (CONTINUED)

B Nom Tube OD, in	Solder Tube OD, in	C Hex, in Nom	D[c] Drill in	D[c] Drill mm	D_1[c] Drill in	D_1[c] Drill mm	D_2[e] Dia in ±0.0010	D_2[e] Dia mm ±0.025	L in ±0.03	L mm ±0.8	M in ±0.03	M mm ±0.8
5/8	1/8	7/8	0.500	12.70	0.094	2.39	0.1290	3.277	1.56	39.6	1.41	35.8
5/8	3/16	7/8	0.500	12.70	0.156	3.96	0.1915	4.864	1.56	39.6	1.41	35.8
5/8	1/4	7/8	0.500	12.70	0.188	4.78	0.2540	6.452	1.56	39.6	1.41	35.8
5/8	5/16	7/8	0.500	12.70	0.250	6.35	0.3165	8.039	1.56	39.6	1.41	35.8
5/8	3/8	7/8	0.500	12.70	0.312	7.92	0.3790	9.627	1.56	39.6	1.41	35.8
5/8	1/2	7/8	0.500	12.70	0.438	11.13	0.5040	12.802	1.62	41.1	1.41	35.8
5/8	3/4	7/8	0.500	12.70	0.688	17.48	0.7540	19.152	1.88	47.8	1.41	35.8
5/8	7/8	1	0.500	12.70	0.781	19.84	0.8790	22.327	2.00	50.8	1.50	38.1

B Nom Tube OD, in	Solder Tube OD, in	M_1 in ±0.03	M_1 mm ±0.8	O in	O mm	S in	S mm	S_1[c] Max in	S_1[c] Max mm	T[d] Ref in	T[d] Ref mm	Y Dia Min in	Y Dia Min mm
5/8	1/8	0.84	21.3	0.31	7.9	0.31	7.9	1.07	27.2	0.37	9.4	0.22	5.6
5/8	3/16	0.84	21.3	0.31	7.9	0.31	7.9	1.07	27.2	0.37	9.4	0.28	7.1
5/8	1/4	0.84	21.3	0.31	7.9	0.31	7.9	1.07	27.2	0.37	9.4	0.34	8.6
5/8	5/16	0.84	21.3	0.31	7.9	0.31	7.9	1.07	27.2	0.37	9.4	0.40	10.2
5/8	3/8	0.84	21.3	0.31	7.9	0.31	7.9	1.07	27.2	0.37	9.4	0.48	12.2
5/8	1/2	0.91	23.1	0.38	9.7	0.38	9.7	1.07	27.2	0.37	9.4	0.60	15.2
5/8	3/4	1.16	29.5	0.62	15.7	0.62	15.7	0.81	20.6	0.37	9.4	0.86	21.8
5/8	7/8	1.38	35.1	0.75	19.0	0.75	19.0	0.94	23.9	0.37	9.4	0.98	24.9

B Nom Tube OD, in	Solder Tube OD, in	C Hex, in Nom	D[c] Drill in	D[c] Drill mm	D_1[c] Drill in	D_1[c] Drill mm	D_2[e] Dia in ±0.0010	D_2[e] Dia mm ±0.025	L in ±0.03	L mm ±0.8	M in ±0.03	M mm ±0.8
3/4	1/8	1-1/16	0.625	15.88	0.094	—	0.1290	3.277	1.75	44.4	1.62	41.1
3/4	3/16	1-1/16	0.625	15.88	0.156	—	0.1915	4.864	1.75	44.4	1.62	41.1
3/4	1/4	1-1/16	0.625	15.88	0.188	—	0.2540	6.452	1.75	44.4	1.62	41.1
3/4	5/16	1-1/16	0.625	15.88	0.250	—	0.3165	8.039	1.75	44.4	1.62	41.1
3/4	3/8	1-1/16	0.625	15.88	0.312	—	0.3790	9.627	1.75	44.4	1.62	41.1
3/4	1/2	1-1/16	0.625	15.88	0.438	—	0.5040	12.802	1.81	46.0	1.62	41.1
3/4	5/8	1-1/16	0.625	15.88	0.547	—	0.6290	15.977	1.94	49.3	1.62	41.1
3/4	3/4	1-1/16	0.625	15.88	0.781	—	0.8790	22.327	2.19	55.6	1.62	41.1

B Nom Tube OD, in	Solder Tube OD, in	M_1 in ±0.03	M_1 mm ±0.8	O in	O mm	S in	S mm	S_1[c] Max in	S_1[c] Max mm	T[d] Ref in	T[d] Ref mm	Y Dia Min in	Y Dia Min mm
3/4	1/8	0.94	23.9	0.31	7.9	0.31	7.9	1.22	31.0	0.43	10.9	0.22	5.6
3/4	3/16	0.94	23.9	0.31	7.9	0.31	7.9	1.22	31.0	0.43	10.9	0.28	7.1
3/4	1/4	0.94	23.9	0.31	7.9	0.31	7.9	1.22	31.0	0.43	10.9	0.34	8.6
3/4	5/16	0.94	23.9	0.31	7.9	0.31	7.9	1.22	31.0	0.43	10.9	0.40	10.2
3/4	3/8	0.94	23.9	0.31	7.9	0.31	7.9	1.22	31.0	0.43	10.9	0.48	12.2
3/4	1/2	1.00	25.4	0.38	9.7	0.38	9.7	1.22	31.0	0.43	10.9	0.60	15.2
3/4	5/8	1.12	28.4	0.50	12.7	0.50	12.7	1.22	31.0	0.43	10.9	0.74	18.8
3/4	7/8	1.38	35.1	0.75	19.0	0.75	19.0	0.97	24.6	0.43	10.9	0.98	24.9

[a] For flare dimensions shown on Figs. 24 and 25 but not covered in Table 5, see corresponding dimensions for the specified Tube OD in Table 3.

[b] Where thread relief undercut is used, last thread shall be chambered 1/2 to 1 pitch long from G diameter and dimension J may be reduced by an amount equal to 1/2 pitch.

[c] At manufacturer's option through passages in fittings shown in Fig. 24 may conform with the smaller diameter specified or be counterbored to the larger diameter from appropriate end for depth S.

[d] Minimum design thickness, not subject to inspection.

[e] ID of solder cup shall not be out of round by more than 0.003 in (0.08 mm).

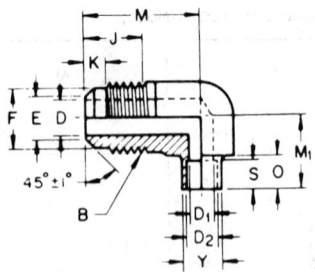

◊FIG. 25—FLARE TO
SOLDER 90 DEG
REDUCING ELBOW
(010204) (ES)

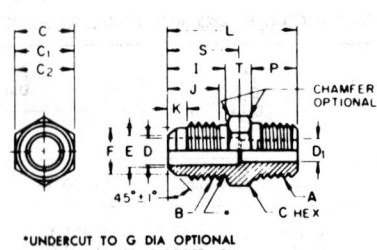

◊FIG. 26—REDUCING CONNECTOR
(HALF UNION)
(010102) (U1)

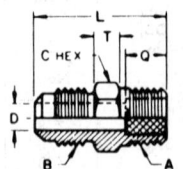

◊FIG. 27—FUSIBLE
REDUCING CONNECTOR
(HALF UNION)
(010163) (FU)

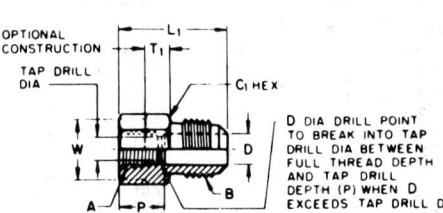

◊FIG. 28—INTERNAL THREAD REDUCING
CONNECTOR (HALF UNION)
(010103) (U3)

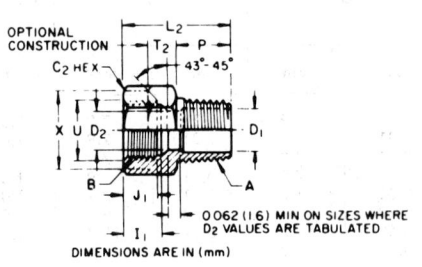

◊FIG. 29—INTERNAL FLARE TO
EXTERNAL PIPE REDUCING
ADAPTOR (010106) (U5)

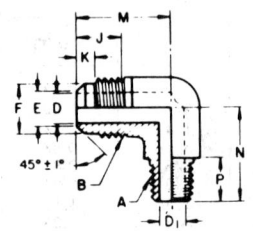

◊FIG. 30—90 DEG REDUCING ELBOW
(010202) (E1)

NOTE: UNSPECIFIED DETAIL WITH RESPECT TO DIMENSIONS, TOLERANCES, CONTOUR, MATERIAL, WORKMANSHIP, ETC., MUST CONFORM TO GENERAL SPECIFICATIONS FOR REFRIGERATION TUBE FITTINGS. THE DIMENSIONAL DESIGNATIONS IN FIGS. 1 AND 30 SHALL APPLY TO CORRESPONDING FEATURES OF OTHER FIGURES ON THIS PAGE UNLESS SHOWN OTHERWISE. THE ILLUSTRATIONS ON THIS PAGE APPLY TO TABLE 6. CODES SHOWN IN BRACKETS ADJACENT TO FIGURE NUMBERS REPRESENT RESPECTIVE FITTING IDENTIFICATION IN ACCORDANCE WITH SAE J846 (FIRST NUMBER) AND ANSI B70.1 (SECOND NUMBER).

◊ TABLE 6—DIMENSIONS OF REDUCING CONNECTORS, REDUCING ADAPTORS, AND REDUCING ELBOWS (FIGS. 26–32)[a]

B Nom Tube OD, in	A Dryseal Pipe Thread NPTF[b]	C Hex, in Nom	C_1 Hex, in Nom	C_2 Hex, in Nom	D^d Drill in	D^d Drill mm	D_1^d Drill in	D_1^d Drill mm	D_2 Drill in	D_2 Drill mm	L in ±0.03	L mm ±0.8	L_1 in ±0.03	L_1 mm ±0.8	L_2 in ±0.03	L_2 mm ±0.8	M in ±0.03	M mm ±0.8	M_1 in ±0.03	M_1 mm ±0.8	M_2 in ±0.03	M_2 mm ±0.8
3/16	1/4	9/16	11/16	9/16	0.125	3.18	0.312	7.92	0.188	4.78	1.19	30.2	1.19	30.2	0.94	23.9	0.81	20.6	0.91	23.1	0.71	18.0
3/16	3/8	11/16	13/16	11/16	0.125	3.18	0.406	10.31	0.188	4.78	1.25	31.8	1.22	31.0	0.84	21.3	0.88	22.4	0.97	24.6	0.75	19.0
3/16	1/2	7/8	1	7/8	0.125	3.18	0.562	14.27	0.188	4.78	1.50	38.1	1.44	36.6	1.06	26.9	0.97	24.6	1.06	26.9	0.80	20.3
3/16	3/4	1-1/16	1-1/4	1-1/16	0.125	3.18	0.750	19.05	0.188	4.78	1.62	41.1	1.50	38.1	1.19	30.2	1.06	26.9	1.22	31.0	0.84	21.3

B Nom Tube OD, in	N in ±0.03	N mm ±0.8	N_1 in ±0.03	N_1 mm ±0.8	N_2 in ±0.03	N_2 mm ±0.8	P in	P mm	Q^e in	Q^e mm	S^d Max in	S^d Max mm	T^f Ref in	T^f Ref mm	T_1 Min in	T_1 Min mm	T_2 Min in	T_2 Min mm	W^g in +0.00/-0.02	W^g mm +0.0/-0.5	X^g in +0.00/-0.02	X^g mm +0.0/-0.5
3/16	0.97	24.6	0.62	15.8	0.86	21.8	0.56	14.2	—	—	0.66	16.8	0.18	4.6	0.24	6.1	0.21	5.3	0.69	17.5	0.56	14.2
3/16	1.00	25.4	0.62	15.8	0.95	24.1	0.56	14.2	—	—	0.69	17.5	0.24	6.1	0.30	7.6	0.24	6.1	0.81	20.6	0.69	17.5
3/16	1.28	32.5	0.81	20.6	1.17	29.7	0.75	19.0	—	—	0.91	23.1	0.30	7.6	0.37	9.4	0.30	7.6	1.00	25.4	0.88	22.4
3/16	1.38	35.1	0.81	20.6	1.20	30.5	0.75	19.0	—	—	0.97	24.6	0.43	10.9	0.49	12.4	0.43	10.9	1.25	31.8	1.06	26.9

B Nom Tube OD, in	A Dryseal Pipe Thread NPTF	C Hex, in Nom	C_1 Hex, in Nom	C_2 Hex, in Nom	D^d Drill in	D^d Drill mm	D_1^d Drill in	D_1^d Drill mm	D_2 Drill in	D_2 Drill mm	L in ±0.03	L mm ±0.8	L_1 in ±0.03	L_1 mm ±0.8	L_2 in ±0.03	L_2 mm ±0.8	M in ±0.03	M mm ±0.8	M_1 in ±0.03	M_1 mm ±0.8	M_2 in ±0.03	M_2 mm ±0.8
1/4	1/4	9/16	11/16	5/8	0.188	4.78	0.312	7.92	0.250	6.35	1.25	31.8	1.25	31.8	1.03	26.2	0.91	23.1	0.97	24.6	0.75	19.0
1/4	3/8	11/16	13/16	11/16	0.188	4.78	0.406	10.31	0.250	6.35	1.31	33.3	1.28	32.5	0.94	23.9	0.94	23.9	1.03	26.2	0.79	20.1
1/4	1/2	7/8	1	7/8	0.188	4.78	0.562	14.27	0.250	6.35	1.56	39.6	1.50	38.1	1.06	26.9	1.03	26.2	1.12	28.4	0.85	21.6
1/4	3/4	1-1/16	1-1/4	1-1/16	0.188	4.78	0.750	19.05	0.250	6.35	1.69	42.9	1.56	39.6	1.19	30.2	1.12	28.4	1.28	32.5	0.89	22.6

(Table 6 continued on next page)

TABLE 6—DIMENSIONS OF REDUCING CONNECTORS, REDUCING ADAPTORS, AND REDUCING ELBOWS (FIGS. 26–32)[a] (CONTINUED)

B Nom Tube OD, in	N in ±0.03	N mm ±0.8	N_1 in ±0.03	N_1 mm ±0.8	N_2 in ±0.03	N_2 mm ±0.8	P in	P mm	Q^e in	Q^e mm	S^d Max in	S^d Max mm	T^f Ref in	T^f Ref mm	T_1 Min in	T_1 Min mm	T_2 Min in	T_2 Min mm	W^g in +0.00/−0.02	W^g mm +0.0/−0.5	X^g in +0.00/−0.02	X^g mm +0.0/−0.5
1/4	0.94	23.9	0.66	16.8	0.86	21.8	0.56	14.2	0.66	16.8	0.66	16.8	0.18	4.6	0.24	6.1	0.24	6.1	0.69	17.5	0.62	15.7
1/4	1.03	26.2	0.66	16.8	0.95	24.1	0.56	14.2	0.69	17.5	0.69	17.5	0.24	6.1	0.30	7.6	0.24	6.1	0.81	20.6	0.69	17.5
1/4	1.28	32.5	0.84	21.3	1.17	29.7	0.75	19.0	—	—	0.91	23.1	0.30	7.6	0.37	9.4	0.30	7.6	1.00	25.4	0.88	22.4
1/4	1.38	35.1	0.84	21.3	1.20	30.5	0.75	19.0	—	—	0.97	24.6	0.43	10.9	0.49	12.4	0.43	10.9	1.25	31.8	1.06	26.9

B Nom Tube OD, in	A Dryseal Pipe Thread NPTF	C Hex, in Nom	C_1 Hex, in Nom	C_2 Hex, in Nom	D^d Drill in	D^d Drill mm	D_1^d Drill in	D_1^d Drill mm	D_2 Drill in	D_2 Drill mm	L in ±0.03	L mm ±0.8	L_1 in ±0.03	L_1 mm ±0.8	L_2 in ±0.03	L_2 mm ±0.8	M in ±0.03	M mm ±0.8	M_1 in ±0.03	M_1 mm ±0.8	M_2 in ±0.03	M_2 mm ±0.8
5/16	1/4	9/16	11/16	11/16	0.219	5.56	0.312	7.92	0.312	7.92	1.34	34.0	1.28	32.5	1.09	27.7	0.97	24.6	1.03	26.2	0.81	20.6
5/16	3/8	11/16	13/16	11/16	0.219	5.56	0.406	10.31	0.312	7.92	1.38	35.1	1.31	33.3	1.00	25.4	1.00	25.4	1.09	27.7	0.85	21.6
5/16	1/2	7/8	1	7/8	0.219	5.56	0.562	14.27	0.312	7.92	1.62	41.1	1.53	38.9	1.12	28.4	1.09	27.7	1.19	30.2	0.91	23.1
5/16	3/4	1-1/16	1-1/4	1-1/16	0.219	5.56	0.750	19.05	0.312	7.92	1.75	44.4	1.52	40.4	1.19	30.2	1.19	30.2	1.34	34.0	0.95	24.1

B Nom Tube OD, in	N in ±0.03	N mm ±0.8	N_1 in ±0.03	N_1 mm ±0.8	N_2 in ±0.03	N_2 mm ±0.8	P in	P mm	Q^e in	Q^e mm	S^d Max in	S^d Max mm	T^f Ref in	T^f Ref mm	T_1 Min in	T_1 Min mm	T_2 Min in	T_2 Min mm	W^g in +0.00/−0.02	W^g mm +0.0/−0.5	X^g in +0.00/−0.02	X^g mm +0.0/−0.5
5/16	0.94	23.9	0.66	16.8	0.86	21.8	0.56	14.2	—	—	0.67	17.0	0.21	5.3	0.24	6.1	0.24	6.1	0.69	17.5	0.69	17.5
5/16	1.03	26.2	0.66	16.8	0.95	24.1	0.56	14.2	—	—	0.69	17.5	0.24	6.1	0.30	7.6	0.24	6.1	0.81	20.6	0.69	17.5
5/16	1.28	32.5	0.84	21.3	1.17	29.7	0.75	19.0	—	—	0.91	23.1	0.30	7.6	0.37	9.4	0.30	7.6	1.00	25.4	0.88	22.4
5/16	1.38	35.1	0.84	21.3	1.20	30.5	0.75	19.0	—	—	0.97	24.6	0.43	10.9	0.49	12.4	0.43	10.9	1.25	31.8	1.06	26.9

B Nom Tube OD, in	A Dryseal Pipe Thread NPTF	C Hex, in Nom	C_1 Hex, in Nom	C_2 Hex, in Nom	D^d Drill in	D^d Drill mm	D_1^d Drill in	D_1^d Drill mm	D_2 Drill in	D_2 Drill mm	L in ±0.03	L mm ±0.8	L_1 in ±0.03	L_1 mm ±0.8	L_2 in ±0.03	L_2 mm ±0.8	M in ±0.03	M mm ±0.8	M_1 in ±0.03	M_1 mm ±0.8	M_2 in ±0.03	M_2 mm ±0.8
3/8	1/8	5/8	5/8	13/16	0.281	7.14	0.219	5.56	—	—	1.25	31.8	1.12	28.4	1.09	27.7	1.03	26.2	1.06	26.9	0.89	22.6
3/8	3/8	11/16	13/16	13/16	0.281	7.14	0.406	10.31	0.375	9.52	1.44	36.6	1.38	35.1	1.12	28.4	1.06	26.9	1.16	29.5	0.93	23.6
3/8	1/2	7/8	1	7/8	0.281	7.14	0.562	14.27	0.375	9.52	1.69	42.9	1.62	41.1	1.25	31.8	1.16	29.5	1.25	31.8	0.99	25.1
3/8	3/4	1-1/16	1-1/4	1-1/16	0.281	7.14	0.750	19.05	0.375	9.52	1.81	46.0	1.66	42.2	1.19	30.2	1.25	31.8	1.41	35.8	1.03	26.2

B Nom Tube OD, in	N in ±0.03	N mm ±0.8	N_1 in ±0.03	N_1 mm ±0.8	N_2 in ±0.03	N_2 mm ±0.8	P in	P mm	Q^e in	Q^e mm	S^d Max in	S^d Max mm	T^f Ref in	T^f Ref mm	T_1 Min in	T_1 Min mm	T_2 Min in	T_2 Min mm	W^g in +0.00/−0.02	W^g mm +0.0/−0.5	X^g in +0.00/−0.02	X^g mm +0.0/−0.5
3/8	0.91	23.1	0.50	12.7	0.67	17.0	0.38	9.7	—	—	0.75	19.0	0.24	6.1	0.24	6.1	0.30	7.6	0.62	15.7	0.81	20.6
3/8	1.09	27.7	0.69	17.5	0.95	24.1	0.56	14.2	0.69	17.5	0.69	17.5	0.24	6.1	0.30	7.6	0.30	7.6	0.81	20.6	0.81	20.6
3/8	1.28	32.5	0.88	22.4	1.17	29.7	0.75	19.0	—	—	0.91	23.1	0.30	7.6	0.37	9.4	0.30	7.6	1.00	25.4	0.88	22.4
3/8	1.38	35.1	0.88	22.4	1.20	30.5	0.75	19.0	—	—	0.97	24.6	0.43	10.9	0.49	12.4	0.43	10.9	1.25	31.8	1.06	26.9

B Nom Tube OD, in	A Dryseal Pipe Thread NPTF	C Hex, in Nom	C_1 Hex, in Nom	C_2 Hex, in Nom	D^d Drill in	D^d Drill mm	D_1^d Drill in	D_1^d Drill mm	D_2 Drill in	D_2 Drill mm	L in ±0.03	L mm ±0.8	L_1 in ±0.03	L_1 mm ±0.8	L_2 in ±0.03	L_2 mm ±0.8	M in ±0.03	M mm ±0.8	M_1 in ±0.03	M_1 mm ±0.8	M_2 in ±0.03	M_2 mm ±0.8
1/2	1/8	3/4	3/4	15/16	0.406	10.31	0.219	5.56	—	—	1.44	36.6	1.19	30.2	1.25	31.8	1.22	31.0	1.22	31.0	1.06	26.9
1/2	1/4	3/4	3/4	15/16	0.406	10.31	0.312	7.92	—	—	1.62	41.1	1.41	35.8	1.34	34.0	1.22	31.0	1.22	31.0	1.06	26.9
1/2	1/2	7/8	1	15/16	0.406	10.31	0.562	14.27	0.500	12.70	1.81	46.0	1.75	44.4	1.47	37.3	1.28	32.5	1.38	35.1	1.12	28.4
1/2	3/4	1-1/16	1-1/4	1-1/16	0.406	10.31	0.750	19.05	0.500	12.70	1.94	49.3	1.81	46.0	1.38	35.1	1.38	35.1	1.53	38.9	1.16	29.5

(Table 6 continued on next page)

TABLE 6—DIMENSIONS OF REDUCING CONNECTORS, REDUCING ADAPTORS, AND REDUCING ELBOWS (FIGS. 26–32)[a] (CONTINUED)

B Nom Tube OD in	N in ±0.03	N mm ±0.8	N_1 in ±0.03	N_1 mm ±0.8	N_2 in ±0.03	N_2 mm ±0.8	P in	P mm	Q[e] in	Q[e] mm	S[d] Max in	S[d] Max mm	T[f] Ref in	T[f] Ref mm	T_1 Min in	T_1 Min mm	T_2 Min in	T_2 Min mm	W[g] in +0.00/−0.02	W[g] mm +0.0/−0.5	X[g] in +0.00/−0.02	X[g] mm +0.0/−0.5
1/2	1.00	25.4	0.56	14.2	0.76	19.3	0.38	9.7	—	—	0.91	23.1	0.30	7.6	0.30	7.6	0.37	9.4	0.75	19.0	0.94	23.9
1/2	1.19	30.2	0.75	19.0	0.95	24.1	0.56	14.2	—	—	0.91	23.1	0.30	7.6	0.30	7.6	0.37	9.4	0.75	19.0	0.94	23.9
1/2	1.38	35.1	0.94	23.9	1.17	29.7	0.75	19.0	—	—	0.91	23.1	0.30	7.6	0.37	9.4	0.37	9.4	1.00	25.4	0.94	23.9
1/2	1.38	35.1	0.94	23.9	1.20	30.5	0.75	19.0	—	—	0.97	24.6	0.43	10.9	0.49	12.4	0.43	10.9	1.25	31.8	1.06	26.9

B Nom Tube OD in	A Dryseal Pipe Thread NPTF Nom	C Hex, in Nom	C_1 Hex, in Nom	C_2 Hex, in Nom	D[d] Drill in	D[d] Drill mm	D_1[d] Drill in	D_1[d] Drill mm	D_2 Drill in	D_2 Drill mm	L in ±0.03	L mm ±0.8	L_1 in ±0.03	L_1 mm ±0.8	L_2 in ±0.03	L_2 mm ±0.8	M in ±0.03	M mm ±0.8	M_1 in ±0.03	M_1 mm ±0.8	M_2 in ±0.03	M_2 mm ±0.8
5/8	1/8	7/8	7/8	1-1/16	0.500	12.70	0.219	5.56	—	—	1.62	41.1	1.28	32.5	1.41	35.8	1.41	35.8	1.41	35.8	1.23	31.2
5/8	1/4	7/8	7/8	1-1/16	0.500	12.70	0.312	7.92	—	—	1.81	46.0	1.50	38.1	1.53	38.9	1.41	35.8	1.41	35.8	1.23	31.2
5/8	3/8	7/8	7/8	1-1/16	0.500	12.70	0.406	10.31	—	—	1.81	46.0	1.59	40.4	1.56	39.6	1.41	35.8	1.41	35.8	1.23	31.2
5/8	3/4	1-1/16	1-1/4	1-1/16	0.500	12.70	0.750	19.05	0.625	15.88	2.06	52.3	1.91	48.5	1.66	42.2	1.44	36.6	1.66	42.2	1.27	32.3

B Nom Tube OD in	N in ±0.03	N mm ±0.8	N_1 in ±0.03	N_1 mm ±0.8	N_2 in ±0.03	N_2 mm ±0.8	P in	P mm	Q[e] in	Q[e] mm	S[d] Max in	S[d] Max mm	T[f] Ref in	T[f] Ref mm	T_1 Min in	T_1 Min mm	T_2 Min in	T_2 Min mm	W[g] in +0.00/−0.02	W[g] mm +0.0/−0.5	X[g] in +0.00/−0.02	X[g] mm +0.0/−0.5
5/8	1.06	26.9	0.62	15.7	0.79	20.1	0.38	9.7	—	—	1.07	27.2	0.37	9.4	0.37	9.4	0.43	10.9	0.88	22.4	1.06	26.9
5/8	1.25	31.8	0.81	20.6	0.98	24.9	0.56	14.2	—	—	1.07	27.2	0.37	9.4	0.37	9.4	0.43	10.9	0.88	22.4	1.06	26.9
5/8	1.25	31.8	0.81	20.6	0.98	24.9	0.56	14.2	—	—	1.07	27.2	0.37	9.4	0.37	9.4	0.43	10.9	0.88	22.4	1.06	26.9
5/8	1.50	38.1	1.00	25.4	1.20	30.5	0.75	19.0	—	—	0.97	24.6	0.43	10.9	0.49	12.4	0.43	10.9	1.25	31.8	1.06	26.9

B Nom Tube OD in	A Dryseal Pipe Thread NPTF Nom	C Hex, in Nom	C_1 Hex, in Nom	C_2 Hex, in Nom	D[d] Drill in	D[d] Drill mm	D_1[d] Drill in	D_1[d] Drill mm	D_2 Drill in	D_2 Drill mm	L in ±0.03	L mm ±0.8	L_1 in ±0.03	L_1 mm ±0.8	L_2 in ±0.03	L_2 mm ±0.8	M in ±0.03	M mm ±0.8	M_1 in ±0.03	M_1 mm ±0.8	M_2 in ±0.03	M_2 mm ±0.8
3/4	1/8	1-1/16	1-1/16	1-5/16	0.625	15.88	0.219	5.56	—	—	1.81	46.0	1.38	35.1	1.62	41.1	1.62	41.1	1.59	40.4	1.41	35.8
3/4	1/4	1-1/16	1-1/16	1-5/16	0.625	15.88	0.312	7.92	—	—	2.00	50.8	1.56	39.6	1.75	44.4	1.62	41.1	1.59	40.4	1.41	35.8
3/4	3/8	1-1/16	1-1/16	1-5/16	0.625	15.88	0.406	10.31	—	—	2.00	50.8	1.66	42.2	1.78	45.2	1.62	41.1	1.59	40.4	1.41	35.8
3/4	3/4	1-1/16	1-1/4	1-5/16	0.625	15.88	0.750	19.05	—	—	2.19	55.6	1.97	50.0	1.78	45.2	1.59	40.4	1.78	45.2	1.41	35.8

B Nom Tube OD in	N in ±0.03	N mm ±0.8	N_1 in ±0.03	N_1 mm ±0.8	N_2 in ±0.03	N_2 mm ±0.8	P in	P mm	Q[e] in	Q[e] mm	S[d] Max in	S[d] Max mm	T[f] Ref in	T[f] Ref mm	T_1 Min in	T_1 Min mm	T_2 Min in	T_2 Min mm	W[g] in +0.00/−0.02	W[g] mm +0.0/−0.5	X[g] in +0.00/−0.02	X[g] mm +0.0/−0.5
3/4	1.22	31.0	0.69	17.5	0.82	20.8	0.38	9.7	—	—	1.22	31.0	0.43	10.9	0.43	10.9	0.49	12.4	1.06	26.9	1.31	33.3
3/4	1.41	35.8	0.88	22.4	1.01	25.7	0.56	14.2	—	—	1.22	31.0	0.43	10.9	0.43	10.9	0.49	12.4	1.06	26.9	1.31	33.3
3/4	1.41	35.8	0.88	22.4	1.01	25.7	0.56	14.2	—	—	1.22	31.0	0.43	10.9	0.43	10.9	0.49	12.4	1.06	26.9	1.31	33.3
3/4	1.62	41.1	1.06	26.9	1.20	30.5	0.75	19.0	—	—	0.97	24.6	0.43	10.9	0.49	12.4	0.49	12.4	1.25	31.8	1.31	33.3

[a] For flare dimensions shown on Figs. 26–32 but not given in Table 6, see corresponding dimensions for the specified Tube OD in Table 3.
[b] Dryseal American Standard Taper Pipe Thread.
[c] Where thread relief undercut is used, last thread shall be chamfered 1/2 to 1 pitch long from G diameter and dimension J may be reduced by an amount equal to 1/2 pitch.
[d] At manufacturer's option, through passages in fittings shown in Fig. 26 may conform with the smaller diameter specified or be counterbored to the larger diameter from the appropriate end for depth S.
[e] Available with three types of fusible alloys as specified in General Specifications.
[f] Minimum design thickness, not subject to inspection.
[g] Basic dimensions shown shall apply as minimum for bosses. The −0.02 in (0.51 mm) tolerance shall apply only to chamfer diameters on full hexagon versions of fittings shown in Figs. 28 and 29.

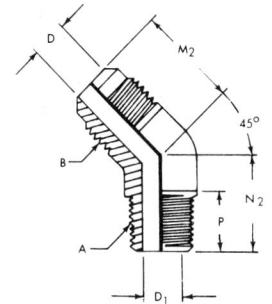

ɸFIG. 31—45 DEG REDUCING ELBOW
(010302) (E5)

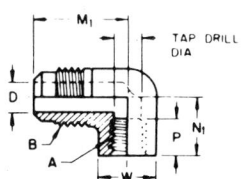

ɸFIG. 32—90 DEG INTERNAL PIPE
THREAD REDUCING ELBOW
(010203) (E3)

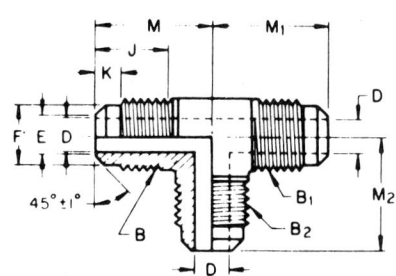

ɸFIG. 33—THREE-WAY REDUCING TEE
(010401) (TR2)

NOTE: UNSPECIFIED DETAIL WITH RESPECT TO DIMENSIONS, TOLERANCES, CONTOUR, MATERIAL, WORKMANSHIP, ETC., MUST CONFORM TO GENERAL SPECIFICATIONS FOR REFRIGERATION TUBE FITTINGS. THE ILLUSTRATIONS ON THIS PAGE APPLY TO TABLES 7-10. CODES SHOWN IN BRACKETS ADJACENT TO FIGURE NUMBERS REPRESENT RESPECTIVE FITTING IDENTIFICATION IN ACCORDANCE WITH SAE J846 (FIRST NUMBER) AND ANSI B70.1 (SECOND NUMBER).

ɸ TABLE 7—END TO CENTER DIMENSIONS OF FLARE TO FLARE ENDS ON REDUCING TEES[a]

B and B_1 Tube OD, of Run, in	End to Center ±0.03 in ±0.8 mm	B_2, Nominal Flare Sizes of Branch, in												
		3/16		1/4		5/16		3/8		1/2		5/8		3/4	
		in	mm	in	mm	in	mm	in	mm	in	mm	in	mm	in	mm
3/16	M or M_1	0.75	19.0	0.75	19.0	0.78	19.8	0.84	21.3	0.91	23.1	0.97	24.6	1.06	26.9
	M_2	0.75	19.0	0.88	22.4	0.91	23.1	1.06	26.9	1.22	31.0	1.41	35.8	1.66	42.2
1/4	M or M_1	0.88	22.4	0.88	22.4	0.84	21.3	0.91	23.1	0.97	24.6	1.03	26.2	1.12	28.4
	M_2	0.75	19.0	0.88	22.4	0.91	23.1	1.06	26.9	1.22	31.0	1.41	35.8	1.66	42.2
5/16	M or M_2	0.91	23.1	0.91	23.1	0.91	23.1	0.97	24.6	1.03	26.2	1.09	27.7	1.19	30.2
	M_2	0.78	19.8	0.84	21.3	0.91	23.1	1.06	26.9	1.22	31.0	1.41	35.8	1.66	42.2
3/8	M or M_1	1.06	26.9	1.06	26.9	1.06	26.9	1.06	26.9	1.09	27.7	1.16	29.5	1.25	31.8
	M_2	0.84	21.3	0.91	23.1	0.97	24.6	1.06	26.9	1.22	31.0	1.41	35.8	1.66	42.2
1/2	M or M_1	1.22	31.0	1.22	31.0	1.22	31.0	1.22	31.0	1.22	31.0	1.28	32.5	1.38	35.1
	M_2	0.91	23.1	0.97	24.6	1.03	26.2	1.09	27.7	1.22	31.0	1.41	35.8	1.66	42.2
5/8	M or M_1	1.41	35.8	1.41	35.8	1.41	35.8	1.41	35.8	1.41	35.8	1.41	35.8	1.50	38.1
	M_2	0.97	24.6	1.03	26.2	1.09	27.7	1.16	29.5	1.28	32.5	1.41	35.8	1.66	42.2
3/4	M or M_1	1.66	42.2	1.66	42.2	1.66	42.2	1.66	42.2	1.66	42.2	1.66	42.2	1.66	42.2
	M_2	1.06	26.9	1.12	28.4	1.19	30.2	1.25	31.8	1.38	35.1	1.50	38.1	1.66	42.2

[a] For flare and pipe thread dimensions shown on Figs. 33–35, see corresponding dimensions for specified Tube OD and specified pipe thread size in Table 3. For passage diameters, see Tables 9 and 10.

ɸ TABLE 8—END TO CENTER DIMENSIONS OF FLARE TO PIPE ENDS ON REDUCING TEES[a]

B, B_1 or B_2 Tube OD, in	End to Center ±0.03 in ±0.8 mm	A_1 Dryseal Taper Thread, NPTF[b], in									
		1/8		1/4		3/8		1/2		3/4	
		in	mm	in	mm	in	mm	in	mm	in	mm
3/16	M_3	0.75	19.0	0.81	20.6	0.88	22.4	0.97	24.6	1.06	26.9
	N	0.75	19.0	0.97	24.6	1.00	25.4	1.28	32.5	1.38	35.1
1/4	M_3	0.81	20.6	0.91	23.1	0.94	23.9	1.03	26.2	1.12	28.4
	N	0.78	19.8	0.94	23.9	1.03	26.2	1.23	32.5	1.38	35.1
5/16	M_3	0.91	23.1	0.97	24.6	1.00	25.4	1.09	27.7	1.19	30.2
	N	0.78	19.8	0.94	23.9	1.03	26.2	1.28	32.5	1.38	35.1
3/8	M_3	1.03	26.2	1.00	25.4	1.06	26.9	1.16	29.5	1.25	31.8
	N	0.91	23.1	1.06	26.9	1.09	27.7	1.28	32.5	1.38	35.1
1/2	M_3	1.22	31.0	1.22	31.0	1.22	31.0	1.28	32.5	1.38	35.1
	N	1.00	25.4	1.19	30.2	1.12	28.4	1.38	35.1	1.38	35.1
5/8	M_3	1.41	35.8	1.41	35.8	1.41	35.8	1.41	35.8	1.44	36.6
	N	1.06	26.9	1.25	31.8	1.25	31.8	1.38	35.1	1.50	38.1
3/4	M_3	1.62	41.1	1.62	41.1	1.62	41.1	1.62	41.1	1.59	40.4
	N	1.22	31.0	1.41	35.8	1.41	35.8	1.50	38.1	1.62	41.1

[a] For flare and pipe thread dimensions shown on Figs. 33–35, see corresponding dimensions for specified Tube OD and specified pipe thread size in Table 3. For passage diameters, see Tables 9 and 10.
[b] Dryseal American Standard Taper Pipe Thread.

TABLE 9—PASSAGE DIAMETERS THROUGH FLARE ENDS

Nom Tube OD, in	D^a Dia Drill	
	in	mm
3/16	0.125	3.18
1/4	0.188	4.78
5/16	0.219	5.56
3/8	0.281	7.14
1/2	0.406	10.31
5/8	0.500	12.70
3/4	0.625	15.88

[a] At manufacturer's option, through passages in tees shown in Figs. 33–35 having varying diameters at opposite ends may conform to the smaller diameters specified or be counterbored to the larger diameter from the appropriate end for a minimum depth equivalent to the maximum end to center length of that end, plus 1/2 the maximum passage through branch plus 0.01 in (0.3 mm).

TABLE 10—PASSAGE DIAMETERS THROUGH PIPE THREAD ENDS

Nom Pipe Size, in	D^a Dia Drill	
	in	mm
1/8	0.219	5.56
1/4	0.312	7.92
3/8	0.406	10.31
1/2	0.562	14.27
3/4	0.750	19.05

[a] At manufacturer's option, through passages in tees shown in Figs. 33–35 having varying diameters at opposite ends may conform to the smaller diameters specified or be counterbored to the larger diameter from the appropriate end for a minimum depth equivalent to the maximum end to center length of that end, plus 1/2 the maximum passage through branch, plus 0.01 in (0.3 mm).

TO DETERMINE CORRECT END TO CENTER LENGTHS ON TEES, EACH 90 DEG MUST BE FIGURED SEPARATELY AS AN ELBOW AND THE LARGER OF THE TWO BRANCH LENGTHS SHALL APPLY. SEE EXAMPLES BELOW.

EXAMPLE: Find lengths for 5/8 x 1/2 x 3/8 in three way tee.
1. From Table 7 obtain values for each 90 deg separately.
2. Use larger of two M_2 dimensions as found.

EXAMPLE: Find lengths for 5/8 x 3/8 x 1/2 in right angle two way tee.
1. From Tables 7 and 8 obtain values for each 90 deg separately.
2. Use the larger dimension M_2 or M_3 as found.

FROM TABLE 7

B Nominal Tube OD, in	End to Center	B_2	
		3/8	
		in	mm
5/8	M	1.41	35.8
	M_2	1.16	29.5

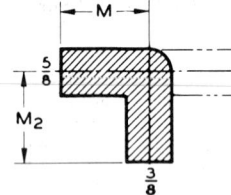

FROM TABLE 7

B Nominal Tube OD, in	Center End to Center	B_2	
		1/2	
		in	mm
5/8	M	1.41	35.8
	M_2	1.28	32.5

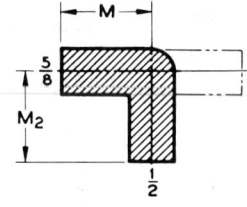

FROM TABLE 7

B_1 Nominal Tube OD, in	End to Center	B_2	
		3/8	
		in	mm
1/2	M_1	1.22	31.0
	M_2	1.09	27.7

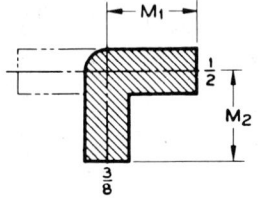

FROM TABLE 8

B_2 Nominal Tube OD, in	End to Center	A	
		3/8	
		in	mm
1/2	M_3	1.22	31.0
	N	1.12	28.4

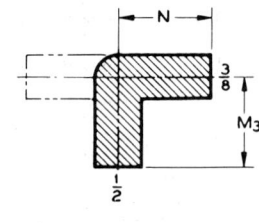

Result:

Dimension		in	mm
M	=	1.41	35.8
N	=	1.12	28.4
M_2	=	1.28	32.5

Result:

Dimension		in	mm
M	=	1.41	35.8
M_1	=	1.22	31.0
M_2	=	1.16	29.5

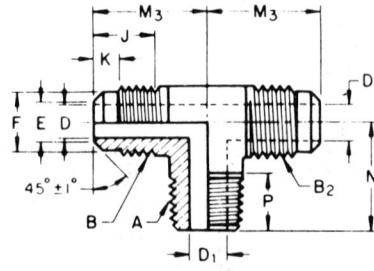

φFIG. 34—TWO-WAY REDUCING TEE
(010425) (T1)

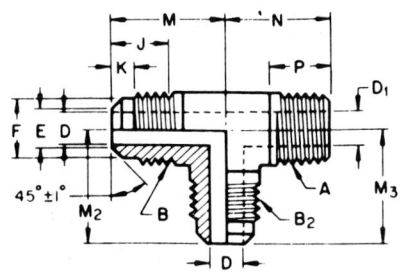

φFIG. 35—RIGHT ANGLE TWO WAY REDUCING TEE
(010424) (T3)

NOTE: UNSPECIFIED DETAIL WITH RESPECT TO DIMENSIONS, TOLERANCES, CONTOUR, MATERIAL, WORKMANSHIP, ETC., MUST CONFORM TO GENERAL SPECIFICATIONS FOR REFRIGERATION TUBE FITTINGS. CODES SHOWN IN BRACKETS ADJACENT TO FIGURE NUMBERS REPRESENT RESPECTIVE FITTING IDENTIFICATION IN ACCORDANCE WITH SAE J846 (FIRST NUMBER) AND ANSI B70.1 (SECOND NUMBER).

21.43

TABLE 11—DIMENSIONS FOR STANDARD SIZES OF SHORT AND LONG FLARE NUTS (FIGS. 36A AND 36B)

Nom Tube OD, in	B Straight Thread Nom Size, in	C Hex, in Nom	D Dia		E Dia Min		I		J Full Thread Min	
			in +0.005 −0.000	mm +0.13 −0.00	in	mm	in	mm	in	mm
3/16	3/8 −24	1/2	0.192	4.88	0.41	10.4	0.28	7.1	0.22	5.6
1/4	7/16−20	5/8	0.255	6.48	0.47	11.9	0.34	8.6	0.27	6.9
5/16	1/2 −20	11/16	0.317	8.05	0.47	11.9	0.38	9.7	0.30	7.6
3/8	5/8 −18	13/16	0.380	9.65	0.59	15.0	0.44	11.2	0.34	8.6
1/2	3/4 −16	15/16	0.505	12.83	0.75	19.0	0.53	13.5	0.44	11.2
5/8	7/8 −14	1- 1/16	0.630	16.00	0.94	23.9	0.66	16.8	0.55	14.0
3/4	1-1/16−14	1- 5/16	0.755	19.18	1.12	28.4	0.78	19.8	0.67	17.0

Nom Tube OD, in	L		L₁		T		U Dia				X	
	in ±0.03	mm ±0.8	in +0.09 −0.00	mm +2.3 −0.0	in ±0.03	mm ±0.8	Min in	mm	Max in	mm	in +0.00 −0.03	mm +0.0 −0.8
3/16	0.88	22.4	0.53	13.5	0.38	9.7	0.39	9.9	0.41	10.4	0.50	12.7
1/4	0.94	23.9	0.59	15.0	0.44	11.2	0.45	11.4	0.47	11.9	0.62	15.7
5/16	0.94	23.9	0.62	15.7	0.44	11.2	0.51	13.0	0.53	13.5	0.69	17.5
3/8	1.06	26.9	0.69	17.5	0.50	12.7	0.64	16.3	0.67	17.0	0.81	20.6
1/2	1.19	30.2	0.81	20.6	0.56	14.2	0.77	19.6	0.80	20.3	0.94	23.9
5/8	1.44	36.6	0.94	23.9	0.75	19.0	0.90	22.9	0.93	23.6	1.06	26.9
3/4	1.75	44.5	1.12	28.4	1.00	25.4	1.08	27.4	1.11	28.2	1.31	33.3

TABLE 12—DIMENSIONS FOR REDUCING SIZES OF SHORT AND LONG FLARE NUTS (FIGS. 36A AND 36B)

Nom Tube OD, in	Nom Tube OD, in	B Straight Thread Nom Size, in	C Hex, in Nom	D Dia		E Dia Min		I		J Full Thread Min	
				in +0.005 −0.000	mm +0.13 −0.00	in	mm	in	mm	in	mm
1/4	3/16	7/16−20	5/8	0.192	4.88	0.41	10.4	0.34	8.6	0.27	6.9
5/16	3/16	1/2 −20	11/16	0.192	4.88	0.41	10.4	0.38	9.7	0.30	7.6
5/16	1/4	1/2 −20	11/16	0.255	6.48	0.47	11.9	0.38	9.7	0.30	7.6
3/8	1/4	5/8 −18	13/16	0.255	6.48	0.47	11.9	0.44	11.2	0.34	8.6
3/8	5/16	5/8 −18	13/16	0.317	8.05	0.47	11.9	0.44	11.2	0.34	8.6
1/2	3/8	3/4 −16	15/16	0.380	9.65	0.59	15.0	0.53	13.5	0.44	11.2
5/8	3/8	7/8 −14	1- 1/16	0.380	9.65	0.59	15.0	0.66	16.8	0.55	14.0
5/8	1/2	7/8 −14	1- 1/16	0.505	12.83	0.75	19.0	0.66	16.8	0.55	14.0
3/4	1/2	1-1/16−14	1- 5/16	0.505	12.83	0.75	19.0	0.78	19.8	0.67	17.0
3/4	5/8	1-1/16−14	1- 5/16	0.630	16.00	0.94	23.9	0.78	19.8	0.67	17.0

Nom Tube OD, in	L		L₁		T		U Dia				X	
	in ±0.03	mm ±0.8	in +0.09 −0.00	mm +2.3 −0.0	in ±0.03	mm ±0.8	Min in	mm	Max in	mm	in +0.03 −0.00	mm +0.8 −0.0
1/4	0.94	23.9	0.59	15.0	0.44	11.2	0.45	11.4	0.47	11.9	0.62	15.7
5/16	0.94	23.9	—	—	0.44	11.2	0.51	13.0	0.53	13.5	0.69	17.5
5/16	0.94	23.9	0.62	15.8	0.44	11.2	0.51	13.0	0.53	13.5	0.69	17.5
3/8	1.06	26.9	0.69	17.5	0.50	12.7	0.64	16.3	0.67	17.0	0.81	20.6
3/8	1.06	26.9	0.69	17.5	0.50	12.7	0.64	16.3	0.67	17.0	0.81	20.6
1/2	1.19	30.2	0.81	20.6	0.56	14.2	0.77	19.6	0.80	20.3	0.94	23.9
5/8	1.44	36.6	—	—	0.75	19.0	0.90	22.9	0.93	23.6	1.06	26.9
5/8	1.44	36.6	0.94	23.9	0.75	19.0	0.90	22.9	0.93	23.6	1.06	26.9
3/4	1.75	44.4	1.12	28.4	1.00	25.4	1.08	27.4	1.11	28.2	1.31	33.3
3/4	1.75	44.4	1.12	28.4	1.00	25.4	1.08	27.4	1.11	28.2	1.31	33.3

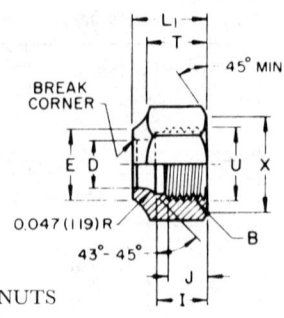

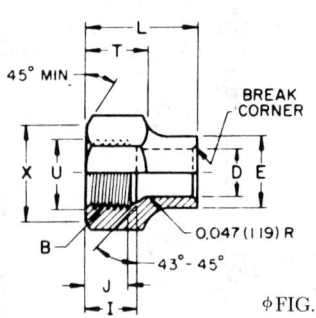

φFIG. 36A—SHORT NUTS
(010166) (NS4)
(010168) (NRS4)

φFIG. 36B—LONG NUTS
(010167) (N4)
(01069) (NR4)

φFIG. 36—FLARE NUTS AND REDUCING FLARE NUTS

φ TABLE 13—DIMENSIONS OF FLARE CAP, FLARE GASKET, AND FLARE SEAL BONNET (FIGS. 37–39)

Nom Tube OD, in	B Straight Thread Nom Size, Class 2B Int	C Hex, in	D Dia		I		J Full Thread Min		L		S Max		T Min	
		Nom	in	mm	in	mm	in	mm	in	mm	in	mm	in	mm
3/16	3/8 –24	1/2	0.12	3.0	0.28	7.1	0.22	5.6	0.47	11.9	0.34	8.6	0.27	6.9
1/4	7/16–20	9/16	0.19	4.3	0.34	8.6	0.27	6.9	0.53	13.5	0.41	10.4	0.40	10.2
5/16	1/2 –20	5/8	0.22	5.6	0.38	9.7	0.30	7.6	0.62	15.7	0.50	12.7	0.40	10.2
3/8	5/8 –18	3/4	0.28	7.1	0.44	11.2	0.34	8.6	0.69	17.5	0.56	14.2	0.46	11.7
1/2	3/4 –16	7/8	0.41	10.4	0.53	13.5	0.44	11.2	0.84	21.3	0.72	18.3	0.52	13.2
5/8	7/8 –14	1-1/16	0.50	12.7	0.66	16.8	0.55	14.0	0.97	24.6	0.84	21.3	0.62	15.7
3/4	1-1/16–14	1-5/16	0.62	15.7	0.78	19.8	0.67	17.0	1.09	27.7	0.97	24.6	0.74	18.8

Nom Tube OD, in	U Dia				V Dia		W Dia Max		X		Y Min	
	Min		Max									
	in	mm	in	mm	±0.03	±0.8	in	mm	±0.02	±0.5	in	mm
3/16	0.39	9.9	0.41	10.4	0.12	3.0	0.31	7.9	0.16	4.1	0.16	4.1
1/4	0.45	11.4	0.47	11.9	0.19	4.8	0.36	9.1	0.22	5.6	0.22	5.6
5/16	0.51	13.0	0.53	13.5	0.22	5.6	0.42	10.7	0.28	7.1	0.28	7.1
3/8	0.64	16.3	0.67	17.0	0.28	7.1	0.55	14.0	0.36	9.1	0.34	8.6
1/2	0.77	19.6	0.80	20.3	0.41	10.4	0.66	16.8	0.47	11.9	0.41	10.4
5/8	0.90	22.9	0.93	23.6	0.50	12.7	0.77	19.6	0.61	15.5	0.44	11.2
3/4	1.08	27.4	1.11	28.2	0.62	15.7	0.95	24.1	0.72	18.3	0.56	14.2

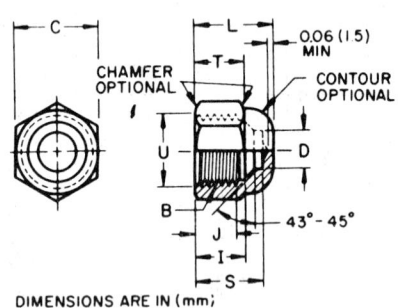

φFIG. 37—FLARE CAP
(010112) (N5)

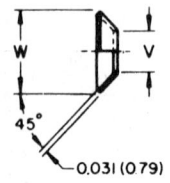

φFIG. 38—FLARE GASKET
(010113) (B2)

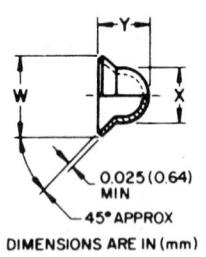

φFIG. 39—FLARE SEAL BONNET
(010114) (B1)

φ TABLE 14—DIMENSIONS OF DRUM ADAPTORS[a] (FIG. 40)

Nom Tube OD, in	A Straight Internal Pipe Thread NPSM[b]	C Hex, in	D Drill		L	
		Nom	in	mm	±0.03 in	±0.8 mm
1/4	1/2	1-1/8	0.188	4.78	1.12	28.4
1/4	3/4	1-1/4	0.188	4.78	1.12	28.4
3/8	3/4	1-1/4	0.281	71.4	1.25	31.8
1/2	3/4	1-1/4	0.406	10.31	1.38	35.1

[a] For flare and dimensions shown on Fig. 40 but not specified in Table 14, see corresponding dimensions for the specified Tube OD in Table 3. Drum adaptor fittings are normally supplied with seal gasket for pipe thread end.
[b] American Standard Straight Pipe Thread for Mechanical Joints.
[c] Where thread relief undercut is used, last thread shall be chamfered 1/2 to 1 pitch long from G diameter and dimension J may be reduced by an amount equal to 1/2 pitch.

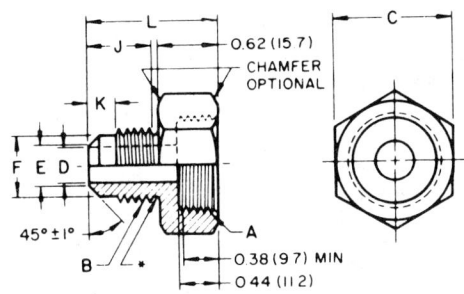

φFIG. 40—DRUM ADAPTOR
(010165) (K)

φ TABLE 15—DIMENSIONS OF FUSIBLE PIPE PLUGS[a] (FIG. 41)

A Dryseal Pipe Thread NPTF[b]	C Hex, in	D Dia		L		T	
	Nom	in	mm	in	mm	in	mm
1/8	7/16	0.219	5.6	0.56	14.2	0.19	4.8
1/4	9/16	0.250	6.4	0.75	19.0	0.19	4.8
3/8	3/4	0.375	9.5	0.78	19.8	0.22	5.6

[a] Plugs are available with three types of fusible alloys as specified in general specifications.
[b] Dryseal American Standard Taper Pipe Thread.

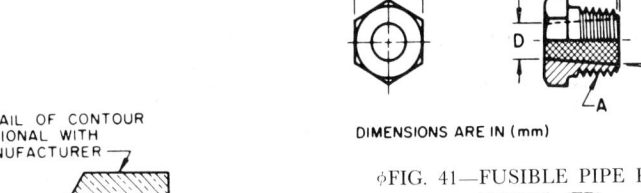

φFIG. 41—FUSIBLE PIPE PLUG
(010164) (FP)

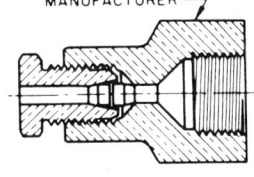

φFIG. 42A—CAPILLARY TUBE CONNECTION ASSEMBLY
(450101BA) (CTN)

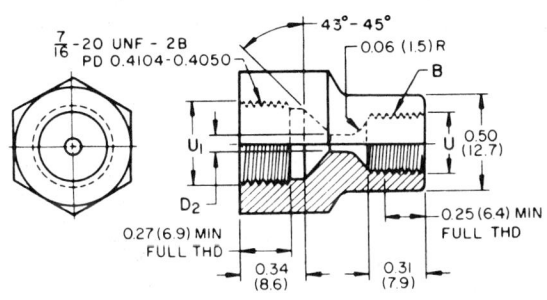

φFIG. 42B—CAPILLARY TUBE BODY
(450101) (CTN)

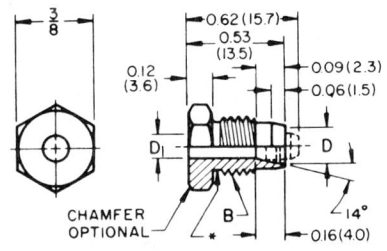

φFIG. 42C—CAPILLARY TUBE COMPRESSION SCREW
(450110) (CTN)

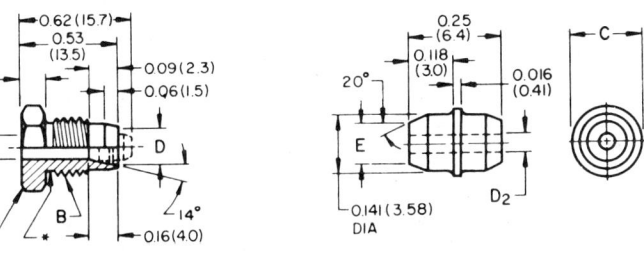

φFIG. 42D—CAPILLARY TUBE SLEEVE
(450115) (CTN)

φ FIG. 42—CAPILLARY TUBE CONNECTION

NOTE: UNSPECIFIED DETAIL WITH RESPECT TO DIMENSIONS, TOLERANCES, CONTOUR, MATERIAL, WORKMANSHIP, ETC., MUST CONFORM TO GENERAL SPECIFICATIONS FOR REFRIGERATION TUBE FITTINGS. CODES SHOWN IN BRACKETS ADJACENT TO FIGURE NUMBERS REPRESENT RESPECTIVE FITTING IDENTIFICATION IN ACCORDANCE WITH SAE J846 (FIRST NUMBER) AND ANSI B70.1 (SECOND NUMBER).

φ TABLE 16—DIMENSIONS OF CAPILLARY TUBE CONNECTIONS (FIG. 42)

Nom Tube OD, in	B Straight Thread Nominal Size, in	C Dia, in		D Dia		D_1 Drill		D_2 Drill		E Dia		G^a Dia		U Dia		U_1 Dia	
		Nom		±0.005	±0.13	±0.005	±0.13	±0.002	±0.05	±0.003	±0.08	+0.000 −0.010	+0.00 −0.25	+0.02 −0.00	+0.5 −0.0	+0.02 −0.00	+0.5 −0.0
				in	mm	in	mm	in	mm	in	mm	in	mm	in	mm	in	mm
0.081	5/16-24	11/64		0.187	4.7	0.104	2.6	0.085	2.16	0.105	2.67	0.250	6.35	0.33	8.4	0.45	11.4
0.093	5/16-24	11/64		0.187	4.7	0.104	2.6	0.097	2.46	0.105	2.67	0.250	6.35	0.33	8.4	0.45	11.4

[a] Where thread relief undercut is used, the last thread shall be chamfered 1/2 to 1 pitch thread long from the G diameter.

HYDRAULIC TUBE FITTINGS—SAE J514 APR80

SAE Standard

Report of the Construction and Industrial Machinery Technical Committee and the Tube, Pipe, Hose, and Lubrication Fittings Committee, approved May 1950, completely revised by the Fluid Connectors and Conductors Technical Committee May 1978, editorial change April 1980. Conforms in general to ANSI B116.1, Hydraulic Tube Fittings.

GENERAL SPECIFICATIONS

Scope—This standard covers complete general and dimensional specifications for 37 deg flared and flareless types of hydraulic tube fittings and the O-ring bosses and related components with which they assemble. Also included are pipe fittings for use in conjunction with these tube fittings. These fittings are intended for general application in hydraulic systems on industrial equipment and commerical products.

These fittings are capable of providing leak proof full flow connections in hydraulic systems operating at working pressures as specified in SAE J1065 "Pressure Ratings for Hydraulic Tubing and Fittings".

Since many factors influence the pressure at which a hydraulic system will or will not perform satisfactorily, the values shown in SAE J1065 should not be construed as a guaranteed minimum.

For any severe applications it is recommended sufficient testing be conducted and reviewed by both the user and fitting manufacturer to assure performance levels will be safe and satisfactory.

The standard is divided into five sections as follows:
Section I—37 deg Flare Tube Fittings.
Section II—Flareless Tube Fittings.
Section III—O-Ring Boss and Plugs.
Section IV—Hydraulic Pipe Fittings (formerly J926).
Section V—Tables for Calculating Dimensions on Special Sizes.

Section I—The 37 deg flared tube fittings shall be as shown in Figs. 1–15 and Tables 1–5. Since the basic design of these fittings is derived from Air Force and Navy Standards for 37 deg flared fittings which meet a performance specification, any future changes should not be detrimental to the design or performance. Dimensions for double and single 37 deg flares on tubing to be used with these fittings are given in Fig. 3 and Table 2 of SAE J533.

Section II—The flareless tube fittings shall be as shown in Figs. 16–32 and Tables 6–10. The basic design of these fittings is derived from existing military standards.

Section III—Specifications for straight thread O-ring bosses into which connector and adjustable styles of hydraulic tube fittings assemble are given in Fig. 33 and Table 11. O-ring boss plugs shall be as shown in Fig. 34 and Table 12. Modification of hexagon chamfers on standard fittings when used in MS 33649 (or superseded AND 10050) type O-ring bosses is covered in Fig. 35 and Table 13. For specifications of O-rings used in conjunction with these fittings, see SAE J515. Assembly instruction for O-ring fittings are shown in Figs. 36 and 37.

Section IV—Hydraulic pipe fittings are shown in Figs. 38–46 and Tables 14–17.

Section V—Tables 18 thru 28 and Fig. 47 and instructions which may be used for determining the overall lengths, leg lengths, and stock sizes applicable to special size combination fittings not covered in the standard dimensional tabulations.

The following general specifications supplement the dimensional data contained in Tables 1–17 with respect to all unspecified detail.

Size Designations—Fitting sizes are designated by the corresponding outside diameter of the tubing for the various types of tube ends and by the corresponding standard nominal pipe size for pipe thread ends.

See SAE J846 for proper coding and call-out.

Dimensions and Tolerances—Except for nominal sizes and thread specifications, dimensions and tolerances are given in both U. S. Customary and SI Units as designated. Tabulated dimensions shall apply to the finished parts, plated or otherwise processed, as specified by the purchasers. The maximum and minimum across flat dimensions shall be within the commercial tolerance of bar or extruded stock from which the fittings are produced. The minimum across corners dimensions of hexagons shall be 1.092 times the nominal width across flats, but shall not result in a side flat width less than 0.43 times the nominal width across flats. The minimum across corners dimensions of external squares shall be 1.25 times the nominal width across flats, but shall not result in a side flat width less than 0.75 times the nominal width across the flats.

Tolerance on all dimensions not otherwise limited shall be ±0.016 in (±0.41 mm). Fitting seats shall be concentric with straight thread pitch diameters within 0.010 in (0.25 mm) full indicator movement (FIM).

Unless otherwise specified, tolerance on hole diameters designated drill in the dimensional tables shall be as tabulated below:

Drill Size Range		Tolerance			
in	mm	in		mm	
		Plus	Minus	Plus	Minus
0.0135–0.246	0.35– 6.25	0.003	0.003	0.08	0.08
0.250 –0.500	6.35–12.70	0.004	0.004	0.10	0.10
0.516 –0.750	13.10–19.05	0.005	0.005	0.13	0.13
0.765 –1.000	19.40–25.40	0.007	0.005	0.18	0.13
1.016 –1.500	25.80–38.10	0.008	0.005	0.20	0.13
1.516 and over	38.50	0.010	0.005	0.25	0.13

Angular tolerance on axis of ends on elbows, tees, and crosses shall be ±2.50 deg for $\frac{1}{8}$–$\frac{3}{8}$ in tube fittings or $\frac{1}{8}$ and $\frac{1}{4}$ pipe fittings; ±1.50 deg for $\frac{1}{2}$–2 in O.D. tube fittings or $\frac{3}{8}$–2 in pipe fittings.

Where so illustrated and not otherwise specified, hexagon corners shall be chamfered 15–20 deg to a diameter equal to the nominal width across flats, with a minus tolerance of 0.016 in (0.41 mm); or where design permits, corners may be chamfered to the diameter of the abutting surface providing the length of chamfer does not exceed that produced by the 20 deg chamfer previously described.

Passages—Where passages in straight fittings are machined from opposite ends, the offset at the meeting point shall not exceed 0.015 in (0.38 mm). The cross-sectional area at the junction of passages in angle fittings shall not be less than that of the smallest passage.

Wall Thickness—Unless otherwise designated, the wall thickness at any point on fittings shall not be less than the thickness established by the specified dimensions, tolerances, and eccentricities for inner and outer surfaces.

Contour—Details of contour shall be optional with manufacturer provided the tabulated dimensions are maintained and serviceability of the fittings is not impaired.

Straight Threads—Unified Standard Class 2A external and Class 2B internal threads with modified minor diameters, where specified, shall apply to plain finish (unplated) fittings of all types. For externally threaded parts with additive finish, the maximum diameters of Class 2A may be exceeded by the amount of the allowance, that is, the basic diameters (Class 2A maximum diameters plus the allowance) apply to an externally threaded part after plating. For internally threaded parts with additive finish, the Class 2B diameters and modified minor diameters apply after plating.

The pitch diameter tolerance shall be the same as the corresponding diameter-pitch combination and class of the Unified fine and 12 thread series. See SAE J475 (ISO R725).

Where external threads are produced by roll threading and body is not undercut, the unthreaded portion of body adjacent to the shoulder may be reduced to the minimum pitch diameter.

External threads shall be chamfered and internal threads shall be countersunk as specified in the dimensional tables.

Thread Eccentricity Tolerances—The various thread elements of Class 2A external and Class 2B internal threads on tube fittings shall be concentric within the limitations specified under General Specifications in SAE J512.

Pipe Threads—Pipe threads, unless there is specific authorization to the contrary, shall conform to the Dryseal American Standard Taper Pipe Thread (NPTF). Specifications are given in detail in SAE J476 (ANSI B1.20.3).

The length of full form external thread shall not be shorter than L_2 plus one pitch (thread).

Where external pipe threads are produced by roll threading, the diameter of the unthreaded shank adjacent to shoulder may be reduced to the E_2 pitch diameter for brass fittings and to the root diameter on steel fittings.

External pipe threads shall be chamfered from the diameters tabulated below to produce the specified length of chamfer or partial thread. Internal pipe threads shall be countersunk 90 deg, included angle, to the diameters tabulated below:

Nominal Pipe Thread Size	External Thread								Internal Thread			
	Chamfer Dia				Length of Chamfer or Partial Thread				Countersink Dia			
	Max		Min		Min		Max		Min		Max	
	in	mm	in	mm	in	mm	in	mm	in	mm	in	mm
1/8	0.32	8.1	0.30	7.6	0.037	0.94	0.055	1.40	0.42	10.7	0.44	11.2
1/4	0.42	10.7	0.40	10.2	0.056	1.42	0.084	2.13	0.55	14.0	0.57	14.5
3/8	0.55	14.0	0.53	13.5	0.056	1.42	0.084	2.13	0.69	17.5	0.71	18.0
1/2	0.68	17.3	0.66	16.8	0.071	1.80	0.107	2.72	0.85	21.6	0.87	22.1
3/4	0.89	22.6	0.87	22.1	0.071	1.80	0.107	2.72	1.06	26.9	1.08	27.4
1	1.12	28.4	1.09	27.7	0.087	2.21	0.130	3.30	1.34	34.0	1.37	34.8
1-1/4	1.46	37.1	1.43	36.3	0.087	2.21	0.130	3.30	1.68	42.7	1.71	43.4
1-1/2	1.70	43.2	1.67	42.4	0.087	2.21	0.130	3.30	1.92	48.8	1.95	49.5
2	2.17	55.1	2.14	54.4	0.087	2.21	0.130	3.30	2.39	60.7	2.42	61.5

Tabulated diameters conform with Appendix A of SAE J476.

Material—Unless otherwise specified, fittings and ferrules shall be made from carbon steel. Flareless type ferrules in Figs 28 and 29 shall be made from SAE 1010, 1112, 1113 or 12L14 steel and cyanide hardened to a depth of 0.0010–0.0019 in (0.03–0.05 mm).

Stainless steel fittings shall be made from AISI Type 300 Series stainless steel of good quality.[1] Flareless type ferrules in Figs. 28 and 29 shall be made from stainless steel of such hardness as to be capable of biting, fully annealed type 304 stainless steel tubing. Unless otherwise specified by the purchaser, stainless steel fittings shall be passivated. Carbon steel and stainless steel fittings fabricated from multiple components must be bonded together with materials having a melting point of not less than 1900°F (1038°C).

37 deg flared type and pipe type brass fittings shall be made from C36000 (CA360) one-half hard barstock or extruded shapes or C37700 (CA377) forgings.

Finish—Unless otherwise specified, carbon steel fittings shall be furnished cadmium or zinc plated to a thickness of 0.0002 in (0.005 mm) minimum followed by a chromate treatment, or with a phosphate coating (oil finished). These parts must meet the requirements of 32 h salt spray test in accordance with ASTM B 117, Method of Salt Spray (Fog) Testing.

Workmanship—Workmanship shall conform to the best commercial practice to produce high quality fittings. Fittings shall be free from all hanging burrs, loose scale and slivers which might become dislodged in usage and all other defects which might affect their serviceability. All sealing surfaces must be smooth except that annular tool marks up to 100 μin (2.5 μm) max shall be permissable.

Assembly Considerations—Use of a compatible lubricant is desirable in assembling dryseal pipe threads on hydraulic tube or pipe fittings to minimize galling and effect a pressure tight seal.

[1] See SAE J405.

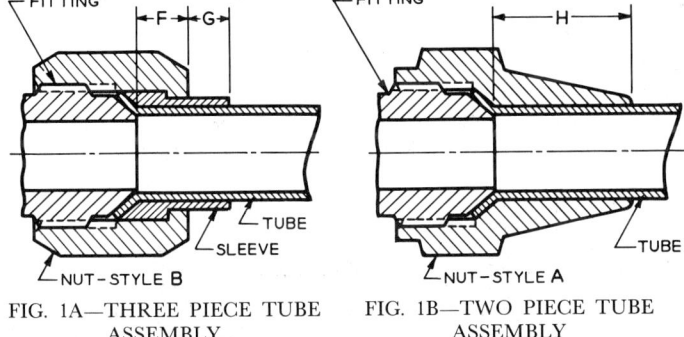

FIG. 1A—THREE PIECE TUBE ASSEMBLY

FIG. 1B—TWO PIECE TUBE ASSEMBLY

FIG. 1—DETAILS OF 37 DEG FLARED HYDRAULIC TUBE FITTING ASSEMBLIES

TABLE 1—DIMENSIONS OF 37 DEG FLARED HYDRAULIC TUBE FITTING ASSEMBLIES (FIG. 1)

Nominal Tube OD	F (Ref)		G (Ref)		H (Ref)	
	in	mm	in	mm	in	mm
1/8	0.19	4.8	0.12	3.0	0.48	12.2
3/16	0.25	6.4	0.12	3.0	0.56	14.2
1/4	0.19	4.8	0.17	4.3	0.59	15.0
5/16	0.30	7.6	0.11	2.8	0.66	16.8
3/8	0.28	7.1	0.19	4.8	0.69	17.5
1/2	0.31	7.9	0.19	4.8	0.81	20.6
5/8	0.38	9.7	0.27	6.9	0.94	23.9
3/4	0.36	9.1	0.25	6.4	1.03	26.2
7/8	0.38	9.7	0.31	7.9	1.19	30.2
1	0.34	8.6	0.41	10.4	1.31	33.3
1-1/4	0.34	8.6	0.39	9.9	1.50	38.1
1-1/2	0.50	12.7	0.55	14.0	1.53	38.9
2	0.55	14.0	0.53	13.5	1.75	44.4

37 DEG FLARED TYPE

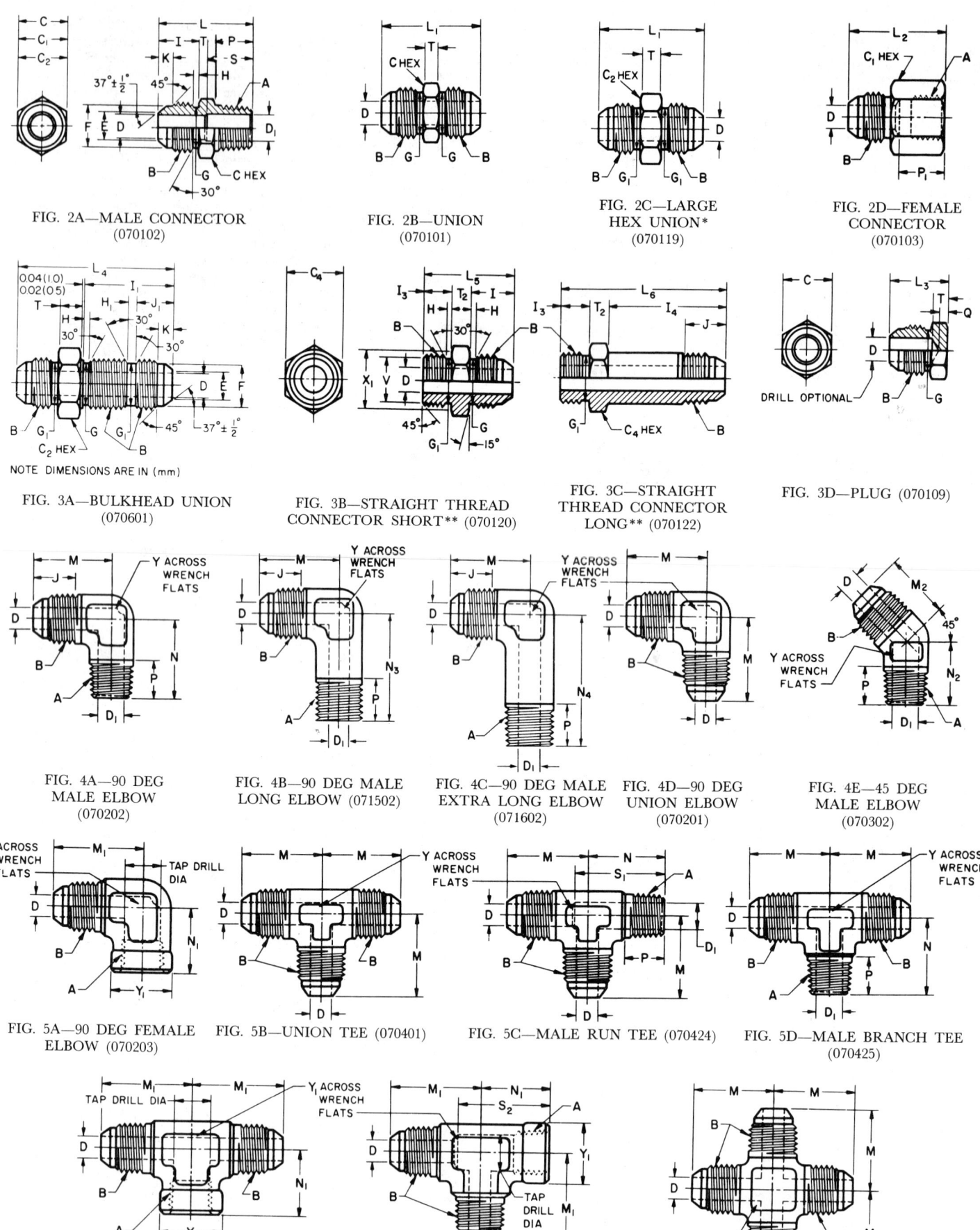

FIG. 2A—MALE CONNECTOR (070102)
FIG. 2B—UNION (070101)
FIG. 2C—LARGE HEX UNION* (070119)
FIG. 2D—FEMALE CONNECTOR (070103)
FIG. 3A—BULKHEAD UNION (070601)
FIG. 3B—STRAIGHT THREAD CONNECTOR SHORT** (070120)
FIG. 3C—STRAIGHT THREAD CONNECTOR LONG** (070122)
FIG. 3D—PLUG (070109)
FIG. 4A—90 DEG MALE ELBOW (070202)
FIG. 4B—90 DEG MALE LONG ELBOW (071502)
FIG. 4C—90 DEG MALE EXTRA LONG ELBOW (071602)
FIG. 4D—90 DEG UNION ELBOW (070201)
FIG. 4E—45 DEG MALE ELBOW (070302)
FIG. 5A—90 DEG FEMALE ELBOW (070203)
FIG. 5B—UNION TEE (070401)
FIG. 5C—MALE RUN TEE (070424)
FIG. 5D—MALE BRANCH TEE (070425)
FIG. 6A—FEMALE BRANCH TEE (070427)
FIG. 6B—FEMALE RUN TEE (070426)
FIG. 6C—CROSS (070501)

TABLE 2—DIMENSIONS OF ALL BODIES AND LOCKNUTS (FIGS. 2A–10C)

Nom Tube OD, in	A Dryseal Pipe Thread SAE J476 (ANSI B1.20.3)	B Thread Size, in SAE J475 (ISO R725) Class 2A Ext Class 2B Int g	C Hex Nom	C₁ Hex Nom	C₂ Hex Nom	C₃ Hex Max	C₄ Hex Nom	D Drill in	D mm	D₁[b] Drill in	D₁ mm	E Dia in ±0.003	E Dia mm ±0.08	F Dia in +0.000/−0.005	F Dia mm +0.00/−0.13	F₁ Dia in ±0.005	F₁ Dia mm ±0.13	F₂ Dia in ±0.005	F₂ Dia mm ±0.13
1/8	1/8-27	5/16-24	7/16	9/16	9/16	7/16	7/16	0.062	1.6	0.188	4.8	0.083	2.11	0.245	6.22	0.195	4.95	—	—
3/16	1/8-27	3/8-24	7/16	5/8	1/2	1/2	1/2	0.125	3.2	0.188	4.8	0.146	3.71	0.307	7.80	0.250	6.35	—	—
1/4	1/8-27	7/16-20	1/2	9/16	11/16	9/16	9/16	0.172	4.4	0.188	4.8	0.193	4.90	0.359	9.12	0.295	7.49	0.250	6.35
5/16	1/8-27	1/2-20	9/16	9/16	3/4	5/8	5/8	0.234	6.0	0.188	4.8	0.255	6.48	0.421	10.69	0.355	9.02	0.312	7.93
3/8	1/4-18	9/16-18	5/8	3/4	13/16	11/16	11/16	0.297	7.5	0.281	7.0	0.318	8.08	0.476	12.09	0.435	11.05	—	—
1/2	3/8-18	3/4-16	13/16	7/8	1	7/8	7/8	0.391	9.9	0.406	10.3	0.426	10.82	0.654	16.61	0.570	14.48	0.500	12.70
5/8	1/2-14	7/8-14	15/16	1-1/8	1-1/8	1	1	0.484	12.3	0.531	13.5	0.539	13.69	0.767	19.48	0.680	17.27	0.625	15.88
3/4	3/4-14	1-1/16-12	1-1/16	1-3/8	1-3/8	1-1/4	1-1/4	0.609	15.5	0.719	18.0	0.664	16.87	0.938	23.82	0.850	21.59	—	—
7/8	3/4-14	1-3/16-12	1-1/4	1-3/8	1-1/4	1-3/8	1-3/8	0.718	18.0	0.719	18.0	0.788	20.02	1.063	27.00	0.975	24.77	—	—
1	1-11-1/2	1-5/16-12	1-3/8	1-5/8	1-5/8	1-1/2	1-1/2	0.844	21.5	0.938	23.8	0.913	23.19	1.188	30.18	1.100	27.94	—	—
1-1/4	1-1/4-11-1/2	1-5/8-12	1-11/16	2	1-7/8	2	1-7/8	1.078	27.5	1.250	31.7	1.147	29.13	1.501	38.12	1.410	35.81	—	—
1-1/2	1-1/2-11-1/2	1-7/8-12	2	2-3/8	2-1/8	2-1/4	2-1/8	1.312	33.0	1.500	38.0	1.381	35.08	1.750	44.45	1.625	41.28	—	—
2	2-11-1/2	2-1/2-12	2-5/8	2-7/8	2-3/4	2-7/8	2-3/4	1.781	45.0	1.938	49.0	1.880	47.75	2.375	60.32	2.240	56.90	—	—

Nom Tube OD, in	G Dia in +0.002/−0.010	G Dia mm +0.05/−0.25	G₁[a] Dia in +0.002/−0.003	G₁ Dia mm +0.05/−0.08	H in +0.015/−0.000	H mm +0.4/−0.0	H₁ in +0.010/−0.000	H₁ mm +0.3/−0.0	I in ±0.015	I mm ±0.4	I₁ in ±0.02	I₁ mm ±0.5	I₂ in ±0.02	I₂ mm ±0.5	I₃ in ±0.005	I₃ mm ±0.13	I₄ in ±0.02	I₄ mm ±0.5	I₅ in ±0.02	I₅ mm ±0.5
1/8	0.250	6.35	0.250	6.35	0.063	1.6	0.125	3.2	0.448	11.4	1.11	28.2	0.92	23.4	0.297	7.54	1.17	29.7	0.61	15.5
3/16	0.313	7.95	0.313	7.95	0.063	1.6	0.131	3.3	0.479	12.2	1.11	28.2	0.92	23.4	0.297	7.54	1.25	31.8	0.61	15.5
1/4	0.364	9.25	0.364	9.25	0.075	1.9	0.140	3.6	0.550	14.0	1.20	30.5	1.02	25.9	0.360	9.14	1.39	35.3	0.70	17.8
5/16	0.427	10.85	0.427	10.85	0.075	1.9	0.140	3.6	0.550	14.0	1.20	30.5	1.02	25.9	0.360	9.14	1.45	36.8	0.70	17.8
3/8	0.482	12.24	0.482	12.24	0.083	2.1	0.156	4.0	0.556	14.1	1.28	32.5	1.09	27.7	0.391	9.93	1.56	39.6	0.77	19.6
1/2	0.660	16.76	0.660	16.76	0.094	2.4	0.187	4.8	0.657	16.7	1.44	36.6	1.25	31.8	0.438	11.13	1.88	47.8	0.88	22.4
5/8	0.773	19.63	0.773	19.63	0.107	2.7	0.219	5.6	0.758	19.3	1.58	40.1	1.39	35.3	0.500	12.70	2.09	53.1	1.02	25.9
3/4	0.945	24.00	0.945	24.00	0.125	3.2	0.234	5.9	0.864	21.9	1.75	44.4	1.56	39.6	0.594	15.09	2.50	63.5	1.16	29.5
7/8	1.070	27.18	1.070	27.18	0.125	3.2	0.234	5.9	0.890	22.6	1.75	44.4	1.56	39.6	0.594	15.09	2.69	68.3	1.16	29.5
1	1.195	30.35	1.195	30.35	0.125	3.2	0.234	5.9	0.911	23.1	1.75	44.4	1.56	39.6	0.594	15.09	2.84	72.1	1.16	29.5
1-1/4	1.507	38.28	1.507	38.28	0.125	3.2	0.234	5.9	0.958	24.3	1.80	45.7	1.61	40.9	0.594	15.09	3.47	88.1	1.16	29.5
1-1/2	1.756	44.60	1.756	44.60	0.125	3.2	0.234	5.9	1.083	27.5	1.81	46.0	1.62	41.1	0.594	15.09	3.88	98.6	1.16	29.5
2	2.381	60.48	2.381	60.48	0.125	3.2	0.234	5.9	1.333	33.9	2.09	53.1	1.91	48.5	0.594	15.09	4.84	122.9	1.16	29.5

Nom Tube OD, in	J Full Thread Min in	J mm	J₁ Full Thread in ±0.005	J₁ mm ±0.13	J₂ Full Thread Min in	J₂ mm	J₃ Full Thread in ±0.005	J₃ mm ±0.13	K in +0.015/−0.000	K mm +0.4/−0.0	K₁ in ±0.02	K₁ mm ±0.5	K₂ in +0.030/−0.015	K₂ mm +0.8/−0.4	L in ±0.02	L mm ±0.5	L₁ in ±0.02	L₁ mm ±0.5	L₂ in ±0.02	L₂ mm ±0.5	L₃ in ±0.02	L₃ mm ±0.5	L₄ in ±0.02	L₄ mm ±0.5
1/8	0.433	11.0	0.385	9.78	0.251	6.4	0.234	5.94	0.177	4.5	0.094	2.4	0.312	7.9	1.11	28.2	1.17	29.7	1.12	28.4	0.70	17.8	1.87	47.5
3/16	0.464	11.8	0.385	9.78	0.282	7.2	0.234	5.94	0.177	4.5	0.094	2.4	0.328	8.3	1.14	29.0	1.23	31.2	1.13	28.7	0.73	18.5	1.90	48.3
1/4	0.535	13.6	0.443	11.25	0.300	7.6	0.281	7.14	0.193	4.9	0.094	2.4	0.344	8.7	1.22	31.0	1.37	34.8	1.19	30.2	0.80	20.3	2.07	52.6
5/16	0.535	13.6	0.443	11.25	0.331	8.4	0.281	7.14	0.193	4.9	0.094	2.4	0.375	9.5	1.22	31.0	1.37	34.8	1.17	29.7	0.80	20.3	2.07	52.6
3/8	0.541	13.7	0.476	12.09	0.333	8.5	0.312	7.92	0.198	5.0	0.094	2.4	0.375	9.5	1.43	36.3	1.41	35.8	1.40	35.6	0.84	21.3	2.18	55.4
1/2	0.642	16.3	0.565	14.35	0.375	9.5	0.344	8.74	0.253	6.4	0.125	3.2	0.422	10.7	1.53	38.9	1.62	41.1	1.56	39.6	0.94	23.9	2.44	62.0
5/8	0.743	18.9	0.623	15.82	0.461	11.7	0.391	9.93	0.266	6.8	0.125	3.2	0.500	12.7	1.89	48.0	1.88	47.8	1.89	48.0	1.10	27.9	2.74	69.6
3/4	0.849	21.6	0.732	18.59	0.469	11.9	0.469	11.91	0.315	8.0	0.125	3.2	0.562	14.3	2.06	52.3	2.16	54.9	2.06	52.3	1.28	32.5	3.09	78.5
7/8	0.875	22.2	0.732	18.59	0.516	13.1	0.469	11.91	0.315	8.0	0.125	3.2	0.578	14.7	2.09	53.1	2.21	56.1	2.06	52.3	1.31	33.3	3.12	79.2
1	0.896	22.8	0.732	18.59	0.563	14.3	0.469	11.91	0.315	8.0	0.125	3.2	0.594	15.1	2.30	58.4	2.25	57.2	2.35	59.7	1.33	33.8	3.14	79.8
1-1/4	0.943	24.0	0.784	19.91	0.563	14.3	0.469	11.91	0.367	9.3	0.125	3.2	0.625	15.9	2.45	62.2	2.43	61.7	2.49	63.2	1.45	36.8	3.31	84.1
1-1/2	1.068	27.1	0.795	20.19	0.661	16.8	0.469	11.91	0.378	9.6	0.125	3.2	0.734	18.6	2.68	68.1	2.75	69.8	2.62	66.5	1.65	41.9	3.52	89.4
2	1.318	33.5	0.878	22.30	0.844	21.4	0.469	11.91	0.461	11.7	0.125	3.2	0.938	23.8	3.11	79.0	3.40	86.4	2.97	75.4	2.05	52.1	4.20	106.7

TABLE 2—DIMENSIONS OF ALL BODIES AND LOCKNUTS (FIGS. 2A–10C) (continued)

Nom Tube OD, in	L_5 in	L_5 mm	L_6 in	L_6 mm	M in	M mm	M_1 in	M_1 mm	M_2 in	M_2 mm	M_3 in	M_3 mm	M_4 in	M_4 mm	M_5 in	M_5 mm	M_6 in	M_6 mm	M_7 in	M_7 mm	M_8 in	M_8 mm	M_9 in	M_9 mm
	±0.02	±0.5	±0.02	±0.5	±0.03	±0.8	±0.03	±0.8	±0.03	±0.8	±0.03	±0.8	±0.03	±0.8	±0.06	±1.5	±0.06	±1.5	±0.03	±0.8	±0.03	±0.8	±0.03	±0.8
1/8	1.06	26.9	1.79	45.5	0.77	19.6	1.00	25.4	0.66	16.8	0.84	21.3	1.42	36.1	0.97	24.6	0.94	23.9	0.91	23.1	0.88	22.4	1.38	35.1
3/16	1.10	27.9	1.87	47.5	0.83	21.1	1.03	26.2	0.66	16.8	0.91	23.1	1.45	36.8	1.00	25.4	0.94	23.9	0.94	23.9	0.88	22.4	1.38	35.1
1/4	1.23	31.2	2.08	52.8	0.89	22.6	1.08	27.4	0.72	18.3	0.97	24.6	1.59	40.4	1.00	25.4	0.94	23.9	1.03	26.2	1.05	26.7	1.53	38.9
5/16	1.23	31.2	2.14	54.4	0.95	24.1	1.08	27.4	0.77	19.6	1.03	26.2	1.62	41.1	1.06	26.9	1.00	25.4	1.09	27.7	1.05	26.7	1.53	38.9
3/8	1.30	33.0	2.31	58.7	1.06	26.9	1.23	31.2	0.83	21.1	1.09	27.7	1.81	46.0	1.25	31.8	1.12	28.4	1.25	31.8	1.14	29.0	1.67	42.4
1/2	1.48	37.6	2.70	68.6	1.25	31.8	1.42	36.1	0.98	24.9	1.36	34.5	2.11	53.6	1.38	35.1	1.28	32.5	1.45	36.8	1.30	33.0	1.94	49.3
5/8	1.70	43.2	3.04	77.2	1.45	36.8	1.64	41.7	1.11	28.2	1.56	39.6	2.39	60.7	1.62	41.1	1.44	36.6	1.70	43.2	1.52	38.6	2.17	55.1
3/4	1.97	50.0	3.61	91.7	1.66	42.2	1.89	48.0	1.28	32.5	1.78	45.2	2.67	67.8	1.75	44.4	1.50	38.1	1.94	49.3	1.73	43.9	2.44	62.0
7/8	1.99	50.5	3.80	96.5	1.73	43.9	1.86	47.2	1.39	35.3	1.86	47.2	2.73	69.3	1.78	45.2	1.62	41.1	2.00	50.8	1.80	45.7	2.50	63.5
1	2.04	51.8	3.98	101.1	1.81	46.0	2.17	55.1	1.47	37.3	1.94	49.3	2.80	71.1	2.00	50.8	1.75	44.4	2.05	52.1	1.86	47.2	2.56	65.0
1-1/4	2.17	55.1	4.69	119.1	2.06	52.3	2.33	59.2	1.59	40.4	2.17	55.1	3.12	79.2	2.31	58.7	2.03	51.6	2.25	57.2	1.91	48.5	2.65	67.3
1-1/2	2.37	60.2	5.17	131.3	2.33	59.2	2.89	73.4	1.78	45.2	2.34	59.4	3.42	86.9	2.59	65.8	2.25	57.2	2.39	60.7	1.91	48.5	2.67	67.8
2	2.78	70.6	6.29	159.8	3.06	77.7	3.30	83.8	2.22	56.4	2.89	73.4	4.11	104.4	3.38	85.9	2.91	73.9	2.89	73.4	1.86	47.2	2.91	73.9

Nom Tube OD, in	N in	N mm	N_1 in	N_1 mm	N_2 in	N_2 mm	N_3 in	N_3 mm	N_4 in	N_4 mm	O in	O mm	P^f in	P^f mm	P_1 Max in	P_1 Max mm	Q Min in	Q Min mm	S^b Max in	S^b Max mm	$S_1^{b,c}$ Min in	$S_1^{b,c}$ Min mm	S_2^c Min in	S_2^c Min mm
	±0.03	±0.8	±0.03	±0.8	±0.03	±0.8	±0.03	±0.8	±0.03	±0.8	±0.02	±0.5	+0.03/−0.00	+0.8/−0.0										
1/8	0.72	18.3	0.66	16.8	0.52	13.2	1.00	25.4	1.28	32.5	0.438	11.1	0.38	9.7	0.46	11.7	0.11	2.8	0.49	12.4	0.79	20.1	0.73	18.5
3/16	0.72	18.3	0.66	16.8	0.52	13.2	1.04	26.4	1.35	34.3	0.500	12.7	0.38	9.7	0.46	11.7	0.11	2.8	0.49	12.4	0.82	20.8	0.76	19.3
1/4	0.78	19.8	0.66	16.8	0.64	16.3	1.17	29.7	1.56	39.6	0.562	14.3	0.38	9.7	0.46	11.7	0.11	2.8	0.49	12.4	0.91	23.1	0.79	20.1
5/16	0.78	19.8	0.66	16.8	0.64	16.3	1.21	30.7	1.63	41.4	0.625	15.9	0.38	9.7	0.46	11.7	0.12	3.0	0.64	16.3	1.11	28.2	0.82	20.8
3/8	1.09	27.7	0.88	22.4	0.86	21.8	1.58	40.1	2.07	52.6	0.688	17.5	0.56	14.2	0.67	17.0	0.12	3.0	0.67	17.0	1.25	31.8	1.07	27.2
1/2	1.22	31.0	1.02	25.9	0.95	24.1	1.82	46.2	2.42	61.5	0.875	22.2	0.56	14.2	0.68	17.3	0.16	4.1	0.68	17.3	1.46	37.1	1.26	32.0
5/8	1.47	37.3	1.23	31.2	1.17	29.7	2.17	55.1	2.87	72.9	1.000	25.4	0.75	19.0	0.90	22.9	0.16	4.1	0.91	23.1	1.75	44.4	1.51	38.4
3/4	1.59	40.4	1.36	34.5	1.20	30.5	2.44	62.0	3.28	83.3	1.188	30.2	0.75	19.0	0.91	23.1	0.19	4.8	0.94	23.9	1.94	49.3	1.71	43.4
7/8	1.69	42.9	1.42	36.1	1.27	32.3	2.65	67.3	3.62	91.9	1.312	33.3	0.75	19.0	0.91	23.1	0.19	4.8	—	—	—	—	1.82	46.2
1	1.97	50.0	1.62	41.1	1.48	37.6	3.01	76.5	4.05	102.9	1.438	36.5	0.94	23.9	1.14	29.0	0.19	4.8	1.13	28.7	2.43	61.7	2.08	52.8
1-1/4	2.38	60.5	1.70	43.2	1.67	42.4	3.69	93.7	5.00	127.0	1.750	44.4	0.97	24.6	1.16	29.5	0.23	5.8	1.20	30.5	2.96	75.2	2.28	57.9
1-1/2	2.64	67.1	2.08	52.8	1.77	45.0	4.10	104.1	5.55	141.0	2.000	50.8	1.00	25.4	1.16	29.5	0.23	5.8	1.27	32.3	3.34	84.8	2.27	57.6
2	3.00	76.2	2.39	60.7	2.11	53.6	4.81	122.2	6.63	168.4	2.625	66.7	1.03	26.2	1.18	30.0	0.33	8.4	1.37	34.8	3.94	100.1	3.33	84.6

Nom Tube OD, in	T^d Ref in	T^d Ref mm	T_1 in	T_1 mm	T_2^d Ref in	T_2^d Ref mm	U Dia in	U Dia mm	V Dia in	V Dia mm	X Dia in	X Dia mm	X_1 Dia in	X_1 Dia mm	Y Forging in	Y Forging mm	Y Barstock Max in	Y Barstock Max mm	Y_1^e in	Y_1^e mm	MM in	MM mm	NN Dia in	NN Dia mm
			±0.02	±0.5			+0.015/−0.000	+0.4/−0.0	±0.005	±0.13	±0.005	±0.13	±0.005	±0.13	±0.02	±0.5			±0.02	±0.5	±0.003	±0.08	±0.015	±0.4
1/8	0.22	5.6	0.22	5.6	0.28	7.1	0.317	8.1	0.250	6.35	0.562	14.27	0.438	11.13	0.312	7.9	—	—	0.562	14.3	0.030	0.76	0.504	12.8
3/16	0.22	5.6	0.22	5.6	0.28	7.1	0.380	9.7	0.310	7.87	0.625	15.88	0.500	12.70	0.375	9.5	—	—	0.562	14.3	0.030	0.76	0.575	14.6
1/4	0.22	5.6	0.25	6.4	0.28	7.1	0.443	11.3	0.365	9.27	0.688	17.48	0.563	14.30	0.438	11.1	0.562	14.3	0.562	14.3	0.035	0.89	0.650	16.5
5/16	0.22	5.6	0.25	6.4	0.28	7.1	0.505	12.8	0.425	10.80	0.750	19.05	0.625	15.88	0.500	12.7	0.625	15.9	0.562	14.3	0.035	0.89	0.722	18.3
3/8	0.25	6.4	0.27	6.9	0.31	7.9	0.567	14.4	0.480	12.19	0.812	20.62	0.688	17.48	0.562	14.3	0.812	20.6	0.750	19.0	0.035	0.89	0.794	20.2
1/2	0.25	6.4	0.31	7.9	0.34	8.6	0.755	19.2	0.660	16.76	1.000	25.40	0.875	22.22	0.750	19.0	0.875	22.2	0.875	22.2	0.041	1.04	1.010	25.7
5/8	0.31	7.9	0.36	9.1	0.40	10.2	0.880	22.4	0.775	19.68	1.125	28.58	1.000	25.40	0.875	22.2	1.125	28.6	1.062	27.0	0.050	1.27	1.155	29.3
3/4	0.38	9.7	0.41	10.4	0.47	11.9	1.067	27.1	0.945	24.00	1.375	34.92	1.250	31.75	1.062	27.0	1.375	34.9	1.312	33.3	0.050	1.27	1.444	36.7
7/8	0.38	9.7	0.41	10.4	0.47	11.9	1.193	30.3	1.070	27.18	1.500	38.10	1.375	34.92	1.188	30.2	1.500	38.1	1.312	33.3	0.050	1.27	1.589	40.4
1	0.38	9.7	0.41	10.4	0.50	12.7	1.315	33.5	1.195	30.35	1.625	41.28	1.500	38.10	1.312	33.3	1.625	41.3	1.625	41.3	0.050	1.27	1.732	44.0
1-1/4	0.46	11.7	0.41	10.4	0.58	14.7	1.630	41.4	1.505	38.23	1.875	47.62	1.875	47.62	1.625	41.3	2.125	54.0	1.875	47.6	0.050	1.27	2.165	55.0
1-1/2	0.53	13.5	0.41	10.4	0.65	16.5	1.880	47.8	1.755	44.58	2.125	53.98	2.125	53.98	1.875	47.6	2.250	57.2	2.562	65.1	0.050	1.27	2.454	62.3
2	0.68	17.3	0.41	10.4	0.81	20.6	2.505	63.6	2.380	60.45	2.750	69.85	2.750	69.85	2.500	63.5	3.250	82.6	2.812	71.4	0.050	1.27	3.160	80.3

[a] O-ring groove undercut must be smooth and free from tool marks.
[b] At manufacturer's option, through passages in Figs. 2A and 5C may conform with the smaller diameter specified or the appropriate end may be counterbored to the larger diameter for depths S and S_1, respectively.
[c] Maximum depth shall be optional with manufacturer providing wall thickness is controlled in compliance with General Specifications.
[d] Minimum design thickness, not subject to inspection.
[e] The basic dimensions shown shall apply as minimum for boss diameters.
[f] P dimension shall be considered minimum for shaped fittings only.
[g] Unified class 2B thread shall apply to swivel nuts Figs. 8C–9C and with minor diameter modified to class 3B limits for locknuts Figs. 8A and 8B.

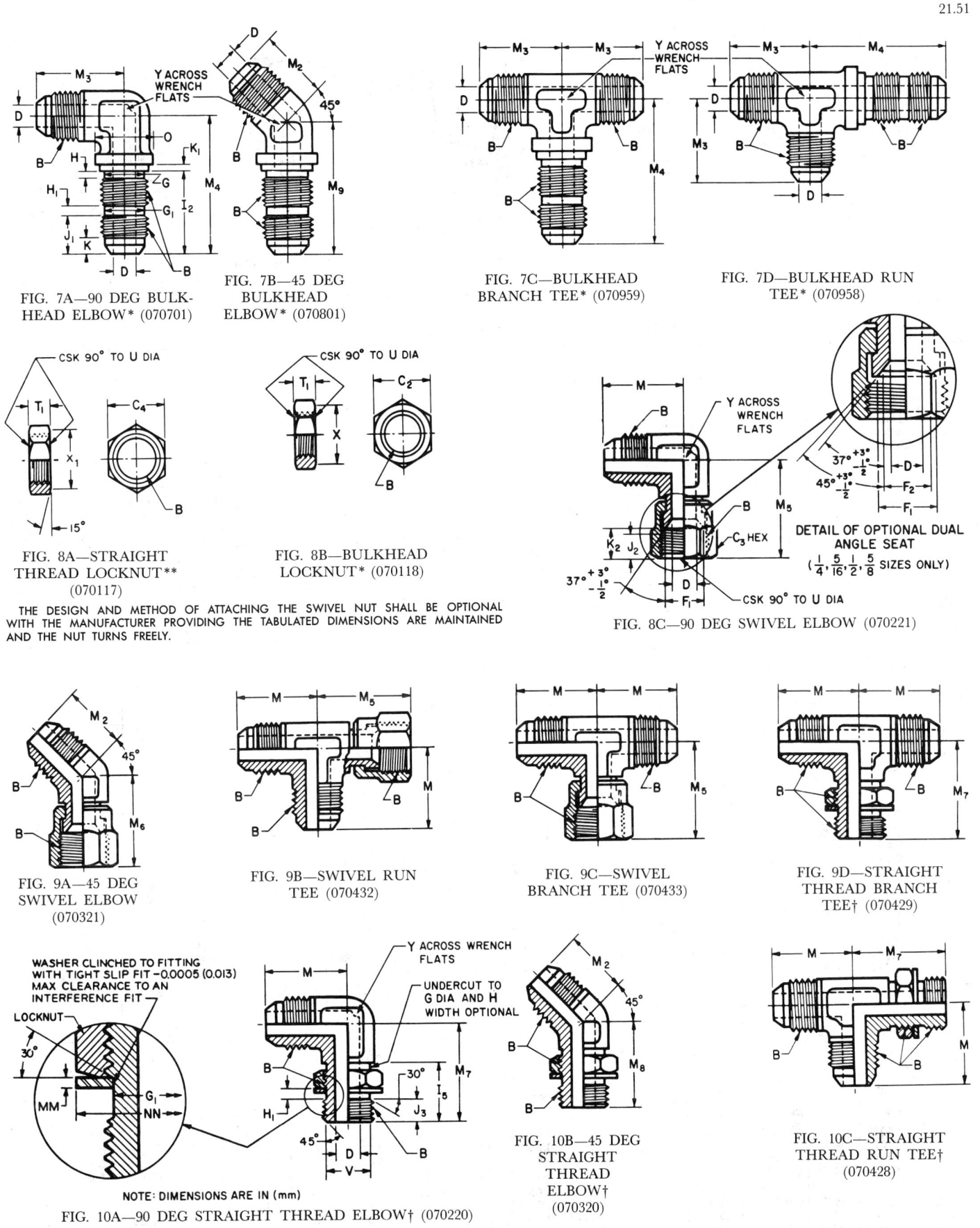

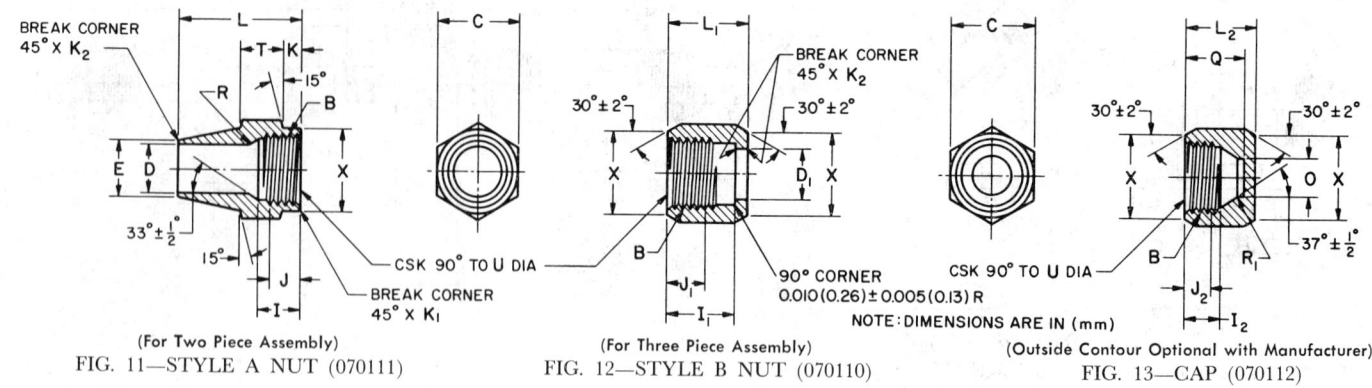

FIG. 11—STYLE A NUT (070111) (For Two Piece Assembly)
FIG. 12—STYLE B NUT (070110) (For Three Piece Assembly)
FIG. 13—CAP (070112) (Outside Contour Optional with Manufacturer)

NOTE: DIMENSIONS ARE IN (mm)

TABLE 3—DIMENSIONS OF STYLE A AND B NUTS AND CAP (FIGS. 11–13)

Nom Tube OD, in	B Thread Size, in SAE J475 (ISO R725) Class 2B Int	Thread Minor Dia		C^a Hex Min Nom	D Dia		D_1 Dia		E Dia		I		I_1		I_2	
		in +0.005 −0.000	mm +0.13 −0.00		in +0.003 −0.000	mm +0.08 −0.00	in +0.003 −0.000	mm +0.08 −0.00	in +0.005 −0.000	mm +0.13 −0.00	in ±0.005	mm ±0.13	in ±0.010	mm ±0.3	in ±0.005	mm ±0.13
1/8	5/16-24	0.272	6.91	3/8	0.130	3.30	0.180	4.57	0.183	4.65	0.282	7.16	0.460	11.7	0.234	5.94
3/16	3/8 -24	0.334	8.48	7/16	0.193	4.90	0.242	6.15	0.246	6.25	0.313	7.95	0.506	12.9	0.266	6.76
1/4	7/16-20	0.388	9.86	9/16	0.255	6.48	0.305	7.75	0.308	7.82	0.375	9.52	0.532	13.5	0.328	8.33
5/16	1/2 -20	0.451	11.46	5/8	0.318	8.08	0.374	9.50	0.371	9.52	0.375	9.52	0.579	14.7	0.344	8.74
3/8	9/16-18	0.508	12.90	11/16	0.380	9.65	0.440	11.18	0.433	11.00	0.385	9.78	0.603	15.3	0.344	8.74
1/2	3/4 -16	0.688	17.48	7/8	0.505	12.83	0.570	14.48	0.558	14.17	0.438	11.13	0.723	18.4	0.375	9.52
5/8	7/8 -14	0.804	20.42	1	0.631	16.03	0.698	17.73	0.694	17.63	0.521	13.23	0.817	20.8	0.469	11.91
3/4	1-1/16-12	0.979	24.87	1-1/4	0.756	19.20	0.834	21.18	0.829	21.06	0.594	15.09	0.868	22.0	0.500	12.70
7/8	1-3/16-12	1.104	28.04	1-3/8	0.881	22.38	0.961	24.41	0.964	24.49	0.594	15.09	0.914	23.2	0.531	13.49
1	1-5/16-12	1.229	31.22	1-1/2	1.006	25.55	1.089	27.66	1.099	27.91	0.625	15.88	0.962	24.4	0.563	14.30
1-1/4	1-5/8 -12	1.541	39.14	2	1.260	32.00	1.347	34.21	1.353	34.37	0.625	15.88	1.017	25.8	0.563	14.30
1-1/2	1-7/8 -12	1.791	45.49	2-1/4	1.510	38.35	1.617	41.07	1.603	40.72	0.708	17.98	1.170	29.7	0.656	16.66
2	2-1/2 -12	2.416	61.37	2-7/8	2.014	51.16	2.167	55.04	2.127	54.03	0.875	22.22	1.462	37.1	0.828	21.03

Nom Tube OD, in	J Full Thread Min		J_1 Full Thread Min		J_2 Full Thread Min		K		K_1		K_2		L		L_1	
	in	mm	in	mm	in	mm	in ±0.02	mm ±0.5	in ±0.005	mm ±0.13	in ±0.005	mm ±0.13	in ±0.02	mm ±0.5	in ±0.01	mm ±0.3
1/8	0.215	5.5	0.246	6.2	0.167	4.2	0.09	2.3	0.010	0.25	0.005	0.13	0.84	21.3	0.540	13.7
3/16	0.246	6.2	0.277	7.0	0.198	5.0	0.09	2.3	0.010	0.25	0.005	0.13	0.94	23.9	0.600	15.2
1/4	0.295	7.5	0.295	7.5	0.245	6.2	0.09	2.3	0.010	0.25	0.005	0.13	1.00	25.4	0.610	15.5
5/16	0.295	7.5	0.326	8.3	0.261	6.6	0.09	2.3	0.010	0.25	0.005	0.13	1.06	26.9	0.670	17.0
3/8	0.297	7.5	0.328	8.3	0.261	6.6	0.09	2.3	0.010	0.25	0.010	0.25	1.09	27.7	0.720	18.3
1/2	0.339	8.6	0.370	9.4	0.276	7.0	0.12	3.0	0.010	0.25	0.010	0.25	1.28	32.5	0.840	21.3
5/8	0.409	10.4	0.456	11.6	0.354	9.0	0.19	4.8	0.010	0.25	0.010	0.25	1.48	37.6	0.970	24.6
3/4	0.464	11.8	0.464	11.8	0.370	9.4	0.19	4.8	0.010	0.25	0.010	0.25	1.66	42.2	1.020	25.9
7/8	0.464	11.8	0.511	13.0	0.401	10.2	0.19	4.8	0.010	0.25	0.010	0.25	1.81	46.0	1.080	27.4
1	0.495	12.6	0.558	14.2	0.433	11.0	0.19	4.8	0.010	0.25	0.010	0.25	1.94	49.3	1.120	28.4
1-1/4	0.495	12.6	0.558	14.2	0.433	11.0	0.25	6.4	0.010	0.25	0.010	0.25	2.19	55.6	1.220	31.0
1-1/2	0.578	14.7	0.656	16.7	0.526	13.4	0.25	6.4	0.010	0.25	0.010	0.25	2.31	58.7	1.410	35.8
2	0.745	18.9	0.839	21.3	0.698	17.7	0.31	7.9	0.015	0.38	0.015	0.38	2.75	69.8	1.740	44.2

Nom Tube OD, in	L_2		O Dia		Q		R Rad		R_1 Rad		T		U Dia		X	
	in ±0.01	mm ±0.3	in ±0.005	mm ±0.13	in ±0.010	mm ±0.3	in ±0.01	mm ±0.3	in ±0.01	mm ±0.3	in ±0.02	mm ±0.5	in +0.015 −0.000	mm +0.4 −0.0	in ±0.01	mm ±0.3
1/8	0.500	12.7	0.062	1.57	0.438	11.1	0.03	0.8	0.03	0.8	0.25	6.4	0.317	8.05	0.36	9.1
3/16	0.562	14.3	0.125	3.18	0.500	12.7	0.03	0.8	0.03	0.8	0.27	6.9	0.380	9.65	0.42	10.7
1/4	0.594	15.1	0.172	4.37	0.531	13.5	0.03	0.8	0.03	0.8	0.33	8.4	0.443	11.25	0.54	13.7
5/16	0.609	15.5	0.234	5.94	0.547	13.9	0.03	0.8	0.03	0.8	0.33	8.4	0.505	12.83	0.60	15.2
3/8	0.625	15.9	0.297	7.54	0.562	14.3	0.05	1.3	0.06	1.5	0.34	8.6	0.567	14.40	0.67	17.0
1/2	0.750	19.0	0.391	9.93	0.625	15.9	0.06	1.5	0.06	1.5	0.45	11.4	0.755	19.18	0.86	21.8
5/8	0.844	21.4	0.484	12.29	0.719	18.3	0.06	1.5	0.06	1.5	0.52	13.2	0.880	22.35	0.98	24.9
3/4	0.906	23.0	0.609	15.47	0.781	19.8	0.08	2.0	0.06	1.5	0.64	16.3	1.067	27.10	1.24	31.5
7/8	0.969	24.6	0.719	18.26	0.844	21.4	0.09	2.3	0.06	1.5	0.69	17.5	1.193	30.30	1.36	34.5
1	1.016	25.8	0.844	21.44	0.859	21.8	0.09	2.3	0.06	1.5	0.73	18.5	1.317	33.45	1.48	37.6
1-1/4	1.062	27.0	1.078	27.38	0.906	23.0	0.09	2.3	0.06	1.5	0.73	18.5	1.630	41.40	1.98	50.3
1-1/2	1.188	30.2	1.312	33.32	1.031	26.2	0.11	2.8	0.06	1.5	0.83	21.1	1.880	47.75	2.24	56.9
2	1.438	36.5	1.781	45.24	1.281	32.5	0.11	2.8	0.06	1.5	0.92	23.4	2.505	63.63	2.86	72.6

[a] Across flat widths must fit standard wrench openings.

NOTES: FIGS. 11–13 APPLY TO TABLE 3. CODES SHOWN IN BRACKETS ADJACENT TO FIGURE NUMBERS REPRESENT RESPECTIVE FITTING IDENTIFICATION IN ACCORDANCE WITH SAE J846.

TABLE 4—DIMENSIONS FOR SLEEVES (FIG. 14) FOR INCH TUBING

Nom Tube OD, in	D Dia		F Dia		I		L		O Dia		R Rad		X Dia	
	in	mm	in	mm	in	mm	in	mm	in	mm	in	mm	in	mm
	+0.003 −0.000	+0.08 −0.00	±0.005	±0.13	±0.02	±0.5	±0.02	±0.5	+0.000 −0.003	+0.00 −0.08	±0.01	±0.3	+0.000 −0.003	+0.00 −0.08
1/8	0.130	3.30	0.205	5.21	0.12	3.0	0.34	8.6	0.172	4.37	0.031	0.8	0.267	6.78
3/16	0.193	4.90	0.267	6.78	0.14	3.6	0.34	8.6	0.234	5.94	0.031	0.8	0.329	8.36
1/4	0.255	6.48	0.315	8.00	0.14	3.6	0.41	10.4	0.297	7.54	0.031	0.8	0.383	9.73
5/16	0.318	8.08	0.375	9.52	0.16	4.1	0.44	11.2	0.366	9.30	0.031	0.8	0.445	11.30
3/8	0.380	9.65	0.441	11.20	0.17	4.3	0.50	12.7	0.432	10.97	0.047	1.2	0.502	12.75
1/2	0.505	12.83	0.589	14.96	0.22	5.6	0.56	14.2	0.562	14.27	0.062	1.6	0.682	17.32
5/8	0.631	16.03	0.705	17.91	0.24	6.1	0.66	16.8	0.690	17.53	0.062	1.6	0.797	20.24
3/4	0.756	19.20	0.880	22.35	0.26	6.6	0.68	17.3	0.826	20.98	0.078	2.0	0.972	24.69
7/8	0.881	22.38	1.005	25.53	0.26	6.6	0.76	19.3	0.953	24.21	0.094	2.4	1.097	27.86
1	1.006	25.55	1.130	28.70	0.28	7.1	0.78	19.8	1.081	27.46	0.094	2.4	1.222	31.04
1-1/4	1.260	32.00	1.412	35.86	0.31	7.9	0.91	23.1	1.339	34.01	0.094	2.4	1.534	38.96
1-1/2	1.510	38.35	1.630	41.40	0.34	8.6	1.12	28.4	1.609	40.87	0.109	2.8	1.784	45.31
2	2.014	51.16	2.195	55.75	0.41	10.4	1.19	30.2	2.159	54.84	0.109	2.8	2.409	61.19

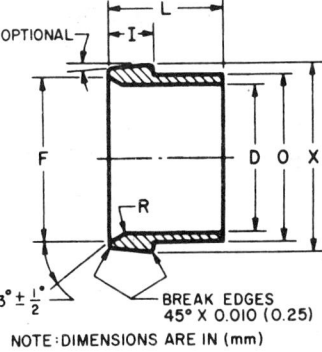

(For Three Piece Assembly)
FIG. 14—SLEEVE (070115)

TABLE 4A—DIMENSIONS FOR SLEEVES (FIG. 14) FOR METRIC TUBING[c]

Nom Tube OD, mm	D Dia		F Dia		I		L		O Dia		R Rad		X Dia		in Size Body[a] and Nut
	in	mm	in	mm	in	mm	in	mm	in	mm	in	mm	in	mm	
	+0.003 −0.000	+0.08 −0.00	±0.005	±0.13	±0.02	±0.5	±0.02	±0.5	+0.000 −0.003	+0.00 −0.08	±0.010	±0.30	+0.000 −0.003	+0.00 −0.08	
6	0.241	6.13	0.315	8.00	0.14	3.6	0.41	10.4	0.297	7.54	0.031	0.80	0.383	9.73	1/4
8	0.320	8.13	0.375	9.52	0.16	4.1	0.44	11.2	0.366	9.30	0.031	0.80	0.445	11.30	5/16
10	0.399	10.13	0.441	11.20	0.17	4.3	0.50	12.7	0.432	10.97	0.047	1.20	0.502	12.75	3/8
12	0.478	12.13	0.589	14.96	0.22	5.6	0.56	14.2	0.562	14.27	0.062	1.60	0.682	17.32	1/2
16	0.636	16.15	0.705	17.91	0.24	6.1	0.66	16.8	0.690	17.53	0.062	1.60	0.797	20.24	5/8
19[b]	0.756	19.20	0.880	22.35	0.26	6.6	0.68	17.3	0.826	20.98	0.078	2.00	0.972	24.69	3/4
20	0.793	20.15	1.005	25.53	0.26	6.6	0.76	19.3	0.953	24.21	0.094	2.40	1.097	27.86	7/8
25	0.990	25.15	1.130	28.70	0.28	7.1	0.78	19.8	1.081	27.46	0.094	2.40	1.222	31.04	1
32	1.270	32.25	1.412	35.86	0.31	7.9	0.91	23.1	1.339	34.01	0.094	2.40	1.534	38.96	1-1/4
38[b]	1.510	38.35	1.630	41.40	0.34	8.6	1.12	28.4	1.609	40.87	0.109	2.80	1.784	45.31	1-1/2
50	1.983	50.36	2.195	55.75	0.41	10.4	1.19	30.2	2.159	54.84	0.109	2.80	2.409	61.19	2

[a] Metric sleeves are used with standard Fig. 12 (070110) tube nuts and standard fitting bodies as shown in Figs. 2A–10C (Table 2) and Fig. 15 (Table 5).
[b] 19 mm and 38 mm are shown only because they use the standard 3/4 and 1-1/2 size Fig. 14 sleeve and there is apparent usage.
[c] In addition to cadmium or zinc plating, sleeves for metric tubing will be dyed light blue for identification.

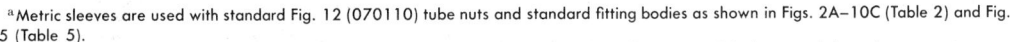

FIG. 15—REDUCING ADAPTER (070123)

TABLE 5—DIMENSIONS[a] OF REDUCING ADAPTER (FIG. 15)

Tube Reduction, in	D Dia Ref		L		O Dia		W		X Dia		Y Dia		Z	
	in	mm	in	mm	in	mm	in	mm	in	mm	in	mm	in	mm
			±0.02	±0.5	+0.000 −0.003	+0.00 −0.08	±0.02	±0.5	+0.000 −0.003	+0.00 −0.08	+0.005 −0.015	+0.13 −0.40	±0.02	±0.5
3/8 x 1/4	0.172	4.4	0.97	24.6	0.432	10.97	0.17	4.3	0.502	12.75	0.441	11.20	—	—
1/2 x 1/4	0.172	4.4	1.00	25.4	0.562	14.27	0.22	5.6	0.682	17.32	0.589	14.96	—	—
1/2 x 3/8	0.297	7.5	1.00	25.4	0.562	14.27	0.22	5.6	0.682	17.32	0.589	14.96	—	—
5/8 x 1/4	0.172	4.4	1.03	26.2	0.690	17.53	0.23	5.8	0.797	20.24	0.705	17.91	—	—
5/8 x 3/8	0.297	7.5	1.03	26.2	0.690	17.53	0.23	5.8	0.797	20.24	0.705	17.91	—	—
3/4 x 1/4	0.172	4.4	1.09	27.7	0.826	20.98	0.27	6.9	0.972	24.69	0.880	22.35	0.41	10.4
3/4 x 3/8	0.297	7.5	1.09	27.7	0.826	20.98	0.27	6.9	0.972	24.69	0.880	22.35	—	—
3/4 x 1/2	0.391	10.0	1.19	30.2	0.826	20.98	0.27	6.9	0.972	24.69	0.880	22.35	—	—
1 x 3/4	0.609	15.5	1.47	37.3	1.081	27.46	0.28	7.1	1.222	31.04	1.130	28.70	—	—

[a] For dimensions shown on Fig. 15 but not specified in above table, see corresponding dimensions for the specified outside diameter in Table 2.

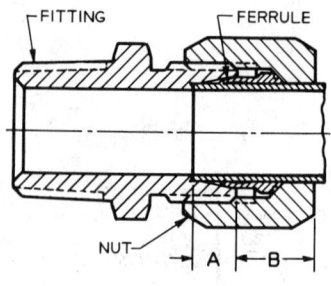

FIG. 16A—ASSEMBLY WITH STYLE A FERRULE (FIG. 28)

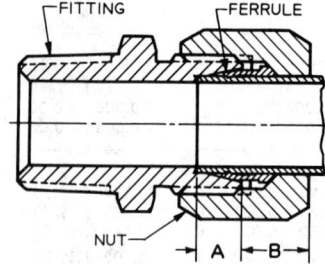

FIG. 16B—ASSEMBLY WITH STYLE B FERRULE (FIG. 29)

FIG. 16—DETAILS OF FLARELESS HYDRAULIC TUBE FITTING ASSEMBLIES

TABLE 6—DIMENSIONS OF FLARELESS HYDRAULIC TUBE FITTING ASSEMBLIES (FIGS. 16 AND 18)

Nominal Tube OD	A (Ref)		B (Ref)		C (Ref)	
	in	mm	in	mm	in	mm
1/8	0.188	4.78	0.31	7.9	0.08	2.0
3/16	0.234	5.94	0.34	8.6	0.12	3.0
1/4	0.234	5.94	0.42	10.7	0.14	3.6
5/16	0.250	6.35	0.42	10.7	0.14	3.6
3/8	0.250	6.35	0.47	11.9	0.14	3.6
1/2	0.305	7.75	0.50	12.7	0.14	3.6
5/8	0.350	8.89	0.53	13.5	0.14	3.6
3/4	0.350	8.89	0.56	14.2	0.14	3.6
7/8	0.350	8.89	0.53	13.5	—	—
1	0.415	10.54	0.66	16.8	0.17	4.3
1-1/4	0.415	10.54	0.72	18.3	0.20	5.1
1-1/2	0.485	12.32	0.72	18.3	0.20	5.1
2	0.485	12.32	0.84	21.3	0.20	5.1

FIG. 17—ENLARGED VIEW OF STYLE A FERRULE BITE

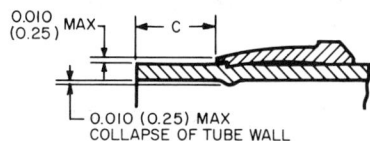

FIG. 18—ENLARGED VIEW OF STYLE B FERRULE BITE

Since the proper make up of the hydraulic flareless tube fittings depends on securing the ferrule to the tube by the cutting action of the ferrule into the tube, the following instructions should be adhered to when assembling and reassembling fittings of the flareless type given in Figs. 19-32 and Tables 7-10.

ASSEMBLY INSTRUCTIONS FOR HYDRAULIC FLARELESS TUBE FITTINGS

These instructions apply to the assembly of hydraulic tube fittings of the flareless type given in Tables 7-10 and Figs. 19-32.

The following instructions should be used to assure proper make-up of the fitting when assembled since the fitting depends on securing the ferrule to the tube by the cutting action of the ferrule into the tube.

1. Cut tube square and burr inside and outside corner (not excessive).

2. Assemble fitting by sliding nut over tubing with open end out. Slide ferrule on tubing with cutting edge out, the large head end should be inside of the nut. Lubricate the ferrule and the threads on the body and nut with oil or petrolatum. Insert tube into fitting.

3. Bottom the tube in the fitting, and tighten the nut until the ferrule just grips the tube. With a little experience, the mechanic can determine this point by feel. If the fittings are bench assembled, the gripping action can be determined by rotating the tube by hand as the nut is drawn down. When the tube can no longer be turned by hand, the ferrule has started to grip the tube.

4. After the ferrule grips the tube, tighten the nut one full turn. This may vary slightly with different tubing materials, but for general practice, it is a good rule for the mechanic to follow.

5. The fittings can now be disassembled for inspection. The two styles of ferrules differ somewhat in inspection even though their principles of make-up and application are similar.

(a) For the ferrule in Fig. 28, the bite or cut into the tube can be readily seen since it is on the lead edge of the ferrule. The bite into the tube should show a definite groove where the ferrule cuts into the tube and peels the metal over the lead edge of the ferrule, see Fig. 17 for further detail.

(b) For the ferrule in Fig. 29, the pilot at the end of the ferrule should be contacting or be within 0.0015 in (0.038 mm) of the tube for hard material or not more than 0.005 in (0.13 mm) on soft material, see Fig. 18 for further detail. This is an indication that the cutting edge has performed its function and has taken a secure bite in the tube. The sleeve should be slightly sprung or arched.

For both styles of ferrules the rounded or lead edge should show a good seat in the fitting, and the head or shoulder end should be collapsed tight against the tube. The ferrules should have no end movement; however, the ferrule may be rotated on the tube due to spring back of the material. The performance of the fitting is not affected if the ferrule rotates.

6. In production, it may be preferable to use a threaded presetting tool to preform the ferrule onto the tubing. The presetting tool is a counterpart of the fitting hardened to provide good wearing properties for repeated usage. When using the presetting tool, the assembly instructions are the same as the presetting tool takes the place of the fitting. Care should be taken to keep the cam surface of the presetting tool free of defects since they would transfer themselves to the ferrule, which would result in improper seating when the fitting is installed.

7. In some installations, it may be necessary to use a mandrel to support the inside of the tube when setting the ferrule. This is only necessary when the tube wall is so thin or so soft that it will not resist the biting action of the ferrule without collapsing. The mandrel in this instance supports the tube and allows the ferrule to bite into the tube without deforming or collapsing. Because the use of a mandrel allows very little give in the tubing, the setting of the ferrule may be made with slightly less turns than described above.

REASSEMBLY INSTRUCTIONS FOR FLARELESS FITTINGS

After disassembly of the fitting joint the flareless fitting can be reassembled by assembling the tube and ferrule into the socket of the fitting and threading the nut onto the fitting.

The operation of assembly up to the point at which the ferrule seats itself in the fitting can usually be accomplished by hand or with the use of a small wrench. If a wrench is required, only low torques are necessary to seat the ferrule.

When the ferrule is seated, an increase in the torque will be quite evident. When this point is reached, draw the nut up approximately $\frac{1}{6}$ of a turn minimum, but not more than $\frac{1}{3}$ of a turn, to complete the tightening operation.

NOTE: Instructions for assembling adjustable and swivel style hydraulic fittings in straight thread O-ring bosses are given immediately following the specifications for straight thread O-ring bosses.

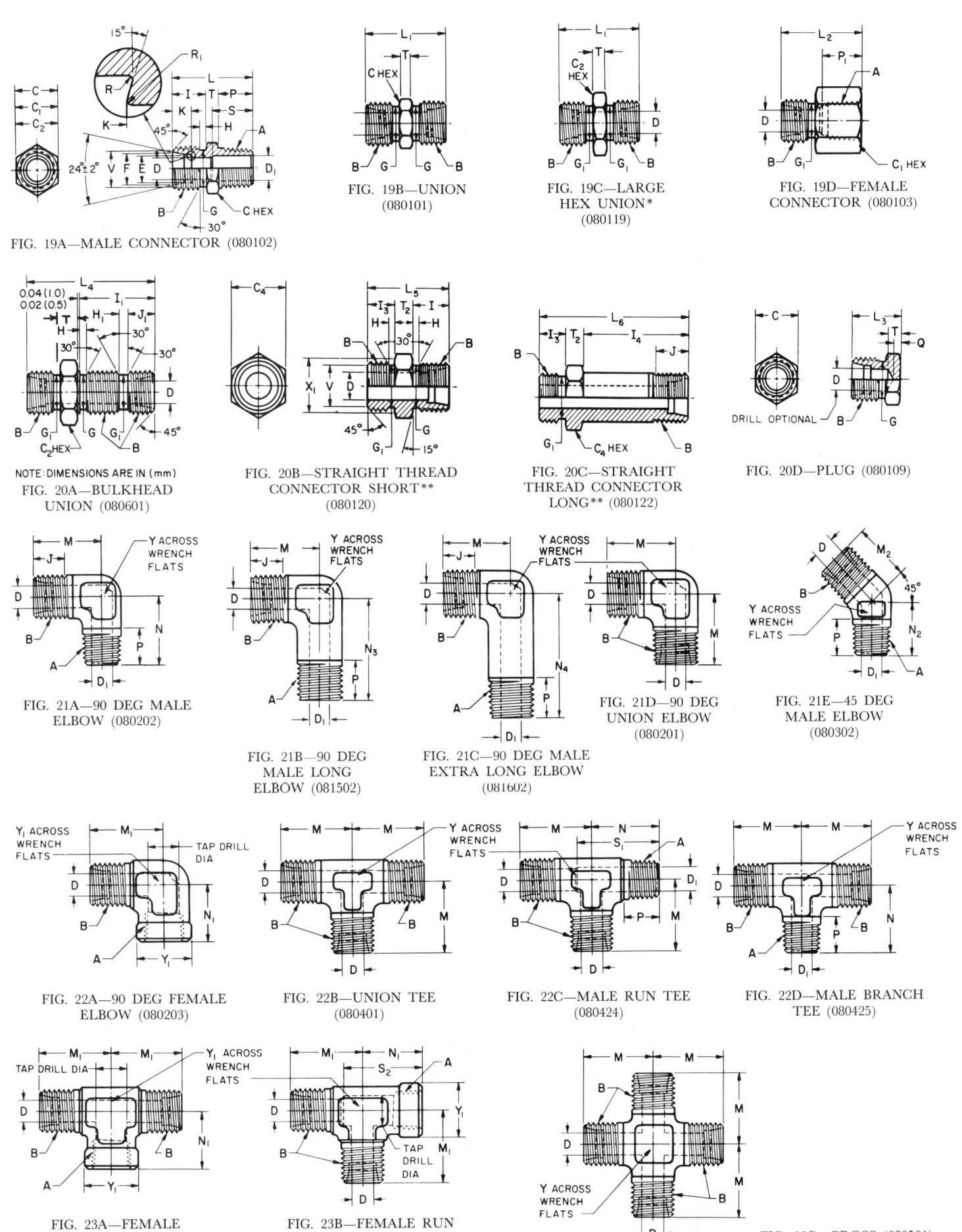

FLARELESS TYPE

TABLE 7—DIMENSIONS OF ALL BODIES AND LOCKNUTS (FIGS. 19A–27C)

Nom Tube OD, in	A Dryseal Pipe Thread, in SAE J476 (ANSI B2.2)	B Thread Size, in SAE J475 (ISO R725) Class 2A Ext Class 2B Int	C Hex Nom, in	C_1 Hex Nom, in	C_2 Hex Nom, in	C_3 Hex Max, in	C_4 Hex Nom, in	D Dia Drill in	D mm	D_1[b] Dia Drill in	D_1 mm	E Dia in (+0.004/−0.000)	E Dia mm (+0.10/−0.00)	F Dia in (+0.004/−0.000)	F Dia mm (+0.10/−0.00)	F_1 Dia in (+0.005/−0.000)	F_1 Dia mm (+0.13/−0.00)
1/8	1/8—27	5/16—24	7/16	9/16	9/16	7/16	7/16	0.093	2.4	0.188	4.8	0.135	3.43	0.189	4.80	0.170	4.32
3/16	1/8—27	3/8—24	7/16	9/16	5/8	1/2	1/2	0.125	3.2	0.188	4.8	0.196	4.98	0.267	6.78	0.250	6.35
1/4	1/8—27	7/16—20	1/2	9/16	11/16	9/16	9/16	0.203	5.2	0.188	4.8	0.261	6.63	0.319	8.10	0.320	8.13
5/16	1/8—27	1/2—20	9/16	9/16	3/4	5/8	5/8	0.234	6.0	0.188	4.8	0.324	8.23	0.382	9.70	0.380	9.65
3/8	1/4—18	9/16—18	5/8	3/4	13/16	11/16	11/16	0.281	7.0	0.281	7.0	0.386	9.80	0.441	11.20	0.445	11.30
1/2	3/8—18	3/4—16	13/16	7/8	1	7/8	7/8	0.422	10.7	0.406	10.3	0.514	13.06	0.601	15.27	0.573	14.55
5/8	1/2—14	7/8—14	15/16	1-1/8	1-1/8	1	1	0.500	12.7	0.531	13.5	0.641	16.28	0.727	18.47	0.715	18.16
3/4	3/4—14	1-1/16—12	1-1/8	1-3/8	1-3/8	1-1/4	1-1/4	0.656	16.6	0.719	18.0	0.766	19.46	0.852	21.64	0.835	21.21
7/8	3/4—14	1-3/16—12	1-1/4	1-3/8	1-1/2	1-3/8	1-3/8	0.718	18.0	0.719	18.0	0.891	22.63	0.977	24.82	0.949	24.10
1	1—11-1/2	1-5/16—12	1-3/8	1-5/8	1-5/8	1-1/2	1-1/2	0.875	22.2	0.938	23.8	1.016	25.81	1.102	27.99	1.060	26.92
1-1/4	1-1/4—11-1/2	1-5/8—12	1-11/16	1-7/8	2	1-7/8	1-7/8	1.093	27.8	1.250	31.7	1.270	32.26	1.355	34.42	1.325	33.66
1-1/2	1-1/2—11-1/2	1-7/8—12	2	2-3/8	2-1/8	2-1/4	2-1/8	1.344	34.1	1.500	38.0	1.520	38.61	1.604	40.74	1.590	40.39
2	2—11-1/2	2-1/2—12	2-5/8	2-7/8	2-3/4	2-7/8	2-3/4	1.813	46.0	1.938	49.0	2.022	51.36	2.108	53.54	2.095	53.21

Nom Tube OD, in	G Dia in (+0.002/−0.010)	G mm (+0.05/−0.30)	G_1[a] Dia in (+0.002/−0.003)	G_1 mm (+0.05/−0.08)	H in (+0.015/−0.000)	H mm (+0.4/−0.0)	H_1 in (+0.010/−0.000)	H_1 mm (+0.3/−0.0)	I in ±0.015	I mm ±0.4	I_1 in ±0.02	I_1 mm ±0.5	I_2 in ±0.02	I_2 mm ±0.5	I_3 in ±0.005	I_3 mm ±0.13	I_4 in ±0.02	I_4 mm ±0.5	I_5 in ±0.02	I_5 mm ±0.5	J Full Thread Min in	J mm
1/8	0.250	6.35	0.250	6.35	0.063	1.6	0.125	3.2	0.375	9.5	1.02	25.9	0.83	21.1	0.297	7.54	1.08	27.4	0.61	15.5	0.360	9.1
3/16	0.313	7.95	0.313	7.95	0.063	1.6	0.131	3.3	0.422	10.7	1.06	26.9	0.88	22.4	0.297	7.54	1.19	30.2	0.61	15.5	0.407	10.3
1/4	0.364	9.25	0.364	9.25	0.075	1.9	0.140	3.6	0.453	11.5	1.12	28.4	0.94	23.9	0.360	9.14	1.33	33.8	0.70	17.8	0.438	11.1
5/16	0.427	10.85	0.427	10.85	0.075	1.9	0.140	3.6	0.453	11.5	1.12	28.4	0.94	23.9	0.360	9.14	1.39	35.3	0.70	17.8	0.438	11.1
3/8	0.482	12.24	0.482	12.24	0.083	2.1	0.156	4.0	0.469	11.9	1.17	29.7	0.98	24.9	0.391	9.93	1.48	37.6	0.77	19.6	0.454	11.5
1/2	0.660	16.76	0.660	16.76	0.094	2.4	0.187	4.7	0.562	14.3	1.31	33.3	1.12	28.4	0.438	11.13	1.80	45.7	0.88	22.4	0.547	13.9
5/8	0.773	19.63	0.773	19.63	0.107	2.7	0.219	5.6	0.625	15.9	1.45	36.8	1.27	32.3	0.500	12.70	2.03	51.6	1.02	25.9	0.610	15.5
3/4	0.945	24.00	0.945	24.00	0.125	3.2	0.234	5.9	0.688	17.5	1.56	39.6	1.38	35.1	0.594	15.09	2.42	61.5	1.16	29.5	0.673	17.1
7/8	1.070	27.18	1.070	27.18	0.125	3.2	0.234	5.9	0.688	17.5	1.56	39.6	1.38	35.1	0.594	15.09	2.61	66.3	1.16	29.5	0.673	17.1
1	1.195	30.35	1.195	30.35	0.125	3.2	0.234	5.9	0.688	17.5	1.56	39.6	1.38	35.1	0.594	15.09	2.77	70.4	1.16	29.5	0.673	17.1
1-1/4	1.507	38.28	1.507	38.28	0.125	3.2	0.234	5.9	0.688	17.5	1.56	39.6	1.38	35.1	0.594	15.09	3.39	86.1	1.16	29.5	0.673	17.1
1-1/2	1.756	44.60	1.756	44.60	0.125	3.2	0.234	5.9	0.688	17.5	1.56	39.6	1.38	35.1	0.594	15.09	3.81	96.8	1.16	29.5	0.673	17.1
2	2.381	60.48	2.381	60.48	0.125	3.2	0.234	5.9	0.688	17.5	1.77	45.0	1.58	40.1	0.594	15.09	4.78	121.4	1.16	29.5	0.673	17.1

Nom Tube OD, in	J_1 Full Thread in ±0.005	J_1 mm ±0.13	J_2 Full Thread min in	J_2 mm	J_3 Full Thread in ±0.005	J_3 mm ±0.13	K in (+0.015/−0.005)	K mm (+0.40/−0.13)	K_1 in ±0.02	K_1 mm ±0.5	K_2 in (+0.04/−0.02)	K_2 mm (+1.0/−0.5)	L in ±0.02	L mm ±0.5	L_1 in ±0.02	L_1 mm ±0.5	L_2 in ±0.02	L_2 mm ±0.5	L_3 in ±0.02	L_3 mm ±0.5
1/8	0.290	7.37	0.297	7.5	0.234	5.94	0.188	4.78	0.094	2.4	0.172	4.4	1.04	26.4	1.02	25.9	1.05	26.7	0.63	16.0
3/16	0.340	8.64	0.297	7.5	0.234	5.94	0.234	5.94	0.094	2.4	0.178	4.5	1.09	27.7	1.11	28.2	1.08	27.4	0.68	17.3
1/4	0.355	9.02	0.300	7.6	0.281	7.14	0.234	5.94	0.094	2.4	0.219	5.6	1.12	28.4	1.18	30.0	1.09	27.7	0.71	18.0
5/16	0.365	9.27	0.331	8.4	0.281	7.14	0.250	6.35	0.094	2.4	0.219	5.6	1.12	28.4	1.18	30.0	1.08	27.4	0.71	18.0
3/8	0.445	11.30	0.333	8.5	0.312	7.92	0.250	6.35	0.094	2.4	0.265	6.7	1.34	34.0	1.24	31.5	1.31	33.3	0.75	19.0
1/2	0.495	12.57	0.375	9.5	0.344	8.74	0.305	7.75	0.125	3.2	0.250	6.4	1.44	36.6	1.42	36.1	1.47	37.3	0.85	21.6
5/8	0.545	13.84	0.461	11.7	0.391	9.93	0.350	8.89	0.125	3.2	0.304	7.7	1.75	44.4	1.61	40.9	1.76	44.7	0.97	24.6
3/4	0.545	13.84	0.469	11.9	0.469	11.91	0.350	8.89	0.125	3.2	0.354	9.0	1.88	47.8	1.81	46.0	1.89	48.0	1.10	27.9
7/8	0.545	13.84	0.469	11.9	0.469	11.91	0.350	8.89	0.125	3.2	0.281	7.1	1.88	47.8	1.81	46.0	1.86	47.2	1.10	27.9
1	0.545	13.84	0.438	11.1	0.469	11.91	0.415	10.54	0.125	3.2	0.250	6.4	2.07	52.6	1.81	46.0	2.13	54.1	1.10	27.9
1-1/4	0.545	13.84	0.438	11.1	0.469	11.91	0.415	10.54	0.125	3.2	0.265	6.7	2.18	55.4	1.89	48.0	2.22	56.4	1.18	30.0
1-1/2	0.545	13.84	0.438	11.1	0.469	11.91	0.485	12.32	0.125	3.2	0.289	7.3	2.28	57.9	1.96	49.8	2.23	56.6	1.25	31.8
2	0.545	13.84	0.438	11.1	0.469	11.91	0.485	12.32	0.125	3.2	0.281	7.1	2.46	62.5	2.11	53.6	2.31	58.7	1.40	35.6

Nom Tube OD, in	L_4 in ±0.02	L_4 mm ±0.5	L_5 in ±0.02	L_5 mm ±0.5	L_6 in ±0.02	L_6 mm ±0.5	M in ±0.03	M mm ±0.8	M_1 in ±0.03	M_1 mm ±0.8	M_2 in ±0.03	M_2 mm ±0.8	M_3 in ±0.03	M_3 mm ±0.8	M_4 in ±0.03	M_4 mm ±0.8	M_5 in ±0.06	M_5 mm ±1.5	M_6 in ±0.06	M_6 mm ±1.5
1/8	1.71	43.4	0.99	25.1	1.70	43.2	0.77	19.6	0.83	21.1	0.64	16.3	0.80	20.3	1.33	33.8	0.98	24.9	0.95	24.1
3/16	1.80	45.7	1.04	26.4	1.81	46.0	0.83	21.1	0.83	21.1	0.64	16.3	0.92	23.4	1.41	35.8	1.02	25.9	0.95	24.1
1/4	1.89	48.0	1.13	28.7	2.02	51.3	0.89	22.6	0.89	22.6	0.70	17.8	0.95	24.1	1.52	38.6	1.05	26.7	0.98	24.9
5/16	1.89	48.0	1.13	28.7	2.08	52.8	0.95	24.1	0.95	24.1	0.73	18.5	1.05	26.7	1.55	39.4	1.14	29.0	1.05	26.7
3/8	1.98	50.3	1.21	30.7	2.23	56.6	1.05	26.7	1.05	26.7	0.83	21.1	1.08	27.4	1.70	43.2	1.28	32.5	1.20	30.5
1/2	2.22	56.4	1.38	35.1	2.62	66.5	1.25	31.8	1.23	31.2	0.98	24.9	1.33	33.8	1.97	50.0	1.47	37.3	1.33	33.8
5/8	2.48	63.0	1.57	39.9	2.98	75.7	1.42	36.1	1.42	36.1	1.08	27.4	1.52	38.6	2.27	57.7	1.61	40.9	1.48	37.6
3/4	2.72	69.1	1.79	45.5	3.53	89.7	1.58	40.1	1.58	40.1	1.27	32.3	1.64	41.7	2.48	63.0	1.77	45.0	1.61	40.9
7/8	2.72	69.1	1.79	45.5	3.72	94.5	1.62	41.1	1.62	41.1	1.27	32.3	1.70	43.2	2.55	64.8	1.80	45.7	1.58	40.1
1	2.72	69.1	1.82	46.2	3.91	99.3	1.73	43.9	1.73	43.9	1.36	34.5	1.73	43.9	2.61	66.3	1.86	47.2	1.70	43.2
1-1/4	2.80	71.1	1.90	48.3	4.61	117.1	1.89	48.0	2.08	52.8	1.45	36.8	2.02	51.3	2.89	73.4	2.08	52.8	1.80	45.7
1-1/2	2.87	72.9	1.97	50.0	5.10	129.5	2.02	51.3	2.58	65.5	1.52	38.6	2.20	55.9	3.17	80.5	2.20	55.9	1.92	48.8
2	3.23	82.0	2.13	54.1	6.23	158.2	2.45	62.2	2.64	67.1	1.83	46.5	2.39	60.7	3.77	95.8	2.52	64.0	2.05	52.1

TABLE 7—DIMENSIONS OF ALL BODIES AND LOCKNUTS (FIGS. 19A-27C) (continued)

Nom Tube OD, in	M_7 in ±0.03	M_7 mm ±0.8	M_8 in ±0.03	M_8 mm ±0.8	M_9 in ±0.03	M_9 mm ±0.8	N in ±0.03	N mm ±0.8	N_1 in ±0.03	N_1 mm ±0.8	N_2 in ±0.03	N_2 mm ±0.8	N_3 in ±0.03	N_3 mm ±0.8	N_4 in ±0.03	N_4 mm ±0.8	O in ±0.02	O mm ±0.5	P[f] in +0.03/−0.00	P[f] mm +0.8/−0.0
1/8	0.91	23.1	0.86	21.8	1.28	32.5	0.72	18.3	0.66	16.8	0.52	13.2	1.00	25.4	1.28	32.5	0.438	11.1	0.38	9.7
3/16	0.94	23.9	0.86	21.8	1.33	33.8	0.72	18.3	0.66	16.8	0.52	13.2	1.04	26.4	1.35	34.3	0.500	12.7	0.38	9.7
1/4	1.03	26.2	1.05	26.7	1.45	36.8	0.78	19.8	0.66	16.8	0.64	16.3	1.17	29.7	1.56	39.6	0.562	14.3	0.38	9.7
5/16	1.09	27.7	1.05	26.7	1.45	36.8	0.78	19.8	0.66	16.8	0.64	16.3	1.21	30.7	1.63	41.4	0.625	15.9	0.38	9.7
3/8	1.25	31.8	1.14	29.0	1.56	39.6	1.09	27.7	0.88	22.4	0.86	21.8	1.58	40.1	2.07	52.6	0.688	17.5	0.56	14.2
1/2	1.45	36.8	1.30	33.0	1.81	46.0	1.22	31.0	1.02	25.9	0.95	24.1	1.82	46.2	2.42	61.5	0.875	22.2	0.56	14.2
5/8	1.70	43.2	1.52	38.6	2.05	52.1	1.47	37.3	1.23	31.2	1.17	29.7	2.17	55.1	2.87	72.9	1.000	25.4	0.75	19.0
3/4	1.94	49.3	1.73	43.9	2.25	57.2	1.59	40.4	1.36	34.5	1.20	30.5	2.44	62.0	3.28	83.3	1.188	30.2	0.75	19.0
7/8	2.00	50.8	1.80	45.7	2.31	58.7	1.69	42.9	1.42	36.1	1.27	32.3	2.65	67.3	3.62	91.9	1.312	33.3	0.75	19.0
1	2.05	52.1	1.86	47.2	2.38	60.5	1.97	50.0	1.62	41.1	1.48	37.6	3.01	76.5	4.05	102.9	1.438	36.5	0.94	23.9
1-1/4	2.25	57.2	1.91	48.5	2.41	61.2	2.38	60.5	1.70	43.2	1.67	42.4	3.69	93.7	5.00	127.0	1.750	44.4	0.97	24.6
1-1/2	2.39	60.7	1.91	48.5	2.42	61.5	2.64	67.1	2.08	52.8	1.77	45.0	4.10	104.1	5.55	141.0	2.000	50.8	1.00	25.4
2	2.89	73.4	1.86	47.2	2.58	65.5	3.00	76.2	2.39	60.7	2.11	53.6	4.81	122.2	6.63	168.4	2.625	66.7	1.03	26.2

Nom Tube OD, in	P_1 Max in	P_1 Max mm	Q Min in	Q Min mm	R Rad Max in	R Rad Max mm	R_1 Rad Max in	R_1 Rad Max mm	S^b Max in	S^b Max mm	$S_1^{b,c}$ Min in	$S_1^{b,c}$ Min mm	S_2^c Min in	S_2^c Min mm	T^d Ref in	T^d Ref mm	T_1 in ±0.02	T_1 mm ±0.5	T_2^d Ref in	T_2^d Ref mm
1/8	0.46	11.7	0.11	2.8	0.010	0.3	0.005	0.1	0.49	12.4	0.81	20.6	0.75	19.0	0.22	5.6	0.22	5.6	0.28	7.1
3/16	0.46	11.7	0.11	2.8	0.015	0.4	0.005	0.1	0.49	12.4	0.82	20.8	0.76	19.3	0.22	5.6	0.22	5.6	0.28	7.1
1/4	0.46	11.7	0.11	2.8	0.015	0.4	0.005	0.1	0.55	14.0	1.07	27.2	0.80	20.3	0.22	5.6	0.25	6.4	0.28	7.1
5/16	0.46	11.7	0.12	3.0	0.015	0.4	0.010	0.3	0.55	14.0	1.11	28.2	0.82	20.8	0.22	5.6	0.25	6.4	0.28	7.1
3/8	0.67	17.0	0.12	3.0	0.015	0.4	0.010	0.3	—	—	—	—	1.06	26.9	0.25	6.4	0.27	6.9	0.31	7.9
1/2	0.68	17.3	0.16	4.1	0.015	0.4	0.010	0.3	0.67	17.0	1.50	38.1	1.27	32.3	0.25	6.4	0.31	7.9	0.34	8.6
5/8	0.90	22.9	0.16	4.1	0.015	0.4	0.010	0.3	0.91	23.1	1.76	44.7	1.52	38.6	0.31	7.9	0.36	9.1	0.40	10.2
3/4	0.91	23.1	0.19	4.8	0.015	0.4	0.010	0.3	0.94	23.9	1.95	49.5	1.73	43.9	0.38	9.7	0.41	10.4	0.47	11.9
7/8	0.91	23.1	0.19	4.8	0.015	0.4	0.010	0.3	—	—	—	—	1.82	46.2	0.38	9.7	0.41	10.4	0.47	11.9
1	1.14	29.0	0.19	4.8	0.015	0.4	0.010	0.3	1.13	28.7	2.45	62.2	2.10	53.3	0.38	9.7	0.41	10.4	0.50	12.7
1-1/4	1.16	29.5	0.23	5.8	0.025	0.6	0.015	0.4	1.20	30.5	2.98	75.7	2.29	58.2	0.46	11.7	0.41	10.4	0.58	14.7
1-1/2	1.16	29.5	0.23	5.8	0.025	0.6	0.015	0.4	1.27	32.3	3.36	85.3	2.80	71.1	0.53	13.5	0.41	10.4	0.65	16.5
2	1.18	30.0	0.33	8.4	0.025	0.6	0.015	0.4	1.37	34.8	3.95	100.3	3.34	84.8	0.68	17.3	0.41	10.4	0.81	20.6

Nom Tube OD, in	U Dia in +0.015/−0.000	U Dia mm +0.4/−0.0	V Dia in ±0.005	V Dia mm ±0.13	X Dia in ±0.005	X Dia mm ±0.13	X_1 Dia in ±0.005	X_1 Dia mm ±0.13	Y in ±0.02	Y mm ±0.5	Y_1^e in ±0.02	Y_1^e mm ±0.5	MM in ±0.003	MM mm ±0.08	NN Dia in ±0.015	NN Dia mm ±0.4
1/8	0.317	8.1	0.250	6.35	0.562	14.27	0.438	11.13	0.312	7.9	0.562	14.3	0.030	0.76	0.504	12.8
3/16	0.380	9.7	0.310	7.87	0.625	15.88	0.500	12.70	0.375	9.5	0.562	14.3	0.030	0.76	0.575	14.6
1/4	0.443	11.3	0.365	9.27	0.688	17.48	0.563	14.30	0.438	11.1	0.562	14.3	0.035	0.89	0.650	16.5
5/16	0.505	12.8	0.425	10.80	0.750	19.05	0.625	15.88	0.500	12.7	0.562	14.3	0.035	0.89	0.722	18.3
3/8	0.567	14.4	0.480	12.19	0.812	20.62	0.688	17.48	0.562	14.3	0.750	19.0	0.035	0.89	0.794	20.2
1/2	0.755	19.2	0.660	16.76	1.000	25.40	0.875	22.22	0.750	19.0	0.875	22.2	0.041	1.04	1.010	25.7
5/8	0.880	22.4	0.775	19.68	1.125	28.58	1.000	25.40	0.875	22.2	1.062	27.0	0.050	1.27	1.155	29.3
3/4	1.067	27.1	0.945	24.00	1.375	34.92	1.250	31.75	1.062	27.0	1.312	33.3	0.050	1.27	1.444	36.7
7/8	1.193	30.3	1.070	27.18	1.500	38.10	1.375	34.92	1.188	30.2	1.312	33.3	0.050	1.27	1.589	40.4
1	1.317	33.5	1.195	30.35	1.625	41.28	1.500	38.10	1.312	33.3	1.625	41.3	0.050	1.27	1.732	44.0
1-1/4	1.630	41.4	1.505	38.23	1.875	47.62	1.875	47.62	1.625	41.3	1.875	47.6	0.050	1.27	2.165	55.0
1-1/2	1.880	47.8	1.755	44.58	2.125	53.98	2.125	53.98	1.875	47.6	2.562	65.1	0.050	1.27	2.454	62.3
2	2.505	63.6	2.380	60.45	2.750	69.85	2.750	69.85	2.500	63.5	2.812	71.4	0.050	1.27	3.160	80.3

[a] O-ring groove undercut must be smooth and free from tool marks.
[b] At manufacturer's option, through passages in Figs. 19A and 22C may conform with the smaller diameter specified or the appropriate end may be counterbored to the larger diameter for depths S and S_1, respectively.
[c] Maximum depth shall be optional with manufactuer providing wall thickness is controlled in compliance with General Specifications.
[d] Minimum design thickness, not subject to inspection.
[e] The basic dimensions shown shall apply as minimum for boss diameters.
[f] P dimension shall be considered minimum for shaped fittings only.

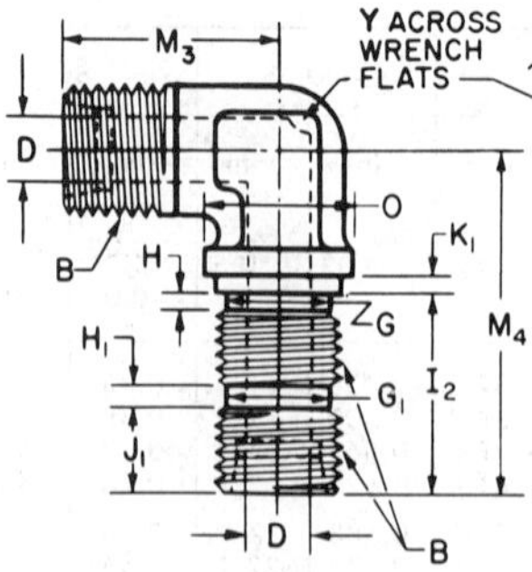

21.58

FIG. 24A—90 DEG BULKHEAD ELBOW* (080701)

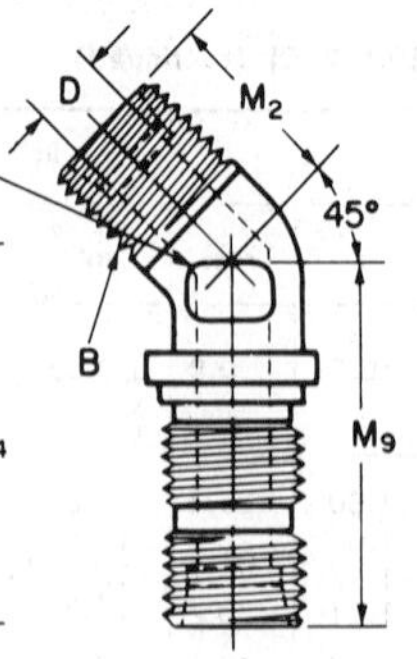

FIG. 24B—45 DEG BULKHEAD ELBOW* (080801)

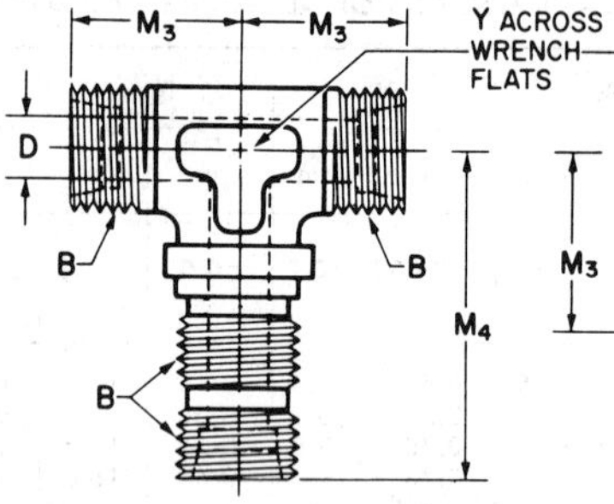

FIG. 24C—BULKHEAD BRANCH TEE* (080959)

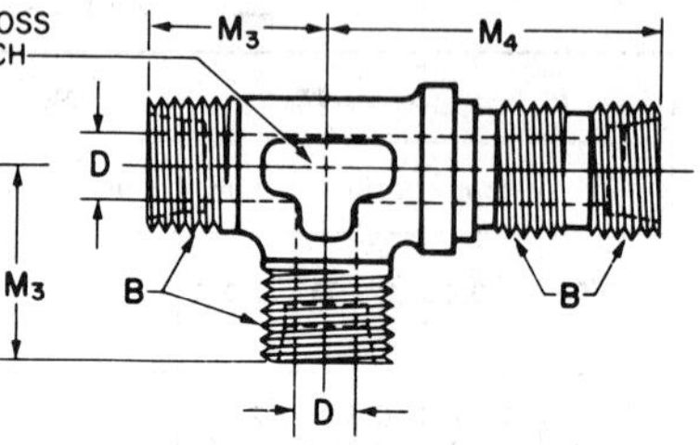

FIG. 24D—BULKHEAD RUN TEE* (080958)

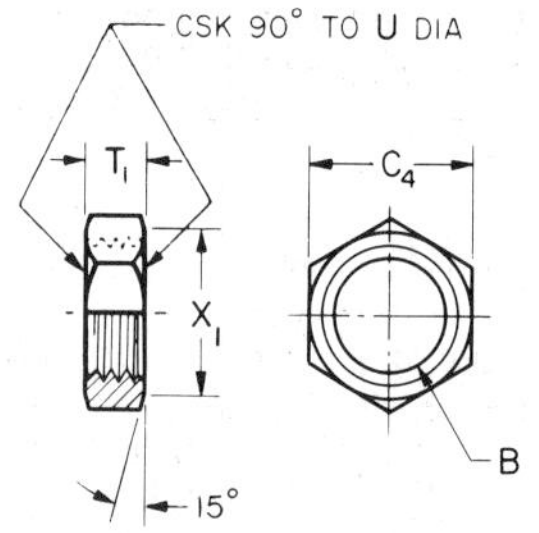

ed. FIG. 25A—STRAIGHT THREAD LOCKNUT** (080117)

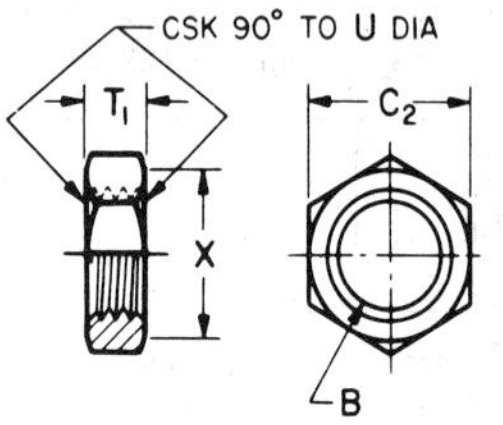

FIG. 25B—BULKHEAD LOCKNUT* (080118)

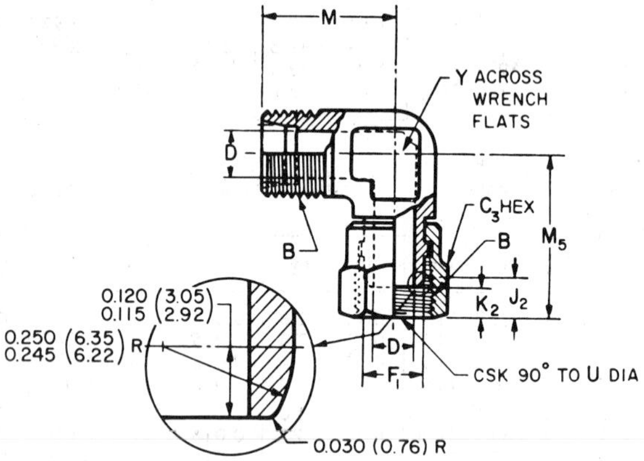

NOTE: DIMENSIONS ARE IN (mm)

FIG. 25C—90 DEG SWIVEL ELBOW (080221)

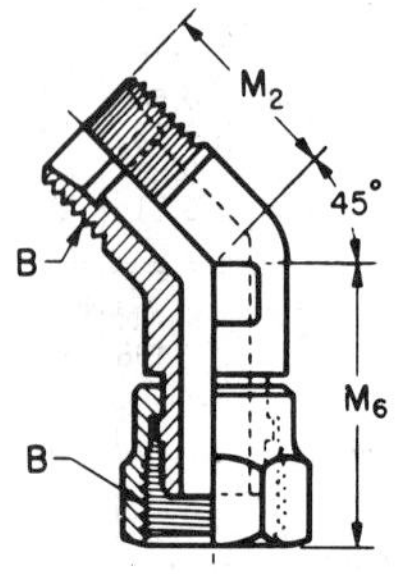

FIG. 26A—45 DEG SWIVEL ELBOW (080321)

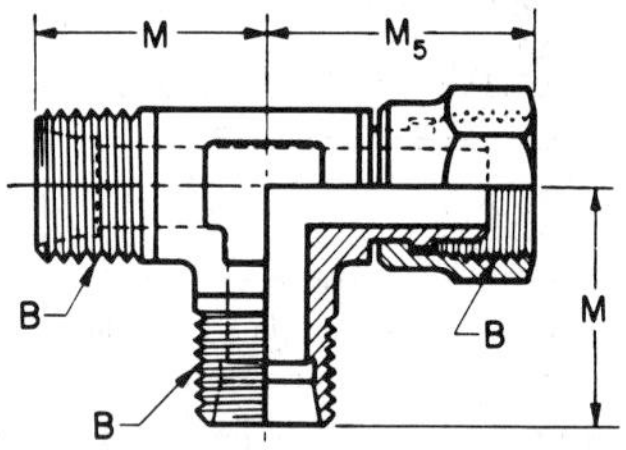

FIG. 26B—SWIVEL RUN TEE (080432)

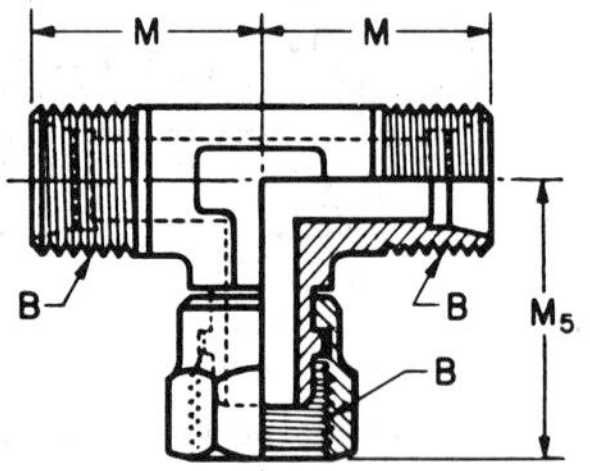

FIG. 26C—SWIVEL BRANCH TEE (080433)

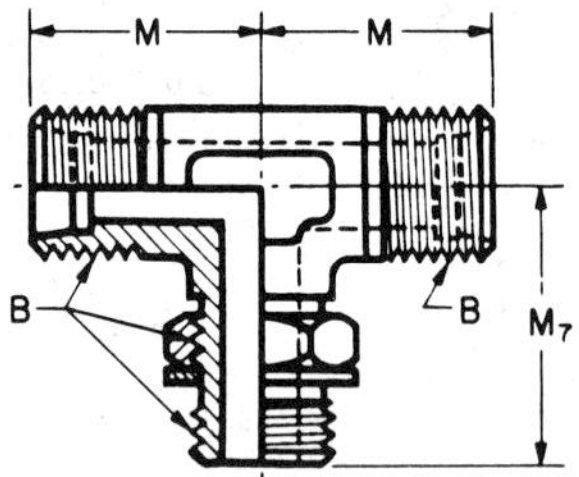

FIG. 26D—STRAIGHT THREAD BRANCH TEE† (080429)

THE DESIGN AND METHOD OF ATTACHING THE SWIVEL NUT SHALL BE OPTIONAL WITH THE MANUFACTURER PROVIDING THE TABULATED DIMENSIONS ARE MAINTAINED AND THE NUT TURNS FREELY.

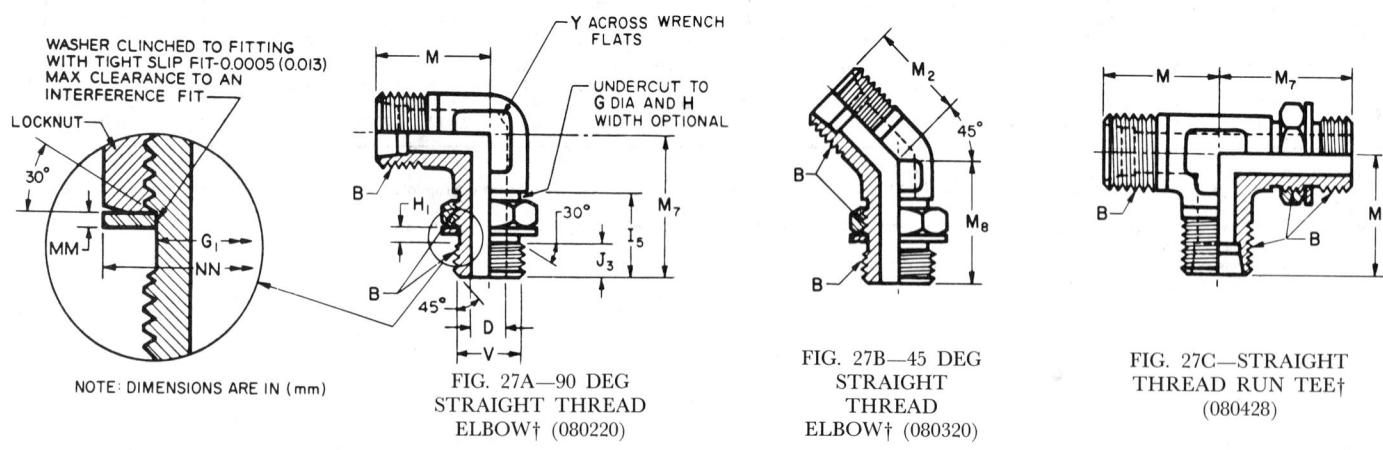

NOTE: DIMENSIONS ARE IN (mm)

FIG. 27A—90 DEG STRAIGHT THREAD ELBOW† (080220)

FIG. 27B—45 DEG STRAIGHT THREAD ELBOW† (080320)

FIG. 27C—STRAIGHT THREAD RUN TEE† (080428)

NOTES: UNSPECIFIED DETAIL WITH RESPECT TO DIMENSIONS, TOLERANCES, CONTOURS, MATERIAL, WORKMANSHIP, ETC., MUST CONFORM TO GENERAL SPECIFICATIONS FOR HYDRAULILC TUBE FITTINGS. THE DIMENSIONAL DESIGNATIONS FOR TUBE ENDS IN FIGS. 19A THRU 27C, FOR SWIVEL ENDS IN FIGS. 25C THRU 26C, FOR O-RING BOSS ENDS IN FIGS. 20B AND 20C, AND FOR ADJUSTABLE STRAIGHT THREAD ENDS IN FIGS. 26D THRU 27C SHALL APPLY TO CORRESPONDING ENDS OF OTHER FIGURES ON THIS AND PRECEDING PAGE UNLESS SHOWN OTHERWISE. FIGS. 19A THRU 27C ON THIS AND PRECEDING PAGE APPLY TO TABLE 7. CODES SHOWN IN BRACKETS ADJACENT TO FIGURE NUMBERS REPRESENT RESPECTIVE FITTING IDENTIFICATION IN ACCORDANCE WITH SAE J846.
*INTENDED FOR USE WITH O-RING SEALS IN MS 33649 (OR SUPERSEDED AND 10050) BOSSES.
** MODIFICATION OF 1/8-1 IN SIZES IN THESE TYPES OF FITTINGS FOR USE WITH MS 33649 (OR SUPERSEDED AND 10050) BOSSES IS SHOWN IN FIG. 35 AND TABLE 13.
† IF DESIRED BY THE PURCHASER AND SO SPECIFIED, THESE FITTINGS MAY BE FURNISHED WITH LARGE HEXAGON LOCKNUT SHOWN IN FIG. 25B.

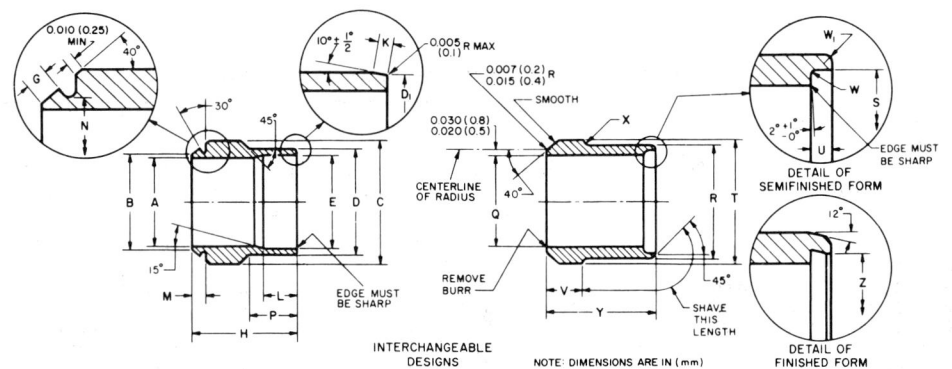

FIG. 28—STYLE A FERRULE (080115A)   FIG. 29—STYLE B FERRULE (080115B)

NOTE: FIGS. 28 AND 29 APPLY TO TABLE 8. CODES SHOWN IN BRACKETS ADJACENT TO FIGURE NUMBERS REPRESENT RESPECTIVE FITTING IDENTIFICATION IN ACCORDANCE WITH SAE J846.

TABLE 8—DIMENSIONS OF FERRULES (FIGS. 28 AND 29)

Nom Tube OD, in	A[a] Dia		B Dia		C[a] Dia		D[a] Dia		D_1 Dia Ref		E[a] Dia		G Ref		H	
	in	mm	in	mm	in	mm	in	mm	in	mm	in	mm	in	mm	in	mm
	+0.003 −0.000	+0.008 −0.00	+0.000 −0.008	+0.00 −0.20	+0.005 −0.000	+0.13 −0.00	+0.000 −0.003	+0.00 −0.08			+0.003 −0.000	+0.08 −0.00			±0.003	±0.08
1/8	0.130	3.30	0.150	3.81	0.237	6.02	0.178	4.52	0.171	4.3	0.140	3.56	0.027	0.69	0.288	7.32
3/16	0.193	4.90	0.212	5.38	0.307	7.80	0.249	6.32	0.242	6.1	0.205	5.21	0.027	0.69	0.329	8.36
1/4	0.255	6.48	0.275	6.98	0.367	9.32	0.312	7.92	0.303	7.7	0.268	6.81	0.027	0.69	0.363	9.22
5/16	0.318	8.08	0.338	8.59	0.430	10.92	0.375	9.52	0.366	9.3	0.330	8.38	0.027	0.69	0.367	9.32
3/8	0.380	9.65	0.400	10.16	0.492	12.50	0.440	11.18	0.431	10.9	0.393	9.98	0.029	0.74	0.393	9.98
1/2	0.505	12.83	0.535	13.59	0.663	16.84	0.587	14.91	0.577	14.7	0.521	13.23	0.044	1.12	0.429	10.90
5/8	0.631	16.03	0.661	16.79	0.780	19.81	0.713	18.11	0.703	17.9	0.647	16.43	0.042	1.07	0.442	11.23
3/4	0.756	19.20	0.786	19.96	0.925	23.50	0.838	21.29	0.828	21.0	0.772	19.61	0.050	1.27	0.475	12.06
7/8	0.881	22.38	0.911	23.14	1.040	26.42	0.963	24.46	0.953	24.2	0.897	22.78	0.052	1.32	0.475	12.06
1	1.006	25.55	1.036	26.31	1.187	30.15	1.088	27.64	1.078	27.4	1.022	25.96	0.054	1.37	0.475	12.06
1-1/4	1.260	32.00	1.289	32.74	1.446	36.73	1.341	32.06	1.331	33.8	1.275	32.38	0.062	1.57	0.475	12.06
1-1/2	1.510	38.35	1.539	39.09	1.694	43.03	1.590	40.39	1.580	40.1	1.524	38.71	0.062	1.57	0.475	12.06
2	2.014	51.16	2.039	51.79	2.210	56.13	2.094	53.19	2.084	52.9	2.026	51.46	0.070	1.78	0.509	12.93

Nom Tube OD, in	K		L		M		N Dia		P				Q[b] Dia		R[b] Dia	
	in	mm	in	mm	in	mm	in	mm	Min		Max		in	mm	Min	
									in	mm	in	mm			in	mm
	+0.003 −0.002	+0.08 −0.05	+0.015 −0.000	+0.40 −0.00	+0.0000 −0.006	+0.00 −0.15	+0.000 −0.005	+0.00 −0.13					+0.003 −0.000	+0.08 −0.00		
1/8	0.020	0.51	0.078	1.98	0.046	1.17	0.152	3.86	0.125	3.18	0.130	3.30	0.129	3.28	0.175	4.44
3/16	0.020	0.51	0.078	1.98	0.047	1.19	0.218	5.54	0.125	3.18	0.130	3.30	0.190	4.83	0.252	6.40
1/4	0.025	0.64	0.109	2.77	0.049	1.24	0.285	7.24	0.156	3.96	0.161	4.09	0.255	6.48	0.309	7.85
5/16	0.025	0.64	0.125	3.18	0.049	1.24	0.352	8.94	0.166	4.22	0.171	4.34	0.318	8.08	0.373	9.47
3/8	0.025	0.64	0.125	3.18	0.049	1.24	0.418	10.62	0.151	3.84	0.156	3.96	0.380	9.65	0.431	10.95
1/2	0.030	0.76	0.162	4.11	0.068	1.73	0.555	14.10	0.240	6.10	0.260	6.60	0.506	12.85	0.585	14.86
5/8	0.030	0.76	0.181	4.60	0.064	1.63	0.681	17.30	0.227	5.77	0.247	6.27	0.633	16.08	0.698	17.73
3/4	0.030	0.76	0.181	4.60	0.076	1.93	0.807	20.50	0.250	6.35	0.270	6.86	0.758	19.25	0.836	21.23
7/8	0.030	0.76	0.181	4.60	0.083	2.11	0.931	23.65	0.255	6.48	0.275	6.98	0.883	22.43	0.961	24.41
1	0.030	0.76	0.187	4.75	0.083	2.11	1.056	26.82	0.244	6.20	0.264	6.71	1.008	25.60	1.086	27.58
1-1/4	0.030	0.76	0.187	4.75	0.083	2.11	1.309	33.25	0.225	5.72	0.245	6.22	1.260	32.00	1.339	34.01
1-1/2	0.030	0.76	0.187	4.75	0.086	2.18	1.559	39.60	0.225	5.72	0.245	6.22	1.511	38.38	1.589	40.36
2	0.030	0.76	0.187	4.75	0.092	2.34	2.059	52.30	0.238	6.05	0.258	6.55	2.014	51.16	2.092	53.14

Nom Tube OD, in	R[b] Dia Max		S[b] Dia Min		S[b] Dia Max		T[b] Dia		U Min		U Max		V Min		V Max	
	in	mm	in	mm	in	mm	in	mm	in	mm	in	mm	in	mm	in	mm
							±0.005	±0.13								
1/8	0.178	4.52	0.154	3.91	0.160	4.06	0.203	5.16	0.015	0.38	0.023	0.58	0.078	1.98	0.084	2.13
3/16	0.256	6.50	0.227	5.77	0.233	5.92	0.312	7.93	0.020	0.51	0.028	0.71	0.078	1.98	0.084	2.13
1/4	0.313	7.95	0.286	7.26	0.292	7.42	0.359	9.12	0.024	0.61	0.032	0.81	0.096	2.44	0.102	2.59
5/16	0.377	9.58	0.350	8.89	0.356	9.04	0.422	10.72	0.024	0.61	0.032	0.81	0.096	2.44	0.102	2.59
3/8	0.435	11.05	0.408	10.36	0.414	10.52	0.484	12.29	0.024	0.61	0.032	0.81	0.116	2.95	0.122	3.10
1/2	0.589	14.96	0.556	14.12	0.562	14.27	0.625	15.88	0.024	0.61	0.032	0.81	0.116	2.95	0.122	3.10
5/8	0.702	17.83	0.669	16.99	0.675	17.14	0.750	19.05	0.024	0.61	0.032	0.81	0.116	2.95	0.122	3.10
3/4	0.840	21.34	0.807	20.50	0.813	20.65	0.875	22.23	0.024	0.61	0.032	0.81	0.116	2.95	0.122	3.10
7/8	0.965	24.51	0.932	23.67	0.938	23.83	1.000	25.40	0.024	0.61	0.032	0.81	0.116	2.95	0.122	3.10
1	1.090	27.69	1.057	26.85	1.063	27.00	1.125	28.58	0.024	0.61	0.032	0.81	0.116	2.95	0.122	3.10
1-1/4	1.343	34.11	1.310	33.27	1.316	33.43	1.406	35.71	0.024	0.61	0.032	0.81	0.116	2.95	0.122	3.10
1-1/2	1.593	40.46	1.560	39.62	1.566	39.78	1.656	42.06	0.024	0.61	0.032	0.81	0.116	2.95	0.122	3.10
2	2.096	53.24	2.063	52.40	2.069	52.55	2.188	55.58	0.024	0.61	0.032	0.81	0.153	3.89	0.159	4.04

TABLE 8—DIMENSIONS OF FERRULES (FIGS. 28 AND 29) (continued)

Nom Tube OD, in	W Rad				W₁ Rad		X Rad		Y				Z Dia			
	Min		Max						Min		Max		Min		Max	
	in	mm	in	mm	in	mm	in	mm	in	mm	in	mm	in	mm	in	mm
1/8	0.003	0.08	0.006	0.15	0.007	0.2	0.010	0.3	0.275	6.98	0.281	7.14	0.148	3.76	0.156	3.96
3/16	0.003	0.08	0.006	0.15	0.007	0.2	0.020	0.5	0.275	6.98	0.281	7.14	0.221	5.61	0.229	5.82
1/4	0.003	0.08	0.006	0.15	0.007	0.2	0.020	0.5	0.333	8.46	0.339	8.61	0.280	7.11	0.288	7.32
5/16	0.003	0.08	0.006	0.15	0.007	0.2	0.020	0.5	0.333	8.46	0.339	8.61	0.344	8.74	0.352	8.94
3/8	0.003	0.08	0.006	0.15	0.007	0.2	0.020	0.5	0.372	9.45	0.378	9.60	0.402	10.21	0.410	10.41
1/2	0.008	0.20	0.014	0.36	0.010	0.3	0.020	0.5	0.372	9.45	0.378	9.60	0.550	13.97	0.558	14.17
5/8	0.008	0.20	0.014	0.36	0.010	0.3	0.020	0.5	0.412	10.46	0.418	10.62	0.663	16.84	0.671	17.04
3/4	0.008	0.20	0.014	0.36	0.010	0.3	0.020	0.5	0.412	10.46	0.418	10.62	0.801	20.35	0.809	20.55
7/8	0.008	0.20	0.014	0.36	0.010	0.3	0.020	0.5	0.412	10.46	0.418	10.62	0.926	23.52	0.934	23.72
1	0.008	0.20	0.014	0.36	0.010	0.3	0.020	0.5	0.412	10.46	0.418	10.62	1.051	26.70	1.059	26.90
1-1/4	0.008	0.20	0.014	0.36	0.010	0.3	0.020	0.5	0.412	10.46	0.418	10.62	1.304	33.12	1.312	33.32
1-1/2	0.008	0.20	0.014	0.36	0.010	0.3	0.020	0.5	0.412	10.46	0.418	10.62	1.554	39.47	1.562	39.67
2	0.008	0.20	0.014	0.36	0.010	0.3	0.020	0.5	0.450	11.43	0.456	11.58	2.057	52.25	2.065	52.45

[a] These diameters, A, C, D, and E must be concentric within 0.005 in (0.13 mm).
[b] These diameters, Q, R, S, and T must be concentric within 0.005 in (0.13 mm).

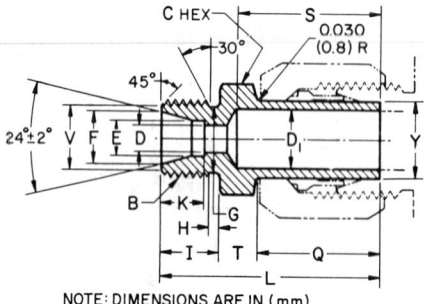

FIG. 30—REDUCER (080123)

TABLE 9—DIMENSIONS[a] OF REDUCERS (FIG. 30)

Tube Reduction, in	B Thread Size, in SAE J475 (ISO R725) Class 2A Ext	C Hex Nom, in	D Dia Drill		D₁ Dia Drill		L		Q		S		T[b] Ref		Y Dia	
							in	mm	in	mm	in	mm				
			in	mm	in	mm	±0.02	±0.5	±0.02	±0.5	±0.02	±0.5	in	mm	±0.003	±0.08
3/8 x 1/4	7/16–20	1/2	0.203	5.2	0.250	6.3	1.61	40.9	0.88	22.4	0.92	23.4	0.22	5.6	0.375	9.52
1/2 x 1/4	7/16–20	9/16	0.203	5.2	0.375	9.5	1.73	43.9	1.00	25.4	1.07	27.2	0.22	5.6	0.500	12.70
1/2 x 3/8	9/16–18	5/8	0.281	7.0	0.375	9.5	1.77	45.0	1.00	25.4	1.10	27.9	0.25	6.4	0.500	12.70
5/8 x 1/4	7/16–20	11/16	0.203	5.2	0.500	12.7	1.85	47.0	1.09	27.7	1.14	29.0	0.25	6.4	0.625	15.88
5/8 x 3/8	9/16–18	11/16	0.281	7.0	0.500	12.7	1.86	47.2	1.09	27.7	1.20	30.5	0.25	6.4	0.625	15.88
5/8 x 1/2	3/4 –16	13/16	0.422	10.7	0.500	12.7	1.96	49.8	1.09	27.7	1.20	30.5	0.25	6.4	0.625	15.88
3/4 x 1/4	7/16–20	13/16	0.203	5.2	0.625	15.9	1.92	48.8	1.16	29.5	1.14	29.0	0.25	6.4	0.750	19.05
3/4 x 3/8	9/16–18	13/16	0.281	7.0	0.625	15.9	1.93	49.0	1.16	29.5	1.20	30.5	0.25	6.4	0.750	19.05
3/4 x 1/2	3/4 –16	13/16	0.422	10.7	0.625	15.9	2.03	51.6	1.16	29.5	1.20	30.5	0.25	6.4	0.750	19.05
3/4 x 5/8	7/8 –14	15/16	0.500	12.7	0.625	15.9	2.15	54.6	1.16	29.5	1.26	31.8	0.31	7.9	0.750	19.05
1 x 1/2	3/4 –16	1-1/16	0.422	10.7	0.844	21.4	2.05	52.1	1.12	28.4	1.10	27.9	0.31	7.9	1.000	25.40
1 x 5/8	7/8 –14	1-1/16	0.500	12.7	0.844	21.4	2.11	53.6	1.12	28.4	1.10	27.9	0.31	7.9	1.000	25.40
1 x 3/4	1-1/16–12	1-1/16	0.656	16.6	0.844	21.4	2.24	56.9	1.12	28.4	1.17	29.7	0.38	9.7	1.000	25.40
1-1/4 x 5/8	7/8 –14	1-3/8	0.500	12.7	1.031	26.2	2.22	56.4	1.16	29.5	1.14	29.0	0.38	9.7	1.250	31.75
1-1/4 x 3/4	1-1/16–12	1-3/8	0.656	16.6	1.031	26.2	2.29	58.2	1.16	29.5	1.20	30.5	0.38	9.7	1.250	31.75
1-1/4 x 1	1-5/16–12	1-3/8	0.875	22.2	1.031	26.2	2.28	57.9	1.16	29.5	1.23	31.2	0.38	9.7	1.250	31.75

[a] For dimensions shown on Fig. 30 but not specified in above table, see corresponding dimensions for the specified tube outside diameter in Table 7.
[b] Minimum design thickness, not subject to inspection.

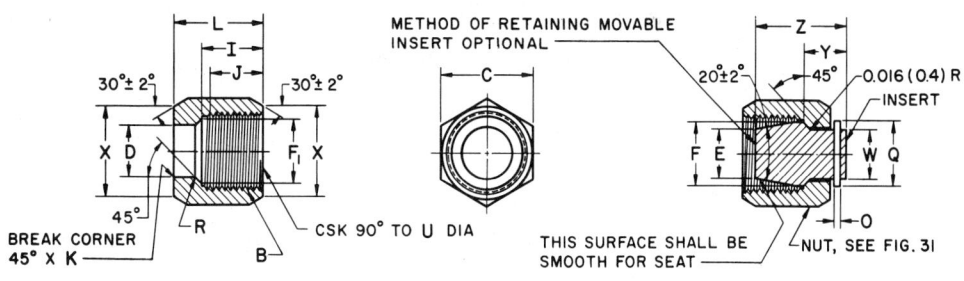

FIG. 31—NUT (080110) FIG. 32—CAP ASSEMBLY (080112)

NOTE: DIMENSIONS ARE IN (mm)

TABLE 10—DIMENSIONS OF NUT AND CAP ASSEMBLY (FIGS. 31 AND 32)

Nom Tube OD, in	B Thread Size, in SAE J475 (ISO R725) Class 2B Int	C Hex Nom, in	D Dia in (+0.003/−0.000)	D Dia mm (+0.08/−0.00)	E Dia in (+0.005/−0.000)	E Dia mm (+0.13/−0.00)	F Dia in (±0.02)	F Dia mm (±0.5)	F₁ Dia in (±0.005)	F₁ Dia mm (±0.13)	I in (+0.015/−0.000)	I mm (+0.40/−0.00)	J Full Thread Min in	J Full Thread Min mm	K in	K mm
1/8	5/16−24	3/8	0.130	3.30	0.126	3.20	0.22	5.6	—	—	0.388	9.86	0.328	8.3	0.005	0.1
3/16	3/8 −24	7/16	0.193	4.90	0.187	4.75	0.28	7.1	—	—	0.459	11.66	0.406	10.3	0.005	0.1
1/4	7/16−20	9/16	0.255	6.48	0.252	6.40	0.34	8.6	—	—	0.537	13.64	0.469	11.9	0.005	0.1
5/16	1/2 −20	5/8	0.318	8.08	0.315	8.00	0.41	10.4	—	—	0.552	14.02	0.483	12.3	0.005	0.1
3/8	9/16−18	11/16	0.380	9.65	0.377	9.58	0.47	11.9	—	—	0.568	14.43	0.500	12.7	0.010	0.3
1/2	3/4 −16	7/8	0.505	12.83	0.503	12.78	0.62	15.7	—	—	0.599	15.21	0.516	13.1	0.010	0.3
5/8	7/8 −14	1	0.631	16.03	0.630	16.00	0.75	19.0	—	—	0.677	17.20	0.578	14.7	0.010	0.3
3/4	1-1/16−12	1-1/4	0.756	19.20	0.755	19.18	0.88	22.4	—	—	0.677	17.20	0.562	14.3	0.010	0.3
7/8	1-3/16−12	1-3/8	0.881	22.38	0.875	22.22	1.00	25.4	—	—	0.677	17.20	0.562	14.3	0.010	0.3
1	1-5/16−12	1-1/2	1.006	25.55	1.005	25.53	1.16	29.5	—	—	0.677	17.20	0.562	14.3	0.010	0.3
1-1/4	1-5/8 −12	2	1.260	32.00	1.257	31.93	1.41	35.8	—	—	0.645	16.38	0.531	13.5	0.010	0.3
1-1/2	1-7/8 −12	2-1/4	1.510	38.35	1.507	38.28	1.66	42.2	1.446	36.73	0.639	16.23	0.515	13.1	0.010	0.3
2	2-1/2 −12	2-7/8	2.014	51.16	2.000	50.80	2.19	55.6	1.695	43.05	0.615	15.62	0.500	12.7	0.010	0.3
									2.285	58.04						

Nom Tube OD, in	L in (±0.02)	L mm (±0.5)	O in (±0.016)	O mm (±0.4)	Q Max in	Q Max mm	R Rad in	R Rad mm	U Dia in (+0.015/−0.000)	U Dia mm (+0.4/−0.0)	W Dia in (+0.004/−0.000)	W Dia mm (+0.10/−0.00)	X Dia in (±0.01)	X Dia mm (±0.3)	Y in (±0.02)	Y mm (±0.5)	Z in (±0.02)	Z mm (±0.5)
1/8	0.53	13.5	0.05	1.3	0.25	6.4	0.031	0.8	0.317	8.1	0.125	3.18	0.36	9.1	0.25	6.4	0.56	14.2
3/16	0.61	15.5	0.05	1.3	0.25	6.4	0.031	0.8	0.380	9.7	0.181	4.60	0.42	10.7	0.28	7.1	0.62	15.7
1/4	0.70	17.8	0.05	1.3	0.38	9.7	0.031	0.8	0.443	11.3	0.246	6.25	0.54	13.7	0.30	7.6	0.70	17.8
5/16	0.72	18.3	0.05	1.3	0.50	12.7	0.031	0.8	0.505	12.8	0.310	7.87	0.60	15.2	0.30	7.6	0.72	18.3
3/8	0.75	19.0	0.05	1.3	0.50	12.7	0.031	0.8	0.567	14.4	0.372	9.45	0.67	17.0	0.31	7.9	0.75	19.0
1/2	0.84	21.3	0.06	1.5	0.62	15.7	0.047	1.2	0.755	19.2	0.498	12.65	0.86	21.8	0.41	10.4	0.91	23.1
5/8	0.92	23.4	0.06	1.5	0.75	19.0	0.047	1.2	0.880	22.4	0.625	15.88	0.98	24.9	0.34	8.6	0.92	23.4
3/4	0.97	24.6	0.06	1.5	0.88	22.4	0.047	1.2	1.067	27.1	0.750	19.05	1.24	31.5	0.41	10.4	0.91	23.1
7/8	1.00	25.4	0.06	1.5	1.00	25.4	0.047	1.2	1.193	30.3	0.875	22.22	1.36	34.5	0.41	10.4	0.94	23.9
1	1.05	26.7	0.09	2.3	1.25	31.8	0.062	1.6	1.317	33.5	1.000	25.40	1.48	37.6	0.58	14.7	1.06	26.9
1-1/4	1.05	26.7	0.12	3.0	1.50	38.1	0.062	1.6	1.630	41.4	1.250	31.75	1.98	50.3	0.62	15.7	1.17	29.7
1-1/2	1.03	26.2	0.16	4.1	1.75	44.4	0.062	1.6	1.880	47.8	1.500	38.10	2.24	56.9	0.70	17.8	1.28	32.5
2	1.12	28.4	0.16	4.1	2.25	57.2	0.062	1.6	2.505	63.6	2.000	50.80	2.86	72.6	0.69	17.5	1.25	31.8

STRAIGHT THREAD O-RING BOSS
(For Swivel and Adjustable Style Fittings)

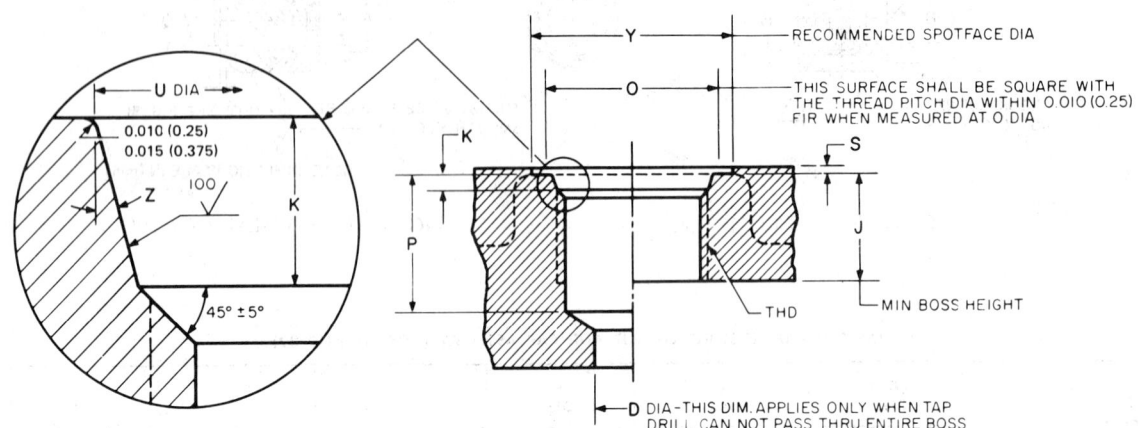

FIG. 33—STRAIGHT THREAD O-RING BOSS

TABLE 11—STRAIGHT THREAD O-RING BOSS DIMENSIONS (FIG. 33)

Nominal Tubing OD[e]		Millimetre		Thread Size SAE J475 (ISO R725), in	D Min Dia		J Full Thread Depth, Min		K +0.015 −0.000	+0.4 −0.0
in										
Nominal	Decimal	mm	Decimal in Equiv.		in	mm	in	mm	in	mm
1/8	0.125	—	—	5/16-24UNF-2B	0.062	1.6	0.390	10.0	0.074	1.9
3/16	0.1875	4	0.1575	3/8-24UNF-2B	0.125	3.2	0.390	10.0	0.074	1.9
		5	0.1968							
1/4	0.250	6	0.2362	7/16-20UNF-2B	0.172	4.4	0.454	11.5	0.093	2.4
5/16	0.3125	8	0.3150	1/2-20UNF-2B	0.234	6.0	0.454	11.5	0.093	2.4
3/8	0.375	10	0.3937	9/16-18UNF-2B	0.297	7.5	0.500	12.7	0.097	2.5
1/2	0.500	12	0.4724	3/4-16UNF-2B	0.391	10.0	0.562	14.3	0.100	2.5
5/8	0.625	16	0.6299	7/8-14UNF-2B	0.484	12.5	0.656	16.7	0.100	2.5
3/4	0.750			1-1/16-12UN-2B	0.609	16.0	0.750	19.0	0.130	3.3
7/8	0.875	20	0.7874	1-3/16-12UN-2B	0.719	18.0	0.750	19.0	0.130	3.3
1	1.000	25	0.9842	1-5/16-12UN-2B	0.844	21.0	0.750	19.0	0.130	3.3
1-1/4	1.250	32	1.2598	1-5/8-12UN-2B	1.078	27.0	0.750	19.0	0.132	3.3
1-1/2	1.500	40	1.5748	1-7/8-12UN-2B	1.312	33.0	0.750	19.0	0.132	3.3
2	2.000	50	1.9685	2-1/2-12UN-2B	1.781	45.0	0.750	19.0	0.132	3.3

Nominal Tubing OD		O Min Dia		P[d] Min		S[b,c] Max		U[a] Dia +0.005 −0.000	+0.13 −0.00	Y[c]		Z ±1 deg
in												
Nominal	Decimal	in	mm	in	mm	in	mm	in	mm	in	mm	
1/8	0.125	0.438	11	0.468	12.0	0.062	1.6	0.358	9.1	0.672	17	12
3/16	0.1875	0.500	13	0.468	12.0	0.062	1.6	0.421	10.7	0.750	19	12
1/4	0.250	0.563	15	0.547	14.0	0.062	1.6	0.487	12.4	0.828	21	12
5/16	0.3125	0.625	16	0.547	14.0	0.062	1.6	0.550	14.0	0.906	23	12
3/8	0.375	0.688	18	0.609	15.5	0.062	1.6	0.616	15.6	0.969	25	12
1/2	0.500	0.875	22	0.688	17.5	0.094	2.4	0.811	20.6	1.188	30	15
5/8	0.625	1.000	26	0.781	20.0	0.094	2.4	0.942	23.9	1.344	34	15
3/4	0.750	1.250	32	0.906	23.0	0.094	2.4	1.148	29.2	1.625	41	15
7/8	0.875	1.375	35	0.906	23.0	0.094	2.4	1.273	32.3	1.765	45	15
1	1.000	1.500	38	0.906	23.0	0.125	3.2	1.398	35.5	1.910	49	15
1-1/4	1.250	1.875	48	0.906	23.0	0.125	3.2	1.713	43.5	2.270	58	15
1-1/2	1.500	2.125	54	0.906	23.0	0.125	3.2	1.962	49.8	2.560	65	15
2	2.000	2.750	70	0.906	23.0	0.125	3.2	2.587	65.7	3.480	88	15

[a] Diameter U shall be concentric with thread pitch diameter within 0.005 in (0.13 mm) FIM, and shall be free from longitudinal and spiral tool marks. Annular tool marks up to 100 μin (2.5 μm) max shall be permissible.
[b] Maximum recommended spotface depth to permit sufficient wrench grip for proper tightening of the fitting or locknut.
[c] If face of boss is on a machined surface, dimensions Y and S need not apply.
[d] Tap drill depths given require use of bottoming taps to produce the specified full thread lengths. Where standard taps are used, the tap drill depths must be increased accordingly.
[e] Nominal tubing OD is shown for the standard inch sizes and the conversion to equivalent millimeter sizes. Figures are for reference only, as any boss can be used for a tubing size depending upon other design criteria.

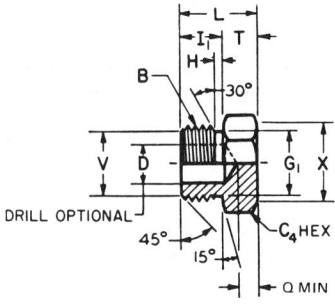

FIG. 34A—HEXAGON HEAD
O-RING BOSS PLUG
(090109A)

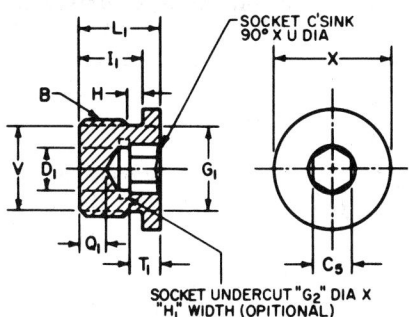

FIG. 34B—HEXAGON SOCKET
O-RING BOSS PLUG
(090109B)

TABLE 12—DIMENSIONS[a] OF O-RING BOSS PLUGS (FIGS. 34A–34B)

Nom Tube OD, in	B Thread Size, in SAE J475 (ISO R725) Class 2A Ext, in	C_4 Hex Nom	C_5 Hex Socket		D Drill		D_1 Dia		G_1[b] Dia		G_2 Dia Optional		H		H_1 Optional	
			in	mm	in	mm	in	mm	in	mm	in	mm	in	mm	in	mm
			+0.005 −0.000	+0.13 −0.00			+0.005 −0.000	+0.13 −0.00	+0.002 −0.003	+0.05 −0.08	+0.010 −0.000	+0.25 −0.00	+0.015 −0.000	+0.4 −0.0	±0.020	±0.5
1/8	5/16–24	7/16	0.125	3.18	0.093	2.4	0.125	3.18	0.250	6.35	0.156	4.0	0.063	1.6	0.063	1.6
3/16	3/8 –24	1/2	0.125	3.18	0.125	3.2	0.125	3.18	0.313	7.95	0.156	4.0	0.063	1.6	0.063	1.6
1/4	7/16–20	9/16	0.188	4.78	0.203	5.2	0.188	4.78	0.364	9.25	0.234	5.9	0.075	1.9	0.094	2.4
5/16	1/2 –20	5/8	0.188	4.78	0.234	5.9	0.188	4.78	0.427	10.85	0.234	5.9	0.075	1.9	0.094	2.4
3/8	9/16–18	11/16	0.250	6.35	0.297	7.5	0.250	6.35	0.482	12.24	0.297	7.5	0.083	2.1	0.094	2.4
1/2	3/4 –16	7/8	0.313	7.95	0.422	10.7	0.313	7.95	0.660	16.76	0.380	9.7	0.094	2.4	0.094	2.4
5/8	7/8 –14	1	0.375	9.52	0.500	12.7	0.375	9.52	0.773	19.63	0.443	11.3	0.107	2.7	0.094	2.4
3/4	1-1/16–12	1-1/4	0.563	14.30	0.656	16.7	0.563	14.30	0.945	24.00	0.661	16.8	0.125	3.2	0.125	3.2
7/8	1-3/16–12	1-3/8	0.563	14.30	0.718	18.2	0.563	14.30	1.070	27.18	0.661	16.8	0.125	3.2	0.125	3.2
1	1-5/16–12	1-1/2	0.625	15.88	0.875	22.2	0.625	15.88	1.195	30.35	0.740	18.8	0.125	3.2	0.125	3.2
1-1/4	1-5/8 –12	1-7/8	0.750	19.05	1.093	27.8	0.750	19.05	1.507	38.28	0.875	22.2	0.125	3.2	0.125	3.2
1-1/2	1-7/8 –12	2-1/8	0.750	19.05	1.344	34.1	0.750	19.05	1.756	44.60	0.875	22.2	0.125	3.2	0.125	3.2
2	2-1/2 –12	2-3/4	0.750	19.05	1.813	46.1	0.750	19.05	2.381	60.48	0.875	22.2	0.125	3.2	0.125	3.2

Nom Tube OD, in	I_1		L		L_1		Q Min		Q_1 Min		T[c] Ref		T_1 Min Hexagon Depth		V Dia		X Dia		U Dia	
	in	mm	in	mm	in	mm	in	mm	in	mm	in	mm	in	mm	in	mm	in	mm	in	mm
	±0.005	±0.13	±0.02	±0.5	±0.02	±0.5									±0.005	±0.13	±0.005	±0.13	±0.010	±0.3
1/8	0.297	7.54	0.60	15.2	0.38	9.7	0.06	1.5	0.11	2.8	0.28	7.1	0.125	3.18	0.250	6.35	0.438	11.13	0.156	4.0
3/16	0.297	7.54	0.60	15.2	0.38	9.7	0.08	2.0	0.11	2.8	0.28	7.1	0.125	3.18	0.310	7.87	0.500	12.70	0.156	4.0
1/4	0.360	9.14	0.67	17.0	0.45	11.4	0.10	2.5	0.12	3.0	0.28	7.1	0.156	3.96	0.365	9.27	0.563	14.30	0.219	5.6
5/16	0.360	9.14	0.67	17.0	0.45	11.4	0.12	3.0	0.12	3.0	0.28	7.1	0.156	3.96	0.425	10.80	0.625	15.88	0.219	5.6
3/8	0.391	9.93	0.73	18.5	0.48	12.2	0.16	4.1	0.12	3.0	0.31	7.9	0.188	4.77	0.480	12.19	0.688	17.48	0.297	7.5
1/2	0.438	11.13	0.80	20.3	0.56	14.2	0.22	5.6	0.15	3.8	0.34	8.6	0.188	4.77	0.660	16.76	0.875	22.22	0.375	9.5
5/8	0.500	12.70	0.93	23.6	0.63	16.0	0.25	6.4	0.15	3.8	0.41	10.4	0.250	6.35	0.775	19.68	1.000	25.40	0.438	11.1
3/4	0.594	15.09	1.09	27.7	0.75	19.0	0.25	6.4	0.15	3.8	0.47	11.9	0.313	7.95	0.945	24.00	1.250	31.75	0.656	16.7
7/8	0.594	15.09	1.09	27.7	0.75	19.0	0.25	6.4	0.15	3.8	0.47	11.9	0.313	7.95	1.070	27.18	1.375	34.92	0.656	16.7
1	0.594	15.09	1.12	28.4	0.75	19.0	0.25	6.4	0.19	4.8	0.50	12.7	0.375	9.52	1.195	30.35	1.500	38.10	0.734	18.6
1-1/4	0.594	15.09	1.20	30.5	0.75	19.0	0.25	6.4	0.19	4.8	0.58	14.7	0.375	9.52	1.505	38.23	1.875	47.62	0.906	23.0
1-1/2	0.594	15.09	1.27	32.3	0.75	19.0	0.25	6.4	0.25	6.4	0.65	16.5	0.375	9.52	1.755	44.58	2.125	53.98	0.906	23.0
2	0.594	15.09	1.43	36.3	0.75	19.0	0.30	7.6	0.25	6.4	0.81	20.6	0.375	9.52	2.380	60.45	2.750	69.85	0.906	23.0

[a] Modification of 1/8–1 in sizes for use with MS 33649 (or superseded AND 10050 bosses is shown in Fig. 35 and Table 13.
[b] O-ring groove undercut must be smooth and free of tool marks.
[c] Minimum design thickness, not subject to inspection.

MODIFICATION OF HEXAGONS ON STANDARD FITTINGS TO ACCOMMODATE MS 33649 (OR SUPERSEDED AND 10050) BOSSES

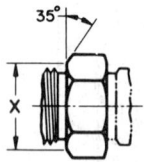

FIG. 35—MODIFIED HEXAGON CHAMFER

TABLE 13—DIMENSIONS OF MODIFIED CHAMFER (FIG. 35)

Nominal Tube OD	X Dia in (+0.000 / −0.010)	X Dia mm (+0.00 / −0.25)	Nominal Tube OD	X Dia in (+0.000 / −0.010)	X Dia mm (+0.00 / −0.25)
1/8	0.433	11.00	1/2	0.870	22.10
3/16	0.495	12.57	5/8	0.995	25.27
1/4	0.558	14.17	3/4	1.245	31.62
5/16	0.620	15.75	7/8	1.370	34.80
3/8	0.683	17.35	1	1.495	37.97

When 37 deg flared fittings shown in Figs. 3B and 3C, flareless fittings shown in Figs. 20B and 20C, or O-ring boss plugs shown in Fig. 34, in sizes from 1/8–1 in inclusive, are to be used with MS 33649 (or superseded AND 10050) type straight thread O-ring bosses the chamfer on the bearing face of the hexagon of these fittings shall be modified as shown in Fig. 35.

ASSEMBLY INSTRUCTIONS FOR ADJUSTABLE STYLE FITTINGS IN STRAIGHT THREAD O-RING BOSSES

These instructions apply to the assembly of hydraulic fittings of the 37 deg flared type shown in Figs. 9D, 10A, 10B, 10C, and 8A, and flareless type shown in Figs. 26D, 27A, 27B, 27C, and 25A, and hydraulic O-rings, Fig. 1 of SAE J515.

1. Lubricate the O-ring by coating with a light oil or petrolatum and install in the groove adjacent to the face of the metal back-up washer which is assembled at the extreme end of the groove as shown in Fig. 36B.

2. Install the fitting into the SAE straight thread boss (Fig. 33) until the metal back-up washer contacts the face of the boss as shown in Fig. 36C.

3. Position the fitting by turning out (counterclockwise) up to a maximum of one turn. Holding the pad of the fitting with a wrench, tighten the locknut and washer against the face of the boss as shown in Fig. 36D.

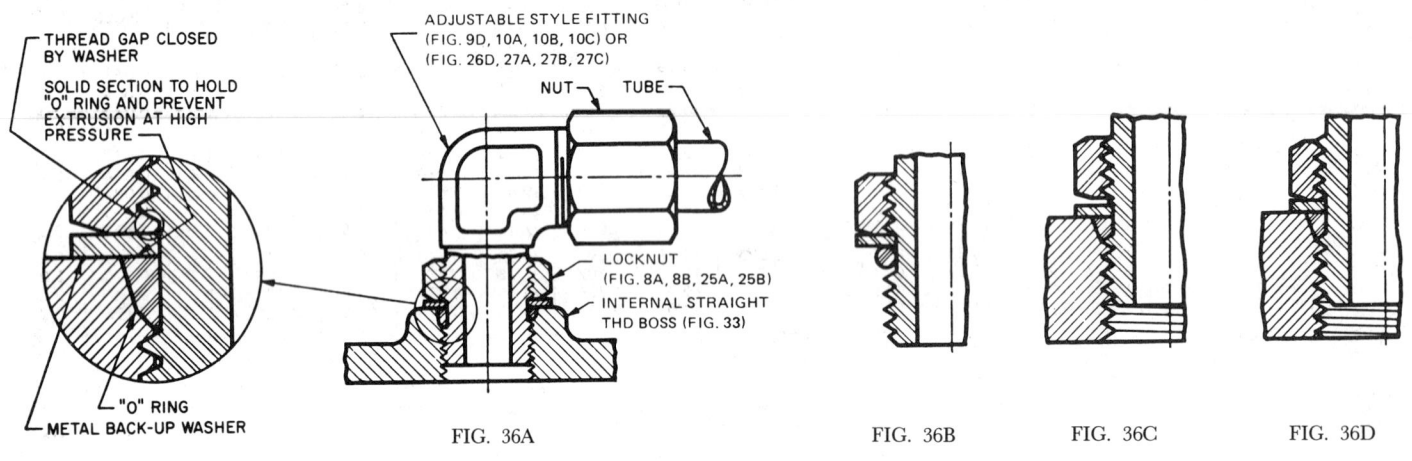

FIG. 36—ADJUSTABLE STYLE FITTINGS

ASSEMBLY INSTRUCTIONS FOR SWIVEL STYLE AND O-RING BOSS FITTINGS IN STRAIGHT THREAD O-RING BOSSES

These instructions apply to the assembly of hydraulic fittings of the 37 deg flared type shown in Figs. 3B and 3C, flareless type shown in Figs. 20B and 20C, and O-ring boss plugs shown in Fig. 34.

1. Lubricate O-ring by coating with a light oil or petrolatum and install in the O-ring groove on the fitting.

2. Screw fitting into the straight thread boss and tighten hexagon against the face of the boss as shown in Fig. 37C.

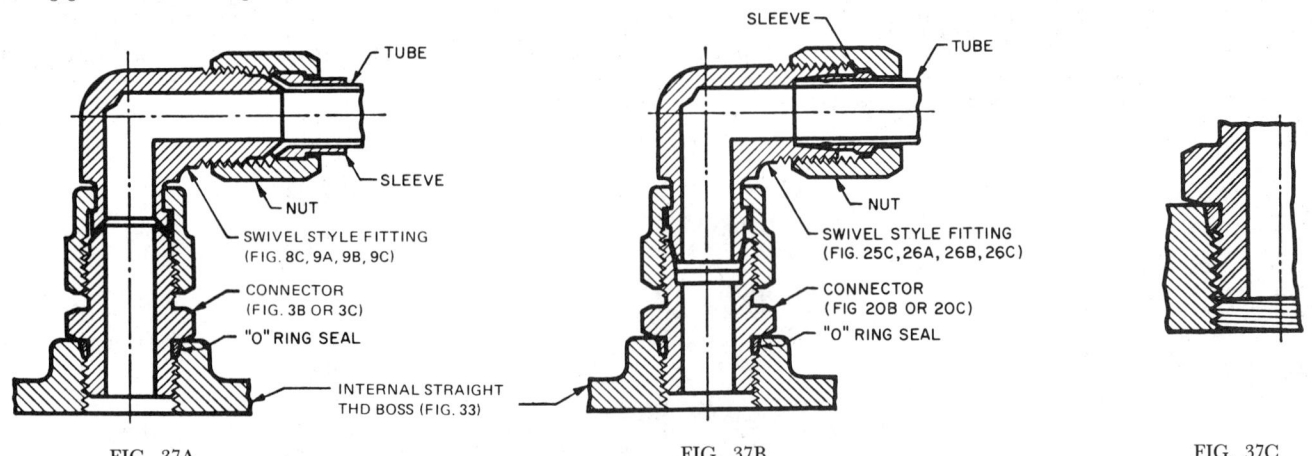

FIG. 37—SWIVEL STYLE FITTINGS

NOTES: UNSPECIFIED DETAIL WITH RESPECT TO DIMENSIONS, TOLERANCES, CONTOURS, MATERIAL AND WORKMANSHIP MUST CONFORM TO GENERAL SPECIFICATIONS FOR HYDRAULIC PIPE FITTINGS. CODES SHOWN IN BRACKETS ADJACENT TO FIGURE NUMBERS REPRESENT RESPECTIVE FITTING IDENTIFICATION IN ACCORDANCE WITH SAE J846.

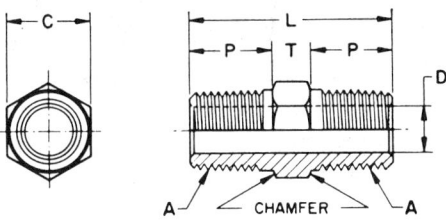

FIG. 38—HEXAGON PIPE NIPPLE
(140137)

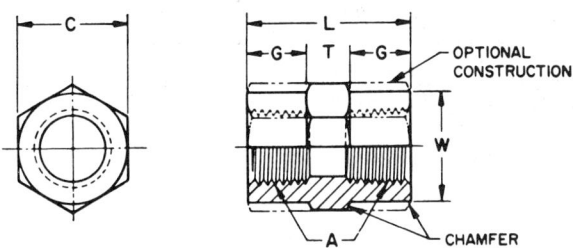

FIG. 39—HEXAGON PIPE COUPLING
(140138)

TABLE 14—DIMENSIONS OF HEXAGON PIPE NIPPLES (FIG. 38)

A Dryseal Pipe Thread SAE J476 (ANSI B1.20.3)	C Hex	D Drill		L		P		T^a Ref	
				in ±0.02	mm ±0.5	in +0.03 −0.00	mm +0.8 −0.0		
Nom	Nom	in	mm					in	mm
1/8–27	7/16	0.188	4.8	1.06	26.9	0.38	9.7	0.22	5.6
1/4–18	5/8	0.281	7.0	1.45	36.8	0.56	14.2	0.25	6.4
3/8–18	3/4	0.406	10.3	1.45	36.8	0.56	14.2	0.25	6.4
1/2–14	7/8	0.531	13.5	1.89	48.0	0.75	19.0	0.31	7.9
3/4–14	1-1/8	0.719	18.0	1.96	49.8	0.75	19.0	0.38	9.7
1 –11-1/2	1-3/8	0.938	23.8	2.34	59.4	0.94	23.9	0.38	9.7
1-1/4–11-1/2	1-3/4	1.250	31.7	2.48	63.0	0.97	24.6	0.46	11.7
1-1/2–11-1/2	2	1.500	38.0	2.61	66.3	1.00	25.4	0.53	13.5
2 –11-1/2	2-1/2	1.938	49.0	2.82	71.6	1.03	26.2	0.68	17.3

[a] Minimum Design Thickness, Not Subject To Inspection.

TABLE 15—DIMENSIONS OF HEXAGON PIPE COUPLING (FIG. 39)

A Dryseal Pipe Thread SAE J476 (ANSI B1.20.3)	C Hex	G Ref[a]		L		T Ref Min		W Dia	
				in ±0.02	mm ±0.5	in	mm	in +0.00 −0.02	mm +0.0 −0.5
Nom	Nom	in	mm						
1/8–27	5/8	0.25	6.4	0.75	19.0	0.22	5.6	0.625	15.88
1/4–18	3/4	0.43	10.9	1.13	28.7	0.25	6.4	0.750	19.05
3/8–18	7/8	0.43	10.9	1.13	28.7	0.25	6.4	0.875	22.22
1/2–14	1-1/8	0.56	14.2	1.50	38.1	0.31	7.9	1.125	28.58
3/4–14	1-3/8	0.56	14.2	1.53	38.9	0.38	9.7	1.375	34.92
1 –11-1/2	1-5/8	0.69	17.5	1.89	48.0	0.38	9.7	1.625	41.28
1-1/4–11-1/2	2	0.69	17.5	1.93	49.0	0.46	11.7	2.000	50.80
1-1/2–11-1/2	2-3/8	0.69	17.5	1.93	49.0	0.53	13.5	2.375	60.32
2 –11-1/2	2-7/8	0.63	16.0	1.96	49.8	0.68	17.3	2.875	73.03

[a] Reference Dimension, Not Subject To Inspection.

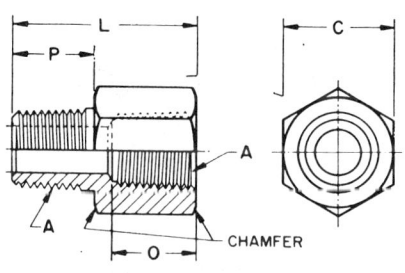

FIG. 40—ADAPTER
(140139)

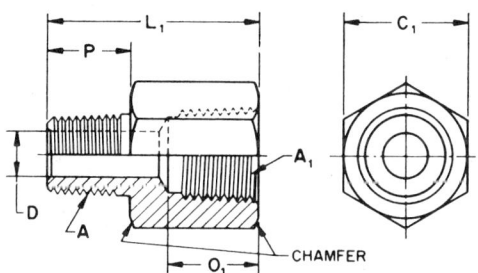

FIG. 41—INCREASE ADAPTERS
(140139)

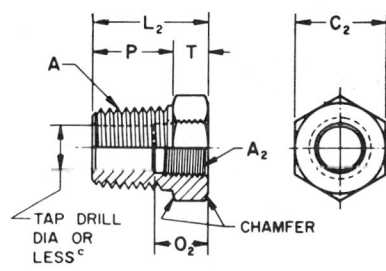

FIG. 42—REDUCER BUSHING
(140140)

TABLE 16—DIMENSIONS OF ADAPTERS, INCREASE ADAPTERS, AND REDUCER BUSHINGS (FIGS. 40, 41, and 42)

Dryseal Pipe Thread SAE J476 (ANSI B1.20.3)			C Hex	C_1 Hex	C_2 Hex	D Drill		L		L_1		L_2		O Min		O_1 Min		O_2^a Min		P		T^b Ref	
A Adapter	A × A_1 Increase Adapter	A × A_2 Reducer Bushing	Nom	Nom	Nom	in	mm	in ±0.02	mm ±0.5	in ±0.02	mm ±0.5	in ±0.02	mm ±0.5	in	mm	in	mm	in	mm	in +0.03 −0.00	mm +0.08 −0.00	in	mm
1/8–27	1/8 × 1/4	—	5/8	3/4	—	0.188	4.8	1.04	26.4	1.21	30.7	—	—	0.38	9.7	0.56	14.2	—	—	0.38	9.7	—	—
1/4–18	1/4 × 3/8	1/4 × 1/8	3/4	7/8	5/8	0.281	7.0	1.39	35.3	1.44	36.6	0.85	21.6	0.58	14.7	0.58	14.7	0.38	9.7	0.56	14.2	0.25	6.4
3/8–18	3/8 × 1/2	3/8 × 1/4	7/8	1-1/8	3/4	0.406	10.3	1.44	36.6	1.68	42.7	0.85	21.6	0.58	14.7	0.75	19.0	0.56	14.2	0.56	14.2	0.25	6.4
1/2–14	1/2 × 3/4	1/2 × 3/8	1-1/8	1-3/8	7/8	0.531	13.5	1.87	47.5	1.93	49.0	1.10	27.9	0.75	19.0	0.77	19.6	0.58	14.7	0.75	19.0	0.31	7.9
3/4–14	3/4 × 1	3/4 × 1/2	1-3/8	1-5/8	1-1/8	0.719	18.0	1.93	49.0	2.18	55.4	1.17	29.7	0.77	19.6	0.94	23.9	0.75	19.0	0.75	19.0	0.38	9.7
1 –11-1/2	1 × 1-1/4	1 × 3/4	1-5/8	2	1-3/8	0.938	23.8	2.37	60.2	2.46	62.5	1.36	34.5	0.94	23.9	0.94	23.9	0.77	19.6	0.94	23.9	0.38	9.7
1-1/4–11-1/2	1-1/4 × 1-1/2	1-1/4 × 1	2	2-3/8	1-3/4	1.250	31.7	2.49	63.2	2.50	63.5	1.47	37.3	0.94	23.9	0.94	23.9	0.94	23.9	0.97	24.6	0.46	11.7
1-1/2–11-1/2	1-1/2 × 2	1-1/2 × 1-1/4	2-3/8	2-7/8	2	1.500	38.0	2.53	64.3	2.63	66.8	1.57	39.9	0.94	23.9	0.97	24.6	0.94	23.9	1.00	25.4	0.53	13.5
2 –11-1/2	—	2 × 1-1/2	2-7/8	—	2-1/2	1.938	49.0	2.66	67.6	—	—	1.75	44.5	0.97	24.6	—	—	0.94	23.9	1.03	26.2	0.68	17.3

[a] Beyond Tap Drill Depth O_2, Hole May Be Reduced Below Tap Drill Diameter, but Shall Not Be Less Than D Diameter in Corresponding External Pipe Size Adapter.
[b] Minimum Design Thickness, not Subject To Inspection.

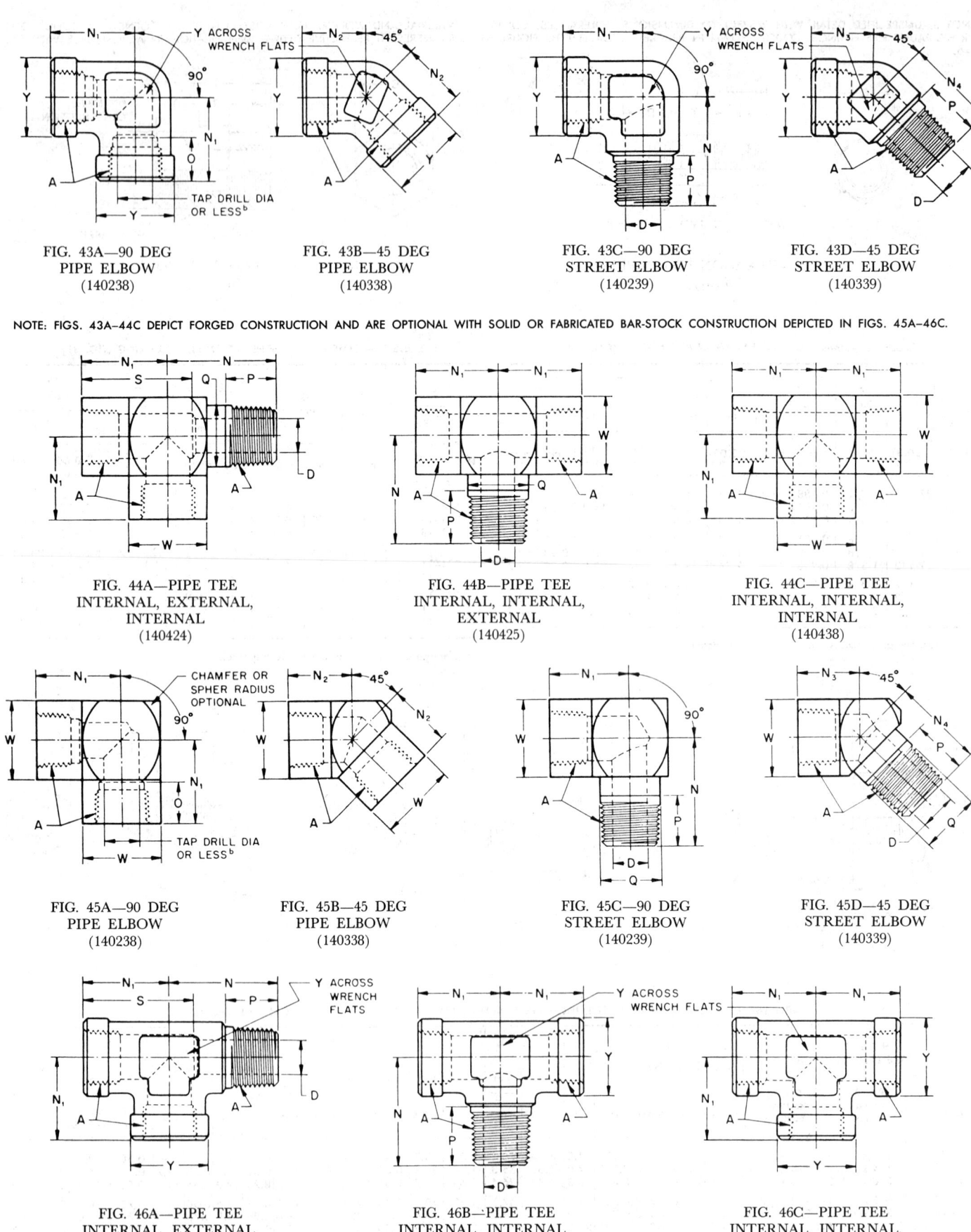

FIG. 43A—90 DEG PIPE ELBOW (140238)
FIG. 43B—45 DEG PIPE ELBOW (140338)
FIG. 43C—90 DEG STREET ELBOW (140239)
FIG. 43D—45 DEG STREET ELBOW (140339)

NOTE: FIGS. 43A–44C DEPICT FORGED CONSTRUCTION AND ARE OPTIONAL WITH SOLID OR FABRICATED BAR-STOCK CONSTRUCTION DEPICTED IN FIGS. 45A–46C.

FIG. 44A—PIPE TEE INTERNAL, EXTERNAL, INTERNAL (140424)
FIG. 44B—PIPE TEE INTERNAL, INTERNAL, EXTERNAL (140425)
FIG. 44C—PIPE TEE INTERNAL, INTERNAL, INTERNAL (140438)

FIG. 45A—90 DEG PIPE ELBOW (140238)
FIG. 45B—45 DEG PIPE ELBOW (140338)
FIG. 45C—90 DEG STREET ELBOW (140239)
FIG. 45D—45 DEG STREET ELBOW (140339)

FIG. 46A—PIPE TEE INTERNAL, EXTERNAL, INTERNAL (140424)
FIG. 46B—PIPE TEE INTERNAL, INTERNAL, EXTERNAL (140425)
FIG. 46C—PIPE TEE INTERNAL, INTERNAL, INTERNAL (140438)

NOTES: UNSPECIFIED DETAIL WITH RESPECT TO DIMENSIONS, TOLERANCES, CONTOURS, MATERIAL AND WORKMANSHIP MUST CONFORM TO GENERAL SPECIFICATIONS FOR HYDRAULIC PIPE FITTINGS. THE DIMENSIONAL DESIGNATIONS FOR TAP DRILL AND NOTES SHOWN ON FIGS. 6 AND 13 SHALL APPLY TO CORRESPONDING FEATURES OF FIGS. 7–12 AND ed. FIGS. 45B–46C UNLESS SHOWN OTHERWISE.

TABLE 17—DIMENSIONS OF FORGED AND BARSTOCK TYPES OF PIPE ELBOWS, STREET ELBOWS, AND PIPE TEES (FIGS. 43A–46C)

A Dryseal Pipe Thread SAE J476 (ANSI B1.20.3)	D^a Drill		N		N₁		N₂		N₃		N₄		O^a Min		P		Q Dia Min		S^b Min		W Square or Diameter		Y^c	
	in	mm	in ±0.03	mm ±0.8	in ±0.03	mm ±0.8	in ±0.03	mm ±0.8	in ±0.03	mm ±0.8	in ±0.03	mm ±0.8	in	mm	in +0.03 −0.00	mm +0.8 −0.0	in	mm	in	mm	in	mm	in ±0.02	mm ±0.5
1/8–27	0.188	4.8	0.78	19.8	0.66	16.8	0.50	12.7	0.47	11.9	0.72	18.3	0.38	9.6	0.38	9.6	0.44	11.2	0.94	23.9	0.62	15.7	0.56	14.2
1/4–18	0.281	7.0	1.09	27.7	0.88	22.4	0.69	17.5	0.62	15.7	1.05	26.7	0.56	14.2	0.56	14.2	0.56	14.2	1.14	29.0	0.75	19.0	0.75	19.0
3/8–18	0.406	10.3	1.22	31.0	1.02	25.9	0.75	19.0	0.72	18.3	1.06	26.9	0.58	14.7	0.56	14.2	0.68	17.3	1.33	33.8	0.88	22.3	0.88	22.4
1/2–14	0.531	13.5	1.47	37.3	1.23	31.2	0.94	23.9	0.91	23.1	1.34	34.0	0.75	19.0	0.75	19.0	0.88	22.4	1.62	41.1	1.12	28.4	1.06	26.9
3/4–14	0.719	18.0	1.59	40.4	1.36	34.5	1.00	25.4	0.97	24.6	1.38	35.1	0.77	19.6	0.75	19.0	1.06	26.9	1.86	47.2	1.38	35.0	1.31	33.3
1 11-1/2	0.938	23.8	1.97	50.0	1.62	41.1	1.19	30.2	1.12	28.4	1.72	43.7	0.94	23.8	0.94	23.8	1.38	35.1	2.23	56.6	1.62	41.1	1.62	41.1
1-1/4–11-1/2	1.250	31.7	2.38	60.5	1.70	43.2	1.44	36.6	1.63	41.4	1.80	45.7	0.94	23.8	0.97	24.6	1.69	42.9	2.47	62.7	2.00	50.8	1.88	47.8
1-1/2–11-1/2	1.500	38.0	2.64	67.1	2.08	52.8	1.46	37.1	1.69	42.9	2.06	52.3	0.94	23.8	1.00	25.4	1.90	48.3	2.97	75.4	2.38	60.5	2.56	65.0
2 –11-1/2	1.938	49.0	3.00	76.2	2.39	60.7	1.59	40.4	2.19	55.6	2.15	54.6	0.97	24.6	1.03	26.2	2.38	60.5	3.52	89.4	2.88	73.2	2.81	71.4

^aBeyond Tap Drill Depth 0, Hole May Be Reduced to Below Tap Drill Diameter, but Shall Not Be Less Than D Drill for Corresponding Size. See Figs. 43A and 45A.
^bMaximum Depth Shall Be Optional With Manufacturer Provided That Strength of Fitting Is Not Impaired.
^cThe Basic Dimension Shown Shall Apply As Minimum for Boss Diameter.

TABLES FOR CALCULATING DIMENSIONS ON SPECIAL SIZES

Tables 18–28 present various factors to be used in determining the dimensions applicable to special size combination fittings not contained in SAE J514.

In Tables 19, 23, 24, and 25, no factors are given for extreme combination sizes because of differences in factors due to method of manufacture. These extreme conditions are rare and it is suggested a manufacturer be contacted for the proper dimension.

No factors are given for bulkhead or swivel ends as combinations are not generally specified for these fittings.

Tables 26–28 present the minimum stock size acceptable for the various machining ends.

For any non-standard size fitting, be it a connector, 45 or 90 deg elbow, tee, or cross, one end is always standard, conforming to the J514 tables of dimensions. Considering this end to be the largest on the fitting, it may then be used as a basis for establishing the stocksize and length (either overall or end to center) for all other ports by deducting factors equivalent to the reduction in machining requirements from the appropriate standard lengths.

The factors applicable to the various end configurations and size reductions tabulated in the tables were determined on the following basis:

Those pertaining to lengths were derived by maintaining the standard hexagon thickness for straight fittings and the standard centerline to machining start for shaped fittings.

The factors shown in Tables 18, 20, 22, 24, and 25 were derived by subtracting the standard machining length required for the smaller end from that required for the larger standard end and rounding the result to a two place decimal.

The factors given in Tables 19, 21, and 23 were derived by subtracting the standard machining length plus an allowance of 1½ pitches (threads) for imperfect thread length required for the smaller end from the same value required for the larger end and rounding the result to a two place decimal.

The minimum allowable stock size for the various types of ends are also tabulated for reference purposes.

EXAMPLES:

For a 37 deg flared male connector (Fig. 2A) with ½ tube OD and ¾ NPTF, the overall length would be determined as follows:

 2.06 = L overall length for ¾ tube to ¾ NPTF from Table 2
−0.21 = Factor from Table 18 for ½ machining on ¾ size fitting
 1.85 = Overall length for the non-standard male connector

Since the ¾ NPTF is the larger machining, the hexagon width of 1⅛ from Table 27 would apply.

For a 37 deg flared 90 deg male elbow (Fig. 4A) with ½ OD tube and ¾ NPTF, the end to center length M would be derived as follows:

 1.66 = M dimension for standard ¾ 37 deg end
−0.24 = Factor from Table 19 for ½ machining on ¾ size fitting
 1.42 = End to center length

Since the ¾ NPTF is the standard end, the N end to center dimension would remain 1.59 as shown in Table 2. The wrench flat size would be as shown by the Y column in Table 2 for the ¾ tube OD.

TOLERANCES: The following tolerances apply to non-standard sizes.

Overall length of straight fittings = ±0.02 in (0.5 mm)
Centerline to end on shaped fittings = ±0.06 in (1.5 mm)

TABLE 18—FACTORS FOR 37 DEG FLARED END ON STRAIGHT FITTINGS (FIGS. 2A-2D, 3A-3C, AND 15)

Nominal Tube OD	\multicolumn{26}{c	}{Standard Machining Size}	Nominal Tube OD
	3/16		1/4		5/16		3/8		1/2		5/8		3/4		7/8		1		1-1/4		1-1/2		2			
	in	mm	in	mm	in	mm	in	mm	in	mm	in	mm	in	mm	in	mm	in	mm	in	mm	in	mm	in	mm	
1/8	0.03	0.8	0.10	2.5	0.10	2.5	0.11	2.8	0.21	5.3	0.31	7.9	0.42	10.7	0.44	11.2	0.46	11.7	0.51	13.0	0.64	16.3	0.89	22.6	1/8
3/16	—	—	0.07	1.8	0.07	1.8	0.08	2.0	0.18	4.6	0.28	7.1	0.39	9.9	0.41	10.4	0.43	10.9	0.48	12.2	0.60	15.2	0.85	21.6	3/16
1/4	—	—	—	—	0.00	0.0	0.01	0.3	0.11	2.8	0.21	5.3	0.31	7.9	0.34	8.6	0.36	9.1	0.41	10.4	0.53	13.5	0.78	19.8	1/4
5/16	—	—	—	—	—	—	0.01	0.3	0.11	2.8	0.21	5.3	0.31	7.9	0.34	8.6	0.36	9.1	0.41	10.4	0.53	13.5	0.78	19.8	5/16
3/8	—	—	—	—	—	—	—	—	0.10	2.5	0.20	5.1	0.31	7.9	0.33	8.4	0.36	9.1	0.40	10.2	0.53	13.5	0.78	19.8	3/8
1/2	—	—	—	—	—	—	—	—	—	—	0.10	2.5	0.21	5.3	0.23	5.8	0.25	6.4	0.30	7.6	0.43	10.9	0.68	17.3	1/2
5/8	—	—	—	—	—	—	—	—	—	—	—	—	0.11	2.8	0.13	3.3	0.15	3.8	0.20	5.1	0.33	8.4	0.58	14.7	5/8
3/4	—	—	—	—	—	—	—	—	—	—	—	—	0.03	0.8	—	—	0.05	1.3	0.09	2.3	0.22	5.6	0.47	11.9	3/4
7/8	—	—	—	—	—	—	—	—	—	—	—	—	—	—	—	—	0.02	0.5	0.07	1.8	0.19	4.8	0.44	11.2	7/8
1	—	—	—	—	—	—	—	—	—	—	—	—	—	—	—	—	—	—	0.05	1.3	0.17	4.3	0.42	10.7	1
1-1/4	—	—	—	—	—	—	—	—	—	—	—	—	—	—	—	—	—	—	—	—	0.13	3.3	0.38	9.7	1-1/4
1-1/2	—	—	—	—	—	—	—	—	—	—	—	—	—	—	—	—	—	—	—	—	—	—	0.25	6.4	1-1/2

TABLE 19—FACTORS FOR 37 DEG FLARED END ON SHAPE FITTINGS (FIGS. 4A-4E, 5A-5D, 6A-6C, 7A-7D, 8C, 9A-9D, AND 10A-10C)

Nominal Tube OD	\multicolumn{26}{c	}{Standard Machining Size}	Nominal Tube OD
	3/16		1/4		5/16		3/8		1/2		5/8		3/4		7/8		1		1-1/4		1-1/2		2			
	in	mm	in	mm	in	mm	in	mm	in	mm	in	mm	in	mm	in	mm	in	mm	in	mm	in	mm	in	mm	
1/8	0.03	0.8	0.11	2.8	0.11	2.8	0.13	3.3	—	—	—	—	—	—	—	—	—	—	—	—	—	—	—	—	1/8
3/16	—	—	0.08	2.0	0.08	2.0	0.10	2.5	0.21	5.3	—	—	—	—	—	—	—	—	—	—	—	—	—	—	3/16
1/4	—	—	—	—	0.00	0.0	0.01	0.3	0.13	3.3	0.24	6.1	—	—	—	—	—	—	—	—	—	—	—	—	1/4
5/16	—	—	—	—	—	—	0.01	0.3	0.13	3.3	0.24	6.1	0.36	9.1	—	—	—	—	—	—	—	—	—	—	5/16
3/8	—	—	—	—	—	—	—	—	0.11	2.8	0.23	5.8	0.35	8.9	0.38	9.7	—	—	—	—	—	—	—	—	3/8
1/2	—	—	—	—	—	—	—	—	—	—	0.12	3.0	0.24	6.1	0.27	6.9	0.29	7.4	0.22	5.6	—	—	—	—	1/2
5/8	—	—	—	—	—	—	—	—	—	—	—	—	0.12	3.0	0.15	3.8	0.17	4.3	0.22	5.6	—	—	—	—	5/8
3/4	—	—	—	—	—	—	—	—	—	—	—	—	0.03	0.8	—	—	0.05	1.3	0.09	2.3	0.22	5.5	—	—	3/4
7/8	—	—	—	—	—	—	—	—	—	—	—	—	—	—	—	—	0.02	0.5	0.07	1.8	0.19	4.8	0.44	11.2	7/8
1	—	—	—	—	—	—	—	—	—	—	—	—	—	—	—	—	—	—	0.05	1.3	0.17	4.3	0.42	10.7	1
1-1/4	—	—	—	—	—	—	—	—	—	—	—	—	—	—	—	—	—	—	—	—	0.13	3.3	0.38	9.7	1-1/4
1-1/2	—	—	—	—	—	—	—	—	—	—	—	—	—	—	—	—	—	—	—	—	—	—	0.25	6.4	1-1/2

TABLE 20—FACTORS FOR FLARELESS STRAIGHT FITTINGS (FIGS. 19A-19D, 20A-20C, AND 30)

Nominal Tube OD	\multicolumn{14}{c	}{Standard Machining Size}
	3/16		1/4		5/16		3/8		1/2		5/8		3/4 thru 2	
	in	mm	in	mm	in	mm	in	mm	in	mm	in	mm	in	mm
1/8	0.05	1.3	0.08	2.0	0.08	2.0	0.10	2.5	0.19	4.8	0.25	6.4	0.31	7.9
3/16	—	—	0.03	0.8	0.03	0.8	0.05	1.3	0.14	3.6	0.20	5.1	0.26	6.6
1/4	—	—	—	—	0.00	0.0	0.02	0.5	0.11	2.8	0.17	4.3	0.23	5.8
5/16	—	—	—	—	—	—	0.02	0.5	0.11	2.8	0.17	4.3	0.23	5.8
3/8	—	—	—	—	—	—	—	—	0.09	2.3	0.15	3.8	0.21	5.3
1/2	—	—	—	—	—	—	—	—	—	—	0.06	1.5	0.12	3.0
5/8	—	—	—	—	—	—	—	—	—	—	—	—	0.06	1.5
3/4 to 1-1/2	—	—	—	—	—	—	—	—	—	—	—	—	0.00	0.0

TABLE 21—FACTORS FOR FLARELESS SHAPED FITTINGS (FIGS. 21A-21E, 22A-22D, 23A-23C, 24A-24D, 25C, 26A-26D, AND 27A-27C)

	Nominal Tube OD	\multicolumn{2}{c}{3/16}	\multicolumn{2}{c}{1/4}	\multicolumn{2}{c}{5/16}	\multicolumn{2}{c}{3/8}	\multicolumn{2}{c}{1/2}	\multicolumn{2}{c}{5/8}	\multicolumn{2}{c}{3/4 thru 2}
		in	mm	in	mm	in	mm	in	mm	in	mm	in	mm	in	mm
Reduced Machining Size	1/8	0.05	1.3	0.10	2.5	0.10	2.5	0.12	3.0	0.23	5.8	0.30	7.6	0.38	9.7
	3/16	—	—	0.05	1.3	0.05	1.3	0.07	1.8	0.18	4.6	0.25	6.4	0.33	8.4
	1/4	—	—	—	—	0.00	0.0	0.02	0.5	0.13	3.3	0.20	5.1	0.28	7.1
	5/16	—	—	—	—	—	—	0.02	0.5	0.13	3.3	0.20	5.1	0.28	7.1
	3/8	—	—	—	—	—	—	—	—	0.11	2.8	0.18	4.6	0.26	6.6
	1/2	—	—	—	—	—	—	—	—	—	—	0.07	1.8	0.15	3.8
	5/8	—	—	—	—	—	—	—	—	—	—	—	—	0.08	2.0
	3/4 to 1-1/2	—	—	—	—	—	—	—	—	—	—	—	—	0.00	0.0

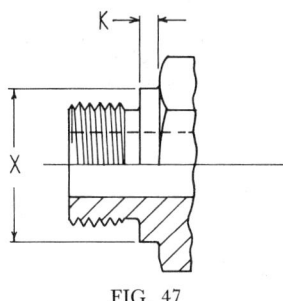

FIG. 47

ed. **TABLE 22—FACTORS FOR NON-ADJUSTABLE STRAIGHT THREAD ENDS (FIGS. 3B, 3C, 20B, AND 20C)**

Nominal Tube OD	3/16		1/4		5/16		3/8		1/2		5/8		3/4		7/8		1		1-1/4		1-1/2		2		Nominal Tube OD	X Diameter ±0.01 in	±0.3 mm	K ±0.02 in	±0.5 mm
	in	mm	in	mm	in	mm	in	mm	in	mm	in	mm	in	mm	in	mm	in	mm	in	mm	in	mm	in	mm					
1/8	0.00	0.0	−0.06	−1.5	+0.03	+0.8	0.00	0.0	−0.05	−1.3	−0.11	−2.8	−0.21	−5.3	−0.21	−5.3	−0.21	−5.3	−0.21	−5.3	−0.21	−5.3	−0.21	−5.3	1/8	0.50	12.7	0.09	2.3
3/16	—	—	−0.06	−1.5	0.00	0.0	−0.05	−1.3	−0.11	−2.8	−0.21	−5.3	−0.21	−5.3	−0.21	−5.3	−0.21	−5.3	−0.21	−5.3	−0.21	−5.3	−0.21	−5.3	3/16	0.56	14.2	0.09	2.3
1/4	—	—	—	—	0.00	0.0	−0.03	−0.8	+0.01	+0.3	−0.05	−1.3	−0.14	−3.6	−0.14	−3.6	−0.14	−3.6	−0.14	−3.6	−0.14	−3.6	−0.14	−3.6	1/4	0.63	16.0	0.09	2.3
5/16	—	—	—	—	—	—	−0.03	−0.8	+0.01	+0.3	−0.05	−1.3	−0.14	−3.6	−0.14	−3.6	−0.14	−3.6	−0.14	−3.6	−0.14	−3.6	−0.14	−3.6	5/16	0.69	17.5	0.09	2.3
3/8	—	—	—	—	—	—	—	—	−0.04	−1.0	+0.01	+0.3	−0.11	−2.8	−0.11	−2.8	−0.11	−2.8	0.11	2.8	0.11	−2.0			3/8	0.75	19.0	0.09	2.3
1/2	—	—	—	—	—	—	—	—	—	—	−0.06	−1.5	−0.03	−0.8	−0.03	−0.8	−0.03	−0.8	−0.03	−0.8	−0.03	−0.8	−0.03	−0.8	1/2	0.94	23.9	0.13	3.3
5/8	—	—	—	—	—	—	—	—	—	—	—	—	−0.09	−2.3	+0.03	+0.8	+0.03	+0.8	+0.03	+0.8	+0.03	+0.8	+0.03	+0.8	5/8	1.06	26.9	0.13	3.3
3/4	—	—	—	—	—	—	—	—	—	—	—	—	0.00	0.0	+0.13	+3.3	+0.13	+3.3	+0.13	+3.3	+0.13	+3.3	+0.13	+3.3	3/4	1.31	33.3	0.13	3.3
7/8	—	—	—	—	—	—	—	—	—	—	—	—	—	—	0.00	0.0	+0.13	+3.3	+0.13	+3.3	+0.13	+3.3	+0.13	+3.3	7/8	1.44	36.6	0.13	3.3
1	—	—	—	—	—	—	—	—	—	—	—	—	—	—	—	—	+0.16	+4.1	+0.16	+4.1	+0.16	+4.1	+0.16	+4.1	1	1.56	39.6	0.16	4.1
1-1/4	—	—	—	—	—	—	—	—	—	—	—	—	—	—	—	—	—	—	+0.16	+4.1	+0.16	+4.1	+0.16	+4.1	1-1/4	1.88	42.8	0.16	4.1
1-1/2	—	—	—	—	—	—	—	—	—	—	—	—	—	—	—	—	—	—	—	—	—	—	+0.16	+4.1	1-1/2	2.13	54.1	0.16	4.1

Fittings involving values to the right of and above the line through the table shall include a turned shoulder of diameter "X" and thickness "K" as shown in Fig. 47. This turned shoulder permits the reduced straight thread end to seat in a standard boss spotface.

For some combinations, additional length will be required to accommodate this feature. Therefore, all factors designated as plus in above table should be added to standard length dimensions to obtain applicable values.

TABLE 23—FACTORS FOR ADJUSTABLE STRAIGHT THREAD ENDS (FIGS. 9D, 10A-10C, 26D, AND 27A-27C)

	Nominal Tube OD	3/16		1/4		5/16		3/8		1/2		5/8		3/4 thru 2	
		in	mm	in	mm	in	mm	in	mm	in	mm	in	mm	in	mm
Reduced Machining Size	1/8	0.00	0.0	0.11	2.8	0.11	2.8	0.17	4.3	—	—	—	—	—	—
	3/16	—	—	0.11	2.8	0.11	2.8	0.17	4.3	0.30	7.6	0.35	8.9	—	—
	1/4	—	—	—	—	0.00	0.0	0.06	1.5	0.19	4.8	0.35	8.9	—	—
	5/16	—	—	—	—	—	—	0.06	1.5	0.19	4.8	0.35	8.9	—	—
	3/8	—	—	—	—	—	—	—	—	0.13	3.3	0.29	7.4	0.45	11.4
	1/2	—	—	—	—	—	—	—	—	—	—	0.16	4.1	0.32	8.1
	5/8	—	—	—	—	—	—	—	—	—	—	—	—	0.16	4.1
	3/4 to 1-1/2	—	—	—	—	—	—	—	—	—	—	—	—	0.00	0.0

TABLE 24—FACTORS FOR ALL MALE PIPE ENDS
(FIGS. 2A, 4A–4C, 4E, 5C, 5D, 19A, 21A–21C, 21E, 22C AND 22D)

Nominal Pipe Size	Reduced Pipe Size	1/4 in	1/4 mm	3/8 in	3/8 mm	1/2 in	1/2 mm	3/4 in	3/4 mm	1 in	1 mm	1-1/4 in	1-1/4 mm	1-1/2 in	1-1/2 mm	2 in	2 mm
	1/8	0.19	4.8	0.19	4.8	0.38	9.7	0.38	9.7	—	—	—	—	—	—	—	—
	1/4	—	—	0.00	0.0	0.19	4.8	0.19	4.8	0.38	9.7	—	—	—	—	—	—
	3/8	—	—	—	—	0.19	4.8	0.19	4.8	0.38	9.7	0.41	10.4	—	—	—	—
	1/2	—	—	—	—	—	—	0.00	0.0	0.19	4.8	0.22	5.6	0.25	6.4	—	—
	3/4	—	—	—	—	—	—	—	—	0.19	4.8	0.22	5.6	0.25	6.4	0.28	7.1
	1	—	—	—	—	—	—	—	—	—	—	0.03	0.8	0.06	1.5	0.09	2.3
	1-1/4	—	—	—	—	—	—	—	—	—	—	—	—	0.03	0.8	0.06	1.5
	1-1/2	—	—	—	—	—	—	—	—	—	—	—	—	—	—	0.03	0.8

TABLE 25—FACTORS FOR ALL FEMALE PIPE ENDS (FIGS. 2D, 5A, 6A, 6B, 19D, 22A, 23A, AND 23B)

Nominal Pipe Size	Reduced Pipe Size	1/4 in	1/4 mm	3/8 in	3/8 mm	1/2 in	1/2 mm	3/4 in	3/4 mm	1 in	1 mm	1-1/4 in	1-1/4 mm	1-1/2 in	1-1/2 mm	2 in	2 mm
	1/8	0.21	5.3	0.22	5.6	0.44	11.2	0.45	11.4	—	—	—	—	—	—	—	—
	1/4	—	—	0.01	0.3	0.23	5.8	0.24	6.1	0.47	11.9	0.48	12.2	—	—	—	—
	3/8	—	—	—	—	0.22	5.6	0.23	5.8	0.46	11.7	0.48	12.2	0.26	6.6	—	—
	1/2	—	—	—	—	—	—	0.01	0.3	0.24	6.1	0.26	6.6	0.26	6.6	—	—
	3/4	—	—	—	—	—	—	—	—	0.23	5.8	0.25	6.4	0.25	6.4	0.27	6.9
	1	—	—	—	—	—	—	—	—	—	—	0.02	0.5	0.02	0.5	0.04	1.0
	1-1/4	—	—	—	—	—	—	—	—	—	—	—	—	0.00	0.0	0.02	0.5
	1-1/2	—	—	—	—	—	—	—	—	—	—	—	—	—	—	0.02	0.5

TABLE 26—MINIMUM STOCK SIZE FOR TUBE ENDS

Nominal Tube OD	Hexagon Width Minimum	Width Over Forged Pads Minimum
1/8	3/8	0.31
3/16	7/16	0.38
1/4	1/2	0.44
5/16	9/16	0.50
3/8	5/8	0.56
1/2	13/16	0.75
5/8	15/16	0.88
3/4	1-1/8	1.06
7/8	1-1/4	1.19
1	1-3/8	1.31
1-1/4	1-11/16	1.62
1-1/2	2	1.88
2	2-5/8	2.50

TABLE 27—MINIMUM STOCK SIZE FOR MALE PIPE ENDS

Nominal Pipe Size	Hexagon Width Minimum	Width Over Forged Pads Minimum
1/8	7/16	0.44
1/4	9/16	0.56
3/8	3/4	0.75
1/2	7/8	0.88
3/4	1-1/8	1.06
1	1-3/8	1.31
1-1/4	1-11/16	1.62
1-1/2	2	1.88
2	2-1/2	2.50

TABLE 28—MINIMUM STOCK SIZE FOR FEMALE PIPE ENDS

Nominal Pipe Size	Hexagon Width Minimum	Width Over Forged Pads Minimum
1/8	9/16	0.56
1/4	3/4	0.75
3/8	7/8	0.88
1/2	1-1/8	1.06
3/4	1-3/8	1.31
1	1-5/8	1.62
1-1/4	2	1.88
1-1/2	2-3/8	2.50
2	2-7/8	2.81

HYDRAULIC HOSE FITTINGS—SAE J516 SEP79

SAE Standard

Report of the Construction and Industrial Machinery Technical Committee, Nonmetallic Materials Committee, and Tube, Pipe, Hose, and Lubrication Fittings Committee, approved January 1952, last revised by the Tube, Pipe, Hose, and Lubrication Fittings Committee December 1975, editorial change September 1979.

GENERAL SPECIFICATIONS

Scope—This standard is intended to document general and dimensional specifications for the various types and styles of hose fittings used in conjunction with hydraulic hoses specified in SAE J517 (January, 1952), in hydraulic systems on construction and industrial equipment and commercial products. Also included are adapter unions. The basic design and dimensional features of the fittings shown are derived from existing industry practice and include the following types:

Male Dryseal pipe thread type hose fittings shall be as shown in Figs. 1A–1C for the respective styles, and in Tables 1A–1F for the applicable hoses and sizes.

Male straight thread O-ring boss type hose fittings shall be as shown in Figs. 2A and 2B for the respective styles, and in Tables 2A–2C for the applicable hoses and sizes.

Male 37-deg flared type hose fittings shall be as shown in Figs. 3A–3C for the respective styles, and in Tables 3A–3F for the applicable hoses and sizes.

Female 37-deg flared type hose fittings shall be as shown in Figs. 4A–7B for the respective styles and shapes, and in Tables 4A–4F for the applicable hoses and sizes.

Male 45-deg flared type hose fittings shall be as shown in Figs. 8A and 8B for the respective styles, and in Tables 5A–5D for the applicable hoses and sizes.

Female 45-deg flared type hose fittings shall be as shown in Figs. 9A–12B for the respective styles and shapes, and in Tables 6A–6D for the applicable hoses and sizes.

Male flareless type hose fittings shall be as shown in Figs. 13A and 13B for the respective styles, and in Tables 7A and 7B for the applicable hoses and sizes.

Female flareless type hose fittings shall be as shown in Figs. 14A–14C for the respective styles, and in Tables 8A and 8B for the applicable hoses and sizes.

4-bolt split flange type hose fittings shall be as shown in Figs. 15A–21C for the respective styles and shapes, and in Tables 9A–9D for the applicable hoses and sizes.

Adapter unions shall be as shown in Figs. 22–25 and Table 10.

It is recommended that where step sizes or additional types of fittings are required, they be designed to conform with the specifications of this standard insofar as they may apply.

The following general specifications shall supplement the dimensional data contained in the tables with respect to all unspecified detail.

Size Designations—The hose fitting size is generally designated by the fractional inch nominal hose inside diameter together with the nominal pipe or straight thread size or nominal split flange size for the respective fitting types. However, these sizes may also be designated by their dash sizes as follows:

The hose dash size is equivalent to the number of sixteenth inch increments in the hose inside diameter, except in the case of SAE 100R5 hose where it is equivalent to the number of sixteenth inch increments in the outside diameter of tubing having approximately the same inside diameter as the hose.

The pipe thread dash size is the number of sixteenth inch increments in the nominal pipe thread size.

The O-ring boss, 37-deg and 45-deg flared, and flareless type thread dash sizes correspond to the number of sixteenth inch increments in the outside diameter of the tubing with which they are designed to be used.

The 4-bolt split flange dash size is the number of sixteenth inch increments in the nominal flange size.

Fitting Identification—Permanently attached style fittings that are not assembled to hose by fitting manufacturers and all field attachable styles of hose fittings shall be permanently and legibly marked to identify the hose size and type on which they are designed to be used.

Dimensions and Tolerances—Tabulated dimensions shall apply to the finished parts, plated or otherwise processed, as specified by the purchaser. Dimensions over external contour of shell portion of fittings shown in the tables reflect the maximum envelope of products available. Dimensions W and Z are indicative of the minimum diameter hole through which the hose fitting will pass. Details of internal construction of the attaching portion of fittings are not specified and shall be optional with the manufacturer, providing the fittings, properly assembled onto the appropriate hose, will not be the point of failure when the assemblies are subjected to the various tests specified in SAE J517 (January, 1952). In the case of field attachable styles of fittings this requirement shall apply to a minimum of one reuse as well.

The maximum and minimum across flat dimensions shall be within the commercial tolerance of bar stock from which the fittings are produced. Formed or upset hexagon contours shall fit standard wrench size openings. The minimum across corners dimensions of external hexagons shall be 1.092 times the nominal width across flats, but shall not result in a side flat width of less than 0.43 times the nominal width across flats.

Tolerance on all dimensions not otherwise limited shall be ±0.016 in. Fitting seats shall be concentric with straight thread pitch diameters within 0.010 in full indicator reading (FIR).

Passages—The tabulated D dimensions reflect the minimum bore at any point through the fitting. Where passages in straight fittings are machined from opposite ends, the offset at the meeting point shall not exceed 0.015 in. On angle fittings, the cross sectional area at the junction of fluid passages shall not be less than that of the smallest passage.

Contour—Details of contour shall be optional with the manufacturer, providing the tabulated dimensions are maintained and serviceability of fittings is not impaired. The wrench clearance dimension Y, where specified, represents the width of the fitting hexagon T plus sufficient clearance in the shell portion of fitting adjacent to the hexagon to provide adequate space for application of a standard wrench to the hexagon without interfering with mating components during assembly.

Straight Threads—Unified Standard Class 2A external and Class 2B internal threads shall apply to plain finish (unplated) fittings having straight threads designated B. For externally threaded parts with additive finish, the maximum diameters of Class 2A may be exceeded by the amount of the allowance, that is, the basic diameters (Class 2A maximum diameters plus the allowance) shall apply to an externally threaded part after plating. For internally threaded parts with additive finish, the Class 2B diameters apply after plating.

The pitch diameter tolerance shall be the same as the corresponding diameter-pitch combination and class of the Unified fine and 12-thread series. See SAE J475 (June, 1964).

Where external threads are produced by roll threading and the body is not undercut, the unthreaded portion adjacent to the shoulder may be reduced to the minimum pitch diameter.

External threads shall be chamfered and internal threads shall be countersunk as specified in the illustrations and dimensional tables.

Thread Eccentricity Tolerances—The various thread elements of Class 2A external and Class 2B internal threads on hose fittings shall be concentric within the limitations specified under General Specifications in SAE J512 (June, 1912).

MALE DRYSEAL PIPE THREAD TYPE

NOTES: UNSPECIFIED DETAIL WITH RESPECT TO DIMENSIONS, TOLERANCES, CONTOURS, MATERIAL, WORKMANSHIP, AND SO ON, MUST CONFORM TO GENERAL SPECIFICATIONS FOR HYDRAULIC HOSE FITTINGS. THE DIMENSIONAL DESIGNATIONS ON THE FIRST FIGURE IN EACH GROUP SHALL APPLY TO ALL OTHER FIGURES IN THAT GROUP EXCEPT AS SHOWN OTHERWISE. CODES SHOWN IN BRACKETS ADJACENT TO FIGURE NUMBERS REPRESENT RESPECTIVE FITTING IDENTIFICATION, WITH XX SUBSTITUTED FOR THE HOSE TYPE CODE DEPICTED IN BRACKETS AT END OF RESPECTIVE TABLE TITLES, IN ACCORDANCE WITH SAE J846.

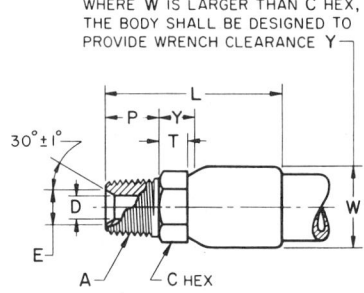

FIG. 1A—PERMANENTLY ATTACHED STYLE (1501XX)

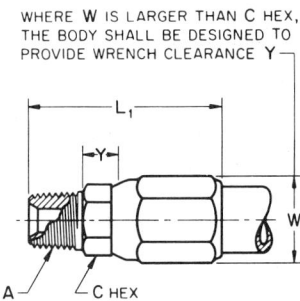

FIG. 1B—FIELD ATTACHABLE SCREW STYLE (1601XX)

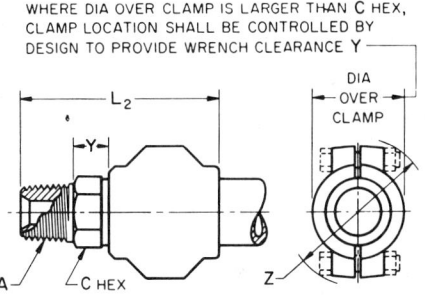

FIG. 1C—FIELD ATTACHABLE SEGMENT CLAMP STYLE (1701XX)

FIG. 1—MALE DRYSEAL PIPE THREAD TYPE HOSE FITTINGS

TABLE 1A—DIMENSIONS OF MALE DRYSEAL PIPE THREAD TYPE HOSE FITTINGS FOR USE ON SAE 100R1 HYDRAULIC HOSE (FIGS. 1A-1C) (CODE 42)

Nominal SAE 100R1 Hose ID	Hose Dash Size	A Dryseal Taper Thread[a] NPTF	Pipe Thread Dash Size	C Min	D Dia Min	E Dia ±0.010	L Max	L_1 Max	L_2 Max	P Min	T Min	W Max	Y Min	Z Max
3/16	-3	1/8	-2	7/16	0.09	0.281	2.16	2.16	—	0.38	0.19	0.75	0.37	—
3/16	-3	1/4	-4	9/16	0.09	0.344	2.25	2.38	—	0.56	0.22	0.75	0.41	—
1/4	-4	1/8	-2	7/16	0.11	0.281	2.31	2.24	—	0.38	0.19	0.87	0.37	—
1/4	-4	1/4	-4	9/16	0.11	0.344	2.49	2.42	—	0.56	0.22	0.87	0.41	—
5/16	-5	1/4	-4	9/16	0.20	0.344	2.49	2.47	—	0.56	0.22	0.94	0.41	—
5/16	-5	3/8	-6	11/16	0.20	0.469	2.49	2.47	—	0.56	0.25	0.94	0.50	—
3/8	-6	1/4	-4	9/16	0.21	0.344	2.66	2.56	—	0.56	0.22	1.09	0.41	—
3/8	-6	3/8	-6	11/16	0.26	0.469	2.66	2.56	—	0.56	0.25	1.09	0.50	—
3/8	-6	1/2	-8	7/8	0.26	0.625	2.77	2.78	—	0.75	0.31	1.09	0.56	—
13/32	-6.5	3/8	-6	11/16	0.28	0.469	2.66	2.56	—	0.56	0.25	1.16	0.50	—
13/32	-6.5	1/2	-8	7/8	0.28	0.625	2.77	2.81	—	0.75	0.31	1.16	0.56	—
1/2	-8	3/8	-6	11/16	0.32	0.469	2.66	2.91	—	0.56	0.25	1.23	0.50	—
1/2	-8	1/2	-8	7/8	0.38	0.625	2.85	3.09	3.09	0.75	0.31	1.23	0.56	2.50
5/8	-10	1/2	-8	7/8	0.48	0.625	3.15	3.53	—	0.75	0.31	1.37	0.56	—
5/8	-10	3/4	-12	1-1/16	0.50	0.813	3.15	3.53	—	0.75	0.38	1.37	0.61	—
3/4	-12	3/4	-12	1-1/16	0.61	0.813	3.61	3.53	3.40	0.75	0.38	1.59	0.61	3.00
1	-16	1	-16	1-3/8	0.78	1.031	3.84	4.22	3.97	0.94	0.38	2.02	0.61	3.59
1-1/4	-20	1-1/4	-20	1-11/16	1.01	1.344	4.31	4.70	4.38	0.97	0.44	2.46	0.94	4.00
1-1/2	-24	1-1/2	-24	2	1.25	1.625	4.59	5.00	4.68	1.00	0.50	2.74	0.94	4.50
2	-32	2	-32	2-7/16	1.74	2.063	5.16	5.50	5.50	1.03	0.69	3.47	1.12	5.12

[a] Dryseal American Standard Taper Pipe Thread.

TABLE 1B—DIMENSIONS OF MALE DRYSEAL PIPE THREAD TYPE HOSE FITTINGS FOR USE ON SAE 100R2 HYDRAULIC HOSE (FIGS. 1A-1C) (CODE 43)

Nominal SAE 100R2 Hose ID	Hose Dash Size	A Dryseal Taper Thread[a] NPTF	Pipe Thread Dash Size	C Min	D Dia Min	E Dia ±0.010	L Max	L_1 Max	L_2 Max	P Min	T Min	W Max	Y Min	Z Max
3/16	-3	1/8	-2	7/16	0.09	0.281	2.19	2.65	—	0.38	0.19	0.90	0.37	—
3/16	-3	1/4	-4	9/16	0.09	0.344	2.31	2.84	—	0.56	0.22	0.90	0.41	—
1/4	-4	1/8	-2	7/16	0.11	0.281	2.31	2.44	—	0.38	0.19	1.00	0.37	—
1/4	-4	1/4	-4	9/16	0.11	0.344	2.45	2.69	2.50	0.56	0.22	1.00	0.41	2.13
3/8	-6	1/4	-4	9/16	0.21	0.344	2.53	2.88	—	0.56	0.22	1.19	0.41	—
3/8	-6	3/8	-6	11/16	0.26	0.469	2.66	2.88	2.63	0.56	0.25	1.19	0.50	2.47
3/8	-6	1/2	-8	7/8	0.26	0.625	2.77	3.07	—	0.75	0.31	1.19	0.56	—
1/2	-8	3/8	-6	11/16	0.32	0.469	2.73	3.00	—	0.56	0.25	1.38	0.50	—
1/2	-8	1/2	-8	7/8	0.38	0.625	2.91	3.22	3.10	0.75	0.31	1.38	0.56	2.47
5/8	-10	1/2	-8	7/8	0.48	0.625	3.19	3.75	—	0.75	0.31	1.56	0.56	—
5/8	-10	3/4	-12	1-1/16	0.50	0.813	3.19	3.75	3.40	0.75	0.38	1.56	0.61	2.88
3/4	-12	3/4	-12	1-1/16	0.61	0.813	3.61	3.75	3.40	0.75	0.38	1.75	0.61	3.00
7/8	-14	1	-16	1-3/8	0.72	1.031	3.70	4.46	—	0.94	0.38	1.81	0.61	—
1	-16	1	-16	1-3/8	0.78	1.031	3.68	4.52	4.25	0.94	0.38	2.06	0.61	3.60
1-1/4	-20	1-1/4	-20	1-11/16	1.01	1.344	5.62	5.22	5.00	0.97	0.44	2.63	0.94	4.50
1-1/2	-24	1-1/2	-24	2	1.25	1.625	6.37	5.50	5.40	1.00	0.50	2.94	0.94	5.00
2	-32	2	-32	2-7/16	1.72	2.063	5.16	6.07	6.30	1.03	0.69	3.50	1.12	5.25

[a] Dryseal American Standard Taper Pipe Thread.

TABLE 1C—DIMENSIONS OF MALE DRYSEAL PIPE THREAD TYPE HOSE FITTINGS FOR USE ON SAE 100R3 HYDRAULIC HOSE (FIGS. 1A-1C) (CODE 44)

Nominal SAE 100R3 Hose ID	Hose Dash Size	A Dryseal Taper Thread[a] NPTF	Pipe Thread Dash Size	C Min	D Dia Min	E Dia ±0.010	L Max	L₁ Max	L₂ Max	P Min	T Min	W Max	Y Min	Z Max
1/4	-4	1/8	-2	7/16	0.11	0.281	2.31	2.18	—	0.38	0.19	0.88	0.37	—
1/4	-4	1/4	-4	9/16	0.11	0.344	2.49	2.38	—	0.56	0.19	0.88	0.37	—
3/8	-6	1/4	-4	9/16	0.26	0.344	2.52	2.56	—	0.56	0.22	1.06	0.41	—
3/8	-6	3/8	-6	11/16	0.26	0.469	2.52	2.56	—	0.56	0.22	1.06	0.41	—
1/2	-8	3/8	-6	11/16	0.36	0.469	2.63	2.56	—	0.56	0.25	1.31	0.50	—
1/2	-8	1/2	-8	7/8	0.38	0.625	2.88	3.03	2.63	0.75	0.31	1.31	0.56	2.50
3/4	-12	3/4	-12	1-1/16	0.61	0.813	3.61	3.12	2.88	0.75	0.31	1.66	0.61	3.00
1	-16	1	-16	1-3/8	0.78	1.031	3.84	3.70	3.38	0.94	0.31	2.02	0.61	3.25
1-1/4	-20	1-1/4	-20	1-11/16	1.01	1.344	4.08	—	4.00	0.97	0.38	2.02	0.94	3.94

[a] Dryseal American Standard Taper Pipe Thread

TABLE 1D—DIMENSIONS OF MALE DRYSEAL PIPE THREAD TYPE HOSE FITTINGS FOR USE ON SAE 100R4 HYDRAULIC HOSE (FIGS. 1A AND 1C) (CODE 45)

Nominal SAE 100R4 Hose ID	Hose Dash Size	A Dryseal Taper Thread[a] NPTF	Pipe Thread Dash Size	C Min	D Dia Min	E Dia ±0.010	L Max	L₂ Max	P Min	T Min	W Max	Y Min	Z Max
3/4	-12	3/4	-12	1-1/16	0.61	0.813	2.63	3.61	0.75	0.31	1.73	0.61	3.00
1	-16	1	-16	1-3/8	0.78	1.031	2.88	3.84	0.94	0.31	2.01	0.61	3.25
1-1/4	-20	1-1/4	-20	1-11/16	1.01	1.344	3.03	4.12	0.97	0.38	2.31	0.94	3.94
1-1/2	-24	1-1/2	-24	2	1.25	1.625	3.56	4.50	1.00	0.50	2.60	0.94	4.34
2	-32	2	-32	2-7/16	1.72	2.063	4.00	4.88	1.03	0.56	3.18	1.12	4.94

[a] Dryseal American Standard Taper Pipe Thread.

TABLE 1E—DIMENSIONS OF MALE DRYSEAL PIPE THREAD TYPE HOSE FITTINGS FOR USE ON SAE 100R5 HYDRAULIC HOSE (FIGS. 1A AND 1B) (CODE 46)

Nominal SAE 100R5 Hose ID	Hose Dash Size	A Dryseal Taper Thread[a] NPTF	Pipe Thread Dash Size	C Min	D Dia Min	E Dia ±0.010	L Max	L₁ Max	P Min	T Min	W Max	Y Min
3/16	-4	1/8	-2	7/16	0.12	0.281	1.82	1.80	0.38	0.19	0.79	0.37
3/16	-4	1/4	-4	9/16	0.12	0.344	2.00	2.06	0.56	0.19	0.79	0.41
1/4	-5	1/4	-4	9/16	0.16	0.344	2.04	2.06	0.56	0.19	0.87	0.41
5/16	-6	1/4	-4	9/16	0.23	0.344	2.07	2.19	0.56	0.19	0.94	0.41
13/32	-8	3/8	-6	11/16	0.35	0.469	2.16	2.54	0.56	0.25	1.08	0.50
1/2	-10	1/2	-8	7/8	0.45	0.625	2.40	2.94	0.75	0.31	1.31	0.56
5/8	-12	3/4	-12	1-1/16	0.54	0.813	2.44	3.30	0.75	0.38	1.52	0.61
7/8	-16	3/4	-12	1-1/16	0.72	0.813	2.70	3.06	0.75	0.38	1.73	0.61
7/8	-16	1	-16	1-3/8	0.81	1.031	2.89	3.19	0.94	0.38	1.73	0.61
1-1/8	-20	1-1/4	-20	1-11/16	1.04	1.344	3.04	3.47	0.97	0.44	2.02	0.94
1-3/8	-24	1-1/2	-24	2	1.28	1.625	3.38	3.55	1.00	0.50	2.31	0.94
1-13/16	-32	2	-32	2-7/16	1.75	2.063	3.79	4.10	1.03	0.69	2.89	1.12

[a] Dryseal American Standard Taper Pipe Thread.

TABLE 1F—DIMENSIONS OF MALE DRYSEAL PIPE THREAD TYPE HOSE FITTINGS FOR USE ON SAE 100R6 HYDRAULIC HOSE (FIGS. 1A AND 1B) (CODE 47)

Nominal SAE 100R6 Hose ID	Hose Dash Size	A Dryseal Taper Thread[a] NPTF	Pipe Thread Dash Size	C Min	D Dia Min	E Dia ±0.010	L Max	L₁ Max	P Min	T Min	W Max	Y Min
3/16	-3	1/8	-2	7/16	0.09	0.281	1.18	1.15	0.38	0.12	0.63	0.37
3/16	-3	1/4	-4	9/16	0.09	0.344	1.25	1.39	0.56	0.17	0.63	0.41
1/4	-4	1/8	-2	7/16	0.11	0.281	1.18	1.25	0.38	0.12	0.75	0.37
1/4	-4	1/4	-4	9/16	0.11	0.344	1.31	1.45	0.56	0.17	0.75	0.41
1/4	-4	3/8	-6	11/16	0.11	0.469	1.31	—	0.56	0.25	0.75	0.50
5/16	-5	1/4	-4	9/16	0.20	0.344	1.31	1.45	0.56	0.12	0.81	0.41
5/16	-5	3/8	-6	11/16	0.20	0.469	1.31	1.52	0.56	0.25	0.81	0.50
3/8	-6	1/4	-4	9/16	0.21	0.344	1.62	1.54	0.56	0.12	1.00	0.41
3/8	-6	3/8	-6	11/16	0.26	0.469	1.62	1.54	0.56	0.25	1.00	0.50
1/2	-8	3/8	-6	11/16	0.36	0.469	1.69	1.63	0.56	0.25	1.12	0.50
1/2	-8	1/2	-8	7/8	0.38	0.625	1.81	1.88	0.75	0.31	1.12	0.56
5/8	-10	1/2	-8	7/8	0.48	0.625	2.06	—	0.75	0.31	1.06	0.56
5/8	-10		-12	1-1/16	0.50	0.813	2.06	—	0.75	0.38	1.06	0.61

[a] Dryseal American Standard Taper Pipe Thread.

MALE STRAIGHT THREAD O-RING BOSS TYPE

NOTES: UNSPECIFIED DETAIL WITH RESPECT TO DIMENSIONS, TOLERANCES, CONTOURS, MATERIAL, WORKMANSHIP, AND SO ON, MUST CONFORM TO GENERAL SPECIFICATIONS FOR HYDRAULIC HOSE FITTINGS. THE DIMENSIONAL DESIGNATIONS ON THE FIRST FIGURE IN EACH GROUP SHALL APPLY TO ALL OTHER FIGURES IN THAT GROUP EXCEPT AS SHOWN OTHERWISE. CODES SHOWN IN BRACKETS ADJACENT TO FIGURE NUMBERS REPRESENT RESPECTIVE FITTING IDENTIFICATION, WITH XX SUBSTITUTED FOR THE HOSE TYPE CODE DEPICTED IN BRACKETS AT END OF RESPECTIVE TABLE TITLES, IN ACCORDANCE WITH SAE J846.

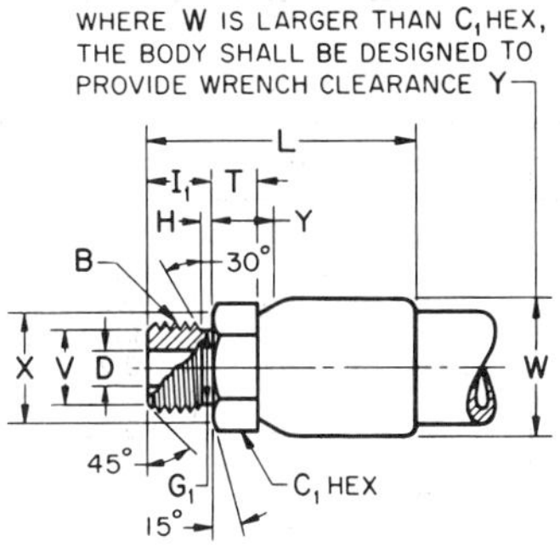

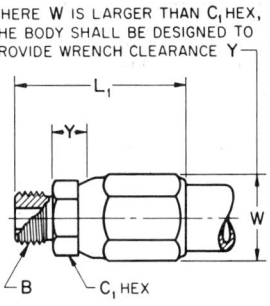

FIG. 2A—PERMANENTLY ATTACHED STYLE (1801XX)

FIG. 2B—FIELD ATTACHABLE SCREW STYLE (1901XX)

FIG. 2—MALE STRAIGHT THREAD O-RING BOSS TYPE HOSE FITTINGS

TABLE 2A—DIMENSIONS OF MALE STRAIGHT THREAD O-RING BOSS TYPE HOSE FITTINGS FOR USE ON SAE 100R1 HYDRAULIC HOSE[a] (FIG. 2) (CODE 42)

Nominal SAE 100R 1 Hose ID	Hose Dash Size	B Nominal Straight Thread Size	Thread Dash Size	C_1 Min	D Dia Min	L Max	L_1 Max	T Min	W Max	Y Min
5/16	-5	9/16-18	-6	11/16	0.20	2.47	2.47	0.31	0.94	0.41
3/8	-6	9/16-18	-6	11/16	0.26	2.54	2.54	0.31	1.09	0.41
3/8	-6	3/4-16	-8	7/8	0.26	2.69	2.69	0.34	1.09	0.50
13/32	-6.5	3/4-16	-8	7/8	0.26	2.69	2.69	0.34	1.16	0.50
1/2	-8	3/4-16	-8	7/8	0.38	2.76	3.00	0.34	1.23	0.50
1/2	-8	7/8-14	-10	1	0.38	2.86	3.10	0.31	1.23	0.56
5/8	-10	1-1/16-12	-12	1-1/4	0.50	3.20	3.58	0.38	1.37	0.69
3/4	-12	1-1/16-12	-12	1-1/4	0.57	3.20	3.58	0.38	1.59	0.69
3/4	-12	1-3/16-12	-14	1-3/8	0.61	3.31	3.69	0.38	1.59	0.69
1	-16	1-5/16-12	-16	1-1/2	0.78	3.72	4.24	0.50	2.02	0.75

[a] For dimensions shown on Fig. 2, but not specified in Table 2A, see corresponding dimensions for respective straight thread size in Fig. 34A and Table 12 of SAE J514. For dimensions of mating bosses, see Fig. 33 and Table 11 of SAE J514.

TABLE 2B—DIMENSIONS OF MALE STRAIGHT THREAD O-RING BOSS TYPE HOSE FITTINGS FOR USE ON SAE 100R2 HYDRAULIC HOSE[a] (FIG. 2) (CODE 43)

Nominal SAE 100R2 Hose ID	Hose Dash Size	B Nominal Straight Thread Size	Thread Dash Size	C_1 Min	D Dia Min	L Max	T Min	W Max	Y Min
3/8	-6	9/16-18	-6	11/16	0.26	2.54	0.31	1.19	0.41
3/8	-6	3/4-16	-8	7/8	0.26	2.69	0.31	1.19	0.50
1/2	-8	3/4-16	-8	7/8	0.38	2.76	0.25	1.38	0.50
1/2	-8	7/8-14	-10	1	0.38	2.86	0.31	1.38	0.56
5/8	-10	1-1/16-12	-12	1-1/4	0.50	3.20	0.38	1.56	0.69
3/4	-12	1-1/16-12	-12	1-1/4	0.57	3.50	0.38	1.75	0.69
3/4	-12	1-3/16-12	-14	1-3/8	0.61	3.50	0.38	1.75	0.69
7/8	-14	1-3/16-12	-14	1-3/8	0.72	3.67	0.38	1.81	0.69
7/8	-14	1-5/16-12	-16	1-1/2	0.72	3.72	0.50	1.81	0.75
1	-16	1-5/16-12	-16	1-1/2	0.78	4.00	0.37	2.06	0.75
1-1/4	-20	1-5/8-12	-20	1-7/8	1.01	5.25	0.37	2.63	0.94

[a] For dimensions shown in Fig. 2, but not specified in Table 2B, see corresponding dimensions for respective straight thread size in Fig. 34A and Table 12 of SAE J514. For dimensions of mating bosses, see Fig. 33 and Table 11 of SAE J514.

TABLE 2C—DIMENSIONS OF MALE STRAIGHT THREAD O-RING BOSS TYPE HOSE FITTINGS FOR USE ON SAE 100R5 HYDRAULIC HOSE[a] (FIG. 2) (CODE 46)

Nominal SAE 100R5 Hose ID	Hose Dash Size	B Nominal Straight Thread Size	Thread Dash Size	C_1 Min	D Dia Min	L Max	L_1 Max	T Min	W Max	Y Min
3/16	-4	7/16-20	-4	9/16	0.12	1.87	1.78	0.17	0.79	0.37
1/4	-5	1/2-20	-5	5/8	0.16	1.90	1.90	0.17	0.87	0.41
5/16	-6	9/16-18	-6	11/16	0.23	1.96	2.09	0.24	0.94	0.41
13/32	-8	3/4-16	-8	7/8	0.35	2.10	2.49	0.25	1.08	0.50
1/2	-10	7/8-14	-10	1	0.45	2.21	2.76	0.31	1.24	0.56
5/8	-12	1-1/16-12	-12	1-1/4	0.54	2.30	3.12	0.38	1.52	0.69
7/8	-16	1-5/16-12	-16	1-1/2	0.81	2.54	2.78	0.38	1.73	0.69
1-1/8	-20	1-5/8-12	-20	1-7/8	1.04	2.66	2.92	0.47	2.02	0.69

[a] For dimensions shown on Fig. 2, but not specified in Table 2C, see corresponding dimensions for respective straight thread size in Fig. 34A and Table 12 of SAE J514. For dimensions of mating bosses, see Fig. 33 and Table 11 of SAE J514.

MALE 37-DEG FLARED TYPE

NOTES: UNSPECIFIED DETAIL WITH RESPECT TO DIMENSIONS, TOLERANCES, CONTOURS, MATERIAL, WORKMANSHIP, AND SO ON, MUST CONFORM TO GENERAL SPECIFICATIONS FOR HYDRAULIC HOSE FITTINGS. THE DIMENSIONAL DESIGNATIONS ON THE FIRST FIGURE IN EACH GROUP SHALL APPLY TO ALL OTHER FIGURES IN THAT GROUP EXCEPT AS SHOWN OTHERWISE. CODES SHOWN IN BRACKETS ADJACENT TO FIGURE NUMBERS REPRESENT RESPECTIVE FITTING IDENTIFICATION, WITH XX SUBSTITUTED FOR THE HOSE TYPE CODE DEPICTED IN BRACKETS AT END OF RESPECTIVE TABLE TITLES, IN ACCORDANCE WITH SAE J846.

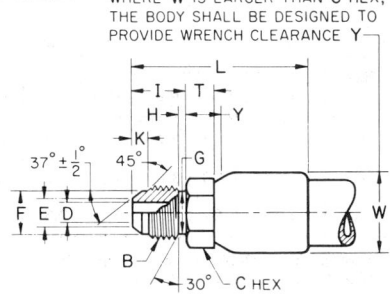
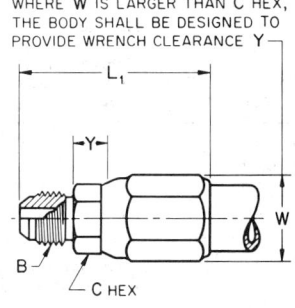
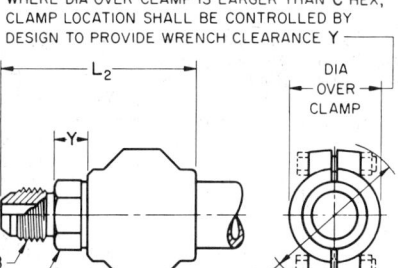

FIG. 3A—PERMANENTLY ATTACHED STYLE (2101XX)

FIG. 3B—FIELD ATTACHABLE SCREW STYLE (2201XX)

FIG. 3C—FIELD ATTACHABLE SEGMENT CLAMP STYLE (2301XX)

FIG. 3—MALE 37-DEG FLARED TYPE HOSE FITTINGS

TABLE 3A—DIMENSIONS OF MALE 37-DEG FLARED TYPE HOSE FITTINGS FOR USE ON SAE 100R1 HYDRAULIC HOSE[a] (FIG. 3) (CODE 42)

Nominal SAE 100R1 Hose ID	Hose Dash Size	B Nominal Straight Thread Size	Thread Dash Size	C Min	D Dia Min	L Max	L_1 Max	L_2 Max	T Min	W Max	Y Min	Z Max
3/16	-3	7/16-20	-4	7/16	0.09	2.33	2.33	—	0.22	0.75	0.37	—
1/4	-4	7/16-20	-4	7/16	0.11	2.48	2.41	—	0.22	0.87	0.37	—
1/4	-4	1/2-20	-5	1/2	0.11	2.48	2.41	—	0.22	0.87	0.41	—
1/4	-4	9/16-18	-6	9/16	0.11	2.49	2.41	—	0.22	0.87	0.41	—
5/16	-5	9/16-18	-6	9/16	0.20	2.49	2.44	—	0.22	0.94	0.41	—
3/8	-6	9/16-18	-6	9/16	0.26	2.52	2.56	—	0.22	1.09	0.41	—
3/8	-6	3/4-16	-8	3/4	0.26	2.63	2.66	—	0.25	1.09	0.50	—
13/32	-6.5	3/4-16	-8	3/4	0.26	2.67	2.66	—	0.25	1.16	0.50	—
1/2	-8	3/4-16	-8	3/4	0.38	2.70	3.08	—	0.25	1.23	0.50	—
1/2	-8	7/8-14	-10	7/8	0.38	2.85	3.18	3.18	0.25	1.23	0.56	2.50
5/8	-10	1-1/16-12	-12	1-1/16	0.50	3.20	3.58	—	0.31	1.37	0.61	—
3/4	-12	1-1/16-12	-12	1-1/16	0.57	3.72	3.58	3.45	0.31	1.59	0.61	3.00
3/4	-12	1-3/16-12	-14	1-1/16	0.61	3.75	3.67	3.54	0.38	1.59	0.61	3.00
1	-16	1-5/16-12	-16	1-5/16	0.78	3.81	4.19	3.75	0.38	2.02	0.69	3.63
1-1/4	-20	1-5/8-12	-20	1-5/8	1.01	4.44	4.73	4.41	0.46	2.46	0.84	4.00

[a] For dimensions shown on Fig. 3A, but not specified in Table 3A, see corresponding dimensions for respective straight thread sizes in Fig. 5A and Table 3 of SAE J514.

TABLE 3B—DIMENSIONS OF MALE 37-DEG FLARED TYPE HOSE FITTINGS FOR USE ON SAE 100R2 HYDRAULIC HOSE[a] (FIG. 3) (CODE 43)

Nominal SAE 100R2 Hose ID	Hose Dash Size	B Nominal Straight Thread Size	Thread Dash Size	C Min	D Dia Min	L Max	L_1 Max	L_2 Max	T Min	W Max	Y Min	Z Max
3/16	-3	7/16-20	-4	7/16	0.09	2.39	2.58	—	0.19	0.90	0.37	—
1/4	-4	7/16-20	-4	7/16	0.11	2.48	2.63	—	0.19	1.00	0.37	—
1/4	-4	1/2-20	-5	1/2	0.11	2.48	2.63	2.49	0.19	1.00	0.41	2.13
1/4	-4	9/16-18	-6	9/16	0.11	2.56	2.63	—	0.22	1.00	0.41	—
3/8	-6	9/16-18	-6	9/16	0.26	2.66	2.81	2.60	0.22	1.19	0.41	2.47
3/8	-6	3/4-16	-8	3/4	0.26	2.69	2.94	2.73	0.25	1.19	0.50	2.47
1/2	-8	3/4-16	-8	3/4	0.38	2.83	3.08	—	0.25	1.38	0.50	—
1/2	-8	7/8-14	-10	7/8	0.38	3.92	3.18	3.05	0.25	1.38	0.56	2.47
5/8	-10	1-1/16-12	-12	1-1/16	0.50	3.29	3.80	—	0.31	1.56	0.61	—
3/4	-12	1-1/16-12	-12	1-1/16	0.57	3.72	3.80	3.45	0.31	1.75	0.61	3.00
3/4	-12	1-3/16-12	-14	1-3/16	0.61	3.75	3.89	—	0.38	1.75	0.61	—
7/8	-14	1-3/16-12	-14	1-3/16	0.72	3.65	4.25	—	0.38	1.81	0.61	—
7/8	-14	1-5/16-12	-16	1-5/16	0.72	3.67	4.27	—	0.38	1.81	0.69	—
1	-16	1-5/16-12	-16	1-5/16	0.78	3.84	4.38	4.22	0.38	2.06	0.69	3.60
1-1/4	-20	1-5/8-12	-20	1-5/8	1.01	5.62	5.35	5.58	0.46	2.63	0.84	4.50

[a] For dimensions shown on Fig. 3A, but not specified in Table 3B, see corresponding dimensions for respective straight thread sizes in Fig. 5A and Table 3 of SAE J514.

TABLE 3C—DIMENSIONS OF MALE 37-DEG FLARED TYPE HOSE FITTINGS FOR USE ON SAE 100R3 HYDRAULIC HOSE[a] (FIG. 3) (CODE 44)

Nominal SAE 100R3 Hose ID	Hose Dash Size	B Nominal Straight Thread Size	Thread Dash Size	C Min	D Dia Min	L Max	L_1 Max	T Min	W Max	Y Min
1/4	-4	7/16-20	-4	7/16	0.11	2.48	2.38	0.22	0.88	0.37
1/4	-4	1/2-20	-5	1/2	0.11	2.48	2.38	0.22	0.88	0.41
3/8	-6	9/16-18	-6	9/16	0.26	2.52	2.56	0.22	1.06	0.41
3/8	-6	3/4-16	-8	3/4	0.26	2.62	2.63	0.25	1.06	0.50
1/2	-8	3/4-16	-8	3/4	0.35	2.73	3.00	0.25	1.31	0.50
1/2	-8	7/8-14	-10	7/8	0.35	2.89	3.06	0.25	1.31	0.56
3/4	-12	1-1/16-12	-12	1-1/16	0.57	3.72	3.36	0.31	1.63	0.61
1	-16	1-5/16-12	-16	1-5/16	0.78	3.81	4.06	0.38	2.00	0.69

[a] For dimensions shown on Fig. 3, but not specified in Table 3C, see corresponding dimensions for respective thread sizes in Fig. 5A and Table 3 of SAE J514.

TABLE 3D—DIMENSIONS OF MALE 37-DEG FLARED TYPE HOSE FITTINGS FOR USE ON SAE 100R4 HYDRAULIC HOSE[a] (FIG. 3) (CODE 45)

Nominal SAE 100R4 Hose ID	Hose Dash Size	B Nominal Straight Thread Size	Thread Dash Size	C Min	D Dia Min	L Max	T Min	W Max	Y Min
3/4	-12	1-1/16-12	-12	1-1/16	0.57	3.72	0.31	1.73	0.61
1	-16	1-5/16-12	-16	1-5/16	0.78	3.81	0.31	2.01	0.69
1-1/4	-20	1-5/8-12	-20	1-5/8	1.01	4.20	0.31	2.31	0.84

[a] For dimensions shown on Fig. 3, but not specified in Table 3D, see corresponding dimensions for respective straight thread sizes in Fig. 5A and Table 3 of SAE J514.

TABLE 3E—DIMENSIONS OF MALE 37-DEG FLARED TYPE HOSE FITTINGS FOR USE ON SAE 100R5 HYDRAULIC HOSE[a] (FIG. 3) (CODE 46)

Nominal SAE 100R5 Hose ID	Hose Dash Size	B Nominal Straight Thread Size	Thread Dash Size	C Min	D Dia Min	L Max	L_1 Max	T Min	W Max	Y Min
3/16	-4	7/16-20	-4	7/16	0.12	1.87	2.06	0.19	0.79	0.37
1/4	-5	1/2-20	-5	1/2	0.16	1.90	2.06	0.19	0.87	0.41
5/16	-6	9/16-18	-6	9/16	0.23	1.94	2.18	0.22	0.94	0.41
13/32	-8	3/4-16	-8	3/4	0.35	2.14	2.63	0.25	1.08	0.50
1/2	-10	7/8-14	-10	7/8	0.45	2.29	2.94	0.25	1.24	0.56
5/8	-12	1-1/16-12	-12	1-1/16	0.54	2.50	3.41	0.31	1.52	0.61
7/8	-16	1-5/16-12	-16	1-5/16	0.81	2.74	3.16	0.38	1.73	0.69
1-1/8	-20	1-5/8-12	-20	1-5/8	1.04	2.91	3.28	0.46	2.02	0.84
1-3/8	-24	1-7/8-12	-24	1-7/8	1.28	3.35	3.62	0.50	2.31	1.00
1-13/16	-32	2-1/2-12	-32	2-1/2	1.75	3.98	4.40	0.62	2.89	1.13

[a] For dimensions shown on Fig. 3, but not specified in Table 3E, see corresponding dimensions for respective thread sizes in Fig. 5A and Table 3 of SAE J514.

TABLE 3F—DIMENSIONS OF MALE 37-DEG FLARED TYPE HOSE FITTINGS FOR USE ON SAE 100R6 HYDRAULIC HOSE[a] (FIG. 3) (CODE 47)

Nominal SAE 100R6 Hose ID	Hose Dash Size	B Nominal Straight Thread Size	Thread Dash Size	C Min	D Dia Min	L Max	T Min	W Max	Y Min
1/4	-4	7/16-20	-4	7/16	0.11	1.38	0.19	0.75	0.37
1/4	-4	1/2-20	-5	1/2	0.11	1.38	0.19	0.75	0.41
3/8	-6	9/16-18	-6	9/16	0.26	1.62	0.22	1.00	0.41
3/8	-6	3/4-16	-8	3/4	0.26	1.72	0.25	1.00	0.50
1/2	-8	3/4-16	-8	3/4	0.38	1.75	0.25	1.12	0.50
5/8	-10	7/8-14	-10	7/8	0.48	2.00	0.25	1.19	0.56

[a] For dimensions shown on Fig. 3, but not specified in Table 3F, see corresponding dimensions for respective straight thread sizes in Fig. 5A and Table 3 of SAE J514.

FEMALE 37-DEG FLARED TYPE

NOTES: UNSPECIFIED DETAIL WITH RESPECT TO DIMENSIONS, TOLERANCES, CONTOURS, MATERIAL, WORKMANSHIP, AND SO ON, MUST CONFORM TO GENERAL SPECIFICATIONS FOR HYDRAULIC HOSE FITTINGS. THE DIMENSIONAL DESIGNATIONS ON THE FIRST FIGURE IN EACH GROUP SHALL APPLY TO ALL OTHER FIGURES IN THAT GROUP EXCEPT AS SHOWN OTHERWISE. THE DESIGN OF AND METHOD OF ATTACHING SWIVEL NUT SHALL BE OPTIONAL WITH MANUFACTURER PROVIDING THE DIMENSIONS SHOWN ARE MAINTAINED AND NUT TURNS FREELY. CODES SHOWN IN BRACKETS ADJACENT TO FIGURE NUMBERS REPRESENT RESPECTIVE FITTING IDENTIFICATION, WITH XX SUBSTITUTED FOR THE HOSE TYPE CODE DEPICTED IN BRACKETS AT END OF RESPECTIVE TABLE TITLES, IN ACCORDANCE WITH SAE J846.

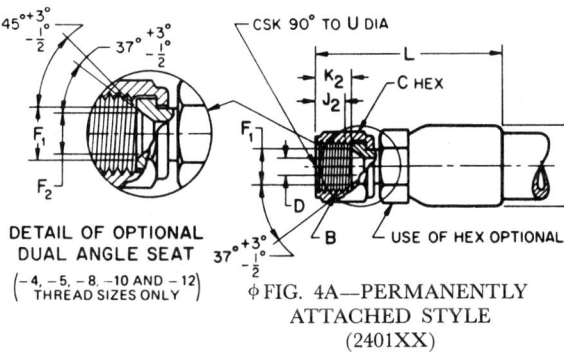
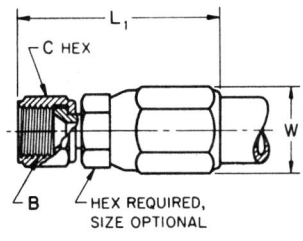

FIG. 4A—PERMANENTLY ATTACHED STYLE (2401XX)

FIG. 4B—FIELD ATTACHABLE SCREW STYLE (2501XX)

FIG. 4C—FIELD ATTACHABLE SEGMENT CLAMP STYLE (2601XX)

FIG. 4—FEMALE 37-DEG FLARED TYPE HOSE FITTINGS

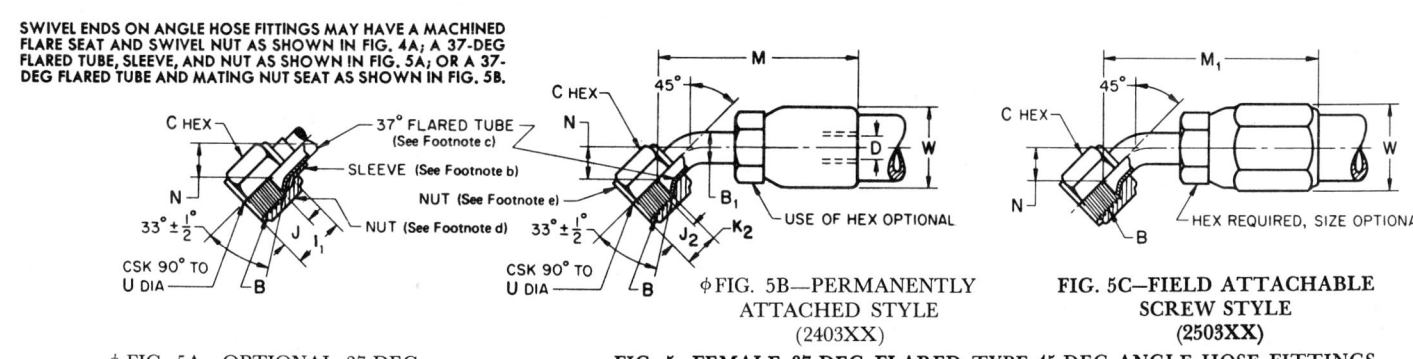

SWIVEL ENDS ON ANGLE HOSE FITTINGS MAY HAVE A MACHINED FLARE SEAT AND SWIVEL NUT AS SHOWN IN FIG. 4A; A 37-DEG FLARED TUBE, SLEEVE, AND NUT AS SHOWN IN FIG. 5A; OR A 37-DEG FLARED TUBE AND MATING NUT SEAT AS SHOWN IN FIG. 5B.

FIG. 5A—OPTIONAL 37-DEG FLARE SWIVEL END WITH SLEEVE

FIG. 5B—PERMANENTLY ATTACHED STYLE (2403XX)

FIG. 5C—FIELD ATTACHABLE SCREW STYLE (2503XX)

FIG. 5—FEMALE 37-DEG FLARED TYPE 45-DEG ANGLE HOSE FITTINGS

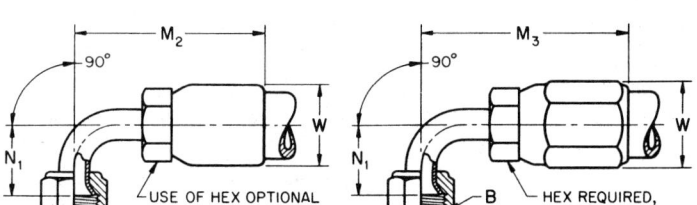
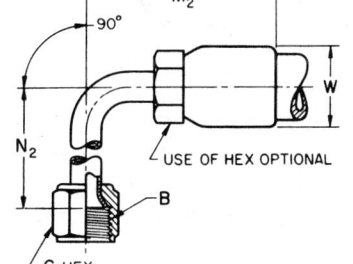
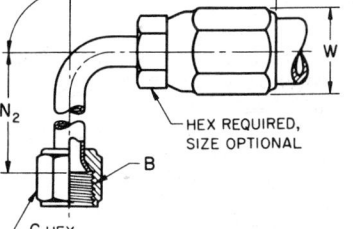

FIG. 6A—PERMANENTLY ATTACHED STYLE (2414XX)

FIG. 6B—FIELD ATTACHABLE SCREW STYLE (2514XX)

FIG. 6—FEMALE 37-DEG FLARED TYPE 90-DEG SHORT DROP ANGLE HOSE FITTINGS

FIG. 7A—PERMANENTLY ATTACHED STYLE (2415XX)

FIG. 7B—FIELD ATTACHABLE SCREW STYLE (2515XX)

FIG. 7—FEMALE 37-DEG FLARED TYPE 90-DEG LONG DROP ANGLE HOSE FITTINGS

TABLE 4A—DIMENSIONS OF FEMALE 37-DEG FLARED TYPE HOSE FITTINGS FOR USE ON SAE 100R1 HYDRAULIC HOSE[a] (FIGS. 4–7) (CODE 42)

Nominal SAE 100R1 Hose ID	Hose Dash Size	B Nominal Straight Thread Size	Thread and Tube Dash Size	B_1 Tube OD[b,c] Ref	C Nom	D Dia Min	L Max	L_1 Max	L_2 Max	M Max
3/16	-3	7/16-20	-4	—	9/16	0.09	2.38	2.56	—	—
1/4	-4	7/16-20	-4	1/4	9/16	0.11	2.61	2.62	—	2.66
1/4	-4	1/2-20	-5	5/16	5/8	0.11	2.67	2.62	—	2.71
1/4	-4	9/16-18	-6	—	11/16	0.11	2.68	2.66	—	—
5/16	-5	9/16-18	-6	3/8	11/16	0.20	2.68	2.66	—	2.94
3/8	-6	9/16-18	-6	3/8	11/16	0.26	2.71	2.78	—	3.00
3/8	-6	3/4-16	-8	1/2	7/8	0.26	2.81	2.91	—	3.38
13/32	-6.5	3/4-16	-8	1/2	7/8	0.26	2.81	2.91	—	3.38
1/2	-8	3/4-16	-8	1/2	7/8	0.38	2.88	3.24	—	3.38
1/2	-8	7/8-14	-10	5/8	1	0.38	3.07	3.27	3.22	3.63
5/8	-10	1-1/16-12	-12	3/4	1-1/4	0.50	3.48	3.68	—	4.19
3/4	-12	1-1/16-12	-12	3/4	1-1/4	0.57	3.86	3.68	3.50	4.57
3/4	-12	1-3/16-12	-14	7/8	1-3/8	0.61	4.04	3.81	3.68	4.50
1	-16	1-5/16-12	-16	1	1-1/2	0.78	4.04	4.28	4.28	4.85
1-1/4	-20	1-5/8-12	-20	—	2	1.01	4.56	4.96	—	—

TABLE 4A—DIMENSIONS OF FEMALE 37-DEG FLARED TYPE HOSE FITTINGS FOR USE ON SAE 100R1 HYDRAULIC HOSE[a] (FIGS. 4–7) (CODE 42) *(continued)*

Nominal SAE 100R1 Hose ID	Thread and Tube Dash Size	M_1 Max	M_2 Max	M_3 Max	M_4 Max	N ±0.06	N_1 ±0.06	N_2 ±0.06	W Max	Z Max
3/16	-4	—	—	—	—	—	—	—	0.75	—
1/4	-4	2.92	2.54	2.63	2.63	0.33	0.68	1.80	0.87	—
1/4	-5	2.97	2.54	2.67	2.67	0.36	0.77	1.80	0.87	—
1/4	-6	—	—	—	—	—	—	—	0.87	—
5/16	-6	3.26	2.69	2.91	2.91	0.39	0.85	2.18	0.94	—
3/8	-6	3.33	2.75	3.03	3.03	0.39	0.85	2.18	1.09	—
3/8	-6	3.55	3.06	3.29	3.29	0.55	1.09	2.43	1.09	—
13/32	-8	3.55	3.06	3.29	3.29	0.55	1.09	2.43	1.16	—
1/2	-8	3.85	3.06	3.56	3.56	0.55	1.09	2.43	1.20	—
1/2	-10	4.12	3.25	3.78	3.78	0.63	1.23	2.57	1.20	2.50
5/8	-12	4.53	3.80	4.22	4.22	0.78	1.82	3.73	1.37	—
3/4	-12	4.53	4.52	4.28	4.28	0.78	1.82	3.73	1.59	3.00
3/4	-14	4.92	4.13	4.55	4.55	0.84	2.00	3.93	1.59	3.00
1	-16	5.78	4.84	5.20	5.20	0.89	2.14	4.33	2.02	3.59
1-1/4	-20	—	—	—	—	—	—	—	2.46	—

[a] For dimensions shown on Figs. 4–7, but not specified in Table 4A, see corresponding dimensions for respective straight thread size in Fig. 11C and Table 3 of SAE J514.
[b] Sleeves shown in Fig. 5A shall conform with Fig. 17 and Table 5 of SAE J514 for respective tube OD referenced above, except on some sizes and shapes where sleeve length must be shortened.
[c] Flares shall conform with single or double 37-deg flares specified in Fig. 3 and Table 2 of SAE J533 for respective tube OD referenced above.
[d] Nuts shown in Fig. 5A shall conform with Fig. 15 and Table 4 of SAE J514 for respective straight thread size.
[e] For dimensions shown on Figs. 5–7, but not specified in Table 4A, see corresponding dimensions for respective straight thread size in Fig. 11C and Table 3 of SAE J514.

Pipe Threads—Taper pipe threads designated A in the illustrations and dimensional tables shall conform to the Dryseal American Standard Taper Pipe Thread (NPTF). Specifications are given in detail in SAE J476.

The length of full form external thread shall not be shorter than L_2 plus one pitch (thread).

Where external pipe threads are produced by roll threading, the diameter of the unthreaded shank adjacent to shoulder may be reduced to the E_2 pitch diameter.

Straight internal pipe threads designated A_1 in the illustrations and tables shall conform with free fitting American Standard Straight Pipe Threads for Mechanical Joints (NPSM) as specified in USA Standard, USAS B2.1, of latest issue.

External pipe threads shall be chamfered from the diameters tabulated below to produce the specified length of chamfered or partial thread. Internal pipe threads shall be countersunk 90 deg, included angle, to the diameters specified below:

Material—Fittings shall be made from materials of good quality, capable of withstanding the stresses resulting from hydraulic pressures equal to the minimum burst pressure of the applicable hose size and type to which they are assembled without failure.

Finish—Unless otherwise specified, fittings made from carbon steel shall be furnished cadmium or zinc plated to a thickness of 0.0002 in minimum, or with a phosphate coating (oil finished). On plated fittings, a subsequent

Nominal Pipe Thread Size	External Thread				Internal Thread	
	Chamfer Diameter[a]		Length of Chamfered or Partial Thread		Countersink Diameter[a]	
	Max	Min	Max	Min	Max	Min
1/8	0.32	0.30	0.055	0.037	0.44	0.42
1/4	0.42	0.40	0.084	0.056	0.57	0.55
3/8	0.55	0.53	0.084	0.056	0.71	0.69
1/2	0.68	0.66	0.107	0.071	0.87	0.85
3/4	0.89	0.87	0.107	0.071	1.08	1.06
1	1.12	1.09	0.130	0.087	1.37	1.34
1-1/4	1.46	1.43	0.130	0.087	1.71	1.68
1-1/2	1.70	1.67	0.130	0.087	1.95	1.92
2	2.17	2.14	0.130	0.087	2.41	2.39

[a] Tabulated diameters conform with Appendix A, SAE J476.

TABLE 4B—DIMENSIONS OF FEMALE 37-DEG FLARED TYPE HOSE FITTINGS FOR USE ON SAE 100R2 HYDRAULIC HOSE[a] (FIGS. 4–7) (CODE 43)

Nominal SAE 100R2 Hose ID	Hose Dash Size	B Nominal Straight Thread Size	Thread and Tube Dash Size	B_1 Tube OD[b,c] Ref	C Nom	D Dia Min	L Max	L_1 Max	L_2 Max	M Max
3/16	-3	7/16-20	-4	—	9/16	0.09	2.44	2.67	—	—
1/4	-4	7/16-20	-4	1/4	9/16	0.11	2.61	2.88	—	2.80
1/4	-4	1/2-20	-5	5/16	5/8	0.11	2.67	2.94	2.60	2.90
1/4	-4	9/16-18	-6	—	11/16	0.11	2.68	2.94	—	—
3/8	-6	9/16-18	-6	3/8	11/16	0.26	2.72	3.16	2.79	3.12
3/8	-6	3/4-16	-8	1/2	7/8	0.26	2.93	3.16	2.81	3.47
1/2	-8	3/4-16	-8	1/2	7/8	0.38	3.02	3.33	—	3.68
1/2	-8	7/8-14	-10	5/8	1	0.38	3.09	3.38	3.32	4.02
5/8	-10	1-1/16-12	-12	3/4	1-1/4	0.50	3.48	4.08	—	4.52
3/4	-12	1-1/16-12	-12	3/4	1-1/4	0.57	3.86	4.08	3.73	4.62
3/4	-12	1-3/16-12	-14	7/8	1-3/8	0.61	4.04	4.14	—	4.76
7/8	-14	1-3/16-12	-14	7/8	1-3/8	0.72	3.90	4.50	—	4.50
7/8	-14	1-5/16-12	-16	1	1-1/2	0.72	3.94	4.54	—	4.75
1	-16	1-5/16-12	-16	1	1-1/2	0.78	4.19	4.54	4.49	5.52
1-1/4	-20	1-5/8-12	-20	—	2	1.01	6.56	5.47	5.25	—
1-1/2	-24	1-7/8-12	-24	—	2-1/4	1.25	6.72	5.85	5.75	—
2	-32	2-1/2-12	-32	—	2-7/8	1.72	7.00	6.53	6.76	—

Nominal SAE 100R2 Hose ID	Thread and Tube Dash Size	M_1 Max	M_2 Max	M_3 Max	M_4 Max	N ±0.06	N_1 ±0.06	N_2 ±0.06	W Max	Z Max
3/16	-4	—	—	—	—	—	—	—	0.90	—
1/4	-4	3.11	2.54	3.00	3.00	0.33	0.68	1.80	0.89	—
1/4	-5	3.04	2.62	2.80	2.80	0.36	0.77	1.80	1.00	2.13
1/4	-6	—	—	—	—	—	—	—	1.00	—
3/8	-6	3.42	2.83	3.32	3.32	0.39	0.85	2.18	1.19	2.47
3/8	-8	3.57	3.16	3.38	3.57	0.55	1.09	2.43	1.19	2.47
1/2	-8	3.95	3.37	3.65	3.65	0.55	1.09	2.43	1.24	—
1/2	-10	4.21	3.63	3.87	3.87	0.63	1.23	2.57	1.24	2.47
5/8	-12	4.51	4.13	4.41	4.41	0.78	1.82	3.73	1.56	—
3/4	-12	4.81	4.52	4.41	4.76	0.78	1.82	3.73	1.75	3.00
3/4	-14	5.10	4.25	4.73	4.73	0.84	2.00	3.93	1.75	—
7/8	-14	5.10	4.25	4.95	4.95	0.84	2.00	3.93	1.79	—
7/8	-16	5.78	4.50	5.20	5.20	0.89	2.14	4.33	1.79	—
1	-16	5.78	5.25	5.31	5.50	0.89	2.14	4.33	2.05	3.60
1-1/4	-20	—	—	—	—	—	—	—	2.63	4.50
1-1/2	-24	—	—	—	—	—	—	—	2.94	5.00
2	-32	—	—	—	—	—	—	—	3.50	5.25

[a] For dimensions shown on Figs. 4–7, but not specified in Table 4B, see corresponding dimensions for respective straight thread size in Fig. 11C and Table 3 of SAE J514.
[b] Sleeves shown in Fig. 5A shall conform with Fig. 17 and Table 5 of SAE J514 for respective tube OD referenced above, except on some sizes and shapes where sleeve length must be shortened.
[c] Flares shall conform with single or double 37-deg flares specified in Fig. 3 and Table 2 of SAE J533 for respective tube OD referenced above.
[d] Nuts shown in Fig. 5A shall conform with Fig. 15 and Table 4 of SAE J514 for respective straight thread size.
[e] For dimensions shown on Figs. 5–7, but not specified in Table 4B, see corresponding dimensions for respective straight thread size in Fig. 11C and Table 3 of SAE J514.

chromate treatment shall be optional with the manufacturer. These parts must meet the requirements of 32 hr salt spray test in accordance with ASTM B 117, Method of Salt Spray (Fog) Testing.

Workmanship—Workmanship shall conform to the best commercial practice to produce high quality fittings. Fittings shall be free from all hanging burrs, loose scales, and slivers which might become dislodged in usage, sharp edges, and all other defects that might affect their serviceability. All sealing surfaces must be smooth except that annular tool marks up to 100 μ in maximum shall be permissible.

Assembly Considerations—Use of a compatible lubricant in assembling Dryseal pipe threads on hose fittings may be desirable to minimize galling and effect a pressure tight seal.

TABLE 4C—DIMENSIONS OF FEMALE 37-DEG FLARED TYPE HOSE FITTINGS FOR USE ON SAE 100R3 HYDRAULIC HOSE[a] (FIGS. 4–7) (CODE 44)

Nominal SAE 100R3 Hose ID	Hose Dash Size	B Nominal Straight Thread Size	Thread Dash Size	C Nom	D Dia Min	L Max	L_1 Max	L_2 Max	W Max	Z Max
1/4	-4	7/16-20	-4	9/16	0.11	2.61	2.50	—	0.88	—
1/4	-4	1/2-20	-5	5/8	0.11	2.67	2.50	—	0.88	—
3/8	-6	9/16-18	-6	11/16	0.26	2.78	2.77	—	1.06	—
3/8	-6	3/4-16	-8	7/8	0.26	2.79	2.75	—	1.06	—
1/2	-8	3/4-16	-8	7/8	0.38	2.93	3.10	—	1.31	—
1/2	-8	7/8-14	-10	1	0.38	3.04	3.25	—	1.31	—
3/4	-12	1-1/16-12	-12	1-1/4	0.57	3.86	3.38	—	1.66	—
1	-16	1-5/16-12	-16	1-1/2	0.78	4.04	3.75	3.68	2.00	3.25

[a] For dimensions shown on Figs. 4–7, but not specified in Table 4C, see corresponding dimensions for respective straight thread size in Fig. 11C and Table 3 of SAE J514.

TABLE 4D—DIMENSIONS OF FEMALE 37-DEG FLARED TYPE HOSE FITTINGS FOR USE ON SAE 100R4 HYDRAULIC HOSE[a] (FIGS. 4–7) (CODE 45)

Nominal SAE 100R4 Hose ID	Hose Dash Size	B Nominal Straight Thread Size	Thread Dash Size	C Nom	D Dia Min	L Max	W Max
3/4	-12	1-1/16-12	-12	1-1/4	0.57	3.86	1.73
1	-16	1-5/16-12	-16	1-1/2	0.78	4.04	2.01
1-1/4	-20	1-5/8-12	-20	2	1.01	4.20	2.31

[a] For dimensions shown on Figs. 4–7 but not specified in Table 4D, see corresponding dimensions for respective straight thread size in Fig. 11C and Table 3 of SAE J514.

TABLE 4E—DIMENSIONS OF FEMALE 37-DEG FLARED TYPE HOSE FITTINGS FOR USE ON SAE 100R5 HYDRAULIC HOSE[a] (FIGS. 4-7) (CODE 46)

Nominal SAE 100R5 Hose ID	Hose Dash Size	B Nominal Straight Thread Size	Thread and Tube Dash Size	B_1 Tube OD[b,c] Ref	C Min	D Dia Min	L Max	L_1 Max	M_1 Max	M_3 Max	M_4 Max	N ±0.06	N_1 ±0.06	N_3 ±0.06	W Max
3/16	-4	7/16-20	-4	1/4	9/16	0.12	2.09	2.12	2.45	2.35	2.35	0.33	0.68	1.80	0.79
1/4	-5	1/2-20	-5	5/16	5/8	0.16	2.21	2.18	2.60	2.54	2.54	0.36	0.77	1.80	0.87
5/16	-6	9/16-18	-6	3/8	11/16	0.23	2.28	2.39	2.75	2.65	2.65	0.39	0.85	2.18	0.94
13/32	-8	3/4-16	-8	1/2	7/8	0.35	2.52	2.88	3.34	3.19	3.19	0.55	1.09	2.43	1.08
1/2	-10	7/8-14	-10	5/8	1	0.45	2.66	3.16	3.66	3.47	3.47	0.63	1.23	2.57	1.25
5/8	-12	1-1/16-12	-12	3/4	1-1/4	0.54	2.77	3.62	4.34	4.29	4.29	0.78	1.82	3.73	1.52
7/8	-16	1-5/16-12	-16	1	1-1/2	0.81	3.18	3.50	4.16	4.08	4.08	0.89	2.14	4.33	1.73
1-1/8	-20	1-5/8-12	-20	1-1/4	2	1.04	3.32	3.88	4.15	4.08	4.08	1.10	2.58	5.28	2.02
1-3/8	-24	1-7/8-12	-24	—	2-1/4	1.28	3.80	3.98	—	—	—	—	—	—	2.31
1-13/16	-32	2-1/2-12	-32	—	2-7/8	1.75	4.50	4.76	—	—	—	—	—	—	2.89

[a] For dimensions shown on Figs. 4-7, but not specified in Table 4E, see corresponding dimensions for respective straight thread size in Fig. 11C and Table 3 of SAE J514.
[b] Sleeves shown in Fig. 5A shall conform with Fig. 17 and Table 5 of SAE J514 for respective tube OD referenced above, except on some sizes and shapes where sleeve length must be shortened.
[c] Flares shall conform with single or double 37-deg flares specified in Fig. 3 and Table 2 of SAE J533 for respective tube OD referenced above.
[d] Nuts shown in Fig. 5A shall conform with Fig. 15 and Table 4 of SAE J514 for respective straight thread size.
[e] For dimensions shown on Figs. 5-7, but not specified in Table 4E, see corresponding dimensions for respective straight thread size in Fig. 11C and Table 3 of SAE J514.

TABLE 4F—DIMENSIONS OF FEMALE 37-DEG FLARED TYPE HOSE FITTINGS FOR USE ON SAE 100R6 HYDRAULIC HOSE[a] (FIGS. 4-7) (CODE 47)

Nominal SAE 100R6 Hose ID	Hose Dash Size	B Nominal Straight Thread Size	Thread Dash Size	C Nom	D Dia Min	L Max	L_1 Max	W Max
1/4	-4	7/16-20	-4	9/16	0.11	1.50	1.56	0.75
1/4	-4	1/2-20	-5	5/8	0.11	1.50	1.56	0.75
5/16	-5	1/2-20	-5	5/8	0.20	1.58	1.69	0.88
5/16	-5	9/16-18	-6	11/16	0.20	1.58	1.46	0.88
3/8	-6	9/16-18	-6	11/16	0.26	1.78	1.84	1.00
3/8	-6	3/4-16	-8	7/8	0.26	1.81	1.84	1.00
1/2	-8	3/4-16	-8	7/8	0.38	1.81	1.88	1.12
1/2	-8	7/8-14	-10	1	0.38	1.91	—	1.12
5/8	-10	7/8-14	-10	1	0.48	2.26	2.32	1.19

[a] For dimensions shown on Figs. 4-7, but not specified in Table 4F, see corresponding dimensions for respective straight thread size in Fig. 11C and Table 3 of SAE J514.

---MALE 45-DEG FLARED TYPE---

NOTES: UNSPECIFIED DETAIL WITH RESPECT TO DIMENSIONS, TOLERANCES, CONTOURS, MATERIAL, WORKMANSHIP, AND SO ON, MUST CONFORM TO GENERAL SPECIFICATIONS FOR HYDRAULIC HOSE FITTINGS. THE DIMENSIONAL DESIGNATIONS ON THE FIRST FIGURE IN EACH GROUP SHALL APPLY TO ALL OTHER FIGURES IN THAT GROUP EXCEPT AS SHOWN OTHERWISE. CODES SHOWN IN BRACKETS ADJACENT TO FIGURE NUMBERS REPRESENT RESPECTIVE FITTING IDENTIFICATION, WITH XX SUBSTITUTED FOR THE HOSE TYPE CODE DEPICTED IN BRACKETS AT END OF RESPECTIVE TABLE TITLES, IN ACCORDANCE WITH SAE J846.

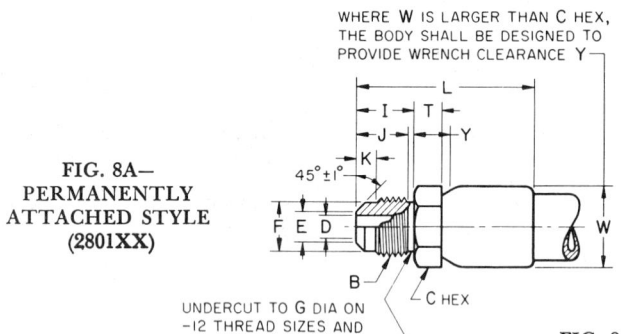
FIG. 8A—PERMANENTLY ATTACHED STYLE (2801XX)

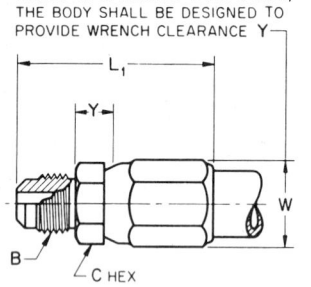
FIG. 8B—FIELD ATTACHABLE SCREW STYLE (2901XX)

FIG. 8—MALE 45-DEG FLARED TYPE HOSE FITTINGS

TABLE 5A—DIMENSIONS OF MALE 45-DEG FLARED TYPE HOSE FITTINGS FOR USE ON SAE 100R1 HYDRAULIC HOSE[a] (FIG. 8) (CODE 42)

Nominal SAE 100R1 Hose ID	Hose Dash Size	B Nominal Straight Thread Size	Thread Dash Size	C Min	D Dia Min	L Max	L_1 Max	T Min	W Max	Y Min
1/4	-4	7/16-20	-4	7/16	0.11	2.37	2.37	0.22	0.87	0.37
1/4	-4	1/2-20	-5	1/2	0.11	2.37	2.42	0.22	0.87	0.41
5/16	-5	5/8-18	-6	5/8	0.20	2.44	2.52	0.22	0.94	0.41
3/8	-6	5/8-18	-6	5/8	0.26	2.50	2.63	0.22	1.09	0.41
3/8	-6	3/4-16	-8	3/4	0.26	2.85	2.63	0.25	1.09	0.50
13/32	-6.5	3/4-16	-8	3/4	0.26	2.85	2.75	0.25	1.16	0.50
1/2	-8	3/4-16	-8	3/4	0.38	2.85	3.08	0.25	1.23	0.50
1/2	-8	7/8-14	-10	7/8	0.38	2.96	3.20	0.25	1.23	0.56
5/8	-10	1-1/16-14	-12	1-1/16	0.50	3.20	3.60	0.31	1.37	0.61
3/4	-12	1-1/16-14	-12	1-1/16	0.57	3.20	3.60	0.31	1.59	0.61

[a] For dimensions shown on Fig. 8, but not specified in Table 5A, see corresponding dimensions for respective straight thread size in Fig. 1A and Table 2 of SAE J512.

TABLE 5B—DIMENSIONS OF MALE 45-DEG FLARED TYPE HOSE FITTINGS FOR USE ON SAE 100R3 HYDRAULIC HOSE[a] (FIG. 8) (CODE 44)

Nominal SAE 100R3 Hose ID	Hose Dash Size	B Nominal Straight Thread Size	Thread Dash Size	C Min	D Dia Min	L Max	T Min	W Max	Y Min
1/4	-4	7/16-20	-4	7/16	0.11	1.88	0.22	0.88	0.37
1/4	-4	1/2-20	-5	1/2	0.11	1.88	0.22	0.88	0.41
3/8	-6	5/8-18	-6	5/8	0.26	2.31	0.22	1.06	0.41
3/8	-6	3/4-16	-8	3/4	0.26	2.38	0.25	1.06	0.50
1/2	-8	3/4-16	-8	3/4	0.38	2.50	0.25	1.31	0.50
1/2	-8	7/8-14	-10	7/8	0.38	2.63	0.25	1.31	0.56
3/4	-12	1-1/16-14	-12	1-1/16	0.57	2.81	0.31	1.66	0.61

[a] For dimensions shown on Fig. 8, but not specified in Table 5B, see corresponding dimensions for respective straight thread size in Fig. 1A and Table 2 of SAE J512.

TABLE 5C—DIMENSIONS OF MALE 45-DEG FLARED TYPE HOSE FITTINGS FOR USE ON SAE 100R5 HYDRAULIC HOSE[a] (FIG. 8) (CODE 46)

Nominal SAE 100R5 Hose ID	Hose Dash Size	B Nominal Straight Thread Size	Thread Dash Size	C Min	D Dia Min	L Max	L_1 Max	T Min	W Max	Y Min
3/16	-4	7/16-20	-4	7/16	0.12	1.82	2.00	0.19	0.79	0.37
1/4	-5	1/2-20	-5	1/2	0.16	1.91	2.06	0.19	0.87	0.41
5/16	-6	5/8-18	-6	5/8	0.23	2.01	2.25	0.22	0.94	0.41
13/32	-8	3/4-16	-8	3/4	0.35	2.23	2.72	0.25	1.08	0.50
1/2	-10	7/8-14	-10	7/8	0.45	2.41	3.06	0.25	1.24	0.56
5/8	-12	1-1/16-14	-12	1-1/16	0.54	2.57	3.55	0.31	1.52	0.61

[a] For dimensions shown on Fig. 8, but not specified in Table 5C, see corresponding dimensions for respective straight thread size in Fig. 1A and Table 2 of SAE J512.

TABLE 5D—DIMENSIONS OF MALE 45-DEG FLARED TYPE HOSE FITTINGS FOR USE ON SAE 100R6 HYDRAULIC HOSE[a] (FIG. 8) (CODE 47)

Nominal SAE 100R6 Hose ID	Hose Dash Size	B Nominal Straight Thread Size	Thread Dash Size	C Min	D Dia Min	L Max	T Min	W Max	Y Min
3/16	-3	3/8-24	-3	3/8	0.09	1.25	0.15	0.63	0.35
3/16	-3	7/16-20	-4	7/16	0.09	1.25	0.15	0.63	0.35
1/4	-4	3/8-24	-3	3/8	0.11	1.35	0.15	0.75	0.35
1/4	-4	7/16-20	-4	7/16	0.11	1.38	0.15	0.75	0.35
1/4	-4	1/2-20	-5	1/2	0.11	1.38	0.15	0.75	0.35
5/16	-5	1/2-20	-5	1/2	0.20	1.40	0.15	0.88	0.35
5/16	-5	5/8-18	-6	5/8	0.20	1.47	0.19	0.88	0.40
3/8	-6	5/8-18	-6	5/8	0.26	1.60	0.19	1.00	0.40
3/8	-6	3/4-16	-8	3/4	0.26	1.72	0.19	1.00	0.43
1/2	-8	3/4-16	-8	3/4	0.38	1.78	0.19	1.12	0.43
5/8	-10	7/8-14	-10	7/8	0.48	2.00	0.25	1.19	0.50

[a] For dimensions shown on Fig. 8, but not specified in Table 5D, see corresponding dimensions for respective straight thread size in Fig. 1A and Table 2 of SAE J512.

FEMALE 45-DEG FLARED TYPE

NOTES: UNSPECIFIED DETAIL WITH RESPECT TO DIMENSIONS, TOLERANCES, CONTOURS, MATERIAL, WORKMANSHIP, AND SO ON, MUST CONFORM TO GENERAL SPECIFICATIONS FOR HYDRAULIC HOSE FITTINGS. THE DIMENSIONAL DESIGNATIONS ON THE FIRST FIGURE IN EACH GROUP SHALL APPLY TO ALL OTHER FIGURES IN THAT GROUP EXCEPT AS SHOWN OTHERWISE. THE DESIGN OF AND METHOD OF ATTACHING SWIVEL NUT SHALL BE OPTIONAL WITH MANUFACTURER PROVIDING THE DIMENSIONS SHOWN ARE MAINTAINED AND NUT TURNS FREELY. CODES SHOWN IN BRACKETS ADJACENT TO FIGURE NUMBERS REPRESENT RESPECTIVE FITTING IDENTIFICATION, WITH XX SUBSTITUTED FOR THE HOSE TYPE CODE DEPICTED IN BRACKETS AT END OF RESPECTIVE TABLE TITLES, IN ACCORDANCE WITH SAE J846.

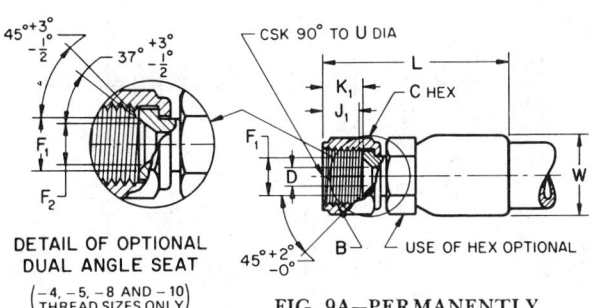

FIG. 9A—PERMANENTLY ATTACHED STYLE (3001XX)

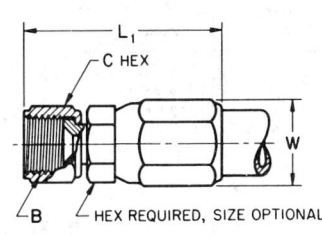

FIG. 9B—FIELD ATTACHABLE SCREW STYLE (3101XX)

FIG. 9—FEMALE 45-DEG FLARED TYPE HOSE FITTINGS

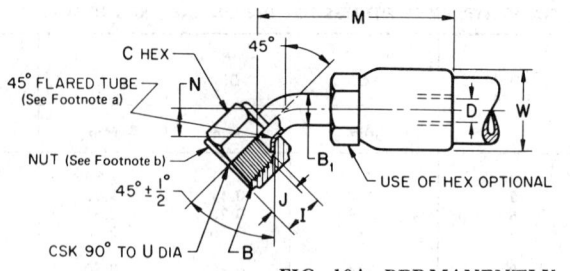

FIG. 10A—PERMANENTLY ATTACHED STYLE (3003XX)

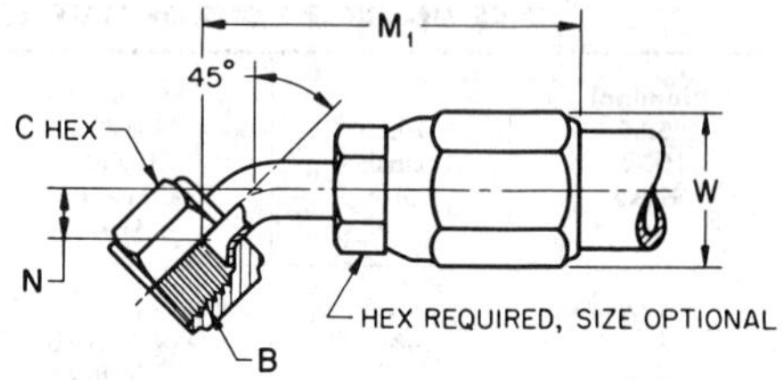

FIG. 10B—FIELD ATTACHABLE SCREW STYLE (3103XX)

FIG. 10—FEMALE 45-DEG FLARED TYPE 45-DEG ANGLE HOSE FITTINGS

SWIVEL ENDS ON ANGLE FITTINGS MAY HAVE A MACHINED FLARE SEAT AND SWIVEL NUT AS SHOWN IN FIG. 9A, OR A 45-DEG FLARED TUBE AND MATING NUT SEAT AS SHOWN IN FIG. 10A.

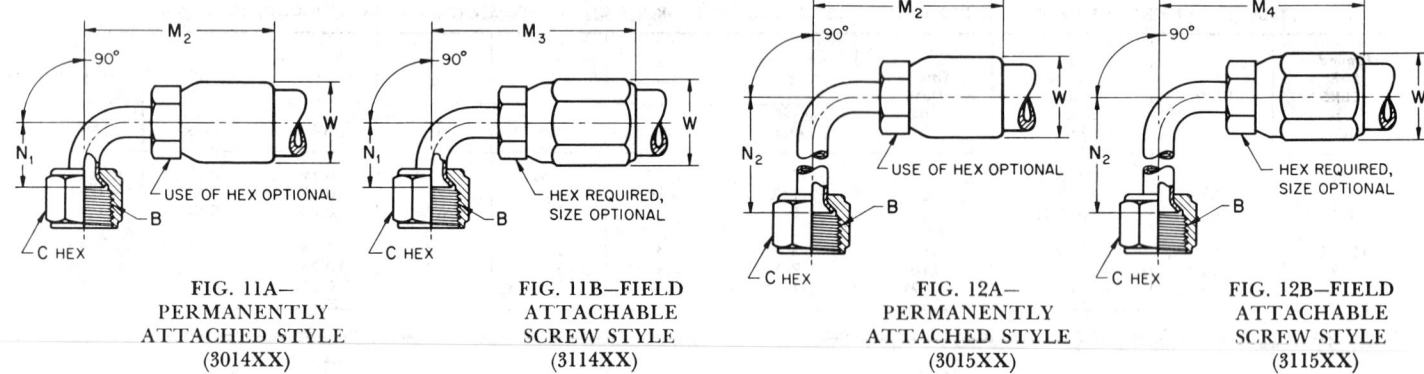

FIG. 11A—PERMANENTLY ATTACHED STYLE (3014XX)

FIG. 11B—FIELD ATTACHABLE SCREW STYLE (3114XX)

FIG. 11—FEMALE 45-DEG FLARED TYPE 90-DEG SHORT DROP ANGLE HOSE FITTINGS

FIG. 12A—PERMANENTLY ATTACHED STYLE (3015XX)

FIG. 12B—FIELD ATTACHABLE SCREW STYLE (3115XX)

FIG. 12—FEMALE 45-DEG FLARED TYPE 90-DEG LONG DROP ANGLE HOSE FITTINGS

TABLE 6A—DIMENSIONS OF FEMALE 45-DEG FLARED TYPE HOSE FITTINGS FOR USE ON SAE 100R1 HYDRAULIC HOSE (FIGS. 9–12) (CODE 42)

Nominal SAE 100R1 Hose ID	Hose Dash Size	B Nominal Straight Thread Size	Thread and Tube Dash Size	B_1 Tube OD[a] Ref	C Nom	D Dia Min	F_1 Dia ±0.010	F_2 Dia ±0.005	J_1 Full Thread Min	K_1 +0.015 −0.000	L Max	L_1 Max
1/4	-4	7/16-20	-4	0.250	9/16	0.11	0.295	0.250	0.30	0.344	2.61	2.62
1/4	-4	1/2-20	-5	0.312	5/8	0.11	0.357	0.312	0.33	0.375	2.67	2.62
5/16	-5	5/8-18	-6	0.375	3/4	0.20	0.470	—	0.36	0.422	2.57	2.66
3/8	-6	5/8-18	-6	0.375	3/4	0.26	0.470	—	0.36	0.422	2.69	2.78
3/8	-6	3/4-16	-8	0.500	7/8	0.26	0.572	0.500	0.38	0.422	2.81	2.85
13/32	-6.5	3/4-16	-8	0.500	7/8	0.26	0.572	0.500	0.38	0.422	2.81	2.91
1/2	-8	3/4-16	-8	0.500	7/8	0.38	0.572	0.500	0.38	0.422	2.88	3.13
1/2	-8	7/8-14	-10	0.625	1	0.38	0.680	0.625	0.46	0.500	3.07	3.27
5/8	-10	1-1/16-14	-12	0.750	1-1/4	0.50	0.850	—	0.47	0.562	3.48	3.68
3/4	-12	1-1/16-14	-12	0.750	1-1/4	0.57	0.850	—	0.47	0.562	3.86	3.68

TABLE 6A—DIMENSIONS OF FEMALE 45-DEG FLARED TYPE HOSE FITTINGS FOR USE ON SAE 100R1 HYDRAULIC HOSE (FIGS. 9–12) (CODE 42)

Nominal SAE 100R1 Hose ID	Thread and Tube Dash Size	M Max	M_1 Max	M_2 Max	M_3 Max	M_4 Max	N ±0.06	N_1 ±0.06	N_2 ±0.06	U Dia +0.015 −0.000	W Max
1/4	-4	2.66	2.92	2.54	2.63	2.63	0.33	0.68	1.80	0.443	0.87
1/4	-5	2.71	2.97	2.54	2.67	2.67	0.36	0.77	1.80	0.505	0.87
5/16	-6	2.94	3.26	2.69	2.91	2.91	0.39	0.85	2.18	0.630	0.94
3/8	-6	3.00	3.33	2.75	3.03	3.03	0.39	0.85	2.18	0.630	1.09
3/8	-8	3.38	3.55	3.06	3.29	3.29	0.55	1.09	2.43	0.755	1.09
13/32	-8	3.38	3.55	3.06	3.29	3.29	0.55	1.09	2.43	0.755	1.16
1/2	-8	3.38	3.85	3.06	3.56	3.56	0.55	1.09	2.43	0.755	1.23
1/2	-10	3.63	4.12	3.25	3.78	3.78	0.63	1.23	2.57	0.880	1.23
5/8	-12	4.19	4.53	3.80	4.22	4.22	0.78	1.82	3.73	1.067	1.37
3/4	-12	4.57	4.53	4.52	4.28	4.28	0.78	1.82	3.73	1.067	1.59

ɸ [a] Flares shall conform with single or double 45-deg flares specified in Fig. 2 and Table 1 of SAE J533 for respective tube OD referenced above.
ɸ [b] For dimensions shown on Figs. 10–12, but not specified in Table 6A, see corresponding dimensions for respective straight thread size in Fig. 4 and Table 3 of SAE J512.

TABLE 6B—DIMENSIONS OF FEMALE 45-DEG FLARED TYPE HOSE FITTINGS FOR USE ON SAE 100R3 HYDRAULIC HOSE (FIGS. 9–12) (CODE 44)

Nominal SAE 100R3 Hose ID	Hose Dash Size	B Nominal Straight Thread Size	Thread Dash Size	C Nom	D Dia Min	F_1 Dia ±0.010	F_2 Dia ±0.005	J_1 Full Thread Min	K_1 +0.015 -0.000	L Max	U Dia +0.015 -0.000	W Max
1/4	-4	7/16-20	-4	9/16	0.11	0.295	0.250	0.30	0.344	2.06	0.443	0.88
1/4	-4	1/2-20	-5	5/8	0.11	0.357	0.312	0.33	0.375	2.06	0.505	0.88
3/8	-6	5/8-18	-6	3/4	0.26	0.470	—	0.38	0.422	2.41	0.630	1.06
3/8	-6	3/4-16	-8	7/8	0.26	0.572	0.500	0.38	0.422	2.63	0.755	1.06
1/2	-8	3/4-16	-8	7/8	0.38	0.572	0.500	0.38	0.422	2.70	0.755	1.31
1/2	-8	7/8-14	-10	1	0.38	0.680	0.625	0.46	0.500	2.88	0.880	1.31
3/4	-12	1-1/16-14	-12	1-1/4	0.57	0.850	—	0.47	0.562	3.03	1.067	1.66

TABLE 6C—DIMENSIONS OF FEMALE 45-DEG FLARED TYPE HOSE FITTINGS FOR USE ON SAE 100R5 HYDRAULIC HOSE (FIGS. 9–12) (CODE 46)

Nominal SAE 100R5 Hose ID	Hose Dash Size	B Nominal Straight Thread Size	Thread and Tube Dash Size	B_1 Tube OD[a] Ref	C Nom	D Dia Min	F_1 Dia ±0.010	F_2 Dia ±0.005	J_1 Full Thread Min	K_1 +0.015 -0.000	L Max	L_1 Max
3/16	-4	7/16-20	-4	0.250	9/16	0.12	0.295	0.250	0.30	0.344	2.09	2.12
1/4	-5	1/2-20	-5	0.312	5/8	0.16	0.357	0.312	0.33	0.375	2.21	2.18
5/16	-6	5/8-18	-6	0.375	3/4	0.23	0.470	—	0.38	0.422	2.28	2.38
13/32	-8	3/4-16	-8	0.500	7/8	0.35	0.572	0.500	0.38	0.422	2.52	2.88
1/2	-10	7/8-14	-10	0.625	1	0.45	0.680	0.625	0.46	0.500	2.66	3.16
5/8	-12	1-1/16-14	-12	0.750	1-1/4	0.54	0.850	—	0.47	0.562	2.77	3.58

Nominal SAE 100R5 Hose ID	Thread Dash Size	M_1 Max	M_3 Max	M_4 Max	N ±0.006	N_1 ±0.006	N_2 ±0.006	U Dia +0.015 -0.000	W Max
3/16	-4	2.45	2.35	2.95	0.33	0.68	1.80	0.443	0.79
1/4	-5	2.60	2.54	2.54	0.36	0.77	1.89	0.505	0.87
5/16	-6	2.75	2.69	2.69	0.39	0.85	2.18	0.630	0.94
13/32	-8	3.34	3.19	3.19	0.55	1.00	2.43	0.755	1.08
1/2	-10	3.66	3.47	3.47	0.63	1.23	2.57	0.880	1.24
5/8	-12	4.34	4.29	4.29	0.78	1.82	3.73	1.067	1.52

[a] Flares shall conform with single or double 45-deg flares specified in Fig. 2 and Table 1 of SAE J533 for respective tube OD referenced above.
[b] For dimensions shown on Figs. 10–12, but not specified in Table 6C, see corresponding dimensions for respective straight thread size in Fig. 4 and Table 3 of SAE J512.

TABLE 6D—DIMENSIONS OF FEMALE 45-DEG FLARED TYPE HOSE FITTINGS FOR USE ON SAE 100R6 HYDRAULIC HOSE (FIGS. 9–12) (CODE 47)

Nominal SAE 100R6 Hose ID	Hose Dash Size	B Nominal Straight Thread Size	Thread Dash Size	C Nom	D Dia Min	F_1 Dia ±0.010	F_2 Dia ±0.005	J_1 Full Thread Min	K_1 +0.015 -0.000	L Max	L_1 Max	U Dia +0.015 -0.000	W Max
3/16	-3	7/16-20	-4	9/16	0.09	0.295	0.250	0.30	0.344	1.42	—	0.443	0.63
1/4	-4	7/16-20	-4	9/16	0.11	0.295	0.250	0.30	0.344	1.50	1.56	0.443	0.75
1/4	-4	1/2-20	-5	5/8	0.11	0.357	0.312	0.33	0.375	1.50	—	0.505	0.75
5/16	-5	1/2-20	-5	5/8	0.11	0.357	0.312	0.33	0.375	1.58	1.69	0.505	0.88
5/16	-5	5/8-18	-6	3/4	0.20	0.470	—	0.38	0.422	1.60	—	0.630	0.88
3/8	-6	5/8-18	-6	3/4	0.26	0.470	—	0.38	0.422	1.78	1.84	0.630	1.00
1/2	-8	3/4-16	-8	7/8	0.38	0.572	0.500	0.38	0.422	1.84	1.88	0.755	1.12
5/8	-10	7/8-14	-10	1	0.48	0.680	0.625	0.46	0.500	2.23	2.32	0.880	1.19

---- MALE FLARELESS TYPE ----

NOTES: UNSPECIFIED DETAIL WITH RESPECT TO DIMENSIONS, TOLERANCES, CONTOURS, MATERIAL, WORKMANSHIP, AND SO ON, MUST CONFORM TO GENERAL SPECIFICATIONS FOR HYDRAULIC HOSE FITTINGS. THE DIMENSIONAL DESIGNATIONS ON THE FIRST FIGURE IN EACH GROUP SHALL APPLY TO ALL OTHER FIGURES IN THAT GROUP EXCEPT AS SHOWN OTHERWISE. CODES SHOWN IN BRACKETS ADJACENT TO FIGURE NUMBERS REPRESENT RESPECTIVE FITTING IDENTIFICATION, WITH XX SUBSTITUTED FOR THE HOSE TYPE CODE DEPICTED IN BRACKETS AT END OF RESPECTIVE TABLE TITLES, IN ACCORDANCE WITH SAE J846.

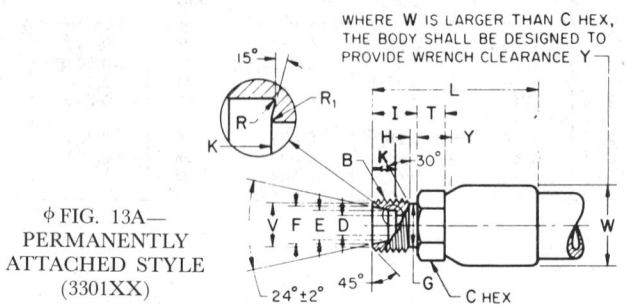
φ FIG. 13A— PERMANENTLY ATTACHED STYLE (3301XX)

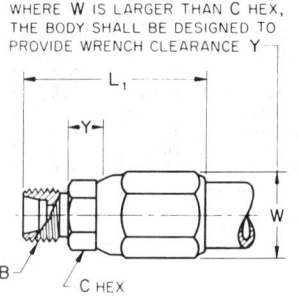
FIG. 13B— FIELD ATTACHABLE SCREW STYLE (3401XX)

FIG. 13— MALE FLARELESS TYPE HOSE FITTINGS

φ TABLE 7A—DIMENSIONS OF MALE FLARELESS TYPE HOSE FITTINGS FOR USE ON SAE 100R1 HYDRAULIC HOSE[a] (FIG. 13) (CODE 42)

Nominal SAE 100R1 Hose ID	Hose Dash Size	B Nominal Straight Thread Size	Thread Dash Size	C Min	D Dia Min	L Max	L_1 Max	T Min	W Max	Y Min
1/4	-4	7/16-20	-4	7/16	0.11	2.38	2.28	0.22	0.87	0.37
5/16	-5	9/16-18	-6	9/16	0.20	2.36	2.36	0.22	0.94	0.41
3/8	-6	9/16-18	-6	9/16	0.26	2.43	2.42	0.22	1.09	0.41
3/8	-6	3/4-16	-8	3/4	0.26	2.54	2.54	0.25	1.09	0.50
13/32	-6.5	3/4-16	-8	3/4	0.26	2.54	2.54	0.25	1.16	0.50
1/2	-8	3/4-16	-8	3/4	0.38	2.61	2.91	0.25	1.23	0.50
1/2	-8	7/8-14	-10	7/8	0.38	2.67	2.97	0.25	1.23	0.56
5/8	-10	1-1/16-12	-12	1-1/16	0.50	3.03	3.47	0.31	1.37	0.61
3/4	-12	1-1/16-12	-12	1-1/16	0.57	3.55	3.47	0.31	1.59	0.61
1	-16	1-5/16-12	-16	1-5/16	0.78	3.59	4.00	0.38	2.02	0.69

φ [a] For dimensions shown on Fig. 13, but not specified in Table 7A, see corresponding dimensions for respective straight thread size in Fig. 19A and Table 7 of SAE J514.

φ TABLE 7B—DIMENSIONS OF MALE FLARELESS TYPE HOSE FITTINGS FOR USE ON SAE 100R2 HYDRAULIC HOSE[a] (FIG. 13) (CODE 43)

Nominal SAE 100R2 Hose ID	Hose Dash Size	B Nominal Straight Thread Size	Thread Dash Size	C Min	D Dia Min	L Max	L_1 Max	T Min	W Max	Y Min
1/4	-4	7/16-20	-4	7/16	0.11	2.38	2.53	0.19	1.00	0.37
3/8	-6	9/16-18	-6	9/16	0.26	2.43	2.72	0.22	1.19	0.41
3/8	-6	3/4-16	-8	3/4	0.26	2.54	2.85	0.25	1.19	0.50
1/2	-8	3/4-16	-8	3/4	0.38	2.61	2.96	0.25	1.38	0.50
1/2	-8	7/8-14	-10	7/8	0.38	2.67	3.02	0.25	1.38	0.56
5/8	-10	1-1/16-12	-12	1-1/16	0.50	3.03	3.63	0.31	1.56	0.61
3/4	-12	1-1/16-12	-12	1-1/16	0.57	3.55	3.63	0.31	1.75	0.61
1	-16	1-5/16-12	-16	1-5/16	0.78	3.59	4.15	0.38	2.06	0.69

φ [a] For dimensions shown on Fig. 13, but not specified in Table 7B, see corresponding dimensions for respective straight thread size in Fig. 19A and Table 7 of SAE J514.

---- FEMALE FLARELESS TYPE ----

NOTES: UNSPECIFIED DETAIL WITH RESPECT TO DIMENSIONS, TOLERANCES, CONTOURS, MATERIAL, WORKMANSHIP, AND SO ON, MUST CONFORM TO GENERAL SPECIFICATIONS FOR HYDRAULIC HOSE FITTINGS. THE DIMENSIONAL DESIGNATIONS ON THE FIRST FIGURE IN EACH GROUP SHALL APPLY TO ALL OTHER FIGURES IN THAT GROUP EXCEPT AS SHOWN OTHERWISE. THE DESIGN OF AND METHOD OF ATTACHING SWIVEL NUT SHALL BE OPTIONAL WITH MANUFACTURER PROVIDING THE DIMENSIONS SHOWN ARE MAINTAINED AND NUT TURNS FREELY. CODES SHOWN IN BRACKETS ADJACENT TO FIGURE NUMBERS REPRESENT RESPECTIVE FITTING IDENTIFICATION, WITH XX SUBSTITUTED FOR THE HOSE TYPE CODE DEPICTED IN BRACKETS AT END OF RESPECTIVE TABLE TITLES, IN ACCORDANCE WITH SAE J846.

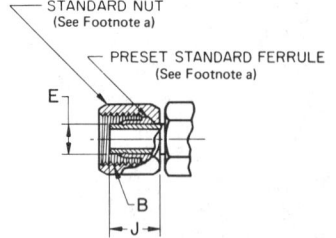
FIG. 14A—OPTIONAL PRESET FERRULE END CONSTRUCTION

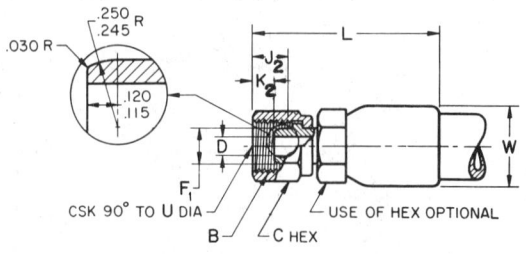
FIG. 14B—PERMANENTLY ATTACHED STYLE (3601XX)

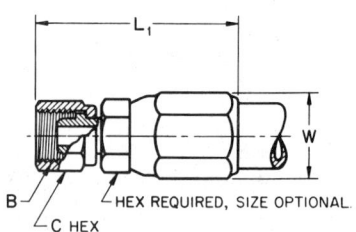
FIG. 14C—FIELD ATTACHABLE SCREW STYLE (3701XX)

FIG. 14—FEMALE FLARELESS TYPE HOSE FITTINGS

TABLE 8A—DIMENSIONS OF FEMALE FLARELESS TYPE HOSE FITTINGS FOR USE ON SAE 100R1 HYDRAULIC HOSE[b] (FIG. 14) (CODE 42)

Nominal SAE 100R1 Hose ID	Hose Dash Size	B Nominal Straight Thread Size	Thread Dash Size	C Nom	D Dia Min	E Dia ±0.003	J Ref	L Max	L_1 Max	W Max
1/4	-4	7/16-20	-4	9/16	0.11	0.250	0.654	2.42	2.62	0.87
5/16	-5	9/16-18	-6	11/16	0.20	0.375	0.720	2.57	2.66	0.94
3/8	-6	9/16-18	-6	11/16	0.26	0.375	0.720	2.69	2.78	1.09
3/8	-6	3/4-16	-8	7/8	0.26	0.500	0.805	2.81	2.91	1.09
13/32	-6.5	3/4-16	-8	7/8	0.26	0.500	0.805	2.81	2.91	1.16
1/2	-8	3/4-16	-8	7/8	0.38	0.500	0.805	2.88	3.24	1.23
1/2	-8	7/8-14	-10	1	0.38	0.625	0.380	3.07	3.27	1.23
5/8	-10	1-1/16-12	-12	1-1/4	0.50	0.750	0.910	3.48	3.63	1.37
3/4	-12	1-1/16-12	-12	1-1/4	0.57	0.750	0.910	3.48	3.63	1.59
1	-16	1-5/16-12	-16	1-1/2	0.78	1.000	1.075	3.94	4.28	2.02

[a] Optional end construction depicted in Fig. 14 shall consist of a standard ferrule preset unto a tubular nipple to retain standard flareless tube nut. Ferrules shall conform to Fig. 28 or 29 and Table 8 of SAE J514 for nominal tube OD corresponding to E diameter above; and nuts shall conform to Fig. 31 and Table 10 for corresponding straight thread size.
[b] For dimensions shown on Fig. 14B, but not specified in Table 8A, see corresponding dimensions for respective straight thread size in Fig. 25C and Table 7 of SAE J514.

TABLE 8B—DIMENSIONS OF FEMALE FLARELESS TYPE HOSE FITTINGS FOR USE ON SAE 100R2 HYDRAULIC HOSE[b] (FIG. 14) (CODE 43)

Nominal SAE 100R2 Hose ID	Hose Dash Size	B Nominal Straight Thread Size	Thread Dash Size	C Nom	D Dia Min	E Dia ±0.003	J Ref	L Max	L_1 Max	W Max
1/4	-4	7/16-20	-4	9/16	0.11	0.250	0.654	2.48	2.67	1.00
3/8	-6	9/16-18	-6	11/16	0.26	0.375	0.720	2.69	3.00	1.19
3/8	-6	3/4-16	-8	7/8	0.26	0.500	0.805	2.81	3.12	1.19
1/2	-8	3/4-16	-8	7/8	0.38	0.500	0.805	2.88	3.33	1.38
1/2	-8	7/8-14	-10	1	0.38	0.625	0.880	3.07	3.38	1.38
5/8	-10	1-1/16-12	-12	1-1/4	0.50	0.750	0.910	3.48	4.08	1.56
3/4	-12	1-1/16-12	-12	1-1/4	0.57	0.750	0.910	3.48	4.08	1.75
1	-16	1-5/16-12	-16	1-1/2	0.78	1.000	1.075	3.94	4.54	2.06

[a] Optional end construction depicted in Fig. 14 shall consist of a standard ferrule preset unto a tubular nipple to retain standard flareless tube nut. Ferrules shall conform to Fig. 28 or 29 and Table 8 of SAE J514 for nominal tube OD corresponding to E diameter above; and nuts shall conform to Fig. 31 and Table 10 for corresponding straight thread size.
[b] For dimensions shown on Fig. 14B, but not specified in Table 8D, see corresponding dimensions for respective straight thread size in Fig. 25C and Table 7 of SAE J514.

---4-BOLT SPLIT FLANGE TYPE---

NOTES: UNSPECIFIED DETAIL WITH RESPECT TO DIMENSIONS, TOLERANCES, CONTOURS, MATERIAL, WORKMANSHIP, AND SO ON, MUST CONFORM TO GENERAL SPECIFICATIONS FOR HYDRAULIC HOSE FITTINGS. THE DIMENSIONAL DESIGNATIONS ON THE FIRST FIGURE IN EACH GROUP SHALL APPLY TO ALL OTHER FIGURES IN THAT GROUP EXCEPT AS SHOWN OTHERWISE. CODES SHOWN IN BRACKETS ADJACENT TO FIGURE NUMBERS REPRESENT RESPECTIVE FITTING IDENTIFICATION, WITH XX SUBSTITUTED FOR THE HOSE TYPE CODE DEPICTED IN BRACKETS AT END OF RESPECTIVE TABLE TITLES, IN ACCORDANCE WITH SAE J846.

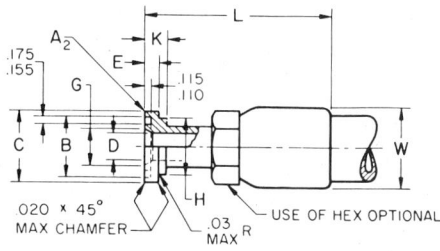

FIG. 15A—PERMANENTLY ATTACHED STYLE (3901XX)

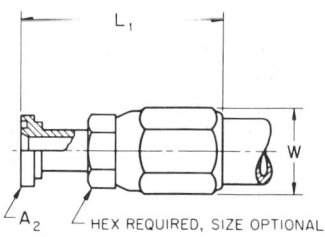

FIG. 15B—FIELD ATTACHABLE SCREW STYLE (4001XX)

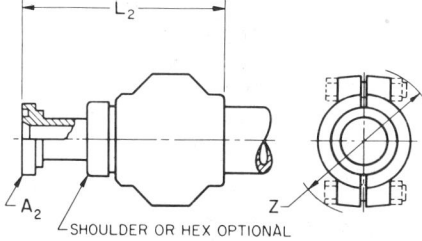

FIG. 15C—FIELD ATTACHABLE SEGMENT CLAMP STYLE (4101XX)

FIG. 15—4-BOLT SPLIT FLANGE TYPE HOSE FITTINGS

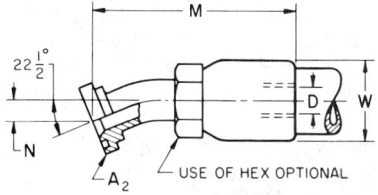

FIG. 16A—PERMANENTLY ATTACHED STYLE (3910XX)

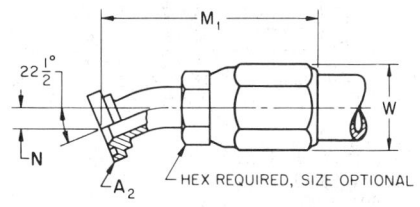

FIG. 16B—FIELD ATTACHABLE SCREW STYLE (4010XX)

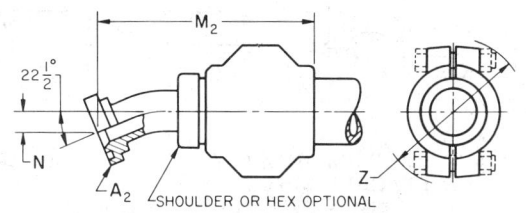

FIG. 16C—FIELD ATTACHABLE SEGMENT CLAMP STYLE (4110XX)

FIG. 16—4-BOLT SPLIT FLANGE TYPE 22½-DEG ANGLE HOSE FITTINGS

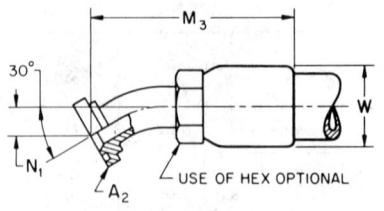

FIG. 17A—PERMANENTLY
ATTACHED STYLE
(3911XX)

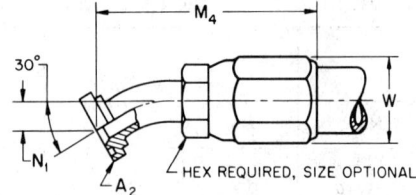

FIG. 17B—FIELD ATTACHABLE
SCREW STYLE
(4011XX)

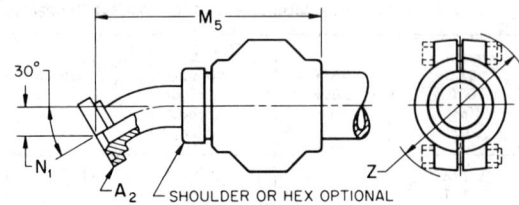

FIG. 17C—FIELD ATTACHABLE
SEGMENT CLAMP STYLE
(4111XX)

FIG. 17—4-BOLT SPLIT FLANGE TYPE 30-DEG ANGLE HOSE FITTINGS

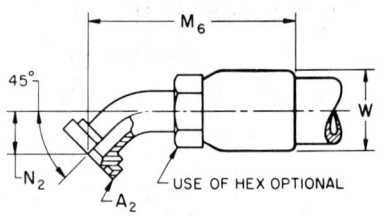

FIG. 18A—PERMANENTLY
ATTACHED STYLE
(3903XX)

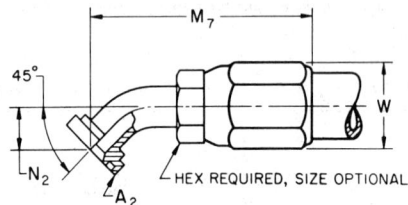

FIG. 18B—FIELD ATTACHABLE
SCREW STYLE
(4003XX)

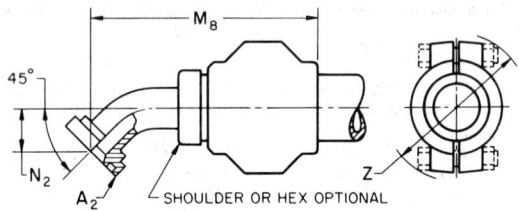

FIG. 18C—FIELD ATTACHABLE
SEGMENT CLAMP STYLE
(4103XX)

FIG. 18—4-BOLT SPLIT FLANGE TYPE 45-DEG ANGLE HOSE FITTINGS

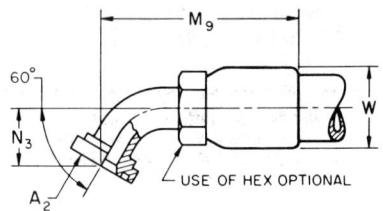

FIG. 19A—PERMANENTLY
ATTACHED STYLE
(3912XX)

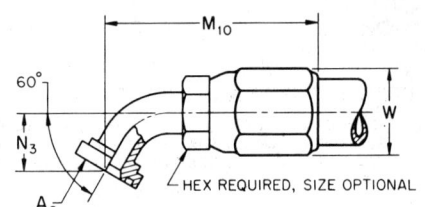

FIG. 19B—FIELD ATTACHABLE
SCREW STYLE
(4012XX)

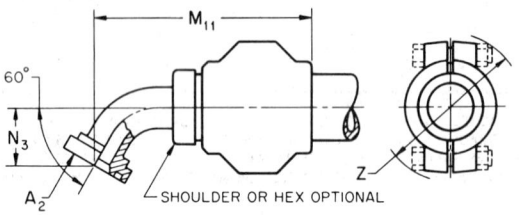

FIG. 19C—FIELD ATTACHABLE
SEGMENT CLAMP STYLE
(4112XX)

FIG. 19—4-BOLT SPLIT FLANGE TYPE 60-DEG ANGLE HOSE FITTINGS

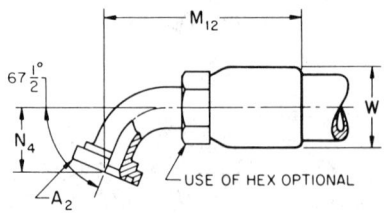

FIG. 20A—PERMANENTLY
ATTACHED STYLE
(3913XX)

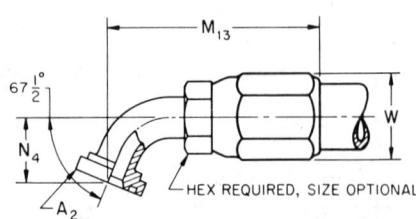

FIG. 20B—FIELD ATTACHABLE
SCREW STYLE
(4013XX)

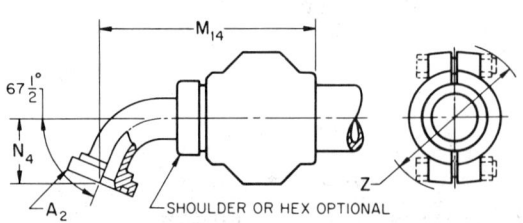

FIG. 20C—FIELD ATTACHABLE
SEGMENT CLAMP STYLE
(4113XX)

FIG. 20—4-BOLT SPLIT FLANGE TYPE 67½-DEG ANGLE HOSE FITTINGS

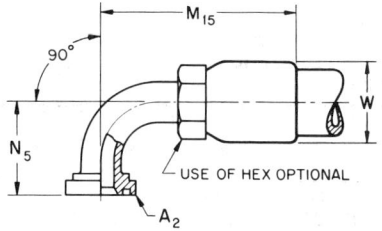
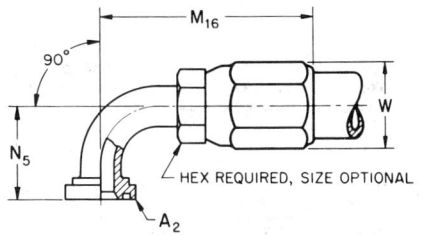
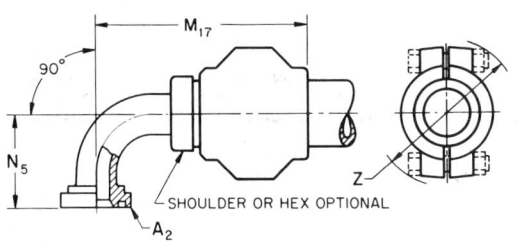

FIG. 21A—PERMANENTLY ATTACHED STYLE (3902XX)

FIG. 21B—FIELD ATTACHABLE SCREW STYLE (4002XX)

FIG. 21C—FIELD ATTACHABLE SEGMENT CLAMP STYLE (4102XX)

FIG. 21—4-BOLT SPLIT FLANGE TYPE 90-DEG ANGLE HOSE FITTINGS

TABLE 9A—DIMENSIONS OF 4-BOLT SPLIT FLANGE TYPE HOSE FITTINGS FOR USE ON SAE 100R1 HYDRAULIC HOSE[a] (FIGS. 15A-21C) (CODE 42)

Nominal SAE 100R1 Hose ID	Hose Dash Size	A_2 Nominal Flange Size	Flange Dash Size	D Dia Min	L Max	L_1 Max	L_2 Max	M Max	M_1 Max	M_2 Max	M_3 Max	M_4 Max	M_5 Max	M_6 Max	M_7 Max	M_8 Max	M_9 Max
1/2	-8	1/2	-8	0.38	3.00	3.68	3.38	3.36	—	—	3.39	—	3.36	3.39	3.91	3.39	3.36
5/8	-10	3/4	-12	0.50	3.50	—	—	3.94	3.94	—	3.94	3.94	—	4.06	—	—	3.84
3/4	-12	3/4	-12	0.57	3.50	3.97	4.00	3.94	3.94	3.94	3.94	3.94	3.94	4.06	4.68	4.06	3.84
1	-16	1	-16	0.78	3.50	4.28	4.38	4.20	4.66	4.66	4.20	4.66	4.66	4.34	5.51	4.81	4.38
1-1/4	-20	1-1/4	-20	1.01	4.18	4.38	4.38	4.40	5.13	5.13	4.82	5.13	5.13	5.00	5.44	5.44	4.69
1-1/2	-24	1-1/2	-24	1.25	4.50	4.75	4.38	5.72	5.72	5.72	5.72	5.72	5.72	5.97	5.97	5.97	6.13
2	-32	2	-32	1.72	5.38	5.06	5.00	6.75	7.44	7.44	6.75	7.44	7.44	7.60	7.75	7.75	8.03

Nominal SAE 100R1 Hose ID	Flange Dash Size	M_{10} Max	M_{11} Max	M_{12} Max	M_{13} Max	M_{14} Max	M_{15} Max	M_{16} Max	M_{17} Max	N ±0.12	N_1 ±0.12	N_2 ±0.12	N_3 ±0.12	N_4 ±0.12	N_5 ±0.12	W Max	Z Max
1/2	-8	—	3.36	3.36	—	3.36	3.30	3.56	3.30	0.38	0.50	0.78	1.05	1.25	1.63	1.23	2.50
5/8	-12	4.09	—	3.84	4.09	—	3.69	3.94	—	0.38	0.53	0.84	1.16	1.31	2.13	1.37	—
3/4	-12	4.09	4.09	3.84	4.09	4.09	3.69	4.28	3.94	0.44	0.58	1.00	1.41	1.60	2.13	1.59	3.00
1	-16	4.88	4.88	4.38	4.88	4.88	4.25	5.20	4.75	0.44	0.62	1.06	1.50	1.75	2.38	2.02	3.59
1-1/4	-20	5.69	5.69	4.69	5.69	5.69	4.81	5.34	5.34	0.50	0.72	1.15	1.66	1.90	2.62	2.46	4.00
1-1/2	-24	6.34	6.34	6.13	6.34	6.34	6.13	6.00	6.00	0.63	0.88	1.41	2.00	2.00	3.12	2.74	4.50
2	-32	8.25	8.25	8.03	8.25	8.25	7.63	8.00	8.00	0.88	1.25	2.00	2.88	3.25	4.50	3.47	5.12

[a] For dimensions of flanged head shown on Fig. 15A, but not specified in Table 9A, see corresponding dimensions for respective nominal flange size in Fig. 3 and Table 1 of SAE J518.

TABLE 9B—DIMENSIONS OF 4-BOLT SPLIT FLANGE TYPE HOSE FITTINGS FOR USE ON SAE 100R2 HYDRAULIC HOSE[a] (FIGS. 15A-21C) (CODE 43)

Nominal SAE 100R2 Hose ID	Hose Dash Size	A_2 Nominal Flange Size	Flange Dash Size	D Dia Min	L Max	L_1 Max	L_2 Max	M Max	M_1 Max	M_2 Max	M_3 Max	M_4 Max	M_5 Max	M_6 Max	M_7 Max	M_8 Max	M_9 Max
1/2	-8	1/2	-8	0.38	3.61	3.78	3.50	3.50	3.78	3.46	3.50	—	3.56	3.62	4.00	3.59	3.62
1/2	-8	3/4	-12	0.38	3.25	3.88	3.50	4.38	—	3.56	3.56	—	3.62	3.62	—	3.65	3.62
5/8	-10	3/4	-12	0.50	2.78	—	—	4.38	—	—	4.48	—	—	4.57	—	—	4.60
3/4	-12	3/4	-12	0.57	3.81	4.25	3.66	4.00	4.44	4.04	4.12	4.54	4.14	4.20	4.68	4.28	4.25
3/4	-12	1	-16	0.57	3.41	4.25	3.66	4.00	4.60	4.20	4.12	4.70	4.30	4.20	4.80	4.40	4.25
7/8	-14	1	-16	0.72	3.25	—	—	—	—	—	—	—	—	—	—	—	—
1	-16	1	-16	0.78	4.28	4.74	4.03	4.40	4.86	4.64	4.54	4.80	4.80	4.72	5.45	4.94	4.75
1	-16	1-1/4	-20	0.78	4.34	4.74	4.32	4.46	4.86	4.76	4.62	4.92	4.92	4.72	5.45	5.04	4.78
1-1/4	-20	1-1/4	-20	1.01	5.38	5.35	5.13	6.06	5.62	5.62	6.25	6.08	5.84	6.56	6.47	5.98	6.75
1-1/4	-20	1-1/2	-24	1.01	5.38	5.35	5.13	6.06	—	5.68	6.25	—	5.84	6.56	—	5.98	6.75
1-1/2	-24	1-1/2	-24	1.25	6.12	5.61	5.31	6.98	6.52	6.52	7.10	6.80	6.80	7.33	7.00	7.00	7.38
1-1/2	-24	2	-32	1.25	6.12	5.61	5.31	6.98	—	6.52	7.10	—	6.80	7.33	—	7.00	7.38
2	-32	2	-32	1.72	5.44	6.37	6.40	7.32	8.12	8.42	7.60	8.40	8.70	7.80	8.60	8.90	8.28

Nominal SAE 100R2 Hose ID	Flange Dash Size	M_{10} Max	M_{11} Max	M_{12} Max	M_{13} Max	M_{14} Max	M_{15} Max	M_{16} Max	M_{17} Max	N ±0.12	N_1 ±0.12	N_2 ±0.12	N_3 ±0.12	N_4 ±0.12	N_5 ±0.12	W Max	Z Max
1/2	-8	3.66	3.62	3.52	3.66	3.56	3.38	3.65	3.50	0.38	0.50	0.78	1.05	1.25	1.63	1.38	2.62
1/2	-12	—	3.65	3.56	—	3.59	3.38	—	3.50	0.38	0.53	0.84	1.16	1.31	1.69	1.38	2.62
5/8	-12	—	—	4.56	—	—	4.45	—	—	0.38	0.53	0.84	1.16	1.31	2.13	1.56	—
3/4	-12	4.68	4.28	4.19	4.65	4.25	4.06	4.40	4.00	0.44	0.58	1.00	1.41	1.60	2.13	1.75	3.12
3/4	-16	4.80	4.40	4.19	4.75	4.35	4.06	4.50	4.10	0.44	0.58	1.00	1.41	1.60	2.13	1.75	3.12
7/8	-16	—	—	—	—	—	—	—	—	—	—	—	—	—	—	1.81	—
1	-16	5.20	5.09	4.84	5.10	5.06	4.62	5.14	4.90	0.44	0.62	1.06	1.50	1.75	2.38	2.06	3.60
1	-20	5.20	5.12	4.84	5.10	5.09	4.62	5.14	5.00	0.47	0.66	1.09	1.56	1.81	2.38	2.06	3.60
1-1/4	-20	6.10	6.10	6.75	6.00	6.00	6.50	6.33	6.00	0.50	0.72	1.15	1.66	1.90	2.62	2.63	4.50
1-1/4	-24	—	6.10	6.75	—	6.00	6.03	6.50	6.33	0.53	0.75	1.22	1.75	2.00	2.69	2.63	4.50
1-1/2	-24	7.06	7.06	7.32	6.96	6.96	7.30	7.01	6.90	0.63	0.88	1.41	2.00	3.00	3.12	2.94	5.00
1-1/2	-32	—	7.06	7.32	—	6.96	7.30	—	6.90	0.63	0.88	1.41	2.00	2.25	3.12	2.94	5.00
2	-32	9.08	9.38	8.18	8.98	9.28	7.80	8.62	8.90	0.88	1.25	2.00	2.88	3.25	4.50	3.50	5.88

[a] For dimensions of flanged head shown on Fig. 15A, but not specified in Table 9B, see corresponding dimensions for respective nominal flange size in Fig. 3 and Table 1 of SAE J518.

φ **TABLE 9C—DIMENSIONS OF 4-BOLT SPLIT FLANGE TYPE HOSE FITTINGS FOR USE ON SAE 100R4 HYDRAULIC HOSE[a]**
(FIGS. 15A, 15C, 16A, 16C, 17A, 17C, 18A, 18C, 19A, 19C, 20A, 20C, 21A AND 21C) (CODE 45)

Nominal SAE 100R4 Hose ID	Hose Dash Size	A_2 Nominal Flange Size	Flange Dash Size	D Dia Min	D Dia Max	L Max	L_2 Max	M Max	M_2 Max	M_3 Max	M_5 Max	M_6 Max	M_8 Max	M_9 Max
3/4	-12	3/4	-12	0.57	3.24	3.00	3.31	3.41	3.41	3.50	3.53	3.63	3.56
1	-16	1	-16	0.78	3.53	3.25	3.69	4.13	3.81	4.25	4.06	4.41	4.06
1-1/4	-20	1-1/4	-20	1.01	4.38	3.81	4.00	4.68	4.13	4.88	4.50	5.19	4.56
1-1/2	-24	1-1/2	-24	1.25	5.02	4.13	5.19	5.19	5.44	5.50	5.62	5.75	5.66
2	-32	2	-32	1.69	5.38	4.82	6.44	7.00	6.69	7.19	7.03	7.50	7.49
2-1/2	-40	2-1/2	-40	2.25	5.00	4.94	5.50	6.00	5.84	6.38	6.53	7.03	6.94
3	-48	3	-48	2.69	6.12	5.50	7.25	6.81	7.66	7.31	8.50	8.06	9.00

Nominal SAE 100R4 Hose ID	Flange Dash Size	M_{11} Max	M_{12} Max	M_{14} Max	M_{15} Max	M_{17} Max	N ±0.12	N_1 ±0.12	N_2 ±0.12	N_3 ±0.12	N_4 ±0.12	N_5 ±0.12	W Max	Z Max
3/4	-12	3.68	3.53	3.63	3.41	3.50	0.44	0.58	1.00	1.41	1.60	2.13	1.73	3.00
1	-16	4.50	4.03	4.44	3.94	4.38	0.44	0.62	1.06	1.50	1.75	2.38	2.01	3.25
1-1/4	-20	5.44	4.53	5.19	4.75	5.09	0.50	0.72	1.15	1.66	1.90	2.62	2.31	3.94
1-1/2	-24	6.13	5.68	6.00	5.28	5.75	0.63	0.88	1.41	2.00	2.00	3.12	2.60	4.34
2	-32	8.00	7.44	8.00	7.50	7.75	0.88	1.25	2.00	2.88	3.25	4.50	3.18	4.94
2-1/2	-40	7.44	7.06	7.56	7.00	7.50	0.62	0.94	1.72	2.62	3.12	4.62	3.93	5.50
3	-48	8.56	9.12	8.81	9.09	8.70	0.72	1.12	2.03	3.12	3.72	5.50	4.46	6.19

[a] For dimensions of flanged head shown on Fig. 15A, but not specified in Table 9C, see corresponding dimensions for respective nominal flange size in Fig. 3 and Table 1 of SAE J518.

TABLE 9D—DIMENSIONS OF 4-BOLT SPLIT FLANGE TYPE HOSE FITTINGS FOR USE ON SAE 100R5 HYDRAULIC HOSE[a] (FIGS. 15B, 16B, 17B, 18B, 19B AND 21B) (CODE 46)

Nominal SAE 100R5 Hose ID	Hose Dash Size	A_2 Nominal Flange Size	Flange Dash Size	D Dia Min	D Dia Max	L_1 Max	M_1 Max	M_4 Max	M_7 Max	M_{10} Max	M_{16} Max	N ±0.12	N_1 ±0.12	N_2 ±0.12	N_3 ±0.12	N_5 ±0.12	W Max
13/32	-8	1/2	-8	0.35	2.82	3.55	—	3.50	—	2.94	0.50	—	1.00	—	1.62	1.08
5/8	-12	3/4	-12	0.54	3.22	3.99	4.08	4.07	4.21	3.80	0.50	0.69	1.00	1.62	2.12	1.52
7/8	-16	1	-16	0.81	2.83	3.61	3.38	3.86	3.92	3.61	0.50	0.50	1.12	1.64	2.38	1.73
1-1/8	-20	1-1/4	-20	1.04	3.64	3.88	4.48	4.16	4.28	4.08	0.50	0.96	1.12	1.64	2.50	2.02
1-3/8	-24	1-1/2	-24	1.28	4.07	4.08	4.02	4.44	4.79	4.48	0.50	0.58	1.13	2.00	2.75	2.31
1-13/16	-32	2	-32	1.75	4.98	4.64	4.74	5.24	—	5.45	0.50	0.65	1.25		3.25	2.89

[a] For dimensions of flanged head shown on Fig. 15A, but not specified in Table 9D, see corresponding dimensions for respective nominal flange size in Fig. 3 and Table 1 of SAE J518.

ADAPTER UNIONS

NOTES: UNSPECIFIED DETAIL WITH RESPECT TO DIMENSIONS, TOLERANCES, CONTOURS, MATERIAL, WORKMANSHIP, AND SO ON, MUST CONFORM TO GENERAL SPECIFICATIONS FOR HYDRAULIC HOSE FITTINGS. THE DIMENSIONAL DESIGNATIONS ON THE FIRST FIGURE IN EACH GROUP SHALL APPLY TO ALL OTHER FIGURES IN THAT GROUP EXCEPT AS SHOWN OTHERWISE. THE DESIGN OF AND METHOD OF ATTACHING SWIVEL NUT SHALL BE OPTIONAL WITH MANUFACTURER PROVIDING THE DIMENSIONS SHOWN ARE MAINTAINED AND NUT TURNS FREELY. CODES SHOWN IN BRACKETS ADJACENT TO FIGURE NUMBERS REPRESENT RESPECTIVE FITTING IDENTIFICATION IN ACCORDANCE WITH SAE J846.

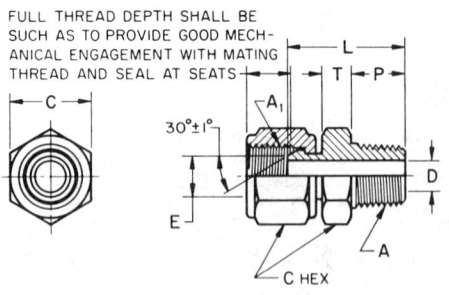

FIG. 22—FEMALE ADAPTER UNION (MALE) (140130)

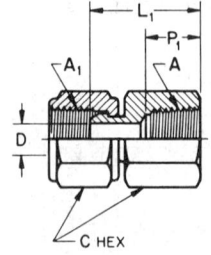

FIG. 23—FEMALE ADAPTER UNION (FEMALE) (140131)

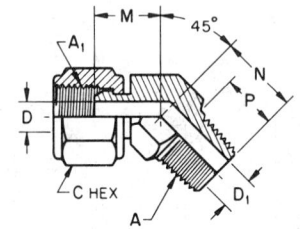

FIG. 24—45-DEG FEMALE ADAPTER UNION (MALE) (140330)

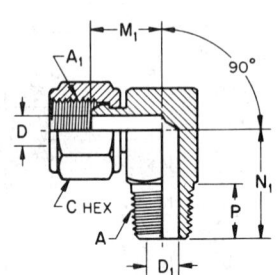

FIG. 25—90-DEG FEMALE ADAPTER UNION (MALE) (140230)

TABLE 10—DIMENSIONS OF ADAPTER UNIONS[a] (FIGS. 22-25)

A Dryseal Taper Thread[b] NPTF	Pipe Thread Dash Size	A₁ Straight Pipe Thread[c] NPSM	C		D Dia		D₁ Dia	E Dia ±0.010	I ±0.06	L₁ ±0.06	M ±0.06	M₁ ±0.06	N ±0.06	N₁ ±0.06	P +0.03 −0.00	P₁		T
			Nom		Max	Min	Max									Max		Min
1/8	-2	1/8	9/16		0.19	0.11	0.188	0.234	0.94	0.88	0.50	0.69	0.62	0.94	0.38	0.46		0.18
1/4	-4	1/4	11/16		0.23	0.21	0.281	0.297	1.25	1.25	0.69	0.75	0.94	1.19	0.56	0.67		0.24
3/8	-6	3/8	7/8		0.36	0.32	0.406	0.422	1.31	1.31	0.75	0.94	1.06	1.50	0.56	0.68		0.30
1/2	-8	1/2	1		0.48	0.44	0.531	0.562	1.62	1.50	0.88	1.00	1.31	1.81	0.75	0.90		0.37
3/4	-12	3/4	1-1/4		0.66	0.61	0.719	0.750	1.75	1.62	1.00	1.19	1.50	2.00	0.75	0.91		0.37
1	-16	1	1-1/2		0.88	0.78	0.938	0.969	2.00	2.00	1.06	1.44	1.50	2.38	0.94	1.14		0.43
1-1/4	-20	1-1/4	1-7/8		1.16	1.01	1.250	1.281	2.06	2.00	1.25	1.62	1.75	2.62	0.97	1.16		0.49
1-1/2	-24	1-1/2	2-1/8		1.38	1.25	1.500	1.500	2.19	2.00	1.38	1.75	1.94	2.81	1.00	1.16		0.49
2	-32	2	2-5/8		1.88	1.72	1.938	2.000	2.38	2.12	1.50	2.31	2.00	3.12	1.03	1.18		0.61

[a]Fitting users should allow clearance for maximum limits specified for Figs. 22-25 in Table 10 for installation.
[b]Dryseal American Standard Taper Pipe Thread.
[c]American Standard Straight Pipe Thread for Mechanical Joints.

HOSE PUSH-ON FITTINGS—SAE J1231

SAE Standard

Report of Fluid Conductors and Connectors Technical Committee approved May 1978.

1. General Specifications

2. Scope—This standard covers complete general and dimensional specifications for hose push-on fittings. These fittings are intended for general applications in low pressure automotive and hydraulic systems on automotive, industrial, and commercial products. The fittings shown in Figs. 1–3 and Table 2 are intended for use with low pressure hose and shall be retained by hose clamps as specified in SAE J536.

3. The following general specifications supplement the dimensional data for all types of fittings contained in Tables 1 and 2 with respect to all unspecified detail.

3.1 Size Designations—Fitting sizes are designated by the corresponding inside diameter of the hose I.D. for the various sizes of formed tube ends and by the corresponding standard nominal pipe size for pipe thread ends.

3.2 Dimensions and Tolerances—Except for nominal sizes and thread specifications, dimensions and tolerances are given in both U. S. Customary and SI units as designated. Tabulated dimensions shall apply to the finished parts, plated or otherwise processed, as specified by the purchaser. Unless otherwise specified, the maximum and minimum across flats dimensions shall be within the commercial tolerance of bar or extruded stock from which the fittings are produced. The minimum across corners dimensions of hexagons shall be 1.092 times the nominal width across flats, but shall not result in a side flat width less than 0.43 times the nominal width across flats.

Unless otherwise specified, tolerance on hole diameters designated drill in the dimensional tables shall be as tabulated below:

Drill Size Range		Tolerance on Hole Diameter			
		plus		minus	
in	mm	in	mm	in	mm
0.0135–0.1850	0.343–4.699	0.003	0.08	0.002	0.05
0.1875–0.2480	4.762–6.299	0.004	0.10	0.002	0.05
0.2500–0.7500	6.350–19.050	0.006	0.15	0.003	0.08
0.7579–1.0000	19.251–25.400	0.007	0.18	0.004	0.10

Tolerance on all dimensions not otherwise limited shall be ±0.010 in (0.25 mm). Angular tolerance on axis of ends on elbows shall be ±2.50 deg for sizes up to and including 3/8 in and ±1.50 deg for sizes larger than 3/8 in. The F diameter shall be concentric with the E diameter within 0.010 in (0.25 mm) full indicator reading (FIR).

Where so illustrated and not otherwise specified, hexagon corners shall be chamfered 30 ± 5 deg to a diameter equal to the nominal width across flats, with a tolerance of −0.016 in (0.41 mm); or where design permits, corners may be chamfered to the diameter of the abutting surface provided the length of chamfer does not exceed that produced by the 30 deg chamfer previously described.

3.3 Passages—Where passages in straight fittings are machined from opposite ends, the offset at the meeting point shall not exceed 0.015 in (0.38 mm). The cross sectional area at the junction of passages in angle fittings shall not be less than that of the smallest passage. Where the passage is specified as a maximum, the minimum shall be no less than the minimum diameter of the smallest passage in the fitting.

3.4 Wall Thickness—Unless otherwise designated, the wall thickness at any point on fittings shall not be less than the thickness established by the specified dimensions, tolerances, and eccentricities for inner and outer surfaces.

3.5 Contour—Details of contour shall be optional with manufacturer provided the tabulated dimensions are maintained and serviceability of the fittings is not impaired. Wrench flats on elbows shall be optional. Where extruded or forged shapes are reduced to conserve material, the wall thickness, unless otherwise specified, shall not be less than the respective minimum values tabulated below:

Nominal Pipe Thread Size, in	Wall Thickness, min			
	Extruded Shape[a]		Forged Shape	
	in	mm	in	mm
1/8	0.04	1.0	0.060	1.52
1/4	0.05	1.3	0.075	1.90
3/8	0.06	1.5	0.090	2.29
1/2	0.08	2.0	0.100	2.54
3/4	0.08	2.0	0.100	2.54
1	0.08	2.0	0.120	3.05

[a]Applies to reduction to one plane of shape only.

3.6 Pipe Threads—Pipe threads, unless there is specific authorization to the contrary, shall conform to the Dryseal American Standard Taper Pipe Thread (NPTF). At purchaser's option, the pipe thread on push-on fittings may be shortened in conformity with the SAE Short Dryseal Taper Pipe Thread (PTF-SAE Short). Specifications for pipe threads are given in detail in SAE J476.

The length of full form external thread shall not be shorter than L, plus one pitch (thread) for Dryseal NPTF and L, for Dryseal PTF-SAE Short, except that where thread is cut through into a relieved body or undercut on the fitting, the minimum full thread length may be reduced by one pitch (thread).

Where external pipe threads are produced by roll threading, the diameter of the unthreaded portion of shank adjacent to shoulder may be reduced to the E, basic diameter.

The tube fitting dimensions tabulated herein are based on length of the Dryseal American Standard Taper Pipe Thread (NPTF), it being the consensus of manufacturers and users that trouble-free assembly cannot be assured unless a full length is used.

External pipe threads shall be chamfered from the diameters shown in Table 1 to produce the specified length of chamfered or partial thread.

3.7 Material and Manufacture—Push on fittings shall be made from brass, steel, or multiple component braze design, as specified by the purchaser and, at manufacturer's option and in accordance with his process of manufacture, may be milled from the bar or forged. Brass shall be SAE UNS C36000 (half-hard), SAE UNS C34500, or SAE UNS C35000 for bar or extruded stock and SAE UNS C37700 for forgings.

21.90

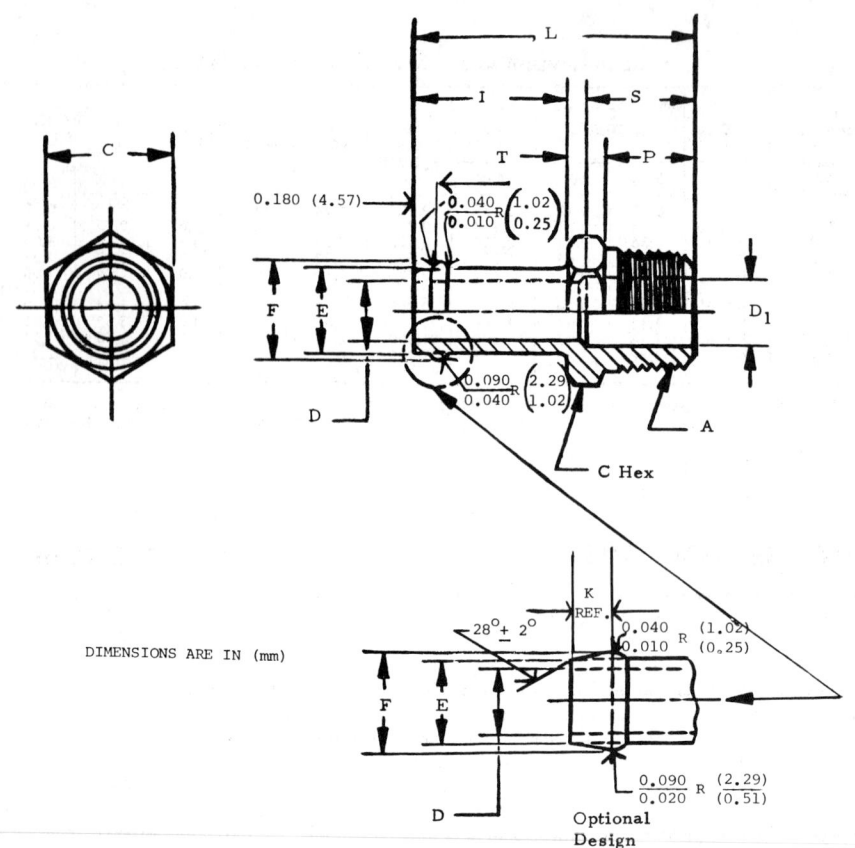

DIMENSIONS ARE IN (mm)

FIG. 1—CONNECTOR (430160)

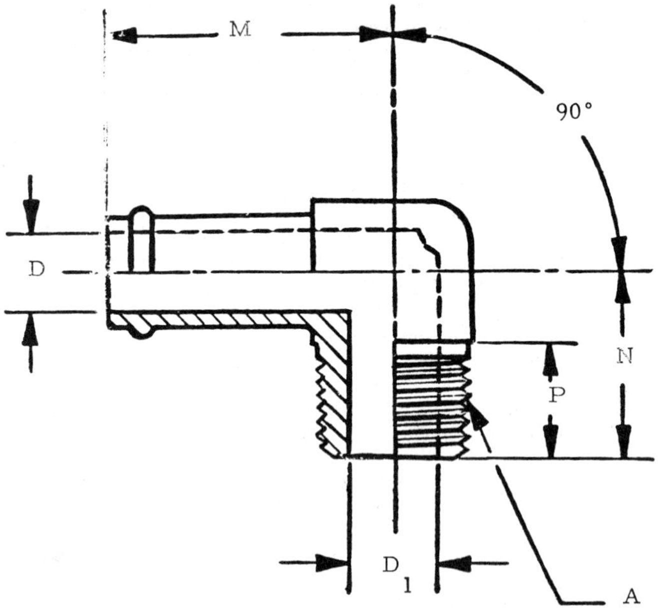

FIG. 2—90° ELBOW (430260)

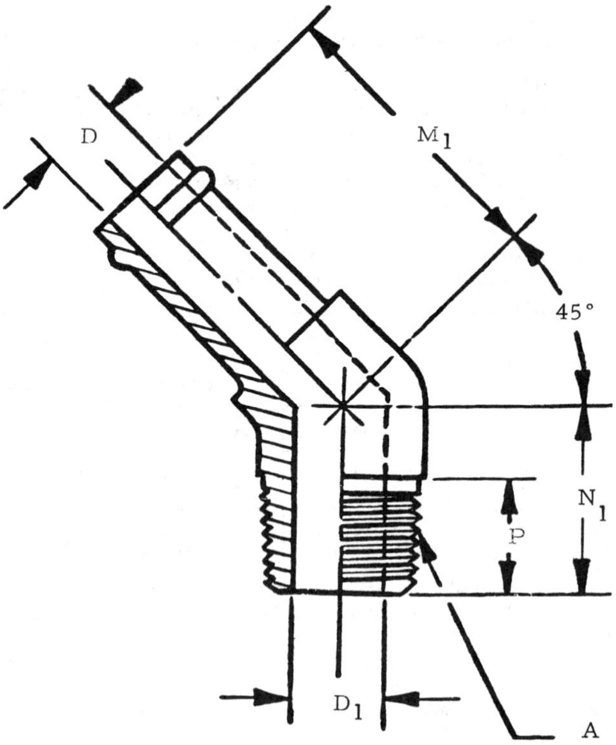

FIG. 3—45° ELBOW (430360)

NOTES: UNSPECIFIED DETAIL WITH RESPECT TO DIMENSIONS, TOLERANCES, CONTOURS, MATERIAL, WORKMANSHIP, ETC. MUST CONFORM TO GENERAL SPECIFICATIONS FOR HOSE PUSH-ON FITTINGS. THE DIMENSIONAL DESIGNATIONS ON THE FIRST FIGURE SHALL APLY TO ALL OTHER FIGURES. CODES SHOWN IN BRACKETS ADJACENT TO FIGURE NUMBERS REPRESENT RESPECTIVE FITTING IDENTIFICATION IN ACCORDANCE WITH SAE J846.

3.8 Finish—Unless otherwise specified by the purchaser, steel fittings shall be furnished cadmium or zinc plated to a thickness of 0.0002 in (0.005 mm) minimum. These parts must meet the requirements of a 32 h salt spray test in accordance with ASTM B 117. At manufacturer's option, plated fittings may be given a subsequent chromate treatment.

3.9 Workmanship—Workmanship shall conform to the best commercial practice to produce high quality fittings. Fittings shall be free from all hanging burrs, loose scale, and slivers which might become dislodged in usage and all other defects which might affect their serviceability. All sealing surfaces must be smooth except that annular tool marks up to 100 μin (2.5 μm) maximum shall be permissible. Fittings manufactured from multiple component braze design shall not leak under 125 psi (862 kPa) or vacuum test of 25 in (84 kPa) of Hg.

3.10 Assembly Considerations—Where it is not objectionable from a function or production standpoint, the use of compatible lubricant or sealant may be desirable in assembling Dryseal pipe threads on hose push-on fittings to minimize galling and effect a pressure-tight seal.

TABLE 1

Nominal Pipe Thread Size, in	External Thread							
	Chamfer Dia[a]				Length of Chamfered or Partial Thread			
	max		min		min		max	
	in	mm	in	mm	in	mm	in	mm
1/8	0.32	8.1	0.30	7.6	0.037	0.94	0.055	1.40
1/4	0.42	10.7	0.40	10.2	0.056	1.42	0.084	2.13
3/8	0.55	14.0	0.53	13.5	0.056	1.42	0.084	2.13
1/2	0.68	17.3	0.66	16.8	0.071	1.80	0.107	2.72
3/4	0.89	22.6	0.87	22.1	0.071	1.80	0.107	2.72
1	1.12	28.4	1.09	27.7	0.087	2.21	0.130	3.30

[a]Tabulated diameters conform with Appendix A, SAE J476.

TABLE 2—DIMENSIONS OF HOSE PUSH-ON CONNECTOR, 90 DEG ELBOW AND 45 DEG ELBOW (FIGS. 1–3)

Nom. Hose I.D. in	A Dryseal Taper Thread NPTF[a] in	C Nom in	D[c] Dia Drill		D₁[c] Dia Drill		E Dia ±0.005		F Dia ±0.005		I		K[e] Ref	
			in	mm	in	mm	in	mm	in	mm	in	mm	in	mm
3/16	1/8	7/16	0.125	3.18	0.219	5.56	0.188	4.78	0.220	5.59	1.110	28.19	0.080	2.03
1/4	1/8	7/16	0.188	4.78	0.219	5.56	0.250	6.35	0.290	7.37	1.130	28.70	0.100	2.54
1/4	1/4	9/16	0.188	4.78	0.344	8.74	0.250	6.35	0.290	7.37	1.130	28.70	0.100	2.54
5/16	1/8	7/16	0.250	6.35	0.219	5.56	0.312	7.92	0.360	9.14	1.130	28.70	0.100	2.54
5/16	1/4	9/16	0.250	6.35	0.344	8.74	0.312	7.92	0.360	9.14	1.130	28.70	0.100	2.54
3/8	1/8	1/2	0.297	7.54	0.219	5.56	0.375	9.52	0.430	10.92	1.130	28.70	0.100	2.54
3/8	1/4	9/16	0.297	7.54	0.344	8.74	0.375	9.52	0.430	10.92	1.130	28.70	0.100	2.54
3/8	3/8	11/16	0.297	7.54	0.406	10.31	0.375	9.52	0.430	10.92	1.130	28.70	0.100	2.54
1/2	1/4	5/8	0.406	10.31	0.344	8.74	0.500	12.70	0.560	14.22	1.160	29.46	0.120	3.05
1/2	3/8	11/16	0.406	10.31	0.406	10.31	0.500	12.70	0.560	14.22	1.160	29.46	0.120	3.05
1/2	1/2	7/8	0.406	10.31	0.562	14.27	0.500	12.70	0.560	14.22	1.160	29.46	0.120	3.05
1/2	3/4	1-1/16	0.406	10.31	0.750	19.05	0.500	12.70	0.560	14.22	1.160	29.46	0.120	3.05
5/8	3/8	3/4	0.500	12.70	0.406	10.31	0.625	15.88	0.690	17.53	1.160	29.46	0.120	3.05
5/8	1/2	7/8	0.500	12.70	0.562	14.27	0.625	15.88	0.690	17.53	1.160	29.46	0.120	3.05
5/8	3/4	1-1/16	0.500	12.70	0.750	19.05	0.625	15.88	0.690	17.53	1.160	29.46	0.120	3.05
3/4	3/8	15/16	0.625	15.88	0.406	10.31	0.750	19.05	0.820	20.83	1.160	29.46	0.120	3.05
3/4	1/2	7/8	0.625	15.88	0.562	14.27	0.750	19.05	0.820	20.83	1.160	29.46	0.120	3.05
3/4	3/4	1-1/16	0.625	15.88	0.750	19.05	0.750	19.05	0.820	20.83	1.160	29.46	0.120	3.05
1	1/2	1-1/8	0.844	21.44	0.562	14.27	1.000	25.40	1.060	26.92	1.160	29.46	0.120	3.05
1	3/4	1-1/8	0.844	21.44	0.750	19.05	1.000	25.40	1.060	26.92	1.160	29.46	0.120	3.05
1	1	1-3/8	0.844	21.44	0.937	23.80	1.000	25.40	1.060	26.92	1.160	29.46	0.120	3.05

L[h]		M		M₁		N[b]		N₁[b]		P[b]		S[b,c] max		T[d] Ref	
in ±0.03	mm ±0.8	in ±0.03	mm ±0.8	in ±0.03	mm ±0.8	in ±0.03	mm ±0.8	in ±0.03	mm ±0.8	in ±0.03	mm ±0.8	in	mm	in	mm
1.67	42.4	1.41	35.8	1.37	34.8	0.71	18.0	0.67	17.0	0.38	9.7	0.46	11.7	0.18	4.6
1.69	42.9	1.43	36.3	1.37	34.8	0.74	18.8	0.67	17.0	0.38	9.7	0.44	11.2	0.18	4.6
1.90	48.3	1.50	38.1	1.47	37.3	0.92	15.4	0.79	20.1	0.56	14.2	0.61	15.5	0.21	5.3
1.69	42.9	1.43	36.3	1.34	34.0	0.78	19.8	0.71	18.0	0.38	9.7	1.21	30.7	0.18	4.6
1.90	48.3	1.50	38.1	1.39	35.3	0.96	24.4	0.84	21.3	0.56	14.2	0.64	16.3	0.21	5.3
1.69	42.9	1.43	36.3	1.31	33.3	0.81	20.6	0.78	19.8	0.38	9.7	1.21	30.7	0.18	4.6
1.90	48.3	1.50	38.1	1.39	35.3	0.99	25.1	0.91	23.1	0.56	14.2	0.60	15.2	0.21	5.3
1.93	49.0	1.57	39.9	1.50	38.1	0.99	25.1	0.84	21.3	0.56	14.2	0.60	15.2	0.24	6.1
1.93	49.0	1.53	38.9	1.36	34.5	1.03	26.2	0.99	25.1	0.56	14.2	1.26	32.0	0.21	5.3
1.96	49.8	1.60	40.6	1.43	36.3	1.03	26.2	0.65	16.5	0.56	14.2	—	—	0.24	6.1
2.15	54.6	1.68	42.7	1.56	39.6	1.22	31.0	1.15	29.2	0.75	19.0	0.84	21.3	0.24	6.1
2.21	56.1	1.79	45.5	1.64	41.7	1.22	31.0	1.09	27.7	0.75	19.0	0.85	21.1	0.30	7.6
1.96	49.8	1.60	40.6	1.38	35.1	1.09	27.7	1.03	26.2	0.56	14.2	1.40	35.6	0.24	6.1
2.15	54.6	1.68	42.7	1.49	37.8	1.28	32.5	1.26	32.0	0.75	19.0	0.88	22.4	0.24	6.1
2.21	56.1	1.79	45.5	1.58	40.1	1.28	32.5	1.19	30.2	0.75	19.0	0.87	22.1	0.30	7.6
1.96	49.8	1.60	40.6	1.30	33.0	1.15	29.2	1.13	28.7	0.56	14.2	1.34	34.0	0.24	6.1
2.15	54.6	1.68	42.7	1.41	35.8	1.34	34.0	1.33	33.8	0.75	19.0	1.28	32.5	0.24	6.1
2.21	56.1	1.79	45.5	1.49	37.8	1.34	34.0	1.28	32.5	0.75	19.0	0.90	22.9	0.30	7.6
2.15	54.6	1.68	42.7	1.26	32.0	1.47	37.3	1.59	40.4	0.75	19.0	1.22	31.0	0.24	6.1
2.21	56.1	1.79	45.5	1.33	33.8	1.47	37.3	1.53	38.9	0.75	19.0	1.33	33.8	0.30	7.6
2.43	61.7	1.82	46.2	1.55	39.4	1.66	42.2	1.56	39.6	0.94	23.9	1.14	29.0	0.33	8.4

[a]Dryseal American Standard Pipe Thread. See General Specifications.
[b]Where SAE Short Pipe Thread is authorized by purchaser, dimensions L, N, N₁, and S are reduced in accordance with the reduction of pipe thread length. See SAE J476 Tables 3 and 4.
[c]At manufacturer's option thru passages in fitting shown in Fig. 1 may conform with the smaller diameter specified or be counterbored to the larger diameter for the appropriate end for depth S.
[d]Minimum design thickness, not subject to inspection.
[e]For reference purposes only, not intended for inspection.

HYDRAULIC HOSE—SAE J517 JUN80

Report of the Construction and Industrial Machinery Technical Committee, Nonmetallic Materials Committee, and Tube, Pipe, Hose, and Lubrication Fittings Committee, approved January 1952, last revised by the Fluid Conductors and Connectors Technical Committee June 1980.

GENERAL SPECIFICATIONS

Scope—This standard covers general, dimensional, and performance specifications for several varieties of hose intended for general application in hydraulic systems on construction and industrial equipment and commercial products.

These general specifications are common to and shall supplement the detailed specifications for the SAE 100R1, SAE 100R2, SAE 100R3, SAE 100R4, SAE 100R5, SAE 100R6, SAE 100R7, SAE 100R8, SAE 100R9, SAE 100R10, SAE 100R11, and SAE 100R12 hydraulic hoses set forth in the following sections.

Hose Application—Hydraulic hose has a finite life and factors which will reduce hose life are:
 (1) Flexing the hose to less than the specified minimum bend radius.
 (2) Twisting, pulling, kinking, crushing, or abrading the hose.
 (3) Operating above or below the hose operating temperature range.
 (4) Exposing the hose to surge pressures above the maximum operating pressure.

Surge, a rapid and transient rise in pressure, will not be indicated on many common pressure gages and can best be measured on electronic measuring instruments with a high-frequency response.

Surge pressures higher than the maximum operating pressure will shorten hose life and should be taken into account by the hydraulic designer.

Size Designations—Hose sizes are normally designated by the nominal hose inside diameter expressed in fractions of inches. The hose size may also be designated by a dash number which, except for SAE 100R5 hose, represents the number of sixteenth inch increments in the hose inside diameter. In the case of the SAE 100R5 hose, the dash number represents the number of sixteenth inch increments in the outside diameter of tubing having approximately the same inside diameter as that of SAE 100R5 hose. See dimensional tables for the respective hoses.

Hose Identification—The entire length of hoses shall be legibly marked, parallel to the longitudinal axis, with a stripe or stripes showing the respective SAE hose specification number (including type designation where applicable) and the fractional nominal hose inside diameter size repeated at intervals of not less than 18.0 in. (460 mm). The hose dash size number designation may also be included at manufacturer's option. Additionally, a colored yarn shall be incorporated into the wall of the hose identifying the manufacturer. The color shall be as designated by the Rubber Manufacturers Association.

Fittings—Fittings for use with the SAE 100R1 through SAE 100R6 hydraulic hoses are covered in SAE J516.

Tests—Unless otherwise agreed upon between the manufacturer and purchaser, tests for evaluating conformance of product with specifications shall be on the basis of Qualification Tests and Inspection Tests set forth in this standard. Tests may be conducted by the manufacturer, the purchaser, or both, as decreed by the purchaser. The tests, sampling, and criteria applicable to both test classifications are given in the detailed specifications and tables for each hose. All tests shall be conducted in accordance with the procedures in SAE J343.

Retests and Rejection—In the event of failure of one or more samples to meet any of the tests specified, the material shall be resampled and retested. Twice the number of specimens designated under initial test procedure shall be selected from the lot in question for such retests, and failure of any of the retested samples shall be cause for rejection.

SUMMARY OF SAE J517 100R-SERIES HOSE MAXIMUM OPERATING PRESSURE PSI/MPa

Nominal Hose I.D. Size In	100R1	100R2	100R3	100R4	100R5	100R6	100R7	100R8	100R9	100R10	100R11	100R12
3/16	3000/20.7	5000/34.5	1500/10.3		3000/20.7	500/3.4	3000/20.7	5000/34.5		10 000/68.9	12 500/86.2	
1/4	2750/19.0	5000/34.5	1250/8.6		3000/20.7	400/2.8	2750/19.0	5000/34.5		8750/60.3	11 250/77.6	
5/16	2500/17.2	4250/29.3	1200/8.3		2250/15.5	400/2.8	2500/17.2					
3/8	2250/15.5	4000/27.6	1125/7.8			400/2.8	2250/15.5	4000/27.6	4500/31.0	7500/51.7	10 000/68.9	4000/27.6
13/32	2250/15.5				2000/13.8							
1/2	2000/13.8	3500/24.1	1000/6.9		1750/12.1	400/2.8	2000/13.8	3500/24.1	4000/27.6	6250/43.1	7500/51.7	4000/27.6
5/8	1500/10.3	2750/19.0	875/6.0		1500/10.3	350/2.4	1500/10.3	2750/19.0				
3/4	1250/8.6	2250/15.5	750/5.2	300/2.1			1250/8.6	2250/15.5	3000/20.7	5000/34.5	6250/43.1	4000/27.6
7/8	1125/7.8	2000/13.8			800/5.5							
1	1000/6.9	2000/13.8	565/3.9	250/1.7			1000/6.9	2000/13.8	3000/20.7	4000/27.6	5000/34.5	4000/27.6
1 1/8					625/4.3							
1 1/4	625/4.3	1625/11.2	375/2.6	200/1.4					2500/17.2	3000/20.7	3500/24.1	3000/20.7
1 3/8					500/3.4							
1 1/2	500/3.4	1250/8.6		150/1.0					2000/13.8	2500/17.2	3000/20.7	2500/17.2
1 13/16					350/2.4							
2	375/2.6	1125/7.8		100/0.7					2000/13.8	2500/17.2	3000/20.7	2500/17.2
2 1/2				62/0.4								
3				56/0.4								

Minimum burst of 100R hoses is 4 times operating pressure

TABLE 1—DIMENSIONS AND SPECIFICATIONS FOR SAE 100R1 HOSE

Nominal SAE 100R1 Hose ID Size, in	Hose Dash Size	Hose ID						Reinforcement Dia				Hose OD Type A			
		Basic		Tolerance				Max		Min		Max		Min	
				Plus		Minus									
		in	mm	in	mm	in	mm	in	mm	in	mm	in	mm	in	mm
3/16	-3	0.188	4.8	0.023	0.6	0.008	0.2	0.398	10.1	0.352	8.9	0.531	13.5	0.469	11.9
1/4	-4	0.250	6.4	0.023	0.6	0.008	0.2	0.461	11.7	0.416	10.6	0.656	16.7	0.594	15.1
5/16	-5	0.312	7.9	0.023	0.6	0.008	0.2	0.523	13.3	0.477	12.1	0.719	18.3	0.656	16.7
3/8	-6	0.375	9.5	0.023	0.6	0.008	0.2	0.617	15.7	0.571	14.5	0.812	20.6	0.750	19.0
13/32	-6.5	0.406	10.3	0.031	0.8	0.015	0.4	0.648	16.4	0.602	15.3	0.844	21.4	0.781	19.8
1/2	-8	0.500	12.7	0.031	0.8	0.015	0.4	0.750	19.0	0.688	17.5	0.938	23.8	0.875	22.2
5/8	-10	0.625	15.9	0.031	0.8	0.015	0.4	0.875	22.2	0.812	20.6	1.062	27.0	1.000	25.4
3/4	-12	0.750	19.0	0.031	0.8	0.015	0.4	1.031	26.2	0.969	24.6	1.219	31.0	1.156	29.4
7/8	-14	0.875	22.2	0.031	0.8	0.015	0.4	1.156	29.4	1.094	27.8	1.344	34.1	1.281	32.5
1	-16	1.000	25.4	0.040	1.0	0.015	0.4	1.344	34.1	1.281	32.5	1.547	39.3	1.453	36.9
1-1/4	-20	1.250	31.8	0.047	1.2	0.015	0.4	1.641	41.7	1.547	39.3	1.875	47.6	1.750	44.4
1-1/2	-24	1.500	38.1	0.047	1.2	0.015	0.4	1.891	48.0	1.797	45.6	2.125	54.0	2.000	50.8
2	-32	2.000	50.8	0.047	1.2	0.015	0.4	2.438	61.9	2.312	58.7	2.688	68.3	2.562	65.1

Nominal SAE 100R1 Hose ID Size, in	Hose OD Type AT		Cover Thickness[a] Type AT				Min Burst Pressure		Proof Pressure		Max Operating Pressure		Min Bend Radius[b]	
	Max		Max		Min									
	in	mm	in	mm	in	mm	psi	MPa	psi	MPa	psi	MPa	in	mm
3/16	0.494	12.5	0.060	1.52	0.030	0.76	12000	82.7	6000	41.4	3000	20.7	3.50	89
1/4	0.557	14.1	0.060	1.52	0.030	0.76	11000	75.8	5500	37.9	2750	19.0	4.00	102
5/16	0.619	15.7	0.060	1.52	0.030	0.76	10000	68.9	5000	34.5	2500	17.2	4.50	114
3/8	0.713	18.1	0.060	1.52	0.030	0.76	9000	62.0	4500	31.0	2250	15.5	5.00	127
13/32	0.744	18.9	0.060	1.52	0.030	0.76	9000	62.0	4500	31.0	2250	15.5	5.50	140
1/2	0.846	21.5	0.060	1.52	0.030	0.76	8000	55.2	4000	27.6	2000	13.8	7.00	178
5/8	0.971	24.7	0.060	1.52	0.030	0.76	6000	41.4	3000	20.7	1500	10.3	8.00	203
3/4	1.127	28.6	0.060	1.52	0.030	0.76	5000	34.5	2500	17.2	1250	8.6	9.50	241
7/8	1.252	31.8	0.060	1.52	0.030	0.76	4500	31.0	2250	15.5	1125	7.8	11.00	279
1	1.440	36.6	0.060	1.52	0.030	0.76	4000	27.6	2000	13.8	1000	6.9	12.00	305
1-1/4	1.766	44.8	0.080	2.03	0.040	1.02	2500	17.2	1250	8.6	625	4.3	16.50	419
1-1/2	2.047	52.0	0.100	2.54	0.050	1.27	2000	13.8	1000	6.9	500	3.4	20.00	508
2	2.594	65.9	0.100	2.54	0.050	1.27	1500	10.3	750	5.2	375	2.6	25.00	635

[a] Cover thickness shall be measured by means of a dial indicator depth gage having a rounded foot placed parallel to the hose, bridging a groove obtained by stripping a 0.50-1.00 in (12.5-25.4 mm) width of cover from the hose. A mandrel should be placed in the hose bore to insure freedom from misalignment.

[b] Bend radius measured at inside of bend.

STEEL WIRE REINFORCED, RUBBER COVERED HYDRAULIC HOSE (SAE 100R1)

Scope—This specification covers hose for use with petroleum and water base hydraulic fluids within a temperature range of −40 to +200°F (−40 to +93°C). Operating temperatures in excess of +200°F (+93°C) may materially reduce the life of the hose. Maximum operating pressure, minimum bend radius, and other performance data are specified in Table 1.

It should be noted that the detailed specifications which follow shall be supplemented by the general specifications given at the beginning of this standard.

Nominal Hose ID, in	Concentricity, FIR			
	ID to OD		ID to Reinforcement	
	in	mm	in	mm
Up to 1/4, incl	0.030	0.8	0.017	0.4
Over 1/4 to 7/8, incl	0.040	1.0	0.024	0.6
Over 7/8	0.050	1.3	0.031	0.8

Dimensions—Dimensions and tolerances applicable to this hose are given in Table 1. The inside diameter of hose shall be concentric with outside diameter of hose and the outer surface of the reinforcement within the following limits:

Hose Construction

Type A—This hose shall consist of an inner tube of oil resistant synthetic rubber, a single wire braid reinforcement, and an oil and weather resistant synthetic rubber cover. A ply or braid of suitable material may be used over the inner tube and/or over the wire reinforcement to anchor the synthetic rubber to the wire.

Type AT—This hose shall be of the same construction as Type A, except having a cover designed to assemble with fittings which do not require removal of the cover or a portion thereof.

Qualification Tests—For qualification to this specification, hose and/or hose assemblies made therefrom shall conform to the following tests and requirements:

1. **Dimensional Check Test** (all samples)—Shall conform to dimensions in Table 1 and these detailed specifications.
2. **Proof Test** (all samples)—Shall not leak at the proof pressure.
3. **Change in Length Test** (one sample)—Shall not exceed +2% to −4% change when pressurized to operating pressure.
4. **Burst Test** (one 18.00 in. (460 mm) assembly)—Shall not leak or fail below the minimum burst pressure.
5. **Leakage Test** (two 12.00 in. (300 mm) assemblies)—Shall not leak or fail.
6. **Cold Bend Test** (one assembly)—Shall exhibit no cover cracks or leakage. Exposure shall be at −40°F (−40°C).
7. **Oil Resistance Test**—After 70 h immersion at 212°F (100°C) in

ASTM No. 3 oil, the volume change of hose inner tube and cover specimens shall be between 0% and +100%.

8. Ozone Resistance Test (two samples)—After 70 h exposure in an atmosphere comprised of 50 parts ozone per 100 million parts of air at an ambient temperature of 100°F (38°C), specimens shall not show evidence of cracking or deterioration when viewed with seven-power magnification while still in a stressed condition.

9. Impulse Test (four unaged assemblies)—Hose assemblies, when tested at 125% of operating pressure for hose sizes 1 in. (25.4 mm) nominal ID and smaller and 100% of operating pressure for hose sizes 1¼ in. (31.8 mm) nominal ID and larger, with 200°F (93°C) circulating petroleum base test fluid, shall withstand a minimum of 150,000 cycles without leakage or other malfunction.

10. Visual Examination (all samples).

Inspection Tests—Inspection tests listed below shall be performed on two samples representing each lot of 500–10,000 ft (150–3000 m) of bulk hose or 100–10,000 assemblies. Lots of less than 500 ft (150 m) of hose or 100 assemblies need not be subjected to these tests if a lot has been tested and met the requirements within the previous 12 month period. Requirements shall be same as for corresponding Qualification Tests:
1. Dimensional Check Test.
2. Proof Test.
3. Change in Length Test.
4. Burst Test.

In addition all hose and/or hose assemblies made therefrom shall be subjected to visual examination.

HIGH PRESSURE, STEEL WIRE REINFORCED, RUBBER COVERED HYDRAULIC HOSE (SAE 100R2)

Scope—This specification covers hose for use with petroleum and water base hydraulic fluids within a temperature range of −40 to +200°F (−40 to +93°C). Operating temperatures in excess of +200°F (+93°C) may materially reduce the life of the hose. Maximum operating pressure, minimum bend radius, and other performance data are specified in Table 2.

It should be noted that the detailed specifications which follow shall be supplemented by the general specifications given at the beginning of this standard.

Dimensions—Dimensions and tolerances applicable to this hose are given in Table 2. The inside diameter of hose shall be concentric with outside diameter of hose and the outer surface of the reinforcement within the following limits:

Nominal Hose ID, in	Concentricity, FIR			
	ID to OD		ID to Reinforcement	
	in	mm	in	mm
Up to 1/4, incl	0.030	0.8	0.021	0.5
Over 1/4 to 7/8, incl	0.040	1.0	0.028	0.7
Over 7/8	0.050	1.3	0.035	0.9

TABLE 2—DIMENSIONS AND SPECIFICATIONS FOR SAE 100R2 HOSE

Nominal SAE 100R2 Hose ID Size, in	Hose Dash Size	Hose ID Basic		Hose ID Tolerance Plus		Hose ID Tolerance Minus		Reinforcement Dia Max		Reinforcement Dia Min		Hose OD Types A and B Max		Hose OD Types A and B Min	
		in	mm	in	mm	in	mm	in	mm	in	mm	in	mm	in	mm
3/16	-3	0.188	4.8	0.023	0.6	0.008	0.2	0.461	11.7	0.416	10.6	0.656	16.7	0.594	15.1
1/4	-4	0.250	6.4	0.023	0.6	0.008	0.2	0.523	13.3	0.477	12.1	0.719	18.3	0.656	16.7
5/16	-5	0.312	7.9	0.023	0.6	0.008	0.2	0.586	14.9	0.539	13.7	0.781	19.8	0.719	18.3
3/8	-6	0.375	9.5	0.023	0.6	0.008	0.2	0.681	17.3	0.633	16.1	0.875	22.2	0.812	20.6
1/2	-8	0.500	12.7	0.031	0.8	0.015	0.4	0.812	20.6	0.750	19.0	1.000	25.4	0.938	23.8
5/8	-10	0.625	15.9	0.031	0.8	0.015	0.4	0.938	23.8	0.875	22.2	1.125	28.6	1.062	27.0
3/4	-12	0.750	19.0	0.031	0.8	0.015	0.4	1.094	27.8	1.031	26.2	1.281	32.5	1.219	31.0
7/8	-14	0.875	22.2	0.031	0.8	0.015	0.4	1.219	31.0	1.156	29.4	1.406	35.7	1.344	34.1
1	-16	1.000	25.4	0.040	1.0	0.015	0.4	1.406	35.7	1.344	34.1	1.609	40.9	1.516	38.5
1-1/4	-20	1.250	31.8	0.047	1.2	0.015	0.4	1.797	45.6	1.703	43.2	2.062	52.4	1.938	49.2
1-1/2	-24	1.500	38.1	0.047	1.2	0.015	0.4	2.047	52.0	1.953	49.6	2.312	58.7	2.188	55.6
2	-32	2.000	50.8	0.047	1.2	0.015	0.4	2.547	64.7	2.453	62.3	2.812	71.4	2.688	68.3

Nominal SAE 100R2 Hose ID Size, in	Hose OD Types AT and BT Max		Cover Thickness[a] Types AB and BT Max		Cover Thickness[a] Types AB and BT Min		Min Burst Pressure		Proof Pressure		Max Operating Pressure		Min Bend Radius[b]	
	in	mm	in	mm	in	mm	psi	MPa	psi	MPa	psi	MPa	in	mm
3/16	0.557	14.1	0.060	1.52	0.031	0.79	20000	137.9	10000	68.9	5000	34.5	3.50	89
1/4	0.619	15.7	0.060	1.52	0.031	0.79	20000	137.9	10000	68.9	5000	34.5	4.00	102
5/16	0.682	17.3	0.060	1.52	0.031	0.79	17000	117.2	8500	58.6	4250	29.3	4.50	114
3/8	0.777	19.7	0.060	1.52	0.031	0.79	16000	110.3	8000	55.2	4000	27.6	5.00	127
1/2	0.908	23.1	0.060	1.52	0.031	0.79	14000	96.5	7000	48.3	3500	24.1	7.00	178
5/8	1.034	26.3	0.060	1.52	0.031	0.79	11000	75.8	5500	37.9	2750	19.0	8.00	203
3/4	1.190	30.2	0.060	1.52	0.031	0.79	9000	62.0	4500	31.0	2250	15.5	9.50	241
7/8	1.315	33.4	0.060	1.52	0.031	0.79	8000	55.2	4000	27.6	2000	13.8	11.00	279
1	1.531	38.9	0.085	2.16	0.042	1.07	8000	55.2	4000	27.6	2000	13.8	12.00	305
1-1/4	1.953	49.6	0.100	2.54	0.050	1.27	6500	44.8	3250	22.4	1625	11.2	16.50	419
1-1/2	2.203	56.0	0.100	2.54	0.050	1.27	5000	34.5	2500	17.2	1250	8.6	20.00	508
2	2.703	68.6	0.100	2.54	0.050	1.27	4500	31.0	2250	15.5	1125	7.8	25.00	635

[a] Cover thickness shall be measured by means of a dial indicator depth gage having a rounded foot placed parallel to the hose, bridging a groove obtained by stripping a 0.50-1.00 in (12.5-25.4 mm) width of cover from the hose. A mandrel should be placed in the hose bore to insure freedom from misalignment.
[b] Bend radius measured at inside of bend.

Hose Construction—The hose shall consist of an inner tube of oil resistant synthetic rubber, steel wire reinforcement according to hose type as detailed below, and an oil and weather resistant synthetic rubber cover. A ply or braid of suitable material may be used over the inner tube and/or over the wire reinforcement to anchor the synthetic rubber to the wire.

Type A—This hose shall have two braids of wire reinforcement.

Type B—This hose shall have two spiral plies and one braid of wire reinforcement.

Type AT—This hose shall be of the same construction as Type A, except having a cover designed to assemble with fittings which do not require removal of the cover or a portion thereof.

Type BT—This hose shall be of the same construction as Type B, except having a cover designed to assemble with fittings which do not require removal of the cover or a portion thereof.

Qualification Tests—For qualification to this specification, hose and/or hose assemblies made therefrom shall conform to the following tests and requirements:

1. **Dimensional Check Test** (all samples)—Shall conform to dimensions in Table 2 and these detailed specifications.
2. **Proof Test** (all samples)—Shall not leak at the proof pressure.
3. **Change in Length Test** (one sample)—Shall not exceed +2% to −4% change when pressurized to operating pressure.
4. **Burst Test** (one 18.00 in. (460 mm) assembly)—Shall not leak or fail below the minimum burst pressure.
5. **Leakage Test** (two 12.00 in. (300 mm) assemblies)—Shall not leak or fail.
6. **Cold Bend Test** (one assembly)—Shall exhibit no cover cracks or leakage. Exposure shall be at −40°F (−40°C).
7. **Oil Resistance Test**—After 70 h immersion at 212°F (100°C) in ASTM No. 3 oil, the volume change of hose inner tube and cover specimens shall be between 0% and +100%.
8. **Ozone Resistance Test** (two samples)—After 70 h exposure in an atmosphere comprised of 50 parts ozone per 100 million parts of air at an ambient temperature of 100°F (38°C), specimens shall not show evidence of cracking or deterioration when viewed with seven-power magnification while still in a stressed condition.
9. **Impulse Test** (four unaged assemblies)—Hose assemblies, when tested at 133% of operating pressure with 200°F (93°C) circulating petroleum base test fluid, shall withstand a minimum of 200,000 cycles without leakage or other malfunction.
10. **Visual Examination** (all samples).

Inspection Tests—Inspection tests listed below shall be performed on two samples representing each lot of 500–10,000 ft (150–300 m) of bulk hose or 100–10,000 assemblies. Lots of less than 500 ft (150 m) of hose or 100 assemblies need not be subjected to these tests if a lot has been tested and met the requirements within the previous 12 month period. Requirements shall be same as for corresponding Qualification Tests:

1. Dimensional Check Test.
2. Proof Test.
3. Change In Length Test.
4. Burst Test.

In addition all hose and/or hose assemblies made therefrom shall be subjected to visual examination.

DOUBLE FIBER BRAID (NONMETALLIC), RUBBER COVERED HYDRAULIC HOSE (SAE 100R3)

Scope—This specification covers hose for use with petroleum and water base hydraulic fluids within a temperature range of −40 to +200°F (−40 to +93°C). Operating temperatures in excess of +200°F (+93°C) may materially reduce the life of the hose. Maximum operating pressure, minimum bend radius, and other performance data are specified in Table 3.

It should be noted that the detailed specifications which follow shall be supplemented by the general specifications given at the beginning of this standard.

TABLE 3—DIMENSIONS AND SPECIFICATIONS FOR SAE 100R3 HOSE

Nominal SAE 100R3 Hose ID Size, in	Hose Dash Size	Hose ID						Hose OD			
		Basic		Tolerance				Max		Min	
				Plus		Minus					
		in	mm	in	mm	in	mm	in	mm	in	mm
3/16	-3	0.188	4.8	0.025	0.6	0.010	0.3	0.531	13.5	0.469	11.9
1/4	-4	0.250	6.4	0.025	0.6	0.010	0.3	0.594	15.1	0.531	13.5
5/16	-5	0.312	7.9	0.025	0.6	0.010	0.3	0.719	18.3	0.656	16.7
3/8	-6	0.375	9.5	0.025	0.6	0.010	0.3	0.781	19.8	0.719	18.3
1/2	-8	0.500	12.7	0.030	0.8	0.010	0.3	0.969	24.6	0.906	23.0
5/8	-10	0.625	15.9	0.030	0.8	0.010	0.3	1.094	27.8	1.031	26.2
3/4	-12	0.750	19.0	0.030	0.8	0.010	0.3	1.281	32.5	1.219	31.0
1	-16	1.000	25.4	0.030	0.8	0.010	0.3	1.547	39.3	1.453	36.9
1-1/4	-20	1.250	31.8	0.045	1.1	0.015	0.4	1.812	46.0	1.688	42.9

Nominal SAE 100R3 Hose ID Size, in	Min Burst Pressure		Proof Pressure		Max Operating Pressure		Min Bend Radius[a]	
	psi	MPa	psi	MPa	psi	MPa	in	mm
3/16	6000	41.4	3000	20.7	1500	10.3	3.00	76
1/4	5000	34.5	2500	17.2	1250	8.6	3.00	76
5/16	4800	33.1	2400	16.5	1200	8.3	4.00	102
3/8	4500	31.0	2250	15.5	1125	7.8	4.00	102
1/2	4000	27.6	2000	13.7	1000	6.9	5.00	127
5/8	3500	24.1	1750	12.1	875	6.0	5.50	140
3/4	3000	20.7	1500	10.3	750	5.2	6.00	152
1	2250	15.5	1125	7.8	565	3.9	8.00	203
1-1/4	1500	10.3	750	5.2	375	2.6	10.00	254

[a] Bend radius measured at inside of bend.

Dimensions—Dimensions and tolerances applicable to this hose are given in Table 3. The inside diameter and outside diameter of the hose shall be concentric within the following limits:

Nominal Hose ID, in	Concentricity, FIR	
	in	mm
Up to 1/4, incl	0.030	0.8
Over 1/4 to 3/4, incl	0.040	1.0
Over 3/4	0.050	1.3

Hose Construction—The hose shall consist of an inner tube of oil resistant synthetic rubber, two braids of suitable textile yarn, and an oil and weather resistant synthetic rubber cover.

Qualification Tests—For qualification to this specification, hose and/or hose assemblies made therefrom shall conform to the following tests and requirements:

1. Dimensional Check Test (all samples)—Shall conform to dimensions in Table 3 and these detailed specifications.

2. Proof Test (all samples)—Shall not leak at the proof pressure.

3. Change in Length Test (one sample)—Shall not exceed +2-4% change when pressurized to operating pressure.

4. Burst Test (one 18.00 in. (460 mm) assembly)—Shall not leak or fail below the minimum burst pressure.

5. Leakage Test (two 12.00 in. (300 mm) assemblies)—Shall not leak or fail.

6. Cold Bend Test (one assembly)—Shall exhibit no cover cracks or leakage. Exposure shall be at −40°F (−40°C).

7. Oil Resistance Test—After 70 h immersion at 212°F (100°C) in ASTM No. 3 oil, the volume change of hose inner tube and cover specimens shall be between 0% and +100%.

8. Ozone Resistance Test (two samples)—After 70 h exposure in an atmosphere comprised of 50 parts ozone per 100 million parts of air at an ambient temperature of 100°F (38°C), specimens shall not show evidence of cracking or deterioration when viewed with seven-power magnification while still in a stressed condition.

9. Impulse Test (four unaged assemblies)—Hose assemblies, when tested at 133% of operating pressure, with 200°F (93°C) circulating petroleum base test fluid, shall withstand a minimum of 200,000 cycles without leakage or other malfunction.

10. Visual Examination (all samples).

Inspection Tests—Inspection tests listed below shall be performed on two samples representing each lot of 500–10,000 ft (150–3000 m) of bulk hose or 100–10,000 assemblies. Lots of less than 500 ft (150 m) of hose or 100 assemblies need not be subjected to these tests if a lot has been tested and met the requirements within the previous 12 month period. Requirements shall be same as for corresponding Qualification Tests:

1. Dimensional Check Test.
2. Proof Test.
3. Change In Length Test.
4. Burst Test.

In addition all hose and/or hose assemblies made therefrom shall be subjected to visual examination.

WIRE INSERTED HYDRAULIC SUCTION HOSE (SAE 100R4)

Scope—This specification covers hose for use in low pressure and vacuum applications with petroleum and water base hydraulic fluids within a temperature range of −40 to +200°F (−40 to +93°C). Operating temperatures in excess of 200°F (93°C) may materially reduce the life of the hose. Maximum operating pressure, minimum bend radius, and other performance data are specified in Table 4.

It should be noted that the detailed specifications which follow shall be supplemented by the general specifications given at the beginning of this standard.

Dimensions—Dimensions and tolerances applicable to this hose are given in Table 4.

Hose Construction—The hose shall consist of an inner tube of oil resistant synthetic rubber, a reinforcement consisting of a ply or plies of woven or braided textile fibers with a suitable spiral of body wire, and an oil and weather resistant synthetic rubber cover.

Qualification Tests—For qualification to this specification, hose and/or hose assemblies made therefrom shall conform to the following tests and requirements:

1. Dimensional Check Test (all samples)—Shall conform to dimensions in Table 4.

2. Proof Test (all samples)—Shall not leak at the proof pressure.

3. Burst Test (one 18.00 in. (460 mm) assembly)—Shall not leak or fail below the minimum burst pressure.

4. Cold Bend Test (one assembly)—Shall exhibit no cover cracks or leakage. Exposure shall be at −40°F (−40°C).

5. Oil Resistance Test—After 70 h immersion at 212°F (100°C) in ASTM No. 3 oil, the volume change of hose inner tube and cover specimens shall be between 0% and +100%.

6. Ozone Resistance Test (two samples)—After 70 h exposure in an atmosphere comprised of 50 parts ozone per 100 million parts of air at an ambient temperature of 100°F (38°C), specimens shall not show evidence of cracking or deterioration when viewed with seven-power magnification while still in a stressed condition.

7. Resistance to Vacuum Test (one sample)—After exposure for 5 min at 25 in. Hg (absolute pressure of 17 kPa), there shall be no evidence of hose blistering or collapse.

8. Visual Examinations (all samples).

Inspection Tests—Inspection tests listed below shall be performed on two samples representing each lot of 500–10,000 ft (150–3000 m) of bulk hose or 100–10,000 assemblies. Lots of less than 500 ft (150 m) of hose or 100 assemblies need not be subjected to these tests if a lot has been tested and met the requirements within the previous 12 month period. Requirements shall be same as for corresponding Qualification Tests:

1. Dimensional Check Test.
2. Proof Test.
3. Vacuum Test.
4. Burst Test.

In addition all hose and/or hose assemblies made therefrom shall be subjected to visual examination.

TABLE 4—DIMENSIONS AND SPECIFICATIONS FOR SAE 100R4 HOSE

Nominal SAE 100R4 Hose ID Size, in	Hose Dash Size	Hose ID						Hose OD Max		Min Burst Pressure		Proof Pressure		Max Operating Pressure		Min Bend Radius[a]	
		Basic		Tolerance													
				Plus		Minus											
		in	mm	in	mm	in	mm	in	mm	psi	MPa	psi	MPa	psi	MPa	in	mm
3/4	-12	0.750	19.0	0.031	0.8	0.031	0.8	1.375	34.9	1200	8.3	600	4.1	300	2.1	5.00	127
1	-16	1.000	25.4	0.031	0.8	0.031	0.8	1.625	41.3	1000	6.9	500	3.4	250	1.7	6.00	152
1-1/4	-20	1.250	31.8	0.047	1.2	0.047	1.2	2.000	50.8	800	5.5	400	2.8	200	1.4	8.00	203
1-1/2	-24	1.500	38.1	0.047	1.2	0.047	1.2	2.250	57.2	600	4.1	300	2.1	150	1.0	10.00	254
2	-32	2.000	50.8	0.062	1.6	0.062	1.6	2.750	69.9	400	2.8	200	1.4	100	0.7	12.00	305
2-1/2	-40	2.500	63.5	0.062	1.6	0.062	1.6	3.250	82.6	250	1.7	125	0.9	62	0.4	14.00	356
3	-48	3.000	76.2	0.062	1.6	0.062	1.6	3.750	95.3	225	1.5	112	0.8	56	0.4	18.00	457
3-1/2	-56	3.500	88.9	0.062	1.6	0.062	1.6	4.250	107.9	180	1.2	90	0.6	45	0.3	21.00	533
4	-64	4.000	101.6	0.062	1.6	0.062	1.6	4.750	120.7	140	1.0	70	0.5	35	0.2	24.00	610

[a] Bend radius measured at inside of bend.

SINGLE WIRE BRAID, TEXTILE COVERED HYDRAULIC HOSE (SAE 100R5)

Scope—This specification covers hose for use with petroleum and water base hydraulic fluids within a temperature range of −40 to +200°F (−40 to +93°C). Operating temperatures in excess of +200°F (+93°C) may materially reduce the life of the hose. Maximum operating pressure, minimum bend radius, and other performance data are specified in Table 5.

It should be noted that the detailed specifications which follow shall be supplemented by the general specifications given at the beginning of this standard.

Dimensions—Dimensions and tolerances applicable to this hose are given in Table 5. The inside diameter and outside diameter of the hose shall be concentric within the following limits:

Nominal Hose ID, in	Concentricity, FIR	
	in	mm
Up to 13/32, incl	0.020	0.6
Over 13/32	0.030	0.8

Hose Construction—The hose shall consist of an inner tube of oil resistant synthetic rubber and two textile braids separated by a high tensile steel wire braid. All braids are to be impregnated with an oil and mildew resistant synthetic rubber compound.

Qualification Tests—For qualification to this specification, hose and/or hose assemblies made therefrom shall conform to the following tests and requirements:

1. **Dimensional Check Test** (all samples)—Shall conform to dimensions in Table 5 and these detailed specifications.
2. **Proof Test** (all samples—Shall not leak at the proof pressure.
3. **Change in Length Test** (one sample)—Shall not exceed +2% to −4% change when pressurized to operating pressure.
4. **Burst Test** (one 18.00 in. (460 mm) assembly)—Shall not leak or fail below the minimum burst pressure.
5. **Leakage Test** (two 12.00 in. (300 mm) assemblies)—Shall not leak or fail.
6. **Cold Bend Test** (one assembly)—Shall exhibit no leakage. Exposure ɸ shall be at −40°F (−40°C).
7. **Oil Resistance Test**—After 70 h immersion at 212°F (100°C) in ASTM No. 3 oil, the volume change of hose inner tube specimens shall be between 0% and +100%.
8. **Impulse Test** (four unaged assemblies)—Hose assemblies, when tested at 125% of operating pressure for hose size 7/8 in (22.2 mm) nominal ID and smaller and 100% of operating pressure for hose sizes 1 1/8 in (28.6 mm) nominal ID and larger, with 200°F (93°C) circulating petroleum base test fluid, shall withstand a minimum of 150,000 cycles for hose sizes 7/8 in (22.2 mm) and smaller and a minimum of 100,000 cycles for hoses sizes 1 1/8 in (28.6 mm) nominal ID and larger, without leakage or other malfunction. Hose sizes 1 1/8 in nominal ID and larger shall be tested straight.
9. **Visual Examination** (all samples).

Inspection Tests—Inspection tests listed below shall be performed on two samples representing each lot of 500–10,000 ft (150–3000 m) of bulk hose or 100–10,000 assemblies. Lots of less than 500 ft (150 m) of hose or 100 assemblies need not be subjected to these tests if a lot has been tested and met the requirements within the previous 12 month period. Requirements shall be same as for corresponding Qualification Tests:

1. Dimensional Check Test.
2. Proof Test.
3. Change In Length Test.
4. Burst Test.

In addition all hose and/or hose assemblies made therefrom shall be subjected to visual examination.

TABLE 5—DIMENSIONS AND SPECIFICATIONS FOR SAE 100R5 HOSE

Nominal SAE 100R5 Hose ID Size, in	Hose Dash Size	Hose ID					Hose OD				Min Burst Pressure		Proof Pressure		Max Operating Pressure		Min Bend Radius[a]	
		Basic		Tolerance			Max		Min									
				Plus		Minus												
		in	mm	in	mm	in	mm	in	mm	in	mm	psi	MPa	psi	MPa	psi	MPa	in	mm
3/16	-4	0.188	4.8	0.026	0.7	0.000	0.0	0.539	13.7	0.500	12.7	12000	82.7	6000	41.4	3000	20.7	3.00	76
1/4	-5	0.250	6.4	0.031	0.8	0.000	0.0	0.601	15.3	0.562	14.3	12000	82.7	6000	41.4	3000	20.7	3.38	86
5/16	-6	0.312	7.9	0.031	0.8	0.000	0.0	0.695	17.6	0.656	16.7	9000	62.0	4500	31.0	2250	15.5	4.00	102
13/32	-8	0.406	10.3	0.031	0.8	0.000	0.0	0.789	20.0	0.743	18.9	8000	55.2	4000	27.6	2000	13.8	4.62	117
1/2	-10	0.500	12.7	0.039	1.0	0.000	0.0	0.945	24.0	0.899	22.8	7000	48.3	3500	24.1	1750	12.1	5.50	140
5/8	-12	0.625	15.9	0.042	1.1	0.000	0.0	1.101	28.0	1.055	26.8	6000	41.4	3000	20.7	1500	10.3	6.50	165
7/8	-16	0.875	22.2	0.042	1.1	0.000	0.0	1.266	32.2	1.203	30.6	3200	22.1	1600	11.0	800	5.5	7.38	187
1-1/8	-20	1.125	28.6	0.047	1.2	0.000	0.0	1.531	38.9	1.469	37.3	2500	17.2	1250	8.6	625	4.3	9.00	229
1-3/8	-24	1.375	34.9	0.047	1.2	0.000	0.0	1.781	45.2	1.719	43.7	2000	13.8	1000	6.9	500	3.4	10.50	267
1-13/16	-32	1.812	46.0	0.047	1.2	0.000	0.0	2.266	57.6	2.172	55.2	1400	9.7	700	4.8	350	2.4	13.25	337

[a] Bend radius measured at inside of bend.

SINGLE FIBER BRAID, (NONMETALLIC), RUBBER COVERED HYDRAULIC HOSE (SAE 100R6)

Scope—This specification covers hose for use with petroleum and water base hydraulic fluids within a temperature range of −40 to +200°F (−40 to +93°C). Operating temperatures in excess of +200°F (+93°C) may materially reduce the life of the hose. Maximum operating pressure, minimum bend radius, and other performance data are specified in Table 6.

It should be noted that the detailed specifications which follow shall be supplemented by the general specifications given at the beginning of this standard.

Dimensions—Dimensions and tolerances applicable to this hose are given in Table 6. The inside diameter and outside diameter of the hose shall be concentric within the following limits:

Nominal Hose ID, in	Concentricity, FIR	
	in	mm
Up to 1/4, incl	0.030	0.8
Over 1/4	0.040	1.0

Hose Construction—The hose shall consist of an inner tube of oil resistant *ed.* synthetic rubber, one braided ply of suitable textile yarn, and an oil and weather resistant synthetic rubber cover.

Qualification Tests—For qualification to this specification, hose and/or hose assemblies made therefrom shall conform to the following tests and requirements:

1. Dimensional Check Test (all samples)—Shall conform to dimensions in Table 6 and these detailed specifications.

2. Proof Test (all samples)—Shall not leak at the proof pressure.

3. Change in Length Test (one sample)—Shall not exceed +2% to −4% change when pressurized to operating pressure.

4. Burst Test (one 18.00 in. (460 mm) assembly)—Shall not leak or fail below the minimum burst pressure.

5. Leakage Test (two 12.00 in. (300 mm) assemblies)—Shall not leak or fail.

6. Cold Bend Test (one assembly)—Shall exhibit no cover cracks or leakage. Exposure shall be at −40°F (−40°C).

7. Oil Resistance Test—After 70 h immersion at 212°F (100°C) in ASTM No. 3 oil, the volume change of hose inner tube and cover specimens shall be between 0% and +100%.

8. Ozone Resistance Test (two samples)—After 70 h exposure in an atmosphere comprised of 50 parts ozone per 100 million parts of air at an ambient temperature of 100°F (38°C), specimens shall not show evidence of cracking or deterioration when viewed with seven-power magnification while still in a stressed condition.

9. Visual Examination (all samples).

Inspection Tests—Inspection tests listed below shall be performed on two samples representing each lot of 500–10,000 ft (150–3000 m) of bulk hose or 100–10,000 assemblies. Lots of less than 500 ft (150 m) of hose or 100 assemblies need not be subjected to these tests if a lot has been tested and met the requirements within the previous 12 month period. Requirements shall be same as for corresponding Qualification Tests:

1. Dimensional Check Test.
2. Proof Test.
3. Change In Length Test.
4. Burst Test.

In addition all hose and/or hose assemblies made therefrom shall be subjected to visual examination.

TABLE 6—DIMENSIONS AND SPECIFICATIONS FOR SAE 100R6 HOSE

Nominal SAE 100R6 Hose ID Size, in	Hose Dash Size	Hose ID Basic		Hose ID Tolerance Plus		Hose ID Tolerance Minus		Hose OD Max		Hose OD Min		Min Burst Pressure		Proof Pressure		Max Operating Pressure		Min Bend Radius[a]	
		in	mm	in	mm	in	mm	in	mm	in	mm	psi	MPa	psi	MPa	psi	MPa	in	mm
3/16	-3	0.188	4.8	0.025	0.6	0.010	0.3	0.469	11.9	0.406	10.3	2000	13.8	1000	6.9	500	3.4	2.00	51
1/4	-4	0.250	6.4	0.025	0.6	0.010	0.3	0.531	13.5	0.469	11.9	1600	11.0	800	5.5	400	2.8	2.50	64
5/16	-5	0.312	7.9	0.025	0.6	0.010	0.3	0.594	15.1	0.531	13.5	1600	11.0	800	5.5	400	2.8	3.00	76
3/8	-6	0.375	9.5	0.025	0.6	0.010	0.3	0.656	16.7	0.594	15.1	1600	11.0	800	5.5	400	2.8	3.00	76
1/2	-8	0.500	12.7	0.030	0.8	0.010	0.3	0.812	20.6	0.750	19.0	1600	11.0	800	5.5	400	2.8	4.00	102
5/8	-10	0.625	15.9	0.030	0.8	0.010	0.3	0.938	23.8	0.875	22.2	1400	9.7	700	4.8	350	2.4	5.00	127

[a]Bend radius measured at inside of bend.

TABLE 7—DIMENSIONS AND SPECIFICATIONS FOR SAE 100R7 HOSE

Nominal SAE 100R7 Hose ID Size, in	Hose Dash Size	Hose ID Basic		Hose ID Tolerance Plus		Hose ID Tolerance Minus		Max Hose OD		Min Burst Pressure		Proof Pressure		Max Operating Pressure		Min Bend Radius[a]	
		in	mm	in	mm	in	mm	in	mm	psi	MPa	psi	MPa	psi	MPa	in	mm
3/16	-3	0.188	4.8	0.015	0.4	0.008	0.2	0.423	10.7	12000	82.7	6000	41.1	3000	20.7	3.50	89
1/4	-4	0.250	6.4	0.015	0.4	0.008	0.2	0.513	13.0	11000	75.8	5500	37.9	2750	19.0	4.00	102
5/16	-5	0.312	7.9	0.015	0.4	0.008	0.2	0.590	15.0	10000	68.9	5000	34.5	2500	17.2	4.50	114
3/8	-6	0.375	9.5	0.015	0.4	0.008	0.2	0.700	17.8	9000	62.0	4500	31.0	2250	15.5	5.00	127
1/2	-8	0.500	12.7	0.020	0.5	0.015	0.4	0.860	21.8	8000	55.2	4000	27.6	2000	13.8	7.00	178
5/8	-10	0.625	15.9	0.025	0.6	0.015	0.4	0.990	24.6	6000	41.4	3000	20.7	1500	10.3	8.00	203
3/4	-12	0.750	19.0	0.030	0.8	0.015	0.4	1.100	27.9	5000	34.5	2500	17.2	1250	8.6	9.50	241
1	-16	1.000	25.4	0.030	0.8	0.015	0.4	1.420	36.8	4000	27.6	2000	13.8	1000	6.9	12.00	305

[a]Bend radius measured at inside of bend.

TABLE 8—DIMENSIONS AND SPECIFICATIONS FOR SAE 100R8 HOSE

Nominal SAE 100R8 Hose ID Size, in	Hose Dash Size	Hose ID Basic		Hose ID Tolerance Plus		Hose ID Tolerance Minus		Max Hose OD		Min Burst Pressure		Proof Pressure		Max Operating Pressure		Min Bend Radius[a]	
		in	mm	in	mm	in	mm	in	mm	psi	MPa	psi	MPa	psi	MPa	in	mm
3/16	-3	0.188	4.8	0.020	0.5	0.008	0.2	0.575	14.6	20000	137.9	10000	68.9	5000	34.5	3.50	89
1/4	-4	0.250	6.4	0.020	0.5	0.008	0.2	0.660	16.8	20000	137.9	10000	68.9	5000	34.5	4.00	102
3/8	-6	0.375	9.5	0.020	0.5	0.008	0.2	0.800	20.3	16000	110.3	8000	55.2	4000	27.6	5.00	127
1/2	-8	0.500	12.7	0.020	0.5	0.015	0.4	0.970	24.6	14000	96.5	7000	48.3	3500	24.1	7.00	178
5/8	-10	0.625	15.9	0.025	0.6	0.015	0.4	1.175	29.8	11000	75.8	5500	37.9	2750	19.0	8.00	203
3/4	-12	0.750	19.0	0.030	0.8	0.015	0.4	1.300	33.0	9000	62.0	4500	31.0	2250	15.5	9.50	241
1	-16	1.000	25.4	0.030	0.8	0.015	0.4	1.520	38.6	8000	55.2	4000	27.6	2000	13.8	12.00	305

[a]Bend radius measured at inside of bend.

THERMOPLASTIC HYDRAULIC HOSE (SAE 100R7)

Scope—This specification covers thermoplastic hose for use with petroleum, water base, and synthetic hydraulic fluids within a temperature range of −40 to +200°F (−40 to +93°C). Operating temperatures in excess of +200°F (+93°C) may materially reduce the life of the hose. Maximum operating pressure, minimum bend radius, and other performance data are specified in Table 7.

It should be noted that the detailed specifications which follow shall be supplemented by the general specifications given at the beginning of this standard.

Dimensions—Dimensions and tolerances applicable to this hose are given in Table 7. The inside diameter and outside diameter of the hose shall be concentric within the following limits:

Nominal Hose ID, in	Concentricity, FIR	
	in	mm
Up to 1/4, incl	0.030	0.8
Over 1/4 to 3/4, incl	0.040	1.0
Over 3/4	0.050	1.3

Hose Construction—The hose shall consist of a thermoplastic inner tube resistant to hydraulic fluids with suitable synthetic fiber reinforcement and a hydraulic fluid and weather resistant thermoplastic cover.

Fitting Compatibility—Fittings for thermoplastic hose may not necessarily be interchangeable. Therefore, it is recommended that fittings and hose be properly matched. Fitting and/or hose manufacturers should be consulted for recommendations.

Qualification Tests—For qualification to this specification, hose and/or hose assemblies made therefrom shall conform to the following tests and requirements:

1. **Dimensional Check Test** (all samples)—Shall conform to dimensions in Table 7 and these detailed specifications.
2. **Proof Test** (all samples)—Shall not leak at the proof pressure.
3. **Change in Length Test** (one sample)—Shall not exceed ±3% change when pressurized to operating pressure.
4. **Burst Test** (one 18.00 in. (460 mm) assembly)—Shall not leak or fail at the minimum burst pressure.
5. **Leakage Test** (two 12.00 in. (300 mm) assemblies)—Shall not leak or fail.
6. **Cold Bend Test** (one assembly)—Shall exhibit no cracks or leakage. φ Exposure shall be at −40°F (−40°C).
7. **Oil Resistance Test**—After 70 h immersion at 212°F (100°C) in ASTM No. 3 oil, the volume change of the hose inner tube and cover specimens shall be between −15% and +35%.
8. **Ozone Resistance Test** (two samples)—After 70 h exposure in an atmosphere comprised of 50 parts ozone per 100 million parts of air at an ambient temperature of 100°F (38°C), specimens shall not show evidence of cracking or deterioration when viewed with seven-power magnification while still in a stressed condition.
9. **Impulse Test** (four unaged assemblies)—Hose assemblies, when tested at 125% of operating pressure, with 200°F (93°C) circulating petroleum base test fluid, shall withstand a minimum of 150,000 cycles without leakage or other malfunction.
10. **Electrical Conductivity Test**—The maximum leakage shall not exceed 50 μA when subjected to 75 kV/ft (75 kV/305 mm) for 5 min. (This test shall not be applicable to hose with pin pricked outer cover.)
11. **Visual Examination** (all samples).

Inspection Tests—Inspection tests listed below shall be performed on two samples representing each lot of 500–10,000 ft (150–3000 m) of bulk hose or 100–10,000 assemblies. Lots of less than 500 ft (150 m) of hose or 100 assemblies need not be subjected to these tests if a lot has been tested and met the requirements within the previous 12 month period. Requirements shall be same as for corresponding Qualification Tests:

1. Dimensional Check Test.
2. Proof Test.
3. Change In Length Test.
4. Burst Test.

In addition all hose and/or hose assemblies made therefrom shall be subjected to visual examination.

HIGH PRESSURE THERMOPLASTIC HYDRAULIC HOSE (SAE 100R8)

Scope—This specification covers thermoplastic hose for use with petroleum, water base, and synthetic hydraulic fluids within a temperature range of −40 to +200°F (−40 to +93°C). Operating temperatures in excess of +200°F (+93°C) may materially reduce the life of the hose. Maximum operating pressure, minimum bend radius, and other performance data are specified in Table 8.

It should be noted that the detailed specifications which follow shall be supplemented by the general specifications given at the beginning of this standard.

Dimensions—Dimensions and tolerances applicable to this hose are given in Table 8. The inside diameter and outside diameter of the hose shall be concentric within the following limits:

Nominal Hose ID, in	Concentricity, FIR	
	in	mm
Up to 1/4, incl	0.030	0.8
Over 1/4 to 3/4, incl	0.040	1.0
Over 3/4	0.050	1.3

Hose Construction—The hose shall consist of a thermoplastic inner tube resistant to hydraulic fluids with suitable synthetic fiber reinforcement and a hydraulic fluid and weather resistant thermoplastic cover.

Fitting Compatibility—Fittings for thermoplastic hose may not necessarily be interchangeable. Therefore, it is recommended that fittings and hose be properly matched. Fitting and/or hose manufacturers should be consulted for recommendations.

Qualification Tests—For qualification to this specification, hose and/or hose assemblies made therefrom shall conform to the following tests and requirements:

1. **Dimensional Check Test** (all samples)—Shall conform to dimensions in Table 8 and these detailed specifications.
2. **Proof Test** (all samples)—Shall not leak at the proof pressure.
3. **Change in Length Test** (one sample)—Shall not exceed ±3% change when pressurized to operating pressure.
4. **Burst Test** (one 18.00 in. (460 mm) assembly)—Shall not leak or fail at the minimum burst pressure.
5. **Leakage Test** (two 12.00 in. (300 mm) assemblies)—Shall not leak or fail.
6. **Cold Bend Test** (one assembly)—Shall exhibit no cracks or leakage. φ Exposure shall be at −40°F (−40°C).
7. **Oil Resistance Test**—After 70 h immersion at 212°F (100°C) in ASTM No. 3 oil, the volume change of the hose inner tube and cover specimens shall be between −15 and +35%.
8. **Ozone Resistance Test** (two samples)—After 70 h exposure in an atmosphere comprised of 50 parts ozone per 100 million parts of air at an ambient temperature of 100°F (38°C) specimens shall not show evidence of cracking or deterioration when viewed with seven-power magnification while still in a stressed condition.
9. **Impulse Test** (four unaged assemblies)—Hose assemblies, when tested at 133% of operating pressure, with 200°F (93°C) circulating petroleum base test fluid, shall withstand a minimum of 200,000 cycles without leakage or other malfunction.
10. **Electrical Conductivity Test**—The maximum leakage shall not exceed 50 μA when subjected to 75 kV/ft (246 kV/m) for 5 min. (This test shall not be applicable to hose with pinpricked outer cover.)
11. **Visual Examination** (all samples).

Inspection Tests—Inspection tests listed below shall be performed on two samples representing each lot of 500–10,000 ft (150–3000 m) of bulk hose or 100–10,000 assemblies. Lots of less than 500 ft (150 m) of hose or 100 assemblies need not be subjected to these tests if a lot has been tested and met the requirements within the previous 12 month period. Requirements shall be same as for corresponding Qualification Tests:

1. Dimensional Check Test.
2. Proof Test.
3. Change in Length Test.
4. Burst Test.

In addition, all hose and/or hose assemblies made therefrom shall be subjected to visual examination.

HIGH PRESSURE, 4-SPIRAL STEEL WIRE REINFORCED, RUBBER COVERED HYDRAULIC HOSE (SAE 100R9)

Scope—This specification covers hose for use with petroleum and water base fluids within a temperature range of −40 to +200°F (−40 to +93°C). Operating temperatures in excess of +200°F (+93°C) may materially reduce the life of the hose. Maximum operating pressure, minimum bend radius, and other performance data are specified in Table 9.

It should be noted that the detailed specifications which follow shall be supplemented by the general specifications given at the beginning of this standard.

Dimensions—Dimensions and tolerances applicable to this hose are given in Table 9. The inside diameter of hose shall be concentric with outside diameter of hose and the outer surfaces of the reinforcement within the following limits:

Nominal Hose ID, in	Concentricity, FIR			
	ID to OD		ID to Reinforcement	
	in	mm	in	mm
Up to 1/4, incl	0.030	0.8	0.021	0.5
Over 1/4 to 7/8, incl	0.040	1.0	0.028	0.7
Over 7/8	0.050	1.3	0.035	0.9

Construction

Type A—This hose shall consist of an inner tube of oil resistant synthetic rubber, 4-spiral plies of wire wrapped in alternating directions, and an oil and weather resistant synthetic rubber cover. A ply or braid of suitable material may be used over the inner tube and/or over the wire reinforcement to anchor the synthetic rubber to the wire.

Type AT—This hose shall be of the same construction as Type A, except having a cover designed to assemble with fittings which do not require removal of the cover or a portion thereof.

Qualification Tests—For qualification to this specification, hose and/or hose assemblies made therefrom shall conform to the following tests and requirements:

1. **Dimensional Check Test** (all samples)—Shall conform to dimensions in Table 9 and these detailed specificatons.
2. **Proof Test** (all samples)—Shall not leak at the proof pressure.
3. **Change in Length Test** (one sample)—Shall not exceed +2%, −4% change when pressurized to operating pressure.
4. **Burst Test** (one 18.00 in. (460 mm) assembly)—Shall not leak or fail below the minimum burst pressure.
5. **Leakage Test** (two 12.00 in. (300 mm) free length assemblies)—Shall not leak or fail.
6. **Cold Bend Test** (one assembly)—Shall exhibit no cover cracks or leakage. Exposure shall be at −40°F (−40°C).
7. **Oil Resistance Test**—After 70 h immersion at 212°F (100°C) in ASTM No. 3 oil, the volume change of hose inner tube and cover specimens shall be between 0% and +100%.
8. **Ozone Resistance Test** (two samples)—After 70 h exposure in an atmosphere comprised of 50 parts ozone per 100 million parts of air at an ambient temperature of 100°F (38°C) specimens shall not show evidence of cracking or deterioration when viewed with seven-power magnification while still in a stressed condition.
9. **Impulse Test** (four unaged assemblies)—Hose assemblies when tested at 133% of operating pressure with 200°F (93°C) circulating petroleum base test fluid shall withstand a minimum of 200,000 cycles for sizes 3/8 and 1/2 in. (9.5 and 12.7 mm), and 300,000 cycles for all other sizes without leakage or other malfunction.
10. **Visual Examination** (all samples).

Inspection Tests—Inspection tests listed below shall be performed on two samples representing each lot of 500–10,000 ft (150–3000 m) of bulk hose or 100–10,000 assemblies. Lots of less than 500 ft (150 m) of hose or 100 assemblies need not be subjected to these tests if a lot has been tested and met the requirements within the previous 12 month period. Requirements shall be same as for corresponding Qualification Tests:

1. Dimensional Check Test.
2. Proof Test.
3. Change in Length Test.
4. Burst Test.

In addition, all hose and/or hose assemblies made therefrom shall be subjected to visual examination.

TABLE 9—DIMENSIONS AND SPECIFICATIONS OF SAE 100R9 HOSE

Nominal SAE 100R9 Hose ID Size, in	Hose Dash Size	Hose ID						Reinforcement Dia				Hose OD Type A			
		Basic		Tolerance				Max		Min		Max		Min	
				Plus		Minus									
		in	mm	in	mm	in	mm	in	mm	in	mm	in	mm	in	mm
3/8	-6	0.375	9.5	0.023	0.6	0.008	0.2	0.710	18.0	0.664	16.9	0.875	22.2	0.812	20.6
1/2	-8	0.500	12.7	0.031	0.8	0.015	0.4	0.828	21.0	0.766	19.4	1.000	25.4	0.938	23.8
3/4	-12	0.750	19.0	0.031	0.8	0.015	0.4	1.109	28.2	1.047	26.6	1.266	32.2	1.203	30.6
1	-16	1.000	25.4	0.040	1.0	0.015	0.4	1.422	36.1	1.360	34.5	1.609	40.9	1.515	38.5
1-1/4	-20	1.250	31.8	0.047	1.2	0.015	0.4	1.797	45.6	1.703	43.3	2.062	52.4	1.938	49.2
1-1/2	-24	1.500	38.1	0.047	1.2	0.015	0.4	2.047	52.0	1.953	49.6	2.312	58.7	2.188	55.6
2	-32	2.000	50.8	0.047	1.2	0.015	0.4	2.608	66.2	2.515	63.9	2.875	73.0	2.750	69.9

Nominal SAE 100R9 Hose ID Size, in	Hose OD Type AT Max		Cover Thickness[a] Type AT				Min Burst Pressure		Proof Pressure		Max Operating Pressure		Min Bend Radius[b]	
			Max		Min									
	in	mm	in	mm	in	mm	psi	MPa	psi	MPa	psi	MPa	in	mm
3/8	0.831	21.1	0.062	1.6	0.031	0.8	18000	124.1	9000	62.0	4500	31.0	5.00	127
1/2	0.958	24.3	0.078	2.0	0.039	1.0	16000	110.3	8000	55.2	4000	27.6	7.00	178
3/4	1.255	31.9	0.078	2.0	0.039	1.0	12000	82.7	6000	41.4	3000	20.7	9.50	241
1	1.594	40.5	0.094	2.4	0.042	1.1	12000	82.7	6000	41.4	3000	20.7	12.00	305
1-1/4	1.997	50.7	0.109	2.8	0.050	1.3	10000	68.9	5000	34.5	2500	17.2	16.50	419
1-1/2	—	—	—	—	—	—	8000	55.2	4000	27.6	2000	13.8	20.00	508
2	—	—	—	—	—	—	8000	55.2	4000	27.6	2000	13.8	26.00	660

[a] Cover thickness shall be measured by means of a dial indicator depth gage having a rounded foot placed parallel to the hose, bridging a groove obtained by stripping a 0.50-1.00 in (12.5-25.4 mm) width of cover from the hose. A mandrel should be placed inside the hose bore to insure freedom from misalignment.
[b] Bend radius measured at inside of bend.

HEAVY DUTY, 4-SPIRAL STEEL WIRE REINFORCED, RUBBER COVERED HYDRAULIC HOSE (SAE 100R10)

Scope—This specification covers hose for use with petroleum and water base fluids within a Temperature Range of −40 to +200°F (−40 to +93°C). Operating temperatures in excess of +200°F (+93°C) may materially reduce the life of the hose. Maximum operating pressure, minimum bend radius, and other performance data are specified in Table 10.

It should be noted that the detailed specifications which follow shall be supplemented by the general specifications given at the beginning of this standard.

Dimensions—Dimensions and tolerances applicable to this hose are given in Table 10. The inside diameter of hose shall be concentric with outside diameter of hose and the outer surface of the reinforcement within the following limits:

Nominal Hose ID, in	Concentricity, FIR			
	ID to OD		ID to Reinforcement	
	in	mm	in	mm
Up to 1/4, incl	0.030	0.8	0.021	0.5
Over 1/4 to 7/8, incl	0.040	1.0	0.028	0.7
Over 7/8	0.050	1.3	0.035	0.9

Construction

Type A—This hose shall consist of an inner tube of oil resistant synthetic rubber, 4-spiral plies of heavy wire wrapped in alternating directions, and an oil and weather resistant synthetic rubber cover. A ply or braid of suitable material may be used over the inner tube and/or over the wire reinforcement to anchor the synthetic rubber to the wire.

Type AT—This hose shall be of the same construction as Type A, except having a cover designed to assemble with fittings which do not require removal of the cover or a portion thereof.

Qualification Tests—For qualification to this specification, hose and/or hose assemblies made therefrom shall conform to the following tests and requirements:

1. **Dimensional Check Test** (all samples)—Shall conform to dimensions in Table 10 and these detailed specifications.
2. **Proof Test** (all samples)—Shall not leak at the proof pressure.
3. **Change in Length Test** (one sample)—Shall not exceed +2%, −4% when pressurized to operating pressure.
4. **Burst Test** (one 18.00 in. (460 mm) assembly)—Shall not leak or fail below the minimum burst pressure.
5. **Leakage Test** (two 12.00 in. (300 mm) free length assemblies)—Shall not leak or fail.
6. **Cold Bend Test** (one assembly)—Shall exhibit no cover cracks or leakage. Exposure shall be at −40°F (−40°C). On sizes larger than 3/4 in. (19 mm), tube and cover samples may be substituted for the hose bending test.
7. **Oil Resistance Test**—After 70 h immersion at 212°F (100°C) in ASTM No. 3 oil, the volume change of hose inner tube and cover specimens shall be between 0% and +100%.
8. **Ozone Resistance Test** (two samples)—After 70 h exposure in an atmosphere comprised of 50 parts ozone per 100 million parts of air at an ambient temperature of 100°F (38°C) specimens shall not show evidence of cracking or deterioration when viewed with seven-power magnification while still in a stressed condition.
9. **Impulse Test** (four unaged assemblies)—Hose assemblies when tested at 133% of operating pressure with 200°F (93°C) circulating petroleum base test fluid shall withstand a minimum of 400,000 cycles without leakage or other malfunctions.

Hose sizes 3/16, 1/4, and 3/8 in. (4.8, 6.4 and 9.5 mm) are not usually impulsed as these sizes are not recommended for systems with conventional hydraulic surges.

10. **Visual Examination** (all samples).

Inspection Tests—Inspection tests listed below shall be performed on two samples representing each lot of 500–10,000 ft (150–3000 m) of bulk hose or 100–10,000 assemblies. Lots of less than 500 ft (150 m) of hose or 100 assem-

TABLE 10—DIMENSIONS AND SPECIFICATIONS FOR SAE 100R10 HOSE

Nominal SAE 100R10 Hose ID Size, in	Hose Dash Size	Hose ID						Reinforcement Dia Max		Hose OD					
		Basic		Tolerance						Type A				Type AT	
				Plus		Minus				Max		Min		Max	
		in	mm	in	mm	in	mm	in	mm	in	mm	in	mm	in	mm
3/16	-3	0.188	4.8	0.023	0.6	0.008	0.2	0.625	15.9	0.781	19.8	0.719	18.3	—	—
1/4	-4	0.250	6.4	0.023	0.6	0.008	0.2	0.687	17.4	0.844	21.4	0.781	19.8	—	—
3/8	-6	0.375	9.5	0.023	0.6	0.008	0.2	0.812	20.6	0.969	24.6	0.906	23.0	—	—
1/2	-8	0.500	12.7	0.039	1.0	0.008	0.2	0.969	24.6	1.125	28.6	1.062	27.0	—	—
3/4	-12	0.750	19.0	0.047	1.2	0.000	0.0	1.281	32.5	1.469	37.3	1.406	35.7	1.450	36.8
1	-16	1.000	25.4	0.063	1.6	0.000	0.0	1.594	40.5	1.797	45.6	1.703	43.3	1.790	45.5
1-1/4	-20	1.250	31.8	0.063	1.6	0.000	0.0	1.859	47.2	2.062	52.4	1.938	49.2	2.060	52.3
1-1/2	-24	1.500	38.1	0.063	1.6	0.000	0.0	2.109	53.6	2.312	58.7	2.188	55.6	2.310	58.7
2	-32	2.000	50.8	0.070	1.8	0.000	0.0	2.640	67.1	2.844	72.2	2.719	69.1	2.840	72.1

Nominal SAE 100R10 Hose ID Size, in	Cover Thickness[a]				Min Burst Pressure		Proof Pressure		Max Operating Pressure		Min Bend Radius[b]	
	Type AT											
	Max		Min									
	in	mm	in	mm	psi	MPa	psi	MPa	psi	MPa	in	mm
3/16	—	—	—	—	40000	275.8	20000	137.9	10000	68.9	4.00	102
1/4	—	—	—	—	35000	241.3	17500	120.6	8750	60.3	5.00	127
3/8	—	—	—	—	30000	206.8	15000	103.4	7500	51.7	6.00	152
1/2	—	—	—	—	25000	172.4	12500	86.2	6250	43.1	8.00	203
3/4	0.078	2.0	0.039	1.0	20000	137.9	10000	68.9	5000	34.5	11.00	279
1	0.094	2.4	0.047	1.2	16000	110.3	8000	55.2	4000	27.6	14.00	356
1-1/4	0.109	2.8	0.054	1.4	12000	82.7	6000	41.4	3000	20.7	18.00	457
1-1/2	0.109	2.8	0.054	1.4	10000	68.9	5000	34.5	2500	17.2	22.00	559
2	0.109	2.8	0.054	1.4	10000	68.9	5000	34.5	2500	17.2	28.00	711

[a]Cover thickness shall be measured by means of a dial indicator depth gage having a rounded foot placed parallel to the hose, bridging a groove obtained by stripping a 0.50-1.00 in (12.5-25.4 mm) width of cover from the hose. A mandrel should be placed in the hose bore to insure freedom from misalignment.
[b]Bend radius measured at inside of bend.

blies need not be subjected to these tests if a lot has been tested and met the requirements within the previous 12 month period. Requirements shall be same as for corresponding Qualification Tests:
1. Dimensional Check Test.
2. Proof Test.
3. Change in Length Test.
4. Burst Test.

In addition, all hose and/or hose assemblies made therefrom shall be subjected to visual examination.

HEAVY DUTY, 6-SPIRAL STEEL WIRE REINFORCED, RUBBER COVERED HYDRAULIC HOSE (SAE 100R11)

Scope—This specification covers hose for use with petroleum and water base fluids within a Temperature Range of −40 to +200°F (−40 to +93°C).

Operating temperatures in excess of +200°F (+93°C) may materially reduce the life of the hose. Maximum operating pressure, minimum bend radius, and other performance data are specified in Table 11.

It should be noted that the detailed specifications which follow shall be supplemented by the general specifications given at the beginning of this standard.

Dimensions—Dimensions and tolerances applicable to this hose are given in Table 11. The inside diameter of hose shall be concentric with outside diameter of hose and the outer surface of the reinforcement within the following limits:

Construction—This hose shall consist of an inner tube of oil resistant synthetic rubber, 6-spiral plies of heavy wire wrapped in alternating directions and an oil and weather resistant synthetic rubber cover. A ply or braid of suitable material may be used over the inner tube and/or over the wire reinforcement to anchor the synthetic rubber to the wire.

Qualification Tests—For qualification to this specification, hose and/or hose assemblies made therefrom shall conform to the following tests and requirements:

1. **Dimensional Check Test** (all samples)—Shall conform to dimensions in Table 11 and these detailed specifications.
2. **Proof Test** (all samples)—Shall not leak at the proof pressure.
3. **Change in Length Test** (one sample)—Shall not exceed +2%, −4% change when pressurized to operating pressure.
4. **Burst Test** (one 18.00 in. (460 mm) assembly)—Shall not leak or fail below the minimum burst pressure.
5. **Leakage Test** (two 12.00 in. (300 mm) free length assemblies)—Shall not leak or fail.
6. **Cold Bend Test** (one assembly)—Shall exhibit no cover cracks or ɸ leakage. Exposure shall be at −40°F (−40°C). On sizes larger than 3/4 in. (19 mm), tube and cover samples may be substituted for the hose bending test. ɸ
7. **Oil Resistance Test**—After 70 h immersion at 212°F (100°C) in ASTM No. 3 oil, the volume change of hose inner tube and cover specimens shall be between 0% and +100%.
8. **Ozone Resistance Test** (two samples)—After 70 h exposure in an atmosphere comprised of 50 parts ozone per 100 million parts of air at an ambient temperature of 100°F (38°C) specimens shall not show evidence of cracking or deterioration when viewed with seven-power magnification while still in a stressed condition.
9. **Impulse Test** (four unaged assemblies)—Hose assemblies when tested at 133% of operating pressure with 200°F (93°C) circulating petroleum base

Nominal Hose ID, in	Concentricity, FIR			
	ID to OD		ID to Reinforcement	
	in	mm	in	mm
Up to 1/4, incl	0.030	0.8	0.021	0.5
Over 1/4 to 7/8, incl	0.040	1.0	0.028	0.7
Over 7/8	0.050	1.3	0.035	0.9

TABLE 11—DIMENSIONS AND SPECIFICATIONS FOR SAE 100R11 HOSE

Nominal SAE 100R11 Hose ID Size, in	Hose Dash Size	Hose ID						Reinforcement Dia				Hose OD			
		Basic		Tolerance				Max		Min		Max		Min	
				Plus		Minus									
		in	mm	in	mm	in	mm	in	mm	in	mm	in	mm	in	mm
3/16	-3	0.188	4.8	0.023	0.6	0.008	0.2	0.750	19.1	0.688	17.5	0.906	23.0	0.844	21.4
1/4	-4	0.250	6.4	0.023	0.6	0.008	0.2	0.812	20.6	0.750	19.1	0.969	24.6	0.906	23.0
3/8	-6	0.375	9.5	0.023	0.6	0.008	0.2	0.938	23.8	0.875	22.2	1.094	27.8	1.031	26.2
1/2	-8	0.500	12.7	0.039	1.0	0.008	0.2	1.094	27.8	1.031	26.2	1.250	31.8	1.188	30.2
3/4	-12	0.750	19.0	0.047	1.2	0.000	0.0	1.406	35.7	1.344	34.1	1.594	40.5	1.531	38.9
1	-16	1.000	25.4	0.062	1.6	0.000	0.0	1.734	44.0	1.641	41.7	1.953	49.6	1.859	47.2
1-1/4	-20	1.250	31.8	0.062	1.6	0.000	0.0	1.984	50.4	1.891	48.0	2.219	56.4	2.094	53.2
1-1/2	-24	1.500	38.1	0.062	1.6	0.000	0.0	2.234	56.7	2.140	54.4	2.469	62.7	2.344	59.5
2	-32	2.000	50.8	0.070	1.8	0.000	0.0	2.797	71.0	2.703	68.6	3.031	77.0	2.906	73.8

Nominal SAE 100R11 Hose ID Size, in	Min Burst Pressure		Proof Pressure		Max Operating Pressure		Min Bend Radius[a]	
	psi	MPa	psi	MPa	psi	MPa	in	mm
3/16	50000	344.7	25000	172.4	12500	86.2	4.00	102
1/4	45000	310.3	22500	155.1	11250	77.6	5.00	127
3/8	40000	275.8	20000	137.9	10000	68.9	6.00	152
1/2	30000	206.8	15000	103.4	7500	51.7	8.00	203
3/4	25000	172.4	12500	86.2	6250	43.1	11.00	279
1	20000	137.9	10000	68.9	5000	34.5	14.00	356
1-1/4	14000	96.5	7000	48.3	3500	24.1	18.00	457
1-1/2	12000	82.7	6000	41.4	3000	20.7	22.00	559
2	12000	82.7	6000	41.4	3000	20.7	28.00	711

[a] Bend radius measured at inside of bend.

test fluid shall withstand a minimum of 400,000 cycles without leakage or other malfunctions.

Hose sizes 3/16, 1/4, 3/8, and 1/2 in. (4.8, 6.4, 9.5, and 12.7 mm) are not usually impulsed as these sizes are not recommended for systems with conventional hydraulic surges.

10. Visual Examination (all samples).

Inspection Tests—Inspection tests listed below shall be performed on two samples representing each lot of 500–10,000 ft (150–3000 m) of bulk hose or 100–10,000 assemblies. Lots of less than 500 ft (150 m) of hose or 100 assemblies need not be subjected to these tests if a lot has been tested and met the requirements within the previous 12 month period. Requirements shall be same as for corresponding Qualification Tests:
1. Dimensonal Check Test.
2. Proof Test.
3. Change in Length Test.
4. Burst Test.

In addition, all hose and/or hose assemblies made therefrom shall be subjected to visual examination.

HEAVY DUTY, HIGH IMPULSE, 4-SPIRAL WIRE REINFORCED, RUBBER COVER HYDRAULIC HOSE (SAE 100R12)

Scope—This specification covers hose for use with petroleum and water base fluids within a temperature range of −40°F to +250°F (−40°C to +121°C). Operating temperatures in excess of +250°F (+121°C) may materially reduce the life of the hose. Maximum operating pressure, minimum bend radius, and other performance data are specified in Table 12.

It should be noted that the detailed specification which follows shall be supplemented by the general specifications given at the beginning of this standard.

Dimensions—Dimensions and tolerances applicable to this hose are given in Table 12. The inside diameter of hose shall be concentric with outside diameter of hose and the outer surface of the reinforcement within the following limits:

Nominal Hose ID, in	Concentricity, FIR			
	ID to OD		ID to Reinforcement	
	in	mm	in	mm
Up to 3/4, incl	0.040	1.0	0.028	0.7
Over 3/4	0.050	1.3	0.035	0.9

Hose Construction—This hose shall consist of an inner tube of oil resistant synthetic rubber, four spiral plies of heavy wire wrapped in alternating directions, and an oil and weather resistant synthetic rubber cover. A ply or braid of suitable material may be used over or within the inner tube and/or over the wire reinforcement to anchor the synthetic rubber to the wire.

Qualification Tests—For qualifications to this specification, hose and/or hose assemblies made therefrom shall conform to the following tests and requirements:

1. Dimensional Check Test (all samples)—Shall conform to the dimensions in Table 12 and these detailed specifications.

2. Proof Test (all samples)—Shall not leak at proof pressure.

3. Change in Length Test (one sample)—Shall not exceed ±2% change when pressurized to operating pressure.

4. Leakage Test (two 12.00 in (300 mm) free hose length assemblies)—Shall not leak or fail.

5. Burst Test (one 18.00 in (460 mm) free hose length assembly)—Shall not leak or fail below minimum burst pressure.

6. Cold Bend Test (one assembly)—Shall exhibit no cover cracks or leakage. Exposure shall be at −40°F (−40°C).

7. Oil Resistance Test—After 70 h immersion at 250°F (121°C) in ASTM No. 3 oil, the volume change of the hose inner tube shall be between 0% and 100% and cover specimens shall be between 0% and 125%.

8. Ozone Resistance Test (two samples)—After 70 h exposure in an atmosphere comprised of 50 parts ozone per 100 million parts of air at an ambient temperature of 100°F (38°C), specimens shall show no evidence of cracking or deterioration when viewed with seven-power magnification while still in a stressed condition.

9. Impulse Test (four unaged assemblies)—Hose assemblies when tested at 133% of the maximum operating pressure per Table 12 with 250°F (121°C) circulating petroleum base fluid shall withstand a minimum of 500 000 cycles without leakage or other malfunction.

10. Visual Examination (all samples).

Inspection Tests—Inspection tests listed below shall be performed on two samples representing each lot of 500 to 10 000 ft (150 to 3000 m) of bulk hose or 100 to 10 000 hose assemblies. Lots of less than 500 ft (150 m) of hose or 100 assemblies need not be subjected to these tests if a lot has been tested and met the requirements within the previous 12 month period. Requirements shall be the same as for the corresponding Qualification Tests:
1. Dimensional Check Test.
2. Proof Test.
3. Change in Length.
4. Burst Test.

In addition, all hose and/or hose assemblies made therefrom shall be subjected to visual examination.

TABLE 12—DIMENSIONS AND SPECIFICATIONS FOR SAE 100R12 HOSE

Nominal SAE 100R12 Hose ID	Hose Dash Size	Hose ID						Reinforcement OD				Hose OD			
		Basic		Tolerance				Max		Min		Max		Min	
				Plus		Minus									
		in	mm	in	mm	in	mm	in	mm	in	mm	in	mm	in	mm
3/8	−6	0.375	9.5	0.023	0.6	0.008	0.2	0.700	17.8	0.631	16.0	0.828	21.0	0.743	18.9
1/2	−8	0.500	12.7	0.031	0.8	0.015	0.4	0.846	21.5	0.764	19.4	0.966	24.6	0.884	22.5
3/4	−12	0.750	19.0	0.031	0.8	0.015	0.4	1.120	28.4	1.059	26.9	1.241	31.5	1.179	29.9
1	−16	1.000	25.4	0.040	1.0	0.015	0.4	1.405	35.7	1.344	34.1	1.542	39.2	1.448	36.8
1-1/4	−20	1.250	31.8	0.047	1.2	0.015	0.4	1.777	45.1	1.683	42.7	1.912	48.6	1.788	45.4
1-1/2	−24	1.500	38.1	0.047	1.2	0.015	0.4	2.032	51.6	1.938	49.2	2.167	55.0	2.043	51.9
2	−32	2.000	50.8	0.047	1.2	0.015	0.4	2.553	64.8	2.459	62.5	2.688	68.3	2.564	65.1

Nominal SAE 100R12 Hose ID	Minimum Burst Pressure		Proof Pressure		Maximum Operating Pressure		Minimum Bend Radius[a]	
	psi	MPa	psi	MPa	psi	MPa	in	mm
3/8	16 000	110.3	8000	55.2	4000	27.6	5.00	127
1/2	16 000	110.3	8000	55.2	4000	27.6	7.00	178
3/4	16 000	110.3	8000	55.2	4000	27.6	9.50	241
1	16 000	110.3	8000	55.2	4000	27.6	12.00	305
1-1/4	12 000	82.7	6000	41.4	3000	20.7	16.50	419
1-1/2	10 000	69.0	5000	34.5	2500	17.2	20.00	508
2	10 000	69.0	5000	34.5	2500	17.2	25.00	635

[a] Minimum bend radius measured at the inside of hose bend.

SELECTION, INSTALLATION, AND MAINTENANCE OF HOSE AND HOSE ASSEMBLIES—SAE J1273 SEP79

SAE Recommended Practice

Report of the Fluid Conductors and Connectors Technical Committee, approved September 1979.

1. Scope—Hose (also includes hose assemblies) has a finite life and there are a number of factors which will reduce its life.

This recommended practice is intended as a guide to assist system designers and/or users in the selection, installation, and maintenance of hose. The designers and users must make a systematic review of each application and then select, install, and maintain the hose to fulfill the requirements of the application. The following are general guidelines and are not necessarily a complete list.

WARNING: IMPROPER SELECTION, INSTALLATION, OR MAINTENANCE MAY RESULT IN PREMATURE FAILURES, BODILY INJURY, OR PROPERTY DAMAGE.

2. Selection—The following is a list of factors which must be considered before final hose selection can be made.

2.1 Pressure—After determining the system pressure, hose selection must be made so that the recommended maximum operating pressure is equal to or greater than the system pressure. Surge pressures higher than the maximum operating pressure will shorten hose life and must be taken into account by the hydraulic designer.

2.2 Suction—Hoses used for suction applications must be selected to insure the hose will withstand the negative pressure of the system.

2.3 Temperature—Care must be taken to insure that fluid and ambient temperatures, both static and transient, do not exceed the limitations of the hose. Special care must be taken when routing near hot manifolds.

2.4 Fluid Compatibility—Hose selection must assure compatibility of the hose tube, cover, and fittings with the fluid used. Additional caution must be observed in hose selection for gaseous applications.

2.5 Size—Transmission of power by means of pressurized fluid varies with pressure and rate of flow. The size of the components must be adequate to keep pressure losses to a minimum and avoid damage to the hose due to heat generation or excessive turbulence.

2.6 Routing—Attention must be given to optimum routing to minimize inherent problems.

2.7 Environment—Care must be taken to insure that the hose and fittings are either compatible with or protected from the environment to which they are exposed. Environmental conditions such as ultraviolet light, ozone, salt water, chemicals, and air pollutants can cause degradation and premature failure and, therefore, must be considered.

2.8 Mechanical Loads—External forces can significantly reduce hose life. Mechanical loads which must be considered include excessive flexing, twist, kinking, tensile or side loads, bend radius, and vibration. Use of swivel type fittings or adapters may be required to insure no twist is put into the hose. Unusual applications may require special testing prior to hose selection.

2.9 Abrasion—While a hose is designed with a reasonable level of abrasion resistance, care must be taken to protect the hose from excessive abrasion which can result in erosion, snagging, and cutting of the hose cover. Exposure of the reinforcement will significantly accelerate hose failure.

2.10 Proper End Fitting—Care must be taken to insure proper compatibility exists between the hose and coupling selected based on the manufacturer's recommendations substantiated by testing to industry standards such as SAE J517d (November, 1976).

2.11 Length—When establishing proper hose length, motion absorption, hose length changes due to pressure, as well as hose and machine tolerances must be considered.

2.12 Specifications and Standards—When selecting hose, government, industry, and manufacturers' specifications and recommendations must be reviewed as applicable.

2.13 Hose Cleanliness—Hose components vary in cleanliness levels. Care must be taken to insure that the assemblies selected have an adequate level of cleanliness for the application.

2.14 Electrical Conductivity—Certain applications require that hose be non-conductive to prevent electrical current flow. Other applications require the hose to be sufficiently conductive to drain off static electricity. Hose and fittings must be chosen with these needs in mind.

3. Installation—After selection of proper hose, the following factors must be considered by the installer.

3.1 Pre-Installation Inspection—Prior to installation, a careful examination of the hose must be performed. All components must be checked for correct style, size, and length. In addition, the hose must be examined for cleanliness, I.D. obstructions, blisters, loose cover, or any other visible defects.

3.2 Follow Manufacturers' Assembly Instructions

3.3 Minimum Bend Radius—Installation at less than minimum bend radius may significantly reduce hose life. Particular attention must be given to preclude sharp bending at the hose/fitting juncture.

3.4 Twist Angle and Orientation—Hose installations must be such that relative motion of machine components produces bending of the hose rather than twisting.

3.5 Securement—In many applications, it may be necessary to restrain, protect, or guide the hose to protect it from damage by unnecessary flexing, pressure surges, and contact with other mechanical components. Care must be taken to insure such restraints do not introduce additional stress or wear points.

3.6 Proper Connection of Ports—Proper physical installation of the hose requires a correctly installed port connection while insuring that no twist or torque is put into the hose.

3.7 Avoid External Damage—Proper installation is not complete without insuring that tensile loads, side loads, kinking, flattening, potential abrasion, thread damage, or damage to sealing surfaces are corrected or eliminated.

3.8 System Check Out—After completing the installation, all air entrapment must be eliminated and the system pressurized to the maximum system pressure and checked for proper function and freedom from leaks.

NOTE: Avoid potential hazardous areas while testing.

4. Maintenance—Even with proper selection and installation, hose life may be significantly reduced without a continuing maintenance program. Frequency should be determined by the severity of the application and risk potential. A maintenance program should include the following as a minimum.

4.1 Hose Storage—Hose products in storage can be affected adversely by temperature, humidity, ozone, sunlight, oils, solvents, corrosive liquids and fumes, insects, rodents, and radioactive materials. Storage areas should be relatively cool and dark and free of dust, dirt, dampness, and mildew.

4.2 Visual Inspection—Any of the following conditions requires replacement of the hose:

(a) Leaks at fitting or in hose. (Leaking fluid is a fire hazard.)
(b) Damaged, cut, or abraded cover. (Any reinforcement exposed.)
(c) Kinked, crushed, flattened, or twisted hose.
(d) Hard, stiff, heat cracked, or charred hose.
(e) Blistered, soft, degraded, or loose cover.
(f) Cracked, damaged, or badly corroded fittings.
(g) Fitting slippage on hose.

4.3 Visual Inspection—The following items must be tightened, repaired, or replaced as required:

(a) Leaking port conditions.
(b) Clamps, guards, shields.
(c) Remove excessive dirt buildup.
(d) System fluid level, fluid type, and any air entrapment.

4.4 Functional Test—Operate the system at maximum operating pressure and check for possible malfunctions and freedom from leaks.

NOTE: Avoid potential hazardous areas while testing.

4.5 Replacement Intervals—Specific replacement intervals must be considered based on previous service life, government or industry recommendations, or when failures could result in unacceptable down time, damage, or injury risk.

TESTS AND PROCEDURES FOR SAE 100R SERIES HYDRAULIC HOSE AND HOSE ASSEMBLIES—SAE J343 JUN80

SAE Standard

Report of the Tube, Pipe, Hose, and Lubrication Fittings Committee, approved June 1968, last revised by the Fluid Conductors and Connectors Technical Committee May 1978, editorial change June 1980.

Scope—This standard is intended to establish uniform methods for the testing and performance evaluation of the SAE 100R Series of hydraulic hose and hose assemblies. The specific tests and performance criteria applicable to each variety of hose and/or assemblies made therefrom are set forth in the respective specifications of SAE J517.

Test Procedures—The test procedures described in the current issue of ASTM D 380, Standard Methods of Testing Rubber Hose, shall be followed. However, in cases of conflict between the ASTM specifications and those described below, the latter shall take precedence.

Standard Tests

Dimensional Check Test—The hose shall conform to all dimensions tabulated in the applicable specification.

Reinforcement diameter and finished outside diameter measurements shall be made by calculation from measurement of the outside circumference. Use of a flexible tape graduated to read the diameter directly shall be acceptable.

Inside diameter measurements shall be made by means of suitable expanding ball or telescoping gages.

Concentricity shall be measured both over the reinforcement and the finished outside diameter using either a dial indicator gage or a micrometer. The foot of the measuring instrument contacting the inside of the hose shall be rounded to conform to the curvature of the hose. The readings shall be taken at 90 deg intervals around the hose and acceptability based on the total variation between high and low readings.

Inside and outside diameter measurements shall be made at a minimum distance of 1.00 in (25.4 mm) and concentricity measurements at a minimum distance of 0.50 in (12.7 mm), back from the ends of the hose.

Proof Test—Hose and/or hose assemblies shall be hydrostatically tested to the specified proof pressure for a period of not less than 30 s nor more than 60 s. There shall be no indication of failure or leakage.

Change in Length Test—Measurements for the determination of elongation or contraction shall be conducted on a previously untested, unaged hose assembly having at least 12.00 in (300 mm) length of free hose between couplings. The hose assembly shall be attached to the pressure source and pressurized to the specified pressure for a period of 30 s, after which time the pressure shall be released. After allowing the hose to restabilize for a period of 30 s following pressure release, reference marks 10.00 in (250 mm) apart shall be accurately placed on the hose outer cover, midway between the hose couplings. The hose shall then be repressurized to the specified pressure for a period of 30 s, after which time, while the hose is pressurized, the distance between the reference marks shall be measured. This length shall be the *final length*.

The change in length shall be computed using the following formula:

$$\% \text{ Change} = \frac{(\text{Final length} - \text{Original length}) \, 100}{\text{Original length}}$$

$(-\%)$ Change = contraction
$(+\%)$ Change = elongation

Burst Test—Hose and/or hose assemblies on which the end fittings have been attached not over 30 days shall be subjected to a hydrostatic pressure increased at a constant rate so as to attain the specified minimum burst pressure within a period of not less than 15 s nor more than 30 s. There shall be no leakage, hose burst, or indication of failure below the specified minimum burst pressure.

Cold Bend Test—Hose and/or hose assemblies shall be subjected to the specified temperature for 24 h in a straight position. After this time and while still at the specified temperature, the sample shall be evenly and uniformly bent over a mandrel having a diameter equal to twice the minimum specified bend radius. Bending shall be accomplished within a period of not less than 8 s nor more than 12 s.

Hoses of less than 1 in (25.4 mm) nominal inside diameter shall be bent 180 deg over the mandrel and hoses of 1 in (25.4 mm) nominal inside diameter and larger shall be bent 90 deg over the mandrel.

After bending, the sample shall be allowed to warm to room temperature, then visually examined for cover cracks and subjected to the proof test. There shall be no cover cracks or leakage. (In lieu of the bending test, hoses of 1 in (25.4 mm) nominal inside diameter and over may be considered acceptable if samples of tube and cover pass the Low Temperature Test on Tube and Cover of ASTM D 380.)

Oil Resistance Test—After 70 h immersion in ASTM No. 3 oil at the designated temperature, the volume change of specimens taken from the hose inner tube and cover shall be within the specified limits.

Ozone Resistance Test—The cover compound shall be tested in accordance with the latest issue of ASTM D 622, procedure 9, and ASTM D 1149. Where space limitations prohibit the use of a hose, specimen cover stock tested in accordance with ASTM D 518, procedure B, may be substituted.

Impulse Test—Impulse testing shall be conducted with unaged hose assemblies and, where the individual standard requires, also with aged assemblies.

The test assemblies shall be impulsed on suitable equipment with the hose bent to the minimum bend radius. Hoses of less than 1 in (25.4 mm) nominal inside diameter shall be bent either 90 or 180 deg and hoses of 1 in (25.4 mm) nominal inside diameter and over shall be bent 90 deg. Individual standards may designate impulsing in a straight position on specific sizes.

The test assembly free length of hose measured between couplings shall be computed using the following:

$$90 \text{ deg bend free length} = \frac{\pi(\text{Min bend radius})}{2} + 2 \text{ (hose OD)}$$

$$180 \text{ deg bend free length} = \pi(\text{Min bend radius}) + 2 \text{ (hose OD)}$$

$$\text{Straight free length} = 14\text{–}18 \text{ in (356–457 mm)}.$$

Where aged samples are required refer to the individual standards.

The test fluid shall be circulated through the assemblies at the specified temperature with a tolerance of $\pm 5°F$ ($\pm 3°C$). The impulse rate shall be 30–75 cpm at the specified pressure. Circulation of the test fluid shall be at a rate which will maintain uniform bore temperature. Cooling or heating of the test chamber shall not be permitted.

The impulse pressure curve must fall entirely within the shaded area of Fig. 1 and should conform as closely as possible to the curve as shown. Unless failure occurs first, the impulse test shall continue for the specified number of cycles.

It is recommended the test fluid be changed frequently to prevent breakdown.

Leakage Test—Unaged hose assemblies on which the end fittings have been attached not over 30 days shall be subjected to a hydrostatic pressure equal to 70% of the specified minimum burst pressure for a period of 5–5.5 min and then reduced to zero after which the 70% of minimum burst pressure shall be reapplied for another 5 min. There shall be no leakage or evidence of failure. This test is to be considered a destructive test and sample shall be destroyed.

Visual Examination of Product—All bulk hose shall be visually inspected to see that the hose identification has been properly applied and all assemblies shall be inspected to see that the correct fittings are properly installed.

Electrical Conductivity Test (for thermoplastic hose only)—Hose assemblies having a free length of 6 ± 0.5 in (152 ± 13 mm) without fluid and capped to prevent entry of moisture shall be exposed to a minimum of 85% relative humidity at $75 \pm 5°F$ ($24 \pm 3°C$) for a period of 168 h. Surface moisture shall be removed prior to testing.

Conditioned assemblies shall have one end fitting attached to the lead from a source of 60 Hz sinusoidal, 37.5 kV (rms) electricity. This lead shall be suspended by dry fabric strings so that the hose hangs free, at least 2 ft (600 mm) from any extraneous objects. The lower end of the hose shall be connected to ground through a 1000–1 000 000 Ω resistor, keeping the resistor near the end of the hose. A suitable a-c voltmeter shall be connected across the resistor, using a fully shielded cable with the shielding well grounded. Thirty-seven and one-half kV shall be applied to the specimen for 5 min and a current reading taken. This current shall not exceed the value specified.

Resistance to Vacuum Test—The hose shall not blister nor show any other indication of failure when subjected to the specified vacuum for a period of 5 min. Where practicable, one end of the hose shall be equipped with a transparent cap and electric light to permit visual examination for failure. Where the length or size of the hose precludes visual examination, failure shall be determined by inability to pass through the hose a ball or cylinder 0.250 in (6.4 mm) less in diameter than the bore of hoses of $\frac{1}{2}$ in (12.7 mm) nominal inside diameter and larger. For hoses under $\frac{1}{2}$ in (12.7 mm) nominal inside diameter, a ball or cylinder 0.125 in (3.2 mm) smaller in diameter than the bore shall be used.

Cubical Expansion Test—Cubical expansion tests shall be run in accordance with the current issue of ASTM D 380.

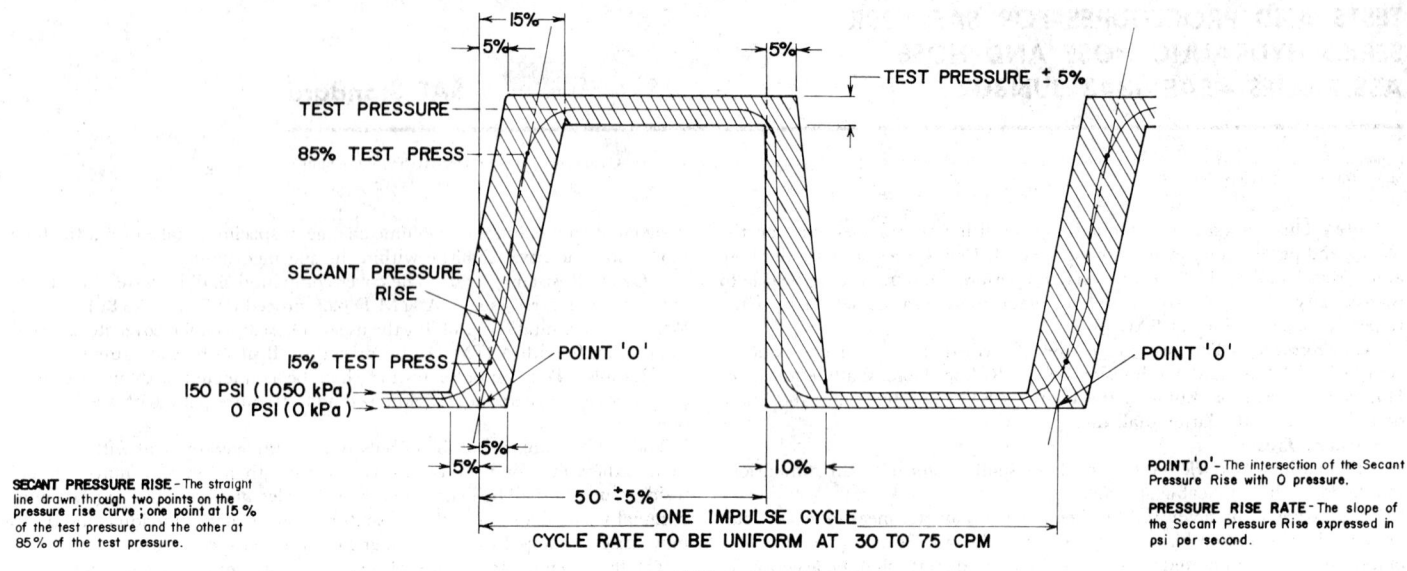

FIG. 1
CYCLE RATE TO BE UNIFORM AT 30-75 CPM

FLEX-IMPULSE TEST PROCEDURE FOR HYDRAULIC HOSE ASSEMBLIES—SAE J1405

SAE Information Report

Report of Fluid Conductors and Connectors Technical Committee approved January 1979.

Scope—The procedures contained herein have been developed to establish a uniform method for comparative impulse testing of hydraulic hose assemblies with and without flexing in order to determine the effect of flexing on ultimate life of hose. The test method minimizes variables to give a comparison between flexing and non-flexing. Basic impulse test parameters are to be in accordance with SAE J343 except as modified to incorporate flexing. This test is not a requirement for SAE J517 hose.

Test Procedure—For optimum validity of comparison, test specimens should be cut from a continuous length of hose with alternate samples along the length designated for flexing and non-flexing impulse test.

Those specimens designated for non-flexing should be tested in accordance with SAE J343. Those specimens designated for flexing are to be made up with free hose length in accordance with the following formula:

Free hose length = 4.142 (min. bend radius) + 3.57 (hose O.D.)

Performance of the flex-impulse test requires a supplementary rig capable of moving one test manifold in a continuous circular pattern as shown in Fig. 1. This manifold is geared so that the center lines of the hose fittings at hose attachment stay parallel at all times. A variable drive is provided, and the number of revolutions per minute are to be controlled to 36 ± 2% of the impulse cycles per minute. This maintains a proportionality between the number of cycles of flexing and impulse and assures that the test specimen is in a different configuration on each succeeding impulse.

The vertical centerline of a stationary manifold is positioned a distance *A* from the center of revolution of the revolving manifold. This distance was determined empirically such that the test specimen is subjected to back bending motion near each fitting with the radius of bend at that point being greater than the aplicable SAE minimum bend radius. However, when the revolving manifold reaches the position nearest the stationary manifold, the bend radius inside the loop is smaller than the applicable SAE minimum bend radius.[1] Distance *A* is calculated with the following formula:

1.75 (min. bend radius) + hose O.D.

Specimens for flex-impulse testing should be mounted with straight end fitting on the rig as described above using care to avoid imparting twist to the hose. (Angular fittings may be used, provided they are installed in such a position to assure the hose travel and geometry of Fig. 1). A like number of samples, preferably not less than three, should be tested simultaneously and should be run to failure.

To accelerate completion of the test for comparative purposes, a pressure based on actual burst values of the hose is recommended, with flexing and non-flexing specimens to be tested at the same pressure. Suggested procedure is to first determine the average burst strength for the test length of hose and from this calculate the impulse test pressure as 35% of average burst. If this test procedure does not produce failures within the desired range, a higher or lower percentage may be used.

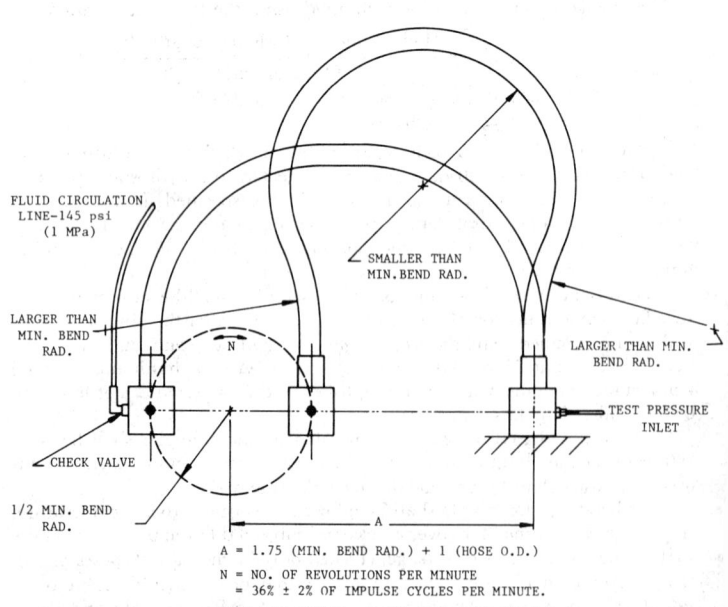

A = 1.75 (MIN. BEND RAD.) + 1 (HOSE O.D.)
N = NO. OF REVOLUTIONS PER MINUTE
  = 36% ± 2% OF IMPULSE CYCLES PER MINUTE.
FREE HOSE LENGTH = 4.142 (MIN. BEND RAD.) + 3.57 (HOSE O.D.)

FIG. 1

---

[1] Violation of the minimum bend radius for this test does not imply that such violation is recommended in applications.

TESTS AND PROCEDURES FOR HIGH-TEMPERATURE TRANSMISSION OIL HOSE, LUBRICATING OIL HOSE AND HOSE ASSEMBLIES—SAE J1019

SAE Standard

Report of Tube, Pipe, Hose, and Lubrication Fittings Committee approved April 1973.

1. Scope—This standard is intended to establish uniform methods for testing and evaluation of hose and hose assemblies for use in: high-temperature transmission oil systems and high-temperature lubricating oil systems.

2. Test Procedures—The test procedures described in the current issue of ASTM D 380, Standard Methods of Testing Rubber Hose, shall be followed where applicable. However, in cases of conflict between the ASTM specifications and those described below, the latter shall take precedence.

3. Qualification Tests—For qualification to this specification, hose and/or assemblies made therefrom shall conform to the following tests and requirements:

3.1 Dimensional Check Test—The hose shall conform to all dimensions tabulated in the required customer specification.

Reinforcement diameter and finished outside diameter measurements shall be made by calculation from measurement of the outside circumference. Use of a flexible tape graduated to read the diameter directly shall be acceptable.

Inside diameter measurements shall be made by means of suitable expanding ball or telescoping gages.

Concentricity shall be measured both over the reinforcement and the finished outside diameter using either a dial indicator gage or a micrometer. The foot of the measuring instrument contacting the inside of the hose shall be rounded to conform to the curvature of the hose. The readings shall be taken at 90 deg intervals around the hose and acceptability based on the total variation between high and low readings.

Inside and outside diameter measurements shall be made at a minimum distance of 1.00 in. (25 mm) and concentricity measurements at a minimum distance of 0.50 in. (12.5 mm) back from the ends of the hose.

3.2 Proof Test—Hose and/or hose assemblies shall be hydrostatically tested to the proof pressure specified by the end user for a period of not less than 30 s nor more than 60 s. There shall be no indication of failure or leakage.

3.3 Change in Length Test—Measurements for the determination of elongation or contraction shall be conducted on a previously untested, unaged hose assembly having at least 12.00 in. (300 mm) length of free hose between hose couplings. The hose assembly shall be attached to the pressure source and pressurized to the operating pressure specified by the end user for a period of 30 s, after which time the pressure shall be released. After allowing the hose to restabilize for a period of 30 s following pressure release, reference marks 10.00 in. (250 mm) apart shall be accurately placed on the hose outer cover, midway between the hose couplings. The hose shall then be repressurized to the specified pressure for a period of 30 s, after which time, while the hose is pressurized, the distance between the reference marks shall be measured. This length shall be the "final length."

The change in length shall be computed using the following formula:

$$\% \text{ change} = \frac{(\text{Final length} - \text{Original length}) \, 100}{\text{Original length}}$$

(−%) change = contraction
(+%) change = elongation

3.4 Burst Test—Hose and/or hose assemblies on which the end fittings have been attached not over 30 d shall be subjected to a hydrostatic pressure increased at a constant rate so as to attain the minimum burst pressure specified by the end user within a period of not less than 15 s nor more than 30 s. There shall be no leakage, hose burst, or indication of failure below the specified minimum burst pressure.

3.5 Cold Flexibility Test—Hose and/or hose assemblies shall be subjected to −40 +0, −5°F (−40 +0, −3°C) for 24 h in a straight position. After this time and while still at the specified temperature, the sample shall be evenly and uniformly bent over a mandrel having a diameter equal to twice the minimum specified bend radius. Bending shall be accomplished within a period of not less than 8 s nor more than 12 s.

Hoses of less than 1 in. (25 mm) nominal inside diameter shall be bent 180 deg over the mandrel and hoses of 1 in. (25 mm) nominal inside diameter and larger shall be bent 90 deg over the mandrel

After flexing, the sample shall be allowed to warm to room temperature, then visually examined for cover cracks and subjected to the proof test. There shall be no cover cracks or leakage. (In lieu of the flexing test, hoses of 1 in. (25 mm) nominal inside diameter and larger may be considered acceptable if samples of tube and cover pass the Low Temperature Test on Tube and Cover of ASTM D 380.)

3.6 Ozone Resistance Test—The cover compound shall be tested in accordance with the latest issue of ASTM D 518, Procedure B—Exposure of Looped Test Specimens, and ASTM D 1149. (Does not apply to fabric covered hose.)

3.7 High-Temperature Circulation Test—Test assemblies having a minimum free length of 14 in. (355 mm) of hose between couplings shall be mounted on a circulating oil test unit in a straight configuration.

The ambient temperature shall be 200 ±20°F (93 ±11°C) and the oil temperature 300 ±5°F (149 ±3°C) between the inlet and outlet.

Either transmission Type A fluid, or lubricating oil to Mil-L-2104C as specified shall be circulated through the hose assemblies.

The circulating pressure should be 50–100 psi (345–690kPa).

Entrained air in the lines must be kept to a minimum and caution must be exercised so that the free hose in the test assemblies not be in contact with the heating elements and located in such a manner as to permit good air circulation.

The test fluid should be changed every 375 ±25 h. Tests are to be run continuously except for oil change and addition or removal of samples. All shutdown time is to be recorded.

After 750 ±5 h, the test assemblies shall be removed, the oil drained, and hose allowed to cool for a minimum of 4 h.

The aged samples shall then be bent around a mandrel having a diameter 12 times the inside diameter of the hose and the cover examined (if rubber covered). No visible cracking will be permitted. The time required to bend the hose around the mandrel shall be between 8 and 12 s.

The assemblies shall then be subjected to a pressure test in a straight position for a period of not less than 30 s nor more than 60 s as follows:

Transmission oil assemblies: 500 psi (3.4 MPa)
Lubricating oil assemblies: 200 psi (1.4 MPa)

There shall be no leakage through the hose or at the hose fitting juncture.

4. Inspection Tests—Inspection tests and lot sizes for inspection shall be negotiated.

HYDRAULIC FLANGED TUBE, PIPE, AND HOSE CONNECTIONS, 4-BOLT SPLIT FLANGE TYPE—SAE J518c

SAE Standard

Report of Construction and Industrial Machinery Technical Committee and Tube, Pipe, Hose, and Lubrication Fittings Committee approved February 1952 and last revised by Tube, Pipe, Hose, and Lubrication Fittings Committee May 1972.

GENERAL SPECIFICATIONS

Scope—This standard covers complete general and dimensional specifications for the flanged heads and split flange clamp halves applicable to 4-bolt split flange type tube, pipe, and hose connections with appropriate references to the O-ring seals and attaching components used in their assembly. Also included are recommended port dimensions and port design considerations.

The flanged heads specified are incorporated into fittings having suitable means for attachment to tubes, pipes, or hoses to provide connection ends. These connections are intended for application in hydraulic systems, on industrial and commercial products, where it is desired to avoid the use of threaded connections.

Flanged heads shall be as specified in Fig. 3 and Table 1. Split flange clamp halves shall be as specified in Fig. 4 and Table 1. Port dimensions and spacing shall be as specified in Fig. 5 and Table 2.

O-ring seals, having nominal dimensions as indicated in Table 1, are used in conjunction with these connections. They shall conform to the seals specified in SAE J120, Table on Dimensions and Tolerances.

Bolts for use with these connections shall be of the sizes and lengths indicated in Table 1. They shall conform with the finished hexagon bolts specified in SAE J105. They shall be of SAE Grade 5 material or better as specified in SAE J429. Socket head cap screws of SAE Grade 5 material or better are acceptable.

Lock washers, if used, shall be in accordance with the light spring lock washers specified in SAE J489, Dimensions of Light, Medium, Heavy, Extra Heavy, and Hi Collar Spring Lock Washers, and of sizes applicable to the corresponding bolts.

The following general specifications supplement the dimensional data contained in Table 1 with respect to all unspecified detail.

TABLE 1A—DIMENSIONS OF HYDRAULIC FLANGED CONNECTIONS, STANDARD PRESSURE SERIES (CODE 61)

Nominal Flange Size, in	Flange Dash Size	A Dia Max (in)	A Dia Max (mm)	B Dia (in)	B Dia (mm)	C Dia ±0.010 in	C Dia ±0.25 mm	D Dia ±0.010 in	D Dia ±0.25 mm	E ±0.005 in	E ±0.13 mm	F ±0.005 in	F ±0.13 mm
1/2	-8	0.50	13	1.005-1.000	25.53- 25.40	1.188	30.18	1.219	30.96	0.265	6.73	0.245	6.22
3/4	-12	0.75	19	1.255-1.250	31.88- 31.75	1.500	38.10	1.531	38.89	0.265	6.73	0.245	6.22
1	-16	1.00	25	1.565-1.560	39.75- 39.62	1.750	44.45	1.781	45.24	0.315	8.00	0.295	7.49
1-1/4	-20	1.25	32	1.755-1.750	44.58- 44.45	2.000	50.80	2.031	51.59	0.315	8.00	0.295	7.49
1-1/2	-24	1.50	38	2.125-2.115	53.98- 53.72	2.375	60.33	2.406	61.09	0.315	8.00	0.295	7.49
2	-32	2.00	51	2.500-2.490	63.50- 63.25	2.812	71.42	2.844	72.24	0.375	9.53	0.355	9.02
2-1/2	-40	2.50	64	3.005-2.995	76.33- 76.07	3.312	84.12	3.344	84.94	0.375	9.53	0.355	9.02
3	-48	3.00	76	3.625-3.615	92.08- 91.82	4.000	101.60	4.031	102.39	0.375	9.53	0.355	9.02
3-1/2	-56	3.50	89	4.115-4.095	104.52-104.01	4.500	114.30	4.531	115.09	0.442	11.23	0.422	10.72
4	-64	4.00	102	4.615-4.595	117.22-116.71	5.000	127.00	5.031	127.79	0.442	11.23	0.422	10.72
5	-80	5.00	127	5.615-5.595	142.62-142.11	6.000	152.40	6.031	153.19	0.442	11.23	0.422	10.72

Nominal Flange Size, in	G Dia Max (in)	G Dia Max (mm)	H Dia Max (in)	H Dia Max (mm)	J Dia ±0.010 in	J Dia ±0.25 mm	K Ref (in)	K Ref (mm)	L ID Ref (in)	L ID Ref (mm)	M OD Ref (in)	M OD Ref (mm)	N Dia Ref (in)	N Dia Ref (mm)	O-Ring Size No.
1/2	0.56	14	0.94	24	0.955	24.26	0.50	13	0.734	18.64	1.012	25.70	0.139	3.53	210
3/4	0.81	21	1.25	32	1.265	32.13	0.56	14	0.984	24.99	1.262	32.05	0.139	3.53	214
1	1.06	27	1.50	38	1.515	38.48	0.56	14	1.296	32.92	1.574	39.98	0.139	3.53	219
1-1/4	1.31	33	1.70	43	1.720	43.69	0.56	14	1.484	37.69	1.762	44.75	0.139	3.53	222
1-1/2	1.56	40	1.98	50	2.000	50.80	0.62	16	1.859	47.22	2.137	54.28	0.139	3.53	225
2	2.06	52	2.45	62	2.470	62.74	0.62	16	2.234	56.74	2.512	63.80	0.139	3.53	228
2-1/2	2.56	65	2.92	74	2.950	74.93	0.69	18	2.734	69.44	3.012	76.50	0.139	3.53	232
3	3.06	78	3.55	90	3.580	90.93	0.75	19	3.359	85.32	3.637	92.38	0.139	3.53	237
3-1/2	3.56	90	4.00	102	4.030	102.36	0.88	22	3.859	98.02	4.137	105.08	0.139	3.53	241
4	4.06	103	4.50	114	4.530	115.06	1.00	25	4.359	110.72	4.637	117.78	0.139	3.53	245
5	5.06	129	5.50	140	5.530	140.46	1.12	28	5.359	136.12	5.637	143.18	0.139	3.53	253

Nominal Flange Size, in	O (in)	O (mm)	P ±0.03 in	P ±0.8 mm	Q ±0.010 in	Q ±0.25 mm	R (in)	R (mm)	S Rad (in)	S Rad (mm)	T Dia ±0.010 in	T Dia ±0.25 mm	U (in)	U (mm)	V (in)	V (mm)
1/2	2.16-2.09	54.9- 53.1	0.86	21.8	1.500	38.10	0.31	8	0.31	8	0.344	8.74	0.50	13	0.75	19
3/4	2.59-2.53	65.8- 64.3	0.98	24.9	1.875	47.63	0.40	10	0.34	9	0.406	10.31	0.56	14	0.88	22
1	2.78-2.72	70.6- 69.1	1.11	28.2	2.062	52.37	0.48	12	0.34	9	0.406	10.31	0.62	16	0.94	24
1-1/4	3.16-3.09	80.3- 78.5	1.39	35.3	2.312	58.72	0.56	14	0.41	10	0.469	11.91	0.56	14	0.88	22
1-1/2	3.72-3.66	94.5- 93.0	1.58	40.1	2.750	69.85	0.67	17	0.47	12	0.531	13.49	0.62	16	1.00	25
2	4.06-3.94	103.1-100.1	1.86	47.2	3.062	77.77	0.81	21	0.47	12	0.531	13.49	0.62	16	1.03	26
2-1/2	4.56-4.44	115.8-112.8	2.09	53.1	3.500	88.90	0.96	24	0.50	13	0.531	13.49	0.75	19	1.50	38
3	5.38-5.25	136.7-133.4	2.53	64.3	4.188	106.38	1.18	30	0.56	14	0.656	16.66	0.88	22	1.62	41
3-1/2	6.06-5.94	153.9-150.9	2.70	68.6	4.750	120.65	1.34	34	0.62	16	0.656	16.66	0.88	22	1.12	28
4	6.44-6.31	163.6-160.3	2.95	74.9	5.125	130.18	1.49	38	0.62	16	0.656	16.66	1.00	25	1.38	35
5	7.31-7.19	185.7-182.6	3.52	89.4	6.000	152.40	1.78	45	0.62	16	0.656	16.66	1.12	28	1.62	41

Nominal Flange Size, in	Bolt Dimensions Thread	Bolt Dimensions Length (in)	Bolt Dimensions Length (mm)	W ±0.010 in	W ±0.25 mm	X ±0.010 in	X ±0.25 mm	Max Working Pressure psi	Max Working Pressure MPa	Recommended Bolt Torque lb-in	Recommended Bolt Torque N·m
1/2	5/16-18	1-1/4	32	0.750	19.05	0.344	8.74	5000	34.5	175- 225	20- 25
3/4	3/8-16	1-1/4	32	0.938	23.83	0.438	11.13	5000	34.5	250- 350	28- 40
1	3/8-16	1-1/4	32	1.031	26.19	0.515	13.08	5000	34.5	325- 425	37- 48
1-1/4	7/16-14	1-1/2	38	1.156	29.36	0.594	15.09	4000	27.6	425- 550	48- 62
1-1/2	1/2-13	1-1/2	38	1.375	34.93	0.703	17.86	3000	20.7	550- 700	62- 79
2	1/2-13	1-1/2	38	1.531	38.89	0.844	21.44	3000	20.7	650- 800	73- 90
2-1/2	1/2-13	1-3/4	44	1.750	44.45	1.000	25.40	2500	17.2	950-1100	107-124
3	5/8-11	1-3/4	44	2.094	53.19	1.219	30.96	2000	13.8	1650-1800	186-203
3-1/2	5/8-11	2	51	2.375	60.33	1.375	34.93	500	3.4	1400-1600	158-181
4	5/8-11	2	51	2.562	65.07	1.531	38.89	500	3.4	1400-1600	158-181
5	5/8-11	2-1/4	57	3.000	76.20	1.812	46.02	500	3.4	1400-1600	158-181

Size Designation—4-bolt split flange connection sizes are designated by the nominal flange size which corresponds to the maximum inside diameter of the hole through the flanged head.

Dimensions and Tolerances—Tabulated dimensions and tolerances shall apply to the finished parts, plated or otherwise processed, as specified by the purchaser. Tolerances on all dimensions for flanged heads, split flange clamp halves and ports not otherwise limited shall be ±0.016 in. (±0.4 mm).

Material—Flanged heads shall be made of steel. Split flange clamp halves shall be made from a material with the following properties:

Standard series—1/2 in (-8) size	Minimum yield, 32,000 psi (221 MPa) Minimum elongation, 3%
All other sizes	Minimum yield, 60,000 psi (414 MPa) Minimum elongation, 3%
High pressure series—all sizes	Minimum yield, 48,000 psi (331 MPa) Minimum elongation, 3%

Finish—Unless otherwise specified by the purchaser, the flanged heads and split flange clamp halves shall be furnished with the following finishes:
1. Cadmium or zinc plated to a thickness of 0.0002 in. (0.005 mm) minimum followed by a chromate treatment. These parts must meet the requirements of a 32 h salt spray test in accordance with ASTM B 117, Method of Salt Spray (Fog) Testing.
2. Phosphate coated (oil finish). These parts must meet the requirements of a 16 h salt spray test in accordance with ASTM B 117.

Bolts shall be finished with a phosphate coating (oil finished). These parts must meet the requirements of a 16 h salt spray test in accordance with ASTM B 117. Lock washers shall have a plain (natural) finish.

Workmanship—Workmanship shall conform to the best commercial practice to produce high quality connection components. Connection components shall be free from all hanging burrs, loose scale, and slivers which might become dislodged in usage and all other defects which might affect their serviceability. All sealing surfaces must be smooth except that annular tool marks up to 100 μin. (3 μm) max shall be permissible.

TABLE 1B—DIMENSIONS OF HYDRAULIC FLANGED CONNECTIONS, HIGH PRESSURE SERIES (CODE 62)

Nominal Flange Size, in	Flange Dash Size	A Dia Max		B Dia		C Dia		D Dia		E		F	
						±0.010 in	±0.25 mm	±0.010 in	±0.25 mm	±0.005 in	±0.13 mm	±0.005 in	±0.13 mm
		in	mm	in	mm								
1/2	-8	0.50	13	1.005-1.000	25.53-25.40	1.250	31.75	1.281	32.54	0.305	7.75	0.285	7.24
3/4	-12	0.75	19	1.255-1.250	31.88-31.75	1.625	41.28	1.656	42.06	0.345	8.76	0.325	8.26
1	-16	1.00	25	1.565-1.560	39.75-39.62	1.875	47.63	1.906	48.41	0.375	9.53	0.355	9.02
1-1/4	-20	1.25	32	1.755-1.750	44.58-44.45	2.125	53.98	2.156	54.76	0.405	10.29	0.385	9.78
1-1/2	-24	1.50	38	2.125-2.115	53.98-53.72	2.500	63.50	2.531	64.29	0.495	12.57	0.475	12.07
2	-32	2.00	51	2.500-2.490	63.50-63.25	3.125	79.38	3.156	80.16	0.495	12.57	0.475	12.07

Nominal Flange Size, in	G Dia Max		H Dia Max		J Dia		K Ref		L ID Ref		M OD Ref		N Dia Ref		O-Ring
					±0.010 in	±0.25 mm									Size No.
	in	mm	in	mm	in	mm	in	mm	in	mm	in	mm	in	mm	
1/2	0.56	14	0.94	24	0.970	24.64	0.56	14	0.734	18.64	1.012	25.70	0.139	3.53	210
3/4	0.81	21	1.25	32	1.280	32.51	0.69	18	0.984	24.99	1.262	32.05	0.139	3.53	214
1	1.06	27	1.50	38	1.530	38.86	0.81	21	1.296	32.92	1.574	39.98	0.139	3.53	219
1-1/4	1.31	33	1.72	44	1.750	44.45	1.00	25	1.484	37.69	1.762	44.75	0.139	3.53	222
1-1/2	1.56	40	2.00	51	2.030	51.56	1.19	30	1.859	47.22	2.137	54.28	0.139	3.53	225
2	2.06	52	2.62	67	2.660	67.56	1.50	38	2.234	56.74	2.512	63.80	0.139	3.53	228

Nominal Flange Size, in	O		P		Q		R		S Rad		T Dia		U		V	
			±0.03 in	±0.8 mm	±0.010 in	±0.25 mm					±0.010 in	±0.25 mm				
	in	mm	in	mm	in	mm	in	mm	in	mm	in	mm	in	mm	in	mm
1/2	2.25-2.19	57.2-55.6	0.89	22.6	1.594	40.49	0.32	8	0.31	8	0.344	8.74	0.62	16	0.88	22
3/4	2.84-2.78	72.1-70.6	1.14	29.0	2.000	50.80	0.43	11	0.41	10	0.406	10.31	0.75	19	1.12	28
1	3.22-3.16	81.8-80.3	1.33	33.8	2.250	57.15	0.51	13	0.47	12	0.469	11.91	0.94	24	1.31	33
1-1/4	3.78-3.72	96.0-94.5	1.48	37.6	2.625	66.68	0.59	15	0.56	14	0.531	13.49	1.06	27	1.50	38
1-1/2	4.50-4.38	114.3-111.3	1.83	46.5	3.125	79.38	0.68	17	0.66	17	0.656	16.66	1.19	30	1.69	43
2	5.31-5.19	134.9-131.8	2.20	55.9	3.812	96.82	0.84	21	0.72	18	0.781	19.84	1.44	37	2.06	52

Nominal Flange Size, in	Bolt Dimensions			W		X		Maximum Recommended Working Pressure		Recommended Bolt Torque Range	
	Thread	Length		±0.010 in	±0.25 mm	±0.010 in	±0.25 mm	psi	MPa	lb-in	N·m
		in	mm								
1/2	5/16-18	1-1/4	32	0.797	20.24	0.359	9.12	6000	41.4	175-225	20-25
3/4	3/8-16	1-1/2	38	1.000	25.40	0.469	11.91	6000	41.4	300-400	34-45
1	7/16-14	1-3/4	44	1.125	28.58	0.547	13.89	6000	41.4	500-600	56-68
1-1/4	1/2-13	1-3/4	44	1.312	33.32	0.625	15.88	6000	41.4	750-900	85-102
1-1/2	5/8-11	2-1/4	57	1.562	39.67	0.719	18.26	6000	41.4	1400-1600	158-181
2	3/4-10	2-3/4	70	1.906	48.41	0.875	22.23	6000	41.4	2400-2600	271-294

21.110

NOTES: UNSPECIFIED DETAIL WITH RESPECT TO DIMENSIONS, TOLERANCES, CONTOURS, MATERIAL, WORKMANSHIP, ETC., MUST CONFORM TO GENERAL SPECIFICATIONS FOR HYDRAULIC FLANGED TUBE, PIPE, AND HOSE CONNECTIONS, 4-BOLT SPLIT FLANGE TYPE. DIMENSIONS IN FIGS. 1-4 APPLY TO TABLE 1. CODES SHOWN IN BRACKETS ADJACENT TO FIGURE NUMBERS REPRESENT RESPECTIVE FLANGED CONNECTION IDENTIFICATION, WITH XX SUBSTITUTED FOR THE PRESSURE RATING CODE DEPICTED IN RESPECTIVE SUBHEADINGS OF TABLE 1, IN ACCORDANCE WITH SAE J846.

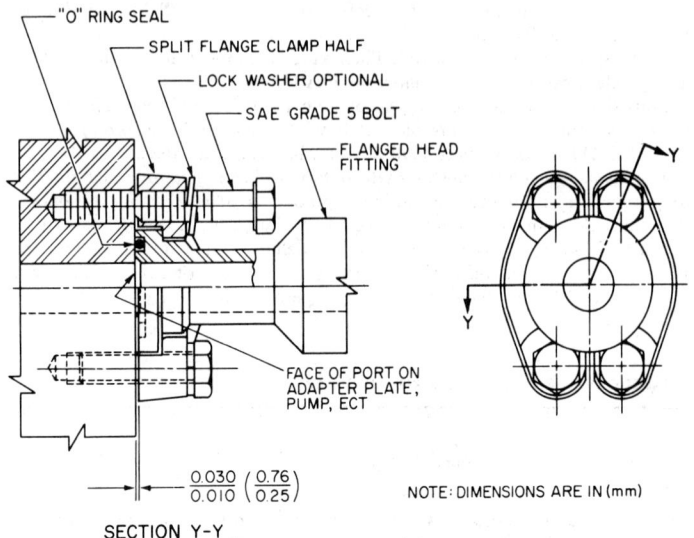

FIG. 1—ASSEMBLED SPLIT FLANGED CONNECTION

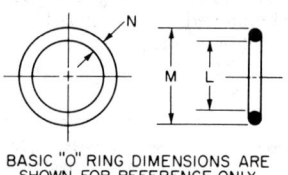

FIG. 2—O-RING SEAL

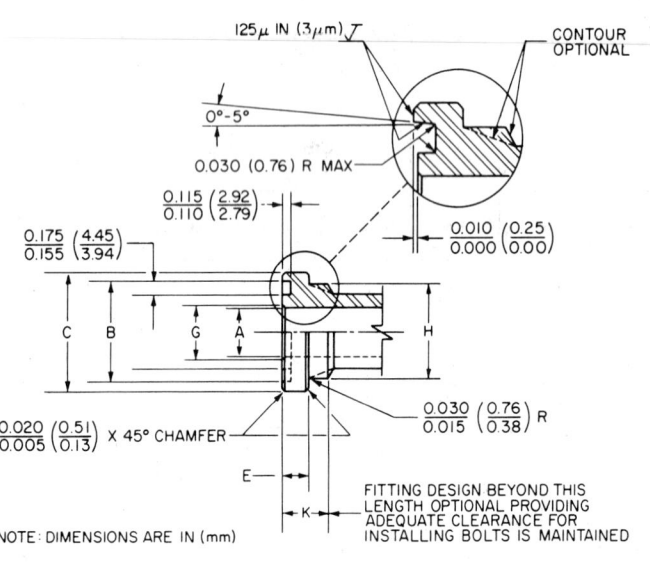

FIG. 3—FLANGED HEAD

FIG. 4—SPLIT FLANGE CLAMP HALF (1101XX)

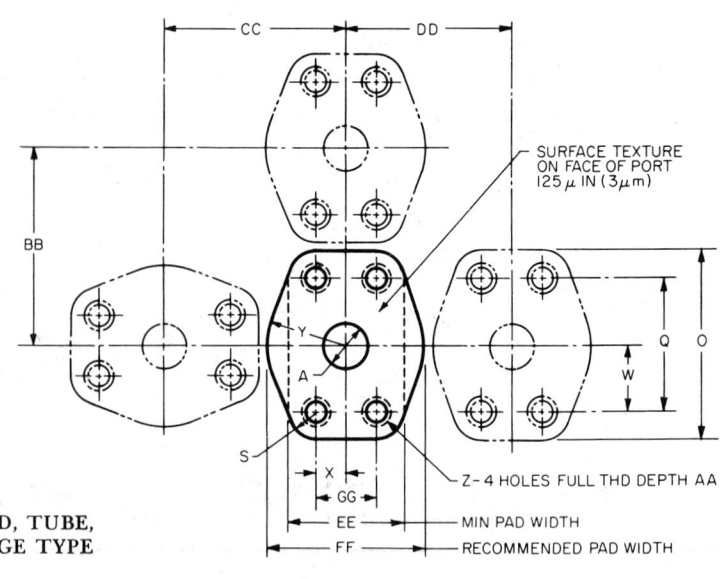

FIG. 5—PORT DIMENSIONS FOR HYDRAULIC FLANGED, TUBE, PIPE, AND HOSE CONNECTIONS, 4-BOLT SPLIT FLANGE TYPE

TABLE 2A—PORT DIMENSIONS FOR BOLTED FLANGE CONNECTIONS, STANDARD PRESSURE SERIES

Nominal Flange Size, in	Flange Dash Size	A Dia +0.00 -0.06 in	A Dia +0.0 -1.5 mm	O in	O mm	FF in	FF mm	Q ±0.010 in	Q ±0.25 mm	GG ±0.010 in	GG ±0.25 mm	S Rad in	S Rad mm	W in	W mm
1/2	-8	0.50	12.7	2.12	54	1.81	46	1.500	38.10	0.688	17.48	0.31	8	0.75	19
3/4	-12	0.75	19.1	2.56	65	2.06	52	1.875	47.63	0.875	22.23	0.34	9	0.94	24
1	-16	1.00	25.4	2.75	70	2.31	59	2.062	52.37	1.031	26.19	0.34	9	1.03	26
1-1/4	-20	1.25	31.8	3.12	79	2.88	73	2.312	58.72	1.188	30.18	0.41	10	1.16	29
1-1/2	-24	1.50	38.1	3.69	94	3.25	83	2.750	69.85	1.406	35.71	0.47	12	1.38	35
2	-32	2.00	50.8	4.00	102	3.81	97	3.062	77.77	1.688	42.88	0.47	12	1.53	39
2-1/2	-40	2.50	63.5	4.50	114	4.28	109	3.500	88.90	2.000	50.80	0.50	13	1.75	44
3	-48	3.00	76.2	5.31	135	5.16	131	4.188	106.38	2.438	61.93	0.56	14	2.09	53
3-1/2	-56	3.50	88.9	6.00	152	5.50	140	4.750	120.65	2.750	69.85	0.62	16	2.38	60
4	-64	4.00	101.6	6.38	162	6.00	152	5.125	130.18	3.062	77.77	0.62	16	2.56	65
5	-80	5.00	127.0	7.25	184	7.12	181	6.000	152.40	3.625	92.08	0.62	16	3.00	76

Nominal Flange Size, in	X in	X mm	Y Rad in	Y Rad mm	Z Thread UNC-2B	AA Min in	AA Min mm	BB[a] Min in	BB[a] Min mm	CC[a] Min in	CC[a] Min mm	DD[a] Min in	DD[a] Min mm	EE Min in	EE Min mm
1/2	0.34	9	0.91	23	5/16-18	0.94	24	2.22	56	2.06	52	1.91	49	1.31	33
3/4	0.44	11	1.03	26	3/8-16	0.88	22	2.66	68	2.41	61	2.16	55	1.62	41
1	0.52	13	1.16	29	3/8-16	0.88	22	2.84	72	2.62	67	2.41	61	1.88	48
1-1/4	0.59	15	1.44	37	7/16-14	1.12	28	3.22	82	3.09	78	2.97	75	2.12	54
1-1/2	0.70	18	1.62	41	1/2-13	1.06	27	3.78	96	3.56	90	3.34	85	2.50	64
2	0.84	21	1.91	49	1/2-13	1.06	27	4.09	104	4.00	102	3.91	99	3.00	76
2-1/2	1.00	25	2.14	54	1/2-13	1.19	30	4.59	117	4.50	114	4.38	111	3.50	89
3	1.22	31	2.58	66	5/8-11	1.19	30	5.41	137	5.34	136	5.25	133	4.19	106
3-1/2	1.38	35	2.75	70	5/8-11	1.31	33	6.09	155	5.84	148	5.59	142	4.69	119
4	1.53	39	3.00	76	5/8-11	1.19	30	6.47	164	6.28	160	6.09	155	5.19	132
5	1.81	46	3.56	90	5/8-11	1.31	33	7.34	186	7.28	185	7.22	183	6.19	151

[a]Dimensions BB, CC, and DD provide 0.06 in (1.5 mm) clearance between flanges, dimensionally on the high limit, when the same size flanges are used on adjacent ports. These dimensions do not apply when more than one size of flanges are used on adjacent ports.

TABLE 2B—PORT DIMENSIONS FOR BOLTED FLANGE CONNECTIONS, HIGH PRESSURE SERIES

Nominal Flange Size, in	Flange Dash Size	A Dia +0.00 -0.06 in	A Dia +0.0 -1.5 mm	O in	O mm	FF in	FF mm	Q ±0.010 in	Q ±0.25 mm	GG ±0.010 in	GG ±0.25 mm	S Rad in	S Rad mm	W in	W mm
1/2	-8	0.50	12.7	2.22	56	1.88	48	1.574	40.49	0.718	18.24	0.31	8	0.80	20
3/4	-12	0.75	19.1	2.81	71	2.38	60	2.000	50.80	0.937	23.80	0.41	10	1.00	25
1	-16	1.00	25.4	3.19	81	2.75	70	2.250	57.15	1.093	27.76	0.47	12	1.12	28
1-1/4	-20	1.25	31.8	3.75	95	3.06	78	2.625	66.68	1.250	31.75	0.56	14	1.31	33
1-1/2	-24	1.50	38.1	4.44	113	3.75	95	3.125	79.38	1.437	36.50	0.66	17	1.56	40
2	-32	2.00	50.8	5.25	133	4.50	114	3.812	96.82	1.750	44.45	0.72	18	1.91	49

Nominal Flange Size, in	X in	X mm	Y Rad in	Y Rad mm	Z Thread UNC-2B	AA Min in	AA Min mm	BB[a] Min in	BB[a] Min mm	CC[a] Min in	CC[a] Min mm	DD[a] Min in	DD[a] Min mm	EE Min in	EE Min mm
1/2	0.36	9	0.94	24	5/16-18	0.81	21	2.34	59	2.22	56	2.09	53	1.50	38
3/4	0.47	12	1.19	30	3/8-16	0.94	24	2.94	75	2.75	70	2.59	66	1.88	48
1	0.55	14	1.38	35	7/16-14	1.06	27	3.31	84	3.16	80	2.97	75	2.12	54
1-1/4	0.62	16	1.53	39	1/2-13	1.00	25	3.88	99	3.56	90	3.25	83	2.38	60
1-1/2	0.72	18	1.88	48	5/8-11	1.38	35	4.56	116	4.25	108	3.97	101	2.75	70
2	0.88	22	2.25	57	3/4-10	1.50	38	5.38	137	5.03	128	4.72	120	3.38	86

[a]Dimensions BB, CC, and DD provide 0.06 in (1.5 mm) clearance between flanges, dimensionally on the high limit, when the same size flanges are used on adjacent ports. These dimensions do not apply when more than one size of flanges are used on adjacent ports.

HYDRAULIC "O" RING—SAE J515a

SAE Standard

Report of Tube, Pipe, Hose, and Lubrication Fittings Committee approved January 1956 and last revised June 1962. Editorial change June 1967.

TYPE 1—PETROLEUM BASE AND NONFLAMMABLE WATER BASE HYDRAULIC FLUIDS

Compliance with—Compound SB915 shown in SAE Standard, Specifications for Elastomer Compounds for Automotive Applications—SAE J14, and ASTM D 735 Suffix letters B, E1, E3, and Z[1].

General Service—High Pressure Applications Pneumatics, Industrial Lubricating Oil, Hydraulic Oils and Gasoline.

Temperature Range	−30[1] to +250 F
Shore Hardness A	90 ±5
Elongation	100% min
Tensile	1500 psi min
Compound	Buna N

Lubrication—When assembling Type 1 "O" rings with "O" ring style fittings the "O" ring shall be coated with a light oil or petroleum before assembly.

[1] Special Requirement: Low temperature test to conform to ASTM D 736, Method of Test for Low-Temperature Brittleness of Rubber and Rubber-Like Materials, at the low temperature limit specified.

TYPE 2—NONFLAMMABLE PHOSPHATE ESTER BASE HYDRAULIC FLUIDS

Identification—Identify with nonpermanent orange ink around the periphery.

General Service—High pressure application of nonflammable hydraulic fluids of the phosphate ester base type.

Temperature Range	−40 to +212 F
Shore Hardness A	88 ±5
Elongation	100%
Tensile	1000 psi min
Compound	Butyl or equivalent resistant compound

Lubrication—When assembling Type 2 "O" rings with "O" ring style fittings lubricate the "O" ring with the fluid used in the system.

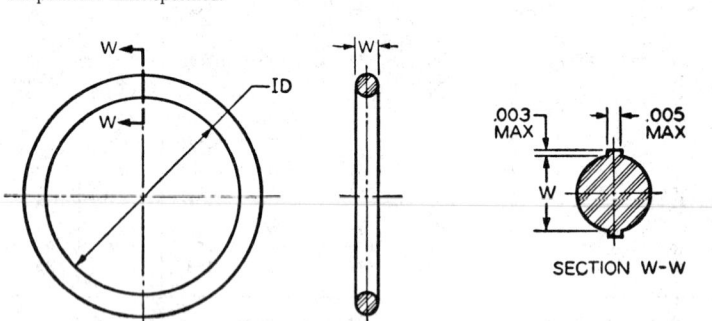

FIG. 1—HYDRAULIC "O" RING

TABLE 1—"O" RING DIMENSIONS, IN. (FIG. 1)

Tubing OD[a]	Dash No.[a]		W, Width, Dia	Inside Dia
	Type 1	Type 2		
1/8	2-1	2-2	0.064 ±0.003	0.239 ±0.005
3/16	3-1	3-2	0.064 ±0.003	0.301 ±0.005
1/4	4-1	4-2	0.072 ±0.003	0.351 ±0.005
5/16	5-1	5-2	0.072 ±0.003	0.414 ±0.005
3/8	6-1	6-2	0.078 ±0.003	0.468 ±0.005
1/2	8-1	8-2	0.087 ±0.003	0.644 ±0.005
5/8	10-1	10-2	0.097 ±0.003	0.755 ±0.005
3/4	12-1	12-2	0.116 ±0.004	0.924 ±0.006
7/8	14-1	14-2	0.116 ±0.004	1.048 ±0.006
1	16-1	16-2	0.116 ±0.004	1.171 ±0.006
1-1/4	20-1	20-2	0.118 ±0.004	1.475 ±0.010
1-1/2	24-1	24-2	0.118 ±0.004	1.720 ±0.010
2	32-1	32-2	0.118 ±0.004	2.337 ±0.010

[a] Tubing OD specified as means of designating which seal is to be used in a given size "O" ring boss assembly.

FORMED TUBE ENDS FOR HOSE CONNECTIONS—SAE J962b

SAE Standard

Report of Tube, Pipe, Hose and Lubrication Fittings Committee approved June 1966 and last revised June 1976. Approved by the American National Standards Institute November, 1976.

GENERAL SPECIFICATIONS

Scope—This SAE Standard covers the dimensional and general specifications applicable to those formed tube end configurations suitable for hose connections made with or without hose clamps (see SAE J536) in relatively low pressure applications.

Dimensions and Tolerances—Dimensions in this standard are based on, and unless designated otherwise, are specified in inches with SI equivalents shown adjacent to respective inch dimensions or designated mm in the text and tables in accordance with SAE J916. Tabulated dimensions shall apply to finished ends, plated or otherwise processed. Dimensions specified apply to metal tubing having a nominal wall thickness of 0.028–0.035 in (0.71–0.89 mm). Forming of tube having a wall thickness outside this range may require adjustment of dimensions. Tolerance on all dimensions not otherwise specified shall be ±0.010 in (±0.25 mm).

φ *Workmanship*—Formed tube ends shall be free from burrs, cracks, sharp edges, irregularities in diameters and any other defects affecting serviceability.

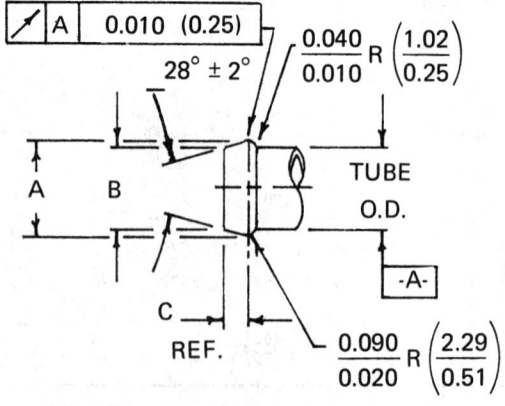

φ FIG. 1—STYLE A

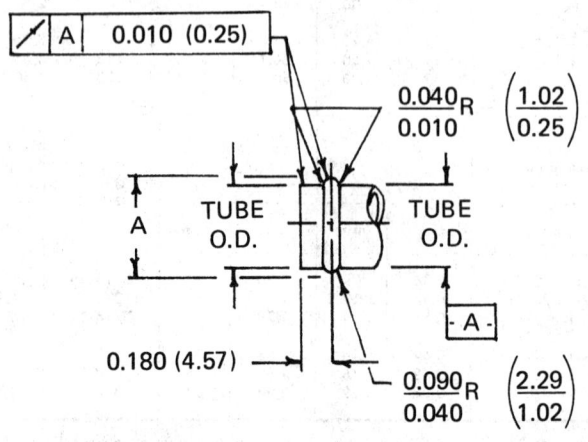

φ FIG. 2—STYLE B

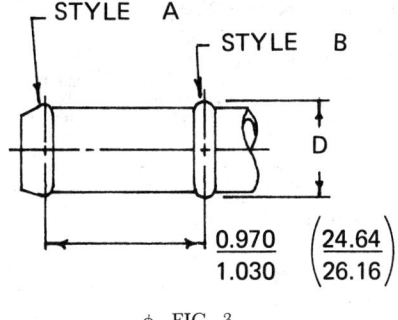

FIG. 3

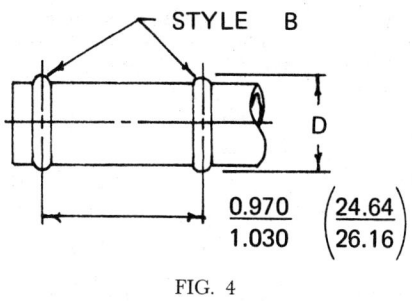

FIG. 4

TABLE 1—DIMENSIONS OF FORMED TUBE ENDS (FIGS. 1–4)

Nominal Tube OD	Outside Dia, A		End Dia, B		End to Center, C Ref[a]		D Dia ±0.020(0.51)	
in	in	mm	in	mm	in	mm	in	mm
3/16	0.220	5.59	0.180	4.57	0.080	2.03	0.220	5.59
1/4	0.290	7.37	0.240	6.10	0.100	2.54	0.290	7.37
5/16	0.360	9.14	0.310	7.87	0.100	2.54	0.360	9.14
3/8	0.430	10.92	0.380	9.65	0.100	2.54	0.430	10.92
7/16	0.490	12.45	0.440	11.18	0.100	2.54	0.490	12.45
1/2	0.560	14.22	0.500	12.70	0.120	3.05	0.560	14.22
9/16	0.620	15.75	0.560	14.22	0.120	3.05	0.620	15.75
5/8	0.690	17.53	0.630	16.00	0.120	3.05	0.690	17.53

[a]For reference purposes only, not intended for inspection.

FLARES FOR TUBING—SAE J533b

SAE Standard

Report of Parts and Fittings Technical Committee approved February 1947 and last revised by Tube, Pipe, Hose, and Lubrication Fittings Committee January 1972.

GENERAL SPECIFICATIONS

Scope—This SAE Standard covers specifications for 37 deg and 45 deg single and double flares for tube ends intended for use with 37 deg flared tube fittings and 45 deg flared or inverted flared tube fittings, respectively.

Dimensions—Dimensions in this standard are based on and, unless designated otherwise, are specified in inches, with metric equivalents shown in parentheses located adjacent to respective inch dimensions on illustrations or designated mm in text and tables, in accordance with SAE J916.

Single and double 45 deg flares shall conform to the dimensions specified in Fig. 2 and Table 1.

Single and double 37 deg flares shall conform to the dimensions specified in Fig. 3 and Table 2.

The following general specifications supplement the dimensional data with respect to unspecified detail and apply to both 37 deg and 45 deg flares for tubing.

Burring Prior to Flaring—To assure producing satisfactory flares, it may be necessary to perform burring operations on the tube end prior to flaring. Smoothly breaking the inside corner before single flaring ferrous, and some nonferrous tubing, is normally required to eliminate the cutoff burr which might otherwise create leakage paths across a substantial portion of the flare. Smoothly breaking the outside corner prior to single flaring, or both outside and inside corners prior to double flaring, shall be permissible on any tube material to minimize splitting.

Inasmuch as the specified dimensions shall prevail, whether or not the corners are broken, the quality of the finished flare shall be the only criterion applied to the burring operation.

Concentricity—Flare seat shall be concentric with tube outside diameter within 0.015 in. (0.38 mm) full indicator reading (FIR). To promote uniformity in checking concentricity of flare seat to the tube outside diameter, it is recommended the gaging method depicted in Fig. 1 and the following procedure, or equivalent means be used.

1. Mount tube in precision collet, dividing head, or equivalent rotational centering and clamping device with the rear of flare not more than 0.12 in. (3.04 mm) ahead of the collet. A minimum straight length of tube behind the flare of 1.00 in. (25.4 mm), or twice the tube outside diameter, whichever is greater, must be available for mounting purposes.

2. Place stylus of indicator gage on the coined portion of flare seat.

3. Rotate the mounted tube through full 360 deg revolution.

4. Read full indicator reading occurring over the 360 deg of rotation.

Workmanship—Flares shall be free from loose scale, burrs, slivers, and cracks. Seating surfaces shall be smooth and free from nicks, pit marks, and any other defects that prevent sealing.

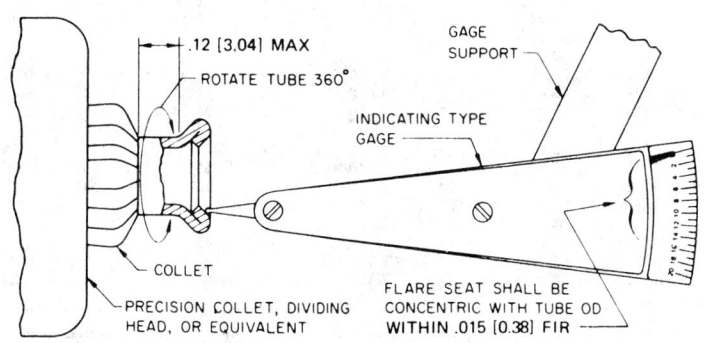

FIG. 1—TYPICAL FLARE CONCENTRICITY GAGE

TABLE 1—DIMENSIONS OF SINGLE AND DOUBLE 45-DEG FLARES FOR TUBING[a] (FIG. 2)

Nominal Tube OD		A Single Flare Diameter				A_1 Double Flare Diameter				B Single Flare Radius		B_1 Double Flare Radius		C Double Flare Coined Seat Length		D[b] Single Flare Wall Thickness		D_1[b] Double Flare Wall Thickness	
in	mm	in		mm		in		mm		±0.01 in	±0.25 mm	±0.01 in	±0.25 mm	in Min	mm Min	in Max	mm Max	in Max	mm Max
		Max	Min	Max	Min	Max	Min	Max	Min										
1/8	3.18	0.181	0.171	4.59	4.35	0.213	0.198	5.41	5.03	0.02	0.51	0.04	1.02	0.040	1.02	0.035	0.88	0.025	0.63
3/16	4.76	0.249	0.239	6.32	6.08	0.280	0.265	7.11	6.74	0.02	0.51	0.04	1.02	0.040	1.02	0.035	0.88	0.028	0.71
1/4	6.35	0.325	0.315	8.25	8.01	0.360	0.345	9.14	8.77	0.02	0.51	0.04	1.02	0.040	1.02	0.049	1.24	0.035	0.83
5/16	7.94	0.404	0.388	10.26	9.86	0.425	0.410	10.79	10.42	0.02	0.51	0.04	1.02	0.062	1.57	0.049	1.24	0.035	0.88
3/8	9.52	0.487	0.471	12.36	11.97	0.500	0.485	12.70	12.32	0.02	0.51	0.04	1.02	0.062	1.57	0.065	1.65	0.049	1.24
7/16	11.11	0.561	0.545	14.24	13.85	0.570	0.555	14.47	14.10	0.02	0.51	0.04	1.02	0.062	1.57	0.065	1.65	0.049	1.24
1/2	12.70	0.623	0.607	15.82	15.42	0.640	0.625	16.25	15.88	0.02	0.51	0.04	1.02	0.062	1.57	0.083	2.10	0.049	1.24
9/16	14.29	0.676	0.660	17.17	16.77	0.712	0.697	18.08	17.71	0.02	0.51	0.04	1.02	0.062	1.57	0.083	2.10	0.049	1.24
5/8	15.88	0.748	0.732	18.99	18.60	0.772	0.757	19.60	19.23	0.02	0.51	0.04	1.02	0.062	1.57	0.095	2.41	0.049	1.24
3/4	19.05	0.916	0.900	23.26	22.86	0.912	0.897	23.16	22.79	0.02	0.51	0.04	1.02	0.062	1.57	0.109	2.76	0.049	1.24
7/8	22.22	1.041	1.025	26.44	26.04	—	—	—	—	0.02	0.51	—	—	—	—	0.109	2.76	—	—
1	25.40	1.157	1.141	29.38	28.99	—	—	—	—	0.02	0.51	—	—	—	—	0.120	3.04	—	—

[a] It is not the intent of this standard to define the appropriateness of fittings to be used in conjunction with the flares specified. Considerations such as the effects of wall thickness on working pressures, length of thread engagements, etc., shall be the responsibility of the user. See SAE J514.

[b] Recommended maximum nominal wall thickness of tubing normally considered suitable for flaring to the above specifications.

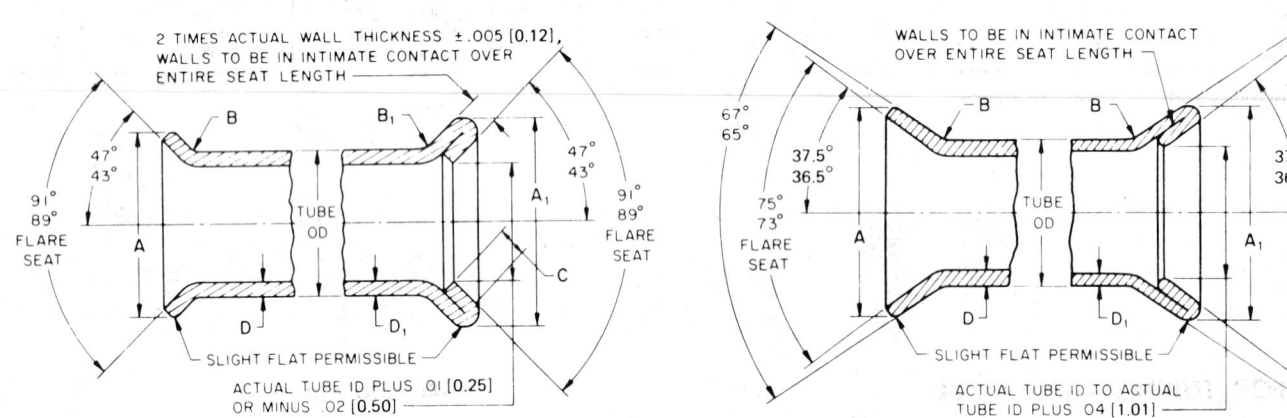

FIG. 2—SINGLE AND DOUBLE 45-DEG FLARES FOR TUBING

FIG. 3—SINGLE AND DOUBLE 37-DEG FLARES FOR TUBING

TABLE 2—DIMENSIONS OF SINGLE AND DOUBLE 37-DEG FLARES FOR TUBING[a] (FIG. 3)

Nominal Tube OD		A Single Flare Diameter				A_1 Double Flare Diameter				B Radius		D[b] Single Flare Wall Thickness		D_1[b] Double Flare Wall Thickness	
in	mm	in		mm		in		mm		±0.02 in	±0.5 mm	in Max	mm Max	in Max	mm Max
		Max	Min	Max	Min	Max	Min	Max	Min						
1/8	3.18	0.200	0.180	5.08	4.58	0.200	0.180	5.08	4.58	0.03	0.8	0.035	0.88	0.025	0.63
3/16	4.76	0.280	0.260	7.11	6.61	0.280	0.260	7.11	6.61	0.03	0.8	0.035	0.88	0.028	0.71
1/4	6.35	0.360	0.340	9.14	8.64	0.360	0.340	9.14	8.64	0.03	0.8	0.065	1.65	0.035	0.88
5/16	7.94	0.430	0.400	10.92	10.16	0.430	0.400	10.92	10.16	0.03	0.8	0.065	1.65	0.035	0.88
3/8	9.52	0.490	0.460	12.44	11.69	0.490	0.460	12.44	11.69	0.04	1.0	0.065	1.65	0.049	1.24
1/2	12.70	0.660	0.630	16.76	16.01	0.660	0.630	16.76	16.01	0.06	1.5	0.083	2.10	0.049	1.24
5/8	15.88	0.790	0.760	20.06	19.31	0.790	0.760	20.06	19.31	0.06	1.5	0.095	2.41	0.049	1.24
3/4	19.05	0.950	0.920	24.13	23.37	0.950	0.920	24.13	23.37	0.08	2.0	0.109	2.76	0.049	1.24
7/8	22.22	1.070	1.040	27.17	26.42	1.070	1.040	27.17	26.42	0.08	2.0	0.109	2.76	0.065	1.65
1	25.40	1.200	1.170	30.48	29.72	1.200	1.170	30.48	29.72	0.09	2.3	0.120	3.04	0.065	1.65
1-1/8	28.58	1.380	1.350	35.05	34.29	1.380	1.350	35.05	34.29	0.09	2.3	0.120	3.04	0.065	1.65
1-1/4	31.75	1.510	1.480	38.35	37.60	1.510	1.480	38.35	37.60	0.09	2.3	0.120	3.04	0.065	1.65
1-1/2	38.10	1.730	1.700	43.94	43.18	1.730	1.700	43.94	43.18	0.11	2.8	0.120	3.04	0.065	1.65
1-3/4	44.45	2.110	2.080	53.59	52.84	2.110	2.080	53.59	52.84	0.11	2.8	0.120	3.04	0.065	1.65
2	50.80	2.360	2.330	59.94	59.19	2.360	2.330	59.94	59.19	0.11	2.8	0.134	3.40	0.065	1.65

[a] It is not the intent of this standard to define the appropriateness of fittings to be used in conjunction with the flares specified. Considerations such as the effects of wall thickness on working pressures, length of thread engagements, etc., shall be the responsibility of the user.

[b] Recommended maximum nominal wall thickness of tubing normally considered suitable for flaring to the above specifications.

SEAMLESS LOW CARBON STEEL TUBING ANNEALED FOR BENDING AND FLARING—SAE J524 JAN80

SAE Standard

Report of the Tube, Pipe, Hose, and Lubrication Fittings Committee, approved January 1954, last revised by the Fluid Conductors and Connectors Technical Committee January 1980.

Scope—This standard covers cold drawn and annealed seamless low carbon steel pressure tubing intended for use as hydraulic lines and in other applications requiring tubing of a quality suitable for flaring and bending.

Manufacture—The tubing shall be cold drawn to size and after forming shall be annealed in such a manner as to produce a finished product which will meet all requirements of this standard.

Dimensions and Tolerances—The tolerances applicable to tubing outside diameter are shown in Table 1. The wall thickness shall not vary more than ±10% for tubing having 0.50 in (12.7 mm) or larger, nominal inside diameter nor more than ±15% for tubing having a smaller nominal inside diameter. Tubing outside diameter and wall thickness shall be as specified by purchaser.

Quality—Lengths of finished tubing shall be reasonably straight and have smooth ends free from burrs. Tubing shall be free from scale and injurious defects and have a workmanlike finish. Surface imperfections, such as handling marks, die marks, or shallow pits, shall not be considered injurious defects provided the imperfections are within the tolerances specified for diameter and wall thickness. The removal of such surface imperfections is not required.

The inside of tubing shall be clean and free from any contamination that cannot be removed readily by cleaning agents normally used in manufacturing.

Material—Tubing shall be made from low carbon steel conforming to the following chemical composition:

Element	φ Cast or Heat Analysis[a] % by Weight
Carbon	0.18 max
Manganese	0.30 to 0.60
Phosphorus	0.040 max
Sulfur	0.050 max

[a] Check analysis tolerance shall be as specified in SAE J409, Table 1.

Mechanical Properties—The finished tubing shall have mechanical properties as tabulated below:

Yield Strength, min	φ 25,000 psi (170 MPa)
Ultimate Strength, min	45,000 psi (310 MPa)
φ Elongation in 2 in (50 mm), min	35%[a]
Hardness (Rockwell B), max	65[b]

[a] For tubing having nominal outside diameter of 0.375 in (9.5 mm) or less, and/or wall thicknesses of 0.035 in (0.9 mm) or less, a minimum elongation of 25% is permissible.
[b] The hardness test shall not be required on tubing with a nominal wall thickness of less than 0.065 in (1.65 mm). Such tubing shall meet all other mechanical properties and performance requirements.

Performance Requirements—The finished tubing shall satisfactorily meet the following performance tests. Test specimens shall be taken from tubing which has not been subjected to cold working after the anneal of the finished sized tubing.

Flattening Test—A section approximately 3 in (75 mm) in length, cut φ from the finished tubing, shall not crack or show any flaws when flattened between parallel plates to a distance equal to three times the wall thickness of the section under test. Superficial ruptures resulting from minor surface imperfections shall not be considered cause for rejection.

Expansion Test—A test specimen shall be taken from every shipment or every 1500 ft (460 m), whichever is smaller, of finished tubing and subjected to expansion over a hardened tapered plug having a slope of 0.1:1.0 until the φ outside diameter has been expanded 25% without evidence of cracking or flaws.

Pressure Proof Test—Unless otherwise specified, tubing supplied under this standard shall have been tested hydrostatically, with no evidence of failure, at a pressure which will subject the material to a yield stress of 20,000 psi (140 MPa). Test pressures shall be as determined by Barlow's φ formula for thin hollow cylinders under tension:

$$P = \frac{2TS}{D}$$

where: D = outside diameter of tubing, in (mm)
P = hydrostatic pressure, psi (MPa)
S = allowable unit stress of material = 20,000 psi (140 MPa)   φ
T = minimum wall thickness of tubing, in (mm)   φ

No tube shall be tested beyond a hydrostatic pressure of 5000 psi (35 MPa), φ unless so specified.

Nondestructive Electric Test—In lieu of the hydrostatic test, when mutually agreed upon by the purchaser and manufacturer, all tubing shall be tested by passing it through an electric eddy current tester which is capable of detecting defects that would prevent the tubing from passing the hydrostatic pressure proof test.

Corrosion Protection—The inside and outside of the finished tubing shall be protected against corrosion during shipment and normal storage. If a corrosion preventive compound is applied, it shall be such that after normal storage periods it can readily be removed by cleaning agents normally used in manufacturing.

φ TABLE 1—TUBING OUTSIDE DIAMETER TOLERANCES

Nominal Tubing OD[a,b]		OD Tolerance ±	
in	mm	in	mm
Up to 1.00	Up to 25.4	0.004	0.10
Over 1.00 to 1.50 inclusive	Over 25.4 to 38.1 inclusive	1.006	0.15
Over 1.50 to 2.00 inclusive	Over 38.1 to 50.8 inclusive	0.008	0.20
Over 2.00 to 3.50 inclusive	Over 50.8 to 88.9 inclusive	0.010	0.25

[a] The actual outside diameter shall be the average of the maximum and minimum outside diameters as determined at any one cross-section through the tubing.
[b] Refer to SAE J514 for nominal tubing outside diameters to be used in conjunction with standard hydraulic tube fittings.

WELDED AND COLD DRAWN LOW CARBON STEEL TUBING ANNEALED FOR BENDING AND FLARING—SAE J525 JAN80

SAE Standard

Report of the Tube, Pipe, Hose, and Lubrication Fittings Committee, approved April 1958, last revised by the Fluid Conductors and Connectors Technical Committee January 1980.

Scope—This standard covers cold worked and annealed electric resistance welded single wall low carbon steel pressure tubing intended for use as hydraulic lines and in other applications requiring tubing of a quality suitable for flaring and bending.

Manufacture—The tubing shall be made from a single strip of steel shaped into a tubular form, the edges of which are joined and sealed by electric resistance welding. After forming and welding, the tubing shall be normalized and subjected to a cold working operation that shall result in a 15% minimum reduction in cross sectional area, of which at least 8% shall consist of a reduction in wall thickness. Subsequent to cold working, the tubing shall be annealed in such a manner as to produce a finished product which will meet all requirements of this standard. Tubing that has been pickled to remove scale shall be suitably treated to eliminate any embrittlement induced by the pickling process.

Dimensions and Tolerances—The tolerances applicable to tubing outside diameter, inside diameter, and wall thickness are shown in Table 1. Tubing shall be subject to any two of the tolerances specified, as designated by the purchaser.

Quality—Lengths of finished tubing shall be reasonably straight and have smooth ends free from burrs. Tubing shall be free from scale and injurious defects and have a workmanlike finish. Surface imperfections such as handling marks, die marks, or shallow pits shall not be considered injurious defects provided the imperfections are within the tolerances specified for diameter and wall thickness. The removal of such surface imperfections is not required. There shall be no dimensional indications of the presence of the weld.

The inside of tubing shall be clean and free from any contamination that cannot be readily removed by cleaning agents normally used in manufacturing.

Material—Tubing shall be made from low carbon steel conforming to the following chemical composition.

Element	ϕ Cast or Heat Analysis[a] % by Weight
Carbon	0.18 max
Manganese	0.30–0.60
Phosphorus	0.040 max
Sulfur	0.050 max

[a] Check analysis tolerance shall be as specified in SAE J409, Table 3.

Mechanical Properties—The finished tubing shall have mechanical properties as tabulated below:

Yield Strength, min	ϕ 25,000 psi (170 MPa)
Ultimate Strength, min	45,000 psi (310 MPa)
ϕ Elongation in 2 in (50 mm), min	35%[a]
Hardness (Rockwell B scale), max	65[b]

[a] For tubing having nominal outside diameter of 0.375 in (9.5 mm) or less, and/or wall thicknesses of 0.035 in (0.9 mm) or less, a minimum elongation of 25% is permissible.
[b] The hardness test shall not be required on tubing with a nominal wall thickness of less than 0.065 in (1.65 mm). Such tubing shall meet all other mechanical properties and performance requirements.

Performance Requirements—The finished tubing shall satisfactorily meet the following performance tests. Test specimens shall be taken from tubing which has not been subjected to cold working after the anneal of the finished sized tubing.

ϕ **Flattening Test**—A section approximately 3 in (75 mm) in length, cut from the finished tubing, shall not crack or show any flaws when flattened between parallel plates to a distance equal to three times the wall thickness of the section under test. Superficial ruptures resulting from minor surface imperfections shall not be considered cause for rejection.

Reverse Flattening Test—A test specimen shall be taken from every shipment or every 1500 ft (460 m), whichever is smaller, of finished tubing and split longitudinally 90 deg on each side of the weld. The section containing the weld shall be opened and flattened with the weld at the point of maximum bend. There shall be no evidence of cracks or lack of penetration or overlaps resulting from flash removal in the weld.

Refer to ASTM A 370, Methods and Definitions for Mechanical Testing of Steel Products, paragraph T5(B), reverse flattening test.

Expansion Test—A test specimen shall be taken from every shipment or every 1500 ft (460 m), whichever is smaller, of finished tubing and subjected to expansion over a hardened tapered plug having a slope of 0.1 : 1.0 until the ϕ outside diameter has been expanded 25% without evidence of cracking or flaws.

Pressure Proof Test—Unless otherwise specified, tubing supplied under this standard shall have been tested hydrostatically, with no evidence of failure, at a pressure which will subject the material to a yield stress of 20,000 psi (140 MPa). Test pressures shall be as determined by Barlow's ϕ formula for thin hollow cylinders under tension:

$$P = \frac{2TS}{D}$$

where: D = outside diameter of tubing, in (mm)
P = hydrostatic pressure, psi (MPa)
S = allowable unit stress of material = 20,000 psi (140 MPa)   ϕ
T = minimum wall thickness of tubing, in (mm)

No tube shall be tested beyond a hydrostatic pressure of 5000 psi (35 MPa) ϕ unless so specified.

Nondestructive Electric Test—In lieu of the hydrostatic test, where mutually agreed upon by the purchaser and manufacturer, all tubing shall be tested by passing it through an electric eddy current tester which is capable of detecting defects that would prevent the tubing from passing the hydrostatic pressure proof test.

Corrosion Protection—The inside and outside of the finished tubing shall be protected against corrosion during shipment and normal storage. If a corrosion preventive compound is applied, it shall be such that after normal storage periods it can readily be removed by cleaning agents normally used in manufacturing.

ϕ **TABLE 1—TUBING OUTSIDE DIAMETER AND WALL THICKNESS TOLERANCE**

Nominal Tubing OD[a,b]		Tolerance ±				
		OD		ID		Wall Thickness (%)
in	mm	in	mm	in	mm	
Up to 0.38	Up to 9.5	0.002	0.05	0.002	0.05	15
Over 0.38 to 0.63 inclusive	Over 9.5 to 15.9 inclusive	0.0025	0.06	0.0025	0.06	10
Over 0.63 to 2.00 inclusive	Over 15.9 to 50.8 inclusive	0.003	0.08	0.003	0.08	10
Over 2.00 to 2.50 inclusive	Over 50.8 to 63.5 inclusive	0.004	0.10	0.004	0.10	10
Over 2.50 to 3.00 inclusive	Over 63.5 to 76.2 inclusive	0.005	0.13	0.005	0.13	10
Over 3.00 to 4.00 inclusive	Over 76.2 to 101.6 inclusive	0.006	0.15	0.006	0.15	10

[a] The actual outside diameter shall be the average of the maximum and minimum outside diameters as determined at any one cross section through the tubing.
[b] Refer to SAE J514 for nominal tubing outside diameters to be used in conjunction with standard hydraulic tube fittings.

WELDED LOW CARBON STEEL TUBING—SAE J526 JAN80

SAE Standard

Report of the Tube, Pipe, Hose, and Lubrication Fittings Committee, approved January 1952, last revised by the Fluid Conductors and Connectors Technical Committee January 1980.

Scope—This standard covers welded single wall low carbon steel tubing intended for general automotive applications and other similar uses.

Manufacture—The tubing shall be made from a single strip of steel shaped into a tubular form, the edges of which are joined and sealed by a suitable butt welding process. After welding, the bead shall be removed from the outside to provide a smooth round surface and the tubing shall be processed in such a manner as to produce a finished product which will meet all requirements of this standard.

Dimensions and Tolerances—The standard nominal diameters and the applicable dimensions and tolerances are shown in Table 1.

Quality—Finished tubing shall be clean, smooth, and round, both inside and outside; and shall be free from scale and injurious defects. A slight weld bead and splatter on the inside surface shall be permissible but must be held to the minimum consistent with good welding practice. Surface imperfections such as handling marks, die marks, or shallow pits shall not be considered injurious defects provided such imperfections are within the tolerances specified for diameter and wall thickness.

The inside of tubing shall be clean and free from any contamination which will impair the processing or serviceability of the tubing.

ϕ *Material*—Tubing shall be made from low carbon steel, such as UNS G10100.

Mechanical Properties—The finished tubing shall have mechanical properties as tabulated below:

Yield Strength, min (0.2% offset)	ϕ 25,000 psi (170 MPa)
Tensile Strength, min	42,000 psi (290 MPa)
ϕ Elongation in 2 in (50 mm)	14–40%
Hardness (Rockwell 30 T scale), max	65

Performance Requirements—The finished tubing shall satisfactorily meet the following performance tests. As designated therein, test specimens having minimum lengths equivalent to two times the tubing outside diameter or 2 in (50 mm), whichever is greater, shall be taken from tubing which has not been ϕ subjected to cold working after the final processing of the finished sized tubing.

Flaring Test—A test specimen having squared and deburred ends shall withstand being double flared at one end to the requirements of SAE J533 ϕ without evidence of splitting or flaws. The test specimen shall be held firmly and squarely in the die and the punch, while being forced down gradually, shall be guided parallel to the axis of the tubing.

Hardness Test—The hardness test shall not be required, it being recognized that hardness will be satisfactory if the tubing meets all other mechanical properties and performance requirements set forth in this SAE Standard.

Bending Test—The finished tubing shall withstand bending on a centerline radius equal to three times the tubing outside diameter without undue reduction of area or flattening where proper bending fixtures are used.

Pressure Proof Test—Unless otherwise specified, the finished tubing shall withstand a hydrostatic proof test, with no evidence of failure, at a pressure ϕ which will subject the material to a yield stress of 20,000 psi (140 MPa). Test pressures shall be as determined by Barlow's formula for thin hollow cylinders under tension:

$$P = \frac{2TS}{D}$$

where: D = outside diameter of tubing, in (mm)
P = hydrostatic pressure, psi (MPa)
ϕ   S = allowable unit stress of material = 20,000 psi (140 MPa)
T = minimum wall thickness of tubing, in (mm)

ϕ No tube shall be tested beyond a hydrostatic pressure of 5000 psi (35 MPa) unless so specified.

Nondestructive Electric Test—In lieu of the hydrostatic test, where mutually agreed upon by the purchaser and manufacturer, all tubing shall be tested by passing it through an electric eddy current tester which is capable of detecting defects that would prevent the tubing from passing the hydrostatic pressure proof test.

ϕ **TABLE 1—TUBING DIMENSIONS AND TOLERANCES**[a]

Nominal Tubing OD		Outside Diameter[b] Tolerance ±			Wall Thickness[c]			
					Basic		Tolerance ±	
in	mm	Basic	in	mm	in	mm	in	mm
0.125	3.18	0.125	0.002	0.05	0.025	0.64	0.005	0.13
0.188	4.76	0.188	0.003	0.08	0.028	0.71	0.005	0.13
0.250	6.35	0.250	0.003	0.08	0.028	0.71	0.003	0.08
0.312	7.94	0.312	0.003	0.08	0.028	0.71	0.003	0.08
0.375	9.52	0.375	0.003	0.08	0.028	0.71	0.003	0.08
0.438	11.11	0.437	0.004	0.10	0.030	0.76	0.003	0.08
0.500	12.70	0.500	0.004	0.10	0.030	0.76	0.003	0.08
0.500	12.70	0.500	0.004	0.10	0.035	0.89	0.0035	0.09
0.562	14.29	0.562	0.004	0.10	0.030	0.76	0.003	0.08
0.625	15.88	0.625	0.004	0.10	0.035	0.89	0.0035	0.09

[a] Other sizes may be specified by agreement between the supplier and the user.
[b] The actual outside diameter shall be the average of the maximum and minimum outside diameters as determined at any one cross section through the tubing.
[c] The tolerances listed represent the maximum permissible deviation at any point.

Corrosion Protection—The inside and outside of the finished tubing shall be protected against corrosion during shipment and normal storage. If a corrosion preventive compound is applied, it shall be such that after normal storage periods it can readily be removed by cleaning agents normally used in manufacturing.

BRAZED DOUBLE WALL LOW CARBON STEEL TUBING—SAE J527 JAN80

SAE Standard

Report of the Tube, Pipe, Hose, and Lubrication Fittings Committee, approved January 1952, last revised by the Fluid Conductors and Connectors Technical Committee January 1980.

Scope—This standard covers brazed double wall low carbon steel tubing intended for general automotive applications and other similar uses.

Manufacture—The tubing shall be made from a single or double strip of steel shaped into the form of a double wall tubing, the seams of which are secured and sealed by copper brazing in a controlled atmosphere. The braze shall be uniform with no evidence of a bead on either the inside or outside of the tubing. The tubing shall be processed in such a manner as to produce a finished product which will meet all requirements of this standard.

Dimensions and Tolerances—The standard nominal diameters and the applicable dimensions and tolerances are shown in Table 1.

Quality—Finished tubing shall be clean, smooth, and round, both inside and outside; and shall be free from scale and injurious defects. Surface imperfections such as handling marks, die marks, or shallow pits shall not be considered injurious defects provided such imperfections are within the tolerances specified for diameter and wall thickness.

The inside of tubing shall be clean and free from any contamination which will impair the processing or serviceability of the tubing.

ϕ **Material**—Tubing shall be made from low carbon steel, such as UNS G10100.

Mechanical Properties—The finished tubing shall have mechanical properties as tabulated below:

Yield Strength, min (0.2% offset)	ϕ 25,000 psi (170 MPa)
Tensile Strength, min	42,000 psi (290 MPa)
ϕ Elongation in 2 in (50 mm)	14–40%
Hardness (Rockwell 30 T scale), max	65

Performance Requirements—The finished tubing shall satisfactorily meet the following performance tests. Test specimens shall be taken from tubing which has not been subjected to cold working after the final processing of the finished sized tubing.

Flaring Test—A test section cut from the finished tubing, have squared and deburred ends, shall withstand being double flared at one end to the dimensions shown in SAE J533. The test section shall be held firmly and squarely in the die and the punch, while being forced down, shall be guided parallel to the axis of the tubing. The flare shall exhibit no evidence of splitting or flaws except that a separation of the outer lap joint with area A (Fig. 1) shall be permissible providing it does not exceed 0.12 in (3.1 mm) in length and is confined to the outer thickness only. Seam separation shall not be permissible in the following areas:

AREA B—The flare seat, defined as the surface within the 90 deg included angle. Conical surface shall be smooth and free from cracks or other irregularities which could cause leaks after assembly.

ϕ **TABLE 1—TUBING DIMENSIONS AND TOLERANCES**[a]

Nominal Tubing OD		Outside Diameter[a] Tolerance ±			Wall Thickness			
					Basic		Tolerance[b]	
in	mm	Basic	in	mm	in	mm	in	mm
0.125	3.18	0.125	0.002	0.05	0.025	0.64	0.005	0.13
0.188	4.76	0.188	0.003	0.08	0.028	0.71	0.003	0.08
0.250	6.35	0.250	0.003	0.08	0.028	0.71	0.003	0.08
0.312	7.94	0.312	0.003	0.08	0.028	0.71	0.003	0.08
0.375	9.52	0.375	0.003	0.08	0.028	0.71	0.003	0.08
0.438	11.11	0.437	0.004	0.10	0.030	0.76	0.003	0.08
0.500	12.70	0.500	0.004	0.10	0.035	0.89	0.0035	0.09
0.562	14.29	0.562	0.004	0.10	0.035	0.89	0.0035	0.09
0.625	15.88	0.625	0.004	0.10	0.035	0.89	0.0035	0.09

[a] The actual outside diameter shall be the average of the maximum and minimum outside diameters as determined at any one cross section through the tubing.
[b] The tolerances listed represent the maximum permissible deviation at any point.

AREA C—The surface beyond the length of the double thickness created by the flare.

Bending Test—The finished tubing shall withstand bending on a centerline radius equal to three times the tubing outside diameter without undue reduction of area or flattening where proper bending fixtures are used.

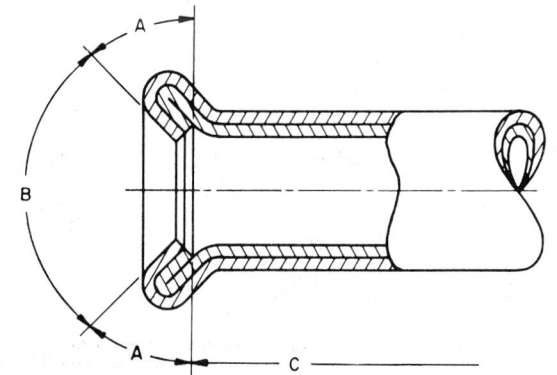

FIG. 1

Pressure Proof Test—Unless otherwise specified, the finished tubing shall withstand a hydrostatic proof test, with no evidence of failure, at a pressure ɸ which will subject the material to a yield stress of 20,000 psi (140 MPa) Test pressures shall be determined by Barlow's formula for thin hollow cylinders under tension:

$$P = \frac{2TS}{D}$$

where: D = outside diameter of tubing, in (mm)
P = hydrostatic pressure, psi (MPa)
ɸ S = allowable unit stress of material = 20,000 psi (140 MPa)
T = minimum wall thickness of tubing, in (mm)

No tube shall be tested beyond a hydrostatic pressure of 5000 psi (35 MPa) ɸ unless so specified.

Nondestructive Electric Test—In lieu of the hydrostatic test, where mutually agreed upon by the purchaser and manufacturer, all tubing shall be tested by passing it through an electric eddy current tester which is capable of detecting defects that would prevent the tubing from passing the hydrostatic pressure proof test.

Corrosion Protection—The inside and outside of the finished tubing shall be protected against corrosion during shipment and normal storage. If a corrosion preventive compound is applied, it shall be such that after normal storage periods it can readily be removed by cleaning agents normally used in manufacturing.

WELDED FLASH CONTROLLED LOW CARBON STEEL TUBING NORMALIZED FOR BENDING, DOUBLE FLARING, AND BEADING—SAE J356 JAN80

SAE Standard

Report of the Tube, Pipe, Hose, and Lubrication Fittings Committee, approved July 1968, last revised by the Fluid Conductors and Connectors Technical Committee January 1980.

Scope—This standard covers normalized electric resistance welded flash controlled single wall low carbon steel pressure tubing intended for use as pressure lines and in other applications requiring tubing of a quality suitable for bending, double flaring, beading, and brazing.

Manufacture—The tubing shall be made from a single strip of steel shaped into a tubular form, the edges of which are joined and sealed by electric resistance welding. After forming and welding, the outside flash shall be removed to provide a smooth surface. The inside flash shall be of uniform contour, free from saw tooth peaks, and controlled in height by seam welding techniques or by cutting, but not by hammering or rolling. The inside flash height shall conform to the following:

ɸ

Nominal Wall Thickness		Nominal Tubing Outside Diameter			
		Thru 1.000 (25.4 mm)		Over 1.000 (25.4 mm)	
		Maximum Flash Height[a]			
in	mm	in	mm	in	mm
Thru 0.035	0.90	0.005	0.13	0.010	0.25
Over 0.035 thru 0.065	0.90 thru 1.65	0.008	0.20	0.010	0.25
Over 0.065	1.65	0.010	0.25	0.010	0.25

[a] For tubes having an ID greater than 0.312 in (8 mm), the height of the inside weld flash shall be measured with a ball micrometer having an 0.156 ± 0.016 in (3.96 ± 0.41 mm) radius on the anvil or ball point. For tubes having an ID 0.312 in (8 mm) or less, screw thread micrometers shall be used. The height of the flash shall be the difference between the thickness of the tubing wall at the point of maximum height of the flash and the average of the wall thickness measured at points adjacent to both sides of the flash.

The tubing shall be normalized to produce a finished product which will meet all requirements of this standard.

Dimensions and Tolerances—The tolerances applicable to tubing outside diameter are shown in Table 1. The tolerances applicable to tubing wall thickness are shown in Table 2. Particular attention shall be given to areas adjacent to the weld to insure against thin spots and/or sharp indentations.

Quality—Lengths of finished tubing shall be reasonably straight and have smooth ends free from burrs. Finished tubing shall be free from scale and injurious imperfections and shall have a workmanlike finish. Outside surface imperfections such as handling marks, straightening marks, light die marks, or shallow pits shall not be considered injurious, provided the imperfections are not detrimental to the function of the tubing. The removal of such surface imperfections shall not be required.

The inside surface shall be free of weld splatter, pits, and all other injurious imperfections detrimental to the function of the tubing.

Material—Tubing shall be made from low carbon hot or cold rolled steel conforming to the chemical composition shown in Table 3. If rimmed steel is used, it shall be single strand. The steel shall be made by the open hearth, basic oxygen, or electric furnace process. A ladle analysis of each heat shall be made to determine the percentages of the elements specified. The chemical composition thus determined shall be reported to the purchaser, or his representative, if requested, and shall conform to the requirements specified. If a check analysis is required, the tolerances shall be as specified in SAE J409, Table 3.

Mechanical Properties—The finished tubing shall have mechanical properties as tabulated below:

Yield Strength, min	ɸ 25,000 psi (170 MPa)
Ultimate Strength, min	45,000 psi (310 MPa)
ɸ Elongation in 2 in (50 mm), min	35%[a]
Hardness (Rockwell B), max	65[b]

[a] For tubing having nominal outside diameter of 0.375 in (9.5 mm) or less, and/or wall thicknesses of 0.035 in (0.9 mm) or less, a minimum elongation of 25% is permissible.
[b] The hardness test shall not be required on tubing with a nominal wall thickness of less than 0.065 in (1.65 mm). Such tubing shall meet all other mechanical properties and performance requirements.

Performance Requirements—The finished tubing shall satisfactorily meet the following performance tests. As designated therein, test specimens having minimum length equivalent to two times the tubing outside diameter or 2 in (50 mm), whichever is greater, shall be taken from finished tubing, as manu- ɸ factured. All tests shall be conducted at room temperature.

Flattening Test—A test specimen shall be taken from every shipment or every 1500 ft (460 m), whichever is smaller, of finished tubing and flattened between parallel plates to a distance equal to three times the actual wall thickness of the specimen under test without any cracking or flaws. The weld shall be placed at 90 deg from the direction of applied force. Superficial ruptures resulting from minor surface imperfections shall not be considered cause for rejection.

Reverse Flattening Test—A test specimen shall be taken from every shipment or every 1500 ft (460 m), whichever is smaller, of finished tubing and split longitudinally 90 deg on each side of the weld. The section containing the weld shall be opened and flattened with the weld at the point of maximum bend. There shall be no evidence of cracks or metal flaking, or lack of weld penetration or overlaps resulting from flash control or flash removal in the weld.

Expansion Test—A test specimen shall be taken from every shipment or every 1500 ft (460 m), whichever is smaller, of finished tubing and subjected to expansion over a hardened tapered plug having a slope of 0.1:1.0 until the ɸ outside diameter has been expanded 25% without evidence of cracking or flaws.

The tubing shall be capable of being double flared as shown in SAE J533 without evidence of cracking or flaws. Refer to footnote b of Table 1 for tubing

ɸ **TABLE 1—TUBING OUTSIDE DIAMETER TOLERANCE**

Nominal Tubing OD[a,b]		Tolerance ±	
in	mm	in	mm
Thru 0.375	9.50	0.0025	0.06
Over 0.375–0.625	9.50–15.88	0.003	0.08
Over 0.625–1.125	15.88–28.57	0.0035	0.09
Over 1.125–2.000	28.57–50.80	0.005	0.13
Over 2.000–2.500	50.80–63.50	0.006	0.15
Over 2.500–3.000	63.50–76.20	0.008	0.20
Over 3.000–3.500	76.20–88.90	0.009	0.23
Over 3.500–4.000	88.90–101.60	0.010	0.25

ɸ [a] OD measurements shall be taken at least 2.0 in (50 mm) from the end of the tubing.
[b] Refer to SAE J514 for nominal tubing OD to be used in conjunction with standard hydraulic tube fittings and SAE J533 for recommended max nominal wall thickness for double flaring.

TABLE 2—TUBING WALL THICKNESS TOLERANCES, IN

Nominal Wall Thickness[a]	Nominal Tubing Outside Diameter					
	Thru 1.000		Over 1.000 thru 2.000		Over 2.000 thru 4.000	
	Tolerance[c]					
	Plus[b]	Minus	Plus[b]	Minus	Plus[b]	Minus
0.028	0.002	0.003	0.003	0.003	0.004	0.003
0.035	0.002	0.004	0.003	0.004	0.004	0.004
0.049	0.002	0.005	0.003	0.005	0.004	0.005
0.065	0.004	0.006	0.005	0.008	0.006	0.008
0.083	0.004	0.006	0.006	0.008	0.007	0.008
0.095	0.004	0.006	0.006	0.010	0.007	0.010
0.109	0.004	0.006	0.008	0.010	0.009	0.010
0.120	0.004	0.008	0.008	0.010	0.009	0.010
0.134	0.004	0.008	0.008	0.010	0.009	0.010
0.148	—	—	0.008	0.011	0.009	0.011
0.165	—	—	0.008	0.011	0.009	0.011
0.180	—	—	0.008	0.011	0.009	0.011
0.203	—	—	0.008	0.012	0.009	0.012
0.220	—	—	0.008	0.012	0.009	0.012
0.238	—	—	0.013	0.018	0.014	0.018
0.259	—	—	0.013	0.020	0.014	0.020

[a] For intermediate wall thicknesses, the tolerance for the next heavier wall thickness shall apply.
[b] Plus tolerances include allowance for crown on flat rolled steel.
φ [c] Millimeter conversions of the inch tolerances are:

in	mm	in	mm	in	mm
0.002	0.05	0.007	0.18	0.012	0.30
0.003	0.08	0.008	0.20	0.013	0.33
0.004	0.10	0.009	0.23	0.014	0.36
0.005	0.13	0.010	0.25	0.018	0.46
0.006	0.15	0.011	0.28	0.020	0.51

TABLE 3—CHEMICAL REQUIREMENTS

Element	φ Cast or Heat Analysis, Wgt %
Carbon	0.18 max
Manganese	0.30 thru 0.60
Phosphorus	0.04 max
Sulfur	0.05 max

OD and wall thickness subject to this capability requirement. Double flaring tests shall not be required.

Hardness Test—One hardness test shall be made on a specimen from each production lot of tubing. The hardness test shall be made on the inside surface of the specimen. The hardness test shall not be required on tubing with a nominal wall thickness less than 0.065 in (1.7 mm). Such tubing shall meet all other mechanical properties and performance requirements.

Tensile Test—One tension test, in accordance with ASTM A 370, shall be made on a specimen from each production lot of tubing. If the percentage of elongation of the test specimen is less than that specified and/or any part of the fracture is more than 0.75 in (19 mm) from the center of the gage length, as indicated by scribe marks on the specimen before testing, a retest shall be allowed.

Pressure Proof Test—Unless otherwise specified, the finished tubing shall withstand a hydrostatic proof test, with no evidence of failure, at an actual pressure of 5000 psi (35 MPa) or at a fiber stress of 20,000 psi (140 MPa) φ whichever is less. Test pressures shall be determined by the following formula:

$$P = \frac{2TS}{D}$$

where: P = hydrostatic test pressure, psi (MPa) (5000 psi (35 MPa) max) φ
T = allowable minimum wall thickness of tubing, in (mm)
S = 20,000 psi (140 MPa) allowable fiber stress (80% of min yield φ strength)
D = nominal outside diameter of tubing, in (mm)

Nondestructive Electric Test—In lieu of the hydrostatic test, when mutually agreed upon by the purchaser and manufacturer, all tubing shall be tested by passing it through an electric eddy current tester which is capable of detecting defects that would prevent the tubing from passing the hydrostatic pressure proof test.

Test Specimens—Test specimens for mechanical tests shall be smooth on the ends and free from flaws. If any test specimen exhibits burrs, flaws or defective machining, before testing, it may be discarded and another specimen may be selected.

Test Certificate—A certificate of compliance to the performance requirements shall be furnished to the purchaser by the producer if requested in the purchase agreement.

Cleanliness—The inside and outside surfaces of the finished tubing shall be commercially bright, clean and free from grease, oxide scale, carbon deposits and any other contamination that cannot be readily removed by cleaning agents normally used in manufacturing plants.

Corrosion Protection—The inside and outside surfaces of the finished tubing shall be protected against corrosion during shipment and normal storage. If a corrosion preventive compound is applied, it shall be such that after normal storage periods, it can be readily removed by cleaning agents normally used in manufacturing plants.

PRESSURE RATINGS FOR HYDRAULIC TUBING AND FITTINGS—SAE J1065

SAE Information Report

Report of Tube, Pipe, Hose and Lubrication Fittings Committee approved January 1974.

1. Scope—This report is intended to provide design guidance in the selection of steel tubing and related tube fittings for general hydraulic system applications. The information presented herein is based on tubing products which conform to SAE J524, SAE J525, and SAE J356, and is subject to due consideration being given to the following limitations:

1.1 Since many factors influence the pressure at which a hydraulic system will or will not perform satisfactorily, this report should not be used as a "standard" nor "specification," and the values shown herein should not be construed as "guaranteed" minimum.

1.1.1 Within the fluid power industry, many criteria are used for determining the pressure capability of tubing. Consideration is given to specified minimum yield or fiber stress factors, to calculated yield or burst pressures, and to yield or burst pressures determined by actual test. Also, varying design factors are applied, commensurate with the total system conditions. Thus, it is impractical to set down specific allowable working pressures that will satisfy all design criteria. It is considered desirable, however, to provide guidelines on the subject such as are published in this report.

1.1.2 Factors such as the thinning of tube walls due to forming operations, shock loads, and vibration characteristics of the system must also be considered.

2. Hydraulic Tubing—Three normally acceptable reference working pressures (psi) for each combination of diameter and wall thickness of commonly used tubing calculated using three generally accepted formulations and reflecting two popular design stress factors are presented in Tables 1 and 2. The designer, therefore, may select the desired tubing on the basis of the value which bests suits the intended application and satisfies any preference he may have regarding formulation.

2.1 Formulas—The formulas from which the three values tabulated vertically opposite each tube size were derived are respectively:

2.1.1 The Barlow formula: $P = \dfrac{2ST}{D}$

2.1.2 The Boardman formula: $P = \dfrac{2ST}{D - 0.8T}$

2.1.3 The Lamé formula: $P = S\left(\dfrac{D^2 - d^2}{D^2 + d^2}\right)$

where: D = nominal outside diameter of tubing, in.
d = nominal inside diameter of tubing, in.
P = hydrostatic working pressure, psi
S = allowable fiber stress of material, psi
T = nominal wall thickness of tubing, in.

2.2 The values shown in Table 1, reflecting a design factor of approximately 4:1, are based on an allowable fiber stress of 12,500 psi which is equivalent to 50% of the minimum yield point and approximately 28% of the minimum ultimate strength of the tubing.

2.3 The values shown in Table 2, reflecting a design factor of approximately 3:1, are based on an allowable fiber stress of 17,000 psi which is equivalent to 68% of the minimum yield point and approximately 38% of the minimum ultimate strength of the tubing.

3. *Hydraulic Tube Fittings*—When properly assembled in conjunction with appropriate tubing selected from respective tables (flared where applicable, in accordance with SAE J533) the hydraulic tube fittings specified in SAE J514, are capable providing leak-proof full-flow connections in hydraulic systems operating at working pressures designated in the following:

3.1 At a design factor of approximately 4:1, these fittings should be suitable for use with tubings selected from Table 1 which have a reference working pressure of 3000 psi or less.

3.2 At a design factor of approximately 3:1, these fittings should be suitable for use with tubings selected from Table 2 which have a reference working pressure of 5000 psi or less.

3.3 These fittings are also capable of higher working pressures in hydraulic systems wherein conditions will permit the use of smaller tube sizes, increased wall thickness (37 deg flared tube fittings are limited in this regard) and/or reduction of the design factor. Prior to such use, however, it is recommended sufficient testing be conducted and reviewed by both the user and fittings manufacturers to assure performance levels will be satisfactory.

TABLE 1—REFERENCE WORKING PRESSURES AT APPROXIMATELY 4:1 DESIGN FACTOR, PSI

Nominal Tube OD, in		See Note[a]	Nominal Tube Wall Thickness, in											
			0.028	0.035	0.049	0.065	0.083	0.095	0.109	0.120	0.134	0.148	0.156	0.188
1/8	0.125	1	5600	7000										
		2	6800	9000										
		3	6650	8450										
3/16	0.188	1	3750	4650										
		2	4250	5500										
		3	4250	5450										
1/4	0.250	1	2800	3500	4900	6500								
		2	3100	3950	5800	8200								
		3	3100	3950	5750	7800								
5/16	0.312	1	2250	2800	3900	5200								
		2	2400	3100	4500	6250								
		3	2450	3100	4500	6150								
3/8	0.375	1	1850	2350	3250	4350	5550	6350						
		2	2000	2500	3650	5050	6700	7950						
		3	2000	2550	3650	5000	6550	7600						
1/2	0.500	1		1750	2450	3250	4150	4750	5450	6000				
		2		1850	2650	3650	4800	5600	6600	7450				
		3		1850	2700	3650	4800	5550	6450	7200				
5/8	0.625	1		1400	1950	2600	3300	3800	4350	4800				
		2		1450	2100	2850	3700	4350	5050	5650				
		3		1500	2100	2850	3750	4350	5050	5600				
3/4	0.750	1		1150	1650	2150	2750	3150	3650	4000				
		2		1200	1700	2350	3050	3500	4100	4600				
		3		1200	1750	2350	3050	3550	4150	4600				
7/8	0.875	1		1000	1400	1850	2350	2700	3100	3400				
		2		1050	1450	1950	2550	2950	3450	3850				
		3		1050	1500	2000	2600	3000	3500	3900				
1	1.000	1		875	1200	1600	2050	2350	2700	3000	3350	3700		
		2		900	1250	1700	2200	2550	3000	3300	3750	4200		
		3		900	1300	1750	2250	2600	3000	3350	3800	4200		
1-1/8	1.125	1			1100	1450	1850	2100	2400	2650	3000	3300		
		2			1150	1500	1950	2250	2650	2900	3300	3700		
		3			1150	1550	2000	2300	2650	2950	3300	3700		
1-1/4	1.250	1			1000	1300	1650	1900	2200	2400	2700	2950	3100	3750
		2			1000	1350	1750	2000	2350	2600	2950	3250	3450	4250
		3			1000	1350	1750	2050	2350	2650	2950	3300	3500	4300
1-1/2	1.500	1				1100	1400	1600	1800	2000	2250	2450	2600	3150
		2				1100	1450	1650	1950	2150	2400	2700	2850	3500
		3				1150	1450	1700	1950	2150	2450	2700	2850	3500
1-3/4	1.750	1				925	1200	1350	1550	1700	1900	2100	2250	2700
		2				950	1250	1400	1650	1800	2050	2250	2400	2950
		3				950	1250	1450	1650	1850	2050	2300	2400	2950
2	2.000	1				800	1050	1200	1350	1500	1650	1850	1950	2350
		2				850	1050	1250	1400	1600	1750	1950	2100	2550
		3				850	1100	1250	1450	1600	1800	2000	2100	2550
2-1/4	2.250	1				700	900	1050	1200	1350	1500	1650	1750	2100
		2				750	950	1100	1250	1400	1550	1750	1850	2250
		3				750	950	1100	1250	1400	1600	1750	1850	2250

NOTE: WALL THICKNESSES HAVING VALUES SHOWN TO RIGHT OF BOLD LINE ARE NOT NORMALLY CONSIDERED SUITABLE FOR 37 DEG SINGLE FLARING TO SAE J533.

[a] Pressure values listed opposite numbers 1, 2, and 3 for each tube OD were derived from the Barlow, Boardman, and Lamé formulas, respectively, with 12,500 psi allowable stress factor. See paragraph 2.1 pressures for tube sizes not shown above may be calculated using these formulas.

TABLE 2—REFERENCE WORKING PRESSURES AT APPROXIMATELY 3:1 DESIGN FACTOR, PSI

Nominal Tube OD, in		See Note[a]	Nominal Tube Wall Thickness, in										
			0.028	0.035	0.049	0.065	0.083	0.095	0.109	0.120	0.134	0.148	0.156	0.185
1/8	0.125	1	7600	9500										
		2	9300	12250										
		3	9050	11500										
3/16	0.188	1	5100	6350										
		2	5750	7450										
		3	5800	7400										
1/4	0.250	1	3800	4750	6650	8850								
		2	4200	5350	7900	11150								
		3	4200	5400	7800	10650								
5/16	0.312	1	3050	3800	5350	7050								
		2	3300	4200	6100	8500								
		3	3300	4200	6100	8350								
3/8	0.375	1	2550	3150	4450	5900	7550	8600						
		2	2700	3450	4950	6850	9150	10800						
		3	2750	3450	5000	6850	8950	10350						
1/2	0.500	1		2400	3350	4400	5650	6450	7400	8150				
		2		2500	3600	4950	6500	7600	9000	10100				
		3		2550	3650	4950	6500	7550	8800	9750				
5/8	0.625	1		1900	2650	3550	4500	5150	5950	6550				
		2		2000	2850	3850	5050	5900	6900	7700				
		3		2000	2900	3900	5100	5900	6900	7650				
3/4	0.750	1		1600	2200	2950	3750	4300	4950	5450				
		2		1650	2350	3150	4150	4800	5600	6250				
		3		1650	2350	3200	4150	4850	5600	6250				
7/8	0.875	1		1350	1900	2550	3250	3700	4250	4650				
		2		1400	2000	2700	3500	4050	4700	5250				
		3		1400	2000	2700	3500	4100	4750	5250				
1	1.000	1		1200	1650	2200	2800	3250	3700	4100	4550	5050		
		2		1200	1750	2350	3000	3500	4050	4500	5100	5700		
		3		1250	1750	2350	3050	3550	4100	4550	5150	5750		
1-1/8	1.125	1			1500	1950	2500	2850	3300	3650	4050	4450		
		2			1550	2050	2650	3100	3550	3950	4500	5000		
		3			1550	2100	2700	3100	3600	4000	4500	5050		
1-1/4	1.250	1			1350	1750	2250	2600	2950	3250	3650	4050	4250	5100
		2			1400	1850	2400	2750	3200	3550	4000	4450	4700	5800
		3			1400	1850	2400	2800	3200	3550	4000	4500	4750	5850
1-1/2	1.500	1				1450	1900	2150	2450	2700	3050	3350	3550	4250
		2				1550	1950	2250	2600	2900	3250	3650	3850	4750
		3				1550	2000	2300	2650	2950	3300	3700	3900	4750
1-3/4	1.750	1				1250	1600	1850	2100	2350	2600	2900	3050	3650
		2				1300	1700	1950	2250	2450	2750	3100	3250	4000
		3				1300	1700	1950	2250	2500	2800	3100	3300	4050
2	2.000	1				1100	1400	1600	1850	2050	2300	2500	2650	3200
		2				1150	1450	1700	1950	2150	2400	2650	2850	3450
		3				1150	1450	1700	1950	2150	2450	2700	2850	3500
2-1/4	2.250	1				975	1250	1450	1650	1800	2000	2250	2350	2840
		2				1000	1300	1500	1700	1900	2150	2350	2500	3050
		3				1000	1300	1500	1750	1900	2150	2400	2500	3050

NOTE: WALL THICKNESSES HAVING VALUES SHOWN TO RIGHT OF BOLD LINE ARE NOT NORMALLY CONSIDERED SUITABLE FOR 37 DEG SINGLE FLARING TO SAE J533.

[a] Pressure values listed opposite numbers 1, 2, and 3 for each tube OD were derived from the Barlow, Boardman, and Lamé formulas, respectively, with 17,000 psi allowable stress factor.

See paragraph 2.1 pressures for tube sizes not shown above may be calculated using these formulas.

SEAMLESS COPPER TUBE—SAE J528 JAN80   SAE Standard

Report of the Tube, Pipe, Hose, and Lubrication Fittings Committee, approved January 1953, last revised by the Fluid Conductors and Connectors Technical Committee January 1980.

Scope—This standard covers minimum requirements for soft (061) annealed seamless copper tube intended for automotive and general purposes. (Comparable specifications are ASTM B 75 and ANSI H23.3. Other copper tube is covered in SAE J463.)

Manufacture—The tube shall be cold drawn to size and after forming shall be annealed in such a manner as to produce a finished product which will meet all requirements of this standard.

Dimensions and Tolerances—Tube furnished to this standard shall conform to the dimensional tolerances shown in Table 1 for the size of tube specified by the purchaser. (Standard nominal sizes are listed.)

Quality—The finished tube shall be clean, smooth, and round, free from internal and external mechanical imperfections, and shall have a bright appearance.

Material—Unless otherwise specified by purchaser, tube shall be made from any one of the materials listed in Table 2. (UNS C12200 is most commonly used.) Average grain size of the tube shall be 0.040 mm, minimum.

◆ TABLE 1—TUBING DIMENSIONS AND TOLERANCES

Nominal Tubing OD		Outside Diameter[a]				Wall Thickness			
		Basic		Tolerance		Basic		Tolerance[b]	
in	mm	in	mm	±in	±mm	in	mm	±in	±mm
1/8	3.18	0.125	3.18	0.0020	0.05	0.030	0.76	0.0030	0.08
3/16	4.76	0.188	4.78	0.0020	0.05	0.030	0.76	0.0025	0.063
1/4	6.35	0.250	6.35	0.0020	0.05	0.030	0.76	0.0025	0.063
5/16	7.95	0.312	7.92	0.0020	0.05	0.032	0.81	0.0025	0.063
3/8	9.53	0.375	9.52	0.0020	0.05	0.032	0.81	0.0025	0.063
1/2	12.70	0.500	12.70	0.0020	0.05	0.032	0.81	0.0025	0.063
5/8	15.88	0.625	15.88	0.0020	0.05	0.035	0.89	0.0025	0.063
3/4	19.05	0.750	19.05	0.0025	0.063	0.035	0.89	0.0025	0.063

[a] The actual outside diameter shall be the average of the maximum and minimum outside diameters as determined at any one cross section through the tubing.
[b] The tolerances listed represent the maximum permissible deviation at any point.

Mechanical Properties—Tube shall conform to the following:

Ultimate Strength (Tensile), min	◆ 30,000 psi (205 MPa)
Yield Strength (Tensile), min[a]	◆ 9,000 psi (62.0 MPa)

[a] At 0.5% extension under load.

Expansion Test—Samples of tube (selected from sections which have not been subjected to cold working after anneal of the finished sized tube) shall be cut square and deburred. These shall be expanded on a hardened and ground tapered steel pin having an included angle of 60 deg until the outside diameter is increased 40%. Care should be taken to keep the axes of the pin and the tube in line during the expansion operation. The test may be made in a die to restrict the expansion to 40%. The expanded tube shall show no cracking or rupture visible to the unaided eye.

◆ *Hydrostatic Test*—Unless otherwise specified, tube shall show no evidence of weakness or defects when subjected to an internal hydrostatic pressure sufficient to subject the material to a fiber stress of 6000 psi (40 MPa) determined by the following formula for thin, hollow cylinders under tension. The tube need not be tested at a hydrostatic pressure of over 1000 psi (7 MPa) unless so specified.

$$P = \frac{2St}{D - 0.8t}$$

where: P = hydrostatic pressure, psi (MPa)
t = minimum thickness of tube wall, in (mm)
D = basic outside diameter of tube, in (mm)
S = allowable stress of the material = 6000 psi (40 MPa) ◆

Embrittlement Test—The tube is expected to pass the following test al- ◆ though the actual performance of the test is not required under this specification unless specifically stipulated by the purchaser:

(a) Heat the cleaned or degreased specimens for 20 min minimum at a temperature of 850 ± 25°C (1562 ± 45°F) in a furnace in which the atmosphere is at least 10% of hydrogen by volume. Then quench the specimens immediately and rapidly in water or in the same atmosphere with minimum contact with air.

(b) Polish and etch if desired, cross-sectional test specimens taken transverse to, and bounded by, an original surface of the material. Examine the prepared surface microscopically under illumination at a magnification of 75–200 diameters inclusive. Specimens shall show no passing or open grain structure characteristic of embrittlement.

◆ TABLE 2—CHEMICAL COMPOSITION, WEIGHT %

SAE Alloy No.[a]	UNS No.[b]	Similar ASTM Copper No.[c]	Copper, min	Phosphorus	Arsenic
CA102	C10200	102 (was OF)	99.95	—	—
CA120	C12000	120 (was DLP)	99.90	0.004–0.012	—
CA122	C12200	122 (was DHP)	99.90	0.015–0.040	—
—	—	142 (was DPA)	99.40	0.015–0.040	0.15–0.50

[a] SAE J463.
[b] Unifed Numbering System.
[c] ASTM B 75 and ANSI H23.3.

METALLIC AIR BRAKE SYSTEM TUBING AND PIPE—SAE J1149

SAE Standard

This standard was formerly designated SAE J844 approved June 1963 and completely revised July 1976.

Scope—This SAE standard covers minimum requirements for two *types* of metallic tubing and pipe as used in automotive air brake systems. It includes material and performance specifications, corrosion precautions, and installation recommendations. Copper tubing is designated *Type 1*, and galvanized steel pipe *Type 2*.

Corrosion Precautions—In the design and selection of air brake system components, adequate provision shall be made to control corrosion due to galvanic coupling of widely dissimilar metals and alloys when such materials used for tubing, pipe, fittings, and attaching or supporting parts are in intimate contact with each other. Also, adequate provision shall be made to protect the tubing, pipe, and fittings from oxygen concentration cell type of corrosion. Where soft nonmetallic cushions are used to prevent metal-to-metal contact between supporting components and the tubing, pipe, and fittings, the cushioning material shall be such that it will not absorb and retain significant amounts of water.

Installation Recommendations—The tubing or pipe installed in air brake systems shall be supported in such a manner as to minimize fatigue conditions. Metal-to-metal contact should be avoided by the use of soft nonmetallic cushions at points of support to control chafing and fretting. Tubing or pipe shall be protected against road hazards either by installation in a protected location or by providing adequate shielding at vulnerable areas. Protective loom, where used, shall be both water and acid resistant.

Specifications
Type 1—Copper Tubing
Scope—This material specification covers the minimum requirements for seamless annealed copper tubing that shall be used for automotive air brake lines.
Manufacture—The tubing shall be seamless cold drawn to size and bright annealed as a final operation in such a manner as to produce a finished product which will meet all requirements of this standard.

Dimensions and Tolerances—The finished tubing shall conform to the dimensions and tolerances shown in Table 1, for the nominal diameter specified by the purchaser.

Quality—The finished tubing shall be clean, smooth, and round, free from internal and external mechanical imperfections, corrosion, scale, seams, and cracks.

Material—The tubing shall be made from phosphorized, low residual phosphorus copper conforming to SAE J463, UNS C12200 which has the following chemical composition:

Element	Ladle Analysis % by Weight
Copper	99.90 min
Phosphorus	0.015–0.040

Mechanical Properties—The finished tubing shall have mechanical properties as tabulated below:

Yield Strength psi (MPa) min[a]	Tensile Strength psi (MPa) max	Elongation in 2 in (50 mm), % min Tubing OD	
		3/4 in (19 mm) and smaller	Over 3/4 in (19 mm)
9000 (62)	30 000 (210)	30	40

[a] At 0.5% extension under load.

TABLE 1—DIMENSIONS AND TOLERANCES OF AIR BRAKE TUBING

Nominal Tubing OD (in)	Outside Diameter[a]				Wall Thickness (min)	
	Specified		Tolerance ±			
	in	mm	in	mm	in	mm
1/4	0.250	6.35	0.002	0.05	0.0295	0.75
5/16	0.312	7.92	0.002	0.05	0.0295	0.75
3/8	0.375	9.52	0.002	0.05	0.0295	0.75
7/16	0.437	11.10	0.002	0.05	0.0455	1.160
1/2	0.500	12.70	0.002	0.05	0.0455	1.160
5/8	0.625	15.87	0.002	0.05	0.0455	1.160
3/4	0.750	19.05	0.0025	0.06	0.0455	1.160
1	1.000	25.40	0.0025	0.06	0.0455	1.160

[a] The actual outside diameter shall be the average of the maximum and minimum outside diameters as determined at any one cross section through the tubing.

Grain Size—The tubing shall be furnished in either of two temper conditions with grain size as tabulated below:

Temper	Grain Size, mm
Light Annealed	0.015–0.040
Soft Annealed	0.040 min

Performance Requirements—The finished tubing shall satisfactorily meet the following performance tests. Test specimens shall be taken from tubing which has not been subjected to cold working after the anneal of the finished sized tubing.

Flaring Test—A test section cut from the finished tubing, having squared and deburred ends, shall withstand being flared at one end over a polished tapered mandrel of 60 deg included angle until the actual average outside diameter is increased 40% without evidence of splitting or flaws. The axis of the mandrel and axis of the tubing shall be kept parallel during the flaring process and the test may be made in a die to restrict the expansion to 40%.

Pressure Proof Test—Unless otherwise specified, tubing supplied under this standard shall withstand, with no evidence of failure, a hydrostatic proof test at a pressure equivalent to a yield stress of 9000 psi (62 MPa). The test pressures shall be as determined from Barlow's formula for thin hollow cylinders under tension:

$$P = \frac{2\,TS}{D}\left[\frac{2000\,TS}{D}\right]$$

where:

D = outside diameter of tubing, in (mm)
P = hydrostatic pressure, psi (kPa)
S = allowable unit stress of material = 9000 psi (62 MPa)
T = minimum wall thickness of tubing, in (mm)

The test pressure at a yield strength of 9000 psi (62 MPa) for the minimum wall thicknesses allowed are given in Table 2.

Air Pressure Test—Each length of finished tubing shall be tested at the maximum operating air pressure, as specified by the purchaser. The tubing shall show no leakage at the test pressure. An electric eddy current test may be substituted for the air pressure test, providing the rejection limits are such that the hydrostatic and air pressure requirements can be guaranteed.

Identification—Tubing shall be permanently and legibly marked at intervals not greater than 15 in (381 mm) with the words *Air Brake*.

Methods of Test—All tests to determine conformance with the foregoing specifications shall be conducted in accordance with the following ASTM Standards:

Chemical Analysis—See ASTM E 62, Method of Test for Antimony in Copper and Copper Base Alloys.

TABLE 2—HYDROSTATIC TEST PRESSURES FOR AIR BRAKE TUBING

Nominal Tubing OD, in	Hydrostatic Test Pressure		Nominal Tubing OD, in	Hydrostatic Test Pressure	
	psi	kPa		psi	kPa
1/4	2100	14 500	1/2	1600	11 000
5/16	1700	11 700	5/8	1300	8950
3/8	1400	9650	3/4	1000	6900
7/16	1800	12 400	1	800	5500

Grain Size—See ASTM E 79, Methods for Estimating the Average Grain Size of Wrought Copper and Copper Base Alloys.

Tensile—See ASTM E 8, Methods of Tension Testing of Metallic Materials.

Type 2—Galvanized Steel Pipe

Scope—This material specification covers the minimum requirements for pipe that shall be used in automotive air brake lines.

Specifications—Welded or seamless steel pipe shall be Schedule 40, Zinc Coated (galvanized by the hot dip process), and manufactured in accordance with ASTM A 120, Specification for Black and Hot-Dipped Zinc-Coated (Galvanized) Welded and Seamless Steel Pipe for Ordinary Uses.

Dimensions and Tolerances—The finished pipe shall conform to the dimensions and tolerances listed for the several nominal sizes in Table 3.

Pipe Threads—Both ends of lengths of pipe shall be threaded after coating, unless there is specific authorization to the contrary, to conform to Dryseal American Standard Taper Thread (NPTF). Specifications for pipe threads are given in detail in SAE J476.

Mechanical Properties—The steel in the finished pipe, including the weld, shall have mechanical properties as tabulated below:

Yield Strength, psi (MPa), min	Elongation in 2 in (50 mm), %
25 000 (170)	14–40

Pressure Proof Tests Per Test Method ASTM A 370, Supplement II

Hydrostatic Test—Unless otherwise specified, each length of pipe shall be tested at the mill to a hydrostatic pressure of 700 psi (4850 kPa). For nominal sizes over 1 in (25.4 mm) dia see ASTM A 120.

Nondestructive Electric Test—In lieu of the hydrostatic test, if mutually agreeable to purchaser and the manufacturer, each pipe may be tested by passing it through an electric eddy current tester which is capable of detecting defects 0.062 in (1.57 mm) in length and one-half the wall thickness, or defects of any length completely penetrating the wall. Such tests shall be made on the welded seam and the adjacent metal affected thereby.

Corrosion Protection—The inside and outside surfaces of the pipe shall be coated with zinc by the hot dip process. The coating shall weigh at least 2.0 oz/ft^2 (610 g/m^2) of total surface. Tests to determine whether product meets this requirement shall be conducted in accordance with ASTM A 120.

Bending—Pipe shall be used for essentially straight runs; however, generous curves having a radius in excess of 20 times the outside diameter shall be permitted. In no case shall heat be used to facilitate bending of pipe.

TABLE 3A—DIMENSIONS AND TOLERANCES OF PIPE FOR AIR BRAKE USE, in

Nominal Pipe Size	Outside Diameter			Inside Diameter (Ref)	Wall Thickness		Threads per in	Weight per ft,[a] lb ±5%
	Specified	Tolerance			Specified	min		
		Plus	Minus					
1/8	0.405	0.016	0.031	0.269	0.068	0.060	27	0.24
1/4	0.540	0.016	0.031	0.364	0.088	0.077	18	0.42
3/8	0.675	0.016	0.031	0.493	0.091	0.080	18	0.57
1/2	0.840	0.016	0.031	0.622	0.109	0.095	14	0.85
3/4	1.050	0.016	0.031	0.824	0.113	0.099	14	1.13
1	1.315	0.016	0.031	1.049	0.133	0.116	11.5	1.68

[a] Nominal Weight Plain Ends.

TABLE 3B—DIMENSIONS AND TOLERANCES OF PIPE FOR AIR BRAKE USE (mm)

Outside Diameter			Inside Diameter (Ref)	Wall Thickness		Threads per in	Nominal Weight Plain Ends kg/m ± 5%	Nominal in Size
mm Specified	Tolerance			Specified	min			
	min	max						
10.29	9.50	10.67	6.83	1.73	1.52	27	0.36	1/8
13.72	12.93	14.10	9.25	2.24	1.96	18	0.63	1/4
17.14	16.36	17.53	12.52	2.31	2.03	18	0.85	3/8
21.34	20.55	21.72	15.80	2.77	2.41	14	1.27	1/2
26.67	25.88	27.05	20.93	2.87	2.51	14	1.68	3/4
33.40	32.61	33.78	26.64	3.38	2.95	11.5	2.50	1

NONMETALLIC AIR BRAKE SYSTEM TUBING—SAE J844d

SAE Standard

Report of Tube, Pipe, Hose, and Lubrication Fittings Committee approved June 1963 and completely revised July 1976.

1. Scope[1]—This standard covers the minimum requirements for nonmetallic tubing as manufactured for use in air brake systems. Non-reinforced products are designated type A and reinforced products type B. It is not intended to cover tubing for any portion of the system which operates below $-40°F$ ($-40°C$), above $+200°F$ ($+93°C$), above a maximum working gage pressure of 150 psi (1030 kPa), or in an area subject to attack by battery acid. This tubing is intended for use in the brake system for connections which maintain a basically fixed relationship between components during vehicle operation. *Coiled tube assemblies* required for those installations where flexing occurs are covered by this standard and SAE J1131 to the extent of setting minimum requirements on the essentially straight tube and tube fitting connections which are used in the construction of such assemblies.[2]

2. Installation and Assembly Recommendations

2.1 End Fittings—End fittings are to be assembled to the tubing in accordance with the fitting manufacturer's recommendations. The fitting may be of the design shown in SAE J246, or any other design suitable for use with nonmetallic air brake tubing. Performance test requirements for nonmetallic air brake assemblies are covered in SAE J1131.

2.2 Non-Coiled Tubing—Non-coiled tubing should not be used in flexing applications such as frame to axle.

2.3 Support and Routing—When installed in a vehicle this tubing shall be routed and supported so as to:
  (a) Eliminate chafing, abrasion, kinking, or other mechanical damage.
  (b) Minimize fatigue conditions.
  (c) Be protected against road hazards by installation in a protected location or by providing adequate shielding at vulnerable areas.
  (d) Not to be exposed to temperatures, internal or external, over $+200°F$ ($+93°C$) or below $-40°F$ ($-40°C$).
  (e) Not to be exposed to attack by battery acid.
  (f) Avoid excessive sag.

3. Identification—Air brake tubing shall be labeled in a contrasting color with the legend repeated every 15 in (380 mm) or less along the entire length of tubing in legible block capital letters.

The following minimum information, in the order listed, is required. Additional information and/or another lay line may be added, if necessary.
  (a) Airbrake
  (b) SAE J844
  (c) Type, A or B
  (d) Nominal, tubing O.D. in fractions of in—1/4, 3/8, 1/2, etc. (6.4, 9.5, 12.7 mm)
  (e) Tubing manufacturer's identification

4. Manufacture—The tubing shall be manufactured to comply with the requirements outlined in this standard.

5. Construction—Type A tubing shall consist of a single wall extrusion of 100% virgin nylon (polyamide) containing additives which provide heat and light resistance. Type B tubing shall consist of a core extrusion of 100% virgin nylon (polyamide) containing additives which provide heat resistance. This core shall be reinforced with polyester braid or equivalent, and covered with a protective jacket of 100% virgin nylon (polyamide) containing additives which provide heat and light resistance. The protective covering shall be bonded to the core through the interstices of the braid. The inner core and outer jacket shall be of contrasting colors.

[1] See SAE J1149 for Metallic Air Brake System Tubing and Pipe.
[2] Federal regulations covering designed requirements and accepted applications for coiled tube assemblies are set forth in 49CFR393.45.

6. Dimensions and Tolerances—The tubing shall conform to dimensions shown in Table 1 under all conditions of moisture. Conformance with this requirement shall be determined on samples which have been subjected to $230°F$[3] ($110°C$) for $4 h$[4] in a circulating air oven, and on separate samples which have been immersed in boiling water for 2 h. Dimensional tests shall be made after samples have been returned to room temperature for $\frac{1}{2}$ to 3 h.

7. Mechanical Properties—The tubing shall conform to the mechanical properties shown in Table 2, when tested according to the methods outlined in this standard.

8. Performance Requirements—The tubing shall satisfactorily meet the following performance tests (see Footnotes 3, 4, 5, and 6).

8.1 Leak Test[6]—The tubing manufacturer shall subject each continuous length of tubing to test at a gage pressure of 200 psi (1380 kPa) with an appropriate gas for a period of time sufficient to determine the presence of any leaks. Defective sections shall be cut off and scrapped. The remaining tubing shall be recoupled at the points where defective sections were removed and again subjected to the 200 psi (1380 kPa) pressure test. The procedure shall be repeated until all sections of tubing designated for distribution to users have successfully withstood the test.

8.2 Moisture Absorption[5]—Expose sample of tubing for 24 h in a circulating air oven at $230°F$ ($110°C$). Remove from oven, weigh immediately and expose for 100 h at 100% relative humidity and $75°F$ ($24°C$). Within 5 min from humidity conditioning, wipe surface moisture from both the interior and exterior surfaces of the tubing and re-weigh. Moisture absorption shall not exceed 2% by weight.

8.3 Ultraviolet Resistance[5]—Place sample of tubing on a turntable 17 in (430 mm) in diameter, rotating at 33 ± 3 rpm, with a RS-4* sunlamp or equivalent centrally located 9 in (230 mm) above the table. Expose for 1200 h using a new bulb that has been seasoned for 50 h prior to test. Do not permit temperature of tubing to exceed $120°F$ ($49°C$) during the test (a fan cooling unit may be utilized). Immediately following this exposure, subject the tubing to the impact test shown in Fig. 1. Tubing shall show no evidence of cracks. Subject tubing to Room Temperature Burst Test as specified in 8.10. Tubing shall withstand no less than 80% of the burst pressure shown in Table 2.

[3] All test temperatures specified may vary by $\pm 5°F$ ($\pm 3°C$).
[4] All times are minimum unless otherwise specified.
[5] Normally considered a *Qualification Test*.
[6] Normally considered an *Inspection Test* conducted on each lot of tubing; and where a lot is defined as "the output of one production shift of one size and color of tubing."

TABLE 2—MECHANICAL PROPERTIES

Type of Tubing	Nominal Tubing OD	Minimum Burst Pressure at 75°F(24°C)		Minimum Bend Radius		Maximum Stiffness	
		psi	MPa	in	mm	lbf	N
A	1/8	1000	6900	0.37	9.4	1	4.4
A	1/4	1200	8300	1.00	25.4	2	8.9
A	5/16	1000	6900	1.25	31.8	6	27.0
B	3/8	1400	9700	1.50	38.1	8	36.0
B	1/2	950	6600	2.00	50.8	20	89.0
B	5/8	900	6200	2.50	63.5	50	222.0
B	3/4	800	5500	3.00	76.2	80	356.0

TABLE 1—DIMENSIONS AND TOLERANCES

Type of Tubing	Nominal Tubing OD	Outside Diameter				Inside Diameter		Wall Thickness			
		Max		Min		Basic		Basic		Tolerances	
		in	mm	in	mm	in	mm	in	mm	in	mm
A	1/8	0.128	3.25	0.122	3.10	0.079	2.01	0.023	0.58	±0.003	±0.08
A	1/4	0.253	6.43	0.247	6.27	0.170	4.32	0.040	1.02	±0.003	±0.08
A	5/16	0.316	8.03	0.308	7.82	0.232	5.89	0.040	1.02	±0.004	±0.10
B	3/8	0.379	9.63	0.371	9.42	0.251	6.38	0.062	1.57	±0.004	±0.10
B	1/2	0.505	12.83	0.495	12.57	0.376	9.55	0.062	1.57	±0.004	±0.10
B	5/8	0.630	16.00	0.620	15.75	0.441	11.20	0.092	2.34	±0.005	±0.13
B	3/4	0.755	19.18	0.745	18.92	0.566	14.38	0.092	2.34	±0.005	±0.13

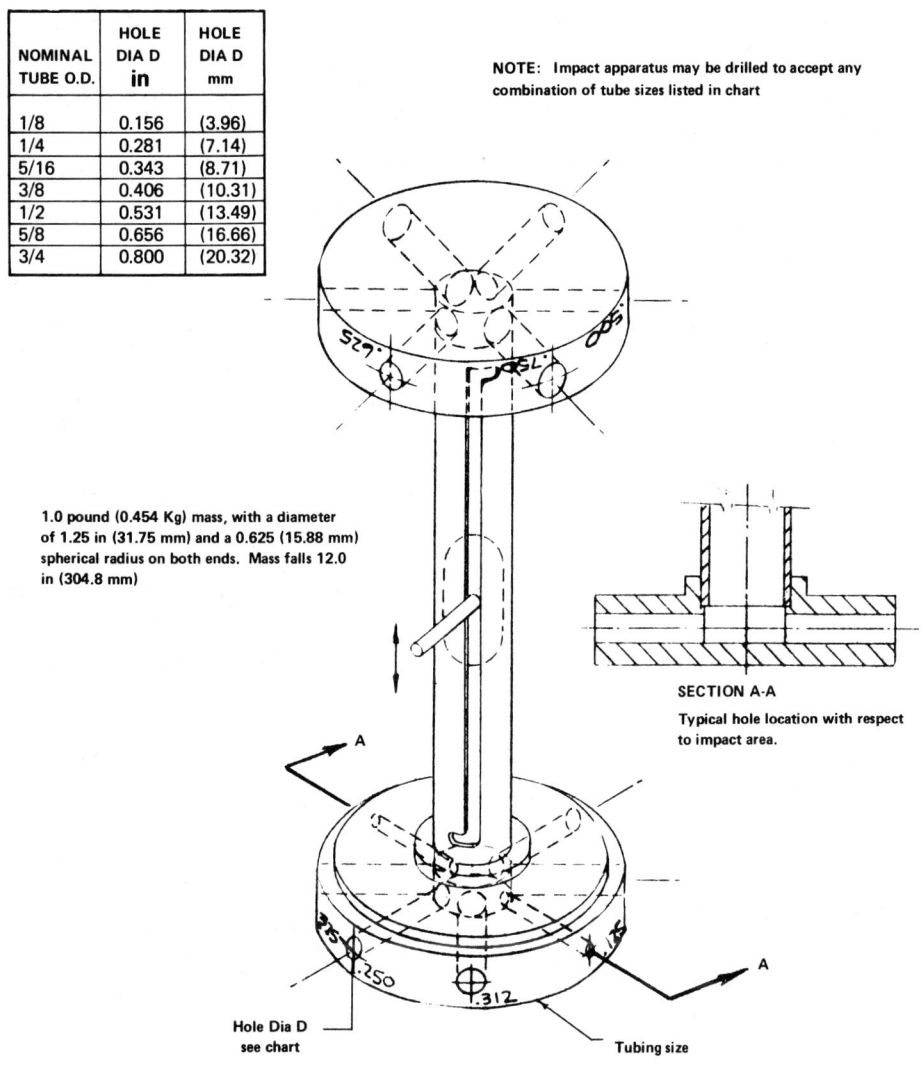

NOMINAL TUBE O.D.	HOLE DIA D in	HOLE DIA D mm
1/8	0.156	(3.96)
1/4	0.281	(7.14)
5/16	0.343	(8.71)
3/8	0.406	(10.31)
1/2	0.531	(13.49)
5/8	0.656	(16.66)
3/4	0.800	(20.32)

FIG. 1—TYPICAL NYLON TUBING IMPACT APPARATUS

*RS-4 sunlamp is manufactured by General Electric Company[7]
Cuyahoga Lamp Plant
Nela Park
Noble Road
Cleveland, OH 44112
*RS-4 sunlamp is available from George W. Gates Co., Inc.
P.O. Box 216
Hempsted Turnpike and Lucille Ave.
Franklin Square
Long Island, NY 11010

The RS-4 sunlamp is a 100 W; 3010 lm mercury arc lamp with an outer glass jacket which eliminates wavelengths below 285 manometres.

8.4 Cold Temperature Flexibility[5]—Expose sample of tubing for 24 h in a circulating air oven at 230°F (110°C). Remove from oven and within 30 min expose for 4 h at −40°F (−40°C). Also expose a mandrel at −40°F (−40°C) having a diameter equal to 12 times the nominal diameter of the tubing. (In order to obtain uniform temperatures the tubing and mandrel may be supported by a nonmetallic surface during the entire period of test.) Immediately following this exposure, bend tubing 180 deg over the mandrel, accomplishing the bending motion within a period of 4—8 s. The tubing shall show no evidence of fracture.

8.5 Heat Aging[5]—Three separate heat aging tests shall be conducted; each phase shall be run on separate tubing samples. After the completion of each phase, tubing shall show no evidence of fracture or kinking. Subject tubing to Room Temperature Burst Test as specified in 8.10. Tubing shall withstand 80% of the burst pressure shown in Table 2.

[7] The manufacturer and distributor of the sunlamp is listed due to the fact that at the present time this is the only known supplier.

Phase 1—Bend samples of tubing 180 deg around a mandrel having a diameter equivalent to twice the minimum bend radius specified in Table 2. While in this position, expose tubing and mandrel for 72 h in an air circulating oven at 230°F (110°C). Remove from oven and permit tubing to return to 75°F (24°C) while still on the mandrel. Within 30 min after stabilization at 75°F (24°C), return the tubing to a straight position in a minimum of 4 s, then rebend (against the set) 180 deg around the mandrel, accomplishing the bending motion within a period of 4–8 s.

Phase 2—Expose samples of tubing for 72 h in a circulating air oven at 230°F (110°C). Remove from oven and permit tubing to return to 75°F (24°C). Within 30 min after stabilization at 75°F (24°C), subject tubing to the Impact Test shown in Fig. 1.

Phase 3—Immerse samples of tubng in boiling water for 2 h. Remove from water and permit to return to 75°F (24°C). Within 30 min after stabilization at 75°F (24°C), subject tubing to the Impact Test shown in Fig. 1.

8.6 Resistance to Zinc Chloride[5]—Bend tubing to the minimum bend radius shown in Table 2. While in this position, immerse in a 50% (by weight) aqueous solution of zinc chloride for 200 h at 75°F (24°C). Remove from solution. Tubing shall show no evidence of cracking on the outside diameter.
NOTE: Fresh, anhydrous zinc chloride should be used to make up a concentration of 50% (by weight) aqueous solution (specific gravity of 1.576 or a Baume rating of 53° at 60°F (15.6°C)).

8.7 Resistance to Methyl Alcohol[5]—Bend tubing to the minimum bend radius shown in Table 2. While in this position, immerse in 95% methyl alcohol for 200 h at 75°F (24°C). Remove from solution. Tubing shall show no evidence of cracking.

8.8 Stiffness[5]—Use samples 11 in (280 mm) long. Insert a rod of suitable size into the tubing to maintain a straight position within ±0.125 in (3.2 mm). Expose tubing and rod for 24 h in a circulating air oven at 230°F

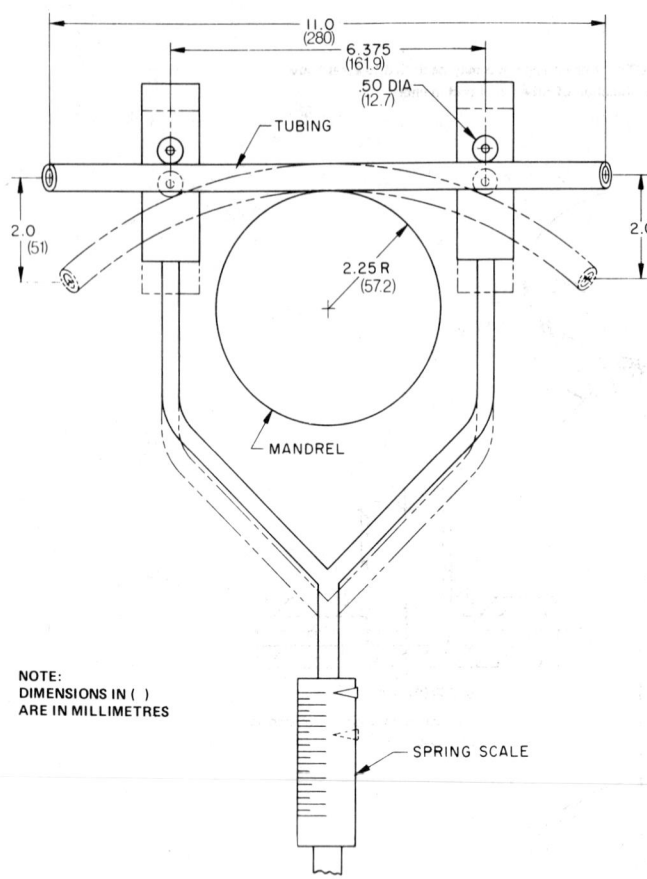

FIG. 2—STIFFNESS TEST APPARATUS

(110°C). Remove from oven and permit tubing and rod to return to 75°F (24°C). Within 30 min after stabilization at 75°F (24°C), remove rod and subject tubing to Stiffness Test shown in Fig. 2. Tubing shall require no more force than specified in Table 2 to deflect 2 in (51 mm).

8.9 Boiling Water Stabilization and Burst Test[5]—Immerse tubing in boiling water for 2 h. Remove from water and subject to the Room Temperature Burst Test as specified in 8.10. Tubing shall withstand no less than 80% of the burst pressure shown in Table 2.

8.10 Room Temperature Burst Test[6]—Tubing shall be stabilized (for ½–3 h at 75°F (24°C) and tested by increasing pressure at a constant rate to reach the specified minimum burst pressure in Table 2 within a time period of 3—15 s. Tubing that bursts below the pressure specified in Table 2 shall be rejected.

8.11 Cold Temperature Impact[6]—Condition tubing by exposing one half the samples for 24 h at 230°F (110°C) in a circulating air oven, and one half the samples in boiling water for 2 h; then expose all the samples to −40°F (−40°C) for 4 h. Also, expose impact test apparatus, shown in Fig. 1, to −40°F (−40°C). While tubing and apparatus are at this cold temperature (approx. −40°F), subject tubing to impact as specified. The tubing shall show no evidence of cracks. After impact testing, permit tubing to return to 75°F (24°C). Within 30 min after stabilization at 75°F (24°C), subject tubing to Room Temperature Burst Test as specified in 8.10. Tubing shall withstand at least 80% of the burst pressure shown in Table 2. Sample size shall be 10 specimens per lot. In the event of any failures a second sample from the same lot consisting of 20 specimens shall be tested. If another failure occurs, the lot shall be rejected.

8.12 Adhesion Test[6]

8.12.1 This test applies only to the reinforced products, Type B.

8.12.2 CONDITION—This test shall be conducted at 75°F (24°C) ambient temperature.

8.12.3 PROCEDURE AND REQUIREMENTS—Cut a strip of tubing into a 0.25 in (6.0 mm) wide helical coil equal in length to 5 times the circumference of the tubing. Bend the helical coil in reverse of coiling so as to expose the braid gap between the outer jacket and core tube section. Start by working a sharp knife blade into the braid gap to initiate separation, and then attempt to separate the outer jacket from the core tube at the braid interstices. The bonded surface (excluding the braided area) between the outer jacket and core section shall be inseparable for the entire test sample length.

8.13 Heat Aging Adhesion Test[5]

8.13.1 PROCEDURE—Subject samples to Phase 1 of the Heat Aging Test Procedure per Paragraph 8.5.

8.13.2 REQUIREMENTS—After completion of the Phase 1 procedure, the tubing shall meet the requirements of 8.12, Adhesion Test.

PERFORMANCE REQUIREMENTS FOR SAE J844d NONMETALLIC TUBING AND FITTING ASSEMBLIES USED IN AUTOMOTIVE AIR BRAKE SYSTEMS—SAE J1131

SAE Standard

Report of the Tube, Pipe, Hose, and Lubrication Fittings Committee approved January 1976.

1. Scope—This standard is intended to establish uniform methods of testing SAE J844d tubing and fitting assemblies as used in automotive air brake systems.

This standard also establishes minimum qualifications for tensile and pressure capabilities, vibrational durability under cyclic temperatures, serviceability, and fitting compatibility requirements. The specific tests and performance criteria applicable to the tubing are set forth in SAE J844d.

NOTE: The test values contained in this performance standard are for test purposes only. For environmental and usage limitations see SAE J844d. Fittings—A type of fitting for use with SAE J844d nonmetallic tubing is included in SAE J246b, however, it is not intended to restrict or preclude the use of other designs of fittings that comply with this standard.

2. Tension Tests

2.1 Description—Both hot and cold tensile tests shall be conducted with different unaged assemblies (fittings attached within 30 days of test date). Tests consist of subjecting the assembly to increasing tension load in a suitable testing machine until the specified force values or elongation percentages are obtained.

2.2 Apparatus—A tension testing machine with suitable indicating device shall be used for the tension test. The fixtures for holding the test specimens shall be arranged so that the tubing and fittings have a straight center line corresponding to the direction of the machine pull. The lower part of the fixture shall be equipped with a container of sufficient dimensions to submerge the required length of tubing in water. A means of heating the water to boiling shall be provided.

2.3 Test Specimens—Obtain tubing specimens from current production stock and cut to a length sufficient to obtain 6 in ± 0.25 in (152 mm) of tubing between end fittings after assembly. Assemble fittings to the tubing using the manufacturer's recommendations.

2.4 Procedure

2.4.1 HOT PULL—Place the test specimen in the tensile machine with the lower fitting and $4 \text{ in}^{+0.25}_{-0.0}$ in (102 mm) of tubing submerged below the surface of the water such that the outside diameter is exposed to the water. Bring the water to a boil and continue boiling for $5 \text{ min}^{+0.5}_{-0.0}$ min. Apply load at a rate of pull of 1 in (25 mm) per min.

2.4.2 CONDITIONED PULL TEST—Soak test specimen in air at −40 ± 5°F (−40 ± 3°C) for $30 \text{ min}^{+0.5}_{-0.0}$ min, normalize at room temperature, and submerge in boiling water for 15 min. Repeat for a total of four complete cycles. Allow the test specimen to normalize at room temperature for 30 min. Conduct the tensile test within 30 min after the normalizing period while at ambient temperature of 75 ± 5°F (24 ± 3°C). Apply load at a rate of pull of 1 in (25 mm) per min.

2.5 Requirements—The test specimen shall elongate 50%, that is, 6 in (152 mm) increased to 9 in (229 mm), or shall withstand the load shown in the following table, without causing separation from the fitting.

Nominal Tubing OD		Tensile Load	
in	mm[a]	lb	N
1/8	3.2	15	67
1/4	6.4	50	222
5/16	7.9	75	334
3/8	9.5	150	667
1/2	12.7	200	890
5/8	15.9	325	1446
3/4	19.0	350	1557

[a] For Reference Only.

3. Vibration Test

3.1 Description—This test is designed to evaluate the effects of vibration under varying ambient temperatures on a tubing and fitting assembly. Leakage rate is used to gage acceptability.

3.2 Apparatus—Equipment capable of vibrating one end of the test specimen at 600 cpm through 0.5 in (12.7 mm) displacement in a plane perpendicular to the tube while the other end is held rigid. The distance between the static and vibrating heads is to be such that when the assembly is displaced 0.5 in, no parallel pull to the longitudinal axis of the assembly will occur. The equipment must be capable of controlling the ambient air temperature between $-40 \pm 5°F$ ($-40 \pm 3°C$) and $220 \pm 5°F$ ($104 \pm 3°C$) and of applying 120 ± 10 psig (827 ± 69 kPa) dry air to the test lines. A mass flow meter capable of determining air leakage shall be provided.

3.3 Test Specimens—Cut tubing specimens to a length sufficient to obtain 18 in (457 mm) between fittings after assembly. Assemble identical fittings to the tubing using the manufacturer's recommendations. Fitting attaching nuts are not permitted to be retightened during the test.

3.4 Procedure—Allowing 0.5 in (12.7 mm) slack, mount the lines straight in the vibration machine. Oscillate one end of the lines at 600 ± 20 cycles per min through a total stroke of 0.5 in (12.7 mm) for a total of $1\,000\,000 \,^{+50\,000}_{-0.0}$ cycles, while maintaining an internal pressure of 120 ± 10 psig (827 ± 69 kPa) using dry air. Starting at $220 \pm 5°F$ ($104 \pm 3°C$), vary the ambient air temperature from $220 \pm 5°F$ ($104 \pm 3°C$) to $-40 \pm 5°F$ ($-40 \pm 3°C$) at 250,000 vibration cycle intervals (approximately 7 h intervals). Using a mass flow meter observe for fitting leakage during and after the test. Check nut tightness after completing the test.

3.5 Requirements—The test specimen is considered a failure if leakage exceeds 50 cm³/min at $-40 \pm 5°F$ ($-40 \pm 3°C$) or 25 cm³/min at $70 \pm 5°F$ ($21 \pm 3°C$). The fitting is considered a failure if the attaching nut becomes loose. This is defined as follows:
1. Record the initial tightening torque.
2. At the conclusion of the test, attempt to tighten the nut further by applying 20% of the initial tightening torque in the tightening direction. Do not apply a higher torque and do not apply any torque or force in the loosening direction.
3. If the nut moves at all under the 20% torque, it shall be defined as a loose nut and failure of the test. Record the movement in degrees to reach 20%.

4. Proof and Burst Pressure Test

4.1 Description—This test is intended to evaluate fitting retention at proof pressure (50% of minimum burst) and at minimum burst pressure as listed in the latest issue of SAE J844.

4.2 Apparatus—The test apparatus consists of a suitable source of hydraulic pressure and the necessary gages and piping.

4.3 Test Specimen—Cut tubing specimens to obtain 12 in (305 mm) between fittings after assembly. Assemble fittings to the tubing using the manufacturer's recommendations.

4.4 Procedure—Plug one end of the test specimen and mount in the apparatus with the end unrestrained. Apply proof pressure at room temperature, $75 \pm 5°F$ ($24 \pm 3°C$) to the test specimen and hold for 30 s. Increase pressure at a constant rate so as to reach the specified minimum burst pressure within a time period of 3—15 s.

4.5 Requirements—Fittings shall not separate from the tubing nor shall the assembly visibly leak at less than specified minimum burst pressure.

5. Serviceability Test

5.1 Description—This test is intended to evaluate the effects of repeated assembly and disassembly of a tubing and fitting assembly. Leakage rate is used to gage acceptability.

5.2 Apparatus—The test apparatus consists of a suitable source of pneumatic pressure and the necessary gages and piping. A mass flow meter capable of determining air leakage shall be provided.

5.3 Test Specimens—Cut tubing specimens to obtain 12 in (305 mm) between fittings after assembly. Assemble fittings to the tubing using manufacturer's recommendations.

5.4 Procedure—The tubing and fitting connections shall be disassembled and reassembled for a minimum of five times. After the fifth reassembly, pressurize the test specimens with air to 120 ± 10 psig (827 ± 69 kPa) at room temperature, $75 \pm 5°F$ ($24 \pm 3°C$), and check for leakage.

5.5 Requirements—Leakage rate must not exceed 25 cm³/min.

6. Fitting Compatability Test

6.1 Description—This test is intended to evaluate the effects of high and low temperatures on fitting performance.

6.2 Apparatus—The test apparatus consists of a suitable source of hydraulic pressure, 450 ± 10 psig (3103 ± 69 kPa), and necessary gages and piping in environmental test chambers at $200 \pm 5°F$ ($93 \pm 3°C$) and $-40 \pm 5°F$ ($-40 \pm 3°C$).

6.3 Test Specimens—Cut tubing specimens to obtain 12 in (305 mm) between fittings after assembly. Assemble fittings to the tubing using manufacturer's recommendations.

6.4 Procedure—Fill test specimens with hydraulic fluid and subject to $200 \pm 5°F$ ($93 \pm 3°C$) and atmospheric pressure for $24 \,^{+1}_{-0.0}$ h then apply internal pressure of 450 ± 10 psig (3103 ± 69 kPa) for 5 min while maintaining temperature of $200 \pm 5°F$ ($93 \pm 3°C$). Reduce to atmospheric pressure and permit test specimens to return to room temperature; then subject test specimens to $-40 \pm 5°F$ ($-40 \pm 3°C$) and atmospheric pressure for 24 h. Apply internal pressure of 450 ± 10 psig (3103 ± 69 kPa) for $5 \,^{+0.5}_{-0.0}$ min while maintaining temperature of $-40 \pm 5°F$ ($-40 \pm 3°C$).

6.5 Requirements—Tubing shall not rupture or disconnect from the fittings.

FUEL INJECTION TUBING—SAE J529b

SAE Standard

Report of Engine Committee and Tube, Pipe, Hose, and Lubrication Fittings Committee approved January 1955 and last revised August 1969. Editorial change August 1973.

Scope—This standard covers tubing intended for use as high pressure fuel injection lines on a range of engines requiring tubing no larger than 1/4 in. (6.35 mm) OD. The material and construction of the tubing shall be cold drawn annealed low carbon seamless steel of a quality suitable for cold swaging, cold upsetting, and cold bending.

Manufacture—The tubing shall be cold drawn from steel billets which, after piercing, have had the internal surface conditioned to remove all hot mill fissures or other defects. After forming, the tubing shall be annealed in such a manner as to prevent formation of scale on the inside surface and produce a finished product which will meet all requirements of this standard.

Dimensions, Tolerances, and Color Code—The basic outside diameter and inside diameter tubing sizes with their tolerances and color code are shown in Table 1.

Quality—The outside and inside surfaces of finished tubing shall be free from scale, rust, seams, laps, laminations, deep pits, or other injurious defects. The inside surface of the tubing shall be finished to ensure a smooth bore of accurate size with no fissures, crevices, or other imperfections deeper than 0.005 in. (0.13 mm). There shall be no more than five imperfections between 0.003–0.005 in. (0.08–0.13 mm) deep per tube cross section.

The above examination for ID finish shall be performed at a minimum 50X magnification of the tubing cross section.

The inside of tubing shall be clean and free from any contamination which will impair the processing or serviceability of the tubing.

Material—Tubing shall be made from nonaging low carbon steel, produced by the open hearth, electric furnace, or basic oxygen process, conforming to the following chemical composition:

Element	Ladle Analysis[a] % by Weight
Carbon	0.08–0.13
Manganese	0.30–0.60
Phosphorus	0.04 max
Sulfur	0.05 max
Silicon	0.10 max

[a] Check analysis tolerance shall be as specified in SAE J409, Table 1.

Mechanical Properties—The finished tubing shall have mechanical properties as tabulated below:

Tensile Strength[a]	45,000–55,000 psi (310–379 MPa)
Elongation in 2 in (51 mm), %, min	30
Hardness (Rockwell B scale), max	65

[a] Per ASTM A 370, Supplement II, Section T-2.

Performance Requirements—The finished tubing shall satisfactorily meet the following performance tests. Standard sampling techniques are available. If these are not used, it is recommended that a minimum of three lengths of tubing be picked at random from each lot of 100 mill lengths or less, and preferably from tubing annealed in the same furnace at the same time. Test specimens shall be taken from tubing which has not been subjected to cold working after the anneal of the finished sized tubing.

Bending Test—The tubing shall withstand cold bending through an angle of 180 deg over a 1/4 in. (6.4 mm) diameter rod without showing other than superficial outside surface ruptures. ID finish shall not deviate from the description under the paragraph on "Quality."

Upsetting Test—The tubing shall withstand cold upsetting from a length of 1/2 to 1/4 in. (12.7–6.4 mm) without showing other than superficial outside surface ruptures.

Inspection—The failure of any specimen to pass any of the foregoing chemical, mechanical property, or performance requirements should be cause for rejection of the entire lot represented.

Corrosion Protection—The inside and outside of the finished tubing shall be protected against corrosion during shipment and normal storage. If a corrosion preventive compound is applied, it shall be such that after normal storage periods it can readily be removed by cleaning agents normally used in manufacturing.

Installation Considerations—The following recommendations should be considered in the design and fabrication of fuel injection tubing systems.

Minimum Bend Radii—In forming tubing the bend radii, as measured at the centerline of the tubing, should not be less than 3/4 in. (19 mm) to ensure good bends under average conditions. With proper tools and care, smaller radii bends are possible.

TABLE 1—TUBING DIMENSIONS AND TOLERANCES, AND COLOR CODE

Nominal Tubing OD, in	Outside Diameter							Inside Diameter						Color Code for Basic ID
	Basic		Tolerance					Basic		Tolerance				
			Standard			Optional				Standard		Optional		
			Plus		Minus	Plus		Minus						
	in	mm	in	mm	in	mm	in	mm	in	mm	±in	±mm	±in	±mm	
1/4	0.250	6.35	0.005	0.13	0.000	0.002	0.05	0.000	0.063	1.60	0.0025	0.064	0.001	0.025	Red
	0.250	6.35	0.005	0.13	0.000	0.002	0.05	0.000	0.067	1.70	0.0025	0.064	0.001	0.025	Black
	0.250	6.35	0.005	0.13	0.000	0.002	0.05	0.000	0.078	1.98	0.0025	0.064	0.001	0.025	Yellow
	0.250	6.35	0.005	0.13	0.000	0.002	0.05	0.000	0.084	2.13	0.0025	0.064	0.001	0.025	Blue
	0.250	6.35	0.005	0.13	0.000	0.002	0.05	0.000	0.093	2.36	0.0025	0.064	0.001	0.025	White

NOTE: In this chart, the color code identifies the basic tubing ID. The color code shall be applied to the OD of the tubing in the form of an intermittent stripe (gaps not longer than 12 in (304 mm)), using Dykem dye or other suitable marking material.

AUTOMOTIVE PIPE FITTINGS—SAE J530a

SAE Standard

Report of Parts and Fittings Committee approved February 1948 and last revised by Tube, Pipe, Hose, and Lubrication Fittings Committee December 1973.

GENERAL SPECIFICATIONS

Scope—This standard includes complete general and dimensional specifications for those types of pipe fittings commonly used in the automotive and other mass production industries where the use of lubricants or sealers is objectionable. The automotive pipe fittings shown in Figs. 1–17 and Tables 2–7 are intended for general automotive and similar applications involving low or medium pressures or in conjunction with automotive tube fittings in piping systems.

Dimensions and Tolerances—Except for nominal sizes and thread specifications, dimensions and tolerances are given in both U. S. customary and SI units as designated. Tabulated dimensions shall apply to the finished fittings, plated or otherwise processed, as specified by the purchaser. Unless otherwise specified, maximum and minimum across flats dimensions shall be within the commercial tolerance of bar or extruded stock from which the fittings are produced. The minimum across corner dimensions of external hexagons shall be 1.092 times the nominal width across flats, but shall not result in a side flat width less than 0.43 times the nominal width across flats. The minimum across corner dimensions of external squares shall be 1.25 times the nominal width across flats, but shall not result in a side flat width less than 0.75 times the nominal width across flats. Unless otherwise specified, tolerance on hole diameters designated drill in the dimensional tables shall be as tabulated below:

Drill Size Range		Tolerance on Hole Diameter			
		Plus		Minus	
in	mm	in	mm	in	mm
0.0135 thru 0.1850	0.343 thru 4.699	0.003	0.08	0.002	0.05
0.1875 thru 0.2480	4.762 thru 6.299	0.004	0.10	0.002	0.05
0.2500 thru 0.7500	6.350 thru 19.050	0.006	0.15	0.003	0.03
0.7579 thru 1.0000	19.25 thru 25.400	0.007	0.18	0.004	0.10

Tolerance on all dimensions not otherwise limited shall be ±0.010 in. (0.25 mm). Angular tolerance on axis of ends on elbows and tees shall be ±2.50 deg for sizes up to and including 3/8 in, and ±1.50 deg for sizes larger than 3/8 in.

Wall Thickness—Unless otherwise designated, the wall thickness at any point on fittings shall not be less than the thickness established by the specified dimensions, tolerances, and eccentricities for inner and outer surfaces.

Contour—Details of contour shall be optional with the manufacturer provided the tabulated dimensions are maintained and serviceability of the fittings is not impaired. Wrench flats on elbows and tees shall be optional. Where extruded or forged shapes are reduced to conserve material, the wall thickness, unless otherwise specified, shall not be less than the respective minimum values tabulated below:

Nominal Pipe Thread Size, in	Wall Thickness, min[a]		Nominal Pipe Thread Size, in	Wall Thickness, min[a]	
	in	mm		in	mm
1/16	0.04	1.0	1/2	0.08	2.0
1/8	0.04	1.0	3/4	0.08	2.0
1/4	0.05	1.3	1	0.08	2.0
3/8	0.06	1.5			

[a] Applies to reduction to one plane only on extruded shapes.

Passages—Where passages in straight fittings are machined from opposite ends, the offset at the meeting point shall not exceed 0.015 in. (0.38 mm). The cross-sectional area at the junction of passages in angle fittings shall not be less than that of the smaller passage.

Pipe Threads—The pipe threads, unless there is specific authorization to the contrary, shall conform with the Dryseal American Standard Taper Pipe Thread (NPTF). At purchaser's option, the pipe thread may be shortened in conformity with the SAE Short Dryseal Taper Pipe Thread (PTF-SAE Short).

Specifications for pipe threads are given in detail in SAE J476. The pipe fitting dimensions tabulated herein are based on length of the Dryseal American Standard Taper Pipe Thread (NPTF), it being the consensus of manufacturers and users that trouble-free assembly and pressure tight joints without lubricant or sealer cannot be assured unless a full-length thread is used.

TABLE 1

Nominal Pipe Thread Size, in	External Thread								Internal Thread			
	Chamfer Diameter[a]				Length of Chamfered or Partial Thread				Countersink Diameter[a]			
	Max		Min		Min		Max		Min		Max	
	in	mm	in	mm	in	mm	in	mm	in	mm	in	mm
1/16	0.23	5.8	0.21	5.3	0.037	0.94	0.056	1.42	0.33	8.4	0.35	8.9
1/8	0.32	8.1	0.30	7.6	0.037	0.94	0.056	1.42	0.42	10.7	0.44	11.2
1/4	0.42	10.7	0.40	10.2	0.056	1.42	0.083	2.11	0.55	14.0	0.57	14.5
3/8	0.55	14.0	0.53	13.5	0.056	1.42	0.083	2.11	0.69	17.5	0.71	18.0
1/2	0.68	17.3	0.66	16.8	0.071	1.80	0.107	2.72	0.85	21.6	0.87	22.1
3/4	0.89	22.6	0.87	22.1	0.071	1.80	0.107	2.72	1.06	26.9	1.08	27.4
1	1.12	28.4	1.09	27.7	0.087	2.21	0.130	3.30	1.34	3.40	1.37	34.8

[a] Tabulated diameters conform with Appendix A, SAE J476.

AUTOMOTIVE PIPE FITTINGS

NOTES: UNSPECIFIED DETAIL WITH RESPECT TO DIMENSIONS, TOLERANCES, CONTOURS, MATERIAL, WORKMANSHIP, ETC., MUST CONFORM TO GENERAL SPECIFICATIONS FOR AUTOMOTIVE PIPE FITTINGS. CODES SHOWN IN BRACKETS ADJACENT TO FIGURE NUMBERS REPRESENT RESPECTIVE FITTING IDENTIFICATION IN ACCORDANCE WITH SAE J846.

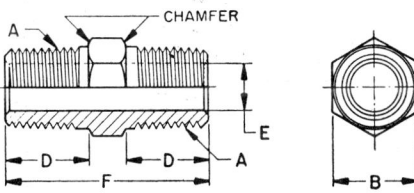

FIG. 1—HEXAGON NIPPLE (130137)

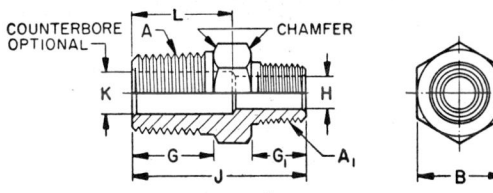

FIG. 2—HEXAGON REDUCER NIPPLE (130137)

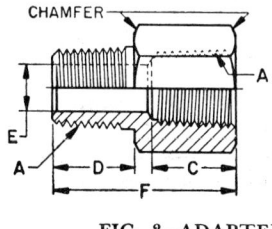

FIG. 3—ADAPTER (130139)

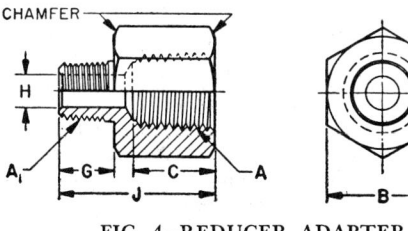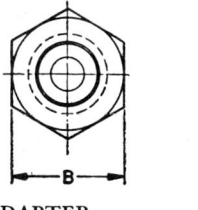

FIG. 4—REDUCER ADAPTER (130139)

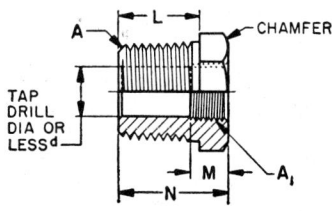

FIG. 5—REDUCER BUSHING (130140)

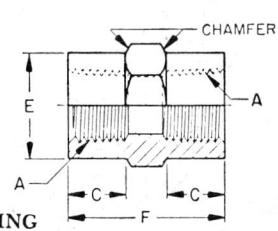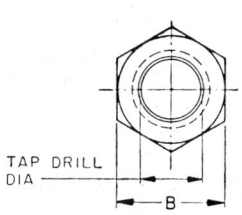

FIG. 6—COUPLING (130138)

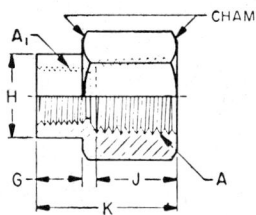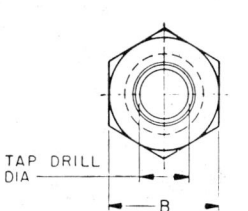

FIG. 7—REDUCER COUPLING (130138)

NOTES: UNSPECIFIED DETAIL WITH RESPECT TO DIMENSIONS, TOLERANCES, CONTOURS, MATERIAL, WORKMANSHIP, ETC., MUST CONFORM TO GENERAL SPECIFICATIONS FOR AUTOMOTIVE PIPE FITTINGS. CODES SHOWN IN BRACKETS ADJACENT TO FIGURE NUMBERS REPRESENT RESPECTIVE FITTING IDENTIFICATION IN ACCORDANCE WITH SAE J846.

TABLE 2—DIMENSIONS OF HEXAGON NIPPLES AND REDUCER NIPPLES (FIG. 1 AND 2)

Dryseal Taper Thread NPTF[a], in		All Nipples				Nipples					
A Hexagon Nipples	A x A₁ Hexagon Reducer Nipples	B Hexagon Width				D Shoulder Length[b]		E Drill Dia		F Overall Length[b]	
		Max		Min							
		in	mm	in	mm	in	mm	in	mm	in	mm
1/16-27	—	0.316	8.03	0.310	7.87	0.38	9.7	0.141	3.58	0.94	23.9
1/8-27	1/8 x 1/16	0.440	11.18	0.434	11.02	0.38	9.7	0.219	5.56	0.97	24.6
1/4-18	1/2 x 1/8	0.566	14.38	0.558	14.17	0.56	14.2	0.312	7.92	1.38	35.1
3/8-18	3/8 x 1/8	0.692	17.58	0.684	17.37	0.56	14.2	0.438	11.13	1.41	35.8
—	3/8 x 1/4	0.692	17.58	0.684	17.37	—	—	—	—	—	—
1/2-14	1/2 x 3/8	0.879	22.33	0.871	22.12	0.75	19.0	0.562	14.27	1.81	46.0

Dryseal Taper Thread NPTF[a], in A x A₁ Hexagon Reducer Nipples	Reducer Nipples											
	G Shoulder Length[b]		G₁ Shoulder Length[b]		H Drill Dia[c]		J Overall Length[b]		Counterbore			
									K Max Dia[c]		L Max Depth[b,c]	
	in	mm	in	mm	in	mm	in	mm	in	mm	in	mm
1/8 x 1/16	0.38	9.7	0.38	9.7	1.041	3.58	0.97	24.6	0.223	5.66	0.47	11.9
1/2 x 1/8	0.56	14.2	0.38	9.7	0.219	5.56	1.19	30.2	0.310	8.08	0.69	17.5
3/8 x 1/8	0.56	14.2	0.38	9.7	0.219	5.56	1.22	31.0	0.444	11.28	0.69	17.5
3/8 x 1/4	0.56	14.2	0.56	14.2	0.312	7.92	1.41	35.8	0.444	11.28	0.69	17.5
1/2 x 3/8	0.75	19.0	0.56	14.2	0.438	11.13	1.62	41.1	0.568	14.43	0.91	23.1

[a] Dryseal American Standard Taper Pipe Thread. See General Specifications.
[b] Where SAE Short Pipe Thread is authorized by purchaser, dimensions D, F, G, G₁, J, and L are reduced in accordance with reduction of pipe thread length. See General Specifications.
[c] At manufacturers option, through passages may conform with the smaller diameter specified or be counterbored to the larger diameter for the depth specified.

TABLE 3—DIMENSIONS OF ADAPTERS AND REDUCER ADAPTERS (FIGS. 3 AND 4)

Dryseal Taper Thread NPTF[a], in		All Adapters						Adapters						Reducer Adapters					
A Adapters	A x A₁ Reducer Adapters	B Hexagon Width				C Tap Drill Depth[b,c]		D Shoulder Length[b]		E Dia Drill		F Overall Length[b]		G Shoulder Length[b]		H Dia Drill		J Overall Length[b]	
		Max		Min		Min													
		in	mm	in	mm	in	mm	in	mm	in	mm	in	mm	in	mm	in	mm	in	mm
1/16-27	—	0.440	11.18	0.434	11.02	0.38	9.7	0.38	9.7	0.141	3.58	0.84	21.3	—	—	—	—	—	—
1/8-27	1/8 x 1/16	0.566	14.38	0.558	14.17	0.38	9.7	0.38	9.7	0.219	5.56	0.88	22.4	0.38	9.7	0.141	3.58	0.84	21.3
1/4-18	1/4 x 1/8	0.754	19.15	0.746	18.95	0.56	14.2	0.56	14.2	0.312	7.92	1.25	31.8	0.38	9.7	0.219	5.56	1.06	26.9
3/8-18	3/8 x 1/4	0.879	22.33	0.871	22.12	0.56	14.2	0.56	14.2	0.438	11.13	1.25	31.8	0.56	14.2	0.312	7.92	1.25	31.8
1/2-14	1/2 x 3/8	1.068	27.13	1.058	26.87	0.75	19.0	0.75	19.0	0.562	14.27	1.66	42.2	0.56	14.2	0.438	11.13	1.47	37.3
3/4-14	3/4 x 1/2	1.380	35.05	1.370	34.80	0.75	19.0	0.75	19.0	0.750	19.05	1.69	42.9	0.75	19.0	0.562	14.27	1.69	42.9
1-11-1/2	1 x 3/4	1.630	41.40	1.620	41.15	0.94	23.9	0.94	23.9	0.938	23.82	2.06	52.3	0.75	19.0	0.750	19.05	1.88	47.8

[a] Dryseal American Standard Taper Pipe Thread. See General Specifications.
[b] Where SAE Short Pipe Thread is authorized by purchaser, dimensions C, F, G, and J are reduced in accordance with reduction of pipe thread length. See General Specifications.
[c] Tap drill depths given require use of bottoming taps to produce standard full thread lengths. See General Specifications.

AUTOMOTIVE PIPE FITTINGS—CAST TYPE

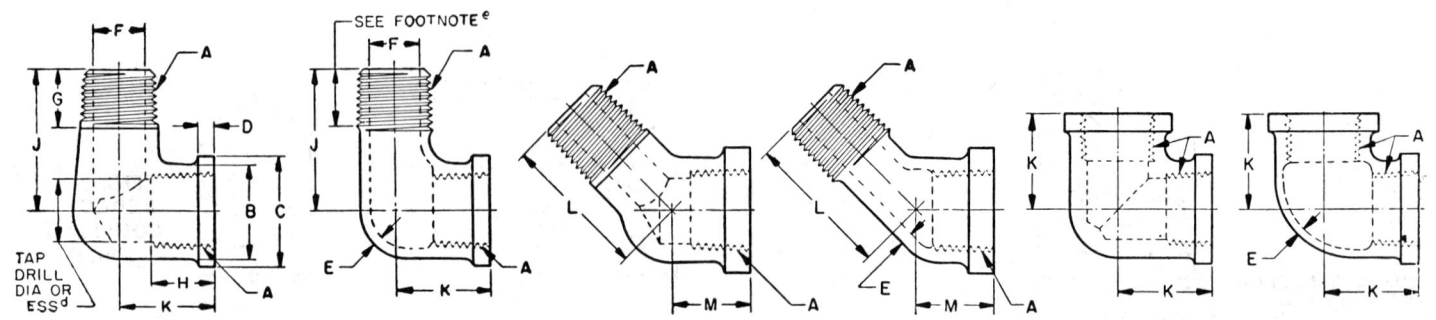

FIG. 8—90 DEG STREET ELBOWS (130239)   FIG. 9—45 DEG STREET ELBOWS (130339)   FIG. 10—90 DEG PIPE ELBOWS (130238)

21.131

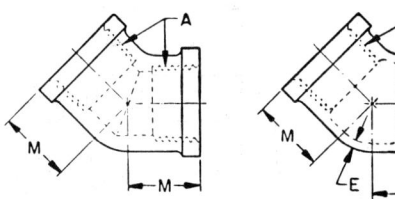

FIG. 11—45 DEG PIPE ELBOWS
(130338)

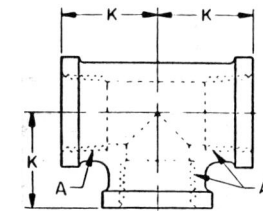

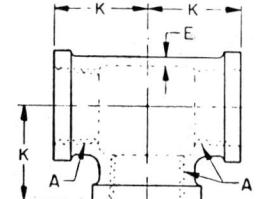

FIG. 12A—INTERNAL, INTERNAL, INTERNAL TEES
(130438)

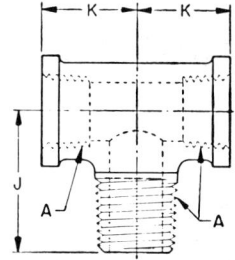

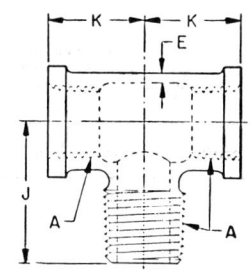

FIG. 12B—INTERNAL, INTERNAL, EXTERNAL TEES
(130425)

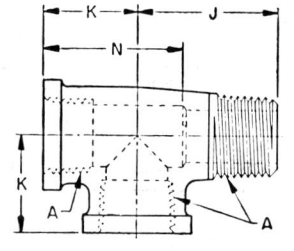

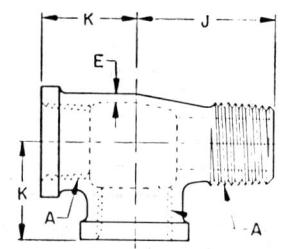

FIG. 12C—INTERNAL, EXTERNAL, INTERNAL TEES
(130424)

NOTES: UNSPECIFIED DETAIL WITH RESPECT TO DIMENSIONS, TOLERANCES, CONTOURS, MATERIAL, WORKMANSHIP, ETC., MUST CONFORM TO GENERAL SPECIFICATIONS FOR AUTOMOTIVE PIPE FITTINGS. THE DIMENSIONAL DESIGNATIONS ON THE FIRST FIGURE IN EACH GROUP SHALL APPLY TO ALL OTHER FIGURES IN THAT GROUP EXCEPT AS SHOWN OTHERWISE. CODES SHOWN IN BRACKETS ADJACENT TO FIGURE NUMBERS REPRESENT RESPECTIVE FITTING IDENTIFICATION IN ACCORDANCE WITH SAE J846.

TABLE 4—DIMENSIONS OF REDUCER BUSHINGS (FIG. 5)

Dryseal Taper Thread NPTF[a], in	B Hexagon Width				C Tap Drill Depth[b,c] Min		D Shoulder Length[b]		E Hole Dia[d] Min		F Overall Length[b]	
	Max		Min									
$A \times A_1$	in	mm	in	mm	in	mm	in	mm	in	mm	in	mm
1/8 x 1/16	0.440	11.18	0.434	11.02	0.38	9.7	0.38	9.7	0.139	3.53	0.56	14.2
1/4 x 1/8	0.566	14.38	0.558	14.17	0.38	9.7	0.56	14.2	0.217	5.51	0.75	19.0
3/8 x 1/8	0.692	17.58	0.684	17.37	0.38	9.7	0.56	14.2	0.217	5.51	0.75	19.0
3/8 x 1/4	0.754	19.15	0.746	18.95	0.56	14.2	0.56	14.2	0.309	7.85	0.75	19.0
1/2 x 1/8	0.879	22.33	0.871	22.12	0.38	9.6	0.75	19.0	0.217	5.51	1.00	25.4
1/2 x 1/4	0.879	22.33	0.871	22.12	0.56	14.2	0.75	19.0	0.309	7.85	1.00	25.4
1/2 x 3/8	0.879	22.33	0.871	22.12	0.56	14.2	0.75	19.0	0.435	11.05	1.00	25.4
3/4 x 1/4	1.130	28.70	1.120	28.45	0.56	14.2	0.75	19.0	0.309	7.85	1.00	25.4
3/4 x 3/8	1.130	28.70	1.120	28.45	0.56	14.2	0.75	19.0	0.435	11.05	1.00	25.4
3/4 x 1/2	1.130	28.70	1.120	28.45	0.75	19.0	0.75	19.0	0.559	14.20	1.00	25.4
1 x 1/2	1.442	36.63	1.432	36.37	0.75	19.0	0.94	23.9	0.559	14.20	1.31	33.3
1 x 3/4	1.442	36.63	1.432	36.37	0.75	19.0	0.94	23.9	0.747	18.98	1.31	33.3

[a] Dryseal American Standard Pipe Thread. See General Specifications.
[b] Where SAE Short Pipe Thread is authorized by purchaser, dimensions C, D, and F are reduced in accordance with reduction of pipe thread length. See General Specifications.
[c] Tap drill depths given require use of bottoming taps to produce standard full thread lengths. See General Specifications.
[d] At manufacturer's option, hole may conform to tap drill diameter or may be reduced beyond tap drill depth C, but in no case shall it be smaller than E diameter specified.

AUTOMOTIVE PIPE FITTINGS—EXTRUDED OR BAR STOCK TYPE

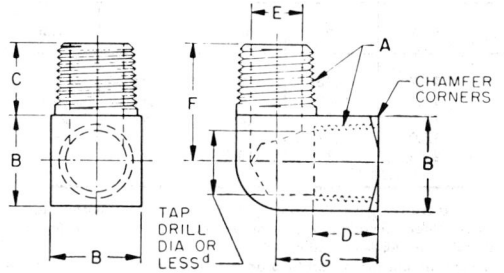

FIG. 13—90 DEG STREET ELBOW
(130239)

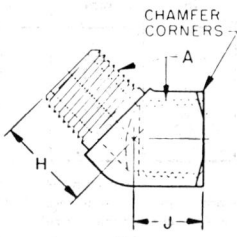

FIG. 14—45 DEG STREET ELBOW
(130339)

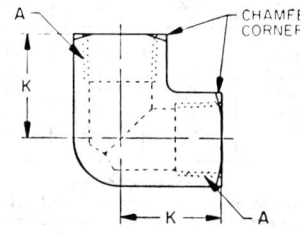

FIG. 15—90 DEG PIPE ELBOW
(130238)

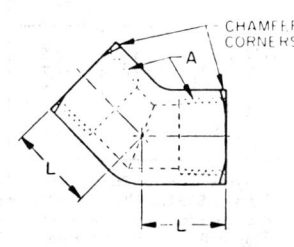

FIG. 16—45 DEG PIPE ELBOW
(130338)

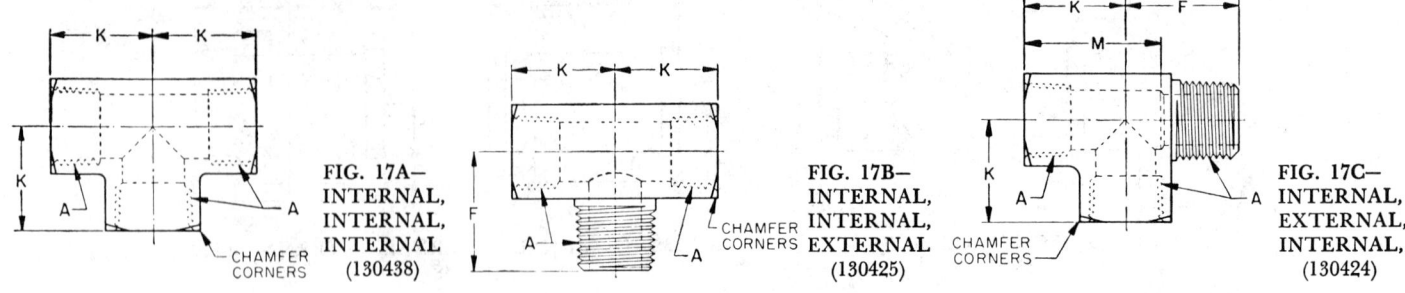

FIG. 17A— INTERNAL, INTERNAL, INTERNAL (130438)

FIG. 17B— INTERNAL, INTERNAL, EXTERNAL (130425)

FIG. 17C— INTERNAL, EXTERNAL, INTERNAL (130424)

FIG. 17—TEES

NOTES: UNSPECIFIED DETAIL WITH RESPECT TO DIMENSIONS, TOLERANCES, CONTOURS, MATERIAL, WORKMANSHIP, ETC., MUST CONFORM TO GENERAL SPECIFICATIONS FOR AUTOMOTIVE PIPE FITTINGS. THE DIMENSIONAL DESIGNATIONS ON THE FIRST FIGURE IN EACH GROUP SHALL APPLY TO ALL OTHER FIGURES IN THAT GROUP EXCEPT AS SHOWN OTHERWISE. CODES SHOWN IN BRACKETS ADJACENT TO FIGURE NUMBERS REPRESENT RESPECTIVE FITTING IDENTIFICATION IN ACCORDANCE WITH SAE J846.

TABLE 5—DIMENSIONS OF COUPLINGS AND REDUCER COUPLINGS (FIGS. 6 AND 7)

Dryseal Taper Thread NPTF[a], in		All Couplings				Couplings						Reducer Couplings							
A Coupling	A x A$_1$ Reducer Coupling	B Hexagon Width				C Shoulder Length[b]		E Min Body Dia		F Overall Length[b]		G Shoulder Length[b]		H Min Body Dia		J Min Tap Drill Depth[b,c]		K Overall Length[b]	
		Max		Min															
		in	mm	in	mm	in	mm	in	mm	in	mm	in	mm	in	mm	in	mm	in	mm
1/16-27	—	0.440	11.18	0.434	11.02	0.28	7.1	0.44	11.2	0.75	19.0	—	—	—	—	—	—	—	—
1/8-27	1/8 x 1/16	0.566	14.38	0.558	14.17	0.27	6.9	0.56	14.2	0.75	19.0	0.31	7.9	0.50	12.7	0.38	9.7	0.78	19.8
1/4-18	1/4 x 1/8	0.754	19.15	0.746	18.95	0.44	11.2	0.75	19.0	1.12	28.4	0.31	7.9	0.56	14.2	0.56	14.2	0.97	24.6
3/8-18	3/8 x 1/8	0.879	22.33	0.871	22.12	0.42	10.7	0.88	22.4	1.12	28.4	0.25	6.4	0.56	14.2	0.56	14.2	0.94	23.9
—	3/8 x 1/4	0.879	22.33	0.871	22.12	—	—	—	—	—	—	0.47	11.9	0.75	19.0	0.56	14.2	1.16	29.5
1/2-4	1/2 x 1/8	1.068	27.13	1.058	26.87	0.59	15.0	1.06	26.9	1.50	38.1	0.25	6.4	0.56	14.2	0.75	19.0	1.19	30.2
—	1/2 x 1/4	1.068	27.13	1.058	26.87	—	—	—	—	—	—	0.34	8.6	0.75	19.0	0.75	19.0	1.25	32.5
—	1/2 x 3/8	1.068	27.13	1.058	26.87	—	—	—	—	—	—	0.44	11.2	0.88	22.4	0.75	19.0	1.38	35.1

[a] Dryseal American Standard Taper Pipe Thread. See General Specifications.
[b] Where SAE Short Pipe Thread is authorized by purchaser, dimensions C, F, G, J, and K are reduced in accordance with reduction of pipe thread length. See General Specifications.
[c] Tap drill depths given require use of bottoming taps to produce standard full thread length. See General Specifications.

TABLE 6—DIMENSIONS OF CAST TYPE STREET ELBOWS, PIPE ELBOWS, AND PIPE TEES (FIGS. 8-12)

A Dryseal Taper Thread NPTF[a], in	B Min Body Dia		C Min Collar Dia		D Min Collar Thickness		E Min Wall Thickness		F Drill Dia[f]		G Turned Length[b]		H Min Tap Drill Depth[b,c]		J Center to End[b]			
															Max		Min	
	in	mm	in	mm	in	mm	in	mm	in	mm	in	mm	in	mm	in	mm	in	mm
1/16-27	0.44	11.2	0.53	13.5	0.12	3.0	0.08	2.0	0.141	3.58	0.38	9.7	0.38	9.7	0.84	21.3	0.78	19.8
1/8-27	0.56	14.2	0.67	17.0	0.14	3.6	0.08	2.0	0.219	5.56	0.38	9.7	0.38	9.7	0.95	24.1	0.89	22.6
1/4-18	0.72	19.3	0.81	20.6	0.16	4.1	0.08	2.0	0.312	7.92	0.56	14.2	0.56	14.2	1.15	29.2	1.07	27.2
3/8-18	0.88	22.4	1.00	25.4	0.17	4.3	0.09	2.3	0.438	11.13	0.56	14.2	0.56	14.2	1.30	33.0	1.20	30.5
1/2-14	1.03	26.2	1.17	29.7	0.19	4.8	0.09	2.3	0.562	14.27	0.75	19.0	0.75	19.0	1.56	39.6	1.44	36.6

A Dryseal Taper Thread NPTF[a], in	K Center to End[b]				L Center to End[b]				M Center to End[b]				N Drill Depth	
	Max		Min		Max		Min		Max		Min			
	in	mm	in	mm	in	mm	in	mm	in	mm	in	mm	in	mm
1/16-27	0.53	13.5	0.47	11.9	0.72	18.3	0.66	16.8	0.44	11.2	0.38	9.7	0.66	16.8
1/8-27	0.58	14.7	0.52	13.2	0.81	20.6	0.75	19.0	0.45	11.4	0.39	9.9	0.75	19.0
1/4-18	0.76	19.3	0.68	17.3	0.92	23.4	0.84	21.3	0.60	15.2	0.52	13.2	0.97	24.6
3/8-18	0.88	22.4	0.78	19.8	0.97	24.6	0.87	22.1	0.67	17.0	0.57	14.5	—	—
1/2-14	1.08	27.4	0.96	24.4	1.12	28.4	7.00	25.4	0.84	21.3	0.72	18.3	—	—

[a] Dryseal American Standard Taper Pipe Thread. See General Specifications.
[b] Where SAE Short Pipe Thread is authorized by purchaser, dimensions G, H, J, K, L, and M are reduced in accordance with reduction of pipe thread length. See General Specifications.
[c] Tap drill depths given require use of bottoming taps to produce standard full thread length. See General Specifications.
[d] Hole diameters may be reduced beyond tap drill depth H, but shall not be less than F specified for corresponding size. (See Fig. 8.)
[e] Minimum pipe thread length where body is relieved or undercut shall not be shorter than L$_2$ plus one turn (thread) full thread. Thread length may be reduced one pitch (thread) if thread is cut through into relief or undercut. See SAE J476 and Fig. 8.
[f] 1/16, 1/8, and 1/4 in size cast fittings are generally produced from solid castings and have drilled passage holes. 3/8 and 1/2 in size cast fittings are generally produced with cored passage holes and may have internal minimum full thread length of 0.36 and 0.43 in (9.1 and 10.9 mm), respectively.

TABLE 7—DIMENSIONS OF EXTRUDED AND FORGED TYPE STREET ELBOWS, PIPE ELBOWS, AND PIPE TEES (FIGS. 13-17)

A Dryseal Taper Thread NPTF[a], in	B Body Size		C Turned Length[b]		D Min Tap Drill Depth[b,c]		E Drill Dia[d]		F Center to End[b]		G Center to End[b]	
									±0.03	±0.8	±0.03	±0.8
	in	mm	in	mm	in	mm	in	mm	in	mm	in	mm
1/16-27	7/16	11.11	0.38	9.7	0.38	9.7	0.141	3.58	0.59	5.0	0.45	11.4
1/8-27	9/16	14.29	0.38	9.7	0.38	9.7	0.219	5.56	0.66	16.8	0.48	12.2
1/4-18	11/16	17.46	0.56	14.2	0.56	14.2	0.312	7.92	0.91	23.1	0.72	18.3
3/8-18	13/16	20.64	0.56	14.2	0.56	14.2	0.438	11.13	0.97	24.6	0.78	19.8
1/2-14	1	25.40	0.75	19.0	0.75	19.0	0.562	14.27	1.25	31.8	1.03	26.2
3/4-14	1-1/4	31.75	0.75	19.0	0.75	19.0	0.750	19.05	1.38	35.1	1.12	28.4
1-11-1/2	1-1/2	38.10	0.94	23.9	0.94	23.9	0.938	23.82	1.69	42.9	1.41	35.8

A Dryseal Taper Thread NPTF[a], in	H Center to End[b]		J Center to End[b]		K Center to End[b]		L Center to End[b]		M Drill Depth	
	±0.03	±0.8	±0.03	±0.8	±0.03	±0.08	±0.03	±0.8		
	in	mm	in	mm	in	mm	in	mm	in	mm
1/16-27	0.47	11.9	0.38	9.7	0.50	12.7	0.44	11.2	0.66	16.8
1/8-27	0.50	12.7	0.38	9.7	0.55	14.0	0.45	11.4	0.75	19.0
1/4-18	0.72	18.3	0.56	14.2	0.78	19.8	0.66	16.8	1.03	26.2
3/8-18	0.78	19.8	0.56	14.2	0.84	21.3	0.69	17.5	1.17	29.7
1/2-14	1.00	25.4	0.75	19.0	1.09	27.7	0.91	23.1	1.48	37.6
3/4-14	1.06	26.9	0.75	19.0	1.16	29.5	0.94	23.9	1.66	42.2
1-11-1/2	1.34	34.0	0.94	23.9	1.52	38.6	1.19	30.2	2.12	53.8

[a] Dryseal American Standard Taper Pipe Thread. See General Specifications.
[b] Where SAE Short Pipe Thread is authorized by purchaser, dimensions C, D, F, G, H, J, K, and L are reduced in accordance with reduction of pipe thread length. See General Specifications.
[c] Tap drill depths given require use of bottoming taps to produce standard full thread length. See General Specifications.
[d] Hole diameter may be reduced beyond tap drill depth D but shall not be less than E specified for corresponding size. (See Fig. 13.)

However, the tap drill depths and the overall lengths specified in the tables for fittings with internal taper pipe threads are not consistent with the tap drill depths and the overall thread lengths of the Dryseal American Standard Taper Pipe Threads (NPTF) specified in Table A2, Appendix A of SAE J476. The full-length Dryseal American Standard Taper Pipe Taps specified in Table B2, Appendix B of SAE J476 cannot be used as the tap drill depths and overall lengths of the fittings have been reduced to the minimum required by bottoming taps to produce standard full thread length. The deviations described herein are peculiar to automotive pipe fittings. As special tooling is required, caution should be exercised in specifying the deviations for any other products. External pipe threads shall be chamfered from the diameters tabulated in Table 1 to produce the specified length of chamfered or partial thread. Internal pipe threads shall be countersunk 90 deg included angle, to the diameters shown in Table 1.

Material and Manufacture—Pipe fittings may be made from cast iron, malleable iron, steel, stainless steel, brass, or aluminum alloy as specified by the purchaser, by casting, forging, milling from the bar, or upsetting from a grade of material free from defects which will affect their serviceability. However, all varieties and sizes of pipe fittings may not be currently available in the aforementioned materials. Nipples, adapters, bushings, and couplings are generally available in brass and steel. Cast elbows and tees are generally available in malleable iron for sizes 1/4 in. and over and in brass. Extruded and forged elbows and tees are generally available in brass and steel.

Finish—Unless otherwise specified by the purchaser, steel fittings shall be furnished cadmium or zinc plated to a thickness of 0.0002 in. (0.05 mm) minimum. These parts must meet the requirements of a 32 h salt spray test in accordance with ASTM B117. At manufacturer's option, plated fittings may be given a subsequent chromate treatment.

Workmanship—Workmanship shall conform to the best commercial practice to produce, high-quality fittings. Fittings shall be free from all hanging burrs, loose scale, and slivers which might become dislodged in usage and all other defects which might affect serviceability.

AUTOMOTIVE PIPE, FILLER, AND DRAIN PLUGS—SAE J531a

SAE Standard

Report of Parts and Fittings Committee approved February 1948 and last revised by Tube, Pipe, Hose, and Lubrication Fittings Committee December 1973.

GENERAL SPECIFICATIONS

Scope—This standard includes complete general and dimensional specifications for those types of pipe, filler, and drain plugs (shown in Figs. 1-6 and Tables 2-4) commonly used in automotive and related industrial applications.

Dimensions and Tolerances—Except for nominal sizes and thread specifications, dimensions and tolerances are given in both U. S. customary and SI units as designated. Tabulated dimensions shall apply to the finished plugs, plated, hardened, or otherwise processed, as specified by the purchaser. The minimum across corner dimensions of external hexagons shall be 1.092 times the nominal width across flats, but shall not result in a side flat width less than 0.43 times the nominal width across flats. The minimum across corner dimensions of external squares shall be 1.25 times the nominal width across flats, but shall not result in a side flat width less than 0.75 times the nominal width across flats. At maximum material condition, the radii at corners of hexagon and square sockets in broached and upset plugs shall not exceed 0.005 in. (0.13 mm). Tolerance on dimensions not otherwise limited shall be ±0.010 in. (0.25 mm).

Pipe Threads—The pipe threads on automotive pipe plugs, unless there is specific authorization to the contrary, shall conform with the Dryseal American Standard Taper Pipe Thread (NPTF) and be gaged accordingly. The automotive pipe plug dimensions are based on the length of the NPTF thread and are intended for assembly with all types of Dryseal taper and straight internal threads. It is the consensus of manufacturers and users that trouble-free assembly and pressure tight joints without lubricant or sealer cannot be assured unless a full length thread is used. The pipe threads on automotive filler and drain plugs, unless there is specific authorization to the contrary, shall conform with the Dryseal SAE Short Taper Pipe Thread (PTF-SAE Short) and be gaged accordingly. The automotive filler and drain plug dimensions are based on the length of the (PTF-SAE Short) thread and are primarily intended for assembly with Dryseal American Standard Taper (NPTF) or Dryseal American Standard Intermediate Straight (NPSI) internal pipe threads in installations where it is desirable to limit the entry of the small end of the plug. Limitations on other application of this thread are explained in SAE J476.

External pipe threads shall be chamfered or rounded from the diameters tabulated in Table 1 to produce a length of chamfered or partial thread as specified. The threads on countersunk headless types of plugs shall be chamfered on both ends to the dimensions shown.

TABLE 1

Nominal Dryseal Pipe Thread Size, in	Chamfer Dia at Small End of Plugs of All Types[a]				Chamfer Dia at Large End of Countersunk Headless Plugs				Length of Chamfered or Partial Threads			
	Max		Min		Max		Min		Max		Min	
	in	mm	in	mm	in	mm	in	mm	in	mm	in	mm
1/16	0.23	5.8	0.21	5.3	0.25	6.4	0.23	5.8	0.056	1.42	0.037	0.94
1/8	0.32	8.1	0.30	7.6	0.34	8.6	0.32	8.1	0.056	1.42	0.037	0.94
1/4	0.42	10.7	0.40	10.2	0.45	11.4	0.43	10.9	0.083	2.11	0.056	1.42
3/8	0.55	14.0	0.53	13.5	0.58	14.7	0.56	14.2	0.083	2.11	0.056	1.42
1/2	0.68	17.3	0.66	16.8	0.72	18.3	0.70	17.8	0.107	2.72	0.071	1.80
3/4	0.89	22.6	0.87	22.1	0.93	23.6	0.91	23.1	0.107	2.72	0.071	1.80
1	1.12	28.4	1.09	27.7	1.17	29.7	1.14	29.0	0.130	3.30	0.087	2.21

[a] Tabulated diameters conform with Appendix A, SAE J476.

Related specifications covering blank sizes, dies, chasers, and gages are shown in SAE J476.

Material and Manufacture—Plugs may be made from low carbon steel, cast iron, malleable iron, brass, bronze, or aluminum alloy as specified by purchaser, by casting, milling from the bar, or upsetting from a grade of material free of defects which will affect their serviceability.

Finish—Unless otherwise specified by the purchaser, steel plugs shall be furnished cadmium or zinc plated to a thickness of 0.0002 in. (0.05 mm) min. These parts must meet the requirements of a 32 h salt spray test in accordance with ASTM B 117. At the manufactuerer's option, plated plugs may be given a subsequent chromate treatment.

Workmanship—Workmanship shall conform to the best commercial practice to produce high quality parts. Plugs shall be free from all hanging burrs, loose scale, and slivers which might become dislodged in usage and all other defects which might affect their serviceability.

SQUARE AND HEXAGON HEAD

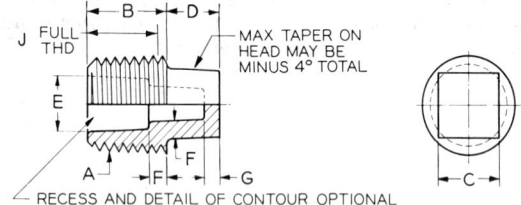

FIG. 1A—SQUARE INSIDE HEAD PIPE PLUGS (130109A)

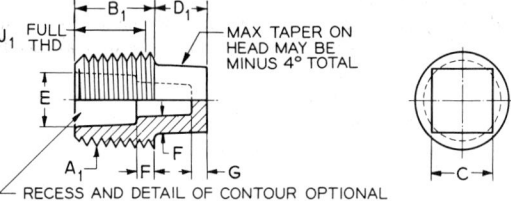

FIG. 1B—SQUARE INSIDE HEAD FILLER AND DRAIN PLUGS[1] (130109B)

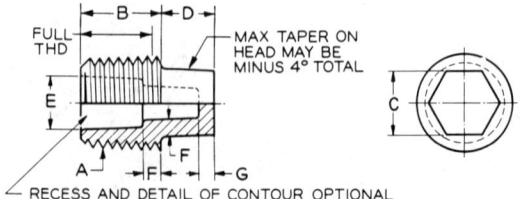

FIG. 1C—HEXAGON INSIDE HEAD PIPE PLUGS (130109C)

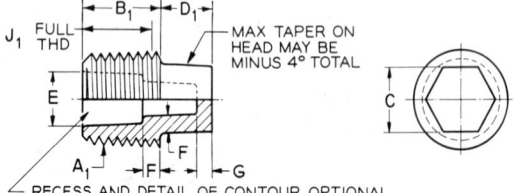

FIG. 1D—HEXAGON INSIDE HEAD FILLER AND DRAIN PLUGS[1] (130109D)

FIG. 1—SQUARE AND HEXAGON INSIDE HEAD PLUGS
CODES SHOWN IN BRACKETS ADJACENT TO FIGURE NUMBERS REPRESENT RESPECTIVE FITTING IDENTIFICATION IN ACCORDANCE TO SAE J846.

TABLE 2—DIMENSIONS OF SQUARE AND HEXAGON INSIDE HEAD PIPE, FILLER AND DRAIN PLUGS (FIGS. 1A-1D)[a]

A Dryseal Thread NPTF, in	A₁ Dryseal Thread PTF-SAE Short, in	B Body Length[b]		B₁ Body Length[b]		C Head Width		D Head Height Square Head		D₁ Head Height Hex Inside Head		E Recess Dia, Max			
												Ferrous		Nonferrous	
		in	mm	in	mm	in	mm	in	mm	in	mm	in	mm	in	mm
1/16-27	1/16-27	0.330	8.38	0.290	7.37	0.214	5.44	0.178	4.52	0.163	4.14	—	—	—	—
		0.350	8.89	0.310	7.87	0.221	5.61	1.93	4.90	0.178	4.52				
1/3-27	1/8-27	0.330	8.38	0.290	7.37	0.276	7.01	0.240	6.10	0.225	5.72	—	—	—	—
		0.350	8.89	0.310	7.87	0.283	7.19	0.255	6.48	0.240	6.10				
1/4-18	1/4-18	0.495	12.57	0.445	11.30	0.370	9.40	0.280	7.11	0.260	6.60	—	—	—	—
		0.525	13.34	0.475	12.06	0.377	9.58	0.300	7.62	0.280	7.11				
3/8-18	3/8-18	0.495	12.57	0.445	11.30	0.428	10.87	0.310	7.87	0.285	7.24	0.31	7.9	0.36	9.1
		0.525	13.34	0.475	12.06	0.440	11.18	0.335	8.51	0.310	7.87				
1/2-14	1/2-14	0.660	16.76	0.590	14.99	0.553	14.05	0.380	9.65	0.350	8.89	0.38	9.7	0.53	13.5
		0.700	17.78	0.630	16.00	0.565	14.35	0.410	10.41	0.380	9.65				
3/4-14	3/4-14	0.670	17.02	0.600	15.24	0.615	15.62	0.440	11.18	0.410	10.41	0.56	14.2	0.72	18.3
		0.710	18.03	0.640	16.26	0.627	15.93	0.470	11.94	0.440	11.18				
1-11-1/2	1-11-1/2	0.830	21.08	0.750	19.05	0.803	20.40	0.500	12.70	0.460	11.68	0.75	19.0	0.93	23.6
		0.870	22.10	0.790	20.07	0.815	20.70	0.540	13.72	0.500	12.70				

A Dryseal Thread NPTF, in	A₁ Dryseal Thread PTF-SAE Short, in	Wall Thickness, Min								J Full Thread Length		J₁ Full Thread Length	
		F				G							
		Ferrous		Nonferrous		Ferrous		Nonferrous					
		in	mm	in	mm	in	mm	in	mm	in	mm	in	mm
1/16-27	1/16-27	—	—	—	—	—	—	—	—	0.50	7.6	0.26	6.6
1/3-27	1/8-27	—	—	—	—	—	—	—	—	0.30	7.6	0.27	6.9
1/4-18	1/4-18	—	—	—	—	—	—	—	—	0.46	11.7	0.41	10.4
3/8-18	3/8-18	0.13	3.3	0.11	2.8	0.13	3.3	0.08	2.0	0.46	11.7	0.41	10.4
1/2-14	1/2-14	0.16	4.1	0.12	3.0	0.16	4.1	0.09	2.3	0.61	15.5	0.53	13.5
3/4-14	3/4-14	0.18	4.6	0.13	3.3	0.18	4.6	0.10	2.5	0.62	15.7	0.55	14.0
1-11-1/2	1-11-1/2	0.20	5.1	0.14	3.6	0.20	5.1	0.11	2.8	0.77	19.6	0.69	17.5

[a] Warning—Automotive filler and drain plugs are primarily intended for installations where it is desirable to limit the entry of the small end of the plug. See General Specifications.

[b] Length B may be reduced one (p) thread if the thread is cut through at head corners.

[1] WARNING — AUTOMOTIVE FILLER AND DRAIN PLUGS ARE PRIMARILY INTENDED FOR INSTALLATION WHERE IT IS DESIRABLE TO LIMIT THE ENTRY OF THE SMALL END OF THE PLUG SEE GENERAL SPECIFICATIONS.

SQUARE AND HEXAGON COUNTERSUNK HEADLESS

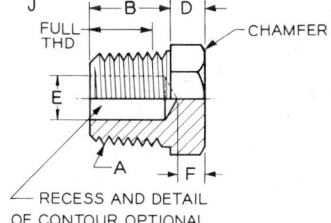
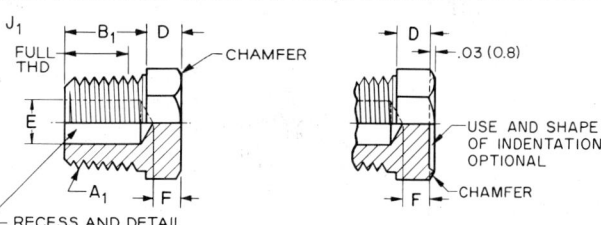

FIG. 2A—HEXAGON OUTSIDE HEAD PIPE PLUGS (130109E)

FIG. 2B—HEXAGON OUTSIDE HEAD FILLER AND DRAIN PIPE PLUGS[1] (130109F)

NOTE: DIMENSIONS ARE IN (mm)

CODES SHOWN IN BRACKETS ADJACENT TO FIGURE NUMBERS REPRESENT RESPECTIVE FITTING IDENTIFICATION IN ACCORDANCE TO SAE J846.

TABLE 3—DIMENSIONS OF HEXAGON OUTSIDE HEAD PIPE, FILLER, AND DRAIN PLUGS[a] (FIGS. 2A AND 2B)

A Dryseal Thread NPTF, in	A_1 Dryseal Thread PTF-SAE Short, in	B Shoulder Length		B_1 Shoulder Length		C Hex (Nom)	D Head Height		E Recess Dia, Max		F Wall Thickness, Min		J Full Thread		J_1 Full Thread	
		in	mm	in	mm	in	in	mm	in	mm	in	mm	in	mm	in	mm
1/16-27	1/16-27	0.37	9.4	0.32	8.1	5/16	0.151 0.162	3.84 4.11	0.10	2.5	0.09	2.3	0.30	7.6	0.26	6.6
1/8-27	1/8-27	0.37	9.4	0.32	8.1	7/16	0.181 0.194	4.60 4.93	0.16	4.1	0.12	3.0	0.30	7.6	0.27	6.9
1/4-18	1/4-18	0.56	14.2	0.49	12.4	9/16	0.181 0.194	4.60 4.93	0.25	6.4	0.12	3.0	0.46	11.7	0.41	10.4
3/8-18	3/8-18	0.56	14.2	0.49	12.4	11/16	0.212 0.227	5.38 5.77	0.38	9.7	0.16	4.1	0.46	11.7	0.41	10.4
1/2-14	1/2-14	0.75	19.0	0.64	16.3	7/8	0.212 0.227	5.38 5.77	0.50	12.7	0.16	4.1	0.61	15.5	0.53	13.5
3/4-14	3/4-14	0.75	19.0	0.65	16.5	1-1/16	0.304 0.323	7.72 8.20	0.69	17.5	0.19	4.8	0.62	15.7	0.55	14.0
1-11-1/2	1-11-1/2	0.94	23.9	0.81	20.6	1-5/16	0.304 0.323	7.72 8.20	0.88	22.4	0.19	4.8	0.77	19.6	0.69	17.5

[a] Warning—Automotive filler and drain plugs are primarily intended for installations where it is desirable to limit the entry of the small end of the plug. See General Specifications.

[b] Length B may be reduced one (p) thread if thread is cut through at head corners.

[1] WARNING—AUTOMOTIVE FILLER AND DRAIN PLUGS ARE PRIMARILY INTENDED FOR INSTALLATION WHERE IT IS DESIRABLE TO LIMIT THE ENTRY OF THE SMALL END OF THE PLUG. SEE GENERAL SPECIFICATIONS.

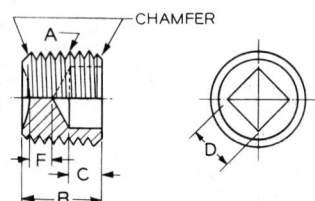 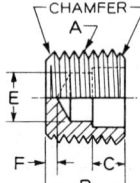 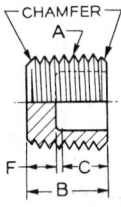

FIG. 3A—UPSET (130109G)   FIG. 3B—BROACHED (130109H)   FIG. 3C—CAST (130109J)

FIG. 3—SQUARE COUNTERSUNK HEADLESS PIPE PLUGS (NPTF)

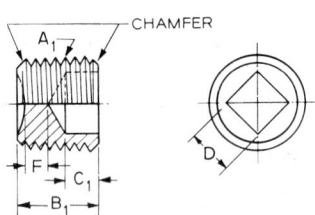 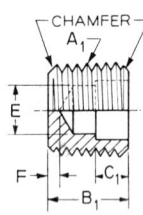 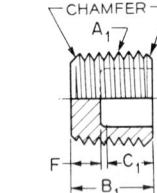 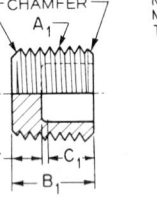

FIG. 4A—UPSET (130109K)   FIG. 4B—BROACHED (130109L)   FIG. 4C—CAST (130109M)

FIG. 4—SQUARE COUNTERSUNK HEADLESS FILLER AND DRAIN PLUGS (PTF)[1]

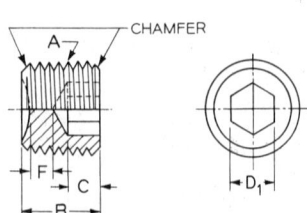 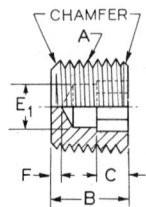 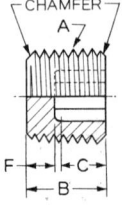 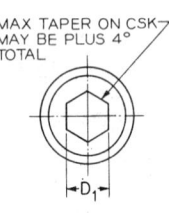

FIG. 5A—UPSET (130109N)   FIG. 5B—BROACHED (130109P)   FIG. 5C—CAST (130109R)

FIG. 5—HEXAGON COUNTERSUNK HEADLESS PIPE PLUGS (NPTF)

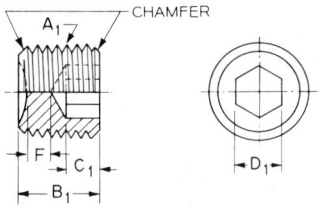
FIG. 6A—UPSET (130109S)

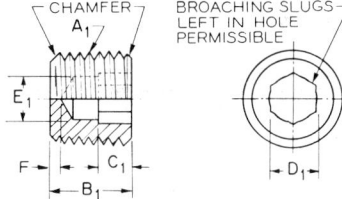
FIG. 6B—BROACHED (130109T)

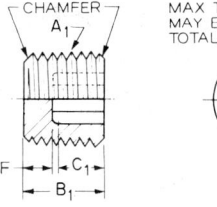
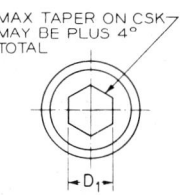
FIG. 6C—CAST (130109U)

FIG. 6—HEXAGON COUNTERSUNK HEADLESS FILLER AND DRAIN PLUGS (PTF)[1]

CODES SHOWN IN BRACKETS ADJACENT TO FIGURE NUMBERS REPRESENT RESPECTIVE FITTING IDENTIFICATION IN ACCORDANCE TO SAE J846.

[1] WARNING — AUTOMOTIVE FILLER AND DRAIN PLUGS ARE PRIMARILY INTENDED FOR INSTALLATION WHERE IT IS DESIRABLE TO LIMIT THE ENTRY OF THE SMALL END OF THE PLUG. SEE GENERAL SPECIFICATIONS.

TABLE 4—DIMENSIONS OF SQUARE AND HEXAGON COUNTERSUNK HEADLESS PIPE PLUGS AND HEADLESS FILLER AND DRAIN PLUGS[a] (FIGS. 3-6)

A Dryseal Thread NPTF, in	A_1 Dryseal Thread PTF-SAE Short, in	B Body Length[b]		B_1 Body Length[b]		C Socket Depth, Min		C_1 Socket Depth, Min		Broached or Upset D Socket Width[c]		Cast D_1 Socket Width[c]	
		in	mm	in	mm	in	mm	in	mm	in	mm	in	mm
1/16-27	1/16-27	0.290	7.37	0.250	6.35	0.12	3.0	0.09	2.3	0.130	3.30	0.13	3.3
		0.310	7.87	0.270	6.86					0.135	3.43	0.14	3.6
1/8-27	1/8-27	0.290	7.37	0.260	6.60	0.12	3.0	0.09	2.3	0.192	4.88	0.19	4.8
		0.310	7.87	0.280	7.11					0.197	5.00	0.21	5.3
1/4-18	1/4-18	0.445	11.30	0.395	10.03	0.19	4.8	0.16	4.1	0.255	6.48	0.26	6.6
		0.475	12.06	0.425	10.80					0.260	6.60	0.28	7.1
3/8-18	3/8-18	0.445	11.30	0.395	10.03	0.19	4.8	0.16	4.1	0.319	8.10	0.32	8.1
		0.475	12.06	0.425	10.80					0.324	8.23	0.35	8.9
1/2-14	1/2-14	0.590	14.99	0.520	13.21	0.25	6.4	0.19	4.8	0.382	9.70	0.38	9.7
		0.630	16.00	0.560	14.22					0.387	9.83	0.41	10.4
3/4-14	3/4-14	0.600	15.24	0.530	13.46	0.31	7.9	0.19	4.8	0.508	12.90	0.51	13.0
		0.640	16.26	0.570	14.48					0.513	13.03	0.54	13.7
1-11-1/2	1-11-1/2	0.750	19.05	0.670	17.02	0.38	9.7	0.25	6.4	0.508	12.90	0.51	13.0
		0.790	20.07	0.710	18.03					0.513	13.03	0.54	13.7

A Dryseal Thread NPTF, in	A_1 Dryseal Thread PTF-SAE Short, in	Broached or Upset D_1[c] Socket Width		Cast D_1 Socket Width		Broached or Upset						Cast			
						E Hole Dia		E_1 Hole Dia				F Wall Thickness			
								Steel		Nonferrous				Ferrous		Nonferrous	
						Max		Max		Max		Min		Min		Min	
		in	mm	in	mm	in	mm	in	mm	in	mm	in	mm	in	mm	in	mm
1/16-27	1/16-27	0.156	3.96	0.143	3.63	0.161	4.09	0.161	4.09	0.06	1.5	0.06	1.5	0.06	1.5
		0.161	4.09												
1/8-27	1/8-27	0.188	4.78	0.209	5.31	0.193	4.90	0.193	4.90	0.06	1.5	0.08	2.0	0.06	1.5
		0.193	4.90												
1/4-18	1/4-18	0.250	6.35	0.278	7.06	0.255	6.48	0.261	6.63	0.06	1.5	0.10	2.5	0.07	1.8
		0.255	6.48												
3/8-18	3/8-18	0.313	7.95	0.345	8.76	0.318	8.08	0.323	8.20	0.06	1.5	0.13	3.3	0.08	2.0
		0.318	8.08												
1/2-14	1/2-14	0.375	9.52	0.410	10.41	0.381	9.68	0.386	9.80	0.09	2.3	0.16	4.1	0.09	2.3
		0.380	9.65												
3/4-14	3/4-14	0.563	14.30	0.553	14.05	0.570	14.48	0.570	14.48	0.09	2.3	0.18	4.6	0.10	2.5
		0.568	14.43												
1-11-1/2	1/11-1/2	0.625	15.88	0.553	14.05	0.633	16.08	0.633	16.08	0.12	3.0	0.20	5.1	0.11	2.8
		0.630	16.00												

[a] Warning—Automotive filler and drain plugs are primarily intended for installation where it is desirable to limit the entry of the small end of the plug. See General Specifications.
[b] Thread must be full or complete thread for length B and B_1.
[c] Tabulated limits shall be maintained for a distance equal to one-half the specified socket depth width at top and bottom portions of socket may slightly exceed maximum.

[1] WARNING — AUTOMOTIVE FILLER AND DRAIN PLUGS ARE PRIMARILY INTENDED FOR INSTALLATION WHERE IT IS DESIRABLE TO LIMIT THE ENTRY OF THE SMALL END OF THE PLUG. SEE GENERAL SPECIFICATIONS.

AUTOMOTIVE STRAIGHT THREAD FILLER AND DRAIN PLUGS—SAE J532a

SAE Standard

Report of Parts and Fittings Committee approved February 1918 and last revised by Tube, Pipe, Hose, and Lubrication Fittings Committee December 1973.

GENERAL SPECIFICATIONS

Scope—This standard includes complete general and dimensional specifications for those types of filler and drain plugs (shown in Figs. 1–7 and Tables 1, 3, and 4) having straight threads which are commonly used with gaskets or seals in automotive and related industrial applications.

Dimensions and Tolerances—Except for nominal sizes and thread specifications, dimensions and tolerances are given in both U. S. customary and SI units as designated. Tabulated dimensions shall apply to the finished plugs, plated, hardened, or otherwise processed, as specified by the purchaser. The minimum across corner dimensions of external hexagons shall be 1.092 times the nominal width across flats, but shall not result in a side flat width less than 0.43 times the nominal width across flats. The minimum across corner dimensions of external squares shall be 1.25 times the nominal width across flats, but shall not result in a side flat width less than 0.75 times the nominal width across flats. The diameter of the washer face on hexagon outside head plugs shall be equal to 95% of the maximum width across flats within a tolerance of ±5%. At maximum material condition, the radii at corners of hexagon and square sockets in broached or upset plugs shall not exceed 0.005 in. (0.13 mm).

Tolerance on dimensions not otherwise limited shall be ±0.010 in. (0.25 mm).

Straight Threads—Unified standard Class 2A external and Class 2B internal threads shall apply to inch sizes of plain finish (unplated) plugs and holes into which they assemble. For externally threaded parts with additive finish, the maximum diameters of Class 2A may be exceeded by the amount of the allowance, that is, the basic diameters (Class 2A maximum diameter plus the allowance) shall apply after plating. The pitch diameter tolerance for special diameter-pitch combinations shall be based on diameter, pitch, and a length of engagement of 9 times the pitch. See SAE J475.

For metric sizes of plugs and mating holes, threads shall conform with SAE J548.

For convenient reference, the data generally required to specify threads is given in Table 1 for both the plugs and mating holes. (Inasmuch as threads are normally produced and gaged with equipment made to the respective measurement system, conversion of size designations and dimensions to other measurement systems is considered unnecessary.)

External threads shall be chamfered or rounded from the diameters tabulated in Table 2 to produce a length of chamfered or partial thread as specified.

Material and Manufacture—Plugs may be made from low carbon steel, cast iron, malleable iron, brass, bronze, or aluminum alloy as specified by purchaser, by casting, milling from the bar, or upsetting from a grade of material free of defects which will affect their serviceability.

Finish—Unless otherwise specified by the purchaser, steel plugs shall be furnished cadmium or zinc plated to a thickness of 0.002 in. (0.05 mm) minimum. These parts must meet the requirements of a 32 h salt spray test in accordance with ASTM B 117. At manufacturer's option, plated plugs may be given a subsequent chromate treatment.

Workmanship—Workmanship shall conform to the best commercial practice to produce high quality parts. Plugs shall be free from all hanging burrs, loose scale, and slivers which might become dislodged in usage and all other defects which might affect their serviceability.

SQUARE AND HEXAGON HEAD

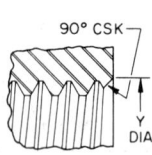
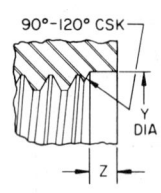

FIG. 1—RECOMMENDED HOLE DATA

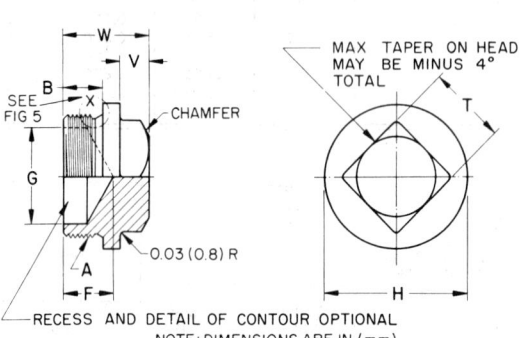

FIG. 3—SQUARE INSIDE HEAD PLUG (420109B)

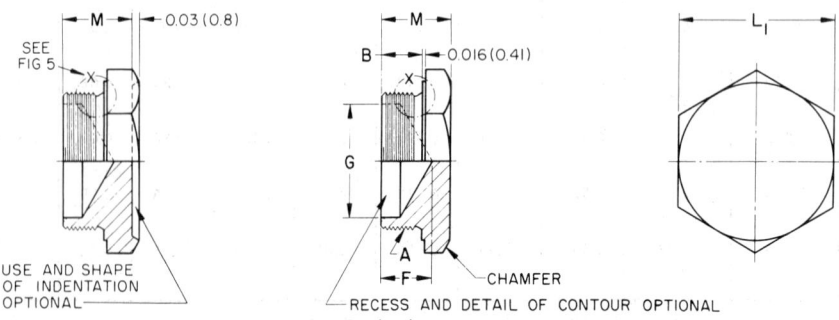

FIG. 2—HEXAGON OUTSIDE HEAD PLUG (420109A)

CODES SHOWN IN BRACKETS ADJACENT TO FIGURE NUMBERS REPRESENT RESPECTIVE FITTING IDENTIFICATION IN ACCORDANCE WITH SAE J846.

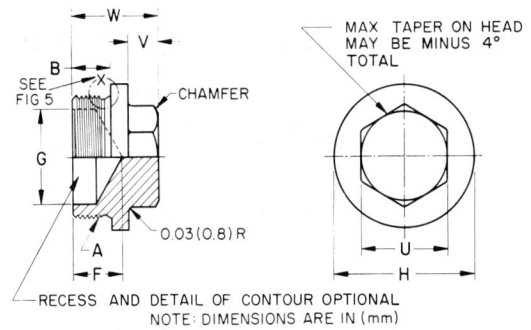

FIG. 4—HEXAGON INSIDE HEAD PLUG (420109C)

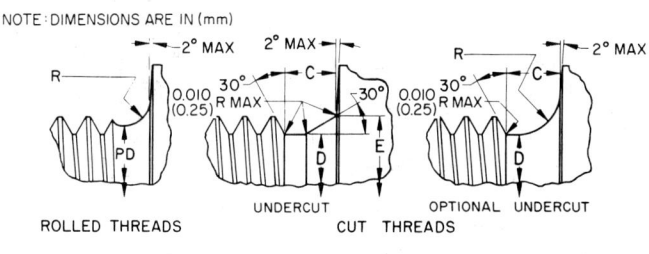

FIG. 5—DETAIL X ENLARGED

TABLE 1—STRAIGHT THREAD SIZES (EXTERNAL AND INTERNAL) (FIGS. 1-7)

Nom Size, in	Series Designation	External Thread				Internal Thread							
		Pitch Diameter				Pitch Diameter				Minor Diameter			
		Max		Min		Max		Min		Max		Min	
		in	mm	in	mm	in	mm	in	mm	in	mm	in	mm
5/16-24	UNF	0.2843		0.2806		0.2902		0.2854		0.277		0.267	
3/8-24	UNF	0.3468		0.3430		0.3528		0.3479		0.340		0.330	
1/2-20	UNF	0.4662		0.4619		0.4731		0.4675		0.457		0.446	
5/8-18	UNF	0.5875		0.5828		0.5949		0.5889		0.578		0.565	
3/4-16	UNF	0.7079		0.7029		0.7159		0.7094		0.696		0.682	
7/8-18	UNS	0.8375		0.8329		0.8449		0.8389		0.828		0.815	
1-18	UNS	0.9625		0.9578		0.9701		0.9639		0.953		0.940	
1-1/4-18	UNEF	1.2124		1.2075		1.2202		1.2139		1.203		1.190	
1-1/2-18	UNEF	1.4624		1.4574		1.4704		1.4639		1.452		1.440	
1-3/4-16	UN	1.7078		1.7025		1.7163		1.7094		1.696		1.682	
2-16	UN	1.9578		1.9524		1.9664		1.9594		1.946		1.932	
Metric Thread Sizes													
10 x 1	—		9.335		9.238		9.446		9.350		8.954		8.844
14 x 1.25	—		13.155		13.048		13.297		13.188		12.962		12.499
18 x 1.5	—		16.980		16.853		17.153		17.026		16.426		16.266

─────────── SQUARE AND HEXAGON SOCKET HEAD ───────────

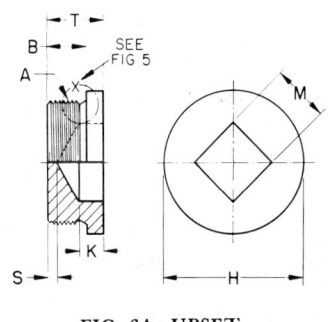

FIG. 6A—UPSET
(420109D)

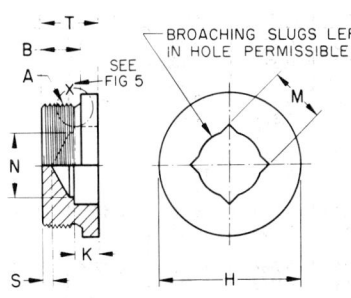

FIG. 6B—BROACHED
(420109E)

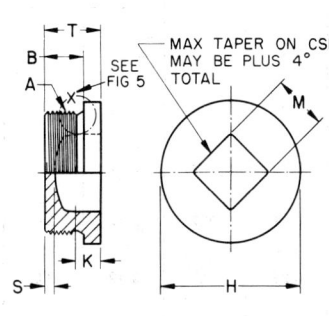

FIG. 6C—CAST
(420109F)

FIG. 6—SQUARE SOCKET HEAD PLUGS

CODES SHOWN IN BRACKETS ADJACENT TO FIGURE NUMBERS REPRESENT RESPECTIVE FITTING IDENTIFICATION IN ACCORDANCE WITH SAE J846.

21.140

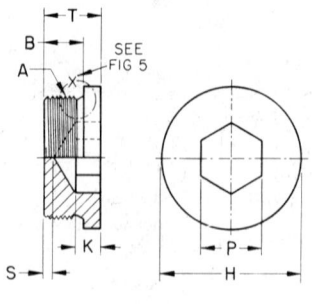 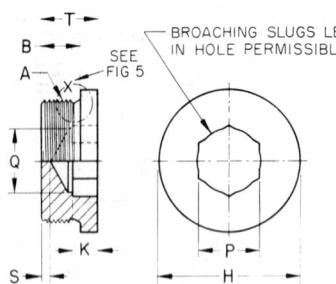 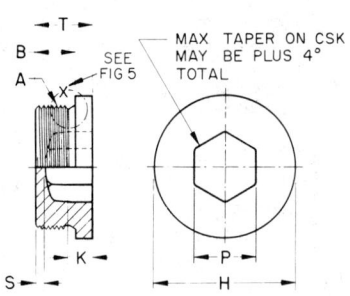

FIG. 7A—UPSET (420109G)   FIG. 7B—BROACHED (420109H)   FIG. 7C—CAST (420109J)

FIG. 7—HEXAGON SOCKET HEAD PLUGS

TABLE 2

Nom Size, in	Series Desig-nation	External Thread								Internal Thread			
		Chamfer Dia				Length of Chamfer or Partial Thread				Y CSK or C'Bore Dia		Z C'Bore Depth	
		Max		Min		Max		Min		Basic		Basic	
		in	mm	in	mm	in	mm	in	mm	in	mm	in	mm
5/16-24	UNF	0.25	6.4	0.24	6.1	0.052	1.32	0.031	0.79	0.34	8.6	0.042	1.07
3/8-24	UNF	0.31	7.9	0.30	7.6	0.052	1.32	0.031	0.79	0.40	10.2	0.042	1.07
1/2-20	UNF	0.43	10.9	0.42	10.7	0.062	1.57	0.038	0.97	0.53	13.5	0.050	1.27
5/8-18	UNF	0.54	13.7	0.52	13.2	0.069	1.75	0.042	1.07	0.66	16.8	0.056	1.42
3/4-16	UNF	0.66	16.8	0.64	16.3	0.078	1.98	0.047	1.19	0.78	19.8	0.062	1.57
7/8-18	UNS	0.79	20.1	0.77	19.6	0.069	1.75	0.042	1.07	0.91	23.1	0.056	1.42
1-18	UNS	0.92	23.4	0.90	22.9	0.069	1.75	0.042	1.07	1.03	26.2	0.056	1.42
1-1/4-18	UNEF	1.17	29.7	1.15	29.2	0.069	1.75	0.042	1.07	1.28	32.5	0.056	1.42
1-1/2-18	UNEF	1.42	36.1	1.40	35.6	0.069	1.75	0.042	1.07	1.53	38.9	0.056	1.42
1-3/4-16	UN	1.66	42.2	1.64	41.7	0.078	1.98	0.047	1.19	1.78	45.2	0.062	1.57
2-16	UN	1.91	48.5	1.89	48.0	0.078	1.98	0.047	1.19	2.03	51.6	0.062	1.57
Metric Thread Sizes													
10 x 1	—	0.33	8.4	0.32	8.1	0.049	1.25	0.030	0.75	0.42	10.8	0.039	1.00
14 x 1.25	—	0.48	12.2	0.47	11.9	0.061	1.56	0.037	0.94	0.58	14.8	0.049	1.25
18 x 1.5	—	0.62	15.7	0.60	15.2	0.074	1.88	0.044	1.12	0.74	18.8	0.059	1.50

TABLE 3—DIMENSIONS OF HEXAGON OUTSIDE HEAD AND SQUARE, AND HEXAGON INSIDE HEAD FILLER AND DRAIN PLUGS (FIGS. 2-5)

Nom Size, in	B Shoulder Length		C Relief Width		D Relief Dia		E Pilot Dia		F Recess Depth		G Recess Dia, Max		H Flange Dia		R Fillet Radius[a]	
															Approx	
	in	mm	in	mm	in	mm	in	mm	in	mm	in	mm	in	mm	in	mm
5/16	0.31	7.9	0.09	2.3	0.252 / 0.245	6.40 / 6.22	0.328 / 0.321	8.33 / 8.15	—	—	—	—	0.56	14.2	0.042	1.07
3/8	0.31	7.9	0.09	2.3	0.314 / 0.307	7.98 / 7.80	0.391 / 0.384	9.93 / 9.75	—	—	—	—	0.62	15.7	0.042	1.07
1/2	0.34	8.6	0.09	2.3	0.428 / 0.421	10.87 / 10.69	0.516 / 0.509	13.11 / 12.93	0.41	10.4	0.25	6.4	0.75	19.0	0.050	1.27
5/8	0.38	9.7	0.12	3.0	0.545 / 0.537	13.84 / 13.64	0.641 / 0.633	16.28 / 16.08	0.47	11.9	0.38	9.7	0.88	22.4	0.056	1.42
3/4	0.38	9.7	0.12	3.0	0.660 / 0.651	16.76 / 16.54	0.766 / 0.757	19.46 / 19.23	0.47	11.9	0.50	12.7	1.00	25.4	0.062	1.57
7/8	0.41	10.4	0.12	3.0	0.793 / 0.785	20.14 / 19.94	0.891 / 0.883	22.63 / 22.43	0.50	12.7	0.56	14.2	1.12	28.4	0.056	1.42
1	0.44	11.2	0.12	3.0	0.918 / 0.910	23.32 / 23.11	0.016 / 1.008	25.81 / 25.60	0.56	14.2	0.69	17.5	1.25	31.8	0.056	1.42
1-1/4	0.47	11.9	0.12	3.0	1.167 / 1.159	29.64 / 29.44	1.266 / 1.258	32.16 / 31.95	0.59	15.0	0.94	23.9	1.50	38.1	0.056	1.42

(Table continued on next page)

TABLE 3—DIMENSIONS OF HEXAGON OUTSIDE HEAD AND SQUARE, AND HEXAGON INSIDE HEAD FILLER AND DRAIN PLUGS (FIGS. 2-5)

Nom Size, in	B Shoulder Length		C Relief Width		D Relief Dia		E Pilot Dia		F Recess Depth		G Recess Dia, Max		H Flange Dia		R Fillet Radius[a] Approx	
	in	mm	in	mm	in	mm	in	mm	in	mm	in	mm	in	mm	in	mm
1-1/2	0.50	12.7	0.12	3.0	1.417 / 1.409	35.99 / 35.79	1.516 / 1.508	38.51 / 38.30	0.69	17.5	1.12	28.4	1.75	44.4	0.056	1.42
1-3/4	0.56	14.2	0.12	3.0	1.657 / 1.648	42.09 / 41.86	1.766 / 1.757	44.86 / 44.63	0.75	19.0	1.38	35.1	2.00	50.8	0.062	1.57
2	0.56	14.2	0.12	3.0	1.907 / 1.898	48.44 / 48.21	2.016 / 2.007	51.21 / 50.98	0.75	19.0	1.62	41.1	2.25	57.2	0.062	1.57
Metric Thread Sizes																
10	0.31	7.9	0.09	2.3	0.336 / 0.329	8.53 / 8.36	0.410 / 0.403	10.41 / 10.24	—	—	—	—	0.62	15.7	0.039	1.00
14	0.34	8.6	0.09	2.3	0.480 / 0.473	12.19 / 12.01	0.567 / 0.560	14.40 / 14.22	0.41	10.4	0.31	7.9	0.81	20.6	0.049	1.25
18	0.38	9.7	0.12	3.0	0.623 / 0.614	15.82 / 15.60	0.724 / 0.715	18.39 / 18.16	0.47	11.9	0.44	11.2	0.94	23.9	0.059	1.50

Nom Size, in	R_1 Fillet Radius[a] Approx		Hex Outside Head				Square and Hexagon Inside Head							
			L Hex Width		M Overall Length		T Square Size		U Hex Size		V Square and Hex Height		W Overall Length	
	in	mm	in	mm	in	mm	in	mm	in	mm	in	mm	in	mm
5/16	0.062	1.57	0.565 / 0.551	14.35 / 14.00	0.50	12.7	0.221 / 0.214	5.61 / 5.44	0.283 / 0.276	7.19 / 7.01	0.22	5.6	0.66	16.8
3/8	0.062	1.57	0.627 / 0.612	15.93 / 15.54	0.50	12.7	0.283 / 0.276	7.19 / 7.01	0.315 / 0.304	8.00 / 7.72	0.25	6.4	0.69	17.5
1/2	0.075	1.90	0.752 / 0.737	19.10 / 18.72	0.53	13.5	0.377 / 0.370	9.58 / 9.40	0.440 / 0.428	11.18 / 10.87	0.25	6.4	0.72	18.3
5/8	0.083	2.11	0.877 / 0.860	22.28 / 21.84	0.62	15.7	0.440 / 0.428	11.18 / 10.87	0.502 / 0.489	12.75 / 12.42	0.28	7.1	0.78	19.8
3/4	0.094	2.39	1.002 / 0.983	25.45 / 24.97	0.62	15.7	0.502 / 0.490	12.75 / 12.45	0.565 / 0.551	14.35 / 14.00	0.28	7.1	0.81	20.6
7/8	0.083	2.11	1.127 / 1.106	28.63 / 28.09	0.66	16.8	0.565 / 0.553	14.35 / 14.05	0.627 / 0.612	15.93 / 15.54	0.31	7.9	0.88	22.4
1	0.083	2.11	1.252 / 1.230	31.80 / 31.24	0.75	19.0	0.627 / 0.615	15.93 / 15.62	0.815 / 0.798	20.70 / 20.27	0.31	7.9	0.91	23.1
1-1/4	0.083	2.11	1.502 / 1.477	38.15 / 37.52	0.78	19.8	0.815 / 0.803	20.70 / 20.40	1.002 / 0.983	25.45 / 24.97	0.34	8.6	1.00	25.4
1-1/2	0.083	2.11	1.752 / 1.725	44.50 / 43.82	0.88	22.4	0.940 / 0.928	23.88 / 23.57	1.127 / 1.106	28.63 / 28.09	0.38	9.7	1.12	28.4
1-3/4	0.094	2.39	2.002 / 1.974	50.85 / 50.14	0.94	23.9	1.127 / 1.115	28.63 / 28.32	1.252 / 1.230	31.80 / 31.24	0.44	11.2	1.25	31.8
2	0.094	2.39	2.252 / 2.232	57.20 / 56.69	0.94	23.9	1.315 / 1.302	33.40 / 33.07	1.502 / 1.477	38.15 / 37.52	0.44	11.2	1.31	33.3
Metric Thread Sizes														
10	0.059	1.50	0.627 / 0.612	15.93 / 15.54	0.50	12.7	0.283 / 0.276	7.19 / 7.01	0.315 / 0.304	8.00 / 7.72	0.25	6.4	0.69	17.5
14	0.074	1.88	0.814 / 0.799	20.68 / 20.29	0.53	13.5	0.377 / 0.370	9.58 / 9.40	0.440 / 0.428	11.18 / 10.87	0.25	6.4	0.72	18.3
18	0.089	2.25	0.940 / 0.921	23.88 / 23.39	0.62	15.7	0.502 / 0.490	12.75 / 12.45	0.565 / 0.551	14.35 / 14.00	0.28	7.1	0.81	20.6

[a] See detail X in Fig. 5.

TABLE 4—DIMENSIONS OF SQUARE AND HEXAGON SOCKET HEAD FILLER AND DRAIN PLUGS (FIGS. 5-7)

Nom Size, in	B Shoulder Length		C Relief Width		D Relief Dia		E Pilot Dia		H Flange Dia		K Socket Depth, Min		S Wall Thickness, Min		T Overall Length	
	in	mm	in	mm	in	mm	in	mm	in	mm	in	mm	in	mm	in	mm
5/16	0.31	7.9	0.09	2.3	0.252 / 0.245	6.40 / 6.22	0.328 / 0.321	8.33 / 8.15	0.56	14.2	0.12	3.0	0.12	3.0	0.44	11.2
3/8	0.31	7.9	0.09	2.3	0.314 / 0.307	7.98 / 7.80	0.391 / 0.384	9.93 / 9.75	0.62	15.7	0.12	3.0	0.12	3.0	0.44	11.2
1/2	0.34	8.6	0.09	2.3	0.428 / 0.421	10.87 / 10.69	0.516 / 0.509	13.11 / 12.93	0.75	19.0	0.19	4.8	0.12	3.0	0.47	11.9
5/8	0.38	9.7	0.12	3.0	0.545 / 0.537	13.84 / 13.64	0.641 / 0.633	16.28 / 16.08	0.88	22.4	0.19	4.8	0.12	3.0	0.50	12.7
3/4	0.38	9.7	0.12	3.0	0.660 / 0.651	16.76 / 16.54	0.766 / 0.757	19.46 / 19.23	1.00	25.4	0.25	6.4	0.12	3.0	0.52	13.2

(Table continued on next page)

TABLE 4—DIMENSIONS OF SQUARE AND HEXAGON SOCKET HEAD FILLER AND DRAIN PLUGS (FIGS. 5-7) (continued)

Nom Size, in	B Shoulder Length		C Relief Width		D Relief Dia		E Pilot Dia		H Flange Dia		K Socket Depth, Min		S Wall Thickness, Min		T Overall Length	
	in	mm	in	mm	in	mm	in	mm	in	mm	in	mm	in	mm	in	mm
7/8	0.41	10.4	0.12	3.0	0.793 / 0.785	20.14 / 19.94	0.891 / 0.883	22.63 / 22.43	1.12	28.4	0.25	6.4	0.12	3.0	0.56	14.2
1	0.44	11.2	0.12	3.0	0.918 / 0.910	23.32 / 23.11	1.016 / 1.008	25.81 / 25.60	1.25	31.8	0.25	6.4	0.12	3.0	0.59	15.0
1-1/4	0.47	11.9	0.12	3.0	1.167 / 1.159	29.64 / 29.44	1.266 / 1.258	32.16 / 31.95	1.50	38.1	0.31	7.9	0.12	3.0	0.66	16.8
1-1/2	0.50	12.7	0.12	3.0	1.417 / 1.409	35.99 / 35.79	1.516 / 1.508	38.51 / 38.30	1.75	44.4	0.38	9.7	0.12	3.0	0.75	19.0
1-3/4	0.56	14.2	0.12	3.0	1.657 / 1.648	42.09 / 41.86	1.766 / 1.757	44.86 / 44.63	2.00	50.8	0.38	9.7	0.12	3.0	0.81	20.6
2	0.56	14.2	0.12	3.0	1.907 / 1.898	48.44 / 48.21	2.016 / 2.007	51.21 / 50.98	2.25	57.2	0.38	9.7	0.12	3.0	0.88	22.4
Metric Thread Sizes																
10	0.31	7.9	0.09	2.3	0.336 / 0.329	8.53 / 8.36	0.410 / 0.403	10.41 / 10.24	0.62	15.7	0.12	3.0	0.12	3.0	0.44	11.2
14	0.34	8.6	0.09	2.3	0.480 / 0.473	12.19 / 12.01	0.567 / 0.560	14.40 / 14.22	0.81	20.6	0.19	4.8	0.12	3.0	0.47	11.9
18	0.38	9.7	0.12	3.0	0.623 / 0.614	15.82 / 15.60	0.724 / 0.715	18.39 / 18.16	0.94	23.9	0.25	6.4	0.12	3.0	0.53	13.5

Nom Size, In	R Fillet Radius[a] Approx		R₁ Fillet Radius[a] Approx		Square Socket						Hexagon Socket					
					Broached or Upset				Cast		Broached or Upset				Cast	
					M[a] Socket Width		N Hole Dia, Max		M Socket Width		P[a] Socket Width		Q Hole Dia, Max		P Socket Width	
	in	mm	in	mm	in	mm	in	mm	in	mm	in	mm	in	mm	in	mm
5/16	0.042	1.07	0.062	1.57	0.130 / 0.135	3.30 / 3.43	0.143	3.63	0.13 / 0.14	3.3 / 3.6	0.156 / 0.161	3.96 / 4.09	0.161	4.09	0.156 / 0.161	3.96 / 4.09
3/8	0.042	1.07	0.062	1.57	0.192 / 0.197	4.88 / 5.00	0.209	5.31	0.19 / 0.21	4.8 / 5.3	0.188 / 0.193	4.78 / 4.90	0.193	4.90	0.188 / 0.193	4.78 / 4.90
1/2	0.50	1.27	0.075	1.90	0.255 / 0.260	6.48 / 6.60	0.278	7.06	0.26 / 0.28	6.6 / 7.1	0.250 / 0.255	6.35 / 6.48	0.255	6.48	0.250 / 0.255	6.35 / 6.48
5/8	0.056	1.42	0.083	2.11	0.319 / 0.324	8.10 / 8.23	0.345	8.76	0.32 / 0.35	8.1 / 8.9	0.313 / 0.318	7.95 / 8.08	0.318	8.08	0.313 / 0.318	7.95 / 8.08
3/4	0.062	1.57	0.094	2.39	0.382 / 0.387	9.70 / 9.83	0.410	10.41	0.38 / 0.41	9.7 / 10.4	0.375 / 0.380	9.53 / 9.65	0.381	9.68	0.375 / 0.380	9.53 / 9.65
7/8	0.056	1.42	0.083	2.11	0.413 / 0.418	10.49 / 10.62	0.443	11.25	0.41 / 0.44	10.4 / 11.2	0.500 / 0.505	12.70 / 12.83	0.506	12.85	0.500 / 0.505	12.70 / 12.83
1	0.056	1.42	0.083	2.11	0.413 / 0.418	10.49 / 10.62	0.443	11.25	0.41 / 0.44	10.4 / 11.2	0.563 / 0.568	14.30 / 14.43	0.570	14.48	0.563 / 0.568	14.30 / 14.43
1-1/4	0.056	1.42	0.083	2.11	0.508 / 0.513	12.90 / 13.03	0.553	14.05	0.51 / 0.54	13.0 / 13.7	0.625 / 0.630	15.88 / 16.00	0.633	16.08	0.625 / 0.630	15.88 / 16.00
1-1/2	0.056	1.42	0.083	2.11	0.632 / 0.637	16.05 / 16.18	0.679	17.25	0.63 / 0.66	16.0 / 16.8	0.750 / 0.755	19.05 / 19.18	0.757	19.23	0.750 / 0.760	19.05 / 19.30
1-3/4	0.062	1.57	0.094	2.39	0.759 / 0.764	19.28 / 19.41	0.820	20.83	0.76 / 0.80	19.3 / 20.3	0.875 / 0.880	22.23 / 22.35	0.883	22.43	0.875 / 0.885	22.23 / 22.48
2	0.062	1.57	0.094	2.39	0.884 / 0.889	22.45 / 22.58	0.945	24.00	0.89 / 0.93	22.6 / 23.6	1.125 / 1.130	28.58 / 28.70	1.134	28.80	1.125 / 1.135	28.58 / 28.83
Metric Thread Sizes																
10	0.039	1.00	0.059	1.50	0.192 / 0.197	4.88 / 5.00	0.209	5.31	0.19 / 0.21	4.8 / 5.3	0.188 / 0.193	4.78 / 4.90	0.193	4.90	0.188 / 0.193	4.78 / 4.90
14	0.049	1.25	0.074	1.88	0.255 / 0.260	6.48 / 6.60	0.278	7.06	0.26 / 0.28	6.6 / 7.1	0.250 / 0.255	6.35 / 6.48	0.255	6.48	0.250 / 0.255	6.35 / 6.48
18	0.059	1.50	0.089	2.25	0.382 / 0.387	9.70 / 9.83	0.410	10.41	0.38 / 0.41	9.7 / 10.4	0.375 / 0.380	9.53 / 9.65	0.381	9.68	0.375 / 0.380	9.53 / 9.65

[a] See detail X in Fig. 5.

LUBRICATION FITTINGS—SAE J534c

SAE Standard

Report of Parts and Fittings Committee approved January 1949 and last revised by Tube, Pipe, Hose, and Lubrication Fittings Committee October 1973.

GENERAL SPECIFICATIONS

Scope—This SAE Standard covers complete general and dimensional specifications for the various types of lubrication fittings and related threaded components intended for general application in the automotive and allied fields.

Designations—Lubrication fittings are designated by the type and size of the threaded ends and the configuration of the fitting.

Dimensions and Tolerances—Except for nominal sizes and thread designations, dimensions and tolerances are given in both U. S. customary and SI units, as designated. Tabulated dimensions shall apply to the finished parts, plated or otherwise processed, as specified by the purchaser. Tolerance on all dimensions not otherwise limited shall be ±0.01 in. (±0.3 mm). The maximum and minimum across flats dimensions shall be within the commercial tolerance of bar or extruded stock from which the fittings are produced. The minimum across corners dimensions of hexagons shall be 1.092 times the nominal width across flats, but shall not result in a side flat width less than 0.43 times the nominal width across flats.

Check Valve—All the standard hydraulic lubrication fittings contained herein are supplied with ball check valves. Fittings without valves are not recommended by the lubrication fitting industry.

Contour—Details of contour shall be optional with the manufacturer, provided the tabulated dimensions are maintained and serviceability of the fittings is not impaired.

Pipe Threads—The pipe threads on fittings, unless there is specific authorization to the contrary, shall conform with the specifications given in detail in SAE J476 for the designated thread series, except that external thread crests may have greater maximum truncation due to manufacturing practices. Experience has shown that the crest of the threads on lubrication fittings, intended for use with grease, does not have to conform to Dryseal American Standard Form to function satisfactorily. The deviations from standard Dryseal practice are peculiar to lubrication fittings and as special considerations are involved, it is not advisable to use them in any other application of pipe thread practice.

External pipe threads shall be chamfered from a diameter (rounded to a two-place decimal) obtained by subtracting 0.016 in. (0.41 mm) from the minimum minor diameter at the small end, with a minus tolerance on the diameter of 0.02 in. (0.5 mm), to produce a length of chamfered or partial thread equivalent to 1 to 1½ times pitch (rounded to a three-place decimal). See Appendix A of SAE J476.

Internal pipe threads shall be countersunk 90 deg included angle to a diameter (rounded to a two-place decimal) obtained by adding 0.016 in. (0.41 mm) to the maximum major diameter at the large end with a plus tolerance on the diameter of 0.02 in. (0.5 mm). See Appendix A of SAE J476.

Recommended assembly considerations for the various combinations of Dryseal pipe threads are given under the respective standard thread series and the paragraph headed Limitation of Assembly, Appendix D, in SAE J476.

¼-28 Taper Thread—External taper threads designated SAE-LT shall be Unified Standard Form ¼-28 with 0.75 in. (19.0 mm), ±0.06 in. (±1.5 mm), diametral taper per 12.00 in. (304.8 mm) of length. The pitch diameter measured at start of thread on small end shall be 0.2257–0.2224 in. (5.733–5.649 mm).

Threads shall be chamfered 0.036–0.054 in. (0.91–1.37 mm) long from a diameter of 0.20 in. (5.1 mm) with a tolerance of −0.02 in. (−0.5 mm).

It is recommended that SAE-LT taper threads be assembled into ¼-28 UNF, Class 3B, straight threaded holes having a modified maximum minor diameter of 0.2152 in. (5.466 mm) to insure 75% minimum thread height.

Special Thread Forming Threads—The ¼-28 special taper thread forming thread and the ⅛-27 pipe special thread forming thread, where specified, shall conform to the dimensions specified in Fig. 1 and Table 1. Fittings employing these threads may be driven or spun into unthreaded holes of diameters recommended and they are generally either marked or colored to provide ready identification.

Material and Manufacture—Unless otherwise specified, fittings shall be made from steel standard with the manufacturer. At the manufacturer's option, caps for water pump fittings may be made from brass, steel, or aluminum.

The greasing end of fittings shall be hardened. They shall have a case depth of 0.005–0.009 in. (0.13–0.23 mm) and minimum hardness of 83 on the Rockwell 15N scale. The threaded end on special thread forming fittings shall also be hardened.

Finish—Steel fittings shall have a minimum plating thickness of 0.002 in. (0.005 mm) of cadmium or zinc. Zinc plated fittings shall have a supplementary treatment other than organic coating and both cadmium and zinc plated fittings shall withstand a minimum 50 h salt spray test in accordance with ASTM B 117, Method of Salt Spray (Fog) Testing, before showing red rust on external surfaces.

Workmanship—Fittings shall be free from burrs, loose scale, sharp edges, and all other defects that might affect their serviceability.

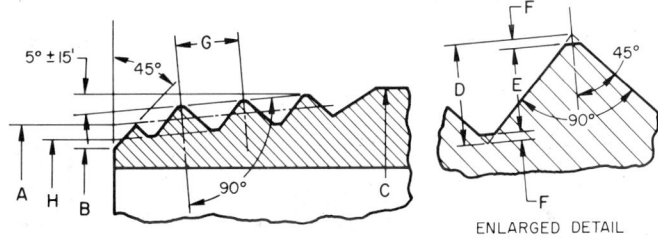

FIG. 1—SPECIAL THREAD FORMING THREADS

TABLE 1—DIMENSIONS OF SPECIAL THREAD FORMING THREADS (FIG. 1)

Nominal Thread Size	A Pitch Dia at Small End		B Chamfer Dia		C Shank Dia		D Height of Sharp V Thread		E Height of Truncated Thread		F Height of Truncated at Crest and Root		G Pitch		H Root Dia at Small End		Recommended Hole Dia[a]	
	in	mm	in	mm	in	mm	in	mm	in	mm	in	mm	in	mm	in	mm	in	mm
¼-28 Spl Taper	0.2226	5.654	0.20	5.1	0.259	6.58	0.0178	0.452	0.0168	0.427	0.0027	0.069	0.0357	0.907	0.208	5.28	0.235	5.97
	0.2156	5.476	0.18	4.6	0.253	6.43			0.0124	0.315	0.0005	0.013			0.201	5.11	0.230	5.84
⅛-27 Spl Pipe	0.3571	9.070	0.33	8.4	0.403	10.24	0.0185	0.470	0.0175	0.445	0.0029	0.074	0.0370	0.940	0.342	8.69	0.380	9.65
	0.3501	8.892	0.31	7.9	0.396	10.06			0.0127	0.323	0.0005	0.013			0.335	8.51	0.373	9.47

[a] It may be desirable to deviate slightly from specified diameters to obtain optimum performance in specific mating materials. Fitting manufacturers should be consulted.

NOTE: Effective thread length shall be as shown and specified in Figs. 3A, 4A, and 4B and Table 2.

21.144

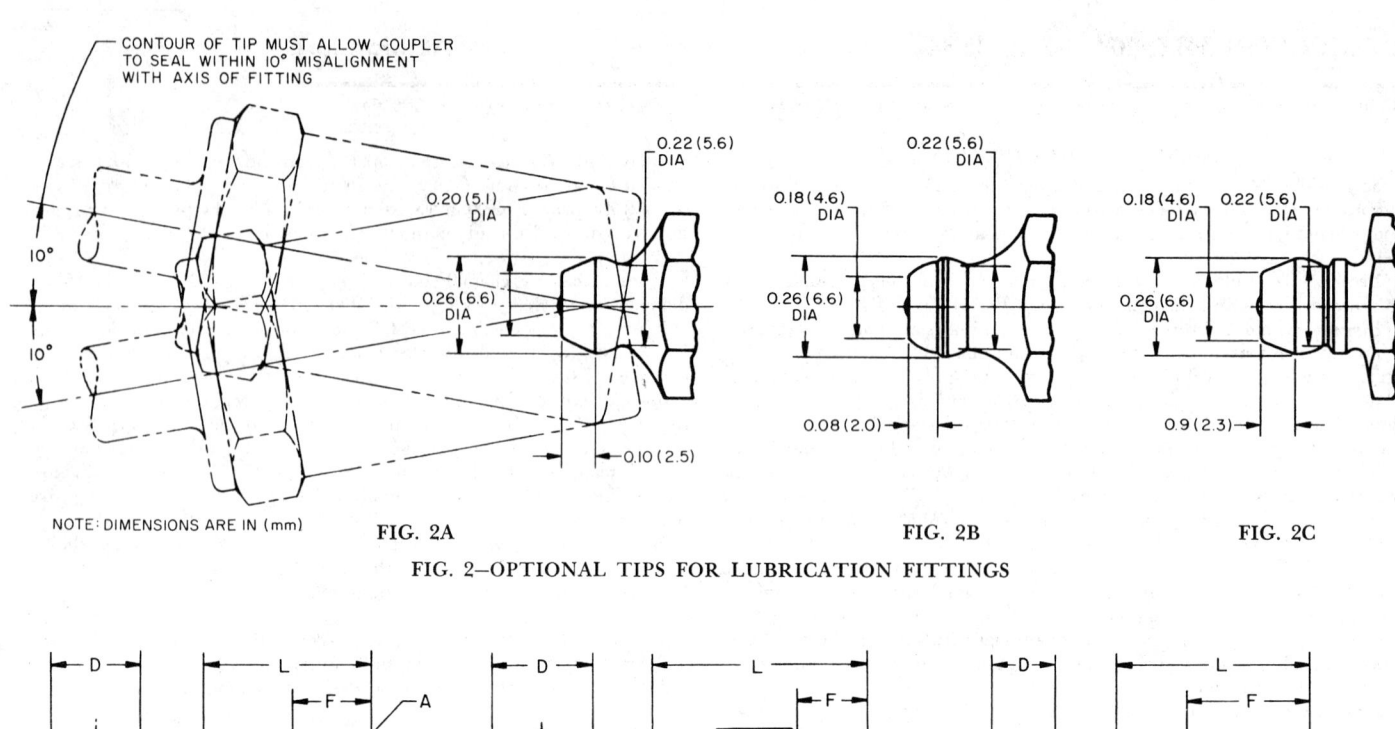

FIG. 2—OPTIONAL TIPS FOR LUBRICATION FITTINGS

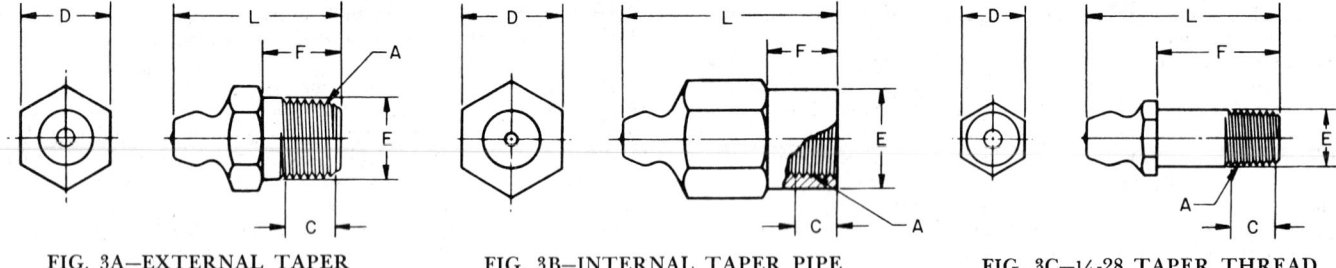

FIG. 3A—EXTERNAL TAPER PIPE THREAD FIG. 3B—INTERNAL TAPER PIPE THREAD FIG. 3C—¼-28 TAPER THREAD

FIG. 3—STRAIGHT FITTINGS

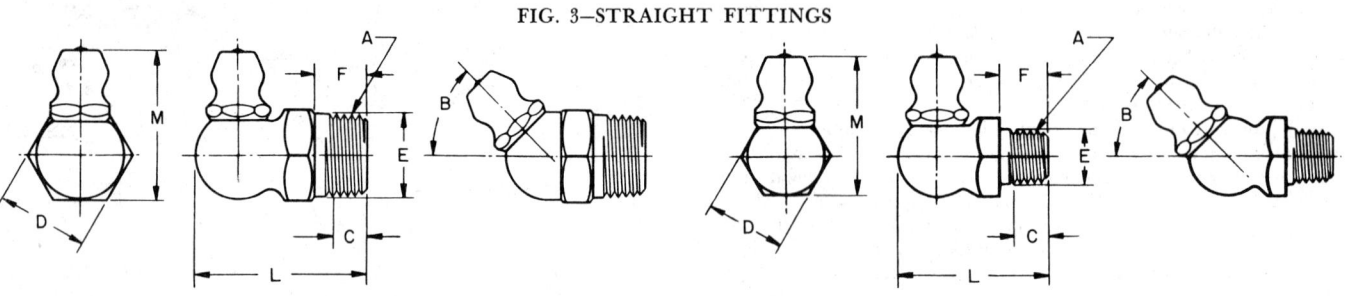

FIG. 4A—TAPER PIPE THREAD FIG. 4B—¼-28 TAPER THREAD

FIG. 4—ELBOW FITTINGS

TABLE 2—DIMENSIONS OF STRAIGHT AND ELBOW FITTINGS (FIGS. 3 AND 4)

Type	A Thread	B Angle, ±3 deg	C Effective Thread Length, min		D Hex Width Across Flats, Nom, in		E Shank Dia		F Shank Length ±0.03		L Overall Length ±0.04		M Overall Height ±0.04	
			in	mm	in	mm	in	mm	in	mm	in	mm	in	mm
													±0.04	±1.0
Straight Fittings	1/8-27 Dryseal-PTF special extra short	—	0.18	4.6	7/16		0.40	10.2	0.28	7.1	0.66	16.8	—	—
	1/8-27 Dryseal-PTF special short	—	0.22	5.6	7/16		0.40	10.2	0.76	19.3	1.26	32.0	—	—
	1/8-27 Dryseal-PTF special short	—	0.22	5.6	7/16		0.40	10.2	1.28	32.5	1.76	44.7	—	—
	1/8-27 Dryseal-PTF special short	—	0.22	5.6	7/16		0.40	10.2	2.18	55.4	2.62	66.5	—	—
	1/8 pipe special thread forming	—	0.14	3.6	7/16		0.40	10.2	0.24	6.1	0.62	15.7	—	—
	1/8-27 Dryseal-NPTF internal thread	—	0.28	7.1	1/2		0.48	12.2	0.32	8.1	1.00	25.4	—	—
	1/4-28 taper thread (SAE-LT)	—	0.10	2.5	5/16		0.26	6.6	0.18	4.6	0.54	13.7	—	—
	1/4-28 taper thread (SAE-LT)	—	0.20	5.1	5/16		0.26	6.6	0.34	8.6	0.68	17.3	—	—
	1/4-28 taper thread (SAE-LT)	—	0.20	5.1	5/16		0.26	6.6	0.62	15.7	0.94	23.9	—	—
	1/4-28 special taper thread forming	—	0.10	2.5	5/16		0.256	6.50	0.20	5.1	0.55	14.0	—	—

(Continued)

TABLE 2—DIMENSIONS OF STRAIGHT AND ELBOW FITTINGS (FIGS. 3 AND 4)

Type	A Thread	B Angle, ±3 deg	C Effective Thread Length, min		D Hex Width Across Flats, Nom, in	E Shank Dia		F Shank Length		L Overall Length		M Overall Height	
								in ±0.03	mm ±0.8	in ±0.04	mm ±1.0	in ±0.04	mm ±1.0
			in	mm		in	mm						
Elbow Fittings	1/8-27 Dryseal-PTF special short	30	0.22	5.6	7/16	0.40	10.2	0.30	7.6	0.90	22.9	0.56	14.2
	1/8-27 Dryseal-PTF special short	30	0.22	5.6	7/16	0.40	10.2	1.26	32.0	2.10	53.3	0.56	14.2
	1/8 pipe special thread forming	30	0.14	3.6	7/16	0.40	10.2	0.20	5.1	0.86	21.8	0.56	14.2
	1/8-27 Dryseal-PTF special short	45	0.22	5.6	7/16	0.40	10.2	0.30	7.6	0.86	21.8	0.64	16.3
	1/4-28 taper thread (SAE-LT)	45	0.10	2.5	3/8	0.26	6.6	0.20	5.1	0.82	20.8	0.58	14.7
	1/4-28 taper thread (SAE-LT)	45	0.20	5.1	3/8	0.26	6.6	0.30	7.6	0.94	23.9	0.58	14.7
	1/4-28 special taper thread forming	45	0.10	2.5	3/8	0.256	6.50	0.19	4.8	0.80	20.3	0.58	14.7
	1/8-27 Dryseal-PTF special short	65	0.22	5.6	7/16	0.40	10.2	0.30	7.6	0.86	21.8	0.72	18.3
	1/8-27 Dryseal-PTF special short	65	0.22	5.6	7/16	0.40	10.2	0.56	14.2	1.18	30.0	0.72	18.3
	1/8 pipe special thread forming	65	0.14	3.6	7/16	0.40	10.2	0.20	5.1	0.78	19.8	0.72	18.3
	1/8-27 Dryseal-PTF special short	90	0.22	5.6	7/16	0.40	10.2	0.30	7.6	0.84	21.3	0.72	18.3
	1/8-27 Dryseal-PTF special short	90	0.22	5.6	7/16	0.40	10.2	1.26	32.0	1.82	46.2	0.72	18.3
	1/8 pipe special thread forming	90	0.14	3.6	7/16	0.40	10.2	0.20	5.1	0.76	19.3	0.72	18.3
	1/4-28 taper thread (SAE-LT)	90	0.10	2.5	3/8	0.26	6.6	0.20	5.1	0.76	19.3	0.66	16.8
	1/4-28 special taper thread forming	90	0.10	2.5	3/8	0.256	6.50	0.19	4.8	0.76	19.3	0.66	16.8
	1/8-27 Dryseal-PTF special short	105	0.22	5.6	7/16	0.40	10.2	0.30	7.6	1.06	26.9	0.76	19.3

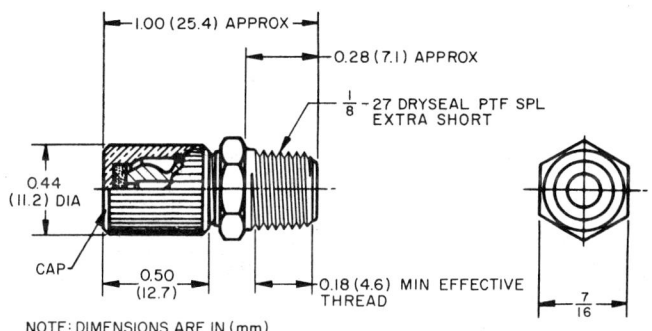

FIG. 5—WATER PUMP FITTING

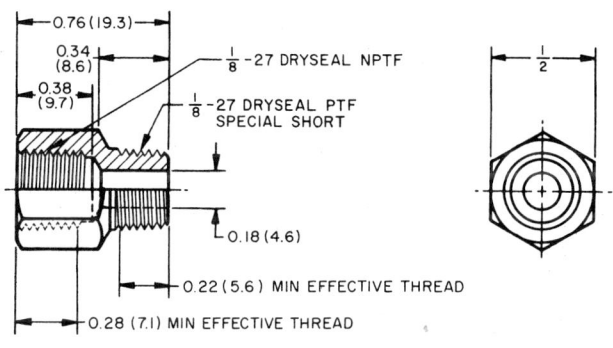

FIG. 6—EXTENSION

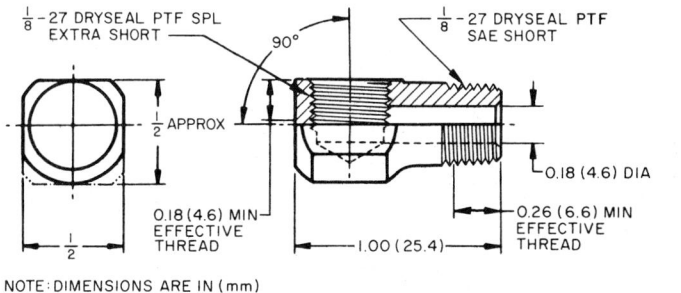

FIG. 7A

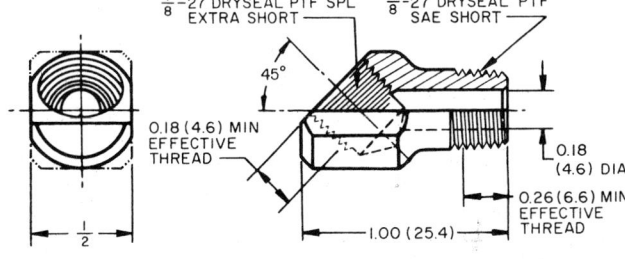

FIG. 7B

FIG. 7—ELBOW ADAPTERS

NOTE: UNSPECIFIED DETAIL WITH RESPECT TO DIMENSIONS, TOLERANCES, CONTOURS, MATERIAL, WORKMANSHIP, ETC., MUST CONFORM TO GENERAL SPECIFICATIONS FOR LUBRICATION FITTINGS.

ELECTRICAL EQUIPMENT AND LIGHTING

ELECTRICAL EQUIPMENT AND LIGHTING

22 Equipment

Nomenclature—Automotive Electrical Systems—SAE J831	22.01
Ignition System Nomenclature and Terminology—SAE J139	22.01
Ignition System Measurements Procedure—SAE J973a	22.02
Storage Batteries—SAE J537j	22.04
Life Test for Automotive Storage Batteries—SAE J240a	22.12
Grounding of Storage Batteries—SAE J538a	22.13
Voltage Drop for Starting Motor Circuits—SAE J541a	22.13
Voltages for Diesel Electrical Systems—SAE J539a	22.13
Starting Motor Mountings—SAE J542c	22.14
Starting Motor Pinions and Ring Gears—SAE J543c	22.15
Generator Mountings—SAE J545b	22.15
Starting Motor and Generator Curves—SAE J544b	22.17
Low Temperature Cranking Load Requirements of an Engine—SAE J1253	22.18
Electrical Generating System (Alternator Type) Performance Curve and Test Procedure—SAE J56	22.19
Magneto Mountings—SAE J546	22.20
Spark Plugs—SAE J548d	22.21
Preignition Rating of Spark Plugs—SAE J549a	22.27
† Limits and Methods of Measurement of Radio Interference Characteristics of Vehicles and Devices (20–1000 MHz)—SAE J551 JUN79	22.27
External Electromagnetic Radiation Suppressors—SAE J552a	22.34
Electric Fuses (Cartridge Type)—SAE J554b	22.35
* Blade Type Electric Fuses—SAE J1284 DEC79	22.37
Circuit Breakers—SAE J553c	22.40
Circuit Breaker—Internal Mounted—Automatic Reset—SAE J258	22.42
Ignition Switch—SAE J259	22.42
* Automobile, Truck, Truck-Tractor, Trailer, and Motor Coach Wiring—SAE J1292 JUN80	22.43
Automotive Printed Circuits—SAE J771c	22.45
High Tension Ignition Cable—SAE J557	22.49
Low Tension Wiring and Cable Terminals and Splice Clips—SAE J163	22.51
Fusible Links—SAE J156a	22.51
Five Conductor Electrical Connectors for Automotive Type Trailers—SAE J895	22.52
Four- and Eight-Conductor Rectangular Electrical Connectors for Automotive Type Trailers—SAE J1239	22.54
Seven-Conductor Electrical Connector for Truck-Trailer Jumper Cable—SAE J560b	22.57
Seven Conductor Jacketed Cable for Truck-Trailer Connections—SAE J1067	22.58
Electrical Terminals—Blade Type—SAE J858a	22.59
† Electrical Terminals—Pin and Receptacle Type—SAE J928 JUN80	22.60
† Electrical Terminals—Eyelet and Spade Type—SAE J561 JUN80	22.61
Nonmetallic Loom—SAE J562a	22.65
Electric Windshield Wiper Switch—SAE J112a	22.66
Electric Windshield Washer Switch—SAE J234	22.66
Electric Blower Motor Switch—SAE J235	22.67
6- and 12-Volt Cigar Lighter Receptacles—SAE J563b	22.67
Electromagnetic Susceptibility Procedures for Vehicle Components (Except Aircraft)—SAE J1113a	22.68
† Battery Cable—SAE J1127 JUN80	22.79
Low Tension Primary Cable—SAE J1128	22.81
Recommended Environmental Practices for Electronic Equipment Design—SAE J1211	22.86
Glossary of Automotive Electronic Terms—SAE J1213	22.102
* Performance Requirements for the Automotive Audio Cassette—SAE J1274 JUN80	22.107
* Testing Methods for Audio Cassettes—SAE J1275 JUN80	22.108

* New
† Technical revision § Editorial change

23 Lighting

Terminology—Motor Vehicle Lighting—SAE J387	23.01
Lighting Identification Code—SAE J759c	23.01
Headlamp Beam Switching—SAE J564c	23.02
Semiautomatic Headlamp Beam Switching Devices—SAE J565c	23.03
Lamp Bulb Retention System—SAE J567c	23.04
Connectors and Plugs—SAE J856	23.07
Dimensional Specifications for Sealed Beam Headlamp Units—SAE J571d	23.09
142 mm x 200 mm Sealed Beam Headlamp Unit—SAE J1132	23.16
Dimensional Specifications for General Service Sealed Lighting Units—SAE J760a	23.18
Lamp Bulbs and Sealed Beam Headlamp Units—SAE J573g	23.20
§ Service Performance Requirements and Test Procedures for Motor Vehicle Lamp Bulbs—SAE J1049 AUG79	23.23
Tests for Motor Vehicle Lighting Devices and Components—SAE J575 JUN80	23.24
Service Performance Requirements for Motor Vehicle Lighting Devices and Components—SAE J256b	23.26
Plastic Materials for Use in Optical Parts Such as Lenses and Reflectors of Motor Vehicle Lighting Devices—SAE J576d	23.28
Plastic Material for Use in Housings of Motor Vehicle Lighting Devices—SAE J29	23.28
Color Specification for Electric Signal Lighting Devices—SAE J578d	23.29
Sealed Beam Headlamp Units for Motor Vehicles—SAE J579c	23.31
Service Performance Requirements for Sealed Beam Headlamp Units for Motor Vehicles—SAE J32a	23.32
§ Sealed Beam Headlamp Assembly—SAE J580 AUG79	23.33
Auxiliary Driving Lamps—SAE J581a	23.35
Fog Lamps—SAE J583d	23.36
Auxiliary Low Beam Lamp—SAE J582a	23.36
Motorcycle Headlamps—SAE J584b	23.37
* Motorcycle Auxiliary Front Lamps—SAE J1306 JUN80	23.39
Motorcycle and Motor Driven Cycle Electrical System (Maintenance of Design Voltage)—SAE J392	23.39
Motorcycle Turn Signal Lamps—SAE J131a	23.40
Tail Lamps (Rear Position Lamps)—SAE J585e	23.40
Stop Lamps—SAE J586d	23.41
Mechanical Stop Lamp Switch—SAE J249	23.42
License Plate Lamps—SAE J587f	23.43
Turn Signal Lamps—SAE J588f	23.44
Headlamp-Turn Signal Spacing—SAE J1221	23.45
Supplemental High Mounted Stop and Rear Turn Signal Lamps—SAE J186a	23.49
Turn Signal Switch—SAE J589b	23.50
Side Turn Signal Lamps—SAE J914b	23.50
Cornering Lamps—SAE J852b	23.51
Turn Signal Flashers—SAE J590e	23.52
Service Performance Requirements for Turn Signal Flashers—SAE J1055	23.52
Vehicular Hazard Warning Signal Flasher—SAE J945b	23.54
Service Performance Requirements for Vehicular Hazard Warning Flashers—SAE J1056	23.55
Warning Lamp Alternating Flashers—SAE J1054	23.56
Service Performance Requirements for Warning Lamp Alternating Flashers—SAE J1104	23.56
Spot Lamps—SAE J591b	23.58
Parking Lamps (Front Position Lamps)—SAE J222a	23.59
Clearance, Side Marker, and Identification Lamps—SAE J592f	23.59
Backup Lamps—SAE J593e	23.60
Backup Lamp Switches—SAE J1076	23.60
Headlamp Switch—SAE J253	23.61
Reflex Reflectors—SAE J594f	23.61
Emergency Warning Device—SAE J774c	23.62
Flashing Warning Lamps for Authorized Emergency, Maintenance and Service Vehicles—SAE J595b	23.63
Hazard Warning Signal Switch—SAE J910b	23.64
360 Deg Emergency Warning Lamp—SAE J845	23.64
School Bus Red Signal Lamps—SAE J887a	23.65
School Bus Stop Arm—SAE J1133a	23.66
Lighting Inspection Code—SAE J599d	23.66
Headlamp Testing Machines—SAE J600a	23.68
Headlamp Aiming Device for Mechanically Aimable Sealed Beam Headlamp Units—SAE J602c	23.69
Flasher Test Equipment—SAE J823c	23.72

22 Equipment

NOMENCLATURE—AUTOMOTIVE ELECTRICAL SYSTEMS—SAE J831

SAE Standard

Report of Electrical Equipment Committee approved June 1962.

Purpose and Scope—The purpose of this standard is to define terms relating to automotive electrical systems to facilitate clear understanding as to their meaning and promote uniformity of nomenclature in engineering discussions, technical papers, and specifications.

Generator, Electric DC, Mechanical Rectification—An electric d-c generator with mechanical rectification is a device which transforms mechanical power into direct current electrical power (d-c) by means of a field structure for excitation, a generating winding, and a segmented commutator for rectification, all of which are contained in an integral package.

Generator, Electric DC, Diode Rectification—An electric d-c generator with diode rectification is a device which transforms mechanical power into direct current electrical power (d-c) by means of a field structure for excitation, a generating winding, and a network of diodes for rectification, all of which are contained in an integral package.

Generator, Electric AC—An electric a-c generator or alternator is a device which transforms mechanical power into alternating current electrical power (a-c) by means of a field structure for excitation and a generating winding. The frequency of the alternating current is determined by the speed of the device.

Output at Idle—Generator "output at idle" on vehicles is the current from the charging system which is available at engine idle speed. The current may be used to supply the connected electrical loads, to charge the battery, or may be divided between the electrical loads and the battery in an infinite number of combinations.

IGNITION SYSTEM NOMENCLATURE AND TERMINOLOGY—SAE J139

SAE Recommended Practice

Report of Electrical Equipment Committee approved January 1970.

1. Purpose—To provide a person interested in presenting technical information, reports, papers, etc., in the field of ignition with standardized terminology and definitions. Ignition measurements, procedures, and instrumentation are covered by SAE J973 and AIR 84A.

2. Ignitions Systems Definitions

2.1 Kettering Ignition System—Inductive system commonly used for internal combustion engines. Employs induction coil, breaker contacts, capacitor, and suitable power supply such as a battery.

2.2 Semiconductor Ignition System—An ignition system for internal combustion engines employing the use of solid-state semiconductors for switching purposes.

2.2.1 INDUCTIVE SYSTEM—An ignition system which stores its primary energy in an inductor or coil.

2.2.2 CAPACITOR DISCHARGE SYSTEM—An ignition system which stores its primary energy in a capacitor.

2.2.3 SUB HEADINGS

2.2.3.1 *Breaker Triggered*—A semiconductor system which utilizes conventional breaker contacts to time and trigger the system.

2.2.3.2 *Breakerless*—A semiconductor system which does not use mechanical breaker contacts for timing or triggering purposes, but retains the distributor for distribution of the secondary voltage.

2.2.3.4 *Distributorless*—A semiconductor ignition system which does not utilize breaker contacts to time or trigger the system nor does it utilize a distributor for distribution of the secondary voltage.

3. Parameter Definitions

3.1 Secondary Available Voltage (Evaluation)—Instrumentation per SAE J973.

3.1.1 OPEN CIRCUIT—The secondary output voltage delivered by the ignition system with a 50 mmf secondary load, measurement to be taken at the supply voltage specified after the system has been temperature stabilized.

3.2 Secondary Available Voltage (on Engine)—Secondary available voltage is the voltage that is available for firing the spark plug. Data for available voltage curve are obtained by making a connection from the secondary coil terminal to the input of a cathode ray oscilloscope (through a voltage divider), then disconnecting the longest spark plug lead from its plug and observing the voltage developed on that lead. The plug end of this lead is taped to prevent flashover to grounded areas, and the lead is arranged in about the same position with respect to engine-grounded areas as it would be if connected to the spark plug. Unless otherwise specified, the longest lead is selected because the higher electrostatic capacity of that lead will give it the lowest available voltage.

Readings are made at constant engine speed, at various increments from idle to speed. At each check point, the speed is held for a sufficient length of time for the observer to be reasonably sure that he has observed the maximum, most usual, and minimum voltages that are likely to occur. Usually, only the minimum values are plotted.

3.3 Required Voltage (Engine or Vehicle)—The voltage required to fire the spark plugs. Requirement data are obtained using the same instrumentation as for available voltage runs, but with all spark plug leads connected. Values plotted are the maximum voltage observed viewing all cylinders simultaneously. Requirements are usually determined with the engine operating at full load (wide-open throttle) and under a variety of part-load conditions to ascertain the maximum required voltages.

3.4 Ignition Reserve (Engine or Vehicle)—The difference between the minimum available and maximum required voltages. An adequate ignition "safety factor" is important if an engine is to be reasonably free from troubles caused by moisture or dirt losses, "leaky" secondary leads, and fouled spark plugs.

3.5 Rise Time—The time required, in microseconds, for the output voltage wave form to rise from 10% of maximum to 90% maximum output.

Measurement to be made under the same conditions as the secondary available voltage.

3.6 Average Current—The input current to an ignition system as measured on a d-c ammeter. In most instances the engine speed must also be specified.

3.7 Peak Coil Current—Applies to inductive systems only, and is the peak current flowing through the ignition coil primary winding at the instant the contacts open.

3.8 Contact Current—The peak current flowing through the contacts of a contact-triggered system at the instant the contacts open.

3.9 Spark Duration—The length of time a spark is established across a spark gap or the length of time current flows in a spark gap. Measured per SAE J973.

3.10 Primary Voltage—The peak magnitude of the first half cycle of the voltage induced across the induction coil primary winding.

3.11 Timing Lag—The number of engine degrees retard in timing caused by electrical lag in the system. Generally stated as number of engine degrees per 1000 rpm engine speed.

3.12 Cut-In Speed—The minimum speed at which the ignition system will operate properly. Distributor and/or engine rpm should be specified.

3.13 Stored Energy—The amount of energy stored in the primary of the ignition system.

3.13.1 INDUCTIVE SYSTEM

$$W_p = \tfrac{1}{2} Li^2$$

where: W_p = energy stored in the primary field, joules
L = primary inductance, henries
i = current flowing in the primary winding, amp

3.13.2 CAPACITOR DISCHARGE SYSTEM

$$W_p = \tfrac{1}{2} CE^2$$

where: W_p = energy stored in the primary capacitor, joules
C = primary capacitance, farads
E = peak primary voltage, V

4. Semiconductor Terminology—Terms used in the semiconductor field in general and which also apply to ignition are omitted because they are covered by EIA, JEDEC, and NEMA Standard RS-245, and also by tentative IEEE standards.

IGNITION SYSTEM MEASUREMENTS PROCEDURE—SAE J973a

SAE Recommended Practice

Report of Electrical Equipment Committee approved October 1966 and last revised January 1973.

1. Introduction—This procedure is intended to provide any technical person or group interested in ignition system design and/or evaluation with the specific equipment, conditions, and methods which will produce test results definitive and reproducible for his own work and yet sufficiently standardized to be acceptable to other groups working on battery ignition systems for automotive engines.

2. D-C Source—The source of d-c voltage to be used in ignition system measurements shall be a variable d-c power supply having a 10–90% transient recovery time of not more than 50 μs over the load range encountered in use. It must have no more than 10 mV variation in average voltage from no load to full ignition system load and no more than 50 mV peak-to-peak ripple over the same load range. This power supply shall be shunted by a suitably tapped automotive-type lead acid battery and be positioned immediately adjacent to the test area so that the source impedance of a vehicle is simulated as closely as possible.

3. Ignition System Definition—The ignition system as defined for the tests tabulated in this report shall consist of:

(a) A coil. This can be the conventional induction coil or an air or magnetic core transformer.

(b) A coil external resistor or resistors if the coil being tested requires an external resistor.

(c) A distributor. This is defined as any device which incorporates a timing mechanism, a spark advance mechanism or mechanisms, and a spark distribution mechanism, all of which have a proper angular interrelationship to themselves and, through a mechanical drive, to the engine.

(d) High voltage, metal conductor ignition cables: coil to distributor—18 in. (455 mm) long, distributor to spark gap—24 in. (610 mm) long. Metal conductor cables are specified to eliminate the varying effects of the different kinds of cable with high impedance conductors. Resistance per foot, as well as inductance of spark plug cables built to suppress radiation, can be quite different from manufacturer to manufacturer.

(e) Any auxiliary switching means implicit with the system being tested such as a transistorized control unit.

The above devices shall be interconnected as the manufacturer recommends or similar to the conventional system illustrated in Fig. 1.

4. System Load—The load connected to the ignition system shall be a multigap spark gap test stand, each gap being individually variable, the number of gaps used being the same as the number of towers on the distributor cap. Using an 8-cylinder distributor as an example, seven gaps will be set to fire at a nominal 12 kV, the remaining gap will be opened to the point where it never can fire. Attached to the nonfiring gap, by not less than 1 ft (305 mm) of secondary ignition cable, will be a high quality (dissipation factor of 3% or less), high voltage, 50 pico farad capacitor (this can be a section of shielded ignition cable) to simulate the capacitance of the cables and spark plugs as normally encountered on a vehicle, and at suitable times a low voltage coefficient (0.0005%/V max), noninductive approximately 10 W, 1.0 MΩ resistor. The resistor simulates lead or carbon fouled spark plugs.

For certain tests, as designated in paragraph 5, the capacitive and resistive loads will be directly connected to the coil high voltage tower with the coil not firing.

5. Measurements to be made

5.1 Group A

5.1.1 AVAILABLE VOLTAGE AT SPARK PLUG—This measurement is fundamental to spark ignition. Comparing available voltage to voltage required to fire spark plugs (in a given engine) determines the adequacy of the ignition system. (See Fig. 2A.)

5.1.2 PEAK COIL PRIMARY CURRENT—This measurement indicates energy into the coil ($E = \tfrac{1}{2} Li^2$) and must be controlled to insure adequate distributor contact life. (See Fig. 2B.)

5.1.3 AVERAGE COIL PRIMARY CURRENT—This measurement determines the average current draw of the system with respect to the d-c source (alternator generator, battery, etc.).

5.1.4 SPARK DURATION—Within limits, this measurement is indicative of the igniting capability of a spark under marginal fuel conditions. It also is an indication of the amount of erosion which will occur on spark plug electrodes due to electrical means. Because of the complexity of both of these areas, however, experience is required to use this information effectively. (See Fig. 2C.)

5.1.5 SPARK VOLTAGE—This is the instantaneous voltage observed across the spark gap halfway through the discharge. (See Fig. 2E.)

5.1.6 SPARK CURRENT—This is the instantaneous current from the secondary winding of the ignition coil flowing through the spark gap after breakdown. (See Fig. 2E.)

5.1.7 SPARK ENERGY—This is the inductive portion of energy dissipated in the spark after breakdown. It is calculated as shown:

$$E_{spark} = \frac{V_a(t_f - t_0)(i_f + i_0)}{2}$$

where: t_0 and i_0 = initial values of time and current of the spark after breakdown
t_f and i_f = final values of time and current of the spark after breakdown
V_a = spark voltage at $(t_f - t_0)/2$

5.2 Group B

5.2.1 COIL SECONDARY VOLTAGE RISETIME—This measurement is an indication of the ability of an ignition system to fire shunted (fouled) spark plugs. The shorter the risetime, the less system energy is lost across the fouled shunt and the more voltage is available to fire the plug. (See Fig. 2A.)

5.2.2 COIL PRIMARY INDUCED VOLTAGE—This measurement is useful with respect to distributor contact life on conventional ignition systems and is a measure of the stress on a semiconductor power switch in inductive energy storage ignition systems. (See Fig. 2D.) This measurement is not applicable to capacitor discharge ignition systems.

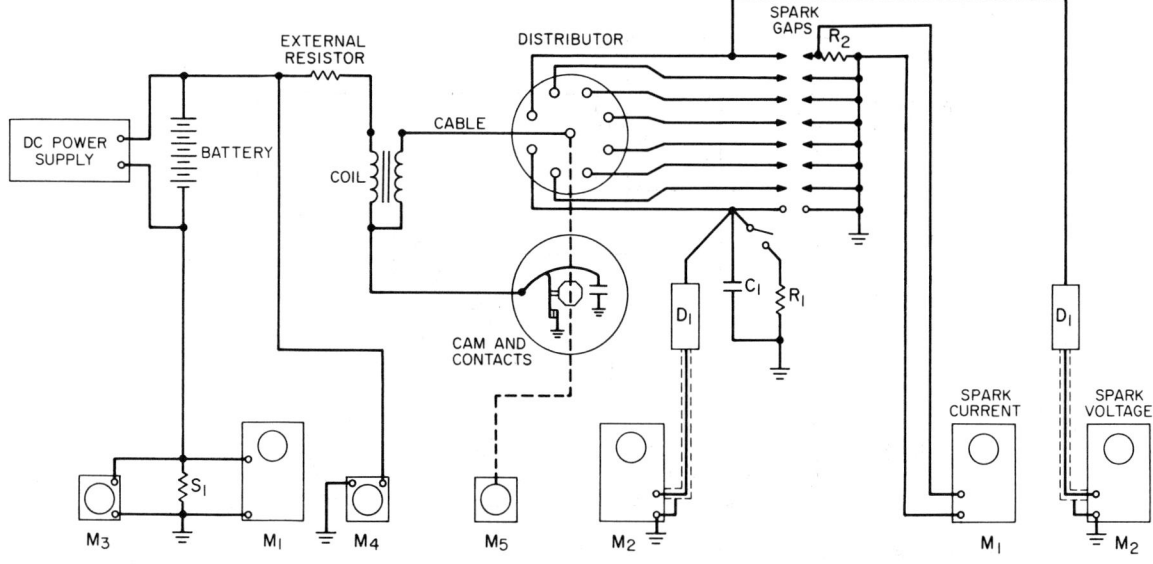

M_1 = CURRENT MEASURING OSCILLOSCOPE
M_2 = VOLTAGE MEASURING OSCILLOSCOPE
M_3 = DC AMMETER
M_4 = DC VOLTMETER
M_5 = TACHOMETER

C_1 = SHUNT CAPACITY ($C_1 + D_1$ = 50 MMF)
S_1 = 0.1 OHM METER SHUNT
D_1 = VOLTAGE DIVIDER
R_1 = 1.0 MEGOHM, 10 WATT
R_2 = SECONDARY CURRENT SENSING RESISTOR 100 OHM, 1/2 WATT - CARBON ±1%

FIG. 1

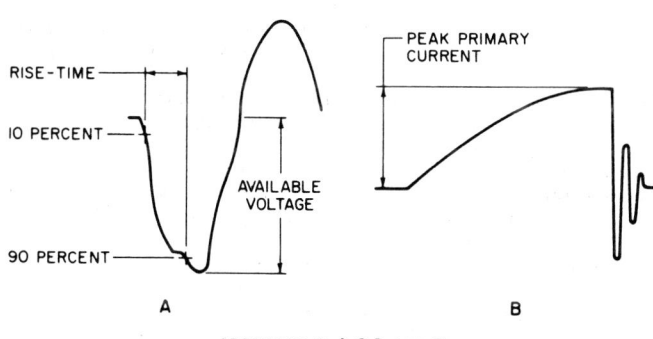

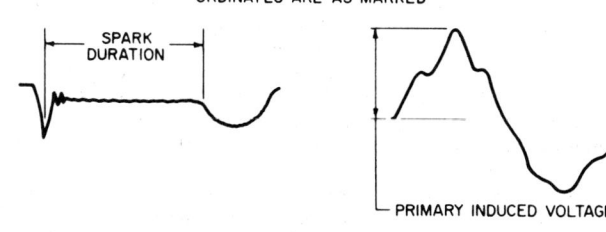

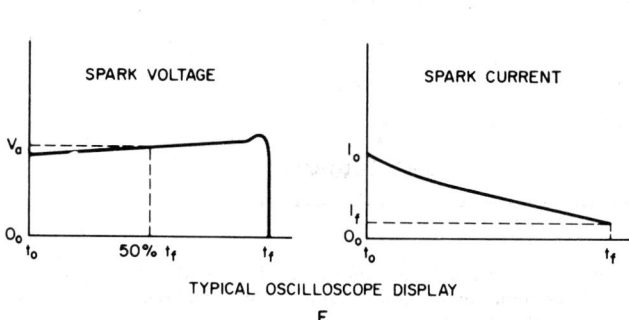

TYPICAL OSCILLOSCOPE DISPLAY
E

FIG. 2

6. Test Equipment

6.1 A voltage divider and oscilloscope for measuring high voltage as defined in SAE AIR 84 should be used to measure available voltage, risetime, and spark duration.

6.2 An oscilloscope with a maximum risetime of 0.035 μs and with a minimum band pass of 10MC (Ref. Tektronix 535A with a type L plug-in unit) with its input connected across a noninductive meter shunt which is in series with the coil primary for peak coil primary current measurements. The sensing resistor shall not have a resistance greater than 0.1 Ω. The oscilloscope must have a minimum deflection sensitivity of 50 mV/cm.

6.3 A good quality d-c ammeter of the permanent magnet-moving coil type should be used for average coil primary current measurements. The meter range selected should easily allow reading resolutions of at least 0.1 A.

6.4 The same oscilloscope required in paragraph 6.2 should be used to measure primary induced voltage.

6.5 A good quality d-c voltmeter with an input resistance of at least 1000 Ω/V and with sufficient resolution to easily indicate differences of 0.1 V. To achieve this resolution the full scale deflection should be appropriate to the voltage rating of the ignition system being tested.

6.6 A distributor drive stand and attached tachometer which will have:
(a) An eccentricity between the mounting fixture and drive of 0.003 in. (0.076 mm) maximum.
(b) A continuously variable speed adjustment with a total speed variation between 15 and 3500 rpm possible.
(c) Speed stability within 5% at any given speed.
(d) A tachometer accurate within 3% of indicated speed and independent of the electrical portion of the ignition system.

7. Procedures

7.1 Group A Tests—The conventional circuit arrangement as shown in Fig. 1 with instrumentation in place, or modified with an auxiliary switching unit connected as the manufacturer intended, can be used to measure available voltage, peak primary coil current, average primary coil current, spark duration, spark voltage, and spark current at the distributor speeds and input voltages listed in Table 1.

The calculation described in paragraph 5.1.7 plus the procedure described here determines the inductive portion of the spark energy dissipated in a 12 kV spark gap under the conditions shown in Table 1. Spark gap currents and voltages can be measured and spark energy calculated equally well under other conditions and with different spark gaps. This procedure can be used in relating the effective amount of spark energy required to ignite a given fuel mixture.

If 6V ignition systems are to be tested, divide the primary voltages listed in Table 1 by two; for 24V systems, multiply by two.

Allow the ignition system to soak at least 1 h at the temperatures listed in Table 1 before beginning tests. Before any readings are recorded at any of the

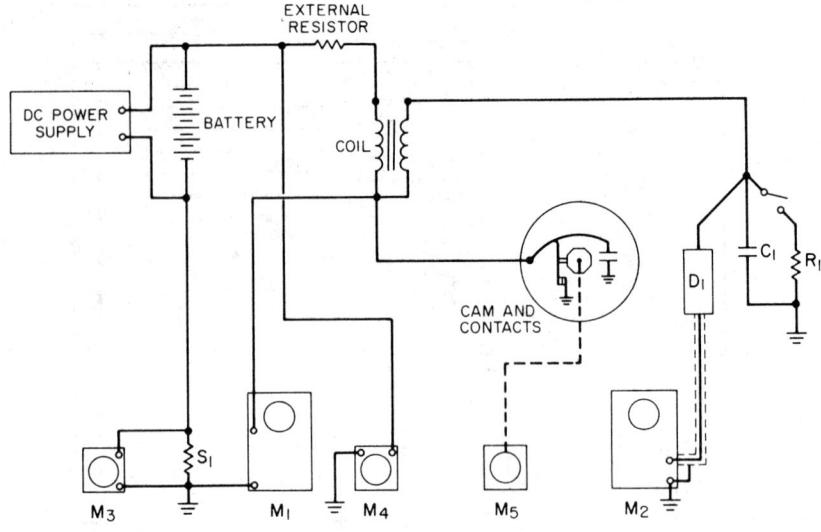

FIG. 3

- M_1 = PRIMARY VOLTAGE MEASURING OSCILLOSCOPE
- M_2 = SECONDARY VOLTAGE MEASURING OSCILLOSCOPE
- M_3 = DC AMMETER
- M_4 = DC VOLTMETER
- M_5 = TACHOMETER
- C_1 = SHUNT CAPACITY ($C_1 + D_1$ = 50 MMF)
- S_1 = 0.1 OHM METER SHUNT
- D_1 = VOLTAGE DIVIDER
- R_1 = 1.0 MEGOHM, 10 WATT

TABLE 1

Distributor rpm	Primary Volts	Environment Temperature °F	Environment Temperature °C	Operating Condition
20	5.0	−20 ±2	−29 ±1	Cold starting
30	5.0	−20 ±2	−29 ±1	Cold starting
40	5.0	−20 ±2	−29 ±1	Cold starting
50	11.0	80 ±5	27 ±3	Hot starting
60	11.0	80 ±5	27 ±3	Hot starting
70	11.0	80 ±5	27 ±3	Hot starting
250	14.0	80 ±5	27 ±3	Running
500	14.0	80 ±5	27 ±3	Running
750	14.0	80 ±5	27 ±3	Running
1000	14.0	80 ±5	27 ±3	Running
1250	14.0	80 ±5	27 ±3	Running
1500	14.0	80 ±5	27 ±3	Running
1750	14.0	80 ±5	27 ±3	Running
2000	14.0	80 ±5	27 ±3	Running
2250	14.0	80 ±5	27 ±3	Running
2500	14.0	80 ±5	27 ±3	Running
2750	14.0	80 ±5	27 ±3	Running
3000	14.0	80 ±5	27 ±3	Running

test points, the system should be allowed to come to a thermally stable operating condition (typically, this takes about 2 min).

Output voltage amplitudes vary due to contact arcing and other small but accumulative factors. It is recommended that the minimum peak amplitude be recorded. This represents the level which can be guaranteed by the system under test.

The voltage divider lead would have to be connected to a firing spark gap for spark duration measurements and this gap set carefully to fire at $12 \pm \frac{1}{2}$ kV. To secure firing voltages stability of this magnitude, special gaps and/or arrangements are usually required. Firing across a surface may help stability. Firing a gap under pressure using a dry inert gas and spherical electrodes also helps.

When environmental equipment is used to control ambient test temperatures, care must be taken that wire and/or cable lengths and, consequently, impedances do not affect test results.

During simulated starting tests, the system shall be operated under conditions simulating vehicle application; that is, if primary resistor in series with coil is normally bypassed during vehicle cranking, resistor should be bypassed for this portion of bench tests.

7.2 Group B Tests—The circuit arrangement shown in Fig. 3 is appropriate to measure the coil's primary induced voltage and secondary voltage. When the 1.0 MΩ resistor is connected, it is also appropriate to measure the risetime of the secondary voltage. The distributor and spark gaps are dispensed within these tests, as the waveform irregularities they introduce add nothing to the results and make stabilized patterns on the oscilloscope difficult to achieve. Oscillograph M_1 is used to measure primary induced voltage in this case. These measurements should be made at an ambient temperature of $80 \pm 5°F$ ($27 \pm 3°C$), a distributor speed of 1000 rpm, and a primary voltage of 14 V. Primary induced voltage test results are usually more meaningful if compared to secondary voltage values measured simultaneously. A satisfactory ratio of secondary voltage to primary induced voltage should be established by each group making these tests if they wish to insure that neither contacts nor semiconductors are overstressed.

Because risetime is measured between 10 and 90% of the peak voltage amplitude, it is usually easier to photograph the oscillograph waveform than to attempt to read this figure directly. Most manufacturers of oscilloscopes furnish compatible cameras for this purpose.

φSTORAGE BATTERIES—SAE J537j — SAE Standard

Report of Electrical Equipment Division approved January 1914 and completely revised by Electrical Equipment Committee June 1978.

1. Applications—This SAE Standard applies to lead-acid types of storage batteries used in motor vehicles, motorboats, tractors, and automotive industrial applications equipped with regulated charging systems.

2. Intent of Standard—This standard serves as a guide for testing procedures and as a publication providing a record of current production batteries, their ratings, and container description. Any battery with planned significant usage may be submitted for consideration. Any battery out of original equipment production for five years should be deleted.

2.1 The ratings submitted are to be based on the procedures shown in this standard. The ratings submitted must be of a level that when any subsequent significant sample is tested, in accordance with this standard, that at least 90% of the batteries will meet the ratings. The choice of 90% compliance recognizes that batteries consist of many plates and require chemical-electrical formation procedures and small variations in either can affect the performance of individual batteries.

2.2 The ratings and container description listed are provided as a record of current battery usage. It is recommended that any potential suppliers and users establish compliance agreements prior to any sample evaluation.

3. *Electrical Tests Testing Procedure*—Individual battery performance values are to be determined by the procedures outlined under Sampling, Conditioning, and Sequence of Tests. Battery classifications, dimensions, and ratings are given in the tables.

3.1 *Sampling*—Compliance determination samples shall be randomly selected from normal production. The age of the samples from date of manufacture shall be not less than 10 nor more than 60 days.

3.2 Conditioning

3.2.1 CHARGING—The battery shall be charged at a rate equal to $1/100$ of the 0°F (−17.8°C) Cold Cranking discharge rate until all cells are gassing freely and the charge voltage and specific gravity of electrolyte are constant over three successive readings taken at 1 h intervals.

3.2.2 ELECTROLYTE TEMPERATURE—During the period of charging the electrolyte temperature shall be maintained between 60 and 100°F (16 and 43°C).

3.2.3 SPECIFIC GRAVITY—The fully charged initial specific gravity at the end of the first conditioning charge shall be adjusted to 1.265 ± 0.005 corrected to 80°F (26.7°C). For dry charged batteries, specific gravity at the end of the first conditioning charge shall be 1.265 ± 0.005.

3.2.4 SEQUENCE OF TESTS

1. Charge battery according to methods given under Conditioning, and repeat this before each discharge.
2. Perform Charge Rate Acceptance Test.

TABLE 1—BATTERY CLASSIFICATIONS, RATINGS AND DIMENSIONS

6 Volt		Ref. No.	Electrical Values Cold Cranking Test		Reserve Capacity Min	Maximum Overall Dimensions						Over Charge Life Units
SAE No.	Assembly Fig. No.		at 0°F (−17.8°C) A	at −20°F (−28.9°C) A		Length		Width		Height		
						mm	in	mm	in	mm	in	
6 Volt Batteries												
1-475	2	1M1	475	380	159	231	9.13	181	7.13	231	9.13	4
1-545	2	1M1A	545	460	185	231	9.13	181	7.13	231	9.13	4
2-520	2	1M2	520	410	192	263	10.38	181	7.13	231	9.13	4
2-560	2	1H2	560	480	220	263	10.38	181	7.13	238	9.38	7
2-650	2	1H2A	650	545	245	263	10.38	181	7.13	238	9.38	7
2-775	2	1H2B	775	610	295	263	10.38	181	7.13	238	9.38	7
2E-595	5	8H2A	595	510	210	492	19.38	104	4.13	238	9.38	5
3EH-830	5	8T2	830	675	340	492	19.38	110	4.34	248	9.77	10
4EH-880	5	ST3	880	700	420	492	19.38	127	5.00	248	9.77	13
2N-495	1	4M2	495	420	170	254	10.00	141	5.57	228	9.00	4
4-700	2	1H4	700	570	275	333	13.13	181	7.13	238	9.35	9
4-720	2	1H4A	720	590	280	333	13.13	181	7.13	238	9.38	9
4-860	2	1T4	860	750	380	330	13.03	178	7.04	241	9.52	18
5D-800	2	2H5	800	675	340	349	13.75	181	7.13	238	9.39	10
7D-900	2	6T3A	900	650	430	428	16.88	193	7.63	276	10.88	13
7DS-900	2		900	650	430	405	15.94	193	7.63	276	10.88	13

TABLE 1—BATTERY CLASSIFICATIONS, RATINGS AND DIMENSIONS

12 Volt		Ref. No.	Electrical Values Cold Cranking Test		Reserve Capacity Min	Maximum Overall Dimensions						Over Charge Life Units
SAE No.	Assembly Fig. No.		at 0°F (−17.8°C) A	at −20°F (−28.9°C) A		Length		Width		Height		
						mm	in	mm	in	mm	in	
12 Volt Batteries												
3EE-290	9	13M2	290	230	85	490	19.32	110	4.35	225	8.88	7
3ET-425	9-C	13TC2	425	340	120	490	19.32	110	4.35	249	9.82	9
20H-235	10-C	9HCO	235	170	45	198	7.82	173	6.82	238	9.38	4
21-325	10-J		325	250	68	208	8.19	173	6.81	222	8.77	7
22R-290	11	10MO	290	215	72	228	8.99	173	6.84	227	8.97	5
22R-290/1	11-L-J		290	210	65	227	8.96ª	174	6.86	214	8.44	8
22R-350	11-L-J		350	270	88	227	8.96ª	174	6.86	214	8.44	10
24R-440	11-L-J		440	320	120	260	10.35ª	174	6.86	227	8.94	14
24R-455	11-J		455	340	135	261	10.27	173	6.81	228	8.97	15
22NF-245	11-F	18M1	245	185	52	239	9.44	139	5.50	226	8.91	5
22F-260	11-F-J	17MJ1D	260	190	50	241	9.50	173	6.82	214	8.46	5
22F-305	11-F-J	17MJ1C	305	210	75	241	9.50	172	6.79	207	8.17	5
24-255	10	9M3A	255	190	60	260	10.25	173	6.82	225	8.88	5
24-285	10-J-C	9M3C	285	220	75	260	10.25	173	6.82	225	8.88	5
24-305	10	9M3B	305	210	75	260	10.25	173	6.82	225	8.88	5
24-305/1	10	9M3F	305	230	68	260	10.25	173	6.82	225	8.88	5
24H-365	10	9H3A	365	280	98	260	10.25	171	6.75	238	9.38	7
24-375	10-C	9MC3A	375	300	86	260	10.25	171	6.82	225	8.88	7
24T-380	10-J	9TJ3	380	290	113	260	10.25	173	6.82	247	9.75	9
24-385/1	10-J	9MJ3D	385	280	95	260	10.25	172	6.79	220	8.67	7
24-385/2	10-J	9MJ3K	385	305	110	260	10.25	172	6.79	222	8.76	8
24-410	10-J	9MJ3G	410	310	110	260	10.25	173	6.82	226	8.90	8
24R-350	11	10M3	350	280	99	261	10.30	173	6.84	227	8.97	6
27-360	10-J	9MJ6A	360	280	110	306	12.04	173	6.82	225	8.88	9
27-440	10-C	9MC6	440	350	102	304	12.00	171	6.75	222	8.75	9

ª Add 0.5 in for Lifting Ledges.

TABLE 1—BATTERY CLASSIFICATIONS, RATINGS AND DIMENSIONS

12 Volt SAE No.	Assembly Fig. No.	Ref. No.	Electrical Values Cold Cranking Test at 0°F (-17.8°C) A	at -20°F (-28.9°C) A	Reserve Capacity Min	Length mm	Length in	Width mm	Width in	Height mm	Height in	Over Charge Life Units
27H-435	10	9H5	435	340	125	297	11.72	173	6.82	238	9.38	9
27-500	10		500	400	140	306	12.06	173	6.81	222	8.75	10
27-620	10		620	496	162	305	12.00	173	6.81	223	8.75	—
27R-430	11	10M7	430	320	125	305	12.01	173	6.81	227	8.95	9
27R-455	11	10H7	455	355	136	305	12.01	173	6.81	232	9.15	14
27HF-425	11-F	17H3A	425	320	136	317	12.50	173	6.82	232	9.15	15
27HF-435	11-F	17H3	435	340	125	317	12.50	173	6.82	232	9.15	9
29NF-290	11-F	18M3	290	235	80	330	13.00	141	5.56	228	9.00	4
30H-460	10	9H9	460	330	158	342	13.50	173	6.82	238	9.38	13
30H-580	10	9H9A	580	480	175	342	13.50	172	6.81	233	9.21	13
31-475	18		475	375	130	333	13.13	173	6.80	239	9.41	—
32N-350	11	11M6	350	280	115	361	14.25	139	5.50	226	8.91	9
53-210	14	14M2A	210	155	40	331	13.07	121	4.79	211	8.32	4
60-360	12	15M4A	360	280	110	331	13.07	159	6.27	225	8.88	9
71-275	17-J		275	210	60	208	8.19	179	7.05	216	8.51	—
71-350	17-J		350	270	80	208	8.19	179	7.05	216	8.51	—
72-275/1	17-J	22MJ1A	275	210	60	231	9.10	184	7.27	222	8.77	5
73-430	17-J		430	330	100	231	9.10	179	7.05	216	8.51	—
74-335	17-J	22MJ2A	335	270	98	260	10.25	184	7.27	222	8.77	8
74-410	17-J	22MJ2B	410	310	110	260	10.25	184	7.27	222	8.77	8
74-455	17-J	22MJ2C	455	360	140	260	10.25	184	7.27	222	8.77	9
74-465	17-J		465	375	125	261	10.28	179	7.05	216	8.51	—
U1-160	10	23LO	160	110	23	198	7.80	133	5.25	187	7.38	3
U1-200	10	23LOA	200	150	32	198	7.80	133	5.25	187	7.38	3

TABLE 1—BATTERY CLASSIFICATIONS, RATINGS AND DIMENSIONS

12 Volt SAE No.	Assembly Fig. No.	Ref. No.	Electrical Values Cold Cranking Test at 0°F (-17.8°C) A	at -20°F (-28.9°C) A	Reserve Capacity Min	Length mm	Length in	Width mm	Width in	Height mm	Height in	Over Charge Life Units
4D-640	8	20T4A	640	450	285	539[b]	21.25[b]	222	8.75	276	10.88	9
4D-800	8	20T4B	800	640	310	539[b]	21.25[b]	222	8.75	276	10.88	9
8D-900[a]	8	20T8A	900	650	430	539[b]	21.25[b]	282	11.13	276	10.88	13
		21T1 21T2 1H2C 1H4B 9H9B	See SAE J930a									

[a] Ratings for batteries recommended for motorcoach and bus service are for double insulation. When double insulation is used in other types, deduct 15% from the rating values for cold cranking.
[b] Dimensions over handles.

3. Conduct the Reserve Capacity Test at 25 A and 80°F (26.7°C).
4. Conduct the Cold Cranking Test at 0°F (−17.8°C) at the discharge rate specified in Table 1 for 0°F (−17.8°C) test.
5. Repeat step 3.[1]
6. Conduct the Cold Cranking Test at −20°F (−28.9°C) and at the discharge rate specified in Table 1 for −20°F (−28.9°C) test.
7. Repeat step 3.[1]
8. Perform life tests.

3.2.5 New batteries may require extra conditioning, not afforded by test event 1, in determining its true reserve capacity, therefore, the highest reserve capacity test value obtained for each battery in test events 3, 5, and 7 shall be used as the reserve capacity performance of that battery.

3.3 Activation Test of Dry Charged Battery—Dry charged batteries when activated according to manufacturer's instructions shall meet the minimum standards for wet batteries.

3.4 Charge Rate Acceptance Test—This test is to reveal the ability of a new, previously untested wet battery (or activated dry charged battery) to accept a charge under conditions existing in the voltage-regulated electrical system of a vehicle at 30°F (−1.1°C) with the battery in a partially discharged condition.

1. Charge battery according to method given under Conditioning.
2. Discharge battery at 25 ± 0.25 A from a starting electrolyte temperature of 80 ± 5°F (27 ± 3°C) for a time equal to 0.8 times the Reserve Capacity rating shown in Table 1.
3. Place battery in cold box until electrolyte of a center cell reaches 30 ± 2°F (−1.1 ± 1.1°C).
4. With battery in 30 ± 2°F (−1.1 ± 1.1°C) ambient, charge at a constant potential equivalent to 2.4 V per cell and record charge rate in amperes at 10 min. This rate in amperes shall be taken as the Charge Rate Acceptance.

NOTE: Minimum requirement in Table 1 is 2% of the 0°F (−17.8°C) Cold Cranking Rate.

3.5 Reserve Capacity Rating—The fully charged battery at a temperature of 80 ± 5°F (27 ± 3°C) is discharged at 25 ± 0.25 A to a terminal voltage equivalent to 1.75 V per cell measured under load.[2] The Reserve

---

[1] To determine compliance with ratings, steps 5 and 7 are not required if the Reserve Capacity Rate is made in step 3.

[2] Record temperature of a center cell at the time the end voltage equivalent to 1.75 V per cell is reached.

Capacity is defined as the time of discharge in minutes. All results shall be corrected to 80°F (27°C) standard. This shall be accomplished by maintaining the battery electrolyte temperature at the end of discharge at 80 ± 1°F (27 ± 0.6°C) or by applying the following correction factor:

$$M_c = M_r[1 - 0.01(T_f - 80)]$$

where: M_c = corrected minutes
M_r = minutes run
T_f = temperature at end of discharge, °F[3]
0.01 = temperature correction factor

3.6 Cold Cranking Test (30 s Test)—The following test is a measure of the cranking capability of a battery at the rating temperature.[4]

1. Place the battery in an air ambient held at the rating temperature[4] ±2°F (±1.1°C) until the temperature of a center cell reaches the rating temperature ±1°F (±0.6°C).

2. With the battery in an air ambient at the rating temperature, discharge the battery at the rating current[5] shown in Table 1 for 30 s. The rating current shall be held constant ±1% throughout the discharge. Measure battery terminal voltage under load at the end of 30 s.

3. The acceptance criterion for this test is that the battery terminal voltage at the end of 30 s shall be equivalent to 1.2 V per cell or greater.

3.7 Cold Cranking (90 s Test for CIM Starting)—The following test is a measure of the cranking capability of a battery at the rating temperature.[4]

1. Place the battery in an air ambient held at the rating temperature[4] ±2°F (±1.1°C) until the temperature of a center cell reaches the rating temperature ±1°F (±0.6°C).

2. With the battery in an air ambient at the rating temperature, discharge the battery at the rating current[5] shown in Table 1 for 90 s. The rating current shall be held constant ±1% throughout the discharge. Measure battery terminal voltage under load at the end of 90 s.

3. The acceptance criterion for this test is that the battery terminal voltage at the end of 90 s shall be equivalent to 1.0 V per cell or greater.

3.8 Life Test—See SAE J240a (January, 1972).

3.9 Overcharge Life Test[6]—This test was designed to provide more overcharge than that covered in the Life Test. A charge of 495 A·h is given, followed by a discharge capacity check, thereby representing one weeks testing on a continuous basis or one complete life unit. The procedure for a life unit is as follows:

1. 495 A·h (1782 kC) of continuous charge is given at 4.5 A, with batteries in a water bath as specified under Test Equipment and Specifications.

2. The battery is given a 48 h stand on open circuit in the water bath.

3. A 150 A discharge check rate is given at the battery temperature obtained in step 2 to an end voltage equivalent to 1.20 V per cell or a minimum discharge time of 30 s whichever occurs first.

4. Repeat steps 1–3 without separate discharge.

5. The overcharge life test shall be considered completed when the battery fails to meet step 3. The battery shall meet the minimum life unit (weeks) specified in Table 1. The number of life units (weeks) on test shall be computed to the last week in which the minimum time of 30 s has been equaled or exceeded; that is, the week in which the battery failed shall not be counted in reporting the life units.

6. Water shall be added daily during charge to restore electrolyte level to normal. See Note A Section 4.6.

3.10 Vibration Test—This vibration test is to determine the ability of a battery to withstand vibration forces without suffering mechanical damage, loss of capacity or electrolyte or without developing internal or external leaks.

3.10.1 PROCEDURE

1. Four batteries shall be selected as a sample for this test.

2. Bring batteries to full charge and pressure test each cell within 2 psi (14 kPa) air pressure for 5 s. Maximum pressure loss shall be 0.1 psi (0.68 kPa) on a closed system which has the pressure supply blocked off.

3. Place one or more of the batteries at 80 ± 5°F (27 ± 3°C) on vibration machines recommended for this test. The battery plates shall be oriented parallel to the axis of the rotating shaft of the machine.

4. The batteries shall be firmly held down by a hold-down bearing on the battery in similar fashion as to that encountered in vehicle applications.

5. The electrolyte shall be at the level recommended by the manufacturer.

6. The batteries shall be vibrated for a minimum of 2 h at an acceleration of 5 g (49 m/s^2) and a frequency of 30–35 Hz. Each 2 h of vibration shall represent one unit of vibration.

---

[3] Test is not valid if T_f is outside temperature of 70–90°F (21–32°C).
[4] Rating temperature for purpose of this test is either 0 or −20°F (−17.8 or −28.9°C) as shown in Table 1.
[5] When a battery is built with double insulation for heavy duty service, see footnote c of Table 1.
[6] For 12 V batteries rated at more than 180 min Reserve Capacity and all 6 V batteries use two times the ampere rates and two times the ampere-hour (kC) values specified for the Overcharge Life Test.

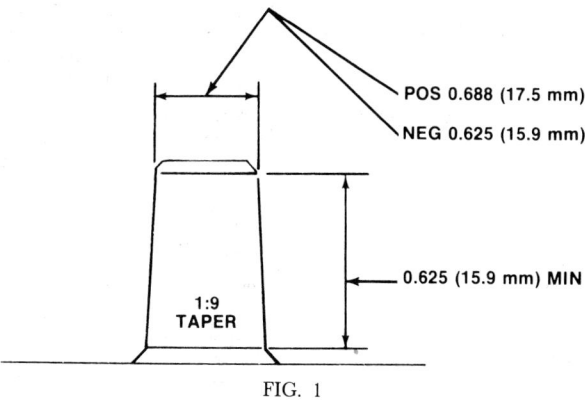

FIG. 1

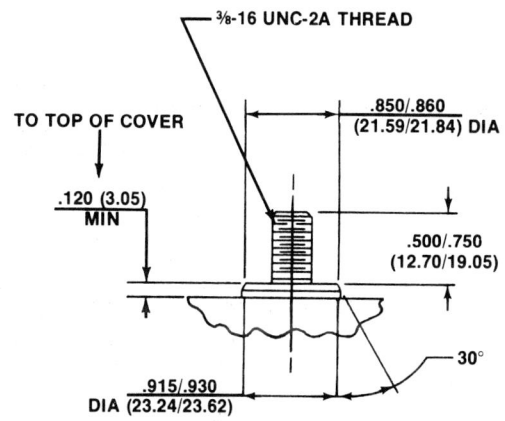

NOTE: DIMENSIONS ARE IN (mm)

CAUTION: Stud length, cable eyelet thickness and terminal nut must be compatible to insure reliable connections. Consult battery supplier for specific stud length.

FIG. 2

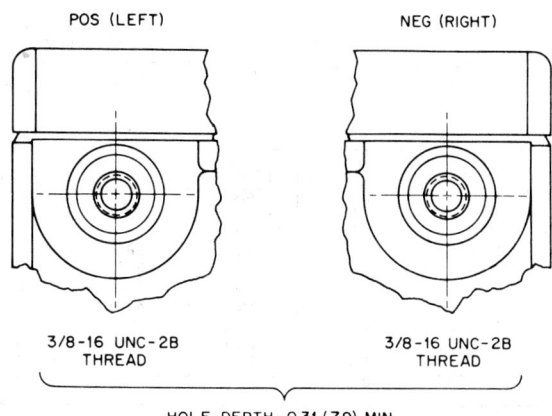

FIG. 3

TABLE 1A—BATTERY CLASSIFICATIONS, RATINGS AND DIMENSIONS FOR NEW BATTERIES

Temporary SAE No.	Volts	Assembly Fig. No.	Ref. Application	Recommended Electrical Values Cold Cranking Test		Reserve Capacity Min	Maximum Overall Dimensions						O. C. Unit
				at 0°F (-17.8°C) A	at -20°F (-28.9°C) A		Length		Width		Height		
							mm	in	mm	in	mm	in	
T54-310	12	19-K	Ford	310	220	60	186	7.34	154	6.04	212	8.36	—
T55-380	12	19-K	Ford	380	275	75	218	8.60	154	6.04	212	8.36	—
T56-450	12	19-K	Ford	450	330	90	254	10.02	154	6.04	212	8.36	—
TFG10-600	12	10	Prestolite	600	455	170	333	13.10	180	7.12	248	9.75	—
TFG18-600	12	18	Prestolite	600	455	170	333	13.10	180	7.12	249	9.79	—
TFG10-625	12	10	Prestolite	625	470	170	333	13.10	180	7.12	248	9.75	—
TFG18-625	12	18	Prestolite	625	470	170	333	13.10	180	7.12	249	9.79	—
T25-430	12	10-J	Chrysler	430	270	100	222	8.77	170	6.67	224	8.82	—
T24-440	12	10-J	Chrysler	440	350	102	260	10.25	173	6.81	222	8.77	—
T61-310	12	20-K	Ford	310	220	60	192	7.57	160	6.30	225	8.86	—
T62-380	12	20-K	Ford	380	275	75	225	8.87	160	6.30	225	8.86	—
T63-450	12	20-K	Ford	450	330	90	258	10.14	160	6.30	225	8.86	—
T64-475	12	20-K	Ford	475	355	120	296	11.64	160	6.30	225	8.86	—

TABLE 1B—BATTERIES FOR DELETION AFTER FIVE YEARS

SAE No.	Last Used	Assembly Fig. No.	Ref. No.	Electrical Values Cold Cranking Test		Reserve Capacity Min	Maximum Overall Dimensions						Over Charge Life Units
				at 0°F (-17.8°C) A	at -20°F (-28.9°C) A		Length		Width		Height		
							mm	in	mm	in	mm	in	
3-650	1975	2	1H3	650	545	245	298	11.75	181	7.13	238	9.38	7
3-550	1975	2	1M3	550	440	222	298	11.75	181	7.13	231	9.13	7
4-600	1975	2	1M4	600	470	256	333	13.13	181	7.13	231	9.13	9
2E-475	1975	5	8H2	475	400	160	492	19.38	104	4.13	238	9.38	5
22F-275	1975	11-F-J	17MJ1B	275	210	60	241	9.50	173	6.82	214	8.46	5
24-240	1975	10	9M3	240	170	52	260	10.25	173	6.82	225	8.88	5
24H-300	1975	10	9H3	300	225	70	260	10.25	173	6.82	238	9.38	5
24-315	1974	10	9MC3B	315	235	81	260	10.25	171	6.75	225	8.88	6
24-335	1975	10-J	9MJ3F	335	270	98	260	10.25	173	6.82	225	8.88	8
24-385	1975	10	9M3D	385	280	95	260	10.25	173	6.82	225	8.88	7
24H-410	1975	10	9H3B	410	300	110	260	10.25	173	6.82	238	9.38	8
24-430	1975	10	9M3E	430	340	105	260	10.25	173	6.82	225	8.88	8
27-410	1975	10-J	9MJ6	410	310	140	306	12.04	173	6.82	225	8.88	9
27H-430	1975	10	9H6	430	310	130	305	12.03	173	6.82	238	9.40	11
27-455	1975	10-J	9MJ6B	455	360	140	306	12.04	173	6.82	225	8.88	9

TABLE 1B—BATTERIES FOR DELETION AFTER FIVE YEARS

SAE No.	Last Used	Assembly Fig. No.	Ref. No.	Electrical Values Cold Cranking Test		Reserve Capacity Min	Maximum Overall Dimensions						Over Charge Life Units
				at 0°F (-17.8°C) A	at -20°F (-28.9°C) A		Length		Width		Height		
							mm	in	mm	in	mm	in	
29NF-330	1975	11-F	18M3A	330	280	100	330	13.00	139	5.50	228	9.00	8
60-310	1975	12	15M4	310	250	83	331	13.07	159	6.27	225	8.88	7
60-330	1975	12	15M4B	330	250	90	331	13.07	159	6.27	225	8.88	6
72-275	1975	17-J	22MJ1	275	210	60	231	9.10	184	7.27	210	8.27	5
74-285	1975	17-J	22MJ2	285	220	75	260	10.25	184	7.27	222	8.77	5
77-360	1973	17-J	22MJ3	360	280	110	305	12.04	184	7.27	222	8.77	9
77-410	1975	17-J	22MJ3A	410	310	140	305	12.04	184	7.27	222	8.77	9
77-455	1975	17-J	22MJ3B	455	360	142	305	12.04	184	7.27	222	8.77	9
U1R-160	1975	11	24LO	160	110	23	198	7.80	133	5.25	187	7.38	3
U1R-200	1975	11	24LOA	200	150	32	198	7.80	133	5.25	187	7.38	3
4B-425	1975	8	20T8	425	340	250	539[b]	21.25[b]	282	11.13	276	10.88	15
6D-800	1975	8	20T6A	800	555	350	539[b]	21.25[b]	254	10.00	276	10.88	10

[a] Ratings for batteries recommended for motorcoach and bus service are for double insulation. When double insulation is used in other types, deduct 15% from the rating values for cold cranking.
[b] Dimensions over handles.

TABLE 2—IDENTIFICATION SELECTION CHART

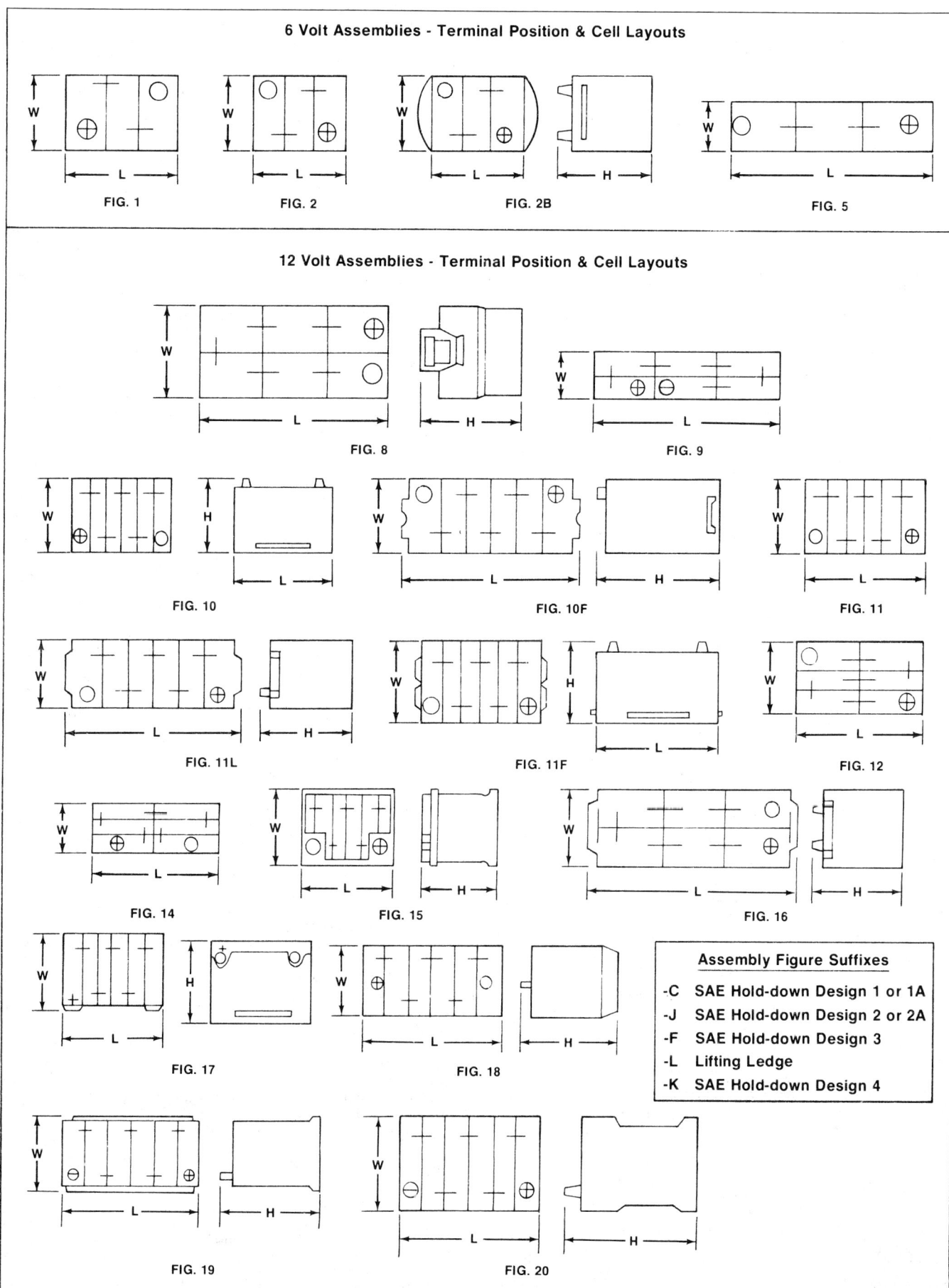

22.10

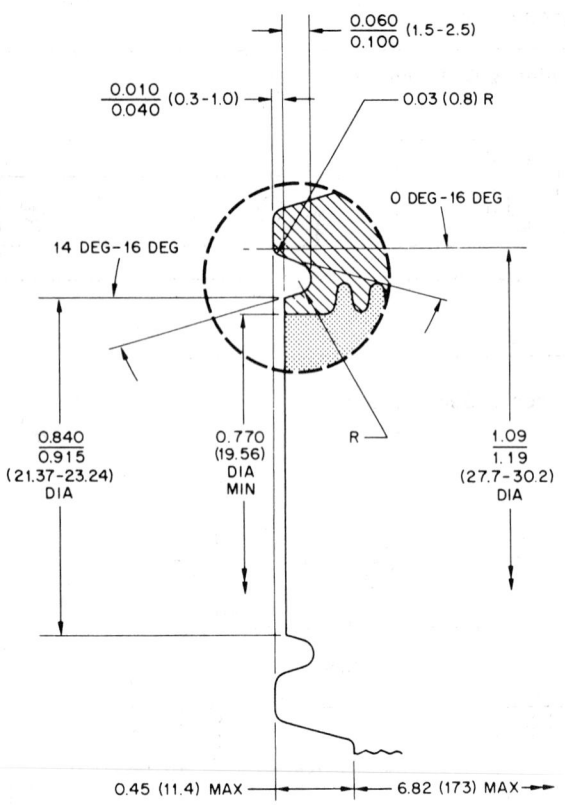

FIG. 3A—SIDE TERMINAL DIMENSIONS

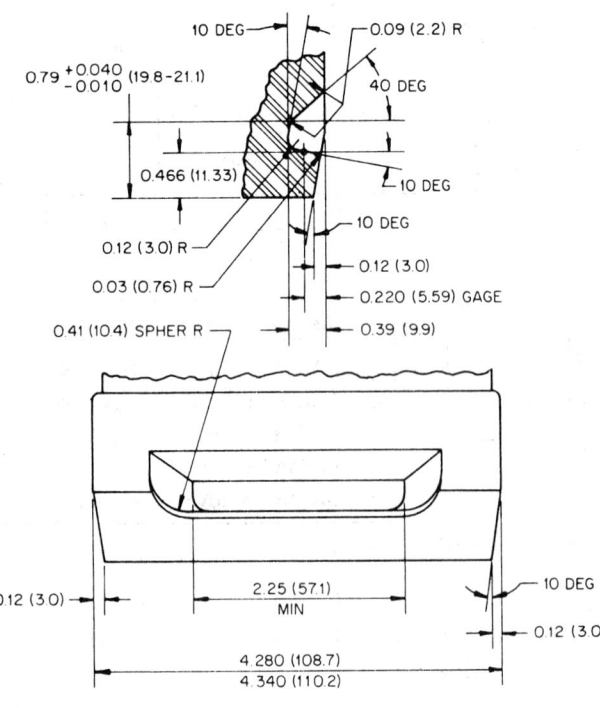

FIG. 4—HOLD-DOWN DESIGN 1—DESIGN "A" FOR BATTERIES APPROXIMATELY 4.35 IN WIDE WITH RECESSES IN ENDS

7. During vibration, there shall be no electrolyte losses.

8. After vibration, each cell must maintain 2 psi (14 kPa) of air pressure for 5 s with a maximum loss of 0.1 psi (0.68 kPa) of pressure as tested in item 2.

9. Immediately discharge the battery at 80 ± 5°F (27 ± 3°C) at the Cold Cranking rate for 0°F (−17.8°C) shown in Table 1. The 80°F (27°C) battery must meet the minimum voltage at 30 s.

10. The battery will be rated at the number of units it can survive while meeting the requirements of the preceding paragraph and on external and internal examination shall have no mechanical defects or leaks.

4. Test Equipment and Specifications

4.1 Tank—The tank shall be of sufficient size and design to permit a minimum of 1 in (25 mm) battery clearance on all sides including the bottom, and a battery immersion depth of 6 in (150 mm). Two battery strip supports are to be used. The width of the contact surface between the battery and each battery support shall be no greater than 1.5 in (40 mm). Means shall be provided to thermostatically control the water temperature at 100 ± 5°F (38 ± 3°C).

4.2 Vibration Machine—U. S. Army Ordnance battery vibration machine as shown on drawing No. D7070340, or L.A.B. Reliability vibration test machine type ARV-30x40-400, or exact equivalent.

4.3 Double Insulated Batteries—Double insulation is defined for the purpose of this SAE Standard as the use of a retaining sheet of porous or perforated material between the positive plate and the customary single separator.

4.4 Location of Battery Parts—The location and polarity of the terminal posts and the position of handles, when used, shall be as shown in Table 2.

4.5 Type Designations and Markings—Type letters, numbers, or symbols, which shall enable the user to determine ratings from the manufacturer's catalogs, shall be stamped or molded on the case or cell connectors or on a name plate permanently attached to the end or side.

4.6 Terminal Posts—Polarity shall be plainly marked as follows: The positive terminal shall be identified by Pos, P, or + on the terminal or on the cover near the terminal. See Figs. 1, 2, 3, and 3A.

NOTE A: This is a standard test, which due to its requirements, is more severe for smaller capacity batteries resulting in a reduced number of life cycles expected for those batteries.

5. Battery Identification Numbers
Procedure for selecting identification numbers for new batteries.

5.1 Select from the plan views[7] shown in Table 2, the cell and terminal arrangement of the new battery. This figure number, along with the appropriate suffix for hold-down design will become the Assembly Figure Number.

5.2 If overall battery dimensions clearly fall in an established group size,[8] the SAE number for Table 1A will have the symbol T followed by the group size followed by the 0°F (−17.8°C) Cold Cranking Rating. If any battery in Tables 1, 1A, or 1B has the same combination of group size and Cold Cranking Rating, add a slant line and Fig. 1, 2, etc. to the new battery as appropriate to designate the next similar battery.

5.3 If there is no fit to an established group size, the new temporary SAE number in Table 1A will consist of the symbol TFG followed by the assembly figure number from Table 2 followed by the 0°F Cold Cranking Rating. Should there be no appropriate assembly figure available, a new figure will be added to Table 2.

5.4 The Storage Battery Subcommittee will act on requests for additions or deletions to Tables 1, 1A, or 1B. All new batteries will appear first in Table 1A.

6. Battery Container Design for Bottom Hold-Down

6.1 Batteries which have either ledges or recesses for the hold-down shall be designed as shown in hold-down Designs 1, 1A, 2, 2A, 3, or 4. (Unless specified, all dimensions ±0.01 in (±0.3 mm), all angles ±1 deg, all radii ±0.03 in (±0.8 mm).) See Figs. 4–9.

6.2 Batteries which have recesses in the ends for the hold-down shall be of the designs shown in hold-down Designs 1 and 1A. Batteries having this feature will have the letter C added after the assembly figure number.

6.3 Batteries which have recesses in the sides for the hold-down shall be of the design shown in hold-down Designs 2, 2A, or 4. Those with Designs 2 and 2A will have the letter J and with Design 4 will have the letter K added after the assembly figure number.

6.4 Batteries which have ledges on the ends for the hold-down shall be of the design shown in hold-down Design 3. This design is used on Figs. 17 and 18 in Table 2.

---

[7] Cell and terminal plans serve for batteries with brand names on either front or back. For example: Fig. 9 in Table 2 can be turned 180 deg.

[8] Battery group size and assembly figure numbers are the designation established by Battery Council International (BCI) for reference in the BCI Replacement Battery Data Book. Except for temporary listings in Table 1A SAE battery numbers start with the BCI group size.

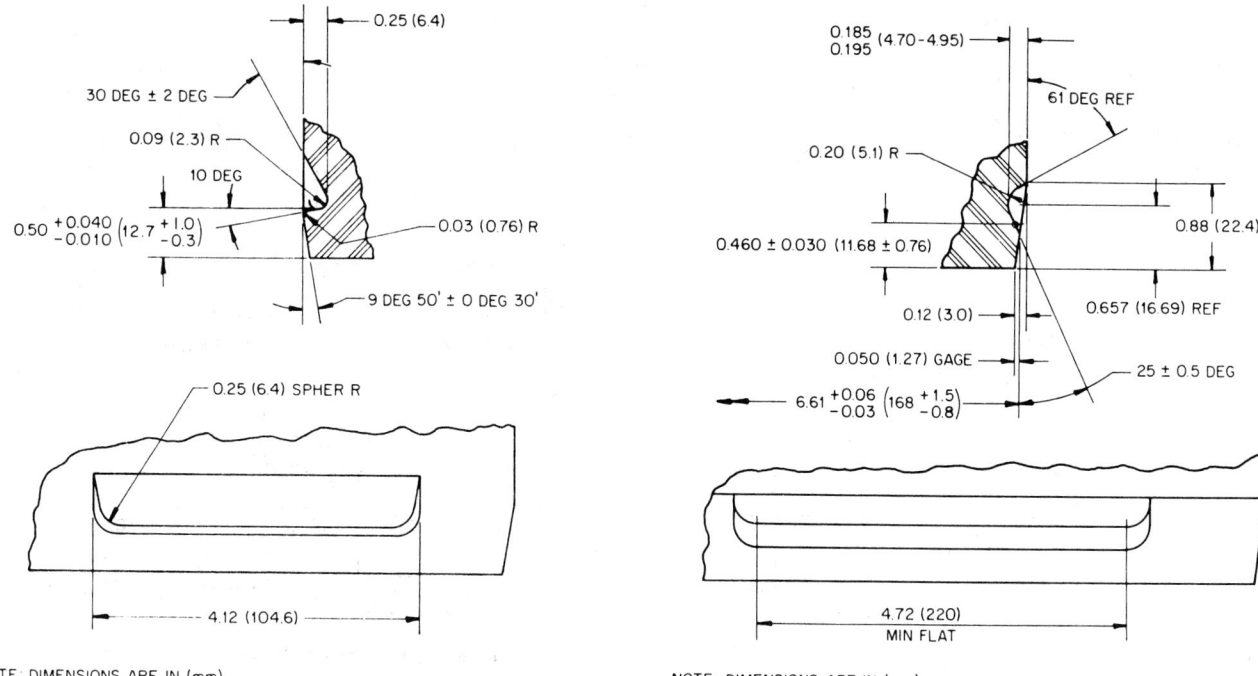

FIG. 5—HOLD-DOWN DESIGN 1A—DESIGN "B" FOR BATTERIES APPROXIMATELY 6.75 IN WIDE WITH RECESSES IN ENDS

FIG. 6—HOLD-DOWN DESIGN 2—DESIGN FOR BATTERIES WITH RECESSES IN SIDES FOR HOLD-DOWN

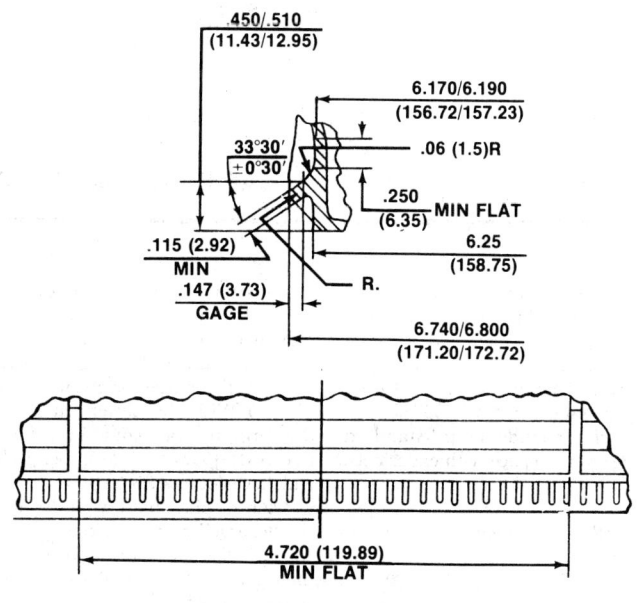

FIG. 7—HOLD-DOWN DESIGN 2A—DESIGN FOR BATTERIES WITH RECESSES IN SIDES FOR HOLD-DOWN

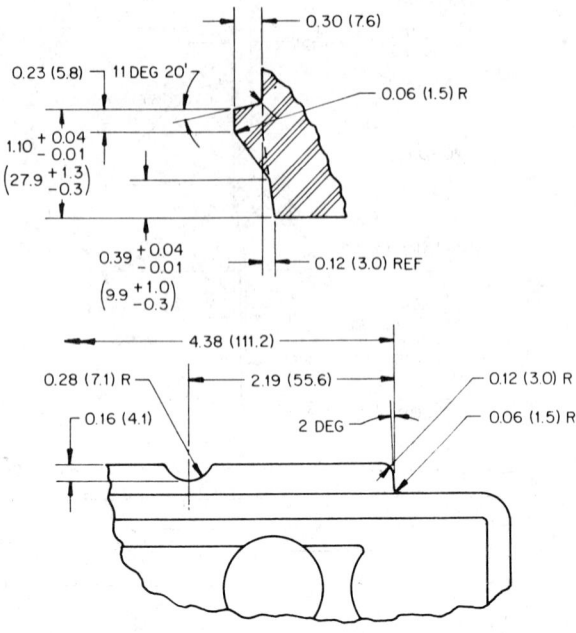

FIG. 8—HOLD-DOWN DESIGN 3—DESIGN FOR BATTERIES WITH LEDGES ON ENDS FOR HOLD-DOWN

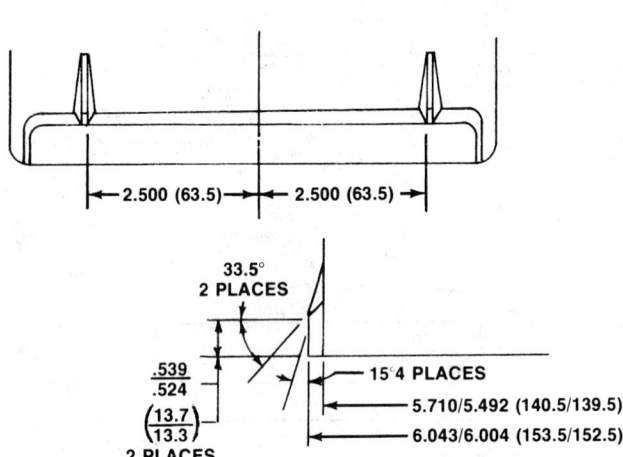

FIG. 9—HOLD-DOWN DESIGN 4—DESIGN FOR BATTERIES WITH RECESSES IN SIDES FOR HOLD-DOWN

LIFE TEST FOR AUTOMOTIVE STORAGE BATTERIES—SAE J240a

SAE Recommended Practice

Report of Electrical Equipment Committee approved May 1971 and last revised January 1972.

1. Scope—This SAE Recommended Practice applies to 12 V, automotive storage batteries of 180 min or less reserve capacity. This life test simulates automotive service where the battery operates in a voltage regulated charging system. It subjects the battery to charge and discharge cycles resulting in failure modes comparable to those encountered in automotive service. Other performance and dimensional information is contained in SAE J537.

This SAE Recommended Practice is intended as a guide toward standard practice, but may be subject to frequent change to keep pace with experience and technical advances.

2. Testing Procedure

2.1 The battery is tested in a water bath maintained at 105 ± 5 F (40.6 ± 2.8 C).

2.2 The test cycle is performed as follows:
Discharge: 2 min $\pm$ 1 s at 25 ± 0.05 A.
Charge:
   (a) Maximum voltage—14.8 ± 0.03 V.
   (b) Maximum rate—25 ± 0.05 A.
   (c) Time—10 min $\pm$ 3 s.

2.3 Battery is continuously cycled for 100 h (example: Monday noon until 4:00 p.m. Friday).

A switching delay of not more than 10 s is permitted from termination of charge to start of discharge and termination of discharge to start of charge.

2.4 The battery is given a 60 h stand on open circuit in the water bath.

2.5 With the battery at the temperature obtained in paragraph 2.4, discharge at a rate equal to its 0 F (-17.8 C) cold cranking rate in amperes (see SAE J537) to 1.20 V per cell, or a minimum discharge time of 30 s, whichever occurs first.

2.6 Replace battery on the life test without a separate recharge. Start on the "charge" portion of the cycle.

2.7 The life test shall be considered completed when the battery fails to maintain 1.2 V per cell for a minimum of 30 s on the manual discharge (paragraph 2.5).

The point of failure shall be determined by plotting the 30 s discharge voltage values.

2.8 Water should be added as required during the cycling portion of the test.

GROUNDING OF STORAGE BATTERIES—SAE J538a — SAE Standard

Report of Electrical Equipment Committee approved December 1955 and last revised June 1963. Reaffirmed without change January 1973.

The negative side of the storage battery shall be securely and adequately grounded so that the voltage drop to the starting motor is held within the limits established in SAE J541.

VOLTAGE DROP FOR STARTING MOTOR CIRCUITS—SAE J541a — SAE Recommended Practice

Report of Electrical Equipment Division approved January 1932 and last revised by Electrical Equipment Committee October 1967. Reaffirmed without change February 1975.

The starting motor circuits in motor vehicles, excluding motors, relays, and solenoids, shall be designed so that the difference between the voltage at the storage battery terminals and the starting motor terminals including connections shall not exceed those shown in Table 1. The voltage drop per hundred amps ($v_d/100_a$) is defined with a normal circuit temperature of 68 F.

TABLE 1—VOLTAGE DROP

System Voltage	($V_d/100_a$)	Use
6V	0.12V	Light and Medium Duty
12V	0.200V	Light and Medium Duty
24V	0.400V	Light and Medium Duty
12V	0.12V	Heavy Duty
24V—32	0.200V	Heavy Duty

VOLTAGES FOR DIESEL ELECTRICAL SYSTEMS—SAE J539a — SAE Recommended Practice

Report of Electrical Equipment Division approved January 1939 and last revised September 1976.

This SAE Recommended Practice is intended to apply to lamps, batteries, heaters, radios, and similar equipment for operation with mobile or automotive diesel engines. Twenty-four V systems have long been used for heavy duty services because 24 V permit operating 12 V systems in series-parallel. Thirty-two V systems have been used for marine, railroad-car lighting, and other uses.

Generators, storage batteries, starting motors, lighting, and auxiliary electrical equipment shall be for nominal system ratings of 12, 24, or 32 V as determined by the power requirements of the application. It is recommended that no intermediate voltages be considered except that a 30 V system may be used when cranking requirements permit and no lighting or auxiliary electrical equipment is involved.

The combination of a 24 V starting motor and two 12 V batteries connected in series for cranking is considered practical where it can be adapted to the installation. The batteries are reconnected in parallel for charging from a 12 V generator/alternator and for operating lights and other auxiliary equipment, or charged separately and used individually for lights and other electrical equipment.

STARTING MOTOR MOUNTINGS—SAE J542c

SAE Recommended Practice

Report of Electrical Equipment Division approved August 1917 and completely revised by Electrical Equipment Committee December 1978.

The purpose of this report is to provide standardized dimensions for mounting starting motors.

It is recommended that a full register diameter having a minimum depth of 0.100 in (2,54) be provided in the flywheel housing to insure proper control of gear center distance and clearance between pitch diameters. The clearance between the starting motor pilot diameter and the register diameter in the flywheel housing should be 0.001 in (0,03) minimum to 0.010 in (0,25) maximum.

Text noted with an asterisk in Figs. 1, 2, and 3, should not exceed root radius of pinion in order to provide clearance for the flywheel.

The face of the starting motor mounting flange should be relieved at its junction with the pilot diameter to avoid mounting interference with flywheel housing.

Dimensional units—inch (millimeter).

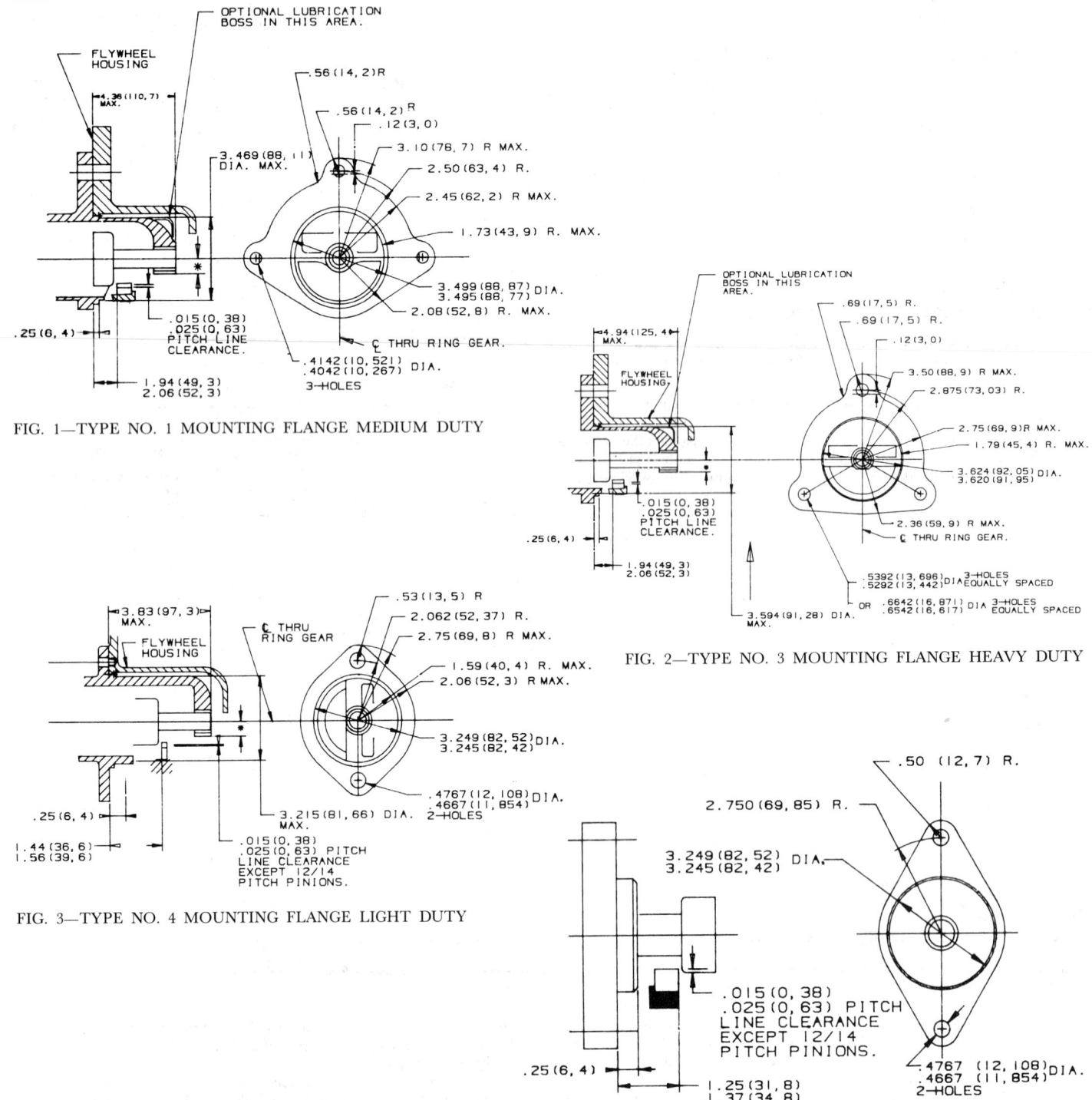

FIG. 1—TYPE NO. 1 MOUNTING FLANGE MEDIUM DUTY

FIG. 2—TYPE NO. 3 MOUNTING FLANGE HEAVY DUTY

FIG. 3—TYPE NO. 4 MOUNTING FLANGE LIGHT DUTY

FIG. 4—TYPE NO. 5 MOUNTING FLANGE LIGHT DUTY

NOTE: TYPE NO. 2 MOUNTING FLANGE SAME AS NO. 3 EXCEPT 0.4142 (10,521)/0.4042 (10,267) MOUNTING HOLES.

STARTING MOTOR PINIONS AND RING GEARS—SAE J543c   SAE Standard

Report of Electrical Equipment Division approved August 1917 and last revised by Electrical Equipment Committee April 1976.

The following table and illustrations are to be used as a guide in establishing starting motor pinions and ring gear designs. Consult the gear manufacturer for detail dimensions.

Ring Gear Design—Ring gears of 10/12 pitch and finer are normally not chamfered. Gears coarser than 10/12 pitch should be chamfered in accordance with Fig. 1.

Ring Gear and Pinion Installation—Backlash is necessary for free meshing and running of the pinion with the ring gear. Backlash may be obtained by increasing the center distance as shown in Fig. 2 or by reducing the tooth thickness.

Ring Gear Hardness—Hardness range for typical ring gears after assembly is:

8/10 pitch and coarser Rockwell C45-52
10/12 pitch and finer Rockwell C48-55

Center Distance—The formula for calculating center distance (C.D.) is:

$$C.D. = \frac{\text{No. Ring Gear Teeth (Blank)}^a + \text{No. Pinion Teeth (Blank)}^a}{2 \times \text{Diametral Pitch}^b} + \Delta C^c$$

where:

a = the number of teeth is equal to the number used to determine blank size. A blank is a disk or cylinder of such size as to relate to a standard gear of standard addendum, dedendum, and given number of teeth. To increase tooth strength and improve cranking ratio, many pinion gears are cut on an oversize blank (example: 10 teeth on 11 tooth blank). In this example, 11 would be used for the number of pinion teeth in calculating center distance.

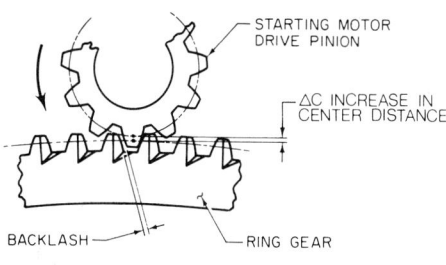

FIG. 2

b = for fractional diametral pitch (example: 8/10 pitch), use the numerator (8 in this example) for center distance calculation.

c = ΔC is the increase in center distance to obtain backlash. See Fig. 2. If backlash is obtained by reducing tooth thickness, omit ΔC from the C.D. formula.

Pinion and Gear Data		Pinion Data[a]			
Diametral Pitch[b]	Pressure Angle, deg	No. of Teeth	Maximum Root Dia,[c] in	Maximum Outside Dia, in	Maximum Circular Tooth Thickness at Theoretical Pitch Dia, in	Layout Pitch Dia, in

Diametral Pitch[b]	Pressure Angle, deg	No. of Teeth	Max Root Dia,[c] in	Max Outside Dia, in	Max Circular Tooth Thickness, in	Layout Pitch Dia, in
12/14	12	9	0.695	1.016	0.161	0.873
12	14½	9	0.698	1.017	0.167	0.874
12	14½	9	0.648	1.027	0.167	0.874
10	14½	9	0.812	1.200	0.194	1.000
10	14½	10	0.839	1.300	0.194	1.100
10/12	20	9	0.789	1.167	0.188	1.000
10/12	20	9	0.821	1.167	0.198	1.000
10/12	20	10	0.889	1.267	0.187	1.100
10/12	20	10	0.942	1.307	0.204	1.140
10/12	20	11	0.989	1.367	0.187	1.200
10/12	20	12	1.089	1.467	0.187	1.300
8/10	20	9	1.000	1.450	0.245	1.250
8/10	20	10	1.109	1.556	0.235	1.356
8/10	20	10	1.125	1.562	0.237	1.375
8/10	20	10	1.125	1.575	0.245	1.375
8/10	20	11	1.250	1.700	0.245	1.500
8/10	20	12	1.375	1.825	0.237	1.625
8/10	20	12	1.375	1.825	0.245	1.625
8/10	20	13	1.393	1.825[d]	0.196	1.625
8/10	20	13	1.525	1.950	0.244	1.750
6/8	20	11	1.521	2.083[d]	0.262	1.833
6/8	20	11	1.688	2.240	0.317	2.000
6/8	20	12	1.688	2.240[d]	0.262	2.000

[a] Dimensions are for maximum metal conditions. Tolerances will result in increased clearances.
[b] The two diametral pitch gear data are based on Fellows stub tooth system.
[c] If larger root diameter is desired, consult gear manufacturers.
[d] Standard blank. All others oversize blanks.

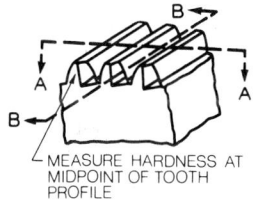

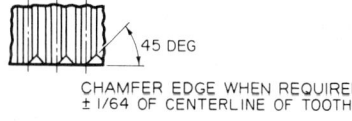

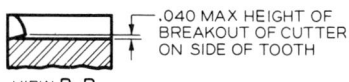

FIG. 1

GENERATOR MOUNTINGS—SAE J545b   SAE Standard

Report of Electrical Equipment Division approved August 1917 and last revised by Electrical Equipment Committee August 1968. Reaffirmed without change January 1970. Editorial change October 1971.

GENERATOR-ELECTRIC, D-C, MECHANICAL RECTIFICATION (COMMUTATOR TYPE)

Generator Ratings and Terminals—Consult the manufacturer for generator ratings and for terminal dimensions and location.[1]

Flange Type Generator—The envelope dimensions for flange type generators are shown in Table 1 and Fig. 1. Flange details A and B can be used with a gear drive. Flange A provides adjustment for a chain drive.

Hinge Mounted Type Generator—The envelope dimensions for hinge mounted type generators are shown in Table 2 and Fig. 2. If a ventilating fan is required, consult manufacturer for fan and pulley dimensions.

[1] For information on electrical charging systems for construction and industrial machinery, see SAE J180.

TABLE 1—TYPICAL FLANGE TYPE GENERATOR DIMENSIONS, IN.

Item	A, dia	B, dia	C	D	E	F	G	H	J	K
1	4.56	0.7495-0.7500	8.75	0.42	0.50	5.38	0.38	0.50	0.31	0.23
2	4.62	0.7495-0.7500	9.97	0.50	0.25	5.38	0.38	0.50	0.31	0.25
3	5.00	0.7495-0.7500	10.62	0.56	0.22	5.50	0.41	0.56	0.33	0.28
4	5.06	0.7495-0.7500	9.53	0.42	0.50	5.38	0.38	0.50	0.31	0.23
5	5.12	0.7495-0.7500	10.22	0.50	0.25	5.38	0.38	0.50	0.31	0.25
6	5.56	0.7495-0.7500	10.38	0.42	0.56	5.38	0.38	0.50	0.31	0.23
7	5.62	0.8740-0.8745	12.06	0.50	0.22	5.38	0.38	0.50	0.31	0.25

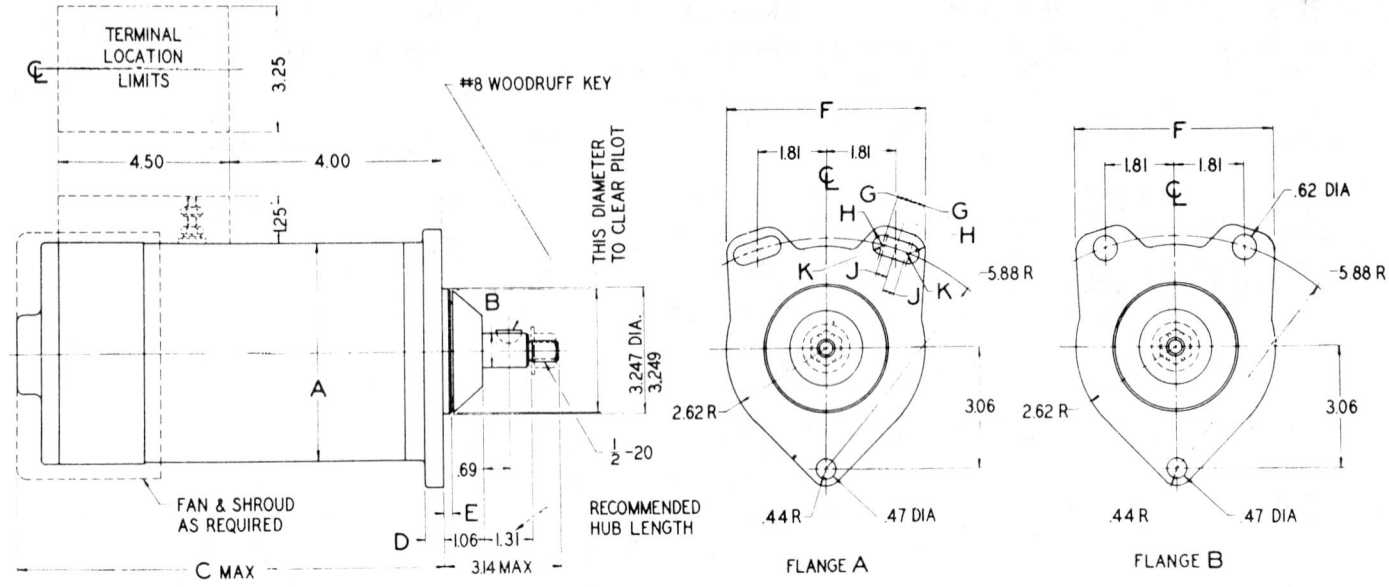

FIG. 1—FLANGE TYPE GENERATOR DIMENSIONS

TABLE 2—TYPICAL HINGE MOUNTED TYPE GENERATOR DIMENSIONS, IN.

Item	A	B	C	D	E	F	G	H	J, Thread
1	4.56	6.69	8.06	2.31	0.88	0.45	0.6690–0.6684	0.98	0.669-20
2	4.62	7.12	8.28	2.30	0.81	0.41	0.6690–0.6684	1.00	0.669-20
3	4.81	6.80	7.58	2.17	0.94	0.54	0.6693–0.6690	0.70	0.669-20
4	5.00	7.13	9.07	2.32	1.23	0.80	0.6691–0.6686	0.62	1/2-20
5	5.06	7.25	8.62	2.38	1.00	0.45	0.6690–0.6684	0.93	0.669-20
6	5.12	7.25	8.73	3.14	1.31	0.66	0.6690–0.6684	1.00	1/2-20
7	5.56	7.81	9.16	2.44	1.06	0.52	0.7500–0.7495	0.91	0.669-20
8	5.75	9.03	11.03	3.39	1.31	0.70	0.8745–0.8740	1.19	5/8-18

Item	K[a]	L	M	N	P, deg	R	S, Thread	T, deg	U	V
1	5	5.50	3.38	3.19	17	0.33	5/16-18	0	0.44	0.49
2	5	5.59	3.06	3.06	3	0.32	5/16-18	13	0.41	0.50
3	6	5.60	3.44	3.54	13	0.45	3/8-16	0	0.61	—
4	5	5.88	3.57	3.50	21	0.45	5/16-18	0	0.39	0.17
5	5	6.00	3.62	3.50	21	0.33	5/16-18	0	0.44	0.41
6	5	6.06	3.88	3.50	8	0.32	5/16-18	13	0.44	0.50
7	8	6.50	3.88	3.88	21	0.45	3/8-16	0	0.44	0.38
8	8	6.56	3.88	3.88	8	0.45	3/8-16	13	0.56	0.56

[a] Standard woodruff key.

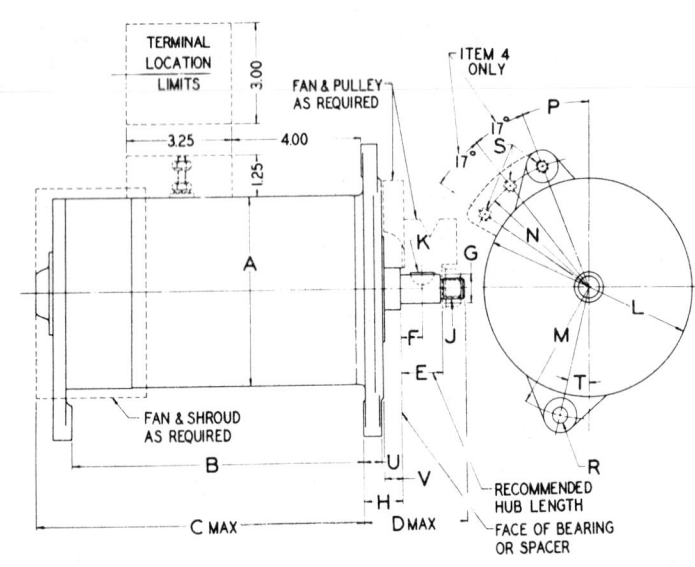

FIG. 2—HINGE MOUNTED TYPE GENERATOR DIMENSIONS

GENERATOR-ELECTRICAL, D-C, DIODE RECTIFICATION (ALTERNATOR TYPE)

Scope—This SAE Standard covers the two basic mounting configurations as shown in Figs. 3 and 4.

Generator Ratings, Terminals, and Clearances—Consult the manufacturer for ratings, terminal locations, and detail dimensions.[1] Dimensions shown do not provide for terminal clearance or possible body protrusions in local areas.

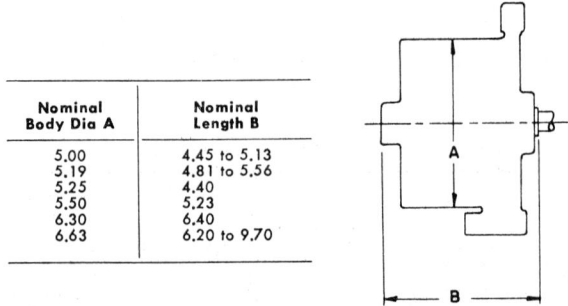

Nominal Body Dia A	Nominal Length B
5.00	4.45 to 5.13
5.19	4.81 to 5.56
5.25	4.40
5.50	5.23
6.30	6.40
6.63	6.20 to 9.70

FIG. 3—TYPE 1

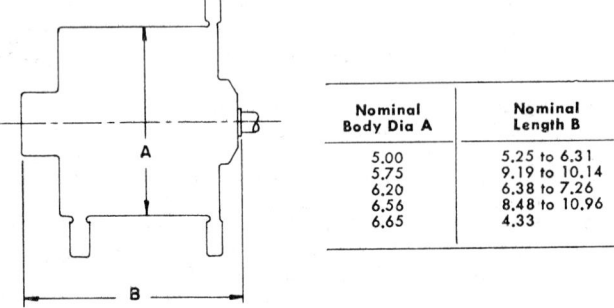

Nominal Body Dia A	Nominal Length B
5.00	5.25 to 6.31
5.75	9.19 to 10.14
6.20	6.38 to 7.26
6.56	8.48 to 10.96
6.65	4.33

FIG. 4—TYPE 2

STARTING MOTOR AND GENERATOR CURVES—SAE J544b

SAE Recommended Practice

Report of Electrical Equipment Division approved January 1945 and last revised by Electrical Equipment Committee July 1966. Editorial change February 1967. Reaffirmed without change January 1970.

A METHOD OF CORRECTING STARTING MOTOR SPEED FOR TEMPERATURES AND VOLTAGES OTHER THAN AT CALIBRATION

The following information may be obtained from a starting motor calibration curve at a given motor current (I) and ambient temperature (t_1):

1. Motor speed (N_1)
2. Motor terminal voltage (E_1)
3. Motor resistance (R_1)

Since the speed of a series cranking motor is directly proportional to the back electromotive force, the back emf (E_{b1}) necessary to produce the above conditions may be found from the following relationship:

4. $(E_{b1}) = (E_1) - (IR_1)$

The resistance of the cranking motor must be corrected for any ambient temperature other than that at the time of calibration. Therefore:

5. $(R_2) = (K)(R_1)$

where (K) is the resistance correction factor for an ambient temperature change from (t_1) to (t_2).

If the motor terminal voltage at (t_2) is (E_2), then the back emf (E_{b2}) at (t_2) is as follows:

6. $(E_{b2}) = (E_2) - (IR_2)$

The change in motor speed will be directly proportional to change in back emf. Therefore:

7. $(N_2) = \dfrac{(E_{b2})}{(E_{b1})}(N_1)$

where (N_2) is the motor speed at (E_2) and (t_2) for a given motor current (I).

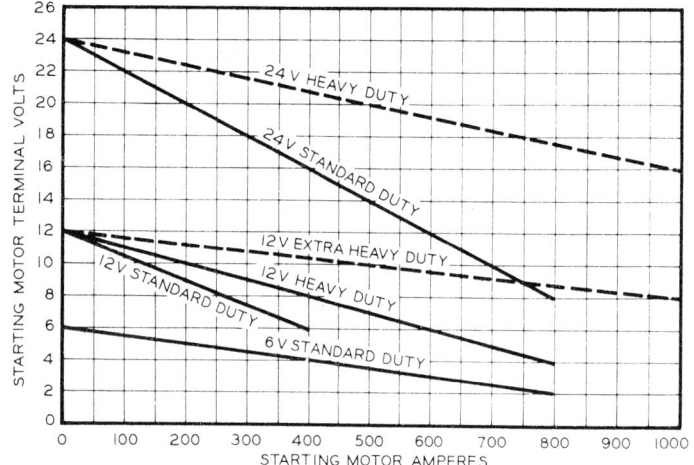

FIG. 1—TERMINAL VOLTAGE CURVES FOR STARTING MOTOR TESTS

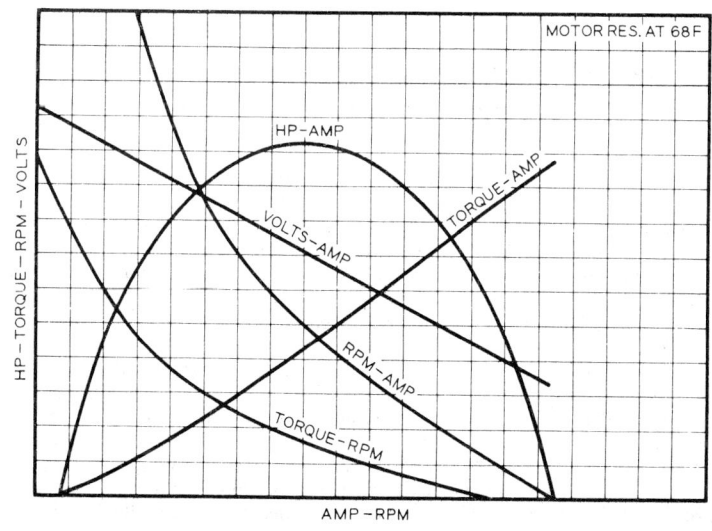

FIG. 2—STANDARD FORM FOR STARTING MOTOR CHARACTERISTIC CURVES

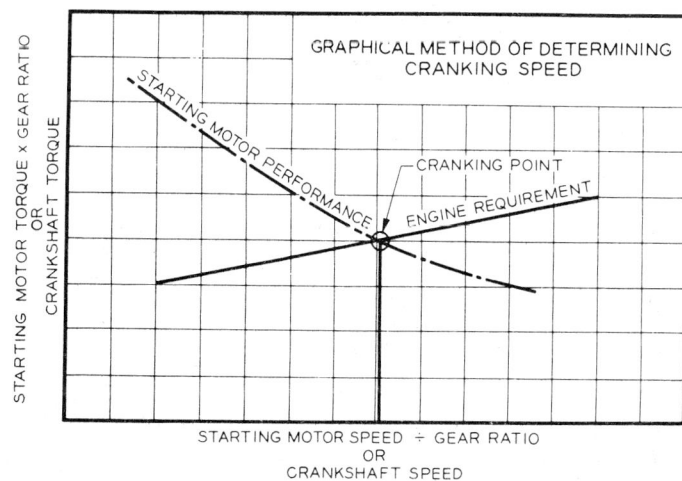

FIG. 3—METHOD OF DETERMINING CRANKING SPEEDS

GENERATOR-ELECTRIC, D-C, DIODE RECTIFICATION (ALTERNATOR TYPE)

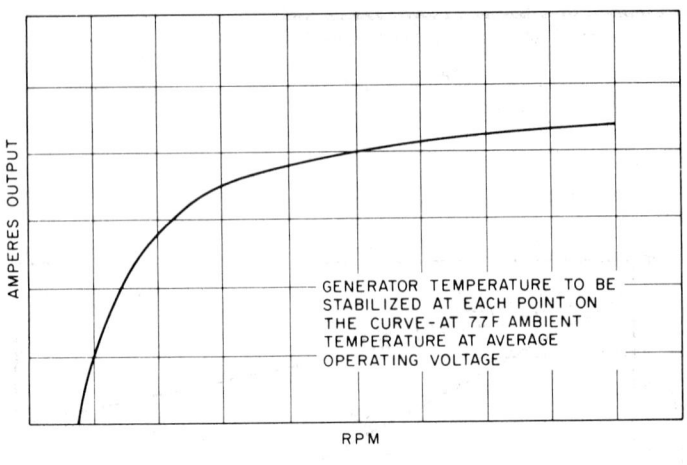

GENERATOR-ELECTRIC, D-C, MECHANICAL RECTIFICATION (COMMUTATOR TYPE)

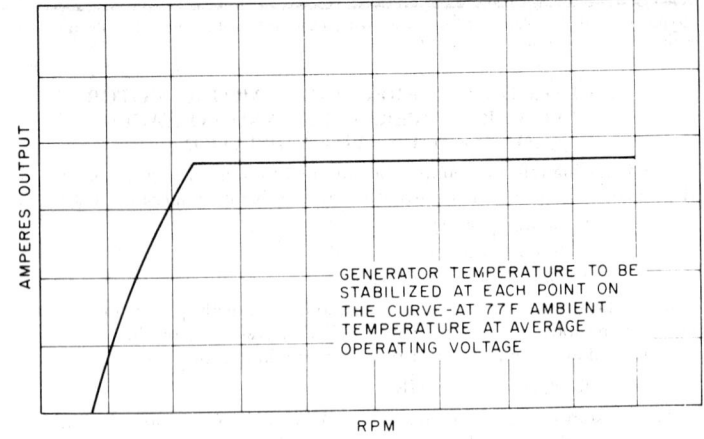

ELECTRICAL SYSTEMS WITH LEAD-ACID BATTERIES ARE BASED ON THE VOLTAGES SHOWN IN THE TABLE.

Cells	Basic System Voltage	Average Operating Voltage	Cells	Basic System Voltage	Average Operating Voltage
3	6	7.0	15	30	35.0
6	12	14.0	16	32	37.5
12	24	28.0			

AMPERE OUTPUT IS TOTAL OUTPUT LESS CURRENT REQUIRED FOR FIELD EXCITATION.
CONSULT THE MANUFACTURER FOR GENERATOR OUTPUT PERFORMANCE AND REGULATOR CHARACTERISTICS.

FIG. 4—STANDARD FORM FOR GENERATOR PERFORMANCE CURVES

LOW TEMPERATURE CRANKING LOAD REQUIREMENTS OF AN ENGINE—SAE J1253

SAE Recommended Practice

Report of Cranking Motor Subcommittee approved February 1979.

1. Purpose—The electrical cranking system components, which include the battery, cables, and cranking motor, must be carefully selected to provide the necessary speed to start an engine under the most severe climatic conditions for which the system is intended. Engine cranking loads increase with cold temperatures, therefore, the initial selection of these components needs to consider low temperature engine torque requirements. To insure an adequate electrical cranking system is obtained, it is important that proper test procedures are used for obtaining the cranking load requirements of the engine.

2. Procedure—The following test procedure is recommended for obtaining low temperature cranking torque requirements:

2.1 Engine Preparation

2.1.1 The engine to be tested should be equipped with all accessories that provide parasitic loads, such as power steering pump, automatic transmission, etc.

2.1.2 The engine, if new, should be run in to stabilize friction loads—equivalent to 1000 miles or 20 h at 2000 engine rpm.

2.1.3 The engine is winterized with anti-freeze solution for the temperature at which the test will be run.

2.14 The engine oil selected for the low temperature test should be representative of the high limit viscosity for the SAE grade recommended by the engine manufacturer for the operating temperature range. Sufficient oil of the same viscosity should be obtained for the complete test program so variations in test results can be minimized.

2.1.5 Fuel dilution of the engine oil will reduce its viscosity, therefore, to avoid this possibility, the cranking test is run without fuel in the carburetor, or with fuel system cut off.

2.1.6 To prepare the engine for test, the engine is warmed up and oil drained hot. This procedure should be repeated two times to assure complete change of oil when oil grade change is made. The oil filter is changed for the final fill. When the same grade of oil is used for other test temperatures and/or additional test days, the engine warm up procedure is repeated and only one drain is required.

2.1.7 Install a thermocouple in the center of the greatest mass of oil so soak temperatures can be monitored.

2.1.8 Equip engine with necessary instrumentation to provide cranking speed, battery voltage, cranking motor voltage, and current data. (The cranking speed can be determined from oscillographic current or voltage traces by calculation of the time span between the current or voltage peaks caused by the cylinder compression loads. The mean cranking speed is obtained over two consecutive revolutions. The mean torque is obtained by measuring the mean cranking current over the same period and calculated as described in paragraph 2.2.)

2.1.9 Prior to starting the cold soak period, warm up the engine for approximately 5-10 min to circulate oil, run carburetor bowl dry and disable ignition or cut off fuel system for diesels and adjust throttle plate to the idle position.

2.1.10 The engine with the calibrated motor is soaked at the test temperature until stabilization, which can be monitored by the oil thermocouple.

2.2 Cranking Motor Preparation—The cranking motor is used to measure the engine cranking torque. To minimize performance variances, a new cranking motor should be *run in* until the motor performance becomes stabilized prior to calibration which is determining the speed, torque, and current under load using a standard SAE terminal voltage curve (Ref. Fig. 1 of SAE

J544b, Starting Motor and Generator Curves (January, 1970)) unless otherwise specified.

After completion of the cranking load tests, a re-calibration curve should be run to verify initial performance.

NOTE: Since torque is proportional to cranking motor current, determination of engine torque can be calculated by obtaining the cranking motor running torque, corresponding to the cranking motor current measured at the test temperature from the performance characteristics of the calibrated cranking motor, and multiplying this value by the proper flywheel ring gear to cranking motor pinion gear ratio.

2.3 Cranking Load Tests

2.3.1 A sufficient number of cranking tests should be run to obtain a curve of average torque versus average engine speed over an approximate range of 30–120 engine rpm for gasoline engines, 50–150 rpm for direct injection diesel engines, and 120–220 rpm for small indirect injection diesel engines.

2.3.2 To obtain the range of speeds required to plot the torque curve, various battery capacities or a regulated DC power supply that can simulate desired battery conditions, are used to supply the appropriate cranking motor terminal voltages. The batteries are not required to be cold soaked but should be maintained at full charge.

2.3.3 The cranking time for each test should be approximately 10 s with readings between 5 and 10 s used as the plotting points. Allow a minimum of 30 min additional soak time before performing the next cranking test.

2.3.4 Using the test data, calibrated cranking motor performance characteristics and engine ring gear to cranking motor pinion gear ratio, calculate the engine torque requirements for each test speed and plot an engine torque requirement curve as shown in Fig. 3 of SAE J544b, Starting Motor and Generator Curves (January, 1970).

2.3.5 It should be noted that since gear efficiencies have been neglected, the torque measured is not true engine torque but that as seen by the cranking motor. However, it provides suitable design and application data for determining cranking motor requirements.

2.3.6 Once the engine torque requirement curve has been determined and the speed required to start the engine is known, cranking motor performance requirements for the engine application can be determined.

ELECTRICAL GENERATING SYSTEM (ALTERNATOR TYPE) PERFORMANCE CURVE AND TEST PROCEDURE—SAE J56

SAE Recommended Practice

Report of Electrical Equipment Committee approved September 1978.

1. Purpose—The purpose of this SAE Recommended Practice is to provide a standard test procedure for the development of alternator output performance, and a standard form for plotting performance curves.

2. Instrumentation

2.1 Voltmeter and ammeter should have an accuracy of ±0.5% of full scale.

2.2 Speed sensing device should have an accuracy of ±1%.

3. Test Procedure

3.1 Alternator test stand with suitable instrumentation is shown in Fig. 1.

3.2 The test conditions require that the alternator output voltage be maintained at the applicable test voltage level below, with the field current to be supplied by the alternator.

3.3 Output performance curve data is obtained by driving the alternator at room ambient, 24 ± 3°C (75.0 ± 5°F) at various discrete speeds, and adjusting the load to maintain constant output test voltage. Temperature stabilization must be attained at each discrete speed.

3.4 When the alternator is normally used with a built in or external solid state type regulator having inherent voltage drop, the performance test should be run with the regulator adjusted or re-connected for *full field current* condition, or an equivalent voltage drop should be connected in the field circuit if the test is performed without a regulator.

3.5 Deviations from the above procedure should be noted on the performance curve.

4. Curve—The output performance curve is made by plotting output current (amps) as the ordinate versus alternator speed (rpm) as the abscissa. (See Fig. 2.)

	Voltage Levels				
Basic System Battery Voltage	6.0	12.0	24.0	30.0	32.0
	7.0	14.0	28.0	35.0	37.5

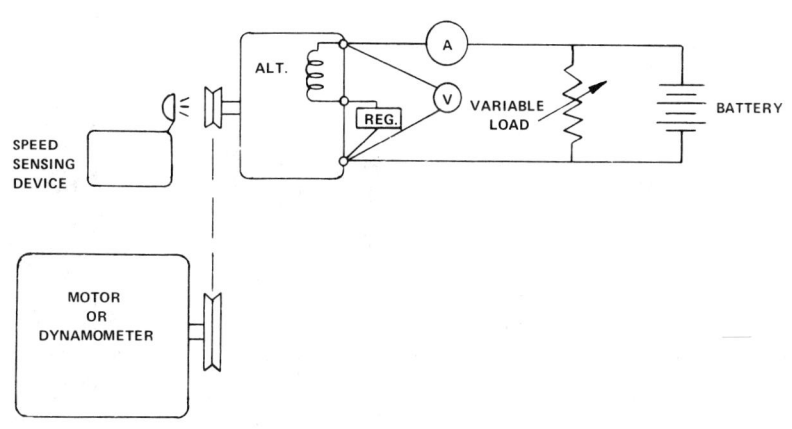

FIG. 1—TYPICAL TEST SETUP FOR OUTPUT PERFORMANCE

ALTERNATOR OUTPUT CURVE

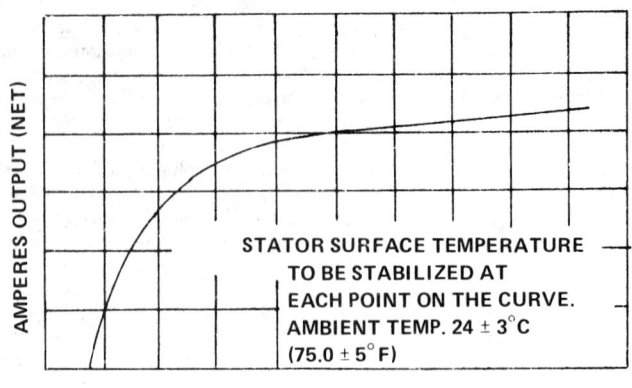

FIG. 2—STANDARD FORM FOR PERFORMANCE CURVE

MAGNETO MOUNTINGS—SAE J546

SAE Standard

Report of Miscellaneous and Electrical Equipment Divisions approved January 1913 and last revised by Electrical Equipment Committee September 1952. Reaffirmed without change February 1970.

For aircraft magneto mountings, see SAE Aerospace Standards, AS 12 and AS 13, published separately.

When the bolt centerline of the flange mounting is horizontal, the adjusting slot shall be to the left when facing the driven end of the magneto. See Fig. 1.

Fractional dimensions may vary $\pm 1/32$, unless otherwise specified.

Angles may vary ± 2 deg, unless otherwise specified.

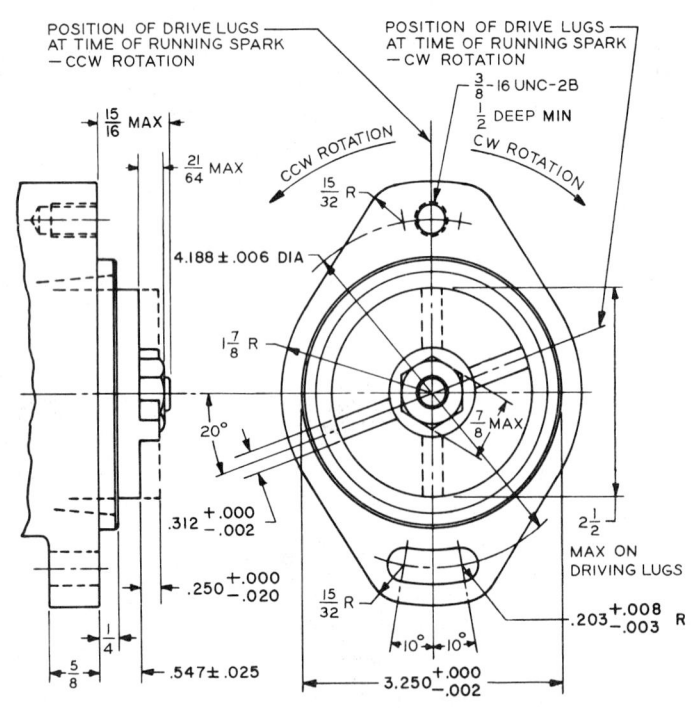

FIG. 1—FLANGE TYPE MAGNETO MOUNTING

SPARK PLUGS—SAE J548d

SAE Standard

Reports of Miscellaneous Division approved January 1915, Aircraft Division approved March 1918, Motorcycle Division approved August 1919, Electrical Equipment Division approved May 1930, and completely revised by Electrical Equipment Committee July 1977.

Scope and Purpose—This SAE Standard applies only to spark plugs used for ground vehicles and stationary engines.

This standard is intended to serve as a guide to dimensions common to the majority of current production spark plugs and future applications. It is not the intent of this standard to prohibit the manufacture of spark plugs having dimensions differing from those presented. Many applications exist which require specialized or nonstandard spark plugs. It is recommended that this standard be used in spark plug design and engine applications wherever possible. Whenever design situations arise that prevent the use of one of these standard spark plugs, a spark plug manufacturer should be contacted for guidance.

The figures and tables show typical configurations of unshielded and shielded spark plug designs, their dimensional characteristics, installation, threaded hole, and spark plug thread sizes.

Thread Gages—In order to keep the wear on threading tools within permissible limits, the threads on the spark plug GO (ring) gage shall be truncated to the maximum minor diameter of the spark plug, and the tapped hole GO (plug) gage to the minimum major diameter of the tapped hole. The plain plug gage for checking the minor diameter of the tapped hole shall be the minimum specified.

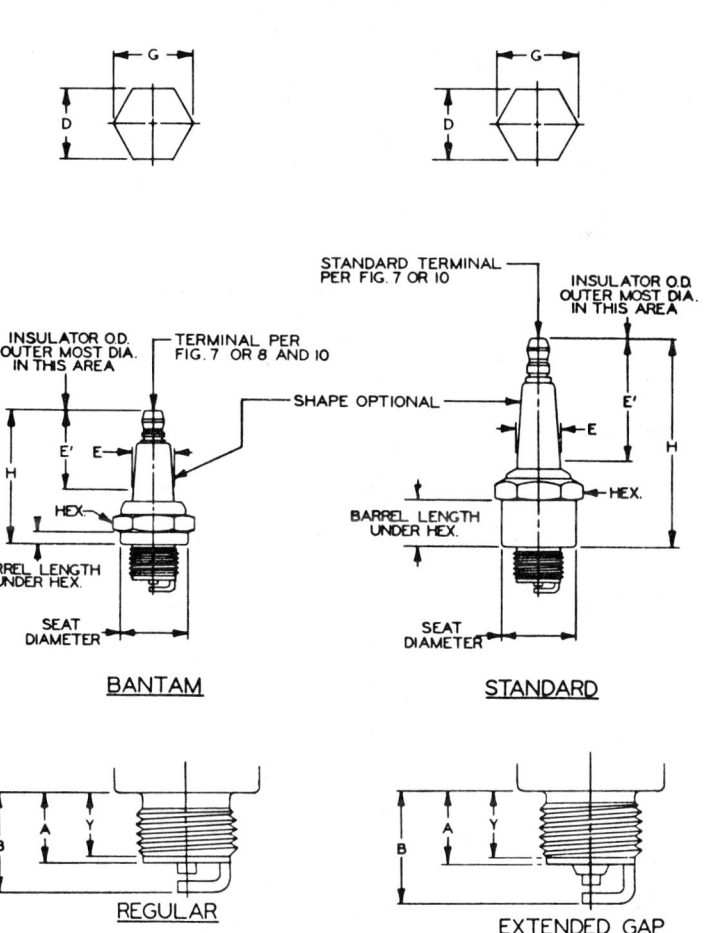

FIG. 1—FLAT SEAT SPARK PLUGS (SEE TABLE 1)

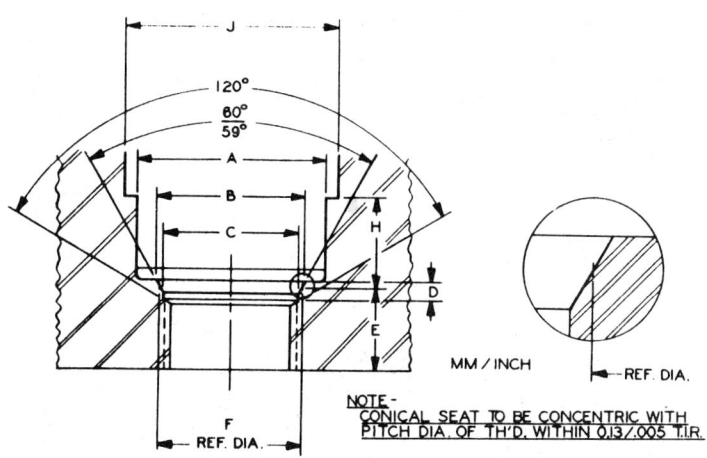

FIG. 3—CONICAL SEAT SPARK PLUGS (SEE TABLE 3)

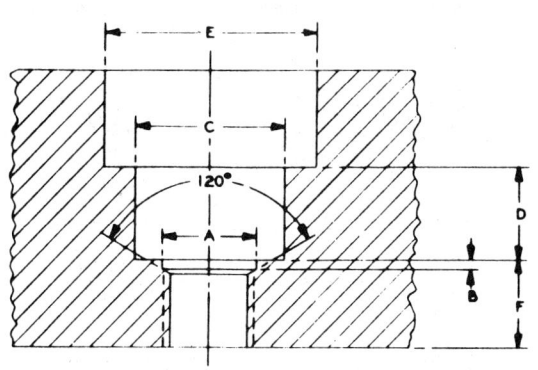

FIG. 2—HEAD COUNTERBORE—FLAT SEAT (SEE TABLE 2)

FIG. 4—HEAD COUNTERBORE—CONICAL SEAT (SEE TABLE 4)

TABLE 1—FLAT SEAT SPARK PLUGS (FIG. 1), mm (in)

Dimension	M14 x 1.25 Standard Installed Height				M14 x 1.25 Bantam Installed Height
	Short Reach	Intermediate Reach	Normal Reach	Long Reach	
A, plug reach, nom[a,b]	9.53 (0.375)	11.13 (0.438)	12.70 (0.500)	19.00 (0.748)	9.53 (0.375)
B, plug depth regular, max[a,b]	13.60 (0.535)	14.27 (0.562)	17.80 (0.700)	24.40 (0.960)	13.60 (0.535)
B, plug depth, extended gap, max[a,b]	17.20 (0.677)	17.45 (0.687)	21.00 (0.830)	27.00 (1.060)	16.00 (0.630)
Y, threaded length, nom[a]	9.02 (0.355)	10.13 (0.399)	11.71 (0.461)	18.00 (0.709)	9.02 (0.355)
Hex size:					
D, across flats	20.80–20.40 (0.819–0.803)	20.80–20.40 (0.819–0.803)	20.80–20.40 (0.819–0.803)	20.80–20.40 (0.819–0.803)	19.00–18.80 (0.748–0.740)
G, across corners, min	23.00 (0.905)	23.00 (0.905)	23.00 (0.905)	23.00 (0.905)	21.01 (0.827)
Barrel length under hex, min	11.40[c] (0.450)[c]	11.20 (0.440)	10.00 (0.394)	10.00 (0.394)	2.92 (0.115)
Seat dia max	20.80 (0.820)	20.80 (0.820)	20.80 (0.820)	20.80 (0.820)	19.00 (0.748)
Insulator, max E, dia	12.70 (0.500)	12.70 (0.500)	12.70 (0.500)	12.70 (0.500)	12.70 (0.500)
E^1, length	33.00 (1.300)	33.00 (1.300)	33.00 (1.300)	33.00 (1.300)	24.10 (0.950)
H, installed height, max	68.00 (2.677)	68.00 (2.677)	68.00 (2.677)	68.00 (2.677)	46.00 (1.810)

Dimension	M12 x 1.25 Standard Installed Height		M10 x 1.0 Standard Installed Height	
	Normal Reach	Long Reach	Normal Reach	Long Reach
A, plug reach, nom[d]	12.70 (0.500)	19.00 (0.748)	12.70 (0.500)	19.00 (0.748)
B, plug depth regular, max[d]	15.80 (0.622)	22.35 (0.880)	15.80 (0.622)	22.35 (0.880)
B, plug depth (extended gap) max[d]	19.00 (0.748)	25.00 (0.984)	19.00 (0.748)	25.00 (0.984)
Y, threaded length, nom[d]	11.71 (0.461)	18.00 (0.709)	11.71 (0.461)	18.00 (0.709)
Hex size:				
D, across flats	17.50–17.20 (0.689–0.677)	17.50–17.20 (0.689–0.677)	15.95–15.62 (0.628–0.615)	15.95–15.62 (0.628–0.615)
G, across corner, min	19.51 (0.768)	19.51 (0.768)	17.65 (0.695)	17.65 (0.695)
Barrel length under hex, min	6.00 (0.236)	6.00 (0.236)	6.00 (0.236)	6.00 (0.236)
Seat dia, max	17.50 (0.689)	17.50 (0.689)	15.95 (0.628)	15.95 (0.628)
Insulator dia, max:				
E, dia	11.00 (0.433)	11.00 (0.433)	11.00 (0.433)	11.00 (0.433)
E^1, length	33.00 (1.300)	33.00 (1.300)	33.00 (1.300)	33.00 (1.300)
H, installed height, max	61.00 (2.400)	61.00 (2.400)	61.00 (2.400)	61.00 (2.400)

Note: See Table 5 for installation torque and gasket thickness.

[a] After tightening the spark plugs with a torque of 38 N·m (28 lb-ft) (threads clean and dry), the gasket must be 1.02–1.52 mm (0.045–0.057 in) thickness. If the thickness of the gasket is outside these dimensions, a corresponding adjustment to dimensions A, B, and Y must be made.
[b] Dimensions A and B may be adjusted for some types of plugs.
[c] Where barrel length under hex requires 11.40 mm (0.450 in), check with engine manufacturer.
[d] After tightening the spark plugs with a torque of 15 N·m (11 lb-ft) (threads clean and dry), the gasket must be 0.97–1.22 mm (0.038–0.048 in) thickness. If the thickness of the gasket is outside these dimensions, a corresponding adjustment to dimensions A, B, and Y must be made.

TABLE 2—HEAD COUNTERBORE—FLAT SEAT (FIG. 2) mm (in)

M Thread Size and Reach	A Min	B Nom	C Min	D Max	E Min	F Min
M14 x 1.25 Bantam installed height	14.25 (0.561)	1.24 (0.049)	20.00 (0.787)	2.03 (0.080)	29.00 (1.142)	8.46 (0.333)
M14 x 1.25 Standard installed height						
Short reach	14.25 (0.561)	1.24 (0.049)	22.00 (0.866)	10.21 (0.402)	31.00 (1.220)	8.46 (0.333)
Intermediate reach	14.25 (0.561)	1.24 (0.049)	22.00 (0.866)	10.21 (0.402)	31.00 (1.220)	10.03 (0.395)
Normal reach	14.25 (0.561)	1.24 (0.049)	22.00 (0.866)	9.00 (0.354)	31.00 (1.220)	11.63 (0.458)
Long reach	14.25 (0.561)	1.24 (0.049)	22.00 (0.866)	9.00 (0.354)	31.00 (1.220)	18.00 (0.709)
M12 x 1.25 installed height						
Normal reach	12.24 (0.482)	1.24 (0.049)	18.49 (0.728)	5.00 (0.197)	29.00 (1.142)	11.63 (0.458)
Long reach	12.24 (0.482)	1.24 (0.049)	18.49 (0.728)	5.00 (0.197)	29.00 (1.142)	18.00 (0.709)
M10 x 1.00 installed height						
Normal reach	10.37 (0.408)	1.00 (0.039)	17.02 (0.670)	5.00 (0.197)	25.00 (0.984)	11.63 (0.458)
Long reach	10.37 (0.408)	1.00 (0.039)	17.02 (0.670)	5.00 (0.197)	25.00 (0.984)	18.00 (0.709)

TABLE 3—CONICAL SEAT SPARK PLUG DIMENSIONS (FIG. 3), mm (in)

Dimension	M14 x 1.25 Bantam Installed Height — Short Reach	M14 x 1.25 Standard Installed Height — Normal Reach	M14 x 1.25 Standard Installed Height — Long Reach
F, reference dia	14.81 (0.583)	14.81 (0.583)	14.81 (0.583)
A, plug reach, nom[a]	7.80 (0.307)	11.00 (0.433)	17.50 (0.689)
B, plug depth regular, max[a]	11.90 (0.470)	16.30 (0.640)	22.90 (1.020)
B, plug depth extended gap, max[a]	14.00 (0.550)	19.30 (0.760)	25.80 (1.020)
Y, threaded length, nom	7.29 (0.287)	10.21 (0.402)	16.48 (0.649)
Hex size:			
D, across flats	15.95–15.62 (0.628–0.615)	15.95–15.62 (0.628–0.615)	15.95–15.62 (0.628–0.615)
G, across corners, min	17.65 (0.695)	17.65 (0.695)	17.65 (0.695)
Barrel length under hex, min[a]	3.00 (0.120)	8.40 (0.330)	8.40 (0.330)
Seat dia	15.95–15.49 (0.628–0.610)	15.95–15.49 (0.628–0.610)	15.95–15.49 (0.628–0.610)
Insulator max:			
E, dia	11.00 (0.433)	11.00 (0.433)	11.00 (0.433)
E¹, length	24.00 (0.940)	33.00 (1.300)	33.00 (1.300)
H, Installed height, max	38.00 (1.500)	63.00 (2.480)	63.00 (2.480)

[a] Dimensions A and B may be adjusted for some types of plugs.

Dimension	M14 x 1.25 Bantam Installed Height — Special Long Reach	M18 × 1.5 Standard Installed Height — Normal Reach	M18 × 1.5 Standard Installed Height — Special Long Reach
F, reference dia	14.81 (0.583)	19.00 (0.748)	19.00 (0.748)
A, plug reach, nom[a,b]	17.3 (0.681)	10.90 (0.429)	23.83 (0.938)
B, plug depth, regular, max[a,b]	22.90 (0.900)	18.00 (0.709)	30.23 (1.190)
B, plug depth, extended gap, max[a,b]	25.80 (1.020)	20.00 (0.790)	32.00 (1.260)
Y, seat to end of thread, nom	16.49 (0.649)	10.21 (0.402)	22.80 (0.898)
J, plug reach, unthreaded barrel, nom	8.36 (0.329)		12.97 (0.511)
Hex Size:			
D, across flats	15.95–15.62 (0.628–0.615)	20.80–20.40 (0.819–0.803)	20.80–20.40 (0.819–0.803)
G, across corners, min	17.65 (0.695)	23.00 (0.906)	23.00 (0.906)
Barrel length under hex	9.10 (0.358)	13.49 (0.531)	14.48 (0.570)
Seat dia	15.95–15.49 (0.628–0.610)	20.19–19.89 (0.795–0.783)	20.19–19.89 (0.795–0.783)
K, Unthreaded dia below conical seat	13.94 max (0.549)		17.93 max (0.706)
Insulator, max			
E, dia	11.00 (0.433)	12.70 (0.500)	12.70 (0.500)
E¹, length	33.00 (1.300)	33.00 (1.300)	33.00 (1.300)
H, installed height	63.00 (2.480)	68.00 (2.680)	63.00 (2.680)

[a] Lengths are to the reference diameter. See Fig. 3.
[b] Dimension A may be increased for some types of plugs.

TABLE 4—HEAD COUNTERBORE—CONICAL SEAT (FIGS. 4 and 5) mm (in)

Thread Size and Reach	A Min	B Tolerance	C Tolerance	D Tolerance	E Min	F Nom	G[a] Max	H[a] Max	J Min
M14, short	17.50 (0.689)	15.39–15.09 (0.606–0.594)	14.40–14.25 (0.567–0.561)	2.29–2.03 (0.090–0.080)	7.82 (0.308)	14.81 (0.583)	[b]	2.03 (0.080)	25.00 (0.984)
M14, normal	17.50 (0.689)	15.39–15.09 (0.606–0.594)	14.40–14.25 (0.567–0.561)	2.29–2.03 (0.090–0.080)	11.25 (0.443)	14.81 (0.583)	[b]	7.10 (0.280)	25.00 (0.984)
M14, long	17.50 (0.689)	15.39–15.09 (0.606–0.594)	14.40–14.25 (0.567–0.561)	2.29–2.03 (0.090–0.080)	17.55 (0.691)	14.81 (0.583)	[b]	7.10 (0.280)	25.00 (0.984)
M18, normal	22.00 (0.866)	19.81–19.56 (0.780–0.770)	18.39–18.24 (0.724–0.718)	2.29–2.03 (0.090–0.080)	11.20 (0.441)	19.00 (0.748)	[b]	9.00 (0.354)	31.00 (1.220)
M14, special long	17.50 (0.690)	15.50–15.00 (0.610–0.591)	14.35–14.22 (0.565–0.560)	2.29–2.03 (0.090–0.080)	[b]	14.81 (0.583)	18.05 (0.710)		28.00 (1.100)
M18, special long	22.35 (0.880)	19.81–19.56 (0.780–0.770)	18.42–18.29 (0.725–0.720)	2.29–2.03 (0.090–0.080)		19.13 (0.753)	24.16 (0.951)	11.18 (0.440)	28.45 (1.120)

[a] Lengths are to the reference diameter. See Figs. 4 and 5 as applicable.
[b] Dimension not applicable.

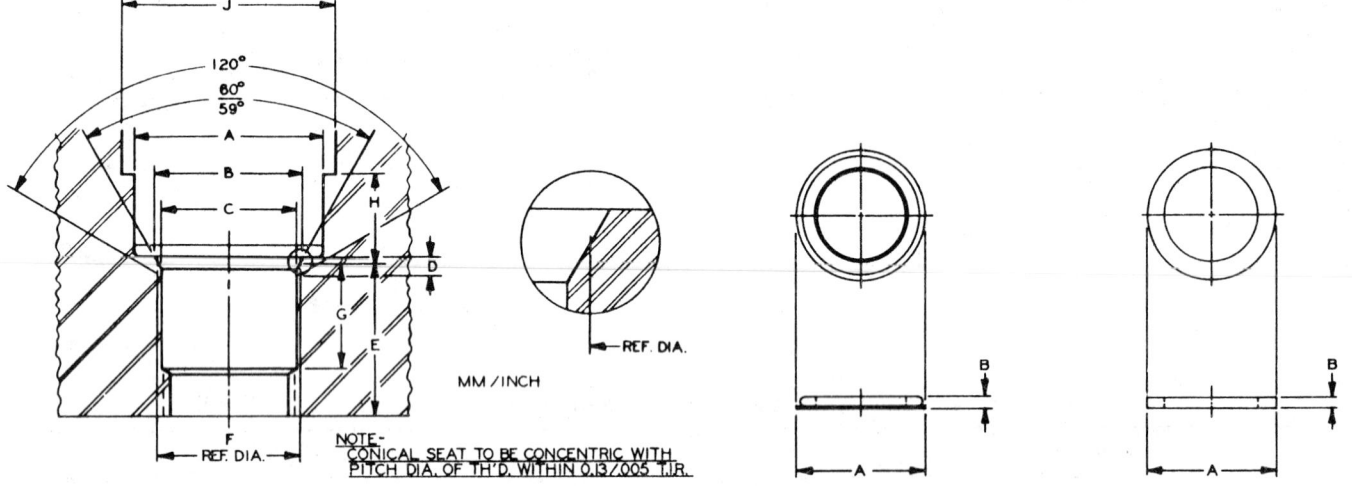

FIG. 5—HEAD COUNTERBORE—CONICAL SEAT (SPECIAL LONG REACH) (SEE TABLE 4)

FIG. 6—GASKETS (SEE TABLE 5)

TABLE 5—INSTALLATION TORQUE AND COMPRESSED GASKET DIMENSIONS (SEE FIG. 6) mm (in)

Plug Size	CAST IRON HEADS			ALUMINUM HEADS		
	Installation Torque N·m (lb-ft)	Compressed Gasket Thickness mm (in) B dim	Gasket OD Maximum mm (in) A dim	Installation Torque N·m (lb-ft)	Compressed Gasket Thickness mm (in) B dim	Gasket OD Maximum mm (in) A dim
M14, folded steel	35–40 (26–30)	1.14–1.45 (0.045–0.057)	20.83 (0.820)	20–30 (15–22)	1.14–1.50 (0.045–0.059)	20.83 (0.820)
solid copper	35–40 (26–30)	1.37 (0.054)	20.83 (0.820)	20–30 (15–22)	1.37 (0.054)	20.83 (0.820)
M18 folded steel	43–52 (32–38)	1.22–1.45 (0.048–0.057)	25.15 (0.990)	38–46 (28–34)	1.09–1.40 (0.043–0.055)	25.15 (0.990)
solid copper	43–52 (32–38)	1.96 (0.077)	24.59 (0.968)	38–46 (28–34)	2.01 (0.079)	24.51 (0.965)
M12 folded steel	15–25 (11–18)	1.00–1.45 (0.040–0.057)	16.51 (0.650)	15–25 (11–18)	1.00–1.45 (0.040–0.057)	16.51 (0.650)
M10 folded steel	10–15 (7–11)	1.00–1.45 (0.040–0.057)	14.73 (0.580)	10–15 (7–11)	1.00–1.37 (0.040–0.054)	14.73 (0.580)
M14 conical seat (gasketless)	9–20 (7–15) service[a]			9–20 (7–15) service[a]		
M18 conical seat (gasketless)	20–27 (15–20) service[a]			20–27 (15–20) service[a]		

[a] Consult with engine manufacturer for original installation in engines.
Without torque wrench, 1/16 turn after finger tight.

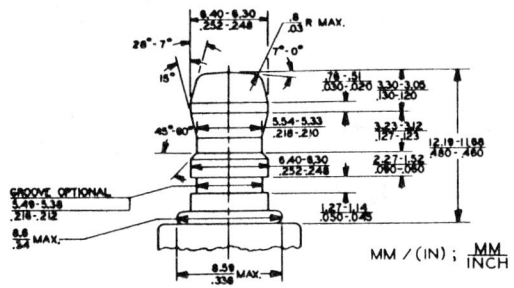

FIG. 7—SOLID POST TYPE

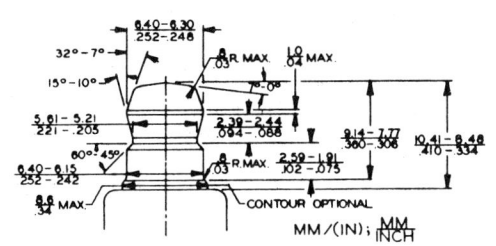

FIG. 8—ALTERNATE BANTAM SPARK PLUG SOLID POST TYPE

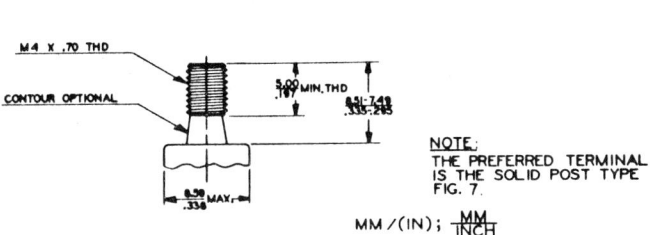

FIG. 9—THREADED POST (SEE TABLE 6)

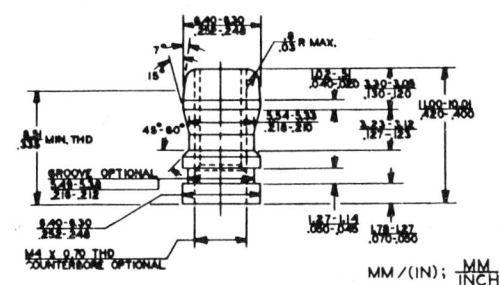

FIG. 10—NUT TYPE FOR THREADED POST (SEE TABLE 6)

TABLE 6—SPARK PLUG AND TAPPED HOLE THREAD SIZES, mm (in)
Installation and Terminals

Size		Major Dia		Pitch Dia		Minor Dia		Go Thread Ring Gage Minor Dia	Go Thread Plug Gage Major Dia	Plain Plug Gage Minor Dia Min
		Max	Min	Max	Min	Max	Min			
M18 x 1.5	Plug	17.955 (0.7069)	17.803 (0.7009)	16.980 (0.6685)	16.853 (0.6635)	16.053 (0.6320)		16.053 (0.6320)		
	Hole		18.039 (0.7102)	17.153 (0.6753)	17.026 (0.6703)	16.426 (0.6467)	16.266 (0.6404)		18.039 (0.7102)	16.266 (0.6404)
M14 x 1.25	Plug	13.868 (0.5460)	13.741 (0.5410)	13.104 (0.5159)	12.997 (0.5117)	12.339 (0.4858)		12.339 (0.4858)		
	Hole		14.034 (0.5525)	13.297 (0.5235)	13.188 (0.5192)	12.692 (0.4997)	12.499 (0.4921)		14.034 (0.5525)	12.499 (0.4921)
M12 x 1.25	Plug	11.862 (0.4670)	11.735 (0.4620)	11.100 (0.4370)	10.998 (0.4330)	10.211 (0.4020)		10.211 (0.4020)		
	Hole		11.935 (0.4699)	11.242 (0.4426)	11.138 (0.4385)	10.559 (0.4157)	10.366 (0.4081)		11.935 (0.4699)	10.366 (0.4081)
M10 x 1.0	Plug	9.974 (0.3927)	9.794 (0.3856)	9.324 (0.3671)	9.212 (0.3627)	8.747 (0.3444)		8.747 (0.3444)		
	Hole		10.000 (0.3937)	9.500 (0.3740)	9.350 (0.3681)	9.153 (0.3604)	8.917 (0.3511)		10.000 (0.3937)	8.917 (0.3511)
7/8-18	Plug	22.225 (0.8750)	22.017 (0.8668)	21.295 (0.8384)	21.191 (0.8343)	20.493 (0.8068)		20.493 (0.8068)		
	Hole		22.225 (0.8750)	21.412 (0.8430)	21.308 (0.8389)	20.851 (0.8209)	20.698 (0.8149)		22.225 (0.8750)	20.698 (0.8149)
M4 x 0.7	Term.	3.944 (0.1553)	3.804 (0.1498)	3.489 (0.1374)	3.399 (0.1338)	3.085 (0.1215)		3.085 (0.1215)		
	Nut		4.000 (0.1575)	3.663 (0.1442)	3.545 (0.1396)	3.422 (0.1347)	3.242 (0.1276)		4.000 (0.1575)	

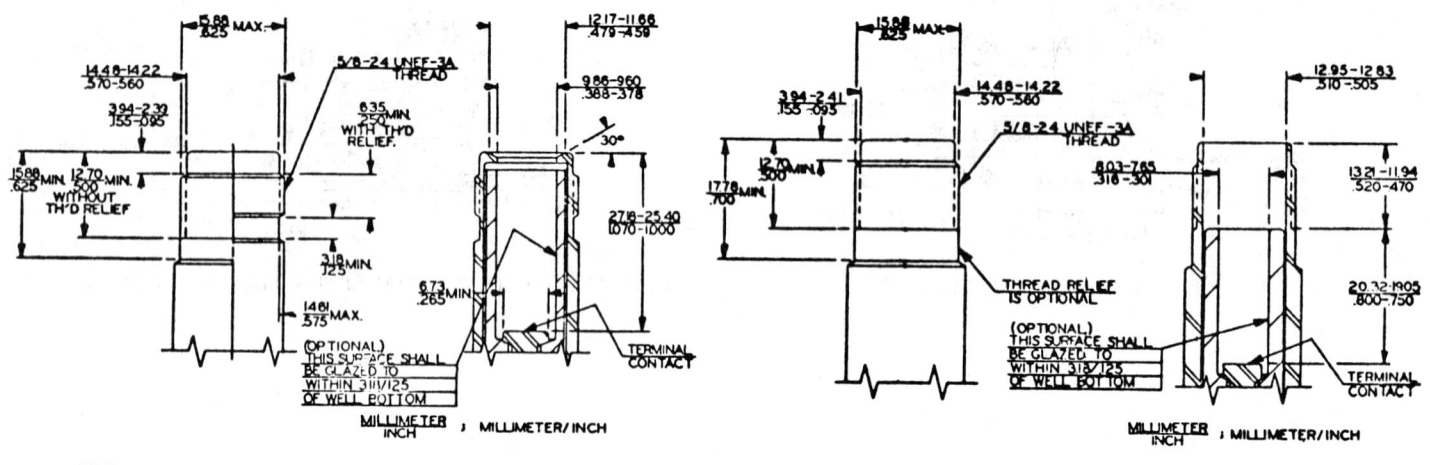

FIG. 11—5/8-24 THREADED 3/4 IN WELL DEPTH

FIG. 12—5/8-24 THREADED 1 IN WELL DEPTH

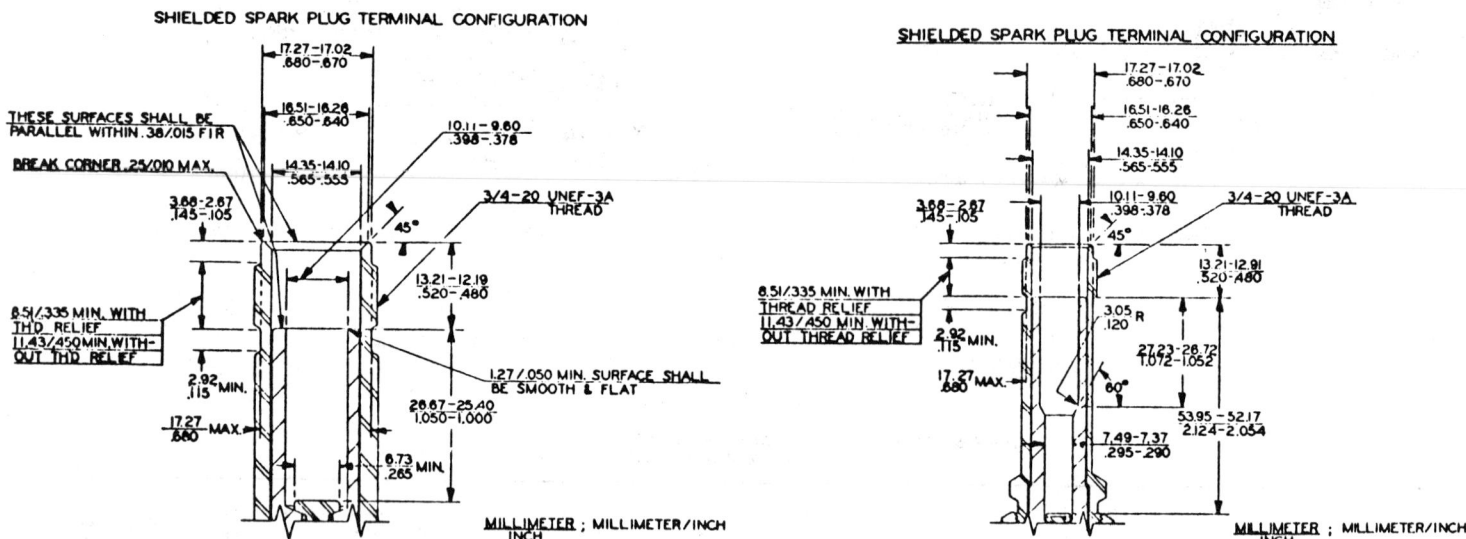

FIG. 13—3/4-20 THREADED 1 IN WELL DEPTH

FIG. 14—3/4-20 THREADED 2 1/8 IN WELL DEPTH

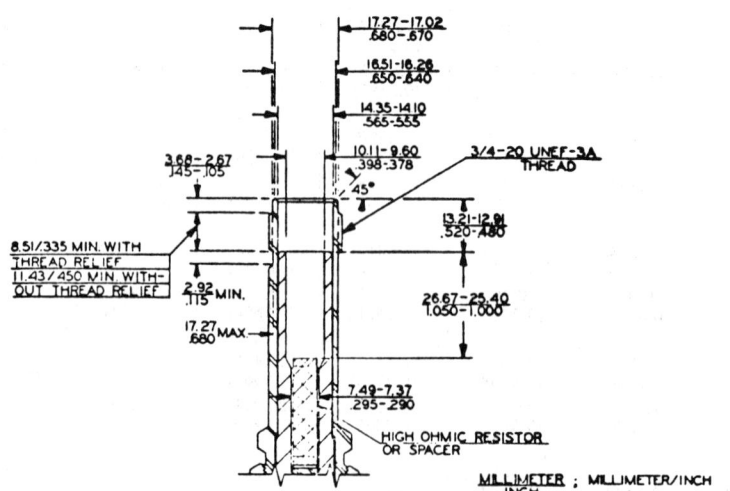

FIG. 15—3/4-20 THREADED 1 IN WELL DEPTH WITH INTERNAL RESISTOR OR SPACER

PREIGNITION RATING OF SPARK PLUGS—SAE J549a

SAE Recommended Practice

Report of Electrical Equipment Committee approved December 1947 and last revised by Propulsion Division and Electrical Equipment Committee October 1970.

Purpose—The purpose of this SAE Recommended Practice is to describe the equipment and procedures used in obtaining preignition ratings of spark plugs.

The spark plug preignition ratings obtained with the equipment and procedure specified herein are useful for comparative purposes and are not to be considered as absolute values since different numerical values may be obtained in different laboratories.

Equipment—SAE 17.6 engine[1] with the cylinder barrel having knurled and chemically treated surface and compression piston rings chromium plated.

Speed—The nominal speed is to be 2700 rpm, but is not to be over 2765 rpm when firing, nor 2670 rpm when motoring.

Compression Ratio—5.6/1.

Spark Advance—30 deg before top dead center (btdc) for nonaviation plugs, 40 deg btdc for aviation plugs or nonaviation plugs that cannot be rated at 30 deg btdc.

Ignition Source—Magneto.

Spark Plug Installation—The thread in the spark plug hole opening should conform in size and length to the standards established by SAE for the rating engine. SAE recommended torque values should be used when installing plugs in the engine. Reducer bushing or adapters should not be used.

Fuel—98% 1 deg benzol, 2% SAE 60 uncompounded aviation oil, with 3.0 cc/gal tetraethyl lead.

Fuel Injection Timing—The fuel injection pump port shall begin to close 60 ±5 deg of crankshaft angle atdc on intake stroke.

Fuel Circulation Rate—$\frac{1}{2} \pm \frac{1}{4}$ gal/min.

Fuel Injection Pump—The gallery pressure of the fuel injection pump is to be 15 ±2 psi.

Fuel Pressures—Injection: 1500 ±100 psi.

Mixture Strength—The mixture strength is that which gives maximum thermal plug temperature.

Inlet Air Temperature—225 ±5 F.

[1] See SAE AS 840, Manual for SAE 17.6 Cubic Inch Spark Plug Rating Engine, Including Maintenance and Overhaul.

Inlet Air Humidity—75 ±25 grains of moisture per pound of dry air.

Coolant—The coolant should be distilled water plus 1 gram of borax per gal as an inhibitor. City water may be used if total dissolved and suspended solids do not exceed 120 ppm.

Jacket Inlet Temperature—With integral head engine: 265 ±5 F; with insert head engine: 190 ±2 F.

Coolant Flow—5 ±½ gal/min.

Crankcase Oil—Oil is to be SAE 60 uncompounded aviation oil.

Oil Pressure—In main bearings, 95 ±5 psi; in valve gear, 15 psi minimum.

Oil Temperature—190 ±10 F.

Oil Quantity—Oil level is maintained at the center of the oil level sight glass.

Operating Conditions—The plug rating is that indicated mean effective pressure (imep) value obtained on the engine at a point when the supercharge pressure is 1 in. Hg below the preignition point. The following steps are recommended to attain this point.

1. The supercharge pressure is increased in 4 in. Hg increments until preignition occurs as indicated by a rapid rise in the thermal plug temperature. At each setting, the mixture strength is adjusted such that a maximum thermal plug temperature is obtained.

2. When preignition occurs, the fuel supply is instantly cut off and the supercharge pressure is decreased 2 in. Hg, at which point the fuel is turned on and again adjusted for maximum thermal plug temperature. This condition should be held for 3 min or until preignition again occurs.

3. If preignition occurs after step 2, the supercharge pressure should be reduced in 1 in. Hg increments until stable engine operation is attained.

4. If, after step 2, stable engine operation is obtained, the supercharge pressure should be increased 1 in. Hg, again adjusting for optimum thermal plug temperature until stable engine operation for 3 min is obtained or preignition again occurs. If preignition occurs, the supercharge pressure should be reduced 1 in. Hg until stable operation for 3 min is obtained.

Friction torque should be measured at supercharge pressure 1 in. Hg below the preignition point and within 30 sec after the engine ceases to fire.

Engine power output may be checked using a plug that has a rating point at least 50 imep above the plugs that are to be rated.

LIMITS AND METHODS OF MEASUREMENT OF RADIO INTERFERENCE CHARACTERISTICS OF VEHICLES AND DEVICES (20–1000 MHz)—SAE J551 JUN79

SAE Standard

Report of the Electrical Equipment Committee, approved December 1947, last revised June 1979. Rationale statement available.

1. Foreword—The limits recommended in SAE J551 were based on television tests conducted in 1944 and 1945. Because measurement equipment was not available, initial publication of the SAE Standard was not possible until 1958. Antenna distance was 50 ft corresponding to the expected distance of TV receivers from the roadside and spectrum coverage was 30–400 MHz.

The SAE J551 limits were recommended to CISPR[1] at its 1958 Plenary Session. Measurements were also made with Land Mobile personnel during 1958 and 1959. The 1961 Plenary Session of CISPR chose different limits, a measurement distance of 10 m, a frequency range from 40–250 MHz, and a means of statistical analysis to ensure 80% compliance at an 80% confidence, given in CISPR Recommendation No. 18.

Ongoing investigation in the early sixties revealed that TV problems had been largely resolved, hence, emphasis was shifted to the Land Mobile Radio Service. Concurrently, the limit of SAE J551 was realized to be an absolute one, at variance with the statistical nature of RFI. Interest of governmental authorities in international standardization lead to investigation of the CISPR standard (18/1) which used a measurement distance of 10 m. The SAE RFI Subcommittee concluded that the levels of interference permitted by J551 and the CISPR limit (40–250 MHz) closely paralleled. Therefore, the CISPR limit was adopted, but extended to cover the broader range of 20–1000 MHz in SAE J551a (1967), and with an 80/80 conformance note added to recognize the inherent variability of RFI. To secure agreement with CISPR, a 20 dB correlation factor relating peak and quasi-peak methods of measurement was developed in 1965 tests. SAE J551c added coverage of small internal combustion engines used on non-automotive applications (1973).

[1] International Special Committee on Radio Interference.

The 1976 revision, SAE J551d, was designed to minimize ambiguities, and assist in reducing test variability by codifying test conditions and procedures. A method of data analysis was provided, and calibration methods specified. To assist the occasional user, liberal explanatory notes and a bibliography provide information not intrinsically part of the basic specification.

SAE J551e provided added coverage of marine applications together with a section on definitions. Equipment accuracy requirements reflect state-of-the-art capability and new specifications are included for plotters used with scanning receivers. Test procedures for use with electric vehicles are under development and hopefully will be available for the next updating of this standard.

The *f* revision amplified historic information in the Foreword, provided additional definitions, provided requirements for scanning plotters, added provision for simultaneous measurements on left and right sides of the vehicle, clarified site soil/pavement requirements, and furnished preliminary information for the test of electric vehicles. Provision for testing in an anechoic chamber was also included for the first time.

SAE J551 JUN79, revised in June, 1979, updated the expected repeatability of the RFI measurement system. Testing of marine engines separately (not installed in a marine vehicle) was clarified.

Experience worldwide indicates that vehicles meeting the SAE limit are compatible with radio services generally. Nevertheless the increasing importance of Land Mobile, the increasing utilization of the radio spectrum as a whole, and the need for further documentation of the limit have prompted a series of tests conducted by the MVMA beginning in 1972 and extending to the present (mid-1979).

It was expected that data from these tests would be available for consideration during the development of J551f and would provide a more firm base for

RFI requirements. Delays in the testing have occurred and this information is now due in time for the *g* revision.

The SAE Radio Interference Subcommittee continues to consider other methods of measurement. It is the long term desire of the RFI Subcommittee to cooperate in assuring that there is a truly international (versus a North American) standard. For this reason, continued and increased emphasis will be placed on cooperation with CISPR (International Special Committee on Radio Interference) seeking to secure worldwide agreement. Methods of measurement which are departures from the existing method will be channeled through the international organization for refinement while under consideration by the Subcommittee.

This SAE Standard was developed by the SAE Radio Interference Subcommittee, which also serves as a source of expertise to the Technical Advisor to the U. S. National Committee of the International Electrotechnical Commission for the work of CISPR.

2. Purpose—This SAE Standard provides test procedures and recommended levels to assist engineers in the measurement of electromagnetic radiation from a vehicle or other device powered by an internal combustion engine or electric motor (excluding aircraft). Adherence to the recommended level will minimize the degradation effects of large populations of potential interference sources on communication services or equipment susceptible to radio frequency interference. Procedures are included to measure the radiation from a single vehicle or device. From these measurements predictions may be made of the radiation levels of individual vehicles or devices in a production run.

3. Scope—This SAE Standard covers the measurement of Impulse Electric Field Strength radiated over the frequency range of 20–1000 MHz from a vehicle or other device powered by an internal combustion engine or electric motor. Operation of all engines (main and auxiliary) on a vehicle or other device is included. All equipment normally operating when the engine is running is also included, except operator-controlled equipment, which is excluded. The recommended limit applies only to complete vehicles or devices in their final manufactured form. Vehicle mounted rectifiers used for battery charging in electric vehicles are included in this specification when operated in their charging mode.

4. Definitions—(See also paragraphs 6.2.2 and 9.1).

NOTE: The definitions in paragraphs 4.5–4.10, inclusive, were added for use in a later revision.

4.1 Vehicle—A self-propelled machine (excluding aircraft and rail vehicles and boats over 10 m in length). Vehicles may be propelled by an internal combustion engine, electrical means, or both. Vehicles include, but are not limited to, mopeds, agricultural tractors, snowmobiles, and small motorboats.

4.2 Device—A machine equipped with an internal combustion engine but not self-propelled. Devices include, but are not limited to, chain saws, irrigation pumps, and air compressors.

4.3 Impulse Electric Field Strength—The root-mean-square value of the sinusoidally varying radiated electric field producing the same peak response in a bandpass system, antenna and bandpass filter, produced by the unknown impulse electric field.

4.4 Shall—Conformance with the specific recommendation is mandatory and deviation is not permitted. The use of *shall* is not qualified by the fact that compliance with the standard is considered voluntary.

4.5 Impulse—An impulse is a noise transient having a frequency spectrum that is instantaneously uniform over a specified frequency band.

4.6 Impulsive Noise—A series of impulses.

4.7 Spectrum Amplitude—The vector sum of the voltages produced by an impulse in a given impulse bandwidth, divided by the impulse bandwidth.

4.8 Impulse Rate—The number of impulses per second (i/s).

4.9 Peak Field Strength—The highest instantaneous value, at a given frequency, of the spectrum amplitude field strength expressed in dB (μV/m)/kHz.

4.10 Peak Spectrum Amplitude—The highest instantaneous value of the Spectrum Amplitude. It is measured at scanning rates no greater than 8.33 times the IF impulse bandwidth ((in MHz) of the receiver being used) per second.

5. Measurements

5.1 Impulse Electric Field Strength—Impulse Electric Field Strength shall be expressed in units of decibles above one microvolt per meter per kilohertz bandwidth. The relationship expressing impulse electric field strength to the measurement system is:

$$F = R + AF + T \quad \text{(Equation 1)}$$

where: F = Impulse electric field strength dB as above $(1\ \mu V/m)/kHz$
R = Instrument reading dB above $1\ \mu V/kHz$ as discussed in paragraph 6.1.2
AF = Antenna factor, defined in paragraph 6.2.2 or 6.2.3
T = Transmission line factor, defined in paragraph 6.3

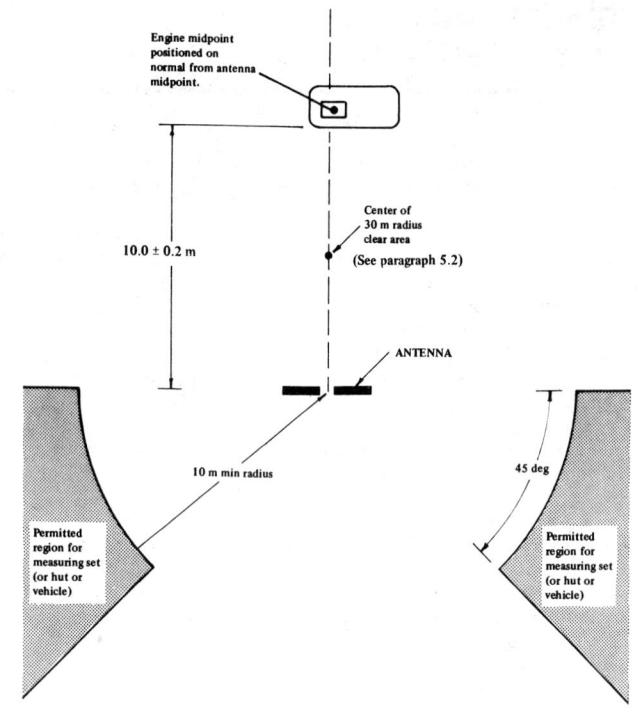

FIG. 1—PLAN VIEW OF SITE AND EQUIPMENT SHOWING POSITION OF ANTENNA AND MEASURING EQUIPMENT RELATIVE TO VEHICLE OR DEVICE

NOTE: Quasi-peak measurements may be made in lieu of peak measurements, if the proper bandwidth factor, limit (20 dB lower than SAE), and equipment are used as specified in CISPR Publication 12. The factor to convert from dB above $(1\ \mu V/m)/kHz$ to dB above $(1\ \mu V/m)/MHz$ is $+60$ dB.

5.2 Source Orientation—(See Figs. 1–4).

5.2.1 VEHICLE ORIENTATION—The vehicle shall be oriented as shown. Measurements shall be taken from both the left and right sides of the vehicle.

5.2.2 DEVICE ORIENTATION—The figures show one typical measuring position. As a minimum, measurements shall be taken at 90 deg increments, in a plan view, around the device with the engine midpoint positioned on the normal from the antenna midpoint.

For devices which normally operate in a multiplicity of positions, measurements shall be made on all six faces of an imaginary cube surrounding the device. Hand held devices shall be positioned on a non-conducting structure 1 m above ground level. Once the position(s) yielding significant contributions to the characteristic reading of the device have been determined, the remaining position(s) need not be measured.

5.2.3 MARINE ORIENTATION—Figs. 3 and 4 show typical measuring positions. As a minimum, measurements shall be taken on all four sides (port, bow, starboard, and stern) with the engine midpoint positioned on the normal from the antenna midpoint. Once the position yielding maximum radiation has been established on a particular marine installation, only that position need be measured.

5.3 Polarization—Both vertical and horizontal components of impulse electric field strength shall be measured at each vehicle or device position.

5.4 Frequency Range—Measurements shall be made over the frequency range of 20–1000 MHz. For analysis, this range shall be divided into a minimum of 15 bands with approximately three bands in each octave of 2:1 frequency ratio. For bands that include the frequency range of 75–400 MHz (that is, where the recommended limit is not constant), the highest frequency in each band shall be no greater than one third above the lowest frequency in that band. Each band shall be scanned to determine the maximum radiation level for that band.

Example:

20–26 MHz	Over 80–100 MHz	Over 300–400 MHz
Over 26–34	Over 100–130	Over 400–525
Over 34–45	Over 130–170	Over 525–700
Over 45–60	Over 170–225	Over 700–850
Over 60–80	Over 225–300	Over 850–1000

5.5 Observation Time—Either manual or automatic frequency scanning may be used, so long as the observation time at each frequency gives reasonable confidence that the higher peak impulses have been observed.

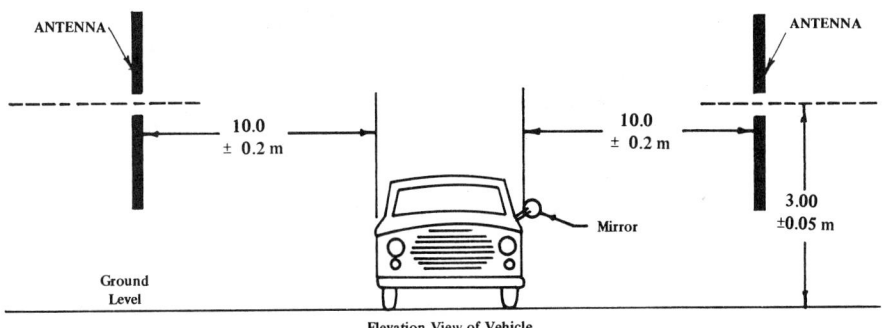

(a) Dipole Antenna In Position To Measure Vertical Component Of Radiation

NOTE: The horizontal distances shown are from the center of the dipole to the nearest peripheral part of the vehicle.

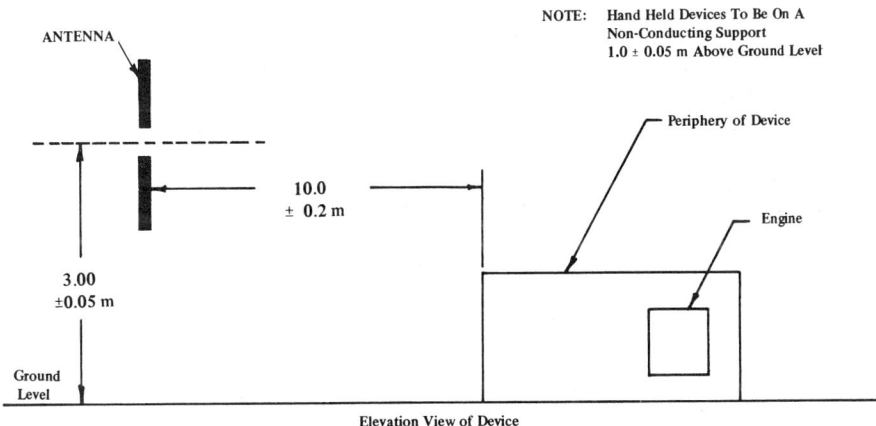

(b) Dipole Antenna In Position To Measure Vertical Component Of Radiation

NOTE: The horizontal distances shown are from the center of the dipole to the nearest peripheral part of the device.

FIG. 2—ELEVATION VIEW OF SITE AND EQUIPMENT SHOWING POSITION OF ANTENNA AND MEASURING EQUIPMENT RELATIVE TO VEHICLE OR DEVICE

6. Equipment

6.1 Measuring Instrument

6.1.1 TYPE—The measuring instrument shall be a receiver capable of detecting the peak of the envelope of the response of a bandpass filter to an impulse type signal over the specified frequency range.

6.1.2 INDICATING DEVICE—The measuring instrument shall have an indicating device (meter, numerical display, graphical display, etc.) to determine the peak response as defined in paragraph 6.1.1. This peak response is a function of the effective bandwidth of the measuring system (usually the impulse bandwidth of the tuned IF circuits of the measuring instrument). Hence the effects of bandwidth shall be included in the calibration of the indicating device (see paragraph 8.1.2). The indicating device readings shall be in units of decibels above one microvolt per kilohertz (dB above $1\ \mu V/kHz$). The impulse bandwidth of the measuring instrument shall not exceed 10% of the frequency at which measurements are made.

6.1.3 ACCURACY—The measuring system, excluding source, shall be able to measure impulse electric field strength over the frequency range 20–1000 MHz with a maximum uncertainty of ± 5 dB. The frequency uncertainty shall be $\pm 3\%$.

NOTE: To insure that the measurements defined in this standard are within the stated tolerances, consideration should be given to all pertinent measuring equipment characteristics such as: frequency and amplitude stability, image rejection, cross-modulation, overload levels, selectivity, time constants, and signal/noise ratio as well as those affecting antenna and lead-in cable.

6.1.4 SCANNING PLOTTERS—The sine wave signal frequency response at 1.25 cm (0.5 in) peak-to-peak shall not be down by more than 3 dB at 10 Hz from the 1 Hz response.

6.2 Antenna

6.2.1 STANDARD ANTENNA—The standard antenna for these measurements is the balanced half-wavelength resonant dipole tuned to the arithmetic midpoint of the bands used. (See paragraph 5.4 for examples of bands.) The reference point is the center of the two dipole elements. In the frequency range 20–30 MHz, the physical length of the resonant dipole prohibits its use in the vertical plane at 3 m height. For that reason, a dipole tuned to 30 MHz shall be used to obtain the horizontal and vertical components of field intensity over the frequency range 20–30 MHz and the 30 MHz antenna factor shall be used.

6.2.2 ANTENNA FACTOR—The factor relating the field strength to the loaded

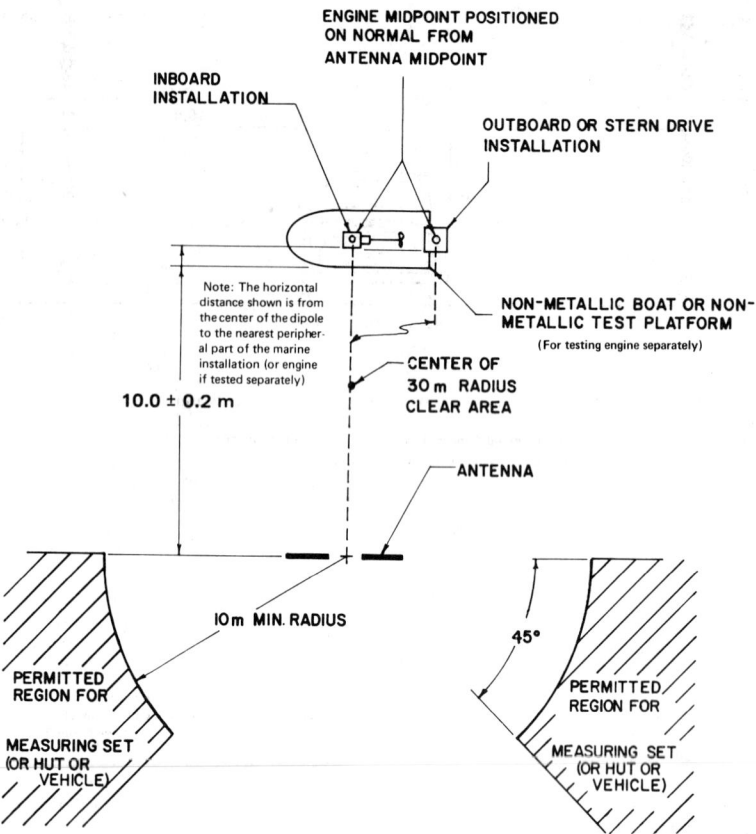

φ FIG. 3—PLAN VIEW OF MARINE TEST SITE SHOWING TESTING SET (OR HUT OR VEHICLE) AND WATER BASED MARINE INSTALLATION

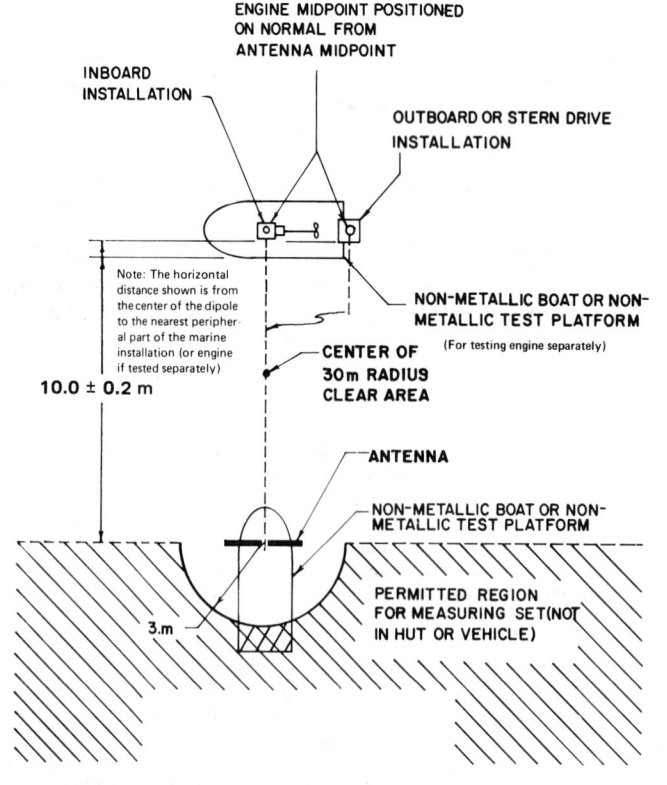

φ FIG. 4—PLAN VIEW OF MARINE TEST SITE SHOWING WATER OR LAND BASED TESTING SET AND WATER BASED MARINE INSTALLATION

antenna terminal voltage[2] at the reference point of the antenna is called the antenna factor, designated AF expressed in dB. The antenna factor shall include the effects of baluns, impedance matching devices, any mismatch losses, and operation off the resonant frequency of the antenna.

NOTE: This factor is a function of frequency and is usually provided by manufacturers of resonant dipoles. Knowledge of the antenna factor for free space operation for resonant dipoles is sufficient accuracy for purposes of this standard. Greater accuracy can be obtained by knowing the antenna factor for the particular resonant dipole being used in the test environment. A method for determining antenna factor is described in SAE ARP 958.

6.2.3 ALTERNATE ANTENNAS—Although linearly polarized antennas are recommended, any receiving antenna is permitted if its output can be normalized to the output of the standard antenna in the actual test environment. The antenna factor for the alternate antenna is then the antenna factor for the standard antenna (resonant dipole) minus the gain (dB) of the alternate antenna over the standard antenna.

NOTE: If a complex receiving antenna system is used, having a phase center which moves with frequency, care must be exercised in its use to assure that the data are essentially the same as that which would have been measured using the standard antenna. (See paragraph 8.1.3 for recommended test method.)

6.2.4 ANTENNA SUPPORT STRUCTURE—Electrical interaction between the antenna elements and the antenna support/guy system shall be avoided.

6.2.5 AUXILIARY ANTENNA—For simultaneous left and right measurement in automatic test systems or for reference antenna purposes, the auxiliary antenna shall be located symmetrically opposite the antenna illustrated in Fig. 2.

NOTE: To minimize possible interaction when active, the antennas may be operated in opposite polarization modes to each other. Example: For simultaneous recording, one antenna would be in the horizontal mode while the other would be in the vertical mode and vice versa.

6.3 Transmission Line—The transmission line factor (loss) as a function of frequency shall be known. The factor is designated T and is:

$$T = 20 \log_{10} \left(\frac{\text{input voltage}}{\text{output voltage}} \right) dB \qquad \text{(Equation 2)}$$

NOTE: It is recommended that the transmission line be double braided or solid shielded coaxial cable to achieve proper shielding. It is preferable that transmission line loss and mismatch errors be accounted for by including the cable in the measuring instrument calibration. When this is done T is dropped from the equation for F in paragraph 5.1.

Theoretical considerations of antenna and feed line geometry demand that the feed line not interact electrically with the antenna elements. One acceptable feed line geometry for dipole antenna is to route the feed line horizontally rearward for a distance of 6 m at a height of 3 m before descending to ground level or below. Other geometries are acceptable if they can be shown not to affect the measurements, or if the effects can be included in equipment calibration.

6.4 Reference Equipment

6.4.1 REFERENCE IMPULSE GENERATOR—The impulse generator output level (dB above 1 μV/kHz) shall be known to within ± 1.0 dB. (See paragraph 8.1.3.2 for criteria relating to impulse generator output voltage requirements.)

NOTE: For convenience of testing, the standard reference instrument should be a broadband impulse generator capable of producing a uniform spectrum to within ± 3.0 dB in the frequency range 20–1000 MHz.

6.4.2 REFERENCE ANTENNA—The prime function of the reference antenna is to provide a repeatable RF field for the comparison of an alternate antenna to the standard dipole antenna and for system repeatability measurements. For ease in measurement and to assure freedom from variation due to antenna adjustment, it is recommended that broadband antennas of the type shown as examples below be used.

Frequency Range	Example Antenna Type	Example Mil. Drwg.
20–200 MHz	φ Biconical with Balun	ES-F-201286 ES-OL-176439
200–1000 MHz	Conical Logarithmic Spiral	62J4040

(Reference: MIL. Std. 462, Notice 3, Pg. 7)

φ **6.5 Repeatability**—The repeatability of the measurements system shall be established, and periodically checked to detect variability; the input/output characteristics of the measuring instrument (paragraph 8.1.2) shall be checked at shorter intervals of time.

NOTE: In view of the variations in ground conductivity and other factors which influence repeatability, it is reasonable to expect a standard deviation not to exceed 3 dB in the measurements made of an impulse electric field (paragraph 8.1.4) within the range 20–1000 MHz.

---

[2] As this is a voltage ratio, calculations to convert to decibels should be made using a factor of 20.

7. Test Conditions

7.1 Site and Equipment Location

7.1.1 SURFACE VEHICLES AND DEVICES (Figs. 1 and 2)

7.1.1.1 *Outdoor Test Site*—The test site shall be a clear area free from electromagnetic reflecting surfaces within a circle of minimum radius 30 m measured from a point midway between the vehicle or device and antenna. The measuring set, test hut, or vehicle in which the measuring set is located, may be within the test site, but only within the permitted regions shown in Fig. 1.

7.1.1.2 *Indoor Anechoic Chamber*—As an alternate to an outdoor test site, an indoor anechoic chamber may be used provided the antenna-to-vehicle or device separation and the antenna height (Fig. 1) is maintained. The chamber walls and ceiling must be lined with RF absorbing material to minimize reflections from these surfaces. The chamber must be designed so that the overall measurement accuracy (paragraph 6.1.3) is maintained over the frequency range specified (paragraph 5.4). The chamber floor shall have electrical constants (dielectric constant, conductivity, etc.), approximating the average surface of an outdoor test site.

NOTE: Indoor anechoic chamber may be considered a marked improvement over outdoor testing because:
1. All-weather capability.
2. Controlled stable electromagnetic environment.
3. Consistent and known ground plane characteristics.
4. Improved repeatability of measurements.

7.1.2 MARINE VEHICLES (FIGS. 3 AND 4)—The test site shall be a clear area φ free from electromagnetic reflecting surfaces within a circle of minimum radius 30 m measured from a point midway between the engine under test and the antenna. The center of the antenna shall be 3 ± 0.05 m above average water level. When tested separately, inboard, stern drive, and outboard engines shall be installed in a non-metallic boat or other non-metallic platform.

7.1.2.1 *Land Based Testing Set*—When the test equipment is on land, the test hut or vehicle in which the measuring set is located may be within the test site, but only within the permitted regions shown in Fig. 3. If the measuring set is not in a hut or vehicle, it may be located within the test site, but only within the permitted region shown in Fig. 4.

7.1.2.2 *Water Based Testing Set*—The measuring set shall be installed in a non-metallic boat or non-metallic test platform which may be within the test site, but only within the permitted region shown in Fig. 4.

7.1.3 ELECTRIC VEHICLES—Electrically propelled vehicles shall be tested by operating them with the drive wheels raised off the road surface. Testing shall consist of full tests on four sides while operating the vehicle throughout its speed range. A preliminary test shall be conducted to determine that speed which is the worst case condition and the full test then conducted at that speed. The vehicle shall conform to the limit of this standard at all expected steady state speeds.

If an electric vehicle has an internal means of recharging the battery, the vehicle shall also be tested with the charger operating and the electric vehicle otherwise inoperative.

NOTE: This requirement represents current state-of-the-art (1978). Nevertheless, the requirement has not been tested in practical usage and is to be regarded as provisional in nature. Before starting the test, it is imperative to determine if the drive system will be damaged by operation in the unloaded state.

7.1.4 SITE AND EQUIPMENT LOCATION (GENERAL)—Testing shall be conducted over ground surfaces natural to the vehicle or device being tested, or if tested over other surfaces, such tests shall be normalized to natural surface levels. For example, typical asphalt, concrete surfaces, or sandy loam surfaces are considered natural for automotive vehicles, and either salt or nonsalt water for marine vehicles.

NOTE: Theoretically, correlation of test results from different sites could be improved if an extensive ground plane were constructed at each site. In practice, it has not been shown that variability is, in fact, improved.

Site requirements mentioned above are made considering that:

(a) Practical testing requirements involve remote sites where construction of an artificial ground plane would be restrictive.

(b) That the purpose of this standard is to minimize the degradation effects of large populations of potential interference sources on communication services operating in their normal environment.

7.2 Moisture—During testing, all surfaces, other than those normally in contact with water, shall be dry. Immersible devices shall be tested at their normal operating depth.

NOTE: Dew or light moisture may seriously affect readings obtained on devices incorporating plastic enclosures.

7.3 Ambient Noise—Ambient noise and interference exclusive of radio, TV, radar, and similar carriers at the test site shall not be of a magnitude sufficient to interfere with measurements from the test vehicle or device.

7.4 Source Operation

7.4.1 TEMPERATURE—The engine shall be at normal operating temperature.

7.4.2 ENGINE SPEED—The engine speed may be set by adjustment of the engine idle speed screw. If this method is not used, care shall be exercised to insure that the speed setting mechanism shall not affect the radio frequency radiation from the engine.

7.4.2.1 *Variable Speed Engines*—During each measurement the engine shall be run at 1500 r/min ± 10%. For engines that cannot operate at 1500 r/min, the engine shall be run at the minimum r/min recommended by the manufacturer at which the engine will operate in a steady-state condition for the duration of the test.

7.4.2.2 *Fixed Speed Engines*—Engines in this category shall be operated at their normal speed during the test.

7.4.3 AUXILIARY ENGINES—Auxiliary engines shall be operated in their normal intended manner and tested in the same manner as the main engine of the vehicle or device.

NOTE: Dependent upon the location of auxiliary engines, this requirement may dictate multiple tests of the vehicle or device with the several engines positioned in front of the antenna on successive tests.

7.4.4 IMMERSIBLE DEVICES—Immersible devices shall be immersed to their normal operating depth. Control cables, if required for normal operation, shall be attached in their proper relation to the engine.

7.4.5 OTHER EQUIPMENT—All operator-controlled equipment not required while the engine is running should be turned OFF. If OFF is not possible (as in the case of some blower motors), the equipment shall be operated in the mode requiring the least current draw (LO instead of MED or HI).

8. Procedure

8.1 Equipment Calibration

8.1.1 FREQUENCY CALIBRATION—Frequency accuracy shall be determined using standard techniques.

8.1.2 INDICATING DEVICE CALIBRATION—The calibration of the measuring instrument shall be accomplished in one of two ways:

8.1.2.1 *Preferred Method*—This method may be used for any indicating device whether previously calibrated or not. Use the reference impulse generator (paragraph 6.4.1) and appropriate RF attenuator or impulse generator output level controls to calibrate the indicating device directly in dB above 1 µV/kHz by injecting the impulse signal into the measuring instrument's RF input port and noting the indicating device readings at various known impulse signal levels. A scale overlay, look-up table, or other suitable technique can then be used to facilitate the determination of readings in dB above 1 µV/kHz (R in Equation 1). The calibration shall be accomplished for at least one frequency in each band described in paragraph 5.4 to determine whether the indicating device calibration is a function of frequency. If so, the appropriate calibration shall be used at each frequency.

8.1.2.2 *Acceptable Method*—This method may be used if the indicating device has been previously calibrated in microvolts or decibels above 1 µV. The measuring instrument impulse bandwidth shall be known. Manufacturer's specifications are acceptable if the unit is in proper condition. If not, the impulse bandwidth shall be measured as discussed in SAE ARP 1267. Calculate a conversion factor using the following relationship:

$$B = 20 \log_{10} (1 \text{ kHz/measuring instrument impulse bandwidth, kHz}) \quad \text{(Equation 3)}$$

The indicating device scale shall be modified by an overlay, look-up table, or other suitable means to account for the bandwidth conversion B. If the indicating device is calibrated in dB above 1 µV, B should be added to the existing scale values to determine the new scale in dB above 1 µV/kHz so that R may be determined in Equation 1. If the indicating device is calibrated in µV, the quotient within the brackets of the definition of B should be multiplied by the scale values to obtain µV/kHz. Then the dB calculation shall be performed to provide R in dB above 1 µV/kHz for use in Equation 1. This calibration procedure shall be used for each frequency to be measured.

8.1.3 ALTERNATE ANTENNA FACTOR DETERMINATION—If an alternate antenna (see paragraph 6.2.3) is used, the antenna factor shall be determined by a substitution technique in the intended test environment. The substitution standard shall be the standard antenna (paragraph 6.2.1). The radiated field to be measured for the substitution technique is generated by the reference antenna and the impulse generator as specified in paragraph 6.4.1.

8.1.3.1 *Test Geometry*—The alternate antenna shall be located at its intended test position. When substitution occurs, the standard antenna shall be placed so that its reference point is at the same place that the reference point for the alternate antenna normally occupies. The reference antenna shall be 10 m in horizontal distance from the alternate antenna reference point (taking the place of the nearest vehicle periphery in Fig. 1) and shall be 1 m high.

8.1.3.2 *Reference Impulse Electric Field Amplitude*—For accurate measurements to be made, the reference impulse electric field shall be at least 3 dB above the least measureable field of the measuring system. A value of at least 10 dB is preferred.

NOTE: Experience indicates that an impulse generator that meets paragraph 6.4.1 and has a nominal 100 dB µV/MHz level can produce a field of approximately 10 dB (µV/m)/kHz at the receiving antenna when an impedance matching attenuator of 10 dB is used at the output of the generator. This field strength varies depending on reference antenna losses and radiation characteristics and on propagation anomalies. This approximate number is provided so that the required sensitivities and tolerable losses in the measuring system can be estimated so that this antenna factor determination can be performed.

8.1.3.3 *Test Procedure*—The procedure to be used is to measure the reference field with the standard antenna positioned as in paragraph 8.1.3.1 to obtain a meter reading (usually voltage). Then the alternate antenna is substituted and a second reading is taken.

The antenna factor for the alternate antenna is calculated as discussed in paragraph 6.2.3. This procedure should be conducted for both horizontal and vertical polarizations to determine whether a separate antenna factor is required for the two cases.

NOTE: The standard antenna factor may be assumed to be the same for both cases.

8.1.3.4 *Frequencies*—The number of frequencies at which antenna factor values are required depends on the alternate antenna being evaluated. If the gain as a function of frequency of the alternate antenna is well behaved, at least one frequency for each band in paragraph 5.4 shall be evaluated. If it is not well behaved, enough frequencies must be evaluated to describe the function adequately.

8.1.4 COMPLETE SYSTEM VERIFICATION—The complete measurement system comprised of antenna, transmission cable, measuring instrument, and readout devices shall be verified by measuring an impulse electric field established with a wideband impulse generator and antenna(s) described in paragraph 6.4. This verification shall be made on a periodic basis so that any change in system performance can be detected.

8.2 Ambient Measurement
—In order to evaluate the effect of extraneous noise, RF carriers, etc. on the test results, both vertical and horizontal measurements shall be made, both immediately before and immediately after a set of source measurements. The ambient measurements shall be made as described in Section 3, except that the source vehicles or device engine(s) shall be inoperative.

8.3 Source Measurement
—Make the measurements described in Section 5 under the test condition specified in Section 7.

9. Assessment of Results

9.1 Characteristic Reading
—A characteristic reading is used for the purpose of comparison with the recommended limit. The characteristic reading for each band shall be the maximum measurement obtained for that band for both polarizations and for all specified measurement positions of the vehicle or device as specified in Section 5 and shall be compared to the limit at the arithmetic mean frequency of the band. Known ambient carriers shall be ignored in determining characteristic readings.

9.2 SAE Recommended Limit
—The recommended limit is contained in Appendix A.

9.3 Conformance of a Group
—An individual test may be regarded as being representative of the performance of a group built in substantial conformity to the sample tested. If additional information about the group is desired, several samples may be tested and statistical analysis applied to the characteristic readings. It is recommended that the statistical analysis be conducted with the criteria that 80% conformance is desired with 80% confidence. (See Section 10 for suitable references.)

10. References

Statistics

1. Albert H. Bowker and Gerald J. Lieberman, "Engineering Statistics." Second Edition, Prentice-Hall Inc.

2. D. B. Owen, "Factors for One-Sided Tolerance Limits and for Variables Sampling Plans." Sandia Corporation Monograph, SCR-607 (March 1963).

3. D. B. Owen, "A Survey of Properties and Applications of the Noncentral t-Distribution." Technometrics, Vol. 10 (1968), pp. 445–478.

4. Edwin L. Crow, Francis A. Davis, Margaret W. Maxfield, "Statistics Manual." U. S. Naval Ordnance Test Station, China Lake, CA, (1955).

5. George J. Resnikoff and Gerald J. Liebermann, "Tables of the Non-Central t-Distribution." Stanford University Press, 1957.

6. Mary Gibbons Natrella, "Experimental Statistics." National Bureau of Standards Handbook 91, Issued August 1, 1963, Reprinted October 1966 with corrections.

7. Roy H. Wampler, "One-Sided Tolerance Limits for the Normal Distribution, P = .80, γ = .80." Journal of Research of the National Bureau of Standards Section B, Vol. 80B, No. 3, (July–September, 1976).

8. N. L. Johnson and F. C. Leone, "Statistics and Experimental Design I." New York: John Wiley & Sons, 1964, pp. 298–348.

9. H. J. Larson, "Introduction to Probability Theory and Statistical Inference." New York: John Wiley & Sons, 1969.

10. "The Statistical Considerations in the Determination of Limits of Radio Interference." International Electrotechnical Commission (International Special Committee on Radio Interference—CISPR) Report No. 48

(Document CISPR (Secretariat) 952E, September 1973 as approved at West Long Branch meeting 1973).

Antennas, Electronics, and Fields

1. Henry Jasik, "Antenna Engineering Handbook." First Edition, McGraw-Hill Book Co., Inc. 1961.
2. Frederick Emmons Terman, "Electronic and Radio Engineering." Fourth Edition, McGraw-Hill Book Co., Inc., 1951.
3. Frederick Emmons Terman and Joseph Mayo Pettit, "Electronic Measurements." Second Edition, McGraw-Hill Book Co., Inc., 1952.
4. Simon Ramo, John R. Whinnery, and Theodore Van Duzer, "Fields and Waves in Communication Electronics." John Wiley & Sons, Inc., 1965.
5. Ezra B. Larsen, "Calibration of Radio Receivers to Measure Broadband Interference." NBSIR 73-335, Final Report, Phase I (September 1973).

APPENDIX A
PERFORMANCE LEVEL

1. SAE Recommended Limit—The recommended limit is the curve defined in Fig. A-1. The coordinates of the end points of the straight line segments on a semilogarithmic (decade) graph are (20, 12), (75, 12), (400, 23), and (1000, 23). The abscissa is frequency in MHz and the ordinate is impulse electric field strength in dB above $(1\ \mu V/m)/kHz$. To meet the intent of this standard, the characteristic reading for each band[3] shall fall at or below the limit at the arithmetic mean frequency of that band.

[3]"Bands" and "characteristic reading" are defined elsewhere in this standard.

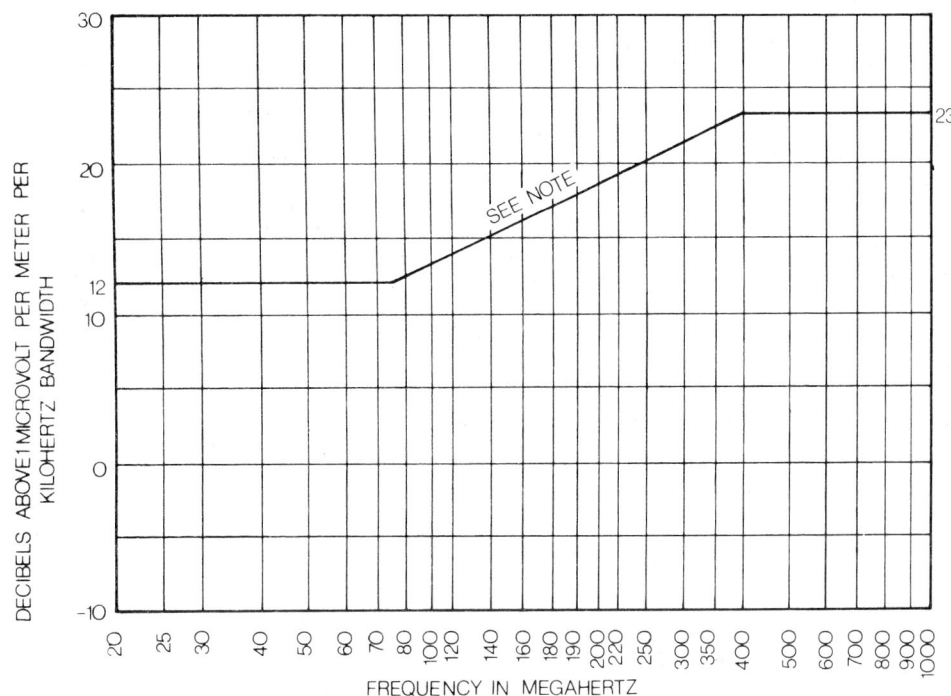

FIG. A-1—SAE RECOMMENDED LIMIT BASED ON MEASUREMENT OF IMPULSE ELECTRIC FIELD STRENGTH; FOR 120 kHz BANDWIDTH ADD 41.58 dB
NOTE: FOR GREATER ACCURACY, THE VALUE OF THE LIMIT IN THE SLOPING PORTION MAY BE CALCULATED BY THE EQUATION:

$$\text{LIMIT} = 12 + 15.13 \log_{10}\left(\frac{\text{Frequency in MHz}}{75}\right) \text{dB above } (1\ \mu V/m)/kHz$$

EXTERNAL ELECTROMAGNETIC RADIATION SUPPRESSORS—SAE J552a

SAE Recommended Practice

Report of Electrical Equipment Committee approved January 1951 and last revised December 1969.

1. *General*—This recommended practice lists the desirable characteristics as well as test procedures to be used in evaluating ignition radio noise suppressors of the lumped resistor type. More specifically, they are resistive devices used in conjunction with the high voltage portion of the ignition system to suppress electromagnetic radiation generated due to the high voltage avalanche associated with arc discharges.

2. *Recommended Practices*—The use of these resistive devices (such as shown in Fig. 1) in conjunction with nonmetallic resistance core ignition cable is not recommended, since the assembly of these devices onto the cable is apt to damage the conductive core and impair the reliability of the cable.

When such a unit having screw connectors is used with metallic conductor ignition cable, it is recommended practice to seal the suppressor against any detrimental environment such as oil, dirt, salt, or water.

3. *Specifications*

3.1 Resistance—Generally, three values are specified: 5,000 ohms, +50%, −25%; 10,000 ohms, +50%, −25%; and 15,000 ohms +50%, −25%. Other resistances and tolerances may be used for specific applications.

3.2 Resistor Types—There are two basic types of resistors used for such devices: carbon composite and metallic wire-wound. The carbon composite exhibits a substantial negative voltage coefficient at constant temperature; whereas the wire-wound units have a negligible voltage coefficient, but have high distributive inductance. If the resistor is of the wire-wound type, its resistance value can be measured by means of a d-c ohmeter. If the resistor is known to be of the carbon composition type or its construction is unknown, then the resistance measuring procedure outlined in paragraph 4.1 must be followed in order to determine its correct value.

Both types of resistors should be capable of meeting the requirements of paragraph 4.2.

3.3 Flashover—Suppressor units shall have an insulating covering which will prevent flashover from the installed suppressor to ground, from end to end of the suppressor, or between adjacent windings of wire-wound units.

4. *Test Conditions and Procedures*

4.1 Resistance Measurements—All resistance values for carbon composition resistors (or any resistor of unknown composition) shall be determined by using a 5000 V pulse test. (These resistance values obtained on carbon type units will not agree with values as read on a low voltage bridge or ohmeter.) The test can be made with any device capable of providing a 5000 V pulse of short enough duration and slow enough repetition rate that the suppressor dissipation rating is not exceeded. The preferred method, as outlined in Federal Specification W-S-506-A, paragraph 4.5.2.2, is as follows:

"Test equipment to produce short duration, high peak voltage pulses shall be used. Each resistor shall be tested in a resistance bridge circuit. A pulse shall be applied across the bridge so that the pulse across the resistor has the following parameters:

Pulse repetition rate: 4 ±1 pulse/sec
Pulse duration: 100 microseconds (maximum)
Pulse peak: 5000 ±25 V
Pulse rise time: 10–15 microseconds"

An oscilloscope shall be used to determine bridge balance at the voltage peak. The peak voltage applied to the suppressor may be measured by an oscilloscope in conjunction with a voltage divider and voltage calibrator, or by other means capable of providing a true peak voltage indication. Any discontinuity or irregularities observed in the voltage pulse waveform are indications of unacceptable internal or external voltage flashover. (See Fig. 2 for resistance test circuit.)

4.2 Flashover Measurements

4.2.1 The following flashover test and detection procedures are to determine if the insulating capability of the suppressor resistor insertion is sufficient.

A suppressor element should be tested in the specific application for which it is intended. A unit which passes the required tests for one application cannot be assumed satisfactory for others.

4.2.2 ENVIRONMENT—If environmental test equipment is available, the recommended conditions for flashover tests would be at 200 F (93 C), 95%

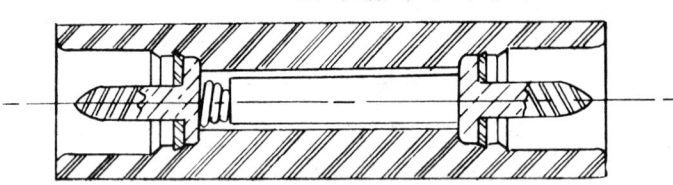

FIG. 1—THREADED SUPPRESSOR RESISTOR

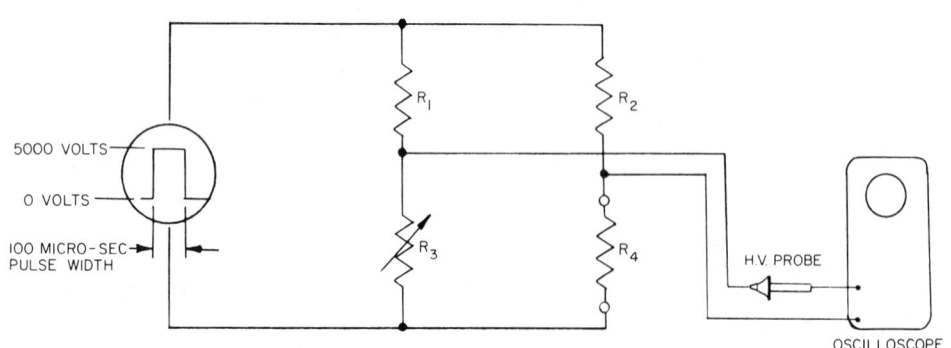

FIG. 2—RESISTANCE MEASUREMENT CIRCUIT

relative humidity, and atmospheric pressure. However, if this kind of equipment is not available, a test at room temperature and humidity can be substituted although test data may be somewhat less consistent and reliable.

4.2.3 IGNITION SYSTEM—These tests should be run using the ignition system in which the suppressor is actually used. System description should be included when qualifying data.

4.2.4 TEST VOLTAGE—The test voltage shall be determined by the maximum output voltage the ignition system can produce without exterior flashover to ground or interior flashover occurring within such components as the coil and distributor cap. The value of this voltage can be determined by the circuit arrangement illustrated in Fig. 3 using the high voltage measuring procedure as defined in SAE AIR 84. Determination of this voltage shall be made without the suppressor in the circuit. Test voltage shall be maintained for a period of 5 min with no internal or external voltage flashover occurring during this time. (All environmental seals such as distributor nipples and spark plug covers should be in place during this and the final test. When the spark plug is part of this test, the shell should be cut back so that flashover does not occur.) Flashover shall be detected by observing the characteristics of the voltage waveform on an oscilloscope connected across the high voltage source as shown in Fig. 3.

4.2.5 TESTS—The suppressor should be inserted into the ignition system described in paragraph 4.2.1 at the appropriate place, and in proper physical relationship to surrounding points at ground potential. With the output voltage adjusted to the previously determined value, if internal and/or external flashover should occur, the suppressor is not acceptable. Test voltage shall be maintained for a period of 5 min.

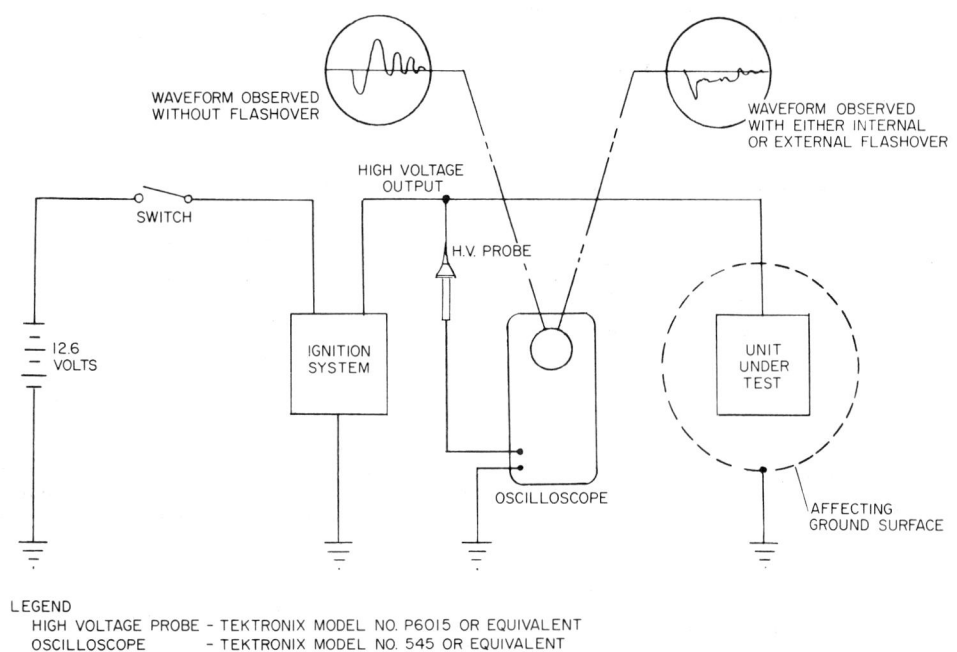

FIG. 3—CIRCUIT FOR RADIO NOISE SUPPRESSOR FLASHOVER TEST

⏀ ELECTRIC FUSES (CARTRIDGE TYPE)—SAE J554b   SAE Standard

Report of Electrical Equipment Division approved January 1914 and completely revised by Electrical Equipment Committee May 1978.

1. *Scope*—The fuses shown are for use in motor vehicles, boats, and trailers to protect electrical wiring and equipment. This standard is for the construction shown and is not intended to restrict the design and use of other configurations and materials capable of meeting the vehicle requirements.

2. *Definition*—A fuse is a device designed to open the electrical circuit when subjected to overcurrents that could damage the circuit or equipment. This action is to be nonreversible, and the fuse is intended to be replaced after the circuit malfunction has been corrected.

3. *Materials*—The fuses shown shall have clear glass tubes. End caps shall be of brass, copper, or other copper alloy and shall be plated with nickel, cadmium, or other suitable material having satisfactory electrical and corrosion protective properties.

4. *Construction*—Fuse caps shall be tightly attached to the glass tube and the ends shall be square and free of solder externally. Fuse elements shall be clearly visible through the glass tube. Fuses shall be capable of being passed through a tubular gage having a length as long as the fuse and having a uniform inside diameter of 0.258–0.259 in (6.55–6.58 mm). Preferred and other fuse dimensions are shown in Fig. 1.

5. *Application*

5.1 General—This standard applies to fuses of all lengths. However, the fuse derating chart shown in Fig. 2 applies specifically to the preferred length of 1.250 in (31.75 mm). The fuse manufacturer should be contacted for recommendations on other available lengths. (See Fig. 1.)

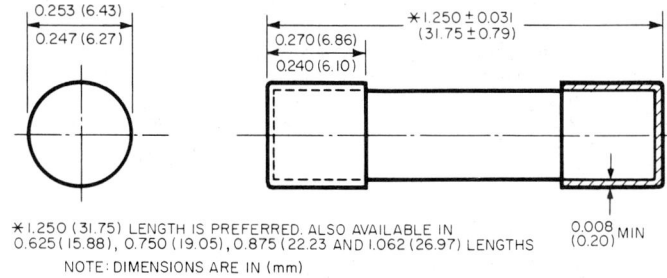

*1.250 (31.75) LENGTH IS PREFERRED. ALSO AVAILABLE IN 0.625 (15.88), 0.750 (19.05), 0.875 (22.23) AND 1.062 (26.97) LENGTHS
NOTE: DIMENSIONS ARE IN (mm)

FIG. 1—FUSE DIMENSIONS

5.2 Ampere Rating—This standard covers ampere ratings up to and including 30 A. Preferred ampere ratings are shown in Table 1. These ratings are determined at 75°F (24°C) ambient temperature. Approximate capacity change with respect to temperature is shown in Fig. 3 for all length fuses. The use of fuses in ambient temperatures beyond the limits shown is not recommended without thorough testing experimentally in the vehicle. It is further recommended that fuses not be loaded to 100% of the adjusted capacity, according to ambient temperature, due to electrical system variances. See

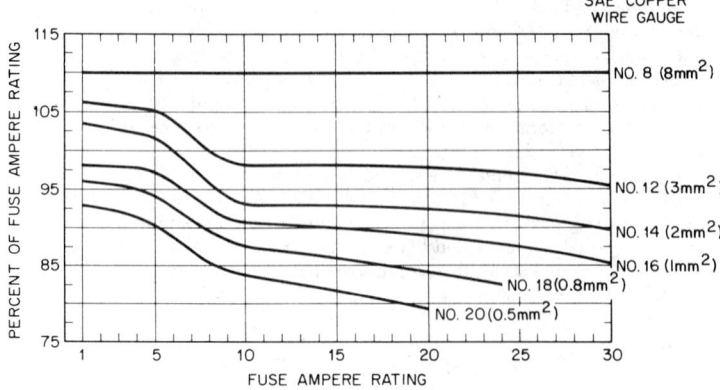

FIG. 2—FUSE DERATING FOR VARIOUS WIRE SIZES (1.250 IN (31.75 MM) LENGTH)

TABLE 1—FUSE COLOR CODES

Ampere Rating	Color	Ampere Rating	Color
1	Dark green	9	Orange
2	Gray	10	Red
2-1/2	Purple	14	Black
3	Violet	15	Light blue
4	Pink	20	Yellow
5	Tan	25	White
6	Gold	30	Light green
7-1/2	Brown		

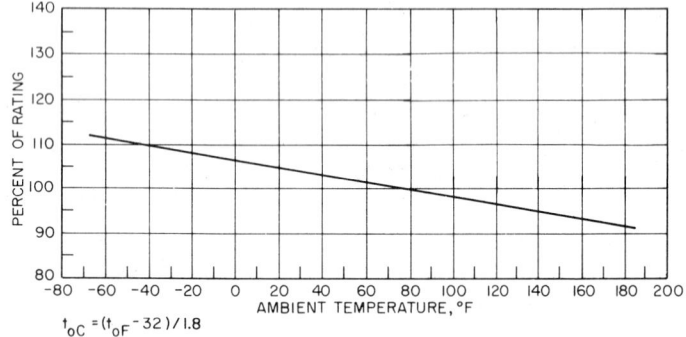

$t_{°C} = (t_{°F} - 32)/1.8$

FIG. 3—EFFECT OF AMBIENT TEMPERATURE ON AMPERE RATING OF SAE SPECIFICATION FUSES

Fig. 2 for additional deratings when fuses are used on cable gage sizes other than the test gage wire.

5.3 Voltage Rating—Fuses shall be capable of interrupting any voltage up to and including 32 VDC.

5.4 Maximum Voltage Drop—The maximum voltage drop (in millivolts) at rated current across the fuse only, shall be as shown in Fig. 4, when measured across the fuse from ferrule to ferrule.

6. Performance—Tests shall be conducted within an ambient temperature range of 75 ± 9°F (24.0 ± 5°C) except for the overcurrent test which is to be conducted at 75 ± 2°F (23.9 ± 1.2°C).

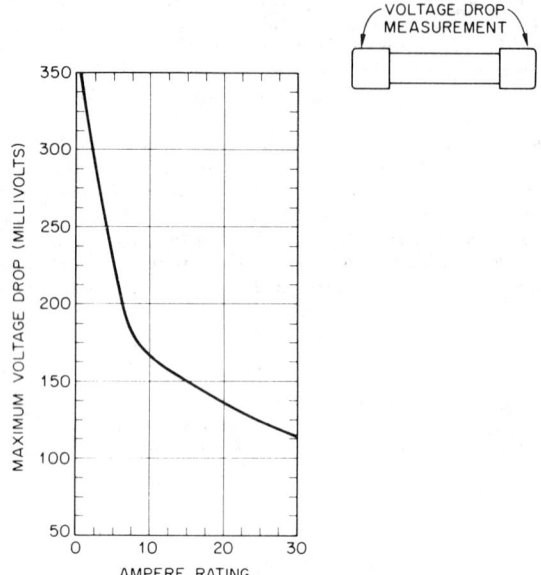

FIG. 4—MAXIMUM VOLTAGE DROP OF FUSE AT RATED CURRENT

6.1 Ampere Rating Tests—Fuses shall carry 110% of rated current continuously for 4 h, shall open at 135% of rated current in less than 1 h, and shall open at 200% of rated current in less than 10 s.

6.2 Cycling Test—Fuses shall perform satisfactorily for a 50 000 cycle load test. Each cycle shall consist of applying 70% of the rated current carrying capacity for 10 s followed by a 10 s interval of no applied current.

6.3 Vibration Test—Fuses shall perform satisfactorily after undergoing the following tests: Suitably mounted samples shall be subjected to a simple harmonic motion having an amplitude of 0.03 in (0.8 mm) travel (0.06 in (1.5 mm) maximum total excursion). The frequency shall be varied uniformly between the limits of 10 and 55 Hz. The entire range of 10–55 Hz and returning to 10 Hz shall be traversed in approximately 1 min. This motion shall be applied for a period of 2 h in each of the three mutually perpendicular directions (total of 6 h).

6.4 Procedure—The fuses, with the exceptions noted in the vibration test, shall be mounted horizontally. When testing two or more fuses in series, the fuses shall be mounted no less than 6 in (152 mm) apart and with no less than 24 in (609 mm) of interconnecting cable. All electrical tests shall be made using SAE No. 8[1] gage copper wire, and with fuse clip attachment terminals that have a maximum voltage drop of 4 mV per ampere when measured between points located on the wire 3 in (76 mm) from the attachment terminals. This determination shall be made by using a solid copper dummy 0.250 in (6.35 mm) in diameter and 1.250 ± 0.005 in (31.75 ± 0.13 mm) long, with suitably plated ferrules, installed in the fuse clips.

6.5 Marking—Fuses shall be permanently and legibly marked on the end caps with the ampere rating and the manufacturer's name or trademark. In addition, the ampere rating may be marked on the glass using numerals that are 0.150–0.200 in (4.0–5.0 mm) in height, or the fuses may be color coded with a permanent stripe around the interior or exterior of the glass tube. If a color stripe is used, the fuse element must still be clearly visible through the glass tube. This color coding shall be as shown in Table 1.

[1] Conductor cross section area to be not less than 15 105 cir mil (7.65 mm²).

BLADE TYPE ELECTRIC FUSES—SAE J1284 DEC79   SAE Standard

Report of the Electrical Equipment Committee, approved December 1979. Rationale statement available.

PART 1—DESIGN PARAMETERS

1. Scope—The fuses shown are for use in motor vehicles, boats, and trailers to protect electrical wiring and equipment. This standard is for the construction shown and is not intended to restrict the design and use of other configurations and materials capable of meeting the vehicle requirements.

2. Definition—A fuse is a device designed to interrupt the electrical circuit when subjected to overcurrents. This action is to be nonreversible, and the fuse is intended to be replaced after the circuit malfunction has been corrected.

3. Materials—The fuses shall have transparent non-conductive bodies capable of withstanding vehicle environmental conditions as set forth in this standard. Terminals shall have a suitable finish which will assure corrosion protection and satisfactory mechanical and electrical properties.

4. Construction—Fuse terminals shall be tightly attached to the fuse body. Fuse elements shall be clearly visible through the body. Typical overall dimensions are shown in Fig. 1.

5. Marking (Initially and After Environmental Exposure)—Fuses shall be marked on the fuse body with the amperage, voltage rating, and the manufacturer's name or trademark. In addition, the fuse may be color coded provided that the element remains clearly visible through the non-conductive body. All color coding shall be as shown in Table 1. Marking shall be legible at the conclusion of all tests set forth in this standard.

PART 2—PERFORMANCE REQUIREMENTS

1. Ampere Rating—This standard covers ampere ratings up to and including 30 A. Preferred ampere ratings are shown in Table 1. These ratings are determined at 24°C ambient temperature using a test procedure detailed in Part 3, Section 1 of this standard.

The specific ampere capacity of the fuses is a function of the particular electrical system being utilized. To aid in determining the actual capacity change, several factors should be considered by the application engineer.

1.1 Wire—Fig. 2 represents the approximate capacity change due to cable sizes other than 5 mm^2.

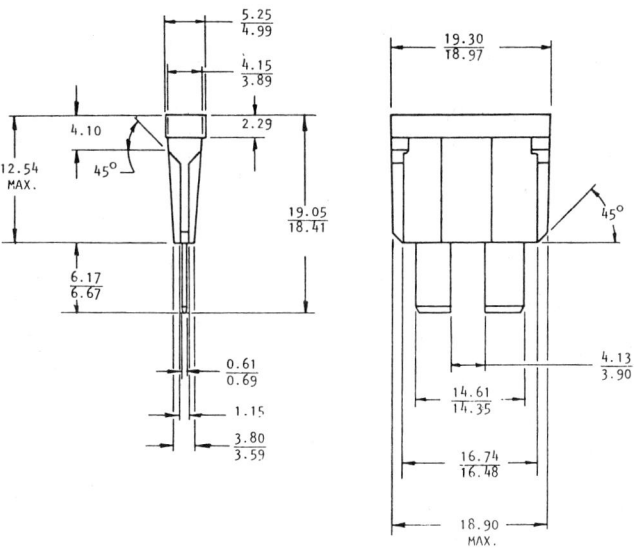

NOTE: ALL DIMENSIONS ARE IN MILLIMETERS

FIG. 1—FUSE DIMENSIONS

TABLE 1—FUSE COLOR CODES

Ampere Rating	Color
3	Violet
4	Pink
5	Tan
7-1/2	Brown
10	Red
15	Light Blue
20	Yellow
25	Natural (White)
30	Light Green

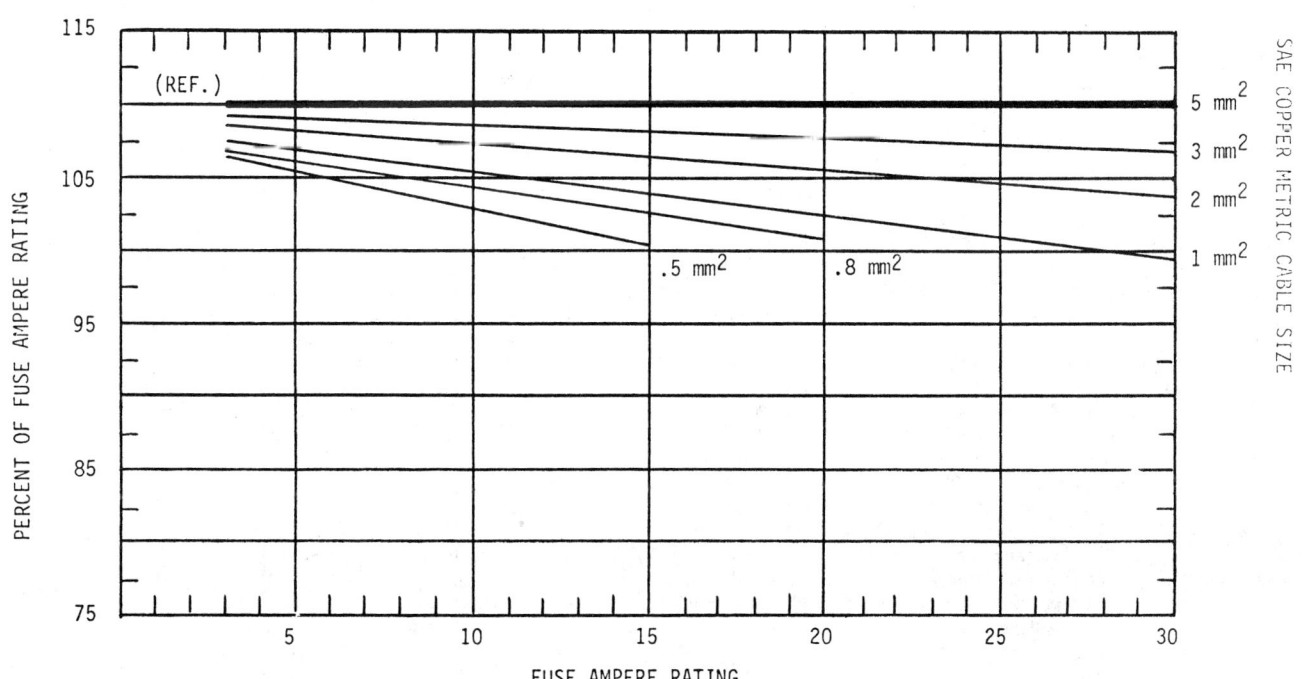

NOTE: THIS CURVE BASED ON 110 PER CENT OF RATED CURRENT USING 5 mm^2 COPPER CABLE AT 24°C, TEST IN STANDARD TEST MODULE

FIG. 2—CHANGE IN FUSE CURRENT CARRYING CAPACITY FOR VARIOUS CABLE SIZES AT 24°C

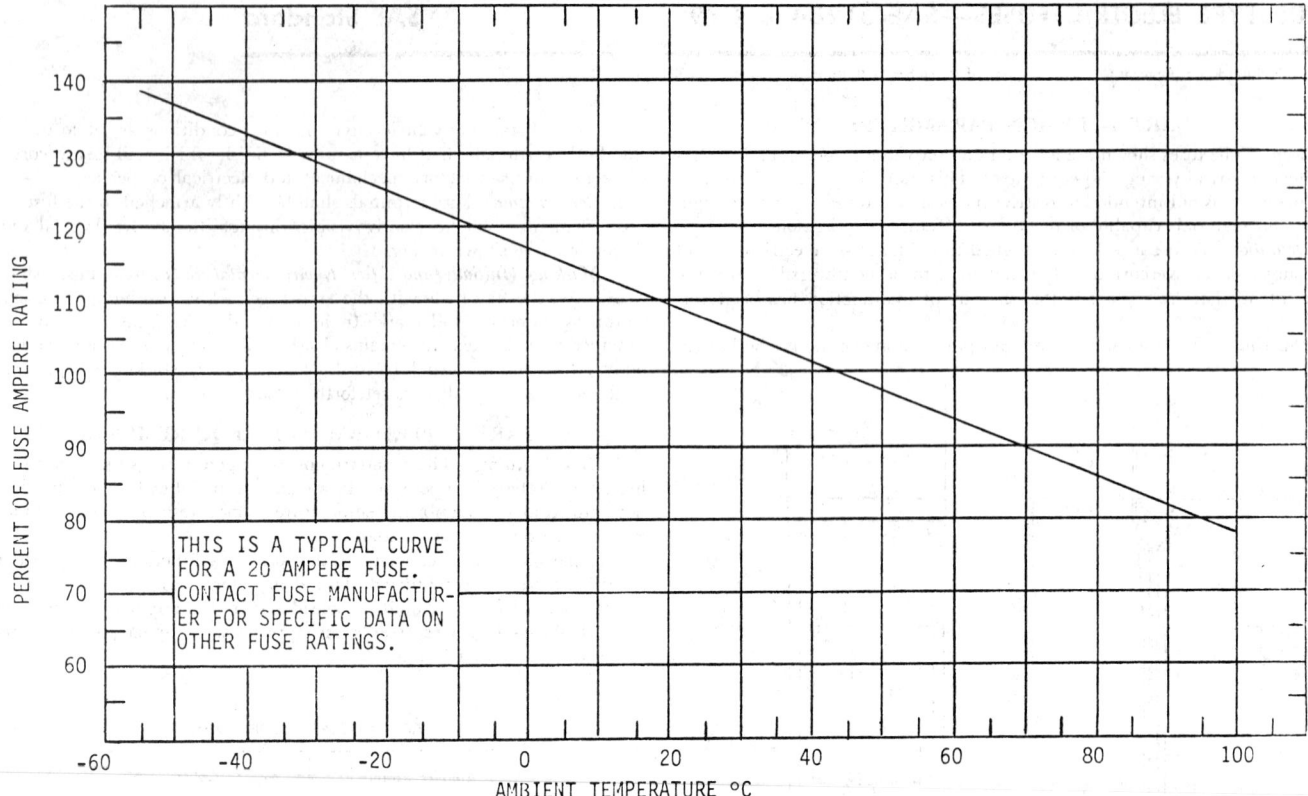

NOTE: THIS CURVE BASED ON 110 PER CENT OF RATED CURRENT USING 5 mm² COPPER CABLE, TESTED IN STANDARD TEST MODULE

FIG. 3—CHANGE IN FUSE CURRENT CARRYING CAPACITY WITH CHANGES IN AMBIENT TEMPERATURE

1.2 Temperature—Fig. 3 represents the approximate capacity change with respect to ambient temperature. The use of fuses in ambient temperatures beyond the limits shown is not recommended.

1.3 Loading—It is recommended that fuses not be loaded above 100% of their adjusted capacity based on ambient temperatures and the use of cable sizes other than 5 mm².

In addition, it is further recommended that actual performance be verified through testing experimentally in the vehicle.

2. Voltage Rating—Fuses shall be capable of interrupting at any voltage up to and including 32 V.

3. Maximum Voltage Drop—The maximum voltage drop (in millivolts) at rated current across the fuse only shall be as shown in Fig. 4.

4. Performance—Tests shall be conducted within an ambient temperature range of 19–29°C.

4.1 Ampere Rating Tests—Fuses shall carry 110% of rated current continuously for a minimum of 100 h; shall open in not less than 0.75 s or more than 1800 s at 135% of rated current; shall open in not less than 0.15 s or more than 5 s at 200% of rated current; and shall not open in less than 0.080 s at 350% of rated current.

4.2 Current Cycling Test—Fuses shall meet the requirements of Part 2, paragraph 4.1 after current cycling for a minimum of 250 000 cycles.

4.3 Transient Current Cycling—Fuses shall meet the requirements of Part 2, paragraph 4.1 after a minimum of 50 000 cycles of transient current cycling.

5. Vibration Test—Fuses shall meet the requirements of Part 2, paragraph 4.1 after undergoing six (6) h of vibration conditioning.

6. Environmental Exposure—Fuses shall meet the requirements of Part 2, paragraph 4.1 after sequential exposure to dust and accelerated aging conditioning.

PART 3—TEST PROCEDURE

1. Procedure—The fuses, with the exceptions noted in the vibration and accelerated aging tests, shall be mounted horizontally. When testing two or more fuses in series, the fuses shall be mounted no less than 150 mm apart and with no less than 600 mm of interconnecting cable, except as noted for the transient current cycling test. All electrical tests shall be made using 5 mm² copper cable and a standard test module (as shown in Fig. 5) or a suitable equivalent. The interface voltage drop ($V_{CD} - V_{AB}$) of the fixture should not exceed 2 mV/A. The total voltage drop (V_{EF}) should not exceed 4 mV/A. The voltage checks shall be made using a solid copper dummy with dimensions as shown in Fig. 6.

2. Voltage Drop—The voltage drop (in millivolts) at rated current across the

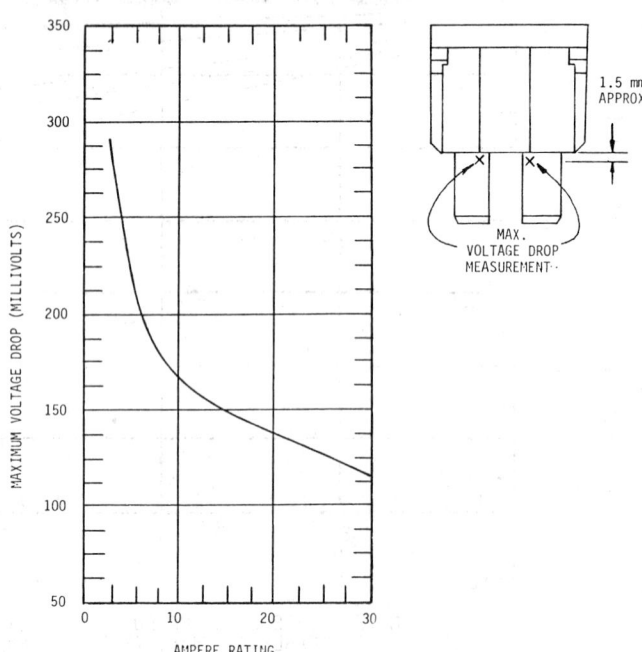

FIG. 4—MAXIMUM VOLTAGE DROP OF FUSE AT RATED CURRENT

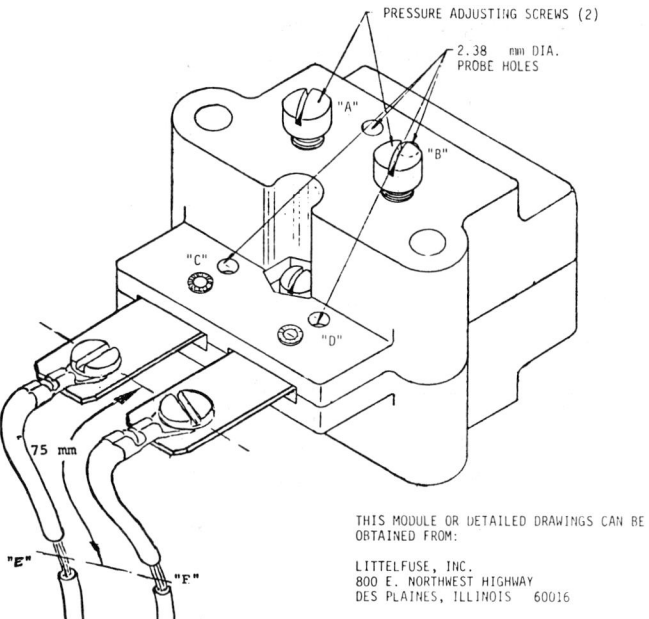

FIG. 5—STANDARD TEST MODULE

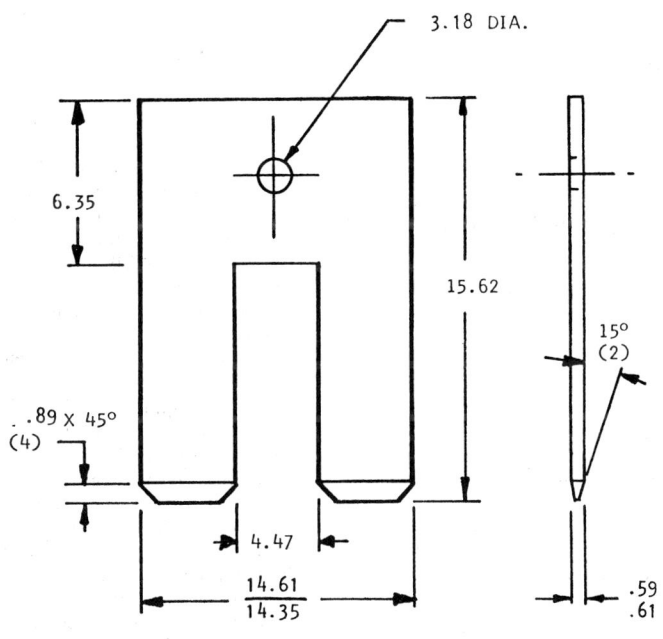

NOTE: ALL DIMENSIONS ARE IN MILLIMETERS
COPPER ALLOY #CA110

FIG. 6—DUMMY SLUG

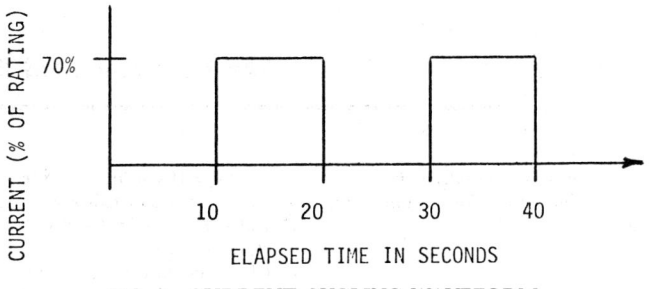

FIG. 7—CURRENT CYCLING WAVEFORM

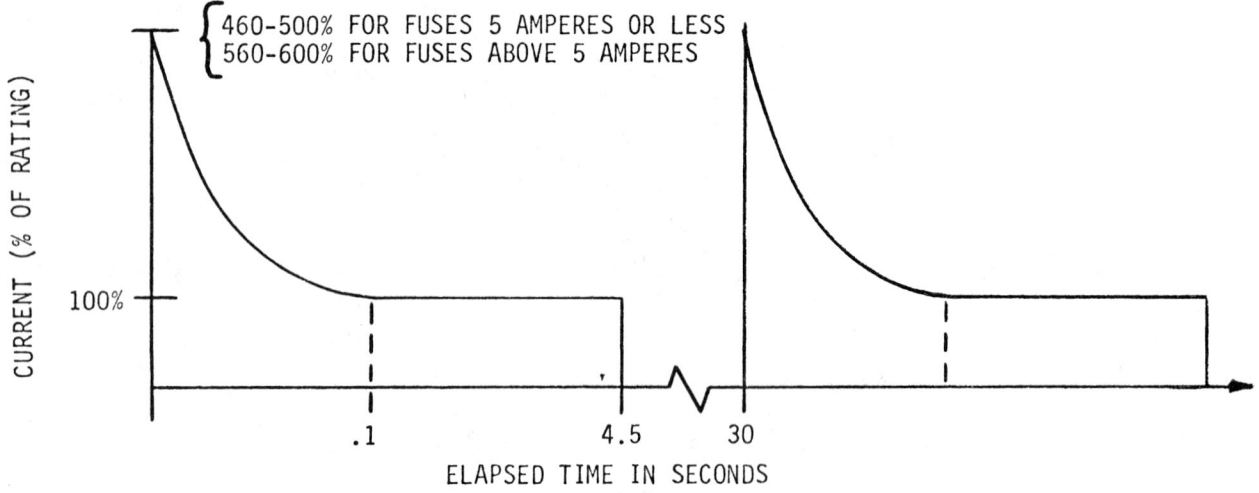

FIG. 8—TRANSIENT CURRENT CYCLING WAVEFORM

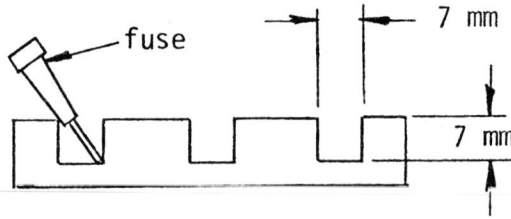

FIG. 10—TRAY (TYPICAL)

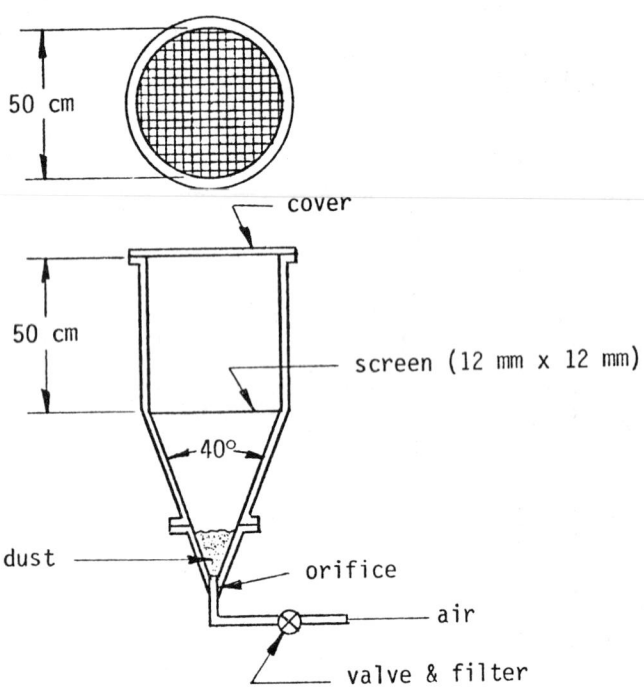

FIG. 9—DUST CHAMBER (TYPICAL)

fuse only shall be measured at the indicated points across the fuse from terminal to terminal as shown in Fig. 4.

3. Current Cycling Test

3.1 Resistors should be employed as load(s) to adjust the current to 70% of the fuse rating as shown in Fig. 7.

3.2 The test system voltage can be any convenient voltage up to and including 32 V (regulation ±0.2 V).

4. Transient Current Cycling Test

4.1 Lamps bulbs (EX. 1157, 194, and 168) should be employed as loads to adjust the initial peak transient current to the percent of fuse rating as shown in Fig. 8 and the initial steady-state current to 100% of the fuse rating. The test system voltage should be 14 ± 0.2 V.

4.2 It is acceptable if current levels decrease below initial stated values during current cycling as a result of normal lamp aging. Lamp bulbs that burn out must be replaced.

5. Vibration Test—Suitably mounted samples shall be subjected to a simple harmonic motion having an amplitude of 0.76 mm travel (1.52 mm total excursion). The frequency shall be varied uniformly between the limits of 10–55 Hz. The entire range of 10–55 Hz and returning back to 10 Hz shall be traversed in approximately 1 min. The motion shall be applied for a period of 2 h in each of three mutually perpendicular directions (total of 6 h). (Testing to be performed without current passing through fuse.)

6. Environmental Exposure

6.1 Dust—Fuses shall be placed in a dust chamber, Fig. 9, unmounted and lying on their side. The chamber shall contain about 1 kg of coarse grade dust conforming to SAE J726a. The dust shall be agitated for 3 s every 20 min by compressed air. The air shall be free of oil and moisture, at a gauge pressure of 5.6 ± 0.06 kg/cm^2 entering through an orifice 1.5 ± 1 mm in diameter. The total exposure time to dust shall be 5 h.

6.2 Accelerated Aging—A total of 15 cycles as follows:
16 h at 95–99% relative humidity and $37 \pm 1°C$[1]
2 h at $-40 \pm 1°C$
2 h at $70 \pm 1°C$
4 h at room ambient

[1]Fuses shall be placed in the humidity chamber on a tray, Fig. 10, with the blades down. Samples shall remain in the humidity chamber when testing cannot be completed in a 24 h period, for example, weekends.

CIRCUIT BREAKERS—SAE J553c

SAE Recommended Practice

Report of Electrical Equipment Committee approved November 1951 and last revised May 1972.

1. Scope—This SAE Recommended Practice covers the requirements for typical externally mounted in-line type automotive circuit breakers as follows:

1.1 12 V, Type I and Type II circuit breakers.

1.2 24 V, Type I circuit breakers.

2. Definitions

2.1 Type I Circuit Breaker is a cycling or continuously self-resetting unit.

2.2 Type II Circuit Breaker is a noncycling unit which remains open as long as the overload exists and resets when the overload no longer exists.

3. Reference Standards—SAE J573.

4. Basic Test Setup

4.1 Unless otherwise specified, the circuit breaker shall be tested in an ambient temperature of 77 ± 3 F (25 ± 1.7 C) in still air.

4.2 During the test of 12 V circuit breakers, the breakers shall be connected to a pair of 3 ft (0.9 m) leads using the wire size listed in Table 1.

4.3 During the test of 24 V circuit breakers, the breakers shall be connected to a pair of 6 ft (1.8 m) leads using the wire size listed in Table 1.

5. Current Rating Test

5.1 The circuit breaker shall continuously carry 100% of its rated current for 1 h after which the millivolt drop across the circuit breaker terminals shall not exceed the value specified in Fig. 1.

5.2 The circuit breaker shall trip at 125% of its rated current within 1 h.

6. Ambient Temperature Test

6.1 To detect the opening and closing temperatures, the circuit breaker shall be installed in a variable temperature controlled oven and connected in series with:

(a) Trade No. 53 bulb and a 12 V battery for 12 V circuit breakers.

(b) Trade No. 1820 bulb and a 24 V battery for 24 V circuit breakers.

6.2 For current ratings of 10 A or less, the contacts shall not open at less than 180 F (82.2 C) and shall not reclose at less than 130 F (54.4 C).

6.3 For circuit breakers with current ratings in excess of 10 A, the contacts shall not open at less than 240 F (112.2 C) and shall not reclose at less than 180 F (82.2 C).

7. Effective Current Test for Type I

7.1 The maximum effective value of current passed through the circuit breaker on overload shall not be greater than 125% of its rated value.

7.2 A suggested test method is as follows:

(a) Connect the circuit breaker with its leads to a fully charged battery (as described in paragraphs 4 and 10) and in series with a variable resistor load.

(b) Bare ½ in. (12.7 mm) of one lead in the middle of its length and solder a thermocouple at this point.

(c) With the circuit breaker bypassed, adjust the current to 125% of the circuit breaker rating and at the end of 10 min, record the temperature of the wire.

(d) Allow the wire with the attached thermocouple to cool to room temperature.

(e) With the circuit breaker not bypassed, adjust the current in the test circuit to 200% of the circuit breaker rating and at the end of 10 min, record the temperature of the wire.

(f) The circuit breaker meets the requirements of paragraph 7.1, if the temperature recorded in step (e) does not exceed that recorded in step (c).

8. Maximum Value of Current for Type II
After opening when subjected to 200% rated current, the current passed through the circuit breaker shall not exceed 1 A. This current shall be measured after 10 min with 14 V applied.

9. Hi-Pot Test for Type I and Type II
To insure that when a circuit breaker is functioning under overload another unsafe electrical path does not occur, an a-c "Hi-Pot" test is required as follows:

(a) Set up the test circuit described in paragraph 7.2 (a).

(b) Adjust the current to 400% of the circuit breaker rating and allow the circuit breaker to function for 10 min.

(c) During the first 5 min apply 440 V a-c from one terminal to the cover.

(d) During the final 5 min apply 440 V a-c from the other terminal to the cover.

(e) During Steps (c) and (d) there shall be no continuity between terminal and cover.

10. Life Test Setup
Using the basic test setup described in paragraph 4, the leads shall be connected as follows:

(a) *For 12 V Circuit Breakers*—To a fully charged 12 V battery the equivalent of 70 A-h capacity. Means shall be provided for maintaining 14 V at the battery terminals during the open circuit portions of the cycle.

(b) *For 24 V Circuit Breakers*—To a fully charged 24 V battery the equivalent of 35 A-h capacity. Means shall be provided for maintaining 28 V at the battery terminals during the open circuit portions of the cycle.

11. Life Test—Test 1

11.1 The circuit breaker shall be subjected to a 30 min short circuit test using the test setup described in paragraph 10.

11.2 At the conclusion of paragraph 11.1, the circuit breaker shall continuously carry 80% of its rated current for 1 h after which the millivolt drop shall not exceed the value specified in Fig. 1 for this 80% of rated current.

11.3 At the conclusion of paragraph 11.2, the circuit breaker shall be reconnected as in paragraph 11.1 and then tested to failure. Ultimate failure shall always result in an open circuit and shall not cause damage to connecting wires.

12. Life Test—Type II

12.1 Using the test setup described in paragraph 10, the circuit breaker shall be subjected to a 20 cycle life test.

The "on" time of each cycle shall be 60 s during which time the circuit breaker must open.

The "off" time of each cycle shall be of sufficient duration to allow the circuit breaker to reclose.

The "on" time of the twentieth cycle shall be 24 h with 11.3 V applied. During this time, the circuit breaker must remain open.

12.2 Repeat paragraph 12.1, but exclude the 24 h "on" requirement on the 20th cycle.

12.3 At the conclusion of paragraph 12.2, the circuit breaker shall continuously carry 80% of its rated current for 1 h after which the millivolt drop shall not exceed the value specified in Fig. 1 for this 80% of rated current.

13. General Requirements

13.1 Circuit breakers shall have suitable identification for voltage and current ratings.

13.2 Fig. 2 depicts a typical design but is not intended to be restrictive. Size, shape, and means of termination may vary providing the performance requirements of this recommended practice are met.

TABLE 1a

Ampere Rating	SAE Wire Size
5–10	16
Greater than 10 to 15 incl.	14
Greater than 15 to 30 incl.	12
Greater than 30 to 40 incl.	10
Greater than 40 to 50 incl.	8

a This table is to be used for test purposes only. It is not a guide for the selection of a particular wire size in conjunction with a circuit breaker rating.

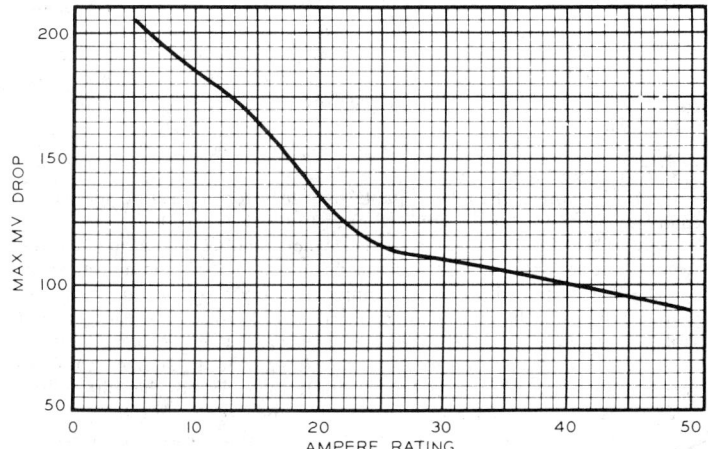

FIG. 1—MAXIMUM MILLIVOLT DROP VERSUS AMPERE RATING

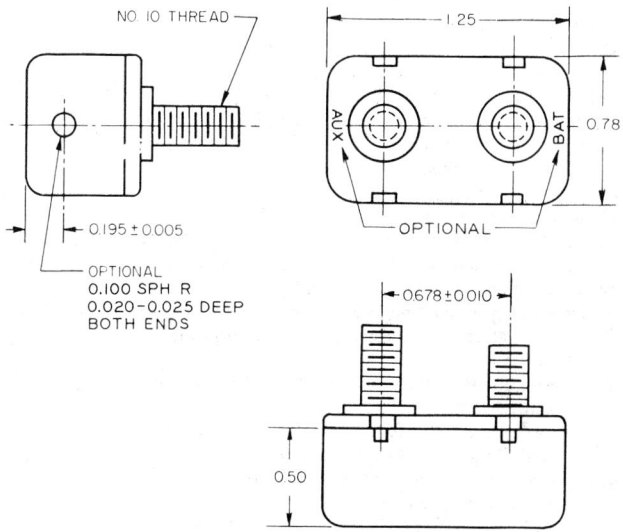

FIG. 2—TYPICAL AUTOMOTIVE CIRCUIT BREAKER

CIRCUIT BREAKER—INTERNAL MOUNTED—AUTOMATIC RESET—SAE J258

SAE Recommended Practice

Report of Electrical Equipment Committee approved June 1971.

1. Definition—An internally mounted circuit breaker is one mounted within an automotive switch or other automotive device for protection against overload of the wiring. In a given application, this same circuit breaker may be designed to protect against overload of the electrically operable devices.

2. Test Procedure

2.1 The circuits containing the internally mounted circuit breakers shall have 3 ft leads connected to their terminals. The wire, size, and mating terminals specified must be the same as the intended application or, if there are a variety of applications, then the minimum wire size with which the circuit breaker will meet the requirements of this recommended practice.

2.2 The *circuits* containing the internally mounted circuit breakers shall be allowed a 0.20 V drop at rated current, when tested individually, in addition to the voltage drop allowed in the switch or other device specification.

3. Requirements—In a switch or other device with multiple circuit breakers, the following requirements are for testing each circuit breaker individually, with the design load applied to the other circuit breakers:

3.1 Continuously carry 100% of its rated current (as specified by the manufacturer for 1 h at 75 ± 3 F (24 ± 1.6 C).

3.2 After a 1 h soak at 125 ± 3 F (51.6 ± 1.6 C) and no load, the circuit breaker shall carry 80% of its rated current for 1 h at that temperature.

3.3 Effective current value shall not exceed 140% of its rated current when tested according to procedure as outlined in SAE J553 for Type I.

3.4 The circuit breaker must fail safe (open circuit) when tested as outlined in the life test of SAE J553 for 30 min or until destruction.

IGNITION SWITCH—SAE J259

SAE Recommended Practice

Report of Electrical Equipment Committee approved June 1971.

1. Definition—An ignition switch is that part of an electrical system by which the operator of a vehicle causes the ignition system to function. In addition to ignition circuitry, the switch may have circuits which control electrical accessories, engine starting, warning indicator lamp checking, etc.

2. Temperature Test

2.1 To insure basic function, the switch shall be manually cycled for 10 cycles at design electrical load at 75 ± 10 F (24 ± 5.5 C), $165 +0, -5$ F ($74 +0, -2.8$ C), and $-25 +5, -0$ F ($-32 +2.8, -0$ C) after a 1 h exposure at each of these temperatures. The switch shall be electrically and mechanically operable during each of these cycles.

2.2 This same switch shall be used for the endurance test described below.

3. Endurance Test Setup

3.1 The switch shall be set up to operate its design electrical load.

3.2 The test shall be set up to operate the switch for the prescribed number of complete cycles.

One complete cycle shall consist of sequencing through each position (with dwell in each position) and return without dwell in intermediate positions to the initial position.

The test equipment shall be so arranged as to provide the following mechanical time requirements:

Travel Time—0.1–0.5 s (time from one position to the next).
Dwell Time—1.0–2.0 s (time in each position).

3.3 During the test the switch shall be operated at:

6.4 V d-c for a 6 V system
12.8 V d-c for a 12 V system
25.6 V d-c for a 24 V system

These voltages shall be measured at the input termination on the switch.

The power supply shall not generate any adverse transients not present in motor vehicles and shall comply with the following specifications:

(a) *Output Current*—Capable of supplying the continuous current of the design electrical load.

(b) *Regulation*

Dynamic—The output voltage at the supply shall not deviate more than 1.0 V from zero to maximum load (including inrush current) and should recover 63% of its maximum excursion within 100 ms.

Static—The output voltage at the supply shall not deviate more than 2% with changes in static load from zero to maximum (not including inrush current), and means shall be provided to compensate for static input line voltage variations.

(c) *Ripple Voltage*—Maximum 300 mV peak to peak.

4. Endurance Requirements

4.1 The switch shall be capable of satisfactory operating for 25,000 complete cycles at a temperature of 75 ± 10 F (24 ± 5.5 C).

4.2 When the switch has a position which operates accessories only, the switch shall be cycled to that position during 25% of the endurance cycles.

4.3 The voltage drop from the input terminal(s) to the corresponding output terminal(s) shall be measured before and after the endurance test and shall not exceed 0.30 V (the average of three consecutive readings) at 10.0 A except in circuits such as warning indicator lamp checking and/or grounding type circuits which do not exceed 1 A load. In these circuits, the voltage drop shall not exceed 0.50 V (the average of three consecutive readings) at 1.0 A. If wiring is an integral part of the switch, the voltage drop measurement shall be made including 3 in. of wire on each side of the switch; otherwise, measurement shall be made at switch terminals.

AUTOMOBILE, TRUCK, TRUCK-TRACTOR, TRAILER, AND MOTOR COACH WIRING—SAE J1292 JUN80

SAE Recommended Practice

Report of the Electrical Equipment Committee, approved June 1980.

[This SAE Recommended Practice combines, revises, and replaces two previous recommended practices: SAE J555a and SAE J556.]

1. Scope—This SAE Recommended Practice covers the application of primary wiring distribution system harnesses to automotive, truck, and similar type vehicles. This is written principally for new vehicles but is also applicable to rewiring and service. It covers the areas of performance, operating integrity, efficiency, economy, uniformity, facility of manufacturing, and service. This practice applies to wiring systems of less than 50 V.

2. General Section

2.1 Definition—The systems of installation known as two wire or single wire are to be designated respectively as *insulated-return* and *ground-return* systems. Installations in which the frame and/or body of the vehicle are used as part of the return circuit are considered as *ground-return* systems.

2.2 Insulated Cable—All insulated cable shall conform to SAE Standards J1127 and J1128.

2.2.1 CONDUCTORS

2.2.1.1 All conductors are to be constructed in accordance with SAE J1127 and J1128 except when good engineering practice dictates special strand constructions.

2.2.1.2 Conductor materials and stranding other than copper can be used if all applicable requirements for physical, electrical, and environmental conditions are met as dictated by the end application.

2.2.2 CONDUCTOR INSULATION—Physical and dimensional values of conductor insulation are to be in conformance with the requirements of SAE J1127 and J1128 except when good engineering practice dictates special conductor insulations.

2.3 Insulated Cable Application

2.3.1 Select cable insulation in accordance with the vehicle's working environment. Consideration is given to physical and environmental factors such as flexing, heat, cold, bend, oil and fuel contact, dielectric, abrasion, short circuit, and pinch resistance among others.

NOTE—Most vehicle working environments permit the use of a thermoplastic insulated, SAE type GPT, general purpose cable. A cable of this type is generally used in static (non-flexing) applications when nominal abrasion, heat, cold, oil, dielectric, short circuit, and pinch resistance properties are desired.

2.3.2 Where vehicle working environments for cable require additional physical and environmental characteristics, upgraded insulations such as SAE types HDT, GPB, HDB, STS, HTS, and SXL shall be used as the severity of the applications dictate.

2.3.3 Specific continuous duty temperature limitations for each SAE cable type shall be observed. The total of the ambient temperature plus cable temperature rise, due to current flow, should not exceed the continuous duty guideline temperatures as shown in Table 1, unless extensive testing and/or evaluation has indicated that higher temperatures can be tolerated.

In addition, the maximum continuous duty temperature rating for any wire insulation shall be determined by an accelerated aging test conducted in accordance with ASTM D 573, with the samples of insulation being removed from the finished wire and aged 168 h. The test temperature shall be 30°F (17°C) above the intended rated temperature. Tensile strength after aging shall be not less than 80% of the original tensile strength. The elongation after aging shall be at least 50% of the original elongation.

NOTE—Heavier conductors may be required to protect the carrying of current in wire bundles when all conductors are carrying maximum current. Temperature rise tests of the conductor bundle should be run to determine the proper conductor size and insulation.

Resistance wire low tension cable may be used to limit the voltage applied to electrical devices. Since the nature of the wire is to limit the voltage applied to electrical devices, the distance of the device from the power source and the current demand of the device will determine the materials used. Because every application is different, no materials, conducting or insulating, can be specifically described as standard; thus the conductor and insulating materials must be carefully chosen for each application by the design engineer. It is desirable to identify resistance wire by printing the words *resistance wire* on the conductor.

Extreme care shall be used by the design engineer in choosing resistance wire as a conducting material to satisfy the current demand of the device and not create a temperature rise in the conductor that would deteriorate the insulating material even though the device is left on continuously.

Circuits using resistance wire shall be carefully placed in the vehicle so that their temperature rise will not create a hazard to, or malfunction of, any part of the vehicle. A general design guide would be that the conductor be required to dissipate no more than 5 W per insulated conductor foot.

2.3.4 FUSIBLE LINKS—A special section of low tension cable designed to open circuit when subjected to an extreme current overload shall conform to SAE J156.

2.3.5 It is desirable to color code each conductor in an electrical circuit to facilitate manufacture and service of a wire assembly. It is further desirable for all motor vehicle manufacturers to assign and use similar color code identifications for commonly used electrical circuits to promote ease of circuit analysis in service among the various manufacturers.

2.3.5.1 When feasible, each circuit shall conform to a recommended color code by category of equipment as shown in Table 2. Otherwise, the color code may be a solid color (basic) and/or a basic color with secondary color stripes, dots, or hashes.

2.3.5.2 Secondary color markings to be applied as to be visible throughout the entire length of the wire, or at each end of a lead.

2.3.5.3 Color combinations for special circuits not shown on Table 2 are to be selected by the user. As special circuit functions become standard with manufacturers, they shall be added to the recomended Color Code by category and shown in Table 2.

NOTE—It is desirable for the wire of any one circuit to be of uniform color code throughout the circuit regardless of the number of connections. A circuit is assumed to be continous until it can be interrupted by a relay or switch contacts, or when it reaches a load (such as bulbs, motors, etc.). Fusible links may differ in color from the circuits they are protecting as it could be advantageous to identify fusible link wire gauge size by insulation color.

2.3.5.4 Each circuit in the same wire assembly shall be distinguished from one another in some manner such as color code, or some substantial difference in insulation diameter (that is, two or more gauge sizes).

2.4 Conductor Termination

2.4.1 All stranded conductor stripped ends are to be fitted with terminals (exception—splices). Solid, precisely shaped conductors whose ends are the termination shall not have this fitting.

2.4.2 All terminal attachments to conductors shall conform with the physical and electrical performance requirements of SAE J163.

2.4.3 As a general practice, all terminations have integral and functional insulation grips, except where other secondary applications preclude their use. Special applications without insulation grips may be employed where other means of relieving strain are provided.

2.4.4 A terminal shall be attached to a conductor by a simple mechanical *crimp-type* process that will conform to the intent of paragraph 2.4.2. For maximum reliability and surety of connection, the *crimp* may also be soldered, swaged, brazed, or welded in a workmanlike manner. Care shall be taken to minimize wicking of solder in a stranded wire to avoid impairment of the strain relief or cable flexing.

2.4.5 CIRCUIT GROUNDING—Ground terminal lugs shall be solder dipped, cadmium, tin, or zinc plated. Ground terminals shall be accessible for service. A serrated paint cutting terminal may be utilized to make proper contact on

TABLE 1

SAE Cable (Ref. SAE J1128)	Temperature[a]
Type GPT, HDT, GPD, HDB	194°F (90°C)
Type STS, HTS	221°F (105°C)
Type SXL	275°F (135°C)

[a]Recommended maximum continuous duty temperature (ambient plus rise).

TABLE 2—CIRCUIT COLOR CODE—BASIC CIRCUITS (AUTOMOTIVE ONLY)

Function	Color
Left rear stop and turn	Yellow
Right rear stop and turn	Dark green
Auxiliary	Blue
Tail, side marker, license	Brown
Ground	White

NOTE—The above code is identical to the color code adopted for automotive type trailers—SAE J895.

painted surfaces. Ground terminal devices shall be cadmium, tin, or zinc plated. In special cases, plating may not be required for lugs and/or attaching devices.

Ground return connections shall be made to the vehicle structure, frame, or engine. In cases where the engine or body is mounted on rubber or other insulation, proper ground shall be provided.

2.4.6 Terminations used shall comply with the requirements of SAE J561, ring and spade types; SAE J858, blade type; and SAE J928, pin and receptacle type. Secondary applications will dictate the use of special terminations for special use or application.

Note—Terminations may be plated with a conductive and corrosion resistant material such as tin or silver to upgrade the current carrying capacity and to improve their resistance to corrosion.

2.5 Conductor Splicing

2.5.1 Conductors shall be mechanically crimped, soldered, swaged, brazed, or welded with other conductors to form a wire splice. All wire splices shall conform with the electrical specifications for splices per SAE J163.

2.5.2 Splices shall be mechanically secure to withstand all fabrication installation and vehicle environment abuse. The splice must be insulated.

2.6 Terminal and Connector Function

2.6.1 Single terminations shall be used only where there is no possibility of misconnections in assembly or service except when special applications may require otherwise.

2.6.2 Multiple terminal connect-disconnect connector bodies shall be used at all points where two or more conductors are terminated and where there is a possibility of misconnection in fabrication, assembly, or service; secondary applications may require a deviation from this practice.

2.6.3 All connections shall be designed to maintain surety of connections while subjected to vibration, shock, and the extreme temperatures that are normal environmental conditions for motor vehicles. Surety may be accomplished by employing the use of integral-molded lock devices, terminal to terminal interferences (detents), secondary locking clips, or attaching devices.

2.6.4 All multiple connect-disconnect connector bodies shall be polarized to prevent incorrect assembly unless circuitry permits use of a nonpolarized connector.

2.6.5 Connections shall be located in clean, dry areas when possible. Connections shall be designed to maintain circuit integrity regardless of environmental conditions (such as high humidity, road splash, rain, drainage, earth particles, fuels, lubricants, high and low temperatures, and solvent).

2.7 Conductor Grouping

2.7.1 Conductors are to be grouped together into multiple conductor assemblies whenever possible.

2.7.2 The number of wiring assemblies and electrical connections per vehicle shall be kept to a minimum with overlay or option wiring used only when justified by the economics of fabrication, vehicle installation, and service.

2.8 Wire Assembly Construction

2.8.1 Conductors are to be grouped, where practical, in cable or harness form.

Note—Suitable material such as braided cotton, braided paper and cotton, braided vinyl/nylon, flexible plastic conduit, friction or thermoplastic tape, extruded rubber and thermoplastic jackets, or woven loom may be used to form the assembly.

2.8.2 Wiring harness covering shall be adequate to protect the harness in the vehicle routing environment and shall furnish protection during all phases of vehicle assembly and operation.

Note—A general guideline to be used in the selection of coverings is specified in Table 3.

2.9 Wire Assembly Installation and Protection

2.9.1 Wiring and related devices shall be installed in a workmanlike manner, mechanically and electrically secure. Devices, lamps, and so forth requiring periodic service shall be serviceable and accessible by providing wire length sufficient to reasonably accomplish this.

2.9.2 In general, wire routing shall be such that maximum protection is provided by the vehicle sheet metal and structural components. Smooth protective channels especially designed for wiring and built into the vehicle body structure should be used when practicable. Avoid areas of excessive heat, vibration, and abrasion.

Note—Extra protection (such as braid, loom, conduit, etc.) should be provided when these areas cannot be avoided (Ref. Table 3).

2.9.3 All parts of the electrical system shall be adequately protected against corrosion.

2.9.4 If significant vibration levels exist, the edges of all metal members through which cables and harnesses pass shall be deburred, flanged, rolled, or bushed with suitable grommets. Suitable tubing or conduit over cables may be substituted for grommets if properly secured. Clips for retaining cables and harnesses shall be securely attached to body or frame member and cable or harness. Clips also assist in locating and routing at assembly.

2.9.5 Wiring shall be located to afford protection from road splash, stones, abrasion, grease, oil, and fuel. Wiring exposed to such conditions shall be further protected by either, or a combination of, the use of heavy wall thermoplastic insulated cable, (see SAE Standard J1128, Low Tension Primary Cable) additional tape application, plastic sleeving or conduit, nonmetallic loom, or metallic or other suitable shielding or covering.

2.9.6 Where cables must flex between moving parts, the last supporting clip shall be securely mounted and secure the cable in a permanent manner.

2.9.7 Wiring fasteners shall be non-conductive unless the wiring or fastener involved is provided with extra heavy outer covering such as nonmetallic conduit, tape, or dip.

Note—Overlay or option wiring should be routed in the same fasteners with standard wiring where practical, or should be fastened to the standard wiring with plastic straps or other mechanical means.

2.9.8 Electrical apparatus with integral wiring shall be supplied with grommets or other suitable mechanical fasteners for strain relief.

2.10 Wiring Overload Protective Devices

2.10.1 The current to all low-tension circuits, except starting motor and ignition circuits, shall pass through short circuit protective devices connected to the battery feed side of switches. Headlight systems shall be independently protected. Circuit protection shall be accomplished by utilizing fuses, circuit breakers, or fusible links which conform to SAE Standards.

2.10.2 The protective device shall be selected to prevent wire damage when subjected to extreme current overload.

2.10.2.1 *Fuses*—Fuse sizes shall be selected using guidelines presented in SAE J554, Electric Fuses.

2.10.2.2 *Circuit Breakers*—Fail-safe automatic reset circuit breakers shall be employed when it is necessary to quickly re-establish circuit continuity when that portion of the wiring has been subjected to an overload condition. Non-cycling type circuit breakers will not reset until the overload is removed, (unless they are the non-cycling manual-reset type). Circuit breakers shall conform to SAE J553 and SAE J258.

2.10.2.3 *Fusible links* shall be employed when heavy feed circuits exceed the continuous working limits of the fuses or circuit breakers. The link of wire, acting like a fuse, shall conform to the guidelines presented in SAE J156, Fusible Links.

3. Truck, Truck-Tractor Section

3.1 The following SAE Recommended Practice relates to wiring for exterior lamps, exclusive of head lamps, of commercial vehicles 80 in (203 cm) or more in width. Except as noted, the wiring system shall conform to the guidelines of Section 2.

3.1.1 Lamp—A lamp is a complete lighting unit. All lamps shall meet the requirements of SAE Standard J575, Lighting Equipment for Motor Vehicles. Lamps with pigtails not in excess of 12 in (30 cm) long shall have a minimum of 16-gauge wire; pigtails in excess of 12 in (30 cm) long shall have wire gauge conforming to the wiring requirements of the vehicle.

3.1.2 Wire Size—To minimize voltage drops, the feed wire size for all circuits shall be a minimum of 12-gauge; branches or taps not in excess of 50 ft (15.2 m) in length shall be 14-gauge. The ground wire for insulated-return systems shall be equal to the respective feed wire. The main ground wire shall be a minimum of 10-gauge.

Note—In many cases 4 or 6 gauge may be required.

3.1.3 Design Voltage of Lamps—Reference SAE Standard J573, Lamp Bulb and Sealed Units for design voltage values applicable to various bulbs.

3.1.4 Truck-tractors shall conform to Section 2 and the following:

3.1.4.1 *Circuit Identification*—It is desirable to follow the SAE Recommended Practice J1067, Seven Conductor Jacketed Cable for Truck and Trailer Connections, for coding of truck-tractor jumper cable throughout the circuit. Where impractical, the coding is to be followed to a junction block or harness terminating point where visual inspection will identify the circuit coding change. The coding may also be numbers and/or letters printed on the

TABLE 3

Type	Wire Harness Covering	General Application
1	Vinyl Plastic Tape—0.007 in (0.18 mm)	Primarily used for grouping cables into wire harnesses. Wiring not subject to damage from scuffing or scrubbing on rough metal edges.
2	Friction Tape, Cotton and Kraft Paper Braid	Generally optional; improved scuff and scrub resistance.
3	Vinyl/Nylon Braid	Improved abrasion resistance.
4	Non-Metallic Loom (Woven Asphalt, Impregnated Loom, Extruded Vinyl Plastic, or Elastomeric Tubing)	Improved scuff and abrasion resistance.
5	Rigid and Flexible Conduit	For maximum abrasion resistance and/or positive positioning for clearance to moving or heated vehicle components.

wire insulation. Whatever coding system is chosen, the system shall facilitate in harness manufacturing and in service.

3.1.4.2 *Circuit Termination*—Wiring for trailer circuits shall terminate in:
(a) A connector socket conforming to SAE Recommended Practice J560, Seven-Conductor Electrical Connector for Truck-Trailer Jumper Cable, or
(b) A jumper cable with cable plug conforming to SAE J560.

4. Trailer Section

4.1 Trailers shall conform to Section 2 and the following:

4.1.1 WIRING—All wiring shall be installed in:
(a) Suitable conduit and boxes,
(b) Structure of the trailer, and
(c) Housings and/or raceways which provide equal protection.

Wiring shall be protected from stones, excess dirt, ice, moisture, chafing, and so forth, that will result in harmful effects.

All wiring for legally required lights shall be serviceable in a manner permitting removal and reinstallation from outside the trailer.

4.1.2 GROUNDING—The trailer shall be grounded to the tractor through the jumper cable.

NOTE—Contact of the trailer king pin or apron plate with the lower coupler or *grounding through the lower coupler* is not to be considered as providing a tractor-to-trailer ground.

4.1.3 MARKING—The voltage of the lighting system shall be permanently or semi-permanently marked in a legible manner on a mounting surface, in proximity to the electrical connector receptacle. Preferably, the marking shall be in amber reflective letters.

4.1.4 TRAILER CONNECTOR SOCKET—The trailer connector socket for receiving the jumper cable plug shall conform to SAE J560.

4.1.5 CIRCUIT PROTECTION—Circuit protection independent of truck-tractor system shall be provided. Trailer circuit protective devices shall conform to SAE Standard J554, Electric Fuses or SAE Recommended Practice J553, Circuit Breakers and shall be located near the trailer wiring connector socket and be readily accessible for service.

5. Motor Coach Section

5.1 Motor coaches shall conform to Section 2 and the following:

5.1.1 WIRING—Where practical, wiring is to be located within the structure of the coach where it will not be subjected to damage by road splash, stones, grease, oil, fuel, or abrasion.

Wiring so located that it will be subjected to more than normal wear or hard usage shall be equipped with a means of disconnecting from the main harness and be easily removable for replacement or repair.

Wiring connections to lights mounted on the coach body shall be accessible from outside, with the light removed or through an access door in an interior trim panel.

6. Storage Battery Cables

6.1 Definition—Battery cables provide the link between the battery(s) and the balance of the starting/charging circuit. Items that dictate the design are:

6.1.1 ROUTING—Routing shall be established with the following guidelines:

6.1.1.1 Areas of excessive heat, abrasion, and vibration are to be avoided. Extra protection (such as loom, conduit, tubing, heat shield, etc.) shall be provided when these areas cannot be avoided.

6.1.1.2 Grommets or ferrules and nipples shall be provided when routed through holes in the frame or sheet metal.

6.1.1.3 Support at intervals of approximately 24 in (61 cm). Insulated or nonconductive supports shall be used.

6.1.1.4 Provide strain relief for the battery and starter motor terminals as close to terminals as practical.

6.1.1.5 Tailor such that the cables are not too loose nor too tight, considering engine rocking due to torque changes.

6.1.2 VOLTAGE DROP—Voltage drop for starting motor circuits as recommended in SAE J541, determines the maximum drop allowed for the total cranking circuits from the battery to the starter motor and the return to the battery.

6.1.3 CABLE SIZE—Cable size is determined by knowing the system parameters and subtracting their fixed resistances (such as connections, starter solenoid, ground path other than the battery cable, etc.) from the total specified in paragraph 6.1.2. This remaining resistance is the maximum allowed for the battery cables.

6.1.4 CABLE CONSTRUCTION—Cable construction is determined from the environment in which the battery cables must survive.

6.1.4.1 *Core Stranding*—Core stranding of conventional cable can be either bunched, concentric stranded, or rope lay. Bunched or concentric will suffice in most applications, except those requiring higher flex life. For larger cable sizes, rope stranding is needed for routing purposes as well. Battery strap is available for extreme flex requirements and restricted space or routing problems. Reference SAE J1127.

6.1.4.2 *Insulation*—Insulation provides electrical as well as environmental protection for the core. Polyvinyl chloride (PVC) can be used in most applications; cross-linked polyethylene, hypalon, neoprene, etc., may be needed for added protection against short circuit, high temperature, abrasion, etc. Reference SAE J1127 and Table 1.

6.1.4.3 *Terminals*—Terminals provide the connection to the battery, starter solenoid, junction blocks, switches, and grounding locations. A multitude of different types and styles are available for the variety of cable sizes. Also available are sleeves and covers which provide additional circuit and corrosion protection. Reference SAE J561 and SAE J163.

AUTOMOTIVE PRINTED CIRCUITS—SAE J771c — SAE Standard

Report of Electrical Equipment Committee approved June 1961 and last revised April 1975.

Scope—This report relates to recommendations and specifications governing the classification, composition, test procedures, and properties of printed circuits commonly used to replace cable in automotive low voltage systems. It is not applicable to miniature circuits for solid state devices, high impedance or high voltage functions.

Base Materials—The base insulating materials fall into three categories: Phenolic Laminates, Plastic Molding Materials, and Flexible Plastic Films of the type which will meet the following flame retardancy requirement.

Flame Retardancy—Flame resistance of the material used for base or overlay shall be equal to or better than phenolic laminates, acrylonitrile butadiene, styrene, or polyester films from polyethylene terephthalate, respectively, as measured by ASTM D 635, Tests for Flammability of Rigid Plastics Over 0.127 cm (0.050 in.) in Thickness, or ASTM D 568, Test for Flammability of Plastics 0.127 cm (0.050 in.) and Under in Thickness.

1. PHENOLIC LAMINATES—When plastic laminate base material is used, the quality characteristics and tolerances must conform to the following:
(a) Paper laminate thoroughly impregnated with thermosetting phenolic resin binder and properly cured.
(b) The material shall be opaque unless otherwise specified.
(c) Thickness shall be 0.062 in. (1.58 mm) unless otherwise specified. (Refer to item E under Design Considerations.)
(d) Flexural strength shall not be less than 13,000 psi (89.6 MPa) with the grain or less than 11,000 psi (75.8 MPa) across the grain (ASTM D 229, Testing Rigid Sheet and Plate Materials Used for Electrical Insulation, and ASTM D 790, Method of Test for Flexural Properties of Plastics).

2. PLASTIC MOLDING MATERIALS—When plastic molding material is used, the quality characteristics and tolerances must conform to the following:

(a) The material must be opaque in cross section of 0.062 ± 0.005 in. (1.58 ± 0.13 mm) thickness.
(b) Minimum physical properties:
 (1) Impact strength: 1.3 ft-lb/in. (69.4 N·m/m) notch ¼ in. (6.35 mm) bar sample notches at 73°F (22.8°C) (ASTM D 256, Methods of Test for Impact Resistance of Plastics and Electrical Insulating Materials).
 (2) Flexural strength: 11,220 psi (77.36 MPa) (ASTM D 790).
 (3) Tensile strength: 7750 psi (53.4 MPa) (ASTM D 638, Method of Test for Tensile Properties of Plastics).
 (4) Heat distortion temperature 225°F (107.2°C) at 264 psi (1.8 MPa) (ASTM D 648, Method of Test for Deflection Temperature of Plastics Under Load).

The materials in paragraphs 1 and 2 shall not be subjected to temperatures in excess of 225°F (107.2°C).

3. FLEXIBLE PLASTIC FILM—When flexible plastic film is used, the quality characteristics and tolerances must conform to the following:
(a) Tensile strength: 17,000 psi (117.2 MPa) (minimum), ASTM D 882, Methods of Test for Tensile Properties of Thin Plastic Sheeting (Method A-100% elongation per minute).
(b) Tensile modulus: 450,000 psi (3.1 GPa) (minimum) ASTM D 882 (Method A-100% elongation per minute).
(c) Tear initiation: 2000 lb/in. (226 N·m) (minimum) Graves, ASTM D 1004, Method of Test for Tear Resistance of Plastic Film and Sheeting.
(d) Thermal coefficient of linear expansion: 15×10^{-6} (maximum) 70–120°F (21–49°C).
(e) Moisture absorption, water immersion: less than 0.5%, 24 h.
(f) Dielectric strength: 500 V/mil (19.7 kV/mm) minimum per ASTM

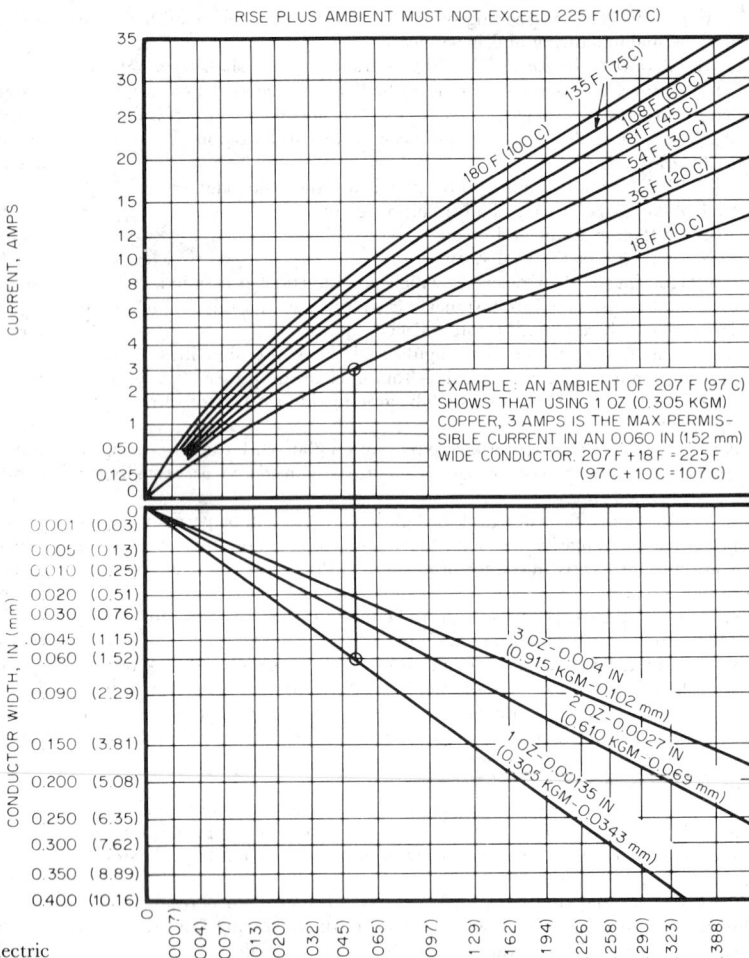

FIG. 1—CONDUCTOR THICKNESS AND WIDTH

FOR USE IN DETERMINING CURRENT CARRYING CAPACITY AND SIZES OF COPPER CONDUCTORS FOR VARIOUS TEMPERATURE RISES ABOVE AMBIENT. CURVES ARE DERATED A NOMINAL 10% TO ALLOW FOR VARIATIONS IN PRODUCTION TECHNIQUES

D 149, Methods of Test for Dielectric Breakdown Voltage and Dielectric Strength of Electrical Insulating Materials at Commercial Power Frequencies.

(g) Melting point: more than 350°F (177°C).

(h) Thickness to be 0.003 in. (0.08 mm) minimum.

Copper

1. Printed circuit grade electrodeposited copper foil with:

   (a) 99.8% minimum purity per ASTM E 53-48.

   (b) Maximum oxygen content 0.1% per ASTM E 53-48.

   (c) Minimum elongation of 10%, both transverse and longitudinal. Measurement shall be made using a 2 in. (50.80 mm) gage length with an extension rate of 2 in./min (0.85 mm/s).

   (d) Tensile strength: Longitudinal—25,000 psi (172 MPa) minimum, transverse—25,000 psi (172 MPa) minimum.

   (e) Bulge height: For test method see Appendix. Bulge height shall be 0.180 in. (4.57 mm) minimum.

   (f) Resistivity: The resistivity shall not exceed $0.15940 \, \Omega/m^2$ at 68°F (20°C).

   (g) The foil shall be free of any lead inclusions or other foreign materials.

   (h) The foil shall have an antitarnish treatment which is compatible with the subsequent bonding or lamination process and which will offer no additional significant increase in the electrical resistance between the foil surface and any component.

2. Thicknesses employed may be 0.00135 in. (0.034 mm), which is 1 oz/ft² (0.305 kg/m²); 0.0027 in. (0.069 mm), which is 2 oz/ft² (0.610 kg/m²); or 0.004 in. (0.102 mm), which is 3 oz/ft² (0.915 kg/m²). The choice of thickness is made on the basis of the thinnest copper foil that will meet the current-carrying requirements for any specific applications (see Fig. 1). Discoloration of the copper faces shall not be considered objectionable.

Coating—On phenolic laminate or molded plastics, a scratch resistant protective insulating coating film shall be bonded to all foil side surfaces which will not require contact with other components. The coating film or overlay used on circuits with flexible plastic base must be firmly bonded such that the fold endurance (see Physical Test Requirements) will be maintained.

Design Considerations

1. Practical Tolerances on Critical Dimensions

   (a) Hole diameters:

   Up to 0.500 ± 0.004 in. (12.7 ± 0.10 mm)
   0.500 to 1.000 ± 0.005 in. (12.7–25.4 ± 0.13 mm)
   Over 1.000 ± 0.006 in. (25.4 ± 0.15 mm)

   For slots and notches, consider both length and width as hole diameters.

   (b) Hole-to-hole centerline tolerance: ±0.005 in. (±0.13 mm). Add ±0.001 in. (±0.03 mm) for every inch (25.4 mm) over 1 in. (25.4 mm).

   (c) Line width and spacing tolerance: ±0.010 in. (±0.25 mm). Line width tolerances do not include nicks, pin holes, and scratches. These imperfections are acceptable provided the line is not reduced by more than 20% in a local area.

   (d) Warpage on phenolic laminate: Measured according to ASTM D 709, Specification for Laminated Thermosetting Materials; 0.025 in./in. (0.025 mm/mm). Closer warp tolerances may limit the selection of raw materials or make necessary unusual manufacturing operations or shipping procedures.

   (e) Panel thickness (0.062 in. or 1.58 mm): Tolerance with copper foil:

oz/ft²	kg/m²	No. of Sides	Tolerance	
			in	mm
1	0.305	1	±0.0055	±0.14
1	0.305	2	±0.0060	±0.15
2	0.610	1	±0.0060	±0.15
2	0.610	2	±0.0065	±0.16

   (f) Bond strength: Normal peel strength testing method to be used: a suitable universal testing machine (Fig. 2) pulling the foil from the laminate within 5 deg (0.083 rad) of the perpendicular to the face of the sample. Jaws must grip foil across entire foil width. Convert results to pounds per inch width (newton/metre). Bond strength to base material 4 lb/in. (0.45 N·m) of width minimum (6 lb/in. (0.68 N·m) of width average) on 10 samples, one line each.

(g) Registration of circuit pattern to part: ±0.015 in. (0.38 mm).

2. Good Practice (line widths, spacings, etc.[1])

(a) Minimum width of conductor: 0.060 in. (1.52 mm). (See Figs. 1 and 3 for special application and typical pattern.)

(b) Minimum spacing between conductors: 0.060 in. (1.52 mm).

(c) Minimum distance between copper conductors and edge of board: 0.060 in. (1.52 mm).

(d) Minimum distance from a hole to the edge of the board should equal the diameter of the hole and should never be less than the thickness of the base material.

(e) Where conductor pad diameter is to be used for circuit mounting and grounding conductor, diameter should exceed screw head diameter by a minimum of 0.090 in. (2.29 mm). Conductor fillet radii should be as large as possible, and in no case less than 0.03 in. (0.76 mm) R.

(f) Maximum distance from edge of lamp socket hole to copper foil pad on phenolic laminate or plastic molded bases shall be 0.030 in. (0.76 mm) (Fig. 4). Maximum distance from edge of lamp socket hole to copper foil pad on flexible plastic films shall be 0.065 in. (1.65 mm) (Fig. 5).

(g) A radius shall be applied to all corners, notches, and slots, and shall not be less than 0.03 in. (0.76 mm) R.

(h) Additional layout considerations (Figs. 3–5).

Test Requirements

1. Electrical

(a) Dielectric: 200,000 Ω minimum resistance must exist between any two conductors after conditioning for 24 h at 170°F (77°C), and 100% relative humidity, when using a low voltage type resistance meter. Moisture present on sample surface to be wiped off or blown off with compressed air prior to the resistance check.

(b) Millivolt drop: Tests to be conducted at ambient room temperature of 75 ± 5°F (24 ± 2.8°C) in draft free area.

(1) Maximum permissible millivolt drop from any component contacting the foil on any printed circuit board or flexible plastic film shall be 5.0 mV at 2.0 A before conditioning.

(2) Maximum millivolt drop from any component to the foil after one humidity cycle of 100 h at 100% relative humidity at 100°F (37.8°C), must not exceed 10.0 mV at 2.0 A.

(c) Continuity of all electrical connections and circuits must be maintained after 25 thermal shock cycles. For printed circuits used in applications where the maximum ambient temperature does not exceed 175°F (80°C), each cycle is to consist of:

15 min at −30 ± 3°F (−34.5 ± 1.7 °C)
15 min at 75 ± 5°F (24 ± 2.8°C)
15 min at 175 ± 3°F (80 ± 1.7°C)
15 min at 75 ± 5°F (24 ± 2.8°C)

For printed circuits used in applications where the maximum ambient temperature is greater than 175°F (80°C) but does not exceed 250°F (121°C), the third 15 min of each cycle will be at 250 ± 3°F (121 ± 1.7°C).

2. Physical—No fractures of the conductors permitted on flexible circuits or on rigid circuits properly attached to the supporting structure after conditioning separate samples as follows:

(a) After 50 heat cycles in test requirement 1(c).

(b) After two humidity cycles. One cycle to consist of 100 h at 100% relative humidity at 100°F (37.8°C), followed by 24 h at room ambient conditions.

(c) After being subjected to five continuous cycles as follows: 100 ± 5°F (37.8 ± 2.8°C) at 95–100% relative humidity for 2 h. Temperature then to be raised gradually to 160 ± 5°F (71 ± 2.8°C) at the same relative humidity over a 1 h period and then maintained for an additional 3 h. The temperature is then to be lowered gradually to 100 ± 5°F (37.8 ± 2.8°C), while at the 95–100% relative humidity, over a period of 6 h. This would constitute one cycle. After completion of five cycles, parts examination should be initiated following 1 h minimum rest at room ambient conditions. If the printed circuit being tested is used in an application where the maximum ambient temperature is greater than 175°F (80°C) but does not exceed 250°F (121°C), at the completion of the preceding five cycles it shall be subjected immediately to an

[1] The 0.060 in. (1.52 mm) minimum conductor widths and spacing refer to printed circuits commonly used to replace cable in automotive wiring. Miniature circuits for solid-state devices, etc., will normally have special requirements and specifications determined by the function.

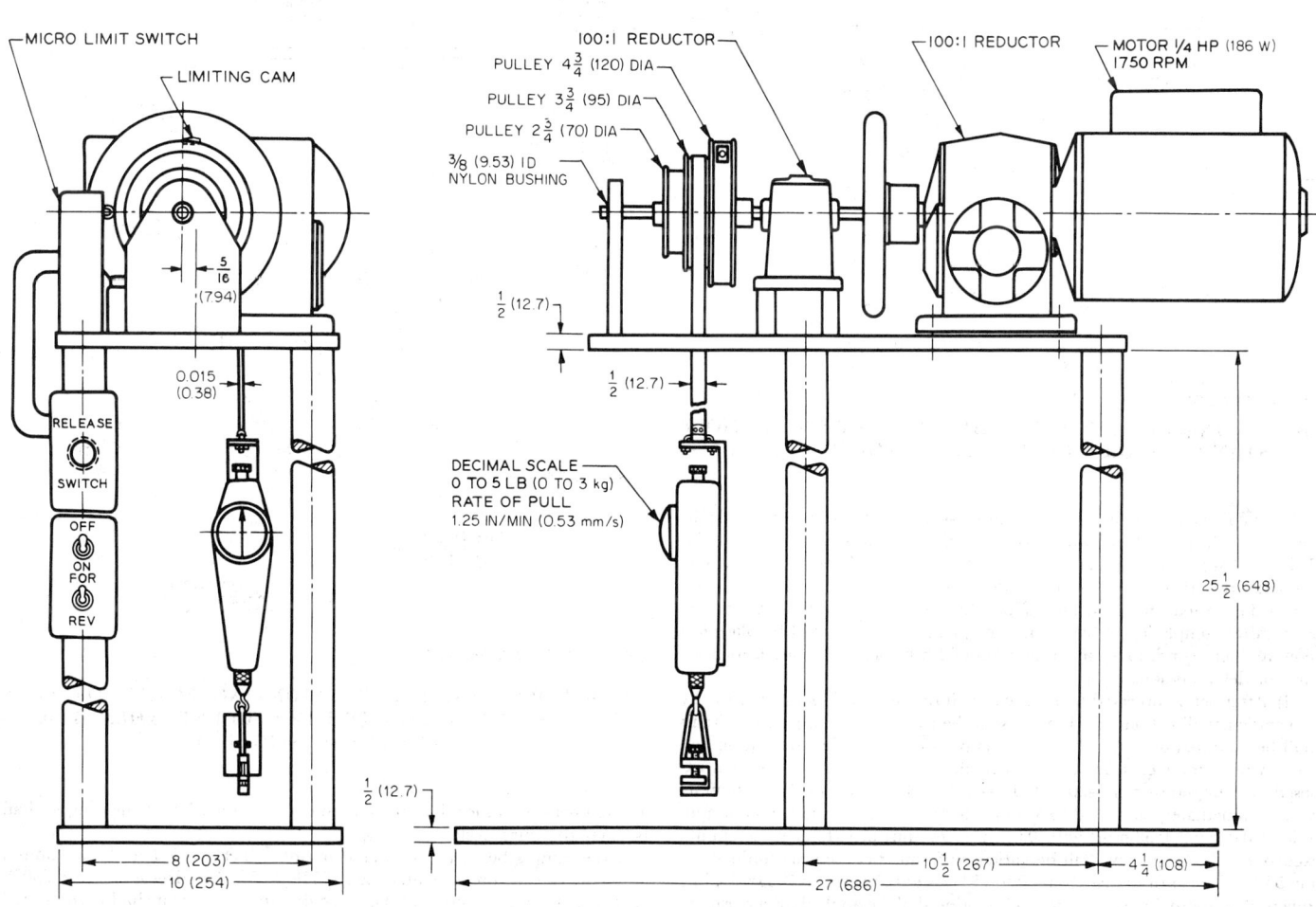

FIG. 2—BOND STRENGTH TESTER

FIG. 3—CONDUCTOR PATTERNS: RECOMMENDED TYPICAL SECTION, WIDE CONDUCTOR LINES AND FILLETS IN AREA OF PADS AND LANDS

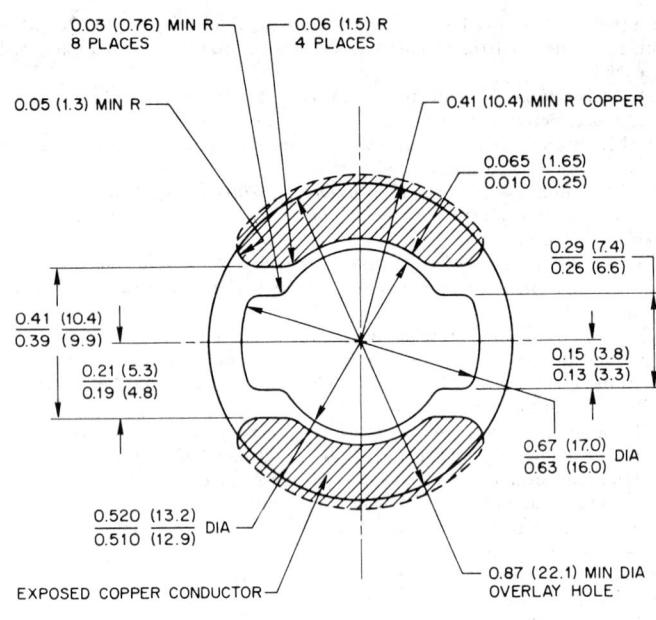

FIG. 5—½ IN (12.7 mm) PRINTED CIRCUIT LAMP SOCKET HOLE STANDARD FOR FLEXIBLE PLASTIC FILM

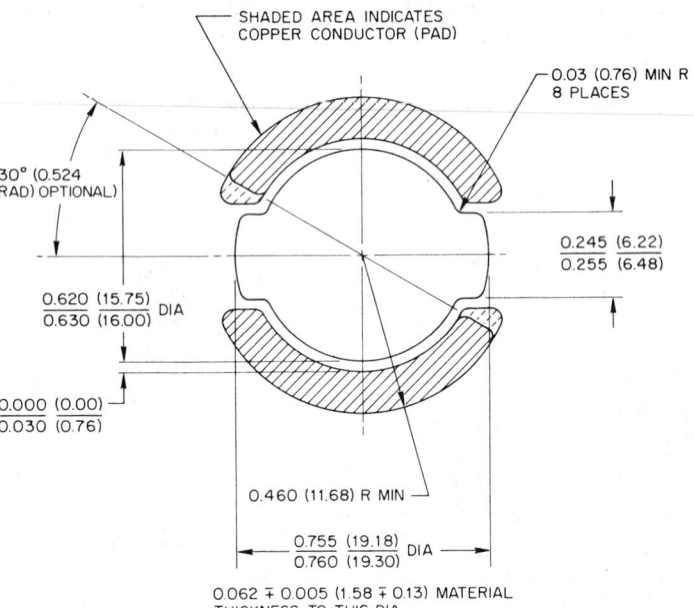

FIG. 4—⅝ IN (15.9 mm) PRINTED CIRCUIT LAMP SOCKET HOLE STANDARD FOR RIGID PLASTIC OR PHENOLIC BASE

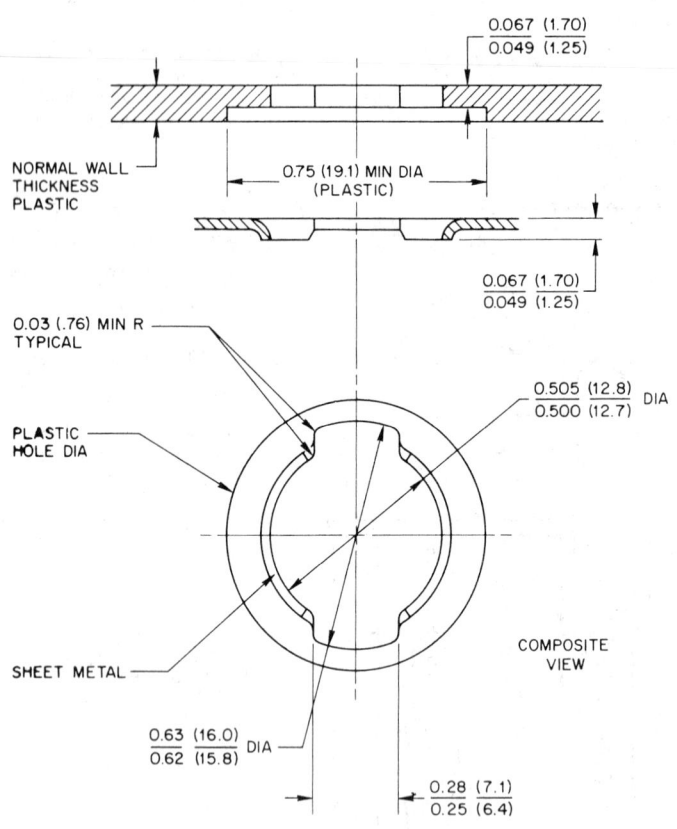

FIG. 6—HOLE REQUIRED IN MOUNTING PLATE FOR ½ IN (12.7 mm) PRINTED CIRCUIT LAMP SOCKET WHEN BASE IN FLEXIBLE PLASTIC FILM

additional five continuous cycles as follows: $-30 \pm 3°F$ ($-34.5 \pm 1.7°C$) for one hour. Temperature then to be raised gradually to $250 \pm 3°F$ ($121 \pm 1.7°C$) over a 3 h period and then maintained for one hour. The temperature is then to be lowered gradually to $-30 \pm 3°F$ ($-34.5 \pm 1.7°C$) over a 3 h period and then maintained for 1 h. This would constitute one cycle. After completion of five cycles, the printed circuit would be placed in room ambient conditions for a minimum of 1 h before examination for fractures or delamination.

(d) After being subjected to one cycle of vibration in each of three mutually perpendicular directions. Each cycle shall be as follows: The printed circuit shall be mounted or fastened as it normally would be in the application and contain all components and attachment devices. This assembly is then exposed to a temperature of $-30 \pm 3°F$ ($-34.5 \pm 1.7°C$) for 1 h. The assembly is then immediately subjected to a simple harmonic motion having an amplitude of 0.015 in. (0.38 mm) with 0.03 in. (0.76 mm) peak-to-peak maximum excursion. The frequency shall be varied uniformly between the limits of 10 and 55 Hz. The entire range from 10 to 55 Hz and returning to 10 Hz shall be traversed in approximately 1 min. This motion shall be applied for a period of 30 min. During this portion of the test the sample shall not be subjected to any temperature other than room ambient. Examination of the sample shall be after the third cycle.

If the sample being tested is used in an application where the maximum ambient temperature is greater than 175°F (80°C) but does not exceed 250°F (121°C), at the completion of the preceding three cycles, it shall be subjected immediately to an additional three cycles. These cycles will be similar to the

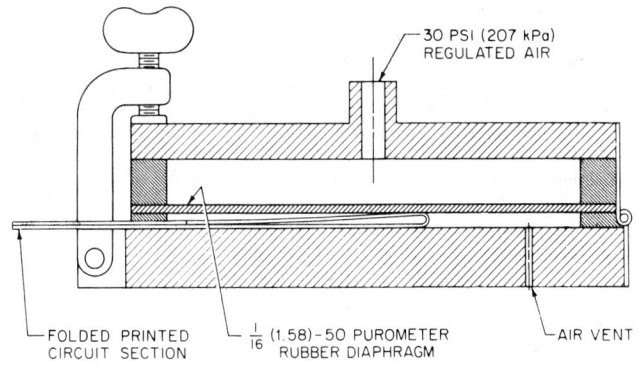

FIG. 7—FOLD TESTER FIXTURE

first three except the conditioning temperature shall be $250 \pm 3°F$ ($121 \pm 1.7°C$) instead of $-30°F$ ($-34.5°C$). Examination of the sample shall be after all six cycles are completed.

(e) Method for determining ductility of copper material intended for use in printed circuits: See Appendix.

APPENDIX
METHOD FOR DETERMINING DUCTILITY OF COPPER MATERIAL INTENDED FOR USE IN FLEXIBLE PLASTIC FILM PRINTED CIRCUITS

1. *Apparatus*—The apparatus used shall conform to that described in ASTM D 2210-64 Modified, Standard Method of Test for Bursting Strength of Leather, with Bulge Diameter of 1.250 in. (32 mm), so that the output shaft turns at 10 rpm.

2. *Test Specimens*—From the samples selected to determine the ductility property, cut at least three specimens $3\frac{1}{2}$ in. (90 mm) square from sections located approximately at equal distances across the width of the material.

3. *Test Procedure*:

3.1 Set the bulge height indicator to zero using a flat rigid sheet between the plates.

3.2 Clamp the specimen securely between the plates of the circular hydraulic bulge tester with the uncoated (bright) side of the material facing downward.

3.3 Admit the hydraulic fluid to the bulging chamber between the plates.

3.4 Observe and record the maximum bulge height at burst.

HIGH TENSION IGNITION CABLE—SAE J557 — SAE Standard

Report of Electrical Equipment Division approved July 1921 and last revised by Electrical Equipment Committee June 1961. Reaffirmed without change January 1968.

Scope—The specifications contained in this report cover high tension cable used in motor vehicles or tractor engine ignition systems.

1. Specification Types

Type HTLR—High Tension, Low Resistance Conductor, Rubber Insulation, Synthetic Sheath.

Type HTHR—High Tension, High Resistance Conductor, Rubber Insulation, Synthetic Sheath.

Type HTS—High Tension, Metallic Conductor, Rubber Insulation, Synthetic Sheath.

Type HTB—High Tension, Metallic Conductor, Rubber Insulation, Braided Covering.

Type HTT—High Tension, Metallic Conductor, Thermoplastic Insulation.

2. General Specifications

Conductors—TYPE HTLR AND TYPE HTHR—The conductor shall consist of a resistor element capable of producing a specified designed resistance in the finished cable.

TYPE HTS, TYPE HTB AND TYPE HTT—The metallic conductor shall consist of 7 or 19 strands of wire having a lay twisted to the left as viewed from the observer. The diameter of the stranded group shall not exceed 0.060 in. or be less than 0.030 in. The individual wires shall be treated so that there will be no interaction between the wire and insulating compound.

Insulation—A homogeneous insulating compound shall be applied directly over the conductor. The insulating compound on Types HTS, HTB and HTT shall adhere closely to, but shall strip readily from the wires of the conductor, leaving them reasonably clean. The thickness of the insulating compound shall be such that it will meet the applicable tests for each cable type.

Inner Braid—When specified, an open braid of suitable textile shall be applied over the insulation.

Sheath—When specified, a synthetic material shall be applied concentrically over the insulation or over the inner braid, if used. It shall be so compounded as to meet the tests specified. Adjacent layers of cable, when wound on reel or coil or packed in sets, shall not stick to one another at any temperature under 121 F.

Outer Braid—When specified, a protective braid shall be applied over the insulation compound. The braid shall be so treated as to be resistant to heat, oil, water and gasoline. Adjacent layers of cable, when wound on a reel or when packed in sets, shall not stick to one another at any temperature under 105 F.

Size—The outside diameter of the cable shall be within 0.270 to 0.285 in. Cable may be slightly flattened or oval in section, but the average of two diameters measured 90 deg apart at any section must be within the values specified.

Marking—When specified, the manufacturer's name and/or manufacturer's identification, the quarter, year, and cable type shall be legibly and reasonably permanently marked on the outside of the cable at intervals not exceeding 2 ft.

3. Cable Construction

Type HTLR—High Tension, Low Resistance Conductor, Rubber Insulation, Synthetic Sheath.

Construction

(a) Conductor—See Section 2, Conductors.
(b) Insulation—See Section 2, Insulations.
(c) Inner Braid—(when specified) See Section 2, Inner Braid.
(d) Sheath—See Section 2, Sheath.
(e) Size—See section 2, Marking.
(f) Marking—See Section 2, Marking.

Tests—The cable shall meet the following tests:

    Life Cycle   —See Test No. 1
    High Temperature—See Test No. 2
    Low Temperature —See Test No. 3
    Hot Oil Test   —See Test No. 4
    Resistance   —See Test No. 9

Type HTHR—High Tension, High Resistance Conductor, Rubber Insulation, Synthetic Sheath.

Construction

(a) Conductor—See Section 2, Conductors.
(b) Insulation—See Section 2, Insulation.
(c) Inner Braid—(when specified) See Section 2, Inner Braid.
(d) Sheath—See Section 2, Sheath.
(e) Size—See Section 2, Size.
(f) Marking—See Section 2, Marking.

Tests—The cable shall meet the following tests:

    Life Cycle   —See Test No. 1
    High Temperature—See Test No. 2
    Low Temperature —See Test No. 3
    Hot Oil Test   —See Test No. 4
    Resistance   —See Test No. 9

Type HTS—High Tension, Metallic Conductor, Rubber Insulation, Synthetic Sheath.

Construction

(a) Conductor—See Section 2, Conductors.
(b) Insulation—See Section 2, Insulation.
(c) Inner Braid—(when specified) See Section 2, Inner Braid.
(d) Sheath—See Section 2, Sheath.
(e) Size—See Section 2, Size.
(f) Marking—See Section 2, Marking.

Tests—The cable shall meet the following tests:

Life Cycle	—See Test No. 1
High Temperature	—See Test No. 2
Low Temperature	—See Test No. 3
Hot Oil Test	—See Test No. 4

Type HTB—High Tension, Metallic Conductor, Rubber Insulation, Braided Covering.

Construction

(a) Conductor—See Section 2, Conductors.
(b) Insulation—See Section 2, Insulation.
(c) Outer Braid—See Section 2, Outer Braid.
(d) Sheath—See Section 2, Sheath.
(e) Size—See Section 2, Size.
(f) Marking—See Section 2, Marking.

Tests—The cable shall meet the following tests:

High Potential	—See Test No. 5
Corona	—See Test No. 6
Low Temperature	—See Test No. 7
Life Cycle	—See Test No. 8

Type HTT—High Tension, Metallic Conductor, Thermoplastic Insulation.

Construction

(a) Conductor—See Section 2, Conductors.
(b) Insulation—See Section 2, Insulation.
(c) Inner Braid—(when specified) See Section 2, Inner Braid.
(d) Sheath—See Section 2, Sheath.
(e) Size—See Section 2, Size.
(f) Marking—See Section 2, Marking.

Tests—The cable shall meet the following tests:

Life Cycle	—See Test No. 1
High Temperature	—See Test No. 2
Low Temperature	—See Test No. 3

4. Tests—All electrical tests shall be made with 60 cycle alternating current. The voltage specified shall be mean effective values (root-mean square). Voltage shall be increased from zero to the prescribed test value at a uniform rate of rise approximately but not exceeding 3 kv per sec or as otherwise specified.

1. Life Cycle Test—One end of a suitable length of cable shall be secured to a ½-in. diameter mandrel. A weight of 10 lb shall be attached to the free end. The mandrel and cable specimen shall then be rotated. Five complete turns shall be wound around the mandrel, coils touching. The cable shall then be unwound and rewound in the opposite direction. The complete winding cycle shall be performed twice.

The specimen shall then be wound on a 1-in. dia mandrel so that there are five turns spaced ¾ in. apart. At the time of winding on the 1-in. mandrel, one end of the cable shall be firmly attached to the mandrel and a 5-lb weight shall be attached to the outside of the other end of the specimen. The mandrel and specimen shall then be placed in a snug fitting, belled metal shield. The specimen with mandrel and shield shall be subjected to the following tests:

(a) Heat 5 hr at 250 ± 3 F. Remove from the oven and immerse in water maintained at 120 ± 3 F for a period of 18 hr. Remove and drain for 30 min. Subject specimen to 15,000 v applied between conductor and sheath for 30 min. Follow by:

(b) Immerse in SAE 30 oil and maintain at 194 ± 5 F for 18 hr. Remove and drain 30 min. Subject specimen to 15,000 v applied between conductor and sheath for 30 min. Follow by:

(c) Immerse in kerosene for a period of 18 hr at room temperature. Drain for 4 hr. After completion of tests, the insulation shall not crack, rupture, show excessive swelling or other evidence of damage.

2. High Temperature Test—Suitable lengths of cable shall be suspended straight in a ventilated oven maintained at a temperature of 250 ± 3 F for 48 hr. The specimen shall then be removed from the oven, allowed to cool to room temperature, then wrapped 360 deg. around a ½-in. mandrel. Cracking of the outer jacket shall constitute failure.

3. Low Temperature Test—Suitable lengths of straight specimens shall be placed in a cold chamber maintained at −30 F for 24 hr. Without removing from the cold chamber, wrap specimen 360 deg around a 3-in. diameter mandrel which has also been subjected to the same time and temperature as the specimen. Cracking of the outer jacket shall constitute failure.

4. Hot Oil Test—Twenty-inch specimens shall be immersed in SAE 30 oil maintained at a temperature of 250 ± 5 F for a 40-hr period. The ends of the specimens shall protrude 3 in. above the oil and be 1 in. apart. The oil shall be circulated during the test. Remove from oil and allow to cool to room temperature, wipe free of oil and wrap center portion 360 deg around a ½-in. dia mandrel against curvature of bend. Excessive swelling, cracking or rupture of jacket shall constitute failure.

5. High Potential Test—One inch of insulation shall be removed from each end of an 18-in. length of the cable and the exposed ends of the conductor twisted together. The loop so formed shall be immersed in water at 68 F (20 C) ± 5 F so that 4 in. of each leg of the loop protrudes above the surface of the solution. After 24-hr immersion and while still immersed, a potential of 20,000 v shall be applied between the conductor of the cable and the water and shall be maintained for 20 min. The cable shall withstand this test without failure.

6. Corona Test—A 5-ft length of cable, having a 5-lb weight attached to the free end, shall be wound 10 turns on a metal rod 1 in. in diameter, with adjacent turns touching. The cable shall then be secured to prevent slipping or release of tension. A potential of 10,000 v shall be applied between the conductor of the cable and the metal rod and maintained for 30 min. The potential shall then be raised to 20,000 v and maintained for 2 hr.

This test shall be started with cable and metal rod at a temperature of 68 F (20 C) ± 5 F. The cable shall withstand this test without failure.

7. Low Temperature Test—A sample of the cable 5 ft in length, shall be stretched straight on a rack and subjected to a temperature of −10 F (−23.3 C) ± 5 F for 12 hr. The cable shall then be wrapped around a 3-in. mandrel, with cable and mandrel at a temperature of −10 F. Cracking of the outer covering of the cable shall constitute failure.

8. Life Cycle Test—One end of a 5-ft specimen shall be firmly attached to suitable cylindrical mandrel with a diameter of ½ in.; a weight of 5 lb shall be firmly attached to the end of the specimen. The specimen shall be wound in a clockwise direction on the mandrel with the coils touching. It shall then be unwound and rewound in a counter-clockwise direction so that the coils again touch. This procedure shall be repeated until no less than two bends in each direction have been imposed on the specimen. The specimen shall then be removed from the ½-in. and wound on a 1-in. mandrel so that there are five turns spaced ¾ in. apart. At the time of winding on the 1-in. mandrel, one end of the cable shall be firmly attached to the mandrel and the 5-lb weight attached to the outside of the other end of the specimen. After winding, a close fitting brass sleeve with belled ends shall be placed over the wound mandrel. The test specimen shall then be subjected to the procedure and conditions specified in Table 1 and shall not be removed from the 1-in. dia mandrel until they have been completed.

9. Resistance Test—The resistance of Type HTLR and Type HTHR shall be within the limits shown in Table 2 when measured with a standard commercial multi-purpose meter with a low voltage power source by using ½ in. probes on each lead and a 13 in. piece of cable. Probes must be carefully inserted ½ in. into the conductor. The cable should then be tested as follows:

(a) Measure resistance of cable after 50 lb tension.

(b) Measure resistance of cable after being wound and unwound five turns on a ½-in. dia mandrel. Wind cable under 5 lb tension.

TABLE 1—OPERATIONS TO BE PERFORMED AND CONDITIONS TO BE MAINTAINED

Time Consumed, hr	First 24-Hr Period	Second 24-Hr Period	Third 24-Hr Period	3-1/2-Hr Period
3	Rest	Rest	Drain	Drain
1/2	Corona Test 15 kv	Corona Test 15 kv	Corona Test 15 kv	Corona Test 15 kv
1/2	Rest	Rest	Rest	—
20	Bake in oven at 176 ±20 F.	Soak in 5% sodium chloride solution at room temperature, ends protruding not more than 8 in. nor less than 4 in.	Soak in solution of 1 part kerosene,[a] 1 part motor fuel, 2 parts engine SAE 30 oil, ends protruding not more than 8 in. nor less than 4 in.	—

[a] Kerosene conforming with Federal Specification VV-K-211.

TABLE 2—RESISTANCE OF TYPES HTLR AND HTHR CABLES

Cable	Resistance - Ohms per Ft	
	Low Limit	High Limit
Type HTLR	3000	7000
Type HTHR	6000	12000

LOW TENSION WIRING AND CABLE TERMINALS AND SPLICE CLIPS—SAE J163

SAE Recommended Practice

Report of Electrical Equipment Committee approved January 1974.

Scope—This SAE Recommended Practice covers the application requirements for terminals and splice clips attached to stranded low tension wiring and cable as shown in J878 and J558. In addition, it covers maximum voltage drop limits for friction type connections.

Use of Terminals—Friction (quick disconnect) type brass connections should be used only where the maximum temperature (environmental ambient plus rise due to current), measured at the center of the terminal surface, does not exceed the capabilities of the physical properties of the material. Maximum temperatures for terminal materials other than brass should also be determined prior to using so as to be compatible with the physical properties of these materials.

Electrical connections and splices of standard types must be protected, as application dictates, from moisture, salt, soil accumulation, acid, or corrosive vapor which will deteriorate the connection beyond the limits of this recommended practice.

Performance Requirements (Electrical)—Terminals or splice clips shall be attached to wire or cable in such a manner that, following the humidity test, the voltage drop across the attachment shall not exceed the values in Table 1. Friction connections (terminal to terminal) shall be such that following four repeated insertions and the humidity test, the voltage drop across the connection shall not exceed the values in Table 2. For a terminal to be acceptable, all specimens tested must meet the requirements.

Test Procedure—Tests shall be conducted at $73 \pm 5°F$ ($23 \pm 3°C$). Test samples shall consist of terminals or splice clips attached to 12 in. (305 mm) of wire. It is suggested that at least 10 specimens of each wire size be subjected to each test.

Voltage Drop Test—Measurements shall be made after the temperature of the specimen has stabilized (2 h under test load).

Measurements across a wire to terminal attachment shall be made between the center of the conductor grip and a point on the cable core 3 in. (76.2 mm) behind the conductor grip. Probe point on the cable core shall be stripped and solder dapped. For preinsulated terminals, the measurements shall be made between a point in front of the conductor grip within $1/16$ in. (1.6 mm) of the end of the insulation and a point on the cable core 3 in. (76.2 mm) behind the conductor grip. The voltage drop across the attachment is defined as the difference between this reading and the voltage drop through the 3 in. (76.2 mm) of wire.

Measurements across a splice clip connection shall be made between points on cable cores 3 in. (76.2 mm) behind the center of the conductor grip crimp. Probe points on the cable core shall be stripped and solder dapped. The voltage drop across the splice is defined as the difference between this reading and the voltage drop through the 6 in. (152 mm) of wire. Measurements shall be made across each combination of conductor pairs in the splice. The current value shall be selected according to the smaller gage cable in the cable pair being measured.

Measurements across a friction connection shall be made between the centers of the conductor grips of two joined line connector type terminals and from the center of the conductor grip of a line terminal to a similar point on a joined fixed terminal.

For preinsulated terminals, the voltage drop across the wire to terminal attachment plus the 3 in. (76.2 mm) of cable shall be determined first per the above procedure. Then the terminals shall be connected and measurements across the connection shall be made from the stripped solder dapped points on the cable cores of two joined line connector type terminals and from the same point on the cable core to a point on a joined fixed terminal equivalent to that used for uninsulated terminals. The voltage drop across the friction connection is defined as the difference between this reading and the previous measurement(s) for the same specimen(s) (wire drop plus attachment drop).

Humidity Test—The humidity test shall consist of 100 h at 95–100% relative humidity at $100 \pm 5°F$ ($38 \pm 3°C$). (Demineralized water shall be used.) Specimens shall be prepared as follows:

1. Mounted at least 1 in. (25.4 mm) apart on test boards.
2. Placed in the humidity cabinet with the axis of each specimen in a horizontal plane and such that it has all surfaces of the terminal completely exposed and not in contact with other objects.
3. Removed from the cabinet after the prescribed exposure period and allowed to dry 24 h at room condition before final MVD test.

TABLE 1—WIRE TO TERMINAL OR WIRE TO WIRE (SPLICE CLIP METHOD) VOLTAGE DROP (AFTER HUMIDITY TEST)

Wire/Cable (SAE Gage)	Test Current, A	Drop, mV	
		Uninsulated Terminal	Preinsulated Terminal
20	5	3	3.5
18	10	5	5.5
16	15	8	9
14	20	10	11
12	30	15	17
10	40	20	22
8	50	25	—
6	60	15	—
4	70	18	—
2	80	20	—
0	90	23	—
00	100	25	—

TABLE 2—FRICTION VOLTAGE DROP (AFTER HUMIDITY TEST AND FOUR INSERTIONS)

Wire/Cable (SAE Gage)	Test Current, A	Drop, mV
20	5	7.5
18	10	15
16	15	22.5
14	20	30
12	30	45
10	40	60
8	50	75

FUSIBLE LINKS—SAE J156a

SAE Standard

Report of Electrical Equipment Committee approved February 1970 and last revised April 1977.

1. Scope—This SAE Recommended Practice covers the details, use, and design evaluation testing of fusible links for motor vehicle electrical wiring protection. The specifications as listed are known good practice and are not intended to restrict new materials or construction.

2. Definition—A fusible link is a special section of low tension cable designed to open the circuit when subjected to an extreme current overload. Its purpose is to minimize wiring system damage when such an overload occurs accidentally in those circuits protected by the fusible link.

3. General Specifications

3.1 Conductors—Conductors shall conform to the specifications shown in Table 1 of SAE J1128.

3.2 Insulation—The insulating material shall meet the requirements shown in SAE J1128 Type HTS. A special insulation with a tensile strength of 1000 psi (6900 kPa) minimum and STS wall may also be used.

3.3 Wire Size—The fusible link must be of a smaller wire size than any connecting cable in the circuits being protected. Wire sizes are to be determined experimentally with the vehicle wiring system based on the type of harness wire insulation, circuit loads, and physical locations. This may be done either in the vehicle or with an equivalent laboratory set up.

3.4 Length—The length of each fusible link for effective protection is to be determined in the same manner as for the wire size.

3.5 Location—Fusible links shall be located such that any fumes generated during their destruction will not cause undue discomfort to any passenger, and no damage will occur to adjacent components, combustible material, or other circuits.

3.6 Terminations—The conductor and insulation at each end of a fusible link shall be securely fastened to its termination. If spliced to a connecting cable, the splice joint must either be welded or mechanically secured and soldered. The splice must then be properly insulated.

3.7 Identification—Each fusible link shall be permanently marked with the wire size and identification that it is a fusible link. After a link has fused and opened the circuit, sufficient identification shall still be present to establish this information for replacement.

4. Testing—Design evaluation testing is to be conducted to verify the ability of a specified fusible link to conduct the maximum design load of the electrical circuit and to ascertain that the link will open *under extreme current overload* without causing damage to the protected wiring, associated harness, or adjacent components.

4.1 Charging Circuit Protection—Fusible links located in circuits which conduct battery charging currents are to be tested either in the vehicle or in a duplicating laboratory set up. The specified generator and battery should be operating at maximum charge current and the battery shall have been completely discharged before the test began. Electrical accessory loads are to be such as would cause maximum current through the fusible link that could occur in a vehicle. In a laboratory set up, the generator may be duplicated by an equivalent current producing source.

At the start of the test, the generator temperature is to be $75 \pm 5°F$ ($24 \pm 3°C$) and the battery electrolyte temperature $110 \pm 5°F$ ($43 \pm 3°C$). The test shall be conducted for at least 5 min after the maximum fusible link core temperature attainable in the vehicle can be reached. After the test is completed, the fusible link insulation shall show no deterioration due to heat.

4.2 Short Circuit Protection—Fusible links are to be tested for design evaluation by grounding the conductor of the protected circuit at a point which is the most electrically remote from the fusible link. The point selected must not have any intervening circuit protecting devices, such as fuses or circuit breakers between it and the fusible link. Under extreme current overload conditions, the fusible link must open the circuit within a period of time such that no damage to the protected wiring, associated harness, or adjacent components occurs. After the link has opened, there shall be no exposed conductor in a location to cause subsequent short circuiting of the battery or generator.

4.3 Observations and Conclusions—At the conclusion of each test, visual inspection of the wiring is to be made. Other than opened fusible link sections, there shall be no evidence of cable insulation deformation or damage regardless of the type of insulation used. Any fusible link tested for maximum design current load shall show no insulation deterioration after the test.

FIVE CONDUCTOR ELECTRICAL CONNECTORS FOR AUTOMOTIVE TYPE TRAILERS—SAE J895

SAE Recommended Practice

Report of Electrical Equipment Committee approved June 1964. Editorial change January 1966.

Scope—This SAE Recommended Practice covers the wiring and connector standards for nonpassenger carrying trailers, SAE Classes 1–3[1], with circuit loads not to exceed 7.5 amp per circuit. It provides the lighting circuits of these trailers with a universal connecting device, standard circuit coding and protection for the wiring from hazards and shorts.

Receptacle—The receptacle shall be of the design as shown in Fig. 1 and shall be attached to the towing vehicle as follows:

(a) White—Ground to frame
(b) Brown—Spliced to tail and license light circuit
(c) Yellow—Spliced to left turn and stop circuit
(d) Green—Spliced to right turn and stop circuit
(e) Blue—Auxiliary

[1] See SAE J684.

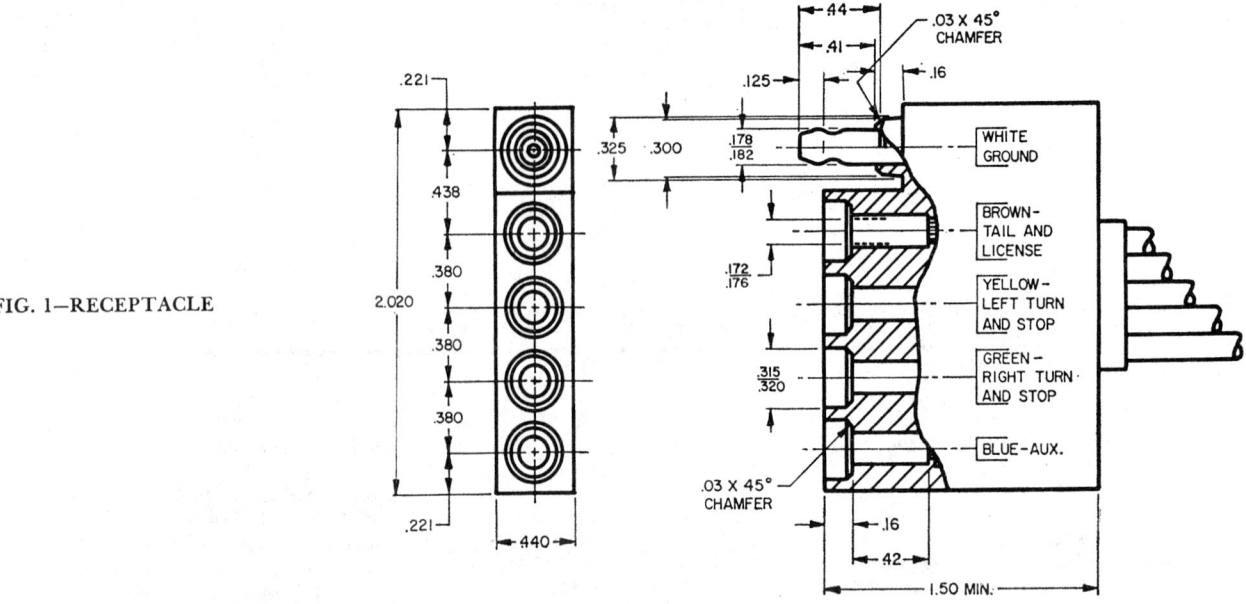

FIG. 1—RECEPTACLE

The receptacle leads shall be attached to the vehicle wiring harness in a workmanlike manner, mechanically and electrically secure. Further, a well insulated strain relief shall be provided between the receptacle and the towing vehicle wiring harness connections so that there will be no strain on the vehicle harness in the event of an abnormal pull on the receptacle. The receptacle shall be placed in a location where it will not be exposed to road hazards either when connected or loose. The receptacle leads must be properly routed and protected against damage from cutting and pinching where they leave the vehicle body. No receptacle leads shall be smaller than 16 gage (single) or smaller than 18 gage (in multiconductor cables) heavy duty SAE insulated automotive primary wire.

Plug—The plug shall be as shown in Fig. 2, and wiring, shall be attached to the trailer so that the wires have the maximum protection against road splash, stones, abrasion, grease, oil, and fuel. The wiring shall be secured to the trailer frame at intervals not greater than 18 in. so that the wiring does not shift or sag. The circuits used shall be color coded as follows:

(a) White—Ground
(b) Brown—The tail and license light
(c) Yellow—Left turn and stop light
(d) Green—Right turn and stop light
(e) Blue—Auxiliary

No plug leads shall be smaller than 16 gage (single) or smaller than 18 gage (in multiconductor cables) heavy duty SAE insulated automotive primary wire. Extra insulation should be provided between the strain relief at the trailer hitch and the wiring assembly so that an abnormal pull on the plug will not damage the wiring.

APPENDIX

Material Requirements—The receptacle and plug shall be made of an insulating material such that they can be processed to provide the spacing and splash protection indicated in Figs. 1 and 2. The material used shall have a hardness of Shore "A" 50 minimum and shall be compatible with the insulation used on the wire leads and/or jacket over the leads. Where thermoplastic materials are used the hardness shall not exceed Shore "A" 70. The jacket on the leads shall be as deemed necessary to adequately protect the wiring. The metal pins and sockets shall be of the size and type shown in Figs. 3 and 4 made of either brass or bronze and suitably coated to protect against corrosion. The coating shall be smooth so as not to bind when the parts are engaged.

Assembly Requirements—The plug and receptacle assembly shall disengage with a minimum force of 3 lb per circuit and a maximum of 7 lb per circuit, except the ground circuit which can be 12 lb max. The mechanical force requirement of disengagement does not preclude the requirement for good electrical connections between the male and female connectors of the circuits.

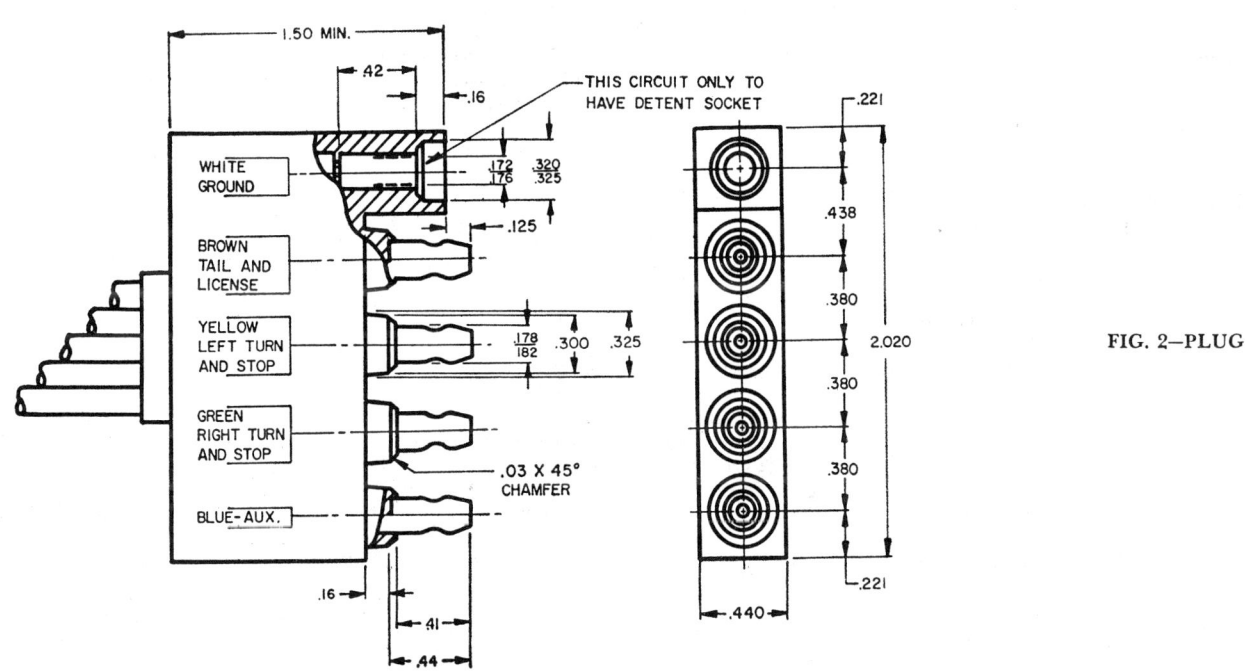

FIG. 2—PLUG

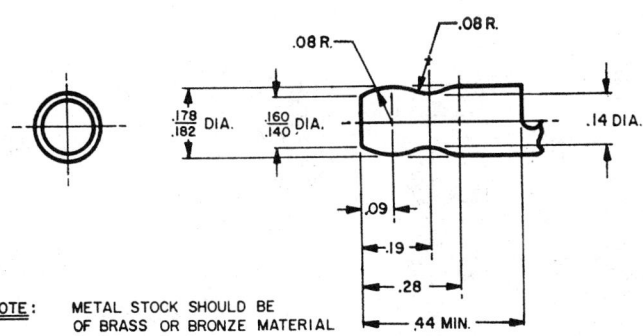

FIG. 3—PIN

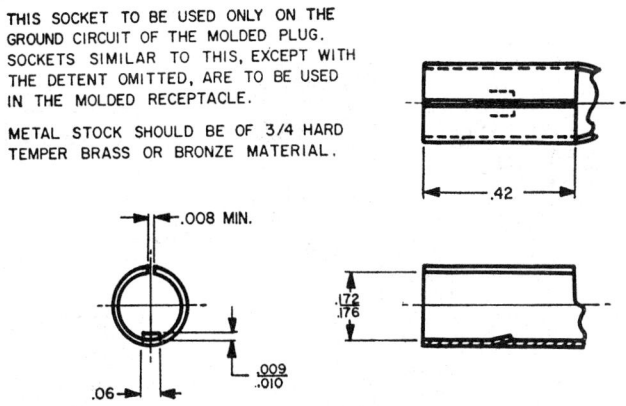

FIG. 4—SOCKET

FOUR- AND EIGHT-CONDUCTOR RECTANGULAR ELECTRICAL CONNECTORS FOR AUTOMOTIVE TYPE TRAILERS—SAE J1239

SAE Recommended Practice

Report of Electrical Equipment Committee approved June 1978.

1. Scope—This SAE Recommended Practice covers the wiring and rectangularly shaped connector standards for all types of trailers whose gross weight does not exceed 10 000 lb (4540 kg). These trailers are grouped in SAE classes 1 through 4 as delineated in SAE J684e (August, 1974), with running light circuit loads not to exceed 7.5 amps per circuit. This recommended practice provides circuits for lighting, electric brakes, trailer battery charging, and an auxiliary circuit color coding and protection for the wiring from hazards or short circuits. Color coding is compatible with SAE J560b (September, 1974) and ISO (1724-1975(E)).

2. Receptacle—The receptacle shall be of the configuration and design dimensions shown in Fig. 1 for four circuits and as shown in Fig. 2 for eight circuits.

2.1 The four-circuit receptacle (Fig. 1) shall be color coded and attached to the towing vehicle as follows:

2.1.1 WHITE—Ground to frame (SAE wire size 16 or metric size 1.2 minimum).

2.1.2 BROWN—Spliced to tail and license lamp circuit.

2.1.3 YELLOW—Spliced to left turn and stop circuit.

2.1.4 GREEN—Spliced to right turn and stop circuit.

2.2 The eight-circuit receptacle (Fig. 2) shall be color coded and shall be attached to the towing vehicle as follows:

Left Bank of Receptacles:

2.2.1 RED—Independent stop.

2.2.2 BLUE—Brake circuit spliced to controller of brake.

2.2.3 OPTIONAL—Auxiliary (see Fig. 2, Note 1).

2.2.4 ORANGE—Battery charge circuit—connect to battery positive terminal through separate fuse or circuit breaker.

Right Bank of Receptacles:

2.2.5 WHITE—Direct to battery negative (SAE wire size 12 or metric size 3.0 minimum).

2.2.6 BROWN—Spliced to tail and license lamp circuit.

2.2.7 YELLOW—Spliced to left turn and stop lamp circuit.

2.2.8 GREEN—Spliced to right turn and stop lamp circuit.

2.3 The receptacle leads shall be attached to the vehicle wiring harness, brake controller, or battery in a workmanlike manner, mechanically and electrically secure.

2.4 The receptacle leads must be properly routed and protected against damage from cutting and pinching where they leave the vehicle body. No receptacle leads designated for lighting shall be smaller than SAE wire size 16 or metric size 1.2 if a single conductor, or smaller than SAE wire size 18 or metric size 0.8 if a multi-conductor cable.

2.5 No receptacle leads for brake circuits shall be smaller than SAE wire size 14 or metric size 2.0 and no circuits shall be smaller than SAE wire size 12 or metric size 3.0 for trailer battery charge circuit or battery return circuit.

The gauge of conductors for the auxiliary circuit shall be sized to provide at least the minimum ampacity for the load it will service with a voltage drop not exceeding 3%. The receptacle shall be placed in a location where it will not be exposed to road hazards when disconnected from trailer.

3. Plug—The plug shall be of the configuration and design dimensions shown in Fig. 1 for four circuits and as shown in Fig. 2 for eight circuits.

3.1 The four circuit plug (Fig. 1) shall be color coded and attached to the trailer harness as follows:

3.1.1 WHITE—Ground to frame (SAE wire size 16 or metric wire size 1.0 minimum).

3.1.2 BROWN—Spliced to tail and license lamp circuit.

3.1.3 YELLOW—Spliced to left turn and stop circuit.

3.1.4 GREEN—Spliced to right turn and stop lamp circuit.

3.2 The eight-circuit plug (Fig. 2) shall be color coded and attached to the trailer harness as follows:

Right Bank of Receptacles:

3.2.1 RED—Independent stop.

3.2.2 BLUE—Brake circuit spliced to controller of brake circuit.

3.2.3 OPTIONAL—Auxiliary (see Fig. 2, Note 1).

3.2.4 ORANGE—Battery charge circuit—connect to trailer battery positive terminal through separate fuse or circuit breaker.

Left Bank of Receptacles:

3.2.5 WHITE—Ground to frame and trailer battery negative terminal.

3.2.6 BROWN—Spliced to tail and license lamp circuit.

3.2.7 YELLOW—Spliced to left turn and stop lamp circuit.

3.2.8 GREEN—Spliced to right turn and stop lamp circuit.

4. Wiring

4.1 Exposed trailer wiring shall be run in conduits or secured at intervals not greater than 18 in (457 mm) to stop lateral movement and prevent rubbing or chafing.

4.2 So far as practicable wiring should be located to afford protection from road splash, stones, or abrasion. Wiring exposed to such conditions shall be further protected by the use of—or combination of—additional tape covering, plastic sleeving, non-metallic, or other suitable shielding or covering.

5. Appendix

5.1 Material Requirements

5.1.1 If the receptacles and plugs are fabricated of either compression molded or extruded plastic, the plastic material shall be stabilized for protection against exposure to ultra-violet light.

5.1.1.2 The hardness of a molded or extruded plastic receptacle or plug shall fall within the limits of Shore *A* 50 as minimum and Shore *A* 70 as a maximum.

5.1.2 The metal pins and sockets shall be of the size and type shown in Figs. 3 and 4.

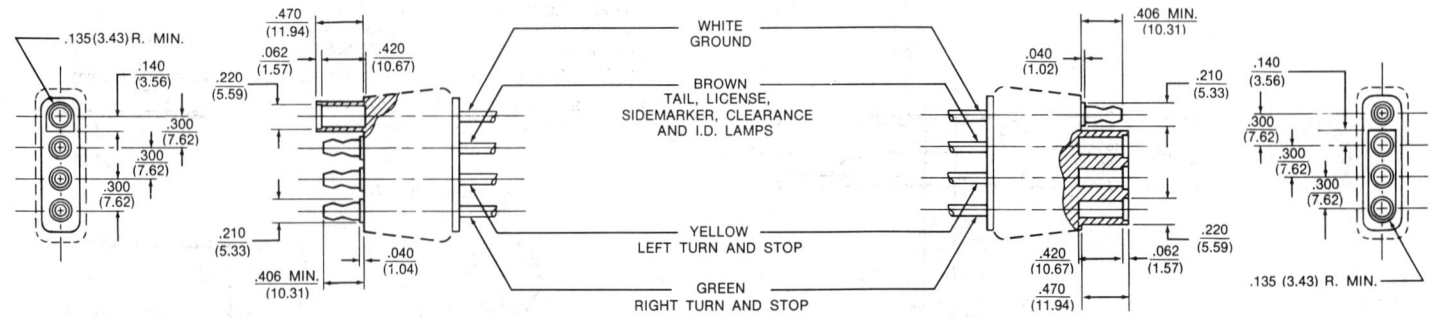

FIG. 1

22.55

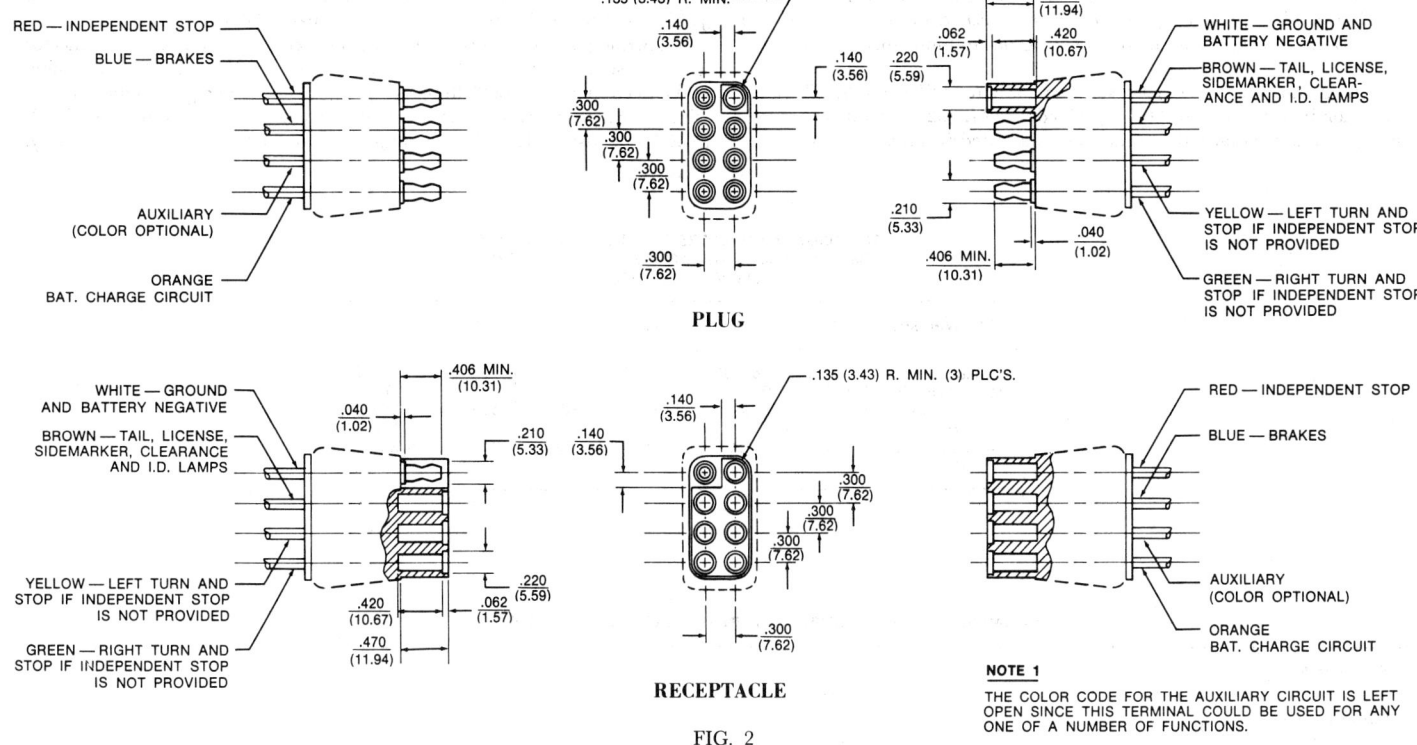

FIG. 2

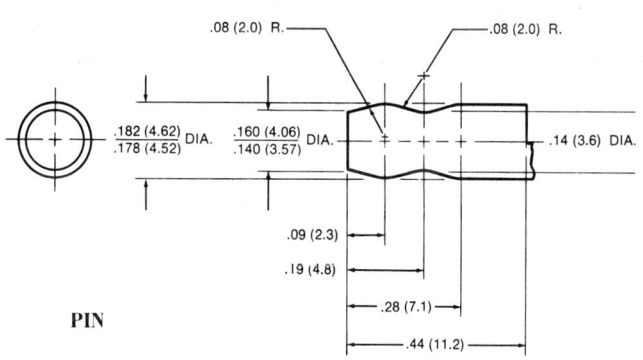

PIN

FIG. 3

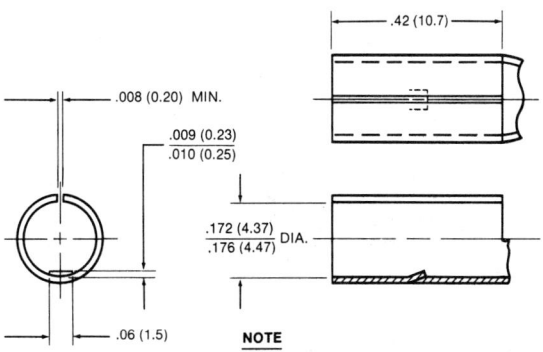

SOCKET

NOTE

THIS SOCKET TO BE USED ONLY ON THE GROUND CIRCUIT OF THE MOLDED PLUG. SOCKETS SIMILAR TO THIS, EXCEPT WITH THE DETENT OMITTED, ARE TO BE USED IN THE MOLDED RECEPTACLE.

METAL STOCK SHOULD BE OF ¾ HARD TEMPER BRASS OR BRONZE MATERIAL.

FIG. 4

5.1.2.1 Pins and receptacles shall be fabricated from brass or bronze and suitably coated to protect against corrosion. Finished surfaces of plugs and interior walls of sockets shall be smooth so as not to bind when the parts are engaged.

5.1.2.2 Pins and receptacles shall conform to *TYPE 1—PIN TERMINALS*, nominal diameter 0.180(5) specified as SAE J928a (April, 1970). Detailed pin and receptacle dimensions are illustrated in Figs. 3 and 4.

5.2 All wire and insulation shall conform to the requirements of SAE J1128 (November, 1975), Low Tension Primary Cable.

6. *Assembly Requirements*—The plug and receptacle of a 4-way connector assembly shall disengage with a minimum force of 5 lb (22.2 N) per assembly and a maximum force of 20 lb (89.0 N) per assembly. The plug and receptacle of an 8-way connector assembly shall disengage with a minimum of 8 lb (35.6 N) per assembly and a maximum force of 30 lb (133.6 N) per assembly.

STRANDED CONDUCTORS FOR 12 VOLT CIRCUITS
(PRIMARY CABLE DATA ABSTRACTED FROM J1128)
3% VOLTAGE DROP

SAE Wire Size	20	18	16	14	12	10
Stranding	7×28	16×30	19×29	19×27	19×25	19×23
Metric Wire Size	0.5	0.8	1.0	2.0	3.0	5.0
Min Cond Area Cir Mil	1072	1537	2336	3702	5833	9343
Min Cond Area mm²	0.508	0.760	1.12	1.85	2.91	4.65

MAXIMUM LENGTH OF CONDUCTOR IN FEET FROM POWER SOURCE TO LOAD

SAE Wire Size	20		18		16		14		12		10	
Circuit Current in AMPS	ft	m	ft	m	ft	m	ft	m	ft	m	ft	m
1	36.4	11.09	52.3	15.94	78.0	23.77						
2	18.2	5.55	26.1	7.96	39.0	11.89	63.0	19.20	99.0	30.17		
3	12.2	3.72	17.4	5.30	26.0	7.92	42.0	12.80	66.0	20.12		
4	9.1	2.77	13.1	3.99	19.5	5.94	31.5	9.60	49.5	15.09	78.8	24.02
5	7.3	2.22	10.4	3.17	15.6	4.75	25.2	7.68	39.6	12.07	63.0	19.20
6	6.1	2.65	8.7	2.65	13.0	3.96	21.0	6.40	33.0	10.06	52.5	16.00
7	5.2	1.58	7.4	2.26	11.1	3.38	18.0	5.49	28.2	8.60	45.0	13.72
8			6.5	1.98	9.8	2.99	15.8	4.82	24.8	7.56	39.4	12.00
9			5.8	1.77	8.6	2.62	14.0	4.27	22.0	6.71	35.0	10.67
10			5.2	1.58	7.8	2.38	12.6	3.84	19.8	6.04	31.5	9.60
15					5.2	1.58	8.4	2.56	13.2	4.02	21.0	6.40
20							6.3	1.92	9.9	3.02	15.8	4.82
20									6.6	2.01	10.5	3.20

FIG. 5

SEVEN-CONDUCTOR ELECTRICAL CONNECTOR FOR TRUCK-TRAILER JUMPER CABLE—SAE J560b

SAE Standard

Report of Electrical Equipment Committee approved January 1951 and last revised September 1974.

1. *Scope*—This SAE Standard covers the dimensions and minimum design requirements of the jumper cable plug and receptacle to achieve interchangeability with electrical connectors of different manufacture. These connectors are used with the cable described in SAE J1067. (Refer to SAE J702 for mounting location.)

2. *Definitions*

    2.1 The receptacle consists of the connector socket, its housing, and a cover which latches the male plug in place. The socket contains the male contacts. See Fig. 1.

    2.2 The cable plug, cylindrical in shape and having an index key for ease of assembly, is attached to the end of the electric jumper cable. The cable plug houses the female contacts. See Fig. 1.

    2.3 The wire color code refers to the color of insulation on the conductors.

3. *Wiring Circuits*—The function and color code of each circuit is shown in Table 1. The location of each circuit is shown in Figs. 2 and 3.

4. *General Requirements*

    4.1 All electric current-carrying parts of the receptacle socket and cable plug shall be made of a copper alloy. Insulating materials shall not fracture during mating and removal of the plug from the receptacle at −40°F (−40°C) and must not deform at 180°F (82.2°C).

    4.2 Male contacts in the receptacle socket may be made of solid or split construction. Female contacts in the cable plug shall be spring loaded radially in order to maintain proper contact for interchangeability.

    4.3 The receptacle socket and cable plug shall be so constructed that the "WHT" terminal shall accommodate at least a No. 8 gage wire and all other terminals at least a No. 10 gage wire.

    4.4 For identification purposes, the number of this standard shall appear on the plug and receptacle, in a visible location when they are installed on a vehicle.

5. *Receptacle Requirements*—Fig. 2 shows receptacle socket dimensions and minimum design requirements. The receptacle shall be provided with a weathertight cover attached to the housing or connector socket, and provided with a latching device which shall make engagement with the back end of the index key on the cable plug. The receptacle cover shall be so constructed that it will latch properly without interference to a cable plug, regardless of the plug's length, when such a plug is properly engaged with the connector socket.

6. *Cable Plug Requirements*

    6.1 Fig. 3 shows cable plug dimensions and minimum design requirements.

    6.2 The force required to connect or disconnect a new cable plug and a new receptacle socket from the same manufacturer shall not exceed 50 lb (222 N).

    6.3 An assembled cable plug and trailer jumper cable shall be so constructed that they will not be damaged by resisting a straight pull of 150 lb

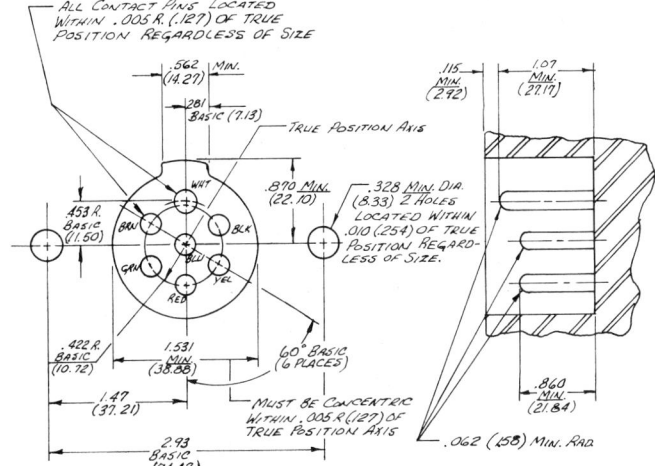

FIG. 2—RECEPTACLE SOCKET

TABLE 1—WIRING CIRCUITS

Conductor Identification	Wire Color	Lamp and Signal Circuits
Wht	White	Ground return to towing vehicle
Blk	Black	Clearance, side marker, and identification lamps
Yel	Yellow	Left-hand turn signal and hazard signal
Red	Red	Stoplamps and antiwheel lock devices
Grn	Green	Right-hand turn signal and hazard signal
Brn	Brown	Tail and license plate lamps
Blu	Blue	Auxiliary

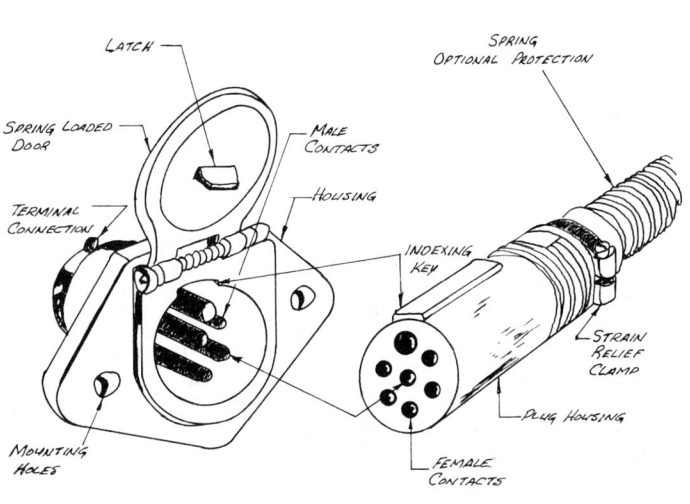

FIG. 1—SEVEN CONDUCTOR ELECTRICAL CONNECTOR

FIG. 3—CABLE PLUG

(667 N) applied to the jumper cable. All cable plugs shall incorporate a strain relief device to ease the tension on the electrical connections between the female contacts and the cable conductors.

7. **Performance Requirements**—The cable plug and receptacle shall be tested in the following sequence:
 (a) Measure the voltage drop across each circuit.
 (b) Perform 2500 coupling cycles.
 (c) Conduct salt spray test (paragraph 7.3.1).
 (d) Measure voltage drop and check for grounds and shorts between circuits.
 (e) Conduct salt spray test (paragraph 7.3.2).
 (f) Check for grounds and shorts between circuits.
 (g) Perform 2500 additional coupling cycles.
 (h) Measure voltage drop across each circuit.

 7.1 Physical Requirement—Couple and uncouple the plug and receptacle 5000 cycles (coupling and uncoupling is one cycle).

 7.2 Voltage Drop—Connect a length of SAE J1067 type cable to the plug and receptacle terminals, then connect the mating parts. Measure the voltage drop across the completed assembly for each circuit at a convenient point on the wire at least 1 in. (25.4 mm) from the terminal. The voltage drop for each circuit not including the wire, must not exceed 3 mV/A initially and at the completion of the coupling and salt spray tests.

 7.3 Salt Spray—After 2500 coupling cycles, the socket and receptacle shall be subjected to the following salt spray tests. (No current to be applied during test, see paragraphs 7.3.1 and 7.3.2.)

 7.3.1 With the socket inserted fully into the receptacle and with the assembly mounted in a normal truck-trailer position, subject the normally exposed portion of the assembly to a 48 h salt spray test per ASTM B 117.

 Four hours (drying time) after this test, check each circuit for voltage drop, shorting between circuits, and grounding to the housing before removing the socket from the receptacle. 70 V d-c shall be used to check for grounding and shorting between circuits. No grounds or shorts are permissible.

 7.3.2 With the receptacle mounted in a normal position with the cover closed and with the open end of the plug pointed down (the terminal openings may be covered), subject the uncoupled units to an additional 48 h salt spray test per ASTM B 117.

 Four hours (drying time) after this test, check each circuit for shorting between circuits and grounding to the housing. 70 V d-c shall be used to check for grounding and shorting between circuits. No grounds or shorts are permissible.

SEVEN CONDUCTOR JACKETED CABLE FOR TRUCK TRAILER CONNECTIONS—SAE J1067

SAE Standard

Report of Electrical Equipment Committee approved October 1973.

1. Scope—This standard covers the minimum construction requirements and the configuration of the conductors in the cable to connect electrically a tractor to a trailer and/or trailer to trailer. This cable is used with the connector described in SAE J560.

 2. Construction
  2.1 Conductor
  2.1.1 The conductors shall be made with tinned annealed copper wire according to ASTM B-33. Steel strands may be added to increase flexibility life.
  2.1.2 The conductor stranding and lay shall be as shown in Table 1.
  2.2 Insulation
  2.2.1 The insulation on the No. 8 wire shall be a layer of .001 in. (.0254 mm) white polyethylene terephthalate film helically wrapped around the conductor with 1/3–1/2 lap.
  2.2.2 The insulation on the No. 10 wire and No. 12 wires shall be as specified in Table 2. The colors of the insulating compounds shall be as shown in Fig. 1. The nominal wall thickness of the insulation shall be 0.032 in. (.813 mm).
  2.3 Cabling—The conductors shall be cabled together with a maximum lay of 6 in. (152.4 mm). The configuration of the conductors shall be as shown in Fig. 1. A suitable filler may be applied in twisting so as to fill the interstices between the conductors to produce a circular cross section. Use of a suitable separator over the conductors is required.

TABLE 1—CONDUCTORS

SAE Wire[a] Size	No. of Wires	Nominal Size of Strand		Lay in.	Conductor Area Cir Mils	Max Dia of Stranded Conductor, in.
		AWG	in.			
12	65	30	.010 (.254 mm)	1.5 (38.1 mm)	6487	.100 (2.54 mm)
10	105	30	.010 (.254 mm)	1.5 (38.1 mm)	10479	.125 (3.18 mm)
8	168 or	30	.010 (.254 mm)	2.0 (50.8 mm)	16414	.175 (4.45 mm)
	427	34	.0063 (.160 mm)	2.0 (50.8 mm)		

[a]SAE wire size numbers indicate that the circular mil area of the stranded conductor approximates the circular mil area of American Wire Gage for equivalent gage size.

TABLE 2—INSULATION COMPOUND[a]

Recovery, in.	5 in. (127.0 mm) str, 1/2 (12.7 mm)
Elongation, %	250
Tensile strength, psi (MPa)	600 (4.12)
Oxygen-bomb aged (96 h at 70°C, 300 psi) elongation	70% of orig. min
Oxygen-bomb aged (96 h at 70°C, 300 psi) tensile	70% of orig. min

[a]Underwriters' Laboratories requirements for Class 3 Rubber—UL 62.

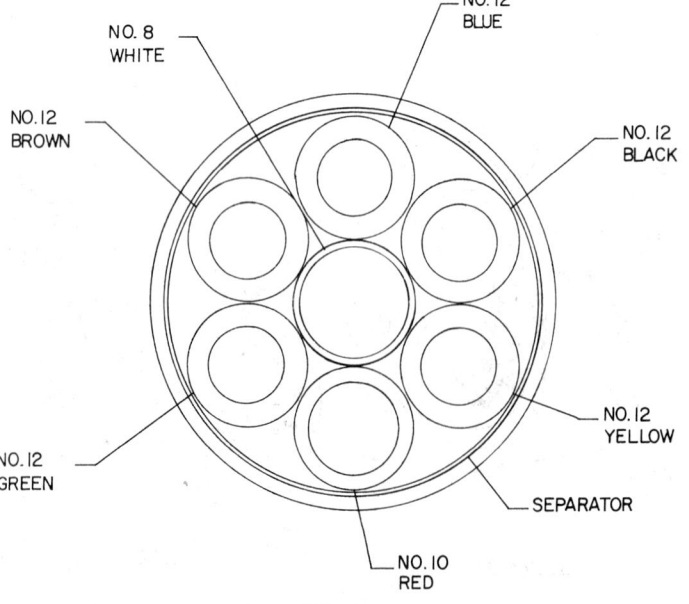

FIG. 1

2.4 Jacket—A 0.062 in. (1.57 mm) minimum thickness jacket shall be applied over the conductors and separator. The jacket compound shall meet the specifications in Table 3. The color of the jacket shall be red.

2.5 Cable Diameter and Finish—The finished outside diameter of the cable shall be 0.690 ± 0.020 in. (17.53 ± 0.508 mm). The finish of the cable shall be smooth and free from defects. Adjacent layers must not stick together when wound around a spool at any temperature below 120°F (49°C).

3. Tests

3.1 The cable shall be free from open circuits or twisted conductor splices.

3.2 Cold Test—A suitable length of cable shall be subjected to −20°F (−29°C) for a period of 4 h. While still at this temperature, it shall be bent 360 deg around a 3 in. (76.2 mm) diameter mandrel. The jacket shall not crack.

TABLE 3—JACKET COMPOUND[a]

Recovery, in.	6 in. (152.4 mm) str, 3/8 (9.53 mm)
Elongation, %	300
Tensile strength, psi (MPa)	1500 (10.35)
Oxygen-bomb aged (96 h at 70°C, 300 psi (2.06 MPa)) elongation	70% of orig. min[b]
Oxygen-bomb aged (96 h at 70°C, 300 psi (2.06 MPa)) tensile	70% of orig. min[b]
Air-oven aged (168 h at 70°C) tensile and elongation	70% of orig. min
Oil immersed (18 h at 121°C) tensile and elongation	60% of orig. min

[a]Underwriters' Laboratories requirements for Class 15 Neoprene—UL 62.
[b]65% of result with unaged specimens if sum of tensile and elongation percentages is at least 140.

ELECTRICAL TERMINALS BLADE TYPE—SAE J858a

SAE Recommended Practice

Report of Electrical Equipment Committee approved June 1963 and last revised August 1969.

Blade terminals listed in this SAE Recommended Practice may be used for terminating wire ends, or for terminating circuits on devices other than wire.

When blade terminals are used for terminating wire, the temper of the terminals shall be sufficiently soft to permit the terminals being assembled to the wire and not show any fracture or cracks which would impair the strength of the assembly.

Terminals may be applied to wire by crimping, welding, swaging, soldering, or any combination at conductor grip.

Insulation grips must be used on all terminals, or some external means of relieving strain shall be provided.

When assembled to wire, the terminals shall fit, and securely grip, the conductor and when applicable, the insulation.

When blade terminals are used to terminate circuits on devices, they shall be of a temper that will permit the terminating section to be formed and attached to the device without fracturing or cracking. The temper should be high enough to resist displacement of the terminal and consequent misalignment to the mating connector.

TYPE 1A BLADE TERMINAL WITH DEPRESSION FOR USE WITH MATING SINGLE CONNECTORS

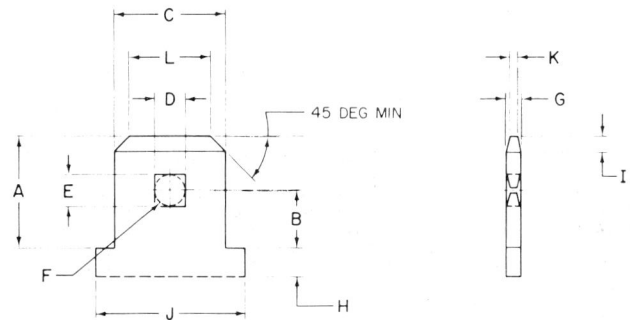

NOTES 1, 2, 3, 4 APPLY

SAE No.	Width	A	B	C	D	E	F	G	H	I	J	K	L
	5/16	0.308–0.318	0.160 ±0.003	0.304–0.320	0.080–0.100	0.070–0.080	0.070–0.080	0.032 ±0.001	0.075 min	0.030 ±0.010	0.369–0.383	0.013–0.019	0.240–0.260
	1/4	0.307–0.317	0.160 ±0.003	0.244–0.252	0.080–0.100	0.070–0.080	0.070–0.080	0.032 ±0.001	0.075 min	0.030 ±0.010	0.294–0.322	0.013–0.019	0.178–0.198
	3/16	0.245–0.260	0.140 ±0.003	0.183–0.192	0.055–0.075	0.045–0.055	0.045–0.055	0.020 ±0.001	0.075 min	0.030 ±0.010	0.240–0.270	0.006–0.012	0.117–0.138

TYPE 1B BLADE TERMINAL WITH HOLE FOR USE WITH MATING SINGLE CONNECTORS

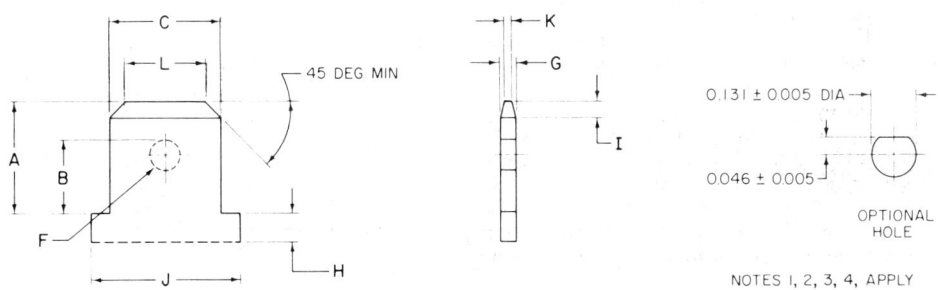

NOTES 1, 2, 3, 4, APPLY

SAE No.	Width	A	B	C	D	E	F	G	H	I	J	K	L
	5/16	0.308–0.318	0.194–0.200	0.304–0.320	—	—	0.090–0.096	0.032 ±0.001	0.075 min	0.030 ±0.010	0.369–0.383	0.013–0.019	0.240–0.260
	1/4	0.307–0.317	0.194–0.200	0.244–0.252	—	—	0.090–0.096	0.032 ±0.001	0.075 min	0.030 ±0.010	0.294–0.322	0.013–0.019	0.178–0.198
	3/16	0.245–0.260	0.150–0.160	0.183–0.192	—	—	0.055–0.065	0.020 ±0.001	0.075 min	0.030 ±0.010	0.240–0.270	0.006–0.012	0.117–0.138

TYPE 1C BLADE TERMINAL WITHOUT HOLE FOR USE WITH MATING MULTIPLE CONNECTOR PLUG

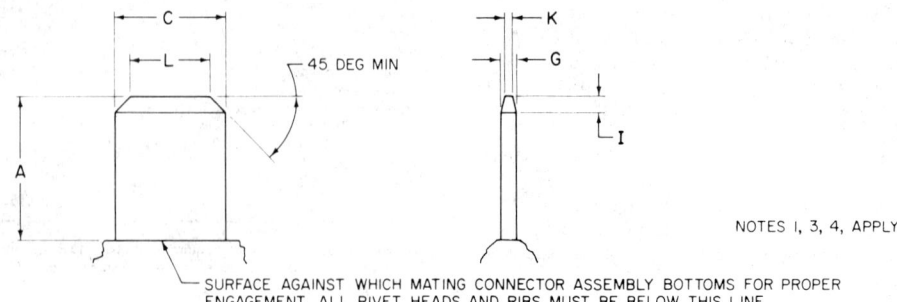

NOTES 1, 3, 4, APPLY

SURFACE AGAINST WHICH MATING CONNECTOR ASSEMBLY BOTTOMS FOR PROPER ENGAGEMENT, ALL RIVET HEADS AND RIBS MUST BE BELOW THIS LINE

SAE No.	Width	A	B	C	D	E	F	G	H	I	J	K	L
	5/16	0.354–0.364	—	0.304–0.320	—	—	—	0.032 ±0.001	—	0.030 ±0.010	—	0.013–0.019	0.240–0.260
	1/4	0.353–0.363	—	0.244–0.252	—	—	—	0.032 ±0.001	—	0.030 ±0.010	—	0.013–0.019	0.178–0.198
	3/16	0.295–0.310	—	0.183–0.192	—	—	—	0.020 ±0.001	—	0.030 ±0.010	—	0.006–0.012	0.117–0.138

TYPE 2 BLADE TERMINAL

NOTES 1, 2, 3, 4, 5 APPLY

SAE No.	Width	A	B	C	D	E	F	G	H	I	J	K	L
	5/16	0.308–0.318	0.194–0.200	0.304–0.320	—	—	0.090–0.096	0.032 ±0.001	0.175 min	0.030 ±0.010	0.369–0.383	0.013–0.019	0.240–0.260
	1/4	0.266–0.296	0.194–0.200	0.244–0.252	—	—	0.090–0.096	0.032 ±0.001	0.175 min	0.030 ±0.010	0.294–0.322	0.013–0.019	0.178–0.198
	3/16	0.245–0.260	0.150–0.160	0.183–0.192	—	—	0.055–0.065	0.020 ±0.001	0.100 min	0.030 ±0.010	0.240–0.270	0.006–0.012	0.117–0.138

NOTES:
1. 45 DEG BEVEL NEED NOT BE A STRAIGHT LINE IF WITHIN THE CONFINES SHOWN.
2. H MINIMUM DIMENSION INDICATES THE AMOUNT OF SHANK NECESSARY ON TERMINAL TO CLEAR MATING PARTS. ALL PROTRUDING RIBS OR HOLD DOWN RIVETS MUST BE BELOW THIS LINE.
3. ALL PORTIONS OF TERMINAL SHOWN SHALL BE FLAT AND FREE FROM OBJECTIONABLE BURRS OR RAISED PLATEAUS. ANY HOLES OR DEPRESSION FOR DETENTS MUST BE FREE FROM RAISES OR BURRS.
4. TERMINALS CAN BE MADE FROM ANY SUITABLE MATERIAL. ANY PLATING MUST BE SMOOTH OR EVEN AND NOT HAVE A SURFACE THAT WILL INDUCE ADDITIONAL DRAG WHEN MATING PARTS ARE ENGAGED. PLATING SHALL NOT INCREASE THE TERMINAL DIMENSIONS OVER 0.0005 IN.
5. HOLE MAY BE OMITTED ACCORDING TO USE.

ELECTRICAL TERMINALS—PIN AND RECEPTACLE TYPE—SAE J928 JUN80

SAE Standard

Report of the Electrical Equipment Committee, approved July 1965, completely revised June 1980.

1. Scope—This SAE Standard covers general requirements and terminal interface dimensions of various sizes of pin and receptacle type terminals.

2. General Requirements—The pin and receptacle type terminals listed in this SAE Standard may be used for terminating wire ends, or for terminating circuits on devices other than wire. Performance requirements for low tension wire terminals are specified in SAE J163 (January, 1974).

Terminals shall be free from burrs, corrosion, or any foreign matter, and shall be of a temper that will permit attachment to wires or circuits on devices without fracturing or cracking.

Terminals may be applied to wire by crimping, welding, swaging, soldering, or any combination thereof at the conductor grip. Insulation grips shall be used on all terminals assembled to 14 gage (2 mm²) and smaller insulated wire except where usage provides other means of relieving strain.

The type, thickness, and finish of the metal used in fabricating these terminals may vary according to the end product use. The dimensions shown in Tables 1 and 2 are included to assure proper fits between manufacturing sources.

Pin terminals fabricated from rod or bar stock must provide suitable stepped internal diameters to fit the wire conductors and insulation consistent with the method by which they are attached.

Insertion and removal forces are also variables that can be adjusted to fit the end use. It is recommended, however, that single connections with indentures should not exceed 15 lb (67 N) and multiple connections without indentures should not exceed 7 lb (31 N) per connection.

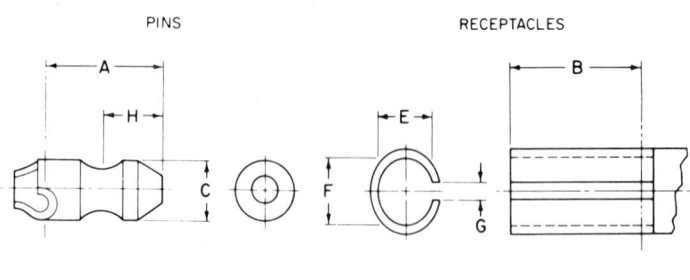

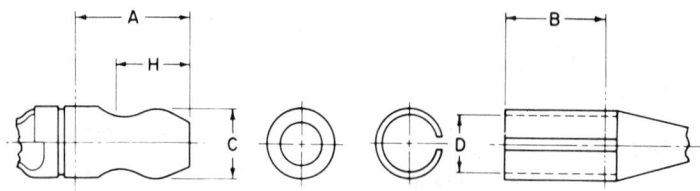

FIG. 1—TYPE I PIN TERMINALS

22.61

TABLE 1—TYPE I PIN TERMINALS

Nominal Dia		SAE Wire Size Range	A Min[a]		B Min[a]		C		D Nominal		E Nominal		F Nominal		G Nominal		H	
in	mm		in	mm	in	mm	in	mm	in	mm	in	mm	in	mm	in	mm	in	mm
0.156	3.96	20–12 (0.5–3.0 mm²)	0.34	8.7	0.34	8.7	0.159–0.155	4.04–3.94	0.150	3.81	0.147	3.73	0.181	4.60	0.034	0.86	0.219	5.56
0.180	4.57	20–10 (0.5–5.0 mm²)	0.40	10.2	0.40	10.2	0.182–0.178	4.62–4.52	0.174	4.42	—	—	—	—	—	—	0.190	4.83

[a]Minimum insertion length.
NOTE: Detent Female—When a female detent is required, the detent of the receptacle must match the H dimension of the pin.

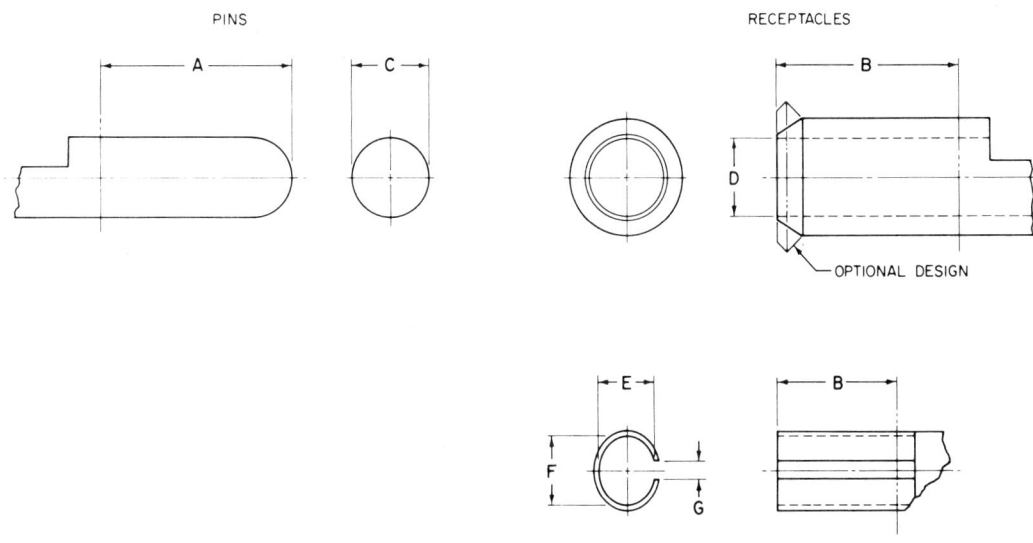

FIG. 2—TYPE II PIN TERMINALS

TABLE 2—TYPE II PIN TERMINALS

Nominal Dia		SAE Wire Size Range	A Min[a]		B Min[a]		C		D Nominal		E Nominal		F Nominal		G Nominal	
in	mm		in	mm	in	mm	in	mm	in	mm	in	mm	in	mm	in	mm
0.086	2.18	20–14 (0.5–2.0 mm²)	0.17	4.4	0.17	4.4	0.086–0.083	2.18–2.11	0.080	2.03	—	—	—	—	—	—
0.093	2.36	20–14 (0.5–2.0 mm²)	0.21	5.4	0.21	5.4	0.093–0.091	2.36–2.31	0.086	2.18	0.086	2.18	0.134	3.40	0.022	0.56
0.156	3.96	20–12 (0.5–3.0 mm²)	0.20	5.1	0.20	5.1	0.159–0.154	4.04–3.91	0.152	3.86	—	—	—	—	—	—

[a]Minimum insertion length.

ELECTRICAL TERMINALS—EYELET AND SPADE TYPE—SAE J561 JUN80

SAE Standard

Report of the Electrical Equipment Division, approved August 1918, completely revised by the Electrical Equipment Committee June 1980.

1. Scope—This SAE Standard covers general requirements and dimensions of various sizes of eyelet and spade type terminals.

2. General Requirements—The eyelet and spade type terminals listed in this SAE Standard may be used for terminating wire ends, or for terminating circuits on devices other than wire. Performance requirements for low tension wire terminals are specified in SAE J163 (January, 1974).

Terminals shall be free from burrs, corrosion, or any foreign matter, and shall be of a temper that will permit attachment to wires or circuits on devices without fracturing or cracking.

Terminals may be applied to wire by crimping, welding, swaging, soldering, or any combination thereof at the conductor grip. Insulation grips shall be used on all terminals assembled to 8 gage (8 mm²) and smaller insulated wire except where usage provides other means of relieving strain.

Materials should be of copper, brass, or other copper alloys. Minimum metal thickness is the nominal thickness shown less a standard strip stock tolerance. Thickness is based on SAE CA260 (UNS C26000) brass conductivity and may be adjusted for use with other materials.

TABLE 1—STUD OR SCREW AND HOLE OR SLOT SIZES

SAE No.	Stud or Screw Size		Hole or Slot Size for Eyelet or Spade, A	
	Nominal	Max	Min	Max
1	4	0.1120	0.123	0.129
2	6	0.1380	0.144	0.150
3	8	0.1640	0.170	0.176
4	10	0.1900	0.201	0.207
5	1/4	0.2500	0.279	0.285
6	5/16	0.3125	0.342	0.348
7	3/8	0.3750	0.404	0.410
8	7/16	0.4375	0.466	0.476
9	1/2	0.5000	0.528	0.538

TABLE 1A—METRIC STUD OR SCREW AND HOLE OR SLOT SIZES

SAE No.	Metric Stud or Screw Size		Hole or Slot Size for Eyelet or Spade, A	
	Nominal	Max	Min	Max
1M	M3	3.00 mm	3.20 mm	3.40 mm
2M	M4	4.00 mm	4.20 mm	4.40 mm
3M	M5	5.00 mm	5.30 mm	5.50 mm
4M	M6	6.00 mm	6.30 mm	6.50 mm
5M	M8	8.00 mm	8.40 mm	8.60 mm
6M	M10	10.00 mm	10.50 mm	10.70 mm
7M	M12	12.00 mm	12.55 mm	12.85 mm
8M	M14	14.00 mm	14.65 mm	14.95 mm
9M	M16	16.00 mm	16.75 mm	17.05 mm

NOTE: Table 1A is for reference purposes only.

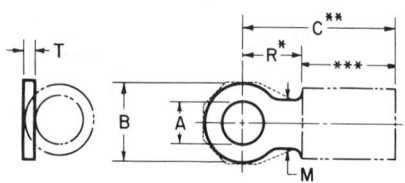
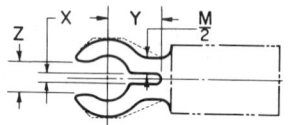

SNAP-ON

* CLEARANCE TO NEAREST OBSTRUCTION
** PRE-INSULATED TERMINALS MAY EXCEED THIS DIMENSION
*** DETAIL DESIGN IS MANUFACTURER'S OPTION

TABLE 2—STRAIGHT-TYPE EYELET AND SNAP-ON EYELET TERMINALS

SAE No.	Screw Size	A Min in	A Min mm	A Max in	A Max mm	B Min in	B Min mm	C Max in	C Max mm	M Min in	M Min mm	R Min in	R Min mm	T Nom in	T Nom mm	X in	X mm	Y in	Y mm	Z in	Z mm
To Use on SAE No. 18 and No. 20 (0.8 mm² and 0.5 mm²) Wire																					
A001	4	0.123	3.13	0.129	3.27	0.24	6.1	0.63	16.0	0.15	3.9	0.19	4.9	0.0253	0.643	—	—	—	—	—	—
A002	6	0.144	3.66	0.150	3.81	0.24	6.1	0.66	16.7	0.15	3.9	0.25	6.4	0.0253	0.643	—	—	—	—	—	—
A003	8	0.170	4.32	0.176	4.47	0.34	8.7	0.70	17.7	0.15	3.9	0.30	7.7	0.0253	0.643	0.06	1.5	0.18	4.6	0.140	3.56
A004	10	0.201	5.11	0.207	5.25	0.34	8.7	0.70	17.7	0.15	3.9	0.30	7.7	0.0253	0.643	0.06	1.5	0.29	7.4	0.150	3.81
A005	1/4	0.279	7.09	0.285	7.23	0.43	11.0	0.85	21.5	0.15	3.9	0.37	9.4	0.0253	0.643	—	—	—	—	—	—
A006	5/16	0.342	8.69	0.348	8.83	0.43	11.0	0.85	21.5	0.15	3.9	0.43	11.0	0.0253	0.643	—	—	—	—	—	—
To Use on SAE No. 14 and No. 16 (2.0 mm² and 1.0 mm²) Wire																					
B101	4	0.123	3.13	0.129	3.27	0.24	6.1	0.76	19.3	0.17	4.4	0.19	4.9	0.0285	0.724	—	—	—	—	—	—
B102	6	0.144	3.66	0.150	3.81	0.24	6.1	0.76	19.3	0.17	4.4	0.25	6.4	0.0285	0.724	—	—	—	—	—	—
B103	8	0.170	4.32	0.176	4.47	0.34	8.7	0.82	20.8	0.17	4.4	0.30	7.7	0.0285	0.724	0.06	1.5	0.18	4.6	0.140	3.56
B104	10	0.201	5.11	0.207	5.25	0.34	8.7	0.82	20.8	0.17	4.4	0.30	7.7	0.0285	0.724	0.06	1.5	0.29	7.4	0.150	3.81
B105	1/4	0.279	7.09	0.285	7.23	0.43	11.0	0.88	22.3	0.17	4.4	0.37	9.4	0.0285	0.724	—	—	—	—	—	—
B106	5/16	0.342	8.69	0.348	8.83	0.55	14.0	1.04	26.4	0.18	4.6	0.43	11.0	0.0285	0.724	—	—	—	—	—	—
B107	3/8	0.404	10.27	0.410	10.41	0.55	14.0	1.04	26.4	0.18	4.6	0.50	12.7	0.0285	0.724	—	—	—	—	—	—
To Use on SAE No. 10 and No. 12 (5.0 mm² and 3.0 mm²) Wire																					
B203	8	0.170	4.32	0.176	4.47	0.34	8.7	0.96	24.3	0.24	6.1	0.30	7.7	0.0403	1.024	0.06	1.5	0.18	4.6	0.140	3.56
B204	10	0.201	5.11	0.207	5.25	0.34	8.7	0.96	24.3	0.24	6.1	0.30	7.7	0.0403	1.024	0.06	1.5	0.29	7.4	0.150	3.81
B205	1/4	0.279	7.09	0.285	7.23	0.50	12.7	1.04	26.4	0.24	6.1	0.37	9.4	0.0403	1.024	—	—	—	—	—	—
B206	5/16	0.342	8.69	0.348	8.83	0.68	17.3	1.16	29.4	0.24	6.1	0.43	11.0	0.0403	1.024	—	—	—	—	—	—
B207	3/8	0.404	10.27	0.410	10.41	0.68	17.3	1.16	29.4	0.24	6.1	0.50	12.7	0.0403	1.024	—	—	—	—	—	—
B208	7/16	0.466	11.84	0.476	12.09	0.68	17.3	1.16	29.4	0.30	7.7	0.62	15.8	0.0403	1.024	—	—	—	—	—	—
B209	1/2	0.528	13.42	0.538	13.66	0.68	17.3	1.16	29.4	0.30	7.7	0.62	15.8	0.0403	1.024	—	—	—	—	—	—
To Use on SAE No. 8 (8.0 mm²) Wire																					
B304	10	0.201	5.11	0.207	5.25	0.34	8.7	1.13	28.7	0.24	6.1	0.30	7.7	0.0453	1.151	—	—	—	—	—	—
B305	1/4	0.279	7.09	0.285	7.23	0.50	12.7	1.13	28.7	0.24	6.1	0.37	9.4	0.0453	1.151	—	—	—	—	—	—
B306	5/16	0.342	8.69	0.348	8.83	0.68	17.3	1.26	32.0	0.24	6.1	0.43	11.0	0.0453	1.151	—	—	—	—	—	—
B307	3/8	0.404	10.27	0.410	10.41	0.68	17.3	1.26	32.0	0.24	6.1	0.50	12.7	0.0453	1.151	—	—	—	—	—	—
B308	7/16	0.466	11.84	0.476	12.09	0.68	17.3	1.38	35.0	0.30	7.7	0.62	15.8	0.0453	1.151	—	—	—	—	—	—
B309	1/2	0.528	13.42	0.538	13.66	0.68	17.3	1.38	35.0	0.30	7.7	0.62	15.8	0.0453	1.151	—	—	—	—	—	—

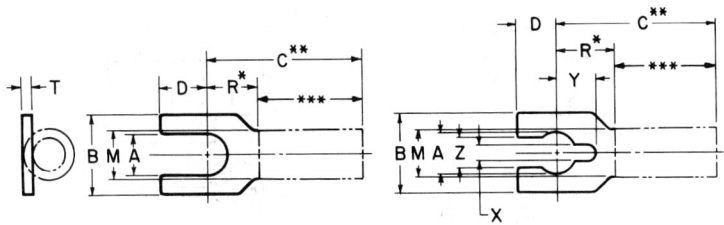

* CLEARANCE TO NEAREST OBSTRUCTION
** PRE-INSULATED TERMINALS MAY EXCEED THIS DIMENSION
*** DETAIL DESIGN IS MANUFACTURER'S OPTION

SNAP-ON

TABLE 3—STRAIGHT-TYPE SPADE TERMINALS

SAE No.	Screw Size	A Min		A Max		B Min		C Max		D		M Min		R Min		T Nom		X		Y		Z in ±0.005	Z mm ±0.13
		in	mm	in	mm	in	mm	in	mm	in	mm	in	mm	in	mm	in	mm	in	mm	in	mm		
To Use on SAE No. 18 and No. 20 (0.8 mm² and 0.5 mm²) Wire
H003	8	0.170	4.32	0.176	4.47	0.37	9.4	0.82	20.8	0.25	6.4	0.15	3.9	0.30	7.7	0.0253	0.643	0.06–0.10	1.5–2.5	0.15–0.21	3.8–5.3	0.148	3.76
H004	10	0.201	5.11	0.207	5.25	0.37	9.4	0.82	20.8	0.25	6.4	0.15	3.9	0.30	7.7	0.0253	0.643	0.06–0.10	1.5–2.5	0.15–0.21	3.8–5.3	0.175	4.44
To Use on SAE No. 14 and No. 16 (2.0 mm² and 1.0 mm²) Wire
H103	8	0.170	4.32	0.176	4.47	0.37	9.4	0.82	20.8	0.25	6.4	0.17	4.4	0.30	7.7	0.0285	0.724	0.06–0.10	1.5–2.5	0.15–0.21	3.8–5.3	0.148	3.76
H104	10	0.201	5.11	0.207	5.25	0.37	9.4	0.82	20.8	0.25	6.4	0.17	4.4	0.30	7.7	0.0285	0.724	0.06–0.10	1.5–2.5	0.15–0.21	3.8–5.3	0.175	4.44
To Use on SAE No. 10 and No. 12 (5.0 mm² and 3.0 mm²) Wire
H203	8	0.170	4.32	0.176	4.47	0.37	9.4	0.82	20.8	0.25	6.4	0.24	6.1	0.30	7.7	0.0403	1.024	0.06–0.10	1.5–2.5	0.15–0.21	3.8–5.3	0.148	3.76
H204	10	0.201	5.11	0.207	5.25	0.37	9.4	0.82	20.8	0.25	6.4	0.24	6.1	0.30	7.7	0.0403	1.024	0.06–0.10	1.5–2.5	0.15–0.21	3.8–5.3	0.175	4.44
To Use on SAE No. 8 (8.0 mm²) Wire
H304	10	0.201	5.11	0.207	5.25	0.37	9.4	1.13	28.7	0.25	6.4	0.24	6.1	0.30	7.7	0.0453	1.151	0.06–0.10	1.5–2.5	0.15–0.21	3.8–5.3	0.175	4.44

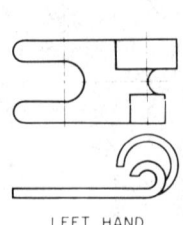 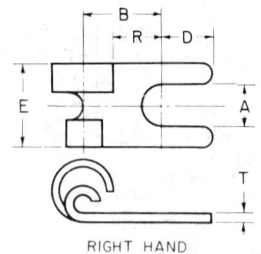 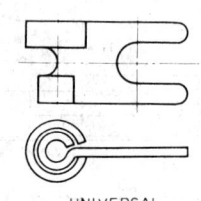 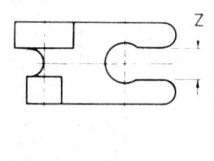

LEFT HAND | RIGHT HAND | UNIVERSAL | SNAP-ON

TABLE 4—SIDE-TYPE SPADE TERMINALS

SAE No.	Screw Size	A Min in	A Min mm	A Max in	A Max mm	B Min in	B Min mm	D in	D mm	E in	E mm	R Min in	R Min mm	T Nom in	T Nom mm	Z in ±0.005	Z mm ±0.13
	To Use on SAE No. 14 (2.0 mm²) Wire and Smaller
M103	8	0.170	4.32	0.176	4.47	0.34	8.7	0.25	6.4	0.38	9.7	0.30	7.7	0.0285	0.724	0.148	3.76
M104	10	0.201	5.11	0.207	5.25	0.34	8.7	0.25	6.4	0.38	9.7	0.30	7.7	0.0285	0.724	0.175	4.44
	To Use on SAE No. 10 and No. 12 (5.0 mm² and 3.0 mm²) Wire
M203	8	0.170	4.32	0.176	4.47	0.37	9.4	0.25	6.4	0.38	9.7	0.30	7.7	0.0403	1.024	0.148	3.76
M204	10	0.201	5.11	0.207	5.25	0.37	9.4	0.25	6.4	0.38	9.7	0.30	7.7	0.0403	1.024	0.175	4.44
	To Use on SAE No. 8 (8.0 mm²) Wire
M304	10	0.201	5.11	0.207	5.25	0.40	10.2	0.25	6.4	0.38	9.7	0.30	7.7	0.0453	1.151	0.175	4.44

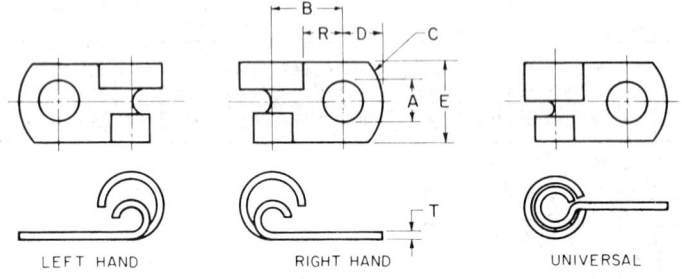

LEFT HAND | RIGHT HAND | UNIVERSAL

TABLE 5—SIDE-TYPE EYELET TERMINALS

SAE No.	Screw Size	A Min in	A Min mm	A Max in	A Max mm	B Min in	B Min mm	C Max in	C Max mm	D in	D mm	E in	E mm	R Min in	R Min mm	T Nom in	T Nom mm
	To Use on SAE No. 14 (2.0 mm²) Wire and Smaller
K101	4	0.123	3.13	0.129	3.27	0.27	6.9	0.38	9.6	0.155	3.94	0.31	7.9	0.19	4.9	0.0285	0.724
K102	6	0.144	3.66	0.150	3.81	0.27	6.9	0.38	9.6	0.155	3.94	0.31	7.9	0.25	6.4	0.0285	0.724
K103	8	0.170	4.32	0.176	4.47	0.34	8.7	0.38	9.6	0.19	4.8	0.38	9.7	0.30	7.7	0.0285	0.724
K104	10	0.201	5.11	0.207	5.25	0.34	8.7	0.38	9.6	0.19	4.8	0.38	9.7	0.30	7.7	0.0285	0.724
K105	1/4	0.279	7.09	0.285	7.23	0.37	9.4	0.63	16.0	0.25	6.4	0.50	12.7	0.37	9.4	0.0285	0.724
K106	5/16	0.342	8.69	0.348	8.83	0.43	11.0	0.63	16.0	0.31	7.9	0.62	15.7	0.43	11.0	0.0285	0.724
K107	3/8	0.404	10.27	0.410	10.41	0.47	12.0	0.76	19.3	0.375	9.52	0.75	19.0	0.50	12.7	0.0285	0.724
	To Use on SAE No. 10 and No. 12 (5.0 mm² and 3.0 mm²) Wire
K203	8	0.170	4.32	0.176	4.47	0.34	8.7	0.38	9.6	0.19	4.8	0.38	9.7	0.30	7.7	0.0403	1.024
K204	10	0.201	5.11	0.207	5.25	0.34	8.7	0.38	9.6	0.19	4.8	0.38	9.7	0.30	7.7	0.0403	1.024
K205	1/4	0.279	7.09	0.285	7.23	0.43	11.0	0.63	16.0	0.25	6.4	0.50	12.7	0.37	9.4	0.0403	1.024
K206	5/16	0.342	8.69	0.348	8.83	0.49	12.5	0.63	16.0	0.31	7.9	0.62	15.7	0.43	11.0	0.0403	1.024
K207	3/8	0.404	10.27	0.410	10.41	0.49	12.5	0.76	19.3	0.375	9.52	0.75	19.0	0.50	12.7	0.0403	1.024
	To Use on SAE No. 6 and No. 8 (13.0 mm² and 8.0 mm²) Wire
K304	10	0.201	5.11	0.207	5.25	0.40	10.2	0.63	16.0	0.19	4.8	0.36	9.7	0.30	7.7	0.0453	1.151
K305	1/4	0.279	7.09	0.285	7.23	0.47	12.0	0.76	19.3	0.375	9.52	0.75	19.0	0.37	9.4	0.0453	1.151
K306	5/16	0.342	8.69	0.348	8.83	0.55	14.0	0.76	19.3	0.375	9.52	0.75	19.0	0.43	11.0	0.0453	1.151
K307	3/8	0.404	10.27	0.410	10.41	0.60	15.3	0.76	19.3	0.375	9.52	0.75	19.0	0.50	12.7	0.0453	1.151
K308	7/16	0.466	11.84	0.476	12.09	0.68	17.3	0.76	19.3	0.375	9.52	0.75	19.0	0.62	15.8	0.0453	1.151

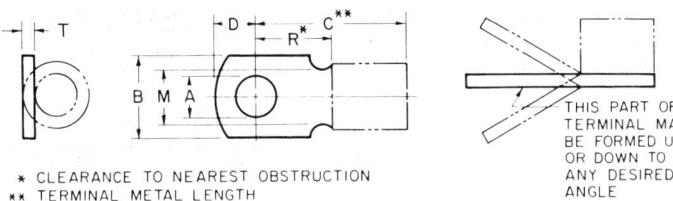

* CLEARANCE TO NEAREST OBSTRUCTION
** TERMINAL METAL LENGTH

THIS PART OF TERMINAL MAY BE FORMED UP OR DOWN TO ANY DESIRED ANGLE

TABLE 6—END-TYPE STARTING-CABLE TERMINALS

SAE No.	Screw Size	A Min		A Max		B Min		C Max		D		M Min		R Min		T Nom	
		in	mm	in	mm	in	mm	in	mm	in	mm	in	mm	in	mm	in	mm
To Use on SAE No. 4 and No. 6 (19.0 mm² and 13.0 mm²) Wire																	
N405	1/4	0.279	7.09	0.285	7.23	0.55	14.0	1.26	32.0	0.34–0.38	8.6–9.7	0.30	7.7	0.62	15.8	0.0720	1.829
N406	5/16	0.342	8.69	0.348	8.83	0.68	17.3	1.26	32.0	0.34–0.38	8.6–9.7	0.30	7.7	0.62	15.8	0.0720	1.829
N407	3/8	0.404	10.27	0.410	10.41	0.68	17.3	1.26	32.0	0.34–0.38	8.6–9.7	0.30	7.7	0.62	15.8	0.0720	1.829
N408	7/16	0.466	11.84	0.476	12.09	0.68	17.3	1.26	32.0	0.34–0.38	8.6–9.7	0.30	7.7	0.62	15.8	0.0720	1.829
N409	1/2	0.528	13.42	0.538	13.66	0.68	17.3	1.26	32.0	0.34–0.38	8.6–9.7	0.30	7.7	0.62	15.8	0.0720	1.829
To Use on SAE No. 1 and No. 2 (40.0 mm² and 32.0 mm²) Wire																	
N505	1/4	0.279	7.09	0.285	7.23	0.55	14.0	1.69	42.9	0.34–0.38	8.6–9.7	0.49	12.5	0.62	15.8	0.0808	2.052
N506	5/16	0.342	8.69	0.348	8.83	0.68	17.3	1.69	42.9	0.34–0.38	8.6–9.7	0.49	12.5	0.62	15.8	0.0808	2.052
N507	3/8	0.404	10.27	0.410	10.41	0.68	17.3	1.69	42.9	0.34–0.38	8.6–9.7	0.49	12.5	0.62	15.8	0.0808	2.052
N508	7/16	0.466	11.84	0.476	12.09	0.68	17.3	1.69	42.9	0.34–0.38	8.6–9.7	0.49	12.5	0.62	15.8	0.0808	2.052
N509	1/2	0.528	13.42	0.538	13.66	0.68	17.3	1.69	42.9	0.34–0.38	8.6–9.7	0.49	12.5	0.62	15.8	0.0808	2.052
To Use on SAE No. 00 and No. 0 (62.0 mm² and 50.0 mm²) Wire																	
N605	1/4	0.279	7.09	0.285	7.23	0.55	14.0	1.76	44.7	0.34–0.38	8.6–9.7	0.55	14.0	0.62	15.8	0.0808	2.052
N606	5/16	0.342	8.69	0.348	8.83	0.68	17.3	1.76	44.7	0.34–0.38	8.6–9.7	0.55	14.0	0.62	15.8	0.0808	2.052
N607	3/8	0.404	10.27	0.410	10.41	0.68	17.3	1.76	44.7	0.34–0.38	8.6–9.7	0.55	14.0	0.62	15.8	0.0808	2.052
N608	7/16	0.466	11.84	0.476	12.09	0.68	17.3	1.76	44.7	0.34–0.38	8.6–9.7	0.55	14.0	0.62	15.8	0.0808	2.052
N609	1/2	0.528	13.42	0.538	13.66	0.68	17.3	1.76	44.7	0.34–0.38	8.6–9.7	0.55	14.0	0.62	15.8	0.0808	2.052
To Use on SAE No. 0000 and No. 000 (103.0 mm² and 81.0 mm²) Wire																	
N705	1/4	0.279	7.09	0.285	7.23	0.74	18.8	1.76	44.7	0.37–0.51	9.4–13.0	0.68	17.3	0.74	18.8	0.0907	2.304
N706	5/16	0.342	8.69	0.348	8.83	0.74	18.8	1.76	44.7	0.37–0.51	9.4–13.0	0.68	17.3	0.74	18.8	0.0907	2.304
N707	3/8	0.404	10.27	0.410	10.41	0.74	18.8	1.76	44.7	0.37–0.51	9.4–13.0	0.68	17.3	0.74	18.8	0.0907	2.304
N708	7/16	0.466	11.84	0.476	12.09	0.74	18.8	1.76	44.7	0.37–0.51	9.4–13.0	0.68	17.3	0.74	18.8	0.0907	2.304
N709	1/2	0.528	13.42	0.538	13.66	0.74	18.8	1.76	44.7	0.37–0.51	9.4–13.0	0.68	17.3	0.74	18.8	0.0907	2.304

NONMETALLIC LOOM—SAE J562a

SAE Recommended Practice

Report of Electrical Equipment Division approved March 1922 and last revised by Electrical Equipment Committee September 1968.

General Data—Nonmetallic flexible loom is recommended for use as an insulated covering giving mechanical protection over insulated wire, metal tubing, or other parts requiring a water-, oil-, and acid-proof covering resistant to fire or abrasion. It is also recommended for use as a covering for copper or other metal tubing to prevent crystallization and to eliminate rattles.

Construction—The loom shall be of single-wall construction, the material used to be strictly nonmetallic and of sufficient mechanical strength so that when formed or woven into a tubing it shall pass the tests for the size specified. Finished loom shall be free from obstruction and shall permit easy introduction of the maximum size wire or other part for which it is normally suited. Loom in any length shall slip freely over a polished mandrel 12 in. long and equal in diameter to the minimum inside diameter specified. The dimensions of the standard sizes are listed in the accompanying table.

TABLE 1—DIMENSIONS OF NONMETALLIC LOOM, IN.

Nominal Size	Inside Dia		Outside Dia		Nominal Size	Inside Dia		Outside Dia	
	Min	Max	Min	Max		Min	Max	Min	Max
5/32	0.156	0.176	0.245	0.265	5/8	0.625	0.645	0.785	0.805
3/16	0.187	0.207	0.287	0.307	11/16	0.687	0.707	0.847	0.867
1/4	0.250	0.270	0.350	0.370	3/4	0.750	0.770	0.934	0.954
5/16	0.312	0.332	0.412	0.432	13/16	0.812	0.832	0.996	1.016
3/8	0.375	0.395	0.505	0.525	7/8	0.875	0.895	1.079	1.099
7/16	0.437	0.457	0.567	0.587	15/16	0.937	0.957	1.141	1.161
1/2	0.500	0.520	0.630	0.650	1	1.000	1.020	1.204	1.224
9/16	0.562	0.582	0.722	0.742					

Saturation

Fire Resistant Loom—The loom shall be of such construction that when the asphaltic compound or equivalent water, acid, and fire resisting compound is applied, it will thoroughly impregnate the outside and lightly impregnate the inside of the loom.

Oil Proof Loom—The loom shall be thoroughly impregnated with a gum saturator or its equivalent. The saturator, when dry, shall be free from tackiness and gummy deposits. This impregnation is introduced to prevent absorption of moisture, oil or gasoline, to bind the material together to give the required wall strength and to prevent fraying.

Finish—For a fire resistant loom the outer surface shall be thoroughly covered by an asphaltic or equivalent water, acid, and fire resisting compound. Over the asphaltic or equivalent water, acid, and fire resisting compound, the loom may be coated with a thin coating of good paraffin wax or equivalent. For an oil-proof loom the outer surface shall be thoroughly covered with at least two coats of black pyroxylin lacquer or its equivalent, producing a good luster and a good bond to the fabric. The lacquer must be thoroughly dried before wrapping or boxing the loom for shipment. The use of heavy finishes or saturators to give artificial appearance is not permitted. The pyroxylin must be sufficiently plasticized so that it will not crack on a piece of loom kept three months at room temperature and then bent back sharply upon itself. Loom with finish of a higher luster than can be obtained with two coats of lacquer as specified above shall be considered special and should be covered by other specifications when required.

Tests—A 6 in. piece of loom totally immersed in water at 70 F for 24 hr and then blown out with a mild air current immediately after removing the water shall not have an increase in weight of more than 35%. The wall must not collapse when the loom is bent to a radius of five times the inside diameter at 70 F. The compound and finish must not crack open in this test.

The material in the wall of the loom shall not crack or break when a 3 in. length is flattened between two steel plates. When the inside diameter of loom is $3/8$ in. or smaller, the distance between plates is to be $11/64$ in. When the inside diameter of loom is over $3/8$ in., the distance between plates is to be $9/32$ in. The finish shall not show excessive cracking when loom is subjected to this test.

The polished mandrels used for checking the inside diameters shall show no sticking or discoloration up to 150 F.

Loom shall be capable of standing a tension test for 5 min without breaking or opening at any point as required in Table 2.

Additional Test for Oil Proof or Fire Resistant Loom—When a piece of loom is totally immersed in an equal mixture of cylinder oil, kerosene, and gasoline at 70 F for 5 min and then subjected to a temperature not exceeding 250 F for 1 hr, the saturating compound must not drip from the loom nor the finish show any appreciable defects.

TABLE 2—TENSILE REQUIREMENTS OF NONMETALLIC LOOM[a]

Loom, Nominal Size, in.	Min Tensile Requirement, lb
3/16	75
1/4	85
5/16 and 3/8	100
7/16 or larger	150

[a] The test piece shall be 6 in. long between supports.

Flame-resisting qualities shall be incorporated in the saturation or finish or both for the loom to pass the following test: The loom shall not convey fire nor support combustion for more than 1 min after five 15-sec applications of a standard test flame with intervals of 15 sec between applications. A standard test flame is the blue flame, about 5 in. high, produced by a $1/2$ in. bunsen burner fed with ordinary illuminating gas at normal pressure. The loom shall be held vertically with either the lower or the upper end thoroughly sealed to prevent the passage of air, and the flame must be applied horizontally.

ELECTRIC WINDSHIELD WIPER SWITCH—SAE J112a — SAE Recommended Practice

Report of Electrical Equipment Committee approved July 1969 and last revised May 1971. Editorial change October 1977.

1. Definition—An electric windshield wiper switch is that part of an electric windshield wiper system by which the operator of a vehicle causes the windshield wipers to function.

2. Reference Standards

2.1 If the switch employs an internal circuit breaker(s), the circuit breaker shall comply with the requirements of SAE J258 (June, 1971).

2.2 If the switch employs the integral exterior circuit breaker, the circuit breaker shall comply with the requirements of SAE J553c (May, 1972).

3. Temperature Test

3.1 To insure basic function, the switch shall be manually cycled for 10 cycles at design electrical load at 75 ± 10 F (24 ± 5.5 C), 165, +0, −5 F (74, +0, −2.8 C), and −25, +5, −0 F (−32, +2.8, −0 C) after a 1 h exposure at each of these temperatures. The switch shall be electrically and mechanically operable during each of these cycles.

3.2 This same switch shall be used for the endurance test described below.

4. Endurance Test Setup

4.1 The switch shall be set up to operate its design electrical load.

4.2 The test shall be set up to operate the switch for the prescribed number of completed cycles.

One complete cycle shall consist of sequencing through each position (with dwell in each position) and return without dwell in intermediate positions to the initial position.

The test equipment shall be so arranged as to provide the following mechanical time requirements:

Travel Time—0.1–0.5 s (time from one position to the next).

NOTE: If the switch employs a rheostat, the travel time through the rheostat segment in each direction shall be 1.0–3.0 s.

Dwell Time—0.50–1.0 s (time in each position).

NOTE: After switching to OFF, if a motor is used, sufficient dwell time shall be provided to allow the motor to park. The dwell time in OFF can then be greater (if required) than the dwell time range indicated above.

4.3 During the test the switch shall be operated at 6.4 V d-c for a 6 V system, 12.8 V d-c for a 12 V system, or 25.6 V d-c for a 24 V system.

These voltages shall be measured at the input termination on the switch.

The power supply shall not generate any adverse transients not present in motor vehicles and shall comply with the following specifications:

(a) Output Current—Capable of supplying the continuous current of the design electrical load.

(b) Regulation

Dynamic—The output voltage at the supply shall not deviate more than 1.0 V from zero to maximum load (including inrush current) and should recover 63% of its maximum excursion within 100 ms.

Static—The output voltage at the supply shall not deviate more than 2% with changes in static load from zero to maximum (not including inrush current), and means shall be provided to compensate for static input line voltage variations.

(c) Ripple Voltage—Maximum 300 mV peak to peak.

5. Endurance Requirements

5.1 The switch shall be capable of satisfactorily operating for 10,000 complete cycles at a temperature of 75 ± 10 F (24 ± 5.5 C) followed by 1 h ON in low position at 75 ± 10 F (24 ± 5.5 C).

5.2 The average voltage drop from the input terminal(s) to the corresponding average output terminal(s) shall be measured before and after the endurance test and after the soak tests and shall not exceed 0.30 (excluding rheostat), the average of three consecutive readings, at design load. If wiring is an integral part of the switch, the voltage drop measurement shall be made including 3 in. (76 mm) of wire on each side of the switch; otherwise, measurement shall be made at switch terminals.

6. Combination Windshield Wiper and Washer Switch—The same combination switch shall be used for the test of each function. If the washer and wiper functions are mechanically coordinated, the functions shall be tested simultaneously. The wiper switch shall meet the requirements of this recommended practice. The washer switch shall meet the requirements of SAE J234 (May, 1971).

ELECTRIC WINDSHIELD WASHER SWITCH—SAE J234 — SAE Recommended Practice

Report of Electrical Equipment Committee approved May 1971. Editorial change October 1977.

1. Definition—An electric windshield washer switch is that part of an electric windshield washer system by which the operator of a vehicle causes the windshield washers to function.

2. Temperature Test

2.1 To insure basic function, the switch shall be manually cycled for 10 cycles at design electrical load at 75 ± 10 F (24 ± 5.5 C), 165 +0, −5 F (74, +0, −2.8 C), and −25, +5, −0 F (−32, +2.8, −0 C) after a 1 h exposure at each of these temperatures. The switch shall be electrically and mechanically operable during each of these cycles.

2.2 This same switch shall be used for the endurance test described below.

3. Endurance Test Setup

3.1 The switch shall be set up to operate its design electrical load.

3.2 The test shall be set up to operate the switch for the prescribed number of completed cycles.

One complete cycle shall consist of sequencing through each position (with dwell in each position) and return without dwell in intermediate positions to the initial position.

The test equipment shall be so arranged as to provide the following mechanical time requirements:

Travel Time—0.1–0.5 s (time from one position to the next).

Dwell Time—1.0–2.0 s (time in each position).

3.3 During the test the switch shall be operated at 6.4 V d-c for a 6 V system, 12.8 V d-c for a 12 V system, or 25.6 V d-c for a 24 V system.

These voltages shall be measured at the input termination on the switch. The power supply shall not generate any adverse transients not present in motor vehicles and shall comply with the following specifications:

(a) Output Current—Capable of supplying the continuous current of the design electrical load.

(b) Regulation

Dynamic—The output voltage at the supply shall not deviate more than 1.0 V from zero to maximum load (including inrush current) and should recover 63% of its maximum excursion within 100 ms.

Static—The output voltage at the supply shall not deviate more than 2% with changes in static load from zero to maximum (not including inrush current), and means shall be provided to compensate for static input line voltage variations.

(c) Ripple Voltage—Maximum 300 mV peak to peak.

4. Endurance Requirements

4.1 The switch shall be capable of satisfactorily operating for 10,000 complete cycles at a temperature of 75 ± 10 F (24 ± 5.5 C).

4.2 The average voltage drop from the input terminal(s) to the corresponding output terminal(s) shall be measured before and after the endurance test and shall not exceed 0.30 V (the average of three consecutive readings) at design load. If wiring is an integral part of the switch, the voltage drop measurement shall be made including 3 in. of wire on each side of the switch; otherwise, measurement shall be made at switch terminals.

5. Combination Windshield Wiper and Washer Switch—The same combination switch shall be used for the test of each function. If the washer and wiper functions are mechanically coordinated, the functions shall be tested simultaneously. The washer switch shall meet the requirements of this recommended ϕ practice. The wiper switch shall meet the requirements of SAE J112a (November, 1971).

ELECTRIC BLOWER MOTOR SWITCH—SAE J235

SAE Recommended Practice

Report of Electrical Equipment Committee approved May 1971.

1. Definition—An electric blower motor switch is that part of a blower system by which the operator of a vehicle causes a blower motor to function.

2. Temperature Test

2.1 To insure basic function, the switch shall be manually cycled for 10 cycles at design electrical load at 75 ± 10 F (24 ± 5.5 C), 165, +0, −5 F (74, +0, −2.8 C), and −25, +5, −0 F (−32, +2.8, −0 C) after a 1 h exposure at each of these temperatures. The switch shall be electrically and mechanically operable during each of these cycles.

2.2 This same switch shall be used for the endurance test described below.

3. Endurance Test Setup

3.1 The switch shall be set up to operate its design electrical load.

3.2 The test shall be set up to operate the switches for the prescribed number of complete cycles.

One complete cycle shall consist of sequencing through each position (with dwell in each position) and return without dwell in intermediate positions to the initial position.

The test equipment shall be so arranged as to provide the following mechanical time requirements:

Travel Time—0.1–0.5 s (time from one position to the next).

Dwell Time—0.5–1.0 s (time in each position).

3.3 During the test the switch shall be operated at 6.4 V d-c for a 6 V system, 12.8 V d-c for a 12 V system, or 25.6 V d-c for a 24 V system.

These voltages shall be measured at the input termination on the switch.

The power supply shall not generate any adverse transients not present in motor vehicles and shall comply with the following specifications:

(a) Output Current—Capable of supplying the continuous current of the design electrical load.

(b) Regulation

Dynamic—The output voltage at the supply shall not deviate more than 1.0 V from zero to maximum load (including inrush current) and should recover 63% of its maximum excursion within 100 ms.

Static—The output voltage at the supply shall not deviate more than 2% with changes in static load from zero to maximum (not including inrush current), and means shall be provided to compensate for static input line voltage variations.

(c) Ripple Voltage—Maximum 300 mV peak to peak.

4. Endurance Requirements

4.1 The switch shall be capable of satisfactorily operating for 10,000 complete cycles at a temperature of 75 ± 10 F (24 ± 5.5 C) followed by 1 h ON in low position at 75 ± 10 F (24 ± 5.5 C).

4.2 The average voltage drop from the input terminal(s) to the corresponding average output terminal(s) shall be measured before and after the endurance test and shall not exceed 0.30 V (the average of three readings) at design load. If wiring is an integral part of the switch, the voltage drop measurement shall be made including 3 in. of wire on each side of the switch; otherwise, measurement shall be made at switch terminals.

SIX- AND TWELVE-VOLT CIGAR LIGHTER RECEPTACLES—SAE J563b

SAE Standard

Report of Electrical Equipment Committee approved June 1949 and last revised May 1978.

This SAE Standard covers the basic cigar lighter receptacle, which may optionally incorporate overload protective devices.

Fig. 1 and Table 1 show the cigar lighter receptacle and its pertinent dimensions.

Lighter plugs shall be permanently marked either "6 volt" or "12 volt".

Manufacturers of accessory plugs to energize devices such as trouble lamps and razors are cautioned to conform to the provisions given in this SAE Standard.

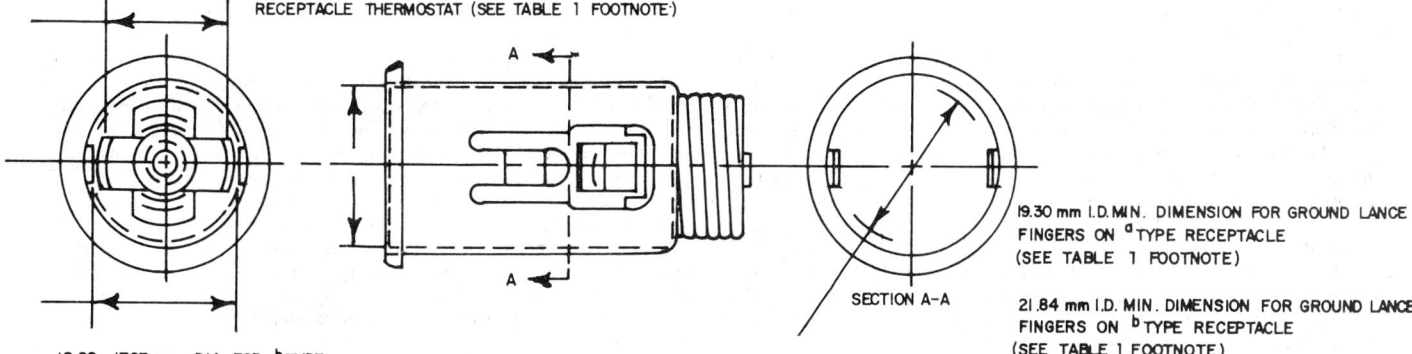

ϕ FIG. 1—CIGAR LIGHTER RECEPTACLE (TYPICAL)

Live contact of any accessory plug shall be made with the center stud; the
ϕ contacting member must not be less than 3.56 mm in diameter and have a
ϕ minimum spherical radius of 2.54 mm at the contacting end.

To prevent damage, the body of an accessory plug back of the contact end should be large enough to serve as a centering guide when it is inserted into the receptacle. The contact end of any accessory plug should have sufficient taper so as not to interfere with the receptacle bimetal fingers. Any ground contact fingers should be so designed as not to interfere with, distort, or catch on the grounding lances in the receptacle when the plug is rotated.

Wiring capacity and overload protection, as provided by the automotive industry to the 12-V lighter receptacle, restricts accessory plug devices to a maximum current draw of 8.0 amp.

Accessory plugs shall be permanently marked "6 volt" or "12 volt" in letters
ϕ 4.7 mm high minimum.

ϕ **TABLE 1—DIAMETERS OF RECEPTACLE AND LIGHTER PLUG**

Volts	Inside Dia, B, mm	Plug Body Dia, mm
6	21.34–21.46	21.08–21.23
12	20.93–21.01[a]	20.73–20.88
12	21.41–21.51[b]	21.13–21.23

[a] Receptacles providing bimetal finger contact to the *outer* periphery of the heating element cup.
[b] Receptacles providing bimetal finger contact to the *inner* periphery of the heating element cup.

ELECTROMAGNETIC SUSCEPTIBILITY PROCEDURES FOR VEHICLE COMPONENTS ϕ (EXCEPT AIRCRAFT)—SAE J1113a

SAE Recommended Practice

Report of Subcommittee on EMI Standards and Test Methods approved April 1975 and completely revised by Electronic Systems Committee June 1978.

1. Introduction

1.1 Scope—This SAE Recommended Practice establishes uniform laboratory measurement techniques for the determination of the susceptibility to undesired electromagnetic sources of electrical, electronic, and electromechanical ground-vehicle components. It is intended as a guide toward standard practice but may be subject to frequent change to keep pace with experience and technical advances, and this should be kept in mind when considering its use.

1.2 Measurement Philosophy—The need for measurement of the susceptibility of vehicle electronic components to electromagnetic sources has recently become more critical as more electronic components are used in safety-related vehicle application. Electronic and electrical equipment may be susceptible to temporary or permanent malfunctions when subjected to electromagnetic sources, either of a transient or steady-state nature.

Electromagnetic interference (EMI) may be either transient, intermittent, or continuous in nature arising from sources such as transmitters or other equipment located either on board, or adjacent to the vehicle, or from component parts of the vehicle ignition or electrical power systems.

This recommended practice sets forth uniform procedures for establishing the susceptibility levels of individual vehicle components. It does not, however, set limits on levels of EMI in which vehicle components must perform.

The most direct method of specifying the EMI environment limits is to measure the actual fields, voltages, currents, and impedances around the component or system of interest under all hazardous conditions. This will, of course, require a large enough sample of installations to determine possible variations.

Another approach to setting limits on levels of EMI would be to establish maximum levels of electric and magnetic field strength with the interconnecting wire and field impedance levels expected for all points within the vehicle, from sources of maximum power levels expected for on-board transmitters, transient sources, and external transmitters. These constitute the ambient level of EMI. Data of this type are being collected and may be supplied later as a separate report on EMI characteristics and test limits.

It is recommended that a statistically valid number of components be tested using procedures adopted as standard by the testing organization. If there is no such standard procedure, one given in Appendix A may be used for all sections of this procedure, except where destructive testing results (Section 5 and possibly Section 4). For destructive testing, such as load-dump simulation of Section 5, consult a handbook on statistical methods for details of the Karber Method or the Bruceton (stair-step) Method of sensitivity measurements. These methods eliminate the effects of cumulative degradation which often occurs during destructive testing.

It is suggested that only those portions of this recommended practice which are critical to the particular use of the component under test be applied, rather than subject the component to the provisions of the entire document. Thus, if the particular component under test is known to be susceptible mainly to transients, but otherwise well protected against conducted and radiated EMI, then only Section 5 need be applied. Or if susceptibility to radiated energy is known to be a primary cause of malfunctions, then only Sections 6 through 8 need be applied.

A list of suitable equipment is provided as an aid in Appendix B. No approval or disapproval of any manufacturer is intended, either by inclusion or exclusion from this list.

Caution must be exercised in many portions of this procedure where high voltages or intense fields may be present. (The present maximum allowable field for human exposure is 10 mW/cm^2 averaged over 6 min, which is equivalent to about 194 V/m electric field strength in the far field.)

1.3 Definitions and Terminology—The following definitions apply to the terms indicated as they are used in this recommended practice:

1.3.1 AMBIENT LEVEL—Those levels of radiated and conducted signal and noise existing at a specified test location and time when the test sample is in operation. Atmospherics, interference from other sources, and circuit noise or other interference generated within the measuring set compose the *ambient level*.

1.3.2 CONDUCTED EMISSION—Desired or undesired electromagnetic energy which is propagated along a conductor. Such an emission is called *conducted interference* if it is undesired.

1.3.3 ELECTROMAGNETIC COMPATIBILITY (EMC)—Is the condition that enables equipment, subsystems, and systems (electronic, chemical, biological, etc.) to function without degradation from electromagnetic sources and without degrading the electromagnetic environment; *i.e.,* it is the condition which allows the coexistence of different electromagnetic sources without significant change in performance of any one in the presence of any or all the others.

1.3.4 EMISSION—Electromagnetic energy propagated from a source by radiation or conduction.

1.3.5 EQUIPMENT UNDER TEST (EUT)—The device or system whose susceptibility is being checked.

1.3.6 FIELD DECAY (VOLTAGE)—The exponentially decaying negative voltage transient such as developed by an automotive alternator when the field excitation is suddenly removed, as when the ignition switch is turned off.

1.3.7 FIELD STRENGTH—The term *field strength* shall be applied to either the electric or the magnetic component of the field, and may be expressed as V/m or A/m. When measurements are made in the far field and in free space, the power density in W/m^2 may be obtained from field strengths approximately as $(V/m)^2/377$ or $(A/m)^2 \times 377$.

1.3.8 GROUND PLANE—A metal sheet or plate used as a common unipotential reference point for circuit returns and electrical or signal potentials.

1.3.9 INTERFERENCE EMISSION—Any undesirable electromagnetic emission.

1.3.10 LOAD DUMP (VOLTAGE)—The exponentially decaying positive voltage transient developed by an automotive alternator when disconnected suddenly from its load, while operating without a storage battery or with a discharged storage battery. Removal of the load, the resulting transient, or both in combination are commonly referred to as *alternator load dump*.

1.3.11 RADIATED EMISSION—Radiation- and induction-field components in space. (For the purpose of this document, induction fields are classed together with radiation fields.)

1.3.12 SPURIOUS EMISSION—Any unintentional electromagnetic emission from a device.

1.3.13 SUSCEPTIBILITY—The characteristic of an object that results in undesirable responses when subjected to electromagnetic energy.

1.3.14 TEST PLAN—The specific document that details all tests and limits for the particular device in question. Several general outlines exist to assist the test designer, e.g., Appendix D of NAVAIR 5335.

2. Conducted Susceptibility, 30 Hz–50 kHz—All Input Leads Including DC and AC Power

2.1 Purpose—This section covers the requirements for determining the susceptibility characteristics from 30 Hz–50 kHz of automotive electronic equipment, sub-systems, and systems to EMI injected onto all input leads, including signal and power.

2.2 Measurement Philosophy—Power-source RF impedance seen by a given type of electronic equipment is dependent upon each particular installation. The effect on the equipment of a powerline RF voltage depends upon this varying impedance, and would render susceptibility measurements meaningless unless the impedance is also measured or controlled. In order to compare measurements made at various locations, powerline impedance seen by the equipment at test frequencies shall be controlled by a shunt capacitor; this also helps to assure adequate voltage at power terminals of the test sample.

At these frequencies, the impedances seen by control and signal leads are generally under the control of the system designer and should be known. Hence, the design values of such impedances shall be used in test measurements to simulate actual performance, as well as to provide ready comparison of measurement results made at various locations.

2.3 Grounding and Shielding—For the stated test frequencies, there are no special grounding and shielding requirements. However, the requirements of Section 3 may be utilized here, if expedient.

2.4 Apparatus—Test apparatus shall be as follows:

(a) Audio Oscillator—30 Hz–50 kHz.

(b) Audio Power Amplifier—50 W or greater with output impedance equal to, or less than, 2.0 Ω (capable of delivering 50 W into a 0.5 Ω resistive load connected across an isolation-transformer secondary).

(c) Measuring Instrument—Calibration Oscilloscope, VTVM or EMI Meter.

(d) A 100 μF Capacitor—required as a shunt to stabilize power-source impedance and to help obtain sufficient test voltage.

(e) Isolation Transformers—secondaries shall be capable of handling power current flow without saturation of core.

(f) DC and AC Power Supplies.

2.5 Test Setup and Procedures—The test setup is shown in Fig. 1 for signal-input circuits (no shunt capacitor) and DC power inputs; and in Fig. 2 for AC power inputs. The procedure is as specified below (see paragraph 2.6):

(a) The EUT shall be connected as shown in Figs. 1 and 2.

(b) The audio oscillator shall be tuned through the required frequency range (30 Hz–50 kHz) with the output progressively adjusted toward the maximum level. Monitor the EUT for (1) malfunction, (2) degradation of performance, or (3) deviation of parameters beyond tolerances indicated in the equipment specifications or approved test plan. The types of failure and their associated susceptibility threshold values shall be recorded.

(c) To determine the susceptibility of power leads, the required power supply voltage applied to the test sample shall be measured and maintained within the specified tolerance as indicated in the equipment specification or approved test plan during the test.

2.6 Notes—If the output impedance of the signal source looking into the secondary terminals of the isolation transformer is unknown, measurement shall be as follows:

(a) Apply a signal to the primary of the transformer and measure the open-circuit secondary voltage (V_{oc}).

(b) Connect a known load R_L, across the secondary and measure the closed-circuit secondary voltage (V_{cc}).

(c) The impedance shall be calculated as follows:

$$Z = \frac{R_L(V_{oc} - V_{cc})}{V_{cc}}$$

(d) Repeat the above procedure at one frequency per decade from 30 Hz–50 kHz (including 30 Hz and 50 kHz).

(e) The measured impedance shall be less than, or equal to, 0.5 Ω on powerlines, and within tolerances of the designed values for signal inputs.

3. Conducted Susceptibility, 50 kHz–100 MHz—All Input Leads, Including DC and AC Power

3.1 Purpose—This section covers the requirements for determining the susceptibility characteristics from 50 kHz–100 MHz of automotive electronic equipment, subsystems, and systems to EMI injected onto all input leads, including signal and power.

3.2 Measurement Philosophy—Power-source RF impedance seen by a given type of electronic equipment is dependent upon each particular installation. The effect on the EUT of a powerline RF voltage depends upon this varying impedance, and would render susceptibility measurements meaningless unless the impedance is also measured or controlled. In order to compare

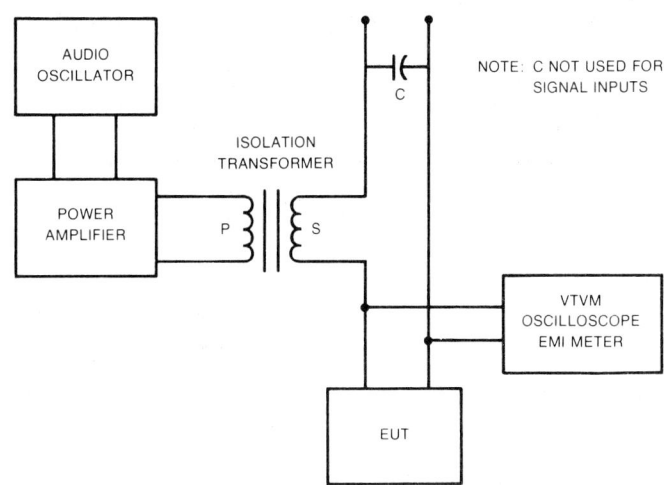

FIG. 1—TEST SETUP FOR MEASURING CONDUCTED SUSCEPTIBILITY, 30 Hz–50 kHz, SIGNAL OR DC POWER INPUT

measurements made at various locations, powerline RF impedance seen by the equipment shall be controlled by line-impedance stabilization networks.

3.3 Grounding and Shielding—To achieve uniform measurement conditions at RF requires that certain grounding practices be followed. Ground requirements are that equipment:

(a) be centered on a metallic ground plane having the following minimum dimensions:

(1) Thickness—1.5 mm (0.060 in) aluminum, copper, or brass sheet.

(2) Surface Area—1 m² (10.8 ft²) or underneath entire equipment plus 0.5 m² (5.4 ft²), whichever is larger.

(3) Width—0.5 m (20 in).

(b) be bonded to the ground plane at its most-sensitive input-terminal(s') ground point(s).

(c) not otherwise be grounded, unless required in installation instructions. The line-impedance stabilization networks shall be bonded to the ground plane as close as possible to the EUT ground. No shielding is to be used other than that called out in installation instructions.

3.4 Power Input Lead Test

3.4.1 APPARATUS—Test apparatus shall be as follows:

(a) Signal Source—A 50 Ω output-impedance source with an output of 100 V or greater into a matched load.

(b) One of the following to measure RF voltage:

(1) Calibrated Oscilloscope.

(2) VTVM.

(3) EMI Meter.

(4) Spectrum Analyzer.

(c) Line-Impedance Stabilization Networks (LISN's)—as specified in Figs. 3A and 4A with 50 Ω resistive RF terminations. When using an LISN, caution should be exercised to avoid load-current limiting due to series inductance in the LISN. This limiting may occur when loads switch between high- and

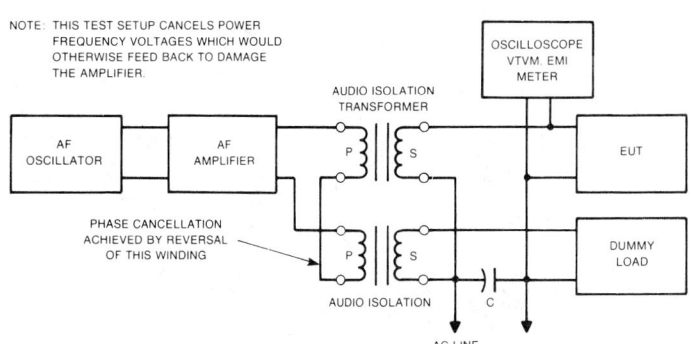

FIG. 2—TEST SETUP FOR MEASURING CONDUCTED SUSCEPTIBILITY, 30 Hz–50 kHz, AC INPUT

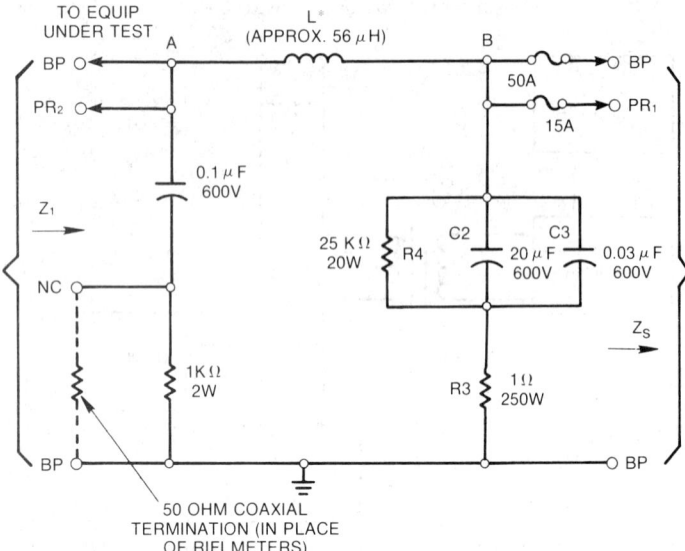

- Z_1 - IMPEDANCE PRESENTED TO THE EQUIPMENT WHEN CONNECTED FOR MEASUREMENTS.
- Z_S - IMPEDANCE OF THE POWER SOURCE USED.
- BP - HEAVY DUTY BINDING POSTS (MFR. STANDARD ELECTRIC TIME CO.)
- PR_1 - POWER RECEPTACLE, 115 V, 15 A (3-WIRE POLARIZED, TWIST LOCK, MALE BASE)
- PR_2 - POWER RECEPTACLE, 115 V, 15 A (3-WIRE NON-POLARIZED, "U" SHAPED GROUNDING SLOT.)
- NC - TYPE "N" CONNECTOR (UG-58/U) PANEL MOUNTING.
- L* - COIL - 26 TURNS OF NO. 6-600 VOLTS INSULATED WIRE (STRANDED), WOUND ON 5-½" (14 cm) DIAMETER COIL FORM.
- CASE: 17-½" L x 17-½" W x 8-¾" H (44.4 cm L x 44.4 cm W x 22.2 cm H) BRASS (DIVIDED IN TWO SECTIONS BY A BRASS PLATE 17-½" x 8-⅝" x 1/16" (44.4 cm x 21.9 cm x 1.6 mm) THICK.)
- NOTE: DUAL LINE STABILIZATION NETWORK CONSISTS OF TWO OF THE ABOVE NETWORKS.

FIG. 3A—LINE IMPEDANCE STABILIZATION NETWORK (LISN), 50 kHz–5 MHz

low-impedance states. Use of an LISN may then result in increased susceptibility.

(d) Test-Source Injection Networks illustrated in Fig. 5.
(e) Power Supply—DC and AC.

3.4.2 TEST SETUP AND PROCEDURE—The test setup is shown in Fig. 6. The procedure is as follows:

(a) Each control and signal lead shall be loaded with a terminating impedance. At these frequencies, however, the impedances as seen by the control and signal leads may no longer be determined by the system designer, due to uncontrollable stray impedances. It may be possible to simulate these impedances with a simple capacitor and inductor added to the actual leads if the frequency in MHz does not greatly exceed $\frac{300}{20\pi \ell}$, where ℓ is the characteristic lead length in metres. Above that frequency, the test designer should design his test plan and setup to given uniform results.

(b) The EUT shall be connected as shown in Fig. 6, observing the grounding and shielding requirements of paragraph 3.3.

(c) Signal sources and measuring instrumentation shall be connected to an LISN through test-source injection networks. Care shall be exercised to insure sufficiently short leads on the injection networks and LISN's to minimize loss of signal due to series inductance and shunt capacitance. A current probe on the injection lead right next to the EUT can be used to monitor the signal. For signal-source and measuring-instrument impedance equal to 50 Ω, use the signal-injection network of Fig. 5A. For a signal-source impedance of 50 Ω and a high-impedance measuring instrument, use the signal-injection network of Fig. 5B. Note the corresponding attenuation factors.

(d) Increase the level of the test signal while continuously scanning through the required frequency range (50 kHz–100 MHz). Tests shall be conducted at not less than three frequencies per octave representing the maximum susceptibilities within that octave. Monitor the equipment under test for (1) malfunction, (2) degradation of performance, or (3) deviation of parameters beyond tolerances indicated in the equipment specification or approved test plan (see paragraph 3.6). Record the highest level before degradation was observed.

(e) See paragraph 2.5(c).

3.5 Control and Signal Lead Test

3.5.1 APPARATUS—Test apparatus shall be as follows:
(a) Signal Sources—as for powerline measurements.
(b) Measuring Instruments—as for powerline measurements.
(c) Test-Source Injection Networks—see Fig. 5.

3.5.2 TEST SETUP AND PROCEDURE

(a) The test setup is shown in Fig. 7. Note that the LISN remains in the powerline circuit and its RF injection terminal is loaded with 50 Ω.

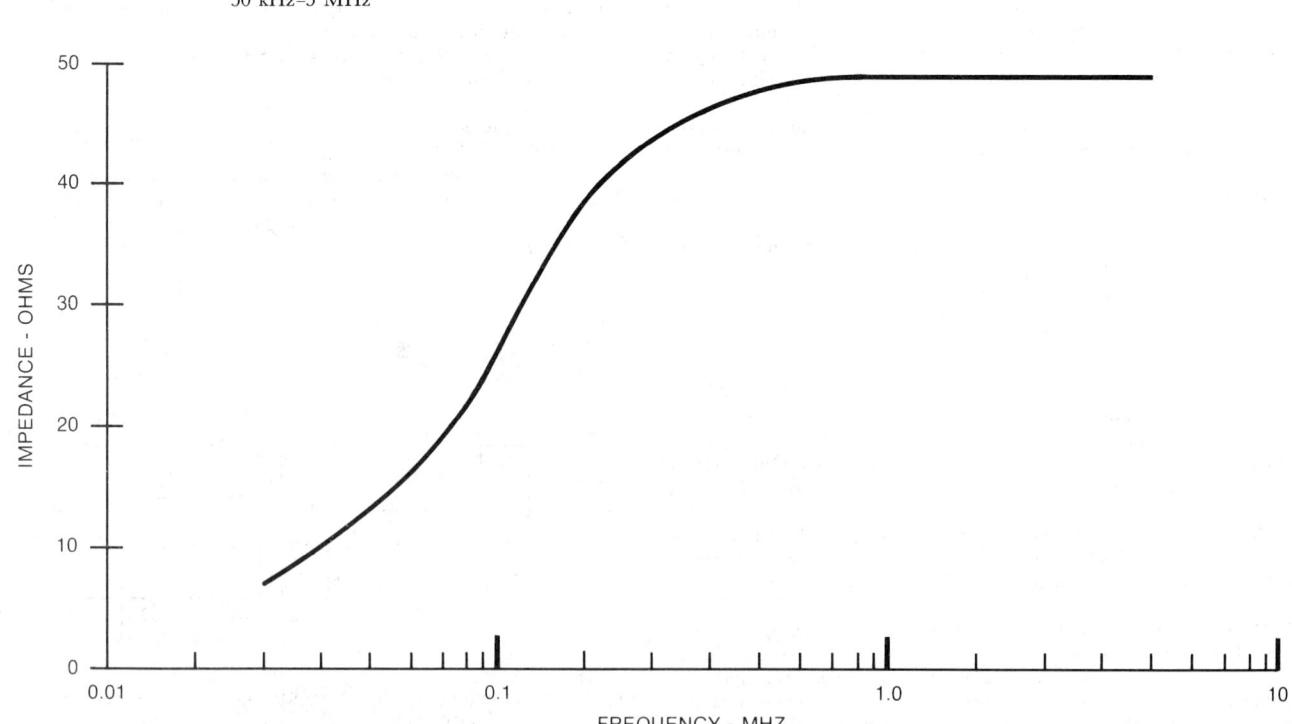

FIG. 3B—LINE IMPEDANCE STABILIZATION NETWORK REFERENCE IMPEDANCE, 50 kHz–5 MHz

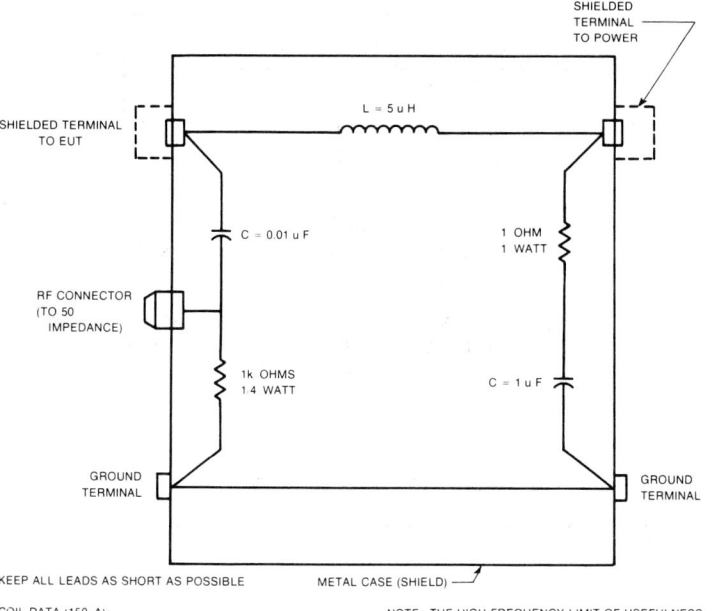

FIG. 4A—LINE IMPEDANCE STABILIZATION NETWORK (LISN), 5–100 MHz

(b) Each control and signal terminal not under test is loaded with its terminating impedance and test signals are injected into the test terminal as indicated in Fig. 7. At these frequencies, however, the impedances seen by control and signal leads may no longer be determined by the system designer. It may be possible to simulate the stray impedance as described in paragraph 3.4.2(a). Care shall be exercised to insure sufficiently short leads on the injection networks and LISNs to minimize loss of signal due to series inductance and shunt capacity. A current probe on the injection lead right next to the EUT can be used to monitor the signal.

(c) Increase the level while continuously scanning through the required frequency range (50 kHz–100 MHz). Tests shall be conducted at not less than three frequencies per octave representing the maximum susceptibilities within that octave. Monitor the EUT for: (1) malfunction, (2) degradation of performance, or (3) deviation of parameters beyond tolerances indicated in the equipment specification or approved test plan. (See paragraph 3.6.) The values at which these occur shall be recorded.

3.6 Notes

(a) Each LISN shall be tested over the range for which it is designed. The impedance should be within 20% of the curves in Figs. 3B and 4B. If any discrepancies occur, then the network should be modified, e.g., by adding ferrites to inductor leads to increase impedance at higher frequencies.

(b) Unless otherwise required in the equipment specifications or approved test plan, the test signals shall be modulated according to the following rules:

(1) Test samples with audio channels/receivers.

AM receivers: Modulate 30% with 1000-Hz tone.

FM receivers: Modulate with 1000-Hz signal using 10-kHz deviation.

SSB receivers: Use no modulation.

Other equipments: Same as for AM receivers.

(2) Test samples with video channels other than receivers. Modulate 90–100% with pulse of duration 2/BW and repetition rate equal to BW/1000, where BW is the video bandwidth (Hz).

(3) Digital equipment—Use pulse modulation with pulse duration and repetition rate equal to that used in the equipment under test.

(4) Non-tuned equipment—Amplitude modulate 30% with 1000-Hz tone or as otherwise specified in the test plan.

4. Conducted Susceptibility, Repetitive Spike, Power Leads

4.1 Purpose—This section describes the requirement to determine equipment susceptibility to spike interference on all AC and DC input power leads.

4.2 Measurement Philosophy—Installed equipment is powered from sources which contain, in addition to the desired electrical voltage, transients with peak values many times this value, due to the release of stored energy during the operation of relays and other loads connected to the source. This test is designed to determine the capability of equipment to withstand such transient over-voltages.

4.3 Apparatus—Test apparatus shall be as follows:

(a) Spike Generator—Characteristics as follows:

Approximately 1-μs rise time and approximately 10-μs fall time (see Fig. 8).

Pulse repetition rate of 3–10 pps.

Voltage output of 0–600 V (or as otherwise specified in the test plan) into a 0.5 Ω with a decaying exponential of less than 20% overshoot (see Fig. 8).

Output for parallel or series injection into power lines (see Figs. 9 and 10).

An output transformer may be required for series injection. If so, adequate current capacity of 25 A or more should be available.

Source impedance (Z_o) of 0.5 Ω or less.

Both polarities of pulse.

(b) Capacitor—Two 10-μf feedthrough capacitors.

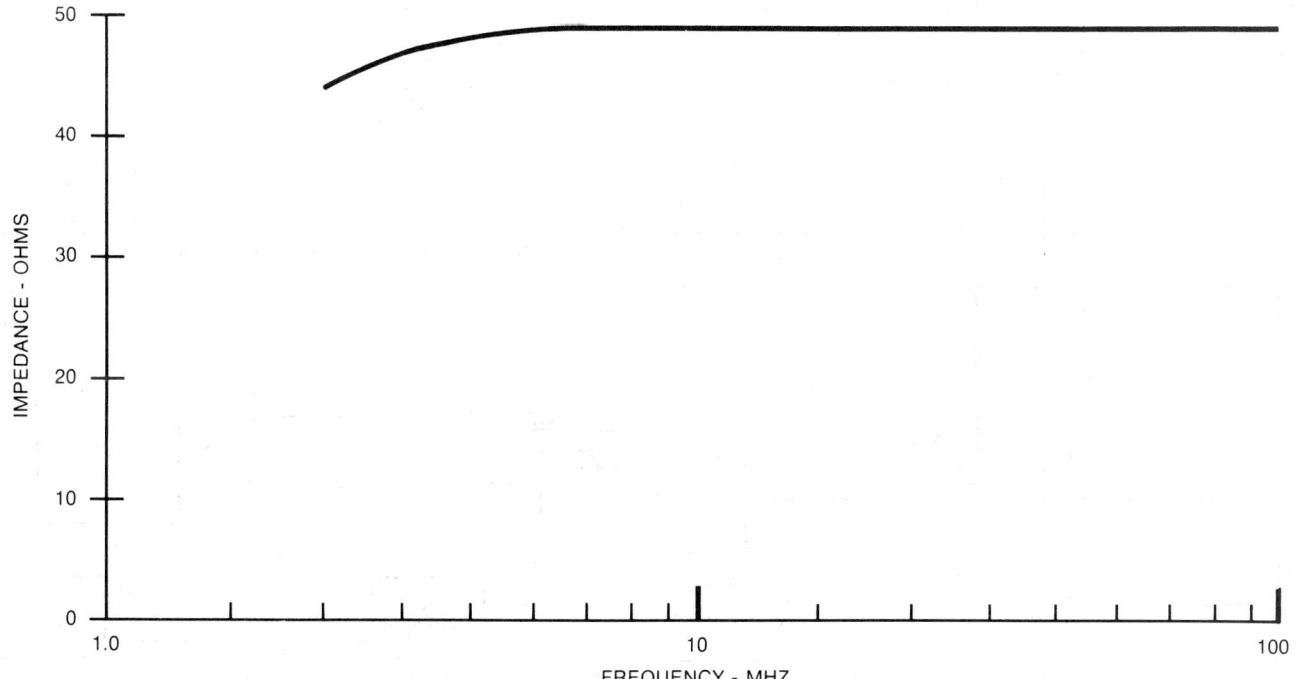

FIG. 4B—LINE IMPEDANCE STABILIZATION NETWORK REFERENCE IMPEDANCE, 5–100 MHz

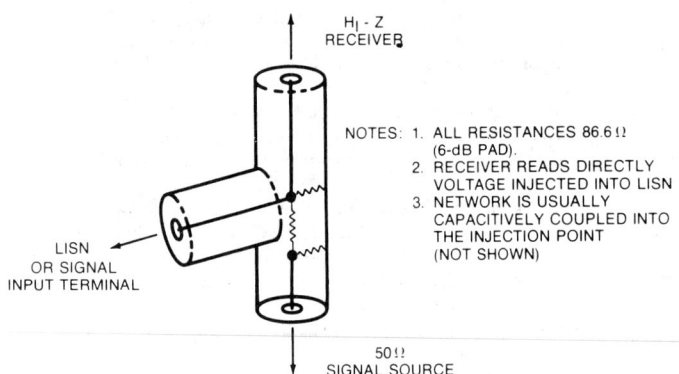

FIG. 5—TEST SOURCE INJECTION NETWORK

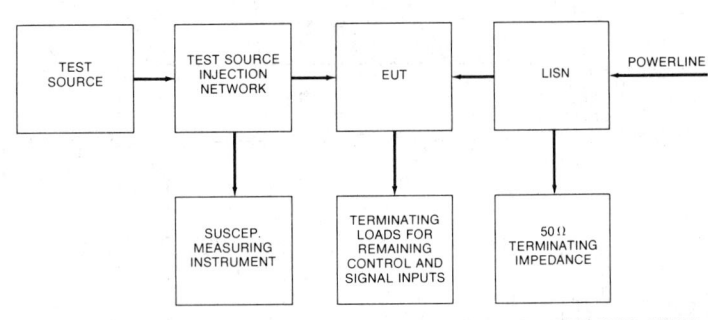

FIG. 7—EQUIPMENT BLOCK DIAGRAM FOR MEASURING CONDUCTED SUSCEPTIBILITY 50 kHz–100 MHz, CONTROL AND SIGNAL INPUTS

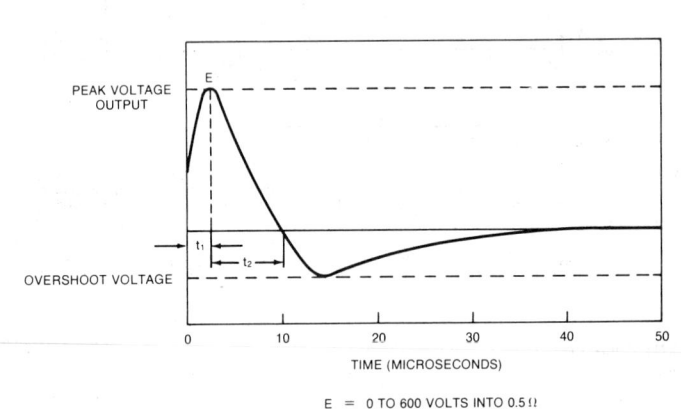

$E = $ 0 TO 600 VOLTS INTO 0.5 Ω
$t_1 \simeq 1\,\mu s$
$t_2 \simeq 10\,\mu s$

$$\frac{\text{OVERSHOOT VOLTAGE}}{\text{PEAK VOLTAGE OUTPUT}} \leq 20\%$$

FIG. 8—SPIKE PARAMETERS

(c) Oscilloscope—Any oscilloscope with 10-MHz bandwidth and adequate sweep rates is acceptable.

(d) Inductor—A 20-μH inductor capable of carrying 20 A continuously (No. 14 wire, 4 in (102 mm) coil of 2 in (51 mm) diameter, 32 turns).

4.4 Test Setup and Procedure—Test procedure shall be as follows:

(a) Connect EUT and test instrumentation as shown in Fig. 9 or Fig. 10.

(b) For DC power leads, both series and shunt test methods shall be used. For AC power leads, only the series test method is used.

(c) Measure the applied spike amplitude, rise time, duration, and repetition rate using a calibrated oscilloscope across the input terminals of the EUT.

(d) Synchronization shall be used to position the spike to specific EUT signal conditions which will produce maximum susceptibility.

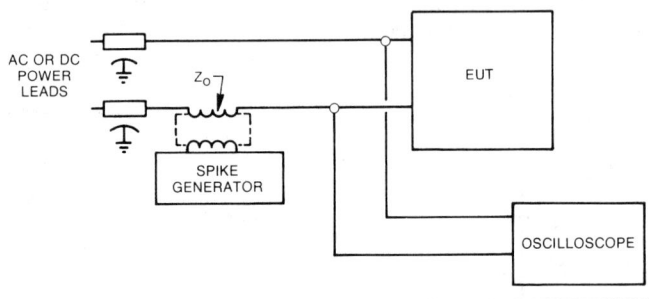

FIG. 9—TEST SETUP FOR CONDUCTED SUSCEPTIBILITY, SPIKE, POWER LEADS, INJECTION IN SERIES

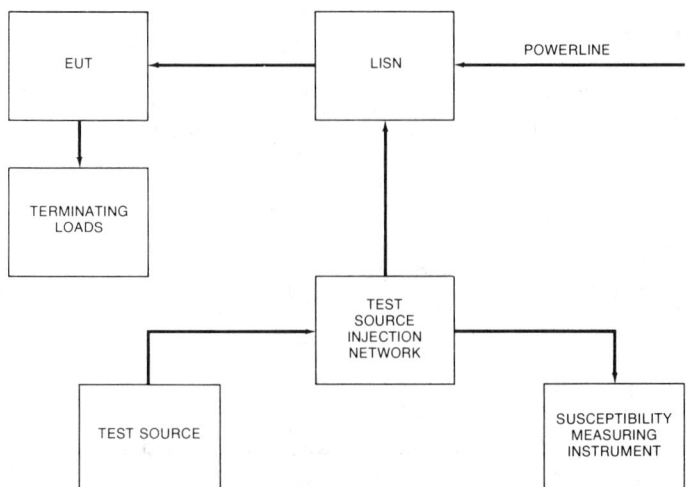

FIG. 6—EQUIPMENT BLOCK DIAGRAM FOR MEASURING CONDUCTED SUSCEPTIBILITY 50 kHz–100 MHz, POWERLINE ONLY

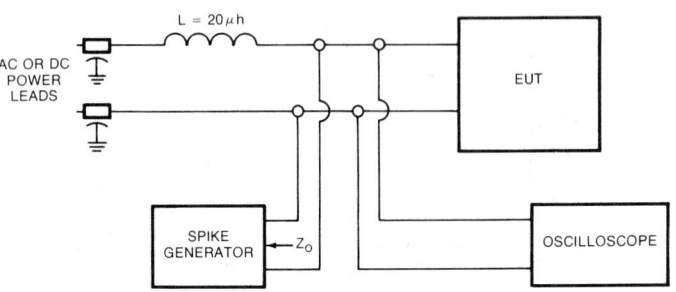

FIG. 10—TEST SETUP FOR CONDUCTED SUSCEPTIBILITY, SPIKE, POWER LEADS, INJECTION IN PARALLEL

(e) Alternately, positive and negative, single and repetitive (6–10 pps) spikes shall be applied to the EUT ungrounded input lines for a period not less than 10 min or as specified in the test plan. On equipment employing gated circuitry, the spike shall be synchronized to occur within the time frame of the gate.

(f) Slowly increase the spike amplitude until (1) malfunction, (2) degradation of performance, or (3) deviation of parameters beyond specifications occurs. Adjust the amplitude as necessary to determine the susceptibility threshold level.

(g) Record the threshold levels, repetition rate, phase position on the AC waveform, and time occurrence on circuit gates, for both safe failure and unsafe failure regions as applicable.

4.5 Notes

(a) As described in paragraph 1.2, a statistical method which eliminates the effect of progressive degradation may have to be used.

5. Conducted Susceptibility, Load Dump and Field Decay (Voltage) Transients, Power Leads

5.1 Purpose—This section describes the requirements to determine the capability of various electrical devices to withstand load-dump and field-decay transients which normally occur in motor vehicles.

5.2 Measurement Philosophy—Field-decay transients occur normally in automotive electrical systems when the ignition switch is turned off. This resulting sudden removal of field excitation causes a negative pulse on the order of 50–100 V to propagate through the electrical system.

The load-dump transient occurs infrequently, but it can occur suddenly and repetitively in certain vehicles under specific circumstances. The most common circumstances are: (1) a loose or corroded battery connection which causes the load (battery) to be suddenly removed from the charging circuit, or (2) sudden disconnection of a jumper battery from a vehicle with a dead battery. Either of these conditions results in a large positive voltage pulse (on the order of 75–125 V) propagating throughout the vehicle electrical system, often with catastrophic results on components of the system.

The vehicle effects of load-dump and field-decay transients depend on the impedance of the electrical system at the moment the transient occurs, the electrical location of the components relative to the alternator, characteristics of the components, and their operating temperatures.

The intent of this section is to provide a standard means of simulating non-repetitive transients in order to: (1) verify that a given component will survive a certain known transient level, or (2) to actually measure, by testing to malfunction or failure, the susceptibility level of the component. In this latter case special statistics may be required. (See paragraph 1.2.)

5.3 Apparatus—Test apparatus shall be as follows:

(a) One-shot pulse generator capable of delivering a voltage pulse of 50–150 V, positive or negative, with an exponential decay extending over an adjustable time base from 100–500 ms, across a resistive load of 10 Ω. The rise time shall lie in the range of 5–10 ms. Parts lists and schematic diagrams for typical pulse generators are included in paragraph 5.6. Higher voltages and longer durations may be required in some environments.

(b) DC voltmeter and an oscilloscope for monitoring the power source and pulse generator.

(c) Equipment as may be required to perform a functional test of the component following transient application.

5.4 Test Setup and Procedure

(a) Connect a known functional EUT as shown in Fig. 11.

(b) Ensure that the ambient temperature and supply voltage V_s are maintained as required by the appropriate test specification.

(c) Set up the pulse generator to provide the polarity, amplitude, and pulse duration as specified in the appropriate test specification.

(d) Generate the test pulse.

(e) Perform the appropriate functional test to determine whether failure has occurred and record results.

(f) Repeat steps (a) through (e) for additional EUTs of the same test components until the statistical requirements of the test specification have been met.

5.5 Notes

(a) If the susceptibility level is to be determined, care must be used to eliminate the effects of cumulative deterioration such as dielectric *punch through* in semi-conductor devices.

(b) When testing to a specified level, unnoticed failures may occur which may be detected only by running life-cycle tests and comparing the results of tested components against those of untested components.

5.6 Test Equipment Description—Figs. 12 and 13 give information for building typical pulse generators. The parts lists and waveforms are shown with each schematic. If desired, the circuits of Figs. 12 and 13 may be combined into a single unit. The pulse durations shown are typical, but other pulse widths can be obtained by selecting appropriate values for C1, R6, R7, and R8.

R7 and R8 also establish pulse amplitude up to a maximum of V_c. The initial voltage V_c is set by adjusting the DC supply voltage to the pulse generator when S1 is in the position shown. The initial value of V_c can be any reasonable voltage as long as the rating of C1 is not exceeded.

The pulse is generated when S1 is switched.

6. Radiated Susceptibility, 30 Hz–15 kHz, Magnetic Field

6.1 Purpose—This section covers requirements to determine the magnetic-field susceptibility of equipment, subsystems, and systems whose largest dimension is less than 60 cm (2 ft) in the frequency range from 30 Hz–15 kHz.

6.2 Measurement Philosophy—In the low-frequency range, some devices that are susceptible to magnetic fields are not so susceptible to the equivalent free-space electric fields or to real electromagnetic power (an elementary example is a compass). For such items, a separate measurement of magnetic-field susceptibility is warranted.

6.3 Apparatus—Test apparatus shall be as follows:

(a) Hemholtz coil—1.8-m (6-ft) diameter, spacing 90 cm (3 ft), to produce a uniform field (±10%) throughout a 60-cm (2-ft) cube; to be self resonant at, or above, 15 kHz and having a known coil factor, say 0.63 G/A.

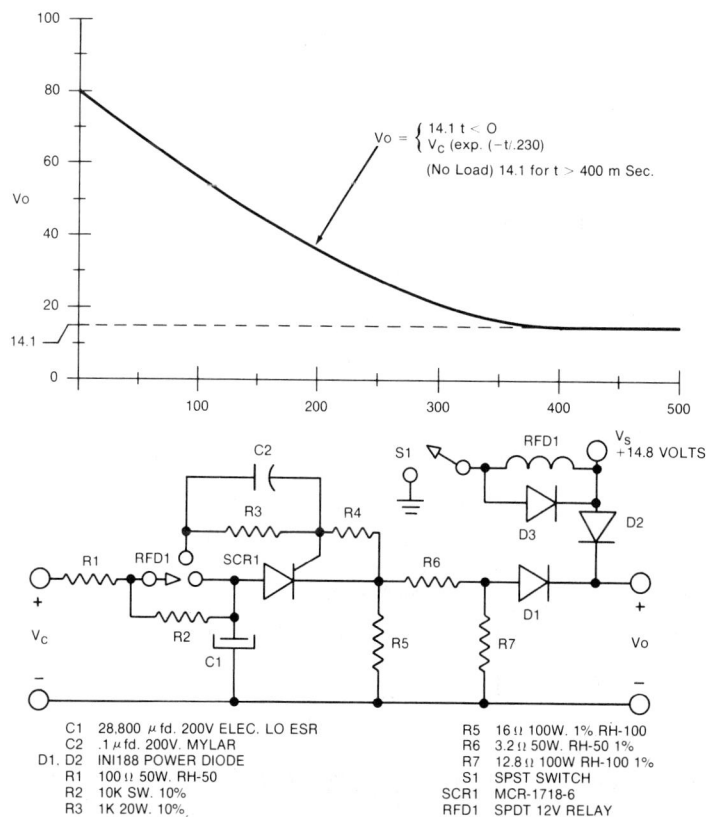

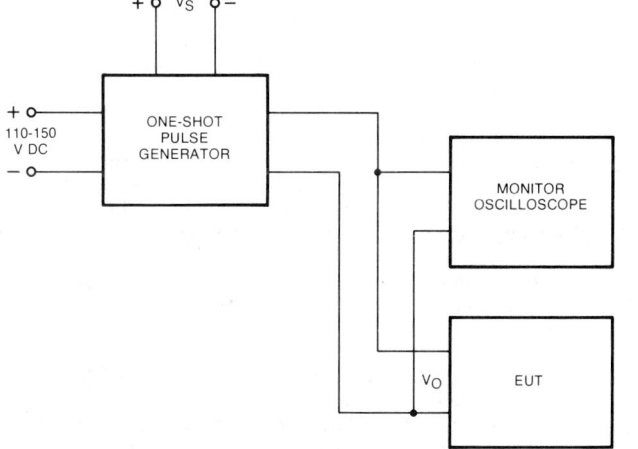

FIG. 11—TEST SETUP FOR CONDUCTED SUSCEPTIBILITY, LOAD DUMP AND FIELD DECAY, POWER LEADS

FIG. 12—TYPICAL LOAD DUMP TRANSIENT SIMULATOR

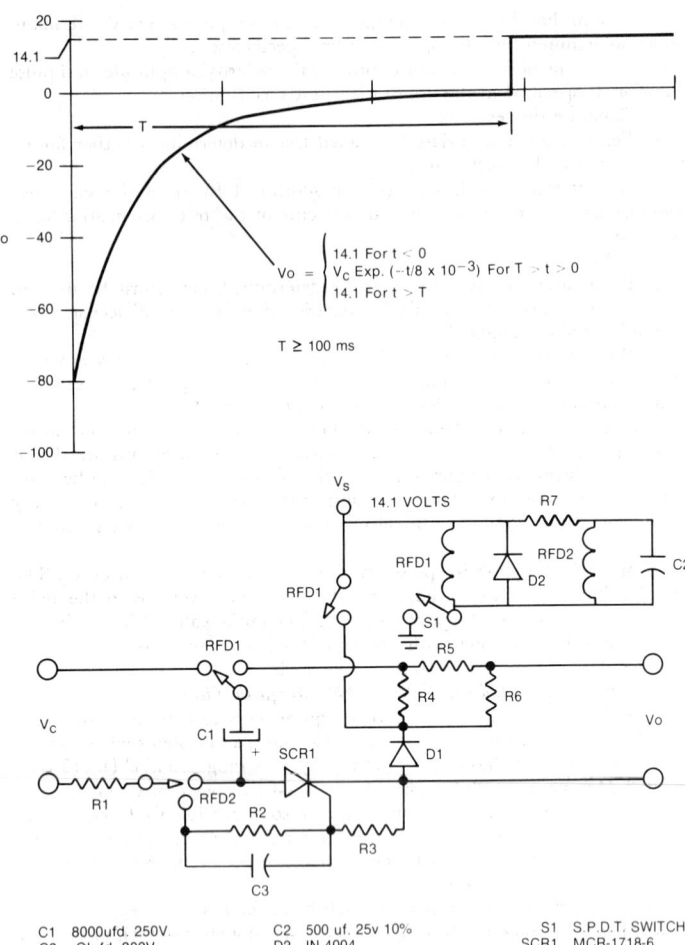

FIG. 13—TYPICAL FIELD DECAY TRANSIENT SIMULATOR

(b) Audio oscillator—30 Hz–15 kHz.
(c) Audio power amplifier—50 W or greater with output impedance equal to, or less than, 2.0 Ω.
(d) Current monitor—0–30 A, 30 Hz–15 kHz.

6.4 Test Setup and Procedure—A Helmholtz-coil setup consists of two identical coaxial loops spaced approximately one radius apart to establish a uniform field in the central region when excited by an audio source as in Fig. 14. The procedure is:

(a) Place the operating EUT in the central region of the Helmholtz coil.

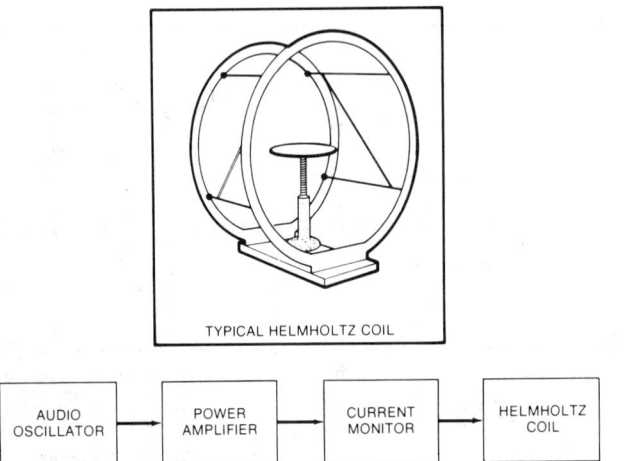

FIG. 14—TEST SETUP FOR RADIATED SUSCEPTIBILITY, 30 Hz–15 kHz, MAGNETIC FIELD

(b) Tune the audio oscillator through the required frequency range (30 Hz–15 kHz) with the output progressively adjusted toward maximum level. Monitor the EUT for (1) malfunction, (2) degradation of performance, or (3) deviation of parameters beyond tolerances indicated in the equipment specifications or approved test plan. The types of failure and their associated susceptibility threshold values of magnetic field (current times coil factor) shall be recorded.

6.5 Notes—Under certain circumstances it may be desirable or necessary to use a non-sinusoidal oscillator. One case where it might be warranted is checking susceptibility to fields next to a transformer since the saturation characteristics of the core may generate both fundamentals and harmonics.

7. Radiated Susceptibility, 14 kHz–200 MHz, Electric Field

7.1 Purpose—This section covers requirements for the determination of electric-field susceptibility of equipment, subsystems, and systems (whose largest dimension is less than 15 cm) in the frequency range 14 kHz–200 MHz.

7.2 Measurement Philosophy—A TEM transmission cell is a rectangular adaptation of a coaxial line which sets up a region of uniform electric and magnetic fields in a traveling wave of essentially free-space impedance. The EUT is exposed to this electromagnetic source, but only the electric-field component is monitored. This technique also prevents disturbance to equipment not under test since the RF field source and EUT are completely self-contained within the electromagnetic enclosure.

7.3 Apparatus—The test apparatus shall consist of the following:

(a) Signal Source—Any commercially available signal source, power amplifier, and general-purpose amplifier capable of supplying at least 100 W of modulated and unmodulated power to develop the susceptibility levels specified in the test plan shall be used, provided the following requirements are met: Frequency accuracy shall be within ±2%. Harmonics and spurious outputs shall not be more than −30 dB referred to the fundamental power.

(b) RF Voltmeter—A commercially available RF voltmeter capable of measuring 100 V over the frequency range 14 kHz–200 MHz.

(c) Termination—One 100 W, 50 Ω load.

(d) Frequency Counter—A frequency counter capable of measuring frequencies up to 200 MHz.

(e) TEM Transmission Cell—A transverse electromagnetic transmission cell is shown in Figs. 15 and 16. Typical dimensions for cells are given in Table 1. The dimension for a cell suggested for use in the frequency range 14 kHz–200 MHz is underlined on the table.

(f) Low-Pass Filter—Cutoff at 200 MHz, with the signal down 60 dB at frequencies greater than 240 MHz.

(g) Signal Samplers or Monitor Tees—Frequency and RF-voltage monitoring equipment.

(h) Monitors—Required test equipment to monitor the operation of the test sample.

7.4 Test Setup and Procedure

(a) The test setup should be as shown in Fig. 17A (or Fig. 17B (see Section 7.4(f))).

(b) The EUT may be placed, if normally ungrounded, midway between the bottom of the cell and the center septum; or if normally grounded, on the bottom of the cell centered with respect to width and length as shown in Fig. 16.

(c) Fields should be generated as required. The field strength, E_v, is determined by

$$E_v = \frac{V_{rf}}{d} \quad \text{volts/meter}$$

where V_{rf} is the input voltage to the cell in volts and d is the cell bottom-to-septum separation in meters.

(d) The EUT shall be operated and monitored by leads from the device to shielded feed-through connectors mounted on the bottom outer shield, as shown in Fig. 16. The leads shall be oriented to obtain minimum interaction with the cell test field.

(e) The EUT shall be oriented in each orthogonal plane within the cell to determine maximum susceptibility.

(f) The entire frequency range from 14 kHz–200 MHz shall be scanned. Tests shall be conducted at not less than three frequencies per octave representing the maximum susceptibilities within that octave. In addition, tests shall also be made at the EUTs critical frequencies (local oscillator frequency, intermediate frequency, and others) as specified in the test plan. The threshold of susceptibility shall be determined by increasing the amplitude of the test field until degradation in performance is observed.

7.5 Notes

(a) Unless otherwise required in the equipment specification or approved test plan, the test signals shall be modulated according to the following rules:

(1) EUTs with audio channels/receivers:

AM receivers: Modulate 30% with 1000-Hz tone.

FM receivers: When monitoring signal-to-noise ratio, modulate with 1000-Hz signal using 10-kHz deviation. When monitoring receiver quieting, use no modulation.

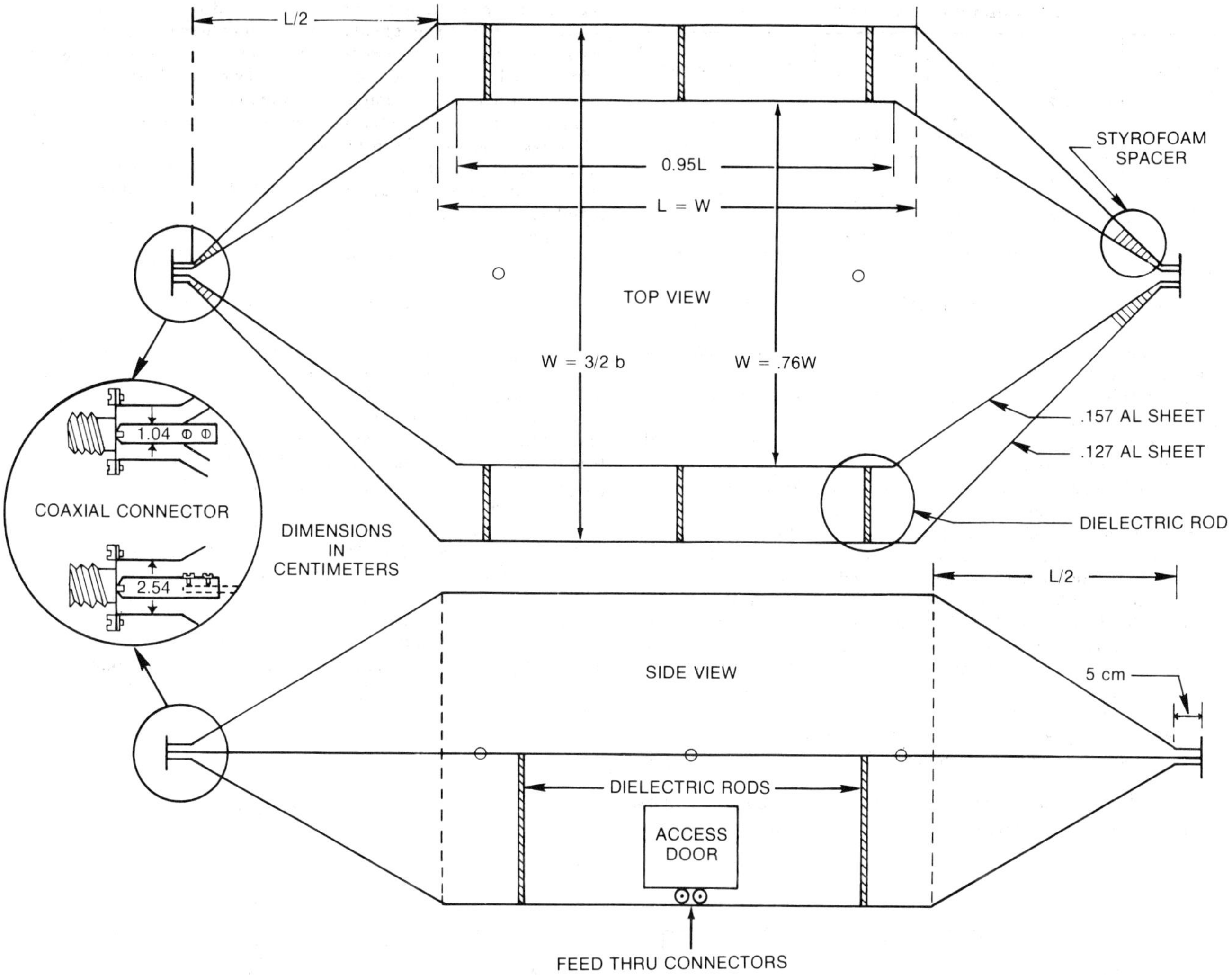

FIG. 15—DESIGN FOR RECTANGULAR TEM TRANSMISSION CELL

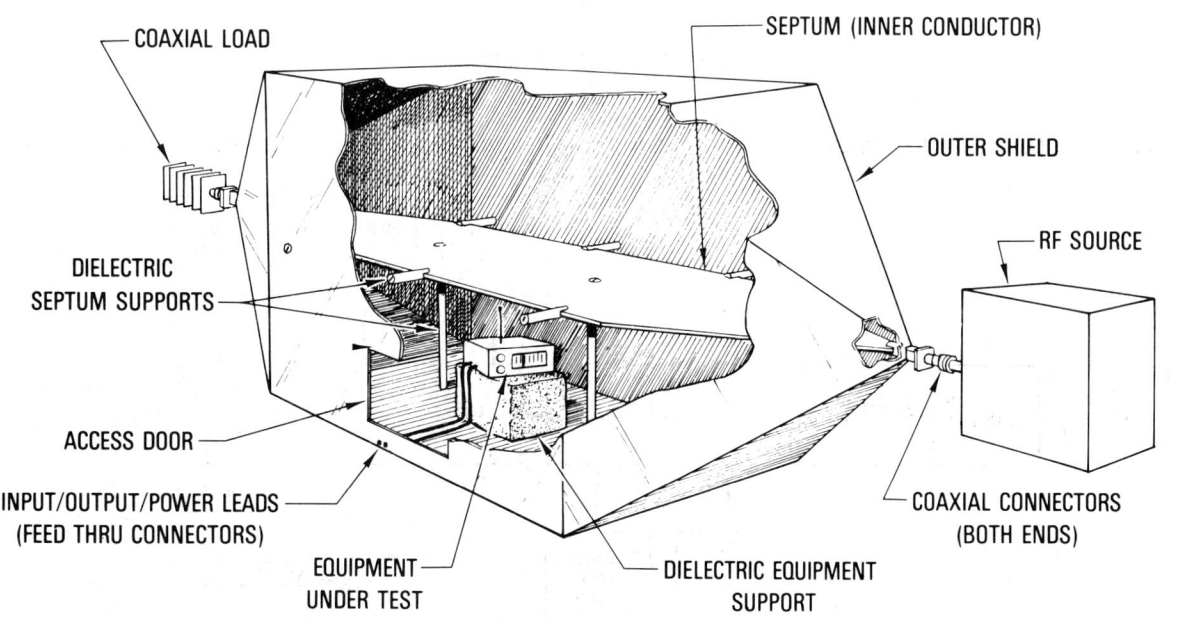

FIG. 16—CUT-AWAY VIEW OF TEM CELL BEING USED FOR EM SUSCEPTIBILITY TESTING. FIGURE SHOWS PLACEMENT OF EUT AND ASSOCIATED INPUT, OUTPUT, AND MONITORING LEADS INSIDE THE CELL

TABLE 1—TEM CELL DIMENSIONS

Recommended Upper Frequency (MHz)	Plate Separation b/2 (cm)	Center Septum		Cutoff/Multimode Frequency f_{11} (MHz)
		w (cm)	t (cm)	
100	60	136	0.157	150
200	30	68	0.157	300
300	20	45.3	0.157	450
500	14	31.7	0.157	644

Other Equipment: Same as for AM receivers.

(2) EUTs with video channels other than receivers—Modulate 90–100% with pulse duration of 2/BW and repetition rate equal to BW/1000, where BW is the video bandwidth.

(3) Digital equipment—Use pulse modulation with pulse duration(s) and repetition rate(s) equal to those used in the EUT.

(4) Non-tuned equipment—Amplitude-modulate 30% with 1000-Hz tone, or as otherwise required in the test plan.

(b) This procedure also exposes the EUT to magnetic fields, and thus the cell may be calibrated for use in determining magnetic-field susceptibility.

(c) Test samples of any size could be tested using a TEM cell modeled from Table 1 to meet the criteria that the size of the EUT be less than L/3 x W/3 x b/6. (These dimensions are considered a maximum to prevent excessive impedance loading and test-field perturbation when inserting the EUT into the cell.) Thus, a small EUT could be tested at higher frequencies in small cells, and a large EUT could be tested at lower frequencies in large cells. The procedure for testing in large cells is the same as for testing in small cells but a larger signal source (higher power) and an appropriate high-power termination (50 Ω) are required.

(d) The useful upper frequency for the cell may be reduced 20–30% from the cutoff/multimode frequency given in Table 1 by the loading effect of the EUT.

(e) Test samples that exceed the ⅓-linear dimension criterion could be tested in the cell, bearing in mind that excessive loading of the cell reduces the accuracy in determining the test field. This effect occurs because the sample tends to short out the test field in the region between the plates, increasing the vertically polarized test field. The error, however, can be corrected by measuring the field in the region above and below the EUT using miniature E-field probes and making an appropriate correction.

(f) Field measurement errors may result from resonances in the EUT. These resonances can be detected at frequencies above 1 MHz by using a bi-directional coupler with RF power monitors in place of the monitor tee and an RF voltmeter as shown in Fig. 17B. The bi-directional coupler and power monitors are used to measure both forward and reflected power at the cell's input giving an indication of the system SWR which may be associated with resonances of the EUT. The field strength E_v is then given as

$$E_v = \frac{\sqrt{P_n R_c}}{d} \quad \text{volts/meter,}$$

where P_n is the net power flow through the cell, R_c is the real part of the cell's complex impedance and is approximately equal to 50 Ω, and d is as defined earlier.

(g) Every effort should be made to match conditions between the TEM cell and actual operating conditions. This includes, if possible, matching (1) lead lengths, (2) lead impedances, and (3) lead exposure to RF fields. If not possible, then a more modest goal of minimizing lead effects by minimizing currents on them would be in order. The lead effect could then be obtained through the use of the conducted susceptibility test.

8. Radiated Susceptibility, 200–1000 MHz, Electric Field

8.1 Purpose—This section covers the requirements for the determination of electric-field susceptibility of equipment, subsystems, and systems in the frequency range 200–1000 MHz.

8.2 Measurement Philosophy—In this frequency range (200–1000 MHz), TEM cells become too small to test many types of equipment. However, RF-absorbing material becomes effective and measurements can be made

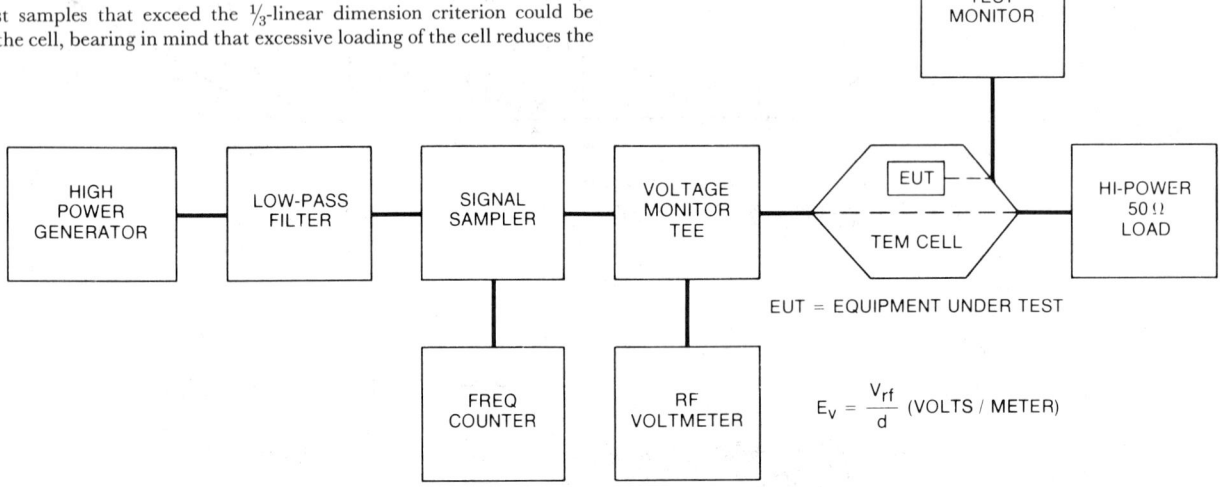

FIG. 17A—BLOCK DIAGRAM OF SYSTEM FOR SUSCEPTIBILITY TESTING OF EQUIPMENT (USED TYPICALLY BELOW 1 MHz)

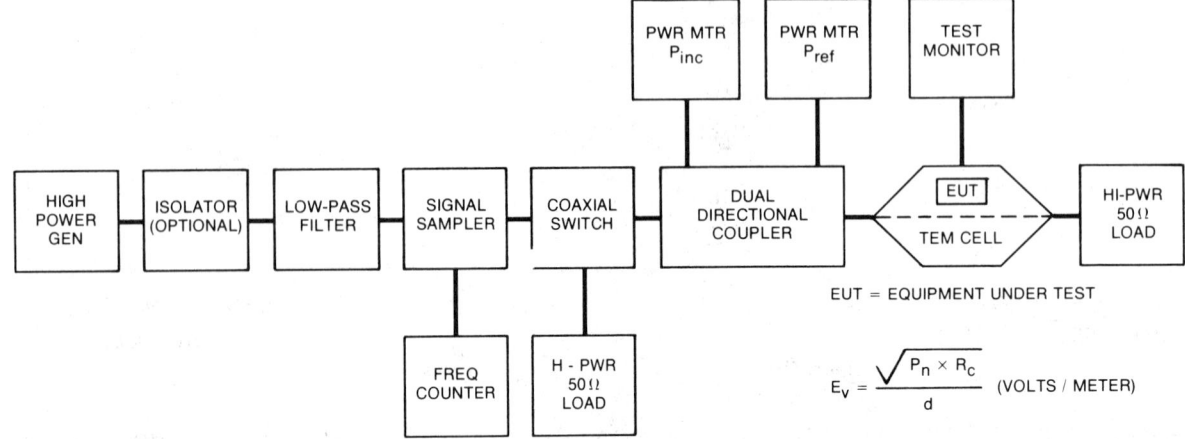

FIG. 17B—BLOCK DIAGRAM OF SYSTEM FOR SUSCEPTIBILITY TESTING OF EQUIPMENT (1 MHz–200 MHz)

within an RF shielded enclosure provided RF-absorbing material is used strategically to make it essentially anechoic.

8.3 Apparatus—The test apparatus shall consist of the following:

(a) *Signal Source*—Any commercially available signal source, power amplifier, and general-purpose amplifier capable of supplying the necessary modulated and unmodulated power to develop the susceptibility levels specified in the test plan may be used, provided the following requirements are met: Frequency accuracy shall be within ±2%. Harmonics and spurious outputs shall not be more than −30 dB referred to the fundamental power.

(b) *EMI Meter or Spectrum Analyzer*.

(c) *Antennas*—Commercial conical logarithmic spirals may be used.

(d) *Output Monitor*—Appropriate instrumentation to monitor the performance of the EUT shall be used.

(e) *Source Antenna Hood*—A hood for shielding the transmitting antenna can be obtained either by special order from RF absorber materials manufacturers (for example as listed in Appendix B) or by private construction. The hood is typically made from $\frac{1}{8}$ in (3.2 mm) aluminum in the form of a rectangle 30 in (0.75 m) square by 40 in (1.0 m) long. The hood, with its end plate, are lined with ferrite absorbing material and mounted on a cart with casters as shown in Figs. 18 and 19.

It may be feasible to develop an antenna, such as a cavity backed spiral that would require a smaller hood; however, such an antenna, capable of transmitting high field strengths (up to 200 V/m) is not commercially available at the present time.

8.4 Test Setup and Procedure

(a) Leakage tests shall be performed in a shielded room which provides adequate attenuation so that the external field strengths do not exceed FCC limits. If a shielded anechoic enclosure covering the 200 MHz–1 GHz frequency range is available, the tests can be performed without a hooded antenna.

(b) In general, the EUT should be placed 1 m from the transmitting antenna. When a large EUT is to be immersed in a field, the transmitting antenna shall be placed at a distance sufficient to allow the entire EUT to fall within the 3-dB beamwidth of the transmitted field. If this is not feasible, the sample may be tested in segments, each segment being equal in dimension to the 6-dB beamwidth of the antenna radiation characteristic. The antenna shall normally be centered in a plane parallel to the absorbing wall behind the EUT.

(c) Fields should be generated, as required, with the specified antenna. Care should be taken so that the test equipment is not affected by the test signals. It may be necessary to place the test equipment, except for the antenna, outside the shielded enclosure.

FIG. 19—UHF HOODED ANTENNA

(d) As suggested in Table 2, the wall behind the EUT should be covered with a microwave absorber material which provides at least 20-dB attenuation of reflected power at 200 MHz.

(e) The specified field strength shall be established with attendant polarization prior to the actual testing by substituting a field-measuring antenna in place of the EUT and by adjusting and recording the transmitter power required to obtain a specified field intensity from the transmitting antenna. (This calibration may be used for all subsequent testing, provided that exactly the same EUT location in the shielded enclosure is used.) The volume surrounding the EUT should be probed with the field measuring antenna to verify a uniform field exists within the test area.

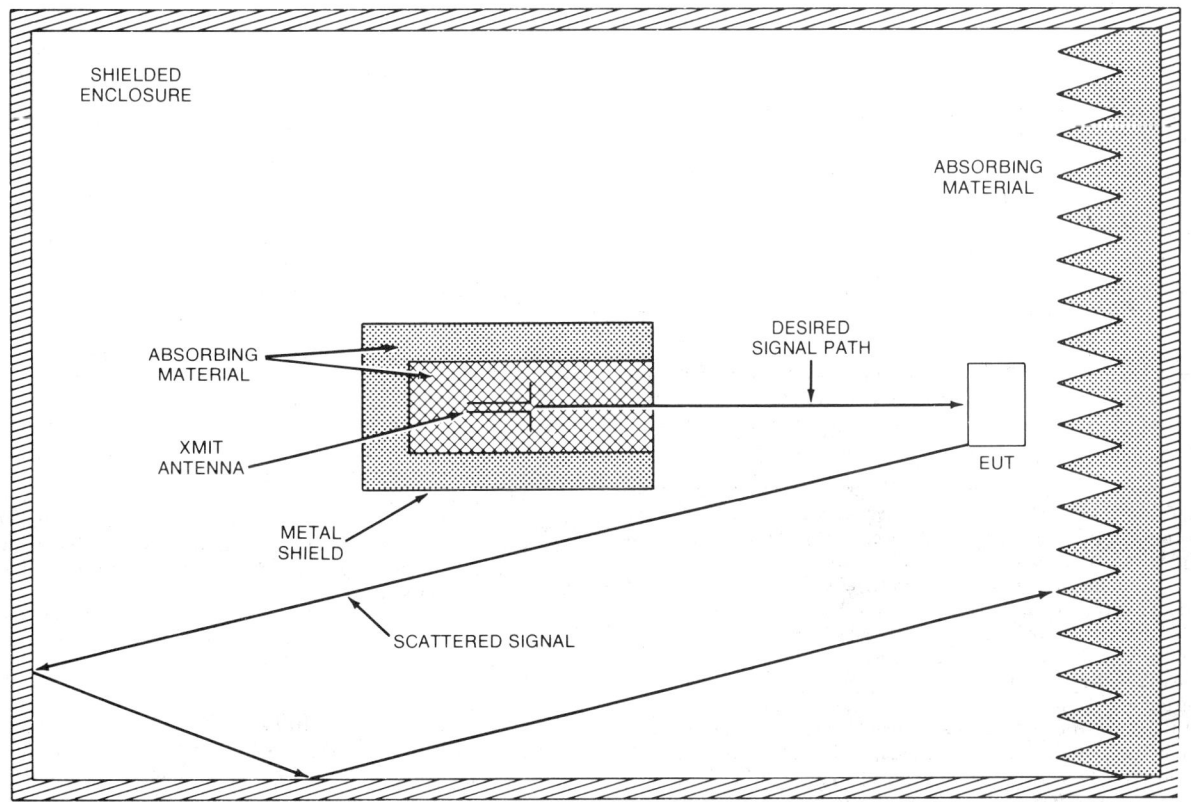

FIG. 18—DIAGRAM OF HOODED ANTENNA MEASUREMENT SETUP IN A SHIELDED ENCLOSURE

TABLE 2—RECOMMENDED MINIMUM THICKNESS OF RF ABSORBER FOR COVERING ENCLOSURE WALL BEHIND TEST SAMPLE

Frequency MHz	Absorber Thickness in (cm)
200	36 (91)
300	24 (61)
500–1000	18 (46)

(f) The EUT shall be oriented in each orthogonal plane within the shielded enclosure to determine maximum susceptibility.

(g) The entire frequency range from 200–1000 MHz shall be scanned. Tests shall be conducted at not less than three frequencies per octave, representing the maximum susceptibilities within that octave. In addition, tests shall also be made at the EUT critical frequencies (local oscillator frequency, intermediate frequency, and others) as specified in the test plan. Determine the threshold of susceptibility by increasing the test signal until degradation of performance is observed.

8.5 Notes

(a) Unless otherwise required in the equipment specification or approved plan, the test signals shall be modulated according to the following rules:

(1) Test samples with audio channels/receivers.

AM receivers: Modulate 30% with 1000-Hz tone.

FM receivers: When monitoring signal-to-noise ratio, modulate with 1000-Hz signal using 10-kHz deviation. When monitoring receiver quieting, use no modulation.

Other equipment: Same as for AM receivers.

(2) EUTs with video channels other than receivers. Modulate 90–100% with a pulse of duration 2/BW and repetition rate equal to BW/1000 where BW is the video bandwidth.

(3) Digital Equipment. Use pulse modulation with pulse duration(s) and repetition rate(s) equal to those used in the equipment.

(4) Non-Tuned Equipment. Amplitude modulate 30% with 1000-Hz tone or as otherwise required in the test plan.

(b) At frequencies approaching 200 MHz and a separation distance of 1 m, the EUT approaches the reactive near field of the antenna. Use of these distances is standard practice despite possible errors caused by them.

(c) The EUT configuration shall be as close to the actual operating situation as possible in terms of surrounding metal structure, lead length, and terminating impedances.

APPENDIX A
STATISTICAL DETERMINATION OF SUSCEPTIBILITY

The susceptibility level (L), below which a given percentage of a population will fall, can be calculated with certain confidence from:

$$L = \bar{x}_n + ks_n$$

where: $\bar{x}_n$ = Arithmetic mean value of the measured susceptibility values, x_i, of n units

$$s_n^2 = \text{Sample variance} = \sum_{i=1}^{n} \frac{(x_i - \bar{x}_n)^2}{n-1}$$

k = A factor derived from the noncentral t-distribution for a given confidence that no more than a given percentage of the units will exceed the calculated level using data from n units tested.

n = Number of tested units.

The above statistic assumes that the readings are normally distributed. That can quickly be verified by plotting the points on normal probability paper and comparing to a straight line. It is necessary to check the data either as original physical values, e.g., watts, or dB values, e.g., dBm, as indicated above to be sure of the validity of the method. Usually, however, the arguments or original physical quantities will tend to have that normal distribution. The dB values, as shown below, will then only be normally distributed if the variation between values of x_i is small. In this case, if there are large variations, it will be better to use the arguments for calculation of mean and standard deviation. The reason for this is the average of the dB values is the dB value of the geometric mean of the arguments, not the arithmetic mean.

The table below gives k factors versus sample size (n) from 3–12, for the condition that, with 80% confidence, 80% of the population will fall below the level calculated from n data points. Those k factors are based on the noncentral t-distribution for one-sided tolerance limits.

TABLE A.1—k VALUES

n	3	4	5	6	7	8	9	10	11	12
k	2.04	1.69	1.52	1.42	1.35	1.30	1.27	1.24	1.21	1.20

The examples that follow illustrate the use of the formulas and the table of k factors in determining the susceptibility levels for two sets of 10-sample data.

TABLE A.2—EXAMPLE 1

The susceptibility values (x_i) of one set of 10 units are as follows:

x_i	$x_i - \bar{x}$	$(x_i - \bar{x})^2$
64 dB	0.1 dB	0.001 (dB)²
62	−1.9	3.61
63	−0.9	0.81
65	1.1	1.21
65	1.1	1.21
64	0.1	0.01
65	1.1	1.21
63	−0.9	0.81
64	0.1	0.01
64	0.1	0.01
$\Sigma x_i = 639$ dB	$\bar{x}_{10} = \frac{\Sigma x_i}{10} = \frac{639}{10} = 63.9$ dB	$\Sigma(x_i - \bar{x})^2 = 8.90$ (dB)²

The standard deviation s_{10} of the 10 units follows from:

$$s_{10}^2 = \frac{(x_i - \bar{x})^2}{9} = \frac{8.90}{9} = 0.989; \; s_{10} \approx 0.99 \text{ dB}$$

With s = 10, k = 1.24 and it follows that

$$L = \bar{x}_{10} + k \cdot s_{10} = 63.9 + 1.24 \cdot 0.99 = 65.1 \text{ dB}$$

TABLE A.3—EXAMPLE 2

The susceptibility values (x_i) of another set of 10 units are as follows:

x_i	$x_i - \bar{x}$	$(x_i - \bar{x})^2$
62.5 dB	1.5 dB	2.25 (dB)²
63	−1.0	1.00
65	1.0	1.00
61	−3.0	9.00
65	1.0	1.00
62	−2.0	4.00
65	1.0	1.00
66	2.0	4.00
65.5	1.5	2.25
65	1.0	1.00
$\Sigma x_i = 640.4$ dB	$\bar{x}_{10} = \frac{\Sigma x_i}{10} = \frac{640}{10} = 64$ dB	$\Sigma(x_i - \bar{x})^2 = 26.50$ (dB)²

The standard deviation s_{10} of the 10 units follows from:

$$s_{10}^2 = \frac{\Sigma(x_i - \bar{x})^2}{9} = \frac{26.50}{9} = 2.94; \; s_{10} \approx 1.71 \text{ dB}$$

With n = 10, k = 1.24 and it follows that

$$L = \bar{x}_{10} + k \cdot s_{10} = 64 + 1.24 \cdot 1.71 = 66.1 \text{ dB}$$

Although both examples have essentially the same mean, the data in example 2 show a higher susceptibility level (L) than example 1 because of example 2's greater standard deviation. Note that, in both examples, the variance of the data (and the sample size) causes the calculated susceptibility level (L) to be higher than the highest individual measured value.

The k factors for higher confidence levels or higher acceptance levels are available in any good statistical handbook, some of which are listed below. Higher confidence or acceptance levels will require the testing of more samples, and this will result in an increase in the cost of the testing program. The minimum number of samples required can be determined from inspection of the k values: that sample size where the k values begin to converge to the asymptotic value will define the sample size. For example, for 90% confidence, that 95% of a population will fall below the calculated value, the minimum number of samples tested should be about 30.

REFERENCES

M. G. Natrella, "Experimental Statistics." NBS Handbook 91.
N. L. Johnson and S. Kotz, "Continuous Univariant Distributions." Vol. 2.
Resnikoff, "Non-Central t Distribution."
D. B. Owen, "Handbook of Statistical Tables."

APPENDIX B—EQUIPMENT LIST

No approval or disapproval of any manufacturer is intended, either by inclusion or exclusion from this list.

LEVEL MEASURING	MODEL	APPLICABLE SECTION
Oscilloscopes		2.4, 3.4.1, 3.5.1, 4.3, 5.3
VM		
Hewlett Packard	410C (W/11042A*)	2.4, 3.4.1, 3.5.1, 7.3*
Boonton Electronics	92C	3.4.1, 3.5.1, 7.3, 8.3
EMI Meters		
Electro-Metrics	EMC-10	2.4
	EMC-25	3.4.1, 3.5.1, 7.3, 8.3
Singer	NM-7	2.4
	NM 17/27 & NM 37/57	3.4.1, 3.5.1, 7.3, 8.3
Spectrum Analysers		7.3, 8.3
Current Monitor		
Electro-Metrics	PCL-10 & PCA-10	6.3
Tektronics	CT-5 & P6042	6.3
Pearson Electronics	411	6.3
GENERATORS		
AF Signal		2.4, Parts of 3.4.1 and 3.5.1, 6.3, Part of 7.3
RF Signal		
Marconi Instruments	TF2008	Parts of 3.4.1 and 3.5.1, Part of 7.3
PF Power		
ALLTECH	445 or 446 w/plug-ins	8.3
MCL	15122 w/plug-ins	8.3
Spike		
Solar Electronics	7054-1	4.3
AMPLIFIERS		
AF Power		2.4, 6.3
RF Power		
IFI	M404	3.4.1, 3.5.1, 7.3
RF Power Labs	220-1k60L	3.4.1, 3.5.1, 7.3
Amplifier Research	100L or 1000L	3.4.1, 3.5.1, 7.3
TEST CHAMBERS		
Screen Room		
Emerson & Cuming		8.3
All-Shield Enclosures, Inc.		8.3
TEM Cell		
IFI	CC-102	7.3
Helmholtz Coil		
Electro-Mechanics	5001	6.3
ASSESSORY MATERIAL		
Feedthrough Capacitors		
Solar Electronics	6512-106R	4.3
RF Absorbing Material		
Emerson & Cuming	VHP 36 or VHP 26	8.3
Rantec	—	8.3
Antenna		
Electro-Metric	LCA-25	8.3
Singer	93490-1	8.3
Hooded Antenna Absorbing Material		
Emerson & Cuming	NZ-41 & AN-77	8.3

BATTERY CABLE—SAE J1127 JUN80   SAE Standard

Report of the Electrical Equipment Committee, approved November 1975, last revised June 1980.

1. Scope—This standard covers battery cables intended for use at 50 volts or less in surface vehicle wiring. Requirements for cable sizes 6 thru 4/0 previously contained in SAE Standard, Low Tension Cable SAE J558 are included. Cable sizes 20 through 4 previously specified in SAE J558 are now included in SAE J1128.

2. Specification Types

2.1 Type SGT—Starter or ground. Thermoplastic insulated.

2.2 Type SGR—Starter or ground. Synthetic rubber insulated.

2.3 Type SGX—Starter or ground. Cross linked polyethylene insulated.

3. General Specifications:

3.1 Conductors—Conductors shall be bunched, concentric or rope stranded as specified in the Appendix and shall be annealed copper wire in accordance with ASTM B-3. When tin or alloy coated wires are used they shall withstand the continuity test as specified under Strand coating test (paragraph 5.1). A separator shall be used between the uncoated conductor and the synthetic rubber insulation. When coated conductors are used no separator is required. The cross sectional area of stranded conductors shall not be less than the values specified in Table 1.

TABLE 1—CONDUCTORS

SAE Wire[a] Size	Metric Wire[b] Size, mm²	Minimum Conductor Area For Finished Cable	
		Cir Mil	mm²
6	13.0	25910	12.1
4	19.0	37360	18.3
2	32.0	62450	31.1
1	40.0	77790	38.1
0	50.0	98980	48.3
2/0	62.0	125100	59.8
3/0	81.0	158600	77.6
4/0	103.0	205500	98.5

[a] SAE wire size number indicates that the cross sectional area of the conductors approximate the area of American Wire Gauge for equivalent sizes.
[b] Metric wire size is the approximate nominal area of the stranded conductor.
Metric dimensions are not direct conversion from circular mils.
See Appendix for various individual conductor constructions and nominal strand diameters.

Minimum conductor area for circular mils is based on 98% of total minimum strand area as specified in ASTM B-3. Minimum conductor area for mm² is based on 98% of minimum strand area. Before processing into finish cable, the minimum strand area for metric strands shall not vary from the specified nominal by more than 1% expressed to the nearest .001 mm.

3.2 Insulation—Insulation shall be homogeneous in character and shall be placed concentrically within commercial tolerances about the conductor. Insulation shall adhere closely to, but strip readily from, the conductors leaving them reasonably clean and in suitable condition for terminating.

Insulation thickness shall be in accordance with the appropriate table for the various cable types. Variations in insulation wall thickness are permissible due to eccentricity. However, the minimum wall thickness at any cross section of a test specimen shall not be less than 70% of the nominal wall thickness of insulation specified in the appropriate table for the various cable types. The minimum wall thickness shall be measured with a pin dial micrometer exerting a force of 0.245 N with a mass of 0.025 kg and using a 0.043 in (1 mm) maximum diameter pin.

4. Cable Type Requirements

4.1 Construction—The conductors and insulation shall be as specified in paragraph 3 for each type of cable.

4.2 Test Requirements—The test requirements for each type of cable shall be as indicated in Table 3.

4.3 Dimensions—The nominal wall thickness and maximum overall diameter of finished cable shall be as specified in Table 4.

5. Tests

5.1 Strand Coating—Tin test shall be conducted on strands prior to stranding and shall be conducted per ASTM B-33. Alloy coated wire shall conform to ASTM B-189.

5.2 Physical Properties (Insulation)—Test samples of insulation that have been removed from the conductors shall be used. The conductor may be stretched for greater ease in removing it from the insulation. The sample may be tested as tube, slit-tube forms, or as dumbbells. The sample shall have marks placed upon it 2 in (50 mm) apart. The sample shall then be stretched at the rate of 20 in (508 mm) per min. The tensile strength shall be calculated upon the original cross section of the test sample before stretching. Physical tests shall be made at room temperature of $75 \pm 5°F$ ($21.1°C$). For the purpose of these tests care must be used in cutting and obtaining samples of uniform cross section.

5.3 High Temperature—One in (25 mm) of insulation shall be removed from each end of a 24 in (610 mm) sample of finished cable. The sample shall be suspended around a cylindrical mandrel with a weight attached to each end of the sample. This condition shall be maintained in a circulating air oven. The mandrel size, weight, temperature and time shall be as specified in Table 5.

At the end of the above conditioning period the sample shall be removed from the oven and allowed to cool to room temperature. When cool the weights shall be removed and the sample bent in the reverse direction around the mandrel at a rate not to exceed one complete turn per minute. The sample shall then be subjected to the Dielectric test as specified in paragraph 5.4.

5.4 Dielectric Test—One in (25 mm) of insulation shall be removed from each end of a 24 in (610 mm) sample of finished cable and the two ends connected together. The loop thus formed shall be immersed in water con-

TABLE 2—INSULATION PROPERTIES

Cable Types	Min. Tensile Strength		Min. Elongation Percent
	psi	kPa	
SGT	1600	11000	200
SGR	1000	6900	150
SGX	1500	10000	150

TABLE 3—TEST REQUIREMENTS

Tests	Cable Types		
	SGT	SGR	SGX
Strand Coating	x	x	x
Physical Properties	x	x	x
High Temperature	x	x	x
Dielectric	x		x
Cold Bend	x	x	x
Flame	x		x
Oil Absorption	x	x	x
Abrasion Resistance	x		x

TABLE 5—HIGH TEMPERATURE TEST

Cable Type Test Conditions	SGT SGR $120\ h/250 \pm 2°F\ (121°C)$				SGX $168\ h/302 \pm 3°F\ (150°C)$			
SAE Wire Size	Mandrel		Weight		Mandrel		Weight	
	in	mm	lb	kg	in	mm	lb	kg
6	10	254	6	2.72	10	254	6	2.72
4	10	254	6	2.72	10	254	6	2.72
2	10	254	6	2.72	10	254	6	2.72
1	10	254	6	2.72	10	254	6	2.72
0	10	254	10	4.54	10	254	10	4.54
2/0	10	254	10	4.54	10	254	10	4.54
3/0	10	254	10	4.54	—	—	—	—
4/0	10	254	10	4.54	—	—	—	—

Note: Metric dimensions and weights are direct conversion from inches and pounds.

TABLE 4—DIMENSIONS[b]

SAE Wire Size	Metric Wire Size, mm²	SGT				SGR				SGX			
		Nom. Wall		Max. Dia.		Nom. Wall		Max. Dia.		Nom. Wall		Max. Dia.	
		in	mm	in	mm	in	mm	in	mm	in	mm	in	mm
6[a]	13.0	.060	1.52	.340	7.36	.047	1.19	.340	7.36	.043	1.09	.300	6.49
4[a]	19.0	.065	1.65	.420	8.86	.047	1.19	.420	8.86	.065	1.65	.420	8.86
2	32.0	.065	1.65	.505	12.74	.065	1.65	.505	12.74	.065	1.65	.505	12.74
1	40.0	.065	1.65	.557	13.86	.065	1.65	.557	13.86	.065	1.65	.557	13.86
0	50.0	.065	1.65	.600	14.95	.065	1.65	.600	14.95	.065	1.65	.600	14.95
2/0	62.0	.065	1.65	.655	16.15	.065	1.65	.655	16.15	.065	1.65	.655	16.15
3/0	81.0	.078	1.98	.750	18.67	.078	1.98	.750	18.67	—	—	—	—
4/0	103.0	.078	1.98	.810	19.73	.078	1.98	.810	19.73	—	—	—	—

[a] The 6 and 4 gage wall thickness can be the same as for GPT.
[b] Metric dimensions are not direct conversion from inches.

taining 5% salt by weight at room temperature so that not more than 6 in (152 mm) of each end of the sample protrudes above the solution. After being immersed for five hours and while still immersed the sample shall withstand the application of 1000 V (rms) at 60 Hz between the conductor and the solution for 1 min without puncture of the insulation.

5.5 Cold Bend Test—One in (25 mm) of insulation shall be removed from each end of a 24 in (610 mm) sample of finished cable. The temperature of the sample shall be lowered at a rate of 122°F (50°C) per minute until the specified temperature is reached. This temperature shall be maintained for three hours. While the sample is still at this low temperature it shall be wrapped around a mandrel for 180 deg at a uniform rate of one turn in 10 s. The temperature and mandrel size shall be as specified in Table 6. Either a revolving or stationary mandrel may be used. When a revolving mandrel is used fasten one end of the sample to the mandrel. The sample shall then be subjected to the dielectric test as specified in paragraph 5.4.

5.6 Flame Test—A bunsen burner having a ½ in (13 mm) inlet, a nominal bore of ⅜ in (10 mm), a length of approximately 4 in (102 mm) above the primary inlets, equipped with a wing top flame spreader having a 1/16 x 2 in (1 x 51 mm) opening fitted to the top of the burner shall be used. A 24 in (610 mm) sample of finished cable shall be suspended taut in a horizontal position within a partial enclosure which allows a flow of sufficient air for complete combustion but is free from drafts. The top of a 2 in (51 mm) gas flame with an inner cone one-third its height shall then be applied to the center of the suspended cable. The time of application of the flame shall be 30 s for SAE wire 6 through 4/0. After removal of the bunsen burner flame the sample shall not continue to burn for more than 30 s.

5.7 Oil Absorption Test—One in (25 mm) of insulation shall be removed from each end of a 24 in (610 mm) sample of finished cable. The sample shall be immersed to within 1½ in (38 mm) from the ends of the insulator in a liquid containing equal parts of kerosene and SAE 10W engine oil at a temperature of 118–122°F (48–50°C) for a period of at least 20 h. The outside diameter of the cable shall not increase more than 15%. The sample shall then be bent around a 10 in (254 mm) mandrel and then subjected to the Dielectric Test, paragraph 5.4.

5.8 Abrasion Resistance—One in (25 mm) of the insulation shall be removed from one end of a 36 in (914 mm) sample of finished cable. The sample shall then be placed taut, without stretching between the cable clamps as shown in military specification MIL-T-5438. Using the weight support bracket and weight specified in Table 8. The sample shall then be subjected to the abrasion test. After each reading the sample shall be moved 2 in (51 mm) and rotated clockwise 90 deg. Eight readings shall be obtained for each sample. Obtain an average by calculating the arithmetic mean of all readings. Discard all readings above the arithmetic mean and average the remaining readings. The average shall define the abrasion resistance of the cable under ϕ test.

ϕ **TABLE 7—ABRASION TEST (REQUIREMENTS)**

SAE Wire Size	Minimum Resistance—in (mm) of Tape			
	SGT		SGX	
	in	mm	in	mm

Previous abrasion data invalid. Corrected values to be established at a later date.

TABLE 8—ABRASIONS TEST (CONDITIONS)

SAE Wire Size	SGT SGX		
	Br	lb	kg
6	C	4.25	1.93
4	C	4.25	1.93
2	C	4.25	1.93
1	C	4.25	1.93
0	C	4.25	1.93
2/0	C	4.25	1.93
3/0	C	4.25	1.93
4/0	C	4.25	1.93

APPENDIX—RECOMMENDED CONDUCTOR CONSTRUCTIONS (AWG STRANDS)

SAE Wire Size	Class I No. Strands/AWG Size (in)	Class II No. Strands/AWG Size (in)
6	37/21 (.0285)	7 x 19/27 (.0142)
4	61/22 (.0253)	7 x 19/25 (.0179)
2	127/23 (.0226)	7 x 19/23 (.0226)
1	127/22 (.0253)	7 x 37/25 (.0179)
0	127/21 (.0285)	7 x 37/24 (.0201)
2/0	127/20 (.0320)	7 x 37/23 (.0226)
3/0	—	7 x 37/22 (.0253)
4/0	—	19 x 22/23 (.0226)

TABLE 6—COLD BEND TEST

Cable Type Test Conditions	SGT SGR SGX 3 h/−40°F (−40°C)			
SAE Wire Size	Mandrel		Weight	
	in	mm	lb	kg
6	10	254	6	2.72
4	10	254	6	2.72
2	10	254	6	2.72
1	18	457	6	2.72
0	18	457	10	4.53
2/0	18	457	10	4.53
3/0	18	457	10	4.53
4/0	18	457	10	4.53

Note: Metric dimensions and weights are direct conversion from inches and pounds.

RECOMMENDED CONDUCTOR CONSTRUCTIONS (METRIC STRANDS)

SAE Wire Size	Metric Size, mm^2	Class I No. Strands/mm Size	Class II No. Strands/mm Size
6	13.0	37/.66	—
4	19.0	61/.63	—
2	32.0	127/.57	7 x 19/.57
1	40.0	127/.63	7 x 19/.63
0	50.0	127/.71	7 x 19/.71
2/0	62.0	127/.79	7 x 19/.79
3/0	81.0	—	7 x 37/.63
4/0	103.0	—	7 x 37/.71

Note: Stranding other than those shown above for both SAE and metric wire sizes are acceptable providing they meet the minimum conductor area specified in Table 1.

LOW TENSION PRIMARY CABLE—SAE J1128

SAE Standard

Report of Electrical Equipment Committee approved November 1975.

1. Scope—This standard covers low tension primary cable intended for use at 50 V or less in surface vehicle wiring. Requirements for cable sizes 20 through 4 previously contained in SAE Standard, Low Tension Cable—SAE J558 and SAE Recommended Practice, Low Tension Cable Thermosetting Insulation—SAE J878, are included. Cable sizes 6 through 4/0 previously specified in SAE J558 are now included in SAE J1127.

2. Specification Types
 2.1 Type GPT—General Purpose, thermoplastic insulated.

2.2 **Type HDT**—Heavy Duty, thermoplastic insulated.
2.3 **Type GPB**—General Purpose, thermoplastic insulated, braided.
2.4 **Type HDB**—Heavy Duty, thermoplastic insulated, braided.
2.5 **Type STS**—Standard Duty, synthetic rubber insulated.
2.6 **Type HTS**—Heavy Duty, synthetic rubber insulated.
2.7 **Type SXL**—Standard Duty, crosslinked polyethylene insulated.

3. *General Specifications*

3.1 **Conductors**—Conductors shall be bunched, concentric, or rope stranded as specified in the Appendix and shall be annealed copper wire in accordance with ASTM B-3. When tin or alloy coated wires are used they shall withstand the continuity test as specified under strand coating test (paragraph 5.1). A separator shall be used between the uncoated conductor and the synthetic rubber insulation. When coated conductors are used no separator is required. The cross sectional area of stranded conductors shall not be less than the values specified in Table 1.

Minimum conductor area for circular mils is based on 98% of total minimum strand area as specified in ASTM B-3. Minimum conductor area for mm^2 is based on 98% of minimum strand area. Before processing the final cable, the minimum strand area for metric strands shall not vary from the specified nominal by more than 1% expressed to the nearest .001 mm.

3.2 **Insulation**—Insulation shall be homogeneous in character and shall be placed concentrically within commercial tolerances about the conductor. Insulation shall adhere closely to, but strip readily from, the conductors leaving them reasonably clean and in suitable condition for terminating.

Insulation thickness shall be in accordance with the appropriate table for the various cable types. Variations in insulation wall thickness are permissible due to eccentricity. However, the minimum wall thickness at any cross section of a test specimen shall not be less than 70% of the nominal wall thickness of insulation specified in the appropriate table for the various cable types. The minimum wall thickness shall be measured with a pin dial micrometer exerting a force of 0.245 N with a mass of 0.025 kg and using a 0.043 in (1 mm) maximum diameter pin.

The physical properties of the insulation shall be as shown in Table 2.

3.3 **Braid**

3.3.1 GENERAL PURPOSE BRAID—When the construction includes a braided covering a closely woven braid of cotton or other fibrous material shall be applied over the insulation. All braided coverage shall be thoroughly covered or saturated with a firmly adhering compound that will present a finished appearance. Adjacent layers of cable when wound on a reel shall not stick to one another at any temperature under 140°F (40°C).

3.3.2 HEAVY DUTY BRAID—The braid shall be made up of one-half .028 in (0.7 mm) paper twine and one-half cotton yarn. The cable shall be braided so that all of the paper twine shall be in one direction and the cotton yarn in the opposite direction. The braid shall be finished or coated with a compound as specified in paragraph 3.3.1.

4. *Cable Type Requirements*

4.1 **Construction**—The conductors, insulation and braid shall be as specified in paragraph 3 for each type of cable.

4.2 **Test Requirements**—The test requirement for each type of cable shall be as indicated in Table 3.

4.3 **Dimensions**—The nominal wall thickness and maximum overall diameter of finished cable shall be as specified in Table 4.

5. *Tests*

5.1 **Strand Coating**—Tin test shall be conducted on strands prior to stranding and shall be conducted per ASTM B-33. Alloy coated wire shall conform to ASTM B-189.

TABLE 1—CONDUCTORS

SAE Wire[a] Size	Metric Wire[b] Size, mm^2	Minimum Conductor Area For Finished Cable	
		Cir Mil[a]	mm^{2b}
20	0.5	1072	.508
18	0.8	1537	.760
16	1.0	2336	1.12
14	2.0	3702	1.85
12	3.0	5833	2.91
10	5.0	9343	4.65
8	8.0	14810	7.23
6	13.0	25910	12.1
4	19.0	37360	18.3

[a] SAE wire size number indicates that the cross sectional area of the conductors approximate the area of American Wire Gauge for equivalent sizes.
[b] Metric wire size is the approximate nominal area of the stranded conductor. Metric dimensions are not direct conversion from circular mils.
See Appendix for various individual conductor constructions and nominal strand diameters.

TABLE 2—INSULATION PROPERTIES

Cable Types	Min Tensile Strength		Min Elongation
	psi	kPa	Percent
GPT HDT	2300	15860	125
GPB HDB	1000	6900	100
STS HTS	1600	11030	250
SXL	1500	10340	150

5.2 **Physical Properties (Insulation)**—Test samples of insulation that have been removed from the conductors shall be used. The conductor may be stretched for greater ease in removing it from the insulation. The sample may be tested as tube, slit-tube forms, or as dumbbells. The sample shall have marks placed upon it 2 in (50 mm) apart. The sample shall then be stretched at the rate of 20 in (508 mm) per min. The tensile strength shall be calculated upon the original cross section of the test sample before stretching. Physical tests shall be made at room temperature of $75 \pm 5°F$ (21.1°C). For the purpose of these tests care must be used in cutting and obtaining samples of uniform cross section.

5.3 **High Temperature**—One in (25 mm) of insulation shall be removed from each end of a 24 in (610 mm) sample of finished cable. The sample shall be suspended around a cylindrical mandrel with a weight attached to each end of the sample. This condition shall be maintained in a circulating air oven. The mandrel size, weight, temperature and time shall be as specified in Table 5.

At the end of the above conditioning period the sample shall be removed from the oven and allowed to cool to room temperature. When cool the weights shall be removed and the sample bent in the reverse direction around the mandrel specified in Table 5 for the bend test at a rate not to exceed one complete turn per minute. The sample shall then be subjected to the Dielectric test as specified in paragraph 5.4.

5.4 **Dielectric Test**—One in (25 mm) of insulation shall be removed from each end of a 24 in (610 mm) sample of finished cable and the two ends twisted together. The loop thus formed shall be immersed in water containing 5% salt by weight at room temperature so that not more than 6 in (152 mm) of each end of the sample protrudes above the solution. After being immersed for five hours and while still immersed the sample shall withstand the application of 1000 V at 60 Hz between the conductor and the solution for one minute without puncture of the insulation.

5.5 **Cold Bend Test**—One in (25 mm) of insulation shall be removed from each end of a 24 in (610 mm) sample of finished cable. The temperature of the sample shall be lowered at a rate of 122°F (50°C) per minute until the specified temperature is reached. This temperature shall be maintained for three hours. While the sample is still at this low temperature it shall be wrapped around a mandrel for 180 deg at a uniform rate of one turn in 10 s. The temperature and mandrel size shall be as specified in Table 6. Either a revolving or stationary mandrel may be used. When a revolving mandrel is used fasten one end of the sample to the mandrel. The sample shall then be subjected to the Dielectric test as specified in paragraph 5.4.

5.6 **Flame Test**—A bunsen burner having a ½ in (13 mm) inlet, a nominal bore of ⅜ in (10 mm), a length of approximately 4 in (102 mm) above the primary inlets, equipped with a wing top flame spreader having a 1/16 x 2 in (1 x 51 mm) opening fitted to the top of the burner shall be used. A 24 in (610 mm) sample of finished cable shall be suspended taut in a horizontal position within a partial enclosure which allows a flow of sufficient air for complete combustion but is free from drafts. The top of a 2 in (51 mm) gas flame with an inner cone one-third its height shall then be applied to the center of the suspended cable. The time of application of the flame shall be

TABLE 3—TEST REQUIREMENTS

Tests	Cable Types						
	GPT	HDT	GPB	HDB	STS	HTS	SXL
Strand Coating	x	x	x	x	x	x	x
Physical Properties	x	x	x	x	x	x	x
High Temperature	x	x	x	x	x	x	x
Dielectric	x	x	x	x	x	x	x
Cold Bend	x	x	x	x	x	x	x
Flame	x	x					x
Oil Absorption	x	x	x	x			
Overload					x	x	x
Short Circuit					x	x	x
Pinch	x	x					
Abrasion Resistance	x	x					

TABLE 4—DIMENSIONS[a]

SAE Wire Size	Metric Wire Size, mm²	GPT Nom Wall		GPT Max Dia		HDT Nom Wall		HDT Max Dia		GPB Nom Wall		GPB Max Dia	
		in	mm	in	mm	in	mm	in	mm	in	mm	in	mm
20	0.5	.023	.58	.095	2.34	.036	.91	.120	2.95	.022	.56	.155	2.92
18	0.8	.023	.58	.100	2.50	.037	.94	.130	3.24	.022	.56	.135	3.43
16	1.0	.023	.58	.115	2.84	.040	1.02	.145	3.58	.022	.56	.145	3.68
14	2.0	.023	.58	.125	3.18	.041	1.04	.165	4.19	.022	.56	.165	4.19
12	3.0	.026	.66	.150	3.81	.046	1.17	.190	4.83	.027	.69	.195	4.95
10	5.0	.031	.79	.185	4.67	.046	1.17	.215	5.42	.031	.79	.230	5.84
8	8.0	.037	.94	.245	6.10	.055	1.40	.280	6.98	.037	.94	.301	7.65
6	13.0	.043	1.09	.305	7.72	.060	1.52	.340	8.60	.047	1.19	.360	9.14
4	19.0	.044	1.12	.375	9.32	.068	1.73	.420	10.44	.047	1.19	.437	11.10

SAE Wire Size	Metric Wire Size, mm²	HDB Nom Wall		HDB Max Dia		STS and SXL Nom Wall		STS and SXL Max Dia		HTS Nom Wall		HTS Max Dia	
		in	mm	in	mm	in	mm	in	mm	in	mm	in	mm
20	0.5					.029	.74	.110	2.71	.036	.91	.125	3.08
18	0.8	.022	.56	.155	3.94	.030	.76	.120	3.00	.037	.94	.135	3.37
16	1.0	.022	.56	.170	4.32	.032	.81	.135	3.33	.040	1.02	.150	3.70
14	2.0	.022	.56	.190	4.83	.035	.89	.155	3.94	.041	1.04	.165	4.19
12	3.0	.027	.69	.230	5.84	.037	.94	.180	4.57	.046	1.17	.200	5.08
10	5.0	.031	.79	.255	6.48	.041	1.04	.210	5.30	.048	1.22	.225	5.67
8	8.0					.043	1.09	.245	6.10	.055	1.40	.270	6.73
6	13.0												
4	19.0					.055	1.40	.335	8.47	.062	1.57	.350	8.85

[a] Metric dimensions are not direct conversion from inches.

15 s for SAE wire size 10 through 20 and 30 s for SAE wire size 8 and larger. After removal of the bunsen burner flame the sample shall not continue to burn for more than 30 s.

5.7 Oil Absorption Test—One in of insulation shall be removed from each end of a 24 in (610 mm) sample of finished cable. The sample shall be immersed to within 1½ in (38 mm) from the ends of the insulator in a liquid containing equal parts of kerosene and SAE 10W engine oil at a temperature of 118–122°F (48–50°C) for a period of at least 20 h. The outside diameter of the cable shall not increase more than 15%. The sample shall then be bent around a mandrel as specified in Table 7, and then subjected to the Dielectric Test, paragraph 5.4.

5.8 Overload Test—In an ambient temperature of 75 ± 5°F (23.9°C) a 60 in (1524 mm) sample cable suspended in air or lying on a transite table top shall be subjected to an overload current sufficient to raise the conductor temperature to 400 ± 3°F (204°C) and to hold it there for a period of 30 min (thermocouple to be inserted into sample conductor stranding 18 in (457 mm) from one end). After the overload test cut 18 in (457 mm) from each end of the cable and discard. The remaining 24 in (610 mm) portion which was in

TABLE 5—HIGH TEMPERATURE TEST

Cable Type Test Conditions	GPT HDT GPB HDB 120 h/250 ± 2°F (121°C)				STS HTS 120 h/275 ± 3°F (135°C)				SXL 240 h/302 ± 3°F (150°C)			
SAE Wire Size	Mandrel		Weight		Mandrel		Weight		Mandrel		Weight	
	in	mm	lb	kg	in	mm	lb	kg	in	mm	lb	kg
20	4.5	114	1.0	0.45	4.5	114	1.0	0.45	0.50	12.7	1.0	0.45
18	4.5	114	1.0	0.45	4.5	114	1.0	0.45	0.50	12.7	1.0	0.45
16	6.5	165	1.0	0.45	6.5	165	1.0	0.45	0.50	12.7	1.0	0.45
14	6.5	165	1.0	0.45	6.5	165	1.0	0.45	0.75	19.0	1.0	0.45
12	6.5	165	3.0	1.36	6.5	165	3.0	1.36	0.75	19.0	3.0	1.36
10	10.0	254	3.0	1.36	10.0	254	3.0	1.36	1.00	25.4	3.0	1.36
8	10.0	254	3.0	1.36	10.0	254	3.0	1.36	2.00	50.8	3.0	1.36
6	10.0	254	6.0	2.72	10.0	254	6.0	2.72	3.00	76.2	6.0	2.72
4	10.0	254	6.0	2.72	10.0	254						

Bend Test Mandrel Dia.

Cable Type	GPT HDT GPB HDB		STS HTS		SXL	
SAE Wire Size	in	mm	in	mm	in	mm
20	3	76.2	3	76.2	1.0	25.4
18	3	76.2	3	76.2	1.0	25.4
16	3	76.2	3	76.2	1.0	25.4
14	6	152.4	6	152.0	3.0	76.2
12	6	152.4	6	152.0	3.0	76.2
10	6	152.4	6	152.0	3.0	76.2
8	6	152.4	6	152.0	6.0	152.4
6	10	254.0	10	254.0	6.0	152.4
4	10	254.0				

Note: Metric dimensions and weights are direct conversion from inches and pounds.

TABLE 6—COLD BEND TEST

Cable Type Test Conditions	GPT HDT GPB HDB 3 h/−40°F (−40°C)				STS HTS 3 h/−40°F (−40°C)				SXL 3 h/−60°F (−51°C)			
	Mandrel		Weight		Mandrel		Weight		Mandrel		Weight	
SAE Wire Size	in	mm	lb	kg	in	mm	lb	kg	in	mm	lb	kg
20	3.0	76	1.0	0.45	3.0	76	1.0	0.45	1.0	25.4	1.5	0.68
18	3.0	76	1.0	0.45	3.0	76	1.0	0.45	1.0	25.4	1.5	0.68
16	3.0	76	1.0	0.45	3.0	76	1.0	0.45	1.0	25.4	1.5	0.68
14	6.0	152	1.0	0.45	6.0	152	1.0	0.45	3.0	76.2	3.0	1.36
12	6.0	152	3.0	1.36	6.0	152	3.0	1.36	3.0	76.2	5.0	2.27
10	6.0	152	3.0	1.36	6.0	152	3.0	1.36	3.0	76.2	5.0	2.27
8	6.0	152	3.0	1.36	6.0	152	3.0	1.36	6.0	152.0	5.0	2.27
6	10.0	254	6.0	2.72	10.0	254	6.0	2.72	6.0	152.0	7.0	3.17
4	10.0	254	6.0	2.72								

the center of the original 60 in (1524 mm) shall then be subjected to the Dielectric test, paragraph 5.4.

5.9 Short Circuit Test (18 gage only)—Using six 36 in (914 mm) lengths and one 48 in (1219 mm) length of 18 gage cable strip 1 in (25.4 mm) of insulation from each end of the 48 in (1219 mm) length. Twist the six 36 in (914 mm) lengths around the 48 in (1219 mm) length with approximately a 4 in (102 mm) lay. Position so that 6 in (152 mm) of the 48 in (1219 mm) cable extends beyond each end. Tape into position using woven glass tape with 1/3 lap. Apply a constant 55 amp current to the center conductor of the bundle for three minutes. Turn off current and allow bundle to cool. Disconnect power source and test for short circuits between all conductors. Use 1000 V (rms) test voltage. There shall be no shorting of conductors and when the glass tape is removed from the bundle the individual wires shall be readily separated without tearing the insulation on the individual cables. This test is conducted to check the thermosetting properties of the insulation.

5.10 Pinch Test—One in (25 mm) of insulation shall be removed from one end of a 36 in (914 mm) sample of finished cable. The sample shall then be placed taut without stretching across a 1/8 in (3 mm) diameter steel bar and be subjected to the force of a weighted steel anvil. Increasing weight shall be applied to the steel anvil at a rate of 5 lb (2.27 kg) per min with a lever advantage of 10 at the moment the insulation is pinched through the 1/8 in (3 mm) diameter rod will contact the conductor and the test shall stop. The weight in the receptacle shall then be recorded. After each reading the sample shall be moved 2 in (51 mm) and rotated clockwise 90 deg. Four readings shall be obtained for each sample. Obtain an average by calculating the arithmetic mean of all those readings. The average shall define the pinch resistance of the cable under test.

5.11 Abrasion Resistance—One in (25 mm) of the insulation shown shall be removed from one end of a 36 in (914 mm) sample of finished cable. The sample shall then be placed taut, without stretching between the cable clamps as shown in military specification MIL-T-5438. Using the weight support bracket and weight specified in Table 10. The sample shall then be subjected to the abrasion test. After each reading the sample shall be moved 2 in (51 mm) and rotated clockwise 90 deg. Eight readings shall be obtained for each sample. Obtain an average by calculating the arithmetic mean of all readings. Discard all readings above the arithmetic mean and average the remaining readings. The average shall define the abrasion resistance of the cable under test. Values for individual cables are shown in Table 9.

Note: The Pinch Test apparatus shall be equivalent to that shown in Fig. 1. The minimum values for each cable type and size shall be as shown in Table 8.

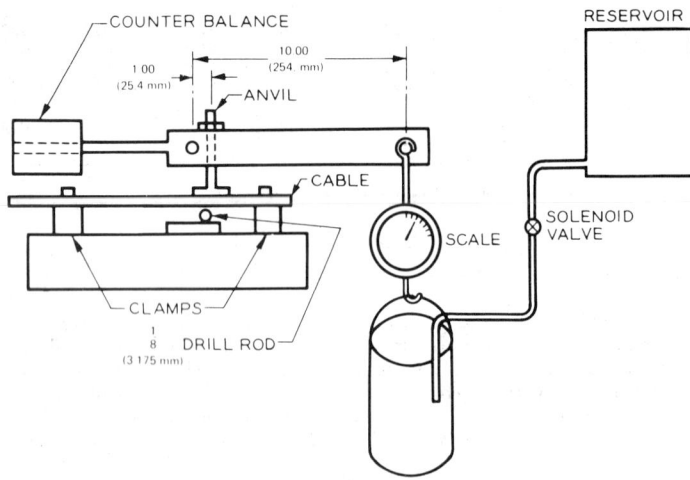

FIG. 1—PINCH TEST APPARATUS

6. Reference Information

6.1 Color Code

6.1.1 RECOMMENDED COLORS—The color of the cables should match as closely as possible the following colors as set forth by "The Color Association of the U.S. Inc., 9th Edition."

6.1.2 STRIPES—Where additional color combinations are required it is possible to apply color stripes. The stripes shall be applied longitudinally along the cable. Black or white stripes are recommended but other colors may be specified.

TABLE 7—OIL ABSORPTION TEST

Cable Type	GPT HDT GPB HDB		STS HTS		SXL	
	Mandrel		Mandrel		Mandrel	
SAE Wire Size	in	mm	in	mm	in	mm
20	3.0	76.0	3.0	76.0	1.0	25.4
18	3.0	76.0	3.0	76.0	1.0	25.4
16	3.0	76.0	3.0	76.0	1.0	25.4
14	6.0	152.0	6.0	152.0	3.0	76.0
12	6.0	152.0	6.0	152.0	3.0	76.0
10	6.0	152.0	6.0	152.0	3.0	76.0
8	6.0	152.0	6.0	152.0	6.0	152.0
6	10.0	254.0	10.0	254.0	6.0	152.0
4	10.0	254.0				

Note: Metric dimensions and weights are direct conversion from inches and pounds.

TABLE 8—PINCH TEST

Minimum Pinch Resistance

Cable Type	GPT		HDT		STS		HTS		SXL	
SAE Wire Size	lb	kg	lb	kg	lb	kg	lb	kg	lb	kg
20	5	2.3	9	4.1	8	3.6	10	4.5	11	5.0
18	6	2.7	10	4.5	8	3.6	10	4.5	16	7.2
16	6	2.7	13	5.9	8	3.6	10	4.5	18	8.2
14	8	3.6	15	6.8	8	3.6	10	4.5	20	9.0
12	8	3.6	18	8.2	8	3.6	10	4.5	25	11.0
10	10	4.5	24	10.9	8	3.6	10	4.5	30	14.0
8	11	5.0	32	14.5	8	3.6	10	4.5	35	16.0
6	15	6.8	43	19.5	8	3.6	10	4.5	40	18.0
4	27	12.2	54	24.5						

Note: Metric dimensions and weights are direct conversion from inches and pounds.

TABLE 9—ABRASION TEST (REQUIREMENTS)

Cable Type	\multicolumn{10}{c}{Minimum Resistance—Inches of Tape}
	GPT		HDT		STS		HTS		SXL	
SAE Wire Size	in	mm	in	mm	in	mm	in	mm	in	mm
20	18	457	16	406	18	457	30	762	22	559
18	21	533	22	559	21	533	35	889	27	686
16	22	559	29	737	22	559	40	1016	30	762
14	12	305	15	381	12	305	18	457	14	356
12	16	406	20	508	16	406	22	559	18	457
10	20	508	27	686	20	508	30	762	24	610
8	25	635	46	1168	25	635	35	889	39	991
6	25	635	50	1270	25	635	60	1524	39	991
4	30	762	60	1524						

Note: Metric dimensions are direct conversion from inches.

TABLE 10—ABRASION TEST (CONDITIONS)

Cable Type	\multicolumn{15}{c}{Test Conditions}
	GPT			HDT			STS			HTS			SXL		
SAE Wire Size	Br	lb	kg	Br	lb	kg	Br	lb	kg	Br	lb	kg	Br	lb	kg
20	A	1	0.45	B	3	1.36	A	1	0.45	B	1	0.45	A	1	0.45
18	A	1	0.45	B	3	1.36	A	1	0.45	B	1	0.45	A	1	0.45
16	A	1	0.45	B	3	1.36	B	1	0.45	B	1	0.45	B	1	0.45
14	B	3	1.36	B	4.25	1.93	B	3	1.36	B	3	1.36	B	3	1.36
12	B	3	1.36	B	4.25	1.93	B	3	1.36	B	3	1.36	B	3	1.36
10	B	3	1.36	B	4.25	1.93	B	3	1.36	B	3	1.36	B	3	1.36
8	B	3	1.36	C	4.25	1.93	B	3	1.36	C	3	1.36	B	3	1.36
6	C	4.25	1.93	C	4.25	1.93	C	4.25	1.93	C	4.25	1.93	C	4.25	1.93
4	C	4.25	1.93	C	4.25	1.93									

Note: Metric weights are direct conversion from pounds.

TABLE 11—TECA COLORS 9TH EDITION

Color	Nom.	Dark	Light
White	70003	70004	
Red	70180	70082	70179
Pink	70098	70099	70097
Orange	70072	70041	70071
Yellow	70205	70068	70067
Lt Green	70062	70063	70061
Dk Green	70065	70066	70064
Lt Blue	70143	70144	70142
Dk Blue	70086	70087	70085
Purple	70135	70164	70134
Tan	70093	70094	70092
Brown	70107	70108	70106
Gray	70152	70153	70185
Black	None	—	—

APPENDIX

Recommended Conductor Constructions (AWG Strands)

SAE Wire Size	Class III No. Strands/ AWG Size (in)	Class IV No. Strands/ AWG Size (in)
20	7/28 (.0126)	
18	16/30 (.0100)	65/36 (.0050)
16	19/29 (.0113)	
14	19/27 (.0142)	
12	19/25 (.0179)	
10	19/23 (.0226)	
8	19/21 (.0285)	
6	37/21 (.0285)	7 x 19/27 (.0142)
4	61/22 (.0253)	7 x 19/25 (.0179)

Recommended Conductor Constructions (Metric Strands)

SAE Wire Size	Metric Wire Size mm^2	Class III No. Strands/ mm Size
20	0.5	7/.31
18	0.8	19/.23
16	1.0	19/.28
14	2.0	19/.36
12	3.0	19/.45
10	5.0	19/.57
8	8.0	19/.71
6	13.0	37/.66
4	19.0	61/.63

Note: Stranding other than those shown above for both SAE and metric wire sizes are acceptable providing they meet the minimum conductor area specified in Table 1.

RECOMMENDED ENVIRONMENTAL PRACTICES FOR ELECTRONIC EQUIPMENT DESIGN—SAE J1211

SAE Recommended Practice

Report of the Electronic Systems Committee approved June 1978. Editorial change November 1978.

1. Purpose—This guideline is intended to aid the designer of automotive electronic systems and components by providing material that may be used to develop environmental design goals.

2. Scope—The climatic, dynamic, and electrical environments from natural and vehicle-induced sources that influence the performance and reliability of automotive electronic equipment are included. Test methods that can be used to simulate these environmental conditions are also included in this document.

The information is applicable to vehicles that meet all the following conditions and are operated on roadways:

2.1 Front engine rear wheel drive vehicles.
2.2 Vehicles with reciprocating gasoline engines.
2.3 Coupe, sedan, and hard top vehicles.

Part of the information contained herein is not affected by the above conditions and has more universal application. Careful analysis is necessary in these cases to determine applicability.

3. Application

3.1 Environmental Data and Test Method Validity—The information included in the following sections is based upon test results achieved by major North American automobile manufacturers and automobile original equipment suppliers. Operating extremes were measured at test installations normally used by manufacturers to simulate environmental extremes for vehicles and original equipment components. They are offered as a design starting point. Generally, they cannot be used directly as a set of operating specifications because some environmental conditions may change significantly with relatively minor physical location changes. This is particularly true of vibration, engine compartment temperature, and electromagnetic compatability. Actual measurements should be made as early as practical to verify these preliminary design baselines.

The proposed test methods are either currently used for laboratory simulation or are considered to be a realistic approach to environmental design validation. They are not intended to replace actual operational tests under adverse conditions. The recommended methods, however, describe standard cycles for each type of test. The designer must specify the number of cycles over which the equipment should be tested. The number of cycles will vary depending upon equipment, location, and function. While the standard test cycle is representative of an actual short term environmental cycle, no attempt has been made to equate this cycle to an acceleration factor for reliability or durability. These considerations are beyond the scope of this guideline.

3.2 Organization of Test Methods and Environment Extremes Information—The data presented in this document is contained in Sections 4 and 5. Section 4, Environmental Factors and Test Methods, describes the 11 major characteristics of the expected environment that have an impact on the performance and reliability of automotive electronic systems. These descriptions are titled:

3.2.1 Temperature.
3.2.2 Humidity.
3.2.3 Salt Spray Atmosphere.
3.2.4 Immersion and Splash (Water, Chemicals, and Oils).
3.2.5 Dust, Sand, and Gravel Bombardment.
3.2.6 Altitude.
3.2.7 Mechanical Vibration.
3.2.8 Mechanical Shock.
3.2.9 Factors Affecting the Automotive Electrical Environment.
3.2.10 Steady State Electrical Characteristics.
3.2.11 Transient, Noise, and Electrostatic Characteristics.

They are organized to cover three facets of each factor:
(a) Definition of the factor.
(b) Description of its effect on control, performance, and long term reliability.
(c) A review of proposed test methods for simulating environmental stress.

Section 5, Environmental Extremes by Location, summarizes the anticipated limit conditions at five general control sites:
(a) Underhood
  1. Engine
  2. Bulkhead—dash panel
(b) Chassis
(c) Exterior
(d) Interior
  1. Instrument Panel
  2. Floor
  3. Rear Deck
(e) Trunk

3.3 Combined Environments—The automotive environment consists of many natural and induced factors. Combinations of these factors are present simultaneously. In some cases, the effect of a combination of these factors is much more serious than the effect of exposing samples to each environmental factor in series. For example, the suggested test method for humidity includes both high and low temperature exposure. This combined environmental test is very important to compmonents whose proper operation is dependent on seal integrity. Temperature and vibration is a second combined environmental test that can be significant to some components. During design analysis a careful study should be made to determine the possibility of design susceptibility to a combination of environmental factors that could occur at the planned mounting location. If the possibility of susceptibility exists, a combined environmental test should be considered.

3.4 Test Sequence—The optimum test sequence is a compromise between two considerations:

3.4.1 The order in which the environmental exposures will occur in operational use.

3.4.2 A sequence that will create a total stress on the sample that is representative of operation stress.

The first consideration is impossible to implement in the automotive case, since exposures occur in a random order. The second consideration prompts the test designer to place the more severe environments last. Many sequences that have been successful follow this general philosophy, except that temperature cycle is placed first in order to condition the sample mechanically.

4. Environmental Factors and Test Methods

4.1 Temperature

4.1.1 DEFINITION—Thermal factors are probably the most pervasive environmental hazard to automotive electronic equipment. Sources for temperature extremes and variations include:

4.1.1.1 The vehicle's climatic environment, including the diurnal and seasonal cycles. Additionally, variations in climate by geographical location must be considered. In the most adverse case, the vehicle that spends the winter in Canada may be driven in the summer in the Arizona desert. Temperature variations due to this source range from -40–85°C (-40–185°F).

4.1.1.2 Heat sources and sinks generated by the vehicle's operation. The major sources are the engine and drive train components, including the brake system. Very wide variations are to be found during operation. For instance, temperatures on the surface of the engine can range from the cooling system's 88–650°C (190–1200°F) on the surface of the exhaust system. This category also includes conduction, convection, and radiation of heat due to various modes of vehicle operation.

4.1.1.3 Self-heating of the equipment due to its own internal dissipation. A design review of the worst case combination of peak ambient temperature (due to 4.1.1.1 and 4.1.1.2 above), minimized heat flow away from the equipment and peak applied steady state voltage should be conducted.

4.1.1.4 Vehicle operational mode and actual mounting location. Measurements should be made at the actual mounting site during the following vehiclular conditions while subjected to the maximum heat generated by adjacent equipment and at the maximum ambient environment:

4.1.1.4.1 Engine start.
4.1.1.4.2 Engine idle.
4.1.1.4.3 Engine high speed.
4.1.1.4.4 Engine turn off—prior history important.
4.1.1.4.5 Various engine/road load conditions.

4.1.1.5 Ambient conditions before installation due to storage and transportation extremes. Shipment in unheated aircraft cargo compartments may lower the minimum storage (non-operating) temperature to −50°C (−58°F).

The thermal environmental conditions that are a result of these conditions can be divided into three categories:

4.1.1.5.1 Extremes—The ultimate upper and lower temperatures the equipment is expected to experience.

4.1.1.5.2 Cycling—The cumulative effects of temperatures cycling within the limits of the extremes.

4.1.1.5.3 Shock—Rapid change of temperature. Fig. 1 illustrates one form

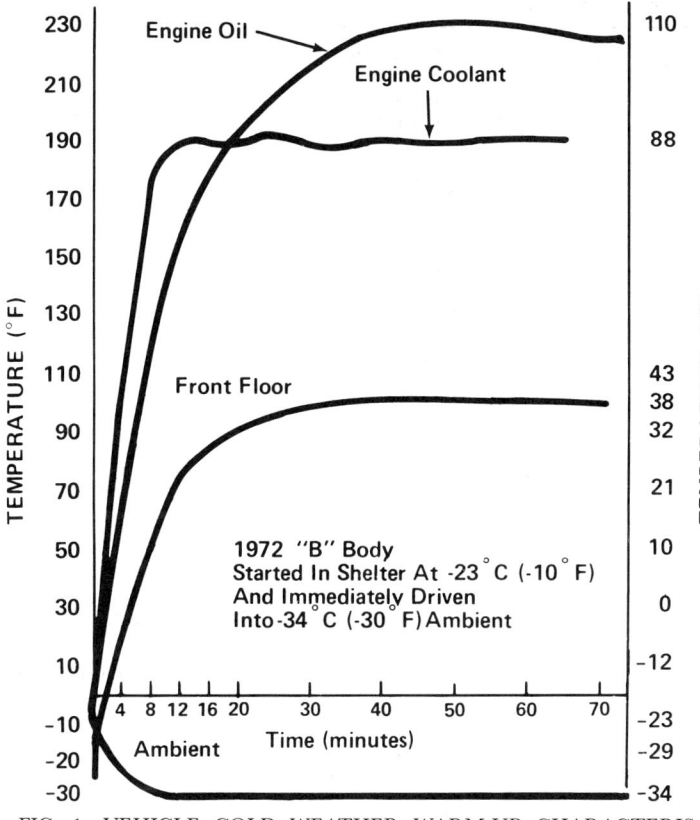

FIG. 1—VEHICLE COLD WEATHER WARM-UP CHARACTERISTICS

4.1.2.4 Seal failures, including the *breathing* action of some assemblies, due to temperature-induced dimensional variation which permit intrusion of liquid or vapor borne contaminants.

4.1.2.5 Failure of circuit components due to direct mechanical stress caused by differential thermal expansion.

4.1.2.6 The acceleration of chemical attack on interconnects, due to temperature rise, can result in progressive degradation of circuit components, printed circuit board conductors, and solder joints.

In addition to this, high temperature extremes can cause a malfunction by:

4.1.2.7 Exceeding the dissociation temperature of surrounding polymer or other packaging components.

4.1.2.8 Carbonization of packaging materials with eventual progressive failure of the associated passive or active components. This is possible in cases of extreme overtemperature. In addition, non-catastropic failure is possible due to electrical leakage in the resultant carbon paths.

4.1.2.9 Changes in active device characteristics with increased heat including changes in gain, impedance, collector-base leakage, peak blocking voltage, collector-base junction second breakdown voltage, etc. with temperature.

4.1.2.10 Changes in passive device characteristics such as permanent or temporary drift in resistor value and capacitor dielectric constants with increased temperature.

4.1.2.11 Changes in interconnect and relay coil performance due to the conductivity temperature coefficient of copper.

4.1.2.12 Changes in the properties of magnetic materials with increasing temperature, including Curie point effects and loss of *permanent* magnetism.

4.1.2.13 Dimensional changes in packages and components leading to separation of subassemblies.

4.1.2.14 Changes in the strength of soldered joints due to changes in mechanical characteristics of the solder.

Further, low temperature extremes can cause failure due to:

4.1.2.15 The severe mechanical stress caused by ice formation in moisture bearing voids or cracks.

4.1.2.16 The very rapid and extreme internal thermal stress caused by applying maximum power to semi-conductor or other components after extended cold soak under aberrant operating conditions such as 24-V battery jumper starts.

4.1.3 RECOMMENDED TEST METHODS

4.1.3.1 *Temperature Cycle Test*—A recommended thermal cycle profile is shown in Fig. 2 and recommended extreme temperatures in Table 1. The test method of Fig. 2A, a 24-h cycle, offers longer stabilization time and permits a convenient room ambient test period. Fig. 2B, and 8-h cycle, provides more temperature cycles for a given test duration. It is applicable only to modules whose temperatures will reach stabilization in a shorter cycle time. Stabilization should be verified by actual measurements. Thermocouples, etc.

Separate or single test chambers may be used to generate the temperature environment described by the thermal cycles. By means of circulation, the air temperature should be held to within ±2.8°C (±5°F) at each of the extreme temperatures. The test specimens should be placed in such a position, with respect to the air stream, that there is substantially no obstruction to the flow of air across the specimen. If two test specimens are used, care must be exercised to assure that the test samples are not subjected to temperature

of vehicle operation which induces thermal shock. Thermal shock is also induced when equipment at elevated temperature is exposed to sudden rain or road splash.

The automotive electronic equipment designer is urged to develop a systematic, analytic method for dealing with steady state and transient thermal analysis. The application of many devices containing semi-conductors will be temperature limited. For this reason, the potential extreme operating conditions for each application must be scrutinized to avoid later field failure.

4.1.2 EFFECT ON PERFORMANCE—The damaging effects of thermal shock and thermal cycling include:

4.1.2.1 Cracking of printed circuit board or ceramic substrates.

4.1.2.2 Thermal stress or fatigue failures of solder joints.

4.1.2.3 Delamination of printed circuit board and other interconnect system substrates.

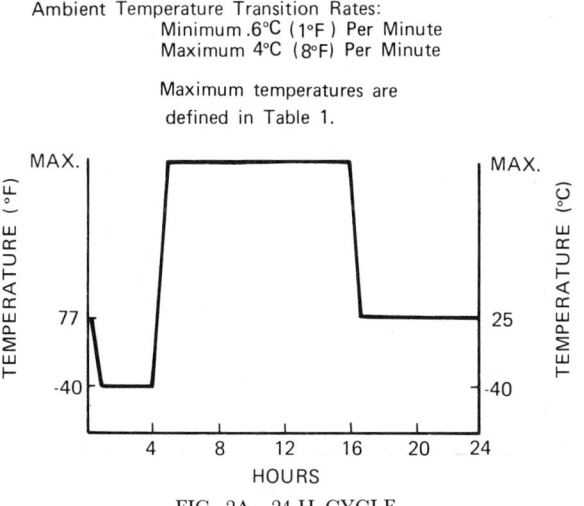

FIG. 2A—24-H CYCLE

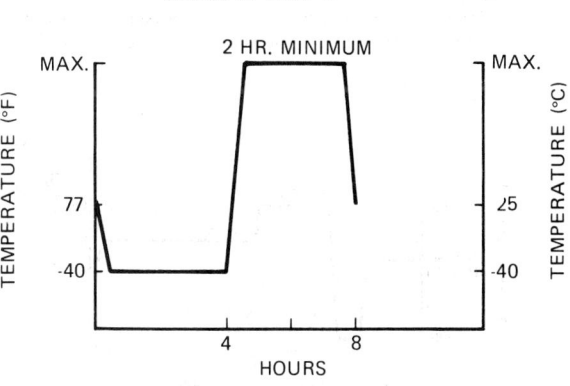

FIG. 2B—8-H CYCLE

FIG. 2—RECOMMENDED THERMAL CYCLES

TABLE 1—RECOMMENDED TEMPERATURE EXTREMES

Location		Maximum Temperature
Chassis	—Isolated Areas	+ 85°C (+185°F)
	—Exposed to Heat Sources	+121°C (+250°F–1200°F)
	—Exposed to Oils	+177°C (+350°F)
Exterior		+121°C (+250°F)
Underhood	—Dash Panel	140°C (285°F)
	—Engine (Typical)	150°C (300°F)
	—Choke Housing	205°C (400°F)
	—Starter Cable Near Manifold	205°C (400°F)
	—Exhaust Manifold	650°C (1200°F)
Interior	—Floor	85°C (185°F)
	—Rear Deck	107°C (225°F)
	—Instrument Panel (Top)	113°C (236°F)
	—Instrument Panel (Other)	85°C (185°F)
Door Interior	—No data available	
Trunk		85°C (185°F)
Minimum Temperature		**−40°C (−40°F)**

transition rates greater than that defined in Fig. 2. Direct heat conduction from the temperature chamber heating element to the specimen should be minimized.

Electrical performance should be measured under the expected operational minimum and maximum extremes of excitation, input and output voltage and load at both the cold and hot temperature extremes. These measurements will provide insight into electrical variations with temperature.

Thermal shock normally expected in the automotive environment is simulated by the maximum rates of change shown on the recommended thermal cycle profile shown in Fig. 2. The proper thermal shock cycle should be determined by analysis of component power dissipation, expected rate of temperature change at its location in the system and the overall ambient operating temperature. In general, thermal shock is most severe when equipment is operated intermittently in low temperature environments. The effects of thermal shock include cracking and delamination of substrates, seal failures, wire bond breaks, and operating characteristic changes.

Thermal stress is caused by repeating cycling through the thermal profile of Fig. 2. The number of cycles required is a function of the equipment application. Functional electrical testing during temperature transitions or immediately after temperature transitions, is a means of detecting poor electrical connections. The effects of thermal stress are similar to thermal shock but are caused by fatigue.

NOTE: Although uniform oven temperatures are desirable, in some vehicle environments the only means of heat removal may be by special heat sinks or by free convection to surrounding air. It may be necessary to use conductive heat sinks with independent temperature controls in the former case and baffles or slow speed air stirring devices in the latter to simulate such conditions in the laboratory. (See Section 3.)

4.1.4 RELATED SPECIFICATIONS—A generally accepted method for small part testing is defined in MIL-STD-202E, Method 102A, Temperature Cycling. The short dwell periods at extreme temperature are satisfactory where temperature stabilization has been verified by actual measurements, thermocouples, etc.

4.2 Humidity

4.2.1, 4.2.2 DEFINITION AND EFFECTS ON PERFORMANCE—Both primary and secondary humidity sources exist in the vehicle. In addition to the primary source, externally applied ambient humidity, the cyclic thermal-mechanical stresses caused by operational heat sources, introduce a variable vapor pressure on the seals. Temperature gradients set up by these cycles can cause the dew point to travel from locations inside the equipment to the outside and back, resulting in additional stress on the seal.

The actual relative humidity in the vehicle depends on location due to operational heat sources, trapped vapors, air conditioning, and cool-down effects. Recorded data indicates an extreme condition of 98% relative humidity at 38°C (100°F).

Primary failure modes include corrosion of metal parts due to galvanic and electrolytic action, as well as corrosion due to interaction with water and due to adverse pH changes. Other failure modes include changes in electrical properties, surface bridging between circuits, and decomposition of organic matter.

4.2.3 RECOMMENDED TEST METHODS—The most common way to determine the effect of humidity on electronic equipment is to overtest and examine any failures for relevance to the more moderate actual operating conditions. Three general test methods are recommended. The most common is an active temperature humidity cycling under accelerated conditions. The second is a 10-day soak at 95% relative humidity and 38°C (100°F) temperature. A third method is an 8–24 h exposure at 103.4 kPa gauge pressure (15 psig) in a pressure vessel. This is a quick and effective method of uncovering defects in plastic encapsulated semi-conductors.

There are many acceptable accelerated humidity test cycles, including MIL-STD-202E, Method 103B; however, the test cycles in Fig. 3 are recommended as the most useful.

An optional frost condition may be incorporated during one of these humidity cycles (Fig. 3A). Electrical performance should be continuously monitored during these frost cycles to note erratic operation. Heat-producing and moving parts may require altering the frost condition portions of the cycle to allow a period of non-operation and induced frosting.

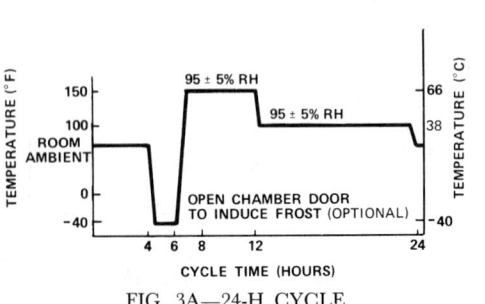

FIG. 3A—24-H CYCLE

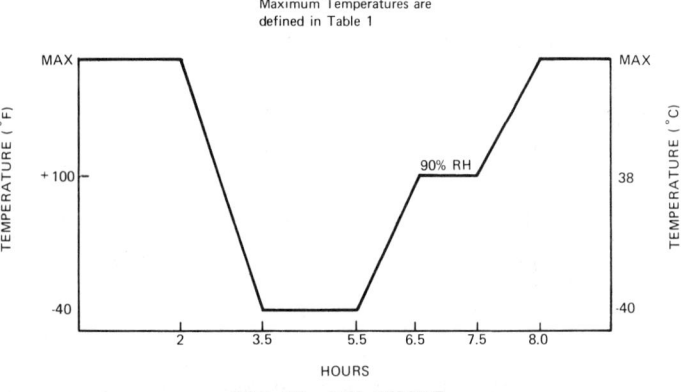

FIG. 3B—8-H CYCLE

FIG. 3—RECOMMENDED HUMIDITY CYCLES

The 10-day soak is normally conducted with equipment non-operating. Equipment that operates with standby voltage excitation and a low current drain when the ignition is off is a significant exception. Examples of this type include seat belt interlocks and electronic clocks. Samples of such equipment should be tested with normal standby conditions. Accelerated humidity effects should be expected under the conditions of high temperature, high humidity, and excitation voltage.

4.2.4 RELATED SPECIFICATIONS—A number of related humidity specifications are recommended for review and reference. The first: MIL-STD-810B, Method 507, Procedure 1, Humidity; is a system-oriented test method. The second; a modified version of MIL-STD-202E, Method 103B, Humidity (Steady State); is intended to evaluate materials. The third; MIL-STD-202E, Method 106D, Moisture Resistance; is a procedure for testing small parts.

4.3 Salt Atmosphere

4.3.1, 4.3.2 DEFINITION AND EFFECT ON PERFORMANCE—Electronic equipment mounted on the chassis, exterior, and underhood are often exposed to a salt spray environment. In coastal regions, the salt is derived from sea breezes and in colder climates, from road salt. Although salt spray is generally not found in the interior and trunk of the vehicle, it is advisable to evaluate the potential effects of saline solutions on the floor area as the result of transfer from the outside environment by vehicle operators, passengers, and transported equipment.

Failure modes due to salt spray are generally the same as those associated with water and water vapor. However, corrosion effects and alteration of conductivity are accelerated by the presence of saline solutions and adverse changes in pH.

4.3.3 RECOMMENDED TEST METHODS—The recommended test method for measuring susceptibility of electronic equipment to salt spray is the American Society for Testing and Materials (ASTM) Standard Method of Salt Spray (Fog) Testing-Number B 117-73.

The test consists of exposing the electronic equipment to a solution of 5 parts salt to 95 parts water atomized at a temperature of 35°C (95°F). The equipment being tested should be exposed to the salt spray for a period of from 24-96 h. The actual exposure time must be determined by analysis of the specific mounting location. When the tests have been concluded, the test specimens should be gently rinsed in clean running water, about 38°C (100°F) to remove salt deposits from their surface and then immediately dried. Drying should be done with a stream of clean, compressed dry air at about 241.3-275.8 kPa gauge pressure (35-40 psig). The equipment should then be tested under nominal conditions of voltage and load throughout the test.

NOTE: The Pascal (Pa) is the designated SI (metric) unit for pressure and stress. It is equivalent to 1 N/m^2.

Where leakage resistance values are critical, appropriate measurements in both the wet and dry states may be necessary.

4.3.4 RELATED SPECIFICATIONS—ASTM B 117-73, Salt Spray (Fog) Testing, is the recommended test method.

4.4 Immersion and Splash (Water, Chemicals, and Oils)

4.4.1 DEFINITION—Electronic equipment mounted on or in the vehicle is exposed to varying amounts of water, chemicals, and oil. A list of potential environmental chemicals and oils includes:

Engine Oils and Additives
Transmission Oil
Rear Axle Oil
Power Steering Fluid
Brake Fluid
Axle Grease
Washer Solvent
Gasoline
Anti-Freeze Water Mixture
Degreasers
Soap and Detergents
Steam
Battery Acid
Water and Snow
Salt Water
Waxes
Freon
Spray Paint
Ether
Vinyl Plasticizers
Undercoating Material

The modified chemical characteristics of these materials when degraded or contaminated should also be considered.

4.4.2 EFFECT OF PERFORMANCE—Loss of the integrity of the container can result in corrosion or contamination of vulnerable internal components. The chemical compatibility can be determined by laboratory chemical analysis. Devices that may be immersed in fluids for a long period, such as sensors, should be subjected to laboratory life tests in these fluids.

4.4.3 RECOMMENDED TEST METHODS—The equipment designer should first determine whether the parts must withstand complete immersion or splash, and which fluids are likely to be present in the application. Immersion and splash tests are generally performed following other environmental tests because this sequence will tend to aggravate any incipient defects in seals, seams, and bushings which might otherwise escape notice.

Splash testing should be done with the equipment mounted in its normal operating position with all drain holes, if used, open. The sample is subjected to precipitation of 0.25 cm (0.1 in)/min delivered at an angle 45 deg below and above the sample with a nozzle having a solid cone spray.

During immersion testing, most commonly utilizing water as the fluid, the equipment ordinarily is not operated due to setup logistics and techniques of testing. Electrical tests should, therefore, be performed immediately before and after this test. In this test, the electronic equipment in its normal exterior package is immersed in tap water at about 18°C (65°F). The test sample should be completely covered by the wter. The sample is first positioned in its normal mounting orientation. It remains in this position for 5 min and then is rotated 180 deg. It should remain in that position for 5 min and then be rotated 90 deg about the other axis where it remains for 5 min. Immediately after removal, the sample should be exposed to some temperature below freezing until the entire mass is below freeezing. The sample is then returned to room temperature, air dried, functionally tested, and inspected for damage.

More severe tests such as combined temperature, pressure, and continuous fluid contact must be considered for equipment subjected to extreme environments as in the case of exposure to coolant water, brake fluid, and transmission oil. Caution must be used in specifying combined tests as they may be unrealistically severe for many applications.

4.5 Dust, Sand, and Gravel Bombardment

4.5.1 DEFINITION—Dust is a significant environment for chassis, underhood, and exterior-mounted devices; and can be a long-term problem in interior locations, such as under the dash and seats. Sand, primarily windblown, is an important environmental consideration for chassis, exterior, and underhood. Bombardment by gravel is significant for chassis and exterior-mounted equipment.

4.5.2 EFFECT ON PERFORMANCE—Exposure to fine dust can cause problems with moving parts, form conductive bridges, and act as an absorbent material for the collection of water vapor. Some electromechanical components may be able to tolerate fine dust, but larger particles may affect or totally inhibit their mechanical action. While the exposure in desert areas is severe, exposure to a reasonable amount of road dust is common to all areas.

4.5.3 RECOMMENDED TEST METHODS—Dust, sand, and gravel bombardment tests should be at room temperature and the sample need not be operating, although functional tests should be performed prior to and after testing.

Dust conforming to that defined in SAE J726b (November, 1976), coarse grade should be used. If this dust packs or seals openings in the test sample or if the sample contains exposed mechanical elements, the following alternate dust mixture may be used:

J726b Coarse or Equivalent	70%
120 Grit Aluminum Oxide	30%

Components should be placed in a dust chamber with sufficient dry air movement to maintain a concentration of 0.88 g/m^3 (0.025 g/ft^3) for a period of 24 h.

An alternate method is to place the sample about 15 cm (6 in) from one wall in a 3-ft cubical box. The box should contain (10 lb) 4.54 Kgm of fine powdered cement in accordance with ASTM C150-56, specification for Portland Cement. At intervals of 15 min, the dust must be agitated by compressed air or fan blower. Blasts of air for a 2-s period in a downward direction assure that the dust is completely and uniformly diffused throughout the entire cube. The dust is then allowed to settle. The cycle is repeated for 5 h.

The recommended test for susceptibility of equipment to damage from gravel bombardment is SAE J400 (July, 1968), Recommended Practice Test for Chip Resistance of Surface Coatings. This document is intended to detect susceptibility of surface coatings to chipping, but the basic test equipment and procedures are useful for evaluation of the electronic equipment. The test consists of exposing the test sample to bombardment by gravel 0.96-1.6 cm ($\frac{3}{8}$-$\frac{5}{8}$ in) in diameter for a period of approximately 2 min. The sample is positioned about 35 cm (13$\frac{3}{4}$ in) from the muzzle of the gravel source. 470 cm^3, (approximately 1 pt) of gravel (250-300 stones) is delivered under a pressure of 483 kPa gauge pressure (70 psig) over an approximate 10-s period. The process is repeated 12 times for a total exposure of 2 min. Judgment must be used in determining which sides should be exposed to the bombardment. Certainly all forward-facing surfaces not shielded by other parts are included. In many cases, the bottom and sides should also be exposed.

4.5.4 RELATED SPECIFICATIONS—Three specifications are referenced. The first: MIL-STD-202E, Method 110A, Sand and Dust, is a piece part test and is included for information and comparison. The second is SAE J726b (November, 1976), Air Cleaner Test Code, which defines the recommended dust. It also describes some test apparatus. The third specification is SAE J400 (July, 1968), Test for Chip Resistance for Surface Coatings, which is recommended in part for a gravel bombardment guide. Continued integrity at the conclusion of the exposure is the passing criteria.

4.6 Altitude

4.6.1 DEFINITION—With the exception of air shipment of unenergized controls, operation in the vehicle should follow the anticipated operating limits. Completed controls are expected to be stressed over these limits of absolute pressure:

Condition	Altitude	Atmospheric Pressure
Operating	3.7 km (12 000 ft)	62.1 kPa absolute pressure (9 psia)
Non-operating	12.2 km (40 000 ft)	18.6 kPa absolute pressure (2.7 psia)

4.6.2 EFFECT ON PERFORMANCE—With increased altitude the following effects are generally observed:

4.6.2.1 Reduction in convection heat transfer efficiency.

4.6.2.2 Change in mechanical stress on packages which have internal cavities. The reference cavity of an absolute pressure sensor is an example of this.

4.6.2.3 A very noticeable reduction in the high voltage breakdown characteristics of systems with electrically stressed insulator, conductor or air surfaces; this may result in setup of surface tracking with eventual component failure.

4.6.3 RECOMMENDED TEST METHODS—The recommended test method is to operate equipment during the thermal cycles described in the Temperature Test Section, but with the added parameter of 62.1 kPa absolute pressure (9 psia) pressure. The equipment should operate under maximum load. Failure effects will be similar to those experienced with thermal cycle and shock. Non-operating tests should be done at a minimum temperature of −51°C (−60°F) if possible.

4.7 Mechanical Vibration

4.7.1 DEFINITION—Vibration, which is prevalent whenever the vehicle engine or suspension system is in motion, is a key factor in the automotive environment. The intensity varies from low severity at smooth engine idle to extreme severity when traversing rough roads at high speed. Vibration also varies with location. Detailed data is included in Figs. 11–18.

4.7.2 EFFECT ON PERFORMANCE—A number of modes of degradation or failure are possible under applied vibration. A partial list includes:

4.7.2.1 Loss of wiring harness electrical connection due to improper connector design and/or assembly.

4.7.2.2 Excitation of tuned mass harmonic vibration within the equipment which eventually leads to failure due to metal fatigue at stress concentration points.

4.7.2.3 Failure of mounting structure due to the added acceleration forces acting on the mass of the equipment.

4.7.2.4 Mechanical flexure at seal and other interface areas which promotes the intrusion of other environmental factors, such as moisture, in a manner similar to the phenomena described under temperature cycling effects.

4.7.2.5 Temporary abberation of equipment performance due to acceleration forces on control component masses. Two examples illustrate this:

4.7.2.5.1 Sensor measurement error due to motion of the sense element. An example of this is a pressure sensor which gives incorrect information under some applied frequencies due to the mass of a diaphragm-spring mechanism.

4.7.2.5.2 False operation of electromechanical components—e.g., a relay whose contacts close or open, due to vibratory movement of its armature's mass.

The designer should be particularly alert to failures which are intermittent or which cause faulty operation during applied vibration. Many malfunctions of this type revert to normal operation after the vibration excitation is removed. It is, therefore, recommended that electronic performance tests be conducted during vibration tests for those functions which must perform under this condition. In most cases this is only practical under laboratory simulation of the road test situation.

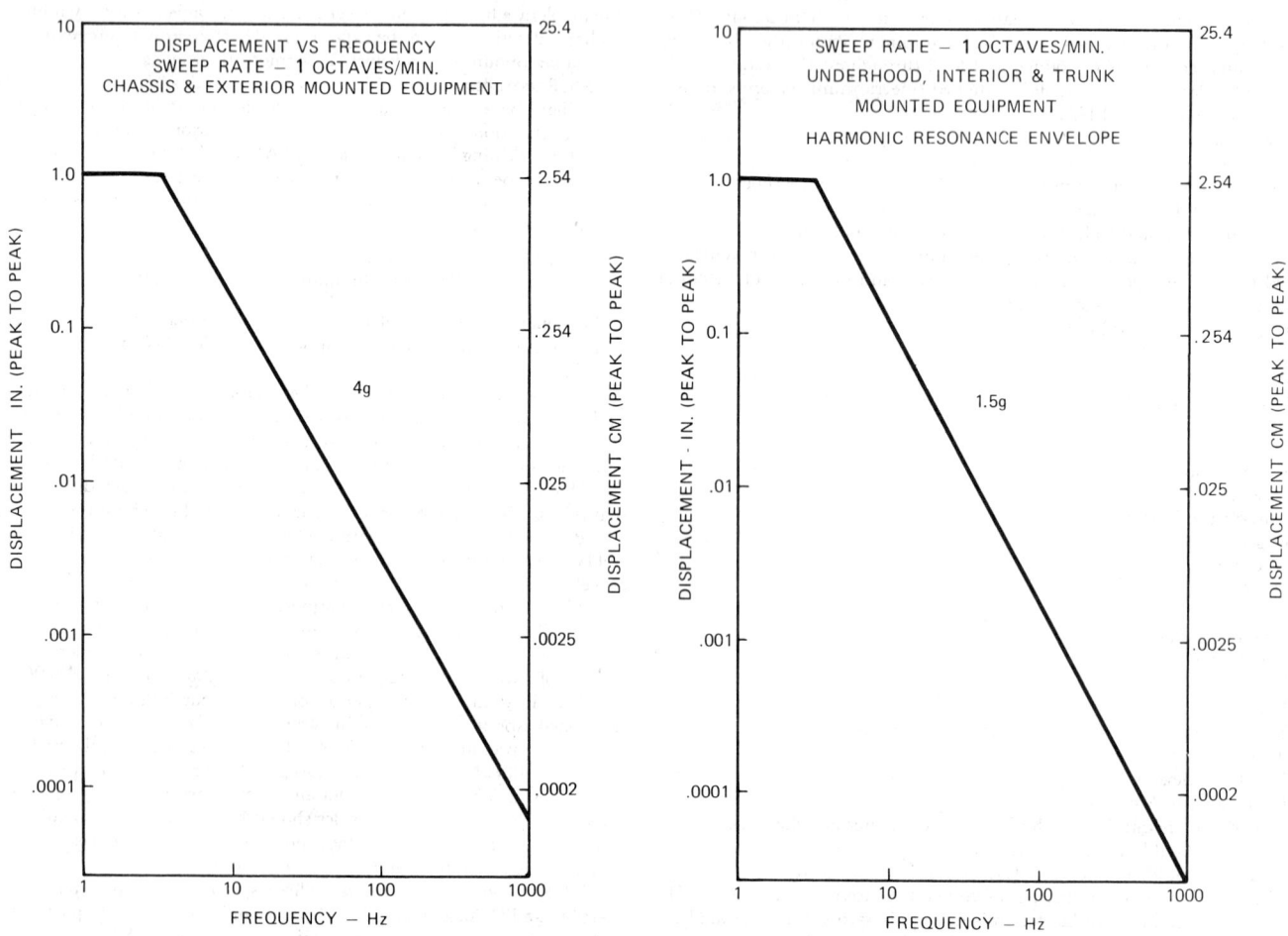

FIG. 4—RECOMMENDED RESONANT SEARCH VIBRATION PROFILE

4.7.3 RECOMMENDED TEST METHODS—A typical test for this environmental factor has been operation of a test vehicle over a group of severe road test track conditions. These include surfaces described as the Belgium Block Road, the Hop, the Tramp, the Square Block Test Course, and other complex surfaces. These courses are excellent test beds for complete transportation packages installed in the vehicle. Unfortunately, they are relatively inconvenient for electronic control evaluation during the design phase. In many cases electronic equipment exhibit intermittent or degraded performance during vibration and returns to normal operation when the excitation is removed.

Failure of electronic equipment in a vibratory environment may be the result of fixed frequency or random vibrations. Current practice within the industry is to conduct a resonant search up to 1000 Hz and then dwell at the major resonances if they are applicable to the operating environment spectrum.

Fig. 4 shows the recommended amplitude and sweep rate for this search. This profile is primarily gravity unit-oriented. A second recommended procedure is to sweep from 10–55 Hz and return in 1 min at an amplitude determined by measurements taken at the proposed mounting location. The test is conducted in each of three mutually perpendicular planes.

Experience has shown that in some cases random vibration may be a valuable approach in uncovering electronic equipment failure modes. While random testing is more difficult and costly, consideration should be given to this approach where required.

In the time sweep and resonant dwell, vibration must be conducted in each of three mutually perpendicular axes. Test duration must be determined by the equipment designer.

4.7.4 RELATED SPECIFICATIONS—Three specifications are referenced. The first, MIL-STD-202E, Method 201A, Vibration, and the second, MIL-STD-202E, Method 204C, Vibration, High Frequency, are concerned with sine vibration and offer procedural details and information on resonant dwell periods. The third, MIL-STD-202E, Method 214, Random Vibration, offers similar information on the random vibration approach.

4.8 Mechanical Shock

4.8.1, 4.8.2 DEFINITION AND EFFECT ON PERFORMANCE—The automotive shock environment is logically divided into four classes:
Shipping and handling shock.
Installation shock.
Operational shock.
Crash shock.

Shipping and Handling Shocks—These are similar to those encountered in non-automotive applications.

Installation Shock—It is common production-line practice to lift and carry equipment by its harness. Therefore, it is recommended that the harness design assure for secure fastening and suitable strain relief.

Operational Shocks—The shocks encountered during the life of the vehicle that are caused by curbs, pot holes, etc. can be very severe. These vary widely in amplitude, duration and number, and the test condition can only be generally simulated.

Crash Shock—This is included as an operating environment for safety systems. The operational requirements of these systems are limited to longitudinal shock at the present time.

4.8.3 RECOMMENDED TEST METHODS

Bench Handling Shock—The component shall be placed on a solid wooden bench top at least 3.4 cm (1⅝ in) thick. The test shall be performed as follows: using one edge as a pivot, lift the opposite edge of the component until one of the following conditions shall first occur:

(a) The component forms a 45 deg angle with bench top.
(b) The lifted edge is just below the balance point. The component shall be allowed to drop to the bench top. Repeat using other practical edges of the same face as pivot points. The procedure is then repeated with the component resting on other faces until it has been dropped on each face that the component might normally be placed when bench handling or servicing.

Transit Drop Test—The drop shall be from a height of 122 cm (48 in) onto a solid 5 cm (2 in) thick plywood base backed by concrete or a rigid steel frame with the test sample properly installed in its shipping container. The drop shall be performed on each face, edge, and corner.

Installation Shock Test—A recommended test is to support the device and the far end of the harness at the same elevation, then release the device. Care should be taken to prevent the equipment from striking another object during this test. The drop should be repeated and the harness terminals or strain relief area inspected for damage.

Operation Shock—With the possible exception of collision, the most severe shock anticipated after production line installation is encountered when driving over complex road surfaces. The complex profile that was used to derive this test profile consists of a rise in the roadway followed by a depression or dip. Upon leaving the dip at 48 km/h (30 mph), the vehicle will often become airborne. The severe shock is experienced when the vehicle returns to the roadway. Fig. 5 shows the shock measured on a steering column just below

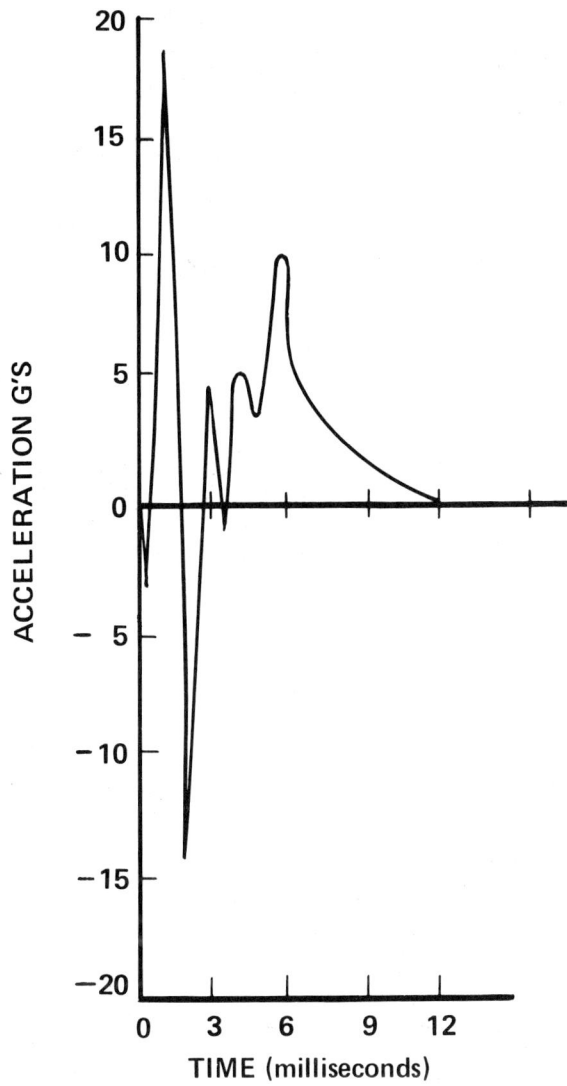

FIG. 5—OPERATIONAL SHOCK PROFILE

the steering wheel. The accelerometer was mounted with its sensitive axis perpendicular to the axis of the column and in the vertical plane.

While this location is not typical of component mounting locations, it probably represents the most severe operational shock environment. This information is provided for guidance only; there are no generally accepted test procedures at the present time.

Crash Shock Test—Only limited and preliminary data on the effects of crash shock on the electronic equipment environment are available. However, a representative deceleration profile for a 48 km/h (30 mph) barrier crash is shown in Fig. 6. The following factors vary with each installation and should be considered in pretest analysis:

(a) Equipment mass.
(b) Mounting system.
(c) Structure of the associated vehicle (crush distance, rate of collapse, etc.)
(d) Particular engine package.
(e) Direction of crash.

4.8.4 RELATED SPECIFICATIONS—Two specifications are recommended for consideration. The first, MIL-STD-202E, Method 203B, Random Drop, is designed to uncover failures that may result from the repeated random shocks that occur in shipping and handling. It is an endurance test. The second, MIL-STD-202E, Method 213B, Shock (Specified Pulse), is intended to measure the effect of known or generally accepted shock pulse shapes. It is intended that operational shock be reduced into a standard pulse shape to achieve a repeatable test method.

4.9 General Automotive Electrical Environment

Factors unique to the automobile that make the vehicular environment more severe than that encountered in most electrical equipment applications are:

Interaction with other vehicular electronic/electrical systems.
Voltage variations.

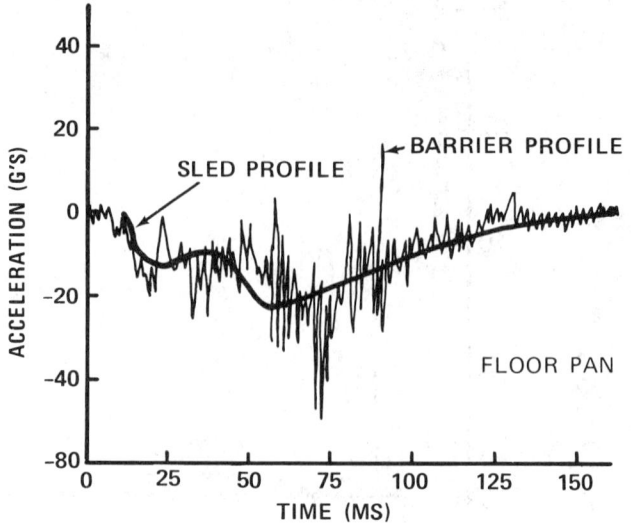

FIG. 6—48 KM/H (30 MPH) BARRIER AND SLED SHOCK PROFILES

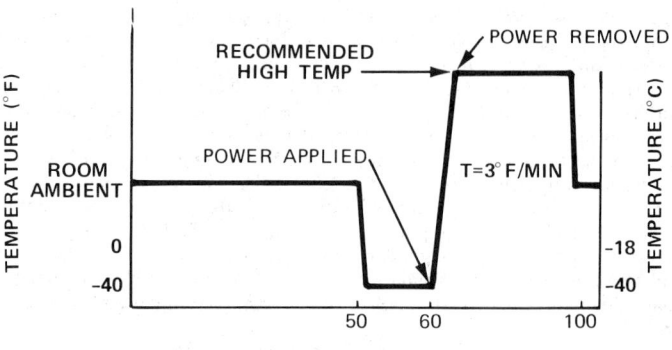

FIG. 7—COMBINED THERMAL AND ELECTRICAL STRESS PROFILE

Customer added equipment.
Lack of maintenance.
Complex external electromagnetic fields.

Discussion of the electrical environment falls into two categories:

(a) Electrical, Steady-State—Including variations in applied vehicle DC voltages with a characteristic frequency of below 1 Hz.

(b) Electrical, Transient, and Noise—Including all noise and high voltage transient with characteristic frequencies above 1 Hz.

These conditions are discussed in Sections 4.10 and 4.11 respectively.

4.10 Steady State Electrical Characteristics

4.10.1 DEFINITION—A normally operating vehicle will maintain supply voltages ranging from +11–+16 VDC. However, under certain conditions, the voltage may fall to approximately 9 VDC. This might happen in an idling vehicle which has a heavy electrical load (lights and air conditioning) and a partially depleted battery. Therefore, depending upon the application, the designer/user may wish to specify the +9–+16 VDC range. For specific equipment such as those that must be functioning during engine start, voltage may be specified as appropriate. Cold starting with a partially depleted battery charge at −40°C (−40°F) can reduce the nominal 12 V voltage to between 4.5 and 6 VDC.

Another condition affecting the DC voltage supply is developed when the vehicle voltage regulator fails, causing the alternator to drive the system 18 V. Extended 18 V operation will eventually cause boil-off of the battery electrolyte resulting in voltages as high as 75–130 V. Other charging system failures could result in lower than normal battery voltages. The general steady state voltage regulation characteristics are shown in Table 2.

Emergency starts by garages and emergency road services sometimes utilize 24 V sources, and there have been reports of 36 V being used for this purpose. High voltages such as these are applied for up to 5 min and sometimes even with reverse polarity. The use of voltages which are above the vehicle system voltage can damage components in a vehicular electrical system, and the higher the voltage, the greater the likelihood of damage. A designer cannot cope with ever-increasing excitation potentials, and the above values usually are not a part of his design criteria. The possibility of the use of voltages above system voltage is included here for information only.

4.10.2 EFFECT ON PERFORMANCE—Equipment that must operate during the starting condition is generally designed to perform with slight degradation over a wide range of voltage. The designer is alerted to the possibility of failure from a combination of voltage and temperature variation. Over-voltage and high temperature, both from the external environment and internal dissipation, may cause excessive heat and result in failure. Under-voltage will probably result in degraded or non-performance. Conditions must be carefully examined to determine the true temperature and excitation voltage of the equipment.

4.10.3 RECOMMENDED TEST METHODS—Critical automotive equipment is performance-tested for operation within predetermined limits. Samples are also subjected to combinations of temperatures and supply voltage variation which are designed to represent the worst case stresses on control components. A typical cycle for this form of test is shown in Fig. 7.

The voltage applied and removed at the two points shown in Fig. 7 is generally 16 V, the maximum normal voltage. If the test is performed for the high voltage *booster battery* start condition of 24 V, a narrower temperature range is used. This is a destructive test which is often used as an indication of basic design environmental capability. The number of cycles expected before failure, the actual limit values for temperature and voltage, and the period of each cycle are dependent on the design goals for the equipment being considered.

Samples of finished units are generally tested for extended operation at the peak voltage/temperature combination expected at the equipment's location. In the absence of actual temperature measurements, the values in Table 1 are recommended. These tests often run for extended periods and are particularly stringent for equipment in the underhood environment.

4.11 Transient Noise and Electrostatic Characteristics

4.11.1, 4.11.2 DEFINITION AND EFFECT ON PERFORMANCE—Four principal types of transients are encountered on automobile wire harnesses. These are load dump, inductive switching transients, alternator field decay, and mutual coupling. Generally, they occur singly, but there are cases where the latter two could occur simultaneously. EMC characteristics vary considerably with type of vehicle and wiring harness. The equipment user and/or designer should determine the actual values of peak voltages, peak current, source impedance, repetition rate, frequency of occurrence at the interface between his equipment and the electrical distribution system, then design and test the electronic equipment to withstand values consistent with the expected use. Table 3 summarizes typical transient characteristics.

TABLE 2—AUTOMOTIVE VOLTAGE REGULATION CHARACTERISTICS

Condition	Voltage
Normal operating vehicle	16 V max
	14.2 V nominal
	9 V min[a]
Cold Cranking at −40°C (−40°F)	4.5–6.0 V
Jumper Starts	+24 V
Reverse Polarity	−12 V
Charging System Failure	<9–18 V
Battery Electrolyte Boil-Off	75–130 V

[a] See Section 4.10.1 for a definition of normal voltage.

TABLE 3—AUTOMOTIVE TRANSIENT VOLTAGE CHARACTERISTICS

Type	Max Amplitude (V)	Characteristic	Remarks
Load Dump	120	$106\epsilon^{-t/0.188} + 14$	Damage potential
Inductive Load Switching	−286	$-300\epsilon^{-t/0.001} + 14$ followed by +80 Volt excursion	Logic Errors
Alternator Field Decay	−90	$-90\epsilon^{-t/0.038}$	Occurs at Shutdown Only
Mutual Coupling	214	$+200\epsilon^{-t/0.001} + 14$	Logic Errors

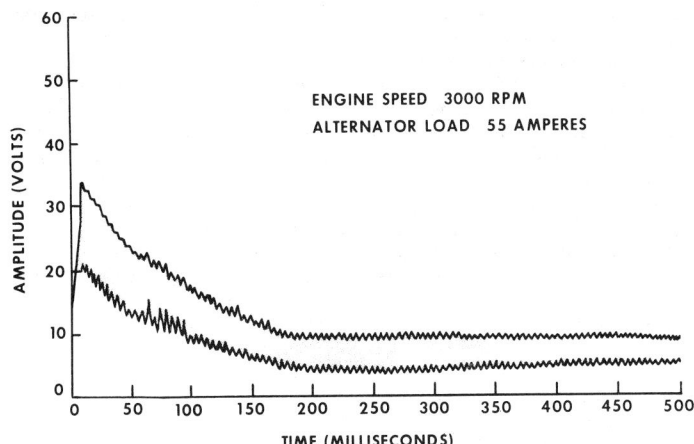

FIG. 8—LOAD DUMP TRANSIENT

TABLE 4—SUMMARY OF AUTOMOTIVE ELECTRICAL CONTINUOUS NOISE CHARACTERISTICS

Type	Max Amplitude	Duration	Repetition Rate	Remarks
Normal Accessory Noise	1.5 V Peak	Frequency	50 Hz–10 kHz	Total Pulse Height is 3 V-PP
Normal Ignition Pulses	3 V Peak	10–15 μs	Dependent on engine speed	Total Pulse Height is 6 V-PP
Abnormal Ignition Pulses	75 V Peak	~90 μs	Dependent on engine speed	
Transceiver Feedback	15–20 mV	Carrier	Frequency	Sinusoid

Load Dump Transient—Load dump occurs when the alternator load is abruptly reduced. This sudden reduction in current causes the alternator to generate a positive voltage spike. The worst case load dump is caused by disconnecting a discharged battery when the alternator is operated at rated load. Using the discharged battery load to create the load dump creates the worst situation for two reasons:

(a) The battery normally acts like a capacitor and absorbs transient energy when it is in the circuit.

(b) The partially discharged battery forms the single greatest load on the alternator and, therefore, disconnecting it creates the greatest possible step load change.

This transient may be the most severe encountered in the automobile and can result in component damage. In the practical case, it is most often initiated by defective battery terminal connections. Transient voltages of as high as 125 V or more have been reported with rise times of approximately 100 μs. Reports of decay time vary from 100 μs–4.5 s. The long duration decay occurs during vehicle turn off with a disconnected or dry vehicle battery. However, even the shortest time (100 ms) is relatively long, requiring that significant energy must be dissipated. Fig. 8 shows oscillograms of more typical load dump transients.

The load dump transient contains considerable electrical energy which must be safely dissipated to prevent damage to electronic equipment. This transient occurs randomly in time appearing as individual or repetitive pulses at random unknown rates due to vibration.

Inductive Load Switching Transient—Inductive transients are caused by solenoid, motor field, air conditioning clutch, and ignition system switching. These occur during vehicle operation whenever an inductive accessory is turned off. The severity is dependent on the magnitude of switched inductive load and line impedance. Unfortunately, measurements to date have not been taken with standardized procedures and were most probably observed with different loads.

These transients generally take the form of a large negative peak, followed by the smaller damped positive excursion. The highest reported by the data acquisition task force is $-300/+80$ V with an effective duration of 320 ms. Transients of this nature may cause component damage or introduce logic or functional computational errors.

Alternator Field Decay Transient—This is a special case of the inductive load switching transient. It is a negative pulse caused by alternator field decay and may occur when the field is disconnected from the battery as the ignition switch is turned to the *off* position. The amplitude is dependent on the voltage regulator cycle and load at the time of shutdown, varying from -40 to -100V and a duration of 200 ms.

Coupling—Coupling is not, strictly speaking, a generator of transients, but a mechanism which is capable of introducing transients into circuits not directly connected to the transient source. There are three general coupling modes in the automobile: magnetic, capacitive, and conducted. Briefly, the automobile coupling problems are caused by long harnesses, nonshielded conductors, and common ground return impedances. Long harnesses are one of the principal coupling media that distribute transients throughout the automobile (Ref. 2). When a number of wires are bundled into a harness and a step change in current or voltage occurs, inductive or capacitive coupling, between the conductor experiencing the change and the other wires, can result.

Other Effects—It is possible that inductive switching of certain solenoids and the alternator decay transient condition occur simultaneously. This hypothesis would account for the higher voltage transients that have been reported, but not explained. Measurement of 600 V transients on engine shutdown have been reported. Also to be considered are noise suppression capacitors that are sometimes placed on the fuse block, and some accessories that are applied to quiet interference on the entertainment radio. In some cases, these capacitors may form tuned circuits with automotive inductive loads, causing high voltage transient conditions.

Certain devices, with high levels of stored energy, such as coasting permanent magnetic motors, may maintain line voltage for a finite interval of time after the ignition is shut off. Some equipment may perform in an unsatisfactory mode of operation under such conditions.

NOTE: Direct conduction through common circuits constitutes the most frequent path by which transients are introduced into electronic equipment.

Electrical Noise—Noise will normally have a repetition rate which is dependent on the characteristics of the interferring device or engine speed. There are four general types as summarized in Table 4. A typical oscillogram of automotive electrical noise is shown in Fig. 9.

Normal Accessory Noise—Generally, the normal compliment of accessories contributes less than 1.5 V peak over a frequency range of 50 Hz–10 kHz.

Normal Ignition Pulses—Normal ignition pulses can cause 3 V peak pulses of 10–15 μs duration at a repetition rate dependent on engine speed.

Abnormal Ignition Pulses—Normally, the battery acts as a low impedance path to ground for the voltage pulse caused by the primary and secondary windings of the ignition coil. If the battery is disconnected, the repetitive voltage pulses will increase to a significant amplitude. Under this condition, there have been reports of voltages as high as 75 V peak and 90 ms duration with the repetition rate dependent on engine speed. The energy level is substantial and component damage is possible.

Since this condition can occur simultaneous with load dump, consideration should be given to testing both conditions together.

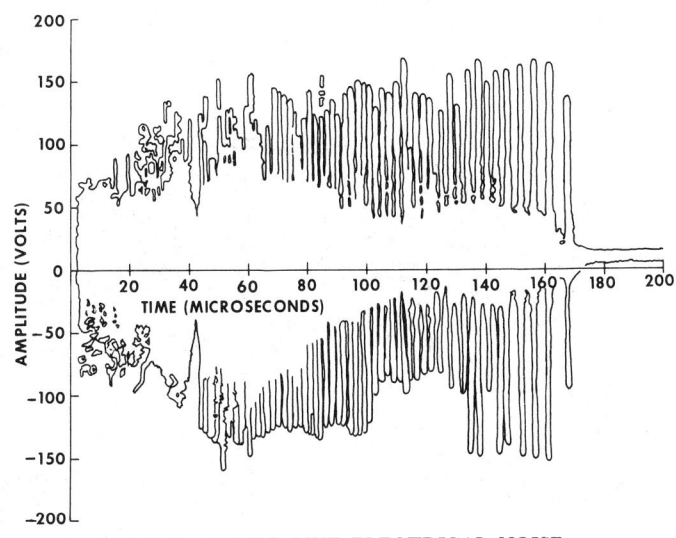

FIG. 9—POWER LINE ELECTRICAL NOISE

Transceiver Feedback—Some automotive transceivers feedback energy to the power line at carrier frequency when the transmitter is keyed. These potentials are small, 15–20 mV peak, and are mentioned here only because they are at a predictable frequency.

Electrostatic Discharge—The electrostatic charge stored by the human body and then discharged into a device may cause operating anomalies. Recent investigations indicate that discharging a 300 pF capacitor that has been charged to a potential of 15 kV through a 5 kΩ resistor is adequate to simulate this effect.

External Sources of Radiated Energy—The vehicle is exposed to radiated energy from a multitude of sources which have the potential to disrupt normal system operation.

A more detailed discussion of these transient and noise effects is available in Ref. 3 and 4.

NOTE: The mechanisms governing the introduction of transients into an electronic assembly or its interrelated components are very complex. The equipment designer/packager must, therefore, be familiar with the configuration of the total vehicle electrical system, e.g., wire routing, shielding, grounding, filtering and decoupling practices and equipment locations.

5. Environmental Extremes by Location—This section quantifies guidelines for the extreme operating conditions for five major in-vehicle equipment mounting sites:

(a) Underhood
  1. Engine Compartment
  2. Bulkhead—dash panel
(b) Chassis
(c) Exterior
(d) Interior
  1. Instrument Panel
  2. Floor
  3. Rear Deck
(e) Trunk

The physical locations of these sites are given in Fig. 10. Each site (denoted by shaded section) is individually discussed together with the following detail:

(a) A table listing extremes of temperature; humidity; salt spray; sand, dust, and gravel; oil and chemical; mechanical shock and vibration and electrical steady and transient; operating conditions.

(b) Comments germane to other operating conditions of interest.

(c) Charts and other information pertaining to the vibration environment.

This section contains data from environmental measurements made by North American vehicle manufacturers or automotive original equipment suppliers. Decisions concerning each environmental factor and the test methods used to determine equipment performance and durability, should only be arrived at after examining the information in Section 4 of this report. In addition, the designer should be satisfied, by referring to pertinent test data, that the particular application falls within the described operating extremes. See Section 3.

5.1 Underhood—Engine—Caution should be exercised in applying electronics equipment in the underhood region because of the wide range of environments. Data is summarized in Table 5.

5.1.1 TEMPERATURE—Equipment in the vicinity of the exhaust system may experience temperature peaks that are beyond the survival limits of many insulation materials and electronic components.

Investigators have found that the lowest peak temperature areas are often forward in the lower compartment, near the interior or exterior radiator support hardware. The exterior has the disadvantage of being subject to more splash with resultant potential for moisture intrusion, corrosion, or thermal shock.

The heat flow temperature control mechanism for typical engine-mounted equipment relies heavily on the conduction of heat via the engine mass rather than convection via fins projecting into the airflow. Equipment thermally interlocked by conduction with the engine, has two advantages during normal operation:

1. During engine operation, the upper temperature limit is set by the coolant peak temperature, which is in turn controlled by the thermostat.

2. The time rate of change of temperature is limited by the combined engine and coolant system thermal mass.

5.1.2 PEAK TEMPERATURE (HEAT SOAK) TEST—The temperature profile varies widely with individual engine/body combinations. Therefore, it is impossible to specify all possible operating conditions. Generally, worst case temperature operating conditions should be obtained by instrumenting a proposed location for the following operating conditions:

5.1.2.1 The largest engine installation expected in that body style.
5.1.2.2 Peak ambient temperature.
5.1.2.3 Air condition ON.

The vehicle is driven at highway speed for about 20 min and then parked. Underhood temperatures are monitored for the *heat soak* conditions as the thermal energy stored in the engine system is released in the absence of underhood airflow. Design modifications which contribute thermal energy to the underhood area, such as secondary air thermal reactors or catalytic reactors, should be in place and operating for this test.

Test procedures of this type have revealed that the region to the rear of the engine compartment, and the locations near radiated and conducted heat from the exhaust/reactor manifold tend to be much higher in temperature.

Present control practice has limited the location of electronic equipment to temperature situations similar to those shown for the intake manifold, although operation in the vicinity of the alternator heat source will probably add about 10°C (18°F) to the peak 121°C (250°F) shown for the intake manifold. Some experimenters expect the temperature near the radiator support structure to be no higher than 100°C (212°F).

Consideration should also be given to heat flow into the engine compartment from the front wheel suspension/brake and tire combination. Some consideration has been given to electronic equipment thermally interlocked with the engine cooling system, although the high pressure-temperature combination experienced during coolant boil-off may cause unacceptable catastrophic failure.

Rate of temperature change with time is also a consideration in this area, since cold starts will result in very rapid changes, as shown in Fig. 1.

5.1.3 VIBRATION—Vibration profiles recorded on the intake manifold are shown in Fig. 11, together with the equivalent power spectral density profiles.

5.2 Underhood—Dash Panel—Data is summarized in Table 6.

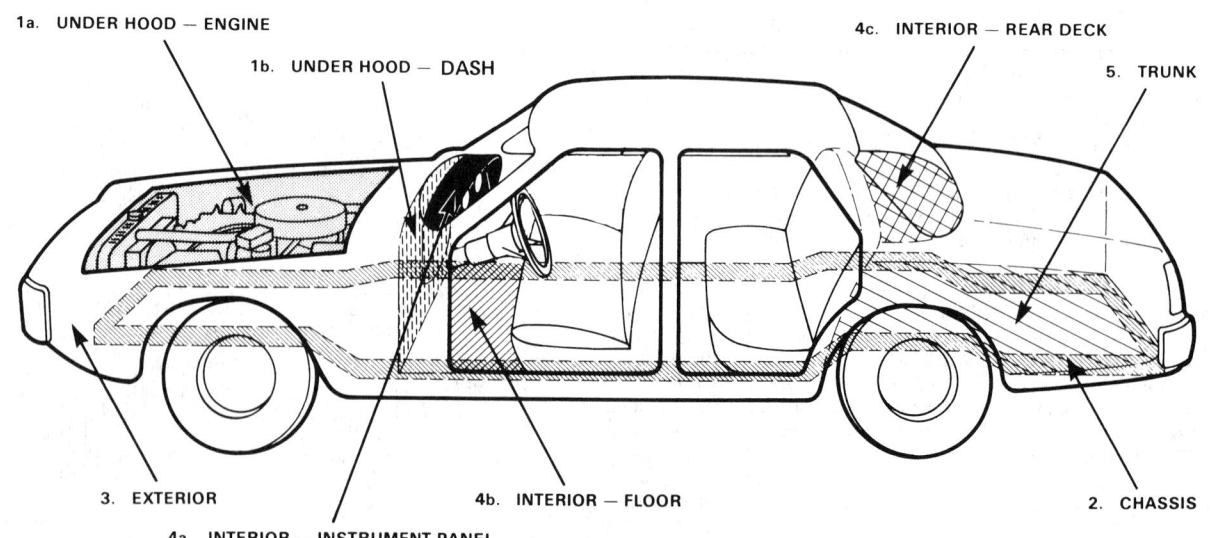

FIG. 10—VEHICLE ENVIRONMENTAL ZONES

UNDERHOOD-ENGINE ENVIRONMENTAL EXTREME DATA

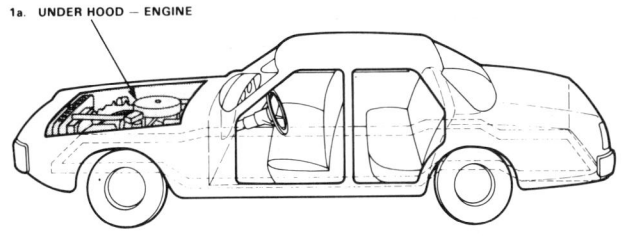

1a. UNDER HOOD – ENGINE

	TEMPERATURE			HUMIDITY (%RH)			SALT SPRAY	IMMERSION	SAND, DUST & GRAVEL	OIL & CHEMICAL	MECHANICAL SHOCK & VIBRATION	ELECTRICAL	
	LOW	HIGH	SLEW RATE	HIGH	LOW	FROST						STEADY-STATE	TRANSIENT
Choke Housing	-40°C (-40°F)	204°C (400°F)		95% at 38°C (100°F)	0	yes	Sect. 4.3	Splash present	Sect. 4.5	Sect. 4.4	Figure 11	Table 2 & Table 4	Table 3
Exhaust Manifold	-40°C (-40°F)	649°C (1200°F)	-7°C/Min. (20°F/Min.)										
Intake Manifold	-40°C (-40°F)	121°C (250°F)											

TABLE 5

5.2.1 TEMPERATURE—Temperature conditions are similar to the Underhood-Engine intake manifold, except that the primary method of heat flow is convection rather than conduction, and the resultant temperature slew rate is less. Equipment in this area generally relies heavily on convection due to the relatively low thermal conduction characteristics and unpredictable thermal interface between the equipment and the dash panel sheet metal. The rate of change in temperature is therefore set by the thermal mass of the equipment itself, and heat flow due to air movement in its vicinity rather than conduction via the mounting surface. Thermal shock due to the impact of cold mud, slush, etc., is not likely in the upper dash panel location. However, consideration should be given to melted snow and ice leakage from the hood/windshield area.

The majority of investigators have experienced peak temperatures of 121°C (250°F), although one data source expects this to be 140°C (285°F). Of course, locations on the dash panel near or just above the exhaust manifold(s) which is at 649°C (1200°F), will experience higher temperatures. The effects of underhood exhaust processing components (catalytic reactors, etc.) will also raise the peak temperatures.

5.2.2 HUMIDITY—This condition is similar to the associated engine condition, with the peak value shown in Table 6. The possibility of snow and ice intrusion, with hot ethylene glycol and water mixtures, due to cooling system failure, should also be considered.

5.2.3 SALT SPRAY—This condition is often a factor, particularly on the lower outboard portions where the dash panel joins the forward floor pan. Driving through salt slush can cause the entrance of salt spray through the radiator. The spray is then delivered to the engine compartment at high velocity by the fan. Spray due to this source is impacted on the dash panel, except for areas shielded by the engine or other underhood components.

5.2.4 IMMERSION—Not generally required.

5.2.5 SAND, DUST, AND GRAVEL—Gravel is not generally a problem, except at the lower dash panel near the transition into the forward floor pan.

5.2.6 OILS AND CHEMICALS—Commonly encountered components (with and without contaminants) are:

    Engine oils and additives (hot and cold)
    Brake fluid
    Gasoline
    Ethylene glycol and water (hot and cold)
    Water and snow
    Waxes
    Transmission oil
    Windshield washer solvent
    Degreasers and cleaning compounds
    Detergents
    Battery acid
    Steam
    Freon

5.2.7 MECHANICAL SHOCK AND VIBRATION—Vibration profiles are shown in Fig. 12.

5.3 **Chassis**—Data is summarized in Table 7.

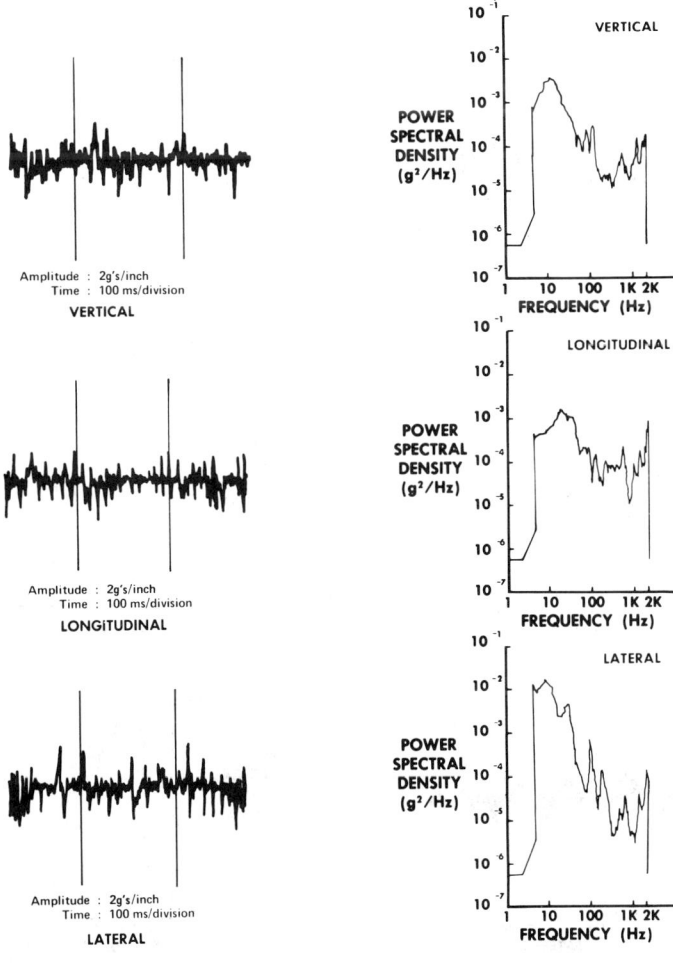

FIG. 11—ENGINE INTAKE MANIFOLD VIBRATION MEASUREMENTS

UNDERHOOD-DASH PANEL ENVIRONMENTAL DATA

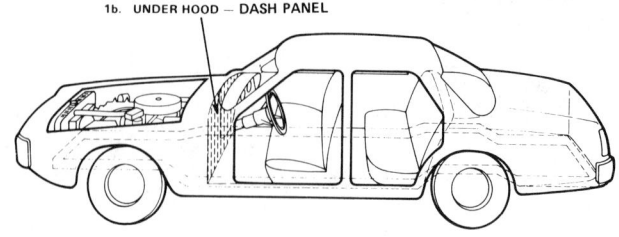

	TEMPERATURE			HUMIDITY (%RH)			SALT SPRAY	IMMERSION	SAND, DUST & GRAVEL	OIL & CHEMICAL	MECHANICAL SHOCK & VIBRATION	ELECTRICAL	
	LOW	HIGH	SLEW RATE	HIGH	LOW	FROST						STEADY-STATE	TRANSIENT
Normal	−40°C (−40°F)	121°C (250°F)	open	95% at 38°C (100°F)			Sect. 4.3	no	Sect. 4.5	Sect. 4.4	Figure 12	Tables 2 & 4	Table 3
Extreme	−40°C (−40°F)	141°C (285°F)	open	80% at 66°C (150°F)				no					

TABLE 6

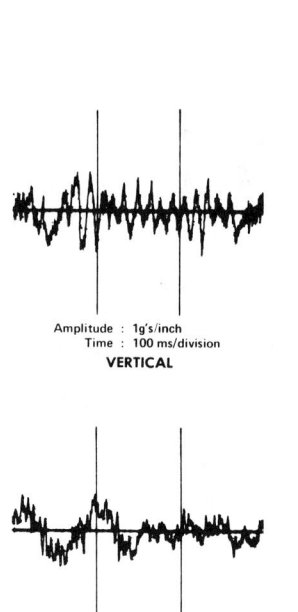

RECORDED DATA

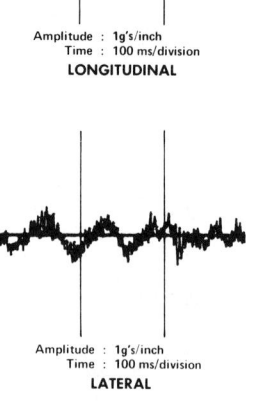

EQUIVALENT P.S.D.

FIG. 12—PLENUM VIBRATION MEASUREMENTS

5.3.1 TEMPERATURE—The heat sources encountered in the chassis area include (in rank of decreasing surface temperature):

Source	Peak Temperature
a. Exhaust/catalytic reactor system	649°C (1200°F)
b. Brake system/tires and transmission/differential drivetrain components	177°C (350°F)
c. Engine	121°C (250°F)
d. Vehicle ambient peak temperature	85°C (185°F)

The practical limitations of equipment components (with the possible exception of sensors) will restrict the designer to locations with the peak temperatures given in *c* and *d* above. Again, the designer is urged to check his particular installation for the actual peak temperatures experienced under operating conditions.

5.3.2 HUMIDITY—As shown in Table 7.

5.3.3 SALT SPRAY—With the exception of a few shielded locations, all chassis components are subject to heavy salt spray.

5.3.4 IMMERSION—Typical chassis components are subject to immersion.

5.3.5 DUST, SAND, AND GRAVEL—All chassis components in line with the wheel track that are not shielded are subject to continuous bombardment during vehicle operation on gravel roads. In nontrack aligned portions of the chassis, some bombardment will be experienced by equipment mounted on forward-facing chassis surfaces. All chassis components are subject to heavy dust and sand environments.

5.3.6 OILS AND CHEMICALS—The chassis is subject to all of the oils and chemicals listed in Section 4.4.

5.3.7 MECHANICAL, SHOCK, AND VIBRATION—Vibration data collected on the frame bumper attachment, frame transmission mount, frame crossmember and wheel backplate with equivalent power spectral density profiles are shown in Figs. 13–15.

5.3.8 ELECTRICAL—Steady State (Refer to Section 4.10.1 for further information—Three operating conditions are recognized:

a. Normal starting and running   9–16 V[1]
b. Cold starting   4.5–6 V
c. Booster battery starting   24 V

5.3.9 ELECTRICAL—TRANSIENT—This condition varies, depending upon the electrical distance of the equipment from the battery and the nearness of transient sources (e.g., inductive motors, solenoids, the alternator). Typical data is shown in Table 7. (Refer to Section 4.11 for further information.)

5.4 Exterior—The exterior consists of all outward and external vehicle surfaces above the chassis. This includes the forward grille area and potential mounting areas just above the bumpers. Data is summarized in Table 8.

[1] See Section 4.10.1 for a definition of normal voltage.

CHASSIS ENVIRONMENTAL EXTREME DATA

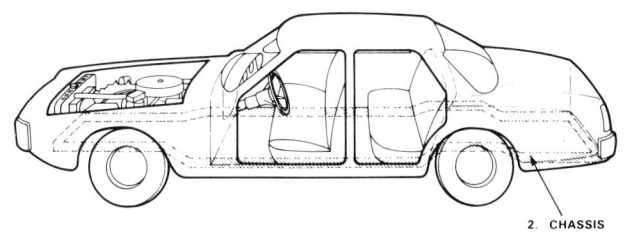

2. CHASSIS

	TEMPERATURE			HUMIDITY (%RH)			SALT SPRAY	IMMERSION	SAND, DUST & GRAVEL	OIL & CHEMICAL	MECHANICAL SHOCK & VIBRATION	ELECTRICAL	
	LOW	HIGH	SLEW RATE	HIGH	LOW	FROST						STEADY-STATE	TRANSIENT
Isolated	-40°C (-40°F)	85°C (185°F)	NA	98% at 38°C (100°F)	0	yes							
Near Heat Source	-40°C (-40°F)	121°C (250°F)	NA	66°C (150°F)	0	yes	Sect. 4.3	Sect. 4.4	Sect. 4.5	Sect. 4.4	Figures 13, 14 & 15	Table 2 & 4	Table 3
At Drive Train High Temp Location	-40°C (-40°F)	177°C (350°F)	NA	80%	0	yes							

TABLE 7

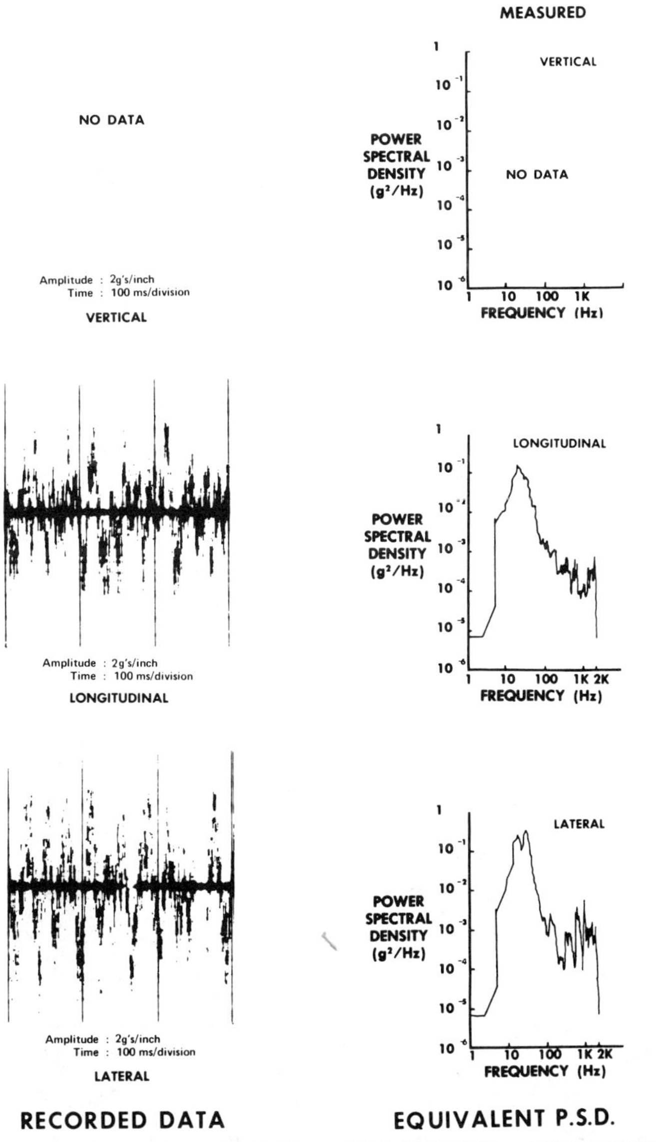

FIG. 13—FRAME BUMPER ATTACHMENT VIBRATION MEASUREMENTS

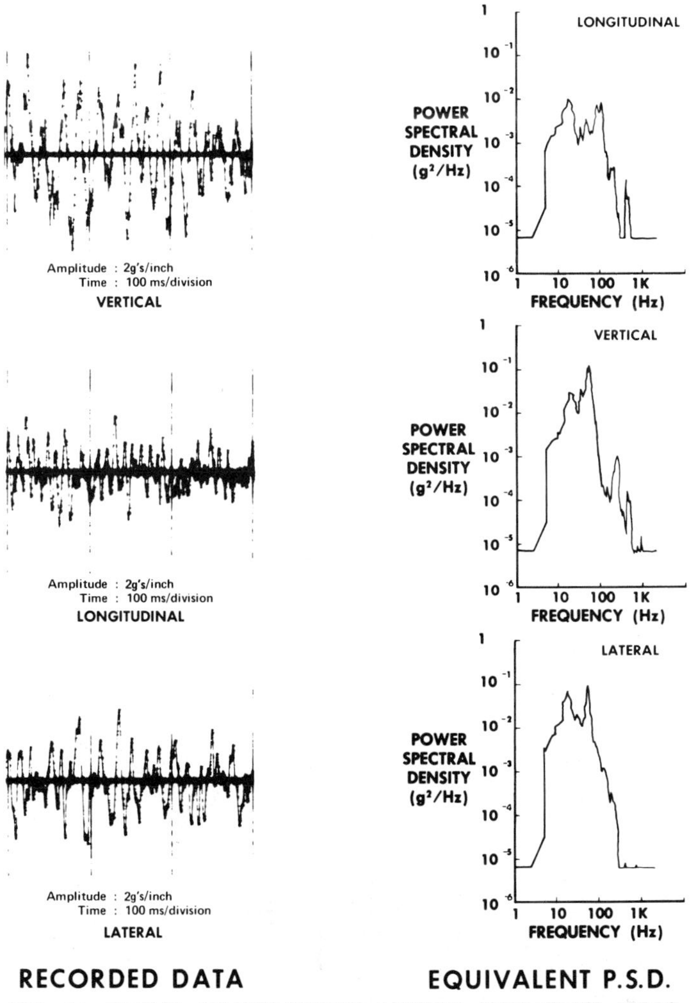

FIG. 14—FRAME TRANSMISSION MOUNT VIBRATION MEASUREMENTS

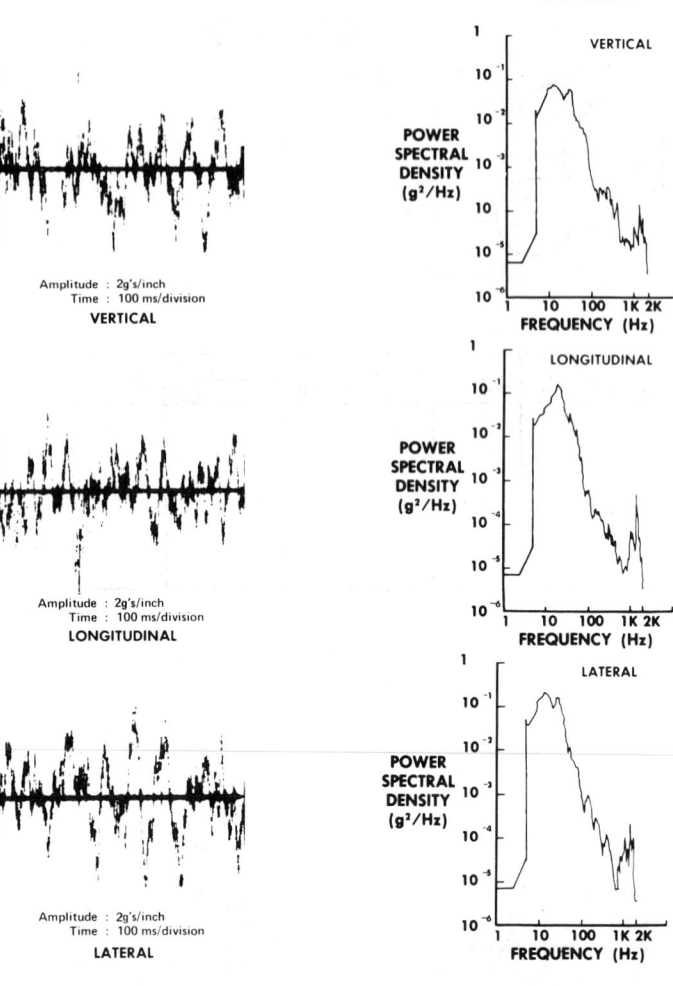

FIG. 15—FRAME CROSS MEMBER VIBRATION MEASUREMENTS

5.4.1 Temperature—Since all surfaces are away from internal vehicle heat sources, the temperature is primarily controlled by the climatic ambient conditions. These are discussed in Section 4.1 and shown in Table 8. Thermal shock due to splash or immersion, particularly on the front of the vehicle, should be anticipated.

5.4.2 Humidity—Shown in Table 8.

5.4.3 Salt Spray—Most exterior surfaces are subject to heavy salt spray, with the possibility of crystalline salt buildup in some grill areas.

5.4.4 Immersion—Equipment mounted approximately below the vehicle axle line are possibly subject to occasional immersion. Components above this line experience splash.

5.4.5 Gravel, Dust, and Sand—Components on the front of the vehicle are subject to bombardment from the vehicle ahead. Sand and dust impinges on all surfaces.

5.4.6 Oils and Chemicals—Environmental chemicals include:

    Road tar
    Anti-freeze/water mixture
    Soaps and detergents
    Steam
    Salt spray
    Washer solvent
    Degreasers
    Waxes
    Water and snow

5.4.7 Mechanical Shock and Vibration—Data collected at the wheel back plate is shown in Fig. 16, and center pillar data is shown in Fig. 17.

5.4.8 Electrical—Steady State—Three operating conditions are recognized:

a.	Normal starting and running	9–16 V[1]
b.	Cold starting	4.5–6 V
c.	Booster battery starting	24 V

5.4.9 Electrical—Transient—This condition appears to vary widely, depending upon the electrical distance of the equipment from the battery and the nearness of transient sources (e.g., inductive motors, solenoids, the alternator). Typical values are shown in Table 8.

5.5 Interior—Instrument Panel—This includes the top of the dashboard and the near vertical section carrying the instruments and steering wheel. Data is shown in Table 9.

5.5.1 Temperature—Two temperature conditions are traceable to the climatic vehicle environment. Components not in direct sunlight experience temperatures from −40–85°C (−40–185°F). Components on the top surface of the instrument panel experience a greater heat buildup when closed vehicles are parked in the bright sun. Heat radiated incident sunlight and re-radiated energy from the windshield cause the temperature to build to 113°C (235°F) in this region. Heat due to underdash components, such as radio or heater, is also a contributing factor.

5.5.2 Humidity—As shown in Table 9. A tightly closed vehicle with wet upholstery experiences very high internal humidity at high temperature.

5.5.3 Salt Spray—Not generally a problem at the instrument panel.

5.5.4 Immersion—Not anticipated, although liquid spills are possible on the upper dash surface.

[1]See Section 4.10.1 for a definition of normal voltage.

EXTERIOR ENVIRONMENTAL DATA

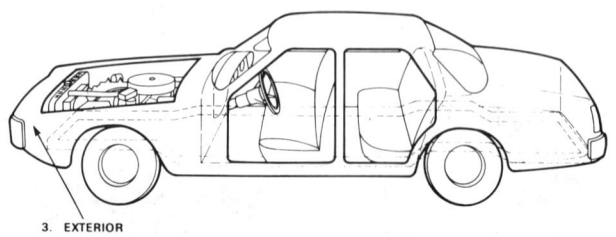

3. EXTERIOR

	TEMPERATURE			HUMIDITY (%RH)			SALT SPRAY	IMMERSION	SAND, DUST & GRAVEL	OIL & CHEMICAL	MECHANICAL SHOCK & VIBRATION	ELECTRICAL	
	LOW	HIGH	SLEW RATE	HIGH	LOW	FROST						STEADY-STATE	TRANSIENT
Normal	−40°C (−40°F)	85°C (185°F)	NA	95% at 38°C (100°F)	0	yes	Sect. 4.3	Sect. 4.4	Sect. 4.5	Sect. 4.4	Figure 16 & 17	Table 2 & 4	Table 3

TABLE 8

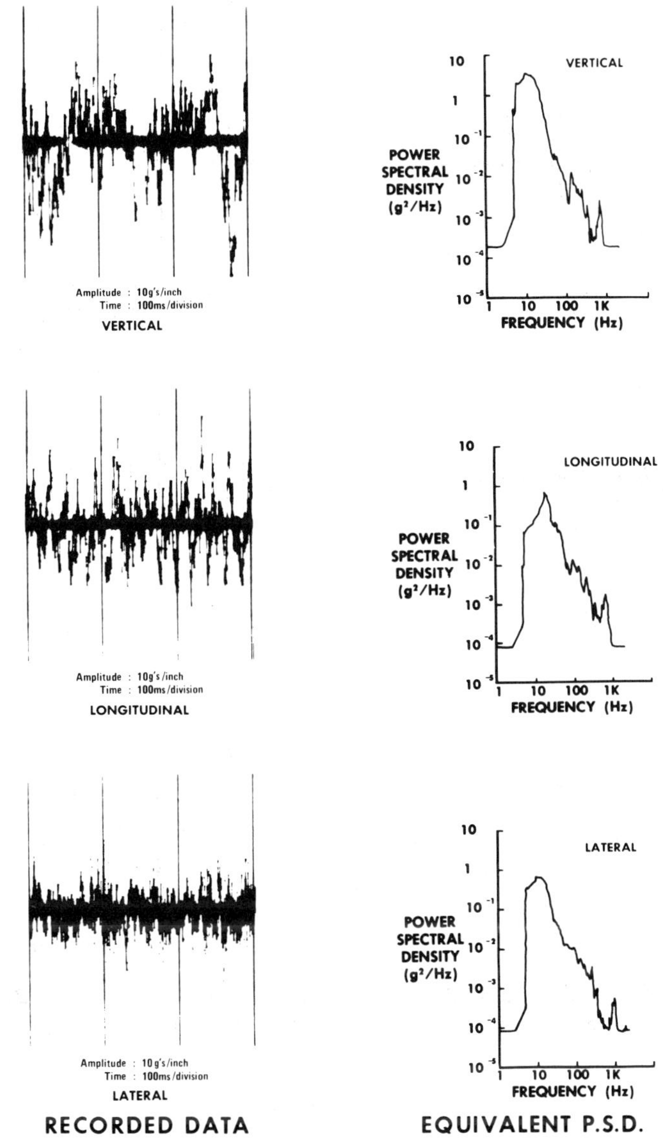

FIG. 16—WHEEL BACK PLATE VIBRATION MEASUREMENTS

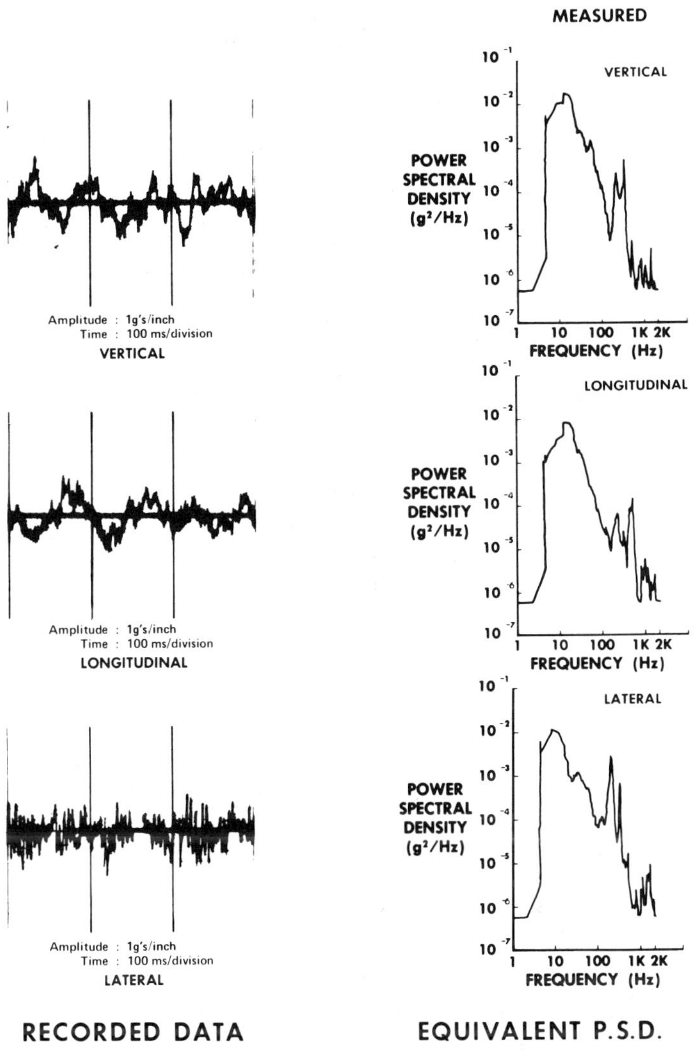

FIG. 17—CENTER PILLAR VIBRATION MEASUREMENTS

INTERIOR - INSTRUMENTAL PANEL ENVIRONMENTAL DATA

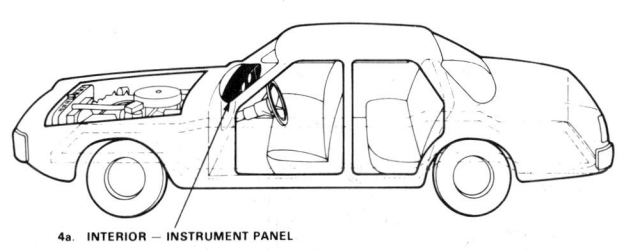

4a. INTERIOR – INSTRUMENT PANEL

	TEMPERATURE			HUMIDITY (%RH)			SALT SPRAY	IMMERSION	SAND, DUST & GRAVEL	OIL & CHEMICAL	MECHANICAL SHOCK & VIBRATION	ELECTRICAL	
	LOW	HIGH	SLEW RATE	HIGH	LOW	FROST						STEADY-STATE	TRANSIENT
Nominal	-40°C (-40°F)	85°C (185°F)	NA	98% at 38°C (100°F)	0	yes	no	Partial	Dust only	Sect. 4.4	Figure 18	Table 2 & 4	Table 3
Top Surface	-40°C (-40°F)	113°C (235°F)	NA	80% at 66°C (150°F)									

TABLE 9

5.5.5 GRAVEL, SAND, AND DUST—Gravel not anticipated. Coatings of sand and dust are expected on all horizontal surfaces.

5.5.6 OILS AND CHEMICALS—Mainly cleaning agents: waxes, soaps, and detergents.

5.5.7 MECHANICAL VIBRATION—As shown in Fig. 18.

5.5.8 ELECTRICAL—STEADY STATE—Three operating conditions are recognized:

    a. Normal starting and running   9–16 V[1]
    b. Cold starting   4.5–6 V
    c. Booster battery starting   24 V

5.5.9 ELECTRICAL—TRANSIENT—This condition appears to vary widely, depending upon the electrical distance of the equipment from the battery and the nearness of transient sources (e.g., inductive motors, solenoids, the alternator). Typical data is shown in Table 9.

5.6 Interior—Floor—This covers all approximately horizontal surfaces, including the floor beneath the front seat(s), the footrest areas in front of the seat(s) and beneath the dashboard, and the interior surfaces of the drive tunnel. Data is shown in Table 10.

5.6.1 TEMPERATURE—As shown in Table 10. Higher temperatures may be experienced directly over drivetrain components (transmission, etc.) and the exhaust system (including catalytic converters) although data is not available at this time.

5.6.2 HUMIDITY—As shown in Table 10, standing water is possible in depressions due to rain entry through open windows or leaking body seals. Also, water is carried into the vehicle by wet garments or packages.

5.6.3 SALT SPRAY—The water entry discussed in humidity section may also be a saturated salt solution.

5.6.4 IMMERSION—Immersion is possible as discussed in humidity section.

5.6.5 GRAVEL, DUST, AND SAND—Gravel bombardment is not a condition, although a buildup of dust and sand is common.

5.6.6 OILS AND CHEMICALS—Contaminants include the following:
    Engine oils and additives (tracked in on occupant's shoes)
    Cleaning solvents
    Water and snow
    Gasoline
    Salt water

5.6.7 MECHANICAL SHOCK AND VIBRATION—No data available at this time. Similar to conditions shown for the transmission mounts in the chassis section.

5.6.8 ELECTRICAL—STEADY STATE—Three operating conditions are recognized:

    a. Normal starting and running   9–16 V[1]
    b. Cold starting   4.5–6 V
    c. Booster battery starting   24 V

5.6.9 ELECTRICAL—TRANSIENT—This condition appears to vary widely, depending upon the electrical distance of the control site from the battery and the nearness of transient sources (e.g., inductive motors, solenoids, the alternator).

5.7 Interior—Rear Deck—This area includes horizontal surface extending from the top of the rear seat to the body work just below the bottom edge of the backlight. Data is shown in Table 11.

5.7.1 TEMPERATURE—The major heat source is climatic incident radiant energy from direct sunlight and sunlight reflected from the backlight. The peak temperature of 104°C (220°F) is slightly less than that given for the upper dashboard surface because of the absence of heat sources beneath the panel.

[1] See Section 4.10.1 for a definition of normal voltage.

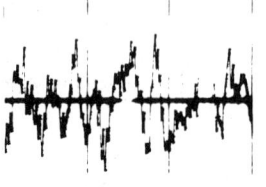
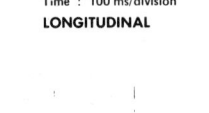
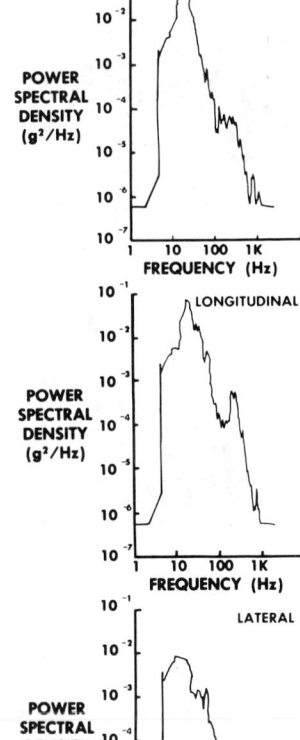

RECORDED DATA   **EQUIVALENT P.S.D.**

FIG. 18—INSTRUMENT PANEL VIBRATION MEASUREMENTS

5.7.2 HUMIDITY—As shown in Table 11.

5.7.3 SALT SPRAY—Not expected.

5.7.4 IMMERSION—Not present.

5.7.5 SAND AND DUST—Light coatings of sand and dust are present.

5.7.6 OILS AND CHEMICALS—Only cleaning agents expected.

5.7.7 MECHANICAL SHOCK AND VIBRATION—No data available at this time. The condition is similar to conditions shown for the dashboard, with the exception that vibration at the vehicle's rear is a function of high variable trunk, rear seat, and bumper loading conditions.

5.7.8 ELECTRICAL—STEADY STATE—Three operating conditions are recognized:

INTERIOR - FLOOR ENVIRONMENTAL DATA

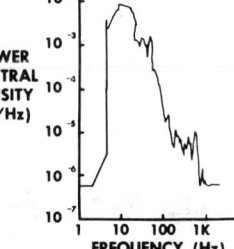

4b. INTERIOR – FLOOR

TEMPERATURE			HUMIDITY (%RH)			SALT SPRAY	IMMERSION	SAND, DUST & GRAVEL	OIL & CHEMICAL	MECHANICAL SHOCK & VIBRATION	ELECTRICAL	
LOW	HIGH	SLEW RATE	HIGH	LOW	FROST						STEADY-STATE	TRANSIENT
-40°C (-40°F)	85°C (185°F)	NA	98%RH 38°C (100°F)	0	–	NO	NO	Sect. 4.5	Sect. 4.4	Not measured	Table 2 & 4	Table 3

TABLE 10

INTERIOR - REAR DECK ENVIRONMENTAL DATA

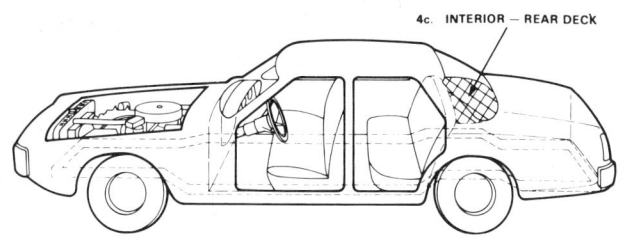

TEMPERATURE			HUMIDITY (%RH)			SALT SPRAY	IMMERSION	SAND, DUST & GRAVEL	OIL & CHEMICAL	MECHANICAL SHOCK & VIBRATION	ELECTRICAL	
LOW	HIGH	SLEW RATE	HIGH	LOW	FROST						STEADY-STATE	TRANSIENT
-40°C (-40°F)	104°C (220°F)	NA	98% at 38°C (100°F) ----- 80% at 66°C (150°F)	0	no	no	no	Sect. 4.5	Sect. 4.4	Not Measured	Table 2 & 4	Table 3

TABLE 11

a. Normal starting and running   9–16 V[1]
b. Cold starting   4.5–6 V
c. Booster battery starting   24 V

5.7.9 ELECTRICAL—TRANSIENT—This condition appears to vary widely, depending upon the electrical distance of the equipment from the battery and the nearness of transient sources (e.g., inductive motors, solenoids, the alternator). Care should be taken to measure transients due to electrical rear window lifts, if present.

5.8 Trunk—Environmental data is defined in Table 12.

5.8.1 TEMPERATURE—The anticipated temperature limits are fairly similar to those expected for the interior. However, the thinner insulation may increase temperatures in some areas of the trunk floor due to radiated and conducted exhaust system heat. The inside of the trunk lid may also experience higher temperatures than those given due to direct sunlight heating.

5.8.2 HUMIDITY—The presence of stored liquids or wet clothing, etc., makes this the highest humidity enclosed volume in the car. Condensation is possible on all surfaces.

5.8.3 SALT SPRAY—A standing saturated salt solution is possible on the compartment floor.

5.8.4 IMMERSION—Standing liquid on the floor is anticipated unless the area is equipped with drains.

5.8.5 SAND AND DUST—A heavy buildup of sand and dust is anticipated.

5.8.6 OILS AND CHEMICALS—Spillage of all of the chemicals listed in Section 4.4 is possible.

5.8.7 MECHANICAL SHOCK AND VIBRATION—No data available at this time. Conditions are similar to those shown for the dashboard, with the exception

[1] See Section 4.10.1 for a definition of normal voltage.

that vibration at the vehicle's rear is a function of highly variable trunk, rear seat, and rear bumper loading conditions.

5.8.8 ELECTRICAL—STEADY STATE—This condition appears to vary widely, depending upon the electrical distance of the equipment from the battery and the nearness of transient sources (e.g., inductive motors, solenoids, the alternator). Typical values are shown in Table 12.

5.9 Environmental Extremes Summary—Table 13 summarizes the information provided in Sections 5.1–5.8.

ACKNOWLEDGMENTS

The subcommittee acknowledges the contribution of Mr. J. R. Morgan. Mr. Morgan provided oscillograms of the load dump transient and electrical noise, and critiqued the electrical section of this document.

REFERENCES

1. O. T. McCarter, "Environmental Guidelines for the Designer of Automotive Electronic Components." Paper 740017 presented at SAE Automotive Engineering Congress, Detroit, March 1974.
2. J. R. Morgan, "Transients in the Automotive Electrical System." Motorola CER-114, 1973.
3. G. B. Andrews, "Control of the Automotive Electrical Environment." Paper 730045 presented at SAE Automotive Engineering Congress, Detroit, January 1973.
4. SAE J1113a, Electromagnetic Susceptibility Procedures for Vehicle Components (Except Aircraft) (June, 1978).

TRUNK & ENVIRONMENTAL DATA

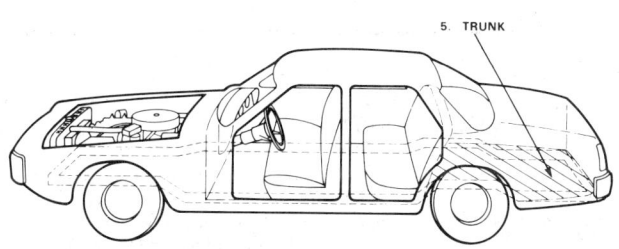

TEMPERATURE			HUMIDITY (%RH)			SALT SPRAY	IMMERSION	SAND, DUST & GRAVEL	OIL & CHEMICAL	MECHANICAL SHOCK & VIBRATION	ELECTRICAL	
LOW	HIGH	SLEW RATE	HIGH	LOW	FROST						STEADY-STATE	TRANSIENT
-40°C (-40°F)	85°C (185°F)	NA	98% at 38°C (100°F) ----- 80% at 66°F	0	yes	no	no	Sect. 4.5	Sect. 4.4	Not Measured	Tables 2 & 4	Table 3

TABLE 12

TABLE 13—ENVIRONMENTAL EXTREME SUMMARY

Location	Temperature			Humidity (%RH)			Salt Spray	Immersion	Sand, Dust, and Gravel	Oil and Chemical	Mechanical Shock and Vibration	Electrical	
	Low	High	Slew Rate	High	Low	Frost						Steady-State	Transient
1. Underhood—Engine Choke Housing Exhaust Manifold Intake Manifold	−40°F −40°C −40°F −40°C −40°F −40°C	400°F 204°C 1200°F 649°C 250°F 121°C	20°F/min −7°C/min	95% at 100°F 38°C	0	yes	Section 4.3	splash present	sand and dust	↑	Fig. 11	↑	↑
Underhood—Firewall Normal	−40°F −40°C	250°F 121°C	open	95% at 100°F 38°C	0		Section 4.3	no	sand and dust		Fig. 12		
Extreme	−40°F −40°C	285°F 141°C	open	80% at 150°F 66°C	0			no					
2. Chassis Isolated	−40°F −40°C	185°F 84°C	NA	98% at 100°F 38°C	0	yes	yes	yes	yes			See Section 4.10 and Tables 2 and 4	
Near Heat Source	−40°F −40°C	250°F 121°C	NA	80% at 150°F 66°C	0	yes		yes	yes		Figs. 13, 14, and 15		See Section 4 and Table 3
At Drivetrain High Temp Locations	−40°F −40°C	350°F 177°C	NA						yes				
3. Exterior Normal	−40°F −40°C	235°F 113°C	NA	95% at 100°F 38°C	0	yes	yes	yes	yes	See Section 4.4	Figs. 16 and 17		
4. Interior Instrument Panel	−40°F −40°C	185°F 84°C	NA	98% at 100°F 38°C	0	yes	no	no	dust only		Fig. 18		
Top		225°F 113°C											
Floor	−40°F −40°C	185°F 84°C	NA	98% at 100°F 38°C 80% at 150°F 66°C	0	no	no		dust and sand		Not measured		
Rear Deck	−40°F −40°C	220°F 104°C	NA	98% at 100°F 38°C	0	no	no		dust and sand		Not measured		
5. Trunk	−40°F −40°C	185°F 84°C	NA	98% at 100°F 38°C	0	yes	no	no	sand and dust	↓	Not measured	↓	↓

GLOSSARY OF AUTOMOTIVE ELECTRONIC TERMS—SAE J1213

Information Report

Report of Electronic Systems Committee approved June 1978.

Active Element—A component capable of producing power gain such as a transistor, tunnel diode, thyristor, etc. Also active device, active component.

Active Filter—A device employing passive network elements and amplifiers used for transmitting or rejecting signals in certain frequency ranges or for controlling the relative output of signals as a function of frequency. (GRAF)

Ambient Conditions—The conditions (pressure, temperature, etc.) of the surrounding medium.

American Wire Gauge—Abbreviated AWG. System of numerical designations for wire size, based on specified ranges of circular mil area. American Wire Gauge starts with 4/0 (0000) at the largest size going to 3/0 (000), 2/0 (00), 1/0 (0), 1, 2, and up to 40 and beyond for the smallest sizes.

Ampere (A)—The standard unit for measuring the strength of an electric current. The rate of flow of a charge in a conductor or conducting medium of one coulomb per second.

Amplifier—A device, circuit, or component which produces as an output an enlarged reproduction of the essential features of its input.

Amplitude Modulation (AM)—Modulation in which the amplitude of a wave is the characteristic subject to variation. (GRAF)

Anode—The positive pole (+) in batteries, galvanic cells, or plating apparatus. In diodes, the positive lead.

Analog—Of or pertaining to the general class of devices or circuits in which the output varies as a continuous function of the input.

Analog Computer—A computer which represents numerical quantities as electrical and physical variables and manipulates these variables in accomplishing solutions to mathematical problems.

And Gate—A combinational logic element such that the output channel is in its one state, if and only if each input channel is in its *one* state.

Avalanche Breakdown—In a semiconductor diode, a nondestructive breakdown caused by the cumulative multiplication of carriers through field-induced impact ionization. (GRAF)

Avalanche Diode—Also called breakdown diode. A silicon diode that has a high ratio of reverse-to-forward resistance until avalanche breakdown occurs. After breakdown the voltage drop across the diode is essentially constant and is independent of the current. Used for voltage regulating and voltage limiting. Originally called zener diode before it was found that the Zener effect had no significant role in the operation of diodes of this type.

Bandwidth—The range within the limits of a band. The least frequency interval of a wave form. The range of frequencies of a device, within which its performance, with respect to some characteristic, conforms to a specified standard.

Bit—The smallest element of information in binary language. A contraction of BInary digiT. These characters in system (computer) language signify "on" and "off" (1 and 0). Word length, memory capacity, etc. can be expressed in number of "bits".

B-Multiplier—See Darlington Amplifier.

Barrier Layer—See Depletion Layer.

Base—(transistor) A region that lies between an emitter and a collector of a transistor and into which minority carriers are injected. (IEEE)

Base Resistance—Resistance in series with the base lead in the common T equivalent circuit of a transistor. (GRAF)

Battery—A DC voltage source which converts chemical, nuclear, thermal or solar energy into electrical energy. (GRAF)

Bidirectional Diode-Thyristor—A two-terminal thyristor having substantially the same switching behavior in the first and third quadrants of the principal voltage-current characteristic. (IEEE)

Bias—To influence or dispose to one direction, as, for example, with a direct voltage or with a spring. (IEEE)

Binary—A characteristic or property involving a selection, choice, or condition, in which there are but two possible alternatives.

Bipolar—Having to do with a device in which both majority and minority carriers are present. In connection with ICs, the term describes a specific type of construction; bipolar and MOS are the two most common types of IC construction. (GRAF)

Bleeder Resistor—A resistor used to draw a fixed current. Also used to discharge a filter capacitor after the circuit is de-energized. (GRAF)

Boolean Algebra—The Algebra of Logic named for mathematician George Boole using alphabetic symbols to stand for logical variables and "zero" and "one" to represent states. AND, OR, NOT, are the three basic logic operations in this algebra. NAND and NOR are combinations of the three basic operations.

Breakdown Voltage—See dielectric strength.

Capacitance (C)—In a system of conductors and dielectrics, that property which permits the storage of electrically separated charges when potential differences exist between the conductors. Its value is expressed as the ratio of a quantity of electricity to a potential difference (Q/V). (IEEE)

Capacitor—(condenser) A device consisting of two electrodes separated by a dielectric, which may be air, for introducing capacitance into an electric circuit. (IEEE)

Carrier—An AC voltage having a frequency suitably high to be modulated by electrical signals.

Cascade—An arrangement of two or more similar circuits or amplifying stages in which the output of one provides the input of the next. (GRAF)

Cathode—A general name for any negative electrode. (GRAF)

Cathode-Ray Tube—An electron-beam tube in which the beam can be focused to a small cross section on a luminescent screen and varied in position and intensity to produce a visible pattern. (IEEE)

Chip—A single substrate on which all the active and passive elements of an electronic circuit have been fabricated. A chip is not ready for use until it is packaged and provided with terminals for connection to the outside world. Also called a die. (GRAF)

Chip Sets—A term describing the microprocessor chip in addition to RAMs, ROMs, and interface I/O devices. Chip sets, mounted on a board, are also referred to as the CPU portion of the microcomputer.

Clock—A device that generates periodic signals used for synchronization. (IEEE)

Coil—See Inductor.

Collector—(transistor) A region through which primary flow of charge carriers leaves the base. (IEEE)

Common Collector Amplifier—A transistor amplifier in which the collector element is common to both the input and output circuit. Also known as an emitter-follower and a grounded-collector amplifier. (GRAF)

Common Mode Rejection—A measure of how well a differential amplifier ignores a signal which appears simultaneously and is in phase at both input terminals. Also, called in-phase rejection. (GRAF)

Complementary MOS—Pertaining to n- and p-channel enhancement-mode devices fabricated compatibly on a silicon chip and connected into push-pull complementary digital circuits. These circuits offer low quiescent power dissipation and potentially high speeds, but they are more complex than circuits in which only one channel type is used. Abbreviated CMOS. (GRAF)

Computer—Any device capable of accepting information, applying prescribed processes to the information, and supplying the results of the process.

Conductivity—The ability to transmit heat or electricity. Electrical conductivity is expressed in terms of the current per unit of applied voltage. The reciprocal of resistivity.

Constant Current Source—A regulated source which acts to keep its output current constant in spite of changes in load, line, or temperature while the output voltage changes by whatever amount is necessary to maintain the constant output current. (GRAF)

Continuous Rating—The rating applicable to specified operation for a specific uninterrupted length of time.

Central Processing Unit (CPU)—The section of a computer that contains the arithmetic, logic, and control circuits. In some systems it may also include the memory unit and the operator's console. Also called main frame.

Creep—A change in output occurring over a specific time period while the input and all environmental conditions are held constant.

Critical Damping—The value of damping which provides the most rapid transient response without overshoot. Operation between underdamping and overdamping. (GRAF)

Current Density—The amount of electric current passing through a given cross-sectional area of a conductor. (GRAF)

Damping—The transitory decay of the amplitude of a free oscillation of a system, associated with energy loss from the system. (IEEE)

Damping Ratio—The ratio of the degree of actual damping to the degree of damping required for critical damping. (GRAF)

Darlington Amplifier—A transistor circuit which, in its original form, consists of two transistors in which the collectors are tied together and the emitter of the first transistor is directly coupled to the base of the second transistor. Therefore, the emitter current of the first transistor equals the base current of the second transistor. This connection of two transistors can be regarded as a compound transistor with three terminals. (GRAF)

Darlington Pair—See Darlington Amplifier.

D'Arsonval Current—A high-frequency, low voltage current of comparatively high amperage. (GRAF)

Decibel—One-tenth of a bel, the number of decibels denoting the ratio of the two amounts of power being ten times the logarithm to the base 10 of this ratio. NOTE: The abbreviation dB is commonly used for the term decibel. With P_1 and P_2 designating two amounts of power and n the number of decibels denoting their ratio.

$$n = 10 \log_{10} [P_1/P_2] \text{ decibel}$$

When the conditions are such that ratios of currents or ratios of voltages (or analogous quantities in other field) are the square roots of the corresponding powers ratios, the number of decibels by which the corresponding powers differ is expressed by the following equations:

$$n = 20 \log_{10} [I_1/I_2] \text{ decibel}$$

$$n = 20 \log_{10} [V_1/V_2] \text{ decibel}$$

where I_1/I_2 and V_1/V_2 are the given current and voltage ratios, respectively. By extension, these relations between numbers of decibels and ratios of currents or voltages are sometimes applied where these ratios are not the square roots of the corresponding power ratios; to avoid confusion, such usage should be accompanied by a specific statement of this application. Such extensions of the term described should preferably be avoided. (IEEE)

Delay—(1) The amount of time by which an event is retarded. (2) The amount of time by which a signal is delayed. NOTE: It may be expressed in time (milliseconds, microseconds, etc.) or in number of characters (pulse times, word times, major cycles, minor cycles, etc.). (IEEE)

Delay Line—(electronic computers) (1) Originally, a device utilizing wave propagation for producing a time delay of a signal. (2) Commonly, any real or artificial transmission line or equivalent device designed to introduce delay. (IEEE)

Depletion Layer—In a semi-conductor, the region in which the mobile-carrier charge density is insufficient to neutralize the net fixed charge density of donors and acceptors. (GRAF)

Dielectric Constant—The property that determines the electrostatic energy stored per unit volume for unit potential gradient. NOTE: This numerical value usually is given relative to a vacuum. (IEEE)

Dielectric Strength—(material) (electric strength) (breakdown strength) The potential gradient at which electric failure or breakdown occurs. To obtain the true dielectric strength the actual maximum gradient must be considered, or the test piece and electrodes must be designed so that uniform gradient is obtained. The value obtained for the dielectric strength in practical tests will usually depend on the thickness of the material and on the method and conditions of test. (IEEE)

Digital Computer—A computer that processes information in numerical form. Electronic digital computers generally use binary or decimal notation and process information by repeated high speed use of the fundamental arithmetic processes of addition, subtraction, multiplication, and division.

Digital-to-Analog (D/A) Converter—A device which transforms digital data into analog data by translating digital magnitude to equivalent voltage level.

Diode—(electronic tube) A two electrode electron tube containing an anode and a cathode. (semi-conductor) A semi-conductor device having two terminals and exhibiting a non-linear voltage-current characteristic; in more restricted usage; a semi-conductor device that has the asymmetrical voltage-current characteristic exemplified by a single p-n junction. (IEEE)

Diode Transistor Logic—Abbreviated DTL. A logic circuit that uses diodes at the input to perform the electronic logic function that activates the circuit transistor output. In monolithic circuits, the DTL diodes are a positive level logic and function or a negative level or function. The output transistor acts as an inverter to result in the circuit becoming a positive NAND or a negative NOR function. (GRAF)

DIP—Abbreviation for dual in-line package. (GRAF)

Dipole Antenna—Any one of a class of antennas producing the radiation pattern approximating that of an elementary electronic dipole. NOTE: Common usage considers a dipole to be a metal radiating structure that supports a line current distribution similar to that of a thin straight wire, a half wavelength long, so energized that the current has two nodes, one at each of the far ends. (IEEE)

Drain (D)—In a field effect transistor, the element that corresponds to the collector of a transistor. (GRAF)

Drift—An undesired change in output over a period of time, which change is not a function of the input.

Duty Cycle—The ratio of the time "On" of a device or system divided by the total cycle time (i.e., "On" plus time "Off"). For a device that normally runs intermittently rather than continuously: the amount of time a device operates as opposed to its idle time.

Electromagnetic Compatability (EMC)—The ability of electronic communications equipment, sub-systems, and systems to operate in their intended environments without suffering or causing unacceptable degradation of performance as a result of unintentional electromagnetic radiation or response. (GRAF)

Electromagnetic Interference (EMI)—Electromagnetic phenomena which, either directly or indirectly, can contribute to the degradation in performance of an electronic receiver or system. (GRAF)

Electron—One of the natural elementary constituents of matter. It carries a negative electric charge of one electronic unit. (GRAF)

Electromotive Force (emf)—The force which may cause current to flow when there is a difference of potential between two points. (GRAF)

Electromagnetic Waves—The radiant energy produced by the oscillation of an electric charge. (GRAF)

Electrolumenescence—Lumenescence resulting from a high-frequency discharge through a gas or from application of an alternating current to a layer of phosphor. (GRAF)

Emitter—A region from which charge carriers that are minority carriers in the base are injected into the base. (IEEE)

Emitter-Coupled Logic—Nonsaturated bipolar logic in which the emitters of the input logic transistors are coupled to the emitter of a reference transistor. The basic gate circuit employs a long-tailed pair. Abbreviated ECL. (GRAF)

Exclusive OR—A logic operator having the property that if P is a statement and Q is a statement, then P exclusive ORQ is true if either but not both statements are true, false if both are true or both are false. (IEEE)

Feedback—The recycling of a portion of the output to the input of a system. Systems employing feedback are called closed-loop systems.

Feedback Amplifier—An amplifier that uses a passive network to return a portion of the output signal to modify the performance of the amplifier. (GRAF)

Ferrites—Chemical compounds of iron oxide and other metallic oxides combined with ceramic material. They have ferromagnetic properties but are poor conductors of electricity. Hence they are useful where ordinary ferromagnetic materials (which are good electrical conductors) would cause too great a loss of electrical energy. (GRAF)

Ferromagnetic Material—Material whose relative permeability is greater than unity and depends upon the magnetizing force. A ferromagnetic material usually has relatively high values of relative permeability and exhibits hysteresis. (IEEE)

FET—(field-effect transistor) A semi-conductor device in which the resistance between the source and drain terminals depends on a field produced by a voltage applied to the gate terminal. (GRAF)

Filter—A selective network of resistors, inductors, or capacitors which offers comparatively little opposition to certain frequencies or to direct current, while blocking or attenuating other frequencies. (GRAF)

Flat Pack—A flat, rectangular integrated circuit or hybrid-circuit package with coplanar leads. (GRAF)

Flux (Magnetic)—The sum of all the lines of force in a magnetic field crossing a unit area per unit time.

Flux Density—Flux per unit area perpendicular to the direction of the flux. (GRAF)

Forward Voltage—(V_F) The voltage across a semi-conductor diode associated with the flow of forward current. The p-region is at a positive potential with respect to the n-region.

Frequency Modulated Output—fm (frequency modulation) A scheme for modulating a carrier frequency in which the amplitude remains constant but the carrier frequency is displaced in frequency proportionally to the amplitude of the modulating signal. A frequency modulation broadcast system is practically immune to atmospheric and man-made interference.

Frequency Response—A measure of how the gain or loss of a circuit, device or system varies with the frequencies applied to it. Also, the portion of the frequency spectrum which can be sensed by a device within specified limits of error.

Gain—Any increase in power when a signal is transmitted from one point to another. Usually expressed in decibels. (GRAF)

GCS—Abbreviation for gate controlled switch. (GRAF)

Gate—(1) A device or element that, depending upon one or more specified inputs, has the ability to permit or inhibit the passage of a signal. (2) (electronic computers) (a) A device having one output channel and one or more input channels, such that the output channel state is completely determined by the contemporaneous input channel states, except during switching transients. (b) A combinational logic element having at least one input channel. (c) An AND gate. (d) An OR gate. (3) In a field effect transistor, the electrode that is analogous to the base of a transistor or the grid of a vacuum tube. (GRAF)

Hall Effect—The development of a transverse electric potential gradient in a current carrying conductor or semi-conductor upon the application of a magnetic field.

Hardware—(1) Mechanical, magnetic, electrical, or electronic devices; physical equipment (contrasted with software). (2) Particular circuits of functions built into a system. (IEEE)

Harmonic Distortion—The production of harmonic frequencies at the output by the nonlinearity of a system when a sinusoidal input is applied. (GRAF)

Heat Sink—A mounting base, usually metallic, that dissipates, carries away, or radiates into the surrounding atmosphere the heat generated within a semi-conductor device. (GRAF)

Hertz—The unit of frequency, one cycle per second. (IEEE)

High-Threshold Logic—Abbreviated HTL. Logic with a high noise margin, used primarily in industrial applications. It closely resembles DTL, except that in HTL a reverse-biased emitter junction is used as a threshold element operating as a zener diode. A typical noise margin is 6 volts with a 15-volt supply. (GRAF)

Hole—In the electronic valence structure of a semi-conductor, a mobile vacancy which acts like a positive electronic charge with a positive mass. (GRAF)

Hole Conduction—The apparent movement of a hole to the more negative terminal in a semi-conductor. Since the hole is positive, this movement is equivalent to a flow of positive charges in that direction.

Hybrid Circuit—A circuit which combines the thin-film and semi-conductor technologies. Generally, the passive components are made by thin-film techniques, and the active components by semi-conductor techniques. (GRAF)

Hysteresis—The difference between the response of a unit or system to an increasing and a decreasing signal. Hysteretical behavior is characterized by inability to *retrace* exactly on the reverse swing a particular locus of input/output conditions. (GRAF)

Impedance (Z)—The total opposition offered by a component or circuit to the flow of alternating or varying current. Impedance is expressed in ohms and is similar to the actual resistance in a direct current circuit. Impedance may be computed as $Z = E/I$, where E is the applied AC voltage and I is the resulting alternating current flow in the circuit.

Inductance—The property of an electric circuit by which a varying current in it produces a varying magnetic field that induces voltage in the same circuit or in a nearby circuit—measured in henrys.

Inductor—A device consisting of one or more associated windings, with or without a magnetic core, for introducing inductance into an electric circuit. (IEEE)

Integrated Circuit—A combination of interconnected circuit elements inseparably associated on or within a continuous substrate. NOTE: To further define the nature of an integrated circuit, additional modifiers may be prefixed. Examples are: (1) dielectric-isolated monolithic integrated circuit.

(2) beam lead monolithic integrated circuit. (3) silicon-chip tantalum thin-film hybrid integrated circuit. (IEEE)

Input Impedance—The impedance a transducer presents to a source. The effective impedance *seen looking into* the input terminals of an amplifier; circuit details, signal level, and frequency must be specified. (GRAF)

Insulator—A high resistance device that supports or separates conductors to prevent a flow of current between them or to other objects. (GRAF)

Inverse Voltage—The effective voltage across a rectifier during the half-cycle when current does not flow. (GRAF)

Ion Implantation—A method of semi-conductor doping in which impurities that have been ionized and accelerated to a high velocity penetrate the semi-conductor surface and become deposited in the interior. (GRAF)

JFET—Abbreviation for Junction Field Effect Transistor.

Jump—(electronic computation) (1) To (conditionally or unconditionally) cause the next instruction to be obtained from a storage location specified by an address part of the current instruction when otherwise it would be specified by some convention. (2) An instruction that specifies a jump. (IEEE)

Latch—A feedback loop used in a symmetrical digital circuit (such as a flip-flop) to retain a state. (GRAF)

Large-Scale-Integration—Abbreviated LSI. (1) The simultaneous achievement of large area circuit chips and optimum density of component packaging for the express purpose of cost reduction by maximization of the number of system interconnections made at the chip level. (2) Monolithic digital ICs with a typical complexity of 100 or more gates or gate-equivalent circuits. The number of gates per chip used to define LSI depends on the manufacturer. The term sometimes describes hybrid ICs built with a number of MSI or LSI chips. (GRAF)

Lead Frame—A metal frame that holds the leads of a plastic encapsulated package (DIP) in place before encapsulation and is cut away after encapsulation. (GRAF)

Light-Emitting Diode—A pn junction that emits light when biased in the forward direction. (GRAF)

Linearity—The relationship between two quantities when a change in a second quantity is directly proportionate to a change in the first quantity. Also, deviation from a straight-line response to an input signal. (GRAF)

Logic—A mathematical approach to the solution of complex situations by the use of symbols to define basic concepts. In computers and information-processing networks, the systematic method that governs the operations performed on the information, usually with each step influencing the one that follows. (GRAF)

Magneto Resistive Effect—The change in the resistance of a conductor or semi-conductor due to the application of a magnetic field.

Memory—(electronic computation) See Storage.

Metalization—The deposition of a thin-film pattern of a conductive material onto a substrate to provide interconnection of electronic components or to provide conductive pads for interconnections. (GRAF)

Microcomputer—A complete system capable of performing minicomputer functions, through a much lower power range. It is a combination of the chip sets; inferface I/O along with the auxiliary circuits, power supply, and control console.

Micron—A unit of length equal to 10^{-6} metre.

Microprocessor—The digital processor on a chip which performs arithmetic logic and control logic. It is the basic building block of a microcomputer system.

Minority Carrier—The less predominate carrier in a semi-conductor. Electrons are the minority carriers in P-type semi-conductors since there are fewer electrons than holes. Holes are the minority carriers in N-types since they are outnumbered by electrons. (GRAF)

Monolithic—An integrated circuit which is built on a single slice of silicon substrate.

MOS—Abbreviation for Metal Oxide Semi-Conductor.

MNOS—Abbreviation for Metal-Nitride-Oxide Semi-Conductor. (GRAF)

Multiplexing—The process of combining several measurements for transmission over the same signal path. There are two widely used methods of multiplexing; time division, and frequency division. Time division utilizes the principle of time sharing among measurement channels. Frequency division utilizes the principle of frequency sharing among information channels where the data from each channel are used to modulate sinusoidal signals called subcarriers so that the resultant signal representing each channel contains only frequencies in a restricted narrow frequency range. Multiplex radio transmission, for instance, is the simultaneous transmission of two signals over a common carrier wave. (GRAF)

NAND Gate—A combination of a *not* function and an *and* function in a binary circuit that has two or more inputs and one output. (GRAF)

Negative Feedback—(degeneration) A process by which a part of the output signal of an amplifying circuit is fed back to the input. (GRAF)

NMOS (N-Type MOS)—MOS devices made on P-type silicon substrates where the active carriers are electrons flowing between N-type source and drain contacts.

Noise—Unwanted disturbances superposed on a useful signal that tend to obscure its information content.

Non-Volatile Memory—Electronic memory which is not lost during power off conditions.

NOR Gate—An *or* gate followed by an inverter to form a binary circuit in which the output is logic zero if any of the inputs is one, and vice versa. (GRAF)

NPN Transistor—A transistor with a P-type base and N-type collector and emitter. (GRAF)

Null—A condition (typically a condition of balance) which results in a minimum absolute value of output. Often specified as the calibration point when the least error can be tolerated by the associated control system.

N-Type Material—A crystal of pure semi-conductor material to which has been added an impurity so that electrons serve as the majority charge carriers. (GRAF)

Ohm—The unit of resistance. One ohm is the value of resistance through which a potential of one volt will maintain a current of one ampere. (GRAF)

Operational Amplifier—An amplifier that performs various mathematical operations. Also called OP-AMP. (GRAF)

OR Gate—A multiple-input gate circuit whose output is energized when any one or more of the inputs is in a prescribed state. Used in digital logic.

Oscillator—An electronic device which generates alternating current power at a frequency determined by the values of certain constants in its circuits. (GRAF)

Parallel Processing—Pertaining to the simultaneous execution of two or more sequences of instructions by a computer having multiple arithmetic or logic units. (IEEE)

Permeability—(μ) The measure of how much better a given material is than air as a path for magnetic lines or force. It is equal to the magnetic induction (B) in gausses, divided by the magnetizing force (H) in oersteds. (GRAF)

Phase Angle—(1) *general.* The measure of the progression of a periodic wave in time or space from a chosen instant or position. NOTES: (a) The phase angle of a field quantity, or of voltage or current, at a given instant of time at any given plane in a waveguide is [wt − Bz + θ], when the wave has a sinuosoidal time variation. The term waveguide is used here in its most general sense and includes all transmission lines; for example, rectangular waveguide, coaxial line, strip line, etc. The symbol B is the imaginary part of the propagation constant for that waveguide, propagation is in the +z direction, and θ is the phase angle when z = t = 0. At a reference time t = 0 and at the plane z, the phase angle [−Bz + θ] will be represented by ϕ. (b) Phase angle is obtained by multiplying the phase by 360 degrees or by 2π radians. (2) *current transformer.* The angle between the current leaving the identified secondary terminal and the current entering the identified primary terminal. NOTE: This angle is conveniently designated by the Greek letter beta (β) and is considered positive when the secondary current leads the primary current. (3) *potential [voltage] transformer.* The angle between the secondary voltage from the identified to the unidentified terminal and the corresponding primary voltage. NOTE: This angle is conveniently designated by the Greek letter gamma (γ) and is considered positive when the secondary voltage leads the primary voltage. (4) (*instrument transformer*) Phase displacement, in minutes, between the primary and secondary values. (IEEE)

Photocell—*photoelectric cell* (1) A solid-state photosensitive electron device in which use is made of the variation of the current-voltage characteristic as a function of incident radiation. (2) A device exhibiting photovolatic or photo-conductive effects. (IEEE)

Piezoelectric—The property of certain crystals, which: (1) produce a voltage when subjected to a mechanical stress, (2) undergo mechanical stress when subject to a voltage. (GRAF)

Plasma—A gas made up of charged particles. NOTE: Usually plasmas are neutral, but not necessarily so, as, for example, the space charge in an electron tube. (IEEE)

PMOS (P-Type MOS)—MOS devices made on an N-type silicon substrate where the active carriers are holes flowing between P-type source and drain controls.

PNP Transistor—A transistor consisting of two P-type regions separated by an N-type region. (GRAF)

PNPN Diode—A semi-conductor device which may be regarded as a two transistor structure with two separate emitters feeding a common collector. (GRAF)

Potential—The difference in voltage between two points of a circuit. Frequently one point is assumed to be ground which has zero potential. (GRAF)

Positive Feedback—regeneration The process by which the amplification is increased by having part of the power in the output returned to the input in order to reinforce the input power. (GRAF)

Potting—An embedding process for parts that are assembled in a container or can into which the insulating material is poured, with the container remaining an integral part as the outer surface of the finished unit. (GRAF)

P-Type Material—A semi-conductor material that has been doped with an excess of acceptor impurity atoms, so that free holes are produced in the material. (GRAF)

PROM—An acronym for Programmable Read Only Memory. An electronic memory which may be permanent (non-volatile) or semi-permanent (erasable electronically or with ultra-violet light) and therefore able to be reprogrammed one or more times.

RAM—An acronym for Random Access Memory. A memory that has stored information immediately available when addressed regardless of the previous memory address location. As the memory words can be selected in any order, there is equal access time to all.

rfi—radio-frequency interference Radio frequency energy of sufficient magnitude to have an influence on the operation of other electronic equipment. (GRAF)

Rectifier—A device which, by virtue of its asymmetrical conduction characteristic, converts an alternating current into a unidirectional current. (GRAF)

Regulated Power Supply—A unit which maintains a constant output voltage or current for changes in line voltage, output load, ambient temperature or time. (GRAF)

Regulation—overall, power supplies The maximum amount that the output will change as a result of the specified change in line voltage, output load, temperature, or time. NOTE: Line regulation, load regulation, stability, and temperature coefficient are defined and usually specified separately. (IEEE)

Relay—An electric device that is designed to interpret input conditions in a prescribed manner and after specified conditions are met to respond to cause contact operation or similar abrupt change in associated electric control circuits. NOTES: (1) inputs are usually electric, but may be mechanical, thermal, or other quantities. Limit switches and similar simple devices are not relays. (2) a relay may consist of several units, each responsive to specified inputs, the combination providing the desired performance characteristic. (IEEE)

Resistivity—The measure of the resistance of a material to electric current either through its volume or on a surface. (GRAF)

Resistor—A device the primary purpose of which is to introduce resistance into an electric circuit. NOTE: Resistor as used in electric circuits for purposes of operation, protection, or control, commonly consists of an aggregation of units. Resistors as commonly supplied consist of wire, metal ribbon, cast metal, or carbon compounds supported by or imbedded in an insulation medium. The insulating medium may enclose and support the resistance material as in the case of the porcelain-tube type, or the insulation may be provided only at the points of support as in the case of heavy-duty ribbon or cast iron grids mounted in metal frames. (IEEE)

Resistor-Capacitor-Transistor Logic—Abbreviated RCTL. A logic circuit design that employs a resistor and a speedup capacitor in parallel for each input of the gate. A transistor's base is connected to one end of the RC network. A positive voltage on the RC input will energize the transistor and turn it on, so that the output voltage is nearly zero volts. This circuit is a positive NOR or negative NAND when NPN transistors are used in the circuit. (GRAF)

Resistor-Transistor Logic—Abbreviated RTL. A form of logic that has a resistor as the input component that is coupled to the base of an NPN transistor. As in RCTL, the transistor is an inverting element that produces the positive NOR gate or the negative NAND gate function. (GRAF)

Resist Plating—Any material which, when deposited on a conductive area, prevents the areas underneath from being plated. (GRAF)

Resonant Frequency—The frequency at which a given system or object will respond with maximum amplitude when driven by an external sinusoidal force of constant amplitude. (GRAF)

Rise Time—The time required for the leading edge of a pulse to rise from 10–90% of its final value. It is proportionate to the time constant and is a measure of the steepness of the wavefront. Also, the measured length of time required for an output voltage of a digital circuit to change from a low voltage level (0) to a high voltage level (1) after the change has started. (GRAF)

ROM—An acronym for Read Only Memory. A memory which permits the reading of a predetermined pattern of *Zeros* and *Ones*. This predetermined information is stored in the ROM at the time of its manufacture. A ROM is analogous to a dictionary where a certain address results in predetermined information output.

Saturation—A circuit condition whereby an increase in the driving or input signal no longer produces a change in the output. (GRAF)

Saturation Voltage—Generally, the voltage excursion at which a circuit self-limits (i.e., is unable to respond to excitation in a proportional manner). (GRAF)

Schottky Barrier—A simple metal to semi-conductor interface that exhibits a nonlinear impedance. (GRAF)

Semi-conductor—An electronic conductor, with resistivity in the range between metals and insulators, in which the electric-charge-carrier concentration increases with increasing temperature over some temperature range. NOTE: Certain semi-conductors possess two types of carriers, namely, negative electrons and positive holes.

Semi-Conductor Controlled Rectifier (SCR)—An alternate name used for the reverse-blocking triode-thyristor. NOTE: The name of the actual semi-conductor material (selenium, silicon, etc.) may be substituted in place of the word *semi-conductor* in the name of the components. (IEEE)

Sensitivity—Measure of the ability of a device or circuit to react to a change in some input. Also, the minimum or required level of an input necessary to obtain rated output.

Serial-Parallel—Pertaining to processing that includes both serial and parallel processing, such as one that handles decimal digits serially but handles the bits that comprise a digit in parallel. (IEEE)

Serial Transmission—data transmission, telecommunication Used to identify a system wherein the bits of a character occur serially in time. Implies only a single transmission channel. Also called serial by bit. (IEEE)

Shift Register—(1) A logic network consisting of a series of memory cells such that a binary code can be caused to shift into the register by serial input to only the first cell. (2) A register in which the stored data can be moved to the right or left.

Signal—(1) A visual, audible, or other indication used to convey information. (2) The intelligence, message, or effect to be conveyed over a communication system. (3) A signal wave; the physical embodiment of a message. (4) *computing systems.* The event or phenomenon that conveys data from one point to another. (5) *control, industrial control.* Information about a variable that can be transmitted in a system. (IEEE)

Signal Generator—A shielded source of voltage or power, the output level and frequency of which are calibrated, and usually variable over a range. NOTE: The output of known waveform is normally subject to one or more forms of calibrated modulation. (IEEE)

Signal-To-Noise-Ratio—The ratio of the value of the signal to that of the noise. NOTES: (a) This ratio is usually in terms of peak values in the case of impulse noise and in terms of the root-mean-square values in the case of the random noise. (b) Where there is a possibility of ambiguity, suitable definitions of the signal and noise should be associated with the terms; as, for example: peak-signal to peak-noise ratio; root-mean-square signal to root-mean square noise ratio; peak-to-peak signal to peak-to-peak noise ratio, etc. (c) This ratio may be often expressed in decibels. (d) This ratio may be a function of the bandwidth of the transmission system. (IEEE)

Silicon-On-Sapphire—Pertaining to the technology in which monocrystalline silicon films are epitaxially deposited onto a single-crystal sapphire substrate to form a structure for the fabrication of dielectrically isolated elements. Abbreviated SOS. (GRAF)

Software—(1) Computer programs, routines, programming languages and systems. (2) The collection of related utility, assembly, and other programs, that are desirable for properly presenting a given machine to a user. (3) Detailed procedures to be followed, whether expressed as programs for a computer or as procedures for an operator or other person. (4) Documents, including hardware manuals and drawings computer-program listings and diagrams, etc. (5) Items such as those in (1), (2), (3), and (4) as contrasted with hardware. (IEEE)

Solid-State—Pertaining to circuits and components using semi-conductors. (See solid-state devices.) (GRAF)

Solid State Device—Any element that can control current without moving parts, heated dilaments, or vacuum gaps. All semi-conductors are solid-state devices, although not all solid-state devices are semi-conductors (e.g., transformers). (GRAF)

Solid State Relay—A relay constructed exclusively of solid-state components.

Source(s) (or Source Electrode)—In a field effect transistor, the electrode that is analogous to the emitter of a transistor or the cathode of a vacuum tube. (GRAF)

Source Impedance—The impedance which a source of energy presents to the input terminal of a device. (GRAF)

Stability—The ability of a component or device to maintain its nominal operating characteristics after being subjected to changes in temperature, environment, current, and time. (GRAF)

Steady-State—A condition in which circuit values remain essentially constant, occurring after all initial transients or fluctuating conditions have settled down. (GRAF)

Storage—electronic computation (1) The act of storing information. (2) Any device in which information can be stored, sometimes called a memory device. (3) In a computer, a section used primarily for storing information. Such a section is sometimes called a memory or store (British). NOTES: (a) The physical means of storing information may be electrostatic, ferroelectric, magnetic, acoustic, optical, chemical, electronic, electric, mechanical, etc., in nature. (b) Pertaining to a device in which data can be entered, in which it can be held, and from which it can be retrieved at a later time. (IEEE)

Substrate—The supporting material on or in which the parts of an integrated circuit are attached or made. (GRAF)

Thermal Resistor—An electronic device which makes use of the change in resistivity of a semi-conductor with changes in temperature. (GRAF)

Thermal Runaway—A condition in which the dissipation in a transistor or other device increases so rapidly with higher temperature that the temperature keeps on rising. (GRAF)

Thermistor—A solid-state semi-conducting device, the electrical resistance of which varies with the temperature. Its temperature coefficient of resistance is high, nonlinear, and negative. (GRAF)

Thermocouple—Also called thermal junction. A device for measuring temperature where two electrical conductors of dissimilar metals are joined at the point of heat application and a resulting voltage difference, directly proportional to the temperature, is developed across the free ends and is measured potentiometrically. (GRAF)

Thick-Film—Pertaining to a film pattern usually made by applying conductive and insulating materials to a ceramic substrate by a silk-screen process. Thick films can be used to form conductors, resistors, and capacitors. (GRAF)

Thin-Film—A film of conductive or insulating material, usually deposited by sputtering or evaporation, that may be made in a pattern to form electronic components and conductors on a substrate or used as insulation between successive layers of components. (GRAF)

Thyristor—A bistable semi-conductor device comprising three or more junctions that can be switched from the *off* state to the *on* state or vice versa, such switching occurring within at least one quadrant of the principal voltage current characteristic. (IEEE)

Transducer—A device by means of which energy can flow from one or more transmission systems or media to one or more other transmission systems or media. NOTE: The energy transmitted by these systems or media may be of any form (for example, it may be electric, mechanical, or acoustical), and it may be of the same form or different forms in the various input and output systems or media. (IEEE)

Transformer—A device consisting of a winding with tap or taps, or two or more coupled windings with or without a magnetic core for introducing mutual coupling between electric circuits. (IEEE)

Transient—A phenomenon caused in a system by a sudden change in conditions, and which persists for a relatively short time after the change. Also a momentary surge on a signal or power line. It may produce false signals or triggering impulses and cause insulation or component breakdowns or failures. (GRAF)

Transistor—An active semi-conductor device with three or more terminals. (IEEE)

Transistor-Transistor Logic—Abbreviated TTL or T^2L. Also called multi-emitter transistor logic. A logic-circuit design similar to DTL, with the diode inputs replaced by a multiple emitter transistor. In a four-input DTL gate, there are four diodes at the input. A four-input TTL gate will have four emitters of a single transistor as the input element. TTL gates using NPN transistors are positive-level NAND gates or negative-level NOR gates. (GRAF)

Triac—A five-layer NPNPN device that is equivalent to two SCRs connected in antiparallel with a common gate. It provides switching action for either polarity of applied voltage and can be controlled in either polarity from the single gate electrode. (GRAF)

Unijunction Transistor—A three terminal semi-conductor device exhibiting stable open-circuit, negative resistance characteristics. (GRAF)

VAR—Abbreviation for Volt Ampere Reactive. The unit of reactive power, as opposed to real power in watts. One VAR is equal to one reactive volt-ampere. (GRAF)

Varistor—A two-electrode semi-conductor device with a voltage-dependent nonlinear resistance that drops markedly as the applied voltage is increased. (GRAF)

Volatile Memory—An electronic memory (RAM) which temporarily stores data that is lost when the power is turned off.

Volt—The unit of voltage or potential difference in SI units. The volt is the voltage between two points of a conducting wire carrying a constant current of one ampere, when the power dissipated between these points is one watt. (IEEE)

Waveguide—(1) Broadly, a system of material boundaries capable of guiding electro-magnetic waves. (2) More specifically, a transmission line comprising a hollow conducting tube within which electromagnetic waves may be propagated or a solid dielectric or dielectric filled conductor for the same purpose. (3) A system of material boundaries or structures for guiding transverse-electromagnetic mode, often and originally a hollow metal pipe for guiding electromagnetic waves. (IEEE)

Watt—A unit of the electric power required to do work at the rate of one joule per second. It is the power expended when one ampere of direct current flows through a resistance of one ohm. (GRAF)

Zener Diode—A two layer device that, above a certain reverse voltage (the Zener value), has a sudden rise in current. If forward-biased, the diode is an ordinary rectifier. But, when reversed-biased, the diode exhibits a typical knee, or sharp break, in its current-voltage graph. The voltage across the device remains essentially constant for any further increase of reverse current, up to the allowable dissipation rating. The Zener diode is a good voltage regulator, over voltage protector, voltage reference, level shifter, etc. True Zener breakdown occurs at less than six volts. (See also Avalanche Diode.) (GRAF)

Zener Effect—A reverse current breakdown due to the presence of a high electrical field at the junction of a semi-conductor or insulator. (GRAF)

PERFORMANCE REQUIREMENTS FOR THE AUTOMOTIVE AUDIO CASSETTE— SAE J1274 JUN80

SAE Recommended Practice

Report of the Electrical Equipment Committee, approved June 1980.

1. Scope—This performance document for the audio cassette was developed by qualified engineers from the automotive industry, the prerecorded tape industry, and the tape and cassette manufacturers. It is based upon sound engineering principles supported by laboratory tests and field experience. The values given are supported by all of these industries and considered necessary to establish customer acceptance and satisfaction.

2. Purpose—This document lists the performance values for the test procedure listed in SAE J1275, Testing Methods for Audio Cassettes.

3. Cross Reference—The performance requirements listed in Section 4 of this document are cross referenced to the appropriate paragraph numbers in Section 6 of J1275, Testing Methods for Audio Cassettes, by showing the test procedure paragraph in parentheses immediately following the test value paragraph number. For example 4.1 (6.1) means that the results listed in paragraph 4.1 are obtained from the test procedure listed in paragraph 6.1 of J1275, and so on.

4. Test Values

4.1 (6.1) **Laboratory Conditions**—These are given in paragraph 6.1 in J1275.

4.2 (6.2) **Breakout Lugs**—The deflection at 2 N force shall not exceed 0.5 mm (0.02 in). The force to break the lug shall be greater than 3 N. When the lug breaks, it shall break cleanly so as to provide the minimum hole size shown on drawing.

4.3 (6.3) **Yield and/or Tensile Strength**—A change in length of not more than 0.2% shall be acceptable and the tape should not break or deform.

4.4 (6.4) **Hub Anchorage**—The hub tape or leader shall not break or separate to be acceptable.

4.5 (6.5) **Splice Strength**—The increase in the gap between the tape and the leader shall not increase more than 0.13 mm (0.005 in).

4.6 (6.6) **Friction Torque**—The maximum torque required to move the tape and both reels shall be 2.7×10^{-3} N·m (0.38 oz·in) when measured at nearly full take up reel and with no hold back torque applied to the supply reel. This same measurement made with a hold back torque of 0.8×10^{-3} N·m (0.113 oz·in) applied to the nearly empty supply reel shall not exceed 5.5×10^{-3} N·m (0.76 oz·in).

4.7 (6.7) Pressure Pad—The pressure due to the pressure pad shall be in the range of 0.005—0.015 N/mm^2 (0.73–2.18 psi).

4.8 (6.8) Mechanical Noise—The difference in the noise levels shall not exceed 8 dB in the play mode and 25 dB in the fast forward or fast reverse mode.

4.9 (6.9) Magnetic Properties of Tapes

4.9.1 (6.9.1) DEFINITIONS—The definitions of the various terms are shown in J1275.

4.9.2 (6.9.2) SENSITIVITY MEASUREMENTS—The output of the sample tape being tested shall be not more than 4 dB below Level Set Reference Cassette at both 315 Hz and 5 kHz.

4.9.3 (6.9.3) FREQUENCY RESPONSE—The frequency response shall be within 4 dB of the Level Set Reference Cassette.

4.9.4 (6.9.4) UNIFORMITY—The uniformity of the 315 Hz signal shall not exceed 2 dB. For the 5 kHz signal the uniformity shall not exceed 3 dB.

4.9.5 (6.9.5) PRERECORDED CASSETTE OUTPUT LEVEL—The output level for a prerecorded cassette shall be -13 vu.

4.10 (6.10) Life Test Cycle—The definition of a life test cycle is defined in J1275.

4.11 (6.11) Environmental Conditions Test—After exposure to the following environmental tests listed in J1275, the cassette shall be capable of meeting all of the test requirements contained in paragraphs 4.2–4.9.5 and may be so measured. The wow and flutter must be measured per paragraph 5.4.3 after each test and shall be less than 0.4%.

Paragraph	Test
6.11.1.1	High Temperature Storage
6.11.2.1	Low Temperature Storage
6.12	Humidity
6.14.2	Non-Operating Vibration
6.16	Thermocycling

4.12 (6.11.1.2) High Temperature Operating Test—The cassette shall operate satisfactorily and wow and flutter shall be less than 0.4%.

4.13 (6.11.2.2) Low Temperature Operating Test—The cassette shall operate satisfactorily and the wow and flutter after 3 min of operation shall be less than 0.6%.

4.14 (6.14.1) Operating Vibration—The cassette shall operate satisfactorily and the wow and flutter shall not be more than 1%.

4.15 (6.15) Life Test—The life expectancy of a cassette shall be such that the mean time to failure at a 90% confidence level is the 100 cycles described in paragraph 6.10, J1275.

4.16 (6.13) Drop Test—After the four drops specified in paragraph 6.13, J1275, the cassette shall not be cracked or disassembled and shall function satisfactorily in a player.

TESTING METHODS FOR AUDIO CASSETTES— SAE J1275 JUN80

SAE Recommended Practice

Report of the Electrical Equipment Committee, approved June 1980.

1. Introduction—With the technical advances made in magnetic tape coatings and record and reproducing heads, a larger and larger portion of recorded entertainment is being put into audio cassettes. Various forms of educational subjects and information are appearing. All of these will become an increasingly important part of the information and entertainment systems in the automobile. As the number of audio cassettes in use in the automobile grows, the consumer demand for a high performance, reliable audio cassette will increase. A large amount of effort has gone into standardizing the cassette by various groups. However, only a small effort has been made toward the special requirements needed in the automobile. This recommended practice covers the physical dimensions, test methods, and procedures; and SAE J1274 covers the performance requirements. This is the first issue of this SAE Recommended Practice on the audio cassette and it is planned to revise and reissue it again in about two years.

2. Scope—This recommended practice covers the mechanical properties, mechanical dimensions, test conditions, and procedures needed to evaluate the audio cassette for automotive use. In addition, a set of cassette player/recorder interface properties which ensures satisfactory operation of the cassette in the automobile is included.

3. Applicable Documents—Reference is made to the following documents covering the audio cassette and associated player/recorder.

SAE J1274
(Available from SAE, Warrendale, PA 15096)
ANSI Y14.5-1973
ANSI C16.5-1961 (IEEE Std. 152-1953)
ANSI S4.3-1972 (IEEE Std. 193-1971)
ANSI/EIA RS-399A
(ANSI reports available from American National Standards Institute, 1430 Broadway, New York, NY 10018)
DIN Standard 45500
DIN Calibration Tape 45513/6
DIN Calibration Tape PES 12/C 521
(DIN Standards and Tapes available from Deutsches Institut fur Normung eV, 8 Burggrafenstrasse, 1000 Berlin 30, West Germany)
Federal Standard W-C-1684
(Available from Naval Publications and Forms Center, 5801 Tabor Avenue, Philadelphia, PA 19120)
IEC-94A and Amendments
IEC 368-1972
(IEC Recommendations available from ANSI)
ISO 370-1975 Standard
(Available from ANSI)
ITA-A-101, Audio Cassette
ITA-A-102, Audio Cassette and Associated Hardware
ITA-A-104, Duplication Guidelines
(ITA Specifications available from ITA, 10 West 66th Street, New York, NY 10023)

Appreciation is expressed to ITA which extended their permission to use parts of their specification and to EIA for the use of their drawings on the cassette.

4. Physical Dimensions—Figs. 1–3 inclusive (Figs. 1A–3A inclusive for English dimensions) contain the dimensional requirements for the cassette.

4.1 Tape Winding—Magnetic coated surface of the tape shall face outward from the open end of the cassette.

4.2 Breakout Lugs—Breakout lugs shown in Fig. 3 (Fig. 3A for English dimensions) when removed shall provide protection against accidental erasure of the relevant side. Dimensions for the lug holes are shown on Fig. 1 (Fig. 1A for English dimensions). The space between the lug and hole edges shall not exceed 1.0 mm (0.04 in).

4.3 Materials—All materials for the housing and other parts which are within the tape path shall be non-magnetic and shall not require lubrication or maintenance after manufacture.

4.4 Shield—Each audio cassette shall be provided with an effective magnetic shield positioned directly behind the pressure pad.

4.5 Support Planes—The support planes for the audio cassette are shown in Fig. 2 (Fig. 2A for English dimensions) by the hatched areas. These planes determine the location of the cassette in the player.

4.6 Tape Path—The unobstructed area for the tape to travel from the supply hub to the take-up hub is shown in Figs. 2 and 2A. The openings reserved for player/recorder components are shown in Fig. 4. These openings shall be unobstructed over the full height of the openings. Most cassettes are constructed in such a way that there is a small gap between two halves of the cassette. If this gap exists, steps shall be taken to minimize the possibility of the tape becoming trapped in this gap.

4.7 Window Area (Window Optional)—The maximum window area shall be as shown in Fig. 3 (Fig. 3A for English dimensions). The maximum allowable cassette thickness (required, for example, to accomodate marks indicating amount of wound and unwound tape) shall be as shown for each support plane. The window area, if provided, shall be covered with a clear, rigid material sealed to the housing to exclude dust and dirt and shall not interfere with the tape motion.

4.8 Tape Width—The tape shall have a width of $3.81 +0.00, -0.05$ mm ($0.150 +0.00, -0.002$ in).

4.9 Tape Length and Thickness—For pre-recorded tapes, the tape length shall be determined by program material. For blank tapes, the cassette shall contain a sufficient length of tape such that when playing one side at 0.0476 m/s (1.875 in/s) the measured playing time will not be less than the playing time per side or one half of the total playing time as specified on the label. Tape thickness shall be 11–20 μ (433–788 μ in).

4.10 Splices—Splices will not be allowed between the ends of the mag-

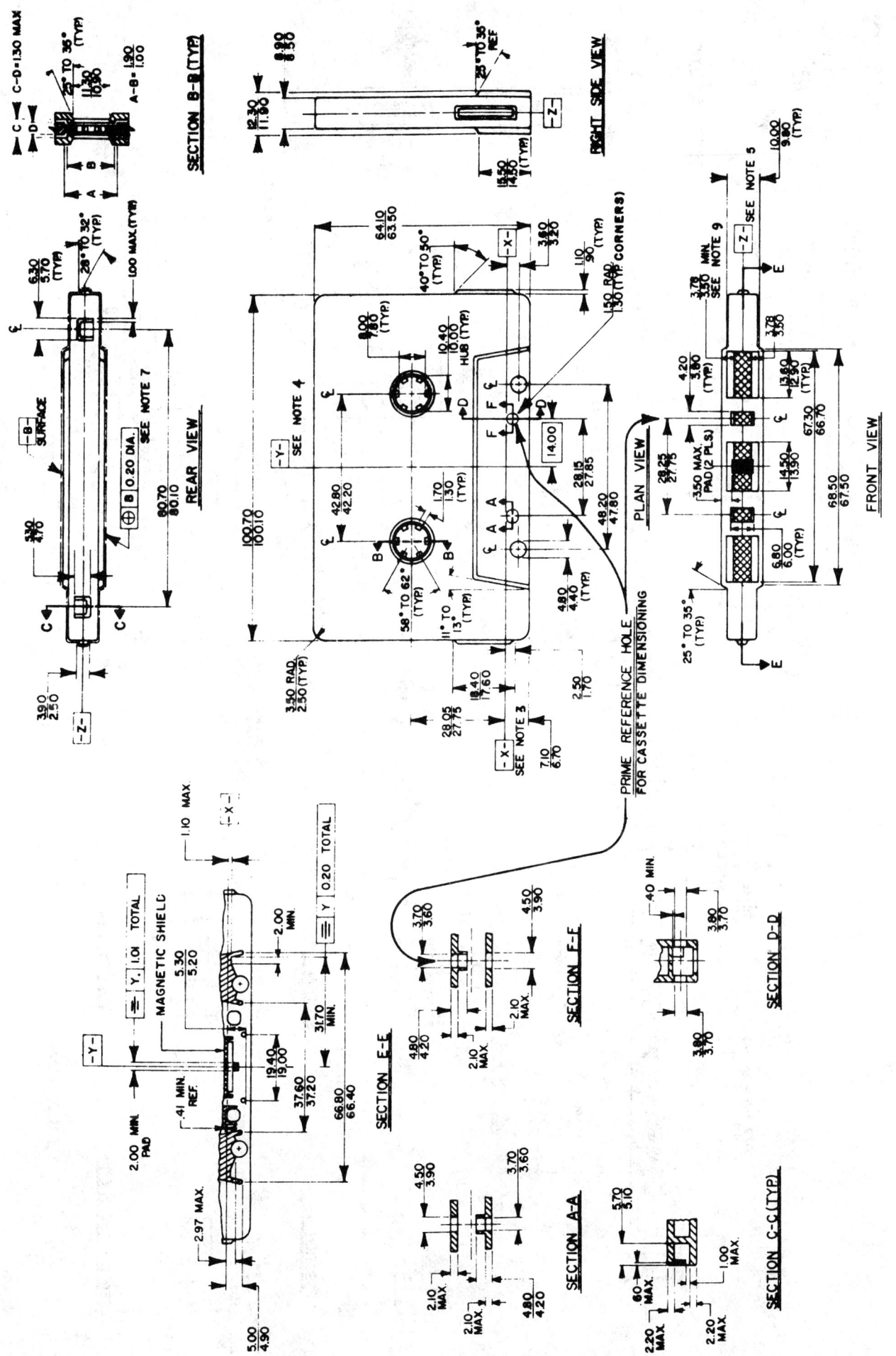

FIG. 1—COMPACT CASSETTE DIMENSIONS (MILLIMETERS)

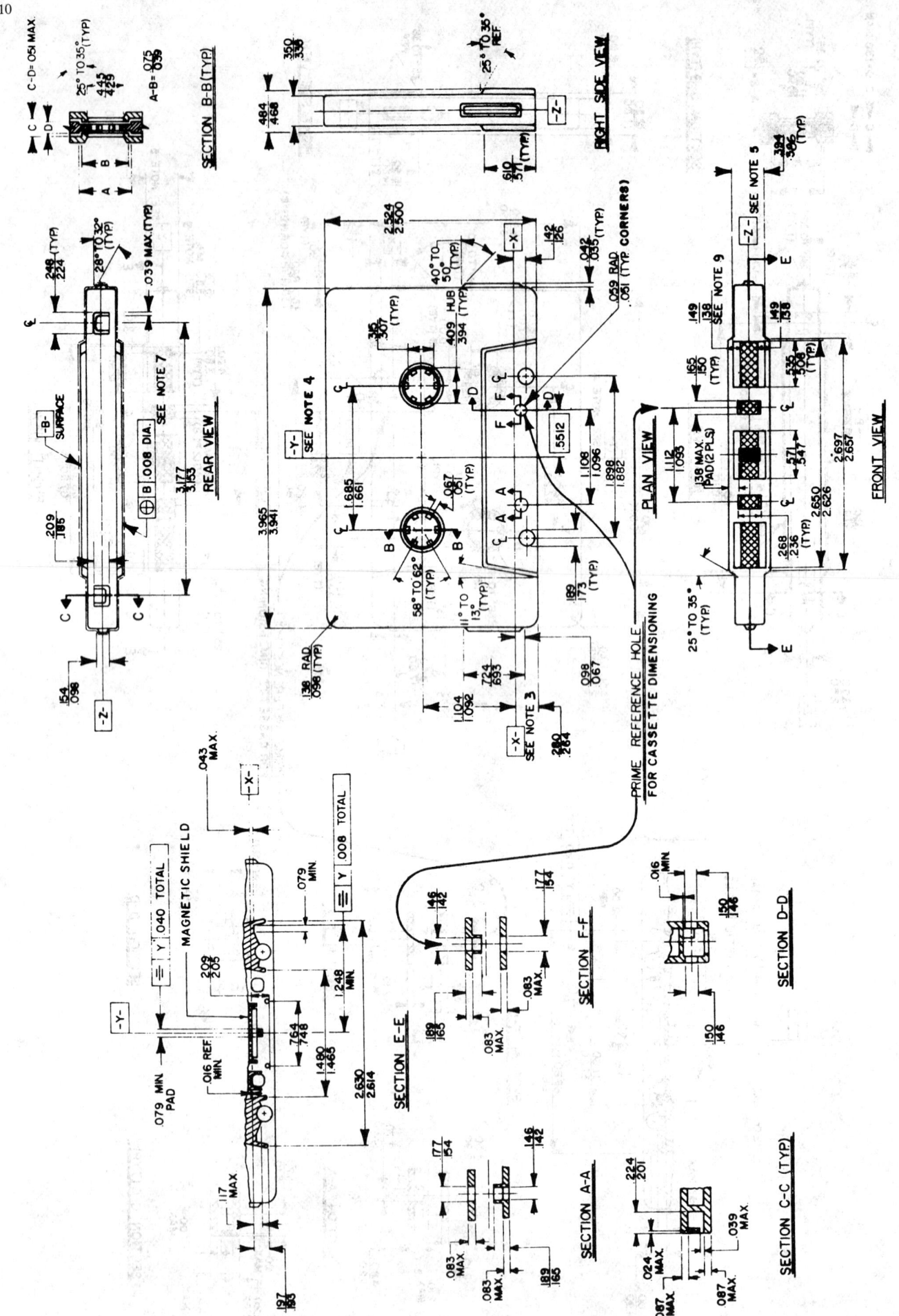

FIG. 1A—COMPACT CASSETTE DIMENSIONS (INCHES)

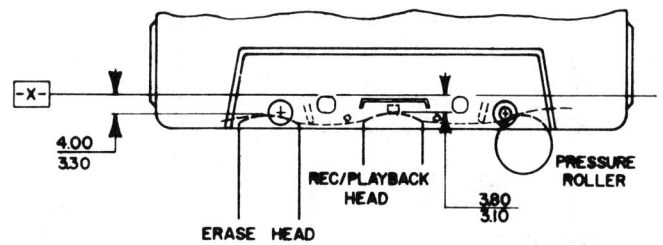

SUGGESTED HEAD PENETRATION

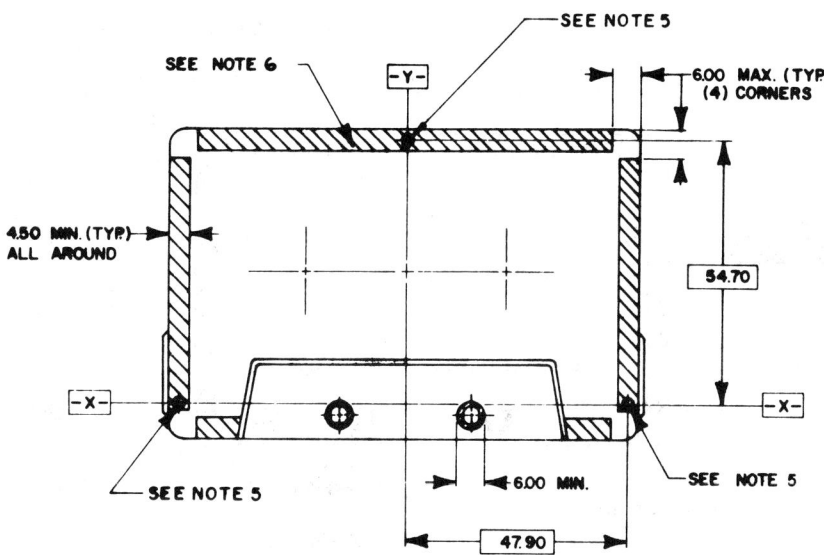

LOCATION, DIMENSIONS AND PARALLELISM OF SUPPORT PLANES

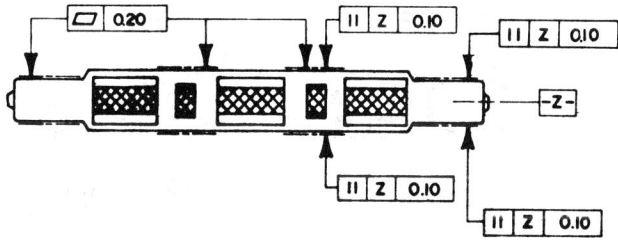

MILLIMETERS

FIG. 2

netic tape. For the splice between the leader and the tape it is recommended that the splice gap not exceed 0.4 mm ($\frac{1}{64}$ in) so that the adjacent layers of tape on the reel will not stick to the splice area.

4.11 Labels

4.11.1 The area for the label is shown in Fig. 3 (Fig. 3A for English dimensions). The maximum allowable depression in the thickness of the cassette in the label area is given for each support plane.

4.11.2 Labels shall be positioned within the recessed surface of the housing. Direct printing on the cassette housing is allowed; however, paragraphs 4.11.4 to 4.11.6 inclusive shall still apply.

4.11.3 Labels shall securely adhere to the housing and not loosen or curl after environmental and physical test conditions.

4.11.4 Inks and colors on labels shall not smear or come off under normal use and when subjected to the environmental and physical test conditions.

4.11.5 For blank tapes, labels shall specify either the total playing time or the playing time per side as well as the speed at which the tape must be played to obtain this playing time.

4.11.6 The label on blank tapes shall provide a blank area for consumer information which will readily accept lead pencil and ball point pen.

4.11.7 The label or the container insert shall contain a warning against exposing the cassettes to high temperatures from direct sunlight on both pre-recorded and blank tape cassettes.

4.12 Workmanship—The cassette shall be clean, free of cracks and molding flash, and have as small as possible a gate break off scar located in a non-critical area. A visual examination of the tape shall not show evidence of excessive tension that can occur during manufacture of the cassette.

4.13 Recording Format—The location of recorded information on the tape shall be as shown in Fig. 7. Tracks 1 and 2 shall be recorded for simultaneous reproduction for one direction of travel and shall be reproduced when playing Side 1 or Side A when inserted into the player with that side upward. Side 2 or Side B shall play in a similar manner using tracks 3 and 4.

5. Cassette Player Interface

5.1 Introduction—The performance and life of any cassette is highly dependent upon the player in which it is operating. A player not meeting certain performance standards will not play a standard cassette as defined in paragraph 5.2.1 in a satisfactory manner. Conversely, a player meeting these standards may not play a substandard cassette in a satisfactory manner. For this reason it is necessary to define certain minimum player interface standards so that the combined result is acceptable. This section describes the characteristics of a player/recorder that will successfully play/record a cassette.

5.2 Definitions

5.2.1 A standard cassette is one which meets the requirements of this document (J1275) and SAE Recommended Practice J1274.

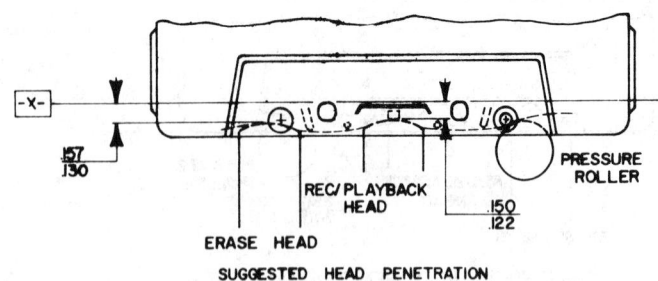

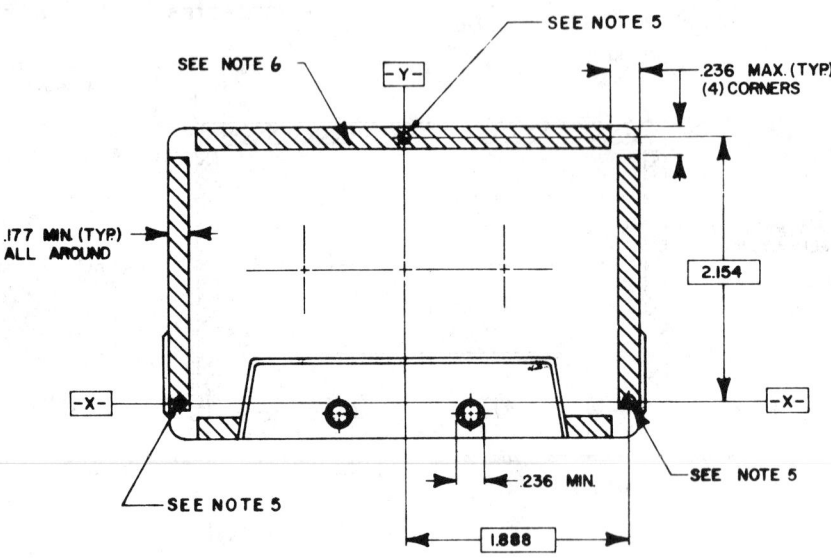

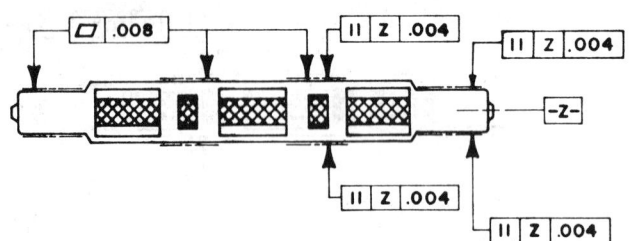

INCHES

FIG. 2A

5.2.2 A recorder for audio cassettes designates an instrument suitable for either recording or reproducing, or both on said cassette. While very few recorders are currently used in automobiles, references to recording are included because the user may record a cassette on other equipment and use it in the automobile.

5.2.3 POSITIONS—Positions as described in this document (that is, normal, horizontal, vertical, up, down, left, right, and so forth) shall be from an observer viewpoint with the cassette in such a position that the relevant label area is in the horizontal plane facing upward and tape head recess area facing the observer.

5.2.4 HEAD POSITIONS—Pictorial definitions of Azimuth (1), Height (2), Tilt (3), Tangency (4), and Contact (5) are shown in Fig. 5.

5.3 Player/Recorder Mechanical Requirements

5.3.1 The player shall satisfactorily accept the cassette as described herein.

5.3.2 SUPPORT PLANES—The player shall be so designed that the cassette is positioned and supported while operating at the support planes, prime reference line and reference hole as defined herein.

5.3.3 TAPE HEAD INSERTION—The tape head insertion shall be in accordance with the dimensions contained herein and shown in Fig. 2.

5.3.4 The deviation of azimuth of the head gap (refer to Fig. 5) when measured with an 8 kHz tape shall be within 3 dB from maximum output.

5.3.5 The surface of the heads in contact with the tape shall be within 2 deg perpendicular to the cassette support plane and within 2 deg of each other.

5.3.6 The spindle axis (refer to Fig. 6) shall be perpendicular to the reference support plane within 2 deg.

5.3.7 The guides, capstan, and pressure roller (refer to Fig. 6) shall be perpendicular to the support plane within 1 deg.

5.3.8 Any other member inserted into any part of the tape path shall be within 2 deg perpendicular to the cassette support plane. Fig. 4 defines the areas where this is permitted.

5.3.9 DRIVE TRAIN LAYOUT—The transport layout showing major interface points as shown in Fig. 6. This figure indicates the required position of the left and right hand spindles and their relationship to the capstan and prime reference line of the cassette.

5.3.10 TORQUE VALUES

5.3.10.1 *Running Torque*—The running torque shall be between 2.7×10^{-3} to 5.5×10^{-3} N·m (0.354 to 0.779 oz·in).

5.3.10.2 *Fast Forward and Rewind Modes*—The torque for fast forward and rewind modes shall be 5.5×10^{-3} to 9.0×10^{-3} N·m (0.778 to 1.27 oz·in).

5.4 Player/Recorder Electrical Requirements

5.4.1 TAPE SPEED—The standard tape speed for the player shall be 0.0476 m/s (1.875 in/s) −1, +3%.

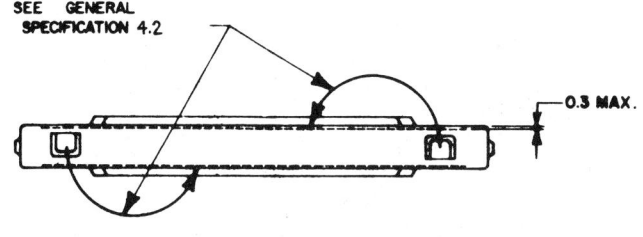

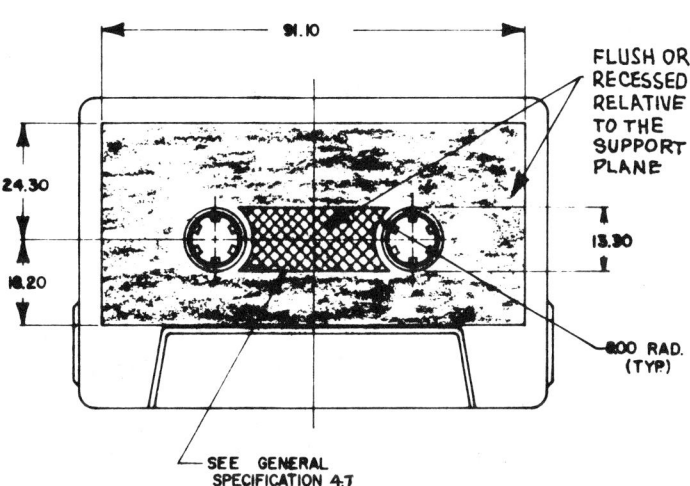

FIG. 3

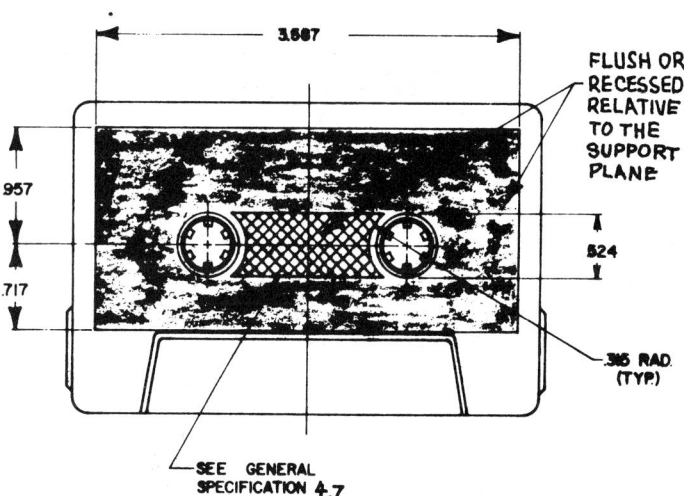

FIG. 3A

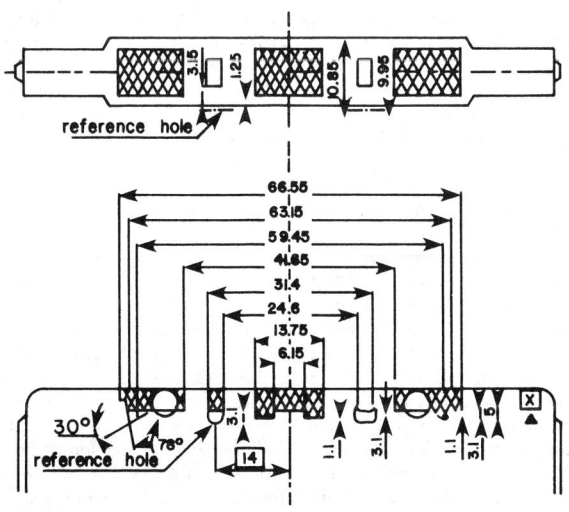

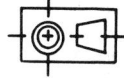

FIG. 4—INTERFACE WITH EQUIPMENT

The hatched areas, containing no other elements than the magnetic tape and the pressure pad (center hole), are available for inserting magnetic heads and other equipment elements. The dimensions of these areas have been derived from Fig. 2.

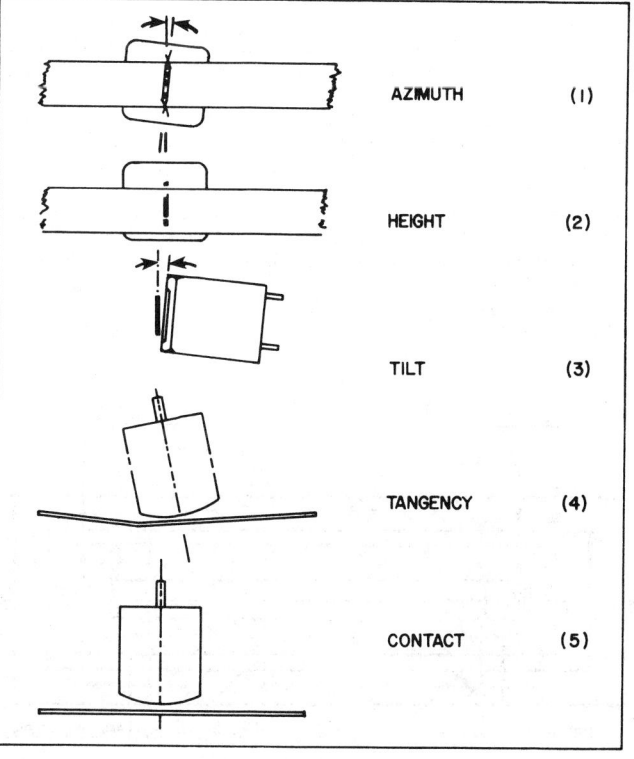

FIG. 5

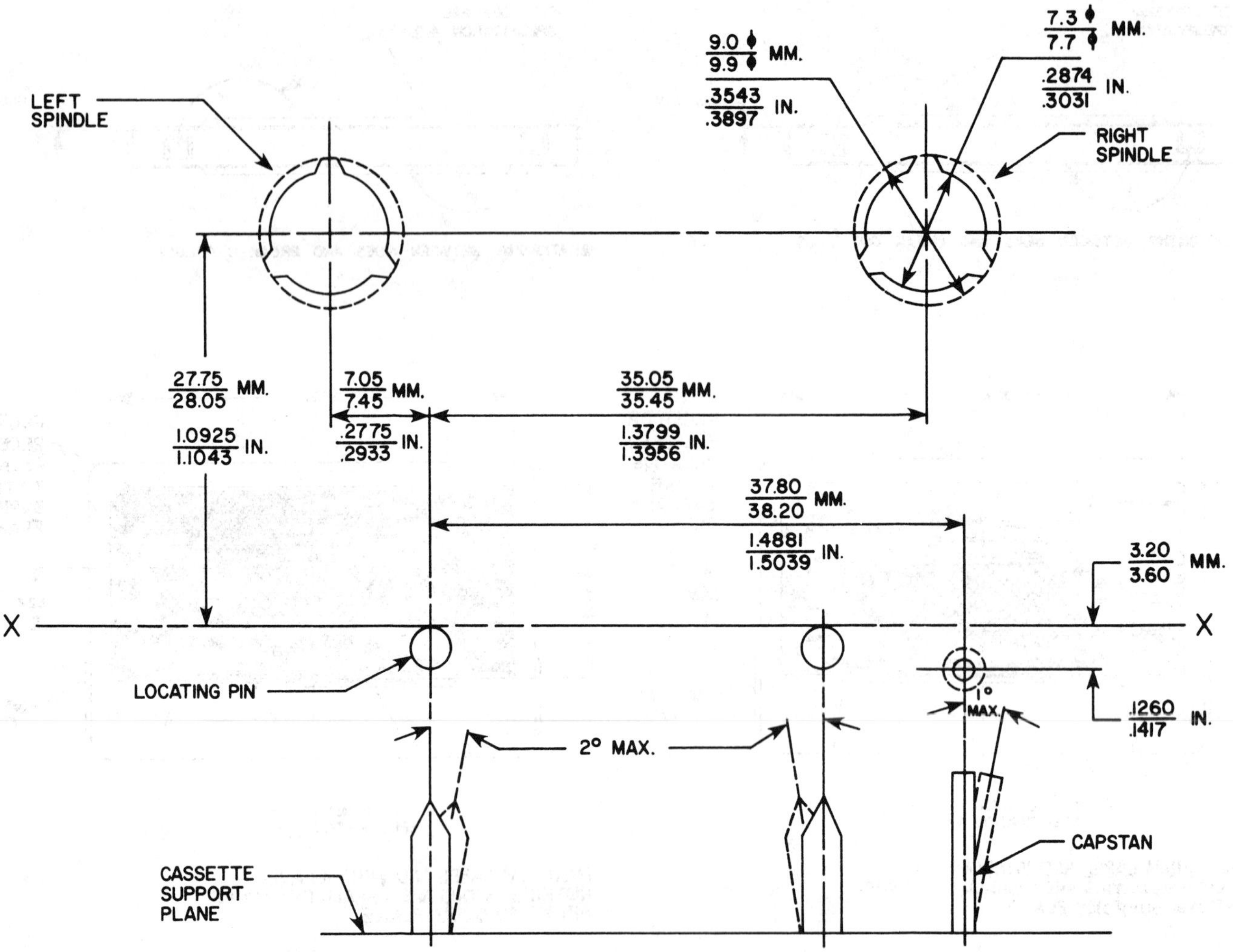

FIG. 6—TRANSPORT LAYOUT OF MAJOR INTERFACE POINTS

5.4.2 OPERATING VOLTAGE—The player shall operate without degraded performance between 11.0 and 16.0 V as measured at the input to the player.

5.4.3 WOW AND FLUTTER—The wow and flutter when measured according to ANSI S4.3 1972 or IEC 368-1972 shall be 0.3% or less.

5.4.4 SIGNAL TO NOISE RATIO—The signal to noise ratio on stereo shall be 40 dB minimum when measured by procedure contained in DIN 45500.

5.4.5 CROSSTALK—The crosstalk from opposite direction tracks shall be 45 dB minimum.

5.4.6 STEREO SEPARATION—The stereo separation between two stereo pairs shall be 20 dB minimum between 500 Hz and 6300 Hz.

5.4.7 FREQUENCY RESPONSE—The normal frequency shall be within 3 dB from 40–10 000 Hz.

6. Test Procedures

6.1 Unless otherwise specified, test shall be run at normal laboratory temperatures and humidities of 25 ± 5°C and a relative humidity of 20–50%.

6.2 **Breakout Lugs**—Apply a force to the center of the lug. Measure the amount of deflection of the lug at 2N and the force at which the lug breaks.

6.3 **Yield and/or Tensile Strength of Tape**—A section of tape 100 ± 2 mm long shall be measured before, during, and after applying a force of 2 N (0.44 lbf) for at least 2 min. Any change in length from before applying the force to that during and after shall be noted.

6.4 **Hub Anchorage**—The hub anchorage strength shall be measured by applying a force of 10 N (2.2 lbf) to the tape or leader that has been pulled out of the reproducing head opening. The force shall be applied tangent to the front face.

6.5 **Splice Strength**—The gap between the leader and the tape shall first be measured. A force of 2 N (0.44 lbf) shall be applied between leader and the tape for 5 min and the tape leader gap remeasured.

6.6 **Frictional Torque**—A torque tester, such as the Information Terminals Model M-400, or Minnetech Labs Model MTM 200 series, or equivalent, shall be used to perform this test. The cassette shall be inserted into the tester such that the nearly full hub is on the takeup side. The torque tester shall be started, and with no holdback torque applied to the supply hub, the average torque reading on the meter shall be noted. The foregoing test shall be repeated with a holdback torque of 0.8×10^{-3} N·m (0.113 oz·in) applied to the supply hub. The tape shall then be wound on the opposite hub and the foregoing tests repeated. After measuring the above listed torques, place the cassette in a player and play it for 5 min followed by 10 s of fast forward then 5 s of fast rewind. The torques shall again be measured. This play cycle shall be done with sufficient tape on each hub to prevent it from reaching the end and stopping. This measurement shall be repeated with the cassette turned over.

6.7 **Pressure Pad**—The pressure exerted by the pressure pad shall be measured on an Information Terminals Model M-400 or equivalent at both minimum and maximum head insertions (see Fig. 2).

6.8 **Mechanical Noise**—Mechanical noise of the cassette shall be measured using a sound level meter such as Bruel and Kjaer type 2209 or equiva-

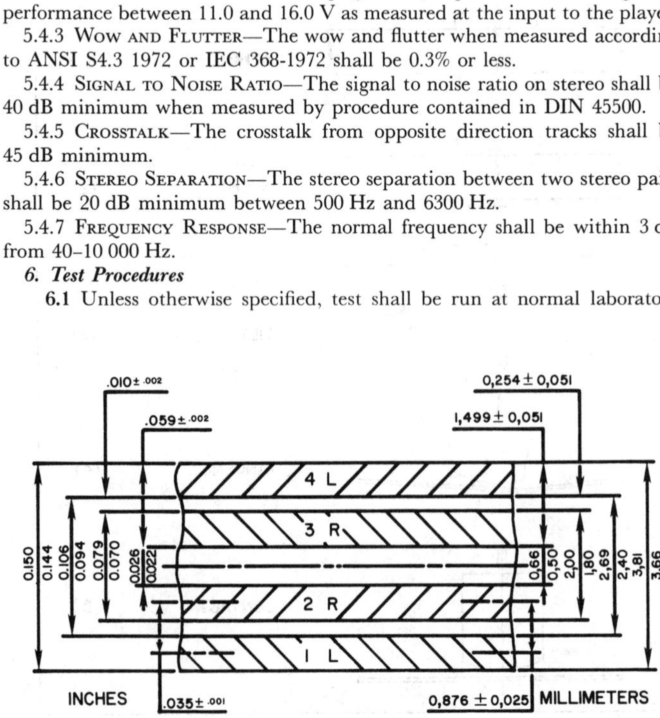

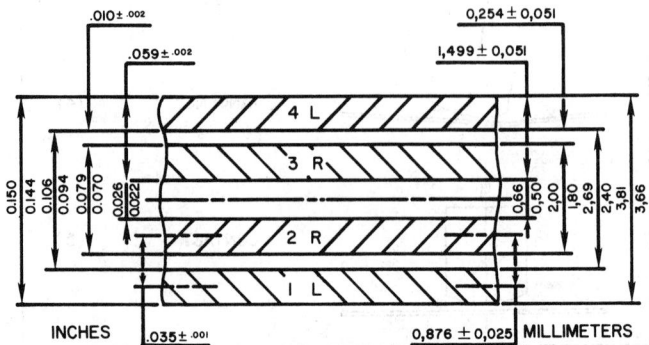

FIG. 7—CASSETTE TWO CHANNEL SOUND FORMAT

lent on the A scale with the microphone 50 ± 10 mm from the cassette or top of the player. This measurement may be done with the cassette in either a cassette player or in a torque measuring machine. The noise generated by the cassette driving machine shall first be measured using a cassette without tape and hubs in the play, fast forward, and rewind modes. This measurement shall be made in an environment that is at least 6 dB below the noise level produced by the machine but need not be below 50 dB. The noise made with a cassette in place, shall then be measured for play, fast forward, and fast reverse mode and the difference between the two measurements noted.

6.9 Magnetic Properties of Tapes

6.9.1 DEFINITIONS

6.9.1.1 *Level Set Reference Cassette*—The level set reference cassette is defined as the BASF DIN Calibration Tape 45513/6 4.75 (Fe) in a standard BASF cassette. Tape flux is 250 nWb/m.

6.9.1.2 *Reference Level*—The reference level is defined as that output level obtained using the Level Set Reference Tape. (See paragraph 6.9.1.1.)

6.9.1.3 *Blank Reference Cassette*—The blank reference cassette is defined as the blank portion of the DIN Calibration Tape PES 12/C 521 (see paragraph 6.9.1.1) or equivalent in a standard cassette housing.

6.9.1.4 *Equalization*—The equalization shall be set such that a tape flux falling at 6.0 dB per octave above the transition frequency of 1325 Hz (120 μs time constant) produces a constant output voltage.

6.9.1.5 *Bias*—The value of the recording bias shall be set for the maximum output voltage at 1000 Hz.

6.9.2 SENSITIVITY MEASUREMENT—With the recording level set to produce the same output as the Level Set Reference cassette (-20 dB section) when using the Blank Reference cassette record a 315 Hz signal (long wave length sensitivity) and a 5 kHz signal (short wave length sensitivity). Play back these sections and compare the output to the Level Set Reference.

6.9.3 FREQUENCY RESPONSE—The frequency response shall be measured by recording both the tape to be measured and a Blank Reference cassette and the response compared between these two tapes at frequencies from 100 Hz to 10 kHz.

6.9.4 UNIFORMITY—Uniformity shall be measured by recording a 315 Hz signal for 65-85 s at a standard speed of 4.76 cm/s and at 20 dB below standard reference level. A similar section of tape shall be recorded with a 5 kHz signal at 20 dB below standard reference level. These sections shall be played back and the uniformity of the output measured with a graphic level recorder whose frequency response is flat between DC and 2 Hz and is not more than 6 dB down at 6 Hz. Both right and left tracks and both sides of the tape shall be measured.

6.9.5 PRERECORDED CASSETTE OUTPUT LEVEL—The output of prerecorded tapes shall be measured with a meter calibrated to read vu (volume units) as defined in ANSI C16.5-1961 (IEEE Standard 152-1953). The output of the player shall be adjusted to read 0 vu when a Level Set cassette is reproducing a 315 Hz signal. The prerecorded tape is then played and the average output in volume units measured for normal program material. Extra loud or soft passages will be excluded from the estimate.

6.10 Life Test Cycle—A single life test cycle shall consist of playing the cassette from beginning to end and rewinding back to the beginning.

6.11 Temperature Test

6.11.1 HIGH TEMPERATURE

6.11.1.1 The high temperature storage test shall consist of a 4 h soak at 75°C $+0$, -3°C (167°F $+0$, -5°F) followed by 2 h at room temperature.

6.11.1.2 The high temperature operating test shall be conducted at 60°C $+0$, -3°C (140°F $+0$, -5°F) after a 2 h soak at that temperature.

6.11.2 LOW TEMPERATURE

6.11.2.1 The low temperature storage test will consist of 4 h at -40°C $+0$, -3°C (-40°F $+0$, -5°F) followed by 2 h at room temperature.

6.11.2.2 The low temperature operating test will be conducted at -20°C -0, $+3$°C (-4°F -0, $+5$°F) after 2 h soak at that temperature.

6.12 Humidity—The humidity tests shall consist of a 24 h soak at 45 ± 5°C (113 ± 9°F) and a relative humidity of 95 ± 5%. The cassette shall be allowed to stabilize at normal conditions for at least 2 h before testing.

6.13 Drop Test—A drop test shall consist of four drops of a cassette, without packing or case, with the specified amount of tape from a 0.75 m height (29.8 in) onto a concrete floor or equal. The first two drops shall be on alternate ends, the second two drops shall be on alternate faces. The ends are defined as the short sides and the faces are the surfaces where the labels are attached.

6.14 Vibration

6.14.1 OPERATING VIBRATION TESTS—In this test the cassette shall be positioned in the player and the player shall be operating. The vibration shall be sinusoidal from 10–55 Hz in 1 min with a peak acceleration of 1.4 g. Tests shall consist of a minimum of 3 min of vibration in the vertical plane with the player in the horizontal plane.

6.14.2 NON-OPERATING VIBRATION—The non-operating vibration tests shall consist of two, one quarter hour tests, one quarter hour with the cassette horizontal and one quarter hour with the cassette vertical with vibration in the vertical plane. During this test the hubs may be anchored. The vibration shall be from 10–55–10 Hz in 1 min with a peak acceleration of 7 g.

6.15 Life Tests—The life tests shall consist of at least 100 cycles of operation using the cycle described in paragraph 6.10. The cassette shall be turned over after every 25 ± 5 cycles. This test shall be conducted under normal conditions as described in paragraph 6.1. The Friction Torque may be measured at anytime during the test. However, it should be checked at least twice during the test and at the conclusion of the test.

6.16 Thermocycling Test—The thermocycling tests shall consist of 12 cycles in which temperature is varied from -20°C $+0$, -3°C to $+60$°C $+0$, -3°C. The average rate of change of temperature between -20 and $+60$°C shall be between 2 and 6°C/min. The chamber shall remain at the extreme temperatures for approximately 1 h on each cycle.

6.17 Test Sequences—A total of 16 cassettes are required for qualification testing. These are to be divided into four groups and each group subjected to only the tests for that group. The tests are to be run in the order listed in each group. Where less than the number of cassettes listed for the group are to be tested, the number to be tested is so indicated by the test.

Group 1—1 Cassette—Test paragraphs:
  (a) 4. Physical Dimensions
  (b) 6.2 Breakout Lugs
  (c) 6.7 Pressure Pad

Group 2—6 Cassettes—Test paragraphs:
  (a) 6.8 Mechanical Noise
  (b) 6.15 Life Test. More than 6 units may be needed to establish life criteria

Group 3—6 Cassettes divided into three subgroups of two each—Test paragraphs:
  (a) 6.11 Temperature Tests—2 cassettes. Perform operating temperature tests, paragraphs 6.11.1.2 and 6.11.2.2. Then perform storage temperature tests, paragraphs 6.11.1.1 and 6.11.2.1
  (b) 6.12 Humidity—2 cassettes
  (c) 6.14.2 Non-Operating Vibration—Use the same two cassettes after either Temperature or Humidity test
  (d) 6.16 Thermocycling—2 cassettes
  (e) 6.13 Drop Test—Use 2 remaining cassettes after either temperature or humidity testing that were not used in Non-Operating Vibration test
  (f) 6.2 Breakout Lugs—Use the same two cassettes after Thermocycling to measure breaking force of lugs

Group 4—3 Cassettes—Test paragraphs:
  (a) 6.9.2 Sensitivity Measurement
  (b) 6.9.3 Frequency Response
  (c) 6.9.4 Uniformity
The above tests apply only to blank tape cassettes.
  (d) 6.9.5 Prerecorded Output Level
  (e) 6.4 Hub Anchorage
  (f) 6.5 Splice Strength
  (g) 6.3 Yield and/or Tensile Strength

General Drawing Notes

1. For explanation of drafting symbols used refer to American National Standards Institute, Inc. Standard Drafting Practice Y 14.5 (also see Appendix).

2. For method of converting inches to millimeters and vice versa as used in this standard refer to International Organization for Standardization ISO Recommendation R370 reference number ANSI Z210.1-1973.

3. The X Datum Line is defined as a line through the intersections of the centerlines and back edge of the two rectangular locating holes in the same plane.

4. The Y Datum Line—The compact cassette is symmetrically disposed about the Y Datum Line, which for the purpose of this standard is defined as a line perpendicular to the X Datum Line.

5. The Z Datum Plane is defined as exactly equidistant between the upper and lower support planes, which are defined between the respective three control points.

6. Shaded areas designate controlled support planes. Flatness, parallelism and dimensional requirements refer to these support planes only.

7. Each surface feature, regardless of feature size, must be within 0.2 mm (0.008 in) dia. of true position as referenced to the corresponding surface feature of the opposite side (half).

8. The compact cassette is symmetrically disposed about the Y Datum Line, and datum features are symmetrically disposed about their respective centerlines or edges to within half of the indicated coordinate tolerances, unless otherwise noted.

9. The tape guide limits within the compact cassette are indicated by the cross hatched area.

10. TYP (TYPICAL) indicates that the dimension so marked applies to identical undimensioned features disposed throughout the drawing.

APPENDIX
DRAFTING SYMBOLS

|.XXX| dimensions are basic or true position dimensions and, as such, are absolute and without tolerance, and are intended as a reference dimension only in the establishment of a particular datum.

|⊕| B ⓢ | .XXX Dia. ⓢ| each surface feature above and below the associated plane must be within .XXX dia. of true position, regardless of feature size, of its corresponding feature. In the case of holes, the axis of the holes (either surface) must be within a cylindrical tolerance zone .XXX dia. This .XXX dia. tolerance zone would be perpendicular to the associated datum plane. In the case of features other than holes, the boundary edges are to be considered to be within .XXX dia. tolerance zone.

|≡| -Y- | .XXX TOTAL| the central plane of the dimensional feature must be within a tolerance zone defined by two parallel planes .XXX apart. The midpoint of these planes is considered to be the associated datum.

|▱| .XXX| the support plane, as designated by Figs. 2, 2A, and note 5.5, must be flat within a tolerance zone as defined by two parallel planes .XXX apart, that is, all surface irregularities of these support planes must be confined within the boundaries of the aforementioned tolerance zone.

|11| -Z- | .XXX| the lower support planes, shown in Figs. 2, 2A, and note 5.5, must be parallel to the Z Datum Plane within a tolerance zone defined by two parallel planes .XXX apart and parallel to the Z Datum Plane.

23 Lighting

TERMINOLOGY—MOTOR VEHICLE LIGHTING—SAE J387

SAE Information Report

Report of Lighting Committee approved March 1969.

Scope—The following are definitions of common terms used in SAE Standards and Recommended Practices pertaining to motor vehicle lighting.

Bulb—An indivisible assembly which contains a source of light and which is normally used in a lamp.

Lamp—A divisible assembly which contains a bulb or other light source and sometimes an optical system such as a lens and/or a reflector, and which provides a lighting function.

Device—Any piece of equipment or mechanism designed to serve a special purpose or perform a special function.

Unit—An indivisible assembly which provides a mechanical, electrical, or lighting function, for example, sealed beam unit or flasher.

Light—Visible radiant energy.

LIGHTING IDENTIFICATION CODE—SAE J759c

SAE Recommended Practice

Report of Lighting Committee approved November 1960 and last revised January 1975.

1. Definition—A lighting identification code is a series of standardized markings for lighting devices which a manufacturer or a supplier may use to mark his product to indicate the SAE Lighting Standard or Standards to which the device is designed to conform. The code is not intended to limit the manufacturer or supplier in applying other markings to the devices.

THE SOCIETY OF AUTOMOTIVE ENGINEERS DOES NOT APPROVE PRODUCTS; HENCE, THE USE OF MARKINGS IN ACCORDANCE WITH THIS CODE SHOULD NOT BE INTERPRETED TO MEAN THAT A DEVICE SO MARKED HAS SAE APPROVAL.

2. Requirements

2.1 Code Location—The identification code should be permanently marked on the lens or body where it can be observed with the device mounted in its normal position on the vehicle, except that headlamps, turn signal switches, hazard warning switches, and flashers may have the markings located where they can be observed by removing other parts. This exception is granted because many of these devices are not visible as installed on the vehicle. However, when these same devices are externally mounted, the markings must be visible. The manufacturer's identification and model designation (or part number) is required on both the housing and lens of separable devices.

2.2 Size of Markings—Identification numerals and letters shall be at least $\frac{1}{8}$ in (3 mm) high, except that raised molded markings 0.08 in (2 mm) high may be used on lenses containing less than 2 in^2 (13 cm^2) of area. The smaller markings are permitted when they are raised, molded markings (not stamped, etched, or lettered in indelible ink).

2.3 Flashers—Flashers, because of their small size, may use 0.08 in (2 mm) high markings and these may be permanently stamped, etched, or lettered in indelible ink. The markings required on flashers are:

(1) The manufacturer's identification and model number (or part number).
(2) The appropriate SAE Standard or Recommended Practice.
   EXAMPLE: Flasher Co. 200
         SAE J590e

2.4 Multicompartment and Multiple Lamp Arrangements—Multicompartment lamps or individual lamps designed to use more than one compartment or lamp to comply with the appropriate SAE requirements shall be marked to indicate the number of compartments or lamps that are to be operated together to meet this requirement. Lamps which do not carry a number within a circle or parentheses to indicate that more than one lamp is to be used are intended to be used alone unless exceptions are noted in the appropriate SAE Standard or Recommended Practices. See example in paragraph 3.

2.5 Content of Identification Code—The identification code should consist of a series of letters and numbers in the following sequence:

2.5.1 SAE.

2.5.2 A number within a circle or parentheses indicating the number of compartments in a multicompartment device or the number of separate lamps when more than one compartment or one lamp is needed to satisfy the requirements of the applicable SAE Standard. No number is required for a single compartment device.

2.5.3 One or more letters identifying the function or functions to which the device is designed to conform. Multipurpose devices shall be marked to cover each function for which the device was originally designed to conform and may be used to carry out one or more of these functions. Table 1 lists the identifying designations for SAE Standards.

2.5.4 The last two numbers of a year which means that the code letters refer to SAE Standards and Recommended Practices listed in the SAE Handbook current in the year indicated, or the applicable requirements of Federal Motor Vehicle Safety Standard 108 specified for the device function in the year indicated. To denote that a function meets the requirements of FMVSS 108, but not the current SAE standard, a dash line shall be placed under the function letter. Example: AIR̲ST 75.

Devices marked prior to the adoption of this Recommended Practice need not be remarked.

2.6 Content of Manufacturer's Identification—The manufacturer's lettered identification and model designation (or part number). Pictorial trade-

TABLE 1—LETTERS INDICATING DEVICE FUNCTIONS

A	Reflex Reflectors—Class A
D	Motorcycle and motor driven cycle turn signal lamps
E	Side turn signal lamps—vehicles 30 ft (9.1 m) or more in length
E2	Side turn signal lamps—vehicles less than 30 ft (9.1 m) in length
F	Fog lamps
H	Sealed beam headlamps (marking applies to housing or unit)
I	Turn signal lamps—Class A
I2	Turn signal lamps—Class A spaced less than 4 in (102 mm) from headlamp (Reference Docket 69-19; Notice 9)
J590e	Turn signal flasher
J945b	Hazard warning signal flasher
J1054	Warning lamp alternating flasher
K	Cornering lamps
L	License plate lamps
M	Motorcycle and motor driven cycle headlamps—motorcycle type
N	Motorcycle and motor driven cycle headlamps—motor driven cycle type
O	Spot lamps
P	Parking lamps
P2	Clearance or side marker or identification lamps
PC	Combination clearance and side marker lamps
Q	Turn signal operating units—Class A
QB	Turn signal operating units—Class B
QC	Vehicular hazard warning signal operating unit
R	Backup lamps
S	Stop lamps
T	Tail lamps
U	Supplemental high mounted stop and turn signal lamps
V	Liquid burning emergency flares
W	Warning lamps for emergency, maintenance and service vehicles
W2	Warning lamps for school buses
W3	360 deg emergency warning lamps
W4	Emergency warning device
X	Electric emergency lanterns
Y	Driving lamps
Z	Auxiliary low beam lamps

For uniformity, the order of marking should be in alphabetical sequence beginning with the first letter of the alphabet where applicable. EXAMPLE: SAE AIST 75.

Devices marked prior to the adoption of this recommended practice need not be remarked.

marks are permitted but they must be in addition to the required lettered identification. The lettered identification must be reproducible on a typewriter or computer. The manufacturer's identification does not have to be located in sequence with the identification code markings but may be located on a different area of the lens or body.

3. Examples

Examples of code and manufacturer's identification markings:

SAE AIST 75

XYZ Corp. 400

Translated, this means SAE Standards current in 1975 for stop lamps, tail lamps, turn signals and reflectors. The manufacturer is the XYZ Corp. (the word corporation is shown for example only, since this is generally not used) and their model number is Model 400.

SAE A (2) I (2) S (3) T 75

Translated, this means a three-compartment or three-lamp arrangement with two of the compartments needed to meet optically the stop and turn signal requirements, and all three of the compartments needed to meet the tail lamp requirements. If, for example, each of the compartments fully complied with the tail lamp requirements, the *3* preceding the letter *T* would be omitted. This would signify that all of the compartments were not necessary to meet the tail lamp requirement and that each individual compartment met it as a separate single function. If the SAE Standards referenced in the examples were not revised and would still be current in the year 1976, for example, the manufacturer could change the marking on this device to 76, if he so desired. This coding change regarding the year might also be made if the SAE Standards have been revised in the interim and the device meets the new requirements.

HEADLAMP BEAM SWITCHING—SAE J564c

SAE Recommended Practice

Report of Lighting Division approved January 1934 and last revised by Lighting Committee June 1971. Editorial change October 1977.

1. Definition

1.1 A headlamp beam switching device is a driver controlled switch used to select the high or low beam headlamp circuit.

1.2 The device may incorporate an auxiliary circuit for override of the semiautomatic beam switching control.

1.3 An auxiliary beam selection switch may be used in conjunction with the beam switching device to select between two or more upper or lower beam systems.

ϕ 2. *Reference Standards*—The following sections from SAE J575f (April, 1975) are a part of this recommended practice:

Section B—Sample for Test

Section C—Lamp Bulbs

Section D—Laboratory Facilities

3. Temperature Test

3.1 To insure basic function, the switch shall be manually cycled for 10 cycles at design electrical load: 75 ± 10 F, (24 ± 5.5 C); 165 +0, −5 F (74 +0, −2.8 C); −25 +5, −0 F (−32 +2.8, −0 C). This to be done after a 1 h exposure at each of these temperatures. The switch shall be electrically and mechanically operable during each of these cycles.

3.2 For each of the 1 h temperature exposures, design electrical load shall be connected to both high and low beam circuits with the switch initially set in the low beam position. After 1 h exposure at each temperature, the switch shall be cycled without interruption of the headlight current in excess of 0.10 s.

3.3 This same switch shall be used for the endurance test described in paragraph 5.

4. *General Requirements*—The switch shall be designed so that the headlight circuits are never maintained open.

5. Endurance Test Setup

5.1 The switch shall be operated with the maximum load stated by the switch manufacturer.

5.2 The test should be set up to operate the switch for the prescribed number of cycles.

One cycle shall consist of the following sequence of positions:

high beam—low beam—high beam

The switch shall be tested at a rate not to exceed 45 cycles/minute.

5.3 During the test, the system or systems specified for the switch shall be used. The switch shall be operated at 6.4 V d-c for a 6 V system, 12.8 V d-c for a 12 V system, or 25.6 V d-c for a 24 V system, measured at the input termination of the switch. The power supply shall not generate any adverse transients not present in motor vehicles and shall comply with the following specifications:

(a) Output Current—Capable of supplying a continuous current of the design load and inrush currents as required by the bulb load complement.

(b) Regulation—

Dynamic—The output voltage shall not deviate more than 1.0 V from zero to maximum load (including inrush current) and should recover 63% of its maximum excursion within 5 ms.

Static—The output voltage shall not deviate more than 2% with changes in static load from zero to maximum (not including inrush current), and means shall be provided to compensate for static input line voltage variations.

(c) Ripple Voltage—Maximum 300 mV, peak to peak.

6. Endurance Requirements

6.1 The switch shall be subjected to an endurance test of 100,000 cycles.

6.2 The voltage drop from the input terminals to the output terminals of each circuit shall be measured at the beginning of the test and at intervals of 25,000 cycles.

This voltage drop shall not exceed 0.250 V at maximum design load for loads up to and including 15 A.

For loads in excess of 15 A, an additional voltage drop of 0.015 V per additional ampere will be allowed, but in no case shall the total voltage drop exceed 0.400 V.

If wiring is an integral part of the switch, the voltage drop measurement shall be made including 3 in. of wire on each side of the switch; otherwise, measurement would be made at the switch terminals.

7. *Semiautomatic Beam Control Switch*—The switch shall meet all the requirements above except as follows:

Endurance Requirements—Semiautomatic beam control switch. The switch shall be subjected to 50,000 cycles. One cycle shall consist of the following:

High Beam—Override Mechanism—Low Beam

Low Beam—Override Mechanism—High Beam

The dwell time in each position shall be evenly divided. The cycle rate shall not exceed 30 cycles/minute.
8. Installation Requirements
8.1 The following requirements apply to the device as used on the vehicle and are not part of the laboratory test requirements and procedures.

8.2 Upper Beam Indicator—An indicator shall be provided to show when the upper beam of the headlamp system is on. The indicator should consist of a red light, with a minimum area equivalent to that of a $3/16$ in. diameter circle, plainly visible to drivers of all heights under normal driving conditions when headlights are required.

SEMIAUTOMATIC HEADLAMP BEAM SWITCHING DEVICES—SAE J565c

SAE Recommended Practice

Report of Lighting Committee approved August 1954 and last revised February 1972.

1. Definition—A semiautomatic headlamp beam switching device is one which provides either automatic or manual control of beam switching at the option of the driver. When the control is automatic, the headlamps switch from the upper beam to the lower beam when illuminated by the headlamps on an approaching car and switch back to the upper beam when the road ahead is dark. When the control is manual, the driver may obtain either beam manually regardless of the condition of lights ahead of the vehicle.

2. Operation Instructions—A set of operating instructions shall be included to permit a driver to operate the device correctly. Items to be covered are:

2.1 How to turn the automatic control on and off.

2.2 How to adjust sensitivity. A sensitivity control shall be provided for the driver.

2.3 Any other specific instructions applicable to the particular device.

3. The following sections from SAE J575 are a part of this recommended practice:

3.1 Section B—Samples for Test

3.2 Section D—Laboratory Facilities

4. Test Requirements—The device shall be adjusted for sensitivity in accordance with the manufacturer's instructions.

4.1 Aim—The device shall be mounted and operated in the laboratory in the same environment as that encountered on the vehicle, that is, tinted glass, grille work, etc.

4.2 Sensitivity—The device shall switch to the lower beam in accordance with the "dim" limits shown in Table 1. The device shall switch back to the upper beam within the limits shown in the "hold" column. The sensitivity test should be made at 13 V input to the device. This voltage will be used for all tests unless otherwise specified.

To provide more complete information on sensitivity throughout the required vertical and horizontal angles, a set of constant foot-candle curves shall be made at "dim" sensitivities of 17, 25, and 100 cd at 100 ft (30.5 m). The curves shall be examined to determine that there are no sensitivity voids within the test angles shown in Table 1.

4.3 Voltage Regulation—With the device adjusted for sensitivity in accordance with paragraphs 4.1 and 4.2, the H-V "dim" sensitivity shall be between 8 and 25 cd at 11 and 15 V input to the device.

4.4 Manual Override of Automatic Control—The device shall include a means convenient to the driver for switching to the opposite beam from the one provided.

With the device set up as in paragraphs 4.1 and 4.2, the test light shall be turned on to cause the device to be on lower beam. The manufacturer's instructions shall be followed to cause the device to override the test light and switch to upper beam.

In a similar manner, the test light shall be turned off to cause the device to be on upper beam. Again, the manufacturer's instructions shall be followed to cause the device to switch to the lower beam.

4.5 Warmup Test—If the warmup time of the device exceeds 10 s, it should maintain the headlamps on lower beam during warmup. The device shall be adjusted in accordance with paragraphs 4.1 and 4.2 and then checked for warmup with the test light at 25 cd to determine compliance.

4.6 Fail-Safe Test—In the event that there is a failure of the automatic control (light sensor) portion of the system, means must be provided for switching to low beam.

4.7 Temperature Test—The device shall be exposed for 1 h in a temperature corresponding to that at the device mounting location. For a device mounted in the passenger compartment or the engine compartment, this temperature is 210 F (98.9 C); mounted elsewhere, the temperature shall be 150 F (65.6 C). After the high temperature exposure, the H-V "dim" sensitivity shall be between 8 and 25 cd over the temperature range of -30 F (-34.4 C) to $+100$ F (37.8 C).

4.8 Dust Test—The device shall be adjusted in accordance with paragraphs 4.1 and 4.2 and then subjected to the dust test, Section G of SAE J575. After the photo-unit lens is wiped clean, the H-V "dim" sensitivity shall be between 8 and 25 cd.

4.9 Corrosion Test—The device shall be adjusted in accordance with paragraphs 4.1 and 4.2 and then all components which are located outside the driver compartment of the vehicle shall be subjected to corrosion testing in accordance with Section H of SAE J575 with the device not operating. (Water should not be allowed to collect on any connector socket.) After the test, the H-V "dim" sensitivity shall be between 8 and 25 cd.

4.10 Vibration Test—The device shall be adjusted in accordance with paragraphs 4.1 and 4.2 and then shall be subjected to vibrations of 5 g constant acceleration as follows:

(a) The device shall be mounted in proper vehicle position and vibrated for 0.5 h in each of three directions: vertical, horizontal and parallel to the vehicle axis, and horizontal and normal to the vehicle axis.

(b) The vibration frequency shall be varied from 30 Hz to 200 Hz and back to 30 Hz cycles per second over a period of approximately 1 min.

(c) The device shall be operating during vibration test.

At the conclusion of the test, the H-V "dim" sensitivity shall be between 8 and 25 cd and the mechanical aim of the photo-unit shall not change more than 0.25 deg.

4.11 Sunlight Test—The device shall be exposed for 1 h in bright noonday sunlight (5000 ft-c (5100 lx) minimum illumination with a clear sky) with the photo-unit aimed as it would be on a car and facing an unobstructed portion of the horizon in the direction of the sun. After being rested for 1 h in normal room light at room temperature, the H-V "dim" sensitivity shall be between 8 and 25 cd.

4.12 Durability Test—The device shall be adjusted in accordance with paragraphs 4.1 and 4.2 and then subjected to the following cycle test:

(a) The photo-unit shall be actuated by a 60 cd light at 100 ft (30.5 m) (or equivalent) which is cycled on and off 4 times per minute.

(b) The device shall be operated at 13 V input for 90 min on and 30 min off for 200 h operating time.

After resting for 2 h at room temperature in a lighted area of 50–150 ft-c, (51–153 lx) the H-V "dim" sensitivity shall be between 8 and 25 cd.

4.13 Automatic Dimming Indicator—There shall be a convenient means of informing the driver when the device is controlling the headlights automatically. The manufacturer's instructions shall be followed to determine the means of indication.

4.14 Upper Beam Indicator—The device shall not affect the function of the upper beam indicator light.

4.15 Return to Upper Beam—With the device adjusted as in paragraphs 4.1 and 4.2, a light of 100 ft-c (102 lx) intensity shall be impressed on the lens of the device for 10 s. The light shall then be extinguished. The device must return to upper beam within 2 s.

4.16 Lens Accessibility—The device lens shall be accessible for cleaning when the unit is mounted on a vehicle.

4.17 Mounting Height—The center of the device lens shall be mounted no less than 24 in. (610 mm) above the road surface.

TABLE 1—OPERATING LIMITS
(Candlepower at 100 Ft (30.5 m))

Test Position, deg	Dim cd	Hold cd
H—V	15—Adjust	1.5 min to 3.75 max
H—2L	25 max	1.5 min
H—4L	40 max	1.5 min
H—6L	75 max	1.5 min
H—2R	25 max	1.5 min
H—5R	150 max to 40 min	1.5 min
1D—V	30 max	1.5 min
1U—V	30 max	1.5 min

LAMP BULB RETENTION SYSTEM—SAE J567c

SAE Standard

Report of Electrical Equipment Division approved August 1915 and last revised by Lighting Committee December 1970.

1. Scope—This SAE Standard covers the performance and functional requirements of the lamp bulb retention system applicable for use in motor vehicles.

2. Definition—The lamp bulb retention system is a device which retains a lamp bulb in its intended application and provides electrical continuity.

3. Requirements

3.1 The lamp bulb retention system shall accept and provide for the retention and removal of the maximum and minimum bulb gages. See Table 1 for bulb gages.

3.2 The bulb retention system shall provide required electrical connections.

3.3 Bulb retention systems employing multiple contacts shall have them spaced so that they will not contact (electrically insulated from) each other or short to ground.

3.4 When the bulb retention system is assembled in its intended application, the insertion and rotational forces required to lock the maximum bulb gage in its final seating position shall not exceed the values shown below:

Maximum Insertion Force
 B and C base sockets—14 lb.
 A base sockets—8 lb.

Maximum Torque
 C base sockets—5 in.-lb.

TABLE 1—MAXIMUM AND MINIMUM BULB RETENTION SYSTEM GAGES

Dimension[a]	C-2 Base Max Gage	C-2 Base Min Gage	B-2 Base Max Gage	B-2 Base Min Gage	B-1 Base Max Gage	B-1 Base Min Gage
A	0.6025 +0.0005/−0.0000	0.5925 +0.0000/−0.0005	0.6025 +0.0005/−0.0000	0.5925 +0.0000/−0.0005	0.6025 +0.0005/−0.0000	0.5925 +0.0000/−0.0005
C	0.6680 +0.0005/−0.0000	0.6425 +0.0000/−0.0005	0.6680 +0.0005/−0.0000	0.6425 +0.0000/−0.0005	0.6680 +0.0005/−0.0000	0.6425 +0.0000/−0.0005
E	0.0800 +0.0005/−0.0000	0.0740 +0.0000/−0.0005	0.0800 +0.0005/−0.0000	0.0740 +0.0000/−0.0005	0.0800 +0.0005/−0.0000	0.0740 +0.0000/−0.0005
F	0.3160 +0.0005/−0.0000	0.2490 +0.0000/−0.0005	0.3160 +0.0005/−0.0000	0.2490 +0.0000/−0.0005	0.3160 +0.0005/−0.0000	0.2490 +0.0000/−0.0005
G	0.0500 +0.0000/−0.0050	0.0150 +0.0000/−0.0005	0.0500 +0.0000/−0.0050	0.0150 +0.0000/−0.0005	0.0500 +0.0000/−0.0050	0.0150 +0.0000/−0.0005
N	0.2050 +0.0010/−0.0000	0.1770 +0.0000/−0.0010	0.2050 +0.0010/−0.0000	0.1770 +0.0000/−0.0010	0.2050 +0.0010/−0.0000	0.1770 +0.0000/−0.0010
P	0.1330 +0.0020/−0.0000	0.1170 +0.0000/−0.0020	0.0000 +0.0010/−0.0000	0.0000 +0.0000/−0.0010	0.0000 +0.0010/−0.0000	0.0000 +0.0000/−0.0010
S	0.2670 +0.0005/−0.0000	0.2670 +0.0000/−0.0005	0.2670 +0.0005/−0.0000	0.2670 +0.0000/−0.0005	NA	NA
T	0.0600 +0.0050/−0.0000	0.0600 +0.0050/−0.0000	0.0600 +0.0050/−0.0000	0.0600 +0.0050/−0.0000	0.0600 +0.0050/−0.0000	0.0600 +0.0050/−0.0000
D	—	0.0250 +0.0000/−0.0005	—	0.0250 +0.0000/−0.0005	—	0.0250 +0.0000/−0.0005
K	—	0.3500 +0.0000/−0.0005	—	0.3500 +0.0000/−0.0005	—	0.3500 +0.0000/−0.0005
L	0.6450 +0.0005/−0.0000	—	0.6450 +0.0005/−0.0000	—	0.6450 +0.0005/−0.0000	—
H	0.1700 +0.0010/−0.0000	0.1380 +0.0000/−0.0010	0.1700 +0.0010/−0.0000	0.1380 +0.0000/−0.0010	0.1700 +0.0010/−0.0000	0.1380 +0.0000/−0.0010
U	—	—	—	—	—	—

Dimension[a]	A-1 Base Max Gage	A-1 Base Min Gage	Wedge Base Max Gage	Wedge Base Min Gage
A	0.3660 +0.0005/−0.0000	0.3570 +0.0000/−0.0005	0.1750 +0.0005/−0.0000	0.1450 +0.0000/−0.0005
C	0.4320 +0.0005/−0.0000	0.4040 +0.0000/−0.0005	0.2400 +0.0005/−0.0000	0.1450 +0.0000/−0.0005
E	0.0670 +0.0005/−0.0000	0.0610 +0.0000/−0.0005	0.3740 +0.0005/−0.0000	0.3740 +0.0000/−0.0005
F	0.2550 +0.0005/−0.0000	0.1800 +0.0000/−0.0005	0.1200 +0.0005/−0.0000	NA
G	0.0500 +0.0000/−0.0050	0.0150 +0.0000/−0.0005	0.1600 +0.0005/−0.0000	0.1300 +0.0000/−0.0005
N	0.1700 +0.0010/−0.0000	0.1470 +0.0000/−0.0010	0.0650 +0.0000/−0.0005	—
P	0.0000 +0.0010/−0.0010	0.0000 +0.0010/−0.0010	0.0950 +0.0005/−0.0000	0.0750 +0.0000/−0.0005
S	NA	NA	—	—
T	0.0600 +0.0050/−0.0000	0.0600 +0.0050/−0.0000	—	—
D	—	0.0250 +0.0000/−0.0005	—	—
K	—	0.1800 +0.0000/−0.0005	—	—
L	0.4100 +0.0005/−0.0000	—	—	—
H	0.1310 +0.0005/−0.0000	0.0950 +0.0000/−0.0010	—	—
U	0.2520 +0.0010/−0.0000	—	—	—

[a] See Figs. 1 and 2.

B base sockets—5 in.-lb.
A base sockets—2 in.-lb.

NOTE: Bulb retention systems designed to be removed from their intended application for bulb service may be checked while removed.

3.5 When the bulb retention system is assembled in its intended application, the B and C base bulb retention systems shall provide a minimum bulb support as measured with the bulb support gage. (See Table 2 for bulb support gages and Fig. 3 for gage.)

NOTES:

1. Bulb retention systems designed to be removed from their intended application for bulb service may be checked while removed.

2. Bulb retention systems which provide alternative equivalent bulb supporting means may be used and need not be checked with the bulb support gage.

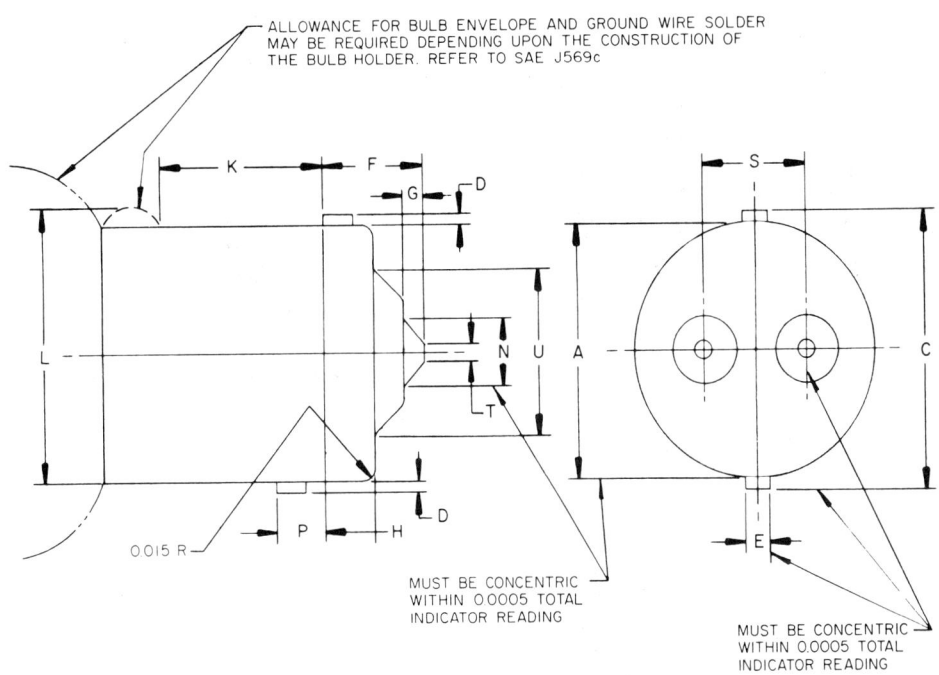

FIG. 1—C-2, B-2, B-1, AND A-1 BASE GAGES

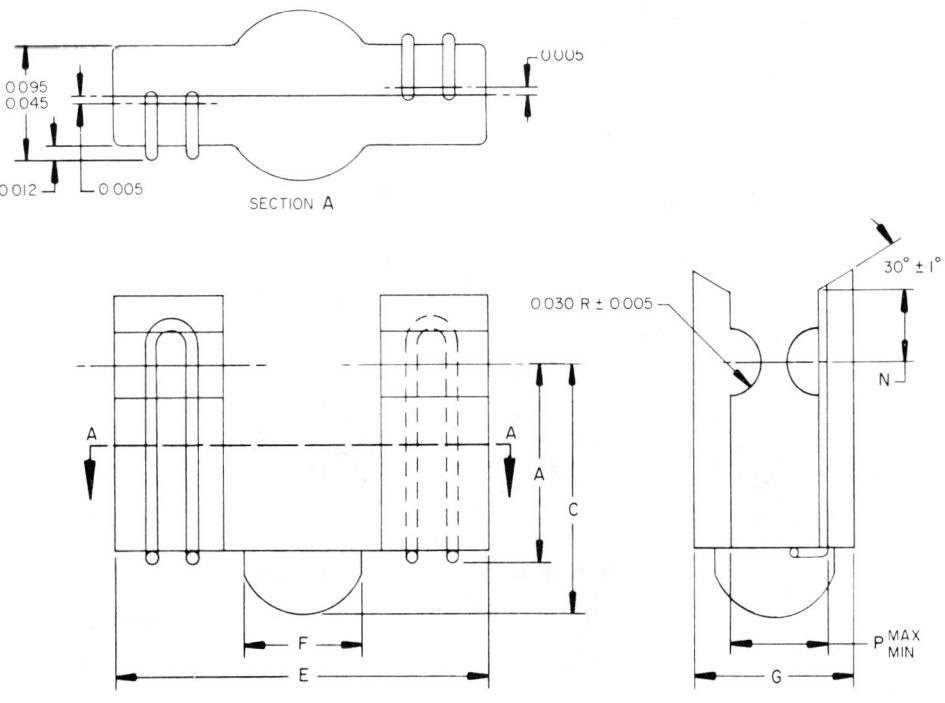

FIG. 2—WEDGE BASE GAGES

TABLE 2—BULB SUPPORT GAGE (SEE FIG. 3)

Dimension	B-1 and B-2 Base Bulb Support Gage	C-2 Base Bulb Support Gage	Dimension	B-1 and B-2 Base Bulb Support Gage	C-2 Base Bulb Support Gage
A	0.5925 +0.0005/−0.0000	0.5925 +0.0005/−0.0000	N	0.191 +0.005/−0.000	0.191 +0.005/−0.000
B	0.750 +/−0.010	0.750 +/−0.010	P	0.4080 +0.0005/−0.0000	0.4080 +0.0005/−0.0000
C	0.6425 +0.0005/−0.0000	0.6425 +0.0005/−0.0000	R	0.744 +0.000/−0.010	1.041 +0.000/−0.010
D	0.0250 +0.0005/−0.0000	0.0250 +0.0005/−0.0000	S	0.187D S/F	0.187D S/F
E	0.0740 +0.0005/−0.0000	0.0740 +0.0005/−0.0000	T	0.250D S/F	0.250D S/F
F	0.2490 +0.0005/−0.0000	0.2590 +0.0005/−0.0000	U	0.035 +0.0000/−0.0050	0.035 +0.000/−0.005
G	0.000 +/−0.001	0.117 +0.002/−0.000	V	0.6095 +0.0005/−0.0000	0.6095 +0.0005/−0.0000
H	0.1380 +0.0005/−0.0000	0.1380 +0.0005/−0.0000	W	0.060 +0.005/−0.000	0.060 +0.005/−0.000
J	1.188 +0.005/−0.000	1.750 +0.005/−0.000	X	B-1 0.000; B-2 0.136 +0.000/−0.003	0.136 +0.000/−0.003
K	0.270 +0.000/−0.005	0.270 +0.000/−0.005	Y	B-1 0.000; B-2 0.2670 +0.005/−0.000	0.2670 +0.005/−0.000
L	0.1250 +0.0005/−0.0000	0.1250 +0.0005/−0.0000			
M	0.030R +/−0.001	0.030R +/−0.001			

NOTE: Dim. P − L = 0.283 in. Dim.

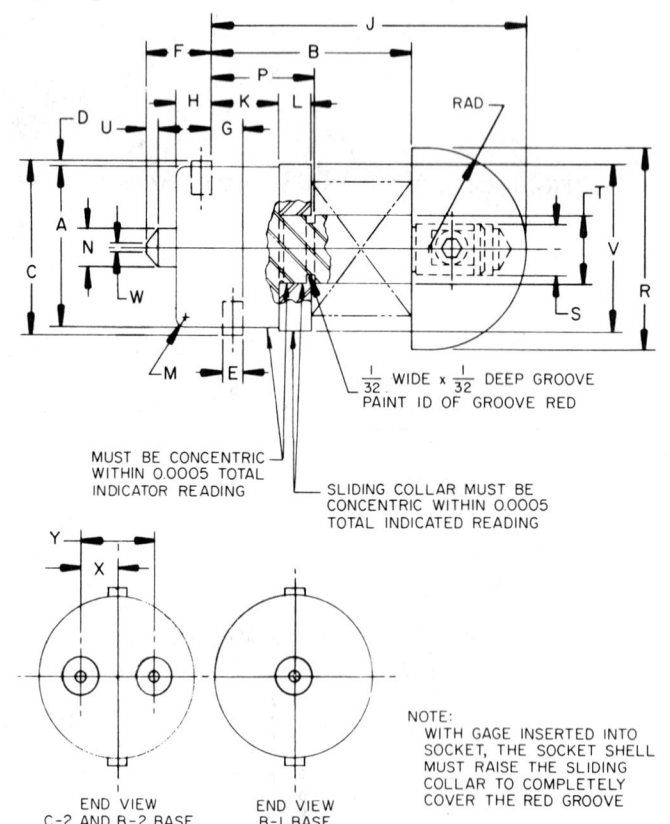

FIG. 3—BULB SUPPORT GAGE

CONNECTORS AND PLUGS—SAE J856

SAE Standard

23.07

Report of Electric Equipment Division approved June 1936 and last revised by Lighting Committee April 1963.

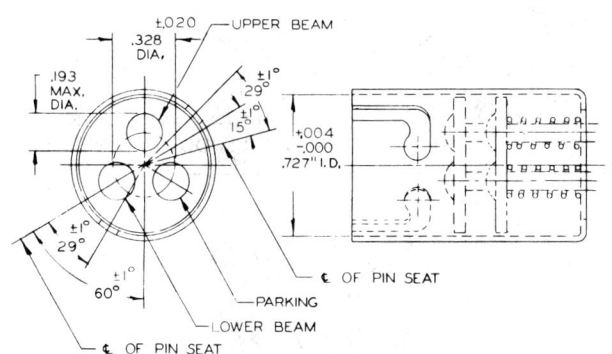

THREE WAY OFFSET PIN, LARGE FOR PLUG BASE TYPE SEE FIG. 11
FIG. 1—SOCKET, PLUG, THREE WAY OFFSET PIN, LARGE

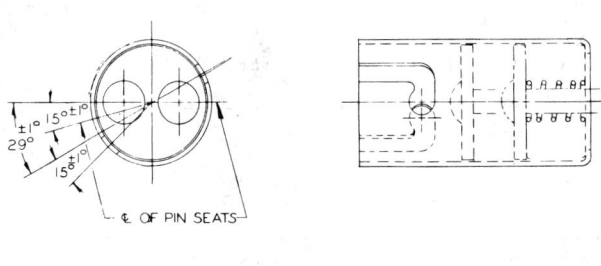

DOUBLE CONTACT OFFSET PIN FOR PLUG BASE TYPE SEE FIG. 8
FIG. 2—SOCKET, PLUG, DOUBLE CONTACT OFFSET PIN

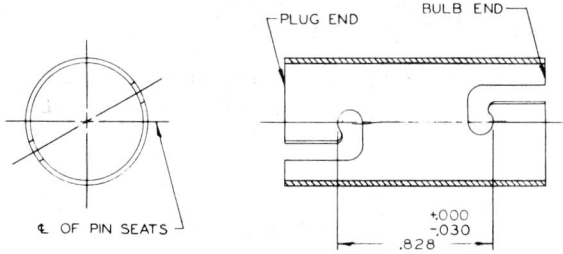

FOR PLUG BASE TYPE SEE FIG. 9, FOR BULB BASE TYPES B-1 AND B-2
FIG. 3—SOCKET, PLUG-BULB, DOUBLE END, SHORT[1]

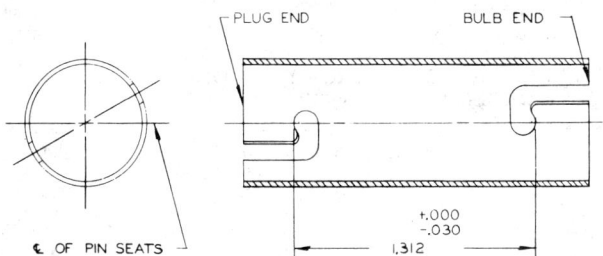

FOR PLUG BASE TYPE SEE FIG. 10, FOR BULB BASE TYPES B-1 AND B-2
FIG. 4—SOCKET, PLUG-BULB, DOUBLE END, LONG[1]

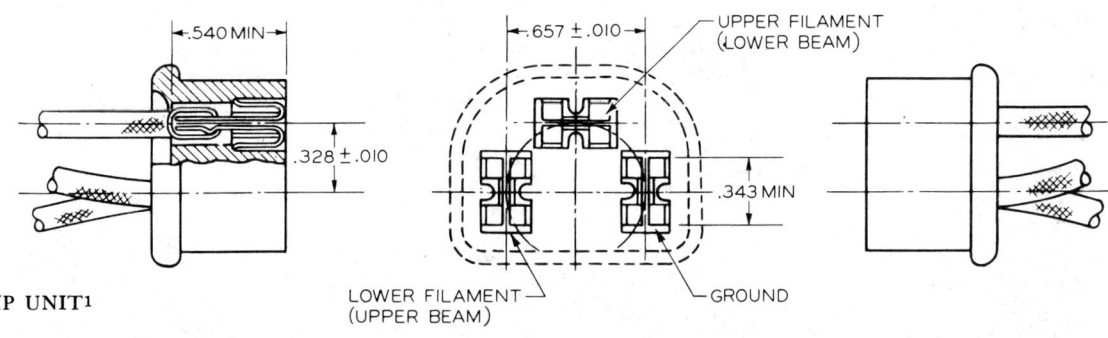

FIG. 5—CONNECTOR, SEALED BEAM HEADLAMP UNIT[1]

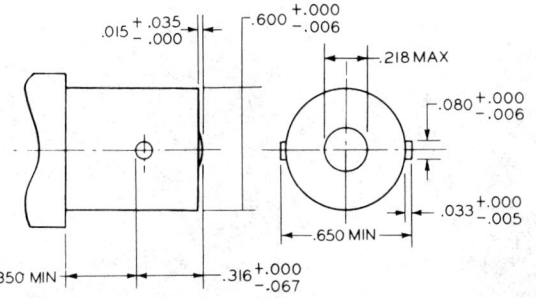

FIG. 6—PLUG, SINGLE CONTACT

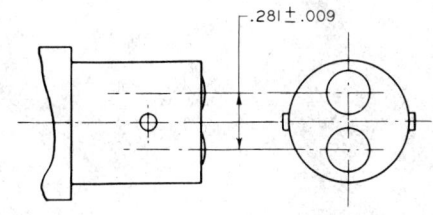

FIG. 7—PLUG, DOUBLE CONTACT

[1] All dimensions for the following are the same as those given for Figs. 1 and 2, except where otherwise indicated.

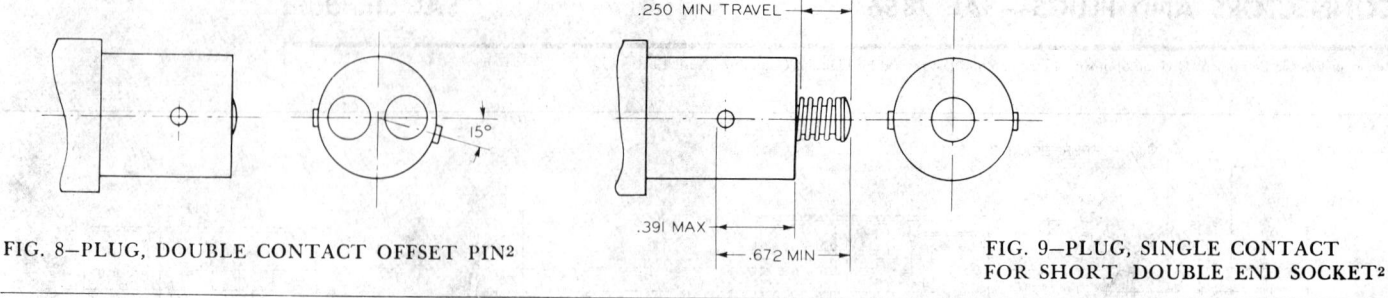

FIG. 8—PLUG, DOUBLE CONTACT OFFSET PIN[2]

FIG. 9—PLUG, SINGLE CONTACT FOR SHORT DOUBLE END SOCKET[2]

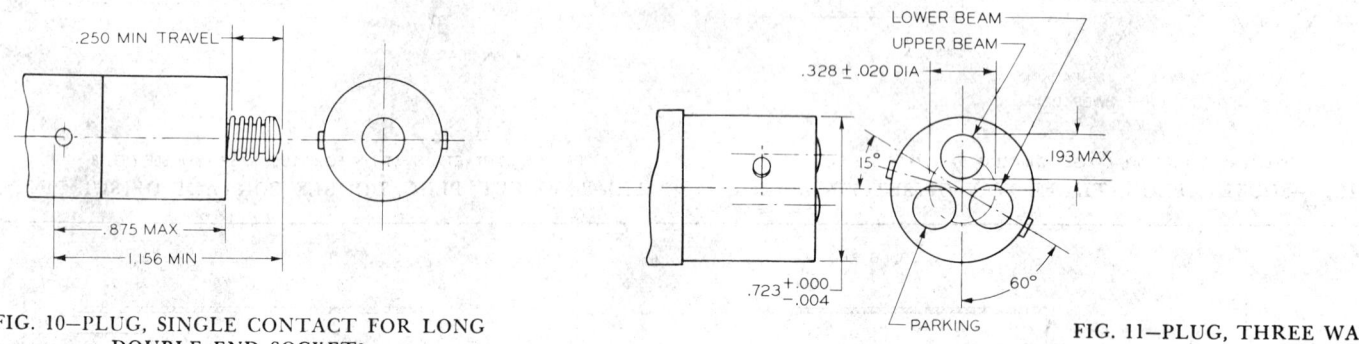

FIG. 10—PLUG, SINGLE CONTACT FOR LONG DOUBLE END SOCKET[2]

FIG. 11—PLUG, THREE WAY OFFSET PIN, LARGE[2]

[2] All dimensions for the following are the same as those given for Figs. 6 and 7, except where otherwise indicated.

DIMENSIONAL SPECIFICATIONS FOR SEALED BEAM HEADLAMP UNITS—SAE J571d

SAE Standard

Report of Lighting Division approved February 1940 and last revised by Lighting Committee February 1975. Editorial change June 1976.

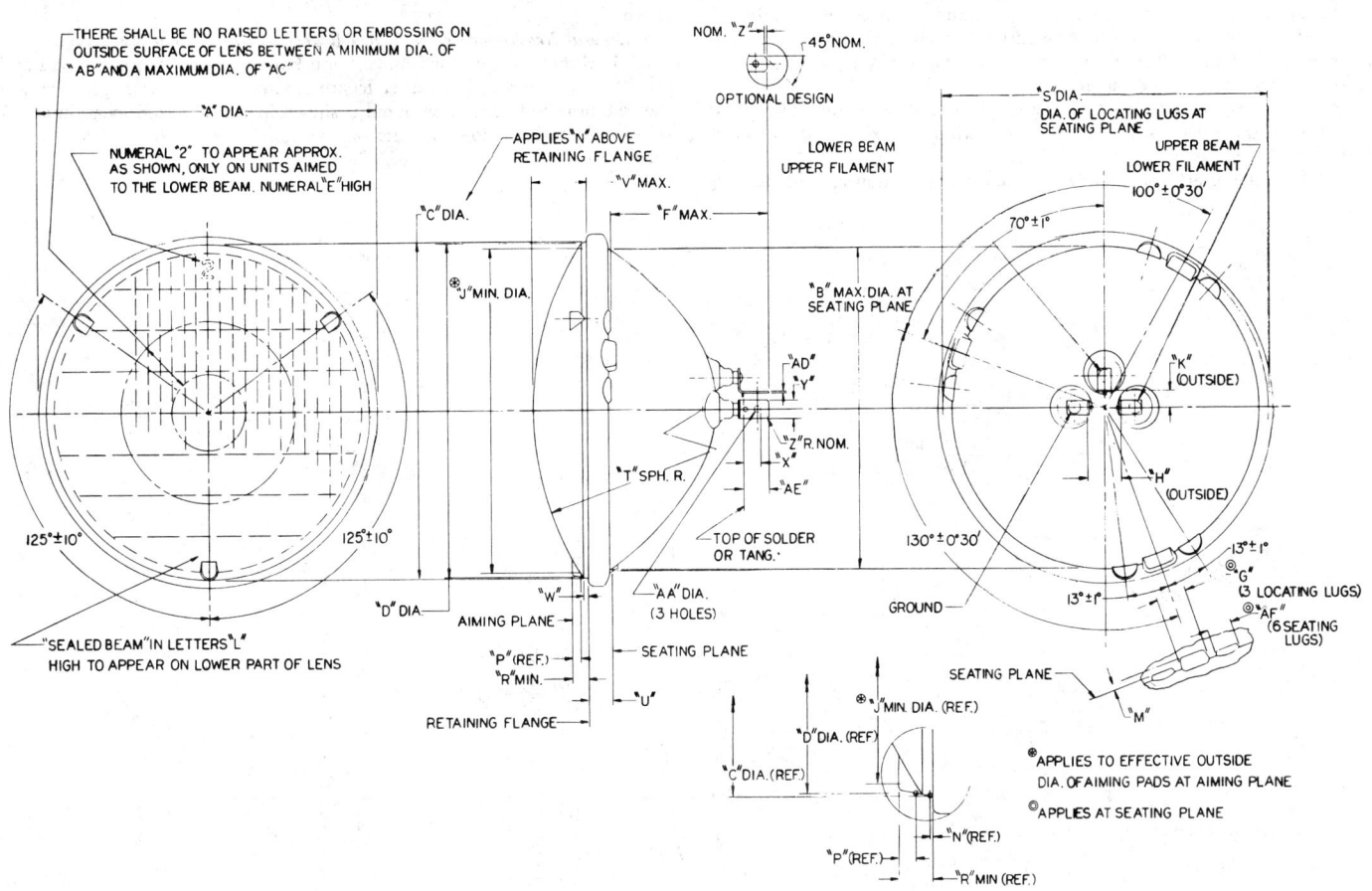

DIMENSIONS

Letter	in	mm	Letter	in	mm
A	7.031 +0.000/−0.109	178.58 +0/−2.76	S	6.770 +0.000/−0.040	171.95 +0/−1.01
B	6.380	162.05	T	6.000 +0.250/−0.000	152.40 +6.35/−0
C	6.687 +0.000/−0.030	169.84 +0/−0.76	U	0.500 ±0.040	12.70 ±1.01
D	6.595−6.675	167.52−169.54	V	1.150	29.21
E	3/8 ±1/32	9.52 ±0.79	W	0.078 +0.062/−0.000	1.98 +1.57/−0
F	3.500	88.90	X	0.345 +0.060/−0.000	8.76 +1.52/−0
G	0.575 +0.000/−0.025	14.60 +0/−0.63	Y	0.304 +0.016/−0.000	7.72 +0.40/−0
H	0.670 +0.035/−0.000	17.01 +0.88/−0	Z	0.06	1.52
J	6.450	163.83	AA	0.120 +0.010/−0.000	3.04 +0.25/−0
K	0.333 +0.020/−0.000	8.45 +0.50/−0	AB	1.50	38.10
L	1/4 ±1/32	6.35 ±0.79	AC	3.60	91.44
M	0.106 +0.100/−0.000	2.69 +2.54/−0	AD	0.030 ±0.002	0.76 ±0.05
N	0.030	0.76	AE	0.535 +0.000/−0.070	13.58 +0/−1.77
P	0.180	4.57	AF	0.50 ±0.25	12.70 ±6.35
R	0.350	8.89			

FIG. 1—SEALED BEAM HEADLAMP UNIT, 7 IN (178 mm) DIAMETER

1. Scope—This standard applies to dimensions for design guidance to assure interchangeability of sealed units, mounting rings or lamp bodies, and retaining rings. It is not intended that this standard be used for compliance. Compliance shall be based on tests specified in other related standards.

2. General Requirements for Sealed Beam Headlamp Units (Figs. 1, 3, 4, 6, and 7).

2.1 For 7 in. (178 mm) and 5¾ in. (146 mm) units there shall be no raised letters or embossing on outside surface of lens between a minimum diameter of 1.50 in. (38.10 mm) and a maximum diameter of 3.60 in. (91.44 mm) about the center of the lens.

2.2 For 4 x 6½ in. (100 x 165 mm) units there shall be no raised letters or embossing on the outside surface of the lens within a maximum diameter of 2.8 in. (71.12 mm) about the center of the lens.

2.3 Manufacturer's name and/or trade mark shall appear on lens.

2.4 "Sealed Beam" in letters ¼ ± ¹⁄₃₂ in. (6.35 ± 0.71 mm) high shall appear on lens.

2.5 The face of letters, numbers, or other symbols molded on the surface of the lens, shall not be raised more than 0.020 in. (0.508 mm).

2.6 Trade number and voltage shall be marked on the unit.

2.7 Aiming pad designs may vary but shall meet limiting dimensions as shown.

3. General Requirements for Sealed Beam Headlamp Mounting Rings.

3.1 Mounting rings for 7 in. (178 mm) and 5¾ in. (146 mm) Type 1 and Type 2 (Figs. 2 and 5) have three locating notches, one of which is narrower than the others. The narrow notch is shown in the preferred location at a 7 o'clock orientation. However, the narrow notch may be located at 2 or 11 o'clock.

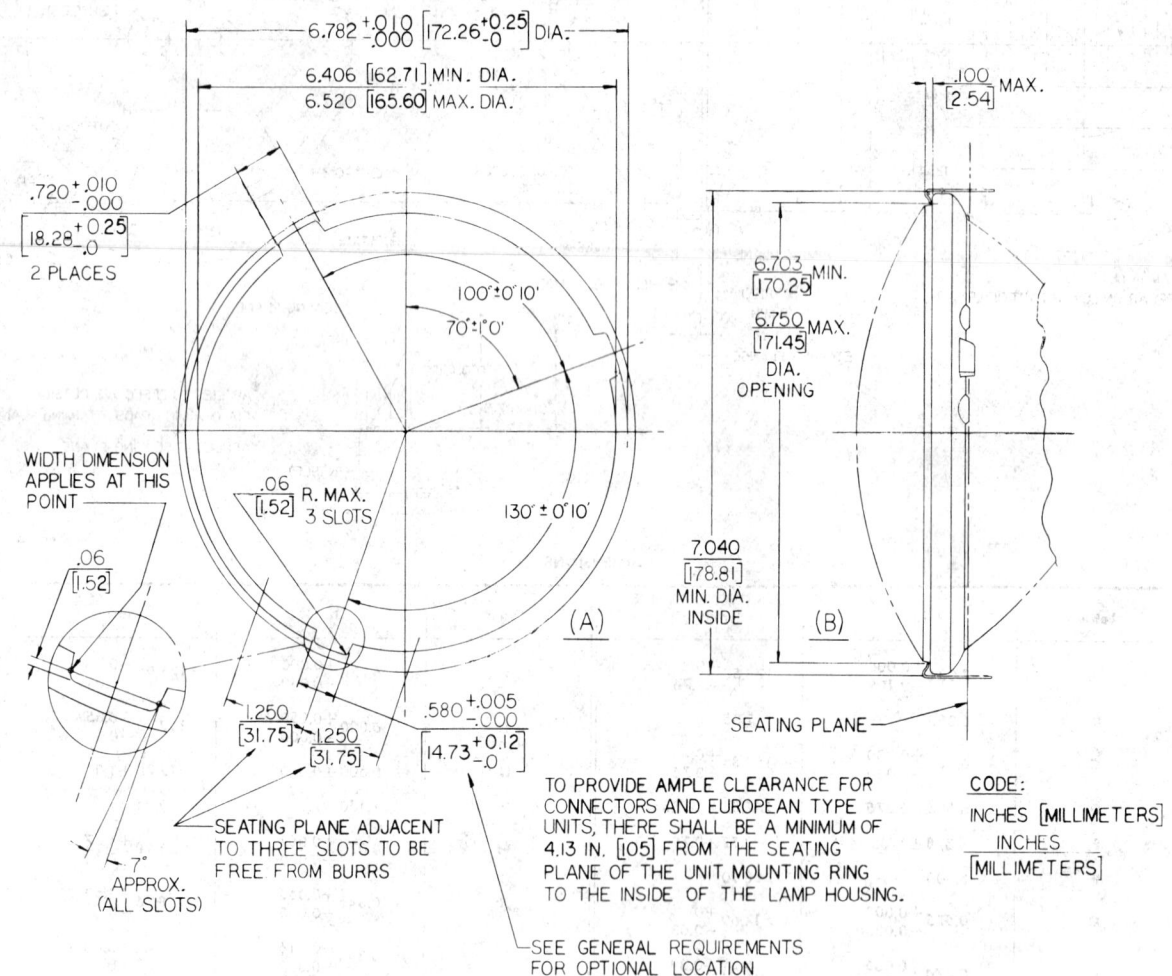

FIG. 2—(A) FRONT VIEW OF SLOT OR NOTCHES FOR 7 IN (178 mm) DIAMETER SEALED BEAM HEADLAMP MOUNTING RING OR LAMP BODY; (B) SEALED BEAM HEADLAMP UNIT RETAINING RING

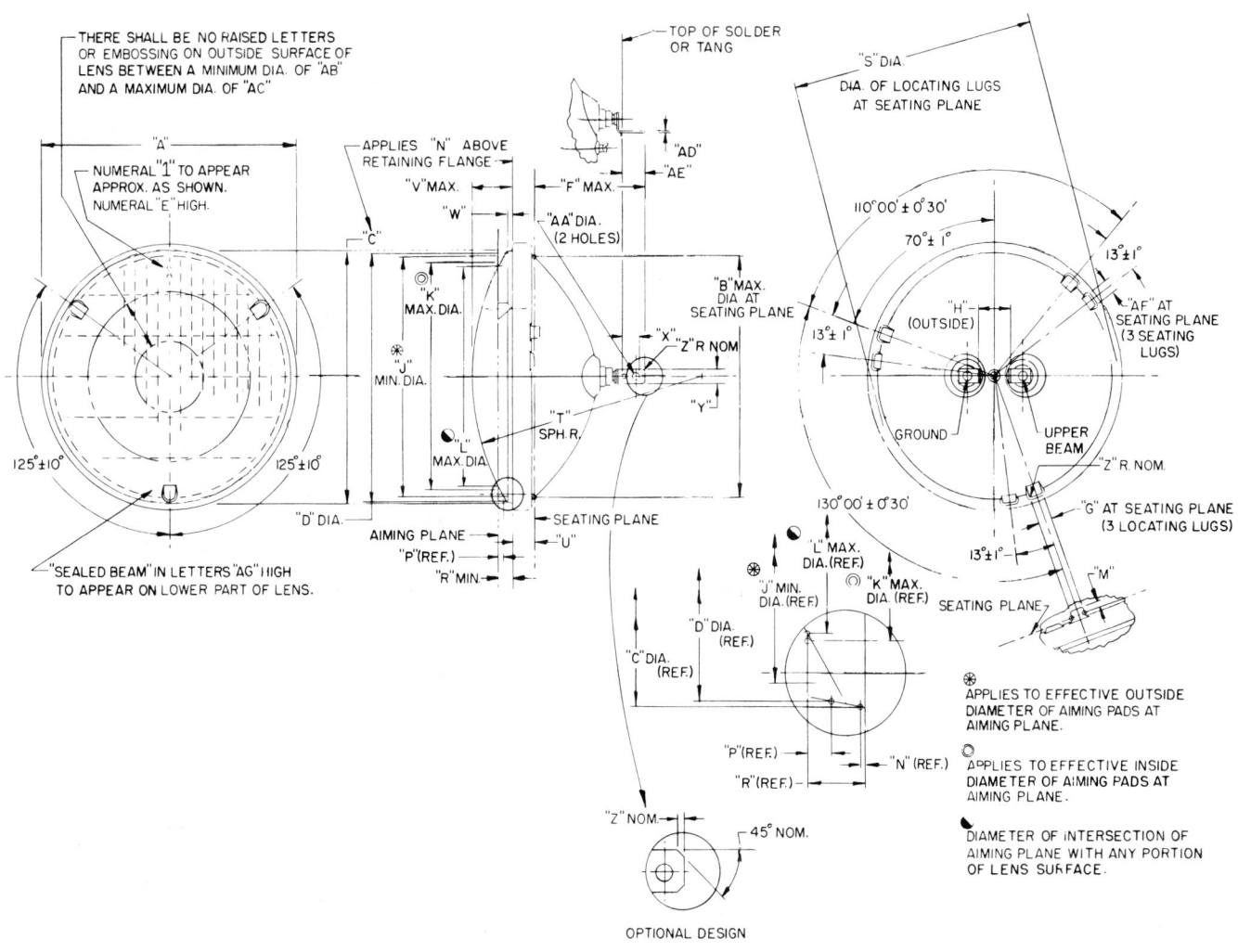

DIMENSIONS

Letter	in	mm	Letter	in	mm
A	5.700 +0.000/−0.100	144.78 +0/−2.54	T	5.06 ±0.12	128.52 ±3.04
B	5.120	130.04	U	0.500 ±0.040	12.70 ±1.01
C	5.355 +0.000/−0.030	136.01 +0/−0.76	V	0.92	23.36
D	5.280−5.340	134.11−135.63	W	0.078 +0.062/−0.000	1.98 +1.57/−0
E	1/4 ±1/32	6.35 ±0.79	X	0.345 +0.060/−0.000	8.76 +1.52/−0
F	2.60	66.04	Y	0.304 +0.016/−0.000	7.72 +0.40/−0
G	0.312 ±0.010	7.92 ±0.25	Z	0.06	1.52
H	0.670 +0.035/−0.000	17.01 +0.88/−0	AA	0.120 +0.010/−0.000	3.04 +0.25/−0
J	5.060	128.52	AB	1.50	38.10
K	4.57	116.07	AC	3.60	91.44
L	4.53	115.06	AD	0.030 ±0.002	0.76 ±0.05
M	0.100 +0.050/−0.000	2.54 +1.27/−0	AE	0.535 +0.000/−0.070	13.58 +0/−1.77
N	0.030	0.76			
P	0.165	4.19	AF	0.31 ±0.12	7.87 ±3.04
R	0.320	8.12			
S	5.440 +0.000/−0.040	138.17 +0/−1.01	AG	5/32 ±1/32	3.96 ±0.79

FIG. 3—TYPE 1 SEALED BEAM HEADLAMP UNIT, 5¾ IN (146 mm) DIAMETER

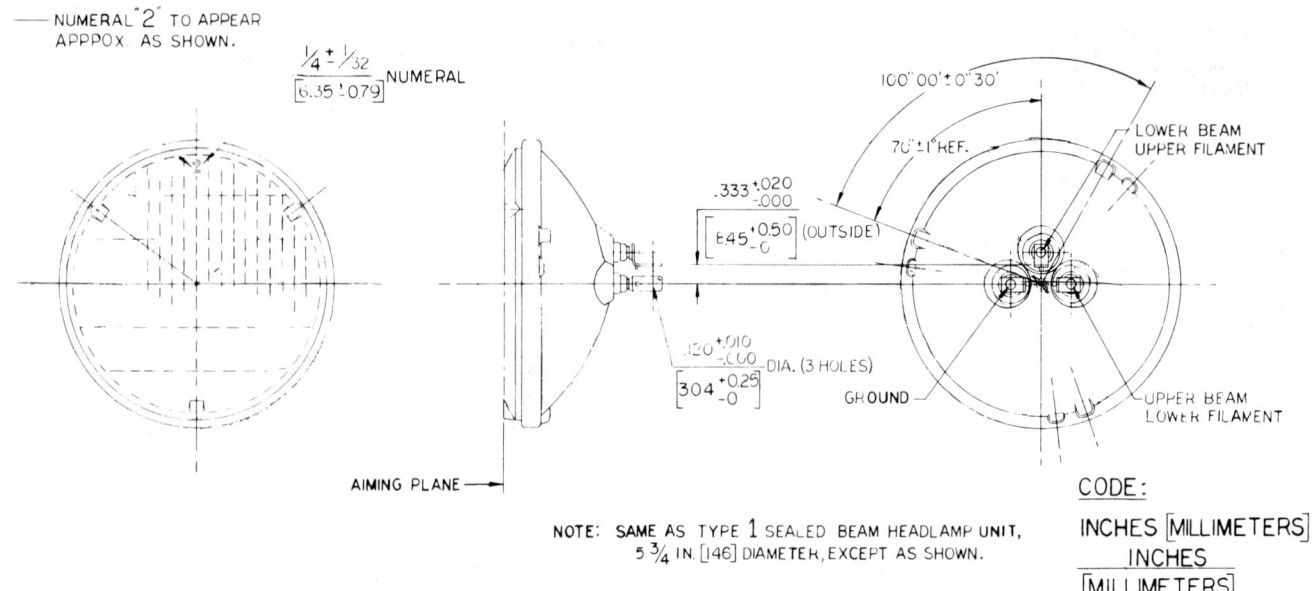

FIG. 4—TYPE 2 SEALED BEAM HEADLAMP UNIT, 5¾ IN (146 mm) DIAMETER

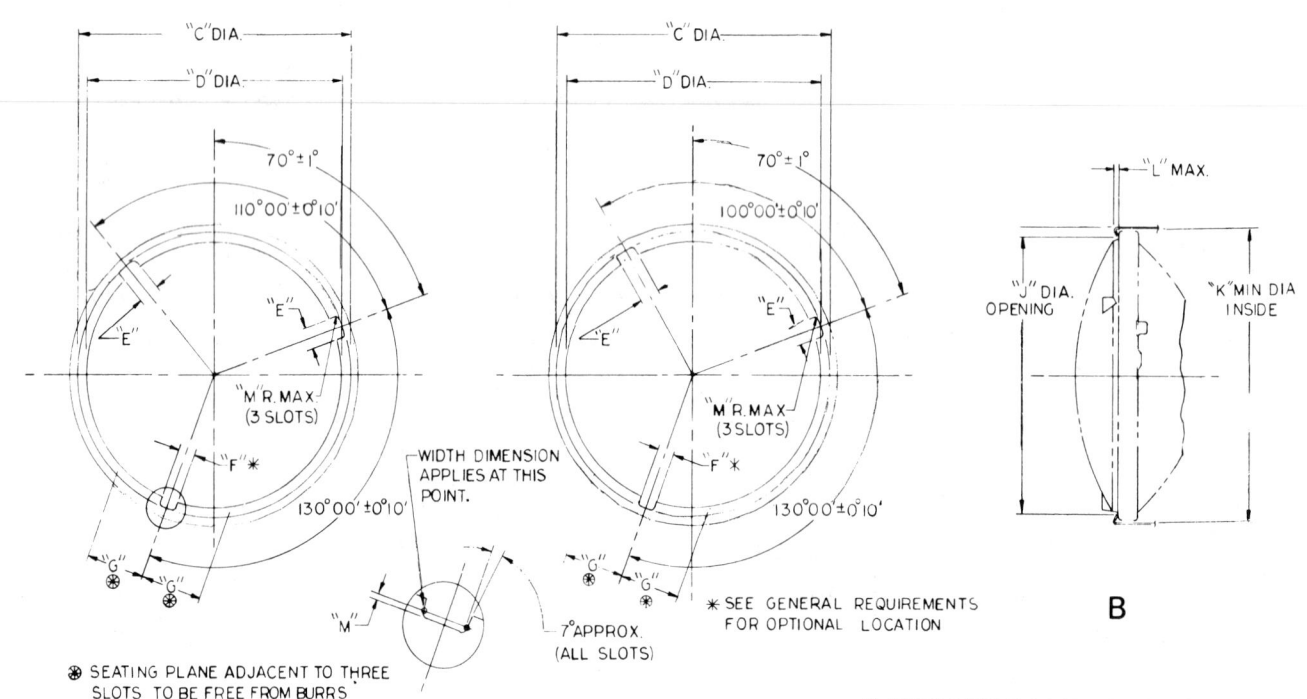

DIMENSIONS

Letter	in	mm	Letter	in	mm
C	5.450 +0.010/−0.000	138.43 +0.25/−0	G	1.20	30.48
D	5.250 − 5.140	133.35 − 130.55	J	5.400 − 5.360	137.16 − 136.14
E	0.410 +0.010/−0.000	10.41 +0.25/−0	K	5.710	145.03
F	0.330 +0.005/−0.000	8.38 +0.12/−0	L	0.100	2.54
			M	0.06	1.52

FIG. 5—(A) FRONT VIEW OF SLOTS OR NOTCHES FOR 5¾ IN (146 mm) DIAMETER SEALED BEAM HEADLAMP MOUNTING RING OR LAMP BODY; (B) SEALED BEAM HEADLAMP UNIT RETAINING RING

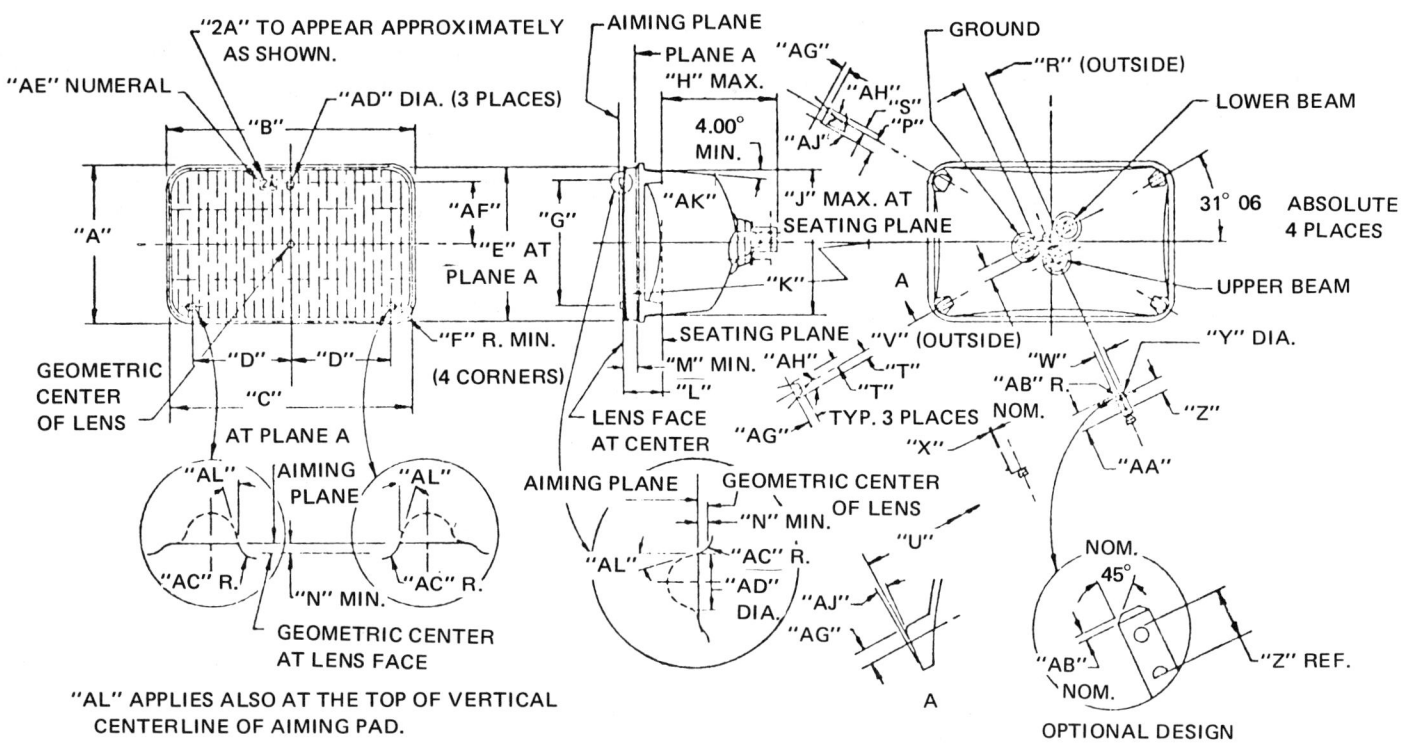

Letter	in	mm	Letter	in	mm	Letter	in	mm
A	4.200 +0.030/−0.170	106.68 +0.76/−4.32	P	0.313 +0.015/−0.010	7.95 +0.38/−0.25	AA	0.535 +0.000/−0.071	13.58 +0/−1.80
B	6.580 +0.030/−0.170	167.13 +0.76/−4.32	R	0.669 +0.035/−0.000	17.02 +0.88/−0.00	AB	0.060 ± 0.020	1.5 ± 0.5
						AC	0.060 ± 0.020	1.5 ± 0.5
C	6.440 ± 0.030	163.58 ± 0.76	S	0.122 ± 0.015	3.10 +0.38/−0.25	AD	0.200 ± 0.010	5.08 ± 0.25
D	2.700 ± 0.020	68.58 ± 0.51	T	0.167 ± 0.010	4.24 ± 0.25	AE	0.250 ± 0.030	6.35 ± 0.76
E	4.060 ± 0.030	103.12 ± 0.76	U	3.640 ± 0.010	92.47 ± 0.25	AF	1.660 ± 0.010	42.16 ± 0.25
F	0.540	13.71	V	0.335 +0.020/−0.000	8.5 +0.5/−0	AG	0.160 ± 0.010	4.06 ± 0.25
G	3.320 ± 0.030	84.33 ± 0.76				AH	15.00° ± 3.00°	
H	3.350	85.09	W	0.304 +0.016/−0.000	7.72 +0.40/−0.00	AJ	3.33° min	
J	4.01	101.85	X	0.030 ± 0.002	0.76 ± 0.05	AK	1.56° max	39.6 max
K	50.000 +0.500/−2.00	1270.0 +13.0/−50.8	Y	0.120 +0.010/−0.000	3.05 +0.25/−0	AL	16° max	
L	1.375 ± 0.040	34.93 ± 1.02	Z	0.345 +0.059/−0.000	8.76 +1.50/−0	AM	32° min	
M	0.420	10.68						
N	0.020	0.51						

FIG. 6—TYPE 2A SEALED BEAM HEADLAMP UNIT 4 × 6½ IN (100 × 165 mm) RECTANGULAR UNIT

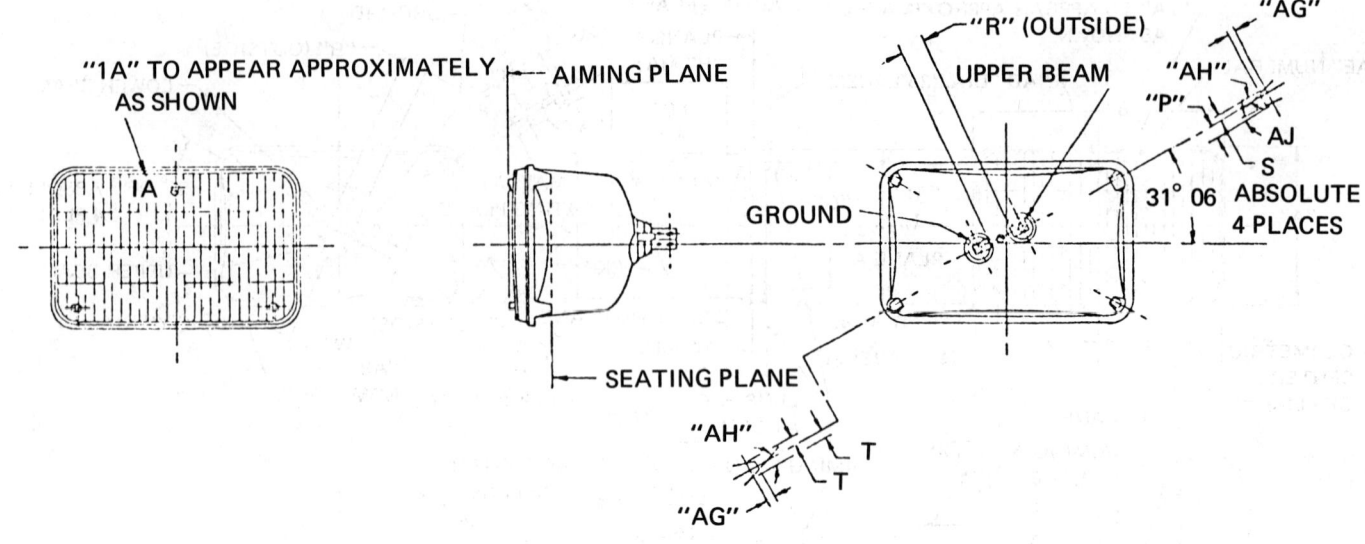

Letter	in	mm
G	0.160 ± 0.010	4.064 ± 0.25
R	0.669 +0.035/−0.000	17.0 +0.9/−0
AH	15°00′ ± 3°00′	
T	0.167 ± 0.0100	4.24 ± 0.25
AJ	4°20′ ± 1°00′	
P	0.313 +0.015/−0.010	7.95 +0.38/−0.25
S	0.122 +0.015/−0.010	3.10 +0.38/−0.25

FIG. 7—TYPE 1A SEALED BEAM HEADLAMP UNIT 4 × 6½ (100 × 165 mm) RECTANGULAR UNIT

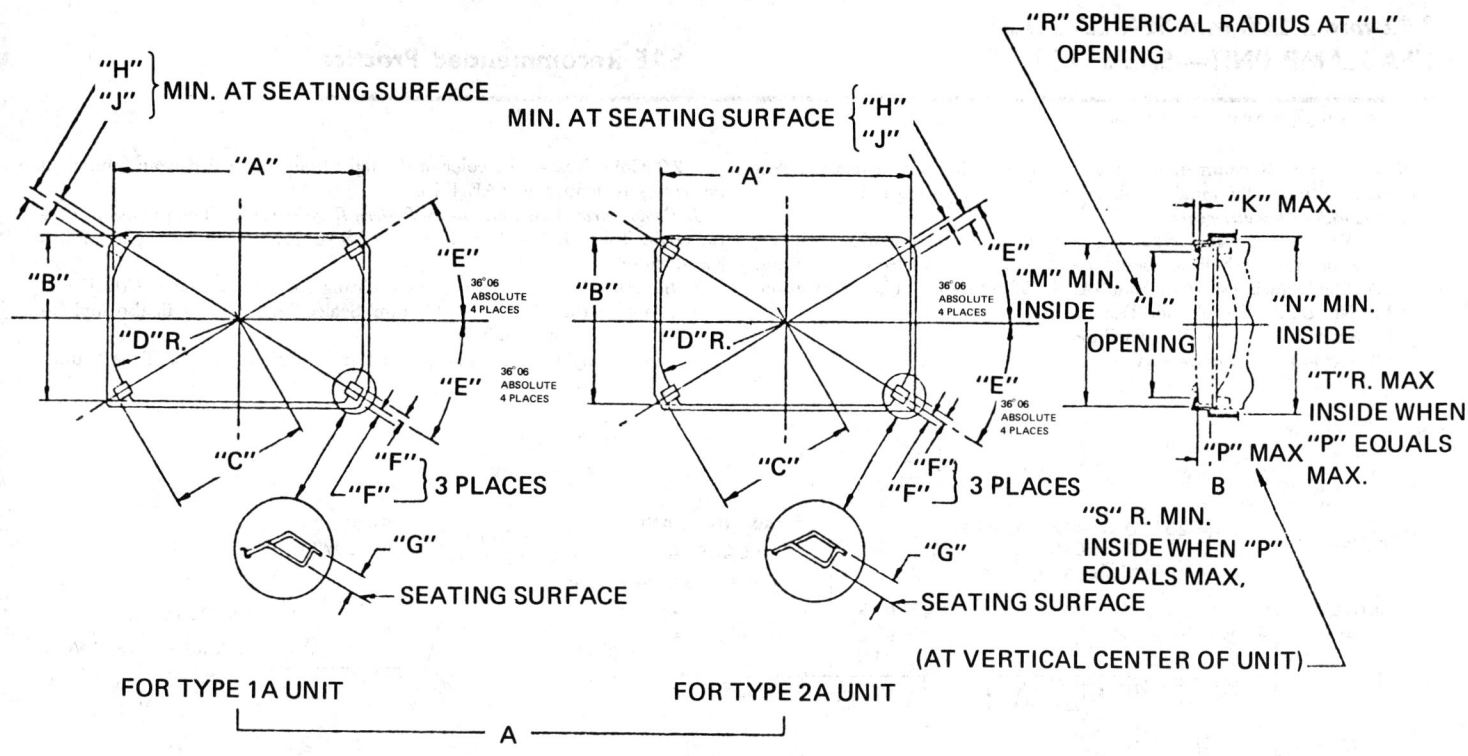

FIG. 8(A)—FRONT VIEW OF SLOTS OR NOTCHES FOR 4 × 6½ IN (100 × 165 mm) RECTANGULAR HEADLAMP MOUNTING RING OR LAMP BODY; (B)—RECTANGULAR SEALED BEAM HEADLAMP UNIT RETAINING RING.

142 mm x 200 mm SEALED BEAM HEADLAMP UNIT—SAE J1132

SAE Recommended Practice

Report of Lighting Committee approved January 1976.

1. *Scope*—This Recommended Practice covers design parameters of a rectangular, mechanically aimable, 2-beam sealed headlighting unit.
2. *Laboratory Requirements*
   2.1 *Test Voltage*—In conducting tests to this Recommended Practice, the sealed beam unit shall be operated at 12.8 V for 12 V electrical systems.
   2.2 The following sections from SAE J575f are part of this specification:
   2.2.1 SECTION 2—Samples for Test
   2.2.2 SECTION 3—Laboratory Facilities
   2.2.3 SECTION 4.6—Photometry

2.3 *Color Test*—The color of the light from this sealed beam unit shall be white, as defined in SAE J578a.
3. *Photometric Design and Beam Pattern Requirements*—The provisions contained in Table 1, Fig. 2 and Fig. 3 of SAE J579c shall be a part of this specification.
4. *Inspection*—The provisions of Lighting Inspection Code, SAE J599d, which apply to the 7 in Type 2 (178 mm) Sealed Beam Unit shall also apply to the 142 mm x 200 mm unit.
5. *Service Performance*—The photometric requirements for Type 2 units

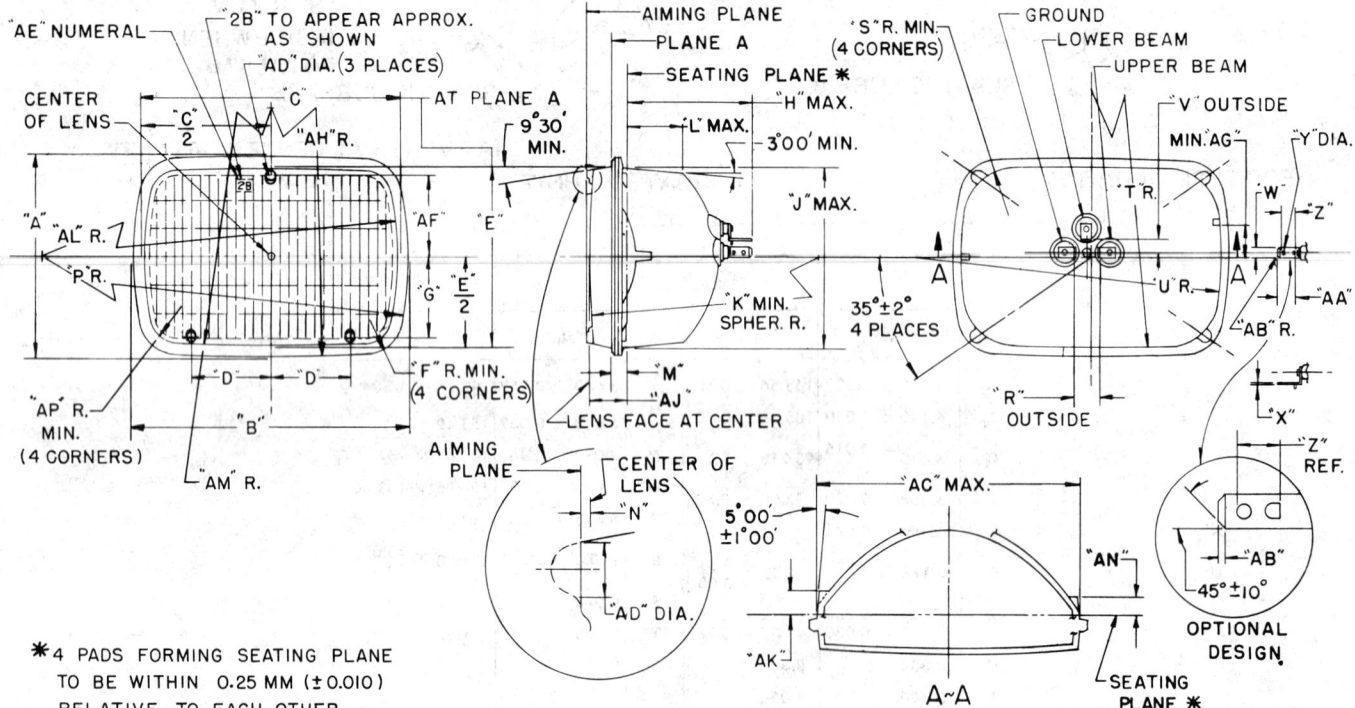

LETTER	MM	INCH
A	142.0 +0.8/-4.2	5.591 +0.032/-0.165
B	200.0 +0.8/-4.2	7.874 +0.032/-0.165
C	189.73 +0.30/-0.80	7.470 +0.012/-0.032
D	64.0 ±1.0	2.520 ±0.039
E	131.73 +0.30/-0.80	5.186 +0.012/-0.032
F	25.5	1.004
G	59.6 ±1.0	2.346 ±0.039
H	107.0	4.213
J	132.23	5.206
K	1200.0	47.244
L	49.0	1.929
M	11.1 ±1.0	0.437 ±0.039
N	0.5 +4.0/-0.5	0.020 +0.157/-0.020

LETTER	MM	INCH
P	254.0 ±5.0	10.000 ±0.197
R	17.0 +0.9/-0	0.669 +0.035/-0.000
S	42.7	1.681
T	2401.5 ±13.0	94.547 ±0.512
U	249.0 ±5.0	9.803 ±0.197
V	8.5 +0.5/-0	0.335 +0.020/-0.000
W	7.72 +0.40/-0	0.304 +0.016/-0.000
X	0.76 ±0.05	0.030 ±0.002
Y	3.05 +0.25/-0	0.120 +0.010/-0.000
Z	8.76 +1.50/-0	0.345 +0.059/-0.000
AA	13.58 +0/-1.80	0.535 +0.000/-0.071

LETTER	MM	INCH
AB	1.52 ±0.50	0.060 ±0.020
AC	190.23	7.489
AD	5.0 ±1.0	0.197 ±0.039
AE	6.35 ±0.80	0.250 ±0.032
AF	60.5 ±1.0	2.382 ±0.039
AG	19.7	0.776
AH	2406.5 ±13.0	94.744 ±0.512
AJ	26.7 +4.0/-1.0	1.051 +0.157/-0.039
AK	16.0 +2.0/-1.0	0.630 +0.079/-0.039
AL	250.0 +0/-25.0	9.843 +0.000/-0.984
AM	2402.0 +0/-775.0	94.567 +0.000/-30.512
AN	12.0 +2.0/-1.0	0.472 +0.079/-0.039
AP	20.4	0.803

FIG. 1—TYPE 2B SEALED BEAM HEADLAMP 142 x 200 mm (5.6 x 7.9 in) RECTANGULAR UNIT

and contained in Table 1 of SAE J32a shall be considered a part of this Recommended Practice.

6. Electrical Specifications

6.1 Design wattage (maximum wattage shall not exceed design wattage by more than 7.5%; there is no minimum spec.).

    U.B.   65 watts
    L.B.   55 watts

6.2 Average design life (test at 14 V):

    U.B.   150 h
    L.B.   320 h

7. Dimensional Requirements—See Figs. 1 and 2.

8. General Requirements

8.1 There shall be no raised letters or embossing on the outside surface of the lens within a maximum diameter of 71.12 mm (2.8 in) about the center of the lens.

8.2 The provisions of Sections 2.3 to 2.7 (inclusive) of SAE J571d shall also apply to the 142 mm x 200 mm unit.

NOTE: Upon approval of this Recommended Practice, Dimensional Specifications for Headlamp Aimer Locating Plate, and contained in SAE J602c, will be revised to include the 142 mm x 200 mm unit.

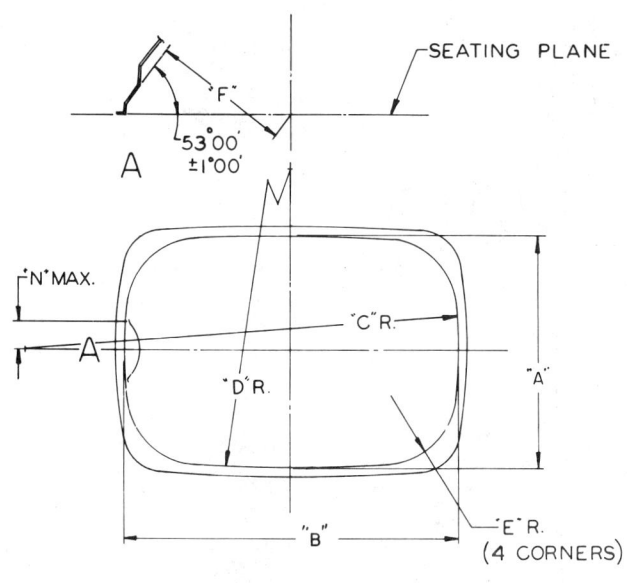

MOUNTING RING (A)

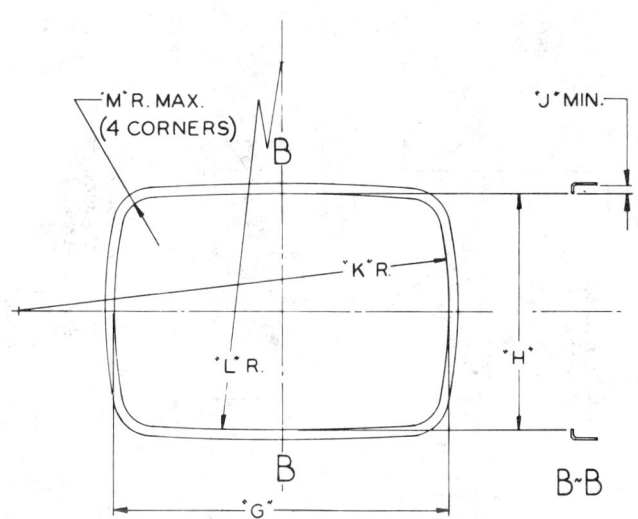

RETAINING RING (B)

LETTER	MM	INCH
A	132.9 ± 0.5	5.232 ± 0.020
B	191.0 ± 0.5	7.520 ± 0.020
C	250.0 ± 5.0	9.843 ± 0.197
D	2400.0 ± 50.0	94.488 ± 1.969
E	41.0 ± 2.0	1.614 ± 0.079
F	79.90 ± 0.40	3.146 ± 0.016

LETTER	MM	INCH
G	190.42 ± 0.30	7.497 ± 0.012
H	132.42 ± 0.30	5.213 ± 0.012
J	5.34	0.210
K	250.0 $^{+30.0}_{-0}$	9.843 $^{+1.181}_{-0.000}$
L	2402.0 $^{+2250.0}_{-0}$	94.567 $^{+88.583}_{-0.000}$
M	20.4	0.803
N	19.0	0.748

FIG. 2—(A) FRONT VIEW OF MOUNTING RING OR LAMP BODY FOR 142 x 200 mm (5.6 x 7.9 in) RECTANGULAR UNIT; (B) RETAINING RING

DIMENSIONAL SPECIFICATIONS FOR GENERAL SERVICE SEALED LIGHTING UNITS—SAE J760a

SAE Recommended Practice

Report of Lighting Committee approved March 1961 and last revised December 1974.

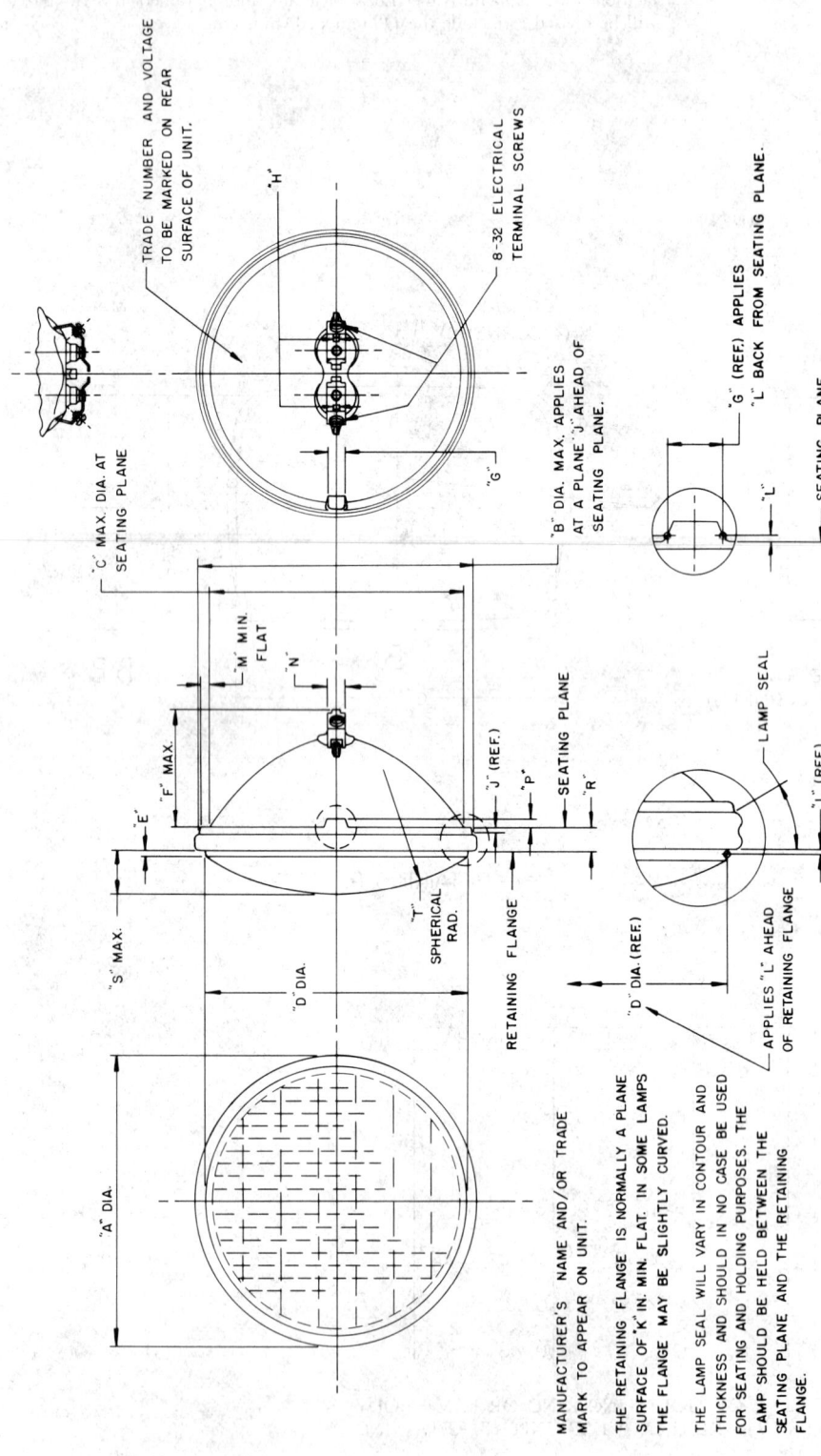

LETTER	INCH	MM
A	5.70 +.00 / -.10	144.8 +0.0 / -2.5
B	5.475	139.06
C	5.100	129.54
D	5.265 +.000 / -.030	133.7 +0.00 / -0.76
E	.078 +.062 / -.000	2.00 +1.50 / -0.00
F	2.50	63.5
G	.440 +.000 / -.025	11.17 +0.00 / -0.63
H	.125 ±.010	3.17 ±0.25
J	.110	2.79
K	.062	1.57
L	.030	0.76
M	.125	3.18

LETTER	INCH	MM
N	.375 ±.010	9.52 ±0.25
P	.135 +.030 / -.000	3.43 +0.76 / -0.00
R	.53 ±.04	13.5 ±1.0
S	.92	23.36
T	5.00 ±.12	127.0 ±3.0

FIG. 1—DIMENSIONS OF 5 3/4 IN (146 mm) DIAMETER SEALED LIGHTING UNIT

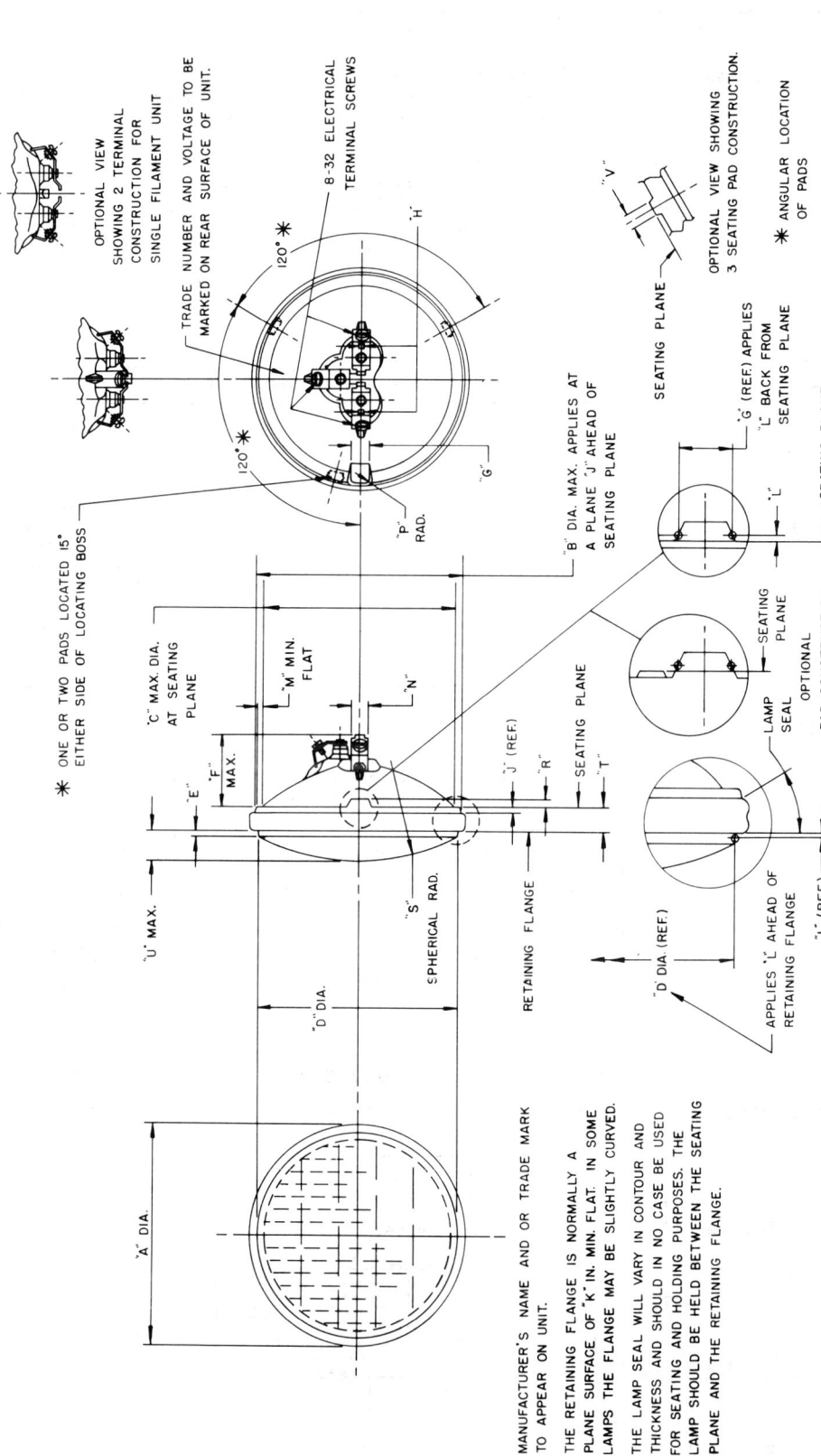

FIG. 2—DIMENSIONS OF 4½ IN (114 mm) DIAMETER SEALED LIGHTING UNIT

LAMP BULBS AND SEALED BEAM HEADLAMP UNITS—SAE J573g

SAE Standard

Report of Lighting Division approved March 1918 and last revised by Lighting Committee December 1976. Rationale statement available.

1. Scope—Many of the lighting devices on motor vehicles are required and are essential to operation on the highway. To maintain lighting performance, it is important that the bulb and sealed unit types employed be readily available, when needed, throughout the country in normal service channels. Therefore, this SAE Standard lists an assortment of *current popular types*, together with their *design characteristics*, which are recommended for use wherever practicable. It is recognized that because of constantly changing and improving technology, the list may be incomplete. Also, instances may arise in the design of some devices which require the employment of other types while achieving the desired performance.

Some of the design characteristics in this standard are listed solely for the sake of standardization and have no bearing on how lamp bulbs perform in lighting devices on the highway. A condensed list of specifications and their applicable tolerances is presented in SAE J1049.

2. Definition

2.1 Accurate Rated Bulbs—A bulb operated at design mean spherical candela (Table 2) and having its filament(s) within ±0.010 in (0.25 mm) of nominal design position. This applies to Nos. 1156, 1157, and 1157NA only. (See Fig. 1 for the spacing between the major and minor filaments of Nos. 1157 and 1157NA.) Rated bulbs shall be seasoned at rated voltage for 1% of their design life or 10 h maximum.

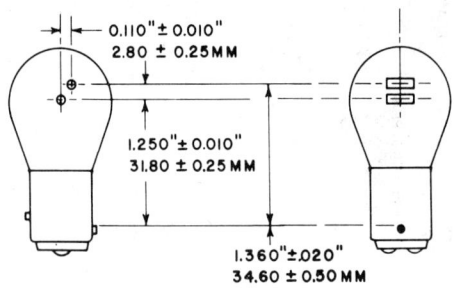

AXIAL ALIGNMENT TOLERANCE:
NARROW VIEW ± 0.010" (±0.25mm)
WIDE VIEW ± 0.020" (±0.50mm)

(USE DIMENSIONS OF LOWER FILAMENT FOR 1156 BULB)

ϕFIG. 1—FILAMENT LOCATION FOR RATED BULBS

TABLE 1—BULB DIMENSIONS (SEE FIG. 2)

Bulb	Base	Max Bulb Dia (D)		Max Exposed Length (L)	
		in	mm	in	mm
G—3-1/2	A—1	0.46	11.7	0.70	17.8
T—1-3/4	W—1	0.23	5.8	0.60	15.2
T—3-1/4	W—2	0.40	10.3	0.81	20.7
T—3-1/4	A—1	0.43	11.0	0.94	23.9
G—4-1/2	A—1	0.59	15.0	0.84	21.4
G—6	B—1, B—2	0.75	19.0	1.19	30.2
B—6	B—1, B—2	0.78	19.7	1.47	37.3
S—8	B—1, B—2	1.04	26.5	1.8	45.0
S—8	C—2	1.04	26.5	1.8	45.0

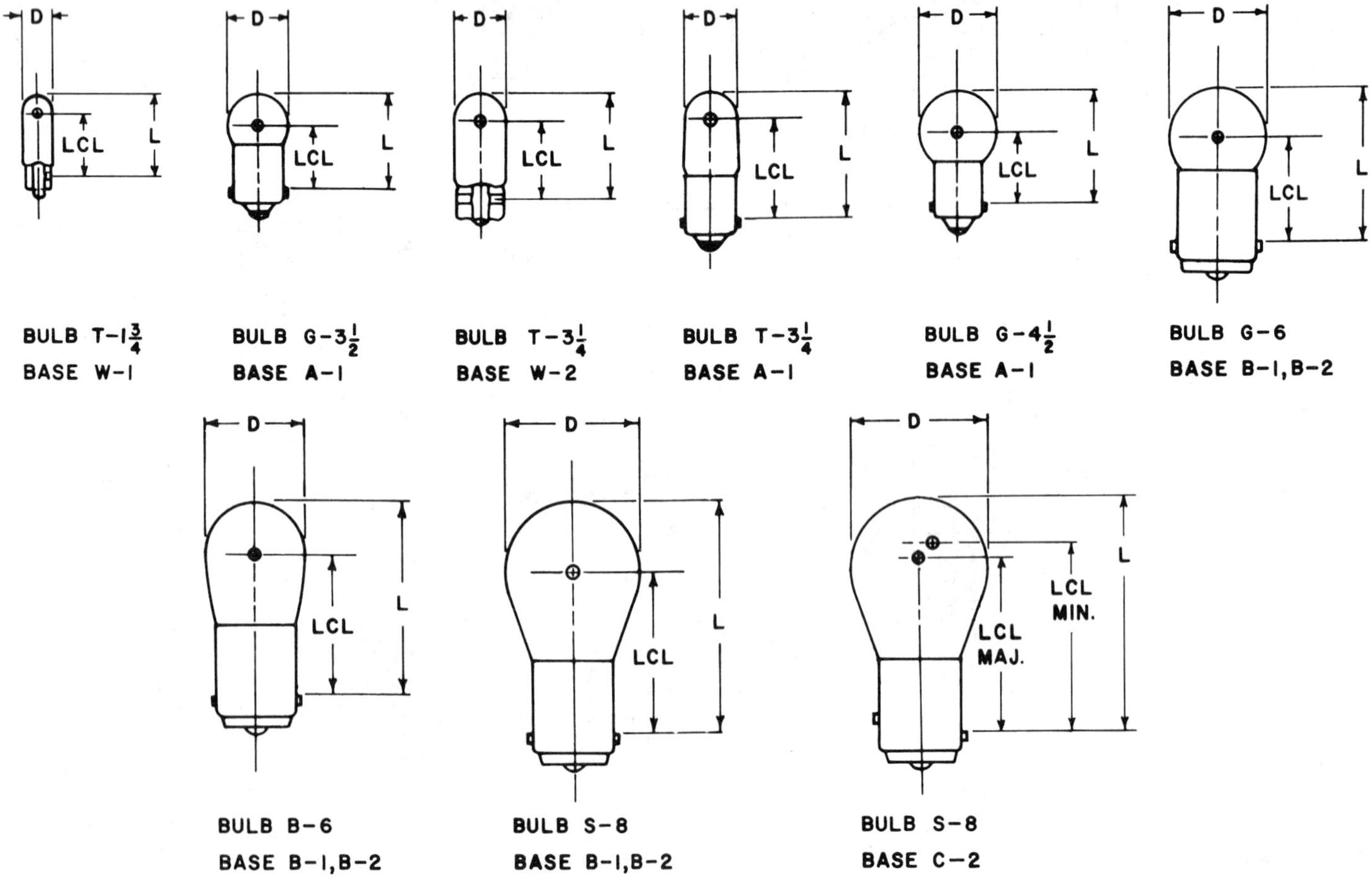

BULB T-1¾ BASE W-1
BULB G-3½ BASE A-1
BULB T-3¼ BASE W-2
BULB T-3¼ BASE A-1
BULB G-4½ BASE A-1
BULB G-6 BASE B-1, B-2

BULB B-6 BASE B-1, B-2
BULB S-8 BASE B-1, B-2
BULB S-8 BASE C-2

ϕFIG. 2—BULB TYPES

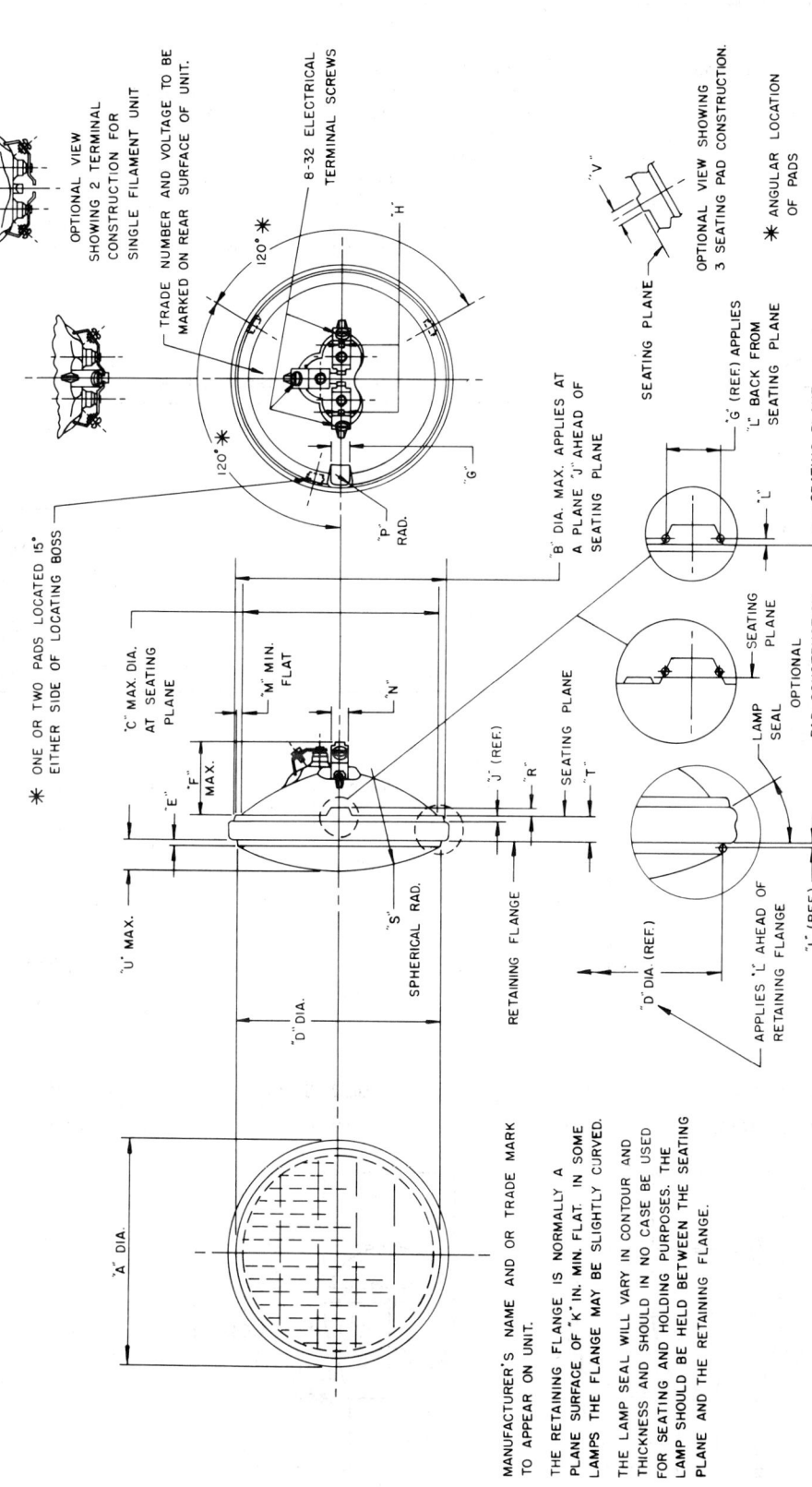

FIG. 2—DIMENSIONS OF 4½ IN (114 mm) DIAMETER SEALED LIGHTING UNIT

LAMP BULBS AND SEALED BEAM HEADLAMP UNITS—SAE J573g

SAE Standard

Report of Lighting Division approved March 1918 and last revised by Lighting Committee December 1976. Rationale statement available.

1. Scope—Many of the lighting devices on motor vehicles are required and are essential to operation on the highway. To maintain lighting performance, it is important that the bulb and sealed unit types employed be readily available, when needed, throughout the country in normal service channels. Therefore, this SAE Standard lists an assortment of *current popular types,* together with their *design characteristics,* which are recommended for use wherever practicable. It is recognized that because of constantly changing and improving technology, the list may be incomplete. Also, instances may arise in the design of some devices which require the employment of other types while achieving the desired performance.

Some of the design characteristics in this standard are listed solely for the sake of standardization and have no bearing on how lamp bulbs perform in lighting devices on the highway. A condensed list of specifications and their applicable tolerances is presented in SAE J1049.

2. Definition

2.1 **Accurate Rated Bulbs**—A bulb operated at design mean spherical candela (Table 2) and having its filament(s) within ±0.010 in (0.25 mm) of nominal design position. This applies to Nos. 1156, 1157, and 1157NA only. (See Fig. 1 for the spacing between the major and minor filaments of Nos. 1157 and 1157NA.) Rated bulbs shall be seasoned at rated voltage for 1% of their design life or 10 h maximum.

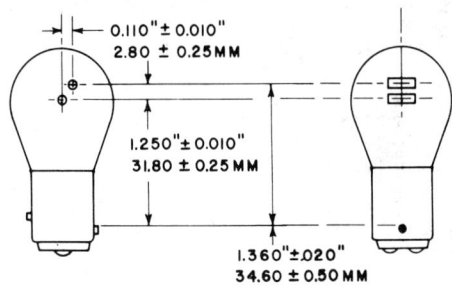

0.110"± 0.010"
2.80 ± 0.25MM
1.250"± 0.010"
31.80 ± 0.25MM
1.360"±.020"
34.60 ± 0.50MM
AXIAL ALIGNMENT TOLERANCE:
NARROW VIEW ± 0.010" (±0.25mm)
WIDE VIEW ± 0.020" (±0.50mm)

(USE DIMENSIONS OF LOWER FILAMENT FOR 1156 BULB)

ϕFIG. 1—FILAMENT LOCATION FOR RATED BULBS

ϕ TABLE 1—BULB DIMENSIONS (SEE FIG. 2)

Bulb	Base	Max Bulb Dia (D)		Max Exposed Length (L)	
		in	mm	in	mm
G—3-1/2	A—1	0.46	11.7	0.70	17.8
T—1-3/4	W—1	0.23	5.8	0.60	15.2
T—3-1/4	W—2	0.40	10.3	0.81	20.7
T—3-1/4	A—1	0.43	11.0	0.94	23.9
G—4-1/2	A—1	0.59	15.0	0.84	21.4
G—6	B—1, B—2	0.75	19.0	1.19	30.2
B—6	B—1, B—2	0.78	19.7	1.47	37.3
S—8	B—1, B—2	1.04	26.5	1.8	45.0
S—8	C—2	1.04	26.5	1.8	45.0

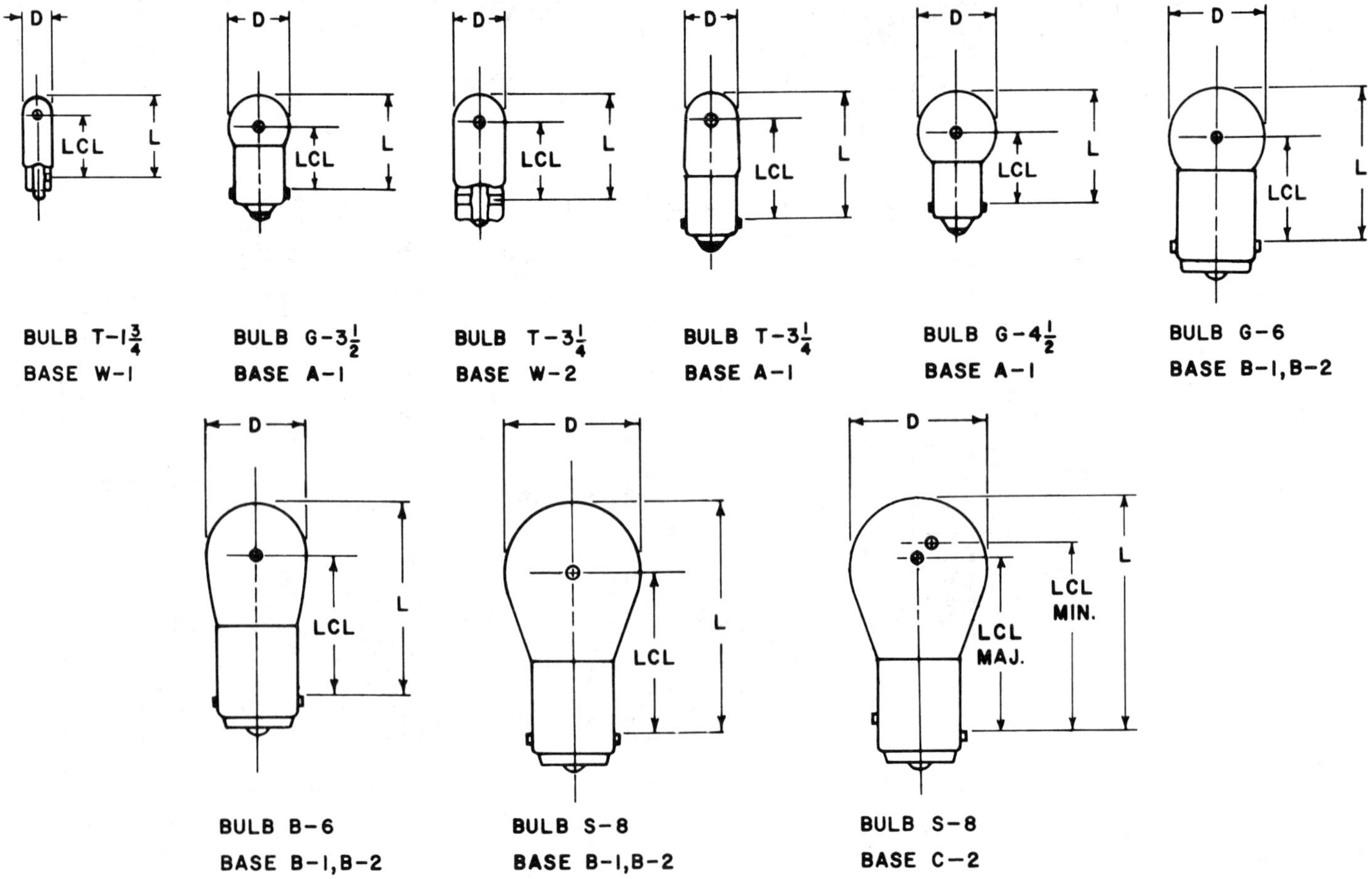
D
LCL
L
BULB T-1 3/4
BASE W-1
BULB G-3 1/2
BASE A-1
BULB T-3 1/4
BASE W-2
BULB T-3 1/4
BASE A-1
BULB G-4 1/2
BASE A-1
BULB G-6
BASE B-1,B-2
BULB B-6
BASE B-1,B-2
BULB S-8
BASE B-1,B-2
LCL MIN.
LCL MAJ.
BULB S-8
BASE C-2

ϕFIG. 2—BULB TYPES

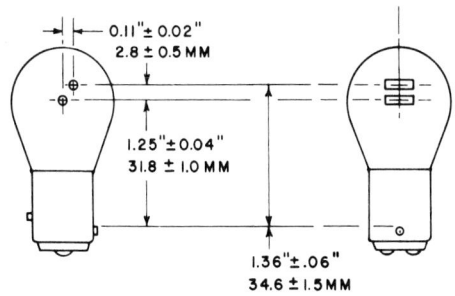

(USE DIMENSIONS OF LOWER FILAMENT FOR 1156 BULB)

FIG. 3—BULB FILAMENT DESIGN LOCATION

TABLE 2—TYPICAL LAMP BULBS FOR MOTOR VEHICLES

Typical Service[a]	Trade No.	Design					Filament Data						Bulb Type[b]	Base Data	
		Mean Spherical Candela	cd Tol, ±%	Volts	Design Amps	Amp Tol, ±%	Rated Average Lab Life, h	Type[b]	Light Center Length (LCL)		LCL Tolerance		Axial Alignment Tolerance			Type[b]	Designation
									in	mm	±in	±mm	±in	±mm			
C	74	0.7	30	14.0	0.1	15	500	C—2F	0.40	10.2	0.04	1.0	0.04	1.0	T—1-3/4	W—1	Sub-Min Wedge
C	53	1	20	14.4	0.12	10	1000	C—2V	0.50	12.7	0.09	2.3	0.09	2.3	G—3-1/2	A—1	Min Bay
C, M	57	2	20	14.0	0.24	10	500	C—2V	0.56	14.2	0.09	2.3	0.09	2.3	G—4-1/2	A—1	Min Bay
C, M	1895	2	20	14.0	0.27	10	1500	C—2F	0.56	14.2	0.09	2.3	0.09	2.3	G—4-1/2	A—1	Min Bay
T, P, M, L	67	4	15	13.5	0.59	8	2000	C—2R	0.81	20.6	0.09	2.3	0.09	2.3	G—6	B—1	SC Bay
T, P, M, L	97	4	15	13.5	0.69	8	2000	C—2V	0.81	20.6	0.09	2.3	0.09	2.3	G—6	B—1	SC Bay
C	161	1	20	14.0	0.19	10	1500	C—2F	0.56	14.2	0.09	2.3	0.09	2.3	T—3-1/4	W—2	Wedge
C	168	3	20	14.0	0.35	10	1500	C—2F	0.56	14.2	0.09	2.3	0.09	2.3	T—3-1/4	W—2	Wedge
C, M	194	2	20	14.0	0.27	10	1500	C—2F	0.56	14.2	0.04	1.0	0.06	1.5	T—3-1/4	W—2	Wedge
D, S, B	1156	32	10	12.8	2.10	5	600	C—6	1.25	31.8	0.04	1.0	0.04	1.0	S—8	B—1	SC Bay[d]
P, S, T, D	1157	32	10	12.8	2.10	5	600	C—6	1.25	31.8	0.04	1.0	0.04	1.0	S—8	C—2	DC Bay[d]
		3	12	14.0	0.59	8	2000	C—6	—c	—c	—c	—c	—c	—c			
D, M, P	1157NA	24	30	12.8	2.10	5	600	C—6	1.25	31.8	0.04	1.0	0.04	1.0	S—8	C—2	DC Bay[d]
		2.2	30	14.0	0.59	8	2000	C—6	—c	—c	—c	—c	—c	—c			

[a] Letter designations are defined as follows: B—backup; C—indicator; D—turn signal; M—marker, clearance, identification; P—parking; S—stop; T—tail; L—license.
[b] See Figs. 2, 4, and 5.
[c] See Fig. 3 for filament spacing and light center length.
[d] Plane of pins with respect to filament is 90 ± 15 deg. On remaining types filament orientation is random.

TABLE 3—TYPICAL SEALED BEAM UNITS

Type of Service[a]	Trade No.	Design		Rated Average Lab Life, h at 14.0 V	Max. Amps at Design Volts	Filament Type	Bulb Type	Dimensional Specification[c]	Terminals	
		Watts	Volts[b]						No.	Type
H	4000	37.5-60	12.8-12.8	200-320	3.14-5.02	C—6/C—6	PAR 46	Fig. 4	3	Lugs
H	4001	37.5	12.8	200	3.14	C—6	PAR 46	Fig. 3	2	Lugs
H	5001	50	12.8	200	4.20	C—6	PAR 46	Fig. 3	2	Lugs
HX	4006	37.5	12.8	240	3.14	C—6	PAR 46	Fig. 3	2	Lugs
HX	4040	37.5-60	12.8-12.8	200-320	3.14-5.02	C—6/C—6	PAR 46	Fig. 4	3	Lugs
H	6014	60-50	12.8-12.8	200-320	5.02-4.20	C—6/C—6	PAR 56	Fig. 1	3	Lugs
HX	6015	60-50	12.8-12.8	200-320	5.02-4.20	C—6/C—6	PAR 56	Fig. 1	3	Lugs
HX	6016	60-50	12.8-12.8	300-500	5.02-4.20	C—6/C—6	PAR 56	Fig. 1	3	Lugs
H	6052	65-55	12.8-12.8	150-320	5.46-4.62	C—6/C—6	142 x 200 mm	J1132	3	Lugs
H	4651	50	12.8	200	4.20	C—6	PAR 4 x 6.5	Fig. 7	2	Lugs
H	4652	40-60	12.8-12.8	200-320	3.36-5.02	C—6/C—6	PAR 4 x 6.5	Fig. 6	3	Lugs

[a] Letter designations are defined as follows: H—sealed beam headlamp; X—heavy-duty; S—spot; f—fog.
[b] All lamps designed for use on 12 V circuits are life tested at 14 V, in general, the life at average service is longer.
[c] See SAE J571.

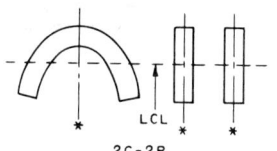 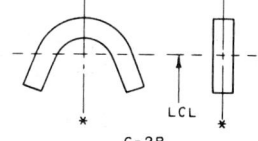 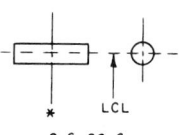 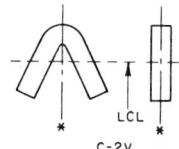 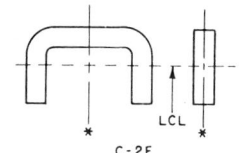

2C-2R   C-2R   C-6, CC-6   C-2V   C-2F

———— PARALLEL WITH BULB AXIS
- - - - - DENOTES LCL REF

FIG. 4—FILAMENT TYPES

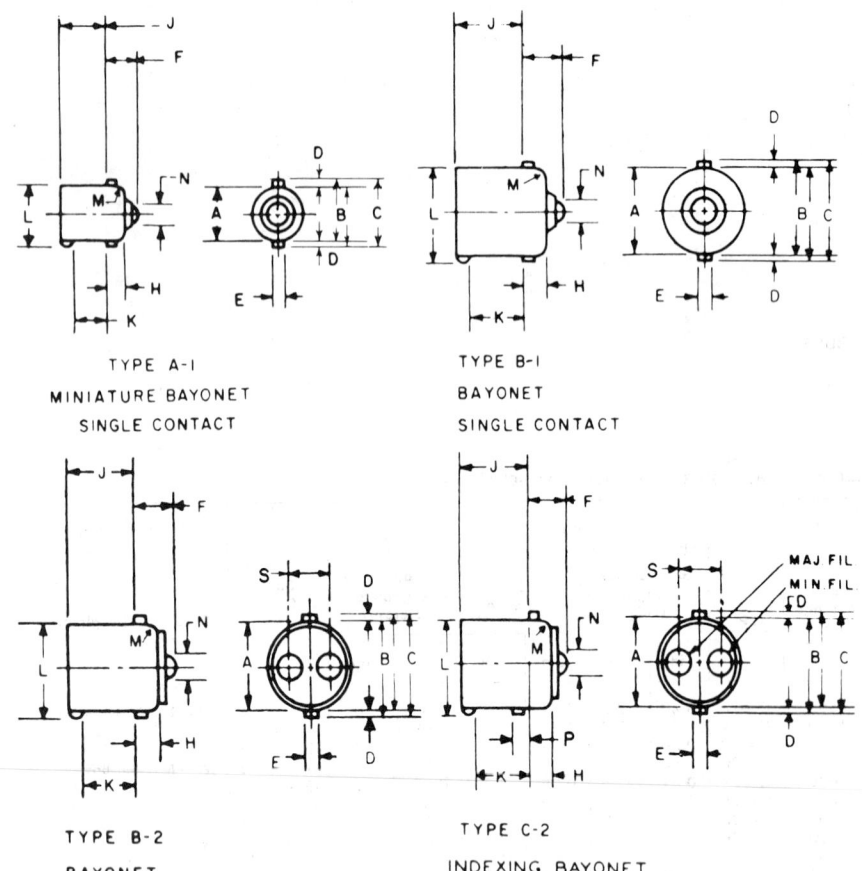

FIG. 5—BASE TYPES

TABLE 4—BASE DIMENSIONS[a] (SEE FIG. 5)

Dimension	Miniature (A-1)				Bayonet (B-1, B-2, C-2)			
	in		mm		in		mm	
	min	max	min	max	min	max	min	max
A[b]	0.357	0.366	9.07	9.30	0.5925	0.6025	15.05	15.30
B	0.384	0.400	9.75	10.16	0.616	0.636	15.65	16.15
C	—	0.432	—	10.97	—	0.668	—	16.97
D	0.025	—	0.64	—	0.025	—	0.64	—
E	0.059	0.067	1.5	1.7	0.071	0.087	1.8	2.2
F	0.180	0.255	4.57	6.48	0.249	0.316[c]	6.32	8.02[c]
H	0.095	0.131	2.41	3.33	0.138	0.170	3.51	4.32
J	0.300	—	7.62	—	0.492	—	12.50	—
K	0.180	—	4.57	—	0.350	—	8.89	—
L	—	0.41	—	10.4	—	0.64	—	16.3
M	0.03 nom	—	0.8 nom	—	0.03 nom	—	0.8 nom	—
N	0.16 nom	—	4 nom	—	0.19 nom	—	4.8 nom	—
P	—	—	—	—	0.117	0.133	2.97	3.38
S	—	—	—	—	0.255	0.279	6.48	7.09

[a] Apply to base on complete lamp bulbs.
[b] Both minimum and maximum to be measured with a ring gage. Applies to all parts of base shell except within 1/8 in (3 mm) from the bulb and base junction.
[c] On bases B-2 and C-2, heights of solder contacts are to be within 0.02 in (0.5 mm) of each other.

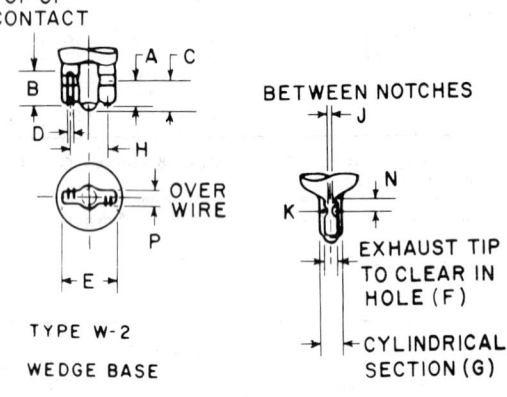

FIG. 6—WEDGE BASE DIMENSIONS

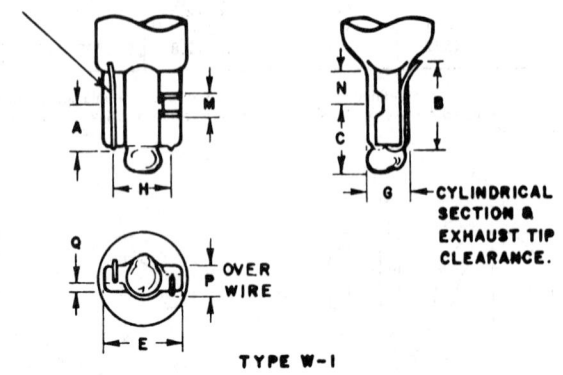

FIG. 7—SUBMINIATURE WEDGE BASE DIMENSIONS

TABLE 5—WEDGE BASE DIMENSIONS (SEE FIG. 6)

Dimension	Type W-2			
	in		mm	
	min	max	min	max
A[a]	0.135	0.175	3.43	4.45
B	0.190	—	4.83	—
C	—	0.250	—	6.35
D	0.06 nom		1.5 nom	
E	0.350	0.374	8.89	9.50
F	—	0.12	—	3.0
G	—	0.16	—	4.1
H	0.22 nom	—	5.6 nom	—
J	0.03 nom		0.8 nom	
K[b]	0.03R nom	—	0.8R nom	—
P	0.075	0.095	1.90	2.40
N	0.065	—	1.65	—

[a] To be measured on longest side only with the wire in intimate contact with the bottom of the glass wedge.
[b] Optional construction, radius under wire not required.

TABLE 6—SUB-MINIATURE WEDGE BASE DIMENSIONS (SEE FIG. 7)

Dimension	Type W-1			
	in		mm	
	min	max	min	max
A[a]	0.08	0.12	2.0	3.0
B	0.12	0.20	3.0	5.1
C	—	0.20	—	5.1
E	0.19	0.20	4.7	5.1
G	—	0.12	—	3.0
H	0.13 nom	—	3.3 nom	—
P	0.07	0.09	1.7	2.2
N	0.065	—	1.65	—
M	0.06 nom	—	1.5 nom	—
Q	0.02 nom	—	0.5	—

[a] To be measured on longest side only with the wire in intimate contact with the bottom of the glass wedge.

SERVICE PERFORMANCE REQUIREMENTS AND TEST PROCEDURES FOR MOTOR VEHICLE LAMP BULBS—SAE J1049 AUG79

SAE Recommended Practice

Report of the Lighting Committee, approved August 1973, completely revised August 1979, editorial change August 1979.

1. *Scope*—This recommended practice covers service performance tests, test methods, and requirements applicable to lamp bulbs covered by SAE J573g (December, 1976). It is intended to supplement the engineering design requirements provided in SAE J573g (December, 1976) by establishing test procedures and requirements for service evaluation of lamp bulbs.

2. *Samples for Test*—Test samples shall be new, unused lamp bulbs fabricated from production processes.

3. *Requirements*—The test samples shall comply with the following requirements:

3.1 Candela—Seasoned bulbs shall be measured at design volts in a properly calibrated photometer in accordance with accepted photometric procedures.

An acceptable seasoning schedule at rated volts is 1% of Rated Average Lab Life as shown in Table 1 or 10 h maximum, whichever is shorter. For lamp bulbs not listed in Table 1, use the manufacturer's published design life for Rated Average Lab Life.

3.2 Physical Dimensions

3.2.1 Table 1 lists candela values and filament relationship dimensions (LCL, AA, base pin rotation) of lamp bulbs.

3.2.2 Table 2 lists the base dimensions considered important for metal-based lamps to insure that lamp bulbs will perform satisfactorily in a bulb retaining device (socket) made in accordance with SAE J567c (December, 1970).

3.2.3 Table 4 lists the base dimensions considered important for subminiature wedge base (Type W-1) lamps to insure that lamp bulbs will perform satisfactorily in a bulb retaining device (socket) made in accordance with SAE J567c (December, 1970).

3.2.4 Table 3 lists the base dimensions considered important for wedge base (Type W-2) lamps to insure that lamp bulbs will perform satisfactorily in a bulb retaining device (socket) made in accordance with SAE J567c (December, 1970).

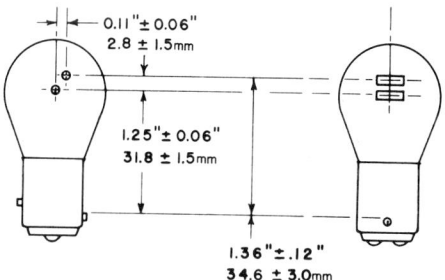

(USE DIMENSIONS OF LOWER FILAMENT FOR 1156 BULB)

FIG. 1—FILAMENT LOCATION

TABLE 1—SERVICE PERFORMANCE REQUIREMENTS FOR MOTOR VEHICLE LAMP BULBS

Trade No.	Mean Spherical Candela	cd Tol, ±%	Design Volts	Design Amps	Amp Tol, ±%	Rated Average Lab Life, h	Tolerance Data						Base Pin Rotation with Respect to Filament, deg
							Light Center Length (LCL)		LCL Tolerance		Axial Align Tolerance		
							in	mm	±in	±mm	±in	±mm	
74	0.7	50	14.0	0.1	30	500	0.40	10.2	0.13	3.3	0.13	3.3	—
53	1	30	14.4	0.12	25	1000	0.50	12.7	0.13	3.3	0.13	3.3	Random
57	2	30	14.0	0.24	25	500	0.56	14.2	0.13	3.3	0.13	3.3	Random
1895	2	30	14.0	0.27	20	1500	0.56	14.2	0.13	3.3	0.13	3.3	Random
67	4	20	13.5	0.59	15	2000	0.81	20.6	0.13	3.3	0.13	3.3	Random
97	4	25	13.5	0.69	15	2000	0.81	20.6	0.13	3.3	0.13	3.3	Random
161	1	30	14.0	0.19	20	1500	0.56	14.2	0.13	3.3	0.13	3.3	—
168	3	30	14.0	0.35	20	1500	0.56	14.2	0.13	3.3	0.13	3.3	
194	2	30	14.0	0.27	20	1500	0.56	14.2	0.09	2.3	0.09	2.3	
1156	32	15	12.8	2.10	10	600	1.25	31.8	0.06	1.5	0.08	2.0	90 ± 40
1157	32	15	12.8	2.10	10	600	1.25	31.8	0.06	1.5	0.08	2.0	90 ± 40
	3	20	14.0	0.59	15	2000	—[a]	—[a]	—[a]	—[a]	—[a]	—[a]	
1157NA	24	35	12.8	2.10	10	600	1.25	31.8	0.06	1.5	0.08	2.0	90 ± 40
	2.2	35	14.0	0.59	15	2000	—[a]	—[a]	—[a]	—[a]	—[a]	—[a]	

TABLE 2—IMPORTANT BASE DIMENSIONS FOR METAL-BASED LAMP BULBS, MIN[a]

Dimension	Miniature (A-1)		Bayonet (B-1, B-2, C-2)	
	in	mm	in	mm
A[b]	0.357	9.07	0.5925	15.05
B	0.384	9.75	0.616	15.65
F	0.180	4.57	0.249	6.32

[a] See Fig. 5 of SAE J573g (December, 1976).
[b] To be measured with a ring gage. Applies to all parts of base shell except within 1/8 in (3 mm) from the bulb and base junction.

TABLE 3—IMPORTANT BASE DIMENSIONS FOR WEDGE BASE LAMPS[a]

Dimension	Type W-2			
	in		mm	
	min	max	min	max
A[b]	—	0.175	—	4.45
B	0.190	—	4.83	—
C	—	0.250	—	6.35
P	0.075	—	1.90	—

[a] See Fig. 6 of SAE J573g (December, 1976).
[b] To be measured on longest side only, with the wire in intimate contact with the bottom of the glass wedge.

TABLE 4—IMPORTANT BASE DIMENSIONS FOR SUBMINIATURE WEDGE BASE LAMPS[a]

Dimension	Type W-1			
	in		mm	
	min	max	min	max
A[b]	—	0.12	—	3.0
B	0.12	—	3.0	—
C	—	0.20	—	5.1
P	0.07	—	1.7	—

[a] See Fig. 7 of SAE J573g (December, 1976).
[b] To be measured on longest side only, with the wire in intimate contact with the bottom of the glass wedge.

TESTS FOR MOTOR VEHICLE LIGHTING DEVICES AND COMPONENTS—SAE J575 JUN80

SAE Recommended Practice

Report of the Lighting Division, approved May 1942, completely revised by the Lighting Committee September 1977.

1. Scope—This standard covers standardized basic tests, test methods, and requirements applicable to many of the lighting devices and components covered by SAE Standards, Recommended Practices, and Information Reports.

2. Samples for Tests

2.1 Lighting Devices—Samples submitted for laboratory test shall be representative of the devices as regularly manufactured and marketed. Each sample shall be securely mounted on a test fixture in its designed operating position and shall include all accessory equipment necessary to operate the device in its normal manner.

2.2 Bulbs—Unless otherwise specified, bulbs used in the tests shall be supplied by the laboratory and be representative of standard bulbs in regular production. They shall be selected for accuracy in accordance with specifications in SAE Standard, Lamp Bulbs and Sealed Units—SAE J573f and be operated at their rated mean spherical candlepower, except as otherwise specified. Where special bulbs are specified, they shall be submitted with the devices and the same or similar bulbs used in the tests and operated at their rated mean spherical candlepower, except as otherwise specified.

2.3 Test Fixture—A device specifically designed to support the lighting device in its designed operating position during laboratory testing. This device with the test sample installed shall not have a resonant frequency in the 10–55 Hz range.

3. Laboratory Facilities—The laboratory shall be equipped to test the sample in accordance with the requirements of the SAE Standard or Recommended Practice for the specific device.

4. Test Requirements—The following sections describe individual tests which need not be performed in any particular sequence. The completion of the tests may be expedited by performing the tests simultaneously on separately mounted samples.

4.1 Vibration Test—A sample device as mounted on the test fixture defined in Sections 2.1 and 2.3 shall be securely bolted to the table of the vibration test machine and subjected to vibration according to the following test parameters:

4.1.1 TABLE SIZE—The table or suitable adapter shall be of sufficient size to completely contain the test fixture base with no overhang.

4.1.2 FREQUENCY—Varied from 10–55 and return to 10 Hz at a linear sweep period of 2 min/complete sweep cycle.

4.1.3 WAVE FORM—Sinusoidal with a maximum permissible harmonic distortion as shown in Fig. 3.

4.1.3.1 *Distortion Measurement*—The test machine output wave form shall be measured with an accelerometer, having a flat frequency response ($\pm 5\%$) from 5–2200 Hz, attached to the unloaded test machine table or its adapter plate to measure acceleration in the direction of table travel.

4.1.3.2 *Harmonic Distortion Analysis*—The percent distortion shall be computed by taking the ratio ($\times 100$) of the rms (root mean squared) voltage of the distortion components to the rms voltage of the total signal (distortion plus fundamental) of the accelerometer.

4.1.4 EXCURSION—$1.0 {}^{+0.1}_{-0.0}$ mm peak to peak over the specified frequency range in Section 4.1.2.

4.1.5 DIRECTION OF VIBRATION—Vertical axis of the device as it is mounted on the vehicle.

4.1.6 TEST DURATION—$60 {}^{+1}_{-0}$ min.

4.1.7 ACCEPTANCE REQUIREMENTS—Upon completion of the test the sample device shall be examined. Any device showing rotation, displacement, cracking, or rupture of parts (except bulb(s) and sealed beam unit internal components) which would result in failure of any other tests contained in Section 4 of this standard shall constitute a failure. Cracking or rupture of parts of the device affecting its mounting shall also constitute a failure.

4.2 Moisture Test—A sample device shall be mounted in its normal operating position with all drain holes open and subjected to a precipitation of 2.5 mm of water per minute delivered from a nozzle with a solid cone spray. The centerline of the nozzle shall be directed at an angle of 45 deg downward on the sample device. During the Moisture Test the lamp shall revolve about

TABLE 1—CYCLE TIMES (MIN)

Device	Steady Burn	5 On 5 Off	3 On 12 Off	Steady[a] Flash
License	X			
Clearance & Identification	X			
Side Marker	X			
Tail	X			
Park	X			
Stop		X		
Back-Up		X		
Corner			X	
Turn				X
Illuminating (Fog Lamp, Driving Lamp, etc.)	X			

[a] Flash rate per SAE J590e.

NOTE: Turn signal and/or stop lamps optically combined with tail or park lamps shall be tested at design voltage of major filament with the minor filament steady burning and the major filament cycled as specified. Stop-turn signal lamps optically combined shall be tested as stop lamps only. Back-up lamps shall always be tested separately.

its vertical axis at a rate of 4 rpm. This test shall be continued for 12 h. The water shall then be turned off and the device permitted to drain for 1 h. The device shall then be examined. Moisture accumulation in excess of 2 cc shall constitute a failure.

4.3 Dust Test—A sample device with all drain holes closed shall be mounted in its normal operating position, at least 150 mm from the wall in a cubical box with inside measurements of 900 mm on each side, containing 4.5 kg of fine powdered cement in accordance with ASTM C150–77, Specification for Portland Cement. At intervals of 15 min this dust shall be agitated by compressed air or fan blower(s) by projecting blasts of air for a 2 s period in a downward direction into the dust in such a way that the dust is completely and uniformly diffused throughout the entire cube. The dust is then allowed to settle. This test shall be continued for 5 h. The unit shall be considered to have met the requirements of this test if no dust is found on the interior surfaces, or if the maximum candlepower output is within 10% as compared with the condition after the unit is cleaned inside and out. Where sealed beam units are used the dust test shall not be required.

4.4 Corrosion Test—A sample device shall be subjected to a salt spray (fog) test in accordance with ASTM B117-73, Method of Salt Spray (Fog) Testing, for a period of 50 h, consisting of two periods of 24 h exposure and 1 h drying time each.

Immediately after the device has been subjected to the corrosion test, there shall be no evidence of corrosion which would result in failure of any tests contained in Section 4 of this standard.

4.5 Color Test—Refer to SAE Standard, Color Specification for Electric Lamps—SAE J578b.

4.6 Photometry—The photometric measurement shall be made at a distance between the light source and the point of measurement specified for the lighting device.

When making photometric measurements at specific test points, the candela values between test points shall not be less than the lower specified value of the two closest adjacent test points (on a horizontal or vertical line) for minimum values.

In locating the test points, as designated in the standards and recommended practices, the following nomenclature shall apply: The line formed by the intersection of a vertical plane through the light source of the device and normal to the test screen is designated V. The line formed by the intersection of a horizontal plane through the light source and normal to the test screen is designated H. The point of intersection of these two lines is designated H-V.

The other points on the test screen are measured in terms of degree from these two lines. Degrees to the right (R) and to the left (L) are regarded as being to the right and left of the vertical line when the observer stands behind the lighting device and looks in the direction of the emanating light beam when the device is properly aimed for photometry with respect to the H-V point.

Similarly, the upward angles designated as U and the downward angles designated D, refer to light emanating at angles above and below the horizontal line, respectively.

Example: 4D-3L is a point 4 deg below H and 3 deg to the left of V.
1U-V is a point 1 deg above H and on the line V.

The recommended goniometer configuration which shall be used to position the device when making the photometric measurements at specific test points is shown in Fig. 1. Other systems may be used to achieve equivalent results.

4.7 Out-of-Focus Tests on Unsealed Units—Tests shall be made for each of four out-of-focus filament positions. Where conventional bulbs with two-pin bayonet bases are used, candela tests shall be made with the light source 1.52 mm above, below, ahead, and behind the designed position. If prefocused bulbs are used, the limiting positions at which tests are made shall be 0.51 mm above, below, ahead, and behind the designed position. The minimum values for out-of-design position shall be 80% of minimum requirements. The lamp may be re-aimed for each of the out-of-focus positions of the light source.

4.8 Warpage Test on Devices with Plastic Components—A sample device shall be mounted in its normal position and operated as specified in Table 1 at design voltage in a circulating air oven for 1 h within a temperature range of 46–49°C. After this warpage test has been completed there shall be no evidence of warpage which would result in failure of any test contained in Section 4 of this standard.

APPENDIX

Please refer to SAE Technical Paper 790747, SAE Vibration Test for Motor Vehicle Lighting Devices and Components, for additional information on paragraph 4.1.

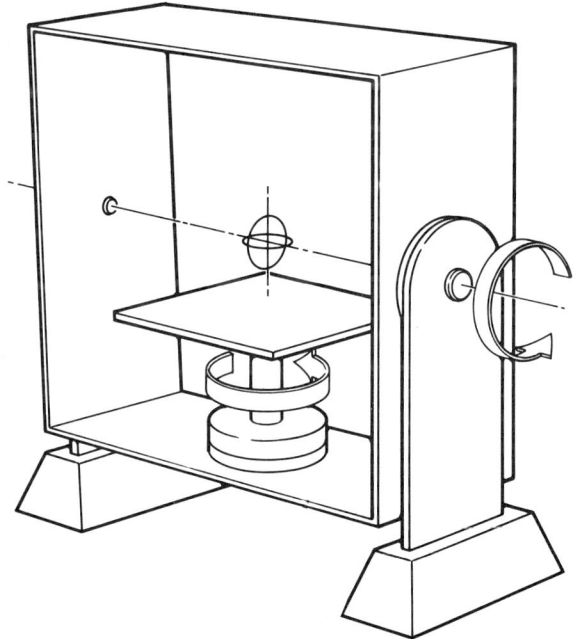

FIG. 1—RECOMMENDED GONIOMETER CONFIGURATION

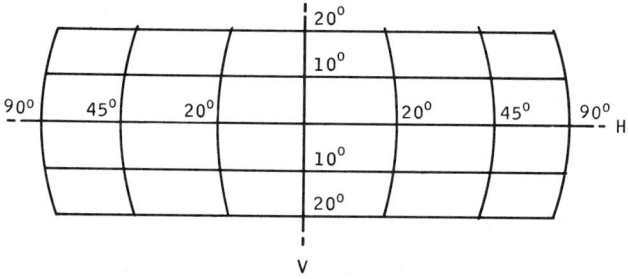

FIG. 2—PROJECTION ON LAMP AS SEEN BY PHOTOCELL

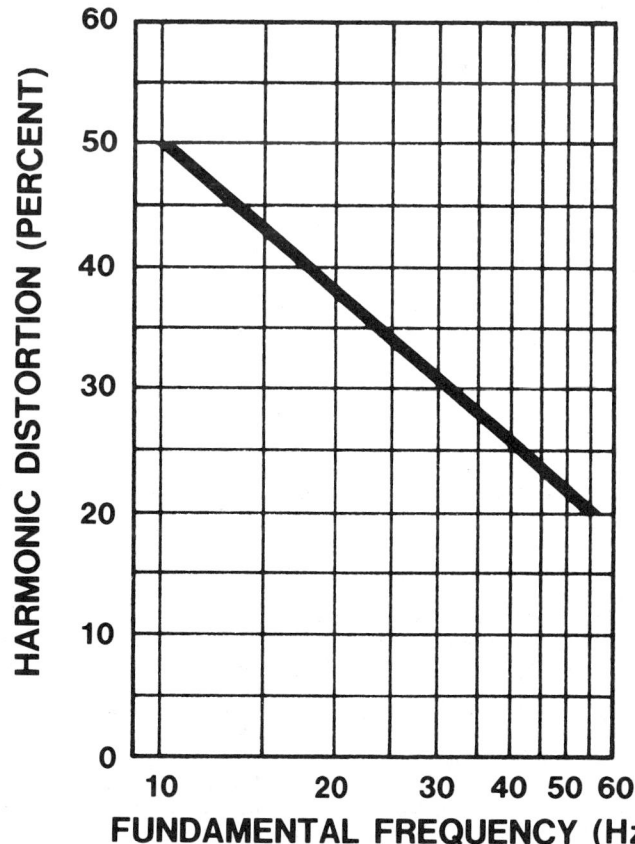

FIG. 3—MAXIMUM PERMISSIBLE VIBRATION WAVE FORM HARMONIC DISTORTION

SERVICE PERFORMANCE REQUIREMENTS FOR MOTOR VEHICLE LIGHTING DEVICES AND COMPONENTS—SAE J256b

SAE Recommended Practice

Report of Lighting Committee approved July 1971 and last revised March 1978. Rationale statement available.

1. Scope—This recommended practice covers service performance tests, test methods, and requirements applicable to lighting devices and components covered by SAE Standards, Recommended Practices, and Information Reports. It is intended to supplement the engineering design test procedures and requirements by establishing test procedures and requirements for service evaluation of many of the lighting devices covered by SAE Standards.

4.3 An adjustment in lamp orientation from design position may be made in determining compliance to the requirements of the appropriate table(s) and applicable footnotes, provided such adjustment does not exceed 3 deg. All zones must comply after final reaim.

4.4 Devices shall be operated at the design voltage. Devices designed for use with both 6 and 12 V systems or 24 and 12 V systems shall be tested with 12 V bulbs.

φ TABLE 1—MINIMUM LUMINOUS INTENSITY REQUIREMENTS,[c,d]

Zones	Test Points, deg	Tail Lamps (Red Lighted Compartments) (see notes b, d)			Stop and Turn Signals (Red Lighted Compartments) (see notes c, d)			Turn Signals (Front Yellow Lighted Compartments) (see notes c, d, e)		Turn Signals (Rear Yellow Lighted Compartments) (see notes c, d)	
		1 Compartment Zone Total	2 Compartment Zone Total	3 Compartment Zone Total	1 Compartment Zone Total[a]	2 Compartment Zone Total	3 Compartment Zone Total	1 Compartment Zone Total[a]	2 Compartment Zone Total	1 Compartment Zone Total	2 Compartment Zone Total
1	20L—5U 20L—H 20L—5D 10L—10U 10L—10D	1.6	2.7	3.8	55	66	80	135	165	85	110
2	10U—V 5U—10L 5U—10R	2.1	3.6	5.5	85	100	115	210	251	140	160
3	10L—H 5L—5U 5L—5D	3.4	5.3	8.0	140	167	195	350	420	225	275
4	5U—V H—5L H—V H—5R 5D—V	9.6	16.5	24.0	380	449	520	950	1130	610	710
5	5R—5U 5R—5D 10R—H	3.4	5.3	8.0	140	167	195	350	420	225	275
6	5D—10L 5D—10R 10D—V	2.1	3.6	5.5	85	100	115	210	251	140	160
7	10R—10U 10R—10D 20R—5U 20R—H 20R—5D	1.6	2.7	3.8	55	66	80	135	165	85	110

[a] Applies to lamps designed to SAE J575d requirements.
[b] Tail lamp candela shall not exceed 120% of maximum values as specified in SAE J585.
[c] Stop and turn lamps candela shall not exceed 120% of maximum values as specified in SAE J586 or J588.
[d] A tolerance of 10% may be applied to each zone.
φ [e] Front turn signal lamps having minimum intensities which are at least 2½ times those shown above for zones 1–7 need not comply with the 4 in (100 mm) spacing requirement with respect to the lighted edge of the low beam headlamp unit.

2. Samples for Test—Test samples shall be new, unused devices fabricated from production tools and assembled by production processes. Bulbs used in tests shall be those supplied with the device and within the design tolerances. If the device to be tested does not contain a bulb(s) and/or socket(s), only those bulb(s) and/or socket(s) approved by the lamp manufacturer shall be used.

3. Requirements

3.1 The lighting devices shall comply with all test requirements specified in applicable SAE Standards unless otherwise noted herein.

3.2 For the device to comply with the photometric requirements, the summation of the candela readings of the specific test points in a zone shall meet the value specified for that zone in the appropriate table(s) and applicable footnotes.

3.3 The measured candela at each test point shall not be less than 60% of the minimum requirements specified in applicable SAE Standards.

4. Photometric Test Procedure

4.1 The photometric measurement shall be made with the device mounted in its normal operating position and at the distance between the light source and the point of measurement as specified in the appropriate SAE Standard.

4.2 A device using a two-filament bulb shall be oriented with respect to its major filament.

TABLE 2—MINIMUM LUMINOUS INTENSITY REQUIREMENTS FOR SIDEMARKER, CLEARANCE, AND IDENTIFICATION LAMPS

Zone	Test Point, deg	Total for Zone cd (see note c)	
		Red[a]	Amber
1	45L[b]—10U 45L[b]—H 45L[b]—10D	0.75	1.86
2	V—10U V—H V—10D	0.75	1.86
3	45R[b]—10U 45R[b]—H 45R[b]—10D	0.75	1.86

[a] Red clearance and identification lamps shall not exceed 120% of the maximum values as specified in SAE J592.
φ [b] The requirements for sidemarkers used on vehicles less than 80 in (2030 mm) wide may be met for inboard test points at a distance of 15 ft (4.6 m) from the vehicle on a vertical plane that is perpendicular to the longitudinal axis of the vehicle and located midway between the front and rear sidemarker lamps.
[c] A tolerance of 10% may be applied to each zone.

TABLE 3—MINIMUM LUMINOUS INTENSITY REQUIREMENTS FOR PARKING LAMPS

Zone	Test Point, deg	Total for Zone, cd (see note a)
1	20L—5U 20L—H 20L—5D 10L—10U 10L—10D	2.8
2	10U—V 5U—10L 5U—10R	2.4
3	10L—H 5L—5U 5L—5D	4.2
4	5U—V H—5L H—V H—5R 5D—V	16.8
5	5R—5U 5R—5D 10R—H	4.2
6	5D—10L 5D—10R 10D—V	2.4
7	10R—10U 10R—10D 20R—5U 20R—H 20R—5D	2.8

[a] Parking lamp candlepower shall not exceed 120% of the maximum values as specified in SAE J222.
[b] A tolerance of 10% may be applied to each zone.

TABLE 4—MINIMUM LUMINOUS INTENSITY REQUIREMENTS FOR BACKUP LAMPS

Zone	Test Point, deg	Total for Zone, cd (see notes a, b)
1a	45L—5U 45L—H 45L—5D	45
2a	30L—H 30L—5D	50
3	10L—10U 10L—5U V—10U V—5U 10R—10U 10R—5U	100
4	10L—H 10L—5D V—H V—5D 10R—H 10R—5D	360
5a	30R—H 30R—5D	50
6a	45R—5U 45R—H 45R—5D	45

[a] When two lamps of the same or symmetrically opposite design are used, the reading along the vertical axis and the averages of the readings for the same angles left and right of vertical for one lamp shall be used to determine compliance with the requirements. If two lamps of differing designs are used, they shall be tested individually and the values added to determine that the combined units meet twice the candela requirements.
[b] Backup lamp candlepower at any test point at H and above shall not exceed 120% of the maximum values specified in SAE J593.
[c] When only one backup lamp is used on the vehicle, it shall be tested to twice the candela requirements.
[d] A tolerance of 10% may be applied to each zone.

TABLE 5—MINIMUM LUMINOUS INTENSITY REQUIREMENTS FOR RED SCHOOL BUS WARNING LAMPS

Zone	Test Point, deg	Total for Zone, cd (see notes b, c, d)
1	30L—H 30L—5D 20L—5U 20L—H 20L—5D	590
2	5U—10L 5U—5L 5U—V 5U—5R 5U—10R	1500
3	H—10L H—5L H—V H—5R H—10R	2400
4	5D—10L 5D—5L 5D—V 5D—5R 5D—10R	1950
5	10D—5L 10D—V 10D—5R	120
6	20R—5U 20R—H 20R—5D 30R—H 30R—5D	590

[a] A tolerance of 10% may be applied to each zone.

PLASTIC MATERIALS FOR USE IN OPTICAL PARTS SUCH AS LENSES AND REFLECTORS OF MOTOR VEHICLE LIGHTING DEVICES—SAE J576d

SAE Standard

Report of Lighting and Nonmetallic Materials Committee approved January 1955. Completely revised and upgraded to Standard Status June 1976.

1. Scope—This SAE Standard provides test methods and requirements to evaluate the suitability of plastic materials intended for optical applications in motor vehicles. The tests of this Standard are intended to determine physical and optical characteristics of the material only and are not intended to cover the performance of plastics when molded and installed in a finished assembly. Performance expectations of such a finished assembly, including its plastic components, are to be based on tests for lighting devices, as specified in SAE Standards and Recommended Practices for motor vehicle lighting equipment.

2. Definitions

2.1 Material—For the purpose of this Standard the term "material" includes type and grade of plastics, composition, and manufacturer's designation (number) and color.

2.1.1 COATED MATERIAL—A coated material is a material as defined in 2.1 which has a coating applied to the outer surface of the finished sample to impart some protective properties. "Coating" includes manufacturer's name, formulation designation (number), and recommendations for application. A trace quantity (100 ppm maximum in wet state) of an optical brightener shall be added to a coating formulation in order to test for presence of the coating.

2.2 Material Exposure

2.2.1 EXPOSED—Materials used in lenses or optical devices exposed to direct sunlight as installed on the vehicle.

2.2.2 SHADED—Materials used in lenses or optical devices exposed to direct sunlight only at angles less than 45 deg above the horizontal as installed on the vehicle, but exposed to other environmental and service factors.

2.2.3 PROTECTED—Material used in inner lenses for optical devices to provide protection as installed on the vehicle.

3. Test Requirements

3.1 Materials to be Tested—These tests shall be made on each material (defined in 2.1 and 2.1.1) offered for use in optical parts employed in motor vehicle lighting devices. A test of one color and formulation shall cover variations in dye concentration but shall not cover changes in dye materials, polymers, or coatings.

3.2 Samples Required

3.2.1 GENERAL—Samples of plastic shall be injection molded into polished molds to produce 3 in (76 mm) diameter discs with two faces flat and parallel. Each exposed surface of the samples should contain a minimum uninterrupted area of 5 in^2 (3226 mm^2).

3.2.2 THICKNESS—Samples shall be furnished in the following thicknesses:

THICKNESS		TOLERANCE	
in	mm	in	mm
0.062	1.57	±0.005	±0.13
0.125	3.18	±0.005	±0.13
0.250	6.35	±0.005	±0.13

3.2.3 NUMBER OF SAMPLES REQUIRED—Outdoor Exposure Test: 1 sample/each thickness/each site × 2 sites for each material = 2 samples/each thickness for each material.

Control: 1 sample/each thickness for each material = 1 sample each.

NOTE: The control sample must be kept properly protected from influences which may change its appearance and properties.

3.3 Outdoor Exposure Test

3.3.1 EXPOSURE SITES—Florida (warm, moist climate) and Arizona (warm climate).

3.3.2 SAMPLE MOUNTING—One sample of each thickness of each material at each test station shall be mounted so that the exposed upper surface of the sample is at an angle of 45 deg to the horizontal, facing south. The exposed surface of the sample shall contain a minimum uninterrupted area of 5 in^2 (3226 mm^2), and they shall be mounted in the open no closer than 12 in (305 mm) to their background.

3.3.3 EXPOSURE TIME AND CONDITIONS—The time of exposure shall be as noted in 3.3.3.1 for each type of material exposed. During the exposure time the samples shall be cleaned once every three months by washing with mild soap or detergent and water, and then rinsing with distilled water. Rubbing shall be avoided.

3.3.3.1 *Exposure Time Based on Material Usage*

Exposed—(Defined by 2.2.1): 3 years
Shaded—(Defined by 2.2.2): 2 years
Protected—(Defined by 2.2.3): 6 consecutive months starting in May

3.4 Luminous Transmittance and Color Measurements—Measurements shall be made in accordance with ASTM E308, Recommended Practice for Spectrophotometry and Description of Color in CIE 1931 System.

4. Material Performance Requirements—A material in the range of thicknesses as stated by the material manufacturer, and as defined by 2.1 or 2.1.1 shall conform to the following conditions.

4.1 Before Exposure to Any Tests—The trichromatic coefficients shall conform with the requirements of SAE J578 in the range of thicknesses stated by the material manufacturer.

4.2 After Outdoor Exposure

4.2.1 LUMINOUS TRANSMITTANCE—The luminous transmittance of the exposed samples using CIE Illuminant A (2856 K) shall not have changed by more than 25% of the luminous transmittance of the unexposed control sample when tested in accordance with ASTM E308.

4.2.2 TRICHROMATIC COEFFICIENTS—The trichromatic coefficients shall conform with the requirements of SAE J578 in the range of thicknesses stated by the material manufacturer.

4.2.3 APPEARANCE—The exposed samples, when compared with the unexposed control samples, shall not show surface deterioration, dimensional changes, color bleeding or delamination.

PLASTIC MATERIAL FOR USE IN HOUSINGS OF MOTOR VEHICLE LIGHTING DEVICES—SAE J29

SAE Recommended Practice

Report of Lighting Committee approved January 1973.

1. Scope—This SAE Recommended Practice provides test methods and requirements to evaluate the suitability of plastic materials intended for use in housings of motor vehicle lighting devices including reflex reflectors. These tests are not intended to cover the performance of plastics when molded and installed in a finished assembly. Performance expectations of such a finished assembly, including its plastics components, are to be based on tests of devices as specified in SAE Standards and Recommended Practices for motor vehicle lighting devices.

2. Definitions

2.1 Material—For the purpose of this recommended practice, the term "material" broadly includes type and grade of plastics, composition, or manufacturer's designation (number) and color.

3. Test Procedures

3.1 Materials to be tested—These tests shall be made of each material to be offered for housings for motor vehicle lighting devices. If a colorant, filler, or stabilizer is satisfactory when tested at a high concentration and a low concentration, any concentration between the two limits is considered satisfactory. New tests will be required if the colorant material is changed, if the polymer is changed, or if additives to the polymer are added or omitted. Supplier should certify that the material supplied is the same, within allowable engineering specifications, as the material tested and approved.

3.2 Types of Tests—Four basic types of tests are employed: outdoor exposure test, heat test, tensile test, and impact test. Tensile and impact tests are to be made after heat and exposure tests.

3.3 Specimens Required

3.3.1 GENERAL—Test specimens shall be tensile and impact test specimens injection molded under optimum conditions for the type of material, using an ASTM specimen mold. The specimen thickness in all cases shall be 0.125 in. (3.2 mm). (Compression molding type materials use compression molded specimens.)

3.3.2 NUMBER OF SPECIMENS REQUIRED

Outdoor Exposure Test—5 tensile specimens and 10 unnotched impact specimens per each site by 2 sites = 30 specimens of each material.

Heat Test—5 tensile specimens and 10 unnotched impact specimens of each material = 15 specimens of each material.

Control—10 impact specimens, unnotched times 2 tests = 20 specimens. 5 tensile specimens times 2 tests = 10 specimens.

Five specimens of each type of test specimen shall be used for appearance comparisons after all testing is completed.

The control specimens shall be stored and they shall be tested at the same time the heat and outdoor exposure specimens are tested. Keep these specimens properly protected from influences which may change their appearance or properties.

3.4 **Outdoor Exposure Test**—(Required only if material is directly exposed to ultraviolet light and unprotected by a paint coating, vacuum metallizing, hot stamp, or electroplate.)

3.4.1 EXPOSURE SITES—Florida (warm, moist climate) and Arizona (warm climate).

3.4.2 SPECIMEN MOUNTING—Five tensile specimens and 10 unnotched impact specimens of each material at each test station shall be mounted at a 45 deg angle to the vertical, facing south. They shall be mounted in the open no closer than 12 in. (305 mm) to their background.

3.4.3 EXPOSURE TIME AND CONDITIONS—The time of exposure shall be 3 years. During this time specimens shall be cleaned at least once every 3 months by washing with mild soap and water, and then rinsing. Rubbing should be avoided.

3.5 **Heat Test**—Aging (ASTM D794) No Load

3.5.1 SPECIMEN MOUNTING—Five tensile specimens and 10 unnotched impact specimens of each material shall be supported in a vertical position with not more than 1 in. (25.4 mm) segments at the bottom retained in the fixtures, leaving ample material uncovered and unsupported.

3.5.2 TEST TIME AND TEMPERATURE—The specimens shall be placed for 240 h in a circulating air oven at:

Class I —175 ± 5°F (79 ± 3°C)
Class II —200 ± 5°F (93 ± 3°C)
Class III—225 ± 5°F (107 ± 3°C)
Class IV—250 ± 5°F (121 ± 3°C)
Class V —275 ± 5°F (135 ± 3°C)

A candidate material shall be tested at the class-temperature below the ASTM D 648 (66 psi or 455 kPa) heat distortion point of the material.

4. *Material Performance Requirements*—A material shall be said to conform to this recommended practice if the following conditions are met:

4.1 **After Outdoor Exposure**

4.1.1 TENSILE STRENGTH—The maximum tensile loss shall be 35% of initial value when measured by ASTM D 638 (speed 0.2 in./min (5.1 mm/min)).

4.1.2 IMPACT STRENGTH—The maximum impact loss shall be 35% of initial value when measured by ASTM D 256 at room temperature and −40°F (−40°C) (Izod Method A). The specimens should be milled before testing to obtain specified notch. High impact materials possessing more than 1.5 ft-lb (2.0 J) per inch (25.4 mm) of notch after test shall also qualify.

4.2 **After Heat Test**

4.2.1 TENSILE STRENGTH—The maximum tensile loss shall be 35% of initial value when measured by ASTM D 638 (speed 0.2 in./min (5.1 mm/min)).

4.2.2 IMPACT STRENGTH—The maximum impact loss shall be 35% of initial value when measured by ASTM D 256 at room temperature and −40°F (−40°C) (Izod Method A). The specimens should be milled before testing to obtain specified notch. High impact materials possessing more than 1.5 ft-lb (2.0 J) per inch (25.4 mm) of notch after test shall also qualify.

COLOR SPECIFICATION FOR ELECTRIC SIGNAL LIGHTING DEVICES—SAE J578d

SAE Standard

Report of Lighting Committee approved January 1942 and last revised September 1978. Rationale statement available.

1. *Scope*—The purpose of this standard is to define and provide for the control of colors employed in motor vehicle external lighting equipment. The specification applies to the overall effective color of light emitted by the device and not to the color of the light from a small area of the lens. It does not apply to any pilot, indicator, or tell-tale lights.

2. *Definitions*—Fundamental definitions of color are expressed by Chromaticity Coordinates according to the CIE (1931) standard colorimetric system. (See Fig. 1.)

2.1 **Red**—The color of light emitted from the device shall fall within the following boundaries:

$y = 0.33$ (yellow boundary)
$y = 0.98 - x$ (purple boundary)

2.2 **Yellow (Amber)**—The color of light emitted from the device shall fall within the following boundaries:

$y = 0.39$ (red boundary)
$y = 0.79 - 0.67x$ (white boundary)
$y = x - 0.12$ (green boundary)

φ 2.2.1 SELECTIVE YELLOW[1]—The color of light emitted from the device shall fall within the following boundaries:

$y = 0.58x + 0.14$ (red boundary)
$y = 1.29x - 0.10$ (green boundary)
$y = 0.97 - x$ (white boundary)

2.3 **White (Achromatic)**—The color of light emitted from the device shall fall within the following boundaries:

$x = 0.31$ (blue boundary) $y = 0.44$ (green boundary)
$x = 0.50$ (yellow boundary) $y = 0.38$ (red boundary)
$y = 0.15 + 0.64x$ (green boundary) $y = 0.05 + 0.75x$ (purple boundary)

φ 2.3.1 WHITE TO YELLOW—The color of light emitted from the device shall fall within one of the following areas:

(a) That defined in paragraph 2.2 Yellow.
(b) That defined in paragraph 2.2.1 Selective Yellow.
(c) That defined in paragraph 2.3 White.
(d) The area between Yellow, Selective Yellow, and White as shown by the dashed lines in Fig. 1.

2.4 **Green**—The color of light emitted from the device shall fall within the following boundaries:

$y = 0.73 - 0.73x$ (yellow boundary)
$x = 0.63y - 0.04$ (white boundary)
$y = 0.50 - 0.50x$ (blue boundary)

2.5 **Blue**—The color of light emitted from the device shall fall within the following boundaries:

2.5.1 RESTRICTED BLUE—This color should be elected when recognition of blue as such is necessary.

$y = 0.07 + 0.81x$ (green boundary)
$x = 0.40 - y$ (white boundary)
$x = 0.13 + 0.60y$ (violet boundary)

2.5.2 SIGNAL BLUE—This color may be elected when, due to other factors, it is not always necessary to identify blue as such.

$y = 0.32$ (green boundary)
$x = 0.16$ (white boundary)
$x = 0.40 - y$ (white boundary)
$x = 0.13 + 0.60y$ (violet boundary)

3. *Method of Color Measurement*—One of the methods listed in the following paragraphs shall be used to check the color of the light from the device for compliance with the color specifications. The device shall be operated at design voltage. Components (bulbs, caps, lenses, and the like) shall be tested in a fixture or manner simulating the intended application.

3.1 **Visual Method**—In this method, the color of the light from the device undergoing the inspection is compared visually with the color of the light from a standard. The standard may consist of a filter or limit glass. In the case of white, CIE Source A is used only as a color reference. The chromaticity coordinates of the color standards shall be as close as possible to the limits listed in the definitions. The color of the standard filters is determined spectrophotometrically.

RED—Red shall not be acceptable if it is less saturated (paler), yellower, or bluer than the limit standards.

YELLOW (AMBER)—Yellow shall not be acceptable if it is less saturated (paler), greener, or redder than the limit standards.

φ [1]Not for use in Turn Signal, Parking, Identification, Clearance, Sidemarker, and School Bus Warning Lamps, as well as yellow reflex reflector applications as required by FMVSS 108.

WHITE—White shall not be acceptable if its color differs materially from that of CIE Source A.

GREEN—Green shall not be acceptable if it is less saturated (paler), yellower, or bluer than the limit standards.

BLUE—Blue shall not be acceptable if it is less saturated (paler), greener, or redder than the limit standards.

In making visual appraisals, the light from the device illuminates a portion of the comparator field. The standard illuminates an immediately adjacent field portion of approximately equal area. It is preferable that the standard field should surround the comparator field, or vice versa. The locations of the standard and test sample shall be adjusted so the comparison fields have equal and uniform luminance (brightness). The test equipment shall be so arranged that light is brought into the comparator field from the full aperture of the device or component.

3.2 Tristimulus Method—In this method, photoelectric receivers, with spectral responses that approximate the CIE standard spectral tristimulus valves, are used to make color measurements. In making these measurements, a sphere may be used to integrate the light from a colored source. The color shift that results from the spectral selectivity of the sphere paint shall be corrected by the use of a filter, correction factor, or appropriate calibration. Color measurement shall be made at H-V.

For those signal light devices that have uniform luminous output and spectral characteristics in all useful directions, no additional precautions need to be taken. Color measurements should be made at as many directions of view as are required to evaluate the color for those directions that apply to the end use of the signal light device. The measured color should be reported for each of these directions. This precaution is necessary because lights that are seen as one color in some directions may be seen as a somewhat different color in other directions. If the fixture under test is placed at the sphere opening, or a short distance from the opening, the sphere will integrate the effects of the light emitted in many directions, with the result that the indication might be called an intermediate color of the light. Measurements of color made in each required direction will reveal the color in that direction and will avoid the erroneous color designation for a spatially integrated measurement.

APPENDIX

A1. Spectroradiometric or Spectrophotometric Method—In the CIE standard colorimetric system, the chromaticity coordinates are computed from the spectral energy distribution curve. This should be regarded as a referee approach rather than the one commonly used.

A2. Precautions—The following are applicable to all methods in determining the chromaticity (color) of light:

(a) Measurements should be made in as many directions as required to define the characteristics of the light.

(b) The lamp should be allowed to reach operating temperature before measurements are made.

If visually the device does not appear to be emitting light with a uniform color, additional precautions should be taken:

(c) The distance between the test instrument and the device under test should be great enough so that further increases in distance will not affect the results.

(d) The entire light emitting surface of the device must be visible from any point on the entrance window of the test instrument.

A3. Color Application—Since blue may be misidentified under some circumstances, it is recommended that blue be used as a secondary rather than a primary signal color.

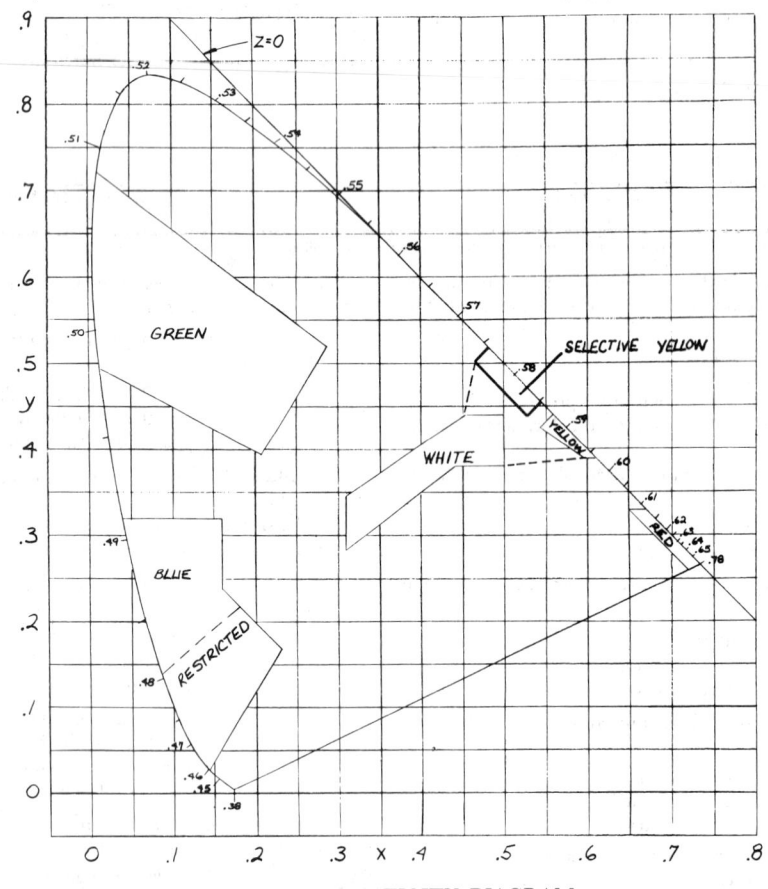

φ FIG. 1—CHROMATICITY DIAGRAM

SEALED BEAM HEADLAMP UNITS FOR MOTOR VEHICLES—SAE J579c

SAE Standard

Report of Lighting Division approved January 1940 and last revised by Lighting Committee December 1974. Editorial change December 1978.

1. Scope—This standard applies to design evaluation of mechanically aimable sealed beam headlamp units for two-beam systems. For service performance requirements and evaluations, see SAE J32.

2. Definitions

2.1 Sealed Beam Unit—An integral and indivisible hermetically sealed optical assembly with the name "Sealed Beam" molded in the lens.

2.2 Upper Beam—A beam intended primarily for distant illumination and for use when not meeting or following other vehicles.

2.3 Lower Beam—A beam intended to illuminate the road ahead of the vehicle when meeting or following another vehicle.

2.4 7 in. (178 mm) Sealed Beam System—A system employing two 7 in. (178 mm) Type 2 sealed beam units.

2.5 7 in. (178 mm) Type 2 Sealed Beam Unit—A 7 in. (178 mm) diameter unit providing an upper and a lower beam. Unit to be aimed to the lower beam.

2.6 5¾ in. (146 mm) Sealed Beam System—A system employing four 5¾ in. (146 mm) sealed beam units: two Type 1 and two Type 2.

2.7 5¾ in. (146 mm) Type 1 Sealed Beam Unit—A 5¾ in. (146 mm) diameter unit having a single filament and used in a four-lamp system to provide the principal portion of the upper beam.

2.8 5¾ in. (146 mm) Type 2 Sealed Beam Unit—A 5¾ in. (146 mm) diameter unit having two filaments and used in a four-lamp system to provide the lower beam and a secondary portion of the upper beam. Unit to be aimed to the lower beam.

2.9 4 x 6½ in. (100 x 165 mm) Sealed Beam System—A system employing four 4 x 6½ in. (100 x 165 mm) sealed beam units: two Type 1A and two Type 2A.

2.10 4 x 6½ in. (100 x 165 mm) Type 1A Sealed Beam Units—A 4 x 6½ in. (100 x 165 mm) rectangular unit having a single filament and used in a four-lamp system to provide the principal portion of the upper beam.

2.11 4 x 6½ in. (100 x 165 mm) Type 2A Sealed Beam Unit—A 4 x 6½ in. (100 x 165 mm) rectangular unit having two filaments and used in a four-lamp system to provide the lower beam and a secondary portion of the upper beam. Unit to be aimed to the lower beam.

2.12 Mechanically Aimable Sealed Beam Unit—A unit having three pads on the face of the lens, forming a mechanical aiming plane used to adjust and inspect the aim of the unit when installed on the vehicle.

2.13 Aiming Plane—A plane through the three aiming pads on the face of the lens.

2.14 Mechanical Axis—A line perpendicular to the aiming plane through the geometric center of the lens.

2.15 H-V Axis—A line from the center of the lens to the intersection of the horizontal and vertical lines on the screen.

3. Laboratory Requirements

3.1 Test Voltage—In conducting tests to this standard, the sealed beam unit shall be operated at 6.4 or 12.8 V for 6 and 12 V electrical systems.

3.2 The following sections from SAE J575 are a part of this standard:

3.2.1 Section B—Samples for Test.
3.2.2 Section D—Laboratory Facilities.
3.2.3 Section J—Photometry. The angular relation between test points for the upper and lower beams is as shown in Fig. 3.

3.3 Color Test—The color of the light from a sealed beam unit shall be white, as defined in SAE J578.

3.4 Beam Pattern Location

3.4.1 BEAM LOCATION—The aiming plane of the sealed beam unit shall be placed parallel to the aiming screen at 25 ft (7.6 m) with the mechanical axis on the H-V axis.

3.4.1.1 *5¾ in. (146 mm) Type 1 Sealed Beam Unit*—The beam shall be photoelectrically aimed so that test points in Fig. 1 designated by the squares have equal intensity and those designated by the triangles have equal intensity. (This will center the high intensity area about the H-V axis.)

3.4.1.1.1 The mechanical axis shall not deviate from the H-V axis more than ±0.3 deg vertically or ±0.6 deg horizontally.

TABLE 1—TEST POINT VALUES FOR 7 IN (178 mm) TYPE 2 SEALED BEAM UNITS

Upper Beam (One 7 in (178 mm) Unit)			Lower Beam (One 7 in (178 mm) Unit)		
Test Points, deg[b]	cd, max	cd, min	Test Points, deg[b]	cd, max	cd, min
2U-V	—	1,000	10U to 90U[a]	125	—
1U-3R and 3L	—	2,000	1U-1-1/2L to L	700	—
H-V	75,000	20,000	1/2U-1-1/2L to L	1,000	—
			1/2D-1-1/2L to L	2,500	—
			1-1/2U-1R to R	1,400	—
H-3R and 3L	—	10,000			
H-6R and 6L	—	3,250	1/2U-1R to 3R	2,700	—
H-9R and 9L	—	1,500	1/2D-1-1/2R	20,000	8,000
H-12R and 12L	—	750	1D-6L	—	750
			1-1/2D-2R	—	15,000
1-1/2D-V	—	5,000			
1-1/2D-9R and 9L	—	1,500	1-1/2D-9L and 9R	—	750
2-1/2D-V	—	2,500	2D-15L and 15R	—	700
2-1/2D-12R and 12L	—	750	4D-4R	12,500	—
4D-V	5,000	—			

[a] From the normally exposed surface of the lens.
[b] A tolerance of ±1/4 deg in location may be allowed for at any test point.

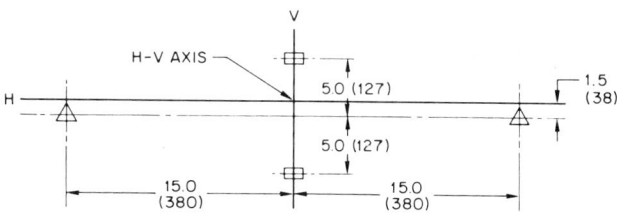

FIG. 1—TEST POINTS ON SCREEN AT 25 FT (7.6 m)

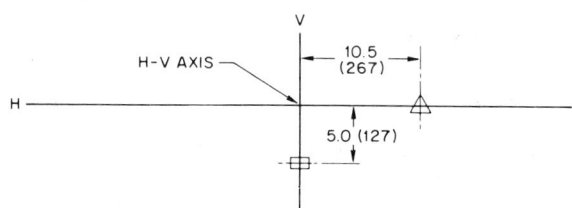

FIG. 2—TEST POINTS ON SCREEN AT 25 FT (7.6 m)

3.4.1.2 *5¾ in. (146 mm) Type 2 and 7 in. (178 mm) Type 2 Sealed Beam Units*—The low beam shall be photoelectrically aimed so that the intensity at the test point designated by the triangle in Fig. 2 is 20% of the maximum beam intensity at the same time that the intensity at the test point designated by the square is 30% of the maximum intensity.

3.4.1.2.1 The mechanical axis shall not deviate from the H-V axis more than ±0.3 deg vertically or ±0.4 deg horizontally.

3.4.1.3 Methods which provide results equivalent to those described in paragraphs 3.4.1.1 and 3.4.1.2 may also be used for beam pattern location evaluation.

3.5 Photometric Design Requirements

3.5.1 PHOTOMETRIC DESIGN REQUIREMENTS—The design requirements established in Fig. 3 and Tables 1 and 2 provide a detailed definition of the headlamp unit light pattern.

3.5.2 TEST PROCEDURES—Photometric tests shall be made with the photometer at a distance of at least 60 ft (18.3 m) from the unit. The unit shall be aimed mechanically by centering the unit on the photometer axis and with the aiming plane on the lens normal to the photometer axis.

3.5.3 DESIGN CANDELA REQUIREMENTS—The beam or beams from the unit shall be designed to conform to the candela specifications in Tables 1 or 2.

3.6 For dimensional requirements, see SAE J571.

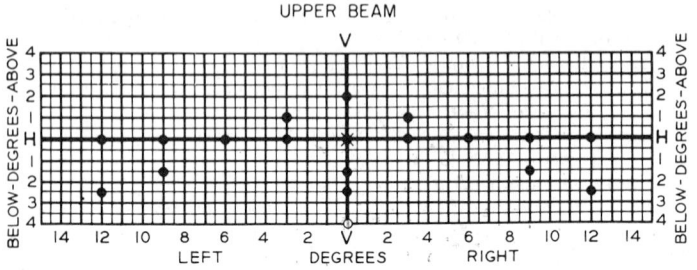

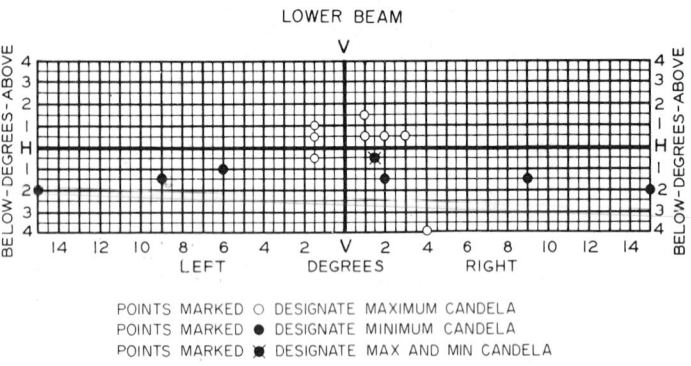

POINTS MARKED ○ DESIGNATE MAXIMUM CANDELA
POINTS MARKED ● DESIGNATE MINIMUM CANDELA
POINTS MARKED ✶ DESIGNATE MAX AND MIN CANDELA

FIG. 3–PHOTOMETRIC DESIGN TEST POINTS

TABLE 2—TEST POINTS FOR 5-3/4 IN (146 mm) AND 4 x 6-1/2 IN (100 x 165 mm) FOUR-LAMP DUAL SEALED BEAM UNITS

Upper Beam (One Type 1 and One Type 2) or (One Type 1A and One Type 2A)					Lower Beam (One Type 2 or 2A Unit)		
Test Points, deg[b]	Type 1 or 1A		Type 2 or 2A		Test Points, deg[b]	cd, max	cd, min
	cd, max	cd, min	cd, max	cd, min			
2U-V	—	750	—	750	10U-90U[a]	125	—
1U-3R and 3L	—	3,000	—	2,000	1U-1-1/2L to L	700	—
H-V	60,000	18,000	15,000	7,000	1/2U-1-1/2L to L	1,000	—
					1/2D-1-1/2L to L	2,500	—
H-3R and 3L	—	12,000	—	3,000	1-1/2U-1R to R	1,400	—
H-6R and 6L	—	3,000	—	2,000			
H-9R and 9L	—	2,000	—	1,000	1/2U-1R to 3R	2,700	—
H-12R and 12L	—	750	—	750	1/2D-1-1/2R	20,000	8,000
					1D-6L	—	750
1-1/2D-V	—	3,000	—	2,000	1-1/2D-2R	—	15,000
1-1/2D-9R and 9L	—	1,250	—	750			
2-1/2D-V	—	1,500	—	1,000	1-1/2D-9L and 9R	—	750
2-1/2D-12R and 12L	—	600	—	400	2D-15L and 15R	—	700
4D-V	5,000	—	2,500	—	4D-4R	12,500	—

[a] From the normally exposed surface of the lens.
[b] A tolerance of ±1/4 deg in location may be allowed for at any test point.

SERVICE PERFORMANCE REQUIREMENTS FOR SEALED BEAM HEADLAMP UNITS FOR MOTOR VEHICLES—SAE J32a

SAE Recommended Practice

Report of Lighting Committee approved January 1973 and last revised July 1974.

1. Scope—This Recommended Practice covers service performance tests, test methods, and requirements applicable to sealed beam headlamp units covered by SAE J579 and J571. It is intended to supplement the dimensional requirements and engineering design test procedures and requirements, and is the means for service evaluation of sealed beam headlamp units.

2. Samples for Test

2.1 Test samples shall be new, unused devices fabricated from production tools and assembled by production processes.

2.2 Test samples shall be seasoned (lighted) at 14 V for 60 min prior to photometry. (For units designed to operate on other than 12 V circuits, check manufacturer for proper seasoning schedule.)

3. Requirements

3.1 The sealed beam headlamp units shall meet the photometric requirements shown in Table 1.

3.2 The sealed beam headlamp units shall meet the dimensional requirements shown in Table 2.

4. Test Procedure

4.1 Units shall be aimed mechanically so that the center of the circle defined by the three aiming pads is on the photometer axis and the aiming plane normal to the photometer axis.

4.2 In conducting tests to this recommended practice, sealed beam units shall be operated at 6.4, 12.8, or 28.0 V for 6, 12, and 24 V units, respectively.

4.3 On the photometric test, a ½ deg reaim shall be permitted at each point.

TABLE 1—PHOTOMETRIC REQUIREMENTS

Test Point	Requirement, cd
Type 1 (5-3/4 in (146 mm))	
1U—V	2,000 min
H—3L	10,000 min
H—3R	10,000 min
H—V	15,000 min
Type 2 (upper beam) (7 in [178 mm])	
1U—V	1,600 min
H—3L	8,000 min
H—3R	8,000 min
H—V	16,000 min
Type 2 (lower beam) (5-3/4 and 7 in [146 and 178 mm])	
1/2 U—1-1/2 L	1,500 max
1/2 U—1R	5,000 max
1/2 D—1-1/2 R	4,500 min

TABLE 2—DIMENSIONAL REQUIREMENTS

	5-3/4 in (146 mm)		7 in (178 mm)	
	in	mm	in	mm
Aiming pad height	0.320 min	8.1 min	0.350 min	8.9 min
Stackup[a]	0.460 min	11.7 min	0.460 min	11.7 min
Aiming ring diameter	5.28-5.34	134.1-135.6	6.59-6.68	167.5-169.5

[a] To be measured using appropriate stackup ring.

SEALED BEAM HEADLAMP ASSEMBLY—SAE J580 AUG79 — SAE Standard

Report of the Lighting Committee, approved March 1960, last revised November 1978, editorial change August 1979.

φ **1. Scope**—This standard applies to the design of sealed beam headlamp
ed. assemblies including the functional parts, except the sealed beam units which
are covered in SAE J571d (June, 1976), J579c (December, 1974), and J1132
(January, 1976).

2. Definitions

2.1 Sealed Beam Headlamp Assembly—A major lighting assembly which includes one or more sealed beam units used to provide general illumination ahead of the vehicle.

2.2 Mounting Ring—The adjustable ring upon which the sealed beam unit is mounted.

2.3 Retaining Ring—The clamping ring that holds the sealed beam unit against the mounting ring.

2.4 Aiming Screws—Horizontal and vertical adjusting screws with self-locking features used to aim and retain the headlamp unit in the proper position.

ed. φ **3. Reference Standards**—The following sections from SAE J575g (September, 1977) are a part of this standard:
- φ 3.1 Section 2—Samples for Test
- φ 3.2 Section 3—Laboratory Facilities
- φ 3.3 Section 4.1—Vibration Test
- φ 3.4 Section 4.4—Corrosion Test

φ **4. Dimensional Specifications**—The mounting ring and retaining ring shall
ed. comply with SAE J571d (June, 1976), Figs. 2, 5, and 8, SAE J1132 (January, 1976) Fig. 2.

5. General Requirements

5.1 Headlamps shall be designed so that they may be inspected and
ed. aimed by mechanical aimers as specified in SAE J602c (December, 1974) without the removal of any ornamental trim rings or other parts.

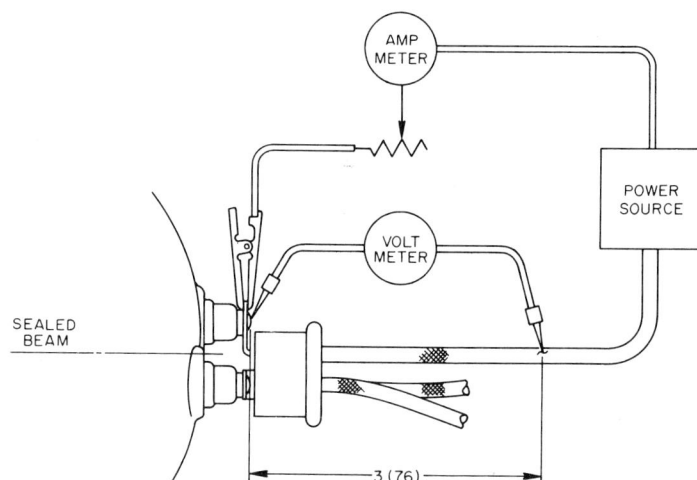

NOTE: DIMENSIONS ARE IN (mm)

FIG. 1—CONNECTOR TEST

UNIT	A DIM	B DIM
SAE 5.75	5.690 (144.53)	5.300 (134.62)
SAE 7.00	7.020 (178.31)	6.640 (168.66)

DIMENSIONS ARE IN (mm)

MACHINE MATERIALS:
DISC, ARM AND BRACE-ALUMINUM-SAE-AA-6061-T6 OR EQUIV
COIL SPRING AND LEVEL CLIP-SPRING STEEL
  SAE 1050-CADMIUM PLATE
WEIGHT AND EYE BOLT ASSEMBLY-STEEL-CADMIUM PLATE
SCREWS-ALUMINUM-MACHINE THREADS
MACHINED DIM ± 0.005
SAE 5.75 AND SAE 7.00 — HEADLAMP TEST FIXTURE

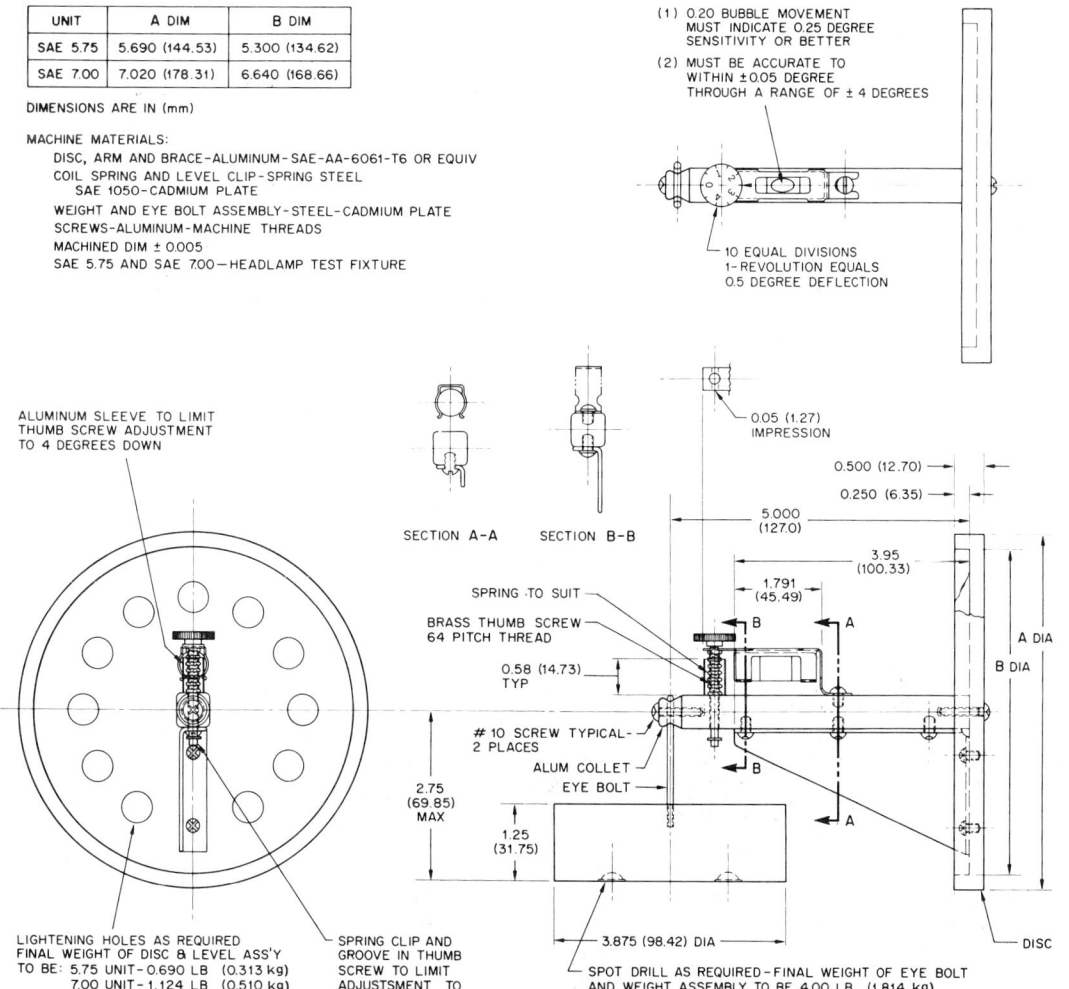

FIG. 2—DEFLECTOMETER

5.2 When in use, a headlamp shall not have any styling ornament or other feature, such as a glass cover or grille, in front of the lens.

6. Design Requirements and Tests—Unless otherwise specified, the following are laboratory tests in which the sealed beam headlamp assembly shall be mounted in design position with the sealed beam unit set at nominal aim (0,0).

6.1 Aiming Adjustment Test—When making the aiming adjustment test, an accurate measurement technique shall be used. A device attached to the sealed beam unit such as a spot projector, or replacing the sealed beam unit with a mirror with a separate light source, or other accurate means can be used.

6.1.1 When the headlamp assembly is tested in the laboratory, a minimum aiming adjustment of ±4.0 deg shall be provided in both the vertical and horizontal planes.

6.1.2 On headlamp assemblies with independent vertical and horizontal aiming screws, the adjustment shall be such that when tested in the laboratory neither the vertical nor horizontal aim shall deviate more than 4 in (100 mm) from horizontal or vertical planes respectively at a distance of 25 ft (7.6 m) through an angle of ±4 deg.

6.1.3 The self-locking devices used to hold aiming screws in position shall continue to operate satisfactorily at least for 20 adjustments on each screw, over a length of screw thread of ±1/8 in (3 mm).

NOTE: Paragraphs 6.1.2 and 6.1.3 are not applicable to lamps with ball and socket or equivalent adjusting means.

6.2 Inward Force Test—The mechanism, including the aiming adjusters, when subjected to an inward force of 50 lb (222 N) directed normal to the headlamp aiming plane and symmetrically about the center of the sealed beam unit face shall meet the following requirements:

6.2.1 The sealed beam unit shall not permanently recede by more than 0.1 in. (2.5 mm).

6.2.2 The aim of the sealed beam unit shall not permanently deviate by more than 1.25 in. (3.2 mm) at a distance of 25 ft (7.6 m).

6.3 Retaining Ring Tests—Positive means shall be provided for holding the sealed beam unit to the mounting ring.

6.3.1 The fastening means shall be deemed adequate if it will withstand and hold the sealed beam unit securely in its proper position at the end of 20 replacements.

6.3.2 When a unit having a flange thickness shown below is secured between the retaining ring and the mounting ring, it shall be held tight enough that it will not rattle.

Flange Thickness

5-3/4 in (146 mm)	0.465 (11.8 mm)
7 in (178 mm)	0.465 (11.8 mm)
4 x 6-1/2 in (100 x 165 mm)	1.24 (31.5 mm)
(142 x 200 mm)	0.398 (10.1 mm)

6.4 Connector Tests—Measure voltage drop as shown in Fig. 1.

6.4.1 The voltage drop shall not exceed 40 mV with a 10 A load.

6.5 Torque Deflection Test—The headlamp assembly to be tested shall be mounted in designed vehicle position and set at nominal aim (0, 0). The sealed unit shall be replaced by the appropriate deflectometer (Figs. 2, 3, and 4). A torque of 20 lb-in (2.25 N·m) shall be applied to the headlamp assembly through the deflectometer and a reading on the thumb wheel shall be taken. The torque shall then be removed and a second reading on the thumb wheel shall be taken.

6.5.1 The difference between the two readings shall not exceed 0.30 deg.

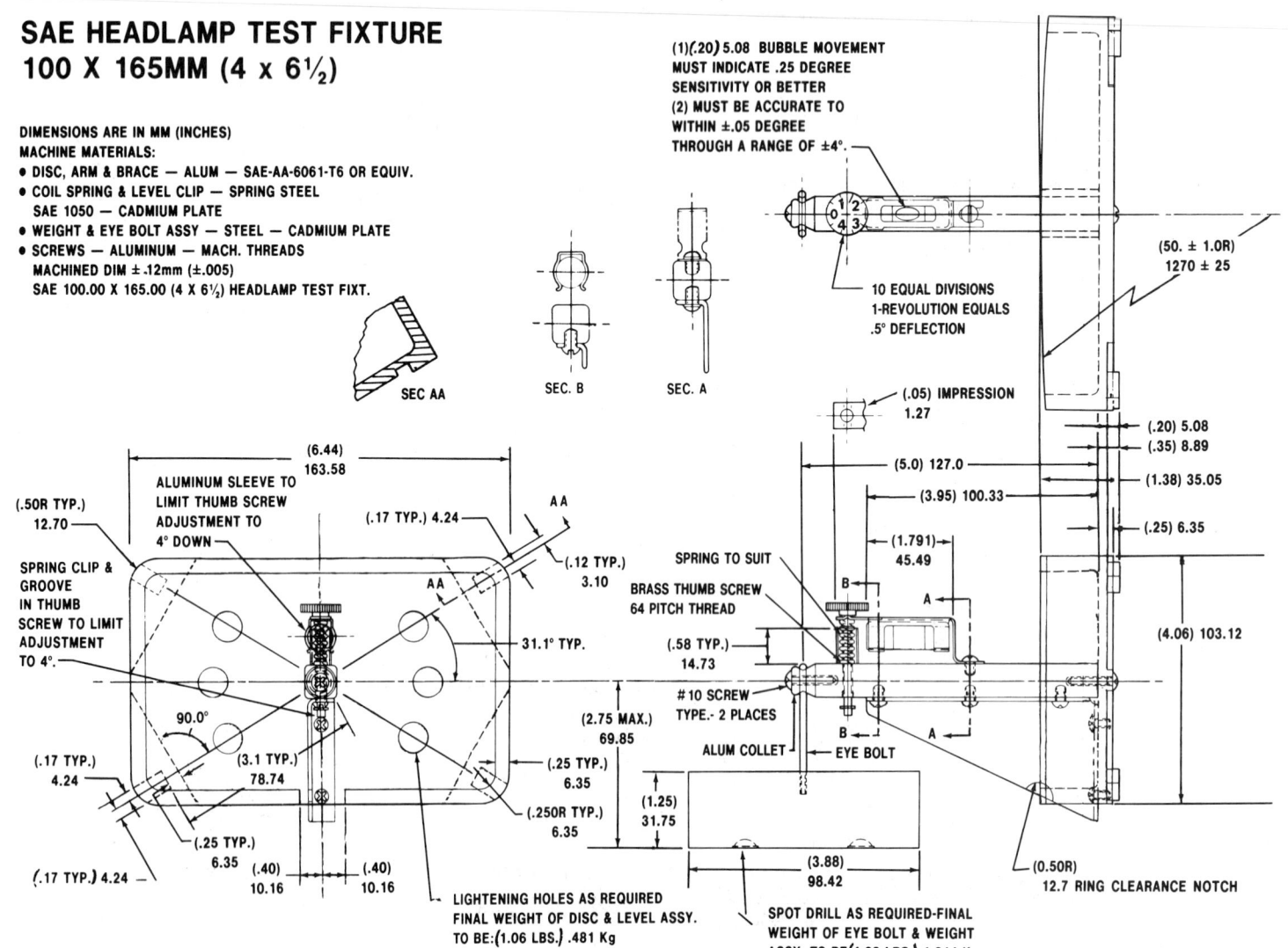

FIG. 3—DEFLECTOMETER

SAE HEADLAMP TEST FIXTURE
142 X 200MM (5.60 x 7.90)

DIMENSIONS ARE IN MM (INCHES)
MACHINE MATERIALS:
- DISC, ARM & BRACE — ALUM — SAE-AA-6061-T6 OR EQUIV.
- COIL SPRING & LEVEL CLIP — SPRING STEEL SAE 1050 — CADMIUM PLATE
- WEIGHT & EYE BOLT ASS'Y — STEEL — CADMIUM PLATE
- SCREWS — ALUMINUM — MACH. THREADS
  MACHINED DIM ±.12mm (±.005)
  SAE (142 X 200) (5.60 X 7.90) HEADLAMP TEST FIXT.

φ FIG. 4—DEFLECTOMETER

AUXILIARY DRIVING LAMPS—SAE J581a   SAE Standard

Report of Lighting Committee approved March 1979. Editorial change June 1979. This report was last reviewed and appeared in the 1971 Handbook. Since the 1971 Handbook, it was deleted and is now being reinstated with revisions in the 1980 Handbook. Rationale statement available.

1. Scope—This SAE Standard is an Engineering Design Standard for auxiliary driving lamps and may also be supplemented by a Service Performance Standard.

2. Definition—An auxiliary driving lamp is a lighting device mounted to provide illumination forward of the vehicle and supplements the upper beam of a standard headlamp system. It is not intended for use alone or with the lower beam of a standard headlamp system.

3. Laboratory Requirements

3.1 The following Sections from SAE J575g (September, 1977) are a part of this standard:

3.1.1 Section 2—Samples for Tests.
3.1.2 Section 2.2—Bulbs.
3.1.3 Section 3—Laboratory Facilities.
3.1.4 Section 4.1—Vibration Test.
3.1.5 Section 4.2—Moisture Test.
3.1.6 Section 4.3—Dust Test.
3.1.7 Section 4.4—Corrosion Test.
3.1.8 Section 4.6—Photometry.
3.1.9 Section 4.7—Out of Focus Tests on Unsealed Units.
3.1.10 Section 4.8—Warpage Test on Devices with Plastic Components.

3.2 Sealed beam units when tested separately need comply only with Section 2, Section 3, Section 4.6, and Section 4.8 of SAE J575g (September, 1977).

3.3 Plastic Materials—Plastic materials used in optical parts shall comply with the requirements set forth in SAE J576d (June, 1976).

3.4 Color Test—The color of the light shall be white as defined in SAE J578c (February, 1977).

3.5 Photometric Requirements

3.5.1 Photometric tests shall be made at a distance of at least 18.3 m from the photometer to the lamp.

3.5.2 Photometric tests shall be made with the filament at the design position. For unsealed units, tests shall also be made at the out-of-focus positions listed in Section 4.7 of SAE J575g (September, 1977).

3.5.3 LAMP AIM FOR THE PHOTOMETRIC TEST—A lamp or sealed beam unit which is designed to be aimed mechanically shall be centered on the photometric axis with the aiming planes normal to that axis. A lamp or sealed unit not designed to be aimed mechanically shall be photoelectrically aimed so that the test points in Fig. 1 designated by the squares have equal intensity and those designated by the triangles have equal intensity (this will center the high intensity zone about the H-V axis).

3.5.4 The lamp shall be designed to conform with the photometric requirements shown in Table 1 for the design filament position and the required

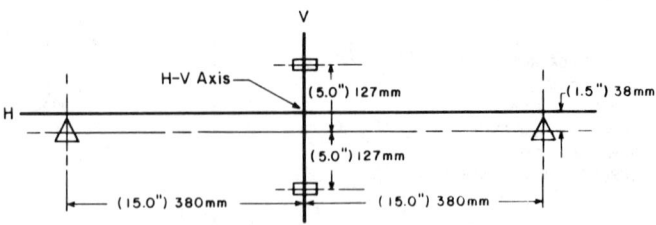

FIG. 1—TEST POINTS ON SCREEN AT 7.6 M

out-of-focus filament positions. An aiming tolerance of ±1/4 deg shall be allowed at each test point.

4. Installation Recommendations

4.1 The following requirements and test procedures apply to the device as used on the vehicle, and are not a part of the laboratory test requirements and procedures.

4.2 Lamp Aim on the Vehicle—Lamp aim adjustments and inspection should be with mechanical aimers if possible. The correct mechanical aim is 0-0, Ref. SAE J599d (December, 1974) and SAE J602c (December, 1974).

4.2.1 If the vehicle mounting or lamp design precludes mechanical aiming, the lamp shall be aimed photometrically (Ref. Section 3.5.3) or visually aimed. The correct visual aim is with the high intensity zone of the beam symmetrical about the H-V axis of the lamp on an aiming screen at 7.6 m (25 ft).

TABLE 1

Photometric Requirements	
Position Deg	Candela, CD
2U—3R and 3L	2000 Min
1U—3R and 3L	5000 Min
H-V	25 000 Min and 50 000 Max
H—3R and 3L	10 000 Min
1D—6R and 6L	3700 Min
2D—6R and 6L	2000 Min
4D-V	5000 Max

FOG LAMPS—SAE J583d   SAE Standard

Report of Lighting Division approved May 1937 and completely revised by Lighting Committee July 1977. Rationale statement available.

1. Scope—This SAE Standard is an engineering design standard for fog lamps and may also be supplemented by a Service Performance Standard.

2. Definition—A fog lamp is a lighting device mounted to provide illumination forward of the vehicle under conditions of rain, snow, dust, or fog. The lamp may be used with a lower beam headlamp or switch controlled in conjunction with the headlamps and used at the drivers discretion with either low or high beam headlamps.

3. Laboratory Requirements

3.1 The following sections from SAE J575 are a part of this standard:

3.1.1 SECTION 2—Samples for Tests
3.1.2 SECTION 2.2—Bulbs
3.1.3 SECTION 3—Laboratory Facilities
3.1.4 SECTION 4.1—Vibration Test
3.1.5 SECTION 4.2—Moisture Test
3.1.6 SECTION 4.3—Dust Test
3.1.7 SECTION 4.4—Corrosion Test
3.1.8 SECTION 4.6—Photometry
3.1.9 SECTION 4.7—Out of Focus Test on Unsealed Units
3.1.10 SECTION 4.8—Warpage Test on Devices with Plastic Components

3.2 Sealed units designed for use as a fog lamp when tested without the other parts of the lamp assembly need only comply with Section 2, Section 3, and Section 4.6 of SAE J575.

3.3 Plastic Materials—Any plastic materials used in optical parts shall comply with the requirements set forth in SAE J576.

3.4 Color Test—The color of the light from a fog lamp shall be white to yellow.

3.5 Photometric Requirements

3.5.1 Photometric tests shall be made at a distance of at least 18.3 m (60 ft) from the point of measurement to the light source.

3.5.2 Photometric test shall be made with the filament at the design position. For unsealed units, tests shall also be made at the out-of-focus positions listed in Section 4.7 of SAE J575.

3.5.3 LAMP AIM—A lamp or sealed beam unit which is designed to be aimed mechanically shall be centered on the photometric axis with the aiming plane normal to that axis. A lamp or sealed unit not designed to be aimed mechanically shall be centered on the photometer axis with the beam aimed downward so that 500 cd is directed at some point on the horizontal between 6L and 6R and balancing the photometric values on the horizontal at 3L and 3R.

3.5.4 The lamp shall be designed to comply with the photometric requirements shown in Table 1 for the design filament position and the required out-of-focus filament position. An aiming tolerance of ±0.25 deg shall be allowed at each test point.

4. Installation Recommendations

4.1 The following requirements and test procedures apply to the device as used on the vehicle, and are not a part of the laboratory test requirements and procedures.

4.2 Lamp Aim on Vehicle—Lamp aim adjustments and inspection should be with mechanical aimers. The correct mechanical aim is 0-0, reference SAE J599 and SAE J602.

4.2.1 If the vehicle mounting or lamp design precludes mechanical aiming, the lamp shall be visually aimed. The *correct* visual aim is with the top of the beam 100 mm (4 in) below the lamp center at 7.6 m (25 ft). The lamp shall be centered laterally about a vertical line directly ahead of the lamp. A higher visual aim may be desired, but the top of the beam should not be higher than the lamp center level at 7.6 m (25 ft).

TABLE 1

Position, deg	Candela, cd
8U—90U	75 max
4U—6L and 6R	125 max
2U—6L and 6R	250 max
1U—6L and 6R	350 max
H—6L — 6R	500 max
1 1/2D—3L and 3R	2000 min—10 000 max
1 1/2D—9L and 9R	1000 min
3D—15L and 15R	1000 min

AUXILIARY LOW BEAM LAMP—SAE J582a   SAE Standard

Report of Lighting Division approved January 1941 and last revised by Lighting Committee January 1973.

1. Scope—This standard provides design requirements and test methods for an auxiliary low beam lamp.

2. Definitions—An auxiliary low beam lamp supplements the lower beam of a standard headlamp system.

3. Laboratory Requirements

3.1 Test Voltage—In conducting tests to this standard lamps or sealed beam units shall be operated at 6.4, 12.8, or 28.0 V for 6, 12, and 24 V systems respectively.

3.2 The following sections from the latest revision of SAE J575 are a part of this standard:

3.2.1 Section B—Samples for Test
3.2.2 Section C—Bulbs
3.2.3 Section D—Laboratory Facilities
3.2.4 Section E—Vibration Test
3.2.5 Section F—Moisture Test
3.2.6 Section G—Dust Test

3.2.7 Section H—Corrosion Test
3.2.8 Section J—Photometry
3.2.9 Section K—Out-of-Focus Tests on Unsealed Units
3.2.10 Section L—Warpage Test on Devices with Plastic Lenses

3.3 Sealed beam units need comply only with Sections B, D, and J of SAE J575.

3.4 Plastic Materials—Plastic materials used in optical parts shall comply with the requirements set forth in SAE J576.

3.5 Color Test—The color of the light shall be white as defined in SAE J578.

3.6 Photometric Requirements

3.6.1 Photometric tests shall be made at a distance of at least 60 ft (18.3 m) from the photometer to the lamp.

3.6.2 Photometric tests shall be made with the filament at the design position. For unsealed units, tests shall also be made at the out-of-focus positions listed in Section K of SAE J575.

3.6.3 A lamp or sealed beam unit which is designed to be aimed mechanically shall be centered on the photometric axis with the aiming plane normal to that axis. A lamp or sealed beam unit not designed to be aimed mechanically shall be centered on the photometer axis with the beam aimed downward and to the right so that 7000 cd and not over that value is directed at 1/2U at some point between 1R and 3R and 5000 cd is directed at 1/2D, 1L.

3.6.4 The lamps shall be designed to conform with the photometric values shown in Table 1 for the design filament position and the required out-of-focus filament positions. For nonsealed beam units, at the time of photometry of the in-focus filament position, the candela values at 1/2D—1L and 1/2U—3R shall be recorded. For each of the out-of-focus positions, specified in Section K SAE J575, the units shall be reaimed photometrically to the two recorded in-focus candelas.

4. Installation Recommendations

4.1 The following requirements and test procedures apply to the device as used on the vehicle, and are not a part of the laboratory test requirements and procedures.

4.2 Lamp Aim on Vehicle—Lamp aim adjustment and inspection should be with mechanical aimers. The correct mechanical aim is 0-0. Reference: SAE J599 and SAE J602.

4.2.1 If vehicle mounting precludes mechanical aiming, the lamp may be visually aimed. The correct visual aim is with the top edge of the high intensity zone 1 in. (25 mm) above horizontal at 25 ft (7.6 mm) and the left edge of the high intensity zone 5 in. (130 mm) left of vertical at 25 ft (7.6 mm).

4.3 Means shall be provided to turn off the auxiliary low beam lamp independently of the lower beam lamps.

4.4 Lamp Mounting—The lamp shall be mounted at the front and to the left side of the center of the vehicle.

TABLE 1

Position, deg	cd, max	cd, min
10U—90U[a]	150	—
1-1/2U—1L to L	800	—
1-1/2U—1R to R	2,000	—
1/2U—1L to L	1,000	—
1/2U—1R to 3R	7,000	—
1/2D—1R to 3R	50,000	15,000
1/2D—1L to L	5,000	—
1D—1R	—	15,000
1D—3R	—	15,000
4D—2R	8,000	—

A tolerance of ±1/4 deg in location may be allowed for at any test point.
[a]From the normally exposed surface of the lens.

MOTORCYCLE HEADLAMPS—SAE J584b

SAE Recommended Practice

Report of Lighting Committee approved January 1949 and last revised by Motorcycle Committee December 1971.

1. Definition—A motorcycle headlamp is a major lighting device used to provide general illumination ahead of the vehicle. For definition and classes of motorcycles, see SAE J213.

2. Laboratory Requirements

2.1 The following sections from SAE J575 are a part of this standard:
2.1.1 Section B—Samples for Test
2.1.2 Section C—Lamp Bulbs
2.1.3 Section D—Laboratory Facilities
2.1.4 Section E—Vibration Test
2.1.5 Section F—Moisture Test
2.1.6 Section G—Dust Test
2.1.7 Section H—Corrosion Test
2.1.8 Section J—Photometric Test Points

NOTE: Where sealed beam units (hermetically sealed) are used and tested separately from the housing, paragraphs 2.1.4—Vibration Test, 2.1.5—Moisture Test, 2.1.6—Dust Test, and 2.1.7—Corrosion Test, shall not be required for sealed beam units.

2.2 Color Test—The light from the lamps shall be white. See SAE J578.

2.3 Aiming Adjustment Tests

2.3.1 A minimum aiming adjustment of ±4 deg shall be provided in both the vertical and horizontal planes.

2.3.2 The mechanism, including the aiming adjustment, shall be so designed as to prevent the unit from receding into the lamp body or housing when an inward force of 50 lbf (224 N) is exerted at the geometric center of the outer surface of the lens.

2.3.3 Headlamps with independent vertical and horizontal aiming adjusting mechanisms.

(a) The headlamp unit mounting shall be provided with independent vertical and horizontal aiming adjustments. The adjustment mechanisms shall be so designed that neither the vertical nor horizontal aim will deviate more than 4.00 in. (101 mm) from the horizontal or vertical planes, respectively, at a distance of 25 ft (7.6 m), through an angle of ±4 deg.

(b) When adjusting screws are employed, they shall be equipped with self-locking devices which operate satisfactorily for a minimum of 10 adjustments on each screw, over a length of screw thread of ±1/8 in. (3 mm).

2.3.4 Headlamps with ball and socket or equivalent adjusting means need not conform with paragraph 2.3.3.

2.4 Clarity of Hot Spot Definition—The geometric center of the high intensity zone of the upper beam of the multiple beam headlamps shall be deemed sufficiently defined for the purpose of service aiming if it can be set by three experienced observers on a vertical screen at 25 ft (7.6 m) within a maximum vertical deviation of ±0.2 deg and within a maximum horizontal deviation of ±0.4 deg. The aim for each observer shall be taken as the average of at least three observations.

2.5 Beam Aim During Photometric Test—The upper beam of a multiple beam headlamp shall be aimed photoelectrically so that the center of the zone of highest intensity falls 0.4 deg vertically below the lamp axis and is centered laterally. The center of the zone of highest intensity shall be established by the intersection of a horizontal plane passing through the point of maximum intensity, and the vertical plane established by balancing the photometric values at 3 deg left and 3 deg right.

The 7 in. (178 mm) Type 2 and 5 3/4 in. (146 mm) dual sealed beam units shall be aimed mechanically by centering the unit on the photometer axis and with the aiming plane through the faces of the pads on the lens normal to the photometer axis.

2.6 Photometric Tests—Shall be made with the photometer at a distance of 60 ft (18.3 m) from the lamp. The bulb or unit shall be operated at its rated voltage during the test.

At-Focus Tests—The light source or sources shall be located in the design position as specified by the manufacturer.

The beams shall meet the candela specifications listed in Tables 1 and 2 except that a tolerance ±1/4 deg in location may be allowed for any test point to allow for variations in readings between laboratories.

3. Optional Systems—One 7 in. (178 mm) Type 2 sealed beam unit or one 5 3/4 in. (146 mm) Type 1 and one 5 3/4 in. (146 mm) Type 2 sealed beam units meeting the requirements of SAE J579 may be used on Class A, B, C, and D motorcycles.

4. Plastic Materials—Any plastic material used in optical parts shall comply with the requirements set forth in SAE J576.

5. Installation Requirements—The following requirements apply to the devices as used on the vehicle and are not part of laboratory test requirements and procedures.

5.1 Beam Switching—The switching of motorcycle headlamps between the upper and the lower beams should be by means of a switch designed and located so that it may be operated conveniently by a simple movement of the driver's hand or foot. The switch shall have no dead point between upper and lower beam switch position.

5.2 Means shall be provided for indicating to the driver that the upper beam is on. The upper beam indicator shall consist of a red or blue light, with a minimum area equivalent to that of a 3/16 in. (5 mm) diameter circle, plainly visible to drivers of all heights under normal driving conditions when headlights are required.

5.3 Semi-automatic headlamp beam switching devices are permitted. See SAE J565.

TABLE 1—UPPER BEAM

Position, deg	Class A and B Motorcycle	Class C and D Motorcycle
	Multiple Beam, cd	Multiple Beam, cd
H—V	10,000 min	2,000 min
1/2D—V	20,000 min	5,000 min
1/2D—3L and 3R	4,000 min	3,000 min
1/2D—6L and 6R	1,000 min	750 min
1D—V	15,000 min	5,000 min
2D—V	5,000 min	3,000 min
3D—V	2,500 min	1,000 min
3D—6L and 6R	750 min	500 min
4D—V	5,000 max	5,000 max
Maximum anywhere	75,000	75,000

TABLE 2—LOWER BEAM

Position, deg	Class A and B Motorcycle	Class C and D Motorcycle
	Multiple Beam, cd	Multiple Beam, cd
1-1/2U—1R to right	1,000 max	1,000 max
1U—1L to left	500 max	500 max
1/2U—1R to 3R	2,000 max	2,000 max
1/2U—1L to left	800 max	800 max
1/2D—1R to right	15,000 max	15,000 max
1/2D—1L to left	2,000 max	2,000 max
2D—3R	3,000 min	2,000 min
2D—3L	2,000 min	1,500 min
2D—6L and 6R	750 min	500 min
4D—4R	12,500 max	12,500 max

MOTORCYCLE AUXILIARY FRONT LAMPS—SAE J1306 JUN80

SAE Recommended Practice

Report of the Motorcycle Committee, approved June 1980. Rationale statement available.

1. Scope—This engineering design specification provides parameters and general requirements for auxiliary front lamps to be used on motorcycles. It may be supplemented by a service performance requirement.

2. Definition—An auxiliary lamp as covered by this specification is a unit, including sealed beam, intended to supplement either the upper or the lower beam from motorcycle headlamps.

3. Laboratory Requirements

3.1 The following sections of SAE J575g are parts of this recommended practice:

Section 2 —Samples for Tests.
Section 2.2—Bulbs.
Section 3 —Laboratory Facilities.
Section 4.1—Vibration Test.
Section 4.2—Moisture Test.
Section 4.3—Dust Test.
Section 4.4—Corrosion Test.
Section 4.6—Photometry.
Section 4.7—Out-of-Focus Tests on Unsealed Units.
Section 4.8—Warpage Test on Devices with Plastic Components.

3.2 Sealed beam units need comply only with Sections 2, 3, and 4.6 of SAE J575.

3.3 *Color Test*—The color of the light from a motorcycle auxiliary front lamp shall be white. (See SAE J578.)

3.4 *Plastic Materials*—Any plastic materials used in exterior optical parts shall comply with the requirements set forth in SAE J576.

3.5 *Photometric Tests*—These shall be made with the photometer at a distance of at least 60 ft (18.3 m) from the lamp.

3.5.1 At-Focus Tests—The light source shall be located in the designed position as specified by the manufacturer.

The beam from the lamp shall be aimed with the left edge of the high intensity zone at a vertical line straight ahead of the lamp center and with the top edge of the high intensity zone at the level of the lamp center at a distance of 25 ft (7.6 m) from the lens.

The beam from the lamp shall meet the photometric specifications listed in Table 1 when it is aimed as specified above.

3.5.2 Out-of-Focus Tests on Unsealed Units—Similar tests shall be made for each of four out-of-focus filament positions, except that the completed distribution may be omitted. Where conventional bulbs with two pin bayonet bases are used, intensity tests shall be made with the light source 0.060 in (1.5 mm) above, below, ahead, and behind the designed position. If prefocused bulbs are used, the limiting positions at which tests are made shall be 0.020 in (0.5 mm) above, below, ahead, and behind the designed position. The beam from the lamp may be reaimed as specified in paragraph 3.5.1 for each of the out-of-focus positions of light source.

4. Installation and Usage Requirements—The following items apply to the device as used on the motorcycle and are not a part of the laboratory test requirements and procedures.

For greatest visibility, with reasonable limitation of glare to approaching drivers, the beam from the lamp shall be aimed in accordance with paragraph 3.5.1. The unit should be turned off when traveling in congested areas in cities. It may be wired so that it can be turned on or off with either beam of the regular headlamps.

TABLE 1—TEST POSITIONS AND LUMINOUS INTENSITY REQUIREMENTS

Position, deg[a]	Max Intensity (cd)
1U-1L to left and above	400
1/2U-1L to left	500
1/2D-1L to left	1000
1-1/2D-1L to left	3000
2U-1R to right and above	1000
1U-1R to right	3000
H-1R to right	7000
1-1/2D-2R to 4R	10 000 min

[a] An aiming tolerance of ±1/4 deg should be allowed on individual points.

APPENDIX

As a matter of information, attention is called to SAE J567 for requirements and gages to be used in socket design for unsealed units.

MOTORCYCLE AND MOTOR DRIVEN CYCLE ELECTRICAL SYSTEM (MAINTENANCE OF DESIGN VOLTAGE)—SAE J392

SAE Recommended Practice

Report of Motorcycle Committee and Lighting Committee approved December 1969. Editorial change November 1971.

1. Purpose—This SAE Recommended Practice provides minimum illumination voltage values for motorcycle and motor driven cycle electrical systems and accompanying test procedures. (NOTE: Wherever the word "motorcycle" appears in the report, it is understood to include "motor driven cycle.")

2. Scope—This recommended practice pertains to both battery-equipped and batteryless motorcycle electrical systems.

3. Test Apparatus

3.1 *Voltmeter*—0–20 V maximum full-scale deflection, accuracy ±1/2% (two voltmeters required).

3.2 *Ammeter*—Capable of carrying full system load current. Accuracy ±3% FS.

3.3 *Means for Measuring Engine RPM*—Accuracy ±3%.

4. Test Procedure

4.1 Install fully charged original equipment battery on the motorcycle (if motorcycle is battery equipped).

4.1.1 Battery temperature to be 80 ± 10 F.

4.2 Connect one voltmeter between the headlamp low beam terminal and the ground; connect the other voltmeter between the tail lamp terminal and the ground.

4.3 Connect the ammeter in series with the battery. (NOTE: Disregard paragraph 4.3 for batteryless machines.)

4.4 Start engine and turn on headlamp(s).

4.4.1 Switch headlamp to the low beam position.

4.4.2 External fan cooling may be applied to the motorcycle engine.

4.5 Run the engine at an rpm equivalent to 30 mph in top gear for 10 minutes.

4.5.1 Record the lowest and highest headlamp voltage and tail lamp voltage observed during the 10 minute period.

4.6 Increase speed to manufacturer's suggested maximum rpm.

4.6.1 Record the highest and lowest headlamp and tail lamp voltages observed during a 5 sec period.

4.7 Run the engine at manufacturer's rated idle speed for 10 minutes.

4.7.1 Record the lowest and highest tail lamp voltage observed during the 10 minute period.

4.7.2 Record the lowest and highest headlamp voltage observed during the 10 minute period.

4.8 Slowly increase the engine speed until generating equipment cancels the system load, indicated by "0" reading on the ammeter. (NOTE: Disregard paragraph 4.8 for batteryless motorcycles.)

4.8.1 Record the engine rpm at ammeter zero point.

5. Test Limits

5.1 Voltages recorded in paragraphs 4.5.1, 4.6.1, and 4.7.1 shall be between 80% and 120% of the rated headlamp design voltage.

5.2 Voltages recorded in paragraph 4.7.2 shall be between 40% and 120% of the rated headlamp design voltage.

5.3 Engine rpm observed in paragraph 4.8.1 shall be less than the motorcycle equivalent speed at 30 mph in top gear operation.

MOTORCYCLE TURN SIGNAL LAMPS—SAE J131a — SAE Recommended Practice

Report of Lighting Committee and Motorcycle Committee approved October 1969 and last revised by Motorcycle Committee December 1971.

1. Definition—Motorcycle turn signal lamps are the signalling elements of a turn signal system which indicate a change in direction by giving flashing lights on the side toward which the turn will be made. (For flashing rate and "on" period, see SAE J590.)

2. General Requirements—The effective projected illuminated area measured on a plane at right angles to the axis of a lamp shall not be less than $3\frac{1}{2}$ in.^{2} (2260 mm^2).

2.1 The following sections from SAE J575 are a part of this standard:
2.1.1 Section B—Samples for Test
2.1.2 Section C—Lamp Bulbs
2.1.3 Section D—Laboratory Facilities
2.1.4 Section E—Vibration Test
2.1.5 Section F—Moisture Test
2.1.6 Section G—Dust Test
2.1.7 Section H—Corrosion Test
2.1.8 Section J—Photometric Test—For turn signal lamps, see Table 1.

2.1.8.1 All beam candlepower measurements shall be made with the incandescent filament of the signal lamp at least 10 ft (3 m) from the photometer screen. The H-V axis shall be taken as parallel to the longitudnal axis of the vehicle.

2.1.9 Section L—Warpage Test on Devices with Plastic Lenses

3. Color Test—The color of the light from turn signal lamps shall be yellow to the rear and yellow to the front of the vehicle. See SAE J578.

4. Plastic Materials—Any plastic materials used in optical parts shall comply with requirements set forth in SAE J576.

5. Installation Requirements—The following requirements apply to the device as used on the vehicle and are not a part of the laboratory requirements and test procedures.

5.1 The filament center of each signal lamp on the front shall be symmetrically spaced a minimum of $4\frac{1}{2}$ in. (114 mm) from the centerline of the vehicle and 4 in. (102 mm) from the inside diameter of the retaining ring of the headlamp unit providing the lower beam. On the rear the symmetrical spacing shall be a minimum of $4\frac{1}{2}$ in. (114 mm) from the centerline of the vehicle to the filament axis of the signal lamp.

5.1.1 When one turn signal is used on each side of the front and rear, visibility of the front turn signal to the front and the rear turn signal to the rear shall not be completely obstructed by any part of the vehicle throughout the photometric test angles.

5.1.2 Signals from lamps mounted on the left side of the vehicle shall be visible through a horizontal angle of 45 deg to the left, and signals from lamps mounted on the right side of the vehicle shall be visible through a horizontal angle of 45 deg to the right. To be considered visible, the lamp must provide a completely unobstructed projected illuminated area of outer lens surface, at least 2 in.^{2} (1290 mm^2) in extent, projected and measured at all angles throughout 45 deg to the longitudinal axis of the vehicle.

5.2 Turn Signal Pilot Indicator—There shall be an illuminated indicator to give the operator a clear and unmistakable indication that the turn signal system is correctly functioning. The illuminated indicator shall consist of one or more lights flashing at the same frequency as the signal lamps, and shall be plainly visible to operators of all heights when seated in normal position in the operator's seat, while driving in bright sunlight. Failure of one or more turn signal lamps to operate shall be indicated by a "steady on" or a "steady off," or by a significant change in the flashing rate of the illuminated indicator.

APPENDIX

As a matter of information attention is called to typical sockets shown in SAE J567.

TABLE 1

Test Points, deg	Min Requirement, cd	Test Points, deg	Min Requirement, cd
10U and 10D { 10L	15	{ 20L	20
10U and 10D { 10R	30	{ 10L	45
	15	{ 5L	120
5U and 5D { 20L	15	H { V	120
{ 10L	40	{ 5R	120
{ 5L	60	{ 10R	45
{ V	90	{ 20R	20
{ 5R	60		
{ 10R	40		
{ 20R	15	Rear only	750 max

φ TAIL LAMPS (REAR POSITION LAMPS)—SAE J585e — SAE Standard

Report of Lighting Division approved March 1918 and completely revised by Lighting Committee September 1977. Rationale statement available.

1. Scope—This engineering design standard provides design parameters and general requirements for tail lamps (rear position lamps). It is intended for use in conjunction with the Service Performance Specification, SAE J256 (July, 1971).

2. Definitions

2.1 Tail Lamps—Lamps used to designate the rear of a vehicle by a steady burning, low intensity light.

2.2 Multiple Compartment Lamp—A device which gives its indication by two or more separately lighted areas which are joined by one or more common parts such as a housing or lens.

2.3 Multiple Lamp Arrangement—An array of two or more separated lamps on each side of the vehicle which operate together to give a signal.

3. Laboratory Requirements

3.1 A multiple compartment lamp or multiple lamps may be used to meet the photometric requirements of a tail lamp. If a multiple compartment or multiple lamps are used and the distance between the optical axes (filament centers) does not exceed 22 in (560 mm) for two compartment or lamp arrangements, and does not exceed 16 in (410 mm) for three compartment or lamp arrangements, then the combination of the compartments or lamps must be used to meet the photometric requirements for the corresponding number of lighted sections (Table 1). If the distance between optical axes exceeds the above dimensions, each compartment or lamp shall comply with the photometric requirements for one lighted section (Table 1).

For vehicles of 80 in (2032 mm) or more in overall width, a maximum of two lamps and/or compartments per side may be mounted closer together than 22 in (560 mm) providing that each compartment and/or lamp meets the single compartment photometric requirements listed in Table 1. Each lamp and/or compartment utilized in this manner shall meet the one lighted section value for all functions for which it is designed.

3.2 The following sections from SAE J575g (March, 1978) are a part of this standard:
3.2.1 SECTION 2—Samples for Test
3.2.2 SECTION 2.2—Bulbs
3.2.3 SECTION 2.3—Test Fixture

3.2.4 SECTION 3—Laboratory Facilities
3.2.5 SECTION 4.1—Vibration Test
3.2.6 SECTION 4.2—Moisture Test
3.2.7 SECTION 4.3—Dust Test
3.2.8 SECTION 4.4—Corrosion Test
3.2.9 SECTION 4.6—Photometry
3.2.10 SECTION 4.8—Warpage Test

3.3 Plastic Materials—Any plastic materials used in optical parts shall comply with the requirements set forth in SAE J576d (June, 1976).

3.4 Color Test—The color of the light from a tail lamp shall be red. (See SAE J578c (February, 1977).)

3.5 If the tail lamp is optically combined with another lamp such as a stop lamp or turn signal and a two-filament bulb is used, the bulb shall have an indexing base and the socket shall be designed so that bulbs with non-indexing bases cannot be used.

3.6 Photometric Requirements

3.6.1 All beam candlepower measurements shall be made with the incandescent filament(s) of the signal lamp(s) at least 10 ft (3 m) from the photometric screen. The H-V axis shall be taken as parallel to the longitudinal axis of the vehicle. When compartments or lamps are photometered together, the H-V axis shall intersect the midpoint between the optical centers (filament).

3.6.2 Beam candlepower measurements of multiple compartment lamp or multiple lamp arrangement shall be made by either of the following methods:

(a) All compartments or lamps may be photometered together provided that a line from the optical axis (filament centers) of each compartment or lamp to the center of the photometer sensing device does not make an angle of more than 0.6 deg with the photometer (H-V) axis.

(b) Each compartment or lamp may be photometered separately by aligning its axis with the photometer and adding the value at each test point.

3.6.3 Table 1 lists design candlepower requirements for a tail lamp.

4. Installation Requirements—The following requirements apply to the device as used on the vehicle and are not part of the laboratory test requirements and procedures.

Visibility of the tail lamp shall not be obstructed by any part of the vehicle throughout the photometric test angles for the lamp, unless the lamp is designed to comply with all photometric and visibility requirements with these obstructions considered. Signal from lamps on both sides of the vehicle shall be visible through a horizontal angle from 45 deg to the left to 45 deg to the right. Where more than one lamp or optical area is lighted on each side of the car, only one such area on each side need comply. To be considered visible, the lamp must provide an unobstructed projected illuminated area of outer lens surface, excluding reflex, at least 2 in^2 (12.5 cm^2) in extent, measured at 45 deg to the longitudinal axis of the vehicle.

APPENDIX

As a matter of information, attention is called to SAE J567c (December, 1970) for requirements and gages to be used in socket design.

TABLE 1—MINIMUM DESIGN CANDLEPOWER REQUIREMENTS

Test Points, deg		Lighted Sections		
		1	2	3
10U and 10D	10L	0.3	0.5	0.7
	V	0.5	1.0	1.5
	10R	0.3	0.5	0.7
5U and 5D	20L	0.3	0.5	0.7
	10L	0.8	1.3	2.0
	5L	1.3	2.0	3.0
	V	1.8	3.0	4.5
	5R	1.3	2.0	3.0
	10R	0.8	1.3	2.0
	20R	0.3	0.5	0.7
H	20L	0.4	0.7	1.0
	10L	0.8	1.3	2.0
	5L	2.0	3.5	5.0
	V	2.0	3.5	5.0
	5R	2.0	3.5	5.0
	10R	0.8	1.3	2.0
	20R	0.4	0.7	1.0
Maximum at H or above (See also note 5)		18	20	25

NOTES:
1. Specifications are based on laboratories using accurate, rated bulbs during testing.
2. Lamps designed for use in both 6V and 12V systems shall be tested with 12V bulbs. Lamps designed to operate on the vehicle through a resistor or equivalent shall be photometered with the listed design voltage of the design source applied across the combination of resistance and filament.
3. A multiple device tail lamp gives its indication by two or more separately lighted sections which may be separate lamps, or areas that are joined by common parts. The photometric values are to apply when all sections which provide the tail signal are considered as a unit except when the dimensions between optical centers exceed those given in paragraph 3.1. For a separate lamp arrangement, where lamps are interchangeable, each lamp should be of approximately the same performance.
4. When a tail lamp is combined with the turn signal lamp or stop lamp, the signal lamp or stop lamp shall not be less than three times the candlepower of the tail lamp at any test point; except that at H-V, H-5L, H-5R, and 5U-V, the signal lamp or stop lamp shall not be less than five times the candlepower of the tail lamp. If a multiple compartment or multiple lamp arrangement is used and the distance between optical axes for both the tail lamp and the turn signal or stop lamp functions is within the dimensions specified in paragraph 3.1, the ratios of the signal lamps or stop lamps to the tail lamp shall be computed with all the compartments or lamps lighted. If a multiple compartment or multiple lamp arrangement is used and the distance between optical axes for one of the functions exceeds the dimensions specified in paragraph 3.1, the ratio shall be computed for only those compartments or lamps where the tail lamp and turn signal or stop lamp are optically combined. When the tail lamp is combined with the turn signal lamp or stop lamp, and the maximum candlepower of the tail lamp is located below horizontal and within an area generated by a 0.5 deg radius around a test point, the ratio for the test point may be computed using the lowest value of the tail lamp candlepower within the generated area.
5. A tail lamp shall not exceed the maximum candlepower at night over any area larger than that generated by a 1/4 deg radius, within a solid cone angle from 20L to 20R and from H to 10U.

STOP LAMPS—SAE J586d   SAE Standard

Report of the Lighting Division approved February 1927 and completely revised by Lighting Committee September 1977. Rationale statement available.

1. Scope—This engineering design standard provides design parameters and general requirements for stop lamps. It is intended for use in conjunction with Service Performance Specification, SAE J256 (July, 1971).

2. Definitions

2.1 Stop Lamps—Lamps giving a steady light to the rear of a vehicle or train of vehicles to indicate the intention of the operator of a vehicle to stop or diminish speed by braking.

2.2 Multiple Compartment Lamp—A device which gives its indication by two or more separately lighted areas which are joined by one or more common parts such as a housing or lens.

2.3 Multiple Lamp Arrangement—An array of two or more separated lamps on each side of the vehicle which operate together to give a signal.

3. Laboratory Requirements

3.1 A multiple compartment lamp or multiple lamps may be used to meet the photometric requirements of a stop lamp. If a multiple compartment or multiple lamps are used and the distance between the optical axes (filament centers) does not exceed 22 in (560 mm) for two compartment or lamp arrangements and does not exceed 16 in (410 mm) for three compartment or lamp arrangements, then the combination of the compartments or lamps must be used to meet the photometric requirements for the corresponding number of lighted sections (Table 1). If the distance between optical axes exceeds the above dimensions, each compartment or lamp shall comply with the photometric requirements for one lighted section (Table 1).

For vehicles of 80 in (2032 mm) or more in overall width, a maximum of two lamps and/or compartments per side may be mounted closer together than 22 in (560 mm) providing that each compartment and/or lamp meets the single compartment photometric requirements listed in Table 1 and has a minimum effective projected luminous lens area of 8 in^2 (50 cm^2). Each lamp and/or compartment utilized in this manner shall meet the one lighted section value for all functions for which it is designed.

3.2 The effective projected luminous lens area of a single compartment lamp measured on a plane at right angles to the axis of a lamp must be at least 8 in^2 (50 cm^2).

3.3 If a multiple compartment lamp or multiple lamps are used to meet the photometric requirements of a stop lamp, the effective projected luminous lens area of each compartment or lamp shall be at least 3.5 in^2 (22 cm^2), provided the combined area is at least 8 in^2 (50 cm^2).

3.4 The following sections from SAE J575g (March, 1978) are a part of this standard:

3.4.1 SECTION 2—Samples for Tests
3.4.2 SECTION 2.2—Bulbs
3.4.3 SECTION 2.3—Test Fixture
3.4.4 SECTION 3—Laboratory Facilities
3.4.5 SECTION 4.1—Vibration Test
3.4.6 SECTION 4.2—Moisture Test
3.4.7 SECTION 4.3—Dust Test

3.4.8 SECTION 4.4—Corrosion Test
3.4.9 SECTION 4.6—Photometry
3.4.10 SECTION 4.8—Warpage Test, except that if the tail lamp and/or side marker lamp are incorporated in the same device as the stop lamp, they shall be operated continuously during the test.

3.5 Plastic Materials—Any plastic materials used in optical parts shall comply with the requirements set forth in SAE J576d (June, 1976).

3.6 Color Test—The color of the light from stop lamps shall be red. (See SAE J578c (February, 1977).)

3.7 If the stop lamp is optically combined with the tail lamp and a two-filament bulb is used, the bulb shall have an indexing base and the socket shall be designed so that bulbs with nonindexing bases cannot be used.

3.8 Photometric Requirements

3.8.1 All beam candlepower measurements shall be made with the incandescent filament(s) of the signal lamp(s) at least 10 ft (3 m) from the photometer screen. The H-V axis shall be taken as parallel to the longitudinal axis of the vehicle. When compartments or lamps are photometered together, the H-V axis shall intersect the midpoint between the optical center (filament).

3.8.2 Beam candlepower measurements of multiple compartment lamp or multiple lamp arrangements shall be made by either of the following methods:

(a) All compartments or lamps may be photometered together provided that a line from the optical center (filament) of each compartment or lamp to the center of the photometer sensing device does not make an angle of more than 0.6 deg with the photometer (H-V) axis.

(b) Each compartment or lamp may be photometered separately by aligning its axis with the photometer and adding the value at each test point.

3.8.3 Table 1 lists design candlepower requirements for a stop lamp.

4. Installation Requirements—The following requirements apply to the device as used on the vehicle and are not part of the laboratory test requirements and procedures.

4.1 Visibility of the stop lamp shall not be obstructed by any part of the vehicle throughout the photometric test angles for the lamp unless the lamp is designed to comply with all photometric and visibility requirements with these obstructions considered. Signal from lamps on both sides of the vehicle shall be visible through a horizontal angle from 45 deg to the left to 45 deg to the right. Where more than one lamp or optical area is lighted on each side of the car, only one such area on each side need comply. To be considered visible, the lamp must provide an unobstructed projected illuminated area of outer lens surface, excluding reflex, at least 2 in^2 (12.5 cm^2) in extent, measured at 45 deg to the longitudinal axis of the vehicle.

4.2 When a stop signal is optically combined with the turn signal, the circuit shall be such that the stop signal cannot be turned on in the turn signal which is flashing.

APPENDIX

As a matter of information, attention is called to SAE J567c (December, 1970) for requirements and gages to be used in socket design.

TABLE 1—MINIMUM DESIGN CANDLEPOWER REQUIREMENTS

Test Points, deg		Lighted Sections		
		1	2	3
10U and 10D	10L	10	12	15
	V	25	30	35
	10R	10	12	15
5U and 5D	20L	10	12	15
	10L	30	35	40
	5L	50	60	70
	V	70	82	95
	5R	50	60	70
	10R	30	35	40
	20R	10	12	15
H	20L	15	18	20
	10L	40	47	55
	5L	80	95	110
	V	80	95	110
	5R	80	95	110
	10R	40	47	55
	20R	15	18	20
Maximum		300	360	420

NOTES:
1. Specifications are based on laboratories using accurate, rated bulbs during testing.
2. Lamps designed for use in both 6V and 12V systems shall be tested with 12V bulbs. Lamps designed to operate on the vehicle through a resistor or equivalent shall be photometered with the listed design voltage of the design source applied across the combination of resistance and filament.
3. A multiple device signaling unit gives its indication by two or more separately lighted sections which may be separate lamps, or areas that are joined by common parts. The photometric values are to apply when all sections which provide the same signal are considered as a unit except when the dimensions between optical centers exceed those given in paragraph 3.1. For a separate lamp arrangement, where lamps are interchangeable, each lamp should be of approximately the same performance.
4. When a tail lamp is combined with the stop lamp, the stop lamp shall be not less than three times the candlepower of the tail lamp at any test point; except that at H-V, H-5L, H-5R, and 5U-V, the stop lamp shall not be less than five times the candlepower of the tail lamp. If a multiple compartment or multiple lamp arrangement is used and the distance between optical axes for both the tail lamp and stop lamp is within the dimensions specified in paragraph 3.1, the ratio of the stop lamp to the tail lamp shall be computed with all the compartments or lamps lighted. If a multiple compartment or multiple lamp arrangement is used and the distance between optical axes for one of the functions exceeds the dimensions specified in paragraph 3.1, the ratio shall be computed for only those compartments or lamps where the tail lamp and stop lamp are optically combined. When the tail lamp is combined with the stop lamp, and the maximum candlepower of the tail lamp is located below horizontal and within an area generated by a 0.5 deg radius around a test point, the ratio for the test point may be computed using the lowest value of the tail lamp candlepower within the generated area.
5. Stop lamps shall not exceed the listed maximum candlepower over any area larger than that generated by a 0.25 deg radius.

MECHANICAL STOP LAMP SWITCH—SAE J249

SAE Recommended Practice

Report of Lighting Committee approved February 1972.

1. Definition—A mechanical stop lamp switch is a mechanically operated device used to energize the stop lamp circuit with operator actuation of the brake pedal.

2. Temperature Test

2.1 To insure basic function, the switch shall be manually cycled for 10 cycles at design electrical load at:

75, ±10 F	(24, ±5.5 C)
165, +0, −5 F	(74, +0, −2.8 C)
−25, +5 F, −0 F	(−32, +2.8 C, −0 C)

This is to be done after a 1 h exposure at each of these temperatures. The switch shall be electrically and mechanically operable during each of these cycles.

2.2 This same switch shall be used for the endurance test described below.

3. Endurance Test Setup

3.1 The switch shall be set up to operate its design electrical load.

3.2 The test shall be set up to operate the switches for the prescribed number of completed cycles.

One complete cycle shall consist of energizing and de-energizing the stop lamps with switch travel as specified by the manufacturer. With the switch exercised through its complete travel, it shall be cycled as follows:

TRAVEL RATE—0.4–0.6 in./s (10.2–15.2 mm/s) (at make and break).
DWELL TIMES—1.0–2.0 s (circuit closed, lamps on).
1.0–2.0 s (circuit open, lamps off).

3.3 During the test the switch shall be operated at 6.4 V d-c for a 6 V system, 12.8 V d-c for a 12 V system, or 25.6 V d-c for a 24 V system. These voltages shall be measured at the input termination on the switch. The power supply shall not generate any adverse transients not present in motor vehicles and shall comply with the following specifications:

(a) *Output Current*—Capable of supplying the continuous current of the design electrical load and inrush current as required by the bulb load complement.

(b) *Regulation*—

Dynamic—The output voltage at the supply shall not deviate more than 1.0 V from 0 to maximum load (including inrush current) and should recover 63% of its maximum excursion within 100 ms.

Static—The output voltage at the supply shall not deviate more than 2%

with changes in static load from 0 to maximum (not including inrush current), and means shall be provided to compensate for static input line voltage variations.

(c) *Ripple Voltage*—Maximum 300 mV, peak to peak.

4. Endurance Requirements

4.1 The switch shall be capable of satisfactorily operating for 250,000 complete cycles at a temperature of 75 ± 10 F (24 ± 5.5 C).

4.2 The voltage drop from the input terminal(s) to the corresponding output terminal(s) shall be measured before and at the end of the endurance test and shall not exceed the following, using the average of three consecutive readings at design load:

0.30 V (4-lamp load or less)
0.40 V (more than a 4 lamp load)

If wiring is an integral part of the switch, the voltage drop measurement shall be made including 3 in. (76.2 mm) of wire on each side of the switch; otherwise, measurement to be made at the switch terminals.

LICENSE PLATE LAMPS—SAE J587f

SAE Standard

Report of Lighting Division approved March 1918 and last revised by Lighting Committee January 1977. Rationale statement available.

1. Scope—This is an Engineering Design Standard which contains design parameters and general requirements for vehicular license plate lamps. It may be augmented by a separate service performance specification.

2. Definition—A license plate lamp is used to illuminate the license plate on the rear of a vehicle. If a tail or a stop lamp is combined with a license plate lamp, the combination should also meet the requirements for these devices.

3. Laboratory Requirements—The following sections from SAE J575 are a part of this standard:

Section 2   —Samples for Test
Section 2.2—Bulbs
Section 2.3—Test Fixture
Section 3   —Laboratory Facilities
Section 4.1—Vibration Test
Section 4.2—Moisture Test
Section 4.3—Dust Test
Section 4.4—Corrosion Test
Section 4.8—Warpage Test on Devices with Plastic Components

4. Plastic Materials—Any plastic materials used in optical parts shall comply with the requirements set forth in SAE J576.

5. Color—The color of the light emitted from license plate lamps shall be white. (See SAE J578.)

6. General Requirements

6.1 The lamp design should be such that, when mounted on a vehicle as intended, the angle between the plane of the license plate and the plane on which the vehicle stands will not exceed 90 ± 15 deg.

6.2 The license plate holder shall be designed and constructed to provide a substantial plane surface on which to mount the plate.

6.3 License plate lamp(s) for vehicles other than motorcycles and motor driven cycles shall be of such size and design as to provide illumination on all parts of a 6 x 12 in (152 x 305 mm) test plate. License plate lamp(s) for motorcycles and motor driven cycles shall be of such size and design as to provide illumination of all parts of a 4 x 7 in (102 x 178 mm) test plate. The light rays shall reach all portions of an imaginary plate of the same size at least 1 in (25.4 mm) ahead of the actual test plate measured perpendicular to the plane of the plate.

6.4 The license plate lamp or lamps for vehicles other than motorcycles and motor driven cycles shall be mounted so as to illuminate the plate from the top or sides.

6.5 When a single lamp is used to illuminate the plate, the lamp and license plate holder shall bear such relation to each other that at no point on the plate will the incident light make an angle of less than 8 deg to the plane of the plate.

6.6 When two or more lamps are used to illuminate the plate, the minimum 8 deg incident light angle shall apply only to that portion of the plate which the particular lamp is designed to illuminate.

6.7 If a clear lens, a lens with partial optical treatment, or no lens is used, the 8 deg requirement set forth in 6.5 and 6.6 shall be measured from the bulb filament.

6.8 If an all-over optical treatment is used on the lens, the 8 deg requirement set forth in 6.5 and 6.6 shall be measured from the geometric center of the optical area.

6.9 All illumination measurements shall be made on a rectangular test plate of clean, white blotting paper mounted on the license plate holder in the position ordinarily taken by the license plate. The face of the test plate shall be 1/16 in (1.6 mm) from the face of the license plate holder.

6.10 For lamps used on vehicles other than motorcycles and motor driven cycles, the test stations shall be located on the face of the test plate as shown in Fig. 1. For lamps used on motorcycles and motor driven cycles, the test stations shall be located on the face of the test plate as shown in Fig. 2.

7. Photometric Test—The illumination of each of the stations on the test plate shall be at least 0.75 ft-c (8.1 lx). The ratio of maximum to minimum illumination shall not exceed 20/1 for the 6 x 12 in (152 x 305 mm) plate and shall not exceed 15/1 for the 4 x 7 in (102 x 178 mm) plate. The average of the two highest and the two lowest illumination values recorded at the eight test stations in the test plate of Fig. 1 shall be taken as maximum and minimum, respectively. The highest illumination value and the average of the two lowest illumination values recorded at the six test stations in the test plate of Fig. 2 shall be taken as maximum and minimum, respectively.

APPENDIX

A1—As a matter of information, attention is called to SAE J567 for requirements and gages to be used in socket design.

A2—Since some license lamps may be mounted in shaded locations, attention is particularly called to the section of SAE J576 which covers exposure time and conditions.

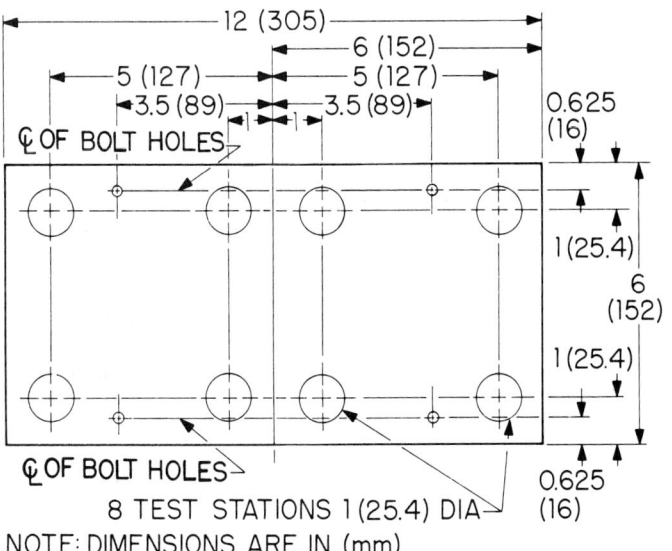

FIG. 1—TEST PLATE FOR VEHICLES OTHER THAN MOTORCYCLES AND MOTOR DRIVEN CYCLES

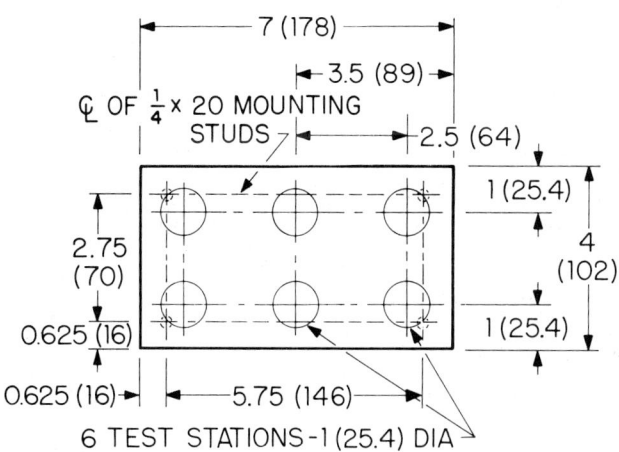

FIG. 2—TEST PLATE FOR MOTORCYCLES AND MOTOR DRIVEN CYCLES

TURN SIGNAL LAMPS—SAE J588f

SAE Standard

Report of Lighting Division approved February 1927 and completely revised by Lighting Committee June 1978. Editorial change November 1978. Rationale statement available.

1. Scope—This engineering design standard provides design parameters and general requirements for turn signal lamps. It is intended for use in conjunction with SAE J256b (March, 1978).

2. Definitions

2.1 Turn Signal Lamps—The signaling elements of a turn signal system which indicate a change in direction by giving a flashing light on the side toward which the turn will be made. (For flashing rate and *on* period, see SAE J590e.)

2.2 Multiple Compartment Lamp—A device which gives its indication by two or more separately lighted areas which are joined by one or more common parts such as a housing or lens.

2.3 Multiple Lamp Arrangement—An array of two or more separated lamps on each side of the vehicle which operate together to give a signal.

3. Laboratory Requirements

3.1 A multiple compartment lamp or multiple lamps may be used to meet the photometric requirements of a turn signal lamp. If a multiple compartment or multiple lamps are used and the distance between the optical axes (filament centers) does not exceed 560 mm (22 in) for two compartment or lamp arrangements and does not exceed 410 mm (16 in) for three compartment or lamp arrangements, then the combination of the compartments or lamps must be used to meet the photometric requirements for the corresponding number of lighted sections (Table 1). If the distance between optical axes exceeds the above dimensions, each compartment or lamp shall comply with the photometric requirements for one lighted section (Table 1).

For vehicles of 2032 mm (80 in) or more in overall width, a maximum of two lamps and/or compartments per side may be mounted closer together than 560 mm (22 in) providing that each compartment and/or lamp meets the single compartment photometric requirements listed in Table 1 and has a minimum effective projected luminous lens area 50 cm² (8 in²). Each lamp and/or compartment utilized in this manner shall meet the one lighted section value for all functions for which it is designed.

3.2 The effective projected luminous area of a single compartment lamp measured on a plane at right angles to the axis of a lamp must be at least 50 cm² (8 in²) for a rear lamp and at least 22 cm² (3.5 in²) for a front lamp.

3.3 If a multiple compartment lamp or multiple lamps are used to meet the photometric requirements of a rear turn signal lamp, the effective projected luminous lens area of each compartment or lamp shall be at least 22 cm² (3.5 in²) provided the combined area is at least 50 cm² (8 in²).

3.4 The flashing signal from a double-faced signal lamp shall not be obliterated when subjected to external light rays from either in front or behind, at any and all angles within the required visibility angles of the signal.

3.5 The following sections from SAE J575g (September, 1977) are a part of this standard:

3.5.1 Section 2—Samples for Test.

3.5.2 Section 2.2—Bulbs.

3.5.2.1 Sealed lighting units, as described in SAE J760a (December, 1974), shall be operated at their design voltage. They shall be seasoned at rated voltage for 1% of design life or 10 h maximum before photometry.

3.5.3 Section 2.3—Test Fixture.

3.5.4 Section 3—Laboratory Facilities

3.5.5 Section 4.1—Vibration Test.

3.5.6 Section 4.2—Moisture Test.

3.5.7 Section 4.3—Dust Test.

3.5.8 Section 4.4—Corrosion Test.

3.5.9 Section 4.5—Color Test.

3.5.9.1 The color of light from the turn signal lamps shall be red or yellow to the rear and yellow to the front of the vehicle (see SAE J578c (February, 1977)).

3.5.10 Section 4.6—Photometry.

3.5.11 Section 4.8—Warpage Test on Devices with Plastic Components.

3.6 Sealed lighting units (as described in SAE J760a (December, 1974), when tested without the other parts of the lamp assembly, need only comply with paragraphs 2, 2.3, 3, 4.5 and 4.6 of SAE J575g (September, 1977).

3.7 Plastic Materials—Any plastic materials used in optical parts shall comply with the requirements set forth in SAE J576d (June, 1976).

3.8 If the turn signal is optically combined with the tail lamp and a two-filament bulb is used, the bulb shall have an indexing base and the socket shall be designed so that bulbs with nonindexing bases cannot be used.

3.9 Photometric Requirements

3.9.1 Rear signals from double-faced turn signal lamps need only meet the photometric requirements in Table 1 from directly to the rear to the left for a left lamp and from directly to the rear to the right for a right lamp. (The intent of the foregoing sentence is to permit the manufacturer to provide glare protection for the driver.)

3.9.2 Photometric measurements shall be made with the incandescent filament(s) of the signal lamp(s) at least 3 m (10 ft) from the photometer screen. The H-V axis shall be taken as parallel to the longitudinal axis of the vehicle. When compartments or lamps are photometered together, the H-V axis shall intersect the midpoint between the optical centers (filament).

3.9.3 Photometric measurements of multiple compartment lamp or multiple lamp arrangements shall be made by either of the following methods:

3.9.3.1 All compartments or lamps may be photometered together provided that a line from the optical center (filament) of each compartment or lamp to the center of the photometer sensing device does not make an angle of more than 0.6 deg with the photometer (H-V) axis.

3.9.3.2 Each compartment or lamp may be photometered separately by aligning its axis with the photometer. The photometric measurement for the entire multiple compartment lamp or multiple lamp arrangement can then be determined by adding the photometric outputs from each individual lamp or component at corresponding test points.

4. Installation Requirements—The following requirements apply to the device as used on the vehicle and are not part of the laboratory test requirements and procedures:

TABLE 1—MINIMUM DESIGN PHOTOMETRIC REQUIREMENTS (cd)

Test Points (Deg)		Front Signals—Yellow						Rear Signals					
		Spacing of Turn Signal From Lighted Edge of Low Beam Unit—See Section 4.2						Red			Yellow		
		(A) 100 mm (4 in) or more			(B) less than 100 mm (4 in)								
Lighted Sections		1	2	3	1	2	3	1	2	3	1	2	3
10U&D	10L&R	25	30	35	60	75	85	10	12	15	15	20	25
	V	60	75	90	150	185	225	25	30	35	40	50	55
5U&D	20L&R	25	30	35	60	75	85	10	12	15	15	20	25
	10L&R	75	88	100	185	220	250	30	35	40	50	55	65
	5L&R	125	150	175	310	375	435	50	60	70	80	100	110
	V	175	205	235	435	510	585	70	82	95	110	130	150
H	20L&R	35	45	50	85	110	125	15	18	20	25	30	35
	10L&R	100	120	140	250	300	350	40	47	55	65	75	90
	5L&R	200	240	275	500	600	685	80	95	110	120	150	175
	V	200	240	275	500	600	685	80	95	110	130	150	175
Maximum—Rear Lamps Only		—	—	—	—	—	—	300	360	420	750	900	900

NOTES:

1. Specifications are based on laboratories using accurate, rated bulbs during testing.
2. Lamps designed for use with both 6V and 12V bulbs shall be tested with 12V bulbs. Lamps designed to operate on the vehicle through a resistor or equivalent shall be photometered with the listed design voltage of the design source applied across the combination of resistance and filament.
3. A multiple device signaling unit gives its indication by two or more separately lighted sections which may be separate lamps, or areas that are joined by common parts. The photometric values are to apply when all sections which provide the same signal are considered as a unit except when the dimensions between optical centers exceed those dimensions given in paragraph 3.1. For a separate lamp arrangement, where lamps are interchangeable, each lamp should be of approximately the same performance.
4. When a tail lamp or parking lamp is combined with the turn signal lamp, the signal lamp shall not be less than three times the candlepower (a) of the tail lamp at any test point, or (b) of the parking lamp at any test point on or above horizontal; except that at H-V, H-5L, H-5R, and 5U-V, the signal lamp shall not be less than five times the candlepower of the tail lamp or parking lamp. If a multiple compartment or multiple lamp arrangement is used and the distance between optical axes for both the tail lamp (parking lamp) and the turn signal is within the dimensions specified in paragraph 3.1, the ratio of the turn signal to the tail lamp (parking lamp) shall be computed with all the compartments or lamps lighted. If a multiple compartment or multiple lamp arrangement is used and the distance between optical axes for one of the functions exceeds the dimensions specified in paragraph 3.1, the ratio shall be computed for only those compartments or lamps where the tail lamp (parking lamp) and turn signal are optically combined. Where the tail lamp is combined with the turn signal lamp, and the maximum candlepower of the tail lamp is located below horizontal and within an area generated by an 0.5 deg radius around a test point, the ratio for the test point may be computed using the lowest value of the tail lamp candlepower within the generated area.
5. Lamps intended for the rear of a vehicle shall not exceed the listed maximum candlepower over any area larger than that generated by an 0.25 deg radius.

4.1 Signal lamps on the front and the rear shall be spaced as far apart laterally as practicable, so that the direction of turn will be clearly understood.

4.2 The minimum design photometric requirements of Column (A) of Table 1 shall apply when the optical axis (filament center) of the front turn signal is at a spacing of at least 100 mm (4 in) from the lighted edge of the headlamp unit providing the lower beam, or from the lighted edge of any additional lamps used to supplement (such as fog lamps and/or auxiliary low beam lamps), or used in lieu of, the lower beam. If such spacing is less than 100 mm (4 in), the minimum design photometric requirements of Column (B) of Table 1 shall apply.

4.3 Visibility of the front signal to the front, and the rear signal to the rear, shall not be obstructed by any part of the vehicle throughout the photometric test angles for the lamps unless the lamp is designed to comply with all photometric and visibility requirements with these obstructions considered. In addition, signals from lamps mounted on the left side of the vehicle shall be visible through a horizontal angle of 45 deg to the left, and signals from lamps mounted on the right side of the vehicle shall be visible through a horizontal angle of 45 deg to the right. To be considered visible, the lamp must provide an unobstructed effective projected illuminated area of outer lens surface, excluding reflex, at least 12.5 cm^2 (2 in^2) in extent measured at 45 deg to the longitudinal axis of the vehicle.

Except that on combinations of vehicles, signal lamps on the rear of other than the rearmost vehicle shall be visible from not less than 20 deg to the left to 45 deg to the left for the left signal and from not less than 20 deg to the right to 45 deg to the right for the right signal.

4.4 When a stop signal is optically combined with the turn signal, the circuit shall be such that the stop signal cannot be turned on if the turn signal is flashing.

4.5 Turn Signal Pilot Indicator

4.5.1 If one right and one left turn signal is not readily visible to the driver, there shall be an illuminated indicator provided to give a clear and unmistakable indication that the turn signal system is activated. The illuminated indicator shall consist of one or more lights flashing at the same frequency as the signal lamps.

4.5.2 If the illuminated indicator is located inside the vehicle, for example in the instrument cluster, it should emit a green colored light and have a minimum area of 18 mm^2 (0.028 in^2).

4.5.3 If the illuminated indicators are located on the outside of the vehicle, for example on the front fenders, they should emit a yellow colored light and have a minimum projected illuminated area of 60 mm^2 (0.1 in^2).

4.5.4 The minimum required illuminated area of the indicators specified in paragraphs 4.5.2 and 4.5.3 shall be visible according to the procedures described in SAE J1050a (January, 1977), Describing and Measuring the Driver's Field of View. The steering wheel shall be turned to a straight-ahead driving position and in the design location for an adjustable wheel or column.

APPENDIX

The appendix contains additional information not considered to be a part of this specification.

A1. Attention is called to SAE J567c (December, 1970) for requirements and gauges to be used in socket design.

A2. SAE J589b (October, 1977), Turn Signal Switch.

HEADLAMP-TURN SIGNAL SPACING—SAE J1221

SAE Information Report

Report of Lighting Committee approved March 1978. Editorial change November 1978.

1. Scope—This information report is being issued to cover the principles and practical problems involved in locating automotive front turn signals with respect to the headlamp unit which contains the low beam; it also records testing and results obtained, which led to new criteria for turn signal location.

2. Background—For a substantial period of time, SAE J588f (November, 1978), Turn Signal Lamps, has provided that, "The optical axis (filament center) of the front turn signal shall be at least 4 in from the inside diameter of the retaining ring of the headlamp unit providing the lower beam." More recently it was recognized that the 4 in requirement could realistically be waived if the turn signal intensity values were at least 2.5 times the otherwise required minimum output. Significant in the foregoing is the fact that optical center and filament center were said to be the same. As long as lamp bulbs were located in the centers of lamps, use of the filament center for the 4 in measurement was convenient and valid. See Fig. 1A. With the increase in bumper size, decrease in front end area, etc., on some car models, a practice was instituted to place the bulb toward one end of the lamp occasionally but to continue to measure off the 4 in requirement to the filament center. See Fig. 1B. Since, in such cases, the optical center had shifted, measurement to the filament center continued to be legal but did not provide for the same lamp performance. To further complicate the situation, dimensional limitations, in many cases, made it extremely difficult to provide a large enough turn signal lamp to produce at least 2.5 times the minimum requirements. Both functional and practical problems, therefore, were in need of better handling.

Early attempts to redefine the optical center included measurement of the unobstructed lamp output at H-V and then moving an opaque shield across the lamp face until the H-V output was reduced to 50% of the original value. Field tests of turn signal lamp performance established the validity of this procedure for determining the point (or line) to be used for the 4 in measurement. Lamp performance, so viewed, became relatively independent of the bulb placement. This method was not adopted, however, since it could only be used after *completion* of design and tooling. What was needed was a method for determining optical center reasonably closely when the vehicle and lamp were still in the design stage.

3. Establishment of Method—Outdoor tests were run on November 1, 1977 which proved the realism of using the geometric centroid of the turn signal lens as the point from which the measurement to the low beam headlamp unit should be made. Test equipment and procedures used are covered in detail in the Appendix to this information report.

4. Test Results—Based upon recognition distance equivalent to that of a lamp whose centroid is 4 in away from the inside edge of the retaining ring of the headlamp unit containing the lower beam, measurement standards are as follows:

4.1 Lamps whose centroids are 4 in (100 mm) or more from the lower beam unit must produce at least the minimum intensity values listed in SAE J588f (November, 1978).[1]

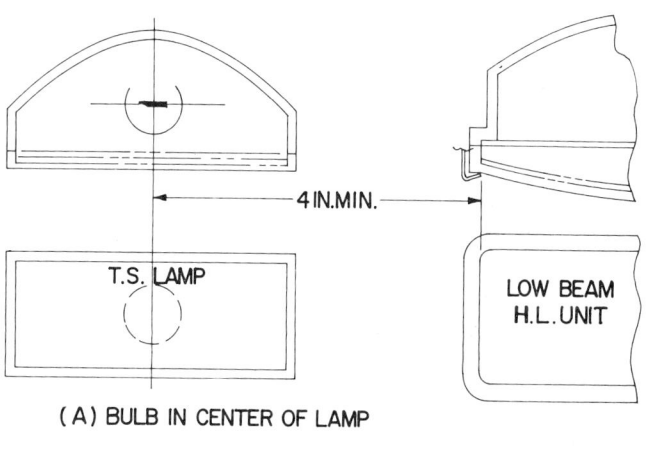

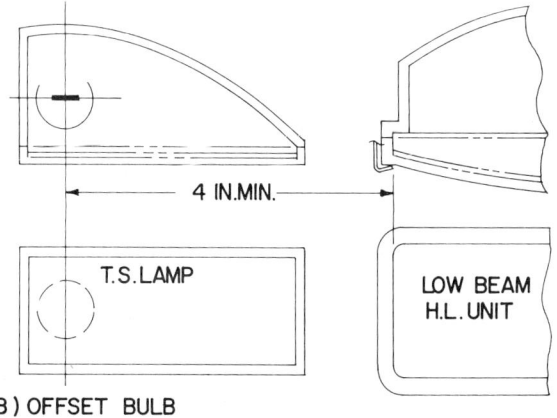

FIG. 1—BULB-HEADLAMP SPACING

4.2 Lamps whose centroids are 3 in (75 mm) and less than 4 in (100 mm) from the lower beam unit must produce at least 1.5 times the minimum intensity values listed in SAE J588f (November, 1978).[2]

[1] Already in effect at time of November 1 test.

[2] Additional, optional criteria established by November 1 test.

4.3 Lamps whose centroids are 2.5 in (60 mm) and less than 3 in (75 mm) from the lower beam unit must produce at least 2 times the minimum intensity values listed in SAE J588f (November, 1978).[2]

4.4 Spacing requirements are waived for those lamps which produce at least 2.5 times the minimum intensity values listed in SAE J588f (November, 1978).[1]

5. *Examples of Measurements* — The illustrations in Fig. 2 show common examples of headlamp-turn signal relative locations. Since all possible configurations cannot be shown, administration of the measurement requirement should be conducted with reason rather than as a narrow, mathematical exercise. In Fig. 2E, for example, a common lamp construction is shown which

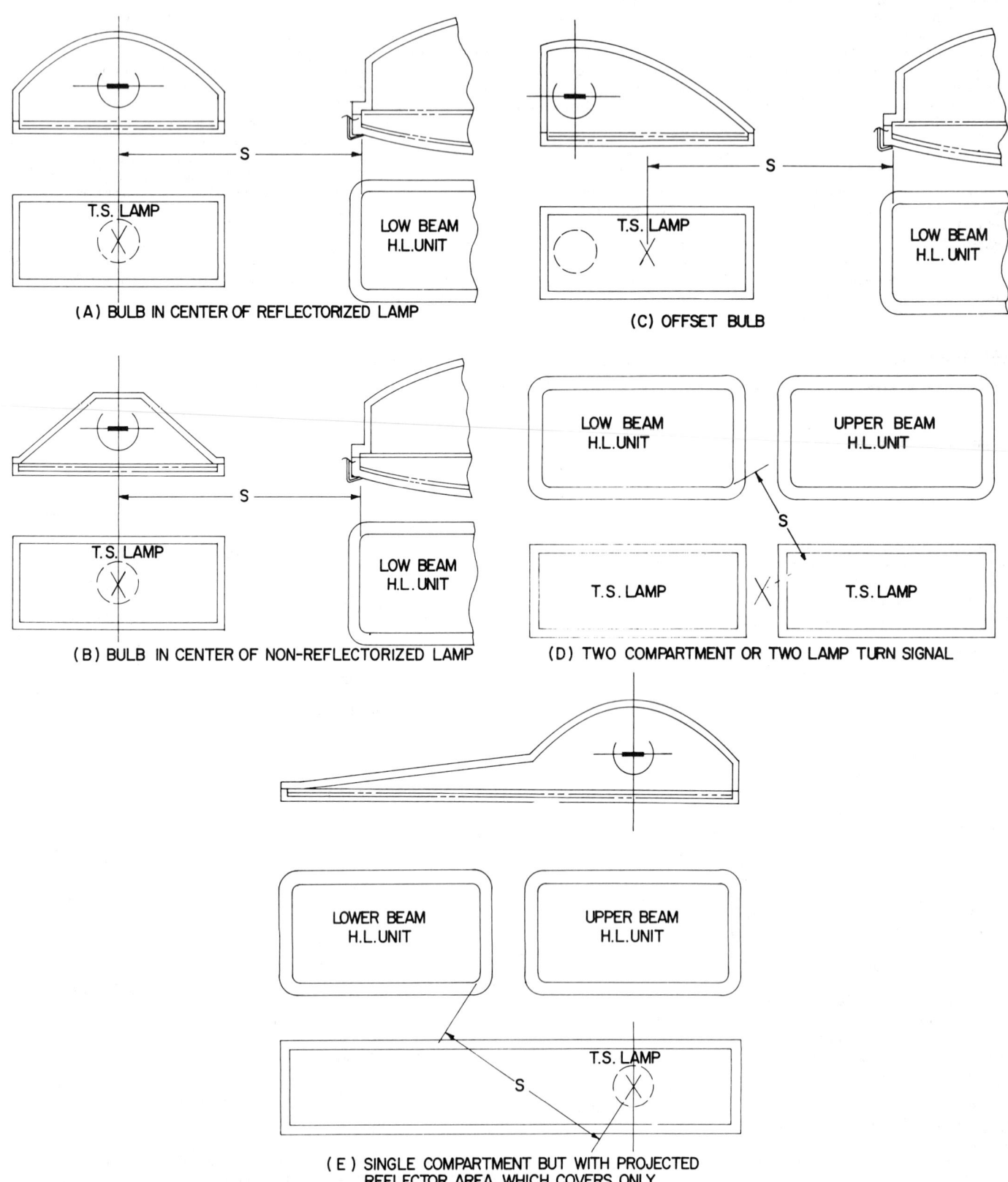

FIG. 2—MEASUREMENT EXAMPLES

illustrates the use of direct filament light for cosmetic purposes. Since the principal mass of light is that coming from the reflector, the centroid should be located on the lens with respect to the effective projected reflector area. Tests on a typical lamp having this type construction have verified this concept by showing that only 5–8% of the total light comes from that portion of the lens (the cosmetic portion) that is outside of the projected reflector area.

Although not shown in Fig. 2, measurements have also indicated the validity of measuring from the lower beam headlamp unit to a point on the lens which is directly in front of the bulb filament, in the case of a non-reflectorized lamp which has Fresnel optics in the lens. S is used in the Fig. 2 examples to show the spacing to points from which measurement should be made; **X** marks the centroid locations.

APPENDIX
FRONT TURN SIGNAL OPTICAL AXIS OBSERVATIONS

A1. Objective—To establish that using the centroid of the lighted portion of the lens designed for the signal function is equivalent to using the 50% H-V point[3] on the lens as the optical axis.

A2. Evaluation—Observers will determine the distance from an approaching car that turn signal operation can be identified. The observers will start timing watches as soon as they can clearly identify that the turn signal lamp is flashing, stop the watch as the turn signal lamp passes their station, and record the elapsed time.

A3. Procedure—(A similar type procedure and lamps used in the SAE lighting observations October 27, 1976 will be used for this observation.)

A3.1 The turn signal lamp will be mounted outboard of the Type 2A sealed unit on the left hand side of the test car.

A3.2 All observations will be made with the test car headlamps on low beam and the turn signal lamp flashing continuously at 90 flashes/min with 50% on time.

[3] 50% H-V point was identified in the second paragraph of Section 2 (Background).

A3.3 A 4-in spacing between the headlamp and either the centroid of the lens or the 50% H-V point on the lens will be used.

A3.4 The observers will be seated in stationary passenger cars spaced at approximately 300 ft (91 m) intervals, along the two land road with the lower beam headlamps illuminating the foreground.

A3.5 The test car will be driven at 20 mph past the observer cars so that all observers will have approximately 600–800 ft (183–244 m) clear view of the test car as it approaches their station.

A4. Test Conditions

A4.1 Yellow turn signal lamps using 1157 NA bulbs will be used at signal intensity of 200 cd uniform within the 10 deg zone.

A4.2 Both symmetrical and offset (non-symmetrical) design lamps shown in this report will be used. Both lamps are the same physical size having 12 in² of lens surface.

A4.3 All observations will be randomized and each condition will be observed two times during the observation period.

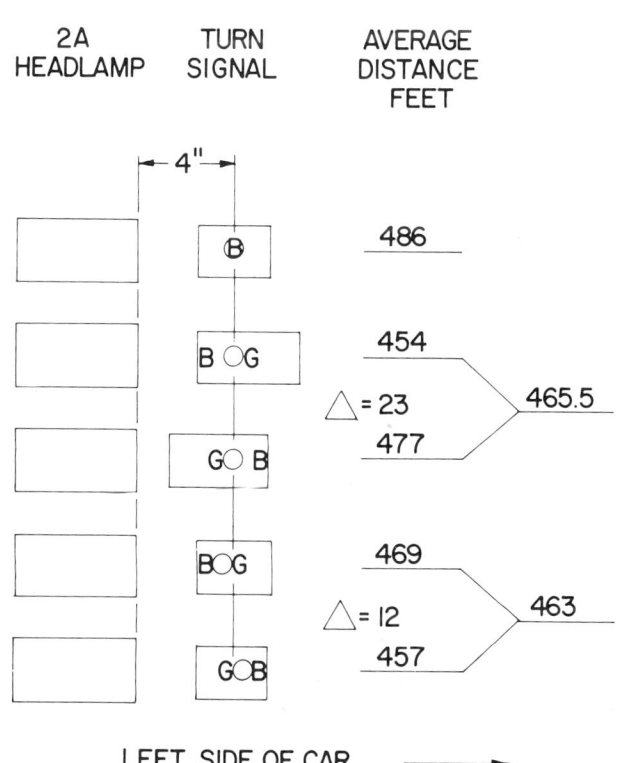

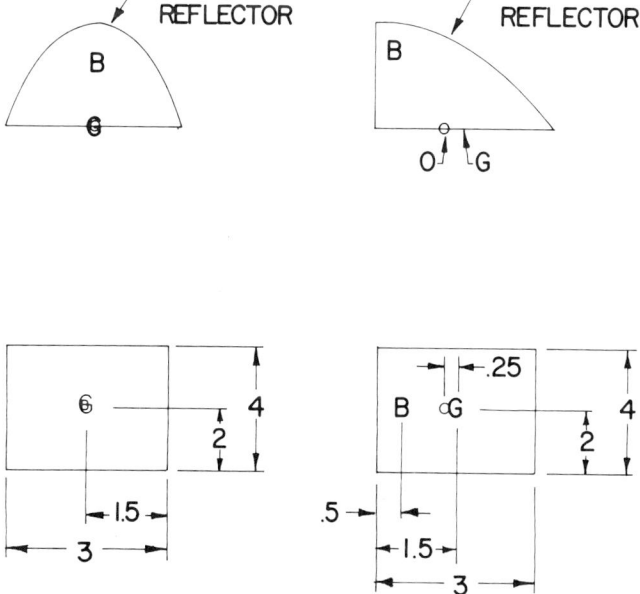

A5. Observation Results—The results of the observations are illustrated. These results are the averages of those determined by 34 observers. The evaluation was made on a straight, concrete test road under a moonlit sky, in a calm ambient temperature of approximately 55°F (13°C).

A6. Observation Conclusions—Using the geometric centroid of the lighted portion of the lens designed for the signal function is somewhat better than using the 50% H-V point on the lens, since recognition distances have lesser differences as a function of bulb placement.

FRONT TURN SIGNAL TO HEADLAMP SPACING OBSERVATIONS

B1. Objective—To determine front turn signal recognition distance relative to spacing and candela output to establish intermediate minimum design candela between the present 200–500 cd requirements relative to spacing between the headlamp and front turn signal.

B2. Evaluation—Observers will determine the distance from an approaching car that turn signal operation can be identified. The observers will start timing watches as soon as they can clearly identify that the turn signal lamp is flashing, stop the watch as the turn signal lamp passes their station, and record the elapsed time.

B3. Procedure—(A similar type procedure and lamps used in the SAE lighting observations October 27, 1976 will be used for this observation.)

B3.1 The turn signal lamp will be mounted outboard of the Type 2A sealed unit on the left hand side of the test car.

B3.2 All observations will be made with the test car headlamps on low beam and the turn signal lamp flashing continuously at 90 flashes/min with 50% on time.

B3.3 The spacing of the turn signal lamp to the headlamp will be varied.

B3.4 The observers will be seated in stationary passenger cars spaced at approximately 300-ft intervals, along the two lane road with the lower beam headlamps illuminating the foreground.

B3.5 The test car will be driven at 20 mph past the observer cars so that all observers will have approximately 600–800 ft clear view of the test car as it approaches their station.

B4. Test Conditions

B4.1 A yellow turn signal lamp of the Symmetrical Design shown in the illustration using an 1157 NA bulb will be used at signal intensities of 200, 300, and 400 cd uniform within the 10 deg zone.

B4.2 Spacing of the signal lamp to the headlamp—2.5, 2.75, 3.0, and

FRONT TURN SIGNAL TO HEADLAMP SPACING SUMMARY

Spacing Used, in	Cd	Average Observed Distance, ft (m)
4.00	200	453 (138)
3.50	300	467 (142)
3.00	300	451 (138)
2.50	300	433 (132)
3.00	400	498 (152)
2.75	400	467 (142)
2.50	400	462 (141)

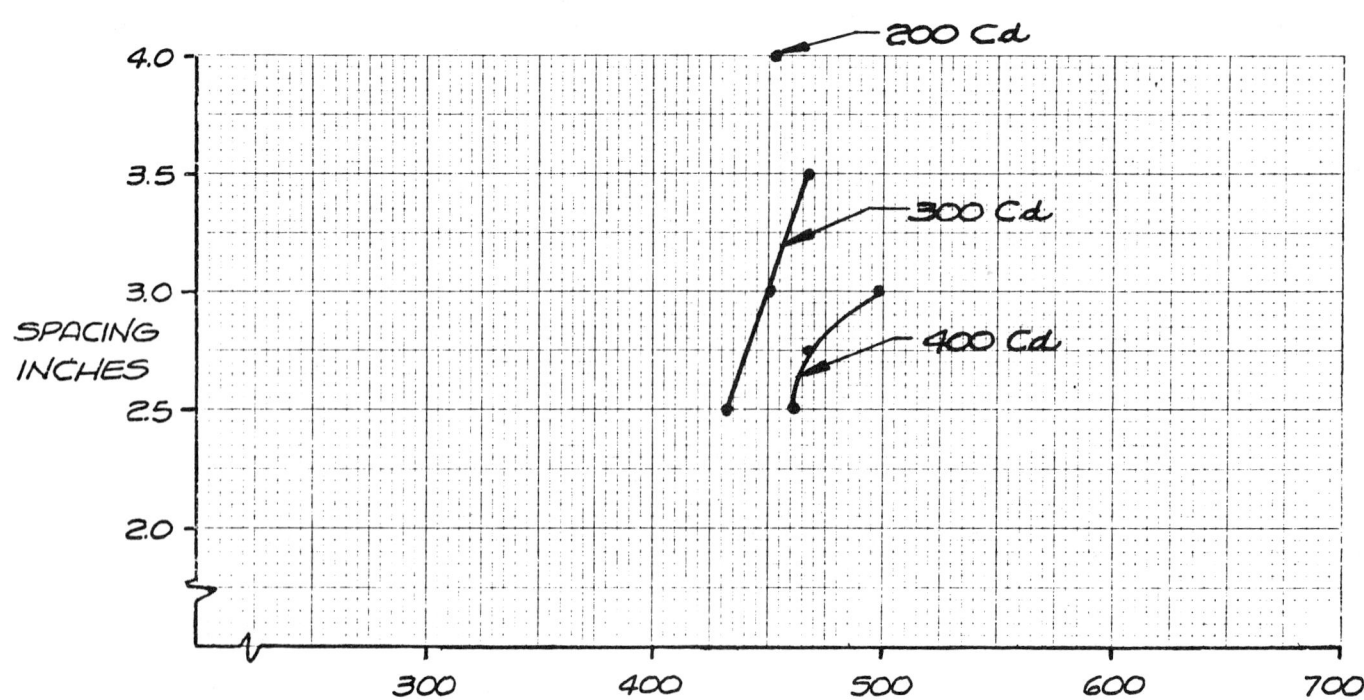

3.5 in for 300 cd and 400 cd observations, and 4 in for the 200 cd observations.

B4.3 All observations will be randomized and each condition will be observed twice during the observation period.

B5. Observation Results—The average findings of 34 observers are shown. The evaluation was made on the straight concrete test road under a moonlit sky, in a calm ambient temperature of approximately 55°F (13°C).

B6. Observation Conclusions—Based on equivalent recognition distance of the lamp at 4 in spacing and emitting 200 cd, the following is concluded:

B6.1 Lamps spaced between 3.0–4.0 in from the lower beam headlamp should emit at least 300 cd, or 1.5 times the minimum candela specified in SAE J588f (November, 1978).

B6.2 Lamps spaced between 2.5–3.0 in from the lower beam headlamp should emit at least 400 cd, or two times the minimum candela specified in SAE J588f (November, 1978).

B6.3 Lamps spaced closer than 2.5 in to the lower beam headlamp should emit at least 500 cd, or 2.5 times the minimum candela specified in SAE J588f (November, 1978).

This relationship is presented in graphical form.

ɸ SUPPLEMENTAL HIGH MOUNTED STOP AND REAR TURN SIGNAL LAMPS—SAE J186a

SAE Recommended Practice

Report of Lighting Committee approved July 1970 and completely revised September 1977. Rationale statement available.

---

1. Scope—This SAE Recommended Practice provides design parameters and general requirements for high mounted lamps intended to supplement stop and/or rear turn signal lamps described in SAE J586c (September, 1977) and J588 (November, 1977).

2. Definitions

2.1 Supplemental high mounted stop and rear turn signal lamps are additional lamps that are mounted high and possibly forward of the rear mounted tail, stop, and turn signal lamps. The supplemental stop and turn signals may be provided by separate lamps or combined in a single lamp.

2.2 The supplemental stop lamp(s) are additional lamp(s) of a stop lamp system giving a steady warning light to the rear of the vehicle and are intended to provide a signal through intervening vehicles to operators of following vehicles.

2.3 The supplemental rear turn signal lamps are additional lamps of a turn signal system, which indicate a change in direction by giving a flashing warning signal on the side toward which the vehicle operator intends to turn and are intended to provide a signal through intervening vehicles to operators of following vehicles.

3. Laboratory Requirements

3.1 The following sections of SAE J575g (March, 1978) are a part of this recommended practice:

3.1.1 SECTION 2—Samples for Test
3.1.2 SECTION 2.2—Bulbs
3.1.3 SECTION 2.3—Test Fixture
3.1.4 SECTION 3—Laboratory Facilities
3.1.5 SECTION 4.1—Vibration Tests
3.1.6 SECTION 4.2—Moisture Test
3.1.7 SECTION 4.3—Dust Test
3.1.8 SECTION 4.4—Corrosion Test
3.1.9 SECTION 4.5—Color Test
  3.1.9.1 *Color Test*—The light from the supplemental stop and/or turn signal lamps shall meet the same color requirements as the required lamps. See SAE J578c (February, 1977).
3.1.10 SECTION 4.6—Photometry
  3.1.10.1 *Photometric Test*—Photometric tests shall be made with the photometer at a distance of at least 10 ft (3 m) from the lamp. In measuring distances and angles, the center of the light emitting area shall be taken as the light source. The lamp axis shall be taken as the horizontal line through the light source and parallel to what would be the longitudinal axis of the vehicle if the lamp were mounted in its normal position on the vehicle.
  3.1.10.2 Photometric requirements for supplemental stop and turn signal lamps are shown in Table 1.
3.1.11 SECTION 4.8—Warpage Test

TABLE 1—MINIMUM DESIGN PHOTOMETRIC REQUIREMENTS

Test Points		Red (cd)	Yellow (cd)
10U	10L	5	8
	V	10	16
	10R	5	8
5U and 5D	10L	10	16
	5L	15	24
	V	15	24
	5R	15	24
	10R	10	16
H	10L	10	16
	5L	15	24
	V	15	24
	5R	15	24
	10R	10	16
Maximum		60[a]	120[a]

[a] The lamp shall not exceed the listed maximum over an area larger than that generated by a 1/4 deg radius within a solid cone angle from 10L to 10R and from 10U to 5D.

3.2 Plastic Materials—Any plastic materials used in optical parts shall comply with the requirements set forth in SAE J576d (June, 1976).

4. Installation Requirements—The following requirements apply to the device as used on the vehicle and are not a part of the laboratory test requirements and procedures.

4.1 Visibility of the signal shall not be obstructed by any part of the vehicle from 10U to 5D and from 10L to 10R unless the lamp is designed to comply with all requirements when the obstruction is considered.

4.2 Supplemental turn signals shall flash simultaneously (not alternately) with the required turn signals.

4.3 No function other than red reflex reflectors shall be combined in the supplemental high mounted stop and/or rear turn signal lamps.

APPENDIX

As a matter of information, attention is called to SAE J567c (December, 1970) for requirements and gages to be used in socket design.

This SAE Recommended Practice is intended as a guide toward standard practice, and it may be supplemented with an SAE Service Performance Standard.

TURN SIGNAL SWITCH—SAE J589b

SAE Standard

Report of Lighting Committee approved September 1950 and last revised June 1971. Editorial change October 1977.

1. Definition

1.1 A turn signal switch is that part of a turn signal system by which the operator of a vehicle causes the turn signal lamps to function.

1.2 A *Class A* turn signal switch may be used on any vehicle but is intended for use on multipurpose passenger vehicles, trucks, and buses that are 80 in. or more wide overall.

1.3 A *Class B* turn signal switch is intended for use in passenger cars, motorcycles, and multipurpose passenger vehicles, trucks, and buses of less than 80 in. overall width.

2. Reference Standards

2.1 The following sections from SAE J575f (April, 1975) are a part of this standard:
Section B—Samples for Test
Section C—Lamp Bulbs
Section D—Laboratory Facilities

2.2 Turn signal pilot indicators—See SAE J588e (September, 1970).

3. Temperature Test

3.1 To insure basic function, the switch shall be manually cycled for 10 cycles at design electrical load at: 75 ± 10 F (24 ± 5.5 C); $165 +0, -5$ F ($74 +0, -2.8$ C); $-25 +5, -0$ F ($-32 +2.8, -0$ C). This to be done after a 1 h exposure at each of these temperatures. The switch shall be electrically and mechanically operable during each of these cycles.

3.2 This same switch shall be used for the endurance test described in paragraph 4.

4. Endurance Test Setup

4.1 The switch shall be operated with the maximum design bulb load stated by the switch manufacturer with the flasher not included in the circuit. Failed bulbs shall be replaced during the test.

4.2 When the switch is provided with a self-canceling mechanism, the test equipment shall be arranged so that the switch can be turned off by the self-canceling mechanism. Provision shall also be made for manual canceling.

4.3 The test shall be set up to operate the switch for the prescribed number of cycles.

One cycle shall consist of the following sequence of positions: off, left turn, off, right turn, off.

The test requirement shall function within the following mechanical timing requirements at a cycle rate of 12–20 cycles/minute:
Travel time—0.1–0.5 s max (time from one position to the next position)
Dwell time—0.4 s min (in each position)

4.4 During the test the switch shall be operated at 6.4 V d-c for a 6 V system, 12.8 V d-c for a 12 V system, or 25.6 V d-c for a 24 V system, measured at the input termination of the switch. The power supply shall not generate any adverse transients not present in motor vehicles and shall comply with the following specifications:

(a) Output current—Capable of supplying a continuous output current of the design load and inrush currents as required by the bulb load complement.

(b) Regulation—
Dynamic—The output voltage shall not deviate more than 1.0 V from zero to maximum load (including inrush current) and should recover 63% of its maximum excursion within 5 ms.
Static—The output voltage shall not deviate more than 2% with changes in static load from zero to maximum (not including inrush current), and means shall be provided to compensate for static input line voltage variations.

(c) Ripple voltage—Maximum 300 mV, peak to peak.

5. Endurance Requirements

5.1 Class A turn signal switches shall be capable of meeting the following endurance requirements:

(a) 165,000 cycles at 75 ± 10 F (24 ± 5.5 C).

(b) When the switch is provided with a self-canceling mechanism it shall be tested as follows: 155,000 cycles of self-canceling followed by 10,000 cycles of manual canceling.

(c) If the turn signal switch includes stop lamp circuitry, the stop lamp circuit shall be fed electrically for the first 100,000 cycles only.

5.2 Class B turn signal switches shall be capable of meeting the following endurance requirements:

(a) 100,000 cycles at 75 ± 10 F (24 ± 5.5 C).

(b) When the switch is provided with a self-canceling mechanism it shall be tested as follows: 95,000 cycles of self-canceling followed by 5000 complete cycles of manual canceling.

(c) If the turn signal switch includes stop lamp circuitry, the stop lamp circuit shall be fed electrically for the first 50,000 complete cycles only.

5.3 If the turn signal switch includes cornering light circuitry which is fed from the headlight switch, the cornering light circuit shall be fed electrically for the first 50,000 cycles only.

5.4 The voltage drop from the input terminal of each circuit to the lamp terminal of each circuit shall be measured at the beginning of the test and at intervals of 25,000 cycles.

This voltage drop shall not exceed:
0.25 V for 2 lamp load (or less) per side
0.30 V for 3 lamp load per side
0.35 V for 4 lamp load per side
0.40 V for 5 lamp load (or greater) per side
before, during, and after the endurance test.

If wiring is an integral part of the switch, the voltage drop measurement is to be made including 3 in. of wire on each side of switch; otherwise, measurement is to be made at switch terminals. Care shall be taken not to include the voltage drop of other devices in the circuit.

6. Combination Turn Signal and Hazard Warning Signal Switches

6.1 The same combination switch shall be used for the test of each function. The turn signal switch function shall meet the requirements of this standard. The hazard warning signal switch function shall meet the requirements of SAE J910b (June, 1971).

6.2 The operating motion of the hazard warning signal switch function shall differ from the actuating motion of the turn signal switch function.

SIDE TURN SIGNAL LAMPS—SAE J914b

SAE Recommended Practice

Report of Lighting Committee approved February 1965 and last revised July 1978.

1. Scope—This engineering design specification provides parameters and general requirements for side turn signal lamps. It may be supplemented by a service performance requirement.

2. Definition—Side turn signal lamps are lighting devices mounted on the side at or near the front of a vehicle and are used as part of the turn signal system to indicate a change in direction by giving a flashing warning signal on the side toward which the vehicle operator intends to turn or maneuver. Side turn signal lamps should not be confused with regular turn signal lamps described in SAE J588, which may be mounted on the side of the vehicle in some vehicle installations.

3. Laboratory Requirements

3.1 The following sections of SAE J575 are a part of this recommended practice:

3.1.1 Section 2—Samples for Test.
3.1.2 Section 2.2—Bulbs.
3.1.3 Section 3—Laboratory Facilities.
3.1.4 Section 4.1—Vibration Test.
3.1.5 Section 4.2—Moisture Test.
3.1.6 Section 4.3—Dust Test.
3.1.7 Section 4.4—Corrosion Test.
3.1.8 Section 4.6—Photometry.
3.1.9 Section 4.8—Warpage Test on Devices with Plastic Components. Procedures shall be the same as required for turn signal lamps.

3.2 Photometric Tests

3.2.1 Photometric tests shall be made with the photometer at a distance no less than 10 ft (3.0 m) from the lamp. The H-V axis shall be taken as normal to the longitudinal axis of the vehicle.

3.2.2 Candela requirements for side turn signal lamps mounted on trucks, buses, and multipurpose vehicles with a length of 30 ft (9.1 m) or more are shown in Table 1.

3.2.3 Candela requirements for side turn signal lamps mounted on passenger cars and buses, trucks, and multipurpose vehicles with a length of less than 30 ft (9.1 m) are shown in Table 2.

3.3 Color Test—The color of the light from a side turn signal shall be yellow. (See SAE J578).

3.4 Plastic Materials—Any plastic materials used in exterior optical parts shall comply with the requirements set forth in SAE J576.

4. Installation Requirements—The following requirements apply to the

TABLE 1—LUMINOUS INTENSITY REQUIREMENTS FOR TRUCKS, BUSES, AND MULTIPURPOSE VEHICLES OF 30 FT (9.1 m) OR MORE IN LENGTH[a,b,c,d]

Position, deg[e]	Candela, min	Position, deg[e]	Candela, min
15U-0	20	H-30L	30
10U-30L	20	H-0	30
5U-85L	40	5D-85L	40
5U-70L	30	5D-70L	30
H-85L	50	10D-30L	20
H-70L	30	15D-0	20

[a] 300 cd maximum at all points.
[b] If a side marker lamp is combined with the side turn signal lamp, the signal lamp shall have no less than 3 times the luminous intensity of the side marker at any test point, except that at 5U-30L, 5U-70L, H-30L, and H-70L, the signal lamp shall have no less than 5 times the luminous intensity of the side marker.
[c] Specifications are based on laboratories using accurate, rated bulbs during testing.
[d] Lamps designed for use in both 6 V and 12 V systems shall be tested with 12 V bulbs.
[e] Substitute right-hand angles for left-hand angles for lamps mounted on the right side of the vehicle.

TABLE 2—LUMINOUS INTENSITY REQUIREMENTS FOR PASSENGER CARS AND MULTIPURPOSE VEHICLES, TRUCKS, AND BUSES OF LESS THAN 30 FT (9.1 m) IN LENGTH[a,b,c,d]

Position, deg[e]	Candela, min	Position, deg[e]	Candela, min
15U-30L	5	H-30L	15
15U-70L	5	H-70L	15
5U-30L	15	5D-30L	5
5U-70L	15	5D-70L	5

[a] If a side marker lamp is combined with the side turn signal lamp, the signal lamp shall have no less than 3 times the luminous intensity of the side marker at any test point, except that at 5U-30L, 5U-70L, H-30L, and H-70L, the signal lamp shall have no less than 5 times the luminous intensity of the side marker.
[b] Specifications are based on laboratories using accurate, rated bulbs during testing.
[c] Lamps designed for use in both 6 V and 12 V systems shall be tested with 12 V bulbs.
[d] 60 cd maximum at all points.
[e] Substitute right-hand angles for left-hand angles for lamps mounted on the right side of the vehicle.

device as used on the vehicle and are not part of the laboratory test procedures.

4.1 Visibility and photometric performance of the signal within the test angles shown in Tables 1 and 2 shall not be obstructed by any portion of the vehicle unless the lamp is designed to comply with all requirements when the obstruction is considered.

4.2 Side turn signal lamps shall flash simultaneously with the required front signal lamps.

4.3 A side turn signal lamp may be combined with a side marker lamp provided the candela output of the individual functions conform to the ratio requirements shown in the footnotes to Tables 1 and 2.

4.4 Mounting Height

4.4.1 Side turn signal lamps shall be mounted on trucks, buses, and multipurpose vehicles with a length of 30 ft (9.1 m) or more at a height of no more than 72 in (1830 mm) or less than 32 in (810 mm).

4.4.2 Side turn signal lamps shall be mounted on passenger cars and multipurpose vehicles, trucks, and buses with a length of less than 30 ft (9.1 m) at a height of no more than 48 in (1220 mm) or less than 27 in (685 mm).

APPENDIX

As a matter of information, attention is called to SAE J567, for requirements and gages to be used in socket design.

CORNERING LAMPS—SAE J852b

SAE Recommended Practice

Report of Lighting Committee approved April 1963 and last revised February 1965.

Definition—Cornering lamps are steady burning lamps used in conjunction with the turn signal system to supplement the headlamps by providing additional illumination in the direction of turn.

Identification Code Designation—K.

General Test Provisions—The following sections from SAE J575 are a part of this recommended practice:

Section B—Samples for Test
Section C—Lamp Bulbs
Section D—Laboratory Facilities
Section E—Vibration Test
Section F—Moisture Test
Section H—Corrosion Test
Section L—Warpage Test: Except that the device shall be cycled 3 minutes on and 12 minutes off until the operating time totals 1 hr.

Dust Tests—Where sealed units are used, the dust test shall not be required. For other type lamps, see Section G of SAE J575.

Color Test—The color of the light from a cornering lamp shall be white or amber (see SAE J578).

Photometric Test—Photometric tests should be made with the photometer at a distance of at least 10 ft from the lamp.

The line on the test screen formed by the intersection of a horizontal plane through the light source of the lamp should be designated as H. The line formed on the test screen by the intersection of a vertical plane through the light source and parallel to vehicles longitudinal axis should be designated as V. The intersection of these two lines should be designated as H-V.

A single lamp mounted on a test stand to simulate mounting on the vehicle should meet the following photometric specifications (test points shown are for a lamp mounted on the left side of the vehicle—right hand angle should be substituted for left angle for a lamp mounted on the right side of the vehicle).

Position, deg	Candlepower	Position, deg	Candlepower
8U-90U—V-L	125 max	H—V to L	500 max
4U—V-L	200 max	2-1/2 D—30 L	300 min
2U—V-L	300 max	2-1/2 D—45 L	500 min
1U—V-L	400 max	2-1/2 D—60 L	300 min

Installation Requirements (These requirements are not a part of the test specifications.)

Cornering lamps are primarily intended to be used during the times that headlamps are required.

Means should be provided to turn on the cornering lamps with the turn signal lamps or by another suitable means and they should turn off when the turn signal lamps are turned off. If the cornering lamps are not turned off automatically, a visible or audible means shall be provided to indicate to the driver when the lamps are on.

TURN SIGNAL FLASHERS—SAE J590e

SAE Standard

Report of Lighting Committee approved March 1960 and last revised July 1975. Editorial change January 1977.

1. Scope—Flasher(s) referred to in this SAE Standard are for nominal 12 V or 6 V d-c circuits and are intended to operate at the design load(s) for the turn signal system as stated by the flasher manufacturer. It is an engineering design standard and is supplemented by SAE J1055 covering service performance requirements for turn signal flashers.

2. Definition—A turn signal flasher is a device which causes all the required signal lamps to flash as long as it is turned on.

3. Sample(s) Required—A flasher(s) representative of those regularly manufactured and marketed shall be submitted for test. The sample(s) shall include means of connection into the lighting circuit if other than the standard sealed beam type plug-in connector is used. The specific ampere design load for fixed-load flashers, or the minimum and maximum design loads for variable-load flashers, and the mounting position (if necessary) shall be specified by the flasher manufacturer.

4. Test Circuitry and Equipment Requirements—See SAE J823. The standard test circuit is shown therein.

5. Pilot Indication—The means of producing the visible pilot indication required in the turn signal system may be incorporated in the flasher. A means of producing an audible signal may be incorporated in the flasher. The "means" shall function satisfactorily under all the test conditions that are applied to the flasher.

6. Performance requirements—The flasher(s) for test shall meet the requirements of paragraphs 6.1, 6.2, and 6.3.

6.1 Starting Time—A normally closed type flasher shall open (turn off) within 1.0 s for a unit designed to operate two signal lamps, or within 1.25 s for a unit designed to operate more than two signal lamps. A normally open type flasher shall complete the first cycle (close the contacts and then open the contacts) within 1.5 s. The time measurement shall start when the voltage is initially applied. The test shall be made in an ambient temperature of $75 \pm 10°F$ ($24 \pm 5.5°C$) and the power source for the test circuit adjusted as specified in SAE J823. For a fixed-load flasher, the test shall be made with the specific ampere design load connected. For a variable-load flasher, the test shall be made with both the minimum and maximum design load. Compliance shall be based on an average of three starts, which shall be separated by a cooling interval of 5 min.

6.2 Voltage Drop—The test shall be made in an ambient temperature of $75 \pm 10°F$ ($24 \pm 5.5°C$) in the standard test circuit and the power source for the test circuit adjusted as specified in SAE J823. For a fixed-load flasher the test shall be made with the specific ampere design load, and for a variable-load flasher the test shall be made with the maximum design load only. The lowest voltage drop across the flasher shall not exceed the following:

0.40 V for two signal lamps
0.45 V for three signal lamps
0.50 V for four or more signal lamps

The voltage drop shall be measured between the input and the load terminals at the flasher and during the "on" period after the flasher has been operating for five consecutive cycles and shall be an average of at least three consecutive cycles.

6.3 Flash Rate and Percent Current "On" Time—The flash rate and the percent current "on" time of a normally closed type flasher shall be within the unshaded portion of the polygon shown in Fig. 1. The flash rate and percent current "on" time of a normally open type flasher shall be within the entire rectangle shown, including the shaded areas. These requirements shall be met

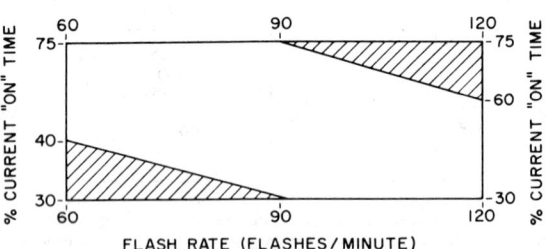

FIG. 1—FLASH RATE (FLASHES PER MINUTE)

by a fixed-load flasher with the specific ampere design load connected, and by a variable-load flasher with both the minimum and maximum design loads, as specified by the flasher manufacturer. Flash rate and percent current "on" time shall be measured after the flasher has been operating for five consecutive cycles and shall be an average of at least three consecutive cycles. The above operating limits shall apply over combinations of bulb voltages and ambient temperatures tabulated:

(a) 12.8 V (or 6.4 V) and $75 \pm 10°F$ ($24 \pm 5.5°C$)
(b) 12.0 V (or 6.0 V) and $0 \pm 5°F$ ($-17 \pm 3°C$)
(c) 15.0 V (or 7.5 V) and $0 \pm 5°F$ ($-17 \pm 3°C$)
(d) 11.0 V (or 5.5 V) and $125 \pm 5°F$ ($50 \pm 3°C$)
(e) 14.0 V (or 7.0 V) and $125 \pm 5°F$ ($50 \pm 3°C$)

6.4 Extreme Temperature Tests—The flasher shall be stabilized in ambient temperatures of $145 \pm 5°F$ ($63 \pm 3°C$) and $-25 \pm 5°F$ ($-32 \pm 3°C$) for a period of 3 h and then operated with the manufacturer's design load(s) and 12.8 V at the bulbs. In each of these tests, the starting time shall not exceed 3.0 s, and the flash rate shall be 60–120 flashes/min after five complete cycles and shall be the average of at least three consecutive cycles.

7. Durability Test Requirements—A flasher(s) conforming to paragraphs 6.1, 6.2, and 6.3(a) shall be subjected to the durability test. The test shall be run on the flasher(s) connected in a standard test circuit specified in SAE J823 and with the power source for the durability test adjusted to apply and maintain 14.0 V (or 7.0 V) according to the flasher rating to the input terminals of the standard test circuit. For a fixed-load flasher, the test shall be made with the specific ampere design load connected. For a variable-load flasher, the test shall be made with the maximum design load only.

Total time: 100 h
Mode of operation: Continuous
Ambient temperature during test: $75 \pm 10°F$ ($24 \pm 5.5°C$)

At the conclusion of the durability test, the flasher(s) shall be tested in the standard test circuit. The power source for the test circuit shall be adjusted as specified in SAE J823 in the paragraph on durability test requirements. For a fixed-load flasher, the test shall be made with the specific ampere design load connected. For a variable-load flasher, the test shall be made with the maximum design load only. The ambient temperature shall be $75 \pm 10°F$ ($24 \pm 5.5°C$).

The flasher(s) tested shall comply with the provisions of paragraphs 6.1, 6.2, and 6.3(a).

SERVICE PERFORMANCE REQUIREMENTS FOR TURN SIGNAL FLASHERS—SAE J1055

SAE Recommended Practice

Report of Lighting Committee approved September 1973. Editorial change January 1977.

1. Scope—This recommended practice covers service performance tests, test procedures and requirements applicable to turn signal flashers. It is intended to supplement the engineering design standard SAE J590 to cover service performance requirements of turn signal flashers.

2. Test Conditions

2.1 Performance Test—Flashers shall be performance tested at the specific design load for fixed-load flashers, or at the minimum and maximum design loads for variable-load flashers, and in the mounting position (if necessary), as specified by the flasher manufacturer.

2.2 Durability Test—Flashers shall be durability tested at the specific design load for a fixed-load flasher, or at the maximum design load for a variable-load flasher, as specified by the flasher manufacturer.

2.3 Test Circuitry and Equipment Requirements—See SAE J823. The standard test circuit is shown therein.

3. Definitions

3.1 Lot—The term "lot" or "batch" shall mean inspection lot, that is, a collection of flashers from which a sample is to be drawn and tested to determine conformance with the acceptability criteria. Each lot shall consist of flashers of a single type manufactured at essentially the same time. Each flasher shall be coded externally to represent the period of manufacture by at least month and year.

3.2 Sample and Sample Size—A sample shall consist of individual flashers drawn from a lot, the individual flashers being selected at random without regard to their quality. The number of flashers in the sample is the sample size.

3.3 Flasher Characteristic—Value of a particular parameter of flasher operation, for example, flash rate, percent current "on" time, starting time, voltage drop.

3.4 Engineering Design Standard—Flasher characteristics as specified in SAE J590.

3.5 Zone—A prescribed range or set of values of a flasher characteristic other than that of the engineering design standard.

3.6 Zone Designation—Each zone shall be designated by the capital letters A, B, C in the order of increasing deviation of the zone from SAE J590.

3.7 Zone A—The range of values of a flasher characteristic which does not alter materially the safety-related aspect of turn signal indication.

3.8 Zone B—The range of values of a flasher characteristic which still provides an adequate safety-related aspect of turn signal indication.

3.9 Zone C—The range of values of a flasher characteristic which does not provide an adequate safety-related aspect of turn signal indication.

3.10 Chance Occurrence—An unusual event.

3.11 Flasher Classification—Each flasher with characteristics outside of the requirements of SAE J590 shall be classified by the zone designation of the flasher characteristic which has the zone designation furthest from the requirement of SAE J590.

Flashers with all characteristics falling within the requirements of SAE J590 shall not be assigned a zone designation.

3.12 Average Laboratory Life—The average number of test hours a sample of flashers remains within SAE J590 engineering design standard or zones A and B. The average shall be based on the sample size tested.

4. Performance Testing

4.1 Select a random group of 32 flashers from the lot. Randomly select a sample of 10 flashers from the group. Submit each of the 10 units to all of the tests of paragraphs 4.2, 4.3, and 4.4.

4.2 Starting Time—The starting time of a normally closed type flasher is the time to open (turn off) after the voltage is applied. The starting time of a normally open type flasher is the time to complete the cycle (close the contacts and then open the contacts) after voltage is applied. The test shall be made in an ambient temperature of 75 ± 10°F (24 ± 5.5°C) and the power source for the test circuit adjusted as specified in SAE J823. For fixed-load flashers, the test shall be made with the specific ampere design load connected. For variable-load flashers, the test shall be made with the minimum and maximum design load connected. Starting time shall be based on a single start. The measured starting times shall be classified by zones in accordance with Table 1.

4.3 Voltage Drop—The test shall be made in an ambient temperature of 75 ± 10°F (24 ± 5.5°C) in the standard test circuit, and the power source for the test circuit adjusted as specified in SAE J823. For fixed-load flashers, the test shall be made with the specific ampere design load connected; and for variable-load flashers, the test shall be made with the maximum design load only. The lowest voltage drop across the flasher shall be measured between the input and the load terminals at the flasher and during the "on" period after the flasher has been operating for five consecutive cycles, and shall be an average of at least three consecutive cycles. Measured voltage drops shall be classified by zones in accordance with Table 2.

4.4 Flash Rate and Percent Current "On" Time—Fixed-load flashers shall be tested with the specific ampere design load connected, and variable-load flashers shall be tested with the minimum design load connected and with the maximum design load connected. Flash rate and percent current "on" time shall be measured after the flasher has been operating for five consecutive cycles, and shall be an average of at least three consecutive cycles. The flashers shall be tested over combinations of bulb voltages and ambient temperatures tabulated:

(a) 12.8 V (or 6.4 V) and 75 ± 10°F (24 ± 5.5°C)
(b) 12.0 V (or 6.0 V) and 0 ± 5°F (−17 ± 3°C)
(c) 15.0 V (or 7.5 V) and 0 ± 5°F (−17 ± 3°C)

TABLE 1—STARTING TIME, s

Type	Turn Signal Lamp Load	SAE J590	Zone A	Zone B	Zone C
Normally closed	2	1.0 max			
Normally open	3 or more Design load(s)	1.25 max 1.5 max	1.7 max	2.0 max	> 2.0

TABLE 2—VOLTAGE DROP, V

Turn Signal Lamp Load	SAE J590	Zone A	Zone B	Zone C
2	0.40 max			
3	0.45 max	0.6 max	0.8 max	> 0.8
4 or more	0.50 max			

(d) 11.0 V (or 5.5 V) and 125 ± 5°F (50 ± 3°C)
(e) 14.0 V (or 7.0 V) and 125 ± 5°F (50 ± 3°C)

Measured flash rates and percent current "on" times shall be classified by zones in accordance with Fig. 1. For normally closed type flashers, the shaded portion of the SAE J590 polygon shall be included in zone A. Where a measured value lies on a classification boundary line, the classification nearest to SAE J590 shall be assigned.

5. Performance Tests Quality Criteria

5.1 Zero flashers in zone C and conformance to the zones A and B requirements of Table 3 shall indicate acceptable quality. If so, submit the sample to the durability test in accordance with paragraph 6.

5.2 Two flashers in zone C shall indicate unacceptable quality. Testing shall be terminated at this point. In the event one flasher is in zone C, and the requirements of Table 3 for zones A and B have been met, the following procedure shall be used to determine whether the zone C result was a chance occurrence. Take another random sample of 10 flashers from the remaining group of 22 flashers and perform the tests in paragraphs 4.2, 4.3, and 4.4. One or more additional flashers in zone C shall indicate unacceptable quality. Testing shall be terminated at this point. Zero additional units in zone C and conformance to the requirements of Table 3 for zones A and B shall mean that the first flasher in zone C was a chance occurrence. Discard the first sample of 10. Submit the second sample of 10 flashers to the durability test in accordance with paragraph 6.

6. Durability Testing—Flashers shall be subjected to the durability test. The durability test shall be run with each flasher connected in a standard test circuit specified in SAE J823 and with the power source for the durability test adjusted to apply 14.0 V (or 7.0 V), according to the flasher rating, to the input terminals of the standard test circuit throughout the test. For fixed-load flashers the durability test shall be made with the specific ampere design load connected, and for variable-load flashers the test shall be made with the maximum design load only. Ambient temperature during test shall be 75 ± 10°F (24 ± 5.5°C) and mode of operation shall be continuous except for two test measurement points which shall be at 25 test hours and, if necessary, at 100 test hours.

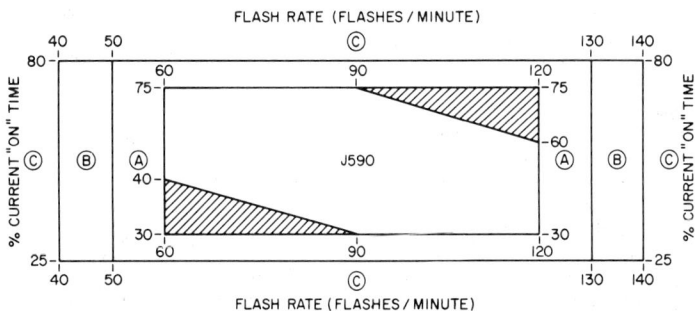

FIG. 1—FLASH RATE (FLASHES PER MINUTE) AND PERCENT CURRENT "ON" TIME CLASSIFICATIONS BY ZONES

TABLE 3—ACCEPTANCE CRITERIA[a]

	Max Quantity Flashers in Zone A	Max Quantity Flashers in Zone B	Max Quantity Flashers in Zone C
For performance testing[b]	3 less zone B	1	0
For durability testing[c]	0 in zone C in 25 h or minimum average laboratory life of 60 h		

[a]Each flasher in the sample shall be classified only by its largest deviation from SAE J590.
[b]See paragraph 5.
[c]See paragraphs 6 and 7.

At each point of measurement each flasher shall be tested in accordance with paragraphs 4.2, 4.3, and 4.4(a) and in the standard test circuit except that for variable-load flashers the test shall be made with the maximum design bulb load only.

7. Durability Test Quality Criteria—Zero flashers in zone C at 25 test hours shall indicate acceptable quality. Two or more flashers in zone C at 25 test hours shall indicate unacceptable quality. In the event one flasher is in zone C at 25 test hours, the remaining flashers shall be run to 100 test hours. Upon conclusion of the 100 h test, an average laboratory life of the tested flashers equal to or greater than 60 h shall indicate acceptable quality. An average laboratory life of less than 60 h shall indicate unacceptable quality. At each point of measurement, each flasher that falls within zone C shall be considered a failure. Each such failed flasher shall be considered to have a test life equal to the number of hours corresponding to the last prior point of measurement for the purpose of computing the average laboratory life of the tested flashers.

VEHICULAR HAZARD WARNING SIGNAL FLASHER—SAE J945b

SAE Standard

Report of Lighting Committee approved February 1966 and last revised September 1973. Upgraded to Standard Status June 1975. Editorial change January 1977.

1. Scope—Flasher(s) referred to in this SAE Standard are for nominal 12 V or 6 V d-c circuits and are required to operate from two signal lamps to the maximum design load including pilot lamps, as stated by the flasher manufacturer. It is an engineering design standard and is supplemented by SAE J1056 covering service performance requirements for vehicular hazard warning flashers.

2. Definition—A vehicular hazard warning signal flasher is a device which, as long as it is turned on, causes all the required signal lamps listed in SAE J910 to flash.

3. Sample(s) Required—A flasher(s) representative of those regularly manufactured and marketed shall be submitted for test. The sample(s) shall include means of connection into the lighting circuit if other than the standard sealed beam type plug-in connector is used. The maximum design load for the hazard warning signal flasher and the mounting position (if necessary) shall be specified by the flasher manufacturer.

4. Test Circuitry and Equipment Requirements—See SAE J823. The standard test circuit is shown therein.

5. Pilot Indication—The means of producing the visible pilot indication required in the hazard warning signal system may be incorporated in the flasher. A means of producing an audible signal also may be incorporated in the flasher. The "means" shall function satisfactorily under all the test conditions that are applied to the flasher.

6. Performance Requirements—The flasher(s) for test shall meet the requirements of paragraphs 6.1, 6.2, and 6.3.

6.1 Starting Time—A flasher having normally closed contacts shall open (turn off) within 1.5 s after voltage is applied. A flasher having normally open contacts shall complete the first cycle (close the contacts and then open the contacts) within 1.5 s after voltage is applied. The test shall be made in an ambient temperature of $75 \pm 10°F$ ($24 \pm 5.5°C$) with the minimum and maximum design load, measured between the input and the load terminals after the flasher has been operating for five consecutive cycles, and shall be an average of at least three consecutive cycles, and shall not exceed 0.45 V.

6.2 Voltage Drop—The test shall be made in an ambient temperature of $75 \pm 10°F$ ($24 \pm 5.5°C$) in the standard test circuit with the maximum design load connected and the power source for the test circuit adjusted as specified in SAE J823. The lowest voltage drop across the flasher during the "on" period, measured between the input and the load terminals after the flasher has been operating for five consecutive cycles, and shall be an average of at least three consecutive cycles, and shall not exceed 0.45 V.

6.3 Flash Rate and Percent Current "On" Time—The flash rate and the percent current "on" time of a normally closed type flasher shall be within the unshaded portion of the polygon shown in Fig. 1. The flash rate and percent current "on" time of a normally open type flasher shall be within the entire rectangle shown including the shaded areas. Flashing rate and percent current "on" time shall be measured after the flasher has been operating for five

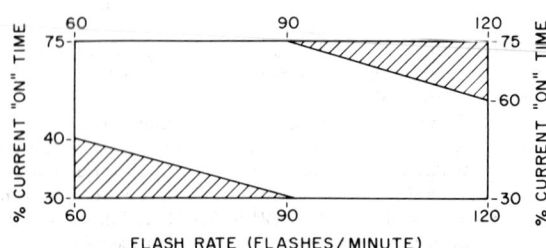

FIG. 1—FLASH RATE (FLASHES PER MINUTE)

consecutive cycles and shall be an average of at least three consecutive cycles. The above operating limits shall apply for loads of two signal bulbs and the maxium design load including pilot lamps, as specified by the flasher manufacturer, over combinations of bulb voltages and ambient temperatures tabulated:

(a) 12.8 V (or 6.4 V) and $75 \pm 10°F$ ($24 \pm 5.5°C$)
(b) 11.0 V (or 5.5 V) and $0 \pm 5°F$ ($-17 \pm 3°C$)
(c) 13.0 V (or 6.5 V) and $0 \pm 5°F$ ($-17 \pm 3°C$)
(d) 11.0 V (or 5.5 V) and $125 \pm 5°F$ ($50 \pm 3°C$)
(e) 13.0 V (or 6.5 V) and $125 \pm 5°F$ ($50 \pm 3°C$)

6.4 Extreme Temperature Tests—The flasher shall be stabilized in ambient temperatures of $145 \pm 5°F$ ($63 \pm 3°C$) and $-25 \pm 5°F$ ($-32 \pm 3°C$) for a period of 3 h and then operated with the manufacturer's design load(s) and 12.8 V at the bulbs. In each of these tests, the starting time shall not exceed 3.0 s, and the flash rate shall be 60–120 flashes/min after five complete cycles and shall be the average of at least three consecutive cycles.

7. Durability Test Requirements—A flasher(s) conforming to paragraphs 6.1, 6.2, and 6.3(a) shall be subjected to the durability test. The test shall be run with the maximum design load connected into the standard test circuit specified in SAE J823 and with the power source for the test circuit adjusted to apply and maintain 13.0 V (or 6.5 V), according to the flasher rating, to the input terminals of the test circuit through the test:

Total time: 36 h
Mode of operation: Continuous
Ambient temperature during test: $75 \pm 10°F$ ($24 \pm 5.5°C$)

At the conclusion of the durability test, the flasher(s) shall be tested in the standard test circuit with the minimum of two signal lamp bulbs and maximum design load including pilot lamps, as specified by the flasher manufacturer. The power source for the test circuit shall be adjusted as specified in SAE J823 in the paragraph on durability test requirements. The ambient temperature shall be $75 \pm 10°F$ ($24 \pm 5.5°C$). The flasher(s) tested shall comply with the provisions of paragraphs 6.1, 6.2, and 6.3(a).

SERVICE PERFORMANCE REQUIREMENTS FOR VEHICULAR HAZARD WARNING FLASHERS—SAE J1056

SAE Recommended Practice

Report of Lighting Committee approved September 1973. Editorial change June 1974. Editorial change January 1977.

1. **Scope**—This recommended practice covers service performance tests, test procedures and requirements applicable to vehicular hazard warning signal flashers. It is intended to supplement engineering design standard SAE J945 to cover service performance requirements of vehicular hazard warning signal flashers.

2. **Test Conditions**

   2.1 **Performance Test**—Flashers shall be performance tested at minimum design load corresponding to two signal lamp bulbs and at maximum design load, and in the mounting position (if necessary), as specified by the flasher manufacturer.

   2.2 **Durability Test**—Flashers shall be durability tested at the maximum design load as specified by the flasher manufacturer.

   2.3 **Test Circuitry and Equipment Requirements**—See SAE J823. The standard test circuit is shown therein.

3. **Definitions**

   3.1 **Lot**—The term "lot" or "batch" shall mean inspection lot, that is, a collection of flashers from which a sample is to be drawn and tested to determine conformance with the acceptability criteria. Each lot shall consist of flashers of a single type manufactured at essentially the same time. Each flasher shall be coded externally to represent the period of manufacture by at least month and year.

   3.2 **Sample and Sample Size**—A sample shall consist of individual flashers drawn from a lot, the individual flashers being selected at random without regard to their quality. The number of flashers in the sample is the sample size.

   3.3 **Flasher Characteristic**—Value of a particular parameter of flasher operation, for example, flash rate, percent current "on" time, starting time, voltage drop.

   3.4 **Engineering Design Standard**—Flasher characteristics as specified in SAE J945.

   3.5 **Zone**—A prescribed range or set of values of a flasher characteristic other than that of the engineering design standard.

   3.6 **Zone Designation**—Each zone shall be designated by the capital letters A, B, C in the order of increasing deviation of the zone from SAE J945.

   3.7 **Zone A**—The range of values of a flasher characteristic which does not alter materially the safety-related aspect of hazard warning signal indication.

   3.8 **Zone B**—The range of values of a flasher characteristic which still provides an adequate safety-related aspect of hazard warning signal indication.

   3.9 **Zone C**—The range of values of a flasher characteristic which does not provide an adequate safety-related aspect of hazard warning signal indication.

   3.10 **Chance Occurrence**—An unusual event.

   3.11 **Flasher Classification**—Each flasher with characteristics outside of the requirements of SAE J945 shall be classified by the zone designation of the flasher characteristic which has the zone designation furthest from the requirement of SAE J945.

   Flashers with all characteristics falling within the requirements of SAE J945 shall not be assigned a zone designation.

   3.12 **Average Laboratory Life**—The average number of test hours a sample of flashers remains within SAE J945 engineering design standard or zones A and B. The average shall be based on the sample size tested.

4. **Performance Testing**

   4.1 Select a random group of 32 flashers from the lot. Randomly select a sample of 10 flashers from the group. Submit each of the 10 units to all of the tests of paragraphs 4.2, 4.3, and 4.4.

   4.2 **Starting Time**—The starting time of a normally closed type flasher is the time to open (turn off) after the voltage is applied. The starting time of a normally open type flasher is the time to complete the cycle (close the contacts and then open the contacts) after voltage is applied. The test shall be made in an ambient temperature of $75 \pm 10°F$ ($24 \pm 5.5°C$) and the power source for the test circuit adjusted as specified in SAE J823. The test shall be made with the minimum and maximum design load connected. Starting time shall be based on a single start. The measured starting times shall be classified by zones in accordance with Table 1.

   4.3 **Voltage Drop**—The test shall be made in an ambient temperature of $75 \pm 10°F$ ($24 \pm 5.5°C$) in the standard test circuit, with maximum design load connected and the power source for the test circuit adjusted as specified in SAE J823. The lowest voltage drop across the flasher during the "on"

TABLE 1—STARTING TIME, s

Type	Vehicular Hazard Warning Signal Design Load	SAE J945	Zone A	Zone B	Zone C
Normally closed or normally open	Rated minimum and rated maximum	1.5 max	2.2 max	3.0 max	> 3.0

period shall be measured between the input and the load terminals after the flasher has been operating for five consecutive cycles, and shall be an average of at least three consecutive cycles. Measured voltage drops shall be classified by zones in accordance with Table 2.

   4.4 **Flash Rate and Percent Current "On" Time**—The flash rate and percent current "on" time shall be tested with the minimum design load connected and with the maximum design load connected. Flash rate and percent current "on" time shall be measured after the flasher has been operating for five consecutive cycles, and shall be an average of at least three consecutive cycles. The flashers shall be tested over combinations of bulb voltages and ambient temperatures tabulated.

   (a) 12.8 V (or 6.4 V) and $75 \pm 10°F$ ($24 \pm 5.5°C$)
   (b) 11.0 V (or 5.5 V) and $0 \pm 5°F$ ($-17 \pm 3°C$)
   (c) 13.0 V (or 6.5 V) and $0 \pm 5°F$ ($-17 \pm 3°C$)
   (d) 11.0 V (or 5.5 V) and $125 \pm 5°F$ ($50 \pm 3°C$)
   (e) 13.0 V (or 6.5 V) and $125 \pm 5°F$ ($50 \pm 3°C$)

Measured flash rates and percent current "on" times shall be classified by zones in accordance with Fig. 1. For normally closed type flashers, the shaded portion of SAE J945 polygon shall be included in zone A. Where a measured value lies on a classification boundary line, the classification nearest to SAE J945 shall be assigned.

5. **Performance Tests Quality Criteria**

   5.1 Zero flashers in zone C and conformance to the zone A and B requirements of Table 3 shall indicate acceptable quality. If so, submit the sample to the durability test in accordance with paragraph 6.

   5.2 Two flashers in zone C shall indicate unacceptable quality. Testing shall be terminated at this point. In the event one flasher is in zone C, and the requirements of Table 3 for zones A and B have been met, the following procedure shall be used to determine whether the zone C result was a chance occurrence. Take another random sample of 10 flashers from the remaining group of 22 flashers and perform the tests in paragraphs 4.2, 4.3, and 4.4. One or more additional flashers in zone C shall indicate unacceptable quality. Testing shall be terminated at this point. Zero additional units in zone C and

TABLE 2—VOLTAGE DROP, V

Vehicular Hazard Warning Signal Design Load	SAE J945	Zone A	Zone B	Zone C
Rated maximum	0.45 max	0.6 max	0.8 max	> 0.8

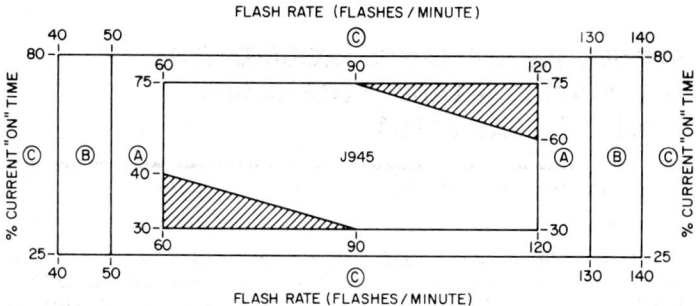

FIG. 1—FLASH RATE (FLASHES PER MINUTE) AND PERCENT CURRENT "ON" TIME CLASSIFICATIONS BY ZONES

TABLE 3—ACCEPTANCE CRITERIA[a]

	Max Quantity Flashers in Zone A	Max Quantity Flashers in Zone B	Max Quantity Flashers in Zone C
For performance testing[b]	3 less zone B	1	0
For durability testing[c]	0 in zone C in 12 h or minimum average laboratory life of 20 h		

[a] Each flasher in the sample shall be classified only by its largest deviation from SAE J945.
[b] See paragraph 5.
[c] See paragraphs 6 and 7.

conformance to the requirements of Table 3 for zones A and B shall mean that the first flasher in zone C was a chance occurrence. Discard the first sample of 10. Submit the second sample of 10 flashers to the durability test in accordance with paragraph 6.

6. Durability Testing—Flashers shall be subjected to the durability test. The durability test shall be run with the maximum design load connected into the standard test circuit specified in SAE J823 and the power source for the test circuit adjusted to apply and maintain 13.0 V (or 6.5 V), according to the flasher rating, to the input terminals of the test circuit throughout the test. Ambient temperature during test shall be 75 ± 10°F (24 ± 5.5°C) and mode of operation shall be continuous except for two test measurement points which shall be made at 12 test hours and, if necessary, at 36 test hours. At each point of measurement, each flasher shall be tested in the standard test circuit with the minimum of two bulbs and maximum design load as specified by the flasher manufacturer. The power source for the test circuit shall be adjusted as specified in SAE J823. The ambient temperature shall be 75 ± 10°F (24 ± 5.5°C). Each flasher shall be measured in accordance with paragraphs 4.2, 4.3, and 4.4(a).

7. Durability Test Quality Criteria—Zero flashers in zone C at 12 test hours shall indicate acceptable quality. Two or more flashers in zone C at 12 test hours shall indicate unacceptable quality. In the event one flasher is in zone C at 12 test hours, the remaining flashers shall be run continuously to 36 test hours. Upon conclusion of the 36 h test, an average laboratory life of the tested flashers equal to or greater than 20 h shall indicate acceptable quality. An average laboratory life of less than 20 h shall indicate unacceptable quality. At each point of measurement, each flasher that falls within zone C shall be considered a failure. Each such failed flasher shall be considered to have a test life equal to the number of hours corresponding to the last prior point of measurement for the purpose of computing the average laboratory life of the tested flashers.

WARNING LAMP ALTERNATING FLASHERS—SAE J1054

SAE Recommended Practice

Report of Lighting Committee approved September 1973. Editorial change January 1977.

1. Scope—Flashers referred to in this SAE recommended engineering design practice are for nominal 12 V circuits and are required to operate the maximum design load per output terminal as stated by the flasher manufacturer. This design practice is intended for use in conjunction with a supplementary Service Performance Standard for Warning Lamp Flashers (under development).

2. Definition—A warning lamp alternating flasher is a device that alternately flashes signal lamps as specified in SAE J887 and/or SAE J595.

3. Identification and Code Marking—The device shall be identified by proper SAE coding and the markings shall include the manufacturer's name, model number, and the maximum rated design load in amperes per load terminal. (Reference: SAE J759.)

4. Pilot Indication—A means of producing a visible pilot indication or an audible signal may be incorporated in the flasher and shall function satisfactorily under all test conditions that are applied to the flasher.

5. Performance Requirements

5.1 *Starting Time*—Starting time is defined as the interval between the instant power is applied to the input terminal and the instant of transfer to the second output terminal. Starting time shall not exceed 1.5 s. Compliance shall be based on an average of three starts separated by a minimum cooling interval of 5 min. The test shall be made in an ambient temperature of 75 ± 10°F (24 ± 5.5°C) at the manufacturer's specified design load at 12.8 V measured at the bulbs.

5.2 *Voltage Drop*—The lowest voltage drop during the "on" period, when operated at the manufacturer's design load with 12.8 V at the bulbs in an ambient temperature of 75 ± 10°F (24 ± 5.5°C), shall not exceed 0.8 V measured between input and each load terminal. The voltage drop shall be measured after the flasher has been operating for five consecutive cycles, and shall be an average of at least three consecutive cycles.

5.3 *Flash Rate and Percent Current On Time*—The flash rate and percent current "on" time of the device when operated at the manufacturer's design load shall be 60–120 flashes/min at each load terminal, with the sum of the percent current "on" time for both output terminals to be 90–110% with a minimum of 30% for either terminal.

The flash rate and percent current on time shall be measured after the flasher has been operating for five consecutive cycles, and shall be an average of at least three consecutive cycles over a combination of bulb voltages and ambient temperatures as follows:

(a) 12.8 V and 75 ± 10°F (24 ± 5.5°C)
(b) 12.0 V and 0 ± 5°F (−18 ± 3°C)
(c) 15.0 V and 0 ± 5°F (−18 ± 3°C)
(d) 11.0 V and 125 ± 5°F (52 ± 3°C)
(e) 14.0 V and 125 ± 5°F (52 ± 3°C)

5.4 *Extreme Temperature Tests*—The flasher shall be stabilized in ambient temperatures of 145 ± 5°F (63 ± 3°C) and −25 ± 5°F (−32 ± 3°C) for a period of 3 h and then operated with the manufacturer's design load and 12.8 V at the bulbs. In each of these tests, the starting time shall not exceed 5 s, and the flash rate shall be 30–150 flashes/min at each load terminal after five complete cycles and shall be the average of at least three consecutive cycles.

6. Durability Test Requirements—The durability test shall be run with the manufacturer's maximum design load per output terminal connected into the standard test circuit adjusted to supply and maintain 13.0 V to the input terminal of the test circuit in an ambient temperature of 75 ± 10°F (24 ± 5.5°C). Flashers shall flash continuously for a total time of 100 h.

After the completion of the durability test, the flasher shall meet the requirements for starting time, voltage drop, flash rate, and percent current "on" time as specified under the performance requirements of paragraphs 5.1, 5.2, and 5.3(a) with 12.8 V at the bulbs and 75 ± 10°F (24 ± 5.5°C).

APPENDIX

As a matter of information, attention is called to test equipment and circuitry as specified in SAE J823.

SERVICE PERFORMANCE REQUIREMENTS FOR WARNING LAMP ALTERNATING FLASHERS—SAE J1104

SAE Recommended Practice

Report of Lighting Committee approved October 1974. Editorial change January 1977.

1. Scope—This recommended practice covers service performance tests, test procedures and requirements applicable to warning lamp alternating flashers. It is intended to supplement the engineering design standard SAE J1054 to cover service performance requirements of warning lamp alternating flashers.

2. Test Conditions

2.1 **Performance Test**—Flashers shall be performance tested at the specified design load and in the mounting position (if necessary), as specified by the flasher manufacturer.

2.2 **Durability Test**—Flashers shall be durability tested at the design load specified by the flasher manufacturer.

2.3 Test Circuitry and Equipment Requirements—As a matter of information, attention is called to test equipment, procedures and circuitry as specified in SAE J823 flasher test equipment.

3. Definitions

3.1 Lot—The term "lot" or "batch" shall mean inspection lot, that is, a collection of flashers from which a sample is to be drawn and tested to determine conformance with the acceptability criteria. Each lot shall consist of flashers of a single type manufactured at essentially the same time. Each flasher shall be coded externally to represent the period of manufacture by at least month and year.

3.2 Sample and Sample Size—A sample shall consist of individual flashers drawn from a lot, the individual flashers being selected at random without regard to their quality. The number of flashers in the sample is the sample size.

3.3 Flasher Characteristic—Value of a particular parameter of flasher operation, for example, Flash Rate, Percent Current "On" Time, Starting Time, Voltage Drop.

3.4 Engineering Design Standard—Flasher characteristics as specified in SAE J1054.

3.5 Zone—A prescribed range or set of values of a flasher characteristic other than that of the Engineering Design Standard.

3.6 Zone Designation—Each zone shall be designated by the capital letters A, B, C in the order of increasing deviation of the zone from SAE J1054.

3.7 Zone A—The range of values of a flasher characteristic which does not alter materially the safety-related aspect of warning signal indication.

3.8 Zone B—The range of values of a flasher characteristic which still provides an adequate safety-related aspect of warning signal indication.

3.9 Zone C—The range of values of a flasher characteristic which does not provide an adequate safety-related aspect of warning signal indication.

3.10 Chance Occurrence—An unusual event.

3.11 Flasher Classification—Each flasher with characteristics outside of the requirements of SAE J1054 shall be classified by the zone designation of the flasher characteristic which has the zone designation furthest from the requirement of SAE J1054. Flashers with all characteristics falling within the requirements of SAE J1054 shall not be assigned a zone designation.

3.12 Average Laboratory Life—The average number of test hours a sample of flashers remains within SAE J1054 Engineering Design Standard or Zones A and B. The average shall be based on the sample size tested.

4. Performance Testing

4.1 Select a random group of 32 flashers from the lot. Randomly select a sample of ten flashers from the group. Submit each of the ten units to all of the tests of paragraphs 2, 3 and 4 below.

4.2 Starting Time—Starting time is defined as the interval between the instant power is applied to the input terminal and the instant of transfer to the second output terminal. Under the unique condition where both load terminals are energized at the instant power is applied to the flasher, then the interval shall be measured to the instant when the first load terminal de-energized is re-energized.

The test shall be made in an ambient temperature of $75 \pm 10°F$ ($24 \pm 5.5°C$) at the manufacturer's specified design load and with 12.8 V at the bulbs. Starting time shall be based on a single start. The measured Starting Times shall be classified by Zones in accordance with Table 1.

4.3 Voltage Drop—The lowest voltage drop during the "On" period, when operated at the manufacturer's design load with 12.8 V at the bulbs in an ambient temperature of $75° \pm 10°F$ ($24 \pm 5.5°C$) shall be measured between the input and each load terminal. The voltage drop shall be measured after the flasher has been operating for five consecutive cycles, and shall be an average of at least three consecutive cycles, and shall be classified by zones in accordance with Table 2.

4.4 Flash Rate and Percent Current "On" Time—The flash rate and percent current "on" time shall be measured after the flasher has been operating for five consecutive cycles, and shall be an average of at least three consecutive cycles over a combination of bulb voltages and ambient temperatures as follows:

TABLE 1—STARTING TIME (TIME IN SECONDS)

Warning Signal Lamp Load	SAE J1054	Zone A	Zone B	Zone C
Specified Design Load	1.5 max	2.2 max	3.0 max	> 3.0

(a) 12.8 V and $75 \pm 10°F$ ($24 \pm 5.5°C$)
(b) 12.0 V and $0 \pm 5°F$ ($-17 \pm 3°C$)
(c) 15.0 V and $0 \pm 5°F$ ($-17 \pm 3°C$)
(d) 11.0 V and $125 \pm 5°F$ ($50 \pm 3°C$)
(e) 14.0 V and $125 \pm 5°F$ ($50 \pm 3°C$)

Measured flash rates and percent current "on" times at each load terminal shall be classified by Zones in accordance with Fig. 1. Where a measured value lies on a classification boundary line, the classification nearest to SAE J1054 shall be assigned.

5. Performance Tests Quality Criteria

5.1 Zero flashers in Zone C and conformance to the Zone A and B requirements of Table 3 shall indicate acceptable quality. If so, submit the sample to Durability Test in accordance with paragraph 6.

5.2 Two flashers in Zone C shall indicate unacceptable quality. Testing shall be terminated at this point. In the event one flasher is in Zone C, and the requirements of Table 3 for Zones A and B have been met, the following procedure shall be used to determine whether the Zone C result was a chance occurrence. Take another random sample of 10 flashers from the remaining group of 22 flashers and perform the tests in paragraphs 4.2, 4.3, and 4.4. Any additional flashers in Zone C shall indicate unacceptable quality. Testing shall be terminated at this point. Conformance of the second sample shall indicate acceptable quality and that the first Zone C flasher was a random chance occurrence. Discard the first sample of ten. Submit the second sample of 10 flashers to Durability Test in accordance with paragraph 6.

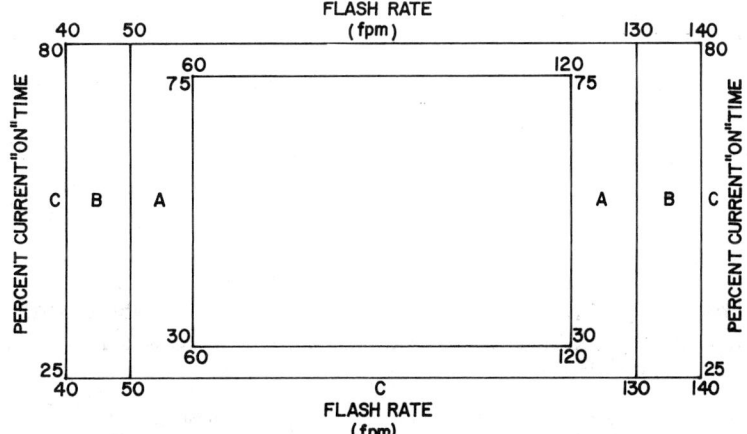

FIG. 1—FLASH RATE (FLASHES PER MINUTE) AND PERCENT CURRENT "ON" TIME CLASSIFICATIONS BY ZONES*

*When tested in accordance with paragraph 4.4(a), the sum of the percent current "On" times for both load terminals shall be 90–110%. Flashers which do not conform to this requirement shall be classified as being in Zone B.

6. Durability Testing—The durability test shall be run with each flasher from the lot selected in paragraph 5 connected in a standard test circuit and with the power source for the durability test adjusted to apply 13.0 V to the input terminals of the standard test circuit throughout the test. The durability test shall be made with the specified design load connected to each output terminal. Ambient temperature during test shall be 75 ± 10°F (24 ± 5.5°C) and mode of operation shall be continuous except for two test measurement points which shall be at 25 test h and, if necessary, at 100 test h. At each point of measurement each flasher shall be tested in accordance with paragraphs 4.2, 4.3, and 4.4(a) under Performance Testing and in the standard test circuit.

7. Durability Test Quality Criteria—Zero flashers in Zone C at 25 test h shall indicate acceptable quality. Two or more flashers in Zone C at 25 test h shall indicate unacceptable quality. In the event one flasher is in Zone C at 25 test h, the remaining flashers shall be run to 100 test h. Upon conclusion of the 100 h test, an average laboratory life of the tested flashers equal to or greater than 60 h shall indicate acceptable quality. An average laboratory life of less than 60 h shall indicate unacceptable quality. At each point of measurement, each flasher that falls within Zone C shall be considered a failure. Each such failed flasher shall be considered to have a test life equal to the number of hours corresponding to the last prior point of measurement for the purpose of computing the average laboratory life of the tested flashers.

TABLE 2—VOLTAGE DROP

Warning Signal Lamp Load	SAE J1054 V	Zone C V
Specified Design Load	0.8 max	> 0.8

TABLE 3—ACCEPTANCE CRITERIA

	Acceptance Criteria[a]		
	Max Qty Flashers in Zone A	Max Qty Flashers in Zone B	Max Qty Flashers in Zone C
For Performance Testing[b]	3 less Zone B	1	0
For Durability Testing[c]	0 in Zone C in 25 h or Minimum Average Laboratory Life of 60 h.		

[a] Each flasher in the sample shall be classified only by its largest deviation from SAE J1054.
[b] See paragraph 5.
[c] See paragraph 6 and 7.

8. Extreme Temperature Operation

8.1 Select a random sample of three flashers, from the remaining sample of 22 (or 12) flashers, which conform to the Engineering Standard when tested in accordance with paragraph 4.4(a). Submit each unit to the tests of paragraph 8.2.

8.2 The flashers shall be stabilized in ambient temperatures of 145 ± 5°F (63 ± 3°C) and −25 ± 5°F (−32 ± 3°C) for a period of 1 h and then operated with the manufacturer's specified design load and with 12.8 V at the bulbs. In each of these tests, for each unit the starting time shall not exceed 5 s and the flash rate shall be 30–150 fpm at each load terminal after five complete cycles and shall be the average of at least three consecutive cycles.

SPOT LAMPS—SAE J591b

SAE Standard

Report of Lighting Committee approved October 1951 and last revised December 1972.

Definition—Spot lamps are lamps which provide a substantially parallel beam of light and which can be aimed at will.

General Requirements—The spot lamp beam pattern should be well defined and substantially round or oval in shape.

The following sections from SAE J575 are a part of this standard.

Section B—Samples for Test
Section C—Lamp Bulbs
Section D—Laboratory Facilities
Section E—Vibration Test
Section F—Moisture Test
Section G—Dust Test—Where sealed units are used, the dust test shall not be required.
Section H—Corrosion Test

Color Test—The color of light from spot lamps should be white.

APPENDIX

As a matter of information, attention is called to typical sockets shown in SAE J567.

PARKING LAMPS (FRONT POSITION LAMPS)—SAE J222a

SAE Standard

Report of Lighting Committee approved December 1970 and last revised January 1977. Rationale statement available.

1. *Scope*—This Engineering design standard provides design parameters and general requirements for parking lamps (front position lamps). It is intended for use in conjunction with the Service Performance Specification, SAE J256.

2. *Definition*—Whether separate or in combination with other lamps, parking lamps (front position lamps) are lamps on both the left and right of the vehicle which show to the front and are intended to mark the vehicle when parked. In addition, these front lamps serve as a reserve front position indicating system in the event of headlamp failure.

3. *Laboratory Requirements*

   3.1 *General Requirements*—The following sections from SAE J575 are a part of this standard:

     Section 2 —Samples for Test
     Section 2.2—Bulbs
     Section 2.3—Test Fixture
     Section 3 —Laboratory Facilities
     Section 4.1—Vibration Test
     Section 4.2—Moisture Test
     Section 4.3—Dust Test
     Section 4.4—Corrosion Test
     Section 4.6—Photometry
     Section 4.8—Warpage Test on Devices with Plastic Components

   3.2 *Plastic Materials*—Any plastic materials used in optical parts shall comply with the requirements set forth in SAE J576.

   3.3 *Color Test*—The color of light from parking lamps (front position lamps) shall be white or yellow (amber). (See SAE J578.)

   3.4 *Photometric Requirements*—Photometric tests shall be made at a distance of at least 4 ft (1.2 m). The H-V axis of a parking lamp (front position lamp) shall be taken as parallel with the longitudinal axis of the vehicle. Candlepower (candela) requirements for parking lamps (front position lamps) are shown in Table 1.

   3.5 *Parking Lamp in Combination*—When a parking lamp (front position lamp) is optically combined with a turn signal and the parking lamp (front position lamp) is connected to be operated with the headlamps, the turn signal shall not be less than three times the candlepower of the parking lamp (front position lamp) at any test point on or above horizontal, except that at H-V, H-5L, H-5R, and 5U-V the turn signal shall not be less than five times the candlepower of the parking lamp (front position lamp). When a two-filament bulb is used, the bulb shall have an indexing base and the socket shall be designed so that bulbs with nonindexing bases cannot be used.

APPENDIX

As a matter of information, attention is called to SAE J567 for requirements and gages to be used in socket design.

TABLE 1—PHOTOMETRIC REQUIREMENTS

Test Point, deg		Min Candle-power (Candela)	Test Point, deg		Min Candle-power (Candela)
10U	10L	0.8	H	5R	3.6
	V	0.8		10R	1.4
	10R	0.8		20R	0.4
5U	20L	0.4	5D	20L	0.4
	10L	0.8		10L	0.8
	5L	1.4		5L	1.4
	V	2.8		V	2.8
	5R	1.4		5R	1.4
	10R	0.8		10R	0.8
	20R	0.4		20R	0.4
H	20L	0.4	10D	10L	0.8
	10L	1.4		V	0.8
	5L	3.6		10R	0.8
	V	4.0			

CLEARANCE, SIDE MARKER, AND IDENTIFICATION LAMPS—SAE J592f

SAE Standard

Report of Lighting Division approved January 1937 and last revised by Lighting Committee September 1977. Rationale statement available.

1. *Scope*—This engineering design standard provides design parameters and general requirements for clearance, side marker, and identification lamps. It is intended for use in conjunction with the Service Performance Specification, SAE J256 (July, 1971).

2. *Definitions*

   2.1 *Clearance Lamps*—Lamps which show to the front or rear of a vehicle, mounted on the permanent structure of the vehicle as near as practicable to the upper left and right extreme edges to indicate the overall width and height of the vehicle.

   2.2 *Side Marker Lamps*—Lamps which show to the side of the vehicle, mounted on the permanent structure of the vehicle as near as practicable to the front and rear edges to indicate the overall length of the vehicle. Additional lamps may also be mounted at intermediate locations on the sides of the vehicle.

   2.3 *Combination Clearance and Side Marker Lamps*—Single lamps which simultaneously fulfill the requirements of clearance and side marker lamps.

   2.4 *Identification Lamps*—Lamps used in groups of three, in a horizontal row, which show to the front or rear or both, having lamp centers spaced not less than 6 in (150 mm) nor more than 12 in (310 mm) apart, mounted on the permanent structure as near as practicable to the vertical centerline and the top of the vehicle to identify certain types of vehicle.

3. *Laboratory Requirements*

   3.1 *General Requirements*—The following sections from SAE J575g (March, 1978) are a part of this standard:

     3.1.1 SECTION 2—Samples for Test
     3.1.2 SECTION 2.2—Bulbs
     3.1.3 SECTION 2.3—Test Fixture
     3.1.4 SECTION 3—Laboratory Facilities
     3.1.5 SECTION 4.1—Vibration Test
     3.1.6 SECTION 4.2—Moisture Test
     3.1.7 SECTION 4.3—Dust Test
     3.1.8 SECTION 4.4—Corrosion Test
     3.1.9 SECTION 4.6—Photometry

TABLE 1—PHOTOMETRIC MINIMUM CANDELA REQUIREMENTS CLEARANCE, SIDE MARKER, IDENTIFICATION LAMPS

Test Points, deg		Clearance, Side Marker, and Identification Lamps	
		Red[a]	Yellow (Amber)
10U	45L[b]	0.25	0.62
	V	0.25	0.62
	45R[b]	0.25	0.62
H	45L[b]	0.25	0.62
	V	0.25	0.62
	45R[b]	0.25	0.62
10D	45L[b]	0.25	0.62
	V	0.25	0.62
	45R[b]	0.25	0.62

[a] The maximum light output for red clearance and identification lamps is 15 cd. When red clearance lamps are optically combined with stop or turn signal lamps, this maximum applies on or above the horizontal.

[b] The requirements for side markers used on vehicles less than 80 in (2032 mm) wide may be met for inboard test points at a distance of 15 ft (4.6 m) from the vehicle on a vertical plane that is perpendicular to the longitudinal axis of the vehicle and located midway between the front and rear side marker lamps.

3.1.10 Section 4.8—Warpage Test on Devices with Plastic Components

3.2 Color Test—The color of light from front clearance lamps, front and intermediate side marker lamps, and front identification lamps shall be yellow (amber). The color of light from rear clearance lamps, rear side marker lamps, and rear identification lamps shall be red. (See SAE J578c (February, 1977).)

3.3 Plastic Materials—Any plastic materials used in optical parts shall comply with the requirements set forth in SAE J576d (June, 1976).

3.4 Photometric Requirements—Photometric tests shall be made at a distance of at least 4 ft (1.2 m). The H-V axis of a clearance or identification lamp shall be taken as parallel with the longitudinal axis of the vehicle. The H-V axis of a side marker lamp shall be taken as normal to the longitudinal axis of the vehicle. The H-V axis of a combination clearance and side marker lamp shall be taken as parallel with the longitudinal axis of the vehicle when checking clearance lamp test points, and normal to this vehicle axis when checking side marker test points. In all cases, the H-V axis shall be taken as parallel to the surface on which the vehicle stands.

Candela requirements for clearance, identification, and side marker lamps are shown in Table 1. Combination clearance and side marker lamps shall comply with both clearance and side marker minimum candela requirements.

ϕ APPENDIX

As a matter of information, attention is called to SAE J567c (December, 1970) for requirements and gages to be used in socket design.

TABLE 1 — PHOTOMETRIC MINIMUM CANDELA REQUIREMENTS[a]

Test Points	45L	30L	10L	V	10R	30R	45R
10U	—	—	10	15	10	—	—
5U	15	—	20	25	20	—	15
H	15	25	50	80	50	25	15
5D	15	25	50	80	50	25	15

[a]The maximum per lamp at H and above shall be 300 cd for a two-lamp system and 500 cd for a single lamp system.

BACKUP LAMPS—SAE J593e

SAE Standard

Report of Lighting Committee approved August 1947 and last revised March 1974.

1. Scope—This SAE Standard provides test methods and engineering requirements for motor vehicle backup lamps.

2. Definition—Backup lamps are devices used to provide illumination behind the vehicle and to provide a warning signal to pedestrians and other drivers when the vehicle is backing up or is about to back up.

3. Laboratory Requirements

3.1 General Requirements—The following sections from SAE J575 are a part of this standard:

3.1.1 Section B—Samples for Test
3.1.2 Section C—Bulbs
3.1.3 Section D—Laboratory Facilities
3.1.4 Section E—Vibration Test
3.1.5 Section F—Moisture Test
3.1.6 Section G—Dust Test
3.1.7 Section H—Corrosion Test
3.1.8 Section J—Photometry

3.2 Sealed units designed for use as a backup lamp when tested without the other parts of the lamp assembly need comply only with sections B, D, and J of SAE J575.

3.3 Color Test—The color of the light from a backup lamp shall be white, in accordance with SAE J578, and normally shall be established by visual appraisal.

A backup lamp may project incidental red, yellow, or white light through reflectors or lenses that are adjacent, close to, or a part of the lamp assembly. If a lamp has portions of its lens which project nonwhite light, that light shall be regarded as incidental if, quantitatively, it does not exceed 20% of the total device output at all specified test points; the lamp shall also meet the photometric requirements of this standard with white light alone.

3.4 Plastic Materials—Any plastic materials used in optical parts shall comply with the requirements set forth in SAE J576.

3.5 Photometric Test

3.5.1 Photometric tests shall be made with the photometer at a distance of at least 10 ft (3 m) from the lamp. The H-V axis shall be taken as parallel to the longitudinal axis of the vehicle.

3.5.2 The light from a single lamp, when used in a two-lamp system, shall meet the photometric requirements shown in Table 1.

3.5.3 When only one backup lamp is used on the vehicle, it shall meet twice the photometric requirements of Table 1.

3.5.4 When two lamps of the same or symmetrically opposite design are used, the reading along the vertical axis and the averages of the readings for the same angles left and right of vertical for one lamp shall be used to determine compliance with the requirements. If two lamps of differing designs are used, they shall be tested individually and the values added to determine that the combined units meet twice the candlepower requirements.

4. Installation Requirements—The following requirements apply to the device as used on the vehicle and are not part of the laboratory test requirements and procedures:

4.1 The backup lamp shall be illuminated only when the ignition switch is energized and reverse gear is engaged.

4.2 Backup lamps shall not be lighted when the vehicle is in forward motion.

4.3 Backup lamps shall be mounted on the rear so that the center of the lens of at least one lamp is visible from any eye point elevation from at least 6 ft to 2 ft (1.8 m to 0.6 m) above the horizontal plane on which the vehicle is standing; and from any position in the area, rearward of a vertical plane perpendicular to the longitudinal axis of the vehicle, 3 ft (0.9 m) to the rear of the vehicle and extending 3 ft (0.9 m) beyond each side of the vehicle.

BACKUP LAMP SWITCHES—SAE J1076

SAE Recommended Practice

Report of Lighting Committee approved February 1974.

1. Scope—This SAE Recommended Practice establishes performance requirements and related test procedures for backup lamp switches which are intended for use in motor vehicles.

2. Definition

2.1 Type "A"—A transmission-mounted backup lamp switch is that device which is mounted in or on the transmission and actuated by a moving part within the transmission that energizes the backup lamps when the transmission is shifted into reverse.

2.2 Type "B"—A backup lamp switch performing the same function as Type "A," except that it is operated by mechanism external of the transmission but not mounted in the passenger compartment.

2.3 Type "C"—A backup lamp switch performing the same function as Type "A," but mounted in the passenger compartment and actuated by movement of the shift mechanism or linkage.

3. Reference Standards—When testing Type "A" units, the transmission fluid shall comply with the following SAE specifications: SAE J308 and SAE J311.

4. Temperature Test

4.1 Types "A" and "B"—The switch shall be installed in a device simulating its vehicle operating conditions. To insure basic function, the switch shall be manually cycled for 10 cycles at design electrical load after 1 h exposure at each of the following ambient temperatures:

°F	°C
75 ± 10	24 ± 5.5
225 (+0, −5)	107 (+0, −2.8)
−25 (+5, −0)	−32 (+2.8, −0)

[a]Due to a potentially large variable in geographical and environmental conditions in a specific vehicle design, this temperature may be exceeded in a given application. If 225 °F (107°C) is exceeded in a given application, it is the responsibility of the switch manufacturer to specify the high temperature test.

The switch shall be electrically and mechanically operable during each of these cycles.

4.2 Type "C"—The temperature test shall be conducted the same as for Types "A" and "B" except the ambient temperatures shall be:

°F	°C
75 ± 10	24 ± 5.5
165 (+0, −5)	74 (+0, −2.8)
−25 (+5, −0)	−32 (+2.8, −0)

4.3 This same switch shall be used for the endurance test described below.

5. Endurance Test Setup

5.1 The switch shall be set up to operate its design electrical load.

5.2 One complete cycle shall consist of energizing and de-energizing the backup lamps, with switch travel as specified by the manufacturer. The test equipment shall function within the following mechanical timing requirements.

Travel time—0.1–0.5 s (time from one position to the next position)

Dwell time—1.0–2.0 s (time in each position)

5.3 During the test, the switch shall be operated at 6.4 V d-c for a 6 V system, 12.8 V d-c for a 12 V system, or 25.6 V d-c for a 24 V system, measured at the input termination of the switch. The power supply shall not generate any adverse transients not present in motor vehicles and shall comply with the following specifications:

(a) Output current—Capable of supplying a continuous output current of the design load and inrush currents as required by the bulb load complement.

(b) Regulation—

Dynamic—The output voltage shall not deviate more than 1.0 V from zero to maximum load (including inrush current) and should recover 63% of its maximum excursion within 100 ms.

Static—The output voltage shall not deviate more than 2% with changes in static load from zero to maximum (not including inrush current) and means shall be provided to compensate for static input line voltage variations.

(c) Ripple voltage—Maximum 0.30 V peak to peak.

6. Endurance Requirements

6.1 The switch shall be capable of operating for 50,000 cycles at 75 ± 10°F (24 ± 5.5°C).

6.2 The voltage drop from the input terminal to the output terminal shall be measured before and after the completion of the endurance test and shall not exceed 0.30 V (the average of three consecutive readings) at design load.

If wiring is an integral part of the switch, the voltage drop measurement shall be made including 3 in. (76 mm) of wire on each side of the switch; otherwise, measurement shall be made at switch terminals.

HEADLAMP SWITCH—SAE J253

SAE Recommended Practice

Report of Lighting Committee approved July 1971.

1. Definition—The headlamp switch is an operator actuated device for control of various vehicle light sources. Primary function is to control headlights, park lights, tail lights, and certain marking lights. A secondary function may be one of control of various accessory and instrument lights. Circuit breaker(s) may be incorporated for circuit overload protection.

2. Temperature Test

2.1 To insure basic function, the switch shall be manually cycled for 10 cycles at design electrical load at: 75 ± 10 F (24 ± 5.5 C); 165 +0, −5 F (74 +0, −2.8 C); −25 +5, −0 F (−32 +2.8, −0 C). This to be done after a 1 h exposure at each of these temperatures. The switch shall be electrically and mechanically operable during each of these cycles.

2.2 This same switch shall be used for the endurance test described below.

3. Endurance Test Setup

3.1 The switch shall be set up to operate its design electrical load. (Both primary and secondary circuit function design electrical loads.)

3.2 The test shall be set up to operate the switches for the prescribed number of completed cycles.

One complete cycle shall consist of sequencing through each position (with dwell in each position) and return without dwell in intermediate positions to the initial position.

The test equipment shall be so arranged as to provide the following mechanical time requirements:

Travel time—0.1–0.5 s (time from one position to the next).

Dwell time—1.0–2.0 s (time in each position).

3.3 During the test the switch shall be operated at 6.4 V for a 6 V system, 12.8 V d-c for a 12 V system, or 25.6 V d-c for a 24 V system. These voltages shall be measured at the input termination on the switch.

The power supply shall not generate any adverse transients not present in motor vehicles and shall comply with the following specifications:

(a) Output current—Capable of supplying the continuous current of the design electrical load and inrush current as required by the bulb load complement.

(b) Regulation—

Dynamic—The output voltage at the supply shall not deviate more than 1.0 V from zero to maximum load (including inrush current) and should recover 63% of its maximum excursion within 100 ms.

Static—The output voltage at the supply shall not deviate more than 2% with changes in static load from zero to maximum (not including inrush current), and means shall be provided to compensate for static input line voltage variations.

(c) Ripple voltage—Maximum 300 mV peak to peak.

4. Endurance Requirements

4.1 The switch shall be capable of satisfactorily operating for 25,000 complete cycles at a temperature of 75 ± 10 F (24 ± 5.5 C) followed by 1 h ON in headlamp position at 75 ± 10 F (24 ± 5.5 C).

4.2 The voltage drop from the input terminal(s) to the corresponding output terminal(s) shall be measured before and after the completion of the endurance test and shall not exceed 0.30 V (the average of three consecutive readings) at design load. These voltage drop readings should exclude the voltage drop across the circuit breaker(s). If wiring is an integral part of the switch, the voltage drop measurement shall be made including 3 in. of wire on each side of the switch; otherwise, measurement to be made at switch terminals.

ϕ REFLEX REFLECTORS—SAE J594f

SAE Standard

Report of Lighting Committee approved July 1971 and completely revised January 1977.

1. Scope—This SAE Standard is an engineering design standard for reflex reflectors. This design standard is intended to be supplemented by an SAE service performance standard for reflex reflectors which is under development.

2. Definition—Reflex reflectors, for the purpose of this specification, include only devices which are used on vehicles to give an indication to an approaching driver by reflected light from the lamps on the approaching vehicle.

3. Requirements

3.1 General—The following sections from SAE J575 are a part of this standard:

3.1.1 SECTION B—Samples for Test

3.1.2 SECTION D—Laboratory Facilities

3.1.3 SECTION E—Vibration Test

3.1.4 SECTION F—Moisture Test—Except that in the case of sealed units there shall be no visible moisture within the unit.

3.1.5 SECTION G—Dust Test

3.1.6 SECTION H—Corrosion Test

3.1.7 SECTION J—Photometry—The reflex reflector shall be set up for testing as shown in Fig. 1. The test distance shall be 100 ft (30.5 m). The

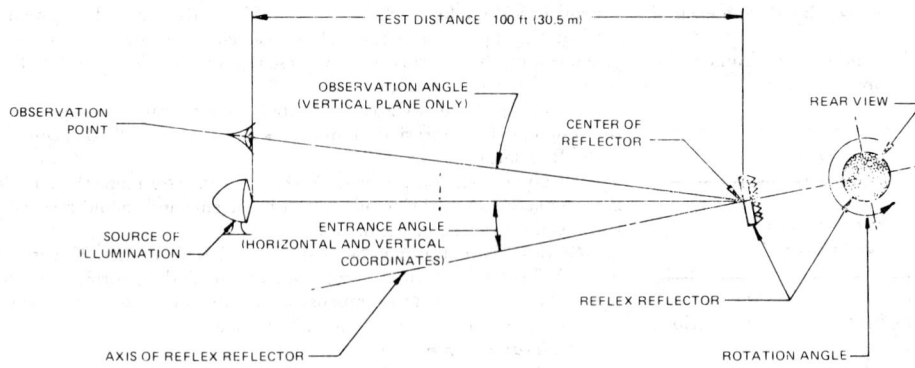

FIG. 1—SETUP FOR TESTING

source of illumination shall be a lamp with a 2 in (50 mm) effective diameter and with a filament operating at 2856 K color temperature. The observation point shall be located directly above the source of illumination. The reflex reflector shall be mounted on a goniometer with the center of the reflex area at the center of rotation and at the same horizontal level as the source of illumination. The H-V axis of reflex reflectors shall be taken as parallel to the longitudinal axis of the vehicle for rear reflectors and perpendicular to a vertical plane passing through the longitudinal axis of the vehicle for side reflectors.

Photometric measurements of reflex reflectors shall be made at various observation angles and entrance angles as shown in Table 1. The observation angle is the angle formed by a line from the observation point to the center of the reflector and a second line from center of the reflector to the source of illumination. The entrance angle is the angle between the axis of the reflex reflector and a line from the center of the reflector to the source of illumination. The entrance angle shall be designated left, right, up, and down in accordance with the position of the source of illumination with respect to the axis of the reflex reflector as viewed from behind the reflector.

Photometric measurements shall be made photoelectrically, and the candlepower which the reflex reflector is projecting toward the observation point shall be determined. Also, the illumination on the reflex reflector from the source of illumination shall be measured in footcandles. The recorded measurement of each test point is the quotient of the projected candlepower divided by the footcandle illumination. Reflex reflectors may have any linear or area dimensions; but, for the photometric test a maximum projected area of 12 in^2 (7740 mm^2) contained within a 10 in (254 mm) diameter circle shall be exposed.

In making photoelectric measurements, the opening to the photocell shall not be more than 0.5 in (13 mm) vertical by 1 in (25 mm) horizontal with the observation point above the source of illumination.

Reflex reflectors, which do not have a fixed rotational position on the vehicle, shall be rotated about their axis through 360 deg to find the minimum candlepower per footcandle which shall be reported for each test point. If the output falls below the minimum requirement at any test point, the reflector shall be rotated ±5 deg about its axis from the angle where the minimum output occurred; and the maximum candlepower per footcandle within this angle shall be reported as a tolerance value.

Reflex reflectors, which, by their design or construction, permit mounting on the vehicle in fixed rotational position, shall be tested in this position. A visual locator, such as the word TOP shall not be considered adequate to establish a fixed rotational position on the vehicle.

If uncolored reflections from the front surface interfere with photometric readings at any test point, the operator shall check 1 deg above, below, right, and left of the test point, and report the lowest reading and location. The latter must meet the minimum requirements for the test point.

3.1.8 Color—Section I—Color Test—The test sample may be either the reflex reflector or a disc of the same material, technique of fabrication, and dye formulation as the reflex reflector. If a disc is used, the thickness should be twice the thickness of the reflector as measured from the face of the lens to the apexes of the reflecting elements.

3.2 Plastic Material Test—See SAE J576

TABLE 1—MINIMUM MILLI-CANDELAS PER INCIDENT LUX FOR A RED REFLEX REFLECTOR[a]

Observation Angle (deg)	Entrance Angle (deg)				
	0 deg	10 deg Up	10 deg Down	20 deg Left	20 deg Right
0.2	420	280	280	140	140
1.5	6	5	5	3	3

[a] Yellow values shall be 2.5 times indicated red values and white values shall be 4 times indicated red values.

TABLE 1A—MINIMUM CANDLEPOWER PER INCIDENT FOOTCANDLE—RED REFLEX REFLECTOR[a]

Observation Angle (deg)	Entrance Angle (deg)				
	0 deg	10 deg Up	10 deg Down	20 deg Left	20 deg Right
0.2	4.5	3.0	3.0	1.5	1.5
1.5	0.07	0.05	0.05	0.03	0.03

[a] Yellow values shall be 2.5 times indicated red values and white values shall be 4 times indicated red values.

EMERGENCY WARNING DEVICE—SAE J774c

SAE Recommended Practice

Report of Lighting Committee approved June 1961 and last revised January 1971.

1. Scope—This SAE Recommended Practice establishes performance requirements and related test procedures for emergency warning devices intended for highway use.

2. Definition—An emergency warning device is a device to be placed on the roadway to warn the driver of an approaching vehicle of a stationary hazard by reflection of light from the lamps of the approaching vehicle at night or by a flag or fluorescent area in the daytime.

3. Laboratory Requirements

3.1 General—The following sections from SAE J575 are a part of this standard with modifications as detailed below. All tests shall be run on a single device. At the conclusion of tests in the order shown, the device shall be photometered and shall meet the minimum specific intensity values listed in Tables 1 or 2.

3.1.1 Section B—Samples for Test

3.1.2 Section E—Vibration Test—The complete device in its opaque container shall be tested in stored position. If a means is not provided to attach device securely to the vehicle, the device in its container shall be vibration tested in a metal box on the test equipment with clearance of 1 in. to the

TABLE 1—MINIMUM CANDLEPOWER PER INCIDENT FOOT-CANDLE—TYPE 1 DEVICES

Observation Angle, deg	Entrance Angle, deg			
	0 Horizontal 0 Vertical	0 Horizontal 5 U & D	0 Vertical 20 L & R	0 Vertical 30 L & R
0.2	50.0	50.0	20.0	5.0
1.5	1.0	1.0	0.4	0.1

TABLE 2—MINIMUM CANDLEPOWER PER INCIDENT FOOT-CANDLE—TYPE 2 DEVICES

Observation Angle, deg	Entrance Angle, deg				
	0	10 Up	10 Down	20 Left	20 Right
0.2	27.0	20.0	20.0	10.0	10.0
1.5	0.14	0.10	0.10	0.06	0.06

closest surface of the device when the device is at rest. The reflex reflectors should show no evidence of surface abrasion at the conclusion of the test.

3.1.3 Section G—Dust Test—Device shall be tested in its functional position. All units shall be subjected to dust whether sealed or not sealed. The "before and after" photometric tests are not required as the photometric performance is evaluated after all mechanical tests are complete.

3.1.4 Section F—Moisture Test—The device shall be tested in functional position. There should be no visible moisture within the device. There should be no visible deterioration of the device that would impair its function.

3.1.5 Section H—Corrosion Test—A reflex reflector unit shall be tested in functional position.

3.2 Color Test—Reflex Reflector—The reflex reflector indication shall be red. See SAE J578. The test sample may be either the reflex reflector or a disc of the same material, technique of fabrication, and dye formulation as the reflex reflector. If a disc is used, the thickness shall be twice the thickness of the reflector as measured from the face of the lens to the apexes of the reflecting elements.

3.3 Color Test—Daytime Warning Area—See SAE J943 for color and test method.

3.4 Plastic Material Test—See SAE J576.

3.5 Photometric Testing—See SAE J594. One emergency warning device shall be tested in accordance with SAE J594 except total values for each side should be as specified in Table 1 for Type 1 devices and Table 2 for Type 2 devices. The total area of the device shall be photometered either in whole or in parts with particular caution regarding beam uniformity. The device shall be tested in functional position. The device shall be securely supported in its functional position during the photometry to prevent relative movement between the axis of the device and the axis of the goniometer.

4. General Requirements

4.1 The Complete Set—A complete emergency warning set may consist of one to three emergency warning devices and a container or rack for storage. During storage the faces of the individual emergency warning devices should be protected from dirt and damage. The emergency warning devices shall be readily extractable from the storage container or rack and should be readily set up in functional position without the use of tools.

4.2 The Emergency Warning Device—Two types of emergency warning devices are described as follows:

4.2.1 A Type 1 warning device shall be an equilateral triangle, each side of which shall display both a daytime and nighttime warning area. The daytime warning shall be a red-orange fluorescent area meeting the color and reflectivity requirements of SAE J943 and shall measure at least 14 in. outside leg length by $1\frac{1}{4}$ in. minimum width. The nighttime warning shall be a retro-reflective area meeting the photometric requirements of Table 1 and shall measure at least 14 in. inside leg length by $\frac{1}{2}$ in. minimum width. With the device in functional position, the lower edge of the horizontal leg of the triangle shall be a minimum of 1 in. above the road surface.

4.2.2 A Type 2 warning device shall consist of a holder with two reflex reflectors on each side, one above the other (two bidirectional reflectors may be used). The effective area of individual reflex reflectors shall be at least 6 in.^{2} The holder shall provide positive means for maintaining the reflector faces perpendicular to the surface on which the warning device rests. With the warning device in functional position, the top of the effective luminous area shall be at least 7 in. above the surface upon which the device rests.

4.3 Wind Test—Emergency warning devices shall be tested for their ability to withstand wind pressure. If flags or flag mounting means are provided, flags shall be included in the Wind Test. One unit in functional position shall be set upon a horizontal brushed concrete surface and subjected to a wind of 40 mph velocity blowing in a horizontal direction. The unit shall be tested with the wind direction perpendicular to the reflector faces, first on one side, second the other side, and then at three intermediate positions. Tipping, turning, or sliding on the supporting surface shall be cause for rejection.

FLASHING WARNING LAMPS FOR AUTHORIZED EMERGENCY, MAINTENANCE AND SERVICE VEHICLES—SAE J595b — SAE Standard

Report of Lighting Committee approved December 1948 and last revised July 1964. Reaffirmed without change May 1972.

Definition—Flashing warning lamps covered in this standard are for use on authorized emergency, maintenance and service vehicles.

Identification Code Designation—The identification code designation is W1.

General Requirements—Warning lamps should have at least 12 sq in. of effective illuminated area. The exposed illuminated area of the lamp need have no word, letter, nor other device intended to identify the signal.

The following sections from SAE J575 are a part of this standard.

Section B—Samples for Test
Section C—Lamp Bulbs
Section D—Laboratory Facilities
Section E—Vibration Test
Section F—Moisture Test
Section G—Dust Test
Section H—Corrosion Test
Section J—Photometric Test—All beam candlepower measurements shall be made with a bar photometer or equivalent, with the filament of the lamp at a distance of at least 10 ft from the photometer screen. The lamp axis shall be taken as the light source parallel to the longitudinal axis of the vehicle. (For minimum candlepower requirements see Table 1.)
Section L—Warpage Test on Devices with Plastic Lenses

Bulb Sockets—See SAE J567.

Color Test—The warning lamp indication shall be red or amber. See SAE J578.

General Warning Signal System Recommendations—(These general recommendations are not a part of the test specifications.)

Each vehicle should be equipped with two flashing lamps on the rear and two flashing lamps on the front.

Front and rear warning lamps should be mounted as high and as far apart as practicable, but in no case should the lateral spacing be less than 3 ft. The location of front warning lamps should be such that they can be clearly distinguished when the headlamps are lighted on the lower beam.

TABLE 1—MINIMUM CANDLEPOWER REQUIREMENTS

Test Points, deg		Candlepower, Min		Test Points, deg		Candlepower, Min	
		Red	Amber			Red	Amber
10U	5L	20	60	H	V	300	900
	V	50	150		5R	200	600
	5R	20	60		10R	75	225
					20R	30	90
5U	20L	20	60				
	10L	50	150	5D	20L	20	60
	5L	100	300		10L	50	150
	V	150	450		V	150	450
	5R	100	300		10R	50	150
	10R	50	150		20R	20	60
	20R	20	60				
				10D	5L	20	60
H	20L	30	90		V	50	150
	10L	75	225		5R	20	60
	5L	200	600				

Visibility of front warning lamps to the front and of rear warning lamps to the rear should be unobstructed by any part of the vehicle 10 deg above to 10 deg below the horizontal and from 45 deg to the right to 45 deg to the left of the centerline of the vehicle.

Warning lamps should flash no slower than 60, nor faster than 120, times per minute under normal operating conditions. The "on" period of the flasher should be between 30 to 75%.

There should be a visible or audible means of giving a clear and unmistakable indication to the driver when the warning lamps are turned "on" and functioning normally.

To improve the effectiveness of the signal, it is recommended that, where practical, the area of the vehicle immediately surrounding the signal be painted black.

HAZARD WARNING SIGNAL SWITCH—SAE J910b

SAE Standard

Report of Lighting Committee approved January 1965 and last revised June 1971. Editorial change October 1977.

1. Definition—A hazard warning signal switch is a driver controlled device which causes at least one turn signal lamp on the left and the right to the front and left and the right to the rear of the vehicle to flash simultaneously to indicate to the approaching driver the presence of a vehicular hazard.

φ *2. Reference Standards*—The following sections from SAE J575f (April, 1975) are a part of this standard:
  Section B—Samples for Test
  Section C—Lamp Bulbs
  Section D—Laboratory Facilities

3. Temperature Test

3.1 To insure basic function, the switch shall be manually cycled for 10 cycles at design electrical load at: 75 ± 10 F (24 ± 5.5 C); 165 +0, −5 F (74 +0, −2.8 C); −25 +5, −0 F (−32 +2.8, −0 C). This to be done after a 1 h exposure at each of these temperatures. The switch shall be electrically and mechanically operable during each of these cycles.

3.2 This same switch shall be used for the endurance test described in paragraph 5.

4. General Requirements

4.1 The hazard warning signal switch function shall operate independently of the ignition switch.

4.2 If the hazard warning signal switch requires the operation of more than one switch, a means shall be provided for actuating all switches simultaneously by a single driver action.

5. Endurance Test Setup

5.1 The switch shall be operated with a maximum design bulb load stated by the switch manufacturer with the flasher not included in the test circuit. Failed bulbs shall be replaced during the test. (In cases where the flasher is an integral part of the assembly, it shall be included.)

5.2 The test shall be set up to operate for the prescribed number of cycles. One cycle shall consist of the following sequence of position: off, on, off. The test equipment shall function within the following mechanical time requirements at a cycle rate of 12–40 cycles/minute:
  Travel time—0.1–0.5 s (time from one position to the next position)
  Dwell time—0.4 s min (in each position)

5.3 During the test the switch shall be operated at 6.4 V d-c for a 6 V system, 12.8 V d-c for a 12 V system, or 25.6 V d-c for a 24 V system, measured at the input of the switch. The power supply shall not generate any adverse transients not present in motor vehicles and shall comply with the following specifications:

  (a) Output current—Capable of supplying a continuous output current of the design load and inrush currents as required by the bulb load complement.
  (b) Regulation—
    Dynamic—The output voltage shall not deviate more than 1.0 V from zero to maximum load (including inrush current) and should recover 63% of its maximum excursion within 5 ms.
    Static—The output voltage shall not deviate more than 2% with changes in static load from zero to maximum (not including inrush current), and means shall be provided to compensate for static input line voltage variations.
  (c) Ripple voltage—Maximum 300 mV, peak to peak.

6. Endurance Requirements

6.1 The switch shall be capable of satisfactorily operating for 10,000 cycles at a temperature of 75 ± 10 F (24 ± 5.5 C) followed by 1 h "on" at a temperature of 75 ± 10 F (24 ± 5.5 C).

6.2 The voltage drop from the input terminal(s) to the corresponding output terminal(s) shall be measured before and after the endurance test and shall not exceed 0.30 V. If wiring is an integral part of the switch, the voltage drop measurement is to be made including 3 in. of wire on each side of switch; otherwise, measurement is to be made at switch terminals. Care shall be taken not to include the voltage drop of other devices in the circuit.

7. Installation Requirements

7.1 The following requirements apply to the device as used on the vehicle and are not part of the laboratory test requirements and procedures.

7.2 *Pilot Indicator Lamps*—For vehicles equipped with right- and left-hand turn signal pilot indicators, both pilots and/or separate pilots shall flash simultaneously while the hazard system is turned on. In vehicles equipped with a single pilot turn signal indicator, a separate hazard warning pilot indicator shall flash and the turn signal pilot may flash while the hazard warning signal switch is turned on. If a separate hazard pilot indicator is used, it shall be red in color and have a minimum area equivalent to a 0.5 in. diameter circle.

8. Combination Turn Signal and Hazard Warning Signal Switch

8.1 The same combination switch shall be used for the test of each function. The hazard warning signal switch shall meet the requirements of this standard. The turn signal switch shall meet the requirements of SAE φ J589b (June, 1971).

8.2 The operating motion of the hazard warning signal switch function shall differ from the actuating motion of the turn signal function.

360 DEG EMERGENCY WARNING LAMP—SAE J845

SAE Recommended Practice

Report of Lighting Committee approved January 1963. Editorial change April 1964. Reaffirmed without change May 1972.

Purpose—The purpose of this SAE Recommended Practice is to provide minimum performance requirements and test procedures for 360 deg emergency warning signal lamps.

Definition—360 deg emergency warning signal lamps are devices for use on authorized emergency vehicles.

General Requirements—This standard specifically pertains to that type of signal device that will project a flashing beam signal throughout 360 deg on the horizontal plane passing through the center of the light source. The device shall project a beam of light on a vertical plane rotating around the horizontal axis of the lamp and shall have a total light spread of at least 10 deg extending a minimum of 5 deg above and below the horizontal axis of the light source.

The flash rate when observed from a fixed position shall be between 60 and 120 flashes per minute. When the flash rate is produced by current interruption, the on period shall be long enough to permit the bulb to come up to full brightness.

The color of the lens when used on vehicles on highway emergency service missions shall be red or amber.

The lamp dome or cover lens shall project a minimum area of 16 sq in. as measured on a vertical plane passing through the center of the lamp.

General Test Provisions—The following sections from SAE J575 are a part of this recommended practice:
  Section B—Samples for Test
  Section C—Lamp Bulbs
  Section D—Laboratory Facilities
  Section E—Vibration Test
  Section F—Moisture Test
  Section G—Dust Test
  Section H—Corrosion Test
  Section L—Warpage Test

Color Test—See SAE J578.

Extreme Temperature Tests

Heat—The device shall be subjected to an ambient temperature of 120 F continuously for 6 hr. From the beginning of the sixth hour to the end of the test, the device shall be lighted and operated at its normal rated voltage. At the end of the sixth hour, the device shall continue to function normally except that for the purpose of this test the flash rate shall not be greater than 130 flashes per minute.

Cold—The device shall be subjected to an ambient temperature of

−25 F for 6 hr. At the end of the sixth hour, the device shall be lighted and operated at its normal rated voltage.

After the unit has been allowed to operate for 3 minutes, the device shall function normally except that for the purpose of this test the flash rate shall not be less than 50 flashes per minute.

Photometric Test

Lamps Flashed by Current Interruption—All candlepower measurements shall be made with the incandescent filament of the signal lamp at least 15 ft or more from the photometer screen.

In photometric tests of lamps which are flashed by current interruption, the lamp shall be mounted so that the horizontal plane through the photometer axis passes through the center of the light source. The vertical axis through the center of the light source shall be perpendicular to this horizontal plane.

The lamp shall be turned about its vertical axis until the photometer indicates minimum candlepower. This shall be the H-V point. With the lamp in this position, the candlepower measured within the angular limits shall not be less than the values specified in the table of minimum photometric values.

Lamps Flashed by Rotation or Oscillation—All candlepower measurements shall be made with the incandescent filament of the signal lamp at least 60 ft or more from the photometer screen.

In photometric tests of lamps which are flashed by rotation, the lamps shall be mounted so that the horizontal plane through the photometer axis passes through the center of the light source of the rotating element. The vertical axis through the center of the light source shall be perpendicular to this horizontal plane.

TABLE 1—PHOTOMETRIC MINIMUM CANDLEPOWER REQUIREMENTS

Test Angles, deg	Candlepower, Min	
	Red	Amber
5 U to 5 D	50	125
2½ U to 2½ D	200	500

The rotating element shall be turned on its vertical axis until the photometer indicates maximum candlepower. This shall be the H-V point. The candlepower measured within the angular limits shall not be less than the values specified in Table 1.

Installation Recommendations—(These recommendations are not part of the test specification.)

1. In view of the wide variety of uses for 360 deg warning lamps, care must be taken to select a lamp with performance characteristics commensurate with the function and speed of the vehicle upon which it will be used.

2. It is desirable that all emergency warning lamps described by this specification be provided with an illuminated switch or pilot indicator to give the driver a clear and unmistakable indication that the emergency warning lamp system is turned on.

3. Emergency warning lamps should be mounted to provide 360 deg visibility at all times.

SCHOOL BUS RED SIGNAL LAMPS—SAE J887a — SAE Standard

Report of Lighting Committee approved July 1964 and last revised February 1975.

1. Definition—School bus red signal lamps are alternately flashing lamps mounted horizontally both front and rear, intended to identify a vehicle as school bus and to inform other users of highway that such vehicle is stopped on highway to take on or discharge school children.

2. Identification Code Designation—The identification code designation will be W2.

3. General Requirements—The effective projected illuminated area measured on a plane at right angles to the axis of the lamp must not be less than 19 in.^{2} (120 cm^2).

3.1 The following sections from SAE J575 are a part of this standard.
3.1.1 SECTION B—Samples for Test
3.1.2 SECTION C—Lamp Bulbs
3.1.3 SECTION D—Laboratory Facilities
3.1.4 SECTION E—Vibration Test
3.1.5 SECTION F—Moisture Test
3.1.6 SECTION G—Dust Test
3.1.7 SECTION H—Corrosion Test
3.1.8 SECTION J—Photometry
3.1.9 SECTION L—Warpage Test on Devices with Plastic Lenses

3.2 Sealed units designed for use as school bus signal lamps, when tested without the other parts of the lamp assembly, need only comply with Sections B, D, and J of SAE J575.

3.3 Color Test—See SAE J578

3.4 Photometric Test

3.4.1 All beam candela measurements shall be made with a bar photometer, or equivalent, with the filament of the lamp at a distance of at least 10 ft (3 m) from the photometer screen. The lamp axis shall be taken as the horizontal line through the light source parallel to what would be the longitudinal axis of the vehicle, if the lamp were mounted in its normal position on the vehicle.

3.4.2 An aiming tolerance of $\pm\frac{1}{2}$ deg vertical and ± 1 deg horizontal should be allowed for manufacturing and laboratory variations.

4. Aiming Provisions—The lamps should be equipped with aiming pads on the lens face suitable for use with mechanical headlamp aimer. The lamp should be designed so that with the aiming plane normal to the photometer axis, the beam shall meet the photometric specifications indicated in Table 1.

5. General Signal System Recommendations—(These general recommendations are not a part of the test specifications.)

5.1 Each vehicle should be equipped with two alternately flashing red lamps on the rear and two alternately flashing red lamps on the front. They should be controlled by a manually actuated switch and shall flash alternately at a rate of 60–120 cycles per min (1–2 Hz). The "on" period of the flasher should be long enough to permit the bulb filament to come up to full brightness. There should be a visible or an audible means of giving a clear and unmistakable indication to the driver when the signal lamps are turned on. Front and rear signal lamps should be spaced as far apart laterally as practical but in no case shall the spacing between the lamps be less than 40 in. (1000 mm).

5.2 The signal lamps should be mounted on the same horizontal center line as high as practical at the front above the windshield and on the same

TABLE 1—MINIMUM PHOTOMETRIC REQUIREMENTS FOR
RED SCHOOL BUS WARNING LAMPS

Test Point		Candela	Test Point		Candela
10U	5L	20			
	V	50			
	5R	20			
5U	20L	150	H	20R	180
	10L	300		30R	30
	5L	300			
	V	300	5D	30L	30
	5R	300		20L	200
	10R	300		10L	300
	20R	150		5L	450
				V	450
H	30L	30		5R	450
	20L	180		10R	300
	10L	400		20R	200
	5L	500		30R	30
	V	600			
	5R	500	10D	5L	40
	10R	400		V	40
				5R	40

horizontal center line as high as practical at the rear so that the lower edge of the lenses are not lower than the top line of the side window openings.

5.3 The visibility of the front signal lamps to the front and of the rear signal lamps to the rear should be unobstructed by any part of the vehicle from 10 deg above to 10 deg below horizontal and from 30 deg to the right to 30 deg to the left of center line of the lamps.

5.4 To improve the effectiveness of the signal, it is recommended that the area of the vehicle immediately surrounding the signal lamp be painted black.

5.5 Lamps should be mounted on the school bus with their aiming plane vertical and normal to the vehicle axis. A suggested tolerance is 1 deg in vertical aim and 2 deg in horizontal aim. If lamps are aimed or inspected with a mechanical headlamp aimer, the graduation settings for aim should be 0 down and 0 sideways. The limits for inspection should be from 5 up to 5 down and from 10 right to 10 left.

APPENDIX

As a matter of information, attention is called to SAE J567 for requirements and gages to be used in socket design and to SAE J571 (Figs. 1 and 3) for aiming pad dimensions and locations. Attention is also called to SAE J602 for mechanical aimers and to SAE J1054 for characteristics of alternating flashers to be used with these signal lamps.

SCHOOL BUS STOP ARM—SAE J1133a

SAE Recommended Practice

Report of Lighting Committee approved April 1976 and last revised by Lighting Committee November 1977. Rationale statement available.

φ **1. Scope**—This engineering design specification provides design parameters and general requirements for School Bus Stop Arms equipped with alternately flashing lights. The stop arm is intended as an optional device which may be used in addition to, but not as a substitute for, school bus signal lamps covered by SAE J887 (July, 1964). Decision for adoption of the recommendations presented in the Appendix is left to governmental units or other bodies who adopt this practice.

2. Definition—A school bus stop arm is an auxiliary device used to signal that a school bus has stopped to load or discharge passengers. It supplements devices specified by SAE J887. It may be operated automatically or manually and shall contain alternately flashing lights.

φ **3. General Requirements**—A stop arm shall have on both the front and rear the word STOP in letters which are at least 150 mm (6 in) in height and have a stroke of at least 20 mm (7/8 in); it also may optionally be reflectorized.

4. Laboratory Requirements

 4.1 The following sections from SAE J575 are a part of this recommended practice:
  4.1.1 Section 2—Samples for Test
  4.1.2 Section 3—Laboratory Facilities
  4.1.3 Section 4.1—Vibration Test
  4.1.4 Section 4.2—Moisture Test
  4.1.5 Section 4.3—Dust Test
  4.1.6 Section 4.4—Corrosion Test
  4.1.7 Section 4.6—Photometry
  4.1.8 Section 4.8—Warpage Test

 4.2 **Plastic Materials**—Any plastic materials used in optical parts shall comply with the requirements of SAE J576.

 4.3 **Color Test**—The color of the light from stop arm lamps shall be red, both to the front and rear and comply with the requirements of SAE J578.

 4.4 **Durability**—The device shall be subjected to a test of 45,000 cycles at a rate not to exceed 10 cpm (0.2 Hz) and at a temperature of 25° ± 3°C. A cycle shall consist of movement from the parked or retracted position to the fully extended position and return to parked position. Failure of the device to operate in the intended electrical or mechanical manner during or at the conclusion of the test shall constitute a failure.

Failure of the light source shall not be considered as failure of the device.

5. Photometric Requirements—Each lamp used shall have a designed light output and distribution to the front and rear as outlined for red lamps in Section 3.9.1 and Table 1 of SAE J588. For service performance, see SAE J256.

6. System Requirements

 6.1 Lamps shall be activated at the commencement of the stop arm extension cycle and deactivated when the stop arm is retracted.

 6.2 Lamps shall flash alternately in the range of 60–120 cpm.

 6.3 The sum of the percent "on" time for both lamps shall be 90%–110%; the minimum for either shall be 30%.

7. Installation Requirements—The following requirements apply to the device as used on the vehicle and are not a part of the laboratory test requirements and procedures.

 7.1 The stop arm shall be installed on the left outside of the bus body and be mounted so as to be seen readily when the arm is in the extended position by motorists approaching from either the front or rear of the bus.

 7.2 Double-faced lamps shall be located in the top and bottommost portions of the stop arm, one above the other.

 7.3 If the device is operated by a manual switch, that switch shall be located so as to be easily accessible to driver.

APPENDIX

A1. As a matter of information, attention is called to SAE J567 for requirements and gages to be used in socket design. Attention is also called to SAE J1054—Warning Lamp Alternating Flashers.

A2. It is recommended that the word STOP be displayed as white letters against a red background.

φ **A3.** It is also recommended that a stop arm have the shape of a regular octagon which is at least 450 mm (18 in) by 450 mm (18 in). The octagon should have a white border at least 12 mm (0.5 in) wide. The maximum extension should not exceed 560 mm (22 in) beyond the left side of the vehicle.

LIGHTING INSPECTION CODE—SAE J599d

SAE Standard

Report of Lighting Division approved January 1937 and last revised by Lighting Committee December 1974.

This code is intended only for the inspection and maintenance of lighting equipment on motor vehicles that are in use.

The original SAE code, adopted in 1937, was drafted for use in preparing Interstate Commerce Commission regulations for trucks and buses in interstate operation under the 1935 Motor-Carrier Act. Subsequently, the SAE code served as a basis for Section 2, Lighting Systems, of the American National Standard Code for Inspection Requirements for Motor Vehicles, ANSI D7-1939. The ANSI inspection requirements for lighting systems were adopted by the Society as the SAE Recommended Practice in January 1940.

1. Definitions

 1.1 **Sealed Beam Unit**—An integral and indivisible optical assembly with the name "Sealed Beam" molded in the lens.

 1.2 **Upper Beam**—A beam intended primarily for distant illumination and for use on the open highway when not meeting other vehicles.

 1.3 **Lower Beam**—A beam intended to illuminate the road ahead of the vehicle without causing undue glare to other drivers.

 1.4 **7 in. (178 mm) Sealed Beam System**—A system employing two 7 in. (178 mm) Sealed Beam units.

1.5 7 in. (178 mm) Type 2 Sealed Beam Unit—A 7 in. (178 mm) diameter unit (with a numeral 2 molded in the lens), which provides an upper and a lower beam. These units are mechanically aimable. NOTE: Original 7 in. (178 mm) Sealed Beam units which can be identified by the absence of "2" on the lens shall be aimed visually on the upper beam.

1.6 $5\frac{3}{4}$ in. (146 mm) Sealed Beam System—A system employing four $5\frac{3}{4}$ in. (146 mm) Sealed Beam units: two Type 1 and two Type 2.

1.7 $5\frac{3}{4}$ in. (146 mm) Type 1 Sealed Beam Unit—A $5\frac{3}{4}$ in. (146 mm) diameter unit having a single filament and used in a four-lamp system to provide the principal portion of the upper beam.

1.8 $5\frac{3}{4}$ in. (146 mm) Type 2 Sealed Beam Unit—A $5\frac{3}{4}$ in. (146 mm) diameter unit having two filaments and used in a four-lamp system to provide the lower beam and a secondary portion of the upper beam.

1.9 $4 \times 6\frac{1}{2}$ in. (100 x 165 mm) Sealed Beam System—A system employing four $4 \times 6\frac{1}{2}$ in. (100 x 165 mm) sealed beam units: two Type 1A and two Type 2A.

1.10 $4 \times 6\frac{1}{2}$ in. (100 x 165 mm) Type 1A Sealed Beam Unit—A $4 \times 6\frac{1}{2}$ in. (100 x 165 mm) rectangular unit having a single filament and used in a four-lamp system to provide the principal portion of the upper beam.

1.11 $4 \times 6\frac{1}{2}$ in. (100 x 165 mm) Type 2A Sealed Beam Unit—A $4 \times 6\frac{1}{2}$ in. (100 x 165 mm) rectangular unit having two filaments and used in a four-lamp system to provide the lower beam and a secondary portion of the upper beam.

1.12 Mechanically Aimable Sealed Beam Unit—A unit having three pads on the face of the lens forming a plane which is intended to be used to adjust and inspect the aim of the unit when installed on the vehicle.

1.13 Symmetrical Beam—A beam in which both sides are symmetrical with respect to the median vertical plane of the beam.

1.14 Asymmetrical Beam—A beam in which both sides are not symmetrical with respect to the median vertical plane of the beam. All lower beams are asymmetrical. NOTE: The inspector should see that the driver understands how to use multiple beam headlamps so as to obtain the best road lighting with minimum glare to other users of the highway.

2. Equipment—It is recommended that mechanically aimable headlamps be aimed and inspected for aim by mechanical aimers. Another aiming and inspection method is by visual means on a screen at a distance of 25 ft (7.6 m) ahead of the headlamps or on the screen of a headlamp testing machine.

2.1 The mechanical aimer used shall conform to the requirements of SAE J602. The device shall be in good repair, calibrated and used according to the manufacturer's instructions.

2.2 If a screen is used, it should be of adequate size with a matte-white surface well shaded from extraneous light and properly adjusted to the floor area on which the vehicle stands. Provision should be made for moving the screen or its vertical centerline so that it can be aligned with the vehicle axis. In addition to the vertical centerline, the screen should be provided with four laterally adjustable vertical tapes and two vertically adjustable horizontal tapes. The four movable vertical tapes should be located on the screen at the left and right limits called for in the specification with reference to centerlines ahead of each headlamp unit. The headlamp centerlines shall be spaced either side of the fixed centerline on the screen by the amount the headlamp units are to the left and right. The horizontal tapes should be located on the screen at the upper and lower limits called for in the specifications with reference to the height of lamp centers and the plane on which the vehicle rests, not the floor on which the screen rests. See Fig. 1.

2.3 The Headlamp Testing Machine used shall conform to the requirements of SAE J600. The device shall be in good repair, calibrated and used according to the manufacturer's instructions.

3. Preparation for Headlamp Aim or Inspection—Before checking beam aim, the inspector shall:

3.1 Remove ice or mud from under fenders.
3.2 See that no tire is noticeably deflated.
3.3 Check car springs for sag or broken leaves.
3.4 See that there is no load in the vehicle other than the driver.
3.5 Check functioning of any "level-ride" control.
3.6 Clean lenses and aiming pads.
3.7 Check for bulb burnout, broken mechanical aiming pads, and proper beam switching.
3.8 Stabilize suspension by rocking vehicle sideways.

4. Headlamp Aim Adjustment for Service Facilities

4.1 The following aim adjustment requirements should apply to dealers, service stations, and others who do headlamp adjusting.

4.2 It is recommended that mechanically aimable headlamps be aimed using mechanical aimers (paragraph 2.1). The aimers shall be calibrated for accuracy and shall be compensated for the level of the floor in the aiming area.

4.3 Mechanical Aiming

4.3.1 The correct mechanical aim for both Type 1 and Type 2 units is 0-0.
4.3.2 If a headlamp being serviced is not so aimed, the aim shall be corrected to 0-0.

4.4 Visual Aiming

4.4.1 The correct visual aim for Type 1 units is with the center of the high intensity zone at horizontal and straight ahead vertically. (See Fig. 2.)

4.4.2 The correct visual aim for Type 2 units is with the top edge of the high intensity zone of the lower beam horizontal and the left edge at vertical. (See Fig. 3.)

4.4.3 If the headlamp being serviced is not so aimed, it should be corrected to the above aim.

5. Headlamp Aim Inspection Limits for Vehicle Inspection Facilities

5.1 The following inspection limits should apply to stations that conduct mandatory inspection of vehicles.

5.2 It is recommended that mechanically aimable lamps be inspected using mechanical aimers (paragraph 2.1). The aimers shall be calibrated for accuracy and shall be compensated for the level of the floor in the inspection area.

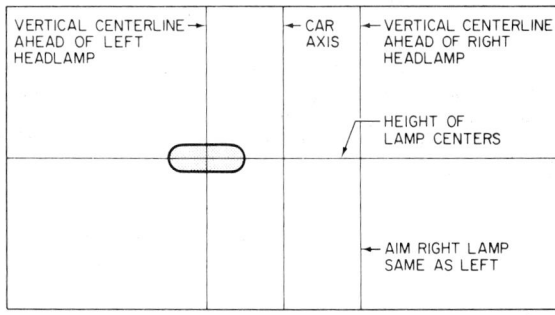

FIG. 2—HOW PROPERLY AIMED UPPER BEAM OF 5¾ IN (146 MM) TYPE 1 AND 7 IN (178 MM) SEALED BEAM (NOT MARKED "2" ON LENS) WILL APPEAR ON THE AIMING SCREEN 25 FT (7.6 M) IN FRONT OF VEHICLE. (SHADED AREA INDICATES HIGH INTENSITY ZONE)

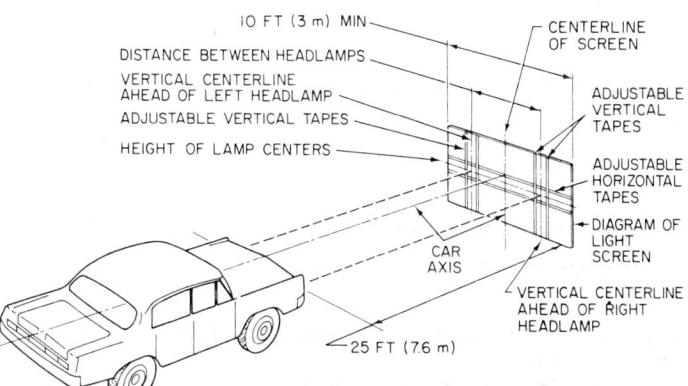

FIG. 1—ALIGNMENT OF HEADLAMP AIMING SCREEN

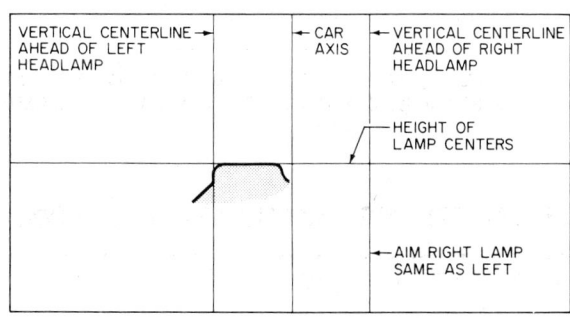

FIG. 3—HOW PROPERLY AIMED LOWER BEAM OF 5¾ IN (146 MM) AND 7 IN (178 MM) TYPE 2 SEALED BEAM WILL APPEAR ON THE AIMING SCREEN 25 FT (7.6 M) IN FRONT OF THE VEHICLE. (SHADED AREA INDICATES HIGH INTENSITY ZONE)

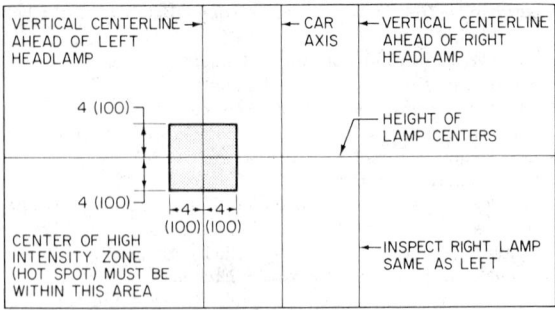

FIG. 4—AIM INSPECTION LIMITS FOR UPPER BEAM OF 5¾ IN (146 MM) TYPE 1 SEALED BEAM AND 7 IN (178 MM) SEALED BEAM UNITS NOT MARKED "2" AT THE TOP OF THE LENS. ALSO, TWO-BEAM LAMPS NOT MARKED SEALED BEAM ON THE LENS

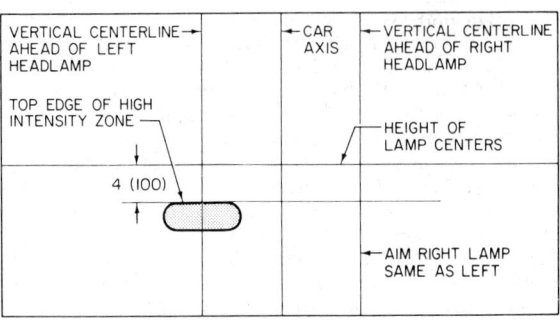

FIG. 6—HOW PROPERLY AIMED FOG LAMP (SYMMETRICAL BEAM) WILL APPEAR ON THE AIMING SCREEN 25 FT (7.6 M) IN FRONT OF VEHICLE. (SHADED AREA INDICATES HIGH INTENSITY ZONE)

5.3 Mechanical Aim Inspection

5.3.1 The mechanical inspection limits for both Type 1 and Type 2 units shall be 4 (100 mm) up to 4 (100 mm) down and 4 (100 mm) left to 4 (100 mm) right.

5.3.2 Failure to meet these limits shall be cause for rejection.

5.4 Visual Aiming

5.4.1 The visual inspection limits for Type 1 units shall be with the center of the high intensity zone from 4 (100 mm) up to 4 (100 mm) down and from 4 (100 mm) left to 4 (100 mm) right based in inches (millimeters) on a screen at 25 ft (7.6 m). (See Fig. 4.)

5.4.2 The visual inspection limits for Type 2 units shall be with the top edge of the high intensity zone from 4 (100 mm) up to 4 (100 mm) down and the left edge of the high intensity zone from 4 (100 mm) left to 4 (100 mm) right based in inches (millimeters) on a screen at 25 ft (7.6 m). (See Fig. 5.)

5.4.3 Failure to meet these limits shall be cause for rejection.

6. Fog Lamps (Symmetrical Beams) Aim Adjustment for Service Facilities

6.1 The following aim adjustment requirements should apply to dealers, service stations, and others who do headlamp adjusting.

6.2 The correct visual aim for fog lamps (symmetrical beams) is with the top edge of the high intensity zone 4 (100) below horizontal and the center of the high intensity zone straight ahead vertically based in inches (millimeters) on a screen at 25 ft (7.6m) (See Fig. 6.)

7. Fog Lamps (Symmetrical Beam) Aim Inspection Limits for Vehicle Inspection Facilities

7.1 The following inspection limits should apply to stations that conduct mandatory inspection of vehicles.

7.2 The visual inspection limits for fog lamps (symmetrical beam) shall be with the top edge of the high intensity zone at horizontal or below and with the center of the high intensity zone from 4 (100) left to 4 (100) right based in inches (millimeters) on a screen at 25 ft (7.6m) (See Fig. 7.)

8. Fog Lamps (Asymmetrical Beam) and Passing Lamps Aim Adjustment and Inspection Limits

8.1 Lamp aim adjustment and inspection is the same as for Type 2 Sealed Beam headlamp units. See paragraph 4.4.2 for adjustment and 5.4.2 for inspection.

9. General Lamp Inspection Other Than Headlamp Aim Inspection—This includes the following types of lamps: head, tail, stop, license, clearance, signal, marker, reflex reflector, and fog. Any of the following defects shall be cause for rejection.

9.1 Any bulb in any lamp which fails to function properly.

9.2 An improperly connected circuit which does not light the proper filaments for the different switch positions.

9.3 A cracked, broken, or missing lens.

9.4 A lens that is rotated, upside down, wrongside out, or is otherwise incorrectly installed. A lens marked "left" or "right", not appropriately installed.

9.5 A separate type lens, the name of which does not correspond with the name stamped on the lamp body, unless it is specifically approved for use with that lamp body.

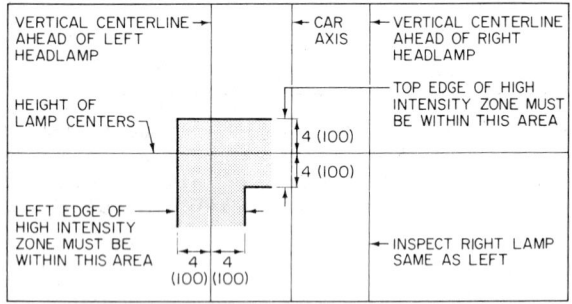

FIG. 5—AIM INSPECTION LIMITS FOR LOWER BEAM OF 5¾ IN (146 MM) TYPE 2 SEALED BEAM AND 7 IN (178 MM) TYPE 2 SEALED BEAM AND FOR AUXILIARY PASSING LAMP

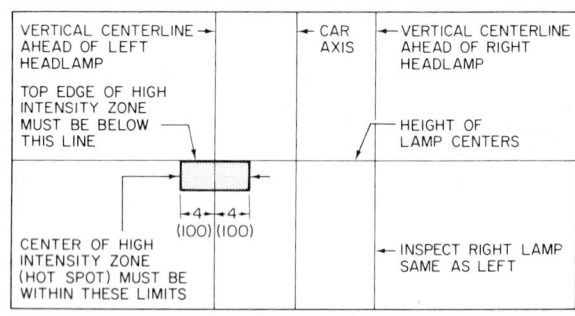

FIG. 7—AIM INSPECTION LIMITS FOR FOG LAMPS (SYMMETRICAL BEAM)

HEADLAMP TESTING MACHINES—SAE J600a

SAE Recommended Practice

Report of Lighting Committee approved December 1952 and last revised November 1963. Editorial change March 1965.

Scope—The purpose of this specification is to provide a laboratory test procedure for headlamp testing machines to determine their ability to aim, or to check the aim, of headlamps, fog lamps, and auxiliary driving and passing lamps, within tolerances prescribed herein. This specification does not apply to aiming devices of the kind covered by SAE Recommended Practice, Headlamp Aiming Device for Mechanically Aimable Sealed Beam Headlamp Units—SAE J602.

Samples for Test—Sample headlamp testing machines submitted for laboratory tests should be representative of the device as regularly manufactured and marketed, excepting in the case of machines using a track, an abbreviated section of track may be supplied for the test. Each sample should include all accessory equipment peculiar to the device and necessary to its service operation and calibration. Full assembly and operating instructions should be provided, including information on how to check accuracy and maintain the device in calibration.

Laboratory Facilities—The laboratory should be equipped with all facilities for accurate screen aiming and all facilities to make accurate physical and optical tests required in this specification, in accordance with established laboratory practice.

General Requirements—The headlamp testing machine should incorporate a fixed track or equivalent for positioning the aiming device in front of the headlamps.

The design of the headlamp testing machine should permit checking the aim of lamps mounted at heights from 12 to 54 in. and spaced up to 48 in. from the center of the motor vehicle.

The device and/or instructions should provide a practical means for a periodic check of its accuracy in the field.

The spirit level or other means provided for indicating vertical aim should be capable of showing at least 0.1 in. deviation with a 0.2 deg (1 in. in 25 ft) change in level.

A vertical aim scale should be provided with numerical graduations in steps, each of which represents 1 in. at 25 ft to provide for variations in aim at least from 4 in. above 0 to 10 in. below 0.

A lateral aim scale should be provided with graduations in steps of not more than 2 in. at 25 ft from straight ahead to at least 6 in. left and right.

The instructions covering use of the headlamp testing machine should include those items in the Preparation for Aiming, Section 2(a) of SAE Recommended Practice, Lighting Inspection Code—SAE J599.

Alignment With Car—Means should be provided in the device for compensating within ±0.05 deg for reasonable variation in floor slope and clearly explained in the operating instructions.

Means should be provided in the device for accurate lateral alignment within 0.1 deg with respect to longitudinal axis of vehicle.

Visual Beam Appraisal—Machines using a photoelectric cell or cells to determine aim should also have a visual screen upon which the beam pattern is projected proportional to its appearance and aim on a screen at 25 ft. Such visual screen should be plainly visible to the operator and should have horizontal and vertical reference lines to permit visual appraisal of the aim of the lamp beam.

Test Procedure—Assuming that the headlamp testing machine complies with the general requirements, it shall be considered acceptable if it complies with additional test requirements as follows:

NOTE: The laboratory should set up the sample headlamp testing machine in accordance with the instructions furnished. At the same time and on the same axis the laboratory should set up a headlamp adjusting screen with adjustable horizontal and vertical reference lines so as to relate the aim obtained on the headlamp testing machine to that obtained with a properly aligned screen.

Aim

1. Symmetrical Beam—Upper beam [all auxiliary driving lamps, all 7 in. sealed beam units (except 7 in. Type 2 units) and all Type 1, 5¾ in. sealed beam units].

A. The headlamp testing machine should permit determining the vertical and horizontal aim of the geometric center of the high intensity zone of the upper beam within the limits set up in these specifications.

(1) A group of sealed beam units, at least of Type 1, which meets SAE specifications but represent the maximum production variations of each lamp manufacturer, should be obtained by the testing laboratory.

Units should be selected by the laboratory which represent the extreme of production variation from the lamp manufacturer's average in their relationship of maximum intensities up, down, right, and left respectively to the geometrical center of the high intensity zone.

(a) The average vertical aim of each lamp obtained on the headlamp testing machine should not vary more than 0.2 deg (1 in. at 25 ft) from the average visual aim of each lamp obtained on the screen by three experienced observers. A minimum of 3 observations should be made by each observer on both the headlamp testing machine and the screen.

(b) The average horizontal aim obtained on each lamp on the headlamp testing machine should not vary left or right more than 0.4 deg (2 in. at 25 ft) from the average visual aim of each lamp obtained on the screen by three experienced observers. A minimum of 3 observations should be made by each observer on both the headlamp testing machine and the screen.

2. Asymmetrical Beam—Auxiliary passing lamps, lower beam 5¾-in. and 7-in. Type 2 sealed beam units.

A. The headlamp testing machine should permit determining the aim of the top and left edge cutoffs of the high intensity zone of the lower beam of Type 2 lamps and of auxiliary passing lamps within the limits set up in these specifications.

(1) A group of sealed beam units, at least of Type 2, which meets SAE specifications but represent the maximum production variations of each lamp manufacturer, should be obtained by the testing laboratory.

Units should be selected by the laboratory which represent the extreme of production variation from the lamp manufacturer's average in their relationship of maximum intensities up, down, right and left to the cut off at the top and left edges.

(a) The average vertical aim of each lamp obtained on the headlamp testing machine should not vary more than 0.2 deg (1 in. at 25 ft) from the average visual aim of each lamp obtained on the screen by three experienced observers. A minimum of 3 observations should be made by each observer on both the headlamp testing machine and the screen.

(b) The average horizontal aim obtained on each lamp on the headlamp testing machine should not vary left or right more than 0.4 deg (2 in. at 25 ft) from the average visual aim of each lamp obtained on the screen by three experienced observers. A minimum of 3 observations should be made by each observer on both the headlamp testing machine and the screen.

3. Symmetrical Fog Lamps

A. The headlamp testing machine should permit determining the vertical aim of the top cutoff of the high intensity zone and the horizontal aim of the geometric center of the high intensity zone within the limits of these specifications.

(1) The laboratory should select at least 3 lamps from a random sampling of lamps obtained from each manufacturer, representative of those currently manufactured and meeting SAE specifications.

(a) The average vertical and horizontal aim of each lamp obtained on the headlamp testing machine should not vary more than 0.4 deg (2 in. at 25 ft) from the average visual aim obtained on the screen by three experienced observers. A minimum of three observations should be made by each observer on both the headlamp testing machine and the screen.

NOTE: Instructions furnished by the manufacturer for aiming lamps should be such that the beam patterns when viewed on the screen will fall within the limits set up in SAE Recommended Practice, Lighting Inspection Code—SAE J599.

HEADLAMP AIMING DEVICE FOR MECHANICALLY AIMABLE SEALED BEAM HEADLAMP UNITS—SAE J602c

SAE Standard

Report of Lighting Committee approved October 1957 and last revised December 1974.

1. Scope—This specification applies to the requirements of a device used in the field and inspection stations to aim the mechanically aimable type of sealed beam headlamp units.

The purpose of this specification is to provide a laboratory test procedure to determine whether the devices under test are capable of accurately positioning sealed beam headlamp units from their aiming pads and maintaining their accuracy in service within the tolerances designated in this specification.

2. Definitions

2.1 Headlamp Aiming Device—A device used to adjust and inspect the aim of mechanically aimable headlamp units consisting of one or more fixtures designed to seat against the three aiming pads (aiming plane) on mechanically aimable headlamp units installed on a vehicle to facilitate accurate aiming of such units, vertically and laterally.

2.2 Mechanically Aimable Headlamp Units—A unit having three pads on the face of the lens forming a mechanical aiming plane used to adjust and inspect the aim of the unit when installed on a vehicle.

2.3 Aiming Plane—A plane through the three aiming pads on the face of the lens.

3. Samples for Test—Sample devices submitted for laboratory tests shall be representative of the devices as regularly manufactured and marketed. Each sample shall include all accessory equipment peculiar to the device. Full assembly and operating instructions shall be provided, including information on how to check accuracy and maintain the device in calibration.

4. Laboratory Facilities—The laboratory shall be equipped with all facilities necessary to make the tests required in this standard.

5. General Requirements

5.1 The device shall be of such design that the seating portion will register only on the three aiming pads on the sealed beam units as covered by SAE J571.

5.2 No part of the device, except those parts (strings, sighting devices, scales, etc.) required for referencing lateral alignment between devices, shall extend beyond the dimensional limits of the headlamp aiming device locating plate (Fig. 1, dimension C and Fig. 2, 4.05 in. (102.9 mm) max. Dia. dimension).

5.3 A device which uses adapters to fit more than one size sealed beam unit shall meet all of the requirements of this recommended practice with and without the adapters.

5.4 The seating plane of the device shall meet the dimensions shown in Fig. 1 or Fig. 2.

5.5 When aiming headlamp units spaced 90 in. (2300 mm) apart, the torque exerted by the device at the aiming plane shall not exceed 18 in.-lb (2.0 N·m) vertically and 12 in.-lb (1.4 N·m) laterally.

5.6 The means of securing the device to the sealed beam unit shall retain the device against the three aiming pads when an axially centered tensile force of 4.0 lb (17.8 N) minimum is applied to the device.

5.7 The device shall be capable of being calibrated and shall have available for immediate use an independent calibration fixture and/or instructions to immediately recalibrate the device.

5.8 If a suction cup is used to retain the device to the headlamp unit, the effective diameter for $5\frac{3}{4}$ in. (146 mm) and 7 in. (178 mm) shall not exceed 3.5 in. (90 mm) and effective diameter for 4 in. x $6\frac{1}{2}$ in. (100 x 165 mm) shall not exceed 2.8 in. (71 mm) when installed.

5.9 Means shall be provided in the device for compensating within ±0.1 deg through a slope range of ±1.5 deg from horizontal. The method for device compensation shall be clearly explained in the operating instructions.

5.10 If the lateral aim is to be accomplished by reference between devices on opposite sides of the vehicle, the means provided for referencing lateral alignment between devices (sight line, string, or equivalent) shall be located as shown in Fig. 1 and Fig. 2.

5.11 The spirit level or other means provided for indicating vertical aim shall be capable of showing at least a 0.1 in. (2.5 mm) deviation with a 1 in. (25 mm)[1] change in level.

5.12 A lateral aim scale shall be provided with graduations in steps of not more than 2 in. (51 mm)[1] from straight ahead to at least 8 in. (203 mm)[1] left and right.

5.13 The instructions covering use of the device shall include those items shown in paragraph 3 of SAE J599.

5.14 The vertical aim scale shall be marked O with the aiming plane vertical.

5.15 The vertical aim scale shall be provided with numerical graduations in steps, each of which represents 1 in. (25 mm)[1] to provide for variations in vertical aim at least from 8 in. (203 mm)[1] above O to 8 in. (203 mm)[1] below O.

6. Test Procedure—Assuming that the devices comply with the general requirements, they shall be considered acceptable if they comply with additional test requirements as follows:

NOTE 1—All tests are to be made in an ambient temperature of 75 ± 5°F (24 ± 3°C) unless otherwise specified.

NOTE 2—If a vertical indication means other than a spirit level is used, an equivalent accuracy shall be maintained.

[1] Represents inches (millimeters) at 25 ft (7.6 m).

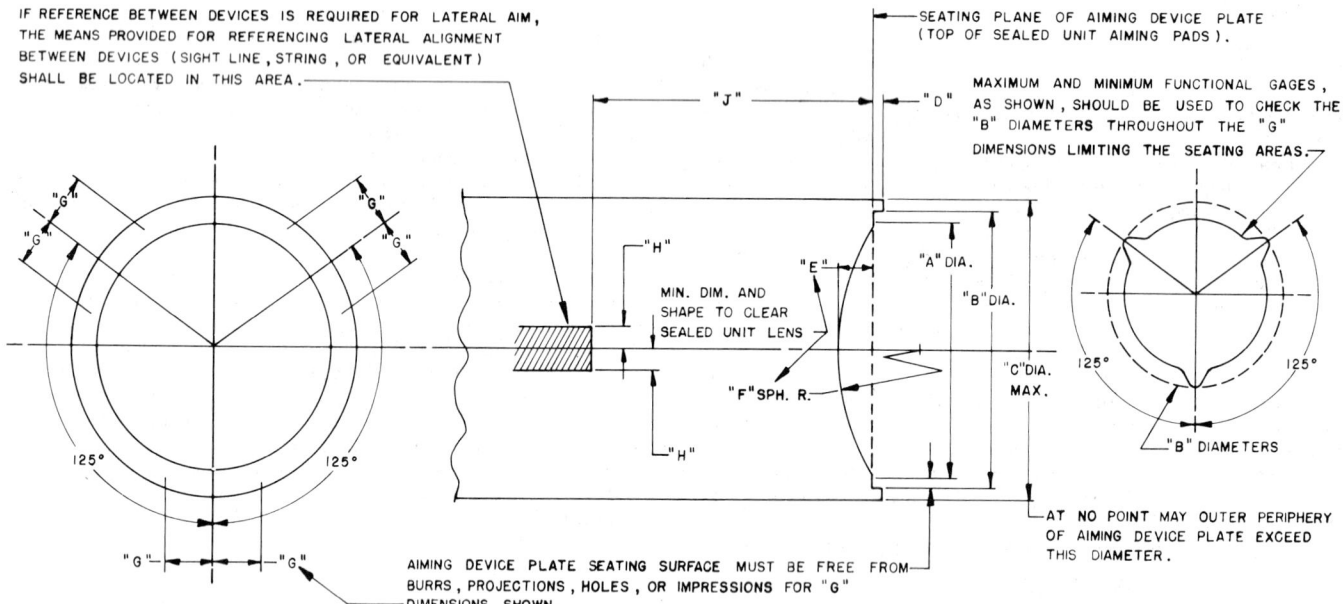

Locating Plate	Unit of Measure	Dimensions											
		A		B		C	D		E	F	G	H	J
		Max	Min	Max	Min	Max	Max	Min	Ref	Ref	Min	Max	Min
5-3/4 in (146 mm)	in	4.830	4.770	5.375	5.345	5.700	0.165	0.145	0.70	4.40	0.70	1.00	9.50
	mm	122.7	121.2	136.5	135.8	144.8	4.19	3.68	17.8	111.8	17.8	25.4	241.3
7 in (178 mm)	in	6.140	6.080	6.710	6.680	7.031	0.180	0.160	0.96	5.60	0.70	1.00	10.25
	mm	156.0	154.4	170.4	169.7	178.6	4.57	4.06	24.4	142.2	17.8	25.4	260.4

FIG. 1—DIMENSIONAL SPECIFICATIONS FOR HEADLAMP AIMING DEVICE LOCATING PLATE

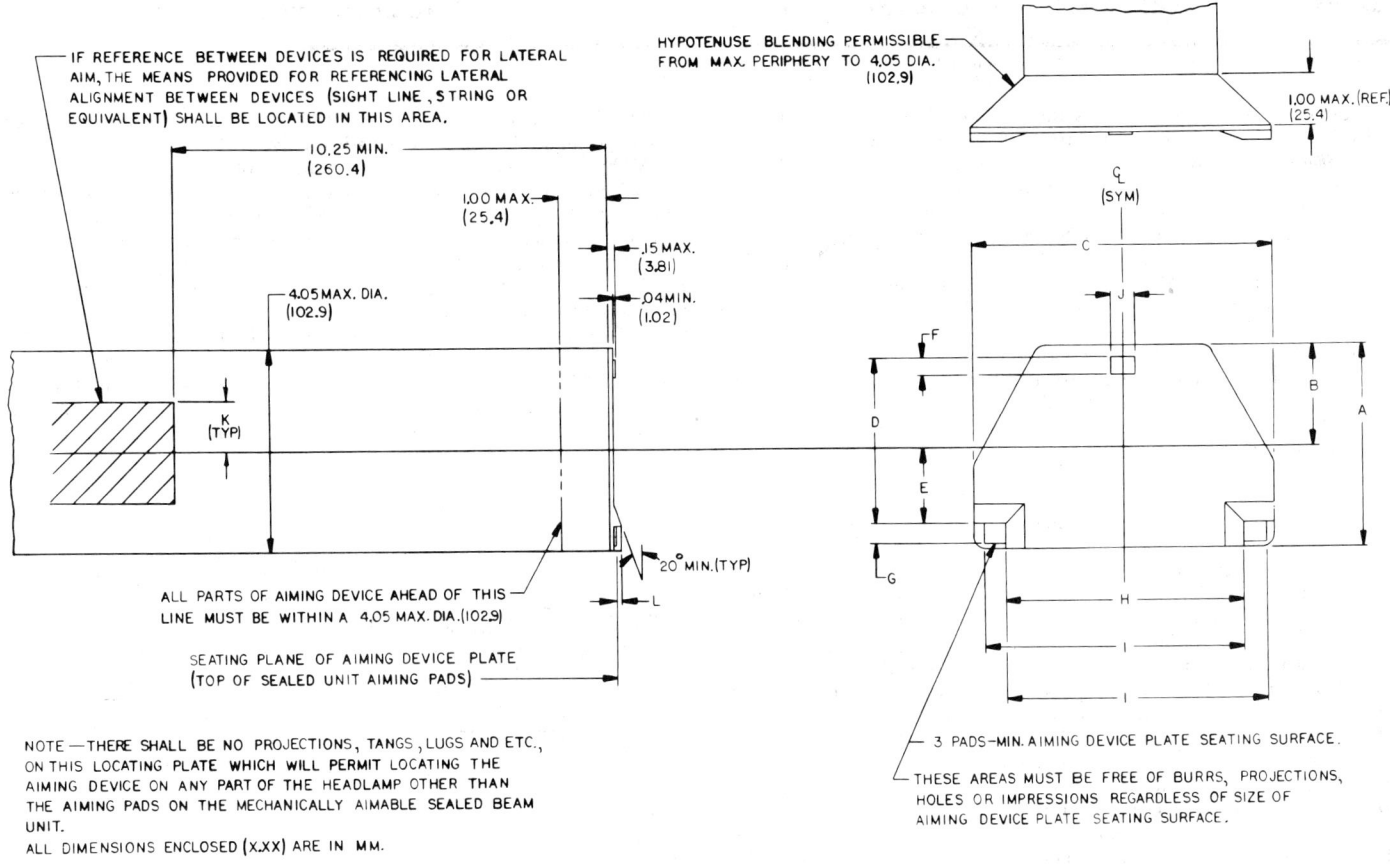

LOCATING PLATE	UNIT OF MEASURE	DIMENSIONS																							
		A		B		C		D		E		F		G		H		I		J		K		L	
		MAX	MIN	MAX	MIN	MAX	MIN	MAX	MIN	MAX	MIN	MAX	MIN	MAX	MIN	MAX	MIN	MAX	MIN	MAX	MIN	MAX	MIN	MAX	MIN
4 x 6½	in.	3.935	3.925	1.975	1.953	6.001	5.991	3.320	3.300	1.550	1.540	.370	.350	.330	.310	5.088	5.018	—	5.421	—	.400	1.000	—	.080	.060
100 x 103	mm	99.95	99.60	50.17	49.66	152.42	152.17	84.33	83.82	39.37	39.12	9.40	8.89	8.38	7.87	129.24	128.98	—	137.20	—	10.16	25.40	—	2.03	1.52

FIG. 2—DIMENSIONAL SPECIFICATIONS FOR HEADLAMP AIMER LOCATING PLATE (4 x 6½)

6.1 With the aiming plane vertical and with the vertical scale on the device set at O, the angle through which the aiming plane must be rotated vertically to center the bubble in the spirit level, or equivalent, shall not exceed 0.5 in. (13 mm).[1]

6.2 With the aiming planes in the same vertical plane and with the means provided for adjusting lateral aim in use, the angle through which the aiming plane must be rotated laterally to indicate straight ahead shall not exceed ±1 in. (25 mm)[1] with the lamps 24 and 90 in. (610 and 2300 mm) apart.

6.3 With the aiming planes initially in the same vertical plane and subsequently toed inward and outward 6 in. (152 mm)[1] and with the means provided for checking lateral aim in use, the error in reading shall not exceed ±1 in. (25 mm)[1] with the lamps 60 in. (1520 mm) apart.

6.4 With the aiming plane vertical and with the vertical scale on the device set at O, the level on the aimer shall be adjusted prior to each of the following tests to center the bubble in the spirit level or equivalent:

6.4.1 Each step on the vertical aim scale shall be checked and in no case shall the variation from correct aim exceed ±0.5 in. (13 mm).[1]

6.4.2 A pair of devices shall be stabilized at 20 ± 5°F (−7 ± 3°C) and then installed on a pair of unlighted sealed beam units spaced 60 in. (1520 mm) apart at the 20°F (−7°C) ambient temperature. After a period of 30 min, the seating portion of the device shall continue to register against the three sealed beam unit aiming pads and, the variation from correct vertical aim shall not exceed ±0.5 in. (13 mm)[1] and the variation from correct lateral aim shall not exceed ±1 in. (25 mm).[1]

6.4.3 A pair of aiming devices shall be stabilized at 100 ± 5°F (38 ± 3°C) and then installed on a pair of lighted sealed beam units spaced 60 in. (1520 mm) apart at the 100°F (38°C) ambient temperature. After a period of 30 min, the seating portion of the device shall continue to register against the three sealed beam unit aiming pads and the variation from correct vertical aim shall not exceed 0.5 in. (13 mm)[1] and the variation from correct lateral aim shall not exceed ±1 in. (25 mm).[1]

6.4.4 A pair of devices shall be exposed with the aiming plane down in a circulating air over to 140 ± 5°F (60 ± 3°C) for 24 h followed by a temperature of −40 ± 5°F (−40 ± 3°C) for 24 h and then permitted to return to room temperature, after which they shall show no visible damage. They shall then be installed on the pair of unlighted sealed beam units spaced 60 in. (1520 mm) apart and the variation from correct vertical aim shall not exceed ±0.5 in. (13 mm)[1] and the variation from correct lateral aim shall not exceed ±1 in. (25 mm).[1]

6.4.5 A sample device shall be exposed to 35 ± 5°F (1.7 ± 3°C) for 1 h and then immediately allowed to free fall onto a concrete floor three times from its normal operating position on a sealed beam unit at a height of 40 in. (1020 mm), after which it shall show no damage that would interfere with the proper recalibration of the device. It shall then be installed in combination with its companion device on a pair of unlighted sealed beam units spaced 60 in. (1520 mm) apart and the variation from correct vertical aim shall not exceed 1 in. (25 mm)[1] and the variation from correct lateral aim shall not exceed 1 in. (25 mm).[1] (This test applies only to devices which are supported by the sealed beam unit.)

6.5 Using the calibration fixture and/or instructions required by paragraph 5.7, the device shall be calibrated and checked for compliance with paragraphs 6.1 and 6.2.

FLASHER TEST EQUIPMENT—SAE J823c

SAE Standard

Report of Lighting Committee approved April 1962 and last revised January 1975.

1. Scope—This standard specifies the test procedure, test circuitry, and instruments required for measuring the performance of flashers.

2. Laboratory Facilities—The laboratory shall be equipped with all of the facilities required to make the tests in this specification, in accordance with established laboratory practice, including the following:

2.1 Means shall be provided to maintain ambient temperatures of $-25°-5°F$ ($-32-3°C$), $0-5°F$ ($-18-3°C$), $75-10°F$ ($24-5.5°C$), $125-5°F$ ($52-3°C$), and $145-5°F$ ($63-3°C$). The chamber atmosphere shall be air.

2.2 Power Supply—Performance Tests—The power supply for testing performance requirements shall not generate any adverse transients not present in motor vehicles and shall comply with the following specifications:

2.2.1 OUTPUT VOLTAGE—Capable of supplying to the input terminals of the standard circuit 11–16 V d-c for 12 V flashers or 5–9 V d-c for 6 V flashers.

2.2.2 OUTPUT CURRENT—Capable of supplying required design current(s) continuously and inrush currents as required by the design bulb load complement.

2.2.3 REGULATION

2.2.3.1 Dynamic—The output voltage shall not deviate more than 1.0 V from 0 to maximum load (including inrush current) and shall recover 63% of its maximum excursion within 100 μ sec.

2.2.3.2 Static—The output voltage shall not deviate more than 2% with changes in static load from 0 to maximum (not including inrush current) nor for static line voltage variations.

2.2.4 RIPPLE VOLTAGE—Maximum 75 mV, peak to peak.

2.3 Power Supply—Durability Tests—The power supply for the durability test requirements shall not generate any adverse transients not present in motor vehicles and shall comply with the following specifications:

2.3.1 OUTPUT VOLTAGE—Capable of supplying, as required, 14 and 13 V (7 and 6.5 V) d-c, according to the flasher rating, to the input terminals of the standard test circuits shown in Figs. 1 and 2.

2.3.2 OUTPUT CURRENT—Capable of supplying a continuous output current of the design load for one flasher times the number of flashers and inrush currents as required by the bulb load complement.

2.3.3 REGULATION

Dynamic: The output voltage shall not deviate more than 1.0 V from 0 to maximum load (including inrush current) and should recover 63% of its maximum excursion within 5 ms.

Static: The output voltage shall not deviate more than 2% with changes in static load from 0 to maximum (not including inrush current), and means shall be provided to compensate for static line voltage variations.

2.3.4 RIPPLE VOLTAGE: Maximum 300 mV, peak to peak.

3. Testing Requirements

3.1 The flashers shall be mounted as specified by the manufacturer if special precautions are required.

3.2 The flashers shall be connected in a standard test circuit as shown in Fig. 1 for turn signal and hazard warning flashers or Fig. 2 for warning lamp alternating flashers using the design load(s) within 0.5% at 12.8 V (6.4 V) as specified by the flasher manufacturer.

3.3 A suitable high impedance measuring device connected to points X-Y in Fig. 1 or to points $X-Y_1$, and to points $X-Y_2$ in Fig. 2 shall be used for measuring flash rate, percent current "on" time, starting time, and voltage drop across the flasher. The measurement of these quantities shall not affect the circuit.

3.4 The resistance at A-B for each load circuit in Fig. 1 or Fig. 2 shall be measured with flasher and bulb loads each shorted out with removable shunt resistances not to exceed 0.005 ohms each.

The effective series resistance in the total circuit (Fig. 1) or in each of the parallel circuits (Fig. 2) between the power supply and bulb sockets (excluding the flasher and bulb loads by using the removable shunt resistances) shall be 0.10–0.01 ohms.

3.5 Adjust the voltage at the bulbs to 12.8 V (6.4 V) as required for testing at C-D in Fig. 1 or C-D and E-F in Fig. 2 with the flasher shorted out by an effective shunt resistance not to exceed 0.005 ohms. The load current shall be held to the rated value for the total flasher design load(s) within 0.5% at 12.8 V (6.4 V) by simultaneously adjusting trimmer resistors, R.

3.6 For testing fixed load flashers at other required voltages, adjust the power supply to provide required voltages at required temperatures at C-D in Fig. 1 or C-D and E-F in Fig. 2 without readjustment of trimming resistors, R.

3.7 For testing variable load flashers, the circuit shall be first adjusted at 12.8 V (6.4 V) at C-D in Fig. 1 or C-D and E-F in Fig. 2 with a minimum required bulb load and the power supply shall be adjusted to provide other required test voltages at required temperatures at C-D in Fig. 1 or C-D and E-F in Fig. 2 without readjustment of trimming resistors, R (each required test voltage shall be set with a minimum bulb load in place). The required voltage tests with a maximum bulb load shall be conducted without readjusting each corresponding power supply voltage previously set with minimum bulb load.

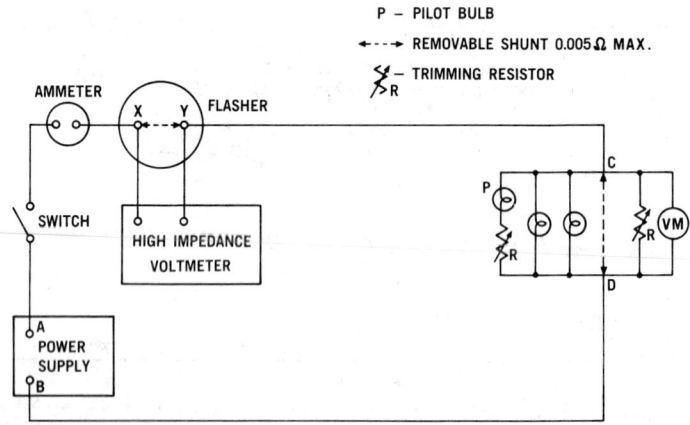

FIG. 1—STANDARD TEST CIRCUIT—TURN SIGNAL AND HAZARD WARNING FLASHERS

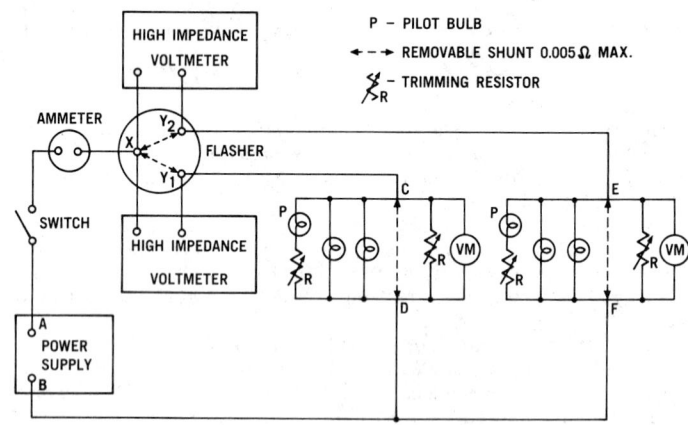

FIG. 2—STANDARD TEST CIRCUIT—WARNING LAMP ALTERNATING FLASHER

NUMERICAL INDEX FOR SAE STANDARDS, RECOMMENDED PRACTICES, AND INFORMATION REPORTS

Each surface vehicle Standard, Recommended Practice, or Information Report has a designation consisting of the letter "J" combined with a number. The letter "J" is combined with a nonsignificant number to eliminate any possible confusion between the report number and the SAE numbers within the reports.

Effective with the 1981 SAE Handbook, subsequent revisions of reports will be indicated by the month and year of revision, for example, J1159 AUG79. This new system will be phased in over a five year period. A lower case "a", "b", etc., appended to the report number indicates successive revisions of older reports.

This index covers reports appearing in the 1981 SAE Handbook and in currently available Handbook Supplements. Also listed are reports which were cancelled or superseded since 1978. Reports which were cancelled or superseded prior to 1978 are not listed in this index. For information regarding these older reports contact the Technical Division or the Publications Divison at SAE Headquarters or SAE Branch Office as follows:

SAE
400 Commonwealth Drive
Warrendale, PA 15096
(412) 776-4841

SAE
Detroit Branch
Suite 206
2100 West Big Beaver
Troy, MI 48084
(313) 649-0420

SAE J4c	Replaced by J117, J114, J140a, J141, J339a, J800c	
SAE J10b	Automotive and Off-Highway Air Brake Reservoir Performance and Identification Requirements	31.129
SAE J14	Specifications for Elastomer Compounds for Automotive Applications	(Cancelled 1980)
SAE J15	Flexible Foams Made From Polymers or Copolymers of Vinyl Chloride	(Cancelled 1978)
SAE J17 OCT79	Latex Foam Rubbers	12.37
SAE J18 DEC79	Sponge- and Expanded Cellular-Rubber Products	12.40
SAE J19	Latex Dipped Goods and Coatings for Automotive Applications	12.45
SAE J20e	Coolant System Hoses	12.98
SAE J29	Plastic Material for Use in Housings of Motor Vehicle Lighting Devices	23.28
SAE J30d	Fuel and Oil Hoses	12.102
SAE J31	Hydraulic Backhoe Lift Capacity	41.214
SAE J32a	Service Performance Requirements for Sealed Beam Headlamp Units for Motor Vehicles	23.32
SAE J33a	Snowmobile Definitions and Nomenclature—General	39.01
SAE J34a	Exterior Sound Level Measurement Procedure for Pleasure Motorboats	35.29
SAE J35	Diesel Smoke Measurement Procedure	25.46
SAE J38	Lift Arm Safety Device for Loaders	41.224
SAE J39	T-Hook Slots for Securement in Shipment of Agricultural Equipment	40.37
SAE J40	SAE J40 series hose replaced by SAE J1400 series	
SAE J43	Axle Rating for Industrial Wheel Loaders and Loader-Backhoes	41.216
SAE J44	Service Brake System Performance Requirements—Snowmobiles	39.10
SAE J45	Brake System Test Procedure—Snowmobiles	39.10
SAE J46 JUN80	Wheel Slip Brake Control System Road Test Code	31.132
SAE J47	Maximum Sound Level Potential for Motorcycles	35.01
SAE J48	Guidelines for Fluid Level Indicators	41.70
SAE J49 APR80	Specification Definitions—Hydraulic Backhoes	41.44
SAE J50a	Windshield Wiper Hose	12.112
SAE J51b	Automotive Air Conditioning Hose	12.112
SAE J52	Steering Wheel Rim Faceform Impact Test Procedure	(Cancelled 1981)
SAE J53	Minimum Performance Criteria for Emergency Steering of Wheeled Earthmoving Construction Machines	41.278
SAE J56	Electrical Generating System (Alternator Type) Performance Curve and Test Procedure	22.19
SAE J57a	Sound Level of Highway Truck Tires	35.21
SAE J58	Flanged 12-Point Screws	15.42
SAE J59	Fuel Injection Nozzle Gaskets	24.100
SAE J60	SAE J60 series rubber replaced by SAE J1600 series	
SAE J63 FEB80	Hole Placement on Dozer End Bits	41.156
SAE J64	Vehicle Identification Numbers—Snowmobile and All-Terrain Vehicles	39.03
SAE J67	Shovel Dipper, Clam Bucket, and Dragline Bucket Rating	41.210
SAE J68 OCT79	Tests for Snowmobile Switching Devices and Components	39.08
SAE J70	SAE J70 series fluid replaced by SAE J1700 series	
SAE J71a	Central Fluid Systems	13.20
SAE J73	Multipurpose Petroleum Base Fluids*	(HS J73)
SAE J75	Motor Vehicle Brake Fluid Container Compatibility	31.48
SAE J76	Handling and Dispensing of Motor Vehicle Brake Fluid	31.49
SAE J77	Service Maintenance of Motor Vehicle Brake Fluid in Motor Vehicle Brake Actuating Systems	31.51
SAE J78 JUN79	Steel Self-Drilling Tapping Screws	5.21
SAE J79	Brake Disc and Drum Thermocouple Installation	31.131
SAE J80	Automotive Rubber Mats	(Cancelled 1981)

*Available as a Related Technical Report. Complete listing and details appear at the end of this index.

Standard	Date	Title	Page
SAE J81	JUN79	Thread Rolling Screws	5.26
SAE J82	JUN79	Mechanical and Quality Requirements for Machine Screws	5.14
SAE J88	SEP79	Exterior Sound Level Measurement Procedure for Earthmoving Machinery	35.54
SAE J89a		Dynamic Cushioning Performance Criteria for Snowmobile Seats	39.12
SAE J90b		Nonmetallic Automotive Gasket Materials	12.80
SAE J92		Snowmobile Throttle Control Systems	39.17
SAE J93		Snowmobile and All-Terrain Vehicle Pulley Guards and Shields	39.14
SAE J94	APR80	Combination Tail and Floodlamp for Industrial Equipment	41.180
SAE J95	APR80	Headlamps for Industrial Equipment	41.180
SAE J96		Flashing Warning Lamp for Industrial Equipment	41.179
SAE J97		Nomenclature for Exhaust System Parts	24.125
SAE J98	APR80	Safety for Industrial Wheeled Equipment	41.264
SAE J99	APR80	Lighting and Marking of Industrial Equipment on Highways	41.181
SAE J100		Passenger Car Glazing Shade Bands	34.35
SAE J101	MAR80	Hydraulic Wheel Cylinders for Automotive Drum Brakes	31.19
SAE J102		Square Head Setscrews	15.41
SAE J103		Lag Screws	15.46
SAE J104		Square and Hex Nuts	15.47
SAE J105		Hex Bolts	15.01
SAE J106	APR80	Soil Type and Strength Classification	41.63
SAE J107	APR80	Operator Controls and Displays on Motorcycles	38.01
SAE J108a		Brake System Road Test Code—Motorcycles	31.114
SAE J109a		Service Brake System Performance Requirements—Motorcycles and Motor-Driven Cycles	31.116
SAE J110c		Seals—Testing of Radial Lip	28.57
SAE J111c		Seals—Terminology of Radial Lip	28.38
SAE J112a		Electric Windshield Wiper Switch	22.66
SAE J113		Hard Drawn Mechanical Spring Wire and Springs	6.04
SAE J114		Seat Belt Assembly Webbing Abrasion Test Requirements	33.12
SAE J115	SEP79	Safety Signs	41.272
SAE J116a		Gas Turbine Engine Test Code	24.34
SAE J117		Dynamic Test Procedure—Type 1 and Type 2 Seat Belt Assemblies	33.04
SAE J118a		Formerly Standard SAE Carbon Steels	2.14
SAE J119a		Fiberboard Crease Bending Test	12.72
SAE J120a		Rubber Rings for Automotive Applications	12.87
SAE J121a		Decarburization in Hardened and Tempered Threaded Fasteners	5.44
SAE J122a		Surface Discontinuities on Nuts	5.40
SAE J123c		Surface Discontinuities on Bolts, Screws, and Studs	5.37
SAE J125		Elevated Temperature Properties of Cast Irons	7.08
SAE J126a		Selecting and Specifying Hot and Cold Rolled Steel Sheet and Strip	4.19
SAE J128		Occupant Restraint System Evaluation—Passenger Cars	33.12
SAE J129		Passenger Car Engine and Transmission Identification Numbers	26.03
SAE J131a		Motorcycle Turn Signal Lamps	23.40
SAE J132		Oil Tempered Chromium-Vanadium Valve Spring Quality Wire and Springs	6.03
SAE J133		Fifth Wheel Kingpin Performance—Commercial Trailers and Semitrailers	37.20
SAE J134	JUN79	Brake System Road Test Code—Passenger Car and Light Duty Truck-Trailer Combinations	31.95
SAE J135a		Service Brake System Performance Requirements—Passenger Car-Trailer Combinations	31.104
SAE J136		Simplified Method for Simulating Glancing Blow Impacts—Motor Vehicles	(Cancelled 1981)
SAE J137c		Lighting and Marking of Agricultural Equipment on Highways	40.39
SAE J138		Film Analysis Guides for Dynamic Studies of Test Subjects	34.129
SAE J139		Ignition System Nomenclature and Terminology	22.01
SAE J140a		Seat Belt Hardware Test Procedure	33.01
SAE J141		Seat Belt Hardware Performance Requirements	33.03
SAE J151		Pressure Relief for Cooling System	24.114
SAE J152		Engine Specification Tag	(Cancelled 1980)
SAE J153		Safety Considerations for the Operator	41.263
SAE J154a		Operator Enclosures Human Factor Design Considerations	41.256
SAE J155		Service Brake System Performance Requirements—Light-Duty Truck	31.92
SAE J156a		Fusible Links	22.51
SAE J157		Oil Tempered Chromium-Silicon Alloy Steel Wire and Springs	6.05
SAE J158a		Automotive Malleable Iron Castings	7.04
SAE J159	OCT79	Crane Load Moment System	41.147
SAE J160	JUN80	Swell, Growth, and Dimensional Stability of Brake Linings	31.60
SAE J161a		Selection of Tires for Multipurpose Passenger Vehicles, Light-, Medium-, and Heavy-Duty Trucks, Buses, Trailers, and SemiTrailers for Normal Highway Service	29.12
SAE J162a		Flywheels for Single Bearing Engine Mounted Power Generators	41.168
SAE J163		Low Tension Wiring and Cable Terminals and Splice Clips	22.51
SAE J164	JUL79	Radiator Caps and Filler Necks	24.128
SAE J165	APR80	Fan Blast Deflectors for Earthmoving Machines	41.223
SAE J167a		Overhead Protection for Agricultural Tractors—Test Procedures and Performance Requirements	40.56
SAE J168a		Protective Enclosures for Agricultural Tractors—Test Procedures and Performance Requirements	(Cancelled 1978)
SAE J169	APR80	Design Guidelines for Air Conditioning Systems for Construction and Industrial Equipment Cabs	41.274
SAE J170a		Measurement of Fuel Evaporative Emissions from Gasoline Powered Passenger Cars and Light Trucks	25.73
SAE J171a		Measurement of Fuel Evaporative Emissions from Gasoline Powered Passenger Cars and Light Trucks Using the Enclosure Technique	25.80
SAE J172		Hard Drawn Carbon Steel Valve Spring Quality Wire and Springs	6.06
SAE J173	JAN80	Specification Definitions—Dozers	41.24
SAE J174		Torque-Tension Test Procedure for Steel Threaded Fasteners	15.76
SAE J175		Wheels—Passenger Cars—Impact Performance Requirements and Test Procedures	30.02
SAE J176a		Dry-Break Fueling Installation for Construction Machinery	41.70
SAE J177a		Measurement of Carbon Dioxide, Carbon Monoxide, and Oxides of Nitrogen in Diesel Exhaust	25.01
SAE J178		Music Steel Spring Wire and Springs	6.09
SAE J179a		Labeling—Disc Wheels and Demountable Rims—Trucks	30.09
SAE J180	APR80	Electrical Charging Systems for Construction and Industrial Machinery	41.184
SAE J182a		Motor Vehicle Fiducial Marks	34.115

*Available as a Related Technical Report. Complete listing and details appear at the end of this index.

Standard	Title	Page
SAE J183 FEB80	Engine Oil Performance and Engine Service Classification	13.01
SAE J184a	Qualifying a Sound Data Acquisition System	35.06
SAE J185	Access Systems for Construction and Industrial Equipment	41.249
SAE J186a	Supplemental High Mounted Stop and Rear Turn Signal Lamps	23.49
SAE J187	Truck Vehicle Identification Numbers	26.05
SAE J188	Power Steering Pressure Hose—High Volumetric Expansion Type	12.107
SAE J189	Power Steering Return Hose—Low Pressure	12.108
SAE J190	Power Steering Pressure Hose—Wire Braid	12.111
SAE J191	Power Steering Pressure Hose—Low Volumetric Expansion Type	12.110
SAE J192 APR80	Exterior Sound Level for Snowmobiles	35.25
SAE J193 SEP79	Ball Stud and Socket Assembly Test Procedure	16.06
SAE J194	Drawbar for Forestry Tractors (Cancelled 1978)	
SAE J195	Automatic Vehicle Speed Control—Motor Vehicles	20.12
SAE J196 JUN80	Electric Tachometer Specification—On Road	20.08
SAE J197	Electric Tachometer Specification—Off Road	20.10
SAE J198	Windshield Wiper Systems—Trucks, Buses, and Multipurpose Vehicles	34.08
SAE J200 APR80	Classification System for Rubber Materials for Automotive Applications	12.01
SAE J201	In-Service Brake Performance Test Procedure Passenger Car and Light-Duty Truck	31.55
SAE J202	Synthetic Skins for Automotive Testing	34.118
SAE J207	Electroplating of Nickel and Chromium on Metal Parts—Automotive Ornamentation and Hardware	11.133
SAE J208d	Safety for Agricultural Equipment	40.44
SAE J209 APR80	Instrument Face Design and Location for Construction and Industrial Equipment	41.258
SAE J210	Wiring Identification System for Industrial and Construction Equipment	41.188
SAE J211 JUN80	Instrumentation for Impact Tests	34.131
SAE J212 JUN80	Brake System Dynamometer Test Procedure—Passenger Car	31.88
SAE J213a	Definitions—Motorcycles	38.01
SAE J214	Hydraulic Cylinder Test Procedure	41.94
SAE J215 JAN80	Continuous Hydrocarbon Analysis of Diesel Emissions	25.16
SAE J216	Passenger Car Glazing—Electrical Circuits	34.35
SAE J217	Stainless Steel 17-7 PH Spring Wire and Springs	6.07
SAE J218a	Passenger Car Identification Terminology	26.02
SAE J220	Crane Boomstop	41.144
SAE J222a	Parking Lamps (Front Position Lamps)	23.59
SAE J223 APR80	Symbols and Color Codes for Maintenance Instructions, Container and Filler Identification	41.71
SAE J224 MAR80	Collision Deformation Classification	34.119
SAE J225 JUN80	Brake System Torque Balance Test Code—Commercial Vehicles	31.123
SAE J226a	Engine Preheaters	24.124
SAE J227a	Electric Vehicle Test Procedure	27.09
SAE J228 JAN80	Carburetor Airflow Reference Standards	24.126
SAE J229 JUN80	Service Brake Structural Integrity Test Procedure—Passenger Car	31.94
SAE J230	Stainless Steel, SAE 30302, Spring Wire and Springs	6.08
SAE J231 APR80	Minimum Performance Criteria for Falling Object Protective Structure (FOPS)	41.235
SAE J232	Industrial Rotary Mowers	41.267
SAE J233 JUN80	Marine Engine Mountings Direct Drive Transmission	42.08
SAE J234	Electric Windshield Washer Switch	22.66
SAE J235	Electric Blower Motor Switch	22.67
SAE J238	Nut and Conical Spring Washer Assemblies	15.74
SAE J240a	Life Test for Automotive Storage Batteries	22.12
SAE J242a	Metric Thread Fuel Injection Tubing Connections	24.105
SAE J243	Methods of Tests for Automotive-Type Sealers, Adhesives, and Deadeners	12.46
SAE J244	Measurement of Intake Air or Exhaust Gas Flow of Diesel Engines	25.06
SAE J245 JUN80	Engine Rating Code—Spark Ignition	24.21
SAE J246 JUN80	Spherical and Flanged Sleeve (Compression) Tube Fittings	21.18
SAE J247 JUN80	Instrumentation for Measuring Acoustic Impulses Within Vehicles	35.04
SAE J248	Crane Overload Indicating System Test Procedure (Cancelled 1978)	
SAE J249	Mechanical Stop Lamp Switch	23.42
SAE J250	Synthetic Resin Plastic Sealers, Nondrying Type	12.45
SAE J253	Headlamp Switch	23.61
SAE J254	Instrumentation and Techniques for Exhaust Gas Emissions Measurement	25.89
SAE J255a	Diesel Engine Smoke Measurement	25.27
SAE J256b	Service Performance Requirements for Motor Vehicle Lighting Devices and Components	23.26
SAE J257 JUN80	Brake Rating Horsepower Requirements—Commercial Vehicles	31.125
SAE J258	Circuit Breaker—Internal Mounted—Automatic Reset	22.42
SAE J259	Ignition Switch	22.42
SAE J260 JUN80	Rear Underride Guard Test Procedure	34.146
SAE J261	Muffler Parts Nomenclature	24.123
SAE J262	Resonator Parts Nomenclature	24.123
SAE J263	Emergency Air Brake Systems—Motor Vehicles and Vehicle Combinations	31.126
SAE J264	Vision Glossary	34.113
SAE J265	Diesel Fuel Injection Nozzle and Holder Assembly (17 mm Nominal Diameter)	24.100
SAE J267a	Wheels/Rims—Trucks—Test Procedures and Performance Requirements	30.05
SAE J268a	Rear View Mirrors—Motorcycles	38.03
SAE J270	Engine Rating Code—Diesel	24.26
SAE J271	Special Quality High Tensile, Hard Drawn Mechanical Spring Wire and Springs	6.04
SAE J272c	Vehicle Identification Number Systems	26.01
SAE J273a	Passenger Car Vehicle Identification Number System	26.02
SAE J274	Rated Spring Capacity	19.15
SAE J275	Test Method for Determining Window Fogging Resistance of Interior Trim Materials	12.61
SAE J276	Steering Frame Lock for Articulated Loaders and Tractors	41.224
SAE J277	Maintenance of Design Voltage—Snowmobile Electrical Systems	39.04
SAE J278	Snowmobile Stop Lamp	39.05
SAE J279	Snowmobile Tail Lamp (Rear Position Lamp)	39.06
SAE J280	Snowmobile Headlamps	39.04
SAE J281	Cast Iron Sealing Rings	28.61
SAE J282	Automotive Gasoline Performance and Information System (Cancelled 1979)	
SAE J283	Test Procedure for Measuring Hydraulic Lift Force Capacity on Agricultural Tractors Equipped with Three-Point Hitch	40.22
SAE J284a	Safety Alert Symbol for Agricultural, Construction and Industrial Equipment	40.43
SAE J285a	Gasoline Dispenser Nozzle Spouts	24.161
SAE J286	Clutch Friction Test Machine (SAE No. 2)—Test Procedure	28.35
SAE J287 FEB80	Driver Hand Control Reach	34.66
SAE J288a	Metallic and Nonmetallic Nonpressure Integral Fuel Tank—Snowmobile	39.16
SAE J291 JUN80	Determination of Brake Fluid Temperature	31.131

*Available as a Related Technical Report. Complete listing and details appear at the end of this index.

SAE J292	Snowmobile and Snowmobile Cutter Lamps, Reflective Devices, and Associated Equipment	39.07
SAE J293	Vehicle Grade Parking Performance Requirements	31.72
SAE J294	Service Brake Structural Integrity Test Procedure—Vehicles Over 10 000 LB (4500 kg) GVWR	31.112
SAE J296	Excavator Hoe Bucket Rating	41.213
SAE J297 APR80	Operator Controls on Industrial Equipment	41.262
SAE J298	Universal Symbols for Operator Controls on Industrial Equipment	41.260
SAE J299 JAN80	Stopping Distance Test Procedure	31.130
SAE J300d	Engine Oil Viscosity Classification	13.05
SAE J301	Effective Dates of Revisions	13.01
SAE J304c	Engine Oil Tests	13.06
SAE J306 OCT79	Axle and Manual Transmission Lubricant Viscosity Classification	13.13
SAE J308c	Axle and Manual Transmission Lubricants	13.12
SAE J310 MAR80	Automotive Lubricating Greases	13.14
SAE J311b	Fluid for Passenger Car Type Automatic Transmissions	13.17
SAE J312 JUN80	Automotive Gasolines	13.22
SAE J313 JUN80	Diesel Fuels	13.28
SAE J314b	Felts—Wool and Part Wool	12.78
SAE J315b	Fiberboard Test Procedure	12.68
SAE J316	Oil Tempered Carbon Steel Spring Wire and Springs	6.01
SAE J317	Range of Anthropometric Measurements for Asian Populations*	(HS J317)
SAE J318	Air Brake Gladhand Service (Control) and Emergency (Supply) Line Couplers—Trucks, Truck-Tractors, and Trailers	31.128
SAE J321b	Tire Guards for Protection of Operator of Earthmoving Haulage Machines	41.223
SAE J322	Nonmetallic Trim Materials—Test Method for Determining the Staining Resistance to Hydrogen Sulfide Gas	12.60
SAE J323	Test Method for Determining Cold Cracking of Flexible Plastic Materials	12.59
SAE J326	Nomenclature—Utility Backhoe	41.212
SAE J327	Ignition Timing	24.41
SAE J328a	Wheels—Passenger Cars—Performance Requirements and Test Procedures	30.01
SAE J331a	Sound Levels for Motorcycles	35.02
SAE J332a	Testing Machines for Measuring the Uniformity of Passenger Car Tires	29.10
SAE J333b	Operator Protection for Wheel Type Agricultural Tractors	(Cancelled 1978)
SAE J334b	Protective Frame for Agricultural Tractors—Test Procedures and Performance Requirements	(Cancelled 1978)
SAE J335b	Multiposition Small Engine Exhaust System Fire Ignition Suppression	24.114
SAE J336a	Sound Level for Truck Cab Interior	35.19
SAE J337	Brake Flange Mounting—On-Highway Vehicles	37.27
SAE J338	Motor Vehicle Instrument Panel Laboratory Impact Test Procedure—Knee-Leg Area	(Cancelled 1981)
SAE J339a	Seat Belt Assembly Webbing Abrasion Test Procedure	33.12
SAE J341a	Truck and Bus Tire Performance Requirements and Test Procedures	29.13
SAE J342	Spark Arrester Test Procedure for Large Size Engines	24.117
SAE J343 JUN80	Tests and Procedures for SAE 100R Series Hydraulic Hose and Hose Assemblies	21.105
SAE J344	Lawn and Garden Tractors Load and Inflation Pressures and Tire Selection Table for Future Design	40.14
SAE J345a	Wet or Dry Pavement Passenger Car Tire Peak and Locked Wheel Braking Traction	29.08
SAE J346	Motor Vehicle Seatback Assembly Laboratory Impact Test Procedure—Head Area	(Cancelled 1981)
SAE J347a	Diesel Fuel Injection Nozzle Holder Assembly (9.5 mm)	24.99
SAE J348	Wheel Chocks	37.36
SAE J349 JUN80	Detection of Surface Imperfections in Ferrous Rods, Bars, Tubes, and Wires	4.42
SAE J350 JAN80	Spark Arrester Test Procedure for Medium Size Engines	24.117
SAE J351	Oil Tempered Carbon Steel Valve Spring Quality Wire and Springs	6.02
SAE J352a	External Ignition-Proofing of Marine Engine Alternators	(Cancelled 1979)
SAE J353	External Ignition-Proofing of Marine Engine Regulators	(Cancelled 1979)
SAE J354a	External Ignition-Proofing of Marine Engine Distributors	(Cancelled 1979)
SAE J355a	External Ignition-Proofing of Marine Engine Cranking Motors	(Cancelled 1979)
SAE J356 JAN80	Welded Flash Controlled Low Carbon Steel Tubing Normalized for Bending, Double Flaring, and Beading	21.118
SAE J357a	Physical and Chemical Properties of Engine Oils	13.08
SAE J358	Nondestructive Tests	4.45
SAE J359a	Infrared Testing	4.47
SAE J360	Motor Vehicle Grade Parking Performance Test Code	31.71
SAE J361a	Test Method for Determining Visual Color Match to Master Specimen for Fabrics, Vinyls, Coated Fiberboards, and Other Automotive Trim Materials	12.75
SAE J362	Vehicle Hood Latch Systems	34.04
SAE J363 FEB80	Filter Base Mounting	24.82
SAE J364a	Vehicle Identification Numbers—Motorcycles	26.03
SAE J365a	Method of Testing Resistance to Scuffing of Trim Materials	12.69
SAE J366b	Exterior Sound Level for Heavy Trucks and Buses	35.18
SAE J367 JUN80	Passenger Car Door System Crush Test Procedure	34.142
SAE J368a	High Strength, Quenched, and Tempered Structural Steels	2.74
SAE J369a	Flammability of Automotive Interior Materials—Horizontal Test Method	12.73
SAE J370	Bolt and Capscrew Sizes for Use in Construction and Industrial Machinery	15.101
SAE J371a	Drain, Fill, and Level Plugs for Earthmoving Machinery	41.171
SAE J372a	LP-Gas Fuel Systems Components	24.113
SAE J373a	Housing Internal Dimensions for Single and Two Plate Spring Loaded Clutches	24.81
SAE J374 JUN80	Passenger Car Roof Crush Test Procedure	34.141
SAE J375a	Radius-of-Load and Boom Angle Measuring System	41.145
SAE J376 APR80	Load Indicating Devices in Lifting Crane Service	41.146
SAE J377	Performance of Vehicle Traffic Horns	35.24
SAE J378c	Marine Engine Wiring	42.16
SAE J379a	Gogan Hardness of Brake Lining	31.58
SAE J380	Specific Gravity of Brake Lining	31.59
SAE J381	Windshield Defrosting Systems Test Procedure—Trucks, Buses, and Multipurpose Vehicles	34.14
SAE J382	Windshield Defrosting Systems Performance Requirements—Trucks, Buses, and Multipurpose Vehicles	34.16
SAE J383	Motor Vehicle Seat Belt Anchorages—Design Recommendations	33.06
SAE J384	Motor Vehicle Seat Belt Anchorages Test Procedure	33.08

*Available as a Related Technical Report. Complete listing and details appear at the end of this index.

Standard	Title	Page
SAE J385	Motor Vehicle Seat Belt Anchorages—Performance Requirements	33.09
SAE J386 APR80	Seat Belts for Construction Machines	41.270
SAE J387	Terminology—Motor Vehicle Lighting	23.01
SAE J388	Dynamic Flex Fatigue Test for Slab Polyurethane Foam	34.29
SAE J389b	Universal Symbols for Operator Controls on Agricultural Equipment	40.47
SAE J390	Dual Dimensioning*	(HS J390)
SAE J391	Definition for Particle Size	4.56
SAE J392	Motorcycle and Motor Driven Cycle Electrical System (Maintenance of Design Voltage)	23.39
SAE J393	Nomenclature—Wheels and Rims—Trucks	30.10
SAE J397b	Deflection Limiting Volume—ROPS/FOPS Laboratory Evaluation	41.234
SAE J398b	Fuel Tank Filler Conditions—Passenger Car Multi-Purpose Passenger Vehicles, and Light-Duty Trucks	24.160
SAE J399a	Anodized Aluminum Automotive Parts	11.28
SAE J400 JUN80	Test for Chip Resistance of Surface Coatings	12.63
SAE J401	Selection and Use of Steels	2.05
SAE J402b	SAE Numbering System for Wrought or Rolled Steel	2.06
SAE J403h	Chemical Compositions of SAE Carbon Steels	2.06
SAE J404j	Chemical Compositions of SAE Alloy Steels	2.09
SAE J405d	Chemical Compositions of SAE Wrought Stainless Steels	2.11
SAE J406c	Methods of Determining Hardenability of Steels	2.15
SAE J407e	Hardenability Bands for Alloy H Steels. (Cancelled 1981)	
SAE J408c	Methods of Sampling Steel for Chemical Analysis	2.69
SAE J409c	Permissible Variations from Specified Ladle Chemical Ranges and Limits for Steels	2.70
SAE J410c	High Strength, Low Alloy Steel	2.72
SAE J411d	Carbon and Alloy Steels	3.01
SAE J412h	General Characteristics and Heat Treatments of Steels	3.05
SAE J413b	Mechanical Properties of Heat Treated Wrought Steels	3.13
SAE J414a	Estimated Mechanical Properties and Machinability of Hot Rolled and Cold Drawn Carbon Steel Bars	3.14
SAE J415j	Definitions of Heat Treating Terms	3.20
SAE J416b	Tensile Test Specimens	4.01
SAE J417b	Hardness Tests and Hardness Number Conversions	4.03
SAE J418a	Grain Size Determination of Steels	4.07
SAE J419	Methods of Measuring Decarburization	4.10
SAE J420b	Magnetic Particle Inspection	4.47
SAE J421b	Cleanliness Rating of Steels by the Magnetic Particle Method	4.37
SAE J422a	Microscopic Determination of Inclusions in Steels	4.39
SAE J423a	Methods of Measuring Case Depth	4.43
SAE J424	Method for Determining Breakage Allowances for Steel Sheets	4.45
SAE J425b	Eddy Current Testing by Electromagnetic Methods	4.48
SAE J426c	Liquid Penetrant Test Methods	4.52
SAE J427b	Penetrating Radiation Inspection	4.52
SAE J428b	Ultrasonic Inspection	4.55
SAE J429 JAN80	Mechanical and Material Requirements for Externally Threaded Fasteners	5.01
SAE J430	Mechanical and Chemical Requirements for Nonthreaded Fasteners	5.18
SAE J431 AUG79	Automotive Gray Iron Castings	7.01
SAE J434c	Automotive Ductile (Nodular) Iron Castings	7.06
SAE J435c	Automotive Steel Castings	7.12
SAE J437a	Selection and Heat Treatment of Tool and Die Steels	8.02
SAE J438b	Tool and Die Steels	8.01
SAE J439a	Sintered Carbide Tools	8.07
SAE J441	Cut Steel Wire Shot	9.05
SAE J442 AUG79	Test Strip, Holder and Gage for Shot Peening	9.05
SAE J443	Procedures for Using Standard Shot Peening Test Strip	9.07
SAE J444a	Cast Shot and Grit Size Specifications for Peening and Cleaning	9.09
SAE J445a	Metallic Shot and Grit Mechanical Testing	9.10
SAE J446a	Preferred Thicknesses for Uncoated Flat Metals (Thru 12 mm)	10.06
SAE J447a	Prevention of Corrosion of Metals*	(HS J447a)
SAE J448a	Surface Texture	10.02
SAE J449a	Surface Texture Control	10.04
SAE J450	Use of the Terms Yield Point and Yield Strength	10.01
SAE J451b	Aluminum Alloys—Fundamentals	11.01
SAE J452c	General Information on SAE Aluminum Casting Alloys	11.05
SAE J453c	Chemical Compositions and Mechanical and Physical Properties of SAE Aluminum Casting Alloys	11.08
SAE J454k	General Data on Wrought Aluminum Alloys	11.11
SAE J457j	Chemical Compositions, Mechanical Property Limits, and Dimensional Tolerances of SAE Wrought Aluminum Alloys	11.27
SAE J459c	Bearing and Bushing Alloys (Information Report)	11.29
SAE J460e	Bearing and Bushing Alloys (Standard)	11.29
SAE J461d	Wrought and Cast Copper Alloys	11.31
SAE J462b	Cast Copper Alloys	11.60
SAE J463d	Wrought Copper and Copper Alloys	11.63
SAE J464c	Magnesium Alloys	11.102
SAE J465c	Magnesium Casting Alloys	11.102
SAE J466c	Magnesium Wrought Alloys	11.107
SAE J467b	Special Purpose Alloys ("Superalloys")	11.111
SAE J468b	Zinc Alloy Ingot and Die Casting Compositions	11.122
SAE J469a	Zinc Die Casting Alloys	11.123
SAE J470c	Wrought Nickel and Nickel Related Alloys	11.123
SAE J471d	Sintered Powder Metal Parts: Ferrous	9.01
SAE J473a	Solders	11.138
SAE J474b	Electroplating and Related Finishes	11.139
SAE J475a	Screw Threads	14.01
SAE J476a	Dryseal Pipe Threads	14.01
SAE J478a	Slotted and Recessed Head Screws	15.07
SAE J479a	Slotted Headless Setscrews	15.40
SAE J481	Round Head Bolts	15.04
SAE J482a	Hexagon High Nuts	15.53
SAE J483a	Crown (Blind, Acorn) Nuts	15.54
SAE J484	Alignment of Nut Slots	15.54
SAE J485	Holes in Bolt and Capscrew Shanks and Slots in Nuts for Cotter Pins	15.79
SAE J487a	Cotter Pins	15.77
SAE J488	Plain Washers	15.68
SAE J489b	Lock Washers	15.68
SAE J490c	Ball Joints	16.08
SAE J491b	Ball Studs and Ball Stud Socket Assemblies	16.01
SAE J492	Rivets and Riveting	15.85
SAE J493	Rod Ends and Clevis Pins	15.84
SAE J494	Grooved Straight Pins	15.80
SAE J495	Straight Pins (Solid)	15.79
SAE J496	Spring Type Straight Pins	15.82
SAE J497	Unhardened Ground Dowel Pins	15.84
SAE J498c	Involute Splines and Inspections	17.01
SAE J499a	Parallel Side Splines for Soft Broached Holes in Fittings	17.01
SAE J500	Serrated Shaft Ends	17.03
SAE J501	Shaft Ends	17.04
SAE J502	Woodruff Keys	17.05
SAE J503	Woodruff Key Slots and Keyways	17.06
SAE J506b	Sleeve Type Half Bearings	24.56
SAE J510c	Leaf Springs for Motor Vehicle Suspension	19.01
SAE J511a	Pneumatic Spring Terminology	19.08
SAE J512 NOV79	Automotive Tube Fittings	21.04

*Available as a Related Technical Report. Complete listing and details appear at the end of this index.

SAE No.	Title	Page
SAE J513f	Refrigeration Tube Fittings	21.26
SAE J514 APR80	Hydraulic Tube Fittings	21.46
SAE J515a	Hydraulic "O" Ring	21.112
SAE J516 SEP79	Hydraulic Hose Fittings	21.72
SAE J517 JUN80	Hydraulic Hose	21.92
SAE J518c	Hydraulic Flanged Tube, Pipe, and Hose Connections, 4-Bolt Split Flange Type	21.107
SAE J520a	Fuel Supply Connections (Cancelled 1978)	
SAE J521b	Fuel Injection Tubing Connections	24.106
SAE J524 JAN80	Seamless Low Carbon Steel Tubing Annealed for Bending and Flaring	21.115
SAE J525 JAN80	Welded and Cold Drawn Low Carbon Steel Tubing Annealed for Bending and Flaring	21.115
SAE J526 JAN80	Welded Low Carbon Steel Tubing	21.116
SAE J527 JAN80	Brazed Double Wall Low Carbon Steel Tubing	21.117
SAE J528 JAN80	Seamless Copper Tube	21.121
SAE J529b	Fuel Injection Tubing	21.127
SAE J530a	Automotive Pipe Fittings	21.128
SAE J531a	Automotive Pipe, Filler, and Drain Plugs	21.134
SAE J532a	Automotive Straight Thread Filler and Drain Plugs	21.138
SAE J533b	Flares for Tubing	21.113
SAE J534c	Lubrication Fittings	21.143
SAE J535 JAN80	Water Connection Flanges	24.83
SAE J536b	Hose Clamps	15.101
SAE J537j	Storage Batteries	22.04
SAE J538a	Grounding of Storage Batteries	22.13
SAE J539a	Voltages for Diesel Electrical Systems	22.13
SAE J541a	Voltage Drop for Starting Motor Circuits	22.13
SAE J542c	Starting Motor Mountings	22.14
SAE J543c	Starting Motor Pinions and Ring Gears	22.15
SAE J544b	Starting Motor and Generator Curves	22.17
SAE J545b	Generator Mountings	22.15
SAE J546	Magneto Mountings	22.20
SAE J548d	Spark Plugs	22.21
SAE J549a	Preignition Rating of Spark Plugs	22.27
SAE J551 JUN79	Limits and Methods of Measurement of Radio Interference Characteristics of Vehicles and Devices (20–1000 MHz)	22.27
SAE J552a	External Electromagnetic Radiation Supressors	22.34
SAE J553c	Circuit Breakers	22.40
SAE J554b	Electric Fuses (Cartridge Type)	22.35
SAE J555a	Truck, Truck-Tractor, Trailer, and Motor Coach Wiring (Cancelled 1981)	
SAE J556	Automobile Wiring (Cancelled 1981)	
SAE J557	High Tension Ignition Cable	22.49
SAE J558a	Low Tension Cable (Cancelled 1978)	
SAE J560b	Seven-Conductor Electrical Connector for Truck-Trailer Jumper Cable	22.57
SAE J561 JUN80	Electrical Terminals—Eyelet and Spade Type	22.61
SAE J562a	Nonmetallic Loom	22.65
SAE J563b	6- and 12-Volt Cigar Lighter Receptacles	22.67
SAE J564c	Headlamp Beam Switching	23.02
SAE J565c	Semiautomatic Headlamp Beam Switching Devices	23.03
SAE J567c	Lamp Bulb Retention System	23.04
SAE J568	Sockets Receiving Prefocus Base Lamps (Cancelled 1978)	
SAE J571d	Dimensional Specifications for Sealed Beam Headlamp Units	23.09
SAE J572a	Dimensional Specifications for Sealed Lighting Unit for Construction and Industrial Machinery	41.178
SAE J573g	Lamp Bulbs and Sealed Beam Headlamp Units	23.20
SAE J575 JUN80	Tests for Motor Vehicle Lighting Devices and Components	23.24
SAE J576d	Plastic Materials for Use in Optical Parts Such as Lenses and Reflectors of Motor Vehicle Lighting Devices	23.28
SAE J577	Vibration Test Machine (Cancelled 1980)	
SAE J578d	Color Specification for Electric Signal Lighting Devices	23.29
SAE J579c	Sealed Beam Headlamp Units for Motor Vehicles	23.31
SAE J580 AUG79	Sealed Beam Headlamp Assembly	23.33
SAE J581a	Auxiliary Driving Lamps	23.35
SAE J582a	Auxiliary Low Beam Lamp	23.36
SAE J583d	Fog Lamps	23.36
SAE J584b	Motorcycle Headlamps	23.37
SAE J585e	Tail Lamps (Rear Position Lamps)	23.40
SAE J586d	Stop Lamps	23.41
SAE J587f	License Plate Lamps	23.43
SAE J588f	Turn Signal Lamps	23.44
SAE J589b	Turn Signal Switch	23.50
SAE J590e	Turn Signal Flashers	23.52
SAE J591b	Spot Lamps	23.58
SAE J592f	Clearance, Side Marker, and Identification Lamps	23.59
SAE J593e	Backup Lamps	23.60
SAE J594f	Reflex Reflectors	23.61
SAE J595b	Flashing Warning Lamps for Authorized Emergency, Maintenance and Service Vehicles	23.63
SAE J596	Electric Emergency Lanterns (Cancelled 1979)	
SAE J597	Liquid Burning Emergency Flares (Cancelled 1979)	
SAE J598 APR80	Sealed Lighting Units for Construction and Industrial Machinery	41.177
SAE J599d	Lighting Inspection Code	23.66
SAE J600a	Headlamp Testing Machines	23.68
SAE J602c	Headlamp Aiming Device for Mechanically Aimable Sealed Beam Headlamp Units	23.69
SAE J603c	Incandescent Lamp Impact Test (Cancelled 1981)	
SAE J604d	Engine Terminology and Nomenclature—General	24.01
SAE J607a	Small Spark Ignition Engine Test Code	24.32
SAE J609a	Mounting Flanges and Power Take-Off Shafts for Small Engines	24.36
SAE J610b	Valve Seat Inserts—Engine	24.38
SAE J614b	Engine and Transmission Dipstick Marking	24.77
SAE J615b	Engine Mountings	24.78
SAE J616c	Engine Foot Mounting (Front and Rear)	41.166
SAE J617c	Engine Flywheel Housings	41.164
SAE J618 JUN80	Flywheels for Single-Plate Spring-Loaded Clutches	24.79
SAE J619 JUN80	Flywheels for Two-Plate Spring-Loaded Clutches	24.80
SAE J620d	Flywheels for Industrial Engines Used with Industrial Power Take-Offs Equipped with Driving-Ring Type Overcenter Clutches and Engine Mounted Marine Gears	41.170
SAE J621d	Industrial Power Take-Offs with Driving Ring Type Overcenter Clutches	41.162
SAE J622 JAN80	Airflow or Vacuum Governor Flanges	24.84
SAE J623a	Automotive Carburetor Flanges	24.86
SAE J624a	Exhaust Flanges—Industrial Engines	41.168
SAE J625a	Fuel Pump Mountings for Diaphragm Type Pumps	24.90
SAE J626b	Diesel Fuel Injection Pump Mountings	24.92
SAE J629b	Diesel Fuel Injection Nozzle Holder Assemblies	24.97
SAE J631b	Radiator Nomenclature	24.132
SAE J634a	Water Thermostat Pockets	24.145
SAE J635a	Fan Hub Bolt-Circles and Pilot Holes	24.145
SAE J636c	V-Belts and Pulleys	18.01
SAE J637b	Automotive V-Belt Drives	18.04
SAE J638	Test Procedure and Ratings for Hot Water	

*Available as a Related Technical Report. Complete listing and details appear at the end of this index.

	Heaters for Motor Vehicles	34.20		mission Mounted Power Take-Offs	28.16
SAE J639	Safety Practices for Mechanical Vapor Compression Refrigeration Equipment or Systems Used to Cool Passenger Compartment of Motor Vehicles	34.22	SAE J705b	Rating of Truck Power Take-Offs	37.32
			SAE J706a	Rating of Winches	37.32
			SAE J708 JUN80	Agricultural Tractor Test Code	40.01
SAE J640b	Symbols for Hydrodynamic Drives	28.01	SAE J709d	Agricultural Tractor Tire Loadings, Torque Factors, and Inflation Pressures	40.05
SAE J641b	Hydrodynamic Drives Terminology	28.02			
SAE J643b	Hydrodynamic Drive Test Code	28.04	SAE J711c	Tire Selection Tables for Agricultural and Light Industrial Machines of Future Design	40.12
SAE J645c	Automotive Transmission Terminology	28.07			
SAE J646b	Planetary Gear(s) Terminology	28.07	SAE J712a	Industrial and Agricultural Disc Wheels	40.32
SAE J647a	Transmissions—Schematic Diagrams	28.08	SAE J714b	Wheel Mounting Elements for Industrial and Agricultural Disc Wheels	40.36
SAE J648a	Automatic Transmission Hydraulic Control Systems—Terminology	28.08			
			SAE J715 MAY80	Three-Point Free-Link Hitch Attachment of Implements to Agricultural Wheeled Tractors	40.19
SAE J649b	Automatic Transmission Functions—Terminology	28.08			
SAE J651c	Passenger Car and Truck Automatic Transmission Test Code	28.10	SAE J716a	Application of Hydraulic Remote Control to Agricultural Tractors and Trailing Type Agricultural Implements	40.24
SAE J652	Truck Transmissions—Test Code (Cancelled 1980)				
SAE J654	Seals—Lathe Cut	28.60	SAE J717 MAY80	Auxiliary Power Take-Off Drives for Agricultural Tractors	40.31
SAE J656g	Automotive Brake Definitions and Nomenclature	31.53			
			SAE J718d	540-RPM Power Take-Off for Farm Tractors (Cancelled 1978)	
SAE J659	Color Code for Location Identification of Combination Linings of Two Different Materials or Two Shoe Brakes	31.64			
			SAE J719d	1000-RPM Power Take-Off for Farm Tractors (Cancelled 1978)	
SAE J660a	Brake Linings (Cancelled 1980)		SAE J720 MAY80	Tractor Belt Speed and Pulley Width	40.32
SAE J661a	Brake Lining Quality Control Test Procedure	31.61			
SAE J662	Brake Block Chamfer	31.64	SAE J721 JUN80	Operating Requirements for Tractors and Power Take-Off Driven Implements	40.30
SAE J663b	Rivets for Brake Linings and Bolts for Brake Blocks	31.64			
			SAE J722 MAY80	Power Take-Off Definitions and Terminology for Agricultural Tractors	40.32
SAE J666	Brake Diaphragm Maintenance—Tractor and Semitrailer	37.29			
SAE J667	Brake Test Code—Inertia Dynamometer	31.73	SAE J725 MAY80	Mounting Brackets and Socket for Warning Lamp and Slow-Moving Vehicles (SMV) Identification Emblem	40.41
SAE J670e	Vehicle Dynamics Terminology*	(HSJ 670e)			
SAE J671	Sound Deadeners and Underbody Coatings	12.56			
SAE J673b	Automotive Safety Glazing	34.31	SAE J726 SEP79	Air Cleaner Test Code	24.67
SAE J674a	Safety Glazing Materials—Motor Vehicles	34.34			
SAE J678e	Speedometers and Tachometers—Automotive	20.03	SAE J727 MAY80	Component Nomenclature—Crawler Tractor	41.16
SAE J680b	Location and Operation of Instruments and Controls In Motor Truck Cabs	37.34	SAE J728a	Component Nomenclature—Scrapers	41.19
			SAE J729 MAY80	Nomenclature—Dozer	41.22
SAE J682	Rear Wheel Splash and Stone Throw Protection	37.34	SAE J731d	Component Nomenclature—Loader	41.25
SAE J683a	Tire Chain Clearance—Trucks, Buses, and Combinations of Vehicles	37.29	SAE J732 FEB80	Specification Definitions—Loaders	41.42
			SAE J733b	Nomenclature—Rippers, Rooters, Scarifiers	41.27
SAE J684f	Trailer Couplings, Hitches and Safety Chains—Automotive Type	36.01	SAE J734c	Component Nomenclature—Dumper Trailer	41.28
			SAE J736a	Mechanical Power Outlet Test Code (Cancelled 1979)	
SAE J685a	Data Plate—Automotive Type Trailers (Cancelled 1981)		SAE J737b	Hole Spacing for Scraper and Bulldozer Cutting Edges	41.156
SAE J686a	Motor Vehicle License Plates	26.07			
SAE J687c	Nomenclature—Truck, Bus, Trailer	37.01	SAE J738b	Scraper and Bulldozer Cutting Edge Cross Sections, Rolled, Cast, Forged or Machined	41.158
SAE J688	Truck Ability Prediction Procedure	37.02			
SAE J689	Approach, Departure, and Ramp Breakover Angles—Passenger Car and Light Duty Trucks	32.03	SAE J739b	Grader Cutting Edge	41.158
			SAE J740b	Countersunk Square Holes for Cutting Edges and End Bits	41.161
SAE J690	Certificates of Maximum Net Horsepower for Motor Trucks and Truck Tractors	37.12	SAE J741b	Capacity Ratings—Scraper, Dumper Body and Trailer Body	41.209
SAE J691	Motor Truck CA Dimensions	37.13			
SAE J693a	Truck Overall Widths Across Dual Tires	37.26	SAE J742 OCT79	Capacity Rating—Loader Bucket	41.211
SAE J694a	Commercial Vehicle Disc Wheels	37.25			
SAE J695b	Turning Ability and Off Tracking—Motor Vehicles	37.14	SAE J743b	Side Booms—Tractor Mounted	41.153
			SAE J744c	Hydraulic Power Pumps	41.79
SAE J697a	Safety Chain of Full Trailers or Converter Dollies	37.24	SAE J745c	Hydraulic Power Pump Test Procedure	41.77
			SAE J746b	Hydraulic Motor Test Procedure	41.87
SAE J699a	Average Vehicle Dimensions for Use in Designing Docking Facilities for Motor Vehicles	37.13	SAE J747b	Hydraulic Control Valve Test Procedure	41.89
			SAE J748	Hydraulic Directional Control Valves, 3000 psi Maximum	41.81
SAE J700b	Fifth Wheel Kingpin—Commercial Trailers and Semitrailers	37.21	SAE J749a	Drawbars—Crawler Tractor	41.208
			SAE J751 APR80	Off-Highway Tire and Rim Classification	41.199
SAE J701a	Truck Tractor Semitrailer Interchange Coupling Dimensions	37.18			
			SAE J752b	Maintenance Interval—Construction Equipment	41.69
SAE J702b	Brake and Electrical Connection Locations—Truck-Tractor and Truck-Trailer	37.28			
			SAE J753 APR80	Lubrication Chart—Construction and Industrial Machinery	41.67
SAE J703a	Fuel Systems—Truck and Truck Tractor (Cancelled 1980)				
SAE J704b	Openings for Six- and Eight-Bolt Truck Trans-		SAE J754a	Lubricant Types—Construction and Industrial Machinery	41.68

*Available as a Related Technical Report. Complete listing and details appear at the end of this index.

Standard	Date	Title	Page
SAE J755	JUN80	Marine Propeller-Shaft Ends and Hubs	42.05
SAE J756	JUN80	Marine Propeller-Shaft Couplings	42.01
SAE J759c		Lighting Identification Code	23.01
SAE J760a		Dimensional Specifications for General Service Sealed Lighting Units	23.18
SAE J762a		Reinforced Plastics for Ground Vehicle Applications*	(HS J762a)
SAE J763		Aging of Carbon Steel Sheet and Strip	3.20
SAE J764	APR80	Loading Ability Test Code—Scrapers	41.108
SAE J765a		Crane Load Stability Test Code	41.142
SAE J770c		Alloy Steel Machinability Ratings	3.24
SAE J771c		Automotive Printed Circuits	22.45
SAE J772a		Clearance Envelopes for Six- and Eight-Bolt Truck Transmission Mounted Power Take-Offs	28.15
SAE J773b		Conical Spring Washers	15.73
SAE J774c		Emergency Warning Device	23.62
SAE J775a		Engine Poppet Valve Materials	2.75
SAE J776f		Hardenability Bands for Carbon H Steels (Cancelled 1981)	
SAE J778a		Formerly Standard SAE Alloy Steels	2.13
SAE J779		Tractor Protection Valve Control	37.31
SAE J780a		Engine Coolant Pump Seals	24.147
SAE J782b		Seating Manual—Motor Vehicles*	(HS J782b)
SAE J783		Influence of Residual Stress on Fatigue of Steel*	(HS 198)
SAE J784a		Residual Stress Measurements by X-Ray Diffraction*	(HS J784a)
SAE J786a		Brake System Road Test Code—Truck, Bus, and Combination of Vehicles	31.105
SAE J787b		Replaced by J383, J384, J385	
SAE J788a		Manual on Design and Application of Leaf Springs*	(HS J788a)
SAE J792a		SAE Manual on Blast Cleaning*	(HS 124)
SAE J793		Evaluation of Methods for Measurement of Residual Stress*	(HS 147)
SAE J795a		Manual on Design and Application of Helical and Spiral Springs*	(HS J795a)
SAE J797		Mobile Radio Telephone for Automotive Use*	(HS 27)
SAE J799		Maintenance of Automotive Engine Cooling Systems*	(HS 40)
SAE J800c		Motor Vehicle Seat Belt Assembly Installations	33.10
SAE J805		Scales for Use with Decimal-Inch Dimensioning*	(HS 156)
SAE J806b		Oil Filter Test Procedure*	(HS J806b)
SAE J808a		Manual on Shot Peening*	(HS 84)
SAE J810	JUN80	Classification of Common Imperfections in Sheet Steel	4.24
SAE J811		Surface Rolling and Other Methods for Mechanical Prestressing of Metals*	(HS 3)
SAE J813		Automotive Air Brake Reservoir Volume	31.127
SAE J814c		Engine Coolants	12.93
SAE J815		Load Deflection Testing of Urethane Foams for Automotive Seating	34.31
SAE J816b		Engine Test Code—Spark Ignition and Diesel	24.08
SAE J817a		Engineering Design Serviceability Guide Lines—Construction and Industrial Machinery	41.280
SAE J818b		Operating Load for Loaders	41.67
SAE J819	MAR80	Engine Cooling System Field Test (Air-to-Boil)	41.106
SAE J820		Crane Hoist Line Speed and Power Test Code	41.141
SAE J821a		Electrical System for Construction and Industrial Machinery	41.182
SAE J823c		Flasher Test Equipment	23.72
SAE J824		Engine Rotation and Cylinder Numbering	24.42
SAE J826	APR80	Devices for Use in Defining and Measuring Vehicle Seating Accommodation	34.22
SAE J827		Cast Steel Shot	9.08
SAE J828	JAN80	Gas Turbine Engine Nomenclature	24.02
SAE J829c		Fuel Tank Filler Cap and Cap Retainer	24.156
SAE J830b		Fuel Injection Equipment Nomenclature	24.91
SAE J831		Nomenclature—Automotive Electrical Systems	22.01
SAE J832		Locating Ledge for Flat Pad Engine Mountings	24.78
SAE J833	JAN80	USA Human Physical Dimensions	41.245
SAE J834a		Passenger Car Rear Vision	34.39
SAE J835		Split Type Bushings—Design and Application	24.63
SAE J836a		Automotive Metallurgical Joining* (Available from American Welding Society)	(HS J836a)
SAE J839b		Passenger Car Side Door Latch Systems	34.02
SAE J840c		Test Procedures for Brake Shoe and Lining Adhesives and Bonds	31.65
SAE J841	MAY80	Operator Controls on Agricultural Equipment	40.46
SAE J843d		Brake System Road Test Code—Passenger Car and Light-Duty Truck	31.75
SAE J844d		Nonmetallic Air Brake System Tubing	21.124
SAE J845		360 Deg Emergency Warning Lamp	23.64
SAE J846	MAY80	Coding System for Identification of Tube, Pipe, and Hose Fittings	21.01
SAE J847		Full Trailer Tow Bar Eye	37.24
SAE J848a		Fifth Wheel Kingpin, Heavy Duty—Commercial Trailers and Semitrailers	37.22
SAE J849b		Connection and Accessory Locations for Towing Doubles Trailers and Multi-Axle Trailers	37.22
SAE J850	JUN80	Barrier Collision Tests	34.136
SAE J851		Recommended Tolerances for Spoke-Type Wheels, Demountable Rims, and Rim Spacers—Trucks and Trailers	37.25
SAE J852b		Cornering Lamps	23.51
SAE J853a		Passenger Car Vehicle Identification Numbers	26.03
SAE J855a		Test Method of Stretch and Set of Textiles and Plastics	12.76
SAE J856		Connectors and Plugs	23.07
SAE J857	JUN80	Roll-Over Tests without Collision	34.137
SAE J858a		Electrical Terminals—Blade Type	22.59
SAE J859		Numbering System for Designating Grades of Cast Ferrous Materials	7.01
SAE J860		Test Method for Measuring Weight of Organic Trim Materials	12.58
SAE J861		Method of Testing Resistance to Crocking of Organic Trim Materials	12.58
SAE J862b		Factors Affecting Automotive Odometer-Speedometer Accuracy	20.01
SAE J863c		Methods of Determining Plastic Deformation in Sheet Metal Stampings	4.14
SAE J864		Surface Hardness Testing with Files	4.02
SAE J866a		Friction Identification System for Brake Linings and Brake Blocks for Motor Vehicles	31.60
SAE J868		Large Size Radiator Filler Necks	24.130
SAE J869a		Component Nomenclature—Two and Four-Wheel Tractors	41.17
SAE J870		Component Nomenclature—Graders	41.21
SAE J871a		Hi-Head Finished Hex Bolts	15.04
SAE J872	APR80	Reserve Tractive Ability Test Code	41.98
SAE J873	APR80	Machine Drag Test Code	41.100
SAE J874	APR80	Center of Gravity Test Code	41.101
SAE J875		Trailer Axle Alignment	37.27
SAE J876a		Extra Wide Single Tire Rim and Wheel Mounting for Rear Axles	37.26
SAE J877		Properties of Low Carbon Steel Sheets and Strip and Their Relationship to Formability	4.17
SAE J878a		Low Tension Cable Thermosetting	

*Available as a Related Technical Report. Complete listing and details appear at the end of this index.

	Insulation . . . . (Cancelled 1978)	
SAE J879b	Motor Vehicle Seating Systems	34.26
SAE J880 MAY80	Brake System Rating Test Code—Commercial Vehicles	31.117
SAE J881a	Lifting Crane Sheave and Drum Sizes	41.156
SAE J882	Test Method for Measuring Thickness of Automotive Textiles and Plastics	12.59
SAE J883	Test Method for Determining Dimensional Stability of Automotive Textile Materials .	12.58
SAE J884 MAY80	Liquid Ballast Table for Drive Tires of Agricultural Machines	40.10
SAE J885 APR80	Human Tolerance to Impact Conditions as Related to Motor Vehicle Design*	(HS J885)
SAE J885a	Human Tolerance to Impact Conditions as Related to Motor Vehicle Design . . . (Cancelled 1981)	
SAE J887a	School Bus Red Signal Lamps	23.65
SAE J891a	Spring Nuts	15.55
SAE J892a	Push-On Spring Nuts	15.63
SAE J893	Vehicle Fuel Consumption Test Code . . (Cancelled 1979)	
SAE J894	Replaced by J1234	
SAE J895	Five Conductor Electrical Connectors for Automotive Type Trailers	22.52
SAE J896	Mounting Flanges for Engine Accessory Drives	41.166
SAE J897	Machine Slope Operation Test Code	41.107
SAE J898 APR80	Control Locations for Construction and Industrial Equipment Design	41.255
SAE J899	Operator's Seat Dimensions—Construction and Industrial Equipment Design	41.254
SAE J900	Crankcase Emission Control Test Code	25.49
SAE J901b	Universal Joints and Driveshafts—Nomenclature and Terminology	28.17
SAE J902b	Passenger Car Windshield Defrosting Systems	34.12
SAE J903c	Passenger Car Windshield Wiper Systems . .	34.06
SAE J905	Fuel Filter Test Method*	(HS J905)
SAE J906	Automotive Safety Glazing Manual*	(HS J906)
SAE J909 APR80	Three-Point Hitch, Implement Quick-Attaching Coupler, Agricultural Tractors	40.17
SAE J910b	Hazard Warning Signal Switch	23.64
SAE J911	Surface Texture Measurement of Cold Rolled Sheet Steel	10.01
SAE J912a	Test Method for Determining Blocking Resistance and Associated Characteristics of Automotive Trim Materials	12.71
SAE J913a	Test Method for Wicking of Automotive Fabrics and Fibrous Materials	12.69
SAE J914b	Side Turn Signal Lamps	23.50
SAE J915	Automatic Transmissions—Manual Control Sequence	28.14
SAE J916 JUN80	Rules for SAE Use of SI (Metric) Units . . .	1.09
SAE J917 JUN80	Marine Push-Pull Control Cables	42.02
SAE J918c	Passenger Car Tire Performance Requirements and Test Procedures	29.01
SAE J919c	Operator Sound Level Measurement Procedure for Earthmoving Machinery—Singular Type Test	35.37
SAE J920a	Format for Construction Equipment Manuals	41.286
SAE J921b	Motor Vehicle Instrument Panel Laboratory Impact Test Procedure—Head Area . (Cancelled 1981)	
SAE J922 NOV79	Turbocharger Nomenclature and Terminology	24.07
SAE J923	Axles—Nomenclature and Terminology . . .	28.25
SAE J924	Thrust Washers	24.65
SAE J925	Minimum Access Dimensions for Construction and Industrial Machinery	41.248
SAE J926b	Hydraulic Pipe Fittings . . . . (Cancelled 1979)	
SAE J927b	Flywheels for Engine Mounted Torque Converters	41.167
SAE J928 JUN80	Electrical Terminals—Pin and Receptacle Type	22.60
SAE J929a	Piston Rings and Pistons	24.42
SAE J930b	Storage Batteries for Construction and Industrial Machinery	41.287
SAE J931a	Hydraulic Power Circuit Filtration	41.92
SAE J932	Definitions for Macrostrain and Microstrain	4.57
SAE J933 JUN79	Mechanical and Quality Requirements for Tapping Screws	5.19
SAE J934a	Vehicle Passenger Door Hinge Systems . . .	34.01
SAE J935	High Strength Carbon and Alloy Die Drawn Steels	3.03
SAE J936	Methods of Residual Stress Measurement* .	(HS J936)
SAE J937b	Service Brake System Performance Requirements—Passenger Car	31.86
SAE J938a	Drop Test for Evaluating Laminated Safety Glass for Use in Automotive Windshields	34.36
SAE J940	Glossary of Carbon Steel Sheet and Strip Terms	3.18
SAE J941e	Motor Vehicle Driver's Eye Range	34.60
SAE J942b	Passenger Car Windshield Washer Systems	34.10
SAE J943a	Slow-Moving Vehicle Identification Emblem	40.40
SAE J944 JUN80	Steering Control System—Passenger Car— Laboratory Test Procedure	34.144
SAE J945b	Vehicular Hazard Warning Signal Flasher . .	23.54
SAE J946d	Seals—Application Guide to Radial Lip . . .	28.43
SAE J947b	Glossary of Fiberboard Terminology	12.67
SAE J948a	Test Method for Determining Resistance to Snagging and Abrasion of Automotive Bodycloth	12.75
SAE J949a	Test Method for Determining Stiffness (Modulus of Bending) of Fiberboards	12.74
SAE J950 APR80	Gradeability Test Code	41.105
SAE J951 FEB80	Florida Exposure of Automotive Finishes . . .	12.66
SAE J953	Passenger Car Backlight Defogging System	34.17
SAE J954	Urethane for Automotive Seating	34.30
SAE J956	Remote and Automatic Control Systems for Construction and Industrial Machinery . .	41.276
SAE J957	Elevating Scrapers—Capacity Rating	41.210
SAE J958	Nomenclature and Dimensions for Crane Shovels	41.34
SAE J959	Lifting Crane, Wire-Rope Strength Factors . .	41.148
SAE J960 JUN80	Marine Control Cable Connection—Engine Clutch Lever	42.03
SAE J961 JUN80	Marine Control Cable Connection—Engine Throttle Lever	42.04
SAE J962b	Formed Tube Ends for Hose Connections . .	21.112
SAE J963	Anthropomorphic Test Device for Use in Dynamic Testing of Motor Vehicles . (Cancelled 1979)	
SAE J964a	Test Procedure for Determining Reflectivity of Rear View Mirrors	34.37
SAE J965	Abrasive Wear*	(HS J965)
SAE J966	Test Procedure for Measuring Passenger Car Tire Revolutions per Mile	29.15
SAE J967d	Calibration Fluid for Diesel Injection Equipment	24.110
SAE J968c	Calibrating Nozzle and Holder Assembly for Diesel Fuel Injection Systems	24.102
SAE J969b	Testing Techniques for Diesel Fuel Injection Systems	24.104
SAE J970a	Test Stands for Diesel Fuel Injection Systems	24.104
SAE J971	Brake Rating Test Code—Commercial Vehicle Inertia Dynamometer	31.124
SAE J972 JUN80	Moving Barrier Collision Tests	34.132
SAE J973a	Ignition System Measurements Procedure . .	22.02
SAE J974 MAY80	Flashing Warning Lamp for Agricultural Equipment	40.38
SAE J975 MAY80	Headlamps for Agricultural Equipment	40.38
SAE J976 MAY80	Combination Tail and Floodlamp for Agricultural Equipment	40.39
SAE J977	Replaced by J211a	
SAE J978	Bumper Jacking Test Procedure—Motor Vehicles	32.01

*Available as a Related Technical Report. Complete listing and details appear at the end of this index.

SAE J979	Bumper Jack Requirements—Motor Vehicles	32.01
SAE J980a	Bumper Evaluation Test Procedure—Passenger Cars	32.02
SAE J981	Master Calibrating Fuel Injection Pump	24.101
SAE J982a	Test Code—Truck, Truck-Tractor, and Trailer Air Service Brake System Pneumatic Pressure and Time Levels	31.127
SAE J983	Crane and Cable Excavator Basic Operating Control Arrangements	41.150
SAE J984 JUN80	Bodyforms for Use in Motor Vehicle Passenger Compartment Impact Development	34.143
SAE J985	Vision Factors Considerations in Rear View Mirror Design	34.38
SAE J986b	Sound Level for Passenger Cars and Light Trucks	35.11
SAE J987 APR80	Crane Structures—Method of Test	41.110
SAE J988	Labeling of Motor Vehicle Brake Fluid Containers	31.48
SAE J989a	Test Procedure for Measuring Carbon Monoxide Concentrations in Vehicle Passenger Compartments	25.53
SAE J990	Nomenclature—Industrial Mowers	41.265
SAE J991 APR80	Soil-Machine Terminology	41.60
SAE J992b	Brake System Performance Requirements—Truck, Bus, and Combination of Vehicles	31.111
SAE J993b	Alloy and Temper Designation Systems for Aluminum	11.02
SAE J994b	Performance, Test, and Application Criteria for Electrically Operated Backup Alarm Devices	35.32
SAE J995 JUN79	Mechanical and Material Requirements for Steel Nuts	5.15
SAE J996 JUN80	Inverted Vehicle Drop Test Procedure	34.140
SAE J997 JAN80	Spark Arrester Test Carbon	24.120
SAE J998	Minimum Requirements for Motor Vehicle Brake Linings	31.63
SAE J999 OCT79	Crane Boom Hoist Disengaging Device	41.147
SAE J1001 APR80	Safety Criteria for Industrial Flail Mowers	41.268
SAE J1002	Seals—Evaluation of Elastohydrodynamic	28.54
SAE J1003	Diesel Engine Emission Measurement Procedure	25.20
SAE J1004	Glossary of Engine Cooling System Terms	24.130
SAE J1006 MAR80	Performance Test for Air Conditioned Agricultural Equipment	40.41
SAE J1008	Exterior Sound Level Measurement Procedure for Self-Propelled Agricultural Field Equipment	35.60
SAE J1010	Emission Control Hose	12.114
SAE J1012 MAR80	Agricultural Equipment Enclosure Pressurization System Test Procedure	40.43
SAE J1013 JAN80	Measurement of Whole Body Vibration of the Seated Operator of Off-Highway Work Machines	40.57
SAE J1014 OCT79	Classification and Nomenclature Towing Winch for Skidders and Crawler Tractors	41.32
SAE J1015	Ton-Mile Per Hour Test Procedure	41.205
SAE J1016	Component Nomenclature—Dumpers	41.29
SAE J1017	Component Nomenclature—Rollers/Compactors	41.30
SAE J1018a	Recreational Trailer Vehicle Identification Number System	(Cancelled 1981)
SAE J1019	Tests and Procedures for High-Temperature Transmission Oil Hose, Lubricating Oil Hose and Hose Assemblies	21.107
SAE J1024 APR80	Fuel-Fired Heaters—Air Heating—For Construction and Industrial Machinery	41.273
SAE J1025	Test Procedures for Measuring Truck Tire Revolutions per Mile	29.15
SAE J1026 SEP79	Braking Performance—Crawler Tractors and Loaders	41.188
SAE J1028	Mobile Crane Working Area Definitions	41.152
SAE J1029 APR80	Lighting and Marking of Construction and Industrial Machinery	41.180
SAE J1030	Maximum Sound Level for Passenger Cars and Light Trucks	35.13
SAE J1032 APR80	Definitions for Machine Availability (Construction and Industrial)	41.58
SAE J1033	Procedure for Measuring Bore Eccentricity and Face Deviation of Flywheels, Flywheel Housings, and Flywheel Housing Adapters	41.163
SAE J1034	Engine Coolant Concentrate—Ethylene-Glycol Type	12.97
SAE J1035 JUN80	Technical Publications for Agricultural Equipment	40.65
SAE J1036	Dimensional Standard for Cylindrical Hydraulic Couplers for Agricultural Tractors	40.26
SAE J1037	Windshield Washer Tubing	12.115
SAE J1038 FEB80	Recommendations for Children's Snowmobile	39.18
SAE J1039	Size Classification for Crawler Tractors	(Cancelled 1980)
SAE J1040c	Performance Criteria for Rollover Protective Structures (ROPS) for Construction, Earthmoving, Forestry, and Mining Machines	41.225
SAE J1041	Brake Test Procedure and Brake Performance Criteria for Agricultural Equipment	40.15
SAE J1042 APR80	Operator Protection for Industrial Equipment	41.233
SAE J1043 APR80	Minimum Performance Criteria for Falling Object Protective Structure (FOPS) for Industrial Equipment	41.236
SAE J1044a	World Manufacturer Identifier	26.06
SAE J1045	Instrumentation and Techniques for Vehicle Refueling Emissions Measurement	25.120
SAE J1046a	Exterior Sound Level Measurement Procedure for Small Engine Powered Equipment	35.47
SAE J1047	Tubing—Motor Vehicle Brake System Hydraulic	31.136
SAE J1048 MAR80	Symbols for Motor Vehicle Controls, Indicators, and Tell-Tales	34.147
SAE J1049 AUG79	Service Performance Requirements and Test Procedures for Motor Vehicle Lamp Bulbs	23.23
SAE J1050a	Describing and Measuring the Driver's Field of View	34.102
SAE J1051 JAN80	Deflection of Seat Cushions for Off-Road Work Machines	41.251
SAE J1052	Motor Vehicle Driver and Passenger Head Position	34.116
SAE J1053	Steel Stamped Nuts of One Pitch Thread Design	15.64
SAE J1054	Warning Lamp Alternating Flashers	23.56
SAE J1055	Service Performance Requirements for Turn Signal Flashers	23.52
SAE J1056	Service Performance Requirements for Vehicular Hazard Warning Flashers	23.55
SAE J1057 NOV79	Identification Terminology of Earthmoving Machines	41.03
SAE J1058	Sheet Steel Thickness and Profile	4.16
SAE J1059	Speedometer Test Procedure	20.07
SAE J1060	Subjective Rating Scale for Evaluation of Noise and Ride Comfort Characteristics Related to Motor Vehicle Tires	29.16
SAE J1061a	Surface Discontinuities on General Application Bolts, Screws, and Studs	5.33
SAE J1062	Snowmobile Passenger Handgrips	39.09
SAE J1063	Cantilevered Boom Crane Structures—Method of Test	41.115
SAE J1065	Pressure Ratings for Hydraulic Tubing and Fittings	21.119
SAE J1066	Recommended Guidelines for Company Metrication Programs in the Metalworking Industry*	(HS J1066)
SAE J1067	Seven Conductor Jacketed Cable for Truck Trailer Connections	22.58
SAE J1069	Oil Change System for Quick Service of Construction and Industrial Equipment	41.69
SAE J1071	Operator Controls for Graders	41.263

*Available as a Related Technical Report. Complete listing and details appear at the end of this index.

SAE No.	Title	Page
SAE J1072	Sintered Tool Materials	8.12
SAE J1073	Spring-Loaded Clutch Spin Test Procedure	24.81
JUN80		
SAE J1074	Engine Sound Level Measurement Procedure	35.61
SAE J1075	Measurement Procedure for Determining a Representative Sound Level at a Construction Site Boundary Location	35.52
SAE J1076	Backup Lamp Switches	23.60
SAE J1077	Measurement of Exterior Sound Level of Trucks with Auxiliary Equipment	35.22
SAE J1078	A Recommended Method of Analytically Determining the Competence of Hydraulic Telescopic Cantilevered Crane Booms*	(HS J1078)
SAE J1079	Overcenter Clutch Spin Test Procedure	41.161
SAE J1081	Chemical Compositions of SAE Experimental Steels	2.12
MAY80		
SAE J1082	Fuel Economy Measurement—Road Test Procedure	24.161
FEB80		
SAE J1083	Unauthorized Starting or Movement of Machines	41.57
FEB80		
SAE J1084	Operator Protective Structure Performance Criteria for Certain Forestry Equipment	41.237
APR80		
SAE J1085a	Test for Dynamic Properties of Elastomeric Isolators	12.34
SAE J1086	Numbering Metals and Alloys	2.01
SAE J1087	One-Way Clutches—Nomenclature and Terminology	28.33
SAE J1088	Test Procedure for Measurement of Small Utility Engine Exhaust Emissions	25.124
SAE J1089	Lateral Impact Test Procedure for Vehicle Interiors	(Cancelled 1981)
SAE J1092	Component Nomenclature—Industrial Tractors (Wheel)	41.17
SAE J1094a	Constant Volume Sampler System for Exhaust Emissions Measurement	25.98
SAE J1096	Measurement of Exterior Sound Levels for Heavy Trucks Under Stationary Conditions	35.20
SAE J1097	Hydraulic Excavator Lift Capacity Rating	41.216
APR80		
SAE J1098	Ton Mile Per Hour Application Practice	41.207
APR80		
SAE J1099	Technical Report on Fatigue Properties	4.57
SAE J1100	Motor Vehicle Dimensions	34.43
JUL79		
SAE J1102	Mechanical and Material Requirements for Wheel Bolts	5.18
SAE J1104	Service Performance Requirements for Warning Lamp Alternating Flashers	23.56
SAE J1105	Performance, Test and Application Criteria of Electrically Operated Forward Warning Horn for Mobile Construction Machinery	35.34
SAE J1106	Laboratory Testing Machines and Procedures for Measuring the Steady State Force and Moment Properties of Passenger Car Tires*	(HS 210)
SAE J1107	Information Report—Laboratory Testing Machines and Procedures for Measuring the Steady State Force and Moment Properties of Passenger Car Tires*	(HS 210)
SAE J1108a	Truck and Truck Tractor VIN Systems	26.07
SAE J1109	Component Nomenclature—Articulated Log Skidder, Rubber-Tired	41.39
APR80		
SAE J1110	Specification Definitions—Articulated, Rubber-Tired Log Skidder	41.47
APR80		
SAE J1111	Component Nomenclature—Skidder-Grapple	41.41
SAE J1112	Specification Definitions—Skidder-Grapple	41.49
APR80		
SAE J1113a	Electromagnetic Susceptibility Procedures for Vehicle Components (Except Aircraft)	22.68
SAE J1114	Fuel Tank Filler Cap and Cap Retainer—Threaded Pressure—Vacuum Type	24.155
SAE J1115	Guidelines for Developing and Revising SAE Nomenclature and Definitions	1.21
SAE J1116	Categories of Off-Road Self-Propelled Work Machines	41.01
JUN80		
SAE J1117	Method of Measuring and Reporting the Pressure Differential-Flow Characteristics of a Hydraulic Fluid Power Valve	41.82
SAE J1118	Hydraulic Valves for Motor Vehicle Brake Systems Test Procedure	31.26
SAE J1119	Steel Products for Rollover Protective Structures (ROPS) and Falling Object Protective Structures (FOPS)	41.238
APR80		
SAE J1120	Spherical Rod Ends	16.12
SEP79		
SAE J1121	Helical Compression and Extension Spring Terminology	19.09
SAE J1122	Helical Springs: Specification Check Lists	19.14
SAE J1123a	Leaf Springs for Motor Vehicle Suspension—Metric Bar Sizes	19.01
SAE J1124	Glossary of Terms Related to Fluid Filters and Filter Testing	24.110
SAE J1127	Battery Cable	22.79
JUN80		
SAE J1128	Low Tension Primary Cable	22.81
SAE J1129	Operator Cab Environment for Heated, Ventilated, and Air Conditioned Construction and Industrial Equipment	41.275
SAE J1130	Determination of Emissions from Gas Turbine Powered Light Duty Surface Vehicles	25.129
SAE J1131	Performance Requirements for SAE J844d Nonmetallic Tubing and Fitting Assemblies Used in Automotive Air Brake Systems	21.126
SAE J1132	142 mm x 200 mm Sealed Beam Headlamp Unit	23.16
SAE J1133a	School Bus Stop Arm	23.66
SAE J1134	SAE Nodal Mount	37.28
SAE J1135	Turbocharger Connections	24.40
SAE J1136	Braking Performance—Rollers and Compactors	41.197
SEP79		
SAE J1137	Hydraulic Valves for Motor Vehicle Brake Systems—Performance Requirements	31.30
SAE J1138	Design Criteria—Driver Hand Controls Location for Passenger Cars, Multi-Purpose Passenger Vehicles, and Trucks (10 000 GVW and Under)	34.99
SAE J1139	Supplemental Information—Driver Hand Controls Location for Passenger Cars, Multi-Purpose Passenger Vehicles, and Trucks (10 000 GVW and Under)	34.101
SAE J1140	Filler Pipes and Openings of Motor Vehicle Fuel Tanks	24.157
MAR80		
SAE J1141	Air Cleaner Elements	24.66
SAE J1142a	Towability Design Criteria—Passenger Cars and Light Duty Trucks	27.01
SAE J1143a	Towed Vehicle/Wrecker Attachment Test Procedure—Passenger Cars and Light Duty Trucks	27.05
SAE J1144a	Towed Vehicle Drivetrain Test Procedure—Passenger Cars and Light Duty Trucks	27.07
SAE J1145a	Emissions Terminology and Nomenclature	25.139
SAE J1146	The Engine Oil Performance and Engine Service Classification Maintenance Procedure	13.34
SAE J1147	Welding, Brazing, and Soldering—Materials and Practices	10.06
SAE J1148	Engine Charge Air Cooler Nomenclature	24.136
SAE J1149	Metallic Air Brake System Tubing and Pipe	21.122
SAE J1150	Terminology for Agricultural Equipment	40.05
SAE J1151	Methane Measurement Using Gas Chromatography	25.142
NOV79		
SAE J1152	Braking Performance—Rubber-Tired Construction Machines	41.189
APR80		
SAE J1153	Hydraulic Master Cylinders for Motor Vehicle Brakes—Test Procedure	31.22
SAE J1154	Hydraulic Master Cylinders for Motor Vehicle Brakes—Performance Requirements	31.26
SAE J1155	Displacement of Pneumatic Brake Actuators	31.130
SAE J1156	Automotive Resistance Spot Welding Electrodes	(HS J1156)
	(Available from American Welding Society)	

*Available as a Related Technical Report. Complete listing and details appear at the end of this index.

ID	Title	Page
SAE J1157	Measurement Procedure for Evaluation of Full-Flow, Light Extinction Smokemeter Performance	25.35
SAE J1158	Specification Definitions—Winches for Crawler Tractors and Skidders	41.53
SAE J1159 AUG 79	Preparation of SAE Technical Reports—Surface Vehicles and Machines: Standards, Recommended Practices, Information Reports	1.06
SAE J1160	Operator Ear Sound Level Measurement Procedure for Snow Vehicles	35.26
SAE J1161 APR80	Operational Sound Level Measurement Procedure for Snow Vehicles	35.27
SAE J1163 JAN80	Determining Operator Seat Location on Off-Road Work Machines	41.252
SAE J1164	Labeling of ROPS and FOPS	41.233
SAE J1165 JUL79	Reporting Cleanliness Levels of Hydraulic Fluids	41.86
SAE J1166a	Operator Station Sound Level Measurement Procedure for Earthmoving Machinery—Work Cycle Test	35.39
SAE J1167	Motorcycle Stop Lamp Switch	38.03
SAE J1168	Motorcycle Bank Angle Measurement Procedure	38.04
SAE J1169	Measurement of Light Vehicle Exhaust Sound Level Under Stationary Conditions	35.16
SAE J1170 JUN80	Rear Power Take-Off for Agricultural Tractors	40.27
SAE J1171	External Ignition Protection of Marine Electrical Devices	42.08
SAE J1172	Engine Flywheel Housings with Sealed Flanges	24.81
SAE J1173	Size Classification and Characteristics of Glass Beads for Peening	9.11
SAE J1174	Operator Ear Sound Level Measurement Procedure for Small Engine Powered Equipment	35.49
SAE J1175	Bystander Sound Level Measurement Procedure for Small Engine Powered Equipment	35.50
SAE J1176	External Leakage Classifications for Hydraulic Systems	41.98
SAE J1177	Hydraulic Excavator Operator Controls	41.221
SAE J1178	Minimum Performance Criteria for Braking Systems for Rubber-Tired Skidders	41.57
SAE J1179	Hydraulic Excavator Digging Forces	41.220
SAE J1180	Telescopic Boom Length Indicating System	41.144
SAE J1183	Recommended Guidelines for Fatigue Testing of Elastomeric Materials and Components	12.29
SAE J1184	Definitions of Acoustical Terms	12.72
SAE J1188	Specifications for Automotive Weld Quality—Resistance Spot Welding	(HS J1188)
	(Available from American Welding Society)	
SAE J1191	High Tension Ignition Cable Assemblies—Marine	42.19
SAE J1193a	Nomenclature and Dimensions for Hydraulic Excavators	41.36
SAE J1194	Roll-Over Protective Structures (ROPS) for Wheeled Agricultural Tractors	40.49
SAE J1195	Cylinder Rod Wiper Seal Ingression Test	41.96
SAE J1196	Specifications for Automotive Frame Weld Quality—Arc Welding	(HS J1196)
	(Available from American Welding Society)	
SAE J1199	Mechanical and Material Requirements for Metric Externally Threaded Steel Fasteners	5.06
SAE J1200	Blind Rivets—Break Mandrel Type	15.95
SAE J1201	Piston Rings and Grooves (Metric)	24.49
SAE J1203	Light Transmittance of Automotive Windshields Safety Glazing Materials	34.34
SAE J1204	Wheels—Recreational and Utility Trailer Test Procedures	30.03
SAE J1205	Performance Requirements for Snap-In Tubeless Tire Valves	29.19
SAE J1206	Methods for Testing Snap-In Tubeless Tire Valves	29.17
SAE J1207	Measurement Procedure for Determination of Silencer Effectiveness in Reducing Engine Intake or Exhaust Sound Level	35.62
SAE J1209	Identification Terminology of Mobile Forestry Machines	41.12
SAE J1211	Recommended Environmental Practices for Electronic Equipment Design	22.86
SAE J1212	Fire Prevention on Forestry Equipment	41.56
SAE J1213	Glossary of Automotive Electronic Terms	22.102
SAE J1214	Tire to Body Clearance Check for Recreational Vehicles	36.03
SAE J1215	Performance Prediction of Roll-Over Protective Structures (ROPS) Through Analytical Methods	41.244
SAE J1216	Test Methods for Metric Threaded Fasteners	5.11
SAE J1220	Rotary-Trochoidal Engine Nomenclature and Terminology	24.03
SAE J1221	Headlamp-Turn Signal Spacing	23.45
SAE J1224 APR80	Braking Performance—Off-Highway Dumpers	41.195
SAE J1225	Development of a Frequency Weighted Portable Ride Meter	40.60
SAE J1227	Assessing Cleanliness of Hydraulic Fluid Power Components and Systems	41.71
SAE J1228 MAR80	Marine Engine Rating Code	42.11
SAE J1229	Truck Identification Terminology	26.04
SAE J1230 OCT79	Minimum Requirements for Wheel Slip Brake Control System Malfunction Signals	31.136
SAE J1231	Hose Push-On Fittings	21.89
SAE J1232	Passenger and Light Truck Tire Traction Device Profile Determination and Classification	37.30
SAE J1233	Format for Commercial Literature Specifications—Off-Road Work Machines	41.15
SAE J1234	Specification Definitions—Off-Road Work Machines	41.10
SAE J1235	Measuring and Reporting the Internal Leakage of a Hydraulic Fluid Power Valve	41.84
SAE J1236	Cast Iron Sealing Rings (Metric)	28.63
SAE J1237	Metric Thread Rolling Screws	5.29
SAE J1238	Rating Lift Cranes on Fixed Platforms Operating in the Ocean Environment*	(HS J1238)
SAE J1239	Four- and Eight-Conductor Rectangular Electrical Connectors for Automotive Type Trailers	22.54
SAE J1240	Flywheel Spin Test Procedure	24.82
SAE J1241	Fuel and Lubricant Tanks for Motorcycles	38.05
SAE J1242	Acoustic Emission Test Methods	4.56
SAE J1243 AUG79	Diesel Emission Production Audit Test Procedure	25.22
SAE J1244	Oil Cooler Nomenclature and Glossary	24.138
SAE J1245 FEB80	Guide to the Application and Use of Engine Coolant Pump Face Seals	24.149
SAE J1247 APR80	Simulated Mountain Brake Performance Test Procedure	31.80
SAE J1250 FEB80	In-Service Brake Performance Test Procedure—Vehicles Over 4500 kg (10 000 lb)	31.57
SAE J1252 OCT79	SAE Wind Tunnel Test Procedure for Trucks and Buses	37.08
SAE J1253	Low Temperature Cranking Load Requirements of an Engine	22.18
SAE J1254	Component Nomenclature—Feller Buncher	41.39
SAE J1255	Specification Definitions—Feller/Buncher	41.54
SAE J1256 JUN80	Fuel Economy Measurement—Road Test Procedure—Cold Start and Warm-Up Fuel Economy	24.193
SAE J1257	Rating Chart for Cantilevered Boom Cranes	41.121
SAE J1258	Automotive Hydraulic Brake System—Metric Connections	31.139
SAE J1259 APR80	Metric Spherical Rod Ends	16.15
SAE J1261 FEB79	Agricultural Tractor Tire Dynamic Indices	40.13
SAE J1262	Sound Level Measurement Procedure for Trenching Equipment	35.57

*Available as a Related Technical Report. Complete listing and details appear at the end of this index.

SAE J1263	Road Load Measurement and Dynamometer Simulation Using Coastdown Techniques	24.170
SAE J1264	Joint RCCC/SAE Fuel Consumption Test Procedure (Short Term-In-Service Vehicle) Type I	37.05
SAE J1265 FEB80	Dozer Capacity	41.159
SAE J1266	Axle Efficiency Test Procedure	28.29
SAE J1267	Leakage Testing	4.50
SAE J1268 JUN80	Hardenability Bands for Carbon and Alloy H Steels	2.24
SAE J1269 JUN80	Rolling Resistance Measurement Procedure for Passenger Car Tires	29.04
SAE J1270 OCT79	The Measurement of Passenger Car Tire Rolling Resistance	29.06
SAE J1271 AUG79	Technical Committee Guideposts	1.01
SAE J1272 OCT79	Felling Head Terminology and Nomenclature	41.51
SAE J1273 SEP79	Selection, Installation, and Maintenance of Hose and Hose Assemblies	21.104
SAE J1274 JUN80	Performance Requirements for the Automotive Audio Cassette	22.107
SAE J1275 JUN80	Testing Methods for Audio Cassettes	22.108
SAE J1276 AUG79	Standardized Fluid for Hydraulic Component Tests	41.86
SAE J1280 JAN80	Determination of Sulfur Compounds in Automotive Exhaust	25.55
SAE J1281 NOV79	Operator Sound Level Exposure Assessment Procedure for Pleasure Motorboats	35.29
SAE J1283 JUN80	Electrical Connector for Auxiliary Starting of Construction, Agricultural, and Off-Road Machinery	41.173
SAE J1284 DEC79	Blade Type Electric Fuses	22.37
SAE J1285 FEB80	Powershift Transmission Fluid Classification	13.19
SAE J1286 APR80	Static Electric Outboard Thrust	42.04
SAE J1287 JUN80	Measurement of Exhaust Sound Levels of Stationary Motorcycles	38.08
SAE J1292	Automobile, Truck, Truck-Tractor, Trailer, and Motor Coach Wiring	22.43
SAE J1293 JAN80	Undervehicle Corrosion Test	4.34
SAE J1294 APR80	Ignition Distributors—Marine	42.15
SAE J1295 JUN80	Identification Terminology—Pipelayers	41.36
SAE J1297 APR80	Alternative Automotive Fuels	13.32
SAE J1298 APR80	Hydraulic Systems Diagnostic Port Sizes and Locations	41.77
SAE J1306 JUN80	Motorcycle Auxiliary Front Lamps	23.39
SAE J1312 JUN80	Procedure for Measuring Basic Highway Vehicle Engine Performance and Fuel Consumption—Spark Ignition and Diesel	24.17
SAE J1401 FEB79	Road Vehicle—Hydraulic Brake Hose Assemblies for Use with Non-Petroleum Base Hydraulic Fluids	31.01
SAE J1402c	Automotive Air Brake Hose and Hose Assemblies	31.07
SAE J1403a	Vacuum Brake Hose	31.11
SAE J1404	Service Brake Structural Integrity Requirements—Vehicles Over 10 000 lb (4500 KG) GVWR	31.72
SAE J1405	Flex-Impulse Test Procedure for Hydraulic Hose Assemblies	21.106
SAE J1500 JUN80	Universal Symbols for Operator Controls	1.22
SAE J1601	Rubber Cups for Hydraulic Actuating Cylinders	31.144
SAE J1603	Rubber Seals for Hydraulic Disc Brake Cylinders (formerly SAE J62)	31.13
SAE J1604	Rubber Boots for Drum Type Hydraulic Brake Wheel Cylinders	31.17
SAE J1605	Brake Master Cylinder Reservoir Diaphragm Gasket (formerly SAE J66)	31.15
SAE J1702f	Motor Vehicle Brake Fluid—Arctic	31.31
SAE J1703 JAN80	Motor Vehicle Brake Fluid	31.39

*Available as a Related Technical Report. Complete listing and details appear at the end of this index.

SUBJECT INDEX

ABRASION
   Abrasive Wear — (HS J965)
   Seat Belt Assembly Webbing Abrasion Test Procedure — SAE J339a ... 33.12
   Seat Belt Assembly Webbing Abrasion Test Requirements — SAE J114 ... 33.12
   Test Method for Determining Resistance to Snagging and Abrasion of Automotive Bodycloth — SAE J948a ... 12.75

ACOUSTICS
   See Also: SOUND
   Acoustic Emission Test Methods — SAE J1242 ... 4.56
   Definitions of Acoustical Terms — SAE J1184 ... 12.72
   Instrumentation for Measuring Acoustic Impulses within Vehicles — SAE J247 JUN80 ... 35.04

ADHESIVES
   Methods of Tests for Automotive-Type Sealers, Adhesives, and Deadeners — SAE J243 ... 12.46
   Test Procedures for Brake Shoe and Lining Adhesives and Bonds — SAE J840c ... 31.65

AGRICULTURAL MACHINERY
   Agricultural Equipment Enclosure Pressurization System Test Procedure — SAE J1012 MAR80 ... 40.43
   Agricultural Tractor Test Code — SAE J708 JUN80 ... 40.01
   Application of Hydraulic Remote Control to Agricultural Tractors and Trailing Type Agricultural Implements — SAE J716a ... 40.24
   Categories of Off-Road Self-Propelled Work Machines — SAE J1116 JUN80 ... 41.01
   Deflection of Seat Cushions for Off-Road Work Machines — SAE J1051 JAN80 ... 41.251
   Determining Operator Seat Location on Off-Road Work Machines — SAE J1163 JAN80 ... 41.252
   Development of a Frequency Weighted Portable Ride Meter — SAE J1225 ... 40.60
   Dimensional Standard for Cylindrical Hydraulic Couplers for Agricultural Tractors — SAE J1036 ... 40.26
   Electrical Connector for Auxiliary Starting of Construction, Agricultural, and Off-Road Machinery — SAE J1283 JUN80 ... 41.173
   Exterior Sound Level Measurement Procedure for Self-Propelled Agricultural Field Equipment — SAE J1008 ... 35.60
   Measurement of Whole Body Vibration of the Seated Operator of Off-Highway Work Machines — SAE J1013 JAN80 ... 40.57
   Multipurpose Petroleum Base Fluids — (HS J73) ... *
   Performance Test for Air Conditioned Agricultural Equipment — SAE J1006 MAR80 ... 40.41
   Slow-Moving Vehicle Identification Emblem — SAE J943a ... 40.40
   T-Hook Slots for Securement in Shipment of Agricultural Equipment — SAE J39 ... 40.37
   Technical Publications for Agricultural Equipment — SAE J1035 JUN80 ... 40.65
   Terminology for Agricultural Equipment — SAE J1150 ... 40.05
   Test Procedure for Measuring Hydraulic Lift Force Capacity on Agricultural Tractors Equipped with Three-Point Hitch — SAE J283 ... 40.22
   Tractor Belt Speed and Pulley Width — SAE J720 MAY80 ... 40.32
   Brake Systems
      Brake Test Procedure and Brake Performance Criteria for Agricultural Equipment — SAE J1041 ... 40.15
   Implements
      Operating Requirements for Tractors and Power Take-Off Driven Implements — SAE J721 JUN80 ... 40.30
      Three-Point Free-Link Hitch Attachment of Implements to Agricultural Wheeled Tractors — SAE J715 MAY80 ... 40.19
      Three-Point Hitch, Implement Quick-Attaching Coupler, Agricultural Tractors — SAE J909 APR80 ... 40.17
   Lamps
      Combination Tail and Floodlamp for Agricultural Equipment — SAE J976 MAY80 ... 40.39
      Flashing Warning Lamp for Agricultural Equipment — SAE J974 MAY80 ... 40.38
      Headlamps for Agricultural Equipment — SAE J975 MAY80 ... 40.38
   Operator Controls
      Operator Controls on Agricultural Equipment — SAE J841 MAY80 ... 40.46
      Universal Symbols for Operator Controls — SAE J1500 JUN80 ... 1.22
      Universal Symbols for Operator Controls on Agricultural Equipment — SAE J389b ... 40.47
   Operator Protective Structures
      Overhead Protection for Agricultural Tractors—Test Procedures and Performance Requirements — SAE J167a ... 40.56
      Roll-Over Protective Structures (ROPS) for Wheeled Agricultural Tractors — SAE J1194 ... 40.49
   Power Take-Off
      Auxiliary Power Take-Off Drives for Agricultural Tractors — SAE J717 MAY80 ... 40.31
      Operating Requirements for Tractors and Power Take-Off Driven Implements — SAE J721 JUN80 ... 40.30
      Power Take-Off Definitions and Terminology for Agricultural Tractors — SAE J722 MAY80 ... 40.30
      Rear Power Take-Off for Agricultural Tractors — SAE J1170 JUN80 ... 40.27
   Safety
      Lighting and Marking of Agricultural Equipment on Highways — SAE J137c ... 40.39
      Safety Alert Symbol for Agricultural, Construction and Industrial Equipment — SAE J284a ... 40.43
      Safety for Agricultural Equipment — SAE J208d ... 40.44
   Tires
      Agricultural Tractor Tire Dynamic Indices — SAE J1261 FEB79 ... 40.13
      Agricultural Tractor Tire Loadings, Torque Factors, and Inflation Pressures — SAE J709d ... 40.05
      Lawn and Garden Tractors Load and Inflation Pressures and Tire Selection Table for Future Design — SAE J344 ... 40.14
      Liquid Ballast Table for Drive Tires of Agricultural Machines — SAE J884 MAY80 ... 40.10
      Tire Selection Tables for Agricultural and Light Industrial Machines of Future Design — SAE J711c ... 40.12
   Wheels
      Industrial and Agricultural Disc Wheels — SAE J712a ... 40.32
      Wheel Mounting Elements for Industrial and Agricultural Disc Wheels — SAE J714b ... 40.36

AIR CLEANERS
   Air Cleaner Elements — SAE J1141 ... 24.66
   Air Cleaner Test Code — SAE J726 SEP79 ... 24.67

AIR CONDITIONING
   Automotive Air Conditioning Hose — SAE J51b ... 12.112
   Design Guidelines for Air Conditioning Systems for Construction and Industrial Equipment Cabs — SAE J169 ... 41.274
   Performance Test for Air Conditioned Agricultural Equipment — SAE J1006 MAR80 ... 40.41
   Safety Practices for Mechanical Vapor Compression Refrigeration Equipment or Systems Used to Cool Passenger Compartment of Motor Vehicles — SAE J639 ... 34.24

ALARMS
   Performance, Test, and Application Criteria for Electrically Operated Backup Alarm Devices — SAE J994b ... 35.32

ALLOYS
   Eddy Current Testing by Electromagnetic Methods — SAE J425b ... 4.48
   Infrared Testing — SAE J359a ... 4.47
   Liquid Penetrant Test Methods — SAE J426c ... 4.52
   Magnetic Particle Inspection — SAE J420b ... 4.47

*Available as a Related Technical Report. Complete listing and details appear at the end of this index.

Nondestructive Tests — SAE J358b **4.45**
Numbering Metals and Alloys — SAE J1086 . . . . . . **2.01**
Penetrating Radiation Inspection — SAE J427b **4.52**
Prevention of Corrosion of Metals — (HS J447a) . . . . *
Second Edition Unified Numbering System Handbook for Metals and Alloys — (HS 1086a) . . . . *
Ultrasonic Inspection — SAE J428b **4.55**

Aluminum
Alloy and Temper Designation Systems for Aluminum — SAE J993b **11.02**
Aluminum Alloys—Fundamentals — SAE J451b **11.01**
Anodized Aluminum Automotive Parts — SAE J399a . **11.28**
Chemical Compositions and Mechanical and Physical Properties of SAE Aluminum Casting Alloys — SAE J453c **11.08**
Chemical Compositions, Mechanical Property Limits, and Dimensional Tolerances of SAE Wrought Aluminum Alloys — SAE J457j **11.27**
General Data on Wrought Aluminum Alloys — SAE J454k **11.11**
General Information on SAE Aluminum Casting Alloys — SAE J452c **11.05**

Copper
Cast Copper Alloys — SAE J462b **11.60**
Wrought and Cast Copper Alloys — SAE J461d **11.31**
Wrought Copper and Copper Alloys — SAE J463d . . . **11.63**

Magnesium
Magnesium Alloys — SAE J464c **11.102**
Magnesium Casting Alloys — SAE J465c **11.102**
Magnesium Wrought Alloys — SAE J466c **11.107**

Nickel
Wrought Nickel and Nickel-Related Alloys — SAE J470c **11.123**

Steel (See Also: STEEL)
Alloy Steel Machinability Ratings — SAE J770c **3.24**
Formerly Standard SAE Alloy Steels — SAE J778a . . . **2.13**
Hardenability Bands for Carbon and Alloy H Steels — SAE J1268 JUN80 . . . . **2.24**
High Strength Carbon and Alloy Die Drawn Steels — SAE J935 . . . . **3.03**

Superalloys
Special Purpose Alloys ("Superalloys") — SAE J467b **11.111**

Zinc
Zinc Alloy Ingot and Die Casting Compositions — SAE J468b **11.122**
Zinc Die Casting Alloys — SAE J469a **11.123**

ALL-TERRAIN VEHICLES
Snowmobile and All-Terrain Vehicle Pulley Guards and Shields — SAE J93 . . . . **39.14**
Vehicle Identification Numbers—Snowmobile and All-Terrain Vehicles — SAE J64 . . . . **39.03**

ALUMINUM ALLOYS
See: ALLOYS

ANTHROPOMETRY
Bodyforms for Use in Motor Vehicle Passenger Compartment Impact Development — SAE J984 JUN80 . . . **34.143**
Describing and Measuring the Driver's Field of View — SAE J1050a **34.102**
Devices for Use in Defining and Measuring Vehicle Seating Accommodation — SAE J826 APR80 . . . . **34.22**
Driver Hand Control Reach — SAE J287 FEB80 . . . . **34.66**
Film Analysis Guides for Dynamic Studies of Test Subjects — SAE J138 . . . . **34.129**
Human Tolerance to Impact Conditions as Related to Motor Vehicle Design — (HS J885) . . . . *
Motor Vehicle Dimensions — SAE J1100 JUL79 . . . . **34.43**
Motor Vehicle Driver's Eye Range — SAE J941e **34.60**
Range of Anthropometric Measurements for Asian Populations — (HS J317) . . . . *
Steering Control System—Passenger Car—Laboratory Test Procedure — SAE J944 JUN80 . . . . **34.144**
Synthetic Skins for Automotive Testing — SAE J202 **34.118**
Vision Factors Considerations in Rear View Mirror Design — SAE J985 . . . . **34.38**

Vision Glossary — SAE J264 . . . . **34.113**

ANTIFREEZE
Engine Coolant Concentrate—Ethylene-Glycol Type — SAE J1034 . . . . **12.97**
Engine Coolants — SAE J814c **12.93**

AUDIO CASSETTES
See: CASSETTES

AXLES
Axle and Manual Transmission Lubricants — SAE J308c **13.12**
Axle and Manual Transmission Lubricant Viscosity Classification — SAE J306 OCT79 . . . . **13.13**
Axle Efficiency Test Procedure — SAE J1266 . . . . . . **28.29**
Axle Rating for Industrial Wheel Loaders and Loader-Backhoes — SAE J43 . . . . **41.216**
Axles—Nomenclature and Terminology — SAE J923 . **28.25**
Brake Flange Mounting—On-Highway Vehicles — SAE J337 . . . . **37.27**
Trailer Axle Alignment — SAE J875 . . . . **37.27**

BACKHOES
See: CONSTRUCTION AND INDUSTRIAL MACHINERY

BACKUP ALARMS
See: ALARMS

BACKUP LAMPS
See: LAMPS

BALL JOINTS
Ball Joints — SAE J490c **16.08**
Spherical Rod Ends — SAE J1120 SEP79 . . . . **16.12**

BALL STUDS
Ball Studs and Ball Stud Socket Assemblies — SAE J491b **16.01**
Ball Stud and Socket Assembly Test Procedure — SAE J193 SEP79 . . . . **16.06**

BATTERIES
Battery Cable — SAE J1127 JUN80 . . . . **22.79**
Grounding of Storage Batteries — SAE J538a **22.13**
Life Test for Automotive Storage Batteries — SAE J240a **22.12**
Storage Batteries — SAE J537j **22.04**
Storage Batteries for Construction and Industrial Machinery — SAE J930b **41.287**

BEARINGS
Bearing and Bushing Alloys (Information Report) — SAE J459c **11.29**
Bearing and Bushing Alloys (Standard) — SAE J460e **11.29**
Sleeve Type Half Bearings — SAE J506b **24.56**

BELTS
See Also: SEAT BELTS
Automotive V-Belt Drives — SAE J637b **18.04**
Tractor Belt Speed and Pulley Width — SAE J720 MAY80 . . . . **40.32**
V-Belts and Pulleys — SAE J636c **18.01**

BOATS
See: MOTORBOATS

BOLTS
See: FASTENERS

BOOMS
A Recommended Method of Analytically Determining the Competence of Hydraulic Telescopic Cantilevered Crane Booms — (HS J1078) . . . . *
Crane Boom Hoist Disengaging Device — SAE J999 OCT79 . . . . **41.147**
Crane Boomstop — SAE J220 . . . . **41.144**
Radius-of-Load and Boom Angle Measuring System — SAE J375a **41.145**
Side Booms—Tractor Mounted — SAE J743b **41.153**
Telescopic Boom Length Indicating System — SAE J1180 . . . . **41.144**

BRAKES
See Also: HYDRAULIC SYSTEMS
Air Brake Gladhand Service (Control) and Emergency (Supply) Line Couplers—Trucks, Truck-Tractors, and Trailers — SAE J318 . . . . **31.128**
Automotive Air Brake Reservoir Volume — SAE J813 **31.127**

---

*Available as a Related Technical Report. Complete listing and details appear at the end of this index.

Automotive and Off-Highway Air Brake Reservoir Performance and Identification Requirements — SAE J10b ... 31.129
Automotive Brake Definitions and Nomenclature — SAE J656g ... 31.53
Automotive Hydraulic Brake System — Metric Connections — SAE J1258 ... 31.139
Brake and Electrical Connection Locations—Truck-Tractor and Truck-Trailer — SAE J702b ... 37.28
Brake Block Chamfer — SAE J662 ... 31.64
Brake Diaphragm Maintenance—Tractor and Semitrailer — SAE J666 ... 37.29
Brake Disc and Drum Thermocouple Installation — SAE J79 ... 31.131
Brake Flange Mounting—On-Highway Vehicles — SAE J337 ... 37.27
Brake Master Cylinder Reservoir Diaphragm Gasket — SAE J1605 ... 31.15
Brake Rating Horsepower Requirements—Commercial Vehicles — SAE J257 JUN80 ... 31.125
Brake System Performance Requirements—Truck, Bus, and Combination of Vehicles — SAE J992b ... 31.111
Braking Performance—Crawler Tractors and Loaders — SAE J1026 SEP79 ... 41.188
Braking Performance—Off-Highway Dumpers — SAE J1224 APR80 ... 41.195
Braking Performance—Rollers and Compactors — SAE J1136 SEP79 ... 41.197
Braking Performance—Rubber-Tired Construction Machines — SAE J1152 APR80 ... 41.189
Emergency Air Brake Systems—Motor Vehicles and Vehicle Combinations — SAE J263 ... 31.126
Hydraulic Master Cylinders for Motor Vehicle Brakes—Performance Requirements — SAE J1154 ... 31.26
Hydraulic Valves for Motor Vehicle Brake Systems—Performance Requirements — SAE J1137 ... 31.30
Hydraulic Wheel Cylinders for Automotive Drum Brakes — SAE J101 MAR80 ... 31.19
Minimum Performance Criteria for Braking Systems for Rubber-Tired Skidders — SAE J1178 ... 41.57
Minimum Requirements for Wheel Slip Brake Control System Malfunction Signals — SAE J1230 OCT79 ... 31.136
Rivets for Brake Linings and Bolts for Brake Blocks — SAE J663b ... 31.64
Rubber Boots for Drum Type Hydraulic Brake Wheel Cylinders — SAE J1604 ... 31.17
Rubber Cups for Hydraulic Actuating Cylinders — SAE J1601 ... 31.144
Rubber Seals for Hydraulic Disc Brake Cylinders — SAE J1603 ... 31.13
Service Brake Structural Integrity Requirements—Vehicles Over 10 000 lb (4500 KG) GVWR — SAE J1404 ... 31.72
Service Brake System Performance Requirements—Light-Duty Truck — SAE J155 ... 31.92
Service Brake System Performance Requirements—Motorcycles and Motor-Driven Cycles — SAE J109a ... 31.116
Service Brake System Performance Requirements—Passenger Car — SAE J937b ... 31.86
Service Brake System Performance Requirements—Passenger Car—Trailer Combinations — SAE J135a ... 31.104
Service Brake System Performance Requirements—Snowmobiles — SAE J44 ... 39.10
Tubing—Motor Vehicle Brake System Hydraulic — SAE J1047 ... 31.136
Vehicle Grade Parking Performance Requirements — SAE J293 ... 31.72

Brake Fluids
Central Fluid Systems — SAE J71a ... 13.20
Determination of Brake Fluid Temperature — SAE J291 JUN80 ... 31.131
Handling and Dispensing of Motor Vehicle Brake Fluid — SAE J76 ... 31.49
Labeling of Motor Vehicle Brake Fluid Containers — SAE J988 ... 31.48

Motor Vehicle Brake Fluid — SAE J1703 JAN80 ... 31.39
Motor Vehicle Brake Fluid—Arctic — SAE J1702f ... 31.31
Motor Vehicle Brake Fluid Container Compatibility — SAE J75 ... 31.48
Service Maintenance of Motor Vehicle Brake Fluid in Motor Vehicle Brake Actuating Systems — SAE J77 ... 31.51

Brake Hoses
Automotive Air Brake Hose and Hose Assemblies — SAE J1402c ... 31.07
Road Vehicle—Hydraulic Brake Hose Assemblies for Use with Non-Petroleum Base Hydraulic Fluids — SAE J1401 FEB79 ... 31.01
Vacuum Brake Hose — SAE J1403a ... 31.11

Brake Linings
Brake Lining Quality Control Test Procedure — SAE J661a ... 31.61
Color Code for Location Identification of Combination Linings of Two Different Materials or Two Shoe Brakes — SAE J659 ... 31.64
Friction Identification System for Brake Linings and Brake Blocks for Motor Vehicles — SAE J866a ... 31.60
Gogan Hardness of Brake Lining — SAE J379a ... 31.58
Rivets for Brake Linings and Bolts for Brake Blocks — SAE J663b ... 31.64
Minimum Requirements for Motor Vehicle Brake Linings — SAE J998 ... 31.63
Specific Gravity of Brake Lining — SAE J380 ... 31.59
Swell, Growth, and Dimensional Stability of Brake Linings — SAE J160 JUN80 ... 31.60
Test Procedures for Brake Shoe and Lining Adhesives and Bonds — SAE J840c ... 31.65

Brake Shoes
Color Code for Location Identification of Combination Linings of Two Different Materials or Two Shoe Brakes — SAE J659 ... 31.64
Test Procedures for Brake Shoe and Lining Adhesives and Bonds — SAE J840c ... 31.65

Brake Tests
Brake System Rating Test Code—Commercial Vehicles — SAE J880 MAY80 ... 31.117
Brake Rating Test Code—Commercial Vehicle Inertia Dynamometer — SAE J971 ... 31.124
Brake System Dynamometer Test Procedure—Passenger Car — SAE J212 JUN80 ... 31.88
Brake System Road Test Code—Motorcycles — SAE J108a ... 31.114
Brake System Road Test Code—Passenger Car and Light-Duty Truck — SAE J843d ... 31.75
Brake System Road Test Code—Passenger Car and Light Duty Truck-Trailer Combinations — SAE J134 JUN79 ... 31.95
Brake System Road Test Code—Truck, Bus, and Combination of Vehicles — SAE J786a ... 31.105
Brake Test Procedure and Brake Performance Criteria for Agricultural Equipment — SAE J1041 ... 40.15
Brake System Test Procedure—Snowmobiles — SAE J45 ... 39.10
Brake System Torque Balance Test Code—Commercial Vehicles — SAE J225 JUN80 ... 31.123
Brake Test Code—Inertia Dynamometer — SAE J667 ... 31.73
Displacement of Pneumatic Brake Actuators — SAE J1155 ... 31.130
Hydraulic Master Cylinders for Motor Vehicle Brakes—Test Procedure — SAE J1153 ... 31.22
Hydraulic Valves for Motor Vehicle Brake Systems Test Procedure — SAE J1118 ... 31.26
In-Service Brake Performance Test Procedure Passenger Car and Light-Duty Truck — SAE J201 ... 31.55
In-Service Brake Performance Test Procedure—Vehicles Over 4500 kg (10 000 lb) — SAE J1250 FEB80 ... 31.57
Motor Vehicle Grade Parking Performance Test Code — SAE J360 ... 31.71
Service Brake Structural Integrity Test Procedure—Passenger Car — SAE J229 JUN80 ... 31.94
Service Brake Structural Integrity Test Procedure—

*Available as a Related Technical Report. Complete listing and details appear at the end of this index.

Vehicles Over 10 000 LB (4500 kg) GVWR — SAE J294 . . . . **31.112**
Simulated Mountain Brake Performance Test Procedure — SAE J1247 APR80 . . . . **31.80**
Stopping Distance Test Procedure — SAE J299 JAN80 **31.130**
Test Code—Truck, Truck-Tractor, and Trailer Air Service Brake System Pneumatic Pressure and Time Levels — SAE J982a **31.127**
Wheel Slip Brake Control System Road Test Code — SAE J46 JUN80 . . . . **31.132**

BRAZING
Automotive Metallurgical Joining — (HS J836a) . . . . *
Welding, Brazing, and Soldering—Materials and Practices — SAE J1147 . . . . **10.06**

BUCKETS
Crane and Cable Excavator Basic Operating Control Arrangements — SAE J983 . . . . **41.150**
Excavator Hoe Bucket Rating — SAE J296 . . . . **41.213**
Capacity Rating — Loader Bucket — SAE J742 OCT79 **41.211**
Shovel Dipper, Clam Bucket, and Dragline Bucket Rating — SAE J67 . . . . **41.210**

BUMPERS
Approach, Departure, and Ramp Breakover Angles—Passenger Car and Light Duty Trucks — SAE J689 **32.03**
Bumper Evaluation Test Procedure—Passenger Cars — SAE J980a **32.02**
Bumper Jack Requirements—Motor Vehicles — SAE J979 . . . . **32.01**
Bumper Jacking Test Procedure—Motor Vehicles — SAE J978 . . . . **32.01**

BUSES
Automobile, Truck, Truck-Tractor, Trailer, and Motor Coach Wiring — SAE J1292 JUN80 . . . . **22.43**
Brake Rating Horsepower Requirements—Commercial Vehicles — SAE J257 JUN80 . . . . **31.125**
Brake Rating Test Code—Commercial Vehicle Inertia Dynamometer — SAE J971 . . . . **31.124**
Brake System Performance Requirements—Truck, Bus, and Combination of Vehicles — SAE J992b **31.111**
Brake System Road Test Code—Truck, Bus, and Combination of Vehicles — SAE J786a **31.105**
Brake System Torque Balance Test Code—Commercial Vehicles — SAE J225 JUN80 . . . . **31.123**
Exterior Sound Level for Heavy Trucks and Buses — SAE J366b **35.18**
Nomenclature—Truck, Bus, Trailer — SAE J687c . . . **37.01**
SAE Wind Tunnel Test Procedure for Trucks and Buses — SAE J1252 OCT79 . . . . **37.08**
School Bus Red Signal Lamps — SAE J887a **23.65**
School Bus Stop Arm — SAE J1133a **23.66**
Truck and Bus Tire Performance Requirements and Test Procedures — SAE J341a **29.13**
Windshield Defrosting Systems Performance Requirements—Trucks, Buses, and Multipurpose Vehicles — SAE J382 . . . . **34.16**
Windshield Defrosting Systems Test Procedure—Trucks, Buses, and Multipurpose Vehicles — SAE J381 . . . . **34.14**
Windshield Wiper Systems—Trucks, Buses, and Multipurpose Vehicles — SAE J198 . . . . **34.08**

BUSHINGS
Bearing and Bushing Alloys (Information Report) — SAE J459c **11.29**
Bearing and Bushing Alloys (Standard) — SAE J460e **11.29**
Split Type Bushings—Design and Application — SAE J835 . . . . **24.63**

CABLES
Battery Cable — SAE J1127 JUN80 . . . . **22.79**
High Tension Ignition Cable — SAE J557 . . . . **22.49**
High Tension Ignition Cable Assemblies—Marine — SAE J1191 . . . . **42.19**
Low Tension Primary Cable — SAE J1128 . . . . **22.81**
Marine Control Cable Connection—Engine Clutch Lever — SAE J960 . . . . **42.03**
Marine Control Cable Connection—Engine Throttle Lever — SAE J961 . . . . **42.04**

Marine Push-Pull Control Cables — SAE J917 JUN80 **42.02**
Seven-Conductor Electrical Connector for Truck-Trailer Jumper Cable — SAE J560b **22.57**
Seven Conductor Jacketed Cable for Truck Trailer Connections — SAE J1067 . . . . **22.58**

CARBURETORS
Automotive Carburetor Flanges — SAE J623a **24.86**
Carburetor Airflow Reference Standards — SAE J228 JAN80 . . . . **24.126**

CASSETTES
Performance Requirements for the Automotive Audio Cassette — SAE J1274 JUN80 . . . . **22.107**
Testing Methods for Audio Cassettes — SAE J1275 JUN80 . . . . **22.108**

CHAINS
Safety Chain of Full Trailers or Converter Dollies — SAE J697a **37.24**
Passenger and Light Truck Tire Traction Device Profile Determination and Classification — SAE J1232 . . . **37.30**
Tire Chain Clearance—Trucks, Buses, and Combinations of Vehicles — SAE J683a **37.29**
Trailer Couplings, Hitches and Safety Chains—Automotive Type — SAE J684f **36.01**

CAST IRON
See: IRON

CIRCUITS
Automotive Printed Circuits — SAE J771c **22.45**

CIRCUIT BREAKERS
Circuit Breakers — SAE J553c **22.40**
Circuit Breaker—Internal Mounted—Automatic Reset — SAE J258 . . . . **22.42**

CLAM BUCKETS
See: BUCKET

CLAMPS
See: FASTENERS

CLUTCHES
Clutch Friction Test Machine (SAE No. 2)—Test Procedure — SAE J286 . . . . **28.35**
Flywheels for Single-Plate Spring-Loaded Clutches — SAE J618 JUN80 . . . . **24.79**
Flywheels for Two-Plate Spring-Loaded Clutches — SAE J619 JUN80 . . . . **24.80**
Housing Internal Dimensions for Single and Two Plate Spring Loaded Clutches — SAE J373a **24.81**
Industrial Power Take-Offs with Driving Ring Type Overcenter Clutches — SAE J621d **41.162**
Marine Control Cable Connection—Engine Clutch Lever — SAE J960 JUN80 . . . . **42.03**
One-Way Clutches—Nomenclature and Terminology — SAE J1087 . . . . **28.33**
Overcenter Clutch Spin Test Procedure — SAE J1079 **41.161**
SAE Nodal Mount — SAE J1134 . . . . **37.28**
Spring-Loaded Clutch Spin Test Procedure — SAE J1073 JUN80 . . . . **24.81**

COATINGS
Electroplating and Related Finishes — SAE J474b . . . **11.139**
Florida Exposure of Automotive Finishes — SAE J951 FEB80 . . . . **12.66**
Prevention of Corrosion of Metals — (HS J447a) . . . . *
Sound Deadeners and Underbody Coatings — SAE J671 . . . . **12.56**
Test for Chip Resistance of Surface Coatings — SAE J400 JUN80 . . . . **12.63**

COLLISIONS
See: IMPACT

COMPACTORS
See: CONSTRUCTION AND INDUSTRIAL MACHINERY

CONNECTORS
See: FASTENERS

CONSTRUCTION AND INDUSTRIAL MACHINERY
See: INDUSTRIAL MACHINERY
Bolt and Capscrew Sizes for Use in Construction and Industrial Machinery — SAE J370 . . . . **15.101**
Categories of Off-Road Self-Propelled Work Machines — SAE J1116 JUN80 . . . . **41.01**

*Available as a Related Technical Report. Complete listing and details appear at the end of this index.

Classification and Nomenclature Towing Winch for Skidders and Crawler Tractors — SAE J1014 OCT79 .... 41.32
Countersunk Square Holes for Cutting Edges and End Bits — SAE J740b 41.161
Definitions for Machine Availability (Construction and Industrial) — SAE J1032 APR80 .... 41.58
Format for Commercial Literature Specifications—Off-Road Work Machines — SAE J1233 .... 41.15
Industrial Power Take-Offs with Driving Ring Type Overcenter Clutches — SAE J621d 41.162
Measurement of Whole Body Vibration of the Seated Operator of Off-Highway Work Machines — SAE J1013 JAN80 .... 40.57
Measurement Procedure for Determining a Representative Sound Level at a Construction Site Boundary Location — SAE J1075 .... 35.52
Multipurpose Petroleum Base Fluids — (HS J73) .... *
Remote and Automatic Control Systems for Construction and Industrial Machinery — SAE J956 .... 41.276
Soil-Machine Terminology — SAE J991 APR80 .... 41.60
Sound Level Measurement Procedure for Trenching Equipment — SAE J1262 .... 35.57
Specification Definitions—Off-Road Work Machines — SAE J1234 .... 41.10
Storage Batteries for Construction and Industrial Machinery — SAE J930b 41.287
Ton Mile Per Hour Application Practice — SAE J1098 APR80 .... 41.207
Unauthorized Starting or Movement of Machines — SAE J1083 FEB80 .... 41.57
Wiring Identification System for Industrial and Construction Equipment — SAE J210 .... 41.188

Backhoes
Axle Rating for Industrial Wheel Loaders and Loader-Backhoes — SAE J43 .... 41.216
Excavator Hoe Bucket Rating — SAE J296 .... 41.213
Hydraulic Backhoe Lift Capacity — SAE J31 .... 41.214
Nomenclature—Utility Backhoe — SAE J326 .... 41.212
Specification Definitions—Hydraulic Backhoes — SAE J49 APR80 .... 41.44

Brake Systems
Braking Performance—Off-Highway Dumpers — SAE J1224 APR80 .... 41.195
Braking Performance—Rubber-Tired Construction Machines — SAE J1152 APR80 .... 41.189

Compactors
Braking Performance—Rollers/Compactors — SAE J1136 SEP79 .... 41.197
Component Nomenclature—Rollers/Compactors — SAE J1017 .... 41.30

Cranes
A Recommended Method of Analytically Determining the Competence of Hydraulic Telescopic Cantilevered Crane Booms — (HS J1078) .... *
Braking Performance—Rubber-Tired Construction Machines — SAE J1152 APR80 .... 41.189
Cantilevered Boom Crane Structures—Method of Test — SAE J1063 .... 41.115
Crane and Cable Excavator Basic Operating Control Arrangements — SAE J983 .... 41.150
Crane Boom Hoist Disengaging Device — SAE J999 OCT79 .... 41.147
Crane Boomstop — SAE J220 .... 41.144
Crane Hoist Line Speed and Power Test Code — SAE J820 .... 41.141
Crane Load Moment System — SAE J159 OCT79 .... 41.147
Crane Load Stability Test Code — SAE J765a 41.142
Crane Structures—Method of Test — SAE J987 APR80 .... 41.110
Lifting Crane, Wire-Rope Strength Factors — SAE J959 .... 41.148
Lifting Crane Sheave and Drum Sizes — SAE J881a 41.150
Load Indicating Devices in Lifting Crane Service — SAE J376 APR80 .... 41.146
Mobile Crane Working Area Definitions — SAE J1028 .... 41.152
Nomenclature and Dimensions for Crane Shovels — SAE J958 .... 41.34
Radius-of-Load and Boom Angle Measuring System — SAE J375a 41.145
Rating Chart For Cantilevered Boom Cranes — SAE J1257 .... 41.121
Rating Lift Cranes on Fixed Platforms Operating in the Ocean Environment — (HS J1238) .... *
Telescopic Boom Length Indicating System — SAE J1180 .... 41.144

Design Dimensions
Access Systems for Construction and Industrial Equipment — SAE J185 .... 41.249
Control Locations for Construction and Industrial Equipment Design — SAE J898 APR80 .... 41.255
Determining Operator Seat Location on Off-Road Work Machines — SAE J1163 JAN80 .... 41.252
Deflection of Seat Cushions for Off-Road Work Machines — SAE J1051 JAN80 .... 41.251
Instrument Face Design and Location for Construction and Industrial Equipment — SAE J209 APR80 .... 41.258
Minimum Access Dimensions for Construction and Industrial Machinery — SAE J925 .... 41.248
Operator Controls on Industrial Equipment — SAE J297 APR80 .... 41.262
Operator Enclosures Human Factor Design Considerations — SAE J154a 41.256
Operator's Seat Dimensions—Construction and Industrial Equipment Design — SAE J899 .... 41.254
Range of Anthropometric Measurements for Asian Populations — (HS J317) .... *
Universal Symbols for Operator Controls — SAE J1500 JUN80 .... 1.22
USA Human Physical Dimensions — SAE J833 JAN80 .... 41.245

DOZERS
Dozer Capacity — SAE J1265 FEB80 .... 41.159
Hole Placement on Dozer End Bits — SAE J63 FEB80 .... 41.156
Hole Spacing for Scraper and Bulldozer Cutting Edges — SAE J737b 41.156
Nomenclature—Dozer — SAE J729 MAY80 .... 41.22
Scraper and Bulldozer Cutting Edge Cross Sections, Rolled, Cast, Forged or Machined — SAE J738b 41.158
Specification Definitions—Dozers — SAE J173 JAN80 .... 41.24

Dumpers
Braking Performance—Off-Highway Dumpers — SAE J1224 APR80 .... 41.195
Braking Performance—Rubber-Tired Construction Machines — SAE J1152 APR80 .... 41.189
Capacity Ratings—Scraper, Dumper Body and Trailer Body — SAE J741b 41.209
Component Nomenclature—Dumpers — SAE J1016 .... 41.29
Component Nomenclature—Dumper Trailer — SAE J734c 41.28
Minimum Performance Criteria for Emergency Steering of Wheeled Earthmoving Construction Machines — SAE J53 .... 41.278

Electrical Systems
Electrical Charging Systems for Construction and Industrial Machinery — SAE J180 APR80 .... 41.184
Electrical Connector for Auxiliary Starting of Construction, Agricultural, and Off-Road Machinery — SAE J1283 JUN80 .... 41.173
Electrical System for Construction and Industrial Machinery — SAE J821a 41.182

Excavators
Braking Performance—Rubber-Tired Construction Machines — SAE J1152 APR80 .... 41.189
Excavator Hoe Bucket Rating — SAE J296 .... 41.213
Hydraulic Excavator Digging Forces — SAE J1179 .... 41.220
Hydraulic Excavator Lift Capacity Rating — SAE J1097 APR80 .... 41.216
Hydraulic Excavator Operator Controls — SAE J1177 .... 41.221
Nomenclature and Dimensions for Hydraulic Excavators — SAE J1193a 41.36

*Available as a Related Technical Report. Complete listing and details appear at the end of this index.

Fuel Systems
  Dry-Break Fueling Installation for Construction Machinery — SAE J176a **41.70**
  Guidelines for Fluid Level Indicators — SAE J48 .... **41.70**

Graders
  Braking Performance—Rubber-Tired Construction Machines — SAE J1152 APR80 .... **41.189**
  Component Nomenclature—Graders — SAE J870 ... **41.21**
  Grader Cutting Edge — SAE J739b **41.158**
  Minimum Performance Criteria for Emergency Steering of Wheeled Earthmoving Construction Machines — SAE J53 .... **41.278**
  Operator Controls for Graders — SAE J1071 .... **41.263**

Heating, Ventilating, and Air Conditioning
  Design Guidelines for Air Conditioning Systems for Construction and Industrial Equipment Cabs — SAE J169 APR80 .... **41.274**
  Fuel-Fired Heaters—Air Heating—For Construction and Industrial Machinery — SAE J1024 APR80 ...... **41.273**
  Operator Cab Environment for Heated, Ventilated, and Air Conditioned Construction and Industrial Equipment — SAE J1129 .... **41.275**

Hydraulic Systems
  Assessing Cleanliness of Hydraulic Fluid Power Components and Systems — SAE J1227 .... **41.71**
  External Leakage Classifications for Hydraulic Systems — SAE J1176 .... **41.98**
  Hydraulic Cylinder Test Procedure — SAE J214 .... **41.94**
  Hydraulic Directional Control Valves, 3000 psi Maximum — SAE J748 .... **41.81**
  Hydraulic Motor Test Procedure — SAE J746b **41.87**
  Hydraulic Power Circuit Filtration — SAE J931a **41.92**
  Hydraulic Power Pump Test Procedure — SAE J745c **41.77**
  Hydraulic Power Pumps — SAE J744c **41.79**
  Hydraulic Systems Diagnostic Port Sizes and Locations — SAE J1298 APR80 .... **41.77**
  Measuring and Reporting the Internal Leakage of a Hydraulic Fluid Power Valve — SAE J1235 .... **41.84**
  Method of Measuring and Reporting the Pressure Differential-Flow Characteristics of a Hydraulic Fluid Power Valve — SAE J1117 .... **41.82**
  Standardized Fluid for Hydraulic Component Tests — SAE J1276 AUG79 .... **41.86**

Lighting Systems
  Combination Tail and Floodlamp for Industrial Equipment — SAE J94 APR80 .... **41.180**
  Dimensional Specifications for Sealed Lighting Unit for Construction and Industrial Machinery — SAE J572a **41.178**
  Flashing Warning Lamp for Industrial Equipment — SAE J96 .... **41.179**
  Headlamps for Industrial Equipment — SAE J95 APR80 .... **41.180**
  Lighting and Marking of Construction and Industrial Machinery — SAE J1029 APR80 .... **41.180**
  Sealed Lighting Units for Construction and Industrial Machinery — SAE J598 APR80 .... **41.177**

Loaders
  Axle Rating for Industrial Wheel Loaders and Loader-Backhoes — SAE J43 .... **41.216**
  Braking Performance—Crawler Tractors and Loaders — SAE J1026 SEP79 .... **41.188**
  Braking Performance—Rubber-Tired Construction Machines — SAE J1152 APR80 .... **41.189**
  Capacity Rating—Loader Bucket — SAE J742 OCT79 **41.211**
  Component Nomenclature—Loader — SAE J731d ... **41.25**
  Lift Arm Safety Device for Loaders — SAE J38 ..... **41.224**
  Minimum Performance Criteria for Emergency Steering of Wheeled Earthmoving Construction Machines — SAE J53 .... **41.278**
  Operating Load for Loaders — SAE J818b **41.67**
  Specification Definitions—Loaders — SAE J732 FEB80 **41.42**
  Steering Frame Lock for Articulated Loaders and Tractors — SAE J276 .... **41.224**

Lubricants
  Lubricant Types—Construction and Industrial Machinery — SAE J754a **41.68**
  Lubrication Chart—Construction and Industrial Machinery — SAE J753 APR80 .... **41.67**
  Oil Change System for Quick Service of Construction and Industrial Equipment — SAE J1069 .... **41.69**

Maintenance
  Engineering Design Serviceability Guide Lines—Construction and Industrial Machinery — SAE J817a **41.280**
  Format for Construction Equipment Manuals — SAE J920a **41.286**
  Maintenance Interval—Construction Equipment — SAE J752b **41.69**
  Symbols and Color Codes for Maintenance Instructions, Container and Filler Identification — SAE J223 APR80 .... **41.71**

Mowers
  Industrial Rotary Mowers — SAE J232 .... **41.267**
  Nomenclature—Industrial Mowers — SAE J990 .... **41.265**
  Safety Criteria for Industrial Flail Mowers — SAE J1001 APR80 .... **41.268**

Operator Protective Structures
  Deflection Limiting Volume—ROPS/FOPS Laboratory Evaluation — SAE J397b **41.234**
  Identification Terminology—Pipelayers — SAE J1295 JUN80 .... **41.36**
  Labeling of ROPS and FOPS — SAE J1164 .... **41.233**
  Lift Arm Safety Device for Loaders — SAE J38 ..... **41.224**
  Minimum Performance Criteria for Falling Object Protective Structure (FOPS) — SAE J231 APR80 ..... **41.235**
  Minimum Performance Criteria for Falling Object Protective Structure (FOPS) for Industrial Equipment — SAE J1043 APR80 .... **41.236**
  Operator Protection for Industrial Equipment — SAE J1042 APR80 .... **41.233**
  Performance Criteria for Rollover Protective Structures (ROPS) for Construction, Earthmoving, Forestry, and Mining Machines — SAE J1040c **41.225**
  Performance Prediction of Roll-Over Protective Structures (ROPS) Through Analytical Methods — SAE J1215 **41.244**
  Safety for Industrial Wheeled Equipment — SAE J98 APR80 .... **41.264**
  Seat Belts for Construction Machines — SAE J386 APR80 .... **41.270**
  Steel Products for Rollover Protective Structures (ROPS) and Falling Object Protective Structures (FOPS) — SAE J1119 APR80 .... **41.238**
  Steering Frame Lock for Articulated Loaders and Tractors — SAE J276 .... **41.224**

Rippers
  Nomenclature—Rippers, Rooters, Scarifiers — SAE J733b **41.27**

Rollers
  Braking Performance—Rollers and Compactors — SAE J1136 SEP79 .... **41.197**
  Component Nomenclature— Rollers/Compactors — SAE J1017 .... **41.30**

Rooters
  Nomenclature—Rippers, Rooters, Scarifiers — SAE J733b **41.27**

Safety
  Lighting and Marking of Industrial Equipment on Highways — SAE J99 APR80 .... **41.181**
  Performance, Test and Application Criteria of Electrically Operated Forward Warning Horn for Mobile Construction Machinery — SAE J1105 .... **35.34**
  Safety Considerations for the Operator — SAE J153 .. **41.263**
  Safety for Industrial Wheeled Equipment — SAE J98 APR80 .... **41.264**
  Safety Signs — SAE J115 SEP79 .... **41.272**
  Seat Belts for Construction Machines — SAE J386 APR80 .... **41.270**

Scarifiers
  Nomenclature—Rippers, Rooters, Scarifiers — SAE J733b **41.27**

Scrapers

*Available as a Related Technical Report. Complete listing and details appear at the end of this index.

Capacity Ratings—Scraper, Dumper Body and Trailer Body — SAE J741b	41.209
Component Nomenclature—Scrapers — SAE J728a	41.19
Elevating Scrapers—Capacity Rating — SAE J957	41.210
Hole Spacing for Scraper and Bulldozer Cutting Edges — SAE J737b	41.156
Loading Ability Test Code—Scrapers — SAE J764 APR80	41.108
Minimum Performance Criteria for Emergency Steering of Wheeled Earthmoving Construction Machines — SAE J53	41.278
Scraper and Bulldozer Cutting Edge Cross Sections, Rolled, Cast, Forged or Machined — SAE J738b	41.158

Sound Levels

Engine Sound Level Measurement Procedure — SAE J1074	35.61
Exterior Sound Level Measurement Procedure for Earthmoving Machinery — SAE J88 SEP79	35.54
Measurement Procedure for Determining a Representative Sound Level at a Construction Site Boundary Location — SAE J1075	35.52
Operator Sound Level Measurement Procedure for Earthmoving Machinery—Singular Type Test — SAE J919c	35.37
Operator Station Sound Level Measurement Procedure for Earthmoving Machinery—Work Cycle Test — SAE J1166a	35.39

Tests

Center of Gravity Test Code — SAE J874 APR80	41.101
Cylinder Rod Wiper Seal Ingression Test — SAE J1195	41.96
Engine Cooling System Field Test (Air-to-Boil) — SAE J819 MAR80	41.106
Gradeability Test Code — SAE J950 APR80	41.105
Hydraulic Control Valve Test Procedure — SAE J747b	41.89
Machine Drag Test Code — SAE J873 APR80	41.100
Machine Slope Operation Test Code — SAE J897	41.107
Reserve Tractive Ability Test Code — SAE J872 APR80	41.98

Tractors

Braking Performance—Crawler Tractors and Loaders — SAE J1026 SEP79	41.188
Braking Performance—Rubber-Tired Construction Machines — SAE J1152 APR80	41.189
Classification and Nomenclature Towing Winch for Skidders and Crawler Tractors — SAE J1014 OCT79	41.32
Component Nomenclature—Crawler Tractor — SAE J727 MAY80	41.16
Component Nomenclature—Industrial Tractors (Wheel) — SAE J1092	41.17
Component Nomenclature—Two and Four—Wheel Tractors — SAE J869a	41.17
Drawbars—Crawler Tractor — SAE J749a	41.208
Side Booms—Tractor Mounted — SAE J743b	41.153
Specification Definitions—Winches for Crawler Tractors and Skidders — SAE J1158	41.53
Steering Frame Lock for Articulated Loaders and Tractors — SAE J276	41.224
Minimum Performance Criteria for Emergency Steering of Wheeled Earthmoving Construction Machines — SAE J53	41.278
Tractor Protection Valve Control — SAE J779	37.31

CONTROLS

Automatic Transmissions—Manual Control Sequence — SAE J915	28.14
Automatic Vehicle Speed Control—Motor Vehicles — SAE J195	20.12
Control Locations for Construction and Industrial Equipment Design — SAE J898 APR80	41.255
Crane and Cable Excavator Basic Operating Control Arrangements — SAE J983	41.150
Design Criteria—Driver Hand Controls Location for Passenger Cars, Multi-purpose Passenger Vehicles, and Trucks (10 000 GVW and Under) — SAE J1138	34.99
Driver Hand Control Reach — SAE J287 FEB80	34.66
Hydraulic Excavator Operator Controls — SAE J1177	41.221
Marine Control Cable Connection—Engine Clutch Lever — SAE J960 JUN80	42.03
Marine Control Cable Connection—Engine Throttle Lever — SAE J961 JUN80	42.04
Marine Push-Pull Control Cables — SAE J917 JUN80	42.02
Operator Controls for Graders — SAE J1071	41.263
Operator Controls on Agricultural Equipment — SAE J841 MAY80	40.46
Operator Controls on Industrial Equipment — SAE J297 APR80	41.262
Operator Controls and Displays on Motorcycles — SAE J107 APR80	38.01
Remote and Automatic Control Systems for Construction and Industrial Machinery — SAE J956	41.276
Supplemental Information—Driver Hand Controls Location for Passenger Cars, Multi-purpose Passenger Vehicles, and Trucks (10 000 GVW and Under) — SAE J1139	34.101
Symbols for Motor Vehicle Controls, Indicators and Tell-Tales — SAE J1048 MAR80	34.147
Tractor Protection Valve Control — SAE J779	37.31
Universal Symbols for Operator Controls — SAE J1500 JUN80	1.22
Universal Symbols for Operator Controls on Agricultural Equipment — SAE J389b	40.47
Universal Symbols for Operator Controls on Industrial Equipment — SAE J298	41.260

COOLING SYSTEMS

Engine Coolant Concentrate—Ethylene-Glycol Type — SAE J1034	12.97
Engine Coolants — SAE J814c	12.93
Engine Cooling System Field Test (Air-to-Boil) — SAE J819 MAR80	41.106
Fan Hub Bolt-Circles and Pilot Holes — SAE J635a	24.145
Glossary of Engine Cooling System Terms — SAE J1004	24.130
Guide to the Application and Use of Engine Coolant Pump Face Seals — SAE J1245 FEB80	24.149
Maintenance of Automotive Engine Cooling Systems — (HS 40)	*
Oil Cooler Nomenclature and Glossary — SAE J1244	24.138
Pressure Relief for Cooling Systems — SAE J151	24.114

COPPER ALLOYS
  See: ALLOYS

CORROSION

Prevention of Corrosion of Metals — (HS J447a)	*
Undervehicle Corrosion Test — SAE J1293 JAN80	4.34

COUPLERS

Air Brake Gladhand Service (Control) and Emergency (Supply) Line Couplers—Trucks, Truck-Tractors, and Trailers — SAE J318	31.128
Connection and Accessory Locations for Towing Doubles Trailers and Multi-Axle Trailers — SAE J849b	37.22
Dimensional Standard for Cylindrical Hydraulic Couplers for Agricultural Tractors — SAE J1036	40.26
Fifth Wheel Kingpin—Commercial Trailers and Semitrailers — SAE J700b	37.21
Fifth Wheel Kingpin, Heavy Duty—Commercial Trailers and Semitrailers — SAE J848a	37.22
Fifth Wheel Kingpin Performance—Commercial Trailers and Semitrailers — SAE J133	37.20
Marine Propeller-Shaft Couplings — SAE J756 JUN80	42.01
Three-Point Hitch, Implement Quick-Attaching Coupler, Agricultural Tractors — SAE J909 APR80	40.17
Trailer Couplings, Hitches and Safety Chains—Automotive Type — SAE J684f	36.01
Truck Tractor Semitrailer Interchange Coupling Dimensions — SAE J701a	37.18

CRANES
  See: CONSTRUCTION AND INDUSTRIAL MACHINERY

CRASHWORTHINESS
  See: IMPACT

CYLINDERS

Cylinder Rod Wiper Seal Ingression Test — SAE J1195	41.96
Engine Rotation and Cylinder Numbering — SAE J824	24.42
Hydraulic Cylinder Test Procedure — SAE J214	41.94

*Available as a Related Technical Report. Complete listing and details appear at the end of this index.

Hydraulic Master Cylinders for Motor Vehicle Brakes—
Performance Requirements — SAE J1154. . . . . . .   **31.26**
Hydraulic Master Cylinders for Motor Vehicle Brakes—
Test Procedure — SAE J1153. . . . .   **31.22**
Hydraulic Wheel Cylinders for Automotive Drum Brakes
— SAE J101 MAR80. . . . .   **31.19**
Rubber Cups for Hydraulic Actuating Cylinders — SAE
J1601 . . . .   **31.144**

DEADENERS
Methods of Tests for Automotive-Type Sealers, Adhesives, and Deadeners — SAE J243 . . . .   **12.46**
Sound Deadeners and Underbody Coatings — SAE
J671 . . . .   **12.56**

DEFOGGING SYSTEMS
Passenger Car Backlight Defogging System — SAE
J953 . . . .   **34.17**

DEFROSTING SYSTEMS
Passenger Car Windshield Defrosting Systems — SAE
J902b **34.12**
Windshield Defrosting Systems Performance Requirements—Trucks, Buses, and Multipurpose Vehicles —
SAE J382 . . . .   **34.16**
Windshield Defrosting Systems Test Procedure—Trucks,
Buses, and Multipurpose Vehicles — SAE J381 . . .   **34.14**

DIESEL ENGINES
See: ENGINES

DIMENSIONING
Dual Dimensioning — (HS J390) . . . .   *
Scales for Use with Decimal-Inch Dimensioning — (HS
156) . . . .   *

DIMENSIONS
Motor Vehicle Dimensions — SAE J1100 JUL79. . . .   **34.43**
Motor Vehicle Fiducial Marks — SAE J182a. **34.115**

DIPPERS
Shovel Dipper, Clam Bucket, and Dragline Bucket Rating
— SAE J67. . . . .   **41.210**

DIPSTICKS
Engine and Transmission Dipstick Marking — SAE
J614b. **24.77**

DISC WHEELS
See: WHEELS

DISTRIBUTORS
Ignition Distributors—Marine — SAE J1294 APR80. .   **42.15**

DOORS
Passenger Car Door System Crush Test Procedure —
SAE J367 JUN80 . . . .   **34.142**

DOZERS
See: CONSTRUCTION AND INDUSTRIAL MACHINERY

DRAWBARS
Drawbars—Crawler Tractor — SAE J749a **41.208**

DRIVESHAFTS
Universal Joints and Driveshafts—Nomenclature and
Terminology — SAE J901b **28.17**

DRIVETRAINS
Towed Vehicle Drivetrain Test Procedure—Passenger
Cars and Light Duty Trucks — SAE J1144a. **27.07**

DUMPERS
See: CONSTRUCTION AND INDUSTRIAL MACHINERY

EARTHMOVING MACHINERY
See Also: CONSTRUCTION AND INDUSTRIAL
MACHINERY
Drain, Fill, and Level Plugs for Earthmoving Machinery
— SAE J371a **41.171**
Fan Blast Deflectors for Earthmoving Machines — SAE
J165 APR80 . . . .   **41.223**
Identification Terminology of Earthmoving Machines —
SAE J1057 NOV79 . . . .   **41.03**
Maintenance Interval—Construction Equipment — SAE
J752b **41.69**
Minimum Performance Criteria for Emergency Steering
of Wheeled Earthmoving Construction Machines —
SAE J53 . . . .   **41.278**
Minimum Performance Criteria for Falling Object Protective Structure (FOPS) — SAE J231 APR80 . . . . .   **41.235**
Operator Station Sound Level Measurement Procedure
for Earthmoving Machinery—Work Cycle Test — SAE
J1166a **35.39**
Performance Criteria for Rollover Protective Structures
(ROPS) for Construction, Earthmoving, Forestry, and
Mining Machines — SAE J1040c. **41.225**
Tire Guards for Protection of Operator of Earthmoving
Haulage Machines — SAE J321b **41.223**

ELASTOMERIC MATERIALS
Classification System for Rubber Materials for Automotive Applications — SAE J200 APR80 . . . .   **12.01**
Recommended Guidelines for Fatigue Testing of Elastomeric Materials and Components — SAE J1183 . . .   **12.29**
Sponge- and Expanded Cellular-Rubber Products — SAE
J18 DEC79 . . . .   **12.40**
Test for Dynamic Properties of Elastomeric Isolators —
SAE J1085a **12.34**

ELECTRICAL SYSTEMS
Automobile, Truck, Truck-Tractor, Trailer, and Motor
Coach Wiring — SAE J1292 JUN80 . . . .   **22.43**
Blade Type Electric Fuses — SAE J1284 DEC79 . . . .   **22.37**
Electrical Charging Systems for Construction and Industrial Machinery — SAE J180 APR80. . . . .   **41.184**
Electrical Generating System (Alternator Type) Performance Curve and Test Procedure — SAE J56 . . . . . .   **22.19**
Electrical System for Construction and Industrial Machinery — SAE J821a **41.182**
Low Temperature Cranking Load Requirements of an
Engine — SAE J1253 . . . .   **22.18**
Maintenance of Design Voltage—Snowmobile Electrical
Systems — SAE J277 . . . .   **39.04**
Motorcycle and Motor Driven Cycle Electrical System
(Maintenance of Design Voltage) — SAE J392 . . . .   **23.39**
Nomenclature—Automotive Electrical Systems — SAE
J831 . . . .   **22.01**
Voltage Drop for Starting Motor Circuits — SAE J541a   **22.13**
Voltages for Diesel Electrical Systems — SAE J539a . .   **22.13**

ELECTRIC VEHICLES
Electric Vehicle Test Procedure — SAE J227a **27.09**

ELECTROMAGNETICS
Eddy Current Testing by Electromagnetic Methods —
SAE J425b **4.48**
Electromagnetic Susceptibility Procedures for Vehicle
Components (Except Aircraft) — SAE J1113a **22.68**
Limits and Methods of Measurement of Radio Interference Characteristics of Vehicles and Devices (20–
1000 MHz) — SAE J551 JUN79 . . . .   **22.27**

ELECTRONICS
Electromagnetic Susceptibility Procedures for Vehicle
Components (Except Aircraft) — SAE J1113a **22.68**
Glossary of Automotive Electronic Terms — SAE J1213   **22.102**
Recommended Environmental Practices for Electronic
Equipment Design — SAE J1211. . . . .   **22.86**

ELECTROPLATING
Electroplating and Related Finishes — SAE J474b . . .   **11.139**
Electroplating of Nickel and Chromium on Metal
Parts—Automotive Ornamentation and Hardware —
SAE J207 . . . .   **11.133**

EMISSIONS
Acoustic Emission Test Methods — SAE J1242 . . . . .   **4.56**
Test Procedure for Measuring Carbon Monoxide Concentrations in Vehicle Passenger Compartments — SAE
J989a **25.53**
Constant Volume Sampler System for Exhaust Emissions
Measurement — SAE J1094a. **25.98**
Continuous Hydrocarbon Analysis of Diesel Emissions
— SAE J215 JAN80 . . . .   **25.16**
Crankcase Emission Control Test Code — SAE J900 . .   **25.49**
Determination of Emissions from Gas Turbine Powered
Light Duty Surface Vehicles — SAE J1130 . . . . . .   **25.129**
Determination of Sulfur Compounds in Automotive Exhaust — SAE J1280 JAN80 . . . .   **25.55**
Diesel Emission Production Audit Test Procedure —
SAE J1243 AUG79 . . . .   **25.22**
Diesel Engine Emission Measurement Procedure — SAE
J1003 . . . .   **25.20**

*Available as a Related Technical Report. Complete listing and details appear at the end of this index.

Diesel Engine Smoke Measurement — SAE J255a ... **25.27**
Diesel Smoke Measurement Procedure — SAE J35 ... **25.46**
Emissions Terminology and Nomenclature — SAE J1145a ... **25.139**
Instrumentation and Techniques for Exhaust Gas Emissions Measurement — SAE J254 ... **25.89**
Instrumentation and Techniques for Vehicle Refueling Emissions Measurement — SAE J1045 ... **25.120**
Measurement of Carbon Dioxide, Carbon Monoxide, and Oxides of Nitrogen in Diesel Exhaust — SAE J177a ... **25.01**
Measurement of Fuel Evaporative Emissions from Gasoline Powered Passenger Cars and Light Trucks — SAE J170a ... **25.73**
Measurement of Fuel Evaporative Emissions from Gasoline Powered Passenger Cars and Light Trucks Using the Enclosure Technique — SAE J171a ... **25.80**
Measurement Procedure for Evaluation of Full-Flow, Light Extinction Smokemeter Performance — SAE J1157 ... **25.35**
Methane Measurement Using Gas Chromatography — SAE J1151 NOV79 ... **25.142**
Test Procedure for Measurement of Small Utility Engine Exhaust Emissions — SAE J1088 ... **25.124**

ENGINES
Engine Charge Air Cooler Nomenclature — SAE J1148 **24.136**
Engine Cooling System Field Test (Air-to-Boil) — SAE J819 MAR80 ... **41.106**
Engine Preheaters — SAE J226a ... **24.124**
Engine Rating Code—Spark Ignition — SAE J245 JUN80 ... **24.21**
Engine Rotation and Cylinder Numbering — SAE J824 **24.42**
Engine Sound Level Measurement Procedure — SAE J1074 ... **35.61**
Engine Terminology and Nomenclature General — SAE J604d ... **24.01**
Engine Test Code—Spark Ignition and Diesel — SAE J816b ... **24.08**
Glossary of Engine Cooling System Terms — SAE J1004 ... **24.130**
Low Temperature Cranking Load Requirements of an Engine — SAE J1253 ... **22.18**
Maintenance of Automotive Engine Cooling Systems — (HS 40) ... *
Marine Engine Rating Code — SAE J1228 MAR80 ... **42.11**
Measurement Procedure for Determination of Silencer Effectiveness in Reducing Engine Intake or Exhaust Sound Level — SAE J1207 ... **35.62**
Passenger Car Engine and Transmission Identification Numbers — SAE J129 ... **26.03**
Procedure for Measuring Basic Highway Vehicle Engine Performance and Fuel Consumption—Spark Ignition and Diesel — SAE J1312 JUN80 ... **24.17**
Rotary-Trochoidal Engine Nomenclature and Terminology — SAE J1220 ... **24.03**

Diesel Engines
Continuous Hydrocarbon Analysis of Diesel Emissions — SAE J215 JAN80 ... **25.16**
Diesel Emission Production Audit Test Procedure — SAE J1243 AUG79 ... **25.22**
Diesel Engine Emission Measurement Procedure — SAE J1003 ... **25.20**
Diesel Engine Smoke Measurement — SAE J255a ... **25.27**
Diesel Smoke Measurement Procedure — SAE J35 ... **25.46**
Engine Rating Code—Diesel — SAE J270 ... **24.26**
Engine Test Code—Spark Ignition and Diesel — SAE J816b ... **24.08**
Measurement of Intake Air or Exhaust Gas Flow of Diesel Engines — SAE J244 ... **25.06**
Testing Techniques for Diesel Fuel Injection Systems — SAE J969b ... **24.104**

Gas Turbine Engines
Gas Turbine Engine Nomenclature — SAE J828 JAN80 **24.02**
Gas Turbine Engine Test Code — SAE J116a ... **24.34**

Small Engines
Bystander Sound Level Measurement Procedure for Small Engine Powered Equipment — SAE J1175 ... **35.50**
Exterior Sound Level Measurement Procedure for Small Engine Powered Equipment — SAE J1046a ... **35.47**
Multiposition Small Engine Exhaust System Fire Ignition Suppression — SAE J335b ... **24.114**
Operator Ear Sound Level Measurement Procedure for Small Engine Powered Equipment — SAE J1174 ... **35.49**
Small Spark Ignition Engine Test Code — SAE J607a **24.32**

EXCAVATORS
See: CONSTRUCTION AND INDUSTRIAL MACHINERY

EXHAUST SYSTEMS
Nomenclature for Exhaust System Parts — SAE J97 ... **24.125**

FABRICS
See: MATERIALS

FALLING OBJECT PROTECTIVE STRUCTURES
See: OPERATOR PROTECTIVE STRUCTURES

FANS
Fan Blast Deflectors for Earthmoving Machines — SAE J165 APR80 ... **41.223**
Fan Hub Bolt-Circles and Pilot Holes — SAE J635a ... **24.145**

FASTENERS
Decarburization in Hardened and Tempered Threaded Fasteners — SAE J121a ... **5.44**
Mechanical and Material Requirements for Externally Threaded Fasteners — SAE J429 JAN80 ... **5.01**
Mechanical and Material Requirements for Metric Externally Threaded Steel Fasteners — SAE J1199 ... **5.06**
Test Methods for Metric Threaded Fasteners — SAE J1216 ... **5.11**
Torque-Tension Test Procedure for Steel Threaded Fasteners — SAE J174 ... **15.76**

Bolts
Bolt and Capscrew Sizes for use in Construction and Industrial Machinery — SAE J370 ... **15.101**
Hex Bolts — SAE J105 ... **15.01**
Hi-Head Finished Hex Bolts — SAE J871a ... **15.04**
Holes in Bolt and Capscrew Shanks and Slots in Nuts for Cotter Pins — SAE J485 ... **15.79**
Mechanical and Material Requirements for Wheel Bolts — SAE J1102 ... **5.18**
Rivets for Brake Linings and Bolts for Brake Blocks — SAE J663b ... **31.64**
Round Head Bolts — SAE J481 ... **15.04**
Surface Discontinuities on Bolts, Screws, and Studs — SAE J123c ... **5.37**
Surface Discontinuities on General Application Bolts, Screws, and Studs — SAE J1061a ... **5.33**

Clamps
Hose Clamps — SAE J536b ... **15.101**

Connectors
Brake and Electrical Connection Locations—Truck-Tractor and Truck-Trailer — SAE J702b ... **37.28**
Connection and Accessory Locations for Towing Doubles Trailers and Multi-Axle Trailers — SAE J849b ... **37.22**
Connectors and Plugs — SAE J856 ... **23.07**
Electrical Connector for Auxiliary Starting of Construction, Agricultural, and Off-Road Machinery — SAE J1283 JUN80 ... **41.173**
Five Conductor Electrical Connectors for Automotive Type Trailers — SAE J895 ... **22.52**
Four- and Eight-Conductor Rectangular Electrical Connectors for Automotive Type Trailers — SAE J1239 **22.54**
Fuel Injection Tubing Connections — SAE J521b ... **24.106**
Hydraulic Flanged Tube, Pipe, and Hose Connections, 4-Bolt Split Flange Type — SAE J518c ... **21.107**
Metric Thread Fuel Injection Tubing Connections — SAE J242a ... **24.105**
Selection, Installation, and Maintenance of Hose and Hose Assemblies — SAE J1273 SEP79 ... **21.104**
Seven-Conductor Electrical Connector for Truck-Trailer Jumper Cable — SAE J560b ... **22.57**
Seven Conductor Jacketed Cable for Truck Trailer Connections — SAE J1067 ... **22.58**

Nuts
Alignment of Nut Slots — SAE J484 ... **15.54**

*Available as a Related Technical Report. Complete listing and details appear at the end of this index.

Crown (Blind, Acorn) Nuts — SAE J483a 15.54
Hexagon High Nuts — SAE J482a 15.53
Holes in Bolt and Capscrew Shanks and Slots in Nuts
   for Cotter Pins — SAE J485 . . . . 15.79
Mechanical and Material Requirements for Steel Nuts —
   SAE J995 JUN79 . . . . 5.15
Nut and Conical Spring Washer Assemblies — SAE
   J238 . . . . 15.74
Push-On Spring Nuts — SAE J892a 15.63
Spring Nuts — SAE J891a 15.55
Square and Hex Nuts — SAE J104 . . . . 15.47
Steel Stamped Nuts of One Pitch Thread Design — SAE
   J1053 . . . . 15.64
Surface Discontinuities on Nuts — SAE J122a 5.40
Rivets
Blind Rivets—Break Mandrel Type — SAE J1200 . . . 15.95
Mechanical and Chemical Requirements for Nonthreaded
   Fasteners — SAE J430 . . . . 5.18
Rivets and Riveting — SAE J492 . . . . 15.85
Rivets for Brake Linings and Bolts for Brake Blocks —
   SAE J663b 31.64
Screws
Bolt and Capscrew Sizes for Use in Construction and
   Industrial Machinery — SAE J370 . . . . 15.101
Flanged 12-Point Screws — SAE J58 . . . . 15.42
Holes in Bolt and Capscrew Shanks and Slots in Nuts
   for Cotter Pins — SAE J485 . . . . 15.79
Lag Screws — SAE J103 . . . . 15.46
Mechanical and Quality Requirements for Machine
   Screws — SAE J82 JUN79 . . . . 5.14
Mechanical and Quality Requirements for Tapping
   Screws — SAE J933 JUN79 . . . . 5.19
Metric Thread Rolling Screws — SAE J1237 . . . . . . 5.29
Screw Threads — SAE J475a 14.01
Slotted and Recessed Head Screws — SAE J478a . . . 15.07
Slotted Headless Setscrews — SAE J479a 15.40
Square Head Setscrews — SAE J102 . . . . 15.41
Steel Self-Drilling Tapping Screws — SAE J78 JUN79 5.21
Surface Discontinuities on Bolts, Screws, and Studs —
   SAE J123c 5.37
Surface Discontinuities on General Application Bolts,
   Screws, and Studs — SAE J1061a 5.33
Thread Rolling Screws — SAE J81 JUN79 . . . . 5.26
Studs
Surface Discontinuities on Bolts, Screws, and Studs —
   SAE J123c 5.37
Surface Discontinuities on General Application Bolts,
   Screws, and Studs — SAE J1061a 5.33
Washers
Conical Spring Washers — SAE J773b 15.73
Lock Washers — SAE J489b 15.68
Nut and Conical Spring Washer Assemblies — SAE
   J238 . . . . 15.74
Plain Washers — SAE J488 . . . . 15.68
Thrust Washers — SAE J924 . . . . 24.65
FATIGUE
Evaluation of Methods for Measurement of Residual
   Stress — (HS 147) . . . . *
Influence of Residual Stress on Fatigue of Steel — (HS
   198) . . . . *
Manual on Shot Peening — (HS 84) . . . . *
Methods of Residual Stress Measurement — (HS J936) *
Recommended Guidelines for Fatigue Testing of Elasto-
   meric Materials and Components — SAE J1183 . . . 12.29
Residual Stress Measurements by X-Ray Diffraction —
   (HS J784a) . . . . *
Surface Rolling and Other Methods for Mechanical Pre-
   stressing of Metals — (HS 3) . . . . *
Technical Report on Fatigue Properties — SAE J1099 4.57
FIBERBOARDS
   See: MATERIALS
FILTERS
Fuel Filter Test Method — (HS J905) . . . . *
Glossary of Terms Related to Fluid Filters and Filter Test-
   ing — SAE J1124 . . . . 24.110
FINISHES
   See: COATINGS
FITTINGS
Automotive Pipe Fittings — SAE J530a 21.128
Automotive Tube Fittings — SAE J512 NOV79 . . . . . 21.04
Coding System for Identification of Tube, Pipe, and
   Hose Fittings — SAE J846 MAY80 . . . . 21.01
Hose Push-On Fittings — SAE J1231 . . . . 21.89
Hydraulic Hose Fittings — SAE J516 SEP79 . . . . . . 21.72
Hydraulic Tube Fittings — SAE J514 APR80 . . . . . . 21.46
Lubrication Fittings — SAE J534c 21.143
Performance Requirements for SAE J844d Nonmetallic
   Tubing and Fitting Assemblies Used in Automotive Air
   Brake Systems — SAE J1131 . . . . 21.126
Pressure Ratings for Hydraulic Tubing and Fittings —
   SAE J1065 . . . . 21.119
Refrigeration Tube Fittings — SAE J513f 21.26
Spherical and Flanged Sleeve (Compression) Tube Fit-
   tings — SAE J246 JUN80 . . . . 21.18
FLANGES
Airflow or Vacuum Governor Flanges — SAE J622
   JAN80 . . . . 24.84
Automotive Carburetor Flanges — SAE J623a 24.86
Brake Flange Mounting—On-Highway Vehicles — SAE
   J337 . . . . 37.27
Exhaust Flanges—Industrial Engines — SAE J624a . . 41.168
Hydraulic Flanged Tube, Pipe, and Hose Connections,
   4-Bolt Split Flange Type — SAE J518c 21.107
Mounting Flanges and Power Take-Off Shafts for Small
   Engines — SAE J609a 24.36
Mounting Flanges for Engine Accessory Drives — SAE
   J896 . . . . 41.166
Water Connection Flanges — SAE J535 JAN80 . . . . 24.83
FLASHERS
Flasher Test Equipment — SAE J823c 23.72
Flashing Warning Lamp for Industrial Equipment —
   SAE J96 . . . . 41.179
Service Performance Requirements for Turn Signal
   Flashers — SAE J1055 . . . . 23.52
Service Performance Requirements for Vehicular Hazard
   Warning Flashers — SAE J1056 . . . . 23.55
Service Performance Requirements for Warning Lamp
   Alternating Flashers — SAE J1104 . . . . 23.56
Turn Signal Flashers — SAE J590e 23.52
Vehicular Hazard Warning Signal Flasher — SAE J945b 23.54
Warning Lamp Alternating Flashers — SAE J1054 . . . 23.56
FLYWHEELS
Engine Flywheel Housings — SAE J617c 41.164
Engine Flywheel Housings with Sealed Flanges — SAE
   J1172 . . . . 24.81
Flywheels for Engine Mounted Torque Converters —
   SAE J927b 41.167
Flywheels for Industrial Engines Used with Industrial
   Power Take-Offs Equipped with Driving-Ring Type
   Overcenter Clutches and Engine Mounted Marine
   Gears — SAE J620d 41.170
Flywheels for Single Bearing Engine Mounted Power
   Generators — SAE J162a 41.168
Flywheels for Single-Plate Spring-Loaded Clutches —
   SAE J618 JUN80 . . . . 24.79
Flywheels for Two-Plate Spring-Loaded Clutches — SAE
   J619 JUN80 . . . . 24.80
Flywheel Spin Test Procedure — SAE J1240 . . . . . . 24.82
Procedure for Measuring Bore Eccentricity and Face De-
   viation of Flywheels, Flywheel Housings, and Flywheel
   Housing Adapters — SAE J1033 . . . . 41.163
FOG LAMPS
   See: LAMPS
FOPS
   See: OPERATOR PROTECTIVE STRUCTURES
FORESTRY MACHINES
Categories of Off-Road Self-Propelled Work Machines —
   SAE J1116 JUN80 . . . . 41.01

---

*Available as a Related Technical Report. Complete listing and details appear at the end of this index.

Fire Prevention on Forestry Equipment — SAE J1212   **41.56**
Format for Commercial Literature Specifications—Off-Road Work Machines — SAE J1233   **41.15**
Identification Terminology of Mobile Forestry Machines — SAE J1209   **41.12**
Labeling of ROPS and FOPS — SAE J1164   **41.233**
Minimum Performance Criteria for Falling Object Protective Structure (FOPS) — SAE J231 APR80   **41.235**
Operator Protective Structure Performance Criteria for Certain Forestry Equipment — SAE J1084 APR80   **41.237**
Performance Criteria for Rollover Protective Structures (ROPS) for Construction, Earthmoving, Forestry, and Mining Machines — SAE J1040c   **41.225**
Specification Definitions—Off-Road Work Machines — SAE J1234   **41.10**
Unauthorized Starting or Movement of Machines — SAE J1083 FEB80   **41.57**

Feller Buncher
Component Nomenclature—Feller Buncher — SAE J1254   **41.39**
Felling Head Terminology and Nomenclature — SAE J1272 OCT79   **41.51**
Specification Definitions—Feller/Buncher — SAE J1255   **41.54**

Skidders
Classification and Nomenclature Towing Winch for Skidders and Crawler Tractors — SAE J1014 OCT79   **41.32**
Component Nomenclature—Articulated Log Skidder, Rubber-Tired — SAE J1109 APR80   **41.39**
Component Nomenclature—Skidder-Grapple — SAE J1111   **41.41**
Minimum Performance Criteria for Braking Systems for Rubber-Tired Skidders — SAE J1178   **41.57**
Specification Definitions—Articulated, Rubber-Tired Log Skidder — SAE J1110 APR80   **41.47**
Specification Definitions—Skidder-Grapple — SAE J1112 APR80   **41.49**
Specification Definitions—Winches for Crawler Tractors and Skidders — SAE J1158   **41.53**

FUEL
Alternative Automotive Fuels — SAE J1297 APR80   **13.32**
Automotive Gasolines — SAE J312 JUN80   **13.22**
Diesel Fuels — SAE J313 JUN80   **13.28**
Effective Dates of Revisions — SAE J301   **13.01**

FUEL ECONOMY
Fuel Economy Measurement—Road Test Procedure — SAE J1082b   **24.161**
Fuel Economy Measurement—Road Test Procedure—Cold Start and Warm-Up Fuel Economy — SAE J1256 JUN80   **24.193**
Joint RCCC/SAE Fuel Consumption Test Procedure (Short Term-In-Service Vehicle) Type I — SAE J1264   **37.05**
Procedure for Measuring Basic Highway Vehicle Engine Performance and Fuel Consumption—Spark Ignition and Diesel — SAE J1312 JUN80   **24.17**
Road Load Measurement and Dynamometer Simulation Using Coastdown Techniques — SAE J1263   **24.170**

FUEL INJECTION
Calibrating Nozzle and Holder Assembly for Diesel Fuel Injection Systems — SAE J968c   **24.102**
Calibration Fluid for Diesel Injection Equipment — SAE J967d   **24.110**
Diesel Fuel Injection Nozzle and Holder Assembly (17 mm Nominal Diameter) — SAE J265   **24.100**
Diesel Fuel Injection Nozzle Holder Assembly (9.5 mm) — SAE J347a   **24.99**
Diesel Fuel Injection Pump Mountings — SAE J626b   **24.92**
Fuel Injection Equipment Nomenclature — SAE J830b   **24.91**
Fuel Injection Nozzle Gaskets — SAE J59   **24.100**
Fuel Injection Tubing — SAE J529b   **21.127**
Fuel Injection Tubing Connections — SAE J521b   **24.106**
Master Calibrating Fuel Injection Pump — SAE J981   **24.101**
Metric Thread Fuel Injection Tubing Connections — SAE J242a   **24.105**
Test Stands for Diesel Fuel Injection Systems — SAE J970a   **24.104**
Testing Techniques for Diesel Fuel Injection Systems — SAE J969b   **24.104**

FUEL SYSTEMS
Dry-Break Fueling Installation for Construction Machinery — SAE J176a   **41.70**
Filler Pipes and Openings of Motor Vehicle Fuel Tanks — SAE J1140 MAR80   **24.157**
Fuel and Lubricant Tanks for Motorcycles — SAE J1241   **38.05**
Fuel Pump Mountings for Diaphragm Type Pumps — SAE J625a   **24.90**
Fuel Tank Filler Cap and Cap Retainer — SAE J829c   **24.156**
Fuel Tank Filler Cap and Cap Retainer—Threaded Pressure-Vacuum Type — SAE J1114   **24.155**
Fuel Tank Filler Conditions—Passenger Car Multi-Purpose Passenger Vehicles, and Light Duty Trucks — SAE J398b   **24.160**
Gasoline Dispenser Nozzle Spouts — SAE J285a   **24.161**
LP-Gas Fuel Systems Components — SAE J372a   **24.113**
Metallic and Nonmetallic Nonpressure Integral Fuel Tank—Snowmobile — SAE J288a   **39.16**

FUSES
Blade Type Electric Fuses — SAE J1284 DEC79   **22.37**
Electric Fuses (Cartridge Type) — SAE J554b   **22.35**
Fusible Links — SAE J156a   **22.51**

GAS TURBINE ENGINES
See: ENGINES

GASKETS
Brake Master Cylinder Reservoir Diaphragm Gasket — SAE J1605   **31.15**
Fuel Injection Nozzle Gaskets — SAE J59   **24.100**
Nonmetallic Automotive Gasket Materials — SAE J90b   **12.80**

GASOLINE
See: FUEL

GEARS
Planetary Gear(s)—Terminology — SAE J646b   **28.07**
Starting Motor Pinions and Ring Gears — SAE J543c   **22.15**

GENERATORS
Electrical Charging Systems for Construction and Industrial Machinery — SAE J180 APR80   **41.184**
Electrical Generating System (Alternator Type) Performance Curve and Test Procedure — SAE J56   **22.19**
Flywheels for Single Bearing Engine Mounted Power Generators — SAE J162a   **41.168**
Generator Mountings — SAE J545b   **22.15**
Starting Motor and Generator Curves — SAE J544b   **22.17**

GLASS
See: MATERIALS

GLAZING
Automotive Safety Glazing — SAE J673b   **34.31**
Automotive Safety Glazing Manual — (HS J906)   *
Drop Test for Evaluating Laminated Safety Glass for Use in Automotive Windshields — SAE J938a   **34.36**
Light Transmittance of Automotive Windshields Safety Glazing Materials — SAE J1203   **34.34**
Passenger Car Glazing—Electrical Circuits — SAE J216   **34.35**
Passenger Car Glazing Shade Bands — SAE J100   **34.35**
Safety Glazing Materials—Motor Vehicles — SAE J674a   **34.34**

GRADERS
See: CONSTRUCTION AND INDUSTRIAL MACHINERY

GRIT
Cast Shot and Grit Size Specifications for Peening and Cleaning — SAE J444a   **9.09**
Metallic Shot and Grit Mechanical Testing — SAE J445a   **9.10**

HARDENABILITY BANDS
Hardenability Bands for Carbon and Alloy H Steels — SAE J1268 JUN80   **2.24**

HEADLAMPS
See: Lamps

HEATERS
Fuel-Fired Heaters—Air Heating—For Construction and

*Available as a Related Technical Report. Complete listing and details appear at the end of this index.

Industrial Machinery — SAE J1024 APR80 . . . . . . **41.273**
Test Procedure and Ratings for Hot Water Heaters for Motor Vehicles — SAE J638 . . . . **34.20**

HINGES
    Vehicle Passenger Door Hinge Systems — SAE J934a   **34.01**

HITCHES
    Test Procedure for Measuring Hydraulic Lift Force Capacity on Agricultural Tractors Equipped with Three-Point Hitch — SAE J283 . . . . **40.22**
    Three-Point Free-Link Hitch Attachment of Implements to Agricultural Wheeled Tractors — SAE J715 MAY80   **40.19**
    Three-Point Hitch, Implement Quick-Attaching Coupler, Agricultural Tractors — SAE J909 APR80 . . . . **40.17**
    Trailer Couplings, Hitches and Safety Chains—Automotive Type — SAE J684f **36.01**

HORNS
    Performance of Vehicle Traffic Horns — SAE J377 . . . **35.24**
    Performance, Test and Application Criteria of Electrically Operated Forward Warning Horn for Mobile Construction Machinery — SAE J1105 . . . . **35.34**

HOSES
    Automotive Air Brake Hose and Hose Assemblies — SAE J1402c **31.07**
    Automotive Air Conditioning Hose — SAE J51b **12.112**
    Coding System for Identification of Tube, Pipe, and Hose Fittings — SAE J846 MAY80 . . . . **21.01**
    Coolant System Hoses — SAE J20e. **12.98**
    Emission Control Hose — SAE J1010 . . . . **12.114**
    Flex-Impulse Test Procedure for Hydraulic Hose Assemblies — SAE J1405 . . . . **21.106**
    Fuel and Oil Hoses — SAE J30d **12.102**
    Hose Clamps — SAE J536b **15.101**
    Hose Push-On Fittings — SAE J1231. . . . . **21.89**
    Hydraulic Flanged Tube, Pipe, and Hose Connections, 4-Bolt Split Flange Type — SAE J518c **21.107**
    Hydraulic Hose — SAE J517 JUN80 . . . . **21.92**
    Hydraulic Hose Fittings — SAE J516 SEP79 . . . . . . **21.72**
    Power Steering Pressure Hose—High Volumetric Expansion Type — SAE J188. . . . . **12.107**
    Power Steering Pressure Hose—Low Volumetric Expansion Type — SAE J191. . . . . **12.110**
    Power Steering Pressure Hose—Wire Braid — SAE J190. . . . . **12.111**
    Power Steering Return Hose—Low Pressure — SAE J189. . . . . **12.108**
    Road Vehicle—Hydraulic Brake Hose Assemblies for Use with Non-Petroleum Base Hydraulic Fluids — SAE J1401 FEB79. . . . . **31.01**
    Selection, Installation, and Maintenance of Hose and Hose Assemblies — SAE J1273 SEP79 . . . . **21.104**
    Tests and Procedures for High-Temperature Transmission Oil Hose, Lubricating Oil Hose and Hose Assemblies — SAE J1019. . . . . **21.107**
    Tests and Procedures for SAE 100R Series Hydraulic Hose and Hose Assemblies — SAE J343 JUN80 . . **21.105**
    Vacuum Brake Hose — SAE J1403a **31.11**
    Windshield Wiper Hose — SAE J50a **12.112**

HOUSINGS
    Engine Flywheel Housings — SAE J617c **41.164**
    Engine Flywheel Housings with Sealed Flanges — SAE J1172. . . . . **24.81**
    Housing Internal Dimensions for Single and Two Plate Spring Loaded Clutches — SAE J373a. **24.81**

HYDRAULIC SYSTEMS
    See Also: **BRAKES**
    Assessing Cleanliness of Hydraulic Fluid Power Components and Systems — SAE J1227 . . . . **41.71**
    Automatic Transmission Hydraulic Control Systems—Terminology — SAE J648a **28.08**
    Automotive Hydraulic Brake System—Metric Connections — SAE J1258 . . . . **31.139**
    Central Fluid Systems — SAE J71a **13.20**
    External Leakage Classifications for Hydraulic Systems — SAE J1176 . . . . **41.98**
    Flex-Impulse Test Procedure for Hydraulic Hose Assemblies — SAE J1405. . . . . **21.106**
    Fluid for Passenger Car Type Automatic Transmissions — SAE J311b **13.17**
    Hydraulic Control Valve Test Procedure — SAE J747b   **41.89**
    Hydraulic Cylinder Test Procedure — SAE J214 . . . . **41.94**
    Hydraulic Directional Control Valves, 3000 psi Maximum — SAE J748. . . . . **41.81**
    Hydraulic Hose — SAE J517 JUN80 . . . . **21.92**
    Hydraulic Master Cylinders for Motor Vehicle Brakes — Performance Requirements — SAE J1154. . . . . . . **31.26**
    Hydraulic Master Cylinders for Motor Vehicle Brakes—Test Procedure—SAE J1153. . . . . **31.22**
    Hydraulic Motor Test Procedure — SAE J746b **41.87**
    Hydraulic "O" Ring — SAE J515a **21.112**
    Hydraulic Power Circuit Filtration — SAE J931a **41.92**
    Hydraulic Power Pump Test Procedure — SAE J745c **41.77**
    Hydraulic Power Pumps — SAE J744c. **41.79**
    Hydraulic Systems Diagnostic Port Sizes and Locations — SAE J1298 APR80 . . . . **41.77**
    Hydraulic Valves for Motor Vehicle Brake Systems—Performance Requirements — SAE J1137. . . . . . . **31.30**
    Hydraulic Valves for Motor Vehicle Brake Systems Test Procedure — SAE J1118. . . . . **31.26**
    Hydraulic Wheel Cylinders for Automotive Drum Brakes — SAE J101 MAR80. **31.19**
    Hydrodynamic Drives Terminology — SAE J641b . . . **28.02**
    Hydrodynamic Drive Test Code — SAE J643b **28.04**
    Measuring and Reporting the Internal Leakage of a Hydraulic Fluid Power Valve — SAE J1235 . . . . **41.84**
    Method of Measuring and Reporting the Pressure Differential-Flow Characteristics of a Hydraulic Fluid Power Valve — SAE J1117 . . . . **41.82**
    Multipurpose Petroleum Base Fluids — (HS J73) . . . . *
    Pressure Ratings for Hydraulic Tubing and Fittings — SAE J1065 . . . . **21.119**
    Reporting Cleanliness Levels of Hydraulic Fluids — SAE J1165 JUL79 . . . . **41.86**
    Road Vehicle—Hydraulic Brake Hose Assemblies for Use with Non-Petroleum Base Hydraulic Fluids — SAE J1401 FEB79 . . . . **31.01**
    Rubber Cups for Hydraulic Actuating Cylinders — SAE J1601 . . . . **31.144**
    Standardized Fluid for Hydraulic Component Tests — SAE J1276 AUG79 . . . . **41.86**
    Symbols for Hydrodynamic Drives — SAE J640b. . . . **28.01**
    Tubing—Motor Vehicle Brake System Hydraulic—SAE J1047. . . . . **31.136**

IGNITION SYSTEMS
    Electrical Generating System (Alternator Type) Performance Curve and Test Procedure — SAE J56. . . . . . **22.19**
    High Tension Ignition Cable — SAE J557 . . . . **22.49**
    High Tension Ignition Cable Assemblies—Marine — SAE J1191 . . . . **42.19**
    Ignition System Nomenclature and Terminology — SAE J139. . . . . **22.01**
    Ignition System Measurements Procedure — SAE J973a **22.02**
    Ignition Switch — SAE J259 . . . . **22.42**
    Ignition Timing — SAE J327 . . . . **24.41**

IMPACT
    Barrier Collision Tests — SAE J850 JUN80 . . . . **34.136**
    Bodyforms for Use in Motor Vehicle Passenger Compartment Impact Development — SAE J984 JUN80 . . . **34.143**
    Collision Deformation Classification — SAE J224 MAR80 . . . . **34.119**
    Human Tolerance to Impact Conditions as Related to Motor Vehicle Design — (HS J885) . . . . *
    Instrumentation for Impact Tests — SAE J211 JUN80 . **34.131**
    Moving Barrier Collision Tests — SAE J972 JUN80 . . **34.132**
    Passenger Car Door System Crush Test Procedure — SAE J367 JUN80 . . . . **34.142**
    Passenger Car Roof Crush Test Procedure — SAE J374 JUN80 . . . . **34.141**
    Steering Control System—Passenger Car—Laboratory Test Procedure — SAE J944 JUN80 . . . . **34.144**

*Available as a Related Technical Report. Complete listing and details appear at the end of this index.

Wheels—Passenger Cars—Impact Performance Requirements and Test Procedures — SAE J175 .... 30.02

INDUSTRIAL MACHINERY
See Also: CONSTRUCTION AND INDUSTRIAL MACHINERY
Exhaust Flanges—Industrial Engines — SAE J624a .. 41.168
Unauthorized Starting or Movement of Machines — SAE J1083 FEB80. .... 41.57
Universal Symbols for Operator Controls on Industrial Equipment — SAE J298 .... 41.260

INSTRUMENT PANELS
Instrument Face Design and Location for Construction and Industrial Equipment — SAE J209 APR80.... 41.258
Location and Operation of Instruments and Controls in Motor Truck Cabs — SAE J680b 37.34

INSULATIONS
Nonmetallic Loom — SAE J562a 22.65

IRON
Automotive Ductile (Nodular) Iron Castings — SAE J434c 7.06
Automotive Gray Iron Castings — SAE J431 AUG79 7.01
Automotive Malleable Iron Castings — SAE J158a .. 7.04
Eddy Current Testing by Electromagnetic Methods — SAE J425b 4.48
Elevated Temperature Properties of Cast Irons — SAE J125 .... 7.08
Magnetic Particle Inspection — SAE J420b 4.47
Nondestructive Tests — SAE J358 JAN80 .... 4.45
Numbering System for Designating Grades of Cast Ferrous Materials — SAE J859 .... 7.01
Penetrating Radiation Inspection — SAE J427b..... 4.52
Prevention of Corrosion of Metals — (HS J447a) .... *

JACKS
Bumper Jack Requirements—Motor Vehicles — SAE J979 .... 32.01

JOINTS
See: UNIVERSAL JOINTS

KEYS
Woodruff Key Slots and Keyways — SAE J503 ..... 17.06
Woodruff Keys — SAE J502 .... 17.05

KINGPINS
Fifth Wheel Kingpin—Commercial Trailers and Semitrailers — SAE J700b 37.21
Fifth Wheel Kingpin, Heavy Duty—Commercial Trailers and Semitrailers — SAE J848a 37.22
Fifth Wheel Kingpin Performance—Commercial Trailers and Semitrailers — SAE J133 .... 37.20

LAMPS
Auxiliary Driving Lamps — SAE J581a 23.35
Auxiliary Low Beam Lamp — SAE J582a. 23.36
Clearance, Side Marker, and Identification Lamps — SAE J592f 23.59
Connectors and Plugs — SAE J856 .... 23.07
Cornering Lamps — SAE J852b 23.51
Lamp Bulb Retention System — SAE J567c 23.04
Lamp Bulbs and Sealed Beam Headlamp Units — SAE J573g 23.20
Lighting Inspection Code — SAE J599d 23.66
Motorcycle Auxiliary Front Lamps — SAE J1306 JUN80.. 23.39
School Bus Red Signal Lamps — SAE J887a 23.65
Service Performance Requirements and Test Procedures for Motor Vehicle Lamp Bulbs — SAE J1049 AUG79 23.23
Snowmobile and Snowmobile Cutter Lamps, Reflective Devices, and Associated Equipment — SAE J292 ... 39.07
Spot Lamps — SAE J591b 23.58
Backup Lamps
Backup Lamp Switches — SAE J1076 .... 23.60
Backup Lamps — SAE J593e 23.60
Fog Lamps
Fog Lamps — SAE J583d 23.36
Headlamps
Dimensional Specifications for General Service Sealed Lighting Units — SAE J760a 23.18
Dimensional Specifications for Sealed Beam Headlamp Units — SAE J571d 23.09
Dimensional Specifications for Sealed Lighting Unit for Construction and Industrial Machinery — SAE J572a 41.178
Headlamp Aiming Device for Mechanically Aimable Sealed Beam Headlamp Units — SAE J602c..... 23.69
Headlamp Beam Switching — SAE J564c 23.02
Headlamps for Industrial Equipment — SAE J95 APR80.. 41.180
Headlamps for Agricultural Equipment — SAE J975 MAY80.. 40.38
Headlamp Switch — SAE J253 .... 23.61
Headlamp Testing Machines — SAE J600a 23.68
Headlamp-Turn Signal Spacing — SAE J1221 ..... 23.45
Lamp Bulbs and Sealed Beam Headlamp Units — SAE J573g 23.20
Motorcycle Headlamps — SAE J584b 23.37
Sealed Beam Headlamp Assembly — SAE J580 AUG79.. 23.33
142 mm x 200 mm Sealed Beam Headlamp Unit — SAE J1132 .... 23.16
Sealed Beam Headlamp Units for Motor Vehicles — SAE J579c 23.31
Sealed Lighting Units for Construction and Industrial Machinery — SAE J598 APR80 .... 41.177
Semiautomatic Headlamp Beam Switching Devices — SAE J565c 23.03
Service Performance Requirements for Sealed Beam Headlamp Units for Motor Vehicles — SAE J32a... 23.32
Snowmobile Headlamps — SAE J280 .... 39.04
License Plate Lamps
License Plate Lamps — SAE J587f 23.43
Parking Lamps
Parking Lamps (Front Position Lamps) — SAE J222a 23.59
Stop Lamps
Mechanical Stop Lamp Switch — SAE J249....... 23.42
Motorcycle Stop Lamp Switch — SAE J1167 ..... 38.03
Snowmobile Stop Lamp — SAE J278 .... 39.05
Stop Lamps — SAE J586d..................... 23.41
Supplemental High Mounted Stop and Rear Turn Signal Lamps — SAE J186a 23.49
Tail Lamps
Combination Tail and Floodlamp for Agricultural Equipment — SAE J976 MAY80 .... 40.39
Combination Tail and Floodlamp for Industrial Equipment — SAE J94 APR80 .... 41.180
Snowmobile Tail Lamp (Rear Position Lamp) — SAE J279 .... 39.06
Tail Lamps (Rear Position Lamps) — SAE J585e 23.40
Turn Signal Lamps
Headlamp Turn Signal Spacing — SAE J1221 ..... 23.45
Motorcycle Turn Signal Lamps — SAE J131a 23.40
Service Performance Requirements for Turn Signal Flashers — SAE J1055....................... 23.52
Side Turn Signal Lamps — SAE J914b........... 23.50
Supplemental High Mounted Stop and Rear Turn Signal Lamps — SAE J186a 23.49
Turn Signal Lamps — SAE J588f 23.44
Turn Signal Switch — SAE J589b 23.50
Warning Lamps
360 Deg Emergency Warning Lamp — SAE J845 ... 23.64
Flashing Warning Lamp for Agricultural Equipment — SAE J974 MAY80 .... 40.38
Flashing Warning Lamps for Authorized Emergency, Maintenance and Service Vehicles — SAE J595b .. 23.63
Flashing Warning Lamp for Industrial Equipment — SAE J96 .... 41.179
School Bus Red Signal Lamps — SAE J887a 23.65
Vehicular Hazard Warning Signal Flasher — SAE J945b 23.54
Warning Lamp Alternating Flashers — SAE J1054 ... 23.56

LATCHES
Passenger Car Side Door Latch Systems — SAE J839b 34.02
Vehicle Hood Latch Systems — SAE J362 .... 34.04

LATEX
See Also: ELASTOMERIC MATERIALS
Latex Dipped Goods and Coatings for Automotive Applications — SAE J19 .... 12.45

---

*Available as a Related Technical Report. Complete listing and details appear at the end of this index.

Latex Foam Rubbers — SAE J17 OCT79 .... 12.37
Nondestructive Tests — SAE J358 JAN80 .... 4.45
Penetrating Radiation Inspection — SAE J427b 4.52
Ultrasonic Inspection — SAE J428b 4.55

LEAKAGE
Leakage Testing — SAE J1267 .... 4.50

LICENSE PLATE LAMPS
See: LAMPS

LICENSE PLATES
Motor Vehicle License Plates — SAE J686a 26.07

LIGHTERS
Six- and Twelve-Volt Cigar Lighter Receptacles — SAE J563b 22.67

LIGHTING
See Also: LAMPS
Color Specification for Electric Signal Lighting Devices — SAE J578d 23.29
Connectors and Plugs — SAE J856 .... 23.07
Dimensional Specifications for General Service Sealed Lighting Units — SAE J760a 23.18
Dimensional Specifications for Sealed Lighting Unit for Construction and Industrial Machinery — SAE J572a 41.178
Lighting and Marking of Agricultural Equipment on Highways — SAE J137c 40.39
Lighting and Marking of Construction and Industrial Machinery — SAE J1029 APR80 .... 41.180
Lighting and Marking of Industrial Equipment on Highways — SAE J99 APR80 .... 41.181
Lighting Identification Code — SAE J759c 23.01
Lighting Inspection Code — SAE J599d 23.66
Plastic Material for Use in Housings of Motor Vehicle Lighting Devices — SAE J29 .... 23.28
Plastic Materials for Use in Optical Parts Such as Lenses and Reflectors of Motor Vehicle Lighting Devices — SAE J576d 23.28
Reflex Reflectors — SAE J594f 23.61
Sealed Lighting Units for Construction and Industrial Machinery — SAE J598 APR80 .... 41.177
Service Performance Requirements for Motor Vehicle Lighting Devices and Components — SAE J256b .. 23.26
Terminology—Motor Vehicle Lighting — SAE J387 .. 23.01
Tests for Motor Vehicle Lighting Devices and Components — SAE J575 JUN80 .... 23.24

LOADERS
See: CONSTRUCTION AND INDUSTRIAL MACHINERY

LOOMS
Nonmetallic Loom — SAE J562a 22.65

LUBRICANTS
See also: OILS
Automotive Lubricating Greases — SAE J310 MAR80 . 13.14
Axle and Manual Transmission Lubricants — SAE J308c 13.12
Axle and Manual Transmission Lubricant Viscosity Classification — SAE J306 OCT79 .... 13.13
Effective Dates of Revisions — SAE J301 .... 13.01
Lubrication Chart—Construction and Industrial Machinery — SAE J753 APR80 .... 41.67
Lubricant Types—Construction and Industrial Machinery — SAE J754a 41.68
Maintenance Interval—Construction Equipment — SAE J752b 41.69
Multipurpose Petroleum Base Fluids — (HS J73) .... *
Powershift Transmission Fluid Classification — SAE J1285 FEB80 .... 13.19
Standardized Fluid for Hydraulic Component Tests — SAE J1276 AUG79 .... 41.86

MAGNESIUM ALLOYS
See: ALLOYS

MARINE EQUIPMENT
External Ignition Protection of Marine Electrical Devices — SAE J1171 .... 42.08
High Tension Ignition Cable Assemblies—Marine — SAE J1191 .... 42.19
Ignition Distributors—Marine — SAE J1294 APR80 .. 42.15
Marine Control Cable Connection—Engine Clutch Lever — SAE J960 JUN80 .... 42.03
Marine Control Cable Connection—Engine Throttle Lever — SAE J961 JUN80 .... 42.04
Marine Engine Mountings Direct Drive Transmission — SAE J233 JUN80 .... 42.08
Marine Engine Wiring — SAE J378c 42.16
Marine Propeller-Shaft Couplings — SAE J756 JUN80 42.01
Marine Propeller-Shaft Ends and Hubs — SAE J755 JUN80 .... 42.05
Marine Push-Pull Control Cables — SAE J917 JUN80 42.02

MASTER CYLINDERS
See Also: BRAKES
Hydraulic Master Cylinders for Motor Vehicle Brakes—Performance Requirements — SAE J1154 .... 31.26
Hydraulic Master Cylinders for Motor Vehicle Brakes—Test Procedure — SAE J1153 .... 31.22

MATERIALS
Flammability of Automotive Interior Materials—Horizontal Test Method — SAE J369a 12.73
Nondestructive Tests — SAE J358 JAN80 .... 4.45
Test Method for Determining Dimensional Stability of Automotive Textile Materials — SAE J883 .... 12.58
Test Method for Determining Resistance to Snagging and Abrasion of Automotive Bodycloth — SAE J948a 12.75
Test Method for Measuring Thickness of Automotive Textiles and Plastics — SAE J882 .... 12.59
Test Method for Wicking of Automotive Fabrics and Fibrous Materials — SAE J913a 12.69
Test Method of Stretch and Set of Textiles and Plastics — SAE J855a 12.76

Felts
Felts—Wool and Part Wool — SAE J314b 12.78

Fiberboards
Fiberboard Crease Bending Test — SAE J119a 12.72
Fiberboard Test Procedure — SAE J315b 12.68
Glossary of Fiberboard Terminology — SAE J947b ... 12.67
Test Method for Determining Stiffness (Modulus of Bending) of Fiberboards — SAE J949a 12.74
Test Method for Determining Visual Color Match to Master Specimen for Fabrics, Vinyls, Coated Fiberboards, and Other Automotive Trim Materials — SAE J361a 12.75

Glass
Automotive Safety Glazing — SAE J673b 34.31
Automotive Safety Glazing Manual — (HS J906) .... *
Drop Test for Evaluating Laminated Safety Glass for Use in Automotive Windshields — SAE J938a 34.36
Light Transmittance of Automotive Windshields Safety Glazing Materials — SAE J1203 .... 34.34
Passenger Car Glazing—Electrical Circuits — SAE J216 34.35
Passenger Car Glazing Shade Bands — SAE J100 ... 34.35
Safety Glazing Materials—Motor Vehicles — SAE J674a 34.34

Plastics
Liquid Penetrant Test Methods — SAE J426c 4.52
Plastic Material for Use in Housings of Motor Vehicle Lighting Devices — SAE J29 .... 23.28
Plastic Materials for Use in Optical Parts Such as Lenses and Reflectors of Motor Vehicle Lighting Devices — SAE J576d 23.28
Reinforced Plastics for Ground Vehicle Applications — (HS 762a) .... *
Test Method for Determining Cold Cracking of Flexible Plastic Materials — SAE J323 .... 12.59
Ultrasonic Inspection — SAE J428b 4.55

Trim Materials
Anodized Aluminum Automotive Parts — SAE J399a . 11.28
Electroplating of Nickel and Chromium on Metal Parts—Automotive Ornamentation and Hardware — SAE J207 .... 11.133
Method of Testing Resistance to Crocking of Organic Trim Materials — SAE J861 .... 12.58
Method of Testing Resistance to Scuffing of Trim Materials — SAE J365a 12.69

*Available as a Related Technical Report. Complete listing and details appear at the end of this index.

Nonmetallic Trim Materials—Test Method for Determining the Staining Resistance to Hydrogen Sulfide Gas — SAE J322 .... 12.60
Test Method for Determining Blocking Resistance and Associated Characteristics of Automotive Trim Materials — SAE J912a 12.71
Test Method for Determining Visual Color Match to Master Specimen for Fabrics, Vinyls, Coated Fiberboards, and Other Automotive Trim Materials — SAE J361a ... 12.75
Test Method for Determining Window Fogging Resistance of Interior Trim Materials — SAE J275 ..... 12.61
Test Method for Measuring Weight of Organic Trim Materials — SAE J860 .... 12.58

METALS
See Also: ALLOYS
Numbering Metals and Alloys — SAE J1086 (ASTM E527) 2.01
Preferred Thicknesses for Uncoated Flat Metals (Thru 12 mm) — SAE J446a 10.06
Residual Stress Measurements by X-Ray Diffraction — (HS J784a) .... *
Second Edition Unified Numbering System Handbook for Metals and Alloys — (HS 1086a) .... *
Sintered Powder Metal Parts: Ferrous — SAE J471d .. 9.01
Surface Rolling and Other Methods for Mechanical Prestressing of Metals — (HS 3) .... *
Surface Texture — SAE J448a 10.02
Surface Texture Control — SAE J449a 10.04

METRIC SYSTEM
Dual Dimensioning — (HS J390) .... *
Recommended Guidelines for Company Metrication Programs in the Metalworking Industry — (HS 1066) .. *
Rules for SAE Use of SI (Metric) Units — SAE J916 JUN80 .... 1.09

MINING MACHINERY
Minimum Performance Criteria for Falling Object Protective Structure (FOPS) — SAE J231 .... 41.235
Performance Criteria for Rollover Protective Structures (ROPS) for Construction, Earthmoving, Forestry, and Mining Machines — SAE J1040c 41.225
Unauthorized Starting or Movement of Machines — SAE J1083 FEB80 .... 41.57

MIRRORS
Passenger Car Rear Vision — SAE J834a 34.39
Rear View Mirrors—Motorcycles — SAE J268a 38.03
Test Procedure for Determining Reflectivity of Rear View Mirrors — SAE J964a 34.37
Vision Factors Considerations in Rear View Mirror Design — SAE J985 .... 34.38

MOTORBOATS
Blade Type Electric Fuses — SAE J1284 DEC79 .... 22.37
Operator Sound Level Exposure Assessment Procedure for Pleasure Motorboats — SAE J1281 NOV79 ... 35.29
Static Electric Outboard Thrust — SAE J1286 APR80 42.04

MOTORCYCLES
Brake System Road Test Code—Motorcycles — SAE J108a 31.114
Definitions—Motorcycles — SAE J213a 38.01
Fuel and Lubricant Tanks for Motorcycles — SAE J1241 .... 38.05
Maximum Sound Level Potential for Motorcycles — SAE J47 .... 35.01
Measurement of Exhaust Sound Levels of Stationary Motorcycles — SAE J1287 JUN80 .... 38.08
Motorcycle and Motor Driven Cycle Electrical System (Maintenance of Design Voltage) — SAE J392 .... 23.39
Motorcycle Auxiliary Front Lamps — SAE J1306 JUN80 .... 23.39
Motorcycle Bank Angle Measurement Procedure — SAE J1168 .... 38.04
Motorcycle Headlamps — SAE J584b 23.37
Motorcycle Stop Lamp Switch — SAE J1167 .... 38.03
Motorcycle Turn Signal Lamps — SAE J131a 23.40
Operator Controls and Displays on Motorcycles — SAE J107 APR80 .... 38.01
Rear View Mirrors—Motorcycles — SAE J268a 38.03
Service Brake System Performance Requirements—Motorcycles and Motor-Driven Cycles — SAE J109a 31.116
Sound Levels for Motorcycles — SAE J331a 35.02
Universal Symbols for Operator Controls — SAE J1500 JUN80 .... 1.22
Vehicle Identification Numbers—Motorcycles — SAE J364a 26.03

MOUNTINGS
Brake Flange Mounting—On-Highway Vehicles — SAE J337 .... 37.27
Diesel Fuel Injection Pump Mountings — SAE J626b 24.92
Engine Foot Mounting (Front and Rear) — SAE J616c 41.166
Engine Mountings — SAE J615b 24.78
Extra Wide Single Tire Rim and Wheel Mounting for Rear Axles — SAE J876a 37.26
Filter Base Mounting — SAE J363 FEB80 .... 24.82
Fuel Pump Mountings for Diaphragm Type Pumps — SAE J625a 24.90
Generator Mountings — SAE J545b 22.15
Locating Ledge for Flat Pad Engine Mountings — SAE J832 .... 24.78
Magneto Mountings — SAE J546 .... 22.20
Marine Engine Mountings Direct Drive Transmission — SAE J233 JUN80 .... 42.08
Mounting Brackets and Socket for Warning Lamp and Slow-Moving Vehicle (SMV) Identification Emblem — SAE J725 MAY80 .... 40.41
Starting Motor Mountings — SAE J542c 22.14
Wheel Mounting Elements for Industrial and Agricultural Disc Wheels — SAE J714b 40.36

MOWERS
See Also: CONSTRUCTION AND INDUSTRIAL MACHINERY

MUFFLERS
Muffler Parts Nomenclature — SAE J261 .... 24.123
Nomenclature for Exhaust System Parts — SAE J97 .. 24.125

NICKEL ALLOYS
See: ALLOYS

NOISE
See: SOUND

NOZZLES
Calibrating Nozzle and Holder Assembly for Diesel Fuel Injection Systems — SAE J968c 24.102
Diesel Fuel Injection Nozzle and Holder Assembly (17 mm Nominal Diameter) — SAE J265 .... 24.100
Diesel Fuel Injection Nozzle Holder Assemblies — SAE J629b 24.97
Diesel Fuel Injection Nozzle Holder Assembly (9.5 mm) — SAE J347a 24.99
Fuel Injection Nozzle Gaskets — SAE J59 .... 24.100
Gasoline Dispenser Nozzle Spouts — SAE J285a 24.161

NUTS
See: FASTENERS

O-RINGS
Hydraulic "O" Ring — SAE J515a 21.112
Hydraulic Tube Fittings — SAE J514 APR80 .... 21.46
Rubber Rings for Automotive Applications — SAE J120a 12.87

OCCUPANT PROTECTION
See: RESTRAINT SYSTEMS AND SEAT BELTS

ODOMETERS
Factors Affecting Automotive Odometer-Speedometer Accuracy — SAE J862b 20.01

OFF-ROAD MACHINES
See Also: CONSTRUCTION AND INDUSTRIAL MACHINERY
Categories of Off-Road Self-Propelled Work Machines — SAE J1116 JUN80 .... 41.01
Electrical Connector for Auxiliary Starting of Construction, Agricultural, and Off-Road Machinery — SAE J1283 JUN80 .... 41.173
Format for Commercial Literature Specifications—Off-Road Work Machines — SAE J1233 .... 41.15
Soil Type and Strength Classification — SAE J106

*Available as a Related Technical Report. Complete listing and details appear at the end of this index.

APR80	41.63
Specification Definitions—Off-Road Work Machines — SAE J1234	41.10
Universal Symbols for Operator Controls — SAE J1500 JUN80	1.22

OILS

Engine Oil Performance and Engine Service Classification — SAE J183 FEB80	13.01
Engine Oil Tests — SAE J304c	13.06
Engine Oil Viscosity Classification — SAE J300d	13.05
Oil Cooler Nomenclature and Glossary — SAE J1244	24.138
Oil Filter Test Procedure — (HS 806b)	*
Physical and Chemical Properties of Engine Oils — SAE J357a	13.08
The Engine Oil Performance and Engine Service Classification Maintenance Procedure — SAE J1146	13.34

OPERATOR PROTECTIVE STRUCTURES

Deflection Limiting Volume—ROPS/FOPS Laboratory Evaluation — SAE J397b	41.234
Fan Blast Deflectors for Earthmoving Machines — SAE J165 APR80	41.223
Inverted Vehicle Drop Test Procedure — SAE J996 JUN80	34.140
Labeling of ROPS and FOPS — SAE J1164	41.233
Lift Arm Safety Device for Loaders — SAE J38	41.224
Minimum Performance Criteria for Falling Object Protective Structure (FOPS) — SAE J231 APR80	41.235
Minimum Performance Criteria for Falling Object Protective Structure (FOPS) for Industrial Equipment — SAE J1043 APR80	41.236
Operator Protection for Industrial Equipment — SAE J1042 APR80	41.233
Operator Protective Structure Performance Criteria for Certain Forestry Equipment — SAE J1084 APR80	41.237
Overhead Protection for Agricultural Tractors—Test Procedures and Performance Requirements — SAE J167a	40.56
Performance Criteria for Rollover Protective Structures (ROPS) for Construction, Earthmoving, Forestry, and Mining Machines — SAE J1040c	41.225
Performance Prediction of Roll-Over Protective Structures (ROPS) Through Analytical Methods — SAE J1215	41.244
Roll-Over Protective Structures (ROPS) for Wheeled Agricultural Tractors — SAE J1194	40.49
Roll-Over Tests Without Collision — SAE J857 JUN80	34.137
Safety Considerations for the Operator — SAE J153	41.263
Safety Criteria for Industrial Flail Mowers — SAE J1001 APR80	41.268
Safety for Industrial Wheeled Equipment — SAE J98 APR80	41.264
Seat Belts for Construction Machines — SAE J386 APR80	41.270
Steel Products for Rollover Protective Structures (ROPS) and Falling Object Protective Structures (FOPS) — SAE J1119 APR80	41.238
Steering Frame Lock for Articulated Loaders and Tractors — SAE J276	41.224
Tire Guards for Protection of Operator of Earthmoving Haulage Machines — SAE J321b	41.223

PARKING LAMPS

See: LAMPS

PARTICLE SIZES

Definition for Particle Size — SAE J391	4.56

PINS

Clevis Pins

Rod Ends and Clevis Pins — SAE J493	15.84

Cotter Pins

Cotter Pins — SAE J487a	15.77
Holes in Bolt and Capscrew Shanks and Slots in Nuts for Cotter Pins — SAE J485	15.79

Dowel Pins

Unhardened Ground Dowel Pins — SAE J497	15.84

Straight Pins

Grooved Straight Pins — SAE J494	15.80
Spring Type Straight Pins — SAE J496	15.82
Straight Pins (Solid) — SAE J495	15.79

PILOT HOLES

Fan Hub Bolt-Circles and Pilot Holes — SAE J635a	24.145

PIPES

Automotive Pipe, Filler, and Drain Plugs — SAE J531a	21.134
Coding System for Identification of Tube, Pipe, and Hose Fittings — SAE J846 MAY80	21.01
Dryseal Pipe Threads — SAE J476a	14.01

PISTON RINGS

Piston Rings and Grooves (Metric) — SAE J1201	24.49
Piston Rings and Pistons — SAE J929a	24.42

PLASTICS

See: MATERIALS

PLUGS

Automotive Pipe, Filler, and Drain Plugs — SAE J531a	21.134
Automotive Straight Thread Filler and Drain Plugs — SAE J532a	21.138
Connectors and Plugs — SAE J856	23.07
Drain, Fill, and Level Plugs for Earthmoving Machinery — SAE J371a	41.171
Preignition Rating of Spark Plugs — SAE J549a	22.27
Spark Plugs — SAE J548d	22.21

POWER TAKE-OFF

Auxiliary Power Take-Off Drives for Agricultural Tractors — SAE J717 MAY80	40.31
Clearance Envelopes for Six- and Eight-Bolt Truck Transmission Mounted Power Take-Offs — SAE J772a	28.15
Flywheels for Industrial Engines Used with Industrial Power Take-Offs Equipped with Driving-Ring Type Overcenter Clutches and Engine Mounted Marine Gears — SAE J620d	41.170
Industrial Power Take-Offs with Driving Ring Type Overcenter Clutches — SAE J621d	41.162
Mounting Flanges and Power Take-Off Shafts for Small Engines — SAE J609a	24.36
Openings for Six- and Eight-Bolt Truck Transmission Mounted Power Take-Offs — SAE J704b	28.16
Operating Requirements for Tractors and Power Take-Off Driven Implements — SAE J721 JUN80	40.30
Power Take-Off Definitions and Terminology for Agricultural Tractors — SAE J722 MAY80	40.32
Rating of Truck Power Take-Offs — SAE J705b	37.32
Rear Power Take-Off for Agricultural Tractors — SAE J1170 JUN80	40.27

PROPELLERS

Marine Propeller-Shaft Couplings — SAE J756 JUN80	42.01
Marine Propeller-Shaft Ends and Hubs — SAE J755 JUN80	42.05

PULLEYS

Automotive V-Belt Drives — SAE J637b	18.04
Snowmobile and All-Terrain Vehicle Pulley Guards and Shields — SAE J93	39.14
Tractor Belt Speed and Pulley Width — SAE J720 MAY80	40.32
V-Belts and Pulleys — SAE J636c	18.01

PUMPS

Diesel Fuel Injection Pump Mountings — SAE J626b	24.92
Fuel Pump Mountings for Diaphragm Type Pumps — SAE J625a	24.90
Guide to the Application and Use of Engine Coolant Pump Face Seals — SAE J1245 FEB80	24.149
Hydraulic Power Pumps — SAE J744c	41.79
Hydraulic Power Pump Test Procedure — SAE J745c	41.77
Master Calibrating Fuel Injection Pump — SAE J981	24.101
Engine Coolant Pump Seals — SAE J780a	24.147

RADIATORS

Large Size Radiator Filler Necks — SAE J868	24.130
Maintenance of Automotive Engine Cooling Systems — (HS 40)	*
Radiator Caps and Filler Necks — SAE J164 JUL79	24.128
Radiator Nomenclature — SAE J631b	24.132

RADIOS

External Electromagnetic Radiation Suppressors — SAE J552a	22.34
Limits and Methods of Measurement of Radio Interfer-	

*Available as a Related Technical Report. Complete listing and details appear at the end of this index.

ence Characteristics of Vehicles and Devices (20–1000 MHz) — SAE J551 JUN79 .... 22.27
Mobile Radio Telephone for Automotive Use — (HS 27) *

RECREATIONAL VEHICLES
Tire to Body Clearance Check for Recreational Vehicles — SAE J1214 .... 36.03

REFLECTORS
Emergency Warning Device — SAE J774c 23.62
Plastic Materials for Use in Optical Parts Such as Lenses and Reflectors of Motor Vehicle Lighting Devices — SAE J576d 23.28
Reflex Reflectors — SAE J594f 23.61
Snowmobile and Snowmobile Cutter Lamps, Reflective Devices, and Associated Equipment — SAE J292 .. 39.07

RESONATORS
Resonator Parts Nomenclature — SAE J262 ...... 24.123

RESTRAINT SYSTEMS
See Also: SEAT BELTS
Motor Vehicle Seat Belt Anchorages Test Procedure — SAE J384 .... 33.08
Motor Vehicle Seat Belt Anchorages—Design Recommendations — SAE J383 .... 33.06
Motor Vehicle Seat Belt Anchorages—Performance Requirements — SAE J385 .... 33.09
Occupant Restraint System Evaluation—Passenger Cars — SAE J128 .... 33.12
Seat Belt Assembly Webbing Abrasion Test Procedure — SAE J339a 33.12
Seat Belt Assembly Webbing Abrasion Test Requirements — SAE J114 .... 33.12

RIMS
Extra Wide Single Tire Rim and Wheel Mounting for Rear Axles — SAE J876a 37.26
Labeling—Disc Wheels and Demountable Rims—Trucks — SAE J179a 30.09
Nomenclature—Wheels and Rims—Trucks — SAE J393 .... 30.10
Off-Highway Tire and Rim Classification — SAE J751 APR80 .... 41.199
Recommended Tolerances for Spoke-Type Wheels, Demountable Rims, and Rim Spacers—Trucks and Trailers — SAE J851 .... 37.25
Wheels/Rims—Trucks—Test Procedures and Performance Requirements — SAE J267a 30.05

RIPPERS
See: CONSTRUCTION AND INDUSTRIAL MACHINERY

RIVETS
See: FASTENERS

ROD ENDS
Metric Spherical Rod Ends — SAE J1259 APR80 ... 16.15
Spherical Rod Ends — SAE J1120 SEP79 .... 16.12

ROLLERS
See: CONSTRUCTION AND INDUSTRIAL MACHINERY

ROLLOVER PROTECTIVE STRUCTURES
See: OPERATOR PROTECTIVE STRUCTURES

ROOTERS
See: CONSTRUCTION AND INDUSTRIAL MACHINERY

ROPS
See: OPERATOR PROTECTIVE STRUCTURES

RUBBER
See: ELASTOMERIC MATERIALS
LATEX

SAFETY
See Also: OPERATOR PROTECTIVE STRUCTURES AND RESTRAINT SYSTEMS
Access Systems for Construction and Industrial Equipment — SAE J185 .... 41.249
Automotive Safety Glazing Manual — (HS J906) .... *
Barrier Collision Tests — SAE J850 JUN80 .... 34.136
Bodyforms for Use in Motor Vehicle Passenger Compartment Impact Development — SAE J984 JUN80 ... 34.143
Collision Deformation Classification — SAE J224 MAR80 .... 34.119
Crane Load Moment System — SAE J159 OCT79 ... 41.147
Handbook of Motor Vehicle, Safety and Environmental Terminology — (HS 215) .... *
Human Tolerance to Impact Conditions as Related to Motor Vehicle Design — (HS J885) .... *
Instrumentation for Impact Tests — SAE J211 JUN80 34.131
Lift Arm Safety Device for Loaders — SAE J38 .... 41.224
Moving Barrier Collision Tests — SAE J972 JUN80 .. 34.132
Rear Underride Guard Test Procedure — SAE J260 JUN80 .... 34.146
Recommendations for Children's Snowmobile — SAE J1038 FEB80 .... 39.18
SAE Documents Referenced in Federal Motor Vehicle Safety Standards — (HS 19) .... *
Safety Alert Symbol for Agricultural, Construction and Industrial Equipment — SAE J284a 40.43
Safety Chain of Full Trailers or Converter Dollies — SAE J697a 37.24
Safety Criteria for Industrial Flail Mowers — SAE J1001 APR80 .... 41.268
Safety for Agricultural Equipment — SAE J208d ... 40.44
Safety for Industrial Wheeled Equipment — SAE J98 APR80 .... 41.264
Safety Glazing Materials—Motor Vehicles — SAE J674a 34.34
Safety Practices for Mechanical Vapor Compression Refrigeration Equipment or Systems Used to Cool Passenger Compartment of Motor Vehicles — SAE J639 34.22
Safety Signs — SAE J115 SEP79 .... 41.272
School Bus Stop Arm — SAE J1133a 23.66
Slow-Moving Vehicle Identification Emblem — SAE J943a 40.40
Steering Control System—Passenger Car—Laboratory Test Procedure — SAE J944 JUN80 .... 34.144
Steering Frame Lock for Articulated Loaders and Tractors — SAE J276 .... 41.224

SCARIFIERS
See: CONSTRUCTION AND INDUSTRIAL MACHINERY

SCHOOL BUSES
See: BUSES

SCRAPERS
See: CONSTRUCTION AND INDUSTRIAL MACHINERY

SCREWS
See: FASTENERS

SEALERS
Methods of Tests for Automotive-Type Sealers, Adhesives, and Deadeners — SAE J243 .... 12.46
Synthetic Resin Plastic Sealers, Nondrying Type — SAE J250 .... 12.45

SEALS
Cast Iron Sealing Rings — SAE J281 .... 28.61
Cast Iron Sealing Rings (Metric) — SAE J1236 ..... 28.63
Cylinder Rod Wiper Seal Ingression Test — SAE J1195 41.96
Guide to the Application and Use of Engine Coolant Pump Face Seals — SAE J1245 FEB80 .... 24.149
Rubber Seals for Hydraulic Disc Brake Cylinders — SAE J1603 .... 31.13
Seals—Application Guide to Radial Lip — SAE J946d 28.43
Seals—Evaluation of Elastohydrodynamic — SAE J1002 .... 28.54
Seals—Lathe Cut — SAE J654 .... 28.60
Seals—Terminology of Radial Lip — SAE J111c ... 28.38
Seals—Testing of Radial Lip — SAE J110c 28.57
Engine Coolant Pump Seals — SAE J780a 24.147

SEAT BELTS
Dynamic Test Procedure—Type 1 and Type 2 Seat Belt Assemblies — SAE J117 .... 33.04
Motor Vehicle Seat Belt Anchorages—Design Recommendations — SAE J383 .... 33.06
Motor Vehicle Seat Belt Anchorages—Performance Requirements — SAE J385 .... 33.09
Motor Vehicle Seat Belt Anchorages Test Procedure —

*Available as a Related Technical Report. Complete listing and details appear at the end of this index.

SAE J384 . . . . **33.08**
Motor Vehicle Seat Belt Assembly Installations — SAE J800c **33.10**
Seat Belt Assembly Webbing Abrasion Test Procedure — SAE J339a **33.12**
Seat Belt Assembly Webbing Abrasion Test Requirements — SAE J114 . . . . **33.12**
Seat Belts for Construction Machines — SAE J386 APR80 . . . . **41.270**
Seat Belt Hardware Performance Requirements — SAE J141 . . . . **33.03**
Seat Belt Hardware Test Procedure — SAE J140a . . . **33.01**

SEATING
Determining Operator Seat Location on Off-Road Work Machines — SAE J1163 JAN80 . . . . **41.252**
Devices for Use in Defining and Measuring Vehicle Seating Accommodation — SAE J826 APR80 . . . . . . **34.22**
Dynamic Cushioning Performance Criteria for Snowmobile Seats — SAE J89a **39.12**
Dynamic Flex Fatigue Test for Slab Polyurethane Foam — SAE J388 . . . . **34.29**
Deflection of Seat Cushions for Off-Road Work Machines — SAE J1051 JAN80 . . . . **41.251**
Load Deflection Testing of Urethane Foams for Automotive Seating — SAE J815 . . . . **34.31**
Motor Vehicle Driver and Passenger Head Position — SAE J1052 . . . . **34.116**
Motor Vehicle Seating Systems — SAE J879b **34.26**
Operator's Seat Dimensions—Construction and Industrial Equipment Design — SAE J899 . . . . **41.254**
Seating Manual—Motor Vehicles — (HS J782b) . . . . *
Urethane for Automotive Seating — SAE J954 . . . . **34.30**

SHOT PEENING
Cast Shot and Grit Size Specifications for Peening and Cleaning — SAE J444a **9.09**
Cast Steel Shot — SAE J827 . . . . **9.08**
Cut Steel Wire Shot — SAE J441 . . . . **9.05**
Manual on Shot Peening — (HS 84) . . . . *
Metallic Shot and Grit Mechanical Testing — SAE J445a **9.10**
Procedures for Using Standard Shot Peening Test Strip — SAE J443 . . . . **9.07**
SAE Manual on Blast Cleaning — (HS 124) . . . . . . *
Size Classification and Characteristics of Glass Beads for Peening — SAE J1173 . . . . **9.11**
Test Strip, Holder and Gage for Shot Peening — SAE J442 AUG79 . . . . **9.05**

SKIDDERS
See: FORESTRY MACHINES

SKINS
Synthetic Skins for Automotive Testing — SAE J202 **34.118**

SMALL ENGINES
See: ENGINES

SMOKEMETERS
Measurement Procedure for Evaluation of Full-Flow, Light Extinction Smokemeter Performance — SAE J1157 . . . . **25.35**

SNOWMOBILES
Brake System Test Procedure—Snowmobiles — SAE J45 . . . . **39.10**
Dynamic Cushioning Performance Criteria for Snowmobile Seats — SAE J89a **39.12**
Exterior Sound Level for Snowmobiles — SAE J192 APR80 . . . . **35.25**
Maintenance of Design Voltage—Snowmobile Electrical Systems — SAE J277 . . . . **39.04**
Metallic and Nonmetallic Nonpressure Integral Fuel Tank—Snowmobile — SAE J288a **39.16**
Operational Sound Level Measurement Procedure for Snow Vehicles — SAE J1161 APR80 . . . . **35.27**
Operator Ear Sound Level Measurement Procedure for Snow Vehicles — SAE J1160 . . . . **35.26**
Recommendations for Children's Snowmobile — SAE J1038 FEB80 . . . . **39.18**
Service Brake System Performance Requirements—Snowmobiles — SAE J44 . . . . **39.10**
Snowmobile and All-Terrain Vehicle Pulley Guards and Shields — SAE J93 . . . . **39.14**
Snowmobile and Snowmobile Cutter Lamps, Reflective Devices, and Associated Equipment — SAE J292 . . **39.07**
Snowmobile Definitions and Nomenclature—General — SAE J33a **39.01**
Snowmobile Headlamps — SAE J280 . . . . **39.04**
Snowmobile Passenger Handgrips — SAE J1062 . . . **39.09**
Snowmobile Stop Lamp — SAE J278 . . . . **39.05**
Snowmobile Tail Lamp (Rear Position Lamp) — SAE J279 . . . . **39.06**
Snowmobile Throttle Control Systems — SAE J92 . . **39.17**
Tests for Snowmobile Switching Devices and Components — SAE J68 OCT79 . . . . **39.08**
Vehicle Identification Numbers—Snowmobile and All-Terrain Vehicles — SAE J64 . . . . **39.03**

SOLDERS
Automotive Metallurgical Joining — (HS J836a) . . . . *
Solders — SAE J473a **11.138**
Welding, Brazing, and Soldering—Materials and Practices — SAE J1147 . . . . **10.06**

SOUND
See Also: ACOUSTICS
Bystander Sound Level Measurement Procedure for Small Engine Powered Equipment — SAE J1175 . . **35.50**
Engine Sound Level Measurement Procedure — SAE J1074 . . . . **35.61**
Exterior Sound Level for Heavy Trucks and Buses — SAE J366b **35.18**
Exterior Sound Level for Snowmobiles — SAE J192 APR80 . . . . **35.25**
Exterior Sound Level Measurement Procedure for Pleasure Motorboats — SAE J34a **35.29**
Exterior Sound Level Measurement Procedure for Earthmoving Machinery — SAE J88 SEP79 . . . . . . **35.54**
Exterior Sound Level Measurement Procedure for Self-Propelled Agricultural Field Equipment — SAE J1008 **35.60**
Exterior Sound Level Measurement Procedure for Small Engine Powered Equipment — SAE J1046a **35.47**
Maximum Sound Level for Passenger Cars and Light Trucks — SAE J1030 . . . . **35.13**
Maximum Sound Level Potential for Motorcycles — SAE J47 . . . . **35.01**
Measurement of Exhaust Sound Levels of Stationary Motorcycles — SAE J1287 JUN80 . . . . **38.08**
Measurement of Exterior Sound Level of Trucks with Auxiliary Equipment — SAE J1077 . . . . **35.22**
Measurement of Exterior Sound Levels for Heavy Trucks Under Stationary Conditions — SAE J1096 . . . . **35.20**
Measurement of Light Vehicle Exhaust Sound Level Under Stationary Conditions — SAE J1169 . . . . **35.16**
Measurement Procedure for Determination of Silencer Effectiveness in Reducing Engine Intake or Exhaust Sound Level — SAE J1207 . . . . **35.62**
Measurement Procedure for Determining a Representative Sound Level at a Construction Site Boundary Location — SAE J1075 . . . . **35.52**
Operational Sound Level Measurement Procedure for Snow Vehicles — SAE J1161 APR80 . . . . **35.27**
Operator Ear Sound Level Measurement Procedure for Small Engine Powered Equipment — SAE J1174 . . **35.49**
Operator Ear Sound Level Measurement Procedure for Snow Vehicles — SAE J1160 . . . . **35.26**
Operator Sound Level Exposure Assessment Procedure for Pleasure Motorboats — SAE J1281 NOV79 . . . **35.29**
Operator Sound Level Measurement Procedure for Earthmoving Machinery—Singular Type Test — SAE J919c **35.37**
Operator Station Sound Level Measurement Procedure for Earthmoving Machinery—Work Cycle Test — SAE J1166a **35.39**
Qualifying a Sound Data Acquisition System — SAE J184a **35.06**
Sound Levels for Motorcycles — SAE J331a **35.02**

*Available as a Related Technical Report. Complete listing and details appear at the end of this index.

Sound Level for Passenger Cars and Light Trucks — SAE J986b 35.11
Sound Level for Truck Cab Interior — SAE J336a . . . 35.19
Sound Level of Highway Truck Tires — SAE J57a . . . 35.21
Sound Level Measurement Procedure for Trenching Equipment — SAE J1262 . . . . 35.57
Surface Vehicle Sound Measurement Procedures — (HS 184) . . . . *

SPARK ARRESTERS
Multiposition Small Engine Exhaust System Fire Ignition Suppression — SAE J335b 24.114
Spark Arrester Test Carbon — SAE J997 JAN80 . . . . 24.120
Spark Arrester Test Procedure for Large Size Engines — SAE J342 . . . . 24.117
Spark Arrester Test Procedure for Medium Size Engines — SAE J350 JAN80 . . . . 24.117

SPARK PLUGS
Preignition Rating of Spark Plugs — SAE J549a 22.27
Spark Plugs — SAE J548d 22.21

SPEEDOMETERS
Automatic Vehicle Speed Control—Motor Vehicles — SAE J195 . . . . 20.12
Factors Affecting Automotive Odometer—Speedometer Accuracy — SAE J862b 20.01
Speedometer Test Procedure — SAE J1059 . . . . 20.07
Speedometers and Tachometers—Automotive — SAE J678e 20.03

SPLINES
Involute Splines and Inspections — SAE J498c 17.01
Parallel Side Splines for Soft Broached Holes in Fittings — SAE J499a 17.01
Serrated Shaft Ends — SAE J500 . . . . 17.03
Shaft Ends — SAE J501 . . . . 17.04

SPRINGS
Hard Drawn Carbon Steel Valve Spring Quality Wire and Springs — SAE J172 . . . . 6.06
Hard Drawn Mechanical Spring Wire and Springs — SAE J113 . . . . 6.04
Music Steel Spring Wire and Springs — SAE J178 . . . 6.09
Oil Tempered Carbon Steel Spring Wire and Springs — SAE J316 . . . . 6.01
Oil Tempered Carbon Steel Valve Spring Quality Wire and Springs — SAE J351 . . . . 6.02
Oil Tempered Chromium-Silicon Alloy Steel Wire and Springs — SAE J157 . . . . 6.05
Oil Tempered Chromium-Vanadium Valve Spring Quality Wire and Springs — SAE J132 . . . . 6.03
Rated Spring Capacity — SAE J274 . . . . 19.15
Special Quality High Tensile, Hard Drawn Mechanical Spring Wire and Springs — SAE J271 . . . . 6.04
Stainless Steel 17-7 PH Spring Wire and Springs — SAE J217 . . . . 6.07
Stainless Steel, SAE 30302, Spring Wire and Springs — SAE J230 . . . . 6.08

Helical Springs
Helical Compression and Extension Spring Terminology — SAE J1121 . . . . 19.09
Helical Springs: Specification Check Lists — SAE J1122 . . . . 19.14
Manual on Design and Application of Helical and Spiral Springs — (HS J795a) . . . . *

Leaf Springs
Leaf Springs for Motor Vehicle Suspension — SAE J510c 19.01
Leaf Springs for Motor Vehicle Suspension—Metric Bar Sizes — SAE J1123a 19.01
Manual on Design and Application of Leaf Springs — (HS J788a) . . . . *

Pneumatic Springs
Pneumatic Spring Terminology — SAE J511a 19.08

STEELS
Aging of Carbon Steel Sheet and Strip — SAE J763 . . 3.20
Alloy Steel Machinability Ratings — SAE J770c 3.24
Automotive Steel Castings — SAE J435c 7.12
Carbon and Alloy Steels — SAE J411d 3.01
Chemical Compositions of SAE Alloy Steels — SAE J404j 2.09
Chemical Compositions of SAE Carbon Steels — SAE J403h 2.06
Chemical Compositions of SAE Experimental Steels — SAE J1081 MAY80 . . . . 2.12
Chemical Compositions of SAE Wrought Stainless Steels — SAE J405d 2.11
Classification of Common Imperfections in Sheet Steel — SAE J810 JUN80 . . . . 4.24
Cleanliness Rating of Steels by the Magnetic Particle Method — SAE J421b 4.37
Definitions for Macrostrain and Microstrain — SAE J932 . . . . 4.57
Definitions of Heat Treating Terms — SAE J415j 3.20
Detection of Surface Imperfections in Ferrous Rods, Bars, Tubes, and Wires — SAE J349 JUN80 . . . . 4.42
Estimated Mechanical Properties and Machinability of Hot Rolled and Cold Drawn Carbon Steel Bars — SAE J414a 3.14
Formerly Standard SAE Alloy Steels — SAE J778a . . . 2.13
Formerly Standard SAE Carbon Steels — SAE J118a . . 2.14
General Characteristics and Heat Treatments of Steels — SAE J412h 3.05
Glossary of Carbon Steel Sheet and Strip Terms — SAE J940 . . . . 3.18
Grain Size Determination of Steels — SAE J418a 4.07
Hardenability Bands for Carbon and Alloy H Steels — SAE J1268 JUN80 . . . . 2.24
Hardness Tests and Hardness Number Conversions — SAE J417b 4.03
High Strength Carbon and Alloy Die Drawn Steels — SAE J935 . . . . 3.03
High Strength, Low Alloy Steel — SAE J410c 2.72
High Strength, Quenched, and Tempered Structural Steels — SAE J368a 2.74
Influence of Residual Stress on Fatigue of Steel — (HS 198) . . . . *
Infrared Testing — SAE J359a 4.47
Liquid Penetrant Test Methods — SAE J426c 4.52
Magnetic Particle Inspection — SAE J420b 4.47
Mechanical Properties of Heat Treated Wrought Steels — SAE J413b 3.13
Method for Determining Breakage Allowances for Steel Sheets — SAE J424 . . . . 4.45
Methods of Determining Hardenability of Steels — SAE J406c 2.15
Methods of Determining Plastic Deformation in Sheet Metal Stampings — SAE J863c 4.14
Methods of Measuring Case Depth — SAE J423a . . . 4.43
Methods of Measuring Decarburization — SAE J419 . . 4.10
Methods of Residual Stress Measurement — (HS J936) *
Methods of Sampling Steel for Chemical Analysis — SAE J408c 2.69
Microscopic Determination of Inclusions in Steels — SAE J422a 4.39
Numbering Metals and Alloys — SAE J1086 . . . . 2.01
Penetrating Radiation Inspection — SAE J427b 4.52
Permissible Variations from Specified Ladle Chemical Ranges and Limits for Steels — SAE J409c 2.70
Preferred Thicknesses for Uncoated Flat Metals (Thru 12 mm) — SAE J446a 10.06
Prevention of Corrosion of Metals — (HS J447a) . . . . *
Properties of Low Carbon Steel Sheets and Strip and Their Relationship to Formability — SAE J877 . . . . 4.17
SAE Numbering System for Wrought or Rolled Steel — SAE J402b 2.06
Selecting and Specifying Hot and Cold Rolled Steel Sheet and Strip — SAE J126a 4.19
Selection and Heat Treatment of Tool and Die Steels — SAE J437a 8.02
Selection and Use of Steels — SAE J401 . . . . 2.05
Sheet Steel Thickness and Profile — SAE J1058 . . . . 4.16
Surface Hardness Testing with Files — SAE J864 . . . 4.02
Surface Texture Measurement of Cold Rolled Sheet Steel

---

*Available as a Related Technical Report. Complete listing and details appear at the end of this index.

— SAE J911 . . . . **10.01**
Technical Report on Fatigue Properties — SAE J1099 **4.57**
Tensile Test Specimens — SAE J416b **4.01**
Tool and Die Steels — SAE J438b **8.01**
Ultrasonic Inspection — SAE J428b **4.55**

STEERING
Minimum Performance Criteria for Emergency Steering of Wheeled Earthmoving Construction Machines — SAE J53 . . . . **41.278**
Power Steering Pressure Hose—Low Volumetric Expansion Type — SAE J191 . . . . **12.110**
Power Steering Pressure Hose—Wire Braid — SAE J190 . . . . **12.111**
Power Steering Return Hose—Low Pressure — SAE J189 . . . . **12.108**
Steering Control System—Passenger Car—Laboratory Test Procedure — SAE J944 JUN80 . . . . **34.144**
Steering Frame Lock for Articulated Loaders and Tractors — SAE J276 . . . . **41.224**

STOP LAMPS
See: LAMPS

STUDS
See: FASTENERS

SUPERALLOYS
See: ALLOYS

SUSPENSIONS
Vehicle Dynamics Terminology — (HS J670e) . . . . . . *

SWITCHES
Backup Lamp Switches — SAE J1076 . . . . **23.60**
Electric Blower Motor Switch — SAE J235 . . . . **22.67**
Electric Windshield Washer Switch — SAE J234 . . . **22.66**
Electric Windshield Wiper Switch — SAE J112a . . . **22.66**
Hazard Warning Signal Switch — SAE J910b **23.64**
Headlamp Beam Switching — SAE J564c **23.02**
Headlamp Switch — SAE J253 . . . . **23.61**
Ignition Switch — SAE J259 . . . . **22.42**
Mechanical Stop Lamp Switch — SAE J249 . . . . . . **23.42**
Semiautomatic Headlamp Beam Switching Devices — SAE J565c **23.03**
Tests for Snowmobile Switching Devices and Components — SAE J68 OCT79 . . . . **39.08**
Turn Signal Switch — SAE J589b **23.50**

TACHOMETERS
Electric Tachometer Specification—Off Road — SAE J197 . . . . **20.10**
Electric Tachometer Specification—On Road — SAE J196 JUN80 . . . . **20.08**
Speedometers and Tachometers—Automotive — SAE J678e **20.03**

TAIL LAMPS
See: LAMPS

TECHNICAL REPORTS
Guidelines for Developing and Revising SAE Nomenclature and Definitions — SAE J1115 . . . . **1.21**
Preparation of SAE Technical Reports—Surface Vehicles and Machines: Standards, Recommended Practices, Information Reports — SAE J1159 AUG79 . . . . . . **1.06**
Technical Committee Guideposts — SAE J1271 AUG79 . . . . **1.01**

TERMINALS
Electrical Terminals Blade Type — SAE J858a **22.59**
Electrical Terminals—Eyelet and Spade Type — SAE J561 JUN80 . . . . **22.61**
Electrical Terminals—Pin and Receptacle Type — SAE J928 JUN80 . . . . **22.60**
Low Tension Wiring and Cable Terminals and Splice Clips — SAE J163 . . . . **22.51**

TEST SUBJECTS
See: ANTHROPOMETRY

TEXTILES
See: MATERIALS

THERMOCOUPLE
Brake Disc and Drum Thermocouple Installation — SAE J79 . . . . **31.131**

THERMOSTATS
Water Thermostat Pockets — SAE J634a **24.145**

THROTTLES
Marine Control Cable Connection—Engine Throttle Lever — SAE J961 JUN80 . . . . **42.04**
Snowmobile Throttle Control Systems — SAE J92 . . . **39.17**

TIRE RIMS
See: RIMS

TIRES
Agricultural Tractor Tire Dynamic Indices — SAE J1261 FEB79 . . . . **40.13**
Agricultural Tractor Tire Loadings, Torque Factors, and Inflation Pressures — SAE J709d **40.05**
Laboratory Testing Machines and Procedures for Measuring the Steady State Force and Moment Properties of Passenger Car Tires — (HS 210) . . . . *
Lawn and Garden Tractors Load and Inflation Pressures and Tire Selection Table for Future Design — SAE J344 . . . . **40.14**
Liquid Ballast Table for Drive Tires of Agricultural Machines — SAE J884 MAY80 . . . . **40.10**
Methods for Testing Snap-In Tubeless Tire Valves — SAE J1206 . . . . **29.17**
Off-Highway Tire and Rim Classification — SAE J751 APR80 . . . . **41.199**
Passenger and Light Truck Tire Traction Device Profile Determination and Classification — SAE J1232 . . . **37.30**
Passenger Car Tire Performance Requirements and Test Procedures — SAE J918c **29.01**
Selection of Tires for Multipurpose Passenger Vehicles, Light-, Medium-, and Heavy-Duty Trucks, Buses, Trailers, and Semitrailers for Normal Highway Service — SAE J161a **29.12**
Performance Requirements for Snap-In Tubeless Tire Valves — SAE J1205 . . . . **29.19**
Rolling Resistance Measurement Procedure for Passenger Car Tires — SAE J1269 JUN80 . . . . **29.04**
Sound Level of Highway Truck Tires — SAE J57a . . . **35.21**
Subjective Rating Scale for Evaluation of Noise and Ride Comfort Characteristics Related to Motor Vehicle Tires — SAE J1060 . . . . **29.16**
Testing Machines for Measuring the Uniformity of Passenger Car Tires — SAE J332a **29.10**
Test Procedure for Measuring Passenger Car Tire Revolutions per Mile — SAE J966 . . . . **29.15**
Test Procedures for Measuring Truck Tire Revolutions per Mile — SAE J1025 . . . . **29.15**
The Measurement of Passenger Car Tire Rolling Resistance — SAE J1270 OCT79 . . . . **29.06**
Tire Selection Tables for Agricultural and Light Industrial Machines of Future Design — SAE J711c **40.12**
Tire to Body Clearance Check for Recreational Vehicles — SAE J1214 . . . . **36.03**
Ton Mile Per Hour Application Practice — SAE J1098 APR80 . . . . **41.207**
Ton-Mile Per Hour Test Procedure — SAE J1015 . . . . **41.205**
Truck and Bus Tire Performance Requirements and Test Procedures — SAE J341a **29.13**
Truck Overall Widths Across Dual Tires — SAE J693a **37.26**
Vehicle Dynamics Terminology — (HS J670e) . . . . . . *
Wet or Dry Pavement Passenger Car Tire Peak and Locked Wheel Braking Traction — SAE J345a **29.08**

TOOL MATERIALS
Selection and Heat Treatment of Tool and Die Steels — SAE J437a **8.02**
Sintered Carbide Tools — SAE J439a **8.07**
Sintered Tool Materials — SAE J1072 . . . . **8.12**
Tool and Die Steels — SAE J438b **8.01**

TOW BARS
Full Trailer Tow Bar Eye — SAE J847 . . . . **37.24**

TOWING
Towability Design Criteria—Passenger Cars and Light Duty Trucks — SAE J1142a **27.01**
Towed Vehicle Drivetrain Test Procedure—Passenger Cars and Light Duty Trucks — SAE J1144a **27.07**

*Available as a Related Technical Report. Complete listing and details appear at the end of this index.

Towed Vehicle/Wrecker Attachment Test Procedure—
Passenger Cars and Light Duty Trucks — SAE
J1143a 27.05

TRACTORS
See: AGRICULTURAL MACHINERY
CONSTRUCTION AND INDUSTRIAL MACHINERY

TRAILERS
Automobile, Truck, Truck-Tractor, Trailer, and Motor
Coach Wiring — SAE J1292 JUN80 .... 22.43
Blade Type Electric Fuses — SAE J1284 DEC79 .... 22.34
Brake System Road Test Code—Passenger Car and
Light Duty Truck-Trailer Combinations — SAE J134
JUN79 .... 31.95
Capacity Ratings—Scraper, Dumper Body and Trailer
Body — SAE J741b 41.209
Connection and Accessory Locations for Towing Doubles
Trailers and Multi-Axle Trailers — SAE J849b 37.22
Fifth Wheel Kingpin—Commercial Trailers and Semitrailers — SAE J700b 37.21
Fifth Wheel Kingpin, Heavy Duty—Commercial Trailers
and Semitrailers — SAE J848a 37.22
Fifth Wheel Kingpin Performance—Commercial Trailers
and Semitrailers — SAE J133 .... 37.20
Five Conductor Electrical Connectors for Automotive
Type Trailers — SAE J895 .... 22.52
Full Trailer Tow Bar Eye — SAE J847 .... 37.24
Nomenclature—Truck, Bus, Trailer — SAE J687c 37.01
Rear Underride Guard Test Procedure — SAE J260
JUN80 .... 34.146
Service Brake System Performance Requirements—Passenger Car—Trailer Combinations — SAE J135a .. 31.104
Seven-Conductor Electrical Connector for Truck-Trailer
Jumper Cable — SAE J560b 22.57
Trailer Axle Alignment — SAE J875 .... 37.27
Trailer Couplings, Hitches and Safety Chains—Automotive Type — SAE J684f 36.01
Truck Tractor Semitrailer Interchange Coupling Dimensions — SAE J701a 37.18

TRANSMISSIONS
Automatic Transmission Functions—Terminology —
SAE J649b 28.08
Automatic Transmission Hydraulic Control Systems—
Terminology — SAE J648a 28.08
Automatic Transmissions—Manual Control Sequence —
SAE J915 .... 28.14
Automotive Transmission Terminology — SAE J645c .. 28.07
Axle and Manual Transmission Lubricants — SAE
J308c 13.12
Axle and Manual Transmission Lubricant Viscosity Classification — SAE J306 OCT79 .... 13.13
Clearance Envelopes for Six- and Eight-Bolt Truck Transmission Mounted Power Take-Offs — SAE J772a .. 28.15
Fluid for Passenger Car Type Automatic Transmissions
— SAE J311b 13.17
Openings for Six- and Eight-Bolt Truck Transmission
Mounted Power Take-Offs — SAE J704b 28.16
Passenger Car and Truck Automatic Transmission Test
Code — SAE J651c 28.10
Passenger Car Engine and Transmission Identification
Numbers — SAE J129 .... 26.03
Powershift Transmission Fluid Classification — SAE
J1285 FEB80 .... 13.19
Transmissions—Schematic Diagrams — SAE J647a .. 28.08

TRUCKS
Approach, Departure, and Ramp Breakover Angles—
Passenger Car and Light Duty Trucks — SAE J689 .. 32.03
Average Vehicle Dimensions for Use in Designing Docking Facilities for Motor Vehicles — SAE J699a 37.13
Brake and Electrical Connection Locations—Truck-Tractor
and Truck-Trailer — SAE J702b 37.28
Categories of Off-Road Self-Propelled Work Machines —
SAE J1116 JUN80 .... 41.01
Certificates of Maximum Net Horsepower for Motor
Trucks and Truck Tractors — SAE J690 ...... 37.12
Joint RCCC/SAE Fuel Consumption Test Procedure

(Short Term-In-Service Vehicle) Type I — SAE J1264 .. 37.05
Location and Operation of Instruments and Controls in
Motor Truck Cabs — SAE J680b 37.34
Motor Truck CA Dimensions — SAE J691 .... 37.13
Nomenclature—Truck, Bus, Trailer — SAE J687c 37.01
Passenger and Light Truck Tire Traction Device Profile
Determination and Classification — SAE J1232 .. 37.30
Rating of Truck Power Take-Offs — SAE J705b 37.32
Rear Underride Guard Test Procedure — SAE J260 .. 34.146
SAE Commercial Vehicle Ability Report Form — (HS
83a) .... *
SAE Nodal Mount — SAE J1134 .... 37.28
SAE Wind Tunnel Test Procedure for Trucks and Buses
— SAE J1252 OCT79 .... 37.08
Tractor Protection Valve Control — SAE J779 ..... 37.31
Truck Ability Prediction Procedure — SAE J688 .... 37.02
Truck Ability Work Sheet Pad — (HS 83) .... *
Truck Overall Widths Across Dual Tires — SAE J693a .. 37.26
Turning Ability and Off Tracking—Motor Vehicles —
SAE J695b 37.14

Brake Systems
Brake Diaphragm Maintenance—Tractor and Semitrailer
— SAE J666 .... 37.29
Brake Flange Mounting—On-Highway Vehicles — SAE
J337 .... 37.27
Brake System Performance Requirements—Truck, Bus,
and Combination of Vehicles — SAE J992b ... 31.111
Brake System Road Test Code—Truck, Bus, and Combination of Vehicles — SAE J786a 31.105
Service Brake System Performance Requirements—
Light-Duty Truck — SAE J155 .... 31.92
Test Code—Truck, Truck-Tractor, and Trailer Air Service
Brake System Pneumatic Pressure and Time Levels —
SAE J982a 31.127

Emissions
Measurement of Fuel Evaporative Emissions from Gasoline Powered Passenger Cars and Light Trucks — SAE
J170a 25.73
Measurement of Fuel Evaporative Emissions from Gasoline Powered Passenger Cars and Light Trucks Using
the Enclosure Technique — SAE J171a 25.80

Sound Levels
Exterior Sound Level for Heavy Trucks and Buses —
SAE J366b 35.18
Maximum Sound Level for Passenger Cars and Light
Trucks — SAE J1030 .... 35.13
Measurement of Exterior Sound Level of Trucks with
Auxiliary Equipment — SAE J1077 .... 35.22
Measurement of Exterior Sound Levels for Heavy Trucks
Under Stationary Conditions — SAE J1096 .... 35.20
Sound Level for Passenger Cars and Light Trucks —
SAE J986b 35.11
Sound Level for Truck Cab Interior — SAE J336a 35.19

Vehicle Identification Numbers
Truck and Truck Tractor VIN Systems — SAE J1108a .. 26.07
Truck Identification Terminology — SAE J1229 .... 26.04
Truck Vehicle Identification Numbers — SAE J187 ... 26.05

Wheels
Commercial Vehicle Disc Wheels — SAE J694a 37.25
Extra Wide Single Tire Rim and Wheel Mounting for
Rear Axles — SAE J876a 37.26
Labeling—Disc Wheels and Demountable Rims—Trucks
— SAE J179a 30.09
Nomenclature—Wheels and Rims—Trucks — SAE
J393 .... 30.10
Recommended Tolerances for Spoke-Type Wheels, Demountable Rims, and Rim Spacers—Trucks and Trailers — SAE J851 .... 37.25
Wheels/Rims—Trucks—Test Procedures and Performance Requirements — SAE J267a 30.05

Windshields
Windshield Defrosting Systems Performance Requirements—Trucks, Buses, and Multipurpose Vehicles —
SAE J382 .... 34.16
Windshield Defrosting Systems Test Procedure—Trucks,

*Available as a Related Technical Report. Complete listing and details appear at the end of this index.

Buses, and Multipurpose Vehicles — SAE J381 ... **34.14**
Windshield Wiper Systems—Trucks, Buses, and Multipurpose Vehicles — SAE J198 .... **34.08**
  Wiring
    Automobile, Truck, Truck-Tractor, Trailer, and Motor Coach Wiring — SAE J1292 JUN80 .... **22.43**
TUBES AND TUBINGS
  Brazed Double Wall Low Carbon Steel Tubing — SAE J527 JAN80 .... **21.117**
  Detection of Surface Imperfections in Ferrous Rods, Bars, Tubes, and Wires — SAE J349 JUN80..... **4.42**
  Flares for Tubing — SAE J533b **21.113**
  Formed Tube Ends for Hose Connections — SAE J962b **21.112**
  Fuel Injection Tubing — SAE J529b **21.127**
  Fuel Injection Tubing Connections — SAE J521b.... **24.106**
  Metallic Air Brake System Tubing and Pipe — SAE J1149 .... **21.122**
  Metric Thread Fuel Injection Tubing Connections — SAE J242a **24.105**
  Nonmetallic Air Brake System Tubing — SAE J844d.. **21.124**
  Performance Requirements for SAE J844d Nonmetallic Tubing and Fitting Assemblies Used in Automotive Air Brake Systems — SAE J1131 .... **21.126**
  Pressure Ratings for Hydraulic Tubing and Fittings — SAE J1065 .... **21.119**
  Seamless Copper Tube — SAE J528 JAN80 .... **21.121**
  Seamless Low Carbon Steel Tubing Annealed for Bending and Flaring — SAE J524 JAN80 .... **21.115**
  Tubing—Motor Vehicle Brake System Hydraulic — SAE J1047 .... **31.136**
  Welded and Cold Drawn Low Carbon Steel Tubing Annealed for Bending and Flaring — SAE J525 JAN80 **21.115**
  Welded Flash Controlled Low Carbon Steel Tubing Normalized for Bending, Double Flaring, and Beading — SAE J356 JAN80 .... **21.118**
  Welded Low Carbon Steel Tubing — SAE J526 JAN80 **21.116**
  Windshield Washer Tubing — SAE J1037 .... **12.115**
TURBOCHARGERS
  Turbocharger Connections — SAE J1135 .... **24.40**
  Turbocharger Nomenclature and Terminology — SAE J922 NOV79 .... **24.07**
TURN SIGNAL LAMPS
  See: LAMPS
UNIVERSAL JOINTS
  Universal Joints and Driveshafts—Nomenclature and Terminology — SAE J901b **28.17**
VALVES
  Engine Poppet Valve Materials — J775a **2.75**
  Hydraulic Control Valve Test Procedure — SAE J747b **41.89**
  Hydraulic Directional Control Valves, 3000 psi Maximum — SAE J748 .... **41.81**
  Hydraulic Valves for Motor Vehicle Brake Systems—Performance Requirements — SAE J1137....... **31.30**
  Hydraulic Valves for Motor Vehicle Brake Systems Test Procedure — SAE J1118 .... **31.26**
  Measuring and Reporting the Internal Leakage of a Hydraulic Fluid Power Valve — SAE J1235 .... **41.84**
  Method of Measuring and Reporting the Pressure Differential-Flow Characteristics of a Hydraulic Fluid Power Valve — SAE J1117 .... **41.82**
  Methods for Testing Snap-In Tubeless Tire Valves — SAE J1206 .... **29.17**
  Performance Requirements for Snap-In Tubeless Tire Valves — SAE J1205 .... **29.19**
  Tractor Protection Valve Control — SAE J779 .... **37.31**
  Valve Seat Inserts—Engine — SAE J610b **24.38**
V-BELTS
  See: BELTS
VEHICLE IDENTIFICATION NUMBERS
  Passenger Car Identification Terminology — SAE J218a **26.02**
  Passenger Car Vehicle Identification Numbers — SAE J853a **26.03**
  Passenger Car Vehicle Identification Number System — SAE J273a **26.02**
  Truck and Truck Tractor VIN Systems — SAE J1108a **26.07**

Truck Vehicle Identification Numbers — SAE J187 ... **26.05**
Vehicle Identification Numbers—Motorcycles — SAE J364a **26.03**
Vehicle Identification Numbers—Snowmobile and All-Terrain Vehicles — SAE J64 .... **39.03**
Vehicle Identification Number Systems — SAE J272c **26.01**
World Manufacturer Identifier — SAE J1044a...... **26.06**
VIN:
  See: VEHICLE IDENTIFICATION NUMBERS
VISCOSITY
  Axle and Manual Transmission Lubricant Viscosity Classification — SAE J306 OCT79 .... **13.13**
  Axle and Manual Transmission Lubricants — SAE J308c **13.12**
  Engine Oil Viscosity Classification — SAE J300d **13.05**
VISION
  Describing and Measuring the Driver's Field of View — SAE J1050a **34.102**
  Motor Vehicle Driver's Eye Range — SAE J941e **34.60**
  Passenger Car Rear Vision — SAE J834a **34.39**
  Rear View Mirrors—Motorcycles — SAE J268a ... **38.03**
  Vision Glossary — SAE J264 .... **34.113**
  Vision Factors Considerations in Rear View Mirror Design — SAE J985 .... **34.38**
WARNING DEVICES
  360 Deg Emergency Warning Lamp — SAE J845 ... **23.64**
  Emergency Warning Device — SAE J774c **23.62**
  Flashing Warning Lamp for Agricultural Equipment — SAE J974 MAY80 .... **40.38**
  Flashing Warning Lamp for Industrial Equipment — SAE J96 .... **41.179**
  Flashing Warning Lamps for Authorized Emergency, Maintenance and Service Vehicles — SAE J595b .. **23.63**
  Hazard Warning Signal Switch — SAE J910b...... **23.64**
  Performance, Test, and Application Criteria for Electrically Operated Backup Alarm Devices — SAE J994b **35.32**
  Safety Signs — SAE J115 SEP79 .... **41.272**
  School Bus Red Signal Lamps — SAE J887a **23.65**
  Vehicular Hazard Warning Signal Flasher — SAE J945b **23.54**
  Warning Lamp Alternating Flashers — SAE J1054 ... **23.56**
WARNING LAMPS
  See: LAMPS
WASHERS
  See: FASTENERS
WEAR
  See Also: ABRASION
  Method of Testing Resistance to Crocking of Organic Trim Materials — SAE J861 .... **12.58**
  Method of Testing Resistance to Scuffing of Trim Materials — SAE J365a **12.69**
  Test Method for Determining Dimensional Stability of Automotive Textile Materials — SAE J883 .... **12.58**
  Test Method for Determining Resistance to Snagging and Abrasion of Automotive Bodycloth — SAE J948a **12.75**
  Test Method for Wicking of Automotive Fabrics and Fibrous Materials — SAE J913a **12.69**
WELDING
  Automotive Metallurgical Joining — (HS J836a) .... *
  Welding, Brazing, and Soldering—Materials and Practices — SAE J1147 .... **10.06**
WHEEL RIMS
  See: RIMS
WHEELS
  Extra Wide Single Tire Rim and Wheel Mounting for Rear Axles — SAE J876a **37.26**
  Mechanical and Material Requirements for Wheel Bolts — SAE J1102 .... **5.18**
  Nomenclature—Wheels and Rims—Trucks — SAE J393 .... **30.10**
  Rear Wheel Splash and Stone Throw Protection — SAE J682 .... **37.34**
  Recommended Tolerances for Spoke-Type Wheels, Demountable Rims, and Rim Spacers—Trucks and Trailers — SAE J851 .... **37.25**
  Wheel Chocks — SAE J348 .... **37.36**

*Available as a Related Technical Report. Complete listing and details appear at the end of this index.

Wheels—Passenger Cars—Impact Performance Requirements and Test Procedures — SAE J175 .... 30.02
Wheels—Passenger Cars—Performance Requirements and Test Procedures — SAE J328a 30.01
Wheels—Recreational and Utility Trailer Test Procedures — SAE J1204 .... 30.03
Wheels/Rims—Trucks—Test Procedures and Performance Requirements — SAE J267a 30.05
Disc Wheels
Commercial Vehicle Disc Wheels — SAE J694a 37.25
Industrial and Agricultural Disc Wheels — SAE J712a 40.32
Labeling—Disc Wheels and Demountable Rims—Trucks — SAE J179a 30.09
Wheel Mounting Elements for Industrial and Agricultural Disc Wheels — SAE J714b 40.36

WINCHES
Classification and Nomenclature Towing Winch for Skidders and Crawler Tractors — SAE J1014 OCT79 .. 41.32
Rating of Winches — SAE J706a 37.32
Specification Definitions—Winches for Crawler Tractors and Skidders — SAE J1158 .... 41.53

WINDSHIELDS
Automotive Safety Glazing — SAE J673b 34.31
Drop Test for Evaluating Laminated Safety Glass for Use in Automotive Windshields — SAE J938a 34.36
Light Transmittance of Automotive Windshields Safety Glazing Materials — SAE J1203 .... 34.34
Passenger Car Glazing—Electrical Circuits — SAE J216 34.35
Passenger Car Glazing Shade Bands — SAE J100 ... 34.35
Safety Glazing Materials—Motor Vehicles — SAE J674a 34.34
Defogging System
Passenger Car Backlight Defogging System — SAE J953 .... 34.17
Defrosting System
Passenger Car Windshield Defrosting Systems — SAE J902b 34.12
Windshield Defrosting Systems Performance Requirements—Trucks, Buses, and Multipurpose Vehicles — SAE J382 .... 34.16
Windshield Defrosting Systems Test Procedure—Trucks, Buses, and Multipurpose Vehicles — SAE J381 ... 34.14
Washers
Electric Windshield Washer Switch — SAE J234 .... 22.66
Passenger Car Windshield Washer Systems — SAE J942b 34.10
Windshield Washer Tubing — SAE J1037 .... 12.115
Wipers
Electric Windshield Wiper Switch — SAE J112a 22.66
Passenger Car Windshield Wiper Systems — SAE J903c 34.06
Windshield Wiper Hose — SAE J50a 12.112
Windshield Wiper Systems—Trucks, Buses, and Multipurpose Vehicles — SAE J198 .... 34.08

WIRE/WIRING
Automobile, Truck, Truck-Tractor, Trailer, and Motor Coach Wiring — SAE J1292 JUN80 .... 22.43
Detection of Surface Imperfections in Ferrous Rods, Bars, Tubes, and Wires — SAE J349 JUN80 ..... 4.42
Electrical System for Construction and Industrial Machinery — SAE J821a 41.182
Five Conductor Electrical Connectors for Automotive Type Trailers — SAE J895 .... 22.52
Hard Drawn Carbon Steel Valve Spring Quality Wire and Springs — SAE J172 .... 6.06
Hard Drawn Mechanical Spring Wire and Springs — SAE J113 .... 6.04
High Tension Ignition Cable — SAE J557 .... 22.49
High Tension Ignition Cable Assemblies—Marine — SAE J1191 .... 42.19
Low Tension Primary Cable — SAE J1128 .... 22.81
Low Tension Wiring and Cable Terminals and Splice Clips — SAE J163 .... 22.51
Marine Engine Wiring — SAE J378c 42.16
Music Steel Spring Wire and Springs — SAE J178 .. 6.09
Oil Tempered Carbon Steel Spring Wire and Springs — SAE J316 .... 6.01
Oil Tempered Carbon Steel Valve Spring Quality Wire and Springs — SAE J351 .... 6.02
Oil Tempered Chromium—Silicon Alloy Steel Wire and Springs — SAE J157 .... 6.05
Oil Tempered Chromium-Vanadium Valve Spring Quality Wire and Springs — SAE J132 .... 6.03
Seven Conductor Jacketed Cable for Truck Trailer Connections — SAE J1067 .... 22.58
Special Quality High Tensile, Hard Drawn Mechanical Spring Wire and Springs — SAE J271 .... 6.04
Stainless Steel, SAE 30302, Spring Wire and Springs — SAE J230 .... 6.08
Stainless Steel 17-7 PH Spring Wire and Springs — SAE J217 .... 6.07
Wiring Identification System for Industrial and Construction Equipment — SAE J210 .... 41.188

WOODRUFF KEYS
See: KEYS

WORLD MANUFACTURER IDENTIFIER (WMI)
World Manufacturer Identifier — SAE J1044a 26.06

YIELD STRENGTH
Use of the Terms Yield Point and Yield Strength — SAE J450 .... 10.01

ZINC ALLOYS
See: ALLOYS

*Available as a Related Technical Report. Complete listing and details appear at the end of this index.

RELATED TECHNICAL REPORTS

The publications listed here are standards, recommended practices, and information reports which are not included in the Handbook. Copies may be ordered by contacting the Publications Division, SAE, 400 Commonwealth Drive, Warrendale, PA 15096.

HS 19—SAE DOCUMENTS REFERENCED IN FEDERAL MOTOR VEHICLE SAFETY STANDARDS—Includes SAE reports specifically referenced in FMVSS, and reports that have been revised or new reports issued since the SS were last revised and are not referenced. Reports generated by other organizations are also listed for the convenience of the user.
$3.00 Mem. $4.00 List

HS 40—MAINTENANCE OF AUTOMOTIVE ENGINE COOLING SYSTEMS—This comprehensive guide to servicing modern truck and passenger-car cooling systems clearly defines limitations of coolants in regard to leakage, evaporation, overheating, and loss of capacity. Interrelations between cooling, fuel, lubrication, and exhaust systems are examined.
$4.00 Mem. $4.95 List

HS J73—MULTIPURPOSE PETROLEUM BASE FLUIDS—SAE Information Report. Specifications for multipurpose petroleum base fluids are given to reflect current industry usage of such materials in agricultural, construction, industrial and other special-purpose vehicles. Covered is a wide range of fluid properties and their suitability for use at various ambient temperatures depending on vehicle design.
$1.50 Mem. $2.50 List

HS 82—TRUCK ABILITY PREDICTION PROCEDURE—SAE J688—Provides practical method for predicting truck performance. By following directions, it is possible to select a truck on the basis of readily available specifications, information provided, and a minimum of calculation.
$3.00 Mem. $4.00 List

HS 83—TRUCK ABILITY WORK SHEET PAD—Gives procedure form for determining grade ability at a given road speed and equivalent acceleration rate. Pad contains 150 work sheets. One copy of HS 83a (which is suitable for reproduction) is supplied with each pad.
$2.00 Mem. $2.75 List

HS 83a—SAE COMMERCIAL VEHICLE ABILITY REPORT FORM—Available as a single sheet for reproduction or may be purchased in quantity.
$.25 Mem. $.50 List

HS 84—MANUAL ON SHOT PEENING—SAE J808a—Intended as a practical aid to engineers, designers, and men in the shop. HS 84 points out some of the possibilities and some of the limitations of the process. Chapters are devoted to a description of the process, the effect of shot peening, shot peening machines, peening media, quality control, fatigue properties, and x-ray diffraction.
$2.00 Mem. $2.75 List

HS 124—SAE MANUAL ON BLAST CLEANING—SAE J792a—Instructs engineers, designers, and shop men in blast cleaning know-how. Discusses blasting abrasives, blast cleaning machines, production procedures, and process specifications.
$2.50 Mem. $3.25 List

HS 184—SURFACE VEHICLE SOUND MEASUREMENT PROCEDURES—This publication includes an extensive rationale statement to accompany SAE Recommended Practice J184a—Qualifying a Sound Data Acquisition System. Recognizing that the use of only a sound level meter is often impractical or insufficient, there is need to ensure that other systems meet the performance requirements of a standard meter. J184a recognizes recent advancements in sound measurement instrumentation and provides the means to qualify the newer systems to existing performance criteria. The rationale behind the modification of the qualification procedure is explained in detail. In addition to complete systems, the requirements for individual instruments and circuit components are also discussed. HS 184 also includes all of the standards and recommended practices found in the sound level section of the SAE Handbook.
$7.25 Mem. $8.95 List

HS 210—LABORATORY TESTING MACHINES AND PROCEDURES FOR MEASURING THE STEADY STATE FORCE AND MOMENT PROPERTIES OF PASSENGER CAR TIRES—SAE J1106 and SAE J1107—These reports describe some basic design requirements and operational procedures associated with equipment for laboratory measurement of tire force and moment properties for the full range of automotive tires. These properties must be known to establish the tires' contribution to vehicle dynamic performance. This HS is a guide for equipment design and test operation so that data from different laboratories can be directly compared.
$4.25 Mem. $5.50 List

HS 215—HANDBOOK OF MOTOR VEHICLE, SAFETY AND ENVIRONMENTAL TERMINOLOGY—Includes alphabetical listings of more than 1500 technical terms as defined within the Motor Vehicle, Safety and Environmental areas of the 1976 SAE Handbook. Available for the first time in an easy to use format, these definitions were compiled to fill the critical need for a single-source reference. HS 215 is one of the most useful Handbook Supplements ever published by the SAE.
$7.95 Mem. $9.50 List

HS J390—DUAL DIMENSIONING—With the rapidly growing use of metric measurement throughout the world, the need for both metric and inch dimensions on a single drawing has led to growing use of dual dimensioning of engineering drawings. This report presents several approaches to displaying and identifying the two different values for each dimension, although only one is shown as preferred.
$3.00 Mem. $5.00 List

HS J447a—PREVENTION OF CORROSION OF METALS—SAE Information Report. A guide to principles of metal corrosion and methods of dealing with its prevention. Materials are described and evaluated. Chemical treatments are covered along with coatings of metal, paints, and ceramics.
$3.00 Mem. $5.00 List

HS J670e—VEHICLE DYNAMICS TERMINOLOGY—SAE Recommended Practice. Defines mechanical vibration—qualitative terminology, vibrating systems, components and characteristics of suspension systems, vibrations of vehicle suspension systems, suspension geometry, tires and wheels, and directional control. (1978)
$4.00 Mem. $5.00 List

HS J762a—REINFORCED PLASTICS FOR GROUND VEHICLE APPLICATIONS—This information report established an engineering approach to selection and use of reinforced plastic materials. Also acquaints engineers with varied classes of fibrous reinforced plastic materials available and their general range of properties and applications.
$3.00 Mem. $5.00 List

HS J782b—SEATING MANUAL—MOTOR VEHICLES—Developed by the SAE Seating Committee, this manual is designed to provide a more uniform system of nomenclature, definition of functional requirements, and test methods for the various materials, components, and manufacturing methods used in automotive seating. The information compiled is for reference for body and trim engineers as well as those who work cooperatively with the engineers.
$3.00 Mem. $4.00 List

HS J784a—RESIDUAL STRESS MEASUREMENTS BY X-RAY DIFFRACTION—This manual brings together the most important aspects of the x-ray techniques. It places emphasis on theoretical aspects but at the same time attempts to show how these theoretical considerations are reduced to a practical technique. Techniques are presented which enable satisfactory measurements of stress even when diffraction lines are diffused. Techniques are based on two-exposure method and use of diffractometers.
$5.50 Mem. $7.00 List

HS J788a—MANUAL ON DESIGN AND APPLICATION OF LEAF SPRINGS—Written as a guide for the designer of leaf spring installations, this manual contains information which will enable the designer to calculate the space required for a leaf spring, to provide suitable attachments, and to determine the elastic and geometric properties of the assembly.
$5.50 Mem. $7.00 List

HS J795a—MANUAL ON DESIGN AND APPLICATION OF HELICAL AND SPIRAL SPRINGS—A concise and simple account of essentials involved in the design and application of helical and spiral springs to help the engineer and designer as a means to better appreciate the nature of spring problems.

$4.00 Mem. $5.25 List

HS J806b—OIL FILTER TEST PROCEDURE—The purpose of this lubrication oil filter test code is to provide means for evaluating the performance characteristics of full-flow oil filters on bench test equipment. This data collected from "in service" applications may be used for establishing standards of performance for filters tested in this manner.

$4.00 Mem. $5.25 List

HS J836a—AUTOMOTIVE METALLURGICAL JOINING—This report is an abbreviated summary of metallurgical joining by welding, brazing, and soldering, intended to reflect current usage in the automotive industry.

$3.00 Mem. $4.00 List

HS J885 APR80—HUMAN TOLERANCE TO IMPACT CONDITIONS AS RELATED TO MOTOR VEHICLE DESIGN—SAE Information Report. This excellent information report provides a wide variety of data on human tolerance to impact resulting from tests of human volunteers, cadavers, and animal subjects. Its purpose is to provide a basis for crash test evaluation and motor vehicle accident reporting and it meets these needs well. It should, however, also be valuable to anyone involved in any type of trauma or accident investigation or product safety evaluation. Definitions are provided for medical terminology used.

$6.50 Mem. $8.00 List

HS J905—FUEL FILTER TEST METHOD—SAE Recommended Practice. This recommended practice provides a complete description of the procedures for testing final fuel filters for both diesel and gasoline engines. Test apparatus and testing materials are described in detail. An effort has been made in developing this test to simulate actual operating conditions.

$3.00 Mem. $5.00 List

HS J906—AUTOMOTIVE SAFETY GLAZING MANUAL—SAE Information Report. Provides a basic guide to characteristics of safety glazing materials and information on usage for those unfamiliar with glazing.

$1.75 Mem. $2.75 List

HS J965—ABRASIVE WEAR—SAE Information Report. Covers fundamentals of abrasive wear phenomena, testing for abrasive wear resistance, and solutions to abrasive wear problems.

$3.25 Mem. $4.50 List

HS J1066—RECOMMENDED GUIDELINES FOR COMPANY METRICATION PROGRAMS IN THE METALWORKING INDUSTRY—This 40 page booklet is divided into four sections: Drawing Practices; Units, Applications, and Terminology; Education and Training Aids; and Machine and Inspection Tools. Each of these sections recommend practices that will promote effective use of metric units, as well as effective handling of the transition period. Lists are included showing supplier of materials and where to obtain additional information.

$5.00 Mem. $6.50 List

HS J1078—A RECOMMENDED METHOD OF ANALYTICALLY DETERMINING THE COMPETENCE OF HYDRAULIC TELESCOPIC CANTILEVERED CRANE BOOMS

$4.00 Mem. $5.00 List

HS 1086a—SECOND EDITION UNIFIED NUMBERING SYSTEM HANDBOOK FOR METALS AND ALLOYS—The UNS provides a means of correlating many nationally used numbering systems currently administered by various societies, trade organizations, governmental bodies, and individual producers and users of metals and alloys, thereby avoiding confusion caused by the use of more than one identification number for the same material; and by the opposite situation, of having the same number assigned to two or more entirely different materials.

$39.00 Mem. $49.00 List

HS J1238—RATING LIFT CRANES ON FIXED PLATFORMS OPERATING IN THE OCEAN ENVIRONMENT—SAE Recommended Practice.

$2.75 Mem. $3.50 List

The following reports, which are out of print, are available in photocopy form. For price and order information contact the Publications Division, SAE, 400 Commonwealth Drive, Warrendale, PA 15096.

HS 3—SURFACE ROLLING AND OTHER METHODS FOR MECHANICAL PRESTRESSING OF METALS—SAE J811
HS 27—MOBILE RADIO TELEPHONE FOR AUTOMOTIVE USE—SAE J797
HS 147—EVALUATION OF METHODS FOR MEASUREMENT OF RESIDUAL STRESS—SAE J793
HS 156—SCALES FOR USE WITH DECIMAL-INCH DIMENSIONING—SAE J805
HS 198—INFLUENCE OF RESIDUAL STRESS ON FATIGUE OF STEEL—SAE J783
HS J317—RANGE OF ANTHROPOMETRIC MEASUREMENTS FOR ASIAN POPULATIONS
HS J936—METHODS OF RESIDUAL STRESS MEASUREMENT

SAE Standards Subscription Service

A subscription to the
SAE STANDARDS SUBSCRIPTION SERVICE
will keep you up to date with all new and revised documents as they are published

SAE STANDARDS SUBSCRIPTION SERVICE

Here is a service that is offered by the SAE that is designed to keep you up to date with all new and revised standards as they are published. Here is what you will receive with this STANDARDS SUBSCRIPTION SERVICE:

1. All **Standards, Recommended Practices,** and **Information Reports** in all HANDBOOK categories as they are published.

2. The complete Safety Standard Service with the annual update of the Index of FMVSS Standards and SAE documents that have been referenced. (HS 19)

3. SAE Technical Committee Standards and Current Project Inventory, which outlines standards in progress.

4. Copies of each new Federal Standard that appears in the Federal Register that pertains to SAE regarding on and off highway transportation, and is issued by the NHTSA, EPA, OSHA.

When this service is used in conjunction with the current edition of the SAE HANDBOOK, it offers a relief from a constant surveillance task of culling the Federal Register and other unnecessary paper work to find the latest standards and reports. Mailings are made at least eight times per year.

TO ORDER CONTACT:
Society of Automotive Engineers, Inc.
400 Commonwealth Drive
Warrendale, PA 15096
(412) 776-4841 Ext. 280

$150.00 Annual Subscription - Domestic
$185.00 Annual Subscription - Foreign

Ref
TL
151
S62
pt.1

NOV 13 1981